Physics for Scientists and Engineers: Foundations and Connections

with Modern Physics

Debora M. Katz

CENGAGE Learning

Australia • Brazil • Mexico • Singapore • United Kingdom • United States

CENGAGE Learning

Physics for Scientists and Engineers: Foundations and Connections with Modern Physics
Debora M. Katz

Product Director: Mary Finch
Product Manager: Rebecca Berardy Schwartz
Managing Developer: Peter McGahey
Senior Content Developer: Susan Dust Pashos
Content Developer: Ed Dodd
Product Assistant: Margaret O'Neill
Senior Marketing Manager: Janet del Mundo
Senior Content Project Manager: Alison Eigel Zade
Art Director and Executive Director of Design: Bruce Bond
Manufacturing Planner: Beverly Breslin
IP Project Manager: Farah J. Fard
IP Analyst: Christine Myaskovsky
Production Service and Compositor: Cenveo® Publisher Services
Photo Researchers: Sharon Donahue (Chapters 1–22), Carly Bergey (Chapters 22–39) and Pandisathya Paul (Chapters 40–43)
Cover Designer: Bruce Bond
Cover Image: United Launch Alliance

Library of Congress Control Number: 2015949144

Student Edition:
ISBN: 978-1-305-25983-6

Loose-leaf Edition:
ISBN: 978-1-337-61652-2

Cengage Learning
20 Channel Center Street
Boston, MA 02210
USA

Cengage Learning is a leading provider of customized learning solutions with employees residing in nearly 40 different countries and sales in more than 125 countries around the world. Find your local representative at **www.cengage.com.**

Cengage Learning products are represented in Canada by Nelson Education, Ltd.

To learn more about Cengage Learning Solutions, visit **www.cengage.com.**

Purchase any of our products at your local college store or at our preferred online store **www.cengagebrain.com.**

To Zak and Jeff
with love.

Printed in the United States of America
Print Number: 01 Print Year: 2017

About the Author

Debora Katz is on the physics faculty at the United States Naval Academy, where she teaches calculus-based introductory physics. Soon after beginning her job at the Academy, she co-authored a book, *The Physics Toolbox*, meant to help students struggling in introductory physics. She was born in Chicago, Illinois and grew up in a suburb of Chicago as well as in Hawaii. She went to Brandeis University in Waltham, Massachusetts, where she earned a Bachelor of Arts degree in physics. She then went to the University of California in Irvine, California, where she earned a Master of Science degree in physics. Finally, she attended the University of Minnesota, earning a doctorate in astrophysics. After her PhD, she immediately began teaching in the physics department at the United States Naval Academy. She lives in Annapolis with her husband Jeff, her son Zak, and their dog Simon.

Contents

Part III Electricity

Preface for the Instructor

When I decided to pursue physics and astrophysics, I didn't think about teaching. I just knew I was interested in the subjects. As a graduate student at the University of Minnesota, however, I won a fellowship that included pedagogical training by well-known figures in physics education research (PER). I learned to take teaching seriously, and when it was time to find a job, I knew I wanted to work at an institution committed to undergraduate education.

Fortunately, I found that job at the United States Naval Academy. I have had the opportunity to teach our year-long calculus-based course not just to physics majors, but also to a varied audience with a diverse set of abilities, interests, learning styles, and preconceptions. In order to engage my students in the process of learning physics, I use many of the techniques that have come out of PER, and over the past two decades my students have taught me how to teach them. I have tried my best to integrate both PER results and the lessons learned from my students into the pages of this text.

One of the major results of PER is that lectures don't work. Students who sit passively in a classroom, listening to a brilliant lecture, retain very little. Fortunately, PER offers many ways to approach teaching. I have used a great number of these approaches: *peer instruction, active learning groups, interactive lectures,* and, most recently, a *flipped classroom.* All of these approaches require students to take an active part in their education, while I act as a coach, urging them to take the necessary steps to learn physics.

Although there is a wide range of physics pedagogy, including lectures in introductory physics, one thing is always true: *students do better if they read their textbook.* So my first goal in writing this textbook was to write a book that students would actually read and value. I know students don't generally like to read a science textbook, and if they do try to read it, they don't get much for their effort. So I looked into what makes other reading material, such as a novel or a magazine article, so effective. One answer is that the human brain enjoys stories about people. Generally, physics textbooks omit stories; science is presented as a series of results, without the process that led to these results or the impact these results have on our lives. Students cannot connect to this traditional, dogmatic approach; they don't see how physics fits into their lives. So, wherever possible, I included the stories of our field. Some of these are stories of discovery, whereas others are stories that show the impact of physics results on the human experience.

Another major result of PER is the insight that students are not blank slates. They come to our classrooms with *preconceptions* about physics, which they have developed over years of observations made during the course of their everyday lives. One of the important jobs of physics education is to help students tie their preconceptions to the appropriate formalisms of physics. For example, students know that seat belts secure passengers in cars, but they are not usually aware of the connection between this fact from their everyday lives and Newton's first law. To make and reinforce these connections, I use dialogues between fictional students (named Avi, Cameron, and Shannon) to highlight and clarify commonly held preconceptions. Our real students are then tasked with critiquing the fictional dialogue, leading them to connect their preconceptions to the correct physics concepts.

As university instructors, our goal is to help our students work their way up to the top of Bloom's taxonomy (see the figure on page xiv), where they create and answer their own questions using physics. Of course, students must start at the bottom and work their way up to the top. For example, at first we want our students to remember and understand Newton's second law. We spend most of our time working in the middle of the pyramid, where we may expect our students to apply Newton's second law to analyze problems. Occasionally, we expect our students to work higher up in Bloom's taxonomy. Often we rely on a laboratory component, which requires students to make judgments, or we find some problems that call for such evaluation. However, there are very few tools at our disposal for helping students to reach the top of Bloom's taxonomy.

But only when students reach the top of Bloom's taxonomy can they see the value in learning physics because it is at the top of the pyramid that students use physics to answer their own questions. So to help students reach the top and to make physics engaging, I use case studies in my classroom and in this book. Case studies relate interesting topics to the concepts, principles, and tools of physics. In my class, students write their own case studies, using physics concepts to understand situations drawn from such subjects as sports, movies, and history. In my textbook, a case study is woven into each chapter, to make

Bloom's pyramid (taxonomy) applied to Newton's second law.

At the lowest level, we want our students to remember the law. A typical physics course spends a lot of time problem-solving (in the middle of the pyramid) by applying Newton's second law to analyze problems. Case Studies allow students to reach the top of the pyramid.

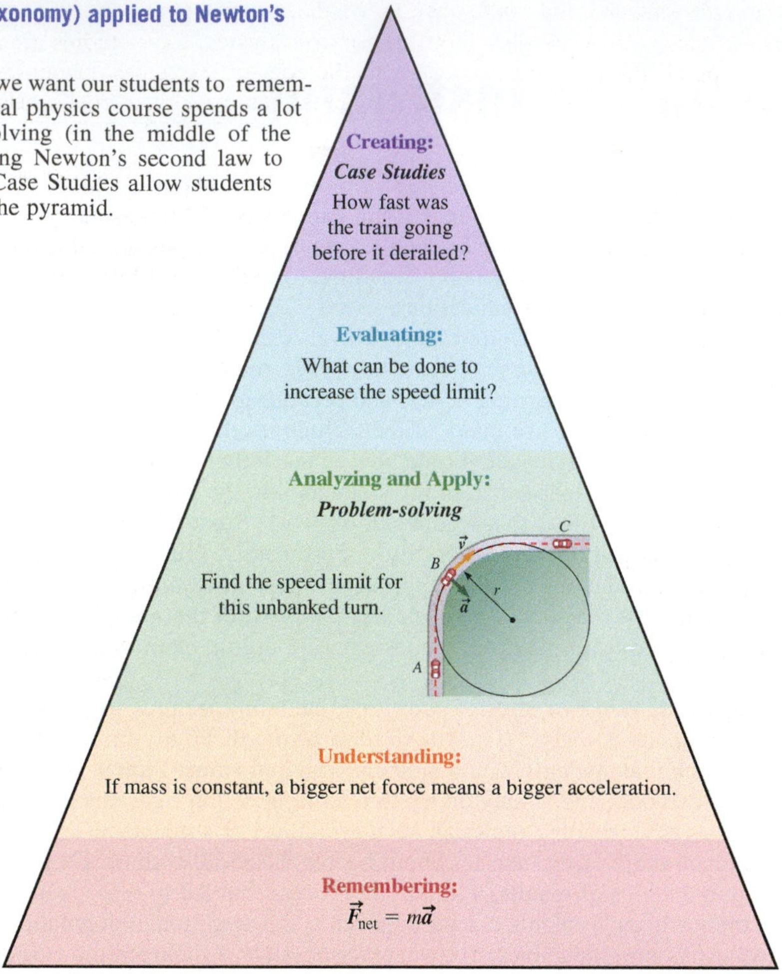

Creating:
Case Studies
How fast was the train going before it derailed?
Evaluating:
What can be done to increase the speed limit?
Analyzing and Apply:
Problem-solving
Find the speed limit for this unbanked turn.
A
B
C
r
Understanding:
If mass is constant, a bigger net force means a bigger acceleration.
Remembering:
$\vec{F}_{\text{net}} = m\vec{a}$

physics more relevant and engaging—for example, Chapter 5 uses a newsworthy train collision to teach Newton's laws of motion.

Case studies fulfill other goals as well. First, because case studies are based on human experience, they also help students realize how the preconceptions they have developed over decades of observation tie into the formalism of physics. Second, case studies are stories—so they are fun to read and memorable. I know you might find it hard to believe, but both my own students and those at other institutions report that they actually *enjoy* the case studies. When was the last time a student said he or she enjoyed reading a science textbook? I hope you'll find this textbook program to be a better learning tool for your students than what has previously been available.

If you have any comments or questions, feel free to contact me at deborakatz@yahoo.com.

How do you engage your students to go beyond the quantitative?

The main goal of *Physics for Scientists and Engineers: Foundations and Connections* is to offer a calculus-based introductory physics textbook designed to assist you in taking your students "beyond the quantitative." Physics Education Research (PER) best practices and the author's extensive classroom experience are leveraged to motivate readers and address the areas where students struggle the most—bridging the gap between abstract language and application, overcoming common preconceptions, and connecting mathematical formalism and physics concepts.

Case studies to motivate students and make abstract concepts concrete.

This text uses case studies to draw readers into the story of physics. Case studies are introduced and revisited throughout the chapters in pedagogy such as the concept exercises, examples, and end-of-chapter problems. Some case studies are based on students' contemporary "real-world" experiences, including events they may read about in the news. These case studies make abstract physics concepts understandable and help bridge the gaps between the key concepts, the formal language, and the mathematics of physics. For example, the case study in Chapter 5 illustrates Newton's laws of motion by examining an actual train collision described in news articles. The author introduces free-body diagrams and Newton's second law, asking students to determine why backward-facing passengers experienced less bodily harm than did the forward-facing passengers. Students apply physics "tools"—Newton's laws and free-body diagrams—to get to the bottom of this real-world example.

FIGURE 5.1 Aerial view of emergency workers helping the injured from a train crash near Los Angeles (April 23, 2002).

CASE STUDY **Train Collision**

On April 23, 2002, a passenger train about 35 miles outside of Los Angeles was hit by a freight train (Fig. 5.1). The accident killed two people and injured more than 260, with all the injured being on the passenger train. Witnesses reported that those people who were seated facing backward suffered little or no injury. News reports said that the passenger train came to a quick stop before the collision and that the impact with the freight train pushed the passenger train 370 ft backward (Fig. 5.2).

One of the most controversial parts of the early reports was how fast the freight train was going at the moment of impact. In Chapter 11, we will reconstruct the accident and estimate the speed of the freight train upon impact. In this chapter, we are concerned only with the following questions:

1. Why did passengers seated facing backward fare better than those who were either standing or seated facing forward?
2. The passenger train was at rest before the collision and was pushed backward. What do those facts tell us about how hard the freight train pushed on the passenger train? Did the passenger train push on the freight train? If so, how hard did it push?

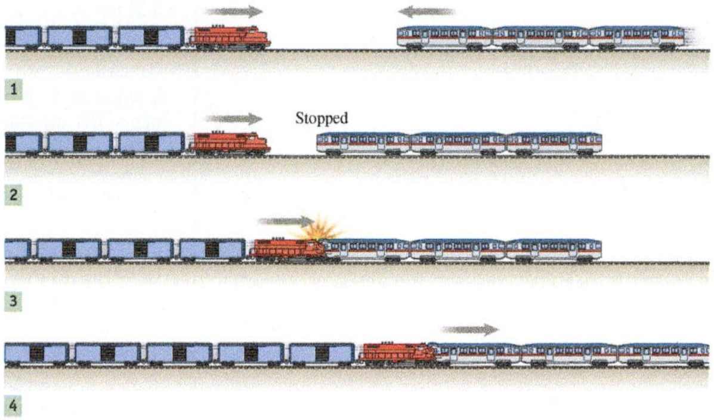

FIGURE 5.2 1 A freight train and a passenger train move toward each other. 2 The passenger train stops. 3 The freight train continues and collides with the passenger train. 4 The passenger train is shoved backward.

Student dialogues to address preconceptions.

Students often come to the classroom with preconceptions (what some call misconceptions). By acknowledging and addressing these preconceptions, we can transform them into building blocks toward proper understanding. Preconceptions are primarily addressed by dialogues between fictional students that allow readers to discover their preconceptions without a sense of failure. Dialogues may be incorporated into case studies, concept exercises, examples, and end-of-chapter problems.

CONCEPT EXERCISE 5.2

CASE STUDY **Train Collision and Newton's First Law**

A group of college students discusses the train collision case study. Use Newton's first law to decide which underlined statements are correct and which are false. Explain your answers.

Shannon: This newspaper says that the people who got really hurt were either standing up or sitting in a forward-facing seat. Those people got thrown forward when the train stopped.

Avi: That's why there are seat belts in cars. If you get into a crash, the force can throw you through the windshield.

Cameron: There is no force that throws you through the windshield. You fly through the windshield because you are already moving and it would take a force to stop you from going forward. *That's* why there's a seat belt.

Avi: That doesn't make sense. Because then you would need a force to stop you from flying through the windshield even when you just stop slowly at a red light.

Cameron: That's right, but when you slow down slowly, you don't need such a big force and the car seat can take care of it.

Shannon: The seat? I don't think a seat can exert a force. It can't move on its own or hold you. That's why the people who were sitting forward on the train were hurt. The people who were sitting backward had the back of the seat to block them.

Two-column format for examples and derivations to connect mathematical formalism and physics concepts.

Research shows that students struggle to make these connections. By presenting many of the **examples and derivations in two columns (what an expert problem-solver thinks on the left and what that expert would write on the board on the right),** students can make the connections between the concept being taught and the mathematical steps to follow. It is like having an instructor within the text: in one column he or she explains the concept, and in the other he or she shows the mathematical steps to follow, just as an instructor would verbally explain a problem in class while simultaneously solving the problem on the board.

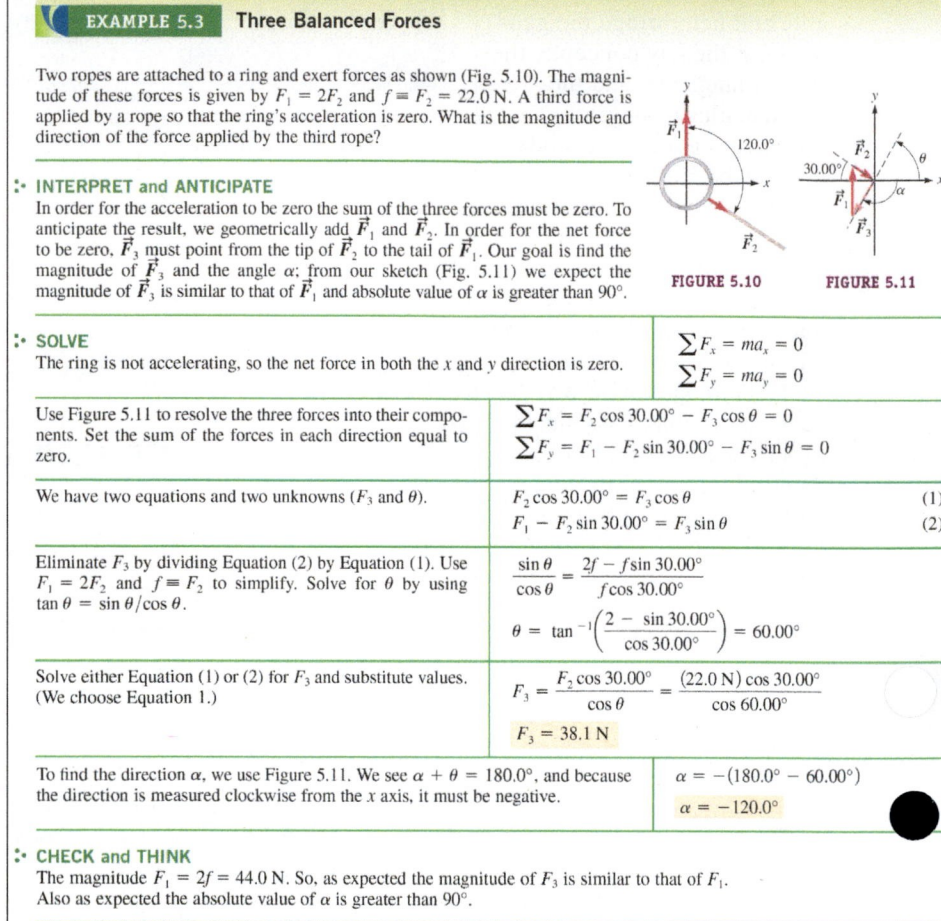

EXAMPLE 5.3 Three Balanced Forces

Two ropes are attached to a ring and exert forces as shown (Fig. 5.10). The magnitude of these forces is given by $F_1 = 2F_2$ and $f \equiv F_2 = 22.0$ N. A third force is applied by a rope so that the ring's acceleration is zero. What is the magnitude and direction of the force applied by the third rope?

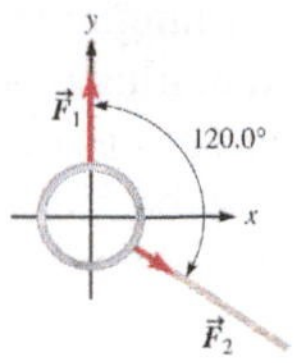

FIGURE 5.10

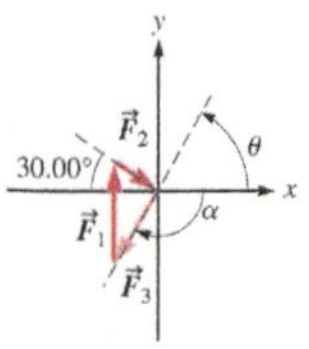

FIGURE 5.11

:• INTERPRET and ANTICIPATE

In order for the acceleration to be zero the sum of the three forces must be zero. To anticipate the result, we geometrically add $\vec{F}_1$ and $\vec{F}_2$. In order for the net force to be zero, $\vec{F}_3$ must point from the tip of $\vec{F}_2$ to the tail of $\vec{F}_1$. Our goal is find the magnitude of $\vec{F}_3$ and the angle α; from our sketch (Fig. 5.11) we expect the magnitude of $\vec{F}_3$ is similar to that of $\vec{F}_1$ and absolute value of α is greater than 90°.

:• SOLVE

The ring is not accelerating, so the net force in both the x and y direction is zero.	$\sum F_x = ma_x = 0$ $\sum F_y = ma_y = 0$
Use Figure 5.11 to resolve the three forces into their components. Set the sum of the forces in each direction equal to zero.	$\sum F_x = F_2 \cos 30.00° - F_3 \cos\theta = 0$ $\sum F_y = F_1 - F_2 \sin 30.00° - F_3 \sin\theta = 0$
We have two equations and two unknowns (F_3 and θ).	$F_2 \cos 30.00° = F_3 \cos\theta$ (1) $F_1 - F_2 \sin 30.00° = F_3 \sin\theta$ (2)
Eliminate F_3 by dividing Equation (2) by Equation (1). Use $F_1 = 2F_2$ and $f \equiv F_2$ to simplify. Solve for θ by using $\tan\theta = \sin\theta/\cos\theta$.	$\frac{\sin\theta}{\cos\theta} = \frac{2f - f\sin 30.00°}{f\cos 30.00°}$ $\theta = \tan^{-1}\left(\frac{2 - \sin 30.00°}{\cos 30.00°}\right) = 60.00°$
Solve either Equation (1) or (2) for F_3 and substitute values. (We choose Equation 1.)	$F_3 = \frac{F_2 \cos 30.00°}{\cos\theta} = \frac{(22.0\text{ N}) \cos 30.00°}{\cos 60.00°}$ $F_3 = 38.1$ N
To find the direction α, we use Figure 5.11. We see $\alpha + \theta = 180.0°$, and because the direction is measured clockwise from the x axis, it must be negative.	$\alpha = -(180.0° - 60.00°)$ $\alpha = -120.0°$

:• CHECK and THINK

The magnitude $F_1 = 2f = 44.0$ N. So, as expected the magnitude of F_3 is similar to that of F_1. Also as expected the absolute value of α is greater than 90°.

Problem-Solving Strategy

Physics is not a spectator sport. Physics students are expected to *do* physics. So, problem-solving is a major component to learning physics. In keeping with the sports analogy, a novice player is given detailed instructions on how to position and move his or her body, but a professional athlete often forgets these details and seems to play the sport naturally. Likewise, physics students need a detailed problem-solving strategy when they first start off. The strategy in this book is streamlined and designed to mimic expert methods. All worked examples in the text have been solved through a three-procedure approach: 1) *Interpret and Anticipate*, 2) *Solve*, and 3) *Check and Think*. (See pages 5 and 6 for more information.) In addition, further problem-solving strategies are provided for specific types of problems. As shown on page xvii, these specific strategies flesh out one or more of the three procedures.

PROBLEM-SOLVING STRATEGY

Applying Newton's Second Law

INTERPRET and ANTICIPATE
Identify the system (often, a single object) that is subject to external force(s). Draw a free-body diagram for that system, making sure that the diagram has all four elements given above. Once the free-body diagram is complete, there are three steps that help in the **SOLVE** procedure when an algebraic or numerical result is required.

SOLVE
Step 1 Apply Newton's second law in component form. Use the coordinate system on the free-body diagram to apply Equation 5.2. You will have one equation for each direction in which there is at least one force.

$$\sum F_x = ma_x \quad \sum F_y = ma_y \quad \sum F_z = ma_z$$

Step 2 Write down any other equations that are relevant to the forces involved. Table 5.1 lists magnitudes of the gravitational force (weight), the spring force, and kinetic friction. If the situation involves one or more of these forces, write down the appropriate equation.

Step 3 Do algebra before substitution. Review your equations. Which parameters are known? Which are unknown? Which do you need to solve for? You may have more equations than you need. Find an algebraic expression for the parameter you need before you substitute any numerical values. This practice makes it easier to find a mistake if you make one and makes it easier for another person to understand your work.

Conceptual Framework

You probably wouldn't be surprised to learn that studies have enumerated the ways physics instructors differ from their students. For example, physicists tend to be *linear* learners, and physics students tend to be *global* learners. This means that physics students need to see the big picture before they can settle into learning the details. To give students a global perspective, each chapter begins with an overview of the material covered in the chapter.

Another difference between physics instructors and their students is that instructors often see physics topics arranged in a conceptual hierarchy, but students have trouble organizing topics. So to help students develop an organization, each chapter begins with a list of the topics arranged into categories: *Underlying Principles, Major Concepts*, *Special Cases*, and *Tools*; an example of this structure is shown below:

Underlying Principles
Newton's three laws of motion

Major Concepts
1. Dynamics
2. Force
3. System
4. Inertia
5. Inertial reference frames

Special Cases: Specific Forces
1. Gravity
2. Hooke's law (spring force)
3. Normal force
4. Tension force
5. Kinetic friction

Tools
Free-body diagrams

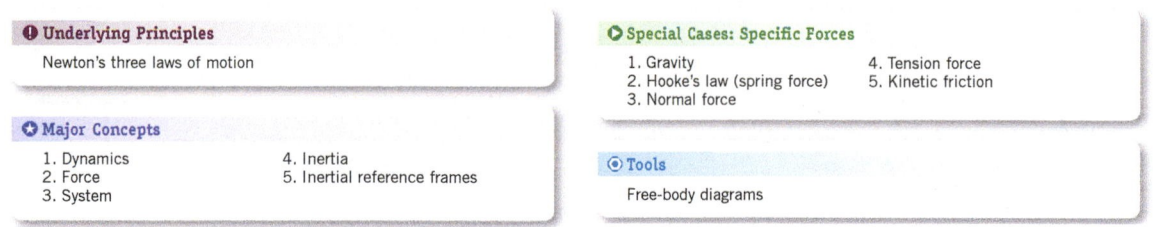

Throughout each chapter, notes in the margins refer to the topics using these categories (see examples below).

NEWTON'S SECOND LAW **Underlying Principle**

FORCE **Major Concept**

GRAVITY **Special Case**

FREE-BODY DIAGRAMS **Tool**

Finally, each chapter has a **Summary** at the end, before the problems and questions. In a traditional textbook, the summary is arranged according to the order in which the topics appear in the chapter. By contrast, the summary in this textbook (see page xviii for an example) is arranged using the same hierarchy found at the beginning of the chapter. This reinforces the conceptual organization by revisiting the concepts found on the chapter's opening page with more complete descriptions. In this way, students can use this information for review before attempting to solve the problems assigned for homework.

SUMMARY

Underlying Principles: Newton's three laws of motion

1. Close translation of **Newton's first law**:

   Every body continues in its state of rest, or of uniform motion in a straight line, unless it is compelled to change that state by forces impressed on it.

   More contemporary statement of Newton's first law:

   If no force acts on an object, then the object cannot accelerate.

   Newton's second law contains his first law:

   The net force on an object is zero if and only if the acceleration of the object is also zero.

2. **Newton's second law**: The acceleration of an object is proportional to the total force acting on it:

$$\vec{F}_{\text{tot}} \equiv \sum \vec{F} = m\vec{a} \quad (5.1)$$

   Newton's second law is often applied in component form:

$$\sum F_x = ma_x \quad \sum F_y = ma_y \quad \sum F_z = ma_z \quad (5.2)$$

3. **Newton's third law**: If two objects A and B interact, the force exerted by A on B is equal in magnitude to the force exerted by B on A but in the opposite direction. Mathematically,

$$\vec{F}_{[\text{B on A}]} = -\vec{F}_{[\text{A on B}]} \quad (5.10)$$

Major Concepts

1. **Dynamics:** a branch of mechanics that focuses on the causes of motion.
2. **Force:** a push or pull; required to make an object accelerate. A force has three properties:
   a. Force does not exist in isolation; it must act on a "subject."
   b. If a force is exerted on a subject, there must be some "source" responsible for that force.
   c. Force is a vector quantity having both magnitude and direction.
3. **System**: any collection of objects.
4. **Inertia**: the tendency of an object to maintain its constant velocity. **Mass**, also known as **inertial mass**, is the intrinsic property of an object that determines its inertia.
5. **Inertial reference frame**: one in which Newton's first law is valid. Inertial reference frames may move with constant velocity, but they do not accelerate. Accelerating frames are known as **noninertial** reference frames, and Newton's first law is not valid in them.

Special Cases

Table 5.1 on page 134 summarizes the five specific forces examined in this chapter.

Tools

Four elements of a **free-body diagram**:

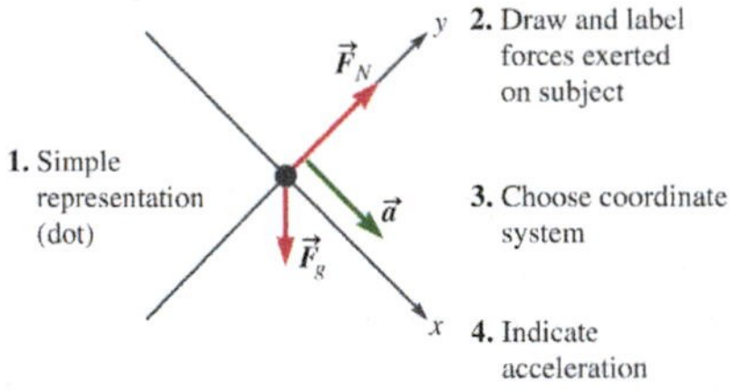
1. Simple representation (dot)
2. Draw and label forces exerted on subject
3. Choose coordinate system
4. Indicate acceleration
$\vec{F}_N$
$\vec{a}$
$\vec{F}_g$
y
x

PROBLEM-SOLVING STRATEGY

Applying Newton's Second Law, $\vec{F}_{\text{tot}} \equiv \sum \vec{F} = m\vec{a}$

:• INTERPRET and ANTICIPATE

Draw a free-body diagram.

:• SOLVE

1. Apply Newton's second law in component form. Write one equation for each direction that has at least one force.
2. Write any other relevant force equations.
3. Do algebra before substituting values.

Problems Set

Over 3,500 problems have been written for this edition. Each part of every problem has been classified as either *Algebraic*, *Conceptual*, *Estimation*, *Graphical*, or *Numerical* in nature. In most traditional textbooks, problems are generally either numerical or algebraic, and in recent years conceptual questions have been added to the end-of-chapter exercises in their own separate section. These books have few, if any, estimation or graphical problems. However, one goal of physics instruction is to teach students to think like physicists. By organizing the questions and problems by topic rather than by type, and by introducing graphical and estimation problems, this textbook teaches students to think more like practicing physicists, who approach a topic by solving a variety of types of problems — *Algebraic*, *Conceptual*, *Estimation*, *Graphical*, or *Numerical*—at once.

Engagement Solutions that Fit Your Teaching Goals and Your Students' Learning Needs

ENHANCED WebAssign

Whether you offer a more traditional lecture-based course or are interested in flipping the classroom, *Physics for Scientists and Engineers: Foundations and Connections* offers print and digital solutions for an active learning environment. The learning resources and analytics in Enhanced WebAssign will provide you with tools that will easily help you check student comprehension before class; create in-class activities and discussions; identify students who need help based on homework results; and compare homework scores with test results. Your students will gain confidence and improve their test scores with ongoing assessments and engagement activities throughout the entire course.

Pre-Lecture

Enhanced WebAssign for *Physics for Scientists and Engineers: Foundations and Connections*. Exclusively from Cengage Learning, Enhanced WebAssign offers an extensive online program for physics to encourage exploration and practice that's so critical for concept mastery. With Enhanced WebAssign you can assign content before the beginning of the class and use powerful analytics to monitor your students' comprehension and engagement. Options include:

- The **Cengage YouBook.** Students should read to succeed. WebAssign has a customizable and interactive eBook, the Cengage YouBook, that lets you tailor the textbook to fit your course and connect with your students. You can remove and rearrange chapters in the table of contents and tailor assigned readings that match your syllabus exactly. Powerful editing tools let you change as much as you'd like—or leave it just like it is.
- **Reading Check Questions.** These questions give your students ample opportunity to test their conceptual understanding before class.
- **PreLecture Explorations (PLE). This option uses** HTML5 interactive simulations enabling students to make predictions, change parameters, and observe results. Each PreLecture Exploration presents an engaging simulation based on a relevant scenario and then asks conceptual and analytic questions, guiding students to a deeper understanding and helping promote a robust physical intuition. This is the perfect resource for a flipped classroom or for professors looking for new ways to increase student engagement and interest in the material prior to lecture.

In-Class Group Discussions and Active Learning

Assign case study problems from the chapter in class or use the text's Student Engagement slides to encourage peer learning and group discussions. The Student Engagement slides, when used with Lecture Tools or any clicker device, will give your students a chance to think critically in the classroom and collaborate in an engaging environment.

Homework

Assessments throughout the semester give you ample opportunity to tailor your lecture and provide targeted help and feedback. Your Enhanced WebAssign course includes:

- **All of the quantitative end-of-chapter problems.**
- **Problem-Solving strategies for when your students need help the most:**
  - **Master It tutorials** help students work through the problems one step at a time.
  - **Watch It solution videos** explain fundamental problem-solving strategies, helping students step through the problem. In addition, instructors can choose to include video hints of problem-solving strategies.

- **Integrated Tutorials (IT)**, written by the author, strengthen students' skills by guiding them through the problem-solving steps identified in their textbook. Tutorials take students through the process, asking questions they will learn to ask themselves when faced with new problems. An example of an Integrated Tutorial in Enhanced WebAssign is shown below.

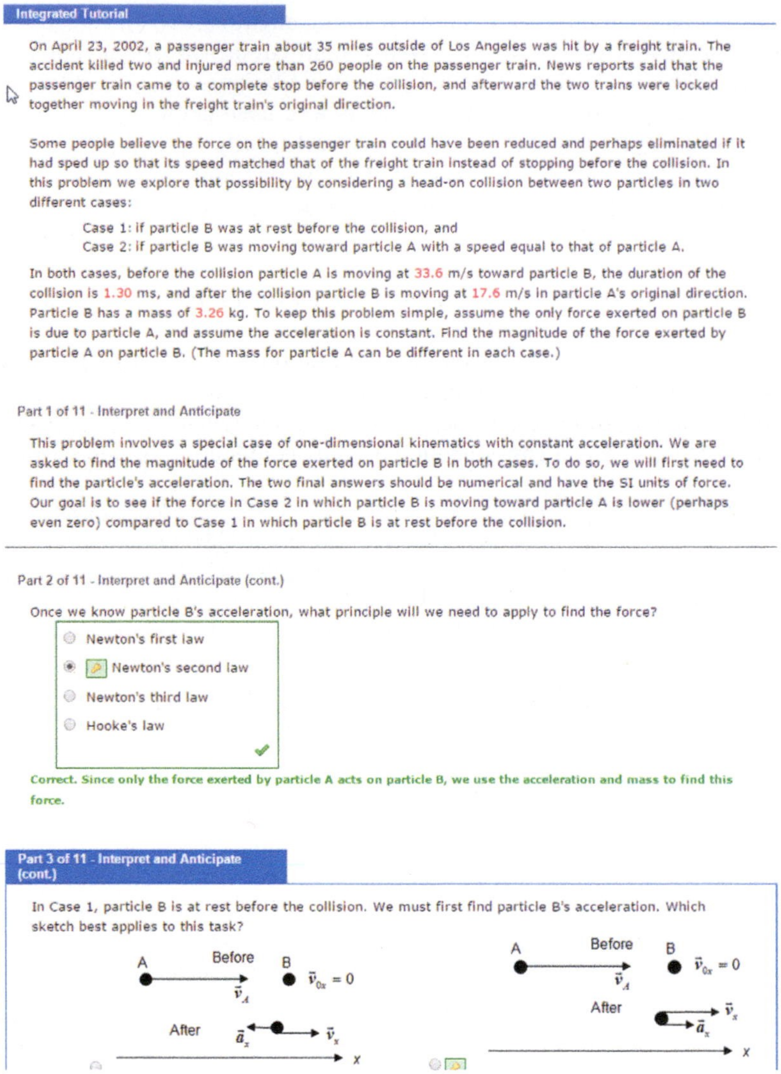

Integrated Tutorial

On April 23, 2002, a passenger train about 35 miles outside of Los Angeles was hit by a freight train. The accident killed two and injured more than 260 people on the passenger train. News reports said that the passenger train came to a complete stop before the collision, and afterward the two trains were locked together moving in the freight train's original direction.

Some people believe the force on the passenger train could have been reduced and perhaps eliminated if it had sped up so that its speed matched that of the freight train instead of stopping before the collision. In this problem we explore that possibility by considering a head-on collision between two particles in two different cases:

Case 1: if particle B was at rest before the collision, and
Case 2: if particle B was moving toward particle A with a speed equal to that of particle A.

In both cases, before the collision particle A is moving at 33.6 m/s toward particle B, the duration of the collision is 1.30 ms, and after the collision particle B is moving at 17.6 m/s in particle A's original direction. Particle B has a mass of 3.26 kg. To keep this problem simple, assume the only force exerted on particle B is due to particle A, and assume the acceleration is constant. Find the magnitude of the force exerted by particle A on particle B. (The mass for particle A can be different in each case.)

Part 1 of 11 - Interpret and Anticipate

This problem involves a special case of one-dimensional kinematics with constant acceleration. We are asked to find the magnitude of the force exerted on particle B in both cases. To do so, we will first need to find the particle's acceleration. The two final answers should be numerical and have the SI units of force. Our goal is to see if the force in Case 2 in which particle B is moving toward particle A is lower (perhaps even zero) compared to Case 1 in which particle B is at rest before the collision.

Part 2 of 11 - Interpret and Anticipate (cont.)

Once we know particle B's acceleration, what principle will we need to apply to find the force?

- Newton's first law
- Newton's second law
- Newton's third law
- Hooke's law

Correct. Since only the force exerted by particle A acts on particle B, we use the acceleration and mass to find this force.

Part 3 of 11 - Interpret and Anticipate (cont.)

In Case 1, particle B is at rest before the collision. We must first find particle B's acceleration. Which sketch best applies to this task?

After Class

Using Enhanced WebAssign, you can administer high-stakes assessments with security measures such as IP restriction, password protection, and LockDown Browser. But before the exam you should make sure your students are ready by assigning a Personalized Study Plan.

- **Personalized Study Plan.** The Personalized Study Plan in Enhanced WebAssign provides chapter and section assessments that show students what material they know and what areas require more work. For items that they answer incorrectly, students can click on links to related study resources such as videos, tutorials, or reading materials. Color-coded progress indicators let them see how well they are doing on different topics. You decide what chapters and sections to include—and whether to include the plan as part of the final grade or as a study guide with no scoring involved.

Please visit **http://www.webassign.net/features/textbooks/katzpse1/details.html** to view an interactive demonstration of Enhanced WebAssign.

Supporting Materials

Please visit **Cengage.com** for information about student and instructor resources for this text, including custom versions and laboratory manuals.

Acknowledgments

Before writing this textbook, I thought books were written by their authors. Period. Of course, they are *written* by their authors, but writing is a small part of *creating* a book. Books are written, edited, reviewed, revised, and tested. Figures are scribbled by the author, marked up and, in many cases, redrawn by the editors before they are sent to professional graphic artists, and then they are further reviewed, tested, and revised. Photos are imagined by the author, then sometimes taken by professional photographers, but often by an editor, or the author's spouse, or a friend or a colleague. Of course, I really want this book to touch the lives of students and instructors, and it is simply impossible to imagine how I could have ever gotten this book into your hands without help. So this book exists and is being read by you in large part due to the tremendous effort of my publishing team, my contributors, my reviewers, my colleagues, my friends, and most of all my family. I cannot possibly thank each individual who touched this book, but I will do my best to mention many here.

First, I want to thank the people at Cengage Learning: Mary Finch, Rebecca Berardy Schwartz, Nicole Hurst, Peter McGahey, Janet del Mundo, Alison Eigel Zade, Cathy Brooks, Bruce Bond, Cate Barr, Brandi Kirksey, Jonathan McDonald, John Wimer, Brendan Killion, Ed Dodd, and especially Susan Dust Pashos. I would also like to thank Irene Nunes, who worked with me in the early stages of development and Chris Hall, who showed great confidence in my abilities.

Second, as I am sure you are aware, textbooks have many components such as the end-of-chapter problems and questions, the online supplements, the student solutions manual, and the instructor's solutions manual. I express my sincere gratitude to the contributors listed below who made these components possible. Most of all, I want to express my deep-felt thanks to Eric S. Mandell for his tremendous fortitude, insight, and creativity.

Contributors of End-of-Chapter Problems

David Bannon, Oregon State University
Andrew Boudreaux, Western Washington University
Juan Cabenela, Minnesota State University, Moorhead
Mark James, Northern Arizona University
Ronald Jodoin, Rochester Institute of Technology
Tom Krause, Towson University
Kingshuk Majumdar, Grand Valley State University
Eric Mandell, Bowling Green State University
Vahe Peroomian, University of California, Los Angeles
Yuri Sikorski, Kettering University
John Stamm, University of Evansville
Brian Utter, James Madison University
Shannon Willoughby, Montana State University

Third, I love to learn, especially about physics and physics education. Writing this book has given me a great opportunity to talk with physics instructors from all around North America. I have learned so much from all of you. I thank all of you, including those who formally reviewed drafts of this book.

Reviewer Consultants

The following dedicated individuals provided valuable long-term guidance for one part of the book (one group of related chapters) through multiple stages of development.

Yildirim Aktas, University of North Carolina at Charlotte
Jason Brown, Clemson University
Andrew Cornelius, University of Nevada, Las Vegas
John DiTusa, Louisiana State University
James Dove, Metropolitan State University of Denver
Scott Dwyer, Rensselaer Polytechnic University
Thomas Hemmick, Stony Brook University
Robert Johnson, University of Pennsylvania
Sally Koutsoliotas, Bucknell University
Jorge Lopez, University of Texas at El Paso
Rafael Lopez-Mobilia, University of Texas at San Antonio
Matthew Mackie, Temple University
James Olsen, Princeton University
Amy Pope, Clemson University
Ramon Ravelo, University of Texas at El Paso
Michael Richmond, Rochester Institute of Technology
Joseph Rothberg, University of Washington
Joseph Scanio, University of Cincinnati
Douglas Sherman, San Jose State University
Chuck Stone, Colorado School of Mines
Catalin Teodorescu, Montgomery College
Jeff Winger, Mississippi State University
Michael Ziegler, Ohio State University

Board of Advisors

These instructors provided guidance during early stages of development, especially while the two-column design for worked examples and derivations was formulated.

Mirela Fetea, University of Richmond
Thomas Keil, late of Worcester Polytechnic Institute
Eric Mandell, Bowling Green State University
Amy Pope, Clemson University
Michael Richmond, Rochester Institute of Technology
Jeff Winger, Mississippi State University

Manuscript Reviewers

Anthony Aguirre, University of California, Santa Cruz
Robert Balogh-Robinson, Marist College
David Bannon, Oregon State University
Arun Bansil, Northeastern University
Peter Barnes, Clemson University
Linda Barton, Rochester Institute of Technology
Dave Besson, University of Kansas
Ken Bolland, Ohio State University
Andrew Boudreaux, Western Washington University
Ryan Case, Elizabethtown Community and Technical College
Amit Chakrabarti, Kansas State University
Raymond Chastain, Louisiana State University
Liao Chen, University of Texas at San Antonio
Krishna Chowdary, Evergreen State College
David Cole, Northern Arizona University
Doug Copely, Sacramento City College
Andrew Cornelius, University of Nevada, Las Vegas
Stephane Coutu, Pennsylvania State University
Grant Denn, Metropolitan State University of Denver
Stephanie Diemel, Shoreline Community College
Susan DiFranzo, Hudson Valley Community College
John DiTusa, Louisiana State University
James Dunne, Mississippi State University
Dipangkar Dutta, Mississippi State University
Mirela Fetea, University of Richmond
Christopher Fischer, University of Kansas
Elena Flitsiyan, University of Central Florida
Rica Sirbaugh French, MiraCosta College
Alejandro Garcia, University of Washington
J. D. Garcia, University of Arizona
Delena Gatch, Georgia Southern University
Christopher Gould, University of Southern California
Benjamin Grinstein, University of California, San Diego
D. Eitan Gross, University of Arkansas
Kevin Haglin, St. Cloud State University
Nathan Harshman, American University
Thomas Hemmick, Stony Brook University
David Heskett, University of Rhode Island
Wendell Hill, University of Maryland, College Park
Jo Hopp, University of Wisconsin—Stout
Yung Huh, South Dakota State University
Anthony Hyder, University of Notre Dame
Robert Jacobsen, University of California, Berkeley
Mark James, Northern Arizona University
John Jaszczak, Michigan Technological University
Ronald Jodoin, Rochester Institute of Technology
Charlie Johnson, University of Pennsylvania
Darrin Johnson, University of Minnesota, Duluth
Philip Johnson, American University
Robert Johnson, University of Pennsylvania
Linda Jones, College of Charleston
Neda Katz, University of Southern California
Thomas Keil, late of Worcester Polytechnic Institute
Yangsoo Kim, Virginia Tech
David Klassen, Rowan University
Igor Kogoutiouk, Minnesota State University, Mankato
Zlatko Koinov, University of Texas at San Antonio
Ichishiro Konno, University of Texas at San Antonio
Dorina Kosztin, University of Missouri, Columbia
Michael Kotlarchyk, Rochester Institute of Technology
Theo Koupelis, University of Wisconsin—Marathon
Sally Koutsoliotas, Bucknell University
Monika Kress, San Jose State University
James LaBelle, Dartmouth College
David Lamp, Texas Tech University
Allen Landers, Auburn University
Pedram Leilabady, University of North Carolina at Charlotte
David Lind, Florida State University
Susannah Lomant, Georgia Perimeter College
Jorge Lopez, University of Texas at El Paso
Rafael Lopez-Mobilia, University of Texas at San Antonio
Daniel Ludwigsen, Kettering University
Kristina Lynch, Dartmouth College
Matthew Mackie, Temple University
Kingshuk Majumdar, Grand Valley State University
Eric Mandell, Bowling Green State University
Andrea Markelz, University at Buffalo
Peter Markowitz, Florida International University
Mark Mattson, James Madison University
Will McElgin, Louisiana State University
David McIntyre, Oregon State University
Marina Milner-Bolotin, University of British Columbia
Drew Milsom, University of Arizona
Rabindra Mohapatra, University of Maryland, College Park
Teruo Morishige, University of Central Oklahoma
Gerhard Muller, University of Rhode Island
Eric Murray, Georgia Institute of Technology
Taha Mzoughi, Kennesaw State University
Craig Ogilvie, Iowa State University
James Olsen, Princeton University
Halina Opyrchal, New Jersey Institute of Technology

Peter Persans, Rensselaer Polytechnic Institute
Doug Petkie, Wright State University
Amy Pope, Clemson University
Richard Quimby, Worcester Polytechnic University
Corneliu Rablau, Kettering University
Gloria Ramos, Citrus College
Ramon Ravelo, University of Texas at El Paso
Michael Richmond, Rochester Institute of Technology
Andreas Riemann, Western Washington University
John Rollino, Rutgers University—Newark
Joseph Rothberg, University of Washington
Baharam Roughani, Kettering University
Dubravka Rupnik, Louisiana State University
Mehmet Sahiner, Seton Hall University
Mahdi Sanati, Texas Tech University
Vladimir Savinov, University of Pittsburgh
Joseph Scanio, University of Cincinnati
Ann Schmiedekamp, Pennsylvania State University—Abington
Ben Shaevitz, Slippery Rock University
Kim Sharp, University of Pennsylvania
Douglas Sherman, San Jose State University
Ethan Siegel, University of Portland
Chandralekha Singh, University of Pittsburgh
Henry Smith, Northeastern University
Bryndol Sones, United States Military Academy
Phillip Sprunger, Louisiana State University
Gay Stewart, University of Arkansas
Chuck Stone, Colorado School of Mines
Jay Strieb, Villanova University
Tad Thurston, Oklahoma City Community College
Ionel Tifrea, California State University, Fullerton
Somdev Tyagi, Drexel University
Brian Utter, James Madison University
Ravi Vadapalli, Texas Tech University
Trina Van Ausdal, Salt Lake Community College
Joan Vogtman, Potomac State College of West Virginia
Keith Warren, North Carolina State University
Laura Weinkauf, Jacksonville State University
Edward Whittaker, Stevens Institute of Technology
Shannon Willoughby, Montana State University
Jeff Winger, Mississippi State University
David Young, Louisiana State University
Michael Ziegler, Ohio State University
Bernard Zygelman, University of Nevada, Las Vegas

Fourth, I thank my friends and colleagues at the United States Naval Academy. You are an inspiration. At the risk of leaving someone out, I must at least mention a few colleagues who helped me with some tricky physics: C. Elise Albert, Peter G. Brereton, Charles A. Edmonson, John P. Ertel, Irene Engle, Jeffrey A. Larsen, Paul T. Mikulski, and Carl E. Mungan. I also thank colleagues who are my role models: Francis David Correll, Robert Shelby, and Donald Treacy.

Fifth, I thank my friends who helped keep me going both emotionally and intellectually: Sallie Gentry, J. Allie Hajian, Milena Higgins, Max and Tess Light, Heidi Manning, Stephen Winchell, and Hui Yang, and the way too many people to name from Ridgley Retreat. Adam Geremia, you've been a great friend, and your photo saved one of my favorite case studies from the chopping block; thanks. I also thank Lawrence Rudnick—my Ph.D. thesis adviser—who taught me how to teach, how to learn, how to write, and how to think. Finally, thanks to Flo and Mary for providing me with a writer's paradise in Northampton.

Most importantly, thanks to my family for support and for allowing me to take the enormous amount of time I needed to work on this book. How can I ever express how grateful I am to Jeff and Zak for allowing me to pursue this dream? Special thanks to Jeff—the best husband on the planet—and also my ghost editor. To the rest of my family: I am not mentioning you by name here because if you read my book carefully, you will find your names throughout the book as my thank-you to you. I love you all!

Sincerely,
Debora

Preface for the Students

If you have taken the time to read this, I am truly proud of and impressed by you! You are off to a great start. Physics is a difficult subject for most students, but *reading your textbook will help you to succeed.* Many textbooks were originally written for previous generations of students. (Now these students are old folks like me.) Although these textbooks have been updated for content, today's students don't find them to be readable. My students easily read 100 pages of history but find it difficult to get through five pages of one of these old physics textbooks. My primary goal in writing this book was to write something that *you* would actually enjoy reading—something that you could learn from.

I may not have met your physics instructor, but I can tell you one thing about him or her. He or she *loves* physics because physics is fun. Physics allows you to answer so many interesting questions. Your instructor wants you to find the joy in applying physics to questions that you care about.

Our job—mine as the author of this book, your instructor's as your mentor, and *yours* as the only person who can teach you anything—is to get you to *think like a physicist*. In other words, we want you to learn how to ask and answer questions using physics. So you will spend a lot of your time in this course working on assigned questions and problems. You must *actively* answer and solve these problems. Physics is not learned passively. But don't worry: you are not alone. Your instructor will find good problems for you to work on, and you will have other people to help you when you're stuck. And you have this textbook. Recent editions of traditional textbooks have begun to include problem-solving strategies. However, many steps in their example problems are still not well explained, and often seem mysterious and opaque to students. I have presented my example problems and major derivations using a two-column format. The column on the right contains the formalism you would typically find in a traditional textbook, or what an instructor might display in a classroom lecture. The left column includes my thoughts. *I have done my best to let you see inside my head while I solve example problems*. Use these two-column examples to see how a physicist thinks.

I have included other features to help you learn physics. For example, it turns out that most physicists (for example, your instructor and me) are *linear learners.* We are perfectly happy learning one detail after another without first seeing the big picture. Most physics students are *global learners.* You probably want the big picture before you are willing to get into the details. So the first page of every chapter begins with an outline and a list of the topics to be covered in the chapter.

Another difference between physics instructors and their students is that instructors see the physics topics arranged in a hierarchy. To them, some topics are more general, and therefore more important, than others. But students have trouble organizing physics topics in this way. So the list of the topics at the beginning of each chapter is arranged into categories. Then throughout each chapter, you will find notes in the margins that refer to the topics using these categories. Finally, each chapter has a summary at the end, before the problems and questions. The summary is arranged using the same hierarchy.

Throughout each chapter are *concept exercises*. These are short exercises you should do to break up the reading. They will help you make sure that you have understood what you have just read. The answers to these *concept exercises* appear near the end of the book, where you will also find the answers to a number of selected end-of-chapter problems and questions.

Students are often surprised to find that physics requires much of the mathematics they have learned in earlier classes. If you have forgotten some of this math, check Appendix A ("Mathematics"). I have included a short review of mathematical topics and formulas. You will find that Appendix B ("Reference Tables") is full of handy lists, such as a list of Greek letters (yes, we use them a lot in physics), and lists of conversion factors and data.

Some of you will find physics to be an easy subject. That is fantastic. Perhaps you will go on to get a Ph.D. in physics, and to solve some very cool problems. Others will find physics a tough subject. I cannot change how you feel about physics, but I hope I have written a book that will help you learn physics—and that you come to find pleasure in studying a subject that has brought me much joy and happiness.

All my best to you!

Debora

PART ONE
Classical Mechanics

Getting Started

1

Key Question

What is physics, and how can I best study it?

Underlying Principle

Physics is the fundamental natural, experimental science.

Major Concepts

1. Theory
2. Scientific evidence
3. Standard unit
4. The SI system
5. Uncertainty and error
6. Significant figures
7. Mass density

Tools

1. Conversion factor
2. Dimensional analysis
3. Order-of-magnitude estimate

How does your brain tell your body to run? Will the Universe expand forever, or will it collapse down to a single point? How do bicycles, airplanes, and rockets work? Why is the sky blue? How did the Earth form? Why can a cockroach survive a fall off a refrigerator? Why do ballerinas and basketball players seem to hover in midair? These questions are just a few of the ones that physics can answer (Fig. 1.1).

1-1 Physics

Physics is a natural science; that is, it deals with natural phenomena, as opposed to a social or political science that deals with human society or human governments. Not all artificial creations are beyond the realm of physics, though. Physicists build spacecraft, fight cancer, and design better bicycles. Physics is also called a physical science. This wording may seem redundant, but it signifies that the goal of physics is to discover the laws governing the *physical* Universe. These laws are not invented by people. For example, not being allowed to drive your car at 100 miles per hour (mph) on an open highway is a law invented by people, but not being able to pedal your ordinary bicycle at 100 mph is determined by the laws of the physical Universe.

There are other natural sciences, such as astronomy, biology, chemistry, and geology. Physics is *the* fundamental natural science because it examines the principles that apply to *all* parts of the physical world, whereas other sciences focus on a more limited part of the physical world. Biology, for example, focuses on living organisms.

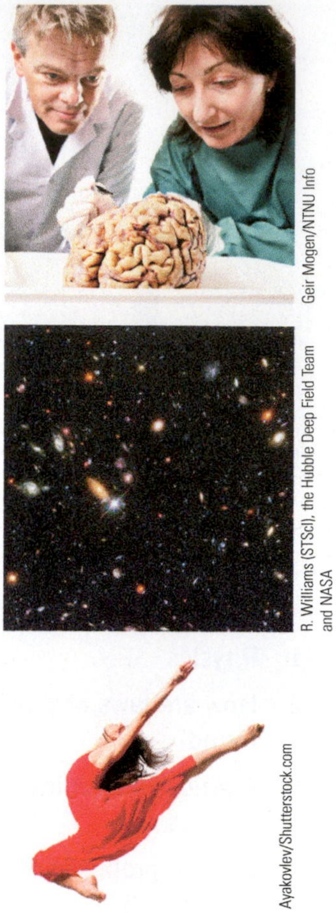

FIGURE 1.1 **A.** How does your brain tell your body to run? **B.** Will the Universe expand forever or recollapse to a single point? **C.** Why do ballerinas seem to hover in the air?

Because physics is such a broad topic, it is helpful to break it into several branches or fields. Each field roughly corresponds to one or two *principles* or *laws* of nature. A **physical principle** or **law** is a rule that governs some behavior or property of the physical Universe. Traditionally, there is a slight distinction between them: Laws are principles that have withstood many experiments and observations. Some principles, however, earn the title of *law* and are later found to have some limitations, but they do not get demoted back to *principle*. Throughout this book, we use the terms *principle* and *law* interchangeably. Understanding the laws of physics is one of your major goals.

You will also learn many *concepts*. A **concept** is an idea that makes it possible to describe the physical world clearly. For example, acceleration (Chapter 2) is a *concept*, but Newton's second law of motion (Chapter 5) is a *principle* that explains how an object accelerates.

This book is divided into six major parts, with each part focusing on just a few principles.

Part I focuses on describing and explaining how objects move. This branch of physics is known as **classical mechanics**. Classical mechanics is governed by Newton's laws of motion.

Part II continues the study of motion. The focus in this part is on more complicated systems such as fluids and more complicated motion such as oscillations and waves. The laws of **thermodynamics** provide the basis for studying complicated systems such as gases.

Part III focuses on **electricity** (how charged particles interact). Electricity is governed by Coulomb's law and Gauss's law.

Part IV shifts to **magnetism**, the study of the interaction between moving charged particles. Magnetism is based on Ampère's law and is so closely related to electricity that physicists consider these two phenomena as one (**electromagnetism**). Faraday's law provides part of the connection between electricity and magnetism.

Part V shows that the connection between electricity and magnetism explains **light** or **radiation**. Maxwell's equations govern electricity, magnetism, and radiation.

Part VI briefly describes two important branches of physics discovered in the 20th century, including **relativity** and **quantum mechanics**. Relativity is divided into two principles. Special relativity describes the motion of objects moving at very high constant speeds. General relativity expands the principles of relativity to include acceleration. Quantum mechanics is important to understanding very small objects, such as **atoms**. Both quantum mechanics and relativity are important to **nuclear physics** and **cosmology** (study of the Universe), and both of these 20th century principles challenge the principles of classical mechanics.

1-2 How Are Laws of Physics Found?

THEORY Major Concept

Physical principles are discovered (not invented) by people. A principle starts off as someone's idea or theory. How do you know when one of your theories is a law of the Universe? A scientific **theory** makes testable predictions. For example, in the early 1500s, Nicolaus Copernicus, a church canon, economist, and physician, theorized that the planets move in circular orbits around the Sun. Copernicus used his theory to make several predictions about the location, brightness, and phase of the planets at certain times. The Copernican theory roughly matched observations of the planets, but it failed to predict their location precisely (Fig. 1.2). This failure means that the theory of circular orbits is not a natural law. In fact, the German mathematician and astronomer Johannes Kepler (1571–1630) showed that elliptical orbits work much better at predicting the location of the planets than do circular orbits (Chapter 7).

Copernicus's model was not original. He revived an ancient Greek theory established by Aristarchus of Samos (310–230 BCE).

This example shows how a scientific theory can be refuted, but not how it can be proved. In fact, no scientific theory can be *proved*. Scientific theories are tested and retested. A theory that holds up under much testing is eventually accepted as a law. Later scientific evidence, however, may refute a theory, even one that was once considered a law.

SCIENTIFIC EVIDENCE Major Concept

Scientific evidence consists of measured observations. In the case of the Copernican theory of planetary orbits, the measured observation is the position of a planet with respect to the background stars (Fig. 1.2). In many branches of science, the observations are made in a laboratory. In that case, a carefully planned and controlled experiment is conducted with the purpose of producing evidence to support or refute a scientific theory. Because theories of physics are often tested in a laboratory, physics is called an *experimental science.*

Once a physical law has been discovered and tested by science, practical applications are often produced by engineers. In the 19th century, physicists discovered the principles of electricity and magnetism. Today, electrical engineers use those principles to design computers, communication devices, and personal music players, just to name a few applications. As we develop an understanding of physics, we will learn how some common and even a few exotic devices work.

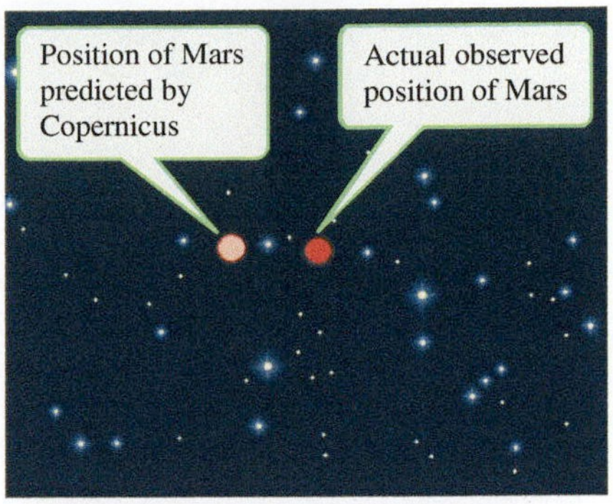

FIGURE 1.2 The Copernican theory of circular planetary orbits was refuted because it failed to predict the position of a planet such as Mars on a particular night and time. Tests of theories are repeated many times to allow for the possibility of observational errors. (The white dots represent fixed background stars.)

1-3 A Guide to Learning Physics

Learning physics is a lot like learning to drive a car. When you learn to drive a car, you learn *concepts* such as *indicating your intention to turn.* Then, you learn traffic *laws* that tell you when to indicate your turn. You *practice* those traffic laws when you drive on the road. The car is usually equipped with *tools* (such as a turn signal indicator) to help you. In physics, you will learn concepts and principles. You will practice physics when you solve problems and perform observations and experiments in a laboratory. The primary tool that will help you is mathematics.

Your everyday experience helped you master driving, and it will make learning physics easier too. Everyday experience is usually very complicated, however, and it is not easy to apply the principles of physics. Take the common experience of walking. What causes you to walk? What principles of physics are involved? Your experience tells you that your muscles cause you to walk, but that doesn't tell you why it is more difficult to walk on some surfaces (sand or ice) than on other surfaces (carpeted floors) or why people cannot walk on water. Part of learning physics is expanding on your experience and connecting that experience to physics principles and to mathematical tools. This book has features that will help you make those connections and use those tools.

Useful Features of this Book

Become familiar with the various useful features of this book.

Concept overview: Every chapter begins with an overview of the concepts, an outline listing the physical principles, new major concepts, and the key questions addressed in the chapter (page 1). The overview also tells you which new tools you will pick up. Finally, it will tell you if any special cases of physical principles or concepts are covered in the chapter. For example, in Chapter 2's overview (page 22), instantaneous acceleration is listed as major concept, and *constant* acceleration is listed as a special case.

Summaries: Every chapter ends with a summary. It is similar to the overview at the beginning of the chapter, but it briefly describes each principle, major concept, mathematical tool, and special case. The summary provides references to particular equations in the main part of the chapter to help you organize your knowledge before a quiz, test, or exam. The summaries are not a substitute for your own notes. The best summaries are made by individual students. Together, the overview at the beginning of a chapter and the summary at the end help you *organize* all the new material you will learn. The details come from your careful reading of the chapter.

Two-column worked examples: In the main body of each chapter are worked examples. Watching an expert solve a physics problem often seems baffling because the expert has trouble explaining what she is thinking. The left column of the worked examples provides those often-unstated thoughts. The column on the right is what the expert would likely write on the board. When you encounter an example, first try to solve the problem without looking at the solution. When you have made your best attempt, read both columns to see how you did. Together, the two columns should help you connect principles and concepts to their mathematical expression. It is okay if you do not get very far on your own. You will still learn a lot by carefully reading the solution. There will be many other opportunities for you to practice your problem-solving skills.

Concept exercises: Physics textbooks are known for being dense, with a lot of information on every page. Throughout each chapter, you will find several concept exercises. Their purpose is to give you a chance to digest some of what you just read. Take a few moments when you encounter a concept exercise to answer the questions. You may find it helpful to work on scratch paper, in your notebook, or even directly in the book. Answers are provided near the end of the book along with those for odd-numbered homework problems. It is very important to read the textbook before the material is discussed in class, and the concept exercises will help make sure your reading is effective.

Case studies: Each chapter has a CASE STUDY to help connect physics principles and concepts to everyday or laboratory experience. Many of the case studies involve discussions between fictional students based on things that real students have said. Use these discussions to uncover your own ideas about physics. You may find that you agree with one or more fictional students.

Some case studies are based on experiments that students like you have performed. Their data help make the explanations throughout the chapter more concrete. Still other case studies are based on historical experiments and discoveries. These help you connect your learning process to the greater scientific endeavor.

Advice from the Author

There are many ways to learn physics and succeed in class. Take some time to think about how you plan to study. If you are a very successful student, you may have never needed to do this kind of planning; university-level physics, however, often requires more effort than other subjects, so it is worth coming up with a plan. Here are some steps that have led many other students to success in physics.

Steps to success in physics

Step 1: On the first day of class, your instructor will probably give you a course outline or syllabus listing each reading assignment. Read each assignment *before* the material is covered in class. Not everything will make sense to you on this first reading. Your goals for the first reading are to (a) become familiar with the material and (b) identify questions or problem areas.

Step 2: Take notes on your reading and in class. Use the concept overview at the beginning of the chapter to help you organize your reading notes. Jot down questions you have and be sure to get your questions answered by your professor, another student, or your teaching assistant.

As you are reading, you may write notes in the textbook itself, on note cards, or in a notebook. Each method has advantages. If you make notes in the textbook, you are likely to find them again even years later. Yes, you should keep your textbook! If you plan to be a scientist or engineer, you need to build up your own personal library of books. That is part of what it means to get a college education. You will refer to this textbook for decades. However, using note cards has the advantage that they can easily be used as flashcards when you review for a test. Finally, if you write your reading notes in a notebook, leave room on each page for your classroom notes so all your notes are kept in one neat place.

Step 3: Working problems is an important part of any physics class. Even if you are not required to turn in homework problems for credit, you should still work several problems after class. After class, reread the material in the textbook. Keep your notes available and make any additions or changes as you reread. Then, work problems. In the next section, we will discuss a problem-solving strategy.

First Case Study

In subsequent chapters, the CASE STUDY often appears near the beginning of the chapter. The case study is referred to throughout the chapter and sometimes even in later chapters. After you read a case study for the first time, think about how you might approach the problem on your own.

CASE STUDY Raisin

Physicists take pride in being able to estimate the value of important quantities. In this case study, we will estimate the *density* of a single raisin from the photo (Fig. 1.3). We will return to this problem several times as we learn about mass, volume, density, and estimation.

FIGURE 1.3 A box of raisins, surrounded by everyday objects (for scale).

1-4 Solving Problems in Physics

Solving problems in physics, whether for homework or on a test, is often the hardest part of the class. Many students do not see the point of solving problems. They believe that they understand everything they have read in the book or have heard in class; they just cannot solve the problems. Many professors believe that problem solving is the primary way to learn physics and certainly one of the best ways to test a student's knowledge of the subject.

Physics professors have been known to say, "Physics is not a spectator sport." They mean that in a physics class you are expected to *do* physics, just as in a physical education class you are expected to *play* the game. By contrast, in your history course you are not expected to *make* history. You learn about history, you discuss history, and you write essays about history. In physics, you are expected to work problems and do experiments in a laboratory using the same sort of skills as a professional physicist.

Our goal is to teach you to think like a physicist. This goal is challenging for you, your instructor, and this textbook's author. Problem solving is an important part of meeting that challenge because when we solve problems our minds are active. Keeping with the sports analogy, problem solving is analogous to practicing. You cannot become a good athlete by watching sports on TV; you must practice on the field.

When you begin a sport, you are given detailed instructions on how to position and move your body. As you become a better player, you often modify what you have learned. In fact, after playing for a long time, an athlete often has trouble remembering what it was like to be a novice. The same is true for solving problems in physics.

To help students develop their problem-solving skills, some expert solvers have developed detailed guidelines, although the experts tend to stray from those guidelines when they work problems. The guidelines we use in this textbook are streamlined and mimic expert methods. Experts devise a strategy, execute a plan, and challenge their results, so in this book, we break problem solving into three procedures: (1) **INTERPRET and ANTICIPATE,** (2) **SOLVE,** and (3) **CHECK and THINK.**

Problem-solving steps

Procedure 1: INTERPRET and ANTICIPATE

The goal of this step is to come up with a **strategy** or plan based on knowing which physical principles are important to the question we must answer. There are two parts to meeting this goal.

The first is to **interpret** the question. What have we been asked to find? What physical principles are relevant to this question? It may be helpful to restate the question in your own words or to ask yourself if you have seen a similar problem before.

The second part is to **anticipate** the result. There are many questions we can ask to develop an expectation, such as the following:

a. Should the result be numerical or algebraic?

b. What are the proper units or what are the expected dimensions? (Units and dimensions are discussed in Sections 1-5 and 1-6.)

c. Can we make an educated guess at an approximate value or order of magnitude (Section 1-8) for the desired quantity?

Often, the interpretation of a problem involves creating some sort of visualization such as a sketch. Students often find **INTERPRET and ANTICIPATE** to be the hardest of the three steps. To help, we'll provide steps and visualization tools (such as diagrams and sketches) in some chapters for specific types of problems.

Procedure 2: SOLVE
In procedure 1, we come up with a strategy or plan. In procedure 2, we **execute** this plan and arrive at an answer. Often, this part involves using mathematical tools. Again, as we learn more physics we'll add some specific steps to this procedure.

Procedure 3: CHECK and THINK
This final procedure is crucial and the most interesting of the three. We must **check** our results. Do we believe that our answers are correct? We "check" by seeing if our expectations from procedure 1 match our results found in procedure 2. If they agree, we may be confident that our answer is correct. Of course, it is possible that we have made mistakes in both procedures, but this comparison is usually a good test.

Once we believe that our answers are correct, we need to stop and think about what we have just learned. What are the **implications** of our results? Often, students are so happy to get to the end of a problem that they forget to reap the benefits of all their hard work. For experts, this part is the most pleasing because at this point major discoveries are made. Examples are *The truck driver was exceeding the speed limit, My results predict the existence of neutron stars,* or *It should be possible to send wireless radio signals.*

Just as it is helpful to watch skilled athletes play when you are learning a new sport, it is helpful to work through expertly solved problems when you are learning physics. It is likely that your physics instructor will demonstrate his or her problem-solving techniques. You will also find many worked examples throughout this book written in a two-column format, using these three procedures. As you solve your own problems, be aware of your own thoughts. Jot down some notes to yourself to help you become aware of what you are thinking.

1-5 Systems of Units

STANDARD UNIT ✪ Major Concept

Physics relies on quantitative measurements. For example, if you are interested in learning about gravity near the surface of the Earth, you might drop a cannonball off the Leaning Tower of Pisa (Fig. 1.4) and measure the time it takes to land. To record your measurement, you need to decide what *standard time unit* to use. A **standard unit** is a precisely defined quantity to which measurements are compared. It is widely believed that the experimenter Galileo Galilei (1564–1642) sometimes used his own heartbeat as a standard time unit. If you use your own heartbeat, you would probably find that it takes roughly four heartbeats for the cannonball to land. Your heartbeat is not a good standard time unit, however, because (1) if anyone else wanted to compare the results of their experiment with yours, they would need to borrow your heart; and (2) your heartbeat is not regular (consistent). If you are sleeping, your heartbeat slows down, and if you exert yourself, it speeds up. A standard unit must be universal (reproducible by all observers) and consistent.

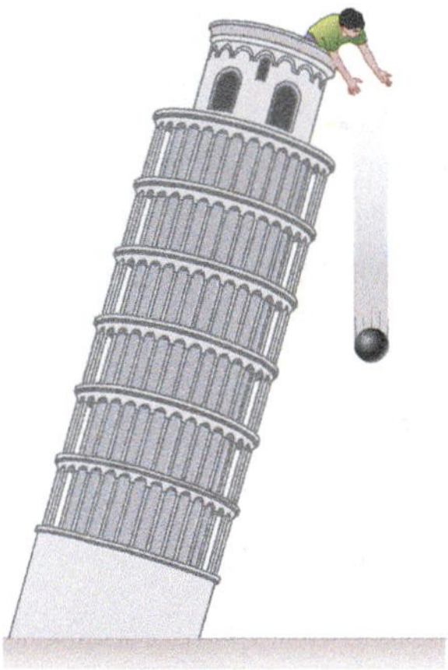

FIGURE 1.4 Galileo is famous for studying the effect of gravity through experimentation. You might be inspired to try your own experiment of dropping a cannonball off a building.

To develop a good system of standard units, an international committee agreed on a set of definitions and standards for comparison of measurements. The resulting unit system, known as the **SI system** (Système International d'Unités), is used worldwide in the scientific community and in this textbook. A complete list of the SI base units is provided in Appendix B. In this section, we consider the SI units for time, length, and mass, three of the fundamental quantities in physics.

Time

THE SI SYSTEM ✪ Major Concept

The SI unit of time is the **second**, abbreviated with a lowercase "s." Units are written in usual plain (roman) type, not in *italics*. The unit of time was formerly based on the rotation period of the Earth, but that standard is not accurate enough for today's experiments. The time standard is now determined by an atomic clock.

Atomic clocks measure the radiation from a particular atom. Atomic radiation has certain characteristic periods. The second is defined in terms of a certain characteristic period of cesium-133's radiation:

> 1 second is the duration of 9,192,631,770 periods of the radiation (corresponding to the transition between hyperfine levels of the ground state) of the cesium-133 atom.

Cesium clocks are kept at several locations, such as the U.S. Naval Observatory in Washington, D.C., and the National Institute of Standards and Technology (NIST) in Boulder, Colorado (Fig. 1.5). The time is broadcast on the Internet and by radio signal, so it is possible to access these clocks from remote locations. This time standard is much more precise than we will need in this textbook, and a typical stopwatch will do when we need to measure time intervals.

FIGURE 1.5 Cesium clock at the National Institute of Standards and Technology in Boulder, Colorado.

Length

The SI unit of length is the **meter** (m). The meter is defined in terms of the speed of light in a vacuum, $c \equiv 299{,}792{,}458\ \text{m/s}$:

1 meter is the distance light travels through empty space in 1/299,792,458 second.

Again, this standard is more precise than we will need, and a meterstick will serve our purposes for measuring length or distance.

Mass

The SI unit of mass is the **kilogram** (kg). The standard is a specific platinum–iridium alloy cylinder kept at the International Bureau of Weights and Measures in Sèvres, France (Fig. 1.6). The mass of that cylinder is defined as 1 kg. In Chapter 5, we discuss the difference between weight and mass. For now, we point out that the **pound** is the U.S. customary unit for weight, not mass.

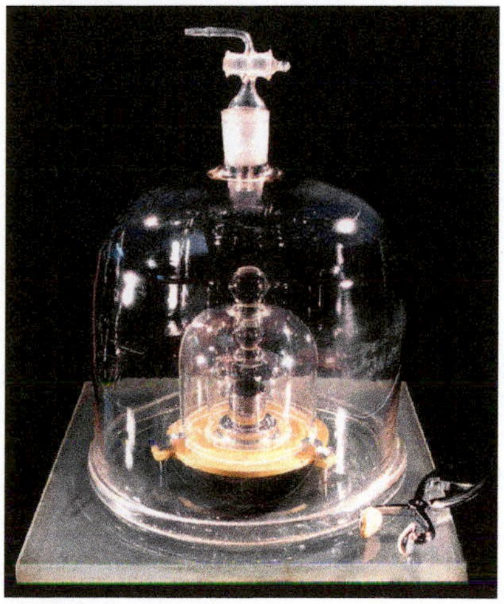

FIGURE 1.6 The mass standard is a platinum–iridium cylinder kept at the International Bureau of Weights and Measures in Sèvres, France.

Scientific Notation and Common Prefixes for SI Units

Many numbers in science are very large or very small. For example, the mass of the Earth is 5,980,000,000,000,000,000,000,000 kg, and the mass of a proton is 0.00000000000000000000000000167262158 kg. In both cases, a long series of zeros is needed as placeholders. All those zeros make the numbers difficult to read. Scientists use **scientific notation** to write numbers in a compact form in terms of powers of 10. So, the mass of the Earth is written as 5.98×10^{24} kg, and the mass of a proton is $1.67262158 \times 10^{-27}$ kg.

The power of 10 used to express a value in scientific notation may also be expressed with a prefix in front of the unit. The most common prefixes are in Table 1.1, and a more extensive list is in Appendix B. For example, $\ell = 9.3 \times 10^{-2}\,\text{m}$ may be expressed as $\ell = 9.3$ cm, where cm stands for centimeters.

TABLE 1.1 Commonly used power-of-10 prefixes.

Power of 10	Prefix	Abbreviation
10^{-12}	pico	Lowercase p
10^{-9}	nano	Lowercase n
10^{-6}	micro	Lowercase Greek μ
10^{-3}	milli	Lowercase m
10^{-2}	centi	Lowercase c
10^{3}	kilo	Lowercase k
10^{6}	mega	Uppercase M
10^{9}	giga	Uppercase G
10^{12}	tera	Uppercase T

CONCEPT EXERCISE 1.1

To practice using the prefixes in Table 1.1, complete the following puns.

a. What do you call 10^6 phones?
b. What do you call 10^{-12} lo?
c. What do you call 2000 mockingbirds?
d. What do you call 0.000001 fish?
e. What do you call 1,000,000,000,000 pins?

Converting Units

Although SI units are the primary standard used in science and engineering, other systems of units are in popular use.

When we know a quantity measured in some other system of units, we often need to convert that measurement to SI units. For example, in U.S. customary units, speed is measured in miles per hour (abbreviated as mph or mi/h). Suppose a speed limit on a highway is 65 mph and we need to convert that speed to SI units (meters per second, abbreviated m/s). To do so, we need a *conversion factor.*

The commonly used metric system is similar to SI, often using the prefixes in Table 1.1, but time may be expressed in hours or minutes.

A **conversion factor** comes from writing equal quantities in terms of a fraction equal to unity. For example, we know that 1 hour equals 3600 seconds:

$$1\ \text{h} = 3600\ \text{s}$$

CONVERSION FACTOR Tool

We find a conversion factor between hours and seconds if we divide each side by 3600 s:

$$\frac{1\text{ h}}{3600\text{ s}} = \frac{3600\text{ s}}{3600\text{ s}}$$

(We could have also come up with a conversion factor by dividing both sides by 1 h instead of 3600 s.) The units may be treated algebraically so that "seconds" cancel out on the right, and the conversion factor is

$$\frac{1\text{ h}}{3600\text{ s}} = 1$$

Because multiplying any number by unity does not change that number's value, we can multiply a measurement by a conversion factor without changing its value, only its units.

To convert 65 mph to the SI unit m/s, we need another conversion factor to convert from miles to meters. From Appendix B,

$$1\text{ mi} = 1609\text{ m}$$

So, the conversion factor is

$$\frac{1609\text{ m}}{1\text{ mi}}$$

Choose the numerator and the denominator of a conversion factor so that the original units cancel out and the desired units remain after you complete the multiplication.

We use both conversion factors to convert 65 mph to SI units:

$$65\text{ mi/h}\left(\frac{1\text{ h}}{3600\text{ s}}\right)\left(\frac{1609\text{ m}}{1\text{ mi}}\right) = 29\text{ m/s}$$

CONCEPT EXERCISE 1.2

CASE STUDY Mass of a Box of Raisins

Find the mass in SI units of the raisins shown in the box in Figure 1.3.

EXAMPLE 1.1 How Far Is a Light-Year?

A light-year (ly) is a unit of distance (length) commonly used in astronomy. A light-year is the distance light travels through a vacuum in 1 year. The speed of light in a vacuum is $c \equiv 299{,}792{,}458$ m/s. Find the number of meters in 1 light-year.

INTERPRET and ANTICIPATE

This example tells us how far light travels through a vacuum in 1 s, and we need the distance light travels in 1 yr. First, we convert 1 yr into seconds. Then, we multiply our answer by the distance light travels in 1 s to find the distance light travels in 1 yr. We expect a numerical result in the form 1 ly = _____ m.

SOLVE There are 365 days in 1 year, 24 hours in 1 day, and 3600 s in 1 hour.	$1\text{ yr}\left(\frac{365\text{ day}}{1\text{ yr}}\right)\left(\frac{24\text{ h}}{1\text{ day}}\right)\left(\frac{3600\text{ s}}{1\text{ h}}\right) = 3.154 \times 10^7\text{ s}$
Because light travels 299,729,458 m in 1 second, we multiply to find the distance light travels in 1 year.	$(3.154 \times 10^7\text{ s})(299{,}729{,}458\text{ m/s}) = 9.45 \times 10^{15}\text{ m}$ $1\text{ ly} = 9.45 \times 10^{15}\text{ m}$ (1)

CHECK and THINK

Our answer has the form that we expected. In the process of solving this example, we found that $1\text{ yr} = 3.154 \times 10^7\text{ s}$, which is an important conversion factor. It is easy to remember if you think of it as $1\text{ yr} \approx \pi \times 10^7\text{ s}$. Of course, $\pi = 3.14159\ldots$, which is a little smaller than 3.154, but often that slight difference is insignificant.

1-6 Dimensional Analysis

There are seven fundamental quantities that are mutually independent. In Part I, we encounter three of these fundamental quantities: time, length, and mass. (We'll learn about the other four in later chapters.) Quantities such as speed, volume, and density are *derived* from fundamental quantities and are known as **derived quantities.** For instance, to derive speed, we divide a length by a time; to derive volume; we multiply three lengths; and to derive a *mass density* (also known simply as *density*), we divide a mass by a volume. To get started and make this section less abstract, let us think about mass density. Density is symbolized by the lowercase Greek letter "rho" ρ. (For a list of Greek letters, see Appendix B.) Mathematically, **density** is expressed as

$$\rho = \frac{m}{V} \tag{1.1}$$

where m is the mass of an object and V is its volume.

We will encounter two meanings of the word *dimension* in physics. One definition of **dimension** is the type or category of a measured quantity. (We won't encounter the other definition until Chapter 2.) For example, distance may be measured in feet, meters, or miles, but all these measurements are units of length. So, the dimension of distance is length. We use the nonitalic uppercase symbols T, L, and M for the fundamental dimensions time, length, and mass, respectively. The dimensions of all the derived quantities in Part I can be written in terms of these three fundamental dimensions.

Dimensional analysis is a method in which the dimensions of a quantity rather than its value or other properties are used to tackle a problem. Physicists often use dimensional analysis to check a result. If your result does not have the expected dimensions, you have made a mistake. (Checking the dimensions of a result is just one way to test your work.) Dimensional analysis may also be used in the first problem-solving step to help interpret a problem.

DIMENSIONAL ANALYSIS Tool

We use the symbol $[\![Q]\!]$ to mean the dimensions (or units) of quantity Q. For example, the phrase *dimensions of density* may be written as $[\![\rho]\!]$.

Some quantities have no dimensions. We say that these quantities are *dimensionless.* For example, the ratio of two lengths is a dimensionless quantity.

Dimensional analysis is based on a few rules:

Rule 1: Dimensions may be treated as algebraic symbols. For example, to find the volume of a rectangular box such as the one in Figure 1.7, we must multiply its length by its width by its height:

$$V = \ell wh$$

The dimensions of volume are L^3:

$$[\![V]\!] = [\![\ell]\!][\![w]\!][\![h]\!] = (\text{L})(\text{L})(\text{L}) = \text{L}^3$$

Rule 2: Quantities can be added or subtracted only if they have the same dimensions.

Rule 3: The terms on both sides of an equation must have the same dimensions.

Rule 4: Trigonometric functions such as sine, cosine, and tangent apply only to (dimensionless) angular quantities, those measured in degrees or radians.

Rule 5: Special functions such as logarithms and exponential functions apply only to dimensionless quantities.

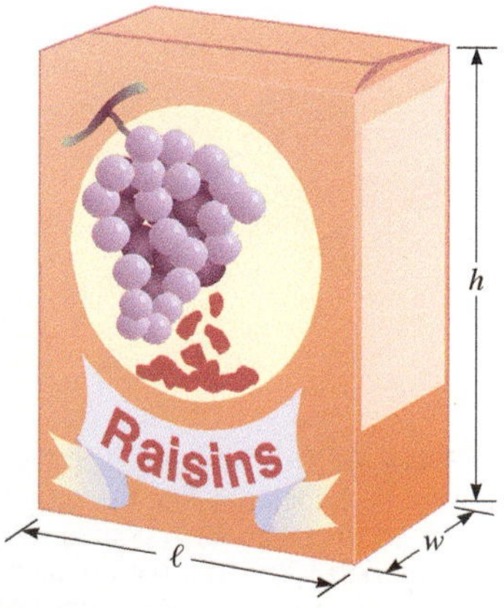

FIGURE 1.7 The volume of the box is $V = \ell wh$.

CONCEPT EXERCISE 1.3

In Einstein's famous equation $E = mc^2$, m stands for mass and c stands for the speed of light. Use this equation to find the dimensions of energy E.

CONCEPT EXERCISE 1.4

Use your results from Concept Exercise 1.3 to find which of the following quantities Q may represent energy. In these expressions, r is the radius of a circle and v is speed.

a. $Q = mv$ **b.** $Q = \frac{1}{2}mv^2$ **c.** $Q = m\frac{v^2}{r}$ **d.** $Q = mvr$

CONCEPT EXERCISE 1.5

CASE STUDY SI Units of Density

The goal of our case study is to find the density of a single raisin. Use dimensional analysis to find the SI units of density. Knowing these units will help us check the answer to the case study.

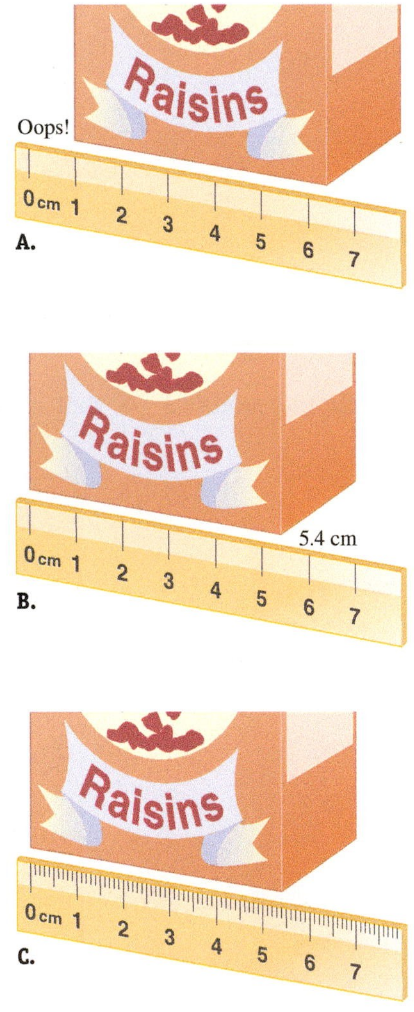

FIGURE 1.8 A. This is a mistake. The edge of the box does not line up with the zero mark on the ruler. **B.** A coarse ruler (few tick marks) leads to a large uncertainty in the measurement. **C.** A finer ruler (many tick marks) reduces the uncertainty in the measurement.

1-7 Error and Significant Figures

Because physics is an experimental science, physicists make a lot of measurements. No measurement is perfect. The terms **uncertainty** or **error** describe the imperfection of measurements. Normally, we think that an "error" is a mistake or fault, such as when a computer message says "Error: Document Failed to Print." A measurement error is not a mistake but rather is a natural result of the measurement process. Mistakes can be fixed. For example, if you are measuring the length ℓ of the raisin box (Fig. 1.8A), you might forget to align the edge of the box with the "0" on the ruler. You can correct that mistake by being careful and paying attention. On the other hand, measurement errors can be reduced but not completely eliminated. When reporting measured quantities, you must estimate the uncertainty of your measurements.

For example, if you are pulled over by a police officer who used a radar gun to measure your speed, the uncertainty in the measurement might make a difference in whether or not you get a ticket. If the speed limit is 65 mph and the radar detector measured your speed at 69 mph with an uncertainty of 1 mph, it is very likely that you were exceeding the speed limit. On the other hand, if the uncertainty in the measurement is 5 mph, the range of the measurement is between 64 and 74 mph, and it is possible that you were driving within the speed limit.

Let's see how to estimate uncertainty. Suppose you try to measure the length of the raisin box with a ruler that only has tick marks for centimeters as shown in Figure 1.8B. You find that the length is between the 5-cm and 6-cm tick marks. You estimate that the length is $\ell = 5.4$ cm, but it may be anywhere in the range of 5.3 to 5.5 cm. You estimate the uncertainty in your measurement to be ± 0.1 cm, and so you report your measurement as $\ell = (5.4 \pm 0.1)$ cm.

You can reduce the error in your measurement by using a ruler that has millimeter tick marks (Fig. 1.8C). You see that the length is between the 5.3-cm and 5.4-cm tick marks. You estimate that the length is $\ell = 5.35$ cm. This time, the uncertainty in your measurement is ± 0.05 cm. You report your measurement as $\ell = (5.35 \pm 0.05)$ cm.

By using the coarse ruler in Figure 1.8B, you can measure length to the nearest tenths place (one place to the right of the decimal), and by using the finer ruler in Figure 1.8C, you can measure the length to the nearest hundredths place. When you report a measurement, the number of digits you use to express the measurement indicates how precisely the measurement is known even if you do not explicitly report the uncertainty. In other words, the number of reported digits, known as the number of **significant figures,** implicitly expresses the uncertainty of the measurement. The last digit you report is uncertain. As another example, if a time is reported as $t = 12.89$ s, the measurement has four significant figures and the last digit is uncertain.

The number of significant figures used to express some quantity must be reported correctly. When using a calculator, it is tempting to write down all the digits returned by the calculator display. This practice, however, is incorrect because it represents our results as being much more certain than they really are. A few rules about significant figures help when making calculations with measured or reported quantities.

SIGNIFICANT FIGURES

Major Concept

There is some disagreement about how to interpret a measurement that is reported without explicitly stating the uncertainty. For example, if a length is reported as 36 cm, it is possible that both digits are known and that the measurement is between 35.5 and 36.5 cm. In this textbook, we will assume the last digit, in this example the "6", is uncertain.

Rule 1: When multiplying or dividing, report the result with the same number of significant figures as the least certain value. For example,

$$\frac{12.3}{4.6} = 2.7$$

because 4.6 has only two significant figures.

Rule 2: When adding or subtracting, the number of decimal places in the result should equal the smallest number of decimal places in any of the given terms. For example,

$$12.34 + 2.006 - 8.9 = 5.4$$

because 8.9 has only one decimal place.

Rule 3: Numbers that are not measured may be considered exact. Irrational numbers such as π and e are known to many significant figures and do not limit your results. For example,

$$\frac{1}{3}(4.56\pi) = 4.78$$

is reported to three significant figures because neither $\frac{1}{3}$ nor π is measured, and our answer is limited only by the three significant figures of 4.56.

Rule 4: It is best to use scientific notation because a zero that acts as a placeholder is not necessarily a significant figure. For example, $m = 390$ kg may have two or three significant figures. To avoid that ambiguity, you may add a decimal point; for example, $m = 390.$ kg has three significant figures. A better way to clarify the number of significant figures is to use scientific notation: $m = 3.90 \times 10^2$ kg has three significant figures, and $m = 3.9 \times 10^2$ kg has two significant figures.

Rule 5: You should keep extra significant figures in intermediate steps when making a calculation, but you should round the final answer to the correct number of significant figures. The extra significant figures in an intermediate result help avoid introducing an error due to rounding a number up or down. This step is particularly important if an intermediate result is a number ending in 5.

Rule 6: When your answer begins with a 1, it is okay to keep one extra significant figure (as long as none of the operands has a leading 1). For example,

$$\frac{5}{4.3} = 1.2$$

CONCEPT EXERCISE 1.6

How many significant figures does each number have? If the number is exact or if the number of significant figures is ambiguous, explain.

a. $\frac{1}{2}$ in the formula $r = \frac{1}{2}d$, where r is radius and d is diameter
b. 105
c. 150
d. 1.50×10^2
e. 1.5×10^2
f. 0.15×10^3

CONCEPT EXERCISE 1.7

Complete the following arithmetic. Report your answers using scientific notation and the correct number of significant figures.

a. $\frac{1}{2}(199) =$
b. $(9.81)\dfrac{4.5}{1.23 \times 10^{-3}} =$
c. $6.789 - 14.1 =$
d. $\dfrac{39.1}{7.75}(0.456 - 1.23) =$

1-8 Order-of-Magnitude Estimates

FIGURE 1.9 It takes longer to eat a nice bunch of grapes than to eat a small box of raisins, but the calorie content is the same.

At lunch with a group of physicists, many topics are discussed: current events, politics, and personal matters. In many ways, the conversation sounds like any lunchtime conversation, but one difference you may notice is that physicists pride themselves on being able to estimate many numerical quantities and will frequently do so at informal gatherings. An **estimate** is not a guess; it is a calculation based on a few roughly known values. For example, if the physicists' conversation turns to the long lines in the cafeteria, it is likely that at least one of them will estimate the total time he will spend in cafeteria lines over the course of his lifetime. His estimate is based on knowing the time he typically waits in line, the number of years he has eaten in cafeterias, and the number of years he expects to eat in cafeterias. Returning to the subject of our CASE STUDY, when comparing their lunches one physicist may remark that it is more efficient to eat raisins instead of grapes because raisins are denser than grapes and provide the same caloric intake. A small box of raisins has 130 calories, which is equivalent to a medium bunch of grapes (Fig. 1.9). Perhaps the physicists will estimate the density of a single raisin and compare it with the density of a single grape.

As a student, part of your job is to learn to think like a physicist, so you, too, must learn to make estimates. Some homework problems will specifically ask you to make an estimate. Estimation is important even if you are expected to find an exact result. Your estimate is another way for you to check your numerical answer.

You may wish to commit some roughly known numbers to memory. Some of these numbers come from common experience, and you may already know them. If remembering such numbers in U.S. customary units is easier for you, it is also worth memorizing a few common conversion factors. To help you think about the sort of values that are worth knowing, consider three categories: your body, your daily experience, and your education.

Your body: You should know several facts about your body besides your height and weight. For example, how long is your thumb? What is the area of your palm? How long is your walking stride? How fast can you run? What is your pulse rate? Table 1.2 provides some typical answers, but it would be handy for you to know these values for your own body.

TABLE 1.2 Typical values associated with a human body.

Quantity	U.S. Customary Units	SI and Metric Units
Length of thumb	2 in.	5 cm
Area of palm	6 in.^{2}	40 cm^2
Height	5–6 ft	1.5–2 m
Weight	110–200 lb	500–1000 N
Mass	4–7 slugs*	50–100 kg
Average stride	1 yd	1 m
Resting heartbeat	60–80 per min	1–1.25 per second
Running speed	6-minute mile (10 mph)	16 km/h

*U.S. customary unit of mass; see Appendix B.

Your daily experience: You should know facts about the things you experience on a regular basis. How much does your car or bicycle weigh? How tall is one story of a typical building? How much does your physics book weigh? What is the area of your cell phone? What is the time between rings on your phone? How long is a matchstick? How long do you spend eating breakfast? What is the area of your campus? Table 1.3 provides some typical numbers, but you should try to make the list specific to your experience, and you should also try to expand the list. Because these facts come from your daily life, you will find them helpful in making estimates for the rest of your life no matter how much or how little physics you do in the future.

TABLE 1.3 Typical values associated with common objects and events.

Quantity	U.S. Customary Units	SI and Metric Units
Top speed of a car	120 mph	200 km/h
Top speed of a *typical* bicycle	30 mph	50 km/h
Weight of a car	1–2 tons	10,000–20,000 N
Mass of a car	70–140 slugs*	1000–2000 kg
Weight of a physics book	10 lb	50 N
Mass of a physics book	0.4 slug	5 kg
Area of a cell phone	5 in.^2	30 cm^2
Length of a matchstick	2 in.	5 cm
Length of a housefly	0.2 in.	0.5 cm
Weight of a quarter	0.2 oz	6×10^{-2} N
Mass of a quarter	4×10^{-4} slug	6 g
Area of a dollar bill	17 in.^2	100 cm^2
Height of a typical story	10 ft	3 m
Area of U.S. Naval Academy	6.5 mi^2	15 km^2
Density of water	62 lb/ft^3	1000 kg/m^3
Density of ice	57 lb/ft^3	9.17×10^2 kg/m^3

*U.S. customary unit of mass; see Appendix B.

Your education: Many calculations you will make in physics involve objects and phenomena outside the range of daily experience. In your school or professional career, however, you may often use certain facts that you learned in an academic setting. Table 1.4 and Appendix B provide some facts to help build up your intuition. As you solve more physics problems, take the time to add new facts to your list.

TABLE 1.4 Approximate values of selected interesting facts.

Quantity	Value in Convenient Units
Age of the Universe	14 billion years
Age of the Earth	4.5 billion years
Time for light to travel from the Sun to the Earth	8 minutes
Time for light to cross the diameter of a proton	3.3×10^{-24} s
Diameter of the Milky Way galaxy	10^5 ly
Size of the smallest visible dust particle	0.1 mm
Size of a living cell	10 μm
Diameter of a hydrogen atom	10^{-10} m
Diameter of a proton	10^{-15} m
Mass of the Milky Way galaxy	10^{42} kg
Mass of an elephant	5×10^3 kg
Mass of a frog	100 g

An estimate is based on roughly known values. To indicate that a result is approximate, we use the mathematical symbol "≈" instead of an equals sign. Suppose you want to know your height in terms of the length of your thumb. If you are 5.5 feet tall,

$$h = 5.5\ \text{ft}\left(\frac{12\ \text{in.}}{1\ \text{ft}}\right) = 66\ \text{in.}$$

$$\left(\frac{h}{\ell_{\text{thumb}}}\right) = \frac{66\ \text{in.}}{2\ \text{in.}} = 33 \approx 3 \times 10$$

Our final estimate is reported to one significant figure because the length of your thumb is only known to one significant figure. So, your height is roughly 30 thumb lengths.

When only the correct power of 10 is known, we say that the value has zero significant figures and the resulting estimate is only expected to provide the correct power of 10. **Order of magnitude** is another term for the power of 10; an estimate based on values with zero significant figures is sometimes referred to as an **order-of-magnitude calculation** or **order-of-magnitude estimate.** The mathematical symbol "$\sim$" is used instead of an equals sign to indicate that a result came from an order-of-magnitude estimate. If we model the Milky Way galaxy as a disk, we can use the diameter provided in Table 1.4 to find an order-of-magnitude estimate for its area:

ORDER-OF-MAGNITUDE ESTIMATE

Tool

$$A = \pi r^2 = \pi\left(\frac{10^5 \text{ ly}}{2}\right)^2 = 8 \times 10^9 \text{ ly}^2 \approx 10 \times 10^9 \text{ ly}^2$$

$$A \sim 10^{10} \text{ ly}^2$$

In this case, because 8 is close to 10, the order-of-magnitude estimate is 10^{10} ly^2, not 10^9 ly^2.

EXAMPLE 1.2 A Great Way to Save Money

A Estimate the money you spend in a school year on your favorite drink purchased from a vending machine or cafe.

FIGURE 1.10 How much money do you spend on your favorite drink in a school year?

wdeon/iStockphoto.com

INTERPRET and ANTICIPATE

You probably know roughly how much money you spend on your favorite drink on a typical school day. After taking into account vacation days and weekends, you can come up with a weekly average. Multiply by the number of weeks you are in school each year to find the money you spend during that time on your favorite drink. Your answer will depend on your particular habits. Our calculation here is based on the habits of an actual physics professor who drinks half a dozen diet sodas during a workday. At home on the weekends, he does not buy soda from vending machines, but during vacations such as spring break, he works five days per week in his office and drinks 8 sodas per weekday. At his university, winter break is 3 weeks long, spring break is 1 week long, each semester is 16 weeks long, and there are two semesters in a school year. We expect a numerical answer of the form $____.

SOLVE

N_{work} is the number of sodas consumed in a typical work week.	$N_{\text{work}} = (6 \text{ sodas/weekday})(5 \text{ weekdays})$ $N_{\text{work}} = 30 \text{ sodas}$
If P is the price of a single soda, $X = PN_{\text{work}}$ is the amount spent on soda in a work week. This particular professor spends \$1.50 on each soda from the vending machine.	$P = \$1.50/\text{soda}$ $X = PN_{\text{work}} = (\$1.50/\text{soda})(30 \text{ sodas})$ $X = \$45 \text{ (work week)}$
Follow a similar procedure to find the amount Y spent on soda during a vacation week.	$N_{\text{vaca}} = (8 \text{ sodas/weekday})(5 \text{ weekdays}) = 40 \text{ sodas}$ $Y = PN_{\text{vaca}} = (\$1.50/\text{soda})(40 \text{ sodas})$ $Y = \$60 \text{ (vacation week)}$
We must find how many weeks this professor spends X and Y on soda during the school year. (Summers don't count because the problem specified "the school year.") Let Z = total amount spent on diet soda.	$Z = (\text{number of work weeks})X + (\text{number of vacation weeks})Y$ (1)
In our professor's case, each semester is 16 weeks long, and there are two semesters in a school year.	$\text{number of work weeks} = (2 \text{ semesters})\left(\frac{16 \text{ weeks}}{\text{semester}}\right)$ $\text{number of work weeks} = 32$

At the professor's school, winter break is 3 weeks long and spring break is 1 week long.	number of vacation weeks = 3 + 1 = 4
Substitute the relevant numbers into Equation (1).	$Z = (32)(\$45) + 4(\$60)$ $Z = \$1680$
Although the cost of a single soda from the vending machine is known to three significant figures, the number of sodas consumed per week is only known to one significant figure. The answer has a leading "1", but none of the operands does, so we keep one extra significant figure.	$Z \approx \$1700$

CHECK and THINK

Our result is in the form we expected. We will think more about the implications after we finish part B.

B About how much money would be saved in one school year if the beverages in part A were purchased from a grocery store?

INTERPRET and ANTICIPATE

We assume the number of sodas consumed per school year remains unchanged, but the cost per soda is lower if the sodas are purchased in a grocery store. We need some everyday experience with grocery store soda prices. There should be a positive savings of the form $_____.

SOLVE Find the number n of sodas the professor in part A consumes in a school year. We said he spent \$1.50 per soda for a total of \$1680 per year.	$n = \dfrac{\$1680}{\$1.50/\text{soda}}$ $n = 1120$ sodas
A package of 12 cans of soda costs around \$4 in a grocery store. To find the total cost z if the professor purchased his sodas at the store, multiply the number of sodas by the cost of a single soda.	$z = 1120\text{ sodas}\left(\dfrac{\$4}{12\text{ sodas}}\right)$ $z = \$373$ $z \approx \$400$
His yearly savings S is found from the difference between Z (vending machine cost) and z (grocery store cost).	$S = Z - z$ $S \approx \$1700 - \$400$ $S \approx \$1300$

CHECK and THINK

As expected, purchasing sodas from the grocery store saves money, but just how much might seem surprising. The cost of one soda purchased in a grocery store is approximately \$4/12 = \$0.33. So, the savings on just a single soda that costs \$1.50 from a vending machine is more than a dollar (\$1.17). That is about a 78% savings, and 78% of \$1700 is \$1300, consistent with our results. The professor has made this calculation and now buys his soda from grocery stores.

EXAMPLE 1.3 A Grape Way to Spend Your Time

Estimate the density of a grape.

INTERPRET and ANTICIPATE

Make this estimate by drawing on everyday experience. For example, you might estimate the number of grapes in a kilogram, about 2.2 pounds. We will approach the problem by estimating the volume and mass of a single grape. Our answer should be numerical, in the form ____ kg/m^3. A grape is mostly water, so we expect our result to be close to the density of water (1000 kg/m^3; see Table 1.3).

Example continues on page 16 ▶

SOLVE Assume the grape is essentially spherical. The diameter of a grape is about half the length of a person's thumb. Use this fact and Table 1.2 to find the radius of a grape.	$r = \frac{2.5\text{ cm}}{2} = 1.25\text{ cm}$ $r = 1.2 \times 10^{-2}\text{ m}$
Find the volume V. (See Appendix A for the volume of a sphere.)	$V = \frac{4}{3}\pi r^3 = \frac{4}{3}\pi(1.2 \times 10^{-2}\text{ m})^3$ $V \approx 7 \times 10^{-6}\text{ m}^3$
Estimate the mass of a grape by holding a grape in the palm of one hand and a quarter in the palm of your other hand. Their weights are about the same, and so are their masses. (Table 1.3 lists the mass of a quarter.)	$m = 6\text{ g} = 6 \times 10^{-3}\text{ kg}$
Use Equation 1.1 to find the density of a grape to one significant figure.	$\rho = \frac{m}{V} = \frac{6 \times 10^{-3}\text{ kg}}{7 \times 10^{-6}\text{ m}^3}$ $\rho \approx 9 \times 10^2\text{ kg/m}^3$

CHECK and THINK

Our result has the expected form and is the same order of magnitude as the density of water. This estimate will be important for the next example. We expect a raisin to be denser than a grape.

EXAMPLE 1.4 CASE STUDY Raisin Your Estimation Skills

Return to the case study (page 5 and Fig. 1.3) and estimate the density of a raisin. Use the mass of the box of raisins found in Concept Exercise 1.2.

INTERPRET and ANTICIPATE

Our job is to gather data from Figure 1.3 to estimate the density of a single raisin. We know the mass of the box (4.25×10^{-2} kg). We need to find the volume of the raisins. If the box is packed tightly, the volume of the box equals the volume of the raisins. We must estimate the volume of the box. We expect to find a numerical solution of the form _____ kg/m³. We expect the density of a raisin to be greater than the density of a grape.

SOLVE Several objects in the photo provide a reference for finding the size of the raisin box. We have decided to use the box of matches, but you may use the cell phone or the battery. Estimate the length ℓ, width w, and height h in terms of the length of a matchstick.	$\ell \approx 1$ match-length $w \approx \frac{1}{3}$ match-length $h \approx 1.3$ match-lengths
According to Table 1.3, one matchstick is about 5 cm or 0.05 m long. Use this estimate to convert ℓ, w, and h to meters.	$\ell \approx 1\,(0.05\text{ m}) \approx 5 \times 10^{-2}\text{ m}$ $w \approx \frac{1}{3}\,(0.05\text{ m}) \approx 1.7 \times 10^{-2}\text{ m}$ $h \approx 1.3\,(0.05\text{ m}) \approx 6.5 \times 10^{-2}\text{ m}$
Calculate the estimated volume. Because the length of a matchstick is only known to one significant figure, the volume should have only one significant figure. We keep an extra significant figure until we have calculated the density.	$V = \ell wh$ $V \approx (5 \times 10^{-2}\text{ m})(1.7 \times 10^{-2}\text{ m})(6.5 \times 10^{-2}\text{ m})$ $V \approx 5.5 \times 10^{-5}\text{ m}^3$
Find the density from Equation 1.1 and the mass, 4.25×10^{-2} kg. Because the volume only has one significant figure, the density estimate is reported to one significant figure.	$\rho = \frac{m}{V}$ (1.1) $\rho \approx \frac{4.25 \times 10^{-2}\text{ kg}}{5.5 \times 10^{-5}\text{ m}^3} \approx 8 \times 10^2\text{ kg/m}^3$

CHECK and THINK

We have a numerical answer in the correct SI units, but our estimated density is lower than we expected. We thought that the density should be greater than the density of a grape. What went wrong? Our estimate was based on the assumption that the box was tightly packed full of raisins, but the manufacturer actually packs the boxes loosely. The raisins in Figure 1.11 have been tightly packed down. (No raisins were removed from the box.) Now we can see that the volume of raisins is about half the volume of the box.

Dr. Jenny Spinner, St. Joseph's University

FIGURE 1.11 A box with tightly packed raisins.

SOLVE

Let's now assume the volume of the box should be half of the old volume estimate.

$$V_{new} \approx \frac{5.5 \times 10^{-5}\ \text{m}^3}{2} \approx 2.8 \times 10^{-5}\ \text{m}^3$$

Use the new volume estimate to find a new estimate for the density of a raisin. By rule 6 on page 13, it is okay to keep one extra significant figure when your answer begins with a 1.

$$\rho_{new} \approx \frac{4.25 \times 10^{-2}\ \text{kg}}{2.8 \times 10^{-5}\ \text{m}^3} \approx 1.5 \times 10^3\ \text{kg/m}^3$$

$$\rho_{new} \approx 1.5 \times 10^3\ \text{kg/m}^3$$

CHECK and THINK

The new estimate fits our expectations: A raisin is denser than a grape.

Welcome to the Beginning of Your Adventure

Physicists love to study all aspects of nature, from the details of subatomic particles to the Universe as a whole. Discoveries made by physicists have been harnessed by engineers who have designed devices and machines that you use every day. Studying physics is an adventurous journey full of uphill struggles with surprising twists and turns. Along the way are beautiful views to see and interesting people to meet. It is an awesome adventure; enjoy it. Many older physicists envy you because you are at the beginning of the road.

SUMMARY

Underlying Principle: Physics is the fundamental natural, experimental science.

Major Concepts

1. Scientific **theories** make testable predictions.
2. **Scientific evidence** comes from measured observations.
3. A **standard unit** is a precisely defined quantity to which measurements are compared.
4. The **SI system** (Système International d'Unités) is a set of standard units used worldwide in the scientific community and in this textbook.
5. The terms **uncertainty** and **error** are used to describe the imperfection of a measurement.
6. The number of reported digits—known as the number of **significant figures**—implicitly expresses the uncertainty of the measurement.
7. **Mass density** is

$$\rho = \frac{m}{V} \tag{1.1}$$

Tools

1. A **conversion factor** comes from writing equal quantities in terms of a fraction equal to unity. Multiplying a quantity by a conversion factor does not change its value, only its units.
2. **Dimensional analysis** is a method in which the dimensions of a quantity rather than its value or other properties are used to attack a problem. It is a good tool for checking or anticipating a result.
3. An **order-of-magnitude estimate** is a calculation based on values with no significant figures. Only the power of 10 is known.

PROBLEM-SOLVING STRATEGY

INTERPRET and ANTICIPATE

Often it is best to start with a sketch. Your goal is to come up with a plan. Ask yourself:

1. Can I restate the problem in my own words?
2. What concepts and physical principles are relevant?
3. Should the result be algebraic or numerical?
4. What should the units or dimensions be?
5. What is the approximate value or order of magnitude?

SOLVE

Execute your plan.

CHECK and THINK

Examine your result by asking:

1. Does my result match my expectation?
2. What is implied by my result?

PROBLEMS AND QUESTIONS

A = algebraic **C** = conceptual **E** = estimation **G** = graphical **N** = numerical

1-5 Systems of Units

1. **N** The average life expectancy in Japan is 81 years. What is this time in SI units?
2. **N** If you live in the United States, you probably know your height in feet and inches. In other countries, metric units are commonly used for measuring such quantities. First, find your height in inches. Then determine your height in **a.** centimeters and **b.** meters
3. **N** The age of the Earth is 4.5 billion years. What is the age of the Earth in the appropriate SI units?
4. **N** How many cubic centimeters (cm^3) are in one cubic meter (m^3)?
5. **N** In a laboratory, it is often convenient to make measurements in centimeters and grams, but SI units are needed for calculations. Convert the following measurements to SI units.
   a. 0.53 cm **b.** 128.92 g **c.** 35.7 cm^3 **d.** 65.7 g/cm^3
6. **C** What do *all* conversion factors have in common?
7. **N** A certain pure 0.9999 gold bullion bar with a mass of 311 g has a volume of 16.1 cm^3. What is the density of gold in appropriate SI units?
8. **N** In Jules Verne's novel, *Twenty Thousand Leagues Under the Sea*, Captain Nemo and his passengers undergo many adventures as they travel the Earth's oceans. **a.** If 1.00 league equals 3.500 km, find the depth in meters to which the crew traveled if they actually went 2.000×10^4 leagues below the ocean surface. **b.** Find the difference between your answer to part (a) and the radius of the Earth, 6.38×10^6 m. (Incidentally, author Jules Verne meant that the total distance traveled, and not the depth, was 20,000 leagues.)
9. **N** The distance to the Sun is 93 million miles. What is the distance to the Sun in the appropriate SI units?
10. **E, N** A popular unit of measure in the ancient world was the cubit (approximately the distance between a person's elbow and the end of the middle finger, when outstretched). Estimate the length of 1 cubit in centimeters. Given that there are about 1.609×10^3 m in 1 mile, how many cubits are there in 1 mile?
11. **N** CASE STUDY On planet Betatron, mass is measured in bloobits and length in bots. You are the Earth representative on the interplanetary commission for unit conversions and find that 1 kg = 0.23 bloobits and 1 m = 1.41 bots. Express the density of a raisin (2×10^3 kg/m^3) in Betatron units.
12. **N** Use your weight in pounds to find your mass in kilograms. (On the Earth, 1 kg weighs roughly 2.2 lb.)
13. **N** A garden snail named Archie, owned by Carl Branhorn of Pott Row, England, covered a 33-cm course in 2.0 min at the 1995 World Snail Racing Championships, held in Longhan, England (Fig. P1.13). Determine Archie's average speed $S_{av} = d/t$ in the appropriate SI units.
14. **C** As part of a biology field trip, you have taken an equal-arm balance (Fig. P1.14) to the beach. Your plan was to measure the masses of various mol-

Rex Features/AP Photo

FIGURE P1.13

lusks, but you forgot to bring along your set of standard gram masses. You notice that the beach is full of pebbles. Although there are variations in color, texture, and shape, you wonder whether you can somehow use the pebbles as a standard mass set. Develop a procedure for assembling a standard mass set from the pebbles on the beach. Describe your procedure step by step so that someone else could follow it.

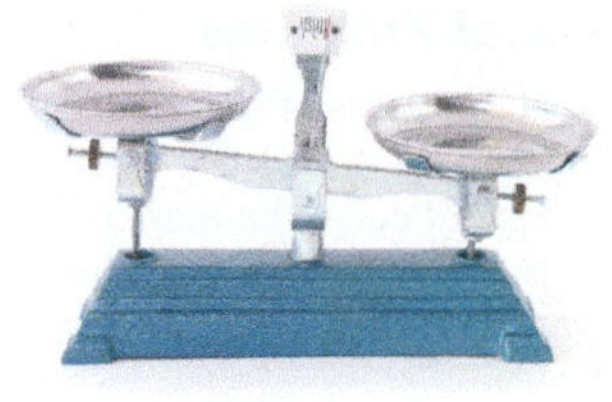

FIGURE P1.14

15. **N** An Olympic-sized swimming pool, with a 660,000-gallon capacity, is filled with a large hose in 10.0 hours. **a.** What is the rate at which the pool is filled, in gallons per minute and **b.** in liters per second? **c.** At this rate, how long would it take to fill a typical backyard pool, with a volume of 40.0 cubic meters?

1-6 Dimensional Analysis

16. **C** Which of the following exaggerated statements nevertheless uses the term "light-year" correctly? **a.** "It will be light-years before I graduate from college." **b.** "Aaron drives so quickly; he must be a dozen light-years in front of us." Explain your answer.

17. **N** The kilogram standard is a circular cylinder whose height and diameter both equal 39.17 mm (Fig. 1.7). What is the density of the alloy used in the standard kilogram?

Problems 18 and 19 are paired.

18. **C** Acceleration a has the dimensions of length per time squared, speed v has the dimensions of length per time, and radius r has the dimension of length. Which of the following expressions may be correct? Explain your answer in each case.

a. $a = vr$ **b.** $a = v/r$ **c.** $a = v^2/r$ **d.** $a = v/r^2$

19. **C** See Problem 18. What are the dimensions of the following combinations of quantities?

a. ma **b.** mv^2/r **c.** mv **d.** mvr

20. **A** In the study of electricity later in this book, you will encounter the intriguing idea that energy can be stored by an electric field present in some region of space. In this case the energy stored is often described in terms of *energy density*, symbolized as u, which is the stored energy per unit volume. **a.** Determine the dimensions of u. *Hint*: Refer back to your results from Concept Exercise 1.3 to find the dimensions of energy. **b.** What are the SI units for u?

21. **A** In subsequent chapters, two different types of a physical quantity called *energy* will be defined. The kinetic energy K of an object is $K = \frac{1}{2}mv^2$. The gravitational potential energy U associated with an object–Earth system can sometimes be expressed as $U = mgy$. In these expressions, m stands for mass, g is the gravitational acceleration with dimensions of length per time squared, v is a quantity called speed that has the dimensions of length per time, and y is a distance with units of length. Show that the two different expressions for energy are consistent in that they have the same dimensions.

22. **C** The symbols for volume flow rate, area, and speed are, respectively, R, A, and v. The SI units of volume flow rate, area, and speed are, respectively, cubic meters per second, square meters, and meters per second. Could $v = RA$? Check by using dimensional analysis.

23. **A** Later in this book, you will study oscillating motion. Such motion is repetitive. Each complete repetition is called a cycle, and the frequency f is the number of cycles per unit time. The SI unit of frequency is s^{-1}. For a block oscillating at the end of a spring, the frequency depends on the mass m of the block, and the stiffness (spring constant) k of the spring. The dimensions of k are mass per time squared. Use dimensional analysis to find the mathematical dependence of f on k and m.

24. **C** What problem might arise if we were to multiply or divide any two numbers that have the same dimension (such as length) but different units (such as meters and feet), while working within a formula or string of calculations, as opposed to multiplying or dividing two numbers that have the same unit of measure?

25. **A** Force is the central concept in classical mechanics. It is measured using the derived SI unit called the newton, represented by N. **a.** Use the relationship $a = F/m$ to determine the dimensions of the newton in terms of mass, length, and time. Here, m stands for mass, a is the acceleration with dimensions of length per time squared, and F is the force. **b.** What is the "fundamental SI unit" of force? (That is, express 1 N in terms of kg, m, and s.) **c.** The force exerted by a spring depends on the length x by which it is stretched or compressed: $F = kx$. The constant of proportionality k, known as the spring constant, depends on the stiffness of the spring and is the force the spring exerts per unit length. Use your results from part (b) to find the fundamental SI units for k.

1-7 Error and Significant Figures

26. **N** Convert 13.7 billion years (the age of the Universe) to the appropriate SI unit. Be sure to report your answer with the correct number of significant figures.

27. **C** How many significant figures does 0.00130 m have?

28. **C** A distance with two significant figures divided by a time interval with three significant figures equals an average speed. How many significant figures does the average speed have?

29. **C** What is the number of significant figures in each of these numbers?

a. 7.913×10^{11} **b.** 0.00643 **c.** 4.1×10^{-4} **d.** 615 ± 3

Problems 30 and 34 are paired.

30. **C** In a laboratory, a researcher measures time using a stopwatch and finds that the time is between 10.53 s and 10.56 s. What should he report for the measured time and its error?

31. **N** Perform the following arithmetic operations, keeping the correct number of significant figures in your answer.

a. The product 56.2×0.154 **b.** The sum $9.8 + 43.4 + 124$
c. The quotient $81.340/\pi$

32. **N** Calculate the result for each of the following cases using the correct number of significant figures.

a. $3.07670 - 10.988$ **b.** $1.0093 \times 10^5 - 9.98 \times 10^4$
c. $\dfrac{5.4423 \times 10^6}{4.008 \times 10^3}$

33. **N** Calculate the result for the following operations using the correct number of significant figures.

$$3.07670 - 10.988 + \frac{\left(\dfrac{5.4423 \times 10^6}{4.008 \times 10^3}\right)}{(1.0093 \times 10^5 - 9.98 \times 10^4)}$$

34. **C** In a laboratory, a researcher fails to start a stopwatch at the beginning of an experiment. Instead he starts it a little late and finds that the time is between 17.89 s and 17.92 s. What should he report for the measurement of the time?

35. **N** Complete the following calculations and report your answer using scientific notation, the correct number of significant figures, and SI units. **a.** Model the Earth as a sphere with a radius of 6378.1 km. Find its volume. **b.** The mass of the Earth is 5.98×10^{24} kg. Find the density of the Earth.

36. **C** A pendulum consists of a bob at the end of a string (Fig. P1.36) In one cycle the bob swings out and back to its original position. The time for one cycle is called "the period," T. Johanna measures the period and finds it is (0.9 ± 0.1) s. Jimmy replaces the bob of the pendulum with a bob that has twice the mass, but is otherwise identical. He finds the period is now (0.95 ± 0.01) s. Jimmy claims that the heavier bob slowed down the pendulum and increased its period. Johanna argues that their measurements are consistent with one another and the period has not changed. Whom do you agree with? Explain.

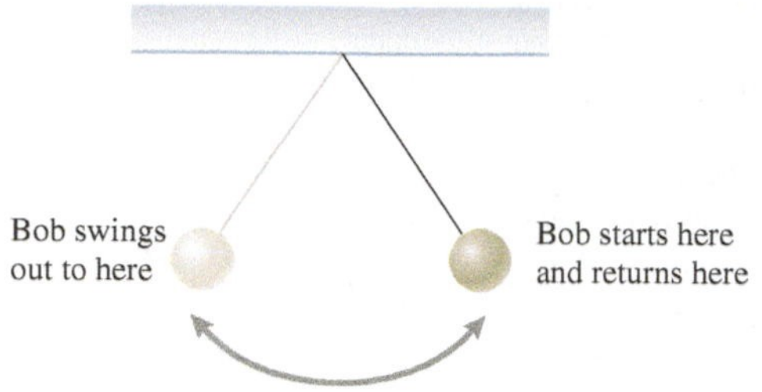

FIGURE P1.36 Problems 36, 37, and 58.

37. **C** In each case (a) through (d), your lab partner has measured the mass of a pendulum bob (Fig. P1.36) and reported the results as shown. Identify and correct the flaw in each case, if one exists. If no correction is possible, explain why you cannot correct the flaw. **a.** $m = 1.23 \pm 0.1$ g **b.** $m = 10.64 \pm 0.03$ g **c.** $m = 18.70 \pm 0.01$ g **d.** $m = 7.6 \pm 0.01$ g

1-8 Order-of-Magnitude Estimates

38. **E** Estimate the area of your room's floor. Measure the area and compare your results.

39. **E** Estimate both the number of hours you will spend studying physics during the current term and the total number of hours the entire class will spend studying physics this term. Show your work and explain your assumptions.

40. **E** Measure the distance from your outstretched thumb to your outstretched pinky finger. Use that information to estimate (a) the top surface area and (b) the volume of your dresser or desk. (Indicate your choice of furniture.)

41. **E** Use your height in SI units to estimate the height of your shower stall or length of your bathtub.

42. **E** Estimate the volume of your physics classroom. Explain your process.

43. **E** Roughly speaking, what are the volume of your lungs and the volume of your stomach? Explain your work.

44. **E** Approximately how much money do you spend each year on entertainment?

45. **E** Estimate the number of living cells in a tiger.

46. **E** How many protons are there in a standard deck of 52 cards?

47. **E** Estimate the time it would take you to hike from Boston to Miami.

48. **E** In 2011, artist Hans-Peter Feldmann covered the walls of a gallery at the New York Guggenheim Museum with 100,000 one-dollar bills (Fig. P1.48). Approximately how much would it cost you to wallpaper your room in one-dollar bills, assuming the bills do not overlap? Consider the cost of the bills alone, not other supplies or labor costs.

FIGURE P1.48

Courtesy Solomon R. Guggenheim Foundation, New York. Photo by David Heald

General Problems

49. **N** The mass of one electron is 9.11×10^{-31} kg, and the mass of one proton is 1.67×10^{-27} kg. How many electrons would it take to equal the mass of one proton?

50. **N** Convert the following distances into SI units. Which is larger?

$$d_1 = 1.02 \text{ km} \qquad d_2 = 102 \times 10^{-5} \text{ Mm}$$

51. **N** A 350-seat rectangular concert hall has a width of 60.0 ft, length of 81.0 ft, and height of 26.0 ft. The density of air is 0.0755 lb/ft^3. **a.** What is the volume of the concert hall in cubic meters? **b.** What is the weight of the air in the concert hall in newtons?

52. **G** Later in this book, you will learn that sound is a wave. The wavelength λ and frequency f of a wave are related by $\lambda f = v$ where v is the speed of the wave. Musicians refer to these different wavelengths or frequencies by their notes (A–G). Use the information in the following table to plot the frequency on the vertical axis and $1/\lambda$ on the horizontal axis. Give a conceptual interpretation and numerical value of the slope on your graph.

Pitch	Wavelength λ (m)	Frequency f (s^{-1})
A	0.7800	440.0
B	0.6949	493.9
C	0.6559	523.2
D	0.5843	587.3
E	0.5206	659.3
F	0.4914	698.5
G	0.4378	784.0

53. **N** Two decorative spheres are carved from the same slab of marble. The radius of the first sphere, which has four times the mass of the second sphere, is 14.5 in. What is the radius of the second sphere in inches?

54. **A** Newton's law of universal gravity $F = G\dfrac{m_1 m_2}{r^2}$ gives the magnitude of the force exerted by two particles with masses m_1 and m_2 on each other. The SI units of force are kg · m/s^2; G is the universal gravitational constant and r is the distance separating the two particles. What are the SI units of G?

55. **A** Two different expressions for finding the magnitude of a quantity called force are $F = ma$ and $F = \dfrac{p_f - p_i}{t_f - t_i}$. The acceleration a has dimensions of length per time squared. Momentum p is equal to mv, where v is speed with the dimensions of length per time. The subscripts f and i refer to the *f*inal and *i*nitial values of those quantities, respectively. Also, m stands for mass and t stands for time. Show that the two different expressions for force are consistent in that they result in a quantity with the same dimensions.

56. The size of a cell in the human body is determined by a need to optimize the exchange rates between the cell's interior and exterior environments. This process requires a suitable surface area-to-volume ratio.

a. **A** Using a sphere of radius r, find an expression for the ratio between the surface area and the volume of a sphere.

b. **G** Cells are typically on the order of micrometers in radius. Plot the ratio A/V as a function of r from part A for values of r ranging from 1 μm to 6 μm.

57. **N** During a visit to New York City, Lil decides to estimate the height of the Empire State Building (Fig. P1.57). She measures the angle θ of elevation of the spire atop the building as 20°. After walking 9.0×10^2 ft closer to the iconic building, she finds the angle to be 25°. Use Lil's data to estimate the height h of the Empire State Building.

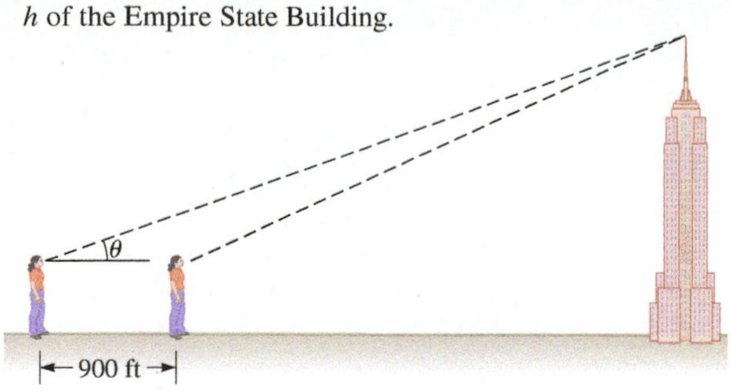

FIGURE P1.57 (Not to scale.)

58. **C, G** You hypothesize that the period of a pendulum is proportional to the length of the string (Fig P1.36). The period is defined as the time for one complete cycle. You conduct some experiments to test your hypothesis. In the first trial, you measure the period to be 0.40 ± 0.08 s. In the second trial, the length of the string is doubled, and you measure the period to be 0.65 ± 0.08 s. All other variables are kept the same in the two trials.

Discuss the extent to which the experimental evidence can be used to support or refute your belief. Are the data consistent with the notion that period is proportional to length? Can the data be used to rule out this possibility? Draw a diagram with a number line to support your answer.

59. **N** You are part of a team in an engineering class that is working on a scale model of a new design for a life vest. You have been asked to find the mass of a piece of foam that will be used for flotation. Because the piece is too bulky to fit on your balance, you break it into two parts. You measure the mass of the first part as 128.3 ± 0.3 g and the second part as 77.0 ± 0.3 g. **a.** What are the maximum and minimum values for the total mass you might reasonably report? **b.** What is the best estimate for the total mass of the foam?
Hint: Propagation of uncertainty is described in Appendix A.

60. **A** Model the human body as three cylinders: a large one for the torso and two small ones for the legs. The height of each cylinder is half the height h of the person. The radius of the large cylinder is r, and the radii of the smaller cylinders are $r/2$. The body has a mass m. Assume the average density of the body is about the same as water, $\rho = 1000\ \text{kg/m}^3$, and the radius of each cylinder is small compared with its height. Show that the body's surface area in meters squared is roughly given by $A = 0.129m^{0.5}h^{0.5}$, where m is in kilograms and h is in meters. *Note*: An empirical fit produces a more accurate estimate of the body's surface area: $A = 0.202m^{0.425}h^{0.725}$.

61. **N** A unit of distance used in astronomy is the parsec (pc): 1 pc = 3.26 ly. The distance to the Earth's next-nearest star, α-Centauri, is 1.3 pc. Find the distance d to α-Centauri in light-years and in meters.

2 One-Dimensional Motion

Key Question

How do physicists describe one-dimensional motion using concepts and mathematics?

Underlying Principles

Kinematics and one-dimensional motion

Major Concepts

1. Translational motion
2. Particle
3. Position
4. Displacement
5. Distance traveled
6. Average and instantaneous velocity
7. Average and instantaneous speed
8. Average and instantaneous acceleration

Special Cases

1. Constant acceleration
2. Free fall

Tools

1. Motion diagrams
2. Vector and scalar quantities
3. Coordinate systems
4. Graphs of position, velocity, and acceleration versus time

Scientific progress is often most exciting when a serendipitous observation leads to a major discovery, which is what happened in 1915 when V. M. Slipher surprised himself and the astronomy community. Measuring the speeds of 15 "spirals" thought to be infant solar systems, Slipher showed that these spirals were moving up to 100 times faster than expected. His observations supported an emerging theory of the Universe. According to this theory, these spirals are whole galaxies outside our own Milky Way galaxy, all moving at high speeds away from one another as the entire Universe expands.

Slipher's experience is not unique. In physics, the most important progress is often made through the careful observation of motion. The study of motion has led to our understanding of many fundamental principles of physics, such as Einstein's theory of relativity and Newton's laws of motion. Newton's laws provide the underpinning for most of the concepts in this book. So, to grasp physics, we must develop a clear and precise description of motion.

Galileo Galilei—whose scientific research during the Italian Renaissance laid the groundwork for Newton's discoveries—divided the study of motion into two parts: kinematics and dynamics. **Kinematics**, the topic of this chapter, is the careful description of motion independent of its cause. **Dynamics** deals with the cause of motion and is the focus of Chapter 5.

KINEMATICS and ONE-DIMENSIONAL MOTION
Underlying Principles

2-1 What Is One-Dimensional Translational Kinematics?

TRANSLATIONAL MOTION and PARTICLE
Major Concepts

The goal of this chapter is to describe motion using the concepts and equations of kinematics. To make this goal easier to achieve, we restrict the motion described in this chapter in two ways. In general, motion can be rotational, vibrational, or translational (Fig. 2.1). **Rotational** motion is spinning motion, such as the motion of a carousel. The back-and-forth motion of a guitar string is an example of **vibrational** motion. The descent of a landing airplane is an example of **translational** motion.

In Chapters 2–11, we restrict our study to objects that are purely translating without rotating or vibrating. A purely translating object may be modeled as a particle. A **particle** is an idealized point that has no spatial extent, no shape, and no internal structure. It has no distinct top or bottom, front or back, or left or right side and therefore cannot rotate and cannot vibrate. Although no real physical object is an ideal particle, many real objects can be approximated, or *modeled*, as particles even if they are not microscopically small or lacking in structure.

Here in Chapter 2, we restrict our study to one-dimensional motion, such as that of a ball falling vertically or a train traveling on a straight track. One-dimensional motion is mathematically simpler than two- and three-dimensional motion. Most of the concepts of kinematics—such as position, displacement, velocity, and acceleration—are mathematically described by vectors. **Vector** quantities ("vectors") have both a magnitude, such as 35 m/s, and a direction, such as northeast. In one-dimensional motion, the direction of any vector is limited to two possibilities, such as east or west, north or south, up or down, or left or right. Mathematically, we can describe these two choices as positive or negative. Our study of physics will also include **scalar** quantities ("scalars") that lack direction; scalars have only magnitude. Examples of scalars are mass, temperature, and time.

A. Rotational motion

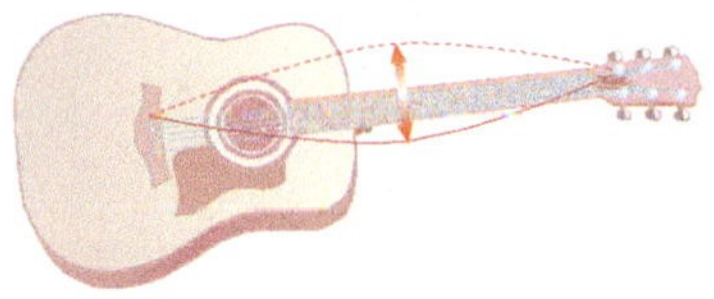

B. Vibrational motion

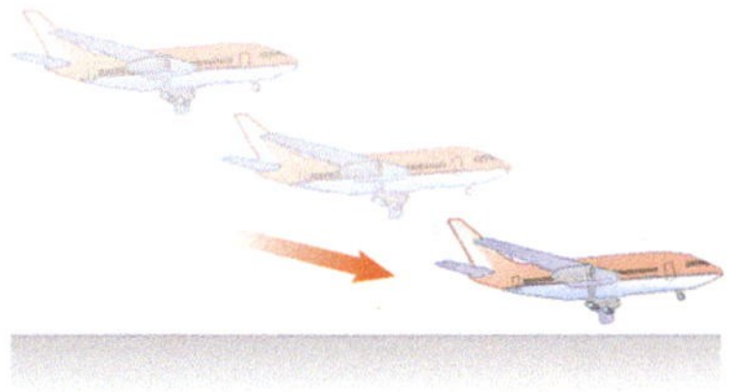

C. Translational motion

FIGURE 2.1 Three types of motion. **A.** The carousel is rotating. **B.** The guitar string is vibrating. **C.** The airplane is translating.

2-2 Motion Diagrams

We observe and describe motion every day. Runners would like to achieve a "4-minute mile," for instance, and the 1968 Ford GT40 can race "from 0 to 100 mph in 8 seconds." Our everyday language is often vague, however. Can a Ford GT40 reach 200 mph in 16 seconds? Although such ambiguity may be okay in some situations, physics requires a carefully defined vocabulary to describe motion. Kinematics can be expressed using words, mathematics, diagrams, and graphs. In this section, we will start with a visual description known as a *motion diagram*.

MOTION DIAGRAM **Tool**

A **motion diagram** is any illustration that shows the location of a particle at regular time intervals. For example, imagine observing the motion of Mars in the night sky. You might start by sketching the region of the night sky around Mars at a particular time. You model Mars as a particle, representing it on your sketch by a dot and labeling its initial location with an A (Fig. 2.2). If you then observed Mars's position at ten-night intervals for a few weeks and recorded and labeled those positions on your sketch, you would end up with a motion diagram of Mars. (In a laboratory course, you may use a camera or motion sensor to make a motion diagram.)

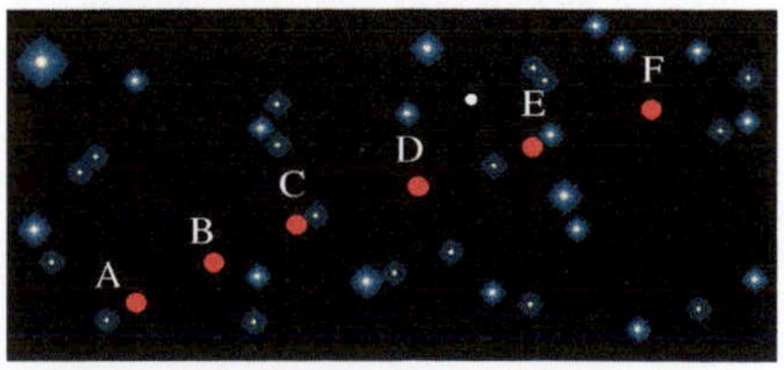

FIGURE 2.2 In this motion diagram, Mars moves from A to F. The time interval between each measurement is ten nights.

A motion diagram contains a lot of information. First, we can see the shape of the particle's path in the plane of the diagram. In Figure 2.2, it is clear that Mars is moving up and to the right in the plane of the sky. Second, because the images are taken at regular time intervals, we can infer something about the particle's speed. A particle at rest is represented by only one dot on a motion diagram because the particle does not move during the time intervals. If a particle maintains its speed, the spacing

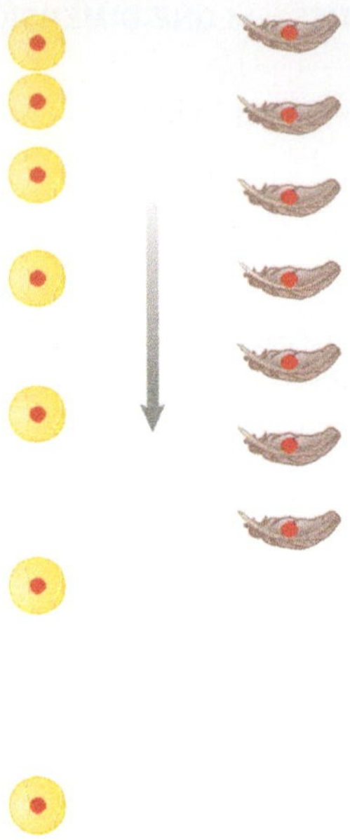

FIGURE 2.3 Motion diagrams of a falling ball and of a falling feather.

between dots is uniform. If the particle is slowing down, the dots get closer together. If the particle is speeding up, the dots get farther apart.

Figure 2.3 shows the motion diagrams for a ball and a feather dropped at the same time. The feather's images are equally spaced, but the ball's images are closer together at the top near the beginning of its fall. The difference in these motion diagrams tells us the feather maintained its speed but the ball sped up as it fell. Because we are using the particle model, it is convenient to represent the object as a dot on a motion diagram as shown in Figure 2.3.

CONCEPT EXERCISE 2.1

In each of the five motion diagrams shown in Figure 2.4, a particle moves in space from position A to position E. For each diagram, describe the motion of the particle as maintaining speed, speeding up, slowing down, or remaining at rest.

FIGURE 2.4

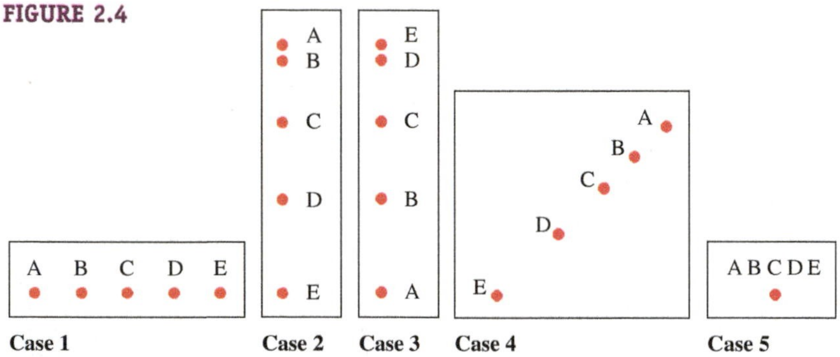

2-3 Coordinate Systems and Position

Motion diagrams alone are not enough to enable us to study kinematics. A precise description of motion requires us to know the location of a particle at particular times. We use a coordinate system to measure location. In this chapter, we need only a one-dimensional **coordinate system** (Fig. 2.5) with two elements: a reference point, called the **origin**, indicated by a zero; and a line passing through the origin called either the **coordinate axis** or just the **axis**, usually labeled x, y, or z. Although the label may be chosen arbitrarily, often a horizontal axis is labeled x and a vertical axis y. One direction along the coordinate axis is chosen to be positive, indicated by an arrow. The opposite direction is negative.

COORDINATE SYSTEM Tool

The **position** of a particle is its location with respect to the chosen coordinate system. Position is a vector whose direction in one dimension is indicated by a positive or negative sign, depending on whether it is on the positive or negative side of the origin. A particle at the origin has a position of zero.

POSITION Major Concept

Introduction to Vectors

VECTORS Tool

We will study four vector quantities in this chapter: position, displacement, velocity, and acceleration. An arrow → above a symbol indicates that the symbol represents a vector quantity. No arrow appears above a symbol representing a scalar quantity.

The overhead arrow points to the right as shown no matter in which direction the vector actually points.

For one-dimensional motion, vectors point in either the positive or negative direction along a single x, y, or z axis. When writing a vector, we must indicate both its

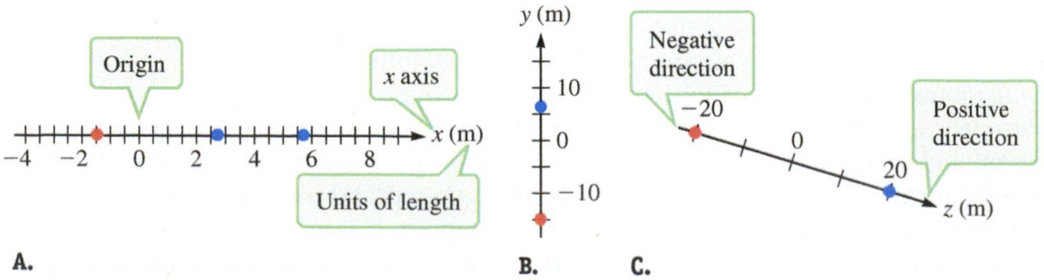

FIGURE 2.5 Three examples of a one-dimensional coordinate system. Each one has an origin; an x, y, or z axis; the distance measured in meters; and an arrow to indicate the positive direction. The labels x, y, and z are an arbitrary choice. In all three cases, the blue particles are at positive positions and the red at negative positions.

magnitude and direction. Mathematically, we use unit vectors to indicate direction. **Unit vectors** always have a magnitude of one and point in the positive direction along a single coordinate axis. The unit vectors $\hat{\imath}$, $\hat{\jmath}$, and $\hat{k}$ point in the positive x, y, and z directions, respectively. Suppose we wish to write a mathematical expression for a one-dimensional vector $\vec{Q}$ pointing in the negative y direction. We use the unit vector $\hat{\jmath}$ to indicate the y axis:

Use a hat ^ to indicate a unit vector.

$$\vec{Q} = -Q\hat{\jmath}$$

where the negative sign indicates that the vector points in the negative direction. The magnitude of the vector is $|-Q| = Q$.

The factor that appears before the unit vector is called the **scalar component**. The scalar component of $\vec{Q} = -Q\hat{\jmath}$ is $Q_y = -Q$. The subscript y indicates a scalar component along the y axis, and so $\vec{Q}$ may be written as

$$\vec{Q} = Q_y\hat{\jmath}$$

The magnitude of a vector is always positive, but the scalar component of a vector may be positive or negative. The scalar component multiplied by the unit vector is called the **vector component**. The vector component of $\vec{Q}$ is $\vec{Q}_y = Q_y\hat{\jmath} = -Q\hat{\jmath}$. In one dimension, the vector component is identical to the vector itself.

Because position is a vector, we must indicate its direction and magnitude. The symbol for position is $\vec{r}$. The vector components of position are written as $\vec{x}$, $\vec{y}$, and $\vec{z}$ and their scalar components as x, y, and z, respectively.

In this book, position is the only vector that does not use subscripts to indicate vector or scalar components.

In one dimension, position has just one vector component and is written in terms of any one of the three vector components $\vec{x}$, $\vec{y}$, or $\vec{z}$:

$$\vec{x} = x\hat{\imath}, \quad \vec{y} = y\hat{\jmath}, \quad \text{or } \vec{z} = z\hat{k}$$

For example, $\vec{r} = -15\hat{\jmath}$ m gives the position of the red particle in Figure 2.5B. Notice that we indicate the SI unit for position (meter) at the end, after the other symbols. The vector component of this position vector is $\vec{y} = -15\hat{\jmath}$ m, and its scalar component is $y = -15$ m. The magnitude of this position is $r = |y| = |-15|$ m $= 15$ m.

CONCEPT EXERCISE 2.2

For each of the following, give the vector component, the scalar component, and the magnitude.

a. $\vec{r} = -2.4\hat{\imath}$ m **b.** $\vec{r} = -2.4\hat{k}$ m **c.** $\vec{z} = -2.4\hat{k}$ m **d.** $x = -2.4$ m

2-4 Position-Versus-Time Graphs

The world does not come with an attached coordinate system; we must *choose* a system that works for a particular problem. There may be more than one good choice of coordinate systems, and choosing a good system takes practice. We consider a **CASE STUDY** to see how coordinate systems are used. This case study is based on experiments two students performed in their physics laboratory.

CASE STUDY An Experiment in Motion

Two students (Crall and Whipple) set up an experiment to study the motion of a small cart along a straight north–south track (Fig. 2.6A). Crall chooses a coordinate system such that the *positive x* direction points north. Crall and Whipple recorded the position of the cart at 0.4-s time intervals in meters from the origin along Crall's axis. Their data are recorded in Table 2.1.

They use these data to make the motion diagram on the left side of Figure 2.6B. We can describe the cart's motion in words as follows: The cart remained at rest at position $\vec{x}_B = 1.64\hat{\imath}$ m for at least the first 0.4 s. It then moved in the negative x direction at a nearly constant speed to position $\vec{x}_P = 0.33\hat{\imath}$ m (point P).

TABLE 2.1 Position data of a laboratory cart at 16 times, using Crall's choice of coordinates.

Label	t (s)	x (m)
A	0.0	1.64
B	0.4	1.64
C	0.8	1.55
D	1.2	1.45
E	1.6	1.35
F	2.0	1.25
G	2.4	1.16
H	2.8	1.06
I	3.2	0.97
J	3.6	0.87
K	4.0	0.78
L	4.4	0.69
M	4.8	0.60
N	5.2	0.51
O	5.6	0.42
P	6.0	0.33

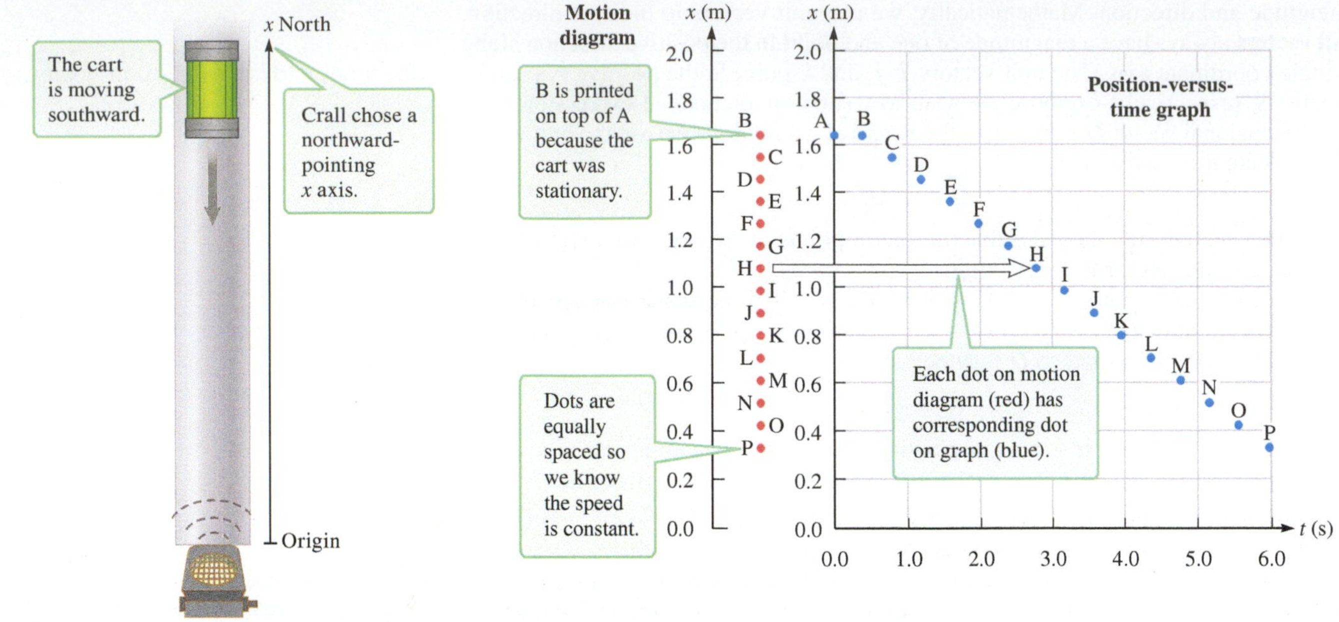

A. B.

FIGURE 2.6 **A.** Overhead view of Crall and Whipple's experimental setup. A sonic ranger (a motion sensor) detects the position of a cart moving along a straight, horizontal track by sending an ultrasonic pulse that echoes back. **B.** The motion diagram (left side) shows the location of the cart every 0.4 s. The position-versus-time graph (right side) is a two-dimensional representation of the same data. Position is plotted on the vertical axis, and time is plotted on the horizontal axis.

Position-Versus-Time Graph

Even with a coordinate system for position imposed on it, a motion diagram is not enough to describe the cart's motion completely. For example, we cannot determine the speed of the cart if we do not know the time interval between each position on the motion diagram. So, when measuring motion, we must also measure the time. We begin by setting an initial time to zero; other times are measured with respect to the initial time. Time is a scalar quantity, and it can be positive or negative. Events that occur *after* the initial time are assigned positive times, and events that occur *before* the initial time are assigned negative times. Like choosing a coordinate system, choosing an initial time takes some care. It is often convenient (but not always practical) to set the initial time to the moment the motion begins.

Crall and Whipple's data in Table 2.1 can be arranged in the position-versus-time graph on the right side of Figure 2.6B. In such a graph, the position's scalar component is plotted on the vertical axis, and time is plotted on the horizontal axis. A complete **position-versus-time graph** has the following features:

POSITION-VERSUS-TIME GRAPH

Tool

1. A vertical position axis labeled *x*, *y*, or *z* corresponding to the chosen coordinate system, with position increasing upward
2. A horizontal time axis labeled *t*, with time increasing to the right
3. Labeled units on both axes (typically "m" for meters and "s" for seconds)
4. Tick marks labeled along both axes

Let's compare a position-versus-time graph to a motion diagram. First, notice that Crall and Whipple's position-versus-time graph (Fig. 2.6B, right) looks like a stretched-out version of the motion diagram (Fig. 2.6B, left). Each dot in a position-versus-time graph represents two pieces of information, the position on the vertical axis and the time on the horizontal axis. In contrast, a motion diagram represents only location.

A position-versus-time graph differs from a motion diagram in one other way. A motion diagram shows the path of the particle, but a position-versus-time graph does not. For example, the position-versus-time graph on the right in Figure 2.6B looks like a particle moving down an incline, but we know that the cart really moved along a horizontal track.

EXAMPLE 2.1 CASE STUDY A Laboratory Cart

Figure 2.7 shows the position-versus-time graph for a cart moving along a straight track like the one used by Crall and Whipple in Figure 2.6A.

A For each labeled point in the position-versus-time graph read off the time and position. Then, describe in words the motion of the cart as it moves through these points.

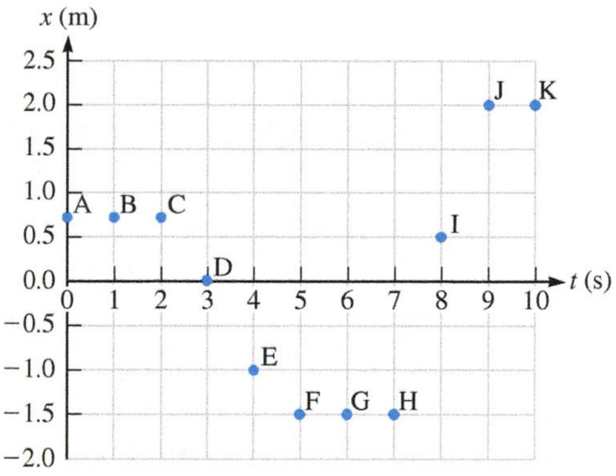

FIGURE 2.7

Point	t (s)	x (m)
A	0	0.75
B	1	0.75
C	2	0.75
D	3	0
E	4	−1.00
F	5	−1.50
G	6	−1.50
H	7	−1.50
I	8	0.50
J	9	2.00
K	10	2.00

INTERPRET and ANTICIPATE

Work your way from A to K on the position-versus-time graph, reading values off the graph as needed.

SOLVE

For each point, read off the position and time. These data are best displayed in a table as at the far right.

Points on graph	*Motion of cart*
Points A through C all have the same position: $\vec{x}_A = \vec{x}_B = \vec{x}_C = 0.75\hat{\imath}$ m.	The cart is at rest at the positive position $\vec{x} = 0.75\hat{\imath}$ m.
Points C through F are successively lower on the x axis.	The cart moves in the negative x direction.
Points F through H are all at the same position: $-1.50\hat{\imath}$ m.	The cart is at rest at $\vec{x} = -1.50\hat{\imath}$ m.
Points H though J are successively higher on the x axis.	The cart moves in the positive x direction.
Points J and K have the same position: $\vec{x}_J = \vec{x}_K = 2.00\hat{\imath}$ m.	The cart is at rest at the positive position $x = 2.00i$ m.

B Create a motion diagram from the position-versus-time graph of Figure 2.7.

INTERPRET and ANTICIPATE

Crall and Whipple's position-versus-time graph looks like a stretched-out motion diagram (Fig. 2.6B). So, we can create a motion diagram from a position-versus-time graph by sliding each point on the graph horizontally onto the position axis, preserving the labeling of the dots.

SOLVE

Figure 2.8 shows the motion diagram on the left and the position-versus-time graph on the right as at the far right.

CHECK and THINK

Compare the motion diagram in Figure 2.8 with the verbal description in part A. Three groups of points (A through C, F through H, and J and K) overlap in the motion diagram, indicating that the cart is at rest, consistent with our previous description. The motion diagram indicates that the cart moves in the negative x direction from C to F and in the positive x direction from H to J, also consistent with our description in part A.

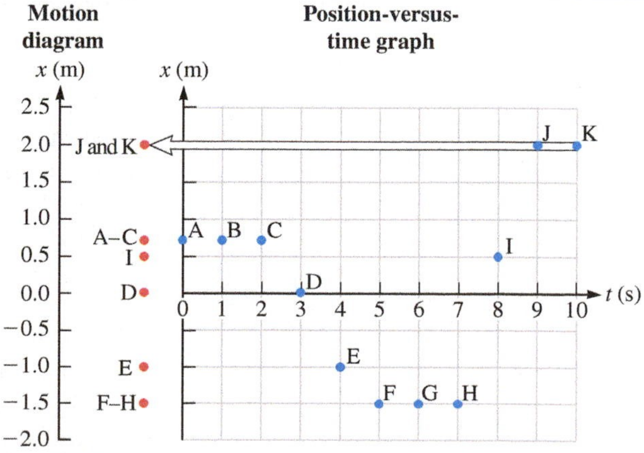

FIGURE 2.8

2-5 Displacement and Distance Traveled

In many ways, kinematics is the study of how certain quantities *change*. We use the language of mathematics to describe kinematics, so we need a mathematical symbol to indicate change. The symbol for change is an uppercase Greek letter delta Δ; the

DISPLACEMENT Major Concept

notation ΔQ means "the change in the quantity Q." This notation is shorthand for writing $Q_f - Q_i$, where the subscripts f and i stand for *f*inal and *i*nitial, respectively. When describing motion, the initial and final times must be *chosen*. The initial time is not necessarily set to zero, and the initial and final times are not necessarily the first and last times the particle moves. They are two relevant moments.

Displacement (Change in Position)

The change in position ($\Delta\vec{r}$) comes up so frequently in kinematics that it is given its own term, ***displacement.*** For one-dimensional motion, the vector component of displacement is

The symbol $\equiv$ does not mean *equals*; instead, it means *is defined to be.*

$$\Delta\vec{x} \equiv (x_f - x_i)\hat{\imath}, \qquad \Delta\vec{y} \equiv (y_f - y_i)\hat{\jmath}, \qquad \text{or } \Delta\vec{z} \equiv (z_f - z_i)\hat{k} \qquad (2.1)$$

depending on the choice of the coordinate axis.

Like position, *change* in position is also a vector quantity. In one dimension, the direction of a displacement vector is indicated by a positive or negative sign. The sign of the displacement, and therefore its direction, are determined by the subtraction in Equation 2.1. Looking back at Crall's coordinate system (Table 2.2), we see that the cart started at position $\vec{x}_i = 1.64\hat{\imath}$ m and ended at $\vec{x}_f = 0.33\hat{\imath}$ m. The cart's displacement between its initial and final positions is therefore

$$\Delta\vec{x} = (x_f - x_i)\hat{\imath} = (0.33 - 1.64)\hat{\imath} = -1.31\hat{\imath}\text{ m}$$

The displacement is negative, meaning that the cart moved in the negative x direction, which in this case is southward toward the origin (Fig. 2.6).

Position depends on the choice of coordinate systems, but displacement does not. As shown in Figure 2.9, Whipple has not chosen the same coordinate system as Crall. Whipple's and Crall's measured positions do not agree (Table 2.2). Crall and Whipple do agree on the cart's displacement between its initial and final positions, however. Using Whipple's coordinate system, we have

$$\Delta\vec{x} = (x_f - x_i)\hat{\imath} = (-1.31 - 0)\hat{\imath} = -1.31\hat{\imath}\text{ m}$$

So, in both coordinate systems, the displacement is the same: 1.31 m southward.

TABLE 2.2 Crall's and Whipple's data

Label	t (s)	Crall's x (m)	Whipple's x (m)
A	0.0	1.64	0
B	0.4	1.64	0
C	0.8	1.55	−0.09
D	1.2	1.45	−0.19
E	1.6	1.35	−0.29
F	2.0	1.25	−0.39
G	2.4	1.16	−0.48
H	2.8	1.06	−0.58
I	3.2	0.97	−0.67
J	3.6	0.87	−0.77
K	4.0	0.78	−0.86
L	4.4	0.69	−0.95
M	4.8	0.60	−1.04
N	5.2	0.51	−1.13
O	5.6	0.42	−1.22
P	6.0	0.33	−1.31

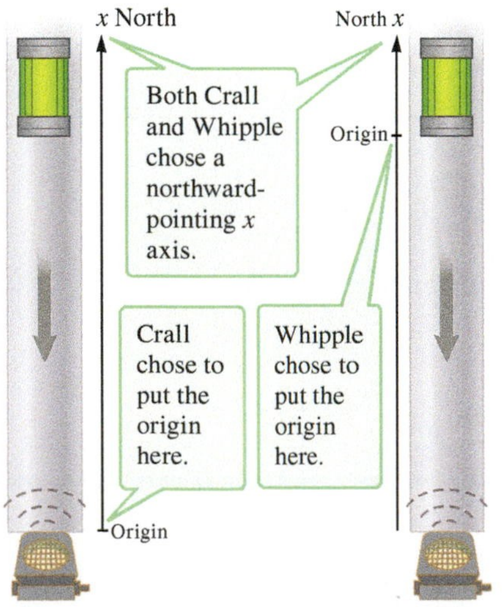

FIGURE 2.9 Crall's and Whipple's position measurements do not agree, but they both measure the same displacement.

Displacement, Translation, and the Particle Model

An object can be modeled as a particle undergoing purely translational motion if every point on the object undergoes exactly the same displacement as every other point on the object. Figure 2.10A shows the motion of a ball in a sled sliding down a slippery slope; each point on the ball has the same displacement. Figure 2.10B shows the same ball rolling down a hill; in this case, the displacement of one eye is greater than the displacement of the other. The sliding ball may be modeled as a particle undergoing pure translational motion, whereas the rolling ball is both rotating and translating, and cannot be modeled as a particle.

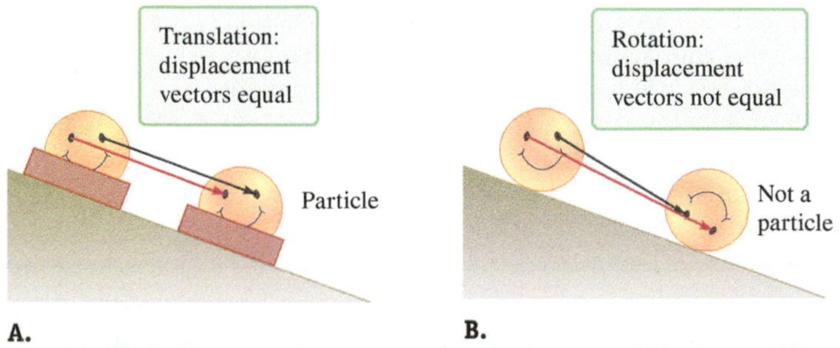

FIGURE 2.10 **A.** A ball in a sled slides down an incline. The black arrow is the displacement vector for one eye and the red arrow is the displacement vector for the other. **B.** The ball rolls down the incline. The displacement of the two eyes in part B is not the same because the object is rotating in addition to translating.

CONCEPT EXERCISE 2.3

Figure 2.11 shows the motion of various objects:

Case 1. A person on a moving Ferris wheel
Case 2. A person on a loop-the-loop roller coaster
Case 3. The tire on a moving bicycle
Case 4. The person on the bicycle

Consider the displacement of points on each object, and decide whether the object is undergoing pure translational motion and therefore may be modeled as a particle.

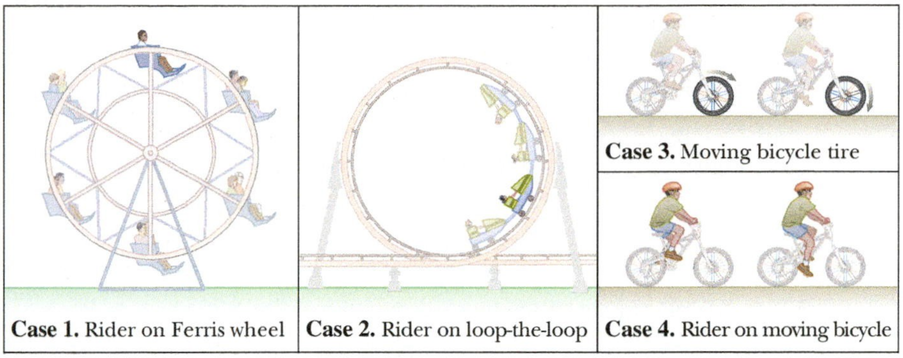

FIGURE 2.11

Distance Traveled

Displacement is not the same thing as distance traveled. The **distance traveled** d by any moving particle is the length of the whole path the particle covers. Distance traveled is a scalar quantity and therefore has no direction. A tiger in a zoo may pace the length of a 20-m enclosure 100 times in a day. Her total distance traveled would be $100 \times 20 \text{ m} = 2000 \text{ m}$, but if she ends at the same position at which she started, her displacement is zero. Distance traveled and the magnitude of displacement are equal if, and only if, the path is a straight line in one direction.

DISTANCE TRAVELED

✪ Major Concept

EXAMPLE 2.2 Block and Spring: Displacement and Distance Traveled

A block on a nearly frictionless table is attached to a spring (Fig. 2.12). The other end of the spring is fixed to a wall. When the block is pulled to the right and then released, it moves back and forth for a long time before coming to rest. A coordinate axis, arbitrarily labeled z, lies along the table; the origin is at the original position of the block, and the positive direction is to the right. The block's position is measured every 0.25 s as given in Table 2.3.

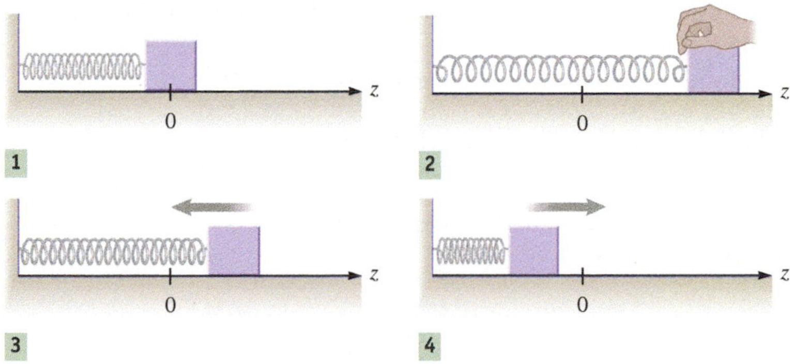

FIGURE 2.12

TABLE 2.3 Position and time data

z (m)	t (s)
0.65	0
0.63	0.25
0.57	0.50
0.48	0.75
0.35	1.00
0.20	1.25
0.05	1.50
−0.12	1.75
−0.27	2.00
−0.41	2.25
−0.52	2.50
−0.60	2.75
−0.64	3.00
−0.65	3.25
−0.61	3.50
−0.53	3.75
−0.42	4.00
−0.29	4.25
−0.14	4.50
0.02	4.75
0.18	5.00
0.33	5.25
0.46	5.50
0.56	5.75
0.62	6.00
0.65	6.25

A Create a position-versus-time graph from these data.

INTERPRET and ANTICIPATE

The block moves back and forth, so we expect our position-versus-time graph to show a repeating curve. The initial time ($t_i = 0$) has already been chosen as the time of the first recorded position (Table 2.3). When working with your own laboratory data, you may make a similarly convenient choice.

Example continues on page 30 ▶

SOLVE

Make a position-versus-time graph with z on the vertical axis and t on the horizontal axis (Fig. 2.13). (It may be helpful to use a spreadsheet program.)

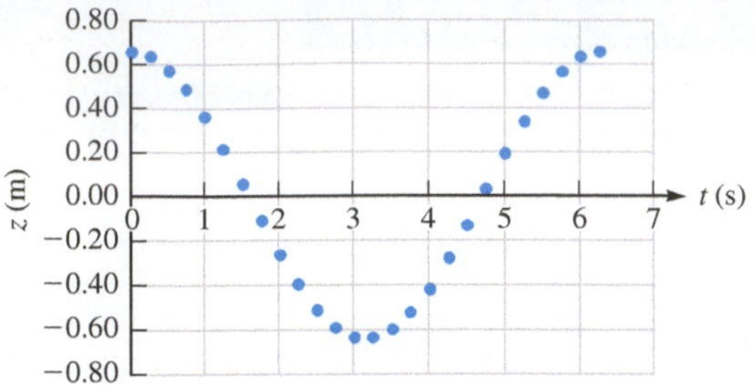

FIGURE 2.13

CHECK and THINK

As expected, Figure 2.13 shows a repeating pattern for the motion.

B What is the displacement of the block during the entire 6.25 s?

INTERPRET and ANTICIPATE

In 6.25 s, the block returned to the original position, so we expect the displacement to be zero.

SOLVE

The initial and final positions are at the same height on the graph. Find the displacement by subtracting.

$$\vec{z}_f = \vec{z}_i = 0.65\hat{k} \text{ m}$$

$$\Delta\vec{z} = \vec{z}_f - \vec{z}_i = (0.65 - 0.65)\hat{k} \text{ m} = 0 \quad \text{as expected.}$$

C What is the displacement of the block during the first 3.25 s?

SOLVE

For this time interval, the final position is negative and the initial position is positive. Find the displacement by subtracting.

$$\vec{z}_f = -0.65\hat{k} \text{ m}$$

$$\vec{z}_i = 0.65\hat{k} \text{ m}$$

$$\Delta\vec{z} = (z_f - z_i)\hat{k} = (-0.65 - 0.65)\hat{k} \text{ m} = -1.30\hat{k} \text{ m}$$

CHECK and THINK

The block moved in the negative z direction during this time interval.

D What is the total distance traveled for the recorded measurement interval?

INTERPRET and ANTICIPATE

Distance is not a vector, so we are not concerned with the algebraic signs of the positions. All we need is the length of the path covered in the recorded 6.25 s.

SOLVE

From part C, we know that the block moved 1.30 m in the first 3.25 s. In the next 3 s, it returned to its original position, so it must have traveled another 1.30 m. The total distance traveled is thus the sum of these two distances.

$$d = (1.30 + 1.30) \text{ m} = 2.60 \text{ m}$$

CHECK and THINK

The block's back-and-forth motion is much like a tiger pacing in her cage. When the block returns to its initial position, its displacement is zero. The distance traveled, however, is not zero!

2-6 Average Velocity and Speed

There are several ways to measure how fast a particle is moving. This section and the next will cover four such measures: average velocity, average speed, instantaneous velocity, and instantaneous speed.

Average Velocity

AVERAGE VELOCITY

Major Concept

Every marathon runner who completes the race undergoes the same displacement, but the one who does it in the shortest time wins. The **average velocity** is the displacement divided by the change in time; for displacement in the x direction,

$$\vec{v}_{av,x} \equiv \frac{\Delta\vec{x}}{\Delta t} = \frac{x_f - x_i}{t_f - t_i}\hat{\imath} \tag{2.2}$$

Similar equations can also be written for the average velocity in the y and z directions; the unit vectors then become $\hat{\jmath}$ and $\hat{k}$, respectively.

where the subscript "av" means that the velocity is averaged over the time interval Δt.

Average velocity is a vector quantity whose magnitude in SI units is measured in meters per second. The change in time in the denominator of Equation 2.2 must always be positive because the final time is always after the initial time. In other words, the sign of the average velocity is determined by the sign of the displacement in the numerator. Therefore, the average velocity points in the same direction as the displacement.

Average Speed

AVERAGE SPEED Major Concept

Sometimes, we are more interested in knowing how fast an object moves without regard to the direction of that motion. **Average speed** S_{av} depends on total distance traveled (not on displacement) such that

$$S_{av} \equiv \frac{d}{\Delta t} \tag{2.3}$$

where d is the total distance traveled in the time interval Δt. Average speed is always positive. Like average velocity, average speed in SI units is measured in meters per second. Because average speed is not a vector quantity, no subscript x, y, or z is needed.

Average speed is important in our everyday experience and is sometimes more meaningful than average velocity. As illustrated in Figure 2.14, the start-to-finish displacement is not necessarily the same for different marathon courses, but all marathon courses are 26.2 miles long. A runner's average *speed* may be (nearly) the same on every course, but the runner's average *velocity* will vary from course to course because the displacement varies. Therefore, the average speed is a better measure of the runner's ability than average velocity, which is more a measure of the shape of the course. Average velocity is a measure of how quickly the displacement changes. The runner who completes the marathon in the shortest time interval has the highest average speed and wins the race.

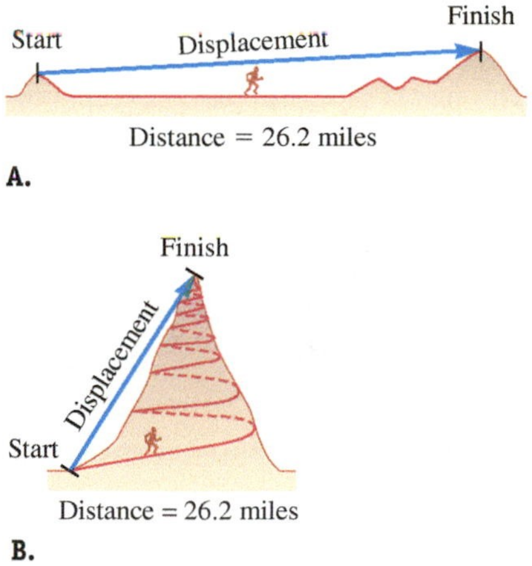

FIGURE 2.14 Two marathon courses. The distance run on each course is 26.2 miles, but the magnitude of the displacement of a runner on course B is less than on course A.

CONCEPT EXERCISE 2.4

The top marathon runners complete the race in around 2 hours, so such an elite marathoner's average speed is around 13.1 mph. To calculate a runner's average velocity in a marathon, we would have to know the displacement, which means knowing the particular course. If the magnitude of the runner's displacement is 20 mi over course A and 12 mi over course B, find the magnitude of the average velocity for each course (Fig. 2.14) and compare your results to the elite marathon average speed.

CONCEPT EXERCISE 2.5

In our everyday experience, we sometimes use the phrase "as the crow flies." Does this expression refer to total distance traveled, or does it refer to magnitude of the displacement?

CONCEPT EXERCISE 2.6

A tiger paces the length of his 20-m enclosure 100 times in 2 hours. He then stops at his initial position. Find his average speed and average velocity over this period. How do they compare with the average speed and average velocity of a tiger that lies asleep in one place for 2 hours?

EXAMPLE 2.3 Motion of a Particle

A The average velocity of a particle measured over a 5.4-s interval is $\vec{v}_{av,x} = 4.9\hat{\imath}$ m/s. Find its displacement during this time interval.

INTERPRET and ANTICIPATE

Because the average velocity points in the positive x direction, we expect the displacement to be in the positive x direction.

SOLVE Solve Equation 2.2 for displacement.	$\vec{v}_{av,x} = \frac{\Delta\vec{x}}{\Delta t}$ (2.2) $\Delta\vec{x} = \vec{v}_{av,x}\Delta t$
Make the appropriate substitutions.	$\Delta\vec{x} = (4.9\hat{\imath}\text{ m/s})(5.4\text{ s}) = 2.6 \times 10^1\hat{\imath}\text{ m}$

CHECK and THINK

The displacement is 26 m in the positive x direction as expected.

B The average velocity of a particle measured over a 5.4-s interval is $\vec{v}_{av,x} = -6.9\hat{\imath}$ m/s. Find its displacement over this time interval.

INTERPRET and ANTICIPATE

This part is similar to part A except now we expect the displacement to be in the negative x direction.

SOLVE As in part A, solve for displacement and make the appropriate substitutions.	$\Delta\vec{x} = \vec{v}_{av,x}\Delta t$ $\Delta\vec{x} = (-6.9\hat{\imath}\text{ m/s})(5.4\text{ s}) = -3.7 \times 10^1\hat{\imath}\text{ m}$

CHECK and THINK

The displacement is 37 m in the negative x direction as predicted.

Average Velocity from a Position-Versus-Time Graph

Graphs are particularly useful tools for presenting data in a visual form. Average velocity may be found from a position-versus-time graph as shown in the following experiment.

EXAMPLE 2.4 CASE STUDY Another Experiment

In a new experiment, Crall and Whipple attached a fan to the cart (Fig. 2.15). They placed the cart near the bottom of an inclined ramp and gave it a push to start it moving up the ramp. Their position-versus-time graph is shown in Figure 2.16. The cart started at $\vec{x} = 0.5\hat{\imath}$ m. It traveled in the positive x direction to a peak position of 1.8 m. Then it traveled in the negative x direction back to a final position of 0.5 m. For clarity on their graph, Crall and Whipple marked the initial, peak, and final positions.

Crall believes that the cart moved faster going up the ramp than it moved going down. Use Crall and Whipple's data from Figure 2.16 to calculate the average velocity for the trip up the ramp and the average velocity for the trip down the ramp. In the **CHECK and THINK** step, compare the magnitude of the average velocity up the ramp to the magnitude of the average velocity down the ramp to decide if Crall is correct.

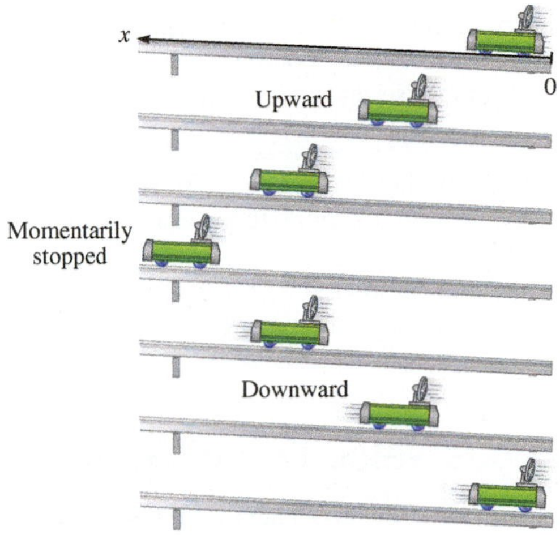

FIGURE 2.15

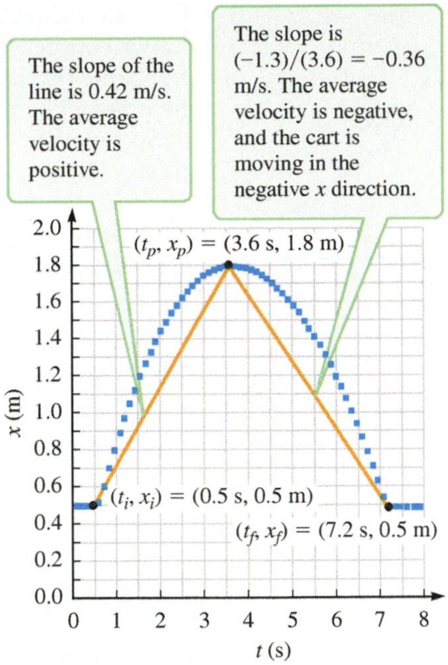

FIGURE 2.16

INTERPRET and ANTICIPATE

In Crall and Whipple's coordinate system, the x axis points up the ramp. So, the cart's displacement is positive when it moves up the ramp and negative when it moves down. We expect the average velocity to be positive when the cart goes up the ramp and negative when it goes down.

SOLVE

Let's find the average velocity for the trip up the ramp (Eq. 2.2). We use the subscript "ip" to indicate the average velocity over the time interval $t_p - t_i$ when the cart moved from its *i*nitial position to its *p*eak position and drop the subscript "x" for simplicity.

$$\vec{v}_{\text{av},\,ip} = \frac{\Delta \vec{x}_{ip}}{\Delta t} = \frac{x_p - x_i}{t_p - t_i}\hat{\imath}$$

$$\vec{v}_{\text{av},\,ip} = \frac{(1.8 - 0.5)\text{m}}{(3.6 - 0.5)\text{s}}\hat{\imath} \tag{1}$$

$$\vec{v}_{\text{av},\,ip} = 0.42\hat{\imath}\ \text{m/s}\ \text{(up the ramp)}$$

We apply a similar process to find the average velocity on the way down the ramp.

$$\vec{v}_{\text{av},\,pf} = \frac{\Delta \vec{x}_{pf}}{\Delta t} = \frac{x_f - x_p}{t_f - t_p}\hat{\imath} = \frac{(0.5 - 1.8)\text{m}}{(7.2 - 3.6)\text{s}}\hat{\imath}$$

$$\vec{v}_{\text{av},\,pf} = -0.36\hat{\imath}\ \text{m/s}\ \text{(down the ramp)}$$

CHECK and THINK

As expected, the average velocity up the ramp is positive and down the ramp is negative.

Crall is interested in whether the cart is faster going up the ramp or down the ramp; to address his interest, we need to compare the magnitude of these average velocities. When we do so, we find that Crall is correct: The cart was faster on the way up the ramp.

$$|\vec{v}_{\text{av},\,ip}| = 0.42\ \text{m/s up the ramp}$$

$$|\vec{v}_{\text{av},\,pf}| = 0.36\ \text{m/s down the ramp}$$

$$|\vec{v}_{\text{av},\,ip}| > |\vec{v}_{\text{av},\,pf}|$$

Let's take another look at how we found the average velocity for the cart in the previous example. The magnitude of the average velocity of the cart on the way up is given by (Eq. 1)

$$\vec{v}_{\text{av},\,ip} = \left[\frac{(1.8 - 0.5)\text{m}}{(3.6 - 0.5)\text{s}}\right]\hat{\imath}$$

Recall that the slope of a line is its rise divided by its run.

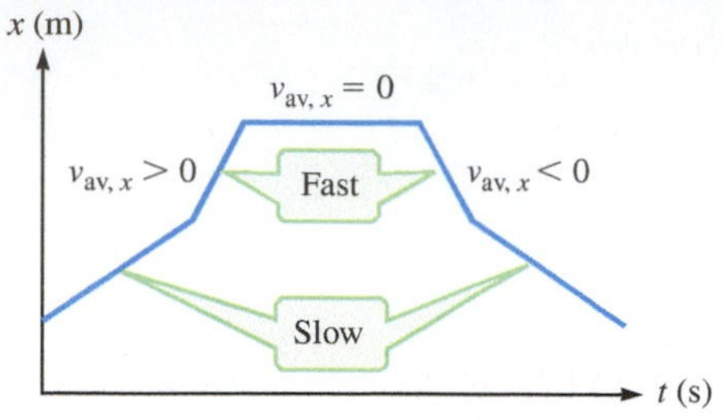

FIGURE 2.17 Finding average velocity from a position-versus-time graph. Steep slope implies a high average velocity. The sign of the slope gives the sign and the direction of the average velocity. A horizontal line implies that the particle is at rest for the time interval during which the position does not change.

The ratio in brackets, [ ], is the same as the slope of the upward-sloping orange line drawn between the initial point and the maximum point in Figure 2.16. On this graph, the rise has units of length (m), the run has units of time (s), and the slope of the orange line is 0.42 m/s.

So, another way to find the average velocity is from the position-versus-time graph. *The slope of the straight line joining two points in a position-versus-time graph is the average velocity over that time interval.* In practice, we inspect a position-versus-time graph to estimate or compare average velocities over different time intervals (Fig. 2.17) using these facts:

1. A steeper slope means a higher average velocity (faster motion).
2. The sign of the slope gives the sign of the average velocity and therefore the direction of motion.
3. A horizontal line has zero rise and therefore zero slope. So, a horizontal line means that the average velocity is zero.

INSTANTANEOUS VELOCITY

Major Concept

2-7 Instantaneous Velocity and Speed

If a particle does not speed up, slow down, or change direction, we say that it is moving at *constant* or *uniform* velocity. In this case, the particle's average velocity is equal to its velocity at every instant in time. The velocity at each instant is called the **instantaneous velocity**, usually referred to simply as **velocity**. Figure 2.18A shows a position-versus-time graph for a particle moving at constant velocity. Because the average and instantaneous velocities are equal in this case, the slope of the line gives both. Further, the slope of the line does not depend on the time interval and is the same between points a and b, b and c, or any pair of points on the position-versus-time graph:

$$\frac{\Delta\vec{x}_{ab}}{\Delta t_{ab}} = \frac{\Delta\vec{x}_{bc}}{\Delta t_{bc}} = \frac{\Delta\vec{x}_{ac}}{\Delta t_{ac}}$$

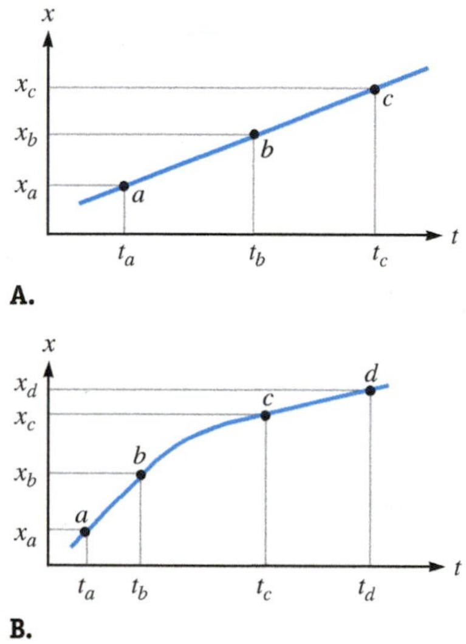

FIGURE 2.18 **A.** The position-versus-time graph has a constant slope, so the velocity is uniform and the average velocity does not depend on the time interval. **B.** The decreasing slope means the particle is slowing down.

If the speed of the particle changes (Fig. 2.18B), the average velocity depends on the time interval during which it is measured. The slope of the line between points a and b is steeper than the slope between c and d. Therefore, the particle had a higher average velocity over the first time interval Δt_{ab} than it did over the second time interval Δt_{cd}:

$$\frac{\Delta x_{ab}}{\Delta t_{ab}} > \frac{\Delta x_{cd}}{\Delta t_{cd}}$$

Instantaneous velocity at one particular time t is obtained in much the same way as average velocity over some time interval Δt, but with a shrinking Δt. Figure 2.19 shows the position-versus-time graph for a particle whose velocity is continuously changing. As the time interval gets shorter, the dashed slanted line in Figure 2.19 becomes a better and better approximation to the velocity at point a (coordinates t_a, x_a). Also, as the time interval approaches zero, the dashed

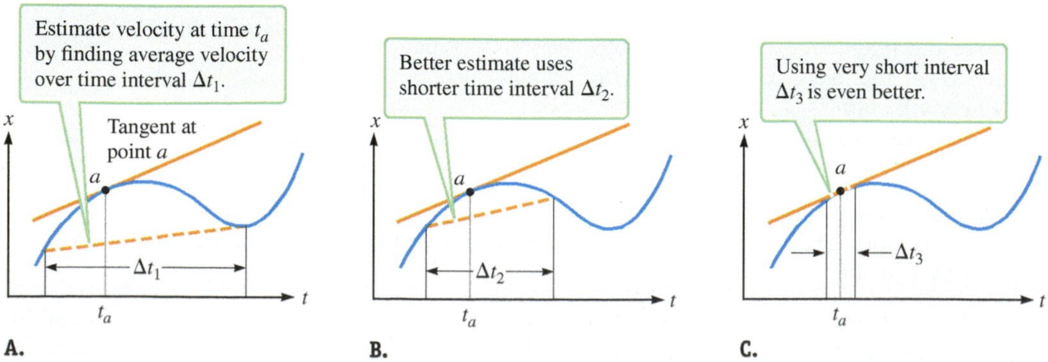

FIGURE 2.19 As Δt gets smaller, the dashed line becomes a better approximation to the tangent to the curve.

slanted line touches the curve at only one point a and is identical to the tangent to the curve at that point (Fig. 2.19C). *The slope of the tangent line at point a is the instantaneous velocity at time t_a.*

Figure 2.19 shows that as Δt approaches zero, the average velocity approaches the instantaneous velocity. So, we find expressions for the instantaneous velocity (in the x direction) by taking the limit of $\vec{v}_{\text{av},x} = \Delta\vec{x}/\Delta t$ (Eq. 2.2) as Δt approaches zero:

$$\vec{v}_x = \lim_{\Delta t \to 0} \frac{\Delta \vec{x}}{\Delta t} = \frac{d\vec{x}}{dt} = \frac{dx}{dt}\hat{\imath} \qquad (2.4)$$

Similar equations can also be written for velocity in the y and z directions.

Like average velocity, instantaneous velocity is a vector quantity. To distinguish the average velocity from instantaneous velocity, we use the additional subscript "av" for the average velocity and no additional subscript for the instantaneous velocity.

Instantaneous Speed

INSTANTANEOUS SPEED

Major Concept

The **instantaneous speed** (or simply the **speed**) is the magnitude of the instantaneous velocity. Speed is a scalar quantity and has no direction. The symbol for speed is v without an arrow.

The definition of *instantaneous* speed is very different from the definition of *average* speed. Both speed and average speed are scalars. Speed is just the magnitude of the velocity vector, but average speed is not simply the magnitude of the average velocity. Average speed is uniquely defined by $S_{\text{av}} \equiv d/\Delta t$ (Eq. 2.3).

In everyday language, we often use speed and velocity interchangeably, but not in physics.

EXAMPLE 2.5 Block and Spring: Velocity and Speed

In this example, we take another look at the block attached to the spring (Fig. 2.12). The block's position as a function of time is given by $\vec{z}(t) = (0.65 \cos \omega t)\hat{k}$, where $\omega = 1.0$ rad/s is a constant. The symbol $\vec{z}(t)$ means "z as a function of t" and not "z multiplied by t," and *rad* is the abbreviation for radians.

A Find the average velocity and average speed over the 6.25-s interval.

INTERPRET and ANTICIPATE

To find the average velocity, we divide the displacement by the time interval. To find the average speed, we must first find the total distance traveled. We expect to find two numerical answers of the form $\vec{v}_{\text{av},z} = \_\_\_\_\_ \hat{k}$ m/s and $S_{\text{av}} = \_\_\_\_\_$ m/s.

SOLVE

Average velocity is given by Equation 2.2 written in terms of z instead of x.	$\vec{v}_{\text{av},z} = \dfrac{\Delta\vec{z}}{\Delta t} \qquad (2.2)$
We already found the displacement was zero in part B of Example 2.2. Therefore the average velocity is zero.	$\Delta\vec{z} = 0$ $\vec{v}_{\text{av},z} = \dfrac{0}{6.25\text{ s}} = 0$
Use Equation 2.3 for average speed.	$S_{\text{av}} = \dfrac{d}{\Delta t} \qquad (2.3)$
We found the total distance traveled in part D of Example 2.2. Now we divide by the total time, 6.25 s.	$d = 2.60$ m $S_{\text{av}} = \dfrac{2.60\text{ m}}{6.25\text{ s}} = 0.416$ m/s

CHECK and THINK

Because the displacement is zero, the average velocity is zero. The distance traveled by the block is not zero, and so its average speed is not zero.

Example continues on page 36 ▶

B Find the velocity and speed at $t = 2.00$ s.

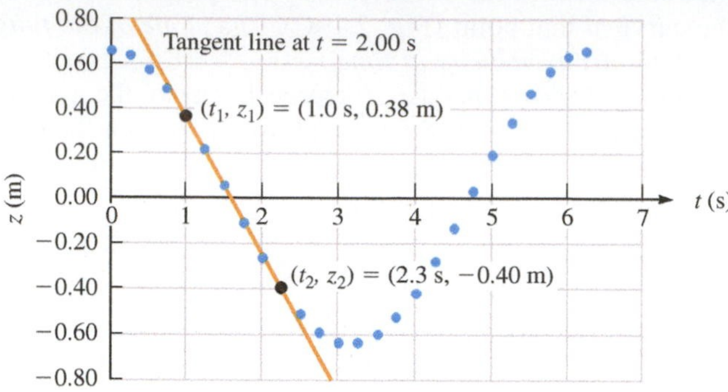

FIGURE 2.20

INTERPRET and ANTICIPATE
In part A, we found the *average* velocity and *average* speed. Here, we are asked for the *instantaneous* velocity and *instantaneous* speed. We see from the position-versus-time graph (Fig. 2.20) that the slope of the tangent line at $t = 2.00$ s is negative, so we expect the velocity to be negative. Instantaneous speed is the magnitude of velocity and will be positive.

Let's further anticipate our results by estimating the slope of the tangent line. Choose two points on the tangent line and find the slope from these two points. For example, $(t_1, z_1) = (1.0 \text{ s}, 0.38 \text{ m})$ and $(t_2, z_2) = (2.3 \text{ s}, -0.40 \text{ m})$.

$$\text{slope} = \frac{z_2 - z_1}{t_2 - t_1} = \frac{(-0.40 - 0.38) \text{ m}}{(2.3 - 1.0) \text{ s}}$$

$$\text{slope} = -0.6 \text{ m/s}$$

SOLVE
Equation 2.4 (written in terms of z) gives the velocity $\vec{v}_z(t)$ as the time derivative of $\vec{z}(t)$.

$$\vec{v}_z(t) = \frac{d\vec{z}}{dt} = \frac{d(0.65 \cos \omega t)\hat{k}}{dt} = 0.65\frac{d(\cos \omega t)}{dt}\hat{k}$$

We need the chain rule (Appendix A).

$$\vec{v}_z(t) = 0.65\frac{d(\omega t)}{dt}\frac{d(\cos \omega t)}{d(\omega t)}\hat{k}$$

Use $d(\cos\theta)/d\theta = -\sin\theta$ to arrive at the velocity at any time t.

$$\vec{v}_z(t) = -(0.65\omega \sin \omega t)\hat{k}$$

We need the velocity at $t = 2.00$ s. Substitute for t and ω. Notice that the "rad" has disappeared from our final answer. Because a radian is defined as the ratio of two lengths (circumference of a circle to its diameter) and is therefore dimensionless, we drop "rad" from our expression whenever it is possible. Here, the velocity's units are simply expressed as meters per second.

$$\vec{v}_z(2.00 \text{ s}) = -(0.65 \text{ m})(1 \text{ rad/s}) \sin[(1 \text{ rad/s})(2.00 \text{ s})]\hat{k}$$

$$\vec{v}_z(2.00 \text{ s}) = -0.59\hat{k} \text{ m/s}$$

The (instantaneous) speed is just the magnitude of the (instantaneous) velocity.

$$|v_z(2.00 \text{ s})| = 0.59 \text{ m/s}$$

CHECK and THINK
As expected, the estimated slope (-0.6 m/s) of the tangent line is close to the velocity we found, and the absolute value of the slope is close to the speed we found.

Displacement from a Velocity-Versus-Time Graph

VELOCITY-VERSUS-TIME GRAPH

Tool

In much the same way that we made a position-versus-time graph, we can make a **velocity-versus-time graph** by plotting instantaneous velocity on the vertical axis and time on the horizontal axis. A velocity-versus-time graph is useful for showing when the particle is speeding up, slowing down, or maintaining a constant velocity.

We can find the displacement of a particle from its velocity-versus-time graph. For example, Figure 2.21 shows velocity-versus-time graphs for two particles. For both particles, the velocity is constant, so the graphs are horizontal lines with $\vec{v}_x = \vec{v}_{\text{av},x}$. Let's find the displacement of each particle during the first 5.00 seconds. According to Equation 2.2, the displacement is

$$\Delta\vec{x} = \vec{v}_{\text{av},x}\Delta t = (v_{\text{av},x}\Delta t)\hat{\imath} \tag{2.5}$$

Notice that the magnitude of the displacement of each particle over the first 5 seconds is the area of the corresponding rectangle. The region between the curve and the horizontal t axis is called "the area under the curve," so the displacement is

$$\Delta\vec{x} = (\text{area under the curve})\hat{\imath}$$

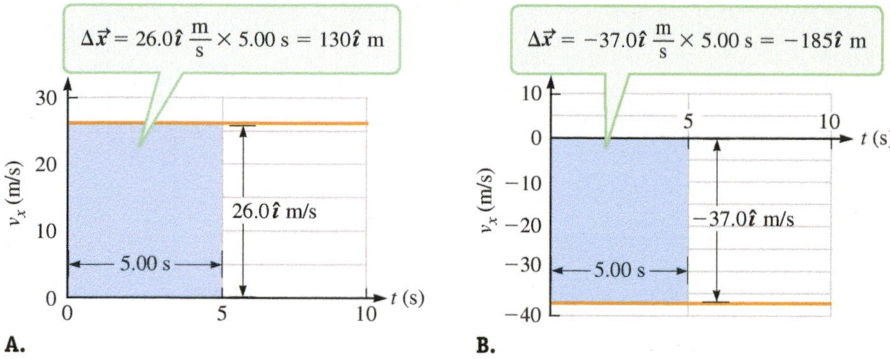

FIGURE 2.21 A. Velocity-versus-time graph for a particle moving at constant velocity $\vec{v}_x = 26.0\hat{\imath}$ m/s. The area of the blue rectangle equals the displacement in the first 5.00-s time interval. The sign of this area and the displacement are positive. **B.** Velocity-versus-time graph for a particle moving at constant velocity $\vec{v}_x = -37.0\hat{\imath}$ m/s. The sign of this area (blue rectangle) and the displacement are negative.

The area of the rectangles in Fig. 2.21 has the units (m/s) × s = m, not m^2. The sign of the displacement comes from the location of the rectangle. If the rectangle is above the t axis, the displacement is positive (Fig. 2.21A). If the rectangle is below the t axis, the displacement is negative (Fig. 2.21B).

2-8 Average and Instantaneous Acceleration

AVERAGE and INSTANTANEOUS ACCELERATION

Major Concepts

Sections 2-3 through 2-7 described motion mathematically using the concepts of position, displacement, speed, and velocity. To complete this mathematical description, we need to explore another concept, acceleration. A particle that is speeding up, slowing down, or changing direction is accelerating.

Average Acceleration

Average acceleration is the change in a particle's velocity over some time interval:

$$\vec{a}_{\text{av},\,x} \equiv \frac{\Delta\vec{v}_x}{\Delta t} = \frac{\Delta v_x}{\Delta t}\hat{\imath} = \frac{v_x(t_f) - v_x(t_i)}{t_f - t_i}\hat{\imath} = \frac{v_{fx} - v_{ix}}{t_f - t_i}\hat{\imath} \tag{2.6}$$

Similar equations can also be written in the y and z directions. The notation v_{fx} means the x component of velocity at the final time, and $v_x(t_f) = v_{fx}$, not velocity multiplied by time.

Average acceleration is a vector whose direction is determined by the subtraction in the numerator of Equation 2.6.

Instantaneous Acceleration

The **instantaneous acceleration**, also known as simply the **acceleration**, is derived from the average acceleration in the same way that instantaneous velocity is derived from average velocity:

$$\vec{a}_x \equiv \lim_{\Delta t \to 0}\frac{\Delta\vec{v}_x}{\Delta t} = \frac{d\vec{v}_x}{dt} = \frac{dv_x}{dt}\hat{\imath} \tag{2.7}$$

In SI units, both average and instantaneous acceleration are measured in meters per second per second, which is usually expressed as meters per second squared (m/s^2). Acceleration is often used in everyday language to describe something that is speeding up or sometimes even something that is increasing, such as accelerating unemployment. In physics, acceleration does not mean that a particle is moving quickly or that the speed is necessarily increasing. Rather, in physics, acceleration refers only to the rate at which the particle's velocity changes. In other words, any particle that is either speeding up or slowing down is accelerating. Furthermore, because velocity is a vector, any particle that changes direction is also accelerating, even if the particle's speed remains constant as the direction of motion changes. This last fact will be very important when we study motion in two and three dimensions.

For one-dimensional motion, a particle's acceleration at any given instant is either in the same direction as its velocity or in the opposite direction (Fig. 2.22). If the acceleration and the velocity are in the same direction, the particle is speeding up. If the acceleration and velocity are in opposite directions, the particle is slowing down, a situation sometimes called **deceleration**. Deceleration does not necessarily indicate motion in the negative direction. Deceleration is really shorthand for the phrase *acceleration in the direction opposite the velocity.*

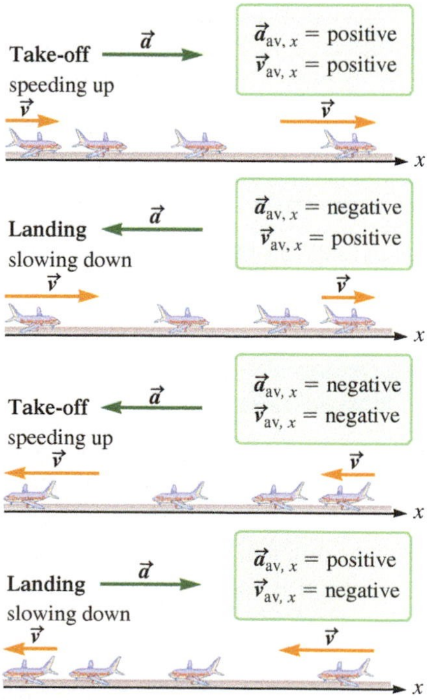

FIGURE 2.22 When the acceleration is in the same direction as the velocity, the plane speeds up. When the acceleration is in the opposite direction to the velocity, the plane slows down.

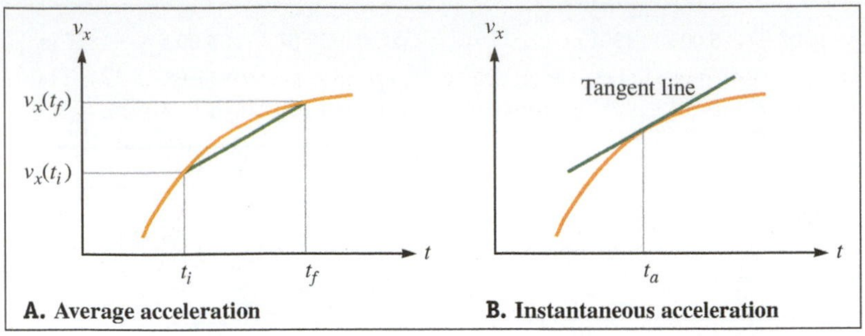

FIGURE 2.23 **A.** The slope of the line between two points gives the average acceleration over the time interval $t_f - t_i$. **B.** The slope of the tangent line gives the instantaneous acceleration at time t_a.

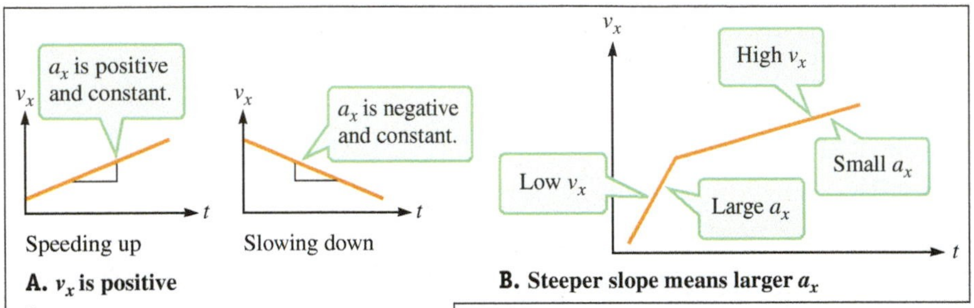

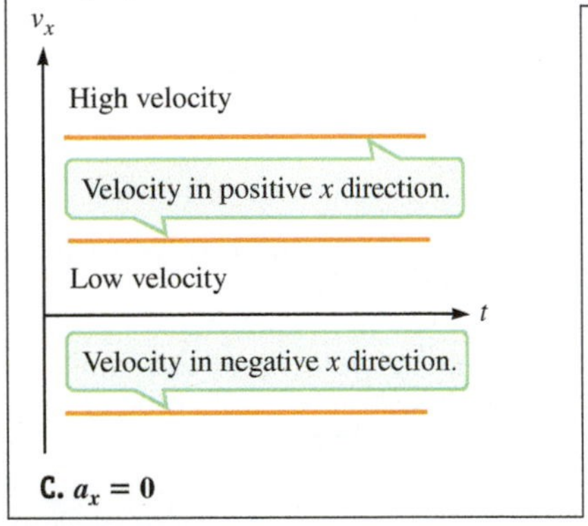

FIGURE 2.24 **A.** The sign of the slope gives the sign and direction of the acceleration. In both cases, the velocity is positive. **B.** The steeper slope means a greater acceleration. **C.** A horizontal line on a velocity-versus-time graph means that the acceleration is zero and the velocity is constant.

ACCELERATION-VERSUS-TIME GRAPH

Tool

Just as average and instantaneous velocity can be found from slopes on a position-versus-time graph, average and instantaneous acceleration can be found from slopes on a velocity-versus-time graph. The slope of the line joining two points on a velocity-versus-time graph is the average acceleration over that time interval (Fig. 2.23A). The slope of the tangent line at a particular point is the instantaneous acceleration at that point (Fig. 2.23B). The slope has dimensions of length per time squared. The sign of the slope gives the direction of the acceleration. If the velocity-versus-time curve is linear, the acceleration is constant (Fig. 2.24A). A steeper slope means a greater acceleration, but not necessarily a high velocity. Figure 2.24B shows velocity versus time for a particle that starts off with a low velocity and a large acceleration but later has a high velocity and a small acceleration. Finally, a horizontal line on a velocity-versus-time graph indicates that the velocity is constant and the acceleration is zero (Fig. 2.24C).

CONCEPT EXERCISE 2.7

Kinematics graphs are great for showing how a quantity depends on time. For example, an **acceleration-versus-time graph** shows how acceleration depends on time.

a. Based on your experience with two other kinematics graphs (position- and velocity-versus-time), describe how to plot an acceleration-versus-time graph.

b. Describe how all three kinematics graphs are connected by slopes of tangent lines (derivatives).

c. If a particle's acceleration is positive and constant, what is the shape of the curve in each of the three kinematics graphs?

EXAMPLE 2.6 Launching a Rocket

Launching a rocket into space is complicated. In this example, we will consider a very simplified scenario, ignoring the effects of the Earth's motion, the effects of the atmosphere, and the details of the rocket's design.

A rocket launched from the surface of the Earth at 11.2 km/s travels along a one-dimensional path (Fig. 2.25). With the origin of the coordinate system at the center of the Earth, the rocket's position for the first few minutes after takeoff is approximately given by

$$y(t) = A + Bt - Ct^2 + Dt^3 \tag{2.8}$$

where A, B, C, and D are constants and

$$A = 6.38 \times 10^6\ \text{m} \qquad C = 4.90\ \text{m/s}^2$$
$$B = 1.12 \times 10^4\ \text{m/s} \qquad D = 5.73 \times 10^{-3}\ \text{m/s}^3$$

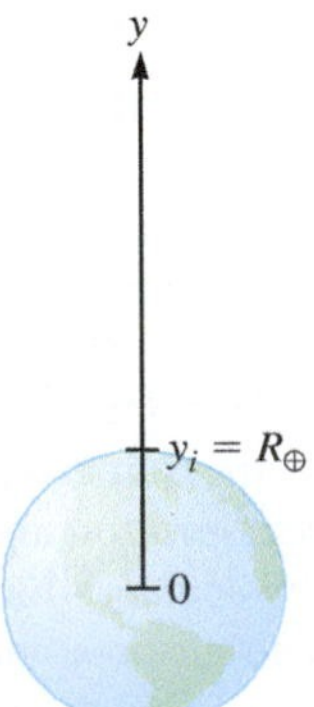

FIGURE 2.25

A Find the initial position y_i of the rocket at $t = 0$ and then find an expression for its displacement $\Delta\vec{y}(t)$.

INTERPRET and ANTICIPATE

The problem statement includes a sketch and a coordinate system (Fig. 2.25). The origin of the y axis is at the center of the Earth, and positive is upward (outward from the center).

SOLVE

To find the rocket's initial position, substitute $t = 0$ into Equation 2.8.

$$y(t) = A + Bt - Ct^2 + Dt^3$$
$$y(0) = A + (B)(0) - (C)(0)^2 + (D)(0)^3$$
$$y_i = A = 6.38 \times 10^6 \text{ m}$$

CHECK and THINK

The rocket is initially on the surface of the Earth, so $y_i = R_\oplus$, where $R_\oplus$ is the radius of the Earth. As listed in reference material (inside cover, Appendix B, or Internet), $R_\oplus = 6.38 \times 10^6$ m, as we found.

SOLVE

The magnitude of the displacement comes from Equation 2.1 written in terms of the vector component of y.

$$\Delta\vec{y}(t) = \vec{y}(t) - \vec{y}_i \quad (2.1)$$
$$\Delta\vec{y}(t) = [(A + Bt - Ct^2 + Dt^3) - (A)]\hat{\jmath}$$
$$\Delta\vec{y}(t) = (Bt - Ct^2 + Dt^3)\hat{\jmath} \quad (1)$$

CHECK and THINK

Let's check that our expression has the correct dimensions.

First, find the dimensions of the constants from their SI units.

$$[\![A]\!] = \text{L} \quad [\![B]\!] = \frac{\text{L}}{\text{T}} \quad [\![C]\!] = \frac{\text{L}}{\text{T}^2} \quad [\![D]\!] = \frac{\text{L}}{\text{T}^3}$$

Second, substitute the dimensions for each constant and variable in Equation (1). As expected, the expression for displacement has the dimensions of length.

$$[\![\Delta y]\!] = \frac{\text{L}}{\text{T}}\text{T} - \frac{\text{L}}{\text{T}^2}\text{T}^2 + \frac{\text{L}}{\text{T}^3}\text{T}^3$$
$$[\![\Delta y]\!] = \text{L}$$

B Derive expressions for $\vec{v}_y(t)$ and $\vec{a}_y(t)$.

SOLVE

Use the same coordinate system (Fig. 2.25). To find an expression for the velocity, take the time derivative of position (Eq. 2.4 written in terms of y).

$$\vec{v}_y = \frac{d\vec{y}}{dt} \quad (2.4)$$
$$\vec{v}_y = \frac{d(A + Bt - Ct^2 + Dt^3)\hat{\jmath}}{dt}$$
$$\vec{v}_y(t) = (B - 2Ct + 3Dt^2)\hat{\jmath} \quad (2)$$

To find an expression for the acceleration, take the time derivative of velocity (Eq. 2.7).

$$\vec{a}_y = \frac{d\vec{v}_y}{dt} \quad (2.7)$$
$$\vec{a}_y = \frac{d(B - 2Ct + 3Dt^2)\hat{\jmath}}{dt}$$
$$\vec{a}_y(t) = (-2C + 6Dt)\hat{\jmath} \quad (3)$$

CHECK and THINK

Use dimensions of the constants we found in part A and substitute the dimensions for each constant and variable in Equation (2). As expected, the expression for velocity has the dimensions of length per time.

$$[\![v_y]\!] = \left(\frac{\text{L}}{\text{T}} - \frac{\text{L}}{\text{T}^2}\text{T} + \frac{\text{L}}{\text{T}^3}\text{T}^2\right)$$
$$[\![v_y]\!] = \frac{\text{L}}{\text{T}}$$

Substitute the dimensions for each constant and variable in Equation (3). As expected, the expression for acceleration has the dimensions of length per time squared.

$$[\![a_y]\!] = \frac{\text{L}}{\text{T}^2} + \frac{\text{L}}{\text{T}^3}\text{T}$$
$$[\![a_y]\!] = \frac{\text{L}}{\text{T}^2}$$

CONSTANT ACCELERATION

Special Case

2-9 Special Case: Constant Acceleration

When we want to describe the motion of some particle, there are three quantities—position, velocity, and acceleration—that we would like to know at all times. In the most general situations, acceleration is *not* a constant (such as the rocket in Example 2.6). The resulting motion can be difficult to describe mathematically, often requiring computer assistance. When the *acceleration is constant*, our task is simplified because the acceleration has the same value at all times. We can describe the position and velocity of the particle at a time t in terms of the particle's position and velocity at an earlier time t_0. Let's assume the particle's motion is one-dimensional along the x axis. Typically, we set $t_0 = 0$. Then, there are five (or six) motion variables:

1. t is the time elapsed since $t_0 = 0$.
2A. $x(t)$ is the position at t.
2B. x_0 is the position at $t_0 = 0$.
3. $v_x(t)$ is the velocity at t.
4. v_{0x} is the velocity at $t_0 = 0$.
5. a_x is the constant acceleration, with the same value at all times.

Usually, the initial and final positions are replaced by the displacement $\Delta x = x(t) - x_0$.

Recall that position depends on the choice of coordinate system, and in any situation, there are many good choices for a coordinate system. The motion of the particle, however, cannot depend on our choice of coordinate system, so kinematic equations cannot depend on that choice. It works out that the mathematical description of motion is in terms of the particle's displacement $(\Delta x = x - x_0)$, which does not depend on our choice of the coordinate system. So, for constant acceleration, there are really just five variables: t, Δx, v_x, v_{0x}, and a_x.

We can derive five equations corresponding to these five variables that can be used to solve all constant-acceleration problems. Throughout these derivations, we will work with the scalar components of the motion variables along an x axis, so direction is indicated simply by the sign of the scalar component. Similar derivations can be done for any other axis.

Velocity as a Function of Time for Constant Acceleration

From Figure 2.24A, we know that the velocity-versus-time graph is linear in the special case of constant acceleration. The tangent at any point on the line coincides with the line itself. Therefore, in the case of constant acceleration a_x, the average acceleration $a_{\text{av},x}$ over any time interval and the instantaneous acceleration $a_x(t)$ at all particular times are equal:

$$a_x(t) = a_x = a_{\text{av},x} = \frac{v_x(t_f) - v_x(t_i)}{t_f - t_i}$$

Here, we use the symbol a_x without (t) to indicate that the acceleration is a constant at all times.

It is conventional practice to set the initial time to zero and replace t_f with t. The initial velocity $v_x(0)$ is written v_{0x}, and the acceleration is

$$a_x = \frac{v_x(t) - v_{0x}}{t}$$

Solve this equation for $v_x(t)$:

$$v_x(t) = v_{0x} + a_x t \qquad (2.9)$$

Equation 2.9 is the first of the five equations and gives velocity as a function of time for constant acceleration. Equation 2.9 is linear, in the form $y = mx + b$, describing either of the lines in Figure 2.24A with a_x equal to the slope (m) and v_{0x} equal to the y intercept (b).

Displacement as a Function of Velocity and Time

In Figure 2.21, we found the displacement from the area under a velocity-versus-time curve in the case of zero acceleration. When there is a nonzero acceleration, we can still find the displacement as the area under the curve. We can break this area into two shapes, the light blue triangle and the dark blue rectangle (Fig. 2.26). The sum of these two areas is the displacement in the time interval t,

$$\Delta x = \tfrac{1}{2}(v_{0x} + v_x)t \qquad (2.10)$$

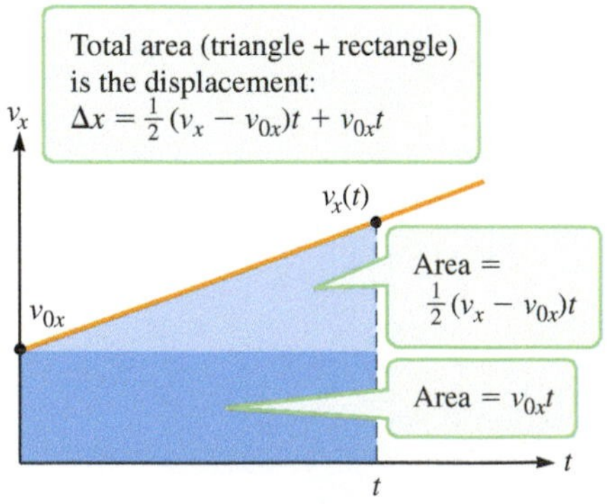

FIGURE 2.26 Velocity-versus-time graph for the special case of constant acceleration. The area under the curve is the displacement in time t.

where we've replaced $v_x(t)$ with v_x to keep the notation compact. This equation is the second of the five.

Before deriving further equations, let's take a closer look at Equation 2.10. Consider the definition of average velocity (Eq. 2.2),

$$v_{\text{av},x} = \frac{\Delta x}{\Delta t} = \frac{\Delta x}{t}$$

if $t_i = 0$ and $t_f = t$. Solving this expression for Δx, we get

$$\Delta x = v_{\text{av},x}\, t$$

Then, substituting this result into $\Delta x = \frac{1}{2}(v_{0x} + v_x)t$ (Eq. 2.10), we find

$$v_{\text{av},x} = \tfrac{1}{2}(v_{0x} + v_x)$$

In words, the average velocity is the arithmetic mean of the initial and final velocities but *only* in the special case of constant acceleration.

The Five Kinematic Equations for Constant Acceleration

Notice that each of the equations we've derived so far only involves four of the five kinematics variables (Δx, t, a_x, v_{0x} and v_x); $v_x(t) = v_{0x} + a_x t$ (Eq. 2.9) does not involve displacement, and $\Delta x = \frac{1}{2}(v_{0x} + v_x)t$ (Eq. 2.10) does not involve acceleration. We can use this information to help us solve problems. For example, if we know acceleration, initial velocity, and final velocity, we can use Equation 2.9 to solve for the time.

If we know acceleration, initial velocity, and time and we need to know the displacement, however, we have to use both Equations 2.9 and 2.10 simultaneously. A more efficient method would be to develop a new equation by eliminating final velocity from the two equations. To do so, substitute Equation 2.9 for v_x in Equation 2.10 and simplify:

$$\Delta x = \tfrac{1}{2}(v_{0x} + v_{0x} + a_x t)t$$

$$\Delta x = v_{0x}t + \tfrac{1}{2}a_x t^2 \qquad (2.11)$$

Equation 2.11 is the third of the five kinematic equations we need for one-dimensional motion with constant acceleration. A similar procedure is used to eliminate either time or initial velocity to obtain

$$\Delta x = v_x t - \tfrac{1}{2}a_x t^2 \qquad (2.12)$$

$$v_x^2 = v_{0x}^2 + 2a_x\Delta x \qquad (2.13)$$

Table 2.4 summarizes these five equations. The derivation of the last two equations—Equation 2.12 in which v_{0x} is missing and Equation 2.13 in which t is missing—is left for homework.

TABLE 2.4 Kinematic equations of motion for constant acceleration.

Equation		Eliminated Variable
$v_x = v_{0x} + a_x t$	(2.9)	Displacement Δx
$\Delta x = \frac{1}{2}(v_{0x} + v_x)t$	(2.10)	Acceleration a_x
$\Delta x = v_{0x}t + \frac{1}{2}a_x t^2$	(2.11)	Final velocity v_x
$\Delta x = v_x t - \frac{1}{2}a_x t^2$	(2.12)	Initial velocity v_{0x}
$v_x^2 = v_{0x}^2 + 2a_x\Delta x$	(2.13)	Final time t

Note: The equations are written in terms of scalar components of the motion variables along the x axis. Similar equations can also be written in the y and z directions.

Using Integral Calculus

Calculus is *the* mathematical tool most widely used by physicists today. Probably one of the first uses of calculus was in the study of motion. You can gain insight into Equation 2.9 by rederiving it using integral calculus. If you have not learned integral calculus, proceed to the **Problem-Solving Strategy: Constant Acceleration** and return to this derivation when you are ready.

(ALTERNATIVE) DERIVATION Equation 2.9

Use integral calculus to derive Equation 2.9, $v_x(t) = v_{0x} + a_x t$ in the special case of constant acceleration.

Start with Equation 2.7 for acceleration in the x direction, then rewrite as a differential equation.	$a_x = \dfrac{dv_x}{dt}$ (2.7) $a_x\,dt = dv_x$

Derivation continues on page 42 ▶

Integrate both sides from an initial time t_i to a final time t_f.	$\int_{t_i}^{t_f} a_x dt = \int_{v_{ix}}^{v_{fx}} dv_x$ (1)
Rewrite using the usual conventions of setting $t_i = 0$, $t_f = t$, $v_{ix} = v_{0x}$, and $v_{fx} = v_x$. Because a_x is constant, it can be pulled outside the integral. (That is why constant acceleration makes the mathematics simpler.)	$a_x \int_0^t dt = \int_{v_{0x}}^{v_x} dv_x$
Carry out the integration and evaluate over the limits of integration.	$a_x t \Big\|_0^t = v_x \Big\|_{v_{0x}}^{v_x}$ $a_x t = v_x - v_{0x}$
Solve for the final velocity.	$v_x = v_{0x} + a_x t$ ✓ (2.9)

COMMENTS

In many situations, the acceleration is not constant. According to Equation (1), however, you can always integrate $a(t)$ over time to find the speed. Sometimes, it is difficult to do by hand, and a computer may be needed.

PROBLEM-SOLVING STRATEGY

Constant Acceleration

As part of the **INTERPRET and ANTICIPATE** procedure, draw a **sketch** that includes a **coordinate system**.

There are four parts to the **SOLVE** procedure:

Step 1 Start by **listing** the six kinematic variables (initial position, final position, initial velocity, final velocity, acceleration, and time). Often, the initial and final positions are combined as the displacement, so your list typically includes five variables. Be sure to include the sign of each value. Write "need" next to any variable you need to solve for and write "not needed" next to any variable that you don't know and don't need to find.

Step 2 Once you have listed the parameters, you may use the constant acceleration equations **(Table 2.4)**. You can avoid doing algebra by **choosing the equation** that does not include the "not needed" variable.

Step 3 Do **algebra** before substituting values.

Step 4 Substitute **values** if appropriate.

EXAMPLE 2.7 Breaking the Sound Speed Barrier

On October 15, 1997, driver Andy Green drove a vehicle dubbed the ThrustSuperSonicCar (ThrustSSC) faster than the speed of sound, officially breaking the sound speed barrier in a land vehicle. The course he drove is 13 miles long. The first 6 miles are used to get the car up to speed. The next mile is timed, and Green's speed over the timed mile broke the record. The last 6 miles are used to slow the car back down. According to the rules, Green had to make two runs within 1 hour. His speed averaged over the two runs of the timed mile was reported as 763.035 mph (1233.704 km/h).

Assume the acceleration was constant over the first and last 6 miles and the velocity was constant over the timed mile. Report your answers to a conservative three significant figures (rather than seven!).

A Find the acceleration for the first 6 miles.

INTERPRET and ANTICIPATE

Make a **simple sketch including a coordinate system**. This sketch applies to all parts of the problem (Fig. 2.27). Because the car is speeding up for the first 6 miles, we expect the acceleration to point in the same direction as the velocity. So, we expect a numerical expression for the acceleration in the form $\vec{a}_x = +\_\_\_\_ \hat{\imath}\ \text{m/s}^2$.

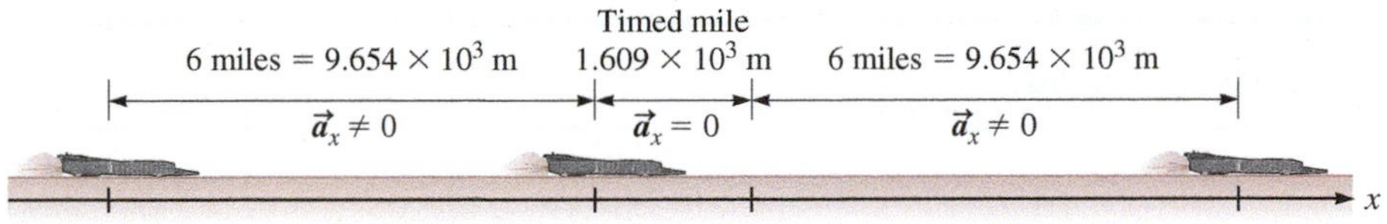

FIGURE 2.27

SOLVE

Convert any values not given in SI units to SI units. It is helpful to write these results on your sketch (Fig. 2.27).

$$6\ \text{mi}\left(\frac{1.609 \times 10^3\ \text{m}}{1\ \text{mi}}\right) = 9.654 \times 10^3\ \text{m}$$

$$1233.704\ \text{km/h}\left(\frac{0.2778\ \text{m/s}}{1\ \text{km/h}}\right) = 3.427 \times 10^2\ \text{m/s}$$

Step 1 **List** the known and unknown variables. Be sure to account for all five.

$\Delta\vec{x} = 9.654 \times 10^3 \hat{\imath}$ m

$\vec{v}_{0x} = 0$

$\vec{v}_x = 3.427 \times 10^2 \hat{\imath}$ m/s

$\vec{a}_x$ needed

t not needed

Step 2 Because time is unknown and not needed (at this point), it is best to **choose the equation** in **Table 2.4** that does not include time.

$$v_x^2 = v_{0x}^2 + 2a_x\Delta x \quad (2.13)$$

Step 3 Do **algebra.** Solve for a_x.

$$a_x = \frac{v_x^2 - v_{0x}^2}{2\Delta x}$$

Step 4 Substitute the given **values.**

$$a_x = \frac{(3.427 \times 10^2\ \text{m/s})^2 - 0^2}{2(9.654 \times 10^3\ \text{m})} = 6.08\ \text{m/s}^2$$

$\vec{a}_x = 6.08\hat{\imath}\ \text{m/s}^2$ Our result has the form we expected.

B How long did the ThrustSSC take to cover the timed mile?

SOLVE

Step 1 Again, **list** the known and unknown variables in SI units. Because we assume the velocity was constant for this part, $a_x = 0$ and $v_{0x} = v_x$. We therefore know four of the five kinematic variables.

$\Delta\vec{x} = 1\hat{\imath}$ mi $= 1.609 \times 10^3 \hat{\imath}$ m

$\vec{v}_{0x} = \vec{v}_x = 3.427 \times 10^2 \hat{\imath}$ m/s

$\vec{a}_x = 0$

t needed

Steps 2 and 3 Because the acceleration is zero, the equations in **Table 2.4** are greatly simplified. Equations 2.9 and 2.13 become $v_{0x} = v_x$, and the other three reduce to $\Delta x = v_x t$. Solve for time.

$$\Delta x = v_x t$$

$$t = \frac{\Delta x}{v_x}$$

Step 4 Substitute the **values.**

$$t = \frac{1.609 \times 10^3\ \text{m}}{3.427 \times 10^2\ \text{m/s}} = 4.70\ \text{s}$$

Example continues on page 44 ▶

CHECK and THINK

To check this result, compare it to the time needed for a car traveling at highway speed to cover 1 mile. Normal highway speed is around 60 mph. A car at 60 mph covers 1 mile in 1 minute, taking 12 to 13 times longer than the ThrustSSC took to cover the same distance. This difference makes sense because the ThrustSSC's record speed was about 12 to 13 times faster than that of a typical car on a highway.

C What was the acceleration for the last 6 miles?

SOLVE

The magnitude of the acceleration has to be the same as for the first 6 miles, but it must point in the negative x direction.	$\vec{a}_x = -6.08\hat{\imath}\ \text{m/s}^2$

CHECK and THINK

The ThrustSSC's velocity is always in the positive x direction, but its acceleration switches direction. When the car is speeding up over the first 6 miles, its acceleration is in the same direction as its velocity. Over the last 6 miles, however, the car is slowing down, so its acceleration is opposite in direction to its velocity.

EXAMPLE 2.8 High-Speed Electrons

Because so much modern technology depends on electricity, moving electrons play an important role in our everyday lives. One way to accelerate an electron is with oppositely charged plates known as electrodes, where one plate carries a positive charge and the other carries a negative charge (Fig. 2.28). We will model the electron as a particle moving with constant acceleration.

An electron initially at rest is released from the negative electrode. The electrodes are 7.5 mm apart. The electron accelerates at $1.5 \times 10^{17}\ \text{m/s}^2$ toward the positive electrode.

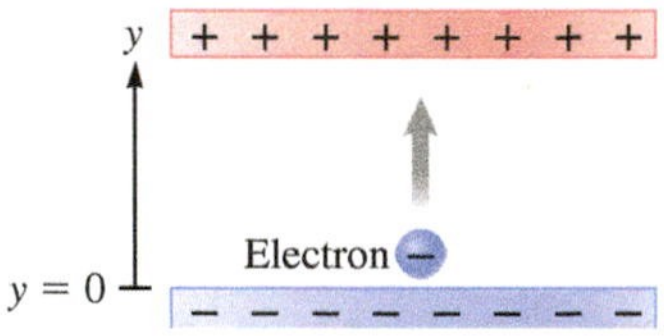

FIGURE 2.28

A Find the velocity of the electron as it strikes the positive electrode.

INTERPRET and ANTICIPATE

A sketch was provided. Because we need to find a vector quantity (velocity), it is important to choose a coordinate system before doing anything else. One good choice in this case is a vertical y axis with its origin at the negative electrode as shown in Figure 2.28. Because the acceleration is constant, we may apply the constant-acceleration equations in Table 2.4. We expect a numerical result in the form $\vec{v}_y = +\_\_\_\_\_\hat{\jmath}\ \text{m/s}$.

SOLVE

Step 1 As in the previous example, **list** the known and unknown variables. Convert them to SI units.	$\Delta\vec{y} = 7.5 \times 10^{-3}\hat{\jmath}\ \text{m}$ $\vec{v}_{0y} = 0$ $\quad\vec{a}_y = 1.5 \times 10^{17}\hat{\jmath}\ \text{m/s}^2$ $\vec{v}_y$ needed $\quad t$ not needed
Step 2 Choose an equation. Equation 2.13 does not involve time and therefore is the best choice to use here. Replace x with y because we chose to describe the motion using a y axis.	$v_y^2 = v_{0y}^2 + 2a_y\Delta y$ (2.13)
Step 3 Do **algebra.** Solve for velocity.	$v_y = \pm(\sqrt{2a_y\Delta y})$
Step 4 Substitute the **values** and choose the positive root because the electron moves in the positive direction of our chosen axis.	$v_y = \sqrt{2(1.5 \times 10^{17}\ \text{m/s}^2)(7.5 \times 10^{-3}\ \text{m})}$ $\vec{v}_y = 4.7 \times 10^7\hat{\jmath}\ \text{m/s}$

CHECK and THINK

The answer is in the form we predicted, but the electron's final speed is five orders of magnitude greater than the ThrustSSC's highest recorded speed, which seems very fast. Consider the electron's very great acceleration, 1.5×10^{17} m/s^2 (typical in the cathode-ray tube of an older television). If the electron accelerated at this rate for just 1 second, its speed would be 1.5×10^{17}m/s. That is nine orders of magnitude greater than the speed of light in a vacuum (3×10^8 m/s). According to Einstein, nothing can exceed the speed of light. So, as we will find in part B, the time for an electron to get to the positive electrode must be much shorter than 1 second.

B Find the time the electron takes to go from the negative to the positive electrode.

INTERPRET and ANTICIPATE

As we concluded in part A, we expect to find $t \ll 1$ s.

SOLVE **Step 1** As always, **list** the known and unknown kinematic variables in SI units. Now we know four of the five variables.	$\Delta\vec{y} = 7.5 \times 10^{-3}\hat{\jmath}$ m $\vec{a}_y = 1.5 \times 10^{17}\hat{\jmath}$ m/s^2 $\vec{v}_{0y} = 0$ t needed $\vec{v}_y = 4.7 \times 10^7\hat{\jmath}$ m/s
Step 2 Any equation in **Table 2.4** that involves time will work. We arbitrarily **choose** Equation 2.9.	$v_y = v_{0y} + a_y t$ (2.9)
Step 3 Do **algebra.** Solve for t.	$t = \dfrac{v_y - v_{0y}}{a_y}$
Step 4 Substitute the **values.**	$t = \dfrac{(4.7 \times 10^7 \text{ m/s}) - 0}{1.5 \times 10^{17} \text{ m/s}^2} = 3.2 \times 10^{-10}$ s $t = 0.32$ ns

CHECK and THINK

As expected, the electron was accelerated for a very short time.

2-10 A Special Case of Constant Acceleration: Free Fall

After the rocket engines in Example 2.6 shut off, the rocket slows down because of the Earth's gravity. When an object accelerates solely under the influence of gravity, it is said to be in **free fall**. The term *free fall* seems to imply that the object must be moving downward, but that is not the case. The rocket, for example, still moves upward after engine shutoff, and the term *free fall* still applies. As long as the object is affected only by gravity, free fall describes its motion after it has been thrown, launched, or dropped. The "fall" refers to the direction of the acceleration. Because this acceleration is due to gravity, it is directed downward toward the center of the Earth.

FREE FALL ▶ **Special Case**

Free fall is an idealization. In real situations, such factors as a planet's atmosphere affect the motion of the object. Often, these effects are much weaker than the effect of gravity, and in these cases, free fall is a good approximation.

When an object rises far above the Earth's surface, its acceleration may drop appreciably. For objects that stay close to the Earth's surface, however, acceleration is nearly constant in free fall, and the magnitude is given by

$$g = 9.81 \text{ m/s}^2$$

The name for "g" is somewhat controversial (more about that in Chapters 5 and 7). For now, we will use the term *free-fall acceleration*. Although we will assume g is

Do **not** refer to "g" as "gravity."

given by the constant above, it is really an approximation. The free-fall acceleration at the Earth's surface varies slightly with altitude and latitude. For example, at the poles, $g \approx 9.83\ \text{m/s}^2$, and at the top of Mount Everest, $g \approx 9.77\ \text{m/s}^2$.

PROBLEM-SOLVING STRATEGY

Free Fall

Free fall near the surface of the Earth is a special case of constant acceleration, and the **Problem-Solving Strategy: *Constant Acceleration*** (p. 42) is applicable. When listing the kinematic variables in step 1, you know that the acceleration is $\pm g$, depending on your choice of coordinate system. (It is often convenient to choose an upward-pointing y axis, and for this choice, $\vec{a}_y = -g\hat{\jmath}$.)

EXAMPLE 2.9 Welcome to the Model Rocket Club

A model rocket is launched straight up at 11.2 m/s. Assume the engine fires very briefly and then burns out. What peak height does the rocket reach?

INTERPRET and ANTICIPATE

This example is similar to Example 2.6, so we can use the same sketch, and by using the same upward-pointing y axis, we can easily compare the two problems. The major difference between them is that the model rocket's initial speed is three orders of magnitude lower than the full-size rocket's initial speed, so we expect the model rocket will not go nearly as high.

SOLVE

Step 1 List the known and unknown kinematic variables. At the peak height, the model rocket's velocity is momentarily zero.

$\Delta\vec{y} = \text{needed}$ $\quad \vec{a}_y = -g\hat{\jmath} = -9.81\hat{\jmath}\ \text{m/s}^2$

$\vec{v}_{0y} = 11.2\hat{\jmath}\ \text{m/s}$ $\quad t$ not needed

$\vec{v}_y = 0$

Step 2 We **choose an equation** that does not involve time (Eq. 2.13).

$$v_y^2 = v_{0y}^2 + 2a_y\Delta y \qquad (2.13)$$

Step 3 Do **algebra.** Solve for displacement.

$$\Delta y = \frac{v_y^2 - v_{0y}^2}{2a_y}$$

Step 4 Substitute **values.**

$$\Delta y = \frac{0^2 - (11.2\ \text{m/s})^2}{2(-9.81\ \text{m/s}^2)}$$

$$\Delta\vec{y} = 6.39\hat{\jmath}\ \text{m}$$

CHECK and THINK

As expected, the model rocket stays near the surface of the Earth. Its displacement is about the height of a two-story building. (Often when checking an answer, it is helpful to consult Appendix B's tables of approximate values.) Think about how this model rocket example differs from the full-size rocket in Example 2.6. The model rocket stays near the surface of the Earth. Its acceleration is essentially constant for its entire flight. Over the course of the full-size rocket's flight, its acceleration decreases according to $\vec{a}_y(t) = (-2C + 6Dt)\hat{\jmath}$. We cannot apply the constant acceleration equations to the full-size rocket, but we can apply them to the model rocket.

EXAMPLE 2.10 Hard Hats Required!

A construction worker is riding an exterior elevator to the top of a skyscraper. The elevator's speed is 4.3 m/s. When the elevator is 72.4 m above the ground, a screwdriver falls off.

A What is the velocity of the screwdriver as it hits the ground?

INTERPRET and ANTICIPATE

If we **choose an upward-pointing *y* axis** as in Example 2.9, we expect the velocity of the screwdriver to be in the negative *y* direction when it hits the ground. Your choice for the origin does not affect your results. Figure 2.29 shows two possible choices.

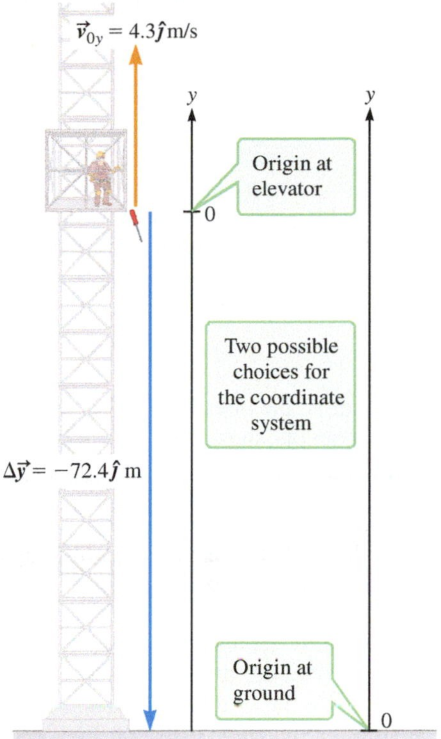

FIGURE 2.29

SOLVE

Step 1 List the known and unknown variables. The screwdriver's initial velocity is the same as the velocity of the elevator, so the screwdriver will actually continue *upward* for a short time. Notice that the displacement does not depend on which origin you choose.	$\Delta\vec{y} = -72.4\hat{\jmath}$ m $\vec{v}_{0y} = +4.3\hat{\jmath}$ m/s $\vec{v}_y$ needed $\vec{a}_y = -g\hat{\jmath} = -9.81\hat{\jmath}$ m/s^2 t not needed
Step 2 Of the equations in **Table 2.4**, Equation 2.13 is the one that contains no time term.	$v_y^2 = v_{0y}^2 + 2a_y\Delta y$ (2.13)
Step 3 Do **algebra.** Solve for v_y. Choose the negative root because as the screwdriver hits the ground, its velocity is downward (in the negative *y* direction in the coordinate system we have chosen).	$v_y = \pm\sqrt{v_{0y}^2 + 2a_y\Delta y}$
Step 4: Substitute the **values.**	$\vec{v}_y = -\sqrt{(4.3\text{ m/s})^2 + 2(-9.81\text{ m/s}^2)(-72.4\text{ m})}\,\hat{\jmath}$ $\vec{v}_y = -37.9\hat{\jmath}$ m/s

CHECK and THINK

The velocity is about 85 mph, or a little faster than typical highway speed.

B How long does it take the screwdriver to reach the ground?

SOLVE

Step 1 As in previous examples, we now know four of the five kinematic variables. We need time. Any of Equations 2.9 through 2.13 will do; we **choose** Equation 2.9.	$v_y = v_{0y} + a_y t$ (2.9)
Step 2 Do **algebra** Solve for time.	$t = \dfrac{v_y - v_{0y}}{a_y}$
Step 3 Substitute the **values.**	$t = \dfrac{-37.9\text{ m/s} - 4.3\text{ m/s}}{-9.81\text{ m/s}^2} = 4.3\text{ s}$

Example continues on page 48 ▶

CHECK and THINK

The model rocket (Example 2.9) and this falling screwdriver problem are very similar. Both required a solution for time after finding some other kinematic variable. In this example, we solved first for v_y and then for t. What if we tried to solve for t first? Our list of known variables would be the same, but we would label t "needed" and v_y "not needed." In that case, choose Equation 2.11, $\Delta y = v_{0y}t + \frac{1}{2}a_y t^2$. Solving for t, we find

$$t = \frac{-v_{0y} \pm \sqrt{v_{0y}^2 + 2a_y \Delta y}}{a_y}$$

We are left with a decision to make. How do we choose between the positive and negative roots of the square-root term? To answer this question, we notice that the expression under the square-root sign is the right side of Equation 2.13, so we can write

$$t = \frac{-v_{0y} \pm \sqrt{v_{0y}^2 + 2a_y \Delta y}}{a_y} = \frac{-v_{0y} \pm \sqrt{v_y^2}}{a_y}$$

As in part A, the direction of $\vec{v}_y$ determines whether we choose the positive root or the negative one. In this example, the correct choice is the negative root, indicating motion in the downward direction.

EXAMPLE 2.11 Movie Stunt

In many movies, a hero jumps off a structure and lands on a moving vehicle. The hero must, of course, watch the moving vehicle and figure out just the right moment to jump.

Superwoman stands on a bridge as a flatbed truck drives toward the bridge at a constant speed of 18 m/s. She steps off the bridge and falls straight down, landing 10.0 m below in the bed of the truck. How far from the bridge was the truck when she stepped off the bridge?

INTERPRET and ANTICIPATE

Make a **simple sketch** that includes two **coordinate axes,** one for Superwoman and one for the truck (Fig. 2.30). The key to solving this problem is to realize that the time it takes Superwoman to fall 10.0 m must be equal to the time it takes the truck to get to the bridge. Once you know that time, use the truck's speed to find the truck's displacement during that time.

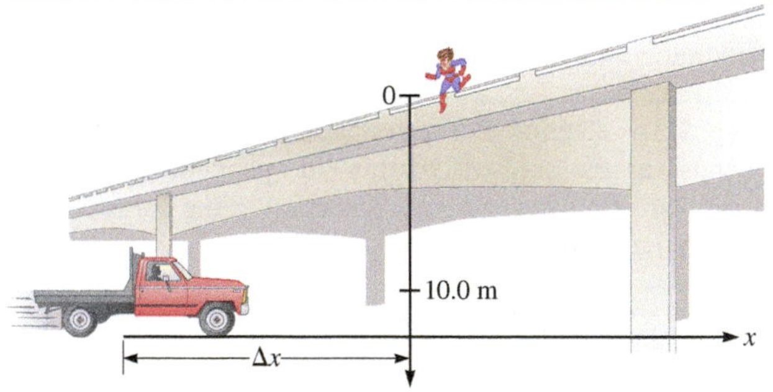

FIGURE 2.30

SOLVE

Step 1 List the known and unknown variables for Superwoman. We have chosen a downward-pointing y axis and for this choice, $\vec{a}_y = g\hat{\jmath}$.	*Superwoman:* $\Delta\vec{y} = 10.0\hat{\jmath}$ m; $\vec{a}_y = g\hat{\jmath} = 9.81\hat{\jmath}$ m/s^2; $\vec{v}_{0y} = 0$; t must be same as for truck; $\vec{v}_y$ not needed
List the known and unknown variables for the truck.	*Truck:* $\Delta\vec{x}$ = needed; $\vec{a}_x = 0$; $\vec{v}_{0x} = \vec{v}_x = 18\hat{\imath}$ m/s (constant); t, same as for Superwoman
Step 2 We **choose an equation** that can be used to find the time it takes Superwoman to fall the 10.0 m. Because we do not know and do not need her final velocity v_y, our best choice is Equation 2.11.	$\Delta y = v_{0y}t + \frac{1}{2}a_y t^2 = 0 + \frac{1}{2}gt^2$ (2.11) $t = \sqrt{\frac{2\Delta y}{g}}$ (1)
Next, we select an equation for Δx for the truck, assuming t is known from Equation (1). Again our **choice** is Equation 2.11.	$\Delta x = v_{0x}t + \frac{1}{2}a_x t^2 = v_{0x}t + 0$ (2.11) $\Delta x = v_{0x}t$

Steps 3 and 4: Do **algebra**. Substitute t from Equation (1) and the given numerical **values** from the problem statement.	$\Delta x = v_{0x}\sqrt{\frac{2\Delta y}{g}} = (18 \text{ m/s})\sqrt{\frac{2(10.0 \text{ m})}{9.81 \text{ m/s}^2}}$ $\Delta\vec{x} = 26\hat{\imath} \text{ m}$

CHECK and THINK

The truck was 26 m or nearly 90 ft from the bridge, or about the distance between adjacent bases in a baseball diamond, when Superwoman stepped off the bridge.

Graphical Solutions

We have seen that graphs of position, velocity, or acceleration versus time may be used in the **INTERPRET and ANTICIPATE** procedure and then in the **CHECK and THINK** procedure. Sometimes, these graphs may be used in the **SOLVE** procedure as shown the next example.

EXAMPLE 2.12 Circus Rehearsal

Two circus performers practice one segment of their act. Horatio stands on a platform 8.0 m above the ground and drops a ball straight down. At the same moment, Amelia uses a spring-loaded device on the ground to launch a dart straight up toward the ball. The dart is launched at 11.5 m/s. How far above the ground and how long after launch does the dart hit the ball? Solve this problem graphically.

INTERPRET and ANTICIPATE

Make a **sketch** and **choose** an upward-pointing y axis with its origin at ground level (Fig. 2.31). To solve this problem graphically, make position-versus-time graphs for the ball and the dart on the same set of axes. To do so, we need position as a function of time for the ball and for the dart. We expect the dart to hit the ball at some point between the ground and the platform.

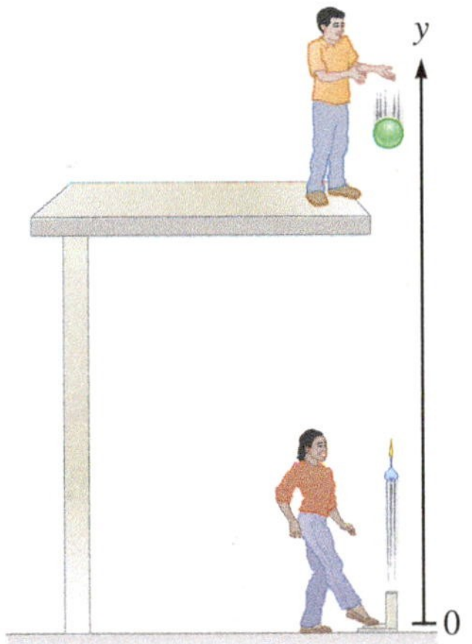

FIGURE 2.31

SOLVE

List the known and unknown variables for the ball. Because we need position and not displacement, we must list initial and final positions separately. The subscript "*B*" stands for ball.	*Ball:* y_{Bf} need equation; v_{Bfy} not needed $y_{Bi} = 8.0 \text{ m}$; $\vec{a}_y = -g\hat{\jmath} = -9.81\hat{\jmath} \text{ m/s}^2$ $v_{Biy} = 0$; t must be part of equation
Because the final velocity is not needed, use Equation 2.11.	$\Delta y_B = v_{Biy} + \frac{1}{2}a_y t^2$ (2.11)
Use the definition of displacement to convert this equation to one for final position of the ball.	$\Delta y_B = y_{Bf} - y_{Bi}$ $y_{Bf} = y_{Bi} + v_{Biy}t + \frac{1}{2}a_y t^2$
Substitute known values.	$y_{Bf} = \left(8.0 - \frac{9.81}{2}t^2\right) \text{ m}$ (1)
Use the same y axis and follow the same procedure for the dart. The subscript "*D*" stands for dart.	*Dart:* y_{Df} need equation; v_{Dfy} unknown $y_{Di} = 0$; $\vec{a}_y = -g\hat{\jmath} = -9.81\hat{\jmath} \text{ m/s}^2$ $v_{Diy} = 11.5 \text{ m/s}$; t must be part of equation
As in the case of the ball, the dart's position comes from Equation 2.11.	$y_{Df} = y_{Di} + v_{Diy}t + \frac{1}{2}a_y t^2$ (2.11) $y_{Df} = \left(11.5t - \frac{9.81}{2}t^2\right) \text{ m}$ (2)

Example continues on page 50 ▶

Plot Equations (1) and (2), the position as a function of time for each object, on the same set of axes (Fig. 2.32). A spreadsheet program is helpful.

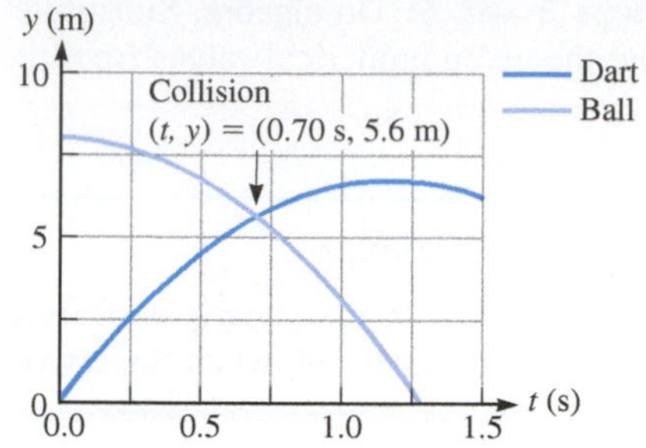

FIGURE 2.32

A close inspection reveals that these curves intersect at $(t, y_f) = (0.70\ \text{s}, 5.6\ \text{m})$.

The dart hits the ball 5.6 m above the ground 0.70 s after the dart is launched.

CHECK and THINK

As expected, the point of intersection $y = 5.6$ m is between the ground $(y = 0)$ and the platform $(y = 8.0\ \text{m})$. We solved this problem graphically, but we could have solved Equations (1) and (2) simultaneously (Problem 88).

SUMMARY

Underlying Principles

1. **Kinematics:** description of motion independent of its cause.
2. **One-dimensional motion:** motion along a single straight line.

Major Concepts

1. **Translational motion:** an object can be treated as a particle undergoing purely translational motion if every point on the object undergoes exactly the same displacement as every other such point.
2. **Particle:** an idealized point with no spatial extent, no shape, and no internal structure.
3. **Position $\vec{r}$** of a particle: its location with respect to a chosen coordinate system. The direction of position in one dimension is indicated by a positive or a negative sign.
4. **Displacement:** the change in position $(\Delta\vec{r})$. The vector component of displacement for each coordinate axis is

$$\Delta\vec{x} \equiv (x_f - x_i)\hat{\imath} \quad \Delta\vec{y} \equiv (y_f - y_i)\hat{\jmath} \quad \Delta\vec{z} \equiv (z_f - z_i)\hat{k} \tag{2.1}$$

5. **Distance traveled** (d): the length of the whole path covered by a moving particle; a scalar quantity.
6. **Average velocity:** the ratio of displacement to change in time.

$$\vec{v}_{\text{av},x} \equiv \frac{\Delta\vec{x}}{\Delta t} = \frac{x_f - x_i}{t_f - t_i}\hat{\imath} \quad \vec{v}_{\text{av},y} \equiv \frac{\Delta\vec{y}}{\Delta t} = \frac{y_f - y_i}{t_f - t_i}\hat{\jmath}$$

$$\vec{v}_{\text{av},z} \equiv \frac{\Delta\vec{z}}{\Delta t} = \frac{z_f - z_i}{t_f - t_i}\hat{k} \tag{2.2}$$

Instantaneous velocity (also known as **velocity**): the average velocity as Δt approaches zero.

$$\vec{v}_x = \lim_{\Delta t\to 0}\frac{\Delta\vec{x}}{\Delta t} = \frac{d\vec{x}}{dt} = \frac{dx}{dt}\hat{\imath} \quad \vec{v}_y = \lim_{\Delta t\to 0}\frac{\Delta\vec{y}}{\Delta t} = \frac{dy}{dt}\hat{\jmath}$$

$$\vec{v}_z = \lim_{\Delta t\to 0}\frac{\Delta\vec{z}}{\Delta t} = \frac{dz}{dt}\hat{k} \tag{2.4}$$

7. **Average speed:** depends on total distance traveled such that

$$S_{\text{av}} \equiv \frac{d}{\Delta t} \tag{2.3}$$

Instantaneous speed (or simply the **speed**): magnitude of the velocity.

Major Concepts—cont'd

8. **Average acceleration:** change in a particle's velocity over some time interval.

$$\vec{a}_{av,x} \equiv \frac{\Delta \vec{v}_x}{\Delta t} = \frac{\Delta v_x}{\Delta t}\hat{\imath} = \frac{v_{xf} - v_{xi}}{t_f - t_i}\hat{\imath}$$

$$\vec{a}_{av,y} \equiv \frac{\Delta v_y}{\Delta t}\hat{\jmath} \quad \vec{a}_{av,z} \equiv \frac{\Delta v_z}{\Delta t}\hat{k} \tag{2.6}$$

Instantaneous acceleration (also known as simply **acceleration**): the average acceleration as Δt approaches zero.

$$\vec{a}_x \equiv \lim_{\Delta t \to 0} \frac{\Delta \vec{v}_x}{\Delta t} = \frac{d\vec{v}_x}{dt} = \frac{dv_x}{dt}\hat{\imath}$$

$$\vec{a}_y \equiv \frac{dv_y}{dt}\hat{\jmath} \quad \vec{a}_z \equiv \frac{dv_z}{dt}\hat{k} \tag{2.7}$$

Special Cases

1. In the special case of **constant acceleration** the acceleration has the same magnitude and direction at all times, greatly simplifying the mathematical analysis.
2. An object accelerating solely under the influence of gravity is said to be in **free fall**. An object in free fall near the surface of the Earth has acceleration of magnitude $g = 9.81 \text{ m/s}^2$.

Tools

1. **Motion diagrams** show the location of a particle at regular time intervals.
2. **Vector** quantities have both magnitude and direction. In one-dimensional motion, a vector can have one of two directions. A **scalar** quantity has a magnitude but not a direction.
3. **Coordinate systems** have two elements in one dimension, (1) a reference point called the **origin** indicated by a zero and (2) a **coordinate axis** (or just **axis**) passing through the origin and usually labeled x, y, or z.
4. A **position-versus-time graph** is made by plotting position on the vertical axis and time on the horizontal axis. **Velocity-versus-time** and **acceleration-versus-time graphs** are similar to position-versus-time graphs except that velocity or acceleration instead of position is plotted on the vertical axis.

GENERAL PROBLEM-SOLVING STRATEGIES

Three visualization tools may help in the **INTERPRET and ANTICIPATE** procedure:

1. Motion diagram
2. Sketch with a coordinate system
3. Position-, velocity-, or acceleration-versus-time graphs

Not all problems involve constant acceleration. When the acceleration changes, you must use calculus (not Table 2.4) to solve the problem.

PROBLEM-SOLVING STRATEGY Special Cases: Constant Acceleration and Free Fall

As part of the **INTERPRET and ANTICIPATE** procedure: Draw a **sketch** that includes a **coordinate system.**

There are four parts to the **SOLVE** procedure:

1. Start by **listing** the six kinematic variables (initial position, final position, initial velocity, final velocity, acceleration, and time). Often, the initial and final positions are combined as the displacement, so your list typically includes five variables. Include the sign of each value. Write "need" next to any variable you need to solve for and write "not needed" next to any variable that you don't know and don't need to find. If the problem involves free fall, remember that the acceleration has a magnitude of $g = 9.81 \text{ m/s}^2$ and points downward toward the center of the Earth.
2. Once you have listed the parameters, you may use the constant acceleration equations (**Table 2.4**). You can avoid doing algebra by choosing the equation that does not include the "not needed" variable.
3. Do **algebra** before substituting values.
4. Substitute **values** if appropriate.

PROBLEMS AND QUESTIONS

A = algebraic **C** = conceptual **E** = estimation **G** = graphical **N** = numerical

2-1 What Is One-Dimensional Translational Kinematics?

1. **C** Is the Moon's motion around the Earth one-dimensional? Explain your answer.

2-2 Motion Diagrams

Problems 2 and 4 are paired.

2. **C** An animal's tracks are frozen in the snow (Fig. P2.2). Can these tracks be used to make a motion diagram? If so, what are the shortcomings of a motion diagram made from these data? If not, why not?

Christopher P. Grant/Shutterstock.com

FIGURE P2.2 Problems 2 and 4.

2-3 Coordinate Systems and Position

Problems 3 and 12 are paired.

3. **G** A particle moves from position A to position C as shown in Figure P2.3. Two different coordinate systems have been chosen. Write expressions for the particle at all three positions using **a.** the top coordinate system, and **b.** the bottom coordinate system.

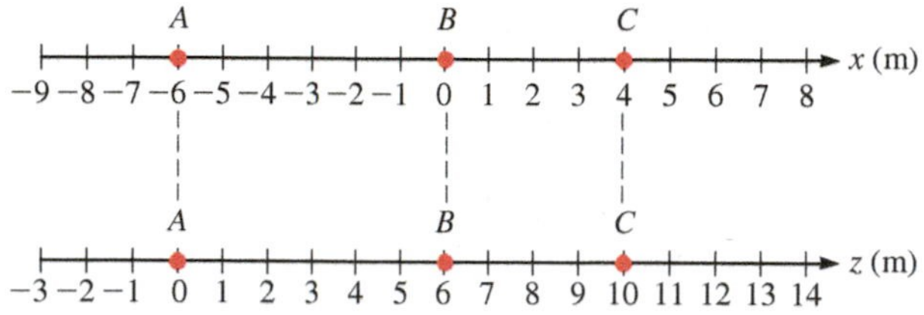

FIGURE P2.3 Problems 3 and 12.

2-4 Position-Versus-Time Graphs

4. **C** An animal's tracks are frozen in the snow (see Fig. P2.2). Can these tracks be used to make a position-versus-time graph? Explain.
5. **N** For each of the following velocity vectors, give the vector component, the scalar component, and the magnitude.
   a. $\vec{v} = 35.0\hat{\jmath}$ m/s **c.** $v_z = -3.50$ m/s
   b. $v_x = 53.0$ m/s **d.** $\vec{v} = -5.30\hat{\imath}$ m/s
6. In the traditional Hansel and Gretel fable, the children drop crumbs of bread on the ground to mark their path through the woods. Unfortunately, the crumbs are eaten by birds, and the children cannot find their way home. In this modern-day problem, the children use a device that releases a drop of food dye once per minute. As long as it does not rain, they can find their way home. As an extra bonus, they make a motion diagram as shown in Figure P2.6.
   a. **C** Describe the motion of the children in words.
   b. **G** Using the coordinate system in Figure P2.6, make a position-versus-time graph. Note any ambiguities you encounter.
   c. **C** Is your position-versus-time graph consistent with your description in part (a)? Explain.
7. **G** After a long and grueling race, two cadets, A and B, are coming into the finish line at the Marine Corps marathon. They move in the same direction along a straight path; the position-versus-time graphs for the runners are shown in Figure P2.7 for a minute near the end of the race. **a.** At time t_1 is the speed of cadet B greater than, less than, or equal to the speed of cadet A? **b.** At time t_2, is cadet B speeding up, slowing down, or moving with constant speed? **c.** From time $t = 0$ to time $t = 60$ s, is the average speed of cadet B greater than, less than, or equal to the average speed of cadet A? Explain.

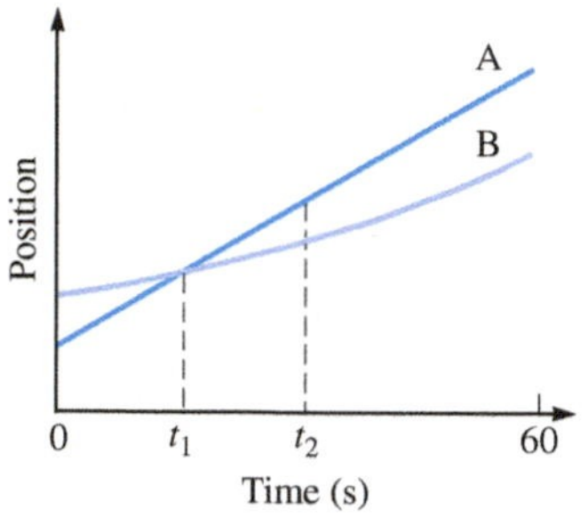

FIGURE P2.7

8. **G** Cassidi Reese (who graduated with a degree in physics) enjoys skydiving in her free time. She volunteered to wear a device during a jump that automatically measures her altitude as a function of time. Assume she maintains a nearly straight path from the plane to her target on the ground during a jump. (We will learn in Chapter 4 that in the absence of air resistance an objected dropped from a plane does not fall along a straight path.) Her data are presented in a position-versus-time graph (Fig. P2.8.) The device automatically stops taking data at an altitude of approximately 500 m. **a.** Draw a sketch of this problem. The position-versus-time graph implies a particular choice of coordinate system. Include this coordinate system on your sketch. **b.** Describe Reese's motion in words. Your description should include when she sped up, slowed down, or maintained

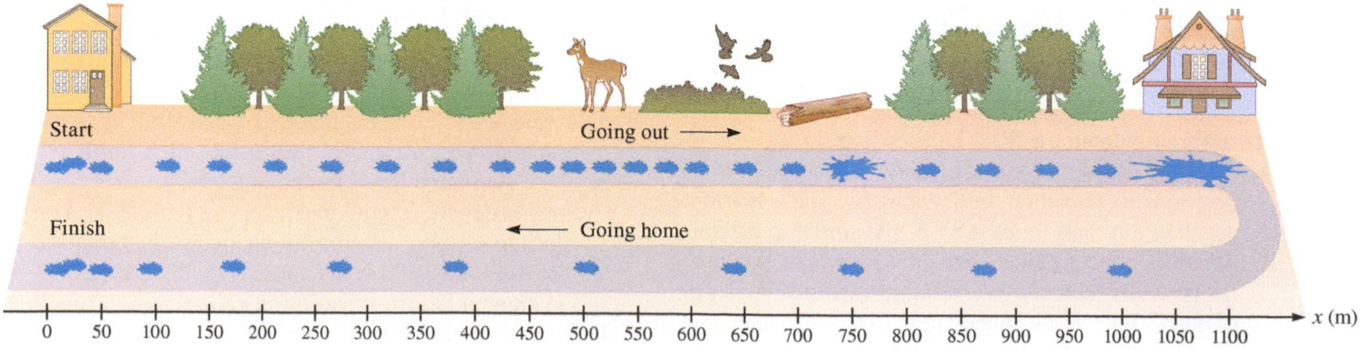

FIGURE P2.6

speed. **c.** For convenience, her altitude was written on the graph every 10 s starting at $t = 0$. Use these data to create a motion diagram for Reese. **d.** Does your motion diagram match your description in part (b)?

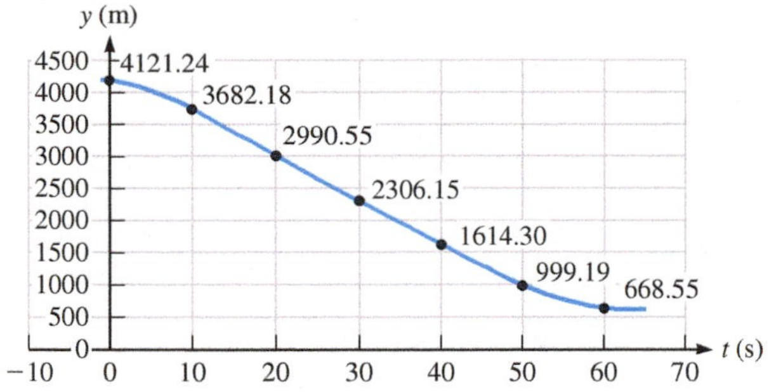

FIGURE P2.8

9. Elisha Graves Otis invented the elevator brake in the mid-1800s, making it possible to build tall skyscrapers with fast elevators. Today's skyscrapers are a large fraction of a mile tall; for example. Taipei 101 in Taiwan has 101 stories and is 515 m (0.32 miles) tall. The top speed of the elevator in the Taipei 101 tower is roughly three times greater than the ascent rate of a commercial jet airplane. The position and time data in the table are based on such an elevator.

a. G Working in SI units, make a position-versus-time graph for the elevator. (You may wish to use a spreadsheet program.)

b. C Describe the motion of the elevator in words.

c. N Find the highest speed of the elevator. When is the elevator going at this speed?

d. C What sort of considerations would the engineers need to make to ensure the comfort of the passengers?

Position (stories)	Time (s)	Position (stories)	Time (s)	Position (stories)	Time (s)
0.0	0	32.7	13	72.0	25
0.3	1	36.0	14	75.3	26
1.1	2	39.3	15	78.5	27
2.5	3	42.5	16	81.6	28
4.4	4	45.8	17	84.2	29
6.8	5	49.1	18	86.5	30
9.8	6	52.4	19	88.2	31
13.1	7	55.6	20	89.6	32
16.4	8	58.9	21	90.0	33
19.6	9	62.2	22	90.0	34
22.9	10	65.5	23	90.0	35
26.2	11	68.7	24	90.0	36
29.5	12				

10. CASE STUDY As shown in Figure 2.9, Whipple chose a coordinate system that was different from Crall's. There are many possible coordinate systems that can be used to analyze the motion of the cart. For example, a third student (Yoon) chose to place the origin of her coordinate system at position A (the cart's initial position) and use a southward-pointing y axis as her positive axis.

a. N Use the distance information given in Table 2.2 to make a new table that gives the cart's position at each of the 16 times according to Yoon's coordinate system. *Hint*: A sketch similar to Figure 2.9 is a good place to start.

b. G Use the data in your table to make a new position-versus-time graph. *Hint*: **CHECK and THINK** about your graph by comparing it with Figure 2.6.

2-5 Displacement and Distance Traveled

11. C When is the distance traveled by a particle along one straight line less than the magnitude of its displacement?

12. G A particle moves from position A to position C as shown in Figure P2.3. Two different coordinate systems have been chosen. Write expressions for the particle's displacement from A to C using **a.** the top coordinate system and **b.** the bottom coordinate system. **c.** Compare your results. Do they make sense? Explain.

13. N A race car travels 825 km around a circular sprint track of radius 1.313 km. How many times did it go around the track?

14. A woman lives on the fourth floor of an apartment building. She works in a high-rise office building 8.5 blocks away from her apartment on the same street. Her office is on the 12th floor. Assume each story of her apartment building is 4.0 m, each story of her office building is 5.5 m, and a block is 146.6 m long.

a. C Sketch her path.

b. E Estimate the distance she travels to work.

c. N Find the magnitude of her displacement.

15. N A train leaving Albuquerque travels 293 miles, due east, to Amarillo. The train spends a couple of days at the station in Amarillo and then heads back west 107 miles where it stops in Tucumcari. Suppose the positive x direction points to the east and Albuquerque is at the origin of this axis. **a.** What is the total distance traveled by the train from Albuquerque to Tucumcari? **b.** What is the displacement of the train for the entire journey? Give both answers in appropriate SI units.

16. E Milwaukee, Wisconsin, and Grand Rapids, Michigan, are on opposite sides of Lake Michigan. Assume they are at the same elevation. The best highway route between the two cities is 266.9 miles long. Use a map to estimate the displacement of a car that travels from Milwaukee to Grand Rapids. Give your answer in miles and in meters. Be sure to include an estimate of the direction.

Problems 17, 18, 19, and 40 are grouped.

17. N The position of a particle attached to a vertical spring is given by $\vec{y} = (y_0 \cos \omega t)\hat{\jmath}$. The y axis points upward, $y_0 = 14.5$ cm, and $\omega = 18.85$ rad/s. Find the position of the particle at **a.** $t = 0$ and **b.** $t = 9.0$ s.
Give your answers in centimeters.

18. A particle is attached to a vertical spring. The particle is pulled down and released, and then it oscillates up and down. Using an upward-pointing y axis, the position of the particle is given by

$$\vec{y} = \left(y_0 \cos \frac{2\pi t}{T}\right)\hat{\jmath}$$

Both y_0 and T are constants in time. The amplitude (y_0) is usually given in meters, and the period (T) is usually in seconds.

a. G Draw a sketch of this situation. Include the y axis.

b. N What is the position of the particle when $t = 0, \frac{1}{2}T, T, \frac{3}{2}T, 2T$ and $\frac{5}{2}T$? Add these positions with clear labels to your sketch.

19. Return to the description of the particle attached to the spring in Problem 17.

a. N Find the displacement of the particle during the time interval from $t = 0$ to $t = 9.0$ s. What does your answer mean?

b. N Find the distance (in centimeters) the particle traveled during this time interval. How can this distance be larger than the magnitude of the displacement?

c. C Many physical systems are modeled by a particle attached to a spring. List some examples of systems that may be modeled by springs. It may be helpful to use the index of this book or the Internet.

2-6 Average Velocity and Speed

20. **C** A particle is constrained such that it can only move in one dimension to the right or to the left. During some fixed interval of time, its average speed was twice the magnitude of its average velocity. Was the particle always moving during this time interval, or was there necessarily a moment when it was at rest? Explain your answer.

21. **N** During a relay race, you run the first leg of the race, a distance of 2.0×10^2 m to the north, in 22.23 s. You then run the same distance back to the south in 24.15 s in the second leg of the race. Suppose the positive y axis points to the north. What is your average velocity **a.** for the first leg of the relay race and **b.** for the entire race?

22. **C** When is the average speed of a particle moving along one straight line less than the magnitude of its average velocity over the same time interval?

Problems 23 through 25 are grouped.

23. Light can be described as a wave or as a particle known as a photon. The speed of light c is 3.00×10^8 m/s.

a. **E** Imagine flipping a switch on the wall that turns on a lamp in the middle of the room. Estimate the time it takes a photon to reach your eye.

b. **N** The Earth–Sun distance is 1.50×10^{11} m. Find the time it takes a photon leaving the surface of the Sun to reach us.

c. **C** Why is it difficult to measure the speed of a photon? Contrast it to measuring the speed of a jogger.

24. **N** Light can be described as a wave or as a particle known as a photon. The speed of light c is 3.00×10^8 m/s. **a.** Sirius is the brightest star in the night sky, 8.18×10^{16} mfrom the Earth. Find the time it takes a photon to reach us from Sirius. Give your answer in years. **b.** A light-year (ly) is the distance that light travels in 1 year. How far from the Earth is Sirius in light-years?

25. During a thunderstorm, a frightened child is soothed by learning to estimate the distance to a lightning strike by counting the time between seeing the lightning and hearing the thunder (Fig. P2.25). The speed v_s of sound in air depends on the air temperature, but assume the value is 343 m/s. The speed of light c is 3.00×10^8 m/s.

a. **E** A child sees the lightning and then counts to eight slowly before hearing the thunder. Assume the light travel time is negligible. Estimate the distance to the lightning strike.

b. **N** Using your estimate in part (a), find the light travel time. Is it fair to neglect the light travel time?

c. **C** Think about how time was measured in this problem. Is it fair to neglect the difference between the speed of sound in cold air (v_s at 0°C = 331.4 m/s) and the speed of sound in very warm air (v_s at 40°C = 355.4 m/s)?

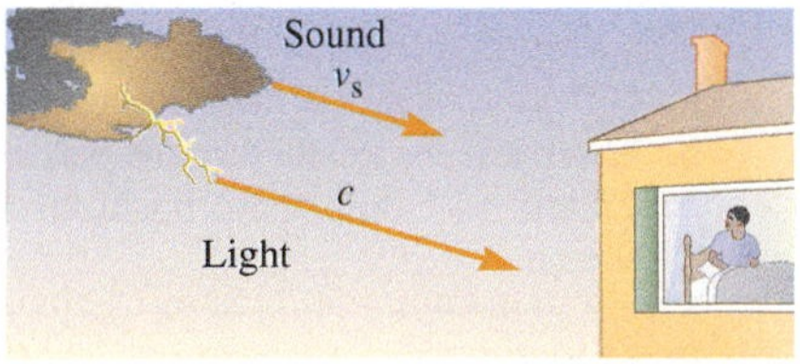

FIGURE P2.25

26. Scientists and engineers must interpret problems from various sources. We can practice this skill anytime we read a newspaper or magazine or browse the Internet. Consider the "Rocket Car" urban legend that can be found on many Internet sites, in which the Arizona Highway Patrol allegedly found the vaporized wreckage of an automobile. The story goes that after some analysis and investigation, it was believed that a former Air Force sergeant attached solid-fuel rockets to his 1967 Chevy Impala and ignited the rockets approximately 3.9 miles from the crash site. The vehicle quickly reached a speed of approximately 275 mph. It continued at this speed for 20 to 25 seconds. The car remained on the highway for 2.6 miles before the driver applied the brakes. The brakes melted and the tires blew out, causing the vehicle to become airborne. It traveled through the air for 1.3 miles before it hit a cliff face 125 feet above the road. Of course, this story was debunked. It is physically implausible, but it can still provide an opportunity to practice analyzing a problem.

a. **C** Draw a sketch of the situation.

b. **C** For the constant-velocity part of the car's motion, identify initial and final positions, the velocity, and the time interval.

c. **N** Calculate the displacement using the position data and then again using the velocity and time data. Are your results consistent?

d. **C** If your results are not consistent, reread the legend and identify possible sources of the discrepancy.

27. **N** The Hawaiian Islands are being formed by the 4.0-in/yr motion of the Pacific Plate over an undersea hot spot in the Earth's crust. New volcanoes are formed as the plate moves, and the age of each volcano can be determined by measuring its distance from Kilauea, the Big Island volcano currently atop the hot spot. Assuming the plate moves at constant speed, what is the age of the Kauai volcano if that island is currently 519 km from Kilauea?

28. When you hear a noise, you usually know the direction from which it came even if you cannot see the source. This ability is partly because you have hearing in two ears. Imagine a noise from a source that is directly to your right. The sound reaches your right ear before it reaches your left ear. Your brain interprets this extra travel time (Δt) to your left ear and identifies the source as being directly to your right. In this simple model, the extra travel time is maximal for a source located directly to your right or left $(\Delta t = \Delta t_{max})$. A source directly behind or in front of you has equal travel time to each ear, so $\Delta t = 0$. Sources at other locations have intermediate extra travel times $(0 \leq \Delta t \leq \Delta t_{max})$. Assume a source is directly to your right.

a. **E** If the speed of sound in air at room temperature is v_s = 343 m/s, find Δt_{max}.

b. **E** Find Δt_{max} if instead you and the source are in seawater at the same temperature, where v_s = 1531 m/s.

c. **C** Why is it difficult to locate the source of a noise when you are under water?

29. **A** In attempting to break one of his many swimming records, Michael Phelps swims the length L of a swimming pool in time t_1 and returns to his starting point in time t_2, completing the lap in world record time. Assume his first lap is in the positive y-direction and use the symbols L, t_1, and t_2. What is Phelps's average velocity during **a.** the first half of this lap and **b.** the second half of the lap? What is his **c.** average velocity and **d.** average speed for the entire lap?

2-7 Instantaneous Velocity and Speed

30. **A** The instantaneous speed of a particle moving along one straight line is $v(t) = ate^{-5t}$, where the speed v is measured in meters per second, the time t is measured in seconds, and the magnitude of the constant a is measured in meters per second squared. What is its maximum speed, expressed as a multiple of a?

31. A particle's velocity is given by $\vec{v}_y(t) = -at\,\hat{\jmath}$, where a = 0.758 m/s² is a constant.

a. **C** Describe the particle's motion. In particular, is it speeding up, slowing down, or maintaining constant speed?

b. **N** Find the particle's velocity at $t = 0$, $t = 10.0$ s, and $t = 5.00$ min.

c. **N** Find the particle's speed at $t = 0$, $t = 10.0$ s, and $t = 5.00$ min.

32. **C** An object initially traveling in the positive x direction undergoes a change in velocity so that, after a finite amount of time passes, it ends up traveling in the negative x direction. Sketch and describe the slope of the position-versus-time graph for this object's motion.

33. **N** Figure P2.33 shows the y-position (in blue) of a particle versus time. **a.** What is the average velocity of the particle during the time interval $t = 1.00$ s to $t = 3.50$ s? **b.** Using the tangent to the curve (shown as the orange line in the figure), what is the instantaneous velocity of the particle at $t = 1.50$ s? **c.** At what time is the velocity of the particle equal to zero?

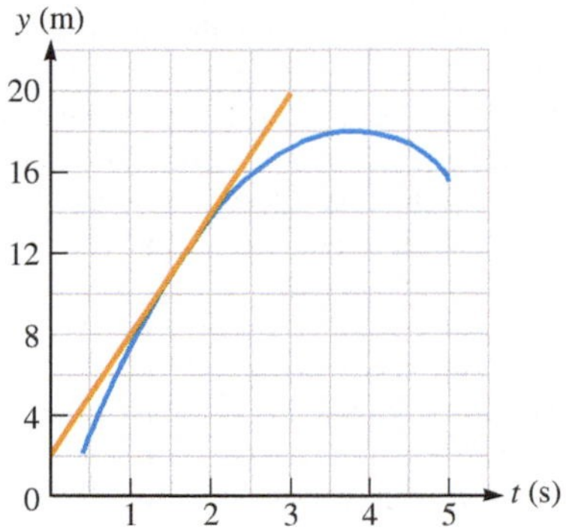

FIGURE P2.33

Problems 34 and 35 are paired.

34. A particle's position is given by $z(t) = -(7.50\ \text{m/s}^2)\ t^2$ for $t \geq 0$.
   a. **A** Find an expression for the particle's velocity as a function of time.
   b. **C** Is the particle speeding up, slowing down, or maintaining a constant speed?
   c. **N** What are the particle's position, velocity, and speed at $t = 6.50$ min?

35. **N** A particle's position is given by $z(t) = -(7.50\ \text{m/s}^2)\ t^2$ for $t \geq 0$. **a.** Find the particle's velocity at $t = 1.50$ s and $t = 3.50$ s. **b.** What is the particle's average velocity during the time interval from $t = 1.50$ s to $t = 3.50$ s?

36. **C** Two sprinters start a race along a straight track at the same time and cross the finish line at the same time. **a.** Are their average velocities necessarily equal? Explain. **b.** Are their instantaneous velocities necessarily *always* equal? Explain. **c.** Are their final velocities necessarily equal? Explain.

2-8 Average and Instantaneous Acceleration

37. **N** An electronic line judge camera captures the impact of a 57.0-g tennis ball traveling at 33.0 m/s with the side line of a tennis court (Fig. P2.37). The ball rebounds with a speed of 20.0 m/s and is seen to be in contact with the ground for 4.00 ms. What is the magnitude of the average acceleration of the ball during the time it is in contact with the ground? Assume one-dimensional motion.

John Evans/Shutterstock.com

FIGURE P2.37

38. **C** During a bungee jump, a student (i) initially moves downward with increasing speed and then (ii) moves downward with decreasing speed, (iii) reaches the turnaround point (that is, the low point of the motion), (iv) moves upward with increasing speed, and finally (v) moves upward with decreasing speed. For each situation (i) through (v), state the direction of the bungee jumper's velocity and acceleration (that is, upward, downward, or zero). *Note*: (i), (ii), (iv), and (v) refer to intervals of motion, whereas (iii) refers to an instant.

39. **C** While studying for your physics test, a friend reminds you that the velocity of an object is zero when the slope on a position-versus-time graph for the object is equal to zero. Your friend also says that the acceleration must also therefore be zero at these times because the slope of the velocity-versus-time graph would also have to be equal to zero. **a.** Has your friend given you good advice for the exam? Explain why or why not. **b.** Construct and describe an example, or case, that would illustrate your response to part (a).

Problems 40 and 17 are paired.

40. As in Problem 17, a particle is attached to a vertical spring. The particle is pulled down and released, and then it oscillates up and down. Using an upward-pointing y axis, the position of the particle is given by

$$\vec{y} = \left(y_0 \cos\frac{2\pi t}{T}\right)\hat{\jmath}$$

Both y_0 and T are constants in time. The amplitude (y_0) is usually given in meters, and the period (T) is usually in seconds.
   a. **G** Draw a sketch of this problem. Include the y axis.
   b. **G** Plot position versus time, extending your graph at least to $t = 2T$.
   c. **A, G** Find $\vec{v}_y(t)$ and plot velocity versus time for the same time interval as in part (b).
   d. **A, G** Find $\vec{a}_y(t)$ and plot acceleration versus time for the same time interval as in part (b).
   e. **C** At what times is speed at a maximum? Where is the particle at these times? Label the locations on a sketch of the physical situation.
   f. **C** At what times is the magnitude of the acceleration at a maximum? Where is the particle at these times? Label the locations on the sketch from part (e).

41. **C** The graph of the scalar component of the position versus time for a particle moving horizontally is a parabola (Fig. P2.41). What can be said about the particle's acceleration?

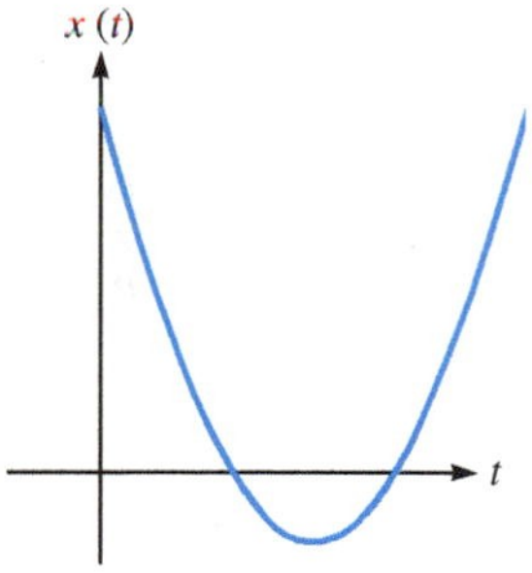

FIGURE P2.41

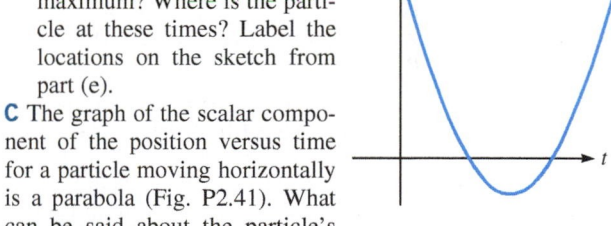

42. **CASE STUDY** Back in the laboratory, Crall gives a cart a quick push up an incline. The cart's speed just after the cart leaves Crall's hand is v_0. The cart rolls up the incline, reaches its highest point, and then rolls back down. When the cart returns to Crall's hand, its speed is once again v_0.
   a. **G** Sketch a velocity-versus-time graph for the cart's motion from just after the initial push to just before it returns. Use a coordinate system in which the positive x direction is upward and parallel to the incline.
   b. **C** At the highest point of the cart's motion, is the acceleration of the cart positive, negative, or zero? Explain.

Problems 43 and 44 are paired.

43. In general, when an object moves through a medium such as water or air, the medium affects the object's motion. If nothing else (like gravity or a motor) acts to counter the effect of the medium, the object will decelerate. Above a certain speed threshold, the speed of the object is given by

$$v_x(t) = \frac{v_0}{bt + C}$$

along an arbitrary x axis. Both b and C are constants in time. The units of b are s^{-1}, and C is unitless. Finally, v_0 is a constant with units of meters per second.

a. C How do you know that the direction of the acceleration is opposite that of the velocity?

b. A Show that the magnitude of its acceleration is given by $a_x = (b/v_0)v_x^2$.

c. C Is the acceleration constant? Can the equations in Table 2.4 be applied to this scenario?

d. C How can a submarine move through the water at constant speed?

44. Consider an object moving through a medium as in Problem 43. Below a certain speed threshold, the speed of the object is given by $v_x(t) = v_{0x}e^{-bt}$ along an arbitrary x axis. The constant b is given in s^{-1}.

a. G Make a graph of velocity versus time.

b. C What is the direction of the acceleration? Explain.

c. A, G Find and graph acceleration as a function of time.

d. A Show that $\vec{a}_x = -b\vec{v}_x$.

45. A computer system, using a preset coordinate system, begins tracking the motion of a high-speed train. The computer system determines the position of the train in that coordinate system, starting at time $t = 0$, and models the motion via the equation

$$\vec{z}(t) = \left(129.1 \text{ m} - \frac{246.3 \text{ m} \cdot \text{s}}{t + 2.0 \text{ s}}\right)\hat{k}$$

a. A Find an expression for the acceleration of the train as a function of time.

b. C Is the train slowing down or speeding up? Explain your answer.

c. C Given the function supplied by the computer system, does the train ever turn around and move in the opposite direction? Explain your answer.

46. In Example 2.6, we considered a simple model for a rocket launched from the surface of the Earth. A better expression for the rocket's position measured from the center of the Earth is given by

$$\vec{y}(t) = \left(R_\oplus^{3/2} + 3\sqrt{\frac{g}{2}}R_\oplus t\right)^{2/3}\hat{\jmath}$$

where $R_\oplus$ is the radius of the Earth $(6.38 \times 10^6 \text{ m})$ and g is the constant acceleration of an object in free fall near the Earth's surface (9.81 m/s^2).

a. A Derive expressions for $\vec{v}_y(t)$ and $\vec{a}_y(t)$.

b. G Plot $y(t)$, $v_y(t)$, and $a_y(t)$. (A spreadsheet program would be helpful.)

c. N When will the rocket be at $y = 4R_\oplus$?

d. N What are $\vec{v}_y$ and $\vec{a}_y$ when $y = 4R_\oplus$?

2-9 Special Case: Constant Acceleration

47. N A uniformly accelerating rocket is found to have a velocity of 15.0 m/s when its height is 5.00 m above the ground, and 1.50 s later the rocket is at a height of 58.0 m. What is the magnitude of its acceleration?

48. C A piece of debris is in space, far from any planets or stars, and is initially moving. The debris is suddenly subject to a constant acceleration. **a.** Is it *possible* the debris is slowing down? **b.** Is it *possible* the debris could ever reverse its motion? Explain your answers.

49. N A driver uniformly accelerates his car such that $\vec{a} = 6.851\hat{\imath} \text{ m/s}^2$. **a.** Assuming he starts from rest, find the velocity of the car after it has accelerated for 4.55 s. **b.** If immediately after that 4.55 s the driver lays off the accelerator, slams on the brakes, and comes to a stop in the subsequent 5.62 s, what is the acceleration he experiences during that time, assuming the acceleration is constant?

50. G Car A and car B travel in the same direction along a straight section of the interstate highway. For the entire interval shown on the velocity-versus-time graph (Fig. P2.50), car A is ahead of car B. **a.** At time t_3, is the magnitude of the acceleration of car A greater than, less than, or equal to that of car B? Explain. **b.** From time t_1 to time t_2, does the distance between cars A and B increase, decrease, or remain constant? Explain.

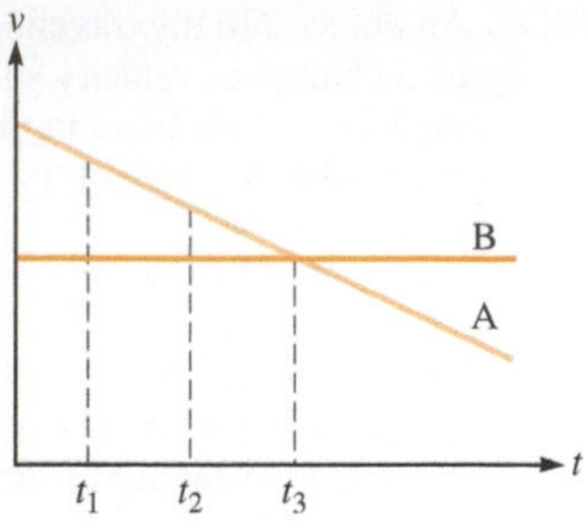

FIGURE P2.50

51. N Accelerating uniformly to overtake a slow-moving truck, a car moving initially at 24.0 m/s covers 68.0 m in 2.50 s. **a.** What is the final speed of the car? **b.** What is the magnitude of the car's acceleration?

52. An object that moves in one dimension has the velocity-versus-time graph shown in Figure P2.52. At time $t = 0$, the object has position $x = 0$.

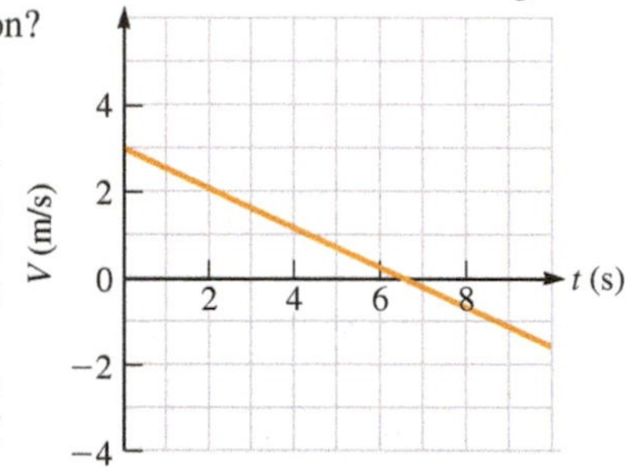

FIGURE P2.52

a. G At time $t = 5$ s, is the acceleration of the object positive, negative, or zero? Explain.

b. G At time $t = 8$ s, is the object speeding up, slowing down, or moving with constant speed? Explain.

c. A Write an expression for the position of the object as a function of time. Explain how you use the graph to obtain your answer.

d. N Use your expression from part (c) to determine the time (if any) at which the object reaches its maximum position. Check your results by examining the graph. *Hint*: To get started with finding the maximum of a function, take the derivative and set it equal to zero.

53. N A particle moves along the positive x axis with a constant acceleration of 3.00 m/s^2 and over time reaches a final speed of 15.0 m/s. **a.** If the particle's initial velocity was $2.00\hat{\imath}$ m/s, what is its displacement during this time, once it reaches its final speed? **b.** What is the distance the particle travels during this time? **c.** If the particle's initial velocity was instead $-2.00\hat{\imath}$ m/s, what is its displacement during this time? **d.** What is the total distance it travels given the initial velocity in part (c)?

54. CASE STUDY Crall and Whipple attached a fan to a cart placed on a level track and then released the cart. They made a position-versus-time graph (Fig. P2.54) and fit a curve to these data such that

$$x = 0.036 \text{ m} + (0.0080 \text{ m/s})t + (0.10 \text{ m/s}^2)t^2$$

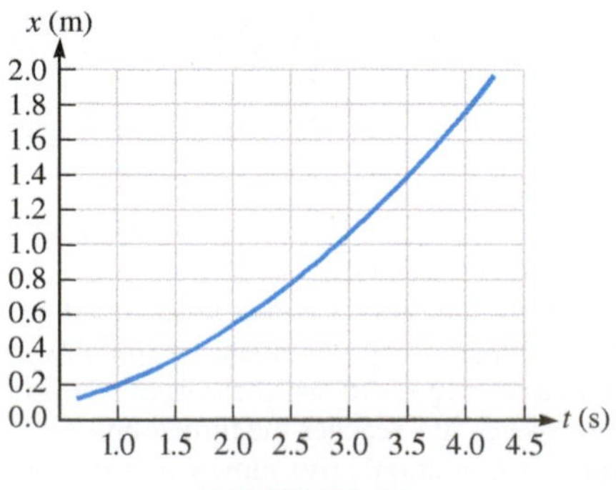

FIGURE P2.54

a. **A, G** Find and graph the velocity as a function of time.
b. **C** What is the shape of the velocity-versus-time graph? What do you expect the acceleration-versus-time graph to look like? Explain.
c. **A, G** Find and graph the acceleration as a function of time.

55. **N** A vehicle moves along the x axis according to the equation $x = 35.0 + 20.0t + 12.0t^2$, where x is in feet and t is in seconds. What is the **a.** position, **b.** velocity, and **c.** acceleration of the vehicle when $t = 4.00$ s?
Give your answer in U.S. customary units.

56. **N** The engineer of an intercity train observes a rock slide blocking the train's path 225.0 m ahead and activates the train's emergency brakes. The train decelerates uniformly at 1.8 m/s^2 for 12.70 s before reaching the rock slide. What is the speed with which the train reaches the rock slide?

2-10 A Special Case of Constant Acceleration: Free Fall

57. **N** A pebble is thrown downward from a 44.0-m-high cliff with an initial speed of 7.70 m/s. How long does it take the pebble to reach the ground?

58. **N** In a cartoon program, Peter tosses his baby, Stewie, up into the air to keep the child entertained. Stewie reaches a maximum height of 0.873 m above the release point. Suppose the positive y axis points upward. **a.** With what initial velocity was Stewie thrown? **b.** How much time did it take Stewie to reach the peak height?

59. **N** Tadeh launches a model rocket straight up from his backyard that takes 4.50 s to reach its maximum altitude. (After launch, the rocket's motion is only influenced by gravity.) **a.** What is the rocket's initial velocity? **b.** What is the maximum altitude reached by the rocket?

60. According to several newspapers, on January 25, 2000, an elevator in the Empire State Building fell 40 stories just after a second passenger boarded on the 44th floor.
a. **C** Draw a sketch and include a coordinate system.
b. **E** Use the data in Appendix B to estimate the displacement of the elevator in SI units.
c. **N** According to at least one newspaper, the fall took 4 s. Find the acceleration, assuming it is constant.
d. **C** Compare your answer to the acceleration of free fall. Does your answer make sense? Explain your reasoning. If needed, explain any sources of discrepancy.

61. **N** In the movie *Star Wars: The Empire Strikes Back*, after being told that Darth Vader is his father, Luke Skywalker falls from a ledge in Cloud City (not on the Earth, so the magnitude of the free-fall acceleration is not necessarily 9.81 m/s^2). Suppose he falls a distance of 28.5 m in 2.5 s. Assuming he starts from rest, answer the following questions. **a.** What is Luke's velocity 2.5 s after he starts to fall? **b.** What is the constant acceleration due to gravity, experienced by Luke, on Cloud City? Use an upward-pointing y axis.

62. **N** A worker tosses bricks one by one to a coworker on a scaffold 5.00 m above his location. Each brick is in flight for 1.80 s. **a.** What is the initial velocity with which the bricks are thrown? **b.** What is the velocity of the bricks just as they are caught?

Problems 63 and 64 are paired.

63. **N** A rock is thrown straight up into the air with an initial speed of 24 m/s at time $t = 0$. Ignore air resistance in this problem. At what times does it move with a speed of 12 m/s? *Note*: There are two answers to this problem.

64. **A** For the rock in Problem 63, determine a symbolic expression in terms of g for the time t when the rock is moving with a speed of $v/2$. *Note*: There are two answers to this problem.

65. **N** A sounding rocket, launched vertically upward with an initial speed of 75.0 m/s, accelerates away from the launch pad at 5.50 m/s^2. The rocket exhausts its fuel, and its engine shuts down at an altitude of 1.20 km, after which it falls freely under the influence of gravity. **a.** How long is the rocket in the air? **b.** What is the maximum altitude reached by the rocket? **c.** What is the velocity of the rocket just before it strikes the ground?

General Problems

66. **N** An object decelerates at a constant rate from an initial velocity of $-7.00\hat{\imath}$ m/s to a final velocity of $10.0\hat{\imath}$ m/s. **a.** What is the object's acceleration if its displacement is $15.0\hat{\imath}$ m? **b.** What would be the object's acceleration if the total distance it traveled were 15.0 m?

67. **N** While strolling downtown on a Saturday afternoon, you stumble across an old car show. As you are walking along an alley toward a main street, you glimpse a particularly stylish Alpha Romeo pass by. Tall buildings on either side of the alley obscure your view, so you see the car only as it passes between the buildings. Thinking back to your physics class, you realize that you can calculate the car's acceleration. You estimate the width of the alleyway between the two buildings to be 4 m. The car was in view for 0.5 s. You also heard the engine rev when the car started from a red light, so you know the Alpha Romeo started from rest 2 s before you first saw it. Find the magnitude of its acceleration.

68. A particle is attached to a vertical spring. The particle is pulled down and released, and then it oscillates up and down. Using an upward-pointing y axis, the position of the particle is given by $\vec{y} = (y_0 \cos \omega t)\hat{\jmath}$. Both y_0 and ω are constants in time.
a. **A** Show that $\vec{a}_y = -\omega^2\vec{y}$.
b. **C** Is the acceleration constant?
c. **C** Can the equations in Table 2.4 be applied to a particle on a spring? Explain.

69. **N** A trooper is moving due south along the freeway at a speed of 21 m/s. At time $t = 0$, a red car passes the trooper. The red car moves with constant velocity of 28 m/s southward. At the instant the trooper's car is passed, the trooper begins to speed up at a constant rate of 2.0 m/s^2. What is the maximum distance ahead of the trooper that is reached by the red car?

70. A dancer moves in one dimension back and forth across the stage. If the end of the stage nearest to her is considered to be the origin of an x axis that runs parallel to the stage, her position, as a function of time, is given by

$$\vec{x}(t) = [(0.02 \text{ m/s}^3)t^3 - (0.35 \text{ m/s}^2)t^2 + (1.75 \text{ m/s})t - 2.00 \text{ m}]\hat{\imath}$$

a. **A** Find an expression for the dancer's velocity as a function of time.
b. **G** Graph the velocity as a function of time for the 14 s over which the dancer performs (the dancer begins when $t = 0$) and use the graph to determine when the dancer's velocity is equal to 0 m/s.

71. **E** The electrical impulse initiated by the nerves in Lina's hand, signaling she has touched a hot stove, travels to her brain as fast as 200 m/s. At this speed, estimate the travel time of this impulse.

72. **C** Two cars leave Seattle at the same time en route to Boston on Interstate 90. The first car moves uniformly the whole way, with constant speed v. The second car travels with constant speed $(v + 1)$ mph for the first half of the distance and travels with constant speed $(v - 1)$ mph for the second half of the distance. Which car gets to Boston first (or is it a tie)? Explain your reasoning.

73. **N** An object begins to move along the y axis and its position is given by the equation $y = 6t^2 - 5t - 2$, with y in meters and t in seconds. **a.** What is the position of the object when it changes its direction? **b.** What is the object's velocity when it returns to its original position?

74. **C** An object starts from rest, traversing a distance d in a time T while moving with constant acceleration along a straight-line path. **a.** After a time $T/2$ has elapsed, has the object traveled a distance greater than, less than, or equal to $d/2$? Explain. **b.** Let v_{av} represent the average velocity of the object. When the object has traveled a distance $d/2$, is its instantaneous velocity greater than, less than, or equal to v_{av}? Explain.

75. **N** The initial velocity of a military jet is 205 m/s eastward. The pilot ignites the afterburners, and the jet accelerates eastward at a constant rate for 1.75 s. The final velocity of the jet is 315 m/s eastward. What was the jet's displacement during the time it was accelerating?

76. Two carts are set in motion at $t = 0$ on a frictionless track in a physics laboratory. The first cart is launched from an initial position of $x = 18.0$ cm with an initial velocity of $11.8\hat{\imath}$ cm/s and a constant acceleration of $-3.40\hat{\imath}$ cm/s^2. The second cart is launched from $x = 20.0$ cm with a constant velocity of $4.30\hat{\imath}$ cm/s.

a. **N** What are the times for which the two carts have equal speeds?

b. **N** What are the speeds of the carts at that time?

c. **N** What are the locations and times at which the carts pass each other?

d. **C** What is the difference between what is asked in parts (a) and (c) of this problem with regard to the times you found?

77. **N** The motion of a spacecraft in the outer solar system is described by the equation $x = 4.00t^2 - 3.00t + 5.00$, where x is in astronomical units (AU) and t is in years. **a.** What is the average speed of the spacecraft between $t = 1.00$ yr and $t = 3.00$ yr? **b.** What is the instantaneous speed of the spacecraft at $t = 1.00$ yr and at $t = 3.00$ yr? **c.** For what time t is the speed of the spacecraft zero?
Report answers in AU/year and years.

78. **C** Cars A and B each move to the right with constant acceleration along a straight road. The velocity vectors of each car are shown in Figure P2.78 for several times separated by equal time intervals. For the entire interval from time t_1 to time t_4, car B is ahead of car A (that is, car B is to the right of car A). **a.** Is the acceleration of car B to the left, to the right, or zero? Explain. **b.** Is the magnitude of the acceleration of car A greater than, less than, or equal to the magnitude of the acceleration of car B? Explain your reasoning. **c.** Is the distance between car A and car B at time t_3 greater than, less than, or equal to the distance between car A and car B at time t_2? Explain.

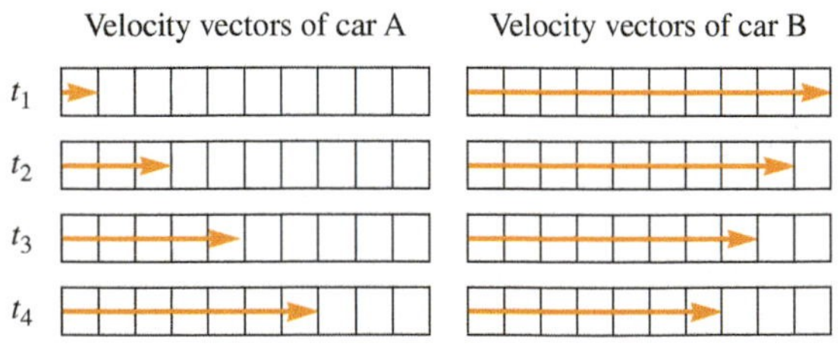

FIGURE P2.78

79. **N** A hydrogen ion moves along the z axis of a bubble chamber according to the equation $z = 5.00t^2 + 4.00t$ where z is in meters and t is in seconds. What is its average velocity during the time interval **a.** $t = 2.50$ to 3.50 s and **b.** $t = 3.50$ to 3.60 s?

80. **N** Trying to determine its depth, a rock climber drops a pebble into a chasm and hears the pebble strike the ground 3.20 s later. **a.** If the speed of sound in air is 343 m/s at the rock climber's location, what is the depth of the chasm? **b.** What is the percentage of error that would result from assuming the speed of sound is infinite?

81. **G** CASE STUDY Similar to that shown in Figure 2.6, Crall and Whipple set up three other experiments (Fig. P2.81). In experiment A, a small fan is attached to a cart. The cart is released from rest and moves along the horizontal track. In experiment B, the cart starts at rest at the bottom of the incline. It is then given a push and moves up the incline. In experiment C, a rubber disk is hung from a spring. The rubber disk is pulled down a few centimeters and then released. The spring causes the disk to oscillate up and down. Pick coordinate systems for each experiment. Be sure to choose an origin and a positive direction. Challenge yourself to think of other possible choices and pick the best coordinate system.

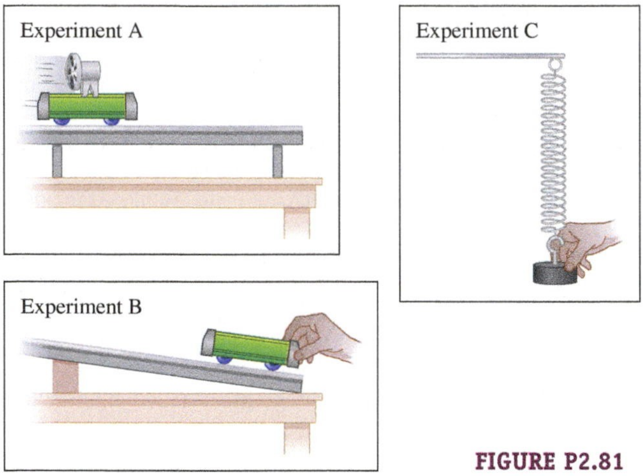

FIGURE P2.81

Problems 82 through 85 are grouped.

82. **A** Write expressions for the average velocity in the y and z directions.

83. **A** Write expressions for the instantaneous velocity in the y and z directions.

84. **A** Write expressions for the average acceleration in the y and z directions.

85. **A** Write expressions for the instantaneous acceleration in the y and z directions.

86. **G** Return to Example 2.6 and calculate the displacement $\Delta\vec{y}(t)$, velocity $\vec{v}_y(t)$, and acceleration $\vec{a}_y(t)$ for $t = 0$, 0.5, 1, 1.5, 2, 2.5, and 3 min. Make graphs of displacement, velocity, and acceleration as functions of time.

87. **N** In 1898, the world land speed record was set by Gaston Chasseloup-Laubat driving a car named Jeantaud. His speed was 39.24 mph (62.78 km/h), much lower than the limit on our interstate highways today. Repeat the calculations of Example 2.7 (acceleration for first 6 miles, time of timed mile, acceleration for last 6 miles) for the Jeantaud car. Compare the results of the ThrustSSC to Jeantaud.

88. **N** In Example 2.12, two circus performers rehearse a trick in which a ball and a dart collide. We found the height and time of the collision graphically. Return to that example, and find height and time by simultaneously solving the equations for the ball and the dart.

89. **A** Use integral calculus to show $\Delta x = v_{0x}t + \frac{1}{2}a_x t^2$ (Eq. 2.11) in the case of constant acceleration.

Vectors

3

Underlying Principles

No new physical principles. This chapter describes mathematical tools and concepts.

Major Concepts

1. Vector and scalar components
2. Resolving a vector into components
3. Vector magnitude and direction
4. Vector addition and subtraction
5. Commutative
6. Associative

Tools

1. Vector manipulation
   a. Parallelogram addition and subtraction
   b. Head-to-tail addition and subtraction
   c. Tail-to-tail subtraction
2. Coordinate systems
   a. Cartesian coordinate system
   b. Right-handed coordinate systems

Key Question

How do you add and subtract vectors?

Each day, we speak our native language, creating sentences that have never been uttered before without thinking about the grammatical rules that govern our speech. Our expectations about speech are based on our native language's grammar. Much of our struggle with learning another language is due to developing a new set of expectations.

Developing new mathematical skills is like learning a language. We have combined scalar quantities using arithmetic and algebraic rules for so long that we often are barely aware of these rules. Now we are going to learn the "language" of **vectors** (quantities with magnitude and direction), including the rules for combining them. Some of these rules will seem familiar from our experience with scalars, whereas other rules will seem new and foreign. The more experience you have with vectors, the more intuitive working with them will become. Vectors such as position, velocity, and acceleration appear frequently in physics.

3-1 Geometric Treatment of Vectors

In Chapter 2, we learned that **vectors** have both a magnitude and a direction, whereas **scalars** have only magnitude. Many quantities in physics—such as position, displacement, average and instantaneous velocity, and average and instantaneous acceleration—are vectors. If motion is one-dimensional, we can use these quantities without learning all the rules of vectors. Vector rules are needed, however, for the description of two-dimensional and three-dimensional motion.

We already know the rules for combining scalars: how to add, subtract, multiply, and divide them. Using these rules allows us to solve scalar equations. Because physical laws are mathematically expressed as equations that involve both vectors and scalars, we must learn to make calculations with vectors. There are two ways to calculate with vectors: geometrically and algebraically. We start with the geometric technique.

The geometric technique will allow you to *anticipate* the result of a problem and *check* your result. Geometric treatment of vectors will also help you think more like a physicist. Recognizing these vectors and coordinate systems is often the first step in *interpreting* a problem.

Because this chapter is dedicated to developing mathematical tools that are used for all vectors, we will often write the rules and relationships in terms of generic vectors designated by uppercase letters such as $\vec{A}$, $\vec{B}$, $\vec{C}$, $\vec{D}$, and $\vec{R}$. To keep the chapter from being too abstract, we will sometimes work with concepts from Chapter 2 such as position, displacement, and average velocity. These vector quantities appear in a number of examples and in the following case study.

CASE STUDY Skydiving

Cassidi Reese, the author's former student who graduated with a degree in physics, likes to skydive and collect data on her dives. Just 10 months after learning to skydive, she celebrated her 300th jump. Achieving such a goal is not easy, and some of her jumps were especially challenging.

On the morning of her 136th jump, Reese boarded a plane in Williamstown, New Jersey, with a group of other skydivers. She happened to be the last jumper out of the plane. The other skydivers were dropped over an empty field, but Reese saw that she was heading for a forest and knew that she had to do something. She took two important measures to avoid landing in a tree. First, she set her parachute in a configuration called "brakes." In this configuration, both her descent velocity and her horizontal velocity were minimized. Second, she opened her parachute when she was at an altitude of 1200 m. (Normally, she would have opened her parachute at an altitude of roughly 750 m.) Reese landed safely in the clearing. Throughout this chapter, we will return to this case study to see what would have happened had Reese deployed her parachute at 750 m. The distance data are given in Figure 3.1. (Reese's motion is influenced by the force exerted by the Earth's gravity as well as by the force exerted by the air, as described in Chapters 5 and 6. In this chapter, we use this case study to practice working with vectors and therefore won't concern ourselves with causes of her motion.)

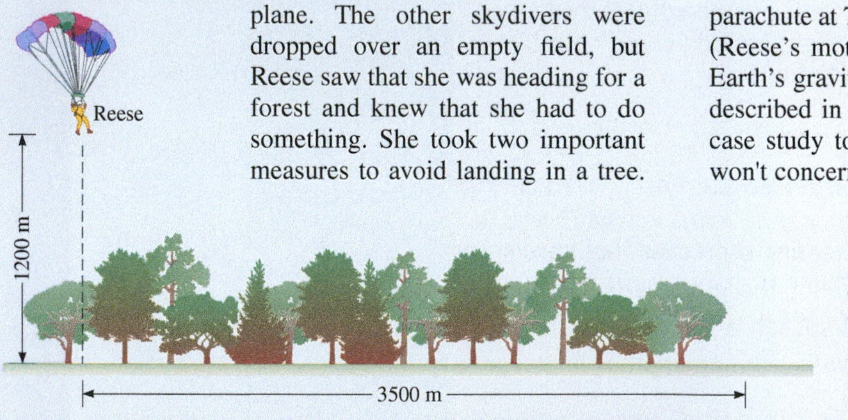
Reese
1200 m
3500 m

FIGURE 3.1 To avoid landing in a forest, Reese set her parachute to "brakes" to slow her travel and she deployed her parachute at 1200 m instead of 750 m. The clearing was 3500 m away from the point just below her when she opened her parachute. She made it to the clearing unharmed, but what would have happened had she not opened her parachute early? (Reese, her parachute, and the trees are not shown to scale.)

Drawing Vectors

Any representation of a vector must reflect both the vector's magnitude and its direction (Fig. 3.2). The length of the arrow represents the magnitude of the vector, and the arrowhead indicates its direction. Two vectors are equal to each other if their magnitudes are equal and they point in the same direction, no matter where the vectors are located. The three vectors in Figure 3.2 are all equal.

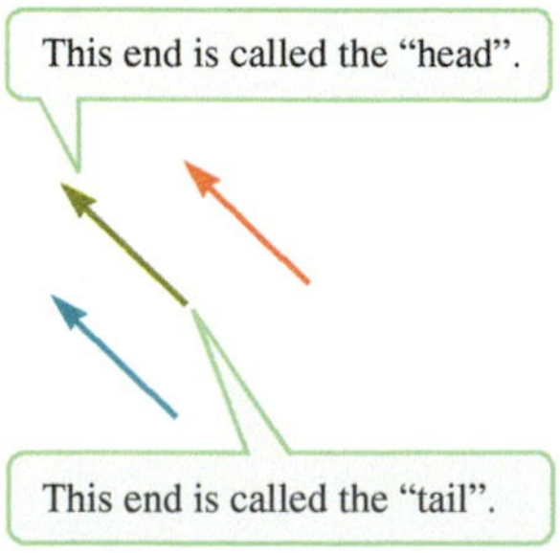
This end is called the "head".
This end is called the "tail".

FIGURE 3.2 A vector may be represented pictorially by an arrow. All three vectors shown are equal.

A vector can be shifted to a new location as long as its magnitude and direction remain unchanged. We will use this property of vectors when we want to combine them geometrically.

Adding Vectors Geometrically

When adding two or more vectors, we must take into account both the magnitude and the direction of each vector. Suppose we need to find the vector $\vec{R}$ resulting from the addition of two other vectors $\vec{A}$ and $\vec{B}$:

$$\vec{R} = \vec{A} + \vec{B} \tag{3.1}$$

The vector $\vec{R}$ is called the ***resultant vector*** or simply the ***resultant***.

The resultant can be found geometrically by either of two methods. The first method is known as ***parallelogram addition*** and is illustrated in Figure 3.3A. The second method of geometric vector addition is known as ***head-to-tail addition*** and is illustrated in Figure 3.3B. Two copies of each vector are used in the parallelogram method, and only one copy of each vector is used in the head-to-tail method.

PARALLELOGRAM VECTOR ADDITION; HEAD-TO-TAIL VECTOR ADDITION

Tools

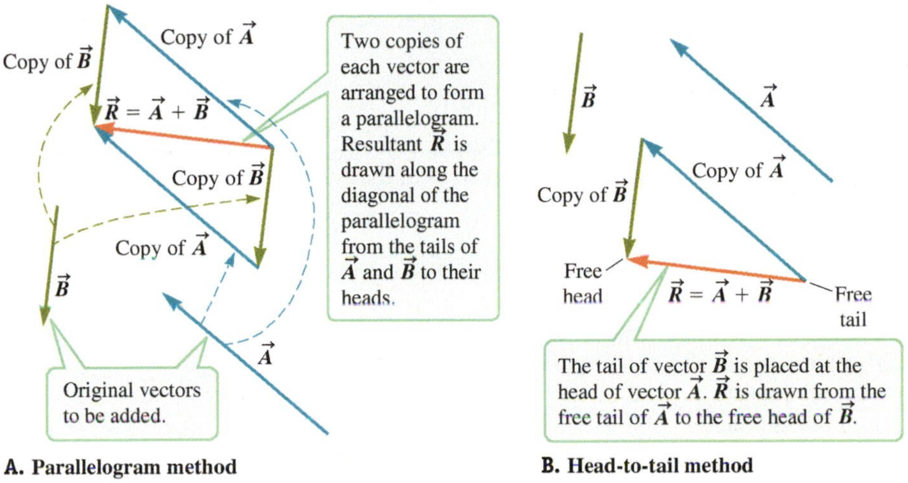

Copy of $\vec{B}$
Copy of $\vec{A}$
$\vec{R} = \vec{A} + \vec{B}$
Copy of $\vec{B}$
Copy of $\vec{A}$
$\vec{B}$
$\vec{A}$
Original vectors to be added.
Two copies of each vector are arranged to form a parallelogram. Resultant $\vec{R}$ is drawn along the diagonal of the parallelogram from the tails of $\vec{A}$ and $\vec{B}$ to their heads.
A. Parallelogram method
$\vec{B}$
$\vec{A}$
Copy of $\vec{A}$
Copy of $\vec{B}$
Free head
$\vec{R} = \vec{A} + \vec{B}$
Free tail
The tail of vector $\vec{B}$ is placed at the head of vector $\vec{A}$. $\vec{R}$ is drawn from the free tail of $\vec{A}$ to the free head of $\vec{B}$.
B. Head-to-tail method

FIGURE 3.3 Two methods for adding vectors geometrically.

Vector addition is **commutative**, meaning that the order of addition does not matter:

$$\vec{A} + \vec{B} = \vec{B} + \vec{A} \tag{3.2}$$

COMMUTATIVE; ASSOCIATIVE

Major Concepts

The two triangles in Figure 3.4 show the head-to-tail addition of vector $\vec{A}$ plus vector $\vec{B}$. The triangle on the right was formed by placing the head of vector $\vec{A}$ at the tail of vector $\vec{B}$. The third leg of the triangle is the resultant vector $\vec{A} + \vec{B}$. In the triangle on the left, the head of vector $\vec{B}$ was placed at the tail of vector $\vec{A}$. The third leg in this triangle is the vector $\vec{B} + \vec{A}$.

We can learn two things from Figure 3.4. First, the vectors $\vec{A} + \vec{B}$ and $\vec{B} + \vec{A}$ are the same length and point in the same direction; therefore, these two vectors are equal, demonstrating that vector addition is commutative. Second, we can see how parallelogram addition is related to the commutative property. Each triangle in Figure 3.4 is half of the parallelogram that we would draw to add vectors $\vec{A}$ and $\vec{B}$ in either order using the parallelogram addition method. Therefore, the vector drawn along the diagonal of the parallelogram from the tails to the heads of these vectors must be equal to both $\vec{A} + \vec{B}$ and $\vec{B} + \vec{A}$.

As you will show in Problem 12, vector addition is also **associative**, meaning that if there are more than two vectors to add, they can be grouped in any order:

$$(\vec{A} + \vec{B}) + \vec{C} = \vec{A} + (\vec{B} + \vec{C}) \tag{3.3}$$

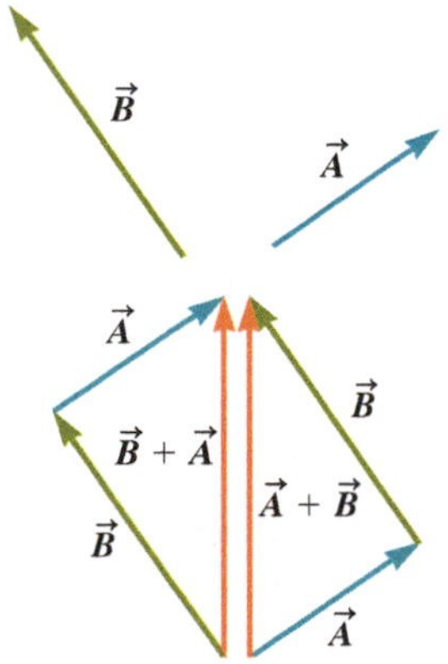

$\vec{B}$
$\vec{A}$
$\vec{A}$
$\vec{B}$
$\vec{B} + \vec{A}$
$\vec{A} + \vec{B}$
$\vec{B}$
$\vec{A}$

FIGURE 3.4 Vector addition is commutative, which means that vectors can be added in any order: $\vec{A} + \vec{B} = \vec{B} + \vec{A}$. This property of vector addition allows us to use parallelogram addition.

Multiplying a Vector by a Scalar

When a vector $\vec{A}$ is multiplied or divided by a positive scalar s, the resultant $\vec{R}$ is a vector that points in the same direction as $\vec{A}$ but differs in magnitude from that of $\vec{A}$ by the factor s:

$$\vec{R} = s\vec{A} \tag{3.4}$$

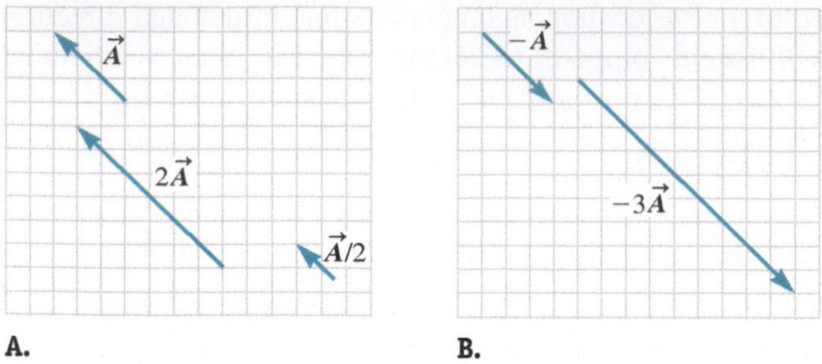

FIGURE 3.5 **A.** When a vector is multiplied or divided by a positive scalar, the resulting vector points in the same direction as the original vector. **B.** When a vector is multiplied (or divided) by a negative scalar, the resulting vector points in the opposite direction.

Figure 3.5A shows vector $\vec{A}$ both multiplied and divided by 2. Both resultant vectors point in the same direction as $\vec{A}$, but the vector $2\vec{A}$ is twice as long as $\vec{A}$ and the vector $\vec{A}/2$ is half as long as $\vec{A}$.

If a vector is multiplied or divided by a negative scalar, its direction reverses. Figure 3.5B shows the same vector $\vec{A}$ multiplied by -1 and -3 to obtain $-\vec{A}$ and $-3\vec{A}$. Both resultant vectors point in the direction opposite to the direction of $\vec{A}$. However, vector $-\vec{A}$ is the same length as $\vec{A}$, so the magnitudes of these two vectors are equal.

Subtracting Vectors Geometrically

Vector subtraction is like vector addition except that one vector is negative: $\vec{R} = \vec{A} - \vec{B}$. We can rewrite the subtraction in the form of an addition:

$$\vec{R} = \vec{A} - \vec{B} = \vec{A} + (-\vec{B}) \tag{3.5}$$

The vector $-\vec{B}$ has the same magnitude as $\vec{B}$ but points in the opposite direction. The vector $-\vec{B}$ is derived from $\vec{B}$ by drawing a vector of the same length but pointing in the opposite direction. Once you have drawn $-\vec{B}$, you can then add $-\vec{B}$ to $\vec{A}$ using either head-to-tail addition or parallelogram addition (Fig. 3.6).

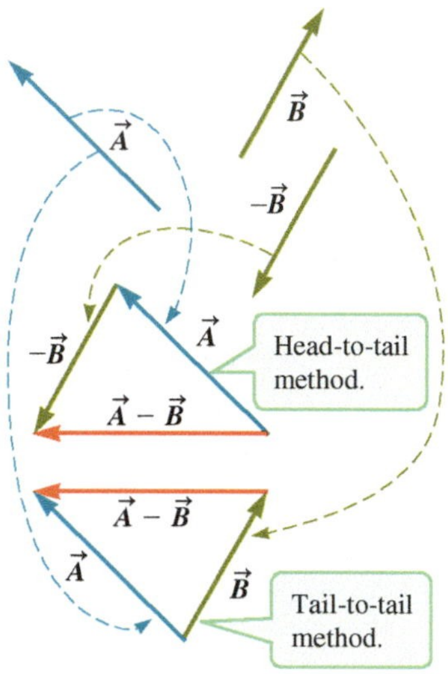

FIGURE 3.6 There are several ways to find $\vec{A} - \vec{B}$ geometrically. As shown in the upper triangle, we may multiply $\vec{B}$ by -1 and then add the result to $\vec{A}$. In the lower triangle, we place $\vec{A}$ and $\vec{B}$ tail to tail, and the resulting vector points from the head of $\vec{B}$ to the head of $\vec{A}$.

An alternative method for geometrically subtracting two vectors $\vec{R} = \vec{A} - \vec{B}$ is **tail-to-tail subtraction.** As shown in the lower triangle in Figure 3.6, the two original vectors $\vec{A}$ and $\vec{B}$ are placed tail to tail, and the resultant vector points from the head of vector $\vec{B}$ to the head of vector $\vec{A}$. This method is often most useful when finding displacement (Example 3.2).

TAIL-TO-TAIL VECTOR SUBTRACTION

Tool

CONCEPT EXERCISE 3.1

The three vectors $\vec{A}$, $\vec{B}$, and $\vec{C}$ in Figure 3.7 all have the same magnitude. Which of these combinations results in a vector of zero magnitude? (More than one choice may be correct.)

a. $\vec{A} - \vec{B}$ **b.** $\vec{B} - \vec{A}$ **c.** $\vec{A} - \vec{C}$ **d.** $\vec{C} - \vec{A}$ **e.** $\vec{A} + \vec{C}$

CONCEPT EXERCISE 3.2

The three vectors $\vec{A}$, $\vec{B}$, and $\vec{C}$ in Figure 3.7 all have the same magnitude.

a. Does $\vec{A} + \vec{C}$ equal $\vec{C} + \vec{A}$? **b.** Does $\vec{A} - \vec{C}$ equal $\vec{C} - \vec{A}$?
c. Is vector subtraction commutative?

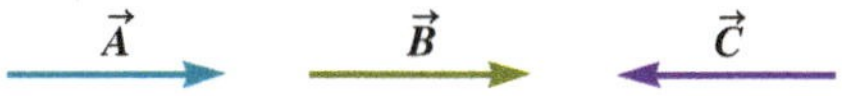

FIGURE 3.7

EXAMPLE 3.1 Adding Vectors Head to Tail

Three vectors $\vec{A}$, $\vec{B}$, and $\vec{C}$ are shown in Figure 3.8. Find the vector $\vec{R} = \vec{A} + \frac{1}{2}\vec{B} - 2\vec{C}$.

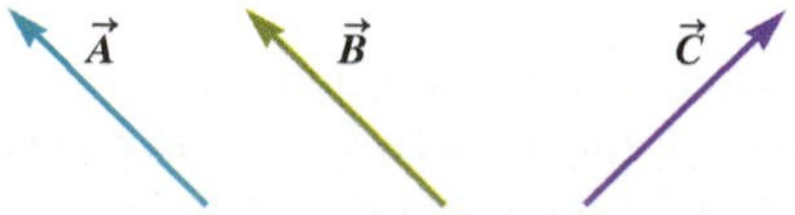

FIGURE 3.8

INTERPRET and ANTICIPATE

As discussed, there are several ways to add and subtract vectors. Because each method produces the same result, the choice is a matter of personal taste. In this example, we choose the head-to-tail method.

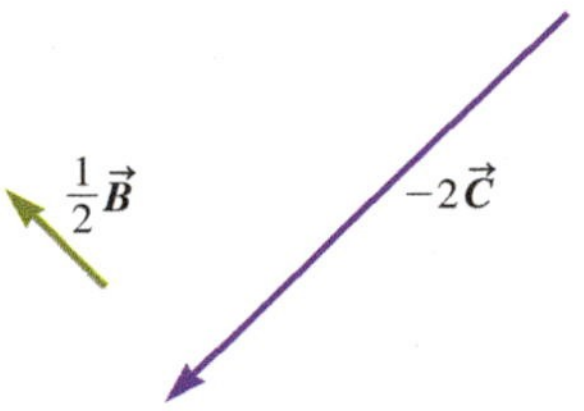

FIGURE 3.9

SOLVE

Regardless of which addition method we choose, our first step must be to multiply vectors $\vec{B}$ and $\vec{C}$ by the appropriate scalars. In Figure 3.9, $\frac{1}{2}\vec{B}$ points in the same direction as $\vec{B}$ but is only half as long. Similarly, $-2\vec{C}$ points in the direction opposite that of $\vec{C}$ and is twice as long as $\vec{C}$.

We arbitrarily choose to add by the head-to-tail method. First, add $\vec{A} + \frac{1}{2}\vec{B}$ by placing the tail of $\frac{1}{2}\vec{B}$ next to the head of $\vec{A}$ (Fig. 3.10). In this problem, the vectors $\vec{A}$, $\frac{1}{2}\vec{B}$, and $(\vec{A} + \frac{1}{2}\vec{B})$ all point in the same direction.

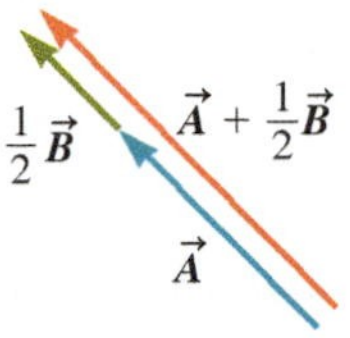

FIGURE 3.10

Finally, we use head-to-tail addition to combine $-2\vec{C}$ and $(\vec{A} + \frac{1}{2}\vec{B})$ as shown in Fig. 3.11.

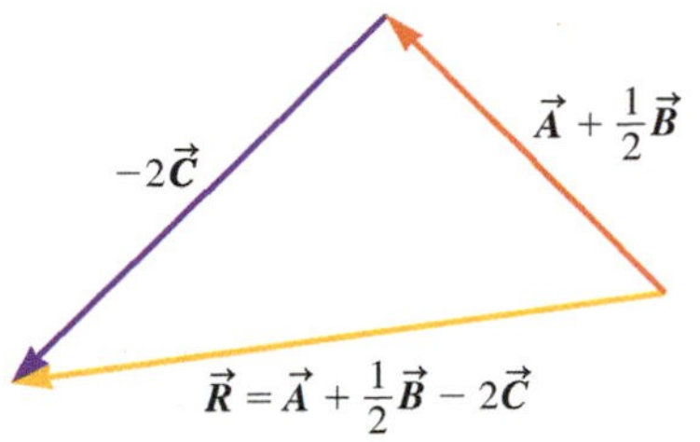

FIGURE 3.11

CHECK and THINK

The resultant vector points to the left and downward. As a check, you can combine the vectors in another order. For example, you could first combine $\frac{1}{2}\vec{B} - 2\vec{C}$ and then add $\vec{A}$. Try it. The resultant should be the same, verifying that vector addition is associative.

Vector addition and subtraction are common in physics. For example, the displacement $\Delta\vec{r}$ (Eq. 2.1) of a particle is found by subtracting its initial position $\vec{r}_i$ from its final position $\vec{r}_f$:

$$\Delta\vec{r} = \vec{r}_f - \vec{r}_i \tag{3.6}$$

The initial and final positions $\vec{r}_i$ and $\vec{r}_f$ need to be measured from a common reference point, the origin of a coordinate system (Section 2-3).

EXAMPLE 3.2 CASE STUDY Cassidi Reese's Displacement

As described previously, Reese opened her parachute when she was directly above a grove of trees, but she landed safely in a nearby field. In Figure 3.12, a reference point has been chosen, and her position $\vec{r}_i$ when she opened the parachute and her position $\vec{r}_f$ when she landed are measured from this reference point. Draw the vector that represents her displacement during this time interval.

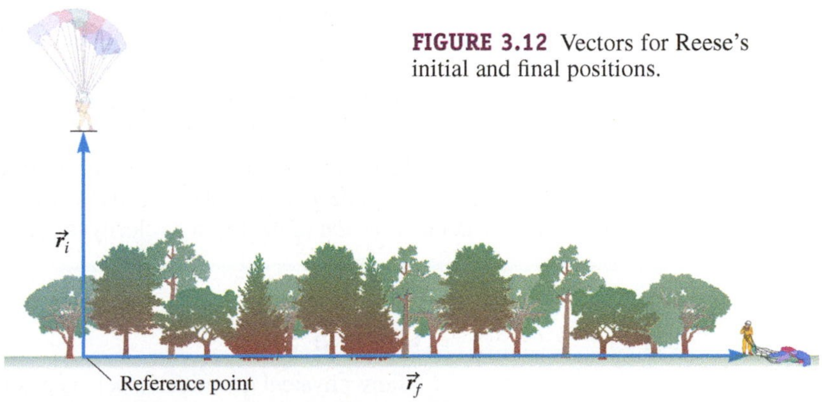

FIGURE 3.12 Vectors for Reese's initial and final positions.

INTERPRET and ANTICIPATE

According to Equation 3.6, to find her displacement, we subtract her initial position vector $\vec{r}_i$ from her final position vector $\vec{r}_f$. There are two ways to do this step geometrically.

$$\Delta\vec{r} = \vec{r}_f - \vec{r}_i$$

Example continues on page 64 ▶

SOLVE
Because the two vectors are already tail to tail, the most convenient subtraction method is tail-to-tail subtraction, in which the displacement vector $\Delta\vec{r}$ points from the head of $\vec{r}_i$ to the head of $\vec{r}_f$ (Fig. 3.13).

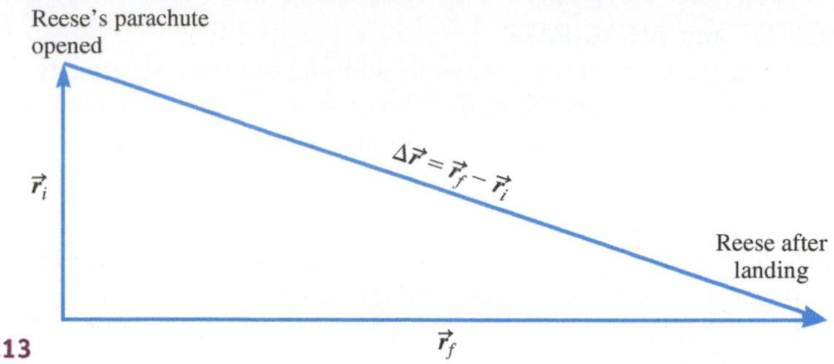

FIGURE 3.13

CHECK and THINK
Displacement is the change in position, so it makes sense that the displacement vector is an arrow that runs from Reese's initial position above the trees to her final position on the ground.

EXAMPLE 3.3 Finding an Acceleration Direction Geometrically

A moon orbits a planet in a circular path. At one instant, the moon's velocity is $\vec{v}_1$; at a later time, its velocity is $\vec{v}_2$ (Fig. 3.14). The average acceleration is given by Equation 2.6:

$$\vec{a}_{av} = \frac{\Delta\vec{v}}{\Delta t}$$

Find the direction of the average acceleration.

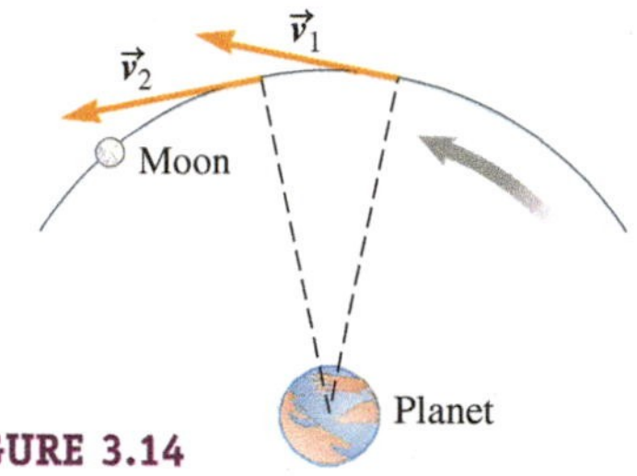

FIGURE 3.14

INTERPRET and ANTICIPATE
The time interval is a scalar. Therefore, the average acceleration must point in the same direction as the change in velocity. If we know the direction of $\Delta\vec{v}$, we also know the direction of $\vec{a}_{av}$.

SOLVE
To determine the direction of $\Delta\vec{v}$, subtract $\vec{v}_2 - \vec{v}_1$ geometrically. We could use tail-to-tail subtraction, but we have arbitrarily chosen to multiply $\vec{v}_1$ by -1 and add the result to $\vec{v}_2$. The resultant vector $\Delta\vec{v}$ runs from the tail of $\vec{v}_2$ to the head of $-\vec{v}_1$. The average acceleration vector is proportional to the change in velocity and therefore points in the same direction.

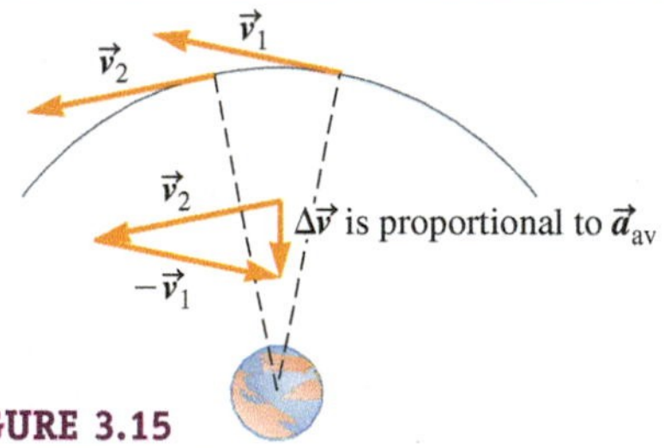

FIGURE 3.15

CHECK and THINK
The two velocity vectors have the same length, so the moon's speed is the same at both instants: $v_1 = v_2$. It might seem surprising that we found a nonzero acceleration, but our answer is correct. In this case, the nonzero acceleration means that the *direction* of the moon's velocity is changing: $\vec{v}_1 \neq \vec{v}_2$.

Using a Scale

Many physical quantities (such as position, displacement, velocity, and acceleration) are vectors that can be represented as arrows and manipulated geometrically. The length of the arrow must be proportional to the magnitude of the vector, but it is often either impossible or not practical to draw an arrow whose length actually equals the magnitude. In such cases, we must use a scale, much like the scale on a map.

For instance, a cartographer may set a scale such that 3 cm on a map equals 10 miles on the Earth, and we write this scale as 3 cm → 10 mi. The procedure for drawing position or displacement vectors is similar to drawing lines on a map. For example, in the CASE STUDY (Fig. 3.1), Reese opened her parachute when her initial position was $r_i = 1200$ m above the ground. By setting a scale such that 1 cm → 400 m, we represented her initial position vector $\vec{r}_i$ by an arrow that is only 3 cm long in Figure 3.12.

Physics involves many vector quantities that do not have the dimensions of length. The scale we use to represent such vectors must take into account the dimensions of the vector. If we want to draw a velocity vector, we need to set a scale so that a unit of length on the paper represents a velocity. For example, we might set a scale such that 1 cm on the paper represents 5 m/s: 1 cm → 5 m/s.

In practice, we draw vectors so as to *anticipate* a result that we want to calculate algebraically. An approximate scale rather than an exact one may be useful. For example, if you need to draw two velocity vectors representing 10 m/s and 5 m/s, you may not need an exact scale, but you should draw one vector approximately twice as long as the other vector.

CONCEPT EXERCISE 3.3

a. You wish to represent free-fall acceleration using a scale such that 1 cm → 1 m/s^2. How long would your vector be?

b. If instead you use a scale such that 0.5 cm → 1 m/s^2, how long would the acceleration vector be? (Give answers to three significant figures.)

3-2 Cartesian Coordinate Systems

In Chapter 2, we combined vector components algebraically and defined a one-dimensional coordinate system for some particular physical situations. By building on those skills, we can study two- and three-dimensional motion.

Axes and Coordinates

A one-dimensional coordinate system requires only one axis, usually labeled x, y, or z. A two-dimensional coordinate system requires two axes (usually x and y), and a three-dimensional coordinate system requires all three axes (x, y, and z). In a **Cartesian coordinate system**, the axes are at right angles to one another. Figure 3.16 shows two two-dimensional coordinate systems. The axes of each system cross at right angles but do not need to be horizontal and vertical. The two axes of each system lie in the same plane and define a two-dimensional space. As before, an arrow on the end of an axis indicates the positive direction.

CARTESIAN COORDINATE SYSTEM

Tool

Two numbers known as **coordinates** specify the location of a point in a two-dimensional system. We write these two coordinates as (x, y). The axes cross at their mutual origin $(x, y) = (0, 0)$. The axes divide up the two-dimensional space into four quadrants labeled with roman numerals. Figure 3.16 shows the relationship between the sign of the coordinates and the four quadrants.

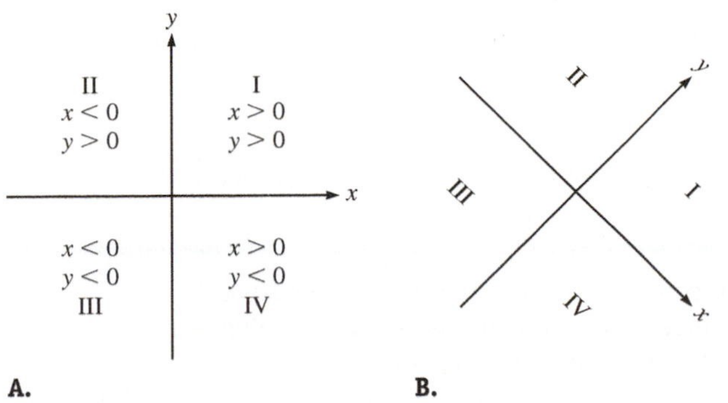

FIGURE 3.16 Two-dimensional Cartesian coordinate systems. **A.** The intersection of the axes divides space into four quadrants. The signs of the x and y coordinates depend on the quadrant as shown. **B.** An x axis might not be horizontal and a y axis might not be vertical, but the axes must be perpendicular to each other.

EXAMPLE 3.4 Reading Coordinates

A pool cue lies on a pool table. The two-dimensional coordinate system shown in Figure 3.17 has been chosen with the origin at the center of the table.

A What are the coordinates of the two ends of the cue?

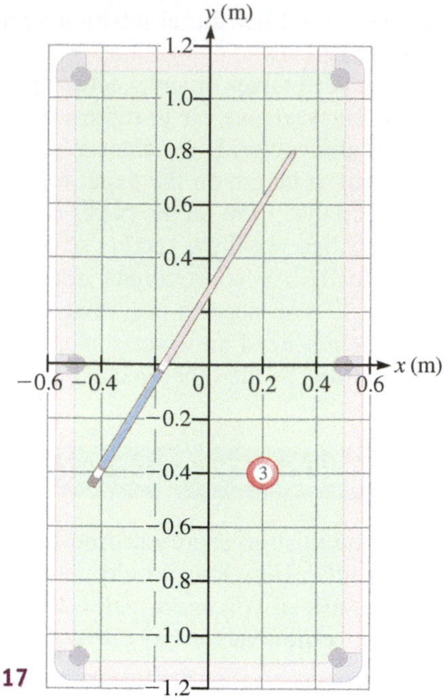

FIGURE 3.17

INTERPRET and ANTICIPATE

In this coordinate system, the x axis is horizontal with the positive direction pointing to the right. The y axis is vertical with the positive direction pointing upward. Both axes are labeled in meters. Figure 3.16 may be used to check the sign of our answer.

SOLVE

Read off the coordinates for each end of the cue to two significant figures. We estimate that the tip (narrow end) is half way between $x = 0.2$ and 0.4 m.	$(x_t, y_t) = (0.30\text{ m}, 0.80\text{ m})$
The bumper (thick end) does not lie on a grid line. So, we must estimate both its x and y coordinates.	$(x_b, y_b) = (-0.45\text{ m}, -0.45\text{ m})$

CHECK and THINK

Because the tip is in quadrant I, both coordinates are positive. The bumper is in quadrant III, so both coordinates are negative.

B How long is the cue?

INTERPRET and ANTICIPATE

Draw two distances labeled Δx and Δy to form a right triangle that has the cue as its hypotenuse (Fig. 3.18). Then, use the Pythagorean theorem to find the length of the cue.

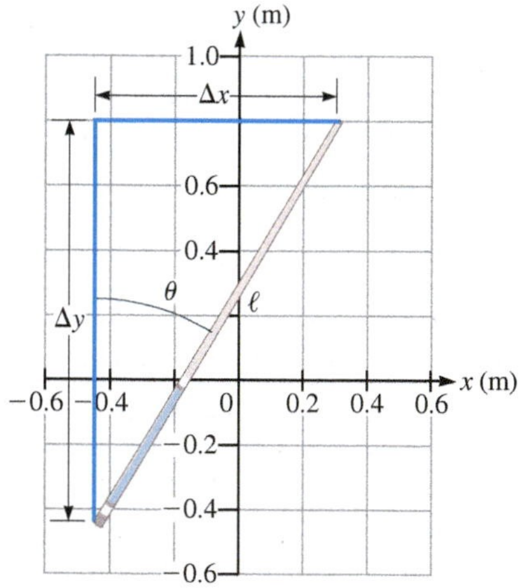

FIGURE 3.18

SOLVE

First, we find the lengths Δx and Δy from the coordinates of the tip and bumper found in part B. Because we plan to square Δx and Δy, it does not matter in which order we do the subtraction.	$\Delta x = x_t - x_b$ $\Delta x = 0.30 - (-0.45)$ $\Delta x = 0.75\text{ m}$ $\Delta y = y_t - y_b$ $\Delta y = 0.80 - (-0.45)$ $\Delta y = 1.25\text{ m}$
To get the length of the cue, use the Pythagorean theorem.	$\ell^2 = \Delta x^2 + \Delta y^2$ $\ell = \sqrt{\Delta x^2 + \Delta y^2}$ $\ell = \sqrt{(0.75\text{ m})^2 + (1.25\text{ m})^2}$ $\ell = 1.46\text{ m}$

CHECK and THINK

If you are familiar with pool, you know that a cue is somewhat shorter than an adult. So, this answer seems reasonable.

C Figure 3.19 shows a different coordinate system imposed on the same pool table we have been working with. Which of the answers above change? Which remain the same? Use this new coordinate system to find the positions of the ends of the cue as well as its length.

INTERPRET and ANTICIPATE

Any coordinate system is *artificially* imposed on a physical situation, meaning that the origin can be anywhere. Because the location of the origin determines the coordinates of any object in the system, the numerical coordinates of an object change as the location of the origin changes. The length of an object cannot depend on the choice of coordinate system. Therefore, the length of the cue must still be 1.46 m.

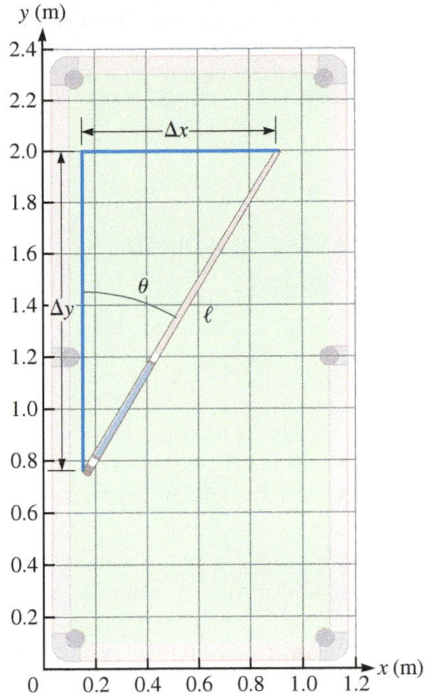

FIGURE 3.19

SOLVE

The new coordinates are read off the figure. In this case, all the coordinates are positive because both ends of the cue are in quadrant I.

For the cue tip:

$$(x_t, y_t) = (0.90 \text{ m}, 2.00 \text{ m})$$

For the cue bumper:

$$(x_b, y_b) = (0.15 \text{ m}, 0.75 \text{ m})$$

Recalculate the length of the cue using the same procedure as in part C but using the new coordinate system.

$$\Delta x = x_t - x_b$$
$$\Delta x = 0.90 - 0.15$$
$$\Delta x = 0.75 \text{ m}$$

$$\Delta y = y_t - y_b$$
$$\Delta y = 2.00 - 0.75$$
$$\Delta y = 1.25 \text{ m}$$

$$\ell = \sqrt{(0.75 \text{ m})^2 + (1.25 \text{ m})^2}$$
$$\ell = 1.46 \text{ m}$$

CHECK and THINK

As expected, the coordinates depend on the choice of coordinate system, but the length of the pool cue does not depend on the coordinate system.

Unit Vectors

As in one dimension, unit vectors specify the direction of vector quantities in two and three dimensions. In the case of one-dimensional vectors, we needed to use only one unit vector ($\hat{\imath}$, $\hat{\jmath}$, or $\hat{k}$) for a given vector quantity. When working with two-dimensional vectors, however, we will need to use two unit vectors simultaneously; and with three-dimensional vectors, all three unit vectors are needed. All facts about unit vectors from Chapter 2 still hold. Unit vectors always have a magnitude of 1. The three unit vectors $\hat{\imath}$, $\hat{\jmath}$, and $\hat{k}$ point along the positive *x*, *y*, and *z* axes, respectively. Like any other vector, a unit vector may be moved to any location without changing it as long as it remains parallel to its original direction.

Right-Handed Coordinate Systems

Figure 3.20 shows a three-dimensional Cartesian coordinate system along with the three unit vectors $\hat{\imath}$, $\hat{\jmath}$, and $\hat{k}$. Because it is a Cartesian system, all three axes cross at right angles. Three-dimensional objects are difficult to represent on a two-dimensional piece of paper. In Figure 3.20, the *x* and *y* axes, meet at right angles; both lie in the plane of the page, just as in the two-dimensional case shown in Figure 3.16. Now imagine poking a pencil through the page, at the point where the *x* and *y* axes intersect. When held perpendicular to the page, the pencil represents the *z* axis.

When drawing three-dimensional situations, it is often better to represent an axis or vector that is perpendicular to the page with either a circled dot ⊙ or a circled cross ⊗. An axis or vector pointing out of the page is represented by ⊙, and one pointing into the page is represented by ⊗. The coordinate system shown in Figure 3.21A is equivalent to the one in Figure 3.20. One way to remember the difference is to picture an archery arrow with tail feathers. If the arrow were pointing at you, you would see the tip as represented by the ⊙ symbol. If the arrow were pointing away from you, you would see the cross made by the tail feathers, resembling the ⊗.

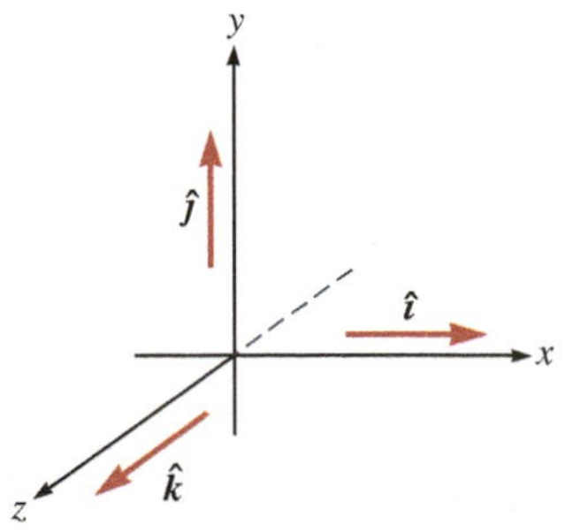

FIGURE 3.20 A three-dimensional coordinate system has *x* and *y* axes in the plane of the page; the *z* axis is perpendicular to the page. The three unit vectors point along the positive *x*, *y*, and *z* axes, respectively.

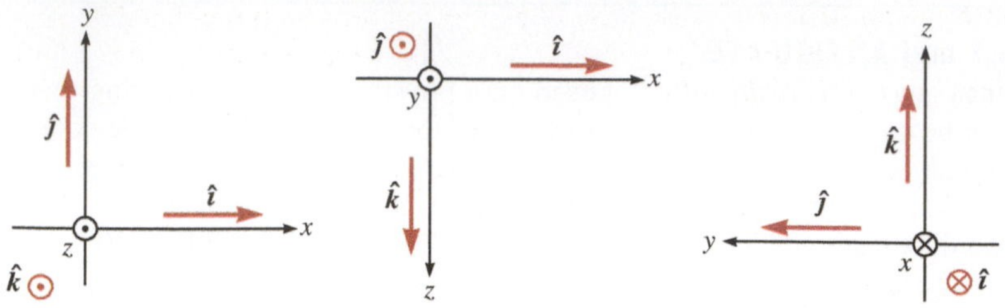

A. **B.** **C.**

FIGURE 3.21 **A.** A right-handed coordinate system equivalent to the one in Figure 3.20. **B.** A right-handed coordinate system with the y axis pointing out of the page. **C.** A right-handed coordinate system with the x axis pointing into the page.

RIGHT-HANDED COORDINATE SYSTEM Tool

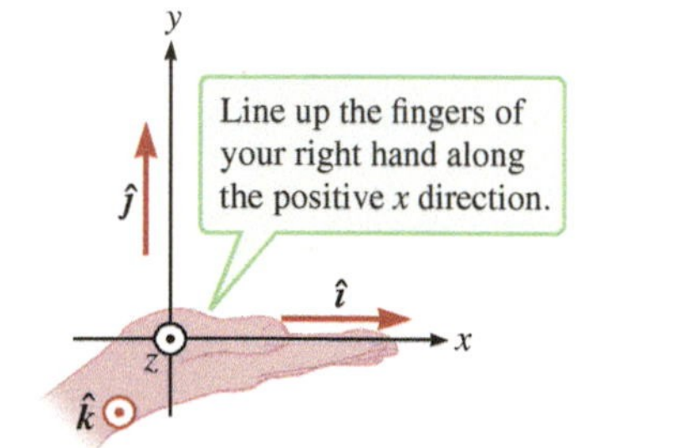

A.

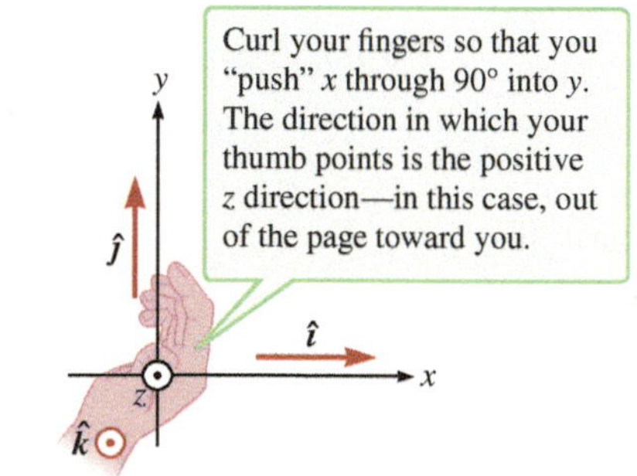

B.

FIGURE 3.22 Finding the z direction in a right-handed coordinate system.

All the coordinate systems used in this book are **right-handed coordinate systems.** Figure 3.21 shows three right-handed x–y–z coordinate systems on a two-dimensional page. In each case, the direction of the z axis (and the unit vector $\hat{k}$) is determined by what is called a *right-hand convention.* Figure 3.22 shows how to use this convention; and it shows that for this coordinate system z points out of the page as it does in Figure 3.21A. Verify for yourself that parts B and C of Figure 3.21 also follow the right-hand convention.

Because the majority of the problems in the first part of this textbook can be analyzed with a two-dimensional Cartesian coordinate system, the rest of this chapter concentrates on two-dimensional vectors.

CONCEPT EXERCISE 3.4

Which coordinate system in Figure 3.23 is right-handed? (More than one choice may be correct.)

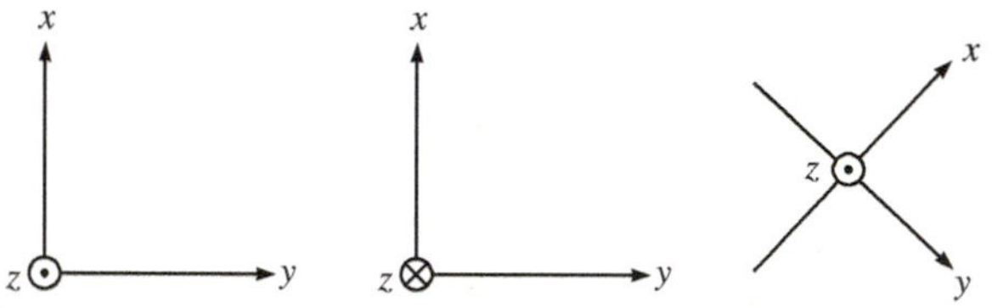

Case 1 **Case 2** **Case 3**

FIGURE 3.23

3-3 Components of a Vector

Vector Components and Scalar Components

Vector quantities can be represented numerically using their components. Vector $\vec{A}$ is shown in a two-dimensional coordinate system (Fig. 3.24). Any vector can be written as the sum of other vectors. In this case, $\vec{A} = \vec{A}_x + \vec{A}_y$, where the vectors $\vec{A}_x$ and $\vec{A}_y$ are the x and y **vector components** of $\vec{A}$. Vector components are either parallel or antiparallel to the coordinate axes. (A vector component antiparallel to a coordinate axis is aligned with the coordinate axis but points in the negative direction.) As is true for all vectors, sliding the vector components to new locations does not change them as long as they remain parallel to their original directions (Fig. 3.24).

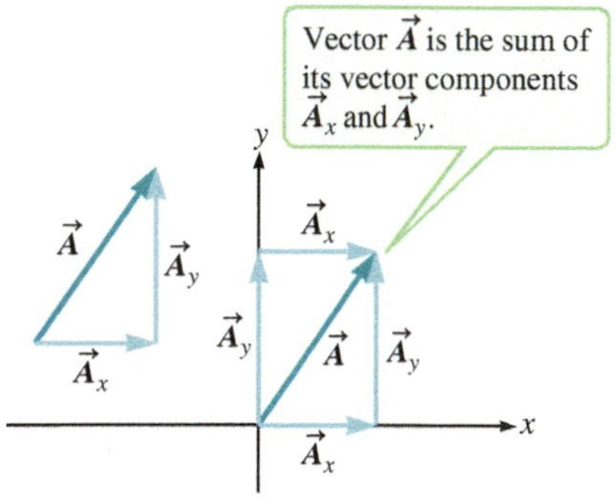

FIGURE 3.24 We may slide the vector $\vec{A}$ and its vector components $\vec{A}_x$ and $\vec{A}_y$ to any location as long as they remain parallel to their original directions.

VECTOR COMPONENTS; SCALAR COMPONENTS Major Concepts

Because vector components are always either parallel or antiparallel to the coordinate axes, they can easily be written in terms of unit vectors. For example, in a three-dimensional coordinate system, the vector components of $\vec{B}$ are

$$\vec{B}_x = B_x\hat{\imath}, \qquad \vec{B}_y = B_y\hat{\jmath}, \qquad \vec{B}_z = B_z\hat{k} \tag{3.7}$$

where B_x, B_y, and B_z are the **scalar components** of $\vec{B}$. Like all scalars, the scalar components of a vector do not have direction. They may be positive or negative. A negative scalar component means that the vector component is pointing in the negative direction along its particular axis.

Any vector is completely specified by either its vector components or its scalar components. For example, the vector $\vec{B}$ may be expressed as a sum of its vector components:

$$\vec{B} = \vec{B}_x + \vec{B}_y + \vec{B}_z$$

Vector $\vec{B}$ can also be written in **component form**, meaning that it is written in terms of its scalar components multiplied by unit vectors:

$$\vec{B} = B_x\hat{\imath} + B_y\hat{\jmath} + B_z\hat{k} \quad (3.8)$$

Vector and scalar components are closely related; if we know one, we can easily figure out the other. For example, if we know the vector component $\vec{A}_x = -3.5\hat{\imath}$, we also know that the scalar component is $A_x = -3.5$. If we know the scalar component $A_y = -7.0$, we know that the vector component is $\vec{A}_y = -7.0\hat{\jmath}$. In this textbook, the term *components* means *scalar* components. Vector components will be specified explicitly when necessary.

Resolving a Vector into Components

RESOLVING A VECTOR INTO COMPONENTS

Major Concept

The process of finding a vector's (vector or scalar) components is known as *resolving the vector into components*. Before learning the mathematical process for resolving a vector, let's build a conceptual understanding. Imagine a vector $\vec{A}$ represented by a real arrow sticking out of the floor at an angle. A vertical screen is placed near the arrow as shown in Figure 3.25. A two-dimensional x–y coordinate system is imposed on the floor and on the screen. We use a light source to find the vector components of the arrow "vector". When we place the light source above the arrow, we see a shadow of the arrow on the floor along the x axis (Fig. 3.25A). This shadow is the x vector component $\vec{A}_x$. When we place the light source to the right of the arrow, we see a shadow of the arrow on the screen along the y axis (Fig. 3.25B). This shadow is the y vector component $\vec{A}_y$.

We can find vector $\vec{A}$'s scalar components mathematically. Figure 3.26A shows vector $\vec{A}$ and its vector components using the conventional arrow representation. The vector components along the x and y axes are the "shadows" of vector $\vec{A}$ projected perpendicular to those axes. These vector components have been assembled to form a right triangle with $\vec{A}$ as the hypotenuse. We can use trigonometry to find the scalar components A_x and A_y in terms of θ and A, where θ is the angle the vector makes with the x axis. For the scalar component A_x, we have

$$\cos\theta = \frac{\text{adjacent}}{\text{hypotenuse}} = \frac{A_x}{A}$$

$$A_x = A\cos\theta \quad (3.9)$$

With a little more trigonometry, we can find the scalar component A_y:

$$\sin\theta = \frac{\text{opposite}}{\text{hypotenuse}} = \frac{A_y}{A}$$

$$A_y = A\sin\theta \quad (3.10)$$

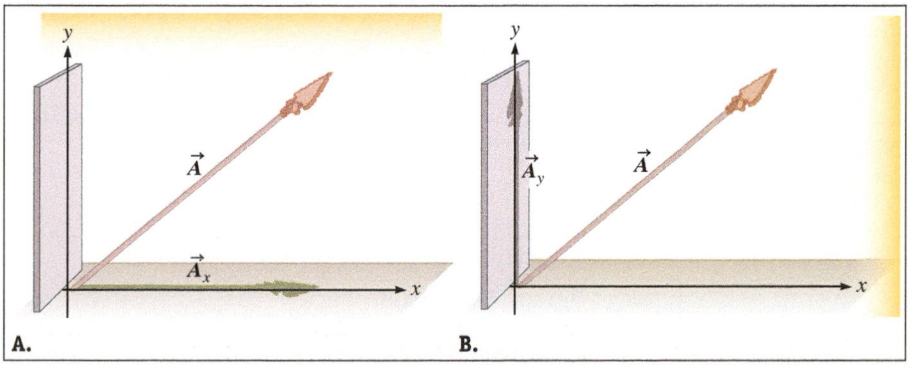

FIGURE 3.25 One way to imagine finding the vector components of $\vec{A}$ is by projecting its shadow onto two perpendicular surfaces. **A.** The x component is the shadow projected onto the horizontal ground. **B.** The y component is the shadow projected onto a vertical screen.

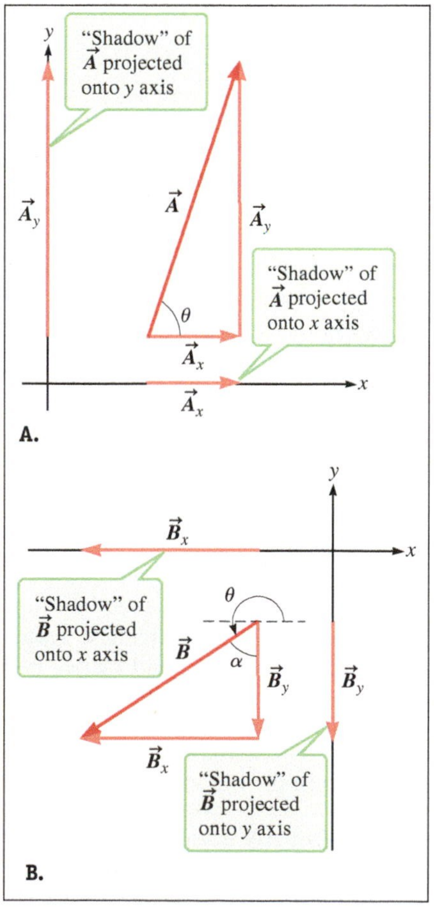

FIGURE 3.26 **A.** The vector components $\vec{A}_x$ and $\vec{A}_y$ are the legs of a right triangle with $\vec{A}$ as the hypotenuse. **B.** The vector components $\vec{B}_x$ and $\vec{B}_y$ are both negative.

We use these scalar components A_x and A_y to write $\vec{A}$ in component form (Eq. 3.8):

$$\vec{A} = A\cos\theta\,\hat{\imath} + A\sin\theta\,\hat{\jmath} \tag{3.11}$$

Vector $\vec{A}$ points up and to the right. Both of its vector components point in the positive direction along their respective axes. As long as the angle θ is measured counterclockwise from the x axis, Equation 3.11 still holds, even if a vector has one or more components that point in the negative direction. For example, the vector components of $\vec{B}$ shown in Figure 3.26B both point in negative directions, and $\vec{B}$ is expressed as

$$\vec{B} = B\cos\theta\,\hat{\imath} + B\sin\theta\,\hat{\jmath}$$

The angle θ is between 180° and 270°, so both cos θ and sin θ are negative.

Although Equation 3.11 can always be used to express a vector in component form, it is often not practical because angles are not always measured counterclockwise from the x axis. It is more common for angles to be expressed as an acute angle measured either from the horizontal or vertical such as the angle α in Figure 3.26B. Start by constructing a right triangle; the legs are the vector components, and the hypotenuse is the vector. Then use trigonometry to express vector $\vec{B}$ in terms of the angle α. Because the angle α is less than 90°, both sin α and cos α are positive. In this case, both vector components are negative, so we must insert a negative sign for each component:

$$\vec{B} = -B\sin\alpha\,\hat{\imath} - B\cos\alpha\,\hat{\jmath}$$

CONCEPT EXERCISE 3.5

Show that if the angle $\theta = 215°$ in Figure 3.26B, then

a. $\alpha = 55°$
b. $\vec{B} = B\cos\theta\,\hat{\imath} + B\sin\theta\,\hat{\jmath} = -B\sin\alpha\,\hat{\imath} - B\cos\alpha\,\hat{\jmath}$

EXAMPLE 3.5 Expressing Vectors in Component Form, with a Twist

Three vectors and a two-dimensional coordinate system are shown in Figure 3.27. The magnitudes of these vectors are $A = 3.6$ m, $B = 2.7$ m, and $C = 3.3$ m.

A Write these three vectors in component form.

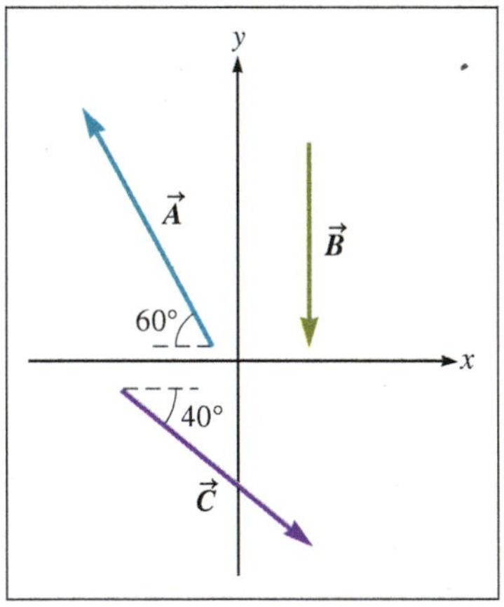

FIGURE 3.27

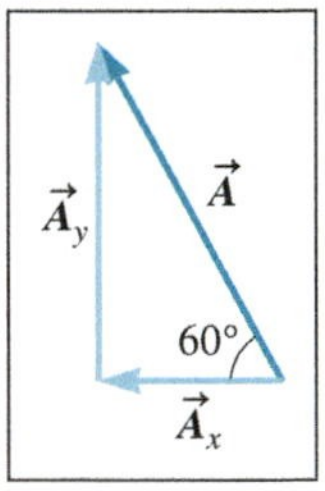

FIGURE 3.28

INTERPRET and ANTICIPATE

To write a vector in component form, we must first find its scalar components and then place the scalar components in front of the appropriate unit vectors.

SOLVE

Draw $\vec{A}_x$ and $\vec{A}_y$ so that together they form a right triangle with $\vec{A}$ as the hypotenuse. The vector components $\vec{A}_x$ and $\vec{A}_y$ are parallel to the x and y axes, respectively (Fig 3.28).

Use trigonometry to find the magnitude of each scalar component. The scalar component A_x must be negative because $\vec{A}_x$ points in the negative x direction.	$A_x = -A\cos 60°$ $A_x = -3.6\cos 60° = -1.8$ m
We can similarly reason that A_y is positive.	$A_y = A\sin 60°$ $A_y = 3.6\sin 60° = 3.1$ m
Place these scalar components in front of the appropriate unit vectors and add the expressions. The dimensional unit appears outside the parentheses because both scalar components are measured in meters.	$\vec{A} = (-1.8\hat{\imath} + 3.1\hat{\jmath})$ m

We see that $\vec{B}$ is antiparallel to the y axis, meaning that it points in the negative y direction. Therefore, $\vec{B}$ is equal to its vector component $\vec{B}_y$ because $\vec{B}_x$ is zero.	$\vec{B}_x = 0$ $\vec{B}_y = -2.7\hat{\jmath}$ m $\vec{B} = (0\hat{\imath} - 2.7\hat{\jmath})$ m $\vec{B} = -2.7\hat{\jmath}$ m
To write $\vec{C}$ in component form, start with a sketch similar to that for $\vec{A}$ (Fig. 3.29).	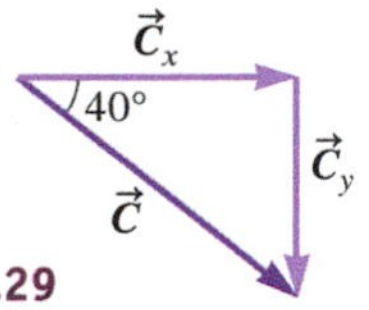 FIGURE 3.29
In this case, C_x is positive and C_y is negative.	$C_x = C\cos 40° = 3.3\cos 40° = 2.5$ m $C_y = -C\sin 40° = -3.3\sin 40° = -2.1$ m $\vec{C} = (2.5\hat{\imath} - 2.1\hat{\jmath})$ m

B Figure 3.30 here shows vector $\vec{B}$ in a new coordinate system with the axes labeled x' and y'. This x'–y' coordinate system is turned 30° counterclockwise with respect to the x–y coordinate system. Write $\vec{B}$ in component form in the x'–y' coordinate system.

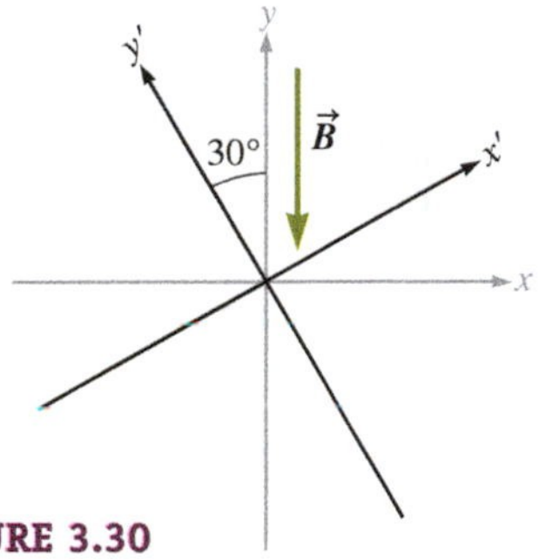

FIGURE 3.30

INTERPRET and ANTICIPATE

In this new coordinate system, $\vec{B}$ is not parallel to an axis, so we must use a procedure similar to that used in finding the components of $\vec{A}$ and $\vec{C}$ in part A. We must find the angle that $\vec{B}$ makes with the new axes to find its components.

SOLVE

Slide $\vec{B}$ leftward and down so that its tail sits on the common origin of the two coordinate systems. This change shows that $\vec{B}$ makes an angle of 30° with the y' axis. Sketch the vector components (Fig. 3.31).

Use trigonometry to find the scalar components $B_{x'}$ and $B_{y'}$ (both negative).	$B_{x'} = -B\sin 30° = -2.7\sin 30° = -1.4$ m $B_{y'} = -B\cos 30° = -2.7\cos 30° = -2.3$ m $\vec{B} = (-1.4\hat{\imath} - 2.3\hat{\jmath})$ m (primed coordinates)

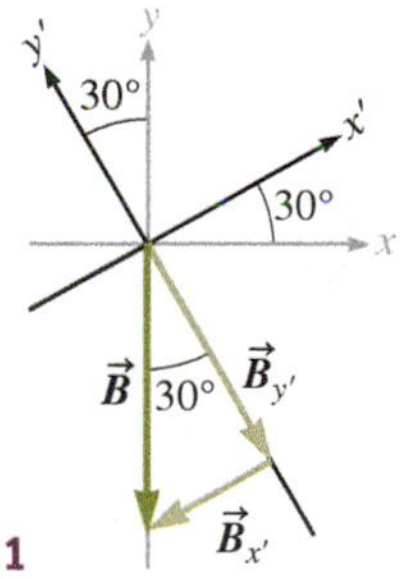

FIGURE 3.31

CHECK and THINK

Many beginners think that the x component of a vector is always found by taking the cosine of some angle and that the y component comes from taking the sine of that angle. In this example, we needed the sine function to find the x component and the cosine function to find the y component. It is best to draw a right triangle and use the general definition of the sine and cosine trigonometric functions (Appendix A) to find vector components. This procedure will give the correct components for any coordinate system, no matter how the angles are drawn.

EXAMPLE 3.6 Using Vector Information to Make a Graphical Representation

A vector $\vec{D}$ is given by $\vec{D} = -3.5\hat{\imath} + 4.0\hat{\jmath}$. Draw a representation of $\vec{D}$ as an arrow on an appropriate coordinate system.

INTERPRET and ANTICIPATE

Because the unit vectors $\hat{\imath}$ and $\hat{\jmath}$ are used, we need a two-dimensional x–y coordinate system.

Example continues on page 72 ▶

SOLVE

Identify the vector components.

$\vec{D}_x = -3.5\hat{\imath}$ $\quad \vec{D}_y = 4.0\hat{\jmath}$

Draw a two-dimensional coordinate system with a suitable scale. The given vector is unitless, so do not show any units on the axes. Draw the vector components $\vec{D}_x$ and $\vec{D}_y$ on the coordinate system, placing the tail of $\vec{D}_y$ at the head of $\vec{D}_x$ (Fig. 3.32). $\vec{D}_x$ is negative, pointing along the negative x axis. $\vec{D}_y$ is positive, pointing parallel to the positive y axis.

To find $\vec{D}$, add $\vec{D}_x + \vec{D}_y$ using the head-to-tail method. The resultant vector is $\vec{D}$, as shown.

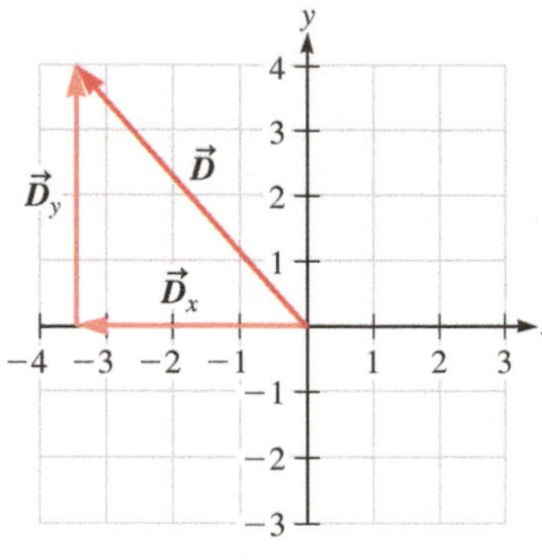

FIGURE 3.32

CHECK and THINK

It makes sense that vector $\vec{D}$ points to the left and up because its x component is negative (left) and its y component is positive (up).

VECTOR MAGNITUDE; VECTOR DIRECTION

Major Concepts

Vector Magnitude and Direction

We have represented vector quantities pictorially and in terms of components. A final way to describe a vector is in terms of its magnitude and direction.

Direction is usually given as an angle in some coordinate system. In the everyday world, this coordinate system is usually the cardinal system used on maps. For example, the phrase "Hurricane Isabelle is heading northeast" means that the hurricane's velocity vector points 45° to the east of north. In physics, the coordinate system used to describe the magnitude and direction of a vector quantity is often a two-dimensional x–y coordinate system. In such a system, we indicate direction as an angle measured from the x axis. The usual convention is that an angle measured counterclockwise from the x axis is considered positive. In Figure 3.33A, the angle between $\vec{A}$ and the x axis is θ, so we say "the direction of $\vec{A}$ with respect to x is θ." If a direction is given as a negative angle, the angle is measured *clockwise* from the x axis. In Figure 3.33B, the direction of $\vec{B}$ is θ measured counterclockwise from the x axis so that $270° < \theta < 360°$. The angle $\alpha = \theta - 360°$ is a negative number.

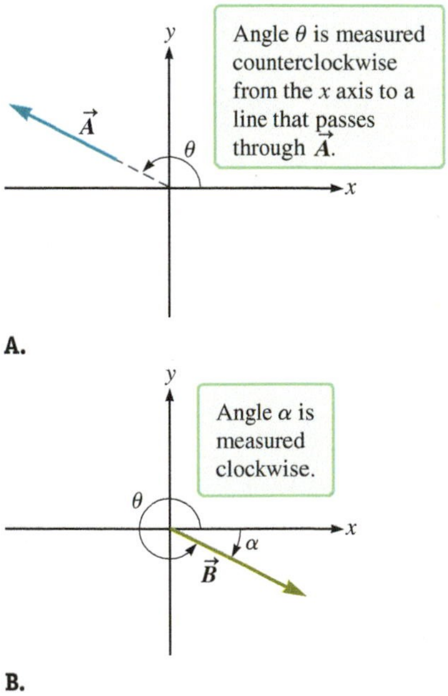

FIGURE 3.33 A. The direction of a two-dimensional vector is given by the positive angle θ measured counterclockwise from the x axis. **B.** The direction may also be given by the negative angle $\alpha = \theta - 360°$ measured clockwise from the x axis.

We can use the scalar components of a vector to find its magnitude and direction. Because $\vec{A}_x$, $\vec{A}_y$, and $\vec{A}$ form a right triangle as in Figure 3.26A, we can use the Pythagorean theorem to find the magnitude A:

$$A^2 = A_x^2 + A_y^2$$

$$A = \sqrt{A_x^2 + A_y^2} \quad (3.12)$$

Because A_x and A_y are squared, their signs do not affect A. Because A is a magnitude, it cannot be negative; therefore, we always choose the positive square root.

The process for finding the magnitude of a three-dimensional vector is similar:

$$A = \sqrt{A_x^2 + A_y^2 + A_z^2} \quad (3.13)$$

We use trigonometry to find the direction θ in terms of A_x and A_y. From Figure 3.26A, we see that

$$\tan\theta = \frac{A_y}{A_x}$$

$$\theta = \tan^{-1}\frac{A_y}{A_x} \quad (3.14)$$

Angles, Inverse Trigonometric Functions, and Vector Directions

You must be careful when using a calculator to take the inverse of a trigonometric function. Because trigonometric functions repeat (are periodic), the inverse of a trigonometric function has an infinite number of solutions. For example, $\tan^{-1} 0 = 0$, 180°, 360°, . . . even though a calculator usually gives only one solution, $\tan^{-1} 0 = 0$.

In a given physics problem, the solution given by a calculator may not be physically reasonable.

The best way to make sure that your answer for a vector's direction is correct is to draw a sketch and estimate the angle. If the angle shown on your calculator is very different from your estimate, you probably need to correct for the geometry of the problem and apply the convention of positive angles measured counterclockwise from the x axis. This convention has been programmed into most calculators.

For example, let us find the direction θ_B of the vector $\vec{B}$ in Figure 3.34. The vector points into quadrant II. So, the angle θ_B must be between 90° and 180°; we estimate that it is about 150°. Reading off $B_x = -3.0$ m and $B_y = 1.5$ m and entering

$$\tan^{-1}\left(-\frac{1.5}{3}\right)$$

into a calculator yields $\tan^{-1}\left(-\frac{1}{2}\right) = -27°$, which is nowhere near our estimate. This answer would mean that the vector points into quadrant IV.

In Figure 3.35, we see that the tangent function repeats every 180°, but a calculator gives only a single solution between −90° and 90°. In other words, the "inverse tangent button" on a calculator cannot tell which of B_x or B_y is negative. To find θ_B, we must take the solution from the calculator and add 180° to return the vector $\vec{B}$ to the proper quadrant (quadrant II; see Fig. 3.16):

$$\theta_B = -27° + 180° = 153°$$

which is very close to our estimate.

The inverse tangent button on a calculator will similarly confuse a vector in quadrant III with one in quadrant I. You must apply the appropriate correction for the geometry of the situation.

One way around this ambiguity produced by our calculators is to sketch the vector along with the angle θ measured from the x axis. If $|\theta| > 90°$, draw an acute angle α between the vector and the x axis as in Figure 3.34. Use the inverse tangent button on your calculator to find α from the magnitudes of the vector components. For the vector in Figure 3.34,

$$\alpha = \tan^{-1}\left(\frac{1.5}{3}\right) = 27°$$

Then use your sketch to find θ. In this case, $\theta_B = 180° - 27° = 153°$.

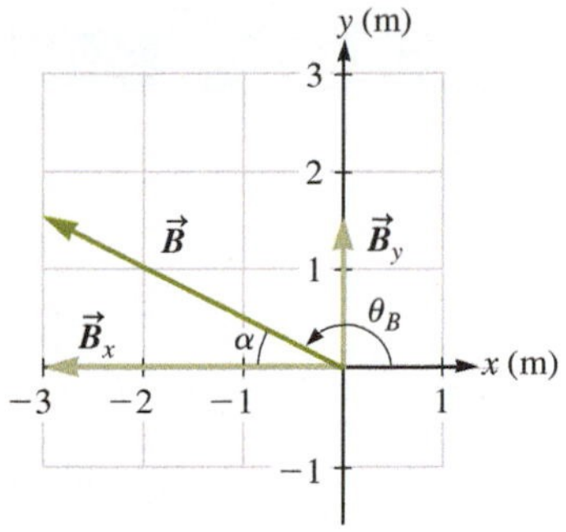

FIGURE 3.34 The direction of a vector is given by the angle measured counterclockwise from the x axis. The Pythagorean theorem and trigonometric functions are used to relate vector components to the magnitude and direction of a vector.

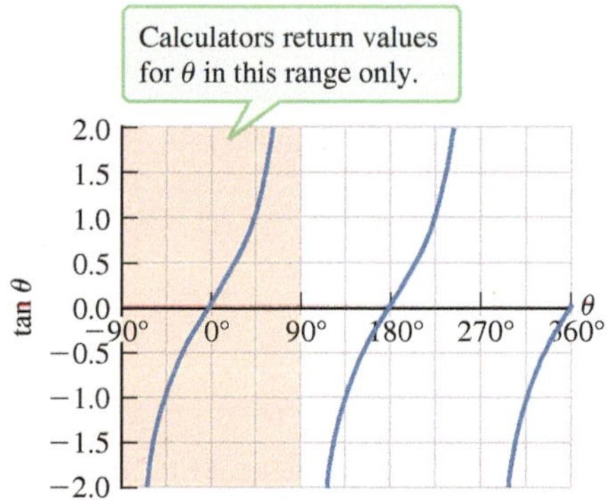

FIGURE 3.35 The tangent function repeats. When you use a calculator to find the inverse tangent, it will only give angles between −90° and 90°.

EXAMPLE 3.7 Translating Unit-Vector Information to Magnitude and Direction

Find the magnitude and direction of the vector $\vec{D} = (-3.5\hat{\imath} + 4.0\hat{\jmath})$ from Example 3.6.

INTERPRET and ANTICIPATE

Our work in Example 3.6 helps us anticipate the answer here. According to our sketch (Fig. 3.32), $\vec{D}$ vector points into quadrant II, so we expect that $90° < \theta < 180°$.

SOLVE The scalar components are the coefficients in front of the unit vectors.	$D_x = -3.5 \qquad D_y = 4.0$
Find the magnitude from Equation 3.12.	$D = \sqrt{D_x^2 + D_y^2}$ $D = \sqrt{(-3.5)^2 + 4.0^2}$ $D = 5.3$
Use Equation 3.14 and a calculator to find the direction.	$\tan^{-1}\dfrac{4.0}{-3.5} = -49°$
CHECK and THINK This result does not meet our expectation that $90° < \theta < 180°$. So, we add 180° to correct the answer returned by the calculator because $\vec{D}$ should point into quadrant II.	$\theta = -49° + 180° = 131°$

EXAMPLE 3.8 CASE STUDY Magnitude and Direction to Miss the Trees

In Example 3.2, we drew a representation of Reese's displacement from the time she set her parachute in the brakes configuration to the moment she landed safely in the field. In this example, we will find the magnitude and direction of her displacement assuming she just barely missed the trees as shown in Figure 3.1. In the next section, we will determine whether she would have survived the jump had she opened her parachute at the usual 750 m instead of at 1200 m.

A Find the magnitude and direction of Reese's displacement if she just barely misses the trees. Use the distance data given in Figure 3.1, page 60.

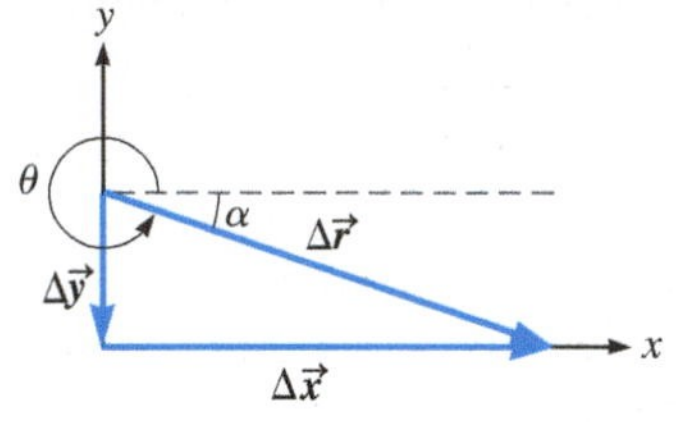

FIGURE 3.36

INTERPRET and ANTICIPATE

Roughly sketch Reese's displacement $\Delta\vec{r}$ in a two-dimensional coordinate system and show the vector components $\Delta\vec{x}$ and $\Delta\vec{y}$ of $\Delta\vec{r}$. Also show the angles θ and α, either of which describes the direction of the displacement vector (Fig. 3.36).

SOLVE

We are given Δx and Δy. Assign Δy a negative value (downward).

$$\Delta x = 3500 \text{ m} \qquad \Delta y = -1200 \text{ m}$$

Modify $A = \sqrt{A_x^2 + A_y^2}$ (Eq. 3.12) to find the magnitude of the displacement.

$$\Delta r = \sqrt{\Delta x^2 + \Delta y^2} \tag{3.12}$$

$$\Delta r = \sqrt{(3500 \text{ m})^2 + (-1200 \text{ m})^2}$$

$$\Delta r = 3.7 \times 10^3 \text{ m}$$

Modify $\theta = \tan^{-1} A_y/A_x$ (Eq. 3.14) to find the direction.

$$\tan^{-1}\frac{\Delta y}{\Delta x} = \tan^{-1}\frac{-1200}{3500} = -19^\circ$$

From our drawing, we see that we have just found α. To find θ, we need to add 360° to our α value.

$$\alpha = -19^\circ$$

$$\theta = 360^\circ + \alpha = 360^\circ - 19^\circ$$

$$\theta = 341^\circ$$

CHECK and THINK

The answer expressed as $\alpha = -19^\circ$ means that Reese fell through an angle of 19° from the horizontal after she set her parachute in the brakes configuration at an altitude of 1200 m.

B Find the magnitude and direction of Reese's displacement if she opens her parachute at an altitude of 750 m instead of 1200 m. Assume she still covers the same horizontal distance as in part A.

INTERPRET and ANTICIPATE

With her parachute closed, Reese falls straight down so that when she opens her parachute the point directly below her is still 3500 m from the edge of the clearing. Her altitude is only 750 m, however.

SOLVE

This situation is like part A except that the y component of displacement has changed.

$$\Delta x = 3500 \text{ m} \qquad \Delta y = -750 \text{ m}$$

As before, modify Equation 3.12 to find the magnitude of the displacement.

$$\Delta r = \sqrt{\Delta x^2 + \Delta y^2} = \sqrt{(3500 \text{ m})^2 + (-750 \text{ m})^2}$$

$$\Delta r = 3.6 \times 10^3 \text{ m}$$

Modify Equation 3.14 to find the direction in terms of α.

$$\alpha = \tan^{-1}\frac{\Delta y}{\Delta x} = \tan^{-1}\left(\frac{-750}{3500}\right) = -12^\circ$$

To find θ we need to add 360° as in part A.	$\theta = 360° + \alpha = 360° - 12°$ $\theta = 348°$

CHECK and THINK

When we return to this CASE STUDY in the next section, we will determine whether or not Reese would have cleared the trees with her chute opening at 750 m.

EXAMPLE 3.9 Translating Magnitude and Direction into Component Form

Jai alai (pronounced "high-lie") is a court game in which players use a long hand-shaped basket strapped to their wrist to propel a ball (Fig. 3.37). The object of the game is to bounce the ball against the wall in such a way that an opponent cannot catch and return the ball. Suppose a jai alai ball collides with a wall at a speed of 69 m/s and a 55° angle from the vertical as shown in Figure 3.38. Choose an x–y coordinate system and write the ball's velocity in component form.

FIGURE 3.37 A jai alai player in action.

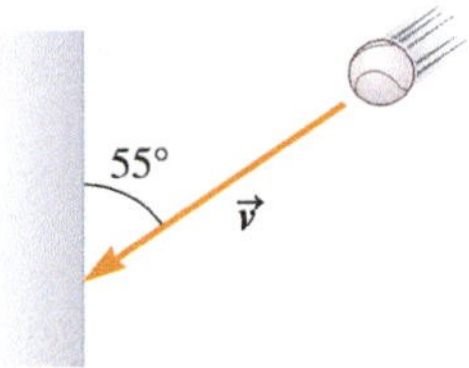

FIGURE 3.38

INTERPRET and ANTICIPATE

Choose a coordinate system and draw a sketch (Fig 3.39). For convenience, place the tail of $\vec{v}$ at the origin and draw the vector components of $\vec{v}$. In anticipation of using $\theta = \tan^{-1}(A_y/A_x)$ (Eq. 3.14), we also show the angle θ measured counterclockwise from the x axis. The velocity vector points in the negative x and y directions, so we expect a numerical result of the form $\vec{v} = (-\_\_\_\_\_\ \hat{\imath} - \_\_\_\_\_\ \hat{\jmath})$ m/s.

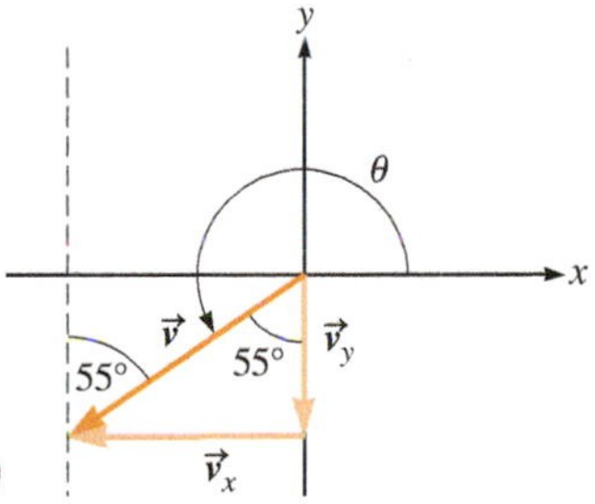

FIGURE 3.39

SOLVE Solve for θ.	$\theta + 55° = 270°$ $\theta = 270° - 55° = 215°$
Trigonometry (Eqs. 3.9 and 3.10) gives the scalar components.	$v_x = v\cos\theta = (69 \text{ m/s})\cos 215° = -57 \text{ m/s}$ $v_y = v\sin\theta = (69 \text{ m/s})\sin 215° = -40 \text{ m/s}$ (two significant figures)
To write the ball's velocity in component form, place the scalar components in front of the appropriate unit vectors.	$\vec{v} = (-57\hat{\imath} - 40\hat{\jmath})$ m/s
CHECK and THINK Both components are negative as we expected. Breaking a velocity into its horizontal and vertical scalar components is an important part of analyzing two-dimensional motion. So, let's use the acute 55° angle and trigonometry to find the components a second time. The velocity vector is the hypotenuse of a right triangle; the x component is opposite the angle, and the y component is adjacent. We can use the sine and cosine functions to find the magnitude of the vector components.	$\sin 55° = \frac{v_x}{v}$ $v_x = v\sin 55°$ $v_x = (69 \text{ m/s})\sin 55° = 57 \text{ m/s}$ $\cos 55° = \frac{v_y}{v}$ $v_y = v\cos 55°$ $v_y = (69 \text{ m/s})\cos 55° = 40 \text{ m/s}$

The negative signs do not come automatically using this method. Instead, we must use our sketch to reason that both components are negative.

3-4 Combining Vectors by Components

In Sections 3-1 and 3-3, we represented vectors pictorially, in component form, and in terms of magnitude and direction. When solving a problem, we often use two or three of these representations. The pictorial representation helps most when planning how to solve a problem, making estimates, and checking a solution. A problem often gives a vector in terms of magnitude and direction, or the final answer may require that form. Complicated calculations are easiest to do using component form, which is the subject of this section. There are three things you need to know about manipulating components.

First, if two vectors are equal to each other, their components are equal. For example, if $\vec{A} = \vec{B}$, then

$$A_x = B_x \tag{3.15}$$

$$A_y = B_y \tag{3.16}$$

These equalities ensure that their magnitudes and directions are equal. To show that the magnitudes are equal, start by finding the magnitude of $\vec{A}$ (Eq. 3.12):

$$A = \sqrt{A_x^2 + A_y^2}$$

Now substitute Equations 3.15 and 3.16 for each component and show that their magnitudes are equal:

$$A = \sqrt{B_x^2 + B_y^2} = B$$

Next, let's find θ_A (the direction of $\vec{A}$) from Equation 3.14:

$$\theta_A = \tan^{-1}\frac{A_y}{A_x}$$

Finally, substitute Equations 3.15 and 3.16 for each component and show that their directions are the same:

$$\theta_A = \tan^{-1}\frac{B_y}{B_x} = \theta_B$$

The second thing you need to know is that when a vector is multiplied (or divided) by a scalar, each of the vector's components is multiplied (or divided) by that scalar. For example, if $\vec{R}$ is the resultant of multiplying $\vec{A}$ by a scalar s, then

$$\vec{R} = s\vec{A} = sA_x\hat{\imath} + sA_y\hat{\jmath} \tag{3.17}$$

The resultant's magnitude is changed by the factor s:

$$R = \sqrt{(sA_x)^2 + (sA_y)^2}$$
$$R = s\sqrt{A_x^2 + A_y^2} = sA$$

but the direction remains unchanged:

$$\tan\theta_R = \frac{sA_y}{sA_x} = \frac{A_y}{A_x} = \tan\theta_A$$

If s is negative, however, the direction of the original vector is reversed.

VECTOR ADDITION

Major Concept

Finally, you need to know that to add or subtract two vectors. you must add or subtract each of their components. For example, if

$$\vec{R} = \vec{A} + \vec{B}$$

the components of $\vec{R}$ are found by adding the components of $\vec{A}$ and $\vec{B}$:

$$R_x = A_x + B_x$$
$$R_y = A_y + B_y$$

The resultant $\vec{R}$ is expressed in component form as

$$\vec{R} = (A_x + B_x)\hat{\imath} + (A_y + B_y)\hat{\jmath} \tag{3.18}$$

VECTOR SUBTRACTION

Major Concept

As an example of the subtraction of two vectors, if

$$\vec{Q} = \vec{A} - \vec{B}$$

the components of $\vec{Q}$ are found by subtracting the components of $\vec{B}$ from those of $\vec{A}$, and $\vec{Q}$ is expressed in component form as

$$\vec{Q} = (A_x - B_x)\hat{\imath} + (A_y - B_y)\hat{\jmath} \tag{3.19}$$

Vector Components of Motion Variables

We are now ready to connect the kinematics we learned in Chapter 2 to the vector mathematics we have learned here. Most of the equations for one-dimensional motion in Chapter 2 were written in terms of vector components. The advantage of using vector components is that it is relatively easy to expand these relationships to three-dimensional motion. Let's do that now for position, displacement, and average velocity.

In Chapter 2, we wrote the position vector $\vec{r}$ in terms of one of three vector components ($\vec{x}$, $\vec{y}$, or $\vec{z}$). The position vector of an object in three-dimensional space is the vector sum of these three vector components:

$$\vec{r} = \vec{x} + \vec{y} + \vec{z} \tag{3.20}$$

By using unit vectors, we write the position vector in terms of its scalar components:

$$\vec{r} = x\hat{\imath} + y\hat{\jmath} + z\hat{k} \tag{3.21}$$

The displacement vector is found by subtracting the initial position vector from the final position vector $\Delta\vec{r} = \vec{r}_f - \vec{r}_i$. This step is done by subtracting each initial component from the corresponding final component (Eq. 3.19):

$$\Delta\vec{r} = (x_f - x_i)\hat{\imath} + (y_f - y_i)\hat{\jmath} + (z_f - z_i)\hat{k} \tag{3.22}$$

Each component in Equation 3.22 is the displacement in one direction (Eq. 2.1):

$$\Delta\vec{r} = \Delta x\hat{\imath} + \Delta y\hat{\jmath} + \Delta z\hat{k} \tag{3.23}$$

or

$$\Delta\vec{r} = \Delta\vec{x} + \Delta\vec{y} + \Delta\vec{z} \tag{3.24}$$

To find the average velocity, divide the displacement vector $\Delta\vec{r}$ by the time interval Δt. The time interval is a scalar, so we just need to divide each component of $\Delta\vec{r}$ (Eq. 3.23) by Δt:

$$\vec{v}_{av} = \frac{\Delta\vec{r}}{\Delta t} = \frac{\Delta x}{\Delta t}\hat{\imath} + \frac{\Delta y}{\Delta t}\hat{\jmath} + \frac{\Delta z}{\Delta t}\hat{k} \tag{3.25}$$

Each component is the average velocity in that particular direction (Eq. 2.2):

$$\vec{v}_{av} = v_{av,x}\hat{\imath} + v_{av,y}\hat{\jmath} + v_{av,z}\hat{k} \tag{3.26}$$

EXAMPLE 3.10 Adding Vectors

Three vectors $\vec{A} = 2\hat{\imath} - 3\hat{\jmath}$, $\vec{B} = 4\hat{\imath} - 2\hat{\jmath}$, and $\vec{C} = -5\hat{\imath} - 7\hat{\jmath}$ are dimensionless. Find vector $\vec{R} = \vec{A} + \vec{B} + \vec{C}$ in component form. Also give the magnitude and direction of $\vec{R}$.

INTERPRET and ANTICIPATE

The best way to anticipate the result of adding three vectors is to sketch the vectors and add them geometrically (Fig. 3.40). The three vectors are drawn head to tail. The resultant vector $\vec{R}$ is drawn from the tail of $\vec{A}$ to the head of $\vec{C}$.

From our sketch, we see that the x component of $\vec{R}$ is positive and small and that the y component is negative and large. We see that angle θ is between 270° and 360°.

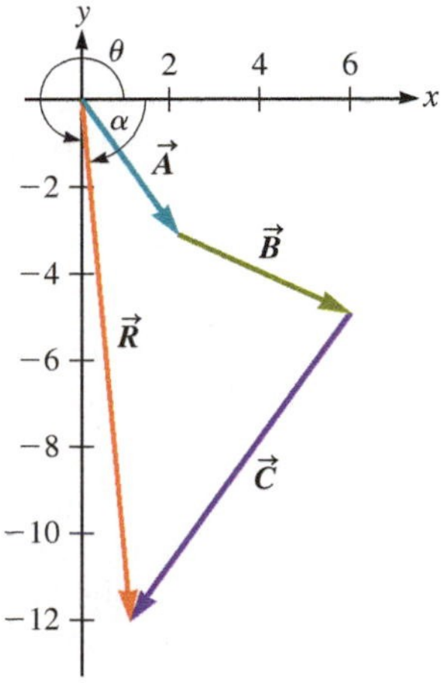

FIGURE 3.40

Example continues on page 78 ▶

SOLVE

According to Equation 3.18, to find the vector sum, we add the x components together and the y components together.

$$R_x = A_x + B_x + C_x = 2 + 4 - 5 = 1$$
$$R_y = A_y + B_y + C_y = -3 - 2 - 7 = -12$$

Writing $\vec{R}$ in component form means adding the vector components of $\vec{R}$.

$$\vec{R} = R_x\hat{\imath} + R_y\hat{\jmath}$$
$$\vec{R} = 1\hat{\imath} - 12\hat{\jmath}$$

CHECK and THINK

As expected, the x component is positive and small, and the y component is negative and large.

SOLVE

The magnitude of the resultant comes from modifying Equation 3.12.

$$R = \sqrt{R_x^2 + R_y^2} = \sqrt{1^2 + (-12)^2} \quad (3.12)$$
$$R = 12 \text{ (two significant figures)}$$

The direction comes from Equation 3.14.

$$\alpha = \tan^{-1}\frac{R_y}{R_x} = \tan^{-1}\frac{-12}{1} = -85° \quad (3.14)$$

Our calculator is giving us angle α instead of angle θ, so we must add 360° to find θ. That is one reason our sketch is so important.

$$\theta - \alpha = 360°$$
$$\theta = \alpha + 360° = -85° + 360°$$
$$\theta = 275°$$

CHECK and THINK

Because the y component is 12 times greater than the x component, the x component has little effect on the magnitude. From our sketch, we would estimate that the magnitude is near 12. Also, we can see that θ is close to 270°, which matches our algebraic results.

EXAMPLE 3.11 CASE STUDY Clearing the Trees in the Brakes Configuration

When Reese saw that she was in danger of hitting the trees, she set her parachute in the brakes configuration (case study, page 60). In this configuration, her velocity was constant at $\vec{v} = (6.0\hat{\imath} - 1.5\hat{\jmath})$ m/s. Had she deployed her parachute at 750 m, would she have cleared the trees?

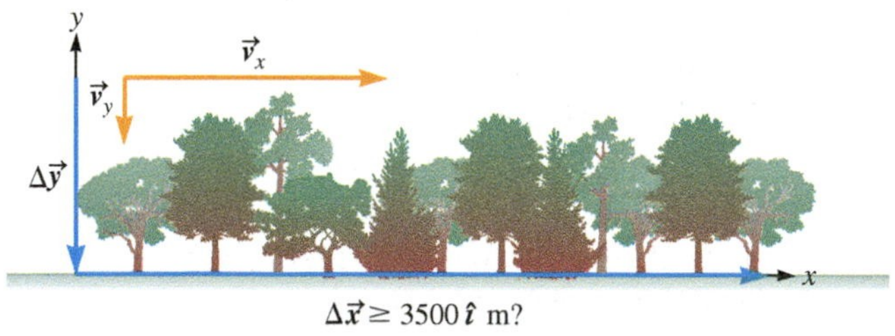

FIGURE 3.41 Suppose Reese releases her parachute at 750 m instead of at 1200 m. Compare with Figures 3.1 and 3.12.

INTERPRET and ANTICIPATE

Our plan is to use Reese's velocity to find her horizontal displacement during the descent. If her horizontal displacement is at least 3500 m ($\Delta x \geq 3500$ m), she would safely clear the trees.

SOLVE

Because Reese's velocity is constant, her average velocity and her instantaneous velocity are equal.

$$\vec{v}_{av} = \vec{v} = (6.0\hat{\imath} - 1.5\hat{\jmath}) \text{ m/s}$$
$$v_{av,x} = 6.0 \text{ m/s}$$
$$v_{av,y} = -1.5 \text{ m/s}$$

If she deploys her parachute at 750 m, then $\Delta y = -750$ m for the jump. Her horizontal displacement Δx depends on time Δt it takes her to descend straight down. Find Δt from the known $v_{av,y}$ and Δy using Equation 3.25.

$$v_{av,y} = \frac{\Delta y}{\Delta t} \quad (3.25)$$
$$\Delta t = \frac{\Delta y}{v_{av,y}} = \frac{-750 \text{ m}}{-1.5 \text{ m/s}} = 500 \text{ s}$$

Use this Δt to find Δx. Equation 3.25 works in this case because the x component of velocity is also constant.	$v_{av,x} = \frac{\Delta x}{\Delta t}$ (3.25) $\Delta x = v_{av,x}\Delta t = (6.0\ \text{m/s})(500\ \text{s})$ $\Delta x = 3000\ \text{m}$ No, she would have hit the trees.

CHECK and THINK

We conclude that Reese would have landed in the trees had she waited and opened her parachute at 750 m. The advantage of deploying her parachute at a higher altitude is that the longer she spends in the air, the larger the value of Δx, assuming that she has the parachute open in the brakes configuration.

SUMMARY

Underlying Principles

No new physical concepts were introduced in the chapter. Instead, we focused on mathematical tools. We have not finished adding all the tools needed to work with vectors, but the rest will be added as we need them. The tools we now have will take care of our needs for the next seven chapters.

Major Concepts

1. Any vector can be written as the sum of its **vector components**: $\vec{A} = \vec{A}_x + \vec{A}_y$. Alternatively, a vector can be written in **component form**, which means that the vector is written in terms of its **scalar components** and the appropriate unit vectors: $\vec{A} = A_x\hat{\imath} + A_y\hat{\jmath}$.
2. The process of finding a vector's (vector or scalar) components is known as **resolving the vector into components**. Trigonometry is used to resolve a vector into components.
3. The **magnitude** of a two-dimensional vector $\vec{A}$ is
$$A = \sqrt{A_x^2 + A_y^2} \quad (3.12)$$
The **direction** of a two-dimensional vector measured counterclockwise from the x axis may be written in terms of its scalar components as
$$\theta = \tan^{-1}\frac{A_y}{A_x} \quad (3.14)$$
4. Vectors may be **added**, $\vec{R} = \vec{A} + \vec{B}$, in terms of their scalar components:
$$\vec{R} = (A_x + B_x)\hat{\imath} + (A_y + B_y)\hat{\jmath} \quad (3.18)$$
Vectors may be **subtracted**, $\vec{Q} = \vec{A} - \vec{B}$, in terms of their scalar components:
$$\vec{Q} = (A_x - B_x)\hat{\imath} + (A_y - B_y)\hat{\jmath} \quad (3.19)$$
5. Vector addition is **commutative**, which means that the order of addition does not matter:
$$\vec{A} + \vec{B} = \vec{B} + \vec{A} \quad (3.2)$$
6. Vector addition is **associative**, which means that if there are more than two vectors to add, they can be grouped in any order:
$$(\vec{A} + \vec{B}) + \vec{C} = \vec{A} + (\vec{B} + \vec{C}) \quad (3.3)$$

Tools

1. Vector manipulation
   a. **Parallelogram addition** is a geometric method for adding vectors $\vec{R} = \vec{A} + \vec{B}$. In this method, two copies of each vector $\vec{A}$ and $\vec{B}$ are used to form a parallelogram. The resultant vector $\vec{R}$ is drawn along the diagonal of the parallelogram from the tails of the two vectors $\vec{A}$ and $\vec{B}$ to their heads.
   b. **Head-to-tail addition** is another geometric method for adding vectors $\vec{R} = \vec{A} + \vec{B}$. In this method, the head of one of the vectors is placed at the tail of the other vector. The resultant vector is drawn from the free tail of the first vector $\vec{A}$ to the free head of the second vector $\vec{B}$.

Tools—cont'd

Geometric **subtraction** $\vec{R} = \vec{A} - \vec{B}$ may be performed by first multiplying $\vec{B}$ by -1 and then applying either the parallelogram or head-to-tail method.

c. **Tail-to-tail subtraction** is an alternative geometric method for subtracting vectors $\vec{R} = \vec{A} - \vec{B}$. The two vectors $\vec{A}$ and $\vec{B}$ are placed tail to tail, and the resultant vector points from the head of the second vector $\vec{B}$ to the head of the first vector $\vec{A}$.

2. Coordinate systems
   a. In a **Cartesian coordinate system**, two or more axes cross at right angles to each other.
   b. All the coordinate systems used in this book are **right-handed coordinate systems.** In a right-handed coordinate system, the direction of $\hat{k}$ is determined by the *right-hand convention* (Fig. 3.22).

PROBLEMS AND QUESTIONS

A = algebraic C = conceptual E = estimation G = graphical N = numerical

3-1 Geometric Treatment of Vectors

1. A velocity vector has a magnitude of 720 m/s. Two students draw arrows representing this vector. Clarisse chooses a scale such that 1 cm $\rightarrow$ 100 m/s.
   a. N What is the length of the arrow that Clarisse draws?
   b. N Francois's arrow is half as long as Clarisse's. What is Francois's scale?
   c. C Is one student's choice better than the other? If so, what makes it a better scale?
2. A young boy throws a baseball through a window.
   a. C Sketch the problem and pick a reference point.
   b. G Use this reference point to draw a vector representing the initial and final position of the ball.
   c. G Draw the displacement vector.
   d. C Which of these vectors change if you pick a different reference point?
3. G Vectors $\vec{A}$ and $\vec{B}$ are perpendicular and have the same nonzero magnitude ($A = B$). If $\vec{C} = \vec{A} + \vec{B}$, what is C, the magnitude of $\vec{C}$? *Hint*: Sketch these vectors.
4. G Vectors $\vec{A}$ and $\vec{B}$ have the same nonzero magnitude ($A = B$), where $\vec{C} = \vec{A} + \vec{B}$ and $\vec{D} = \vec{A} - \vec{B}$. If the magnitudes of $\vec{C}$ and $\vec{D}$ are the same, how are the directions of $\vec{A}$ and $\vec{B}$ related? *Hint*: Sketch these vectors.
5. N Vector $\vec{A}$, with a magnitude of 18 units, points in the positive x direction. Adding vector $\vec{B}$ to vector $\vec{A}$ yields a resultant vector that points in the negative x direction with a magnitude of 6 units. What are the magnitude and direction of vector $\vec{B}$?

Problems 6, 42, and 64 are grouped.

6. G Figure P3.6 shows three vectors. Copy this figure on to your own paper and find $\vec{R} = \vec{F}_1 + \vec{F}_2 + \vec{F}_3$ geometrically.
7. C A student makes a mistake in finding displacement $\Delta\vec{r}$ using head-to-tail addition. Instead of drawing $-\vec{r}_i$, the student draws $-\vec{r}_f$. What, if anything, is wrong with the resulting displacement vector's magnitude or direction?

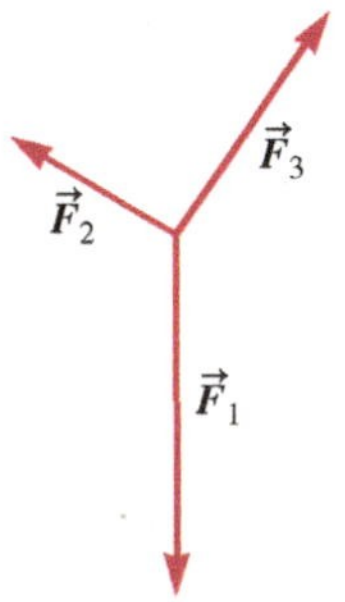

FIGURE P3.6
Problems 6 and 64.

8. The layout of the town of Popperville is a perfectly square grid, with blocks 100 feet long (Fig. P3.8). A cat leaves her house and travels to Mike Mulligan's place 4 blocks west and 3 blocks north of its starting point. After lapping up the milk that Mike always leaves out, the cat travels an additional 1 block north and 3 blocks east to Mrs. McGillicutty's house.
   a. G On your paper, carefully copy the grid of streets that contains the cat's house, Mike Mulligan's house, and Mrs. McGillicutty's house. Then draw and label three displacement vectors on your sketch: $\Delta\vec{r}_1$, the cat's displacement from her house to Mike Mulligan's; $\Delta\vec{r}_2$, the cat's displacement from Mike Mulligan's to Mrs. McGillicutty's; and $\Delta\vec{r}_{tot}$, the cat's total displacement.
   b. A Write an equation to relate the three displacement vectors you have drawn.
   c. C Is the distance the cat traveled greater than, less than, or equal to the magnitude of the cat's total displacement vector? Explain.

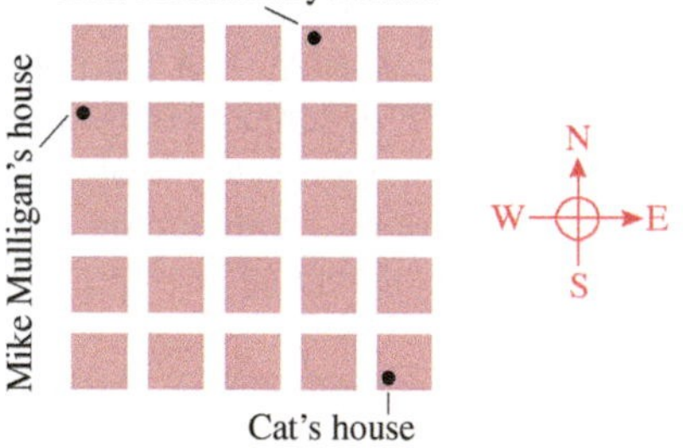

FIGURE P3.8

9. C The magnitude of average velocity is miscalculated by dividing the displacement Δr by $2\Delta t$ instead of Δt. The mistake is carried forward, and an average velocity vector is drawn to scale. What, if anything, is wrong with this vector's magnitude or direction?
10. G An Olympic runner races through the final turn of the track on her way to the finish line. For the following questions, assume the curved section is a semicircle and that the runner moves clockwise around the track as watched from above. **a.** On your paper, draw a sketch of the curved section of the track as seen from above. Mark two points on the track, one to represent the position of the runner at time t_i, just after she has entered the turn, and one to represent her position at time t_f, part of the way through the turn. (You have freedom in your choices.) **b.** Using a point in the interior of the track as an origin, draw vectors to represent the position of the runner at time t_i (vector $\vec{r}_i$) and the position of the runner at time t_f (vector $\vec{r}_f$). **c.** Use your position vectors to find the displacement of the runner.
11. A Three vectors $\vec{A}$, $\vec{B}$, and $\vec{C}$ are related to one another such that $2\vec{A} + 3\vec{B} = 18\vec{C}$. A fourth vector $\vec{D}$ exists such that $6\vec{C} + \vec{D} = \vec{A}$. Determine an expression for $\vec{D}$ in terms of the vectors $\vec{A}$ and $\vec{B}$.

12. **G** Draw three arbitrary vectors $\vec{A}$, $\vec{B}$, and $\vec{C}$. Use these three vectors to demonstrate that vector addition is associative (Eq. 3.3).

13. **C** In Chapter 5, you will study a very important vector, *force*. Each case in Figure P3.13 shows an example of force vectors exerted on an object. These forces are all of the same magnitude F_0. Assume that the forces lie in the plane of the paper. Rank the cases from greatest to smallest according to the magnitude of the total force. *Note*: The total force is the vector sum of the individual forces exerted on the object.

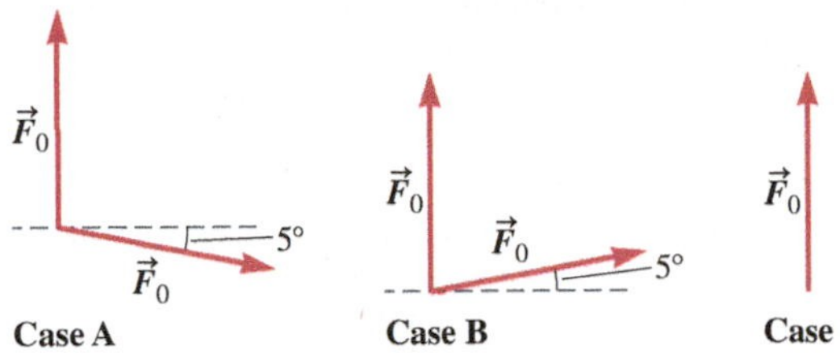

FIGURE P3.13

Problems 14 and 15 are paired.

14. **G** For this problem, you will need a ruler. A classroom clock has a small magnifying glass embedded near the end of the minute hand. The magnifying glass may be modeled as a particle undergoing translational motion along a circular path. Class begins at 7:55 and ends at 8:50 (Fig. P3.14) **a.** Suppose the minute hand of the clock has an actual length of 0.300 m. Use your ruler to find the scale we used when drawing Figure P3.14. Use that scale to draw the displacement vector of the magnifying glass during the 55-min class period. In the **CHECK and THINK** step, be sure to comment on the direction of the displacement. **b.** Use your ruler to find the magnitude of the displacement vector.

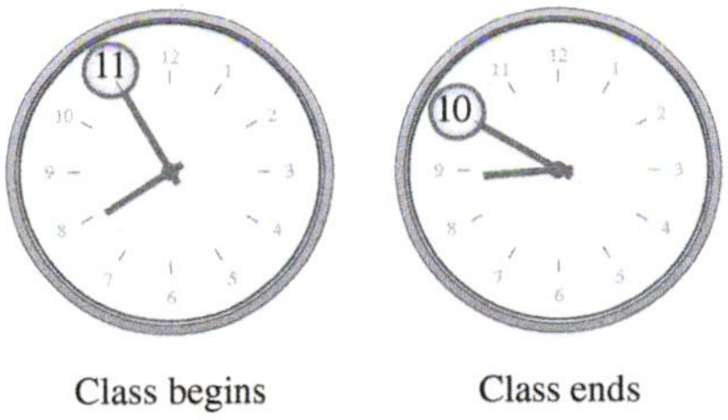

FIGURE P3.14 Problems 14 and 15.

15. Refer to the situation described in Problem 14. Average velocity is the displacement divided by the time interval:

$$\vec{v}_{av} = \frac{\Delta \vec{r}}{\Delta t}$$

a. **N** Graphically find the magnitude of the average velocity of the magnifying glass during the 55-min class period.

b. **C** Describe the direction of the average velocity vector.

16. **G** Vector $\vec{A}$ has a magnitude of 4.50 m and makes an angle of 64.0° with the positive x axis. Vector $\vec{B}$, with a magnitude equal to that of vector $\vec{A}$, points along the negative y axis (Fig. P3.16). Graphically find the magnitude and direction of the resultant vector **a.** $\vec{A} + \vec{B}$, **b.** $\vec{A} - \vec{B}$, **c.** $\vec{B} - \vec{A}$, and **d.** $2\vec{A} - \vec{B}$.

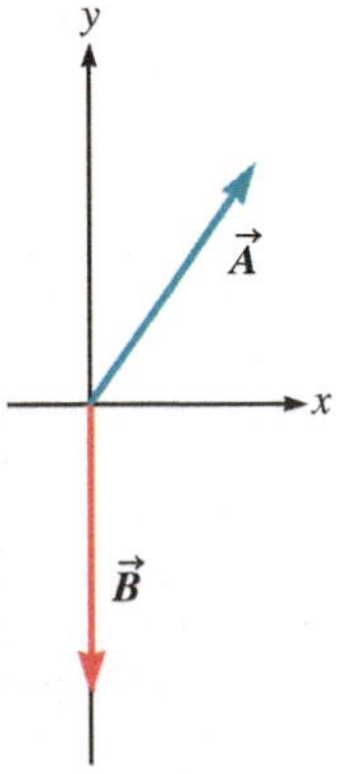

FIGURE P3.16

17. Miguel, an Ultimate Frisbee player, is running three drills (Fig. P3.17). In the first drill, Miguel runs a distance d straight down the field and then makes a 90° turn to the right, running an additional distance d. In the second drill, Miguel runs a distance d straight down the field and makes a 95° turn to the right before running an additional distance d. In the final drill, Miguel runs a distance d straight down the field and makes an 85° turn to the right before running an additional distance d.

a. **C** In which case did Miguel end up farthest from his starting point? In which case did he end up closest to his starting point?

b. **C** In which case(s) does Miguel end up farther away from his starting point than distance d?

c. **N** At what angle will his distance away be exactly d?

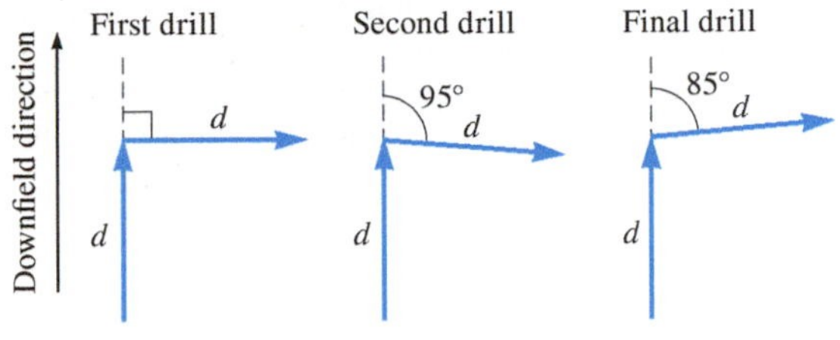

FIGURE P3.17

3-2 Cartesian Coordinate Systems

18. **N** A baseball diamond consists of four plates arranged in a square. Each side of the square is 90 ft (27.43 m) long. Use an $x-y$ coordinate system with the origin at the center of the diamond as shown in Figure P3.18. **a.** What is the position of each plate in this system? **b.** What is the distance from home plate to second base?

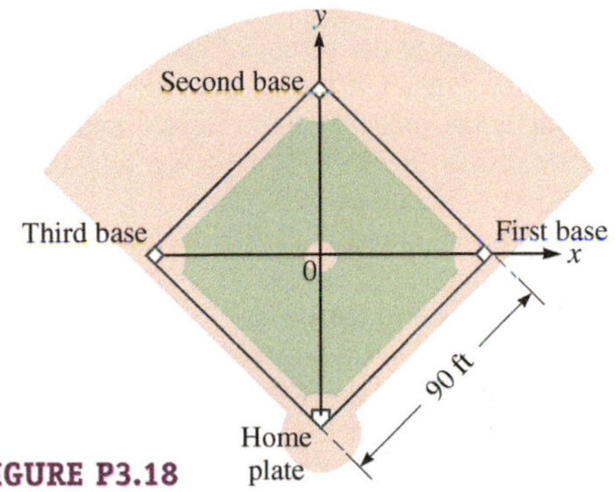

FIGURE P3.18

19. **N** A museum curator is setting up a new display. The show has a particularly large painting that is 6.6 m long. The floor plan for the gallery is shown in Figure P3.19. Can the curator display this large painting in this gallery? If so, on which walls can it hang?

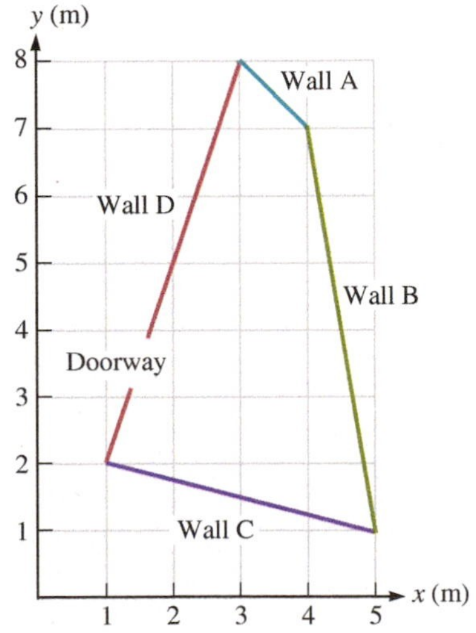

FIGURE P3.19

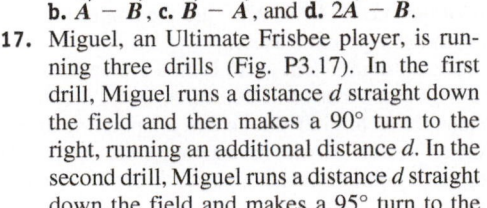

20. A football is thrown along an arc. The ball is released at a height of 6.54 ft (1.99 m) above the ground and, after traveling a horizontal distance of 87.8 ft (26.8 m), is caught by a receiver at a point that is 4.22 ft (1.29 m) above the ground.

a. **G** Sketch the throw from the point of release to the moment the ball is caught (as viewed by an observer from the sideline), placing an x–y coordinate system in the picture with the origin at ground level, directly below where the ball is released.

b. **N** Determine the coordinates of the ball at the point it is released and where it is caught.

c. **N** What is the magnitude of the displacement between the initial and final points in the ball's flight?

21. **N** Two aircraft approaching an aircraft carrier are detected by radar. The first aircraft is at a horizontal distance of 34.3 km in the direction 42.0° north of east and has an altitude of 2.40 km. The second aircraft, at half this altitude, is at a horizontal distance of 42.6 km in the direction 37.0° north of east. What is the distance separating the two aircraft? *Hint*: Use a coordinate system with the x axis pointing east, the y axis pointing north, and the z axis in the vertical direction.

22. **C** Because right-handed people hold the pencil in their right hand, they often mistakenly use their free left hand to draw a coordinate system. In effect, they draw a left-handed coordinate system. Suppose that such a person has drawn the x axis pointing to the right and the y axis pointing toward the top of the page. Which way does the z axis point in a left-handed system?

23. **N** A truck driver delivering office supplies downtown travels 5.00 blocks north, 3.00 blocks east, and finally 2.00 blocks south. **a.** What is the magnitude (in blocks) and the direction of the driver's displacement? **b.** What is the total distance traveled (in blocks) by the truck driver?

24. **C** Instead of closing their right hand so that their fingers push the x axis into the y axis and their thumb points in the direction of the z axis, some people like to hold their right index finger, middle finger, and thumb at right angles to one another as shown in Figure P3.24. Suppose that the z axis is associated with the thumb. With which axis is the index finger associated? With which axis is the middle finger associated?

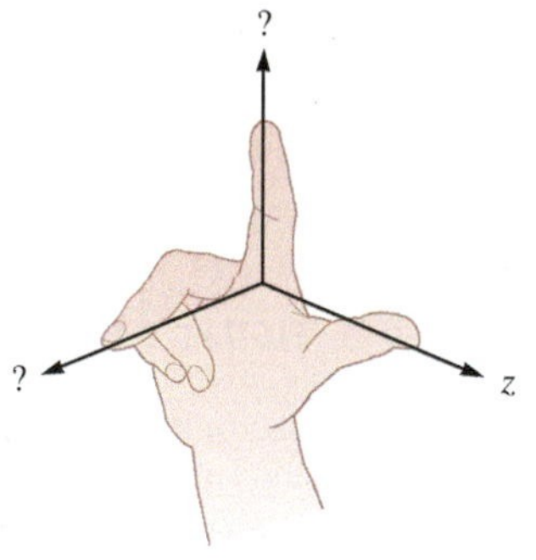

FIGURE P3.24

3-3 Components of a Vector

25. **N** Carolyn rides her bike 40.0° south of west for 5.40 miles as illustrated in Figure P3.25. What is the distance she would have to ride due south and due west to reach the same location?

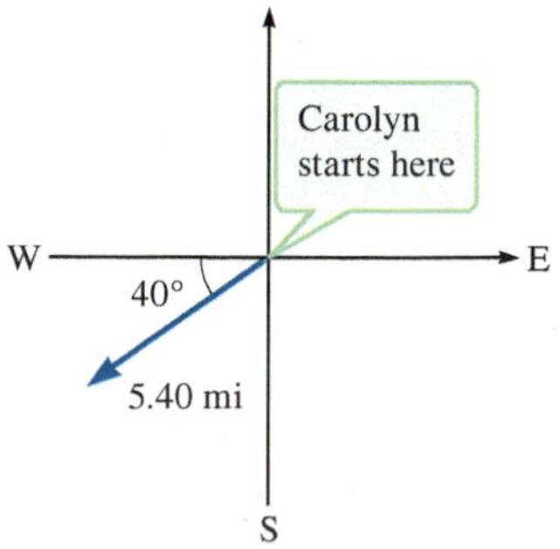

FIGURE P3.25

Problems 26 and 27 are paired.

26. **G** Draw the vector $\vec{A} = 16.5\hat{\imath} - 33.0\hat{\jmath}$ on a coordinate system.

27. **N** Find the magnitude and direction of the vector $\vec{A} = 16.5\hat{\imath} - 33.0\hat{\jmath}$.

28. **N** A vector's x component is twice its y component. What is the acute angle between the two vector components?

Problems 29 and 30 are paired.

29. **N** Vector $\vec{B}$ has a magnitude of 19.45. Its direction measured counterclockwise from the x axis is 127.3°. Find its vector components.

30. **G** Draw the vector in Problem 29 on a coordinate system.

31. **N** A vector points into the first quadrant, and its x and y components are both positive. If its magnitude is equal to twice the magnitude of its x component, what is the angle between the vector and the positive x axis?

32. **N** Find the magnitude of the vector $\vec{B} = -2.00\hat{\imath} + 3.00\hat{\jmath} - 4.00\hat{k}$.

33. **N** Vector $\vec{A}$ has scalar components of 4.00 units in the negative x direction and 2.00 units in the negative y direction. **a.** Write the vector $\vec{A}$ in component form. **b.** What are the magnitude and direction of $\vec{A}$? **c.** Find the vector $\vec{B}$ that when added to $\vec{A}$ yields a resultant vector with an x component 6.00 units in the negative x direction and no y component.

34. A soccer player starts at one end of the field and runs along a straight line making an angle θ with the goal line. She is able to move d meters closer to the opposing goal line before going out of bounds.

a. **G** Make a sketch of this situation that includes the soccer player's displacement vector $\Delta\vec{r}$. Label d and θ on your sketch.

b. **A** Write an expression for the distance the soccer player ran in terms of θ and d.

35. **N** A firecracker explodes into four equal pieces (Fig. P3.35). Given the magnitude and direction of the velocity for each piece and the coordinate system shown, determine the x and y velocity components for each piece of the firecracker.

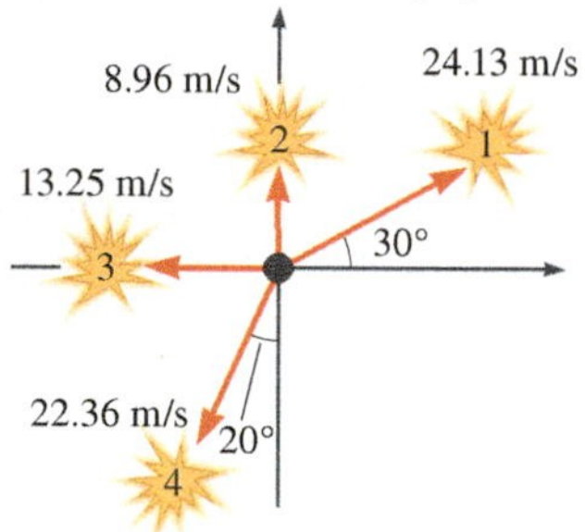

FIGURE P3.35

36. **A** In a soccer match, a breakaway forward is streaking toward the goal line with speed v_0. A defender closes from the side as shown in the motion diagram in Figure P3.36. Notice that the "downfield positions" of the forward and the defender are the same at each instant in time. If the defender moves in a direction making an angle θ with respect to the downfield direction, what is his speed? Answer in terms of the given quantities.

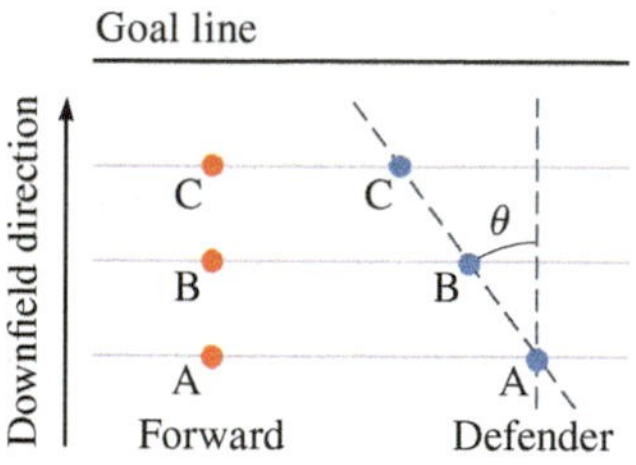

FIGURE P3.36

37. **N** Ezri, a greyhound dog, is running 30.00° west of north with a constant speed of 17.88 m/s. If the positive x axis points to the east and the positive y axis points to the north, determine Ezri's displacement (in component form) during a time interval of 4.000 s.

38. **N** Consider the vector $\vec{A} = 3.00\hat{\imath} + 2.00\hat{\jmath} + 5.00\hat{k}$. **a.** What are the magnitudes of the scalar components of $\vec{A}$? **b.** What is the magnitude of the vector $\vec{A}$? **c.** What are the angles the vector $\vec{A}$ makes with the x, y, and z axes?

39. **N** Vector $\vec{A}$ has an x component of 15.0 m, a y component of −6.00 m, and a z component of −3.00 m. **a.** Write the vector $\vec{A}$ in component form. **b.** Find the vector $\vec{B}$, pointing in the same direction as vector $\vec{A}$, but having one-third its length. **c.** Find the vector $\vec{C}$, pointing in the opposite direction of vector $\vec{A}$, but with three times its length.

3-4 Combining Vectors by Components

40. **G** Figure P3.40 shows a map of Grand Canyon National Park in Arizona. You need a ruler and protractor for this problem. **a.** Paul hikes from Cape Royale to Point Sublime. Find the magnitude and direction of his displacement, ignoring any difference in altitude between the two points. **b.** Lil hikes from Point Sublime to Cape Royale. Find the magnitude and direction of her displacement. Compare your answer with that of part (a).

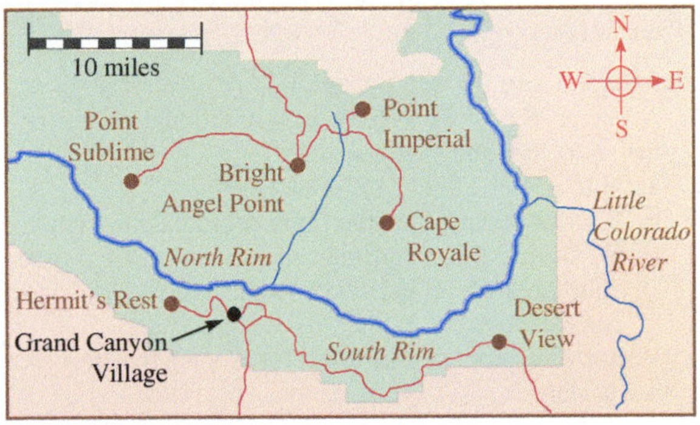

FIGURE P3.40

41. **N** CASE STUDY In the case study, Reese opened her parachute at 1200 m. **a.** Find α_{safe}, the angle below the horizontal at which she would just barely miss the grove of trees and would land safely in the field. **b.** Use the data in Example 3.11 to find the direction of her actual displacement. Comment on her relative safety. (Was her landing too close for comfort, or did she land far from the grove of trees?) Your comment should make reference to your results in part (a).

Problems 6, 42 and 64 are grouped.

42. The same vectors that are shown in Figure P3.6 are shown in Figure P3.42. The magnitudes are $F_1 = 1.90f$, $F_2 = f$, and $F_3 = 1.4f$, where f is a constant.

a. **N** Use the coordinate system shown in Figure P3.42 to find $\vec{R} = \vec{F}_1 + \vec{F}_2 + \vec{F}_3$ in component form in terms of f.

b. **N** If $R_x = 0.33$, what is R_y?

c. **C** Check your result by comparing your answer to that of Problem 6.

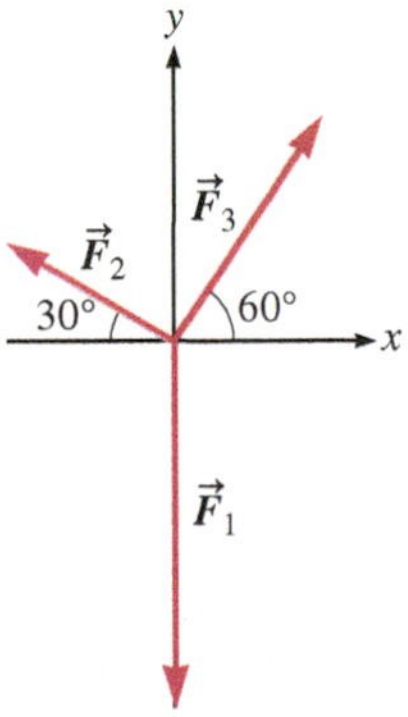

FIGURE P3.42

43. **N** A supertanker begins in Homer, Alaska, sails 125 km west, and then sails 275 km south. In which direction would it need to sail to return in a straight line to where it began?

44. **A** Three vectors are shown in Figure P3.44, but they are not drawn to scale. The sum of the three vectors is $\vec{R} = \vec{F}_1 + \vec{F}_2 + \vec{F}_3$. If $R_y = 0$ and $F_2 = 0.5F_3$, find R_x in terms of F_1.

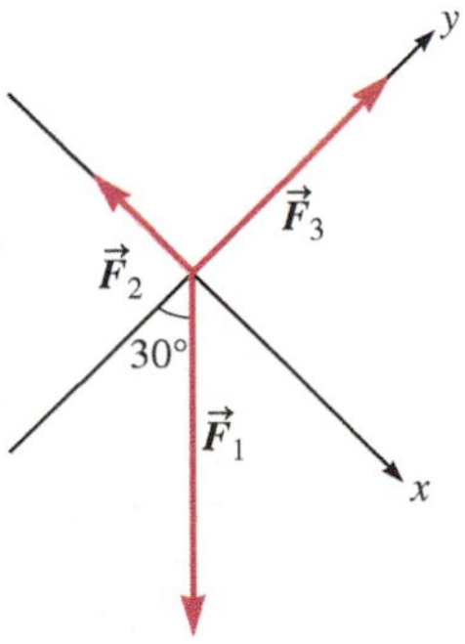

FIGURE P3.44

45. **N** A vector $\vec{A} = (5.20\hat{\imath} - 3.70\hat{\jmath})$ m and a vector $\vec{B} = (1.04\hat{\imath} + B_y\hat{\jmath})$ m are related by a scalar quantity such that $\vec{A} = s\vec{B}$. Determine the value of the y component of the vector $\vec{B}$.

46. **N** Frustrated with her physics class, Janet leaves campus and walks into town to acquire some refreshments. She continues on to a friend's house, where she happens to meet a physics tutor. The tutor agrees to help her, and Janet and the tutor walk to the nearby library to study before the upcoming exam. Janet's three displacements can be described by the following vectors: $\Delta\vec{r}_1 = (625\hat{\imath} - 271\hat{\jmath})$ m, $\Delta\vec{r}_2 = (415\hat{\jmath})$ m, and $\Delta\vec{r}_3 = (-296\hat{\imath} + 181\hat{\jmath})$ m. What is Janet's total displacement?

Problems 47 and 48 are paired.

47. **N** Consider the vectors $\vec{A} = 16.5\hat{\imath} - 33.0\hat{\jmath}$ and $\vec{B} = -2.00\hat{\imath} + 3.00\hat{\jmath} - 4.00\hat{k}$. Find $\vec{R} = \vec{B} - \vec{A}$ in component form.

48. **N** Consider vectors $\vec{A} = 16.5\hat{\imath} - 33.0\hat{\jmath}$, $\vec{B} = -2.00\hat{\imath} + 3.00\hat{\jmath} - 4.00\hat{k}$, and $\vec{R} = \vec{B} - \vec{A}$. Find the magnitude of $\vec{R}$.

49. **A** An airplane leaves city A and flies a distance d_1 due north, landing at city B. The next day, the airplane leaves city B and flies a distance d_2 in a direction making an angle θ to the east of due north, landing at city C. Determine the distance between city A and city C. Express your answer in terms of d_1, d_2, and θ.

50. **N** An aircraft undergoes two displacements. If the first displacement is directed at an angle of 225° with the positive x axis and has magnitude 30.0 miles, and if the resultant displacement is directed at an angle of 75.0° with the positive x axis and magnitude 22.0 miles, what are the magnitude and direction of the second displacement?

51. **N** The resultant vector $\vec{R} = 2\vec{A} - \vec{B} - 2\vec{C}$ has zero magnitude. Vector $\vec{A}$ has an x component of 4.60 m and a y component of -12.1 m, and vector $\vec{B}$ has an x component of -3.00 m and a y component of -4.00 m. What are the x and y components of vector $\vec{C}$? (All of these are two-dimensional vectors.)

52. **A** Three vectors all have the same magnitude. The symbol for the magnitude of each of these vectors is M. The first vector $\vec{A}$ points in the positive x direction. The second vector $\vec{B}$ points in the negative y direction. The third vector $\vec{C}$ points in the positive z direction. These three vectors added together are equal to a fourth vector $\vec{D}$. What is the magnitude of the fourth vector?

53. **N** The two-dimensional vectors $\vec{A}$ and $\vec{B}$ both have magnitudes of 7.00 m. If $\vec{A} + \vec{B} = 4.00\hat{\imath}$ m, what is the angle between $\vec{A}$ and $\vec{B}$?

Problems 54 and 55 are paired.

54. **A** Two birds begin next to each other and then fly through the air at the same elevation above level ground at the same speed v measured in meters per second. One flies northeast, and the other flies northwest. Northeast is exactly halfway between north and east, and northwest is exactly halfway between north and west. After flying for a time t measured in seconds, what is the distance d measured in meters between them? Ignore the curvature of the Earth.

55. **N** Two birds begin next to each other and then fly through the air at the same elevation above level ground at 22.5 m/s. One flies northeast, and the other flies northwest. After flying for 10.5 s, what is the distance between them? Ignore the curvature of the Earth.

56. **N** Vector $\vec{A} = -\hat{\imath} + 2\hat{\jmath} - 5\hat{k}$ and vector $\vec{B} = -5\hat{\imath} - 3\hat{\jmath} - 2\hat{k}$. **a.** What are the scalar components of vector $\vec{R} = \vec{A} + \vec{B}$? What is the magnitude of vector $\vec{R}$? **b.** What are the scalar components of vector $\vec{S} = 3\vec{A} - 2\vec{B}$? What is the magnitude of vector $\vec{S}$?

General Problems

57. **G** A spider undergoes the displacement represented by the vector $\Delta\vec{r}_1$ in Figure P3.57. The spider rests for a while and then undergoes the displacement represented by the vector $\Delta\vec{r}_2$. Each of these

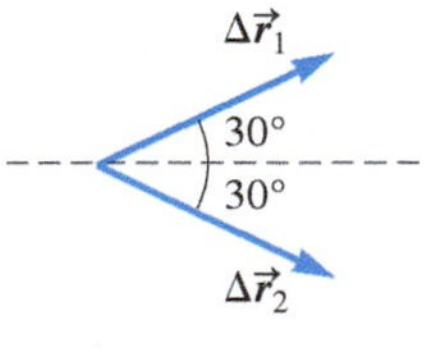

FIGURE P3.57

displacements has magnitude d. After making these two consecutive displacements, is the spider's distance from its original starting point greater than $2d$, equal to $2d$, between d and $2d$, equal to d, or less than d? Explain your reasoning.

Problems 58 and 59 are paired.

58. Peter throws a baseball through a house's window. He stands 3.4 m from the window. The ball starts in his hand 1.1 m above the ground and goes through the window at a point 2.4 m off the ground.

a. G Sketch the problem, including an x–y coordinate system with the origin on the ground directly below the initial position of the ball, the y axis pointing upward, and the x axis pointing toward the house.

b. N What are the initial and final coordinates of the ball?

59. N After finding the initial and final coordinates of the ball in Problem 58, find the displacement of the ball in component form.

60. N Consider the two vectors $\vec{A} = -4\hat{\imath} + 3\hat{\jmath}$ and $\vec{B} = -2\hat{\imath} - 5\hat{\jmath}$. Find the vectors **a.** $\vec{A} + \vec{B}$ and **b.** $\vec{A} - \vec{B}$. Find the quantities **c.** $|\vec{A} + \vec{B}|$ and **d.** $|\vec{A} - \vec{B}|$. **e.** Find the directions of the vectors $\vec{A} + \vec{B}$ and $\vec{A} - \vec{B}$.

61. N Vector $\vec{R}$, with a magnitude of 22.5 units, is directed at an angle of 73.0° below the positive x axis. What are its x and y components?

62. N A glider aircraft initially traveling due west at 85.0 km/h encounters a sudden gust of wind at 35.0 km/h directed toward the northeast (Fig. P3.62). What are the speed and direction of the glider relative to the ground during the wind gust? (The velocity of the glider with respect to the ground is the velocity of the gilder with respect to the wind plus the velocity of the wind with respect to the ground.)

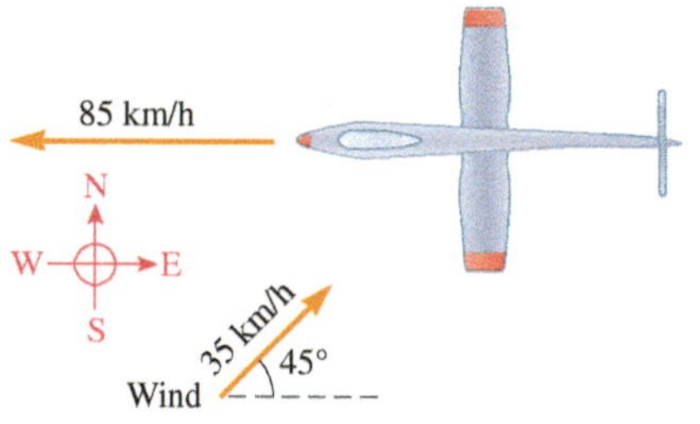

FIGURE P3.62

63. N What are the magnitude and direction of a vector that has an x component of -33.0 units and a y component of -45.0 units?

Problems 6, 42 and 64 are grouped.

64. G Three vectors are shown in Figure P3.6. Copy this figure onto your paper and find $\vec{R} = \vec{F}_2 - \vec{F}_3 - \frac{1}{2}\vec{F}_1$ geometrically.

65. C Vector $\vec{A}$ is in the positive x direction and has magnitude 10. Vector $\vec{B}$ makes an angle of $+\theta$ (measured counterclockwise) from the positive x axis, whereas vector $\vec{C}$ makes an angle of $-\theta$ (measured counterclockwise) from the positive x axis. Vectors $\vec{B}$ and $\vec{C}$ have equal magnitudes. The three vectors $\vec{A}$, $\vec{B}$, and $\vec{C}$ are related according to the equation $\vec{A} = \vec{B} + \vec{C}$. Are the magnitudes of vectors $\vec{B}$ and $\vec{C}$ greater than 5, less than 5, or equal to 5? Explain. If your answer depends on the specific value of θ state that explicitly. *Hint*: Sketch these vectors, including θ.

66. A Using the rules of vector addition, prove that $A = \sqrt{A_x^2 + A_y^2 + A_z^2}$ represents the magnitude of a three-dimensional vector.

67. C A vector $\vec{A}$ and a vector $\vec{B}$ each have the same magnitude but point in different directions. Under what circumstances will the sum of these two vectors result in a vector $\vec{C}$ with the same magnitude as vectors $\vec{A}$ and $\vec{B}$? *Hint*: Draw a sketch.

68. Draw two unit vectors $\hat{\imath}$ and $\hat{\jmath}$.

a. G Geometrically find the sum of these two vectors.

b. N What are the magnitude and direction of the resultant vector?

c. N Write the resultant in component form.

69. C Vector $\vec{A}$ and vector $\vec{B}$ point in different directions. **a.** Under what circumstances are the two vectors related by a scalar quantity as in $\vec{B} = s\vec{A}$? **b.** What must be true about the scalar quantity?

Problems 70 and 71 are paired.

70. A vector $\vec{F} = 3.3\hat{\imath} + 6.3\hat{\jmath}$ is proportional to vector $\vec{A}$ such that $\vec{F} = m\vec{A}$ and m is a scalar.

a. N If $\vec{A} = 1.1\hat{\imath} + 2.1\hat{\jmath}$, what is m?

b. G Draw $\vec{A}$ and $\vec{F}$ on the same coordinate system.

c. N, C Find the magnitude and direction of $\vec{A}$ and $\vec{F}$. Are your answers consistent? Explain how you checked for consistency.

71. Vector $\vec{F}$ is proportional to vector $\vec{A}$ such that $\vec{F} = m\vec{A}$ and m is a scalar.

a. N If $\vec{A} = 2.4\hat{\imath} + 3.0\hat{\jmath}$ and $\vec{F} = 4.0\hat{\imath} + 5.0\hat{\jmath}$, what is m?

b. C Why is it impossible to have $\vec{A} = 2.4\hat{\imath} + 3.0\hat{\jmath}$ and $\vec{F} = 40.0\hat{\imath} + 0.50\hat{\jmath}$, given the relationship between the two vectors?

72. N A chef drops a chopstick on a countertop that is made of square tiles that form a regular grid. She immediately notices that the chopstick spans almost exactly two grids in the horizontal direction. From years of experience, the chef knows that the chopstick has a length of 7.3 grid units. What is the orientation of the chopstick with respect to the horizontal? Answer by specifying an angle in degrees.

73. N A function is given as $f(x) = 3x$, where $x > 0$. Plot this function. What is the angle between the graph of this function and the positive x axis?

Problems 74 and 75 are paired.

74. N A classroom clock has a small magnifying glass embedded near the end of the minute hand. The magnifying glass may be modeled as a particle. Class begins at 7:55 and ends at 8:50. The length of the minute hand is 0.300 m. **a.** Find the average velocity of the magnifying glass at the end of the minute hand using the coordinate system shown in Figure P3.74. Give your answer in component form. **b.** Find the magnitude and direction of the average velocity. **c.** Find the average speed and in the **CHECK and THINK** step, compare to the average velocity.

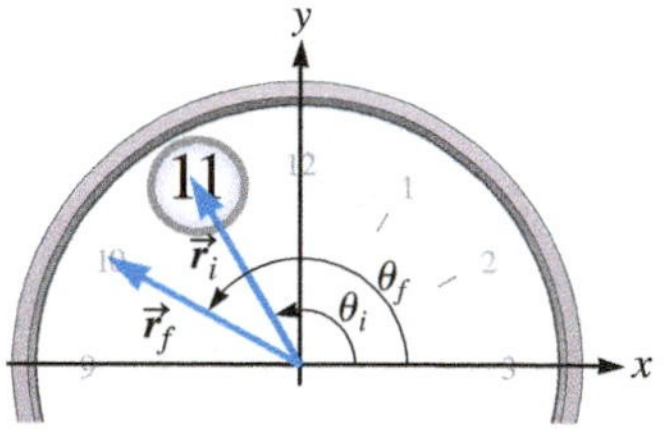

FIGURE P3.74 Problems 74 and 75.

75. C In Problem 74, we found the direction of the average velocity of a magnifying glass at the end of the minute hand on a clock. Now consider the direction of the *instantaneous* velocity when the minute hand points to the 10 and when it points to the 11. If you did Problem 74, compare the direction of the instantaneous velocity vectors to the direction of the average velocity.

Two- and Three-Dimensional Motion

4

Underlying Principles

Two- and three-dimensional kinematics

Major Concepts

1. Position
2. Displacement
3. Average and instantaneous velocity
4. Average and instantaneous acceleration
5. Relative motion and reference frames

Special Cases

1. Projectile motion
   a. Range
   b. Maximum range
2. Uniform circular motion
3. Centripetal acceleration

Key Question

How can the principles of kinematics and the mathematics of vectors be used to describe two- and three-dimensional motion?

Imagine the thrill of zigzagging down a ski slope in the Alps, cliff diving into the crystal blue waters of Hawaii, or playing a game of Ultimate Frisbee in front of the physics building on campus. All these thrills and many more are possible because we live in a three-dimensional world.

Why stop at three dimensions? An even more exciting prospect is that, according to some theories in physics, the Universe may have as many as 11 dimensions! These models are beyond the scope of this book, of course, but it is fun to think that what we are studying here is the basis for such mind-boggling ideas about the Universe.

It would be a lot easier to ace a physics class if our Universe were only one-dimensional because we would have already learned all the kinematics we need in Chapter 2. In such a one-dimensional world, however, we would only be able to move back and forth along a single straight line. Fortunately, our world is not that dull.

TWO- AND THREE-DIMENSIONAL KINEMATICS

❶ Underlying Principle

4-1 What Is Multidimensional Motion?

Except for relative motion, the concepts in this chapter are similar to those we looked at in Chapter 2. This chapter combines the kinematics concepts of Chapter 2 with the mathematical tools of Chapter 3 to study two- and three-dimensional motion. We only study objects that can be modeled as particles, but our particles are no longer restricted to motion along a straight line.

One-dimensional motion occurs along a single straight line, whereas multidimensional motion may be along a curved path or along a path made up of more than one straight line. Figure 4.1 shows motion in one, two, and three dimensions. Notice that the number of axes required corresponds to the number of dimensions.

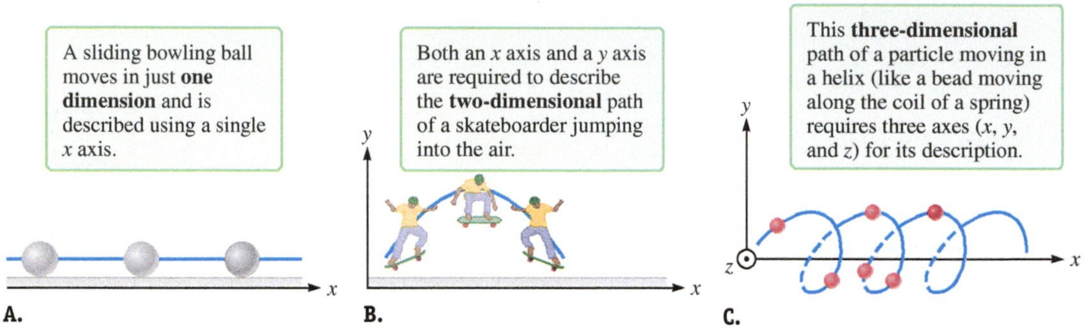

A sliding bowling ball moves in just **one dimension** and is described using a single *x* axis.
x
A.
Both an *x* axis and a *y* axis are required to describe the **two-dimensional** path of a skateboarder jumping into the air.
y
x
B.
This **three-dimensional** path of a particle moving in a helix (like a bead moving along the coil of a spring) requires three axes (*x*, *y*, and *z*) for its description.
y
z
x
C.

FIGURE 4.1 Examples of motion. **A.** One dimension. **B.** Two dimensions. **C.** Three dimensions. In part C, recall that the symbol ⊙ means out of the page; here, the *z* axis points out of the page.

Because many of the situations we study in this book are two-dimensional, we will emphasize two-dimensional motion in this chapter. Many equations are written in the more general three-dimensional form, however, and we will work out several three-dimensional examples.

A powerful way to study multidimensional motion is to break the vector quantities (position, displacement, velocity, and acceleration) into components that describe the motion. Each component can then be manipulated and analyzed separately. This idea is illustrated throughout the chapter as we work on different aspects of the following case study.

CASE STUDY Skateboarding

If you have ever skateboarded or watched skateboarders, you might have noticed that many skateboarding tricks are based on a move known as the *ollie*. As shown in Figure 4.2A, both rider and skateboard fly through the air, with the rider's feet never losing contact with the board. This move, invented by Alan "Ollie" Gelfand of Florida in the late 1970s, was originally used to get over an obstacle.

Before Gelfand invented the ollie, a popular trick at skateboarding competitions was the high jump, now known as the hippy jump. Like the ollie, the hippy involves an obstacle, but instead of both rider and skateboard going over the obstacle, only the rider goes over. The skateboard goes under the obstacle, and the rider lands back on the board on the other side (Fig. 4.2B). This case study focuses on the hippy jump.

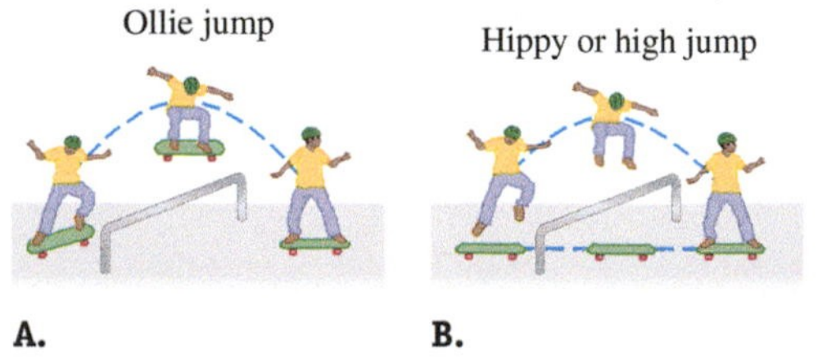

Ollie jump
Hippy or high jump
A.
B.

FIGURE 4.2 **A.** The ollie is a skateboarding trick in which both the rider and the board pass above an obstacle. **B.** In the hippy jump trick, the skateboard passes under the obstacle, and the skater goes over it.

The motion of the skater in either trick is more complicated than simple translational motion. Nevertheless, we treat the skater as a particle. Two students, Avi and Cameron, are asked to model this trick using a small cart on a track (Fig. 4.3A). The cart is equipped with a spring gun that is used to launch a ball. The cart models the skateboard, and the ball models the rider. An obstacle is placed above the track and does not impede the cart's motion. The students can set the launch angle θ and they are to determine at what value of θ the ball must be launched so that it clears the obstacle and lands back in the cart. As we proceed, we will develop the tools needed to figure out which student has the right idea. Here is part of their discussion.

Avi: We need to point the launcher forward. I would guess that θ needs to be less than 90°, maybe around 45°.

Cameron: I think we need to point the launcher straight up.

Avi: That won't work because the ball has to travel a greater distance than the cart. It has to go up in an arc and travel back down. The cart goes only in a straight line. We need to figure out where the cart will be when the ball lands and then aim for that place (Fig. 4.3B). Just imagine that there is a target where the cart will be on the other side of the obstacle. We need to aim for that imaginary target. The motion of the cart doesn't really affect how we aim the launcher.

Cameron: I don't think it works that way. I think it is more like playing tennis on a cruise ship. If you bounce the ball straight up off the face of your racket, it lands right back on your racket (Fig. 4.3C). You don't have to worry about the speed or direction of the ship when you're playing. You just play like you always do.

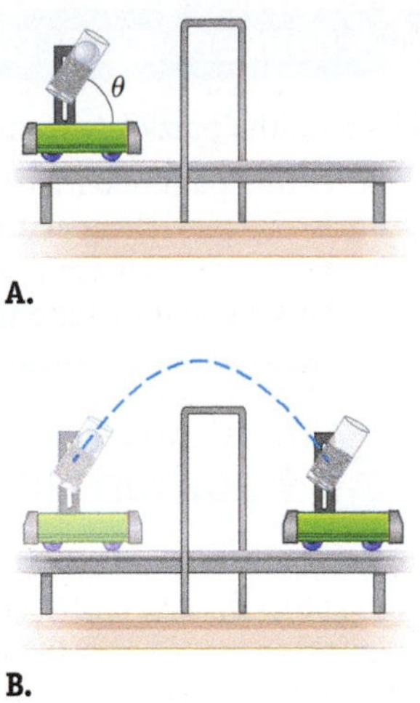

A.

B.

C.

FIGURE 4.3 **A.** Avi and Cameron must model the hippy jump with a cart and spring gun launcher. **B.** Avi suggests launching at an angle smaller than 90°, arguing that the ball must be aimed at a target on the track where the cart will soon be. Essentially, Avi argues to ignore the motion of the cart. **C.** Cameron mentions that if a person on a ship hits a tennis ball straight up, it will land back on the racket. Cameron uses this argument to say that the ball should be launched straight up.

CONCEPT EXERCISE 4.1

CASE STUDY How Many Dimensions?

In each case, determine whether the object is moving in one, two, or three dimensions.

a. The rider in the ollie (Fig. 4.2A).
b. The skateboard in the ollie (Fig. 4.2A).
c. The rider in the hippy (Fig. 4.2B).
d. The skateboard in the hippy (Fig. 4.2B).

4-2 Motion Diagrams for Multidimensional Motion

A motion diagram is an illustration that shows the location of a particle at regular time intervals (Chapter 2). We often use a dot to represent the particle in a motion diagram. As in one-dimensional motion, the spacing between the dots in a diagram for two- or three-dimensional motion indicates the relative speed of the particle. A particle at rest is represented by a single dot. If a particle maintains its speed, the dot spacing is uniform. If the particle is slowing down, the dots get closer together. If the particle is speeding up, the dots get farther apart.

Figure 4.4 shows the motion of three particles. We use the term **nonuniform** to describe the motion of a particle whose speed changes and the term **uniform** to describe the motion when the speed does not change. Whether a particle maintains its speed or not is independent of its path as shown by the examples of one- and two-dimensional motion in Figure 4.4.

We need to choose a coordinate system to analyze motion. Figure 4.4 shows a possible coordinate choice for each motion diagram. (Remember, though, that there are many other good choices that are not shown here.) As usual, we are free to choose both the location of the origin and which direction on each axis is to be the positive direction. In Figure 4.4B, for instance, we have chosen to place the origin at A, but in Figure 4.4C, we have chosen to place the origin at the center of the circular path. One thing we are *not* free to choose about our coordinate system is the minimum number of axes; this number is determined by the dimensionality of the motion.

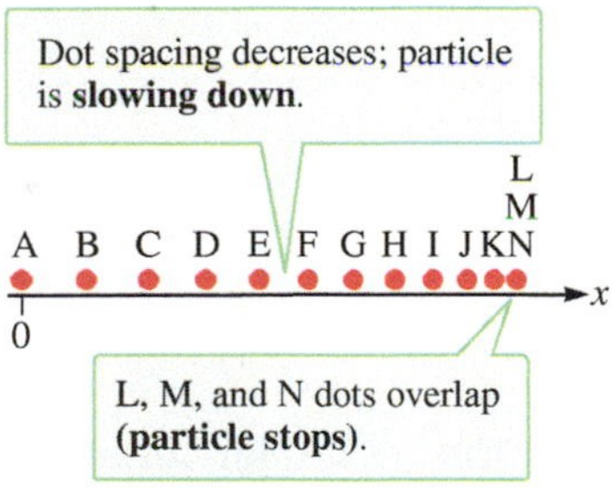

A.

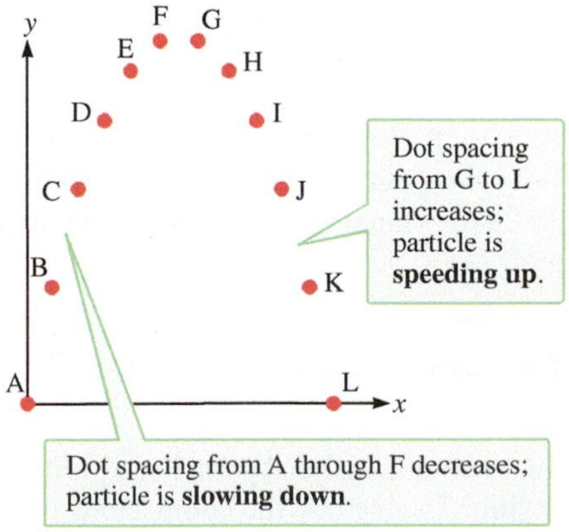

B.

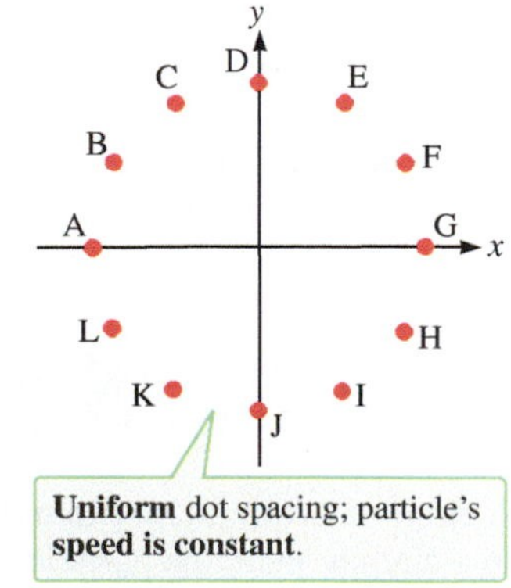

C.

FIGURE 4.4 Motion diagram of a particle moving along **A.** a one-dimensional path, **B.** a two-dimensional parabolic path, and **C.** a two-dimensional circular path. A possible choice of coordinate system for each motion diagram is shown in all three cases.

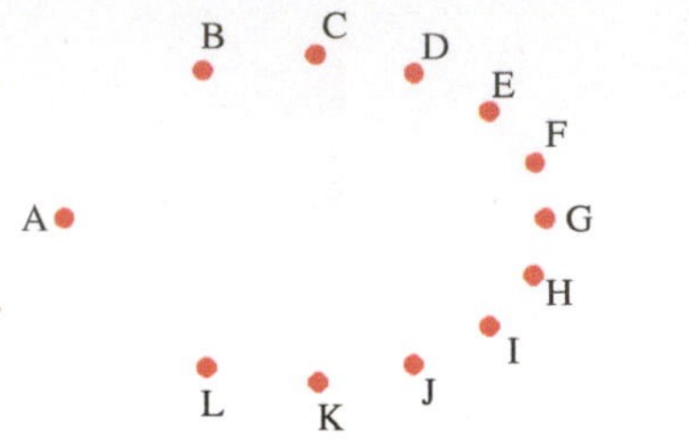

FIGURE 4.5 All dots are in the plane of the page.

CONCEPT EXERCISE 4.2

Based on the particle's motion diagram in Figure 4.5:

a. Is this path one-, two-, or three-dimensional?

b. Is the particle speeding up, slowing down, or maintaining a constant speed? Does your answer change depending on the range of dots you consider? If so, be sure you include this information in your description.

4-3 Position and Displacement

A position vector is used to locate a particle. For example, Figure 4.6 shows the position vectors $\vec{r}_C$ and $\vec{r}_G$ for each motion diagram in Figure 4.4. In each case, the tail of the position vector is at the origin of the chosen coordinate system, and the head is at dot C or dot G. The displacement vector $\Delta\vec{r}_{CG}$ from C to G is also shown in each case. The displacement vector is found by subtracting $\vec{r}_C$ from $\vec{r}_G$ geometrically.

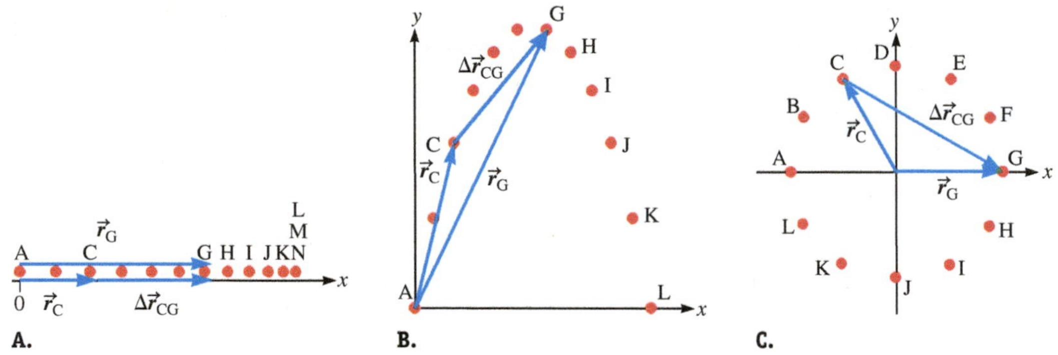

FIGURE 4.6 The same motion diagrams from Figure 4.4, showing position vectors $\vec{r}_C$ and $\vec{r}_G$ for a particle moving along **A.** a one-dimensional path, **B.** a two-dimensional parabolic path, and **C.** a two-dimensional circular path.

Often, the best way to work with multidimensional vectors is in component form. (Eq. 3.8). The position vector is written

POSITION ✪ Major Concept

$$\vec{r} = x\hat{\imath} + y\hat{\jmath} + z\hat{k} \tag{4.1}$$

where x, y, and z are the scalar components.

The displacement $\Delta\vec{r}$ is found by subtracting the initial position vector from the final one:

DISPLACEMENT ✪ Major Concept

$$\Delta\vec{r} = \vec{r}_f - \vec{r}_i = \Delta x\hat{\imath} + \Delta y\hat{\jmath} + \Delta z\hat{k} \tag{4.2}$$

Part of the power of working with components is that each component can be calculated independently. The next example illustrates this procedure.

EXAMPLE 4.1 Mathematical Description of a Circular Path

A circle is an important example of a two-dimensional path. For a particle moving at constant speed in a circular path, the particle's position is given by the components

$$x = A\cos(\omega t) \qquad y = A\sin(\omega t) \tag{4.3}$$

where the numerical values of A and ω are constant for a given circle, but may differ from one circle to another. Suppose their values are $A = 2.0$ m and $\omega = 0.50\pi$ rad/s.

A Show that Equation 4.3 describes a circular path.

INTERPRET and ANTICIPATE

There are at least two ways to find the shape of the path defined by the functions $x(t)$ and $y(t)$. Method 1 is to make a graph of x versus y. Method 2 is to eliminate t between the functions, creating one function of the form $y(x)$. As an illustration, we will use both methods here. Method 1 should produce a circular graph, and method 2 should result in the equation of a circle (Appendix A).

t (s)	x (m)	y (m)
0	2.0	0
1.0	0	2.0
2.0	−2.0	0
3.0	0	−2.0

SOLVE

Method 1. We get numerical values (coordinates) for the points to plot by substituting a range of time values in Equation 4.3 to find values for x and y simultaneously. We show just four times here, but you need many more to draw an adequate circle.

Plot these points on an x–y coordinate system. Connect them with a smooth curve as in Figure 4.7. (The red dots indicate points that are in the table above.) Each tick mark in Figure 4.7 represents 0.5 m along either the x or y axis.

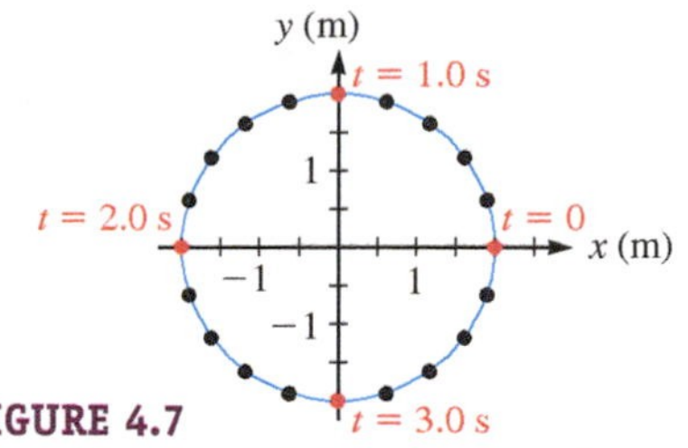

FIGURE 4.7

Method 2. Because the expressions for x and y both involve trigonometric functions, use the trigonometric identity $\cos^2\theta + \sin^2\theta = 1$ to combine the two equations. Square the expressions for x and y.

$$x^2 = A^2 \cos^2(\omega t)$$
$$y^2 = A^2 \sin^2(\omega t)$$

Add these expressions together.

$$x^2 + y^2 = A^2[\cos^2(\omega t) + \sin^2(\omega t)]$$

The term in square brackets equals 1.

$$x^2 + y^2 = A^2$$

CHECK and THINK

The expression we just obtained is the equation of a circle with radius A (Appendix A). So, the constant A in Equation 4.3 is the radius of the circular path. In this case, the radius $A = 2.0$ m, which we can also see on the figure we drew by using method 1.

B Find the position in component form at an initial time $t_i = 0$ and at a final time $t_f = 5.0$ s.

INTERPRET and ANTICIPATE

We expect to find two numerical answers of the form $\vec{r} = (x\hat{\imath} + y\hat{\jmath})$ m. We find the x and y components directly from Equation 4.3. Our sketch can be used to check our answer.

SOLVE

Find the components at $t_i = 0$ by substitution.

$$x_i = A\cos(\omega t_i) = 2.0\text{ m}\cos(0) = 2.0\text{ m}$$
$$y_i = A\sin(\omega t_i) = 2.0\text{ m}\sin(0) = 0$$

Combine these scalar components to find the initial position (Eq. 4.1).

$$\vec{r}_i = x_i\hat{\imath} + y_i\hat{\jmath} = (2.0\hat{\imath} + 0\hat{\jmath})\text{ m}$$
$$\vec{r}_i = 2.0\hat{\imath}\text{ m}$$

Repeat this procedure to find the position at $t_f = 5.0$ s.

$$\omega t_f = (0.50\pi\text{ rad/s})(5.0\text{ s}) = 2.5\pi\text{ rad}$$
$$x_f = (2.0\text{ m})\cos(2.5\pi\text{ rad}) = 0$$
$$y_f = (2.0\text{ m})\sin(2.5\pi\text{ rad}) = 2.0\text{ m}$$
$$\vec{r}_f = x_f\hat{\imath} + y_f\hat{\jmath}$$
$$\vec{r}_f = 2.0\hat{\jmath}\text{ m}$$

CHECK and THINK

It makes sense that the magnitudes r_i and r_f have the same value because as the particle moves along a circular path, its distance from the origin does not change. So, the magnitude of the position vector is always equal to the radius of the circle. Also, our drawing in part A confirms our result for $\vec{r}_i$.

Example continues on page 90 ▶

C What is the particle's displacement over this 5-s interval?

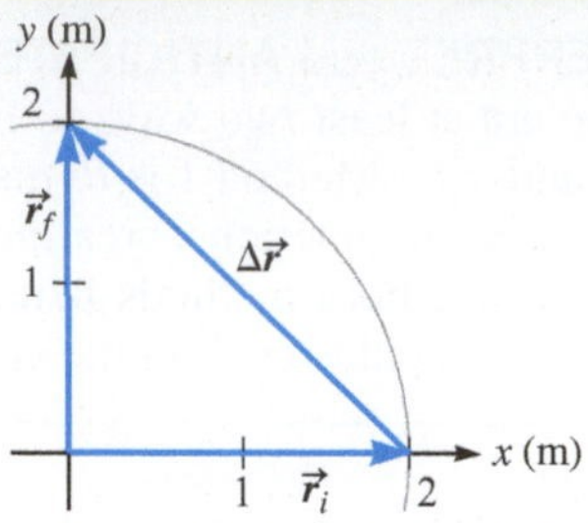

FIGURE 4.8

INTERPRET and ANTICIPATE

We start by getting a rough idea of the displacement geometrically. Because we know both the direction and the magnitude of the two position vectors, we draw these vectors on a coordinate system as in Figure 4.8. Using tail-to-tail subtraction, we find the direction of the displacement vector $\Delta\vec{r} = \vec{r}_f - \vec{r}_i$.

SOLVE

Next, we find an algebraic solution, using Equation 4.2 to calculate the displacement.

$$\Delta\vec{r} = \Delta x\hat{\imath} + \Delta y\hat{\jmath} = (x_f - x_i)\hat{\imath} + (y_f - y_i)\hat{\jmath}$$

$$\Delta\vec{r} = [(0 - 2.0)\hat{\imath} + (2.0 - 0)\hat{\jmath}]\text{ m}$$

$$\Delta\vec{r} = (-2.0\hat{\imath} + 2.0\hat{\jmath})\text{ m}$$

CHECK and THINK

Our result is consistent with $\Delta\vec{r}$ found geometrically. Both show that $\Delta r_x < 0$ and $\Delta r_y > 0$.

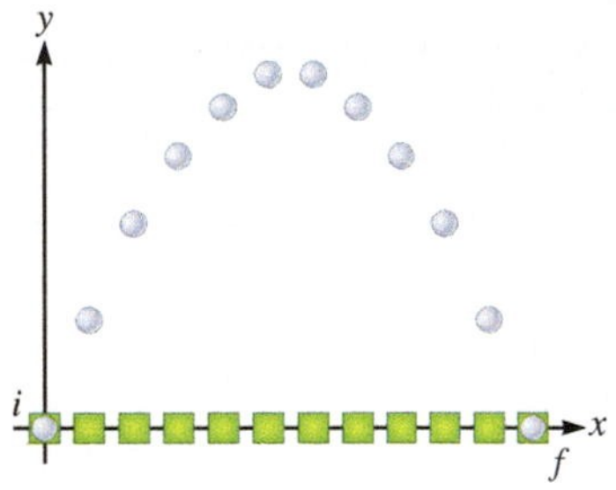

FIGURE 4.9 Avi and Cameron's motion diagram for the ball and cart in the case study. The ball is represented by the circles, and the cart is represented by the squares.

CONCEPT EXERCISE 4.3

CASE STUDY Ball's Displacement

Avi and Cameron have come up with a motion diagram in Figure 4.9 for the ball and cart in the case study, along with their choice of coordinate system. They define the initial position of both objects to be their common location at the moment the ball leaves the cart and the final position to be the location where the ball lands in the cart. If the displacement of the cart is $\Delta\vec{r} = 0.75\hat{\imath}$ m, what is the displacement of the ball?

4-4 Velocity and Acceleration

We saw in Section 2-4 that to find the average velocity of a particle, divide the particle's displacement $\Delta\vec{r}$ by the time interval Δt over which that displacement occurred. So,

AVERAGE VELOCITY Major Concept

$$\vec{v}_{av} = \frac{\Delta\vec{r}}{\Delta t} \tag{4.4}$$

It is often best to write the average velocity in terms of its components:

$$\vec{v}_{av} = v_{av,x}\hat{\imath} + v_{av,y}\hat{\jmath} + v_{av,z}\hat{k} \tag{4.5}$$

Each component is the displacement in one particular direction divided by the time interval (Eq. 2.2). So, the average velocity is written as

$$\vec{v}_{av} = \frac{\Delta x}{\Delta t}\hat{\imath} + \frac{\Delta y}{\Delta t}\hat{\jmath} + \frac{\Delta z}{\Delta t}\hat{k} \tag{4.6}$$

or as

$$\vec{v}_{av} = \frac{x_f - x_i}{t_f - t_i}\hat{\imath} + \frac{y_f - y_i}{t_f - t_i}\hat{\jmath} + \frac{z_f - z_i}{t_f - t_i}\hat{k} \tag{4.7}$$

We also know that instantaneous velocity (usually referred to as velocity) comes from taking the limit of the average velocity as Δt approaches zero (Section 2-7). In other words, we find the velocity $\vec{v}$ from the time derivative of $\vec{r}$:

VELOCITY Major Concept

$$\vec{v} = \frac{d\vec{r}}{dt} \tag{4.8}$$

By taking the limit of each component in Equation 4.6, we find an expression for the velocity in component form:

$$\vec{v} = \frac{dx}{dt}\hat{\imath} + \frac{dy}{dt}\hat{\jmath} + \frac{dz}{dt}\hat{k} \tag{4.9}$$

Each term is the corresponding velocity component:

$$\vec{v} = v_x\hat{\imath} + v_y\hat{\jmath} + v_z\hat{k} \tag{4.10}$$

The average acceleration and the instantaneous acceleration are found in a similar manner (Section 2-8). The average acceleration comes from dividing the change in velocity by the time interval:

AVERAGE ACCELERATION
Major Concept

$$\vec{a}_{\text{av}} = \frac{\Delta\vec{v}}{\Delta t} \tag{4.11}$$

The average acceleration is written in component form as

$$\vec{a}_{\text{av}} = \frac{\Delta v_x}{\Delta t}\hat{\imath} + \frac{\Delta v_y}{\Delta t}\hat{\jmath} + \frac{\Delta v_z}{\Delta t}\hat{k} \tag{4.12}$$

or as

$$\vec{a}_{\text{av}} = a_{\text{av},x}\hat{\imath} + a_{\text{av},y}\hat{\jmath} + a_{\text{av},z}\hat{k} \tag{4.13}$$

The instantaneous acceleration (usually referred to as acceleration) is the time derivative of the velocity:

ACCELERATION
Major Concept

$$\vec{a} = \frac{d\vec{v}}{dt} \tag{4.14}$$

Each component of the acceleration is the time derivative of the velocity component:

$$\vec{a} = \frac{dv_x}{dt}\hat{\imath} + \frac{dv_y}{dt}\hat{\jmath} + \frac{dv_z}{dt}\hat{k} \tag{4.15}$$

So, each term is the corresponding acceleration component:

$$\vec{a} = a_x\hat{\imath} + a_y\hat{\jmath} + a_z\hat{k} \tag{4.16}$$

Because the velocity is the time derivative of position, we can write the acceleration as the second derivative of position:

$$\vec{a} = \frac{d^2x}{dt^2}\hat{\imath} + \frac{d^2y}{dt^2}\hat{\jmath} + \frac{d^2z}{dt^2}\hat{k} \tag{4.17}$$

Working with these quantities in component form is a powerful tool for analyzing multidimensional motion.

EXAMPLE 4.2 Helical Motion

A particle travels along the helical path in Figure 4.10. It is initially at $\vec{r}_i = (3.0\hat{\imath} + 4.0\hat{\jmath} + 2.3\hat{k})$ m, and after 3.0 s its position is $\vec{r}_f = (3.0\hat{\imath} + 4.0\hat{\jmath} + 5.0\hat{k})$ m. The position and displacement vectors are shown. What is the average velocity of the particle during this time interval?

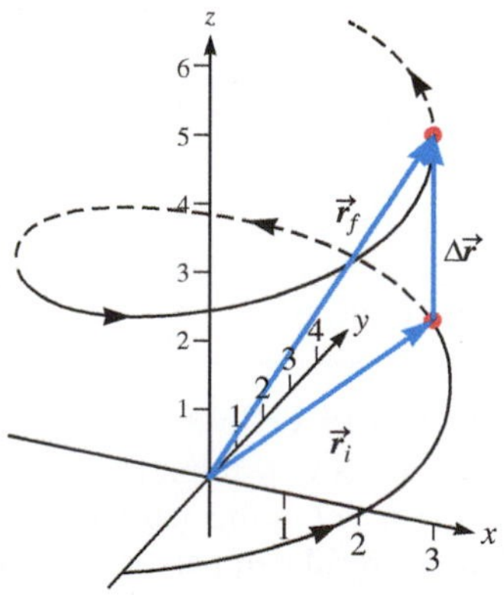

FIGURE 4.10

INTERPRET and ANTICIPATE

From Chapter 2, the average velocity points in the same direction as the displacement, so we expect to find a numerical answer in the form $\vec{v}_{\text{av}} =$ _____ $\hat{k}$ m/s because the displacement is parallel to the z axis.

Example continues on page 92 ▶

SOLVE Substitute values into Equation 4.7, which is the average velocity in component form.	$\vec{v}_{av} = \frac{x_f - x_i}{t_f - t_i}\hat{\imath} + \frac{y_f - y_i}{t_f - t_i}\hat{\jmath} + \frac{z_f - z_i}{t_f - t_i}\hat{k}$ (4.7) $\vec{v}_{av} = \frac{(3.0 - 3.0)\text{ m}}{3.0\text{ s}}\hat{\imath} + \frac{(4.0 - 4.0)\text{ m}}{3.0\text{ s}}\hat{\jmath} + \frac{(5.0 - 2.3)\text{ m}}{3.0\text{ s}}\hat{k}$ $\vec{v}_{av} = 0\hat{\imath} + 0\hat{\jmath} + 0.90\hat{k}\text{ m/s}$ $\vec{v}_{av} = 0.90\hat{k}\text{ m/s}$

CHECK and THINK
As expected, the displacement and the average velocity point in the same direction: straight up along the z axis.

EXAMPLE 4.3 Velocity and Speed in Circular Motion

From Example 4.1, the motion of a particle traveling at a constant speed around a circular path is described by

$$x = A\cos\omega t \qquad y = A\sin\omega t$$

where the radius A and factor ω have specific constant values for a given circular path.

A Find an expression for the velocity of a particle moving along this circular path.

INTERPRET and ANTICIPATE
The expressions for x and y tell us the components of the particle's position vector. According to Equation 4.9, we need to take the time derivative of each component to find velocity.

SOLVE Start with the x component. The constant A does not depend on time, so we pull it outside the derivative. The derivative of the cosine function is negative one multiplied by the sine function (Appendix A).	$\frac{dx}{dt} = \frac{d(A\cos\omega t)}{dt} = A\frac{d(\cos\omega t)}{dt}$ $\frac{dx}{dt} = A(-\omega\sin\omega t) = -A\omega\sin\omega t = v_x$
Now we do the same thing for the y component. The derivative of the sine function is the cosine function.	$\frac{dy}{dt} = \frac{d(A\sin\omega t)}{dt} = A\frac{d(\sin\omega t)}{dt}$ $\frac{dy}{dt} = A(\omega\cos\omega t) = A\omega\cos\omega t = v_y$
Combining these scalar components gives us the velocity in component form.	$\vec{v}(t) = (-A\omega\sin\omega t)\hat{\imath} + (A\omega\cos\omega t)\hat{\jmath}$ (4.18)

B Find an expression for the speed of this particle.

INTERPRET and ANTICIPATE
The speed is the magnitude of velocity.

SOLVE Use the Pythagorean theorem to find the magnitude of a vector (Eq. 3.12). Choose the positive root because magnitude is always positive.	$\lvert\vec{v}\rvert = \sqrt{v_x^2 + v_y^2} = \sqrt{(-A\omega\sin\omega t)^2 + (A\omega\cos\omega t)^2}$ $\lvert\vec{v}\rvert = \sqrt{(A\omega)^2(\sin^2\omega t + \cos^2\omega t)} = \sqrt{(A\omega)^2}$ $\lvert\vec{v}\rvert = A\omega$

CHECK and THINK
This result is very important. Because both A and ω have a constant value for any given circle, we just found that the speed of this particle moving in a circular path is also constant (does not depend on time). By contrast, our results in part A show that the velocity vector *does* depend on time and is therefore not constant. We conclude that only the direction of the velocity changes as a particle travels in uniform circular motion.

EXAMPLE 4.4 Acceleration and Velocity in Circular Motion

Return to the particle traveling at a constant speed around a circular path as described in Example 4.3.

A Find an expression for the particle's acceleration.

INTERPRET and ANTICIPATE

Take the time derivative of the velocity to find acceleration (Eq. 4.15).

SOLVE

We found the velocity in Equation 4.18. Take the time derivative of v_x to find the x component of acceleration.

$$\frac{dv_x}{dt} = \frac{d(-A\omega \sin \omega t)}{dt} = -A\omega \frac{d(\sin \omega t)}{dt}$$
$$a_x = -A\omega^2 \cos \omega t$$

Take the time derivative of v_y to find the y component of acceleration.

$$\frac{dv_y}{dt} = \frac{d(A\omega \cos \omega t)}{dt} = A\omega \frac{d(\cos \omega t)}{dt}$$
$$a_y = -A\omega^2 \sin \omega t$$

The acceleration in component form is found by combining these scalar components (Eq. 4.15).

$$\vec{a}(t) = (-A\omega^2 \cos \omega t)\hat{\imath} - (A\omega^2 \sin \omega t)\hat{\jmath}$$
$$\vec{a}(t) = -A\omega^2(\cos \omega t\, \hat{\imath} + \sin \omega t\, \hat{\jmath}) \quad (4.19)$$

CHECK and THINK

We will consider the implications of this result in part B. Let's just check the units here. Because there are no units associated with trigonometric functions, the units are determined by the coefficient $A\omega^2$ and are $\text{m(rad/s)}^2 = \text{m/s}^2$ as expected for acceleration. (Recall that a radian is defined as the ratio of two lengths, an arc length and the radius of a circle. So, a radian is really dimensionless.)

B In Example 4.1, with $A = 2.0$ m and $\omega = 0.50\pi$ rad/s, we found that the particle's positions at $t = 0$ and 5.0 s are $\vec{r}(0) = 2.0\hat{\imath}$ m and $\vec{r}(5.0 \text{ s}) = 2.0\hat{\jmath}$ m, respectively. Find the velocity and acceleration at these times. Make a sketch showing the relative directions of the position, velocity, and acceleration vectors at $t = 0$ and at $t = 5.0$ s.

INTERPRET and ANTICIPATE

To find the velocity at these two times, substitute values into the expression for $\vec{v}$ (Eq. 4.18). Find acceleration in a similar manner, substituting into the expression for $\vec{a}$ (Eq. 4.19).

SOLVE

It is convenient to find a value for the constant factor $A\omega$, since it appears in both velocity components.

$$A\omega = (2.0 \text{ m})(0.50\pi \text{ rad/s}) = 3.1 \text{ m/s}$$

Substitute $t = 0$ into Equation 4.18 for velocity.

$$\vec{v}(t) = A\omega(-\sin \omega t\, \hat{\imath} + \cos \omega t\, \hat{\jmath})$$
$$\vec{v}(0) = (3.1 \text{ m/s})(-\sin 0\, \hat{\imath} + \cos 0\, \hat{\jmath}) = 3.1\hat{\jmath} \text{ m/s}$$

Repeat for $t = 5.0$ s.

$$\vec{v}(5.0 \text{ s}) = (3.1 \text{ m/s})(-\sin 2.5\pi\, \hat{\imath} + \cos 2.5\pi\, \hat{\jmath})$$
$$\vec{v}(5.0 \text{ s}) = -3.1\hat{\imath} \text{ m/s}$$

The factor $A\omega^2$ appears in both acceleration components.

$$A\omega^2 = (2.0 \text{ m})(0.50\pi \text{ rad/s})^2 = 4.9 \text{ m/s}^2$$

Substitute $t = 0$ into Equation 4.19 for acceleration.

$$\vec{a}(t) = -A\omega^2(\cos \omega t\, \hat{\imath} + \sin \omega t\, \hat{\jmath})$$
$$\vec{a}(0) = -4.9 \text{ m/s}^2(\hat{\imath} + 0\,\hat{\jmath}) = -4.9\hat{\imath} \text{ m/s}^2$$

Repeat for $t = 5.0$ s.

$$\vec{a}(5.0 \text{ s}) = -4.9 \text{ m/s}^2(\cos 2.5\pi\hat{\imath} + \sin 2.5\pi\, \hat{\jmath})$$
$$\vec{a}(5.0 \text{ s}) = -4.9\hat{\jmath} \text{ m/s}^2$$

Example continues on page 94 ▶

Sketch these six vectors on a two-dimensional coordinate system. There is no need to set a scale for the vectors because we are only interested in their relative directions.

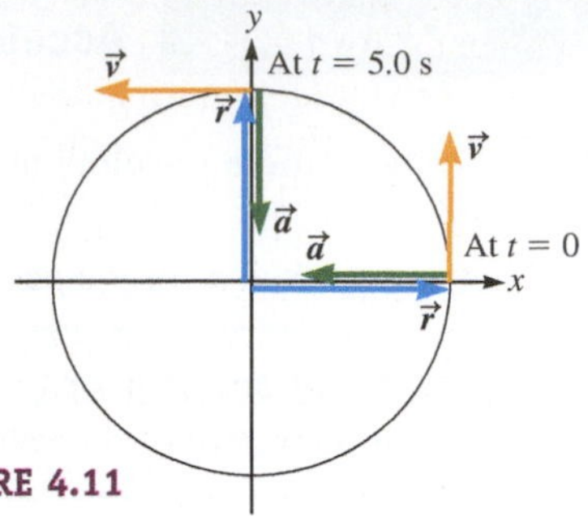

y
$\vec{v}$
At t = 5.0 s
$\vec{r}$
$\vec{a}$
$\vec{a}$
$\vec{v}$
At t = 0
x
$\vec{r}$

FIGURE 4.11

CHECK and THINK

Our sketch shows that the velocity is perpendicular to the acceleration in both instances, as is *always* the case for uniform circular motion (circular motion at constant speed). In fact, whenever the velocity's direction changes while the speed remains constant, the acceleration must be perpendicular to the velocity. We will look more closely at uniform circular motion in Section 4-6.

4-5 Special Case of Projectile Motion

A bullet shot out of a gun, a water balloon dropped out a third-story dormitory window, and a skateboarder doing a hippy are all projectiles. A **projectile** is any particle that is launched or *projected*. The two defining properties of a projectile are (1) that the launching gives it an initial velocity $\vec{v}_0$ and (2) that after it is launched its motion is influenced only by gravity and the projectile is therefore in free fall. For example, once the ball has left a basketball player's hand, the ball is a projectile (Fig. 4.12). The player gives the ball an initial velocity $\vec{v}_0 = \vec{v}_{0x} + \vec{v}_{0y}$, and after that *only the vertical vector component $\vec{v}_y$ is altered by gravity*. The horizontal vector component $\vec{v}_x = \vec{v}_{0x}$ is a constant and therefore never changes from its initial value.

PROJECTILE MOTION ▶ Special Case

Projectile motion is a special case in which (1) there is no acceleration in the x (horizontal) direction and (2) the acceleration in y (vertical) direction is the constant free-fall acceleration g. When the initial velocity's x component is zero ($\vec{v}_{0x} = 0$), the projectile's path is along a straight, vertical line. When the initial velocity's x component is nonzero ($\vec{v}_{0x} \neq 0$), the projectile's path is a parabola in the xy plane.

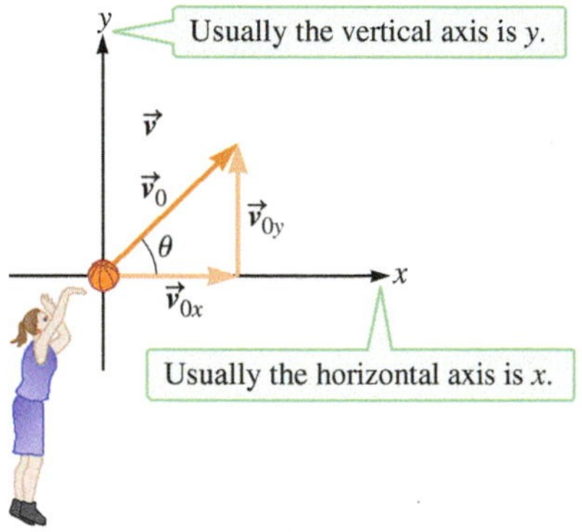

y
Usually the vertical axis is y.
$\vec{v}$
$\vec{v}_0$
$\vec{v}_{0y}$
θ
x
$\vec{v}_{0x}$
Usually the horizontal axis is x.

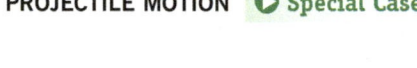

FIGURE 4.12 A basketball player shoots a ball. After the ball leaves her hand, the ball is a projectile.

The Equations for Projectile Motion

A mathematical description of projectile motion means finding equations for the position, velocity, and acceleration as functions of time. We'll continue to use the coordinate system shown in Figure 4.12. We already know the equation for the acceleration because the projectile is in free fall: $\vec{a} = -g\hat{\jmath}$.

Before we get into the mathematical description, let's create a visual description. Figure 4.13 shows the motion diagram of a projectile such as a basketball in a free

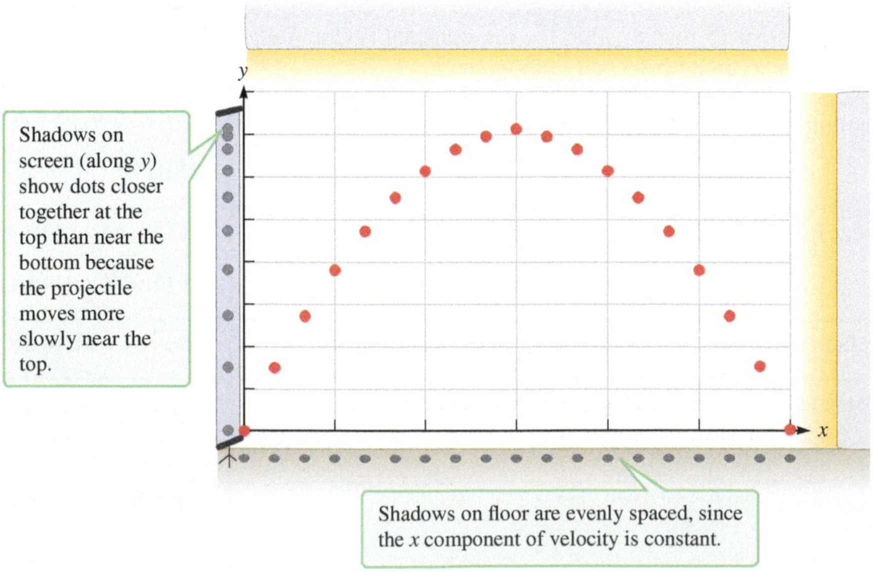

y
Shadows on screen (along y) show dots closer together at the top than near the bottom because the projectile moves more slowly near the top.
x
Shadows on floor are evenly spaced, since the x component of velocity is constant.

FIGURE 4.13 A motion diagram of a projectile. The parabolic path is characteristic of projectile motion. Two lights project the shadow of a projectile. While the shadow on the floor moves at constant velocity, the shadow on the screen moves much like a particle tossed straight up. At the peak, the y component of velocity is zero.

throw or a skateboarder doing an ollie. The projectile is modeled as a particle and is represented by a dot. Imagine using two light sources to break this two-dimensional motion into motion in the x direction and motion in the y direction. One light source is overhead, emitting rays perpendicular to the floor. The other light source is placed in front of the projectile, producing rays perpendicular to a vertical screen behind the projectile. The shadows on the floor are evenly spaced, indicating uniform motion in the x direction. The shadows on the screen are closer together near the top, indicating that the particle moves more slowly near the top. In fact, these shadows look exactly like the motion diagram of a particle in purely vertical free fall.

The shadows in Figure 4.13 show us how to connect the motion of a projectile to its mathematical description. First, because the acceleration is constant in either direction, the five constant acceleration equations (Table 2.4, repeated here) may be applied in both directions. Second, in the x direction, because $a_x = 0$, the five constant equations collapse to two:

$$v_x = v_{0x} \tag{4.20}$$
$$x = x_0 + v_{0x}t \tag{4.21}$$

Third, in the y direction, $\vec{a} = -g\hat{\jmath}$, and so the velocity and position are given by

$$v_y = v_{0y} - gt \tag{2.9}$$
$$y = y_0 + v_{0y}t - \tfrac{1}{2}gt^2 \tag{2.11}$$

Typically, we know the initial velocity $\vec{v}_0$, and we must break it into its components. For example, in Figure 4.12, we have

$$v_{0x} = v_0 \cos\theta \qquad v_{0y} = v_0 \sin\theta \tag{4.22}$$

where the launch angle θ is measured counterclockwise from the x axis.

We combine the above equations to write vector equations for the projectile's motion:

$$\vec{a} = -g\hat{\jmath} \tag{4.23}$$
$$\vec{v} = (v_0 \cos\theta)\hat{\imath} + (v_0 \sin\theta - gt)\hat{\jmath} \tag{4.24}$$
$$\vec{r} = [v_0t \cos\theta + x_0]\hat{\imath} + [(v_0t \sin\theta - \tfrac{1}{2}gt^2) + y_0]\hat{\jmath} \tag{4.25}$$

We often choose to put the origin of the coordinate system at the initial position so that $(x_0, y_0) = (0, 0)$. The projectile's position is then

$$\vec{r} = (v_0t \cos\theta)\hat{\imath} + (v_0t \sin\theta - \tfrac{1}{2}gt^2)\hat{\jmath} \tag{4.26}$$

and we don't need to use Equation 4.25 for this choice.

Projectile motion is a combination of uniform motion in the horizontal direction and free fall in the vertical direction.

TABLE 2.4 Kinematic equations of motion for constant acceleration

Equation	
$v_x = v_{0x} + a_xt$	(2.9)
$\Delta x = \frac{1}{2}(v_{0x} + v_x)t$	(2.10)
$\Delta x = v_{0x}t + \frac{1}{2}a_xt^2$	(2.11)
$\Delta x = v_xt - \frac{1}{2}a_xt^2$	(2.12)
$v_x^2 = v_{0x}^2 + 2a_x\Delta x$	(2.13)

Notes: The equations are written in terms of scalar components of the motion variables along the x axis. Similar equations can also be written in the y and z directions.

EXAMPLE 4.5 CASE STUDY How Do the Shadows Move?

In Concept Exercise 4.3, Avi and Cameron came up with a motion diagram (Fig. 4.9) for the cart and ball in the case study. Imagine placing two light sources and a screen in this drawing so that the moving ball casts shadows on horizontal and vertical surfaces (similar to Fig. 4.13). Describe the path of the shadows.

SOLVE

If we used lights to create shadows of the ball, the shadows would look much like those in Figure 4.13. The vertical shadows would be close together near the top and farther apart near the bottom, indicating the change in velocity along this vertical axis. Because there is no horizontal acceleration in projectile motion, the horizontal component of the ball's velocity is constant. Thus, the horizontal shadows would be evenly spaced. Because the (constant) *horizontal* velocity of the ball is the same as the (constant) velocity of the cart, the ball's horizontal shadows would, at every moment, be projected onto the cart as the cart moves along the track. So, the ball is always directly above the cart. We will need this important conclusion as we wrap up our case-study analysis in Section 4-8.

EXAMPLE 4.6 Parabolic Path

Use Equation 4.26, $\vec{r} = (v_0 t \cos\theta)\hat{\imath} + (v_0 t \sin\theta - \frac{1}{2}gt^2)\hat{\jmath}$, to show that the path of a projectile is a parabola.

INTERPRET and ANTICIPATE

Finding a path means finding y as a function of x. Equation 4.26 provides x and y as functions of time t. From method 2 in Example 4.1, one way to find the path is to eliminate t from the x and y equations.

SOLVE

Use the x component of the projectile's position vector (Eq. 4.26) to obtain an expression for time.

$$x = v_0 t \cos\theta$$

$$t = \frac{x}{v_0 \cos\theta}$$

Substituting this result into the expression for the y component of position allows us to eliminate t between the equations.

$$y = v_0 t \sin\theta - \tfrac{1}{2}gt^2 = \frac{v_0 x \sin\theta}{v_0 \cos\theta} - \tfrac{1}{2}g\left(\frac{x}{v_0 \cos\theta}\right)^2$$

$$y = x \tan\theta - \left(\frac{g}{2v_0^2 \cos^2\theta}\right)x^2 \qquad (4.27)$$

CHECK and THINK

The coefficients a, b, and c determine the geometry—height, width, and center—of a given parabola. Because g is constant, our result tells us that the geometry of a projectile's parabolic path depends only on the initial velocity of the projectile.

This result is the equation of a parabola in the form $y = ax^2 + bx + c$ with

$$a = -\frac{g}{2v_0^2 \cos^2\theta} \qquad b = \tan\theta \qquad c = 0$$

RANGE EQUATION ▶ Special Case

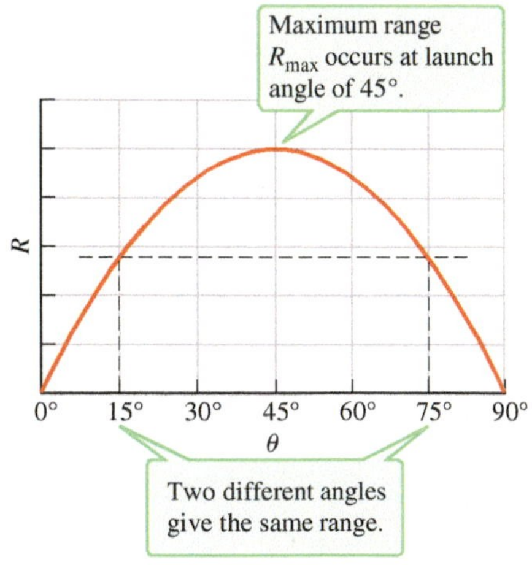
Maximum range R_{max} occurs at launch angle of 45°.
R
0°
15°
30°
45°
60°
75°
90°
θ
Two different angles give the same range.

FIGURE 4.14 Range of a projectile depends on launch angle θ. The maximum range occurs for $\theta = 45°$. For all other ranges, there is a low angle θ_L and a high angle θ_H that both result in the same R. Notice that $\theta_L + \theta_H = 90°$.

Range Equation

A special case of projectile motion occurs when the projectile's final vertical position equals its initial vertical position: $y = y_0$. In this specific case, the displacement is purely horizontal (along x). The **range** of a projectile is the magnitude of its horizontal displacement *when its vertical displacement is zero.* In Problem 82, you will show that the range is given by

$$R = \frac{v_0^2}{g}\sin 2\theta \qquad (4.28)$$

Equation 4.28 (the range equation) is used to find the magnitude of the displacement *only if the displacement along y is zero*: $R = \Delta r$ only if $\Delta y = 0$. (Because the vertical displacement is zero, the horizontal displacement is the total displacement.)

According to the range equation, the range of a projectile launched on the Earth depends on two factors: initial speed v_0 and launch angle θ. Suppose a launcher always gives the projectile the same initial speed v_0. The range of the projectile is then controlled only by the launch angle θ as shown in Figure 4.14. We see that (1) the longest range occurs at $\theta = 45°$ and (2) two possible angles can be used to achieve any range value smaller than the maximum.

Whenever a manufacturer specifies the range for a launch device, the range is understood to mean the maximum value, which is the value at which $\theta = 45°$ (and $\sin 2\theta = 1$):

MAXIMUM RANGE ▶ Special Case

$$R_{max} = \frac{v_0^2}{g} \qquad (4.29)$$

PROBLEM-SOLVING STRATEGY

Two-Dimensional Projectile Motion

There are two parts to the **INTERPRET and ANTICIPATE** procedure:

1. Start with a **sketch** that includes a **coordinate system**. Often, it is best to put the origin of the coordinate system at the position where the projectile is launched.
2. Ask yourself if the vertical displacement is zero ($\Delta y = 0$). If your answer is no, you **cannot** apply the **range equation** (Eq. 4.28).

There are also two parts to the **SOLVE** procedure:

1. Start by **listing the six kinematic variables** (initial position, final position, initial velocity, final velocity, acceleration, and time) for x and y separately. The time t is the same in both lists. Assuming you have chosen a horizontal x axis and a vertical y axis, there are a few things about these parameters that are always true:
   - **a.** $a_x = 0$
   - **b.** $v_x = v_{0x}$
   - **c.** $a_y = \pm g$ (Choose the minus sign if your y axis points upward.)
   - **d.** Trigonometry (Eq. 4.22) may be used to find the initial velocity components when the launch angle θ is known.
2. Once you have listed the parameters, you may use the **constant-acceleration equations** (Table 2.4). In practice, Equations 4.24 through 4.26 (page 111) were derived from the constant-acceleration equations for the special case of two-dimensional projectile motion.

CONCEPT EXERCISE 4.4

A manufacturer claims to have a three-person water-balloon launcher with a 300-yd range (Fig. 4.15), which is a distance equal to the length of three football fields placed end to end!

a. What is the displacement of a water balloon launched at $\theta = 45°$ if it is launched from the ground and returns to the ground? Give your answer in yards.

b. If a water balloon is launched off a building at $\theta = 45°$ and lands on the ground 50 ft below, is its horizontal displacement greater than, less than, or equal to the range?

Courtesy, Main Sauce Productions Ltd.

FIGURE 4.15

CONCEPT EXERCISE 4.5

Does the range of a launcher change if it is used on the Moon? If so, is the range longer or shorter on the Moon than on the Earth?

EXAMPLE 4.7 **A Training Exercise**

A pilot learns how to hit a target by dropping mock bombs on an abandoned building as seen in Figure 4.16. The bomber flies quickly (505 km/h) at a low altitude (1.83×10^3 m). Assume air resistance is negligible.

A How far from the abandoned building should the bomber drop its mock bomb?

INTERPRET and ANTICIPATE

Steps 1 and 2 The problem statement provides a **sketch with a coordinate system**, so we don't need to draw our own. We might be tempted to use the range equation, but we cannot because $\Delta y \neq 0$. Instead, we need to think about the motion of the mock bomb along x and y separately.

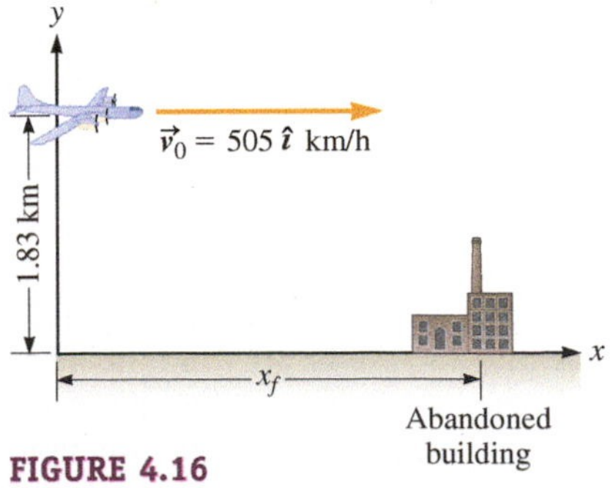

FIGURE 4.16

Example continues on page 98 ▶

SOLVE **Step 1** Start by **listing the six kinematic** quantities for x and y separately. The mock bomb starts inside the plane. Therefore, its initial velocity is the same as the velocity of the plane. In this list, all the variables pertain to the mock bomb. Notice that the initial and final values for both x and y are a function of the coordinate system provided in the problem statement.	$x_i = 0$ $x_f =$ needed $v_x = v_{0x} = 505\ \text{km/h} = 1.40 \times 10^2\ \text{m/s}$ $a_x = 0$ t unknown	$y_i = 1.83 \times 10^3\ \text{m}$ $y_f = 0$ $v_{0y} = 0$ v_y not needed $a_y = -g$ t unknown
Step 2 We start with the displacement along y and solve for the time t the mock bomb takes to fall to the target. The displacement along y comes from modifying Equation 2.11.	$\Delta y = v_{0y}t - \frac{1}{2}gt^2 = 0 - \frac{1}{2}gt^2$ $t = \sqrt{\frac{2\Delta y}{-g}} = \sqrt{\frac{2(-1.83 \times 10^3\ \text{m})}{-9.81\ \text{m/s}}}$ $t = 19.3\ \text{s}$ (1)	
The x component of position comes from the first term in Equation 4.26. This term involves the time of flight (which we just found above) and the x component of initial velocity, which we have in Equation 4.22.	At impact, $x_f = v_0 t \cos\theta = (v_0 \cos\theta)t$ $v_{0x} = v_0 \cos\theta$ (4.22) $x_f = v_{0x}t$	
Substitute values, including the time found in Equation (1).	$x_f = (1.40 \times 10^2\ \text{m/s})(19.3\ \text{s})$ $x_f = 2.70 \times 10^3\ \text{m} = 2.70\ \text{km}$	

CHECK and THINK

It might seem incredible that the mock bomb drops when the plane is 2.70 km (1.7 mi) away from the target. To get a better feel for this seemingly strange fact, think about the shadow of the plane on the ground. As in Figure 4.13, suppose a light directly above the plane gives off rays perpendicular to the ground, projecting the plane's shadow. The mock bomb leaving the plane would always be directly above the plane's shadow. So, the mock bomb and the plane's shadow would be on the abandoned building at the same time.

B Many bomber pilots are trained to fly upward as they drop their bomb. This maneuver allows them to be farther from the explosion. If $\theta = 20.0°$ in Figure 4.17, at what distance from the abandoned building should the mock bomb be dropped?

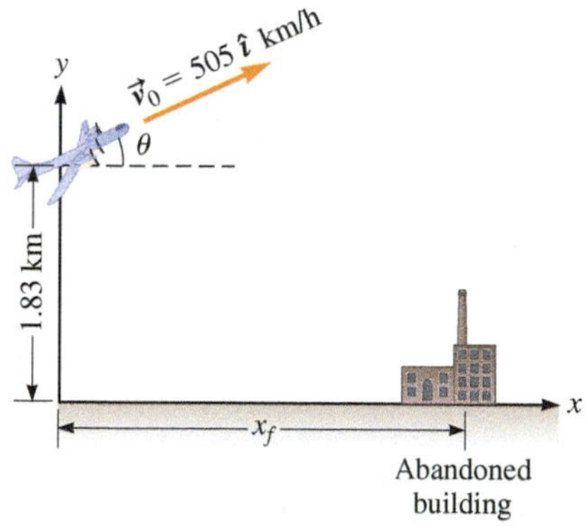

FIGURE 4.17

INTERPRET and ANTICIPATE

The only difference between this question and part A is the mock bomb's initial velocity. It now has a nonzero y component. Use part A to take some shortcuts.

SOLVE Find the new x and y components of the initial velocity (Eq. 4.24). The other parameters are the same as in part A.	$v_{0x} = v_0 \cos\theta = (1.40 \times 10^2\ \text{m/s}) \cos 20.0°$ $v_{0x} = 1.32 \times 10^2\ \text{m/s}$ $v_{0y} = v_0 \sin\theta = (1.40 \times 10^2\ \text{m/s}) \sin 20.0°$ $v_{0y} = 47.9\ \text{m/s}$
As before, find the time it takes the mock bomb to fall from the plane to the target from the displacement along y. Now, however, we must use the quadratic formula to solve for t because the v_{0y} term in our equation is not zero.	$\Delta y = v_{0y}t - \frac{1}{2}gt^2$; rearrange to $(-\frac{1}{2}g)t^2 + (v_{0y})t - \Delta y = 0$ $t = \frac{-v_{0y} \pm \sqrt{v_{0y}^2 - 2g\Delta y}}{-g}$

Substitute values.	$t = \dfrac{-47.9\text{ m/s} \pm \sqrt{(47.9\text{ m/s})^2 - 2(9.81\text{ m/s}^2)(-1.83 \times 10^3\text{ m})}}{-9.81\text{ m/s}^2}$ $t = -15.0\text{ s or } 24.8\text{ s}$
Choose the positive solution for the time and substitute it in the expression for the x component of the mock bomb's displacement. The negative solution occurs before the mock bomb is dropped and is not a physically reasonable solution.	$x = v_{0x}t = (1.32 \times 10^2\text{ m/s})(24.8\text{ s})$ $x = 3.27 \times 10^3\text{ m} = 3.27\text{ km}$

:• CHECK and THINK

Compared with part A, the mock bomb is dropped when the plane is even farther from the target (a little more than 2 mi) because the mock bomb's initial velocity now has an upward component. So, the mock bomb travels above its original altitude and as a result takes longer to fall back to the Earth. In addition, its horizontal velocity is less than in part A. The mock bomb must be dropped sooner to compensate.

4-6 Special Case of Uniform Circular Motion

UNIFORM CIRCULAR MOTION

Special Case

A particle moving along a circular path at constant speed is said to undergo uniform circular motion, as the term suggests. To ancient mathematicians, uniform circular motion was considered perfect and divine. They believed that only heavenly bodies such as the Moon, planets, and stars could exhibit such perfect motion. Today, we do not associate uniform circular motion with perfection, yet it is still an important special case of two-dimensional motion requiring a mathematical description.

Let's start by finding the speed of a particle moving in a circle of radius r. The particle completes one revolution around the circle in a time T known as the **period**. In one period, the distance traveled is the circumference of the circle, $2\pi r$. Because the speed is uniform, it is given by

$$v = \frac{2\pi r}{T} \tag{4.30}$$

Equation 4.30 gives the particle's "translational" or "linear" speed.

Next, let's consider the particle's position as it moves. At one particular instant (Fig. 4.18) the particle's position is

$$\vec{r} = (r\cos\theta)\hat{\imath} + (r\sin\theta)\hat{\jmath} \tag{4.31}$$

where θ is measured counterclockwise from the positive x axis. Notice that the radius r of the circle is also the magnitude of the position vector $\vec{r}$. As the particle travels around the circle, only the direction of $\vec{r}$ changes. Figure 4.19 shows that only θ changes as the particle moves in a circle. So, once we know the radius of the circle, we need only θ to locate the particle.

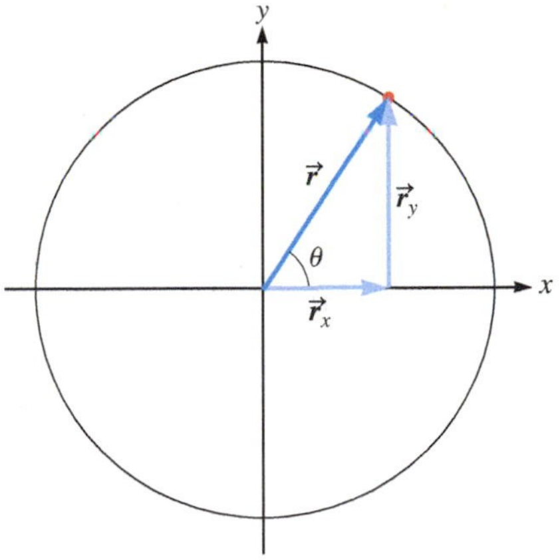

FIGURE 4.18 A particle moves uniformly in a circle. Its position vector at one instant is $\vec{r}$.

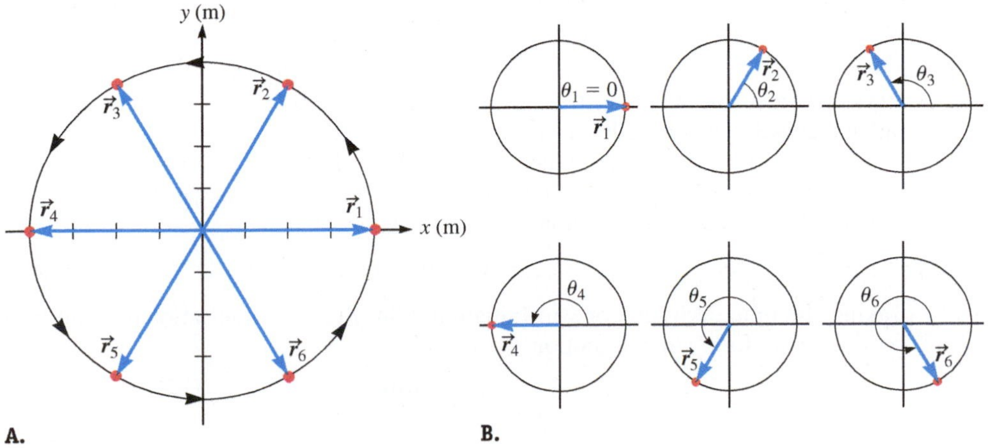

FIGURE 4.19 **A.** A motion diagram of a particle in uniform circular motion. **B.** The same positions shown on separate diagrams. Only the direction θ changes as the particle moves around the circle.

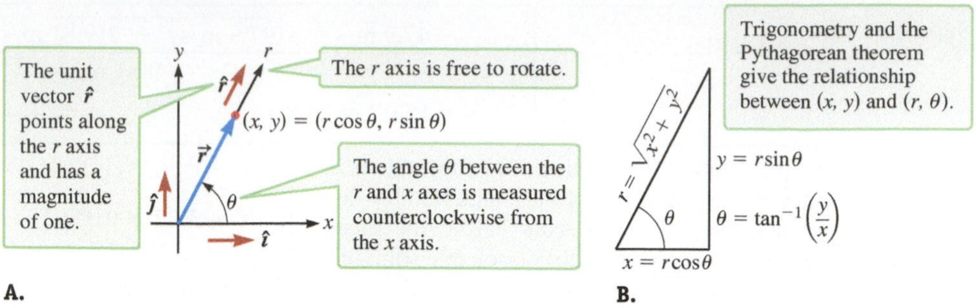
The unit vector $\hat{r}$ points along the r axis and has a magnitude of one.
The r axis is free to rotate.
$(x, y) = (r\cos\theta, r\sin\theta)$
The angle θ between the r and x axes is measured counterclockwise from the x axis.
A.
Trigonometry and the Pythagorean theorem give the relationship between (x, y) and (r, θ).
$r = \sqrt{x^2 + y^2}$
$y = r\sin\theta$
$\theta = \tan^{-1}\left(\frac{y}{x}\right)$
$x = r\cos\theta$
B.

FIGURE 4.20 Plane polar coordinate system.

Polar Coordinate System

We now introduce a coordinate system that will help us exploit the key feature of uniform circular motion—as a particle moves at constant speed along a circular path, only the direction of its position vector changes. As we have seen, two coordinates are needed to describe the position of a particle that moves in two dimensions. In a Cartesian coordinate system, the two coordinates are x and y. As shown in Figure 4.20A, a particle's position can also be given by the magnitude of its position vector $\vec{r}$ and the angle θ. These two coordinates r and θ are called **polar coordinates**, an alternative to Cartesian coordinates. In a polar coordinate system, the r axis is free to rotate. The angle θ is measured counterclockwise from a fixed x axis to the r axis.

By counterclockwise, we mean in the direction that goes from quadrant I to quadrant II to quadrant III to quadrant IV.

The two coordinate systems are related by trigonometry (Fig. 4.20B). To transform from x, y coordinates to r, θ coordinates, use the relationships

$$r = \sqrt{x^2 + y^2} \qquad \text{and} \qquad \tan\theta = \frac{y}{x}$$

and to get from (r, θ) to (x, y), use the relationships

$$x = r\cos\theta \qquad \text{and} \qquad y = r\sin\theta$$

Finally, the polar coordinate system involves two unit vectors, but for the work in this book, we only need to know about one of them. The unit vector $\hat{r}$ points radially outward from the origin as shown in Figure 4.20A.

Linear and Angular Speed

Once we know the radius of the circular path, only the θ component is needed to determine the particle's position at any given moment. We can describe how quickly the particle moves in terms of the change in θ with time. Speed measured in this way is called **angular speed** and is symbolized by the Greek letter omega, ω:

Equation 4.32 gives the same ω described in Example 4.1.

$$\omega = \frac{d\theta}{dt} \tag{4.32}$$

The angle θ is measured in radians, so ω is in radians per second. In one period, the position vector sweeps through an angle $\theta = 2\pi$ rad. Therefore, for a particle in uniform circular motion,

$$\omega = \frac{2\pi}{T} \tag{4.33}$$

We sometimes refer to the instantaneous speed v (Section 2-7) as the **linear** or **translational speed** to distinguish it from the angular speed ω.

Figure 4.21 shows the initial and final positions $\vec{r}_i$ and $\vec{r}_f$ of a particle traveling in uniform circular motion. The position vector swept out an angle θ, and the particle has traveled a distance s along the circle such that

$$s = \theta r \tag{4.34}$$

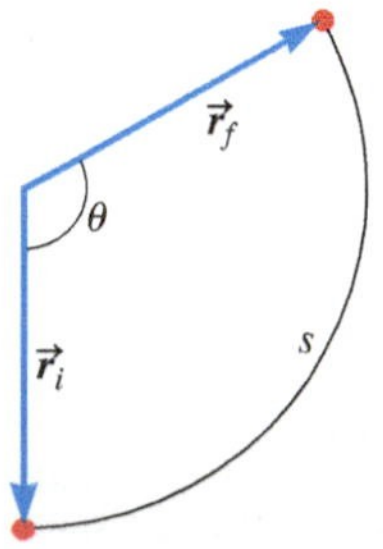
$\vec{r}_f$
θ
$\vec{r}_i$
s

FIGURE 4.21 A particle travels along a portion of a circular path.

Taking the time derivative of s in Equation 4.34 gives the translational speed of a particle in uniform circular motion.

$$v = \frac{ds}{dt} = \frac{d(\theta r)}{dt}$$

Because the radius of the circle r does not change, we can take it outside the derivative. Doing so leaves us with the derivative defining angular speed.

$$\frac{ds}{dt} = r\frac{d\theta}{dt}$$

$$v = r\omega \tag{4.35}$$

Because the radius is measured in meters and the angular speed in rad/s, you might be tempted to report the translational speed v in m·rad/s. The "rad" is dropped, though, because a radian is a ratio of two lengths (Eq. 4.34), and is therefore dimensionless.

CONCEPT EXERCISE 4.6

A particle travels at a uniform linear speed around a circle of radius $r = 0.5$ m, completing one revolution in 12 s.

a. What is the value of the linear speed v?
b. Find the particle's angular speed.
c. What are the angle θ and arc length s swept out by the particle in 4 s?

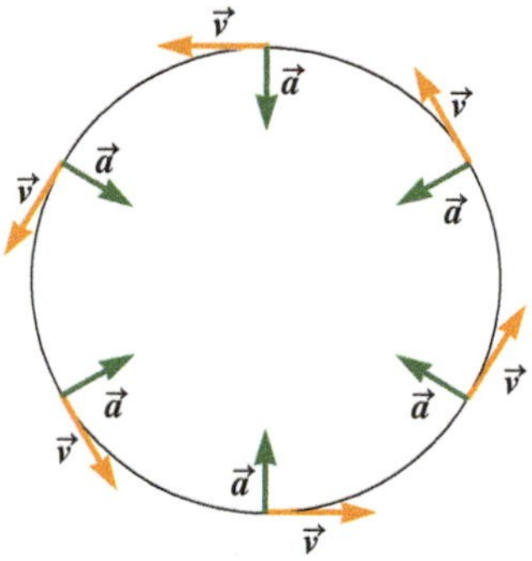

FIGURE 4.22 A particle undergoes uniform circular motion. The particle's velocity is tangent to the circle. Its acceleration points to the center of the circle.

Centripetal Acceleration

The velocity of a particle in uniform circular motion is always tangent to its path (Fig. 4.22). For a particle's velocity to change direction while its speed remains constant, the acceleration must be perpendicular to the velocity (Example 4.4). So, in the case of uniform circular motion, the acceleration always points toward the center of the path and is known as **centripetal acceleration** $\vec{a}_c$. The term *centripetal* comes from a Latin word meaning "center-seeking."

CENTRIPETAL ACCELERATION
Special Case

DERIVATION Centripetal Acceleration

In uniform circular motion, the magnitude of the centripetal acceleration is constant and depends on the particle's speed and the path's radius. We will show that for a particle undergoing uniform circular motion centered on the origin as in Figure 4.23, the centripetal acceleration is given by

$$\vec{a}_c = -\frac{v^2}{r}\hat{r} \tag{4.36}$$

and

$$\vec{a}_c = -\omega^2 r\hat{r} \tag{4.37}$$

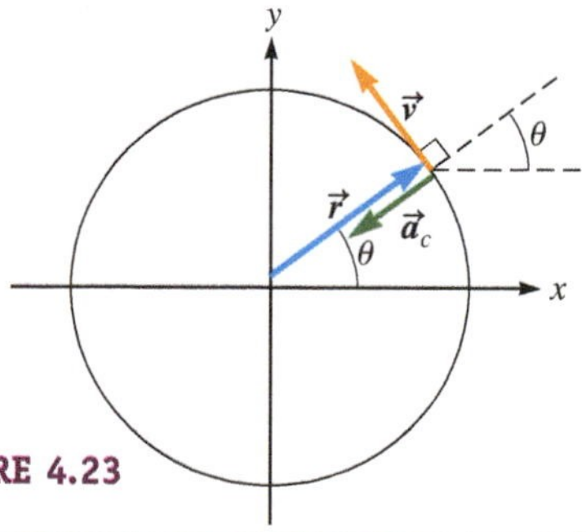

FIGURE 4.23

Consider the arbitrary moment shown in Figure 4.23. Sketch the velocity vector and its components at that moment (Fig. 4.24).

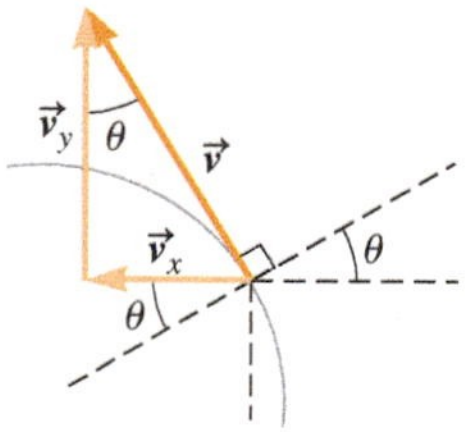

FIGURE 4.24

Write the velocity in component form.

$$\vec{v} = (-v\sin\theta)\hat{\imath} + (v\cos\theta)\hat{\jmath}$$

Take the time derivative (Eq. 4.14) to find the centripetal acceleration. The speed v is constant, but θ is a function of time.

$$\vec{a}_c = \frac{d\vec{v}}{dt} = \frac{d(-v\sin\theta)}{dt}\hat{\imath} + \frac{d(v\cos\theta)}{dt}\hat{\jmath}$$

$$\vec{a}_c = -v\frac{d(\sin\theta)}{dt}\hat{\imath} + v\frac{d(\cos\theta)}{dt}\hat{\jmath}$$

Derivation continues on page 102 ▶

Use the chain rule (Appendix A) to differentiate, then substitute $\omega = d\theta/dt$ (Eq. 4.32).	$\vec{a}_c = -v\frac{d(\sin\theta)}{d\theta}\frac{d\theta}{dt}\hat{\imath} + v\frac{d(\cos\theta)}{d\theta}\frac{d\theta}{dt}\hat{\jmath}$ $\vec{a}_c = -(v\cos\theta)\left(\frac{d\theta}{dt}\right)\hat{\imath} - (v\sin\theta)\left(\frac{d\theta}{dt}\right)\hat{\jmath}$ $\vec{a}_c = -v\omega\cos\theta\hat{\imath} - v\omega\sin\theta\hat{\jmath}$
Replace the angular speed ω using $v = r\omega$ (Eq. 4.35).	$\vec{a}_c = -\frac{v^2}{r}\cos\theta\hat{\imath} - \frac{v^2}{r}\sin\theta\hat{\jmath}$
Ah! At last. The magnitude of the centripetal acceleration is constant and is obtained from the Pythagorean theorem applied to the components of the acceleration.	$a_c = \sqrt{\left(-\frac{v^2}{r}\cos\theta\right)^2 + \left(-\frac{v^2}{r}\sin\theta\right)^2}$ $a_c = \frac{v^2}{r}\sqrt{(\cos\theta)^2 + (\sin\theta)^2}$ $a_c = \frac{v^2}{r}$ (4.38)
The centripetal acceleration always points to the center of the circle. So, using polar coordinates, the centripetal acceleration is in the negative $\hat{r}$ direction. Use $v = \omega r$ to replace v with ω.	$\vec{a}_c = -\frac{v^2}{r}\hat{r}$ ✓ (4.36) $\vec{a}_c = -\omega^2 r\hat{r}$ ✓ (4.37)

EXAMPLE 4.8 Ferris Wheel

Estimate the angular speed ω and the linear speed v of a passenger on a Ferris wheel like the one in Figure 4.25.

FIGURE 4.25

Courtesy, RampantScotland.com

INTERPRET and ANTICIPATE

To find the angular speed, we need to know the period of revolution, and we need the radius of the wheel to find the linear speed. The problem statement gives us none of this information, and so we must come up with our own values based on experience.

SOLVE

To estimate the period, recall from experience that the Ferris wheel is a slow ride. If a ride lasts 2 minutes after all the passengers are seated, a given cart might get to the top four to six times. Therefore, let us assume a passenger makes five revolutions in 2 minutes and from this estimate calculate the period in seconds.	$T = \frac{(2\text{ min})(60\text{ s/min})}{5\text{ rev}} = \frac{120\text{ s}}{5}$ $T \approx 24\text{ s}$
To estimate the angular speed, use Equation 4.33, $\omega = 2\pi/T$	$\omega = \frac{2\pi}{T} = \frac{2\pi}{24\text{ s}}$ $\omega = 0.26\text{ rad/s}$
Because our period value is an estimate, we should not be surprised to find that the period is actually not 24 s but is instead 30 s or even as low as 10 s. Therefore we should report our estimate for ω to one significant figure.	$\omega \approx 0.3\text{ rad/s}$
Now we need to estimate the radius of the Ferris wheel. Figure 4.25 shows a four-story building just behind the Ferris wheel. The building's height is about the same as the radius of the wheel. One story in a typical building is between 2.5 and 4 m high. From the photo, it looks like a medium-size building, so we will assume 3 m per story.	$r = (4\text{ stories})(3\text{ m/story})$ $r = 12\text{ m}$
Now we can find the linear speed. To avoid rounding error, we use two significant figures, but our final estimate should have only one significant figure.	$v = r\omega = (12\text{ m})(0.26\text{ rad/s})$ $v \approx 3\text{ m/s}$

CHECK and THINK

Ferris wheels come in a great range of sizes and speeds. The first Ferris wheel was built for the Chicago World's Fair in 1893. It held 36 cars, each capable of carrying 60 passengers. The wheel's diameter was 250 ft (which is more than 76 m) and the wheel had a period of 9 min (540 s). Passengers experienced a linear speed of only 0.4 m/s, which is slower than walking speed. Still, the ride was so overwhelming that there were reports of a man who became suicidal and tried to get out before the car reached the bottom. The wheel was built as the American response to the Eiffel Tower, which had been built for the 1889 World's Fair in Paris. Unfortunately, after the Chicago World's Fair, the wheel lost money and so was dynamited.

EXAMPLE 4.9 Centrifuge

Centrifuges are used to accelerate a variety of objects, including medical specimens, wet laundry, and even human test subjects. After the centrifuge reaches a set speed, the subject moves in uniform circular motion so that the acceleration is centripetal. A centrifuge used by NASA (Fig. 4.26) to measure the effects of great acceleration on human subjects has a revolution rate of 33.5 revolutions per minute (rpm) and a radius of 8.84 m. Find the centripetal acceleration of the human subject using a polar coordinate system with the origin at the center of the circle. For the **CHECK and THINK** procedure, consider that acceleration around $4g$ to $6g$ can cause a person to have tunnel vision or even lose consciousness in just a few seconds. Would this subject experience such effects?

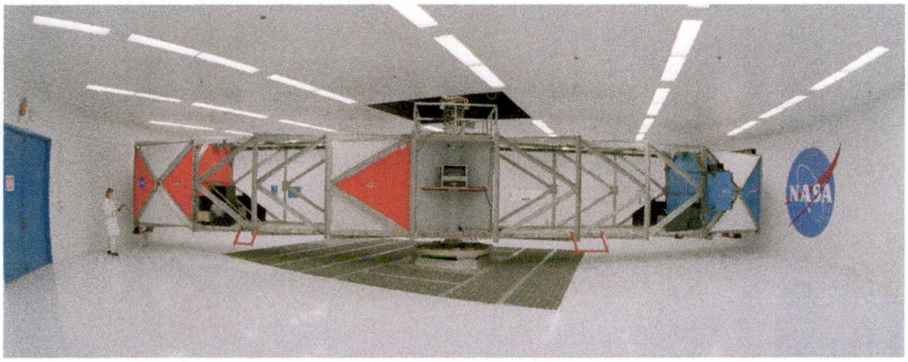

NASA Amesbury

FIGURE 4.26

INTERPRET and ANTICIPATE

Because the centripetal acceleration always points to the center of the circle, we should write our final answer in the form $\vec{a}_c = -(\_\_\_\_\_)\hat{r}$ m/s^2 in plane polar coordinates.

SOLVE First, convert the revolution rate to revolutions per second.	$33.5 \text{ rpm}\left(\frac{1 \text{ min}}{60 \text{ s}}\right) = 0.558 \text{ rev/s}$
Use the revolution rate to find the angular speed. The subject sweeps through 2π rad in each revolution.	$\omega = 2\pi \text{ rad}(0.558 \text{ rev/s}) = 3.51 \text{ rad/s}$
Substitute ω and r into Equation 4.37. The direction is always toward the center of the circle, which is the negative $\hat{r}$ direction.	$\vec{a}_c = -\omega^2 r\hat{r}$ (4.37) $\vec{a}_c = -(3.51 \text{ rad/s})^2(8.84 \text{ m})\hat{r}$ $\vec{a}_c = -109\hat{r} \text{ m/s}^2$

CHECK and THINK

Our expression has the form that we expected. It is common for acceleration to be expressed in terms of the free-fall acceleration—that is, as a number n times the free-fall acceleration: $a_c = ng$. We find n by dividing

$$a_c = \frac{109 \text{ m/s}^2}{9.81 \text{ m/s}^2}g \approx 11g.$$

So, the subject would likely experience tunnel vision or even lose consciousness.

4-7 Relative Motion in One Dimension

RELATIVE MOTION
Major Concept

The word *relative* has so many meanings. Think of such statements as "My uncle is my favorite relative," and "Compare the relative merits of democracy and fascism." A physicist uses the phrase **relative motion** when motion is observed (and measured) from different perspectives. To a physicist, an observer's perspective depends only on the observer's motion: at rest, moving at constant nonzero velocity, or accelerating.

You have probably observed motion from different perspectives. For instance, imagine that you are standing by the side of the road. Your friend—a passenger in a car—passes by at 30 mph and waves. You barely have time to see that he is waving to you because he appears to be moving so quickly. Now imagine riding along on a multilane highway at 60 mph and spotting your friend riding in the next lane, also traveling at 60 mph. This time, he does not appear to speed past you; instead, he seems to be stationary. If your windows were open, you could (aside from a little shouting) converse with him as you would with any passengers in your car.

REFERENCE FRAME
Major Concept

Measurements of motion depend on the observer's perspective. A **reference frame** or simply a **frame** is a coordinate system attached to an observer's particular perspective; in other words, an observer selects a reference frame by choosing a coordinate system that—from her perspective—is at rest. The observer's measurements are made *relative to* or *with respect to* that frame. (These two phrases mean the same thing, and we will use them interchangeably through this discussion.)

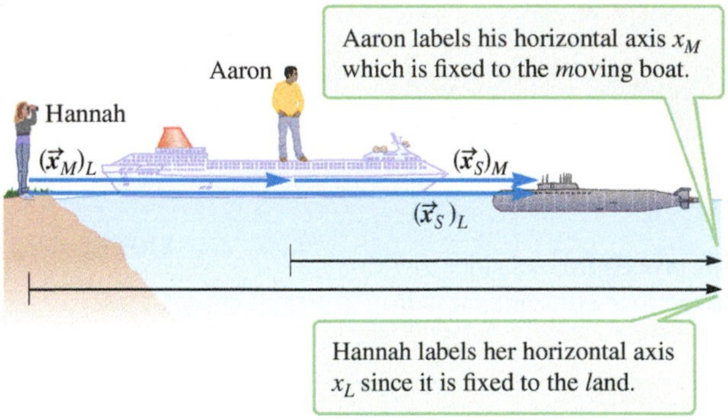

FIGURE 4.27 A submarine is observed from two frames.

The relationships between position, velocity, and acceleration measured in two frames are the foundation for our later study of Einstein's theory of relativity (Chapter 39). Let's use a specific one-dimensional example to derive these relationships. In Figure 4.27, a submarine is observed both by Hannah on the shore and by Aaron on a boat moving at constant velocity. Each of the two observers measures the position of the submarine relative to his or her own frame, and each observer chooses an x axis pointing to the right. Each observer is at rest at the origin of his or her own coordinate system. Hannah's coordinate system is stationary with respect to the shore. Aaron's coordinate system is fixed to the boat and therefore moves with respect to the shore. To distinguish between the two reference frames, Aaron uses an M subscript for *m*oving boat and Hannah uses an L for *l*and. (The choice of M and L will become clear in the next section.) Therefore, the position of the *s*ubmarine relative to Hannah on the *l*and is $(\vec{x}_S)_L$ and relative to Aaron is $(\vec{x}_S)_M$. According to Hannah on the *l*and, the position of Aaron on the *m*oving boat is $(\vec{x}_M)_L$.

The first subscript tells you which object is observed, and the second subscript tells you from which frame. When you are writing, it is common to drop the parentheses and just write a double subscript.

Figure 4.27 shows that these three vectors are related by the expression

$$(\vec{x}_S)_L = (\vec{x}_S)_M + (\vec{x}_M)_L \tag{4.39}$$

We take the time derivative of Equation 4.39 to find the relationship between the submarine velocity measured by Hannah $(\vec{v}_S)_L$ and that measured by Aaron $(\vec{v}_S)_M$:

$$\frac{d(\vec{x}_S)_L}{dt} = \frac{d(\vec{x}_S)_M}{dt} + \frac{d(\vec{x}_M)_L}{dt}$$

$$(\vec{v}_S)_L = (\vec{v}_S)_M + (\vec{v}_M)_L \tag{4.40}$$

where $(\vec{v}_M)_L$ is Aaron's velocity as measured by Hannah on the land.

We take another time derivative to find the acceleration:

$$\frac{d(\vec{v}_S)_L}{dt} = \frac{d(\vec{v}_S)_M}{dt} + \frac{d(\vec{v}_M)_L}{dt}$$

Because Aaron's velocity $(\vec{v}_M)_L$ (measured by Hannah) is constant, its time derivative is zero, so both observers measure the same acceleration for the submarine:

$$(\vec{a}_S)_L = (\vec{a}_S)_M \tag{4.41}$$

EXAMPLE 4.10 Here Comes the Pilot

A pilot is in a boat moving at constant velocity as in Figure 4.27, on his way to a submarine to help the vessel's captain navigate through a difficult channel. A harbormaster stationed on shore observes that the submarine's velocity is $5.3\hat{\imath}$ m/s and that the pilot boat's velocity is $6.5\hat{\imath}$ m/s. According to the pilot, the submarine is initially 5.65×10^2 m away. How long will it take the pilot boat to reach the submarine?

INTERPRET and ANTICIPATE

Use the coordinate system shown in Figure 4.27. We know the velocities of the *s*ubmarine $(\vec{v}_S)_L$ and the pilot boat $(\vec{v}_M)_L$ relative to the harbormaster on the *l*and. If we want to know how long the pilot boat must travel to reach the sub, our first task is to find the velocity $(\vec{v}_S)_M$ of the submarine relative to the pilot boat.

SOLVE

Start with Equation 4.40 and solve for the velocity of the submarine with respect to the pilot boat's frame $(\vec{v}_S)_M$.

$$(\vec{v}_S)_L = (\vec{v}_S)_M + (\vec{v}_M)_L \qquad (4.40)$$

$$(\vec{v}_S)_M = (\vec{v}_S)_L - (\vec{v}_M)_L$$

$$(\vec{v}_S)_M = (5.3\hat{\imath} - 6.5\hat{\imath})\text{ m/s} = -1.2\hat{\imath}\text{ m/s}$$

CHECK and THINK

The negative sign means that, according to the pilot, the submarine is moving in the negative x direction. In other words, to the pilot it looks like the submarine is moving toward him. If that troubles you, think about the situation from the harbormaster's point of view. To her, both the pilot boat and the submarine are moving away from shore. Because the pilot boat has a higher velocity, the harbormaster sees the gap between the pilot boat and the submarine getting smaller.

SOLVE (continued)

We use the submarine's velocity $(\vec{v}_S)_M$ and its initial position $(\vec{x}_S)_M$ measured by the pilot on the boat to find the time it takes the pilot's boat to meet the submarine. To do so, we use $\Delta x = v_{0x}t + \frac{1}{2}a_x t^2$ (Eq. 2.11) with $a = 0$.

When the pilot's boat reaches the submarine, the submarine will be at the origin of his coordinate system $(x_S)_M = 0$. (The time t measured by the harbormaster and by the pilot must be the same.)

$$(\vec{v}_S)_M = \frac{(\Delta\vec{x}_S)_M}{t} = \frac{0 - (x_S)_M}{t}\hat{\imath}$$

$$t = \frac{-(x_S)_M}{(v_S)_M} = \frac{-5.65 \times 10^2\text{ m}}{-1.2\text{ m/s}}$$

$$t = 4.7 \times 10^2\text{ s} = 7.8\text{ min}$$

EXAMPLE 4.11 Does the Speeder Get a Ticket?

A local officer and a state trooper each measure the velocity of a speeding car. The local officer is parked by the side of the road, and the state trooper is pursuing the speeder. According to the local officer, the speeder is going 95.0 mph (42.4 m/s), whereas according to the state trooper, the speeder is going 15.1 mph (6.70 m/s).

A What is the velocity of the trooper relative to the local officer?

INTERPRET and ANTICIPATE

Because this question involves relative motion, we must use two coordinate systems. We use the subscript L for the *l*ocal officer and M for the *m*oving trooper. Figure 4.28 applies to all three parts of the problem.

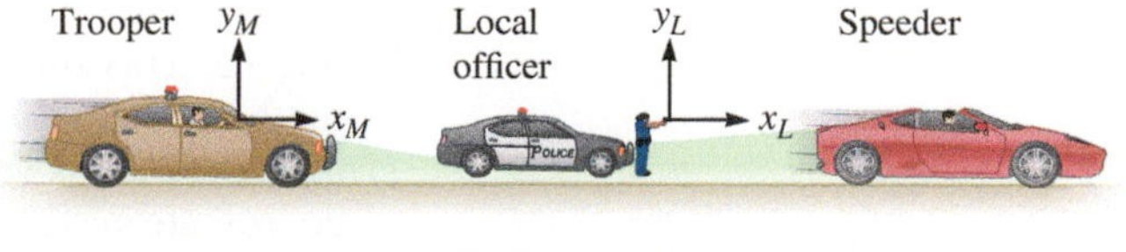

FIGURE 4.28

Example continues on page 106 ▶

SOLVE

We know $(\vec{v}_S)_L$, the speeder's velocity measured by the stationary local officer (42.4 m/s), and $(\vec{v}_S)_M$, the speeder's velocity measured by the moving trooper (6.70 m/s). We can therefore use Equation 4.40 to find $(\vec{v}_M)_L$, the velocity of the trooper measured by the local officer.

$$(\vec{v}_S)_L = (\vec{v}_S)_M + (\vec{v}_M)_L \quad (4.40)$$

$$(\vec{v}_M)_L = (\vec{v}_S)_L - (\vec{v}_S)_M$$

$$(\vec{v}_M)_L = (42.4\hat{\imath} - 6.70\hat{\imath})\text{ m/s}$$

$$(\vec{v}_M)_L = 35.7\hat{\imath}\text{ m/s} = 79.8\hat{\imath}\text{ mph}$$

CHECK and THINK

The stationary officer sees both cars traveling in the same direction. Because the trooper's speed is lower (79.8 mph) than that of the speeder (95.0 mph), the local officer sees the gap between them widen. (Don't worry if you worked in US customary units and found a slight difference in your answer due to a rounding error.)

B The speeder sees the local officer in his rearview mirror and slows down such that 2.5 s after he hits the brakes his speed is, according to the local officer, 60.0 mph (26.8 m/s). What new velocity does the state trooper measure?

SOLVE

We now know $(\vec{v}_M)_L$, velocity of state trooper measured by local officer, from part A. We were just given $(\vec{v}_S)_L$, new velocity of speeder measured by the local officer. Use Equation 4.40 to find $(\vec{v}_S)_M$, velocity of speeder measured by the state trooper.

$$(\vec{v}_S)_L = (\vec{v}_S)_M + (\vec{v}_M)_L \quad (4.40)$$

$$(\vec{v}_S)_M = (\vec{v}_S)_L - (\vec{v}_M)_L$$

$$(\vec{v}_S)_M = (26.8\hat{\imath} - 35.7\hat{\imath})\text{ m/s}$$

$$(\vec{v}_S)_M = -8.93\hat{\imath}\text{ m/s} = -20.0\hat{\imath}\text{ mph}$$

CHECK and THINK

The state trooper now sees the speeder going in the negative x direction; in other words, the speeder seems to be moving toward the state trooper. The local officer still sees both cars moving away, but the gap between them is getting smaller.

C Find the acceleration of the speeder measured by each observer during this 2.5-s interval.

SOLVE

We can use $v_x = v_{0x} + a_x t$ (Eq. 2.9) for either observer. Let's start with the acceleration observed by the local officer. According to $(\vec{a}_S)_L = (\vec{a}_S)_M$ (Eq. 4.41), it should be the same as the acceleration measured by the state trooper.

$$v_x = v_{0x} + a_x t \quad (2.9)$$

$$a_x = \frac{(v_x - v_{0x})_L}{t} = \frac{(26.8 - 42.4)\text{ m/s}}{2.5\text{ s}}$$

$$(\vec{a}_S)_L = -6.25\hat{\imath}\text{ m/s}^2$$

CHECK and THINK

To check, we apply Equation 2.9 to the state trooper's observation and find that the two observers agree on the acceleration.

$$(a_S)_M = \frac{(v_x - v_{0x})_M}{t} = \frac{(-8.93 - 6.70)\text{ m/s}}{2.5\text{ s}}$$

$$(\vec{a}_S)_M = -6.25\hat{\imath}\text{ m/s}^2$$

4-8 Relative Motion in Two Dimensions

RELATIVE MOTION **Major Concept**

We are now ready to consider two-dimensional motion observed from two reference frames. After this discussion, we will be equipped to wrap up the skateboarding case study.

To make the extension to two-dimensional relative motion, we return to the submarine. This time, though, the submarine is in a holding pattern just offshore, and the boat's approach is not along the line joining Hannah to the submarine. Figure 4.29 shows this scene from above. Now each observer must use a two-dimensional coordinate system, with the corresponding axes of the coordinate systems parallel to each other. For us to be able to work out the relationship between these two coordinate systems, they must maintain their relative orientation. In other words, as the boat moves, it may translate, but is not allowed to rotate.

Hannah, on *l*and, measures the position vectors of Aaron on the *m*oving boat $(\vec{r}_M)_L$ and of the *s*ubmarine $(\vec{r}_S)_L$. Aaron measures the position $(\vec{r}_S)_M$ of the submarine, and the three position vectors in Figure 4.29 are related by

$$(\vec{r}_S)_L = (\vec{r}_S)_M + (\vec{r}_M)_L \tag{4.42}$$

As before, the first time derivative gives the velocity:

$$\frac{d(\vec{r}_S)_L}{dt} = \frac{d(\vec{r}_S)_M}{dt} + \frac{d(\vec{r}_M)_L}{dt}$$

$$(\vec{v}_S)_L = (\vec{v}_S)_M + (\vec{v}_M)_L \tag{4.43}$$

We take the second time derivative to find the acceleration:

$$\frac{d(\vec{v}_S)_L}{dt} = \frac{d(\vec{v}_S)_M}{dt} + \frac{d(\vec{v}_M)_L}{dt}$$

As before, the boat is moving at constant velocity, so the time derivative of $(\vec{v}_M)_L$ is zero, and both observers measure the same acceleration:

$$(\vec{a}_S)_L = (\vec{a}_S)_M \tag{4.44}$$

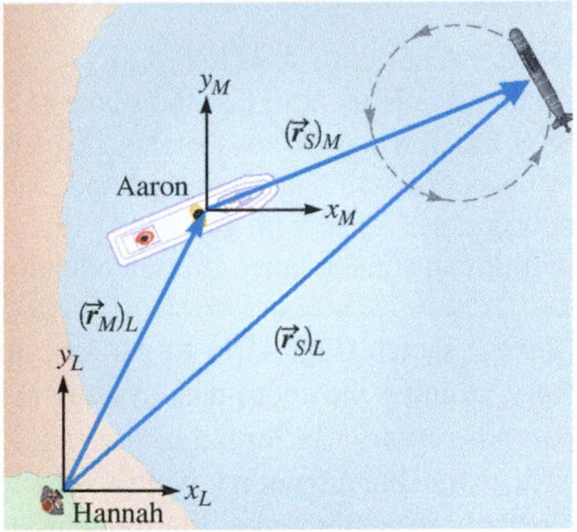

FIGURE 4.29 A submarine is observed from two frames. One frame is attached to the land, and the other frame is attached to the boat and moves relative to the land. Each observer uses a two-dimensional coordinate system. The axes of the two systems remain parallel to each other as the boat moves toward the submarine.

These three equations (4.42, 4.43, and 4.44) are two-dimensional vector equations, and as is often the case, it is helpful to write them in terms of their scalar components:

$$(x_S)_L = (x_S)_M + (x_M)_L \qquad (y_S)_L = (y_S)_M + (y_M)_L \tag{4.45}$$

$$(v_{Sx})_L = (v_{Sx})_M + (v_{Mx})_L \qquad (v_{Sy})_L = (v_{Sy})_M + (v_{My})_L \tag{4.46}$$

$$(a_{Sx})_L = (a_{Sx})_M \qquad (a_{Sy})_L = (a_{Sy})_M \tag{4.47}$$

Don't be intimidated by all the subscripts. The *x* or *y* indicates the component. Think of *S* as the *s*ubject being observed from the *l*ab (*L*) frame and from the *m*oving (*M*) frame with respect to the lab.

We have developed all our equations for relative motion with two observers in mind. It is not necessary for there to be two actual human observers, however. In other words, there can be two reference frames (lab and moving) that are not occupied by human observers. For example, in an airplane, there may be two velocity gauges, ground velocity and air velocity. The ground velocity gauge gives the airplane's velocity relative to the ground, and the air velocity gauge gives its velocity relative to the wind. Although there are likely actual observers on the ground (in the control tower), there is (probably) no observer traveling with the wind. We can still define two reference frames, one attached to the ground ("the lab") and another "attached" to the (moving) wind (Fig. 4.30). The speed of the airplane relative to the wind is called the airspeed, and its speed relative to the ground is its groundspeed.

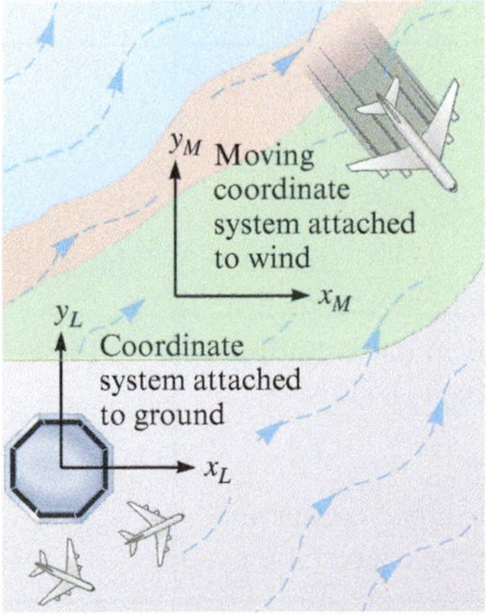

FIGURE 4.30 The speed of the airplane relative to the wind (the airspeed) and its speed relative to the ground (groundspeed) define two reference frames: "the lab" attached to the ground and "moving" attached to the wind.

EXAMPLE 4.12 Groundspeed

An airplane flies at an airspeed of 809 km/h at an angle $\alpha = 20°$ (exactly) south of east. The wind blows at 85.0 km/h toward the northeast with respect to the ground. What are the plane's groundspeed and direction?

Example continues on page 108 ▶

INTERPRET and ANTICIPATE

Airspeed tells us how fast the plane is moving relative to the 85.0-km/h wind; *groundspeed* is the plane's speed as measured by someone standing on the ground. Because the plane's motion is not aligned with the wind's direction, we must use our two-dimensional relative-motion equations.

Make a sketch (Fig. 4.31) to get an estimate for our final results, aligning the coordinate systems with the cardinal (compass) directions. The term *northeast* means exactly 45°north of east, so the direction of the wind's velocity is $\theta = 45°$ counterclockwise from the x axis. Our sketch shows a rough geometric addition of $(\vec{v}_S)_L = (\vec{v}_S)_M + (\vec{v}_M)_L$ as a guide. The vector $(\vec{v}_S)_L$ is slightly longer than $(\vec{v}_S)_M$ (groundspeed a little faster than airspeed), and $(\vec{v}_S)_L$ points more easterly than $(\vec{v}_S)_M$.

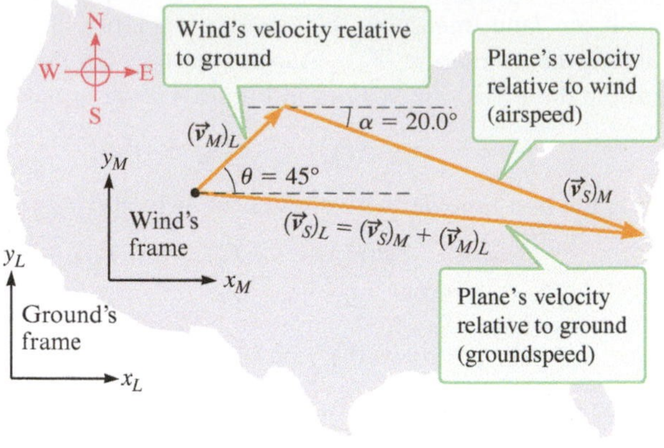

FIGURE 4.31

SOLVE

Convert the given speeds to SI units.

$$(v_M)_L = (85.0\ \text{km/h})\left(\frac{0.278\ \text{m/s}}{1\ \text{km/h}}\right) = 23.6\ \text{m/s}$$

$$(v_S)_M = (809\ \text{km/h})\left(\frac{0.278\ \text{m/s}}{1\ \text{km/h}}\right) = 225\ \text{m/s}$$

Write the velocities $(\vec{v}_M)_L$ and $(\vec{v}_S)_M$ in component form.

$$(\vec{v}_M)_L = (v_{ML}\cos\theta)\hat{\imath} + (v_{ML}\sin\theta)\hat{\jmath}$$

$$v_{ML}\cos 45° = v_{ML}\sin 45° = (23.6\ \text{m/s})\left(\frac{\sqrt{2}}{2}\right) = 16.7\ \text{m/s}$$

$$(\vec{v}_M)_L = (16.7\hat{\imath} + 16.7\hat{\jmath})\ \text{m/s}$$

$$(\vec{v}_S)_M = (v_{SM}\cos\alpha)\hat{\imath} - (v_{SM}\sin\alpha)\hat{\jmath}$$

$$(\vec{v}_S)_M = (211\hat{\imath} - 76.9\hat{\jmath})\ \text{m/s}$$

We find the plane's groundspeed components by adding the wind's velocity and the airspeed velocity.

$$(v_{Sx})_L = (v_{Sx})_M + (v_{Mx})_L \qquad (4.46)$$

$$(v_{Sx})_L = (211 + 16.7)\ \text{m/s} = 227.7\ \text{m/s}$$

$$(v_{Sy})_L = (v_{Sy})_M + (v_{My})_L \qquad (4.46)$$

$$(v_{Sy})_L = (-76.9 + 16.7)\ \text{m/s} = -60.2\ \text{m/s}$$

The groundspeed and direction are found from these components using $v = \sqrt{v_x^2 + v_y^2}$ and $\theta = \tan^{-1}(v_y/v_x)$ (Eqs. 3.12 and 3.14).

$$(v_S)_L = \sqrt{(227.7\ \text{m/s})^2 + (-60.2\ \text{m/s})^2}$$

$$(v_S)_L = 236\ \text{m/s}$$

$$\tan^{-1}\left(\frac{-60.2\ \text{m/s}}{227.7\ \text{m/s}}\right) = -14.8°$$

CHECK and THINK

The airplane is heading 14.8° south of east, which meets our expectation of being slightly more easterly than the 20° south-of-east angle seen by our imaginary wind observer. Also, as expected from our sketch, the groundspeed of 236 m/s is a little higher than the airspeed of 225 m/s.

CONCEPT EXERCISE 4.7

Which speed, groundspeed or airspeed, is needed to determine the flight time of an airplane that flies from city A to city B?

EXAMPLE 4.13 CASE STUDY Who's Right?

Reread the case study (pages 86–87) and determine which student is correct: Cameron, who wants to use a 90° launch angle; or Avi, who wants to use a smaller angle.

INTERPRET and ANTICIPATE

The solution we present here is based on viewing the situation from two reference frames. One is the laboratory frame L, the frame from which Cameron and Avi observe the experiment. The other frame M is attached to the cart. These two coordinate systems are shown in Figure 4.32.

If the cart is not in motion, there is no difference between the two reference frames. If the ball is launched straight up, observers in both frames see the same motion diagram. The ball moves along a single axis (y_L or y_M) from point A to point E and back down to point I.

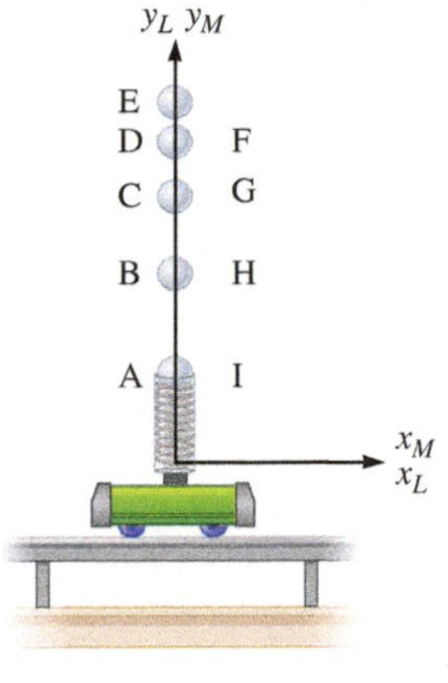

FIGURE 4.32

SOLVE

If the cart is **not** in motion and the ball is launched at an angle smaller than 90° as in Figure 4.33, observers in both frames see the ball move along a parabolic path. They agree that the ball lands in front of the cart at point I.

What changes when the cart (with its frame) is already in motion at constant velocity at the instant the ball is launched? For the observer in the moving cart frame, the answer is simple: Nothing changes. To this observer, it does not matter whether the cart is stationary as the ball is launched or moving with constant velocity during the launch. Either way, the path of the ball in the cart frame is along a straight, vertical line if the launch angle is 90° and along a parabola if the launch angle is smaller than 90°. A straight-up launch brings the ball back to the cart, and a launch at $\theta < 90°$ means that the ball lands at some point in front of the cart.

The ball can land only in one place, of course. So, what the observer on the cart sees must also be what the laboratory observer sees: a 90° launch angle lands the ball in the cart, and a launch angle smaller than 90° lands the ball on the track in front of the cart.

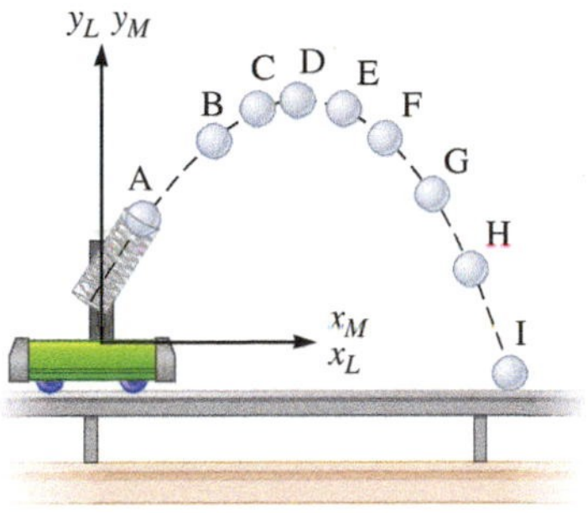

FIGURE 4.33

So, Cameron is correct: The ball must be launched at a 90° angle.

CHECK and THINK

Consider what the laboratory observer sees for a 90° launch from the moving cart. To land back in the cart, the ball must move along a straight, vertical path in the cart frame, yet the ball's path in the laboratory frame cannot be along a straight, vertical line.

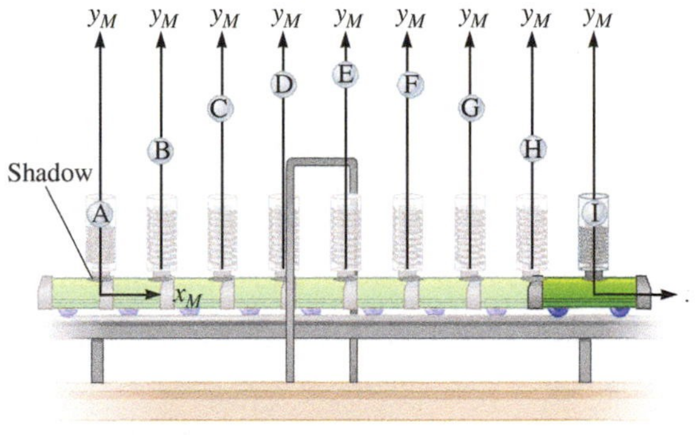

FIGURE 4.34

Figure 4.34 shows how this apparent contradiction is reconciled. The cart moves at constant velocity along the x axis while, from the perspective of the laboratory observer, the ball moves along a two-dimensional parabolic path. If we use an overhead light to project the ball's shadow (Section 4-5) along the x axis, we would find that this shadow is always superimposed on the cart. When the ball reaches point I in its parabolic path, the cart is also at that point, and the ball falls into the cart.

Now consider what the laboratory observer sees when the cart is moving and the launch angle is less than 90°. The motion diagram shows that the ball still moves in a parabolic path, but its velocity along the x axis is now greater than the velocity of the cart (Fig. 4.35). The result of this greater horizontal velocity is that the parabola is wider. The ball's shadow along the x axis is still

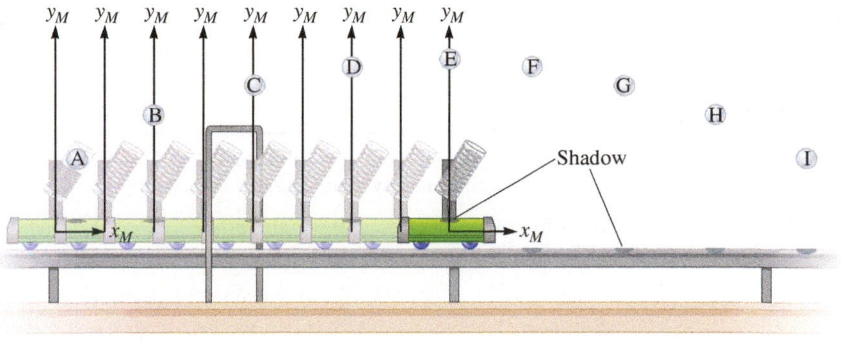

FIGURE 4.35

Example continues on page 110 ▶

moving at a constant velocity, but the magnitude of that velocity is greater than the magnitude of the cart's velocity. The ball's shadow is no longer superimposed on the cart; instead, the shadow is in front of the cart, and the gap between the shadow and the cart widens. When the ball lands on the track, the cart is still far behind.

According to this model, the skateboarder executing the hippy must jump straight up to land on his skateboard. While he is in the air, his shadow will be on top of the skateboard just as the ball's shadow is on the cart.

A Word About Air Resistance

We have just seen that Cameron is correct: The ball must be launched straight up if it is to land back in the cart, whether the cart is moving or not. Still, there is something about Avi's belief that the launch has to be aimed not straight up but rather along the track that agrees with our intuition, which is usually based on our experiences in the everyday world. In this case, our intuition may be based on what we know about throwing lightweight objects such as balloons, foam balls, and feathers. In this chapter, we ignored air, the medium through which all these projectiles move.

In many situations, ignoring air is acceptable, but experience tells us that sometimes the air greatly affects the motion of objects. For example, imagine replacing the metal ball in the case study with a very light foam ball. The air in the laboratory opposes the motion of the foam ball, with the result that its maximum height and range are noticeably shorter than if it were moving in a vacuum. So, like Avi, we might try to compensate for the air's resistance by aiming the launcher at a lower angle. (The air actually opposes the motion of both balls, but the effect is more noticeable in the case of the foam ball.)

In physics, we often think of how things work under ideal conditions—in this case, projectile motion in a vacuum. In the laboratory, we do our best to design experiments and equipment that are a close approximation to ideal conditions. For example, we use objects whose physical dimensions minimize air resistance, and in some laboratories, we may even evacuate a chamber. So, under laboratory conditions, Cameron is correct. In some situations, however, we may need to factor in air resistance, as we will in Chapter 6.

SUMMARY

Underlying Principles: Two- and three-dimensional kinematics

One-dimensional motion is along a single straight line, whereas multidimensional motion may be along a curved path or along a path consisting of two or more straight lines.

Major Concepts

Summary of one- and three-dimensional kinematic quantities.

Quantity	Three Dimensions		One Dimension (along x axis)
1. Position is a vector used to locate a particle with respect to the origin of a particular coordinate system.	$\vec{r} = x\hat{\imath} + y\hat{\jmath} + z\hat{k}$	(4.1)	$\vec{r} = x\hat{\imath}$
2. Displacement $\Delta\vec{r}$ is found by subtracting the initial position vector from the final position vector.	$\Delta\vec{r} = \vec{r}_f - \vec{r}_i = \Delta x\hat{\imath} + \Delta y\hat{\jmath} + \Delta z\hat{k}$ $\Delta x \equiv (x_f - x_i)$ $\Delta y \equiv (y_f - y_i)$ $\Delta z \equiv (z_f - z_i)$	(4.2)	$\Delta\vec{r} = (x_f - x_i)\hat{\imath}$

Major Concepts—cont'd

Quantity	Three Dimensions		One Dimension
3a. Average velocity is the displacement divided by the time interval.	$\vec{v}_{av} = \dfrac{\Delta\vec{r}}{\Delta t}$	(4.4)	$\vec{v}_{av} = \dfrac{\Delta x}{\Delta t}\hat{\imath}$
	$\vec{v}_{av} = \dfrac{\Delta x}{\Delta t}\hat{\imath} + \dfrac{\Delta y}{\Delta t}\hat{\jmath} + \dfrac{\Delta z}{\Delta t}\hat{k}$	(4.6)	
3b. Velocity is the time derivative of $\vec{r}$.	$\vec{v} = \dfrac{d\vec{r}}{dt}$	(4.8)	$\vec{v} = \dfrac{dx}{dt}\hat{\imath}$
	$\vec{v} = \dfrac{dx}{dt}\hat{\imath} + \dfrac{dy}{dt}\hat{\jmath} + \dfrac{dz}{dt}\hat{k}$	(4.9)	
4a. Average acceleration comes from dividing the change in velocity by the time interval.	$\vec{a}_{av} = \dfrac{\Delta\vec{v}}{\Delta t}$	(4.11)	$\vec{a}_{av} = \dfrac{\Delta v_x}{\Delta t}\hat{\imath}$
	$\vec{a}_{av} = \dfrac{\Delta v_x}{\Delta t}\hat{\imath} + \dfrac{\Delta v_y}{\Delta t}\hat{\jmath} + \dfrac{\Delta v_z}{\Delta t}\hat{k}$	(4.12)	
4b. Acceleration is the time derivative of the velocity.	$\vec{a} = \dfrac{d\vec{v}}{dt}$	(4.14)	$\vec{a} = \dfrac{dv_x}{dt}\hat{\imath}$
	$\vec{a} = \dfrac{dv_x}{dt}\hat{\imath} + \dfrac{dv_y}{dt}\hat{\jmath} + \dfrac{dv_z}{dt}\hat{k}$	(4.15)	

5. Relative motion refers to motion observed from different perspectives. An observer's perspective depends only on the observer's motion.

A **reference frame** or simply a **frame** is a coordinate system attached to an observer's particular perspective.

When a relative-motion situation involves two frames, the one that is stationary relative to the Earth is usually called the *laboratory* frame, and the other is the *moving* frame. For relative motion in two dimensions, we write these relationships in terms of their scalar components:

$$(x_S)_L = (x_S)_M + (x_M)_L \qquad (y_S)_L = (y_S)_M + (y_M)_L \tag{4.45}$$

$$(v_{Sx})_L = (v_{Sx})_M + (v_{Mx})_L \qquad (v_{Sy})_L = (v_{Sy})_M + (v_{My})_L \tag{4.46}$$

$$(a_{Sx})_L = (a_{Sx})_M \qquad (a_{Sy})_L = (a_{Sy})_M \tag{4.47}$$

Special Cases

1. Two-dimensional projectile motion is a special case in which (a) there is no acceleration in the x (horizontal) direction and (b) the acceleration in the y (vertical) direction is the constant free-fall acceleration g. A projectile's path is parabolic. The equations of motion for a projectile are

$$\vec{a} = -g\hat{\jmath} \tag{4.23}$$

$$\vec{v} = (v_0 \cos\theta)\hat{\imath} + (v_0 \sin\theta - gt)\hat{\jmath} \tag{4.24}$$

$$\vec{r} = [v_0 t\cos\theta + x_0]\hat{\imath} + [(v_0 t\sin\theta - \tfrac{1}{2}gt^2) + y_0]\hat{\jmath} \tag{4.25}$$

a. The **range** of a projectile is the magnitude of its horizontal displacement when its vertical displacement is zero:

$$R = \frac{v_0^2}{g}\sin 2\theta \tag{4.28}$$

b. The **maximum range** occurs when $\theta = 45°$:

$$R_{max} = v_0^2/g \tag{4.29}$$

2. Uniform circular motion is a particular case of two-dimensional motion in which the path of the particle is a circle and its speed is constant. The magnitude of the position vector is the radius r of the circle.

The particle completes one revolution around the circle in a time T known as the **period**. In one period, the distance traveled is the circumference of the circle, $2\pi r$. So, its speed is given by

$$v = \frac{2\pi r}{T} \tag{4.30}$$

and the centripetal acceleration is

$$a_c = \frac{v^2}{r} = r\omega^2 \text{ toward the center} \tag{4.36 and 4.37}$$

The rate at which θ changes is the **angular speed** of the motion:

$$\omega = \frac{d\theta}{dt} \tag{4.32}$$

For a particle moving in uniform circular motion,

$$\omega = \frac{2\pi}{T} \tag{4.33}$$

The speed and angular speed are related by

$$v = r\omega \tag{4.35}$$

PROBLEM-SOLVING STRATEGY Two-Dimensional Projectile Motion

There are two parts to the **INTERPRET and ANTICIPATE** procedure:

1. Start with a **sketch** that includes a **coordinate system**.
2. If the vertical displacement is zero ($\Delta y = 0$), you can apply the range equation $R = (v_0^2/g)\sin 2\theta$ (Eq. 4.28).

There are also two parts to the **SOLVE** procedure:

1. **List six kinematic variables** for x and y separately.
2. You may use the **constant acceleration equations** (Table 2.4 or Equations 4.23 through 4.25).

PROBLEMS AND QUESTIONS

A = algebraic C = conceptual E = estimation G = graphical N = numerical

4-1 What Is Multidimensional Motion?

1. C In each case, determine if the train is moving in one or two dimensions. Explain your answers. **a.** A train moves in one direction along a flat, straight track. **b.** A train moves in one direction along a flat, straight track and then reverses direction, moving in the opposite direction along the same track. **c.** A train moves in one direction along a straight, downhill track.
2. C In each case, determine whether the object is moving in one, two, or three dimensions. Explain. **a.** The car driving up the hill in Figure P4.2A **b.** The car winding down the mountain in Figure P4.2B

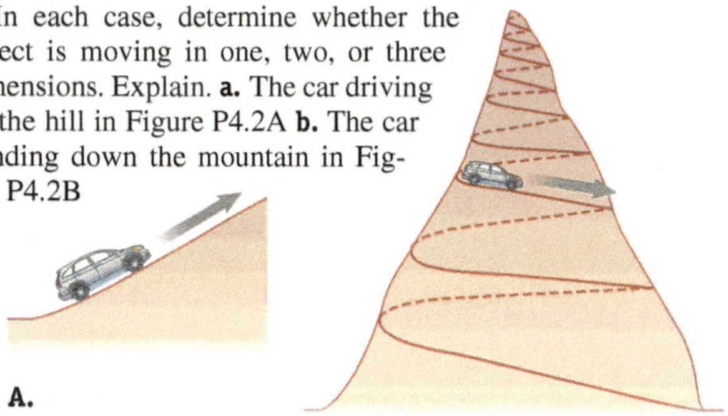

A. B.

FIGURE P4.2

3. C CASE STUDY Imagine an indoor tennis court on a cruise ship moving at constant velocity. Assume the court is oriented so that one player faces the bow (forward) and the other the stern (backward). Does one player have an advantage over the other? If so, which one? Explain your answers.

4-2 Motion Diagrams for Multidimensional Motion

4. G A basketball player dribbles the ball while running at a constant speed straight across the court. Think about the velocity and acceleration of the ball and then sketch a motion diagram for the ball. Explain how you arrived at your sketch.
5. C A motion diagram of a bouncing ball is shown in Figure P4.5. **a.** Does the ball move in one, two, or three dimensions? **b.** For which points is the ball's speed the highest? **c.** Where is the ball at its lowest speeds?

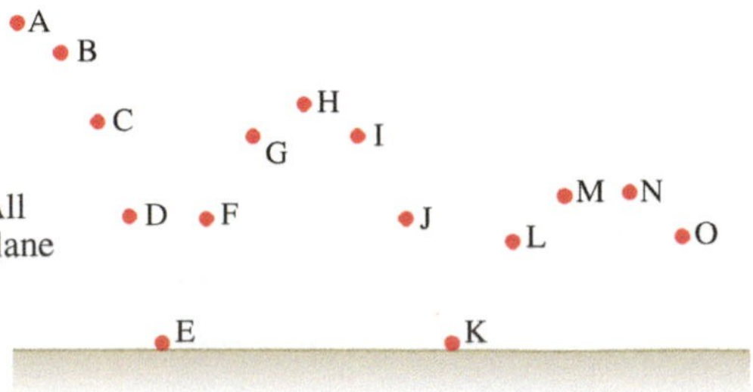

FIGURE P4.5 All dots are in the plane of the page.

6. A ball hangs from a string. The string is kept taut as the ball is displaced to one side and released. The ball swings freely back and forth. This is an example of a simple pendulum.
   a. G Use the data in the accompanying table to create a motion diagram for this ball.
   b. C Does the ball maintain a constant speed? If not, where does it speed up or slow down?

t (s)	x (cm)	y (cm)	t (s)	x (cm)	y (cm)
0	30.90	4.89	11	−29.44	4.43
1	29.44	4.43	12	−25.14	3.21
2	25.14	3.21	13	−18.36	1.70
3	18.36	1.70	14	−9.69	0.47
4	9.69	0.47	15	0.00	0.00
5	0.00	0.00	16	9.69	0.47
6	−9.69	0.47	17	18.36	1.70
7	−18.36	1.70	18	25.14	3.21
8	−25.14	3.21	19	29.44	4.43
9	−29.44	4.43	20	30.90	4.89
10	−30.90	4.89			

4-3 Position and Displacement

7. An ice skater moves along a circular path at constant speed with $\omega = \pi$ rad/s. With the origin located at the center of the circular path, her coordinates at time $t = 0$ are $x = 5.00$ m and $y = 0$.
   a. C Sketch the circular path and include the coordinate system, marking the location of the origin and the initial position of the skater. What is the radius of the circular path?
   b. N Write an expression for the x coordinate of the skater as a function of time.
   c. G Plot the function you found in part (b) with time on the horizontal axis and position x along the vertical axis for all times between $t = 0$ and $t = 6.0$ s.
   d. C After the first time the skater goes around the path, she is then repeating her motion as she goes around again and again. Describe how this repetitive behavior is illustrated in your graph from part (c).

Problems 8 and 26 are paired.

8. Figure P4.8 shows the motion diagram of two balls, one on the left and one on the right. Each ball starts at a point labeled i. The ball on the left is released and falls straight down. At the

same time, the ball on the right is launched horizontally and follows the path shown.

a. C Use the given coordinate system to write the position of points *i*, C, E, G, and K in component form for each ball.

b. N Find the displacement of each ball from *i* to points C, E, G, and K.

c. C Compare your answers for the two balls in part (b). What similarities do you notice?

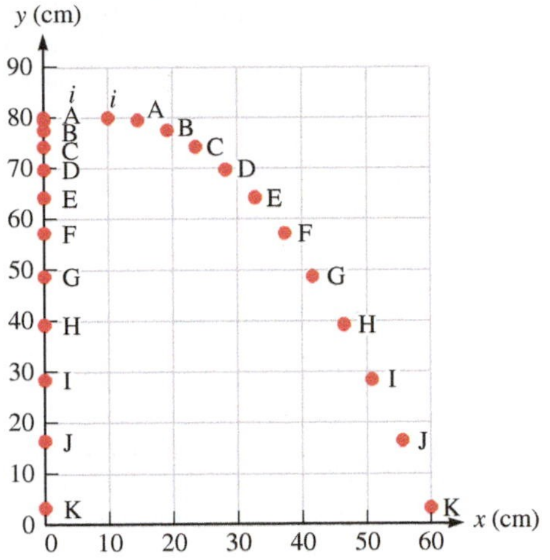

FIGURE P4.8 Problems 8 and 26.

9. A particle moves at constant speed in a circular path, centered about the origin, such that $\omega = 2.0$ rad/s. At some instant, its position is $(x, y) = (3.56, 0.44)$ m.

a. N What is the radius of the circle?

b. C How does your answer change if $\omega = 4.0$ rad/s?

4-4 Velocity and Acceleration

10. N A glider dives toward the ground at a constant velocity of 4.50 m/s and at an angle of 56.0° below the horizontal. If the Sun is directly overhead, what is the speed of the glider's shadow on the level ground below?

11. N An object moves with an initial velocity $\vec{v}_i = 3.00\hat{\jmath}$ m/s and an acceleration $\vec{a} = 2.50\hat{\imath}$ m/s^2. Assume the object is initially at the origin. **a.** What is the position vector of the object as a function of time? **b.** What is the velocity vector of the object as a function of time? **c.** What is the position of the object at time $t = 3.00$ s? **d.** What is the speed of the object at time $t = 3.00$ s?

Problems 12 and 13 are paired.

12. C If a particle's speed is always increasing, what are the possible angles between the particle's velocity and acceleration?

13. C If a particle's speed is always decreasing, what are the possible angles between the particle's velocity and acceleration?

14. N An aircraft flies at constant altitude (with respect to sea level) over the South Rim of the Grand Canyon (Fig. P3.40, page 83). Consider a coordinate system such that the positive *x* axis points to the east, and the positive *y* axis points north. The aircraft's initial position and velocity are 1350 m at an angle of 145° and 60.0 m/s at an angle of 55.0° where both angles are measured counterclockwise with respect to the positive *x* axis. The aircraft's acceleration is 4.0 m/s^2 at an angle of 195° with respect to the positive *x* axis. **a.** What is the velocity of the aircraft after 7.50 s have elapsed? **b.** What is the position vector of the aircraft after 7.50 s have elapsed?

15. N A glider is initially moving at a constant height of 3.59 m. It is suddenly subject to a wind such that its velocity at a later time *t* can be described by the equation $\vec{v}(t) = 15.72\hat{\imath} - 7.88(1 + t)\hat{\jmath} + 0.79t^3\hat{k}$, where $\vec{v}$ and its components are in meters per second, *t* is in seconds, and the *z* axis is perpendicular to the level ground. **a.** What was the initial velocity of the glider? **b.** Write an expression for the acceleration of the glider in component form, when $t = 2.15$ s.

Problems 16 and 17 are paired.

16. N If the vector components of the position of a particle moving in the *xy* plane as a function of time are $\vec{x} = (2.5\ \text{m/s}^2)t^2\hat{\imath}$ and $\vec{y} = (5.0\ \text{m/s}^3)t^3\hat{\jmath}$, at what time *t* is the angle between the particle's velocity and the *x* axis equal to 45°?

17. A If the vector components of a particle's position moving in the *xy* plane as a function of time are $\vec{x} = bt^2\hat{\imath}$ and $\vec{y} = ct^3\hat{\jmath}$, where *b* and *c* are positive constants with the appropriate dimensions such that the components will be in meters, at what time *t* is the angle between the particle's velocity and the *x* axis equal to 45°?

18. C An object is subject to a constant acceleration. Under what circumstances does the object travel (a) in a straight line, (b) along a circular path, or (c) along a curved noncircular path?

19. A The spiral is an example of a mathematical form appearing in nature, from the visible construction of seashells, pinecones, and galaxies to the movement behavior of certain animals. The position of a hungry animal that moves outward along a spiral path, searching for food, can be written as $\vec{r}(t) = A\omega t \cos(\omega t)\hat{\imath} + A\omega t \sin(\omega t)\hat{\jmath}$. Write an expression for the velocity of the animal in component form.

4-5 Special Case of Projectile Motion

20. N A circus performer stands on a platform and throws an apple from a height of 45 m above the ground with an initial velocity $\vec{v}_0$ as shown in Figure P4.20. A second, blindfolded performer must catch the apple. If $v_0 = 26$ m/s, how far from the end of the platform should the second performer stand?

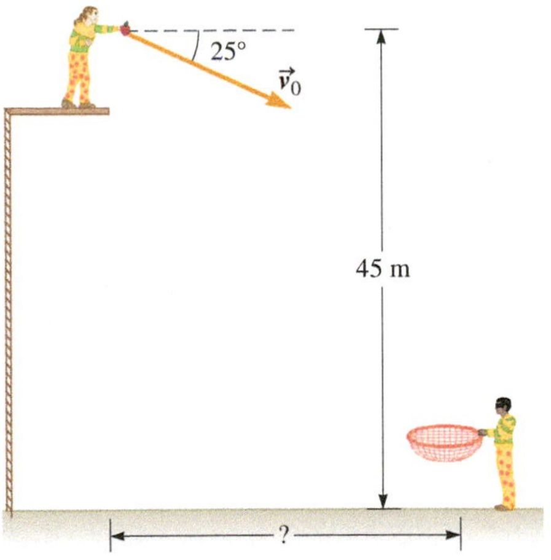

FIGURE P4.20

21. N Anthony carelessly rolls his toy car off a 74.0-cm-high table. The car strikes the floor a horizontal distance of 97.0 cm from the edge of the table. **a.** What was the velocity with which the car left the table? **b.** What was the angle of the car's velocity with respect to the floor just prior to impact?

22. C A physics student stands on a second-story balcony and uses a potato gun to launch a potato horizontally with speed *v*. The potato has flight time *t* and lands on the ground a horizontal distance *d* from the balcony. **a.** If the launch speed of the potato were doubled, would the time of flight increase, decrease, or stay the same? If the flight time changes, would it double or be halved? Explain. **b.** If the launch speed of the potato were doubled, would the horizontal distance increase, decrease, or stay the same? If the horizontal distance changes, would it double or be halved? Explain.

Problems 23 and 24 are paired.

23. **N** During the battle of Bunker Hill, Colonel William Prescott ordered the American Army to bombard the British Army camped near Boston. The projectiles had an initial velocity of 45 m/s at 35° above the horizon and an initial position that was 35 m higher than where they hit the ground. How far did the projectiles move horizontally before they hit the ground? Ignore air resistance.

24. **A** During the battle of Bunker Hill, Colonel William Prescott ordered the American Army to bombard the British Army camped near Boston. The projectiles had an initial velocity of v measured in meters per second at an angle θ above the horizon and an initial position that was h higher than where they hit the ground. How far did the projectiles move horizontally before they hit the ground? Ignore air resistance.

25. **N** A softball is hit with an initial velocity of 29.0 m/s at an angle of 60.0° above the horizontal and impacts the top of the outfield fence 5.00 s later. Assuming the initial height of the softball was 0.500 m above (level) ground, what are the ball's horizontal and vertical displacements?

26. Figure P4.8 shows the motion diagram of two balls. The time interval between images is 0.036 s. Each ball starts at point *i*. The ball on the left is released and falls straight down. At the same time, the ball on the right is launched horizontally and follows the path shown.

a. **A** Write the velocity and acceleration of each ball as a function of time.

b. **N** What is the velocity and acceleration of each ball at *i*, G, and K?

c. **C** Compare your answers for the two balls in part (b). What similarities do you notice?

27. A circus performer throws an apple toward a hoop held by a performer on a platform (Fig. P4.27). The thrower aims for the hoop and throws with a speed of 24 m/s. At the exact moment the thrower releases the apple, the other performer drops the hoop. The hoop falls straight down.

a. **N** At what height above the ground does the apple go through the hoop?

b. **C** If the performer on the platform did not drop the hoop, would the apple pass through it?

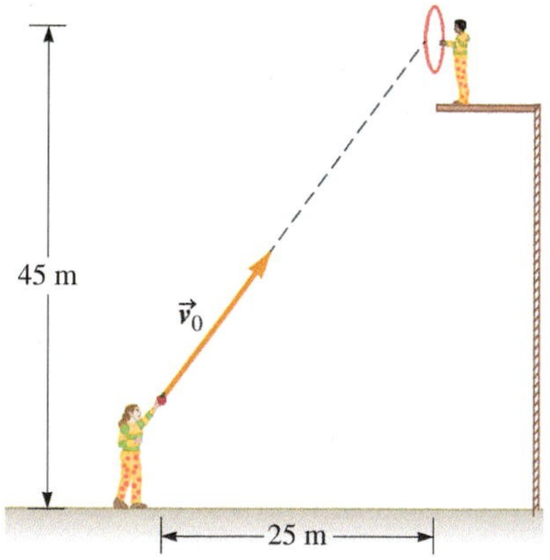

FIGURE P4.27

28. **A** An arrow is fired with initial velocity v_0 at an angle θ from the top of battlements, a height h above the ground. **a.** In terms of h, v_0, θ, and g, what is the time at which the arrow reaches its maximum height? **b.** In terms of h, v_0, θ, and g, what is the maximum height above the ground reached by the arrow?

29. **N** A rock is thrown horizontally off a 56.0-m-high cliff overlooking the ocean, and the sound of the splash is heard 3.60 s later. If the speed of sound in air at this location is 343 m/s, what was the initial velocity of the rock?

30. **A** A projectile is launched up and to the right over flat, level ground. If air resistance is ignored, its maximum range occurs when the angle between its initial velocity and the ground is 45°. Which angles would result in the range being equal to half the maximum?

31. **N** Sienna tosses a ball from the window of her high-rise apartment building with an initial velocity of 6.50 m/s at 25.0° above the horizontal. The ball strikes the ground 4.50 s later. **a.** What is the horizontal distance from the base of the building to the point where the ball strikes the ground? **b.** What is the height from which the ball is thrown? **c.** What is the time it takes for the ball to reach a point 15.0 m below the window where it was thrown?

32. **N** Some cats can be trained to jump from one location to another and perform other tricks. Kit the cat is going to jump through a hoop. He begins on a wicker cabinet at a height of 1.750 m above the floor and jumps through the center of a vertical hoop, reaching a peak height 3.125 m above the floor. **a.** With what initial velocity did Kit leave the cabinet if the hoop is at a horizontal distance of 1.544 m from the cabinet? **b.** If Kit lands on a bed at a horizontal distance of 3.587 m from the cabinet, how high above the ground is the bed?

33. **N** Dock diving is a great form of athletic competition for dogs of all shapes and sizes (Fig. P4.33). Sheba, the American Pit Bull Terrier, runs and jumps off the dock with an initial speed of 9.02 m/s at an angle of 25° with respect to the surface of the water. **a.** If Sheba begins at a height of 0.84 m above the surface of the water, through what horizontal distance does she travel before hitting the surface of the water? **b.** Write an expression for the velocity of Sheba, in component form, the instant before she hits the water. **c.** Determine the peak height above the water reached by Sheba during her jump.

Brian Cahn/Zuma Press/Corbis

FIGURE P4.33

4-6 Special Case of Uniform Circular Motion

34. **C** A graduate student discovers that the only centrifuge in the laboratory is malfunctioning. The angular speed ω was supposed to be adjustable, but now it is stuck on a high angular speed, and the student is running an experiment that requires a low translational speed. Fortunately, the arm is adjustable. What can the student do to the length of the arm to get the correct translational speed?

35. **N** The bola is a traditional weapon used for tripping up or grounding an animal (Fig. P4.35). Once it is set into motion, each ball at the end of the bola can be thought of as a single object in uniform circular motion. Suppose it takes the bola 0.3250 s to traverse a circular path with a radius of 0.8661 m. What is the magnitude of the centripetal acceleration experienced by either ball at the end of the bola?

FIGURE P4.35 A bola is spun in a circle above the hunter, eventually being released and thrown forward.

36. **C** In three different driving tests, a car moves with constant speed v_0. In case 1, the car passes over a mark painted on a horizontal, straight section of road. In case 2, the car passes over a mark painted at the crest of a small hill. In case 3, the car passes over a mark painted at the bottom of a small dip. The hill and the dip are circular in profile, with the same radius (Fig. P4.36). Rank the cases from greatest to least according to the magnitude of the acceleration of the car when it passes the mark. Explain.

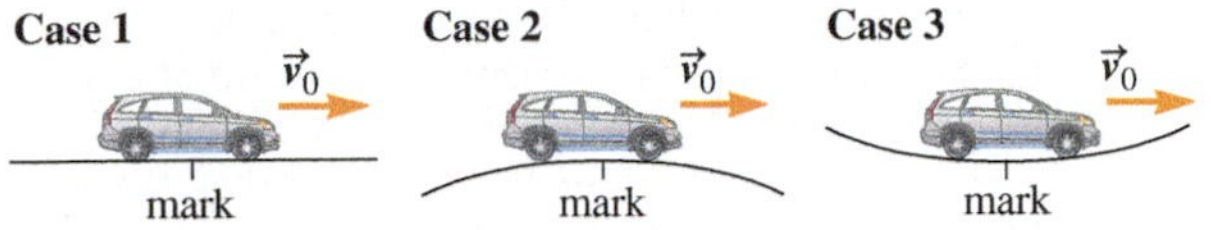

FIGURE P4.36

37. A child swings a tennis ball attached to a 0.750-m string in a horizontal circle above his head at a rate of 5.00 rev/s.

a. **N** What is the centripetal acceleration of the tennis ball?

b. **C** The child now increases the length of the string to 1.00 m but has to decrease the rate of rotation to 4.00 rev/s. Is the speed of the ball greater now or when the string was shorter?

c. **N** What is the centripetal acceleration of the tennis ball when the string is 1.00 m in length?

Problems 38, 39, and 40 are grouped.

38. **A** Two particles A and B move at a constant speed in circular paths at the same angular speed ω. Particle A's circle has a radius that is twice the length of particle B's circle. What is the ratio T_A/T_B of their periods?

39. **A** For particles A and B in Problem 38, what is the ratio v_A/v_B of their translational speeds?

40. **A** For particles A and B in Problem 38, what is the ratio a_A/a_B of (the magnitude of) their centripetal accelerations?

41. **N** Approaching one of the many sharp horizontal turns in the Monaco Grand Prix, an experienced Formula-1 driver slows down from 135 km/h to 55.0 km/h while rounding the bend in 10.0 s. If the driver continues to decelerate at this same rate and the radius of the curve is 15.0 m, what is the acceleration of the car the moment that its speed reaches 55.0 km/h?

42. A pendulum constructed with a bowling ball at the end of a cable 4.00 m in length has an acceleration vector $\vec{a} = (-8.50\hat{\imath} + 24.3\hat{\jmath})\,\text{m/s}^2$ when it is 24.7° past the lowest point in its swing.

a. **G** Sketch the scalar components of the acceleration vector in a vector diagram at this point in the pendulum's motion.

b. **N** What is the magnitude of the radial acceleration of the pendulum at this point?

c. **N** What are the speed and the velocity of the bowling ball at this point?

Problems 43 and 44 are paired.

43. **N** The Moon's orbit around the Earth is nearly circular and has a period of approximately 28 days. Assume the Moon is moving in uniform circular motion. **a.** Find the angular speed of the Moon. **b.** What is its centripetal acceleration?

44. The Earth's orbit around the Sun is nearly circular. Assume the Earth is moving in uniform circular motion.

a. **N** Find the angular speed of the Earth.

b. **N** What is its centripetal acceleration?

c. **C** Compare your answers to parts (a) and (b) with those in Problem 43 for the Moon. Do your answers make sense? Explain.

4-7 Relative Motion in One Dimension

45. **N** Pete and Sue, two reckless teenage drivers, are racing eastward along a straight stretch of highway. Pete is traveling at 98.0 km/h, and Sue is chasing him at 125 km/h. **a.** What is Pete's velocity with respect to Sue? **b.** What is Sue's velocity with respect to Pete? **c.** If Sue is initially 325 m behind Pete, how long will it take her to catch up to him?

46. A state trooper parked near the side of the road sees a car pass a truck. According to the trooper, the car's velocity is 75 mph and the truck's velocity is 58 mph.

a. **G** Draw a motion diagram for these two vehicles as seen by the trooper.

b. **G** Draw a motion diagram of the truck as seen by the car's driver.

c. **N** What is the velocity of the truck according to the car's driver?

d. **C** Are your answers to parts (b) and (c) consistent? Explain.

47. **C** A person might use the relative motion of a train, car, or similar vehicle to perform the apparently superhuman act of throwing a 200-mph fastball. **a.** Explain at least one way to do this trick. **b.** An observer in what reference frame would see this act as superhuman? Could a baseball pitcher ever observe his or her own fastball moving at 200 mph?

48. **C** A brother and sister, Alan and Beth, have just adopted a new dog, Sparky. Alan and Beth walk directly toward each other, each moving with constant speed along a straight-line path. Sparky leaves Alan and runs to Beth. Is the magnitude of Sparky's displacement with respect to Alan greater than, less than, or equal to the magnitude of Sparky's displacement with respect to Beth? Explain your reasoning.

49. **N** A man paddles a canoe in a long, straight section of a river. The canoe moves downstream with constant speed 3 m/s relative to the water. The river has a steady current of 1 m/s relative to the bank. The man's hat falls into the river. Five minutes later, he notices that his hat is missing and immediately turns the canoe around, paddling upriver with the same constant speed of 3 m/s relative to the water. How long does it take the man to row back upriver to reclaim his hat?

50. A trooper drives her car with a constant speed of 20.0 m/s to the east along a straight road. A speeder drives his car to the east on the same road while slowing down at a constant rate. The speeder moves at 16.0 m/s at time $t = 0$, and 8.00 m/s at time $t = 4$ s. For the entire interval from $t = 0$ to $t = 4$ s, the speeder is ahead of the trooper (that is, the speeder is located to the east of the trooper).

a. **N** In the frame of the trooper, what is the speeder's velocity at $t = 2$ s?

b. **C** In the reference frame of the trooper, is the speeder's car speeding up, slowing down, or moving with a constant speed at $t = 2$ s? Explain.

c. **C** The trooper has a laser ranging device that can determine the distance between the two cars. She finds that the distance is d at time $t = 2$ s and $0.9d$ at time $t = 3$ s. At time $t = 4$ s, will the distance between the cars be greater than, less than, or equal to $0.8d$? Explain.

4-8 Relative Motion in Two Dimensions

51. **N** CASE STUDY In the case study, we saw that a skateboarder should jump straight up at an angle $\theta = 90°$ to do a high (hippy) jump over an obstacle and land back on the board. We modeled this trick on a level track. If instead a skateboarder must perform a high jump down a 20.0° incline, what should be the launch angle relative to the board?

52. An ant and a spider each move with constant velocity on a horizontal table. The velocity vectors and positions of the ant and the spider (with respect to the table) at time $t = 0$ s are shown in Figure P4.52.

a. **G** Draw the velocity vector of the ant in the frame of the spider.

b. **C** Is there a time at which the ant and the spider will have the same position? Explain.

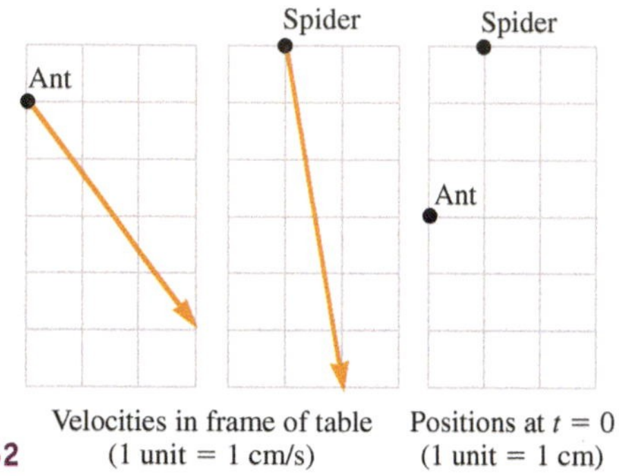

FIGURE P4.52

Problems 53 and 54 are paired.

53. **N** Suppose at one point along the Nile River a ferryboat must travel straight across a 10.3-mile stretch from west to east. At this location, the river flows from south to north with a speed of 2.41 m/s. The ferryboat has a motor that can move the boat forward at a constant speed of 20.0 mph in still water. In what direction should the ferry captain direct the boat so as to travel directly across the river?

54. **N** Return to the ferryboat scenario in Problem 53. If the captain points the boat directly at the target location on the east bank of the river, how far downstream will she be from the target when she lands on the east bank?

55. **N** A jetliner travels with a constant speed of 710.0 km/h relative to the air to a city 880.0 km due south. **a.** What is the time interval required to complete the trip if the jetliner is experiencing a headwind of 55.0 km/h toward the north? **b.** What is the time interval required to complete the trip if the jetliner now experiences a tailwind of the same speed? **c.** What is the time interval required to complete the trip if the jetliner now experiences a crosswind of the same speed relative to the ground towards the west?

56. **C** Avi and Cameron are back in physics class and are working on the following problem:

Car A moves to the east along a straight road as shown in the motion diagram in Figure P4.56. A traffic cone is at rest on the road at the location shown. Car B, not shown on the diagram, is located due south of the traffic cone and is moving to the south, directly away from the cone. Their task is to describe the approximate direction ("south," "southeast," and so forth) of the velocity of car B relative to car A at the following instants: instant 1, when car A is west of the cone; instant 2, when car A is next to the cone; and instant 3, when car A is east of the cone. Consider the discussion that Avi and Cameron have about this problem.

Avi: The velocity of car B relative to car A is to the southeast at instant 1, due south at instant 2, and southwest at instant 3. If car A is picked as an origin, the velocity would follow the line from car A to car B.

Cameron: I agree with your answer for the instant at which car A is next to the cone. At this time, the velocity of car B is due south relative to car A because no east–west movement is seen. The velocity, though, should be southwest at both of the other times. The west–east separation between the cars is decreasing at 1 and increasing the other way at 3.

State whether you agree or disagree with each statement. If you disagree, describe what specifically is incorrect and how it could be corrected.

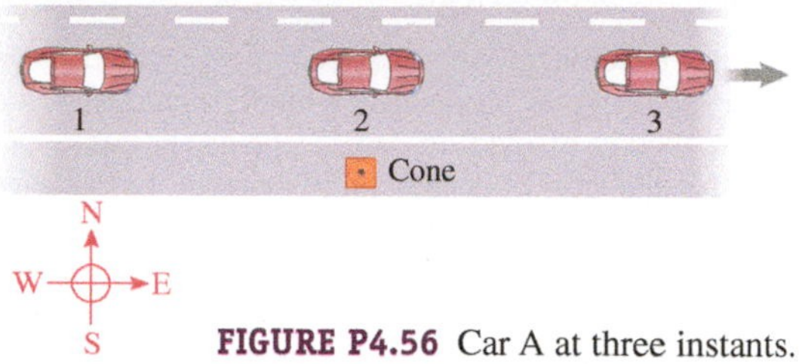

FIGURE P4.56 Car A at three instants.

57. **N** A pair of sunglasses balanced on a car's rearview mirror suddenly drops down as the car accelerates eastward at 2.33 m/s^2. **a.** What is the magnitude and direction of the sunglasses' acceleration as seen by an observer inside the car? **b.** What is the acceleration of the sunglasses as seen by an observer standing outside the car?

58. **N** Two bicyclists in a sprint race begin from rest and accelerate away from the origin of an x–y coordinate system. Miguel's acceleration is given by $(-0.700\hat{\imath} + 1.00\hat{\jmath})$ m/s^2, and Lance's acceleration is given by $(1.20\hat{\imath} + 0.300\hat{\jmath})$ m/s^2. **a.** What is Miguel's acceleration with respect to Lance? **b.** What is Miguel's speed with respect to Lance after 4.50 s have elapsed? **c.** What is the distance separating Miguel and Lance after 4.50 s have elapsed?

General Problems

59. **C** A particle has a nonzero acceleration and a nonzero constant speed at all times. What is the angle between the particle's velocity and acceleration?

60. **N** A golfer hits his approach shot at an angle of 50.0°, giving the ball an initial speed of 38.2 m/s (Fig. P4.60). The ball lands on the elevated green, 5.50 m above the initial position near the hole, and stops immediately. **a.** How much time passed while the ball was in the air? **b.** How far did the ball travel horizontally before landing? **c.** What was the peak height reached by the ball?

FIGURE P4.60

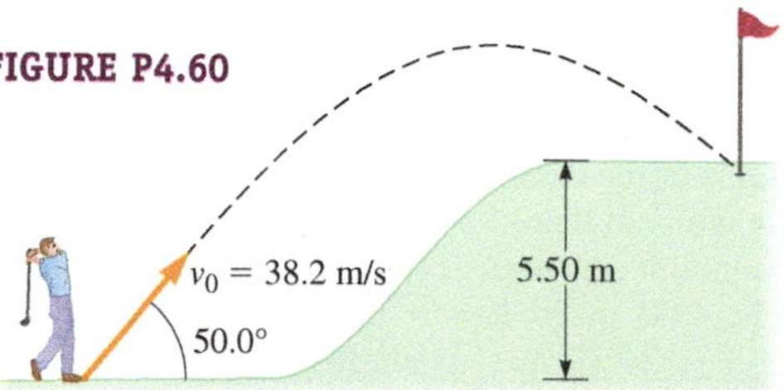

61. **A** You are watching a friend practice archery when he misses the target completely, and the arrow sticks into the ground. Discouraged, your friend asks whether you can help by estimating the speed with which the arrow left the bow. Remembering your physics class, you realize that you can do so by measuring its height above the ground when it was launched, if you assume the arrow was launched horizontally and that the ground is level. For your analysis, you let h represent the arrow's height above the ground when it was launched, v_0 represent the launch speed, and θ represent the angle the arrow makes with the horizontal when it is stuck into the ground. Find an expression for v_0 in terms of h and θ.

Problems 62 and 63 are paired.

62. C A ball starts from rest at the left end of the track shown in Figure P4.62. The ball flies off the right end of the track and follows the parabolic trajectory shown. At each instant 1 and 2, is the direction of the acceleration of the ball upward, downward, to the right, to the left, or is the acceleration zero? Explain.

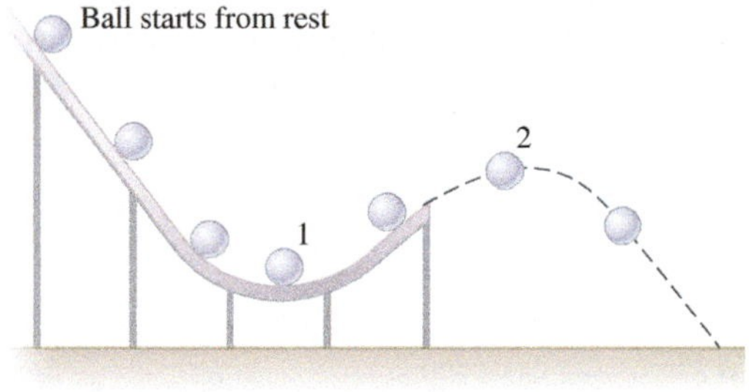

FIGURE P4.62 Problems 62 and 63.

63. N A ball starts from rest at the left end of the track shown in Figure P4.62. The ball flies off the right end of the track and follows the parabolic trajectory shown. Suppose the right end of the track is 1.0 m above the floor and makes an angle of 30.0° with the horizontal. Also suppose the ball leaves the track with a speed of 4.0 m/s. Determine where the ball hits the floor.

64. N David Beckham has lined up for one of his famous free kicks from a point 25.0 m from the goal. He kicks the soccer ball with a speed of 22.0 m/s at 28.0° to the horizontal. The height of the goal's crossbar is 2.44 m. **a.** What is the distance from the crossbar with which the ball will go into the goal or sail over? **b.** Does the soccer ball reach the goal on its way up or on its way down?

Problems 65, 66, and 67 are grouped.

65. G Suppose a particle's position is given by

$$x = At \quad \text{and} \quad y = Bt - Ct^2$$

where A, B, and C are constants. What is the shape of the particle's path?

66. For the particle in Problem 65, $A = 2$ m/s, $B = 4$ m/s, and $C = 4$ m/s^2.

a. G Sketch the particle's path.

b. N Complete the accompanying table.

c. C On your sketch of the particle's path, draw vector components for the velocity and acceleration for the times in the accompanying table. Indicate any components that equal zero.

t (s)	x (m)	y (m)	v_x (m/s)	v_y (m/s)	a_x (m/s^2)	a_y (m/s^2)
0	0			4		
0.5		1			0	
1.0			2			−8

67. A Return to the particle in Problem 65 where A, B, and C are constants. Find expressions for the particle's velocity and acceleration.

68. N Frequently, a weapon must be fired at a target that is closer than the weapon's maximum range. To hit such a target, a weapon has two possible launch angles (Fig. P4.68A): one higher than 45° (θ_H) and one lower than 45° (θ_L). Although the displacement of the projectile is the same for the two angles, a projectile launched at θ_H has a longer flight time and a higher peak position than one launched at θ_L. Usually, some tactical situation makes one angle preferable to the other. For example, if the projectile must go over some nearby object such as a grove of trees, the higher angle may be desirable. A shorter flight time and therefore θ_L are preferable if the target is mobile.

In practice, many weapons are designed to operate either at angles lower than 45° or at angles higher than 45°, but not both. Tanks, for example, often must face mobile targets; to minimize the time the target has to move, tanks fire at low angles. Grenades, on the other hand, are launched at high angles because a soldier launching a grenade is often close to the target, but has no armor plating for protection. The high launch angle allows the soldier to stay out of sight by hiding behind some obstacle, and the longer flight time may make it possible for the soldier to move farther from the exploding grenade.

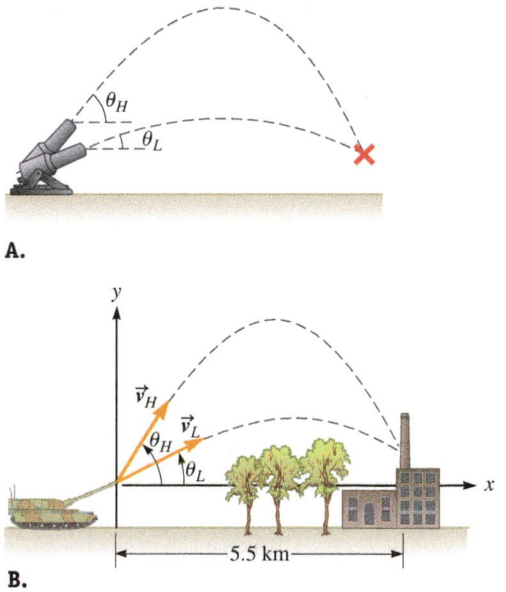

FIGURE P4.68

Imagine an unusual scenario in which a large gun mounted on a vehicle is required to hit an explosives factory (Fig. P4.68B). A huge explosion is expected, and there must be time for the gunner to retreat. A grove of trees provides cover. The maximum range of the gun is 17.6 km, and the maximum speed of the vehicle is 80.0 km/h. **a.** What is the muzzle speed v_0? (Muzzle speed is the speed at which the projectile leaves the barrel of the gun.) **b.** The target is 5.5 km away. Find the low angle θ_L and the high angle θ_H at which the gunner may aim so as to hit the target. **c.** Find the time the projectile takes to hit the target for both angles. **d.** Assume the vehicle retreats at its maximum speed (80.0 km/h) to be as far from the ensuing explosion as possible. How far is it from the factory at the time of the explosion for each launch angle?

Problems 69 and 70 are paired.

69. N A projectile is launched up and to the right over flat, level ground. Its range is 177 m, and its maximum elevation above the ground is 354 m. What was the angle between its initial velocity and the ground? Ignore air resistance.

70. A A projectile is launched up and to the right over flat, level ground. Its range is equal to half of its maximum elevation above the ground. What was the angle between its initial velocity and the ground? Ignore air resistance.

71. N A World War II–era dive bomber is being used to drop food and supplies to outposts on remote mountaintops. During one such drop, the bomber descends with velocity v directed at an angle of 37.0° below the horizontal. The cargo is released at an altitude of 2450 m and reaches its intended drop zone with a displacement Δr of 3850 m. What is the speed of the bomber when it releases its cargo?

72. **C** An observer sitting on a park bench watches a person walking behind a runner. Figure P4.72A is the motion diagram representing what this observer sees. To better reveal the changing distance between runner and walker, five observations (A through E) are shown on five separate lines in Figure P4.72B. To the observer on the bench, both the runner and the walker move to the right, and the gap between them widens. Draw the motion diagram of the runner from the reference frame of the walker.

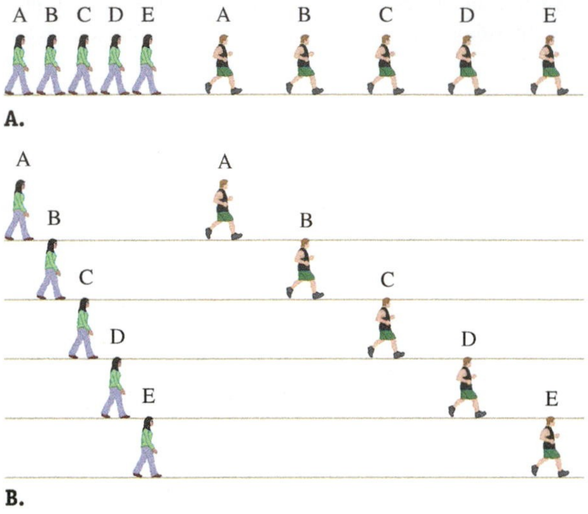

FIGURE P4.72

73. **N** In a dramatic physics demonstration, an apple is suspended 3.45 m above the floor. A pellet gun on the floor is 4.62 m from the point just below the apple. The gun is aimed at the apple; the gun's muzzle speed is 38.3 m/s. At the moment the pellet is shot from the gun, the apple is released. At what height above the floor does the pellet hit the apple?

Problems 74 and 75 are paired.

74. **G** Figure P4.74 is a motion diagram of Mars from positions A to T. The time interval between "dots" is 10 days. Describe the motion of Mars, including changes in Mars's velocity. Also note the fastest and slowest points in Mars's motion. (White dots represent the positions of fixed stars.)

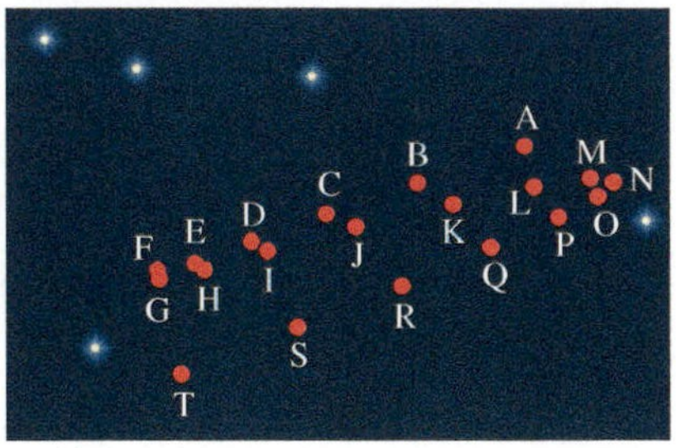

FIGURE P4.74 Problems 74 and 75.

75. **C** Figure P4.74 is a motion diagram of Mars from positions A to T. The time interval between "dots" is 10 days. The diagram is made with respect to an observer on the Earth. Although Mars appears to stop near points F and N, it does not actually stop in its orbit around the Sun. Why does it appear to stop in this motion diagram?

76. A riverboat with a speed in still water of 20 knots (10.3 m/s) travels on a river that has a constant speed of 0.650 m/s.

a. **N** What is the time interval required for the riverboat to travel a distance of 4.00 km upstream and return to its starting point?

b. **N** What would be the time interval required for the same trip in still water?

c. **C** Why does the boat trip take longer when there is a river current?

77. **A** An object located at the origin at $t = 0$ moves with an initial velocity of $(2.00\hat{\imath} + 7.00\hat{\jmath})$ m/s and a nonconstant acceleration of $\vec{a} = (\sqrt{3}t\hat{\imath} - t\hat{\jmath})$ m/s^2. **a.** What is the velocity of the particle as a function of time? **b.** What is the position of the particle as a function of time?

Hint: Integration is required.

78. Stretching his arm to a height of 1.05 m above the ground, Arman mischievously fires his Nerf dart gun at Irene. The dart leaves the gun with initial speed v_0 at an angle of 42.0° above the horizontal. Irene eludes the dart, and it harmlessly strikes the ground.

a. **A** What is the horizontal displacement x of the dart as a function of the initial speed v_i, written as $x(v_i)$, just as the dart strikes the floor?

b. **N** What is x if $v_0 = 0.300$ m/s?

c. **N** What is x if $v_0 = 30.0$ m/s?

d. **A** For small but nonzero values of v_0, show that $x(v_0)$ simplifies because one of its terms dominates.

e. **A** What is the form of $x(v_i)$ for large values of v_0?

79. **N** A circus cat has been trained to leap off a 12-m-high platform and land on a pillow. The cat leaps off at $v_0 = 3.5$ m/s and an angle $\theta = 25°$ (Fig P4.79). **a.** Where should the trainer place the pillow so that the cat lands safely? **b.** What is the cat's velocity as she lands in the pillow?

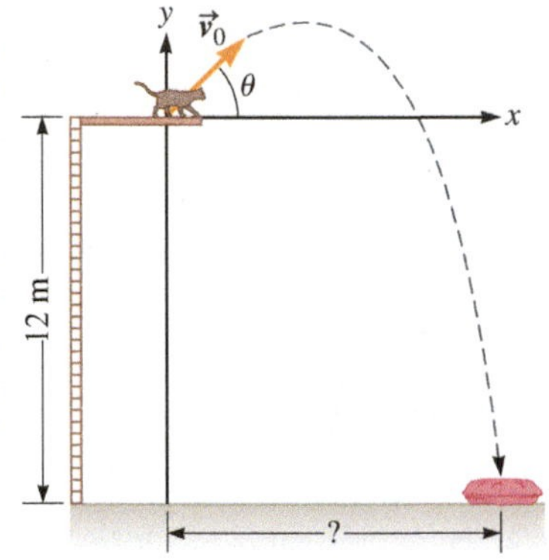

FIGURE P4.79

Problems 80 and 81 are paired.

80. Cosmic rays are high-speed charged particles from space. They are harmful to people, other organisms, and equipment, but fortunately the Earth's upper atmosphere and magnetic field act to shield us from these particles. The trajectory of charged particles trapped by the Earth's atmosphere is complicated. In this problem, we will explore a simpler scenario in which a single charged particle is trapped by a magnetic field and travels in a helical path. The position of this particle is given by

$$\vec{r}(t) = A\cos\omega t\hat{\imath} + A\sin\omega t\hat{\jmath} + Bt\hat{k}$$

where A, B, and ω are constants.

a. **G** Substitute values for t to confirm graphically that the path of the particle is a helix.

b. **A** Find $\vec{v}(t)$ and $\vec{a}(t)$.

c. **C** Describe the z component of $\vec{v}(t)$ and $\vec{a}(t)$.

d. **C** Do $\vec{r}(t)$, $\vec{v}(t)$, and $\vec{a}(t)$ in this problem apply only to charged particles in a magnetic field, or do they apply to any particle moving in a helical path? Explain.

81. An experimentalist in a laboratory finds that a particle has a helical path. The position of this particle in the laboratory frame is given by

$$\vec{r}(t) = R\cos\omega t\hat{\imath} + R\sin\omega t\hat{\jmath} + v_z t\hat{k}$$

where R, v_z, and ω are constants. A moving frame has velocity $(\vec{v}_M)_L = v_z\hat{k}$ relative to the laboratory frame.

a. **C** What is the path of the particle in the moving frame?

b. **A** What is the velocity of the particle as a function of time relative to the moving frame?

c. **A** What is the acceleration of the particle in each frame?

d. **C** How should the acceleration in each frame be related? Does your answer to part (c) make sense? Explain.

82. **A** Derive the range equation $R = \frac{v_0^2}{g}\sin 2\theta$ (Eq. 4.28). *Hint*: What is the vertical displacement when the horizontal displacement equals the range?

Newton's Laws of Motion

5

Underlying Principles

Newton's three laws of motion

Major Concepts

1. Dynamics
2. Force
3. System
4. Inertia
5. Inertial reference frames

Special Cases: Specific Forces

1. Gravity
2. Hooke's law (spring force)
3. Normal force
4. Tension force
5. Kinetic friction

Tools

Free-body diagrams

Key Questions

How is constant velocity maintained?

What causes acceleration?

How do two objects interact?

When the greatest minds in physics today grab a cup of coffee together, they might argue about an 11-dimensional model of our Universe, about the requirements of sending a crewed mission to Mars, or about the way signals are transmitted in the human brain. Scientific argument was also taking place over three centuries ago—although on vastly different topics, of course—in letters exchanged between such great scientists as Isaac Newton, Edmond Halley, and Robert Hooke. The crux of those scientific arguments was likely to be motion and its causes, such as what kept the Moon in its orbit or why apples and coconuts falling from trees have the same acceleration. The research resulting from these 17th-century arguments led to such practical applications as the manufacture of accurate clocks and watches.

17th-century science may seem mundane because clocks just do not seem as exciting as, say, nanomachines. You might have trouble believing that motion was a hot topic then. In fact, the debate among some of those early scientists was bitter and hard fought. Imagine losing your temper over the trajectory of an apple!

The 17th-century study of motion laid the foundation for science today not just because of actual discoveries but also because of the scientific process that developed as these discoveries were made. Without that earlier work, we could not even *begin* to imagine an 11-dimensional Universe, a mission to Mars, or detailed studies of the human brain.

5-1 Our Experience with Dynamics

In Chapters 2 and 4 (kinematics), we learned how to define and describe motion both in words and mathematically. Relative to an observer, objects may be at rest, may move with constant velocity, may accelerate by speeding up or slowing down, or may accelerate without changing speed at all. We did not discuss what *causes* the motion (or lack thereof), however. Why are some objects at rest? Why do others move at constant velocity and still others accelerate? In this chapter, we investigate the answers to these questions as we look at the branch of mechanics known as **dynamics** (the cause of motion).

DYNAMICS ✪ Major Concept

Principia is pronounced prin-kip-pee-ah.

For centuries, dynamics was on the cutting edge of scientific thought in much the same way that certain questions in cosmology and human genetics are today. In 1687, English scientist Isaac Newton (1642–1727) published the *Principia,*[1] a report on his scientific studies in which he explained the causes of motion. Today, his explanations, having been debated and tested innumerable times over the intervening centuries, are largely accepted and are considered laws of nature.[2]

Newton's work in dynamics is summarized in his three laws of motion. Before these three laws were formulated by him and then accepted by the scientific community, many great thinkers reasoned that a cause or a force was required for an object to maintain its motion, even if that motion was constant velocity. This belief seems reasonable, as you can easily see by considering just one simple example: If you push a book across your desktop, the book will come to rest soon after you stop pushing. If you want it to keep moving, you need to keep pushing.

Such common experiences in our everyday lives have helped shape our intuition concerning motion. We observe that to keep something moving, it must be pushed or pulled in some way. Newton's first law of motion, however, says that this belief is not true. It may seem frustrating to find that the laws of physics run counter to our intuition, but we can become more comfortable with physics by seeing how it fits into our experiences. Therefore, an important step in mastering these laws is becoming aware of our intuition and our current conceptions about motion and about forces. We begin this chapter with a case study to help us gain this awareness.

FIGURE 5.1 Aerial view of emergency workers helping the injured from a train crash near Los Angeles (April 23, 2002).

CASE STUDY Train Collision

On April 23, 2002, a passenger train about 35 miles outside of Los Angeles was hit by a freight train (Fig. 5.1). The accident killed two people and injured more than 260, with all the injured being on the passenger train. Witnesses reported that those people who were seated facing backward suffered little or no injury. News reports said that the passenger train came to a quick stop before the collision and that the impact with the freight train pushed the passenger train 370 ft backward (Fig. 5.2).

One of the most controversial parts of the early reports was how fast the freight train was going at the moment of impact. In Chapter 11, we will reconstruct the ac-

[1]The full title is *Philosophiae Naturalis Principia Mathematica,* Latin for *Mathematical Principles of Natural Philosophy.*

[2]In later chapters, we will discuss situations in which the laws of Newtonian mechanics are known to break down.

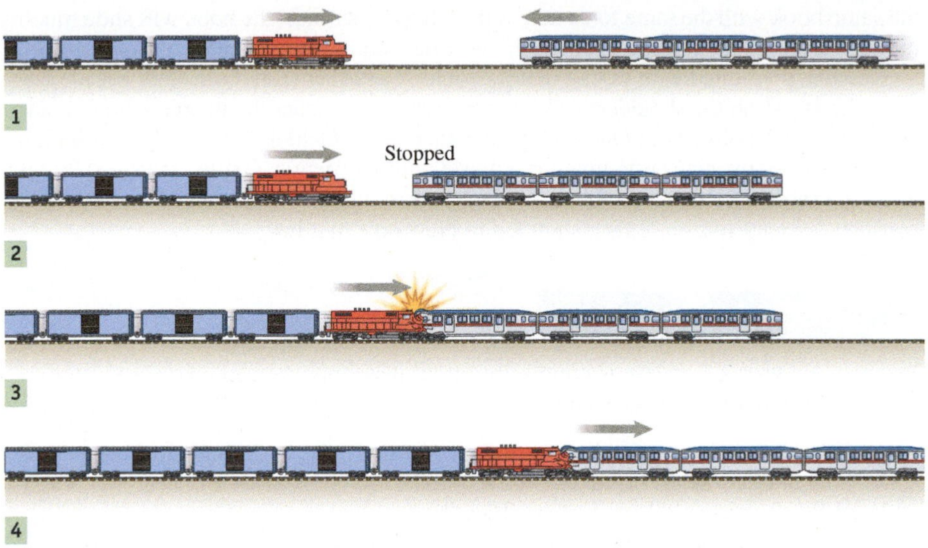

FIGURE 5.2 1 A freight train and a passenger train move toward each other. 2 The passenger train stops. 3 The freight train continues and collides with the passenger train. 4 The passenger train is shoved backward.

cident and estimate the speed of the freight train upon impact. In this chapter, we are concerned only with the following questions:

1. Why did passengers seated facing backward fare better than those who were either standing or seated facing forward?
2. The passenger train was at rest before the collision and was pushed backward. What do those facts tell us about how hard the freight train pushed on the passenger train? Did the passenger train push on the freight train? If so, how hard did it push?

5-2 Newton's First Law

In the years since Newton wrote his laws of motion in Latin, his words have been translated and interpreted many times. Although physicists generally agree on the principles of dynamics, they often do not agree on the best way to state Newton's laws. For this reason, we will state it three times in this chapter and discuss the subtle differences in each statement.

A close translation of the Latin in which **Newton's first law** was originally written is

1ST STATEMENT OF NEWTON'S FIRST LAW ❶ **Underlying Principle**

Every body continues in its state of rest, or of uniform motion in a straight line, unless it is compelled to change that state by forces impressed on it.

Newton's first law says that there is no essential difference between rest and moving in a straight line at constant speed. Both states will continue unchanged unless a force acts on the object. This law is in sharp contrast to the idea that a cause or a force is required to maintain constant velocity.

A book set on a level tabletop will not start sliding until you apply a force to it with your hand. Why does the moving book stop, though? Once it is moving and you have removed your hand, the book has a certain velocity. According to Newton's first law, the book should continue in a straight line at the same speed. Because it slows down and stops, a force must have acted on it and caused this change in motion. The force of friction between the book and the table causes the book to decelerate and eventually come to rest. Because of friction, we might conclude (mistakenly) that rest is the natural state of an object because moving objects eventually stop.

We will study friction in detail in Section 5-7 and in Chapter 6. For now, we will imagine ways to reduce its effects. For example, instead of sliding a book across a clean, dry tabletop, suppose you slide it across a tabletop coated in a slippery wax. If you push the same book with the same force across this slippery surface, the book will slide much farther before coming to rest because the wax has reduced the force of friction.

Newton is credited with saying that he could see farther than most other people because he stood on the shoulders of giants, by which he meant the people who did scientific research before him. One such giant was Galileo Galilei (1564–1642), an Italian scientist whose insight was to imagine a world without friction. Galileo reasoned that if we could reduce friction to nothing, an object set in motion would continue to move at constant velocity forever. This insight led to Newton's first law.

CONCEPT EXERCISE 5.1

Because Newton's first law is counterintuitive, it is important to take some time to think about what the law says and about how and why it differs from our intuition.

a. Why did the unavoidable presence of friction make it difficult for earlier scientists to come to the conclusion expressed in Newton's first law?
b. What is the natural state of an object?
c. How much force does it take to keep an object moving at constant velocity?

CONCEPT EXERCISE 5.2

CASE STUDY Train Collision and Newton's First Law

A group of college students discusses the train collision case study. Use Newton's first law to decide which underlined statements are correct and which are false. Explain your answers.

Shannon: This newspaper says that the people who got really hurt were either standing up or sitting in a forward-facing seat. Those people got thrown forward when the train stopped.

Avi: That's why there are seat belts in cars. If you get into a crash, the force can throw you through the windshield.

Cameron: There is no force that throws you through the windshield. You fly through the windshield because you are already moving and it would take a force to stop you from going forward. *That's* why there's a seat belt.

Avi: That doesn't make sense. Because then you would need a force to stop you from flying through the windshield even when you just stop slowly at a red light.

Cameron: That's right, but when you slow down slowly, you don't need such a big force and the car seat can take care of it.

Shannon: The seat? I don't think a seat can exert a force. It can't move on its own or hold you. That's why the people who were sitting forward on the train were hurt. The people who were sitting backward had the back of the seat to block them.

5-3 Force

In our everyday language, force has many meanings and usages, such as *he forced me to do my homework, the force of evil,* or *an armed force*. As with most terminology, the term *force* is more specific in physics than it is in our everyday usage. We can derive an operational definition of force from Newton's first law.

When Newton wrote the *Principia*, the language and mathematics of physics were not as well defined as they are today. Newton's own statement of his first law (page 121) is somewhat vague for a modern reader. Another way we can grasp Newton's first law is by restating it using the term *accelerate* that we carefully defined in Chapter 2:

2ND STATEMENT OF NEWTON'S FIRST LAW **Underlying Principle**

If no force acts on an object, then the object cannot accelerate.

This statement means that an object's velocity cannot change in either magnitude or direction unless a force is applied. From Newton's first law, we can reason that a **force** is a push or pull that is required to make an object accelerate. Force is a vector quantity; it has both magnitude and direction. Force must therefore be manipulated using the vector algebra we discussed in Chapter 3.

FORCE ✪ Major Concept

Contact Versus Field Forces

In classical mechanics, it is convenient to divide forces into two types. **Contact forces** are forces that require the source to touch the subject. If you want to accelerate your textbook with your hand, you must touch the book and exert a contact force on it. **Field forces** are those that do not require contact between source and subject; instead, field forces can act through empty space. One of the first forces we learn about as infants is gravity. Gravity is a field force; an apple falls from a baby's hand to the ground because of the Earth's gravitational pull on the apple, but the Earth is not in contact with the apple. Another example of a field force is the electric force (Chapter 23). The source of the electric force is charged particles, such as protons and electrons. This force is responsible for such phenomena as lightning and for keeping our electronic equipment running. A force results from an *interaction* between two objects, such as the interaction between your hand and a textbook or the interaction between the Earth and an apple. We may arbitrarily think of one member of the pair involved in the interaction—that is, your hand or the Earth—as the **source** of the force. The other member—that is, the textbook or the apple—is then the **subject** on which the force acts.

Internal Versus External Forces

In addition to distinguishing between contact forces and field forces, we must also distinguish between internal and external forces. Any collection of two or more objects is known as a **system.** An **internal force** is any force that acts *inside a system*, that is, any force exerted by one object in a system on another object in the system. To determine which forces are internal to the system and which are external, we must first decide what we want to call the "system." For example, is the system in Figure 5.3 the whole car? Only some part of it? One or more occupants? Suppose we decide the whole car and its occupants are the system as indicated by the red loop. Then, any force exerted on the car (the system) by anything *outside* the system—such as the road, the Earth, or another car—is an **external force.** Any forces that are inside the system, such as the force exerted by the driver on the steering wheel or the force exerted by one child on the other child—are internal forces. In this chapter, we continue to use the particle model for objects and systems, so we represent the whole car and its occupants as a particle.

SYSTEM ✪ Major Concept

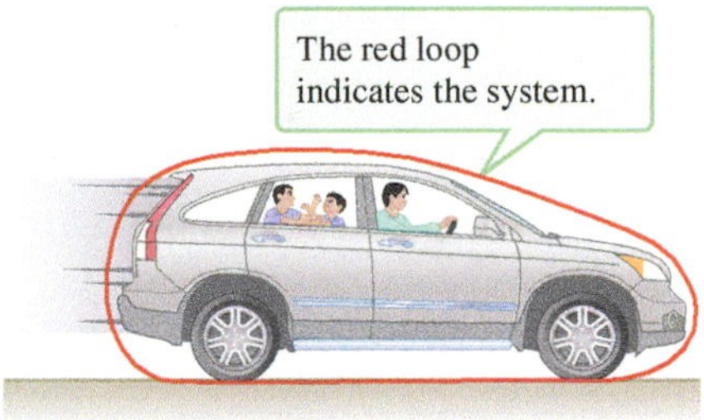

FIGURE 5.3 The people and the car have been chosen as the system (encircled in red). External forces due to the road or the Earth can accelerate the system. Internal forces such as the two children pushing on each other cannot.

CONCEPT EXERCISE 5.3

Shown in Figure 5.4 are four situations in which a force acts on a subject. The subject is labeled in each case.

Case 1. A baseball glove stops a vertically falling baseball.
Case 2. CASE STUDY A freight train collides with a stopped passenger train.
Case 3. A satellite orbits the Earth.

For each case, identify the source of the force and the direction of the force. Then state whether a contact force or a field force is involved.

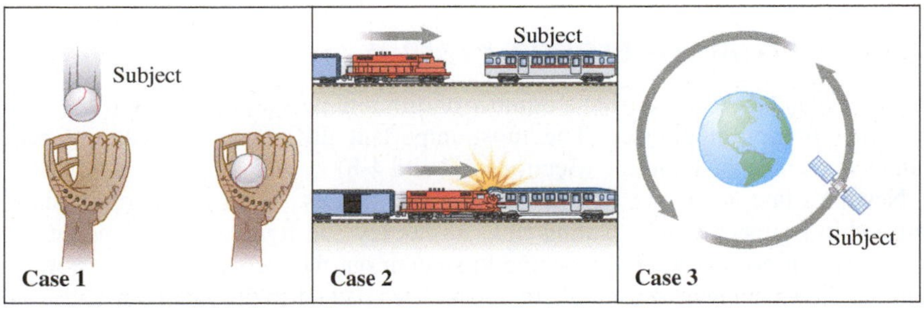

FIGURE 5.4

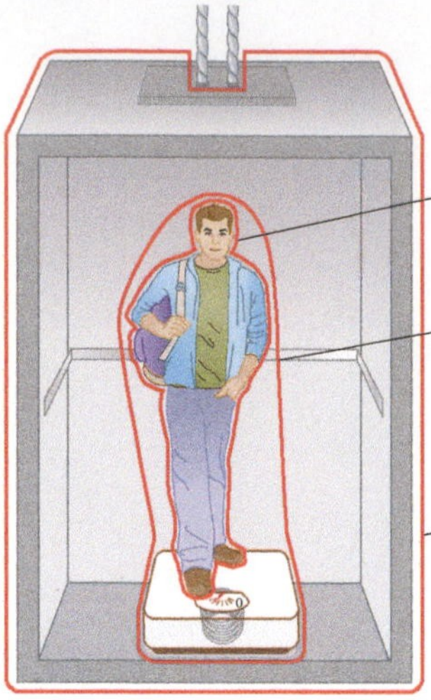

FIGURE 5.5

CONCEPT EXERCISE 5.4

A person stands on a spring scale in an elevator car as shown in Figure 5.5. Which of these sources—the Earth, spring scale, elevator car, and cable—exert an external force if the system consists of:

a. Only the person?
b. The person and the spring scale?
c. The person, the spring scale, and the elevator car?

5-4 Inertial Mass

Newton's first law represents a revolutionary change in human understanding. It implies that what is natural about motion is an object's tendency to maintain constant velocity. This idea is so important that a new word, *inertia*, is required to express the concept. **Inertia** is the tendency of an object to maintain its constant velocity (to resist changes in velocity). Consequently, Newton's first law is often referred to as the *law of inertia*.

INERTIA ✪ Major Concept

Not all objects have the same inertia. It is, for instance, difficult to stop a runaway truck but easy to flip a quarter into the air. It is difficult to change the truck's velocity because the truck has a lot of inertia. The quarter, on the other hand, has little inertia, and it is therefore easy to change its velocity with just your thumb. The truck and quarter are different from each other in many ways—shape, size, and composition, to name just a few—but their difference in mass is what counts when we try to accelerate them. **Mass,** also known as **inertial mass,** is an intrinsic scalar property of any object. Mass measures the object's inertia. In SI units, mass is measured in kilograms. In the U.S. customary system, the unit of mass is the slug, a term we almost never use in everyday conversation.

The more mass an object has, the more difficult it is to accelerate that object. If we apply the same force to objects having different masses, we find that the acceleration of each object is inversely proportional to its mass:

$$a \propto \frac{1}{m}$$

Mass is sometimes confused with weight. Weight is a measure of the gravitational force acting on an object (Section 5-7). The weight of an object can change from place to place, but mass does not. For example, an astronaut weighs less on the Moon than she does on the Earth, but her mass is the same no matter where she is.

CONCEPT EXERCISE 5.5

Often, words that are precisely defined in physics are used loosely in everyday language. A person who cannot seem to get things done is described as having a lot of *inertia*.

a. How does that metaphor fit with the physics definition of inertia?
b. How can that metaphor be misleading?
c. How is the term *massive* used in everyday language?

5-5 Inertial Reference Frames

An object can be observed from a number of different perspectives, which are called *reference frames* in physics. The most important difference between reference frames is their relative motion (Sections 4-7 and 4-8).

Newton's first law is not valid in all reference frames. As an example, consider a cruise ship equipped with a skating rink. As hockey and figure-skating fans know, a skating rink needs a Zamboni machine to smooth out the ice surface every now and then. As shown in Figure 5.6A, a Zamboni has been parked on the ice of our cruise ship

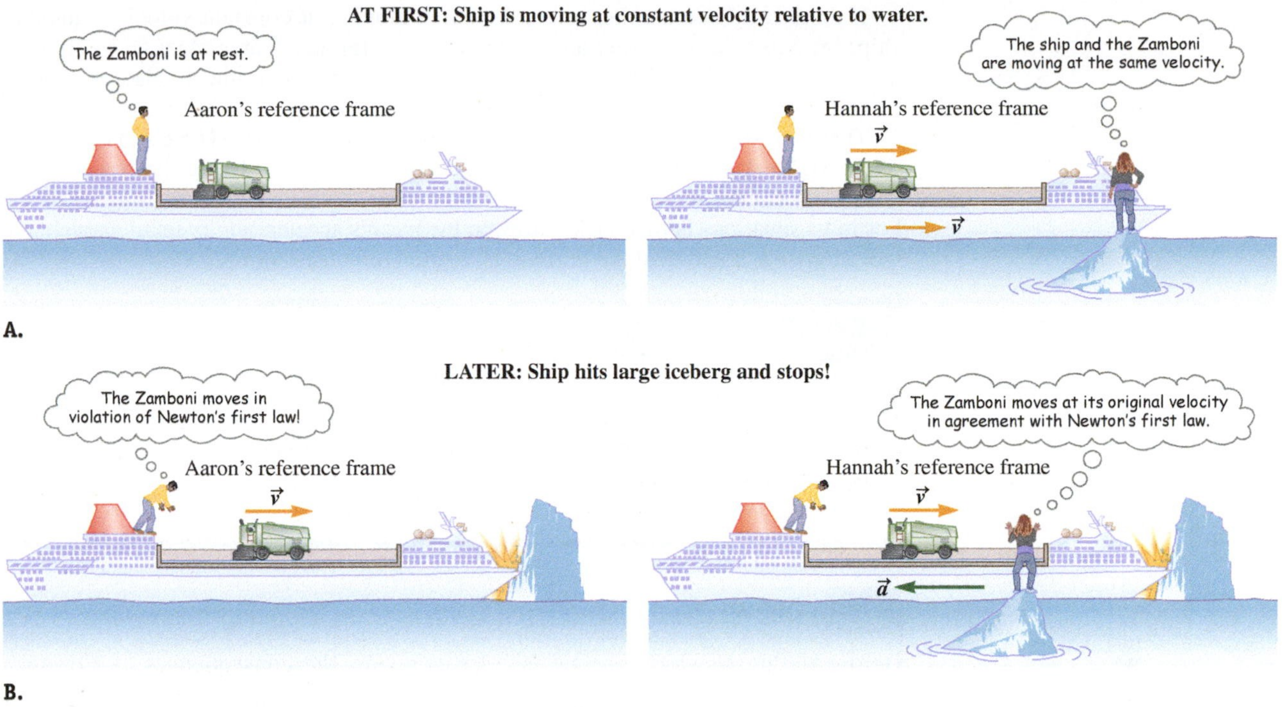

FIGURE 5.6 Two observers see a Zamboni machine on an ice rink aboard a cruise ship. **A.** At first, Aaron on the ship sees the Zamboni at rest. Hannah is on an iceberg, and she sees the Zamboni moving at the same velocity $\vec{v}$ as the ship. **B.** When the ship suddenly stops, Aaron sees the Zamboni move across the ice, and he cannot identify a force that caused the Zamboni's acceleration. Hannah on the iceberg notices that as the ship stops (accelerates to the left), the Zamboni continues moving rightward at its original velocity $\vec{v}$ until it reaches the wall (consistent with Newton's first law).

rink and is at rest relative to the ship. Assume friction between the Zamboni and ice is negligible. Aaron, on the ship looking at the rink, would report that the Zamboni has zero velocity. According to Hannah on a nearby iceberg, however, both the ship and the Zamboni are moving at constant velocity to the right relative to the water.

Now suppose the ship stops suddenly (Fig. 5.6B). Hannah notes that, as the ship's speed decreases to zero, the ship has an acceleration to the left relative to the water. According to Hannah, however, the Zamboni continues to move to the right at its original velocity relative to the water. Hannah can account for the Zamboni's motion in terms of Newton's first law. To her, while the ship is slowing down, the Zamboni continues moving at constant velocity relative to the water because no force acts to change its velocity (until it reaches the wall).

To Aaron, the Zamboni starts with zero velocity but then accelerates toward the rink wall. He cannot identify any force that could cause this acceleration. To him, the Zamboni moves in violation of Newton's first law. According to this law, the Zamboni should remain at rest unless acted upon by a force.

The observers' explanations differ due to their different reference frames. Aaron is in an *accelerating reference frame* (the frame is the ship, which is slowing down), but Hannah is in a stationary reference frame.

INERTIAL REFERENCE FRAMES

Major Concept

Reference frames in which Newton's first law is valid are called **inertial reference frames.** Inertial reference frames may move with constant velocity, but they do not accelerate. Accelerating frames are known as **noninertial reference frames,** and Newton's first law is not valid in them. In our example, the shipboard observer is in a noninertial reference frame once the ship starts slowing down.

Before the ship began decelerating, both observers were in inertial reference frames. Neither of them observed the Zamboni violate Newton's first law. The Zamboni was at rest from the point of view of the shipboard observer, and moving with a nonzero constant velocity according to the point of view of the observer on the iceberg.

As long as no external force acts on an object, it is always possible to find an inertial reference frame in which the velocity of the object is zero. In other words, there is nothing special or "natural" about a state of rest.

If there had been another reference frame moving at constant velocity—another ship, for example—it would also be an inertial reference frame. Any frame moving at constant velocity with respect to an inertial reference frame is also an inertial reference frame.

There is a subtle point about all reference frames on the Earth. The Earth is spinning on its axis and at the same time orbiting the Sun, which is orbiting the center of our galaxy. From Chapter 4, spinning and orbital motion mean that the Earth is accelerating. For most situations in this book, however, the Earth's acceleration is small enough that we can ignore it and treat the Earth as an inertial reference frame.

CONCEPT EXERCISE 5.6

Which of the following reference frames are inertial frames?

a. An airplane cruising in a straight path at constant speed
b. An airplane taking off
c. A car taking a sharp turn

5-6 Newton's Second Law

Before stating Newton's second law, let us review the consequences of Newton's first law:

1. Rest is not a particularly special state; it is merely a special case of constant velocity.
2. A force is required to change an object's velocity.
3. Inertia is the tendency of an object to maintain its constant velocity. An object's inertia is determined by its mass. The more mass (inertia) an object has, the more difficult it is to accelerate that object.

Newton's second law relates the acceleration of an object to the vector sum of all the forces exerted on that object. This vector sum of all those external forces is called either the **total force** or the **net force** on the object. Although Newton did not write his second law using mathematical symbols, today we combine the concepts he presented into one equation:

NEWTON'S SECOND LAW
Underlying Principle

$$\vec{F}_{\text{tot}} \equiv \sum \vec{F} = m\vec{a} \tag{5.1}$$

where $\sum \vec{F}$ is the total external force acting on an object of mass m and $\vec{a}$ is the acceleration of that object. The symbol $\sum$ (a summation sign, represented by an oversized uppercase Greek letter sigma) indicates that we must take the sum of all the forces acting on the object. According to Newton's second law, the total external force acting on a system can accelerate the system, but forces internal to a system cannot.

As discussed in Chapter 3, multiplying the vector $\vec{a}$ by the scalar m results in a vector that points in the same direction as $\vec{a}$. Applying this rule to Newton's second law (Eq. 5.1), shows that the total force on an object points in the same direction as the object's acceleration.

If only one force acts on an object, the object must accelerate in the direction of that force. If more than one force acts on an object, it is possible for the total force to be zero, in which case the acceleration is also zero, which leads to the third way to state Newton's first law:

3RD STATEMENT OF NEWTON'S FIRST LAW
Underlying Principle

The net force on an object is zero if and only if the acceleration of the object is also zero.

Like all other vector equations, Newton's second law can be written in terms of scalar equations with one equation for each scalar component. For a three-dimensional Cartesian coordinate system, Newton's second law is therefore written

$$\sum F_x = ma_x \qquad \sum F_y = ma_y \qquad \sum F_z = ma_z \tag{5.2}$$

Because $\sum \vec{F}$ is the vector sum of all the forces acting on the object, $\sum F_x$ is the sum of all x components of force acting on the object. Further, $\sum F_x$ can cause acceleration only in the x direction. Similar statements can be made for $\sum F_y$ and $\sum F_z$.

Newton's second law is used to define a unit of force. If a single force is applied to a standard mass of 1 kg such that the mass accelerates at 1 m/s^2, the applied force is defined to be 1 newton (1 N):

$$1\ \text{N} \equiv (1\ \text{kg})(1\ \text{m/s}^2) = 1\ \text{kg} \cdot \text{m/s}^2 \tag{5.3}$$

In U.S. customary units, force is measured in pounds:

$$1\ \text{lb} \equiv (1\ \text{slug})(1\ \text{ft/s}^2) = 4.45\ \text{N}$$

Newton's second law is a fundamental principle of classical mechanics. Although it might seem straightforward, its full meaning and subtlety are brought out when it is applied to problems; therefore, much of the rest of this chapter and the next are devoted to studying applications of this law.

CONCEPT EXERCISE 5.7

a. Take a moment to be sure that you understand the distinction between Newton's first two laws. How are they different from each other?

b. According to Newton's second law, what is the acceleration of an object if there are no forces acting on it? Is your answer consistent with Newton's first law?

EXAMPLE 5.1 Electrodes

In Example 2.8 (page 44), an electron initially at rest is released from a negative electrode positioned 7.5 mm away from a positive electrode (Fig. 5.7). Together the two electrodes produce an electric force that gives the electron an acceleration of 1.5×10^{17} m/s^2 toward the positive electrode. No other force is exerted on the electron (mass $m_e = 9.109 \times 10^{-31}$ kg).

A Find the electric force $\vec{F}_E$ acting on the electron.

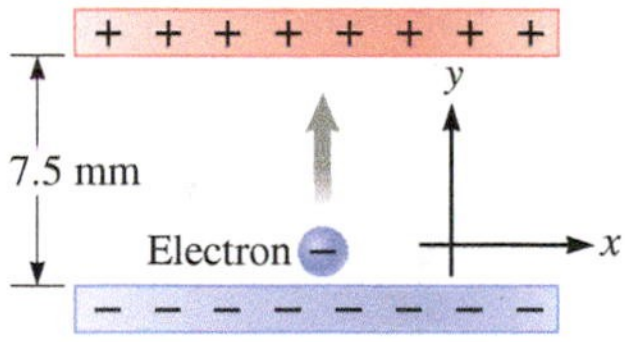

FIGURE 5.7

INTERPRET and ANTICIPATE

Treat the electron as just another particle with mass m_e, accelerating upward. We can use Newton's second law to find the force on a particle whose mass and acceleration we know. Force is a vector quantity, so our answer should have both a magnitude and a direction. The total force and the acceleration point in the same direction. Because only one force (the electric force) acts on the electron, that force must point upward because the acceleration is upward toward the positive electrode. Our answer should be in the vector form $\vec{F} =$ ____ $\hat{\jmath}$ N.

SOLVE

Using Newton's second law in component form, find the scalar component of the electric force. The mass of the particle is the mass of an electron. The only force exerted on the electron is the electric force, so $\sum F_y = F_E$.	$\sum F_y = ma_y$ $m = m_e$ $F_E = m_e a_y$
Substitute values for the mass of the electron and the given acceleration.	$F_E = (9.109 \times 10^{-31}\ \text{kg})(1.5 \times 10^{17}\ \text{m/s}^2)$
Using the definition $1\ \text{N} = 1\ \text{kg} \cdot \text{m/s}^2$, express this value for the electric force ($\vec{F}_E$) in newtons.	$\vec{F}_E = 1.4 \times 10^{-13}\ \hat{\jmath}\ \text{N}$

CHECK and THINK

The calculated force might seem small, but only a small force is required for the given acceleration because the electron's inertial mass is so small.

Example continues on page 128 ▶

B If the electron is replaced by a proton (mass $m_p = 1.673 \times 10^{-27}$ kg) initially at rest and released from the positive electrode, find the velocity of the proton as it strikes the negative electrode. Assume the electric force exerted on the proton has the same magnitude as the force exerted on the electron found in part A but is in the opposite direction (Fig. 5.8).

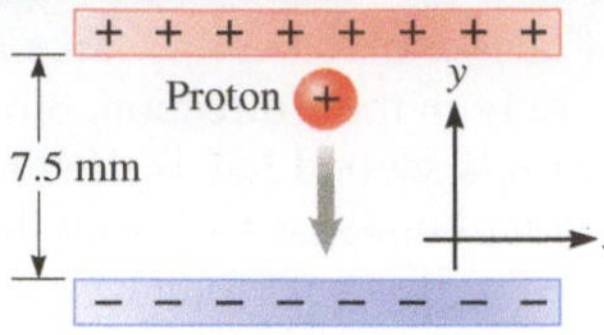

FIGURE 5.8

INTERPRET and ANTICIPATE

There are two differences from part A. First, the force on the particle and its velocity are now downward. Second, we are asked to find velocity instead of force. Our answer should be in the form $\vec{v} = -(\_\_\_)\hat{j}$ m/s. We will use Newton's second law and our answer to part A to find the acceleration, and from that we will use kinematics to find the velocity.

SOLVE

The force on and acceleration of the proton are both in the negative y direction in the coordinate system chosen. The mass is the mass of a proton: $m = m_p$.

$$\sum F_y = ma_y = -F_E$$

$$a_y = \frac{-F_E}{m_p} \quad (1)$$

The acceleration is constant, so we can use the constant acceleration problem-solving strategy (page 42) to find the proton's velocity. We used this strategy for a similar situation in Example 2.8 (pages 44–45).

$$v_y^2 = v_{0y}^2 + 2a_y\Delta y$$

$$v_{0y} = 0 \quad \text{(initially at rest)}$$

$$v_y = \pm\sqrt{2a_y\Delta y} \quad (2)$$

Here we need the negative square root because the velocity is in the negative y direction. Substitute from Equation (1) for a_y.

$$v_y = -\sqrt{2\left(\frac{-F_E}{m_p}\right)\Delta y}$$

Substitute values, remembering that Δy for this proton is negative in our coordinate system.

$$v_y = -\sqrt{2\left(\frac{-1.4 \times 10^{-13}\text{ N}}{1.673 \times 10^{-27}\text{ kg}}\right)(-7.5 \times 10^{-3}\text{ m})}$$

$$\vec{v} = -1.1 \times 10^6\,\hat{j}\text{ m/s}$$

CHECK and THINK

A force of the same magnitude acts on the proton, which is 1800 times more massive than the electron. According to Newton's second law, the proton's acceleration should be about 1800 times smaller. Because the particle's speed depends on the square root of a_y (Eq. 2), the proton's final speed should be about $\frac{1}{\sqrt{1800}}$ times the electron's speed. In Example 2.8, we found the electron's speed is $4.7 \times 10^7\,\hat{j}$ m/s. So, our answer seems reasonable.

$$\frac{v_p}{v_e} = \frac{1.1 \times 10^6}{4.7 \times 10^7} \approx 0.023 \approx \frac{1}{\sqrt{1800}} ✓$$

EXAMPLE 5.2 More Electrodes

Two pairs of electrodes are oriented as shown in Figure 5.9. An electron is at rest in the center of the configuration. When the electrodes are turned on, each pair produces a force on the electron such that

$$\vec{F}_E = (1.8 \times 10^{-13}\,\hat{\imath} + 7.2 \times 10^{-14}\,\hat{j})\text{ N}$$

Find the acceleration of the electron.

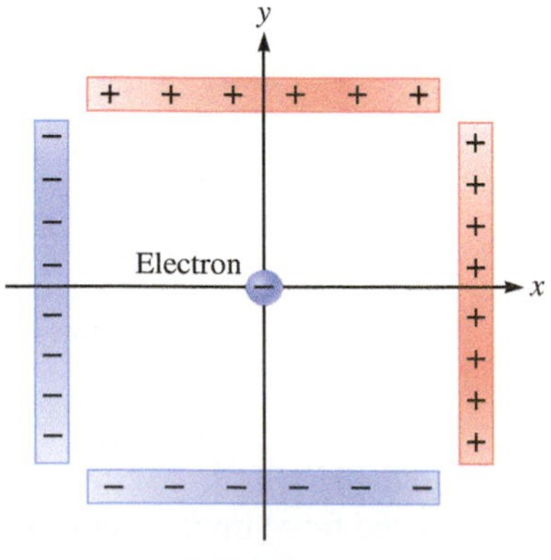

FIGURE 5.9

INTERPRET and ANTICIPATE

From Example 5.1, we can simply treat the electron as a particle on which a two-component force acts. We need to find a magnitude and direction for the electron's acceleration. Because there are forces in both the positive x direction and the positive y direction, we expect the acceleration to have positive x and y components as in $\vec{a} = (\_\_\_\,\hat{\imath} + \_\_\_\,\hat{j})$ m/s^2.

SOLVE

Follow a procedure similar to that used in Example 5.1, but now use Newton's second law twice, once for each component of the force.

$$\sum F_x = ma_x$$
$$\sum F_y = ma_y$$

The mass is the electron's mass. Because there is only one force in each direction, the sums are reduced to a single term each.

$$F_x = m_e a_x$$
$$F_y = m_e a_y$$

Solve both equations for acceleration.

$$a_x = \frac{F_x}{m_e} \quad \text{and} \quad a_y = \frac{F_y}{m_e}$$

Substitute the values. The electron mass is 9.109×10^{-31} kg.

$$a_x = \frac{1.8\times10^{-13}\text{ N}}{9.109\times10^{-31}\text{ kg}} = 2.0\times10^{17}\text{ m/s}^2 \qquad a_y = \frac{7.2\times10^{-14}\text{ N}}{9.109\times10^{-31}\text{ kg}} = 7.9\times10^{16}\text{ m/s}^2$$

Write the solution in component form.

$$\vec{a} = (2.0\,\hat{\imath} + 0.79\,\hat{\jmath}) \times 10^{17}\text{ m/s}^2$$

CHECK and THINK

The final answer is in the form we expected and roughly the same order of magnitude as the acceleration given in Example 5.1.

EXAMPLE 5.3 Three Balanced Forces

Two ropes are attached to a ring and exert forces as shown (Fig. 5.10). The magnitude of these forces is given by $F_1 = 2F_2$ and $f \equiv F_2 = 22.0$ N. A third force is applied by a rope so that the ring's acceleration is zero. What is the magnitude and direction of the force applied by the third rope?

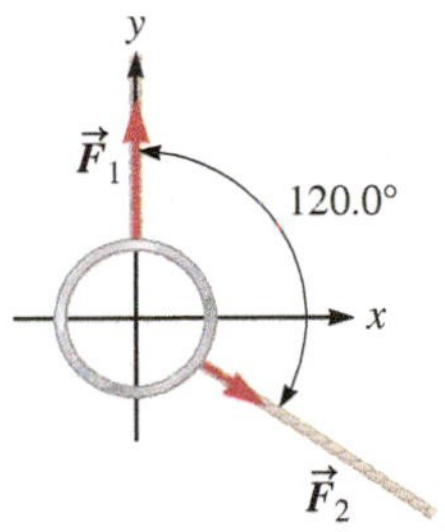

FIGURE 5.10

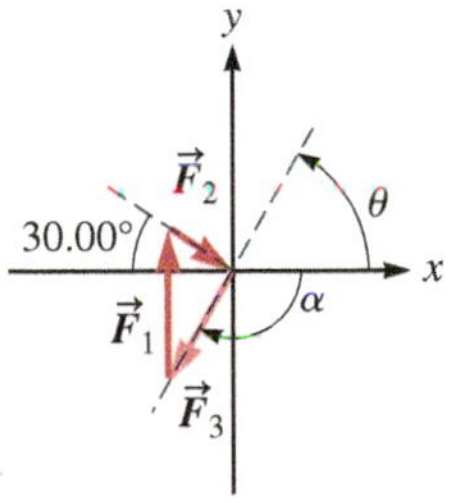

FIGURE 5.11

INTERPRET and ANTICIPATE

In order for the acceleration to be zero the sum of the three forces must be zero. To anticipate the result, we geometrically add $\vec{F}_1$ and $\vec{F}_2$. In order for the net force to be zero, $\vec{F}_3$ must point from the tip of $\vec{F}_2$ to the tail of $\vec{F}_1$. Our goal is find the magnitude of $\vec{F}_3$ and the angle α; from our sketch (Fig. 5.11) we expect the magnitude of $\vec{F}_3$ is similar to that of $\vec{F}_1$ and absolute value of α is greater than 90°.

SOLVE

The ring is not accelerating, so the net force in both the x and y direction is zero.

$$\sum F_x = ma_x = 0$$
$$\sum F_y = ma_y = 0$$

Use Figure 5.11 to resolve the three forces into their components. Set the sum of the forces in each direction equal to zero.

$$\sum F_x = F_2 \cos 30.00° - F_3 \cos\theta = 0$$
$$\sum F_y = F_1 - F_2 \sin 30.00° - F_3 \sin\theta = 0$$

We have two equations and two unknowns (F_3 and θ).

$$F_2 \cos 30.00° = F_3 \cos\theta \qquad (1)$$
$$F_1 - F_2 \sin 30.00° = F_3 \sin\theta \qquad (2)$$

Eliminate F_3 by dividing Equation (2) by Equation (1). Use $F_1 = 2F_2$ and $f \equiv F_2$ to simplify. Solve for θ by using $\tan\theta = \sin\theta/\cos\theta$.

$$\frac{\sin\theta}{\cos\theta} = \frac{2f - f\sin 30.00°}{f\cos 30.00°}$$

$$\theta = \tan^{-1}\left(\frac{2 - \sin 30.00°}{\cos 30.00°}\right) = 60.00°$$

Solve either Equation (1) or (2) for F_3 and substitute values. (We choose Equation 1.)

$$F_3 = \frac{F_2 \cos 30.00°}{\cos\theta} = \frac{(22.0\text{ N})\cos 30.00°}{\cos 60.00°}$$

$$F_3 = 38.1\text{ N}$$

Example continues on page 130 ▶

To find the direction α, we use Figure 5.11. We see $\alpha + \theta = 180.0°$, and because the direction is measured clockwise from the x axis, it must be negative.	$\alpha = -(180.0° - 60.00°)$ $\alpha = -120.0°$

CHECK and THINK

The magnitude $F_1 = 2f = 44.0$ N. So, as expected the magnitude of F_3 is similar to that of F_1. Also as expected the absolute value of α is greater than 90°.

5-7 Some Specific Forces

As you study physics, you will encounter many forces. You should expect to learn three things about each one: its source, its magnitude, and its direction. This section presents a partial list that will make our study of Newton's laws less abstract.

Gravity Near the Earth's Surface

GRAVITY **Special Case**

Gravity, or the **gravitational force** (the two terms are synonymous), is the field force that keeps us on the surface of the Earth and keeps the planets in orbit around the Sun. In general, any two objects that have mass exert an attractive gravitational force on each other. The closer the objects are together, the stronger their gravitational attraction. The general gravitational force between any two bodies is described in Chapter 7. For now, let's focus on the gravitational force exerted by the Earth on objects located near its surface.

From Chapter 2, when an object is in free fall, the only force acting on it is gravity. Experiments show that in the absence of air resistance, all objects in free fall near the surface of the Earth have the same downward acceleration g. Because the Earth's gravitational force $\vec{F}_g$ is the only force acting on such an object, Newton's second law for this situation is simply

$$\vec{F}_{\text{tot}} = \vec{F}_g = m\vec{a} \tag{5.4}$$

If we choose an upward-pointing y axis, we can represent the acceleration as

$$\vec{a} = -g\hat{\jmath} \tag{5.5}$$

Substitute Equation 5.5 into Equation 5.4 to find the gravitational force acting on an object of mass m near the Earth's surface:

$$\vec{F}_g = -mg\hat{\jmath} \tag{5.6}$$

The gravitational force is always attractive; that is, all objects are always pulled toward the center of the Earth. Because of its role in Equation 5.6, g is also known as the **local acceleration due to gravity.**

There is an intimate connection between gravity and what we call *weight.* In this book, we define the **weight** w of any object to be the magnitude of the gravitational force acting on the object:

There are other definitions of weight. Sometimes, weight is defined as a vector quantity, and other definitions are operational in that they involve measuring weight with a scale.

$$w \equiv F_g = mg \tag{5.7}$$

Because the local acceleration due to gravity g is greater at sea level than at high elevations, the weight of an object is greater near sea level.

We can also talk about the weight of an object on another planet. In this case, g is not the local gravitational acceleration near the Earth's surface (9.81 m/s^2), but rather the local gravitational acceleration near the surface of that planet. *Mass,* however, does *not* depend on the local acceleration due to gravity, which means that the mass of an object is the same at sea level, on top of a mountain, or even on another planet.

Because weight is the magnitude of a force, the SI unit for weight is the newton. The U.S. customary unit for weight is the familiar pound. The conversion factors between newtons and pounds are 1 N = 0.225 lb and 1 lb = 4.45 N. Note that the kilogram is *not* a unit of weight; it is the SI unit for mass. A translation can be made, however, between pounds and kilograms: 1 kg = 2.2 lb. This translation is not a true conversion because one side of the equation is a mass and the other side is a weight.

The translation 1 kg = 2.2 lb should be applied only to objects near sea level on Earth, but it is commonly used at other elevations.

An object does not need to be in free fall for gravity to act on it. For instance, even though gravity is acting on you right now, your acceleration is (probably) zero because another force (due to the chair) is also acting on you.

You may be reading your textbook in a moving vehicle, in which case your acceleration is not necessarily zero.

Equation 5.6 gives the gravitational force on any object near the surface of a planet such as the Earth whether the object is in free fall, stationary, or moving with some acceleration other than the free-fall acceleration. If the object is not in free fall, another force besides gravity must be acting on it, and its acceleration must be determined by applying Newton's second law.

EXAMPLE 5.4 A Heavy Space Suit

On the Earth, an astronaut's space suit weighs 1.36×10^3 N (about 300 lb).

A What is the mass of the space suit?

FIGURE 5.12 Edwin E. (Buzz) Aldrin on the Moon, July 1969.

INTERPRET and ANTICIPATE

The weight is a force, so its SI units are newtons as given. We expect to find the mass in kilograms.

SOLVE

Find the mass of the space suit from Equation 5.7.

$$w = mg$$

$$m = \frac{w}{g} = \frac{1.36 \times 10^3 \text{ N}}{9.81 \text{ m/s}^2}$$

$$m = 1.39 \times 10^2 \text{ kg}$$

CHECK and THINK

These units are what we expected.

B The free-fall acceleration near the surface of the Moon is $g_M = 0.16g$. What does the space suit weigh on the Moon?

INTERPRET and ANTICIPATE

We expect the space suit to weigh less on the Moon than it does on the Earth.

SOLVE

The weight of the suit on the Moon is the magnitude of the Moon's gravitational force on it.

$$w_M = mg_M = (1.39 \times 10^2 \text{ kg})(0.16 \times 9.81 \text{ m/s}^2)$$

$$w_M = 2.18 \times 10^2 \text{ N}$$

CHECK AND THINK

The weight of the space suit on the Moon is about one-sixth of its weight on the Earth: $1360 \text{ N}/218 \text{ N} \approx 6$. So would the space suit seem heavy to the astronaut on the Moon? To answer this, imagine walking around on the Earth in your heaviest winter clothes and boots. How much do you think they weigh? Perhaps 5 to 10 lb. Let's say 10 lb. Since you can manage wearing about 10 lb of clothing, let's assume that if the space suit weighs no more than 10 lb on the Moon, the astronaut won't consider it too heavy. Using 1 lb = 4.45 N, we find that 10 lb is about 45 N. The space suit's weight on the Moon is about 4 times heavier (or about 50 lb). So, the suit would feel heavy to the astronaut. (In fact, lower mass suits have been developed for today's astronauts.)

HOOKE'S LAW Special Case

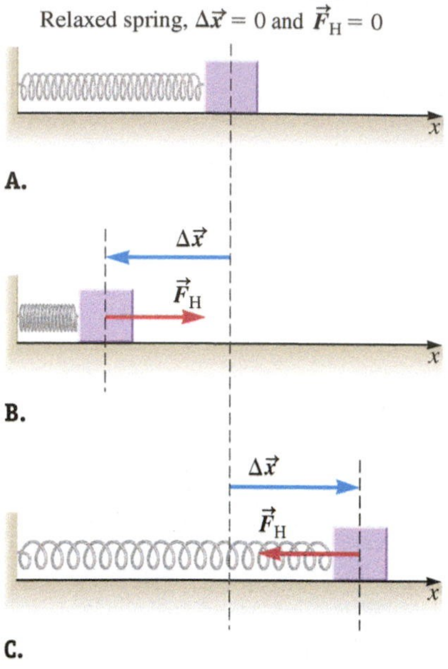

FIGURE 5.13 **A.** The spring is relaxed and does not exert a force. **B.** A compressed spring pushes on the block (contact force). **C.** A stretched spring pulls on the block (contact force).

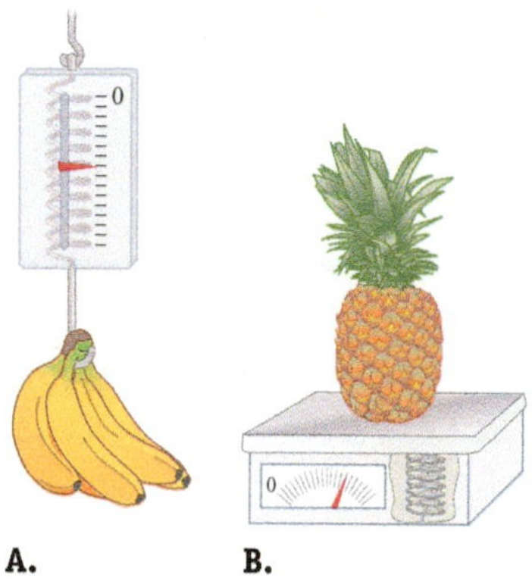

FIGURE 5.14 Spring scales are used to measure the apparent weight of an object. **A.** An object may hang from the spring. **B.** A pan scale employs a main spring (cutaway) connected to a system of levers (not shown).

Spring Force

Figure 5.13 shows a spring fixed on one end and attached to a block on the other end, which is free to move. A *relaxed spring*—that is, one that is neither compressed nor stretched—does not exert a force. When either compressed or stretched from the relaxed state, however, the spring exerts a contact force on the block. Depending on whether the spring is compressed or stretched, the contact force can be either a push or a pull.

The force due to the spring is a **restoring force**; that is, it is directed so as to return ("restore") the spring to its relaxed state. The farther the spring is from being in its relaxed state, the stronger the restoring force. For many springs, the force exerted is proportional to how far the attached block is displaced from the relaxed position. These springs are said to obey **Hooke's law**. For a spring lying along the x axis, Hooke's law is mathematically expressed as

$$\vec{F}_H = -k\Delta\vec{x} \tag{5.8}$$

where the subscript H stands for *H*ooke. The negative sign means that the restoring force is in the direction opposite the direction of the displacement. Similar equations are written for a spring oriented along a y or z axis. The constant k is known as the **spring constant**, a scalar property that differs from one spring to another; it measures the stiffness of the spring. The SI unit of k is newtons per meter (N/m).

Springs are used in many mechanical systems, including such common objects as retractable pens, mattresses, and jewelry clasps. In addition, many forces exerted by agents other than springs can be modeled as spring forces. We will return to springs many times throughout this book.

Spring scales are often used to measure the weight of an object. An object may be hung from a spring scale (Fig. 5.14A) or placed on top (Fig. 5.14B). We define an object's **apparent weight** as the reading of a scale displaying the magnitude of the spring force on that object. Why does the dial reading give the magnitude of the spring force? The force F_H exerted by the spring is proportional to how far the spring is stretched or compressed. The scale manufacturer knows the k value of the spring and designs the scale in such a way that the extension or compression of the spring moves a pointer on the dial to a number that is equal to the magnitude of the force exerted by the spring. We will show in Example 5.8 that the apparent weight (scale reading) does not necessarily equal the true weight of the object (magnitude of the gravitational force exerted on it).

CONCEPT EXERCISE 5.8

Imagine weighing the same bunch of bananas with two different spring scales like the one in Figure 5.14A. Scale 1 has a spring with spring constant k_1. When the bananas are hung from scale 1, the spring stretches Δy_1 from its relaxed position. Therefore, the apparent weight of the bananas is

$$w_{app} = k_1 \Delta y_1$$

Scale 2 has a stiffer spring; its spring constant is $k_2 = 3k_1$. If the same bunch of bananas is hung from scale 2, what is the displacement of its spring Δy_2 in terms of Δy_1?

Normal Force

NORMAL FORCE Special Case

Whenever any object is in contact with a surface, the surface exerts a contact force on the object. The component of this contact force that is perpendicular to the surface is called the **normal force**. (*Normal* is another word for *perpendicular*.) The normal force gets its name because its direction is always perpendicular to the surface.

The magnitude of the normal force varies depending on the situation. The harder the object presses against the surface, the greater the normal force exerted by the surface. Because our intuition tells us that only living organisms can respond to being touched, it is counterintuitive to imagine that an inanimate object like a tabletop can alter the force it exerts. The force exerted by an inanimate object can indeed change, though, and to see how, consider placing a bowling ball on a mattress.

A good spring mattress contains hundreds of springs. If you place a bowling ball on the mattress, the springs compress downward (Fig. 5.15A), and the spring force on the bowling ball is consequently upward. Because the bowling ball is not accelerating, we reason that the spring force on the ball exactly balances the gravitational force on the ball. How far the springs compress depends on how stiff they are. If the mattress were made of springs much stiffer than those shown in Figure 5.15A (that is, if k were higher), the springs, according to $\vec{F}_H = -k\Delta\vec{x}$ (Eq. 5.8), would not compress as far to exert the same spring force. If the mattress springs were very, very stiff, you might not even notice the compression, but they would still be slightly compressed and would exert the same upward force on the bowling ball.

Now imagine placing the bowling ball on a table (Fig. 5.15B). The table is made of molecules held together by molecular bonds that act like very stiff springs. They are slightly compressed by the bowling ball, and as a result they exert an upward force on the ball.

This principle is true of any surface. If a force pushes on a surface, the molecular bonds between the atoms in that surface are compressed, and the surface pushes back with a normal force. If the bowling ball is replaced by a Ping-Pong ball, on either the mattress with its springs or the tabletop with its molecular "springs," the compression will, of course, not be as great. As a result, the normal force exerted by either the mattress or the tabletop on the Ping-Pong ball will be smaller than the normal force on the bowling ball. In other words, the surface can alter the amount of force it exerts, counterintuitive though that may seem.

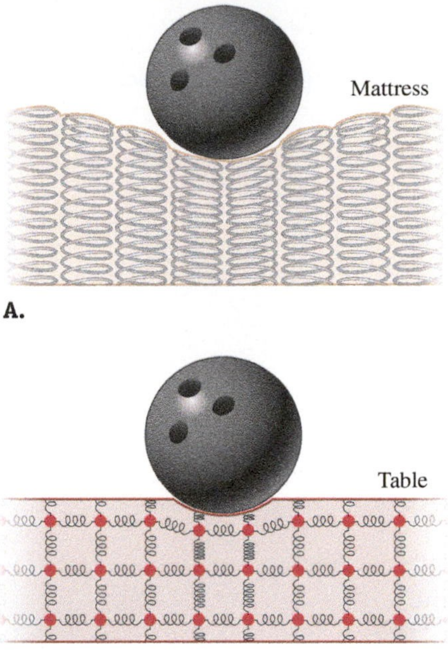

FIGURE 5.15 **A.** The bowling ball is supported by the springs in a mattress. If stiffer (higher k) springs are used, a smaller compression results in the same force on the bowling ball. **B.** The molecular bonds in the table act like very, very stiff springs in a mattress.

Tension Force

A **tension force** $\vec{F}_T$ is exerted by a taut rope, cable, or similar cord such as the rope pulling on the bananas in Figure 5.16. The rope must be in contact with the object to exert a tension force. Because the rope pulls on the object, the tension force is directed along the rope. In Figure 5.16, the rope pulls upward on the bananas. The magnitude of the tension force is called the **tension**.

Like the normal force, the tension force can be modeled as a spring force exerted by molecular bonds between molecules (close-up, Fig. 5.16). In the case of the tension force, a taut rope stretches the molecular bonds. Like very stiff springs, these molecular bonds pull back. If the "springs" are very stiff, the rope does not stretch perceptibly, and we say the rope is *unstretchable*.

In this textbook, we will consider only ropes that are unstretchable and massless, where by "massless" we mean that the mass of any rope in our discussion is so small that we can ignore the effect of gravity on the rope. For such an ideal rope, the tension is the same everywhere in it. Thus, if the rope connects two objects, the magnitude of the tension force on each object is the same.

TENSION FORCE **Special Case**

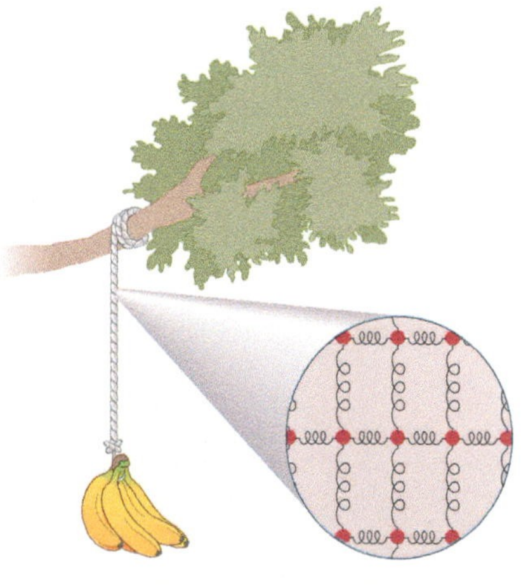

FIGURE 5.16 Bananas hang from a rope. The tension force on the bananas is upward. Molecular bonds in the rope act like very, very stiff springs (close-up). When those bonds are stretched, they pull back.

CONCEPT EXERCISE 5.9

For all three situations, find the magnitude and direction of the tension force(s) exerted on Rochelle. If not enough information is given, say so. Explain your answers.

a. Rochelle and Buddy pull on opposite ends of rope. The tension force exerted on Buddy is 15 N and is directed toward Rochelle.

b. Now, one end of the rope is tied to a sturdy pole and Rochelle pulls on the other end. The tension force on the pole is 15 N directed toward Rochelle.

c. Finally, Rochelle holds one rope in her left hand, and the other end of that rope is pulled by Joe. She holds another rope in her right hand, and the other end of that rope is pulled by Buddy. Both Buddy and Joe experience a 15-N force directed toward Rochelle.

Kinetic Friction

As stated previously, when an object is in contact with a surface, the surface exerts a force on the object. The component of that force perpendicular to the surface is the normal force. The force exerted by the surface may also have a parallel component, which is known as the frictional force. Experiments have shown that friction is a function of the relative motion of the object and the surface. Whenever an object is sliding across a surface, **kinetic friction** is the parallel component of the contact force exerted by the surface on the object. **Static friction** may exert a force on an object that is at rest with respect to the surface. Finally, if an object rolls along a surface, such as a tire on a road, **rolling friction** acts on the object.

KINETIC FRICTION ▶ Special Case

In Chapter 6, we will discuss all three forms of friction in detail. In this chapter, we will mainly consider kinetic friction, which causes a sliding object to slow down and finally stop. The direction of kinetic friction is always parallel to the surface and opposite the direction in which the object is moving relative to the surface.

The magnitude of kinetic friction is proportional to the normal force acting on the object:

$$F_k = \mu_k F_N \tag{5.9}$$

where the constant of proportionality μ_k is known as the **coefficient of kinetic friction.** (The symbol μ is the lowercase Greek letter mu.)

Table 5.1 is a summary of the five forces we have just described. The table mentions the source, magnitude, and direction for each force. Each time you study a new force, be sure that you can identify these three parameters for it.

TABLE 5.1 Summary of the five forces examined in this chapter.

Force	Source	Magnitude	Direction
Gravity ($\vec{F}_g$)	Earth	"Weight," mg	Toward center of the Earth
Hooke's law ($\vec{F}_H$)	Spring	$k\,\Delta x$	Push or pull along spring
Normal ($\vec{F}_N$)	Surface	Varies	Perpendicular to surface
Tension ($\vec{F}_T$)	Rope	"Tension," the same all along the rope	Pull along rope
Kinetic friction ($\vec{F}_k$)	Surface	$\mu_k F_N$	Parallel to surface, opposite to object's $\vec{v}$ relative to surface

5-8 Free-Body Diagrams

This section describes an important technique for applying Newton's second law to solve problems. A good way to begin analyzing such problems is with a drawing. Although a nicely drawn, artistic rendering of the problem may be aesthetically pleasing, it is not only unnecessary but can actually be distracting. Instead, physicists use a type of drawing called a **free-body diagram,** having the following elements:

FREE-BODY DIAGRAMS ⦿ Tool

Element 1. **A simple representation of a subject.** We will continue to use the particle model for all objects until Chapter 13. Until then, represent any object (or system) by a dot.

Element 2. **Clearly labeled vector representations** of all the external forces exerted on the subject. In most cases, the tails of the vectors should be on the subject. To make sure that you don't miss any force exerted on the subject, consult Table 5.1 and decide which of the five forces are present. Make your decision based on whether or not the source of a particular force is present in the situation.

Element 3. A clearly labeled **coordinate system.** Picking a good coordinate system comes with practice. One helpful tip is to choose one axis to be parallel to the subject's acceleration.

Element 4. **An indication of the acceleration.** If relevant, draw a vector arrow indicating the direction of the subject's acceleration, but visually different from the arrows representing the force vectors. For example, you might use a dashed line or a different color. If the subject's acceleration is zero, note that on the diagram.

Drawing free-body diagrams takes practice. You'll get that practice solving problems that involve Newton's second law. In the following problem-solving strategy, drawing a free-body diagram is the primary tool used in the **INTERPRET and ANTICIPATE** procedure.

PROBLEM-SOLVING STRATEGY

Applying Newton's Second Law

INTERPRET and ANTICIPATE

Identify the system (often, a single object) that is subject to external force(s). Draw a free-body diagram for that system, making sure that the diagram has all four elements given above. Once the free-body diagram is complete, there are three steps that help in the **SOLVE** procedure when an algebraic or numerical result is required.

SOLVE

Step 1 Apply Newton's second law in component form. Use the coordinate system on the free-body diagram to apply Equation 5.2. You will have one equation for each direction in which there is at least one force.

$$\sum F_x = ma_x \quad \sum F_y = ma_y \quad \sum F_z = ma_z$$

Step 2 Write down any other equations that are relevant to the forces involved. Table 5.1 lists magnitudes of the gravitational force (weight), the spring force, and kinetic friction. If the situation involves one or more of these forces, write down the appropriate equation.

Step 3 Do algebra before substitution. Review your equations. Which parameters are known? Which are unknown? Which do you need to solve for? You may have more equations than you need. Find an algebraic expression for the parameter you need before you substitute any numerical values. This practice makes it easier to find a mistake if you make one and makes it easier for another person to understand your work.

EXAMPLE 5.5 Sheldon's Morning Routine

Sheldon, of mass 60.4 kg, stands on his bathroom scale. What is the normal force exerted by the scale on Sheldon? What does the scale read (in newtons)?

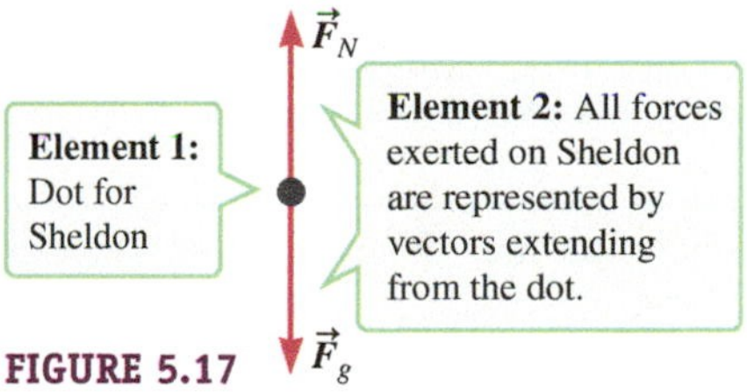

FIGURE 5.17

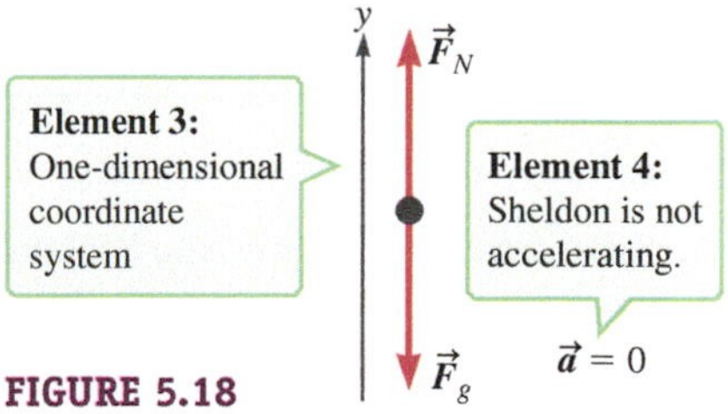

FIGURE 5.18

INTERPRET and ANTICIPATE

Sheldon is the subject, represented by a dot on our free-body diagram (Fig. 5.17). Consider the five forces listed in Table 5.1. There is no rope in this problem, and Sheldon is not moving; so, there is no tension force or kinetic friction exerted on him. He does experience the downward pull of gravity $\vec{F}_g$, and the scale exerts an upward normal force $\vec{F}_N$ on him.

The two forces are vertical, so we only need a one-dimensional coordinate system (Fig. 5.18). We choose an upward-pointing y axis. Sheldon is at rest, so his acceleration is zero as indicated.

The reading on the scale equals the magnitude of the force that the scale exerts. So, in this case, the magnitude of the normal force is the reading on the scale. We expect $\vec{F}_N = +(\_\_\_\_\_\hat{\jmath})$ N because the normal force is upward.

SOLVE

Step 1 Apply Newton's second law in component form. We only need the y-component portion of Newton's second law (Eq. 5.2). In this case, the normal force is in the positive direction, and gravity is in the negative direction. The sum of the two forces is zero because Sheldon is not accelerating.

$$\sum F_y = ma_y \quad (5.2)$$
$$F_N - F_g = 0 \quad (1)$$

Step 2 Write relevant force equations. Table 5.1 lists the magnitude of the gravitational force. There is nothing more we can write about the normal force.

$$w = F_g = mg \quad (2)$$

Step 3 Do algebra. Substitute Equation (2) into Equation (1) and solve for the scalar component of the normal force.

$$F_N - mg = 0$$
$$F_N = mg$$

Example continues on page 136 ▶

The normal force is in the positive y direction.	$\vec{F}_N = mg\hat{\jmath}$
Substitute values. Remember that g is a positive number.	$\vec{F}_N = (60.4\text{ kg})(9.81\text{ m/s}^2)\hat{\jmath}$ $\vec{F}_N = 593\hat{\jmath}\text{ N}$
The reading on the scale is the magnitude of the normal force.	$F_N = 593\text{ N}$

CHECK and THINK

Our answers have the form we expected. It is interesting that Sheldon's weight in U.S. customary units is about 130 lb, which is reasonable. Many physics problems involve the reading on a scale. Keep in mind that the scale's readout is the magnitude of the force it exerts on the object. However, in your laboratory you are likely to find a scale that reports mass in kilograms rather than weight in newtons.

EXAMPLE 5.6 Christa's Backpack

Christa's backpack is full of physics books and hangs from a single strap that loops around a hook as shown in Figure 5.19. The combined mass of the backpack and its contents is 3.14 kg, and the mass of the strap is very much less than this combined mass. Use a free-body diagram and Newton's second law to find the tension F_T in the strap and the total tension force $(\vec{F}_T)_{\text{tot}}$ exerted by the strap on the backpack.

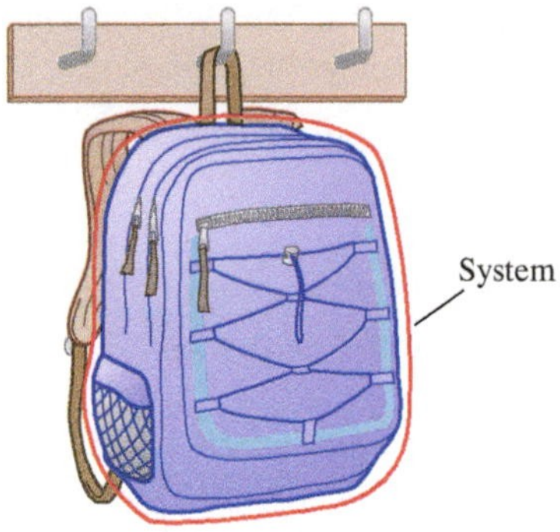

FIGURE 5.19

INTERPRET and ANTICIPATE

Start by identifying the system for which the free-body diagram will be drawn. Because we must find a force exerted on the backpack, the backpack and its contents are the system. We must consider the strap to be external to the system to find the force it exerts.

Review the four other forces in Table 5.1 to determine which are exerted on the system.

No surface in this problem $\rightarrow$ no friction or normal force.
No springs $\rightarrow$ no spring force.
The strap exerts a tension force at each point of contact.

Gravity is acting.
Only gravity and the tension force act on the backpack.

Draw a free-body diagram, including all four elements (Figs. 5.20 and 5.21).

1. **Represent the system**—the backpack and its contents—**by a dot.**

2. **Draw and label the forces.** The tail of each force is on the dot. Gravity points to the center of the Earth (straight down in this problem). The strap attaches vertically to the backpack at two points, so it exerts an upward tension force at both of these points (Fig. 5.20).

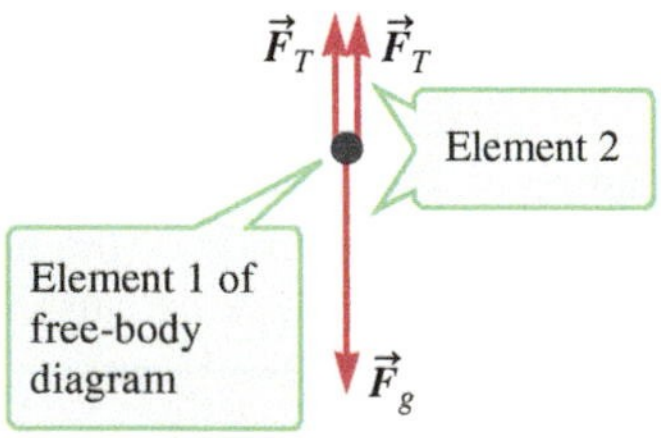

FIGURE 5.20

3. Choose a **coordinate system.** A vertical y axis was chosen because all the forces lie along this axis. An x and a z axis are implied but not drawn because they are not used in this one-dimensional problem (Fig. 5.21).
4. Because the backpack hangs from the hook and does not move, its **acceleration is zero as indicated.**

We expect both numerical answers to have the SI units of newtons. Also, the total tension force points up, so we expect to find an answer in the form $(\vec{F}_T)_{\text{tot}} =$ ____ $\hat{\jmath}$ N.

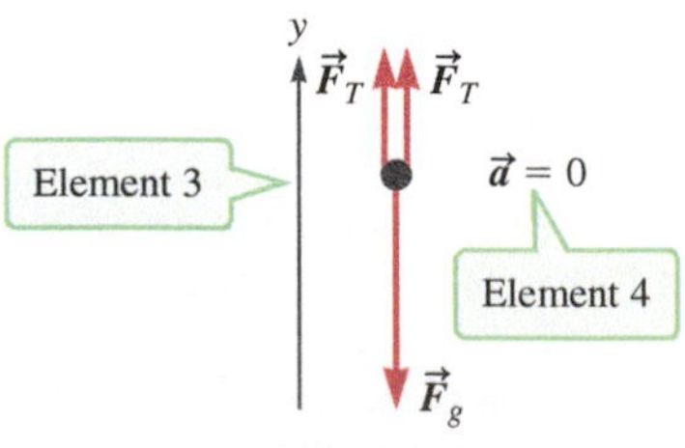

FIGURE 5.21

SOLVE

Step 1 Apply Newton's second law in component form. From our free-body diagram, all forces are along the y axis, so we use just the y-component portion of Newton's second law (Eq. 5.2).	$\sum F_y = ma_y$
Because the acceleration is zero, the ma_y term is zero.	$\sum F_y = 0 \quad (1)$
Now, write each force in the free-body diagram in component form and substitute them into Equation (1). The magnitude of the gravitational force is the weight w. The tension has the same magnitude F_T throughout the single strap. The strap attaches to the backpack in two places, however, so there are two tension forces of magnitude F_T in the positive y direction. The total tension force is the vector sum of these two (Eq. 2).	$\vec{F}_g = -w\hat{\jmath}$ $(\vec{F}_T)_{tot} = 2\vec{F}_T = 2F_T\hat{\jmath} \quad (2)$ $\sum F_y = 2F_T - w = 0 \quad (3)$
Step 2 Write relevant force equations. The weight w of the backpack is the magnitude mg of the Earth's gravitational force on it.	$w = mg \quad (4)$
Step 3 Do algebra. Solve Equations (3) and (4) for the tension F_T by eliminating w from the two equations and isolating F_T.	$\sum F_y = 2F_T - mg = 0$ $F_T = \frac{1}{2}mg$
Substitute values to find the tension in the strap. Notice that the tension is half the weight because the strap is attached at two points.	$F_T = \frac{1}{2}mg = \frac{1}{2}(3.14\ \text{kg})(9.81\ \text{m/s}^2)$ $F_T = 15.4\ \text{N}$
The total tension force exerted on the backpack comes from Equation (2).	$(\vec{F}_T)_{tot} = 2F_T\hat{\jmath} = 2(15.4\ \text{N})\hat{\jmath}$ $(\vec{F}_T)_{tot} = 30.8\hat{\jmath}\ \text{N}$

CHECK and THINK

We just found that the magnitude of the total tension force is equal to the weight of the backpack and its contents, which makes sense because the backpack is not accelerating and so the net force must be zero.

EXAMPLE 5.7 CASE STUDY Train Collision

When the passenger train of our case study came to a sudden stop before the collision, passengers who were in forward-facing seats were thrown forward, and passengers in backward-facing seats were not (Fig. 5.22). Friction due to the seat bottoms is exerted on both passengers. (For the backward-facing passenger, it is static friction; for the forward-facing passenger, it is kinetic friction.) For both passengers, friction is parallel to the seat bottom and points to the right in this case. Draw two free-body diagrams: one for a forward-facing passenger and one for a backward-facing passenger. Assume both passengers have roughly the same mass. Use Newton's laws to explain why the forward-facing passenger fell forward but the backward-facing passenger did not.

FIGURE 5.22

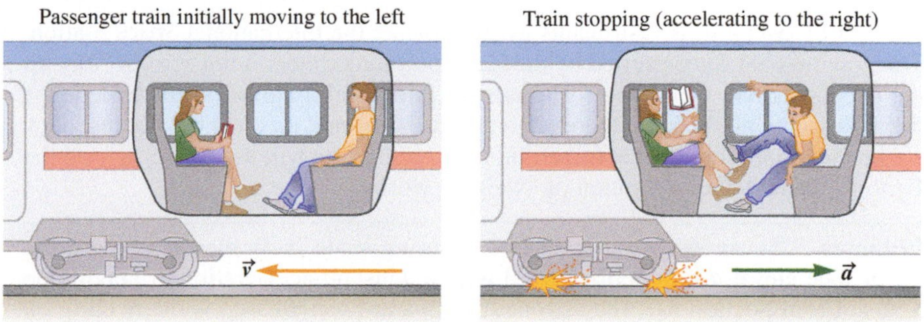

INTERPRET and ANTICIPATE

First, determine which of the five forces discussed in Section 5-7 are exerted on each passenger.

Example continues on page 138 ▶

No tension or spring forces.

Gravity $\vec{F}_g$ acts on both passengers (equal masses assumed, so equal gravitational force on each passenger).

Normal force:

1. $\vec{F}_N$ due to seat bottoms acts on both passengers.
2. Passenger seated facing backward is acted on by additional normal force $\vec{f}_N$ due to the back of her seat.

Friction $\vec{F}_f$ (static or kinetic) due to the seat bottoms points to the right for both passengers.

The two free-body diagrams each contain all four elements (Fig. 5.23). Each passenger is **represented as a dot**. The **accelerations** are indicated. A **coordinate system** has been chosen so that the x axis is parallel to the direction of the accelerations (often a convenient choice). All the **forces** exerted on each passenger have been **drawn and labeled**. There are three forces exerted on the forward-facing passenger and four forces on the backward-facing passenger. Notice that friction due to the seat bottoms points in the same direction as the acceleration (the x direction).

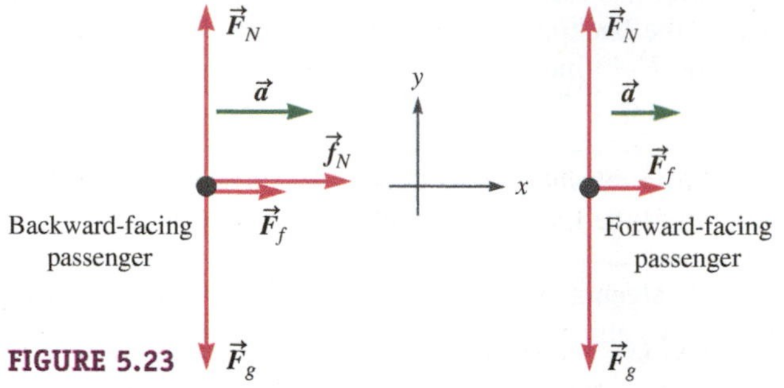

FIGURE 5.23

SOLVE

We can tell why one passenger fell and the other didn't by examining the free-body diagrams. Before the train decelerated suddenly, both passengers were moving at constant velocity in the negative x direction. According to Newton's first law, a force is required to slow these passengers down (to give them the same acceleration as the train). This force must be in the same direction as the train's acceleration, the positive x direction. Only the frictional force due to the seats and the normal force due to the back of the backward-facing passenger's seat point in the x direction. Friction alone is too weak to provide the required acceleration in this case. Therefore, the forward-facing passenger continues to move forward at a speed somewhat less than his original speed. The backward-facing passenger accelerates with the train and remains in her seat because of the additional normal force on her back.

CHECK and THINK

This example explains why seat belts are so important. If you are in a car that decelerates suddenly, friction due to the seat is not strong enough to keep you accelerating along with the car. Inertia will carry you through the windshield if you rely on that friction alone. The seat belt provides the additional force needed to slow you down and keep you in the car.

EXAMPLE 5.8 Astronaut on an Elevator

We hear of "weightless" astronauts in orbiters like the International Space Station. The gravitational force exerted by the Earth on such orbiters is not zero, so why do the occupants feel weightless? The answer is that orbiting astronauts experience weightlessness not because there is no gravitational pull on them but because they are in a noninertial reference frame. In this example, we will use an elevator on the Earth to see how the perception of "weight" works.

In Figure 5.24, an astronaut stands on a spring scale in an elevator car on the Earth. The elevator moves downward with a constant acceleration of magnitude $a < g$. From Section 5-7, the astronaut's apparent weight w_{app} is the reading on the scale.

A Find the astronaut's apparent weight in terms of her true weight w, the magnitude of the elevator's acceleration a, and the free-fall acceleration g.

FIGURE 5.24

INTERPRET and ANTICIPATE

Our free-body diagram for the astronaut (Fig. 5.25) includes all four elements. The **astronaut is represented as a dot.** Only two forces are exerted on the astronaut: gravity $\vec{F}_g$ and the contact force from the spring scale. In this case, the contact force is provided by the spring so that $\vec{F}_N = \vec{F}_H$. These **forces are labeled** on the free-body diagram, and are vertical. We choose an upward-pointing y axis for our **coordinate system.** Finally, the astronaut accelerates downward with the elevator, which is indicated by the downward-pointing **acceleration** vector.

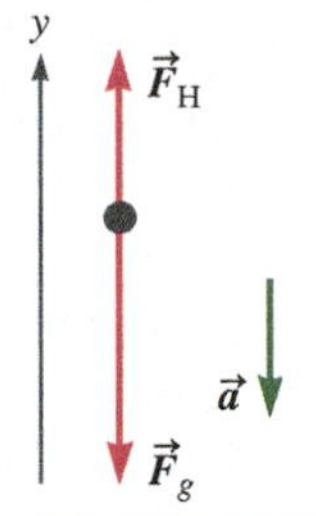

FIGURE 5.25

The reading on the scale dial is the magnitude of the spring force. This reading is also the astronaut's apparent weight, so we can equate these two quantities. We seek an algebraic expression for the apparent weight, which is a positive scalar.

$$w_{app} = F_H \tag{1}$$

SOLVE

Step 1 Apply Newton's second law in component form. Both forces lie along the y axis, so only one equation is needed.

$$\sum F_y = F_H - F_g = ma_y \tag{2}$$

The acceleration is downward as indicated on the free-body diagram.

$$\vec{a}_y = -a\hat{j} \tag{3}$$

Write Newton's second law explicitly in terms of the magnitude of the elevator's acceleration by substituting Equation (3) into Equation (2).

$$\sum F_y = F_H - F_g = -ma \tag{4}$$

Step 2 Write other relevant force equations. The astronaut's true weight is the magnitude of the gravitational force on her (Eq. 5.7), and her mass can be written in terms of this (true) weight. (There is no need to write $F_H = k\Delta y$ because we are not interested in k or Δy.

$$w = F_g = mg \tag{5}$$

$$m = \frac{w}{g} \tag{6}$$

Step 3 Do algebra. Substitute Equations (1) (w_{app} for F_H), (5) (w for F_g), and (6) (w/g for m) into Equation (4), and then solve for w_{app}.

$$w_{app} - w = -\frac{w}{g}a$$

$$w_{app} = w\left(1 - \frac{a}{g}\right) \tag{7}$$

CHECK and THINK

According to Equation (7), as long as the elevator is accelerating downward, the astronaut's apparent weight is less than her true weight. (For an elevator accelerating upward, the minus sign becomes a plus sign and the apparent weight is greater than her true weight.)

How does a person experience "apparent weight"? You do not actually feel the pull of gravity downward; instead, you feel the normal force on your body due to a chair, the floor, or whatever surface is preventing you from falling to the center of the Earth. In this case, the spring scale is acting on the astronaut in this way. She feels lighter because the force due to the scale has decreased.

Because the elevator is accelerating, the astronaut is in a noninertial reference frame. Just as the observer on the ship in Figure 5.6 cannot identify a force that caused the Zamboni to accelerate, the astronaut cannot identify a force that caused her weight to appear to decrease.

B What happens to the astronaut's apparent weight if the elevator is in free fall as seen in Figure 5.26?

FIGURE 5.26

INTERPRET and ANTICIPATE

We expect that something special must happen if the elevator is in free fall. The free-body diagram for the astronaut is now as shown in Figure 5.27. The spring force F_H is no longer acting on the astronaut because she is not in contact with the scale.

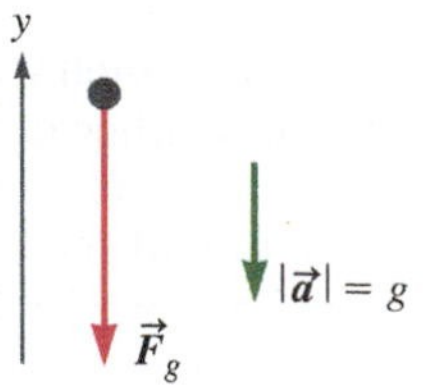

FIGURE 5.27

Example continues on page 140 ▶

SOLVE

If the elevator is in free fall, its acceleration is exactly g. Substitute $a = g$ into Equation (7).

$$w_{app} = w\left(1 - \frac{g}{g}\right) = 0$$

CHECK and THINK

If the elevator is in free fall, the astronaut's apparent weight is zero, and we say that she is "weightless." She does not experience any normal force and therefore does not feel her own weight. The force of the Earth's gravity is not zero, but the force of the spring scale is zero because the astronaut's feet hover above the scale. Therefore, the spring is in its relaxed state, with the pointer at zero on the dial. Similarly, an astronaut in orbit around the Earth is also in free fall, so her apparent weight is zero. Both astronauts feel "weightless."

EXAMPLE 5.9 Pulleys Are Everywhere

Pulleys are found on the end of cranes, on sailboats, and even in the weight room of a workout gym. Ideal pulleys are considered to be massless and frictionless. In the common laboratory setup shown in Figure 5.28, the hanging cylinder of mass m is released from rest. If the friction between the cart of mass M and the track is negligible, find the acceleration of the cart and the tension in the rope.

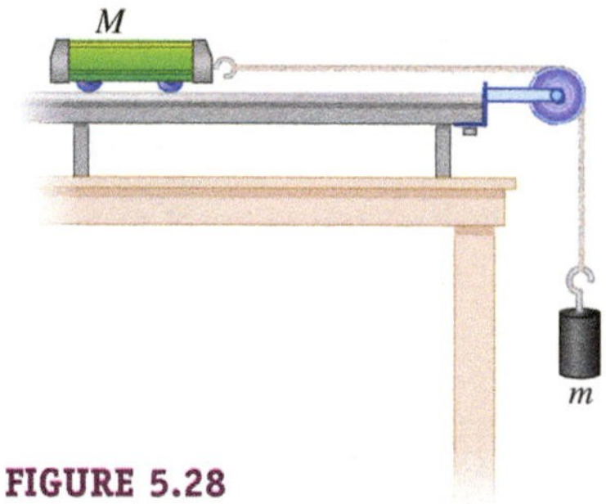

FIGURE 5.28

INTERPRET and ANTICIPATE

Draw one free-body diagram for the cart and another for the hanging cylinder as in Figure 5.29.

1. Each subject is **represented by a dot** (labeled with the mass) in Figure 5.29.
2. **Forces are drawn and labeled on each object.**
   No springs or friction.
   Two forces on the cylinder: gravity $\vec{F}_g$ and the tension force $\vec{F}_T$.
   Three forces on the cart: gravity $\vec{F}_G$, the normal force $\vec{F}_N$, and the tension force $\vec{F}_T$.
   For an ideal pulley, the tension is the same throughout the rope (same symbol $\vec{F}_T$ in both diagrams). We chose an uppercase subscript for gravitational force on the cart and a lowercase subscript for gravitational force on the hanging cylinder.

You can add elements 3 and 4 in any order. Here we'll add element 4 first.

4. The **acceleration** of each subject is indicated. The cart accelerates to the right when the cylinder accelerates downward.
3. **Coordinate systems** were picked to make calculations easy. For the hanging cylinder, the choice may seem unconventional but the taut rope ensures that the acceleration of both the cylinder and the cart has the same magnitude. This choice of coordinates means that the acceleration of both objects is along a positive x axis (to the right for the cart; downward for the cylinder).

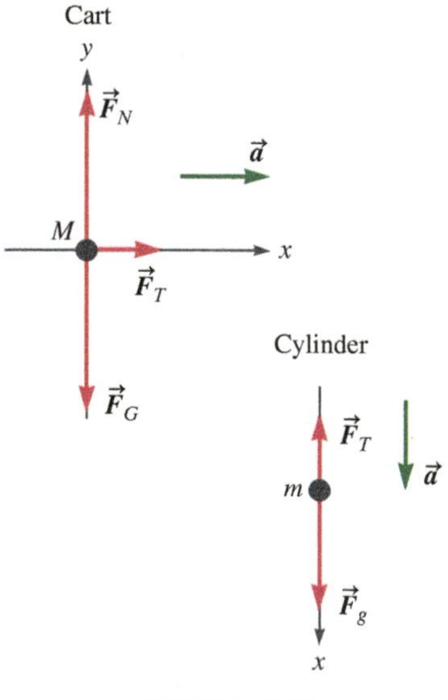

FIGURE 5.29

We expect to find the same algebraic expression for the acceleration of both objects in the form $\vec{a} = (\_\_\_\_)\hat{\imath}$ and an additional algebraic scalar expression for the tension.

SOLVE

Step 1 Apply Newton's second law in component form. We must write two equations for the cart because there are forces along both x and y directions. There is only one equation for the cylinder. The acceleration of the cart along x must equal the acceleration of the cylinder, so we have used the same symbol a for both. The cart has no acceleration in the y direction.

Cart:

$$\sum F_x = F_T = Ma \quad (1)$$

$$\sum F_y = F_N - F_G = 0$$

Hanging cylinder:

$$\sum F_x = F_g - F_T = ma \quad (2)$$

Step 2 Write other relevant force equations. Use Equation 5.7 to write expressions for the weight of each subject.	$F_g = mg$ (3) $F_G = Mg$
Step 3 Do algebra. The y equation for the cart gives no information about the acceleration, so we don't need it. Equations (1) and (2) both have two unknowns, F_T and a. We use these equations along with Equation (3) to solve for the unknowns. Substitute Equations (3) and (1) into Equation (2), and then solve for a.	$mg - Ma = ma$ $mg = Ma + ma = a(M + m)$ $a = \dfrac{mg}{M + m}$ (4)
Write the acceleration in vector notation. Find the tension in the rope by substituting Equation (4) into Equation (1). Remember tension is a scalar.	$\vec{a} = \left(\dfrac{mg}{M + m}\right)\hat{\imath}$ $\quad F_T = \dfrac{Mmg}{M + m}$ (5)

CHECK and THINK

Our expressions for tension and acceleration are in the forms we expected as we can verify by checking their SI units.

Finally, the $\hat{\imath}$ in Equation (5) means that the cart accelerates to the right while the hanging cylinder accelerates downward.

$$[\![a]\!] = \left[\!\!\left[\frac{mg}{M + m}\right]\!\!\right] = \frac{(\text{kg})(\text{m/s}^2)}{\text{kg}} = \text{m/s}^2$$

and

$$[\![F_T]\!] = \left[\!\!\left[\frac{Mmg}{M + m}\right]\!\!\right] = \frac{(\text{kg})(\text{N})}{\text{kg}} = \text{N}$$

5-9 Newton's Third Law

Unlike Newton's first two laws, which deal with forces acting on a single subject, Newton's third law is concerned with the forces between two objects that interact with each other. Conceptually, Newton's third law states that if two objects A and B interact, the force exerted by A on B is equal in magnitude to the force exerted by B on A but in the opposite direction.

We experience Newton's third law all the time. When you are sitting in a chair, your body presses down on the chair, and you can feel the chair pressing back on you. (If the chair did not exert this force on you, you would end up falling through the chair and sitting on the floor. Then you and the *floor* would exert a pair of forces on each other.) As another example, when your arm lifts a heavy backpack upward, you can feel the backpack pulling your arm downward.

We can write Newton's third law mathematically. First, we define $\vec{F}_{[\text{B on A}]}$ as the force exerted *by* object B *on* object A. In the language of Section 5-3, B is the source of the force $\vec{F}_{[\text{B on A}]}$, and A is the subject on which the force acts. Because B exerts a force on object A, it must be true that A also exerts a force on B, and we denote this force by $\vec{F}_{[\text{A on B}]}$. By Newton's third law, these two forces are a pair that must be equal in magnitude and opposite in direction. Mathematically, this idea is expressed as

It is common to drop the word *on* from the subscript so that $\vec{F}_{[\text{B on A}]}$ is written as $\vec{F}_{\text{BA}}$.

$$\vec{F}_{[\text{B on A}]} = -\vec{F}_{[\text{A on B}]} \tag{5.10}$$

NEWTON'S THIRD LAW

Underlying Principle

CONCEPT EXERCISE 5.10

In Concept Exercise 5.3, we identified the source and subject of a force in three situations. Return to those situations and draw a free-body diagram for each subject. In each diagram, draw a loop around the vector arrow representing the force that is part of a third-law pair associated with the source. Label this with the [source on subject] subscript notation.

Sometimes it is hard to see the difference between Newton's second and third laws. There are two important distinctions, however. (1) The third law always applies to *a pair of* interacting objects and to *a pair of* forces; one force acts on one object of the pair, and the other force acts on the other object of the pair. The second law involves all the forces acting *on a single object.* (2) The third law says nothing about acceleration, whereas the second law relates total force exerted on an object to its acceleration.

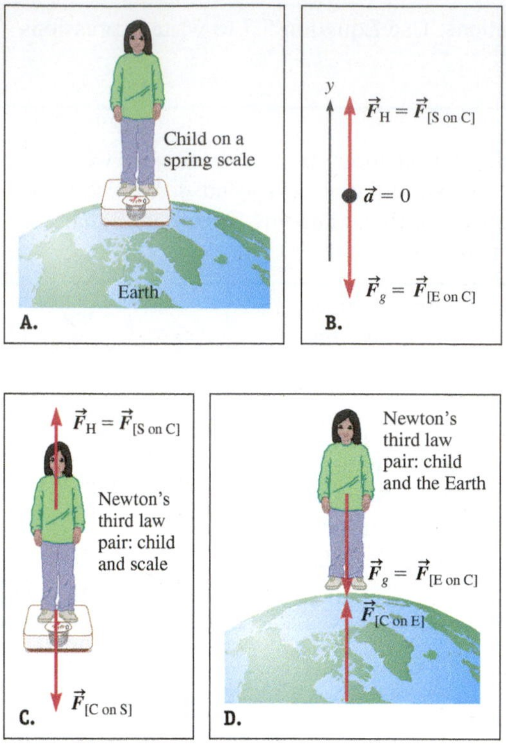

FIGURE 5.30 **A.** A child stands on a spring scale. **B.** A free-body diagram shows all the forces that act on the child, but not the forces that she exerts. According to Newton's second law and because the child is not accelerating, $\vec{F}_{[S\text{ on }C]} = -\vec{F}_{[E\text{ on }C]}$. **C.** The child interacts with the spring scale. According to Newton's third law, $\vec{F}_{[S\text{ on }C]} = -\vec{F}_{[C\text{ on }S]}$. **D.** The child also interacts with the Earth, and, by the third law, $\vec{F}_{[E\text{ on }C]} = -\vec{F}_{[C\text{ on }E]}$.

To further illuminate the distinction between the second and third laws, consider a child standing on a spring scale (Fig. 5.30A). When applying Newton's second law, we use a free-body diagram to represent all the forces acting on the child (Fig. 5.30B). *The forces for which the child is a source are never drawn in the free-body diagram for the child.* The free-body diagram shows two forces acting on her: the spring force $\vec{F}_H$, which in our $\vec{F}_{[\text{source on subject}]}$ notation becomes $\vec{F}_{[S\text{ on }C]}$; and gravity $\vec{F}_g$, which in our third-law notation is $\vec{F}_{[E\text{ on }C]}$. The subscripts stand for *S*pring, *C*hild, and *E*arth.

Parts C and D of Figure 5.30 help illustrate Newton's third law. They are *not* free-body diagrams, differing from a free-body diagram in three ways:

1. Each figure has two objects instead of one.
2. In addition to being acted on by external forces, each object is also the source of a force.
3. Most important, not all the forces exerted on each object are drawn. Instead, each figure shows only interacting force pairs.

In Figure 5.30C, the source of $\vec{F}_{[S\text{ on }C]}$ is the spring, and the subject is the child. This force forms an interaction pair with $\vec{F}_{[C\text{ on }S]}$, the force for which the child is the source and the spring is the subject. Similarly, Figure 5.30D shows the interaction force pair between the Earth and the child, $\vec{F}_{[E\text{ on }C]}$ and $\vec{F}_{[C\text{ on }E]}$. Because the child is not the subject of $\vec{F}_{[C\text{ on }S]}$ or $\vec{F}_{[C\text{ on }E]}$, these two forces do not appear in the free-body diagram for the child (Fig. 5.30B).

In some situations, the distinction between the second and third laws is further blurred because all the forces have the same magnitude. This child on the spring scale is such a case: $F_{[S\text{ on }C]} = F_{[E\text{ on }C]} = F_{[C\text{ on }S]} = F_{[C\text{ on }E]}$, and we must look at both the second law and the third law to see why. According to Newton's second law and because the child is not accelerating (Fig. 5.30B), the two forces acting on her are equal in magnitude:

$$\sum F_y = F_{[S\text{ on }C]} - F_{[E\text{ on }C]} = ma_y = 0$$
$$F_{[S\text{ on }C]} = F_{[E\text{ on }C]}$$

Notice that the two forces have different sources: the spring scale for $F_{[S\text{ on }C]}$ and the Earth for $F_{[E\text{ on }C]}$. Both of these forces are acting on the same subject (the child) as denoted by "on C" in the subscript.

We use Newton's third law to relate the magnitudes of forces that act on a pair of interacting objects (Fig. 5.30C). For the child-spring pair, we have

$$F_{[S\text{ on }C]} = F_{[C\text{ on }S]}$$

and for the child–Earth pair (Fig. 5.30D), we have

$$F_{[E\text{ on }C]} = F_{[C\text{ on }E]}$$

The wording of the subscripts should help you keep track of a third-law force pair. The subscripts should form a pair involving only two objects; for example, E and C in the case of [E on C] and [C on E].

EXAMPLE 5.10 Pushy Skaters

Two ice skaters initially at rest push off from each other as shown in Figure 5.31. The coefficient of kinetic friction between their skates and the ice is $\mu_k = 0.05$. Wilma weighs 6.45×10^2 N, and Mark has a mass of 74.8 kg. At the moment of the push-off, the magnitude of his acceleration is 4.58 m/s^2.

A Find the force with which Wilma pushed Mark.

FIGURE 5.31

INTERPRET and ANTICIPATE

To find the force responsible for Mark's 4.58-m/s^2 acceleration, start with a free-body diagram for him, **represented as usual by a dot** (Fig. 5.32). We have included a two-dimensional **coordinate system** and indicated the direction of his **acceleration**. All the **forces** exerted on him have been **drawn and labeled**. To help keep track of our symbols, we'll use lowercase f for all the forces acting on Mark and uppercase F for those acting on Wilma.

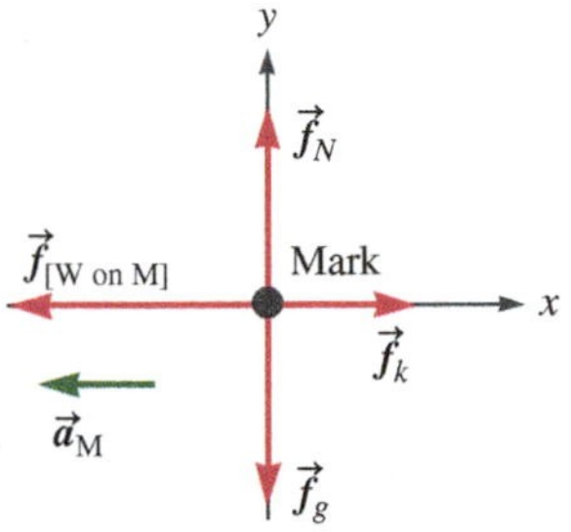

FIGURE 5.32

From our free-body diagram, we expect $\vec{f}_{[\text{W on M}]}$, the force exerted by Wilma on Mark, to be in the negative x direction. We anticipate a numerical answer of the form $\vec{f}_{[\text{W on M}]} = -\_\_\_\_\ \hat{\imath}$ N.

SOLVE

Step 1 Apply Newton's second law in component form. There are forces in both thc x and y directions, so there are two equations.

$$\sum F_x = f_k - f_{[\text{W on M}]} = ma_x \quad (1)$$

$$\sum F_y = f_N - f_g = ma_y \quad (2)$$

Mark is not accelerating in the direction perpendicular to the ice (the y direction), so $a_y = 0$ in Equation (2).

$$f_N - f_g = 0$$

$$f_N = f_g \quad (3)$$

Mark's acceleration (magnitude 4.58 m/s^2) must be entirely in the negative x direction. Write Equation (1) in terms of his acceleration a_{Mx}.

$$\vec{a}_x = -a_{Mx}\hat{\imath} = -4.58\,\hat{\imath} \text{ m/s}^2$$

$$f_k - f_{[\text{W on M}]} = -ma_{Mx} \quad (4)$$

Step 2 Write other force equations. Use Equation 5.7 to express Mark's weight.

$$f_g = mg \quad (5)$$

We also need Equation 5.9 for kinetic friction.

$$f_k = \mu_k f_N \quad (6)$$

Step 3 Do algebra. To find $\vec{f}_{[\text{W on M}]}$, start by eliminating f_N from Equation (6) with Equation (3).

$$f_k = \mu_k f_g \quad (7)$$

Substitute Equation (5) into Equation (7) to write the frictional force in terms of Mark's weight.

$$f_k = \mu_k mg \quad (8)$$

Substitute Equation (8) into Equation (4) and solve for force exerted by Wilma on Mark.

$$\mu_k mg - f_{[\text{W on M}]} = -ma_{Mx}$$

$$f_{[\text{W on M}]} = ma_{Mx} + \mu_k mg$$

$$f_{[\text{W on M}]} = m(a_{Mx} + \mu_k g)$$

Insert numerical values and write the final answer in component form.

$$f_{[\text{W on M}]} = (74.8 \text{ kg})[4.58 \text{ m/s}^2 + (0.05)(9.81 \text{ m/s}^2)]$$

$$\vec{f}_{[\text{W on M}]} = -3.79 \times 10^2\,\hat{\imath} \text{ N}$$

CHECK and THINK

Our answer is in the form we expected.

Example continues on page 144 ▶

B Find Wilma's acceleration.

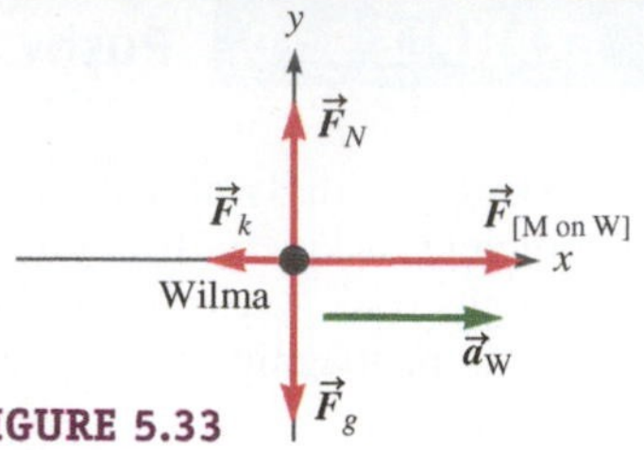

FIGURE 5.33

INTERPRET and ANTICIPATE

To find her acceleration, we need a free-body diagram for Wilma (Fig. 5.33). Her free-body diagram contains all four elements and is similar to Mark's diagram.

SOLVE

Step 1 Apply Newton's second law in component form. The results are similar to Equations (1) and (2) for Mark.

$$\sum F_x = F_{[M\text{ on }W]} - F_k = Ma_{Wx} \quad (9)$$

$$\sum F_y = F_N - F_g = Ma_{Wy} \quad (10)$$

Like Mark, Wilma accelerates only in the x direction, so $a_{Wy} = 0$ in Equation (10).

$$F_N - F_g = 0$$

$$F_N = F_g \quad (11)$$

Step 2 Write other force equations. Use Equation 5.7 to express Wilma's weight. We also need Equation 5.9 for kinetic friction.

$$F_g = Mg \quad (12)$$

$$F_k = \mu_k F_N \quad (13)$$

According to Newton's third law, the force exerted by Mark on Wilma has the same magnitude as the force she exerts on him, but in the opposite direction. We found the force she exerts on him in part A.

$$\vec{F}_{[M\text{ on }W]} = -\vec{f}_{[W\text{ on }M]} = 3.79 \times 10^2\,\hat{\imath}\text{ N} \quad (14)$$

Step 3 Do algebra. Combine Equations (11) and (13) to find an expression for kinetic friction in terms of Wilma's weight.

$$F_k = \mu_k F_g \quad (15)$$

Substitute Equation (15) into Equation (9), and solve for her acceleration a_{Wx}.

$$F_{[M\text{ on }W]} - \mu_k F_g = Ma_{Wx}$$

$$a_{Wx} = \frac{1}{M}(F_{[M\text{ on }W]} - \mu_k F_g)$$

We were given Wilma's weight, but not her mass. Use Equation (12) to express her mass in terms of her weight: $M = F_g/g$.

$$a_{Wx} = \frac{g}{F_g}(F_{[M\text{ on }W]} - \mu_k F_g)$$

$$a_{Wx} = g\left(\frac{F_{[M\text{ on }W]}}{F_g} - \mu_k\right)$$

Substitute numerical values, including Equation (14) for the force Mark exerts on Wilma.

$$a_{Wx} = (9.81\text{ m/s}^2)\left(\frac{3.79 \times 10^2\text{ N}}{6.45 \times 10^2\text{ N}} - 0.05\right)$$

$$\vec{a}_W = 5.27\,\hat{\imath}\text{ m/s}^2$$

CHECK and THINK

According to Newton's *third* law, each skater exerted the same magnitude of force on the other skater. According to Newton's *second* law, the skater with the smaller mass had a higher acceleration. Because the direction of the force on Wilma was opposite the direction of the force on Mark, her acceleration was in the opposite direction, too.

EXAMPLE 5.11 Superman of the Insect World

In 2003, Malcolm Burrows of the University of Cambridge, England, reported that the 6-mm froghopper insect could leap up 70 cm, equivalent to a person leaping up 70 stories.[3] Burrows reported that the best jumps have an acceleration of 408g at an angle of $(58 \pm 2.6)°$. The mass of his froghopper was (12.3 ± 0.7) mg.

A Identify the source of the force that propels the froghopper upward.

[3]Malcom Burrows, "Froghopper Insects Leap to New Heights," *Nature* 424:509 (2003).

SOLVE

To jump, the froghopper must push downward on the ground. By Newton's third law, the ground must therefore push upward on the froghopper, and in the opposite direction as shown in Figure 5.34. The vertical component of the force exerted by the ground is the normal force acting on the insect; the horizontal component is due to static friction, which we will study in detail in Chapter 6.

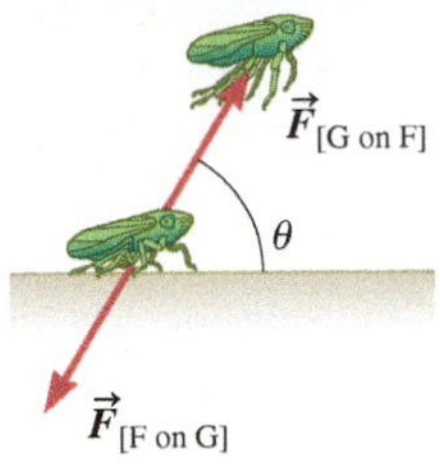

FIGURE 5.34

B Assume a 12.3-mg froghopper leaps at a 60° angle with the ground and accelerates at 408g. What is the magnitude of the force the froghopper exerts on the ground? *Hint*: Make a simplification.

INTERPRET and ANTICIPATE

Start by drawing a free-body diagram for the froghopper including all four elements as in Figure 5.35. The froghopper is **represented by a dot**. The **forces are drawn and labeled**. There are two forces acting on the froghopper: $\vec{F}_{[G\ on\ F]}$ due to the ground and $\vec{F}_g$ due to gravity. The **acceleration** has been indicated, and a **coordinate system** has been chosen so that the y axis lines up with the acceleration (often a good choice).

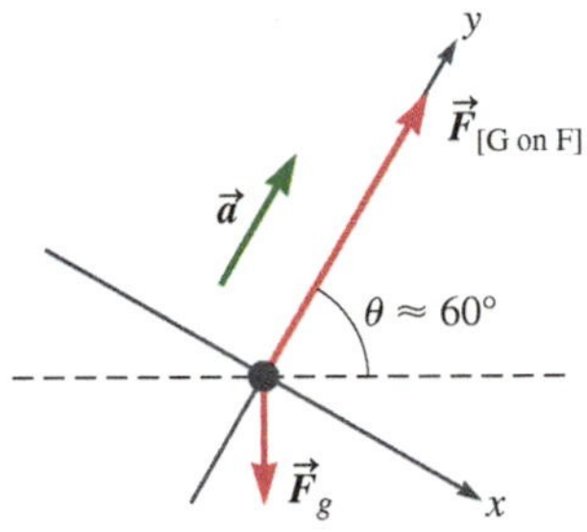

FIGURE 5.35

SOLVE

Step 1 Apply Newton's second law in component form. We could solve this problem exactly by applying Newton's second law in component form along both the x and y axes, but that step is not necessary. Because the froghopper is so light, the gravitational force on it is much smaller than the force exerted by the ground. Let's assume we can ignore the gravitational force for now, and find the force exerted by the ground on the froghopper. If the magnitude of that force is comparable to the froghopper's weight, we will rework the problem and include gravity.

Because we are ignoring gravity, we simply apply Newton's second law in the y direction.	$\sum F_y = ma_y$ $F_{[G\ on\ F]} = ma_y$

Step 2 Write down relevant force equations. There are no other relevant force equations.

Step 3 Do algebra. The force we are looking for is already isolated on the left side of the equation. We only need to substitute the given value of the acceleration.	$F_{[G\ on\ F]} = ma_y = 408mg$ (1) $F_{[G\ on\ F]} = (408)(12.3 \times 10^{-6}\,\text{kg})(9.81\ \text{m/s}^2)$ $F_{[G\ on\ F]} = 4.92 \times 10^{-2}\,\text{N}$

CHECK and THINK

According to Equation (1), the force exerted by the ground on the insect is more than 400 times its weight, so it is reasonable to neglect gravity. By Newton's third law, the force exerted by the froghopper on the ground $(F_{[F\ on\ G]})$ is equal in magnitude to this force but in the opposite direction. In other words, the force exerted by the froghopper is more than 400 times its weight. Imagine exerting more than 60,000 pounds with your legs!

CONCEPT EXERCISE 5.11

A child jumping off the monkey bars at a playground accelerates toward the ground because of the gravitational force exerted on him by the Earth. From Newton's third law, the force exerted by the Earth on the child is equal in magnitude to the force exerted by the child on the Earth. Does the Earth accelerate? Explain.

CONCEPT EXERCISE 5.12

CASE STUDY Train Collision and Newton's Third Law

Avi, Cameron, and Shannon continue to discuss the train collision case study. Use Newton's third law to decide which student is correct. Explain your answer.

Cameron: The passenger train got shoved backward. That means that the freight train did all the pushing. The passenger train couldn't push on the freight train because the passenger train wasn't moving.

Shannon: No. It's not like the freight train exerted all the force. They both exerted a force on each other.

Avi: Shannon's right. Newton's third law says that they have to push on each other. The freight train just pushed harder than the passenger train so that the net force was in the direction the freight train was moving.

Shannon: The force is the same on each train. The passenger train got shoved backward because it was lighter than the freight train.

Avi: I think the freight train pushed harder because it was heavy. Look, the freight train suffered almost no damage and the passenger train was a wreck. The passenger train had to have been hit harder.

5-10 Fundamental Forces

In this chapter, we studied five specific forces: gravity, the spring force, the normal force, the tension force, and kinetic friction. Although we will encounter many other forces throughout this text, the modern theory of physics states that there are only four fundamental forces: gravitational, electromagnetic, strong, and weak. The gravitational force was introduced previously and will be revisited often.

The electromagnetic force is discussed in Parts III and IV of this book. It is the force between electrical charges and is an important part of our daily lives, governing computers, motors, and even human nerve function and thought. In Section 5-7, we found that tension and the normal force can be understood in terms of the molecular bonds between atoms. These bonds are due to electromagnetic forces. In fact, most of the forces we study in this book are really a consequence of the electromagnetic force.

The strong force and the weak force are not studied either in classical mechanics or in electricity and magnetism, so they are briefly mentioned in this book. The strong force keeps protons (and neutrons) together in the nucleus of an atom. Without the strong force, the electromagnetic force between any two protons would cause them to repel each other (with the result that atoms could not form and the Universe would be a very different place). The weak force acts in the nucleus of atoms. Neutrons are unstable outside the nucleus, and the weak force governs how they decay as a result of this instability. The weak force also plays a role in fusion reactions that power the Sun.

SUMMARY

Underlying Principles: Newton's three laws of motion

1. Close translation of **Newton's first law**:
   Every body continues in its state of rest, or of uniform motion in a straight line, unless it is compelled to change that state by forces impressed on it.

   More contemporary statement of Newton's first law:
   If no force acts on an object, then the object cannot accelerate.

   Newton's second law contains his first law:
   The net force on an object is zero if and only if the acceleration of the object is also zero.

Underlying Principles: Newton's three laws of motion—cont'd

2. **Newton's second law**: The acceleration of an object is proportional to the total force acting on it:

$$\vec{F}_{tot} \equiv \sum \vec{F} = m\vec{a} \quad (5.1)$$

Newton's second law is often applied in component form:

$$\sum F_x = ma_x \quad \sum F_y = ma_y \quad \sum F_z = ma_z \quad (5.2)$$

3. **Newton's third law**: If two objects A and B interact, the force exerted by A on B is equal in magnitude to the force exerted by B on A but in the opposite direction. Mathematically,

$$\vec{F}_{[B\text{ on }A]} = -\vec{F}_{[A\text{ on }B]} \quad (5.10)$$

Major Concepts

1. **Dynamics:** a branch of mechanics that focuses on the causes of motion.
2. **Force:** a push or pull; required to make an object accelerate. A force has three properties:
   a. Force does not exist in isolation; it must act on a "subject."
   b. If a force is exerted on a subject, there must be some "source" responsible for that force.
   c. Force is a vector quantity having both magnitude and direction.
3. **System**: any collection of objects.
4. **Inertia**: the tendency of an object to maintain its constant velocity. **Mass**, also known as **inertial mass**, is the intrinsic property of an object that determines its inertia.
5. **Inertial reference frame**: one in which Newton's first law is valid. Inertial reference frames may move with constant velocity, but they do not accelerate. Accelerating frames are known as **noninertial** reference frames, and Newton's first law is not valid in them.

Special Cases

Table 5.1 on page 134 summarizes the five specific forces examined in this chapter.

Tools

Four elements of a **free-body diagram**:

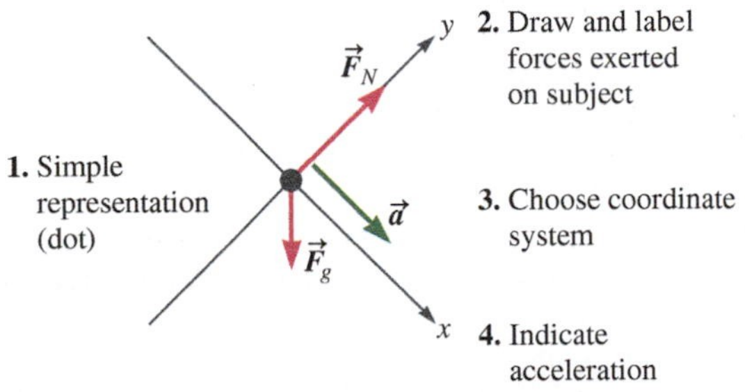

1. Simple representation (dot)
2. Draw and label forces exerted on subject
3. Choose coordinate system
4. Indicate acceleration
y
x

PROBLEM-SOLVING STRATEGY

Applying Newton's Second Law, $\vec{F}_{tot} \equiv \sum \vec{F} = m\vec{a}$

INTERPRET and ANTICIPATE

Draw a free-body diagram.

SOLVE

1. Apply Newton's second law in component form. Write one equation for each direction that has at least one force.
2. Write any other relevant force equations.
3. Do algebra before substituting values.

PROBLEMS AND QUESTIONS

A = algebraic C = conceptual E = estimation G = graphical N = numerical

5-1 Our Experience with Dynamics

1. **C** Why is it easier to lift a very large beach ball than a small lead ball?
2. **C** A heavy bowling ball and a light soccer ball are both rolling toward you at the same speed. Which object is easier to stop with your foot? Which is more likely to break your toe?
3. **C** Imagine pushing two blocks on ice. The light block slides farther than the heavy block. Why?

5-2 Newton's First Law

4. **C** When Julia Child would cook an omelet, she would rapidly jostle the pan back and forth (Fig. P5.4). The egg would slosh

back and forth in the pan as it cooked. Use Newton's laws to explain the egg's motion.

FIGURE P5.4

5. **C** In a science-fiction movie, the hero must get from one spaceship to another, which is several kilometers away. The hero (wearing the appropriate boots) stands on his ship and runs along its surface toward the other ship. When he reaches the edge of his ship, he jumps toward the other ship. Assuming he is far from any major sources of gravity such as planets and friction is negligible. Does he arrive with a lower speed, higher speed or about the same speed at which he left his ship? Explain.
6. **C** A car is traveling directly east for 90 minutes, yet there is no net force on the car during this time. Explain how this is possible.
7. **C** Determine whether at least one force is necessary for the following motions. A particle moves **a.** in a straight line at constant speed, **b.** in a straight line while slowing down and **c.** along an L-shaped path at constant speed.

5-3 Force

8. **C** A bar magnet is used to lift keys that have fallen into a sewer. The keys fly up a few millimeters to meet the bar magnet. Is the magnetic force a contact force or a field force? Explain.
9. **N** Two forces act on an object with $\vec{F}_1 = (7.263\hat{\imath} + 8.889\hat{\jmath})$N and $\vec{F}_2 = (-13.452\hat{\imath} + 7.991\hat{\jmath})$N. What is the magnitude of the net force experienced by the object?
10. **C** An object is subject to a single constant force. Under what circumstances does the object travel in a straight line, along a curved path, or along a circular path?
11. **N** Three forces act on an object with $\vec{F}_1 = (6.03\hat{\imath} - 10.64\hat{\jmath})$N and $\vec{F}_2 = (-3.71\hat{\imath} - 12.93\hat{\jmath})$N. If the net force on the object is zero, what is the unknown force, $\vec{F}_3$?
12. **C** You blow a small piece of paper through the air. Is the force on the paper a contact force or a field force? Explain.
13. **C** A fish swims in the ocean. Does the fish exert a force on the water? If so, is it a field force or a contact force? Explain.

5-4 Inertial Mass

14. **C** Read the following (fictional) dialogue between two people in 2066. Find at least two statements that are incorrect. Explain why they are incorrect.

Europa: I am making Thanksgiving dinner for 12 people. So, what weight turkey do I need?

Dallas: I would suggest about 1 kilogram per person. So, get a 12-kilogram turkey.

Europa: Oh, I forgot to mention that this dinner is for people on the Moon. I plan to shop in the Moon market. How do I find the turkey's weight on the Moon?

Dallas: That's easy. It still weighs 12 kilograms on the Moon.

15. **C** What has more inertia, a bowling ball at rest or the same ball when it is rolling? Explain.
16. **C** Do you have more inertia when you are sitting or when you are running? Explain.
17. **C** Who has more inertia, a child sitting on the sofa or a heavy man jogging?

5-5 Inertial Reference Frames

Problems 18, 19, and 42 are grouped.

18. **C** A ball hanging from a light string or rod can be used as an accelerometer (a device that measures acceleration) as shown in Figure P5.18. What force causes the deflection of the ball? Is the cart in the lower part of the photo an inertial reference frame? How can the ball's deflection be used to find the cart's acceleration? In which direction is the cart accelerating? Explain your answers.

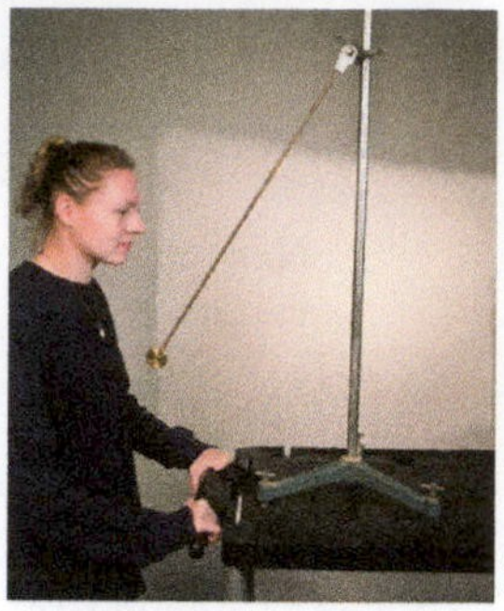

FIGURE P5.18 Problems 18, 19, and 42.

19. **C** Suppose the cart in Figure P5.18 is moving at constant velocity in a strong wind. What force causes the deflection of the ball? Is the cart an inertial reference frame? Explain.
20. **C** You are riding a luxury bus. In front of you is a cup of tea resting on the seat-back tray. Which of the following events may lead to spilled tea in your lap? The bus **a.** remains at rest, **b.** moves at constant velocity, **c.** speeds up or **d.** slows down. Don't worry about other circumstances such as a person knocking your cup over. (More than one choice may be correct.) Explain your answers.

5-6 Newton's Second Law

21. **N** A particle's acceleration is $\vec{a} = (3.45\hat{\imath} - 1.84\hat{\jmath})$ m/s^2, and the total force exerted on the particle is $\vec{F}_{tot} = (10.8\hat{\imath} - 5.76\hat{\jmath})$ N. What is the particle's mass?
22. **N** A particle with mass $m = 4.00$ kg accelerates according to $\vec{a} = (-3.00\hat{\imath} + 2.00\hat{\jmath})$ m/s^2. **a.** What is the net force acting on the particle? **b.** What is the magnitude of this force?
23. **N** The x and y coordinates of a 4.00-kg particle moving in the xy plane under the influence of a net force F are given by $x = t^4 - 6t$ and $y = 4t^2 + 1$, with x and y in meters and t in seconds. What is the magnitude of the force F at $t = 4.00$ s?
24. **C** In the movie *Garden State*, one of the characters uses a bow to launch an arrow straight up into the air. The arrow travels straight up, stops momentarily, and then comes straight down along the same path. The graph of its position versus time is a parabola. What can be said about the sum of the forces on the arrow?
25. **N** The starship *Enterprise* has its tractor beam locked onto some valuable debris and is trying to pull it toward the ship. A Klingon battle cruiser and a Romulan warbird are also trying to recover the item by pulling the debris with their tractor beams as shown in Figure P5.25. **a.** Given the following magnitudes of the tractor beam forces, find the net force experienced by the debris: $F_{Ent} = 7.59 \times 10^6$ N, $F_{Rom} = 2.53 \times 10^6$ N, and $F_{Kling} = 8.97 \times 10^5$ N. **b.** If the debris has a mass of 2549 kg, what is the net acceleration of the debris?

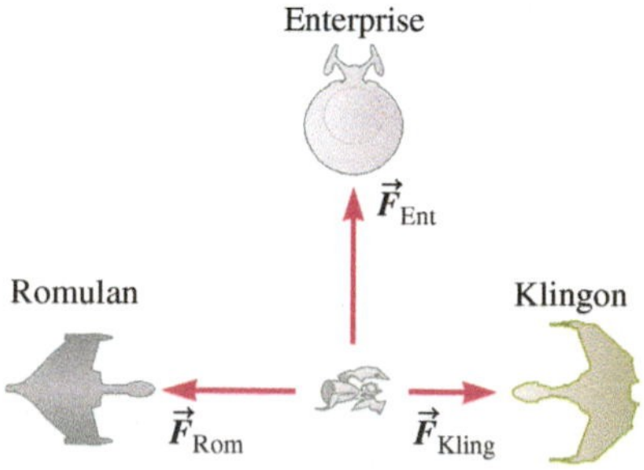

FIGURE P5.25

26. **C** A race car is moving around a circular track at a constant speed of 65 m/s. It continues moving in this manner at this speed for 15 minutes. During this time, what is the direction of the net force exerted on the race car?

27. **N** A particle of mass m_1 accelerates at 4.25 m/s^2 when a force F is applied. A second particle of mass m_2 experiences an acceleration of only 1.25 m/s^2 under the influence of this same force F. **a.** What is the ratio of m_1 to m_2? **b.** If the two particles are combined into one particle with mass $m_1 + m_2$, what is the acceleration of this particle under the influence of the force F?

28. **N** Jim Bob needs to tow his truck home using his neighbor's SUV. The SUV pulls the truck by exerting a horizontal force of 3133 N. If this force alone causes the truck to accelerate at a rate of 2.91 m/s^2, what is the mass of Jim Bob's truck?

29. **N** Two forces $\vec{F}_1 = (62.98\hat{\imath} - 15.80\hat{\jmath})$N and $\vec{F}_2 = (23.66\hat{\imath} - 78.05\hat{\jmath})$N are exerted on a particle. The particle's mass is 14.23 kg. **a.** Find the particle's acceleration in component form. **b.** What are the magnitude and direction of the acceleration?

30. **N** Three forces $\vec{F}_1 = (62.98\hat{\imath} - 15.80\hat{\jmath})$N, $\vec{F}_2 = (23.66\hat{\imath} - 78.05\hat{\jmath})$N, and $\vec{F}_3 = (-86.64\hat{\imath} + 233.4\hat{\jmath})$N are exerted on a particle. The particle's mass is 14.23 kg. Find the particle's acceleration.

31. **N** A hockey stick pushes a 0.160-kg puck with constant force across the frictionless surface of an ice rink. During this motion, the puck's velocity changes from $4.00\hat{\imath}$ m/s to $(6.00\hat{\imath} + 12.00\hat{\jmath})$ m/s in 4.00 s. **a.** What are the scalar components of the force acting on the puck? **b.** What is the magnitude of the force acting on the puck?

Problems 32 and 33 are paired.

32. **N** If the vector components of the position of a particle moving in the xy plane as a function of time are $\vec{x}(t) = \left(2.5\,\frac{\text{m}}{\text{s}^2}\right)t^2\hat{\imath}$ and $\vec{y}(t) = \left(5.0\,\frac{\text{m}}{\text{s}^3}\right)t^3\hat{\jmath}$, when is the angle between the net force on the particle and the x axis equal to 45°?

33. **A** If the vector components of the position of a particle moving in the xy plane as a function of time are $\vec{x}(t) = bt^2\hat{\imath}$ and $\vec{y}(t) = ct^3\hat{\jmath}$, where b and c are positive constants, b has dimensions of length per time squared, and c has dimensions of length per time cubed, when is the angle between the net force on the particle and the x axis equal to 45°?

5-7 Some Specific Forces

34. **N** A 15.0-kg object is in free fall near the surface of the Earth. What is its weight? What is its acceleration? What is the direction of the gravitational force exerted on it? How do your answers change if the same object is at rest on the surface of the Earth?

35. **N** A black widow spider hangs motionless from a web that extends vertically from the ceiling above. If the spider has a mass of 1.5 g, what is the tension in the web?

36. **C** Determine whether each of the following statements is true or false. **a.** An object's weight is always equal to its mass. **b.** The force of tension always pushes. **c.** The magnitude of the sum of the forces on an object is never greater than its weight. Explain.

37. **N** You place tomatoes in the pan of a hanging spring scale and find that they weigh 2.5 lb. You measure the downward displacement of the scale's pan to be 1.5 in. What is the spring constant? Give your answer in lb/in. and in SI units.

38. **C** Kinetic friction is proportional to the normal force (Eq. 5.9). Why should there be an intimate connection between these two forces?

39. **N** A student takes the elevator up to the fourth floor to see her favorite physics instructor. She stands on the floor of the elevator, which is horizontal. Both the student and the elevator are solid objects, and they both accelerate upward at 5.19 m/s^2. This acceleration only occurs briefly at the beginning of the ride up. Her mass is 80.0 kg. What is the normal force exerted by the floor of the elevator on the student during her brief acceleration?

40. A sleigh is being pulled horizontally by a train of horses at a constant speed of 8.05 m/s. The magnitude of the normal force exerted by the snow-covered ground on the sleigh is 6.37×10^3 N.
   a. **N** If the coefficient of kinetic friction between the sleigh and the ground is 0.23, what is the magnitude of the kinetic friction force experienced by the sleigh?
   b. **C** If the only other horizontal force exerted on the sleigh is due to the horses pulling the sleigh, what must be the magnitude of this force?

5-8 Free-Body Diagrams

41. **N** Two blocks are connected by a rope that passes over a massless and frictionless pulley as shown in Figure P5.41. Given that $m_1 = 15.93$ kg and $m_2 = 10.45$ kg, determine the magnitudes of the tension in the rope and the blocks' acceleration.

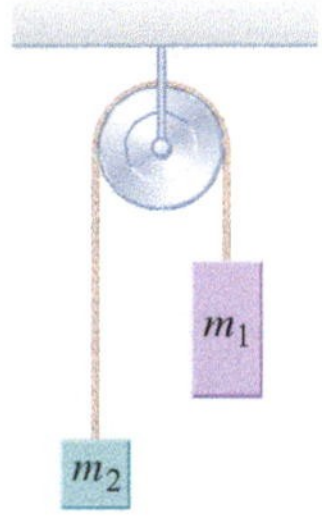

FIGURE P5.41

42. **A** Find an expression for the cart's acceleration in Figure P5.18 in terms of the ball's mass m and the angle θ.

43. **A** A woman uses a rope to pull a block of mass m across a level floor at a constant velocity. The coefficient of kinetic friction between the block and the floor is μ_k. The rope makes an angle θ with the floor. Find an algebraic expression for the tension in the rope in terms of the parameters listed in the problem and any constants.

44. **N** A student working on a school project modeled a trampoline as a spring obeying Hooke's law and measured the spring constant of a certain trampoline as 4617 N/m. If a child of mass 27.0 kg compresses the trampoline vertically by a maximum of 0.25 m, while bouncing up and down, what is the child's acceleration at the moment of maximum compression?

45. **N** One great form of athletic competition for bulldogs, American pit bull terriers, huskies, and many other breeds is the weight pull. Many of these dogs can pull weights two orders of magnitude greater than their own weight! Railsplitter, an American pit bull terrier, can pull a sled that has a total weight of 2215 lb along a horizontal surface. Suppose he pulls with a horizontal force of 3.50×10^3 N while friction is also working against the sled. If the magnitude of the net acceleration of the sled is 0.152 m/s^2, find the value of the coefficient of kinetic friction, μ_k, between the sled and the pulling surface. (The sled is normally on a set of rails.)

46. A heavy crate of mass 50.0 kg is pulled at constant speed by a dockworker who pulls with a 345-N force at an angle θ with the horizontal (Fig. P5.46). The magnitude of the friction force between the crate and the pavement is 212 N.
   a. **G** Draw a free-body diagram of the forces acting on the crate.
   b. **N** What is the angle θ of the rope with the horizontal?
   c. **N** What is the magnitude of the normal force exerted by the pavement on the crate?

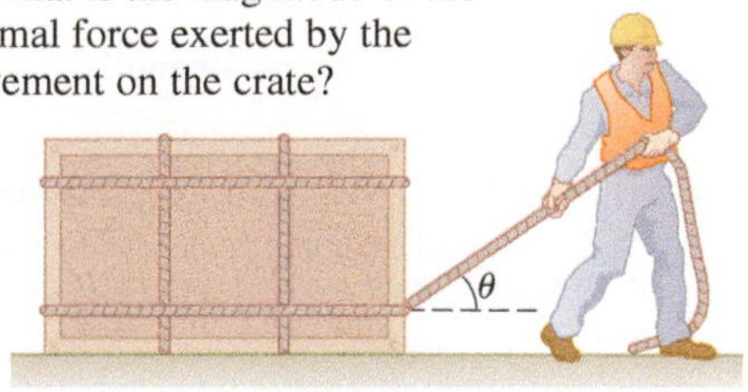

FIGURE P5.46

47. **N** A block with mass m_1 hangs from a rope that is extended over an ideal pulley and attached to a second block with mass m_2 that sits on a ledge. The second block is also connected to a third block with mass m_3 by a second rope that hangs over a second ideal pulley as shown in Figure P5.47. If the friction between the ledge and the second block is negligible, $m_1 = 3.00$ kg, $m_2 = 5.00$ kg, and $m_3 = 8.00$ kg, find the magnitude of the tension in each rope and the acceleration of each block.

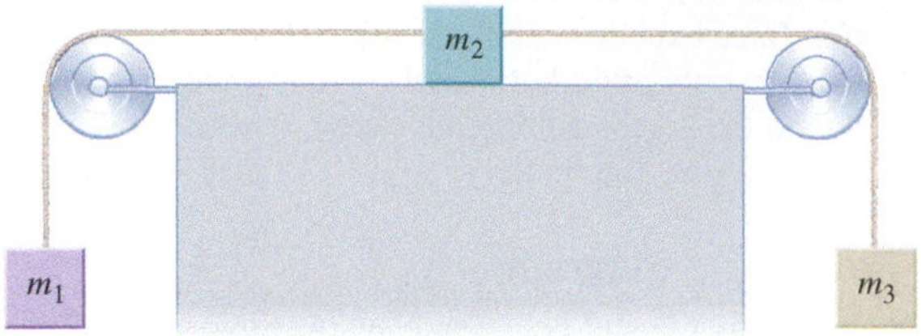

FIGURE P5.47

48. **N** To get in shape, you head to the local gym to exercise by lifting weights. Using a lat machine (Fig. P5.48), you notice that the wire connected to the bar is run around three different pulleys and is connected to a vertical stack of weights on the other end. The weights move up and down as you pull down on the bar. Suppose you set the machine so that the weight to be lifted is 120.0 lb. When you pull down on the bar, exerting a constant force, the acceleration experienced by the weights is 0.250 m/s^2. What tension must exist in the wire connecting the bar and weights under these conditions? Assume the pulleys are massless and frictionless.

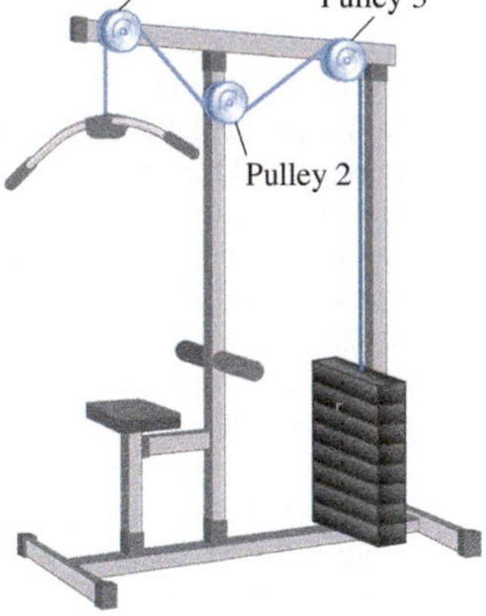

FIGURE P5.48

Problems 49 and 50 are paired.

49. **N** A block with mass m_1 hangs from a rope that is extended over an ideal pulley and attached to a second block with mass m_2 that sits on a ledge slanted at an angle of 20° (Fig. P5.49). Suppose the system of blocks is initially motionless and held still, and then it is released. If $m_1 = 7.00$ kg and $m_2 = 2.00$ kg, find the magnitude of the acceleration of the blocks, assuming there is no friction between the second block and the ledge.

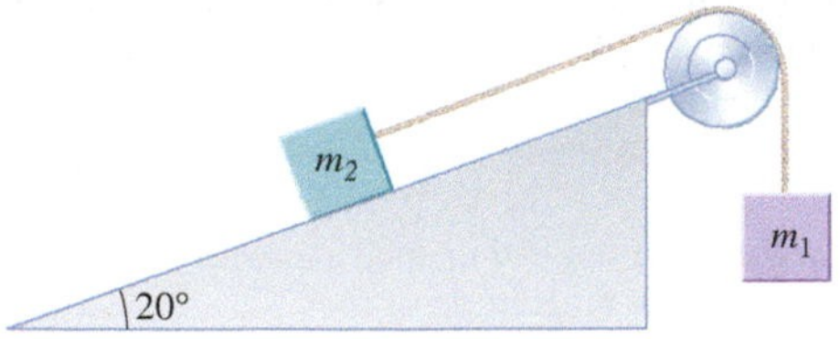

FIGURE P5.49 Problems 49 and 50.

50. **N** Suppose the system of blocks in Problem 49 is initially held motionless and, when released, begins to accelerate. **a.** If $m_1 = 7.00$ kg, $m_2 = 2.00$ kg, and the magnitude of the acceleration of the blocks is 0.134 m/s^2, find the magnitude of the kinetic frictional force between the second block and the ledge. **b.** What is the value of the coefficient of kinetic friction between the block and the ledge?

51. Two objects, $m_1 = 3.00$ kg and $m_2 = 8.50$ kg, are attached by a massless cord passing over a frictionless pulley as shown in Figure P5.51. Assume the horizontal surface is frictionless.
    a. **G** Draw a free-body diagram for each of the two objects.
    b. **N** What is the tension in the cord?
    c. **N** What is the magnitude of the acceleration of the two objects?

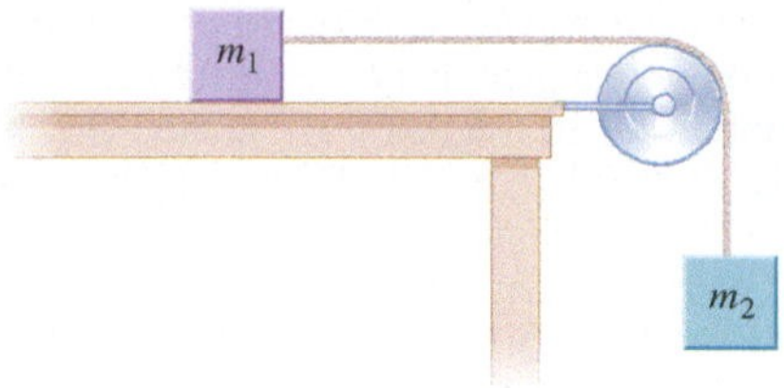

FIGURE P5.51 Problems 51 and 65.

52. A runaway piano starts from rest and slides down a 20.0° frictionless incline 5.00 m in length.
    a. **G** Draw a free-body diagram of the piano.
    b. **N** What is the acceleration of the piano?
    c. **N** What is the speed of the piano at the bottom of the incline?

5-9 Newton's Third Law

53. **C** Does the ground need to exert a force on you for you to jump off the ground, or do you need to exert a force on the ground? If the ground must exert a force on you, is that force greater than the force you exert on the ground?
54. **C** A boxer breaks his hand by punching another boxer's jaw. How is it possible to break your hand when you exert such a force?
55. Paul places a toy block of mass m on a hockey puck of mass M and sets the two in motion at constant velocity on the surface of a frozen lake.
    a. **G** Draw a free-body diagram of the block, identifying each of the forces acting on the block.
    b. **G** Draw a free-body diagram of the hockey puck, identifying each of the forces acting on the puck.
    c. **C** Which are the action and reaction pairs of forces in the block–puck–frozen lake system?
56. **C** A textbook rests on a movable wooden plank that is initially parallel to the ground. **a.** How does the normal force on the book compare to the gravitational force on the book as it rests in the horizontal position? **b.** If you push down on the book, what happens to the magnitude of the normal force as it rests in the horizontal position? **c.** The normal force on the book is part of a third-law interaction pair. Describe the third-law partner of this normal force.
57. An astronaut is stationary while floating in space, a short distance from the safety of her spacecraft. The mass of the astronaut including all her gear is 106.4 kg. Her tether to the craft becomes disconnected, and she needs to get back before she runs out of air. She removes her backpack unit, which has a mass of 30.1 kg and, putting herself between the backpack and the craft, pushes the pack with a force of 212 N directly away from the craft.
    a. **C** Explain how this action returns the astronaut to the craft.
    b. **N** What is the magnitude of the acceleration experienced by the backpack while the force is applied?
    c. **N** What is the magnitude of the acceleration experienced by the astronaut while the force is applied?
58. **C** A talking horse is attached to a wagon. The driver says, "Giddy-up, boy. Let's go!" The horse says, "I can't. When I

pull on the wagon, the wagon pulls back on me with a force of the same magnitude and in the opposite direction. The two forces cancel, so I cannot accelerate the cart." The horse is very clever, but doesn't really understand Newton's laws. Explain where the horse has gone wrong.

59. C When you dive off a cliff into the ocean, the Earth's gravity pulls down on your body, and you accelerate downward. According to Newton's third law, your body's gravity exerts a force on the Earth. In which direction is this force? Why doesn't the Earth accelerate in this direction at 9.81 m/s^2 as a result?

60. N A worker is attempting to lift a 55.0-kg palette of bricks resting on the ground by means of a rope attached to a pulley. **a.** Before the worker pulls on the rope, what is the force exerted by the ground on the palette? **b.** The worker exerts a force of 295 N downward on his end of the rope. What is the force exerted by the ground on the palette? **c.** If the worker doubles the downward force, what is the force exerted by the ground on the palette?

61. C CASE STUDY Return to the dialogue in Concept Exercise 5.12 and review the train collision. The passenger train was stopped on the tracks when it was hit by the freight train. Imagine instead that both trains were moving and both accelerating toward each other. Assume the freight train's acceleration is greater than that of the passenger train. In this case, what can you say about the relative force each train exerts on the other when they collide?

5-10 Fundamental Forces

62. C A concept map is a visual representation of concepts and their connections. Create a concept map including the five specific forces listed in Table 5.1 and the four forces discussed in Section 5.10. Justify the placement of each force.

General Problems

63. N A 75.0-g arrow, fired at a speed of 110 m/s to the left, impacts a tree, which it penetrates to a depth of 12.5 cm before coming to a stop. Assuming the force of friction exerted by the tree is constant, what are the magnitude and direction of the friction force acting on the arrow?

64. C An airplane is moving to the south with a speed of 125 m/s while maintaining a constant altitude. It continues moving in this direction at this speed for 15 seconds. During this time, what is the direction of the net force exerted on the airplane?

65. N A box with mass $m_1 = 6.00$ kg sliding on a rough table with a coefficient of kinetic friction of 0.220 is connected by a massless cord strung over a massless, frictionless pulley to a second box of mass $m_2 = 12.0$ kg hanging from the side of the table (Fig. P5.51). What is the tension in the cord connecting the boxes?

66. C One hundred and twenty-one forces act on an object that is at rest and stays at rest. If one of these forces is removed and stops acting on the object and all the others continue acting on the object with the same magnitudes and in the same directions, what would happen to the object?

67. N A cosmic ray muon with mass $m_\mu = 1.88 \times 10^{-28}$ kg impacting the Earth's atmosphere slows down in proportion to the amount of matter it passes through. One such particle, initially traveling at 2.50×10^8 m/s in a straight line, decreases in speed to 1.50×10^8 m/s over a distance of 1.20 km. **a.** What is the magnitude of the force experienced by the muon? **b.** How does this force compare to the weight of the muon?

68. N A boulder of mass 80.0 kg rests directly on a spring with a spring constant of 6850 N/m. **a.** What is the compression of the spring? **b.** Now, instead, the boulder and the spring are in an elevator accelerating upward at 5.19 m/s^2. What is the compression of the spring in this case?

69. C Only four forces act on an object. They all have the same magnitude. Their directions are north, south, northwest, and northeast. In which direction does the object accelerate?

70. N A 1.50-kg particle initially at rest and at the origin of an x–y coordinate system is subjected to a time-dependent force of $\vec{F}(t) = (3.00t\hat{\imath} - 6.00\hat{\jmath})$ N with t in seconds. **a.** At what time t will the particle's speed be 15.0 m/s? **b.** How far from the origin will the particle be when its velocity is 15.0 m/s? **c.** What is the particle's total displacement at this time?

Problems 71 and 72 are paired.

71. N A block of ice ($m = 15.0$ kg) with an attached rope is at rest on a frictionless surface. You pull the block with a horizontal force of 95.0 N for 1.54 s. **a.** Determine the magnitude of each force acting on the block of ice while you are pulling. **b.** With what speed is the ice moving after you are finished pulling?

72. N Repeat Problem 71, but this time you pull on the block at an angle of 20.0°.

73. A Ezra is pulling a sled, filled with snow, by pulling on a rope attached to the sled. The rope makes an angle θ with respect to the horizontal ground, and the sled is being pulled at a constant speed. If the sled and snow have a total mass of m, the acceleration due to gravity is g, the magnitude of the normal force is F_N, and the coefficient of kinetic friction between the sled and the ground is μ_k, what is the angle θ that the rope makes with the ground in terms of these five quantities?

74. N Starting from rest, a rectangular toy block with mass 300 g slides in 1.30 s all the way across a table 1.20 m in length that Zak has tilted at an angle of 42.0° to the horizontal. **a.** What is the magnitude of the acceleration of the toy block? **b.** What is the coefficient of kinetic friction between the block and the table? **c.** What are the magnitude and direction of the friction force acting on the block? **d.** What is the speed of the block when it is at the end of the table, having slid a distance of 1.20 m?

75. When a 1.50-kg dress hangs midway from a taut clothesline stretched between two poles planted 7.50 m apart, the clothesline is seen to sag 0.0500 m.

a. G Draw a free-body diagram of the dress.

b. N What is the tension produced on the clothesline by the dress? Assume the clothesline is massless.

76. N Jamal and Dayo are lifting a large chest, weighing 207 lb, by using the two rope handles attached to either side. As they lift and hold it up so that it is motionless, each handle makes a different angle with respect to the vertical side of the chest (Fig. P5.76). If the angle between Jamal's handle and the vertical side is 25.0° and the angle between Dayo's handle and the vertical side of the chest is 30.0°, what are the tensions in each handle?

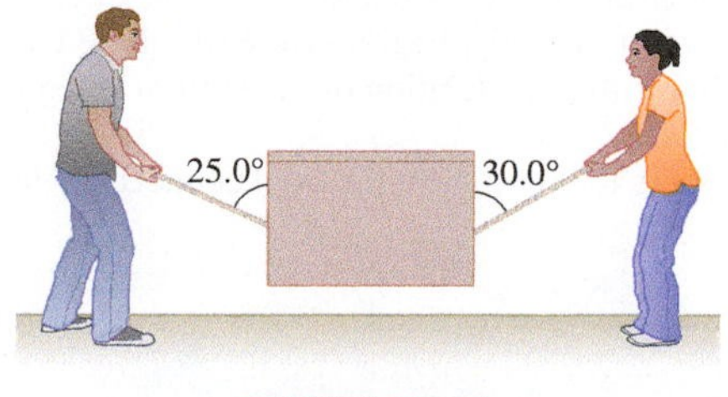

FIGURE P5.76

77. N A heavy chandelier with mass 125 kg is hung by chains in equilibrium from the ceiling of a concert hall as shown in Figure P5.77, with $\theta_1 = 37.0°$ and $\theta_2 = 64.0°$. Assuming the chains are massless, what are the tensions F_{T1}, F_{T2}, and F_{T3} in the three chains?

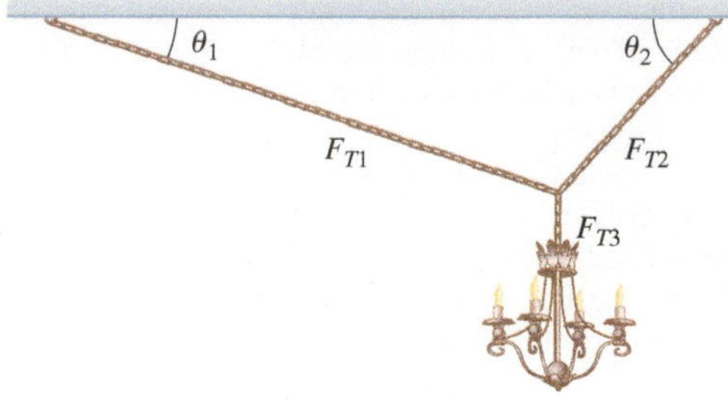

FIGURE P5.77

78. Two children, Raffi and John, sitting on sleds tied together with a massless rope, are being dragged across a frozen river by their playful Siberian husky Rex who is supplying a 112-N horizontal force (Fig. P5.78). The coefficient of friction between the sleds and the ice is 0.08, the combined mass of Raffi and his sled is 42.0 kg, and the combined mass of John and his sled is 51.0 kg.

a. G Draw a free-body diagram for each of the child–sled systems.
b. N What is the acceleration of the system?
c. N What is the tension F_T in the rope connecting the two sleds?

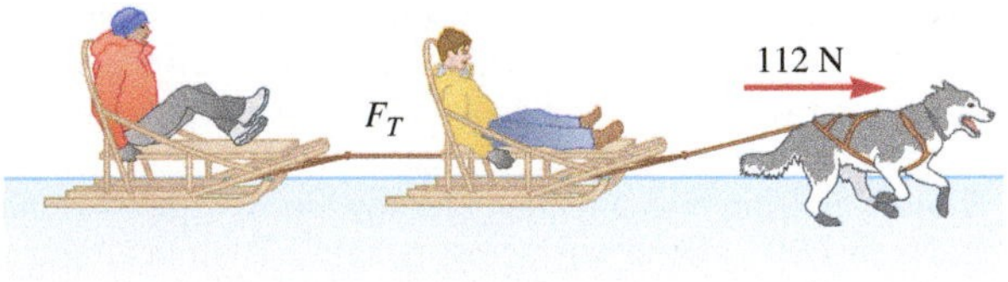

FIGURE P5.78

79. Two boxes with masses $m_1 = 4.00$ kg and $m_2 = 10.0$ kg are attached by a massless cord passing over a frictionless pulley as shown in Figure P5.79. The incline is frictionless, and $\theta = 30.0°$.

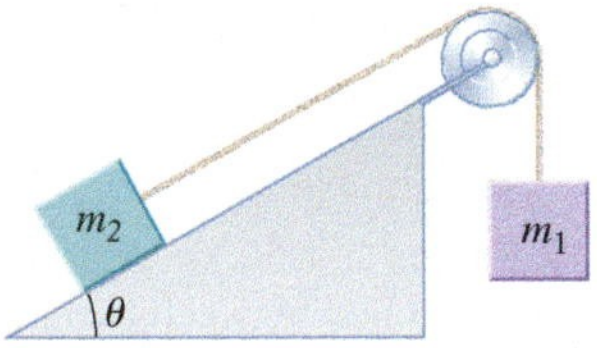

FIGURE P5.79

a. G Draw a free-body diagram for each of the boxes.
b. N What is the magnitude of the acceleration of the boxes?
c. N What is the tension in the cord connecting the boxes?
d. N What is the speed of each of the boxes 3.00 s after the system is released from rest?

80. Two blocks of mass $m_1 = 1.50$ kg and $m_2 = 5.00$ kg are connected by a massless cord passing over a frictionless pulley as shown in Figure P5.80, with $\theta = 35.0°$. The coefficient of kinetic friction between block 1 and the horizontal surface is 0.400, and the coefficient of kinetic friction between block 2 and the inclined surface is 0.330.

a. G Draw a free-body diagram for each of the two blocks.
b. N What is the acceleration of the system when it is released from rest?
c. N What is the tension in the cord connecting the blocks?

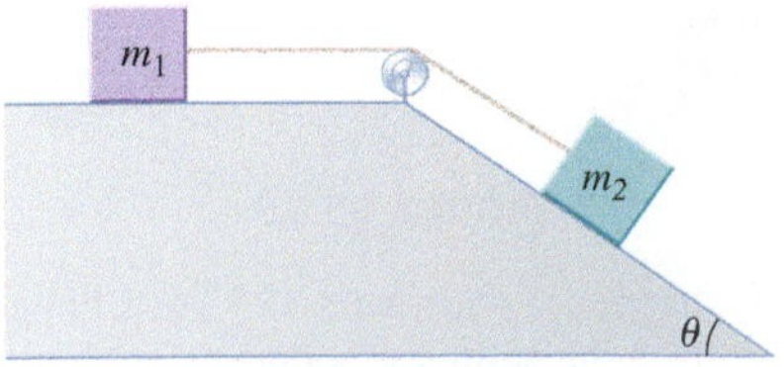

FIGURE P5.80

81. N An aerial demonstration aircraft dives at an angle θ, accelerating from 95.0 m/s to 145 m/s in 7.00 s (Fig. P5.81). The pilot has hung a 0.150-kg keepsake at the end of a string from the hinge on the aircraft's canopy. During the dive, the string of the keepsake is seen to remain perpendicular to the top of the canopy. **a.** What is the angle θ of the aircraft's dive? **b.** What is the tension in the string?

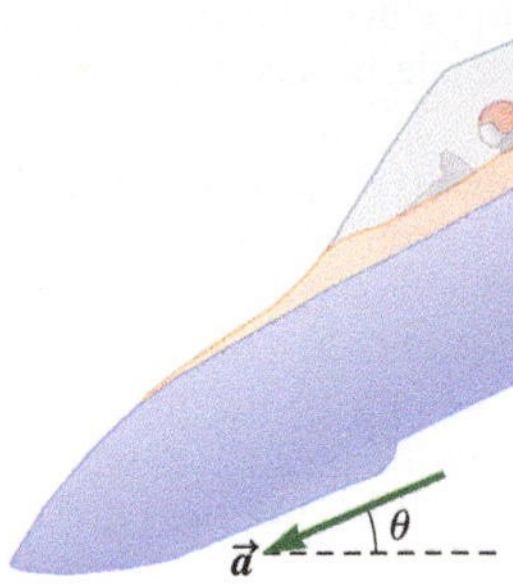

FIGURE P5.81

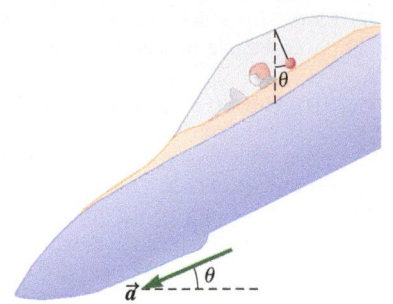

82. N A painter sits on a scaffold that is connected to a rope passing over a pulley. The other end of the rope rests in the hands of the painter who wants to lift the scaffold. She plans to pull downward on the loose end of the rope, thinking that the scaffold will then rise vertically with her along for the ride. The scaffold has a mass of 52 kg, and her mass is 63 kg. The painter pulls downward on the rope with a force of 600.0 N, while she and the scaffold are hanging from the other end above the ground. **a.** What is the net acceleration on the system consisting of the painter and the scaffold? **b.** What is the magnitude of the normal force exerted on the painter by the scaffold?

83. N Three crates with masses $m_1 = 5.45$ kg, $m_2 = 7.88$ kg, and $m_3 = 4.89$ kg are in contact on a frictionless surface. A horizontal force $F = 205$ N is applied to the third crate as shown in Figure P5.83. **a.** What is the magnitude of the contact force between crates 1 and 2? **b.** What is the magnitude of the contact force between crates 2 and 3?

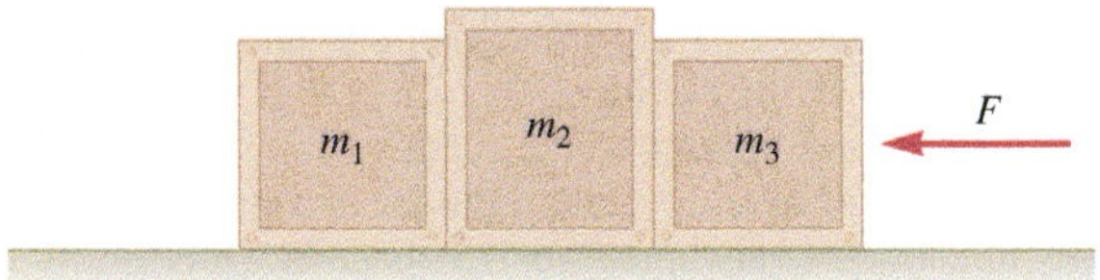

FIGURE P5.83

84. A A small block with mass m is set on the top of an upside-down hemispherical bowl. If the coefficient of static friction between the block and the bowl is μ_s and the block is slowly repositioned at different points down the surface of the bowl, at what angle measured from the vertical will the block begin to slide? Write your answer in terms of the mass, m; the gravitational acceleration on Earth, g; and the coefficient of static friction, μ_s.

Applications of Newton's Laws of Motion

6

Underlying Principle

Newton's laws of motion

Major Concepts

No new major concepts

Special Cases

1. Static friction (revisited)
2. Kinetic friction (revisited)
3. Rolling friction
4. Drag force
5. Terminal speed
6. Centripetal force
7. Nonuniform circular motion

Key Questions

How are Newton's laws of motion applied to complicated situations?

How do we handle friction and air resistance in real situations?

Newton's laws of motion help us understand many different situations. We can use them to figure out how skydivers fly in formation, to set a speed limit on a curved road, and to pick out a good pair of running shoes. Newton's three laws are at the heart of classical mechanics, and we will continue to use their power throughout the next three parts of this book.

In this chapter, we introduce no new principles or major concepts. Instead, we digest, refine, and synthesize the previously introduced principles as we practice applying them to practical situations.

6-1 Newton's Laws in a Messy World

Kinematics enables us to describe motion, and Newton's laws govern the *dynamics* (causes) of motion. In this chapter, we (1) practice using Newton's laws, (2) study a few new forces, and (3) weave kinematics and dynamics together. As a particularly important example of this last point, we revisit circular motion.

In this chapter, we tackle more complicated, realistic, and interesting problems that involve resistive forces such as friction and drag. Part of the reason we focus on resistive forces is that our intuition is based on their presence in our daily lives. We always need to reconcile our intuition with the laws of physics. The following **CASE STUDY** will help you sharpen your intuition. Later, you can compare that intuition with what the laws of physics say about the situation.

CASE STUDY Skydiving in Formation

A group of college students is working on physics homework in the study lounge. An inspirational poster hangs on the wall. It declares "teamwork" under a picture of skydivers, their parachutes not yet deployed, holding hands to form a large circle (Fig. 6.1). Avi notices the picture.

Courtesy, Federation Aeronautique Internationale

FIGURE 6.1 A poster shows skydivers flying in a formation like the one shown here. Avi wonders how this formation is possible.

Avi: Check out that picture. I don't get it. There are about 10 people holding hands. Did they all jump out of the plane together?

Shannon: I don't see how 10 people can jump out of a plane at the same time. Why do you ask?

Avi: Well, if they didn't jump out together, they can't all be together now. Think about it. Their parachutes are closed, which means that they are all in free fall. They each start with the initial velocity of the plane, and then they leave the plane. After that, only gravity acts on them. Each skydiver travels in her or his own parabolic trajectory (Fig. 6.2A). How can they possibly catch up to one another? They should each fall separately and land on the ground in the same order they jumped out of the plane.

Cameron: The problem is that what we learn in class has almost nothing to do with what happens in the real world. Physics works only in ideal conditions, like in our imagination or maybe in a perfect lab. The real world is just too messy for physics. The prof even said that real problems have to be solved on a computer.

Shannon: Okay. I see what Avi means. If all the skydivers are in free fall and if they left the plane separately, they cannot get back together while they are in the air. But, Cameron, just because we study physics under ideal conditions doesn't mean that everything outside of class happens by magic. Physics still holds. We just have to figure out what else is going on that makes a situation more complicated, like maybe the plane was flying at a downward angle instead of flying horizontally (Fig. 6.2B). That way, the second skydiver does not leave the plane until the door is at the level of the first skydiver, who's in free fall. The plane then continues at a slant so that it is always at the same level as the first and second skydivers, and then the third skydiver exits, and so on.

Avi: Shannon, I think you hit on something when you mentioned free fall, but I think it's the opposite. The divers are *not* in free fall. Instead, the air exerts a force on them. Remember that we learned that the range of a projectile is shorter if we take air into account.

Cameron: Okay. I see what you're both saying. The problem is harder than the stuff we did in class where the plane flies level and we ignore air, but the reason we do it that way is because the real world is too hard to study with physics. There is no way to figure it all out. All we do in physics is pretend that everything is a particle, even with two huge trains that are stuck together. Maybe not everything can be modeled as a particle. People jumping out of airplanes are not particles, so maybe we cannot pretend that they are.

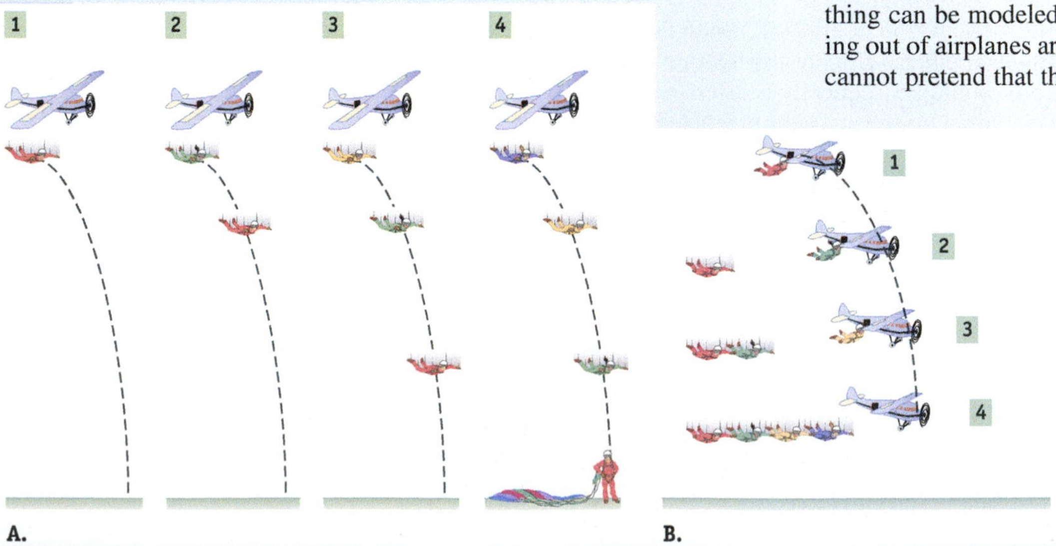

FIGURE 6.2 **A.** Avi explains that each skydiver falls along a separate parabolic path. **1** Only one skydiver has left the plane. **4** The last skydiver has left the plane, and the first one is already on the ground. Avi argues the skydivers never are close enough to hold hands in a circle. **B.** Shannon's explanation of how the skydivers can hold hands is that the plane loses altitude as it flies, remaining close to the skydivers as they exit the plane.

CONCEPT EXERCISE 6.1

CASE STUDY **Skydiving Arguments**

Take a moment to think about each student's argument.

a. Avi argues that the air exerts a force on the skydivers. Does that argument seem possible? If so, how could air help the skydivers create the formation we see in Figure 6.1? As you formulate your answers to these questions, think about this key question: Why do skydivers use parachutes in the first place?

b. According to Shannon's suggestion, the plane's altitude is constantly decreasing so that it is always alongside the divers. Does this situation mean that the plane must be accelerating? If so, what is that acceleration, and what sensation would a person on the plane experience? If not, what can you say about the motion of the plane?

c. Cameron argues that the physics we study in the classroom is good only for ideal conditions and that the world is more complicated than any situation we see in class. Cameron says people jumping out of airplanes are not particles and cannot be treated like particles. Do you agree or disagree? Explain your answer. It may be helpful to return to Chapter 2 to see how a particle was defined.

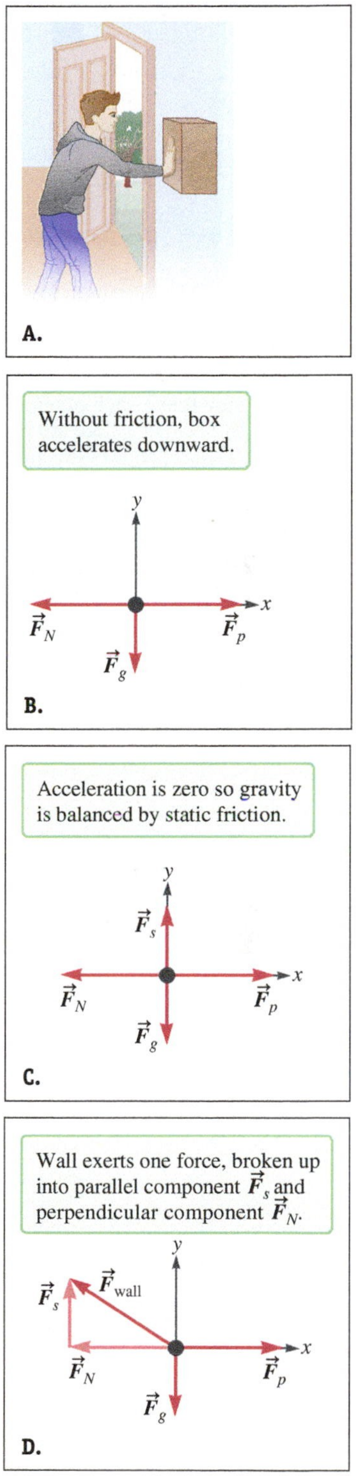
A.
Without friction, box accelerates downward.
y
x
$\vec{F}_N$
$\vec{F}_p$
$\vec{F}_g$
B.
Acceleration is zero so gravity is balanced by static friction.
y
$\vec{F}_s$
x
$\vec{F}_N$
$\vec{F}_p$
$\vec{F}_g$
C.
Wall exerts one force, broken up into parallel component $\vec{F}_s$ and perpendicular component $\vec{F}_N$.
y
$\vec{F}_s$
$\vec{F}_{\text{wall}}$
x
$\vec{F}_N$
$\vec{F}_p$
$\vec{F}_g$
D.

FIGURE 6.3 **A.** A person holds a box against a wall. **B.** If there were no friction, the box would accelerate downward. **C.** The box does not accelerate because static friction supports the box against gravity. **D.** The wall actually exerts one force that is broken into its parallel and perpendicular components for convenience.

6-2 Friction and the Normal Force Revisited

Whenever two surfaces rub against each other, a friction force acts on both of them. Friction can be reduced but never truly eliminated. Our intuition about the nature of motion, an intuition that tells us that a force is required to keep things moving, is based on the presence of friction in most real-world situations. Early researchers studying motion believed that a force or cause was required to keep an object moving, even at constant velocity, because they often observed moving objects slowing down and then stopping. The presence of kinetic friction probably delayed the discovery of Newton's law of inertia. Like Galileo and Newton, we often try to imagine a world without friction. We develop theories based on this ideal condition, and then we test these theories in laboratory experiments designed to minimize friction.

It might sound like friction is some ugly fact of our Universe that we wish would just go away. That is far from the truth, however, for our world would be very weird if friction simply vanished. Without friction, we could not walk, drive a car, pet a dog, or strike a match.

In Figure 6.3A, a person does not support the box from below, but instead presses it against a wall. If there were no friction, the free-body diagram for the box would look like Figure 6.3B. The person pushes with $\vec{F}_p$, and the wall exerts a normal force $\vec{F}_N$. The box does not accelerate in the x direction, so by Newton's second law these two forces have equal magnitudes. The Earth's gravity pulls the box downward, and in the absence of friction, the box would consequently accelerate along the negative y direction. The box is *not* accelerating, however. Therefore, another force must also be acting on the box, one with the same magnitude as the gravitational force but in the positive y direction. It is the static friction force (Fig. 6.3C).

The wall is the source of the normal force and of static friction, so the wall actually exerts a single force up and to the left (Fig. 6.3D). For convenience, we break this force into two components. The component perpendicular to the wall is called the normal force, and the component parallel to the wall is called friction (in this case, static friction). Because the wall really exerts only one force on the box, you may think that as one component increases, the other would, too. In fact, that is true. Friction (the parallel force) is proportional to the normal force.

We observe the large-scale, or macroscopic, effect of the stationary box against the wall, but the normal force and friction underlying this macroscopic observation are due to many microscopic, or small-scale, interactions. These interactions are the electromagnetic force exerted by the molecules in each surface. The number of interactions depends on how much contact the two surfaces have with each other. Even

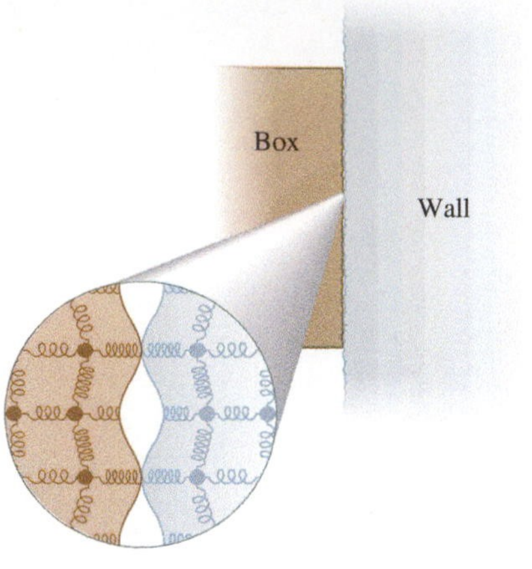

FIGURE 6.4 A close-up of the wall and box show that they touch each other at certain points. In between, there are pockets of no contact. Molecular bonds form at the points of contact. We model these bonds as stiff springs.

surfaces that seem smooth on the macroscopic scale are jagged on the microscopic scale. So, whenever two surfaces come into contact with each other, the surfaces touch only at certain points. As a result, there are, in addition to the contact points, numerous pockets of no contact (Fig. 6.4). So, the *real* contact area is smaller than the *apparent* contact area. The real contact area depends on the normal force. A large normal force means a large real contact area, which means a strong frictional force.

It might seem counterintuitive, but polishing two surfaces increases the magnitude of the friction force between them because polishing increases their real contact area. For example, imagine two highly polished pieces of metal that are kept in a vacuum chamber to make sure that they stay clean. When they are brought into contact with each other, there are very few pockets; consequently, a very high proportion of their surfaces are in contact. It is therefore possible for these two metals to bond, or *cold-weld*, together. We can think of cold-welding as an extreme case of friction. Further, if you want to reduce friction between two highly polished surfaces, it is a good idea to sprinkle the surfaces with small particles such as powder so that their real contact area is reduced.

How can we possibly expect to measure and calculate the electromagnetic forces between all the molecules that interact on the microscopic scale? There are two answers. (1) Usually, we don't. Instead of making all the necessary calculations on the microscopic scale, we run experiments on the macroscopic scale and measure the effects of friction. (2) When we need to think more deeply about friction on the microscopic scale, we use computers to make the large number of required calculations. Of course, the computer calculations involve simulated data as opposed to observed laboratory data. The results of the computer simulations are a deep understanding of how friction works on the microscopic scale and a framework for the macroscopic scale.

Part of what the macroscopic laboratory experiments have shown is that the friction force acts differently depending on the relative motion of the two surfaces: whether they are at rest, sliding past each other, or one rolling across the other. One goal in this chapter is to deepen our understanding of friction in all three of these cases. To help, we use a model of the microscopic interaction between surfaces. When two surfaces come into contact with each other, molecules in one surface form electromagnetic bonds with molecules in the other surface. Through each bond, the surfaces exert a force on one another. So, the more bonds formed between the two surfaces, the stronger the force. The bonds form wherever the surfaces touch each other (Fig. 6.4). As in Chapter 5, we model these microscopic bonds by stiff springs.

Modeling, which is used throughout this book, is an important technique used in physics to understand complex systems. Models are used in other fields as well. For example, a globe is a model for the Earth. Whenever you use a model to understand a complex system, you must keep in mind that the model is simpler than the real systems and so has certain limitations. For example, a globe is usually a smooth spherical model of the Earth. You may find it useful to consult a globe to compare the relative sizes of India and China, but you would not consult a globe to compare the relative climates of the two countries. With this caveat in mind, we model friction.

6-3 A Model for Static Friction

Let's begin with two surfaces at rest with respect to each other such as the box and wall (Fig. 6.3A). When an object is at rest with respect to a surface with which it is in contact, the surface may exert a force on the object that is parallel to the surface. This parallel force is **static friction**. If the object is not accelerating, the static friction force is balanced by the vector sum of any other forces that are applied to the object.

To learn more about static friction, let us combine the results of a macroscopic experiment with a microscopic picture (Fig. 6.5). (This experiment is similar to Example 5.9, so we use the same coordinate systems here.) At first, a box rests on a steel surface, with no string or hook attached. A free-body diagram for the box shows just two forces: gravity $\vec{F}_g$ due to the Earth and the normal force $\vec{F}_N$ due to the surface (Fig. 6.5A). On the microscopic scale, we imagine that the normal force is due to molecular bonds in the steel surface that act like very stiff springs (Section 5-7). The box compresses these bonds slightly, and, like springs, they exert an upward force on the box (Fig. 6.5A, close-up).

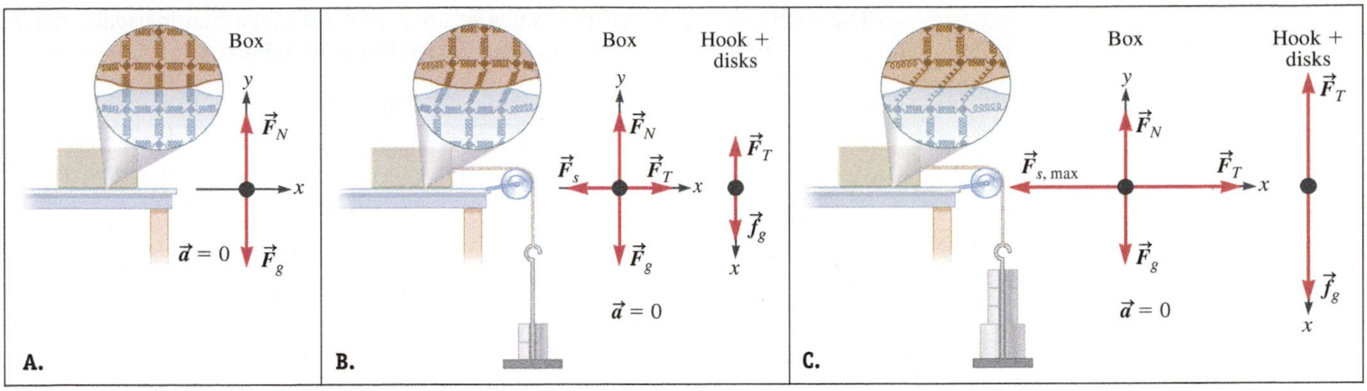

FIGURE 6.5 In this experimental setup, lead disks added to a hook attached by a pulley to a box on a steel surface increase the tension applied to the box. As long as the box does not accelerate, static friction balances this tension. **A.** When two surfaces are in contact, bonds form. These bonds are modeled as springs. **B.** As a force is applied to one of the objects, these bonds are stretched. Like stretched springs, the bonds apply a restoring force. **C.** When the applied force is increased to $\vec{F}_{s,\,max}$, these bonds are stretched to their limit. An increase in the applied force snaps the bonds.

Next, we attach a string and hook to the box. First, we put only a few lead disks on the hook so that the box remains at rest. There are now four forces acting on the box: gravity, the normal force, a tension force $\vec{F}_T$, and static friction force $\vec{F}_s$, (Fig. 6.5B). The normal force and gravity are the same as before (Fig. 6.5A). Because the box is not accelerating, we know from Newton's second law that $F_s = F_T$ (and $F_N = F_g$).

We also draw a free-body diagram for the system made up of the hook and lead disks. There are two forces acting on this system: gravity $\vec{f}_g$ and a tension force. Because the system is not accelerating, its weight must be equal to the tension in the string. So, by adding more disks to the hook, we can increase the tension force exerted on the box.

We find that we can add more lead disks to the hook without accelerating the box. The only way to do so is if the force of static friction F_s increases as the tension F_T increases. Our microscopic model of molecular bonds forming between molecules in the box surface and molecules in the table surface shows how this situation is possible. When there is tension force on the box, the molecular bonds are skewed (Fig. 6.5B, close-up). The bonds pull back (that is, leftward) on the box just like stretched springs would. This pulling by the microscopic bonds is responsible for the static friction force we observe on the macroscopic scale. A stronger tension force exerted on the box means that the bonds stretch farther and, according to Hooke's law, pull back with a greater force. On the macroscopic scale, this increase is described as being an increase in static friction.

Our experience tells us that there must be a limit to static friction. As we add more lead disks, the tension increases and so does static friction. With enough disks added, the box starts to slide. Just before sliding begins, static friction is at its maximum value $F_{s,\,max}$ (Fig. 6.5C). On the microscopic level, this maximum friction level corresponds to the molecular bonds being stretched as far as they can go without breaking (Fig. 6.5C, close-up). Just a little more tension in the string breaks these bonds, and then the box slides.

We reason that the maximum force of static friction must be related to the number of molecular bonds that form between the two surfaces. We already know that the number of bonds between the surfaces depends on how hard the surfaces press on each other (i.e., on the magnitude of the normal force they exert on each other). Mathematically the maximum force of static friction is proportional to the normal force exerted by the two surfaces on each other:

$$F_{s,\,max} = \mu_s F_N \qquad (6.1)$$

The constant of proportionality μ_s is known as the **coefficient of static friction** and depends on the composition and smoothness of the two surfaces. Highly polished metal on metal, for instance, will have more points of contact than wood on wood, so the coefficient of static friction is higher for metal on metal.

STATIC FRICTION **Special Case**

Notice that Equation 6.1 is not a vector equation. *Static friction is always parallel to the two contacting surfaces, whereas the normal force is perpendicular to the surfaces.* For Equation 6.1 to be a vector equation, these two forces would need to be in the same direction. We infer the direction of the static friction force from a free-body diagram.

The coefficient of static friction is a positive, unitless scalar that is usually determined by experiments. Some values for μ_s are found in Table 6.1.

TABLE 6.1 Coefficients of static, kinetic, and rolling friction.

Surfaces	Coefficient of Static Friction μ_s	Coefficient of Kinetic Friction μ_k	Coefficient of Rolling Friction μ_r
Aluminum on steel	0.56 ± 0.08	0.47	
Brass on steel	0.45 ± 0.07	0.42 ± 0.03	
Copper on cast iron	1.08 ± 0.03	0.30 ± 0.01	
Copper on steel	0.53	0.36	
Glass on glass	0.92 ± 0.02	0.4	
Ice on ice	0.10	0.03	
Rubber on dry concrete	1.0	0.8	0.01–0.02
Rubber on wet concrete	0.4 ± 0.2	0.3 ± 0.1	
Steel on steel (dry)	0.72 ± 0.06	0.58 ± 0.02	0.001–0.002
Steel on steel (lubricated)	0.10	0.05	
Synovial joints in humans	0.01	0.006 ± 0.005	
Teflon on Teflon	0.04	0.04	
Teflon on steel	0.04	0.04	
Wood on wood	0.4 ± 0.1	0.2	
Wood on snow	0.11 ± 0.01	0.06	
Waxed ski on snow	0.1	0.04 ± 0.01	

These coefficients are found experimentally. Where possible, an estimate of the margin of error is given along with the actual value. These values are reasonable for classroom use, but if your work requires more precision, you should do your own literature search or experimentation.

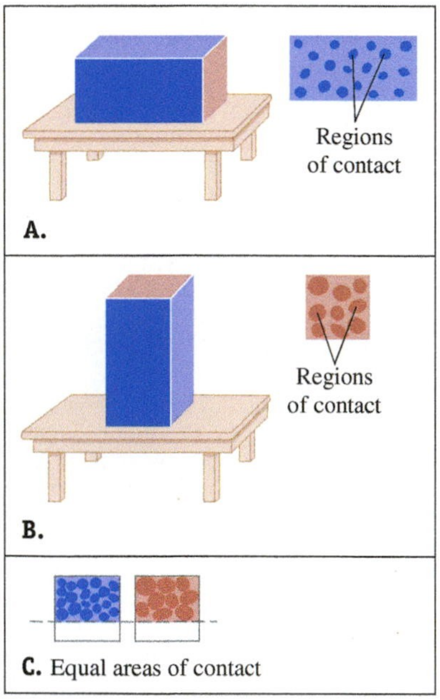

FIGURE 6.6 **A.** When the large side touches the table, there are many small regions of contact. **B.** When the small side touches the table, there are fewer, but larger areas of contact. **C.** Total effective areas of contact are equal.

Experiments have also shown that the magnitude of the *apparent* contact area between the two surfaces does not significantly affect the magnitude of static friction force because the *effective* contact area depends only on the normal force. For example, when the largest side of a rectangular block rests on a table, there are many small regions of contact between the block and the table (Fig. 6.6A). If the block is rotated so that one of its smallest sides rests on the table, there are fewer, but larger regions of contact (Fig. 6.6B). The total effective contact areas (and the normal force) in the two cases are equal (Fig. 6.6C), so friction between the block and the table is the same, too.

We saw that before the static friction force acting on the box in Figure 6.5 reaches its maximum value, the magnitude of that force is equal to the tension in the string. Notice that if no force parallel to the contacting surfaces acts on the box, there is no static friction force acting on the box either (Fig. 6.5A). So, you might guess that the magnitude of the static friction force is equal to the vector sum of all the other forces applied parallel to the contacting surfaces. Although that statement is often true, it is *not* always the case. For example, imagine that our box is at rest on the bed of a flatbed truck (Fig. 6.7A). When the truck is at rest, the only forces acting on the box are gravity and the normal force due to the truck bed. Now imagine that the truck accelerates to the right as shown. If the box remains at rest with respect to the truck, the acceleration of the box relative to the ground must be the same as the acceleration of the truck relative to the ground (Fig. 6.7B). We know from Newton's first law that there must be a force responsible for the box's acceleration, and here that force is static friction (Fig. 6.7C). In this case, there are no other forces acting either parallel or antiparallel to the

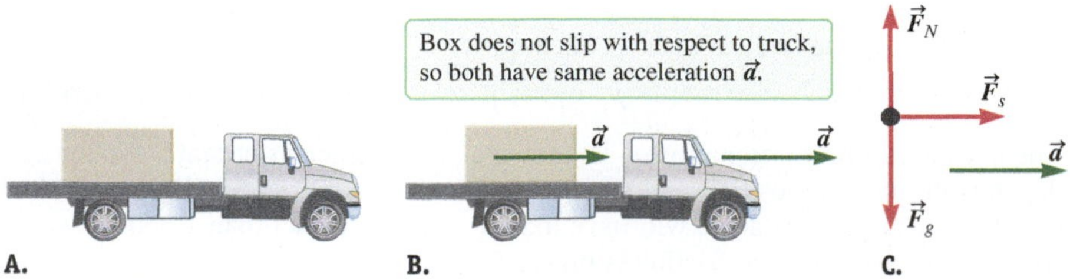

FIGURE 6.7 **A.** A box rests on a flatbed truck. **B.** When the truck accelerates, the box remains at rest with respect to the truck. So, the box must have the same acceleration as the truck with respect to the ground. **C.** A free-body diagram for the box shows that static friction must accelerate the box.

bed surface; thus, it would be incorrect to describe the friction force as being a force opposing the vector sum of other forces acting along the bed's surface.

CONCEPT EXERCISE 6.2

A box rests on a steel surface. Four sides of the box are made of aluminum, and the other two sides are coated with Teflon. Each panel in Figure 6.8 shows a pair of experiments. In each pair, decide which experiment requires the larger tension force to move the box and therefore has the higher maximum static friction force.

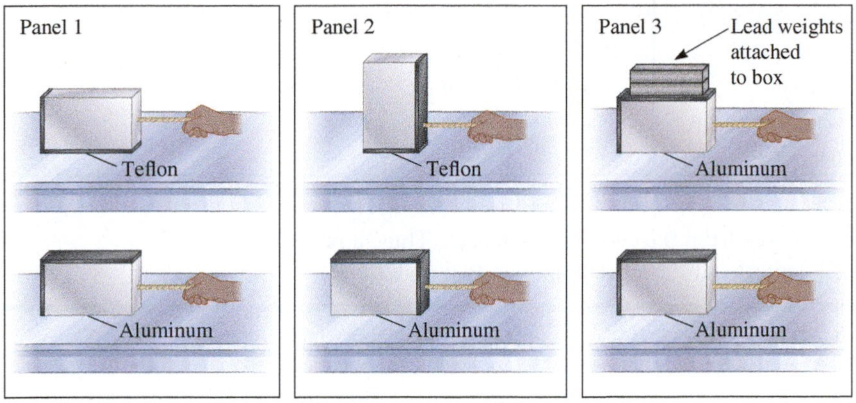

FIGURE 6.8

CONCEPT EXERCISE 6.3

For a particular object and a particular tabletop, the maximum value possible for the magnitude of the static friction force exerted on the object is 15 N. The object and table are at rest, and a free-body diagram for this object is shown in Figure 6.9.

a. If the tension force on the object is zero, what is F_s?
b. If the tension on the object is $F_T = 5$ N, what is F_s?
c. If the tension is increased to $F_T = 15$ N, what is F_s?
d. What happens if the tension is increased to $F_T = 25$ N?

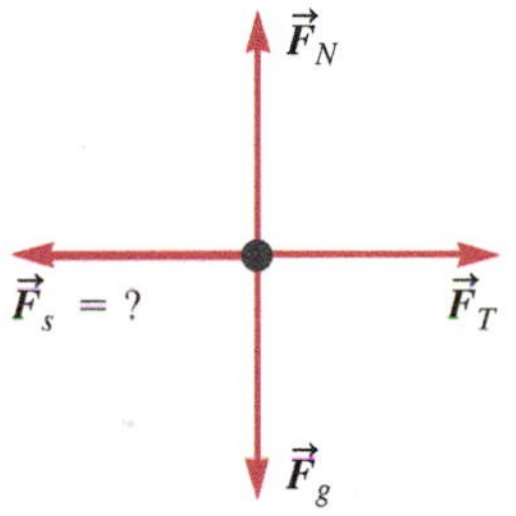

FIGURE 6.9

EXAMPLE 6.1 Testing Running Shoes

The coefficients of friction are usually found experimentally. In this example, we explore an experiment that requires only a protractor to find the coefficient of static friction between two surfaces.

Lisa, a runner, has purchased new shoes and wishes to measure the coefficient of static friction between her shoes and the running track. A higher coefficient of static friction means a faster run (see Concept Exercise 6.7). The coach has extra tiles of track and has given her a piece for her experiment.

She places a shoe on the tile and then slowly tilts the tile, measuring the angle θ at which the shoe just begins to move (Fig. 6.10).

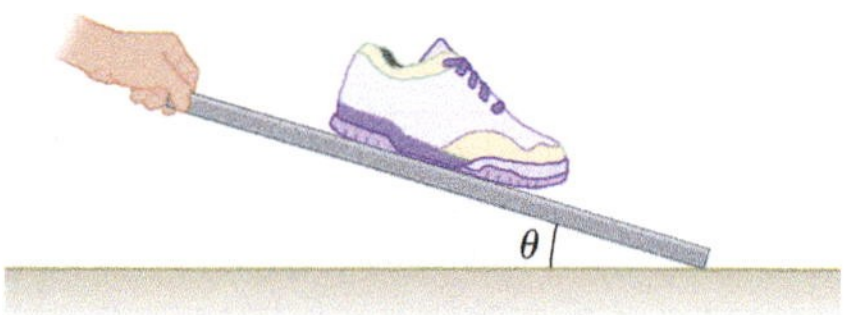

FIGURE 6.10

A Find an expression she can use to determine μ_s in terms of θ.

INTERPRET and ANTICIPATE

Start by drawing a free-body diagram for the shoe. It is often best to choose a coordinate system that has one axis parallel to the incline (Fig. 6.11). We are interested in the moment just before the shoe slips, at which time the acceleration is zero.

Our answer for μ_s should be unitless and algebraic. We expect that the coefficient of static friction depends on the tilt angle. Imagine if Lisa used Velcro between her shoe and the tile; she probably could stand the tile straight up without the shoe slipping.

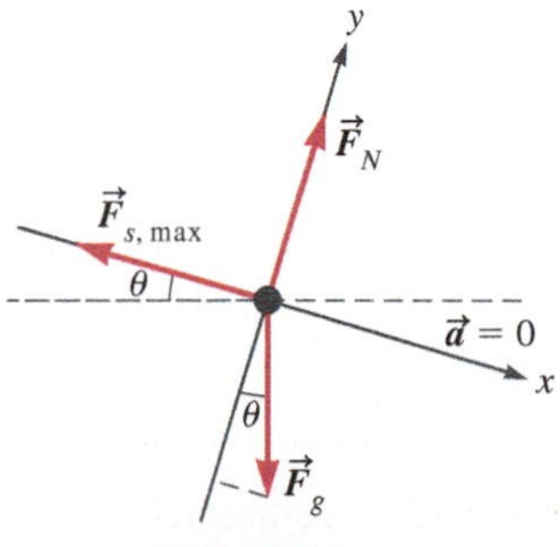

FIGURE 6.11

Example continues on page 160 ▶

SOLVE

Next, apply Newton's second law with the acceleration equal to zero. The gravitational force must be broken into components parallel and perpendicular to the chosen x axis.	$\sum F_x = F_g \sin\theta - F_{s,\max} = 0$ (1) $\sum F_y = F_N - F_g \cos\theta = 0$ (2)
Rewrite $F_{s,\max}$ using $F_{s,\max} = \mu_s F_N$ (Eq. 6.1) and substitute into Equation (1).	$F_g \sin\theta - \mu_s F_N = 0$ (3)
Solve Equations (2) and (3) for μ_s simultaneously. Often, when two equations involve sine and cosine functions, it is helpful to divide one equation by the other.	$\mu_s F_N = F_g \sin\theta$ $F_N = F_g \cos\theta$ $\dfrac{\mu_s F_N}{F_N} = \dfrac{F_g \sin\theta}{F_g \cos\theta}$ $\quad \mu_s = \dfrac{\sin\theta}{\cos\theta}$ $\mu_s = \tan\theta$

CHECK and THINK

Our answer is in the form we expected; as θ increases from 0 to 90°, μ_s increases. Thus, a protractor is the only tool needed to find the coefficient of static friction between two materials.

B If Lisa finds that $\theta = 52°$ at the instant the shoe begins to slip, what is μ_s for her shoes on the track?

We just need to substitute Lisa's measurement into our solution.	$\mu_s = \tan\theta = \tan 52° = 1.3$

CHECK and THINK

This value is a bit higher than μ_s for rubber on dry concrete (Table 6.1), which seems about right.

6-4 Kinetic and Rolling Friction

Let us now move to the kinetic friction between two objects, once again imagining a box on a surface (Fig. 6.12). Once the molecular bonds between the two surfaces are broken and the box begins to move, static friction no longer plays a role, but there is still friction between the two surfaces. Whenever an object is in contact with some surface and slides with respect to that surface, kinetic friction works to stop that motion. The direction of kinetic friction is always parallel to the surface and in the direction opposite the direction of the motion with respect to the surface (Fig. 6.12B).

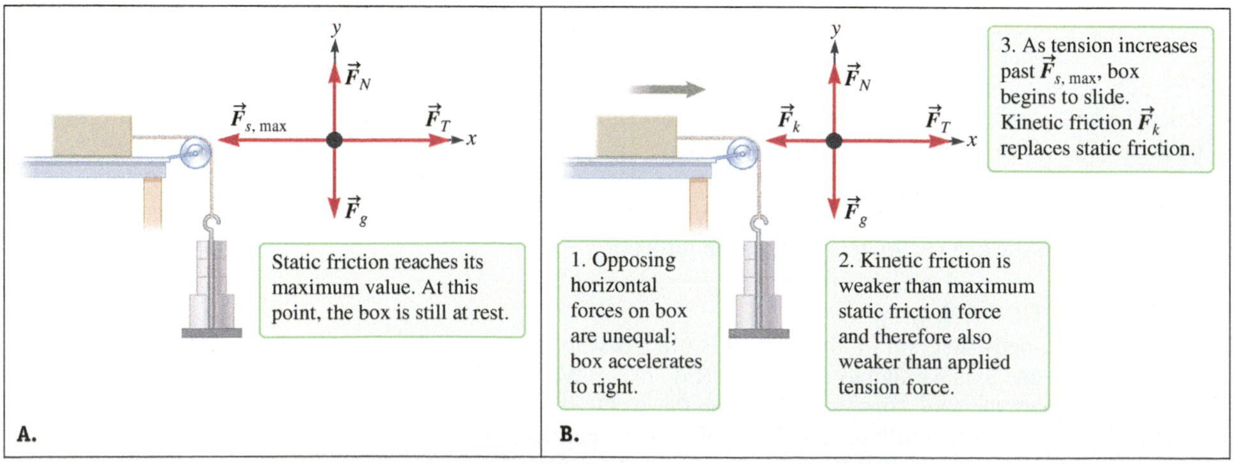

FIGURE 6.12 Experimental setup corresponding to Figure 6.5, but with increased force on the box. **A.** As long as the box does not accelerate, static friction balances the tension in the string. **B.** Once tension exceeds the maximum static friction force, kinetic friction takes over, and the box accelerates to the right.

Model for Kinetic Friction

To learn more about kinetic friction, let us return to our microscopic model. As the sliding surfaces rub against each other, points of contact form weak bonds momentarily. Individually, these bonds are relatively easy to break, but added together they are responsible for kinetic friction on the macroscopic scale. Because it is easier to break the bonds once the surfaces are in motion, the kinetic friction between any two surfaces is always weaker than the maximum value of the static friction force between the surfaces.

Like static friction, kinetic friction must be related to the number of molecular bonds that form between the two surfaces, which depends, in turn, on the composition of the surfaces and on the magnitude of the normal force they exert on each other. According to Equation 5.9, the magnitude of kinetic friction is proportional to the normal force acting on the object:

$$F_k = \mu_k F_N \tag{6.2}$$

KINETIC FRICTION Special Case

where the constant of proportionality μ_k is known as the coefficient of kinetic friction. Like the coefficient of static friction, μ_k is a unitless scalar that depends on the composition of both surfaces and is found experimentally (Table 6.1). Experiments have shown that the speed of an object experiencing kinetic friction does not significantly affect the magnitude of this force. In fact, unlike static friction, kinetic friction is relatively constant.

Like Equation 6.1, Equation 6.2 is *not* a vector equation because $\vec{F}_N$ is perpendicular to the surface and $\vec{F}_k$ is parallel to the surface and in the direction opposite the motion. Because kinetic friction is weaker than the maximum value of static friction, we expect $\mu_s > \mu_k$ for each pair of surfaces in Table 6.1.

Figure 6.12 shows two free-body diagrams for the box as the tension increases. The box remains at rest as long as the static friction force balances the tension in the string. When a tiny bit more weight on the hook increases the tension such that $F_T > F_{s,\max}$, however, the box begins to move (Fig. 6.12B). At that moment, static friction is replaced by kinetic friction. Because $F_k < F_{s,\max}$, we know that $F_k < F_T$. So, the box is accelerated to the right:

$$\sum F_x = ma_x = F_T - F_k$$

$$a_x = \frac{F_T - F_k}{m} > 0$$

Because both F_T and F_k are constant, the acceleration of the box is constant. This information is helpful because it means that we can use the constant-acceleration kinematics equations (Table 2.4).

CONCEPT EXERCISE 6.4

Imagine trying to push a heavy sofa across the room. At first you have difficulty getting the sofa to budge, but once it starts to move your task becomes easier.

a. Does the sofa accelerate for a moment after you get it moving?

b. Use what you now know about static and kinetic friction to explain what is going on. Why won't the sofa move at first? Why do you have to apply a stronger and stronger force to get it to move? Why is the force you must apply to keep the sofa moving less than the force you applied just at the instant motion began?

Sometimes when one surface slides over another, a few strong molecular bonds form at the points of contact. When that happens, there are two possible results. One is that the bonds break almost as soon as they are formed. In this case, the moving object stops only momentarily and then begins sliding again. When this making and breaking of bonds happens repeatedly, the motion is jerky. This jerky motion can lead to squeaky, screeching sounds, such as when a car stops suddenly.

The other possibility is that the strong molecular bonds do not break. Instead, weaker bonds break so that some bits of one surface are left embedded in the other surface, such as tire marks left on the road (Fig. 6.13). Wear and tear on the tire surfaces causes tires to go bald, and roads eventually need to be resurfaced.

FIGURE 6.13 When the wheels locked and slid without rolling, small pieces of the car tires became embedded in the road, leaving skid marks.

CONCEPT EXERCISE 6.5

Why is it important to lubricate such things as car engines, bicycle chains, and pocket knives?

EXAMPLE 6.2 Fun in the Snow

Children are sledding on a hill as illustrated in Figure 6.14. Zak gives Sophia a push to start her off with a speed of 1.7 m/s at a vertical height of 4.6 m above the bottom of the hill. They are using a low-friction sled, and we will assume this sled sliding on snow is like waxed skis on snow. Find the velocity of the sled at the bottom of the hill.

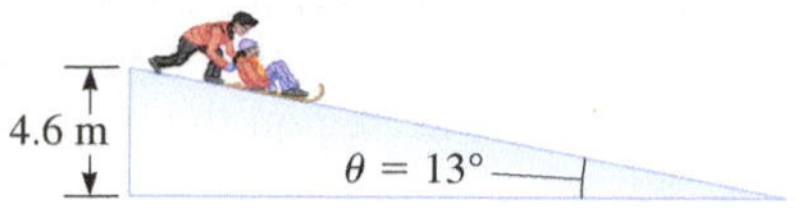

FIGURE 6.14

INTERPRET and ANTICIPATE

This problem requires the synthesis of dynamics and kinematics. We need dynamics to find the acceleration of the sled and kinematics to find the final velocity. We will use the initial velocity of the sled imparted by Zak when we work on the kinematics. Our problem begins just after Zak has lost contact with Sophia.

Start with a free-body diagram for the sled, which has three forces acting on it: gravity, the normal force, and kinetic friction (Fig. 6.15). The acceleration is downhill. We have decided to use an x axis that points in the same direction as the acceleration, as is often best. With this coordinate system our answer should be in the form $\vec{v} =$ ____$\hat{\imath}$ m/s.

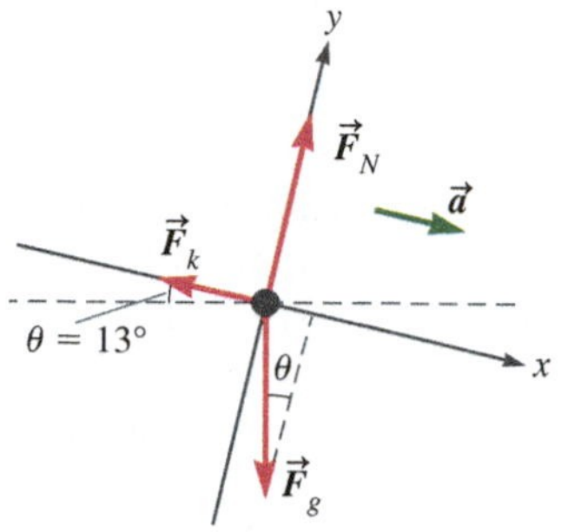

FIGURE 6.15

SOLVE

To find the acceleration, use Newton's second law.

$$\sum F_x = F_g \sin\theta - F_k = ma_x$$
$$\sum F_y = F_N - F_g \cos\theta = 0$$

Substitute $F_k = \mu_k F_N$ (Eq. 6.2) for F_k and mg for F_g.

$$mg\sin\theta - \mu_k F_N = ma_x$$
$$F_N - mg\cos\theta = 0$$

Solve these two equations simultaneously for a_x, eliminating F_N. Notice that mass m cancels out.

$$ma_x = mg\sin\theta - \mu_k mg\cos\theta$$
$$a_x = g(\sin\theta - \mu_k\cos\theta)$$

Substitute values. From Figure 6.14, $\theta = 13°$. Look up μ_k for waxed skies on snow from Table 6.1.

$$a_x = 9.81\ \text{m/s}^2(\sin 13° - 0.04\cos 13°)$$
$$a_x = 1.82\ \text{m/s}^2$$

CHECK and THINK

As a quick check, it makes sense that the sled's acceleration is less than the free-fall acceleration $g = 9.81\ \text{m/s}^2$. Also, we expect the acceleration to be less than it would be if there were no kinetic friction (which would be $a_x = g\sin\theta = 2.2\ \text{m/s}^2$), and it is.

SOLVE

Because the acceleration is constant, we can use Equation 2.9 to find the speed of the sled at the bottom of the hill. Apply trigonometry to Figure 6.15, and find $\Delta x = 4.6/\sin(13)$.

$$v_x^2 = v_{0x}^2 + 2a_x\Delta x \qquad (2.9)$$
$$v_x^2 = (1.7\ \text{m/s})^2 + 2(1.82\ \text{m/s}^2)\left(\frac{4.6\ \text{m}}{\sin 13°}\right) = 77.3\ (\text{m/s})^2$$
$$v_x = \pm 8.8\text{m/s}$$
$$\vec{v} = 8.8\,\hat{\imath}\ \text{m/s}$$

CHECK and THINK

It seems reasonable that the sled is going about five times faster at the bottom of the hill than at the top. This speed is roughly 20 mph, however, which is probably a bit faster than a typical sled. Notice that our answer did not depend on the mass of Sophia or that of the sled.

Rolling Friction

It is probably obvious that the invention of the wheel made life a lot easier. Like objects sliding on a horizontal surface, however, rolling ones are also subject to friction that causes them to slow down and stop.

The friction associated with the rolling motion of one object against a surface is called **rolling friction**, and it is weaker than both kinetic and static friction. That is why it is easier to move a cart with wheels than to move one without wheels.

On the microscopic level, rolling friction is similar to both static and kinetic friction. Consider a ball rolling along a horizontal floor. Where the ball is in contact with the floor, molecular bonds form and then immediately break as the ball rolls over the floor. New bonds form and break repeatedly. The number of bonds that forms depends on the real contact area.

The portion of the ball in contact with the pavement at any instant is slightly flattened, and the portion of the floor in contact with the ball is slightly deformed as well (Fig. 6.16). The more the ball and floor deform, the greater their real contact area. All other things being equal, a greater normal force causes a greater deformation. So, rolling friction depends on the normal force between the two surfaces:

FIGURE 6.16 A ball rolling on a floor causes the ball to flatten and the floor to deform.

$$F_r = \mu_r F_N \tag{6.3}$$

ROLLING FRICTION **Special Case**

Like the other two coefficients of friction, the **coefficient of rolling friction** μ_r is a unitless scalar that is found experimentally (Table 6.1). Like kinetic friction, the direction of the rolling friction force is opposite the direction of motion with respect to the surface. Because the object moves with respect to the surface, the term ***moving friction*** is use to include both kinetic and rolling friction, but not static friction.

CONCEPT EXERCISE 6.6

A car is on an incline at an angle θ to the horizontal (Fig. 6.17). Clearly indicating particular frictional forces, draw a free-body diagram (no need to indicate $\vec{a}$) for the following:

a. The car while parked on the incline
b. The car rolling downhill
c. The car sliding downhill when the surface is icy (so the car's wheels are not rolling)

FIGURE 6.17

CONCEPT EXERCISE 6.7

What forces act on you as you walk across a room? Draw a free-body diagram showing all of them. Which force or forces propel you forward? Why is it more difficult to walk on a slippery surface than on a nonslippery one? Explain how you use Newton's third law to control your motion.

6-5 Drag and Terminal Speed

When an object moves in a *fluid medium* such as air or water, the medium exerts a resistive **drag force** on the object. If you parachute out of an airplane, the drag force exerted by the air on the chute keeps you from free-falling to the Earth. Moving friction and drag are called *resistive forces* because they work to resist the motion of an object relative to the source (the surface or medium).

DRAG FORCE **Special Case**

Like friction, the drag force is a macroscopic effect generated by many microscopic interactions. Drag results from many collisions that take place between molecules in the medium and the object. When an object is moving with respect to a medium, there are more collisions to the object's front side than to the object's back side, and the object tends to slow down. The drag force on an object depends on the density (mass per unit volume) of the medium and also on the cross-sectional area of the object perpendicular to the direction of motion.

FIGURE 6.18 **A.** In a diffuse (low-density) medium, the drag force on the parachute is lower because relatively few molecules of the medium collide with the chute. **B.** In a denser medium, the drag force is greater because here there are more molecules of the medium colliding with the parachute. (Differences in air density have been exaggerated.)

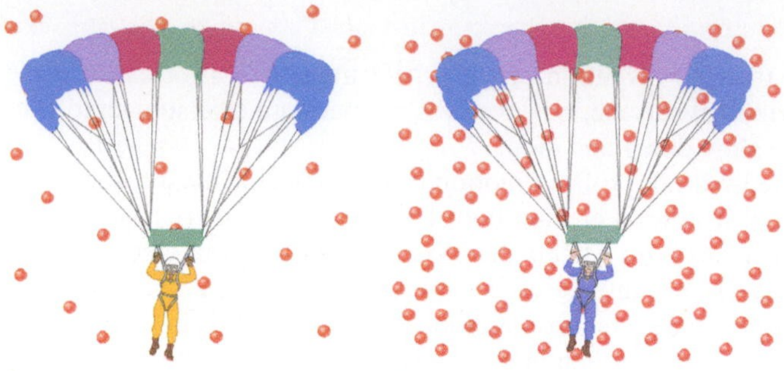

A. B.

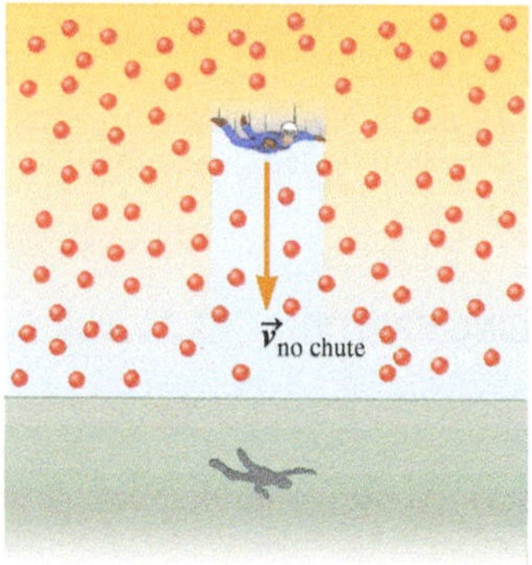

A.

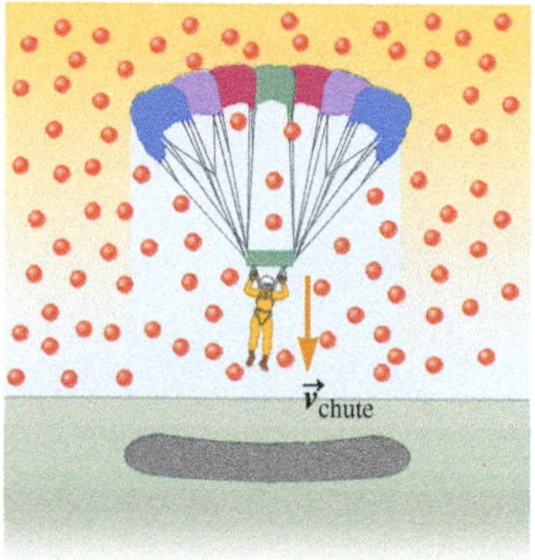

B.

FIGURE 6.19 The cross-sectional area of an object is the area of its shadow projected on a plane that is perpendicular to its velocity. **A.** A skydiver has not yet opened her parachute. Her cross-sectional area is small. **B.** A skydiver has opened his parachute. The larger cross-sectional area of his parachute means that he (and his chute) experience a greater drag force than the other skydiver.

To illustrate density dependence, let us look at two skydivers who have opened their parachutes at different altitudes (Fig. 6.18). The density of air decreases with increasing altitude. So, the skydiver at the higher altitude is surrounded by lower-density air. The skydiver in the more diffuse medium (Fig. 6.18A) experiences fewer collisions with molecules in the medium than the skydiver in the denser medium (Fig. 6.18B). As a result of the higher number of collisions, the skydiver in the denser medium has a larger drag force acting on her.

Now consider how drag force depends on the area of the object. The cross-sectional area of importance is the area perpendicular to the velocity. We can use shadow projections to show how the size of this cross-sectional area is related to drag force magnitude. Imagine that the Sun is directly above a plane when a jump takes place so that its rays are perpendicular to the ground below the skydiver. Figure 6.19 shows that the area of the shadow of a skydiver falling without an open parachute is smaller than the area of a skydiver falling with an open parachute. The latter skydiver experiences a larger drag force because the number of air molecules his parachute encounters is greater than the number encountered by the skydiver who has not yet opened her chute.

CONCEPT EXERCISE 6.8

Figure 6.20 shows four objects moving downward. Find the cross-sectional area of each object for the surface perpendicular to the direction of its motion.

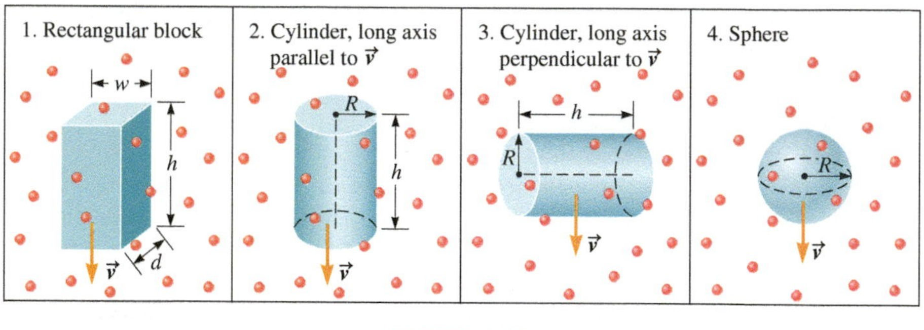

FIGURE 6.20

We are building a mathematical description of drag. So far, we know that the magnitude of the drag force F_D acting on any object is proportional to the density of the medium ρ and the cross-sectional area A of the object:

$$F_D \propto \rho A$$

This magnitude also depends on the relative motion between the object and the medium because a faster relative motion means that each molecule–object collision is stronger. (Think of the difference between getting hit by a fast-moving baseball and a slow-moving one.)

Relative motion between the object and medium is necessary if an object is to experience a drag force, but the object may be at rest with respect to the ground. So, you do not have to jump out of an airplane to experience a drag force; instead, you can just sit on a park bench on a windy day.

The faster an object moves through a medium (or, in our park-bench case, the faster the medium moves past the object), the stronger the drag force on the object. Expressed as a proportionality, we have

$$F_D \propto v^n$$

The constant n is found experimentally; it depends on both the medium and the object. Experiments have shown that $n = 1$ both for small objects (dust particles in air, for instance) moving at low speeds and for large objects moving at typical speeds through liquids (such as a boat in water). In these cases, the drag force is given by

$$\vec{F}_D = -b\vec{v} \tag{6.4}$$

where b is a constant found experimentally with the dimensions of force per speed or mass per time. The negative sign indicates that the direction of the drag force is opposite the velocity's direction.

In this book, we are mainly interested in the drag force exerted on blunt objects moving through air at high speeds. For all such objects (airplanes, skydivers, baseballs, cars), $n = 2$.

One additional factor—the object's shape—affects the drag force on an object moving through air at high speed. For example, car designers who wish to reduce drag use wind tunnels to study the effect of shape on drag. These studies give rise to the tapered, smooth design of energy-efficient automobiles.

We write all these factors—object's shape, density of the medium ρ, object's perpendicular cross-sectional area A, and speed v— into one neat equation:

$$F_D = \tfrac{1}{2}C\rho Av^2 \tag{6.5}$$

The **drag coefficient** C is a unitless number that is found experimentally to be between 0.4 and 1.0. It depends primarily on the shape of the object.

The drag coefficient is not necessarily a constant, but we ignore such complications in this textbook.

Equation 6.5 is not a vector equation, so we cannot use it to determine the direction of the drag force. Instead, we simply state without proof that the direction of the drag force is opposite the direction of the relative velocity.

EXAMPLE 6.3 Ball Tossed

A baseball is tossed straight up (Fig. 6.21). Draw a free-body diagram for the ball at points A, B, and C.

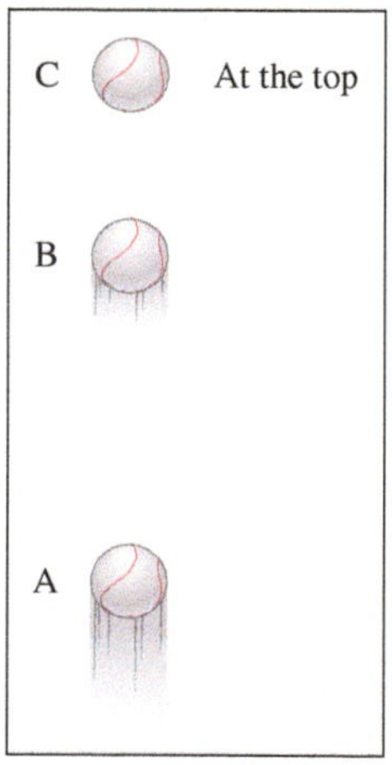

FIGURE 6.21

INTERPRET and ANTICIPATE

Gravity always points to the center of the Earth, and drag points in the direction opposite the direction of the relative velocity. At A and B, the velocity is upward, so the drag force is downward.

SOLVE

The force of gravity is the same at all three points, so we draw three arrows all of the same length (Fig. 6.22).

The drag force on the ball depends on the ball's speed. Because of the downward force of gravity, the ball's speed decreases as it moves from A to C. As a result, the drag force also decreases, so we draw our $\vec{F}_D$ arrow shorter in the free-body diagram at B than in the diagram at A.

At the top of the ball's trajectory, its speed is zero, so there is no drag on the ball at the top (C).

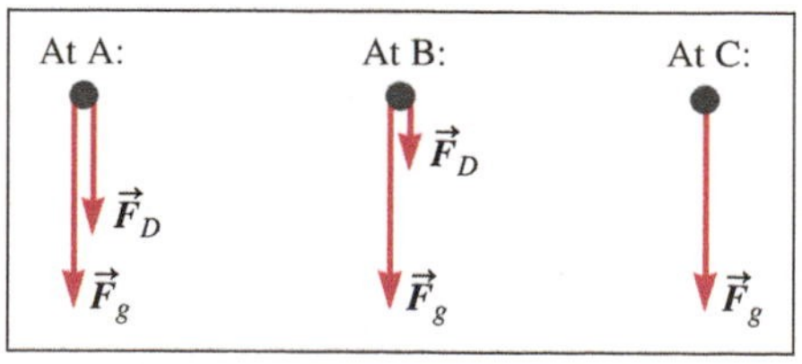

FIGURE 6.22

CHECK and THINK

Let us compare this ball's trajectory with the trajectory of a ball under the ideal free-fall condition (no drag). A ball in free fall without drag has only gravity acting on it. The time for the velocity to reach zero will be longer and the ball's height at the top will be greater if the ball is thrown in a vacuum than if it is thrown in a medium such as air.

TERMINAL SPEED **Special Case**

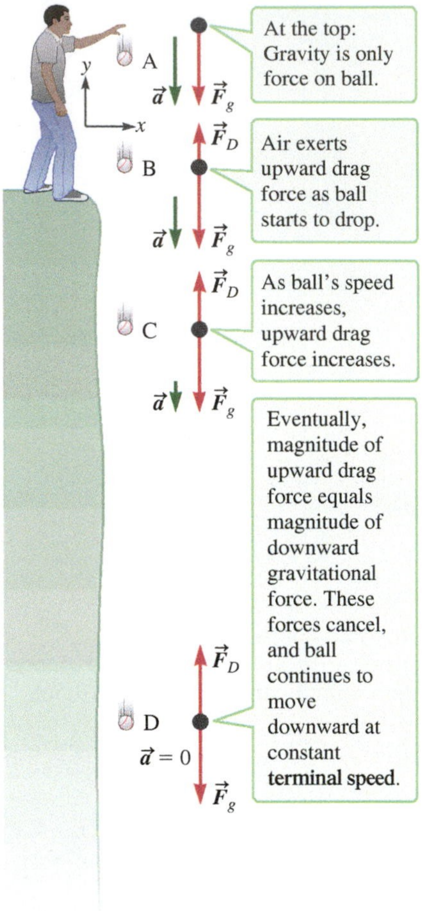

FIGURE 6.23 A ball falls from a great height, reaching terminal speed at position D.

Terminal Speed

When an object falls through the air, the Earth's gravity pulls the object downward, and at the same time the drag force due to the air pushes the object upward. Imagine standing at the edge of a very high cliff (Fig. 6.23). You drop a baseball and then watch as it falls to the base of the cliff far below. As the ball's speed increases, the upward drag force increases. Eventually, the magnitude of the upward drag force equals the ball's weight, and the ball stops accelerating. The ball continues to move downward at a constant speed known as the **terminal speed**. The terminal speed is the highest speed the ball reaches. (If the ball fell in a vacuum instead of through air, no drag force would act on it, and its speed would continue to increase until it hit the ground.)

We use Newton's second law to find an expression for the terminal speed:

$$\sum F_y = F_D - F_g = ma_y$$

At the terminal speed, the acceleration is zero (position D in Fig. 6.23), which means that this expression becomes

$$F_D - F_g = 0$$
$$F_D = F_g$$

Once we substitute for F_D from Equation 6.5 and rearrange to solve for the speed, we end up with

$$v_t = \sqrt{\frac{2mg}{C\rho A}} \tag{6.6}$$

Equation 6.6 confirms our intuition that heavier objects fall faster than light ones. Our intuition is based on observing objects moving not in a vacuum but rather through air. For example, imagine dropping a golf ball and a Ping-Pong ball off the Leaning Tower of Pisa. The two balls are nearly the same shape and size, and they fall through the same air, which means that the density of the medium is the same for both. The only factor that is different is their masses. According to Equation 6.6, the more massive golf ball will reach a higher terminal speed than the Ping-Pong ball and will therefore reach the ground more quickly. This result matches our experience, but what is hard to accept is what happens in a vacuum.

In a vacuum, the only force acting on either ball is gravity. Then, according to Newton's second law,

$$\sum F_y = -F_g = ma_y$$
$$-mg = ma_y$$

The two balls have the same acceleration $a_y = -g$ independent of mass so that if they are both in free fall in a vacuum, they land on the ground at the same time and are moving at the same speed.

CONCEPT EXERCISE 6.9

Do all objects falling through the air reach terminal speed? Explain.

CONCEPT EXERCISE 6.10

a. An object falls through an evacuated space. Which of the three graphs in Figure 6.24 best represents the object's acceleration as a function of time? Which of the three graphs in Figure 6.24 best represents its *speed* as a function of time?

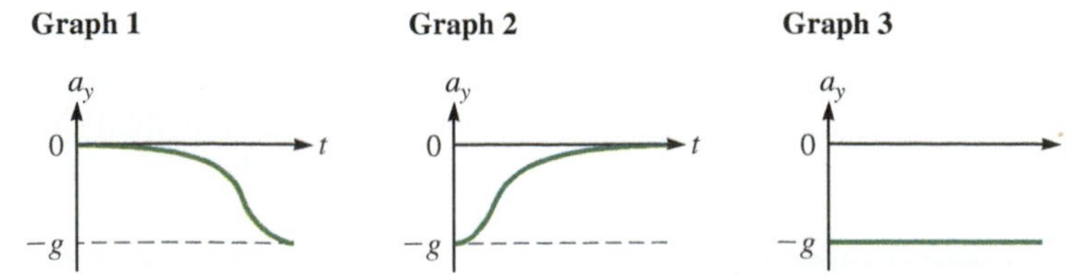

FIGURE 6.24

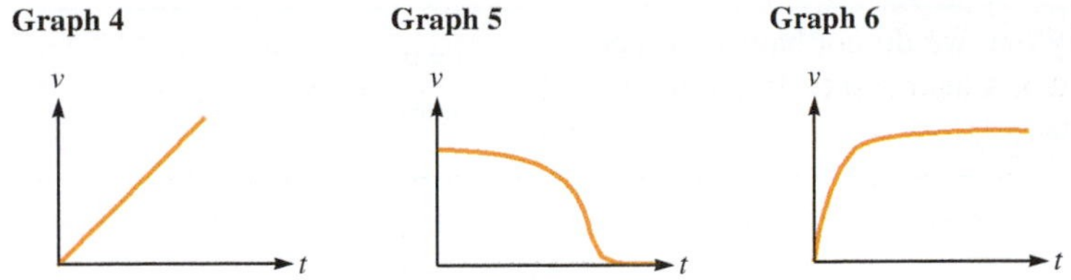

FIGURE 6.25

b. Repeat part (a) for an object that falls through air and at some point reaches its terminal speed.

EXAMPLE 6.4 CASE STUDY Data From a Skydiving Physicist

Cassidi Reese—the skydiver from Chapter 3—agreed to collect jump data for us so that we can study terminal speed. Her data will help us figure out the case study in this chapter. On her data-collecting jump, she varied her speed in the interval between about $t = 6$ s and $t = 65$ s as shown in Figure 6.26.

With her parachute closed, Reese was able to vary her speed by holding her body in different positions as she fell through the air. To increase her speed, she positioned her head downward and held her arms straight back by her sides. To slow down, she oriented herself belly-to-Earth, with her body stretched out and acting almost like a parachute.

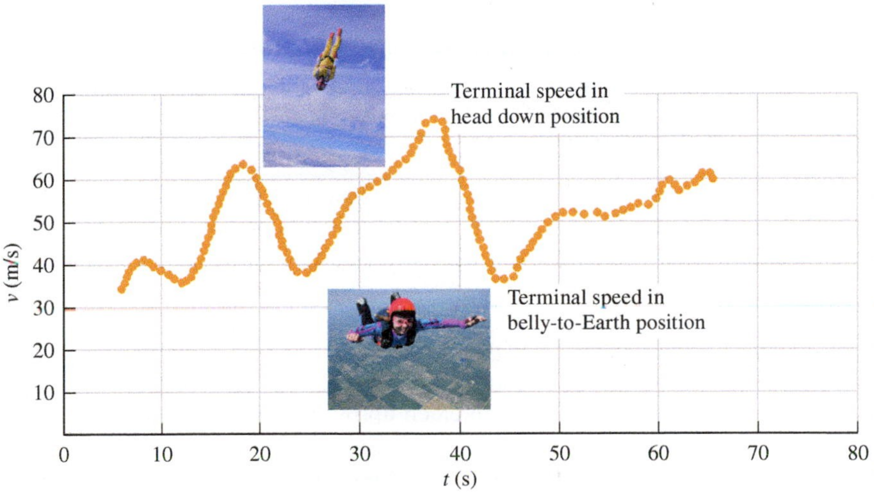

FIGURE 6.26 Reese's speed versus time during a jump in which she varied her body's orientation as shown.

(A) Masterfile Royalty Free; (B) dzphotovideo/iStockphoto.com

A Explain how these two orientations can account for Reese's variations in speed.

With her head pointing downward and her arms at her side, Reese's cross-sectional area (which is equal to the area of her shadow cast on the ground with the Sun directly overhead) is as small as she can make it. In addition, this orientation is tapered and smooth, like a well-designed car. Therefore, this position minimizes both C and A in $v_t = \sqrt{2mg/C\rho A}$ (Eq. 6.6). Because both these factors appear in the denominator, minimizing them maximizes her terminal speed.

In the belly-to-Earth position, Reese's cross-sectional area is large, and her shape is almost like a parachute. In this position, she maximizes C and A, and her terminal speed is consequently minimized.

B Reese and her gear weighed 667 N on the day she collected data. The air density is 1.15 kg/m^3. Use her data to estimate her **effective cross-sectional area** CA at her highest and lowest speeds.

INTERPRET and ANTICIPATE

From our reasoning in part A, we expect that the effective cross-sectional area is largest when Reese is at a lowest terminal speed.

SOLVE

Our first step is to solve Equation 6.6 for CA because this term is the one we are after.

$$v_t = \sqrt{\frac{2mg}{C\rho A}} \quad (6.6)$$

$$CA = \frac{2mg}{\rho v_t^2}$$

Example continues on page 168 ▶

Of the factors on the right in this equation, the only one we do not have a numerical value for is v_t. Because our task is to determine CA at two speeds, we read off Reese's highest and lowest speeds from her data graph.

$$v_{t,\,\text{max}} \approx 73\ \text{m/s}$$
$$v_{t,\,\text{min}} \approx 36\ \text{m/s}$$

We substitute values.

$$(CA)_{\text{min}} = \frac{2(667\ \text{N})}{(1.15\ \text{kg/m}^3)(73\ \text{m/s})^2} = 0.22\ \text{m}^2$$

$$(CA)_{\text{max}} = \frac{2(667\ \text{N})}{(1.15\ \text{kg/m}^3)(36\ \text{m/s})^2} = 0.90\ \text{m}^2$$

CHECK and THINK

Reese can change her effective area by about a factor of four, which allows her to control her speed even before she opens her parachute.

EXAMPLE 6.5 CASE STUDY Flying in Formation

Return to the case study in Section 6-1. Explain how the skydivers in the poster (Fig. 6.1) ended up falling together.

Avi was on to the correct idea. If the skydivers were free-falling, they could not all catch up with one another in flight. Therefore, to all be at the same altitude at the same moment, they would have to step out of the plane together. Because of the drag force exerted by the air, they weren't in free-fall. The drag exerted by the air allows them to join up after exiting the plane individually.

We learned in Example 6.4 that a skydiver can control her speed by changing her effective area. A skydiving team could use this physical fact to join up in the air. Skydivers exiting the plane early on orient their bodies to have a large effective area CA and consequently a low terminal speed. The later skydivers then catch up by holding their bodies in a tapered orientation so as to have a high terminal speed.

A refinement of this technique is to order the skydivers by weight. Equation 6.6, $v_t = \sqrt{2mg/C\rho A}$ tells us that, for a given body orientation, the lighter skydivers have the lower terminal speeds and so should exit the plane before the heavier ones.

Shannon's idea that the plane must fly at a downward slant is not necessary. The plane flies level while the divers jump out of it, one after the other.

Cameron believes that problems like this one are too complicated to be solved using physics alone. In a sense, that is correct. In many cases, our numerical results are based on parameters that usually are determined experimentally, such as the coefficients of friction μ_s, μ_k, and μ_r and the drag coefficient C. There is no analytical basis for giving exact values for these parameters. Once such values are found in a controlled environment, however, they can be used to analyze complicated problems outside the laboratory.

6-6 Centripetal Force

Let us now return to uniform circular motion to show the connection between the kinematics and dynamics of this special case of two-dimensional motion. From Chapter 4, a particle moving in uniform circular motion has a centripetal acceleration. This acceleration points to the center of the circle, and the particle's velocity is tangent to the circle. The velocity and acceleration vectors are perpendicular to each other, which means that the speed remains constant but the direction of the velocity changes.

The constant magnitude of the centripetal acceleration is $a_c = v^2/r$ (Eq. 4.38). Newton's second law tells us that there must be a net force on the object that is responsible for this acceleration. This net force points to the center of the circle and is known as the **centripetal force**. Its magnitude is given by

CENTRIPETAL FORCE **Special Case**

$$F_c = m\frac{v^2}{r} \tag{6.7}$$

If the origin of a polar coordinate system is at the center of the circle, the centripetal force is written as

$$\vec{F}_c = -m\frac{v^2}{r}\hat{r} \tag{6.8}$$

The centripetal force is *not* a *new* force. It is not *generated by* the circular motion of a particle; instead, it is a *requirement of* circular motion. Some physical force (or forces)—gravity, a spring force, the normal force, a tension force, static friction—must act on an object in uniform circular motion in such a way that the net force on the object is perpendicular to the velocity and points to the center of the circular path. Neither drag nor moving friction can generate a centripetal force because they are always directed opposite the velocity.

In the case of uniform circular motion, the net force is the centripetal force, which is always perpendicular to the velocity. So, imagine that the source of the centripetal force were suddenly removed such that there was no net force exerted on the object. Then, according to Newton's first law, the object would continue at the same speed but in a straight line tangent to the point where the object was when the force suddenly vanished.

PROBLEM-SOLVING STRATEGY

When Centripetal Force Is Present

Problems that involve centripetal force are no different from other problems that require us to apply Newton's second law. So, the strategy developed in Section 5-8 works here. Two modifications, however, may be helpful.

INTERPRET and ANTICIPATE

As always, indicate the direction of the object's acceleration. In the case of uniform circular motion, the acceleration is directed toward the center of the circle and is the centripetal acceleration. As in Chapter 5, it is often best to choose to align an axis with the acceleration.

Modification 1 Draw a free-body diagram for one particular instant and **circle the centripetal force(s)**. A free-body diagram should only include forces that are exerted by particular sources. The source of the centripetal force is due to one or more of these sources; there is no separate source of the centripetal force. So, when you draw a free-body diagram, do *not* draw a separate vector for the centripetal force. Instead, circle the force or forces that are parallel (or antiparallel) to the centripetal acceleration. In some cases, a force may have a component that is parallel to the centripetal acceleration and a component that is perpendicular. In that case, circle the force and indicate the parallel component with a short phrase.

SOLVE

Modification 2 The vector sum of the force or forces that you have circled is the centripetal force. So, when you apply Newton's second law, you can expect to set the **sum of the forces equal to the centripetal force** whose magnitude is given by $F_c = m(v^2/r)$ (Eq. 6.7).

CONCEPT EXERCISE 6.11

The following objects are moving in uniform circular motion. Draw a free-body diagram for each object and identify the force responsible for the centripetal acceleration.

Object 1. A person riding on the barrel-of-fun ride (Fig. 6.27, top)
Object 2. The lead object in the laboratory set-up (Fig. 6.27, center)
Object 3. A jogger running on a circular track (Fig. 6.27, bottom)

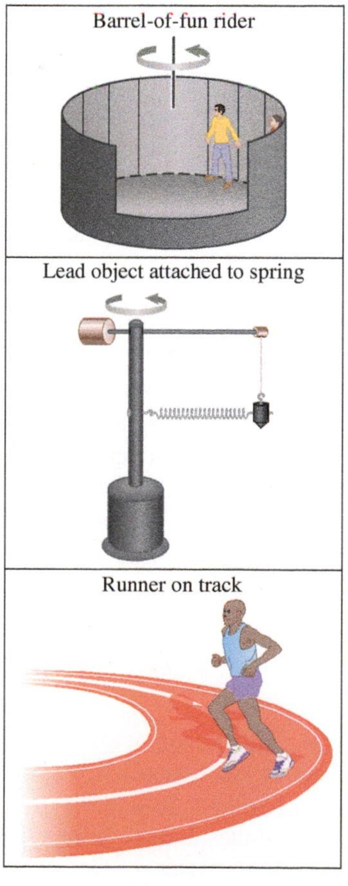

FIGURE 6.27

EXAMPLE 6.6 Particle Accelerator

Particle accelerators are used to study subatomic particles, fundamental building blocks of the Universe. In the type of accelerator known as a *cyclotron*, the particles are accelerated around a circular path, constrained to that path by a magnetic force. The first cyclotron particle accelerator was developed in the early 1930s by American physicists Ernest O. Lawrence and M. Stanley Livingston. In this example, we estimate the strength of the magnetic force in the original cyclotron.

For the purpose of our estimate, we assume a proton is moving in uniform circular motion with a speed of 3.9×10^6 m/s. The diameter of the first cyclotron was about 4.5 in. (1.1×10^{-1} m).

A Find the magnitude of the magnetic force required to keep a proton in uniform circular motion at the specified speed and radius.

INTERPRET and ANTICIPATE

It might seem intimidating to be asked to find a *magnetic* force when we know almost nothing about magnetism. What we need to keep in mind, though, is that this question is really asking us to find the magnitude of the centripetal force. We do not need to know anything about the agent responsible for the centripetal force. Without this force, the proton would travel in a straight line.

SOLVE

Look up the mass of the proton.

$$m_p = 1.67 \times 10^{-27} \text{ kg}$$

We were given the cyclotron diameter, but Equation 6.7 is given in terms of the radius of the motion, which means that we need to calculate the radius and substitute.

$$F = m_p \frac{v^2}{r} \tag{6.7}$$

$$F = (1.67 \times 10^{-27} \text{ kg}) \frac{(3.9 \times 10^6 \text{ m/s})^2}{5.5 \times 10^{-2} \text{ m}}$$

$$F = 4.6 \times 10^{-13} \text{ N}$$

CHECK and THINK

As we expect for any force when we work in SI units, our answer is in newtons.

B Using the same magnetic force, how could you design a cyclotron for a faster proton?

INTERPRET and ANTICIPATE

This question is asking us to examine the relationships among the parameters in $F_c = m(v^2/r)$ (Eq. 6.7). We must find an algebraic result and think about how changing one or more parameters increases the proton's speed.

SOLVE

Because we need to use the same magnetic force and the particle is a proton, both F and m_p are constants. The only other variable in Equation 6.7 that affects particle speed is r. Solve for r in terms of v to see how these two parameters are related.

$$F = m_p \frac{v^2}{r} \tag{6.7}$$

$$r = \left(\frac{m_p}{F}\right) v^2$$

CHECK and THINK

If F and m_p are constants, a faster proton travels in a larger radius, and a larger cyclotron is therefore required. To achieve great speeds, contemporary (synchrotron) particle accelerators use other techniques that are more efficient than the original cyclotrons, and still these accelerators are miles in diameter.

EXAMPLE 6.7 Flat Turn

An engineer is designing a road with a curve in it as in Figure 6.28. The curve is a circular arc of radius 43.3 m, and the road surface is to be concrete. The engineer must figure out the maximum speed limit for this part of the road. So, she must assume the road will be wet sometimes and a driver will attempt to maintain the maximum speed all along the road. Find the maximum speed limit under the wet-road condition.

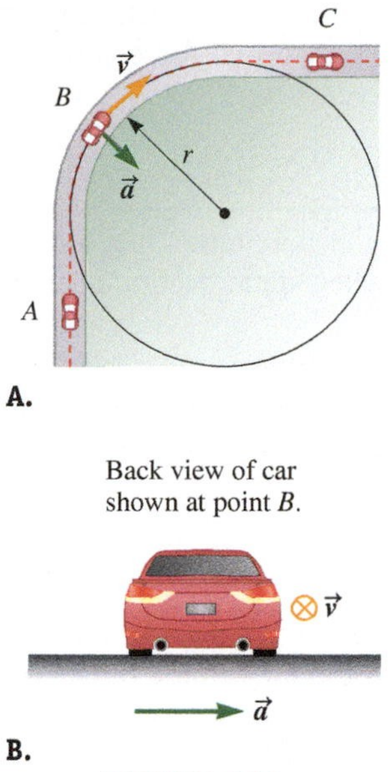

FIGURE 6.28

INTERPRET and ANTICIPATE

This example combines dynamics and kinematics. When the car is in the curve, it is in uniform circular motion. Its velocity is tangent to the curve, and its acceleration is directed toward the center of the circular arc. We need to identify the force or forces that are acting as the centripetal force and then use Equation 6.7, $F_c = m(v^2/r)$, to find the maximum speed at which the car can stay on the road and not fly off (nearly) tangent to the curve. This speed will be the speed limit posted for the bend.

Consider all the forces you know about. Identify which forces act on the car and then which of them may contribute to the centripetal force.

Spring force: There is no spring and so no spring force.

Tension force: There is no rope and therefore no tension force.

Gravity: Because the downward force of gravity has no component directed toward the center of this circle, it does not contribute to the centripetal force.

Normal force: This upward force has no component directed toward the center of the circle and so cannot contribute to the centripetal force.

Rolling friction: Rolling friction always opposes the motion, which in this case means that rolling friction is along the circular path in the counterclockwise direction. Because rolling friction is not directed toward the center, it cannot contribute to the centripetal force. The rolling friction between the tires and the road acts to slow down the car, but the engine compensates for it. Therefore, we ignore rolling friction in this example.

Drag force: The air acts to slow down the car, but the engine compensates for it, meaning that here is another force we can ignore.

Static friction: It may seem surprising, but static friction is the only force that can and does produce the required centripetal force. It may be helpful to think back to the situation with the box in the flatbed truck (Fig. 6.7), where the box was accelerating. We reasoned that because the box was at rest with respect to the flatbed truck, static friction caused the acceleration in the horizontal direction.

You might think that because the car is moving with respect to the road (unlike the box on the flatbed), there can be no static friction. The tires are rolling along the road, and we expect there to be rolling friction. How can static friction also act in this case?

Static friction is parallel to the road surface and is directed toward the center of the circle. Figure 6.28B shows the car at point B, well into the turn. In this figure, the car is heading into the page. The center of the circle is to the right, which is the direction of the centripetal acceleration. The tires do not roll or slip to the right (or left), and static friction is directed to the right.

We are ready to **draw a free-body** diagram for the car **when it is at point *B*** from the perspective of Figure 6.28B. The free-body diagram (Fig. 6.29) includes three forces; there is **no** separate vector drawn for the centripetal force. **Static friction has been circled** because it is parallel to the centripetal acceleration.

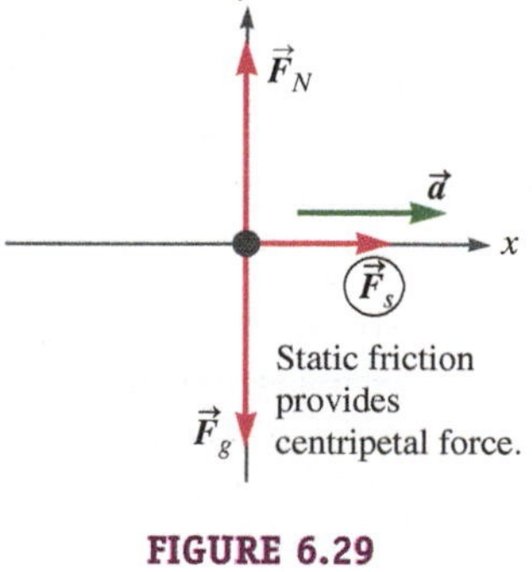

FIGURE 6.29

Example continues on page 172 ▶

SOLVE

Apply Newton's second law. The **total force in the *x* direction equals the centripetal force** $F_c = m(v^2/r)$, Equation 6.7. Because all three forms of friction are proportional to the normal force (Eqs. 6.1–6.3), it is often helpful to solve for F_N when friction is involved.

$$\sum F_x = ma_x$$
$$F_s = m\frac{v^2}{r}$$
$$\sum F_y = F_N - F_g = ma_y = 0$$
$$F_N = F_g = mg$$

To find the maximum speed, consider the maximum possible centripetal force. That is when static friction is at its maximum. Substitute $F_{s,\max} = \mu_s F_N$ (Eq. 6.1).

$$F_{s,\max} = m\frac{v_{\max}^2}{r}$$
$$\mu_s F_N = m\frac{v_{\max}^2}{r}$$
$$\mu_s mg = m\frac{v_{\max}^2}{r}$$
$$v_{\max} = \sqrt{\mu_s gr}$$

Consult Table 6.1 for μ_s for rubber on wet concrete and substitute values.

$$v_{\max} = \sqrt{\mu_s gr}$$
$$v_{\max} = \sqrt{(0.4)(9.81\ \text{m/s}^2)(43.3\ \text{m})}$$
$$v_{\max} = 13.0\,\text{m/s} \approx 29\,\text{mph}$$

CHECK and THINK

There are several important points to be learned from this result. (1) The maximum speed limit does not depend on the mass of the vehicle. There are no separate speed limits by weight class. (2) We found the maximum speed of a vehicle in this turn. If a vehicle is going slower than this speed, the static friction force required to maintain circular motion is smaller than its maximum, and the vehicle safely negotiates the turn. (3) If the car exceeds the speed limit, static friction cannot supply the required centripetal force. As a result, the car slips out of the turn. Kinetic friction takes over, but it is weaker than static friction. With the weaker force acting on it, the car cannot complete the turn and careens off the road. (4) If the road is not wet, $\mu_s = 1.0$, and the car can safely take the turn at $v_{\max} = \sqrt{(1.0)(9.81\ \text{m/s}^2)(43.3\ \text{m})} = 20.6\ \text{m/s} \approx 46\ \text{mph}$. It makes sense that this speed is higher than the speed limit posted.

EXAMPLE 6.8 Banked Turn

An engineer needs to design the curve in Example 6.7 for a higher speed limit. One way to design a road that allows for higher speeds in a curve is to bank the road as shown in Figure 6.30.

We consider the role of the same forces as in Example 6.7: gravity, the normal force, and static friction. Because of the banking, it is possible to take this curve in the absence of static friction. To do so, however, the car must be traveling at one particular speed v_N. Whenever a car takes the curve at any speed other than v_N, static friction comes into play.

A Find an expression for v_N. What is v_N if $\theta = 7.5°$?

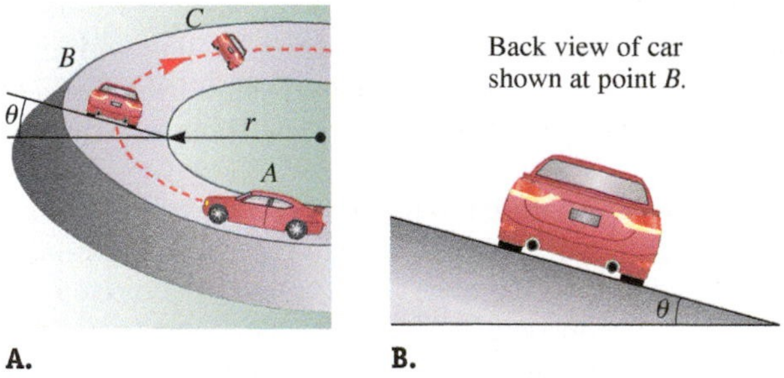

FIGURE 6.30

INTERPRET and ANTICIPATE

There is no static friction in this case from the way v_N is defined above. So, we need only to concern ourselves with gravity and the normal force.

The car does not slide up or down the banked roadway, so the car's circular path is in a horizontal plane. Therefore, the direction of the centripetal acceleration is in that same horizontal plane. It is helpful to choose a coordinate system in which one axis is along the acceleration, such as the x axis in Figure 6.31.

The x component of the normal force is parallel to the centripetal acceleration, so **we circle the normal force and write a phrase near the circle.**

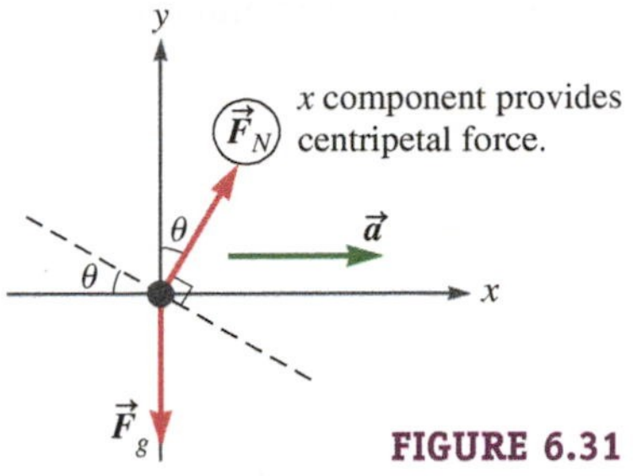
y
x component provides centripetal force.
x
θ
θ

FIGURE 6.31

SOLVE

Apply Newton's second law. **The x component of the normal force acts as the centripetal force.** There is no acceleration in the y direction.

$$\sum F_x = ma_x \qquad \sum F_y = F_N \cos\theta - F_g = ma_y = 0$$

$$F_N \sin\theta = \frac{mv_N^2}{r} \text{ in } x \text{ direction} \quad (1)$$

$$F_N \cos\theta = F_g \text{ in } y \text{ direction} \quad (2)$$

To solve Equations (1) and (2) simultaneously to find v_N, first substitute $F_g = mg$ into Equation (2) to produce Equation (3). Eliminate F_N by dividing Equation (1) by Equation (3). Notice that the mass m cancels.

$$F_N \cos\theta = mg \quad (3)$$

$$\frac{\sin\theta}{\cos\theta} = \frac{mv_N^2}{mgr} = \tan\theta$$

$$v_N = \sqrt{gr\tan\theta}$$

Substitute values.

$$v_N = \sqrt{(9.81 \text{ m/s}^2)(43.3 \text{ m})\tan 7.5°}$$

$$v_N = 7.5 \text{ m/s} \approx 17 \text{ mph}$$

CHECK and THINK

This speed is less than the maximum speed limit we found for the unbanked turn in Example 6.7. This result seems surprising because our experience tells us that banking a turn should allow the car to go faster. It does, once we take static friction into account (part B).

B If the car exceeds v_N, what is the direction of F_s?

INTERPRET and ANTICIPATE

As in Example 6.7, static friction cannot be in the direction in which the car is moving. Static friction must be parallel to the surface, which in this case means the embankment. Thus, static friction may be directed either up or down the embankment. Because the car's speed is greater than v_N, the x component of the normal force alone is not enough to supply the required centripetal force. Static friction directed down the embankment adds to centripetal force as shown in the free-body diagram (Fig. 6.32). So, the **x component of the static friction and the normal force both contribute to the centripetal force.** If friction were directed up the embankment, the centripetal force would be smaller.

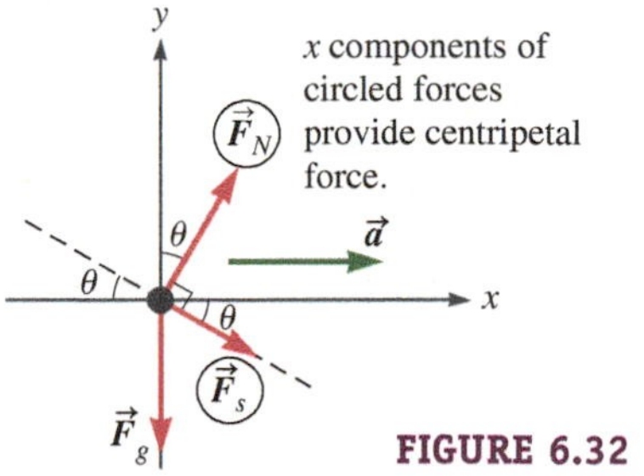
y
x components of circled forces provide centripetal force.
x
θ
θ
θ

FIGURE 6.32

C What is the speed limit of this banked road if $\theta = 7.5°$? (As before, assume the speed limit is set for a wet road and the road surface is concrete.)

SOLVE

Apply Newton's second law to the forces in the free-body diagram. Because we need the highest possible speed, we set static friction to its maximum.

$$\sum F_x = ma_x$$

$$F_N \sin\theta + F_{s,\max}\cos\theta = m\frac{v_{\max}^2}{r}$$

$$\sum F_y = ma_y$$

$$F_N \cos\theta - F_{s,\max}\sin\theta - F_g = 0$$

Example continues on page 174 ▶

In both equations, substitute $\mu_s F_N$ (Eq. 6.1) for static friction $F_{s,\max}$.	$F_N \sin\theta + \mu_s F_N \cos\theta = m\dfrac{v^2_{\max}}{r}$ (1) $F_N \cos\theta - \mu_s F_N \sin\theta = F_g$ (2)
Substitute $F_g = mg$ in Equation (2), and factor F_N in Equation (1). Divide Equation (4) by Equation (3) to eliminate F_N. Notice that mass m cancels out.	$F_N = \dfrac{mg}{\cos\theta - \mu_s \sin\theta}$ (3) $F_N(\sin\theta + \mu_s\cos\theta) = m\dfrac{v^2_{\max}}{r}$ (4) $\dfrac{mg(\sin\theta + \mu_s\cos\theta)}{\cos\theta - \mu_s\sin\theta} = m\dfrac{v^2_{\max}}{r}$ $v_{\max} = \sqrt{\dfrac{gr(\sin\theta + \mu_s\cos\theta)}{\cos\theta - \mu_s\sin\theta}}$ $v_{\max} = \sqrt{\dfrac{(9.81\ \text{m/s}^2)(43.3\ \text{m})(\sin 7.5° + 0.4\cos 7.5°)}{\cos 7.5° - 0.4\sin 7.5°}}$ $v_{\max} = 15\ \text{m/s} \approx 34\ \text{mph}$

As expected, the speed limit on this banked turn is higher than the limit on the unbanked turn (29 mph). The banking angle results in a normal force with a component directed toward the center of the circular path, resulting in a centripetal force that is stronger than the centripetal force due to static friction alone. If the car exceeds this speed limit, it cannot maintain the circular path and will move off the road.

D What is the direction of static friction if $v < v_N$?

If $v < v_N$, the x component of the normal force is greater than the required centripetal force. In this case, static friction reduces the centripetal force. To do so, it must be directed up the embankment as shown in the free-body diagram (Fig. 6.33).

This answer may seem counterintuitive, but think of the extreme case where $v = 0$ so that the car is at rest. Without static friction, gravity would pull the car down the embankment. In fact, if the embankment is very steep, $F_{s,\max}$ may be too small to keep the car from slipping downward, which is true even if the car is moving at some nonzero speed $v < v_N$. So, engineers cannot increase the embankment angle beyond a practical limit.

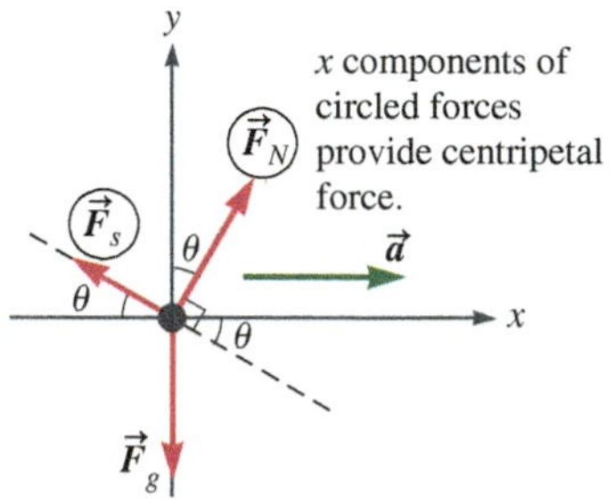

FIGURE 6.33

NONUNIFORM CIRCULAR MOTION

Special Case

Nonuniform Circular Motion

We have seen that if the total force on a particle is constant and perpendicular to its velocity, the particle will travel in uniform circular motion. In this case, the particle's speed is constant, its velocity is tangent to the circle, and its acceleration is directed to the center of the circle.

If a particle's path is circular and the total force is not perpendicular to the velocity, the particle speed changes (Fig. 6.34). This type of circular motion, in which both the magnitude and direction of the velocity change, is called **nonuniform circular motion**. The speed of a particle in nonuniform circular motion changes because the total force has a component parallel to the velocity. (Of course, to maintain a circular path, the total force must also have a component that is centripetal.)

As we have seen, a powerful way to analyze vectors is to break them into components. In this case, instead of using the usual x and y components, we break $\vec{F}_{\text{tot}}$ (or $\vec{a}$)

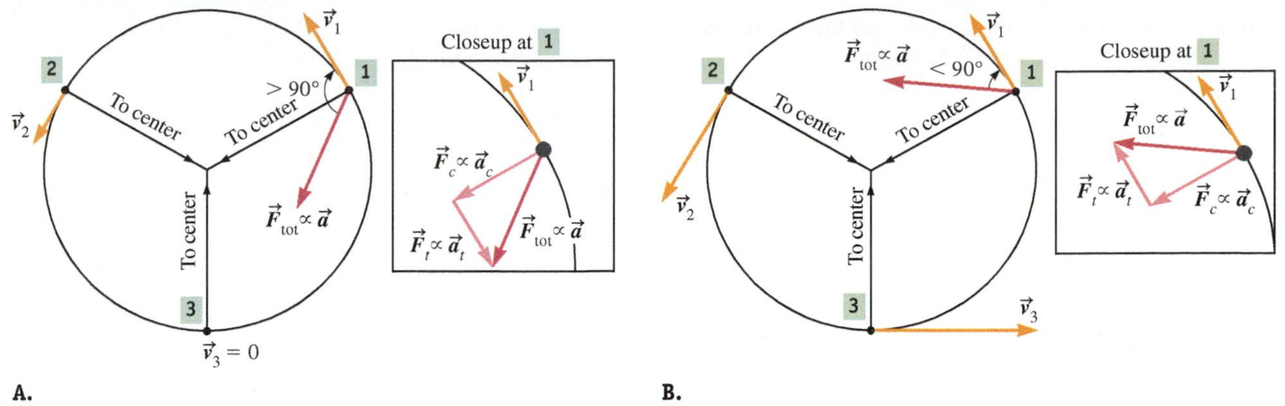

FIGURE 6.34 **A.** A particle travels in a circular path. The net force on it makes an angle greater than 90° with its velocity. The result is that the particle slows down and momentarily stops at 3. **B.** A particle travels in a circular path. The net force on it makes an angle less than 90° with its velocity. The result is that the particle speeds up indefinitely.

into components that are parallel and perpendicular to the velocity vector as shown in the insets of Figure 6.34. The parallel component $\vec{F}_t$ is referred to as the **tangential component** because it is tangent to the circle, and the perpendicular component $\vec{F}_c$ is called the **centripetal component** because it acts toward the center of the motion. The tangential component of the force is responsible for changing the speed of the particle. If this component points in the direction opposite the direction of the velocity as in Figure 6.34A, the particle slows down. If the tangential component points in the direction of the velocity as in Figure 6.34B, the particle speeds up.

The magnitude of the centripetal force F_c and centripetal acceleration a_c are the same as in uniform circular motion:

$$F_c = m\frac{v^2}{r} \quad \text{and} \quad a_c = \frac{v^2}{r}$$

The difference is that in nonuniform circular motion these two quantities are not constant. As the speed changes, the centripetal force and acceleration must also change.

EXAMPLE 6.9 Loop-the-Loop Ride

The loop-the-loop ride is a roller coaster in which the cars travel on the inside of a vertical circular track (Fig. 6.35). After the cars are launched at the bottom *B*, no motor or engine pulls them along the track.

A Treat each car as a particle and assume rolling friction and drag are negligible. Draw a free-body diagram for a car at points (bottom) *B*, (right) *R*, (top) *T*, and (left) *L*. Indicate the acceleration in each case.

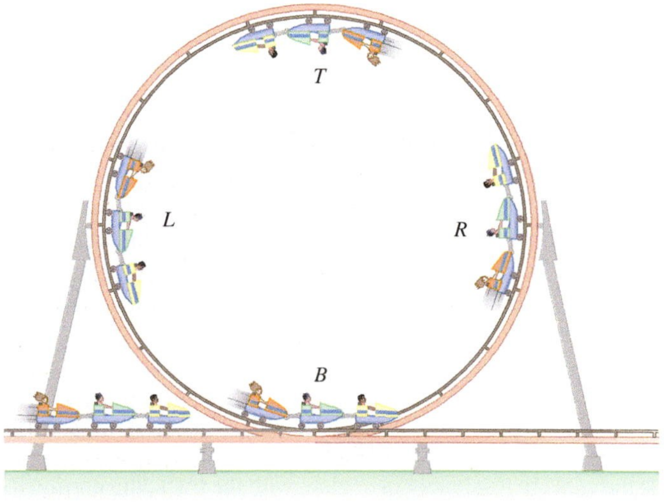

FIGURE 6.35

Example continues on page 176 ▶

Only two forces act on the car: gravity and the normal force due to the track. (Unlike the banked turn in Example 6.8, static friction does not act on a car in the vertical loop-the-loop.) The gravitational force on a car is the same at all four positions. The normal force changes in both magnitude and direction (Fig. 6.36). The direction of the acceleration is estimated by adding the two forces geometrically.

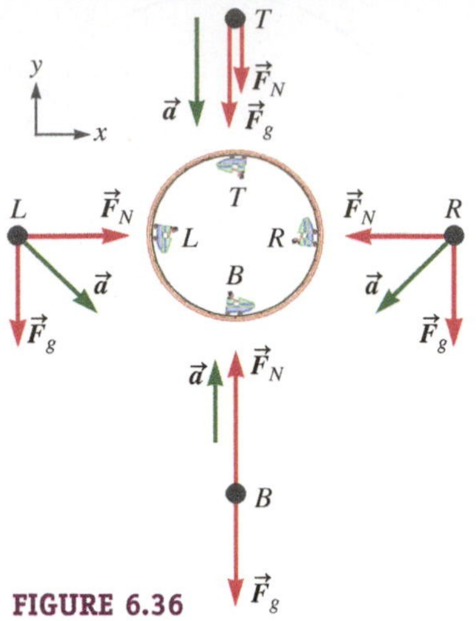

FIGURE 6.36

B Is this uniform circular motion? Explain.

No. Uniform circular motion requires that the velocity and acceleration be perpendicular to each other. In our diagrams, the acceleration is perpendicular to the velocity only at T and B. At all other positions, the acceleration has a tangential component. At R and all other positions on the right half of the track, the tangential component opposes the velocity. Therefore, cars slow down as they move from B to T. On the left side of the track, the tangential acceleration is in the same direction as the velocity. Consequently, cars speed up as they move from T back down to B.

C Write an expression for the magnitude of the centripetal force F_c at each of the four positions in terms of the car's mass m, radius of the loop r, and the speed at the four positions.

INTERPRET and ANTICIPATE

At each position, we must identify the force or forces that point toward the center of the circle. If more than one force points in this direction, we must add them to find the magnitude of the total centripetal force at that point. The top and bottom points (T and B) are the simplest because for those two points both forces are radial.

SOLVE

At T, both the normal force and gravity point to the center of the circle, so they both contribute to the centripetal force.

For the top T:

$$F_c = F_N + F_g = m\frac{v_T^2}{r}$$

At B, both the normal force and gravity affect the centripetal force just as at T. Now, however, gravity points away from the center of the circle, so it is negative.

For the bottom B:

$$F_c = F_N - F_g = m\frac{v_B^2}{r}$$

Gravity is tangent to the circle at R and L. Therefore, only the normal force contributes to the centripetal force at these two points.

For the left L and right R:

$$F_c = F_N = m\frac{v_{L\text{ or }R}^2}{r}$$

CHECK and THINK

The loop-the-loop ride might seem very dangerous because our intuition tells us that a car should fall from T to B. Our analysis seems to confirm our intuition. The net force exerted on a car at point T points straight down. If the velocity were zero at T, the car would indeed fall straight down to B. Because the velocity is tangent to the track at T, however, a car accelerates downward at T, but that does not mean that its motion is straight down. Instead, its motion is downward and *along the track*.

D If the cars are just barely in contact with the track at T, find an expression for their speed at that position.

INTERPRET and ANTICIPATE

If the cars are just barely in contact with the track, the normal force exerted by the track on the cars is approximately zero. Only gravity is responsible for the centripetal force.

$$F_c = F_g$$

$$m\frac{v_T^2}{r} = mg$$

We solve for v_T.

$$v_T = \sqrt{gr}$$

CHECK and THINK

If the cars are moving at any speed higher than $\sqrt{gr}$ when they get to the top, the normal force must contribute to the centripetal force. A problem arises if the cars are moving at a speed less than $\sqrt{gr}$ at the top. If $v = 0$ at the top, the car would fall straight down. If $0 < v < \sqrt{gr}$, the car would follow a parabolic path as it fell to the ground. Next time you plan to board such a ride, estimate the speed of the cars and be sure it greater than $\sqrt{gr}$.

SUMMARY

Underlying Principles: Newton's laws of motion

Special Cases

1. When an object is in contact with a surface and is at rest relative that surface, the **static friction** force on the object is parallel to the surface. If the object is not accelerating, the static friction force is balanced by the vector sum of any other forces that are applied to the object parallel to the surface.

   The maximum force of static friction is given by

   $$F_{s,\max} = \mu_s F_N \tag{6.1}$$

   The **coefficient of static friction** μ_s is a positive, unitless scalar that is usually found through experimentation (Table 6.1).
2. Whenever two surfaces are in contact and sliding relative to each other, **kinetic friction** works against that motion. The direction of kinetic friction on the object is always parallel to the contacting surfaces and in the direction opposite the direction of the object's motion relative to the other surface. The magnitude of kinetic friction is given by

   $$F_k = \mu_k F_N \tag{6.2}$$

   where μ_k is the **coefficient of kinetic friction** and is found experimentally (Table 6.1).
3. The friction force associated with the rolling motion of an object along a surface is called **rolling friction**. The direction of the rolling friction force is opposite the motion of the object relative to the surface. The magnitude of the rolling friction force is given by

   $$F_r = \mu_r F_N \tag{6.3}$$

   The **coefficient of rolling friction** μ_r is a unitless scalar that is found experimentally (Table 6.1).
4. There is a **drag force** on an object when it moves relative to a fluid medium. The direction opposes the object's relative velocity. For small objects moving slowly through air or larger objects moving through water,

   $$\vec{F}_D = -b\vec{v} \tag{6.4}$$

   For blunt objects moving at high speed through air, the magnitude of the drag force is

   $$F_D = \frac{1}{2}C\rho A v^2 \tag{6.5}$$
5. When an object falling through the air stops accelerating, it continues to move downward at a constant speed known as its **terminal speed**:

   $$v_t = \sqrt{\frac{2mg}{C\rho A}} \tag{6.6}$$
6. The net force exerted on object in uniform circular motion points to the center of the circle and is known as the **centripetal force**. Its magnitude is given by

   $$F_c = m\frac{v^2}{r} \tag{6.7}$$

PROBLEM-SOLVING STRATEGY When Centripetal Force Is Present

Problems that involve centripetal force are no different from other problems that require us to apply Newton's second law. Two modifications, however, may be helpful.

INTERPRET and ANTICIPATE

Modification 1 Draw a free-body diagram for one particular instant and **circle the centripetal force(s)**.

SOLVE

Modification 2 The **sum of the forces that are parallel or antiparallel to the centripetal acceleration equals the centripetal force** (Eq. 6.7).

PROBLEMS AND QUESTIONS

A = algebraic C = conceptual E = estimation G = graphical N = numerical

6-1 Newton's Laws in a Messy World

1. **C** In many textbook problems, we ignore certain complications such as friction and drag. The problems contain key words that indicate such a simplification is being used. For example, if a surface is described as "slippery," it means that we can ignore friction. Look at the previous chapters' problem sets. Find five uses of these key words and explain how to interpret each case.
2. **C** CASE STUDY On page 154, Cameron says, "the real world is too hard to study with physics." Do you agree with that statement? If so, explain why. If not, come up with a better statement and explain why it is better.
3. **C** CASE STUDY Are skydivers with their parachutes closed in free fall? Explain.

6-2 Friction and the Normal Force Revisited

4. **C** We often need to model a problem, such as when we model molecular bonds as springs or people as particles. For which circumstances are these examples good models? Under what conditions do they break down?

Problems 5 and 9 are paired.

5. **C** In Figure 6.3, a man holds a box against the wall. The box is at rest. **a.** Compare the normal force exerted by the wall to the force exerted by the man. **b.** If the man reduces his force on the box, what will happen to the normal force? **c.** If the man reduces his force on the box, what will happen to static friction?

FIGURE P6.6

6. **G** Draw a free-body diagram for the burglar, who is shown at rest while sneaking through a chimney in Figure P6.6.
7. **C** The shower curtain rod in Figure P6.7 is called a tension rod. The rod is not attached to the wall with screws, nails, or glue, but is pressed into the wall instead. Explain why the rod remains at rest, supporting the curtain. Explain why the name is misleading and come up with a better name.

Courtesy, Signature Hardware and Gatco Inc.

FIGURE P6.7

6-3 A Model for Static Friction

8. **N, C** A rectangular block has a length that is five times its width and a height that is three times its width. The block's surfaces are all identical except for size. When the block is placed on a horizontal tabletop so that the area in contact with the table is length × width, it is found that a horizontal force of 10.0 N applied to the block is just sufficient to overcome the static friction force and cause the block to move. The block is then knocked over so that the area in contact with the table is length × height. Now, what minimum horizontal force will cause the block to move? Explain.
9. **N** A man exerts a force of 16.7 N horizontally on a box so that it is at rest in contact with a wall as in Figure 6.3. The box weighs 6.52 N. **a.** Find the static friction force exerted on the box, given the forces being applied. **b.** If the coefficient of static friction between the wall and the box is 0.50, find the maximum static friction force that may be exerted on the box. Comment on your results.

Problems 10 and 11 are paired.

10. **C** A makeshift sign hangs by a wire that is extended over an ideal pulley and is wrapped around a large potted plant on the roof as shown in Figure P6.10. When first set up by the shopkeeper on a sunny and dry day, the sign and the pot are in equilibrium. Is it possible that the sign falls to the ground during a rainstorm while still remaining connected to the pot? What would have to be true for that to be possible?

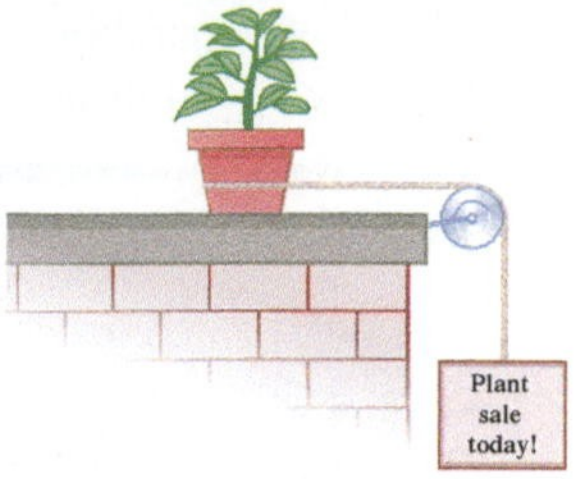

FIGURE P6.10 Problems 10 and 11.

11. **N** In Problem 10, the mass of the sign is 25.4 kg, and the mass of the potted plant is 66.7 kg. **a.** Assuming the objects are in equilibrium, determine the magnitude of the static friction force experienced by the potted plant. **b.** What is the maximum value of the static friction force if the coefficient of static friction between the pot and the roof is 0.572?
12. **C** A heavy box of tools rests on a tabletop. You push on the box with a horizontal force of 100 N and observe that the box does not move. You then push on the box with a horizontal force of 200 N and observe that the box does not move. How does the magnitude of the static friction force compare to the force with which you push in each case?
13. **N** A motorcyclist is traveling at 55.0 mph on a flat stretch of highway during a sudden rainstorm. The rain has reduced the coefficient of static friction between the motorcycle's tires and the road to 0.080 when the motorcyclist slams on the brakes, locking the tires in place. **a.** What is the minimum distance required to bring the motorcycle to a complete stop? **b.** What would be the stopping distance if it were not raining and the coefficient of static friction were 0.630?

Problems 14 and 15 are paired.

14. A small steel I-beam (Fig. P6.14) is at rest with respect to the steel surface of a truck. The truck is accelerating with respect to the road. The mass of the I-beam is 5.8×10^3 kg.
    a. **G** Draw a free-body diagram for the I-beam.
    b. **C** What force or forces accelerate the I-beam with respect to the ground?
    c. **N** The I-beam must remain at rest with respect to the truck. What is the maximum acceleration of the truck? Evaluate your answer.
    d. **G** On the highway, the truck moves with a constant velocity. Draw a free-body diagram for the I-beam. Compare it with your diagram in part (a).

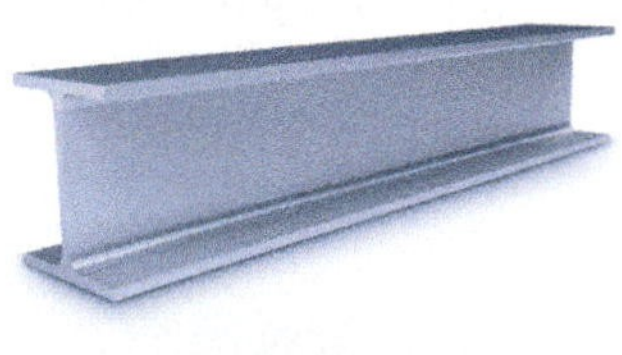

3dfoto/iStockphoto.com

FIGURE P6.14

15. A box is at rest with respect to the surface of a flatbed truck. The coefficient of static friction between the box and the surface is μ_S.
    a. **A** Find an expression for the maximum acceleration of the truck so that the box remains at rest with respect to the truck. Your expression should be in terms of μ_S and g.
    b. **C** How does your answer change if the mass of the box is doubled?

Problems 16 and 17 are paired.

16. **A** A filled treasure chest of mass m with a long rope tied around its center lies in the middle of a room. Dirk wishes to drag the chest, but there is friction between the chest and the floor with a coefficient of static friction μ_s. If the angle between the rope and the floor is θ, what is the magnitude of the tension required to just get the chest moving? Express your answer in terms of m, μ_s, θ, and g.

17. **N** A filled treasure chest ($m = 375$ kg) with a long rope tied around its center lies in the middle of a room. Dirk wishes to drag the chest, but there is friction between the chest and the floor with $\mu_s = 0.52$. If the angle between the rope and the floor is 30.0°, what is the magnitude of the tension required to just get the chest moving?

18. **N** Rochelle holds her 2.80-kg physics textbook by pressing horizontally against both sides of the textbook with the palms of her hands. If the coefficient of static friction between her hands and the textbook is 0.500, what is the minimum force with which she must compress the textbook with each hand to keep it from slipping out of her hands?

Problems 19, 20, and 21 are grouped.

19. **A** A sled and rider have a total mass M. They are at rest on a snowy hill. The coefficient of static friction between the sled and the snow is μ_S. Find an expression for the maximum angle of the hill's slope measured upward from the horizontal in terms of μ_S. How does your answer change if the rider is not on the sled?

20. **N** A sled and rider have a total mass 56.8 kg. They are at rest on a snowy hill. The coefficient of static friction between the sled and the snow is 0.225. What is the maximum angle of the hill's slope measured upward from the horizontal?

21. **C, N** A sled and rider have a total mass 56.8 kg. They are on a snowy hill. The coefficient of static friction between the sled and the snow is 0.225. The angle of the hill's slope measured upward from the horizontal is 19.5°. Are they at rest?

6-4 Kinetic and Rolling Friction

22. **N** You need to design an experiment to measure the free-fall acceleration. You have an inclined plane made of steel and several objects that you may place on the plane. The plane is fixed at 45°. You have a Teflon-coated block with a mass of 0.5 kg, a cart with steel wheels and a mass of 1.5 kg, and a steel box with a mass of 0.75 kg. You have a sonar device to measure the acceleration. You are to place one or more objects on the plane and measure their acceleration along the plane. Does it matter which object or objects you choose? If not, why not? If so, which would you use? Explain your answer.

23. **N** A block with mass $M = 2.00$ kg is placed on an inclined plane. The plane makes an angle of 30° with the horizontal, and the coefficients of kinetic and static friction between the block and the plane are $\mu_k = 0.400$ and $\mu_s = 0.600$, respectively. Will the block slide down the plane, or will it remain motionless? Justify your answer.

24. **C** Lisa measured the coefficient of static friction between two pairs of running shoes and the track in Example 6.1 (page 159). If she wants to have an advantage in a race, which shoes should she wear, the ones with a high coefficient or the ones with the low coefficient of static friction? Explain.

25. **N** An ice cube with a mass of 0.0507 kg is placed at the midpoint of a 1.00-m-long wooden board that is propped up at a 50° angle. The coefficient of kinetic friction between the ice and the wood is 0.133. **a.** How much time does it take for the ice cube to slide to the lower end of the board? **b.** If the ice cube is replaced with a 0.0507-kg wooden block, where the coefficient of kinetic friction between the block and the board is 0.275, at what angle should the board be placed so that the block takes the same amount of time to slide to the lower end as the ice cube does? You may find a spreadsheet program helpful in answering this question.

26. **C** In a pinewood derby race, children release small cars they have constructed from the top of a ramp, record the time it takes to reach the finish line, and rank the winners through several trials (Fig. P6.26). Juan proclaims to his friend, Antonio, that his car will reach the finish line first because he has adjusted his wheels to be more narrow. This adjustment means that the area in contact with the racetrack will be less than normal. Juan suggests that this adjustment will make the magnitude of the friction between his wheels and the track less than that for Antonio's car. Assuming the masses of the cars are the same and air resistance can be ignored, is Juan correct, or does Antonio have a valid counterargument?

Ken Stewart/Zuma Press/Alamy

FIGURE P6.26

27. **A** Curling is a game similar to lawn bowling except it is played on ice and instead of rolling balls on the lawn, stones are slid along ice. A curler slides a stone across a sheet of ice with an initial speed v_i in the positive x direction. The coefficient of kinetic friction between the stone and the curling lane is μ_k. Express your answers in terms of v_i, μ_k, and g only. **a.** What is the acceleration of the stone as it slides down the lane? **b.** What distance does the curling stone travel?

28. **C** Janelle and Jordan need to push their car to the mechanic. Janelle pushes the car while Jordan steers with the car in the neutral gear. (It is free to roll.) Initially, Janelle is pushing the car downhill, later on level ground, and finally uphill to get it into the mechanic's lot. She finds it much easier to push the car downhill than uphill. Assume the angles of elevation θ of the downhill and uphill slopes are the same with respect to the level part of the road. **a.** If the coefficient of rolling friction, μ_r, between the tires and the road is the same on each part of the journey, how does the magnitude of the friction force compare on each part? **b.** Is the friction force always working against Janelle's pushing force on the car? Explain why it is easier to push the car downhill.

Problems 29, 30, and 31 are grouped.

29. **N** A sled and rider have a total mass of 56.8 kg. They are on a snowy hill. The coefficient of kinetic friction between the sled and the snow is 0.195. The angle of the hill's slope measured upward from the horizontal is 19.5°. What is the acceleration of the rider? Is the acceleration greater, less than, or equal to your result if a more massive rider uses the same sled on the same hill? Explain.

30. **N** A sled and rider have a total mass of 56.8 kg. They are on a snowy hill accelerating at $0.7g$. The coefficient of kinetic friction between the sled and the snow is 0.18. What is the angle of the hill's slope measured upward from the horizontal? You may find a spreadsheet program helpful in answering this question.

31. **N** A cart and rider have a total mass of 56.8 kg. The cart is rolling down a hill and accelerating at $0.5g$. The coefficients of kinetic friction and of rolling friction between the cart's wheels and the road are 0.20 and 0.020 respectively. What is the angle of the hill's slope measured upward from the horizontal? You may find a spreadsheet program helpful in answering this question.

6-5 Drag and Terminal Speed

32. **C** Why do you crumple up a sheet of paper before you toss it toward the garbage can?

33. **N** On a particularly hot day at the racetrack, Dale Earnhardt's famous No. 88 Chevy Impala ($m = 1600$ kg), speeding at 200.0 mph, blows a gasket in the engine and begins to decelerate. The car is shifted into neutral and coasts toward the pit lane. Neglecting all other sources of friction and assuming the drag coefficient is 0.330 and the car's frontal area is 2.76 m^2, what is the initial deceleration of the Impala? Report your answer to three significant figures.

34. **C** An intrepid experimenter with two identical objects stands at the edge of a cliff, looking over a canyon. She throws one of the objects directly upward with speed v while holding her hand out over the edge. She throws the second object directly downward with speed v from the same height. Neither object reaches terminal velocity, yet they each have the same speed just before they strike the canyon floor below. Explain how this could be true, considering the drag force on the object during each object's motion.

35. **N** A small sphere of mass $m = 0.500$ kg is dropped from rest into a viscous liquid in which the resistive force on the sphere can be expressed as $\vec{F}_D = -b\vec{v}$, and reaches one-fourth its terminal speed in 3.45 s. **a.** What is the terminal speed of the sphere? **b.** What is the distance traveled by the sphere in 3.45 s?

Problems 36 and 37 are paired.

36. A racquetball with a radius of 0.0285 m is thrown vertically into the air with an initial speed of 3.07 m/s. The drag coefficient of the ball is 0.47, and the density of the air is 1.29 kg/m^3.

a. **N** What is the cross-sectional area of the ball?

b. **N** What is the magnitude of the drag force on the ball the instant it is thrown?

c. **N** What is the magnitude of the drag force on the ball when it reaches the peak?

d. **G** Make a sketch of the magnitude of the drag force versus time from the moment the ball is thrown until it reaches the peak.

37. **N** A racquetball has a radius of 0.0285 m. The drag coefficient of the ball is 0.47, and the density of the air is 1.29 kg/m^3. What would be the terminal speed for the racquetball if it were dropped from a very high cliff, assuming it has a mass of 0.0450 kg?

38. **C** The terminal speed of a penny in air has been found to be between 11 m/s and 32 m/s. A man tosses a penny in the air and fails to catch it. **a.** Does the penny reach its terminal speed before landing at the man's feet? **b.** If the man is on top of the Empire State Building in Manhattan and the penny falls to the ground (roughly 1000 ft below), does it reach its terminal speed? **c.** If a penny falls off a large building, would it bore a hole into the sidewalk below? Explain.

39. **CASE STUDY** Let's take another look at the skydivers flying in formation (Example 6.5, page 168). Assume there are three skydivers of different body types. Let's model each one as a cylinder. Skydiver A's cylinder is 2.00 m tall, has a radius of 0.25 m, and has a total weight of 670 N; skydiver B's cylinder is 1.75 m tall, has a radius of 0.33 m, and has a total weight of 675 N; and skydiver C's cylinder is 1.95 m tall, has a radius of 0.43 m, and has a total weight of 907 N. Assume the drag coefficient is 0.5 when the skydivers are pointing head down and the drag coefficient is 1.0 when the skydivers are oriented belly-to-Earth. The density of air is 1.29 kg/m^3.

a. **N** Find the maximum and minimum terminal speed for each skydiver.

b. **C** If the skydivers wish to fly in formation as in Figure 6.1, in which order should they leave the plane?

c. **N** Imagine that the plane flies in a circular path so that skydivers can exit over the same spot. Suppose that the time between skydivers exiting the plane is 30 s. Assume the skydivers hit terminal speed immediately (not true; actually takes about 10 s). Find the time that the first skydiver must wait after exiting the plane until the last skydiver joins the formation. How would your answer change if you took into account that each diver accelerates for about 10 s just after exiting the plane?

40. In this problem, we compare the drag force to rolling friction on a typical car. The density of air is 1.29 kg/m^3.

a. **E** Estimate the weight and effective area of a typical car. Assume the car has been designed to reduce drag.

b. **G** Use your estimate to plot v^2 versus F_D. Does the shape of your graph make sense? Explain.

c. **E** Estimate rolling friction on this car. Add your estimate to the graph.

d. **G** At what speed does rolling friction equal the drag on the car? For what range of speeds is it okay to ignore drag?

e. **C** How do your answers change if you take into account that at slow speeds drag is proportional to v, not v^2?

41. **N** An inflated spherical beach ball with a radius of 0.3573 m and average density of 10.65 kg/m^3 is being held under water in a pool by Janelle. The density of the water in the pool is 1000.0 kg/m^3. When Janelle releases the ball, it begins to rise to the surface. If the drag coefficient of the ball in the water is 0.470 and the constant upward force on the ball is 1875 N, what will be the terminal speed of the ball as it rises? Ignore the effects of gravity on the ball.

42. **E** **CASE STUDY** In the train collision case study (Chapter 5, page 119), we ignored the drag force on the trains. Estimate the drag on the trains and compare it to the kinetic friction on them. Is it okay to ignore drag? Explain.

Problems 43 and 44 are paired.

43. **N** Your sailboat has capsized! Fortunately, you are no longer aboard the boat. Instead, you are hanging onto the end of a long rope, the other end of which is attached to a Coast Guard helicopter. Model yourself as a particle of mass $M = 55.0$ kg with a diameter equal to 0.500 m. The density of the air is $\rho = 1.29$ kg/m^3. Assume the drag coefficient between you and the air is $C = 0.500$. **a.** First, ignore the drag force due to the air. If the helicopter is flying at a constant speed $v_0 = 35.0$ m/s, what angle will the rope make with the vertical? **b.** Now, consider the drag force due to the air. What angle does the rope make with the vertical given the information in part (a)?

44. You are hanging onto the end of a long rope, the other end of which is attached to a Coast Guard helicopter. Model yourself as a particle of mass $M = 55.0$ kg with a diameter equal to 0.500 m. The density of the air is $\rho = 1.29$ kg/m^3. Assume the drag coefficient between you and the air is $C = 0.500$. Now, at time $t = 0$, the helicopter begins to accelerate horizontally with an acceleration $a = 3.00$ m/s^2.

a. **N** If we ignore the effect of the air drag force, what is the angle of the rope with the vertical?

b. **A** Suppose in this case where the helicopter accelerates we also consider the drag force due to the air. Derive an expression for the angle made by the rope with the vertical as a function of time.

c. **C** Neither in this problem nor Problem 43 has a terminal speed been calculated. Is terminal speed a meaningful concept for this situation? Why or why not?

d. **C** What happens to the tension if the helicopter continues to accelerate? What would be the result for a real rope in this situation?

45. C The **drag coefficient** C in $F_D = \frac{1}{2}C\rho Av^2$ (Eq. 6.5) depends primarily on the shape of the object. You already have developed an intuition about what shapes correspond to a low C by observing the shapes of aerodynamic cars, boats, and even bullets. Which object, a sphere or a cube, would have a larger drag coefficient, assuming they are nearly the same size? Explain your reasoning. What aspect of an object most determines its drag coefficient?

6-6 Centripetal Force

46. N Amir (m = 56 kg) begins from rest and starts to jog around a circular track with a radius of 32.0 m. After a brief 1.56 s, Amir reaches his top speed of 3.20 m/s. What is the magnitude of the centripetal force responsible for keeping Amir on the circular path?

47. N The speed of a 100-g toy car at the bottom of a vertical circular portion of track is 8.00 m/s. If the radius of curvature of this portion of the track is 60.0 cm, what are the magnitude and direction of the force the track exerts on the car? Report your answer to three significant figures.

48. You are riding in a car around a turn. You do not move with respect to the seat. The turn is flat.
  a. G Draw a free-body diagram for you.
  b. C Which force or forces are responsible for the centripetal force on you?
  c. C What happens to you if the driver takes the turn too quickly?

49. N Artificial gravity is produced in a space station by rotating it, so it is a noninertial reference frame. The rotation means that there must be a centripetal force exerted on the occupants; this centripetal force is exerted by the walls of the station. The space station in Arthur C. Clarke's *2001: A Space Odyssey* is in the shape of a four-spoked wheel with a diameter of 155 m. If the space station rotates at a rate of 2.40 revolutions per minute, what is the magnitude of the artificial gravitational acceleration provided to a space tourist walking on the inner wall of the station?

50. N Escaping from a tomb raid gone wrong, Lara Croft (m = 57.0 kg) swings across an alligator-infested river from a 9.00-m-long vine. If her speed at the bottom of the swing is 7.50 m/s and she makes it safely across the river, what is the minimum breaking strength of the vine?

Problems 51 and 52 are paired.

51. N Harry Potter decides to take Pottery 101 as an elective to satisfy his arts requirement at Hogwarts. He sets some clay (m = 3.25 kg) on the edge of a pottery wheel (r = 0.600 m), which is initially motionless. He then begins to rotate the wheel with a uniform acceleration, reaching a final angular speed of 2.400 rev/s in 3.00 s. **a.** What is the speed of the clay when the initial 3.00 s has passed? **b.** What is the centripetal acceleration of the clay initially and when the initial 3.00 s has passed? **c.** What is the magnitude of the constant tangential acceleration responsible for starting the clay in circular motion?

52. C Harry sets some clay (m = 3.25 kg) on the edge of a pottery wheel (r = 0.600 m), which is initially motionless. He then begins to rotate the wheel with a uniform acceleration, reaching a final angular speed of 2.400 rev/s in 3.00 s, while not touching the clay. As a result, the clay is subject to a tangential and centripetal acceleration while it sits on the edge of the wheel. **a.** What force is responsible for the tangential acceleration, and what force is responsible for the centripetal acceleration? **b.** Which of these two forces, tangential or centripetal, will necessarily fail to keep the clay in place on the wheel first? Why?

Problems 53 and 54 are paired.

53. A A small disk of mass m is attached by a rope to a block with a larger mass M through a hole in a table as shown in Figure P6.53. The disk moves in circle of radius R at constant speed, and the block is at rest. Assume friction between the disk and the table is negligible. Find an expression for its speed v_{ucm} in terms of the parameters given and the tension F_T in the rope.

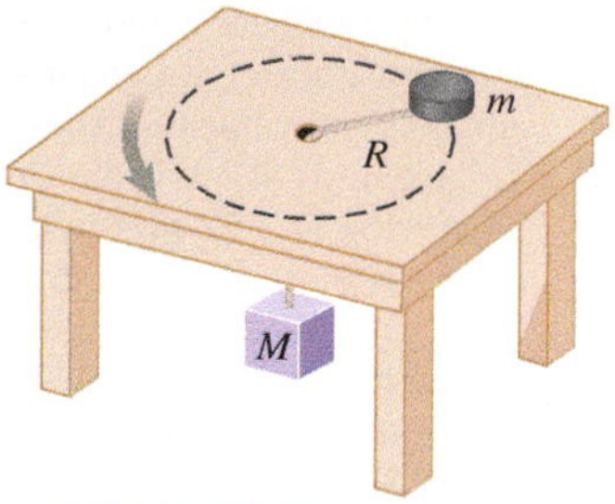

FIGURE P6.53 Problems 53 and 54.

54. The setup in Figure P6.53 is built in a laboratory using a Teflon tabletop and a small Teflon disk of mass m. The disk is attached by a rope to a block with a larger mass M. The small disk starts with a speed v_{ucm} (from Problem 53 in which friction is negligible) and travels in a circular path.
  a. G Draw two free-body diagrams for the disk. One should be from an overhead perspective; the other should be from behind.
  b. C Describe the path of the disk. Explain.
  c. C What happens to the block? Explain.

55. N A cyclist experiences a total horizontal force of 85.0 N when he rounds a horizontal curve on a track at a constant speed of 8.00 m/s. What are the magnitude and direction of the total horizontal force he would experience after he doubled his speed through the curve?

Problems 56, 57, and 58 are grouped.

56. E The Earth's gravity is responsible for the centripetal force on the Moon. Assume the Moon's orbit represents uniform circular motion. Estimate the centripetal force on the Moon.

57. N When a star dies, much of its mass may collapse into a single point known as a black hole. The gravitational force of a black hole on surrounding astronomical objects can be very great. Astronomers estimate the strength of this force by observing the orbits of such objects around a black hole. What is the gravitational force exerted by a black hole on a 1-solar-mass star whose orbit has a 1.4×10^{10} m radius and a period of 5.6 days?

58. N A satellite of mass 16.7 kg in geosynchronous orbit at an altitude of 3.58×10^4 km above the Earth's surface remains above the same spot on the Earth. Assume its orbit is circular. Find the magnitude of the gravitational force exerted by the Earth on the satellite. *Hint*: The answer is not 163 N.

59. N Banked curves are designed so that the radial component of the normal force on the car rounding the curve provides the centripetal force required to execute uniform circular motion and safely negotiate the curve. A car rounds a banked curve with banking angle θ = 22.0° and radius of curvature 150 m. **a.** If the coefficient of static friction between the car's tires and the road is μ_s = 0.400, what is the range of speeds for which the car can safely negotiate the turn without slipping? **b.** What is the minimum value of μ_s for which the car's minimum safe speed is zero? Note that friction points up the incline here.

General Problems

60. C A block lies motionless on a horizontal tabletop. You apply a force $\vec{F}_{app}$ horizontally to the block, but it does not move. What can you say about the relative sizes and magnitudes of $\vec{F}_{app}$, the static friction force between the block and the table, and the kinetic friction force between the block and the table?

61. N A car with a mass of 1453 kg is rolling along a flat stretch of road and eventually comes to a stop due to rolling friction. If the car begins with a speed of 10.0 m/s and the car comes to a stop in 6.88 s, what is the coefficient of rolling friction between the tires and the road?

62. To improve health and train physically, some people choose to run in a pool of water rather than outside or on a treadmill in air. Typical values for the density of air and water are 1.29 kg/m^3 and 1.00×10^3 kg/m^3, respectively. Consider a runner who is moving at 2.57 m/s in air.
   a. **E, N** Assuming the drag coefficient for the human body is approximately 1.00, estimate the cross-sectional area of the person running and determine the resulting drag force on the person while running in air.
   b. **E, N** Again, using your estimate of the cross-sectional area and the same drag coefficient, determine the resulting drag force on the person running at the same speed in water.
   c. **C** Does the speed in part (b) seem achievable while running in water? Explain your answer.

63. **N** A force of 87.0 N in the horizontal direction is required to set a 30.0-kg box, initially at rest on a flat surface, in motion. After the box begins to move, a horizontal force of 64.0 N is required to maintain a constant speed. **a.** What is the coefficient of static friction between the box and the surface? **b.** What is the coefficient of kinetic friction between the box and the surface?

64. A box rests on a surface (Fig. P6.64). A force $\vec{F}_{app}$ is applied to the box in two different ways. In both cases, $\vec{F}_{app}$ has the same magnitude, but in case 1 the force is directed below the horizontal, whereas in case 2 it is directed above the horizontal.
   a. **G** Draw a free-body diagram for both cases.
   b. **C** Now F_{app} is increased in both cases until the box just barely remains at rest. Compare $F_{s,\max}$ for each free-body diagram.
   c. **C** Use your answer to part (b) to find a best way to move a heavy desk. Describe and explain your solution.

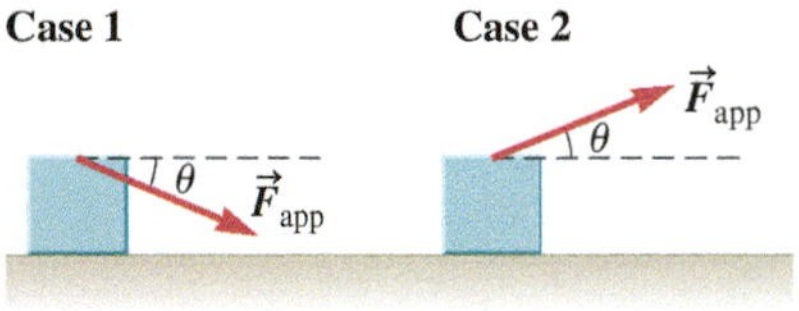

FIGURE P6.64

65. **A** A box of mass m rests on a rough, horizontal surface with a coefficient of static friction μ_s. If a force $\vec{F}_p$ is applied to the box at an angle θ as shown, what is the minimum value of θ for which the box will not move regardless of the magnitude of $\vec{F}_p$?

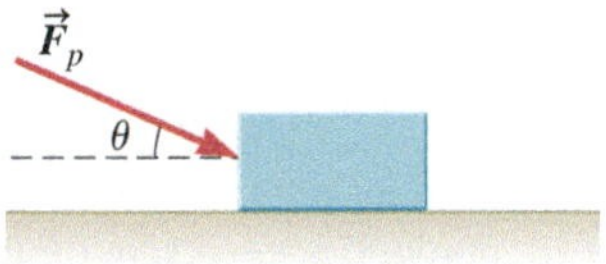

FIGURE P6.65

66. **C** A cylinder of mass M at rest on the end of a string causes a tension F_{T1} in the string. If the cylinder is allowed to swing, the tension in the string when it is vertical is F_{T2}. How do the two tensions F_{T1} and F_{T2} compare in magnitude?

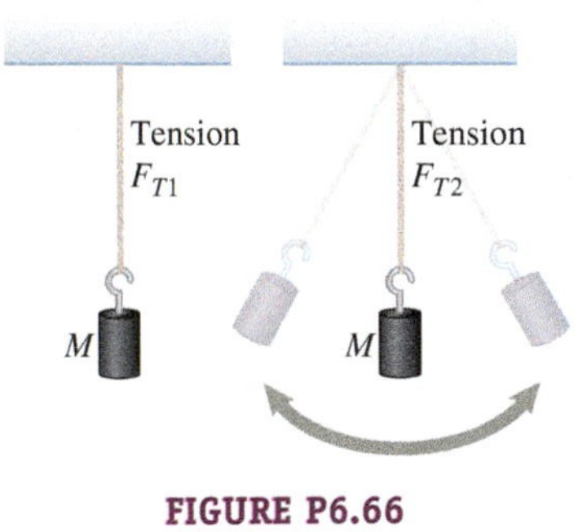

FIGURE P6.66

Problems 67, 70, 71, and 72 are grouped.

67. **A** A block of mass M is placed on a frictionless plane. The plane is inclined at an angle θ, and the block is a distance d from its end. Of course, we would expect the block to slip down the plane. Suppose we revolve the incline around the vertical axis shown in Figure P6.67 instead. At what period of revolution will the block remain in place on the plane?

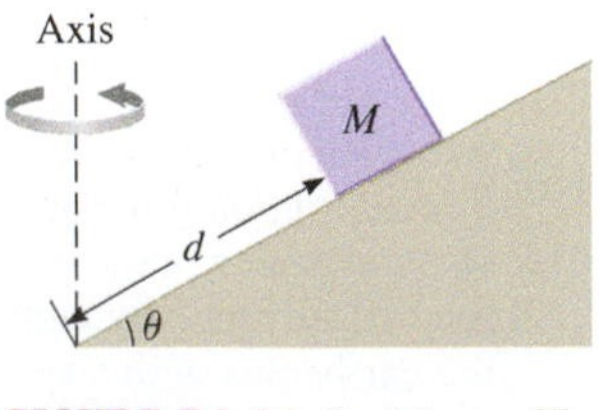

FIGURE P6.67 Problems 67, 71, and 72.

68. **N** Instead of moving back and forth, a conical pendulum moves in a circle at constant speed as its string traces out a cone (Fig. P6.68). One such pendulum is constructed with a string of length $L = 12.0$ cm and bob of mass 0.210 kg. The string makes an angle $\theta = 7.00°$ with the vertical. **a.** What is the radial acceleration of the bob? **b.** What are the horizontal and vertical components of the tension force exerted by the string on the bob?

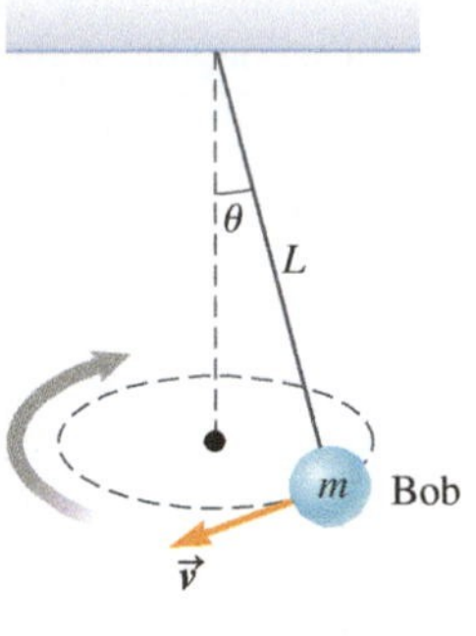

FIGURE P6.68

69. **N** A child is swinging a 325-g ball at the end of a 74.0-cm-long string in a vertical circle. The string can withstand a tension of 12.0 N before breaking. **a.** What is the tension in the string when the ball is at the top of the circle if its speed at that point is 3.40 m/s? **b.** What is the maximum speed the ball can have at the bottom of the circle if the string does not break?

70. **A** Suppose you place a block of mass M on a plane inclined at an angle θ. Show that the mass will slide down the plane if $\mu_s < \tan\,\theta$, where μ_s is the coefficient of static friction between the block and the plane.

71. **A** A block of mass M is placed on a plane, where μ_s is the coefficient of static friction between the block and the plane. The plane is inclined at an angle θ. When the plane is stationary, the block slips down the plane. We can keep the block in place on the incline by revolving the plane around the vertical axis shown in Figure P6.67. In this case, what is the *maximum* period of revolution (that is, the slowest rotation) that will keep the block on the incline a distance d from the lower end?

72. **A** Return to the situation in Problem 71, where the plane rotates around the axis as shown in Figure P6.67. What is the plane's *minimum* rotation period if the block is to remain stationary? *Hint*: Use the solution to Problem 71.

Problems 73 and 74 are paired.

73. **N** A car is driving around a flat, circularly curved road with a radius of 5.00×10^2 m. The mass of the car is 1500 kg, and the coefficient of static friction between its tires and the road is 0.30. **a.** Using the work of Example 6.7 (pages 202–203), what is the maximum speed the car can have without slipping? **b.** What is the magnitude of the centripetal force on the car if it were to travel at the maximum speed around the track?

74. A car is driving around a banked, circularly curved road with a radius of 5.00×10^2 m. The mass of the car is 1500 kg, and the road is banked at an angle of 20°.
   a. **N** Using the results of Example 6.8 (pages 204–205) and ignoring friction, what is the maximum speed the car can have without slipping?
   b. **N** What is the magnitude of the centripetal force on the car if it were to travel at the maximum speed around the track?
   c. **C** Would including friction in this problem increase the maximum speed allowed or decrease it? Explain your answer.

75. **N** Two children, with masses $m_1 = 35.0$ kg and $m_2 = 43.0$ kg, are swinging on a tire swing attached to a tree overhanging a pond. The mass of the tire is negligible. At the lowest point of the swinging motion, the tension in each of the three vertical 4.00-m-long chains supporting the swing is 275 N. **a.** What is the speed of the children at the lowest point of the swinging motion? **b.** What is the force exerted on each child by the tire at the lowest point of the swinging motion?

76. **C** Chris, a recent physics major, wanted to design and carry out an experiment to show that an object's mass determines its inertia. He used an ultrasound device to measure acceleration of a low-friction cart attached to a hanging block to provide the same force on the cart during each run (Fig. P6.76A). Chris varied the mass of the cart by varying the number of lead rods placed in it.

Chris used Newton's second law $\sum F_x = F_T = Ma_x$ to predict his results. He reasoned that because $\vec{F}_T$ is the same for each run, the cart's acceleration should be inversely proportional to its mass:

$$a_x = \frac{F_T}{M} = \frac{\text{constant}}{M} \qquad (1)$$

Chris's goal was to show that his data fit Equation (1). He decided to analyze his results by plotting a_x as a function of $1/M$; Equation (1) predicted that he should get a straight line, passing through the origin with a slope equal to the tension (red line in Fig. P6.76B):

Chris ran several trials for each run, averaged his results and estimated the error. He then plotted his data (green line in Fig. P6.76B). Chris was excited to see that he correctly predicted that the data fell along a straight line:

$$a_x = (0.27\text{ N})\frac{1}{M} - (0.048\text{ m/s}^2)$$

According to the straight-line fit to the data, the slope of the line is 0.27 N, which was close to the weight of the hanging mass and therefore close to the tension in the string. Chris, though, was disappointed to see that the line had a negative intercept. Mathematically, as $M \rightarrow \infty$, $\frac{1}{M} \rightarrow 0$. Chris was confused because he believed that as the mass increased, the cart's acceleration should approach zero. He was quite sure that he did not discover some new property of inertia or mass. After convincing himself that he was not being careless in the laboratory and that his data were correct, he started to search for an explanation for the discrepancy between his prediction and his data. Help Chris find an explanation.

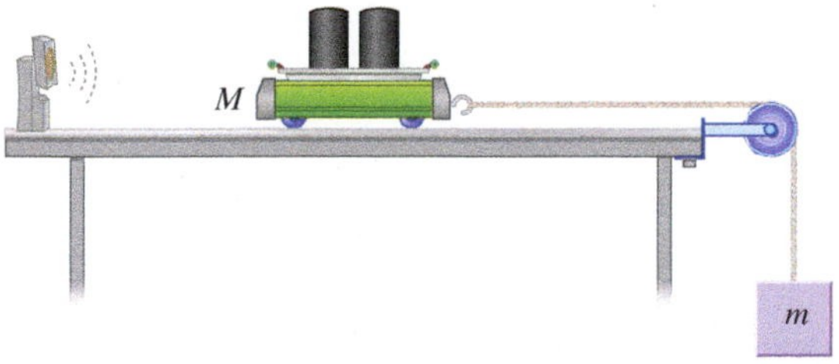

A.

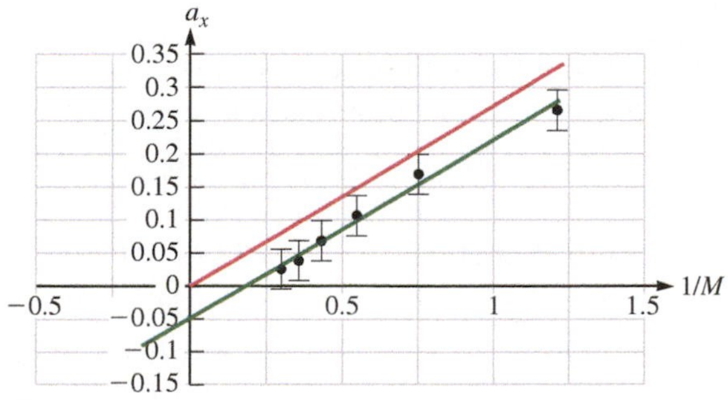

B.

FIGURE P6.76 **A.** Chris's experimental apparatus. **B.** Chris's prediction (red line) and experimental results (green line).

77. **A** Having just crossed the finish line in the 200-km Finland ice marathon with a final straight-line sprint, a skater stands to full height and coasts to a stop by the resistive force of air, which is proportional to the square of his speed v and is given by $F_D = -kmv^2$, where m is the skater's mass and k is a constant. What is the skater's speed at time t after he begins to coast?

78. An aircraft carrier approaching home port cuts its engines and coasts to a full stop at the dock. Upon cutting engine power, the aircraft carrier's speed is seen to be decreasing exponentially, given by $v = v_i e^{-0.050t}$, where t is in seconds.

a. **N** If the aircraft carrier's initial speed is 22 knots (11.3 m/s), what is its speed at $t = 30.0$ s after cutting its engines?

b. **A** What is the acceleration of the aircraft carrier as a function of time?

79. **N** The radius of circular electron orbits in the Bohr model of the hydrogen atom are given by $(5.29 \times 10^{-11}\text{ m})n^2$, where n is the electron's energy level (Fig. P6.79). The speed of the electron in each energy level is $(c/137n)$, where $c = 3 \times 10^8$ m/s is the speed of light in vacuum. **a.** What is the centripetal acceleration of an electron in the ground state ($n = 1$) of the Bohr hydrogen atom? **b.** What are the magnitude and direction of the centripetal force acting on an electron in the ground state? **c.** What are the magnitude and direction of the centripetal force acting on an electron in the $n = 2$ excited state?

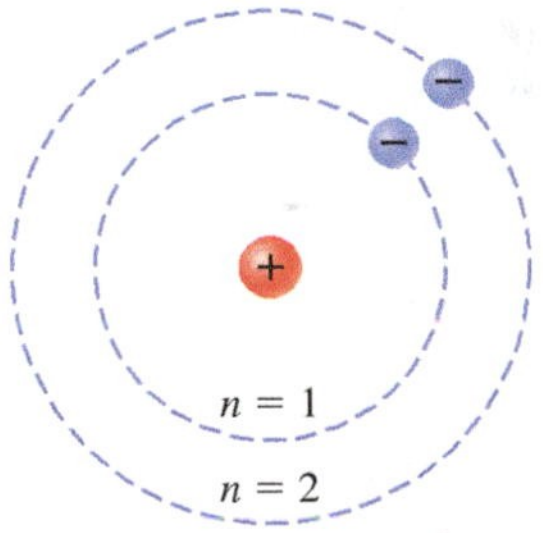

FIGURE P6.79

80. **N** A particle of dust lands 45.0 mm from the center of a compact disc (CD) that is 120 mm in diameter. The CD speeds up from rest, and the dust particle is ejected when the CD is rotating at 90.0 revolutions per minute. What is the coefficient of static friction between the particle and the surface of the CD?

81. **N** Since March 2006, NASA's Mars Reconnaissance Orbiter (MRO) has been in a circular orbit at an altitude of 316 km around Mars (Fig. P6.81). The acceleration due to gravity on the surface of the planet Mars is $0.376g$, and its radius is 3.40×10^3 km. Assume the acceleration due to gravity at the satellite is the same as on the planet's surface. **a.** What is MRO's orbital speed? **B.** What is the period of the spacecraft's orbit?

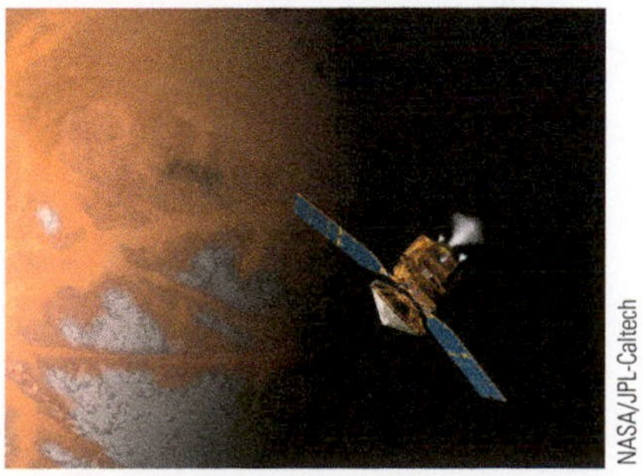

FIGURE P6.81

7 Gravity

Key Questions

How is Newton's law of universal gravity applied in various situations?

What is the scientific process that leads to important developments such as Newton's law of universal gravity?

Underlying Principles

Newton's law of universal gravity

Major Concepts

1. Geocentric model
2. Heliocentric model
3. Empirical laws
4. Inertial mass
5. Gravitational mass
6. Gravitational field

Special Cases

1. Kepler's first law
2. Kepler's second law
3. Kepler's third law

Tools

1. Ellipse
2. Inverse-square law

Most of us cannot remember a time in our lives before we heard about gravity. It seems perfectly logical to us that gravity keeps our feet on the ground and the planets in their orbit. When these notions of gravity—which we take for granted—were discovered, however, they overthrew two millennia of thoughts on the nature of the Universe, and most people found gravity incomprehensible. The powerful effect of studying gravity does not stop there. When Albert Einstein studied gravity, he developed theories that took physics in a whole new direction. Even today, gravity is the most elusive of the fundamental forces.

This chapter has two goals: (1) to study the physical properties of gravity, and (2) to think about the process that led to its discovery. As any historian will say, we learn a lot about the present day by studying the past. So, as we study the history of gravity, let us reflect on the way science operates today and the lessons we can learn from the past.

7-1 A Knowable Universe

Scientists are very much a part of the society in which they live. That society influences the work scientists do; in turn, discoveries made by scientists influence society. Technology, wars, national poverty or affluence, and religious beliefs are among the many factors that impact scientific progress. For one scientist's discovery to become accepted as part of our collective picture of the Universe, that discovery must be supported by other scientists. Scientists, like all other members of society, may have trouble seeing the truth of a discovery because they have personal biases, and sometimes a discovery acceptable in the final analysis is initially rejected.

Today, we take for granted many scientific discoveries that were once highly controversial. We teach our children that the Earth is a planet, for instance, and that, like the other planets in the solar system, it orbits the Sun. When the idea that the Sun (and not the Earth) is the center of the solar system was first put forth, however, it ran counter to societal beliefs at the time. The scientists who put forth this and other equally controversial theories were bold, courageous, and insightful. Once such theories were supported by observation and experimentation, they were accepted and changed society. Newton's law of universal gravity is one such important theory.

This chapter is devoted to one force, gravity, one of the four fundamental forces (Section 5-10).

A closer look at gravity will be a good foundation for our study of the electric force in Chapters 23-29.

By focusing on gravity, Einstein formulated his theory of general relativity (Chapter 39). So, to appreciate general relativity, we must have a good understanding of gravity.

Newton's theory of gravity marked the birth of modern science; accepting his findings required an important shift in how scientists think, what questions they ask, and how they answer them. Part of any engineer's or scientist's job is to develop the critical thinking skills necessary to evaluate new theories, and understanding the process that led to Newton's understanding of gravity will help you hone these skills.

Detlev van Ravenswaay/Photo Researchers, Inc.

FIGURE 7.1 Tycho Brahe's observatory. He made the best-possible naked-eye observations of the sky.

Although we think of it as Newton's work alone, formulating the law of universal gravity involved many scholars and required the collection of accurate data, intellectual honesty, and bold insight in interpreting those data. The data came from astronomical observations. Tycho Brahe (1546–1601) spent two decades, from 1577 until he left Denmark in 1597, making the best-possible naked-eye observations of the sky (Fig. 7.1). At the time Tycho was collecting his data, ideas about the Universe and how it "works" were heavily influenced by early Greek philosophers as well as by Christian theology. Many scholars still accepted the **geocentric model** (*geo-*, "Earth"; *-centric*, "centered") of the solar system, a model described in detail by Ptolemy, a Greek astronomer and mathematician who lived in the second century CE. According to this model, the Sun, the Moon, the stars, and the planets orbited the Earth. Even before Ptolemy's time, however, a few scholars supported a **heliocentric model** (*helio-*, "Sun") of the Universe, a model in which the Earth and other planets orbit the Sun. This heliocentric model was then redeveloped by Nicolaus Copernicus (1473–1543). Both of these early models assumed astronomical bodies move in uniform circular motion. To account for astronomical observations, both models included elaborate combinations of circular motion (Fig. 7.2).

GEOCENTRIC MODEL AND HELIOCENTRIC MODEL

Major Concept

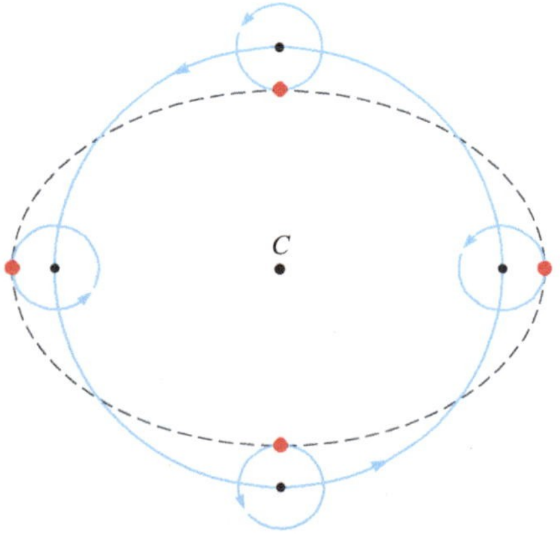

FIGURE 7.2 The Ptolemaic model of the solar system was based on the theory that the planets move in uniform circular motion around the Earth located near the center at *C*. Today, we know that planetary motion is not uniform and not circular.

In his own day, Tycho became famous for his discovery of what he thought was the birth of a new star. (Today, we know that what he observed was actually a star *dying*, but that's another story.) In gratitude for his findings, King Frederick of Denmark gave Tycho an estate and money to establish an astronomical observatory. At the time, most people who studied nature did not believe that quantitative observations were necessary. Tycho, unlike his contemporaries, strived to make the most accurate measurements possible. He observed as often as he could and estimated the magnitude of his measurement *uncertainties,* also called *errors*. He examined his methods so as to find sources of errors and make improvements. His hard work paid off, and his measurements are the best ever made without binoculars or a telescope.

Tycho collected data on the naked-eye planets (Mercury, Venus, Mars, Jupiter, and Saturn), tracking movements over periods of days, months, and years. He then used his data to develop a model of the solar system. Needing the help of a skilled mathematician, he hired Johannes Kepler (1571–1630) to develop a new model, but the two disagreed about it. Tycho's model was a combination of geocentric and heliocentric, and not until Tycho died did Kepler develop a fully heliocentric model.

CONCEPT EXERCISE 7.1

What important experimental skills can we learn from Tycho Brahe?

CASE STUDY Dark Matter and MOND

Today, the field of astrophysics extends far beyond our solar system. We have models of our galaxy, other galaxies, and the Universe as a whole.

There is still much to learn, however. This case study focuses on one controversial issue: the existence of *dark matter* (matter that we have not observed directly). If we haven't observed it, how do we know that dark matter exists? The answer is that we have observed dark matter's influences on ordinary objects such as stars and galaxies. The term *luminous matter* describes objects we can detect directly with our eyes, a telescope or other instrument. So far, the existence of dark matter has only been deduced *indirectly* by observing its interaction with luminous matter.

To get a handle on one way the existence of dark matter is deduced, let's look at the story of the star Sirius, the brightest star in the night sky. In the early 1800s, Sirius was observed to wander back and forth in the sky (Fig. 7.3). From Newton's laws of motion, a force must act on this star to produce this nonuniform motion, and there must be a source of this force. By observing the motion of Sirius, scientists in the early 1800s *indirectly* detected an unseen object.

In 1862, Alvan Clark, an American astronomer and telescope maker, directly observed that Sirius is actually made up of two stars, one much brighter than the other. The brighter one is now called Sirius A; it was the one that was seen to wander back and forth. The fainter one is called Sirius B; it could only be detected directly by Clark's powerful telescope. Although the fainter Sirius B is hard to observe, it provides the gravitational force responsible for the wandering motion observed in Sirius A, the brighter star.

Both Sirius A and B are ordinary matter, not dark matter. The original story of Sirius A and B before Clark's observation, however, illustrates one way that dark matter is indirectly detected. Just as in the case of Sirius B, the existence of dark matter is deduced from the acceleration of luminous matter. According to Newton's second law, a force must be responsible for accelerating the luminous matter. The claim is that the dark matter supplies the necessary gravitational force.

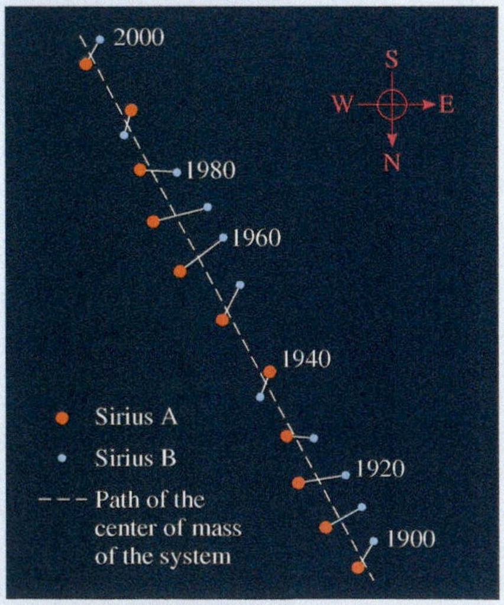

FIGURE 7.3 Observations of Sirius A made at 10-year intervals show that it appears to wobble in the night sky. Initially, astronomers speculated that the gravity of an unseen object caused the wobble. Today, we are able to observe a small secondary star named Sirius B.

There is, however, a major difference between faint objects such as Sirius B and dark matter. Sirius B (before Clark's observation) is an example of *"non-luminous" ordinary matter* that happens to be very faint and hard to detect directly. Such ordinary matter is made up of familiar particles such as protons, neutrons, and electrons that have all been detected directly in laboratories. **Dark matter** is *extraordinary* matter made up of some other types of particles. So far, no laboratory experiments have provided direct evidence for the existence of dark matter, but because of the great deal of indirect evidence, many scientists today are certain that dark matter exists.

Among the many astronomical observations providing evidence for dark matter, we will consider only one greatly simplified piece of evidence, the motion of stars in our own galaxy. Our Milky Way galaxy is a spiral galaxy that from the side looks like a disk with a spherical bulge (Fig. 7.4). The galaxy consists of roughly 200 billion to 400 billion stars and 10^{40} kg of gas. These stars and gas orbit the central bulge much as the eight major planets of our solar system orbit the Sun. For the stars and gas far from the center of the Milky Way, the observed centripetal acceleration is greater than can be accounted for by the gravitational force exerted by the ordinary matter in the galaxy. Scientists hypothesize that dark matter must also be present to account for the

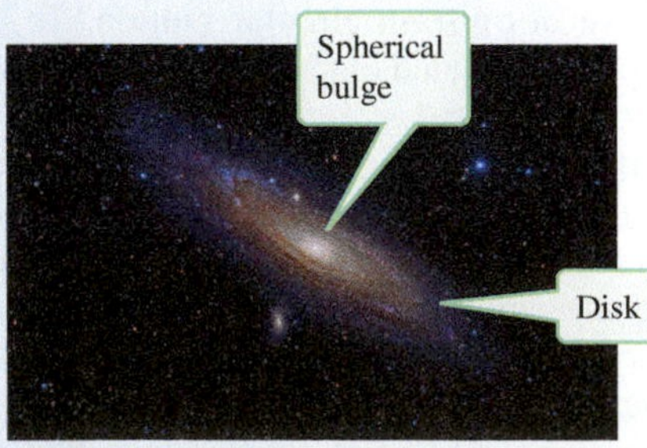

Spitzer/NASA

FIGURE 7.4 The Andromeda galaxy is a spiral galaxy believed to look much like the Milky Way galaxy.

gravitational force required. In this case study, we will use Newton's law of gravity to estimate the dark matter content in our own galaxy.

Because dark matter has yet to be directly observed, some physicists are skeptical that dark matter exists. Instead, they say that Newton's laws must be incorrect. While scientists agree that these laws are valid in many situations, some scientists argue that Newton's laws must be modified to explain phenomena that naturally occur only in astronomical situations. For example, theoretical astrophysicist Mordehai Milgrom has developed a theory of *mo*dified *N*ewtonian *d*ynamics (MOND), a very controversial theory that has not been widely accepted and the details of which are beyond the scope of this book. In this chapter, however, we consider MOND as an alternative to the theory of dark matter as we think about the scientific method.

7-2 Kepler's Laws of Planetary Motion

Kepler began his work on planetary orbits by analyzing the information Tycho had collected on Mars. Some of the data gave a circular orbit for this planet, in agreement with the belief at the time that all heavenly bodies moved in circular orbits. Two of Tycho's Mars measurements, however, could not be fit to a circular orbit. At that time, most mathematicians might have ignored the discrepancy, arguing that all measurements are subject to experimental uncertainty. Kepler, however, realized that even taking into account the uncertainty Tycho estimated for his measurements, these data did not fit a circular model. Intellectually honest in his search for the truth, Kepler boldly tossed out the 2000-year-old assumption of circular orbits. This action was the critical step in his subsequent discovery of his three laws of planetary motion.

Kepler's First Law

KEPLER'S FIRST LAW

Special Case

To understand Kepler's first law, you need to refresh what you know about ellipses. One way to see the difference between an ellipse and a circle is to draw them both. Figure 7.5A shows how to draw a circle. A tack holds a piece of string on one end. The length of the string is r. A pencil is attached to the other end. The pencil traces out a circle because each point on the circle is the same distance r from the tack.

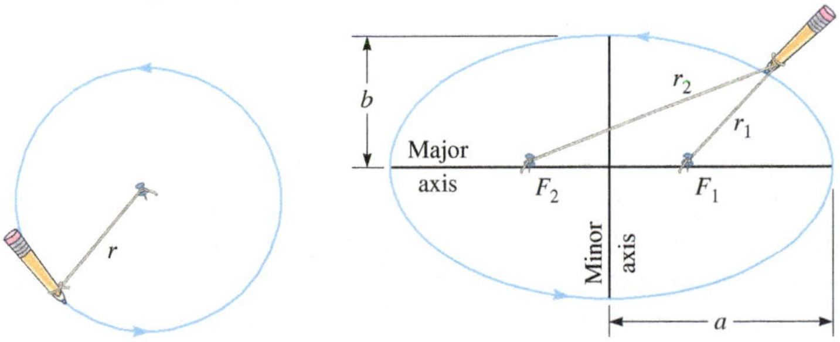

FIGURE 7.5 A. A circle is drawn by attaching a pencil to a string held down by a single tack. Every point on the circle is the same distance r from the center of the circle. **B.** An ellipse is drawn by sliding a pencil against a string that is fixed by two tacks located at the two foci F_1 and F_2. The total distance $r_1 + r_2$ is constant. A circle is a special case of an ellipse.

Two tacks are needed to draw an ellipse. One end of the string is tacked at point F_1, and the other end is tacked at point F_2 (Fig. 7.5B). The points F_1 and F_2 are known as the **foci** (plural of **focus**) of the ellipse. The length of the string $r_1 + r_2$ is greater than the distance between the two tacks. A pencil keeps the string taut. The distances between the pencil and the points F_1 and F_2 are r_1 and r_2, respectively. As the pencil traces an ellipse, r_1 and r_2 change, but the length of the string $r_1 + r_2$ remains constant. So, an **ellipse** is the locus of points for which the sum of the distances from the two foci F_1 and F_2 to the perimeter of the ellipse is constant. If the tacks are brought closer together, the ellipse becomes more like a circle. In fact, if one tack is driven into the top of the other, the ellipse is a circle of radius $(r_1 + r_2)/2$.

ELLIPSE Tool

Figure 7.5B shows the major and minor axes of the ellipse. The **major axis** passes through both foci. The **minor axis** is the perpendicular bisector of the major axis. Either half of the major axis is called a *semimajor axis*, and either half of the minor axis is called a *semiminor axis*. The lengths of the semimajor and semiminor axes are a and b, respectively.

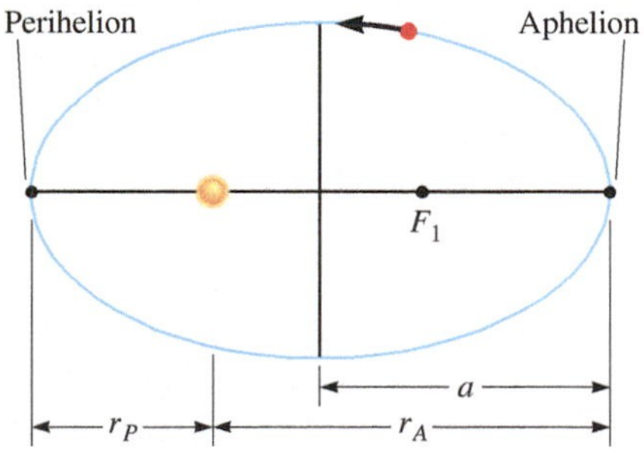

FIGURE 7.6 According to Kepler's first law, planetary orbits are elliptical. The Sun is at one focus, and there is nothing at the other focus.

Kepler's first law, illustrated in Figure 7.6, states that **planetary orbits are ellipses with the Sun at one focus**. The ellipse in Figure 7.6 is obviously not a circle, but the elliptical orbits of most planets are very nearly circular. Also, notice that there is nothing at the other focus.

The closest position of a planet to the Sun is called the planet's **perihelion** (*peri-*, "around, near"; *helios*, "Sun"). The farthest position is the planet's **aphelion** (*api-*, "away from"). The length of the semimajor axis a is the average distance the planet is from the Sun, and this distance can be found from the perihelion and aphelion distances r_P and r_A:

$$a = \frac{r_P + r_A}{2} \quad (7.1)$$

For a circular orbit, $r_A = r_P = a = r$. The Earth's orbit is only slightly elliptical; its perihelion distance, $r_P = 1.48 \times 10^{11}$ m, is nearly equal to its aphelion distance, $r_A = 1.52 \times 10^{11}$ m. The semimajor axis of the Earth's orbit is

$$a = \frac{r_P + r_A}{2} = \frac{1.48 \times 10^{11}\text{ m} + 1.52 \times 10^{11}\text{ m}}{2} = 1.50 \times 10^{11}\text{ m}$$

This average Earth–Sun distance is used to define the **astronomical unit (AU)**: 1 AU $= 1.50 \times 10^{11}$ m. Distances in the solar system are usually measured in astronomical units.

KEPLER'S SECOND AND THIRD LAWS

Special Cases

Put simply, Kepler's second law says that a planet's speed is highest when it is near the Sun.

Kepler's Second Law

For nearly two millennia, scholars believed that the motion of astronomical objects was based on uniform circular motion. Kepler's first law, published in 1609, eliminated the circular part, and Kepler's second law, published at the same time, removed the uniform part. According to **Kepler's second law, a line joining a planet to the Sun sweeps out equal areas in equal time intervals.**

Figure 7.7 illustrates Kepler's second law for a planet shown at 12 positions. Because it is a motion diagram, the time intervals between adjacent positions are equal all around the ellipse. According to Kepler's second law, all the pie-shaped wedges have the same area. For that to be true, the planet's speed must change. The planet moves more slowly near aphelion than it does near perihelion.

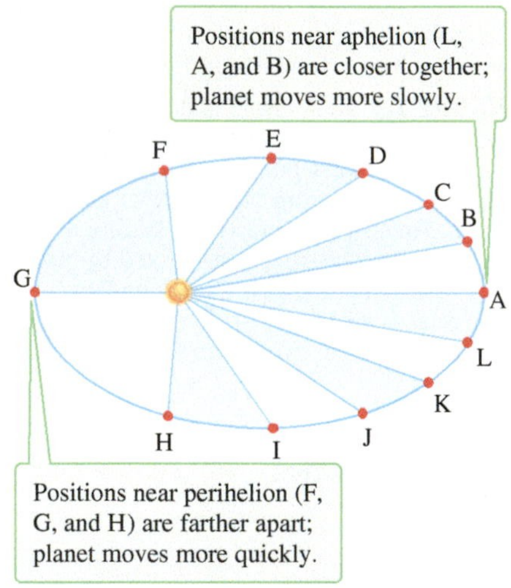

Positions near aphelion (L, A, and B) are closer together; planet moves more slowly.
F
E
D
C
B
G
A
L
K
H
I
J
Positions near perihelion (F, G, and H) are farther apart; planet moves more quickly.

FIGURE 7.7 A motion diagram for a planet orbiting the Sun. According to Kepler's second law, all the pie-shaped wedges have the same area.

Kepler's Third Law

Unlike Kepler's first two laws, which are about the orbit of a single planet, his third law involves all the planets. The time it takes a planet to complete one orbit is known as the planet's **period**. Kepler found that a planet's period depends on its average distance from the Sun. Planets that are closer to the Sun have shorter periods than planets that are farther from the Sun. Mercury, the planet closest to the Sun, has a period of less than three months; Earth's is, of course, one year; and Pluto, a very distant body, has a period of nearly two and a half centuries.

Kepler's third law states that **the square of a planet's period T is proportional to the cube of its semi-major axis**, or $T^2 = Ca^3$. The constant C is the same for every object orbiting the Sun. We can find C easily from the Earth's orbital data. For the Earth, $T = 1$ yr and $a = 1$ AU; therefore,

$$C = \frac{T^2}{a^3} = \frac{(1\text{ yr})^2}{(1\text{ AU})^3} = 1\text{ yr}^2/\text{AU}^3$$

Therefore, if the period T of any object is measured in years and its semimajor axis a in astronomical units, Kepler's third law is expressed mathematically as

$$T^2 = (1\text{ yr}^2/\text{AU}^3)a^3$$

$$T^2_{[\text{yr}]} = a^3_{[\text{AU}]} \quad (7.2)$$

where the subscripts are a reminder that this form of Kepler's third law holds only when these two particular units are used. The constant C with its units is implied in Equation 7.2, so the equation does not imply that time is somehow equivalent to distance. Figure 7.8 shows the elliptical orbit, period, and average distance from the Sun for each major planets and Pluto.

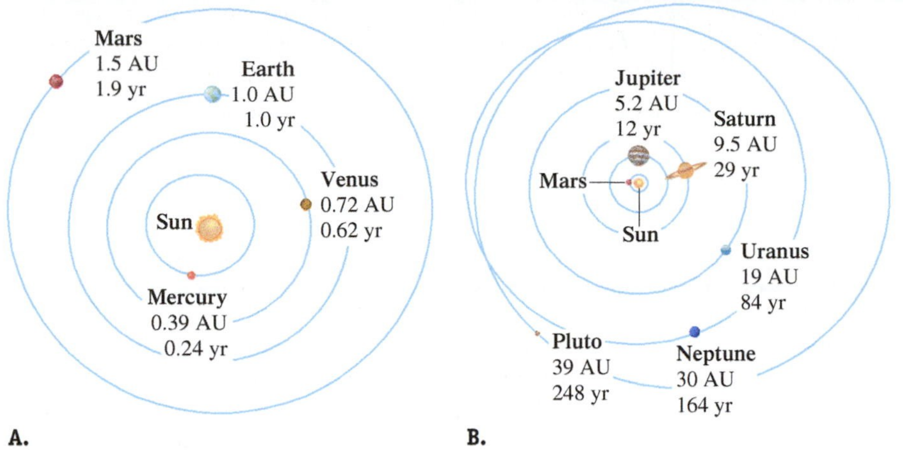

FIGURE 7.8 **A.** The period and semimajor axis of each planet in the inner solar system. **B.** The period and semimajor axis of each planet in the outer solar system and Pluto. (Scales are different in parts A and B.)

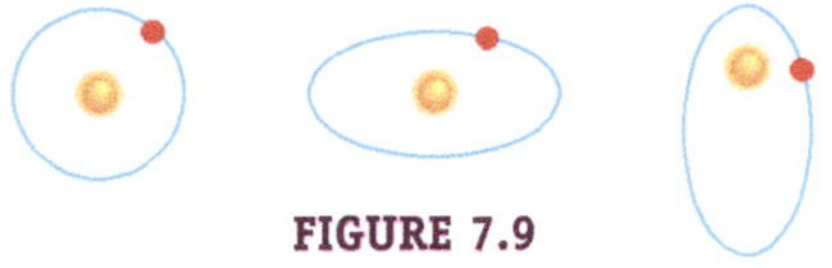

FIGURE 7.9

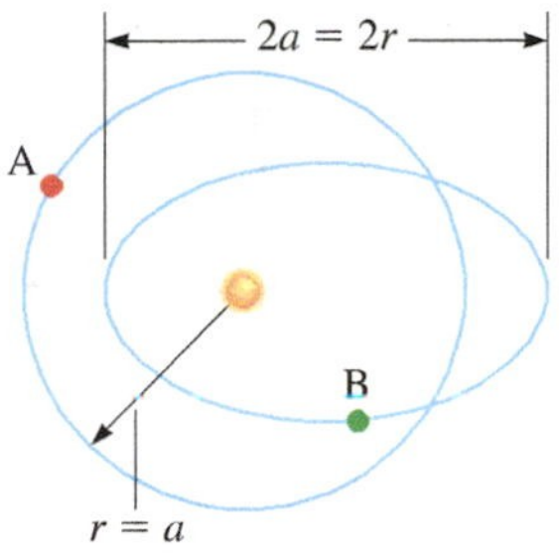

FIGURE 7.10

CONCEPT EXERCISE 7.2

Three possible planetary orbits are shown in Figure 7.9. According to Kepler's first law, which orbits are possible and which are not? Explain.

CONCEPT EXERCISE 7.3

Planet A's orbit is circular with the Sun at the center. Planet B's orbit is elliptical with the Sun at one focus (Fig. 7.10). The radius of planet A's orbit is equal to the semimajor axis of planet B's orbit: $r = a$.

a. Compare the periods of the two planets.
b. Which planet reaches the higher speed? Explain.

EXAMPLE 7.1 The Largest Planet in the Solar System

Use Kepler's third law to find Jupiter's period from its semimajor axis $a = 5.2$ AU.

INTERPRET and ANTICIPATE

An object with a semimajor axis greater than 1 AU should have a period longer than 1 year (Fig. 7.8).

SOLVE

Solve Equation 7.2 for period T.	$T^2_{[\text{yr}]} = a^3_{[\text{AU}]}$
Substitute the semimajor axis value.	$T = (5.2\text{ AU})^{3/2} = 12\text{ yr}$

CHECK and THINK

The period is greater than 1 year as expected.

EXAMPLE 7.2 The Closest Planet to the Sun

Mercury's period is 0.24 yr. Find Mercury's average distance from the Sun.

INTERPRET and ANTICIPATE

An object with a period shorter than 1 year should have a semimajor shorter than 1 AU (Fig. 7.8).

Example continues on page 190 ▶

SOLVE Solve Equation 7.2 for *a*.	$T^2_{[yr]} = a^3_{[AU]}$ $a_{[AU]} = T^{2/3}_{[yr]}$
Substitute the given value for the period.	$a = (0.24\,yr)^{2/3} = 0.39\ AU$

CHECK and THINK

As expected, our result is less than 1 AU.

EMPIRICAL LAWS Major Concept

Kepler's three laws of planetary motion are often described as **empirical laws**. Empirical laws are formulated based solely on data, without any hypothesis presented first, which was certainly true of Kepler's first law. Kepler had no theoretical reason to try fitting an ellipse to Tycho's observations of Mars, and he had not observed any mechanism that causes elliptical motion in the heavens. Kepler's contemporaries theorized that astronomical objects were "perfect" and therefore would exhibit perfect motion, which to them was uniform and circular. People often find it difficult to give up their theories, and such researchers would probably have never tried fitting an ellipse to these data.

It is not entirely correct to say that Kepler's second and third laws are strictly empirical. Kepler theorized that the Sun influences the motion of the planets. He reasoned that the closer a planet is to the Sun, the faster the planet would move. It is true that he did not need this theoretical basis to derive his second or third laws, but it seems likely that his theory helped him discover these laws more easily. It still remains true that Kepler had no single theory that accounts for all three laws; as far as he knew, a different mechanism may have been responsible for each law. It was Newton who developed a theoretical basis for all three laws.

CONCEPT EXERCISE 7.4

Imagine a comet whose period is 500 years.

a. What is its average distance from the Sun?
b. If its perihelion distance is 0.5 AU, what is its aphelion distance?
c. Where does the comet spend most of its time: in the inner solar system, where it is < 2 AU from the Sun, or in the outer solar system, where it is > 2AU from the Sun?

CONCEPT EXERCISE 7.5

Today's employees are rewarded for *thinking outside the box*. In what ways did Kepler think outside the box?

7-3 Newton's Law of Universal Gravity

Because Galileo supported the idea of a heliocentric solar system, the Roman Catholic Church placed him under house arrest.The conflict between the Roman Catholic Church and Galileo was complicated. His house arrest was likely influenced by his personality and by global factors such as the Protestant Reformation. In the 1990s, Pope John Paul II expressed regret for the church's response to Galileo and acquitted him.

Galileo Galilei (1564–1642) used a telescope for astronomical observations. It was the first known time a scientist used a device to explore the heavens beyond the capability of his own senses. Galileo's observations of Venus provided important supporting evidence for Kepler's first law. Galileo found that, like the Earth's moon, Venus goes through phases (new, crescent, quarter, gibbous, and full). (In the geocentric model, Venus is always in a region of space that is between the Earth and the Sun; therefore, the geocentric model predicts that Venus would never go through quarter, gibbous, or full phases.) For an observer on the Earth to see these phases of Venus, it must orbit the Sun and not the Earth.

Isaac Newton (1642–1727) was born the year that Galileo died. By the time he studied Kepler's laws, Newton could assume they were solid. Galileo's observations supported the heliocentric model, and Kepler's laws fit observations of planetary orbits. Newton was thus able to focus on finding a general theory to explain all three laws. One of his remarkable insights was his hypothesis that terrestrial and celestial physical laws are the same. Starting from this idea, Newton theorized that gravity is not limited to objects near the Earth but instead applies to all objects in the Universe. Today, this theory is known as **Newton's law of universal gravity**.

UNIVERSAL GRAVITATIONAL FORCE

Underlying Principle

Like Kepler, Newton theorized that the Sun exerts a force on the planets and that this force is responsible for their kinematics. Newton identified this force as gravity. He used Kepler's second and third laws to derive two properties of gravity. From Kepler's second law, Newton found that the force of gravity must be directed straight toward the Sun. Using Kepler's third law, he mathematically reasoned that the magnitude of the gravitational force F_G felt by any planet depends on the distance r between the Sun and the planet. In particular, Newton found that gravity obeys an *inverse-square law*, which means that the gravitational force on a planet diminishes with the inverse square of its distance from the Sun:

INVERSE-SQUARE LAW

Tool

$$F_G \propto \frac{1}{r^2} \tag{7.3}$$

We will see in later chapters that many other quantities in physics diminish with the inverse square of distance.

Having figured out these two properties of gravity, Newton's next step was an important test of the validity of his thinking. He used his laws of motion and these two properties of gravity that he had inferred from Kepler's second and third laws to derive Kepler's first law. His success in deriving Kepler's first law showed that all three of Kepler's planetary laws stemmed from one physical theory: The Sun exerts a gravitational force on all the planets.

Newton's law of universal gravity has three features:

1. Every particle with nonzero mass in the Universe exerts a gravitational force on every other particle with nonzero mass in the Universe.
2. The gravitational force is attractive. For example, the Sun pulls the Earth toward the Sun, and the Earth pulls the Sun toward the Earth.
3. The magnitude of the gravitational force between two particles is directly proportional to the product of their masses and inversely proportional to the square of their separation.

Mathematically, the law of universal gravity is summarized as

$$F_G = G\frac{m_1 m_2}{r^2} \tag{7.4}$$

where G is the **universal gravitational constant** found through experimentation to be

$$G = 6.673 \times 10^{-11}\ \text{N} \cdot \text{m}^2/\text{kg}^2$$

Figure 7.11A shows two particles separated by a distance r. First, consider particle 1 as the source of the gravitational force. Then, particle 1 exerts a gravitational force $\vec{F}_{[1\text{ on }2]}$ on the subject, particle 2. According to Newton's third law, we could, instead, consider particle 2 to be the source. Then, particle 2 exerts a gravitational force $\vec{F}_{[2\text{ on }1]}$ on the subject, particle 1. Further, from Newton's third law, the magnitude of $\vec{F}_{[1\text{ on }2]}$ is equal to the magnitude of $\vec{F}_{[2\text{ on }1]}$. It doesn't matter how unequal the two masses are; the magnitude of the forces is the same. Particle 1's pull on particle 2 is just as strong as particle 2's pull on particle 1. We can see this equality from Equation 7.4 because the universal gravitational constant G, the separation r, and the product of the two masses $m_1 m_2$ are the same no matter which is considered the source and which the subject.

All the features of universal gravity can be expressed as one vector equation using the polar coordinate system introduced in Section 4-6. Particle 1 is placed at the origin of a polar coordinate system, and $\hat{r}$ points from particle 1 to particle 2 (Fig. 7.11B). With this choice, the gravitational force exerted by particle 1 on particle 2 is

$$\vec{F}_{[1\text{ on }2]} = -F_G\hat{r}$$

$$\vec{F}_{[1\text{ on }2]} = -G\frac{m_1 m_2}{r^2}\hat{r}$$

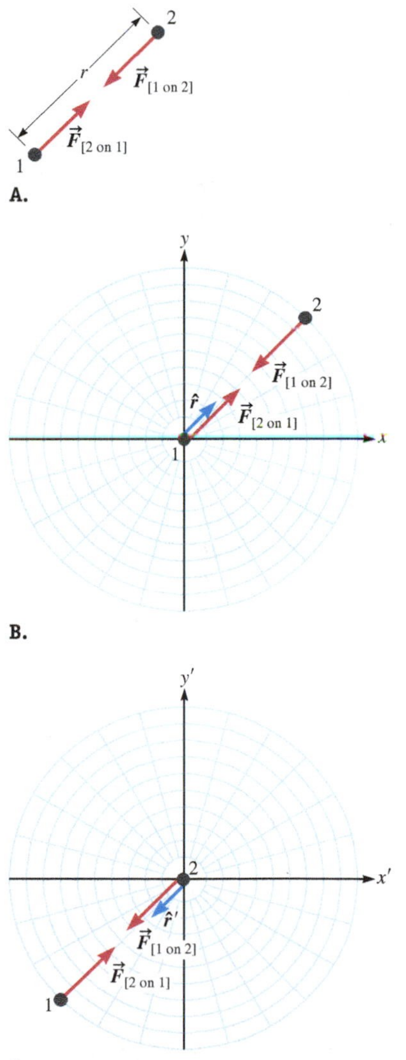

FIGURE 7.11 **A.** The gravitational force particle 1 exerts on particle 2 is equal to the magnitude of the gravitational force exerted by particle 2 on particle 1. **B.** With particle 1 located at the origin of a coordinate system, the unit vector $\hat{r}$ points from particle 1 toward particle 2. **C.** Another choice of coordinate system places particle 2 at the origin. Now unit vector $\hat{r}'$ points from particle 2 toward particle 1.

and the force exerted by particle 2 on particle 1 is

$$\vec{F}_{[2 \text{ on } 1]} = F_G\hat{r} = G\frac{m_1m_2}{r^2}\hat{r}$$

Of course, it is possible to put particle 2 at the origin (Fig. 7.11C). With this coordinate system, we have

$$\vec{F}_{[1 \text{ on } 2]} = F_G\hat{r}' = G\frac{m_1m_2}{r^2}\hat{r}'$$

$$\vec{F}_{[2 \text{ on } 1]} = -F_G\hat{r}' = -G\frac{m_1m_2}{r^2}\hat{r}'$$

where $\hat{r}'$ is the unit vector that points from particle 2 toward particle 1.

The law of universal gravity is stated in terms of particles. Is it possible to treat a large object such as a planet as a particle? The answer depends on both the shape of the source, and distance between the source and the subject. A uniform spherical source may be treated as a particle as long as the subject it acts on is outside the sphere. For example, if we are interested in the gravitational force exerted by the Earth on a comet, we treat the Earth as a particle because it is (nearly) uniform and spherical and the comet is outside the Earth. For a subject outside a sphere, the gravitational force acts as though all the sphere's mass were concentrated at its center. Then, r in Equation 7.4 is measured from the center of the sphere. (This property is due to the inverse-square nature of the law of universal gravity, a feature shared by the electric force.) If, on the other hand, we are interested in a comet that has fallen inside the gaseous Sun, we cannot model the Sun as a particle.

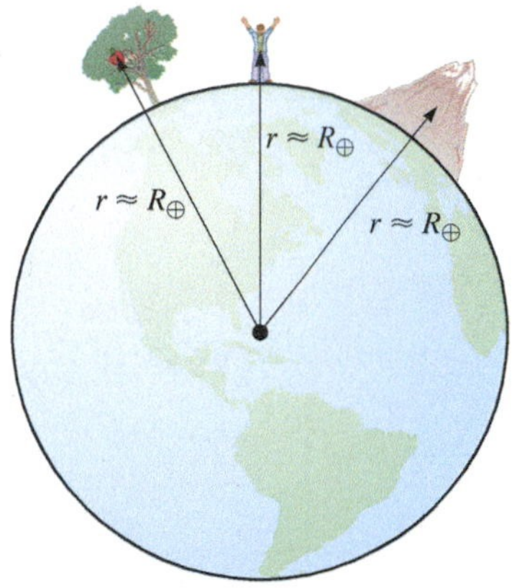

FIGURE 7.12 We can model objects such as apples, people, and mountains on the surface of the Earth as particles.

Nonspherical objects such as galaxies consisting of hundreds of billions of stars may be modeled as particles if the center-to-center distance between them is much greater than their size. The distance between the Earth and any object located on the surface of the Earth is approximated as being the radius of the Earth $r = R_\oplus$ (Fig. 7.12). (Recall that the symbol $\oplus$ stands for "Earth.") We can model objects located on the surface on the Earth as particles as far as gravity is concerned because the distance ($R_\oplus = 6.37 \times 10^6$ m) between the Earth's center and anything on its surface is far greater than the size of the object.

EXAMPLE 7.3 Cavendish's Experiment

With any two nonastronomical objects, the gravitational force is weak, which makes it difficult to measure the universal gravitational constant G. In fact, this constant is the least well determined of all the physical constants.

In 1798, Henry Cavendish used a device designed by John Michell to measure the density of the Earth. His results could have been used to measure G. Today, many physics students use a tabletop apparatus based on Michell's design to do just that.

Figure 7.13 is a schematic of a contemporary "Cavendish" apparatus. Two large spheres of mass M are attached to the frame of the apparatus. Two smaller spheres of mass m are attached to a lightweight rod that is free to rotate. A laser beam is reflected from a mirror attached to the rod, first with the large spheres in one position and then with the large spheres moved to a second position. By measuring the change in the reflected light's path, a student infers the motion of the two small spheres caused by the gravitational force exerted on them by the large spheres. This device is a torsion balance (Chapter 16).

The masses of the spheres are 1.50 kg and 38.3 g. What is the magnitude of the gravitational force between one large sphere and one small one when they are separated by 46.5 mm?

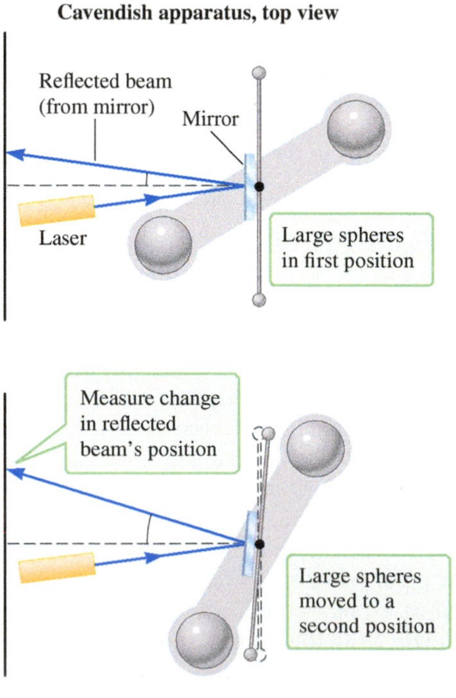

FIGURE 7.13

INTERPRET and ANTICIPATE

This example is a straightforward application of Newton's law of universal gravity (Eq. 7.4). The key is to realize that the separation distance given is actually the center-to-center distance.

Remember to work in SI units.	$F_G = G\frac{m_1m_2}{r^2} = (6.67 \times 10^{-11}\ \text{N}\cdot\text{m}^2/\text{kg}^2)\frac{(1.50\ \text{kg})(38.3 \times 10^{-3}\ \text{kg})}{(46.5 \times 10^{-3}\ \text{m})^2}$ $F_G = 1.77 \times 10^{-9}\ \text{N}$

CHECK and THINK

Gravity is weak. This result is about seven orders of magnitude lower than the weight of a single grain of rice.

EXAMPLE 7.4 A Cluster of Galaxies

Three galaxies are at the vertices of an equilateral triangle, with the distance d between any two of them being much, much greater than their diameters (Fig. 7.14). Use the relative masses indicated to find an expression for the gravitational force on C due to galaxies A and B.

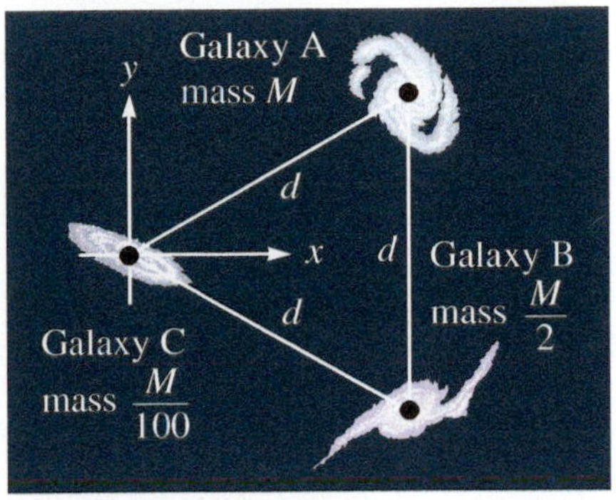

FIGURE 7.14 Not to scale.

INTERPRET and ANTICIPATE

Because the distance between galaxies is much greater than their sizes, we can treat them as particles. A free-body diagram helps interpret the situation.

We choose to put C at the origin of a coordinate system. As shown on our free-body diagram (Fig. 7.15), two gravitational forces act on C: the force exerted by A on C, $\vec{F}_{[\text{A on C}]}$, and the force exerted by B on C, $\vec{F}_{[\text{B on C}]}$. We have sketched the triangular configuration of the galaxies to help find the angles between these forces and the x axis. Because the interior angles of an equilateral triangle are 60°, the angle between each force and the x axis is 30° as shown. We do not need the acceleration, so we do not indicate it on the diagram.

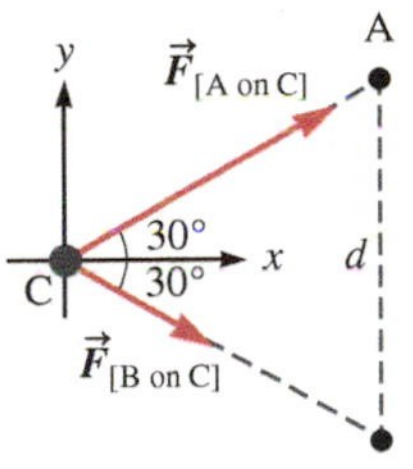

FIGURE 7.15

SOLVE

Start by breaking $\vec{F}_{[\text{A on C}]}$ and $\vec{F}_{[\text{B on C}]}$ into components.	$(F_{[\text{A on C}]})_x = F_{[\text{A on C}]}\cos 30° = \frac{\sqrt{3}}{2}F_{[\text{A on C}]}$ $(F_{[\text{A on C}]})_y = F_{[\text{A on C}]}\sin 30° = \frac{1}{2}F_{[\text{A on C}]}$ $(F_{[\text{B on C}]})_x = F_{[\text{B on C}]}\cos 30° = \frac{\sqrt{3}}{2}F_{[\text{B on C}]}$ $(F_{[\text{B on C}]})_y = -F_{[\text{B on C}]}\sin 30° = -\frac{1}{2}F_{[\text{B on C}]}$
Then, find the total force on C in component form.	$\vec{F}_{\text{tot}} = (\frac{\sqrt{3}}{2}F_{[\text{A on C}]} + \frac{\sqrt{3}}{2}F_{[\text{B on C}]})\hat{\imath} + (\frac{1}{2}F_{[\text{A on C}]} - \frac{1}{2}F_{[\text{B on C}]})\hat{\jmath}$ (1)
We now take into account that $\vec{F}_{[\text{A on C}]}$ and $\vec{F}_{[\text{B on C}]}$ are gravitational forces by applying $F_G = G(m_1m_2/r^2)$ (Eq. 7.4) to find their magnitudes.	$F_{[\text{A on C}]} = G\frac{M(M/100)}{d^2} = \frac{GM^2}{100d^2}$ (2) $F_{[\text{B on C}]} = G\frac{(M/2)(M/100)}{d^2} = \frac{1}{2}F_{[\text{A on C}]}$
Simplify Equation (1) for $\vec{F}_{\text{tot}}$ by substituting $F_{[\text{B on C}]} = \frac{1}{2}F_{[\text{A on C}]}$.	$\vec{F}_{\text{tot}} = (\frac{\sqrt{3}}{2}F_{[\text{A on C}]} + (\frac{\sqrt{3}}{2})\frac{1}{2}F_{[\text{A on C}]})\hat{\imath} + (\frac{1}{2}F_{[\text{A on C}]} - (\frac{1}{2})\frac{1}{2}F_{[\text{A on C}]})\hat{\jmath}$ $\vec{F}_{\text{tot}} = \frac{3\sqrt{3}}{4}F_{[\text{A on C}]}\hat{\imath} + \frac{1}{4}F_{[\text{A on C}]}\hat{\jmath}$
Next, substitute Equation (2) for $F_{[\text{A on C}]}$.	$\vec{F}_{\text{tot}} = \frac{3\sqrt{3}}{4}\left(\frac{GM^2}{100d^2}\right)\hat{\imath} + \frac{1}{4}\left(\frac{GM^2}{100d^2}\right)\hat{\jmath} = \left(\frac{3\sqrt{3}GM^2}{400d^2}\right)\hat{\imath} + \left(\frac{GM^2}{400d^2}\right)\hat{\jmath}$

CHECK and THINK

To check our expression, we first consider the dimensions. In each term, we have GM^2/d^2, which has the dimensions of force.

$$\left[\frac{GM^2}{d^2}\right] = \left(\frac{\mathrm{F \cdot L^2}}{\mathrm{M^2}}\right)(\mathrm{M^2})\left(\frac{1}{\mathrm{L^2}}\right) = \mathrm{F}$$

Next, we see from our free-body diagram that the x components of the forces add together and the y components partially cancel each other. We therefore expect the term in x to be greater than the term in y, as we found.

$$\left(\frac{3\sqrt{3}GM^2}{400d^2}\right) > \left(\frac{GM^2}{400d^2}\right)$$

DERIVATION Kepler's Third Law

Newton derived the law of universal gravity from Kepler's second (equal areas) and third ($T^2 \propto a^3$) laws. In Section 13-7, we will derive Kepler's second law. Here, we derive Kepler's third law for an object in a circular orbit around the Sun,

$$T^2 = \left(\frac{4\pi^2}{GM_\odot}\right)r^3 \quad (7.5)$$

from Newton's laws. We then extend this equation to the general form of Kepler's third law of an object in an elliptical orbit with another object of mass M at one focus:

GENERAL FORM OF KEPLER'S THIRD LAW

Special Case

$$T^2 = \left(\frac{4\pi^2}{GM}\right)a^3 \quad (7.6)$$

Consider a planet in a circular orbit around the Sun. Draw a free-body diagram for this planet at a particular moment (Fig. 7.16). We have chosen a polar coordinate system with $\hat{r}$ directed from the planet to the Sun. The only force acting on it is the gravitational force due to the Sun.

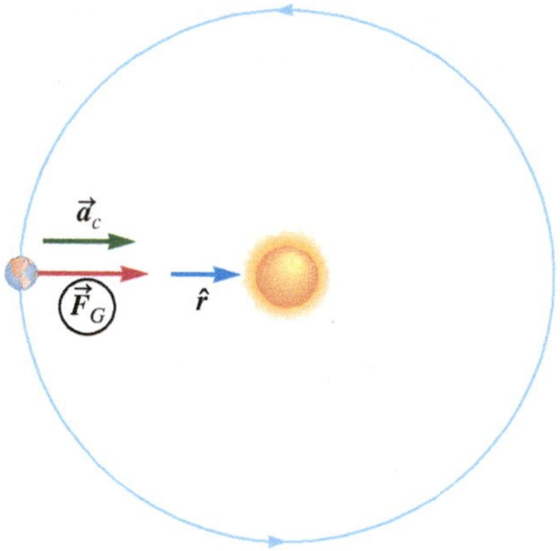

FIGURE 7.16

Gravity depends on the planet's distance to the Sun r, and the centripetal force required depends on the planet's period T. The key to finding the relationship between T and r is that the gravitational force of the Sun provides the centripetal force (circled in the free-body diagram).

Applying Newton's second law shows that the net force exerted on the planet is gravity and that it equals the centripetal force. Substitute Equation 6.7, $F_c = m(v^2/r)$, for F_C and Equation 7.4, $F_G = G(m_1m_2/r^2)$, for F_G.

$$\sum F_r = M_P a_c$$
$$F_G = F_C$$
$$G\frac{M_\odot M_P}{r^2} = M_P\frac{v^2}{r}$$

The planet mass M_P cancels out of this equation, and so does one power of r.

$$v^2 = G\frac{M_\odot}{r} \quad (7.7)$$

We find the planet's speed from Equation 4.30.

$$v = \frac{2\pi r}{T} \quad (4.30)$$

Square that speed and set it equal to Equation 7.7.

$$v^2 = \frac{4\pi^2 r^2}{T^2} = \frac{GM_\odot}{r}$$

Rearrange this equation to arrive at Kepler's third law for a planet in a circular orbit around the Sun.

$$T^2 = \left(\frac{4\pi^2}{GM_\odot}\right)r^3 \; ✓ \quad (7.5)$$

COMMENTS

A more general derivation yields Equation 7.6 which holds when the orbit is elliptical with semimajor axis a and an object of mass M is at one focus. If G and M are expressed in SI units, then T and a are in seconds and meters, respectively.

$$T^2 = \left(\frac{4\pi^2}{GM}\right)a^3 \; ✓ \quad (7.6)$$

Next, let's compare the first equation for Kepler's third law (Eq. 7.2) with this general expression. The term $(4\pi^2/GM)$ is the same for each planet in an elliptical orbit around the Sun, and we arbitrarily use the letter C to represent that term.	$T^2 = Cr^3$
For the Earth, $T = 1$ yr and $r = 1$ AU. Therefore, if T is measured in years and r is measured in astronomical units, then $C = 1\ \text{yr}^2/\text{AU}^3$. So, Equation 7.2 is consistent with Equation 7.6 when time is measured in years and distance in astronomical units.	$T^2_{[\text{yr}]} = a^3_{[\text{AU}]}$ (7.2)

Gravitational and Inertial Mass

It may seem obvious that the gravitational force between any two objects depends on the objects' masses, but there is no theoretical basis for this. That is, no one knows why the gravitational force depends on inertia (mass) and not some other property of the objects, such as volume, density, or composition.

The mass in the law of universal gravity (Eq. 7.4) is referred to as the object's **gravitational mass**. It is the property of the particles that creates a gravitational force between them. As we know, the mass in Newton's second law ($\sum \vec{F}_{\text{tot}} = m\vec{a}$) is the *inertial mass* of an object. Experimental evidence supports the idea that the gravitational mass of any object equals its inertial mass.

GRAVITATIONAL MASS AND INERTIAL MASS

Major Concept

Let us see how this reasoning works. We *assume* the gravitational mass m_{grav} is identical to the inertial mass m_{inert} and then see if this assumption matches experimental evidence. From Equation 7.4, we know that the magnitude of the gravitational force on an object of gravitational mass m_{grav} near the Earth's surface is

$$F_G = G\frac{M_\oplus m_{\text{grav}}}{R_\oplus^2}$$

Suppose our object of inertial mass m_{inert} is in free fall near the Earth's surface. If it is, the only force acting on it is gravity, and we use Newton's second law to find the object's free-fall acceleration g:

$$F_G = m_{\text{inert}}g$$

$$G\frac{M_\oplus m_{\text{grav}}}{R_\oplus^2} = m_{\text{inert}}g$$

Using our assumption that inertial mass and the gravitational mass are identical,

$$m_{\text{inert}} = m_{\text{grav}} \equiv m$$

This assumption leads to

$$G\frac{M_\oplus m}{R_\oplus^2} = mg$$

$$\frac{GM_\oplus}{R_\oplus^2} = g \tag{7.8}$$

Equation 7.8 predicts that g is a constant because $GM_\oplus/R_\oplus^2$ is a constant. We have found a testable prediction for the assumption that $m_{\text{inert}} = m_{\text{grav}} \equiv m$: If the assumption is correct, the gravitational acceleration near the surface of the Earth g has the same value for all objects and does not depend on any property such as size or shape of the object in free fall.

We know from countless experiments carried out over centuries that in the absence of air resistance, all objects near the surface of the Earth fall at the same acceleration g no matter what physical properties they have. Therefore, we infer that the gravitational mass of any object is equal to its inertial mass from the observation that the acceleration due to gravity is constant near the surface of the Earth.

The idea that gravitational mass and inertial mass are identical may seem trivial, but this idea is actually very important and led Einstein to discover his theory of *general relativity* (Chapter 39). As a last thought on this matter, imagine that the gravitational mass did *not* equal the inertial mass. If that were the case, you might

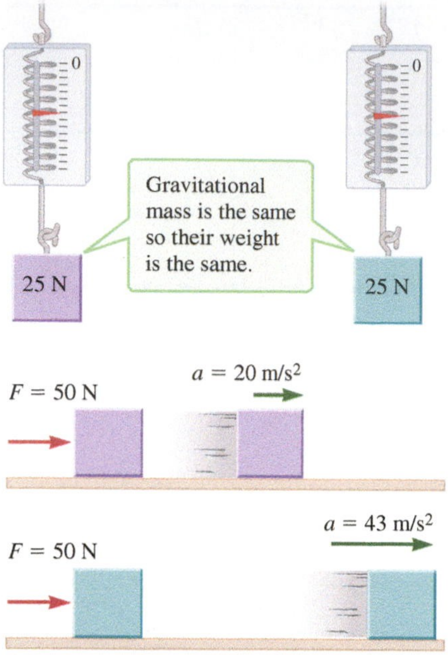

FIGURE 7.17 If gravitational mass and inertial mass were not the same, two objects that have the same weight may not acquire the same acceleration when the same net force is exerted on each.

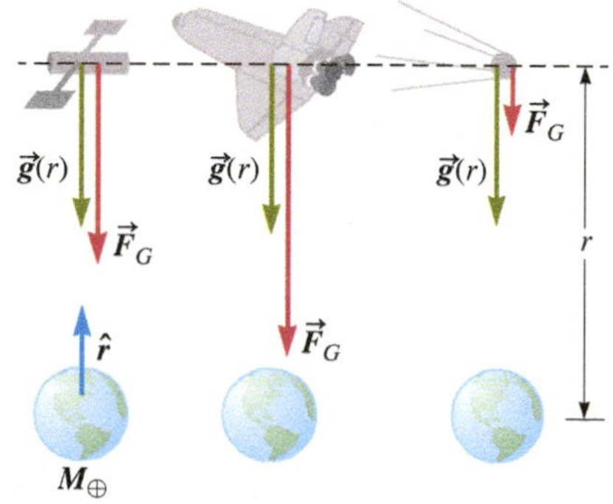

FIGURE 7.18 The gravitational force $\vec{F}_G$ on a spacecraft at a position r depends on the mass of the spacecraft that is actually there. The gravitational field $\vec{g}$ at r is the same no matter which spacecraft is there. In fact, the gravitational field is the same even if nothing is there.

weigh two objects on a spring scale and find that they have identical weights, but when you pushed each object horizontally along a frictionless surface, the same force would *not* produce the same acceleration (Fig. 7.17).

7-4 The Gravitational Field

Imagine the task of putting several spacecraft with different masses into orbit at some distance r from the center of the Earth (Fig. 7.18). One thing you would need to know in such a project is the Earth's gravitational force exerted on each spacecraft when it is in orbit. We find that force from Newton's law of universal gravity (Eq. 7.4), expressed in vector form:

$$\vec{F}_G(r) = -G\frac{M_\oplus m_{SC}}{r^2}\hat{r} \tag{7.9}$$

where m_{SC} is the mass of a particular spacecraft and the unit vector $\hat{r}$ points outward from the Earth toward a spacecraft. You would use Equation 7.9 three times for the three different spacecraft to obtain three different values for $\vec{F}_G$. If the three spacecraft were all at the same distance r from the Earth, the difference in the gravitational force exerted on a particular spacecraft would depend only on the mass of that spacecraft. In Figure 7.18, $\vec{F}_G$ is strongest for the middle spacecraft because it has the greatest mass and is weakest for the rightmost spacecraft because it has the smallest mass.

In this situation, where the Earth is present in all three cases but there are three different spacecraft, it is convenient to think of the Earth as the source of the gravitational force and each spacecraft as a subject of that force. We can rewrite Equation 7.9 as

$$\vec{F}_G(r) = m_{SC}\left(-G\frac{M_\oplus}{r^2}\hat{r}\right) \tag{7.10}$$

The mass m_{SC} is a property of the subject (each spacecraft). The term in parentheses is the same in all three cases independent of the particular spacecraft; it depends on the source (the Earth). For convenience, let's give this term its own mathematical symbol:

$$\vec{g}(r) \equiv -\frac{GM_\oplus}{r^2}\hat{r} \tag{7.11}$$

where $\vec{g}(r)$ depends only on the source. Then, the gravitational force is written in terms of two pieces, one that depends on the subject (m_{SC}) and the other that depends on the source $\vec{g}(r)$:

$$\vec{F}_G(r) = m_{SC}\vec{g}(r)$$

Breaking a field force such as gravity (or the electric force) into pieces—one piece that depends on the subject and the other that depends on the source—is a powerful technique that is used in many situations, and it introduces another concept, the field. The general term *field* refers to a region under the influence of some physical source. So, in this case, the source is the Earth, and the region is the environment around the Earth. Every position in a field is assigned a particular value of some quantity that is influenced by the source. If the source's influence is a vector quantity, the field is a **vector field**, and each position in the field is associated with a particular magnitude and direction. The Earth's gravitational field is a vector field, and Equation 7.11 assigns a magnitude and direction to every position in that field (Fig. 7.19). The term *field* also refers to the mathematical description of the source's influence; in this case, $\vec{g}(r)$. So, Equation 7.11 is known as the *gravitational vector field* of the Earth.

Although the terms *force* and *field* sound alike, they have very different meanings. Unlike the gravitational force, the gravitational field depends only on a source (the Earth) and not on any subject (spacecraft) in the field. The Earth's gravitational field $\vec{g}(r)$ is a measure of how the Earth influences its environment and is a function only of distance from the center as indicated by the (r) in the notation. In Figure 7.18, the gravitational field $\vec{g}(r)$ is the same at the location of each spacecraft. In fact, even if all the spacecraft vanished, $\vec{g}(r)$ would still have the same magnitude and direction because it depends only on the mass $M_\oplus$ of the Earth and the distance r from the center of the Earth.

The word *vector* is often dropped when referring to specific vector fields, so we refer to a *gravitational field*, not a *gravitational* ***vector*** *field*.

This case of the various spacecraft and the Earth illustrates the following important concepts behind a vector field.

1. The field has a **source** that influences its surroundings. We are interested in knowing the gravitational force on a variety of spacecraft, so we choose the Earth as the source of the gravitational field.
2. At each position r, the magnitude and direction of the Earth's gravitational field $\vec{g}(r)$ **only depend on properties of the source**, which is the Earth's mass in this case.
3. **Subjects** are items that may be placed in the source's field. The force exerted on a subject depends on properties of the source and the subject. In this case, the gravitational force exerted on the spacecraft depends on the mass of the spacecraft and the Earth's gravitational field at the spacecraft's location.

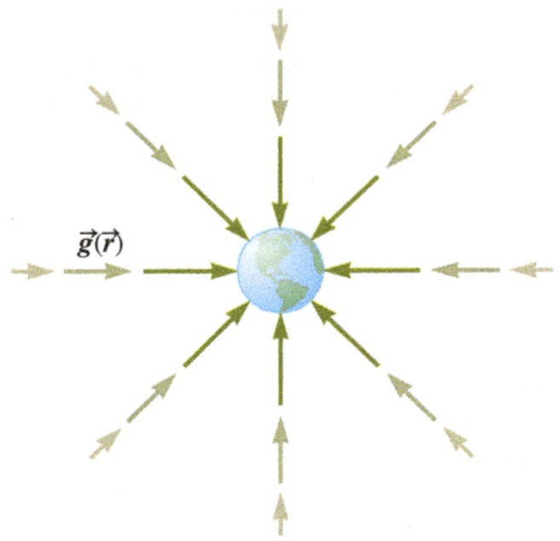

FIGURE 7.19 The Earth's gravitational field points to the center of the Earth at each point. It is strongest near the surface of the Earth, where the vectors are the longest. The gravitational field is equivalent to the local free-fall acceleration at each point.

The dimensions of the gravitational field are force per mass, so it has SI units of newtons per kilogram. Also notice that the gravitational field has the dimensions of acceleration

$$[\![g]\!] = \frac{\text{F}}{\text{M}} = \frac{\text{M} \cdot \text{L}/\text{T}^2}{\text{M}} = \frac{\text{L}}{\text{T}^2}$$

We can think of the Earth's gravitational field at a particular point as being the local free-fall acceleration (which is not true for other fields, such as the electric field). Figure 7.19 shows that the Earth's gravitational *field* is strongest near the surface. So, the free-fall acceleration is higher near the surface than it is at any point out in space. In fact, near the surface of the Earth, the gravitational field has the value of the free-fall acceleration we encountered in Chapter 2 and Equation 7.8:

$$\vec{g}(R_\oplus) = -\frac{GM_\oplus}{R_\oplus^2}\hat{r} \approx -9.8\hat{r} \text{ m/s}^2 \tag{7.12}$$

We use only two significant figures here because the free-fall acceleration $g = 9.81 \text{ m/s}^2$ we have been using is actually slightly less than the value calculated when we substitute into Equation 7.12. The reason for the discrepancy is discussed in Section 7.5.

EXAMPLE 7.5 Apples and the Moon

One of Newton's important contributions to science was the idea that the laws of physics are universal; there is no difference between terrestrial and celestial laws. This idea underlies the law of universal gravity. Newton said that the same gravitational pull by the Earth that causes an apple to fall to the ground is what keeps the Moon in its orbit. Earlier scientists did not necessarily think that the Earth could influence the Moon and would not have used properties of the Moon to better understand the Earth. Newton, however, used the period of the Moon's orbit to estimate the magnitude of the Earth's gravitational *field* at the Moon's distance. This example is based on his estimation.

A Use the period of the Moon's orbit to find the magnitude of the Moon's centripetal acceleration, assuming the orbit is circular.

INTERPRET and ANTICIPATE

This part of the problem is a straightforward application of the concept of uniform circular motion from Chapter 4. We expect a numerical result of the form $a_c =$ ____ m/s².

SOLVE

Because our ultimate task is to find the value of something at the location of the Moon, let us label the Earth–Moon distance $r_{[\text{at Moon}]}$. Apply $\vec{a}_c = -(v^2/r)\hat{r}$ (Eq. 4.36) for centripetal acceleration.	$a_c = \dfrac{v^2}{r_{[\text{at Moon}]}}$ (1)
The period of the Moon's orbit is roughly a month. Find the Moon's speed from its period using $v = 2\pi r/T$ (Eq. 4.30).	$v = \dfrac{2\pi r_{[\text{at Moon}]}}{T}$
Substitute into Equation (1). Notice that one power of $r_{[\text{at Moon}]}$ cancels out.	$a_c = \dfrac{4\pi^2 r_{[\text{at Moon}]}}{T^2}$

Example continues on page 198 ▶

The Earth–Moon distance is $r_{[\text{at Moon}]} = 3.84 \times 10^8$ m, and the period of the Moon's orbit is $T = 27.3$ d $= 2.36 \times 10^6$ s.	$a_c = \dfrac{4\pi^2(3.84 \times 10^8 \text{ m})}{(2.36 \times 10^6 \text{ s})^2}$ $a_c = 2.72 \times 10^{-3} \text{ m/s}^2$

CHECK and THINK

Our result is in the form we expected.

B Now use the mass of the Earth and the Earth–Moon distance to find the magnitude of the Earth's gravitational *field* at the Moon, $g(r_{[\text{at Moon}]})$.

INTERPRET and ANTICIPATE

Let us make the approximation that the only force acting on the Moon is the gravitational force of the Earth. Therefore, (1) the Moon is in free fall, and (2) the Earth's gravity alone provides the centripetal force (Section 6-6) that keeps the Moon in its orbit. So, we expect the centripetal acceleration of the Moon to equal the Earth's gravitational *field* at distance $r_{[\text{at Moon}]}$.

SOLVE

Use Equation 7.11, omitting the unit vector $\hat{r}$ because we are interested in magnitude only.

$$g(r_{[\text{at Moon}]}) = \frac{GM_\oplus}{r^2_{[\text{at Moon}]}} \qquad (7.11)$$

$$g(r_{[\text{at Moon}]}) = \frac{(6.67 \times 10^{-11} \text{N} \cdot \text{m}^2/\text{kg}^2)(5.98 \times 10^{24} \text{ kg})}{(3.84 \times 10^8 \text{ m})^2}$$

$$g(r_{[\text{at Moon}]}) = 2.70 \times 10^{-3} \text{ N/kg} = 2.70 \times 10^{-3} \text{ m/s}^2$$

CHECK and THINK

We found essentially what we expected to find: $a_c \approx g(r_{[\text{at moon}]})$. There is a slight discrepancy (less than 1%) due to our assumptions that the Moon's orbit is a circle centered on the center of the Earth and that only the Earth provides a gravitational force on it. Other objects, such as the Sun, also exert a gravitational force on it.

EXAMPLE 7.6 Your TV Satellite

A *geosynchronous* satellite is one that orbits with the same 24-hour period as the Earth's rotation. Thus, such a satellite is always above the same point on the Earth. Find the altitude of a geosynchronous satellite in a circular orbit.

INTERPRET and ANTICIPATE

As in Example 7.5, we assume the only force acting on the satellite is the Earth's gravity, which must provide a centripetal force.

SOLVE

The magnitude of the centripetal acceleration is equal to the magnitude of the gravitational field at the satellite's position.

$$a_c = g(r)$$

$$\frac{v^2}{r} = \frac{GM_\oplus}{r^2}$$

$$v^2 = \frac{GM_\oplus}{r} \qquad (1)$$

For a geosynchronous satellite to remain always above the same place on the Earth, its period must be 24 hours. Use Equation 4.30 to find the satellite's speed.

$$v = \frac{2\pi r}{T} \qquad (4.30)$$

We eliminate v from Equations (1) and 4.30 and solve for r. Notice that our result looks like Kepler's third law (Eq. 7.5) with the Sun's mass replaced by the Earth's mass.

$$\frac{4\pi^2 r^2}{T^2} = \frac{GM_\oplus}{r}$$

$$r = \left(\frac{GM_\oplus T^2}{4\pi^2}\right)^{1/3}$$

Substitute numerical values.	$r = \left[\frac{(6.67 \times 10^{-11}\ \text{N} \cdot \text{m}^2/\text{kg}^2)(5.98 \times 10^{24}\ \text{kg})(8.64 \times 10^4\ \text{s})^2}{4\pi^2}\right]^{1/3}$ $r = 4.23 \times 10^7\ \text{m}$
This result is the distance from the center of the Earth to the satellite. To find the altitude h of the satellite above the ground, subtract the Earth's radius.	$h = r - R_\oplus$ $h = 4.23 \times 10^7\,\text{m} - 6.37 \times 10^6\,\text{m}$ $h = 3.59 \times 10^7\,\text{m}$

CHECK and THINK

We are often asked to find a numerical result that is not easy to confirm, but in this case it is easy to find the actual altitudes of geosynchronous orbits. For example, according to the National Aeronautics and Space Administration (NASA) and the *Encyclopedia Britannica*, the altitude of a satellite's geosynchronous orbit is about 36,000 km.

So far, we have focused on the Earth as a source of a gravitational field, but, of course, anything with mass creates a gravitational field. We can find the gravitational field of any source that may be modeled as a particle of mass M using

GRAVITATIONAL FIELD OF A PARTICLE
Major Concept

$$\vec{g}(r) \equiv -\frac{GM}{r^2}\hat{r} \tag{7.13}$$

where $\hat{r}$ is a unit vector pointing away from the particle. Equation 7.13 also applies at the surface and outside any spherical source, such as a planet or star; in this case, $\hat{r}$ points away from the center of the spherical source. Equation 7.13 also applies to sources that are not spherical—galaxies, for instance, which might be disk-shaped—when the distance from the source is much greater than the size of the source.

If more than one particle is the source of a gravitational field, the resulting field at any point is found by adding the gravitational fields due to each particle individually. We will learn how to find the gravitational field due to more complicated sources when we study another field force, electricity, in Chapter 24.

EXAMPLE 7.7 SOHO: *So*lar and *H*eli*o*spheric Satellite

SOHO is a space observatory used to study the Sun, orbiting the Sun with the same 1-yr period as the Earth. As it does so, the observatory maintains its position, known as the first Lagrange or L1 point, between the Earth and the Sun. In this example, we will find the distance between the Earth and the L1 point.

A For the moment, assume the Earth and SOHO are stationary relative to the Sun. Find the distance $r_{\oplus S}$ from the Earth at which SOHO remains stationary. Report your answer in meters.

INTERPRET and ANTICIPATE

A sketch as in Figure 7.20 helps interpret this problem. The Earth and SOHO are assumed to be stationary. SOHO is located at a point S between the Earth and the Sun, and S must be the point at which the vector sum of the Earth's gravitational field and the Sun's gravitational field is zero. Because the Sun is much more massive than the Earth, we expect S to be closer to the Earth than to the Sun.

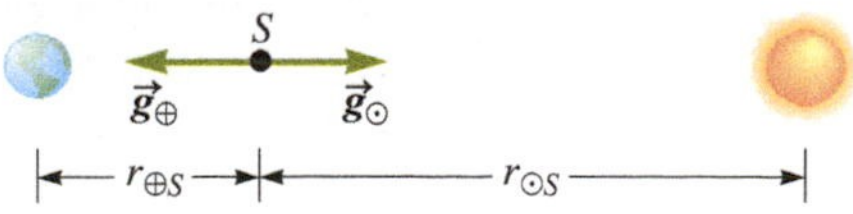

FIGURE 7.20

SOLVE The gravitational fields due to the Earth and the Sun are both given by $\vec{g}(r) \equiv -\frac{GM}{r^2}\hat{r}$ (Eq. 7.13). At S, the magnitudes of the two fields are equal. Let's call the Sun–SOHO distance $r_{\odot S}$ and the Earth–SOHO distance $r_{\oplus S}$.	$\frac{GM_\odot}{r_{\odot S}^2} = \frac{GM_\oplus}{r_{\oplus S}^2}$ (1)

Example continues on page 200 ▶

Because we have two unknowns, $r_{\odot S}$ and $r_{\oplus S}$, we need another equation. This second equation comes from the average distance between the Earth and the Sun, 1 AU.

$$r_{\odot S} + r_{\oplus S} = 1\,\text{AU} \quad (2)$$

:• SOLVE

Solve Equations (1) and (2) simultaneously for $r_{\oplus S}$. Start by simplifying Equation (1).

$$\frac{M_\odot}{r_{\odot S}^2} = \frac{M_\oplus}{r_{\oplus S}^2} \qquad \frac{r_{\odot S}}{r_{\oplus S}} = \sqrt{\frac{M_\odot}{M_\oplus}}$$

$$r_{\odot S} = r_{\oplus S}\sqrt{\frac{M_\odot}{M_\oplus}}$$

Substitute the result into Equation (2).

$$r_{\oplus S}\sqrt{\frac{M_\odot}{M_\oplus}} + r_{\oplus S} = 1\,\text{AU}$$

$$r_{\oplus S} = \frac{1\,\text{AU}}{1 + \sqrt{M_\odot/M_\oplus}}$$

Substitute values and convert the answer to meters. Recall $1\,\text{AU} = 1.50 \times 10^{11}\,\text{m}$.

$$r_{\oplus S} = \frac{1\,\text{AU}}{1 + \sqrt{1.99 \times 10^{30}\,\text{kg}/5.98 \times 10^{24}\,\text{kg}}}$$

$$r_{\oplus S} = 1.73 \times 10^{-3}\,\text{AU} = 2.60 \times 10^{8}\,\text{m}$$

:• CHECK and THINK

As expected, point S is much closer to the Earth (1.73×10^{-3} AU) than to the Sun ($1 - 1.73 \times 10^{-3}\,\text{AU} = 0.998$ AU).

B Now assume SOHO stays aligned with the Earth as both bodies orbit the Sun. The L1 point lies along the imaginary straight line connecting the Earth and the Sun, and this point is the location of SOHO as it orbits. Is $r_{\oplus L}$, the distance between the Earth and L1, the same as the distance $r_{\oplus S}$ found in part A? If not, is the distance $r_{\oplus L}$ longer or shorter than the value for $r_{\oplus S}$ found in part A?

:• INTERPRET and SOLVE

Because it orbits the Sun, SOHO is accelerating. Its centripetal acceleration is the net gravitational field at its position L1; therefore, this field cannot be zero. The centripetal acceleration must point toward the Sun; therefore, the net gravitational field at L1 must point toward the Sun (Fig. 7.21).

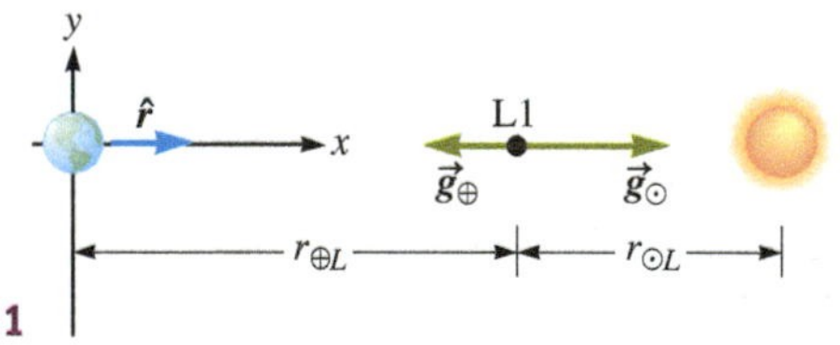

FIGURE 7.21

For the total gravitational field at L1 to point toward the Sun, $g_\odot > g_\oplus$. So, L1 must be closer to the Sun than S. Therefore, $r_{\oplus L} > r_{\oplus S}$.

C According to NASA, the L1 point for SOHO is $r_{\oplus L} = 1.55 \times 10^{9}\,\text{m}$. Confirm this value.

:• INTERPRET and ANTICIPATE

We must develop an appropriate expression and substitute values to show that $r_{\oplus L} = 1.55 \times 10^{9}\,\text{m}$ is correct. The key to finding this appropriate expression comes from realizing that the centripetal acceleration of SOHO must equal the total gravitational field at L1.

:• SOLVE

SOHO's period is same as the Earth's: 1 year. Write the centripetal acceleration in terms of the period.

$$v = \frac{2\pi r_{\odot L}}{T}$$

$$a_c = \frac{v^2}{r_{\odot L}} = \frac{4\pi^2 r_{\odot L}}{T^2} \quad (1)$$

The total gravitational field $(\vec{g}_L)_{\text{tot}}$ at L1 comes from adding the gravitational fields due to the Earth and the Sun at L1. For this vector addition, we choose a coordinate system centered on the Earth as shown in Figure 7.21.

$$(\vec{g}_L)_{\text{tot}} = \frac{GM_\odot}{r_{\odot L}^2}\hat{r} - \frac{GM_\oplus}{r_{\oplus L}^2}\hat{r} \quad (2)$$

As in part A, the average distance between the Earth and the Sun is 1 AU.

$$r_{\oplus L} + r_{\odot L} = 1\text{ AU}$$
$$r_{\odot L} = 1\text{ AU} - r_{\oplus L}$$

Eliminate $r_{\odot L}$ from Equations (1) and (2).

$$a_c = \frac{4\pi^2(1\text{ AU} - r_{\oplus L})}{T^2} \quad (3)$$

$$(\vec{g}_L)_{\text{tot}} = \frac{GM_\odot}{(1\text{ AU} - r_{\oplus L})^2}\hat{r} - \frac{GM_\oplus}{r_{\oplus L}^2}\hat{r} \quad (4)$$

Now substitute values into Equations (3) and (4). If we find that they are equal, we will have confirmed SOHO's distance from the Earth.

$$r_{\oplus L} = 1.55 \times 10^9\text{ m}$$
$$r_{\odot L} = 1\text{ AU} - r_{\oplus L} = 1.48 \times 10^{11}\text{ m}$$

Start with Equation (3) for the centripetal acceleration.

$$a_c = \frac{4\pi^2(1.48 \times 10^{11}\text{ m})}{(3.15 \times 10^7\text{ s})^2} = 5.98 \times 10^{-3}\text{ m/s}^2 \quad (5)$$

Next, find the magnitude of the total gravitational field at the L1 point from Equation (4).

$$(g_L)_{\text{tot}} = (6.67 \times 10^{-11}\text{ N}\cdot\text{m}^2/\text{kg}^2)\left[\frac{1.99 \times 10^{30}\text{ kg}}{(1.48 \times 10^{11}\text{ m})^2} - \frac{5.98 \times 10^{24}\text{ kg}}{(1.55 \times 10^9\text{ m})^2}\right]$$

$$(g_L)_{\text{tot}} = 5.98 \times 10^{-3}\text{ m/s}^2 \quad (6)$$

Compare Equations (5) and (6).

$$(g_L)_{\text{tot}} = a_c = 5.98 \times 10^{-3}\text{ m/s}^2 \checkmark$$

CHECK and THINK

Our work confirms that the L1 point is six times farther from the Earth than S, the (fictitious) stationary position. (Notice that Figures 7.20 and 7.21 are not drawn to scale.)

EXAMPLE 7.8 CASE STUDY Dark Matter in Our Galaxy

To estimate the amount of dark matter in the Milky Way galaxy, let us apply a simple model that describes the galaxy in terms of two components: a central spherical component known as the bulge and a flat component known as the disk in which the arms of the galaxy are embedded (Fig. 7.22). Stars in the disk orbit the bulge. Consider the circular orbit of a star far from the center of the galaxy on the outer edge of the disk, with orbital radius $r = 7.7 \times 10^{20}$ m and linear speed $v = 2.2 \times 10^5$ m/s. Assume we can treat all the matter in the bulge and disk as a particle located at the galactic center. An estimate of the amount of ordinary matter (both luminous matter and nonluminous)—such as stars, gas, and planets—in the galaxy is $M_{\text{ord}} = 1.8 \times 10^{41}$ kg.

Estimate the mass of the dark matter in the Milky Way galaxy.

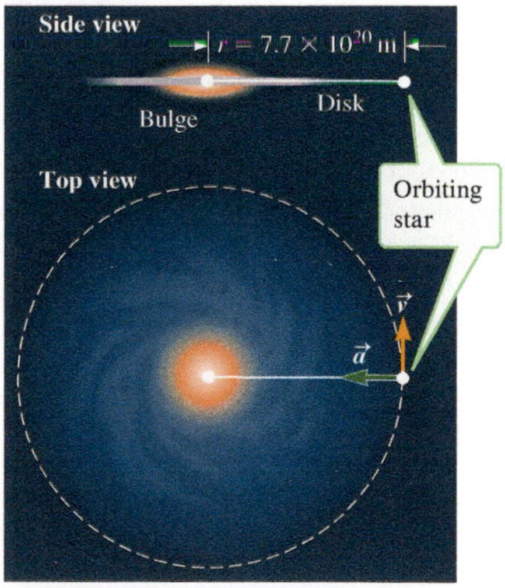

FIGURE 7.22

INTERPRET and ANTICIPATE

As in Examples 7.6 and 7.7, the key to solving this one is to realize that the centripetal force on the star at the edge of the Milky Way is provided by a gravitational force. In this case, that gravitational force is due to all the mass of the bulge and disk of the Milky Way. Once we have found the mass of the Milky Way, we will need to subtract the mass of the ordinary matter to find a numerical estimate for the mass of the dark matter.

SOLVE

Start with the magnitude of the centripetal acceleration of the orbiting star.

$$a_c = \frac{v^2}{r} \quad (4.38)$$

Because the gravitational force of the Milky Way's bulge and disk provides the centripetal force on the orbiting star, the star's centripetal acceleration must equal the gravitation *field* of the Milky Way at the star's location. Because we are modeling the matter in the Milky Way as a particle at the center, we use Equation 7.13, $\vec{g}(r) \equiv -(GM/r^2)\hat{r}$. The mass M_{MW} is the mass of both dark and ordinary matter.

$$g(r) = \frac{GM_{MW}}{r^2}$$

Example continues on page 202 ▶

Set the magnitude of the gravitational field equal to the magnitude of the centripetal acceleration and solve for M_{MW}.	$\frac{v^2}{r} = \frac{GM_{MW}}{r^2} \qquad M_{MW} = \frac{v^2 r}{G}$
Substitute values.	$M_{MW} = \frac{(2.2 \times 10^5\ \text{m/s})^2(7.7 \times 10^{20}\ \text{m})}{6.67 \times 10^{-11}\ \text{N} \cdot \text{m}^2/\text{kg}^2}$ $M_{MW} = 5.6 \times 10^{41}\ \text{kg}$
To find the mass of the dark matter, we must subtract the mass of the ordinary matter.	$M_{DM} = M_{MW} - M_{\text{ord}} = 5.6 \times 10^{41}\ \text{kg} - 1.8 \times 10^{41}\ \text{kg}$ $M_{DM} = 3.8 \times 10^{41}\ \text{kg}$

CHECK and THINK

This result means that $(3.8/5.6) \times 100 = 68\%$ of the Milky Way's matter is dark matter. We just found that there is more dark matter than all the ordinary matter in the Milky Way. To put this value in perspective, imagine standing in a room with 100 people, 68 of whom are invisible. The room might look pretty empty with only 32 people in it, but if you started to walk around, you would bump into people and find that it is actually quite crowded.

Many other indirect observations support the theory that dark matter exists. Like the evidence here, these observations show that much of the Universe is made up of extraordinary dark matter. It seems hard to believe that the kind of ordinary matter (protons, neutrons, and electrons) we are used to seeing everywhere is actually just a small component of the Universe.

EXAMPLE 7.9 CASE STUDY MOND in Our Galaxy

According to Example 7.8, 68% of our own galaxy is made up of extraordinary particles—dark matter—that have never been observed on the Earth. There is no laboratory evidence for the existence of dark matter. Some scientists—Milgrom among them—argue that there is little or no dark matter. So, how do these scientists account for the high speeds of luminous objects such as stars observed in our own Milky Way galaxy? Milgrom argues that Newton's law must be modified to take these observations into account and has developed a new theory, called MOND (*mo*dified *N*ewtonian *d*ynamics). According to this theory, Newton's laws as presented in Chapter 5 are good approximations for objects moving with relatively great acceleration, but the modified version of Newton's laws is necessary for accelerations smaller than 10^{-10} m/s^2. Is the centripetal acceleration of the star in Example 7.8 small enough that MOND (as proposed by Milgrom) plays an important role?

INTERPRET and ANTICIPATE

All the necessary concepts are covered in Chapter 4 about centripetal acceleration.

SOLVE Find the centripetal acceleration of the star (modeled as a particle).	$a_c = \frac{v^2}{r} = \frac{(2.2 \times 10^5\ \text{m/s})^2}{7.7 \times 10^{20}\ \text{m}}$ $a_c = 6.3 \times 10^{-11}\ \text{m/s}^2$

Yes, the resulting acceleration is below the MOND limit.

CHECK and THINK

Let's think about the dark-matter problem in the Milky Way galaxy. The problem stems from the observation of stars orbiting on the galaxy's outer edge. The centripetal acceleration of such stars requires a gravitational force that is too great to be supplied by the amount of luminous matter we observe. Even if we take faint objects such as planets and dim stars into account, nonluminous ordinary matter cannot make up the difference.

There are at least two possible solutions. One is that there must be some type of extraordinary matter (dark matter). Another solution is that Newton's laws do not hold up for the motion of the stars on the galaxy's outer edge.

To sort out which solution, if either, is correct, we must test each one. Tests for dark matter and MOND are beyond the scope of this book, but we can take a moment to think about tests for the solar system models in Galileo's time. When Galileo observed the sky through a telescope, the geocentric model of the solar system was widely accepted. His observations of Venus tested that geocentric model. Because the geocentric model could not account for the phases of Venus he observed, the model failed the test, and the heliocentric model gained acceptance. When there are two different competing theories in science, observations and experiments are used to eliminate one or both of them.

7-5 Variations in the Earth's Gravitational Field

Because the Earth is not a perfect sphere and because its density is not uniform, the gravitational field is not the same everywhere on the planet's surface. Roughly speaking, the Earth's gravitational field is slightly stronger at sea level than at higher elevations. The situation is made more complicated because the Earth is always changing as a result of earthquakes and volcanic eruptions. Such changes mean that the Earth's gravitational field changes over time. Space-based instruments measure these variations in the Earth's gravitational field near its surface.

The gravitational field varies both with elevation and over time, which means that the free-fall acceleration near the Earth's surface depends both on location and on time, but the variation is fairly small. Even though they are small, these variations can affect sensitive experiments. For most of the work we do in this textbook, however, we do not need to worry about these slight variations.

EXAMPLE 7.10 How Much Do You Weigh on Mount Everest?

To see just how much the gravitational field varies on the surface of the Earth, let us calculate it at sea level and on top of the highest mountain, Everest.

Find $\Delta g_{\text{alt}} = g_{\text{sea level}} - g_{\text{Everest}} = g(R_\oplus) - g(R_\oplus + h)$, where the subscript "alt" stands for altitude. To find a precise value, work to four significant figures and use the polar radius of the Earth, $R_{\oplus,\text{pole}} = 6.365 \times 10^6\,\text{m}$, for your sea-level calculation. The height of Mount Everest is $h = 8.848 \times 10^3\,\text{m}$ above the average radius of the Earth, $R_\oplus = 6.376 \times 10^6\,\text{m}$.

INTERPRET and ANTICIPATE

We expect to find a relatively small numerical result for Δg_{alt}.

SOLVE

Use Equation 7.13, $\vec{g}(r) \equiv -(GM/r^2)\hat{r}$, to find the magnitude of the gravitational field at each location.	$g(R_{\oplus,\text{pole}}) = \dfrac{GM_\oplus}{R^2_{\oplus,\text{pole}}}$ $\qquad g(R_\oplus + h) = \dfrac{GM_\oplus}{(R_\oplus + h)^2}$
The numerator is the same in both equations.	$GM_\oplus = (6.673 \times 10^{-11}\,\text{N}\cdot\text{m}^2/\text{kg}^2)(5.976 \times 10^{24}\,\text{kg})$ $GM_\oplus = 3.9878 \times 10^{14}\,\text{N}\cdot\text{m}^2/\text{kg}$
Substitute values. Use $1\,\text{N/kg} = 1\,\text{m/s}^2$ to express g in familiar units.	$g(R_{\oplus,\text{pole}}) = \dfrac{3.9878 \times 10^{14}\,\text{N}\cdot\text{m}^2/\text{kg}}{(6.365 \times 10^6\,\text{m})^2}$ $g(R_{\oplus,\text{pole}}) = 9.843\,\text{m/s}^2$ $g(R_\oplus + h) = \dfrac{3.9878 \times 10^{14}\,\text{N}\cdot\text{m}^2/\text{kg}}{(6.376 \times 10^6 + 8.848 \times 10^3\,\text{m})^2}$ $g(R_\oplus + h) = 9.782\,\text{m/s}^2$
Find the difference in the gravitational field.	$\Delta g_{\text{alt}} = g(R_{\oplus,\text{pole}}) - g(R_\oplus + h) = (9.843 - 9.782)\,\text{m/s}^2$ $\Delta g_{\text{alt}} = 0.061\,\text{m/s}^2$

CHECK and THINK

This difference is small but easily measurable. It means that a 50-kg scientist weighs 492.2 N (110.7 lb) at sea level but only 489.1 N (110.0 lb) on top of Mount Everest.

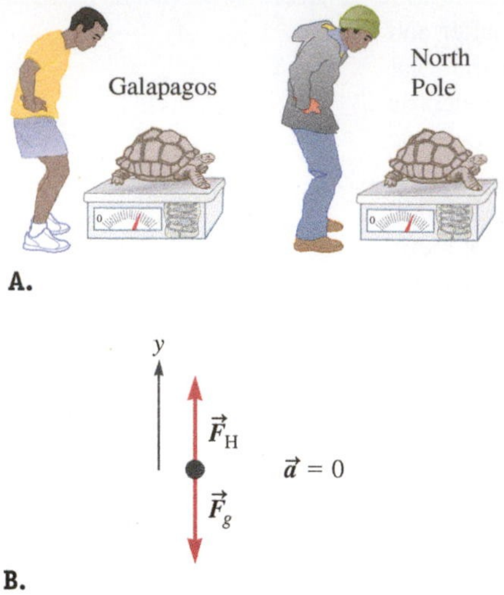

FIGURE 7.23 **A.** Aaron on the Earth weighs a turtle at two locations on Earth. Aaron is in a *non*inertial reference frame. **B.** The free-body diagram drawn by Aaron is the same in both locations.

Here are a few warnings about notation. First, do not confuse G and g. Uppercase G is the universal gravitational constant. Lowercase g without any other symbols is the free-fall acceleration value near the surface of the Earth. The National Institute of Standards and Technology (NIST) defined g to be $9.80665\ \mathrm{m/s^2}$; unless other information is given, you may assume this value. Second, $\vec{g}(r)$ is the notation used for the gravitational field produced by any source. It is a vector quantity, and its SI units are newtons per kilogram. It is numerically equivalent to the local free-fall acceleration when that acceleration is expressed in meters per second squared. Its magnitude is written as $g(r)$, and only near the surface of the Earth is $\vec{g}(R_\oplus) \approx -g\hat{r}$.

The Earth as a Noninertial Reference Frame

There is one more thing we must consider regarding the gravitational field near the Earth's surface. The Earth is a *non*inertial frame. Recall from Chapter 5 that an inertial reference frame is one that is not accelerating. Because the Earth rotates on its axis and orbits the Sun, it is accelerating and is therefore a noninertial frame. Observers in the Earth's noninertial frame may see a particle violate Newton's first law. To Earth-bound observers (at the same altitude), it seems that at the Earth's surface the gravitational field varies with latitude. We refer to the magnitude of the gravitational field measured by an Earth-bound observer as the *effective* gravitational field g_{eff}.

To see how latitude affects our observations of the Earth's gravitational field, let us investigate g_{eff} at the equator and at the North Pole. Imagine Aaron weighing a turtle on a spring scale first in the Galapagos Islands, located at the Earth's equator, and then at the North Pole (Fig. 7.23A). Aaron, being in the noninertial frame of the Earth, draws the same free-body diagram for the turtle at both locations (Figure 7.23B). He identifies two forces, gravity $\vec{F}_g$ and Hooke's spring force $\vec{F}_H$, and he does not observe the turtle accelerating in either location. Applying Newton's second law to the turtle, he says that

$$\sum F_y = F_H - F_g = ma_y = 0$$
$$F_H = F_g$$
$$w_{\text{app}} = w = mg$$

For those who might wonder if turtles are found near the poles: The leatherneck turtle can withstand extreme temperatures and can be found anywhere from the equator to polar regions.

Aaron reasons that the apparent weight w_{app} of the turtle (the value read off the scale) equals the turtle's weight w (magnitude of the Earth's gravitational force on the turtle), and he expects the weight to be the same in the Galapagos as it is at the North Pole. The scale reads slightly less in the Galapagos, however, meaning that the turtle's apparent weight there is slightly less than its apparent weight at the North Pole. To Aaron on the Earth, it therefore appears that the Earth's gravitational field is weaker at the *e*quator than at the *p*ole, or $g_e < g_p$. This apparent difference in the Earth's effective gravitational field results from the observer being in a noninertial frame.

To find a mathematical expression for the effective gravitational field at the equator g_e, we must imagine another observer—Hannah—in an inertial frame hovering above the North Pole watching Aaron weigh the turtle (Fig. 7.24). Her free-body diagram is essentially the same as Aaron's at the North Pole (Fig. 7.23B). (There is one slight difference in that she would see the turtle rotating in place like a record on a turntable. For our purposes here, however, we can ignore this effect and continue to treat the turtle as a particle.) So, both Hannah hovering above the Earth and Aaron standing on the ground find that at the North Pole,

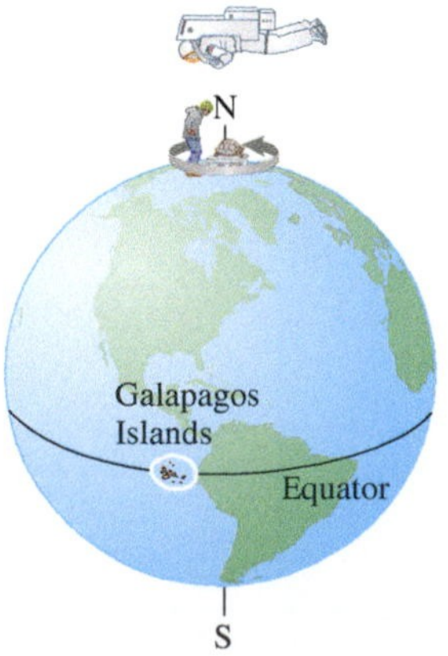

FIGURE 7.24 Hannah in an inertial reference frame above the Earth's North Pole.

$$w_{\text{app}} = w = mg_p \tag{7.14}$$

The major difference between the noninertial and inertial frames comes from observing the turtle in the Galapagos. From Hannah's view above the North Pole, she sees both Aaron and the turtle moving in a circle around the Earth's axis (Fig. 7.25A). Like Aaron, Hannah identifies two forces acting on the turtle, but from her vantage point in an inertial frame, these forces must account for the turtle's centripetal acceleration (Fig. 7.25B). When Hannah applies Newton's second law, she finds that the turtle's apparent weight at the equator equals its weight minus the centripetal force, $m(v^2/R_\oplus)$:

$$\sum F_y = F_H - F_g = ma_y = -m\frac{v^2}{R_\oplus}$$

$$F_H = F_g - m\frac{v^2}{R_\oplus}$$

$$w_{app} = w - m\frac{v^2}{R_\oplus} \qquad (7.15)$$

We can now use Equation 7.15 to find an expression for g_e, the magnitude of the Earth's effective gravitational field at the equator. The weight w is the magnitude of the gravitational force, which must be the same at the North Pole and in the Galapagos:

$$w = mg_p \qquad (7.14)$$

The apparent weight read off the scale by Aaron on the equator is written in term of the gravitational field g_e at the equator:

$$w_{app} = mg_e \qquad (7.16)$$

Substitute Equations 7.14 and 7.16 into Equation 7.15:

$$mg_e = mg_p - m\frac{v^2}{R_\oplus}$$

Removing the common factor m, we see that to Aaron on the equator, the gravitational field appears to be

$$g_e = g_p - \frac{v^2}{R_\oplus} \qquad (7.17)$$

To him, it appears that the Earth's gravitational field at the equator is weaker than at the poles, $g_e < g_p$. To Hannah in an inertial frame (hovering above the North Pole), the gravitational field at the equator is still g, and the reason $w_{app} < w$ is because of the turtle's centripetal acceleration at the equator.

We have found the effective gravitational field g_{eff} at two extremes—the North Pole and the equator. An object at the equator has the highest centripetal acceleration and therefore the lowest effective gravitational field $(g_{eff})_{min} = g_e$ (Eq. 7.17). An object at the North Pole has no centripetal acceleration and therefore the highest effective gravitational field $(g_{eff})_{max} = g_p$.

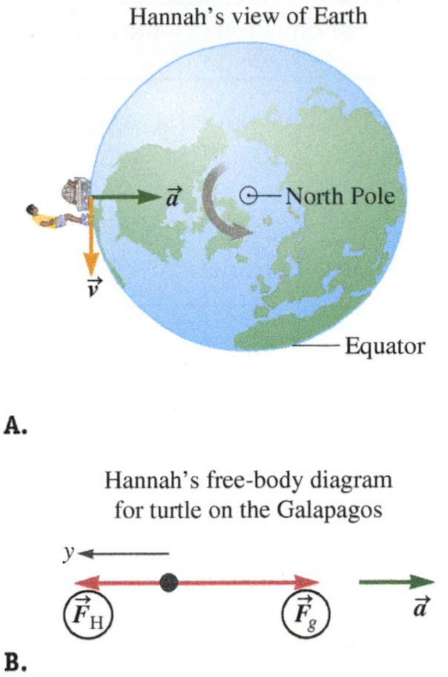

FIGURE 7.25 A. Hannah sees that both Aaron and the turtle in the Galapagos are moving in a circle. Therefore, they are both are accelerating. Aaron in the Galapagos is in a noninertial reference frame. **B.** Free-body diagram for a turtle as drawn by Hannah in an inertial reference frame off the Earth. Hannah sees that the turtle has a centripetal acceleration, but Aaron does not see this centripetal acceleration.

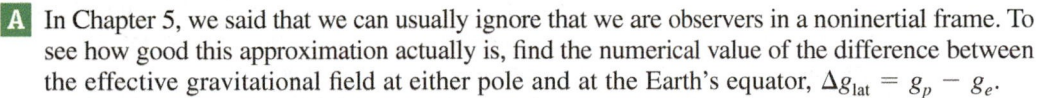

EXAMPLE 7.11 Ignoring the Earth's Rotation

A In Chapter 5, we said that we can usually ignore that we are observers in a noninertial frame. To see how good this approximation actually is, find the numerical value of the difference between the effective gravitational field at either pole and at the Earth's equator, $\Delta g_{lat} = g_p - g_e$.

INTERPRET and ANTICIPATE

As in Example 7.10, we expect to find a small numerical result for Δg_{lat}.

Start with Equation 7.17 to find an expression for Δg_{lat}. Rearrange this equation to get the two g terms together.	$g_e = g_p - \frac{v^2}{R_\oplus}$ $\Delta g_{lat} = g_p - g_e = \frac{v^2}{R_\oplus} \qquad (1)$
The right side of this equation is the centripetal acceleration from Equation 4.38, $a_c = v^2/r$. We do not know the speed v, but $v = 2\pi r/T$ allows us to eliminate v and write this expression in terms of the period T, which we know to be 24 h for the Earth. For $R_\oplus$, we use the Earth's equatorial radius.	$\frac{v^2}{R_\oplus} = \left(\frac{1}{R_\oplus}\right)\left(\frac{4\pi^2 R_\oplus^2}{T^2}\right)$ $\frac{v^2}{R_\oplus} = \frac{4\pi^2 R_\oplus}{T^2} = \frac{4\pi^2(6.387 \times 10^6\ \text{m})}{(8.64 \times 10^4\ \text{s})^2} = 3.38 \times 10^{-2}\ \text{m/s}^2$

Example continues on page 206 ▶

According to Equation (1), the centripetal acceleration is Δg_{lat}.	$\Delta g_{\text{lat}} = 3.38 \times 10^{-2}\ \text{m/s}^2 \text{ or N/kg}$

B Compare this value for Δg_{lat} with the variation in the gravitational field due to altitude Δg_{alt} (Example 7.10). What is the ratio $\Delta g_{\text{alt}}/\Delta g_{\text{lat}}$?

INTERPRET and ANTICIPATE

This part of the problem gives us an opportunity to think about our result in part A.

SOLVE

In Example 7.10, we found the variation in the Earth's gravitational field due to the altitude difference between the North Pole and the top of Mount Everest.

$$\Delta g_{\text{alt}} = 0.061\ \text{N/kg or m/s}^2$$

$$\frac{\Delta g_{\text{alt}}}{\Delta g_{\text{lat}}} = \frac{0.061\ \text{N/kg}}{0.0338\ \text{N/kg}} = 1.8$$

CHECK and THINK

Although the latitude variation Δg_{lat} is only about half the altitude variation, it is easily observable. In fact, this variation is the reason planets are wider at their equators than at their poles.

SUMMARY

Underlying Principles: Newton's law of universal gravity

Newton theorized that gravity is not limited to objects near the Earth, but instead applies to all objects in the Universe. The gravitational force is attractive, and its magnitude is

$$F_G = G\frac{m_1 m_2}{r^2} \tag{7.4}$$

Major Concepts

1. A **geocentric model** was described by Ptolemy, a Greek astronomer and mathematician who lived in the second century CE. According to this model, the Sun, the Moon, the stars, and the planets orbited the Earth.
2. In the **heliocentric model**, the Earth and other planets orbit the Sun.
3. **Empirical laws** are laws formulated based solely on data, without any hypothesis presented first.
4. The mass in Newton's second law ($\sum \vec{F}_{\text{tot}} = m\vec{a}$) is the **inertial mass** of an object. It is a measure of the object's resistance to change its velocity.
5. **Gravitational mass** is the property of the particles that creates a gravitational force between them. Experimental evidence supports the idea that the gravitational mass of any object equals its inertial mass.
6. *Field* is a general term referring to a region under the influence of some physical source. A **gravitational field** $\vec{g}(r)$ depends only on a source and not on any subject in the field. The value of the gravitational field at a particular point is equal to the local free-fall acceleration at that point. Usually, the SI units for the gravitational field are newtons per kilogram (N/kg), but they may also be expressed as meters per second squared (m/s^2). The gravitational field due to any particle with mass M is given by

$$\vec{g}(r) \equiv -\frac{GM}{r^2}\hat{r} \tag{7.13}$$

where $\hat{r}$ is a unit vector pointing away from the particle.

Special Cases

1. **Kepler's first law:** planetary orbits are ellipses with the Sun at one focus.
2. **Kepler's second law:** a line joining a planet to the Sun sweeps out equal areas in equal times.
3. If period T is measured in years and the semimajor axis a is measured in astronomical units (AU), **Kepler's third law** is expressed mathematically as

$$T^2_{[\text{yr}]} = a^3_{[\text{AU}]} \tag{7.2}$$

The general form of Kepler's third law is

$$T^2 = \left(\frac{4\pi^2}{GM}\right)a^3 \tag{7.6}$$

⦿ Tools

1. An **ellipse** is the locus of points for which the sum of the distances from the two foci F_1 and F_2 is constant (Fig. 7.5B). The **major axis** passes through both foci. The **minor axis** is the perpendicular bisector of the major axis. The length of the semimajor and semiminor axis are a and b, respectively.
2. **Inverse-square law:** Newton found that gravity obeys an *inverse-square law*, which means that the gravitational force on a planet varies as the inverse square of its distance from the Sun:

$$F_G \propto \frac{1}{r^2} \tag{7.3}$$

Any quantity that diminishes with the inverse square of distance is said to obey an inverse-square law.

PROBLEMS AND QUESTIONS

A = algebraic C = conceptual E = estimation G = graphical N = numerical

7-1 A Knowable Universe

1. C We use the terms *sunset* and *sunrise*. In what way are these terms misleading?
2. C Briefly describe a contemporary scientific endeavor. How is that scientific project influenced by society? How does it affect society?

7-2 Kepler's Laws of Planetary Motion

3. N For many years, astronomer Percival Lowell searched for a "Planet X" that might explain some of the perturbations observed in the orbit of Uranus. These perturbations were later explained when the masses of the outer planets and planetoids, particularly Neptune, became better measured (*Voyager 2*). At the time, however, Lowell had proposed the existence of a Planet X that orbited the Sun with a mean distance of 43 AU. With what period would this Planet X orbit the Sun?
4. C Many scientific projects are carried out by teams rather than by individuals. Tycho was a great observer, and Kepler was a great mathematician. They had difficulty working together. It is likely that you will be expected to work with others. Come up with strategies for working with people whom you might find difficult.
5. N You are given a string of length 10 cm, two tacks, and a pencil. You are to use these to draw an ellipse as in Figure 7.5B. How far apart must you place the tacks to ensure that the semimajor axis of the ellipse is 5 cm? Explain your answer. If possible, try it yourself.
6. N Io and Europa are two of Jupiter's many moons. The mean distance of Europa from Jupiter is about twice as far as that for Io and Jupiter. By what factor is the period of Europa's orbit longer than that of Io's?

Problems 7 and 8 are paired.

7. Model the Moon's orbit around the Earth as an ellipse with the Earth at one focus. The Moon's farthest distance (apogee) from the center of the Earth is $r_A = 4.05 \times 10^8$ m, and its closest distance (perigee) is $r_P = 3.63 \times 10^8$ m.
   a. N Calculate the semimajor axis of the Moon's orbit.
   b. N How far is the Earth from the center of the Moon's elliptical orbit?
   c. G Use a scale such as 1 cm → 10^8 m to sketch the Earth–Moon system at apogee and at perigee and the Moon's orbit. (The semiminor axis of the Moon's orbit is roughly $b = 3.84 \times 10^8$ m.)
8. Deimos is one of Mars's two moons (Fig. P7.8). Deimos's farthest distance from the center of Mars is $r_A = 2.359 \times 10^7$ m, and its closest distance is $r_P = 2.345 \times 10^7$ m.
   a. G Use a scale such that 1 cm → 10^7 m to sketch the Mars–Deimos system and include the orbit of Deimos.
   b. N Calculate the semimajor axis of Deimos's orbit.
   c. N How far is Mars from the center of Deimos's elliptical orbit? Give your answer in meters and in terms of Mars's radius.

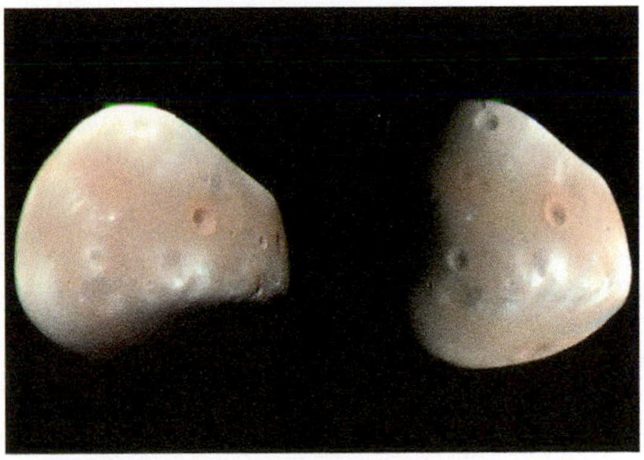

NASA/JPL-Caltech/University of Arizona

FIGURE P7.8

9. N Figure P7.9 is a scaled representation of a planet's orbit. with a semimajor axis of 1.524 AU. **a.** Find the ratio of the aphelion-to-perihelion distance. **b.** Find the perihelion and aphelion distances in astronomical units. **c.** Find the distance the Sun is from the center of the orbit.

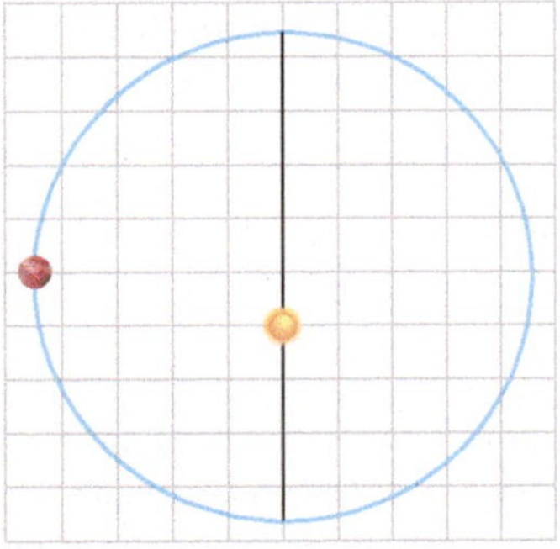

FIGURE P7.9 Problems 9 and 63.

10. C Students often incorrectly assert that the Earth's seasons are caused by its elliptical orbit around the Sun. Why is that assertion incorrect?

11. N One method for estimating the travel time from the Earth to the Moon is to use Kepler's third law to find half the period of an orbit whose perigee is at the Earth and whose apogee is at the Moon. What is the travel time to the Moon, a distance of 3.84×10^5 km from the Earth's center, for a spacecraft initially at an altitude of 200 km in the Earth's orbit, assuming it does not use any means of propulsion?

12. Before Kepler, the prevailing model for the shape of a planetary orbit was that it was circular. Kepler's first law of planetary motion challenges this model, stating planetary orbits are ellipses with the Sun at one focus. Of the eight major planets, Mercury's orbit is the one that differs most from a circular orbit. It has an aphelion distance from the Sun (r_A) of 0.46670 AU and a perihelion distance from the Sun (r_P) of 0.30750 AU.
    a. N Compute the length of the semimajor axis of Mercury.
    b. G The semiminor axis b of the orbit is 0.37883 AU. Using this and the information from part (a), make an accurate sketch of the orbit of Mercury. The sketch should contain not only Mercury's orbit, but also labels for perihelion distance, aphelion distance, major axis, and minor axis. For this sketch, you will need to position the Sun accurately.
    c. C Does the orbit you sketched in part (b) look very different from a circle? Was Copernicus a poor scientist for trying to fit Mercury's orbit with a circle?
    d. C Does the orbit you sketched have the Sun near the center as Copernicus would have thought when he tried to model planetary orbits as circles centered on the Sun?

13. N A massive black hole is believed to exist at the center of our galaxy (and most other spiral galaxies). Since the 1990s, astronomers have been tracking the motions of several dozen stars in rapid motion around the center. Their motions give a clue to the size of this black hole. **a.** One of these stars is believed to be in an approximately circular orbit with a radius of about 1.50×10^3 AU and a period of approximately 30 yr. Use these numbers to determine the mass of the black hole around which this star is orbiting. **b.** What is the speed of this star, and how does it compare with the speed of the Earth in its orbit? How does it compare with the speed of light?

14. N Since 1995, hundreds of extrasolar planets have been discovered. There is the exciting possibility that there is life on one or more of these planets. To support life similar to that on the Earth, the planet must have liquid water. For an Earth-like planet orbiting a star like the Sun, this requirement means that the planet must be within a habitable zone of 0.9 AU to 1.4 AU from the star. The semimajor axis of an extrasolar planet is inferred from its period. What range in periods corresponds to the habitable zone for an Earth-like planet orbiting a Sun-like star?

15. N When Sedna was discovered in 2003, it was the most distant object known to orbit the Sun. Currently, it is moving toward the inner solar system. Its period is 10,500 years. Its perihelion distance is 75 AU. **a.** What is its semimajor axis in astronomical units? **b.** What is its aphelion distance?

7-3 Newton's Law of Universal Gravity

16. C Which is greater, the gravitational force of the Sun on an asteroid or the gravitational force of an asteroid on the Sun? Clearly explain your reasoning without using any formulas. A clever student should be able to answer this question *without* citing Newton's law of universal gravity.

17. N The mass of the Earth is approximately 5.98×10^{24} kg, and the mass of the Moon is approximately 7.35×10^{22} kg. The Moon and the Earth are separated by about 3.84×10^8 m. **a.** What is the magnitude of the gravitational force that the Moon exerts on the Earth? **b.** If Serena is on the Moon and her mass is 25 kg, what is the magnitude of the gravitational force on Serena due to the Moon? The radius of the Moon is approximately 1.74×10^6 m.

Problems 18 and 27 are paired.

18. C Saturn's beautiful rings are composed of millions of particles ranging in size from boulders to tiny dust particles. Does a particle nearer Saturn orbit with the same period as one in the outer part of the rings? Why or why not? Does a particle with a greater mass than a second particle orbit with a different period if they are the same distance from Saturn? Why or why not?

19. N Maria, with mass $m = 48.0$ kg, is in the second row of seats, a distance of 3.00 m from Prof. Karen ($M = 55.0$ kg). What is the magnitude of the gravitational force between Maria and Prof. Karen?

20. A black hole is an object with mass, but no spatial extent. It truly is a particle. A black hole may form from a dead star. Such a black hole has a mass several times the mass of the Sun. Imagine a black hole whose mass is ten times the mass of the Sun.
    a. C Would you expect the period of an object orbiting the black hole with a semimajor axis of 1 AU to have a period greater than, less than, or equal to 1 yr? Explain your reasoning.
    b. N Use Equation 7.6 to calculate this period.

21. N Two spheres of mass $M_1 = 750$ kg and $M_2 = 350$ kg are placed 5.00 m apart. A particle of mass $m = 10.0$ kg is now placed midway between the two spheres. **a.** What is the net gravitational force on the particle due to the two spheres? **b.** At what position between the two spheres should the particle be placed so that the net gravitational force on the particle is zero?

22. E Estimate the magnitude of the gravitational force between the electron and proton in a hydrogen atom.

23. N The Lunar Reconnaissance Orbiter (LRO), with mass $m = 1850$ kg, maps the surface of the Moon from an orbital altitude of 50.0 km. What are the magnitude and direction of **a.** the force the LRO experiences due to the Moon's gravity and **b.** the force exerted by the LRO on the Moon?

Problems 24 and 25 are paired.

24. A Suppose a planet with mass m is orbiting star with mass M and the mean distance between the planet and the star is a. Using Newton's law of universal gravity, derive an algebraic expression for the speed of the planet when it is at the mean distance from the star.

25. N Suppose a planet with a mass of 2.44×10^{25} kg is orbiting a star with a mass of 3.65×10^{31} kg and the mean distance between the planet and the star is 1.12×10^{12} m. Using Newton's law of universal gravity, determine the speed of the planet when it is at the mean distance from the star.

Problems 26 and 41 are paired.

26. N Three billiard balls, the two-ball, the four-ball, and the eight-ball, are arranged on a pool table as shown in Figure P7.26. Given the coordinate system shown and that the mass of each ball is 0.150 kg, determine the gravitational force on the eight-ball due to the other two balls.

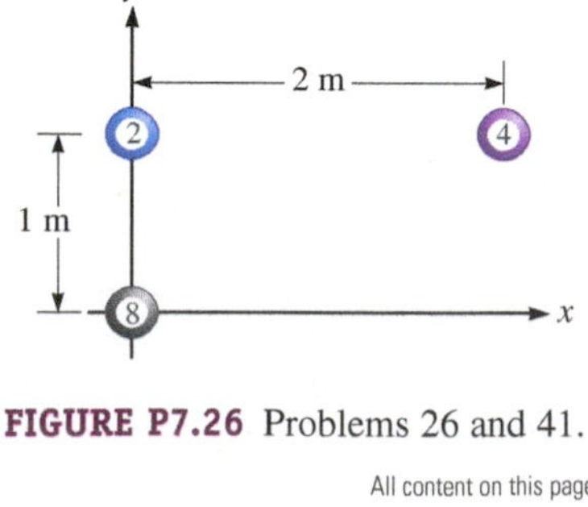

FIGURE P7.26 Problems 26 and 41.

27. **N** Saturn's ring system forms a relatively thin, circular disk in the equatorial plane of the planet. The inner radius of the ring system is approximately 92,000 km from the center of the planet, and the outer edge is about 137,000 km from the center of the planet. The mass of Saturn itself is 5.68×10^{26} kg. **a.** What is the period of a particle in the outer edge compared with the period of a particle in the inner edge? **b.** How long does it take a particle in the inner edge to move once around Saturn? **c.** While this inner-edge particle is completing one orbit abound Saturn, how far around Saturn does a particle on the outer edge move?

28. **N** Three spheres are arranged in the *xy* plane as in Figure P7.28. The first sphere, of mass $m_1 = 12.5$ kg, is located at the origin; the second sphere, of mass $m_2 = 4.50$ kg is located at (−6.00, 0.00) m; and the third sphere, of mass $m_3 = 8.00$ kg is located at (0.00, 5.00) m. Assuming an isolated system, what is the net gravitational force on the sphere located at the origin?

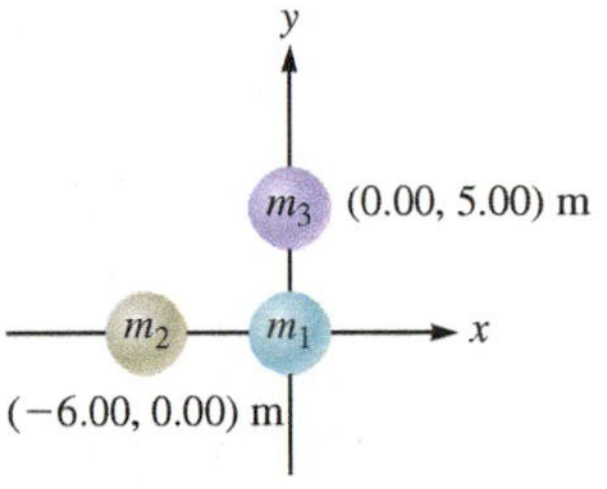

FIGURE P7.28

Problems 29 and 30 are paired.

29. **N** Find the magnitude of the Sun's gravitational force on the Earth when the Earth is at perihelion and at aphelion. The perihelion and aphelion distances are 1.48×10^{11} m and 1.52×10^{11} m, respectively.

30. Andromeda is a spiral galaxy much like the Milky Way. Andromeda's mass is about 2.8×10^{12} times the mass of the Sun and is 2.42×10^{22} m away. The mass of the Milky Way is 2.1×10^{12} times the mass of the Sun.
   a. **E** Estimate the magnitude of the gravitational force between the Milky Way and Andromeda.
   b. **N** Assuming no other forces act on the Milky Way, find its acceleration.

7-4 The Gravitational Field

31. **N** When first detected, near-Earth asteroid 2011 MD was at its closest approach of only 12,000 km above the Earth's surface. What was the asteroid's acceleration due to the Earth's gravity at this point in its trajectory?

32. **C** CASE STUDY To estimate the mass of the dark matter in our galaxy, we need to adopt Newton's hypothesis that there is no difference between celestial and terrestrial laws of physics. Explain why this idea is a necessary part of our dark-matter estimate and of astrophysical study in general.

33. **A** Three particles, each with mass *m*, are located at coordinates (0, *L*), (*L*, 0), and (*L*, *L*) as shown in Figure P7.33. What are the magnitude and direction of the gravitational field at the origin due to the three particles?

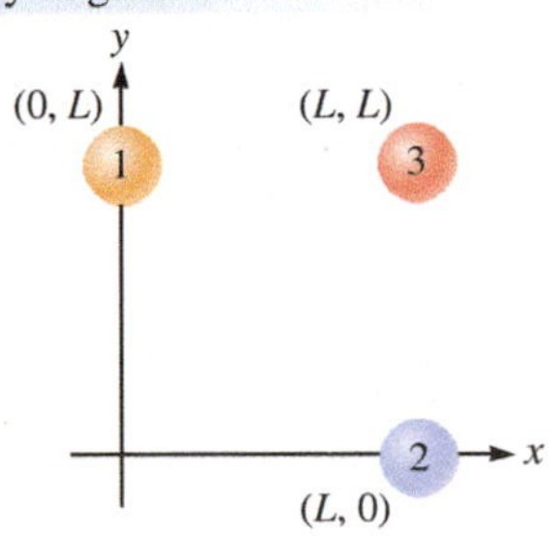

FIGURE P7.33

Problems 34, 35, and 36 are grouped.

34. **C** It is common for students to confuse the terms *force* and *field*. Let us find a way to keep them straight. **a.** What are the SI units of a gravitational force? Is it a vector or a scalar? **b.** What are the SI units of a gravitational field? Give at least two answers. Is it a vector or a scalar?

35. **C** You sit on your chair. You then get off and let your dog sit on your chair. **a.** Is the Earth's gravitational force on you equal to the Earth's gravitational force on your dog? **b.** What happens to the Earth's gravitational field when you are replaced by your dog on the chair? Does it increase, decrease, or stay the same. Explain.

36. **C** In your own words, describe the difference between the terms *gravitational force* and *gravitational field*.

Problems 37, 38, and 39 are grouped.

37. The Sun has a mass of approximately 1.99×10^{30} kg.
   a. **N** Given that the Earth is on average about 1.50×10^{11} m from the Sun, what is the magnitude of the Sun's gravitational field at this distance?
   b. **G** Sketch the magnitude of the gravitational field due to the Sun as a function of distance from the Sun. Indicate the Earth's position on your graph. Assume the radius of the Sun is 7.00×10^8 m and begin the graph there.
   c. **N** Given that the mass of the Earth is 5.97×10^{24} kg, what is the magnitude of the gravitational force on the Earth due to the Sun?

38. The Earth has a mass of approximately 5.97×10^{24} kg.
   a. **N** Given that the Earth is on average about 1.50×10^{11} m from the Sun, what is the magnitude of the Earth's gravitational field at the location that is occupied by the Sun?
   b. **G** Sketch the magnitude of the gravitational field due to the Earth as a function of distance from the Earth.
   c. **N** Given that the mass of the Sun is 1.99×10^{30} kg, what is the magnitude of the gravitational force on the Sun due to the Earth?

39. **C** There are gravitational fields associated with the Earth and the Sun. The magnitude of the gravitational field due to the Earth at the location of the Sun is significantly less than the magnitude of the gravitational field due to the Sun at the location of the Earth. This seems to violate Newton's third law. Explain how Newton's third law is verified when comparing the gravitational force each body exerts on the other.

40. **C** CASE STUDY **MOND** According to Example 7.8 (page 201), 68% of our own galaxy is made up of extraordinary particles—dark matter—that have never been observed on the Earth. In fact, there is no laboratory evidence for the existence of dark matter. So, some scientists say that there is little or no dark matter; instead, they argue that Newton's law must be modified to take these observations into account. This theory is called MOND, for *m*odified *N*ewtonian *d*ynamics. **a.** If the estimated amount of dark matter were much smaller, say a few percent instead of 68% (or more), do you believe that the MOND picture of the Universe would generally be supported? **b.** In what way are MOND proponents like Kepler? Does your answer depend on whether MOND or dark matter is supported by future experiments?

41. **N** Three billiard balls, the two-ball, the four-ball, and the eight-ball, are arranged on a pool table as shown in Figure P7.26. Given the coordinate system shown and that the mass of each ball is 0.150 kg, determine the gravitational field at $x = 2$ m, $y = 0$.

42. **C** Example 7.7 discusses the Lagrange point, L1, located between the Earth and the Sun. Suppose we placed an object in orbit *around* this point. Do you suppose we could obtain a *circular* orbit? Why or why not? *Hint*: At the Earth's distance from the Sun, the Sun's gravitational pull is just sufficient to pull it into an almost circular orbit. If you launch an object into a nearby orbit around the Earth, the object feels a gravitational field due almost entirely to the Earth. What would be the comparable situation at L1?

43. **C** Example 7.6 discusses *geosynchronous* satellites, which remain above a fixed point on the Earth's surface. Suppose you want to put a satellite in orbit so that it always remains above New York City. Is it possible to do so? Why or why not?

44. **C** In the 2009 motion picture *Star Trek*, the antagonist uses a device to convert a planet into a *black hole*, an object with mass

but no spatial extent. Assume the only fictional aspect of this event is the existence of a device capable of turning a planet's mass into a black hole; otherwise, the laws of physics described in this chapter hold. What happens to the orbit of that planet around its star? Carefully explain your reasoning.

45. Figure P7.45 shows a picture of American astronaut Clay Anderson experiencing weightlessness on board the International Space Station.

a. **N** Most people have the misconception that a person in a spacecraft is weightless because he or she is no longer affected by gravity. Show that this premise cannot be true by computing the gravitational field of the Earth at an altitude of 200 km, the typical altitude of a spacecraft in orbit. Compare this result with the gravitational field on the surface of the Earth.

b. **C** Why would astronauts in orbit experience weightlessness even if they are experiencing a gravitational field (and therefore a gravitational force)?

NASA

FIGURE P7.45

7-5 Variations in the Earth's Gravitational Field

46. **G** A simple plumb line may be constructed by hanging a small, dense object such as a lead ball from a lightweight string. It is used to determine a vertical line by holding the open end of the string such that the lead ball hovers just above the ground. Imagine observing three plumb lines from an inertial reference frame. For each location listed, make a sketch and determine if the plumb line is pointing to the center of the Earth. If it is not, indicate on your sketch where it is pointing. Explain your answers. **a.** North Pole **b.** Equator **c.** 45° north latitude

47. The Moon rotates on its axis as it orbits the Earth. The same side of the Moon always faces the Earth.

a. **G** Sketch a motion diagram for the Moon. Include the Earth on your sketch.

b. **N** Find the Moon's rotational period.

c. **N** Compare (by finding $w_{\text{pole}} - w_{\text{equator}}$) the apparent weight of a 1-kg object on the Moon's equator with its apparent weight at one of the Moon's poles. Assume the same elevation.

d. **N** How does your answer compare with what you would expect to find for a 1-kg object's apparent weight measured on the Earth's equator and at one of its poles? Explain your results.

48. **E** Suppose you dropped a small object such as a golf ball from a very tall building such as the Willis Tower in Chicago. Ignoring air drag, estimate the direction and horizontal position of where the ball lands with respect to the point just below where it was dropped.

Problems 49 and 50 are paired.

49. **N** In 1851, Leon Foucault built a small pendulum in his basement that confirmed the theory that the Earth rotates. Today, you can see his much larger pendulum at the Pantheon in Paris, France (Fig. P7.49). To an inertial observer, the pendulum appears to swing back and forth along a single plane. To an observer on the Earth, this plane appears to rotate. **a.** Imagine that such a pendulum is at the North Pole. Through how many degrees would the plane appear to rotate in a single day? **b.** Now imagine that such a pendulum is on the equator. Through how many degrees would the plane appear to rotate in a single day?

Marco Cannizzaro/Shutterstock.com

FIGURE P7.49 Problems 49 and 50. Close-up of the bob of the Foucault pendulum.

50. **N** Consider the pendulum shown in Figure P7.49. To an inertial observer, the pendulum appears to swing back and forth along a single plane. To an observer on the Earth, this plane appears to rotate. The apparent rotation angle θ of the plane in a single day depends on the pendulum's latitude φ: $\theta = 360° \sin \varphi$. The latitude of Paris is 48° 48'. **a.** Confirm that this equation holds at the poles and at the equator. **b.** Find θ for the pendulum in Paris.

General Problems

51. **N** The International Space Station (ISS) experiences an acceleration due to the Earth's gravity of 8.83 m/s^2. What is the orbital period of the ISS?

Problems 52, 55, 56, 57, and 58 are grouped.

52. If you were to calculate the pull of the Sun on the Earth and the pull of the Moon on the Earth, you would undoubtedly find that the Sun's pull is much stronger than that of the Moon, yet the Moon's pull is the primary cause of tides on the Earth. Tides exist because of the *difference* in the gravitational pull of a body (Sun or Moon) on opposite sides of the Earth. Even though the Sun's pull is stronger, the difference between the pull on the near and far sides is greater for the Moon.

a. **A** Let $F(r)$ be the gravitational force exerted on one mass by a second mass a distance r away. Calculate $dF(r)/dr$ to show how F changes as r is changed.

b. **N** Evaluate this expression for $dF(r)/dr$ for the force of the Sun at the Earth's center and for the Moon at the Earth's center.

c. **N** Suppose the Earth–Moon distance remains the same, but the Earth is moved closer to the Sun. Is there any point where $dF(r)/dr$ for the two forces has the same value?

53. **N** Two black holes (the remains of exploded stars), separated by a distance of 10.0 AU (1 AU = 1.50×10^{11} m), attract one another with a gravitational force of 8.90×10^{25} N. The combined mass of the two black holes is 4.00×10^{30} kg. What is the mass of each black hole?

54. **C** Quito is a city in Ecuador located at the equator but in the very high Andes Mountains. How do you expect the apparent weight of a turtle measured in the Galapagos Islands to differ from that measured in Quito?

55. Tides on the Earth are caused by the difference in the Moon's gravitational field at the far side and at the near side of the Earth. The side of the Earth facing the Moon experiences a greater gravitational force than the side facing away from the Moon. The distance from the Earth's center to the Moon's center is 3.84×10^8 m, and the radius of the Earth is 6.37×10^6 m. The mass of the Moon is 7.35×10^{22} kg.

a. **N** Calculate the Moon's gravitational field at the side of the Earth facing the Moon.

b. **N** Calculate the Moon's gravitational field at the side of the Earth facing away from the Moon.

c. **N** Calculate the Moon's gravitational field at the center of the Earth.

d. **G** Sketch the Earth and include the three vectors from parts (a) through (c).

e. **C** Qualitatively explain why there are two tides a day on most places on the Earth due to the Moon. (Because of friction, the high tides do not exactly line up with the Moon.)

56. Consider the Earth and the Moon as a two-particle system.

a. **A** Find an expression for the gravitational field $\vec{g}$ of this two-particle system as a function of the distance r from the center of the Earth. (Do not worry about points inside either the Earth or the Moon.)

b. **G** Plot the scalar component of $\vec{g}$ as a function of distance from the center of the Earth.

57. **N** Consider the Earth and the Moon as a two-particle system. **a.** Find the gravitational field of this two-particle system at the point that is exactly halfway between the Earth and the Moon. **b.** An asteroid of mass 6.69×10^{15} kg is at the point exactly halfway between the Earth and the Moon. What is the magnitude of the gravitational force on it?

58. **N** Consider the Earth and the Moon as a two-particle system. **a.** How far from the center of the Earth is the gravitational field of this two-particle system zero? **b.** Sketch gravitational field vectors $\vec{g}$ along the line joining the Earth and the Moon. Indicate the point at which $\vec{g} = 0$. (Do not consider positions inside either object.)

59. **N** Many science-fiction spacecraft are shown with cylindrical modules that rotate to provide the crew with artificial gravity—a centripetal acceleration that is comparable to the gravitational acceleration on the Earth, 9.81 m/s^2. If one such module is 500 m in diameter, what is the speed of rotation that would provide an artificial gravity equal to that on the Earth's surface? (Report your answer to 2 significant figures in revolutions per minute and radians per second.)

60. You are a planetary scientist studying the atmosphere of Jupiter through a large telescope when you observe an asteroid approaching the planet. This asteroid is large, so you know it is held together by gravity rather than the cohesive forces that hold a large rock together. If the asteroid gets too close to Jupiter, the massive tidal forces will tear it apart, scattering small particles that will add to the ring system. You have calculated the closest distance the asteroid will come to Jupiter. How do you know if the asteroid will survive?

a. **N** A measure of the cohesive gravitational force holding such an asteroid together is the gravitational field on the surface due to the mass of the asteroid. This field is independent of the distance of the asteroid from Jupiter. Calculate the gravitational field at the surface of the asteroid due only to the mass of the asteroid. Assume the asteroid has a diameter of 10,000 km and a density of 1300 kg/m^3.

b. **G** Tidal forces from Jupiter tend to disrupt the asteroid by pulling it apart. The tidal forces depend on the distance between Jupiter and the asteroid. There is a distance between Jupiter and the asteroid known as the *Roche limit* where the tidal forces are balanced by the asteroid's own cohesive gravitational force. If the asteroid is within the Roche limit, it will be torn apart. Figure P7.60 shows Jupiter's gravitational field as a function of distance from its center. By looking at this graph, can you determine an approximate value for the Roche limit for this asteroid in the vicinity of this planet?

c. **C** What will happen to the Roche limit if we consider an asteroid of lower density?

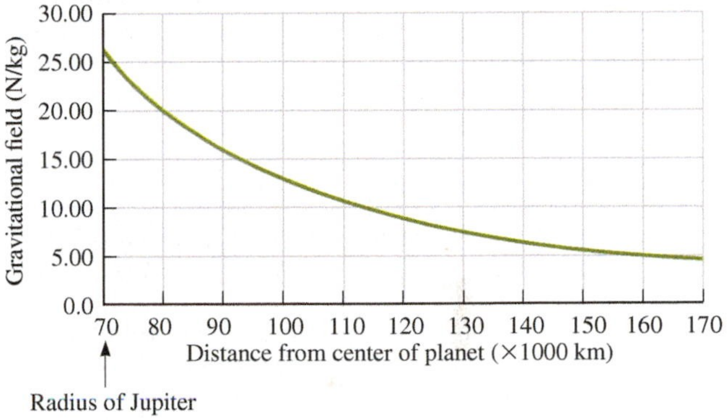

FIGURE P7.60

61 **C** Is it possible to put a geosynchronous satellite above London by placing it in an orbit parallel to the Earth's equator and moving eastward (the same direction as the Earth's rotation)? Why or why not?

62. **CASE STUDY** According to MOND, Newton's second law ($\vec{F}_{tot} = m\vec{a}$) must be modified when we are dealing with very slowly accelerating objects such as stars on the outer edge of the Milky Way. In MOND, if the acceleration of an object is very small (less than about 10^{-10} m/s^2), the net force applied to the object is proportional not to the acceleration as in Newton's second law, but rather to the square of the acceleration: $F_{tot} \propto a^2$.

a. **C** Why has the MOND theory not been tested in a laboratory on the Earth? Explain.

b. **N** Can the theory be tested in a space station? Explain.

63. Planetary orbits are often approximated as uniform circular motion. Figure P7.9 is a scaled representation of a planet's orbit with a semimajor axis of 1.524 AU.

a. **G** Use Figure P7.9 to find the ratio of the Sun's maximum gravitational field to its minimum gravitational field on the planet's orbit.

b. **N** What is the ratio of the planet's maximum speed to its minimum speed?

c. **C** Comment on the validity of approximating this orbit as uniform circular motion.

64. Kepler's second law of planetary motion can be used to relate the ratio of perihelion to aphelion speeds of a planet to the ratio of its perihelion to aphelion distances.

a. **G** Start by sketching the area swept out by the line joining the planet to the Sun during a short period (say 1 day) when the Sun is near aphelion and another area for when the Sun is near perihelion. Based on your sketch, argue that it would be safe to approximate the shapes of those areas as triangles.

b. **A** Write the expression for the "triangular" area swept out by the line connecting the planet to the Sun at aphelion in terms of r_A, v_A, and time t. Write a similar expression for the "triangular" area swept out at perihelion in terms of r_P, v_P, and t. *Note*: The speeds of the planet at aphelion and perihelion are v_A, and v_P, respectively.

c. **A** Show that the ratio of aphelion to perihelion distances, r_A/r_P, is expressible in terms of a ratio of v_A and v_P.

d. **N** The Earth varies in speed during its orbit. The Earth's lowest orbital speed is 29.29 km/s. What is the Earth's highest orbital speed?

65. Newton's version of Kepler's third law offers one of the few ways of determining the mass of objects in space. One form of this relationship is expressed in Equation 7.6. To derive it, however, we assumed the orbit of the less massive object is circular and is centered on the more massive object, which is not completely realistic. In reality, both objects move in an orbit in response to the gravitational force between them! A careful derivation shows that for two masses M_1 and M_2 in orbit around one another with semimajor axis a, the proper expression of Newton's version of Kepler's third law is

$$T^2 = \left[\frac{4\pi^2}{G(M_1 + M_2)}\right]a^3$$

Equation 7.6 differs from this "proper" expression in that we've replaced the mass of a single object with the sum of the masses of both objects.

a. **N** Consider Jupiter, the second most massive object in the solar system. Compute the orbital period of Jupiter around the Sun using this "proper" expression of Newton's version of Kepler's third law and compare the result with the answer you find using Equation 7.6. Do they differ significantly given $M_\odot = 1.98892 \times 10^{30}$ kg, $M_J = 1.8987 \times 10^{27}$ kg, and $a = 7.7857 \times 10^{11}$ m?

b. **C** Would you expect Equation 7.6 to hold for the Moon orbiting the Earth? Clearly explain your reasoning.

c. **C** Would you expect Equation 7.6 to be accurate for two asteroids orbiting each other? Clearly explain your reasoning.

66. **E** Some people insist on believing that the positions of the planets at the time of your birth can have an influence on your life, a subject sometimes referred to as *astrology*. Some astrologers even insist that there is a physical basis in astrology because the gravitational pull of the various planets affects a newborn's body. Estimate the magnitude of the gravitational force on a newborn from the planet Jupiter and compare it with the magnitude of the gravitational force due to the pediatrician who happens to be present during the birth. Which is more likely to have a gravitational "influence" on a newborn child? Is this method a valid scientific test of the gravitational basis of astrology?

67. **A** One intriguing bit of trivia is that for almost any planet made of solid matter, the orbital period of a satellite in a low orbit is about the same, on the order of 90 minutes! Here, a low orbit is one that has a semimajor axis on the order of the planet's radius in size. (Obviously, the orbit must be slightly bigger than the radius of the planet; otherwise, the satellite would hit the ground.) Show that for any planet with a density comparable to the density of the Earth, the orbital period for a low orbit is the same by solving for this period in terms of the planet's density. For simplicity, assume these planets are spherical and of uniform density.

68. **E** In his book *Death from the Skies!*, astronomer Phil Plait describes what would happen if a very massive, very compact object (a *black hole*) were to get near enough to the Earth to destroy it. Estimate at what distance a 5-solar-mass black hole would literally start ripping the surface of the Earth apart. *Note*: Plait estimates the likelihood of this sort of interaction to be on the order of one in one trillion in your lifetime, so it's not really worth losing much sleep over.

69. In his book *A Brief History of Time*, physicist Stephen Hawking popularized the term *Spahettification* to describe what happens to someone who falls feet first into a small but highly massive object. In essence, the gravitational field at the person's feet is sufficiently higher than the gravitational field at the head, and the person gets stretched out like a spaghetti noodle (a phenomenon that would likely be fatal). Imagine that you are 1000 km from an object with a mass of 1 solar mass and you are approximately 1.5 m tall. You want to figure out if you are in danger of being "spaghettified."

a. **A** As a first step, we need to come up with an expression that will allow us to compute the change in gravitational field, $\Delta\vec{g}(d)$, over a small distance ℓ such as a person's height at a distance d from the source of the gravitational field. Show that when $d \gg \ell$, you can write this change as

$$\Delta\vec{g}(d) = \vec{g}\left(d + \frac{\ell}{2}\right) - \vec{g}\left(d - \frac{\ell}{2}\right) \approx -GM\left(\frac{2\ell}{d^3}\right)\hat{r}$$

b. **N** Calculate the difference between the gravitational field of the black hole at your feet and your head if you are falling feet first into the black hole.

c. **C** Is this difference in gravitational field large enough to "spaghettify" you? Clearly explain your reasoning.

70. **N** CASE STUDY If the only force exerted on a star far from the center of the Galaxy ($r = 7.7 \times 10^{20}$ m) is the gravitational force exerted by the ordinary matter ($M_{ord} = 1.8 \times 10^{41}$ kg), find the speed of the star. Assume a circular orbit and assume all the Galaxy's matter is concentrated at the center.

Conservation of Energy

8

Underlying Principle

Conservation of mechanical energy

Major Concepts

1. Energy
2. Kinetic energy
3. Potential energy
4. Isolated system

Special Cases

1. Potential energy
   a. Gravitational potential energy near Earth's surface of the Earth
   b. Universal gravitational potential energy
   c. Elastic potential energy
2. Orbital energies
   a. Circular orbit
   b. Elliptical orbit

Tools

1. Bar chart
2. Energy graph

Key Questions

What is energy?

How does the principle of conservation of mechanical energy simplify complicated situations?

It seems as if nothing stays the same. You go home on a break from college only to find that your room is now a home gym, your favorite science teacher has retired, and your younger brother is taller than you. Change is not unique to the human condition. Our solar system formed out of a cloud of gas, radioactive elements below the Earth's surface decay, and the Sun will eventually die.

So, with everything apparently changing all the time, it is remarkable that philosopher René Descartes (1596–1650) believed that some quantities in the Universe remain constant. He took a very different approach to physics than Newton. Whereas Newton searched to explain changes in motion, Descartes said that the *motion* of the Universe as a whole is conserved, or constant. Today, we describe Descartes's original idea more precisely and say that the *energy* and *momentum* of the Universe are conserved. The notion that some quantities remain constant throughout a process—like the formation of the solar system—is a new approach for us.

8-1 Another Approach to Newtonian Mechanics

In Chapter 5, we introduced the underlying principle of dynamics in the form of Newton's laws of motion. It might seem that all we need to do is continue learning about other specific forces, and then we could study the physics of any phenomenon. In principle, this idea is true because, according to Newton's second law, if we know all the forces acting on a particle, we can find its acceleration. Once we know its acceleration, kinematics (the mathematical description of motion) tells us how to find the particle's velocity and position.

There are, however, many practical situations in which applying Newton's second law does not easily lead to a description of the motion. Fortunately, there is an alternative approach that does not replace Newton's laws (which still govern mechanics); instead, it complements Newton's laws. Some problems are more easily solved using this alternative approach (the conservation approach). The following case study is such a problem.

CASE STUDY Comet Halley

Comet Halley has a period of 76 years. Its first recorded sighting dates from 240 BCE. All its subsequent 30 trips through the inner solar system have been observed from the Earth. *Giotto*, a European Space Agency spacecraft, made history by taking the first close-up pictures of a comet nucleus in 1986.

Since then, other missions have been flown to observe comets as they pass near the Earth. Such missions take careful planning. Because the spacecraft needs to get close to the comet without crashing into it, mission planners must know the comet's orbit precisely (Fig. 8.1A). In this case study, we focus on just one aspect of the Comet Halley's orbit, its speed when *Giotto* flew by it.

Comet Halley had already passed perihelion (closest point to the Sun) and was on its way back out of the inner solar system when *Giotto* made its rendezvous. The spacecraft encountered the comet on March 13, 1986, at a distance $r_G = 0.89$ AU from the Sun and 0.98 AU from the Earth. Our goal is to find Comet Halley's speed when *Giotto* made its rendezvous.

Figure 8.1B illustrates the difficulty in using the force approach. At all positions, the gravitational force exerted on the comet is directed toward the Sun, but the magnitude of that force depends on distance. Because acceleration is proportional to this force, Comet Halley's acceleration depends on position. Also, the relative angle between the acceleration and the velocity depends on position. Using the force approach to find the speed of an object is very complicated when the force, acceleration, and velocity are all changing in both magnitude and direction. The conservation approach, however, greatly simplifies this problem and readily yields the desired quantity.

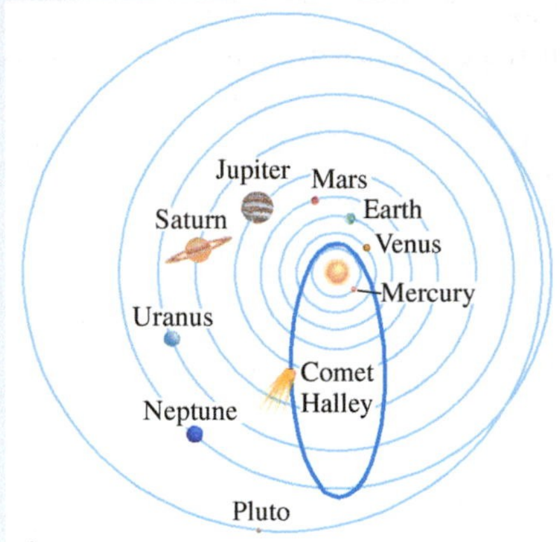

A.

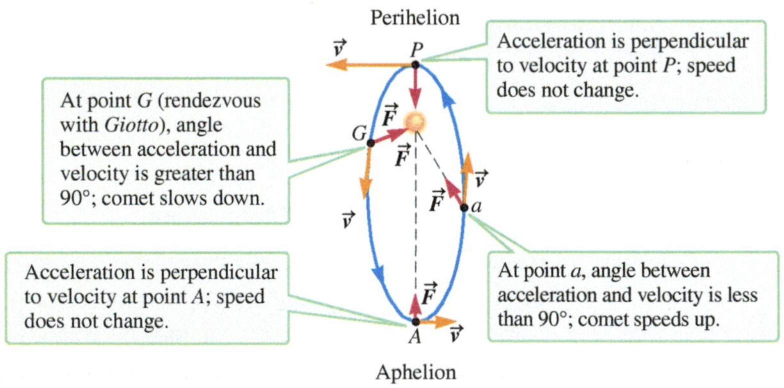

FIGURE 8.1 **A.** Comet Halley's orbit. (Orbits not drawn to scale.) **B.** Gravitational force on the comet and orbital velocity vectors at various points along its orbit.

B.

CONCEPT EXERCISE 8.1

CASE STUDY Comet Halley's Orbital Parameters

Figure 8.1 shows Comet Halley's elliptical orbit.

a. Use the information given in the case study above and Kepler's third law from Chapter 7 to find the semimajor axis a of Comet Halley's orbit.

b. The comet's perihelion distance is $r_P = 0.59$ AU. Find its aphelion distance (farthest point from the Sun).

8-2 Energy

Our first approach to studying mechanics focused on understanding forces and applying Newton's laws of motion (the *force approach*). Our second approach to mechanics is the *conservation approach*, the basic idea of which is that certain quantities remain unchanged during a particular process or activity. In this chapter, we focus on the *conservation of mechanical energy*, so we must start by defining energy.

Newton's first law provides a working definition of force. There is no such law to define energy. *Energy*, like so many terms in physics, also has many everyday usages, some of which can help us to develop our intuition. For example, we might say, "My physics professor has a lot of energy." In this context, *energy* describes the teacher's style: High energy is associated with a lively style, and a low-energy professor has a dull style. This usage fits well into the physics meaning of the word: In physics, the term **energy** describes the state of a particle, object, or system. So, energy is *not* a material substance; instead, it is a scalar quantity that describes the state—in particular, the motion and the configuration—of a system.

ENERGY ✪ Major Concept

For brevity, we use the term *system* to stand for a particle, an object, or a collection of particles and objects.

Kinetic Energy

Kinetic energy K describes the motion of a system. The kinetic energy of a single particle depends on its mass m and speed v:

$$K = \tfrac{1}{2}mv^2 \qquad (8.1)$$

KINETIC ENERGY ✪ Major Concept

To find the kinetic energy of a system that consists of more than one particle, we add up the kinetic energy of each one:

$$K = \frac{1}{2}m_1v_1^2 + \frac{1}{2}m_2v_2^2 + \frac{1}{2}m_3v_3^2 \ldots + \frac{1}{2}m_nv_n^2 = \sum_{i=1}^{n}\frac{1}{2}m_iv_i^2 \qquad (8.2)$$

The SI unit for energy is named for British physicist James Joule. The joule is abbreviated as J and is given by

$$1\,\text{J} = 1\,\text{kg}\cdot\text{m}^2/\text{s}^2 = 1\,\text{N}\cdot\text{m}$$

Because the speed of a particle depends on the reference frame in which speed is measured, the particle's kinetic energy also depends on the choice of reference frame. For example, if a spider sits on the dashboard of a moving car, the spider's kinetic energy is zero according to the driver of the car, but not according to an observer parked on the side of the road.

CONCEPT EXERCISE 8.2

A ball is tossed straight up. What is its kinetic energy at the top of its flight?

CONCEPT EXERCISE 8.3

Can scalar quantities be negative? Can kinetic energy be negative? Explain your answers.

EXAMPLE 8.1 **CASE STUDY** **Comet Halley's Kinetic Energy**

Suppose a comet has the same period (76 years) and mass (1.7×10^{15} kg) as Comet Halley, but it moves in uniform circular motion around the Sun. Find its kinetic energy.

INTERPRET and ANTICIPATE

Knowing the period of the comet, we use Kepler's third law to find the comet's speed. We can then find a numerical value for the comet's kinetic energy. Kinetic energy must always be positive (Concept Exercise 8.3). Always check to see that any kinetic energy you have calculated is positive.

Example continues on page 216 ▶

SOLVE

From Kepler's third law, if the period of the comet in the circular orbit has the same period as that of Comet Halley, its orbital radius must equal the semimajor axis of Comet Halley's orbit.

$$r = a$$

Find the comet's speed from Equation 4.30, with $r = a$.

$$v = \frac{2\pi r}{T} = \frac{2\pi a}{T} \qquad (4.30)$$

Substitute this result into $K = \frac{1}{2}mv^2$ (Eq. 8.1).

$$K = \frac{1}{2}mv^2 = \frac{1}{2}m\left(\frac{2\pi a}{T}\right)^2 = \frac{2\pi^2 ma^2}{T^2}$$

We used Kepler's third law to find a in Concept Exercise 8.1.

$$a = 18 \text{ AU} = 2.7 \times 10^{12} \text{ m}$$

Convert 76 years to seconds and substitute.

$$K = \frac{2\pi^2(1.7 \times 10^{15}\text{ kg})(2.7 \times 10^{12}\text{ m})^2}{(2.4 \times 10^9\text{ s})^2}$$

$$K = 4.2 \times 10^{22}\text{ J}$$

CHECK and THINK

Our kinetic energy result is positive and in the right units. More than 10^{22} J seems like a lot of energy. To check the magnitude of our result, let us calculate the Earth's kinetic energy. The orbital speed of the Earth is about 30 km/s, and its mass is 6×10^{24} kg, so the Earth's kinetic energy is $K_{\oplus} \approx \frac{1}{2}(6 \times 10^{24}\text{ kg})(30 \times 10^3\text{ m/s})^2 \approx 3 \times 10^{33}$ J. The comet is much less massive than the Earth, so it is reasonable that it has much less kinetic energy.

EXAMPLE 8.2 Tossing a Ball Revisited

A ball is tossed upward, leaving the person's hand at speed v_0. It returns to the same height and is caught moving at speed v_0. The ball has mass m. Sketch the ball's kinetic energy as a function of time.

INTERPRET and ANTICIPATE

The ball is in free fall, so we can easily find an expression for its speed as a function of time. Once we know that expression, we can find kinetic energy.

SOLVE

The ball's speed v is given by Equation 2.9. We take upward to be positive. The acceleration is then $a = -g$.

$$v = v_0 + at \qquad (2.9)$$

$$v = v_0 - gt \qquad (1)$$

To find an expression for kinetic energy as a function of time, we substitute Equation (1) into Equation 8.1.

$$K = \tfrac{1}{2}mv^2 = \tfrac{1}{2}m(v_0 - gt)^2 \qquad (2)$$

$$K(t) = \tfrac{1}{2}m(v_0^2 - 2v_0gt + g^2t^2) \qquad (3)$$

Our job now is to sketch Equation (3). There are three *positive* constants: m, v_0, and g. The t^2 term indicates that we have a parabola. We know that at the top of the motion, the velocity is zero, so by Equation (1) $v_0 = gt_{top}$; this is the maximum speed of the ball. From Equation (2), the ball's kinetic energy is at its maximum as it leaves the hand at $t = 0$ and again when it returns at $t = 2t_{top}$ (Fig. 8.2).

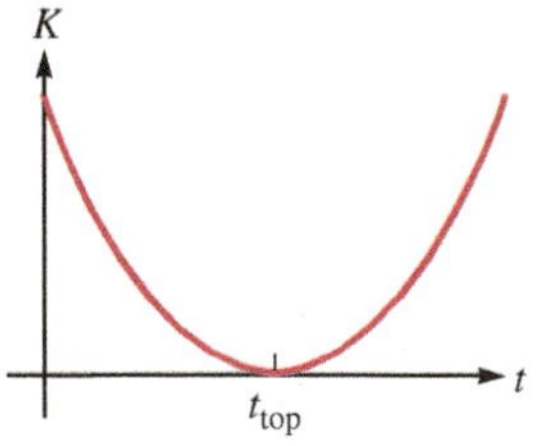

FIGURE 8.2 Kinetic energy versus time

CHECK and THINK

As expected, the kinetic energy is never negative. It is zero when the ball momentarily stops at the top (Fig. 8.2). Contrast this figure to the velocity-versus-time graph, which is positive until the ball reaches the top and is negative when it falls back down (Fig. 8.3).

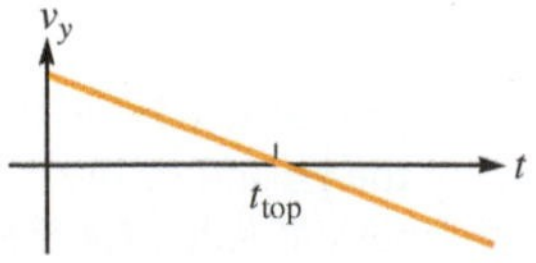

FIGURE 8.3 Velocity versus time

EXAMPLE 8.3 Crall and Whipple's Carts Collide

Crall and Whipple (CASE STUDY Chapter 2) place two carts on a track and set them on a collision course (Fig. 8.4). Cart A has a mass of 0.831 kg, and cart B has a mass of 1.336 kg. Just before they collide, their velocities are $\vec{v}_A = 1.23\hat{\imath}$ m/s and $\vec{v}_B = -0.91\hat{\imath}$ m/s (measured in a frame that is at rest with respect to the laboratory).

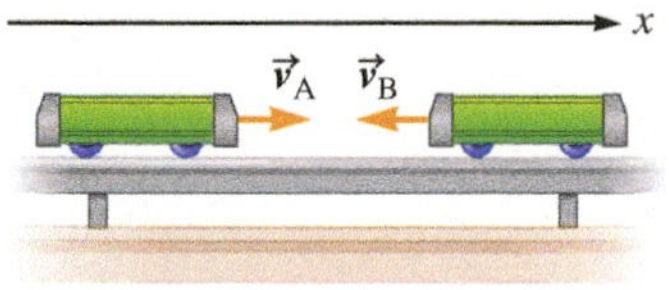

FIGURE 8.4 Two carts are set on a collision course.

A Crall chooses a system that consists of both carts. According to Crall, what is the kinetic energy of the system?

INTERPRET and ANTICIPATE

Because the system consists of two particles, add the kinetic energy of each one to find the kinetic energy of the system.

SOLVE

According to Crall, each cart has nonzero kinetic energy. We add the kinetic energy of each cart.

$$K = \tfrac{1}{2}m_A v_A^2 + \tfrac{1}{2}m_B v_B^2$$

$$K = \tfrac{1}{2}(0.831 \text{ kg})(1.23 \text{ m/s})^2 + \tfrac{1}{2}(1.336 \text{ kg})(-0.91 \text{ m/s})^2$$

$$K = 1.18 \text{ J} \quad \text{in Crall's (lab's) frame}$$

CHECK and THINK

Because kinetic energy only depends on speed, the direction in which the carts are moving does not matter. The kinetic energy of the system according to Crall would be the same whether the carts were moving toward each other, away from each other, or in the same direction.

B Whipple decides to calculate the kinetic energy of cart B as measured by an imaginary observer moving at the same velocity as cart A. What is the kinetic energy measured by this observer?

INTERPRET and ANTICIPATE

This part is similar to part A in that the system has two particles. What is different is the speed of carts in Whipple's chosen frame. In this frame, cart A is at rest, and only cart B is in motion. So, only cart B has nonzero kinetic energy.

SOLVE

We must find the velocity of cart B (the subject) in Whipple's *m*oving reference frame by using Equation 4.46. The relative velocity between the two frames $(v_M)_L$ is the velocity of cart A as observed in the *l*aboratory (Crall's) frame.

$$(\vec{v}_B)_L = (\vec{v}_B)_M + (\vec{v}_M)_L \tag{4.46}$$

$$(\vec{v}_B)_M = (\vec{v}_B)_L - (\vec{v}_M)_L$$

$$(\vec{v}_B)_M = -0.91\hat{\imath} \text{ m/s} - 1.23\hat{\imath} \text{ m/s} = -2.14\hat{\imath} \text{ m/s}$$

Use $K = \frac{1}{2}mv^2$ (Eq. 8.1) to find the kinetic energy, substituting for v the speed seen by the moving observer.

$$K = \tfrac{1}{2}m_B[(v_B)_M]^2 \tag{8.1}$$

$$K = \tfrac{1}{2}(1.336 \text{ kg})(-2.14 \text{ m/s})^2 = 3.06 \text{ J}$$

CHECK and THINK

The kinetic energy measured in the moving frame (Whipple's choice) does not equal the kinetic energy measured in the laboratory frame (Crall's choice). The lesson here is that once you choose a particular reference frame, you must stick with it throughout all relevant calculations. (It is particularly important when working with potential energy, our next topic.)

Potential Energy

POTENTIAL ENERGY
Major Concept

Kinetic energy describes the system's motion, but even a system that has no motion may still have energy associated with the arrangement of the particles in it. This energy, known as **potential energy,** depends on the configuration of a system. Potential energy is only used to describe a system consisting of two or more particles that interact with each other via one or more *internal* forces. That is, forces acting on subjects in the system come only from sources within that system.

Not all types of internal forces can be associated with potential energy. Potential energy can only be associated with internal **conservative** forces. We will define *conservative* and *nonconservative* forces completely (and learn about a third classification) in Chapter 9. Of the forces we have studied so far, only gravity and the spring force (Hooke's law) are conservative.

Potential energy cannot be used to describe a single particle because a single (isolated) particle cannot experience internal forces. To associate potential energy with a system, we must define the system so that it includes both the source(s) and subject(s) of a conservative force or forces.

Consider a two-particle system such as the Earth and a small spacecraft (Fig. 8.5). A coordinate system has been chosen with its origin is at the center of the Earth. Initially, at time t_i, the spacecraft is at $\vec{r}_i = x_i\hat{\imath}$ (Fig. 8.5A). At a later time, $t = t_f$, the spacecraft is at a new position, $\vec{r}_f = x_f\hat{\imath}$, and the two particles are farther apart in this new configuration (Fig. 8.5B). Because the system's configuration has changed, there is a change in potential energy U. The change in potential energy is defined by

$$\Delta U = U_f - U_i \equiv -\int_{x_i}^{x_f} F_x dx \quad (8.3)$$

where F_x is the x component of the force exerted by the Earth on the spacecraft. In this case, that force is gravity, directed along the x axis. The spacecraft's displacement is also along the x axis. Equation 8.3 may be rewritten for displacement along any axis (x, y, z, r) as long as the component of the force along that axis is used.

We find the units of potential energy from Equation 8.3:

$$1\ \text{N}\cdot\text{m} = 1\ \text{kg}\cdot\text{m/s}^2\ \text{m} = 1\ \text{kg}\cdot\text{m}^2/\text{s}^2$$
$$1\ \text{N}\cdot\text{m} = 1\ \text{J}$$

So, the SI unit of energy—both potential and kinetic—is the joule. In the next three sections, we will use Equation 8.3 to derive expressions for the potential energy associated with gravity and the spring force.

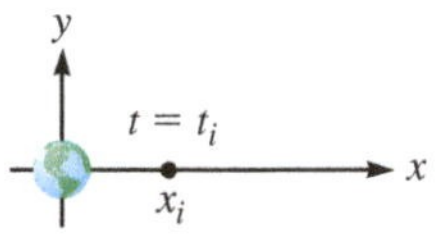

A.

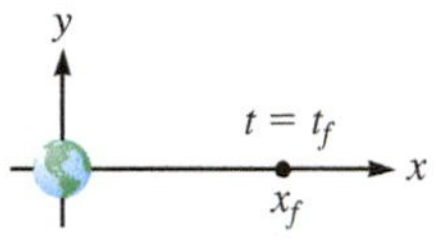

B.

FIGURE 8.5 A two-particle system such as the Earth and a spacecraft (represented by a black dot). **A.** The spacecraft starts at an initial position x_i. **B.** The spacecraft has moved to a final position x_f.

GRAVITATIONAL POTENTIAL ENERGY NEAR EARTH'S SURFACE

Special Case

8-3 Gravitational Potential Energy Near the Earth

A very common system you will encounter consists of the Earth and an object near the Earth. In this section, we'll use Equation 8.3 to find an easily-remembered expression for the change in the system's gravitational potential energy.

DERIVATION Potential Energy Near the Surface of a Planet

We show here that the change in a system's potential energy near the surface of a planet such as the Earth is given by

$$\Delta U = mgy_f - mgy_i \quad (8.4)$$

To calculate the gravitational potential energy of a system involving the Earth, imagine a two-particle system in which one particle is the Earth (Fig. 8.6). For the sake of argument, imagine that the other particle is an apple of mass m. We have chosen to place the origin of the coordinate system on the surface of the Earth. The apple (represented as a dot) moves from y_i to y_f. (The distance shown in Figure 8.6 is exaggerated.)

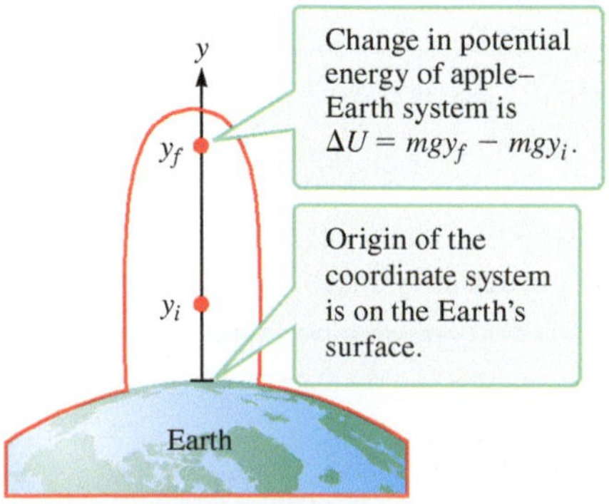

FIGURE 8.6 System = Earth and apple

The Earth's gravitational force on an object near its surface does not depend on distance. So, as the apple is raised from its initial to its final position, the gravitational force on it does not change.	$\vec{F}_g = -mg\hat{j}$
Modify Equation 8.3 to find the change in the gravitational potential energy of the Earth–apple system due to the apple's displacement along the y axis from y_i to y_f.	$\Delta U = -\int_{y_i}^{y_f} F_y dy$ (8.3)
Substitute $F_y = -mg$.	$\Delta U = -\int_{y_i}^{y_f} (-mg) dy$
Because $-mg$ does not depend on y, we pull it outside the integral and integrate.	$\Delta U = mg\int_{y_i}^{y_f} dy = mgy\Big\vert_{y_i}^{y_f}$
Substitute the limits of integration.	$\Delta U = U_f - U_i = mgy_f - mgy_i$ ✓ (8.4)

COMMENTS

Only changes in potential energy ΔU are physically important. The initial (or final) potential energy is set to a convenient value. Choosing the zero point for the potential energy (or equivalently assigning the reference configuration) is an important step in solving problems as discussed below.

Reference Configuration

To make calculations easier, we typically choose a particular configuration to be the *reference configuration* and assign it a potential energy of zero. The potential energy of all other configurations can then be found from a change in potential energy relative to the reference configuration. Choosing the reference configuration in a situation that involves the Earth and a particle near its surface means setting the origin of the coordinate system to a convenient place. For example, if we place the origin of the coordinate system in Figure 8.7 at the initial position of the apple, it becomes the reference configuration, with a potential energy of zero:

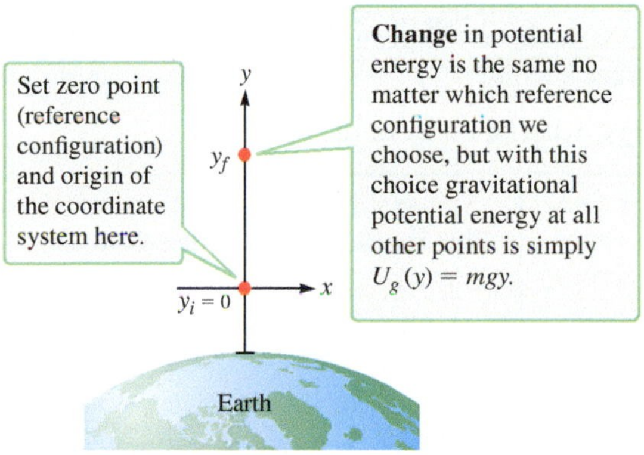

FIGURE 8.7 As in Figure 8.6 the system consists of the apple and the Earth, but here the choice of the origin and reference configuration is more convenient.

$$\Delta U = U_f - 0 = mgy_f - 0$$

The **gravitational potential energy** of any other configuration may now be written as

$$U_g(y) = mgy \qquad (8.5)$$

So, if the apple is at the origin, the potential energy of the Earth–apple system is zero. If the potential energy of the system is greater than zero, the apple is above the origin. If the potential energy is negative, the apple is below the origin. Potential energy is a scalar, so the signs do not indicate direction; rather, they indicate a relative change in potential energy. If the apple is above the origin, the system has *more* energy than if the apple is below the origin.

Path Independence

Only a vertical change in the particle's position results in a change in the potential energy of the Earth–particle system. If the particle moves upward, the gravitational potential energy of the system increases. If the particle's displacement is downward, the potential energy of the system decreases. If the particle's displacement is horizontal, the potential energy remains the same. Furthermore, the change in potential energy only depends on the final and initial configurations of the system, no matter

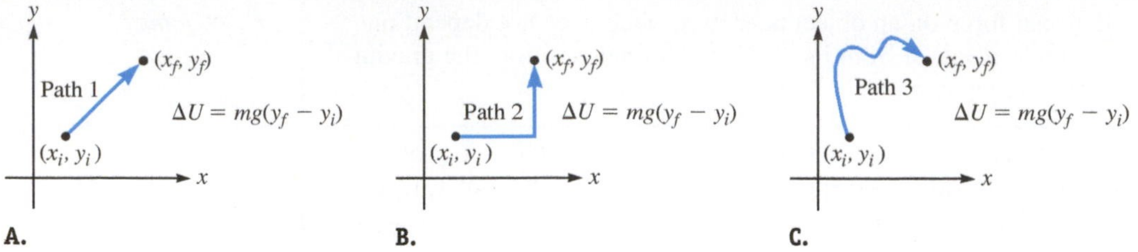

FIGURE 8.8 The change in gravitational potential energy does not depend on the path. The change in potential energy is the same for all three paths because the initial and final positions are the same.

what configurations the system takes in between. We say that the change in potential energy of the system is *path independent*. Figure 8.8 shows three possible paths a particle may take getting from (x_i, y_i) to (x_f, y_f); in all three cases, the change in potential energy is the same: $\Delta U = mg(y_f - y_i)$.

EXAMPLE 8.4 Look Out Below!

A construction worker is repairing the roof of a house. His hammer of mass 0.55 kg slips from his hand and falls through a basketball hoop to the driveway below. The basketball hoop is 3.0 m above the ground and 5.7 m below the roof where the hammer left the worker's hand.

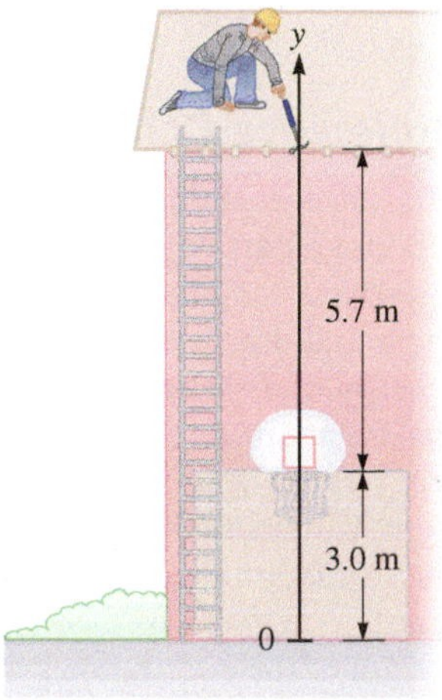

FIGURE 8.9

A To calculate potential energy, what must be in the system?

The system must have at least two objects that interact with each other through one of the conservative forces we know, gravity or the spring force.	The system includes the hammer and the Earth. The force between them is gravity.

B Place the origin of an upward-pointing y axis on the driveway. What is the change in the potential energy of the system?

INTERPRET and ANTICIPATE

Following the directions, we draw an upward-pointing y axis on our sketch (Fig. 8.9).

SOLVE

With this choice for the origin, the reference point for the potential energy of the system is at the driveway. When the hammer is on the *d*riveway, the potential energy of the system is zero.	$U_D = 0$
The potential energy of the system when the hammer is at the *r*oof is given by Equation 8.5.	$U_R = mgy_R$ (8.5)
Find the change in the system's gravitational potential energy due the hammer falling from the roof to the driveway.	$\Delta U = U_D - U_R = 0 - mgy_R = -mgy_R$
Figure 8.9 shows that $y_R = 3.0\text{ m} + 5.7\text{ m} = 8.7\text{ m}$.	$\Delta U = -(0.55\text{ kg})(9.81\text{ m/s}^2)(8.7\text{ m}) = -47\text{ kg m}^2/\text{s}^2$ $\Delta U = -47\text{ J}$

CHECK and THINK

The negative sign means that the Earth–hammer system lost potential energy due to the hammer's fall.

C Now use the basketball hoop as the origin of the y axis and find the change in potential energy.

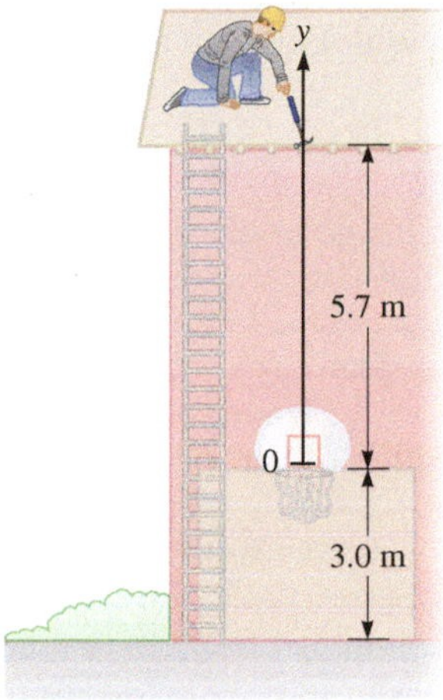

FIGURE 8.10

INTERPRET and ANTICIPATE

In Figure 8.10, we use an upward-pointing y axis with the origin at the basketball hoop. The roof is at a positive position, and the driveway at a negative position. With this choice, the reference point for the potential energy of the system is at the basketball hoop. When the hammer is at the roof, the potential energy of the system is positive, and when the hammer is on the driveway, the potential energy is negative. We expect to find the same change in potential energy.

SOLVE

Use Equation 8.5 to find the potential energy of the system for both configurations: hammer at roof and hammer on driveway.	$U_R = mgy_R$ $U_D = mgy_D$
Find the change in the system's gravitational potential energy.	$\Delta U = U_D - U_R = mgy_D - mgy_R$ $\Delta U = mg(y_D - y_R)$
Our sketch shows that the roof's position is positive and the driveway's is negative: $y_R = +5.7$ m and $y_D = -3.0$ m.	$\Delta U = (0.55\text{ kg})(9.81\text{ m/s}^2)(-3.0\text{ m} - 5.7\text{ m})$ $\Delta U = (0.55\text{ kg})(9.81\text{ m/s}^2)(-8.7\text{ m}) = -47\text{ J}$

CHECK and THINK

There are no surprises here; this result is exactly what we found in part B. With this choice for the origin, neither the final (nor the initial) configuration was set as the reference configuration. Therefore, neither U_D nor U_R was zero making our calculation slightly more complicated. It is helpful to choose either the initial or final configuration as the reference configuration, but it is not necessary.

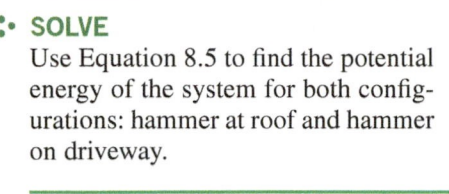

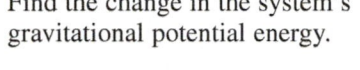

CONCEPT EXERCISE 8.4

In Figure 8.11, a person launches a ball off of a building three times.

Case 1. He drops the ball and it lands at the base of the building.
Case 2. He uses a slingshot to launch the ball in a parabolic path. The ball lands on the ground at the same level as the base of the building.
Case 3. He launches the ball. It flies in a parabolic path and lands on the roof of a building at the same height at which the ball was launched.

Compare the change in potential energy of the Earth–ball system for each case. Explain your answer.

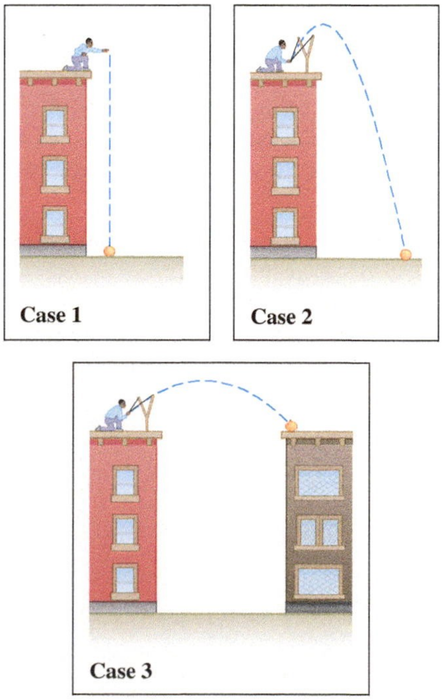

FIGURE 8.11

8-4 Universal Gravitational Potential Energy

Equation 8.5, $U_g(y) = mgy$, can be used to find the gravitational potential energy of any system that consists of a particle near the surface of a large object such as the Earth, the Moon, or another planet. In fact, it can be used whenever the gravitational force exerted on the particle is roughly constant. If it isn't (approximately) constant such as when a rocket is launched from the Earth, however, we need to use an expression for *universal gravitational potential energy*.

DERIVATION Universal Gravitational Potential Energy

Starting with Equation 8.3, we will derive the expression

$$\Delta U = \frac{GMm}{r_i} - \frac{GMm}{r_f} \quad (8.6)$$

for the change in the gravitational potential energy of any system that consists of two particles such as the Earth (of mass M) and a spacecraft (of mass m).

Let's consider a spacecraft (represented by a dot) that moves away from the Earth from $\vec{r}_i$ to $\vec{r}_f$. We have chosen a unit vector $\hat{r}$ that points away from the Earth in the direction of the spacecraft (Fig. 8.12).

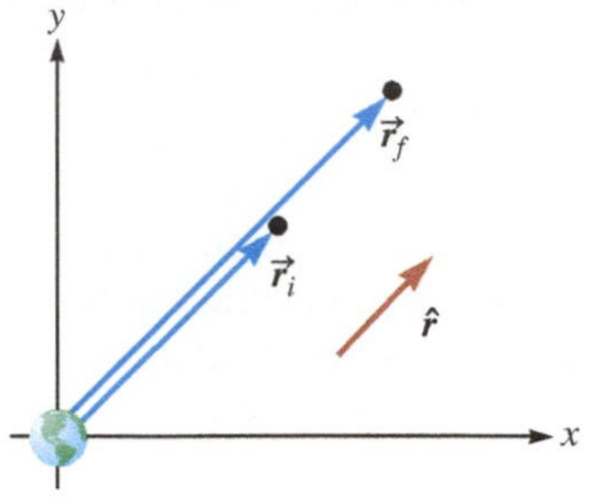

FIGURE 8.12

The gravitational force of the Earth on the spacecraft is given by Newton's law of universal gravity (Eq. 7.4). The subscript r indicates that the force is along the radial axis joining the Earth to the spacecraft.	$\vec{F}_G = F_r\hat{r} = -G\frac{Mm}{r^2}\hat{r}$
Rewrite Equation 8.3 in terms of this radial axis.	$\Delta U = -\int_{r_i}^{r_f} F_r\,dr \quad (8.3)$
Substitute for F_r.	$\Delta U = -\int_{r_i}^{r_f}\left(-G\frac{Mm}{r^2}\right)dr$
Pull the constants outside the integral and integrate. Notice that the negative signs cancel out.	$\Delta U = GMm\int_{r_i}^{r_f}\left(\frac{1}{r^2}\right)dr = GMm\left(-\frac{1}{r}\right)\Big\vert_{r_i}^{r_f}$
Substitute the limits of integration.	$\Delta U = -\frac{GMm}{r_f} - \left(-\frac{GMm}{r_i}\right) = \frac{GMm}{r_i} - \frac{GMm}{r_f}$ ✓ (8.6)

COMMENTS

We showed in the previous section that if a particle moves upward from the surface of the Earth, the gravitational potential energy increases. So we expect $\Delta U > 0$ in this case because the spacecraft moves upward (Fig. 8.12).

The initial position is closer to the origin than the final position.	$r_i < r_f$
Because r_i is in the denominator of the first term and r_f is in the denominator of the second term in Equation 8.6, the first term is larger than the second term.	$\frac{GMm}{r_i} > \frac{GMm}{r_f}$
So, the change in potential energy is positive as predicted.	$\Delta U = \frac{GMm}{r_i} - \frac{GMm}{r_f} > 0$

Reference Configuration for Universal Gravity

If the spacecraft moves away from the Earth, the potential energy of the system increases (Fig. 8.13A). If the spacecraft moves toward the Earth, the system's potential energy decreases (Fig. 8.13B). If spacecraft ends up at the same distance from the Earth at which it started such as on part of a circular orbit, there is no change in the system's gravitational potential energy (Fig. 8.13C).

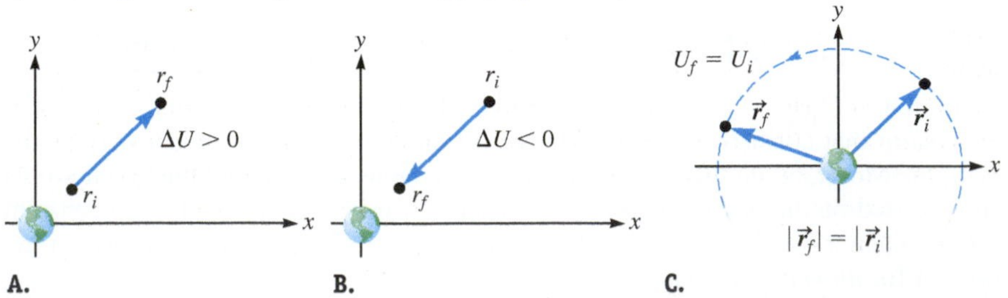

FIGURE 8.13 A system consists of a spacecraft and the Earth.

In principle, the reference configuration can be set to any arbitrary choice such as the spacecraft on the surface of the Earth or in a circular geosynchronous orbit. In practice, because position appears in the denominator, it is best and therefore *usual* to set the reference configuration such that when the two particles are infinitely far apart, the potential energy is zero. So, for a system that consists of two particles 1 and 2 with the reference configuration chosen such that $\lim_{r\to\infty} U = 0$, the gravitational potential energy is

UNIVERSAL GRAVITATIONAL POTENTIAL ENERGY

Special Case

$$U_G(r) = -G\frac{m_1 m_2}{r} \tag{8.7}$$

In any particular problem, you may need $U_g = mgy$ (Eq. 8.5) *or* $U_G = -G(m_1m_2/r)$ (Eq. 8.7), *but not both.* Use $U_g = mgy$ only if the system consists of a large object such as a planet and another particle such as an apple that stays near the surface of the large object.

The subscript G is a reminder that Equation 8.7 is good for particles that interact through (universal) gravity.

EXAMPLE 8.5 Plotting Universal Gravitational Potential Energy

Plot $U_G(r)$ versus r and show with a sketch how it connects to a spacecraft moving directly away from the Earth. Should the change in the system's gravitational potential energy be positive, negative, or zero? Does the graph give the predicted results for ΔU_G?

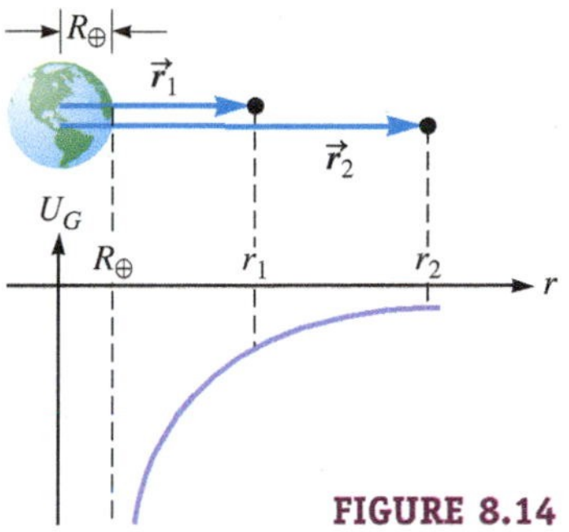

FIGURE 8.14

INTERPRET and ANTICIPATE

We put r on the horizontal axis and $U_G(r)$ on the vertical axis (Fig. 8.14). It may be helpful to use a spreadsheet or similar program. Our sketch shows the Earth and the spacecraft at two different positions $\vec{r}_1$ and $\vec{r}_2$. Because the spacecraft moves away from the Earth, we expect that the system's gravitational potential energy will increase: $\Delta U_G > 0$.

SOLVE

From the graph, $U_G(r_2) > U_G(r_1)$. Therefore the change in potential energy if the spacecraft moves away from the Earth from $\vec{r}_1$ to $\vec{r}_2$ is positive, which means that the gravitational potential energy increases as the spacecraft moves away.

$$\Delta U_G = U_G(r_2) - U_G(r_1)$$

$$U_G(r_2) > U_G(r_1)$$

$$\Delta U_G > 0$$

CHECK and THINK

The graph shows that as $r \to \infty$, the potential energy $U_G \to 0$ as required by our choice of the usual reference configuration.

8-5 Elastic Potential Energy

ELASTIC POTENTIAL ENERGY

Special Case

The other conservative force we study in mechanics is exerted by a spring that obeys Hooke's law. In this section, we derive an expression for the potential energy stored by such a spring, which is known as **elastic potential energy**.

DERIVATION Elastic Potential Energy

Figure 8.15 shows a block on a frictionless, horizontal surface attached to a spring. The other end of the spring is attached to a fixed wall. The system consists of the block, the spring, and the wall. For brevity, we omit the wall and refer to the system as a spring–block system. The block is on a frictionless surface. The block starts at position x_i (Fig. 8.15A). It moves to the left to position x_f, and the spring compresses. The force exerted by the spring on the block (and the wall) is internal to the system, is conservative, and is given by Hooke's law $\vec{F}_H = -k\Delta\vec{x}$ (Eq. 5.8) where $\Delta\vec{x}$ is the displacement of the block from the relaxed (or equilibrium) position. We will show that the change in the system's spring potential energy is given by

$$\Delta U = \tfrac{1}{2}kx_f^2 - \tfrac{1}{2}kx_i^2 \tag{8.8}$$

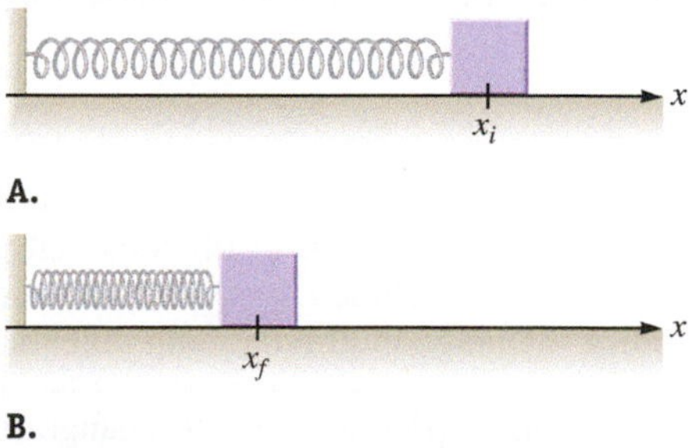

FIGURE 8.15 A system consists of a block, a wall, and a spring. **A.** The block is initially at x_i. **B.** The block moves to x_f.

Derivation continues on page 224 ▶

To find the potential energy of this system, we take the relaxed position to be at the origin and substitute the resulting expression for force into $\Delta U = -\int_{x_i}^{x_f} F_x dx$ (Eq. 8.3).	Hooke's law with origin at relaxed position: $F_x = -kx$ $\Delta U = -\int_{x_i}^{x_f} -kx\,dx$
Simplify and integrate.	$\Delta U = k\int_{x_i}^{x_f} x\,dx = k\frac{x^2}{2}\Big\vert_{x_i}^{x_f}$
Substitute the limits of integration.	$\Delta U = \frac{1}{2}kx_f^2 - \frac{1}{2}kx_i^2$ ✓ (8.8)

COMMENTS

We derived this expression for a horizontal spring lying along the x axis. Our derivation would be the same, however, if the spring were vertical and parallel to the y axis as is common in many problems. (See Example 8.6.)

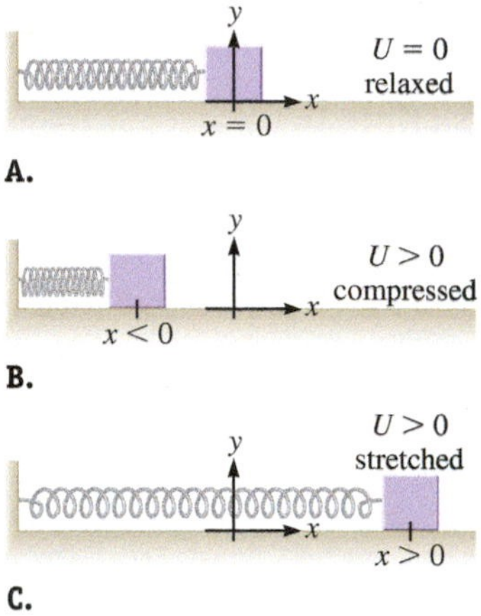

FIGURE 8.16 **A.** The reference configuration and the origin are set to the position of the block when the spring is relaxed. If the spring is then compressed **B.** or stretched **C.**, the elastic potential energy of the system increases.

Reference Configuration for Spring Potential Energy

We usually place the origin of the coordinate system at the relaxed (or equilibrium) position and set this position to the reference configuration. Then, the **elastic potential energy** of the spring–block system is given by

$$U_e = \tfrac{1}{2}kx^2 \tag{8.9}$$

where the spring lies along the x axis. If the spring lies along a different axis such as the y axis, then

$$U_e = \tfrac{1}{2}ky^2 \tag{8.10}$$

where e stands for *e*lastic. Because the component of position is squared, the potential energy associated with any configuration (except the reference configuration) is positive. If the block is at the relaxed position (Fig. 8.16A), the potential energy of the system is zero. If the spring is compressed (Fig. 8.16B) or stretched (Fig. 8.16C), the potential energy is positive. In other words, if the block moves from the relaxed position to any other position, the potential energy increases. However, this statement does *not* mean that all changes in potential energy are positive.

EXAMPLE 8.6 A Vertical Spring

In your laboratory, you are likely to encounter a vertical rather than a horizontal spring because then there is no need to create a frictionless surface. You will probably hang a ball of mass m from the spring of spring constant k. The experiment will begin when the ball is displaced straight downward by a distance h (Fig. 8.17).

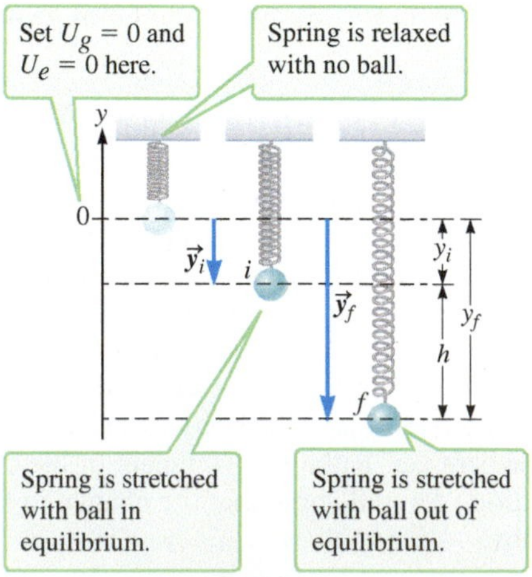

FIGURE 8.17

A To find the change in the system's potential energy, what must be included in the system? (Be sure that there are no external forces exerted on your system.)

It is probably obvious that the system must include the ball and the spring (and the ceiling). It might not be as obvious that the Earth must also be in the system. The Earth must be included, however, because as the ball moves along the vertical axis there is a change in gravitational potential energy.

B Find an expression for the change in the system's potential energy from the equilibrium position (after the ball has been attached to the spring as in Fig. 8.17, middle) to a distance h below equilibrium (Fig. 8.17, right).

INTERPRET and ANTICIPATE

Two conservative forces—gravity and the spring force—are internal to this system. It is convenient to set the reference configuration to the point where the spring is *relaxed* (before the ball is hung from it), so $y = 0$ at the bottom end of the spring (Fig. 8.17, left). The elastic and gravitational potential energies are then both zero when the spring is relaxed. At the *initial* position with the ball in equilibrium attached to the stretched spring (Fig. 8.17, middle), the gravitational potential energy is negative, and the elastic potential energy is positive.

SOLVE

Use $U_g(y) = mgy$ (Eq. 8.5) to find the initial gravitational potential energy and $U_e = \frac{1}{2}ky^2$ (Eq. 8.10) to find the initial elastic potential energy at $\vec{y}_i$.

$$U_{gi} = -mgy_i \qquad (1)$$

$$U_{ei} = \tfrac{1}{2}ky_i^2 \qquad (2)$$

Next, find the potential energies at the final position. According to Figure 8.17, right the final position is given by $\vec{y}_f = -(h + y_i)\hat{\jmath}$.

$$U_{gf} = mg(-y_f) = -mg(h + y_i)$$

$$U_{gf} = -mgh - mgy_i \qquad (3)$$

$$U_{ef} = \tfrac{1}{2}k(-y_f)^2 = \tfrac{1}{2}k(h + y_i)^2$$

$$U_{ef} = \tfrac{1}{2}kh^2 + khy_i + \tfrac{1}{2}ky_i^2 \qquad (4)$$

Find the change in gravitational potential energy by subtracting Equation (1) from Equation (3).

$$\Delta U_g = (-mgh - mgy_i) - (-mgy_i)$$

$$\Delta U_g = -mgh \qquad (5)$$

Find the change in elastic potential energy by subtracting Equation (2) from Equation (4).

$$\Delta U_e = \left(\tfrac{1}{2}kh^2 + khy_i + \tfrac{1}{2}ky_i^2\right) - \tfrac{1}{2}ky_i^2$$

$$\Delta U_e = \tfrac{1}{2}kh^2 + khy_i \qquad (6)$$

Add Equations (5) and (6) to find the change in the system's potential energy (both gravitational and elastic).

$$\Delta U = \Delta U_g + \Delta U_e$$

$$\Delta U = -mgh + \tfrac{1}{2}kh^2 + khy_i \qquad (7)$$

The first term, $-mgh$, is due to change in gravitational potential energy, and the second term, $\frac{1}{2}kh^2$, is due to the change in elastic potential energy. The system's potential energy was not zero at the initial (equilibrium) position, however, so a third term, khy_i, arises. Let's use Newton's second law to deal with this third term. When the ball is in equilibrium (Fig. 8.17, middle), two forces, the spring force and gravity, are exerted on it, and its acceleration is zero (Fig. 8.18).

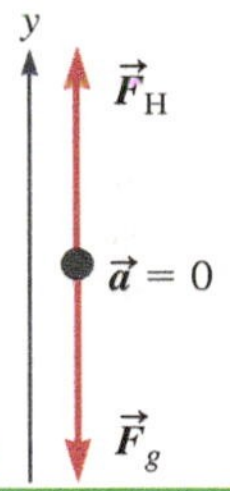

FIGURE 8.18

Applying Newton's second law, we find that the spring force is equal in magnitude to the ball's weight.

$$\sum F_y = F_H - F_g = 0$$

$$F_g = F_H$$

$$mg = ky_i \qquad (8)$$

Substitute Equation (8) into Equation (7).

$$\Delta U = -mgh + \tfrac{1}{2}kh^2 + mgh$$

$$\Delta U = \tfrac{1}{2}kh^2 \qquad (9)$$

CHECK and THINK

Our final expression, Equation (9), does not appear to depend on gravitational potential energy, which is very helpful in the laboratory. If you measure the displacement h of the ball from its equilibrium position, you apparently can ignore changes in gravitational potential energy. Actually, these changes balance out because the system has potential energy in its equilibrium configuration.

8-6 Conservation of Mechanical Energy

Because potential energy can only be associated with a system that consists of two or more particles, the conservation approach requires that we choose a system of two or more particles. In this chapter, we choose systems that include only particles and objects that can be modeled as particles; in the next chapter, we consider more complicated systems.

ISOLATED SYSTEM ✪ **Major Concept**

In this chapter, we also make the special choice of an *isolated system*, one that does not interact with its environment. An **isolated system** does not exert a force on its environment, and the environment does not exert a force on it. For example, if a situation involves an asteroid falling to the Earth, we would not choose the asteroid alone to be the system (Fig. 8.19A). Instead, we should choose a system that includes both the asteroid and the Earth so that gravity is internal to the system (Fig. 8.19B).

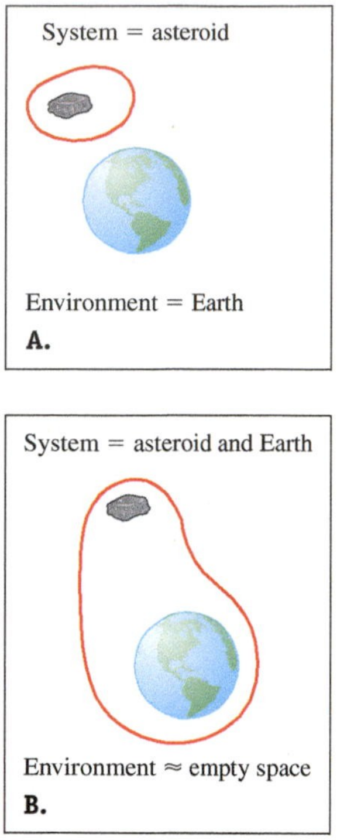

FIGURE 8.19 Two possible system choices. **A.** Only the asteroid is in the system. The system is **not isolated** because the Earth exerts a force on it. **B.** Both the Earth and the asteroid are in the system. The environment is considered to be empty and cannot exert a force on the **isolated** Earth–asteroid system.

An isolated system is an idealization in that almost no real system can be truly isolated. For example, in the Earth–asteroid system in Figure 8.19B, the environment is not truly empty space. The Moon, the Sun, and other planets are in the environment, and they exert gravitational forces on the system. Also, as the asteroid passes through the Earth's atmosphere, molecules interact with the system through the drag force. We can often ignore such environmental effects and consider the system to be isolated, however.

Once we have decided what is in the system, we cannot change that decision partway through our analysis. We then apply the conservation approach to our chosen system. Part of the conservation approach comes from the *conservation of mechanical energy.* The sum of a system's kinetic and potential energy is its **mechanical energy,**

$$E = K + U \tag{8.11}$$

CONSERVATION OF MECHANICAL ENERGY ❶ **Underlying Principle**

According to the **principle of conservation of mechanical energy,** *if only conservative forces are acting, the mechanical energy of an isolated system is conserved.* Put another way, the kinetic energy $K(t)$ and the potential energy $U(t)$ change in time, but the mechanical energy E is a constant that does *not* depend on time. So, it is possible for the system's energy to change forms from kinetic to potential and from potential to kinetic, but the sum of the two is a constant.

The conservation of mechanical energy may be expressed in terms of an initial time and a final time:

$$E_i = E_f \tag{8.12}$$

Another way to express the conservation of mechanical energy is

$$\Delta E = 0 \tag{8.13}$$

Because mechanical energy is the sum of the kinetic and potential energy of the system, we have

$$K_i + U_i = K_f + U_f \tag{8.14}$$

By rearranging terms, we can write Equation 8.14 in terms of the change in energy:

$$(K_f - K_i) + (U_f - U_i) = 0$$

$$\Delta K + \Delta U = 0 \tag{8.15}$$

Equations 8.12 through 8.15 are all mathematical expressions for the conservation of mechanical energy principle. To illustrate the conservation of mechanical energy approach, let us begin with an example we have already considered many times.

EXAMPLE 8.7 A Falling Ball Revisited

A ball of mass m falls straight down and lands in a person's hand (Fig. 8.20). The ball's initial speed is v_i, and it falls with constant acceleration $\vec{a}_y = -g\hat{\jmath}$. Starting with the conservation of mechanical energy, show that

$$v_f^2 = v_i^2 - 2g\Delta y$$

in agreement with our work in Chapter 2.

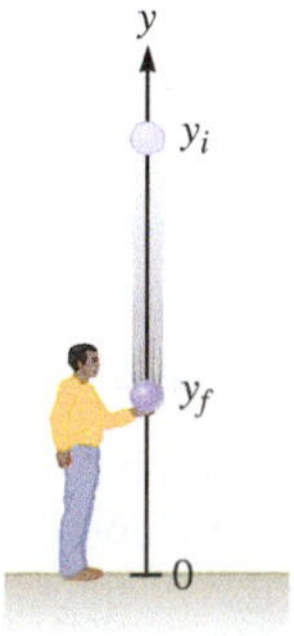

FIGURE 8.20

INTERPRET and ANTICIPATE

The system consists of the ball and the Earth. If we ignore the surrounding air, the system is essentially isolated, and mechanical energy is conserved.

SOLVE

Find an expression for the initial mechanical energy. Use $K = \frac{1}{2}mv^2$ (Eq. 8.1) and $U_g(y) = mgy$ (Eq. 8.5).	$E_i = K_i + U_i = \frac{1}{2}mv_i^2 + mgy_i$ (8.11)
Find a similar expression for the final mechanical energy.	$E_f = K_f + U_f = \frac{1}{2}mv_f^2 + mgy_f$
Mechanical energy is conserved, so set the initial and final mechanical energies equal to each other.	$E_i = E_f$ $\frac{1}{2}mv_i^2 + mgy_i = \frac{1}{2}mv_f^2 + mgy_f$
The mass drops out of this equation. Solve for the square of the final speed.	$v_i^2 + 2gy_i = v_f^2 + 2gy_f$ $v_f^2 = v_i^2 + 2gy_i - 2gy_f = v_i^2 - 2g(y_f - y_i)$ $v_f^2 = v_i^2 - 2g\Delta y$

CHECK and THINK

This result is exactly what we would find if we started with kinematic Equation 2.13 for constant acceleration. The conservation approach does not replace the kinematics and dynamics we learned previously, but it gives us other tools for solving complicated problems.

DERIVATION Principle of Conservation of Mechanical Energy

We started with this familiar example of a free-falling particle to show that conservation of mechanical energy provides an alternative and consistent approach to mechanics. To strengthen that connection, we will use Newton's second law to derive Equation 8.13, $\Delta E = 0$, for any isolated system in which only conservative forces are exerted between the system's particles.

Start with Newton's second law in one dimension along the x axis to make the derivation mathematically simpler.	$F_x = ma_x$
Write the acceleration in terms of its time derivative (Eq. 2.7).	$F_x = m\dfrac{dv_x}{dt}$ (1)
To eliminate time, use the chain rule and Equation 2.4.	$\dfrac{dv_x}{dt} = \dfrac{dx}{dt}\dfrac{dv_x}{dx} = v_x\dfrac{dv_x}{dx}$ (2)
Rewrite Newton's second law by substituting (2) into (1).	$F_x = mv_x\dfrac{dv_x}{dx}$
Solve this differential equation by moving dx to the left side and then integrating.	$F_x dx = mv_x dv_x$ $\int F_x dx = \int mv_x dv_x$
We must make sure that the limits of integration match. The integration is from the initial configuration and speed to the final configuration and speed. At the initial time, the particle's position and speed are x_i and v_i, respectively. Its final position and speed are x_f and v_f.	$\int_{x_i}^{x_f} F_x dx = \int_{v_i}^{v_f} mv_x dv_x$
Integrate and substitute the limits on the right side.	$\int_{x_i}^{x_f} F_x dx = \frac{1}{2}mv^2\Big\vert_{v_i}^{v_f} = \frac{1}{2}mv_f^2 - \frac{1}{2}mv_i^2$
The right side of the equation is the change in kinetic energy (Eq. 8.1). The left side is the negative of the change in potential energy (Eq. 8.3).	$-\Delta U = \Delta K$

Derivation continues on page 228 ▶

Upon rearranging, we find that we have derived Equation 8.13, the conservation of mechanical energy.

$$\Delta K + \Delta U = 0$$
$$\Delta E = 0 \checkmark$$

COMMENTS

This derivation shows that the conservation of mechanical energy may be derived from Newton's second law. So, the conservation approach does not replace the force approach; rather, they complement one another.

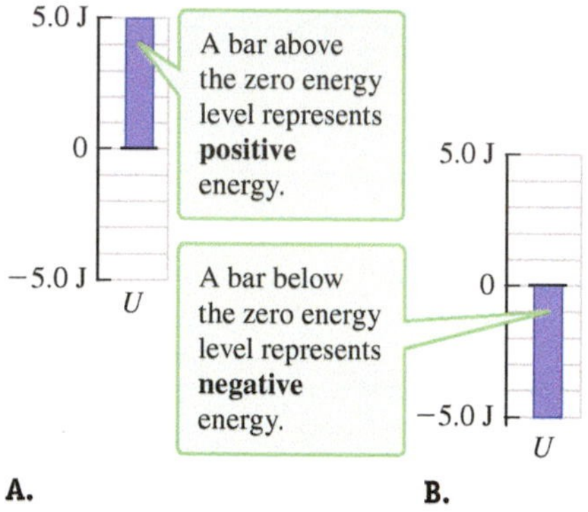

FIGURE 8.21 Bar charts for potential energy.

BAR CHART Tool

8-7 Applying the Conservation of Mechanical Energy

Applying conservation of mechanical energy is a good approach when you are interested in motion at two separate times, and the conservation approach is particularly powerful when the acceleration is *not* constant. In this section, we learn a strategy for applying the conservation of mechanical energy.

When we use the force approach, we choose a coordinate system and draw a free-body diagram. Likewise, using the conservation approach involves choices and diagrams. We must decide what to include in the system and what is in the outside environment. Some choices may make a problem easier to analyze, but the answer cannot depend on our choice.

Two kinds of diagrams help us visualize conservation of mechanical energy. The first is a **bar chart** (Fig. 8.21). (The second is discussed in the next section.) The vertical height of a bar represents the amount of energy. If there is no energy, write "zero" on the horizontal blank. Kinetic energy can never be negative, so it cannot be represented by a bar below the zero energy level. The mechanical energy bar's height (E) must equal the sum of the heights of the kinetic (K) and potential (U) energy bars (Fig. 8.22). If mechanical energy ($E = K + U$) is conserved, the height of its bar must remain unchanged.

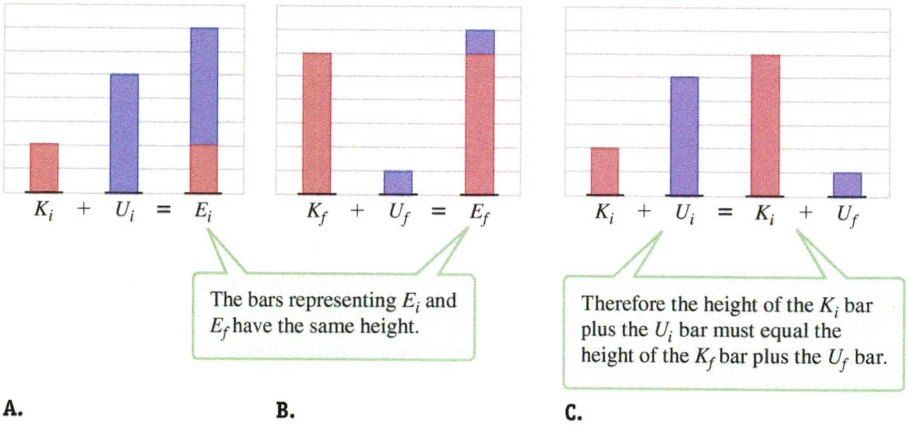

FIGURE 8.22 **A.** The initial mechanical energy bar is the sum of the initial kinetic energy and potential energy bars. **B.** The final kinetic energy and potential energy bars add up to the final mechanical energy bar. **C.** We can combine the initial and final bar charts. The sum of the initial energy bars equals the sum of the final energy bars.

CONCEPT EXERCISE 8.5

Each chart shown in Figure 8.23 has one unknown energy. Use the conservation of mechanical energy to find the missing energy and draw in the appropriate bar. If it is zero, write the word *zero* in place of the bar.

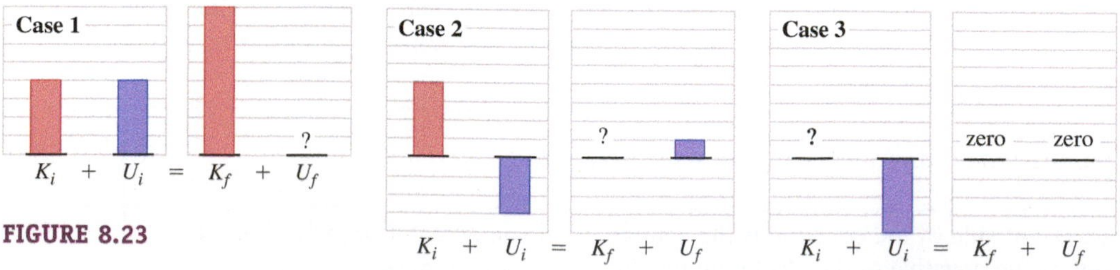

FIGURE 8.23

To solve problems using the principle of conservation of mechanical energy, it is useful to rewrite $E_i = E_f$ (Eq. 8.12). First, the potential energy is associated with conservative forces. Right now, we only know of two conservative forces, gravity and the Hooke's law spring force. So, we write the potential energy as the sum of these two:

$$U = U_g + U_e \quad \text{or} \quad U = U_G + U_e$$

Second, we write mechanical energy in terms of the kinetic and potential energies:

$$K_i + U_{gi} + U_{ei} = K_f + U_{gf} + U_{ef} \tag{8.16}$$

Each term in Equation 8.16 is represented by a bar in a bar chart, and so when applying Equation 8.16 expect three bars on each side of the equal sign.

Remember that we have two expressions for gravitational potential energy, $U_g = mgy$ and $U_G = -G(m_1m_2/r)$ (Eqs. 8.5 and 8.7), and we need only one for any situation. We have used U_g for simplicity in Equation 8.16, but the choice is still implied.

PROBLEM-SOLVING STRATEGY

Applying the Principle of Conservation of Mechanical Energy

INTERPRET and ANTICIPATE

Step 1 Choose an **isolated system** (Section 8-6).

Step 2 Pick a **coordinate system and the reference configuration** (Section 8-6).

Step 3 Identify the configuration of the system at an **initial time and at a final time**. (These times do not necessarily come at the beginning and end of the motion.)

Step 4 Draw a **bar chart** with three bars on each side of the equal sign to visualize the conservation of mechanical energy.

SOLVE

Step 5 Apply the principle of **conservation of mechanical energy** $K_i + U_{gi} + U_{ei} = K_f + U_{gf} + U_{ef}$ (Eq. 8.16).

Step 6 Write down any **other equations** (Eqs. 8.1, 8.5, 8.7, 8.9, and 8.10) that are relevant to the kinetic and potential energies involved.

Step 7 **Do algebra** before substitution.

EXAMPLE 8.8 Escape Speed

To launch a spacecraft so that it never returns, its initial speed must at least be equal to the escape speed of the planet. If a spacecraft is launched at exactly the escape speed, its kinetic energy is zero when it is infinitely far from the planet. Because a spacecraft can never truly be infinitely far away, we assume *very far away* is a good approximation to *infinitely* far away. Find an expression for the escape speed of a planet of mass M and radius R.

INTERPRET and ANTICIPATE

This example is an ideal candidate for solving using the principle of conservation of mechanical energy because we are interested in the spacecraft's motion at two different times: when it is launched and when it is very far away. (It would be difficult to use the force approach because the force exerted on the spacecraft gets weaker as the spacecraft moves away from the Earth.)

Step 1 A suitable **isolated system** consists of the spacecraft and the planet. (As long as any other objects are very far from the system so that their gravitational force is negligible, the system may be considered isolated.)

Step 2 Place the **coordinate system** at the center of the planet as shown in Figure 8.24. The unit vector $\hat{r}$ points outward. Set the **reference configuration** to when the spacecraft is infinitely far away from the planet.

Step 3 Set the **initial time** to when the spacecraft is still on planet and the **final time** to when the spacecraft is very (infinitely) far away.

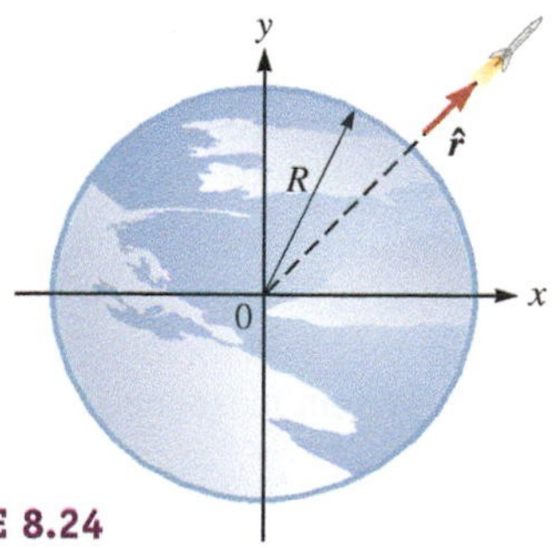

FIGURE 8.24

Example continues on page 230 ▶

Step 4 Draw a bar chart. Initially, the system has positive kinetic energy and negative gravitational potential energy. At the final time, the spacecraft is very far away, and there is no kinetic energy (the spacecraft *just* barely escaped) and no potential energy. (The final gravitational potential energy is zero because infinite separation is the reference configuration.) There is no spring involved in this example, so there can be no elastic potential energy (Fig. 8.25).

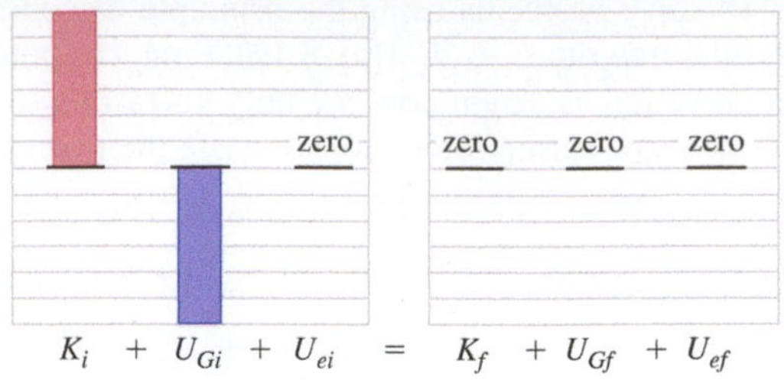

FIGURE 8.25

SOLVE

Step 5 Apply the **conservation of mechanical energy** (Eq. 8.16). The bar chart helps us see which terms are zero. We find that the sum of the initial kinetic and potential energies must equal zero.

$$K_i + U_{Gi} + U_{ei} = K_f + U_{Gf} + U_{ef} \quad (8.16)$$

$$K_i + U_{Gi} + 0 = 0 + 0 + 0$$

$$K_i + U_{Gi} = 0$$

Step 6 Write relevant energy equations. For launch at escape speed, the initial kinetic energy is found from Equation 8.1.

$$K_i = \tfrac{1}{2}mv_{esc}^2 \quad (8.1)$$

The initial gravitational potential energy of the system is given by Equation 8.7. The masses involved are M (mass of the planet) and m (mass of the spacecraft). At launch, the spacecraft begins on the planet's surface at a distance R from the origin. You might be tempted to use $U_g(y) = mgy$ (Eq. 8.5) for the initial gravitational potential energy, but this equation will not be valid over the entire path of the escaping spacecraft. We must use the universal gravitational potential energy.

$$U_{Gi} = -G\frac{Mm}{R} \quad (8.7)$$

Step 7 Do algebra to find the escape speed.

$$\tfrac{1}{2}mv_{esc}^2 - G\frac{Mm}{R} = 0$$

$$v_{esc} = \sqrt{\frac{2GM}{R}} \quad (8.17)$$

CHECK and THINK

Notice that the escape speed of a planet does not depend on the mass of the spacecraft. (The force required to accelerate the spacecraft up to the escape speed does depend on its mass, however.)

Finally, it is convenient to write our result in terms of the free-fall acceleration near the surface of the planet using Equation 7.13, $\vec{g}(r) = -\frac{GM}{r^2}\hat{r}$.

$$g = \frac{GM}{R^2}$$

$$v_{esc} = \sqrt{2gR} \quad (8.18)$$

EXAMPLE 8.9 Dart Gun

Figure 8.26 shows a spring-loaded dart gun used to launch a small ball of mass m straight up. When the spring is relaxed, it comes to the end of the gun barrel. If the spring constant is k and the spring is compressed by y_c, find an expression for the maximum height y_{max} of the ball above the end of the gun barrel.

INTERPRET and ANTICIPATE

Step 1 We choose to include the ball, the spring, the gun, and the Earth in the system. As long as the drag force of the air and the friction due to the gun barrel may be ignored, the **system is isolated**.

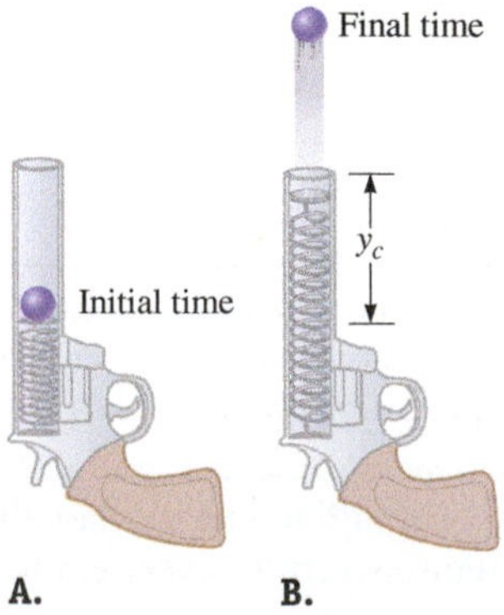

FIGURE 8.26

Step 2 Place the origin of the **coordinate system** at the end of the gun barrel as shown in Figure 8.27.

The **reference configuration** is the one in which the ball is at the end of the gun barrel. The spring is relaxed at that point. The elastic and gravitational potential energies of the system are zero in that configuration.

Step 3 We set the **initial time** to when the ball is on top of a fully compressed spring and the **final time** to when the ball has reached its maximum height.

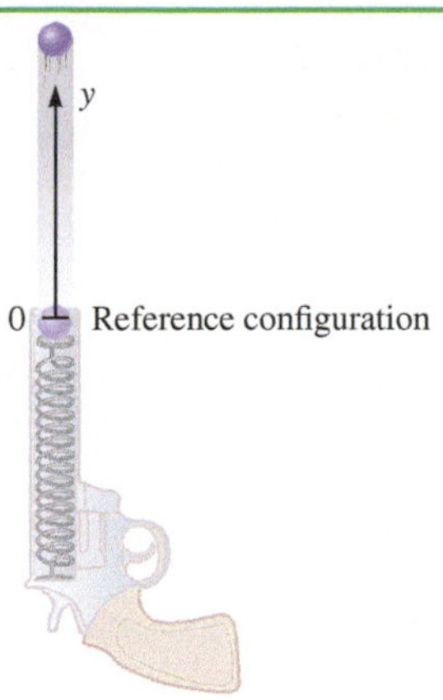

FIGURE 8.27

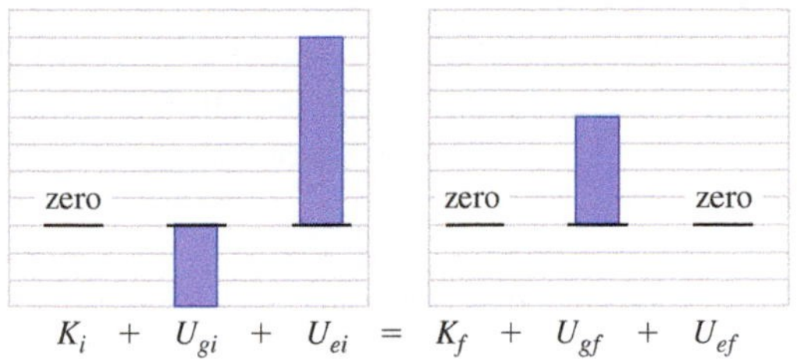

FIGURE 8.28

Step 4 Draw a bar chart. The initial kinetic energy is zero, but the initial gravitational potential energy is negative, and the initial spring potential energy is positive (Fig. 8.28). At the final time, the spring is relaxed and not touching the ball, so the elastic potential energy is zero. The final kinetic energy is zero because the ball momentarily stops at the top of its path. The gravitational potential energy is positive because the ball is above the reference level.

SOLVE

Step 5 Apply the **conservation of mechanical energy** (Eq. 8.16). The bar chart helps us see which terms are zero, and our expression is very simple.

$$U_{gi} + U_{ei} = U_{gf} \tag{1}$$

Step 6 Write relevant energy equations. The gravitational potential energy terms come from $U_g(y) = mgy$ (Eq. 8.5), and the elastic potential energy term comes from $U_e = \frac{1}{2}ky^2$ (Eq. 8.10).

$$U_{gi} = -mgy_c$$
$$U_{gf} = mgy_{max}$$
$$U_{ei} = \tfrac{1}{2}ky_c^2$$

Step 7 Do algebra. Substitute these energies into Equation (1) and solve for y_{max}.

$$-mgy_c + \tfrac{1}{2}ky_c^2 = mgy_{max}$$

$$y_{max} = \frac{k}{2mg}y_c^2 - y_c$$

CHECK and THINK

To see if this expression makes sense, check the dimensions. We expect to find that $[\![y_{max}]\!] = L$.

$$\left[\!\!\left[\frac{k}{2mg}y_c^2\right]\!\!\right] - [\![y_c]\!] = \left[\frac{(F/L)}{F}L^2\right] - L = L$$

As a further check, think about how y_{max} should depend on the various parameters. A stiffer spring (larger k) and a greater compression (greater y_c) should give a greater maximum height. A more massive ball means a lower maximum height. Our result is consistent with all of these.

EXAMPLE 8.10 Roller Coaster

The first roller coaster built in the United States was the Gravity Pleasure Switchback Railway at Coney Island in Brooklyn, New York (Fig. 8.29). The Gravity Pleasure consisted of two identical towers connected by a roller-coaster track. The ride reached a top speed of 6 mph. The cars had to be manually towed to the top of the hills at the beginning of both tracks. The passengers dismounted at the end of the first track, climbed the stairs up the second tower, and then got back in the cars for the return ride.

Most of today's roller coasters have a closed-loop track, and passengers do not dismount until after the ride is completed. Steel Force is a roller coaster located in Dorney Park in Allentown, Pennsylvania. Its tallest hill is 205 ft. Assume rolling friction between the wheels of the car and the track is negligible. Find the maximum speed of a car on Steel Force. For **CHECK and THINK:** How does that speed compare with that of the Gravity Pleasure?

The Granger Collection, New York

FIGURE 8.29 The Gravity Pleasure Switchback Railway (1884).

Example continues on page 232 ▶

INTERPRET and ANTICIPATE

We expect the maximum speed of the car to be at the bottom of the hill.

Step 1 The **isolated system** includes the Earth and the roller-coaster car in the system. As long as rolling friction and drag are negligible, mechanical energy is conserved.

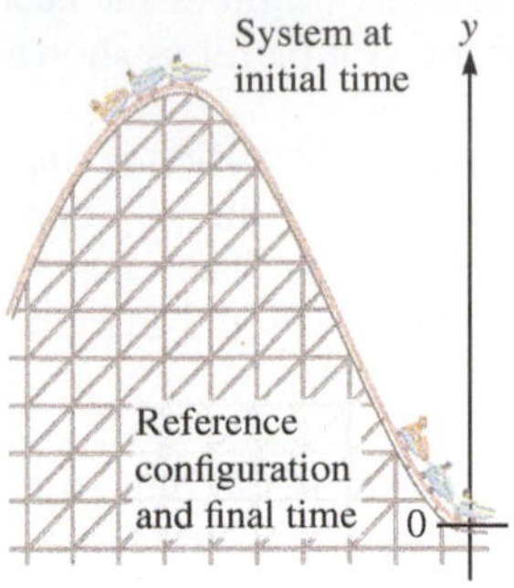

FIGURE 8.30

Step 2 It is convenient to place the origin of the **coordinate system** at the bottom of the track as shown in Figure 8.30. Then the **reference configuration** is the one in which the car is at the bottom of the track. The gravitational potential energy is zero for this configuration.

Step 3 Set the **initial time** to when the car is at the top of the hill and the **final time** to when the car is at the bottom of the hill.

Step 4 Draw a bar chart (Fig. 8.31). We can assume the initial kinetic energy is zero because the car moves very slowly at the top of the hill before it begins its descent. The initial gravitational potential energy is positive. The final kinetic energy is positive. The final gravitational potential energy is zero (at the reference configuration).

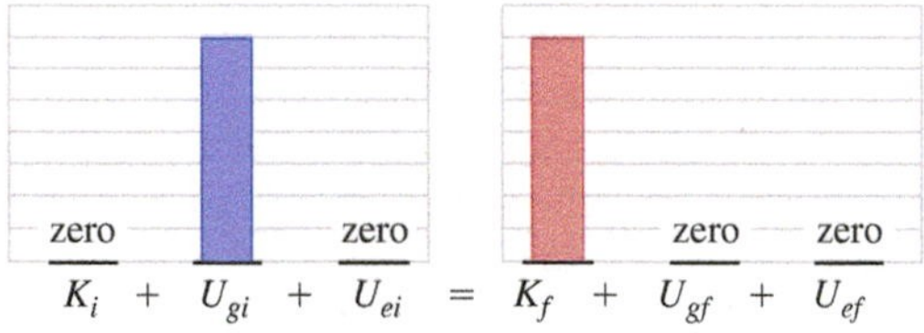

FIGURE 8.31

SOLVE

Step 5 Apply the **conservation of mechanical energy** (Eq. 8.16). Again the bar chart shows that the equation we need is very simple.

$$U_{gi} = K_f \tag{1}$$

Step 6 Write relevant energy equations. In the expression for kinetic energy, v_{max} represents the maximum speed.

$$K_f = \tfrac{1}{2}mv_{max}^2 \tag{8.1}$$

$$U_{gi} = mgy_i \tag{8.5}$$

Step 7 Do algebra. Substitute these energies into Equation (1) and solve for v_{max}. Convert 205 ft to 62.5 meters and substitute.

$$mgy_i = \tfrac{1}{2}mv_{max}^2$$

$$v_{max} = \sqrt{2gy_i} = \sqrt{2(9.81\ \text{m/s}^2)(62.5\ \text{m})}$$

$$v_{max} = 35.0\ \text{m/s}$$

CHECK and THINK

This speed is about 78 mph, or about 13 times the maximum speed attained on the Gravity Pleasure. The Dorney Park website claims that the Steel Force is somewhat slower than 78 mph, which may be because rolling friction and drag play a role in dissipating mechanical energy. In the next chapter, we will learn how to apply the conservation approach to situations that involve nonconservative forces such as friction.

8-8 Energy Graphs

In the previous section, we used energy bar charts to give us a pictorial representation of the conservation of mechanical energy. Another pictorial representation comes from plotting the potential energy. Once a coordinate system and reference configuration have been established, it is possible to write an expression for the potential energy in terms of the relative position between particles in the system (Eqs. 8.5, 8.7, 8.9, and 8.10).

Graphs of potential energy versus relative position, also known as **potential energy curves** (or U curves), give us a way to visualize potential energy (Fig. 8.32). A potential energy curve alone is not enough to visualize the conservation of mechanical energy, however. Instead, we must create an **energy graph** to represent a system's mechanical energy E, potential energy U, and kinetic energy K on a single set of axes.

ENERGY GRAPH ⦿ Tool

The energy graph starts with a U curve to represent the potential energy. Then, because we are only considering systems that have constant mechanical energy, the me-

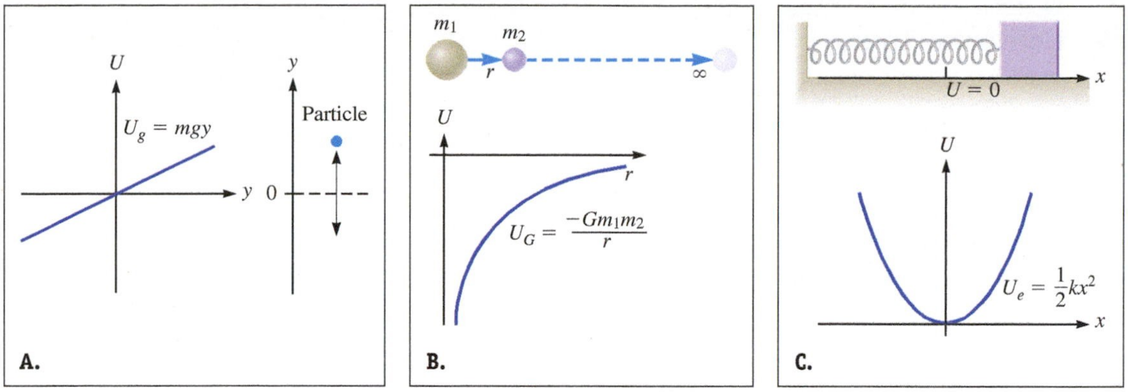
A.
U
$U_g = mgy$
y
y
0
Particle
B.
m_1
m_2
r
∞
U
r
$U_G = \frac{-Gm_1m_2}{r}$
C.
x
U = 0
U
$U_e = \frac{1}{2}kx^2$
x

FIGURE 8.32 Potential energy curves for **A.** gravity near the surface of a large object such as the Earth, **B.** universal gravity, and **C.** Hooke's law.

chanical energy does not depend on the relative position of the particles in the system. So, E is represented by a horizontal line on an energy graph ("the horizontal E line").

We do not actually plot kinetic energy on the energy graph, but we find it in the following way. The kinetic energy is given by (Eq. 8.11)

$$K = E - U$$

So, on the energy graph, kinetic energy is represented by the vertical space between the horizontal E line and the potential energy curve.

To see how to use an energy graph, consider a familiar physical situation: a ball thrown straight upward from the top of a tall building (Fig. 8.33A) with a certain initial velocity. The (nearly) isolated system consists of the ball and the Earth, and we assume mechanical energy of the system is conserved. We start by drawing a potential energy curve. For this situation, it looks like the one in Figure 8.32A. The mechanical energy of the system is constant and is therefore represented by the horizontal line shown in Figure 8.33B. To help see the relationship between the bar charts and the energy graph, we have overlaid the bar charts on the energy graph. Usually, an energy graph does not have such bar charts drawn on it.

To better understand Figure 8.33B, consider two facts about kinetic energy. First, kinetic energy cannot be negative; second, kinetic energy is given by $K = E - U$. Taken together, we reason that a system *cannot* be in a configuration that would have $U > E$ because then the kinetic energy would have to be negative. Figure 8.33B shows that point B corresponds to the highest position of the ball because, for higher positions $y > y_{max}$, the potential energy would be greater than the mechanical energy, $U > E$. In other words, the ball cannot rise higher than the corresponding position y_{max} because that situation would require more mechanical energy than the system possesses. We say that some configurations are *forbidden* to the system. In this case, any position y to the right of $y = y_{max}$ in Fig. 8.33B is not allowed for the ball's given initial velocity.

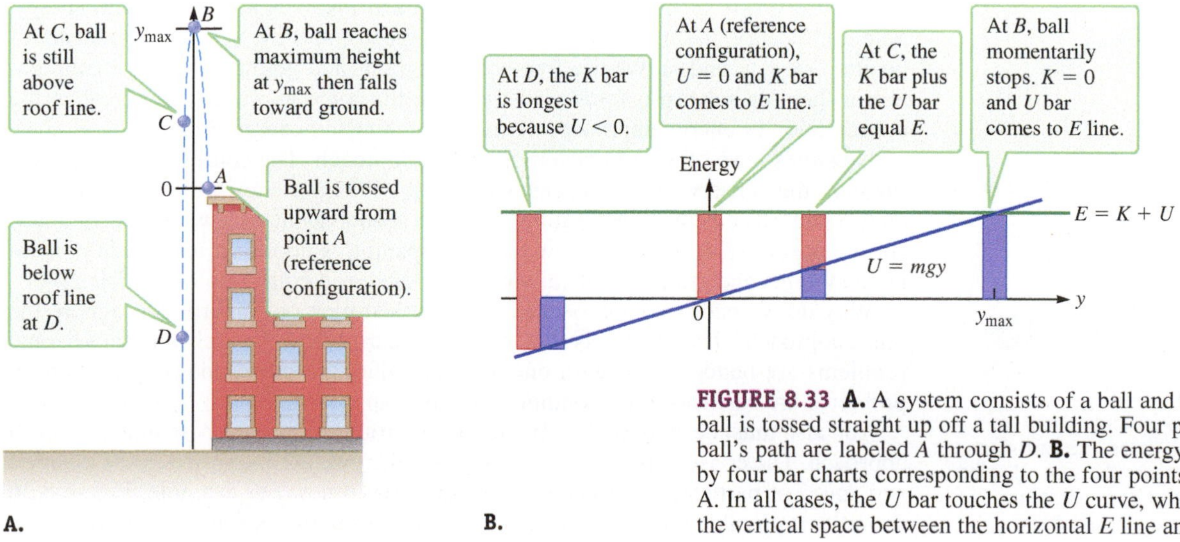
At C, ball is still above roof line.
y_{max}
B
At B, ball reaches maximum height at y_{max} then falls toward ground.
C
0
A
Ball is tossed upward from point A (reference configuration).
Ball is below roof line at D.
D
A.
At D, the K bar is longest because U < 0.
At A (reference configuration), U = 0 and K bar comes to E line.
At C, the K bar plus the U bar equal E.
At B, ball momentarily stops. K = 0 and U bar comes to E line.
Energy
E = K + U
U = mgy
0
y_{max}
y
B.

FIGURE 8.33 **A.** A system consists of a ball and the Earth. The ball is tossed straight up off a tall building. Four points along the ball's path are labeled A through D. **B.** The energy graph is overlaid by four bar charts corresponding to the four points labeled in part A. In all cases, the U bar touches the U curve, while the K bar fills the vertical space between the horizontal E line and the U curve.

EXAMPLE 8.11 Horizontal Spring

A block is attached to a spring and is supported by a frictionless, horizontal surface (Fig. 8.34). The block is displaced such that the spring stretches from its relaxed position to x_s and is then released. Draw a sketch, an energy graph, and bar charts for this example. Then describe the block's allowed positions and motion, connecting your description to the three visual aids.

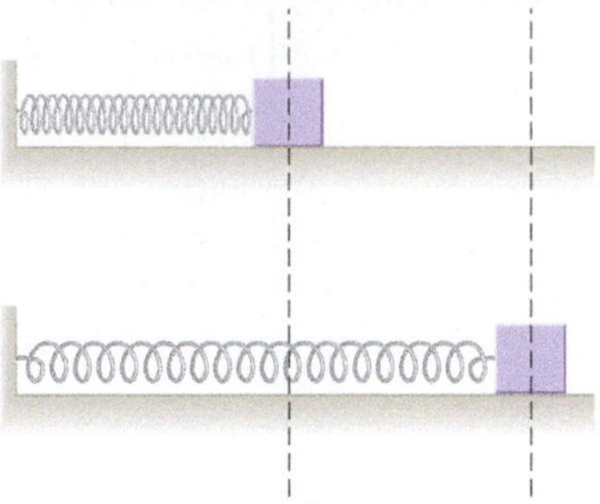

FIGURE 8.34

SOLVE

The potential energy curve is a parabola as in Figure 8.32C. The mechanical energy E is represented by a horizontal line that crosses the potential energy curve in two places, at x_s (representing maximum stretch of the spring) and x_c (representing maximum compression of the spring; Fig. 8.35). When the block is at x_s or x_c, the kinetic energy of the spring–block system is zero. We know that the only allowable configurations of a system have $E \geq U$, so the block must always satisfy $x_c \leq x \leq x_s$. (The system does not have enough mechanical energy to be outside this range.)

Figure 8.35 shows that the block moves from x_s to x_c and back again repeatedly; the bar charts indicate the kinetic and potential energies in three configurations. At x_s, the spring is fully stretched, and the block is momentarily at rest. The kinetic energy is zero. At x_c, K is again zero; now U has the same value that it does at x_s, but the spring is fully compressed. At $x = 0$, the potential energy of the system is zero. All the mechanical energy is in the form of kinetic energy $E = K$. So, the block is at its maximum speed when it passes through the origin.

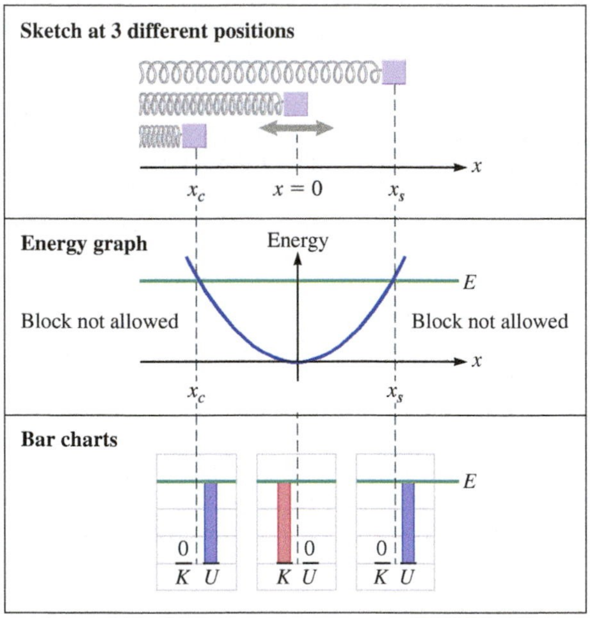

FIGURE 8.35

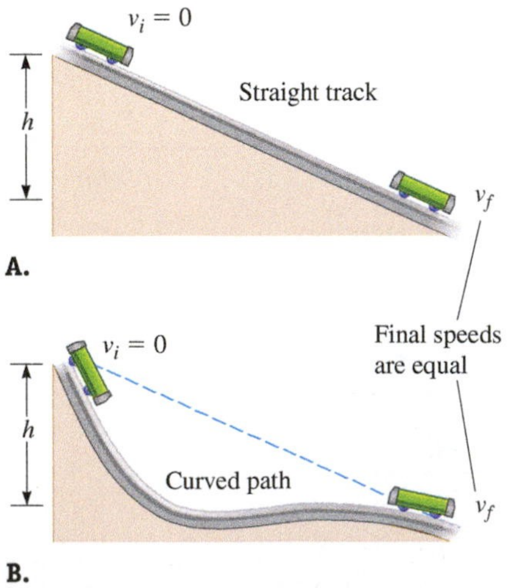

FIGURE 8.36 Two possible paths for a cart to take. In both cases, the cart starts from rest and undergoes the same displacement. The final speed does not depend on the path, but the time elapsed does.

Force Approach Versus Conservation Approach

Let us take a moment to compare the force and conservation approaches. The force approach involves vectors: force, acceleration, velocity, displacement, and position. The conservation of mechanical energy approach involves scalars: kinetic energy, potential energy, the magnitude of velocity (speed), and the scalar components of position (x, y, z, and r). We use free-body diagrams and graphical vector manipulation to help us solve problems visually using the force approach. To visualize the conservation of energy approach, we use bar charts and energy graphs.

Newton's laws of motion govern the force approach. The conservation approach involves the conservation of mechanical energy. We have shown that in the case of an isolated particle, the conservation approach is equivalent to Newton's second law. (In the next chapter, we will show how to expand this approach to include nonisolated systems and systems that cannot be modeled as a collection of particles.)

Why do we need two approaches? Can we just pick our favorite and ignore the other approach? The answer is that you must learn both approaches because some problems are better solved with one approach than the other, and many advanced problems are best solved by combining the two approaches.

You also must be able to identify the sort of problems that are best suited for each approach. The conservation approach is good when we are interested in the position and speed of particles in a system at two different times. For example, Figure 8.36 shows two possible paths a small cart may take. In both cases the cart starts at the same position from rest and ends at the same position. If we want to compare the final speed of the cart in the two cases, the conservation approach works well. We know that the Earth–cart system has the same change in gravitational potential energy, so we conclude that the cart has the same final speed in both cases.

If, however, we want to know which is the quicker path, the conservation approach cannot help. The conservation approach does not help us figure out the time elapsed. So, if we are interested in knowing the elapsed time, we must use the force approach. Because the path in Figure 8.36B is curved, the net force on the cart is not a constant, and it is difficult to find the net force on the cart at every moment. So, applying the force approach would be difficult, but with the help of a computer, we could find the answer.

8-9 Special Case: Orbital Energies

ORBITAL ENERGIES Special Case

In this section, we take a closer look at the energy associated with a particle in orbit around another, much more massive particle at rest such as a comet in orbit around the Sun. We consider this two-particle system to be isolated with only conservative forces acting so that mechanical energy is conserved.

DERIVATION Mechanical Energy for Circular Orbit System

Let us start with the circular orbit of a particle (of mass m) such as a comet around a much more massive particle (of mass M) such as the Sun (Fig. 8.37). The comet maintains a constant distance r from the Sun and a constant speed v_c, where the subscript c stands for circular. We will show that the mechanical energy of the system is given by

$$E = -\frac{1}{2}\left(\frac{GMm}{r}\right) \tag{8.19}$$

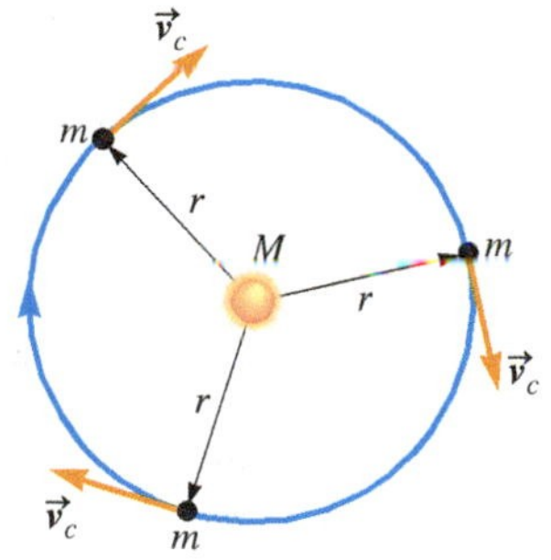

FIGURE 8.37 A two-particle system. A particle of mass m orbits a particle of mass M at constant speed v_c in a circular path of radius r.

For the purpose of this derivation, we neglect the motion of the more massive particle (the Sun). Then, the system's kinetic energy is given by the comet's motion, and K is constant because the comet's mass and speed are constant.	$K = \frac{1}{2}mv_c^2$
The system's potential energy is constant because the masses are constant, and so is the separation r.	$U_G = -\frac{GMm}{r}$ (8.7)
Because the kinetic and potential energies are constant, the mechanical energy is constant, too.	$E = K + U_G = \frac{1}{2}mv_c^2 - \frac{GMm}{r}$ (1)
The centripetal acceleration of the comet is equal to the gravitational field of the Sun at r. So, $\vec{a}_c = -\frac{v^2}{r}\hat{r}$ (Eq. 4.36) equals $\vec{g}(r) = -\frac{GM}{r^2}\hat{r}$ (Eq. 7.13)	$a = g(r)$ $\frac{v_c^2}{r} = \frac{GM}{r^2}$
Cancel out one r.	$v_c^2 = \frac{GM}{r}$ (8.20)
Substitute Equation 8.20 into Equation (1) and simplify to find the mechanical energy of the system.	$E = \frac{1}{2}m\left(\frac{GM}{r}\right) - \frac{GMm}{r}$ $E = -\frac{1}{2}\left(\frac{GMm}{r}\right)$ ✓ (8.19)

COMMENTS

We can take this result a step further by substituting Equation 8.7 for the term in parentheses.

$$E = \tfrac{1}{2}U_G \tag{8.21}$$

What we found is important and remarkable: For a system in which one particle orbits the other in a circular orbit, the mechanical energy is a constant equal to one-half the potential energy. In Problem 59, you will be asked to show that $K = -E$. If these relationships among the potential, kinetic, and mechanical energy don't hold for an orbiting system, the orbit is not circular.

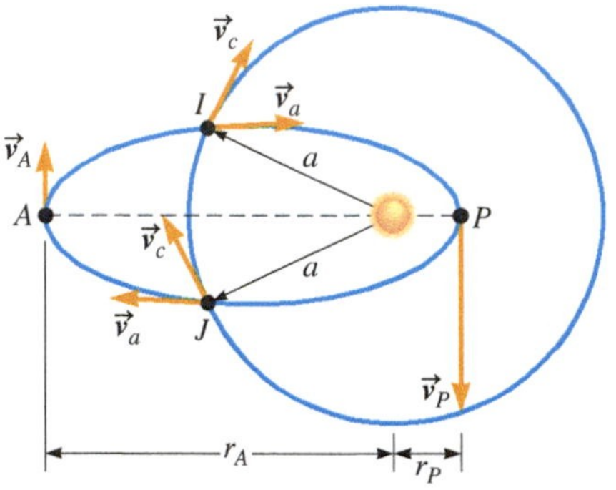

FIGURE 8.38 Comparison of a circular and an elliptical orbit. The semimajor axis of the ellipse is the radius of the circular orbit. At points *I* and *J* on the ellipse, the particle's speed is the same as it would be on the circle. At perihelion *P*, the particle's speed is too high to maintain a circular orbit, and at aphelion *A*, it is too low.

Elliptical Orbits

Let's start by comparing two systems. In one system, a comet has an elliptical orbit around the Sun; in the other, the comet's orbit is circular (Fig. 8.38). For the purpose of our comparison, the radius of the circular orbit is equal to the semimajor axis of the elliptical orbit ($r = a$), and both systems have the same mechanical energy (Eq. 8.19):

$$E = -\frac{1}{2}\frac{GMm}{r} = -\frac{1}{2}\frac{GMm}{a} \tag{8.22}$$

The kinetic and potential energies are constant for a circular orbit, but they are not constant for an elliptical orbit. We can, however, easily find the kinetic and potential energies of the elliptical orbit system when the comet is at point *I* or point *J* (Fig. 8.38). At these locations, the circular orbit crosses the elliptical orbit and $r = a$.

So, at *I* and *J*, the potential energy of the system is

$$U_G = -\frac{GMm}{a} \tag{8.23}$$

The kinetic energy at *I* or *J* is found from the conservation of mechanical energy (with Eqs. 8.22 and 8.23):

$$K = E - U_G = -\frac{1}{2}\frac{GMm}{a} - \left(-\frac{GMm}{a}\right)$$

$$K = \frac{1}{2}\frac{GMm}{a} \tag{8.24}$$

In the case of an elliptical orbit, only the mechanical energy is constant. The equations for potential (Eq. 8.23) and kinetic (Eq. 8.24) energies only hold at *I* and *J* where $r = a$. We can gain insight into how these energies change by examining the comet's speed as it moves in the elliptical orbit.

First, we find the comet's speed v_a at point *I* or *J* from Equation 8.24:

$$K = \frac{1}{2}\frac{GMm}{a} = \frac{1}{2}mv_a^2$$

$$v_a = \sqrt{\frac{GM}{a}} = \sqrt{\frac{GM}{r}} \tag{8.25}$$

where we substituted $r = a$ at these two points. By comparing Equation 8.25 with Equation 8.20, we find that at *I* or *J* (where the two orbits cross), the comet's speed on the elliptical orbit equals the speed it would have on the circular orbit, $v_a = v_c$. We know that the speed v_c of the comet in a circular orbit is constant. From Kepler's second law, however, we know that the speed of the comet in the elliptical orbit varies.

When the comet is at *I* or *J*, $v_a = v_c$, but it does not maintain a circular orbit because its velocity is not perpendicular to the centripetal force provided by the Sun's gravity. So, for a comet at *I* (analogous to point *a* in Fig. 8.1B, p. 214), the Sun's gravity causes the comet to speed up, and for a comet at *J*, gravity causes it to slow down. The comet's velocity is perpendicular to its acceleration only at perihelion and aphelion.

Even at those points, though, the comet cannot maintain a circular orbit. For a comet to maintain a circular orbit of radius *r*, its speed must be given by $v_c = \sqrt{GM/r}$

(Eq. 8.20). The comet in an elliptical orbit moves faster near perihelion than it does in a circular orbit ($v_P > v_c$), so the comet moves away from the Sun. Similarly, when the comet is at aphelion, its speed is too low ($v_A < v_c$) to maintain a circular orbit, so the comet moves toward the Sun. In the next example, we connect the comet's changing speed to the system's changing kinetic and potential energies.

EXAMPLE 8.12 Bar Charts for an Elliptical Orbit

Sketch an elliptical orbit for a comet around the Sun, including points A, P, I, and J as in Figure 8.38. For each of these four points, sketch a bar chart.

INTERPRET and ANTICIPATE

At points I and J, the comet's kinetic and potential energies are equal to what they would be if the comet were in a circular orbit. At perihelion P, the comet is moving faster than v_c, so it has more kinetic energy. Likewise, at aphelion A, the comet is moving slower than v_c, so it has less kinetic energy. The mechanical energy is constant through the entire orbit, so we can find the potential energy at P and A.

SOLVE

The orbit closely resembles Figure 8.38.

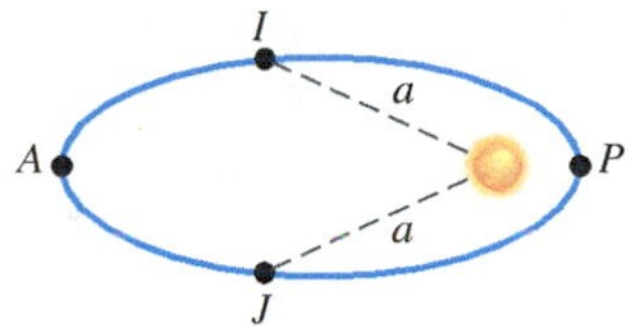

FIGURE 8.39

Bar charts are shown in Figure 8.40. Mechanical energy is constant, so E bars are the same length at all points (A, I, P, and J).

At aphelion *A*, the comet moves too slowly to maintain a circular orbit. The K bar is shorter than the E bar. The comet is farthest from the Sun and the U_G bar is shorter than in other configurations (less negative U_G).

At points *I* and *J*, the bar chart is the same as for a circular orbit.

At perihelion *P*, the comet moves too fast to maintain a circular orbit. The K bar is longer than the E bar. The comet is closer to the Sun than at I or J and U_G is less, so it is shown as a longer (negative) bar.

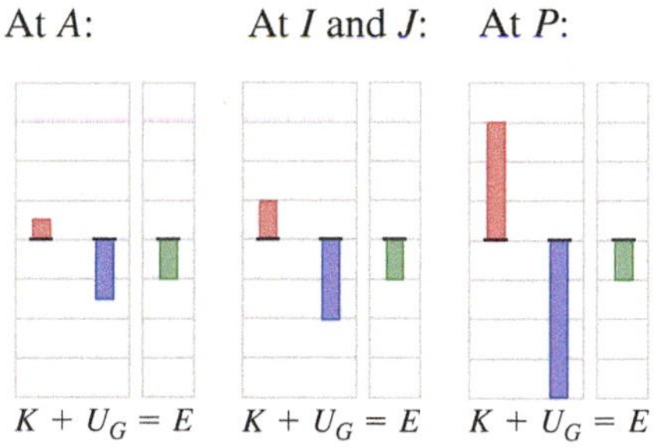

FIGURE 8.40

CHECK and THINK

When working with universal gravity, remember that when the reference configuration is set so that $U_G(\infty) = 0$, the potential energy becomes more negative as the particles move closer together as represented by the long negative U bar at perihelion in Figure 8.40. This change is nonetheless a *decrease* in potential energy, so the kinetic energy must *increase*. (The comet's highest speed is near perihelion.)

EXAMPLE 8.13 CASE STUDY Halley's Speed

Find the speed of Comet Halley when *Giotto* took its pictures. At that time, the comet was at $r_G = 1.33 \times 10^{11}$ m. To check your result, find the comet's speed at perihelion $r_P = 8.83 \times 10^{10}$ m, at aphelion $r_A = 5.27 \times 10^{12}$ m, and at $a = 2.678 \times 10^{12}$ m. (There is some variation in the observed values of the comet's orbital parameters. The values in this example come from the average of a small sample of such measurements. Work with all significant figures given and round your final answer in kilometers per second to two significant figures.)

Example continues on page 238 ▶

INTERPRET and ANTICIPATE

The ellipse in Figure 8.38 works as a sketch of Comet Halley's orbit. The bar charts are similar to Figure 8.40 in Example 8.12. Figure 8.41 is our energy graph. It might seem that we still need the mass of Comet Halley to find its speed, but we will show that the speed of a comet does not depend on its mass.

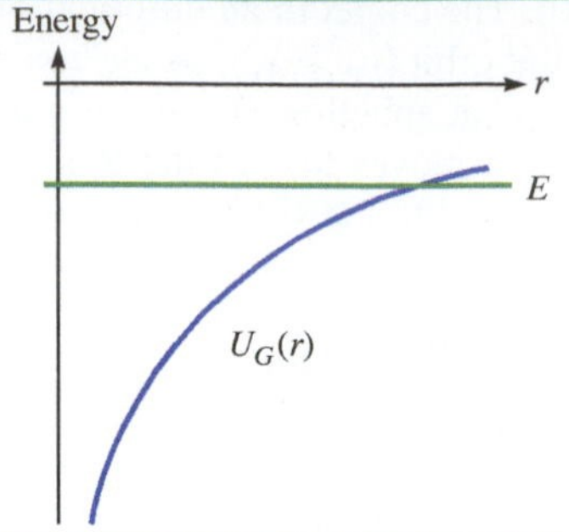

FIGURE 8.41

SOLVE

We need an expression for Comet Halley's speed. We start by writing an expression for its mechanical energy.

$$E = -G\frac{M_\odot m}{2a} \quad (8.22)$$

Now we need to write down the potential energy of the system at some arbitrary position.

$$U_G = -G\frac{M_\odot m}{r} \quad (8.7)$$

We also need an expression for the kinetic energy.

$$K = \tfrac{1}{2}mv^2$$

We know that the potential energy plus the kinetic energy equals the mechanical energy.

$$E = K + U$$

$$-G\frac{M_\odot m}{2a} = \frac{1}{2}mv^2 - G\frac{M_\odot m}{r}$$

Solve this equation for speed. The mass of the comet falls out of the equation.

$$-G\frac{M_\odot}{2a} = \frac{1}{2}v^2 - G\frac{M_\odot}{r}$$

$$v^2 = 2\left(\frac{GM_\odot}{r} - \frac{GM_\odot}{2a}\right)$$

$$v = \sqrt{2GM_\odot\left(\frac{1}{r} - \frac{1}{2a}\right)}$$

Because we need to evaluate this expression for four different values of r, it is useful to substitute for all the other quantities (G, $M_\odot$, and a).

$$v = \sqrt{2(6.673 \times 10^{-11}\,\text{N}\cdot\text{m}^2/\text{kg}^2)(1.989 \times 10^{30}\,\text{kg})\left[\frac{1}{r} - \frac{1}{2(2.678 \times 10^{12}\,\text{m})}\right]}$$

$$v(r) = \sqrt{(2.655 \times 10^{20}\,\text{m}^3/\text{s}^2)\left(\frac{1}{r} - \frac{1}{5.356 \times 10^{12}\,\text{m}}\right)}$$

Now convert AU to m and substitute the four values of r. We have shown only the explicit calculation at r_G; you may verify the other results. We make our calculations with three significant figures and report our final answers to two significant figures.

$$v(r_G) = \sqrt{(2.655 \times 10^{20}\,\text{m}^3/\text{s}^2)\left(\frac{1}{1.33 \times 10^{11}\,\text{m}} - \frac{1}{5.356 \times 10^{12}\,\text{m}}\right)}$$

$$v(r_G) = 4.41 \times 10^4\,\text{m/s} = 44\,\text{km/s}$$

$$v(r_P) = 54\,\text{km/s}$$

$$v(r_A) = 0.92\,\text{km/s}$$

$$v(a) = 7.0\,\text{km/s}$$

CHECK and THINK

These speeds are consistent with the energy graph and bar charts, in Example 8.12, from which we expect $K_P > K_G > K_a > K_A$. So, we expect $v(r_P) > v(r_G) > v(a) > v(r_A)$, just as we found. Let's now think about the value we found; 44 km/s is about 100,000 mph. Imagine trying to meet up with a friend moving at that speed. If you were running just 10 minutes behind schedule, your friend would be nearly 20,000 miles from your rendezvous point at the time you got there. Clearly, a lot of precise calculations must go into trying to rendezvous with a small, fast-moving object like a comet.

SUMMARY

Underlying Principles: Conservation of mechanical energy

If only conservative forces are acting, the mechanical energy ($E = K + U$) of an isolated system is constant:

$$K_i + U_i = K_f + U_f$$

Major Concepts

1. No energy is exchanged between an **isolated system** and its environment.
2. **Energy** describes the motion and configuration of a system.
3. **Kinetic energy** K describes the motion of a system. It depends both on the mass and speed of the system:

$$K = \tfrac{1}{2}mv^2 \tag{8.1}$$

4. **Potential energy** depends on the configuration of the system. For a system that consists of two particles, the change in potential energy is given by

$$\Delta U = U_f - U_i = -\int_{x_i}^{x_f} F_x\,dx \tag{8.3}$$

when one particle moves from x_i to x_f.

Special Cases

1. Potential energy
   a. **Gravity near surface of the Earth:** Gravitational potential energy of the system is given by

   $$U_g(y) = mgy \tag{8.5}$$

   where the reference configuration is at $y = 0$.
   b. For a system that consists of two particles 1 and 2 with the reference configuration chosen such that $\lim_{r\to\infty} U = 0$, the **universal gravitational potential energy** is

   $$U_G(r) = -G\frac{m_1 m_2}{r} \tag{8.7}$$

   c. If the origin of a coordinate system is at the relaxed position of a spring, which is set to the reference configuration, the **elastic potential energy** of the particle–spring system is given by

   $$U_s = \tfrac{1}{2}kx^2 \tag{8.9}$$

2. Conservation of mechanical energy and orbital motion
   a. For a **circular orbit**, there is a special relationship between the mechanical energy, the potential energy, and the kinetic energy of the system, with $E = \frac{1}{2}U_G$ (Eq. 8.21) and $E = -K$.
   b. For an **elliptical orbit**, the mechanical energy is

   $$E = -\frac{1}{2}\frac{GMm}{a} \tag{8.22}$$

Tools

1. A **bar chart** helps visualize the conservation of mechanical energy. The vertical height of the bar represents the energy. A horizontal line is drawn to represent the zero point of the energy. If the energy is positive, the bar is drawn above that horizontal line, and if the energy is negative, the bar is below the line.
2. Another pictorial representation is an **energy graph**, which displays a potential energy curve and a horizontal E line representing the constant mechanical energy. The kinetic energy is not displayed, but is represented by the space between the horizontal E line and the potential energy curve.

PROBLEM-SOLVING STRATEGY Applying the Principle of Conservation of Mechanical Energy

INTERPRET and ANTICIPATE

1. Choose an **isolated system** (Section 8-6).
2. Pick a **coordinate system and the reference configuration.**
3. Identify an **initial and final time.**
4. Draw a **bar chart.**

SOLVE

5. Apply the principle of **conservation of mechanical energy.**
6. Write down any **other energy equations.**
7. Do **algebra** before substitution.

PROBLEMS AND QUESTIONS

A = algebraic C = conceptual E = estimation G = graphical N = numerical

8-1 Another Approach to Newtonian Mechanics

1. C CASE STUDY From Figure 8.1B for Comet Halley, is the comet speeding up, slowing down, or maintaining its speed just before it reaches aphelion? What about just after it passes aphelion? Explain.

8-2 Energy

2. E Estimate the kinetic energy of the following: **a.** An ant walking across the kitchen floor **b.** A baseball thrown by a professional pitcher **c.** A car on the highway **d.** A large truck on the highway

3. N A 70-kg man runs by a seated woman at 4.5 km/h. The woman's mass is 55 kg. **a.** According to the woman, what is the man's kinetic energy? **b.** According to the man, what is the woman's kinetic energy?

4. N An astronaut and his gear (m_A = 90.0 kg) are at rest with respect to his shuttlecraft (m_S = 12,500 kg). The astronaut and the shuttlecraft are initially separated by a distance of 10.0 m. If the astronaut moves farther away from the shuttlecraft to a distance of 20.0 m, what is the change in gravitational potential energy for the astronaut–shuttlecraft system? (*Hint*: Write an expression for the gravitational force between the astronaut and the shuttle using Eq. 8.3.)

5. N A 0.430-kg soccer ball is kicked at an initial speed of 34.0 m/s at an angle of 35.0° to the horizontal. What is the kinetic energy of the soccer ball when it has reached the apex of its trajectory?

6. Both a car (m_c = 1550 kg) and a truck (m_t = 8150 kg) are initially at rest and are each sped up to a final speed of 30 mph.
- **a.** N What is the final kinetic energy of the car?
- **b.** N What is the final kinetic energy of the truck?
- **c.** C Are we able to determine which vehicle was subject to a greater acceleration? If so, explain how. If not, what would we need to know to make that determination?

Problems 7 and 8 are paired.

7. N According to a seated woman, a 67.7-kg man runs with his 32.4-kg dog toward a 20.2-kg child who is running toward the man. The man and dog's speed is 3.50 km/h, and the child's speed is 1.25 km/h. **a.** According to the woman, what is the kinetic energy of the man–dog–child system? **b.** According to the man, what is the kinetic energy of the man–dog–child system?

8. According to a seated woman, a 67.7-kg man runs toward a 20.2-kg child who is running toward the man. The man's speed is 3.50 km/h, and the child's speed is 1.25 km/h.
- **a.** N According to the woman, what is the kinetic energy of the child–man system?
- **b.** C Is it possible to find a reference frame in which the child–man system's kinetic energy is zero? If so, describe the reference frame. If not, explain why not.
- **c.** C If the child and man were running in the same direction, would the kinetic energy of the system increase or would it decrease according to the woman?

9. N Lottery balls are thrown into a hopper and are set into motion by air forced through the container. Each ball has a mass of 0.0050 kg. At one instant in time, ball 1 is moving upward with a speed of 1.5 m/s, and ball 2 is moving downward with a speed of 1.75 m/s. If the kinetic energy of the three-ball system at this instant is 0.0260 J, what must be the speed of ball 3?

10. C If you model the Earth as a particle, can you estimate its potential energy? Why or why not?

8-3 Gravitational Potential Energy Near the Earth

11. N A ball (m = 2.5 kg) is given an initial velocity upwards of 15 m/s. What is the change in the gravitational potential energy of the ball from its initial height to when the ball is at the peak of its motion?

Problems 12, 13, and 14 are grouped.

12. C When you are asked to find the change in potential energy, the contents of the system are implied by the question. For example, in Figure P8.12 several situations are shown in which a ball moves from point A to point B. If you are asked to find the change in gravitational potential energy, what object or objects must be included in the system in each case?

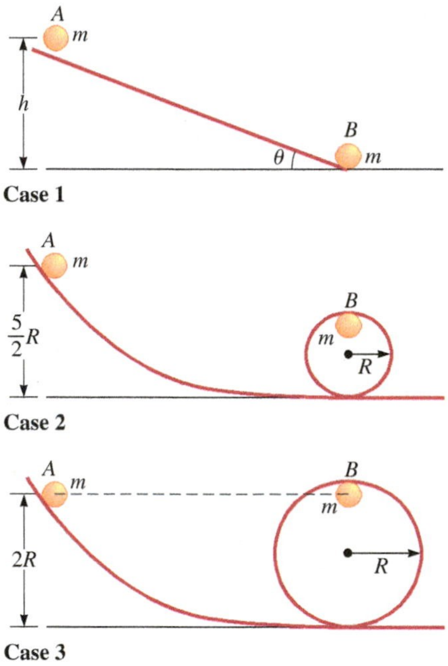

FIGURE P8.12 Problems 12, 13, and 14.

13. A In each situation shown in Figure P8.12, a ball moves from point A to point B. For each case, find an algebraic expression for the change in potential energy in terms of the parameters given in the figure. You can assume that the radius of the ball is negligible.

14. N In each situation shown in Figure P8.12, a ball moves from point A to point B. Use the following data to find the change in the gravitational potential energy in each case. You can assume that the radius of the ball is negligible.
- **a.** h = 1.35 m, θ = 25°, and m = 0.65 kg
- **b.** R = 33.5 m and m = 756 kg
- **c.** R = 33.5 m and m = 756 kg

15. N A wooden block (m = 1.75 kg) slides down an incline from a height 1.50 m above a tabletop and eventually comes to a stop due to friction between the block and the table. What is the change in the block-Earth system's gravitational potential energy?

16. **A** Two blocks of masses m_1 and m_2 are connected by a cord that passes over a pulley as shown in Figure P8.16. They start at the same height, and the hanging block is lowered through a distance h, causing the block on the incline to rise. Calculate the change in gravitational potential energy for each of the blocks.

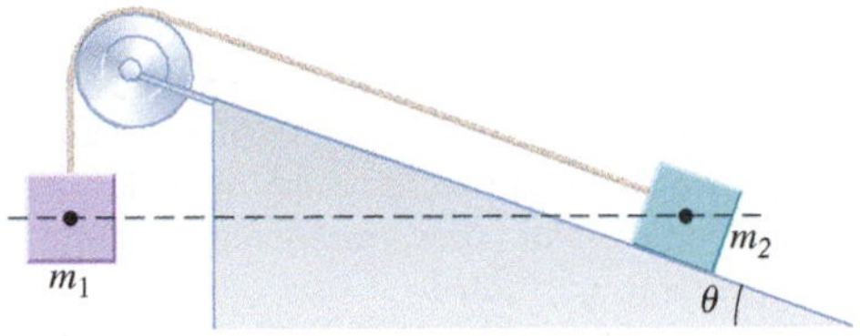

FIGURE P8.16

17. **N** A playground slide is 8.00 ft long and makes an angle of 33.0° with the horizontal. A 55.0-kg child, initially at the top, slides all the way down to the bottom of the slide. **a.** Choosing the bottom of the slide as the reference configuration, what is the system's potential energy when the child is at the top and at the bottom of the slide? What is the change in potential energy as the child slides from the top to the bottom of the slide? **b.** Repeat part (a), choosing the top of the slide as the reference configuration.
18. **N** In Example 8.4, we calculated the change in potential energy of the Earth–hammer system when the reference configuration was set at the driveway and at the basketball hoop. Repeat this calculation, this time setting the reference configuration to the roof at the point of the hammer's release.
19. **N** A ball of mass 0.40 kg hangs straight down on a string of length 15 cm. It is then swung upward, keeping the string taut, until the string makes an angle of 65° with respect to the vertical. Find the change in the gravitational potential energy of the Earth-ball system.
20. **N** A 34.0-cm-long pendulum with a 245-g bob swings through an angle of 180°. Choose the reference configuration as the point where the pendulum string is vertical. Assuming that the pendulum string remains taut, what is the gravitational potential energy of the pendulum bob–Earth system when the pendulum string makes an angle of **a.** 90° with the vertical? and **b.** 40.5° with the vertical? **c.** What is the gravitational potential energy of the pendulum bob–Earth system at the lowest point of the pendulum's motion?

8-4 Universal Gravitational Potential Energy

21. The average distance between the Earth and the Sun is about 1.5×10^{11} m, whereas the average distance between the Earth and Mars is about 2.3×10^{11} m. Assume the usual convention that the gravitational potential energy referenced to the Earth is zero when infinitely far away.
    a. **N** What is the gravitational potential energy of the Earth–Sun system?
    b. **N** What is the gravitational potential energy of the Earth–Mars system?
    c. **C** Describe how these values indicate that the Sun has a greater effect on the Earth's motion than Mars does. What is the most dominant factor that causes the Sun to have the greater effect?
22. **E** Imagine that an object the size of a car is raised to the top of a skyscraper. Estimate the change in the Earth–car system's potential energy. If the car were raised into geosynchronous orbit instead, estimate the change in the system's potential energy. *Hint*: The altitude of geosynchronous orbit is $h = 3.59 \times 10^7$ m; see Example 7.6.

8-5 Elastic Potential Energy

23. One type of toy car contains a spring that is compressed as the wheels are rolled backward along a surface. The spring remains compressed until the wheels are freed and the car is allowed to roll forward. Jose learns that if he rolls the car backward for a greater distance (up to a certain point), the car will go faster when he releases it. The spring compresses 1.00 cm for every 10.0 cm the car is rolled backward.
    a. **N** Assuming the spring constant is 150.0 N/m, what is the elastic potential energy stored in the spring when Jose rolls the car backward 20.0 cm?
    b. **N** What is the elastic potential energy stored in the spring when he rolls the car backward 30.0 cm?
    c. **C** Explain the correlation between the results for parts (a) and (b) and Jose's observations of different speeds.

Problems 24 and 26 are paired.

24. A block is placed on top of a vertical spring, and the spring compresses. Figure P8.24 depicts a moment in time when the spring is compressed by an amount h.
    a. **C** To calculate the change in the gravitational and elastic potential energies, what must be included in the system?
    b. **A** Find an expression for the change in the system's potential energy in terms of the parameters shown in Figure P8.24.
    c. **N** If $m = 0.865$ kg and $k = 125$ N/m, find the change in the system's potential energy when the block's displacement is $h = 0.0650$ m, relative to its initial position.

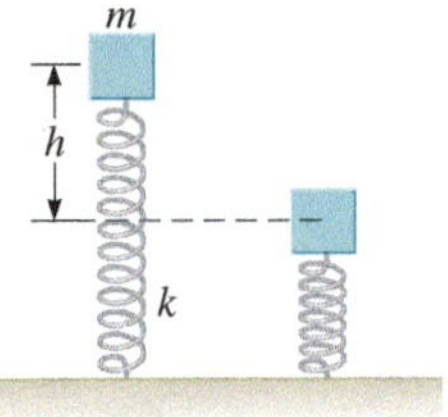

FIGURE P8.24

25. **N** Rubber tends to be nonlinear as an elastic material. Suppose a particular rubber band exerts a restoring force given by $F_x(x) = -Ax - Bx^2$, where the empirical constants are $A = 14$ N/m and $B = 3.3$ N/m^2 so that F_x is in newtons when x is in meters. Calculate the change in elastic potential energy of the rubber band when an external force stretches it from $x = 0$ to $x = 0.20$ m.
26. **N** A block is hung from a vertical spring. The spring stretches ($h = 0.0650$ m) as shown for a particular instant in time in Figure P8.26. Consider the Earth, spring, and block to be in the system. If $m = 0.865$ kg and $k = 125$ N/m, find the change in the system's potential energy between the two times depicted in the figure.

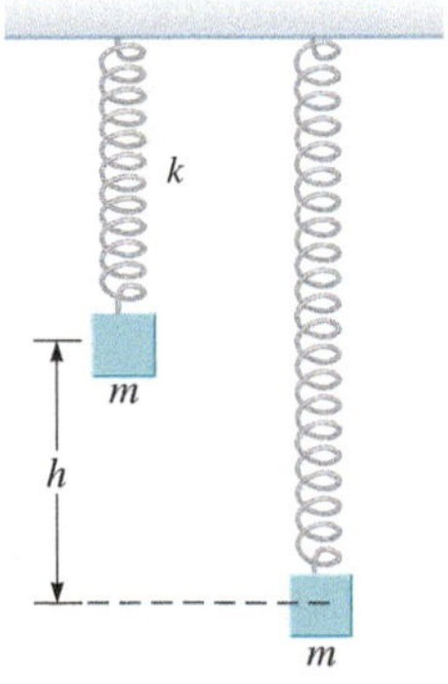

FIGURE P8.26

27. **A** A spring of spring constant k lies along an incline as shown in Figure P8.27. A block of mass m is attached to the spring. The spring compresses, and the block comes to rest as shown. Find an expression for the change in the Earth–block–spring system's potential energy in terms of the parameters given in the figure.

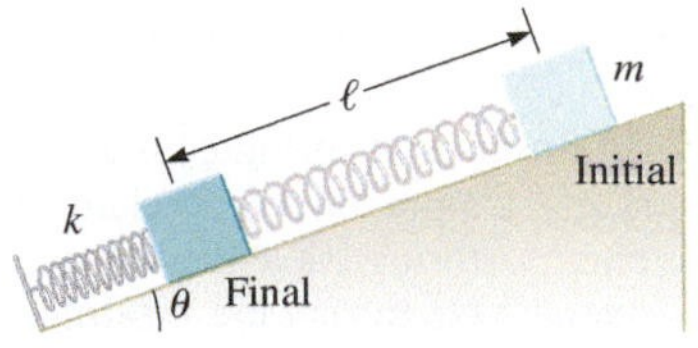

FIGURE P8.27

28. A block on a frictionless, horizontal surface is attached to two springs as shown in Figure P8.28. The block is displaced, compressing one spring and stretching the other.

a. **A** Find an expression for the change in the block–springs system's potential energy in terms of the parameters given in the figure.

b. **C** Is it possible to displace the block in such a way that the system's potential energy does not change?

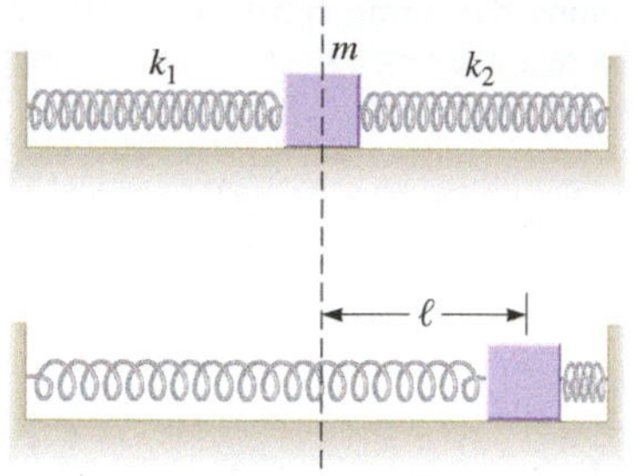

FIGURE P8.28

8-6 Conservation of Mechanical Energy

29. **N** A falcon is soaring over a prairie, flying at a height of 45.0 m with a speed of 12.9 m/s. The falcon spots a mouse running along the ground and dives to catch its dinner. Ignoring air resistance, and assuming the falcon is only subject to the gravitational force as it dives, how fast will the falcon be moving the instant it is 5.00 m above the ground?

30. A stellar black hole may form when a massive star dies. The star collapses down to a single point. Our Sun will not become a black hole because it does not have enough mass, so let us imagine a black hole with eight times the mass of the Sun. A single point has no radius, but a black hole has an effective radius known as the *Schwarzschild radius*. If an object gets inside the Schwarzschild radius, it never escapes. The Schwarzschild radius of an eight-solar-mass black hole is about 24 km.

a. **N** Find the escape speed using the Schwarzschild radius.

b. **C** Compare your answer to the speed of light and speculate on how these objects got the name "black hole."

31. **N** A newly established colony on the Moon launches a capsule vertically with an initial speed of 1.50 km/s. Ignoring the rotation of the Moon, what is the maximum height reached by the capsule?

32. The Flybar high-tech pogo stick is advertised as being capable of launching jumpers up to 6 ft. The ad says that the minimum weight of a jumper is 120 lb and the maximum weight is 250 lb. It also says that the pogo stick uses "a patented system of elastometric rubber springs that provides up to 1200 lbs of thrust, something common helical spring sticks simply cannot achieve (rubber has 10 times the energy storing capability of steel)."

FIGURE P8.32

a. **E** Use Figure P8.32 to estimate the maximum compression of the pogo stick's spring. Include the uncertainty in your estimate.

b. **N** What is the effective spring constant of the elastometric rubber springs? Comment on the claim that rubber has 10 times the energy-storing capability of steel.

c. **N** Check the ad's claim that the maximum height a jumper can achieve is 6 ft.

33. **N** An uncrewed mission to the nearest star, Proxima Centauri, is launched from the Earth's surface as a projectile with an initial speed of 43.0 km/s, just enough for the spacecraft to escape the Earth's gravity and leave the solar system. Ignoring air resistance and the Earth's rotation, what is the speed of the spacecraft when it is more than halfway to the star? Assume we are ignoring the effect of the Sun on the spacecraft.

34. **C** A small ball is tied to a string and hung as shown in Figure P8.34. It is released from rest at position 1, and, during the swing, the string meets a fixed peg as shown. Explain why position 2 at which the ball comes momentarily to rest must be at the same height as position 1.

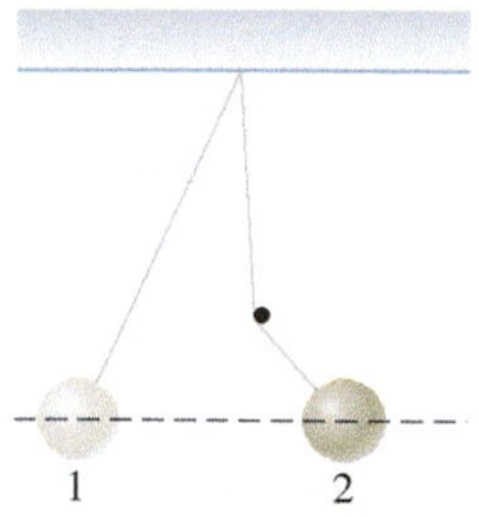

FIGURE P8.34

35. **C** On a late-night TV talk show, the host tosses watermelons from the top of a building with the sole purpose of watching them splatter as they hit the sidewalk several stories below. If the host tosses a watermelon up, down, or horizontally with the same initial speed, it will hit the sidewalk with the same final speed. Explain why.

36. Suppose an astronaut jumps straight up in his full gear, first on the Earth and later on the Moon. Assume his take-off speed is the same in each case.

a. **C** **INTERPRET and ANTICIPATE:** Do you expect him to jump higher on the Earth or on the Moon?

b. **N** Find the ratio of his maximum positions, $y_{\text{max}\,\oplus}/y_{\text{max Moon}}$. Does your result match your expectations? Explain.

37. **N** In the Marvel comic series *X-Men*, Colossus would sometimes throw Wolverine toward an enemy in what was called a *fastball special*. Suppose Colossus throws Wolverine at an angle of 30.0° with respect to the ground (Fig. P8.37). Wolverine is 2.15 m above the ground when he is released, and he leaves Colossus's hands with a speed of 20.0 m/s. **a.** Using conservation of energy and the components of the initial velocity, find the maximum height attained by Wolverine during the flight. **b.** Using conservation of energy, what is Wolverine's speed the instant before he hits the ground?

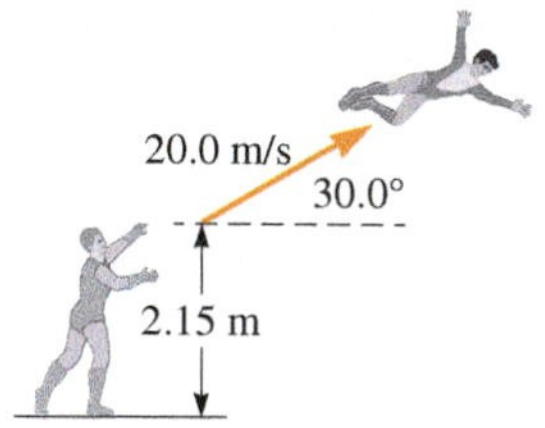

FIGURE P8.37

8-7 Applying the Conservation of Mechanical Energy

38. **N** A game booth at a carnival invites contestants to strike a softball affixed to the bottom end of a lightweight stick 1.25 m long whose opposite end is attached to a frictionless, horizontal axle in such a way that the ball and stick are able to swing around in a full circle. If the ball and stick are initially at rest and hanging straight down, what is the minimum speed the softball must have after it is struck for the contestant to win?

39. **N** Figure P8.39 shows two bar charts. In each, the final kinetic energy is unknown. **a.** Find K_f. **b.** If $m = 2.5$ kg, find v_f.

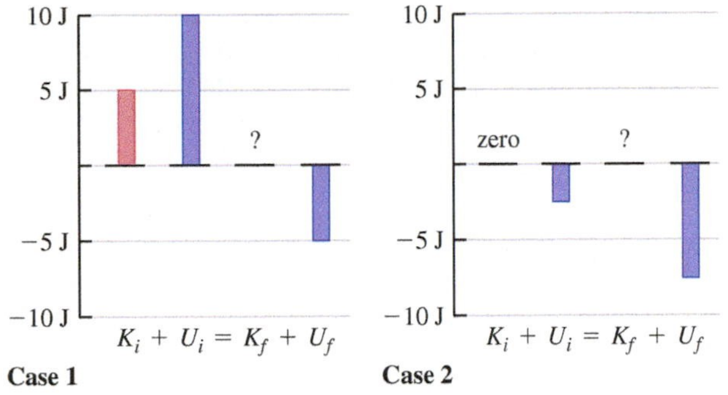

FIGURE P8.39

Problems 40 and 41 are paired.

40. **N** What is the escape speed of the Earth and of the Moon?

41. **N** If a spacecraft is launched from the Moon at the escape speed of the Earth, how fast will the spacecraft be going when it is very far away from the Moon, ignoring the effects of other celestial bodies?

42. **N** A 1.50-kg box rests atop a massless vertical spring with $k = 4250$ N/m that has been compressed by 15.0 cm from its equilibrium position. The box is released and leaves the spring when it reaches its equilibrium position. What is the maximum height the box reaches above its original position?

43. **N** A man unloads a 5.0-kg box from a moving van by giving it an initial speed of 1.3 m/s down a ramp that makes an angle of 20° with the horizontal. It slides with negligible friction for 2.7 m along the ramp before reaching the man's helper. What is the speed of the box as the helper catches it?

44. **N** Starting at rest, Tina slides down a frictionless waterslide with a horizontal section at the bottom that is 4.00 ft above the surface of the swimming pool and strikes the water a distance of 15.0 ft away from the end of the slide. Using conservation of energy, what is Tina's initial height on the waterslide?

Problems 45 and 46 are paired.

45. **N** Karen and Randy are playing with a toy car and track. They set up the track on the floor as shown in Figure P8.45, where the apparatus on the far left is used to launch the car forward by pressing down on the top portion. After the car is launched, it follows the track and continues upward, leaving the track. If the car has a mass of 130 g and it reaches a maximum vertical height of 1.2 m above the floor, what was the speed of the car while it was moving along the floor? Ignore the effects of friction and air resistance.

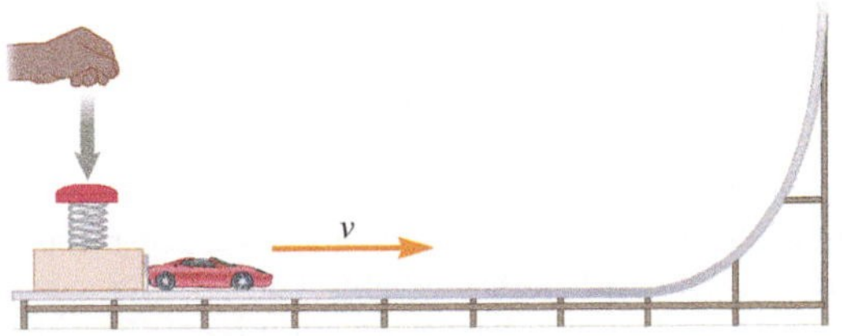

FIGURE P8.45

46. **N** Karen and Randy are playing with a toy car and track. They set up the track on the floor as shown in Figure P8.46, where the apparatus on the far left is used to launch the car forward by pressing down on the top portion. After the car is launched, it follows the track and continues, leaving the track at an angle of 35° with respect to the floor, at a height of 0.35 m above the floor. If the car has a mass of 130 g and it leaves the track with a speed of 3.85 m/s, what is the maximum height reached by the car, relative to the floor? Use conservation of energy to solve this problem. Ignore the effects of friction and air resistance.

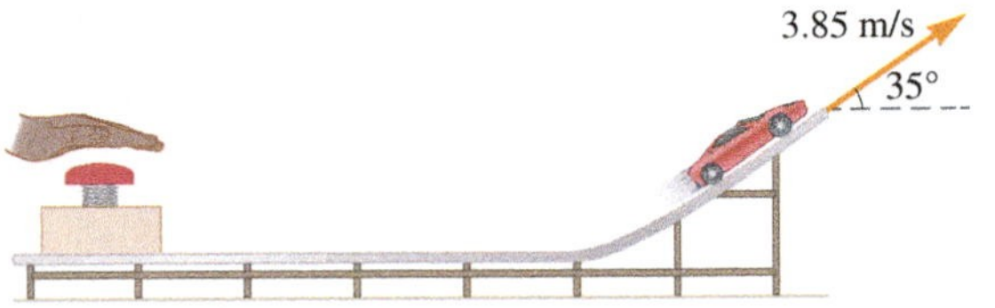

FIGURE P8.46

47. **N** An intrepid physics student decides to try bungee jumping. She obtains a cord that is 9.00 m long and has a spring constant of 5.00×10^2 N/m. When fully suited, she has a mass of 70.0 kg. She looks for a bridge to which she can tie the cord and step off. Determine the minimum height of the bridge that will allow her to stay dry (that is, so that she stops just before hitting the water below). Assume air resistance is negligible.

48. **N** A block of mass $m = 1.50$ kg attached to a horizontal spring with force constant $k = 6.00 \times 10^2$ N/m that is secured to a wall is stretched a distance of 5.00 cm beyond the spring's relaxed position and released from rest. What is the elastic potential energy of the block–spring system **a.** just before the block is released and **b.** when the block passes through the spring's relaxed position? What is the speed of the block **c.** as it passes through the spring's relaxed position and **d.** when it has compressed the spring 2.50 cm beyond its relaxed position?

49. **N** A roller-coaster track is being designed so that the roller-coaster car can safely make it around the circular vertical loop of radius $R = 10.0$ m on the frictionless track. The loop is immediately after the highest point in the track, which is a height h above the bottom of the loop. What is the minimum value of h for which the roller-coaster car will barely make it around the vertical loop?

8-8 Energy Graphs

50. **G** A jack-in-the-box is actually a system that consists of an object attached to the top of a vertical spring (Fig. P8.50). **a.** Sketch the energy graph for the potential energy and the total energy of the spring–object system as a function of compression distance x from $x = -x_{max}$ to $x = 0$, where $-x_{max}$ is the maximum amount of compression of the spring. Ignore the change in gravitational potential energy. **b.** Sketch the kinetic energy of the system between these points—the two distances in part (a)—on the same graph (using a different color).

FIGURE P8.50
Problems 50 and 79.

51. **G** A side view of a half-pipe at a skateboard park is shown in Figure P8.51. Sketch a graph of the gravitational potential energy of the skateboarder–Earth system as a function of position for a skateboarder who travels from the left side of the half-pipe to the right side. Let the leftmost point be where $x = 0$ and the lowest point in the half-pipe be where $U = 0$.

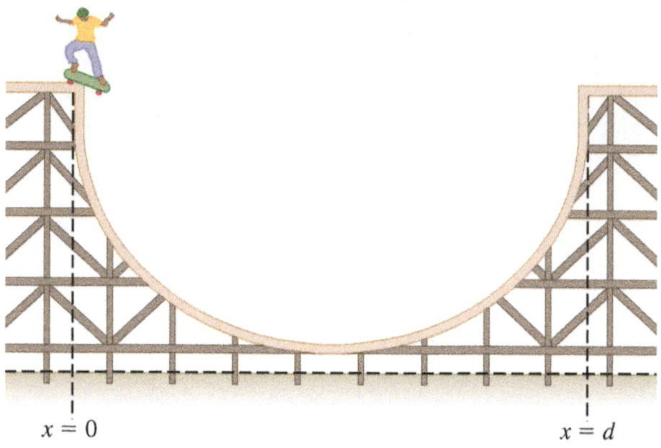

FIGURE P8.51

Problems 52, 53, and 54 are grouped.

52. **G** For this problem, you may wish to use a spreadsheet or similar program to make the following potential energy graphs. Make two graphs of potential energy versus position, where $U = mgy$, on the same set of axes, where you first set $mg = 1$ N and then $mg = 2$ N. Describe your results.

53. **G** For this problem, you may wish to use a spreadsheet or similar program to make the following potential energy graphs. Make two graphs of potential energy versus position, where $U = -GMm/r$, on the same set of axes, where you first set $GMm = 1\ \text{J} \cdot \text{m}$ and then $GMm = 2\ \text{J} \cdot \text{m}$. Describe your results.

54. **G** For this problem, you may wish to use a spreadsheet or similar program to make the following potential energy graphs. Make two graphs of potential energy versus position, where $U = \frac{1}{2}kx^2$, on the same set of axes, where you first set $k = 1$ N/m and then $k = 2$ N/m. Describe your results.

55. A particle moves in one dimension under the action of a conservative force. The potential energy of the system is given by the graph in Figure P8.55. Suppose the particle is given a total energy E, which is shown as a horizontal line on the graph.
   a. **G** Sketch bar charts of the kinetic and potential energies at points $x = 0$, $x = x_1$, and $x = x_2$.
   b. **C** At which location is the particle moving the fastest?
   c. **C** What can be said about the speed of the particle at $x = x_3$?

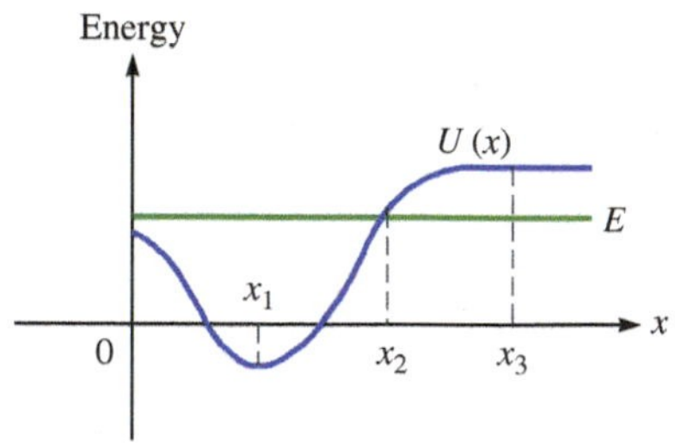

FIGURE P8.55

56. The Sun is powered by nuclear fusion. In each chain reaction, four hydrogen nuclei fuse together to become a single helium nucleus. The total combined mass of the individual hydrogen nuclei before the reaction is greater than the mass of the helium nucleus. The mass lost as a result of this chain reaction is turned into energy according to Einstein's famous equation, $E = mc^2$ (Chapter 39). Consider the first part of the chain reaction in which two hydrogen nuclei (each a single proton) fuse. Figure P8.56A shows the potential energy curve of a two-proton system. Energy in this graph is given in keV, where "eV" is the abbreviation for the unit electron-volt and $1\ \text{eV} = 1.6 \times 10^{-19}$ J. When the protons are more than about 2 fm (2×10^{-15} m) apart, the potential energy is positive. When the protons are close together, the potential energy is negative, and the protons are fused. Figure P8.56B shows a close-up of the potential energy curve when the protons are far apart.
   a. **G** What is the minimum mechanical energy required for the two protons to fuse? Give your answer in keV.
   b. **G** The mechanical energy of an average two-proton system in the Sun is around 2 keV. What is the minimum distance between the protons in such a system?
   c. **C** Explain why it seems impossible for fusion to take place in the Sun. *Note*: Of course, fusion must take place in the Sun. Although it seems impossible according to classical mechanics, quantum mechanics (Part VI) makes fusion possible. According to quantum mechanics, the seemingly impossible is merely improbable, and even events with a low probability of occurrence may still happen. Because there are so many protons in the Sun, there is a good chance that at any moment many of them will fuse.

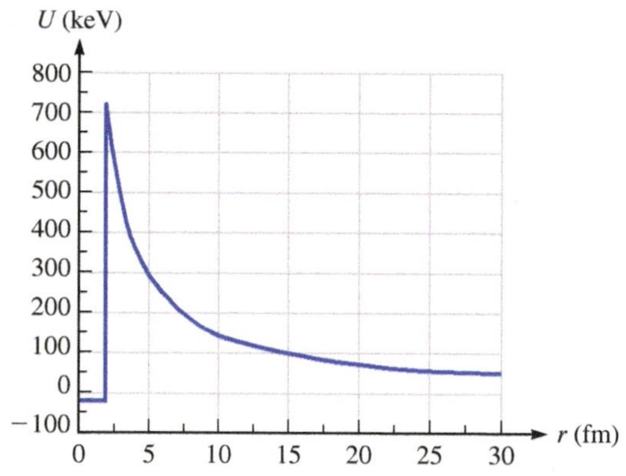

A.

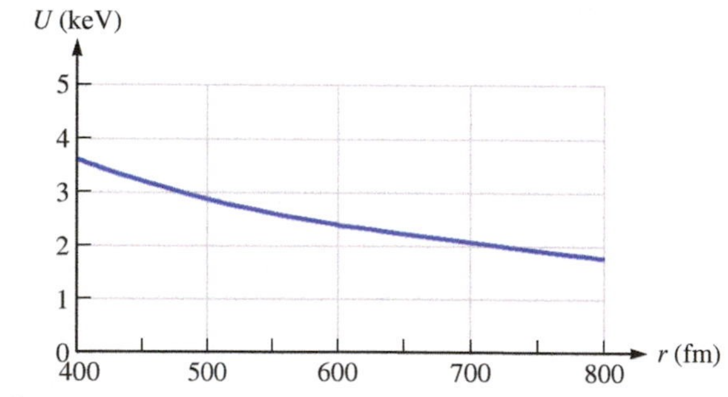

B.

FIGURE P8.56

57. Figure P8.57A shows the potential energy curve for a two-particle system. Particle 1 remains at rest at the origin, and particle 2 ($m = 2.75$ kg) is initially at $\vec{x}_i = 24.0\hat{\imath}$ m with velocity $\vec{v}_i = -3.00\hat{\imath}$ m/s as shown in Figure P8.57B. Report numerical answers to three significant figures.
   a. **N** What is the mechanical energy of this system?

b. **N** What is the kinetic energy of the system when particle 2 is at $\vec{x} = 9.00\hat{\imath}$ m?

c. **C** Will the two particles ever touch? Explain why or why not.

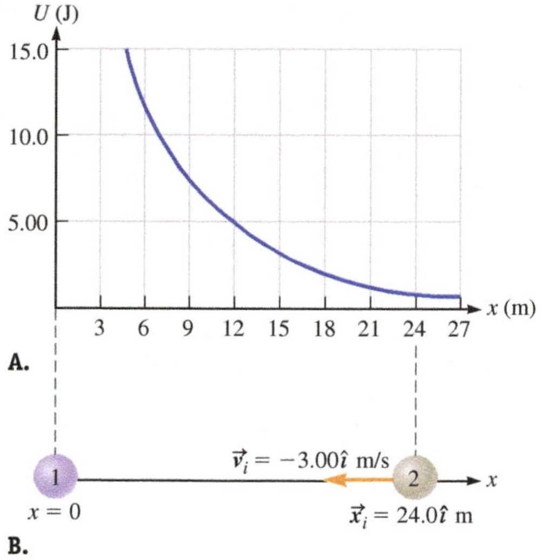

FIGURE P8.57

8-9 Special Case: Orbital Energies

58. **G** Make a bar chart for a two-particle system in which one particle is in a circular orbit around the other as in Figure 8.37 (page 235).

59. **A** Show that for the case of a circular orbit, $E = -K$. *Hint*: First show that $K = -\frac{1}{2}U_G$.

60. **N** Much of the mass of our Milky Way galaxy is concentrated in a central sphere of radius $r = 2$ kpc, where "pc" is the abbreviation for the unit "parsec"; 1 pc = 3.26 ly. Assume the Sun is in a circular orbit of radius $r = 8.0$ kpc around the central sphere of the Milky Way. The Sun's orbital speed is approximately 220 km/s; assume the central sphere is at rest.

a. **N** Estimate the mass in the inner Milky Way. Report your answer in kilograms and in solar masses.

b. **E** What is the escape speed of the Milky Way?

c. **C** **CHECK and THINK:** Do you believe that stars in the Milky Way have been observed to have speeds of 500 km/s? Explain.

61. A stellar black hole may form when a massive star dies. The mass of the star collapses down to a single point. Imagine an astronaut orbiting a black hole having eight times the mass of the Sun. Assume the orbit is circular.

a. **N** Find the speed of the astronaut if his orbital radius is $r = 1$ AU.

b. **N** Find his speed if his orbital radius is $r = 11.8$ km.

c. **C** **CHECK and THINK:** Compare your answers to the speed of light in a vacuum. What would the astronaut's orbital speed be if his orbital radius were smaller than 11.8 km?

62. **C** Astronomers are searching for Planet X in the plane of our solar system. It is expected to be beyond Pluto. Is there more mechanical energy in the Pluto–Sun system or in the Planet X–Sun system? Explain.

63. **N** CASE STUDY Comet Hale-Bopp's closest approach to the Sun is 0.914 AU, and its aphelion, or maximum distance from the Sun, is 371 AU in an elliptical orbit. (The average distance of the Earth from the Sun is 1 AU = 1.496×10^{11} m.) **a.** What is the eccentricity of Comet Hale-Bopp's orbit? **b.** What is the period of this comet's orbit? **c.** The comet's mass is estimated to be 1.30×10^{16} kg. What is the potential energy of the Comet Hale-Bopp–Sun system when it is at its farthest distance from the Sun?

64. **G** CASE STUDY Comet Hale-Bopp's elliptical orbit is described in Problem 63. Draw an energy graph for the Sun–comet system. For points *A*, *P*, *I*, and *J* (Fig. 8.38, page 236), superimpose a bar chart on the energy graph.

General Problems

65. A 50.0-g toy car is released from rest on a frictionless track with a vertical loop of radius *R*. The initial height of the car is $h = 4.00R$.

a. **A** What is the speed of the car at the top of the vertical loop?

b. **N** What is the magnitude of the normal force acting on the car at the top of the vertical loop?

66. **A** In Chapter 23, you will encounter the electric force that exists between two electrically charged objects, given by the formula $F_E = kq_1q_2/r^2$, where *k* is a constant called Coulomb's constant, each *q* represents the amount of excess charge on each object, and *r* is the distance between the objects. Find the change in electric potential energy between two charges that are initially a distance r_1 apart and later a distance r_2 apart.

67. **N** The Earth's perihelion distance (closest approach to the Sun) is $r_P = 1.48 \times 10^{11}$ m, and its aphelion distance (farthest point) is $r_A = 1.52 \times 10^{11}$ m. What is the change in the Sun–Earth's gravitational potential energy as the Earth moves from aphelion to perihelion? What is the change in its gravitational potential energy from perihelion to aphelion?

68. **E, N** After ripping the padding off a chair you are refurbishing, you notice that there are six springs beneath, which are intended to contribute equally in supporting your weight when you sit. You find a tag that indicates that the springs are identical and that each has a spring constant of 1.5×10^3 N/m. What would be the elastic potential energy stored in the six-spring system if you were to sit on the chair?

69. **A** In a classic laboratory experiment, a cart of mass m_1 on a horizontal air track is attached via a string over an ideal pulley to a hanging cylinder of mass m_2. The system is released from rest, and the motion is measured. Find the speed of the cart after the cylinder has descended a distance *H*.

70. **G** A block is attached to a spring, and the block makes contact with a frictionless surface. Sketch a graph of the potential energy of the block–spring system as a function of position along with the corresponding system as in Figure 8.16 (page 224). Use this sketch to show how the block can move to the left from a positive position to a negative position and decrease the elastic potential energy of the system.

Problems 71 and 72 are paired.

71. **N** At the start of a basketball game, a referee tosses a basketball straight into the air by giving it some initial speed. After being given that speed, the ball reaches a maximum height of 4.25 m above where it started. Using conservation of energy, find **a.** the ball's initial speed and **b.** the height of the ball when it has a speed of 2.5 m/s.

72. **A** At the start of a basketball game, a referee tosses a basketball straight into the air by giving it some initial speed. After being given that speed, the ball reaches a maximum height y_{max} above where it started. Using conservation of energy, find expressions for **a.** the ball's initial speed in terms of the gravitational acceleration *g* and the maximum height y_{max} and **b.** the height of the ball when it has speed v in terms of its current height y, the gravitational acceleration g, and the maximum height y_{max}.

73. **N** A rocket carrying a new 950-kg satellite into orbit misfires and places the satellite in an orbit with an altitude of 125 km, well below its operational altitude in low-Earth orbit. **a.** What would be the height of the satellite's orbit if its total energy were 500 MJ greater? What would be the difference in the system's **b.** kinetic energy and **c.** potential energy?

74. **N** In the far future, astronauts travel to the planet Saturn and land on Mimas, one of its 62 moons. Mimas is small compared with the Earth's moon, with mass $M_m = 3.75 \times 10^{19}$ kg and radius $R_m = 1.98 \times 10^5$ m, giving it a free-fall acceleration of $g = 0.0636$ m/s^2. One astronaut, being a baseball fan and having a strong arm, decides to see how high she can throw a ball in this reduced gravity. She throws the ball straight up from the surface of Mimas at a speed of 40 m/s (about 90 mph, the speed of a good major league fastball). **a.** Predict the maximum height of the ball assuming g is constant and using energy conservation. Mimas has no atmosphere, so there is no air resistance. **b.** Now calculate the maximum height using universal gravitation. **c.** How far off is your estimate of part (a)? Express your answer as a percent difference and indicate if the estimate is too high or too low.

75. **N** At 220 m, the bungee jump at the Verzasca Dam in Locarno, Switzerland, is one of the highest jumps on record. The length of the elastic cord, which can be modeled as having negligible mass and obeying Hooke's law, has to be precisely tailored to each jumper because the margin of error at the bottom of the dam is less than 10.0 m. Kristin prepares for her jump by first hanging at rest from a 10.0-m length of the cord and is observed to stretch the rope to a total length of 12.5 m. **a.** What length of cord should Kristin use for her jump to be exactly 220 m? **b.** What is the maximum acceleration she will experience during her jump?

76. **N** Two 67.0-g arrows are fired in quick succession with an initial speed of 80.0 m/s. The first arrow makes an initial angle of 33.0° with the horizontal, and the second arrow is fired straight upward. Assume an isolated system and choose the reference configuration at the initial position of the arrows. **a.** What is the maximum height of each of the arrows? **b.** What is the total mechanical energy of the arrow–Earth system for each of the arrows at their maximum height?

77. **N** A block of mass $m_1 = 4.00$ kg initially at rest on top of a frictionless, horizontal table is attached by a lightweight string to a second block of mass $m_2 = 3.00$ kg hanging vertically from the edge of the table and a distance $h = 0.450$ m above the floor (Fig. P8.77). If the edge of the table is assumed to be frictionless, what is the speed with which the first block leaves the edge of the table?

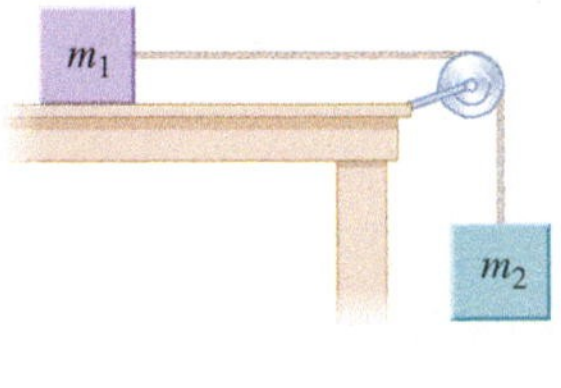

FIGURE P8.77

78. **A** Eric is twirling a ball of mass $m = 0.150$ kg attached to a lightweight string in a vertical circle. If the total energy of the system is conserved, by what factor does the tension on the string increase at the bottom compared with the top of the circle?

79. **N** A jack-in-the-box with a box of mass M and clown of mass $m = 200.0$ g rests on a horizontal table (Fig. P8.50). The clown is attached to the box by means of a lightweight, vertical spring with force constant $k = 1.00 \times 10^2$ N/m. Zara pushes down on the clown with an additional force of 4 mg, further compressing the spring. When Zara removes her hand, the clown jumps upward, and the spring lifts the box off the table. What is the maximum value for the mass M of the box for this action to occur?

Problems 80 and 81 are paired.

80. **C** A system consists of an object and the source of a conservative force. No other forces are exerted on the object. Figure P8.80 shows a graph of the potential energy of the system as a function of the object's position. At what location could the object be in equilibrium? Explain your answer. (The location is an example of what is sometimes called a *stable equilibrium*; explain why.) *Hint*: Considering Eq. 8.3, the force F would be equal to the negative of the derivative of U as a function of position. Use this fact to help answer the question.

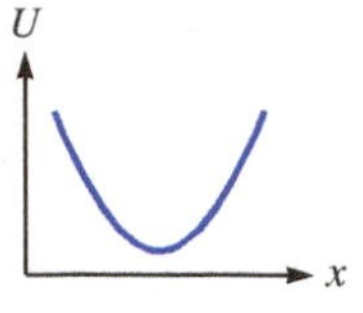

FIGURE P8.80

81. **C** A system consists of an object and the source of a conservative force. No other forces are exerted on the object. Figure P8.81 shows a graph of the potential energy of the system as a function of the object's position. At what location could the object be in equilibrium? Explain your answer. (The location is an example of what is sometimes called an *unstable equilibrium*; explain why.) *Hint*: Considering Eq. 8.3, the force F would be equal to the negative of the derivative of U as a function of position. Use this fact to help answer the question.

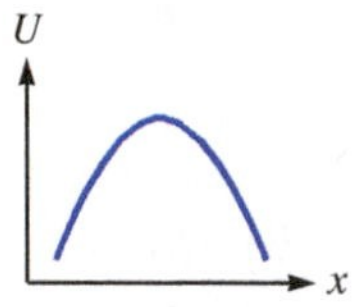

FIGURE P8.81

82. **N** A cluster of grapes is removed from a frictionless, hemispherical bowl 44.0 cm in diameter, leaving behind a single spherical grape of mass 3.00 g initially at rest at the upper edge of the bowl along its horizontal diameter. Choose the bottom of the bowl as the reference configuration where $h = 0$, and answer the following questions as the grape slides to the bottom of the bowl. **a.** What is the gravitational potential energy of the grape–Earth system at the grape's initial position? **b.** What is the kinetic energy of the grape when it reaches the bottom of the bowl? **c.** What is the speed of the grape when it reaches the bottom of the bowl? **d.** What are the potential and kinetic energies of the grape when it reaches a point that is a height $h = 15.0$ cm above the bottom of the bowl?

83. **N** A planet ($m = 5.542 \times 10^{26}$ kg) is in orbit around a star ($M = 1.056 \times 10^{31}$ kg) in another part of our galaxy. **a.** What is the total energy of the planet if its orbit is circular with a radius of 6.35×10^{12} m? **b.** What is the total energy of the planet if its orbit is instead elliptical and if its maximum distance is 7.21×10^{12} m from the star?

84. **N** As of 2012, all of the shuttles in NASA's space shuttle program have been retired. When they once launched, a space shuttle (mass $= 2.03 \times 10^6$ kg) would lift off as the thrust of its three main engines was suddenly augmented by the firing of the solid rocket boosters at $t = 0$ s. Suppose the shuttle's acceleration is given by $a = 3.00t + 0.330t^2 - 0.100t^3$, where a is in meters per second squared and t is in seconds, in the first few seconds after liftoff. What is the change in the kinetic energy of the space shuttle during the first 3.50 s after liftoff in this scenario?

Energy in Nonisolated Systems

9

Underlying Principles

1. Work–kinetic energy theorem
2. Work–mechanical energy theorem
3. Work–energy theorem

Major Concepts

1. Work
2. General expression for work done
3. General expression for change in potential energy
4. Conservative (and nonconservative) force
5. Zero-work force
6. Center of mass
7. Internal energy
8. Dissipative forces
9. Power

Special Cases: Internal Energy

1. Thermal energy
2. Chemical energy

Tools

Dot product (scalar product)

Key Questions

How can the conservation approach be extended to account for energy transfer to and from nonisolated systems?

How are dissipative forces such as moving friction and drag accounted for in the conservation approach?

The Sun is the source of energy that supports our lives. Deep in the Sun's core, nuclear reactions between protons release energy. This energy does not stay in the Sun; instead, it is radiated away in the form of light and other electromagnetic waves (Part V). The Sun's energy lights and heats the solar system, driving the process of photosynthesis that captures that energy in the form of food we eventually eat. Therefore, the Sun is a nonisolated system, transferring energy to its environment. If the Sun's energy were not transferred to the environment, the Earth could not support life at all.

Unlike the conversion of energy from one form to another form—say from potential to kinetic energy—the transfer of energy to or from a system means a net gain or loss of energy. The Sun is actually losing energy, and in about 4.5 billion years, it will no longer be able to produce sufficient nuclear reactions in its core to sustain its present structure. The Sun will then become a white dwarf, the compact remains of a dead star.

This chapter is the second in a trilogy with the same goal of developing a conservation approach to mechanics. In Chapter 8, we focused on the conservation of mechanical energy in isolated systems. In this chapter, we adapt the idea of conservation of energy to account for energy transfer in nonisolated systems.

9-1 Energy Transfer to and from the Environment

For the isolated systems in Chapter 8, no energy is transferred and mechanical energy is conserved. Any real system is not completely isolated, however. Therefore, energy is transferred between the system and its environment. The transfer of energy can lead to spectacular events such as described in the following case study.

CASE STUDY A Meteor Shower

Comet Halley's orbit brings it into the inner solar system every 76 years. Each time a comet comes into the inner solar system, the Sun heats it up and small particles are left behind, forming a debris stream. When the Earth passes through a debris stream, these small particles, known as meteors, interact with the atmosphere. The Earth's atmosphere heats the meteors, which give off light before they evaporate. Meteors are observable without the use of any optical aid and are commonly called *shooting stars*. When the Earth passes through a large debris stream, it is possible to see several shooting stars per second in what is known as a *meteor shower* (Fig. 9.1). Meteor showers are named after the constellation from which the shooting stars seem to originate.

One evening in mid-August, three physics professors—Black, Noir, and Kuro—observed the Perseid meteor shower. On their drive home, Black asked if either of the other two professors knew the size of a piece of the debris. Neither of the other two professors knew the answer, but between the three of them, they knew enough information to estimate an answer. They worked in the dark without a calculator and came up with a good estimate using a conservation approach. In fact, they tried two different approaches and found the same estimate to within a factor of two. By the end of this chapter, we will make our first estimate, and we will make another estimate in Chapter 34.

The basis of their first estimate and ours is that the meteor is the system and the Earth's atmosphere is the environment. The meteor starts with a certain amount of energy that is transferred to the atmosphere. We observe this transfer as a shooting star.

John R. Foster/Photo Researchers, Inc.

FIGURE 9.1 Three meteors are visible in this photo taken during a meteor shower. The meteors appear to come from the same direction in the sky.

Let us compare the Comet Halley case study (Example 8.13) with this meteor case study. In the Halley case study, we included the Sun and comet in the system, which we considered to be isolated. An isolated system's energy is only *converted* or *changed* from one form to another, but no energy is *transferred* into or out of the system. In the meteor case study, the meteor is the system, and the atmosphere is outside the system. The meteor transfers some of its energy to the atmosphere. The meteor loses energy, and the atmosphere gains energy.

9-2 Work Done by a Constant Force

WORK ✪ Major Concept

One way for a system to gain or lose energy is to be acted on by an external force. If the force changes the energy of the system, we say that the force *does work* on the system. **Work** is the amount of energy transferred to (or from) the system by an external force. Work has the dimensions of energy usually expressed as newton meters (N · m), but recall that 1 N · m = 1 J. The use of the word *work* in physics is very different from our common usages such as "the work done by Michelangelo covers the ceiling of the Sistine Chapel" or "I would like to go to a movie, but I have too much work to do." Instead of looking to our intuition for the meaning of "work" in physics, let us turn to three scenarios and develop a mathematical and conceptual sense.

In each case, we will consider a simple system consisting of one object that is modeled as a particle. The system is not isolated. Energy is transferred to or from the system through some external force.

Figure 9.2A shows a hockey player pushing a puck (the system) to his coach, who then stops it. We can think of this action as two separate interactions: first the player and the puck and then the coach and the puck. Because the puck is on ice, we will ignore friction. Because the puck is modeled as a single particle, we cannot associate potential energy with the system. (Recall that potential energy is only associated with a system that consists of two or more particles.) Therefore, any energy gained or lost by the system must appear as a change in the puck's kinetic energy. We will assume in each case that the player or the coach exerts a constant force on the puck as long as his or her stick is in contact with the puck. (Keeping a constant force on the puck is actually quite difficult to do, but the assumption makes our task much easier.)

Parts B and C of Figure 9.2 show free-body diagrams for the puck: one for when the player's stick is in contact with the puck and the other for when the coach's stick is in contact with it. The normal force and the gravitational force are the same in both free-body diagrams, but the *player's* stick exerts a force $\vec{F}_P$ in the positive x direction, and the *coach's* stick exerts a force $\vec{F}_C$ in the negative x direction.

The conservation approach stems from Newton's laws of motion. From both free-body diagrams, there is no acceleration in the y direction because the gravitational force and the normal force cancel out. The sticks, though, cause acceleration in the x direction. We need to look at each stick–puck interaction separately. Our goal is to find the change in kinetic energy in each case and associate it with the work done by the stick.

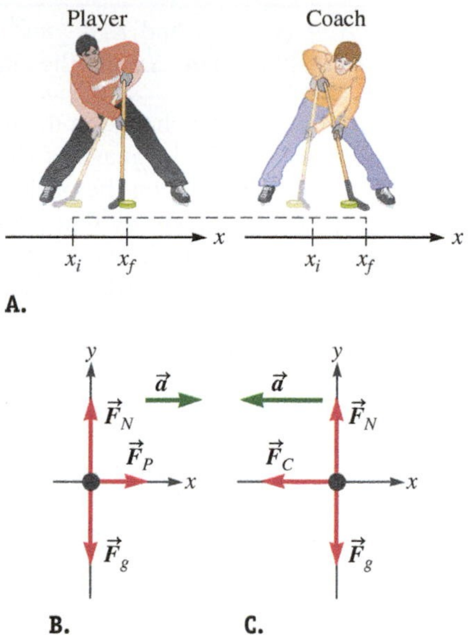

FIGURE 9.2 **A.** A hockey player passes a puck to his coach. **B.** Free-body diagrams show the forces acting on the puck when the player's stick is in contact and then **C.** when the coach's stick is in contact with the puck.

DERIVATION Work Done by a Constant Force in Same Direction as Displacement

We will use the specific example of the player's stick pressing on the puck to show that the work done by a constant force in the x direction is given by

$$W = F_x \Delta x \tag{9.1}$$

where the puck's displacement while in contact with the stick is Δx.

Using the free-body diagram in Figure 9.2B, apply Newton's second law in component form.	$\sum F_x = ma_x$ $F_P = ma_x \quad (1)$
The stick is in contact with the puck from x_i to x_f, and the puck accelerates as it moves in the x direction between these two points. We can write a differential equation describing the puck's motion during the time it is contact with the stick. Substitute $a_x = dv_x/dt$ (Eq. 2.7) into Equation (1), and use the chain rule.	$F_P = ma_x = m\dfrac{dv_x}{dt}$ $F_P = m\dfrac{dv_x}{dx}\dfrac{dx}{dt}$
Use $v_x = dx/dt$ (Eq. 2.4).	$F_P = m\dfrac{dv_x}{dx}v_x$
This differential equation is solved by bringing dx to the left side and integrating from the moment the stick first makes contact with the puck at x_i to when the puck loses contact at x_f. The limits on the left are simply x_i to x_f, but the integral on the right is over dv_x. The limits on the right must be the x component of velocity at x_i and x_f, which we call v_{xi} to v_{xf}.	$F_P dx = mv_x dv_x$ $\int_{x_i}^{x_f} F_P dx = \int_{v_{xi}}^{v_{xf}} mv_x dv_x \quad (9.2)$
The mass of the puck is constant, and we have assumed the stick exerts a constant force on the puck. Therefore, we can pull m and F_P outside the integrals.	$F_P \int_{x_i}^{x_f} dx = m\int_{v_{xi}}^{v_{xf}} v_x dv_x$ $F_P x\Big\|_{x_i}^{x_f} = \frac{1}{2}mv_x^2\Big\|_{v_{xi}}^{v_{xf}}$ $F_P(x_f - x_i) = \frac{1}{2}mv_{xf}^2 - \frac{1}{2}mv_{xi}^2 \quad (9.3)$

Derivation continues on page 250 ▶

We use $\Delta x = (x_f - x_i)$ and $K = \frac{1}{2}mv^2$ (Eqs. 2.1 and 8.1) to rewrite Equation 9.3 in terms of displacement Δx and the change in kinetic energy ΔK.	$F_P \, \Delta x = \Delta K$ (9.4)
Equation 9.4 relates the change in the puck's kinetic energy to the force applied by the player's stick. In the language of Newton's second law, we say that the constant force exerted by the stick on the puck accelerated the puck. In the language of the conservation approach, we say that the **work done by the stick on the puck** changes (increases) the puck's kinetic energy. To write a general expression, we have dropped subscript P (for player) and added a subscript x as a reminder that the stick's force is in the x direction. (Work is a scalar, and no such direction subscript is needed.)	$W_P = \Delta K = F_P \, \Delta x$ $W = F_x \Delta x$ ✓ (9.1)

COMMENTS

Remember that Equation 9.1 was derived in the special case of a constant force exerted on a particle whose displacement is parallel to that force; in fact, it is just our first step in finding a general expression for work.

We now turn our attention to the work done by the coach (Fig 9.2C). We do not need to repeat the derivation we did for the player's interaction with the puck. Let's just see if Equation 9.1 gives a reasonable result for the work done by the coach's stick on the puck. The puck has a positive x component of velocity when it makes contact with the coach's stick, and the stick decelerates the puck. Therefore, the *change* in the puck's kinetic energy is negative ($\Delta K < 0$) as a result of this interaction.

Because the puck loses energy, we expect that the work done by the coach's stick on the puck must be negative. The force exerted by the coach is in the negative x direction: $\vec{F}_C = -F_C\hat{\imath}$. The puck's displacement Δx is in the positive x direction while it is in contact with the coach's stick. So, the work done by the coach's stick on the puck is negative (Eq. 9.1):

$$W_C = -F_C \Delta x$$

which is consistent with what we reasoned: The work done by the coach's stick is negative, which results in a loss in kinetic energy.

WORK–KINETIC ENERGY THEOREM

Underlying Principle

In the two scenarios above, only one force (due either to the hockey player's stick or the coach's stick) does work on the puck. If more than one external force does work on the particle at one time, the total work done is found from the arithmetic sum of the work done by each force. The change in the system's kinetic energy is a result of the total work done. These facts are summarized in the **work–kinetic energy theorem**, which says that the total work done on a particle by all external forces equals the change in the particle's kinetic energy. Mathematically, this theorem is written as

$$W_{\text{tot}} = \Delta K \tag{9.5}$$

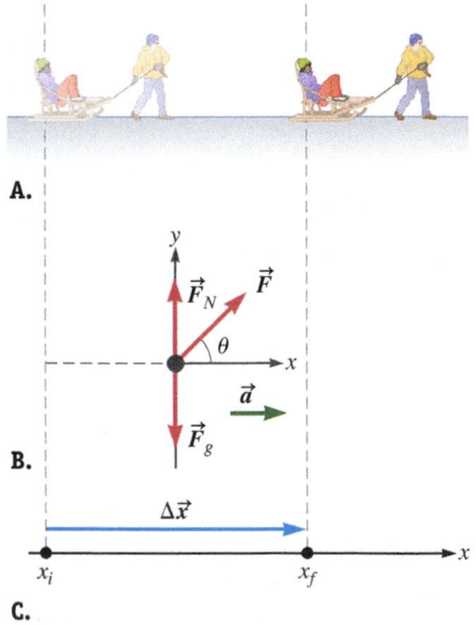

FIGURE 9.3 **A.** Paul pulls a sled. **B.** The free-body diagram shows that the force exerted by Paul makes an angle θ with the x axis. **C.** The displacement of the sled is along the x axis from x_i to x_f.

Now consider the third scenario, where a constant force is applied to a particle whose displacement is not parallel or antiparallel to the force; our goal is to derive a somewhat more general expression for work. Figure 9.3A shows Paul pulling a sled along the ice. As in the case of the puck, we consider the sled (and its contents) to be a one-particle system and friction between the ice and the sled to be negligible. Paul exerts a constant force $\vec{F}$ at an angle θ with respect to the x axis (Fig. 9.3B). The sled's displacement Δx is purely along the x axis (Fig. 9.3C). According to Equation 9.1, we need the x component of Paul's force, $F_x = F\cos\theta$. Then, the work done by him on the sled is

$$W = F_x \Delta x = (F\cos\theta)\Delta x$$

This example of Paul pulling the sled gives us an idea of how to write a more general expression for the work done by a constant force $\vec{F}$ exerted on a particle whose displacement is $\Delta\vec{r}$:

$$W = F\,\Delta r\cos\theta \tag{9.6}$$

where θ is the angle between $\vec{F}$ and $\Delta\vec{r}$, and F and Δr are (positive) magnitudes.

EXAMPLE 9.1 Work Done on a Laboratory Cart

In the common laboratory experiment shown in Figure 9.4, a small cart of mass $m = 0.78\,\text{kg}$ is pulled up a ramp. The experiment is designed so that friction is negligible. Find the work done by each force shown acting on the cart as it is displaced 1.5 m up the ramp. The hanging object weighs 17.8 N, and the tension in the rope is 6.0 N.

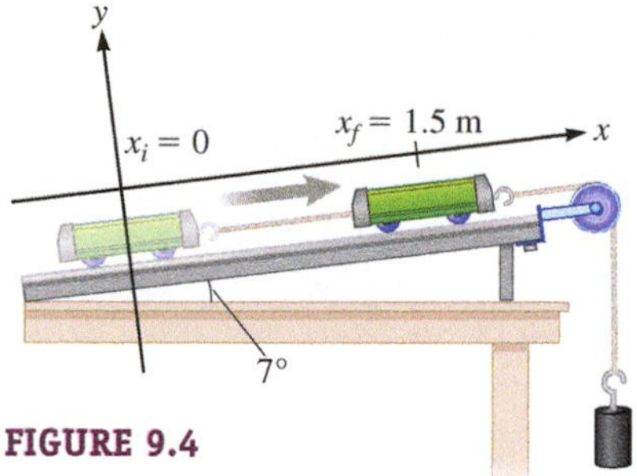

FIGURE 9.4

INTERPRET and ANTICIPATE

The problem statement implies that the system consists of the cart alone modeled as a particle. Our job is to identify each force acting on the cart and then apply $W = F\Delta r\cos\theta$ (Eq. 9.6) once for each force. We expect to find numerical results with SI units of newton meters. Because the weight of the hanging mass is greater than the tension, we expect that the hanging mass accelerates downward and the cart accelerates up the ramp. So, we expect that the kinetic energy of the cart increases and the total work done by all the forces is positive.

Start with a free-body diagram identifying each force acting on the cart. They are gravity, the normal force, and the tension force. When calculating the work done on an object, it is often convenient to indicate the displacement of the object as we have done in Figure 9.5.

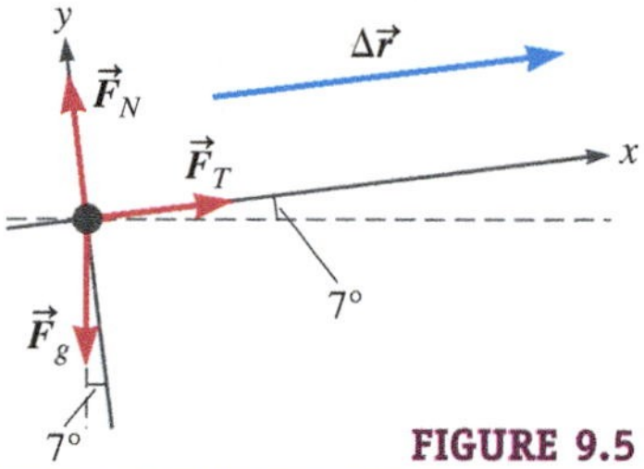

FIGURE 9.5

SOLVE

Apply $W = F\Delta r\cos\theta$ three times. The displacement of the cart is the same ($\Delta\vec{r} = 1.5\hat{\imath}\,\text{m}$) each time, but the angle between the force and the displacement changes for each force. The tension force on the cart is parallel to the cart's displacement, so the angle between $\Delta\vec{r}$ and $\vec{F}_T$ is $\theta_T = 0$.

$$W_T = F_T\Delta r\cos\theta_T = F_T\Delta r\cos 0$$
$$W_T = F_T\Delta r = (6.0\,\text{N})(1.5\,\text{m})$$
$$W_T = 9.0\,\text{N}\cdot\text{m}$$

CHECK and THINK

The tension force and the displacement are parallel, and the work done by the tension on the cart is positive.

SOLVE

The angle between the normal force $\vec{F}_N$ and the cart's displacement $\Delta\vec{r}$ is $\theta_N = 90°$.

$$W_N = F_N\Delta r\cos\theta_N$$
$$W_N = F_N\Delta r\cos 90° = 0$$

CHECK and THINK

The work done by the normal force on the cart is zero because the normal force is perpendicular to the cart's displacement. The cosine part of $W = F\Delta r\cos\theta$ (Eq. 9.6) ensures that the work done by any force that is perpendicular to the displacement is zero.

SOLVE

The angle between the gravitational force $\vec{F}_g$ and the displacement $\Delta\vec{r}$ is $\theta_g = 97°$ (Fig. 9.5).

$$W_g = F_g\Delta r\cos\theta_g = mg\Delta r\cos 97°$$
$$W_g = (0.78\,\text{kg})(9.81\,\text{m/s}^2)(1.5\,\text{m})\cos 97°$$
$$W_g = -1.4\,\text{N}\cdot\text{m}$$

CHECK and THINK

The gravitational force on the cart has both x and y components. Like the normal force, the y component of the gravitational force $\vec{F}_g$ is perpendicular to the displacement and therefore does **no** work on the cart. The x component of the gravitational force is antiparallel (opposite direction) to the displacement and therefore does negative work on the cart. The cosine function in $W = F\Delta r\cos\theta$ selects the vector component of the force that is parallel (or antiparallel) to the displacement. This fact is important in the next section.

To check our work, we find the total work done on the cart by all three external forces. As expected, the total work is positive. According to the work–kinetic energy theorem $W_{tot} = \Delta K$, the change in kinetic energy is positive, which makes sense because the cart's speed increases as it is pulled up the ramp.

$$W_{tot} = W_T + W_N + W_g$$
$$W_{tot} = 9.0\,\text{N}\cdot\text{m} + 0 - 1.4\,\text{N}\cdot\text{m}$$
$$W_{tot} = 7.6\,\text{N}\cdot\text{m}$$

DOT PRODUCT ⦿ Tool

9-3 Dot Product

From Equation 9.6, $W = F\Delta r\cos\theta$, work involves the multiplication of two vectors: force and displacement. In Section 9-2, to build a basic understanding of work, we avoided the mathematics behind multiplying two vectors. To deal with more general cases involving nonconstant forces in up to three dimensions, we now present the necessary mathematics.

There are three types of multiplication involving vectors. The first involves the multiplication of a vector by a scalar. As discussed in Chapter 3, this type of multiplication results in new vector that is parallel or antiparallel to the original vector. The subject of this section is the second type, in which the multiplication of two vectors results in a scalar. This type of vector multiplication is called the **dot product** or **scalar product**. In this section, we will work with generic vectors $\vec{A}$ and $\vec{B}$.

Figure 9.6A shows two vectors $\vec{A}$ and $\vec{B}$ that lie in a plane; φ is the smaller angle between them when they are placed tail to tail. The dot product or scalar product of these two vectors is defined to be

$$D = \vec{A}\cdot\vec{B} \equiv AB\cos\varphi \tag{9.7}$$

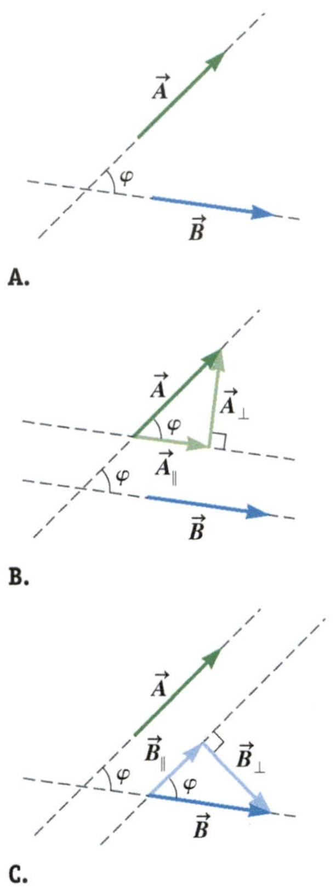

FIGURE 9.6 **A.** Vectors $\vec{A}$ and $\vec{B}$ line in a plane; the angle between them is φ. **B.** Vector $\vec{A}$ is broken into two components, one parallel to $\vec{B}$ and the other perpendicular to $\vec{B}$. **C.** Now vector $\vec{B}$ is broken into components that are parallel and perpendicular to $\vec{A}$.

We can see why both names fit this multiplication. First, the symbol for the dot product is a dot •. (It is *not* correct to write $\vec{A}\vec{B}$.) Second, the result of taking a dot product is a scalar quantity D. The dot product is found by multiplying the magnitudes of the two vectors and the cosine of the angle φ. Because A and B are magnitudes, they must be positive, so the cosine term determines the sign of D. If $\varphi < 90°$, then $D > 0$, and if $90° < \varphi < 180°$, then $D < 0$.

Because D is a scalar quantity, we cannot represent the result of a dot product D with an arrow on a piece of paper, but pictorial representations of the two constituent vectors $\vec{A}$ and $\vec{B}$ help us interpret the role of the dot product. In Figure 9.6B, we have broken vector $\vec{A}$ into two vector components: one perpendicular ($\vec{A}_\perp$) and the other parallel ($\vec{A}_\parallel$) to vector $\vec{B}$. Trigonometry gives us the scalar component $A_\parallel = A\cos\varphi$. Rewriting Equation 9.7 in terms of this parallel scalar component gives

$$D = \vec{A}\cdot\vec{B} = A_\parallel B \tag{9.8}$$

We use a similar procedure to rewrite Equation 9.7 in terms of the component of $\vec{B}$ that is parallel to $\vec{A}$. From Figure 9.6C, we see that $B_\parallel = B\cos\varphi$, and we rewrite Equation 9.7 as:

$$D = \vec{A}\cdot\vec{B} = AB_\parallel \tag{9.9}$$

According to Equations 9.8 and 9.9 the dot product can be thought of either as B multiplied by the component of $\vec{A}$ that is parallel to $\vec{B}$ (Eq. 9.8) or as A multiplied by the component of $\vec{B}$ that is parallel to $\vec{A}$ (Eq. 9.9).

Another way to find the dot product is by using a coordinate system and breaking each vector into components (Fig. 9.7). Then, the dot product of these two vectors is found by multiplying their like components and then adding: $D = A_xB_x + A_yB_y$ (Problem 13). In three dimensions, the dot product may be found using

$$D = \vec{A}\cdot\vec{B} = A_xB_x + A_yB_y + A_zB_z \tag{9.10}$$

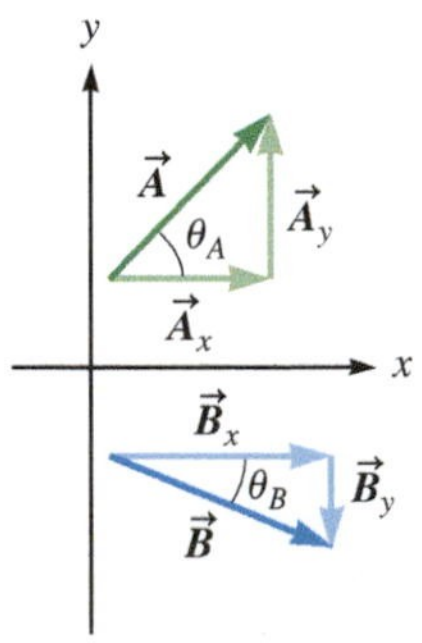

FIGURE 9.7 Once a coordinate system has been chosen, both vectors may be broken into their components.

In addition to knowing how to calculate a dot product, there are three important properties of the dot product. (For homework, you may be asked to prove these properties.)

1. The dot product of a vector with itself gives the square of the magnitude of that vector:

$$\vec{A}\cdot\vec{A} = A^2 \tag{9.11}$$

2. The dot product is *commutative*:

$$\vec{A}\cdot\vec{B} = \vec{B}\cdot\vec{A} \tag{9.12}$$

3. The dot product is *associative*:

$$\vec{A}\cdot(\vec{B} + \vec{C}) = \vec{A}\cdot\vec{B} + \vec{A}\cdot\vec{C} \tag{9.13}$$

We are ready to go back to *work* (pun intended). By comparing work $W = F\Delta r \cos\theta$ (Eq. 9.6) to the dot product $\vec{A}\cdot\vec{B} \equiv AB\cos\varphi$ (Eq. 9.7), we see that the **work done by a constant force** is the dot product of the force and the displacement:

$$W = \vec{F}\cdot\Delta\vec{r} \tag{9.14}$$

We use Equations 9.8 through 9.10 to write work in three other convenient forms:

$$W = F_{\parallel}\Delta r \tag{9.15}$$

$$W = F\Delta r_{\parallel} \tag{9.16}$$

$$W = F_x\Delta x + F_y\Delta y + F_z\Delta z \tag{9.17}$$

EXAMPLE 9.2 Laboratory Cart Revisited

Return to the cart in Example 9.1. Using each of Equations 9.15 through 9.17, find the work done by gravity. Compare your results.

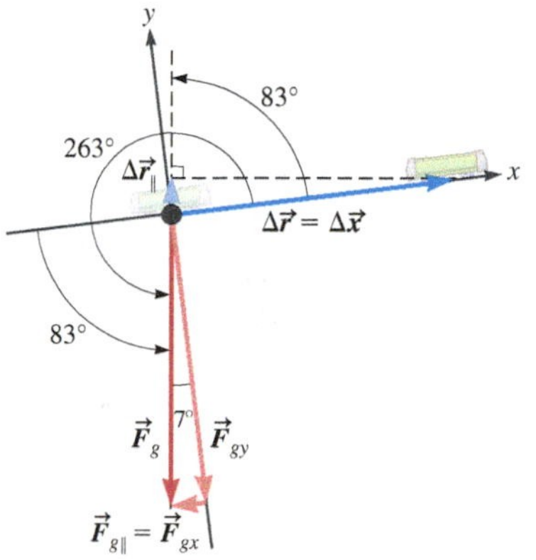

FIGURE 9.8

INTERPRET and ANTICIPATE

In Figure 9.8, we have drawn vectors $\vec{F}_g$ and $\Delta\vec{r}$ on a common coordinate system. This figure will help us find different components of the two vectors as we work each part of the example.

A Use Equation 9.15 to find the work done by gravity.

SOLVE

In all three parts, it is helpful to know the weight of the cart.

$$w = mg = (0.78\text{ kg})(9.81\text{ m/s}^2)$$

$$w = 7.65\text{ N}$$

We need to find $F_{\parallel}$. Figure 9.8 shows that $\vec{F}_{g\parallel}$ is antiparallel to $\Delta\vec{r}$, so $\vec{F}_{g\parallel}$ is negative. Find the magnitude of $\vec{F}_{g\parallel}$ using trigonometry.

$$\vec{F}_{g\parallel} = -w\sin 7^\circ = -(7.65\text{ N})\sin 7^\circ$$

$$\vec{F}_{g\parallel} = -0.93$$

Apply $W = F_{\parallel}\Delta r$ (Eq. 9.15).

$$W = \vec{F}_{g\parallel}\Delta r = (-0.93\text{ N})(1.5\text{ m})$$

$$W = -1.4\text{ N}\cdot\text{m}$$

B Use Equation 9.16 to find the work done by gravity.

SOLVE

We need to find $\Delta r_{\parallel}$. Figure 9.8 shows that $\Delta\vec{r}_{\parallel}$ is antiparallel to $\vec{F}_g$. So, $\Delta r_{\parallel}$ is negative, and its magnitude comes from trigonometry.

$$\Delta r_{\parallel} = -\Delta r\cos 83^\circ$$

$$\Delta r_{\parallel} = -(1.5\text{ m})\cos 83^\circ = -0.18\text{ m}$$

Use $W = F\Delta r_{\parallel}$ (Eq. 9.16) to find work.

$$W = w\Delta r_{\parallel} = (7.65\text{ N})(-0.18\text{ m})$$

$$W = -1.4\text{ N}\cdot\text{m}$$

C Use Equation 9.17 to find the work done by gravity.

SOLVE

We must find the x and y components for both vectors $\vec{F}_g$ and $\Delta\vec{r}$. By inspecting Figure 9.8, we notice that $\vec{F}_{g\parallel} = \vec{F}_{gx}$ and $\Delta\vec{r} = \Delta\vec{x}$.

$$F_{gx} = w\cos(263^\circ) = (7.65\text{ N})\cos(263^\circ) = -0.932\text{N}$$

$$F_{gy} = w\sin(263^\circ) = (7.65\text{ N})\sin(263^\circ) = -7.59\text{N}$$

$$\Delta\vec{r} = \Delta\vec{x} = 1.5\hat{\imath}\text{ m}$$

Example continues on page 254 ▶

Using $W = F_x\Delta x + F_y\Delta y + F_z\Delta z$ (Eq. 9.17), we find work one last time.	$W = F_x\Delta x + F_y\Delta y + F_z\Delta z = F_{gx}\Delta x$ $W = (-0.932\ \text{N})(1.5\ \text{m}) = -1.4\ \text{N}\cdot\text{m}$

CHECK and THINK

We have now found the work done by gravity on the cart in four different ways, three here and one in Example 9.1. All four times, we found the same result.

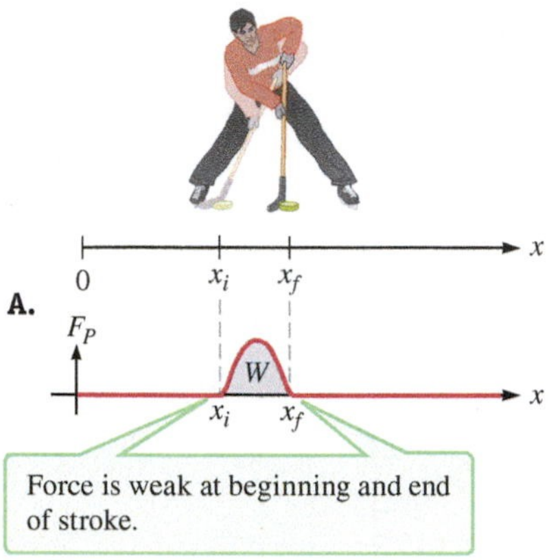

FIGURE 9.9 **A.** A hockey player exerts a nonconstant force on a puck. While the player's stick is in contact with it, the puck is displaced from x_i to x_f. **B.** A plot of the player's force F_P on the puck as a function of the puck's position x looks like a hill because the player's force is strongest in the middle of his stroke. The work done by the player on the puck is the area under the curve.

9-4 Work Done by a Nonconstant Force

We derived our expressions for work by thinking about hockey sticks and a puck. Our assumption was that the hockey sticks exerted a constant force on the puck as long as the sticks were in contact with the puck. Because that assumption was the basis for our derivation, $W = \vec{F}\cdot\Delta\vec{r}$ (Eq. 9.14) holds only if the force is constant. We need to find an expression for work that applies even when the force is *not* constant.

Let us return to the hockey player and the puck (Fig. 9.9A), but this time we will assume the magnitude of the force exerted by the player's stick is a function of position. As before, the stick is in contact with the puck from x_i to x_f, and it exerts a force of varying magnitude on the puck in the positive x direction (Fig. 9.9B).

In our earlier derivation of work (page 249), we assumed the stick exerted a constant force on the puck, so when we got to Equation 9.2, we pulled F_p out of the integral. In this case, F_x is not constant and cannot be pulled out of the integral. We must conclude that work is given by

$$W = \int_{x_i}^{x_f} F_x\,dx \tag{9.18}$$

We interpret this integral as the area under the force-versus-position graph (Fig. 9.9B).

The hockey player's force is parallel to the puck's displacement, so Equation 9.18 is only good for the special case of a nonconstant force that is parallel (or antiparallel) to the displacement of the particle. A more general situation is shown in Figure 9.10, where a nonconstant force acts on an amusement park car that is spiraling up a steep hill. The motion of the car is three-dimensional, so a three-dimensional coordinate system is used. At any instant, the car's infinitesimal displacement $d\vec{r}$ is tangent to its path at that point. As shown in Figure 9.10, the force $\vec{F}$ at point A is neither parallel or antiparallel to the car's displacement $d\vec{r}$.

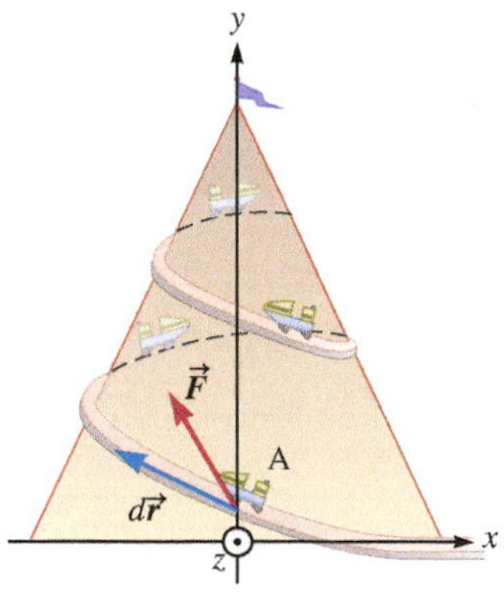

FIGURE 9.10 A nonconstant force acts on an amusement park car that spirals up a steep hill.

In general, the force and infinitesimal displacement has three components, and we write $\vec{F}$ and $d\vec{r}$ in terms of their components:

$$\vec{F} = F_x\hat{\imath} + F_y\hat{\jmath} + F_z\hat{k} \tag{9.19}$$

$$d\vec{r} = dx\hat{\imath} + dy\hat{\jmath} + dz\hat{k}$$

The work dW done by the external force that results in the car's infinitesimal displacement is the dot product $\vec{F}\cdot d\vec{r}$. Because we know the vector components, we use Equation 9.10:

$$dW = \vec{F}\cdot d\vec{r} = F_x dx + F_y dy + F_z dz \tag{9.20}$$

To find the work done by the force over some finite displacement $\Delta\vec{r}$, we integrate (add up) the work dW done over each infinitesimal displacement $d\vec{r}$:

GENERAL EXPRESSION FOR WORK DONE Major Concept

$$W = \int_{r_i}^{r_f} \vec{F}\cdot d\vec{r} = \int_{r_i}^{r_f} (F_x dx + F_y dy + F_z dz) \tag{9.21}$$

Equation 9.21 is the general expression for work. The integral in this equation is sometimes called a *path integral*, and we say that the force is integrated over the path from r_i to r_f.

CONCEPT EXERCISE 9.1

In the three cases shown in Figure 9.11, a force acts on a particle, and the particle is displaced from an initial position to a final position. From the position-versus-force graphs, determine if the force is parallel or antiparallel to the displacement in each case. Explain.

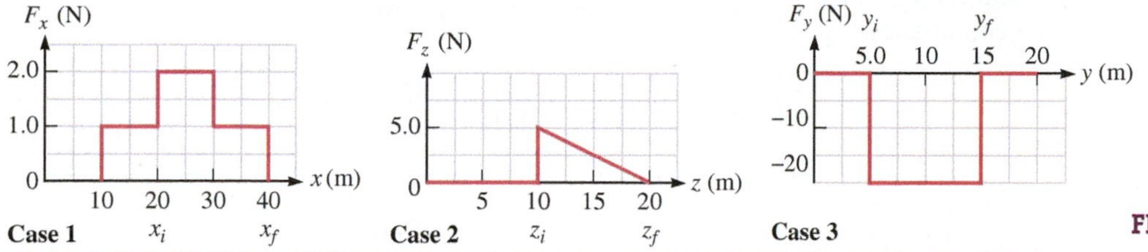

FIGURE 9.11

EXAMPLE 9.3 Work Done by a General Force

A force $\vec{F} = (3x\hat{\imath} - 8y\hat{\jmath})$ N acts on a particle. Find the work done by this force as the particle moves in an xy plane from $(x_i, y_i) = (1, 0)$ m to $(x_f, y_f) = (-3, 8)$ m.

INTERPRET and ANTICIPATE

The force depends on both x and y, so it is two-dimensional and not constant. We expect to find a numerical answer in the form $W =$ _____ N · m.

SOLVE

Write Equation 9.21 in component form. We have two integrals, one for x and one for y.	$W = \int_{r_i}^{r_f} \vec{F} \cdot d\vec{r} = \int_{x_i}^{x_f} F_x dx + \int_{y_i}^{y_f} F_y dy$
Substitute for the force components. The limits on the integrals come from the initial and final positions of the particle.	$W = \int_{1}^{-3} 3x\,dx + \int_{0}^{8} (-8y)dy$
We dropped the units from the integral, but the integrand is in newtons and the limits are in meters.	$W = \frac{3}{2}x^2\Big\vert_1^{-3} - \frac{8}{2}y^2\Big\vert_0^{8} = \frac{3}{2}[(-3)^2 - 1^2] - \frac{8}{2}[8^2 - 0]$ $W = -244\,\text{N} \cdot \text{m}$

CHECK and THINK

Our answer is in the form we expected; the negative sign means that the particle loses energy. Why can't we use a simpler expression for work such as $W = \vec{F} \cdot \Delta\vec{r}$ (Eq. 9.14)? The reason is that as the particle moves, the force acting on it changes. In fact, the x component of the force initially points in the positive direction, but by the end of the path, F_x is negative. Also, the magnitude of the y component of force varies from 0 to 64 N.

EXAMPLE 9.4 CASE STUDY Work Done by Gravity on a Meteor

Model a meteor as a single particle of mass m (Fig. 9.12). External forces such as gravity do work on the particle. Suppose the particle starts very far from a large planet (such as the Earth) with mass M. Find the work done by the gravitational force of the planet on the meteor as the meteor falls along a radial path to a position r measured from the planet's center.

INTERPRET and ANTICIPATE

Whenever we are asked to find the work done by some force, a choice has been made for us; the source of the force must be external to the system. So, here the system is just the meteor and not the planet. We will take the initial position "very far" to mean infinitely far away, $r_i \to \infty$. As the meteor falls to the position $\vec{r}_f = \vec{r}$, gravity does work on it. Because the gravitational force gets stronger as the meteor gets closer to the planet, we must integrate. We expect to find an algebraic expression.

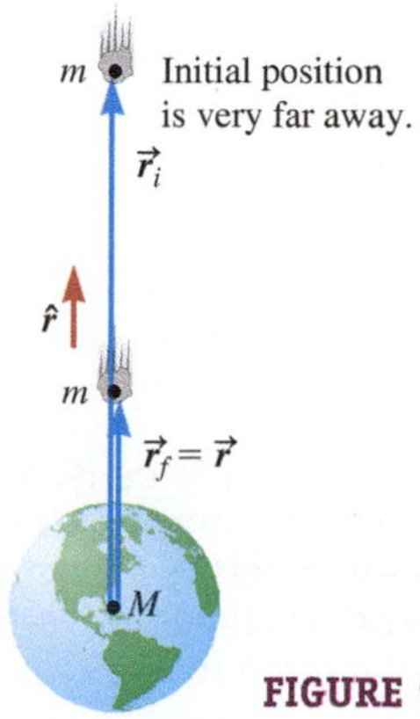

FIGURE 9.12

Example continues on page 256 ▶

SOLVE

Using the unit vector $\hat{r}$ shown, write an expression for the gravitational force exerted on the meteor. This force is toward the planet in the negative $\hat{r}$ direction.

$$\vec{F}_G = F_G\hat{r} = -G\frac{Mm}{r^2}\hat{r}$$

Substitute this expression for force into Equation 9.21. We have left the limits of integration in their general form for now.

$$W = \int_{r_i}^{r_f} \vec{F}\cdot d\vec{r} = \int_{r_i}^{r_f} F_G(\hat{r}\cdot d\vec{r})$$

$$W = \int_{r_i}^{r_f} F_G dr = \int_{r_i}^{r_f}\left(-G\frac{Mm}{r^2}\right)dr$$

Integrate and substitute limits. The final position is simply r, and the initial position is infinitely far away. (Recall that $1/\infty \to 0$.)

$$W = -GMm\left[-\frac{1}{r}\right]\Bigg|_{r_i}^{r_f} = GMm\left[\frac{1}{r_f} - \frac{1}{r_i}\right]$$

$$W = GMm\left[\frac{1}{r} - \frac{1}{\infty}\right] = \frac{GMm}{r} \tag{9.22}$$

CHECK and THINK

We have found an algebraic expression as expected. Equation 9.22 is important and requires further discussion. (1) It is the work done by any large object's gravitational force on a particle that moves toward the large object from very far away. (2) This expression for work is always positive as expected because the force and the displacement of the particle are in the same direction: toward the center of the large object. (3) Equation 9.22 is very similar to Equation 8.7 for the gravitational potential energy of a two-particle system, $U_G(r) = -G(Mm/r)$. The two expressions only differ by a negative sign. We will discuss this difference at length in the next section. (4) Finally, remember that other external forces may do work on the particle.

9-5 Conservative and Nonconservative Forces

From Example 9.4, the expression for the gravitational potential energy of a planet–particle system $U_G(r) = -G(Mm/r)$ differs from the work done by the planet on the particle $W = G(Mm/r)$ only by a negative sign. (This relationship between potential energy and work is true for any conservative force, not just gravity.) The difference comes from deciding what is included in the system. When we choose the contents of a system, our choice affects which forces are internal and which are external. This choice subsequently determines which forces can do work on the system and which forces determine the potential energy of a system. In this section, we see how our choices affect our analysis. Rest assured that our choices cannot change the actual results. It is only our analysis that changes.

For example, suppose a (nonconstant) force is exerted on some particle, and the particle's displacement is parallel to the force such as when a spring exerts a force on a block. If we decide to include the source of that force (the spring) inside the system, then the system may experience changes in its potential energy. In one dimension, the change in potential energy is given by Equation 8.3:

$$\Delta U = U_f - U_i \equiv -\int_{x_i}^{x_f} F_x dx$$

However, if we decide that the source of the force is external to the system, then the source (the spring) does work (Eq. 9.18):

$$W = \int_{x_i}^{x_f} F_x dx$$

When a source of a force is included in a system, there may be a change in the system's potential energy ΔU. If the source is outside the system, however, the force may do work on the system.

When we compare these two choices for what is in the system, we find a general rule: The amount of work done if the source is outside the system equals the negative of the change in energy if the source is included in the system, or

$$W_{\text{ext}} = -\Delta U_{\text{int}} \tag{9.23}$$

where the subscript "ext" is a reminder that to do work the source must be *ext*ernal to the system and the subscript "int" is a reminder that to change the potential energy the source must be *int*ernal. (Often these subscripts are dropped.)

In Chapter 8, we only considered a one-dimension equation for change in potential energy. The most general expression for change in potential energy comes from multiplying Equation 9.21 by −1:

GENERAL EXPRESSION FOR CHANGE IN POTENTIAL ENERGY

Major Concept

$$\Delta U = -\int_{r_i}^{r_f} \vec{F} \cdot d\vec{r} \tag{9.24}$$

We also said in Chapter 8 that potential energy could only be associated with conservative forces such as gravity and the spring force obeying Hooke's law. We can now define a **conservative force** as one that has two properties:

CONSERVATIVE FORCE

Major Concept

1. A change in a conservative force depends only on the change in the (relative) position of the subject.
2. The work done by a conservative force on a subject does not depend on the particular path of the subject.

Let's look at each property separately using universal gravity between the Earth (the source) and a moving meteor (the subject) as a specific example of a conservative force.

The first property might seem puzzling because universal gravity depends not only on the relative position of the source and the subject but also on their masses: $F_G = GMm/r^2$ (Eq. 7.4). Does this dependence on mass mean that gravity isn't a conservative force? The answer is no because as the position r of the subject changes, neither its mass m nor the source's mass M changes. Any change in the gravitational force between them is due only to a change in their relative position.

By contrast, a change in a nonconservative force may depend on some other factors such as the relative velocity of the source and the subject. Moving (kinetic and rolling) friction and drag are **nonconservative** forces because these forces depend on the relative motion between the source and the subject. Take the case of the drag force due to air on an accelerating meteor. As the meteor's speed increases, the drag force increases. So, the change in the drag force does not depend (solely) on the position of the subject; therefore, drag is a nonconservative force.

NONCONSERVATIVE FORCE

Major Concept

The second property says that if a subject moves from an initial position to a final position, the work done on the subject by a conservative force does not depend on the path the subject takes in getting from $\vec{r}_i$ to $\vec{r}_f$. For instance, if a meteor falls along a straight path toward the Earth from $\vec{r}_i$ to $\vec{r}_f$ or takes some roundabout path, the work done by the gravity is the same in either case.

There are two ways to test a force to see if it has this second (path-independent) property. The first test involves calculating the work done by the force on a subject along several different paths. If the work done is not the same along each path, the force is nonconservative. The alternate test involves calculating the work done by the force on a particle along a closed path. If the work done along a closed path does not equal zero, the force is nonconservative.

EXAMPLE 9.5 Gravity Is a Conservative Force

A particle of mass m moves from $\vec{r}_i$ to $\vec{r}_f$ and is acted on by the gravitational force exerted by a planet of mass M as shown in Figure 9.13. Show that the work done by gravity as the particle moves along path 1 (green) equals the work done along path 2 (red) and that gravity (so far) passes the first test of a conservative force.

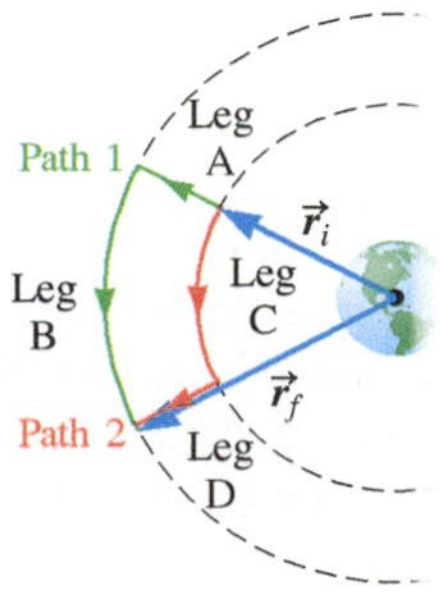

FIGURE 9.13 Two paths from $\vec{r}_i$ to $\vec{r}_f$. Path 1 is in green and path 2 is in red.

INTERPRET and ANTICIPATE

Each path has a straight part and a curved part. So, path 1 is made of leg A (straight part) and leg B (curved part), and path 2 is made of leg C (curved) and leg D (straight). Finding the work done along each whole path means finding the work done along each leg of the path and then adding them together.

Example continues on page 258 ▶

SOLVE

Finding the work done along leg A is similar to our procedure in finding the work done on the meteor falling toward the Earth (Example 9.4). The difference here is that the particle moves directly away from the planet. So, $r_f > r_i$.

$$W = \int_{r_i}^{r_f} \vec{F} \cdot d\vec{r}$$

$$W_A = \int_{r_i}^{r_f} \vec{F}_G \cdot d\vec{r} = \int_{r_i}^{r_f} -F_G \hat{r} \cdot d\vec{r} = -\int_{r_i}^{r_f} F_G dr$$

$$W_A = -\int_{r_i}^{r_f} \left(G\frac{Mm}{r^2} \right) dr$$

$$W_A = GMm\left[\frac{1}{r_f} - \frac{1}{r_i}\right]$$

The work done by gravity as the particle moves along leg B is zero because the force (toward the center) and the displacement (along the circular arc) are perpendicular all along this part of the path. The dot product of perpendicular vectors is zero.

$$W = \int_{r_i}^{r_f} \vec{F} \cdot d\vec{r}$$

$$W_B = 0$$

The total work done by gravity along path 1 is the sum of the work done along each leg.

$$W_1 = W_A + W_B = GMm\left[\frac{1}{r_f} - \frac{1}{r_i}\right] \quad (1)$$

The work done along path 2 is found in a similar way. First, the work along leg C is zero for the same reason that the work done along leg B is zero.

$$W_C = 0$$

Finding the work done by gravity along leg D is just like finding the work along leg A.

$$W_D = GMm\left[\frac{1}{r_f} - \frac{1}{r_i}\right]$$

So, the total work done along path 2 is the sum of the work done along each leg.

$$W_2 = W_C + W_D = GMm\left[\frac{1}{r_f} - \frac{1}{r_i}\right] \quad (2)$$

Compare Equation (1) with Equation (2) to find that the work done by gravity along path 1 equals the work done along path 2.

$$W_2 = W_1$$

CHECK and THINK

Because the work done along each path is the same, we can conclude that gravity *has not failed* this test of a conservative force. We cannot possibly test every path from $\vec{r}_i$ to $\vec{r}_f$ because there are an infinite number of them. So, passing this test is a requirement, but not a sufficient condition, to prove that gravity is a conservative force. Even the second test—that zero work is done around a closed path—is a necessary, but not sufficient, condition. A more sophisticated test exists, but it is beyond the scope of this book. Instead, we state that gravity, the spring force obeying Hooke's law, and the electrostatic force (Chapter 23) all are conservative forces.

EXAMPLE 9.6 A Nonconservative Force

To illustrate the second test for a conservative force, let us return to Paul and his sled to examine two forces that are *not* conservative. The sled is on the ice, and Paul pulls a rope that is attached to the sled (Fig. 9.14A). The tension F_P in Paul's rope is constant. (This constant tension is, of course, difficult for Paul, but it makes our calculations simpler.) The sled is constrained to move in a circle because a second rope attached to it is fixed to a stake in the ice. We assume friction between the sled and ice is low enough to be ignored. As a test, we calculate the work done by the tension force in Paul's rope in one circuit of the sled.

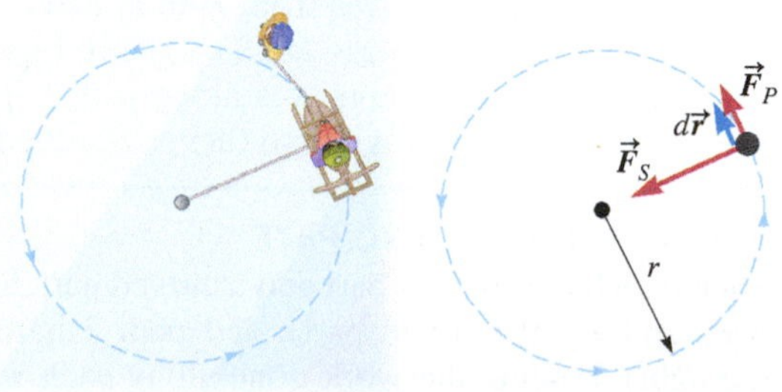

FIGURE 9.14 **A.** Overhead view **B.** Free-body diagram

INTERPRET and ANTICIPATE

We are interested in finding the work done by Paul's rope around the closed circular path, so Paul and his rope are outside the system. The system consists only of the sled and its contents.

SOLVE

Start with the general expression for work (Eq. 9.21). At every point on the circle as shown in Fig. 9.14B, the tension force $\vec{F}_P$ exerted by Paul and the infinitesimal displacement $d\vec{r}$ are parallel ($\varphi = 0$), so the cosine of the angle between them is 1.

$$W = \int_{r_i}^{r_f} \vec{F}_P \cdot d\vec{r} \qquad (9.21)$$

$$W = \int_{r_i}^{r_f} F_P dr \cos \varphi$$

$$W = \int_{r_i}^{r_f} F_P dr \cos 0 = \int_{r_i}^{r_f} F_P dr$$

Because the tension is constant over the entire path, we can pull it outside the integral.

$$W = F_P \int_{r_i}^{r_f} dr$$

The integral is a path integral over a single circular path. The result of such an integral is the length of the path; in this case, it is the circumference of the circle.

$$W = F_P 2\pi r$$

CHECK and THINK

We just found that the work done by the tension force in Paul's rope is *not* zero around a closed path. We conclude that this tension force is nonconservative. Also notice that because the work is positive, the sled's kinetic energy increases. This means that the sled and Paul must be accelerating; a feat that would be difficult for Paul to maintain for an extended period of time.

In the scenario in Example 9.6 in which Paul pulls a sled around a closed, circular path, the sled is constrained to move in a circle by a rope attached to a stake. Figure 9.14B shows the sled at one particular moment along with its infinitesimal displacement and the tension forces exerted on it. We want to calculate the work done by the rope attached to the stake as the sled moves on its circular path. From the figure, we do not need to make any actual calculation because the force $\vec{F}_S$ exerted by the stake's rope is always perpendicular to $d\vec{r}$. Therefore, the work done by the tension force in the stake's rope is always zero. We say that the force $\vec{F}_S$ is a zero-work force. A **zero-work force** may accelerate a particle without doing any work on it. Any centripetal force is a zero-work force. A force may be a zero-work force in some situations and not in others. For instance, gravity—a conservative force—can be a zero-work force when it keeps a particle in a circular orbit. Other examples of zero-work forces are static friction and the normal force. To see why, we need to take another look at the particle model.

ZERO-WORK FORCE

Major Concept

9-6 Particles, Objects, and Systems

So far in this book, we have modeled all sorts of real objects—such as people, cars, and planets—as particles. This model does not work when we are interested in the energy associated with some situations that involve friction and the normal force.

Recall from Chapter 2 that a **particle** is an idealized point that has no spatial extent, no shape, and no internal structure. An object can be treated as a particle undergoing purely translational motion if every point on the object undergoes exactly the same displacement as every other point on the object.

Now let us look at a simple object that cannot be modeled as a single particle. A chain of five metal links is held by an end link and then dropped (Fig. 9.15). The chain falls straight down so that the bottom link lands first. The displacement of the bottom link is less than the displacement of the top link. Because the displacement of each link of the chain is not exactly the same, we say that the chain is a **deformable object** or a **deformable system**, and it cannot be modeled as a particle in this situation.

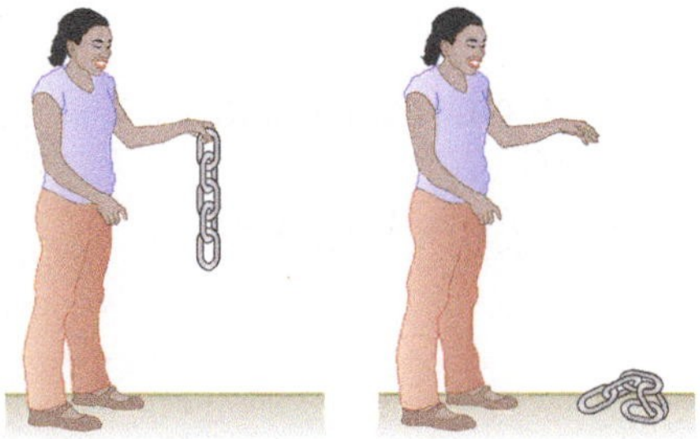

A. B.

FIGURE 9.15 **A.** A chain consisting of five links is held by an end link and then dropped. **B.** Each link lands on the ground, but each link has a different displacement. For example, the bottom link has a smaller displacement than the top link.

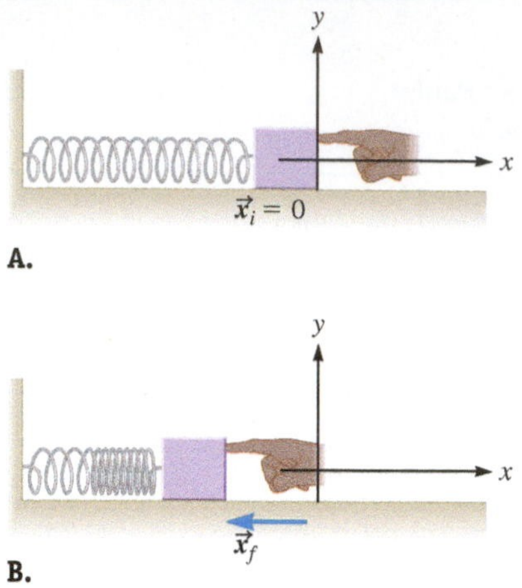

FIGURE 9.16 **A.** A person's hand pushes on the block, and the spring compresses. **B.** The person's hand is outside the block–spring system, so it does work on the system. That work increases the system's potential energy.

Work and Mechanical Energy

Our goal is to incorporate deformable systems into the conservation approach, leading us to extend the work–*kinetic* energy theorem into a work–*mechanical* energy theorem. We start by finding an expression for the work done by a person's hand when compressing a spring (Fig. 9.16). In this case, the deformable system consists of the spring, the wall, and the block.

When the person pushes on the block, the coil attached to the block is displaced, but because the spring is deformable, the coil attached to the wall is not displaced. So, what displacement should we use to calculate the work done by the person on the spring–block system? In this case, we must use the displacement of the *point of application* of the force. The **point of application** is the place in the system where the force is exerted. For a system modeled as a particle, it is simply the location of the particle, but that is not the case for a deformable system. The hand exerts a contact force, so the point of application is the place where the hand is touching the block. The hand has displacement $\Delta\vec{x} = -|x_f - x_i|\hat{\imath}$ (Fig. 9.16).

We also need to know the force exerted by the person to find the work done on the system. According to Newton's second law, as long as the block is not accelerating, the magnitude of the force exerted by the person must be given by Hooke's law, $\vec{F}_H = -|kx|\hat{\imath}$. Because the force is not constant, we must integrate to find the work done by the person. Because the force and the displacement point in the same direction, the angle between them is zero, and the work done by the hand is positive:

$$W = \int_{x_i}^{x_f} (F_H\,dx)(\cos 0) = \int_{x_i}^{x_f} kx\,dx$$

$$W = k\int_{x_i}^{x_f} x\,dx = \tfrac{1}{2}kx^2\Big|_{x_i}^{x_f}$$

$$W = \tfrac{1}{2}k(x_f^2 - x_i^2),\ \text{done by person} \tag{9.25}$$

In this example, there is no friction between the block and the surface, and the block is at rest at both its initial and final positions; so, its kinetic energy does not change. Therefore, the energy transferred from the person's hand to the block–spring system did not go into the system's kinetic energy. Instead, the work done on the system went into changing the system's configuration; the spring was initially relaxed, and in its final configuration it is compressed. So, in this case, the positive work done increased the system's potential energy. In fact, if you compare work done by the external person (Eq. 9.25) to the change in elastic potential energy of spring–block system (Eq. 8.8), you will find that they are equal: $W_{ext} = \Delta U_e$.

In a more general case, the work done can change both the kinetic energy and the potential energy. For example, if the block was initially at rest but moving at the final time, the system's kinetic energy and potential energy would have increased. So, in general, the total work done on a system by external forces in the absence of resistive forces (such as friction and drag) changes the system's mechanical energy; this **work–mechanical energy theorem** is

WORK–MECHANICAL ENERGY THEOREM

Underlying Principle

$$W_{tot} = \Delta E$$

$$W_{tot} = \Delta K + \Delta U \tag{9.26}$$

Center of Gravity and Center of Mass

For a deformable system, we need to know the displacement of the point of application to find the work done by an external force. For the case of a contact force such as a person pushing on a spring–block system, the point of application is the place where the hand makes contact with the system.

In the case of the falling chain (Fig. 9.15), however, the external force is gravity, and gravity is a field force. Gravity pulls each component (each link) of the system downward. So, where is the point of application in that case? The answer is that we define a point called the **center of gravity**, and total gravitational force exerted on the system is modeled as acting on the center of gravity.

When the gravitational field is constant (or nearly so) as it is near the surface of the Earth, the center of gravity is at the same point as the *center of mass*. The **center of mass (CM)** is a point associated with any system such that when a net external force is exerted on the system, the center of mass accelerates according to Newton's second law $\vec{F}_{tot} = m\vec{a}_{CM}$, where m is the system's mass. When that net external force is due to a uniform gravitational field near the surface of the Earth, the center of mass (and equivalently the center of gravity) accelerates downward. (In general, not every component has the same acceleration, so it is possible that only the center of mass has the acceleration $\vec{a}_{CM}$.) Put in terms of energy, the kinetic energy of the center of mass depends on the system's mass and its center-of-mass speed: $K_{CM} = \frac{1}{2}mv_{CM}^2$.

CENTER OF MASS ✪ Major Concept

In the next chapter, we will learn how to calculate the center-of-mass position. For now, you can think of the center of mass as the average location of the system's mass, and you can find it from symmetry. For example, the center of mass of a sphere with uniform density is at the center of the sphere. There does not need to be any matter at the center of mass. In the case of the chain, the center of mass is in the center of the middle link, where there is no metal.

Zero-Work Forces

With the deformable system model in mind, let's take another look at zero-work forces. As an example, consider the simple act of walking across a level floor. To walk, you swing one leg forward and then place your foot down and press against the floor. So, when you walk, you change your shape (deform the system), allowing yourself to press on the floor. By Newton's third law, the floor exerts a force back on you, and a horizontal component accelerates you forward. This horizontal force exerted by the floor is static friction, and its point of application is on the bottom of your shoe. By the definition of *static* friction, however, that point of application has no displacement. So, static friction does zero work on you.

Static friction is not always a zero-work force. For example, when a box sits in a flatbed truck, static friction accelerates the box, matching the truck's acceleration (Fig. 9.17). In this case, the point of application is displaced in the same direction as the static friction force, so we conclude that static friction does positive work on the box.

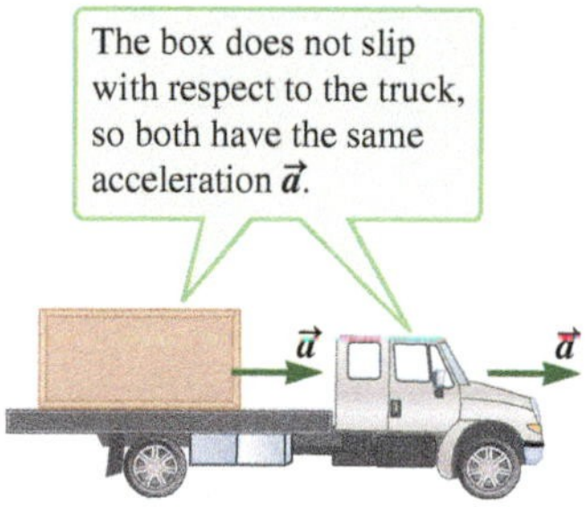
The box does not slip with respect to the truck, so both have the same acceleration $\vec{a}$.
$\vec{a}$
$\vec{a}$

FIGURE 9.17 The point of application of static friction moves in the same direction as the box's displacement, so static friction does positive work in this case.

CONCEPT EXERCISE 9.2

For a person to jump straight up, she must crouch down and quickly extend her legs.

a. What force accelerates her upward?
b. Does this force do work on her? Explain.

CONCEPT EXERCISE 9.3

A man is accelerated upward in an elevator.

a. What force causes him to accelerate?
b. Does this force do work on him? Explain.

9-7 Thermal Energy

So far in our conservation approach to mechanics, we have avoided situations that involve moving friction and drag. In these next two sections, we learn how to include these nonconservative forces.

We experience thermal energy whenever we rub our hands together to warm them up. **Thermal energy** is the internal energy associated with the temperature of a system. It is not really a new form of energy, though; it's just a way to describe the kinetic energy (and potential energy) of the many particles in a system.

THERMAL ENERGY ▶ Special Case

FIGURE 9.18 Globular clusters have hundreds of thousands of stars. The center of mass of a globular cluster orbits the Milky Way much as a planet orbits the Sun. The stars in the globular cluster whiz around the cluster like a swarm of bees.

Globular Cluster Analogy

We can develop our intuition about thermal energy by looking at an analogy (developed by Richard Feynman, an American Nobel Prize–winning physicist). A globular cluster is a roughly spherical group of hundreds of thousands of stars (Fig. 9.18). The center of mass of a globular cluster orbits the center of the Milky Way galaxy much like a planet orbits the Sun. In addition, the stars in the cluster zip around inside the cluster on erratic paths.

First, we model the globular cluster as a system of particles and imagine that some external force (due perhaps to the gravitational attraction of another very large star cluster) does work on our globular cluster. The work W done by this external force is likely to do three things: (1) change the shape of the cluster, (2) accelerate its center of mass, and (3) accelerate the individual stars moving inside the cluster. Each of these changes corresponds to some change in the system's energy as follows: (1) Because potential energy depends on the configuration of the system, a change in the cluster's shape means that its potential energy U changes. (2) The acceleration of the cluster's center of mass means that the cluster's center-of-mass kinetic energy K_{CM} changes. (3) Similarly, the acceleration of the individual stars means that the sum of their kinetic energy K_{stars} changes. So, the work done by the external force equals the change the system's potential energy plus the change in the center-of-mass kinetic energy plus the change in kinetic energy of all the individual stars:

$$W = \Delta U + \Delta K_{CM} + \Delta K_{stars} \tag{9.27}$$

Equation 9.27 is actually just the work–mechanical energy theorem $W_{tot} = \Delta E$ (Eq. 9.26) for this cluster.

Now imagine observing this cluster from a great distance so that we cannot see the individual stars and thus can no longer model the cluster as a system of particles. Instead, we model the cluster as a single deformable object. There are no other changes to the situation; an external force still does work on the cluster. Even from this great distance, we can still observe the cluster deform and its center of mass accelerate. There is a major difference, though. Because we cannot see the individual stars, we cannot observe the change in the individual stars' kinetic energy. It seems that some of the work done on the cluster is missing because $W > \Delta U + \Delta K_{CM}$.

We know, though, that energy cannot be destroyed, so we assume the missing energy is really internal to the system and account for the missing energy this way:

$$W = \Delta U + \Delta K_{CM} + \Delta E_{int} \tag{9.28}$$

INTERNAL ENERGY ✪ **Major Concept**

By comparing Equations 9.27 and 9.28, we equate the change in internal energy ΔE_{int} with the change in kinetic energy of the individual stars ΔK_{stars}. **Internal energy** is a manifestation of potential or kinetic energy on a small scale (compared with the whole system) involving a large number of particles. Imagine trying to add up the kinetic energy of the hundred thousand stars in the cluster; the concept of internal energy makes it possible for us to avoid making all those individual calculations and still apply the conservation approach.

Internal energy comes in several forms. For now, the form of internal energy we are concerned with is thermal energy, so we write the change in internal energy as $\Delta E_{int} = \Delta E_{th}$. In our analogy, the motion of the individual stars is thermal energy. (This wording may seem silly because that motion has nothing to do with the star's temperature, but astronomers have been known to use the phrase *thermal energy* in this way.)

Change in Thermal Energy due to Moving Friction

Suppose we need to calculate the change in thermal energy due to moving friction. Consider a particular example in which a student slides a book across a table from point i to point f along the straight path S (Fig. 9.19). To make our argument simple and clear, we ignore the momentary acceleration at i and the momentary deceleration at f. Four forces—gravity, the normal force, the push of the person, and kinetic friction—act on the book. The person pushes with a constant force $\vec{F}_P$ equal in magnitude to kinetic friction on the book so that the book moves with constant velocity.

At first, let's assume we can treat the book as a particle and find the work done on it by the four forces. Gravity and the normal force do no work because they are perpendicular to the book's displacement. The work done by the person on the book is positive and is given by $W_P = F_P \Delta r$. Now, we might be tempted to argue that the work done by kinetic friction on the book is $W_k = -F_k \Delta r$ because the magnitude of kinetic friction is the same as that of the person's applied force, but in the opposite direction as the book's displacement. Then, the total work W_{tot} done on the book would be zero.

There are two problems with that temptation. First, the book is not a particle, and we know that sliding the book across the table increases the book's thermal energy. (If we touch the bottom of the book, it feels warmer.) So, the total work done on the book cannot be zero[1]. The second problem is that the table's thermal energy also increases. It would be very difficult to separate the book's increase in thermal energy from the table's increase. We avoid this difficulty by cleverly choosing to include both the book and the tabletop in the system. Then, we identify $F_k \Delta r$ as the change in the system's thermal energy. It doesn't matter how much of that thermal energy is in the book versus how much is in the table.

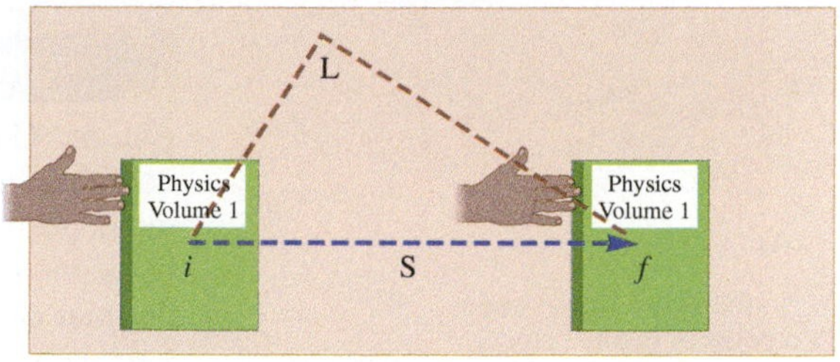

A. Top view

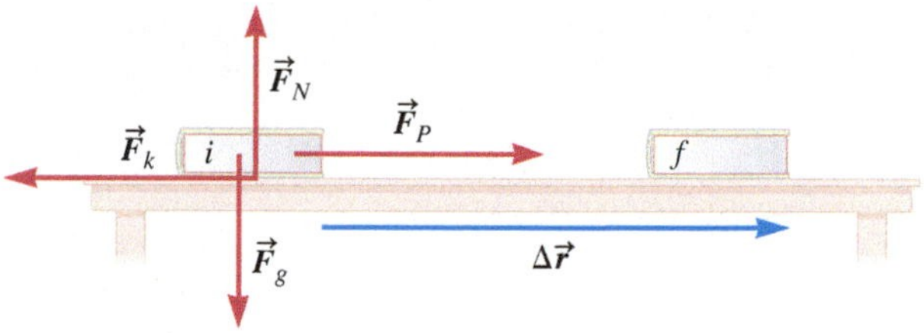

B. Side view

FIGURE 9.19 **A.** A student slides a book across a table along the straight path S. Later we consider a student sliding the book along the L–shaped path L. **B.** Four forces act on the book. The book does not accelerate, so $F_P = F_k$.

So, with this clever choice for the system, the person does positive work on the book–tabletop system. This work does *not* go into the kinetic energy of the system's center of mass. (The book moves at constant velocity.) The work does not go into the potential energy either. (There are no conservative forces in the system.) Instead, the work done by the person goes into the thermal energy of the system. (Both the tabletop and the book get warmer.) Comparing this system to the globular cluster, the increased temperature is analogous to the change in the stars' individual speeds due to (part) of the work done by the external force on the cluster. For the book–tabletop system, the work done by the person equals the change in thermal energy $W_P = \Delta E_{th}$.

THERMAL ENERGY
Special Case

To find the change in thermal energy, you need to know one more thing: that it depends on the path taken by objects in a system. In the book–tabletop system, if the student slides the book along the L-shaped path instead of the straight path S (Fig. 9.19A), the change in the system's thermal energy is greater because path L is longer than path S. The change in thermal energy depends on the total path length s such that

$$\Delta E_{th} = F_k s \tag{9.29}$$

Equation 9.29 is always positive. So, if kinetic (or rolling) friction acts in some system, it always increases the system's thermal energy. Unlike changes in potential or kinetic energy, this increase in thermal energy cannot be converted back to mechanical energy. The increase in thermal energy is at the expense of mechanical energy, and we say that mechanical energy has been dissipated. The thermal energy will eventually leave the system entirely and increase the energy of the environment (Chapter 21). Nonconservative forces (such as moving friction and drag) are known as **dissipative forces** because they reduce the mechanical energy of the system, and this energy is eventually lost by the system.

DISSIPATIVE FORCE
Major Concept

CONCEPT EXERCISE 9.4

Avi passes a book up to Cameron who's on the upper bunk, and Cameron then passes the book back down to Avi. Consider the Earth and the book to be the system. In this round trip, does the system's gravitational energy increase, decrease, or remain the same? Explain.

[1]This argument assumes the only way to transfer energy to a system is through work. In Chapter 21, we will learn about another way to transfer energy to a system. Our argument here is not affected because work is the only energy transfer mechanism possible in this case.

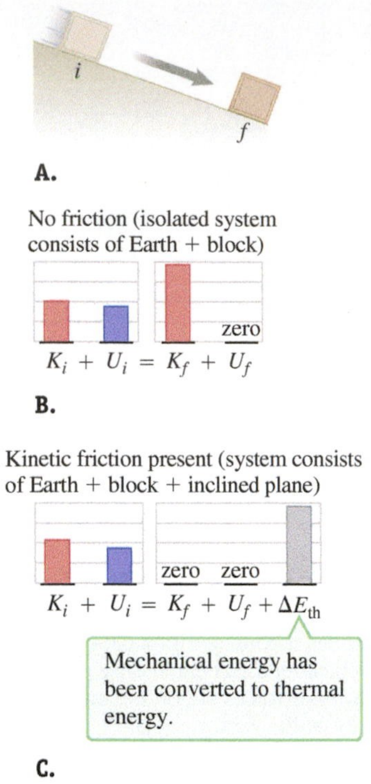
i
f
A.
No friction (isolated system consists of Earth + block)
zero
$K_i + U_i = K_f + U_f$
B.
Kinetic friction present (system consists of Earth + block + inclined plane)
zero zero
$K_i + U_i = K_f + U_f + \Delta E_{th}$
Mechanical energy has been converted to thermal energy.
C.

FIGURE 9.20 **A.** A block slides down a plane. **B.** If kinetic friction is negligible, mechanical energy is conserved. **C.** If kinetic friction causes the block to stop, mechanical energy is not conserved and is instead converted to thermal energy.

CONCEPT EXERCISE 9.5

Avi and Cameron sit on opposite sides of the table as in Figure 9.19. Avi pushes a book to Cameron, Cameron pushes the book back toward Avi, and they repeat this rapid pushing for several rounds before the book is returned to Avi. Consider the book and the tabletop to be the system. At the end of these rounds, does the system's thermal energy increase, decrease, or remain the same? Explain. Compare your answer here with your answer to Concept Exercise 9.4.

9-8 Work–Energy Theorem

Now we are prepared to expand the work–*mechanical energy* theorem to a work–energy theorem that includes energy dissipation. To do so, we use the familiar example of a block sliding on an inclined plane (Fig. 9.20A). First, suppose no friction acts, only the Earth and the block are in the system, and no external forces do work on the system. (Set the reference configuration to the bottom of the inclined plane.) As illustrated by the bar chart (Fig. 9.20B), the mechanical energy is conserved in this isolated system.

Now assume kinetic friction acts between the block and the inclined plane. The new system consists of the block, the Earth, and the inclined plane. As before, the system starts with positive kinetic energy and potential energy, but this time when the system reaches its final configuration, the block has stopped. In the final configuration, the system has no kinetic or potential energy. All the mechanical energy has been converted to thermal energy. A new bar must be included in the bar chart for thermal energy ΔE_{th} (Fig. 9.20C). Mechanical energy ($K + U$) has been dissipated, but the overall energy is conserved in this isolated system.

Now let us add one more complication: that of an external force that does work on the system. Now, the block–Earth–inclined plane system is not isolated. Specifically, work is done by a person who pulls a rope attached to the block (Fig. 9.21A). (The person and the rope are outside the system.) So, we must add a new bar to the chart (Fig. 9.21B). All the energies are the same as they were previously except that the kinetic energy in the final configuration is not zero because the tension in the rope did positive work on the nonisolated system.

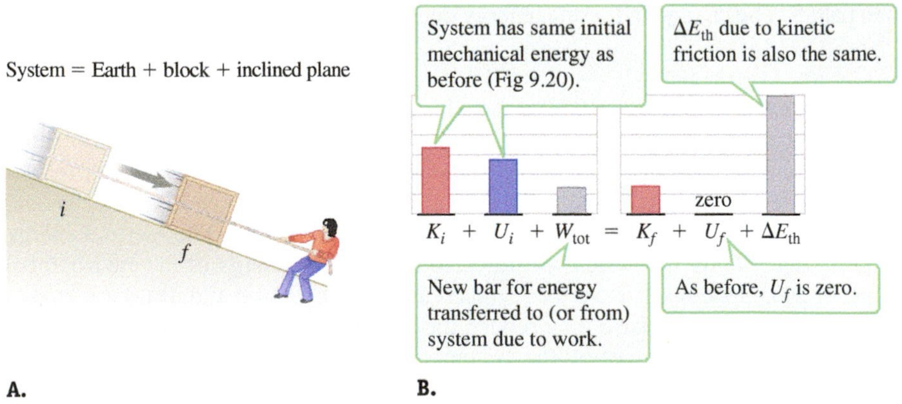
System = Earth + block + inclined plane
i
f
A.
System has same initial mechanical energy as before (Fig 9.20).
ΔE_{th} due to kinetic friction is also the same.
zero
$K_i + U_i + W_{tot} = K_f + U_f + \Delta E_{th}$
New bar for energy transferred to (or from) system due to work.
As before, U_f is zero.
B.

FIGURE 9.21 **A.** A person uses a rope to pull a block down an inclined plane. Kinetic friction acts between the block and the plane. We consider the block, the Earth, and the plane to be inside the system; the person and the rope are outside the system. The rope does work on the system, and kinetic friction increases the thermal energy of the system. **B.** The total height of the three bars on the left equals the total height of the two bars on the right.

We use this bar chart (Fig. 9.21B) to write a general work–energy theorem for a nonisolated system including internal dissipative forces. According to that bar chart, the initial mechanical energy (potential energy plus kinetic energy) plus the energy transferred to or from the system equals the final mechanical energy plus the change in internal energy (thermal energy, in this case.) Mathematically,

$$\begin{pmatrix}\text{initial}\\ \text{mechanical}\\ \text{energy}\end{pmatrix} + \begin{pmatrix}\text{total work,}\\ \text{which is}\\ \text{energy transferred}\end{pmatrix} = \begin{pmatrix}\text{final}\\ \text{mechanical}\\ \text{energy}\end{pmatrix} + \begin{pmatrix}\text{change in}\\ \text{internal}\\ \text{energy}\end{pmatrix}$$

$$E_i + W_{tot} = E_f + \Delta E_{int} \tag{9.30}$$

We write the mechanical energy in terms of the kinetic energy and potential energy:

$$K_i + U_i + W_{tot} = K_f + U_f + \Delta E_{int} \qquad (9.31)$$

WORK–ENERGY THEOREM ❗ **Underlying Principle**

Equation 9.31 is a very general expression for the **work–energy theorem**. Purely for convenience, we rewrite Equation 9.31 explicitly for the two forms of potential energy—gravitational (Eq. 8.5 and 8.7) and elastic (Eq. 8.9)—we are most concerned with in this part of the book. So,

$$K_i + U_{gi} + U_{ei} + W_{tot} = K_f + U_{gf} + U_{ef} + \Delta E_{int} \qquad (9.32)$$

The work–energy theorem involved includes a change in internal energy ΔE_{int}. Internal energy comes in several forms, including chemical energy. Chemical energy is important when we consider certain deformable objects and forces that do no work. For example, a froghopper insect leaps upward by pushing on the ground, and by Newton's third law, the ground pushes back on the froghopper, accelerating the froghopper (Example 5.11). Consider the froghopper and the Earth as the system. The components of the force exerted by the ground are the normal force and static friction. The point of application of these forces has no displacement, so these forces do no work on the froghopper (Section 9-6). No external forces do work on the system, yet the system's mechanical energy increases (Fig. 9.22A). Where does this energy come from? The answer is that the increase in mechanical energy must come from the froghopper's internal energy. The froghopper is a deformable object: To push against the ground and leap into the air, it compresses and quickly releases muscles. The ultimate source of energy that allows an animal to control its muscles is food. Chemical reactions (the digestive process) release energy, and the animal uses that energy (in part) to control its muscles. So, this form of energy is known as *chemical energy*.

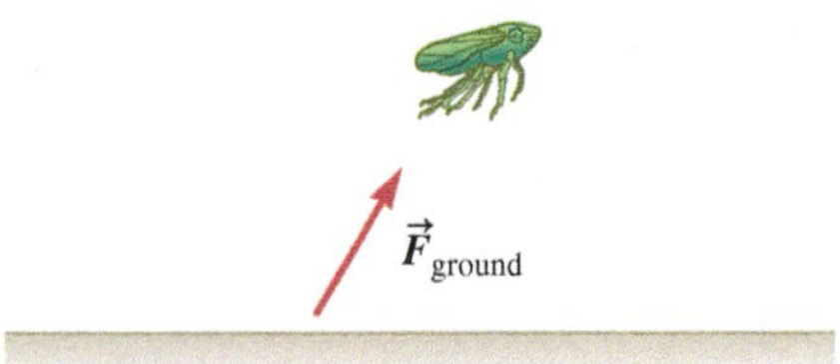

A.

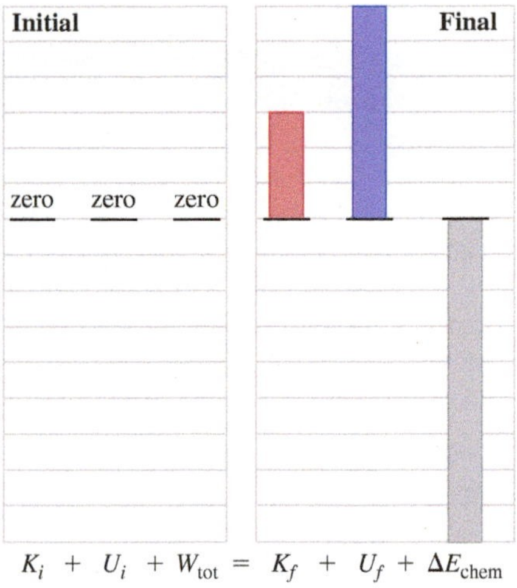

B.

FIGURE 9.22 **A.** A froghopper starts on the ground at rest. The froghopper leaps up, and mechanical energy of the system increases. **B.** The mechanical energy comes from the decrease in internal chemical energy.

CHEMICAL ENERGY ▶ **Special Case**

Living organisms are not the only objects that use chemical energy to gain mechanical energy. For example, a car runs on fossil fuel. Like food in animals, fossil fuel is the source of chemical energy that is converted to mechanical energy. When the insect leaps, the change in its chemical ΔE_{chem} is negative (Fig. 9.22B). So, the source of chemical energy (gasoline or pizza) must be replenished. Table 9.1 lists the rough energy content of some common fuels.

TABLE 9.1 **Energy content in joules.**

Astronomical sources	Joules (J)
Typical supernova (exploding star)	10^{43}
Milky Way's radiation in 1 s	10^{38}
Sun's radiation in 1 year	10^{34}

Terrestrial sources	Joules (J)
Earth's rotation	10^{29}
Volcanic detonation	10^{19}
Largest recorded earthquake	10^{18}
San Francisco earthquake of 1906	10^{17}
Tornado or thunderstorm	10^{15}
Lightning flash	10^{10}

Human needs and everyday phenomena	Joules (J)
1000-MW power station in 1 year	10^{16}
1 lb of uranium-235	3.7×10^{13}
Energy to put space shuttle into orbit	10^{13}
Crossing Atlantic Ocean in a jet plane	10^{12}
1 U.S. gallon of gasoline	1.3×10^{8}
1 lb of coal	1.6×10^{7}
Two-ton truck traveling at highway speed	10^{6}
1 lb of TNT	10^{6}
1 can lighter fluid	10^{6}
1 candy bar, order of french fries, or slice of pizza	10^{6}
1 tablespoon sugar, 1 apple, or 5 crackers	10^{5}
1 AA battery	10^{3}
Major league pitch	10^{2}
Striking a typewriter key	10^{-2}

Adapted from Howard Keller, "Energy Yield of Various Sources," *Physics Teacher* 30:455 (1992).

Remember that we have two expressions for gravitational potential energy, $U_g = mgy$ and $U_G = -G(m_1 m_2/r)$, and we need only one for any situation.

A more detailed study of chemical energy is beyond the scope of this book; instead, here we are primarily interested in thermal energy. So, for convenience in problem solving, we write the work–energy equation in terms of one specific internal energy, thermal energy:

$$K_i + U_{gi} + U_{ei} + W_{\text{tot}} = K_f + U_{gf} + U_{ef} + \Delta E_{\text{th}} \tag{9.33}$$

PROBLEM-SOLVING STRATEGY

Applying the Work–Energy Theorem

The problem-solving strategy for applying the work–energy theorem (Eq. 9.33) is nearly the same as it is when applying the conservation of energy principle. However, now the system no longer needs to be isolated. Also, you need four bars on each side of the chart to account for the four terms on each side of the work–energy theorem.

INTERPRET and ANTICIPATE

Step 1 Choose a **system**. Here are some tips for choosing a system:

- The source of a conservative force (gravity or spring force) is best included in the system. Then, associate potential energy with these forces.
- Dissipative forces such as moving friction increase the thermal energy of two objects; include both objects in the system.
- Sources of other nonconservative forces such as a rope are often best left out of the system. Then, the work done by the force is easily calculated, and it accounts for the transfer of energy from the environment to the system.

Step 2 Pick a **coordinate system and the reference configuration.**

Step 3 Identify an **initial time and a final time.**

Step 4 Draw a **bar chart.**

SOLVE

Step 5 Apply the **work–energy theorem** (Eq. 9.32 or 9.33).

Step 6 Write down any **other equations.**

Step 7 **Do algebra** before substitution.

EXAMPLE 9.7 My Kitten Is Stuck in a Tree!

By examining the same problem three times, this example shows how choosing a system affects our analysis. A girl uses a rope to lower a basket of kittens down from a tree at a constant velocity (Fig. 9.23). She wears leather gloves to protect her hands, and the rope slips through them. Analyze the problem by making bar charts using three different systems: the basket of kittens; the basket of kittens and the Earth; and the basket of kittens, the Earth, the rope, and the girl's leather gloves. To keep the charts simple, leave out the bars for elastic potential energy because there are no springs in this problem.

FIGURE 9.23

INTERPRET and ANTICIPATE

For each system choice, consider whether a force is internal or external. If a force is internal to the system, we need to consider whether it changes the system's potential or thermal energy. If it is external, calculate the work done by the force as it acts on the system. We expect that the total height of the bars on the left should equal the total height of the bars on the right in each case, no matter which system choice has been made. In all three cases, the kinetic energy of the system is unchanged because the basket is lowered at constant velocity. So, the initial kinetic energy bar has the same height as the final kinetic energy bar in all three cases.

A Make a bar chart using only the basket of kittens as the system.

SOLVE

Two external forces act on the basket of kittens: gravity exerted by the Earth and the tension in the rope. We model the basket and its contents as a single particle. A particle has no internal structure, so there can be no potential energy or internal energy. Both gravity and tension do work on the particle.

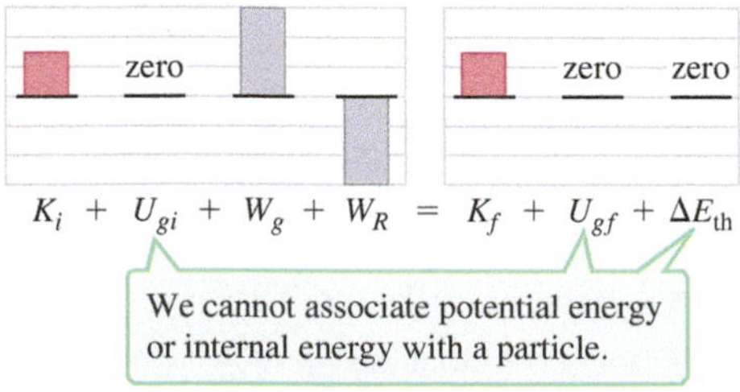
zero
zero
zero
$K_i + U_{gi} + W_g + W_R = K_f + U_{gf} + \Delta E_{\text{th}}$
We cannot associate potential energy or internal energy with a particle.

FIGURE 9.24

The work W_g done by gravity is positive because the force and the displacement are in the same direction. The work W_R done by the tension in the rope is negative because the force and the displacement are in opposite directions. For the total height of the bars on the left of Figure 9.24 to equal the total height of the bars on the right, the bars for W_g and W_R must be the same length.

CHECK and THINK

According to the work–kinetic energy theorem for a particle (Eq. 9.5), the total work done on a particle equals the change in its kinetic energy. Because the basket is lowered at constant velocity, its kinetic energy does not change. Therefore, the total work done by gravity plus tension must equal zero.

B Make a bar chart using the basket of kittens and the Earth as the system.

SOLVE

The rope is still outside the system, so tension still does work on the system. With the Earth inside the system, however, gravity no longer does work. Instead, the gravitational potential energy of the system changes, so we must choose a reference configuration. We choose the reference configuration so that it corresponds to the final configuration of the system just before the basket makes contact with the ground.

The work done by the rope is the same as it was in part A. Earth cannot do work because it is *in* the system. The gravitational potential energy is initially positive, and is zero in the final configuration (Fig. 9.25).

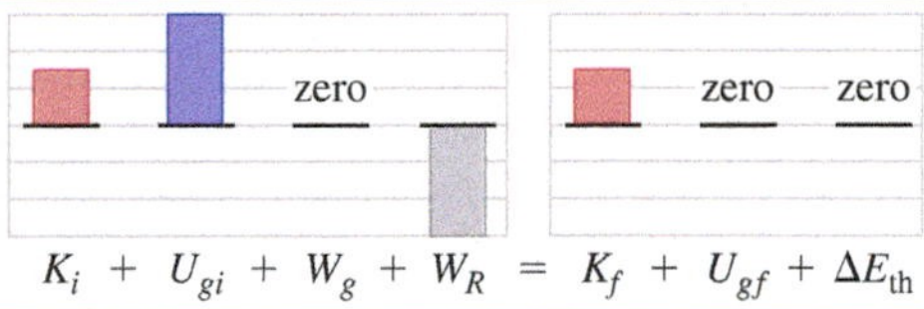

FIGURE 9.25

CHECK and THINK

The total height of the bars on the left equals the total height of the bars on the right. This situation is a special case of the work–mechanical energy theorem (Eq. 9.26) in which the change in kinetic energy is zero, so we have $W_{tot} = \Delta U$. In this particular case, the only work done is by the rope W_R, and the potential energy is gravitational U_g. We have $W_R = U_{gf} - U_{gi} = -U_{gi}$ as seen in the bar chart.

C Make a bar chart using the basket of kittens, the Earth, the rope, and the girl's leather gloves as the system.

SOLVE

With this situation, there are no forces left outside the system to do work. The rope slides through the leather gloves, and kinetic friction acts on both the rope and the gloves. The internal thermal energy of the system increases.

The gravitational potential energy is the same as it was in part B. The rope is in the system, so it cannot do work on the system. Instead, friction between the rope and the leather gloves increases the system's thermal energy (Fig. 9.26).

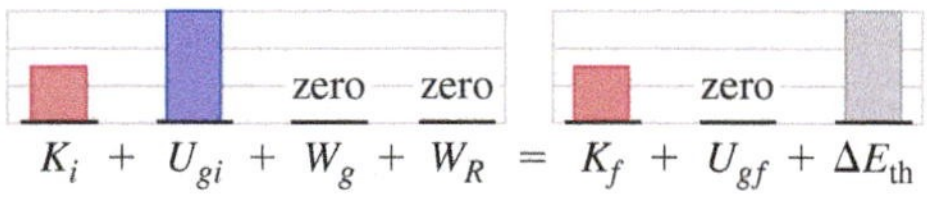

FIGURE 9.26

CHECK and THINK

If there were no friction, we would expect mechanical energy to be conserved, and the decrease in gravitational potential energy would be converted to an increase in kinetic energy. This system includes friction, however, and the decrease in gravitational potential energy goes instead into increasing the thermal energy.

EXAMPLE 9.8 Dart Gun Revisited

In Example 8.9 (page 230), we found an expression for the maximum height of a small ball launched straight up with a spring-loaded dart gun. In that example, we ignored the friction between the ball and the barrel of the gun, and we found that the maximum height of the ball is given by

$$y_{max} = \frac{k}{2mg}y_c^2 - y_c$$

where y_c is the distance by which the spring is initially compressed. Now assume kinetic friction F_k is significant and find an expression for the maximum height of the ball.

Example continues on page 268 ▶

INTERPRET and ANTICIPATE

Step 1 In solving Example 8.9, our (isolated) system consisted of the Earth, the ball, and the spring. Now that we need to take kinetic friction into account, it is best to include the gun barrel in the system. The system is still isolated, but kinetic friction dissipates some mechanical energy and increases the thermal energy. We expect the ball's maximum height to be lower than it was when friction was negligible.

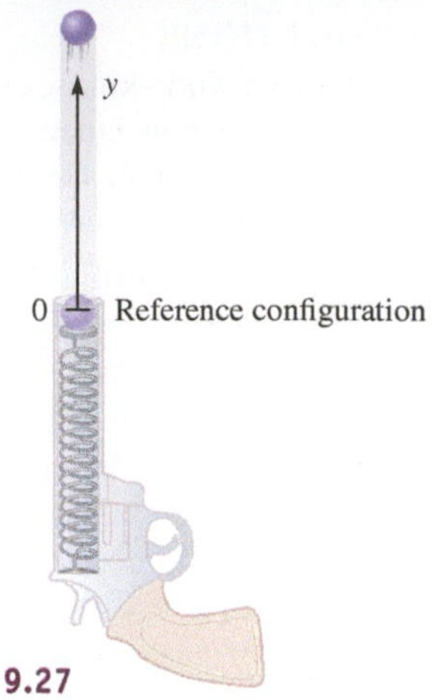

FIGURE 9.27

Step 2 As before, place the origin of the **coordinate system** at the end of the gun barrel in Figure 9.27. The **reference configuration** is the one in which the ball is at the end of the gun barrel. The spring is relaxed at that point. The elastic and gravitational potential energies of the system are zero in that configuration.

Step 3 Also as before, we set the **initial time** to when the ball is on top of a fully compressed spring and the **final time** to when the ball has reached its maximum height.

Step 4 Draw a **bar chart.** The initial and final kinetic energies are both zero. The initial gravitational potential energy is negative, and the initial elastic potential energy is positive. When the system is in its final configuration, the spring is relaxed and has lost contact with the ball, and the gravitational potential energy is positive. The difference between the bar chart in Figure 9.28 and the one in Example 8.9 is that here there is a bar for the increase in the thermal energy of the system.

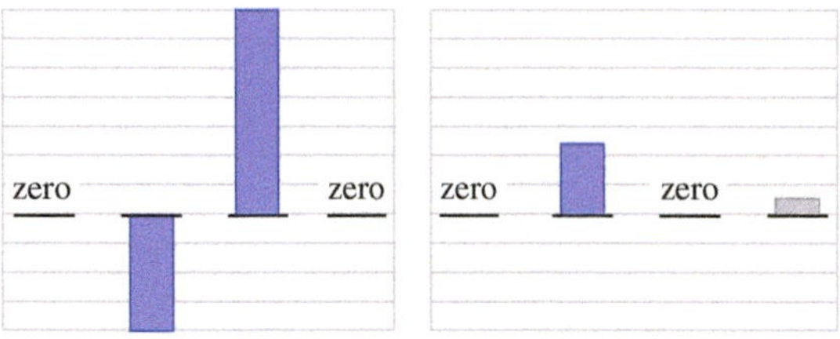

FIGURE 9.28 $K_i + U_{gi} + U_{ei} + W = K_f + U_{gf} + U_{ef} + \Delta E_{th}$

SOLVE

Step 5 Apply the **work–energy theorem** (Eq. 9.33) to write an algebraic expression, substituting zeros as indicated by the bar chart.

$$K_i + U_{gi} + U_{ei} + W_{tot} = K_f + U_{gf} + U_{ef} + \Delta E_{th}$$
$$0 + U_{gi} + U_{ei} + 0 = 0 + U_{gf} + 0 + \Delta E_{th}$$
$$U_{gi} + U_{ei} = U_{gf} + \Delta E_{th} \quad (1)$$

Step 6 Write down **other equations.** Write expressions for potential energy (Eqs. 8.5 and 8.10) and thermal energy (Eq. 9.29). The path length over which friction acts is $s = y_c$.

$$U_{gi} = -mgy_c \qquad U_{ei} = \tfrac{1}{2}ky_c^2$$
$$U_{gf} = mgy_{max} \qquad \Delta E_{th} = F_k y_c$$

Step 7 Do **algebra.** Substitute for these energies in Equation (1) and solve for y_{max}.

$$-mgy_c + \tfrac{1}{2}ky_c^2 = mgy_{max} + F_k y_c$$
$$y_{max} = \frac{k}{2mg}y_c^2 - y_c - \frac{F_k}{mg}y_c$$

CHECK and THINK

The first two terms are the same as what we found for the maximum height of the ball in the absence of friction. Because $F_k y_c/mg$ is subtracted from these terms, the maximum height y_{max} reached by the ball is lower in this case, as expected.

EXAMPLE 9.9 Snow Tubing

At a winter recreation resort, snow tubers at the bottom of the hill hook their tubes to a tow rope. A motor pulls the rope so that tubers move at constant velocity to the top of the hill. (Ignore the momentary acceleration of the tuber when he first attaches his tube to the rope.) The coefficient of kinetic friction between the tube and the snow is $\mu_k = 0.16$. A boy and his tube with a total weight of 415 N are pulled a distance of 255 m up the 18° incline (Fig. 9.29). Consider the Earth, the boy, his tube, and the snow along his path to make up the system. Find the energy transferred from the motor to the system via the work done by the rope.

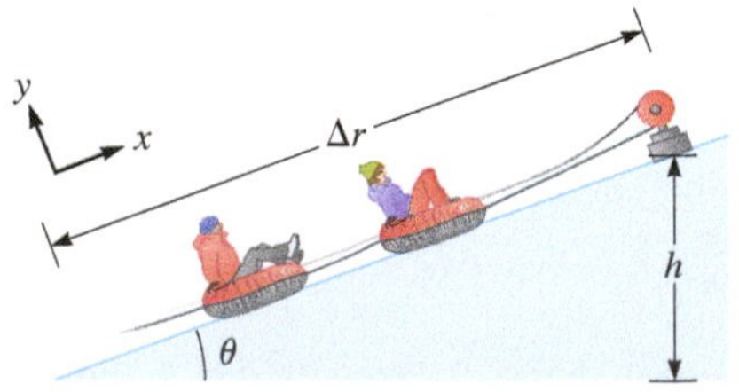

FIGURE 9.29

INTERPRET and ANTICIPATE

Step 1 A **system** has been chosen.

Step 2 A **coordinate system** has been chosen, and we set the **reference configuration** to when the boy is at the bottom of the hill.

Step 3 The **initial time** is when the boy is at the reference configuration, and the **final time** is when he is at the top of the hill.

Step 4 Draw a **bar chart.** The kinetic energy K of the system is unchanged because the tuber moves up at constant velocity. The thermal energy of the system increases due to friction between the tube and the snow. The initial gravitational potential energy is zero. We expect that the motor transfers energy to the system so that the rope must do positive work W_R on the system (Fig. 9.30).

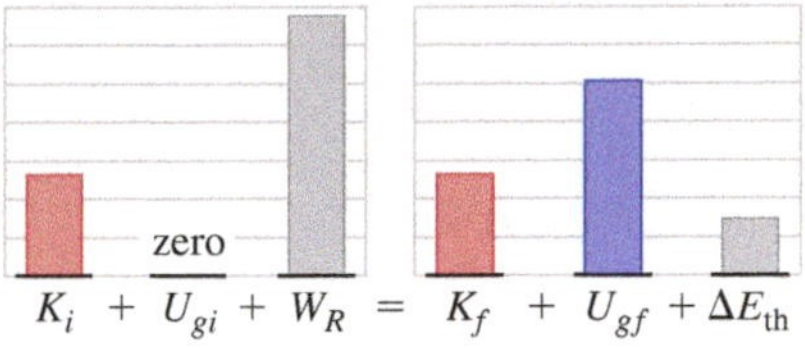

FIGURE 9.30

SOLVE

Step 5 Use the bar chart to apply the **work–energy theorem** (Eq. 9.33). The kinetic energy is constant and cancels out.

$$K_i + U_{gi} + U_{ei} + W_{tot} = K_f + U_{gf} + U_{ef} + \Delta E_{th}$$
$$K + W_R = K + U_{gf} + \Delta E_{th}$$
$$W_R = U_{gf} + \Delta E_{th} \quad (1)$$

Step 6 Write down **other equations.** Equation 8.5 gives the gravitational potential energy of the tuber, whose vertical displacement is h. Using trigonometry, rewrite h in terms of the tuber's given displacement Δr along the slope.

$$U_{gf} = mgh = mg(\Delta r \sin\theta) \quad (2)$$

The change in thermal energy comes from Equation 9.29. Kinetic friction is related to the normal force by Equation 5.9.

$$\Delta E_{th} = F_k \Delta r = \mu_k F_N \Delta r \quad (3)$$

Find the normal force using a free-body diagram (Fig. 9.31).

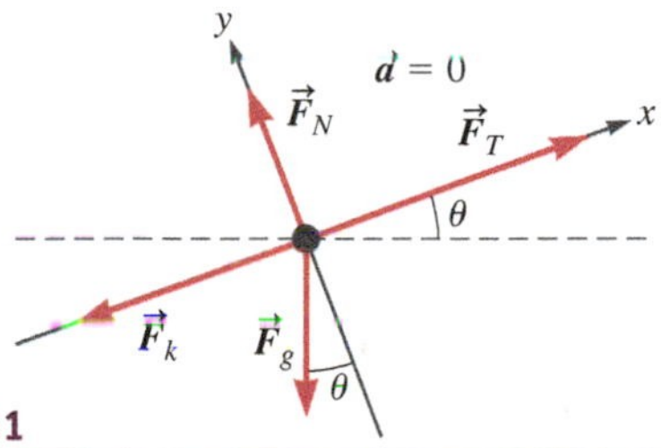

FIGURE 9.31

From Newton's second law and because the tuber is not accelerating, we find the normal force in terms of the weight mg of the tuber and his tube.

$$\sum F_y = F_N - F_g \cos\theta = 0$$
$$F_N = F_g \cos\theta = mg\cos\theta \quad (4)$$

Step 7 **Do algebra.** Substitute Equation (4) into Equation (3).

$$\Delta E_{th} = \mu_k mg\,\Delta r\cos\theta \quad (5)$$

Substitute Equations (2) and (5) into Equation (1).

$$W_R = mg\,\Delta r\sin\theta + \mu_k mg\,\Delta r\cos\theta$$
$$W_R = mg\,\Delta r(\sin\theta + \mu_k\cos\theta)$$
$$W_R = (415\,\text{N})(255\,\text{m})[\sin 18° + (0.16)\cos 18°]$$
$$W_R = 4.9 \times 10^4\,\text{J}$$

CHECK and THINK

As expected, the motor transfers energy to the system, or, in other words, the rope does positive work on the system. It might seem that more than 10,000 J is a lot of energy just to pull a boy up a hill, but it is roughly the energy content of a few french fries (Table 9.1).

9-9 Power

In Example 9.9, we found the energy transferred from a motor to the snow tuber's system. The work done by the rope does not depend on the time it takes the motor to get the tuber up the hill; the same amount of energy is transferred whether the uphill trip takes a long time or a short time. The rate of energy transfer, however, is often important. We would not want a motor that transferred energy so slowly that it took all afternoon to get the tuber up the hill or one that transferred energy so quickly that the trip became unsafe.

POWER ✪ Major Concept

Power is the rate at which energy is transferred into or out of a system or the rate at which energy is converted from one form to another within a system. The word *rate* implies a change in a quantity per unit time. Just as velocity is the rate of change of position, power is the rate of change of energy. We defined the average velocity as the displacement divided by the time interval for that displacement. Similarly, the average power P_{av} transferred to a system is the work W done on the system divided by the time interval Δt over which the energy was transferred:

$$P_{av} = \frac{W}{\Delta t} \tag{9.34}$$

When energy is converted from one form to another within a system, the average power for the conversion is found in a similar way. For example, if we are interested in finding the average power dissipated by friction, we replace W by the change in thermal energy ΔE_{th}:

$$P_{av} = \frac{\Delta E_{th}}{\Delta t} \tag{9.35}$$

Keeping with our velocity analogy, just as instantaneous velocity is simply called *velocity* and is given by the time derivative of position, so instantaneous power is simply called *power* and is given by the time derivative of energy transferred or converted. So, for power transferred to or from a system through work, the instantaneous power is

$$P = \frac{dW}{dt} \tag{9.36}$$

If we are interested in finding the power in converting from one form of energy to another, we take the time derivative of that energy converted. Again for the conversion of mechanical energy to thermal energy, the power dissipated is

$$P = \frac{dE_{th}}{dt} \tag{9.37}$$

Power is a scalar with the dimensions of energy per time. Its SI units are joules per second or newton meters per second. This combination of units is called the *watt*, abbreviated "W," where

Both *work* and *watt* are symbolized by the same letter. Keep them distinct in your mind by remembering that italic uppercase *W* stands for work, and regular (roman, or nonitalic) uppercase W stands for watts.

$$1\ \text{W} = 1\ \text{N}\cdot\text{m/s} = 1\ \text{J/s}$$

We are familiar with the watt as a unit of power. We buy 100-W lightbulbs, the exercise machine reports that we are converting chemical energy to mechanical energy (commonly called *burning* fat) at 400 W, and our regional power plant supplies 10^9 W. In U.S. customary units, power is measured in horsepower (hp), where

$$1\ \text{hp} = 746\ \text{W}$$

In the case of power transferred to or from a system through work, it is often convenient to rewrite Equation 9.36 in terms of external force $\vec{F}$ applied. To do so, substitute $dW = \vec{F}\cdot d\vec{r}$ (Eq. 9.20):

$$P = \frac{\vec{F}\cdot d\vec{r}}{dt} = \vec{F}\cdot\left(\frac{d\vec{r}}{dt}\right)$$

The term in parentheses is velocity (Eq. 4.8), so

$$P = \vec{F}\cdot\vec{v} \tag{9.38}$$

where $\vec{v}$ is the velocity of the point at which the force is applied.

CONCEPT EXERCISE 9.6

Electric companies charge consumers for the amount of energy used in each billing cycle. The companies often express energy in units of kilowatt-hours (kWh), where

$$1 \text{ kWh} = (1000 \text{ W})(3600 \text{ s}) = 3.6 \times 10^6 \text{ J}$$

In Maryland, a major electric supplier charges 4.72 cents per kilowatt-hour. If a Maryland consumer accidentally leaves a 100-W lightbulb on continuously for a 30-day billing cycle, how much would she owe the electric company for her mistake?

EXAMPLE 9.10 Let's Take the Elevator

The Taipei 101 tower shown in Figure 9.32 has the fastest elevators in the world today. The elevator's ascent is so fast that a special pressure-control system was installed. After the express elevator is loaded, it quickly reaches a constant speed of 16.6 m/s. Assume the elevator and its passengers have a mass of 4800 kg and there is a counterweight that reduces the effective weight of the elevator by 40%. Also assume there is a constant 5000-N kinetic friction force acting between the elevator and the shaft. Find the power supplied by the elevator's motor.

FIGURE 9.32

punksid/Shutterstock.com

INTERPRET and ANTICIPATE

This example is similar to Example 9.9 in that a rope is used to transfer energy from a motor to an object that is pulled upward. We include the elevator, the passengers, the Earth, and the elevator shaft in the system. Unlike Example 9.9, however, here we are asked to find the rate at which work is transferred from the motor to the system. We expect to find a numerical result in the form $P = +$____W.

SOLVE

The external force is due to the tension in the rope. So, we need to find the tension force on the elevator. As usual, draw a free-body diagram (Fig. 9.33).

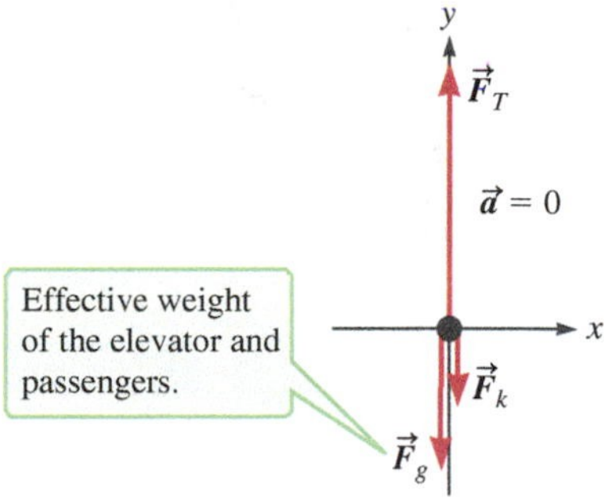

FIGURE 9.33

Apply Newton's second law to find the tension. The elevator moves at constant velocity, so the acceleration is zero. The effective weight is $F_g = 0.60mg$.

$$\sum F_y = F_T - F_g - F_k = 0$$
$$F_T = F_g + F_k = 0.60mg + F_k$$
$$F_T = (0.60)(4800 \text{ kg})(9.81 \text{ m/s}^2) + 5000 \text{ N}$$
$$\vec{F}_T = 3.3 \times 10^4 \hat{\jmath} \text{ N}$$

Power is found using Equation 9.38. Both the velocity and the tension force are upward, so the angle between them is zero.

$$P = \vec{F} \cdot \vec{v} = (3.3 \times 10^4 \text{ N})(16.6 \text{ m/s}) \cos 0° \quad (9.38)$$
$$P = 5.5 \times 10^5 \text{ W}$$

CHECK and THINK

Our result has the form we expected. It is sometimes helpful to convert the power output of a motor into horsepower. In this case, the elevator motor supplies more than 700 hp. This result may seem high until we compare it with something we know, such as a typical car with a 130-hp engine. The elevator's motor puts out as much power as roughly eight car engines. The speed of the Taipei 101 elevator is comparable to the speed limit on a small road, but the elevator is designed to hold five to six times more passengers than a typical car. So, it seems reasonable that the elevator motor would be several times as powerful as a typical car's engine.

EXAMPLE 9.11 Cruising Down the Highway

A car manufacturer reports that when a particular model is driven on a level highway at a constant speed of 60 mph, it has an energy efficiency (gasoline mileage) of 30 miles to 1 gallon of gasoline. One gallon of gasoline yields 1.3×10^8 J. What is the power requirement of the engine for the car to maintain its highway cruising speed? Comment on your results.

INTERPRET and ANTICIPATE

This example is much like converting a measurement from one system of units to another. The interesting part is thinking about our results.

SOLVE

Using the car's speed of 60 mph and its gasoline mileage of 30 miles per gallon, find the rate at which the car consumes gasoline.

$$(60\,\text{mi/h})\left(\frac{1\text{ gal}}{30\text{ mi}}\right) = 2\text{ gal/h}$$

Using the equivalence of 1 gallon of gasoline to 1.3×10^8 J, find the power supplied by the engine.

$$P = (2\text{ gal/h})(1.3 \times 10^8\,\text{J/gal})(1\text{ h}/3600\,\text{s})$$

$$P = 7.2 \times 10^4\,\text{W} \approx 97\text{ hp}$$

CHECK and THINK

We just found that the car's engine must supply about 100 hp just to maintain a constant speed of 60 mph on a level highway. This power is within the capability of a typical car (about 130 hp). The example statement asks us to comment on our results.

What is amazing is that this power is not changing the car's mechanical energy. The car maintains constant speed, so its kinetic energy does not change, and the highway is level, so there is no change in gravitational potential energy of the Earth–car system. The car's engine converts chemical energy, and none of it goes into mechanical energy. Where does it go? The converted energy is dissipated by both rolling friction between the car's tires and the road and by the drag force of the air on the body of the car. Energy is also dissipated by kinetic friction between moving parts inside the car's drivetrain, engine, and other systems. In fact, most of the power supplied by the engine never makes it to the wheels of the car, and is instead dissipated in the car's internal workings. Of course, even more power is required if the system's mechanical energy is increasing, such as when the car accelerates or climbs a hill.

EXAMPLE 9.12 CASE STUDY Size of a Meteor

In the meteor case study, three physics professors estimated the size of a meteor. They worked in the dark without a calculator, no steps were written down, and their estimate was made to one significant figure. One professor had read an article stating that the light given off by a meteor is equivalent to turning on a 100-W lightbulb for 2.5 s. The professors knew that a shooting star results when a meteor travels through the Earth's atmosphere, heating both the meteor and the atmosphere. The meteor gives off light and is vaporized. Three other facts they knew came in handy: $1\,\text{AU} = 1.5 \times 10^{11}\,\text{m}$, $1\,\text{yr} \approx \pi \times 10^7$ s, and the density of water is $\rho = 1000\,\text{kg/m}^3$. Estimate the size (radius) of a meteor.

INTERPRET and ANTICIPATE

The professors decided to include only the meteor in the system. The meteor had some mechanical energy when it entered the Earth's atmosphere, and all that energy was lost in 2.5 s, at which time the meteor was completely vaporized. The professors started with three assumptions. First, the meteor is small, so little energy is needed to vaporize it. They also reasoned that if the meteor has only a small mass, the amount of work done on it by the Earth's gravity during the 2.5-s time interval must also be negligible. Second, they assumed the half of the meteor's energy goes into heating the atmosphere and the other half is radiated away in the form of light. Third, they assumed the meteor's initial speed is twice the speed of the Earth around the Sun. Finally, they assumed the meteor was about twice as dense as water. (Don't worry if you wouldn't come up with all these.)

The bar chart in Figure 9.34 summarizes these assumptions. The meteor starts with kinetic energy K_i. Half of that kinetic energy is transferred in the form of light E_L, and the other half goes into thermal energy E_{th} of the atmosphere.

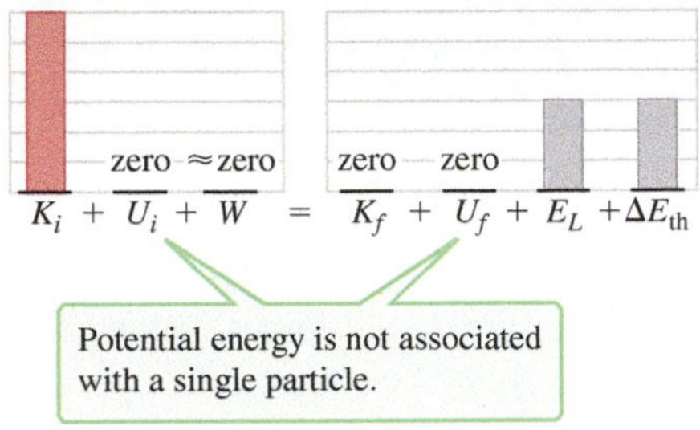

zero ≈zero
zero zero
$K_i + U_i + W = K_f + U_f + E_L + \Delta E_{th}$
Potential energy is not associated with a single particle.

FIGURE 9.34

SOLVE

Our professors estimated the energy that goes into light from the article that one professor read.

$$E_L = P\Delta t = (100\,\text{W})(2.5\,\text{s}) = 250\,\text{J}$$

They assumed the energy that is radiated away in the form of light E_L is half the kinetic energy.

$$E_L = \tfrac{1}{2}K_i = \tfrac{1}{2}\left(\tfrac{1}{2}mv_i^2\right)$$

They solved for the mass of the meteor.

$$m = \frac{4E_L}{v_i^2}$$

The speed of the Earth is its orbital circumference divided by 1 year. It is very handy to memorize the number of seconds in 1 year this way because the π's cancel.

$$v_\oplus = \frac{2\pi a}{t} \approx \frac{2\pi(1.5 \times 10^{11}\,\text{m})}{\pi \times 10^7\,\text{s}}$$

$$v_\oplus \approx \frac{3 \times 10^{11}\,\text{m}}{10^7\,\text{s}} = 3 \times 10^4\,\text{m/s}$$

The professors assumed the meteor's speed was twice the speed of the Earth, giving a (very approximate) mass of the meteor.

$$m = \frac{4E_L}{(2v_\oplus)^2} \approx \frac{4(250\,\text{J})}{4(3 \times 10^4\,\text{m/s})^2}$$

$$m \approx \frac{(300\,\text{J})}{(3 \times 10^4\,\text{m/s})^2} = \frac{1}{3} \times 10^{-6}\,\text{kg} \approx 3 \times 10^{-7}\,\text{kg}$$

If the meteor is roughly spherical, we can find its size if we know its density.

$$m = \rho V = \rho\frac{4\pi R^3}{3} \approx \rho 4R^3$$

$$R \approx \left(\frac{m}{4\rho}\right)^{1/3}$$

The professors assumed the meteor had twice the density of water.

$$R \approx \left[\frac{3 \times 10^{-7}\,\text{kg}}{4(2000\,\text{kg/m}^3)}\right]^{1/3} = \left(\frac{300 \times 10^{-9}\,\text{kg}}{8 \times 10^3\,\text{kg/m}^3}\right)^{1/3}$$

$$R \approx (30 \times 10^{-12}\,\text{m}^3)^{1/3}$$

$$R \approx 3 \times 10^{-4}\,\text{m} = 0.3\,\text{mm}$$

CHECK and THINK

The radius of a meteor is roughly 0.3 mm, about the size of a grain of sand, and this result is about the correct size. A second estimate in Chapter 34 confirms this result.

This estimate shows the power of the conservation approach. We do not need to know the detailed physics of how the meteor vaporizes, gives off light, and heats the atmosphere. To find a reasonably good estimate of a meteor's size, all we need to do is account for the transfer and conversion of energy.

SUMMARY

Underlying Principles

1. **Work–kinetic energy theorem:** The total work done on a particle by all external forces equals the change in the particle's kinetic energy:
$$W_{tot} = \Delta K \qquad (9.5)$$
2. **Work–mechanical energy theorem:** The total work done on a system by external forces in the absence of resistive forces changes the system's mechanical energy:
$$W_{tot} = \Delta K + \Delta U \qquad (9.26)$$
3. **Work–energy theorem:** The equation
$$K_i + U_i + W_{tot} = K_f + U_f + \Delta E_{int} \qquad (9.31)$$
is an extension of the work–mechanical energy theorem that holds when there are internal dissipative forces.

Major Concepts

1. **Work** is the amount of energy transferred to (or from) the system by an external force.
2. A **general expression for work done** is
$$W = \int_{r_i}^{r_f} \vec{F} \cdot d\vec{r} = \int_{r_i}^{r_f} (F_x dx + F_y dy + F_z dz) \qquad (9.21)$$
3. A **general expression for change in potential energy** is
$$\Delta U = -\int_{r_i}^{r_f} \vec{F} \cdot d\vec{r} \qquad (9.24)$$
4. Conservative (and nonconservative) force:
A **conservative force** depends only on the relative position of the source and object. The work done by a conservative force on a particle does *not* depend on the particle's path, and the work done by a **nonconservative force** *does* depend on the particle's path.
5. A **zero-work force** may accelerate a particle without doing any work on it.
6. The **center of mass** is a point associated with any object that moves as though all the mass were concentrated there and all external forces were applied there.
7. **Internal energy** within a system or deformable object is a manifestation of potential or kinetic energy that is on a scale that is too small for us to calculate directly.
8. When **dissipative forces** (nonconservative forces) act in a system, the system's thermal energy increases, and its mechanical energy decreases.
9. **Power** is the rate at which energy is transferred into or out of a system or the rate at which energy is converted from one form to another within a system. The instantaneous power is
$$P = \frac{dW}{dt} \quad (9.36) \qquad \text{and} \qquad P = \vec{F} \cdot \vec{v} \quad (9.38)$$

Special Cases: Internal Energy

1. **Thermal energy** is internal energy associated with the temperature of the system or deformable object. A change in thermal energy due to moving friction is given by
$$\Delta E_{th} = F_k s \qquad (9.29)$$
2. **Chemical energy** is another form of internal energy governed by chemical reactions within the system or deformable object.

Tools

If two vectors $\vec{A}$ and $\vec{B}$ lie in a plane at an angle φ, the **dot product** or **scalar product** of these two vectors is given by
$$D = \vec{A} \cdot \vec{B} \equiv AB \cos \varphi \quad (9.7) \qquad \text{and} \qquad D = \vec{A} \cdot \vec{B} = A_x B_x + A_y B_y + A_z B_z \quad (9.10)$$

PROBLEM-SOLVING STRATEGY Applying the Work–Energy Theorem

INTERPRET and ANTICIPATE

1. Choose a **system**. Here are some tips for choosing a system:
   - Include the source of a conservative force (gravity or spring force).
   - Be sure that dissipative forces are internal.
   - Sources of other nonconservative forces should be external.
2. Pick a **coordinate system and the reference configuration.**
3. Identify an **initial time** and a **final time.**
4. Draw a **bar chart.**

SOLVE

5. Apply the **work–energy theorem** (Eq. 9.32 or 9.33).
6. Write down any **other equations.**
7. **Do algebra** before substitution.

PROBLEMS AND QUESTIONS

A = algebraic C = conceptual E = estimation G = graphical N = numerical

9-1 Energy Transfer to and from the Environment

1. **C** Pick an isolated system for the following scenarios while including the fewest number of objects as possible. **a.** A satellite in orbit around the Earth **b.** An airplane in flight **c.** A truck driving along the road **d.** A person jumping

9-2 Work Done by a Constant Force

2. **E** Estimate the work done by the air drag applied to a parachute when a paratrooper descends from an initial height of 3.00×10^3 ft above the ground.
3. **C** Riders on a merry-go-round experience a force accelerating them toward the center while they are in circular motion. Does this force do any work? Explain your answer.

Problems 4 and 5 are paired.

4. During practice, a hockey player passes the puck to his coach who is 6.5 m away as shown in Figure 9.2. The puck, which has a mass of 2.0 kg, was initially at rest. The hockey player exerts a constant 47.4-N force as his stick pushes the puck 0.25 m. Include only the puck in the system and assume the friction between the ice and the puck is negligible.
   a. **C** Draw a free-body diagram for the puck while it is in contact with the player's stick.
   b. **N** Find the work done by all the forces in your free-body diagram.
   c. **N** What is the speed of the puck as it leaves the player's stick?
   d. **C** In an all-star hockey game, the puck reaches speeds of 100 mph. Use that information to check your results to parts (a) through (c).
5. **N** In Problem 4, what work must the coach do on the puck to stop it?
6. A team of dockworkers does 5.25×10^3 J of total work while pushing a large 435-kg container by applying a constant horizontal force. The container is displaced through a horizontal distance of 15.0 m.
   a. **N** What is the magnitude of the force F applied by the dockworkers?
   b. **C** How would the motion of the crate change if the dockworkers applied a total force greater than F?
   c. **C** How would the motion of the crate change if the dockworkers applied a total force smaller than F?
7. **N** Kerry is pulling a 154-kg sled along a snowy, horizontal path with a 615-N force directed at an angle of 30.0° above the ground. If he pulls the sled over a distance of 30.0 m, how much work has Kerry performed on the sled?
8. **N** A 537-kg trailer is hitched to a truck. Find the work done by the truck on the trailer in each of the following cases. Assume rolling friction is negligible. **a.** The trailer is pulled at constant speed along a level road for 2.30 km. **b.** The trailer is accelerated from rest to a speed of 88.8 km/h. **c.** The trailer is pulled at constant speed along a road inclined at 12.5° for 2.30 km.
9. **N** Consider two objects, one that is moving to the east with a speed of 25 m/s and a second that is initially motionless. If each object has an identical mass and each object is to have a final speed of 25 m/s moving to the west, compare and contrast the necessary net amount of work that must be done on each object. Does one object require a greater amount of work? If so, which one?
10. **N** A helicopter rescues a trapped person of mass $m = 65.0$ kg from a flooded river by lifting the person vertically upward using a winch and rope. The person is pulled 12.0 m into the helicopter with a constant force that is 15% greater than the person's weight. **a.** Find the work done by each of the forces acting on the person. **b.** Assuming the survivor starts from rest, determine his speed upon reaching the helicopter.

9-3 Dot Product

11. **N** Find the dot product of the vectors $\vec{A} = 7.12\hat{\imath} + 2.00\hat{\jmath} - 3.90\hat{k}$ and $\vec{B} = 4.10\hat{\imath} - 11.00\hat{\jmath}$.
12. **N** An object is subject to a force $\vec{F} = (512\hat{\imath} - 134\hat{\jmath})$ N such that 10,125 J of work is performed on the object. If the object travels 25.0 m in the positive x direction while this work is performed, what must be the displacement of the object in the y direction?
13. **A** Show that the dot product of $\vec{A}$ and $\vec{B}$ is given by $D = A_xB_x + A_yB_y$. *Hint*: Refer to Figure 9.7.

Problems 14 and 15 are paired.

14. Vector $\vec{A} = (A\cos\alpha)\hat{\imath} + (A\sin\alpha)\hat{\jmath}$ and vector $\vec{B} = (B\cos\alpha)\hat{\imath} + (B\sin\alpha)\hat{\jmath}$.
    a. **A** What is the angle between vectors $\vec{A}$ and $\vec{B}$?
    b. **G** To check your answer, sketch the vectors on an x–y coordinate system.
15. Vector $\vec{A} = (A\cos 15°)\hat{\imath} + (A\sin 15°)\hat{\jmath}$ and vector $\vec{B} = (B\cos 15°)\hat{\imath} + (B\sin 15°)\hat{\jmath}$.
    a. **N** What is the angle between vectors $\vec{A}$ and $\vec{B}$?
    b. **G** To check your answer, sketch the vectors on an x–y coordinate system.

16. **N** In Figure P9.16, the magnitude of the vectors are $A = 6.00$ and $B = 3.00$. Find $\vec{A} \cdot \vec{B}$ in each case.

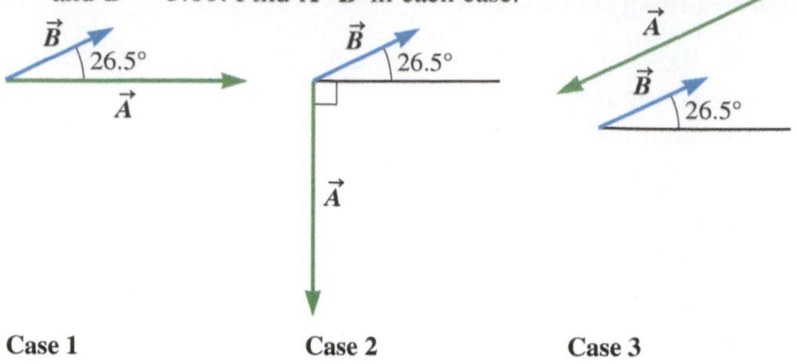

FIGURE P9.16

17. **N** A cart of mass $m = 1.30$ kg travels 2.4 m down a frictionless ramp inclined at 15° relative to a horizontal plane. Find the work done by gravity using **a.** Equation 9.15, **b.** Equation 9.16, and **c.** Equation 9.17.
18. **A** Show that the dot product of a vector with itself gives the square of the magnitude of that vector: $\vec{A} \cdot \vec{A} = A^2$ (Eq. 9.11).
19. **A** Show that the dot product is *commutative*: $\vec{A} \cdot \vec{B} = \vec{B} \cdot \vec{A}$ (Eq. 9.12).
20. **A** Show that the dot product is *associative*: $\vec{A} \cdot (\vec{B} + \vec{C}) = \vec{A} \cdot \vec{B} + \vec{A} \cdot \vec{C}$ (Eq. 9.13).

Problems 21 and 22 are paired.

21. **A** Figure P9.21 shows two vectors $\vec{A}$ and $\vec{B}$. Find the dot product $D = \vec{A} \cdot \vec{B}$ using Equation 9.8.

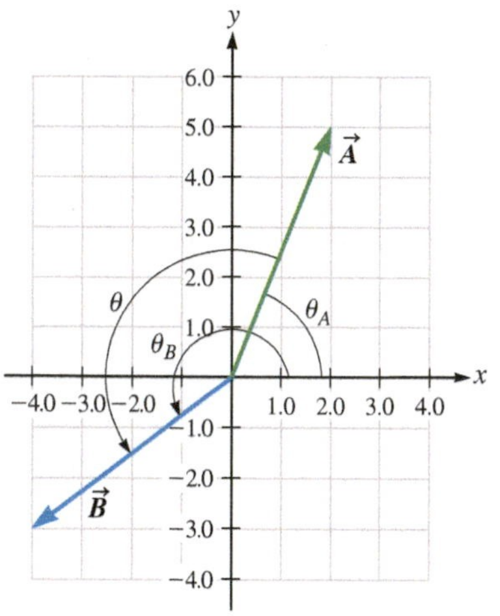

FIGURE P9.21 Problems 21 and 22.

22. **A** Figure P9.21 shows two vectors $\vec{A}$ and $\vec{B}$. Find the dot product $D = \vec{A} \cdot \vec{B}$ using Equation 9.10. If you did Problem 21, compare your results.
23. **N** A constant force of magnitude 4.75 N is exerted on an object. The force's direction is 60.0° counterclockwise from the positive x axis in the xy plane, and the object's displacement is $\Delta\vec{r} = (4.2\hat{\imath} - 2.1\hat{\jmath} + 1.6\hat{k})$ m. Calculate the work done by this force.

9-4 Work Done by a Nonconstant Force

24. **C** In three cases, a force acts on a particle, and the particle is displaced from an initial position to a final position. Figure 9.11 (page 255) shows the position-versus-force graphs, indicating the initial and final positions of the particle in each case. Find the work done by the force on the particle and sketch the force and displacement vectors along with the appropriate axis in each case.
25. **N** An object of mass $m = 5.8$ kg moves under the influence of one force. That force causes the object to move along a path given by $x = 6.0 + 5.0t + 2.0t^2$, where x is in meters and t is in seconds. Calculate the work done by the force on the object from $t = 2.0$ s to $t = 7.0$ s.

Problems 26 and 27 are paired.

26. **N** A nonconstant force is exerted on a particle as it moves in the positive direction along the x axis. Figure P9.26 shows a graph of this force F_x versus the particle's position x. Find the work done by this force on the particle as the particle moves as follows. **a.** From $x_i = 0$ to $x_f = 10.0$ m **b.** From $x_i = 10.0$ to $x_f = 20.0$ m **c.** From $x_i = 0$ to $x_f = 20.0$ m

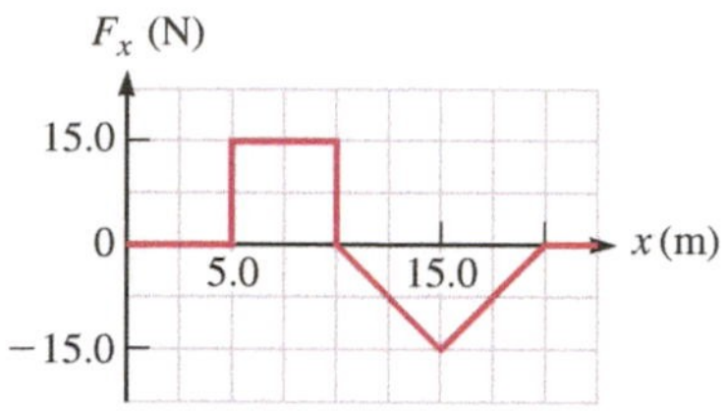

FIGURE P9.26 Problems 26 and 27.

27. **N** If the force in Figure P9.26 is the net force on the particle, find the change in the particle's kinetic energy as it moves as follows. **a.** From $x_i = 0$ to $x_f = 10.0$ m **b.** From $x_i = 10.0$ to $x_f = 20.0$ m **c.** From $x_i = 0$ to $x_f = 20.0$ m
28. A block on a frictionless table is attached to a horizontal spring of spring constant $k = 345$ N/m.
    a. **G** Sketch a graph of the spring force on the block F versus the block's position x. Use the conventional choice of placing the origin at the relaxed position of the spring.
    b. **N** Use your graph to estimate the work done by the spring on the block as the block moves from the relaxed position of the spring, $x_i = 0$, to $x_f = 35.0$ m.
    c. **N** Use Equation 9.18 to find the work done by the spring on the block as the block moves from $x_i = 0$ to $x_f = 35.0$ m.
29. **N** A force $\vec{F} = (4.000x^2\hat{\imath} - 6.000y\hat{\jmath})$ N acts on an object that moves 550.0 m along the direction pointing 22.5° clockwise from the positive y axis. Find the work done by the force on the object as it moves along the path.
30. A particle moves in the xy plane (Fig. P9.30) from the origin to a point having coordinates $x = 7.00$ m and $y = 4.00$ m under the influence of a force given by $\vec{F} = 3y^2\hat{\imath} + x\hat{\jmath}$.
    a. **N** What is the work done on the particle by the force F if it moves along path 1 (shown in red)?

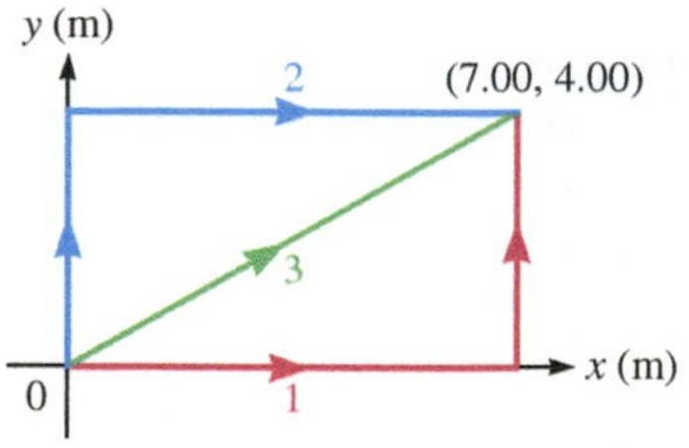

FIGURE P9.30

    b. **N** What is the work done on the particle by the force F if it moves along path 2 (shown in blue)?
    c. **N** What is the work done on the particle by the force F if it moves along path 3 (shown in green)?
    d. **C** Is the force F conservative or nonconservative? Explain.
31. **A** A small object is attached to two springs of the same length ℓ, but with different spring constants k_1 and k_2 as shown in Figure P9.31. Initially, both springs are relaxed. The object is then displaced straight along the x axis from x_i to x_f. Find an expression for the work done by the springs on the object.

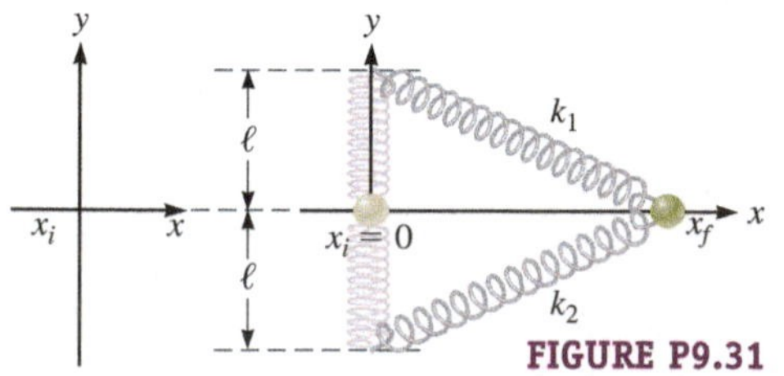

FIGURE P9.31

32. The force acting on a particle is given by $\vec{F}(r) = (A/r^2)\hat{r}$, where A is a positive constant.
   a. C What are the dimensions of A? Can the force be negative?
   b. G Sketch F versus r.
   c. A Find the work done by the force on the particle as it moves from $r_i = r_0$ to $r_f = 2r_0$.
   d. C Check your answer by comparing it with Example 9.5.

9-5 Conservative, Nonconservative, and Zero-Work Forces

33. A Show that the total work done by universal gravity on an object around a closed path is zero.
34. C In each of the situations described, determine whether or not the work done by the man is positive, negative, or zero and briefly explain your reasoning. **a.** A man pushes a child in a stroller, causing it to move horizontally across an ordinary floor, where friction is present. **b.** A man moves a child in a stroller across a frictionless floor so that the man and child glide together at a constant speed. **c.** A man prevents a child in a stroller from accelerating as it travels downhill. **d.** A man pushes the Great Wall of China, yet it remains motionless.
35. C **Review** In Example 9.6, Paul exerts a constant force on a sled that is constrained to move along a circular path. Because the sled is on ice, friction between the sled and the ice is negligible. Is the sled speeding up, slowing down, or maintaining its speed? Is the sled accelerating? If so, draw an overhead sketch and show the sled's acceleration, direction of motion along the path, and forces that are easily drawn from this perspective. If the sled isn't accelerating, show that the sum of the forces exerted on it is zero.

9-6 Particles, Objects, and Systems

36. N A 2.15-g hailstone, which can be modeled as a particle, falls a vertical distance of 145 m at constant speed. What is the work done on the hailstone by **a.** gravity and **b.** air resistance?

Problems 37 and 38 are paired.

37. A 6.0-lb dry concrete block is on a rubber conveyor belt in a factory. The block remains stationary on the conveyor belt as it goes 2.45 m along a 12.0° incline. The conveyor belt maintains a constant speed of 0.95 m/s.
   a. C Draw a free-body diagram for the concrete block.
   b. N Consider the Earth and the block to be the system. Find the work done by any external force on the system.
38. A 6.0-lb dry concrete block is on a rubber conveyor belt in a factory. The block remains stationary on the conveyor belt as it moves 2.45 m horizontally. The conveyor belt maintains a constant speed of 0.95 m/s.
   a. C Draw a free-body diagram for the concrete block.
   b. N Consider the Earth and the block to be the system. Find the work done by any external force on the system.
   c. C Compare your result with your result from Problem 37.
39. N A shopper weighs 3.00 kg of apples on a supermarket scale whose spring obeys Hooke's law and notes that the spring stretches a distance of 3.00 cm. **a.** What will the spring's extension be if 5.00 kg of oranges are weighed instead? **b.** What is the total amount of work that the shopper must do to stretch this spring a total distance of 7.00 cm beyond its relaxed position?
40. A In Figure 9.15, a chain of five links is dropped to the ground. Show that the work done on the whole chain by the Earth's gravity is

$$W = Mg\,\Delta y_{CM}$$

where M is the mass of the whole chain and Δy_{CM} is the displacement of the chain's center of mass. *Hint*: Start by treating each link in the chain as a particle of mass m. (You may wonder if it is okay to model each link as a particle. Each part of a single link has nearly the same displacement. For our discussion here, the slight variation in displacement of a single link is not important, so it is okay to model each individual link as a particle.) Work done on the whole chain is found by adding up the work done on the individual links.

9-7 Thermal Energy

41. A 4.50-kg wooden block slides 2.35 m down a wooden incline at constant velocity. (The coefficient of static friction is 0.400, and the coefficient of kinetic friction is 0.200.)
   a. N What is the angle of the incline?
   b. C When we are considering energy, why is it useful to include the surface of the incline in the system?
   c. N Calculate the increase in the system's thermal energy.
42. C The term *internal energy* could use improvement. In some sense, all energies are internal. Potential energy is the energy associated with the relative position of the particles *in* the system, and kinetic energy is associated with the *motion* of the particles in the system. Come up with a better term for internal energy and explain why it is better.
43. N CASE STUDY Consider a problem similar to the case study, but brought down to the Earth. A 6.0×10^3 kg bus traveling at 90.0 km/h skids to a halt on a wet (horizontal) road, where $\mu_k = 0.30$. How much thermal energy was generated by the friction?
44. C In calculating the work done on a deformable object, we must use the displacement of the point at which the force is applied. Some people choose a system in which the source of friction is external, which is not the choice we use in this book. Explain why it is especially complicated in the case of kinetic friction to find the displacement of the point of application.

Problems 45 and 46 are paired.

45. A A bullet flying horizontally hits a wooden block that is initially at rest on a frictionless, horizontal surface. The bullet gets stuck in the block, and the bullet–block system has a final speed v_f. Find the final speed of the bullet–block system in terms of the mass of the bullet m_b, the speed of the bullet before the collision v_b, the mass of the block m_{wb}, and the amount of thermal energy generated during the collision E_{th}.
46. A A bullet flying horizontally hits a wooden block that is moving before the collision in the same direction as the bullet with an initial speed v_{wb}. The bullet gets stuck in the block, and the bullet–block system has a final speed v_f. Find the final speed of the bullet–block system in terms of the mass of the bullet m_b, the speed of the bullet before the collision v_b, the mass of the block m_{wb}, the speed of the block before the collision v_{wb}, and the amount of thermal energy generated during the collision.
47. N You wish to make a simple amusement park ride in which a steel-wheeled roller-coaster car travels down one long slope, where rolling friction is negligible, and later slows to a stop through kinetic friction between the roller coaster's locked wheels sliding along a horizontal plastic (polystyrene) track. Assume the roller-coaster car (filled with passengers) has a mass of 750.0 kg and starts 80.7 m above the ground. **a. Review** Calculate how fast the car is going when it reaches the bottom of the hill. **b.** How much does the thermal energy of the system change during the stopping motion of the car? **c.** If the car stops in 230.66 m, what is the coefficient of kinetic friction between the wheels and the plastic stopping track?

9-8 Work–Energy Theorem

48. A Show that $K_i + U_{gi} + U_{ei} + W_{tot} = K_f + U_{gf} + U_{ef} + \Delta E_{th}$ (Eq. 9.33) is the same as $K_i + U_i = K_f + U_f$ (Eq. 8.14) in the

case of an isolated system in which only conservative forces act.

49. **A** Show that $K_i + U_{gi} + U_{ei} + W_{tot} = K_f + U_{gf} + U_{ef} + \Delta E_{th}$ (Eq. 9.33) is the same as $W_{tot} = \Delta K + \Delta U$ (Eq. 9.26) in the case of a particle.

Problems 50 and 51 are paired.

50. A small 0.65-kg box is launched from rest by a horizontal spring as shown in Figure P9.50. The block slides on a track down a hill and comes to rest at a distance d from the base of the hill. Kinetic friction between the box and the track is negligible on the hill, but the coefficient of kinetic friction between the box and the horizontal parts of track is 0.35. The spring has a spring constant of 34.5 N/m, and is compressed 30.0 cm with the box attached. The block remains on the track at all times.

a. **C** What would you include in the system? Explain your choice.

b. **N** Calculate d.

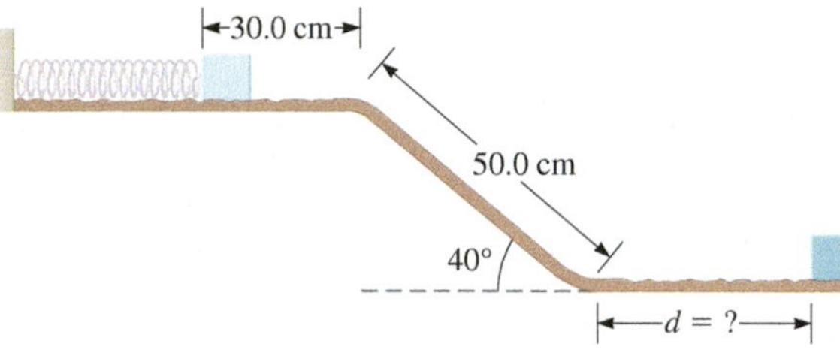

FIGURE P9.50 Problems 50 and 51. (Not to scale.)

51. A small 0.65-kg box is launched from rest by a horizontal spring as shown in Figure P9.50. The block slides on a track down a hill and comes to rest at a distance d from the base of the hill. The coefficient of kinetic friction between the box and the track is 0.35 along the entire track. The spring has a spring constant of 34.5 N/m, and is compressed 30.0 cm with the box attached. The block remains on the track at all times.

a. **C** What would you include in the system? Explain your choice.

b. **N** Calculate d.

c. **C** Compare your answer with your answer to Problem 50 if you did that problem.

52. **N** A horizontal spring with force constant $k = 625$ N/m is attached to a wall at one end and to a block of mass $m = 3.00$ kg at the other end that rests on a horizontal surface. The block is released from rest from a position 3.50 cm beyond the spring's equilibrium position. **a.** If the surface is frictionless, what is the speed of the block as it passes through the equilibrium position? **b.** If the surface is rough and the coefficient of kinetic friction between the box and the surface is $\mu_k = 0.280$, what is the speed of the block as it passes through the equilibrium position?

53. **N** A box of mass $m = 2.00$ kg is dropped from rest onto a massless, vertical spring with spring constant $k = 2.40 \times 10^2$ N/m that is initially at its natural length. How far is the spring compressed by the box if the initial height of the box is 1.75 m above the top of the spring?

54. A person jumps straight upwards.

a. **E** Estimate the change in the person's internal energy.

b. **C** What is the form of the internal energy? What is the ultimate source of that energy?

55. **N** Return to Example 9.9 and use the result to find the tension in the rope.

Problems 56 and 57 are paired.

56. **A** Crall and Whipple design a loop-the-loop track for a small toy car (Fig. P9.56). The car starts at height y_i above the bottom of the loop, goes through the loop of radius R, and then travels along a flat, horizontal track. Rolling friction is negligible. What is the minimum height y_i from which the car can be released so that the car just barely makes it around the loop?

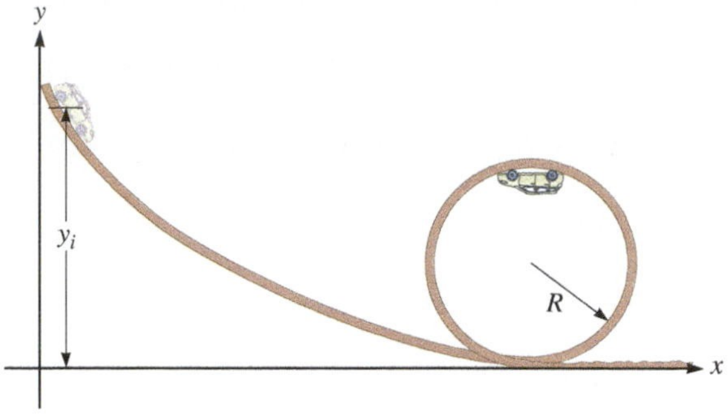

FIGURE P9.56 Problems 56 and 57.

57. **N** Crall and Whipple design a loop-the-loop track for a small toy car. The car starts at height y_i above the bottom of the loop, goes through the loop of radius R, and then travels along a flat, horizontal track before coming to rest. Rolling friction between the horizontal track and the car is significant, but it is negligible along the rest of the track. Assume Crall and Whipple release the car from the minimum height y_i (found in Problem 56). If the coefficient of rolling friction between the car and the horizontal track is 0.30 and the radius of the loop is 0.45 m, how far does the car travel along the horizontal track from the base of the loop before coming to rest?

58. **C** A pendulum is constructed by hanging a 0.500 kg-ball at the end of a 1.00-m-long lightweight, strong string. The other end of the string is attached to a support in the ceiling so that the ball is free to swing back and forth within a plane perpendicular to the ceiling. If this pendulum is displaced from the equilibrium position by 10.0° (with respect to vertical) and allowed to move freely, do you expect the pendulum to swing back and forth, reaching this angle again? Explain your answer.

59. **N** Calculate the force required to pull a stuffed toy duck (mass $m = 1.25$ kg) at a constant velocity of 3.6 m/s horizontally across the floor if the string is 50.0° above the horizontal. The coefficient of kinetic friction between the duck and the floor is 0.70.

9-9 Power

60. **N** What is the power output of a 75.0-kg student that climbs a knotted vertical rope 9.50 m in height at his high school gym at constant speed in 10.0 s?

61. **N** Return to Example 9.10 and the elevators in the Taipei 101 tower. What power is dissipated due to friction between the shaft and the elevator?

62. **E** Estimate the average power required for you to climb two floors in an office building in 10 s.

63. **N** An elevator motor moves a car with six people upward at a constant speed of 2.50 m/s. The mass of the elevator is 8.00×10^2 kg, and the average mass of a person on the elevator is about 80.0 kg. Calculate the electric power that must be delivered to lift the elevator car, assuming half of the necessary power delivered goes into thermal energy.

64. **N** A pail in a water well is hoisted by means of a frictionless winch, which consists of a spool and a hand crank. When Jill turns the winch at her fastest water-fetching rate, she can lift the pail the 25.0 m to the top in 12.2 s. Calculate the average power supplied by Jill's muscles during the upward ascent. Assume the pail of water when full has a mass of 6.82 kg.

65. **N** Figure P9.65A shows a crate attached to a rope that is extended over an ideal pulley. Boris pulls on the other end of the rope with a constant force until the crate has risen a total distance of 6.53 m (Fig. P9.65B). If the crate has a mass of 81.36 kg, what is the average power exerted by Boris, assuming he accomplishes the task in 5.33 s?

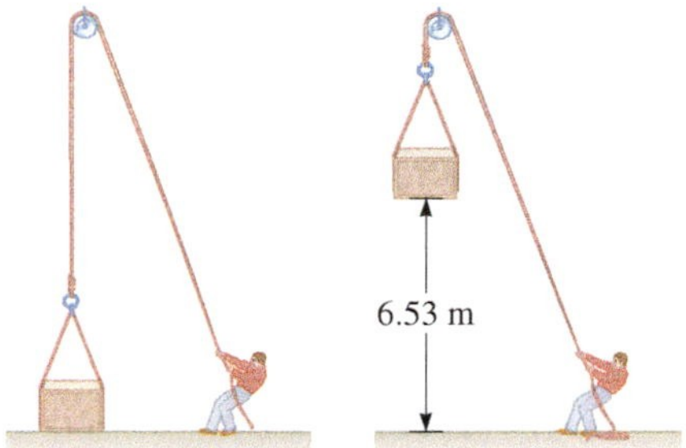

A. **B.**

FIGURE P9.65

66. **E, N** While playing an automobile simulation computer program, you are test-driving a 2011 Ford F-150 pickup truck. The program allows several players to choose a vehicle and drive around in a virtual world, using real physical parameters and characteristics for the simulation. While waiting at a red light, another player pulls up in a 2011 Chevrolet Malibu and challenges you to a race. The race will be over as soon as someone reaches 60 mph. By estimating the average power of each engine in watts and the mass of each vehicle, determine who wins the race. (You may find it helpful to search for information about the engine of each vehicle on the Internet.)

General Problems

67. **N** Use the definition of the scalar dot product to determine the angle between each pair of vectors.
    a. $\vec{A} = -3\hat{\imath} - \hat{\jmath} + 4\hat{k}$ and $\vec{B} = 2\hat{\imath} + 2\hat{\jmath} + 2\hat{k}$
    b. $\vec{A} = \hat{\imath} + 2\hat{\jmath}$ and $\vec{B} = -2\hat{\jmath} - 3\hat{k}$
    c. $\vec{A} = 4\hat{\imath} + 2\hat{k}$ and $\vec{B} = -\hat{\imath} + 5\hat{\jmath} + 3\hat{k}$
68. **C** A physics professor keeps a yo-yo moving in a horizontal circle by its string as it is brought up from a canyon floor to the edge of a tall cliff by an elevator. Determine whether or not the work done on the yo-yo by the each of the following forces is positive, negative, or zero and briefly explain your reasoning. **a.** The elevator (contact force) **b.** The Earth's gravity **c.** The yo-yo string's tension
69. **N** A constant horizontal force of $F = 200.0$ N pushes a crate of mass $m = 25.0$ kg, initially at rest, along a rough, horizontal surface. The crate travels a total distance of 8.00 m, and the coefficient of kinetic friction between the crate and the surface is 0.250. **a.** What is the work done on the crate by the force F? **b.** How much does friction increase the internal energy of the crate–surface system? **c.** What is the change in kinetic energy of the crate? **d.** What is the final speed of the crate?
70. **N** Given the vectors $\vec{A} = \hat{\imath} + \hat{\jmath} + \hat{k}$, $\vec{B} = 4\hat{\imath} - 3\hat{\jmath} - 2\hat{k}$, and $\vec{C} = 2\hat{\imath} + 2\hat{\jmath}$, calculate $\vec{A} \cdot (\vec{C} - \vec{B})$.
71. **N** An object is subject to a nonconstant force $\vec{F} = (6x^3 - 2x)\hat{\imath}$ such that the force is in newtons when x is in meters. Determine the work done on the object as a result of this force as the object moves from $x = 0$ to $x = 1.00 \times 10^2$ m.
72. **E** Estimate the power required for a boxer to jump rope.
73. **N** A mother gently lifts her 4.55-kg baby from the crib at constant speed through a vertical distance of 85.0 cm.
    a. What is the work done on the baby by the mother?
    b. What is the total force exerted on the mother by the baby?
74. Kerry is pulling a 154-kg sled along a snowy, horizontal path with a 615-N force directed at an angle of 30.0° above the ground. He pulls the sled over a distance of 30.0 m, and the coefficient of kinetic friction between the sled and the ground is 0.0612.
    a. C Define the system to be used to account for the change in thermal energy as the sled is moved across the ground.
    b. N How much work is performed by Kerry as he pulls the sled over this distance?
    c. N What is the increase in thermal energy experienced by the system during the motion?
75. **N** A particle of mass $m = 2.50$ kg moving along the x axis from $x = 0$ to $x = 10.0$ m experiences a net conservative force in an isolated system given by $F = 3x - 5$, where F is in newtons and x is in meters. **a.** What is the work done on the particle by the force F? **b.** What is the change in the potential energy of the system during this motion? **c.** If the speed of the particle at the origin is 1.25 m/s, what is its kinetic energy at $x = 10.0$ m?
76. **N** A particle with mass $m = 750$ g is found to be moving with velocity $\vec{v} = (-4.00\hat{\imath} + 3.00\hat{\jmath})$ m/s. From the definition of the scalar product, $v^2 = \vec{v} \cdot \vec{v}$. **a.** What is the particle's kinetic energy at this time? **b.** If the particle's velocity changes to $\vec{v} = (5.00\hat{\imath} - 4.00\hat{\jmath})$ m/s, what is the net work done on the particle?
77. **N** Maria sets up a simple track for her toy block ($m = 0.25$ kg) as shown in Figure P9.77. She holds the block at the top of the track, 0.54 m above the bottom, and releases it from rest. **a.** Neglecting friction, what is the speed of the block when it reaches the bottom of the curve (the beginning of the horizontal section of track)? **b.** If friction is present on the horizontal section of track and the block comes to a stop after traveling 0.75 m along the bottom, what is the magnitude of the friction force acting on the block?

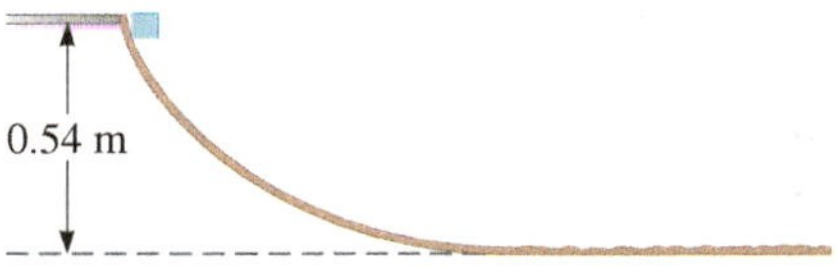

FIGURE P9.77

78. **A** A particle in an isolated system experiences a net conservative force given by $\vec{F} = (Ay - By^2)\hat{\jmath}$, where $\vec{F}$ is in newtons, y is in meters, and A and B are constants, with $U(y = 0) = 0$. **a.** What is the potential energy function $U(y)$ of the system? **b.** What is the change in potential energy of the system if the particle moves from $y = 1.00$ m to $y = 4.00$ m? **c.** What is the change in kinetic energy of the system if the particle moves from $y = 1.00$ m to $y = 4.00$ m?
79. **N** A snowmobile is pulling a sled ($m = 375$ kg) across the snowy tundra (Fig. P9.79). The snowmobile applies a pulling force of 2511 N at an angle of 15° with respect to the ground and pulls the sled over a distance of 50.0 m. **a.** How much work is done by the snowmobile? **b.** If the sled is also being acted on by a constant kinetic friction force of 255 N and it started from rest, what is the final speed of the sled?

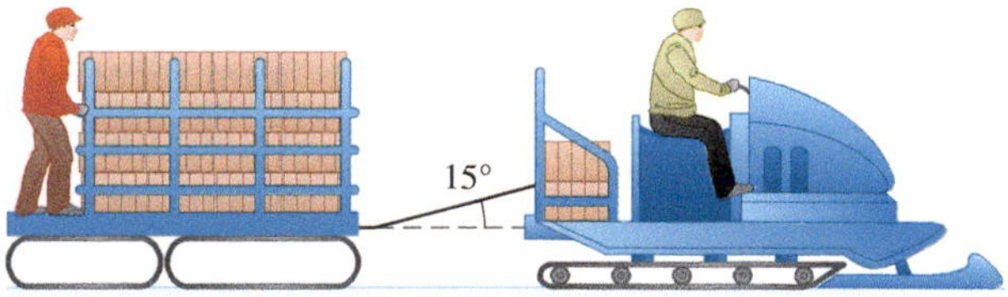

FIGURE P9.79

Substitute this result into Equation 10.1 to find each ball's momentum, using subscript 1 for the 2.5-kg ball and subscript 2 for the 5.0-kg ball. Both results have the form we expected.	$\vec{p}_f = m\vec{v}_f$ (10.1) $\vec{p}_{1f} = -(2.5\,\text{kg})(52\text{ m/s})\hat{\jmath} = -1.3 \times 10^2\hat{\jmath}\text{ kg}\cdot\text{m/s}$ $\vec{p}_{2f} = -(5.0\,\text{kg})(52\text{ m/s})\hat{\jmath} = -2.6 \times 10^2\hat{\jmath}\text{ kg}\cdot\text{m/s}$
SOLVE **Method 2** Equation 10.2 is simplified because only one force—the Earth's gravity—acts on either ball.	$\Sigma\vec{F} = \vec{F}_g = \dfrac{d\vec{p}}{dt}$ (10.2) $-mg\hat{\jmath} = \dfrac{d\vec{p}}{dt}$
Multiply each side by dt and integrate from the instant the ball is released ($t_i = 0$) to the instant the ball lands ($t_f = t$). Because the ball was released from rest, its initial momentum is $\vec{p}_i = 0$. Because $-mg\hat{\jmath}$ is constant, it is pulled outside the integral.	$-(mg\,dt)\hat{\jmath} = d\vec{p}$ $\int_0^t -(mg\,dt)\hat{\jmath} = -(mg)\hat{\jmath}\int_0^t dt = \int_0^{\vec{p}_f} d\vec{p}$ $-(mgt)\hat{\jmath} = \vec{p}_f$
It takes one ball 5.3 s to reach the ground, which we use to find the momentum of each.	$\vec{p}_{1f} = -(2.5\,\text{kg})(9.81\text{ m/s}^2)(5.3\text{ s})\hat{\jmath} = -1.3 \times 10^2\hat{\jmath}\text{ kg}\cdot\text{m/s}$ $\vec{p}_{2f} = -(5.0\,\text{kg})(9.81\text{ m/s}^2)(5.3\text{ s})\hat{\jmath} = -2.6 \times 10^2\hat{\jmath}\text{ kg}\cdot\text{m/s}$

CHECK and THINK

We found the same results using two different methods. As we have come to expect, both balls land with the same velocity. The ball with twice the mass, however, has twice as much momentum. A commonly held belief is that when two different objects are dropped from a great height (in the absence of drag), the heavier one is faster. These sorts of mistaken ideas are usually based on experience that is not identified with the correct physical interpretation. Imagine catching these two balls; clearly, the heavier one is harder to stop. It is incorrect to connect this experience to the speed of the balls. Instead, we should say that it is harder to stop the heavier ball because it has a greater momentum.

10-3 Center of Mass Revisited

When you think about the case of a falling ball, the ideas of momentum and the general form of Newton's second law (Eq. 10.2) don't seem very important. (We studied problems like that in Chapter 2.) The real power of the concept of momentum comes out when we analyze a system of two or more particles. To do so, we need to know how to find a system's center of mass.

The **center of mass (CM)** is a point associated with any system such that when a net external force is exerted on the system, the center of mass accelerates according to Newton's second law (Section 9-6). When a system may be modeled as a particle, the center of mass is just the position of the particle, and there is no need to distinguish the center-of-mass motion from the motion of the rest of the system. For example, we often are able to model a walking person as a particle. However, when a person walks, her arms and legs do not move with the same velocity and acceleration as her center of mass. If we need to know how much energy a person needs to walk a great distance, we must take into account the motion of her arms and legs, as well as her center-of-mass motion.

To locate the center of mass, we start by considering a simple two-particle system (Figure 10.3A).We choose a coordinate system with the origin on particle 1. The x component of particle 2's position is then x_2. For this coordinate system, the x component of the center-of-mass position is defined to be

$$x_{CM} \equiv \left(\frac{m_2}{m_1 + m_2}\right)x_2$$

Because we placed the origin on particle 1, x_2 is the distance between the two particles. The factor $m_2/(m_1 + m_2)$ is a unitless fraction between 0 and 1. If the two particles have equal mass, the fraction is $\frac{1}{2}$, and the center of mass is halfway between the two particles. If $m_2 > m_1$, the fraction is greater than $\frac{1}{2}$, and the center of mass is closer to particle 2. Of course, if $m_2 < m_1$, the center of mass is closer to particle 1.

Sometimes, it is not convenient to place the origin on one of the particles. A change in coordinate system, of course, cannot change the physical location of the center of mass, but it changes that point's coordinates relative to the new origin. Using the coordinate system in Figure 10.3B, the center of mass of the same two-particle system is now

$$x_{CM} = \frac{m_1 x_1 + m_2 x_2}{m_1 + m_2}$$

If there are more than two particles, we find the x component of the center-of-mass position in a similar way. For n particles, we have

$$x_{CM} = \frac{m_1 x_1 + m_2 x_2 + m_3 x_3 + \cdots + m_n x_n}{m_1 + m_2 + m_3 + \cdots + m_n}$$

The sum in the denominator is the total mass M of the system. Using summation notation, we can write a compact expression for the x component of the center-of-mass position:

$$x_{CM} = \frac{1}{M}\sum_{j=1}^{n} m_j x_j \tag{10.3}$$

According to this expression, the center of mass is a *weighted average of position*. The term *weighted average* does not have anything to do with the gravitational force; rather, it is a mathematical description meaning that the average takes into account the relative importance of the quantities being averaged. To understand this term, imagine students standing in a line along one wall of a classroom. You choose an x axis that runs parallel to the wall and whose origin is at one corner, and then you measure the position of each student. The average position of the students is found by adding up the position of each student and dividing by the total number of students. In this case, each student's position is given equal importance, or *weight*, in your calculation. You could instead find a weighted average based on some property of the students such as their physics grades. You would assign "weight" to each student's position by multiplying each position by his or her grade and then find the average of these *weighted* positions. The students with the best grades would then count more when you find the average position. The center of mass is weighted based on the relative mass of each particle in the system; mathematically, the weighting is done by multiplying each term by m_j/M. So, the more massive particles are more important when finding the center-of-mass position.

We started by thinking about two particles lying along a single x axis. A more complicated system may have particles located in two or three dimensions as in Figure 10.3C. Similar expressions are written for the y and z components of the center-of-mass position:

$$y_{CM} = \frac{1}{M}\sum_{j=1}^{n} m_j y_j \qquad z_{CM} = \frac{1}{M}\sum_{j=1}^{n} m_j z_j$$

The position vector of the center of mass is then

$$\vec{r}_{CM} = x_{CM}\hat{\imath} + y_{CM}\hat{\jmath} + z_{CM}\hat{k} \tag{10.4}$$

which is written in compact form as

$$\vec{r}_{CM} = \frac{1}{M}\sum_{j=1}^{n} m_j \vec{r}_j \tag{10.5}$$

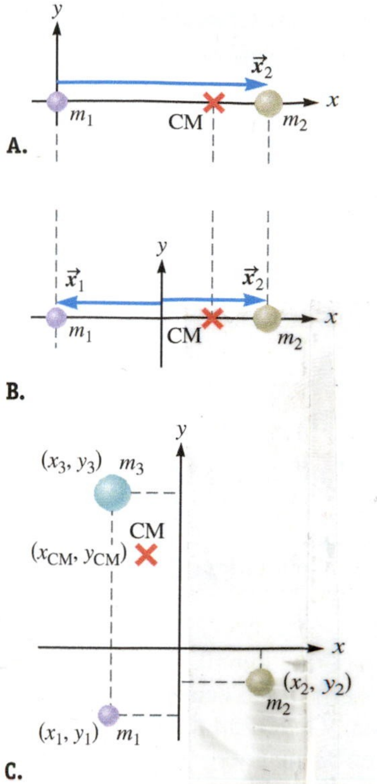

FIGURE 10.3 **A.** A two-particle system with the origin on particle 1. **B.** A two-particle system with the origin between the two particles. **C.** A three-particle system with the origin arbitrarily placed.

$x_{1 \text{ or } 2}$ is the scalar x component of the particle's position, which may be positive or negative. With the origin as shown in Figure 10.3B, x_1 is negative and x_2 is positive.

CENTER OF MASS **Major Concept**

CONCEPT EXERCISE 10.2

We often think of the Earth and the Moon as a two-particle system with the center of mass located at the center of the Earth. Find the center of mass of the Earth–Moon system and comment on the validity of the usual assumption that it is at the center of the Earth. *Hint*: The Earth and the Moon's masses are 5.98×10^{24} kg and 7.36×10^{22} kg, respectively; their center-to-center distance is 3.84×10^{8} m; and the Earth's radius is 6.37×10^{6} m.

EXAMPLE 10.2 A Strange Game of Pool

Two billiard balls of mass 0.170 kg and one bowling ball of mass 4.54 kg are on a pool table. Use the coordinate system in Figure 10.4 to find the center-of-mass position of the three-ball system. Report your answer to two significant figures.

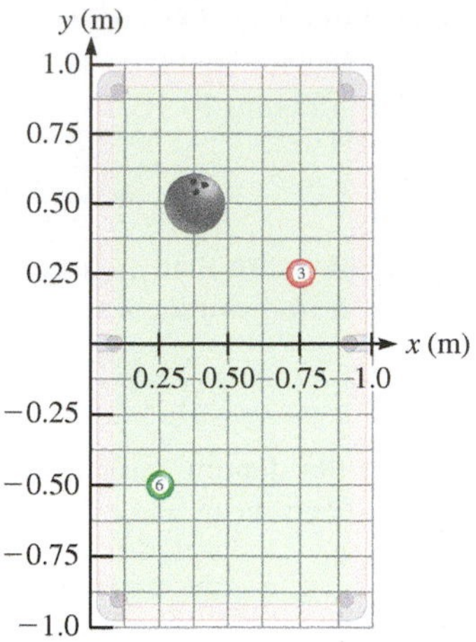

FIGURE 10.4

INTERPRET and ANTICIPATE

The bowling ball is much more massive than the billiard balls, so we expect the center of mass to be close to the bowling ball. Our result should be in the form $\vec{r}_{CM} = (\\_\\_\\_\\_ \hat{\imath} + \\_\\_\\_\\_ \hat{\jmath})$ m.

SOLVE

Read off the position of each ball.

$$\vec{r}_3 = (0.75\hat{\imath} + 0.25\hat{\jmath})\text{ m}$$
$$\vec{r}_6 = (0.25\hat{\imath} - 0.50\hat{\jmath})\text{ m}$$
$$\vec{r}_b = (0.375\hat{\imath} + 0.50\hat{\jmath})\text{ m}$$

Find the total mass of the system.

$$M = 2 \times (0.170\text{ kg}) + 4.54\text{ kg}$$
$$M = 4.88\text{ kg}$$

Use Equation 10.3 to find the x component of the center-of-mass position. Then, find the y component in a similar manner.

$$x_{CM} = \frac{1}{M}\sum_{j=1}^{n} m_j x_j \qquad (10.3)$$

$$x_{CM} = \frac{(0.170\text{ kg})(0.75\text{ m}) + (0.170\text{ kg})(0.25\text{ m}) + (4.54\text{ kg})(0.375\text{ m})}{4.88\text{ kg}} = 0.41\text{ m}$$

$$y_{CM} = \frac{1}{M}\sum_{i=1}^{n} m_i y_i$$

$$y_{CM} = \frac{(0.170\text{ kg})(0.25\text{ m}) + (0.170\text{ kg})(-0.50\text{ m}) + (4.54\text{ kg})(0.50\text{ m})}{4.88\text{ kg}} = 0.46\text{ m}$$

Write the center-of-mass position in component form using Equation 10.4.

$$\vec{r} = x_{CM}\hat{\imath} + y_{CM}\hat{\jmath} \qquad (10.4)$$

$$\vec{r}_{CM} = (0.41\hat{\imath} + 0.46\hat{\jmath})\text{ m}$$

CHECK and THINK

As expected, the center of mass is near the bowling ball.

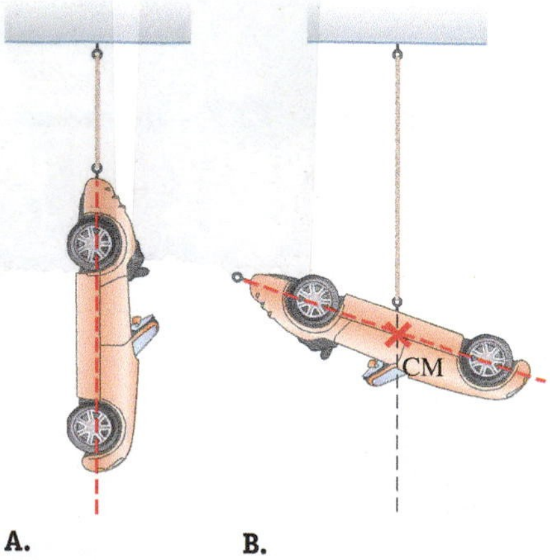

FIGURE 10.5 Finding the car's center of mass by suspending it from two separate points.

Equations 10.3 through 10.5 are useful for finding the center of mass for a system of a small number of particles, but how do we find the center of mass for an extended object such as a car? In that case, we imagine slicing the object into a great number of very small pieces. Each piece is modeled as a particle of mass dm. Then, when we find the weighted average over this great number of slices, the sums become integrals:

$$x_{CM} = \frac{1}{M}\int x\,dm \qquad y_{CM} = \frac{1}{M}\int y\,dm \qquad z_{CM} = \frac{1}{M}\int z\,dm$$

Often, there is no need to actually do these integrals. For an object with uniform density and a regular shape, we exploit the object's symmetry. For example, for a sphere of uniform density, the center of mass is at the center of the sphere. For a donut shape of uniform density, the center of mass is in the center of the hole.

To find the center of mass of an irregularly shaped object or one with nonuniform density, we perform an experiment. External forces applied to an extended object act as though all the mass of the object were concentrated at the center of mass. So, if we suspend an object from a rope, two forces—gravity and the tension force—act on the object. The center of mass must be somewhere along the line that runs through the length of the rope (Problem 14.63).

For example, Figure 10.5A shows a car suspended by a rope; the car's center of mass must be somewhere on the line shown. To find the exact location of the center of mass, we need to suspend the car from a different point (Fig. 10.5B); now we know that the center of mass must be somewhere on the new line. The two lines intersect at one point, and this point must be the center of mass of the car. There is nothing unique or special about the two points from which we suspended the car; any two well-separated points work.

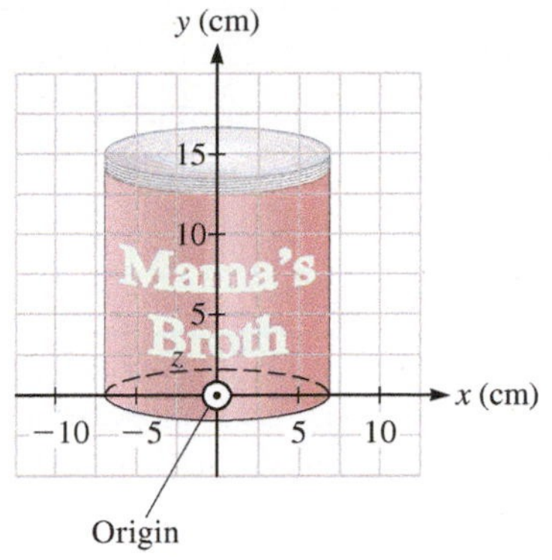

FIGURE 10.6 The z axis points out of the page, directly at the viewer.

CONCEPT EXERCISE 10.3

The car's trunk in Figure 10.5 is empty. What would happen to the center-of-mass position if instead the trunk were full of heavy rocks?

CONCEPT EXERCISE 10.4

Unlike a can of vegetable soup, a can of broth has uniform density. Find the center-of-mass position (x_{CM}, y_{CM}, z_{CM}) for the can of broth in Figure 10.6 assuming it is vacuum sealed so that there is no air bubble. The origin of the coordinate system is in the center of the can's circular base.

10-4 Systems of Particles

Now that we know how to locate the center of mass, we are ready to apply Newton's second law to a system of particles. If an external force is applied to the system, the center of mass accelerates as if all the mass of the system were concentrated at that point. Newton's second law for the system is written as

NEWTON'S SECOND LAW APPLIED TO A SYSTEM WITH CONSTANT MASS
Underlying Principle

$$\sum \vec{F}_{ext} = M\vec{a}_{CM} \quad (10.6)$$

There are two subtle but important facts about Eq. 10.6:

1. It is possible that the net external force may be zero, in which case the center-of-mass acceleration $\vec{a}_{CM}$ is zero. In fact, choosing a system such that the net external force *is* zero is an important part of applying the conservation of momentum principle.
2. The equation $\sum \vec{F}_{ext} = M\vec{a}_{CM}$ involves the center-of-mass acceleration $\vec{a}_{CM}$, which is generally *not* the same as the acceleration of any particle within the system. So, for example, the center of mass may have a constant velocity, but the particles within the system may be accelerating.

EXAMPLE 10.3 Back to the Drawing Board

Zak, a novice artillery enthusiast, decides to launch a projectile at an angle $\theta = 85°$. The total mass of the projectile is $M = 0.45$ kg. Fortunately, he has taken the precaution of launching into an empty farm field, but unfortunately, there is a malfunction. At the peak of the projectile's trajectory, the projectile splits in two sections. Zak finds the back section ($m_1 = 0.15$ kg) 155 m from the launch point (Fig. 10.7). If the launch speed was $v_0 = 125$ m/s and drag is negligible, where should he look for the front section ($m_2 = 0.30$ kg)?

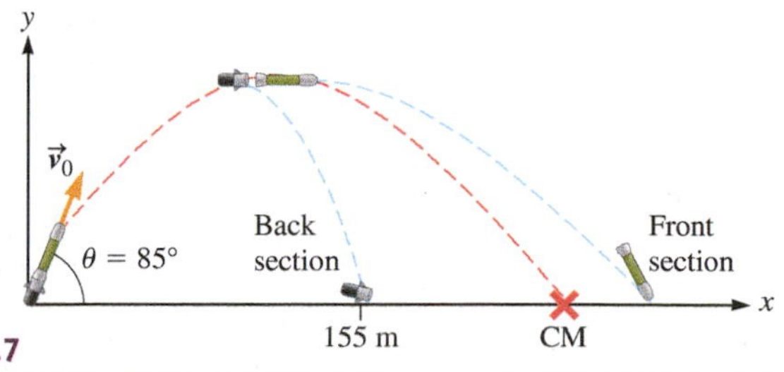

FIGURE 10.7

INTERPRET and ANTICIPATE

If drag is negligible, after launching only gravity acts on the center of mass, which acts as a simple projectile. We can easily find where the center of mass lands, although there will not be any physical material there. Because we know where the back section landed, we can calculate where the front section landed. We expect to find a numerical result of the form $\vec{x}_2 =$ _____ $\hat{\imath}$, and we expect that the front section is farther from the origin than either the center of mass or the back section.

Example continues on page 288 ▶

SOLVE

Find the final center-of-mass position from the range of the projectile (Eq. 4.28).

$$R = \frac{v_0^2}{g}\sin 2\theta \quad (4.28)$$

$$R = \frac{(125\ \text{m/s})^2}{9.81\ \text{m/s}^2}\sin[2(85)°] = 2.76 \times 10^2\ \text{m}$$

$$\vec{x}_{CM} = 2.76 \times 10^2 \hat{\imath}\ \text{m}$$

Model the projectile as a two-particle system. Particle 1 is the back section. The front section is particle 2. Solve Equation 10.3 for $\vec{x}_2$, particle 2's position.

$$x_{CM} = \frac{1}{M}(m_1x_1 + m_2x_2) \quad (10.3)$$

$$m_2x_2 = Mx_{CM} - m_1x_1$$

$$x_2 = \frac{Mx_{CM} - m_1x_1}{m_2}$$

$$x_2 = \frac{(0.45\ \text{kg})(276\ \text{m}) - (0.15\ \text{kg})(155\ \text{m})}{0.30\ \text{kg}}$$

$$\vec{x}_2 = 3.4 \times 10^2 \hat{\imath}\ \text{m}$$

CHECK and THINK

Our answer matches our expectations. It might seem bizarre that if Zak were to go to where the center of mass of the system lands ($\vec{x}_{CM} = 2.76 \times 10^2 \hat{\imath}$ m), he would not find any part of the projectile there. Even if there is no physical matter at the center of mass, it still follows Newton's laws of motion. The front section travels farther if the back section falls off than if the entire projectile remains intact and follows the center-of-mass trajectory. This fact will be helpful when we return to the case study.

Momentum of a System of Particles

The concept of center of mass is also important when we consider the momentum of a system of particles. For a system of n particles, the total momentum is the vector sum of the momentum of each particle and is given by

MOMENTUM OF A SYSTEM OF PARTICLES

Major Concept

$$\vec{p}_{tot} = \sum_{j=1}^{n} \vec{p}_j \quad (10.7)$$

In Problem 81 you will show that the total momentum of a system of particles equals the momentum of the center of mass:

$$\vec{p}_{tot} = \sum_{j=1}^{n} \vec{p}_j = \vec{p}_{CM} \quad (10.8)$$

If the center of mass is not accelerating, there is an inertial reference frame in which the center of mass is at rest ($\vec{v}_{CM} = 0$) known as the *center-of-mass frame*. In the center-of-mass frame, $\vec{p}_{CM} = 0$; therefore, the total momentum of the system is zero ($\vec{p}_{tot} = 0$).

EXAMPLE 10.4 Skating Partners

Consider a two-particle system consisting of two ice skaters gliding on the ice. Paul's mass is 76.0 kg, and his speed is $v_P = 7.30$ m/s; Lil's mass is 52.0 kg, and her speed is $v_L = 6.50$ m/s. The direction of each skater's velocity is shown in Figure 10.8. Friction and drag are negligible, so the skaters' velocities are constant.

A Use the coordinate system shown in Figure 10.8 to find the total momentum of the system with respect to the ice.

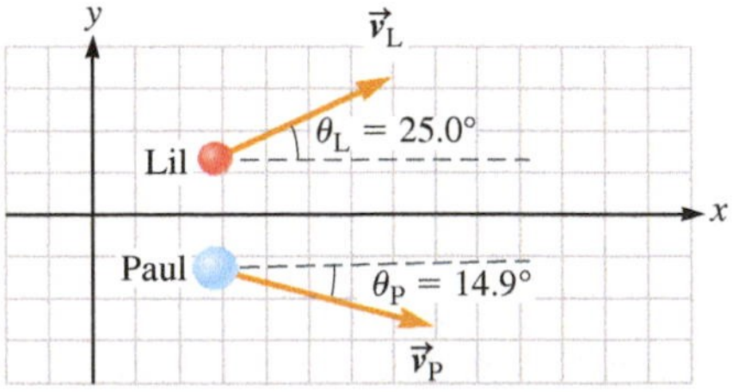

FIGURE 10.8 Velocity vectors for Paul and Lil.

INTERPRET and ANTICIPATE

We will find the momentum of each skater with respect to the ice and then add our results. We can develop our expectations graphically.

Find the magnitude of each skater's momentum by taking the magnitude of $\vec{p} \equiv m\vec{v}$ (Eq. 10.1). The subscript L stands for *L*il, and the subscript P stands for *P*aul.	$p_L = m_L v_L = (52.0\text{ kg})(6.50\text{ m/s}) = 338\text{ kg}\cdot\text{m/s}$ $p_P = m_P v_P = (76.0\text{ kg})(7.30\text{ m/s}) = 555\text{ kg}\cdot\text{m/s}$
Momentum is parallel to velocity. Sketch the two momentum vectors on the coordinate grid (Fig. 10.9). The *y* components of momentum cancel, so we expect the total momentum to be in the form $\vec{p}_{tot} =$ _____ $\hat{\imath}$ kg · m/s.	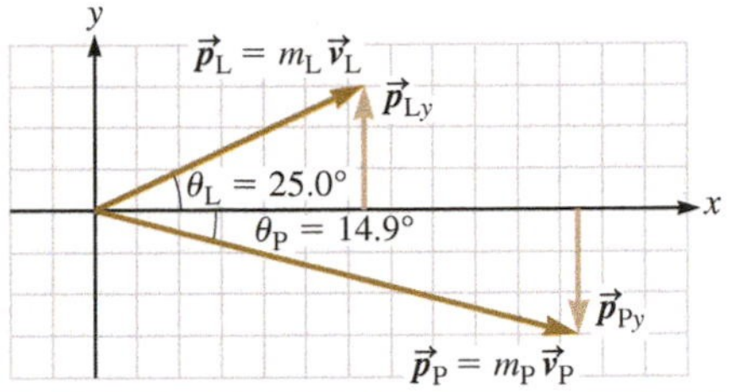 FIGURE 10.9 Momentum vectors for Paul and Lil.

FIGURE 10.9 Momentum vectors for Paul and Lil.

SOLVE

Use the magnitudes found above and the directions given in the figures to write the momentum of each skater in component form.

$$\vec{p}_L = (p_L\cos\theta_L)\hat{\imath} + (p_L\sin\theta_L)\hat{\jmath}$$
$$\vec{p}_L = (338\text{ kg}\cdot\text{m/s})(\cos 25.0°)\hat{\imath} + (338\text{ kg}\cdot\text{m/s})(\sin 25.0°)\hat{\jmath}$$
$$\vec{p}_L = (306\hat{\imath} + 143\hat{\jmath})\text{ kg}\cdot\text{m/s}$$
$$\vec{p}_P = (p_P\cos\theta_P)\hat{\imath} - (p_P\sin\theta_P)\hat{\jmath}$$
$$\vec{p}_P = (555\text{ kg}\cdot\text{m/s})(\cos 14.9°)\hat{\imath} - (555\text{ kg}\cdot\text{m/s})(\sin 14.9°)\hat{\jmath}$$
$$\vec{p}_P = (536\hat{\imath} - 143\hat{\jmath})\text{ kg}\cdot\text{m/s}$$

Now add the two momentum vectors to find the total momentum.

$$\vec{p}_{tot} = (306\text{ kg}\cdot\text{m/s} + 536\text{ kg}\cdot\text{m/s})\hat{\imath} + (143\text{ kg}\cdot\text{m/s} - 143\text{ kg}\cdot\text{m/s})\hat{\jmath}$$
$$\vec{p}_{tot} = 842\hat{\imath}\text{ kg}\cdot\text{m/s}$$

CHECK and THINK

As expected, the total momentum points in the positive *x* direction.

B What is the velocity of the center of mass with respect to the ice?

INTERPRET and ANTICIPATE

The total momentum of the system and the momentum of the center of mass are the same. Therefore, the velocity of the center of mass should point in the same direction as the total momentum. So, we expect to find $\vec{v}_{CM} =$ _____ $\hat{\imath}$ m/s.

SOLVE

Divide the total momentum by the total mass of the system to find the center-of-mass velocity.

$$\vec{p}_{CM} = \vec{p}_{tot} = M\vec{v}_{CM} \qquad (10.8)$$
$$\vec{v}_{CM} = \frac{\vec{p}_{tot}}{M} = \frac{842\hat{\imath}\text{ kg}\cdot\text{m/s}}{(52.0\text{ kg} + 76.0\text{ kg})}$$
$$\vec{v}_{CM} = 6.58\hat{\imath}\text{ m/s}$$

CHECK and THINK

Our result is in the form we expected. Think about the total momentum in the center-of-mass frame. In this frame, the velocity and the momentum of the center of mass are zero. Therefore, the total momentum of the system *relative* to the center of mass is zero. This does not mean that the skaters are at rest in the center-of-mass frame. They are still moving in this frame, but their momenta cancel out.

10-5 Conservation of Momentum

In Chapters 8 and 9, we used the conservation of energy approach to tackle a number of difficult problems. Momentum conservation is an equally powerful concept. Although we can think of momentum transfer from one system to another, often the most powerful and useful way to consider momentum is to define a system for which momentum is conserved. So, we must first figure out what conditions are required for momentum to be conserved.

Unlike energy, momentum does not come in different forms, so we will not be concerned with momentum converting from one form to another.

First, conservation of momentum means that the *total* momentum does not change in time. Second, when any quantity is conserved, the time derivative of that quantity is zero. So, when momentum is conserved,

$$\frac{d\vec{p}_{\text{tot}}}{dt} = 0$$

Third, the total momentum is the same as the momentum of the center of mass $\vec{p}_{\text{tot}} = \vec{p}_{\text{CM}}$ (Eq. 10.8). So, when momentum is conserved, the time derivative of the center of mass momentum is zero:

$$\frac{d\vec{p}_{\text{CM}}}{dt} = 0$$

Next, write the center-of-mass momentum in terms of the system's total mass and the center-of-mass velocity:

$$\frac{d\vec{p}_{\text{CM}}}{dt} = \frac{d(M\vec{v}_{\text{CM}})}{dt} = 0$$

If the total mass of the system is constant, we can pull M outside the derivative:

$$\frac{d\vec{p}_{\text{CM}}}{dt} = M\frac{d(\vec{v}_{\text{CM}})}{dt} = M\vec{a}_{\text{CM}} = 0$$

Finally, when we compare this expression with $\sum \vec{F}_{\text{ext}} = M\vec{a}_{\text{CM}}$ (Eq. 10.6), we find that **when momentum is conserved, the total external force must be zero**:

$$\frac{d\vec{p}_{\text{CM}}}{dt} = M\vec{a}_{\text{CM}} = \sum \vec{F}_{\text{ext}} = 0$$

This last expression is an important finding. The total momentum of a system is conserved if two conditions are met: (1) there is *no* net external force acting on the system, and (2) the mass of the system is constant. Condition 1 does *not* require the system to be isolated. Forces may act on the system as long as the total (net) external force is zero. Of course, the net force on an isolated system is zero, so an isolated system meets condition 1. Condition 2 means that no mass may enter or leave the system. A system that does not lose or gain mass is called a **closed system**. So, we must choose a system that meets both conditions 1 and 2 if we want to apply the conservation of momentum.

CLOSED SYSTEM
Major Concept

Conservation of momentum is best expressed in practice by thinking about the momentum of the system at two times, usually referred to as the *initial* and *final* times. Using this convention for a system for which momentum is conserved, we have

CONSERVATION OF MOMENTUM
Underlying Principle

$$\vec{p}_{i,\text{tot}} = \vec{p}_{f,\text{tot}} \tag{10.9}$$

The initial and final times are chosen to be convenient and do not necessary refer to the beginning or end of motion.

Before we try some examples, we notice that if the total external force on a system is not zero, momentum is not conserved. In such cases, the change in momentum is given by Newton's second law:

NEWTON'S SECOND LAW APPLIED TO A SYSTEM OF PARTICLES
Underlying Principle

$$\sum \vec{F}_{\text{ext}} = \frac{d\vec{p}_{\text{CM}}}{dt} = \frac{d\vec{p}_{\text{tot}}}{dt} \tag{10.10}$$

CONCEPT EXERCISE 10.5

What is the purpose of the ropes attached to the cannon in Figure 10.10?

FIGURE 10.10

Because momentum is a vector, it may be conserved in one, two, or three dimensions. If the vector sum of the external forces on a closed system is zero, momentum is conserved in all three dimensions, and Equation 10.9 may be written in component form as

$$\sum p_{ix} = \sum p_{fx} \qquad \sum p_{iy} = \sum p_{fy} \qquad \sum p_{iz} = \sum p_{fz} \tag{10.11}$$

The summation symbol indicates that we must add the momentum components of all particles in the system to arrive at the total momentum component in each direction.

In some cases, the total external force is zero in only one or two dimensions. In those cases, momentum is only conserved along the directions for which the total external force is zero. For example, in the absence of drag, the only force acting on a projectile near the Earth's surface is gravity along a vertical y axis. Therefore, the momentum of the projectile is not conserved in the y direction, but it is conserved in the x direction. So, for a projectile (moving in the xy plane), $\sum p_{ix} = \sum p_{fx}$, but $\sum p_{iy} \neq \sum p_{fy}$.

EXAMPLE 10.5 Nuclear Decay

An unstable nucleus decays into three particles. Particles 1 and 2 each have mass $2m$, and particle 3 has mass m. We wish to use two detectors to observe particles 1 and 2. In the center-of-mass frame, the nucleus initially has no kinetic energy. When the nucleus decays, the three particles have the velocities shown in Figure 10.11 and total kinetic energy $K = \frac{17}{8}mv^2$. Particles 1 and 2 have the same speed v_α, and particle 3's speed is $\sqrt{2}v$. Find the angles θ_1 and θ_2.

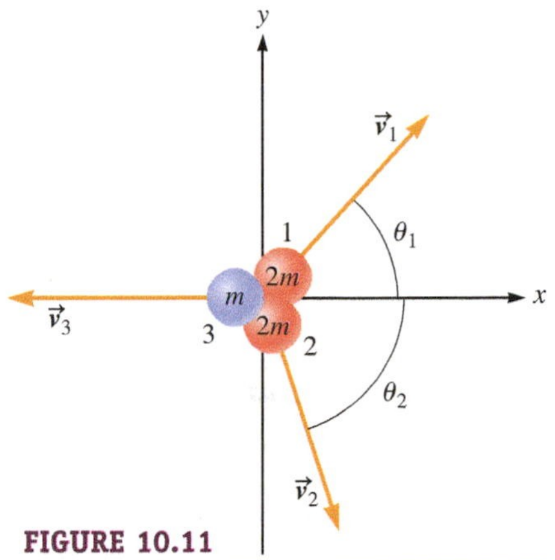

FIGURE 10.11

INTERPRET and ANTICIPATE

If we consider all three particles to make up the system, the system is closed, and no external forces act on it. Therefore, momentum is conserved. In the center-of-mass frame, the system's momentum is zero. The particles travel in the xy plane, so total momentum is conserved in two dimensions. Our result should be numerical values for angles θ_1 and θ_2.

SOLVE

Find the speed v_α of particle 1 in terms of v from the expression for kinetic energy $K = \sum_{i=1}^{n} \frac{1}{2}m_i v_i^2$ (Eq. 8.2).

$$K = \frac{1}{2}m_1v_1^2 + \frac{1}{2}m_2v_2^2 + \frac{1}{2}m_3v_3^2 = \frac{17}{8}mv^2$$

$$K = \frac{1}{2}(2m)v_\alpha^2 + \frac{1}{2}(2m)v_\alpha^2 + \frac{1}{2}m(\sqrt{2}v)^2 = \frac{17}{8}mv^2$$

$$2mv_\alpha^2 = \frac{17}{8}mv^2 - mv^2 = \frac{9}{8}mv^2$$

$$v_\alpha^2 = \frac{9}{16}v^2$$

$$v_\alpha = \frac{3}{4}v \qquad (1)$$

Momentum is conserved in both directions. Let's consider momentum in the y direction first. Initially, the y component of the nucleus's momentum is zero. After the decay, only particles 1 and 2 have momentum in the y direction. For momentum to be conserved, the sum of the two y components of momentum must be zero. We find $\theta_1 = \theta_2$. For convenience, we define this angle as θ.

$$\sum p_{iy} = 0$$

$$\sum p_{fy} = 2mv_\alpha \sin\theta_1 - 2mv_\alpha \sin\theta_2$$

$$\sum p_{iy} = \sum p_{fy}$$

$$0 = 2mv_\alpha \sin\theta_1 - 2mv_\alpha \sin\theta_2$$

$$\theta_1 = \theta_2 \equiv \theta$$

Next, consider momentum in the x direction. Initially, there is no momentum in the x direction, so the final momentum must also be zero.

$$\sum p_{ix} = 0$$

$$\sum p_{fx} = 2(2mv_\alpha \cos\theta) - \sqrt{2}mv$$

$$\sum p_{ix} = \sum p_{fx}$$

$$0 = 4mv_\alpha \cos\theta - \sqrt{2}mv$$

Substitute $v_\alpha = \frac{3}{4}v$ (Eq. 1).

$$4m\left(\frac{3}{4}v\right)\cos\theta = \sqrt{2}mv$$

$$3\cos\theta = \sqrt{2}$$

$$\theta = \cos^{-1}\left(\frac{\sqrt{2}}{3}\right) = 62°$$

$$\theta_1 = \theta_2 = 62°$$

Example continues on page 292 ▶

CHECK and THINK

We found that in Figure 10.11, the angles θ_1 and θ_2 should be equal. In the center-of-mass frame of this closed, isolated system, the total momentum must be zero before and after the decay. The center of mass must remain at rest. The kinetic energy, however, actually increases from zero before the decay to $K = \frac{17}{8}mv^2$. Kinetic energy is not a conserved quantity in this case. Instead, internal energy of the system is converted to kinetic energy. Because this process involves only internal forces, momentum is conserved. As a result, parts of the system move while its center of mass remains at rest.

EXAMPLE 10.6 "Here I Come to Save the Day!"

Imagine that a dastardly villain has planted dynamite on a runaway abandoned mine cart of mass $m_c = 93$ kg. The cart is moving quickly at 45 mph ($v_c = 20$ m/s) along a straight, level track (Fig. 10.12). The hero, who weighs an incredible 750 lbs ($m_h = 340$ kg), jumps off a bridge and falls straight down into the cart. What is the speed of the cart after the hero has made his landing? Assume rolling friction and drag are negligible.

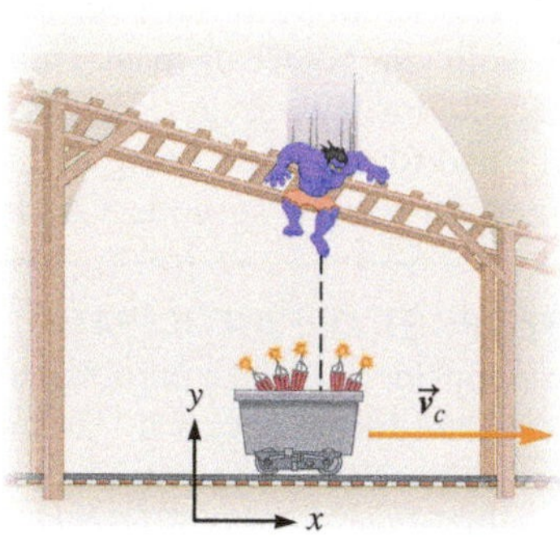

FIGURE 10.12

INTERPRET and ANTICIPATE

Let us consider the cart and the hero to make up the system. The total external force on the system is *not* zero. Before the hero falls, there is a net force due to gravity in the *y* direction. So, momentum is *not* conserved in the *y* direction. Because friction and drag are negligible, there is no net force in the *x* direction. Therefore, momentum is conserved in the *x* direction. We expect to find a numerical result for the speed.

SOLVE

Before the hero lands, only the cart has momentum in the *x* direction.	$p_{ix} = m_c v_c$
After the hero lands in the cart, both he and the cart move with the same velocity $\vec{v}_f$ (in the *x* direction).	$p_{fx} = (m_c + m_h)v_f$
Use the information that momentum is conserved in the *x* direction to find the final speed of the cart–hero system.	$m_c v_c = (m_c + m_h)v_f$ $v_f = \frac{m_c v_c}{m_c + m_h} = \frac{(93\text{ kg})(20\text{ m/s})}{93\text{ kg} + 340\text{ kg}} = 4.3\text{ m/s}$

CHECK and THINK

We just found that the cart slows down by nearly a factor of five when the hero falls into it. This finding makes sense because initially the hero has no momentum in the *x* direction, but the relatively lightweight cart has momentum in the *x* direction. When the hero lands, he acquires an *x* component of momentum. Because the hero is so massive, the cart must slow down considerably when he lands for momentum in the *x* direction to be conserved. This example also illustrates collision, the subject of Chapter 11.

10-6 Case Study: Rockets

When you think of a rocket, you probably imagine a spacecraft being launched. Here we start with an Earth-bound craft that illustrates the physics of a rocket. Imagine Sophia on a light raft in a still pond. She uses a slingshot launcher to shoot large water balloons horizontally off the back of her raft (Fig. 10.13A). To think about conservation of momentum, we must choose a closed system. We include Sophia, the slingshot, the raft, and the water balloons in the system. So, even after she launches a balloon, we will still consider it part of the system. For now, we will also assume no net external force acts on the system, so momentum of the system is conserved. (In Section 10-7, we consider the raft and its contents to be its own separate system distinct from the launched water balloons, and we will also consider the effects of external forces. For now, let's accept this somewhat unrealistic situation in which water does *not* exert a drag force on the raft.) The goal of this section is to derive an expression—known as the *first rocket equation*—for the velocity of the raft.

Imagine that the raft is initially at rest with respect to the pond so that the system's momentum is zero in the pond's frame (*not* depicted in Fig. 10.13). Now imagine that Sophia launches a water balloon in the negative x direction (Fig. 10.13B). The balloon's momentum is in the negative x direction, and for the total momentum of the system to remain zero, the rest of the system (raft and contents) must have momentum in the positive x direction. If Sophia continues to launch her water balloons, her speed will continue to increase. This raft example has the essence of a rocket. The water balloons are the fuel, the launcher is the exhaust system, the raft is the body of the rocket, and Sophia is the astronaut.

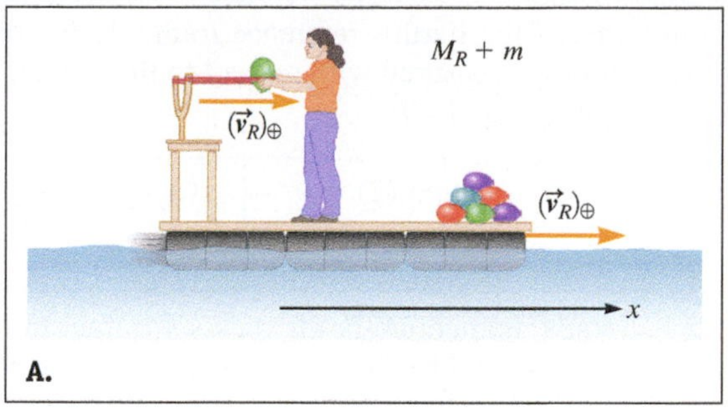

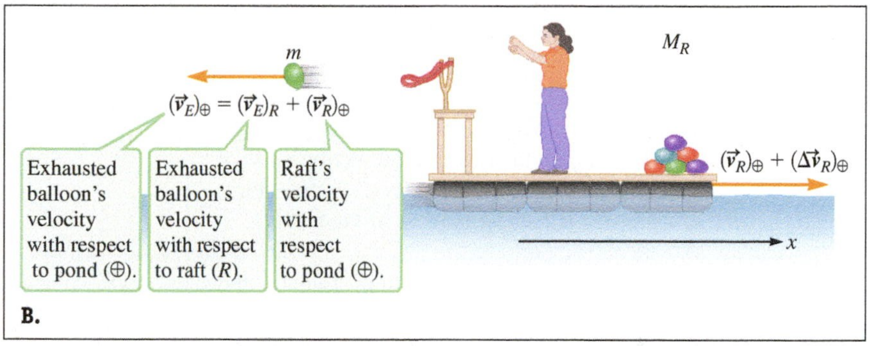

FIGURE 10.13 **A.** Sophia on a raft with water balloons and a launcher. **B.** Sophia launches a water balloon in the negative x direction. To conserve momentum, the raft's velocity in the positive x direction must increase.

Engineers need to know how the change in a rocket's velocity relates to the velocity of its exhausted fuel and to the masses involved (rocket and fuel). This relation is contained in the **first rocket equation**.

DERIVATION First Rocket Equation

Let $(\vec{v}_E)_R$ be the velocity of the *e*xhausted fuel with respect to the *r*ocket. Let $\Delta\vec{v}_R$ be the change in the rocket's velocity with respect to the Earth, ⊕. Here, M_{Ri} is the mass of the rocket (plus the remaining fuel) at some initial time, and M_{Rf} is its mass at some final time. We derive the first rocket equation:

$$\Delta\vec{v}_R = (\vec{v}_E)_R \ln\left(\frac{M_{Rf}}{M_{Ri}}\right) \tag{10.12}$$

FIRST ROCKET EQUATION
Special Case

Let's start with Sophia's rocket. At the initial time (Fig. 10.13A), Sophia has already launched one or more balloons; the *r*ocket has velocity $(\vec{v}_R)_\oplus$ relative to the Earth. She is about to launch a balloon of mass m, so initially the mass of the rocket and its contents is $M_R + m$. The final time occurs just after she has launched the balloon with velocity $(\vec{v}_E)_R$ with respect to the rocket (Fig. 10.13B).

One way to model this system is as just two particles. One particle is the balloon that has just been launched. The other particle is the remainder of the system: the rocket and its contents.

Momentum of this system is conserved.	$\sum \vec{p}_i = \sum \vec{p}_f$ (10.9)
There are two particles in the system: the exhausted balloon (E) and the rocket (R). Collect terms that involve the rocket on the right and terms that involve the balloon on the left.	$\vec{p}_{Ei} + \vec{p}_{Ri} = \vec{p}_{Ef} + \vec{p}_{Rf}$ $\vec{p}_{Ei} - \vec{p}_{Ef} = \vec{p}_{Rf} - \vec{p}_{Ri}$ $-(\vec{p}_{Ef} - \vec{p}_{Ei}) = \vec{p}_{Rf} - \vec{p}_{Ri}$
We find that the change in the rocket's momentum must be equal in magnitude and opposite in direction to the change in the balloon's momentum (with respect to the pond fixed to the Earth).	$-(\Delta\vec{p}_E)_\oplus = (\Delta\vec{p}_R)_\oplus$ (1) (exhaust) (rocket)
Before the balloon is launched, it moves along with the rocket at velocity $(\vec{v}_R)_\oplus$, so its initial momentum is $m(\vec{v}_R)_\oplus$. The balloon's final momentum is $m(\vec{v}_E)_\oplus$.	$(\Delta\vec{p}_E)_\oplus = m(\vec{v}_E)_\oplus - m(\vec{v}_R)_\oplus$ (2)

Derivation continues on page 294 ▶

So far, we have been working in terms of the Earth's reference frame. In the first rocket equation, the speed of the balloon is measured with respect to the rocket, so we switch to the rocket's reference frame (Eq. 4.40).

$$(\vec{v}_E)_\oplus = (\vec{v}_E)_R + (\vec{v}_R)_\oplus \quad (3)$$

Use Equation (3) to eliminate $(\vec{v}_E)_\oplus$ from Equation (2).

$$(\Delta\vec{p}_E)_\oplus = m[(\vec{v}_E)_R + (\vec{v}_R)_\oplus] - m(\vec{v}_R)_\oplus$$
$$(\Delta\vec{p}_E)_\oplus = m(\vec{v}_E)_R \quad (4)$$

Now, we find the change in the balloon's momentum in the rocket's frame. In the rocket's frame, the balloon initially is at rest and has no momentum. When the balloon is launched, its velocity is $(\vec{v}_E)_R$.

$$(\Delta\vec{p}_E)_R = m(\vec{v}_E)_R - 0$$
$$(\Delta\vec{p}_E)_R = m(\vec{v}_E)_R \quad (5)$$

Compare Equations (4) and (5). The balloon's change in momentum is the same in both frames.

$$(\Delta\vec{p}_E)_\oplus = (\Delta\vec{p}_E)_R \quad (10.13)$$

Substitute into Equation (1). We find that the change in the rocket's momentum $(\Delta\vec{p}_R)_\oplus$ (in the positive x direction *relative to the Earth*) is equal to the change in the balloon's momentum, $-(\Delta\vec{p}_E)_R$ (in the negative x direction *relative to the rocket*).

$$(\Delta\vec{p}_R)_\oplus = -(\Delta\vec{p}_E)_R \quad (10.14)$$

Now we need the rocket's change in momentum. Initially, the rocket has velocity $(\vec{v}_R)_\oplus$. After the balloon is launched, the rocket's velocity is $(\vec{v}_R)_\oplus + (\Delta\vec{v}_R)_\oplus$, where $(\Delta\vec{v}_R)_\oplus$ is the increase in the rocket's velocity (relative to the pond).

$$(\Delta\vec{p}_R)_\oplus = M_R[(\vec{v}_R)_\oplus + (\Delta\vec{v}_R)_\oplus] - M_R(\vec{v}_R)_\oplus$$
$$(\Delta\vec{p}_R)_\oplus = M_R(\Delta\vec{v}_R)_\oplus \quad (6)$$

Substitute Equations (5) and (6) into Equation 10.14.

$$M_R(\Delta\vec{v}_R)_\oplus = -m(\vec{v}_E)_R \quad (7)$$

Now we are ready to consider the continuous exhausting of water with the fire hose (Fig. 10.14). Just as in the case of the balloon, the water is exhausted from the hose at $(\vec{v}_E)_R$ with respect to the raft rocket. We imagine breaking up the steady stream of water coming out of the fire hose into small disks of mass dm. Each of these imaginary disks of water is like a very small water balloon. When a small mass of water dm is ejected from the rocket, the rocket's velocity changes by a small amount $(d\vec{v}_R)_\oplus$.

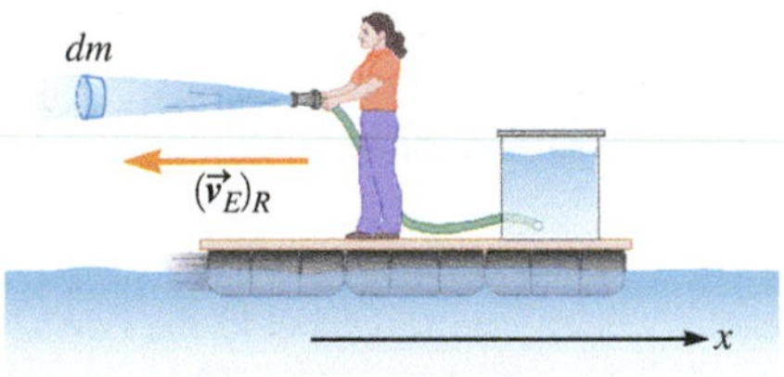

FIGURE 10.14

Rewrite Equation (7) in terms of these differential changes.

$$M_R(d\vec{v}_R)_\oplus = -(dm)(\vec{v}_E)_R \quad (10.15)$$

When a small mass of water dm is ejected, the mass of the rocket and its contents is reduced by dM_R. Therefore, $dM_R = -dm$.

$$M_R(d\vec{v}_R)_\oplus = -(-dM_R)(\vec{v}_E)_R$$
$$(d\vec{v}_R)_\oplus = (\vec{v}_E)_R\left(\frac{dM_R}{M_R}\right) \quad (8)$$

Integrate Equation (8) from some initial time when the rocket's velocity (with respect to the water) was $\vec{v}_{Ri}$ and its mass was M_{Ri} to a final time when the rocket's velocity (with respect to the Earth) and mass are $\vec{v}_{Rf}$ and M_{Rf}. The exhaust velocity of the water with respect to the rocket $(\vec{v}_E)_R$ is constant, so it is pulled outside the integral. The integral on the right results in the natural logarithm function (Appendix A).

$$\int_{\vec{v}_{Ri}}^{\vec{v}_{Rf}} (d\vec{v}_R)_\oplus = \int_{M_{Ri}}^{M_{Rf}} (\vec{v}_E)_R\left(\frac{dM_R}{M_R}\right)$$
$$\vec{v}_{Rf} - \vec{v}_{Ri} = (\vec{v}_E)_R \ln\left(\frac{M_{Rf}}{M_{Ri}}\right)$$
$$\Delta\vec{v}_R = (\vec{v}_E)_R \ln\left(\frac{M_{Rf}}{M_{Ri}}\right) \checkmark \quad (10.12)$$

COMMENTS

Equation 10.12 is the first of two rocket equations. It gives the change in the rocket velocities in terms of its mass. Because the final mass is less than the initial mass of the rocket ($M_{Rf} < M_{Ri}$) and Equation 10.12 involves the natural logarithm of a fraction less than 1, the result is a negative number. So, as expected, the change in the rocket's velocity $(\Delta\vec{v}_R)$ is in the opposite direction to the exhaust velocity $(\vec{v}_E)_R$.

EXAMPLE 10.7 CASE STUDY Rocket Science

Return to the rocket case study (page 282). How would you explain to Shannon that a rocket does not need anything to push against? A spacecraft usually has several small rocket thrusters (Fig. 10.15) that exhaust the fuel at several different angles. Explain how these thrusters might be used to maneuver the spacecraft.

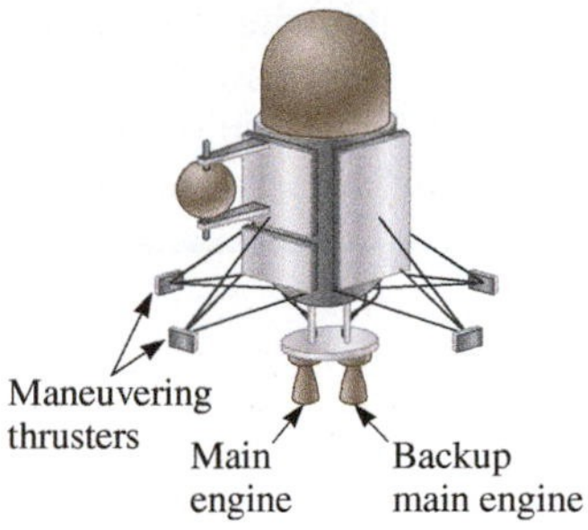

FIGURE 10.15

SOLVE

We assume the spacecraft is far from any massive objects so that gravity can be ignored and there are no other external forces.

If we consider the spacecraft and its exhausted fuel to make up a single system, momentum is conserved because no external forces act on the system. Therefore, to conserve momentum, any change in the exhausted fuel's momentum must result in a change in the spacecraft's momentum, equal in magnitude and opposite in direction: $(\Delta\vec{p}_R)_\oplus = -(\Delta\vec{p}_E)_R$ (Eq. 10.14).

The change in the rocket's velocity is always in the direction opposite to the exhausted fuel's velocity. The small thrusters are used to change the spacecraft's direction. Imagine using the hose on the raft in Figure 10.14 for steering. If Sophia wishes to turn the rocket into the page, she would need to direct the hose out of the page. If resistive forces are very small as they are in space, there is considerable risk of oversteering. Small maneuvering thrusters are about 100 times weaker than the main thrusters used to control the spacecraft.

EXAMPLE 10.8 "How Fast Will We Go?"

A team of engineering students has entered a competition. Each team is provided with a long, thin boat known as a crew shell. Two team members must be on board, and the team that achieves the fastest speed wins. The students may use any mechanical means to power the crew shell. They have decided to use a small commercial pump that ejects water at an impressive speed of 36 m/s. The students mount a 55-gal drum of water on the crew shell. The total mass of the crew shell, pump, empty water drum, and two team members is 157 kg. Ignore resistive forces and find the maximum possible speed of their crew shell on a still pond.

INTERPRET and ANTICIPATE

We include the crew shell and its contents even after the water is ejected in the system. There are no net external forces on the system. Therefore, momentum is conserved. The crew shell starts at rest with respect to the pond. As long as fuel is being exhausted, the shell's speed increases. The shell reaches its maximum speed just as all the fuel is exhausted. We expect to find a numerical result in the form $v_{\text{max}} =$ ______ m/s.

SOLVE

Use the first rocket equation (Eq. 10.12) to find the maximum speed of the crew shell. The initial velocity is zero, and the final velocity is the maximum velocity of the shell.

$$\vec{v}_{Rf} - \vec{v}_{Ri} = (\vec{v}_E)_R \ln\left(\frac{M_{Rf}}{M_{Ri}}\right)$$

$$\vec{v}_{\text{max}} = (\vec{v}_E)_R \ln\left(\frac{M_{Rf}}{M_{Ri}}\right) \quad (1)$$

The initial mass includes the mass of the 55-gal drum of fuel. We convert from gallons to cubic meters and use the density of water to find the mass of this fuel.

$$V_w = 55\text{ gal}\left(\frac{3.786 \times 10^{-3}\text{ m}^3}{1\text{ gal}}\right) = 0.21\text{ m}^3$$

$$M_E = \rho_w V_w = (0.998 \times 10^3\text{ kg/m}^3)(0.21\text{ m}^3)$$

$$M_E = 2.1 \times 10^2\text{ kg}$$

In Equation (1), the final mass M_{Rf} is the mass of the raft's contents not including the mass of the fuel because it has been ejected; the initial mass includes the mass of the fuel.

$$M_{Rf} = 157\text{ kg}$$

$$M_{Ri} = M_{Rf} + M_E = 367\text{ kg}$$

We are only interested in speed, so we take the absolute value of the velocity.

$$v_{\text{max}} = \left|(36\text{ m/s})\ln\left(\frac{157\text{ kg}}{367\text{ kg}}\right)\right|$$

$$v_{\text{max}} = 31.\text{ m/s}$$

Example continues on page 296 ▶

CHECK and THINK

Our answer is in the form we expected, and 31 m/s is about 70 mph, which seems very fast. If we set up this situation as an experiment, the shell's maximum speed would be significantly lower than our answer. Drag plays a major role and cannot be ignored in determining the shell's motion. We will return to this situation in Example 10.10, after we have learned how to account for external forces acting on a rocket system.

10-7 Rocket Thrust: An Open System (Optional)

OPEN SYSTEM Major Concept

In Section 10-6, we considered the rocket and its exhausted fuel to be part of the system. The total mass of the system is a constant. If we decide to choose just the rocket and its contents to be the system, the mass of the rocket decreases as fuel is exhausted. We have then chosen an **open system**, that is, a system that either gains or loses mass.

The mass M_R in this open system is *not* constant; it is the mass of the rocket plus the fuel that has not yet been exhausted. The force exerted by the fuel is known as the **thrust**, and

$$\vec{F}_{\text{thrust}} = (\vec{v}_E)_R \frac{dM_R}{dt} \quad (10.16)$$

where $(\vec{v}_E)_R$ is the velocity of the exhausted fuel with respect to the rocket as defined in Section 10-6.

In Problem 83 you will show that for an open system such as a rocket, the **second rocket equation** is

SECOND ROCKET EQUATION Special Case

$$M_R(\vec{a}_R)_\oplus = \sum \vec{F}_{\text{ext}} + \vec{F}_{\text{thrust}} \quad (10.17)$$

where $(\vec{a}_R)_\oplus$ is the rocket's acceleration observed in the Earth's frame. Because the exhausted fuel is not considered to be in the system, the thrust exerted by the fuel on the rocket is an external force. The first term in Equation 10.17 represents the sum of the *other* external forces such as gravity and drag on the rocket, but not the thrust exerted by the fuel. If no other forces (such as gravity or drag) act on the rocket, the rocket's acceleration is due solely to the thrust exerted on it by the exhausted fuel.

EXAMPLE 10.9 CASE STUDY Rocket Science Revisited

Suppose Shannon is not convinced by the explanation made in Example 10.7. Try another explanation. This time, do *not* include the exhausted fuel in the system. How would you account for the rocket's acceleration?

SOLVE

As before, we assume the spacecraft is far from any massive objects, so gravity can be ignored, and there are no other external forces. If we consider the spacecraft and its contents, but not the exhausted fuel, to be the system, we can use Newton's third law to explain how the spacecraft is maneuvered. The spacecraft's thrusters exert a force on the fuel. By Newton's third law, the fuel must exert a force on the spacecraft of equal magnitude and in the opposite direction. Because there are essentially no resistive forces in space, the force exerted by the fuel on the spacecraft easily causes it to accelerate.

EXAMPLE 10.10 What a Drag!

Let's return to the crew shell competition in Example 10.8 and find a more realistic expectation for the team's maximum speed. The engineering students tested their pump and found that it ejects water at a rate of 23 gal/min. Further tests on the crew shell have shown that the drag force (Eq. 6.5) is given by $F_D = \frac{1}{2}C\rho Av^2 = (19.02 \text{ kg/m})v^2$. Find the terminal speed of the crew shell. (Don't worry about the pump running out of water.)

INTERPRET and ANTICIPATE

The water exhausted provides the thrust and the other external force on the shell is due to the drag force. At the terminal speed, the crew shell is no longer accelerating (Section 6-5), so $(\vec{a}_R)_\oplus = 0$. Therefore, according to the second rocket equation, the drag force must be balanced by the thrust. We expect a lower speed than the one found in Example 10.8.

SOLVE Set $(\vec{a}_R)_\oplus = 0$ in Equation 10.17.	$0 = \sum \vec{F}_{ext} + \vec{F}_{thrust}$ $\sum \vec{F}_{ext} = -\vec{F}_{thrust}$
Because we only need to find the magnitude of the velocity, we only need to work with the magnitude of the forces.	$F_D = F_{thrust} = (19.02\ \text{kg/m})v_t^2$ $v_t = \sqrt{\dfrac{F_{thrust}}{19.02\ \text{kg/m}}}$ (1)
Mass is lost from the crew shell at a rate of 23 gal/min. Use the density of water and several conversion factors to express this rate in kilograms per second.	$\dfrac{dM_R}{dt} = \left(\dfrac{23\ \text{gal}}{1\ \text{min}}\right)\left(\dfrac{1\ \text{min}}{60\ \text{s}}\right)\left(\dfrac{3.786 \times 10^{-3}\ \text{m}^3}{1\ \text{gal}}\right)\left(\dfrac{998\ \text{kg}}{1\ \text{m}^3}\right)$ $\dfrac{dM_R}{dt} = 1.4\ \text{kg/s}$
Use Equation 10.16 to find (the magnitude of) the thrust. The exhaust speed $(v_E)_R$ was given in Example 10.8 as 36 m/s.	$F_{thrust} = \left(\dfrac{dM_R}{dt}\right)(v_E)_R = (1.4\ \text{kg/s})(36\ \text{m/s})$ $F_{thrust} = 50\ \text{N}$
The terminal speed comes from substituting the thrust into Equation (1).	$v_t = \sqrt{\dfrac{F_{thrust}}{19.02\ \text{kg/m}}} = \sqrt{\dfrac{50\ \text{N}}{19.02\ \text{kg/m}}} = 1.6\ \text{m/s}$

CHECK and THINK

Once drag is taken into account, the crew shell is expected to have a much lower maximum speed. Drag slows down the shell by more than one order of magnitude, to just about 4 mph. This result seems reasonable because we know that a crew shell—which is normally rowed—does not get far once the rowers stop rowing.

EXAMPLE 10.11 CASE STUDY "The *Eagle* Has Landed!"

Our modern daily life relies heavily on artificial satellites, but crewed space missions are rare. It is hard for us to imagine that only in the 1960s did we plan such a mission to the Moon, a remarkable act of bravery and ingenuity.

On July 16, 1969, *Apollo 11* was launched from the Kennedy Space Center with three astronauts on board. In lunar orbit on Sunday morning, July 20, Neil Armstrong and Buzz Aldrin climbed into the lunar lander known as the *Eagle*. Michael Collins remained on board the command module, and the *Eagle* of total mass 15,065 kg was released.

Rocket thrusters on the *Eagle* were used to control its descent to the Moon. Although the *Eagle* had overshot the planned landing spot, Armstrong knew that he could still manage to land manually. Aldrin and Armstrong worked without a computer, with Aldrin providing calculations to Armstrong.

The descent (engine) rocket thruster had a maximum thrust of 45,000 N to control the *Eagle*'s deceleration. In addition, the *Eagle* was equipped with sets of smaller thrusters to control its direction (Fig. 10.16). The descent engine had 8212 kg of fuel. It was fired for 756.3 s, and it only had 60 s of fuel remaining when the *Eagle* landed.

FIGURE 10.16 Control thrusters on the *Eagle* were grouped in sets of five (thrusting up, down, forward, backward, and sideways). The descent engine is on the underside of the spacecraft to Aldrin's right in this photo taken by Armstrong.

A Assume the velocity and rate at which mass was ejected from the descent engine were constant at maximum thrust. Use the given data to find dM_R/dt and $(\vec{v}_E)_R$ for the *Eagle*'s descent engine.

Example continues on page 298 ▶

INTERPRET and ANTICIPATE

To find dM_R/dt, calculate how long it would take to expel all the fuel. Assuming the fuel was ejected at a steady rate, divide the mass of the fuel by the time to expel it all to obtain the rate of mass loss. Once we have that result, we find the velocity of the fuel from the thrust.

SOLVE

The descent engine fired for 756.3 s and had 60 s of fuel remaining. Add these figures to find the total time to expel all the fuel.

$$\Delta t = 756.3 \text{ s} + 60 \text{ s} = 816.3 \text{ s}$$

The initial mass of the *Eagle* is greater than its final mass; therefore, dM_R/dt is negative. In the time $\Delta t = 816.3$ s, the *Eagle* would have lost all the mass of the fuel.

$$\frac{dM_R}{dt} = \frac{M_{Rf} - M_{Ri}}{\Delta t} = \frac{-8212 \text{ kg}}{816.3 \text{ s}}$$

$$\frac{dM_R}{dt} = -10.06 \text{ kg/s}$$

To find the velocity of the ejected fuel, use Equation 10.16. Choose a coordinate system with an upward-pointing y axis. Here, $\vec{F}_{\text{thrust}}$ is in the positive y direction (slowing the spacecraft).

$$\vec{F}_{\text{thrust}} = (\vec{v}_E)_R \frac{dM_R}{dt} \quad (10.16)$$

$$(\vec{v}_E)_R = \left(\frac{dM_R}{dt}\right)^{-1} \vec{F}_{\text{thrust}} = \left(-\frac{1}{10.06 \text{ kg/s}}\right)(45000\hat{\jmath} \text{ N})$$

$$(\vec{v}_E)_R = -4.5 \times 10^3 \hat{\jmath} \text{ m/s} = -4.5\hat{\jmath} \text{ km/s}$$

CHECK and THINK

Our numerical results are reasonable. The direction for exhaust velocity makes sense. The thrust must point upward to decelerate the *Eagle*, which would otherwise be in free fall toward the Moon. Therefore, the exhaust must be ejected downward.

B What was the *Eagle*'s acceleration just before it landed on the Moon? Assume the descent engine was at maximum thrust. The local gravitational acceleration near the surface of the Moon is $\vec{g}_{\text{moon}} = -1.67\hat{\jmath} \text{ m/s}^2$.

INTERPRET and ANTICIPATE

Clearly, the Moon's gravity acted on the *Eagle*. Because the Moon has (almost) no atmosphere, there are no other external forces to consider. The engine slowed the descent, so we expect the *Eagle*'s acceleration to be upward.

SOLVE

Solve the second rocket equation (Eq. 10.17) for acceleration.

$$M_R(\vec{a}_R)_\oplus = \sum \vec{F}_{\text{ext}} + \vec{F}_{\text{thrust}} \quad (10.17)$$

$$(\vec{a}_R)_\oplus = \frac{1}{M_R}\left(\sum \vec{F}_{\text{ext}} + \vec{F}_{\text{thrust}}\right) = \frac{1}{M_R}(M_R\vec{g}_{\text{Moon}} + \vec{F}_{\text{thrust}})$$

$$(\vec{a}_R)_\oplus = \left(\vec{g}_{\text{Moon}} + \frac{\vec{F}_{\text{thrust}}}{M_R}\right) \quad (1)$$

The mass M_R needed in Equation (1) is the mass of the *Eagle* just before touching down. We find that information from the initial mass of the lunar lander minus the mass M_E of the fuel that was burned.

$$M_E = (756.3 \text{ s})(10.06 \text{ kg/s}) = 7608 \text{ kg}$$

$$M_R = M_i - M_E = 15065 \text{ kg} - 7608 \text{ kg} = 7457 \text{ kg}$$

$$(\vec{a}_R)_\oplus = \left(-1.67\hat{\jmath} \text{ m/s}^2 + \frac{4.5 \times 10^4\hat{\jmath} \text{ N}}{7457 \text{ kg}}\right) = 4.4\hat{\jmath} \text{ m/s}^2$$

CHECK and THINK

As expected, the acceleration was upward. The acceleration just before landing was about 2.5 times the gravitational acceleration on the surface of the Moon, but less than half of the Earth's gravitational acceleration ($g = 9.81 \text{ m/s}^2$). Astronauts are expected to experience about three g's when launched from the Earth's surface. Because the Moon's gravitational acceleration is about six times weaker than that of the Earth (measured at their surfaces), landing on and taking off from the Moon are much more pleasant rides than doing the same from the Earth.

SUMMARY

Underlying Principles

1. A **general form of Newton's second law** is $\vec{F}_{tot} = d\vec{p}/dt$. This form holds even for systems whose mass changes, and $\vec{F}_{tot} = m\vec{a}$ is only good in the special case of a system with constant mass.
2. The **total momentum of a system is conserved** if two conditions are met: (1) there is no net external force acting on it, with $\sum\vec{F}_{ext} = 0$; and (2) the system is **closed**. Mathematically, conservation of momentum is

$$\vec{p}_{i,\,tot} = \vec{p}_{f,\,tot} \quad (10.9)$$

3. **Newton's second law applied to a system of particles:**

$$\sum\vec{F}_{ext} = \frac{d\vec{p}_{CM}}{dt} = \frac{d\vec{p}_{tot}}{dt} \quad (10.10)$$

If the **system's mass is constant**, then

$$\sum\vec{F}_{ext} = M\vec{a}_{CM} \quad (10.6)$$

Major Concepts

1. **Momentum** is a vector that points in the same direction as a **particle's** velocity:

$$\vec{p} \equiv m\vec{v} \quad (10.1)$$

For a **system of *n* particles, the total momentum** is given by

$$\vec{p}_{tot} = \sum_{j=1}^{n}\vec{p}_j = \vec{p}_{CM} \quad (10.7 \text{ and } 10.8)$$

2. **Center of mass** is a point associated with an object or system that moves as though all of its mass was concentrated there and all the external forces were applied there. It is found from

$$\vec{r}_{CM} = \frac{1}{M}\sum_{j=1}^{n} m_j\vec{r}_j \quad (10.5)$$

The momentum of the center of mass is the total momentum of the system $\vec{p}_{tot} = \vec{p}_{CM}$ (Eq. 10.8).

3. A system that does not lose or gain mass is called a **closed system**. An **open system** is one that either gains or loses mass.

Special Cases: Rockets

1. The **first rocket equation** is

$$\Delta\vec{v}_R = (\vec{v}_E)_R \ln\left(\frac{M_{Rf}}{M_{Ri}}\right) \quad (10.12)$$

2. The **second rocket equation** is

$$M_R(\vec{a}_R)_{\oplus} = \sum\vec{F}_{ext} + \vec{F}_{thrust} \quad (10.17)$$

where the force exerted by the fuel is the thrust:

$$\vec{F}_{thrust} = (\vec{v}_E)_R\frac{dM_R}{dt} \quad (10.16)$$

PROBLEMS AND QUESTIONS

A = algebraic C = conceptual E = estimation G = graphical N = numerical

10-1 A Second Conservation Principle

1. **C** Which of the following questions cannot be answered by applying the conservation of energy principle? Explain your answers. **a.** A ball is tossed up at a given speed. What is the ball's maximum height? **b.** If a ball strikes a wall at a given angle, at what angle does it bounce back? **c.** An object initially at rest explodes into two pieces of equal mass. If the velocity of one piece is measured, what is the velocity of the other piece?

2. **C Review** Chapters 8 through 10 were about the conservation approach. List the general equations in those three chapters that are worth memorizing. Explain your choices.

10-2 Momentum of a Particle

3. **N** Find the momentum in each of the following cases.
   a. A 65-kg person is walking with a velocity of $1.2\hat{\imath}$ m/s.
   b. A 2.3-g bullet is traveling at 720 m/s in the $-\hat{\imath}$ direction.

4. A mother pushes her son in a stroller at a constant speed of 1.52 m/s. The boy tosses a 56.7-g tennis ball straight up at 1.75 m/s and catches it. The boy's father sits on a bench and watches.
   a. N According to the mother, what are the ball's initial and final momenta?
   b. N According to the father, what are the ball's initial and final momenta?
   c. C According to the mother, is the ball's momentum ever zero? If so, when? If not, why not?
   d. C According to the father, is the ball's momentum ever zero? If so, when? If not, why not?
5. N A pitcher throws a 145-g ball horizontally with velocity $\vec{v} = 32.0\ \hat{\imath}$ m/s, and it lands in the catcher's glove 0.450 s later. What is the momentum of the ball the instant before it lands in the glove? Ignore air resistance.
6. E Estimate the magnitude of the momentum of a car on the highway.
7. A An object moves with linear momentum of magnitude p. If the object's mass is m, show that its kinetic energy is given by $K = p^2/2m$.
8. E In comparing a thrown spear versus an arrow shot from a bow, you find that the spear's speed is approximately 35.0 m/s, whereas the arrow's speed is nearly 125 m/s. Treating the spear and the arrow as particles, estimate the mass of the spear and the arrow and, given the speeds discussed here, determine which object would have a greater momentum.
9. N What is the magnitude of the Earth's momentum associated with its orbital motion around the Sun?
10. The velocity of a 10-kg object is given by $\vec{v} = 5t^2\hat{\imath} - (7t + 2t^3)\hat{\jmath}$, where if t is in seconds, then v will be in meters per second.
    a. A What is the net force on the object as a function of time?
    b. N What is the momentum of the object when $t = 15.0$ s?
11. N A particle has a momentum of magnitude 40.0 kg · m/s and a kinetic energy of 3.40×10^2 J. **a.** What is the mass of the particle? **b.** What is the speed of the particle?
12. N **Review** In Example 10.1, two balls are dropped from the top of a building at the same time. One ball has a mass of 2.5 kg, and the other has a mass of 5.0 kg. Drag on the balls is negligible, and they land on the ground 5.3 s after they were released. Using the conservation of energy principle, find the momentum of each ball just as it reaches the ground.
13. N Latoya, sitting on a sled, is being pushed by Dewain on the horizontal surface of a frozen lake. Dewain slips and falls, giving the sled one final push, and the sled comes to rest 9.50 s later. The speed of the sled after the final push is 4.00 m/s, and the combined mass of the sled and Latoya is 32.5 kg. Using a momentum approach, determine the magnitude of the average friction force acting on the sled during this interval.
14. A A baseball is thrown vertically upward. The mass of the baseball is m, and its initial speed is v_i. **a.** What is the momentum of the baseball at its maximum height? **b.** What is the momentum of the baseball when it has traveled three-fourths of the total distance in its up-and-down motion?

10-3 Center of Mass Revisited

15. N Find the center of mass of a system with three particles of masses 1.0 kg, 2.0 kg, and 3.0 kg kept at the vertices of an equilateral triangle of side 1.0 m (Fig. P10.15).

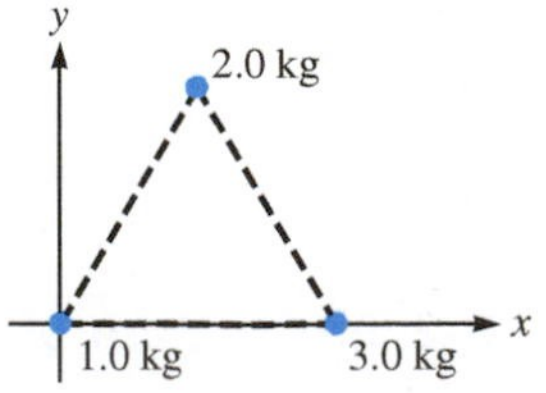

FIGURE P10.15

Problems 16 and 17 are paired.

16. N Fifteen students are lined up along a wall. An x axis has been chosen to run parallel to the wall with the origin in one corner. The positions and (fractional) physics grades of the students are given in Table P10.16. **a.** Find the average of the students' positions. **b.** Find the weighted average of their positions based on their grades. Comment on your results.

TABLE P10.16

x (m)	Grade	Mass (kg)	x (m)	Grade	Mass (kg)
0.25	0.92	54	4.27	0.69	61
0.78	0.95	55	4.65	0.84	56
1.16	0.88	49	5.32	0.75	54
2.04	0.88	69	5.67	0.89	84
2.56	0.78	63	6.10	0.83	78
3.20	0.55	67	6.31	0.82	76
3.62	0.65	65	6.55	0.98	96
3.95	0.46	53			

17. N Fifteen students are lined up along a wall. An x axis has been chosen to run parallel to the wall with the origin in one corner. The positions and masses of the students are given in Table P10.16. **a.** Find the average of the students' positions. **b.** Find the position of their center of mass. Comment on your results.
18. Two metersticks are connected at their ends as shown in Figure P10.18. The center of mass of each individual meterstick is at its midpoint, and the mass of each meterstick is m.
    a. N Where is the center of mass of the two-stick system as depicted in the figure, with the origin located at the intersection of the sticks?
    b. C Can the two-stick system be balanced on the end of your finger so that it remains lying flat in front of you in the orientation shown? Why or why not?

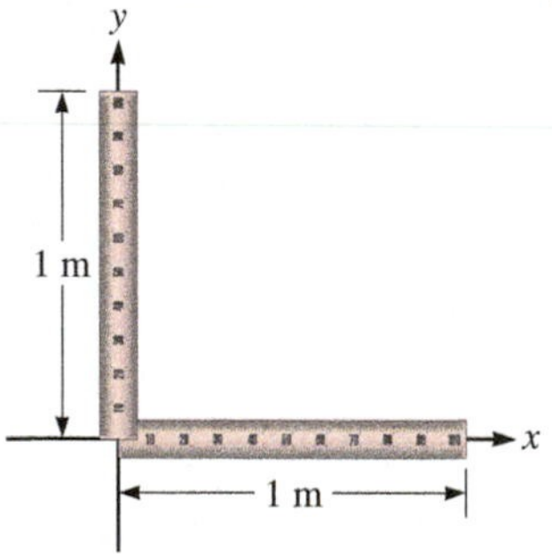

FIGURE P10.18

19. N A boy of mass 25.0 kg is sitting on one side of a seesaw while his older sister, who has a mass of 35.0 kg, is sitting on the other side. Each child is 1.50 m from the center of the seesaw. **a.** Where is the center of mass of the two children relative to the center of the seesaw? **b.** The older sister moves so that the center of mass of the two children is directly above the center of the seesaw. How far did she move and in which direction? Did she move closer to or further from the center?
20. E Estimate the position of your center of mass. Explain your assumptions.
21. N A water molecule, consisting of two hydrogen atoms and one oxygen atom, resides in the plane of an x–y coordinate system. The coordinates of the oxygen atom and the two hydrogen atoms are, respectively, (0, 0), (-7.57×10^{-11} m, 5.86×10^{-11} m) and (7.57×10^{-11} m, 5.86×10^{-11} m). Where is the center of mass of the water molecule located in this coordinate system?
22. The center of mass of a meterstick is at the 50-cm mark. (We are ignoring the thickness of the stick.) You know where it is from the stick's symmetry, and you can confirm it by trying to balance the stick at one point. (You can only get it to balance when you support it at the 50-cm mark.) In this problem, we

find the center of mass of a meterstick by integration. Because we know the answer, this problem is actually about learning the method. So, follow these steps.

a. G Draw the meterstick lying horizontally along an x axis. Place the origin at the 0 mark.

b. G Imagine slicing the stick into small pieces. Each piece has mass dm and length dx. Draw on your sketch one such piece at an arbitrary position.

c. C The linear mass density λ of the stick is its mass M divided by its length L:

$$\lambda = \frac{M}{L}$$

If you had the meterstick in the laboratory, describe what you would measure to find its linear mass density. What are the SI units of λ?

d. A Find an expression for dm in terms of λ and dx.

e. N The stick's center of mass is found by substituting your expression for dm into $x_{CM} = (1/M)\int x\,dm$ and integrating. Do so, being sure to find the appropriate limits for the integration. Check your answer.

Problems 23 and 24 are paired.

23. N When there is a solar eclipse, the Moon is between the Earth and the Sun. Find the center of mass of this three-particle system during a solar eclipse.

24. When the Moon's phase is a "quarter Moon," the Earth, the Moon, and the Sun are at the vertices of a right triangle.

a. N Find the center of mass of the Earth–Moon–Sun system during this phase of the Moon.

b. C If you did Problem 23, use it to check and think about your result.

10-4 Systems of Particles

25. N The distance between Jupiter and one of its moons, Io, is 4.22×10^8 m. What is the location of the center of mass of the Jupiter–Io system if Jupiter's mass is 1.9×10^{27} kg and Io's mass is 8.9×10^{22} kg?

26. A person of mass m stands on a rope ladder that is hanging from a freely floating balloon of mass M. The balloon is initially at rest with respect to the ground. (The buoyant force on the person–balloon system is countering the force of gravity.)

a. C In what direction will the balloon move if the person starts to climb the rope ladder at constant velocity $\vec{v}$ relative to the ladder?

b. A At what speed will the balloon move if the person starts to climb the rope ladder at constant velocity $\vec{v}$ relative to the ladder?

27. N A rugby player with a mass of 65.0 kg is running to the right at a speed of 6.00 m/s toward another player of mass 90.0 kg, who is running in the opposite direction at a speed of 5.00 m/s. What is the total momentum of the two players (both magnitude and direction)?

28. A Your goal here is to show that $\Sigma\vec{F}_{ext} = M\vec{a}_{CM}$ holds for a system of particles. The system's mass M is constant; no particles enter or leave the system. **a.** Show that the velocity of the center of mass is given by

$$\vec{v}_{CM} = (1/M)\sum_{j=1}^{n} m_j\vec{v}_j$$

where m_j and $\vec{v}_j$ are the mass and the velocity of the jth particle, respectively. **b.** Find an expression for the acceleration of the center of mass. Then show that

$$M\vec{a}_{CM} = \sum_{j=1}^{n} m_j\vec{a}_j = \sum_{j=1}^{n}\vec{F}_j$$

where $\vec{F}_j$ is the net force exerted on particle j. **c.** The forces exerted on any particle j include both internal and external forces. Use Newton's third law to argue that the sum of the internal forces exerted on all the particles is zero so that $\sum_{j=1}^{n}\vec{F}_j$ is the sum of only the external forces $\Sigma\vec{F}_{ext}$. Explain your answer.

29. N Two particles with masses 2.0 kg and 4.0 kg are approaching each other with accelerations of 1.0 m/s^2 and 2.0 m/s^2, respectively, on a smooth, horizontal surface (with negligible friction). Find the magnitude of the acceleration of the center of mass of the system.

30. N A billiard player sends the cue ball toward a group of three balls that are initially at rest and in contact with one another. After the cue ball strikes the group, the four balls scatter, each traveling in a different direction with different speeds as shown in Figure P10.30. If each ball has the same mass, 0.16 kg, determine the total momentum of the system consisting of the four balls immediately after the collision.

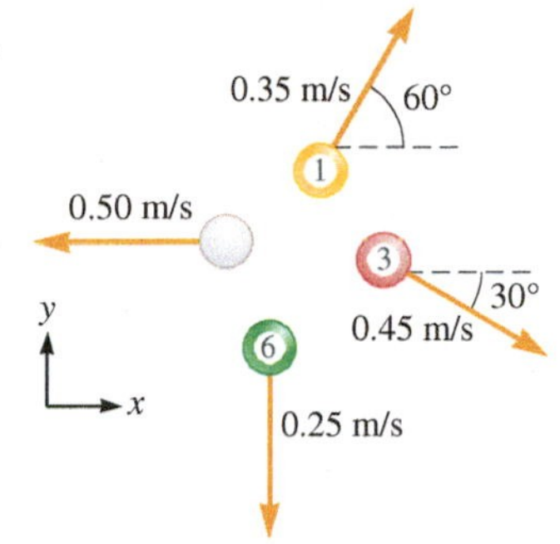

FIGURE P10.30

Problems 31 and 32 are paired.

31. C A crate of mass M is initially at rest on a frictionless, level table. A small block of mass m ($m < M$) moves toward the crate as shown in Figure P10.31. Later, the block and crate are stuck together and are moving with some final speed. The momentum of the block–crate system is the same both before and after the collision. Is the magnitude of the change in momentum of the crate greater than, less than, or equal to the magnitude of the change in the momentum of the block? Explain.

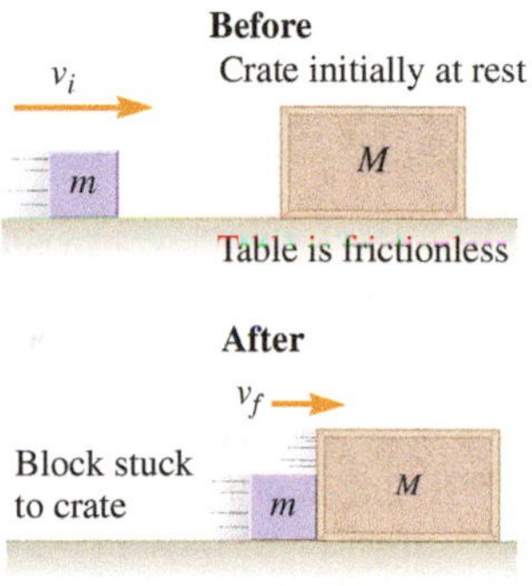

FIGURE P10.31

32. G Shown in the grid in Figure P10.32 are vectors representing the final momenta of the block and the crate in Figure P10.31. Reproduce the grid on your paper and draw the vector that represents the initial momentum of the block m in the space provided.

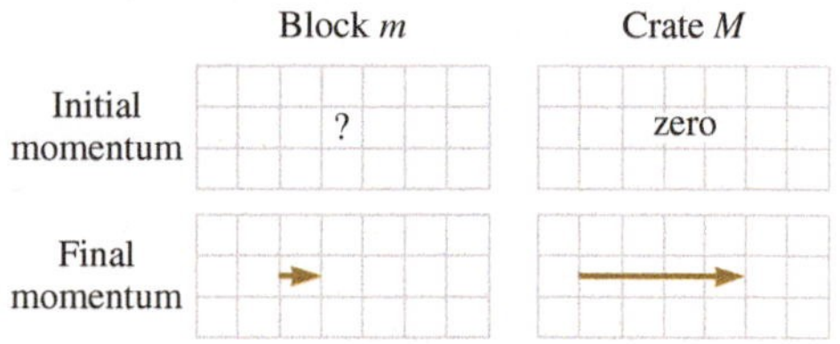

FIGURE P10.32

33. N A particle initially at rest is constrained to move on a smooth, horizontal surface. Another identical particle moving along the surface hits the stationary particle with a speed v. If the particles move together after the collision and their total momentum is the same as the initial momentum of the two-particle system, what is the speed of the combination just after the impact?

34. According to the National Academy of Sciences, the Earth's surface temperature has risen about 1°F since 1900. There is evidence that this *climate change* may be due to human activity. The organizers of World Jump Day argue that if the Earth were in a slightly larger orbit, we could avoid global warming and climate change. They propose that we move the Earth into this new orbit by jumping. The idea is to get people in a particular time zone to jump together. The hope is to have 600 million people jump in a 24-hour period. Let's see if it will work. Consider the Earth and its inhabitants to make up the system.
   a. **E** Estimate the number of people in your time zone. Assume they all decide to jump at the same time; estimate the total mass of the jumpers.
   b. **C** What is the net external force on the Earth–jumpers system?
   c. **N** Assume the jumpers use high-tech Flybar pogo sticks (Fig. P8.32), which allow them to jump 6 ft. What is the displacement of the Earth as a result of their jump?
   d. **C** What happens to the Earth when the jumpers land?

35. **N** Four metal spheres are placed along the x axis. The first sphere, with $m_1 = 1.00$ kg, is located at the origin; the second sphere, with mass $m_2 = 4.00$ kg, is located at $x = -4.00$ m; the third sphere, with mass $m_3 = 3.00$ kg, is located at $x = 2.00$ m; and the final sphere, with mass $m_4 = 6.00$ kg, is located at $x = 4.00$ m. What is the location of the center of mass of the system of the four spheres?

36. **C** In 1970, the carcass of a sperm whale beached near Florence, Oregon. The Oregon Highway Division had the task of removing it. They decided to use half a ton of dynamite to blow it up. The idea was that if the whale carcass were blown into smaller pieces, animals such as seagulls would eat the remains. A television news reporter caught the event on film. Pieces of whale blubber were thrown up to 0.25 mi from the blast site. One chunk of blubber even smashed a car. Include the dynamite and whale carcass in the system. **a.** What is the net external force on the system? **b.** Assuming the dynamite was placed near the whale carcass's center of mass, how was the blubber distributed after the blast? **c.** Would you expect to find the larger chunks of blubber near the blast site, far from the blast site, or randomly distributed? Explain your answer.

10-5 Conservation of Momentum

37. **N** Nicholas, with mass 75.0 kg, jumps vertically upward to a maximum height of 45.0 cm. With what speed does the Earth recoil because of his jump?

38. **C** Usually, we do not walk or even stand on a lightweight boat or raft because of the danger of falling into the water. If you have ever stepped off a small boat onto a dock, however, you have probably noticed that the boat moves away from the dock as you step toward the dock or out of the boat. A heavy dog running on a long lightweight raft presents a similar situation. At first, the raft and the dog are at rest with respect to the water (Fig. P10.38A) so that $\vec{v}_i = 0$. The dog then runs on top of the raft at $\vec{v}_d = v_d\hat{\imath}$ with respect to the water (Fig. P10.38B). The dog has twice the mass of the raft. Find an expression for the velocity of the raft after the dog began running.

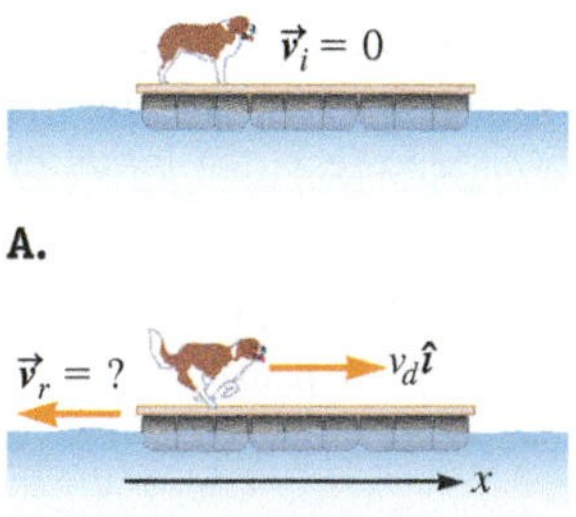

FIGURE P10.38

39. **N** A 175-g billiard ball is shot toward an identical ball at velocity $\vec{v}_i = 6.50\hat{\imath}$ m/s. The identical ball is initially at rest. After the balls hit, one of them travels with velocity $\vec{v}_{1,f} = (1.20\hat{\imath} + 2.52\hat{\jmath})$ m/s. What is the velocity of the second ball after the impact? Ignore effects of friction during this process.

Problems 40 and 41 are paired.

40. There is a compressed spring between two laboratory carts of masses m_1 and m_2. Initially, the carts are held at rest on a horizontal track (Fig. P10.40A). The carts are released, and the cart of mass m_1 has velocity $\vec{v}_1$ in the positive x direction (Fig. P10.40B). Assume rolling friction is negligible.
   a. **C** What is the net external force on the two-cart spring system?
   b. **A** Find an expression for the velocity of cart 2.
   c. **C** Sometimes, mistakes are made in a laboratory. For example, what changes in parts (a) and (b) if the track is not level as shown in Figure P10.40C? Explain your answer.

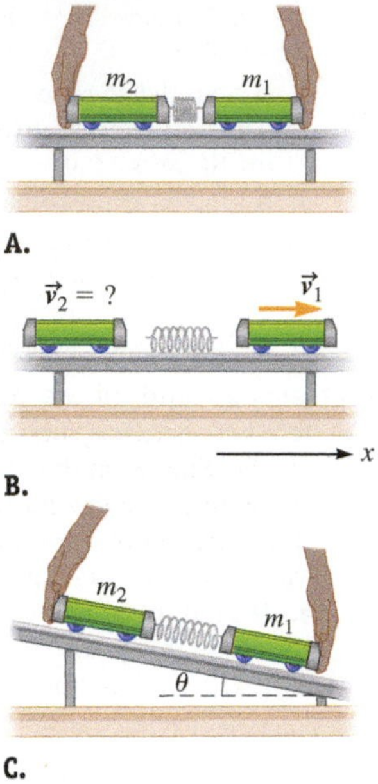

FIGURE P10.40 Problems 40 and 41.

41. **N** There is a compressed spring between two laboratory carts of masses $m_1 = 105$ g and $m_2 = 212$ g. Initially, the carts are held at rest on a horizontal track (Fig. P10.40A). The carts are released, and the cart of mass m_1 has velocity $\vec{v}_1 = 2.035\hat{\imath}$ m/s in the positive x direction (Fig. 10.40B). Assume rolling friction is negligible. **a.** What is the net external force on the two-cart system? **b.** Find the velocity of cart 2.

42. **N** A submarine with a mass of 6.26×10^6 kg contains a torpedo with a mass of 354 kg. The submarine fires the torpedo at an angle of 25° with respect to the horizontal as shown in Figure P10.42. **a.** If the submarine and the torpedo were initially at rest and the torpedo left the submarine with a speed of 89.2 m/s, what is the recoil speed of the submarine? **b.** What is the direction of recoil of the submarine?

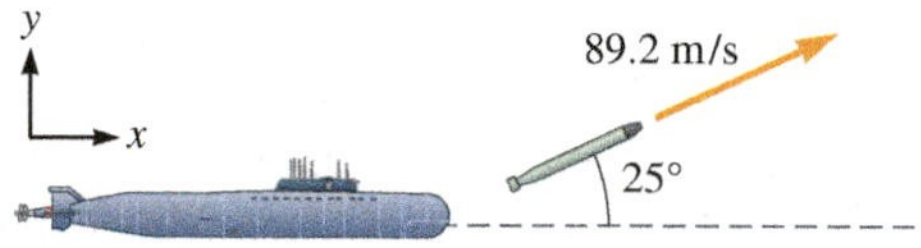

FIGURE P10.42

43. A 44.0-kg child finds himself trapped on the surface of a frozen lake, 10.0 m from the shore. The child slips with each step on the frictionless ice and remains the same distance from the shoreline. Egged on by his parents, he throws a 0.750-kg ball he is carrying toward the center of the lake with a horizontal speed of 1.50 m/s, in the direction opposite that of the shoreline.
   a. **N** Does the act of throwing the ball cause the child to move? If so, what are the speed and the direction of his motion with respect to the Earth?
   b. **C** What are the forces acting on the child when he throws the ball?

Problems 44 and 45 are paired.

44. **C** A model rocket is shot straight up. As it reaches the highest point in its trajectory, it explodes in midair into three pieces

with velocities indicated by the arrows in Figure P10.44, as viewed from directly above the explosion. Rank the mass of each piece in order from smallest to largest and justify your answer.

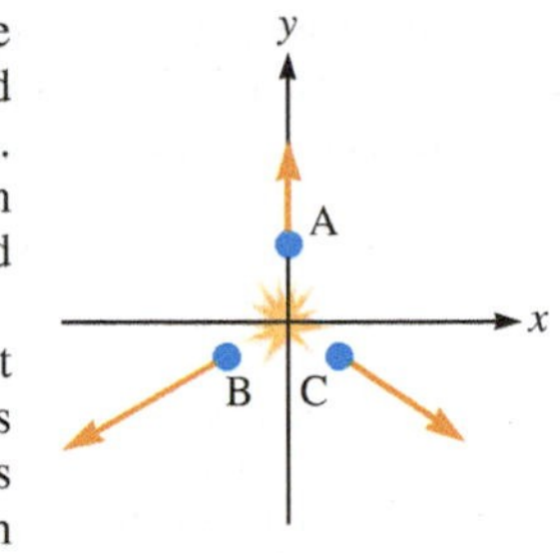

FIGURE P10.44 Problems 44 and 45.

45. **N** A model rocket is shot straight up and explodes at the top of its trajectory into three pieces as viewed from above and shown in Figure P10.44. The masses of the three pieces are $m_A = 100.0$ g, $m_B = 20.0$ g, and $m_C = 30.0$ g. Immediately after the explosion, piece A is traveling at 1.50 m/s, and piece B is traveling at 7.00 m/s in a direction 30° below the negative x axis as shown. What is the velocity of piece C?

Problems 46 and 47 are paired.

46. **C, N** An astronaut finds herself in a predicament in which she has become untethered from her shuttle. She figures that she could get back to her shuttle by throwing one of three objects she possesses in the opposite direction of the shuttle. The masses of the objects are 5.6 kg, 7.9 kg, and 10.3 kg, respectively. She is able to throw the first object with a speed of 15.00 m/s, the second with a speed of 10.0 m/s, and the third with a speed of 7.5 m/s. If the mass of the astronaut and her remaining gear is 75.0 kg, which object should she throw to get back to the shuttle as quickly as possible? Justify your choice.

47. **N** Using the information in Problem 46, determine the final speed of the astronaut with respect to the shuttle if she were to throw each object successively, starting with the least massive and ending with the most massive. Assume that the speeds described are those measured in the rest frame of the astronaut.

10-6 Case Study: Rockets

48. **N** A model rocket consisting of a 60.7-g chassis and a 30.5-g engine is taken on an interstellar mission and fired in outer space. The engine produces 4.95 N of thrust while expending 13.5 g of fuel at a constant rate during a 2.10-s burn. **a.** What is the average exhaust speed of this rocket's engine? **b.** If the rocket starts from rest, what is its final velocity?

Problems 49, 50 and 55 are grouped.

49. **N** A tennis ball machine holds 200 balls and has a mass of 38.5 kg when empty. Each tennis ball has a mass of 56.7 g. A coach has set the machine to launch balls horizontally with a speed of 52.0 m/s. He forgot to set the brakes on the machine. If rolling friction between the machine and the court is negligible and all 200 balls are launched, what is the final speed of the machine?

50. **N** A physics student has set the ball machine in Problem 49 to launch balls horizontally with a speed of 52.0 m/s. The brakes on the machine are set so that it does not move. The student has constructed a "land boat" out of a large grocery cart and a cardboard sail. Each ball is guided into the cart by the cardboard sail. The cart of mass 15.3 kg is initially at rest. If rolling friction between the cart and the floor is negligible and all 200 balls are launched, what is the final speed of the cart?

51. **N** The space shuttle uses its thrusters with an exhaust velocity of 4440 m/s. The shuttle is initially at rest in space and accelerates to a final speed of 1.00 km/s. **a.** What percentage of the initial mass of the shuttle (including the full fuel tank) must be ejected to reach that speed? **b.** If the mass of the shuttle and fuel is initially 1.85×10^6 kg, how much fuel is expelled?

52. **N** A shuttlecraft is initially moving at 94.5 m/s as it travels in deep space. The mass of the shuttle given its current fuel level is approximately 6.33×10^6 kg. When the shuttle fires its rockets to move faster, the speed of the exhaust leaving the rockets is 2750 m/s. What is the mass of the fuel necessary to get the shuttlecraft up to a final speed of 125 m/s?

Problems 53 and 54 are paired.

53. **C** CASE STUDY Shannon is not convinced by the argument in Example 10.7 (page 295), so the conversation continues. Avi and Shannon are sitting on the grass near the softball field.

Shannon: I just don't think that you can propel a rocket or even a boat by pushing something off of it. Catch this. (Shannon throws a softball to Avi). Why am I not moving backward?

Answer Shannon's question.

54. **C** CASE STUDY Shannon and Avi continue to explore the case study (page 282). They sit on skateboards facing each other. Initially, they are both at rest. Shannon, who is very strong, holds a heavy (medicine) ball.

Shannon: I'm going throw this ball to you. You catch it. If the system consists of you, me, and the ball, there are no external forces and momentum is conserved, right? So, I will move backward.

Avi: Yeah, you'll go backward, and so will I.

Shannon: That doesn't make sense. Momentum is already conserved by my going backward. If you go backward too, somehow the system would have to gain momentum. You got that from watching too many cartoons.

Shannon throws the ball to Avi, who catches it. Are either of them moving afterward? If so, who? Explain your answers.

10-7 Rocket Thrust: An Open System (Optional)

55. **N** A coach has set the machine in Problem 49 to launch balls horizontally with a speed of 52.0 m/s at a rate of one ball every 1.50 s. He forgot to set the brake on the machine. If the coefficient of rolling friction between the machine and the court is 0.00300 and all 200 balls are launched out, what is the maximum speed of the machine? (Compare your result with that of Problem 49.)

56. **N** The cryogenic main stage of a rocket has an exhaust speed of 4.21×10^3 m/s and burns liquid hydrogen and liquid oxygen at a combined rate of 317 kg/s. **a.** What is the thrust produced by the rocket's main engine? **b.** If the initial mass of the rocket is 1.10×10^5 kg, what is the initial acceleration of the rocket upon liftoff from the Earth?

57. **N** To lift off from the Moon, a 9.50×10^5 kg rocket needs a thrust larger than the force of gravity. If the exhaust velocity is 4.25×10^3 m/s, at what rate does the exhaust need to be expelled to provide sufficient thrust? The acceleration due to gravity on the Moon is 1.62 m/s^2.

General Problems

58. **N** A system consists of two particles. The first particle has mass $m_1 = 5.00$ kg and a velocity of $(-4.00\hat{\imath} + 2.00\hat{\jmath})$ m/s, and the second particle has mass $m_2 = 2.00$ kg and a velocity of $(2.00\hat{\imath} + 4.00\hat{\jmath})$ m/s. **a.** What is the velocity of the center of mass of this system? **b.** What is the total momentum of this system?

59. **N** A ball of mass $m = 450.0$ g traveling at a speed of 8.00 m/s impacts a vertical wall at an angle of $\theta_i = 45.0°$ below the horizontal (x axis) and bounces away at an angle of $\theta_f = 45.0°$ above the horizontal. What is the average force exerted by the wall on the ball if the ball is in contact with the wall for 250.0 ms?

60. **C** In the original concept, Superman reaches great heights by jumping. In later stories, Superman is able to fly. **a.** In the original concept, is Superman modeled as a rocket? Explain. **b.** Ignore the means by which he levitates in the later stories, and account for how he propels himself.

61. N Two balls with masses $m_1 = 2.0$ g and $m_2 = 3.0$ g move in a plane with velocities $\vec{v}_1 = 6.0\hat{\imath}$ m/s and $\vec{v}_2 = 4.0\hat{\jmath}$ m/s. What is the total momentum of the system?

62. N An astronaut out on a spacewalk to construct a new section of the International Space Station walks with a constant velocity of 2.00 m/s on a flat sheet of metal placed on a flat, frictionless, horizontal honeycomb surface linking the two parts of the station. The mass of the astronaut is 75.0 kg, and the mass of the sheet of metal is 245 kg. **a.** What is the velocity of the metal sheet relative to the honeycomb surface? **b.** What is the speed of the astronaut relative to the honeycomb surface?

63. N Two particles with masses 2.0 kg and 4.0 kg are moving in the same direction with speeds 2.0 m/s and 3.0 m/s, respectively, on a smooth, horizontal surface (with negligible friction). Find the speed of the center of mass of the system.

64. The large slingshot launcher discussed in Section 10-6 exerts a force on the water balloon that is given by the formula $F = 500(1 - t^2)$, where the time is measured in seconds and the calculated force is in newtons.

a. G The slingshot is released and begins exerting a force on the water balloon when $t = 0$. The balloon is considered to be launched when the force on the balloon is zero. Plot the force versus time for the event of launching the balloon. Indicate the time at which the balloon is launched.

b. A Write an expression for the magnitude of the momentum of the water balloon as a function of time during its launch.

c. G Plot the momentum versus time until the moment the balloon is launched.

65. N A racquetball of mass $m = 43.0$ g, initially moving at 30.0 m/s horizontally in the positive x direction, is struck by a racket. After being struck, the ball moves back in the opposite direction at an angle of 30.0° above the horizontal with a speed of 50.0 m/s. What is the average vector force exerted on the racket by the ball if they are in contact for 2.50 ms?

66. N A parcel moving in a horizontal direction with speed $v_0 = 12$ m/s breaks into two fragments of weights 1.0 N and 1.5 N, respectively. The speed of the larger piece remains horizontal immediately after the separation and increases to $v_{1.5} = 25$ m/s. Find the necessary speed and direction of the smaller piece immediately after the separation.

67. N The position of a particle of mass $m_1 = 1.00$ kg is described by the vector $\vec{r}_1 = (2t + 5t^2)\hat{\imath} + 4t\hat{\jmath}$, where t is in seconds and $\vec{r}$ is in meters. The motion of a second particle, of mass $m_2 = 2.50$ kg, is described by the vector $\vec{r}_2 = (3 - 5t)\hat{\imath} + (-t - 2t^2)\hat{\jmath}$. **a.** What is the vector position of the center of mass at $t = 3.00$ s? **b.** What is the velocity of the center of mass at $t = 3.00$ s? **c.** What is the total linear momentum of the system at $t = 3.00$ s?

68. Alika, with mass 45.0 kg, and Armon, with mass 55.0 kg, are both atop their skateboards at rest on a frictionless, horizontal surface. Alika pushes Armon, giving him a speed of 3.00 m/s to the right.

a. N What are the magnitude and the direction of Alika's speed after the push?

b. N What is the amount of potential energy converted by Alika into kinetic energy for the system?

c. C Is momentum conserved in the Alika–Armon system during the push?

d. C Explain your answer to part (c) given that there are large forces acting between Alika and Armon during the push and that their speeds are zero initially and nonzero after the push.

69. N A comet is traveling through space with speed 3.33×10^4 m/s when it encounters an asteroid that was at rest. The comet and the asteroid stick together, becoming a single object with a single velocity. If the mass of the comet is 1.11×10^{14} kg and the mass of the asteroid is 6.66×10^{20} kg, what is the final velocity of their combination?

70. N A ballistic pendulum is used to measure the speed of bullets. It comprises a heavy block of wood of mass M suspended by two long cords. A bullet of mass m is fired into the block horizontally. The block, with the bullet embedded in it, swings upward (Fig. P10.70). The center of mass of the combination rises through a vertical distance h before coming to rest momentarily. In a particular experiment, a bullet of mass 40.0 g is fired into a wooden block of mass 10.0 kg. The block–bullet combination is observed to rise to a maximum height of 20.0 cm above the block's initial height. **a.** What is the initial speed of the bullet? **b.** What is the fraction of initial kinetic energy lost after the bullet is embedded in the block?

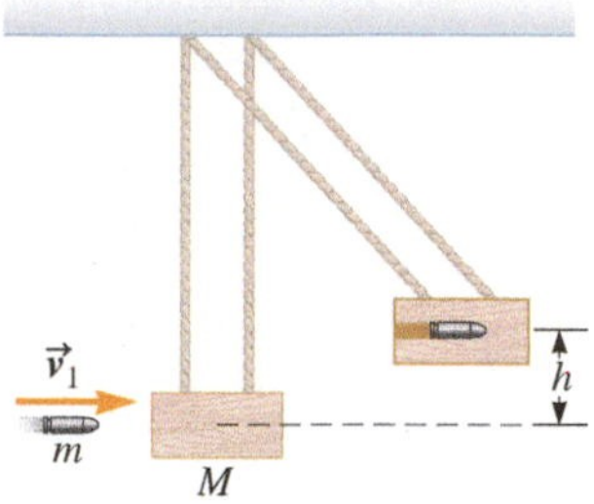

FIGURE P10.70

71. A Rochelle and Sheldon are two astronauts in space tethered together by a strong rope of length L. Initially, Sheldon is holding a large tool, which Rochelle needs. He then throws it to Rochelle, who catches it. Sheldon, Rochelle, and the tool all have the same mass m. Find an expression for the distance traveled by the tool relative to the center of mass of the system.

72. C Physics students at a prominent school were given an assessment test. Consider a winter service truck moving at constant velocity $\vec{v}$ along an icy road. The truck ejects sand out the back at a constant rate $\Delta m/\Delta t$. The sand falls straight down with zero horizontal momentum. The question is, what is the net force on the truck? No students got the correct answer. Most students said that the net force on the truck is zero. Why is this answer incorrect? What is the correct answer?

73. A Joe is teaching Buddy to ice-skate. They each hold an end of a long rope of length L. Initially, they are at rest. Buddy then pulls herself along the rope until she reaches Joe. Buddy's mass is three-fourths of Joe's mass. Find an expression for the distance traveled by Buddy.

74. C Figure P10.74 provides artists with human proportions. Notice that the center of mass moves lower in the body as the person grows. Explain this change. What does it tell you about human proportions as a person grows?

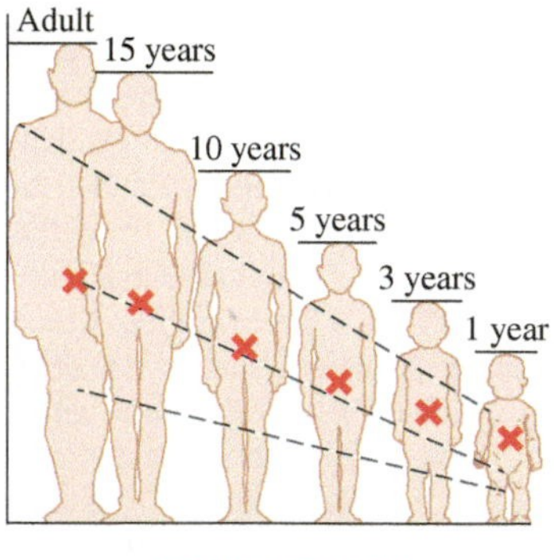

FIGURE P10.74

75. N During the final seconds of the Iditarod race, a sled's harness breaks, sending the team of dogs and the sled and driver careening away from one another on the icy surface. The sled, its driver, and his backpack, with combined mass 275 kg, move with a speed of 3.00 m/s in the positive x direction. In an attempt to catch the harness without slowing down, the driver throws his heavy backpack of mass 20.0 kg in the direction opposite the motion of the sled, giving it a final speed relative to the sled of 5.00 m/s in the opposite direction. **a.** What is the final speed of the sled and driver relative to the ground? **b.** What is the final speed of the backpack relative to the ground?

76. A single-stage rocket of mass 308 metric tons (not including fuel) carries a payload of 3150 kg to low-Earth orbit. The exhaust speed of the rocket's cryogenic propellant is 3.20×10^3 m/s.
 a. **N** If the speed of the rocket as it enters orbit is 8.00 km/s, what is the mass of propellant used during the rocket's burn?
 b. **N** The rocket is redesigned to boost its exhaust speed by a factor of two. What is the mass of propellant used in the redesigned rocket to carry the same payload to low-Earth orbit?
 c. **C** Because the exhaust speed of the redesigned rocket is increased by a factor of two, why is the fuel consumption of the redesigned rocket not exactly half that of the original rocket?

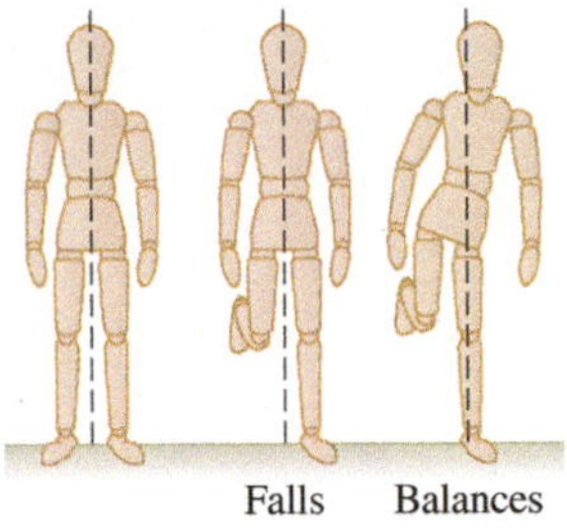

FIGURE P10.77

77. **C** If you try standing on one leg, you will see that your body shifts (Fig. P10.77). Explain why this shift is necessary to prevent you from falling.

78. **N** A light spring is attached to a block of mass $4m$ at rest on a frictionless, horizontal table. A second block of mass m is now placed on the table, in contact with the free end of the spring, and the two blocks are pushed together (Fig. P10.78). When the blocks are released, the more massive block moves to the left at 2.50 m/s. **a.** What is the speed of the less massive block? **b.** If $m = 1.00$ kg, what is the elastic potential energy of the system before it is released from rest?

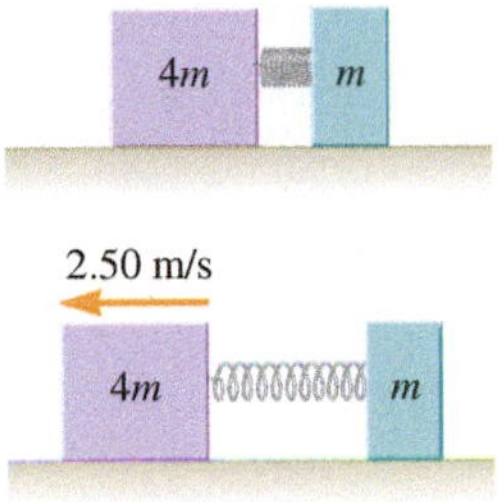

FIGURE P10.78

79. **N** Two blocks of mass 1.00 kg and 0.500 kg are pressed together on a frictionless, horizontal surface, compressing a light spring of force constant 8.1 N/m between them by 6.50 cm (similar to Fig. P10.78). The two blocks are released simultaneously from rest. What is the maximum speed of each block?

80. **C** In Figure P10.80A, a man shows a yoga pose known as the chair pose. Notice that the man leans forward so that he looks like a folding chair. In Figure P10.80B, a woman uses a wall so that she forms a straight-back chair. Why is the wall necessary to create this straight-back chair?

FIGURE P10.80

81. **A** Show that the total momentum of a system of particles equals the momentum of the center of mass:

$$\vec{p}_{\text{tot}} = \sum_{j=1}^{n} \vec{p}_j = \vec{p}_{\text{CM}} \tag{10.8}$$

82. **A** Your friend is making an animated short film, and he consults you on the physics of a scene. In the scene, a penguin and a dog are riding on a model train set. The penguin is in the front car, and the dog is in the second car (Fig. P10.82). The train is moving quickly at speed v_0 when the power is cut. The penguin then pulls the pin connecting its car to the rest of the train. The dog and the remaining cars continue at the same speed v_0 along the track. The penguin and the first car have about one-fifth of the total mass, and the dog and the remaining cars make of four-fifths of the mass. Your friend believes that the penguin will speed up because momentum is conserved and the penguin's part of the system has less mass than the other part of the system. Would such a scene violate laws of physics? To answer this question, find the penguin's speed v_p in terms of v_0 after the cars have been separated. You may ignore friction and drag.

FIGURE P10.82

83. **A** CASE STUDY Show that for an open system such as a rocket, the second rocket equation is

$$M_R(\vec{a}_R)_\oplus = \sum \vec{F}_{\text{ext}} + \vec{F}_{\text{thrust}} \tag{10.17}$$

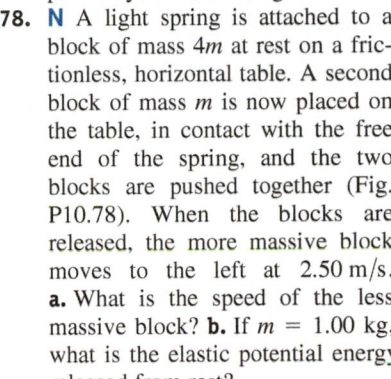

11 Collisions

Key Question

How can the principle of conservation of momentum be applied to the complicated situation of colliding objects?

Underlying Principles

1. Impulse–momentum theorem
2. Conservation of momentum (Chapter 10)

Major Concepts

1. Collision
2. Impulse approximation
3. Impulse
4. Inelastic collision
5. Completely inelastic collision
6. Elastic collision

Special Cases

One-dimensional and two-dimensional collisions

Collisions occur naturally on scales from subatomic particles to whole galaxies. In July 1994, Comet Shoemaker-Levy 9, consisting of more than 20 fragments, crashed into Jupiter. The larger comet fragments left scars in Jupiter's atmosphere that were larger than the Earth and took more than 6 months to heal. Not all collisions are as dramatic. Air molecules colliding with your skin keep you warm, and photons colliding with cells in your eyes allow you to see. Some collisions are deliberate, such as when a car manufacturer tests the safety of a new car or when a physicist studies subatomic particles in an accelerator. In a particularly spectacular example, a spacecraft was designed to collide with a comet so as to create a crater and observe the comet's interior (Fig. 11.1).

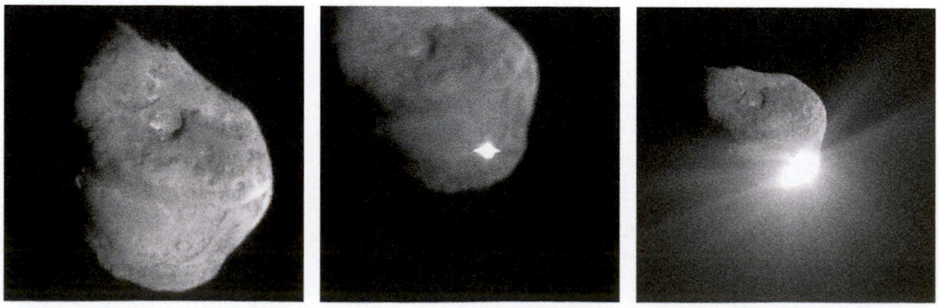

NASA/JPL-Caltech/UMD

FIGURE 11.1 A spacecraft impacts Comet Tempel 1, throwing material into space.

11-1 What Is a Collision?

We have already analyzed a variety of situations using the principles of the conservation of energy and momentum. In this chapter, we apply the conservation approach to *collisions*. Some collisions involve direct contact between the colliding object, as when two cars collide. Other collisions involve "close encounters," as when a spacecraft is sent on a special type of collision course with a planet known as a *gravitational slingshot*, allowing the spacecraft to pick up kinetic energy as it passes near the planet. In physics, a **collision** is an isolated event in which two or more objects exert relatively strong forces on one another for a relatively short time. We usually assume the force between the colliding objects during the time of the collision is much stronger than any other forces exerted on either of them; this assumption is known as the **impulse approximation**. If the force of interaction is a field force such as gravity between a spacecraft and a planet, the objects do not need to touch for them to collide.

COLLISION ✪ Major Concept

IMPULSE APPROXIMATION ✪ Major Concept

CASE STUDY Train Collision Revisited

In Chapter 5, we studied a train collision that took place on April 23, 2002. A passenger train about 35 miles outside of Los Angeles was hit by a freight train (Fig. 5.1). The accident killed two people and injured more than 260. All the injured were on the passenger train. News reports said that the passenger train came to a quick stop before the collision and that the impact with the freight train pushed the passenger train 370 ft backward. One of the most controversial parts of the early reports was the speed of the freight train at the moment of impact. Using physics to reconstruct the events of such an accident is routine in an investigation. In this chapter, we reconstruct the accident and estimate the freight train's speed upon impact.

CONCEPT EXERCISE 11.1

CASE STUDY Forensic Science

Forensic science is the application of scientific techniques and methodology to discover the cause of a particular event such as an accident. Modern forensics have been used to solve historic crimes and mysteries such as the cause of death of King Tutankhamen of ancient Egypt and composer Ludwig van Beethoven in 1827. Forensic evidence is often admitted into a court case. For example, skid marks at an accident scene are used to find the initial speed of a car. Explain why skid marks are left by the car and how they may be used to find the car's initial speed.

11-2 Impulse

You might not usually think of it this way, but catching something can be modeled as a collision. Consider an old picnic game known as the *egg toss*. A team of two players tosses a raw egg back and forth. Each time a player catches the egg without breaking it, he must take a step backward. If the egg breaks, the team is out of the game. The team whose members end up the farthest apart wins. Players seem to know instinctively how to catch the egg gently; no player tries to catch it quickly. Instead, he tries to meet the egg with his fingertips, cupping his hands and moving them backward as the egg stops (Fig. 11.2). This gentle catch means that the egg takes a longer time to stop moving than would a tossed baseball.

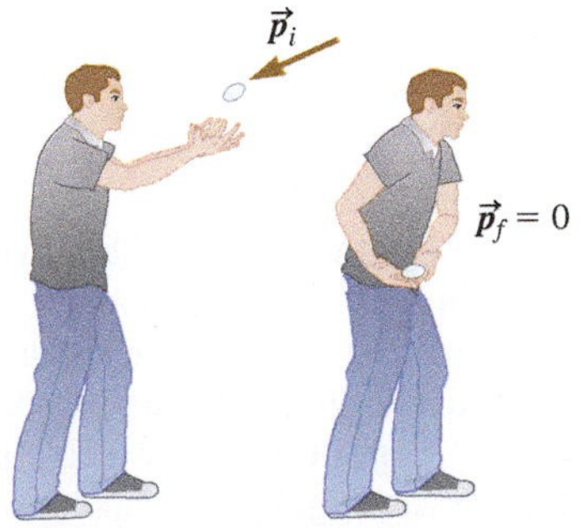

FIGURE 11.2 A player in an egg toss gently catches an egg by allowing it to move from his fingertips to his palm as he moves his hands backward and downward.

EXAMPLE 11.1 How to Win the Egg Toss

The length of time the player spends making the catch actually determines the magnitude of the force he must apply to the egg to make it stop. Show that the average force $\vec{F}_{p,\text{av}}$ exerted by a player (Fig. 11.2) is related to the time Δt he takes to stop the egg and the egg's initial momentum $\vec{p}_i$ by

$$\vec{F}_{p,\text{av}}\Delta t \approx -\vec{p}_i \tag{11.1}$$

INTERPRET and ANTICIPATE

Newton's second law written in terms of momentum (Eq. 10.2) relates change in momentum to force, so it is a good starting point.

$$\sum \vec{F}_{\text{ext}} = \frac{d\vec{p}}{dt}$$

SOLVE

Two forces act on the egg as the player catches it: gravity $\vec{F}_g$ and the player's applied force $\vec{F}_p$.

$$\vec{F}_g + \vec{F}_p = \frac{d\vec{p}}{dt}$$

Multiply each side by dt and integrate from the initial time when the player first makes contact with the egg to the final time when the egg has stopped. Because the gravitational force on the egg is constant, we can pull it outside the integral. The force exerted by the player is not constant, however.

$$\vec{F}_g dt + \vec{F}_p dt = d\vec{p}$$

$$\vec{F}_g \int_{t_i}^{t_f} dt + \int_{t_i}^{t_f} \vec{F}_p dt = \int_{p_i}^{p_f} d\vec{p}$$

The magnitude of the integral $\int_{t_i}^{t_f} \vec{F}_p dt$ is the area under the graph of F_p versus t (Fig. 11.3A). This area is equal to the area of the rectangle of height $F_{p,\text{av}}$ and width Δt, where $F_{p,\text{av}}$ is the magnitude of the average force exerted by the player on the egg during the time interval $\Delta t = t_f - t_i$ (Fig. 11.3B).

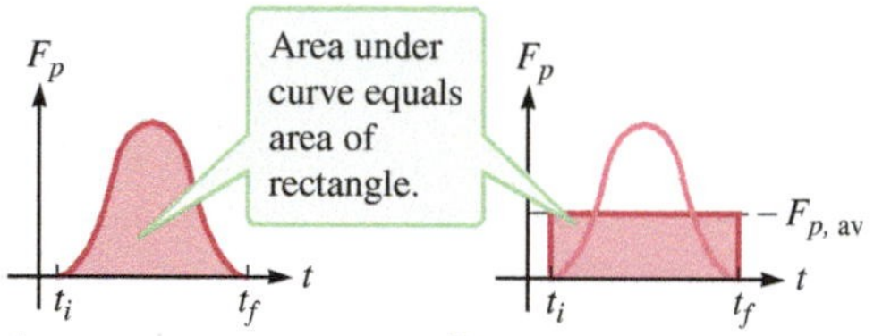

FIGURE 11.3

Complete the integration.

$$\vec{F}_g \int_{t_i}^{t_f} dt + \vec{F}_{p,\text{av}} \int_{t_i}^{t_f} dt = \int_{p_i}^{p_f} d\vec{p}$$

$$\vec{F}_g(t_f - t_i) + \vec{F}_{p,\text{av}}(t_f - t_i) = \vec{p}_f - \vec{p}_i$$

$$\vec{F}_g \Delta t + \vec{F}_{p,\text{av}} \Delta t = \Delta\vec{p}$$

According to the impulse approximation, during the short time of the collision, the force exerted by the colliding objects on each other is much stronger than any other forces exerted on either of them. So, the force exerted by the player on the egg is much greater than gravity. In other words, the change in the egg's momentum is primarily due to the force exerted by the player.

$$\vec{F}_{p,\text{av}}\Delta t \approx \Delta\vec{p} = \vec{p}_f - \vec{p}_i \tag{11.2}$$

The egg has momentum $\vec{p}_i$ when it first makes contact with the player's fingers. The player's job is to stop the egg so that its final momentum $\vec{p}_f$ is zero.

$$\vec{F}_{p,\text{av}}\Delta t \approx 0 - \vec{p}_i = -\vec{p}_i \checkmark \tag{11.1}$$

CHECK and THINK

Equation 11.1 shows how the player uses time to catch the egg gently. The egg has a nonzero initial momentum when it reaches the player's hand; that initial momentum is a result of gravity and how the egg was thrown by his teammate. According to Equation 11.1, the player catching the egg must apply an average force $\vec{F}_{p,\text{av}}$ over a time interval Δt such that combination $\vec{F}_{p,\text{av}}\Delta t$ equals the magnitude of the egg's initial momentum. The player cannot change the initial momentum of the egg, but he can control the force he exerts on the egg by controlling the time over which he makes the catch. If he takes a long time, the average force he exerts on the egg is weak, and the egg will not break. If he stops the egg too quickly, the average force he exerts on the egg will be strong, the egg will break, and he is out of the game.

The egg toss illustrates that the outcome of a collision depends in part on the time interval during which the collision occurs. This relationship between time interval and force is explicitly described by a quantity known as *impulse*. The **impulse** exerted on an object during a time interval Δt is a vector defined as

$$\vec{I} \equiv \int_{t_i}^{t_f} \vec{F}(t)dt = \vec{F}_{av}\Delta t \quad (11.3)$$

IMPULSE ✪ Major Concept

The impulse points in the same direction as the force exerted on the object. Be careful, though; impulse is not the same as force.

Comparing $\vec{I} = \vec{F}_{av}\Delta t$ (Eq. 11.3) with $\vec{F}_{p,av}\Delta t = \Delta\vec{p}$ (Eq. 11.2), we can see that the impulse is also equal to the change in momentum. To put it generally, the total impulse exerted on an object equals the change in the object's momentum:

$$\vec{I}_{tot} = \sum \vec{I} = \Delta\vec{p} \quad (11.4)$$

IMPULSE–MOMENTUM THEOREM ❶ Underlying Principle

Equation 11.4 is referred to as the **impulse–momentum theorem**, and it is another statement of Newton's second law (Problem 12). According to the impulse–momentum *theorem*, the total impulse exerted on an object is in the same direction as the *change* in the object's momentum. According to the impulse *approximation* during a collision, we can usually ignore all the forces except those exerted by one colliding object on the other. So, applying the impulse approximation means that Equation 11.4 reduces to just the impulses exerted by one colliding object on the other(s).

Impulse has both the dimensions of force multiplied by time (Eq. 11.3) and of momentum (Eq. 11.4). In SI units, impulse is given either in N · s or kg · m/s.

CONCEPT EXERCISE 11.2

Why does a coach instruct a gymnast to bend her knees when she lands on the ground?

EXAMPLE 11.2 "Heads Up!"

Two balls, one of mass 2.5 kg and the other of mass 5.0 kg, are dropped from a building at the same time. Drag on the balls is negligible, and they land on the ground 5.3 s after they are released. In Example 10.1, we found that the momentum of the two balls just as they reached the ground was $\vec{p}_1 = -1.3 \times 10^2\hat{\jmath}$ kg · m/s and $\vec{p}_2 = -2.6 \times 10^2\hat{\jmath}$ kg · m/s. Use the same coordinate system as in Figure 10.2 (page 283).

A Find the impulse exerted by the ground on the balls.

INTERPRET and ANTICIPATE

The momenta provided are *initial* momenta for the ball–ground collisions. The ground stops the balls, and their final momenta are zero. The only other force acting on the balls during their collision with the ground is gravity, which (according to the impulse approximation) is weak compared with the force exerted by the ground. We ignore gravity, so, according to the impulse–momentum theorem, the change in each ball's momentum equals the impulse exerted by the ground. We expect to find two numerical results of the form $\vec{I} =$ _____ $\hat{\jmath}$ kg · m/s.

SOLVE

During the collision, the only significant force on each ball is exerted by the ground (subscript "grd"), so Equation 11.4 reduces to one term on the left.

$$\vec{I}_{tot} = \sum \vec{I} = \Delta\vec{p} \quad (11.4)$$

$$\vec{I}_{tot} = \vec{I}_{grd} = \Delta\vec{p}$$

Substitute for the momenta. The final momentum is zero because the balls eventually come to rest on the ground. The initial momenta were found in Example 10.1.

$$\vec{I}_{grd} = \Delta\vec{p} = \vec{p}_f - \vec{p}_i = 0 - \vec{p}_i = -\vec{p}_i$$

$$\vec{I}_{1\,grd} = -\vec{p}_{1\,i} = 1.3 \times 10^2\hat{\jmath} \text{ kg} \cdot \text{m/s}$$

$$\vec{I}_{2\,grd} = -\vec{p}_{2\,i} = 2.6 \times 10^2\hat{\jmath} \text{ kg} \cdot \text{m/s}$$

Example continues on page 310 ▶

CHECK and THINK

The impulse exerted by the ground on the balls points upward as we expected. Notice that the ground exerts twice as much impulse on the ball having twice the mass.

B If it takes 0.005 s for the ground to stop each ball, find the average force exerted by the ground on the 2.5-kg ball. Compare the magnitude of the force exerted by the ground with the ball's weight.

INTERPRET and ANTICIPATE

In part A, we found the impulse. Now that we know the time interval of the collisions, we can find the average force exerted by the ground. We can compare it with the weight of the ball to see if gravity can be ignored as we did in part A. Our results should be in the form $\vec{F}_{av} = \underline{\quad\quad}\hat{\jmath}$ N.

SOLVE Solve $\vec{I} = \vec{F}_{av}\Delta t$ (Eq. 11.3) for the average force.	$\vec{F}_{av} = \dfrac{\vec{I}}{\Delta t}$
Use the impulse found in part A. Recall that $1\ N = 1\ kg \cdot m/s^2$. The subscript "grd" was omitted from the average force for simplicity.	$\vec{F}_{av} = \dfrac{\vec{I}_{grd}}{\Delta t} = \dfrac{1.3 \times 10^2\hat{\jmath}\ kg \cdot m/s}{0.005\ s}$ $\vec{F}_{av} = 2.6 \times 10^4\hat{\jmath}\ N$
Now find the ball's weight.	$w = m_1 g = (2.5\ kg)(9.81\ m/s^2) = 25\ N$
To compare the magnitude of the average force exerted by the ground with the ball's weight, find the ratio F_{av}/w.	$\dfrac{F_{av}}{w} = \dfrac{2.6 \times 10^4\ N}{25\ N} = 1.0 \times 10^3$

CHECK and THINK

The results are in the form we expected. The force exerted by the ground is about 1000 times the ball's weight. Therefore, we are justified in neglecting gravity during these collisions.

11-3 Conservation During a Collision

To analyze a collision with the conservation approach, we choose a system that includes the colliding objects. For objects (modeled as particles) to collide, at least one of the objects must be moving relative to the other(s). So, before the collision, the system has both momentum and kinetic energy. For the collisions we study, usually momentum is conserved (or nearly so), and kinetic energy may or may not be conserved.

Conservation of Momentum During a Collision

If the system of colliding objects is closed (constant mass) and no net external force is exerted on the system, the system's momentum must be conserved (Section 10-5). According to the impulse approximation, the forces exerted by the colliding objects on each other are often much stronger than any external forces; so, it may be acceptable to apply conservation of momentum to a system if the net external force is relatively weak. For example, if we are interested in the collision of an arrow and a tin can, we may ignore the gravitational force exerted by the Earth on the arrow–can system and apply the conservation of momentum principle to the system during the time of the collision. Of course, we must take gravity into account before and after the collision. Throughout this textbook, we assume *the total linear momentum of a system of colliding objects remains constant during the collision.*

So, if the impulse approximation holds, the total momentum of the system before the collision must equal the total momentum of the system after the collision:

$$(\vec{p}_{tot})_i = (\vec{p}_{tot})_f \quad (11.5)$$

CONSERVATION OF MOMENTUM
Underlying Principle

Sometimes a more convenient way to express conservation of momentum is in terms of the system's center of mass. The total momentum of a system is equal to the momentum of the system's center of mass (Eq. 10.8). So, conservation of momentum can be written as

$$(\vec{p}_{\text{CM}})_i = (\vec{p}_{\text{CM}})_f \qquad (11.6)$$

From this equation, we can show that during a collision the center-of-mass velocity is conserved:

$$M(\vec{v}_{\text{CM}})_i = M(\vec{v}_{\text{CM}})_f$$

$$(\vec{v}_{\text{CM}})_i = (\vec{v}_{\text{CM}})_f \qquad (11.7)$$

Equation 11.7 is consistent with what we already know from Chapter 10. A net external force is required to accelerate a system's center of mass. So, if there is no net external force (or if the net external force is negligible) during the collision, the center-of-mass velocity is not changed. The implications of conservation of momentum during a collision are summarized in the concept map shown in Figure 11.4.

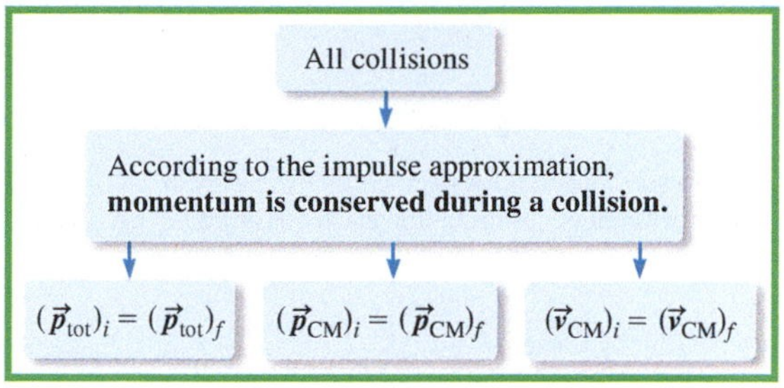

FIGURE 11.4 This concept map shows the implications of the conservation of momentum principle during a collision.

CONCEPT EXERCISE 11.3

When two objects collide, the impulse exerted on object 1 by object 2 is equal in magnitude and opposite and direction to the impulse exerted on object 2 by object 1:

$$\vec{I}_{[1\text{ on }2]} = -\vec{I}_{[2\text{ on }1]} \qquad (11.8)$$

And the change in their momenta is given by:

$$\Delta\vec{p}_1 = -\Delta\vec{p}_2 \qquad (11.9)$$

Which of Newton's three laws justifies these two equations?

Conservation of Kinetic Energy During a Collision

You have probably noticed that when you drop a rubber ball it doesn't quite get back up to the original height after it bounces off the floor. Clearly the Earth–ball system loses some mechanical energy. Where does that energy go?

You may find the answer from another experience. If you have ever played a racket sport (such as squash) with a small rubber ball, you may have noticed that the ball gets very warm after it has been hit around for a few minutes. Some of the ball–wall system's kinetic energy has been converted into thermal energy. The system also loses energy to the surrounding air in the form of sound waves. During any collision, some or even all of the system's kinetic energy is converted into another form or is lost to the environment. In an **inelastic collision**, the system of colliding objects loses kinetic energy. So, the system's initial kinetic energy is greater than the kinetic energy after the collision:

INELASTIC COLLISION
Major Concept

$$K_{\text{tot},\,i} > K_{\text{tot},\,f}$$

The greatest loss of kinetic energy occurs when the colliding objects stick together in what is known as a **completely inelastic collision**. For example, a dart thrown into a dartboard is a completely inelastic collision. However, the objects do not need to be at rest after the collision. For example, if the legendary William Tell actually shot an apple off his son's head with an arrow, the arrow stuck in the apple, and the apple and the arrow moved together after the completely inelastic collision.

COMPLETELY INELASTIC COLLISION
Major Concept

An **elastic collision** is an ideal case in which the system of colliding objects conserves its total kinetic energy:

ELASTIC COLLISION
Major Concept

$$K_{\text{tot},\,i} = K_{\text{tot},\,f} \qquad (11.10)$$

Although no actual collision is ever truly elastic, many collisions may be approximated as elastic.

So, although *linear momentum is always conserved* for a colliding system when external forces can be neglected, *kinetic energy is only conserved if the collision is elastic*. We apply conservation of momentum to both elastic and completely inelastic collisions, but we apply conservation of kinetic energy only to elastic collisions (Fig. 11.5).

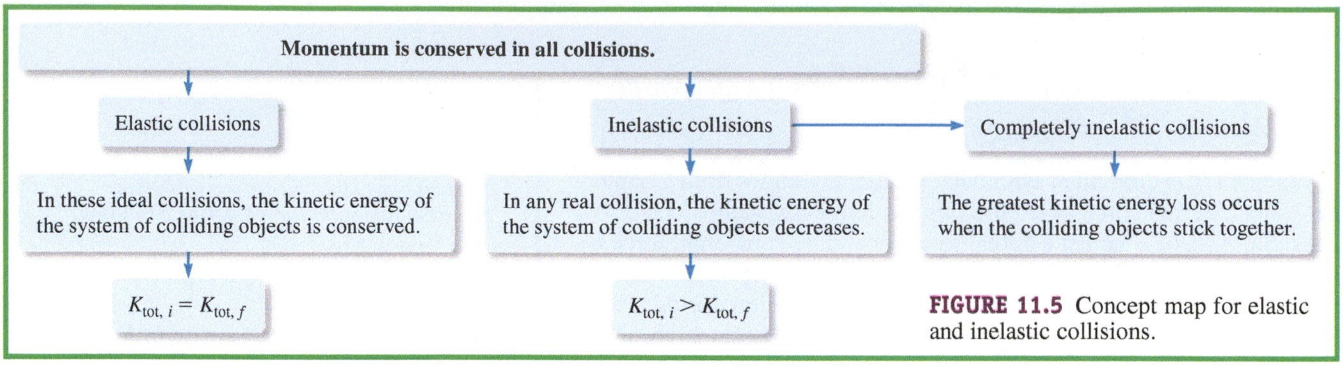
Momentum is conserved in all collisions.
Elastic collisions
Inelastic collisions
Completely inelastic collisions
In these ideal collisions, the kinetic energy of the system of colliding objects is conserved.
In any real collision, the kinetic energy of the system of colliding objects decreases.
The greatest kinetic energy loss occurs when the colliding objects stick together.
$K_{tot,i} = K_{tot,f}$
$K_{tot,i} > K_{tot,f}$

FIGURE 11.5 Concept map for elastic and inelastic collisions.

CONCEPT EXERCISE 11.4

CASE STUDY Trains Stick Together

Two trains collide head-on and stick together. Is the collision elastic, inelastic, or completely inelastic? Explain. Which of Equations 11.5 through 11.10 apply to the two-train system? Explain.

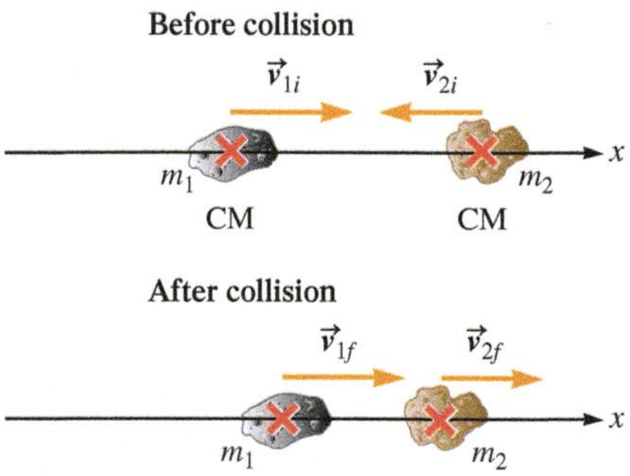
Before collision
$\vec{v}_{1i}$
$\vec{v}_{2i}$
m_1
m_2
CM
CM
x
After collision
$\vec{v}_{1f}$
$\vec{v}_{2f}$
m_1
m_2
x

FIGURE 11.6 Two asteroids collide along the line joining their centers of mass. The collision is one-dimensional.

11-4 Special Case: One-Dimensional Inelastic Collisions

In a one-dimensional collision, the motion of the objects before and after the collision is along a single coordinate axis. So, the objects must collide along the line joining their centers of mass (Fig. 11.6).

Only momentum is conserved in an inelastic collision. For a one-dimensional inelastic collision between two objects along the x axis such as in Figure 11.6, we have

$$\sum \vec{p}_i = \sum \vec{p}_f$$

$$(p_{1i} + p_{2i})\hat{\imath} = (p_{1f} + p_{2f})\hat{\imath}$$

$$(m_1 v_{1i} + m_2 v_{2i})\hat{\imath} = (m_1 v_{1f} + m_2 v_{2f})\hat{\imath} \tag{11.11}$$

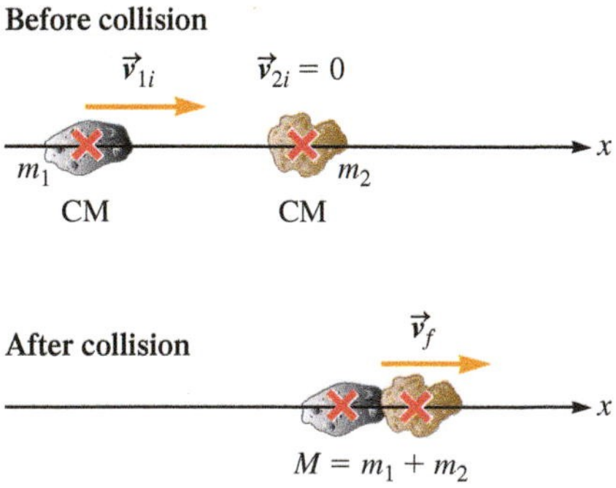
Before collision
$\vec{v}_{1i}$
$\vec{v}_{2i} = 0$
m_1
m_2
CM
CM
x
After collision
$\vec{v}_f$
x
$M = m_1 + m_2$

FIGURE 11.7 A completely inelastic, one-dimensional collision involving a stationary target.

In many collisions, one of the objects is initially at rest. The stationary object is known as the *target*, and the moving object is called the *projectile*. If object 2 in Equation 11.11 is the target, then

$$v_{2i} = 0$$

$$m_1 v_{1i}\hat{\imath} = (m_1 v_{1f} + m_2 v_{2f})\hat{\imath} \tag{11.12}$$

If the collision is *completely* inelastic such as shown in Figure 11.7, the two objects stick together upon collision. After the collision, they act as one large particle of mass $M = m_1 + m_2$ moving at velocity $\vec{v}_f$. Equation 11.12 is simplified in the case of a completely inelastic collision:

COMPLETELY INELASTIC ONE-DIMENSIONAL COLLISION WITH A STATIONARY TARGET

Special Case

$$m_1 v_{1i}\hat{\imath} = (m_1 + m_2)v_f\hat{\imath} = Mv_f\hat{\imath} \tag{11.13}$$

Equations 11.12 and 11.13 are good for certain special cases (Fig. 11.8). When solving an inelastic collision problem, it is best to start with conservation of momentum in a general form (Eqs. 11.5–11.7).

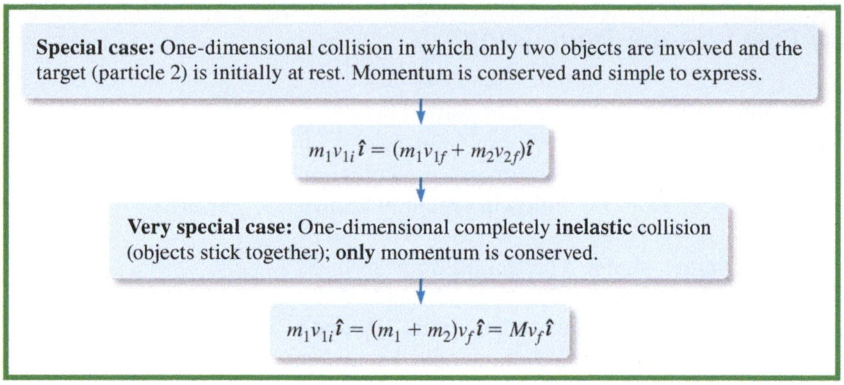

Special case: One-dimensional collision in which only two objects are involved and the target (particle 2) is initially at rest. Momentum is conserved and simple to express.
$m_1 v_{1i}\hat{\imath} = (m_1 v_{1f} + m_2 v_{2f})\hat{\imath}$
Very special case: One-dimensional completely **inelastic** collision (objects stick together); **only** momentum is conserved.
$m_1 v_{1i}\hat{\imath} = (m_1 + m_2)v_f\hat{\imath} = Mv_f\hat{\imath}$

FIGURE 11.8 Concept map for conservation of momentum in the special case of a one-dimensional collision involving only two objects, one of which is initially at rest.

EXAMPLE 11.3 Two Colliding Particles

Figure 11.9 shows a head-on collision between two particles. Particle 1 has three times the mass of particle 2. Initially, particle 1's velocity is $\vec{v}_{1i} = v_0\hat{\imath}$, and particle 2's velocity is $\vec{v}_{2i} = -2v_0\hat{\imath}$. After the collision, particle 2's velocity is $\vec{v}_{2f} = v_0\hat{\imath}$. Find the change in the two-particle system's kinetic energy. In thinking about your result, answer this question: Is the collision elastic, inelastic, or completely inelastic?

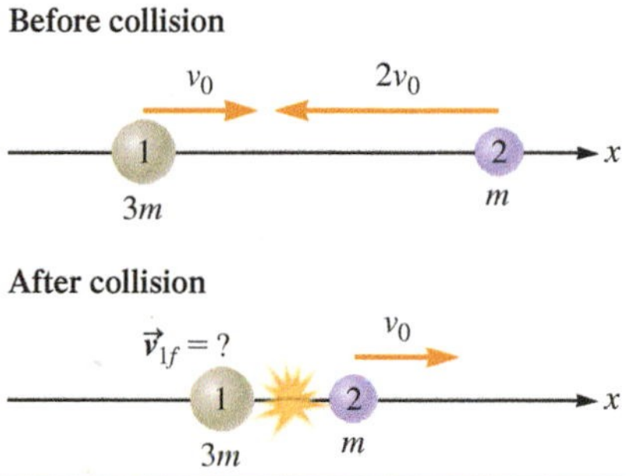

Before collision
v_0
$2v_0$
1
2
x
$3m$
m
After collision
$\vec{v}_{1f} = ?$
v_0
1
2
x
$3m$
m

FIGURE 11.9

INTERPRET and ANTICIPATE

Both particles make up the system. Use conservation of momentum to find particle 1's velocity after the collision and then Equation 8.2 to find the kinetic energy of the system before and after the collision. Our result should be algebraic.

SOLVE

Use the coordinate system shown in Figure 11.9 to find the momentum of the system before and after the collision (Eq. 11.5). (It is reassuring to see that the third equation is the same as Eq. 11.11.)

$$\sum \vec{p}_i = \sum \vec{p}_f \tag{11.5}$$

$$(p_{1i} + p_{2i})\hat{\imath} = (p_{1f} + p_{2f})\hat{\imath}$$

$$(m_1 v_{1i} + m_2 v_{2i})\hat{\imath} = (m_1 v_{1f} + m_2 v_{2f})\hat{\imath}$$

$$(3mv_0 - 2mv_0)\hat{\imath} = (3mv_{1f} + mv_0)\hat{\imath}$$

Solving for v_{1f} shows that particle 1 is at rest after the collision.

$$3v_0 - 2v_0 = 3v_{1f} + v_0$$

$$v_0 = 3v_{1f} + v_0$$

$$v_{1f} = 0$$

Find the total kinetic energy K before and after the collision.

$$K_i = \tfrac{1}{2}m_1 v_{1i}^2 + \tfrac{1}{2}m_2 v_{2i}^2 = \tfrac{1}{2}(3m)v_0^2 + \tfrac{1}{2}m(2v_0)^2$$

$$K_i = \tfrac{3}{2}mv_0^2 + \tfrac{4}{2}mv_0^2 = \tfrac{7}{2}mv_0^2$$

$$K_f = \tfrac{1}{2}m_1 v_{1f}^2 + \tfrac{1}{2}m_2 v_{2f}^2 = 0 + \tfrac{1}{2}mv_0^2 = \tfrac{1}{2}mv_0^2$$

The change in kinetic energy is negative, meaning that the system loses kinetic energy during the collision.

$$\Delta K = K_f - K_i = \tfrac{1}{2}mv_0^2 - \tfrac{7}{2}mv_0^2 = -\tfrac{6}{2}mv_0^2$$

$$\Delta K = -3mv_0^2$$

CHECK and THINK

Because the kinetic energy was not conserved during the collision, the collision must be *inelastic*. (The energy lost may go into the internal energy of each ball or may be lost to the environment through sound.) The collision is not completely inelastic because the particles have two different velocities after the collision, and so cannot be stuck together.

EXAMPLE 11.4 CASE STUDY How Fast Was the Train Going?

We return to the train collision case study from Chapter 5. One of the most important details—the speed of the freight train upon impact—was never reliably reported by witnesses. Initial reports indicated that the freight train was going only about 10 mph at the time of the collision. The goal of this example is to test this estimated train speed from the following reliably known information.

The freight train operator failed to slow down at a yellow signal. The freight train[1] of mass $m_1 = 5.221 \times 10^6$ kg approached a red signal at a speed of 48 mph. The operator applied the emergency brakes after the train passed the red signal. The passenger train of mass $m_2 = 4.1 \times 10^5$ kg was stopped on the tracks.[2] When the trains collided, they locked together and skidded 370 ft.

Two additional assumptions are needed. First, when the emergency brake locks the wheels, the train slides along the track, and kinetic friction (not rolling friction) slows the train. Second, the coefficient of kinetic friction for steel wheels on steel rails varies with weather conditions and lubrication, but we can assume a range of $0.03 \le \mu_k \le 0.4$.

With this information, find the speed of the freight train upon impact. *Hint*: The answer must be given as a range.

INTERPRET and ANTICIPATE

The system consists of the two trains. Figure 11.10 shows the system at three key times: (1) just before the collision, (2) just after the collision, and (3) when the trains finally come to rest. We want to find $\vec{v}_{1i}$ shown before the collision.

During the collision, the net external force on the system is due to kinetic friction. We assume friction is weak compared to the force exerted by the trains on each other; therefore, the momentum of the system is conserved. The two trains stick together, so the collision is completely inelastic.

We can use either the conservation approach or the force approach to find the velocity of the system during the period from just after the collision until the trains come to rest. We choose the conservation approach. (You may use the force approach in Problem 32). There are two major principles used in the solution: conservation of momentum and conservation of energy.

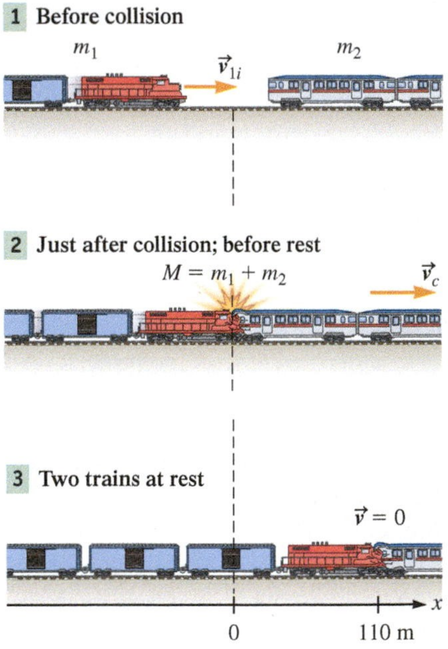

FIGURE 11.10

SOLVE

Let's first use conservation of momentum. Initially, only the freight train has momentum. After the collision, the two trains move together at $\vec{v}_c$. (It is reassuring to find that the penultimate equation is the same as Eq. 11.13.)

$$\sum \vec{p}_i = \sum \vec{p}_f$$

$$(m_1 v_{1i} + m_2 v_{2i})\hat{\imath} = (m_1 v_{1f} + m_2 v_{2f})\hat{\imath}$$

$$m_1 v_{1i}\hat{\imath} = (m_1 + m_2)v_c\hat{\imath}$$

$$m_1 v_{1i}\hat{\imath} = Mv_c\hat{\imath}$$

$$v_{1i} = \frac{M}{m_1}v_c \quad (1)$$

If we know the speed v_c of the two-train system just after the collision, we can use Equation (1) to find the initial velocity of the freight train. We now seek v_c by applying the work–energy theorem, using the 7-step problem-solving strategy on page 266.

Steps 1–4 The **system** consists of the two trains and the track. Our energy bar chart is shown in Figure 11.11. There are no external forces doing work on the system. For this part of the example, we set the **initial time** to the moment of impact, when both trains are moving in the positive x direction. Therefore, the system has kinetic energy initially. There is friction between the track and the trains, so this kinetic energy is converted to thermal energy, and at the **final time** the trains have stopped.

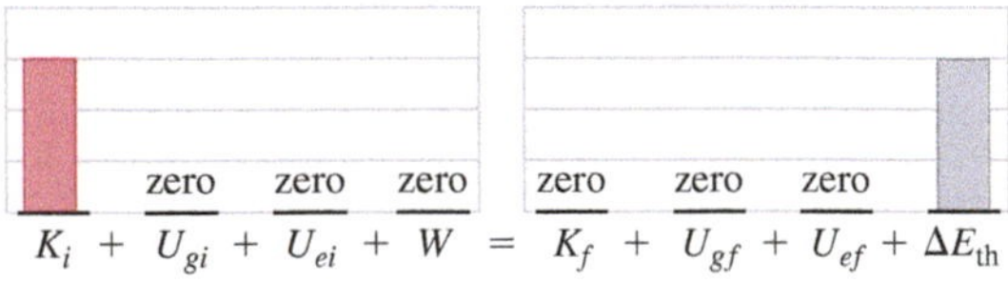

FIGURE 11.11 Energy bar chart for the system.

[1] The spokeswoman for the freight railroad reported the weight of the train to four significant figures.

[2] The passenger train company gave the average weight of a passenger train to two significant figures.

Step 5 Apply the **work–energy theorem** (Eq. 9.33) to write an algebraic expression, using the bar chart as a guide.	$K_i + U_{gi} + U_{ei} + W_{tot} = K_f + U_{gf} + U_{ef} + \Delta E_{th}$ (9.33) $K_i = \Delta E_{th}$ (2)
Step 6 Other equations. The initial kinetic energy is due to the motion of the combined trains moving at v_c. Kinetic friction increases the thermal energy over the stopping distance Δx. The track is level, and only gravity and the normal force are exerted on the system. Therefore, the magnitude of the normal force exerted on the trains equals their weight.	$K_i = \frac{1}{2}Mv_c^2$ (3) $\Delta E_{th} = F_k \Delta x = \mu_k F_N \Delta x$ $\Delta E_{th} = \mu_k Mg\,\Delta x$ (4)
Step 7 Do **algebra.** Find an expression for v_c by substituting Equations (3) and (4) into Equation (2).	$\frac{1}{2}Mv_c^2 = \mu_k Mg\,\Delta x$ $v_c = \sqrt{2\mu_k g\,\Delta x}$ (5)
Substitute Equation (5) in Equation (1) and solve for freight train's speed v_{1i} just prior to the collision.	$v_{1i} = \frac{M}{m_1}\sqrt{2\mu_k g \Delta x}$
To find the range of possible initial freight train speeds, use the extreme values of the kinetic friction coefficient. This estimate is limited to one significant figure. Use 370 ft = 110 m.	$v_{1i}\,(\text{max}) = \left(\frac{5.63 \times 10^6\,\text{kg}}{5.221 \times 10^6\,\text{kg}}\right)\sqrt{2(0.4)(9.81\ \text{m/s}^2)(1.1 \times 10^2\ \text{m})} = 30\ \text{m/s}$ $v_{1i}\,(\text{min}) = \left(\frac{5.63 \times 10^6\,\text{kg}}{5.221 \times 10^6\,\text{kg}}\right)\sqrt{2(0.03)(9.81\ \text{m/s}^2)(1.1 \times 10^2\ \text{m})} = 9\ \text{m/s}$ $(9 \le v_{1i} \le 30)$ m/s
According to the report, the freight train was going 48 mph (21 m/s) when it passed the red signal, so the upper limit we found should be replaced by this figure. (We have to assume the train did not speed up after passing the signal.)	$(9 \le v_{1i} \le 21)$ m/s $(20 \le v_{1i} \le 48)$ mph

CHECK and THINK

The lower limit of 20 mph tells us that the freight train was going at least twice as fast as the first reports indicated. Later reports stated that the freight train exceeded 20 mph at the moment of impact. In one report, it was suggested that the train may not have slowed even after passing the red signal.

11-5 One-Dimensional Elastic Collisions

In Section 11-4, we considered inelastic one-dimensional collisions; in such collisions, only momentum is conserved. If the collision is elastic, kinetic energy is also conserved. Although no collision is truly elastic, some may be modeled as elastic collisions.

In many problems involving the one-dimensional elastic collision of two objects, we often know the mass of each object and two velocities, leaving two velocities unknown. For instance, if we know the initial velocity of each particle $\vec{v}_{1i}$ and $\vec{v}_{2i}$, but the final velocity of each particle $\vec{v}_{1f}$ and $\vec{v}_{2f}$ is unknown (Fig. 11.12), we can use conservation of momentum and conservation of kinetic energy to solve for the two unknown velocities. The algebra leading to $\vec{v}_{1f}$ and $\vec{v}_{2f}$ is messy, but it involves nothing new. To help you solve such problems, we derive expressions for $\vec{v}_{1f}$ and $\vec{v}_{2f}$ here.

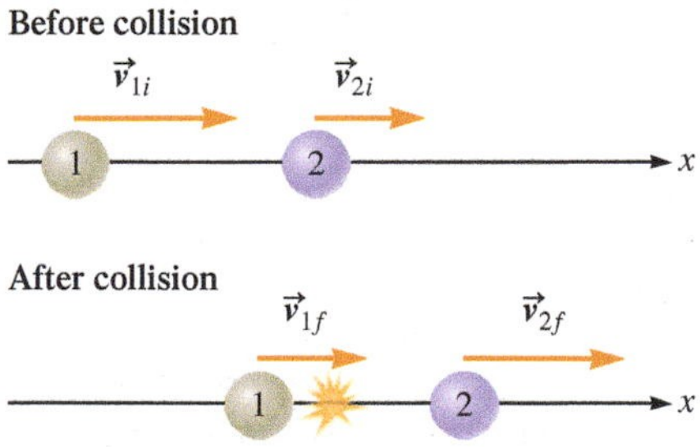

FIGURE 11.12 Two particles collide elastically along the line joining their centers of mass. The collision is one-dimensional.

DERIVATION Particle Velocities After One-Dimensional Elastic Collision

We will show that the velocity of the two particles in Figure 11.12 after their one-dimensional elastic collision is given by

$$\vec{v}_{1f} = \frac{m_1 - m_2}{M}\vec{v}_{1i} + \frac{2m_2}{M}\vec{v}_{2i} \quad (11.14)$$

$$\vec{v}_{2f} = \frac{2m_1}{M}\vec{v}_{1i} + \frac{m_2 - m_1}{M}\vec{v}_{2i} \quad (11.15)$$

FINAL VELOCITIES IN ONE-DIMENSIONAL ELASTIC COLLISION

Special Case

where m_i is the mass of the ith particle and M is the sum of their masses.

Derivation continues on page 316 ▶

Because the collision is elastic, kinetic energy is conserved. Write the kinetic energy of each particle (before and after the collision) in terms of its mass and speed.	$K_{1i} + K_{2i} = K_{1f} + K_{2f}$ $\frac{1}{2}m_1v_{1i}^2 + \frac{1}{2}m_2v_{2i}^2 = \frac{1}{2}m_1v_{1f}^2 + \frac{1}{2}m_2v_{2f}^2$
The factor of $\frac{1}{2}$ cancels out. Collect terms for particle 1 on the left and particle 2 on the right.	$m_1(v_{1i}^2 - v_{1f}^2) = m_2(v_{2f}^2 - v_{2i}^2)$
Factor the difference of the squares as $(a^2 - b^2) = (a - b)(a + b)$.	$m_1(v_{1i} - v_{1f})(v_{1i} + v_{1f}) = m_2(v_{2f} - v_{2i})(v_{2f} + v_{2i})$ (1)
Next, apply conservation of momentum. Both particles have momentum before and after the collision.	$m_1\vec{v}_{1i} + m_2\vec{v}_{2i} = m_1\vec{v}_{1f} + m_2\vec{v}_{2f}$
Again, collect terms for particle 1 on the left and particle 2 on the right.	$m_1\vec{v}_{1i} - m_1\vec{v}_{1f} = m_2\vec{v}_{2f} - m_2\vec{v}_{2i}$ $m_1(v_{1i} - v_{1f})\hat{\imath} = m_2(v_{2f} - v_{2i})\hat{\imath}$ $m_1(v_{1i} - v_{1f}) = m_2(v_{2f} - v_{2i})$ (2)
Use Equation (2) to eliminate $m_2(v_{2f} - v_{2i})$ from Equation (1).	$m_1(v_{1i} - v_{1f})(v_{1i} + v_{1f}) = m_1(v_{1i} - v_{1f})(v_{2i} + v_{2f})$
Divide by the common factor $m_1(v_{1i} - v_{1f})$ to obtain an expression for v_{2f}.	$v_{1i} + v_{1f} = v_{2i} + v_{2f}$ $v_{2f} = v_{1i} + v_{1f} - v_{2i}$ (3)
Solve for v_{1f} by substituting Equation (3) into Equation (2).	$m_1(v_{1i} - v_{1f}) = m_2(v_{1i} + v_{1f} - v_{2i} - v_{2i})$ $(m_1 - m_2)v_{1i} + 2m_2v_{2i} = (m_1 + m_2)v_{1f}$ $v_{1f} = \frac{m_1 - m_2}{m_1 + m_2}v_{1i} + \frac{2m_2}{m_1 + m_2}v_{2i}$ (4)
We find v_{2f} by substituting Equation (4) into Equation (3).	$v_{2f} = v_{1i} + \frac{m_1 - m_2}{m_1 + m_2}v_{1i} + \frac{2m_2}{m_1 + m_2}v_{2i} - v_{2i}$ $v_{2f} = \frac{2m_1}{m_1 + m_2}v_{1i} + \frac{m_2 - m_1}{m_1 + m_2}v_{2i}$ (5)
Finally, Equations (4) and (5) are rewritten in terms of the total mass of the two-particle system, $M = m_1 + m_2$, and as vector equations.	$\vec{v}_{1f} = \frac{m_1 - m_2}{M}\vec{v}_{1i} + \frac{2m_2}{M}\vec{v}_{2i}$ ✓ (11.14) $\vec{v}_{2f} = \frac{2m_1}{M}\vec{v}_{1i} + \frac{m_2 - m_1}{M}\vec{v}_{2i}$ ✓ (11.15)

COMMENTS

Usually, it is best to start any collision problem by writing down the conservation expressions that are valid. For an elastic collision, you would start by writing down an expression for conservation of momentum and another for conservation of kinetic energy. You now have two equations, so you could solve for two unknown quantities. As this derivation shows, however, the algebra involved in finding the two unknowns may be daunting. So, if the problem is a one-dimensional elastic collision, you may wish to start by using Equations 11.14 and 11.5 in that special case.

Stationary Target in an Elastic Collision: Special Cases

Our derivation of the final velocities of the two particles involved in one-dimensional elastic collision allows us to consider the special case of an initially stationary target. In this case, particle 2 (the target) is initially at rest ($v_{2i} = 0$), and Equations 11.14 and 11.15 become

$$\vec{v}_{1f} = \frac{m_1 - m_2}{M}\vec{v}_{1i} \tag{11.16}$$

$$\vec{v}_{2f} = \frac{2m_1}{M}\vec{v}_{1i} \tag{11.17}$$

According to Equation 11.16, particle 1 (the projectile) reverses direction if it has less mass than the target (particle 2) so that $m_1 < m_2$. To further explore how mass affects the final velocity of the particles, imagine a one-dimensional collision between two bumper cars at an amusement park (Fig. 11.13). In all three cases, car 1 is initially moving and car 2 is initially at rest. Their relative mass determines how the cars move after the collision. If the cars have equal mass, then after the collision car 1 is at rest and car 2 is moving at the same speed as car 1 originally had (Fig. 11.13A). If car 1 is much less massive than car 2, after the collision car 1 has a high speed in the reverse direction and car 2 moves slowly in the same direction as car 1's initial velocity (Fig. 11.13B). Finally, if car 1 is much more massive than car 2, after the collision car 1 continues moving with nearly its same velocity and car 2 moves in the same direction at nearly twice the speed as car 1 (Fig. 11.13C).

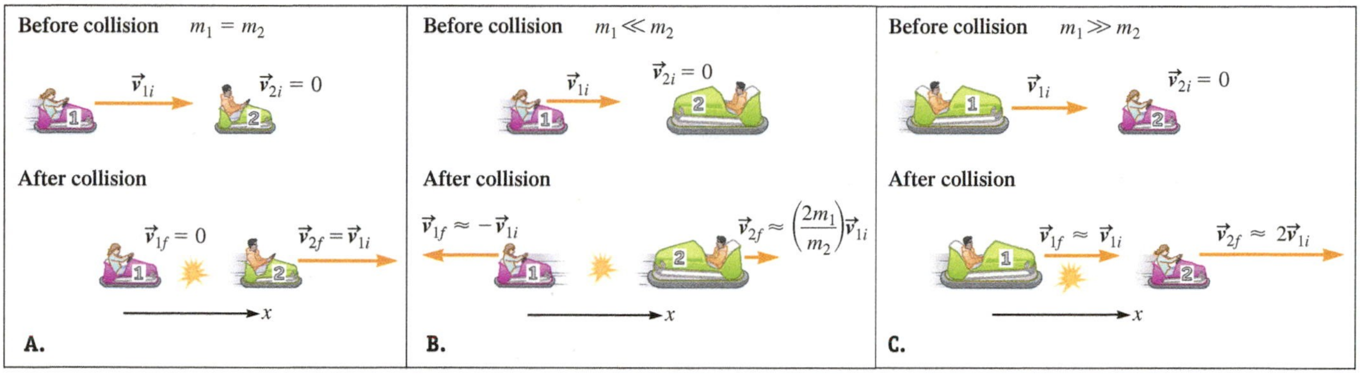

FIGURE 11.13 A one-dimensional collision between two bumper cars in which car 2 is initially at rest in all three cases. **A.** The cars have equal mass. **B.** Car 2 is much more massive than car 1. **C.** Car 1 is much more massive than car 2.

In this subsection, we considered the special case in which the target is initially stationary. As long as the target is not accelerating, it is always possible to find an inertial reference in which the target is initially at rest. As illustrated in the next two examples, using the target's frame is often the best way to tackle a problem.

EXAMPLE 11.5 One-Dimensional Collision in Two Reference Frames

Crall and Whipple set up a laboratory experiment to study a one-dimensional collision between two low-friction carts of mass $m_1 = 2m$ and $m_2 = m$. A small spring on cart 2 ensures that the collision is essentially elastic. Crall decides to work with a reference frame fixed to the laboratory (Fig. 11.14). In the laboratory frame, Crall measures the initial velocities of both carts before the collision: $\vec{v}_{1i} = 3.2\hat{\imath}$ m/s and $\vec{v}_{2i} = 0.8\hat{\imath}$ m/s.

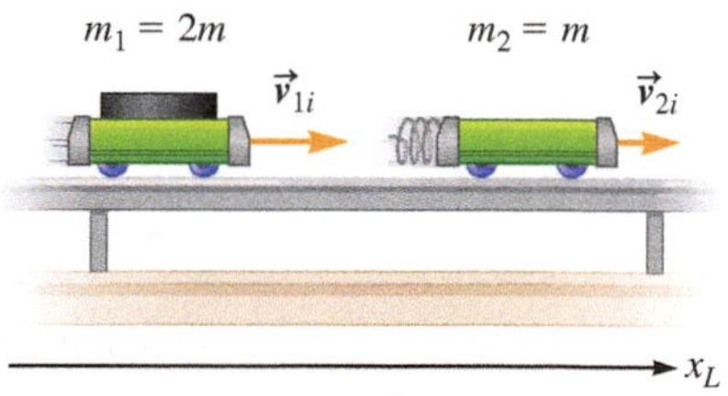

FIGURE 11.14 Crall's laboratory frame, before collision

A Use Crall's measurements to find the velocity of the two carts after the collision in the laboratory reference frame.

INTERPRET and ANTICIPATE

This example is a one-dimensional elastic collision between two objects. Neither object is initially at rest. We expect to find two numerical results of the form $\vec{v}_f =$ _____ $\hat{\imath}$ m/s.

SOLVE

Find the velocities from a straightforward substitution. For cart 1, use Equation 11.14.

$$v_{1f} = \frac{m_1 - m_2}{M}v_{1i} + \frac{2m_2}{M}v_{2i} \qquad (11.14)$$

$$v_{1f} = \frac{2m - m}{3m}v_{1i} + \frac{2m}{3m}v_{2i}$$

$$v_{1f} = \tfrac{1}{3}v_{1i} + \tfrac{2}{3}v_{2i}$$

$$\vec{v}_{1f} = \tfrac{1}{3}(3.2\hat{\imath}\text{ m/s}) + \tfrac{2}{3}(0.8\hat{\imath}\text{ m/s})$$

$$\vec{v}_{1f} = 1.6\hat{\imath}\text{ m/s}$$

Example continues on page 318 ▶

For cart 2, use Equation 11.15.

$$v_{2f} = \frac{2m_1}{M}v_{1i} + \frac{m_2 - m_1}{M}v_{2i} \quad (11.15)$$

$$v_{2f} = \frac{2(2m)}{3m}v_{1i} + \frac{m - 2m}{3m}v_{2i} = \tfrac{4}{3}v_{1i} - \tfrac{1}{3}v_{2i}$$

$$\vec{v}_{2f} = \tfrac{4}{3}(3.2\hat{\imath}\text{ m/s}) - \tfrac{1}{3}(0.8\hat{\imath}\text{ m/s}) = 4.0\hat{\imath}\text{ m/s}$$

CHECK and THINK

According to our results, the carts continue to move in the positive x direction after the collision. Before the collision, cart 1 was moving faster than cart 2. After the collision, cart 2 is moving faster than cart 1.

To check our results, we compare the change in the two cart's momenta. We don't know the value of m, so we leave our expressions in terms of m. As expected, $\Delta\vec{p}_1 = -\Delta\vec{p}_2$ (Eq. 11.9).

$$\Delta\vec{p}_1 = 2m(\vec{v}_{1f} - \vec{v}_{1i}) = 2m(1.6\hat{\imath} - 3.2\hat{\imath})\text{ m/s}$$

$$\Delta\vec{p}_1 = -(3.2\text{ m/s})m\hat{\imath}$$

$$\Delta\vec{p}_2 = m(\vec{v}_{2f} - \vec{v}_{2i}) = m(4.0\hat{\imath} - 0.8\hat{\imath})\text{ m/s}$$

$$\Delta\vec{p}_2 = (3.2\text{ m/s})m\hat{\imath}$$

B Whipple decides to work in the reference frame moving at cart 2's initial velocity (Fig. 11.15). Find the final velocity of the two carts in this frame.

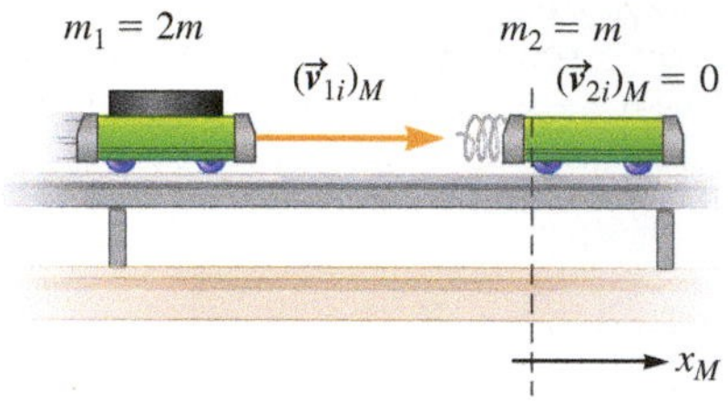

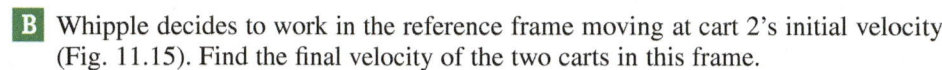

FIGURE 11.15 Whipple's moving frame, before collision

INTERPRET and ANTICIPATE

Whipple's chosen frame is an inertial one that moves at cart 2's *initial* velocity relative to the laboratory. So, cart 2 is initially at rest in Whipple's frame, but its speed is nonzero after the collision. We use the notation from Section 4-7 to indicate the two reference frames: L for *l*aboratory and M for *m*oving. Cart 2 is initially a stationary target in Whipple's frame like the targets in Figure 11.13, so we can use the equations for the special case of a stationary target. Our results should be consistent with part A.

SOLVE

Solve for cart 1's initial velocity $(\vec{v}_{1i})_M$ in Whipple's *moving* frame (Eq. 4.40). The relative velocity between the two frames $(\vec{v}_M)_L$ is cart 2's initial velocity with respect to the laboratory, $\vec{v}_{2i} = 0.8\hat{\imath}$ m/s.

$$(\vec{v}_{1i})_L = (\vec{v}_{1i})_M + (\vec{v}_M)_L \quad (4.40)$$

$$(\vec{v}_{1i})_M = (\vec{v}_{1i})_L - (\vec{v}_M)_L = 3.2\hat{\imath}\text{ m/s} - 0.8\hat{\imath}\text{ m/s}$$

$$(\vec{v}_{1i})_M = 2.4\hat{\imath}\text{ m/s}$$

Cart 1's final velocity in Whipple's reference frame comes from Equation 11.16 in this special case of an elastic one-dimensional collision with a stationary target.

$$(\vec{v}_{1f})_M = \frac{m_1 - m_2}{M}(\vec{v}_{1i})_M \quad (11.16)$$

$$(\vec{v}_{1f})_M = \frac{2m - m}{3m}(\vec{v}_{1i})_M = \tfrac{1}{3}(\vec{v}_{1i})_M$$

$$(\vec{v}_{1f})_M = \tfrac{1}{3}(2.4\hat{\imath}\text{ m/s}) = 0.8\hat{\imath}\text{ m/s}$$

Similarly, cart 2's final velocity in Whipple's frame comes from Equation 11.17.

$$(\vec{v}_{2f})_M = \frac{2m_1}{M}(\vec{v}_{1i})_M = \frac{2(2m)}{3m}(\vec{v}_{1i})_M$$

$$(\vec{v}_{2f})_M = \tfrac{4}{3}(\vec{v}_{1i})_M = \tfrac{4}{3}(2.4\hat{\imath}\text{ m/s})$$

$$(\vec{v}_{2f})_M = 3.2\hat{\imath}\text{ m/s}$$

CHECK and THINK

Use Equation 4.40 to compare the results of parts A and B. The two sets of results are consistent.

$$(\vec{v}_{1f})_L = (\vec{v}_{1f})_M + (\vec{v}_M)_L$$

$$(\vec{v}_{1f})_L = 0.8\hat{\imath}\text{ m/s} + 0.8\hat{\imath}\text{ m/s} = 1.6\hat{\imath}\text{ m/s} \checkmark \quad \text{(lab frame)}$$

$$(\vec{v}_{2f})_L = (\vec{v}_{2f})_M + (\vec{v}_M)_L$$

$$(\vec{v}_{2f})_L = 3.2\hat{\imath}\text{ m/s} + 0.8\hat{\imath}\text{ m/s} = 4.0\hat{\imath}\text{ m/s} \checkmark \quad \text{(lab frame)}$$

Because the target is initially stationary in the Whipple's frame, our calculation using Equations 11.16 and 11.17 is simpler here than it was in part A. This problem-solving skill will help us solve our next example.

EXAMPLE 11.6 A Gravitational Slingshot

In the rocket case study in Chapter 10 (page 282), Cameron suggests that rockets are accelerated by the gravitational attraction of large objects such as planets and moons. It's true. Rocket scientists sometimes set spacecraft on collision courses in which the spacecraft does not crash on the planet's surface. The spacecraft and the planet interact through gravity, a field force that does not require the two objects to touch to collide. We require only that the gravitational force exerted by one object on the other be relatively strong for a relatively short time. When the spacecraft collides with a planet, the spacecraft is accelerated in a process known as a *gravitational slingshot*.

Consider an elastic collision between a spacecraft and a planet (Fig. 11.16A). The spacecraft is initially moving along the positive x axis, and the planet is moving in the negative x direction. During the collision, the spacecraft arcs around the planet so that after the collision, the spacecraft has reversed direction. Of course, the spacecraft's path around the planet is two-dimensional, but because the spacecraft's motion before and after the collision is along a single straight line, we consider it to be a one-dimensional collision. In the solar system's (fixed "laboratory") frame, the initial velocity of the spacecraft is $\vec{v}_{Si} = v_{Si}\hat{\imath}$. We also consider the planet's frame (Fig. 11.16B); the *planet*'s velocity relative to the solar system frame is $\vec{v}_P = -v_P\hat{\imath}$. Find an expression for the spacecraft's final velocity $\vec{v}_{Sf}$ relative to the solar system's frame. Comment on the change in the spacecraft's speed.

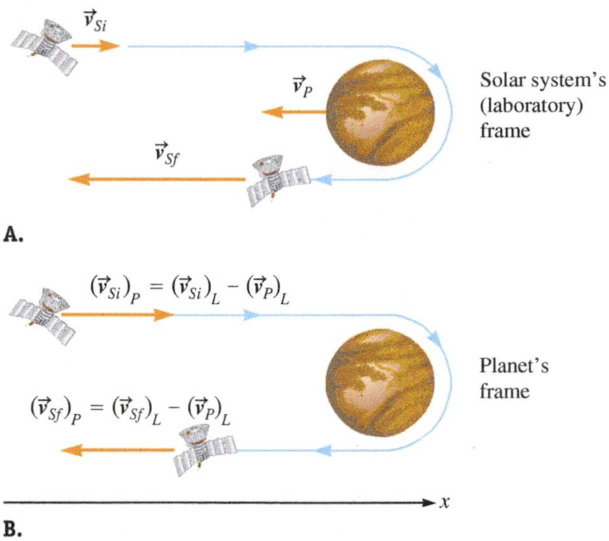

FIGURE 11.16

INTERPRET and ANTICIPATE

The planet is much more massive than the spacecraft, so the planet's velocity $\vec{v}_P$ is essentially unchanged by the collision. To verify that observation, we work in the solar system's (*l*aboratory) frame and use Equation 11.15 (because we have a one-dimensional elastic collision) with $m_1 \ll m_2$.

$$(\vec{v}_P)_f = \frac{2m_1}{M}\vec{v}_{Si} + \frac{m_2 - m_1}{M}(\vec{v}_P)_i \qquad (11.15)$$

$$(\vec{v}_P)_f \approx 0 + \frac{m_2}{m_2}(\vec{v}_P)_i \approx (\vec{v}_P)_i$$

$$(\vec{v}_P)_L = -(v_P\hat{\imath})_L$$

is approximately constant in the laboratory (solar system) frame.

Now consider the planet's (moving) frame. Because the collision does not change the planet's velocity in the planet's own reference frame, its velocity before and after the collision is zero. If we make our calculation using the planet's frame, we have a special case of a one-dimensional elastic collision with a very massive ($m_1 \ll m_2$) stationary target. So, we can use $\vec{v}_{1f} \approx -\vec{v}_{1i}$ (Fig. 11.13B) to find the velocity of the spacecraft after the collision in the planet's frame. We will then transform our answer back to the solar system's frame.

SOLVE

Working in the *p*lanet's frame is very simple. According to Figure 11.13B, the spacecraft's speed is constant, and it simply reverses direction after the collision.

$$(\vec{v}_{Sf})_P \approx -(\vec{v}_{Si})_P \qquad (1)$$

We must transform the spacecraft's initial velocity relative to the solar system frame to the planet's (moving) frame (Eq. 4.40). The subscript L stands for the *l*aboratory (solar system) frame, P stands for *p*lanet, and S stands for spacecraft (the subject).

$$(\vec{v}_{Si})_L = (\vec{v}_{Si})_P + (\vec{v}_P)_L \qquad (4.40)$$

$$(\vec{v}_{Si})_P = (\vec{v}_{Si})_L - (\vec{v}_P)_L$$

$$(\vec{v}_{Si})_P = [v_{Si}\hat{\imath} - (-v_P\hat{\imath})]_L$$

$$(\vec{v}_{Si})_P = [(v_{Si} + v_P)\hat{\imath}]_L \qquad (2)$$

Substitute Equation (2) into Equation (1). The result is the spacecraft's velocity after the collision with respect to the planet's reference frame.

$$(\vec{v}_{Sf})_P \approx -[(v_{Si} + v_P)\hat{\imath}]_L$$

Example continues on page 320 ▶

Now we must transform the spacecraft's final velocity back to the solar system (laboratory) frame (Eq. 4.40). For simplicity, we dropped the subscript L in our final expression.	$(\vec{v}_{Sf})_L = (\vec{v}_{Sf})_P + (\vec{v}_P)_L$ (4.40) $(\vec{v}_{Sf})_L = -[(v_{Si} + v_P)\hat{\imath}]_L + [-v_P\hat{\imath}]_L$ After the gravitational slingshot, the spacecraft's speed (in the solar system's frame) is: $\vec{v}_{Sf} = -(v_{Si} + 2v_P)\hat{\imath}$
CHECK and THINK Finally, find the change in the spacecraft's speed in the solar system's reference frame.	$\Delta v_S = \lvert\vec{v}_{Sf}\rvert - \lvert\vec{v}_{Si}\rvert = (v_{Si} + 2v_P) - v_{Si}$ $\Delta v_S = 2v_P$

The change in spacecraft's speed in the solar system's frame is twice the planet's speed. It seems like the planet-spacecraft's system gains kinetic energy. How is that possible if no net external force is exerted on the system (as it must be if momentum is conserved)? The explanation is that we have an approximation built into our work. The approximation is that the planet's speed and kinetic energy are unchanged by the collision. Although it is a very good approximation, the planet actually loses kinetic energy, which means a very small decrease in its speed (Problem 48).

Finally, in this scenario, the spacecraft reverses direction. Most gravitational slingshots are not simply 180° reversals. Instead, the spacecraft's path is deflected along some other angle, and rocket scientists may plan several slingshots along the trajectory of a single spacecraft.

CONCEPT EXERCISE 11.5

If a spacecraft is headed for the outer solar system, it may require several gravitational slingshots with planets in the inner solar system. If a spacecraft undergoes a head-on slingshot with Venus as in Example 11.6, find the spacecraft's change in speed Δv_S. *Hint*: Venus's orbital period is 1.94×10^7 s, and its average distance from the Sun is 1.08×10^{11} m.

11-6 Two-Dimensional Collisions

Figure 11.17 shows examples of two-dimensional collisions. A cue ball hits the eight-ball on its side, so the collision is not head-on. Afterward, neither ball's motion is along the x axis (Fig. 11.17A). In Figure 11.17B, a car moves in the positive x direction while a truck moves in the positive y direction before the collision. After the collision, the truck and the car are stuck together and move at an angle θ with respect to the x axis. In both situations, two coordinate axes are needed to describe the motion of each system before and after the collision.

The physics of two-dimensional collisions is the same as that for one-dimensional collisions. The momentum of the system is conserved, and kinetic energy is also conserved if the collision is elastic.

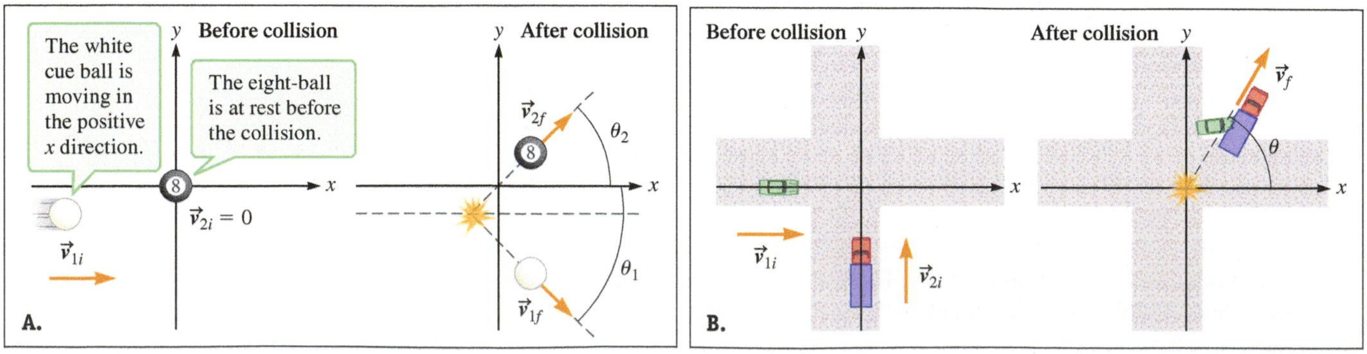

FIGURE 11.17 Two examples of two-dimensional collision. **A.** A white cue ball collides elastically with the side of the eight-ball in a game of pool (billiards). **B.** Two vehicles collide at an intersection. This collision is completely inelastic.

The real difference between one- and two-dimensional collisions is seen when we express conservation of momentum mathematically (Eq. 11.5):

$$\sum \vec{p}_i = \sum \vec{p}_f$$

For a two-dimensional collision, it is often convenient to write the conservation of momentum equation in terms of components:

$$\sum p_{ix} = \sum p_{fx} \qquad \sum p_{iy} = \sum p_{fy} \tag{11.18}$$

Because we have two conservation of momentum equations and not just one, the mathematics behind a two-dimensional elastic collision is more complicated. To keep things manageable, for the rest of this section, we'll focus our attention on two-dimensional collisions that only involve two particles such as those in Figure 11.17.

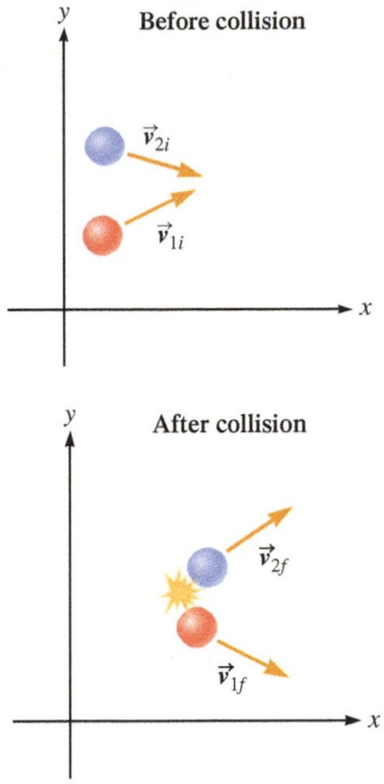

FIGURE 11.18 Two moving particles collide in two dimensions.

Elastic Two-Dimensional Two-Particle Collision

Figure 11.18 shows a two-dimensional elastic collision between two particles. Before and after the collision, both particles are moving. For such a collision, conservation of kinetic energy is rewritten in terms of the particle's speeds:

$$\tfrac{1}{2}m_1 v_{1i}^2 + \tfrac{1}{2}m_2 v_{2i}^2 = \tfrac{1}{2}m_1 v_{1f}^2 + \tfrac{1}{2}m_2 v_{2f}^2 \tag{11.19}$$

Equation 11.19 holds for any elastic collision involving two objects, regardless of the number of dimensions.

For collisions involving two objects, the conservation of momentum (Eq. 11.5) is

$$m_1\vec{v}_{1i} + m_2\vec{v}_{2i} = m_1\vec{v}_{1f} + m_2\vec{v}_{2f} \tag{11.20}$$

Equation 11.20 is good for any collision of any dimensionality involving two objects. It is helpful to rewrite conservation of momentum in terms of the scalar components. These velocity components can be found in the usual way from the angle each vector makes with the x axis. Using the coordinate system shown in Figure 11.18:

$$m_1 v_{1ix} + m_2 v_{2ix} = m_1 v_{1fx} + m_2 v_{2fx} \tag{11.21}$$

$$m_1 v_{1iy} + m_2 v_{2iy} = m_1 v_{1fy} + m_2 v_{2fy} \tag{11.22}$$

Although the physical principles (conservation of momentum and conservation of kinetic energy) of one- and two-dimensional collisions are the same, one thing that makes two-dimensional collisions more complicated is keeping track of all the subscripts. These subscripts account for the particle's label (1 or 2), the time (*i*nitial or *f*inal), and component (x or y).

When solving a problem involving a two-dimensional elastic collision of two objects, there are three independent equations—one from conservation of kinetic energy (Eq. 11.19) and two from conservation of momentum (Eqs. 11.21 and 11.22)—which means that you can solve for three unknown quantities. Often the problem is simplified because one object is initially at rest as is the case in a game of pool, the subject of the next example.

EXAMPLE 11.7 Playing Billiards

Figure 11.17A is an example of a somewhat simplified two-dimensional elastic collision. Often, but not always, a shot in a game of pool may be approximated as an elastic collision between two objects of equal mass $m_1 = m_2$. In addition, only one of the objects (the cue ball) is initially moving, so $\vec{v}_{2i} = 0$. Find the total angle $(\theta_1 + \theta_2)$ between the velocities of the two balls after the collision.

INTERPRET and ANTICIPATE

Both momentum and kinetic energy are conserved in this two-dimensional collision.

Example continues on page 322 ▶

SOLVE

Conservation of kinetic energy (Eq. 11.19) is simplified because the masses are equal ($m_1 = m_2$) and particle 2 is initially at rest ($\vec{v}_{2i} = 0$).

$$\tfrac{1}{2}m_1v_{1i}^2 + \tfrac{1}{2}m_2v_{2i}^2 = \tfrac{1}{2}m_1v_{1f}^2 + \tfrac{1}{2}m_2v_{2f}^2 \qquad (11.19)$$

$$\tfrac{1}{2}m_1v_{1i}^2 + 0 = \tfrac{1}{2}m_1v_{1f}^2 + \tfrac{1}{2}m_1v_{2f}^2$$

$$v_{1i}^2 = v_{1f}^2 + v_{2f}^2 \qquad (1)$$

It is possible to apply conservation of momentum in component form, but here we use the vector form (Eq. 11.20). This expression becomes a simple vector addition of the velocities in the case of billiards.

$$m_1\vec{v}_{1i} + m_2\vec{v}_{2i} = m_1\vec{v}_{1f} + m_2\vec{v}_{2f} \qquad (11.20)$$

$$m_1\vec{v}_{1i} + 0 = m_1\vec{v}_{1f} + m_1\vec{v}_{2f}$$

$$\vec{v}_{1i} = \vec{v}_{1f} + \vec{v}_{2f} \qquad (2)$$

In Figure 11.19, we graphically represent the vector addition in Equation (2). Equation (1) is the Pythagorean theorem applied to the triangle shown. We conclude that $\vec{v}_{1i}$ is the hypotenuse of the triangle and that $\alpha = 90°$.

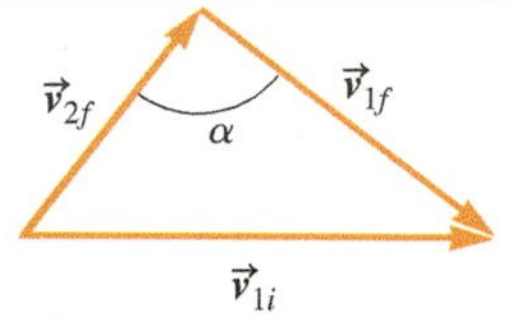

FIGURE 11.19

To compare this triangle to the pool balls after the collision, in Figure 11.20 we slide $\vec{v}_{1f}$ so that its tail touches the tail of $\vec{v}_{2f}$.

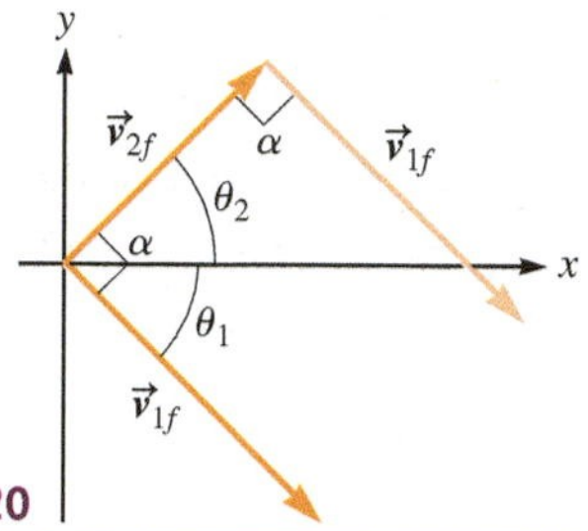

FIGURE 11.20

The angle α is the sum of $\theta_1 + \theta_2$, and it is the angle between $\vec{v}_{1f}$ and $\vec{v}_{2f}$.

$$\theta_1 + \theta_2 = 90° \qquad (11.23)$$

CHECK and THINK

The angle $\alpha = 90°$ holds only for an elastic two-dimensional collision with $m_1 = m_2$ and $\vec{v}_{2i} = 0$. Because many collisions in pool are nearly elastic, $\alpha = 90°$ for many shots throughout the game, and players can predict both where the struck ball will go and where the cue ball will go after the collision. If the collision is head-on, it is a one-dimensional elastic collision between objects of equal mass. In this case, the cue ball stops, and the other ball continues along the same straight line. The angle between the two final velocities isn't well-defined in this case because the cue ball stops.

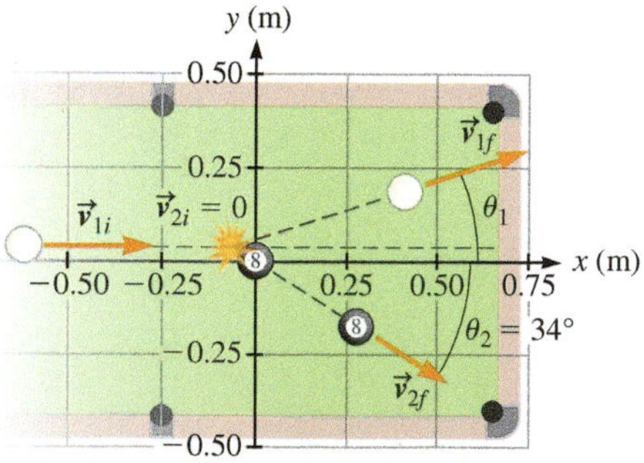

FIGURE 11.21

CONCEPT EXERCISE 11.6

The cue ball hits the eight-ball in a game of pool (Fig. 11.21). Assume the collision is elastic.

a. Will the eight-ball be sunk into a pocket if it is hit at the angle $\theta_2 = 34°$ shown in the figure?

b. If the cue ball is also sunk, the player is said to have scratched. If she scratches when trying to sink the eight-ball, she loses. Did the player lose?

Completely Inelastic Two-Dimensional Two-Particle Collision

In Figure 11.22, the two particles stick together in a completely inelastic collision. Kinetic energy is not conserved, so there is one fewer equation that can be applied. Because both objects stick together, however, they have the same final velocity $\vec{v}_f$, which reduces the number of possible unknown

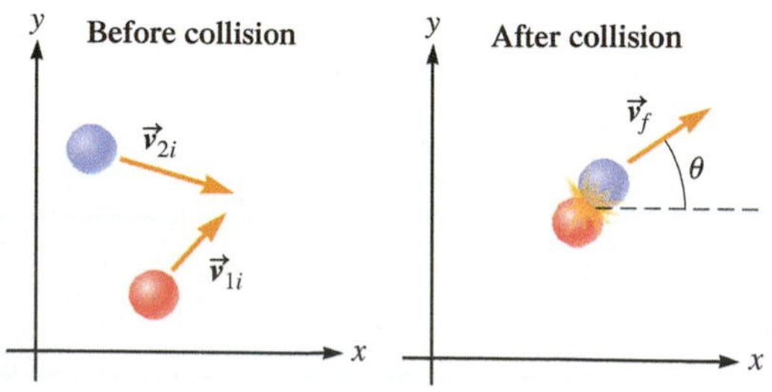

FIGURE 11.22 A two-dimensional completely inelastic collision.

quantities. Further, the conservation of momentum equations are simplified. For the collision shown in Figure 11.22, conservation of momentum in component form is

$$m_1 v_{1ix} + m_2 v_{2ix} = (m_1 + m_2) v_{fx} = M v_{fx} \quad (11.24)$$

$$m_1 v_{1iy} + m_2 v_{2iy} = (m_1 + m_2) v_{fy} = M v_{fy} \quad (11.25)$$

Sometimes, it is convenient to rewrite Equations 11.24 and 11.25 in terms of their common final speed v_f and the angle θ:

$$m_1 v_{1ix} + m_2 v_{2ix} = M v_f \cos\theta \quad (11.26)$$

$$m_1 v_{1iy} + m_2 v_{2iy} = M v_f \sin\theta \quad (11.27)$$

EXAMPLE 11.8 Molecular Collision

A hydrogen molecule (H_2) of mass $m_1 = 3.35 \times 10^{-27}$ kg and an oxygen atom (O) of mass $m_2 = 2.66 \times 10^{-26}$ kg collide and form a water molecule (H_2O) as in Figure 11.23. If the initial speeds of the H_2 molecule and O atom are $v_{1i} = 550$ m/s and $v_{2i} = 378$ m/s, respectively, find the velocity of the H_2O molecule after the collision.

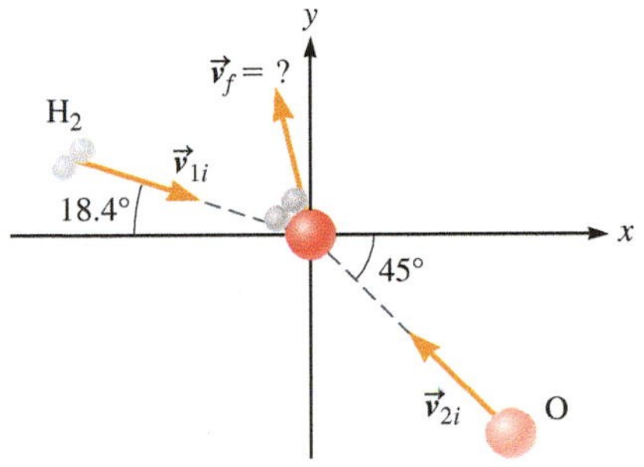

FIGURE 11.23

INTERPRET and ANTICIPATE

The two colliding objects stick together, so the collision is completely inelastic. Therefore, kinetic energy is not conserved, but momentum in each direction is conserved. Our result should be in the form $\vec{v}_f = (\_\_\_\_\,\hat{\imath} + \_\_\_\_\,\hat{\jmath})$ m/s.

SOLVE

Write each object's initial velocity in component form.

$$\vec{v}_{1i} = (v_{1i}\cos 18.4°)\hat{\imath} - (v_{1i}\sin 18.4°)\hat{\jmath} = (522\hat{\imath} - 174\hat{\jmath}) \text{ m/s}$$

$$\vec{v}_{2i} = (-v_{2i}\cos 45°)\hat{\imath} + (v_{2i}\sin 45°)\hat{\jmath} = (-267\hat{\imath} + 267\hat{\jmath}) \text{ m/s}$$

Find the total mass of the two-object system.

$$M = m_1 + m_2 = 3.35 \times 10^{-27} \text{ kg} + 2.66 \times 10^{-26} \text{ kg}$$

$$M = 2.99 \times 10^{-26} \text{ kg}$$

The two particles stick together. We express conservation of momentum in terms of the particles' common final speed v_f and the angle θ (Eqs. 11.24 and 11.25).

$$m_1 v_{1ix} + m_2 v_{2ix} = M v_{fx} \quad (11.24)$$

$$m_1 v_{1iy} + m_2 v_{2iy} = M v_{fy} \quad (11.25)$$

Solve Equations 11.24 and 11.25 for the components of the final velocity.

$$v_{fx} = \frac{m_1 v_{1ix} + m_2 v_{2ix}}{M}$$

$$v_{fx} = \frac{(3.35 \times 10^{-27} \text{ kg})(522 \text{ m/s}) + (2.66 \times 10^{-26} \text{ kg})(-267 \text{ m/s})}{2.99 \times 10^{-26} \text{ kg}} = -179 \text{ m/s}$$

$$v_{fy} = \frac{m_1 v_{1iy} + m_2 v_{2iy}}{M}$$

$$v_{fy} = \frac{(3.35 \times 10^{-27} \text{ kg})(-174 \text{ m/s}) + (2.66 \times 10^{-26} \text{ kg})(267 \text{ m/s})}{2.99 \times 10^{-26} \text{ kg}} = 218 \text{ m/s}$$

Write the result in component form.

$$\vec{v}_f = (-179\hat{\imath} + 218\hat{\jmath}) \text{ m/s}$$

CHECK and THINK

Our answer has the form we expected. Our analysis shows that the water molecule moves up and to the *left*. Our result makes sense because even though both objects have comparable initial speeds, the oxygen atom is about eight times more massive than the hydrogen molecule. Therefore, the oxygen atom's momentum makes up most of the system's initial momentum. After the collision, the motion of the water molecule is nearly in the same direction as the oxygen atom's initial motion.

SUMMARY

Underlying Principles

1. Another statement of Newton's second law is the **impulse–momentum theorem**:
$$\vec{I}_{\text{tot}} = \sum \vec{I} = \Delta \vec{p} \qquad (11.4)$$
2. **Conservation of momentum** (Chapter 10): If the impulse approximation holds, the total momentum of the system before the collision must equal the total momentum of the system after the collision:
$$(\vec{p}_{\text{tot}})_i = (\vec{p}_{\text{tot}})_f \qquad (11.5)$$
Equation 11.5 can be written in terms of the center-of-mass momentum:
$$(\vec{p}_{\text{CM}})_i = (\vec{p}_{\text{CM}})_f \qquad (11.6)$$

Major Concepts

1. A **collision** is an isolated event in which two or more objects exert relatively strong forces on one another for a relatively short time.
2. The **impulse approximation** assumes the force between the colliding objects during the time of the collision is much stronger than any other force exerted on either of them.
3. The **impulse** exerted on an object during a time interval Δt is a vector defined as
$$\vec{I} \equiv \int_{t_i}^{t_f} \vec{F}(t)dt = \vec{F}_{\text{av}}\Delta t \qquad (11.3)$$
4. In an **inelastic collision**, the system of colliding objects loses kinetic energy.
5. In a **completely inelastic collision**, the system of colliding objects loses kinetic energy, and after the collision, the objects are stuck together.
6. An **elastic collision** is an ideal case in which the system of colliding objects conserves its total kinetic energy.

Special Cases

1. Momentum in a completely **inelastic one-dimensional** collision with a stationary target:
$$m_1 v_{1i}\hat{\imath} = (m_1 + m_2)v_f\hat{\imath} = Mv_f\hat{\imath} \qquad (11.13)$$
2. Final velocities in an **elastic one-dimensional** collision:
$$\vec{v}_{1f} = \frac{m_1 - m_2}{M}\vec{v}_{1i} + \frac{2m_2}{M}\vec{v}_{2i} \qquad (11.14)$$
$$\vec{v}_{2f} = \frac{2m_1}{M}\vec{v}_{1i} + \frac{m_2 - m_1}{M}\vec{v}_{2i} \qquad (11.15)$$
3. Conservation of momentum for two-object, two-**dimensional collision**:
$$m_1 v_{1ix} + m_2 v_{2ix} = m_1 v_{1fx} + m_2 v_{2fx} \qquad (11.21)$$
$$m_1 v_{1iy} + m_2 v_{2iy} = m_1 v_{1fy} + m_2 v_{2fy} \qquad (11.22)$$
4. Conservation of kinetic energy for two-object elastic collision:
$$\tfrac{1}{2}m_1 v_{1i}^2 + \tfrac{1}{2}m_2 v_{2i}^2 = \tfrac{1}{2}m_1 v_{1f}^2 + \tfrac{1}{2}m_2 v_{2f}^2 \qquad (11.19)$$

PROBLEMS AND QUESTIONS

A = algebraic C = conceptual E = estimation G = graphical N = numerical

11-1 What Is a Collision?

1. **C** When a spacecraft collides with a planet, it is not necessary for them to actually touch each other, but when a car collides with a truck, the car and the truck must touch. Explain the difference between these two types of collisions.

11-2 Impulse

2. **C** When a person feels that he is about to fall, he will often put out his hand to try to "break the fall." Explain why this natural reaction usually leads to bruises or minor broken bones such as in the wrists instead of major broken bones such as the skull.
3. **N** A tall man walking at 1.25 m/s accidentally bumps his head of mass 3.10 kg on a steel doorjamb. His head stops in 0.010 s. What is the magnitude of the average force exerted by the doorjamb on the man's head? As part of the **CHECK and THINK** step, explain how padding the doorjamb can help reduce this force.
4. **N** A 35.0-kg child steps off a 4.0-ft-high diving board and executes a cannonball jump into a pool. (The child holds her body in a tight ball so that air resistance is negligible as she falls

downward.) What is the impulse exerted by the water on the child? In the **CHECK and THINK** step, explain why the child would be greatly injured if she were to land on the cement edge of the pool instead of in the water.

5. **N** A basketball of mass $m = 625$ g rolls off the hoop's rim, falls from a height of 3.05 m to the court's floor, and then bounces up to a height of 1.40 m. **a.** What are the magnitude and direction of the impulse delivered to the basketball by the floor? **b.** If the ball is in contact with the floor for 0.150 s, what is the average force exerted on the basketball by the floor?

6. **E** Two glass pickle jars fall off the same kitchen table and land on the floor. One jar is empty, and the other is full of (cucumber) pickles (and brine). Estimate the impulse exerted by the floor on each jar. As part of the **CHECK and THINK** step, answer the question: Which is more likely to break? Explain your answer.

7. **N** Sven hits a baseball ($m = 0.15$ kg). He applies an average force of 50.0 N. The ball had an initial velocity of 35.0 m/s to the right and a final velocity of 40.0 m/s to the left as viewed by a fan in the stands. **a.** What is the impulse delivered by Sven's bat to the baseball? **b.** How long is his bat in contact with the ball?

8. **C** A car's air bag helps protect the driver in the case of a collision. **a.** If a driver traveling at a particular speed hits an obstacle and comes to rest, what factors influence the total impulse the driver experiences while coming to rest? **b.** The air bag increases the time over which the collision takes place compared to a collision without an air bag present. How does this increase help protect the occupants of the car?

9. **N** A 65.0-kg driver of a vehicle traveling with a velocity of $11.5\hat{\imath}$ m/s in a parking lot hits a parked car and comes to rest in 0.150 s. **a.** Find the impulse experienced by the driver. **b.** Find the average force acting on the driver during the collision.

10. In a laboratory, a cart collides with a wall and bounces back. Figure P11.10 shows a graph of the force exerted by the wall versus time.

a. **N** Find the impulse exerted by the wall on the cart.

b. **N** What is the average force exerted by the wall on the cart?

c. **N** If the cart has a mass of 0.448 kg, what is its change in velocity?

d. **C** Make a sketch of the situation. Include a coordinate system and explain the significance of the signs in parts (a) through (c).

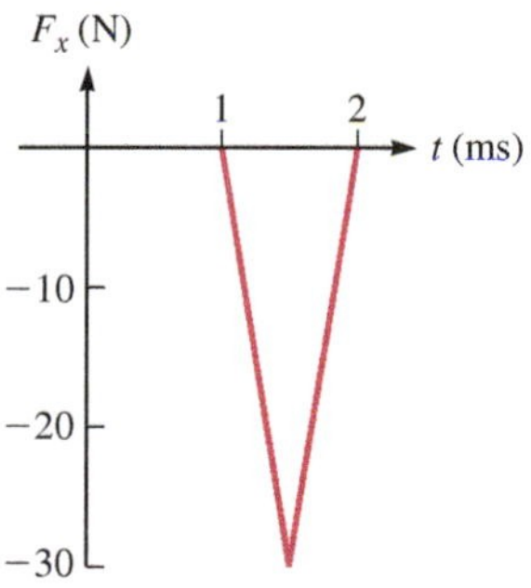

FIGURE P11.10

11. **N** An empty bucket is placed on a scale, and the scale is reset to read 0.00 N. Water falls from a faucet, high overhead, into the bucket without splashing at a rate of 350 mL/s. What is the reading on the scale 4.50 s after the water first hits the bottom of the empty bucket if at that time the falling water travels 1.25 m before hitting the water already in the bucket?

12. **A** Show that Equation 11.4 (the **impulse–momentum theorem)** is another statement of Newton's second law.

13. **C** A crate of mass M is initially at rest on a level, frictionless table. A small block of mass m ($m < M$) moves toward the crate as shown in Figure P11.13. After the collision, the block sticks to the crate. Is the magnitude of the impulse exerted on the crate by the block greater than, less than, or equal to the magnitude of the impulse exerted on the block by the crate during the collision process? Explain.

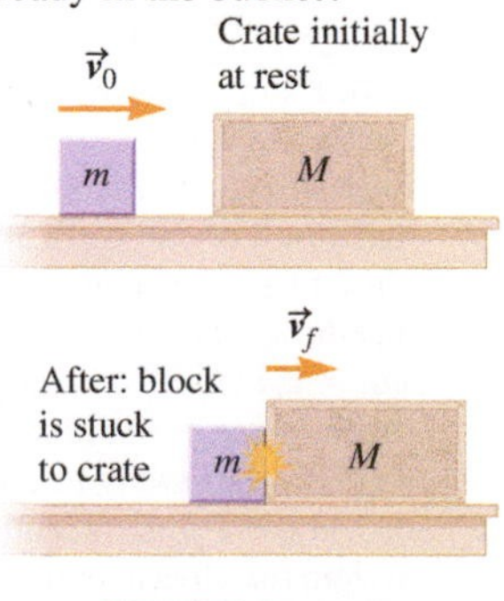

FIGURE P11.13

11-3 Conservation During a Collision

14. **C** Two students, Cameron and Avi, are riding in a small car and have the following discussion.

Avi: If I shoot a cue ball and hit another ball that is just resting there, the cue ball stops and the other ball is propelled at the same speed as the incoming cue ball. So, if a small car rear ends a parked truck, the car would stop, and the truck would be propelled forward at the same speed the car had been traveling.

Cameron: No way, and don't even think of trying that one. The truck has more mass, so it experiences a smaller impulse during the collision. The small car would experience a huge impulse.

Do you agree with either of these students? Explain.

15. **N** Two pucks in a laboratory are placed on an air table. Puck 1 has twice the mass of puck 2. They are pushed toward each other and strike in a head-on collision. Initially, puck 2 is twice as fast as puck 1. **a.** What is the total momentum before the collision? **b.** What is the center-of-mass velocity before the collision? **c.** If the pucks are initially 2.70 m apart, how far did puck 1 move before the collision?

16. **C** A truck collides with a small, empty parked car. Explain your answers to the parts below. **a.** Compare the force exerted by the truck on the car with the force exerted by the car on the truck. **b.** Compare the impulse exerted by the truck on the car with the impulse exerted by the car on the truck. **c.** Compare the change in the truck's momentum with the change in the car's momentum.

Problems 17 and 18 are paired.

17. **N** A comet is traveling through space with a speed of 3.33×10^4 m/s when it collides with an asteroid that was at rest. The comet and the asteroid stick together during the collision process. The mass of the comet is 1.11×10^{14} kg, and the mass of the asteroid is 6.66×10^{20} kg. **a.** What is the speed of the center of mass of the asteroid–comet system before the collision? **b.** What is the speed of the system's center of mass after the collision?

18. **N** For the comet and asteroid in Problem 17, **a.** what is the magnitude of the momentum of the system's center of mass before the collision? **b.** What is the magnitude of the momentum of the system's center of mass after the collision?

Problems 19 and 20 are paired.

19. **A** A skater of mass m standing on ice throws a stone of mass M with speed v in a horizontal direction. Find the distance over which the skater will move in the opposite direction if the coefficient of kinetic friction between the skater and the ice is μ_k.

20. **N** A skater of mass 45.0 kg standing on ice throws a stone of mass 7.65 kg with a speed of 20.9 m/s in a horizontal direction. Find the distance over which the skater will move in the opposite direction if the coefficient of kinetic friction between the skater and the ice is 0.03.

11-4 Special Case: One-Dimensional Inelastic Collisions

21. **N** An object of mass 2.0 kg moving with a velocity of 3.0 m/s collides with another object of mass 1.0 kg moving with a velocity of 4.0 m/s in the same direction. The two objects get stuck together in the collision. What is the velocity of the combination after the collision?

22. **A** In a laboratory experiment, **1** a block of mass M is placed on a frictionless table at the end of a relaxed spring of spring constant k. **2** The spring is compressed a distance x_0 and **3** a small ball of mass m is launched into the block as shown in Figure P11.22. The ball and block stick together and are projected off the table of height h. Find an expression for the horizontal displacement ℓ of the ball–block system from the end of the table until it hits the floor in terms of the parameters given.

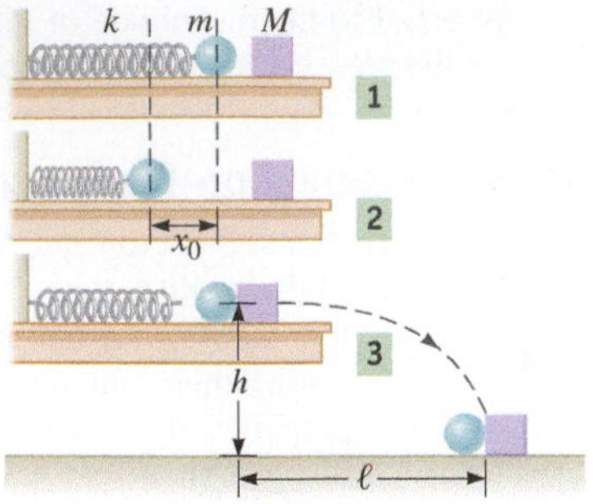

FIGURE P11.22

23. **N** Ezra (m = 25.0 kg) has a tire swing and wants to swing as high as possible. He thinks that his best option is to run as fast as he can and jump onto the tire at full speed. The tire has a mass of 10.0 kg and hangs 3.75 m straight down from a tree branch. Ezra stands back 10.0 m and accelerates to a speed of 3.50 m/s before jumping onto the tire swing. **a.** How fast are Ezra and the tire moving immediately after he jumps onto the swing? **b.** How high does the tire travel above its initial height?

24. **E** A suspicious physics student watches a stunt performed at an ice show. In the stunt, a performer shoots an arrow into a bale of hay (Fig. P11.24). Another performer rides on the bale of hay like a cowboy. After the arrow enters the bale, the bale–arrow system slides roughly 5 m along the ice. Estimate the initial speed of the arrow. Is there a trick to this stunt?

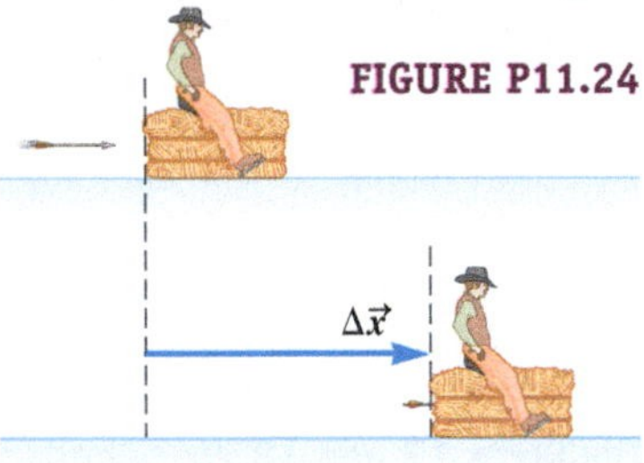

FIGURE P11.24

25. **N** A 2.45-kg ball is shot into a 0.450-kg box that is at rest on a frictionless, horizontal table (Fig. P11.25); after the collision, the ball is embedded in the box. The box is attached to a 0.30-m rope that is attached to the table on the other end. The ball's initial velocity is perpendicular to the rope as shown. If the ball's initial speed before impact is 13.5 m/s, what is the tension in the rope after the collision?

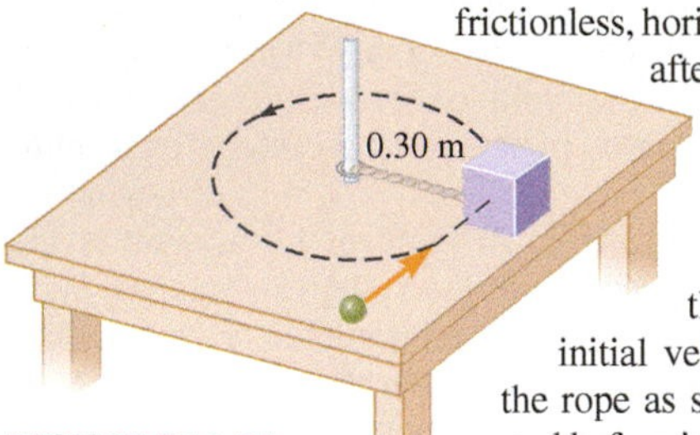

FIGURE P11.25

26. **A** In an attempt to overtake a slow truck of mass $3m$ moving with speed v_T, a car with mass m initially moving with speed v_C in the same direction instead collides and locks bumpers with the truck. **a.** What is the final speed of the car and the truck immediately after the collision? **b.** What is the change in kinetic energy of the car–truck system during the collision?

Problems 27 and 28 are paired.

27. **N** Jeff and Zak have worked out a stunt-show performance involving hovercrafts on a collision course. Just before the collision, Jeff and Zak will safely leap out of the hovercrafts, but they still have a little physics to work out. The two hovercrafts, each with a mass of 966 kg, are hovering at a constant height above the ground and are traveling toward each other as shown in Figure P11.27. The first hovercraft is traveling at 15.0 m/s to the right, and the second is traveling at 20.0 m/s to the left. **a.** If the hovercrafts undergo a completely inelastic collision, what is their final velocity after the collision? (Report to two significant figures.) **b.** How much kinetic energy is lost in the collision? (Report to three significant figures.)

FIGURE P11.27 Problems 27 and 28.

28. **C** The two (unoccupied) hovercrafts, each with a mass of 966 kg, are hovering at a constant height above the ground and are traveling toward each other as shown in Figure P11.27. The first hovercraft is traveling at 15.0 m/s to the right, and the second is traveling at 20.0 m/s to the left. Suppose the two hovercrafts collide, but the collision is not completely inelastic. What bounds can you put on the kinetic energy lost in the collision process? Explain your answer.

Problems 29 and 30 are paired.

29. **A** A dart of mass m is fired at and sticks into a block of mass M that is initially at rest on a rough, horizontal surface. The coefficient of kinetic friction between the block and the surface is μ_k. After the collision, the dart and the block slide a distance D before coming to rest. If the dart were fired horizontally, what would its speed be immediately before impact with the block?

30. **N** A dart of mass m = 10.0 g is fired at and sticks into a block of mass M = 85.0 g that is initially at rest on a rough, horizontal surface. The coefficient of kinetic friction between the block and the surface is 0.400. After the collision, the dart and the block slide a distance of 6.00 m before coming to rest. If the dart were fired horizontally, what would its speed be immediately before impact with the block?

31. **N** A bullet of mass m = 8.00 g is fired into and embeds itself in a large 1.50-kg block of wood, initially at rest. What was the original speed of the bullet if that block with the embedded bullet were moving at a speed of 1.10 m/s immediately after the collision?

32. **N Review** CASE STUDY In Example 11.4, we used the conservation of energy principle to find the velocity of the two-train system just after impact. Now, use Newton's second law to find v_c, the velocity of the system just after the collision. Compare your result with that found in Example 11.4.

33. **A** A bullet of mass m is fired into a ballistic pendulum and embeds itself in the wooden bob of mass M (Fig. P11.33). After the collision, the pendulum reaches a maximum height h above its original position. **a.** Show that the kinetic energy of the system decreases by the factor $m/(m + M)$ immediately after the collision. **b.** What is the change in momentum of the bullet-bob system due to the collision?

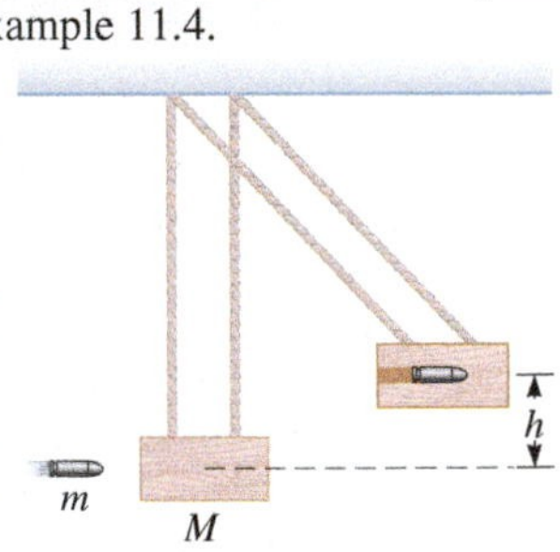

FIGURE P11.33

34. **N** In Examples 8.9 (page 230) and 9.8 (page 267), we found the maximum height of a small ball launched straight up from a spring-loaded dart gun. In this problem, we find the muzzle speed v_m of the dart gun. The dart gun is held horizontally such that its muzzle is very close to an open container of modeling dough (called *Fun Doh*) so that the ball hits the Fun Doh at the muzzle velocity. The ball of mass $m_1 = 2.7 \times 10^{-2}$ kg is fired into the packed Fun Doh container of total mass m_2 = 0.15 kg. The container is suspended from a string so that it can swing as a pendulum bob. The ball becomes embedded in the Fun Doh, and the center of mass of the ball–Fun Doh system rises to a maximum height h = 0.26 m. Assume dissipative forces may be ignored. Find the muzzle speed of the dart gun.

11-5 One-Dimensional Elastic Collisions

35. **N** One object ($m_1 = 0.200$ kg) is moving to the right with a speed of 2.00 m/s when it is struck from behind by another object ($m_2 = 0.300$ kg) that is moving to the right at 6.00 m/s. If friction is negligible and the collision between these objects is elastic, find the final velocity of each.

Problems 36, 37, and 38 are grouped.

36. **C** Particle 1 collides head-on with particle 2, which is initially at rest. After the collision, particle 1 has *reversed* direction and is moving at nearly the same speed it had before the collision. What can you say about the relative mass of the particles and particle 2's velocity after the collision? Explain.

37. **C** Particle 1 collides head-on with particle 2, which is initially at rest. After the collision, particle 1 *continues moving* at nearly its original velocity. What can you say about the relative mass of the particles and particle 2's velocity after the collision? Explain.

38. **C** Particle 1 collides head-on with particle 2, which is initially at rest. After the collision, particle 1 has *stopped*. What can you say about the relative mass of the particles and particle 2's velocity after the collision? Explain.

39. **N** Two objects collide head-on (Fig. P11.39). The first object is moving with an initial speed of 8.00 m/s, and the second object is moving with an initial speed of 10.00 m/s. Assuming the collision is elastic, $m_1 = 5.15$ kg, and $m_2 = 6.25$ kg, determine the final velocity of each object.

FIGURE P11.39

Problems 40 and 41 are paired.

40. **A** Initially, ball 1 rests on an incline of height h, and ball 2 rests on an incline of height $h/2$ as shown in Figure P11.40. They are released from rest simultaneously and collide in the trough of the track. If $m_2 = 4\,m_1$ and the collision is elastic, find an expression for the velocity of each ball immediately after the collision.

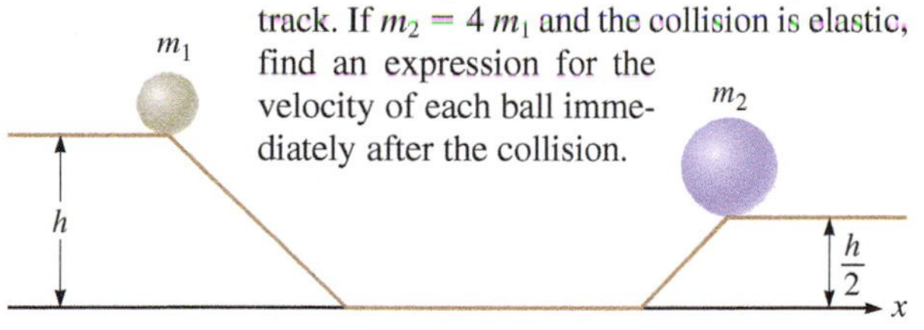

FIGURE P11.40 Problems 40 and 41.

41. **N** Initially, ball 1 rests on an incline of height h, and ball 2 rests on an incline of height $h/2$ as shown in Figure P11.40. They are released from rest simultaneously and collide elastically in the trough of the track. If $m_2 = 4\,m_1$, $m_1 = 0.045$ kg, and $h = 0.65$ m, what is the velocity of each ball after the collision?

42. **N** In an attempt to produce exotic new particles, a proton of mass $m_p = 1.67 \times 10^{-27}$ kg is accelerated to $0.99c$ ($c = 3.00 \times 10^8$ m/s is the speed of light) and crashed into a helium nucleus of mass $m_{\text{He}} = 6.64 \times 10^{-27}$ kg initially at rest. The collision is elastic. **a.** What is the kinetic energy of the helium nucleus after the collision? **b.** What is the kinetic energy of the proton after the collision? (In Chapter 39, we'll learn what Einstein says about making such calculations.)

43. Pendulum bob 1 has mass m_1. It is displaced to height h_1 and released. Pendulum bob 1 elastically collides with pendulum bob 2 of mass m_2 (Fig. P11.43).

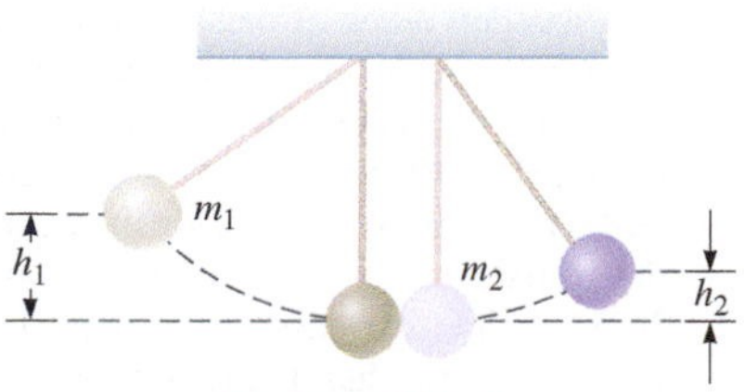

FIGURE P11.43

a. **A** Find an expression for the maximum height h_2 of pendulum bob 2.

b. **N** If $m_2 = 2.5m_1$ and $h_1 = 5.46$ m, what is h_2?

Problems 44, 45, and 46 are grouped.

44. **N** Crall and Whipple observe two carts collide elastically in a laboratory experiment. Cart 1 has a mass of 1.504 kg, and cart 2 has a mass of 1.040 kg. Crall chooses to work in the laboratory frame. Before the collision in his frame, the carts' speeds are $\vec{v}_{1i} = 0.353\hat{\imath}$ m/s and $\vec{v}_{2i} = -0.294\hat{\imath}$ m/s. According to Crall, what are the velocities of the carts after the collision?

45. **N** Refer to Crall and Whipple's experiment in Problem 44. Whipple chooses to do his calculations in a frame moving at cart 2's initial velocity. According to Whipple, what are the velocities of the carts before and after the collision?

46. **N** Refer to Crall and Whipple's experiment in Problem 44. Consider the two carts to make up a single system. What is that system's center-of-mass velocity before and after the collision?

47. **N** A roller-coaster car of mass $m_1 = 8.00 \times 10^2$ kg starts from rest and slides on a frictionless track from point A to point B at a height $h = 10.0$ m below point A, where it collides elastically with a second car of mass $m_2 = 2.00 \times 10^2$ kg, initially at rest. What is the maximum height that the second car rises above point B on the track?

48. **E** In Example 11.6, we found the change in a spacecraft's speed during a gravitational slingshot with a planet. In working that example, we assumed the planet does not lose any kinetic energy. In this problem, we test that assumption. In 1972, *Pioneer 10* was launched. It was the first spacecraft to take close-up images of Jupiter. It was launched from the Earth at a very high speed, fast enough to pass our Moon in just 11 hours. It approached Jupiter at a speed of 14 km/s. The gravitational slingshot with Jupiter increased its speed to 37 km/s. The mass of *Pioneer 10* is 270 kg, and Jupiter's mass is 1.9×10^{27} kg. Find the decrease in Jupiter's kinetic energy and speed as a result of this slingshot. Jupiter's speed is about 1.3×10^4 m/s. Comment on your findings.

49. **N** Two skateboarders, with masses $m_1 = 75.0$ kg and $m_2 = 65.0$ kg, simultaneously leave the opposite sides of a frictionless half-pipe at height $h = 4.00$ m as shown in Figure P11.49. Assume the skateboarders undergo a completely elastic head-on collision on the horizontal segment of the half-pipe. Treating the skateboarders as particles and assuming they don't fall off their skateboards, what is the height reached by each skateboarder after the collision?

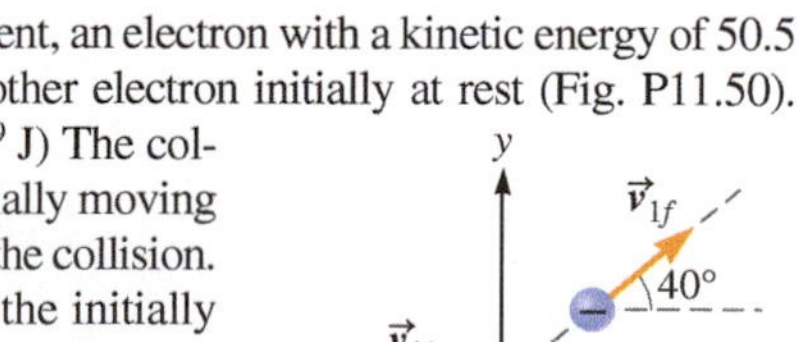

FIGURE P11.49

11-6 Two-Dimensional Collisions

50. In a laboratory experiment, an electron with a kinetic energy of 50.5 keV is shot toward another electron initially at rest (Fig. P11.50). (1 eV $= 1.602 \times 10^{-19}$ J) The collision is elastic. The initially moving electron is deflected by the collision.

a. **C** Is it possible for the initially stationary electron to remain at rest after the collision? Explain.

b. **N** The initially moving electron is detected at an angle of 40.0° from its original path. What is the speed of each electron after the collision?

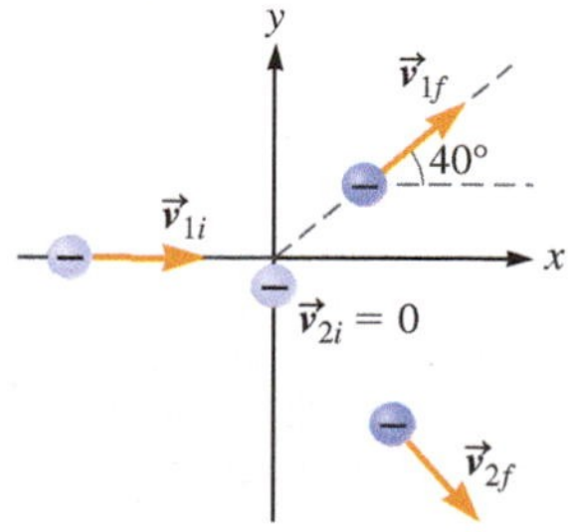

FIGURE P11.50

51. N In Figure P11.51, a cue ball is shot toward the eight-ball on a pool table. The cue ball is shot at the eight-ball with a speed of 8.00 m/s in a direction 30.0° from the y axis. Both balls have the same mass of 0.170 kg. After the balls undergo an elastic collision, the eight-ball travels in the negative x direction into the side pocket. What is the velocity of the cue ball after this collision?

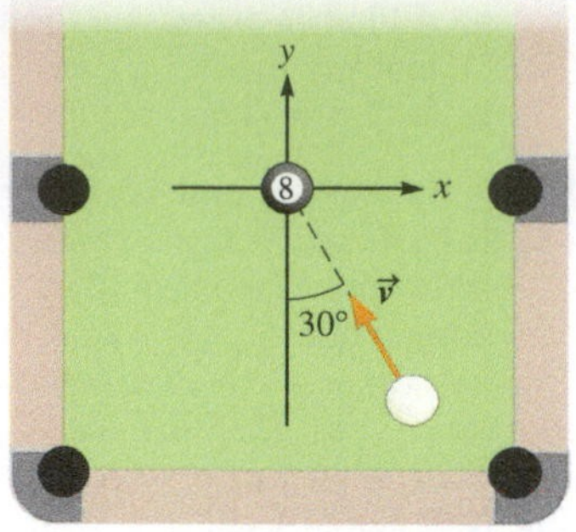

FIGURE P11.51

52. N A proton with an initial speed of 2.00×10^8 m/s in the x direction collides elastically with another proton initially at rest. The first proton's velocity after the collision is 1.64×10^8 m/s at an angle of 35.0° with the horizontal. What is the velocity of the second proton after the collision?

53. N A football player of mass 95 kg is running at a speed of 5.0 m/s down the field as shown in Figure P11.53. A second player of mass 140 kg, running at a speed of 2.5 m/s, tackles the first player so that they move together after the collision. What is the velocity of the two players immediately after the collision?

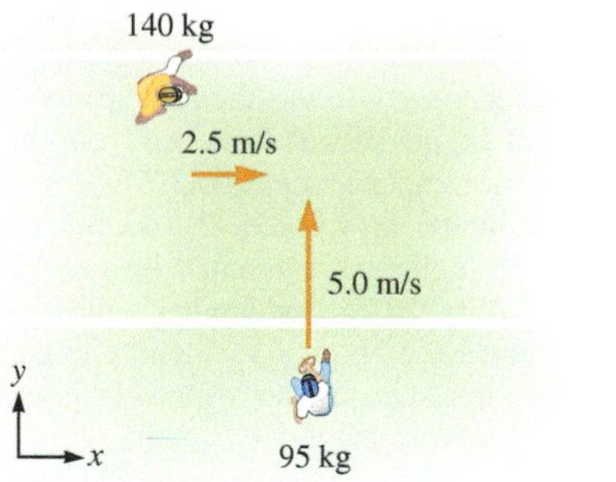

FIGURE P11.53

Problems 54 and 55 are paired.

54. N Two bumper cars at the county fair are sliding toward one another (Fig. P11.54). Initially, bumper car 1 is traveling to the east at 5.62 m/s, and bumper car 2 is traveling 60.0° south of west at 10.00 m/s. After they collide, bumper car 1 is observed to be traveling to the west with a speed of 3.14 m/s. Friction is negligible between the cars and the ground. **a.** If the masses of bumper cars 1 and 2 are 596 kg and 625 kg respectively, what is the velocity of bumper car 2 immediately after the collision? **b.** What is the kinetic energy lost in the collision?

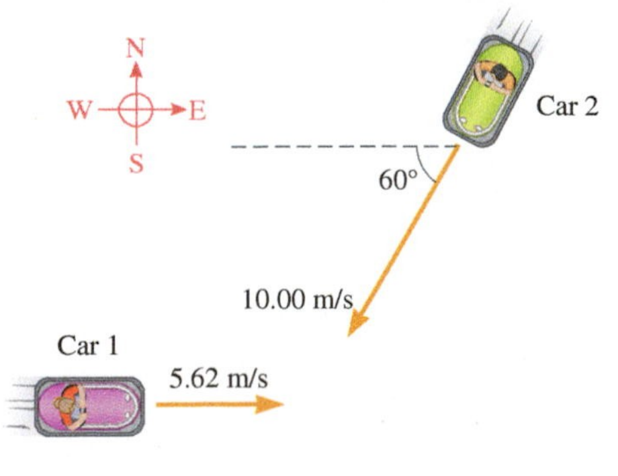

FIGURE P11.54 Problems 54 and 55.

55. Two bumper cars at the county fair are sliding toward one another (Fig. P11.54). Initially, bumper car 1 is traveling to the east at 5.62 m/s, and bumper car 2 is traveling 60.0° south of west at 10.00 m/s. They collide and stick together, as the driver of one car reaches out and grabs hold of the other driver. The two bumper cars move off together after the collision, and friction is negligible between the cars and the ground.
   a. N If the masses of bumper cars 1 and 2 are 596 kg and 625 kg respectively, what is the velocity of the bumper cars immediately after the collision?
   b. N What is the kinetic energy lost in the collision?
   c. C Compare your answers to part (b) from this and Problem 54. Is one answer larger than the other? Discuss and explain any differences you find.

56. N A police officer is attempting to reconstruct an accident in which a car traveling southward with a speed of 23.0 mph collided with another car of equal mass traveling eastward at an unknown speed. After the collision, the two cars coupled and slid at an angle of 60.0° south of east. If the speed limit in that neighborhood is 25.0 mph, should the officer cite the second driver for speeding? Explain your answer.

57. N A bomb explodes into three pieces A, B, and C of equal mass. Piece A flies with a speed of 40.0 m/s, and piece B with a speed of 30.0 m/s at an angle of 90° relative to the direction of A as shown in Figure P11.57. Determine the speed of piece C and the direction of its velocity relative to the direction of piece A.

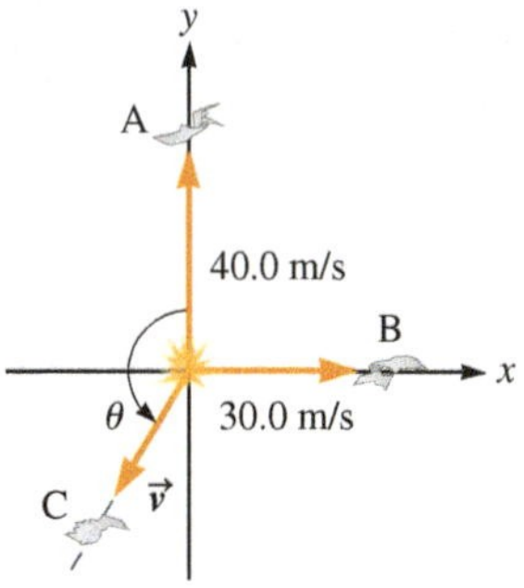

FIGURE P11.57

58. N The spontaneous decay of a heavy atomic nucleus of mass $M = 14.5 \times 10^{-27}$ kg that is initially at rest produces three particles. The first particle, with mass $m_1 = 6.64 \times 10^{-27}$ kg, is ejected in the positive x direction with a speed of 2.50×10^7 m/s. The second particle, with mass $m_2 = 3.00 \times 10^{-27}$ kg, is ejected in the negative y direction with a speed of 2.00×10^7 m/s. **a.** What is the velocity of the third particle? **b.** What is the increase in the kinetic energy of the system after the decay?

General Problems

59. N An object of mass m = 4.00 kg that is moving with a speed of 10.0 m/s collides head-on with another object, and the collision lasts 1.50 s. A graph showing the magnitude of the force during the collision versus time is shown in Figure P11.59, where the force is exerted in the direction opposite the initial velocity. Find the speed of the 4.00-kg mass after collision.

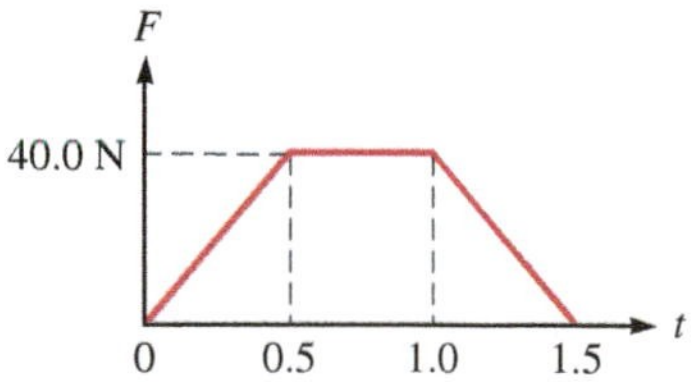

FIGURE P11.59

60. A A wooden block of mass M is initially at rest at the edge of a frictionless table at a height h above the ground. A bullet of mass m is fired horizontally into the block and embeds itself in the block. The block lands a distance d from the edge of the table. Find an expression for the speed of the bullet just before the collision.

61. N A particle of mass m_1 = 1.50 kg with an initial velocity $\vec{v}_1 = 4.50\hat{\jmath}$ m/s has a completely inelastic collision with a second particle of mass m_2 = 3.50 kg with an initial velocity

$\vec{v}_2 = 3.00\hat{\imath}$. What is the velocity of the combined particles immediately after the collision?

62. A person jumping on a trampoline lands at time $t = 0$ and experiences a force given by $F = 280(1.8 - t)t$. The time t is measured in seconds, and the resulting force is in newtons. After 1.8 s, the force is zero, and the person is launched off of the trampoline.

a. G Sketch the magnitude of the force as a function of time.

b. N What is the magnitude of the total impulse in the time interval from 0 to 1.8 s?

63. N In an experiment designed to determine the velocity of a bullet fired by a certain gun, a wooden block of mass $M = 500.0$ g is supported only by its edges, and a bullet of mass $m = 8.00$ g is fired vertically upward into the block from close range below. After the bullet embeds itself in the block, the block and the bullet are measured to rise to a maximum height of 25.0 cm above the block's original position. What is the speed of the bullet just before impact?

64. From what might be a possible scene in the comic book *The X-Men*, the Juggernaut (m_J) is charging into Colossus (m_C) and the two collide. The initial speed of the Juggernaut is v_{Ji} and the initial speed of Colossus is v_{Ci}. After the collision, the final speed of the Juggernaut is v_{Jf} and the final speed of Colossus is v_{Cf} as they each bounce off of the other, heading in opposite directions.

a. A What is the impulse experienced by the Juggernaut?

b. A What is the impulse experienced by Colossus?

c. C In your own words, explain how these impulses must compare with each other and how they are related to the average force each superhero experiences during the collision.

65. N A 110-kg rugby player running east with a speed of 4.00 m/s tackles an 85.0-kg opponent running north with a speed of 3.50 m/s. Assume the tackle is a perfectly inelastic collision. **a.** What is the velocity of the players immediately after the tackle? **b.** What is the amount of mechanical energy lost during the collision?

Problems 66 and 67 are paired.

66. Two pucks in a laboratory are placed on an air table (Fig. P11.66). Puck 2 has four times the mass of puck 1 ($m_2 = 4m_1$). Initially, puck 1's speed is three times puck 2's speed ($v_{1i} = 3v_{2i}$), puck 1's position is $\vec{r}_{1i} = -x_{1i}\hat{\imath}$, and puck 2's position is $\vec{r}_{2i} = -y_{2i}\hat{\jmath}$. The pucks collide at the origin.

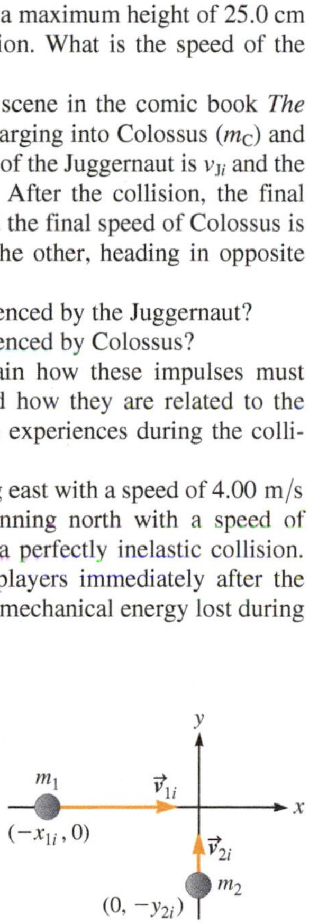

FIGURE P11.66 Problems 66 and 67.

a. G Copy Figure P11.66 and then add $\vec{v}_{CM}$ to your sketch.

b. C Does puck 2 travel a greater distance, lesser distance, or the same distance as puck 1?

c. A Find an expression for y_{2i} in terms of x_{1i}.

d. N If puck 1 moves 1.33 m, how far does puck 2 move before the collision?

67. N Assume the pucks in Figure P11.66 stick together after their collision at the origin. Puck 2 has four times the mass of puck 1 ($m_2 = 4m_1$). Initially, puck 1's speed is three times puck 2's speed ($v_{1i} = 3v_{2i}$), puck 1's position is $\vec{r}_{1i} = -x_{1i}\hat{\imath}$, and puck 2's position is $\vec{r}_{2i} = -y_{2i}\hat{\jmath}$. **a.** Find an expression for their velocity after the collision in terms of puck 1's initial velocity. **b.** What is the fraction K_f/K_i that remains in the system?

68. C Two objects travel toward each other, collide, and are motionless after the collision. Object 1 has a greater speed than object 2. **a.** Can you determine whether the collision was elastic or inelastic? **b.** Can you determine which object has more mass? **c.** Can you determine which object initially had a larger kinetic energy? Explain all your answers.

69. N An object of mass 4.0 kg collides elastically with a second object, initially at rest, and then continues to move in the original direction with one-half its original speed. What is the mass of the second object?

Problems 70 and 71 are paired.

70. N A ball of mass 50.0 g is dropped from a height of 10.0 m. It rebounds after losing 75% of its kinetic energy during the collision process. If the collision with the ground took 0.010 s, find the magnitude of the impulse experienced by the ball.

71. N A ball of mass 0.10 kg that is released from a height of 2.5 m above the ground rebounds to a height 0.65 m. The time of contact of the ball with the ground is 0.010 s. Find the magnitude of the average force exerted by the ball on the ground during the collision.

72. A A pendulum consists of a wooden bob of mass M suspended by a massless rod of length ℓ. A bullet of mass $m \ll M$ is fired horizontally with speed v at the bob and emerges from the bob with speed $v/3$ as shown in Figure P11.72. If the pendulum bob just barely reaches the highest point such that it is able to swing through one complete circle, find an expression for the speed v of the bullet before the collision.

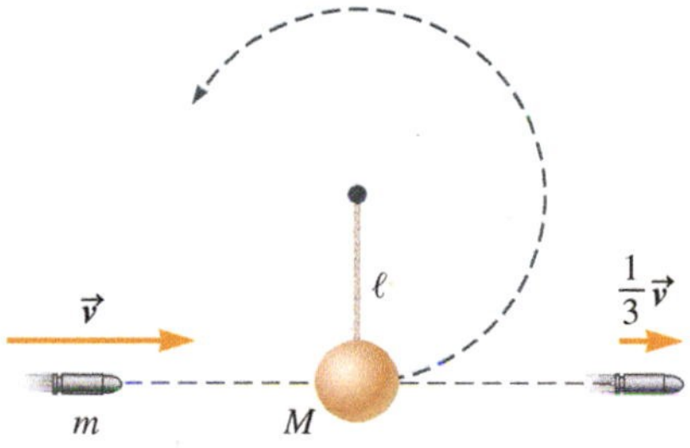

FIGURE P11.72

73. Three runaway train cars are moving on a frictionless, horizontal track in a railroad yard as shown in Figure P11.73. The first car, with mass $m_1 = 1.50 \times 10^3$ kg, is moving to the right with speed $v_1 = 10.0$ m/s; the second car, with mass $m_2 = 2.50 \times 10^3$ kg, is moving to the left with speed $v_2 = 5.00$ m/s, and the third car, with mass $m_3 = 1.20 \times 10^3$ kg, is moving to the left with speed $v_3 = 8.00$ m/s. The three railroad cars collide at the same instant and couple, forming a train of three cars.

a. N What is the final velocity of the train cars immediately after the collision?

b. C Would the answer to part (a) change if the three cars did not collide at the same instant? Explain.

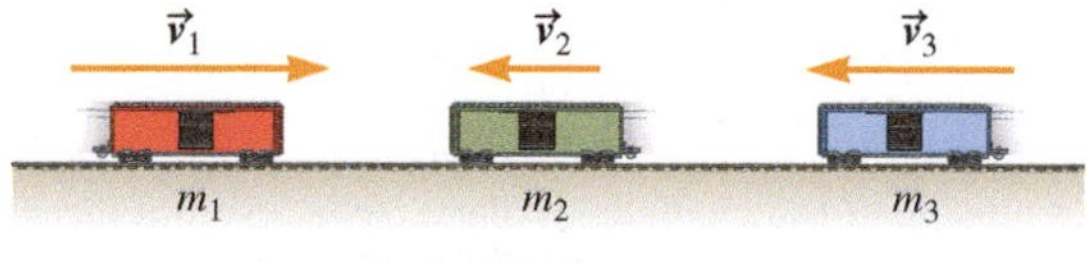

FIGURE P11.73

Problems 74 and 75 are paired.

74. In the early part of the 20th century, Sir Joseph J. Thomson (discoverer of the electron) proposed a "plum pudding" model of the atom. He believed that the positive charge of the atom was spread out like a pudding and that the negative charges (electrons) were embedded in the pudding like plums. His student Ernest Rutherford performed an experiment in 1911 that disproved the plum pudding model. Rutherford fired a beam of alpha particles (helium nuclei) at a thin metal sheet. Because alpha particles are positive, the plum pudding model predicted that they should only be slightly deflected by the positive pudding. Instead, Rutherford found that the alpha particles were greatly deflected, and some even reversed direction completely. Rutherford was surprised by his results and said that it was like firing a bullet at a tissue paper and seeing it bounce back. Rutherford concluded that the positive charge of the atom was not spread out like pudding, but rather was concentrated in the center or *nucleus* of the atom. The alpha particles used in the experiment had an initial speed of 2×10^7 m/s and a mass of 6.7×10^{-27} kg.

a. **N** Assuming the nucleus is initially at rest, the collision is head-on and elastic, and the nucleus is much more massive than the alpha particle, find the final speed of the alpha particle after it collides with the nucleus.

b. **C** Rutherford used gold in his experiment. Check the assumption that the nucleus is much more massive than the alpha particle by finding the speed of the nucleus after the collision.

75. **N** Rutherford fired a beam of alpha particles (helium nuclei) at a thin sheet of gold. An alpha particle was observed to be deflected by 90.0°; its speed was unchanged. The alpha particles used in the experiment had an initial speed of 2×10^7 m/s and a mass of 6.7×10^{-27} kg. Assume the alpha particle collided with a gold nucleus that was initially at rest. Find the speed of the nucleus after the collision.

Problems 76 and 77 are paired.

76. A dramatic (and perhaps unexpected) collision occurs if you hold a tennis ball on top of a basketball and drop them at the same time. After impacting the ground, the tennis ball can be launched much higher than the original height of the two balls. Assume this process can be modeled as an elastic collision of the basketball with the ground followed by a second elastic collision of the basketball with the tennis ball. The basketball (with mass m_b) and tennis ball (with mass m_t) each falls a distance h with an acceleration g before the basketball hits the ground.

a. **A** What is the velocity of each ball the instant before the basketball collides with the ground in terms of the variables specified?

b. **A** Assume the basketball first undergoes an elastic collision with the ground. How fast is the basketball moving after this collision?

c. **A** The basketball then collides elastically with the tennis ball. What is the final velocity of the tennis ball?

d. **N** Assume the basketball has a mass that is eight times that of the tennis ball. What is the ratio of the final speed of the tennis ball to the speed of the balls just before impact?

77. **N** Refer to the dramatic collision described in Problem 76. The basketball and the tennis ball have masses $m_b = 0.45$ kg and $m_t = 57$ g, respectively, and fall from a height of $h = 150$ cm. What is the maximum height reached by the tennis ball after the balls collide with the ground?

78. **E** February 3, 2009, was a very snowy day along Interstate 69 just outside of Indianapolis, Indiana. As a result of the slippery conditions and low visibility (50 yards or less), there was an enormous accident involving approximately 30 vehicles, including cars, tractor-trailers, and even a fire truck. Many witnesses said that people were driving too fast for the conditions and were too close together. In this problem, we explore two rules of thumb for driving in such conditions. The first is to drive at a speed that is half of what it would be in ideal conditions. The other is the "8-second" rule: Watch the vehicle in front of you as it passes some object such as a street sign, and you should pass that same object 8 seconds later. On a dry road, the 8-second rule is replaced by a 3-second rule. **a.** Assume vehicles on a slippery interstate highway follow both rules. What is the distance between the vehicles? **b.** If a driver followed the first rule of thumb, driving at a lower speed, but used the 3-second rule instead of the 8-second rule, what is the distance between the vehicles? How does that distance compare with the visibility on the day of the accident? **c.** Suppose drivers do not follow either rule of thumb for slippery conditions. What is the distance between vehicles? How does that distance compare with the visibility on that day? **d.** Suppose a driver was not obeying either rule of thumb when she sees a tractor-trailer that stopped on the highway. She presses on her brakes, locking the wheels, and her car crashes into the truck. Estimate the magnitude of the impulse exerted on her car. **e.** Estimate the impulse on the car in part (d) had the driver followed both rules of thumb for slippery conditions instead of ignoring them.

Problems 79 and 80 are paired.

79. **N** A cart filled with sand rolls at a speed of 1.0 m/s along a horizontal path without friction. A ball of mass $m = 2.0$ kg is thrown with a horizontal velocity of 8.0 m/s toward the cart as shown in Figure P11.79. The ball gets stuck in the sand. What is the velocity of the cart after the ball strikes it? The mass of the cart is 15 kg.

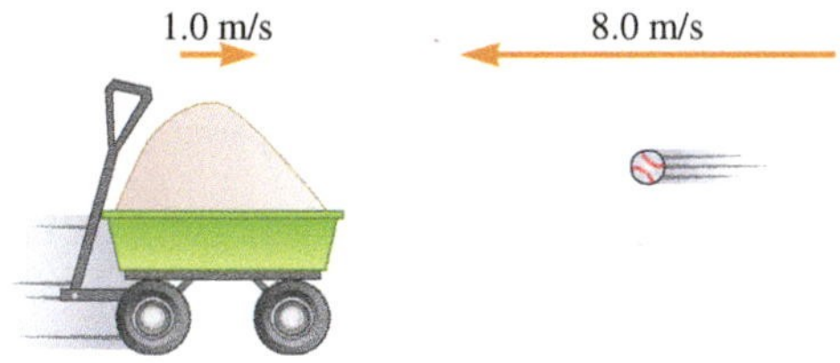

FIGURE P11.79 Problems 79 and 80.

80. Consider the cart, ball, and sand in Problem 79.

a. **N** What are the initial and final kinetic energies of the system? Is the energy conserved in this collision process?

b. **C** What type of collision process is it?

81. A ball of mass m moving in the positive x direction with speed v_i experiences an elastic glancing collision with a second ball of mass $3m$ moving in the opposite direction with the same speed as the first ball. After the collision, the first ball moves in the positive y direction.

a. **A** What are the speeds of the two balls in terms of their initial speeds?

b. **N** What is the angle θ at which the second ball travels after the collision, measured with respect to the original direction of the first ball (the positive x direction)?

82. **A** The force on a particle is given by $\vec{F} = F_{max} \cos(2\pi t/T)\hat{\imath}$, where F_{max} and T are constants. Find expressions for the impulse on the particle during each of the following intervals. **a.** $0 < t < T/4$ **b.** $T/4 < t < 3T/4$ **c.** $0 < t < T$

83. **A** Consider a system of two colliding objects, and derive the conservation of momentum (Equation 11.5) by considering the impulse on each object.

84. **N** Return to Example 11.8 (page 323), and find the system's center-of-mass velocity before and after the collision.

Rotation I: Kinematics and Dynamics

12

Underlying Principles

Newton's second law in rotational form

Major Concepts

1. Rigid object
2. Rotation axis
3. Fixed axis
4. Angular kinematic quantities
   a. Angular position
   b. Angular displacement
   c. Angular velocity
   d. Angular acceleration
5. Torque
6. Rotational inertia

Tools

1. Right-hand rule
2. Cross product (vector product)

Key Questions

How do we describe the kinematics of a rotating object?

How can we apply Newton's second law to a rotating object?

From the earliest times, rotational motion of wheels and other objects has played a major role in culture and technology. Rotating waterwheels may be used to transfer energy from a river, the rotation of gears on a bicycle allows a rider to pedal up steep mountains more easily (Fig. 12.1), and the rotation of a lever enables a person to lift very heavy objects. In this chapter, we explore **rotational kinematics** (the study of rotational motion independent of its cause) and **rotational dynamics** (the cause of rotational motion).

FIGURE 12.1 Wheels and rotational motion have played a major role in human endeavors since ancient times. Even without powered machinery, gears allow people to pedal easily up steep hills.

12-1 Rotation Versus Translation

In Chapter 2, we treated an object as a particle undergoing purely translational motion if every point of the object underwent exactly the same displacement. Figure 12.2A shows Zak riding a Ferris wheel from an initial position to a final position. As the Ferris wheel goes around, Zak's chair pivots so that he is always facing the same direction. All points associated with Zak have the same displacement as he moves between an initial position and a final position. Thus, Zak's motion is translational, and he may be approximated as a particle.

Compare his motion with that of Sophia on a witch's wheel in Figure 12.2B. Sophia's chair does not pivot, so when she is at the top of the wheel, she is upside

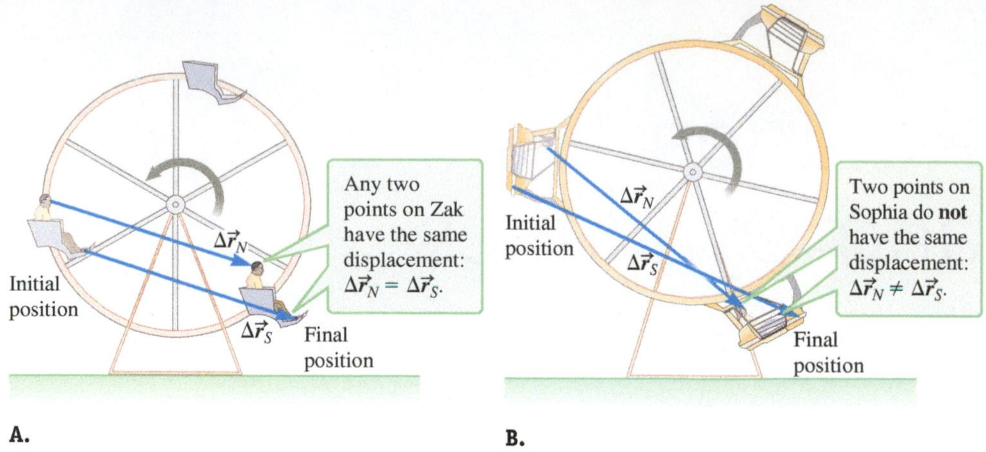

FIGURE 12.2 A. When Zak rides a Ferris wheel, the orientation of his body with respect to the ground remains the same; he is always facing the same direction. His motion is translational. In a given time interval, the displacement of his nose equals the displacement of his shoe: $\Delta\vec{r}_N = \Delta\vec{r}_S$. He can be modeled as a particle. **B.** When Sophia rides a witch's wheel, the orientation of her body with respect to the ground changes. Her motion is not translational; in a given time interval, the displacement of her nose does not equal the displacement of her shoe: $\Delta\vec{r}_N \neq \Delta\vec{r}_S$. She cannot be modeled as a particle.

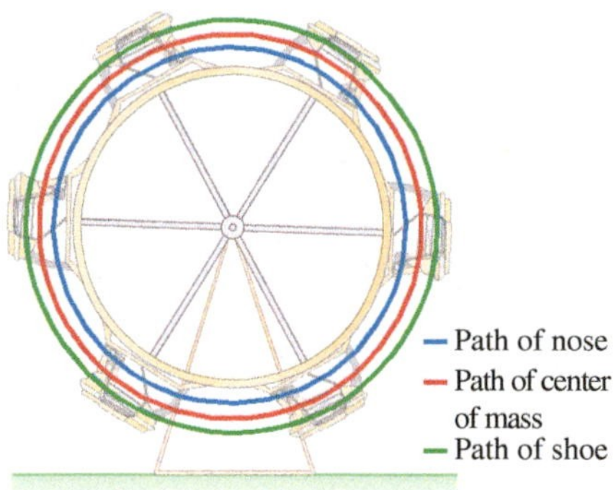

FIGURE 12.3 Sophia on the witch's wheel ride may be modeled as a rotating rigid object. Points such as her shoe, center of mass, and nose travel along arcs of concentric circles.

RIGID OBJECT; ROTATION AXIS; FIXED AXIS

✪ Major Concepts

down. The displacements of her nose and shoe are unequal: $\Delta\vec{r}_N \neq \Delta\vec{r}_S$. Because *not* every point associated with Sophia has the same displacement, her motion is *not* translational, and we *cannot* approximate her as a particle. Instead, we consider the path of several points associated with her. Figure 12.3 shows the path of her nose, shoe, and center of mass. Each point is traveling along the arc of a circle, and all the circles have a common center at the center of the witch's wheel. We model Sophia as a *rotating rigid object*.

A **rigid object** is an object that moves with all its components locked together so that its shape does not change. When we are considering a real object, we need to decide if it can be modeled as a rigid object. A test-crash dummy on the witch's wheel is likely to maintain its shape and so is well modeled as a rigid object, but can we model a live woman moving her arms and legs as a rigid object? If the internal movement of her arms and legs is small compared with the overall motion, we may be safe in modeling her as a rigid object. A typical rider on an amusement park ride is usually well modeled as a rigid object. When an ice skater pulls her arms inward as she spins, however, the rigid object model fails to fit her motion.

If a rigid object is rotating, every particle in the object moves in a circular path centered on the same axis, known as the **rotation axis**. In the case of Sophia on the witch's wheel, the rotation axis runs through the center of the wheel and is perpendicular to the page. In this chapter, we limit our discussion to rotations around a **fixed axis**, that is, a rotation axis that does not move relative to the observer. Sophia's rotation axis is fixed, but the rotation axis of a bicycle wheel rolling down the road is not fixed. We discuss rolling motion in Chapter 13. An axis may be fixed in one reference frame but not in another. If we observe the bicycle wheel in a frame moving along with the bicycle, the axis of rotation is fixed.

For the object's motion to be considered a rotation, the particles in the object do not need to make a complete circle. The following case study illustrates such rotations.

CASE STUDY Ancient Megaliths

Ancient peoples built megalithic monuments of impressive size and mass with primitive technology. Three examples are shown in Figure 12.4.

- The great trilithons of Stonehenge in England were constructed approximately 4600 years ago (Fig. 12.4A). A large, upright stone in a trilithon weighs about 40 tons and stands about 25 ft above the ground.
- Queen Hatshepsut's obelisk was erected approximately 3500 years ago in Egypt (Fig. 12.4B). The largest Egyptian obelisks are made of a single block of granite; they weigh about 400 tons and stand 100 ft tall.
- Moai statues on Easter Island were created roughly 400 to 1000 years ago (Fig. 12.4C). There are nearly 900 standing Easter Island Moai, weighing an average of 14 tons and standing about 14 ft tall.

Today, it is hard to imagine creating any megalithic structure without the use of modern construction equipment. Because moving a 400-ton granite obelisk safely would be difficult even with all the technological advantages of today, some people believe that it was *impossible* for ancient people to accomplish such feats with their technology. It has even been suggested that extraterrestrial aliens visited ancient peoples and constructed the megaliths.[1]

Science usually proceeds by testing the simplest theories first. A claim that an extraterrestrial workforce created the ancient monuments is not supported by other evidence such as written records, and this assertion dismisses the much simpler explanation that ancient people built their own megaliths using tools and teamwork. Archeological evidence from artifacts and written records supports this simpler theory.

Many of the details of ancient construction were unrecorded, or perhaps the records have been lost. So, archeologists must develop theories and test them by using a team of contemporary workers to build replicas of ancient monuments using primitive technology. Of course, even when a modern workforce successfully builds a replica of an ancient monument, it does not prove how the ancient workforce built it. In this case study, we explore how ancient people may have used the Earth's gravity and levers to raise an obelisk or an upright stone in a trilithon and how levers may have been used to rotate a Moai.

FIGURE 12.4 **A.** A trilithon is made of three large stones. Together they form a square arch. **B.** An obelisk is a single large stone. **C.** Easter Island has about 900 standing Moai such as the four intact statues shown here.

CONCEPT EXERCISE 12.1

Figure 12.5 shows two rotating objects. Indicate the rotation axis in each case and determine in which reference frame the rotation axis is fixed. Explain.

Case 1. The Earth's daily rotation on its axis.

Case 2. A bowling ball.

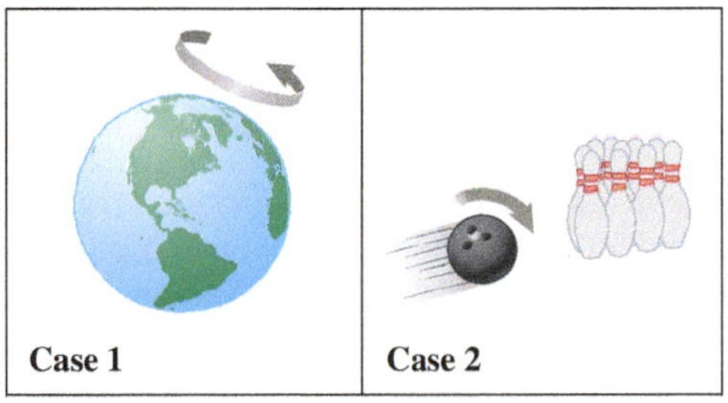

FIGURE 12.5

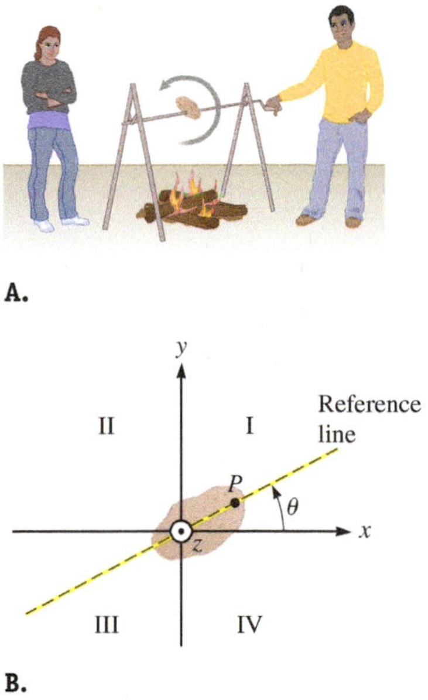

FIGURE 12.6 **A.** Aaron rotates a potato. **B.** The potato rotates counterclockwise in the *xy* plane.

12-2 Rotational Kinematics

As we did for translational motion, we begin our study of rotational motion with kinematics, that is, the mathematical description of rotational motion. We will use analogies with translational kinematics (Chapters 2 and 4) to help us.

Figure 12.6A shows a potato being cooked over an open fire. A stake has been driven through the potato, and it slowly rotates as the stake is turned. As in translational kinematics, we must choose a coordinate system (right-handed in this book). Figure 12.6B shows our choice. The z axis is the rotation axis, aligned with the stake. With this choice, the potato rotates in the xy plane. The coordinate system in Figure 12.6B is right-handed because it follows the rule from Chapter 3: Line up the fingers of your right hand along the positive x axis and curl your fingers so that you "push" the x axis through 90° into the positive y axis; then, the direction in which your thumb points is the direction of the positive z axis (Fig. 3.22). Although you may orient the rotation axis in any convenient way, in this book we will usually align the z axis with the rotation axis.

We will sometimes include the adjective *translational* for clarity when describing a quantity, but if no adjective is explicitly used, *translational* is implied.

[1]See, for example, Erich von Däniken's best-selling book *Chariots of the Gods?*

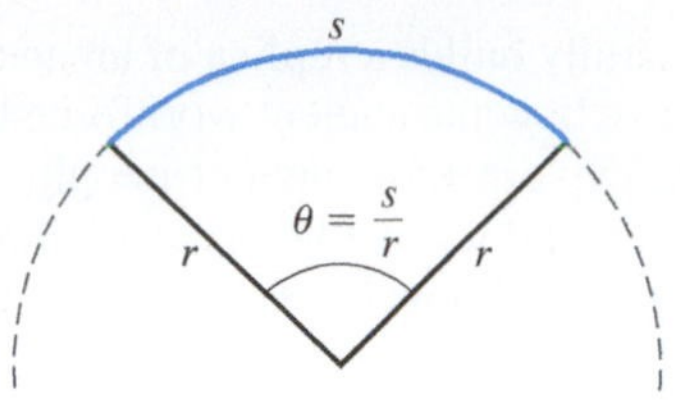

FIGURE 12.7 The arc length s and radius r are measured in the same length units (such as meters).

To describe the rotational motion of the potato, we need to choose a **reference line** fixed on the object and perpendicular to the rotation axis. If the rotation axis is aligned with the z axis, the reference line is parallel to the xy plane. The reference line rotates with the object. For the potato in Figure 12.6B, we have picked a reference line that passes through the rotation axis and a spot ("an eye") on the potato labeled P.

Angular Position and Angular Displacement

ANGULAR POSITION

Major Concept

Polar coordinates are used to describe rotational motion (Chapter 4). The **angular position** θ of the reference line is measured counterclockwise from the positive x axis as shown in Figure 12.6B. The angular position θ is negative if it is measured clockwise from the positive x axis.

In rotational kinematics, we specify the angular position in radians because it is simpler to work with radians than degrees or revolutions. Consider the angle θ shown in Figure 12.7. The circle has radius r, and s is the arc length subtended by θ; so, in radians, θ is

$$\theta = \frac{s}{r} \tag{12.1}$$

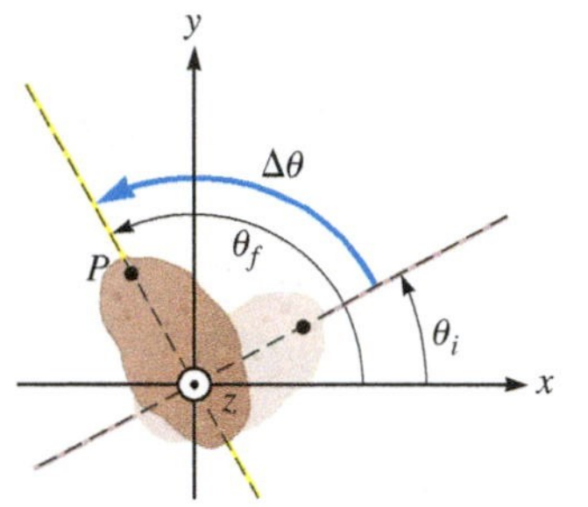

FIGURE 12.8 The angular displacement $\Delta\theta$ of a potato.

From Equation 12.1, an angle in radians is the ratio of two lengths and is therefore dimensionless. For example, if $s = 5.0\text{ m}$ and $r = 2.5\text{ m}$, then $\theta = s/r = 5.0\text{ m}/2.5\text{ m} = 2.0$ according to Equation 12.1. We write $\theta = 2.0$ "rad" to clarify that the angle is measured in radians and *not* in degrees or revolutions. Equation 12.1 does not hold if θ is measured in degrees or revolutions; in rotational kinematics, we convert all angles to radians. Because a radian comes from a dimensionless ratio, we often drop the "rad" when radian measure is understood from the context.

The angular position $\theta(t)$ is a function of time for a rotating object. If, for example, at $t = 0$ the reference line were aligned with the positive x axis, then $\theta(0) = 0$. When the potato completes one revolution so that the reference line is once again lined up with the positive x axis, the angular position is not zero, but 2π rad.

ANGULAR DISPLACEMENT

Major Concept

A change in the angular position is known as the **angular displacement**,

$$\Delta\theta = \theta_f - \theta_i \tag{12.2}$$

where (as usual) i and f stand for *i*nitial and *f*inal, respectively. Figure 12.8 shows the angular displacement of the potato. Equation 12.2 does not require angular displacement to be in radians, but subsequent kinematic equations will.

Although angular displacement has both magnitude and direction, it does not obey the commutative rule of vector addition. So, angular displacement is not a vector, and we should not represent its direction with a straight-line arrow. Instead, we refer to the direction of angular displacement with the terms *clockwise* and *counterclockwise*. These terms are consistent with the sign convention for angular position, so a counterclockwise angular displacement is positive. If $\theta_f > \theta_i$, the angular displacement is positive, and the net rotation of the object is counterclockwise as shown in Figure 12.8.

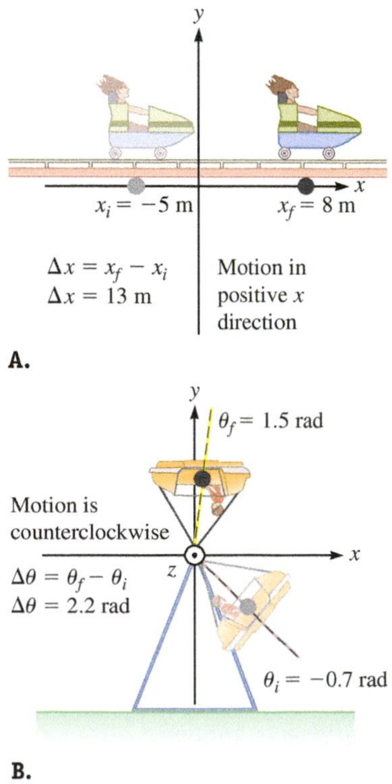

FIGURE 12.9 An analogy between one-dimensional motion along an x axis and rotational motion around a z axis.

In this book, we focus on objects that rotate around a single rotation axis, similar to restricting translational kinematics to one dimension. A single axis makes the mathematics considerably simpler and allows us to concentrate on a conceptual understanding of rotation. When restricted to a single rotation axis, angular displacement *is* commutative. For example, an object may be rotated around a single axis by 1.5 rad and then by 0.75 rad, or it may be rotated by 0.75 rad and then by 1.5 rad with the same result.

Let's take a closer look at the analogy with one-dimensional translational kinematics. Figure 12.9A shows Charlotte riding in a car that is restricted by a track to move in one dimension while Figure 12.9B shows Jim riding in a gondola that is restricted by its axle to rotate about a single, fixed axis. We'll compare the choice of coordinate system, model, measured position, and displacement for the two different riders.

Coordinate system: For one-dimensional translational motion, we typically choose to align one coordinate axis with the direction of movement. If the motion is horizontal, we often choose the direction of motion to be the x axis as shown in Figure 12.9A. For motion around a single rotation axis, we typically align the z axis with the rotation axis. Then, the object's rotation is in the xy plane as shown in Figure 12.9B.

Model: Charlotte may be modeled as a particle and therefore may be represented by a single dot on the x axis. Jim is modeled as a rigid object, and a reference line is drawn on the coordinate system to represent him.

Measured position: Charlotte's position x is positive if she is to the right of the origin. Charlotte's initial position is shown as $x_i = -5$ m. By analogy, Jim's angular position θ is positive if θ is measured counterclockwise from the positive x axis. Measuring θ is different from measuring x. Once a coordinate system has been established, there is only one way to measure x, but θ may be measured either clockwise or counterclockwise. Jim's initial position is measured clockwise and is therefore negative: $\theta_i = -0.7$ rad. (Had it been measured counterclockwise, it would have been $2\pi - 0.7$ rad.) Linear or translational position x is analogous to angular position θ.

The direction in which we measure angular positions is chosen to be consistent with direction of angular displacement.

Displacement: Charlotte's motion may be either in the positive or the negative x direction. If $x_f > x_i$, her displacement is positive and her motion is in the positive x direction as shown in Figure 12.9A. Jim's rotation around the z axis may be either counterclockwise or clockwise in the xy plane. If $\theta_f > \theta_i$, his angular displacement is positive and his rotation is counterclockwise as shown in Figure 12.9B. Translational displacement Δx is analogous to angular displacement $\Delta\theta$.

CONCEPT EXERCISE 12.2

In Figure 12.10, the angle θ subtended is 15.5°, and the radius is 0.25 m. Find the arc length and explain why it would be simpler to find it if the angle were measured in radians instead of degrees.

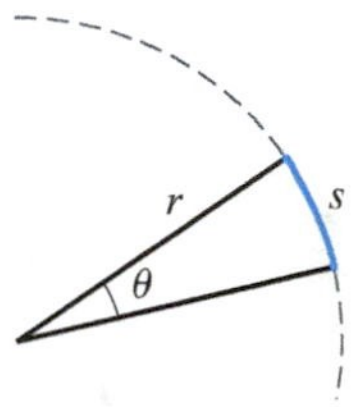

FIGURE 12.10

CONCEPT EXERCISE 12.3

Table 12.1 gives the angular position of a rotating bottle (Fig. 12.11) at five instants in time.

a. Sketch a motion diagram for the bottle along with a coordinate system. It is not necessary to draw the actual bottle; instead, draw the reference line at each time corresponding to positions A through E.

b. What is the angular displacement from A to E? Is the rotation clockwise or counterclockwise?

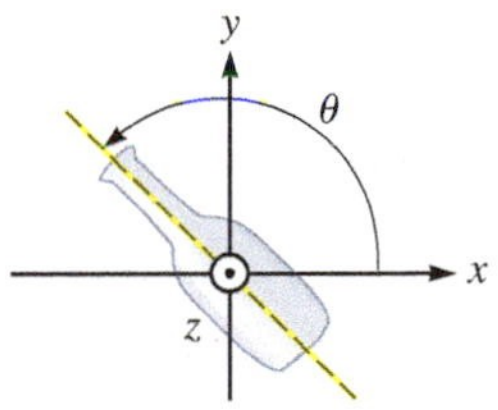

FIGURE 12.11

TABLE 12.1 Angular position of the rotating bottle at various time instants.

Position	t (s)	θ (rad)
A	0	0.75
B	2	3.45
C	4	5.95
D	6	8.25
E	8	10.05

Angular Velocity

The average angular speed and instantaneous angular speed are analogous to average and instantaneous translational speed. Average and instantaneous angular speed are represented by the lowercase Greek letter omega, ω. The **average angular speed** ω_{av} over a time interval Δt is the angular displacement divided by the time interval:

Neither angular displacement nor average angular speed is a vector.

$$\omega_{av} = \frac{\theta_f - \theta_i}{t_f - t_i} = \frac{\Delta\theta}{\Delta t} \qquad (12.3)$$

The **instantaneous angular speed** comes from taking the limit of Equation 12.3 as the time interval goes to zero:

$$\omega = \lim_{\Delta t \to 0} \frac{\Delta\theta}{\Delta t} = \frac{d\theta}{dt} \qquad (12.4)$$

If we measure θ in radians, the SI units of ω_{av} and ω are radians per second, or rad/s.

The instantaneous angular speed is the magnitude of the **instantaneous angular velocity** vector. (We usually drop the word "instantaneous") Just as the direction of $\vec{v}$ refers to the specific coordinate system we choose, so does the direction of $\vec{\omega}$.

ANGULAR VELOCITY

Major Concept

It might seem convenient to use the terms *clockwise* or *counterclockwise* to describe the direction of angular velocity, but those terms are too ambiguous as shown by the two observers of the rotating potato in Figure 12.12A. Aaron is in the background, and Hannah is in the foreground. To Hannah, the potato appears to be rotating counterclockwise, but to Aaron, the potato is rotating clockwise.

RIGHT-HAND RULE Tool

There is an unambiguous way to describe the direction of the angular velocity. An observer uses the **right-hand rule** convention to find the direction of $\vec{\omega}$ by curling the fingers of his or her right hand in the same sense as the rotation. The observer's right thumb then points in the direction of $\vec{\omega}$ (Fig. 12.12B). Try it for yourself; you'll find that your right thumb points out of the page. Aaron and Hannah would also agree that $\vec{\omega}$ points out of the page.

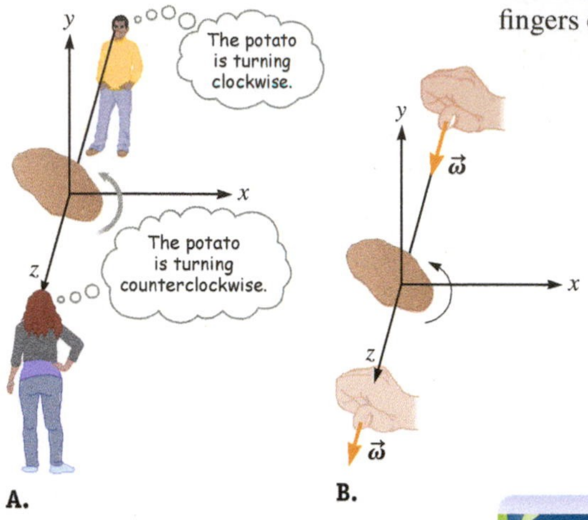

FIGURE 12.12 **A.** Aaron and Hannah watch a rotating potato. **B.** An observer uses the right-hand rule convention to find the direction of $\vec{\omega}$ by curling the fingers of his or her right hand in the same sense as the rotation. The observer's right thumb then points in the direction of $\vec{\omega}$.

We have chosen a particular coordinate system as shown in Figure 12.12. Because we have decided to align the rotation axis with the z axis, $\vec{\omega}$ will point in either the positive or negative z direction. We found that $\vec{\omega}$ points out of the page, which corresponds to the positive z direction in our coordinate system. The potato's motion is in the xy plane, but its angular velocity points in the positive z direction. Although the right-hand rule convention might seem counterintuitive, with practice it becomes second nature. The right-hand rule for angular velocity provides an unambiguous, consistent way to describe rotational motion.

CONCEPT EXERCISE 12.4

What is the average angular speed of the rotating bottle in Concept Exercise 12.3 over the 8-second time interval?

Angular Acceleration

If an object's translational speed changes, the object must be accelerating. Likewise, if an object's rotation speeds up or slows down, the object has a nonzero angular acceleration. The **average angular acceleration** $\vec{\alpha}_{av}$ is given by

$$\vec{\alpha}_{av} = \frac{\vec{\omega}_f - \vec{\omega}_i}{t_f - t_i} = \frac{\Delta\vec{\omega}}{\Delta t} \tag{12.5}$$

ANGULAR ACCELERATION

Major Concept

The **instantaneous angular acceleration** (also known simply as **angular acceleration**) $\vec{\alpha}$ is found by taking the limit of $\vec{\alpha}_{av}$ as Δt goes to zero:

$$\vec{\alpha} = \lim_{\Delta t \to 0} \frac{\Delta\vec{\omega}}{\Delta t} = \frac{d\vec{\omega}}{dt} \tag{12.6}$$

If an object's rotation is speeding up, then $\vec{\alpha}$ points in the same direction as $\vec{\omega}$. If the object is slowing down, then $\vec{\alpha}$ and $\vec{\omega}$ point in opposite directions. Because we are only studying rotations about a single stationary rotation axis, we are only concerned with angular accelerations that are parallel or antiparallel to the angular velocity.

Let us return to Charlotte riding on the track and Jim riding on the wheel (Fig. 12.9) to see how angular velocity and angular acceleration fit in the analogy between rotation and one-dimensional translation.

(Angular) velocity: In Figure 12.13, Charlotte is represented by a dot. Her motion and initial velocity $\vec{v}_i$ are in the positive x direction (Fig. 12.13A, left). Jim is represented by a reference line rotating counterclockwise about the z axis in the xy plane. According to the right-hand rule, Jim's initial angular velocity $\vec{\omega}_i$ is in the positive z direction as shown. Translational velocity $\vec{v}$ is analogous to angular velocity $\vec{\omega}$.

(Angular) acceleration: If Charlotte speeds up so that $v_f > v_i$ (Fig. 12.13B, left), her acceleration $\vec{a}$ must be in the same direction as her velocity. In this case,

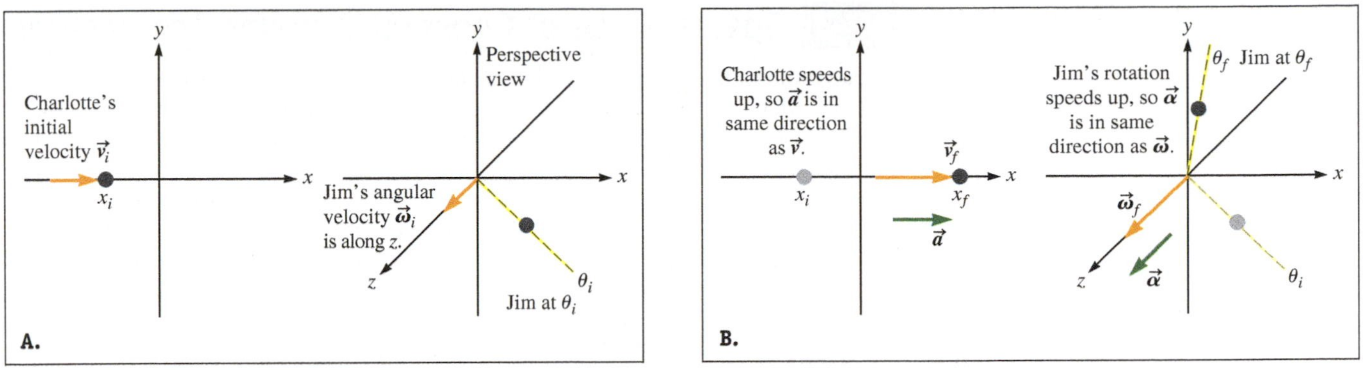

FIGURE 12.13 **A.** An analogy between translational velocity and angular velocity. **B.** An analogy between translational acceleration and angular acceleration. In both parts, Jim's situation is shown in perspective with the z axis extending from upper right to lower left.

$\vec{a}$ is in the positive x direction. Now imagine that Jim's rotation speeds up: $\omega_f > \omega_i$. The increase in his angular speed means that his angular acceleration $\vec{\alpha}$ is in the same direction as his angular velocity. In this case, $\vec{\alpha}$ is in the positive z direction. Translational acceleration $\vec{a}$ is analogous to angular acceleration $\vec{\alpha}$.

CONCEPT EXERCISE 12.5

What are the directions of $\vec{\omega}$ and $\vec{\alpha}$ for the rotating bottle in Concept Exercise 12.3?

EXAMPLE 12.1 "As the World Turns"

Treat the Earth as a rigid object rotating on a fixed axis and consider a time interval $\Delta t = 12$ h. Find the magnitude of the Earth's angular displacement $\Delta\theta$, angular speed ω, and angular acceleration α during that time interval. The Earth's rotation is very nearly uniform.

INTERPRET and ANTICIPATE

Every 24 hours, the Earth completes one revolution. We expect to find numerical answers in the form $\Delta\theta =$ _____ rad, $\omega =$ _____ rad/s, and $\alpha =$ _____ rad/s^2.

SOLVE

The uniform rotation of the Earth means that in half its period it completes half a revolution.

$$\Delta\theta = \tfrac{1}{2}\text{ rev} = \tfrac{1}{2}(2\pi\text{ rad}) = \pi\text{ rad}$$

The Earth's angular speed is constant. The instantaneous angular speed (Eq. 12.4) equals the average angular speed (Eq. 12.3).

$$\omega = \omega_{av} = \frac{\Delta\theta}{\Delta t} = \frac{\pi\text{ rad}}{12\text{ h}\cdot(3600\text{ s}/1\text{ h})} \quad (12.3)$$

$$\omega = \frac{\pi\text{ rad}}{4.3\times10^4\text{ s}} = 7.3\times10^{-5}\text{ rad/s}$$

The Earth's constant angular speed means that its angular acceleration is zero.

$$\alpha = 0$$

CHECK and THINK

Uniform rotation makes this example mathematically simple. Solving this example gives us a chance to develop our intuition for angular speed: Once-a-day rotation corresponds to $\omega = 7.3\times10^{-5}$ rad/s.

12-3 Special Case of Constant Angular Acceleration

In one-dimensional translational kinematics (Chapter 2), we considered the special case in which a particle moves with constant acceleration. The special case of $\vec{a}$ = constant for translational motion is analogous to the special case of $\vec{\alpha}$ = constant for rotational motion. The mathematics describing an object rotating at constant angular acceleration is exactly the same as that describing a particle moving along a line at constant acceleration. In Section 2-9, we solved $v = dx/dt$ and $\vec{a} = d\vec{v}/dt$ (Eqs. 2.4 and 2.7, respectively) in the special case of $\vec{a}$ = constant. It is mathematically equivalent to solving $\omega = d\theta/dt$ and $\vec{\alpha} = d\vec{\omega}/dt$ (Eqs. 12.4 and 12.6, respectively) in the special case of $\vec{\alpha}$ = constant, so we do not need to work out the solution again. Instead, all we need to do is rewrite the solutions from Chapter 2 in terms of the variables for rotational motion. The analogy we developed (Figs. 12.9 and 12.13) shows us how to make the conversion from translational variables to rotational variables: $\Delta x \rightarrow \Delta\theta$, $v_x \rightarrow \omega$, and $a \rightarrow \alpha$. Analogous pairs of kinematic equations are listed in Table 12.2. The two left columns of the table are from Chapter 2 and apply to translational motion, and the two right columns apply to rotational motion and come from converting the translational variables to rotational variables.

TABLE 12.2 Translational kinematic equations for constant acceleration (left) and rotational kinematic equations for constant angular acceleration (right).

Translation (along x axis) with $\vec{a}$ = constant		Rotation (single rotation axis) with $\vec{\alpha}$ = constant	
Equation	**Eliminated variable**	**Equation**	**Eliminated variable**
$v_x = v_{0x} + a_x t$ (2.9)	Displacement Δx	$\omega = \omega_0 + \alpha t$ (12.7)	Angular displacement $\Delta\theta$
$\Delta x = \frac{1}{2}(v_{0x} + v_x)t$ (2.10)	Acceleration a_x	$\Delta\theta = \frac{1}{2}(\omega_0 + \omega)t$ (12.8)	Angular acceleration α
$\Delta x = v_{0x}t + \frac{1}{2}a_x t^2$ (2.11)	Final velocity v_x	$\Delta\theta = \omega_0 t + \frac{1}{2}\alpha t^2$ (12.9)	Final angular velocity ω
$\Delta x = v_x t - \frac{1}{2}a_x t^2$ (2.12)	Initial velocity v_{0x}	$\Delta\theta = \omega t - \frac{1}{2}\alpha t^2$ (12.10)	Initial angular velocity ω_0
$v_x^2 = v_{0x}^2 + 2a_x \Delta x$ (2.13)	Final time t	$\omega^2 = \omega_0^2 + 2\alpha\Delta\theta$ (12.11)	Final time t

Recall from Chapter 2 that these five equations are not independent. Only two are independent; the other three are derived by algebraically combining the two independent equations. Writing out these five equations with a single variable eliminated from each often makes problem solving easier. Many physicists only memorize two of the equations (usually Eqs. 12.7 and 12.8) and then solve them simultaneously as needed for any particular situation. It is not necessary to memorize all five of them, but it is important to be able to derive any one of them as needed to solve a problem.

PROBLEM-SOLVING STRATEGY

Constant Angular Acceleration

We only need to slightly modify the steps for one-dimensional constant-acceleration problems. As part of the **INTERPRET and ANTICIPATE** procedure: Draw a sketch that includes a coordinate system. Keep in mind that you will likely need a three-dimensional coordinate system.

There are also two parts to the **SOLVE** procedure:

Step 1 Start by **listing** the six kinematic variables (initial angular position, final angular position, initial angular velocity, final angular velocity, angular acceleration, and time). Often, the initial and final angular positions are combined as the displacement, so your list typically includes five variables. Be sure to include the sign of each value. Write the word *need* next to any variable you need to solve for and write *not needed* next to any variable that you don't know and don't need to find.

Step 2 Once you have listed the parameters, you may use the constant angular acceleration equations **(Table 12.2)**. You can avoid doing algebra by choosing the equation that does not include the unneeded variable.

Step 3 Do **algebra** before substituting values.

Step 4 Substitute **values** if appropriate.

EXAMPLE 12.2 A Spinning Pulsar

Pulsars are astronomical objects that give off very regular pulses of radio waves with periods roughly between 1 millisecond and 1 second. When pulsars were first discovered in 1967 by Jocelyn Bell, their uniform pulses led her to think that she had discovered extraterrestrial intelligence. Today we know that pulsars are rotating compact objects about the size of a city. Two spots on a pulsar emit radio waves, one of them becoming visible to us periodically as the pulsar rotates (Fig. 12.14) much like a light spinning on top of a lighthouse. Astronomers measure the angular speed of a rotating pulsar by observing the period of its radio wave pulses.

A pulsar may form when a star of medium mass dies. On July 4, 1054 CE, Chinese astronomers witnessed the death of a star that led to the formation of the pulsar in the Crab Nebula. Contemporary astronomers have measured the angular speed of the nebula's pulsar as $\omega = 1.89 \times 10^2$ rad/s and have found that the pulsar is slowing down such that $\alpha = 2.39 \times 10^{-9}$ rad/s^2. What was the initial angular speed of the Crab Nebula's pulsar in 1054 CE?

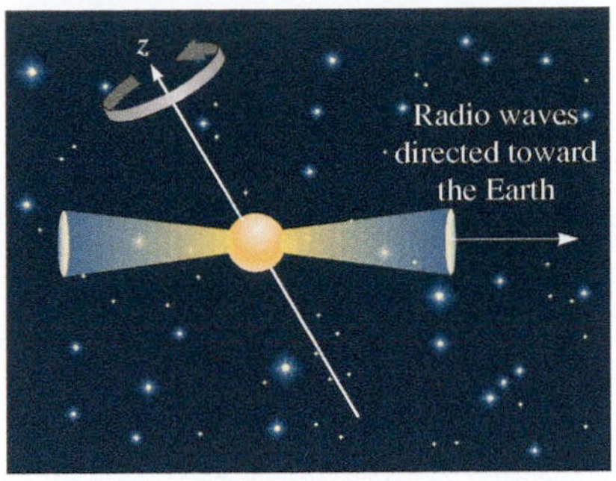

FIGURE 12.14

INTERPRET and ANTICIPATE

We expect a numerical result for ω_0 that is somewhat greater than ω. Because the angular acceleration is so low, however, we should expect that the pulsar's angular speed has not decreased very much since its formation; the exact amount of time that has elapsed since the pulsar was formed is not very significant. So, the date you are working this problem will not matter very much. To be specific, we have decided to work this problem for the year 2015. Work the problem for whatever year you'd like—from the year you were born to 50 years from now—and you will get roughly the same answer. A sketch with a coordinate system is provided as part of the problem statement.

SOLVE

Step 1 List variables. We have assumed the Crab Nebula's pulsar formed 961 years ago. In Figure 12.14, the angular velocity ω is aligned with the positive z axis, and because the pulsar is slowing down, α must be in the negative z direction.

$\Delta\theta =$ not needed $\quad \omega_0 =$ needed

$$\omega = 1.89 \times 10^2 \text{ rad/s}$$
$$\alpha = -2.39 \times 10^{-9} \text{ rad/s}^2$$
$$t = 961 \text{ y} = 3.03 \times 10^{10} \text{ s}$$

Steps 2 and 3 Because $\Delta\theta$ is not needed, use Equation 12.7 to solve for ω_0.

$$\omega = \omega_0 + \alpha t \qquad (12.7)$$
$$\omega_0 = \omega - \alpha t$$

Step 4 Substitute.

$$\omega_0 = (1.89 \times 10^2 \text{ rad/s}) - (-2.39 \times 10^{-9} \text{ rad/s}^2)(3.03 \times 10^{10} \text{ s})$$
$$\omega_0 = 2.61 \times 10^2 \text{ rad/s}$$

CHECK and THINK

As expected, the Crab Nebula's pulsar had a greater angular speed in 1054 CE than it does today.

$$\frac{\omega_0 - \omega}{\omega_0} = \frac{2.61 \times 10^2 \text{ rad/s} - 1.89 \times 10^2 \text{ rad/s}}{2.61 \times 10^2 \text{ rad/s}} = 0.28$$

In the past millennium, the pulsar has only slowed down about 28%. A pulsar's rotation is so nearly uniform that some pulsars have been used to keep time. In fact, the value $\omega = 1.89 \times 10^2$ rad/s was measured in 2004, but even if you are working on this problem in 2019 and you assume that year is when ω was measured, your answer will be the same to three significant figures.

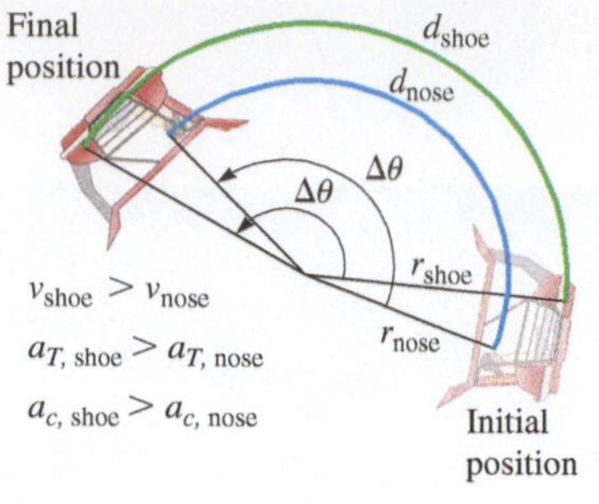

FIGURE 12.15 Sophia in rotation is modeled as a rigid object, so her nose and her shoe have equal angular displacements $\Delta\theta$, angular speeds ω, and angular accelerations α. Her shoe is farther from the axis of rotation than her nose is. Therefore, the translational parameters d, v, a_T, and a_c of her shoe are greater than those for her nose.

12-4 The Connection Between Rotation and Circular Motion

In Section 12-1 we modeled Sophia on a witch's wheel (Fig. 12.2B) as a rigid rotating object. Another way to model her is as a collection of particles, each undergoing circular motion around a common center. In this section, we explore how the rigid-object and collection-of-particles models are related.

Distance Traveled by a Point on a Rotating Object

Take another look at Sophia riding on the witch's wheel (Fig. 12.15). If we model her as a rigid object as she moves from some initial position to a final position, her nose and shoe have equal angular displacements $\Delta\theta$.

Now we instead model Sophia as a collection of particles, each in circular motion. For example, her nose travels along the arc of a circle of radius r_{nose} and her shoe along a circle of radius r_{shoe}. The distance her shoe travels along its circular path is d_{shoe}, and the distance traveled by her nose along its circular path is d_{nose}. From Equation 12.1,

$$\Delta\theta = \frac{d_{shoe}}{r_{shoe}} \quad \text{and} \quad \Delta\theta = \frac{d_{nose}}{r_{nose}}$$

Because the angular displacement $\Delta\theta$ is equal for all of her particles, we can equate these expressions:

$$\frac{d_{shoe}}{r_{shoe}} = \frac{d_{nose}}{r_{nose}}$$

and solve for d_{shoe}:

$$d_{shoe} = d_{nose}\frac{r_{shoe}}{r_{nose}}$$

Sophia's shoe travels along a larger circle than her nose: $r_{shoe} > r_{nose}$, so her shoe travels a longer distance than her nose: $d_{shoe} > d_{nose}$. This particular example illustrates a statement that is true for any rigid rotating object: *A particle far from the rotation axis of a rigid object travels a greater distance than a particle that is closer to the axis.* The distance traveled is directly proportional to r, the distance from the particle to the axis of rotation. Expressed mathematically,

$$d = (\Delta\theta)r \tag{12.12}$$

where $\Delta\theta$ is measured in radians. In SI units, both d and r are measured in meters, showing again that radians are dimensionless.

Translational Velocity of a Point on a Rotating Object

Now let us look at the translational and angular speeds of Sophia's nose and shoe. Sophia moves from her initial position to her final position in a time interval Δt. Her nose and shoe undergo the same angular displacement $\Delta\theta$ in the same time interval. Therefore, her nose and shoe have the same average angular speed $\omega_{av} = \Delta\theta/\Delta t$ (Eq. 12.3), and if we consider an infinitesimal angular displacement $d\theta$ in an infinitesimal time interval dt, her nose and shoe have identical instantaneous angular speeds $\omega = d\theta/dt$ (Eq. 12.4).

You already know that *all particles of a rotating rigid object have the same angular speed* ω, but you may be surprised to learn that not all the particles have the same *translational* speed. Sophia's shoe travels a greater distance than her nose in the same time interval, so her shoe must be moving at a higher average speed than her nose:

$$v_{av,\,shoe} = \frac{d_{shoe}}{\Delta t} \quad \text{and} \quad v_{av,\,nose} = \frac{d_{nose}}{\Delta t}$$

Using $d = r(\Delta\theta)$ (Eq. 12.12) for distance traveled,

$$v_{av,\,shoe} = \frac{r_{shoe}\Delta\theta}{\Delta t} \quad \text{and} \quad v_{av,\,nose} = \frac{r_{nose}\Delta\theta}{\Delta t}$$

but because $r_{\text{shoe}} > r_{\text{nose}}$,

$$v_{\text{av, shoe}} > v_{\text{av, nose}}$$

We extend this result to the more general case of any rotating rigid object. In each case, the average translational speed of a particle of the object is proportional to r, the particle's distance from the rotation axis. Therefore, a particle far from the rotation axis has a higher average translational speed than does a particle close to the axis.

We can extend this idea further to infinitesimal time intervals: The instantaneous translational speed of any particle in a rotating rigid object is proportional to r, the particle's distance from the rotation axis. *Therefore, a point far from the rotation axis has a higher instantaneous translational speed than does a point close to the axis.* Mathematically, we write

$$v = \left(\frac{d\theta}{dt}\right)r = \omega r \tag{12.13}$$

ω is measured in rad/s but v is in m/s, so we drop the "rad" when expressing the units of v.

Equation 12.13 gives the translational speed of any point in a rotating rigid object. The direction of the translational velocity of each point is *tangential* to the circular path of that point at each instant in time.

Tangential Acceleration

Imagine that as Sophia rotates her angular speed changes; perhaps the ride is slowing down. In that case, the *tangential* velocity of both her nose and her shoe must decrease. Because her nose and her shoe have identical angular speeds at each time instant, the nose and the shoe must also have identical *angular* accelerations α, but their *tangential* accelerations are not the same. We find an expression for *tangential* acceleration by differentiating Equation 12.13 with respect to time:

$$a_T = \frac{dv}{dt} = \left(\frac{d^2\theta}{dt^2}\right)r = \left(\frac{d\omega}{dt}\right)r$$

$$a_T = \alpha r \tag{12.14}$$

α is measured in rad/s^2 but we drop the "rad" when writing the units of a_T.

The subscript T stands for *tangent* to the circle. Conceptually, Equation 12.14 says that the tangential acceleration of a particle in a rotating rigid object is proportional to r, the particle's distance from the rotation axis. A particle far from the axis has a greater tangential acceleration than does a particle close to the rotation axis.

Centripetal Acceleration

Recall that when a particle moves in a circle, the direction of its velocity continually changes and the particle must therefore be accelerating toward the center of the circle. The same is true for any point of a rotating rigid object; each particle has a centripetal acceleration given by $a_c = v^2/r$ (Eq. 4.38). We use Equation 12.13 to rewrite v:

$$a_c = \frac{(\omega r)^2}{r} = \omega^2 r \tag{12.15}$$

Equation 12.15 says that *the centripetal acceleration of a particle in a rotating rigid object is proportional to the particle's distance from the rotation axis. A particle far from the axis has a higher centripetal acceleration than does a point close to the rotation axis.* So, at any instant, Sophia's shoe has a greater centripetal acceleration than does her nose.

Rotational Versus Translational Parameters

Equations 12.12 through 12.15 connect the translational kinematics of any particle in a rotating rigid object to the rotational kinematics of the object as a whole. The rotational parameters (angular displacement, angular velocity, and angular acceleration) are useful because every particle in a rotating rigid object has the same $\Delta\theta$, the same ω, and the same α. The translational parameters (d, v, a_T, and a_c) depend on the particular particle's distance r from the axis of rotation.

EXAMPLE 12.3 Two Cities

Treat the Earth as a rigid object rotating on a fixed axis and consider a time interval $\Delta t = 12$ h. Assume the Earth's angular speed is constant.

A Esmeraldas, Ecuador, is nearly on the equator. Find the translational distance traveled d, translational speed v, tangential acceleration a_T, and centripetal acceleration a_c of Esmeraldas during this time interval.

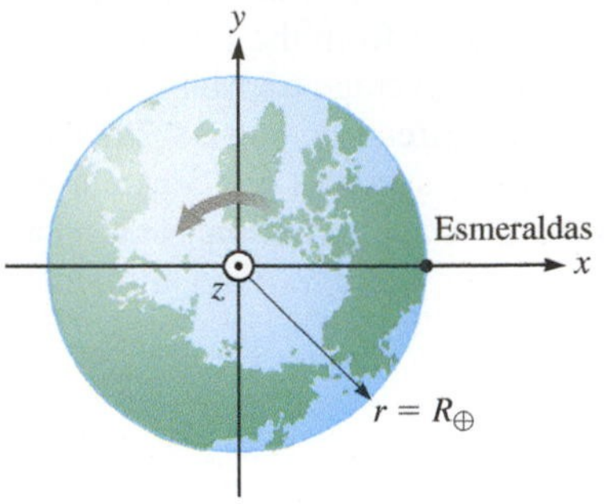

FIGURE 12.16 Looking down from the North Pole.

INTERPRET and ANTICIPATE

We are asked to use rotational parameters for the *entire* Earth to find the translational parameters of one particle, Esmeraldas. The results from Example 12.1 are helpful.

Figure 12.16 shows the Earth from above the north geographic pole. The z axis is aligned with the rotation axis. Esmeraldas is on the equator, so its distance from the rotation axis is equal to the radius of the Earth: $r = R_\oplus$.

SOLVE

The distance traveled comes from Equation 12.12 and the angular displacement found in Example 12.1. We drop "rad" when writing the units of d.

$$d = (\Delta\theta)r \qquad (12.12)$$

$$d = (\Delta\theta)R_\oplus = (\pi\,\text{rad})(6.37 \times 10^6\ \text{m})$$

$$d = 2.00 \times 10^7\ \text{m} = 20{,}000\ \text{km}$$

Use Equation 12.13 and the angular speed found in Example 12.1 to find Esmeraldas's translational speed.

$$v = \omega r \qquad (12.13)$$

$$v = \omega R_\oplus = (7.3 \times 10^{-5}\ \text{rad/s})(6.37 \times 10^6\ \text{m})$$

$$v = 4.7 \times 10^2\ \text{m/s} = 1700\ \text{km/h}$$

Because the Earth's angular acceleration is zero, Esmeraldas's tangential acceleration must be zero: $a_T = \alpha r$ (Eq. 12.14).

$$a_T = \alpha r = (0)R_\oplus$$

$$a_T = 0$$

Esmeraldas's centripetal acceleration is not zero and is given by Equation 12.15.

$$a_c = \omega^2 r \qquad (12.15)$$

$$a_c = \omega^2 R_\oplus = (7.3 \times 10^{-5}\ \text{rad/s})^2(6.37 \times 10^6\ \text{m})$$

$$a_c = 3.4 \times 10^{-2}\ \text{m/s}^2 \approx g/300$$

CHECK and THINK

We continue to develop our intuition for rotational motion. A point on the Earth's equator travels a distance of 20,000 km in half a day or 40,000 km per day with a constant speed of 1700 km/h. The centripetal acceleration of a point on the equator is much less than gravitational acceleration at the surface of the Earth: $a_c \approx g/300$.

B Repeat part A for New Orleans, which is at roughly 30° north latitude.

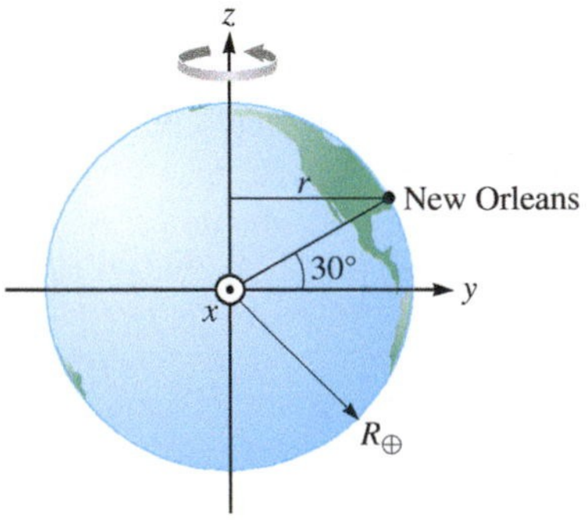

FIGURE 12.17 Looking directly at the equator.

INTERPRET and ANTICIPATE

The difference between this part of the problem and part A is that New Orleans is closer to the axis of rotation than is Esmeraldas.

Figure 12.17 shows the rotation axis aligned with z as in part A, but in this figure, z is up instead of out of the page. From this figure, $r = R_\oplus \cos 30°$. Because $r < R_\oplus$, we expect the distance traveled d, speed v, and centripetal acceleration a_c of New Orleans to have smaller magnitudes than they do for Esmeraldas.

SOLVE

Use the new value of r and follow the same procedure as in part A.

$$r = (6.37 \times 10^6\ \text{m})\cos 30° = 5.52 \times 10^6\ \text{m}$$

Use Equation 12.12 to find the distance traveled by New Orleans in 12 h.	$d = (\Delta\theta)r$ (12.12) $d = (\pi\,\text{rad})(5.52 \times 10^6\text{ m}) = 1.73 \times 10^7\text{ m}$
From Equation 12.13, find the translational speed for New Orleans.	$v = \omega r$ (12.13) $v = (7.3 \times 10^{-5}\text{ rad/s})(5.52 \times 10^6\text{ m}) = 4.0 \times 10^2\text{ m/s}$
Just as in the case of Esmeraldas, New Orleans's tangential acceleration must be zero (Eq. 12.14). Its centripetal acceleration, however, is nonzero and is given by Equation 12.15.	$a_T = 0$ $a_c = \omega^2 r = (7.3 \times 10^{-5}\text{ rad/s})^2(5.52 \times 10^6\text{ m})$ (12.15) $a_c = 2.9 \times 10^{-2}\text{ m/s}^2$

CHECK and THINK

New Orleans's distance traveled, translational speed, and centripetal acceleration are all less than corresponding quantities for Esmeraldas.

Part B gives us a chance to see the difference between rotational and translational parameters. All the Earth's particles such as Esmeraldas and New Orleans have the same value for each rotational parameter—the same $\Delta\theta$, the same ω and the same α— but the values of the translational parameters—d, v, a_T, and a_c—depend on the particle's location. Because Esmeraldas is farther from the rotation axis than is New Orleans, Esmeraldas travels farther in the same time interval, at a higher speed, and with a greater centripetal acceleration than does New Orleans.

Finally, notice that the Earth's rotation is a special case. Because the Earth's rotation is uniform ($\alpha = 0$), all points on the Earth have zero tangential acceleration $a_T = 0$, but that is not the case for a rotating object with $\alpha \neq 0$.

EXAMPLE 12.4 Shift Gears When You Need To

Gears play major role in our everyday lives. Nearly all mechanical devices that move, such as the windshield wipers on your car and the antique clock in the tower on campus, have gears. Because the gears are often hidden inside the device, we cannot see them in action. A bicycle's gears, on the other hand, are easily observed.

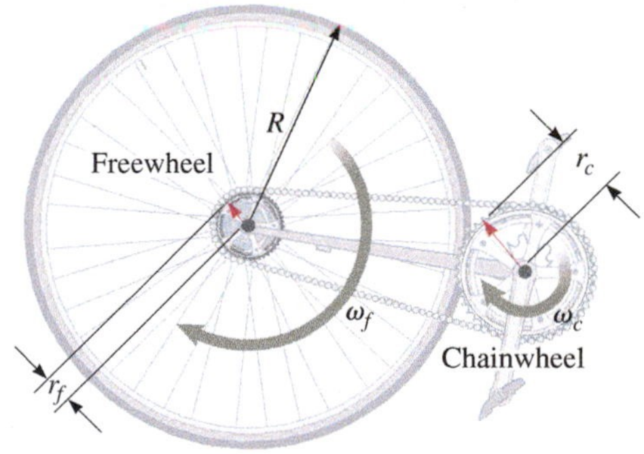

FIGURE 12.18

A bicycle rider moves the bike by pedaling the front gear known as the *chainwheel*. The chainwheel is connected by a chain to the rear gear known as the *freewheel*, which is attached to the rear tire (Fig. 12.18). The angular speed of the rear tire determines the translational speed of the bicycle. A typical rider is most comfortable when he pedals the chainwheel at 80 rpm (revolutions per minute). Most modern bicycles are equipped with several chainwheels and freewheels of varying sizes. The gear ratio γ is defined to be the radius of the chainwheel divided by the radius of the freewheel r_c/r_f. A modern bicycle has many gear ratios available, and the rider may vary the angular speed of the rear tire to maintain his ideal pedaling rate by selecting a gear ratio.

A certain bicycle has a rear tire 27 inches in diameter. Find the bicycle's translational speed if a typical rider selects a gear ratio of **A** $r_c/r_f = 4.82$ and **B** $r_c/r_f = 0.688$. Assume the rider maintains a pedal rate of 80 rpm and the translational speed of a point on the tire's rim equals the bicycle's translational speed.

INTERPRET and ANTICIPATE

The angular speed of the rear tire ω_{tire} determines the translational speed of the bicycle v_{bike}; the faster the tire rotates, the faster the bicycle moves. The freewheel gear is attached to the rear tire. So, the angular speed of the tire must equal the angular speed of the freewheel $\omega_{\text{tire}} = \omega_f$. Now consider the chainwheel, connected to the freewheel by a chain. All the links in the chain have the same translational speed, so the translational speed of a tooth on the freewheel must equal the translational speed of a tooth on the chainwheel: $v_{\text{chain}} = \omega_c r_c = \omega_f r_f$. The freewheel and chainwheel may have different radii, however, so their angular speeds are not necessarily equal. If the chainwheel is larger than the freewheel ($r_c > r_f$), the freewheel will have a greater angular speed ($\omega_f > \omega_c$). Therefore, a high gear ratio makes for a fast bicycle, and we expect that the bicycle's speed in part A is faster than its speed in part B.

Example continues on page 344 ▶

SOLVE The bicycle tire rolls along the ground, so a point on the tire's rim is a distance R from the axis of rotation, where R is the tire's radius. Use $v = \omega r$ (Eq. 12.13) and $\omega_{\text{tire}} = \omega_f$ to write an expression for the bicycle's speed in terms of R.	$v_{\text{bike}} = \omega_f R \quad (1)$
As mentioned above, the translational speed of a tooth on the freewheel must equal the translational speed of a tooth on the chainwheel. This fact gives us an expression for the angular speed of the freewheel.	$v_{\text{chain}} = \omega_c r_c = \omega_f r_f$ $\omega_f = \omega_c \dfrac{r_c}{r_f} \quad (2)$
Substitute Equation (2) into Equation (1).	$v_{\text{bike}} = \omega_c R \dfrac{r_c}{r_f} = \omega_c R\gamma \quad (3)$
Both ω_c and R have the same values for both parts of the problem.	$\omega_c = 80\ \text{rpm}\left(\dfrac{2\pi\ \text{rad}}{1\ \text{rev}}\right)\left(\dfrac{1\ \text{min}}{60\ \text{s}}\right) = 8.4\ \text{rad/s}$ $R = \left(\dfrac{27\ \text{in.}}{2}\right)\left(\dfrac{0.0254\ \text{m}}{1\ \text{in.}}\right) = 0.34\ \text{m}$
For part A, substitute $\gamma = r_c/r_f = 4.82$ into Equation (3).	$v_{\text{bike}} = \omega_c R\gamma = (8.4\ \text{rad/s})(0.34\ \text{m})(4.82)$ $v_{\text{bike}} = 14\ \text{m/s} = 31\ \text{mph}$
For part B, substitute $\gamma = r_c/r_f = 0.688$.	$v_{\text{bike}} = \omega_c R\gamma = (8.4\ \text{rad/s})(0.34\ \text{m})(0.688)$ $v_{\text{bike}} = 2.0\ \text{m/s} = 4.4\ \text{mph}$

CHECK and THINK

As expected, the bicycle's speed is faster for the higher gear ratio in part A. This example illustrates the usefulness of gears. The gears allow the rider to maintain a comfortable pedaling rate even as the terrain changes. More athletic riders often do not make full use of the gears available to them. Instead, they change their pedaling rate to maintain their speed despite any changes in the terrain. These riders are not making use of the mechanical advantage gears add to a bicycle. A story is told that the inventor of the multiple-ratio bicycle gear, Paul de Vivie (1853–1930), who was also known as Velocio, convinced bicycle racers to start using multiple gears by holding a race. A racing champion with a single-gear bicycle rode against an inexperienced rider who rode a bicycle with just three gear ratios. The inexperienced rider won the race.

TORQUE **Major Concept**

12-5 Torque

Imagine trying to loosen a bolt that has rusted into place. If your wrench has a short handle as shown in Figure 12.19A, you might find the task impossible. A longer-handled wrench as in Figure 12.19B makes the job much easier.

This common experience illustrates the factors that cause an object to rotate. The first factor is the magnitude F of the force applied to the handle. If one person cannot get the wrench to budge, he might ask a stronger person to give it a try.

The second factor is the distance r from the axis of rotation to the point of application of the force. In both cases shown in Figure 12.19, the same person applies the same force to rotate the bolt–wrench system. The axis of rotation is the same in both cases (along the axis of the bolt). In part B, the longer handle of the wrench allows the person to apply a force farther from the rotation axis than in part A, allowing him to rotate the bolt–wrench system.

The third factor is the direction in which the force is applied. If you want to rotate the bolt, you know from experience not to pull or push in a direction parallel to the wrench handle. In fact, it is best to apply a force perpendicular to the handle.

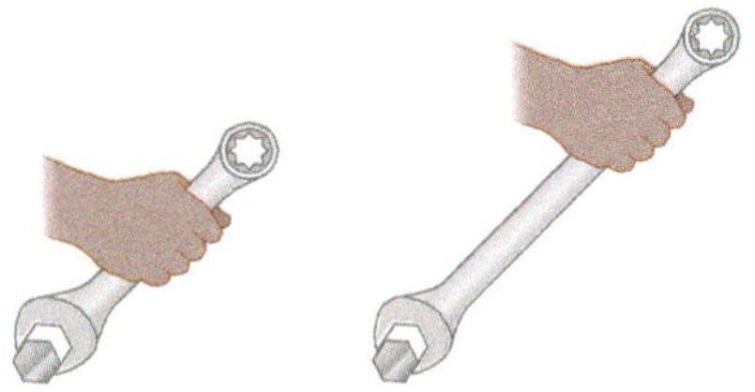

A. **B.**

FIGURE 12.19 Wrenching a bolt.

We combine these three factors into one concept: *torque*, a word coming from the Greek verb meaning "to twist." A **torque** on a rotating object is analogous to a force on a translating object: Just as a single force causes the acceleration of a particle, a single torque causes the angular acceleration of a rigid object.

Figure 12.20 illustrates the similarity and difference between a force and a torque. Figure 12.20A shows a crane lifting a steel beam off the ground. The tension force accelerates the beam upward. Figure 12.20B shows a crane rotating a radio antenna into an upright position. The base of the radio antenna is fixed on one side, forming the rotation axis. The tension force is applied at point P, which is at position $\vec{r}$ from the rotation axis as shown. The angle between $\vec{F}$ and $\vec{r}$ is φ. The torque due to the crane's cable rotates the radio antenna upright.

Torque is symbolized by the lowercase Greek letter τ (pronounced "tau"). There are several mathematical expressions for torque. The expression

$$\tau = rF \sin \varphi \tag{12.16}$$

best shows how the three factors F, r, and φ are combined in the concept of torque. As shown in Figure 12.20B, $\vec{r}$ points from the rotation axis to the point of application P, and the angle φ between $\vec{F}$ and $\vec{r}$ is found by extending $\vec{r}$. Only the magnitudes of force and position enter into Equation 12.16. The SI units of torque are newton meters $(\mathrm{N \cdot m})$.

Equation 12.16 fits our experience of trying to turn a bolt–wrench system. A stronger person applying a larger force F will produce a greater torque on the system. A longer wrench handle r allows a person to apply a greater torque. If a person pulls parallel to the handle, then $\varphi = 0$; so, according to Equation 12.16, the torque is zero. The torque is also zero if the person pushes parallel to the handle because then $\varphi = 180°$ and $\tau = Fr \sin 180° = 0$. In fact, the torque is a maximum for $\varphi = 90°$, which confirms our experience that the best way to turn the bolt–wrench system is to apply a force perpendicular to the handle.

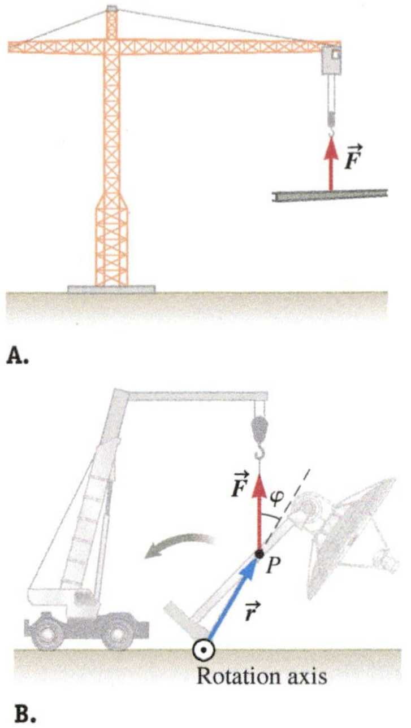

FIGURE 12.20 Analogy between force and torque.

Let's use the example of rotating the radio antenna to look at two other ways to write Equation 12.16. Figure 12.21A shows $\vec{F}$, $\vec{r}$, and φ from Figure 12.20B. We can think of vector $\vec{r}$ as defining a coordinate axis. Vector $\vec{F}$ may then be broken into two components, one parallel to the r axis and the other perpendicular to it. The perpendicular component $\vec{F}_\perp$ is shown in Figure 12.21A. The magnitude of the perpendicular component is found using trigonometry:

$$F_\perp = F \sin \varphi \tag{12.17}$$

Substituting Equation 12.17 into Equation 12.16, we find a new expression for torque:

$$\tau = rF_\perp \tag{12.18}$$

Equation 12.18 says that torque only depends on the component of force that is applied perpendicular to $\vec{r}$ and not on the parallel component.

Another way to express torque comes from looking at the components of $\vec{r}$ relative to $\vec{F}$. As shown in Figure 12.21B, we draw a line along $\vec{F}$ known as the *line of action*. This time, we think of $\vec{F}$ as defining a coordinate axis, and we break $\vec{r}$ into components parallel and perpendicular to that axis. The **moment arm** is defined as the component of $\vec{r}$ that is perpendicular to the force $\vec{F}$. The magnitude of the moment arm is found from trigonometry:

$$r_\perp = r \sin \varphi \tag{12.19}$$

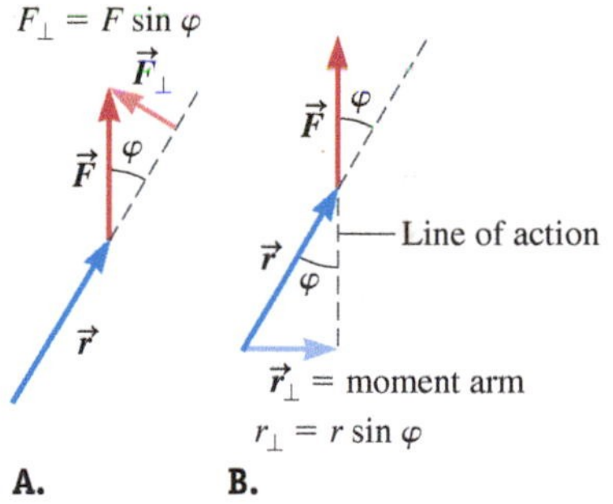

FIGURE 12.21 **A.** The vector $\vec{F}$ may be broken into two components: one parallel to the axis defined by the vector $\vec{r}$ and the other perpendicular to it. We've shown only $\vec{F}_\perp$ here. **B.** The line along $\vec{F}$ is known as the *line of action*. The moment arm is the component of $\vec{r}$ that is perpendicular to line of action.

We find a third expression for torque by substituting Equation 12.19 into Equation 12.16:

$$\tau = r_\perp F \tag{12.20}$$

Equation 12.20 says that the magnitude of the torque is proportional to the moment arm. A longer moment arm produces a greater torque.

CONCEPT EXERCISE 12.6

For each exercise shown in Figure 12.22, how does the moment arm on the dumbbell compare to the length of the person's arm?

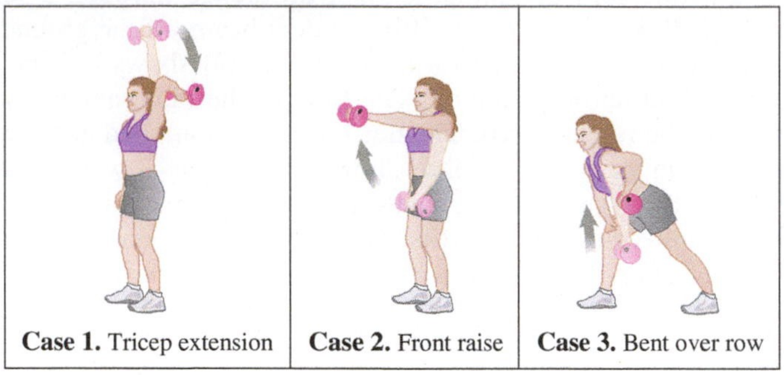

FIGURE 12.22

Recall the first way: A vector multiplied by a scalar results in a vector parallel to the original vector as in $\vec{F}_{tot} = m\vec{a}$. The second way is the dot product between two vectors, yielding a scalar as in $W = \vec{F} \cdot \Delta\vec{r}$.

12-6 Cross Product

Like force, torque is a vector. Equations 12.16, 12.18, and 12.20 give the magnitude of torque, but not its direction. To write a vector equation for torque, we must introduce the third and final way to multiply vectors.

The third type of vector multiplication is known as the *cross product* or *vector product*. We will begin with three generic vectors $\vec{A}$, $\vec{B}$, and $\vec{R}$. Figure 12.23A shows two vectors $\vec{A}$ and $\vec{B}$ that define a single plane. The **cross product** or **vector product** between these two vectors $\vec{A}$ and $\vec{B}$ results in a third vector $\vec{R}$:

VECTOR CROSS PRODUCT Tool

$$\vec{R} = \vec{A} \times \vec{B} \tag{12.21}$$

The magnitude of $\vec{R}$ is given by

$$R = AB \sin \varphi \tag{12.22}$$

where A and B are magnitudes of the vectors $\vec{A}$ and $\vec{B}$, and φ is the angle between the two vectors (Fig. 12.23A); the angle φ must be between 0° and 180°. R is always positive because the magnitudes A and B must be positive, and the sine of an angle between 0 and 180° is positive. R may also be written as $R = A_{\perp}B = AB_{\perp}$, where $A_{\perp} = A \sin \varphi$ and $B_{\perp} = B \sin \varphi$.

The vector resulting from a cross product $\vec{R}$ is always perpendicular to the plane defined by the original two vectors $\vec{A}$ and $\vec{B}$. If $\vec{A}$ and $\vec{B}$ are in the plane of the page as in Figure 12.23, then $\vec{R}$ must either point into or out of the page. The direction of $\vec{R}$ comes from the **right-hand rule** for cross products, a procedure similar to that used in Section 3-2 to find the direction of the z axis in a right-handed coordinate system.

Line up the fingers of your right hand along the first vector in the cross product (in this case, vector $\vec{A}$) and then imagine closing your fingers so that you "push" $\vec{A}$ through the angle φ into the second vector (in this case, vector $\vec{B}$; see Fig. 12.23B). The direction in which your thumb points as you push $\vec{A}$ into $\vec{B}$ is the direction of the resulting vector $\vec{R}$, which in this case is out of the page. Figure 12.23C shows another way to use the right-hand rule. In this procedure, align your right index finger with the first vector $\vec{A}$ and align your right middle finger with the second vector $\vec{B}$. Your right thumb points in the direction of the resulting vector $\vec{R}$.

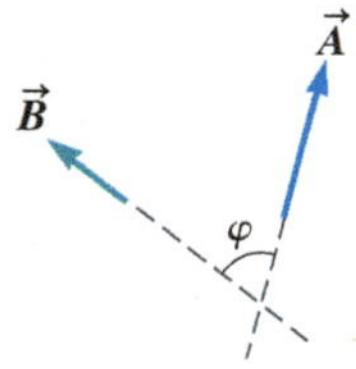

A.

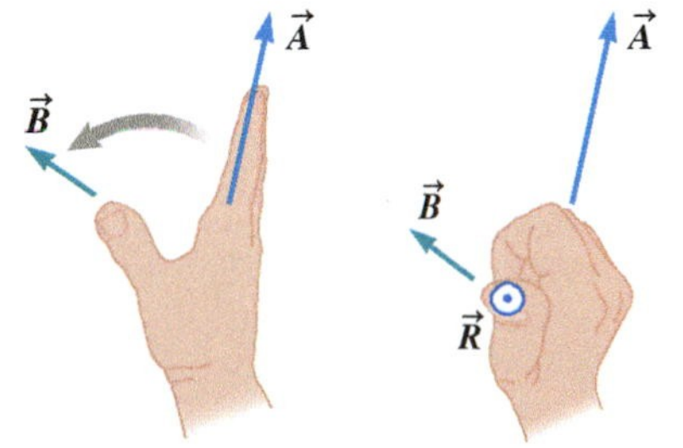

B.

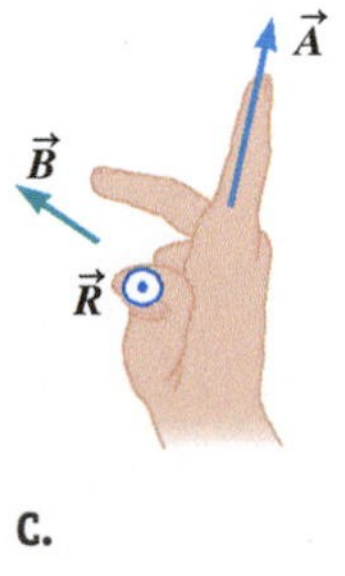

C.

FIGURE 12.23 **A.** Two vectors and $\vec{A}$ and $\vec{B}$ are placed tail to tail and are separated by the angle φ. **B.** We find the direction of the cross product using the right-hand rule. Align the fingers of your right hand with the first vector $\vec{A}$. Then imagine closing your hand such that you push vector $\vec{A}$ through angle φ into vector $\vec{B}$. Your right thumb points in the direction of the resulting vector $\vec{R}$. **C.** Another way to use the right-hand rule is to align the index finger of your right hand with the first vector $\vec{A}$ and your middle finger with the second vector $\vec{B}$. Your right thumb points in the direction of the resulting vector $\vec{R}$.

Switching the order of vectors in the cross product changes the direction of the result. For example, let us find the direction of $\vec{Q} = \vec{B} \times \vec{A}$ using the procedure illustrated in Figure 12.23B. Because $\vec{B}$ is the first vector, align the fingers of your right hand with $\vec{B}$. To close your hand such that you push vector $\vec{B}$ into vector $\vec{A}$, you must orient your hand such that your right thumb goes into the page. (Remember that $0° \le \varphi \le 180°$.) Therefore, $\vec{Q}$ points in the opposite direction to $\vec{R}$ and

$$\vec{A} \times \vec{B} = -\vec{B} \times \vec{A}$$

Thus, the cross product is not commutative and the two vectors must be written in the correct order or you will find the wrong direction for the result.

Now we are ready to write a vector equation for torque. By comparing Equations 12.16 and 12.22, we see that torque may be written in terms of a cross product:

$$\vec{\tau} = \vec{r} \times \vec{F} \quad (12.23)$$

The direction of the torque is found using the right-hand rule for cross products. For example in Figure 12.20B, if you align the fingers of your right hand with $\vec{r}$ and close your fingers into $\vec{F}$, your thumb points out of the page. So, the torque due to the crane on the radio antenna is out of the page.

The direction of torque (just like the direction of angular velocity or angular acceleration) is perpendicular to the plane in which the object rotates. As described in Section 12-2, we restrict our study to rotations around a single, fixed axis usually aligned with the z axis. In that case, a torque will either be in the positive or negative z direction.

EXAMPLE 12.5 Parallelograms and the Cross Product

Two vectors $\vec{B}$ and $\vec{C}$ make up a parallelogram as shown in Figure 12.24. Show that the area A of the parallelogram can be found from the magnitude of the cross product $\vec{A} = \vec{B} \times \vec{C}$.

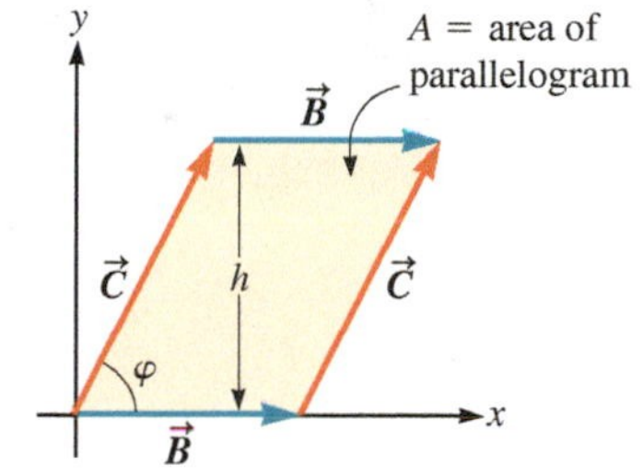

FIGURE 12.24

INTERPRET and ANTICIPATE

Our goal is to show that something we know, the formula for the area of a parallelogram, is equal to another expression, the cross product $\vec{A} = \vec{B} \times \vec{C}$. By working this simple proof, we gain insight into the cross product.

SOLVE

Start with the area of a parallelogram (Appendix A), where b is the length of the "base" and h is the "height" labeled in Figure 12.24.	$A = bh$
From the figure, $b = B$, the magnitude of $\vec{B}$, and $h = C \sin \varphi$. Make the substitutions.	$A = BC \sin \varphi \quad (1)$
Comparing Equation (1) with Equation 12.22, we see that the area A is the magnitude of $\vec{A} = \vec{B} \times \vec{C}$.	$\lvert\vec{A}\rvert = BC \sin \varphi$ ✓

12-7 Rotational Dynamics

Our goal in this section is to find an expression for Newton's second law that is convenient for rotational motion using our analogy with translational motion. In Section 12-5, we found that force is analogous to torque: $\vec{F} \rightarrow \vec{\tau}$. When we want to find the acceleration of an object, we must first find the net force acting on it. By analogy, when we want to find the angular acceleration of an object, we must first find the net torque acting on it.

If more than one torque acts on an object, we find the net torque on the object with vector addition. Restriction to a single, fixed axis greatly simplifies this vector addition. Let's find the total torque acting on the radio antenna in Figure 12.25. Three forces act on the radio antenna: the normal force $\vec{F}_N$ due to the ground, the tension force $\vec{F}_T$ due to the cable, and gravity $\vec{F}_g$ due to the Earth. The normal force acts where the ground touches the antenna along the rotation axis. Because the point of application is on the rotation axis, $\vec{r}_N = 0$, and according to Equation 12.23, the normal force produces zero torque on the antenna. That leaves just the torque due to the crane's cable and the torque due to gravity. The tension force exerted by the crane acts at point P. The vectors $\vec{r}_P$ and

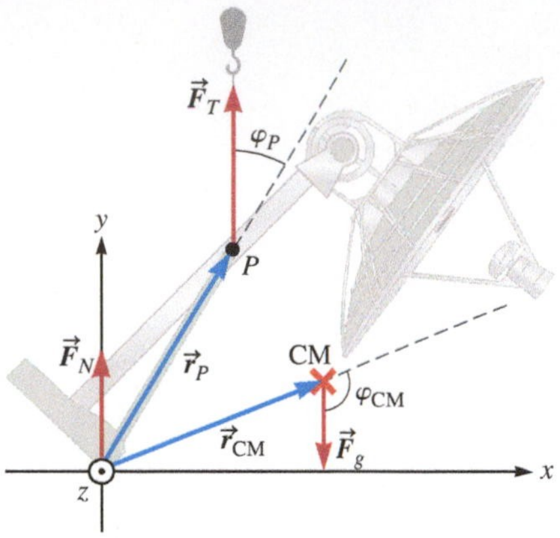

FIGURE 12.25 A crane rotating a radio antenna to a vertical position. The normal force does not exert a torque on the radio antenna because its point of application is on the rotation axis. The torque exerted by the crane is in the positive z direction, and the torque exerted by gravity is in the negative z direction.

$\vec{F}_T$ are shown in Figure 12.25. The torque due to the tension $\vec{\tau}_T$ points out of the page in the positive z direction. The Earth's gravity acts at the antenna's center of mass, which is slightly below the antenna. Applying the right-hand rule to the vectors $\vec{r}_{CM}$ and $\vec{F}_g$ shown in Figure 12.25, we find that the torque due to gravity points into the page (in the negative z direction). So, the total torque acting on the radio antenna is

$$\vec{\tau}_{tot} = \vec{\tau}_N + \vec{\tau}_T + \vec{\tau}_g = 0 + \tau_T\hat{k} - \tau_g\hat{k}$$

Equation 12.16 gives the magnitudes τ_T and τ_g in terms of the variables shown in Figure 12.25:

$$\vec{\tau}_{tot} = (r_P F_T \sin\varphi_P)\hat{k} - (r_{CM} F_g \sin\varphi_{CM})\hat{k}$$

The expression we just found for the total torque exerted on the antenna consists of two terms. The first term is the torque due to the crane, and the second is the torque due to gravity. The direction of the total torque is either in the positive or negative z direction depending on the relative magnitude of these two torques.

Newton's Second Law in Rotational Form

NEWTON'S SECOND LAW IN ROTATIONAL FORM

Underlying Principle

By Newton's second law, the total force acting on a particle is proportional to the particle's acceleration:

$$\vec{F}_{tot} \propto \vec{a}$$

Because the total force and acceleration vectors are proportional, the total force points in the same direction as the particle's acceleration. By analogy, the total torque acting on an object is proportional to the object's angular acceleration:

$$\vec{\tau}_{tot} \propto \vec{\alpha}$$

As in the translational case, because the two vectors are proportional, the total torque points in the same direction as the object's angular acceleration. If we restrict rotational motion to a single, fixed axis aligned with the z axis, the total torque and angular acceleration point in either the positive or negative z direction.

In the case of the antenna (Fig. 12.25), if the torque due to the crane is greater than the torque due to gravity, the angular acceleration is in the positive z direction. However, if the torque due to gravity is the greater one, the angular acceleration is in the negative z direction. If the two torques have the same magnitude, the total torque is zero, and the antenna has zero angular acceleration. It may be rotating at a constant angular speed or be at rest.

The proportionality constant in Newton's second law is inertial mass (Eq. 5.1):

$$\vec{F}_{tot} = m\vec{a}$$

ROTATIONAL INERTIA

Major Concept

Many people refer to rotational inertia as *moment of inertia*. We will not use that phrase in this book because it is too easy to confuse it with other terms.

Inertial mass is the measure of a particle's resistance to acceleration; in other words, mass is a measure of a particle's tendency to maintain its translational velocity. By analogy, **rotational inertia** I is a measure of an object's resistance to angular acceleration; in other words, rotational inertia is a measure of an object's tendency to maintain its angular velocity. Mathematically, Newton's second law for a rotating object is

$$\vec{\tau}_{tot} = I\vec{\alpha} \tag{12.24}$$

where I is the constant of proportionality just as m is in $\vec{F}_{tot} = m\vec{a}$. The SI units of I are found from the units of τ and α.

$$[I] = \frac{[\tau]}{[\alpha]} = \frac{\text{N}\cdot\text{m}}{\text{rad/s}^2} = \frac{\text{kg}\cdot\text{m/s}^2\cdot\text{m}}{\text{rad/s}^2}$$

$$[I] = \text{kg}\cdot\text{m}^2$$

In Chapter 13, we show how to calculate the rotational inertia I, but it is possible to find I experimentally. Imagine applying a single known torque (Eq. 12.23) to an object and measuring its resulting angular acceleration. The torque and angular acceleration vectors point in the same direction, and we find I from Equation 12.24:

$$I = \frac{\tau}{\alpha} \tag{12.25}$$

If more than one torque is applied to the object, then τ in Equation 12.25 is the total torque.

EXAMPLE 12.6 Crall and Whipple Set Things Spinning

Two student researchers, Crall and Whipple, used the laboratory setup shown in Figure 12.26 to measure the rotational inertia of a system consisting of two metal cylinders attached to a light crossbar. The crossbar is mounted on a light axle of radius $R = 0.50$ cm. A lead ball attached to a lightweight string wrapped around the axle rotates the system as the tension in the string provides a constant torque. The tension is measured to be 0.981 N. When the two cylinders are far apart as shown in Figure 12.26, Crall and Whipple measure a constant angular acceleration $\alpha = 0.331\ \text{rad/s}^2$.

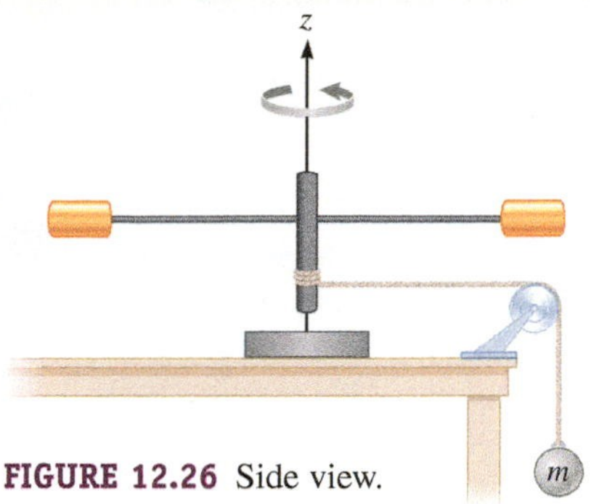

FIGURE 12.26 Side view.

A Find the rotational inertia of the system.

INTERPRET and ANTICIPATE

The only torque acting on the system is due to the tension in the string (Fig. 12.27). Therefore, both the (single) torque and the angular acceleration point in the same direction—the positive z direction—and we can use Equation 12.25 to find the rotational inertia once we find the magnitude of the torque. We should find a numerical result in the form $I =$ _____ $\text{kg} \cdot \text{m}^2$.

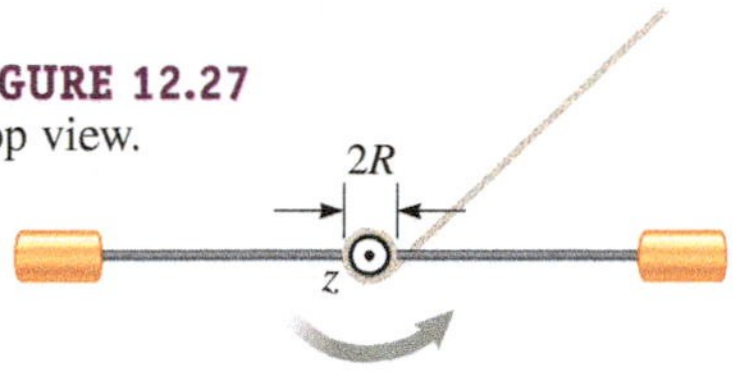

FIGURE 12.27 Top view.

SOLVE

As shown in Figure 12.27, the string is tangent to the axle. Therefore, the tension force in the string is perpendicular to the radius of the axle. This radius is the moment arm in Equation 12.20 with $r_\perp = R$.

$$\tau = r_\perp F$$

$$\tau = RF_T \tag{1}$$

Substitute Equation (1) into Equation 12.25 and the given angular acceleration to find the rotational inertia. We drop the "rad" notation when stating the units of I.

$$I = \frac{\tau}{\alpha} = \frac{RF_T}{\alpha}$$

$$I = \frac{(0.50 \times 10^{-2}\ \text{m})(0.981\ \text{N})}{0.331\ \text{rad/s}^2}$$

$$I = 1.5 \times 10^{-2}\ \text{kg} \cdot \text{m}^2$$

CHECK and THINK

The units are as we expected.

B Whipple moves the two cylinders closer together as shown in Figure 12.28. No other changes are made to the setup. Crall measures the angular acceleration and finds $\alpha = 1.29\ \text{rad/s}^2$. What is the rotational inertia of the system?

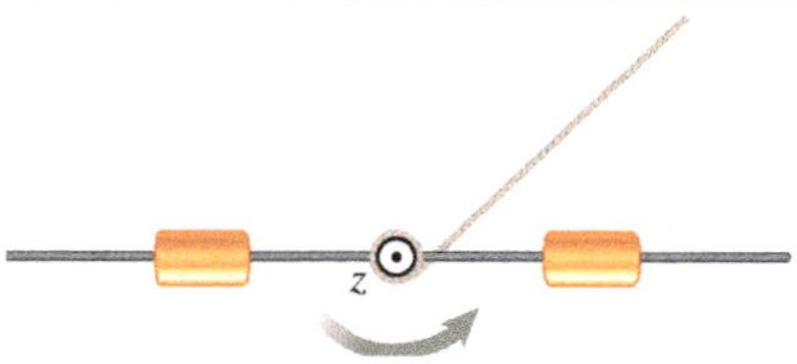

FIGURE 12.28 The cylinders have been moved closer together.

INTERPRET and ANTICIPATE

Because the change to the setup only involves the cylinders, the torque is the same as it was in part A. All we need to do is use the new observed angular acceleration to find the new rotational inertia.

SOLVE

Substitute the new value of α into Equation 12.25.

$$I = \frac{\tau}{\alpha} = \frac{(0.50 \times 10^{-2}\text{m})(0.981\ \text{N})}{1.29\ \text{rad/s}^2}$$

$$I = 3.8 \times 10^{-3}\text{kg} \cdot \text{m}^2$$

CHECK and THINK

We just found that moving the two cylinders closer together reduced the rotational inertia of the system. Would it be easier to stop the rotating system shown in Figure 12.27 or the one shown in Figure 12.28? When the cylinders are moved closer together, the system has a smaller rotational inertia, so a weaker torque would result in the same angular acceleration, and it is easier to stop the system in Figure 12.28. The difference between the systems in Figures 12.27 and 12.28 is analogous to the difference between a bowling ball and a soccer ball. Imagine a bowling ball and a soccer ball moving with the same velocity. It is much easier to stop the moving soccer ball because it has much less inertial mass than does the bowling ball.

EXAMPLE 12.7 CASE STUDY Queen Hatshepsut's Obelisk

Our case study focused on how ancient people constructed megalithic monuments. For example, how did ancient Egyptians raise granite obelisks that were up to 100 ft tall and weighed 400 tons without damaging them? One possibility is that they used gravity to get an obelisk to a near-vertical position on its pedestal and then pulled it up from there. In a PBS *Nova* documentary[2], researchers are shown testing this idea with a 40-ft, 40-ton replica obelisk in a quarry in Massachusetts (Fig. 12.29A). To prevent damage, the obelisk's rotation is primarily controlled by sand, which required the construction of a large sandbox. To understand why the sand is required, let us imagine the same scenario but with no sand (or ropes) as shown in Figure 12.29B. Assume the center of mass is at the center of the obelisk; the obelisk hangs over a pit and pivots on an edge that is about 1 m to left of the center of mass. Find the magnitude of the obelisk's angular acceleration and the magnitude of the tangential acceleration of a point on its base at the moment the obelisk is released. The obelisk's rotational inertia is $I = 4.7 \times 10^5\ \text{kg}\cdot\text{m}^2$.

© PBS NOVA

A.

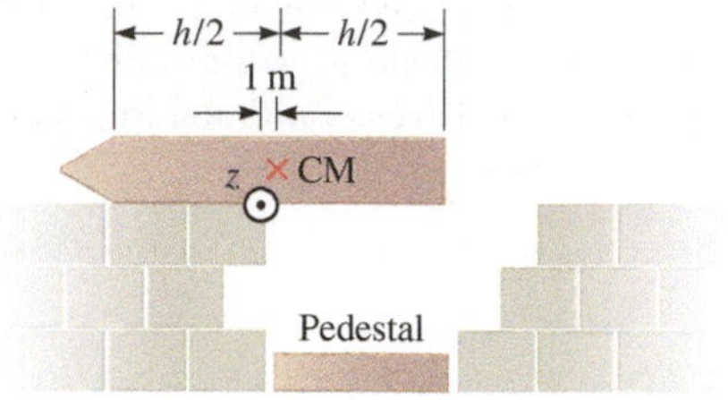

B.

FIGURE 12.29 **A.** In the PBS *Nova* documentary, researchers try to raise a replica obelisk. **B.** Raising an obelisk without sand or ropes.

INTERPRET and ANTICIPATE

When the obelisk is released, the only torque acting on it is due to gravity $\vec{\tau}_g$. There is a normal force acting at the edge of the pit, but it does not contribute a torque because that force is applied along the axis of rotation. With only one torque acting on the obelisk, Equation 12.24 is simplified. Gravitational torque $\vec{\tau}_g$ and $\vec{\alpha}$ both point in the negative z direction. Our results should be numerical in the forms $\alpha =$ _____ rad/s^2 and $a_T =$ _____ m/s^2.

SOLVE

First convert the obelisk's mass and height to SI units.	$m = 3.6 \times 10^4\ \text{kg} \qquad h = 12\ \text{m}$
Use Equation 12.20 to find the torque due to gravity.	$\tau_g = r_\perp F_g = r_\perp mg$
The moment arm $r_\perp = 1$ m. Substitute numerical values.	$\tau_g = (1\ \text{m})(3.6 \times 10^4\ \text{kg})(9.81\ \text{m/s}^2) = 3.5 \times 10^5\ \text{N}\cdot\text{m}$
The angular acceleration comes from $\vec{\tau}_{\text{tot}} = I\vec{\alpha}$ (Eq. 12.24).	$\tau_g = I\alpha$ $\alpha = \dfrac{\tau_g}{I} = \dfrac{3.5 \times 10^5\ \text{N}\cdot\text{m}}{4.7 \times 10^5\ \text{kg}\cdot\text{m}^2} = 0.74\ \text{rad/s}^2$
The tangential acceleration comes from Equation 12.14, where $r = (h/2) + 1$ m.	$a_T = \alpha r$ $a_T = (0.74\ \text{rad/s}^2)(7\,\text{m}) = 5.2\ \text{m/s}^2 \approx \frac{1}{2}g$

CHECK and THINK

The results are in the form we expected. The tangential acceleration is about $\frac{1}{2}g$, which is rather high. Without the sand, the obelisk would rotate too quickly and would probably break when the base hit the pedestal. When damage to the base of the stones was not a concern as it was at Stonehenge (Example 12.8), it is believed that rotation was controlled without using sand.

[2]The PBS television show *Nova* organized teams to build replicas of several famous ancient monuments in two series of documentaries, *Secrets of Lost Empires* (1997) and *Secrets of Lost Empires II* (2000).

Levers

Archimedes (c. 287–212 BCE), an ancient Greek mathematician and inventor, discussed levers in his book *On the Equilibrium of Planes*. Archimedes understood that levers provide a mechanical advantage allowing a person to move very massive objects that would otherwise be impossible to move with just his or her own human strength. A famous quotation is attributed to Archimedes, "Give me a place to stand and I will move the Earth."[3]

A **lever** is a simple machine used to amplify the force exerted on an object. In Figure 12.30A, a load is placed on one end of the arm of the lever, and a force $\vec{F}_{app}$ is applied to the other end as shown. The net torque causes the arm to rotate around the fulcrum or pivot, moving the load.

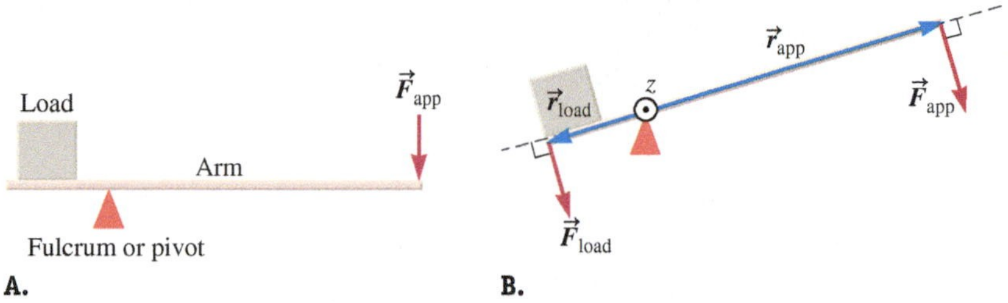

FIGURE 12.30 **A.** The arm of a lever may rotate on a fulcrum. **B.** Two torques act on the lever: one exerted by the load, the other by the applied force. (The normal force exerted by the fulcrum exerts no torque, so it is not included here.)

We can find an expression for $\vec{F}_{app}$ in terms of the parameters shown in Figure 12.30B. Assume the arm rotates at a constant angular speed so that $\vec{\alpha} = 0$. Then, according to Equation 12.24 (Newton's second law for rotation), the net torque is zero. Assume also the weight of the arm is negligible so that there are three forces acting on the arm: the normal force exerted by the fulcrum, the applied force $\vec{F}_{app}$, and the normal force $\vec{F}_{load}$ due to the load. (Remember that the normal force exerted by the load will not always equal its weight.) The rotation axis passes through the fulcrum, so the normal force exerted by the fulcrum does not exert a torque. We use Equation 12.23 (vector definition of torque) to find the torque $\vec{\tau}_{app}$ due to the applied force,

$$\vec{\tau}_{app} = -(r_{app}F_{app}\sin 90°)\hat{k} = -r_{app}F_{app}\hat{k} \tag{12.26}$$

and the torque $\vec{\tau}_{load}$ due to the load,

$$\vec{\tau}_{load} = (r_{load}F_{load}\sin 90°)\hat{k} = r_{load}F_{load}\hat{k} \tag{12.27}$$

Substitute Equations 12.26 and 12.27 into Equation 12.24 and set the expression equal to zero because $\vec{\alpha} = 0$:

$$\vec{\tau}_{tot} = \vec{\tau}_{app} + \vec{\tau}_{load} = I\vec{\alpha}$$

$$\vec{\tau}_{tot} = -r_{app}F_{app}\hat{k} + r_{load}F_{load}\hat{k} = 0$$

Solve for $\vec{F}_{app}$:

$$r_{app}F_{app}\hat{k} = r_{load}F_{load}\hat{k}$$

$$F_{app} = \frac{r_{load}}{r_{app}}F_{load}$$

By Newton's third law, the normal force $\vec{F}_{load}$ exerted by the load on the arm is equal in magnitude to the normal force $\vec{F}_N$ exerted by the arm on the load, so we have

$$F_{app} = \frac{r_{load}}{r_{app}}F_N \tag{12.28}$$

If $r_{app} > r_{load}$ as shown in Figure 12.30B, then $F_{app} < F_N$. In other words, the force F_N exerted by the lever on the load is greater than the applied force F_{app} by a factor of r_{app}/r_{load}. If a person applies a 500-N force and $r_{app} = 3r_{load}$, the force exerted on the load is 1500 N!

[3]Archimedes may never have actually made the statement, but many ancient authors such as Pappus of Alexandria (290–350 CE) and Plutarch (c. 45–120 CE) attribute such a comment to him.

EXAMPLE 12.8 CASE STUDY Building Stonehenge

The upright stones in the largest trilithon at Stonehenge are partially buried in the ground (Fig. 12.31A), unlike Queen Hatshepsut's obelisk, whose base is exposed. Therefore, Stonehenge's ancient builders in Britain did not need to be concerned about slightly damaging the base of the upright. The PBS *Nova* documentary presents a test of how the upright may have been raised. Archeologists believe that pits were first excavated for the upright stones (Fig. 12.31B) and that each upright was then rotated by a torque due to gravity so that the stone was nearly vertical as in Figure 12.31C. An A-frame lever then aided in bringing the stone to vertical (Fig. 12.31D). The A-frame lever is approximately parallel to the stone. The *Nova* team, consisting of 70 people, successfully raised a replica stone. Estimate the number of people that would be required if the A-frame lever were not used. Model the A-frame lever as a simple rod with the fulcrum at the ground and approximately parallel ropes attached at the top and at a distance r_{load} from the bottom, measured along the rod.

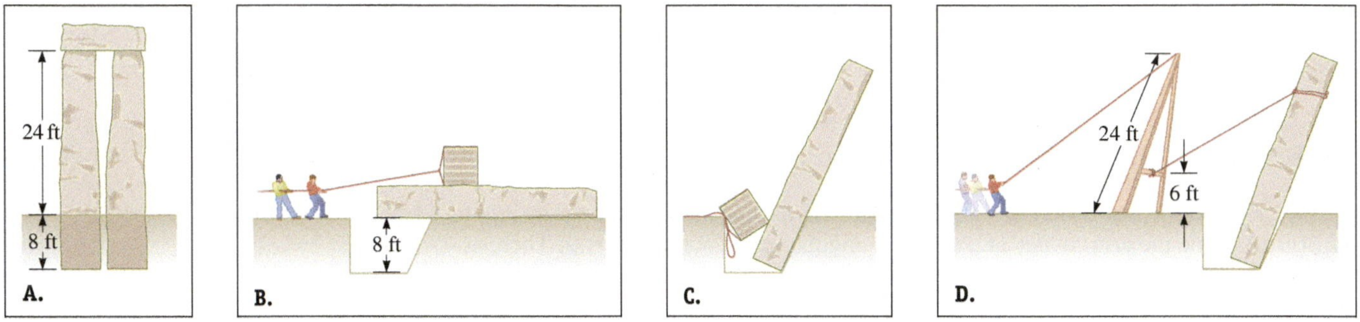

FIGURE 12.31 **A.** The largest stone in Stonehenge supports a trilithon. The stone is about 32 ft tall with 8 ft under ground. **B.** A large stone is slid into place partially over a pit. **C.** A smaller stone is slid across so that a torque due to gravity rotates the large stone into the pit. **D.** The large stone is brought to a vertical position with an A-frame lever.

INTERPRET and ANTICIPATE

We need to find the mechanical advantage the A-frame lever provides. In other words, we need to find by what factor the lever amplifies the force exerted by the people. Then, we assume the size of the team must be increased by that factor if no lever is used.

Start with a coordinate system and a sketch (Fig. 12.32) to identify the torques exerted on the rod representing the A-frame lever. The z axis is on the fulcrum and points out of the page. Two nonzero torques act on the rod: the torque $\vec{\tau}_{app}$ due to the force $\vec{F}_{app}$ exerted by the team and the torque $\vec{\tau}_{load}$ due to the force $\vec{F}_{load}$ exerted by the stone. The torque $\vec{\tau}_{app}$ points in the positive z direction, and the torque $\vec{\tau}_{load}$ points in the negative z direction. When the angular acceleration is zero (so that the stone is just on the verge of moving), the two torques have equal magnitudes.

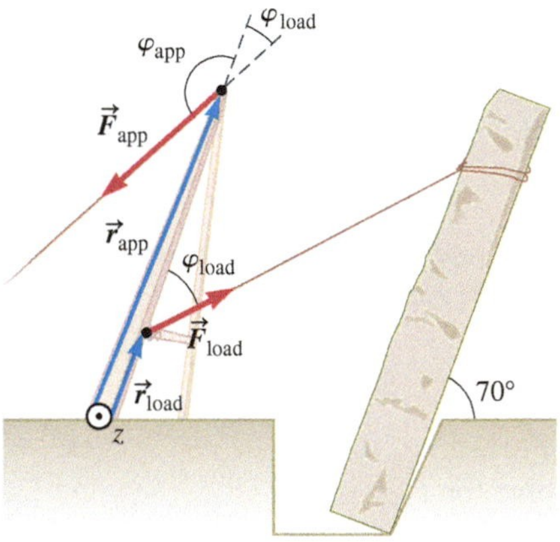

FIGURE 12.32

SOLVE

Assume the ropes are parallel to find the relationship between the angles φ_{app} and φ_{load} and the sine of these angles.

$$\varphi_{app} = 180° - \varphi_{load}$$
$$\sin\varphi_{app} = \sin(180° - \varphi_{load})$$
$$\sin\varphi_{app} = \sin\varphi_{load}$$

Because $\sin\varphi_{app} = \sin\varphi_{load}$, Equation 12.28 applies to this lever problem, but instead of a normal force F_N on the load, we have a tension force F_T. Solve for the ratio F_T/F_{app}. Because we are interested in finding a unitless ratio, we can work in the U.S. customary units of feet (Fig. 21.31D).

$$r_{load}F_T\sin\varphi_{load} = r_{app}F_{app}\sin\varphi_{load}$$
$$F_{app} = \frac{r_{load}}{r_{app}}F_T \quad \text{which resembles Eq. 12.28}$$
$$\frac{F_T}{F_{app}} = \frac{r_{app}}{r_{load}} = \frac{24\text{ ft}}{6\text{ ft}} = 4$$

The A-frame amplifies the force exerted by the team by a factor of four. If no A-frame lever were used, the team would need to be four times larger.

$$\text{size of team using A-frame} = 70\text{ people}$$
$$\text{size of team without A-frame} = 4 \times 70 = 280\text{ people}$$

CHECK and THINK

This example illustrates the power of even simple mechanical tools. Levers fashioned from modest materials such as wood and rope were probably used by many ancient civilizations to construct monuments. No extraterrestrial workforce or tools were required.

We end this chapter by considering another example of ancient engineering. The Moai statues on Easter Island are counted among the seven wonders of the modern world. There is essentially no written record on how the Moai were constructed, transported, and erected, but many surviving partially constructed Moai provide important evidence. A volcanic crater served as the primary quarry (Fig. 12.33A). The Moai's fronts were carved while they lay inside the quarry. Just outside the quarry are incomplete Moai that have been erected facing the sea so that their backs could be carved (Fig. 12.33B). There are broken Moai along the transportation path to the sea. The final Moai stand near the sea facing inland (Fig. 12.33C).

Archeologists have developed and tested theories for each stage of the Moai construction processes. The Easter Islanders probably used stone tools to carve the statues and may have used wooden sleds to transport and erect them. Ramps made out of stones were probably used to raise the Moai to about 45°, and then levers were likely used to lift them to a vertical position. This method is similar to the tools and processes we discussed for raising obelisks in Egypt and stones at Stonehenge. One part of the Moai construction process is unique, however. All the complete Moai face inland, but all the partially complete Moai near the quarry face the sea. The Easter Islanders must have had a way to rotate their statues. One possibility is that a small workforce used levers to rotate the statues while they were attached to their sleds. Another recently described possibility for transport and rotation (thought not accepted by all experts) is that once a statue was nearly vertical, three teams of islanders "walked" the statue with a rocking motion using ropes attached to the statue's head.[4]

Although no experiment can be performed that conclusively refutes the extraterrestrial hypothesis, there is no evidence that uniquely supports the claim that extraterrestrials built the ancient monuments. There is, however, very strong evidence that ancient people used modest tools made out of readily available materials and basic physics to construct megaliths on their own.

[4]Hannah Bloch, "The Riddle of Easter Island," *National Geographic* 222(1), 31-49, July 2012.

A.

B.

C.

FIGURE 12.33 **A.** Stones are quarried near the top of a volcano. **B.** Moai are erected facing the sea so that their backs can be carved. **C.** Moai are transported to the sea, where they are erected facing inland.

SUMMARY

Underlying Principle

Newton's second law in rotational form: $\vec{\tau}_{\text{tot}} = I\vec{\alpha}$ (Eq. 12.24)

Major Concepts

1. A **rigid object** is an object that moves with all its components locked together so that its shape does not change.
2. Every particle in a rotating rigid object moves in a circular path around the **rotation axis**.
3. A **fixed axis** is a rotation axis that does not move relative to the observer.
4. Angular kinematic quantities
   a. The (positive) **angular position** θ of the reference line is measured counterclockwise from the positive x axis.
   b. A change in the angular position is known as the **angular displacement** and is given by Equation 12.2: $\Delta\theta = \theta_f - \theta_i$.

Major Concepts—cont'd

c. **Angular speed** is given by Equation 12.4: $\omega = d\theta/dt$. The direction of **angular velocity** is given by the right-hand rule.

d. **Angular acceleration** is given by Equation 12.6: $\vec{\alpha} = d\vec{\omega}/dt$. If an object's rotation is speeding up, then $\vec{\alpha}$ points in the same direction as $\vec{\omega}$. If the object is slowing down, then $\vec{\alpha}$ and $\vec{\omega}$ point in opposite directions.

5. A **torque** on a rotating object is analogous to a force on a translating object: A force causes the acceleration of a translating object; a torque causes the angular acceleration of a rotating object. Mathematically, torque is given by Equation 12.23: $\vec{\tau} = \vec{r} \times \vec{F}$. The torque's direction is found from the right-hand rule for vector cross products.
6. **Rotational inertia** I is a measure of an object's resistance to have an angular acceleration.

Tools

1. An observer uses the **right-hand rule** convention to find the direction of $\vec{\omega}$ by curling the fingers of his or her right hand in the same sense as the rotation. The observer's right thumb then points in the direction of $\vec{\omega}$. Because we have decided to use the convention that the rotation axis is aligned with the z axis, $\vec{\omega}$ will point in either the positive or negative z direction.
2. The **cross product** or **vector product** between two vectors $\vec{A}$ and $\vec{B}$ results in a third vector $\vec{R}$: $\vec{R} = \vec{A} \times \vec{B}$. The magnitude of R is given by Equation 12.22: $R = AB \sin \varphi$. The direction of $\vec{R}$ comes from the right-hand rule for cross products. Line up the fingers of your right hand along the first vector in the cross product (in this case, vector $\vec{A}$) and then imagine closing your fingers so that you "push" $\vec{A}$ through the angle φ into the second vector (in this case, vector $\vec{B}$); see Figure 12.23B. The direction in which your thumb points as you push $\vec{A}$ into $\vec{B}$ is the direction of the resulting vector $\vec{R}$.

PROBLEMS AND QUESTIONS

A = algebraic C = conceptual E = estimation G = graphical N = numerical

12-1 Rotation Versus Translation

Problems 1 and 2 are paired.

1. **C** Often, we model the Moon as a particle in a circular orbit around the Earth. The same side of the Moon always faces the Earth. Sketch the Moon in its orbit. Explain in what way the particle model is insufficient.
2. **C** Suppose a satellite orbits the Earth such that it is well modeled as a particle. Draw a sketch of it in its orbit. Explain how its motion is different from the Moon's motion around the Earth.

Problems 3 and 4 are paired.

3. **C** An ice skater spins on one foot. Is he rotating? If so, describe his axis of rotation. Explain your answer.
4. **C** A speed ice skater travels around a closed, circular track. Is she rotating? If so, describe her axis of rotation. Explain your answer.

12-2 Rotational Kinematics

5. **N** A ceiling fan is rotating counterclockwise with a constant angular acceleration of π rad/s^2 about a fixed axis perpendicular to its plane and through its center. Assume the fan starts from rest. **a.** What is the angular velocity of the fan after 4.0 s? **b.** What is the angular displacement of the fan after 4.0 s? **c.** How many revolutions has the fan gone through in 4.0 s?
6. **N** As seen from above the Earth's North Pole, the Moon's orbit is counterclockwise. Use a coordinate system with the positive z axis pointing north. Find the magnitude and direction of the Moon's angular velocity. *Hint*: Draw a sketch of the Moon's orbit from this perspective above the North Pole, including the coordinate system.

Problems 7 and 8 are paired.

7. **N** A rotating object's angular position is given by $\theta(t) = (1.54t^2 - 7.65t + 2.75)$ rad, where t is measured in seconds. Find **a.** the object's angular speed when $t = 3.50$ s and **b.** the magnitude of the angular acceleration when $t = 3.50$ s.
8. **N** A rotating object's angular position is given by $\theta(t) = (1.54t^2 - 7.65t + 2.75)$ rad, where t is measured in seconds. **a.** When is the object momentarily at rest? **b.** What is the magnitude of the angular acceleration at that time?
9. Jupiter rotates about its axis once every 9 hours 55 minutes.
   a. **N** What is Jupiter's angular speed of rotation?
   b. **C** What is the effect of this rapid rotation on the shape of the planet?
10. **C** The text of this chapter says that "angular displacement does not obey the commutative rule of vector addition." What does that statement mean? Try the following. Place a book on a table with the front cover facing upward and the spine toward you. Now make two rotations.

    First, perform a 90° rotation (clockwise as seen from above) around the axis perpendicular to the table and through the book. Next, perform a 90° rotation around the horizontal axis parallel to the table and through the book (clockwise as seen from the left side of the book). The book should now be upright and facing you.

Now, return the book to its original position and make these same two rotations, *but in the opposite order*. What result do you get this time? Does angular displacement obey the commutative rule of vector addition? Explain.

Problems 11 and 12 are paired.

11. **N** The long hand on a clock is known as the *minute hand*. It completes one clockwise rotation each hour. Choose a coordinate system such that a z axis points out of the clock. Consider a 15-minute time interval. Find the minute hand's **a.** angular displacement, **b.** angular velocity, and **c.** angular acceleration.

12. The long hand on a clock is known as the *minute hand*. Usually, it completes one clockwise rotation each hour. A particular clock, however, is slowing down at a constant rate of 2.909×10^{-7} rad/s^2. Choose a coordinate system such that a z axis points out of the clock.

a. **C** What is the direction of the minute hand's angular acceleration?

b. **N** Suppose the slow clock and a working clock both read noon at the same moment and at that moment their angular velocities are equal. What does the broken clock read when the working clock reads 1 PM?

13. **E** Assuming an automobile in Los Angeles is driven a total of 12,000 miles per year in commuting to work, what is the approximate number of revolutions the car's tires undergo each year?

Problems 14 and 15 are paired.

14. **N** Use the information about the Crab Nebula's pulsar from Example 12.2 (page 339). **a.** Determine the period of the pulsar's rotation **b.** Determine the number of seconds by which this period increases each second (dT/dt).

15. **N** Example 12.2 gives the rotation rate for the Crab Nebula's pulsar and also the (negative) angular acceleration of this rotation. The rotation rate of this pulsar as well as that of many other pulsars does not show a long, steady decrease, however. There are occasional abrupt *increases* in the rotation rate called *glitches* that are caused by mechanisms in the interior of the pulsar. For the glitch observed in the nebula's pulsar in June 2000, the fractional increase in the rotation rate was $\Delta\omega/\omega = (\omega_{new} - \omega_{orig})/\omega_{orig} = 2.4 \times 10^{-8}$. **a.** What was the rotation rate after the glitch? **b.** The angular acceleration (or spin-down rate, as it is called) also changed after the glitch: $\Delta\alpha/\alpha = 5 \times 10^{-3}$. What was the new angular acceleration? **c.** How long after the glitch did it take the pulsar to return to its pre-glitch rotation rate?

16. **C** A disk rolls up an inclined plane as shown in Figure P12.16, reaches point A, stops there momentarily, and then rolls down the inclined plane. Use the coordinate system shown to determine the direction of the angular velocity and the angular acceleration in each part of the motion as given below. If either one is zero, say so. Explain your answers. **a.** When the disk is going up the incline. **b.** At point A when the disk stops momentarily. **c.** When the disk is rolling down the incline

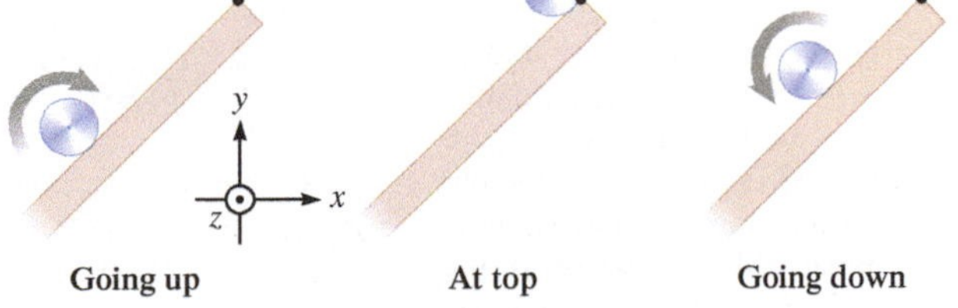

FIGURE P12.16

17. **N** Jeff, running outside to play, pushes on a swinging door, causing its motion to be briefly described by $\theta = t^2 + 0.800t + 2.00$, where t is in seconds and θ is in radians. At $t = 0$ and at $t = 1.50$ s, what are the **a.** angular position, **b.** angular speed, and **c.** angular acceleration of the door?

12-3 Special Case of Constant Angular Acceleration

18. **N** A potter's wheel rotating at 240 rev/min is switched off and rotates through 80.0 revolutions prior to coming to rest. What is the constant angular acceleration of the potter's wheel during this interval?

19. **N** Friction in an old clock causes it to lose 1 minute per hour. Assume the angular acceleration is constant, the positive z axis points out of the face of the clock, and at the beginning of the hour it was at its ideal angular speed. Find the angular acceleration of the minute hand (long hand) and its angular velocity at the end of the hour.

20. **N** A wheel starts from rest and in 12.65 s is rotating with an angular speed of $5.435\,\pi$ rad/s. **a.** Find the magnitude of the constant angular acceleration of the wheel. **b.** Through what angle does the wheel move in 6.325 s?

21. **N** Some people are able to spin a basketball on the tip of a finger. It is often done by balancing the ball on the tip of the finger and brushing the other hand along the side of the ball to cause it to rotate. Suppose the ball begins from rest and reaches a final angular speed of 18.65 rad/s in 1.10 s. **a.** Assuming the ball is subject to a constant angular acceleration, what is the magnitude of the constant acceleration? **b.** Through how many revolutions does the ball rotate during the 1.10 s?

22. **N** Starting from rest, a wheel reaches an angular speed of 15.0 rad/s in 5.00 s. **a.** What is the magnitude of the constant angular acceleration of the wheel? **b.** Through what angle in radians does the wheel rotate during this time interval?

23. **N** A potter's wheel is rotating with an angular velocity of $\vec{\omega} = 24.2\hat{k}$ rad/s. The wheel is slowing down as it is subject to a constant angular acceleration and eventually comes to a stop. **a.** If the wheel goes through 21.0 revolutions as it is slowed to a stop, what is the constant angular acceleration applied to the wheel? **b.** How long does it take for the wheel to come to rest?

Problems 24 and 25 are paired.

24. The angular speed of a wheel is given by $\omega(t) = 72.5 \text{ rad/s} + (9.34 \text{ rad/s}^2)t$.

a. **C** Is the wheel's angular acceleration constant? Explain.

b. **A** Find an expression for the angular acceleration.

25. The angular speed of a wheel is given by $\omega(t) = 72.5 \text{ rad/s} + (9.34 \text{ rad/s}^2)t + (2.24 \text{ rad/s}^3)t^2$.

a. **C** Is the wheel's angular acceleration constant? Explain.

b. **A** Find an expression for the angular acceleration.

26. **N** A wheel is programmed to follow a specific cycle in which it accelerates from rest for 3.40 s and reaches an angular speed of 25.0 rev/s. Then, 3.00 s after reaching this speed, a braking mechanism is engaged, stopping the wheel smoothly in 15.0 s. What is the number of revolutions completed by the wheel during this cycle?

12-4 The Connection Between Rotation and Circular Motion

27. **N** An electric food processor comes with many attachments for blending and slicing food. Assume the motor maintains the same angular speed for all the various attachments. The largest attachment has a diameter of 12.0 cm, and the smallest has a diameter of 5.00 cm. Find the ratio of the translational speeds of points on the edges of each attachment.

28. **C** A truck's tire starts from rest and begins to rotate with a constant angular acceleration. Consider a point on the outer edge of the tire. **a.** Is the tangential acceleration component constant? Explain. **b.** Is the centripetal acceleration component constant? Explain.

29. **N** A bicyclist is testing a new racing bike on a circular track of radius 53.0 m. The bicyclist is able to maintain a constant speed of 24.0 m/s throughout the test run. **a.** What is the angular speed of the bike? **b.** What are the magnitude and direction of the bike's acceleration?

30. **N** Paul is listening to an old-fashioned vinyl record on a turntable. He likes to place coins on the record and watch them spin. A penny is 3.55 cm from the axis of rotation, and a nickel is 8.10 cm from the axis of rotation. The record rotates at 33 rpm (revolutions per minute). What is the tangential speed of each coin?

31. **N** A disk is initially at rest. A penny is placed on it at a distance of 1.0 m from the rotation axis. At time $t = 0$ s, the disk begins to rotate with a constant angular acceleration of 2.0 rad/s^2 around a fixed, vertical axis through its center and perpendicular to its plane. Find the magnitude of the net acceleration of the coin at $t = 1.5$ s.

Problems 32, 33, and 34 are grouped.

32. **N** The blades on modern wind turbines are typically 20 to 40 meters in length. Two different turbines have their blades attached with one end at the center of rotation. Consider two points, each at the outer end of a blade on each of the turbines: one with a blade length of 20.0 m, and the other with a blade length of 40.0 m. Each blade is rotating with a constant angular speed of 3.88 rad/s, and the rotation is considered for 10.0 s. **a.** What is the speed of each point in meters per second? **b.** What is the angular distance traveled by each point? **c.** What is the translational distance traveled by each point? **d.** What is the magnitude of the centripetal acceleration that would be experienced by an object located at each point?

33. **C** Consider again the two wind turbines in Problem 32. **a.** At what point along the 40-m blade would the angular speed of the point be the same as the angular speed at the end of the 20-m blade? Explain your answer. **b.** At what point along the 40-m blade would the translational speed of the point be the same as the speed at the end of the 20-m blade? Explain your answer.

34. **C** Consider again the two wind turbines in Problem 32. **a.** At what point along the 40-m blade would the angular distance traveled by the point be the same as the angular distance traveled by the point at the end of the 20-m blade? **b.** At what point along the 40-m blade would the translational distance traveled by the point be the same as the translational distance traveled by the point at the end of the 20-m blade?

35. **N** In testing an automobile tire for proper alignment, a technician marks a spot on the tire 0.200 m from the center. He then mounts the tire in a vertical plane and notes that the radius vector to the spot is at an angle of 35.0° with the horizontal. Starting from rest, the tire is spun rapidly with a constant angular acceleration of 3.00 rad/s^2. **a.** What is the angular speed of the wheel after 4.00 s? **b.** What is the tangential speed of the spot after 4.00 s? **c.** What is the magnitude of the total accleration of the spot after 4.00 s?" **d.** What is the angular position of the spot after 4.00 s?

Problems 36 and 37 are paired.

36. **N** Two children, each with a mass of 25.0 kg, are at fixed locations on a merry-go-round (a disk that spins about an axis perpendicular to the disk and through its center; Fig. P12.36). One child is 0.75 m from the center of the merry-go-round, and the other is near the outer edge, 3.00 m from the center. With the merry-go-round rotating at a constant angular speed, the child near the edge is moving with translational speed of 12.5 m/s. **a.** What is the angular speed of each child? **b.** Through what angular distance does each child move in 5.0 s? **c.** Through what distance in meters does each child move in 5.0 s? **d.** What is the centripetal force experienced by each child as he or she holds on? Which child has a more difficult time holding on?

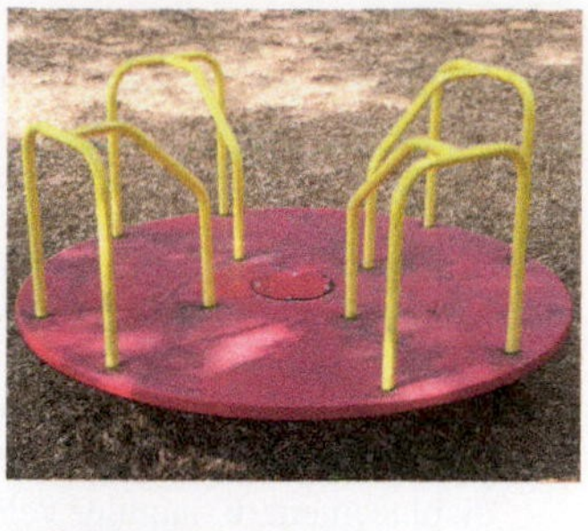

Peter McGahey © Cengage Learning

FIGURE P12.36 Problems 36 and 37. A merry-go-round in a neighborhood park.

37. **N** A merry-go-round at a children's park begins at rest and is pushed by a parent while a child holds onto the ride at the outer edge, 3.00 m from the center. The parent pushes with a constant force for 2.54 s until the child is moving with a speed of 12.5 m/s. What is the magnitude of the net acceleration on the child the instant before the parent stops pushing the merry-go-round?

38. **N** A wheel rotating at a constant rate of 1850 rev/min has a diameter of 17.8 cm. **a.** What is the angular speed of the wheel in radians per second? **b.** What is the tangential speed of a point on the wheel's rim? **c.** What is the centripetal acceleration of a point on the wheel's rim? **d.** What is the total distance traversed by a point on the wheel's rim from $t = 0$ to $t = 3.00$ s?

12-5 Torque

39. **C** Why are doorknobs placed on the edge opposite the hinges?

40. **C** Why would a person with weak hands prefer to have a door handle such as the one shown in Figure P12.40 instead of a round doorknob?

Peter McGahey © Cengage Learning

FIGURE P12.40

41. **N** A student pulls with a 300.0-N force on the outer edge of a door of width 1.51 m that pivots about point P as shown in Figure P12.41. Find the magnitude of the torque applied to the door in each case, where the force is applied in a different direction as shown.

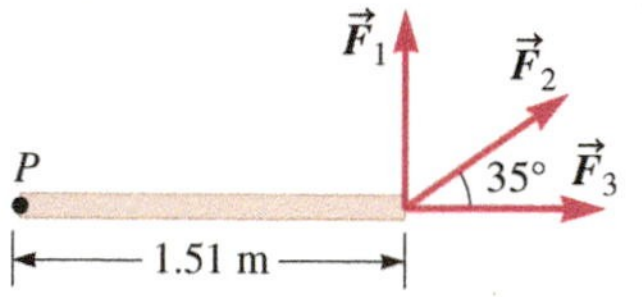

FIGURE P12.41

42. **C** You are working on your car and need to remove a cover in the engine compartment, which requires removing several screws. As you search through your tools, you find that you have two screwdrivers that will fit the screws, but each has a different handle diameter. One of the handles is very thick, whereas the other handle diameter is rather small by comparison. Which screwdriver should you choose to make the job easier for you to accomplish? Explain your choice and explain what quantity most directly applies in describing why the job is "easier" for you.

43. **N** A wheel of inner radius $r_1 = 15.0$ cm and outer radius $r_2 = 35.0$ cm shown in Figure P12.43 is free to rotate about the axle through the origin O. What is the magnitude of the net torque on the wheel due to the three forces shown?

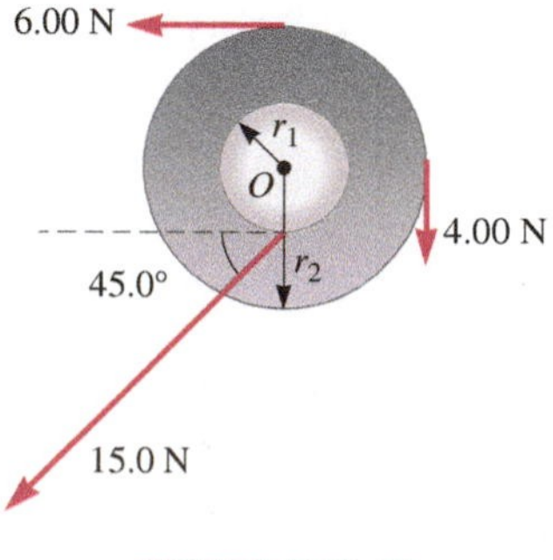

FIGURE P12.43

44. **N** A uniform plank 6.0 m long rests on two supports, 2.5 m apart (Fig. P12.44). The gravitational force on the plank is 100 N. The left end of the plank is 1.5 m to the left of the left support, so the plank is not centered on the supports. A person is standing on the plank half a meter to the right of the right support. The gravitational force on this person is 80.0 N. How far to the right can the person walk before the plank begins to tip?

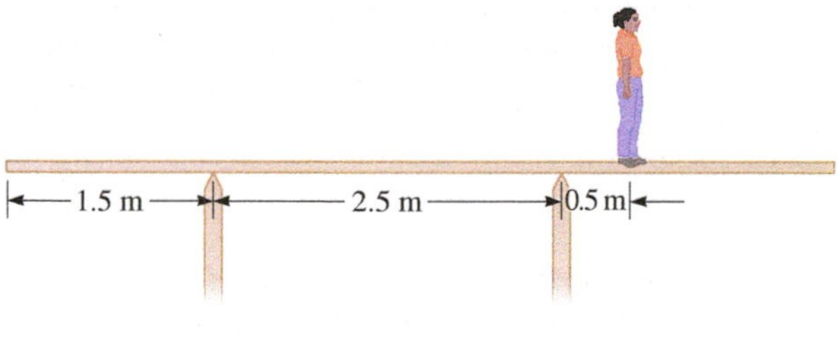

FIGURE P12.44

45. **N** Five forces of equal magnitude are applied to a rod of length L (Fig. P12.45). Rank in order from smallest to largest the magnitudes of the torques produced by these forces around the hinge shown.

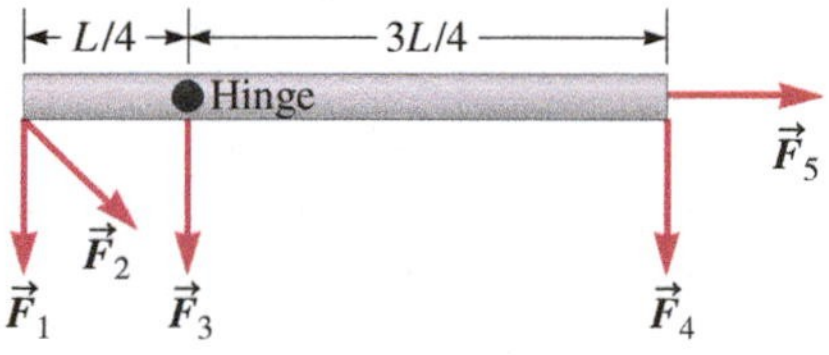

FIGURE P12.45

46. **N** At one point during its swing, a wrecking ball exerts a tension force of 9800 N on its cable, which makes an angle of 30.0° with the horizontal. The crane's 9.00-m-long boom is at an angle of 55.0° with the horizontal. What is the torque exerted by the wrecking ball on the crane about an axis perpendicular to the page and passing through point P shown in Figure P12.46?

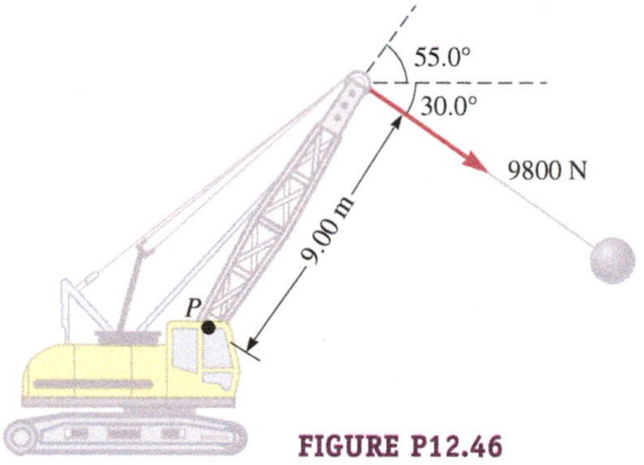

FIGURE P12.46

12-6 Cross Product

47. **N** Find the cross product $\vec{A} \times \vec{B}$ in each case: **a.** $\vec{A} = 15.0\hat{\imath}$ and $\vec{B} = 15.0\hat{\imath}$ **b.** $\vec{A} = 15.0\hat{\imath}$ and $\vec{B} = 15.0\hat{\jmath}$ **c.** $\vec{A} = 15.0\hat{\imath}$ and $\vec{B} = -15.0\hat{\jmath}$

Problems 48 and 49 are paired.

48. **N** The cross product $\vec{A} \times \vec{B} = 42.0\hat{k}$, and the magnitudes of the vectors are $A = 12.0$ and $B = 7.00$. What is the angle between the vectors?

49. **N** The cross product $\vec{A} \times \vec{B} = -42.0\hat{k}$, and the magnitudes of the vectors are $A = 12.0$ and $B = 7.00$. What is the angle between the vectors?

50. **C** Can the dot product and the cross product between two vectors both be zero? Explain.

51. **N** A force $\vec{F} = (2\hat{\imath} + 3\hat{\jmath} + 4\hat{k})$N is applied to a point with position vector $\vec{r} = (3\hat{\imath} + 2\hat{\jmath} + \hat{k})$m. Find the torque due to this force about the axis passing through the origin.

52. **N** Given a vector $\vec{A} = 4.5\hat{\imath} + 4.5\hat{\jmath}$ and a vector $\vec{B} = -4.5\hat{\imath} + 4.5\hat{\jmath}$, determine the magnitude of the cross product of these two vectors, $\vec{A} \times \vec{B}$. *Hint*: Make a sketch of both vectors including a coordinate system.

12-7 Rotational Dynamics

Problems 53 and 54 are paired.

53. **N** A square plate with sides 2.0 m in length can rotate around an axle passing through its center of mass (CM) and perpendicular to its surface (Fig. P12.53). There are four forces acting on the plate at different points. The rotational inertia of the plate is 24 kg · m². Use the values given in the figure to answer the following questions. **a.** What is the net torque acting on the plate? **b.** What is the angular acceleration of the plate?

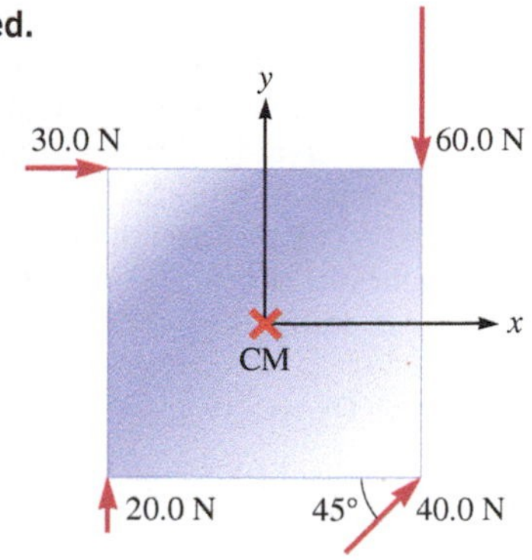

FIGURE P12.53
Problems 53 and 54.

54. **N** Assume the plate in Problem 53 (Fig. P12.53) starts from rest. **a.** What is the angular velocity of the plate after 4.0 s? **b.** Through how many revolutions will the plate rotate in 4.0 s?

55. A disk with a radius of 4.5 m has a 100-N force applied to its outer edge at two different angles (Fig. P12.55). The disk has a rotational inertia of 165 kg · m².
   a. **N** What is the magnitude of the torque applied to the disk in case 1?
   b. **N** What is the magnitude of the torque applied to the disk in case 2?
   c. **N** Assuming the force on the disk is constant in each case, what is the magnitude of the angular acceleration applied to the disk in each case?
   d. **C** Which case is a more effective way of spinning the disk? Describe which quantity you are using to determine "effectiveness" and why you chose that quantity.

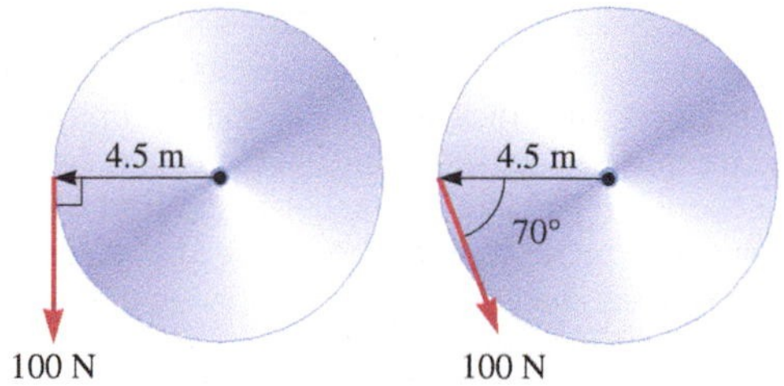

FIGURE P12.55

56. **N** Disc jockeys (DJs) use a turntable in applying their trade, often using their hand to speed up or slow down a disc record so as to produce a desired change in the sound (Fig. P12.56). Suppose DJ Trick wants to slow down a record initially rotating clockwise (as viewed from above) with an angular speed of 33.0 rpm to an angular speed of 22.0 rpm. The record has a rotational inertia of 0.012 kg · m² and a radius of 0.15 m. **a.** What angular acceleration is necessary if he wishes to accomplish this feat in exactly 0.65 s with a constant acceleration?

Peter McGahey © Cengage Learning

FIGURE P12.56

b. How many revolutions does the record go through during this change in speed? **c.** If DJ Trick applies a vertical force with his finger to the edge of the record, with what force must he push so as to slow the record in the above time? Assume the coefficient of kinetic friction between his finger and the record is 0.50, and ignore the mass of the finger.

57. **N** Frictional torque causes a disk to decelerate from an angular speed of 4.00 rad/s at $t = 0$ to 1.00 rad/s at $t = 5.00$ s. The equation describing the angular speed of the wheel during this time interval is given by $d\theta/dt = \omega_0 e^{-bt}$, where b and ω_0 are constants. **a.** What are the values of b and ω_0 during this time interval? **b.** What is the magnitude of the angular acceleration of the disk at $t = 5.00$ s? **c.** How many revolutions does the disk make during the interval $t = 0$ to $t = 5.00$ s?

General Problems

58. **N** A wheel rotates through 45.0 revolutions in 4.00 s, during which time it reaches a final angular speed of 66.0 rad/s. What is the constant angular acceleration of the wheel during this 4.00-s interval?

59. **N** A wheel initially rotating at 85.0 rev/min decelerates at the constant rate of 3.50 rad/s^2 when a braking mechanism is engaged. **a.** What is the time interval required to bring the wheel to a complete stop? **b.** What is the angle in radians through which the wheel rotates in this time interval?

60. A child's pinwheel rotates as the wind passes through it.

a. **N** If the pinwheel rotates from $\theta = 0°$ to $\theta = 90°$ in a time of 0.150 s, what is the average angular velocity of the pinwheel?

b. **N** If the pinwheel rotates from $\theta = 0°$ to $\theta = 180°$ in a time of 0.300 s, what is the average angular velocity of the pinwheel?

c. **N** If the pinwheel rotates from $\theta = 0°$ to $\theta = 270°$ in a time of 0.450 s, what is the average angular velocity of the pinwheel?

d. **N** If the pinwheel rotates from $\theta = 0°$ through one revolution to $\theta = 360°$ in a time of 0.600 s, what is the average angular velocity of the pinwheel?

61. **N** A centrifuge used for training astronauts rotating at 0.810 rad/s is spun up to 1.81 rad/s with an angular acceleration of 0.050 rad/s^2. **a.** What is the magnitude of the angular displacement that the centrifuge rotates through during this increase in speed? **b.** If the initial and final speeds of the centrifuge were tripled and the angular acceleration remained at 0.050 rad/s^2, what would be the factor by which the result in part (a) would change?

Problems 62 and 63 are paired.

62. **C** A disk is rotating around a fixed axis that passes through its center and is perpendicular to the face of the disk. Consider a point on the rim of the disk (point R) and another point halfway between the center and the rim (point H) at one particular instant. **a.** How does the angular speed ω of the disk at point H compare with the angular speed of the disk at point R? **b.** How does the tangential speed of the disk at point H compare with the tangential speed of the disk at point R? **c.** Suppose we pick a point H on the disk at random (by throwing a dart, for example), and we compare the speeds at that point with the speeds at point R. How will the answers to parts (a) and (b) be different? Explain.

63. **C** Consider the rotating disk in Problem 62 at one particular instant. **a.** How does the angular acceleration α of the disk at point H compare with the angular acceleration of the disk at point R? **b.** How does the tangential acceleration a_T of the disk at point H compare with the tangential acceleration of the disk at point R? **c.** Suppose we pick a point H on the disk at random (by throwing a dart, for example), and we compare the accelerations at that point with the accelerations at point R. How will the answers to parts (a) and (b) be different? Explain. **d.** Suppose the disk is rotating at a *constant rate*. Will the answers to parts (a) and (b) change? Explain.

64. **A** A potter's wheel rotates with an angular acceleration $\alpha = 4at^3 - 3bt^2$, where t is the time in seconds, a and b are constants, and α has units of radians per second squared. The initial position of the wheel is θ_0 and the initial angular velocity is ω_0. Obtain expressions for the angular velocity and the angular displacement of the wheel as a function of time.

65. **N** A racing motorcycle attempting to break a track record accelerates uniformly from rest and without slipping and tops 100 mph (44.7 m/s) in 4.99 s. The motorcycle's tires have a radius of 28.9 cm. **a.** What is the number of revolutions through which the motorcycle's tires turn during this time? **b.** What is the angular speed of the tires in revolutions per second after 4.99 s?

Problems 66 and 67 are paired.

66. **N** As seen from above the Earth's North Pole, the Earth's rotation is counterclockwise. In Example 12.1, we assumed the Earth's angular velocity is constant. Actually, the Earth is slowing down at a rate of roughly 0.002 s per century. Choose a coordinate system such that the positive z axis points north. Find the angular acceleration of the Earth, keeping at least eight significant figures in your calculations. Report your answer using two significant figures.

67. **N** As seen from above the Earth's North Pole, the Earth's rotation is counterclockwise. In Example 12.3, we assumed the Earth's angular velocity is constant. Actually, the Earth is slowing down at a rate of roughly 0.002 s per century. Find the magnitude of the tangential acceleration of Esmeraldas and New Orleans.

68. **N** Lara is running just outside the circumference of a carousel, looking for her favorite horse to ride, with a constant angular speed of 1.00 rad/s. Just as she spots the horse, one-fourth of the circumference ahead of her, the carousel begins to move, accelerating from rest at 0.050 rad/s^2. **a.** Taking the time when the carousel begins to move as $t = 0$, when will Lara catch up to the horse? **b.** Lara mistakenly passes the horse and keeps running at constant angular speed. If the carousel continues to accelerate at the same rate, when will the horse draw even with Lara again?

69. **N** The propeller of an aircraft accelerates from rest with an angular acceleration $\alpha = 4t + 6$, where α is in rad/s^2 and t is in seconds. What is the angle in radians through which the propeller rotates from $t = 1.00$ s to $t = 6.00$ s?

70. **G** A ball rolls to the left along a horizontal surface, up the slope, and then continues along a horizontal surface (Fig. P12.70). Sketch the angular speed ω and the magnitude of the angular acceleration α of the ball as functions of time.

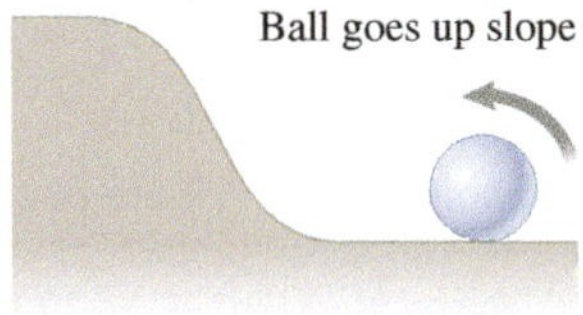

FIGURE P12.70

Problems 71 through 75 are grouped.

71. **A** Three forces are exerted on the disk shown in Figure P12.71, and their magnitudes are $F_3 = 2F_2 = 2F_1$. The disk's outer rim has radius R, and the inner rim has radius $R/2$. As shown in the figure, $\vec{F}_1$ and $\vec{F}_3$ are tangent to the outer rim of the disk, and $\vec{F}_2$ is tangent to the inner rim. $\vec{F}_3$ is parallel to the x axis, $\vec{F}_2$ is parallel to the y axis, and $\vec{F}_1$ makes a 45° angle with the negative x axis. Find expressions for the magnitude of each torque exerted around the center of the disk in terms of R and F_1.

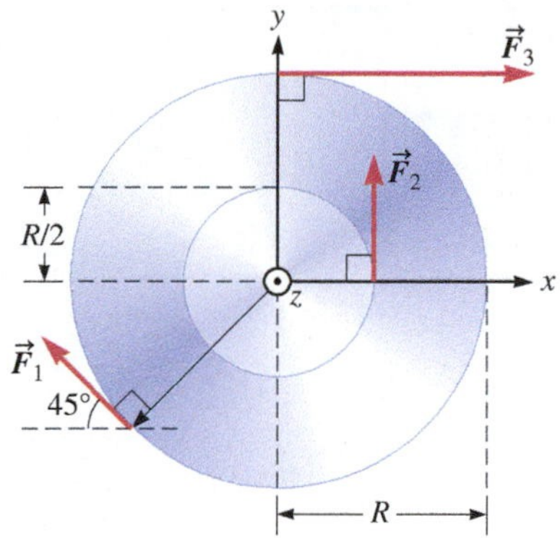

FIGURE P12.71 Problems 71–75.

72. **N** Consider the disk in Problem 71. The disk's outer rim has radius $R = 4.20$ m, and $F_1 = 10.5$ N. Find the magnitude of each torque exerted around the center of the disk.

73. **A** For the disk in Problem 71, find an expression for total torque exerted around the center of the disk in terms of R and F_1.

74. **N** Consider the disk in Problem 71. The disk's outer rim has radius $R = 4.20$ m, and $F_1 = 10.5$ N. The disk's rotational inertia is 0.545 kg · m^2. Find the angular acceleration of the disk.

75. **N** Consider the disk in Problem 71. Three forces are exerted on the disk (Fig. P12.71), with $F_2 = F_1 = 10.5$ N. The disk's outer rim has radius $R = 4.20$ m, and the inner rim has radius $R/2$. The angular acceleration is zero. What is the magnitude of $\vec{F}_3$?

76. Suppose the angular displacement of a rotating object is given by the equation $\Delta\theta(t) = (-5.3 \text{ rad/s}^3)t^3 + (7.0 \text{ rad/s}^2)t^2 + (8.1 \text{ rad/s})t$.

  a. **N** Find a time when the object is momentarily not rotating.

  b. **C** Determine whether the object is beginning to turn clockwise or counterclockwise. (Let clockwise be indicated by a negative angular velocity.)

77. **N** Starting from rest, an automobile accelerates with a uniform tangential acceleration of 2.00 m/s^2 while negotiating a semicircular unbanked curve. Three-fourths of the way into the curve, the car's tires lose traction with the road, causing the car to skid off the curve. What is the coefficient of static friction between the car's tires and the road?

78. **N** Consider a particle on the rim of a disk of radius R rotating around a fixed axis that passes through its center and is perpendicular to the face of the disk. The rim has a tangential acceleration that is constant in magnitude, a_T What is the direction of the total acceleration vector of this particle after it has made three complete rotations, beginning from rest? Give your answer to three significant figures.

Problems 79 through 81 are grouped.

79. **N** A uniform 4.55-kg horizontal rod is fixed on one end as shown in Figure P12.79. The rod is 1.75 m long. A rope is attached to the opposite end, and it makes a 30.0° angle with respect to the rod. What is the magnitude of the torque exerted by the tension in the rope around the fixed end of the rod? Assume the rod is in equilibrium.

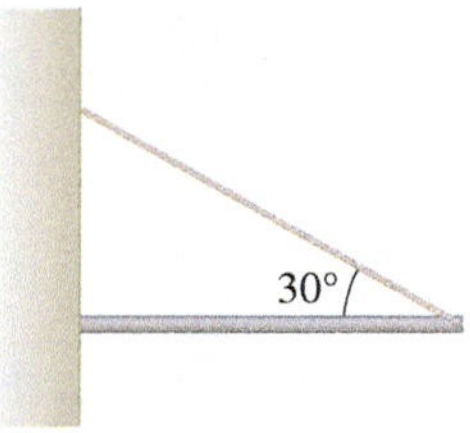

FIGURE P12.79 Problems 79–81.

80. **C** Consider the rod in Problem 79. What is the total torque around any point on the rod?

81. **N** If the rod in Problem 79 is in equilibrium, what is the tension in the rope at the instant shown?

82. **N** As a compact disc (CD) spins clockwise as seen from above, information is read from it, starting with the innermost ring and moving outward. When the information is being read from the innermost ring, the CD's angular speed is $\omega_0 = 52.4$ rad/s. The CD slows down so that when information is read from the outermost ring, $\omega = 20.9$ rad/s. It takes 74 min 33 s to read the music from a particular CD. Find the constant angular acceleration of the CD.

83. **N** A disk-shaped machine part has a diameter of 40.0 cm. Its angular position is given by $\theta = -1.20t^3 + 1.50t^2$, where t is in seconds and θ is in radians. **a.** What is the maximum angular speed of the part during this time interval? **b.** What is the maximum tangential speed of a point halfway to the rim of the part during this time interval? **c.** What is the time t for which the part reverses its direction of motion after $t = 0$? **d.** Through how many revolutions has the part turned when it reaches the time found in part (c)?

84. **E** **CASE STUDY** Wally Wallington is a retired carpenter who believes that he found a way for ancient people to have built Stonehenge. In fact, he has been reconstructing Stonehenge on his own using only simple tools such as wooden levers, stones, sand, water, ropes, and old paint drums. On his own, he has erected concrete columns weighing approximately 20,000 lb. He says that his passion for moving heavy objects with modest equipment started when he was working as a carpenter and had to remove 1200-lb concrete blocks. One solution would have been to break the blocks into smaller pieces and then use wheelbarrows to remove the pieces. Wally thought of a better way. He used levers to move the blocks; he found that the process became very easy with practice. Think of how you might reproduce Wally's original work. Assume he had some reasonably strong lengths of wood and a few old paint drums that he could fill with loose rock, sand, or water. Describe in detail how you might lift a 1200-lb block using a lever.

13 Rotation II: A Conservation Approach

Key Question

When and how can we apply the principles of conservation of energy and angular momentum to rotational motion?

Underlying Principles

1. Conservation of energy
2. Newton's second law
3. Conservation of angular momentum

Major Concepts

1. Rotational inertia
2. Parallel-axis theorem
3. Rotational kinetic energy
4. Work
5. Power
6. Angular momentum

Special Case

Pure rolling

In Chapter 12, we began our study of rotational motion by considering the relatively uncomplicated situation of a rigid object rotating around a fixed axis. Rotational motion, however, can be much more complicated and exciting. First, objects such as Frisbees, bicycle wheels, and somersaulting divers rotate and translate at the same time. Because they translate, their rotation axis must also be in motion. Second, rotating objects are not necessarily rigid: for instance, collapsing stars, spinning figure skaters, and clay thrown on a potter's wheel all change shape as they rotate.

We've faced complicated motion before, and we have found that conservation principles are important tools for analyzing such situations. This chapter has two goals: (1) to find the conditions under which conservation principles may be applied to a rotating object or system and (2) to find the most convenient way to apply these principles.

13-1 Conservation Approach

So far, we have restricted our study of rotational motion to rotation around a fixed axis, analogous to translational motion restricted to one dimension. We also only considered rigid objects, analogous to studying translational motion of objects that can be approximated as particles. For more complicated translational motion such as the elliptical orbit of Comet Halley or the flight of a rocket, we needed the principles of conservation of energy and momentum. These same conservation principles hold for rotational motion, and they give us another way to analyze complicated problems.

CASE STUDY Accounting for All the Energy

The case study in this chapter is based on an experiment a student—Chris—performed in a physics laboratory. Chris applies the principle of conservation of energy to an object that both rotates and translates.

EXAMPLE 13.1 CASE STUDY Loop-the-Loop Laboratory Challenge

Chris's laboratory instructor gives each student an identical steel loop-the-loop setup as shown in Figure 13.1. The instructor has a solid steel ball that rolls on the track. The challenge is to calculate the minimum height h from which the ball can be released so as to just barely keep in contact with the track as it completes the loop. The instructor hands out the ball only when a student is ready to test his or her calculation. Each student will be given three opportunities. The student who has the smallest h in any one of his or her successful trials wins.

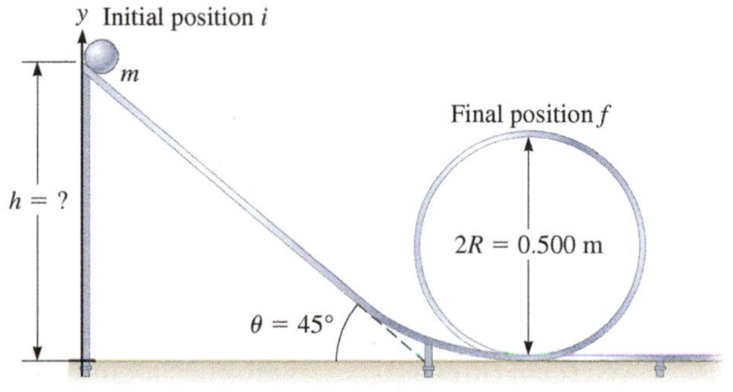

FIGURE 13.1 Experimental setup for a laboratory challenge.

INTERPRET and ANTICIPATE

Chris took a conservation of energy approach. **Step 1:** He included the Earth and the ball in the **system**. **Step 2:** He set the **reference configuration** and chose a **coordinate system**. **Step 3:** He set the **initial** position i to the place where the ball is released and the **final** position f to the top of the loop (Fig. 13.1), ignoring the small horizontal portion of the track. (Compare Chris's work here to Problem 9.56.)

SOLVE

Step 4 Many of the energy terms are zero as shown in Chris's **bar chart** (Fig. 13.2). The ball is released from rest, so $K_i = 0$. There are no external forces to do work on the system; therefore, $W_{\text{tot}} = 0$. Chris ignored internal thermal changes due to friction so that $\Delta E_{\text{int}} = 0$.

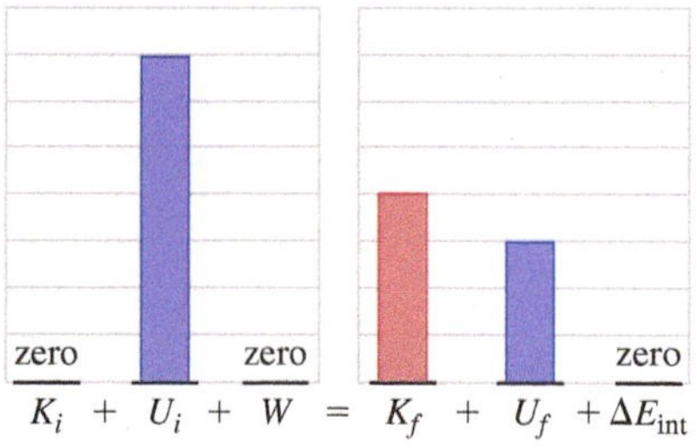

FIGURE 13.2

Step 5 Chris wrote the **conservation of energy equation** using his bar chart.

$$0 + U_i + 0 = K_f + U_f + 0$$

Step 6 Then he used **other equations** to write the conservation of energy equation in terms of the variables shown in Figure 13.1.

$$mgh = \tfrac{1}{2}mv_f^2 + mg(2R) \qquad (13.1)$$

If the ball just barely makes contact with the track at the top of the loop, the only force acting on the ball at that moment is gravity. The ball's centripetal acceleration equals the acceleration due to gravity.

$$a_c = \frac{v_f^2}{R} = g \qquad (13.2)$$

Step 7 **Do algebra.** Chris substituted Equation 13.2 into Equation 13.1 and simplified to find h.

$$mgh = \tfrac{1}{2}mv_f^2 + 2mgR$$

$$mgh = \tfrac{1}{2}mgR + 2mgR = \tfrac{5}{2}mgR$$

$$h = \tfrac{5}{2}R = \tfrac{5}{2}(0.250\,\text{m})$$

$$h = 0.625\text{ m} = 62.5\text{ cm}$$

Example continues on page 362 ▶

CHECK and THINK

When Chris tested his calculation, the ball fell off the track. He figured the problem was that he had ignored rolling friction.

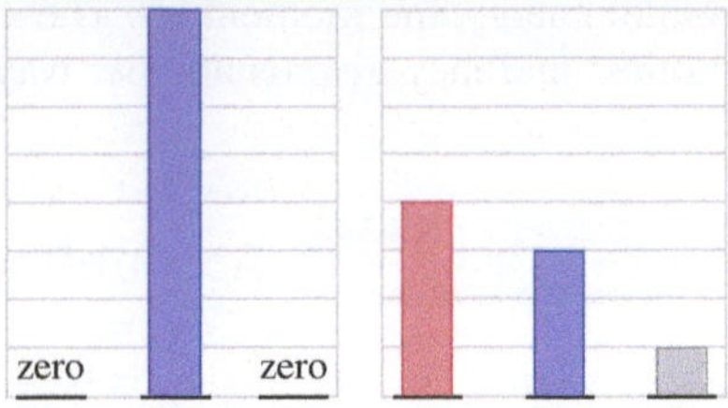

FIGURE 13.3

SOLVE

Step 4 Chris made a new bar chart (Fig. 13.3), taking into account changes in internal thermal energy: $\Delta E_{int} = \Delta E_{th}$.

Step 5 He wrote the **conservation of energy equation.**

$$0 + U_i + 0 = K_f + U_f + \Delta E_{th}$$

Step 6 To find ΔE_{th}, he used $\Delta E_{th} = F_k s$ (Eq. 9.29) with rolling friction substituted for kinetic friction: $F_k \rightarrow F_r$. He assumes that the magnitude of the normal force is $F_N = F_g \cos\theta$.

$$\Delta E_{th} = F_r s = \mu_r F_N s$$
$$\Delta E_{th} = \mu_r (F_g \cos\theta) s$$
$$\Delta E_{th} = \mu_r (mg \cos\theta) s$$

The total distance s traveled from i to f is the distance $h/\sin\theta$ down the incline plus the distance πR around half the circular loop. The normal force is *not* constant as the ball travels over the circular loop. So, Chris's assumption that the normal force is $mg \cos\theta$ over the whole trip means that he is making an estimate.

$$\Delta E_{th} = \mu_r mg \cos\theta\left(\frac{h}{\sin\theta} + \pi R\right)$$
$$\Delta E_{th} = \mu_r mg \cos 45°\left(\frac{h}{\sin 45°} + \pi R\right)$$
$$\Delta E_{th} = \mu_r mg\left(h + \frac{\pi}{\sqrt{2}}R\right) \quad (13.3)$$

Step 7 Do algebra. Now, taking rolling friction into account, Chris substituted ΔE_{th} into the conservation of energy equation and solved for h. He chose the largest value for the coefficient of rolling friction for steel on steel from Table 6.1: $\mu_r = 0.002$.

$$mgh = \tfrac{1}{2}mv_f^2 + 2mgR + \mu_r mg[h + (\pi/\sqrt{2})R]$$
$$mgh = \tfrac{5}{2}mgR + \mu_r mg[h + (\pi/\sqrt{2})R]$$
$$h = \frac{[\frac{5}{2} + \mu_r(\pi/\sqrt{2})]R}{1 - \mu_r} = \frac{[\frac{5}{2} + (0.002)(\pi/\sqrt{2})](0.25\text{ m})}{1 - 0.002}$$
$$h = 0.627\text{ m} = 62.7\text{ cm}$$

CHECK and THINK

This result is only 2 mm higher than Chris's previous calculation, so apparently not much mechanical energy is lost to rolling friction in this case. Chris's instructor gave him the steel ball to test his second calculation, but again the ball fell off the track.

The trouble is that Chris has missed part of the ball's motion. We will return to Chris's experiment to find the missing piece later in this chapter. (His assumption that the normal force is a constant over the whole path does not account for the ball falling off the track.)

CONCEPT EXERCISE 13.1

CASE STUDY When Is Energy Conserved?

Under what conditions is a system's energy conserved? Review Chapter 8 if necessary. Was Chris wrong to think that energy is conserved in the apparatus in Figure 13.1?

ROTATIONAL INERTIA
Major Concept

13-2 Rotational Inertia

In Section 12-7, we introduced rotational inertia as a measure of an object's tendency to maintain its angular velocity. In translational motion, a more massive particle has more inertia, so we expect that a more massive object also has more rotational inertia. It turns out that mass is only one factor that determines an object's rotational inertia.

A clue to the second factor comes from Example 12.6 (page 349). Crall and Whipple measured the rotational inertia of a dumbbell-shaped object (Figs. 12.26–12.28; also Fig. 13.4) and found that when the lead cylinders were close together, the rotational inertia of the dumbbell was reduced. The two factors in an object's rotational inertia are (1) its mass and (2) how its mass is distributed with respect to the rotation axis. An object whose mass is far from the axis of rotation has a larger rotational inertia than an object with the same mass located closer to the rotation axis. Imagine starting to rotate Crall and Whipple's dumbbell; it is easier to rotate the dumbbell when the cylinders are closer to the axis of rotation (Fig. 13.4B).

The rotational inertia of an object that consists of n particles may be found mathematically from

$$I = \sum_{i=1}^{n} m_i r_i^2 \tag{13.4}$$

where m_i is the mass of the ith particle and r_i is that particle's distance from the rotation axis. The SI unit of rotational inertia is kg · m^2.

EXAMPLE 13.2 Crall and Whipple's Dumbbells Revisited

Crall and Whipple measured the rotational inertia of a system of two lead cylinders (Example 12.6, page 349) in two different configurations. Each lead cylinder has a mass of 0.200 kg. For each of the configurations, use Crall and Whipple's experimental results (shown in Fig. 13.4) to determine the distance d between each cylinder's center of mass for the two different configurations. Assume the cross bar and axle are very light compared with the cylinders so that their contribution to the rotational inertia of the object may be ignored.

INTERPRET and ANTICIPATE

The object is modeled as two particles of mass $m = 0.200$ kg. Each particle is a distance r from the rotation axis. Use the rotational inertia experimentally determined by Crall and Whipple to solve for r in Equation 13.4. Our answer should match the configurations shown; that is, we expect $d_a > d_b$.

SOLVE

Determine the distance d_a in Figure 13.4A. Equation 13.4 takes a simple form because both particles have the same mass m and are the same distance r from the rotation axis.	$I = \sum_{i=1}^{n} m_i r_i^2 = m_1 r_1^2 + m_2 r_2^2$ (13.4) $I = mr^2 + mr^2 = 2mr^2$ $r = \sqrt{\dfrac{I}{2m}}$ (13.5)
Crall and Whipple found $I = 1.5 \times 10^{-2}$ kg · m^2 for the object in Figure 13.4A. The distance between the cylinders is twice their distance from the axis of rotation, $d_a = 2r_a$.	$r_a = \sqrt{\dfrac{1.5 \times 10^{-2}\ \text{kg} \cdot \text{m}^2}{2(0.200\ \text{kg})}}$ $r_a = 0.194\ \text{m} = 19.4\ \text{m}$ $d_a = 39\ \text{cm}$
Now determine the distance d_b. For Figure 13.4B, Crall and Whipple found $I = 3.8 \times 10^{-3}$ kg · m^2.	$r_b = \sqrt{\dfrac{3.8 \times 10^{-3}\ \text{kg} \cdot \text{m}^2}{2(0.200\ \text{kg})}}$ $r_b = 9.75 \times 10^{-2}\ \text{m} = 9.75\ \text{cm}$ $d_b = 19\ \text{cm}$

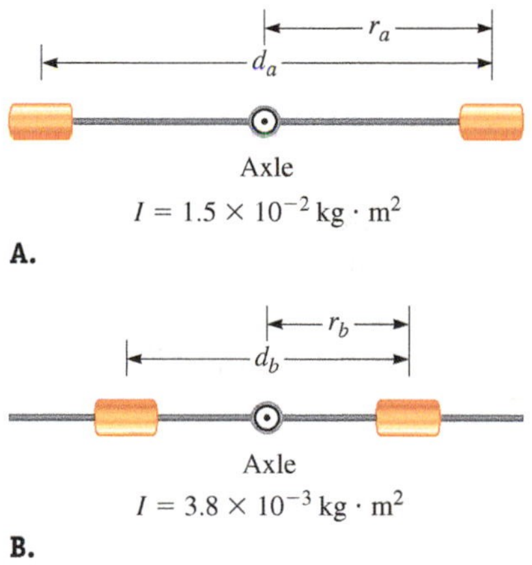

FIGURE 13.4 Crall and Whipple measured the rotational inertia of a dumbbell-shaped object. Rotational inertia depends both on mass and the location of the mass with respect to the axis of rotation.

CHECK and THINK

As expected, $r_a > r_b$ and $d_a > d_b$. In fact, $r_a \approx 2r_b$, and the rotational inertia of the larger dumbbell is about four times greater than that of the small dumbbell even though the two dumbbells are identical in mass. Clearly, the distribution of mass within an object is a major factor in determining the object's rotational inertia.

Rotation Axis Must Be Specified

Because the distribution of an object's mass around the rotation axis is important in determining its rotational inertia, the location of the rotation axis must be specified when reporting a value for I. In other words, it is not good enough to say the rotational inertia of the dumbbell in Figure 13.4A is $I = 1.5 \times 10^{-2}\ \text{kg}\cdot\text{m}^2$ without specifying that the rotation axis is along the axle at the center of the dumbbell (perpendicular to the rod joining the two cylinders). In fact, if a different rotation axis is used, the rotational inertia of the dumbbell is not equal to what Crall and Whipple found experimentally. In the following example, we find the rotational inertia of an object around two different rotation axes.

EXAMPLE 13.3 Rotational Inertia: Same Object, Different Axes

Figure 13.5 shows an object made up of three lead spheres attached to three lightweight, stiff rods. The mass of the rods is negligible compared with the mass of the spheres. Each of the three spheres has mass m, and each lies at the vertex of an equilateral triangle of side ℓ. Find expressions for the rotational inertia of this object around an axis that is perpendicular to the plane of the triangle, (1) first passing through the object's center of mass (Fig. 13.5A) and then (2) passing through a vertex (Fig. 13.5B).

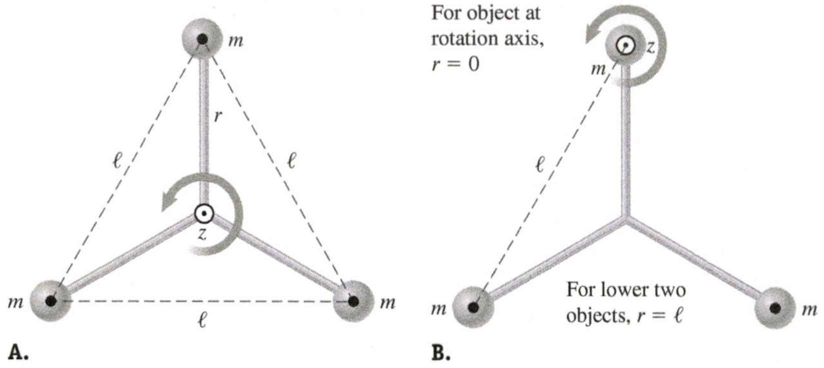

FIGURE 13.5

INTERPRET and ANTICIPATE

To find the rotational inertia around each axis, we need the distance r of each sphere with respect to the rotation axis. In Figure 13.5A, r is the same for each sphere; in Figure 13.5B, however, r is zero for one of the spheres. We expect to find two algebraic expressions for the rotational inertia with dimensions $\text{M}\cdot\text{L}^2$.

SOLVE

In Figure 13.5A, each mass is attached to a rod that runs to the center of mass of the three-sphere system. Using Figure 13.6, we can express the length of each rod in terms of the length ℓ of each side of the equilateral triangle as $r = \ell/\sqrt{3}$.

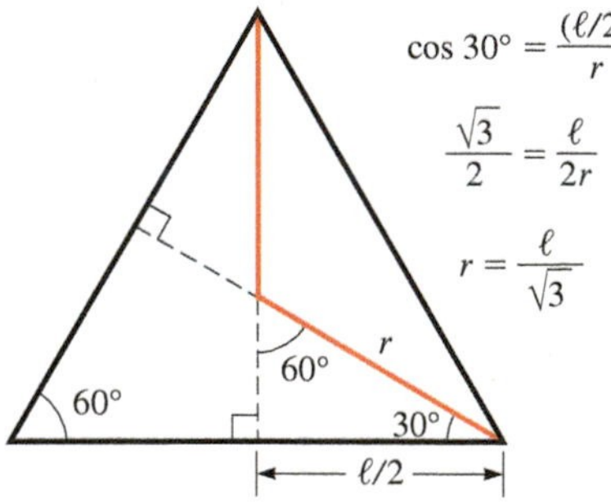

FIGURE 13.6 Geometry for finding r in Figure 13.5A.

The rotational inertia around the center of mass comes from Equation 13.4.

$$I = \sum_{i=1}^{n} m_i r_i^2 = \sum_{i=1}^{3} m_i r_i^2 \qquad (13.4)$$

$$I = m\left(\frac{\ell}{\sqrt{3}}\right)^2 + m\left(\frac{\ell}{\sqrt{3}}\right)^2 + m\left(\frac{\ell}{\sqrt{3}}\right)^2$$

$$I_{\text{CM}} = m\ell^2 \qquad (13.6)$$

In Figure 13.5B, now the axis of rotation passes through one of the three spheres. For that sphere, $r = 0$. For the other two spheres, $r = \ell$.

$$I = \sum_{i=1}^{n} m_i r_i^2 = \sum_{i=1}^{3} m_i r_i^2$$

$$I = m(0)^2 + m\ell^2 + m\ell^2$$

$$I_{\text{vertex}} = 2m\ell^2 \qquad (13.7)$$

CHECK and THINK

The expressions we found both have the correct dimensions: $\text{M}\cdot\text{L}^2$. If the axis of rotation passes through a vertex of the triangle (Fig. 13.5B), the rotational inertia is twice the rotational inertia found when the axis of rotation passes through the center of mass (Fig. 13.5A): $I_{\text{vertex}} = 2I_{\text{CM}}$. This result makes some sense. When we rotate the system around the vertex, the mass is on average farther from the rotation axis than when we rotate the system around the center of mass. Just moving the axis of rotation changes the rotational inertia of the object. So, if you want the system to have a particular angular acceleration, it takes twice the torque to get that angular acceleration if the axis of rotation passes through a vertex instead of through the center of mass.

Rotational Inertia of Continuous Objects

The summation in Equation 13.4 is only practical if we need to calculate the rotational inertia of an object that can be modeled as a small collection of particles. To calculate the rotational inertia of a continuous object like a bicycle tire, a Frisbee, or baseball bat, we must model the continuous object as a large number of infinitesimal pieces of mass dm (Fig. 13.7). Each infinitesimal piece is a perpendicular distance r from the axis of rotation. The rotational inertia of each infinitesimal particle around the axis is

$$dI = r^2 dm$$

To find the rotational inertia of the whole object, we must add up the contribution of each infinitesimal particle using integration:

$$\int dI = \int r^2 dm$$

$$I = \int r^2 dm \tag{13.8}$$

If the mass of the object is uniformly distributed and the shape of the object is simple and symmetric such as a hoop, disk, or straight rod, Equation 13.8 may be solved with a pencil and paper in a few lines of calculation. Table 13.1 gives the rotational inertia of nine symmetric objects around axes that pass through the object's center of mass, found from Equation 13.8. If the object is more complicated, such as a hammer, gymnast, or airplane, it may be necessary to either model the object as a collection of simpler objects or solve Equation 13.8 numerically with the aid of a computer.

In some situations, we may be able to apply a very useful shortcut. Suppose we know the rotational inertia of some object (say a hammer; Fig. 13.8) around an axis through its center of mass, but we need to know the rotational inertia around a dif-

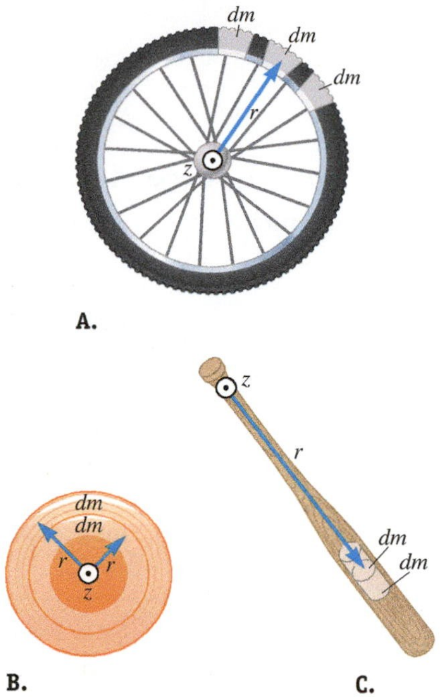

FIGURE 13.7 To find the rotational inertia, we model a continuous object as a large number of particles of mass dm, exploiting the symmetry of the object (if present).

TABLE 13.1 Rotational inertia of various objects.

Hoop around the central axis $I_{CM} = MR^2$ Rotation axis 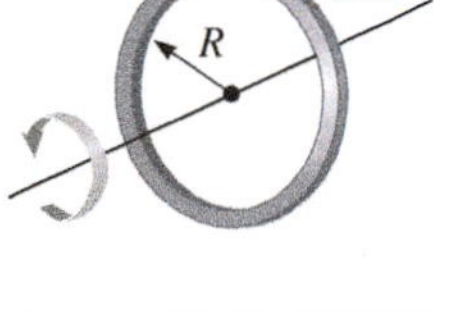	Thin spherical shell around any diameter $I_{CM} = \frac{2}{3}MR^2$ Rotation axis 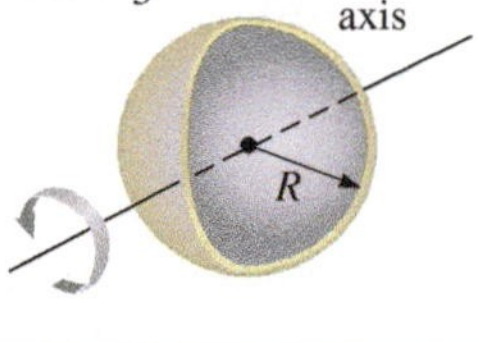	Thin rod around axis through center and perpendicular to rod $I_{CM} = \frac{1}{12}ML^2$ Rotation axis 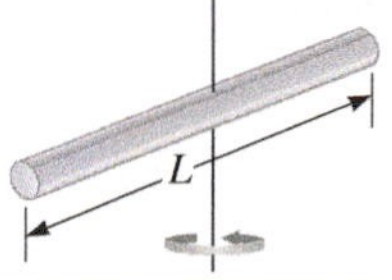
Hollow cylinder around the central axis $I_{CM} = \frac{1}{2}M(R_1^2 + R_2^2)$ Rotation axis 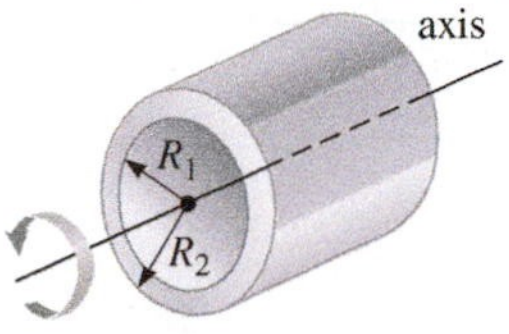	Solid sphere around any diameter $I_{CM} = \frac{2}{5}MR^2$ Rotation axis 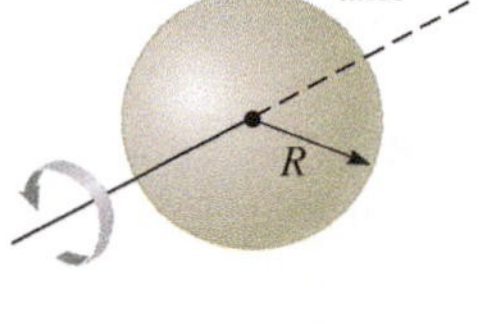	Solid cylinder or disk around a central diameter $I_{CM} = \frac{1}{4}MR^2 + \frac{1}{12}ML^2$ Rotation axis 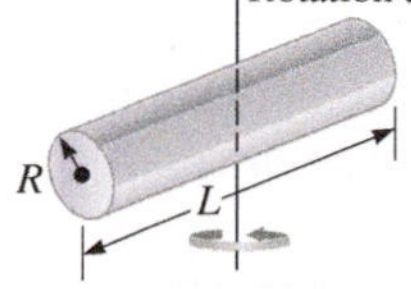
Hoop around any diameter $I_{CM} = \frac{1}{2}MR^2$ Rotation axis 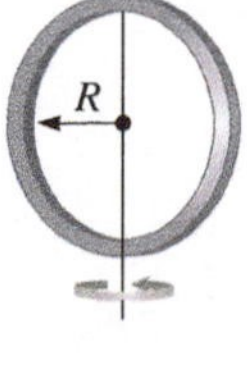	Thin rectangular solid around axis through center $I_{CM} = \frac{1}{12}M(a^2 + b^2)$ Rotation axis 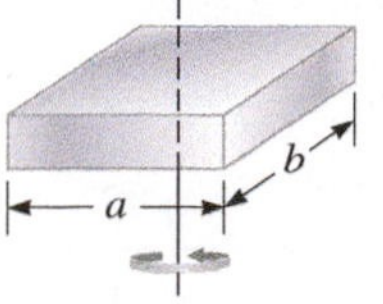	Solid cylinder or disk around the central axis $I_{CM} = \frac{1}{2}MR^2$ Rotation axis 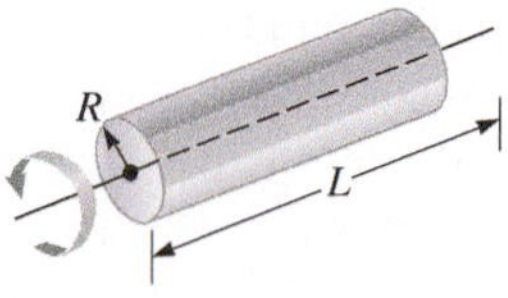

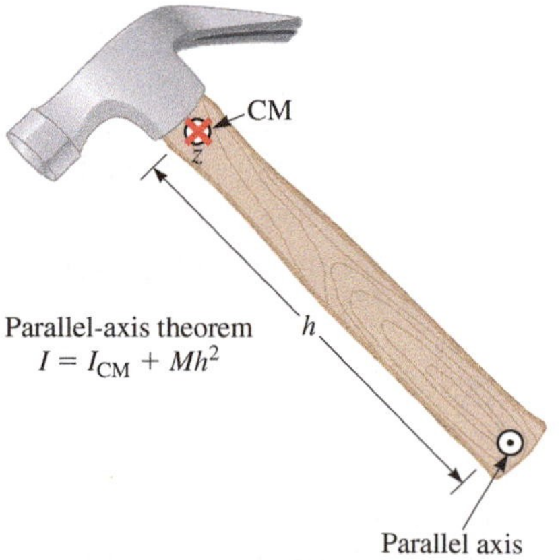

FIGURE 13.8 If you know the rotational inertia I_{CM} around an axis through an object's center of mass (CM), the parallel-axis theorem may be used to find the rotational inertia I around a new axis parallel to the original axis.

ferent axis. If that axis is parallel to the axis that passes through the center of mass, we can use the *parallel-axis theorem* to find the new rotational inertia. If M is the mass of the object, h is the perpendicular distance between the new axis and the axis through the center of mass, and I_{CM} is the rotational inertia around the axis passing through the center of mass, the rotational inertia around the new axis is

PARALLEL AXIS THEOREM

Major Concept

$$I = I_{CM} + Mh^2 \tag{13.9}$$

according to the **parallel-axis theorem**.

EXAMPLE 13.4 Rotational Inertia of a Bicycle Tire

Model a bicycle tire as a hoop of radius R. Use $I = \int r^2 dm$ (Eq. 13.8) to find the rotational inertia of the bicycle tire around the axis.

INTERPRET and ANTICIPATE

Model the object as a large number of infinitesimal pieces (Fig. 13.7A). We expect our results to be consistent with Table 13.1.

SOLVE

The distance from the axis of rotation to each infinitesimal piece equals the radius of the tire: $r = R$.	$I = \int r^2 dm = \int R^2 dm$
Because R is a constant for each piece of mass dm, we can bring R outside the integral.	$I = R^2 \int dm$
Integration over all pieces of mass dm equals the total mass M of the tire.	$I_{tire} = MR^2$

CHECK and THINK

The rotational inertia of the tire equals the rotational inertia of a hoop as shown in Table 13.1. We often need to know the rotational inertia of a hoop or disk around its center of mass. A disk of the same mass and radius has half the rotational inertia of a hoop because the disk's mass is distributed such that on average it is closer to the axis of rotation (Table 13.1, lower right).

EXAMPLE 13.5 Using the Parallel-Axis Theorem

In Example 13.3 (page 364), we found the rotational inertia of an object consisting of three spheres lying at the vertices of an equilateral triangle around an axis perpendicular to the plane of the triangle and passing through the object's center of mass (Fig. 13.5A) is $I_{CM} = m\ell^2$. Now use the parallel-axis theorem to once again find the rotational inertia around an axis passing through a vertex (Fig. 13.5B).

INTERPRET and ANTICIPATE

Start with $I_{CM} = m\ell^2$ (Eq. 13.6) and use the parallel-axis theorem to find the rotational inertia around the axis shown in Figure 13.5B. We expect our result to be consistent with Example 13.3 ($I_{vertex} = 2m\ell^2$).

SOLVE

Figure 13.6 shows that the perpendicular distance between the two axes of rotation is the same as the distance r from one vertex to the center of mass.	$h = r = \dfrac{\ell}{\sqrt{3}}$
Substituting for I_{CM} and h into the parallel-axis theorem (Eq. 13.9) gives the rotational inertia around the axis through the vertex.	$I = I_{CM} + Mh^2$ (13.9) $I_{vertex} = m\ell^2 + 3m\left(\dfrac{\ell}{\sqrt{3}}\right)^2$ $I_{vertex} = m\ell^2 + 3m\left(\dfrac{\ell^2}{3}\right) = m\ell^2 + m\ell^2 = 2m\ell^2$

CHECK and THINK

As expected, this result is exactly what we found in Example 13.3 (Eq. 13.7). Often, the parallel axis theorem makes finding the rotational inertia much easier than using Equation 13.4 or Equation 13.8 directly.

EXAMPLE 13.6 "Swing, Batter!"

Model the baseball bat in Figure 13.7C as a thin rod of length L. Find the rotational inertia around an axis through one end of the bat.

INTERPRET and ANTICIPATE

Table 13.1 lists the rotational inertia of a thin rod around an axis through the center of the rod. We can use that information and the parallel-axis theorem to find the rotational inertia around an axis that runs through one end of the rod. Our result should be an algebraic expression with dimensions $M \cdot L^2$. Because more of the mass is farther from the new axis of rotation, we expect our expression will be greater than the rotational inertia given in Table 13.1 for an axis through the center.

SOLVE

The rotational inertia around the center of mass comes from Table 13.1. The distance h is half the length of the rod, $h = L/2$.

$$I = I_{CM} + Mh^2$$

$$I = \frac{1}{12}ML^2 + M\left(\frac{L}{2}\right)^2$$

$$I = \frac{1}{12}ML^2 + \frac{1}{4}ML^2 = \frac{1}{3}ML^2$$

CHECK and THINK

As expected, the rotational inertia is greater for an axis through the end of the bat: $I > I_{CM}$. Sometimes, a coach will instruct a weak batter to "choke up" on the bat, that is, to move his or her hands closer to the bat's center of mass. This change reduces the rotational inertia so that the bat is easier to rotate or "swing." (Of course, a real bat is more complicated because it is not a uniform rod.)

13-3 Rotational Kinetic Energy

All moving objects have kinetic energy. If the object may be modeled as a translating particle, its kinetic energy is easily found from $K = \frac{1}{2}mv^2$ (Eq. 8.1). If the object is rotating, however, Equation 8.1 is inadequate because different parts of the object move with different translational speeds v around the rotation axis.

ROTATIONAL KINETIC ENERGY

Major Concept

DERIVATION Rotational Kinetic Energy

For any object or system with rotational inertia I rotating at angular speed ω, the rotational kinetic energy K_r is given by

$$K_r = \tfrac{1}{2}I\omega^2 \tag{13.10}$$

To keep our derivation simple, we will show that this expression holds in the case of an object that consists of just two particles. For example, Figure 13.9 shows two small balls attached to a lightweight, stiff rod.

Recall from Chapter 12 that one way to model a rotating object is as a collection of particles moving in concentric circles. So, each ball moves on a circular path around the z axis at the speed indicated in the figure. The total kinetic energy of this two-particle system comes from Equation 8.2.

$$K_r = \sum_{i=1}^{n} \tfrac{1}{2}m_i v_i^2$$

$$K_r = \tfrac{1}{2}m_1 v_1^2 + \tfrac{1}{2}m_2 v_2^2$$

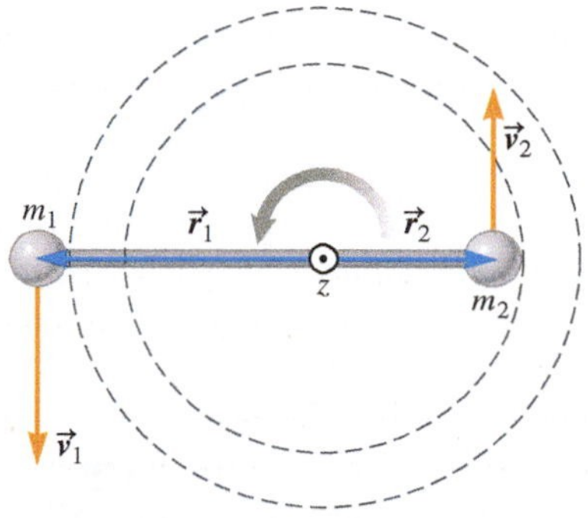

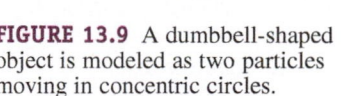

FIGURE 13.9 A dumbbell-shaped object is modeled as two particles moving in concentric circles.

Derivation continues on page 368 ▶

Both balls have the same angular speed ω. Using $v = \omega r$ (Eq. 12.13), we rewrite their translational speeds in terms of their angular speed.	$K_r = \frac{1}{2}m_1(\omega r_1)^2 + \frac{1}{2}m_2(\omega r_2)^2$ $K_r = \frac{1}{2}(m_1 r_1^2 + m_2 r_2^2)\omega^2$ (1)
The term in parentheses in Equation (1) is the rotational inertia of this two-particle object from Equation 13.4.	$I = \sum_{i=1}^{n} m_i r_i^2$ (13.4) $I = m_1 r_1^2 + m_2 r_2^2$ (2)
Substitute Equation (2) into Equation (1) to find the expression for rotational kinetic energy.	$K_r = \frac{1}{2}I\omega^2$ ✓ (13.10)

COMMENTS

We found the rotational kinetic energy of a simple two-particle system, but $K_r = \frac{1}{2}I\omega^2$ (Eq. 13.10) holds for any rotating object or system. It is easy to remember this expression because it is so similar to $K = \frac{1}{2}mv^2$; the inertia m is replaced by the *rotational* inertia I, and the speed v is replaced by the *angular* speed ω.

CONSERVATION OF ENERGY

Underlying Principle

The **conservation of energy** principle is slightly modified to include rotational motion. The initial mechanical energy (translational kinetic energy K, **rotational kinetic energy** K_r, and potential energy U) of the system plus the total work W done on the system by external forces must equal the final mechanical energy of the system plus changes in the system's internal (thermal) energy ΔE_{th}:

$$K_i + K_{ri} + U_i + W = K_f + K_{rf} + U_f + \Delta E_{th}$$

So, in practice, when we apply this modified version of the conservation of energy principle, we must include an extra bar for rotational kinetic energy in our bar chart as we do in the following example.

EXAMPLE 13.7 An Atwood Machine

Pulleys are found on the end of cranes, in the resistance machines in the gym, and in the hand reels on fishing poles. So far, we have only considered very light pulleys, ignoring their rotational inertia. (For instance, see Example 5.9, page 140.) Now, we consider an Atwood machine (Fig. 13.10). A light rope runs over a pulley and connects two objects of masses m and $2m$. The pulley is modeled as a solid disk of radius R and mass $2m$. When the lighter object is displaced upward by an amount Δy, the heavier object is displaced downward by the same distance. Therefore, it is convenient to use a coordinate axis that bends along the rope (similar to the coordinate system used in Example 5.9). The system starts at rest (Fig. 13.10A). Assume there are no energy losses. Find an expression for the angular speed of the pulley at a later time (Fig. 13.10B) in terms of Δy, R, and g.

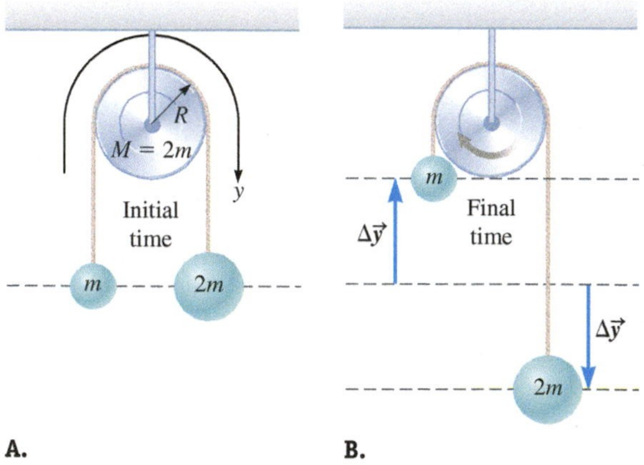

FIGURE 13.10 An Atwood machine.

INTERPRET and ANTICIPATE

Step 1 Using an energy approach, we choose a system including the hanging objects, the pulley, and the Earth. **Steps 2** (choosing a coordinate system) and **3** (indicating the initial and final times) were done for us.

Step 4 **Bar chart.** To save space, the bars for the various components of the system on the left have been combined (Fig. 13.11). On the right, however, we must consider the components (each hanging object and the pulley) separately. As shown in the bar chart, the change in system's thermal energy is zero. No external forces do work on the system. After the system is released, the lighter object rises, and the heavier object falls. The net result is that the system loses gravitational potential energy as demonstrated by adding up the two bars drawn for U_g. Both hanging objects have translational kinetic energy, and the pulley has rotational kinetic energy. Because the hanging masses are attached to each other by the rope, they have the same translational speed v

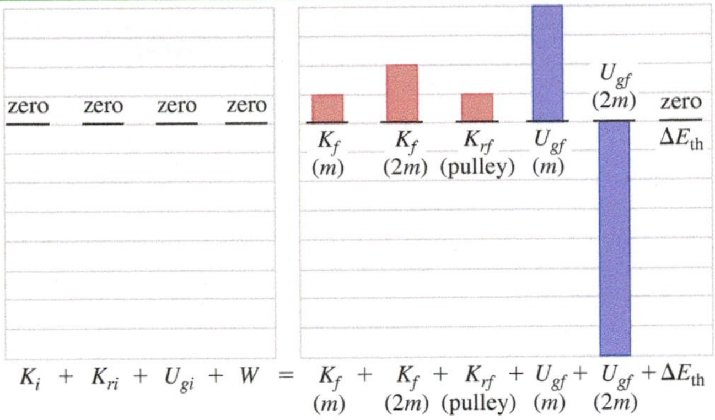

FIGURE 13.11

SOLVE

Step 5 Aided by the bar chart, write an **equation for the conservation of energy.**

$$0 = K_f + K_{rf} + U_{gf}$$

$$0 = \tfrac{1}{2}mv^2 + \tfrac{1}{2}(2m)v^2 + \tfrac{1}{2}I\omega^2 + mg\Delta y - 2mg\Delta y$$

$$0 = \tfrac{3}{2}mv^2 + \tfrac{1}{2}I\omega^2 - mg\Delta y \qquad (1)$$

Step 6 **Other equations.** The rotational inertia of the pulley comes from Table 13.1 (lower right). The translational speed v of the hanging objects is proportional to the angular speed of the pulley (Eq. 12.13).

$$I = \tfrac{1}{2}MR^2 = \tfrac{1}{2}(2m)R^2 = mR^2$$

$$v = \omega R \qquad (12.13)$$

Step 7 **Do algebra.** Substitute for v and I in Equation (1).

$$0 = \tfrac{3}{2}m\omega^2R^2 + \tfrac{1}{2}mR^2\omega^2 - mg\Delta y$$

$$mg\Delta y = 2m\omega^2R^2$$

$$\omega = \frac{1}{R}\sqrt{\frac{g\Delta y}{2}}$$

CHECK and THINK

Does the result make sense? It makes sense that ω depends directly on Δy because as Δy increases, more gravitational potential energy is converted into kinetic energy, increasing ω. Also, if Δy is zero, the hanging blocks are at rest, and the system has no kinetic energy. It also makes sense that ω is inversely proportional to R because it is more difficult to rotate a larger disk, so we expect a larger disk to have a lower angular speed. (As always, you should confirm that the dimensions are correct.)

13-4 Special Case of Rolling Motion

We have studied pure translational motion and pure rotational motion. Rolling motion is a combination of these two. Imagine, for instance, a game of Frisbee. The Frisbee's center of mass translates from the thrower to the catcher; meanwhile, the Frisbee also rotates around an axis passing through its center of mass. The axis is not fixed; rather, it moves at the center-of-mass velocity. There are many other examples of a combination of rotational and translational motion: a gymnast doing a backflip across a balance beam, a stunt plane corkscrewing across the sky, and a graduation hat being tossed high into the air. In this section, we consider one special case of combination motion, rolling without slipping.

Imagine an object such as an automobile or truck tire rolling along a flat, horizontal surface without slipping. If we attach a light to the tire rim, then a long-exposure photo of the tire as it rolls reveals a continuous-motion diagram (Fig. 13.12). The tire's axle is along the rotation axis passing through the tire's center of mass. The center of mass moves along a straight line, parallel to the images formed by the running light and taillight in Figure 13.12. By contrast, the path of the light attached at the rim forms a repeating curve. The shape of this curve comes from the combination of the tire's rotational and translational motion.

FIGURE 13.12 A light attached to the rim of a truck or automobile tire describes a repeating curved pattern as the tire rolls.

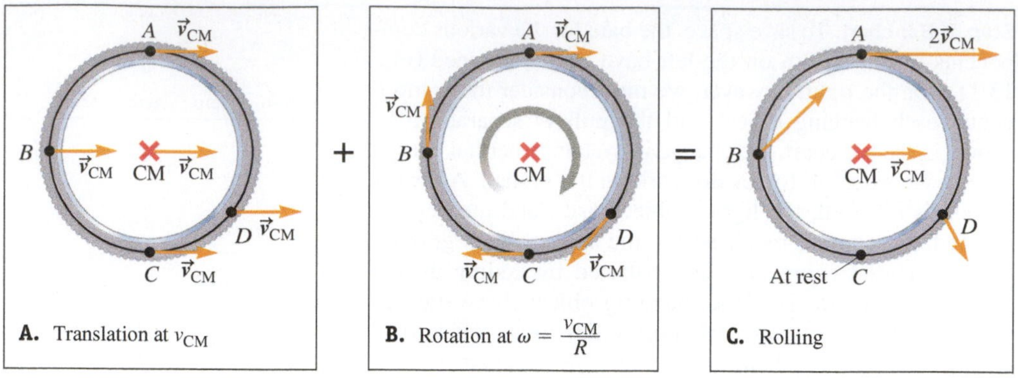

FIGURE 13.13 **A.** Each particle of the tire translates at the same velocity $\vec{v}_{CM}$. **B.** Each particle on the rim rotates around the center of mass at $\omega = v_{CM}/R$. **C.** Pure rolling can be modeled as translational motion plus rotational motion; so the velocity at each point is vector sum of the velocities from parts A and B. Notice that the top of the tire has the highest speed $2v_{CM}$, and the point of contact with the horizontal surface is momentarily at rest.

FIGURE 13.14 The tire and spokes are blurry near the top and are clear near the bottom, showing that the point of contact between the tire and the ground is at rest.

Figure 13.13 further illustrates pure rolling as a special case of motion that combines translation and rotation. Because rolling is a combination of translating (Fig. 13.13A) and rotating (Fig. 13.13B), the velocity of each point comes from adding up the velocity vectors for translating and rotating (Fig. 13.13C). The speed at the center of mass is v_{CM}. At the top of the tire in Figure 13.13C (point A), the speed is $2v_{CM}$, so point A is momentarily the fastest-moving point on the tire. At the bottom, where the tire is in contact with the road (point C), the speed is zero as the photograph in Figure 13.14 shows. Even though we know that the tire continues to roll, zero velocity at point C is another way to say that the tire rolls *without slipping*. If the tire slips as it rolls, the velocity of the point of contact would be nonzero.

Because the tire is rolling without slipping, it is relatively easy to connect the translational and rotational motions mathematically. Figure 13.15 shows how a string may be used to measure the center-of-mass displacement of a rolling tire. After one revolution, the magnitude of the center-of-mass position measured from the origin equals the length of the string. In this case, the string is as long as the circumference of the tire, so $x_{CM} = 2\pi R$.

What happens if the tire does not complete exactly one revolution, but instead rotates through an angle θ? In that case, the center-of-mass position does not equal the circumference of the tire; it equals the arc length s:

$$x_{CM} = s = \theta R \tag{13.11}$$

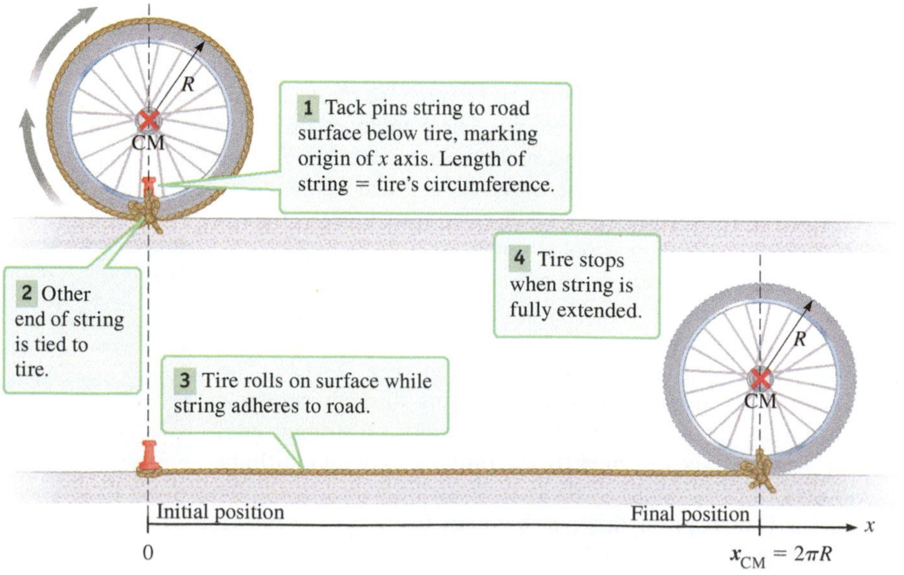

FIGURE 13.15 A string is tied to the tire and tacked to the surface. The tire rolls along the surface through one complete revolution. The tire's center-of-mass displacement equals its circumference, $x_{CM} = 2\pi R$.

The center-of-mass speed v_{CM} comes from taking the time derivative of the center-of-mass position:

$$v_{CM} = \frac{dx_{CM}}{dt} = \frac{d(\theta R)}{dt}$$

Of course, the tire's radius does not change as it rolls, so we bring R outside the derivative and recognize $d\theta/dt$ as the angular speed ω (Eq. 12.4):

$$v_{CM} = R\frac{d\theta}{dt}$$

$$v_{CM} = \omega R \qquad (13.12)$$

PURE ROLLING **Special Case**

In Equation 13.12, v_{CM} is the translational speed of the center of mass parallel to the surface and ω is the angular speed around an axis through the center of mass.

If an object rolls without slipping, Equation 13.12 applies, and we describe the object's motion as **pure rolling**. According to this equation, pure rolling is a special combination of translational and rotational motion in which the center-of-mass speed v_{CM} is determined by the angular speed ω and the radius R of the object.

The magnitude of the center-of-mass acceleration a_{CM} comes from taking the time derivative of Equation 13.12:

$$a_{CM} = \frac{dv_{CM}}{dt} = \frac{d(\omega R)}{dt} = R\frac{d\omega}{dt}$$

where $d\omega/dt$ is the angular acceleration α (Eq. 12.6):

$$a_{CM} = \alpha R \qquad (13.13)$$

If an object's motion is described by a combination of rotation around an axis through its center of mass and translation of the center of mass, its kinetic energy must be given by the sum of Equations 8.1 and 13.10:

$$K = K_{CM} + K_r = \tfrac{1}{2}mv_{CM}^2 + \tfrac{1}{2}I_{CM}\omega^2$$

In the special case of a pure rolling, the center-of-mass speed and the angular speed are connected by Equation 13.12.

EXAMPLE 13.8 CASE STUDY Loop-the-Loop Laboratory Challenge Revisited

Return to the loop-the-loop laboratory challenge in Example 13.1 on page 361 and help Chris find the winning value for h.

INTERPRET and ANTICIPATE

Chris failed to take the rotational kinetic energy of the ball into account. The ball is rolling, so it has translational and rotational kinetic energy (Fig. 13.16). Chris already found that rolling friction did not make much of a difference to his results. We will ignore rolling friction here. So, we start with **Step 4**, making a new **bar chart.**

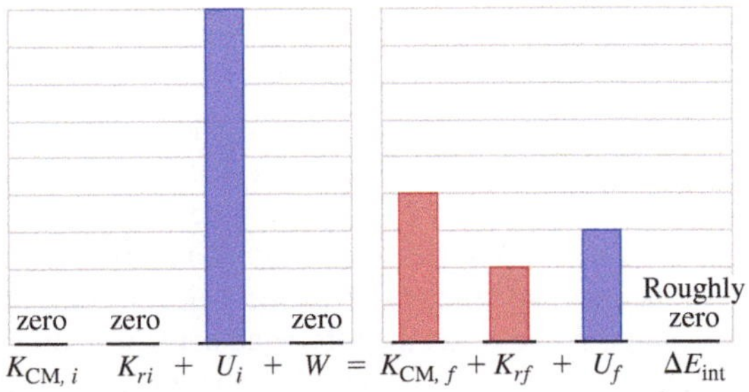

FIGURE 13.16

SOLVE

Step 5 Chris should rewrite the conservation of energy equation using this new bar chart. Refer to Figure 13.1 for the initial and final positions of the ball.

$$K_i + U_i + W_{tot} = K_f + U_f + \Delta E_{int}$$

$$0 + U_i + 0 = K_{CM,f} + K_{rf} + U_f + 0$$

$$mgh = \tfrac{1}{2}mv_f^2 + \tfrac{1}{2}I_{CM}\omega_f^2 + 2mgR \qquad (13.14)$$

Step 6 Other equations. The ball's rotational inertia is found in Table 13.1. Its angular speed at the top of the loop is given by $v_{CM} = \omega r$ (Eq. 13.12) where r is the ball's radius.

$$I_{CM} = \tfrac{2}{5}mr^2 \qquad \omega_f = \frac{v_f}{r}$$

Example continues on page 372 ▶

Step 7 Do algebra. Substitute into Equation 13.14 and solve for h.

$$mgh = \frac{1}{2}mv_f^2 + \frac{1}{2}\left(\frac{2}{5}mr^2\right)\left(\frac{v_f}{r}\right)^2 + 2mgR$$

$$mgh = \tfrac{7}{10}mv_f^2 + 2mgR \qquad (1)$$

Now we need Equation 13.2, a result from Chris's earlier work. If the ball just barely makes contact with the track at the top of the loop, the ball's centripetal acceleration equals the acceleration due to gravity.

$$a_c = \frac{v_f^2}{R} = g \qquad (13.2)$$

$$v_f^2 = gR$$

Substitute for v_f^2 in Equation (1).

$$mgh = \tfrac{7}{10}mgR + 2mgR = (2.7)mgR$$

$$h = 2.7R = (2.7)(0.250\text{ m})$$

$$h = 0.675\text{ m} = 67.5\text{ cm}$$

CHECK and THINK

This new release height is about 5 cm higher than what Chris found previously. He needs to release the ball from this greater height so that the system starts with enough potential energy to complete the loop. Some of the potential energy is converted into the ball's rotational kinetic energy, and the ball successfully makes it around the loop. Without this extra initial potential energy, the ball's center-of-mass speed at the top of the loop is too low, and the ball cannot complete the loop.

Notice that when Chris added the effect of rolling friction to his first answer, he only found a height difference of 2 mm. Because the new difference we found is 25 times greater, we are safe to ignore mechanical energy losses due to friction.

EXAMPLE 13.9 "And the Winner Is . . ."

Imagine a downhill race between four objects each of mass M (Fig. 13.17). One object such as a skier is best modeled as a particle (shown as a small flat object) translating down the hill. The other objects are modeled as a solid cylinder, hoop, and solid sphere each of radius R rolling down the hill without slipping. All four objects start from rest at height h. Assume mechanical energy losses due to friction may be ignored and determine the order in which the objects cross the finish line.

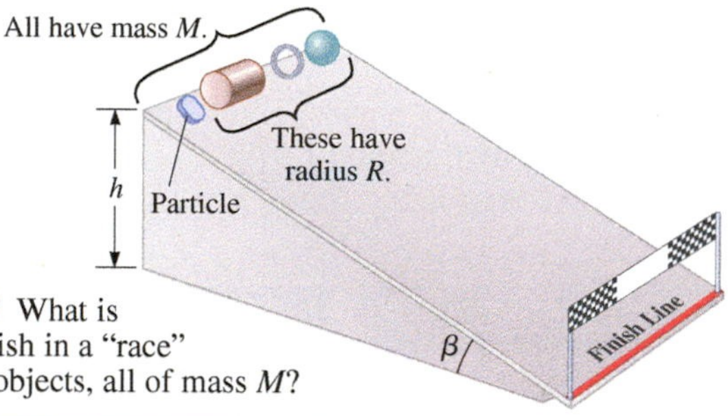

FIGURE 13.17 What is the order of finish in a "race" between these objects, all of mass M?

INTERPRET and ANTICIPATE

The object whose center of mass crosses the finish line in the shortest time interval wins the race. Because all objects start from rest and travel the same distance, the object with the fastest center-of-mass speed as it crosses the finish line wins. For **Step 1**, consider each object to be part of a **system** with the Earth and use conservation of mechanical energy to find the object's center-of-mass speed at the finish line. For **Steps 2 and 3**, set the **reference configuration** to the bottom of the incline, choose an upward-pointing ***y* axis**, set the **initial** time to when an object is at the top of the incline, and set the **final** time to when the object is at the bottom.

Step 4 Each **bar chart** (Fig. 13.18) shows the same initial energy. Because the particle does not rotate, its bar chart has zero rotational kinetic energy. The bar charts for the rolling objects have nonzero rotational kinetic energy.

SOLVE

Step 5 Conservation of energy for the Earth–particle system produces a familiar result (Chapter 8).

$$U_{gi} = K_{\text{CM}}$$

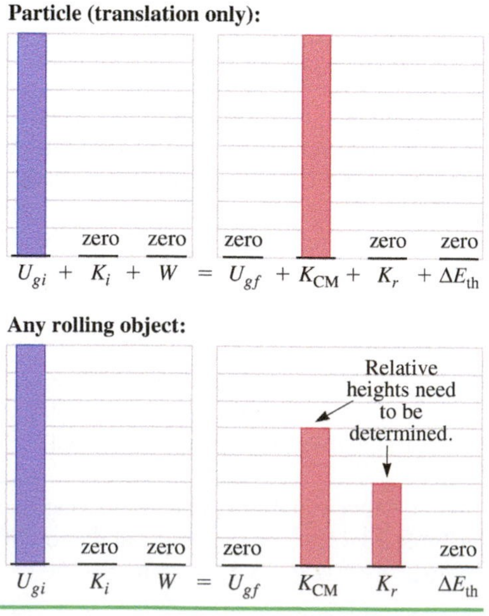

FIGURE 13.18

Steps 6 and 7 Other equations and do algebra. Write the potential energy in terms of the particle's initial height and the kinetic energy in terms of the particle's final speed. Solve for the particle's final speed.	$Mgh = \frac{1}{2}Mv_{CM}^2$ $v_{CM,\,part} = \sqrt{2gh}$ (13.15)
Steps 5 and 6 Conservation of energy and other equations. There are three systems involving a rolling object and the Earth. Conservation of energy for these three systems has the same mathematical form.	$U_{gi} = K_{CM} + K_r$ $Mgh = \frac{1}{2}Mv_{CM}^2 + \frac{1}{2}I_{CM}\omega^2$
From Table 13.1, the rotational inertia of any of these objects may be written as $I_{CM} = (f)MR^2$, where f is a unitless fraction such as $\frac{2}{5}$ in the case of the solid sphere.	$Mgh = \frac{1}{2}Mv_{CM}^2 + \frac{1}{2}(f)MR^2\omega^2$
We use the condition for pure rolling motion (Eq. 13.12, $v_{CM} = \omega R$) to write the expression in terms of v_{CM} and eliminate ω and R.	$Mgh = \frac{1}{2}Mv_{CM}^2 + \frac{1}{2}(f)Mv_{CM}^2$ (13.16) $gh = \frac{1}{2}v_{CM}^2 + \frac{1}{2}(f)v_{CM}^2 = \frac{1}{2}(1+f)v_{CM}^2$ $v_{CM}^2 = \frac{2gh}{(1+f)}$ $v_{CM} = \sqrt{\frac{2gh}{1+f}}$ (13.17)
Find f in Table 13.1 and substitute into Equation 13.17. For a solid cylinder, $f = \frac{1}{2}$.	$v_{CM,\,cyl} = \sqrt{\frac{2gh}{3/2}} = \sqrt{\frac{4}{3}gh} = \sqrt{1.33gh}$
For a hoop, $f = 1$.	$v_{CM,\,hoop} = \sqrt{\frac{2gh}{2}} = \sqrt{gh}$
For a solid sphere, $f = \frac{2}{5}$.	$v_{CM,\,sph} = \sqrt{\frac{2gh}{7/5}} = \sqrt{\frac{10}{7}gh} = \sqrt{1.43gh}$
So, the order in which the objects cross the finish line from first to last is particle, solid sphere, solid cylinder, and hoop.	$v_{CM,\,part} = \sqrt{2gh}$ $v_{CM,\,sph} = \sqrt{1.43gh}$ $v_{CM,\,cyl} = \sqrt{1.33gh}$ $v_{CM,\,hoop} = \sqrt{gh}$ $v_{CM,\,part} > v_{CM,\,sph} > v_{CM,\,cyl} > v_{CM,\,hoop}$

CHECK and THINK

As a quick check, we see that all the expressions have the correct dimensions. A particle does not rotate, so none of the gravitational potential energy goes into rotation; instead, all the initial potential energy goes into translational motion, so the particle wins the race. For objects that rotate, the amount of gravitational potential energy that goes into rotational kinetic energy depends on how close the mass is to the rotation axis. If much of the mass is located farther away from the axis, more of the gravitational potential energy goes into rotational kinetic energy, and less is available for translation. So, the object's center-of-mass speed does *not* depend on its mass or radius; rather, it depends on f, which is determined by the geometry of the object.

13-5 Work and Power

On New Year's Day 1982, the movie *Conan the Barbarian*, starring Arnold Schwarzenegger, opened. His character, Conan, developed his muscles by pushing a very heavy wheel around a vertical axle. Today, *Conan wheels* are used as part of certain strength competitions (Fig. 13.19). Energy is transferred to the wheel by the work done by the athlete. If a system such as a Conan wheel is rotating, work is more conveniently expressed in terms of torque rather than force.

FIGURE 13.19 The contestant is pushing a heavy load around a rotation axis to the left in the photo.

DERIVATION Work in Terms of Torque

Figure 13.20 shows a Conan wheel from above. A force $\vec{F}$ is applied to the wheel's arm at point P, making an angle φ with respect to the arm. We will assume that the wheel moves counterclockwise as shown and derive work in terms of torque:

$$W = \int_{\theta_i}^{\theta_f} \tau d\theta \tag{13.18}$$

We will also show that the power is given by

$$P = \tau\omega \tag{13.19}$$

WORK; POWER ✪ Major Concepts

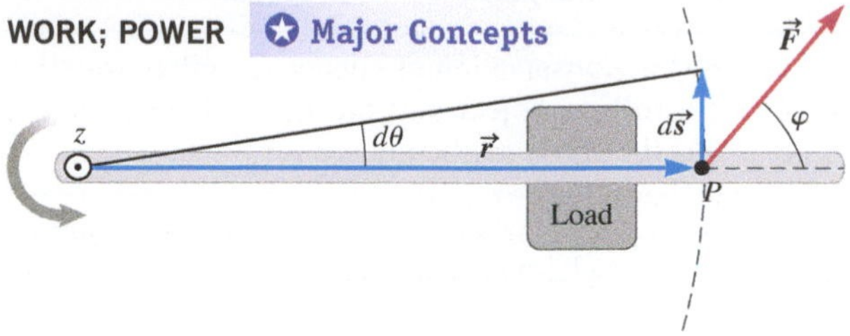

FIGURE 13.20 A person applies a torque to this wheel.

Consider an infinitesimal displacement $d\vec{s}$ of the point of contact P. The work done by the force on the Conan wheel during this infinitesimal displacement comes from Equation 9.20 written in terms of the dot product.	$dW = \vec{F} \cdot d\vec{s}$ (9.20) $dW = F\,ds\cos(90° - \varphi)$ $dW = F\,ds\sin\varphi$ (13.20)
Because the wheel is rotating, angular displacement is much more convenient to work with than translational displacement. The translational displacement ds can be written in terms of the infinitesimal angle $d\theta$ and distance r between point P and the rotation axis ($ds = r\,d\theta$, Eq. 12.12).	$ds = r d\theta$ (13.21)
Substitute Equation 13.21 into Equation 13.20.	$dW = (Fr\sin\varphi)d\theta$
The term $Fr\sin\varphi$ is the torque applied by the person to the Conan wheel ($\tau = Fr\sin\varphi$, Eq. 12.16).	$dW = \tau d\theta$ (13.22)
The work done by the person on the Conan wheel as the wheel rotates from some initial angular position θ_i to a final angular position θ_f comes from integrating Equation 13.22.	$W = \int_{\theta_i}^{\theta_f} \tau d\theta$ ✓ (13.18)

COMMENTS

In keeping with the analogy in Chapter 12 between rotation around a single axis and translation in one dimension, Equation 13.18 is analogous to $W = \int_{x_i}^{x_f} F_x dx$ (Eq. 9.18).

We often need to know the power (the rate at which energy is transferred to the system). To find the power, we divide Equation 13.22 by dt. The resulting Equation 13.19 is analogous to $P = \vec{F} \cdot \vec{v}$ (Eq. 9.38).	$P = \frac{dW}{dt} = \tau\frac{d\theta}{dt}$ $P = \tau\omega$ ✓ (13.19)

Applications: Waterwheels

One of the oldest technological innovations was the wheel. You probably think about how useful the wheel is for transportation, but it has another important use: transferring gravitational potential energy into (rotational) kinetic energy.

Falling water loses gravitational potential energy. That energy may be transferred to a waterwheel (Fig. 13.21) that is then used to operate machinery or generate an electric current. The earliest record of a waterwheel dates from 400 BCE. Ancient waterwheels were mainly used to grind grains into flour. Later waterwheels were used in other industries, for example to operate bellows and hammers in forging iron, to sharpen tools and weapons, and to make textiles. The basic principle of the waterwheel is still in use today in modern hydroelectric power stations (Chapter 32).

There are several major variations in the design of a waterwheel, and many details of these designs contribute to the overall efficiency of a particular device. For example, an efficient waterwheel will have buckets designed to hold the water until the bucket reaches the bottom (as in the next example).

FIGURE 13.21 Waterwheels may be used to transfer energy from moving water to other systems.

EXAMPLE 13.10 What a Grind!

Consider a 12-ft-diameter waterwheel that may have been used to run a textile factory in the United States in the 19th century. The wheel has 24 buckets, each holding 30 gal of water. Assume (somewhat unrealistically) water fills a 30-gal bucket at 10° from the top of the wheel and the water remains in the bucket until it is at 100° from the top, at which point all the water is emptied completely out of the bucket. Friction acts on the waterwheel so that the net torque is zero and the angular speed is a constant 10 rpm. At what rate is energy transferred from the waterfall to the wheel? *Hint*: First find the work done by the water in a single bucket, ignoring the mass of the bucket.

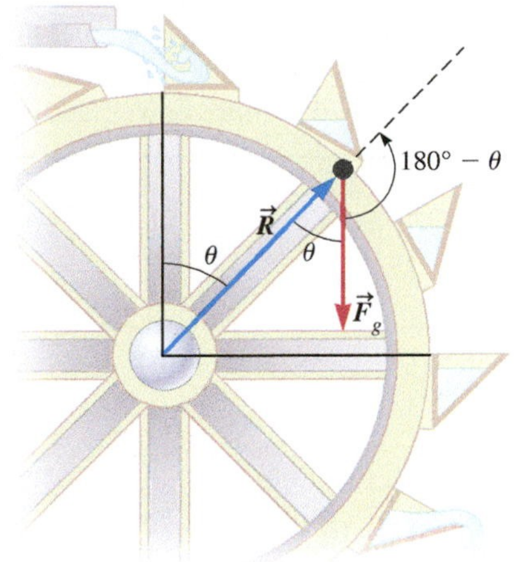

FIGURE 13.22 The black dot represents a bucket of water, where the water has mass m.

INTERPRET and ANTICIPATE

The gravitational force on the water in a full bucket exerts a torque on the waterwheel. A free-body diagram (Fig. 13.22) of a full bucket when its angular position is θ from the top of the waterwheel shows that the torque depends on θ. The torque is zero when the bucket is at the top ($\theta = 0$) and is maximized when the bucket is at $\theta = 90°$.

SOLVE

The magnitude of the torque comes from $\tau = Fr \sin \varphi$ (Eq. 12.16). Use $\sin(180° - \theta) = \sin\theta$.

$$\tau = F_g R \sin(180° - \theta)$$

$$\tau = mgR \sin\theta \qquad (13.23)$$

Because the torque changes as the waterwheel rotates, we must integrate Equation 13.18. Substitute Equation 13.23 for the torque and integrate.

$$W = \int_{\theta_i}^{\theta_f} \tau\, d\theta \qquad (13.18)$$

$$W = \int_{\theta_i}^{\theta_f} mgR \sin\theta\, d\theta = mgR \int_{\theta_i}^{\theta_f} \sin\theta\, d\theta$$

$$W = -mgR \cos\theta \Big|_{\theta_i}^{\theta_f} = mgR(\cos\theta_i - \cos\theta_f)$$

Before evaluating our expression, convert all the given values to SI units and then find the mass of water in the bucket from the density of water. We do not need to convert θ into radians because θ only appears as the input to a trigonometric function.

$$R = \frac{D}{2} = \frac{12\text{ ft}}{2} = 6\text{ ft}\left(\frac{1\text{ m}}{3.28\text{ ft}}\right) = 1.83\text{ m}$$

$$30\text{ gal} \times \frac{1\text{ m}^3}{264\text{ gal}} = 0.114\text{ m}^3$$

$$m = \rho V = (1000\text{ kg/m}^3)(0.114\text{ m}^3)$$

$$m = 1.14 \times 10^2\text{ kg}$$

Substitute for m, R, θ_i, and θ_f.

$$W = mgR(\cos\theta_i - \cos\theta_f)$$

$$W = (1.14 \times 10^2\text{ kg})(9.81\text{ m/s}^2)(1.83\text{ m})(\cos 10° - \cos 100°) = 2.4 \times 10^3\text{ J}$$

During one revolution, a single bucket transfers this energy to the waterwheel. In a single revolution, energy is transferred from the water to the waterwheel by all 24 buckets. So the total work is 24 times the work done by one bucket.

$$W_{tot} = 24W = 24(2.4 \times 10^3\text{ J})$$

$$W_{tot} = 5.8 \times 10^4\text{ J}$$

The power transferred to the waterwheel is given by the total work done by the water in one revolution divided by the period. Find the period from the angular speed.

$$T = \frac{60\text{ s}}{10\text{ rev}} = 6\text{ s}$$

$$P = \frac{W_{tot}}{T} = \frac{5.8 \times 10^4\text{ J}}{6\text{ s}}$$

$$P = 9.6 \times 10^3\text{ W} = 9.6\text{ kW}$$

Example continues on page 376 ▶

CHECK and THINK

Our results provide an estimate of the capabilities of a historic waterwheel. A typical modern house requires between 4 kW and 7 kW. So, the waterwheel in this example would supply enough power for a modern house or apartment. As long as the weather conditions provided water, the waterwheel could provide a steady source of power 24 hours a day, seven days a week. If a factory needed more power, additional waterwheels were added.

For more than two millennia, the waterwheel provided an important means to power machinery. The main alternatives were either human muscles or animal muscles. Also, this sort of waterwheel was a reliable and steady power source. By comparison, consider a person on a stationary exercise bicycle who transfers power from food he has eaten to the kinetic energy of the bicycle. A person in good condition can put out about 100 W for an hour or so but would then need to rest and eat more food. A waterwheel such as the one in Example 13.10 can deliver about 10,000 W without needing a rest.

ANGULAR MOMENTUM
Major Concept

13-6 Angular Momentum

A Segway can be safely maneuvered through busy streets. A unicycle is steered without a steering wheel. Helicopters require two separate rotors. According to Kepler's second law, planets slow down when they are far away from the Sun and speed up when they are close to the Sun. To understand such phenomena, we need to introduce another concept: *angular momentum*. Like linear momentum, angular momentum helps when analyzing complicated motion. Conservation of angular momentum is yet another powerful tool for problem solving in physics.

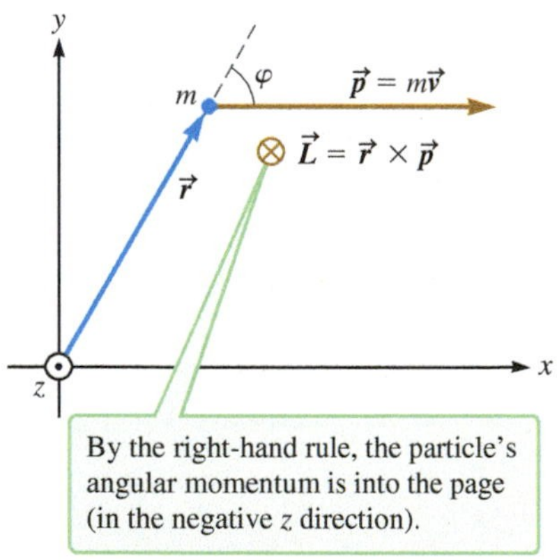

FIGURE 13.23 Definition of angular momentum. A particle does not need to be in orbit around the z axis to have angular momentum around that axis.

Angular Momentum of a Particle

Angular momentum $\vec{L}$ is defined for a particle of mass m moving with translational momentum $\vec{p} = m\vec{v}$:

$$\vec{L} \equiv \vec{r} \times \vec{p} \tag{13.24}$$

where $\vec{r}$ is the position vector of the particle pointing from the origin of the coordinate system to the particle (Fig. 13.23). Angular momentum is defined in terms of the motion of a single particle that does not need to be orbiting the origin of the coordinate system. In fact, the particle does not even need to be moving in a curved path.

Because angular momentum can be determined by using the vector cross product, its direction can be found by using the right-hand rule (Section 12-6). The vector $\vec{L}$ must be perpendicular to the plane containing the vectors $\vec{r}$ and $\vec{p}$. As with any cross product, the magnitude of angular momentum may be found from $R = AB \sin \varphi$ (Eq. 12.22):

$$L = rp \sin \varphi = r_{\perp} p = r p_{\perp} \tag{13.25}$$

where φ is the angle between $\vec{r}$ and $\vec{p}$, $r_{\perp}$ is the component of $\vec{r}$ that is perpendicular to $\vec{p}$, and $p_{\perp}$ is the component of $\vec{p}$ that is perpendicular to $\vec{r}$. The SI units of angular momentum are $kg \cdot m^2/s$.

CONCEPT EXERCISE 13.2

Figure 13.24 shows a particle with momentum $\vec{p}$. Using the coordinate systems shown, determine the direction of the angular momentum of the particle around the origin in each case, and write expressions for $\vec{L}$, using symbols defined in Figure 13.23.

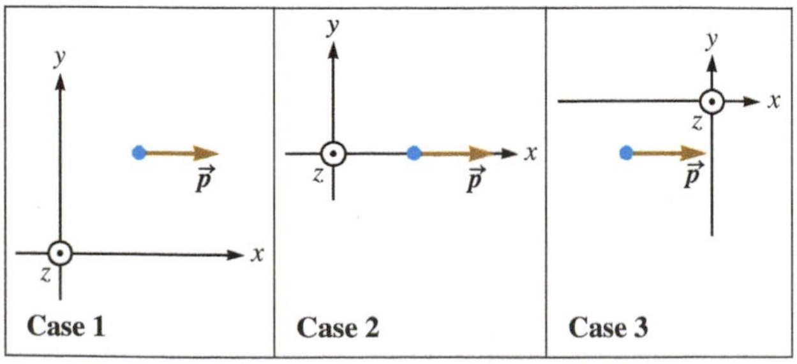

FIGURE 13.24

DERIVATION Angular Momentum of a Rotating Rigid Object

We can find an expression for the angular momentum of a rotating rigid object by considering it to be a collection of particles orbiting the common rotation axis. As an example, consider the dumbbell-shaped object in Figure 13.25. Modeling the dumbbell as two particles moving in concentric circles around the z axis, we will show that its angular momentum is given by

$$\vec{L} = I\vec{\omega} \tag{13.26}$$

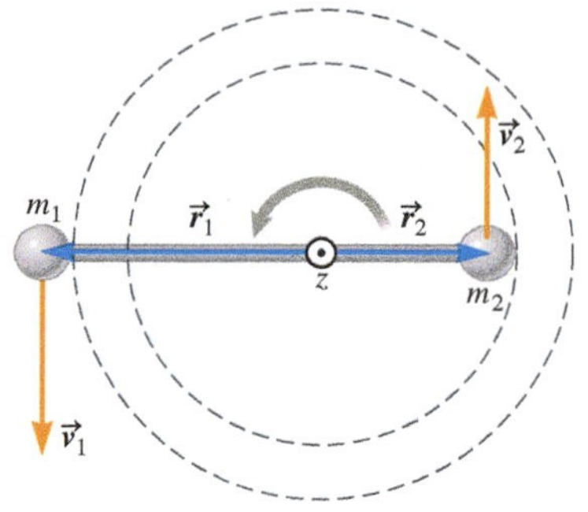

FIGURE 13.25

The angular momentum of each particle is given by $\vec{L} = \vec{r} \times \vec{p}$ (Eq. 13.24).	$\vec{L}_1 = \vec{r}_1 \times \vec{p}_1$ $\vec{L}_2 = \vec{r}_2 \times \vec{p}_2$
By the right-hand rule, both $\vec{L}_1$ and $\vec{L}_2$ point in the positive z direction. From Figure 13.25, $\vec{r}$ is perpendicular to $\vec{p}$ for each particle.	$\vec{L}_1 = r_1 p_1 \hat{k}$ $\vec{L}_2 = r_2 p_2 \hat{k}$
The angular momentum $\vec{L}$ of the entire object is the sum of $\vec{L}_1$ plus $\vec{L}_2$.	$\vec{L} = \vec{L}_1 + \vec{L}_2$ $\vec{L} = (r_1 p_1 + r_2 p_2)\hat{k}$
From $p = mv$ (Eq. 10.1), we can write $\vec{L}$ in terms of the mass and velocity of each particle.	$\vec{L} = (r_1 m_1 v_1 + r_2 m_2 v_2)\hat{k}$
Both particles have the same angular speed ω, related to their translational speeds by $v = \omega r$ (Eq. 12.13).	$\vec{L} = (r_1 m_1 \omega r_1 + r_2 m_2 \omega r_2)\hat{k}$ $\vec{L} = (m_1 r_1^2 + m_2 r_2^2)\omega\hat{k}$
The term in parentheses is the rotational inertia of the dumbbell-shaped object (Eq. 13.4).	$\vec{L} = I\omega\hat{k}$
The angular velocity in Figure 13.25 is in the z direction: $\vec{\omega} = \omega\hat{k}$. Therefore, the dumbbell-shaped object's angular momentum may be written as a vector.	$\vec{L} = I\vec{\omega}$ ✓ (13.26)

COMMENTS

We found Equation 13.26 by calculating the angular momentum around the rotational axis of the dumbbell. The relationship between angular momentum and angular velocity may be more complicated in some situations, but Equation 13.26 holds for any object if $\vec{L}$ is the component of the angular momentum along the rotation axis.[1]

EXAMPLE 13.11 CASE STUDY Comet Halley's Orbital Angular Momentum

Use the results of the case study in Chapter 8 to find the orbital angular momentum of Comet Halley around the Sun **A** when the comet is at perihelion ($r_P = 8.83 \times 10^{10}$ m) and **B** when it is at aphelion ($r_A = 5.27 \times 10^{12}$ m). In Example 8.13, we found the perihelion and aphelion speeds, $v(r_P) = 54.4$ km/s and $v(r_A) = 0.92$ km/s, respectively. Recall that the mass of Comet Halley is $m = 1.7 \times 10^{15}$ kg. (We have provided an extra significant figure here for $v(r_P)$. Use all the significant figures given and report your final results to two significant figures.)

Example continues on page 378 ▶

[1]According to Equation 13.26, $\vec{L}$ and $\vec{\omega}$ are parallel. Sometimes, though, the motion is more complicated, and the two vectors point in different directions. When $\vec{L}$ and $\vec{\omega}$ are not parallel, I is not a scalar, and Equation 13.26 does not hold. This sort of complicated motion is beyond the scope of this textbook.

INTERPRET and ANTICIPATE

The example asks for the *orbital* angular momentum around the Sun. In this case, we can treat Comet Halley as a particle, ignoring its rotational motion around its center of mass. At perihelion and aphelion, the position vector is perpendicular to the velocity vector (Fig. 13.26). Finding the angular momentum for these two positions is straightforward.

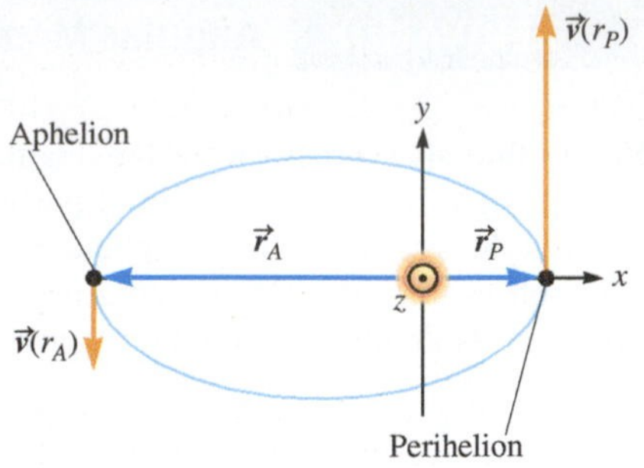

FIGURE 13.26

SOLVE According to the right-hand rule, the angular momentum is in the positive z direction given the coordinate system shown in Figure 13.26.	$\vec{L} = \vec{r} \times \vec{p}$ (13.24) $\vec{L} = rp\hat{k}$ $\vec{L} = rmv\hat{k}$
A Substitute the perihelion values.	$\vec{L}(r_P) = r_P m v_P \hat{k}$ $\vec{L}(r_P) = (8.83 \times 10^{10}\,\text{m})(1.7 \times 10^{15}\,\text{kg})(5.44 \times 10^{4}\,\text{m/s})\hat{k}$ $\vec{L}(r_P) = (8.2 \times 10^{30})\hat{k}\,\text{kg} \cdot \text{m}^2/\text{s}$
B Substitute the aphelion values.	$\vec{L}(r_A) = r_A m v_A \hat{k}$ $\vec{L}(r_A) = (5.27 \times 10^{12}\,\text{m})(1.7 \times 10^{15}\,\text{kg})(9.2 \times 10^{2}\,\text{m/s})\hat{k}$ $\vec{L}(r_A) = (8.2 \times 10^{30})\hat{k}\,\text{kg} \cdot \text{m}^2/\text{s}$

CHECK and THINK

As a quick check, notice that our results have the correct units. A more interesting point is that Comet Halley's angular momentum seems to be the same at perihelion and aphelion. It is *not* a coincidence, as we show in the next section.

EXAMPLE 13.12 The Spinning CD

Find the change in angular momentum of a CD (m = 16.3 g, R = 6.00 cm) as it is played from beginning to end. From Problem 12.82 (page 359), when information is being read from the innermost ring, the CD's angular speed is 52.4 rad/s. The CD slows down so that when information is read from the outermost ring, its angular speed is 20.9 rad/s.

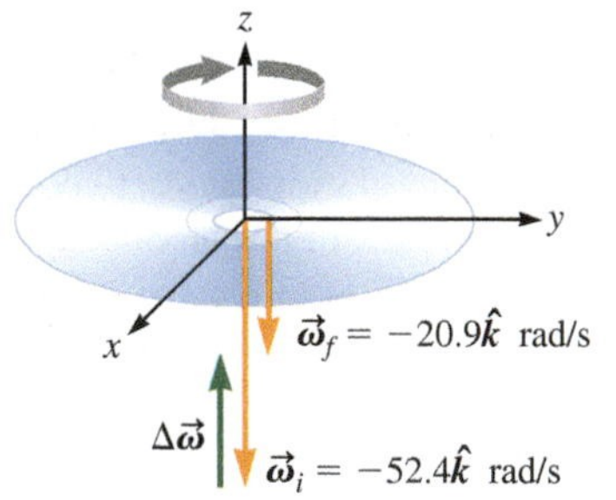

FIGURE 13.27

INTERPRET and ANTICIPATE

Figure 13.27 shows the initial and final angular velocities of the CD. The change in angular velocity is in the positive z direction. Because $\vec{L}$ is parallel to $\vec{\omega}$, we expect the change in angular momentum to be in the positive z direction also.

SOLVE Model the CD as a solid disk and use Table 13.1 to find its rotational inertia.	$I = \frac{1}{2}MR^2 = \frac{1}{2}(16.3 \times 10^{-3}\,\text{kg})(6.00 \times 10^{-2}\,\text{m})^2$ $I = 2.934 \times 10^{-5}\,\text{kg} \cdot \text{m}^2$
The change in angular momentum is derived from $\vec{L} = I\vec{\omega}$ (Eq. 13.26). The rotational inertia does not change as the CD is played.	$\Delta\vec{L} = \Delta(I\vec{\omega}) = I\Delta\vec{\omega}$ $\Delta\vec{L} = I(\vec{\omega}_f - \vec{\omega}_i)$
Substitute the rotational inertia and the angular velocities.	$\Delta\vec{L} = (2.934 \times 10^{-5}\,\text{kg} \cdot \text{m}^2)[-20.9\hat{k}\,\text{rad/s} - (-52.4\hat{k}\,\text{rad/s})]$ $\Delta\vec{L} = (9.24 \times 10^{-4})\hat{k}\,\text{kg} \cdot \text{m}^2/\text{s}$

CHECK and THINK

Because the CD slows down and because its angular velocity is in the negative z direction, the change in its angular momentum is in the positive z direction. In Problem 12.82, you should find that the angular acceleration is also in the positive z direction. Again, it is *not* a coincidence.

13-7 Conservation of Angular Momentum

NEWTON'S SECOND LAW
Underlying Principle

We found in Chapter 10 that when no net force acts on a system, the system's *translational* momentum is conserved, and this principle gave us a way to analyze complicated situations such as the decay of a nucleus. In this section, we will learn when and how to apply the principle of conservation of *angular* momentum. However, to better understand conservation of angular momentum, we consider a system whose angular momentum is changing and derive another expression for Newton's second law.

DERIVATION Another Expression of Newton's Second Law

Another way to express Newton's second law for a system is

$$\frac{d\vec{L}_{tot}}{dt} = \sum \vec{\tau}_{ext} \tag{13.27}$$

where $\vec{\tau}_{ext}$ is the total external torque exerted and $\vec{L}_{tot}$ is the system's total angular momentum. Equation 13.27 holds for a single particle, a collection of particles, or an object.

Our derivation is made simpler if we consider a single particle moving along a straight path (Fig. 13.28). A net force $\vec{F}_{tot}$ acts on the particle so that it speeds up along that path. We will show that for this particle,

$$\frac{d\vec{L}}{dt} = \vec{\tau}_{tot} \tag{13.28}$$

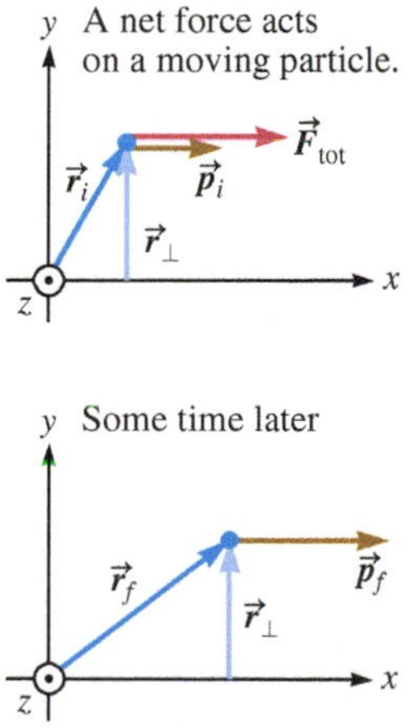

FIGURE 13.28 The particle moves in a straight line, and so $\vec{r}_\perp$ is constant.

Because the particle's speed increases, its translational momentum also increases. The particle's path is along the straight line shown, so $r_\perp$ does not change, but because the translational momentum p increases, the particle's angular momentum L around the z axis increases also.

Take the time derivative of Equation 13.24 to derive an expression for the rate of change of angular momentum.	$\vec{L} = \vec{r} \times \vec{p}$ (13.24) $\frac{d\vec{L}}{dt} = \frac{d(\vec{r} \times \vec{p})}{dt}$
Expanding the right side using the chain rule, we obtain a cross product in each term.	$\frac{d(\vec{r} \times \vec{p})}{dt} = \left(\frac{d\vec{r}}{dt} \times \vec{p}\right) + \left(\vec{r} \times \frac{d\vec{p}}{dt}\right)$
Rewrite the derivative in the first term using Equation 4.8 (definition of translational velocity). Rewrite the derivative in the second term using Equation 10.2 (a general form of Newton's second law).	$\vec{v} = \frac{d\vec{r}}{dt}$ (4.8) $\vec{F}_{tot} = \frac{d\vec{p}}{dt}$ (10.2) $\frac{d(\vec{r} \times \vec{p})}{dt} = (\vec{v} \times \vec{p}) + (\vec{r} \times \vec{F}_{tot})$ $\frac{d(\vec{r} \times \vec{p})}{dt} = [\vec{v} \times (m\vec{v})] + (\vec{r} \times \vec{F}_{tot})$
The first term is zero because $\vec{v}$ and $(m\vec{v})$ are parallel, and the cross product of parallel vectors is zero. The second term is the torque $\vec{\tau}_{tot}$ on the particle due to the net force $\vec{F}_{tot}$ (Eq. 12.23).	$\frac{d(\vec{r} \times \vec{p})}{dt} = 0 + \vec{\tau}_{tot}$

Derivation continues on page 380 ▶

So, a net torque on a particle changes its angular momentum.	$\frac{d\vec{L}}{dt} = \vec{\tau}_{tot}$ ✓	(13.28)

COMMENTS

Equation 13.28 is analogous to $\vec{F}_{tot} = d\vec{p}/dt$ (Eq. 10.2) and is another way to express Newton's second law.

We derived Equation 13.28 for a particle. With two modifications, it also holds for a system of particles or for a rotating object, where $\vec{L}_{tot}$ is the total angular momentum and $\vec{\tau}_{ext}$ is the total external torque acting on the system.	$\frac{d\vec{L}_{tot}}{dt} = \sum \vec{\tau}_{ext}$	(13.27)

Our result $d\vec{L}_{tot}/dt = \sum\vec{\tau}_{ext}$ is analogous to $d\vec{p}_{tot}/dt = \sum\vec{F}_{ext}$ (Eq. 10.10). Also, $d\vec{L}_{tot}/dt = \sum\vec{\tau}_{ext}$ may be applied to a particle. In that case of a single particle, the total angular momentum is $\vec{L}_{tot} = \vec{L}$, and all the torques must be external. So, $\sum\vec{\tau}_{ext} = \vec{\tau}_{tot}$.

CONSERVATION OF ANGULAR MOMENTUM

Underlying Principle

According to $d\vec{L}_{tot}/dt = \sum\vec{\tau}_{ext}$ (Eq. 13.27), if $\sum\vec{\tau}_{ext} = 0$, then $\vec{L}_{tot}$ is constant. Putting this expression into words, **if the net external torque acting on a system is zero, the system's total angular momentum is conserved.**

Recall from Chapters 8 through 11 that a conservation approach works well when we would like to consider a particle or system at two different times, t_i and t_f. If the angular momentum of a particle is conserved, then

$$\vec{r}_i \times \vec{p}_i = \vec{r}_f \times \vec{p}_f \tag{13.29}$$

If the angular momentum of a rotating object (or collection of particles) is conserved, it is convenient to write

$$I_i\omega_i = I_f\omega_f \tag{13.30}$$

by making use of Equation 13.26.

CONCEPT EXERCISE 13.3

In Problem 12.82, (page 359) we found a CD's angular acceleration, and in Example 13.12, we found its change in angular momentum. Use $d\vec{L}_{tot}/dt = \sum\vec{\tau}_{ext}$ (Eq. 13.27) and $\vec{\tau}_{tot} = I\vec{\alpha}$ (Eq. 12.24) to explain why the CD's angular acceleration $\vec{\alpha}$ and change in angular momentum $\Delta\vec{L}$ point in the same direction.

FIGURE 13.29 The rotors on a helicopter rotate in opposite directions. The best way to understand the motion of these vehicles is by considering their angular momentum.

CONCEPT EXERCISE 13.4

Use the principle of conservation of angular to momentum to explain why the rotors on a helicopter such as the one in Figure 13.29 rotate in opposite directions. What would happen if one rotor stopped operating?

CONCEPT EXERCISE 13.5

A simple gyroscope is a spinning disk. Why are gyroscopes used on space telescopes?

EXAMPLE 13.13 Kepler's Second Law Revisited

According to Kepler's second law (Section 7-2), a line joining a planet to the Sun sweeps out equal areas in equal time intervals. Use the principle of conservation of angular momentum to prove Kepler's second law.

INTERPRET and ANTICIPATE

To express this law mathematically, imagine an infinitesimal area dA swept out in a very short time interval dt (Fig. 13.30).

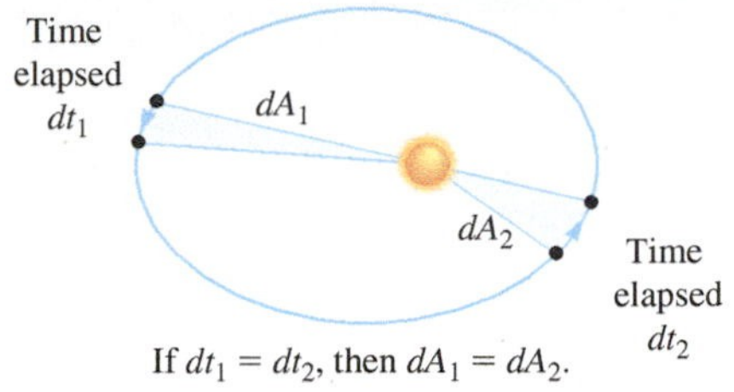

FIGURE 13.30

Kepler's second law says that if $dt_1 = dt_2$, then $dA_1 = dA_2$..	$\frac{dA_1}{dt_1} = \frac{dA_2}{dt_2}$ or $\frac{dA}{dt} = \text{constant}$ (13.31)

SOLVE

Now that we have a mathematical expression for Kepler's second law, we can use physical concepts and mathematics to prove it. Our results in Example 13.11 showed that the angular momentum of Comet Halley is nearly constant as it orbits the Sun. Let us see if a particle such as a planet in orbit around the Sun meets the condition for conserved angular momentum.

We assume the other planets and objects in the solar system exert a negligible force on the planet. The only force then acting on the planet is $\vec{F}_G$ due to the Sun's gravity (Fig. 13.31).

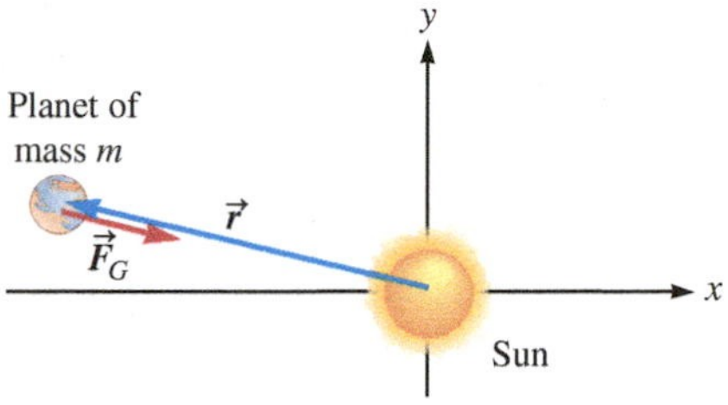

FIGURE 13.31

Because the angle between $\vec{r}$ and $\vec{F}_G$ is 180° at all times, there is no torque acting on the planet, and its angular momentum around the Sun is conserved.	$\vec{\tau} = \vec{r} \times \vec{F}_G$ $\tau = rF_G \sin 180° = 0$
Write an expression for the planet's constant angular momentum.	$\vec{L}_{\text{const}} = \vec{r} \times \vec{p}$ $\vec{L}_{\text{const}} = \vec{r} \times m\vec{v}$ $\vec{L}_{\text{const}} = m(\vec{r} \times \vec{v})$ (13.32)

The infinitesimal area dA is half the area of the parallelogram in Figure 13.32.

From Example 12.5 (page 347), the area of a parallelogram may be written in terms of a cross product. The area of the parallelogram is the magnitude of $\vec{r} \times d\vec{r}$.	$dA = \frac{1}{2}\lvert\vec{r} \times d\vec{r}\rvert$ (13.33)

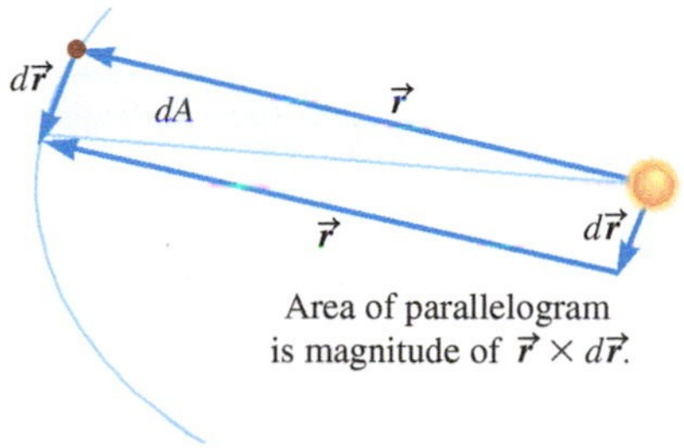

FIGURE 13.32

Write $d\vec{r}$ in terms of the velocity $\vec{v}$ (Eq. 4.8).	$\vec{v} = \frac{d\vec{r}}{dt}$ (4.8) $d\vec{r} = \vec{v}dt$
Substitute $d\vec{r}$ into Equation 13.33.	$dA = \frac{1}{2}\lvert\vec{r} \times \vec{v}dt\rvert$ $dA = \frac{1}{2}\lvert\vec{r} \times \vec{v}\rvert dt$ $\frac{dA}{dt} = \frac{1}{2}\lvert\vec{r} \times \vec{v}\rvert$ (13.34)
Finally, substitute Equation 13.32 into Equation 13.34.	$\frac{dA}{dt} = \frac{1}{2}\left\lvert\frac{\vec{L}_{\text{const}}}{m}\right\rvert$ $\frac{dA}{dt} = \frac{L_{\text{const}}}{2m}$ ✓ (13.35)

CHECK and THINK

Because m and L_{const} are constant for a particular planet, dA/dt is a constant, exactly as Kepler's second law says in Equation 13.31.

EXAMPLE 13.14 More on Spinning Pulsars

Typical neutron stars (the remains of dead stars) have a radius of 10 km. A pulsar is a neutron star that we observe as pulses or flashes of light. Pulsars rotate rapidly with periods roughly between 1 millisecond and 1 second (Example 12.2, page 339). The period of the pulses equals the period of the pulsar's rotation. A pulsar may form when a massive star is unable to produce energy in its core through nuclear fusion. At that time, the core collapses and becomes a pulsar while the outer layers are blown out in a powerful supernova explosion (Fig. 13.33). Before it collapses, the core's radius is roughly the same as that of the Earth (5000 km). Assume a newly formed neutron star's period is 5 ms and none of the core's mass is lost when it collapses. Model the core and the pulsar as solid spheres of uniform density and estimate the rotational period of the core before it collapses.

© NASA/Hubble

FIGURE 13.33 The Crab Nebula, the remains of a star that exploded over 1000 years ago.

INTERPRET and ANTICIPATE

Assume no external forces act on the core during its collapse. Because no net external torque acts on the core, its angular momentum is conserved.

SOLVE

Using Equation 13.30, apply the principle of conservation of angular momentum.

$$I_i\omega_i = I_f\omega_f \qquad (13.30)$$

Write the angular speeds in terms of the period before and after collapse.

$$I_i\left(\frac{2\pi}{T_i}\right) = I_f\left(\frac{2\pi}{T_f}\right)$$

$$T_i = T_f\left(\frac{I_i}{I_f}\right)$$

Use Table 13.1 to find the rotational inertia of a sphere.

$$T_i = T_f\left(\frac{\frac{2}{5}Mr_i^2}{\frac{2}{5}Mr_f^2}\right) = T_f\left(\frac{r_i}{r_f}\right)^2$$

Substitute the given final period (T_f = 5 ms) and radius (r_f = 10 km) and initial core radius (r_i = 5000 km). We do not need to convert from kilometers to meters in this case.

$$T_i = (5 \times 10^{-3}\ \text{s})\left(\frac{5000\ \text{km}}{10\ \text{km}}\right)^2$$

$$T_i = 1.25 \times 10^3\ \text{s}$$

$$T_i \approx 20\ \text{min}$$

CHECK and THINK

Because the rotational inertia is reduced by the collapse of the core, the core's angular speed must increase after collapse to compensate, in accordance with conservation of angular momentum.

EXAMPLE 13.15 Bicycle Tires and Unicycles

At the Smithsonian Air and Space Museum in Washington, D.C., visitors are invited to sit on a stool and hold a rapidly rotating bicycle tire (Fig. 13.34). Initially, the visitor and the stool are at rest, and the bicycle tire is rotating (Fig. 13.34A). The visitor flips the rapidly rotating tire upside down (Fig. 13.34B), which causes the stool and visitor to rotate. In which direction does the visitor rotate? What happens if the visitor flips the tire right side up again?

FIGURE 13.34

INTERPRET and ANTICIPATE

Let's include the visitor, the stool, and the bicycle tire in the system. There is no net external force on the system, so there is no net external torque on it; thus, angular momentum is conserved. Before the visitor flips the bicycle tire, the system's angular momentum is upward in the positive z direction. After she flips the tire, the system's angular momentum must still have the same magnitude and direction.

FIGURE 13.35

SOLVE

Initially (Fig. 13.35), the angular momentum of the system equals the angular momentum of the *b*icycle tire $\vec{L}_i = L_b\hat{k}$.

After the visitor flips the bicycle tire upside down, its angular momentum points in the negative z direction (Fig. 13.36). Because angular momentum is conserved ($\vec{L}_f = \vec{L}_i$), the angular momentum of the whole system must still be in the positive z direction. So, the visitor and the stool must rotate, and their angular momentum must be in the positive z direction and given by $\vec{L}_f = L_v\hat{k} - L_b\hat{k}$.

According to $\vec{L}_v = I\vec{\omega}_v$ (Eq. 13.26), the visitor's angular velocity is in the positive z direction. She rotates counterclockwise when viewed from above.

If the visitor flips the tire right side up again, angular momentum is still conserved. The bicycle tire's angular momentum points in the positive z direction again. All the angular momentum of the system is in the bicycle tire, so the visitor must stop rotating.

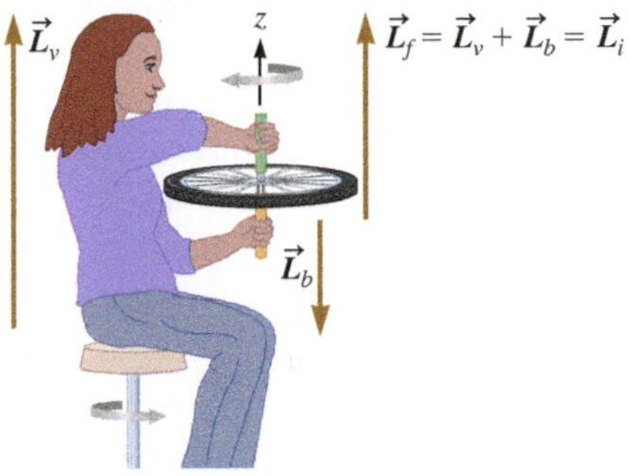

FIGURE 13.36

CHECK and THINK

Our results give us a way to think about how a person might steer a unicycle. Consider the person, the unicycle, and the Earth as the system. No net torque acts on the system. Ignoring the motion of the Earth, Figure 13.37A shows the initial angular momentum of the system with the unicycle in motion: $\vec{L}_i = \vec{L}_{\text{tire}}$. The person leans to his left so that the angular momentum of the tire rotates downward. The total angular momentum must still point to the right (Fig. 13.37B), so the angular momentum of the person must be upward to compensate. The person's angular velocity therefore points upward (parallel to his own angular momentum), and he is able to make a turn. Another way to analyze this situation is to exclude the Earth from the system and calculate the torque done by gravity. This approach is left as a homework problem.

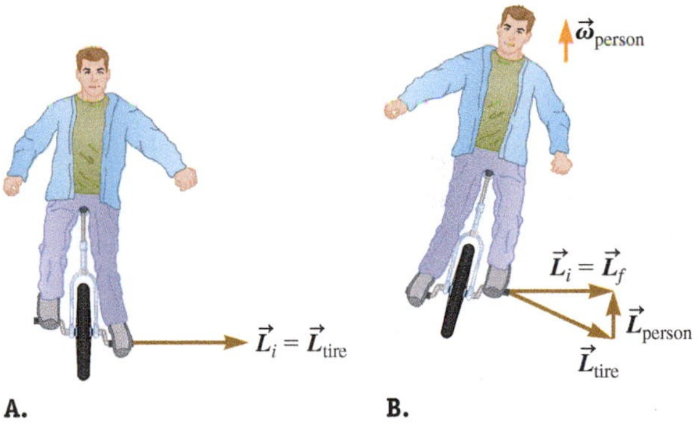

FIGURE 13.37

SUMMARY

Underlying Principles

1. The **conservation of energy** principle is slightly modified to include rotational motion:

$$K_i + K_{ri} + U_i + W = K_f + K_{rf} + U_f + \Delta E_{\text{th}}$$

2. **Newton's second law** can be written in terms of torque and angular momentum:

$$\frac{d\vec{L}_{\text{tot}}}{dt} = \sum \vec{\tau}_{\text{ext}} \quad (13.27)$$

3. A system's **angular momentum is conserved** if the total external torque on the system is zero.

Major Concepts

1. **Rotational inertia** is a measure of an object's tendency to maintain its angular velocity. The rotational inertia of an object that consists of n particles may be found mathematically from

$$I = \sum_{i=1}^{n} m_i r_i^2 \tag{13.4}$$

Rotational inertia of a continuous object is found from

$$I = \int r^2 dm \tag{13.8}$$

Table 13.1 (page 365) provides the result of such integration for a number of simple-shaped objects.

2. According to the **parallel-axis theorem**, if h is the perpendicular distance between a new rotation axis and the axis through the center of mass, the rotational inertia through the new axis is

$$I = I_{CM} + Mh^2 \tag{13.9}$$

where M is the mass of the object.

3. **Rotational kinetic energy** is most easily found in terms of the object's rotational inertia and angular speed:

$$K_r = \tfrac{1}{2} I \omega^2 \tag{13.10}$$

4. **Work** done by an external torque is

$$W = \int_{\theta_i}^{\theta_f} \tau d\theta \tag{13.18}$$

5. **Power** supplied by an external torque is

$$P = \tau\omega \tag{13.19}$$

6. **Angular momentum** $\vec{L}$ is defined for a particle of mass m moving with translational momentum $\vec{p} = m\vec{v}$:

$$\vec{L} \equiv \vec{r} \times \vec{p} \tag{13.24}$$

In general, a system's angular momentum may be written as

$$\vec{L} = I\vec{\omega} \tag{13.26}$$

Special Case

Pure rolling is a special combination of rotational and translation motion in which the object rolls along a surface without slipping. In this special case, the center-of-mass position is given by

$$x_{CM} = s = \theta R \tag{13.11}$$

The center-of-mass speed is given by

$$v_{CM} = \omega R \tag{13.12}$$

The center-of-mass acceleration is given by

$$a_{CM} = \alpha R \tag{13.13}$$

PROBLEMS AND QUESTIONS

A = algebraic C = conceptual E = estimation G = graphical N = numerical

13-1 Conservation Approach

1. C In the following four situations, you do positive work on a system. In other words, energy is transferred from you to a system. In each case, describe what happens to the energy after it enters the system. If there is kinetic energy in the system after the work is performed, is it possible for the speed to be constant in each case? **a.** You pull a sled across a frozen lake. **b.** You pull a sled up a snowy hill. **c.** You pull a string wrapped around a pulley with a fixed axle. **d.** You pull a cart with large wheels across the flat ground.

2. C CASE STUDY When Chris discovers that the ball does not complete the loop-the-loop challenge, he estimates the increase in thermal energy due to rolling friction. He finds that he needs to release the ball from a greater initial height. **a.** Why does it make sense that the ball must be released from a greater height? **b.** Is Chris's value of the increase in thermal energy an overestimate or an underestimate? Explain.

3. C A Frisbee flies across a field. Determine if the system has translational kinetic energy, rotational kinetic energy, neither, or both as determined by the observer in each of the following cases. **a.** The observer watches the flight of the Frisbee across the field from a park bench. **b.** The observer is a dog that runs directly beneath the Frisbee. **c.** The observer is an ant at rest on the Frisbee.

4. C Does a car moving along a highway have rotational kinetic energy? (You may ignore motions within the car.) Explain.

13-2 Rotational Inertia

Problems 5 and 6 are paired.

5. N A system consists of four boxes modeled as particles connected by very lightweight, stiff rods (Fig. P13.5). The system rotates around the z axis, which points out of the page. Each particle has a mass of 5.00 kg. The distances from the z axis to each particle are $r_1 = 32.0$ cm, $r_2 = 16.0$ cm, $r_3 = 17.0$ cm, and $r_4 = 34.0$ cm. Find the rotational inertia of the system around the z axis.

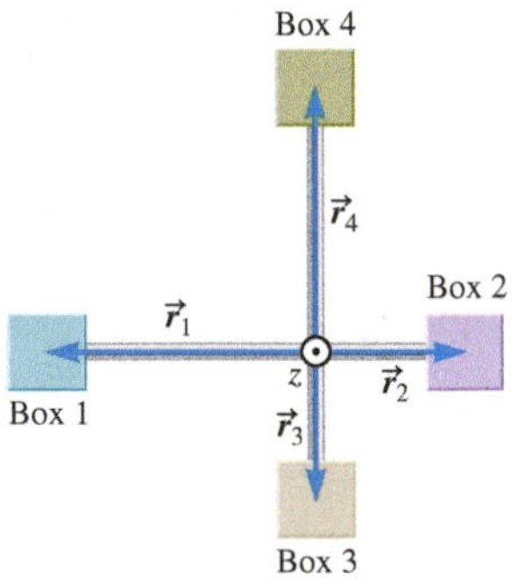

FIGURE P13.5 Problems 5 and 6.

6. N Use the information in Problem 5 to find the rotational inertia of the system around particle 1.

7. N A 12.0-kg solid sphere of radius 1.50 m is being rotated by applying a constant tangential force of 10.0 N at a perpendicular distance of 1.50 m from the rotation axis through the center

of the sphere. If the sphere is initially at rest, how many revolutions must the sphere go through while this force is applied before it reaches an angular speed of 30.0 rad/s?

8. **E** A figure skater clasps her hands above her head as she begins to spin around a vertical axis that passes through her hands, through the top of her head, and through the bottom of her skates. What is the order of magnitude of her rotational inertia around this axis?

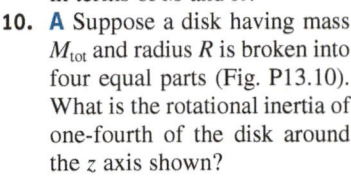

9. **A** A solid sphere of mass M and radius R is rotating around an axis that is tangent to the sphere (Fig. P13.9). What is the rotational inertia of the sphere in this scenario in terms of M and R?

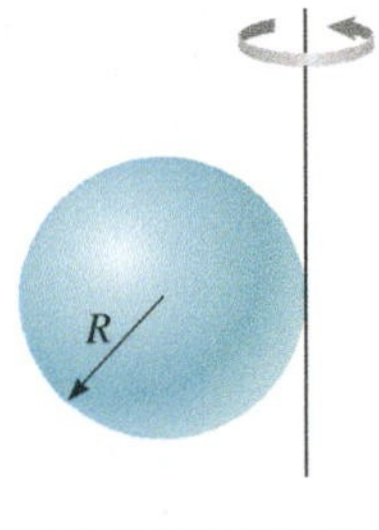

FIGURE P13.9

10. **A** Suppose a disk having mass M_{tot} and radius R is broken into four equal parts (Fig. P13.10). What is the rotational inertia of one-fourth of the disk around the z axis shown?

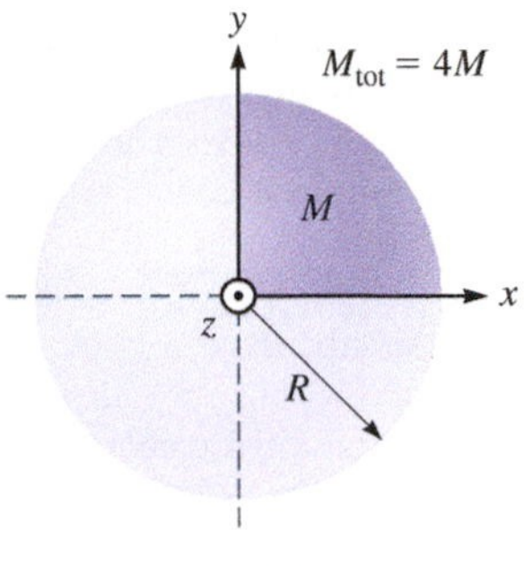

FIGURE P13.10

Problems 11 and 12 are paired.

11. **A** A thin disk of radius R has a nonuniform density $\sigma = 4.5r^2$, where σ has units of kg/m^2 when r is in meters. Derive an expression for the rotational inertia of this disk around an axis through its center and perpendicular to the disk's surface, assuming R is given in meters.

12. **A** Given the disk and density in Problem 11, derive an expression for the rotational inertia of this disk around an axis at the edge of the disk and perpendicular to the disk's surface.

13. **N** A large stone disk is viewed from above and is initially at rest as seen in Figure P13.13. The disk has a mass of 150.0 kg and a radius of 2.000 m. A constant force of 40.0 N is applied tangent to the edge of the disk for 60.0 s, causing the disk to spin around the z axis. **a.** Calculate the angular acceleration of the stone, finding both the direction and magnitude. **b.** What is the final angular velocity of the stone? **c.** Calculate the translational speed for a point on the edge of the stone after 60.0 s.

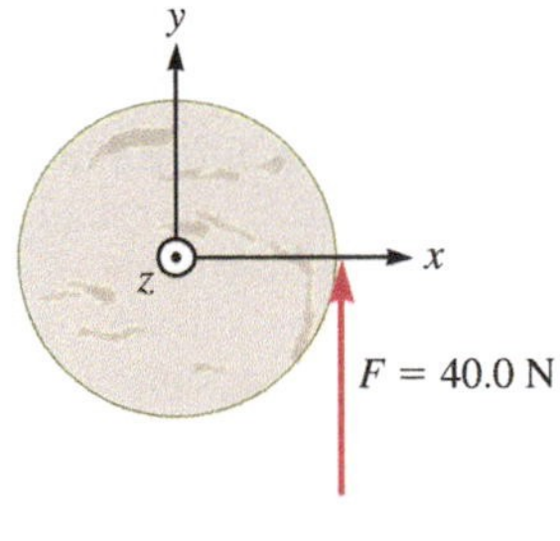

FIGURE P13.13

Problems 14 and 17 are paired.

14. A bike tire has a diameter of 0.775 m and a mass of 3.58 kg. Its mass is concentrated in its rim.

a. **N** Find its rotational inertia around its axle.

b. **N** Find its rotational inertia around its diameter.

c. **C** Which is greater, the rotational inertia around its axle or around its diameter, and why is it greater?

15. **N** A uniform disk of mass M = 3.00 kg and radius r = 22.0 cm is mounted on a motor through its center. The motor accelerates the disk uniformly from rest by exerting a constant torque of 1.00 N · m. **a.** What is the time required for the disk to reach an angular speed of 8.00×10^2 rpm? **b.** What is the number of revolutions through which the disk spins before reaching this angular speed?

16. **N** The net total torque of 50.0 N · m on a wheel rotating around an axis through its center is due to an applied force and a frictional torque at the axle. Starting from rest, the wheel reaches an angular speed of 12.0 rad/s in 5.00 s. At t = 5.00 s, the applied force is removed, and the frictional torque brings the wheel to a stop in 30.0 s. **a.** What is the rotational inertia of the wheel? **b.** What is the magnitude of the frictional torque acting on the wheel? **c.** What is the total number of revolutions the wheel undergoes during this 35.0-s interval?

13-3 Rotational Kinetic Energy

17. **N** A bike tire has a diameter of 0.775 m and a mass of 3.58 kg. Its mass is concentrated in its rim. Assuming the tire rotates with an angular speed of 5.34 rev/s, what is its rotational kinetic energy if it is rotating around its axle?

18. **N** The system shown in Figure P13.18 consisting of four particles connected by massless, rigid rods is rotating around the x axis with an angular speed of 2.50 rad/s. The particle masses are m_1 = 1.00 kg, m_2 = 4.00 kg, m_3 = 2.00 kg, and m_4 = 3.00 kg. **a.** What is the rotational inertia of the system around the x axis? **b.** Using $K_r = \frac{1}{2}I\omega^2$ (Eq. 13.10), what is the total rotational kinetic energy of the system? **c.** What is the tangential speed of each of the four particles? **d.** Considering the system as four particles in motion and using $K = \sum_i \frac{1}{2}mv_i^2$, what is the total kinetic energy of the system? How does this value compare with the result obtained in part (b)?

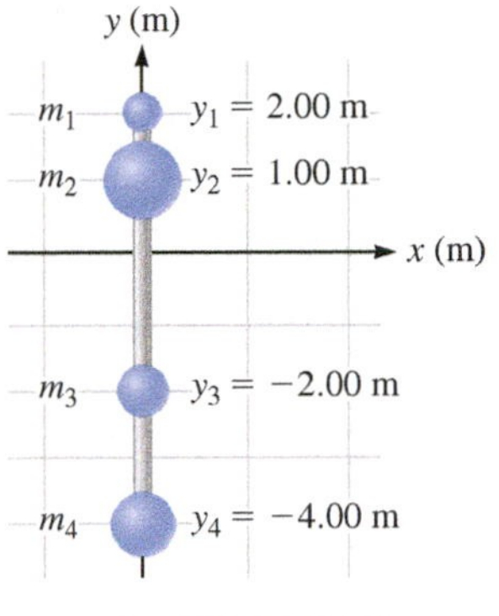

FIGURE P13.18

19. **N** A 10.0-kg disk of radius 2.0 m rotates from rest as a result of a 20.0-N tangential force applied at the edge of the disk. What is the kinetic energy of the disk 4.00 s after the force is applied?

20. **N** Model the Earth as a solid sphere rotating around an axis through its center of mass. Find the rotational kinetic energy of the Earth.

Problems 21 and 22 are paired.

21. **N** A thin, hollow sphere of mass 2.00 kg and radius 0.600 m is rolling on a horizontal surface with a constant angular speed of 60.0 rpm. Find the total kinetic energy of the sphere.

22. **N** In Problem 21, what fraction of the kinetic energy is translational kinetic energy, and what fraction of the kinetic energy is rotational kinetic energy?

23. **N** The 2.00-m-long hour hand of a large clock has a mass of 25.0 kg, and the 3.00-m-long minute hand has a mass of 15.0 kg. Assuming the clock hands can be modeled as thin rods and the clock hands rotate at a constant rate, what is the total rotational kinetic energy of the clock hands around their axis of rotation?

24. **N** When a pitcher throws a baseball or softball, the rotation speed is important when it comes to making the ball "break," or curve away from its expected motion. A baseball that has a mass of 0.143 kg and a radius of 3.65 cm can rotate at a rate of 1.80×10^3 rpm on its way to home plate, causing the ball to curve! Suppose this baseball is rotating around an axis through its center at this extreme rate. What is the rotational kinetic energy of the baseball as it travels toward home plate?

25. **N** In the fall of 2003, Sanyo announced that it could make 10 compact disks (CDs) from one ear of corn. Each CD has a mass of 16.3 g and a radius R = 6.00 cm. From Problem 12.82 (page 359), when information is being read off the innermost ring of a CD, its angular speed is ω_0 = 52.4rad/s. The CD slows down within the player so that when information is read off the outermost ring, ω = 20.9 rad/s. Assume a Sanyo corn CD can be modeled as a solid disk rotating around its central axis. Find the change in its rotational kinetic energy between the beginning and end of its play cycle.

26. **E** A student amuses herself by spinning her pen around her thumb (Fig. P13.26). Estimate the rotational kinetic energy of the pen. She wonders if she can count this amusement as her aerobic activity for the day. Estimate the amount of energy she consumes in spinning the pen for 1 hour. (Compare this amount with chewing gum, which consumes about 11 calorie/h.)

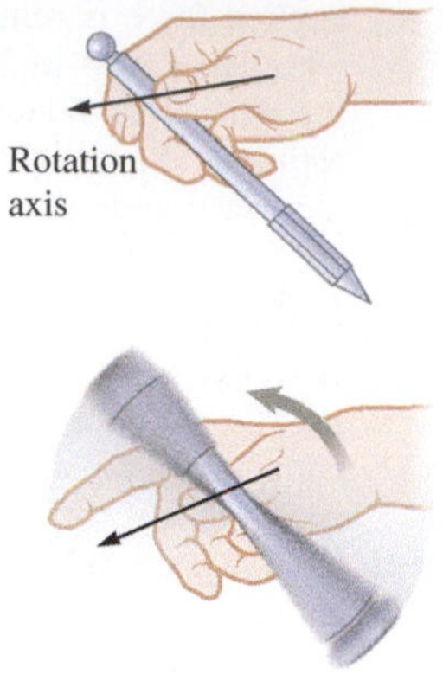

FIGURE P13.26

27. **N** The motion of spinning a hula hoop around one's hips can be modeled as a hoop rotating around an axis not through the center, but offset from the center by an amount h, where h is less than R, the radius of the hoop. Suppose Maria spins a hula hoop with a mass of 0.75 kg and a radius of 0.62 m around her waist. The rotation axis is perpendicular to the plane of the hoop, but approximately 0.40 m from the center of the hoop. **a.** What is the rotational inertia of the hoop in this case? **b.** If the hula hoop is rotating with an angular speed of 13.7 rad/s, what is its rotational kinetic energy?

13-4 Special Case of Rolling Motion

28. **C** A person riding a bicycle on a level road accelerates forward. Consider the person and the bicycle to make up the system and identify the external force that accelerates the system.

29. A uniform hoop and a uniform solid disk are released from rest from a height h on an incline and roll without slipping.

- **a.** **C** Which object reaches the bottom of the incline first? Explain your reasoning.
- **b.** **A** Find expressions for the speed with which each object reaches the bottom of the incline to confirm your answer to part (a).

30. A bicycle and its rider have a mass of 85.6 kg. The bike's tires have a radius of 0.382 m. The mass of each tire is 0.980 kg and is concentrated near each tire's rim. The rider is riding at a constant speed of 5.40 m/s.

- **a.** **N** Find the kinetic energy of the center of mass.
- **b.** **N** Find the rotational kinetic energy of the tires.
- **c.** **C** Compare and comment on your results.

Problems 31 and 32 are paired.

31. **N** Sophia is playing with a set of wooden toys, rolling them off the table and onto the floor. One of the toys is a small sphere with a mass of 0.024 kg and a radius of 0.020 m, and another is a small cylinder that also has a mass of 0.024 kg but a radius of 0.013 m. She rolls each toy so that it has the same translational speed of 0.40 m/s. How much greater is the kinetic energy of the cylinder than the kinetic energy of the sphere?

32. **N** Consider again the two toys in Problem 31 and their translational speed. **a.** Determine the ratio of the rotational kinetic energy to the translational kinetic energy for each toy. **b.** If the sphere and the cylinder had the same angular speed instead of the same translational speed, how would their translational speeds compare?

Problems 33 and 34 are paired.

33. **N** A spring with spring constant 25 N/m is compressed a distance of 7.0 cm by a ball with a mass of 202.5 g (Fig. P13.33). The ball is then released and rolls without slipping along a horizontal surface, leaving the spring at point A. The process is repeated, using a block instead, with a mass identical to that of the ball. The block compresses the spring by 7.0 cm and is also released, leaving the spring at point A. Assume the ball rolls, but ignore other effects of friction. **a.** What is the speed of the ball at point B? **b.** What is the speed of the block at point B?

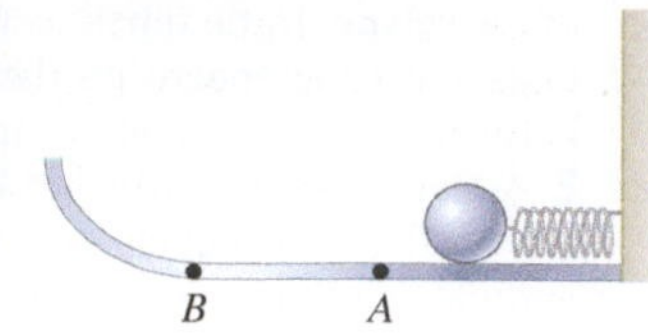

FIGURE P13.33 Problems 33 and 34.

34. **N** Consider the situation described in Problem 33. Each object travels up the incline shown in Figure P13.33 before coming to rest momentarily. Assume the ball rolls, but ignore other effects of friction. **a.** How high above the starting position is the ball when it comes to rest momentarily? **b.** How high above the starting position is the block when it comes to rest momentarily?

35. **C** **CASE STUDY** In the laboratory, you have probably used a small cart on a track. The cart's mass is approximately 0.5 kg, and it rolls on four small wheels. Usually, we ignore the rotational kinetic energy of those four wheels. When Chris did not take the rotational kinetic energy of the ball on the loop-the-loop challenge into account, however, he ran into significant errors. Why is it safe to ignore the kinetic energy of rotation in the case of the cart, but not in case of the ball?

13-5 Work and Power

36. **C** Suppose you wish to get more power out of a waterwheel (Example 13.10, page 375). How could you improve the wheel?

37. **N** A tangential force of 1.5 N acts on an object of mass 2.0 kg that is moving in a circular path of radius 3.0 m. The force causes the object's speed to increase. Find the work performed during one complete revolution of the object.

Problems 38 and 39 are paired.

38. **N** A merry-go-round at a park is subject to a constant torque with a magnitude of 645 N · m as a parent pushes the ride. **a.** How much work is performed by the torque as the merry-go-round rotates through 1.75 revolutions? **b.** If it takes the merry-go-round 4.51 s to go through the 1.75 revolutions, what is the power transferred by the parent to the ride?

39. **N** A parent exerts a torque on a merry-go-round at a park. The torque has a magnitude given by $\tau = 32\theta^2 + 120\theta - 145$, where the torque has units of newton meters when θ is in units of radians. **a.** How much work is performed by the torque as the merry-go-round rotates through 1.75 revolutions, beginning at $\theta_i = 0$? **b.** If it takes the merry-go-round 4.51 s to go through the 1.75 revolutions, what is the power transferred by the parent to the ride?

40. **E** Most people know that Benjamin Franklin studied electricity by flying a kite during a storm. He also studied electricity indoors using a hand-cranked electrical generator (Fig. P13.40). For now, we are interested in the work done by the person in turning the crank. The entire apparatus is about 4.5 ft tall. Assume a person can apply a constant force on the crank's handle. Estimate the work done by a person in turning the crank through one revolution. We'll

The Library Company of Philadelphia

FIGURE P13.40

study this device in detail in Chapter 27. The work done by the person goes into electrifying the ball shown on the top of the generator.

41. **N** Today, waterwheels are not often used to grind food. Instead, we have electrical devices such as blenders, choppers, and mixers. The electric motors in these devices are similar to a waterwheel, but instead of falling water causing the wheel to spin, electricity causes a shaft to spin. The specifications on a particular electric motor reports that at 1.75×10^3 rpm, it puts out 5 hp. What is the corresponding torque in N · m?

13-6 Angular Momentum

42. Model Jupiter's orbit around the Sun as uniform circular motion.
    a. **N** Find the angular momentum of Jupiter's orbital motion.
    b. **N** Although Jupiter is a gas giant, model it as a solid uniform sphere and find the angular momentum of Jupiter's rotational motion.
    c. **C** Compare your results and comment.

43. **N** A buzzard (m = 9.29 kg) is flying in circular motion with a speed of 8.44 m/s while viewing its meal below. If the radius of the buzzard's circular motion is 8.00 m, what is the angular momentum of the buzzard around the center of its motion?

44. **A** An object of mass M is thrown with a velocity v_0 at an angle θ with respect to the horizontal (Fig. P13.44). Find the angular momentum of the object around the origin when the object is at the highest point of its trajectory.

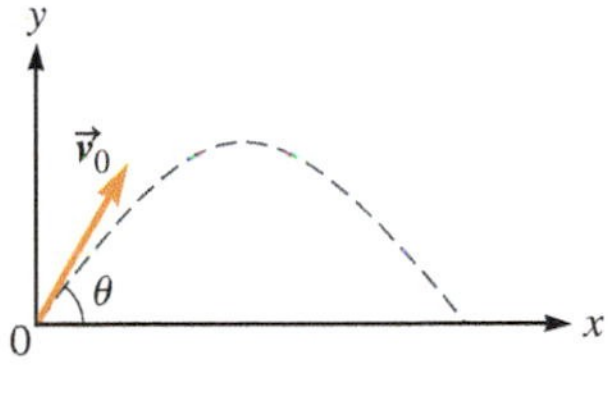

FIGURE P13.44

Problems 45, 46, and 47 are grouped.

45. **N** A thin rod of length 2.65 m and mass 13.7 kg is rotated at an angular speed of 3.89 rad/s around an axis perpendicular to the rod and through its center of mass. Find the magnitude of the rod's angular momentum.

46. **N** A thin rod of length 2.65 m and mass 13.7 kg is rotated at an angular speed of 3.89 rad/s around an axis perpendicular to the rod and through one of its ends. Find the magnitude of the rod's angular momentum.

47. **N** A cylinder of length 2.65 m, radius 0.350 m, and mass 13.7 kg is rotated at an angular speed of 3.89 rad/s around an axis parallel to the length of the cylinder and through its center. Find the magnitude of the cylinder's angular momentum.

48. **N** Two particles of mass m_1 = 2.00 kg and m_2 = 5.00 kg are joined by a uniform massless rod of length ℓ = 2.00 m (Fig. P13.48). The system rotates in the xy plane about an axis through the midpoint of the rod in such a way that the particles are moving with a speed of 3.00 m/s. What is the angular momentum of the system?

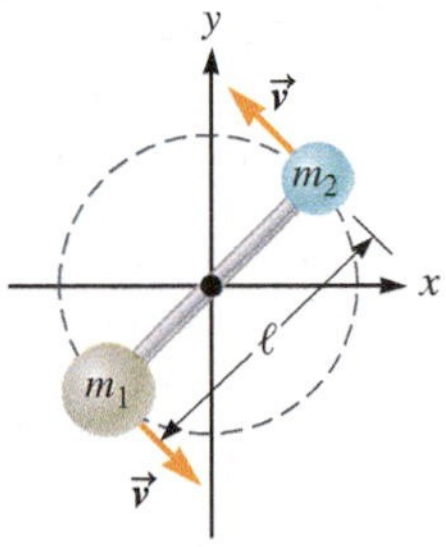

FIGURE P13.48

13-7 Conservation of Angular Momentum

49. **N** A turntable (disk) of radius r = 26.0 cm and rotational inertia 0.400 kg · m^2 rotates with an angular speed of 3.00 rad/s around a frictionless, vertical axle. A wad of clay of mass m = 0.250 kg drops onto and sticks to the edge of the turntable. What is the new angular speed of the turntable?

50. **C** Reanalyze the unicycle's motion in Example 13.15 (page 382). This time, leave the Earth out of the system and explain how the torque exerted by gravity causes the unicycle to turn. Your explanation should involve a diagram.

Problems 51 and 52 are paired.

51. **C** While at a party, a friend proposes that if you sit and rotate freely in an office chair holding two heavy barbells close to your chest and then extend your arms, you will rotate so much faster that you will fall out of the chair. Is your friend correct? If yes, explain. If no, correct your friend as to how the outcome might be achieved in this setting. What can you do or change that would cause the increased rotation to be magnified further?

52. **N** While at a party, a friend challenges you to a game of sorts, where the challenge is to remain sitting throughout. He suggests you hold two 10-kg iron barbells as you sit on a barstool that is free to rotate. Your friend has you extend your arms outward so that the barbells are held away from your body. He then starts the stool rotating. Upon reaching an angular speed of 5.00 rad/s, your friend tells you to pull the barbells back toward your chest. Assuming your body has a rotational inertia of about 8.59 kg · m^2 without the barbells and your arms are about 0.75 m long, what is your angular speed after you pull the barbells so they are at the center of your chest?

53. **N** Two children (m = 30.0 kg each) stand opposite each other on the edge of a merry-go-round. The merry-go-round, which has a mass of 1.80×10^2 kg and a radius of 1.5 m, is spinning at a constant rate of 0.50 rev/s. Treat the two children and the merry-go-round as a system. **a.** Calculate the angular momentum of the system, treating each child as a particle. **b.** Calculate the total kinetic energy of the system. **c.** Both children walk half the distance toward the center of the merry-go-round. Calculate the final angular speed of the system.

54. **A** A disk of mass m_1 is rotating freely with constant angular speed ω. Another disk of mass m_2 that has the same radius is gently placed on the first disk. If the surfaces in contact are rough so that there is no slipping between the disks, what is the fractional decrease in the kinetic energy of the system?

Problems 55, 56, 60, and 61 are grouped.

55. **N** A *month* is the time for the Moon to orbit the Earth. Currently, a month is about 28 days. The same side of the Moon always faces the Earth, so the period of the Moon's rotation equals the period of its orbit. Take into account the orbital motion of the Moon around the Earth and the rotational motion of the Earth and the Moon to find the total angular momentum of the Earth–Moon system. Assume the Moon orbits the Earth in uniform circular motion and the Earth and the Moon are solid, uniform spheres. All the motion is counterclockwise as seen from above the North Pole.

56. In Problem 55, the external torques on the Earth–Moon system are negligible so that the angular momentum of the system is nearly constant. The tidal force between the Earth and the Moon is causing the Earth's rotation to slow down, however. In the distant future, the period of the Earth's rotation will equal the period of the Moon's orbit and rotation. All will be equal to about 50 (24-hour) days.
    a. **C** What must happen to the Moon's orbit for angular momentum to be conserved? (This question may be answered in the **INTERPRET and ANTICIPATE** step.)
    b. **N** What will be the radius of the Moon's orbit at that time?

57. **A** The angular momentum of a sphere is given by $\vec{L} = (-4.59t^3)\hat{\imath} + (6.01 - 1.19t^2)\hat{\jmath} + (6.26t)\hat{k}$, where L has units of kg · m^2/s when t is in seconds. What is the net torque on the sphere as a function of time?

General Problems

58. **A** A ball of mass M is connected to a second ball of mass m by a massless, rigid rod of length L (Fig. P13.58). **a.** Show that for an axis perpendicular to the rod, the minimum rotational inertia of the system is for an axis passing through the center of mass. **b.** What is the minimum rotational inertia of the system for an axis perpendicular to the rod?

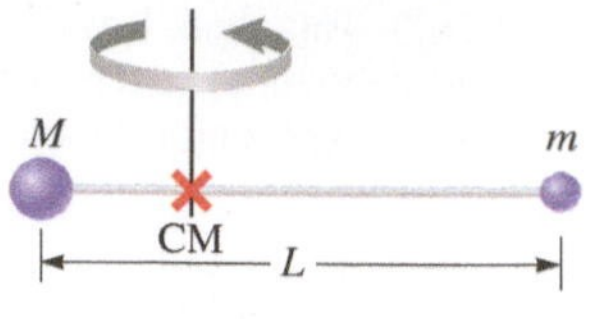

FIGURE P13.58

59. **N** An 80.0-kg disk with a radius of 2.50 m that can rotate around a frictionless axle through its center is initially at rest. A force of 200.0 N can be applied to the disk at any radial distance from the axis of rotation, perpendicular to the radial direction in the plane of the disk. At what radial distance should the force be applied to accelerate the disk so that it completes 3.00 revolutions in 6.00 s?

Problems 55, 56, 60, and 61 are grouped.

60. **N** Find the rotational inertia of the Earth–Moon system around its center of mass.

61. **N** The Earth and the Moon orbit their system's center of mass roughly every 28 days. Find the system's rotational kinetic energy.

62. **A** A rigid rod of mass M and length L is pivoted around point P at one of its ends (Fig. P13.62). Find the speed of the center of mass of the rod when it is vertical if it is released from its horizontal position.

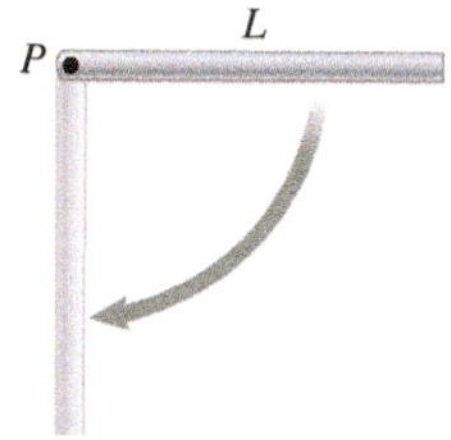

FIGURE P13.62

63. **N** A uniform cylinder of radius r = 10.0 cm and mass m = 2.00 kg is rolling without slipping on a horizontal tabletop. The cylinder's center of mass is observed to have a speed of 5.00 m/s at a given instant. **a.** What is the translational kinetic energy of the cylinder at that instant? **b.** What is the rotational kinetic energy of the cylinder around its center of mass at that instant? **c.** What is the total kinetic energy of the cylinder at that instant?

64. **C** Why does a twirling baton (Fig. P13.64) have a fairly massive rubber tip on its ends (aside from it being a safety feature)?

© sshepard/iStockphoto.com

FIGURE P13.64

65. **A** A thin, spherical shell of mass m and radius R rolls down a parabolic path PQR from height H without slipping (assume $R \ll H$) as shown in Figure P13.65. Path PQ is rough (and so the shell will roll on that path), whereas path QR is smooth, or frictionless (so the shell will only slide, not roll, in this region). Determine the height h above point Q reached by the shell on path QR.

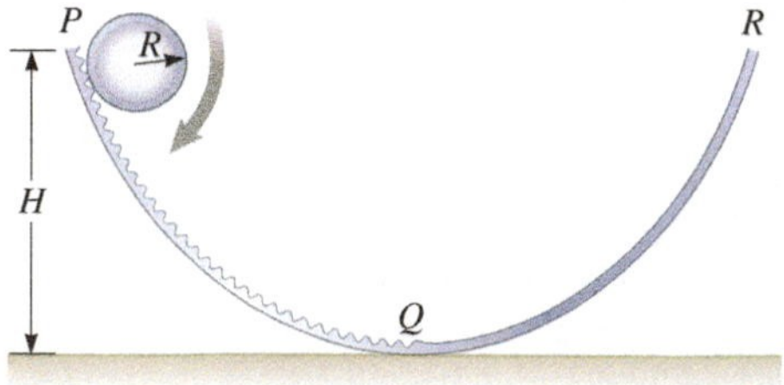

FIGURE P13.65

Problems 66 and 67 are paired.

66. **E** To give a pet hamster exercise, some people put the hamster in a ventilated ball and allow it roam around the house (Fig. P13.66). When a hamster is in such a ball, it can cross a typical room in a few minutes. Estimate the total kinetic energy in the ball–hamster system.

Peter McGahey © Cengage Learning

FIGURE P13.66 Problems 66 and 67.

67. **E** **Review** Estimate the power consumed by a hamster exercising in a ventilated ball (Fig. P13.66). *Hint*: Read Problem 13.66.

68. Some potters prefer to keep their wheel rotating at a constant speed while they work the clay. Suppose you are going to try your hand at some pottery and want to make a tall, cylindrical vase. Should you be worried about the angular speed changing due to a change in shape of the clay? Assume the wheel–clay system is brought to a modest angular speed (120 rpm) while the clay is in the approximate shape of a sphere in the center of the wheel. Estimate and find appropriate values for the masses of the clay and the wheel, the radius of the wheel, and the density of the clay.

a. **E** Given your estimates, what would be the total angular momentum of the initial wheel–clay system?

b. **C** When you shape the clay into a tall, cylindrical vase (one-fourth the radius of the sphere), does the system need to be sped up or slowed down, or should nothing be done to maintain a constant angular speed? (Assume no net external torque is applied as you carefully shape the clay.) Explain.

69. **N** The velocity of a particle of mass m = 2.00 kg is given by $\vec{v} = -5.10\hat{\imath} + 2.40\hat{\jmath}$ m/s. What is the angular momentum of the particle around the origin when it is located at $\vec{r} = -8.60\hat{\imath} - 3.70\hat{\jmath}$ m?

70. **N** A ball of mass M = 5.00 kg and radius r = 5.00 cm is attached to one end of a thin, cylindrical rod of length L = 15.0 cm and mass m = 0.600 kg. The ball and rod, initially at rest in a vertical position and free to rotate around the axis shown in Figure P13.70, are nudged into motion. **a.** What is the rotational kinetic energy of the system when the ball and rod reach a horizontal position? **b.** What is the angular speed of the ball and rod when they reach a horizontal position? **c.** What is the linear speed of the center of mass of the ball when the ball and rod reach a horizontal position? **d.** What is the ratio of the speed found in part (c) to the speed of a ball that falls freely through the same distance?

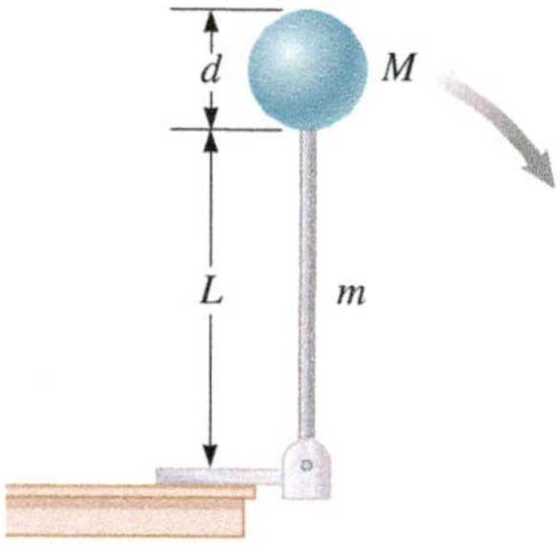

FIGURE P13.70

71. **N** A long, thin rod of mass m = 5.00 kg and length ℓ = 1.20 m rotates around an axis perpendicular to the rod with an angular speed of 3.00 rad/s. **a.** What is the angular momentum of the rod if the axis passes through the rod's midpoint? **b.** What is the angular momentum of the rod if the axis passes through a point halfway between its midpoint and its end?

72. **A** A solid sphere and a hollow cylinder of the same mass and radius have a rolling race down an incline as in Example 13.9 (page 372). They start at rest on an incline at a height h above a horizontal plane. The race then continues along the horizontal plane. The coefficient of rolling friction between each rolling object and the surface is the same. Which object rolls the farthest? (Justify your answer with an algebraic expression.)

73. **N** A uniform disk of mass $m = 10.0$ kg and radius $r = 34.0$ cm mounted on a frictionless axle through its center, and initially at rest, is acted upon by two tangential forces of equal magnitude F, acting on opposite sides of its rim until a point on the rim experiences a centripetal acceleration of 4.00 m/s^2 (Fig. P13.73). **a.** What is the angular momentum of the disk at this time? **b.** If $F = 2.00$ N, how long do the forces have to be applied to the disk to achieve this centripetal acceleration?

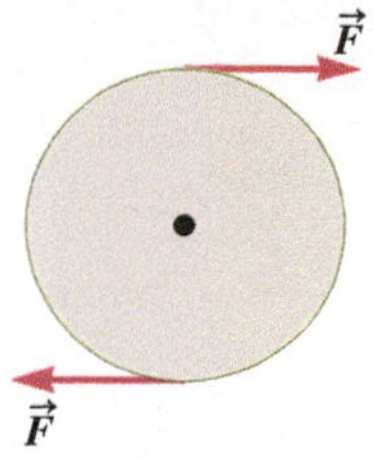

FIGURE P13.73

74. **C** When a person jumps off a diving platform, she imparts some amount of angular momentum to her body. (To see why, consider a person who jumps straight up: She falls straight down. When a person jumps off a diving platform, she must jump at an angle to the vertical, or she would land on the board.) While she is in the air, angular momentum is conserved. Draw a free-body diagram of the diver in the air and use it to explain why angular momentum is conserved during flight. It may be useful to include the axes of rotation for the two common types of spinning moves that divers perform: twists and flips (Fig. P13.74).

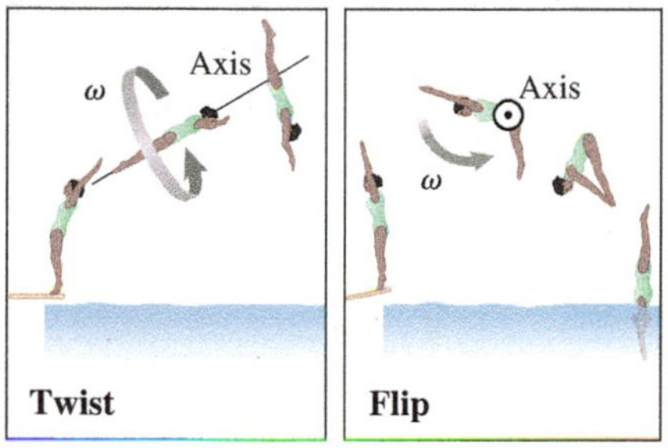

FIGURE P13.74

75. **A** One end of a massless rigid rod of length ℓ is attached to a wooden block of mass M resting on a frictionless, horizontal tabletop, and the other end is attached to the table through a pivot (Fig. P13.75). A bullet of mass m traveling with a speed v in a direction perpendicular to the rod and parallel to the table impacts the block and embeds itself inside. **a.** What is the angular momentum of this system around a vertical axis through the pivot after the collision? **b.** What is the fraction of the bullet's initial kinetic energy that is lost to internal energy during the collision?

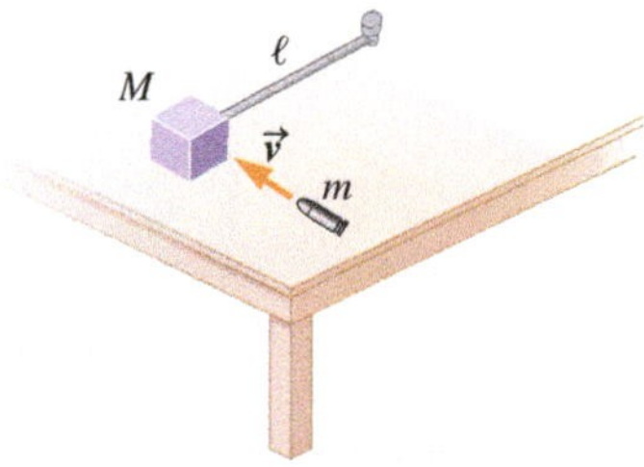

FIGURE P13.75

76. **A** A uniform solid sphere of mass m and radius r is released from rest and rolls without slipping on a semicircular ramp of radius $R \gg r$ (Fig. P13.76). If the initial position of the sphere is at an angle θ to the vertical, what is its speed at the bottom of the ramp?

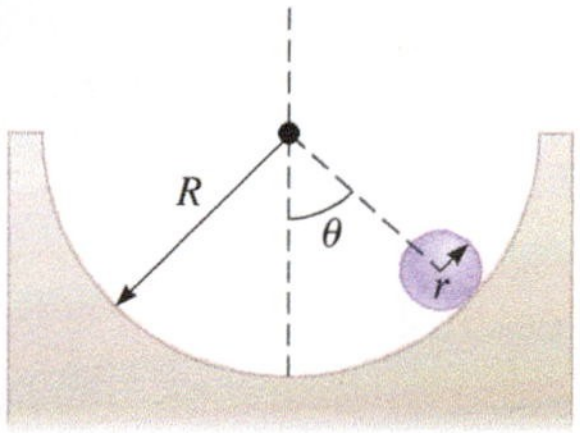

FIGURE P13.76

77. **N** A rod of length 2.0 m and of negligible mass is constrained to move in a vertical plane (Fig. P13.77). The ends of the rod always touch the x and y axes, respectively, as the rod slides downward. Find the coordinates of the instantaneous axis of rotation of the rod when it makes an angle of exactly 60° with the horizontal.

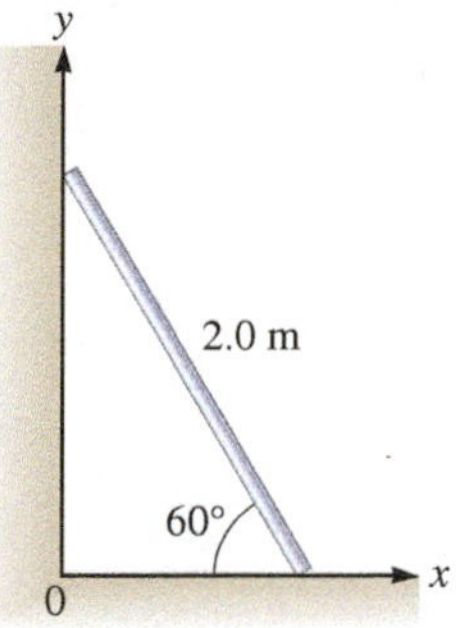

FIGURE P13.77

78. **A** A cam of mass M is in the shape of a circular disk of diameter $2R$ with an off-center circular hole of diameter R is mounted on a uniform cylindrical shaft whose diameter matches that of the hole (Fig. P13.78). **a.** What is the rotational inertia of the cam and shaft around the axis of the shaft? **b.** What is the rotational kinetic energy of the cam and shaft if the system rotates with angular speed ω around this axis?

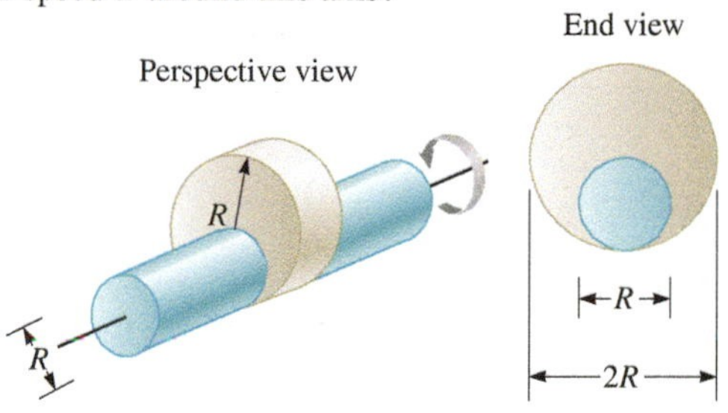

FIGURE P13.78

79. **A** A uniform solid sphere of radius R initially at rest is subjected to an applied force $\vec{F}$ and begins to roll without slipping (Fig. P13.79). **a.** What is the acceleration of the center of mass of the sphere? **b.** What are the magnitude and direction of the force of friction on the sphere? **c.** What is the speed of the center of mass of the sphere after it has traversed a distance d?

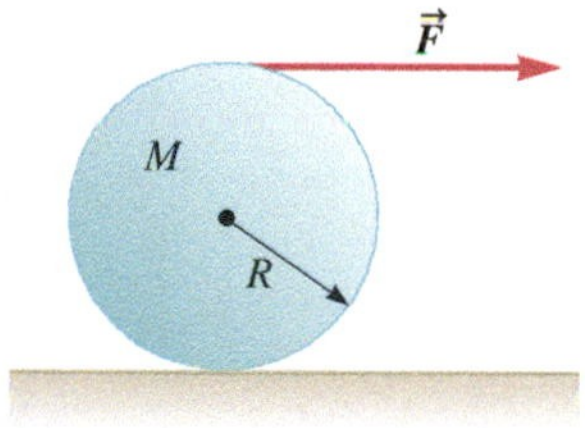

FIGURE P13.79

80. **A** Consider the downhill race in Example 13.9 (page 372). The acceleration of a particle down an incline is $a_{\text{particle}} = g \sin \beta$, where β is the angle the incline makes with the horizontal. Show that the acceleration of a rolling object down an incline is given by

$$a_{\text{CM}} = \frac{a_{\text{particle}}}{1 + f}$$

where f is the unitless fraction described in Example 13.9.

81. **A** A disk of mass M and radius R is released from rest along a smooth (frictionless) inclined plane having an angle of inclination θ as shown in Figure P13.81. The disk will slide instead of roll. Find the magnitude of the angular momentum of the disk around the instantaneous point of contact after a time t from the instant of release. Assume dissipative forces are negligible.

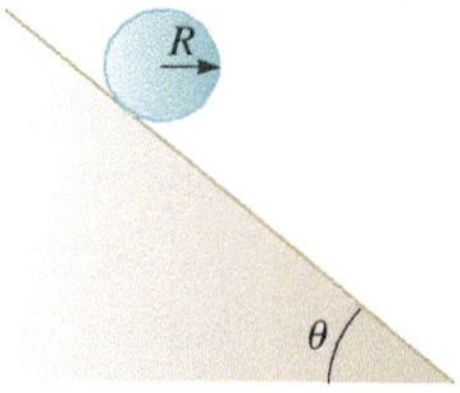

FIGURE P13.81

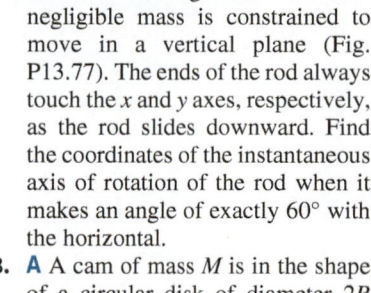

PART TWO

Mechanics of Complex Systems

14 Static Equilibrium, Elasticity, and Fracture

Underlying Principle

Newton's second law applied in the special case of equilibrium

Major Concepts

1. Static equilibrium
   a. Stable static equilibrium
   b. Unstable static equilibrium
   c. Neutral static equilibrium
2. Equilibrium conditions
   a. Force balance condition
   b. Torque balance condition
3. Stress
4. Strain
5. Young's modulus
6. Elastic object and elastic limit
7. Tensile strength
8. Compressive strength
9. Shear modulus

Tools

Cross products

Key Questions

What are the conditions for equilibrium?

How do static objects respond when forces are applied?

© Cedric Weber/Shutterstock.com

A.

© podgorsek/iStockphoto.com

B.

FIGURE 14.1 A. Golden Gate Bridge, San Francisco, California. **B.** Rialto Bridge, Venice, Italy.

Imagine that an engineer is asked to design a roadway to cross a body of water. The engineer has many choices to make. The engineer may choose a suspension bridge like the Golden Gate Bridge in San Francisco or an arch bridge like the Rialto Bridge in Venice (Fig. 14.1) or perhaps a longer and more complex structure such as the Chesapeake Bay Bridge-Tunnel that connects Virginia's eastern shore with the mainland at Virginia Beach near Norfolk. In addition to choosing a type of structure, the engineer has many materials from which to choose: stone, concrete, and steel, just to name a few.

Many structures such as bridges, buildings, and furniture are designed to be stationary, that is, at rest relative to an observer. In this chapter, we take a closer look at the conditions that are necessary for an object to be at rest.

We'll also see how the choice of material affects the structure when forces are applied. Take a classroom chair, for example. When a student sits on the chair, the chair deforms. Bare wooden chairs are less comfortable than cushioned chairs because the wood does not deform as much as does a fabric cushion when compressed.

14-1 What Is Static Equilibrium?

So far, we have studied objects in translational motion, objects in rotational motion, and even objects in combined translational and rotational motion. In this chapter, we study objects that are at rest, that is, not moving relative to the observer. An object that is at rest is said to be in **static equilibrium**. Why do we need a chapter dedicated to objects at rest? The answer is that objects at rest are very common and important; many engineers and architects spend their entire careers designing structures that must maintain static equilibrium despite the application of large forces.

Static equilibrium is a special case of motion. Start by thinking about a ball rolling (without slipping) down an inclined plane and assume rolling friction is negligible. In Section 13-4, we found it convenient to think about rolling as a combination of translational motion and rotational motion. In this case, the ball speeds up. So, both the translational acceleration $\vec{a}$ of its center of mass and its angular acceleration $\vec{\alpha}$ have some nonzero value. Put another way, the ball's translational momentum $\vec{p} = m\vec{v}_{CM}$ and angular momentum $\vec{L} = I\vec{\omega}$ continually change while it is rolling down the incline.

Now consider a special case of motion in which the ball's translational acceleration and angular acceleration are both zero: $\vec{a} = 0$ and $\vec{\alpha} = 0$. For example, if the ball rolls along a level floor instead of down an incline, the ball will not speed up, slow down, or change direction (as long as dissipative forces are negligible). Therefore, the ball's translational momentum and angular momentum are both constant, and the ball is said to be in *equilibrium*. **Equilibrium** is a special case of motion in which an object's translational momentum $\vec{p}$ and angular momentum $\vec{L}$ are both constant.

EQUILIBRIUM
Underlying Principle

Static equilibrium is a special case of equilibrium in which the object's translational and angular momenta equal a particular constant, zero: $\vec{p} = 0$ and $\vec{L} = 0$. Static equilibrium depends on the inertial reference frame from which the object is observed. For example, if you were sitting on a bridge, you would observe that the bridge is in static equilibrium. If you are drifting with a constant velocity on a raft in the water below the bridge, however, you would say that the bridge has a constant, nonzero translational momentum. You would conclude that the bridge was in equilibrium but not in static equilibrium. Throughout this chapter, we will assume the observer is in the reference frame that allows him to observe static equilibrium.

Types of Static Equilibrium

Static equilibrium is classified into three types: stable, unstable, and neutral. To distinguish between these types, you must imagine what would happen if the object in equilibrium were displaced slightly and then released from its new resting position. If, after being released, the forces acting on the object return it to its equilibrium position, the object is in **stable static equilibrium**. For example, a marble ball at rest in the bottom of a bowl is in stable static equilibrium. If the marble is displaced and then released from its new position, gravity returns the marble to its equilibrium position (Fig. 14.2A). In this case, the marble would roll back and forth through its equilibrium position, and rolling friction would cause it to eventually come to rest at the bottom of the bowl. If, after being released, the object moves farther away from its equilibrium position, the object is in **unstable static equilibrium**. For example, a marble at rest on top of an inverted bowl is in unstable static equilibrium. If the marble is displaced and then released, gravity causes it to roll off the bowl, and the marble will not return

STABLE, UNSTABLE, AND NEUTRAL STATIC EQUILIBRIUM
Major Concept

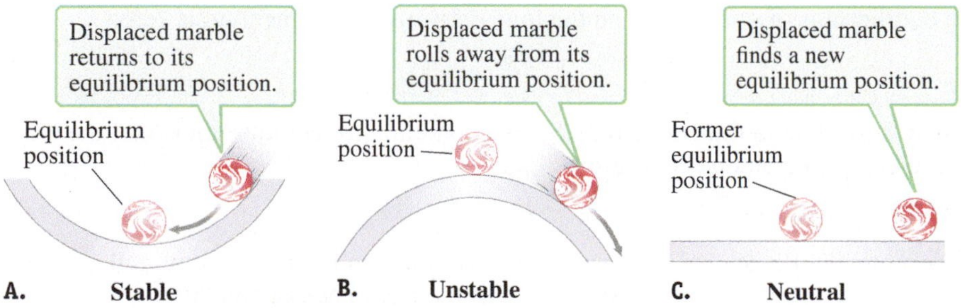

FIGURE 14.2 Types of static equilibrium.

to its equilibrium position (Fig.14.2B). If an object is moved and released from a new position and does not move toward or away from its equilibrium position, the object is in **neutral static equilibrium**. In this case, the forces exerted on the object do not maintain its equilibrium; so, if the object is displaced, it is not restored to its equilibrium position, and it is not forced farther away. For example, a marble at rest on a level surface is in neutral static equilibrium (Fig. 14.2C). If you displace the marble to the right and then release your hold, gravity neither moves the ball back to equilibrium nor farther away. Although the marble is in a new position, there has been no change in the Earth–marble system's gravitational potential energy.

CASE STUDY Designing an Air and Space Exhibit

Besides making sure that a particular design meets the project goals, an engineer must also make sure that the design is safe from collapse, explosion, or other hazards. The designers of the Udvar-Hazy Center—a companion facility to the Smithsonian Air and Space Museum in Washington, D.C.—faced the extraordinary challenge of planning a structure to support 73 hanging aircraft. Their design had to protect the public, the aircraft, and the building. The aircraft must be low enough to be easily seen from the ground and must be accessible by high-lift cherry pickers. To convey a sense of movement and excitement, the aircraft are hung at various angles (Fig. 14.3). To make this design possible, the architects and engineers incorporated important structural elements into the museum. In this case, we explore one such element: the steel cables used to support a single plane. Our goal is to calculate the maximum airplane weight that can be hung from two steel cables without breaking the cables.

FIGURE 14.3 Aircraft at the Udvar-Hazy Center.

CONCEPT EXERCISE 14.1

A rubber duck floats in a bathtub. Imagine moving the duck by a small amount in different directions such as up and down or side to side. Is the floating duck in stable, unstable, or neutral equilibrium?

CONCEPT EXERCISE 14.2

If the rubber duck in Concept Exercise 14.1 is replaced by a small toy boat with an open hull such as a rowboat or canoe, is the boat in stable, unstable, or neutral equilibrium? *Hint*: What if you displace the boat downward or rock it so that water enters the hull?

14-2 Conditions for Equilibrium

For an object to be in equilibrium, two conditions must be met: (1) the object's translational momentum must be constant, and (2) its angular momentum must be constant. We can express these two conditions for equilibrium in mathematical terms. We start with Newton's second law written in terms of momentum:

$$\sum \vec{F} = \vec{F}_{\text{tot}} = \frac{d\vec{p}}{dt} \tag{10.2}$$

In equilibrium, $\vec{p}$ = constant, so the time derivative of momentum is zero:

$$\frac{d\vec{p}}{dt} = 0$$

Then, according to Equation 10.2, the first condition for equilibrium is that the total force exerted on an object is zero:

FORCE BALANCE CONDITION FOR EQUILIBRIUM
✪ Major Concept

$$\vec{F}_{\text{tot}} = 0 \tag{14.1}$$

Conceptually, Equation 14.1 says that for an object to be in equilibrium, the forces must "balance out," so we will refer to it as the **force balance condition**.

For the second equilibrium condition (angular momentum is constant), we start with Newton's second law written in terms of torque and angular momentum:

$$\sum \vec{\tau} = \vec{\tau}_{\text{tot}} = \frac{d\vec{L}}{dt} \quad (13.27)$$

An object in equilibrium has a constant angular momentum $\vec{L}$, so the time derivative of angular momentum is zero:

$$\frac{d\vec{L}}{dt} = 0$$

Then, according to Equation 13.27, the second condition for equilibrium is that the total torque acting on an object is zero:

$$\vec{\tau}_{\text{tot}} = 0 \quad (14.2)$$

TORQUE BALANCE CONDITION FOR EQUILIBRIUM

Major Concept

Conceptually, this second condition for equilibrium says that the torques on the object must "balance out," so we will refer to it as the **torque balance condition**.

These conditions (Eqs. 14.1 and 14.2) are true for all equilibrium situations, static or not. In the next section, we will apply the equilibrium conditions to objects that are in static equilibrium.

Equations 14.1 and 14.2 and are vector equations. It is often practical to apply these equations in their component forms. The condition that the forces must balance may be written in terms of three scalar component equations:

$$\sum F_x = 0 \qquad \sum F_y = 0 \qquad \sum F_z = 0 \quad (14.3)$$

Similarly, the condition that the torques must balance may be written in terms of three scalar component equations:

$$\sum \tau_x = 0 \qquad \sum \tau_y = 0 \qquad \sum \tau_z = 0 \quad (14.4)$$

As with any vector, the subscript indicates the Cartesian component of the torque. For example, $\sum \tau_x$ is the sum of the x components of all the torques.

Cross Product Revisited

CROSS PRODUCTS **Tool**

Torque is the result of a cross product (Eq. 12.23, $\vec{\tau} = \vec{r} \times \vec{F}$, where $\vec{r}$ is the position vector from the rotation axis to the point of application of the force $\vec{F}$). In Section 12-6, we found the magnitude and direction of a vector resulting from a cross product. Now we need to write the result of a cross product in component form because we will need the components of the torque to use Equation 14.4.

Consider vector $\vec{R}$, the cross product of vectors $\vec{A}$ and $\vec{B}$:

$$\vec{R} = \vec{A} \times \vec{B}$$

To write $\vec{R}$ in component form, we start by writing $\vec{A}$ and $\vec{B}$ in component form:

$$\vec{R} = (A_x\hat{\imath} + A_y\hat{\jmath} + A_z\hat{k}) \times (B_x\hat{\imath} + B_y\hat{\jmath} + B_z\hat{k})$$

The cross product is distributive, meaning that

$$\begin{aligned}\vec{R} &= (A_x\hat{\imath} \times B_x\hat{\imath}) + (A_x\hat{\imath} \times B_y\hat{\jmath}) + (A_x\hat{\imath} \times B_z\hat{k}) \\ &+ (A_y\hat{\jmath} \times B_x\hat{\imath}) + (A_y\hat{\jmath} \times B_y\hat{\jmath}) + (A_y\hat{\jmath} \times B_z\hat{k}) \\ &+ (A_z\hat{k} \times B_x\hat{\imath}) + (A_z\hat{k} \times B_y\hat{\jmath}) + (A_z\hat{k} \times B_z\hat{k})\end{aligned}$$

The cross product of parallel vectors is zero (Eq. 12.22 with $\varphi = 0$). Therefore, the three of the terms containing $(\hat{\imath} \times \hat{\imath})$, $(\hat{\jmath} \times \hat{\jmath})$, and $(\hat{k} \times \hat{k})$ are zero:

$$\begin{aligned}\vec{R} &= 0 + (A_x\hat{\imath} \times B_y\hat{\jmath}) + (A_x\hat{\imath} \times B_z\hat{k}) \\ &+ (A_y\hat{\jmath} \times B_x\hat{\imath}) + 0 + (A_y\hat{\jmath} \times B_z\hat{k}) \\ &+ (A_z\hat{k} \times B_x\hat{\imath}) + (A_z\hat{k} \times B_y\hat{\jmath}) + 0\end{aligned}$$

The angle between other pairs of unit vectors is $\varphi = 90°$, so the magnitude of their cross products is $|1||1| \sin 90° = 1$. The directions of their cross products are found by using the right-hand rule:

$$\vec{R} = A_xB_y\hat{k} - A_xB_z\hat{\jmath} - A_yB_x\hat{k} + A_yB_z\hat{\imath} + A_zB_x\hat{\jmath} - A_zB_y\hat{\imath}$$

Regrouping to collect all scalar factors of each unit vector,

$$\vec{R} = (A_yB_z - A_zB_y)\hat{\imath} + (A_zB_x - A_xB_z)\hat{\jmath} + (A_xB_y - A_yB_x)\hat{k} \quad (14.5)$$

Equation 14.5 gives us a way to find the components of $\vec{R}$ directly from the components of $\vec{A}$ and $\vec{B}$:

$$R_x = A_yB_z - A_zB_y \qquad R_y = A_zB_x - A_xB_z \qquad R_z = A_xB_y - A_yB_x \quad (14.6)$$

Applying Equations 14.5 and 14.6 to the equation for torque (Eq. 12.23), we have

$$\vec{\tau} = \vec{r} \times \vec{F} = (r_yF_z - r_zF_y)\hat{\imath} + (r_zF_x - r_xF_z)\hat{\jmath} + (r_xF_y - r_yF_x)\hat{k} \quad (14.7)$$

and

$$\tau_x = r_yF_z - r_zF_y \qquad \tau_y = r_zF_x - r_xF_z \qquad \tau_z = r_xF_y - r_yF_x \quad (14.8)$$

Equation 14.8 allows us to apply the torque balance condition (Eq. 14.4) in component form. For many situations, applying the equilibrium conditions in component form is the most direct way to solve a problem.

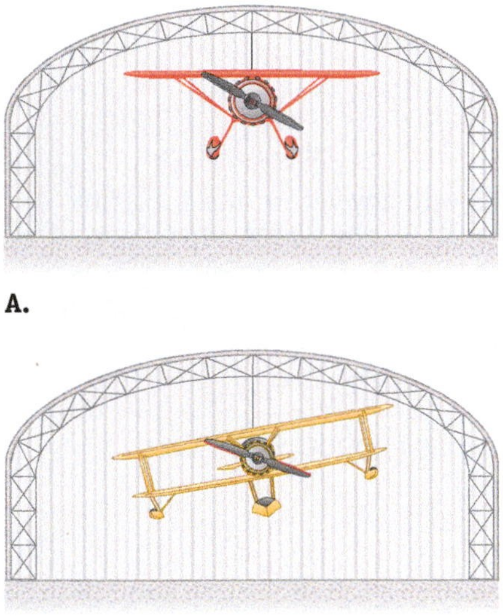

A.

B.

FIGURE 14.4

CONCEPT EXERCISE 14.3

CASE STUDY Hanging a Plane from a Single Point

In the Air and Space Museum, airplanes are usually hung from two or more cables.

a. Is it possible to hang an airplane from a cable attached at a single point such that its wings and fuselage are parallel to the ground as in Figure 14.4A? If so, describe how to do it. If not, why not?

b. Is it possible to hang an airplane from a cable attached at a single point such that its wings are tilted as in Figure 14.4B? If so, describe how to do it. If not, why not?

CONCEPT EXERCISE 14.4

Two people want to sit on opposite ends of an adjustable seesaw.

a. If the two people have equal masses, where should they put the pivot so as to have balanced horizontal seesaw?

b. If one person has more mass than the other person, where should they put the pivot? In the middle? Closer to the heavier person? Closer to the lighter person?

14-3 Examples of Static Equilibrium

In this section, we consider several examples involving objects that are in static equilibrium.

PROBLEM-SOLVING STRATEGY

Static Equilibrium

The strategy we follow is similar to that established in previous chapters (primarily Chapters 5, 6, and 12) when applying Newton's second law.

A modified free-body diagram is helpful for visualizing the problem in the **INTERPRET and ANTICIPATE** step. Because the object is in equilibrium, its acceleration is zero. In Section 5-8, we asked for information about acceleration on the diagram. Here, however, you may wish to keep the diagram clean, so leave the acceleration off the diagram. The modified free-body diagram has five elements.

Element 1. Draw the object as a simple shape. Because we must calculate the torque acting on the object, the object cannot be represented by the particle model. So, represent the object by a simple shape rather than by a dot.

Element 2. As always, your free-body diagram must include all forces acting on the object. You must also **draw the forces at their point of contact.** Recall from Section 9-6 that gravity acts at the object's center of mass, so $\vec{F}_g$ should always be attached to the object's center of mass and point toward the center of the Earth.

Element 3. Choose a rotation axis to calculate torque. Because an object in static equilibrium is at rest (not actually rotating), you may choose any convenient rotation axis. Often, a rotation axis that passes though the point of contact of one of the forces is convenient because that force does not exert a torque around your rotation axis and you will therefore have one torque fewer to calculate. In a situation in which the geometry is complicated—such as when the forces and the points of contact lie in different planes or when the situation has little or no symmetry—you may need to choose more than one rotation axis.

Element 4. It is often convenient to **choose your coordinate system** so that one of the coordinate axes is the same as the rotation axis.

Element 5. Draw the position vector $\vec{r}$ for each force in your free-body diagram. Show each position vector from the rotation axis to the point where the force is exerted on the object. (It is not necessary to draw $\vec{r}$ if the force is on the rotation axis. In that case, $\vec{r} = 0$.) Usually, position vectors are not drawn on a free-body diagram, but they are included on the modified free-body diagram to help calculate torques.

There are three steps in the **SOLVE** procedure:

Step 1 The equilibrium conditions are often best applied in component form. If you calculate the torques using Equation 14.7 or 14.8, it is often helpful to **write every $\vec{F}$ and $\vec{r}$ vector in component form.** Alternatively, you may wish to calculate the torques using $\tau = rF \sin \varphi$ (Eq. 12.16) and the right-hand rule, in which case it may not be necessary to write the vectors in component form. Instead it is helpful to list the magnitude of these vectors if they are known either numerically or algebraically.

Step 2 Finally, **apply the equilibrium condition equations.**

- **a. Apply the force balance condition.**
- **b. Find the torques and apply the torque balance condition.** You will need to find the torques by using either Equation 12.16 ($\tau = rF \sin \varphi$) or Equation 14.7 or 14.8 (in component form) before you apply the torque balance condition. You can use Equations 14.1 (force balance) and 14.2 (torque balance) or apply the conditions in component form Equations 14.3 and 14.4.

Step 3 **Do algebra** as needed to isolate and solve for the unknown quantity you seek before substituting values.

Here is an example of a simple geometry with perpendicular forces and position vectors. The right-hand rule and $\tau = rF \sin \varphi$ (Eq. 12.16) suffice for finding torques.

EXAMPLE 14.1 A Day at the Playground

A granddaughter of mass m_D and her grandfather of mass m_F are on an adjustable seesaw of mass m_s. The seesaw is a board of length ℓ with notches on its underside so that the board can be moved relative to the pivot. For the seesaw to be in equilibrium when the granddaughter and grandfather sit on opposite ends, the pivot must be placed closer to the grandfather than to the granddaughter (Fig. 14.5). Find an expression for the placement of the pivot relative to the seesaw's center of mass.

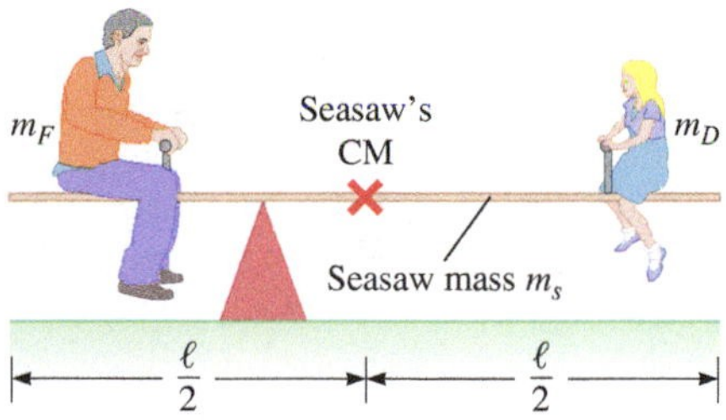

FIGURE 14.5

INTERPRET and ANTICIPATE

Start by creating a modified free-body diagram containing all five elements.

Element 1 Simple shape. The seesaw is the object, represented as a simple thick line (Fig. 14.6). The pivot has been included to help visualize its placement.

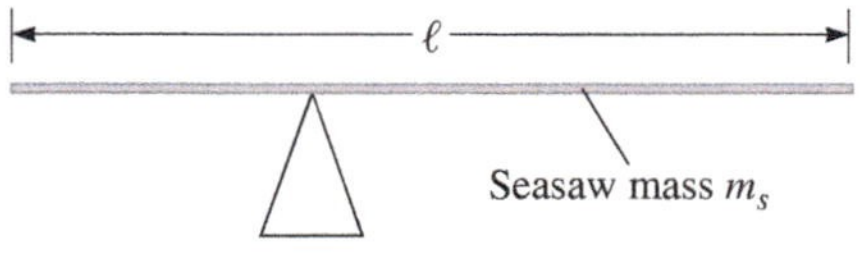

FIGURE 14.6

Element 2 Forces at points of application. Four forces are exerted on the seesaw. Three of them are normal forces: $\vec{F}_F$ due to the grand*f*ather, $\vec{F}_P$ due to the *p*ivot, and $\vec{F}_D$ due to the grand*d*aughter. The fourth force is the gravitational force $\vec{F}_g$ acting at the seesaw's center of mass, which, in this case, is at the seesaw's geometric center. Each force has been drawn at its point of application (Fig. 14.7). At this stage, you do not need to know the relative size of the force vectors.

FIGURE 14.7

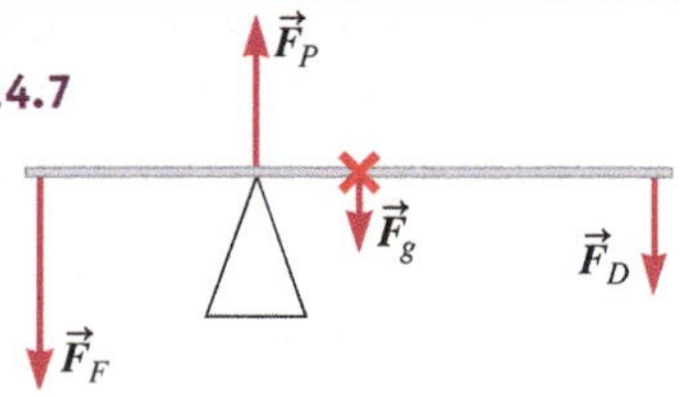

Element 3 A rotation axis. We decided to put the rotation axis through the center of mass so that gravity $\vec{F}_g$ does not exert a torque on the seesaw.

Element 4 Coordinate system. It is best to align one of the axes with the rotation axis. Here we choose a coordinate system with its origin at the center of mass and the z axis aligned with the rotation axis (Fig. 14.8).

FIGURE 14.8

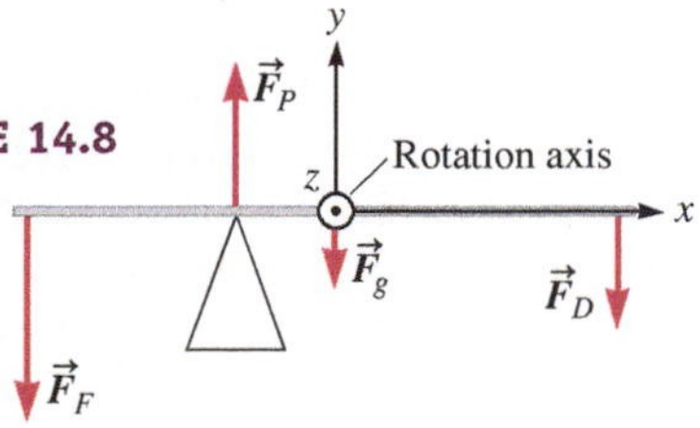

Example continues on page 396 ▶

Element 5 Position vectors. Three position vectors are drawn ($\vec{r}_F$, $\vec{r}_P$, and $\vec{r}_D$) for the three forces ($\vec{F}_F$, $\vec{F}_P$, and $\vec{F}_D$) that exert nonzero torques on the seesaw. The example requires us to find the placement of the pivot; that is, find $\vec{r}_P$ (Fig. 14.9).

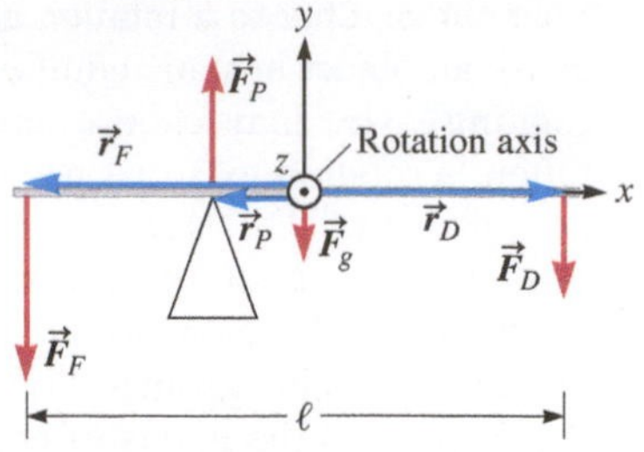

FIGURE 14.9

SOLVE

Step 1 Write each force and position vector in component form. Although the geometry is simple here, we demonstrate the use of unit vectors. Because the seesaw is level, the normal force exerted by each person on the seesaw is equal to the person's weight.

$$\vec{F}_F = -F_F\hat{j} = -m_F g\hat{j} \quad (1)$$

$$\vec{F}_D = -F_D\hat{j} = -m_D g\hat{j} \quad (2)$$

$$\vec{F}_g = -F_g\hat{j} = -m_s g\hat{j} \quad (3)$$

$$\vec{F}_P = +F_P\hat{j} \quad (4)$$

Use Figure 14.9 to write the position vectors in component form. The origin is at the seesaw's center of mass, and each person is at an end, so $r_F = r_D = \ell/2$. We are solving for the pivot's position vector $\vec{r}_P$.

$$\vec{r}_F = -r_F\hat{\imath} = -\frac{\ell}{2}\hat{\imath} \quad (5)$$

$$\vec{r}_D = r_D\hat{\imath} = \frac{\ell}{2}\hat{\imath} \quad (6)$$

$$\vec{r}_{CM} = 0 \quad (7)$$

$$\vec{r}_P = -r_P\hat{\imath} \quad (8)$$

Step 2a Apply the force balance condition. The forces exerted on the seesaw have only y components (Eqs. 1 through 4), so we only need $\sum F_y = 0$ in Equation 14.3.

$$\sum F_y = -F_F - F_D - F_g + F_P = 0 \quad (14.3)$$

$$F_P = F_F + F_D + F_g$$

$$F_P = (m_F + m_D + m_s)g \quad (9)$$

Step 2b Find the torques and apply the torque balance condition. In this case we can use $\tau = rF\sin\varphi$ to find the magnitude of the torque and the right-hand rule to find the direction. Each force is perpendicular to its corresponding position vector (Fig. 14.9), so $\sin\varphi = \sin 90° = 1$ in each case. The magnitude of the torque is just the magnitude of the position multiplied by the magnitude of the force (Eq. 10).

$$\tau = rF\sin\varphi \quad (12.16)$$

$$\tau = rF \quad (10)$$

To find the torque exerted by the grandfather, substitute the magnitude of the force (Eq. 1) and the magnitude of the position (Eq. 5) into Equation (10). The magnitude of a vector is always positive. The right-hand rule shows that the torque is out of the page in the positive z direction.

$$\tau_F = r_F F_F = \left(\frac{\ell}{2}\right)(m_F g)$$

$$\vec{\tau}_F = \left(\frac{\ell}{2}\right)m_F g\hat{k} \quad (11)$$

Next use Equation (10) to find the torque exerted by the granddaughter. Again, we need the magnitude of the force (Eq. 2) and the magnitude of the position (Eq. 6). The right-hand rule shows the torque is into the page, in the negative z direction.

$$\tau_D = r_D F_D = \left(\frac{\ell}{2}\right)(m_D g)$$

$$\vec{\tau}_D = -\left(\frac{\ell}{2}\right)m_D g\hat{k} \quad (12)$$

Now use Equation (10) to find the pivot's torque. The magnitude of the torque τ_P due to the pivot is the magnitude of its position (Eq. 8) multiplied by the magnitude of the force it exerts (Eq. 4). Using the right-hand rule, we find that this torque points into the page in the negative z direction.

$$\tau_P = r_P F_P$$

$$\vec{\tau}_P = -r_P F_P\hat{k} \quad (13)$$

Finally, substitute Equation (9) for F_P in Equation (13).

$$\vec{\tau}_P = -r_P(m_F + m_D + m_s)g\hat{k} \quad (14)$$

The gravitational torque is zero due to our choice of rotation axis.

$$\tau_g = 0$$

because $r_{CM} = 0$ in Equation (7).

We are finally ready to **apply the torque balance condition**. This step is especially easy because all the torques have only z components.

$$\sum\vec{\tau} = \sum\tau_z = 0$$

Step 3 Do algebra. Set the sum of the torques (Eqs. 11, 12, and 14) equal to zero and solve for r_P. The unit vector $\hat{k}$ cancels out.

$$\left(\frac{\ell}{2}\right)m_F g\hat{k} - \left(\frac{\ell}{2}\right)m_D g\hat{k} - r_P(m_F + m_D + m_s)g\hat{k} = 0$$

$$r_P(m_F + m_D + m_s)g = \left(\frac{\ell}{2}\right)m_F g - \left(\frac{\ell}{2}\right)m_D g$$

$$r_P = \left(\frac{\ell}{2}\right)\frac{m_F - m_D}{m_F + m_D + m_s}$$

From Figure 14.9, we see that $\vec{r}_P$ is in the negative x direction. So, the numerator becomes $m_D - m_F$ when we express the vector in component form.

$$\vec{r}_P = \left(\frac{\ell}{2}\right)\frac{m_D - m_F}{m_F + m_D + m_s}\hat{\imath} \quad (15)$$

CHECK and THINK

Our answer tells us where to put the pivot relative to the center of mass so that when the grandfather and granddaughter are on the seesaw, the system is in equilibrium. The dimensions are length, as we would expect for position.

To further check our expression, imagine that the grandfather and granddaughter have the same mass. In that case, we would expect that the pivot would be placed in the middle of the seesaw. Because our coordinate system is at the center of the seesaw, we find $r_P = 0$ if $m_D = m_F$.

However, based on Figure 14.5 we assume that the grandfather is heavier than the granddaughter. So, we expect that he would be closer to the pivot than she is. In other words, we expect $\vec{r}_P$ to point in the negative x direction if $m_F > m_D$ (Fig. 14.9), exactly as we found from Equation (15).

Here is another example showing the use of $\tau = rF \sin \varphi$ and the right-hand rule to find the torques. This method may be especially useful when the unknown quantity is an angle.

EXAMPLE 14.2 Don't Let the Rake Fall Down

A rake is placed against a smooth wall so that it makes an angle θ with the floor of the garage (Fig. 14.10). Assume friction between the wall and the rake is negligible, but the rake's tines are made from a hard rubber that rests on the concrete floor. The coefficient of static friction μ_s for rubber on dry concrete is 1.0. Find the minimum value of θ (other than zero) so that the rake does not slide. The rake's center of mass is one-fourth of the way up from the tines as shown.

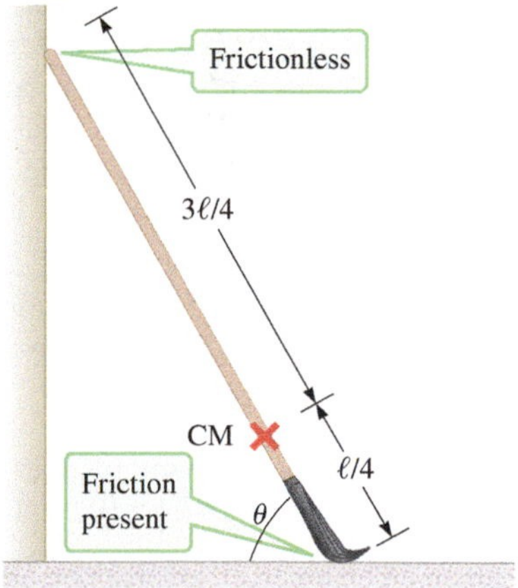

FIGURE 14.10 For a rough floor, what is the smallest angle θ such that the rake won't fall?

INTERPRET and ANTICIPATE

Again, start with a modified free-body diagram that includes all five elements.

Element 1 Simple shape. Represent the rake as a simple thick line.

Element 2 Forces at points of application. Four forces are exerted on the rake: the normal force $\vec{f}_N$ exerted by the wall, the gravitational force $\vec{F}_g$ acting at the center of mass, the normal force $\vec{F}_N$ exerted by the floor, and static friction $\vec{F}_s$ exerted by the floor. Each force has been drawn at its point of contact (Fig. 14.11). But at this stage, you do not need to know the relative size of the force vectors.

Element 3 A rotation axis. We chose a rotation axis through the point of contact between the wall and the rake so that the wall (more specifically, the normal force $\vec{f}_N$) cannot exert a torque on the rake.

Element 4 Coordinate system. We placed the origin at the point of contact with the wall and aligned the z axis with the rotation axis.

Example continues on page 398 ▶

Element 5 Position vectors. Draw position vectors $\vec{r}_{CM}$ and $\vec{r}_F$ for the three forces $\vec{F}_g$, $\vec{F}_N$, and $\vec{F}_s$ that exert nonzero torques around the rotation axis (Fig. 14.11). The forces exerted by the floor—$\vec{F}_N$ and $\vec{F}_s$—act at the same point of contact, so the position vector $\vec{r}_F$ is the same for these two forces.

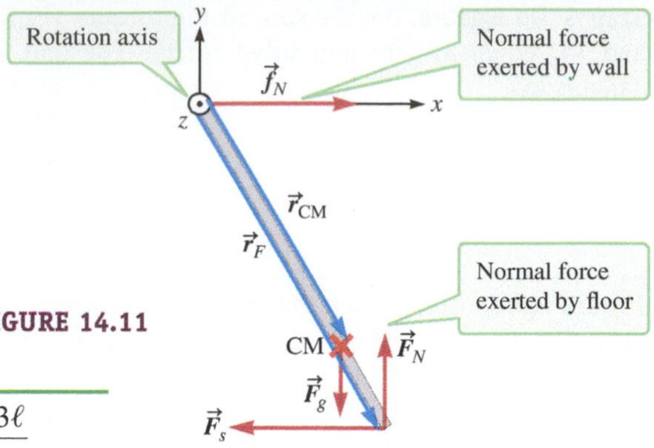

FIGURE 14.11

SOLVE

Step 1 Write each force in component form. Because we plan to use $\tau = rF \sin \varphi$ (Eq. 12.16), we don't need to write the forces in component form.

List the position vectors. Writing the position vectors would be complicated and is not necessary in this example. Instead, we list their magnitudes. There are only two positions: the center-of-mass position and the position of the floor.

$$r_{CM} = \frac{3\ell}{4}$$
$$r_F = \ell$$

Step 2a Apply the force balance condition. Because there are forces in the x and y directions, we must use Equation 14.3 separately for each of them.

$$\sum F_x = f_N - F_s = 0 \qquad \text{(14.3 for } x\text{)}$$
$$f_N = F_s$$
$$\sum F_y = F_N - F_g = 0 \qquad \text{(14.3 for } y\text{)}$$
$$F_N = F_g \qquad (1)$$

Step 2b Find the torques and apply the torque balance condition. When you use $\tau = rF \sin \varphi$, it is helpful to sketch the force and position vectors tail to tail as shown in Figure 14.12. Before calculating the torques, use the right-hand rule and Figure 14.11 to find their directions. Here the torques have only z components.

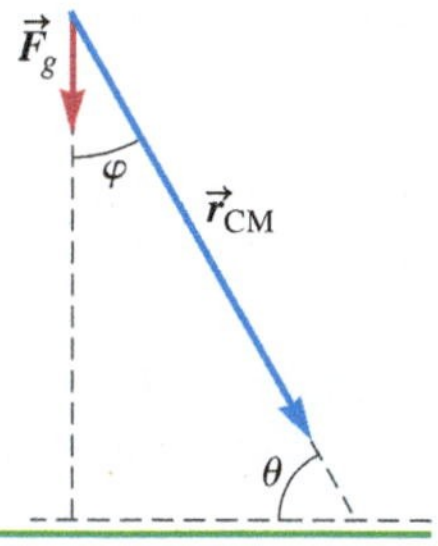

FIGURE 14.12

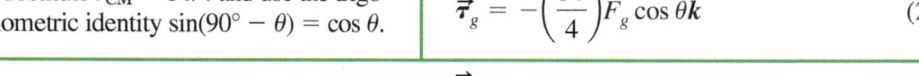

Now calculate each torque. The angle between $\vec{F}_g$ and $\vec{r}_{CM}$ is φ, and from Figure 14.12, $\varphi = 90° - \theta$.

$$\vec{\tau}_g = -F_g r_{CM} \sin \varphi \hat{k}$$
$$\vec{\tau}_g = -F_g r_{CM} \sin (90° - \theta)\hat{k}$$

Substitute $r_{CM} = 3\ell/4$ and use the trigonometric identity $\sin(90° - \theta) = \cos \theta$.

$$\vec{\tau}_g = -\left(\frac{3\ell}{4}\right)F_g \cos \theta \hat{k} \qquad (2)$$

Again, to find the angle φ between $\vec{r}_F$ and $\vec{F}_N$, slide one vector or the other so that their tails are together (Fig. 14.13).

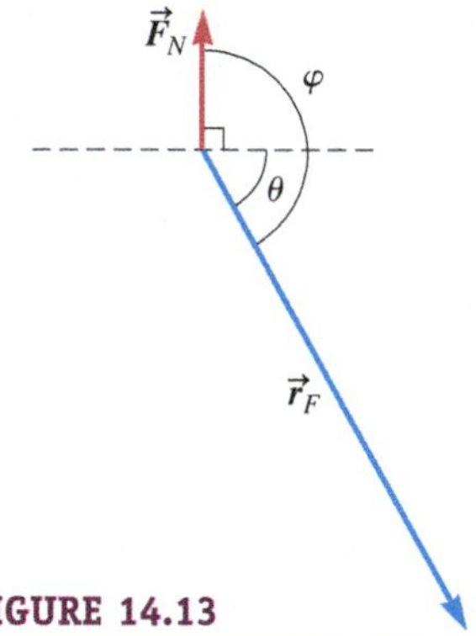

FIGURE 14.13

Find the torque due to the normal force exerted by the floor in terms of θ.

$$\vec{\tau}_N = r_F F_N \sin \varphi \hat{k}$$
$$\vec{\tau}_N = r_F F_N \sin (90° + \theta)\hat{k}$$

Substitute $r_F = \ell$ and use the trigonometric identity $\sin(90° + \theta) = \cos \theta$.

$$\vec{\tau}_N = \ell F_N \cos \theta \hat{k} \qquad (3)$$

The angle between $\vec{F}_s$ and $\vec{r}_F$ is θ, so no separate sketch is needed.

$$\vec{\tau}_s = -\ell F_s \sin \theta \hat{k} \qquad (4)$$

Now we are ready to **apply the torque balance condition.** The torques in this case only have z components.

$$\vec{\tau}_{tot} = \sum \tau_z = 0$$
$$\vec{\tau}_{tot} = \vec{\tau}_g + \vec{\tau}_N + \vec{\tau}_s$$

Step 3 Do algebra. Substitute the expressions for torque (Eqs. 2–4), and solve for θ, the unknown quantity.

$$-\left(\frac{3\ell}{4}\right)F_g \cos \theta + \ell F_N \cos \theta - \ell F_s \sin \theta = 0$$
$$F_s \sin \theta = \left(F_N - \frac{3F_g}{4}\right) \cos \theta$$

By Equation (1), $F_g = F_N$. The minimum angle θ_{min} is possible when static friction is maximum, so $F_s = \mu_s F_N$ (Eq. 6.1) when $\theta = \theta_{min}$.

$$\mu_s F_N \sin \theta_{min} = \left(F_N - \frac{3F_N}{4}\right) \cos \theta_{min} = \frac{F_N}{4} \cos \theta_{min}$$

Solve for $\theta_{\min}$.	$\frac{\sin\theta_{\min}}{\cos\theta_{\min}} = \tan\theta_{\min} = \frac{1}{4\mu_s}$ $\theta_{\min} = \tan^{-1}\left(\frac{1}{4\mu_s}\right)$
Substitute the given value of μ_s.	$\theta_{\min} = \tan^{-1}\left[\frac{1}{4(1.0)}\right] = 14°$

CHECK and THINK

To check our result, imagine that the floor is wet so that the coefficient of static friction between the rake's tines and the floor is reduced. We expect the minimum angle to increase. When the floor is slippery, suppose $\mu_s = 0.5$ instead of 1.0. According to our results, $\theta_{\min}$ would be $\tan^{-1}(0.5) = 27°$, a larger minimum angle, as expected.

Here is an example using a complex geometry. In such a case, Equations 14.7 and 14.8 should be used to find the torques.

EXAMPLE 14.3 CASE STUDY Hanging an Airplane

We consider a somewhat simplified version of the sort of problems solved by engineers and architects who designed the Udvar-Hazy Center. To minimize damage to an airplane, it should be hung from as few cables as possible, and the cables should be connected to support elements in the plane. An airplane has mass $m = 948$ kg and is hung from two steel cables so that the wings make an angle $\theta = 10.0°$ with the horizontal. Cable 1 is attached near the middle of the upper wing, and cable 2 is attached to a wing support near the edge (parts A and B of Fig. 14.14). The placement of cable 2 is determined by the location of the support structure. Cable 1, however, can be attached anywhere along another support structure running across the short length of the wing, so the x component of its position is determined by the desired equilibrium configuration of the plane. Use the coordinate system and data provided in Figure 14.14 to find r_{1x} (the x coordinate of cable 1's point of contact) and the tension in the two cables.

A Find r_{1x}, the x coordinate of cable 1's point of contact.

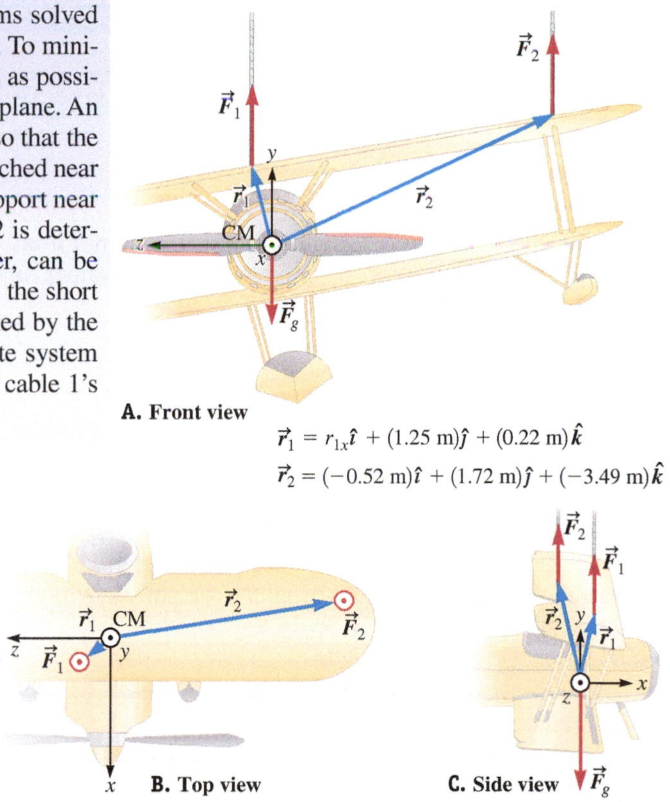

FIGURE 14.14 **A.** Front view of airplane showing a coordinate system with the origin at the center of mass, the x axis out of the page, and the z axis to the left. **B.** Top view looking down from the $+y$ axis. **C.** Side view looking at the fuselage from the $+z$ axis.

INTERPRET and ANTICIPATE

Because the plane is in equilibrium, the sum of the forces and the sum of the torques acting on it are zero (Eqs. 14.1 and 14.2). While Figure 14.14 is somewhat more detailed than you may draw on your own, it provides the free-body diagram with all the elements. The forces are drawn at points of application. There is a rotation axis and coordinate system. The position vectors have been drawn. Three forces act on the plane: gravity $\vec{F}_g$ and the tension forces $\vec{F}_1$ and $\vec{F}_2$ exerted by each cable. Because the origin of the coordinate system is at the center of mass, it is relatively easy to calculate the torques around an axis passing through the center of mass.

SOLVE

Step 1 Write each force in component form. As shown in Figure 14.14, the forces on the airplane have only y components.	$\vec{F}_1 = F_1\hat{\jmath}$ $\quad \vec{F}_2 = F_2\hat{\jmath}$ $\quad F_1$ and F_2 are unknown. $\vec{F}_g = -mg\hat{\jmath}$

Example continues on page 400 ▶

We are given the position vectors $\vec{r}_1$ and $\vec{r}_2$ in component form (Fig. 14.14).	$\vec{r}_1 = r_{1x}\hat{\imath} + (1.25\text{ m})\hat{\jmath} + (0.22\text{ m})\hat{k}$ $\vec{r}_2 = (-0.52\text{ m})\hat{\imath} + (1.72\text{ m})\hat{\jmath} + (-3.49\text{ m})\hat{k}$ r_{1x} is unknown.
Step 2a Apply the force balance condition. We only need to apply Equation 14.3 along the y axis.	$\sum F_y = 0$ (14.3) $F_1 + F_2 - mg = 0$ $F_1 + F_2 = mg$ (1)
Step 2b Find the torques and apply the torque balance condition. Find the total torque around the center of mass. Gravity cannot exert a torque around the center of mass in our chosen coordinate system, so the total torque comes from adding torques $\vec{\tau}_1$ and $\vec{\tau}_2$ exerted by cable 1 and cable 2, respectively.	$\vec{\tau}_{\text{tot}} = \sum \vec{\tau} = 0$ $\sum \vec{\tau} = \vec{\tau}_1 + \vec{\tau}_2 = 0$ (2)

Find the torque due to each cable using Equation 14.7. The tensions $\vec{F}_1$ and $\vec{F}_2$ have only y components, so the F_x and F_z terms disappear.

Find the torque due to each cable.	$\vec{\tau}_1 = \vec{r}_1 \times \vec{F}_1 = -r_{1z}F_1\hat{\imath} + r_{1x}F_1\hat{k}$ $\vec{\tau}_2 = \vec{r}_2 \times \vec{F}_2 = -r_{2z}F_2\hat{\imath} + r_{2x}F_2\hat{k}$
Add the torques and set their sum to zero (Eq. 2).	$(-r_{1z}F_1\hat{\imath} + r_{1x}F_1\hat{k}) + (-r_{2z}F_2\hat{\imath} + r_{2x}F_2\hat{k}) = 0$
Step 3 Do algebra. Collect terms in $\hat{\imath}$ and $\hat{k}$.	$-(r_{1z}F_1 + r_{2z}F_2)\hat{\imath} + (r_{1x}F_1 + r_{2x}F_2)\hat{k} = 0$
In equilibrium, each component of the total torque is zero.	$\tau_x = r_{1z}F_1 + r_{2z}F_2 = 0$ $\tau_z = r_{1x}F_1 + r_{2x}F_2 = 0$
Solve these two equations for F_1.	$F_1 = -\dfrac{r_{2z}}{r_{1z}}F_2$ and $F_1 = -\dfrac{r_{2x}}{r_{1x}}F_2$ (3)
The x coordinate of cable 1's point of contact r_{1x} comes from equating expressions for F_1 in Equation (3). (The values of the position components are given in Fig. 14.14.)	$\dfrac{r_{2z}}{r_{1z}} = \dfrac{r_{2x}}{r_{1x}}$ $r_{1x} = \dfrac{r_{2x}r_{1z}}{r_{2z}}$ $r_{1x} = \dfrac{(-0.52\text{ m})(0.22\text{ m})}{-3.49\text{ m}} = 3.3 \times 10^{-2}\text{ m}$

CHECK and THINK

This result is the answer to part A. Imagine for a moment that the plane were suspended from only one cable. In that case, the cable would be attached directly above the center of mass, and the tension in the cable would equal the weight of the plane. We now expect that because cable 1 is attached almost directly above the center of mass, the tension in cable 1 should be nearly equal to the weight of the plane.

B Find the tension in the two cables.

SOLVE Our work in part A allows us to jump straight into solving this part. The tension in cable 2 comes from substituting Equation (3) into Equation (1).	$-\dfrac{r_{2z}}{r_{1z}}F_2 + F_2 = mg$ $F_2 = \dfrac{mg}{1 - r_{2z}/r_{1z}} = \dfrac{(948\text{ kg})(9.81\text{ m/s}^2)}{1 - (-3.49\text{ m}/0.22\text{ m})}$ $F_2 = 5.50 \times 10^2\text{ N}$ (4)
Finally, the tension in cable 1 comes from substituting Equation (4) into Equation (3).	$F_1 = -\dfrac{r_{2z}}{r_{1z}}F_2 = -\dfrac{(-3.49\text{ m})}{(0.22\text{ m})}(5.50 \times 10^2\text{ N})$ $F_1 = 8.7 \times 10^3\text{ N}$

CHECK and THINK
The plane's weight is $mg = (948 \text{ kg})(9.81 \text{ m/s}^2) = 9.30 \times 10^3$ N. So, the tension F_1 in cable 1 is about 94% of the plane's weight, and the tension in cable 2 is about 6% of the plane's weight. As expected, the tension in cable 1 is nearly equal to the plane's weight.

Here is an example with a complicated geometry. The example has symmetry, however, which helps us predict the outcome and makes the calculation a little easier than usual.

EXAMPLE 14.4 Making a Three-Legged Table

Children often hear interesting stories about a parent's childhood. According to the author's grandmother, when the author's father was about eight years old, he decided to see if their table would stand up on just three of its four legs, so he sawed off one of the legs. Assume the tabletop was a square of side a. Find expressions for the force exerted by the remaining three legs on the tabletop just after the fourth leg was removed. Did the table remain standing up? Model the tabletop without the legs as the system.

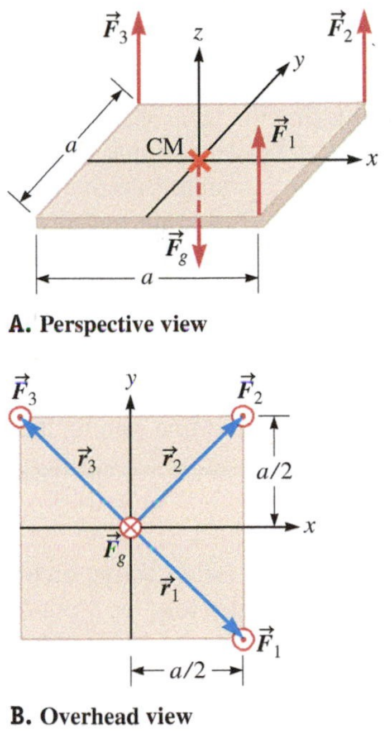

FIGURE 14.15 Representation of a tabletop. One leg of the table has been removed.

INTERPRET and ANTICIPATE
The modified free-body diagram in Figure 14.15 is shown from two perspectives. The **tabletop is represented by a square**. Four forces act on the tabletop: gravity and the three normal forces $\vec{F}_1$, $\vec{F}_2$, and $\vec{F}_3$ exerted by the three remaining legs. These forces are drawn at the **point of application** *(but not to scale)*. The origin of the **coordinate system** is located at the center of mass. We will calculate torques around a **rotation axis** passing through the center of mass and aligned with the z axis. The **position vectors** are drawn. Because of the symmetry in this example, we expect that the expressions for $\vec{F}_1$ and $\vec{F}_3$ are identical.

SOLVE
Step 1 Write each force in component form. The four forces have only z components.

$$\vec{F}_1 = F_1\hat{k}$$
$$\vec{F}_2 = F_2\hat{k}$$
$$\vec{F}_3 = F_3\hat{k}$$
$$\vec{F}_g = -F_g\hat{k}$$

Write the position vectors in component form. The position vectors $\vec{r}_1$, $\vec{r}_2$, and $\vec{r}_3$ have x and y components (Fig. 14.15B). Because the origin of the coordinate system was placed at the center of mass, $\vec{r}_{CM} = 0$.

$$\vec{r}_1 = \tfrac{a}{2}\hat{\imath} - \tfrac{a}{2}\hat{\jmath}$$
$$\vec{r}_2 = \tfrac{a}{2}\hat{\imath} + \tfrac{a}{2}\hat{\jmath}$$
$$\vec{r}_3 = -\tfrac{a}{2}\hat{\imath} + \tfrac{a}{2}\hat{\jmath}$$

Step 2a Apply the force balance condition. We only need the z component of Equation 14.3.

$$\sum F_z = F_1 + F_2 + F_3 - F_g = 0 \quad (14.3 \text{ for } z)$$
$$F_1 + F_2 + F_3 = F_g \quad (1)$$

Step 2b Find the torques. Many of the terms disappear from Equation 14.7 in this example because $F_x = F_y = 0$ and $r_z = 0$ for each torque.

$$\vec{\tau} = (r_yF_z - r_zF_y)\hat{\imath} + (r_zF_x - r_xF_z)\hat{\jmath} + (r_xF_y - r_yF_x)\hat{k} \quad (14.7)$$
$$\vec{\tau} = r_yF_z\hat{\imath} - r_xF_z\hat{\jmath} \quad (2)$$

Use Equation (2) to find the torque exerted by each of the normal forces. The torque exerted by gravity is zero because $\vec{r}_{CM} = 0$.

$$\vec{\tau}_1 = -\tfrac{a}{2}F_1\hat{\imath} - \tfrac{a}{2}F_1\hat{\jmath}$$
$$\vec{\tau}_2 = \tfrac{a}{2}F_2\hat{\imath} - \tfrac{a}{2}F_2\hat{\jmath}$$
$$\vec{\tau}_3 = \tfrac{a}{2}F_3\hat{\imath} + \tfrac{a}{2}F_3\hat{\jmath}$$

Example continues on page 402 ▶

Apply the torque balance condition using Equation 14.2.	$\vec{\tau}_{tot} = \vec{\tau}_1 + \vec{\tau}_2 + \vec{\tau}_3 = 0$ (14.2) $\vec{\tau}_{tot} = \frac{a}{2}(-F_1 + F_2 + F_3)\hat{\imath} + \frac{a}{2}(-F_1 - F_2 + F_3)\hat{\jmath} = 0$
Step 3 Do algebra. Each component of torque must be zero, so we can derive two equations.	$-F_1 + F_2 + F_3 = 0$ $F_1 = F_2 + F_3$ (3) $-F_1 - F_2 + F_3 = 0$ $F_1 + F_2 = F_3$ (4)
Solve Equations (1), (3), and (4) simultaneously for the three normal forces. These forces point upward in the positive z direction (Fig. 14.15A).	$\vec{F}_1 = \vec{F}_3 = \frac{F_g}{2}\hat{k}$ $\vec{F}_2 = 0$

CHECK and THINK

As expected, $\vec{F}_1 = \vec{F}_3$. Because we were able to find real solutions for all the normal forces, it seems that the table should remain standing even after removal of one of its legs. In fact, the father in this story claimed that table remained standing until his mother tried to set the table for dinner.

It seems counterintuitive that the leg across from the missing leg exerts no force on tabletop ($\vec{F}_2 = 0$). This result means that the second leg may also be removed with the same observable effect: The table should remain standing. Our experience, however, tells us that such a two-legged table would be *unstable*. Imagine setting a dinner plate on the three-legged table. As long as the plate is placed on the triangular half of the table that contains all three legs, the table will not tip over. If the plate were placed on the other half of the table, the table would, of course, tip over. Now imagine placing the same plate on a two-legged table. Unless the plate is placed precisely along the line joining the two remaining legs, the table will tip over.

When the forces acting on an object do not lie in a single plane, it may be necessary to consider torques around two separate rotation axes. In the next example, we study a situation in which there is more than one rotation axis.

EXAMPLE 14.5 Designing a Large Floor Fan

Fans are more portable, more reliable, and less expensive to buy and to run than air conditioners, so military bases often use large floor fans to circulate air in a large space such as a gym. (The fan's blades rotate at a constant angular velocity, so there is no net torque on the blades. In this example, consider the force the blades exert on the air.) Because the fans must exert a great force on the air, the air must exert a great force on the fan, similar to the force exerted by the exhaust on a rocket. This force is known as *thrust* (Chapter 10). If the thrust is too great, the fan will topple over. The large fan in Figure 14.16 weighs 20 lb; it has four legs of length $b = 12$ in., and its height (measured from the floor to the center of the fan blades) is $h = 36$ in. Each leg has a small rubber foot at the end so that the floor is only in contact with the end of each leg. Find the maximum thrust that may be exerted without toppling the fan. Give your answer in pounds and in newtons. At the maximum thrust, the front two feet are just lifting off the floor (Problem 58).

FIGURE 14.16

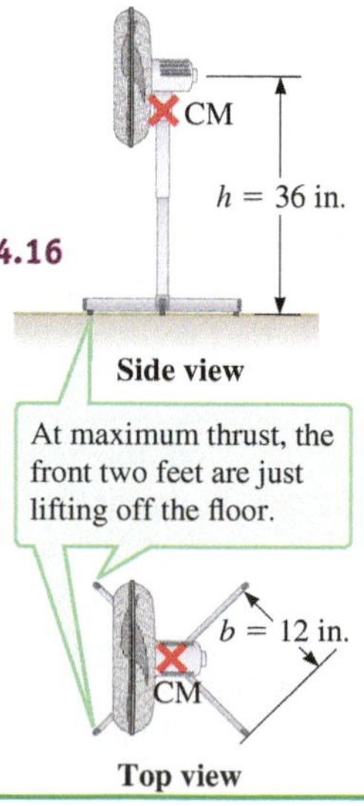

INTERPRET and ANTICIPATE

The modified free-body diagram (Fig. 14.17) shows two views. The fan is **represented by a simple rectangle** with an X-shaped base.

A total of 10 forces act on the fan and are **drawn at their points of application.** Gravity $\vec{F}_g$ acts on the center of mass. The thrust $\vec{F}_{\text{thrust}}$ is a horizontal force whose point of contact we take to be the center of the fan's blades. The eight other forces are due to the four points of contact between the floor and the fan's feet. At each point of contact, the floor exerts a normal force upward ($\vec{F}_{N1}, \vec{F}_{N2}, \vec{F}_{N3}$, and $\vec{F}_{N4}$) and static friction horizontally ($\vec{F}_{S1}, \vec{F}_{S2}, \vec{F}_{S3}$, and $\vec{F}_{S4}$). Static friction must point in the opposite direction to the thrust; otherwise, the fan would glide across the floor.

The origin of the **coordinate system** is at the center of the fan's base. We consider **three rotation axes** aligned with the coordinate axes. The forces act at six points of contact, so we **draw six position vectors.** The position of the center of mass is $\vec{r}_{\text{CM}}$. The thrust acts at $\vec{r}_{\text{thrust}}$. The four points of contact between the floor and the fan's feet are $\vec{r}_1, \vec{r}_2, \vec{r}_3$, and $\vec{r}_4$.

The front two feet are not in contact with the floor, however, so there is no normal force or friction exerted on these feet ($\vec{F}_{N3} = \vec{F}_{N4} = \vec{F}_{S3} = \vec{F}_{S4} = 0$), and there is no need to find $\vec{r}_3$ and $\vec{r}_4$.

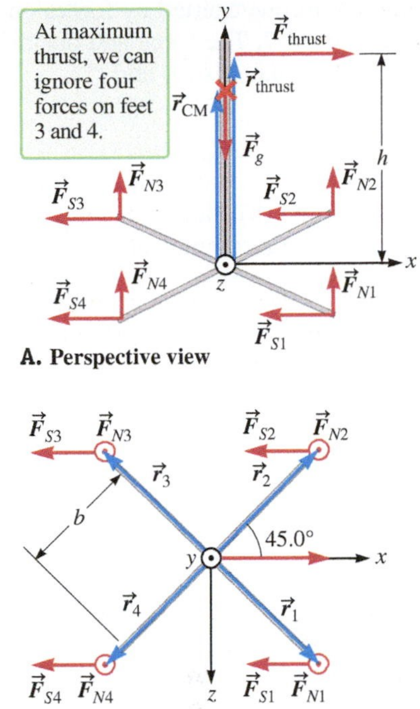

A. Perspective view

B. Overhead view

FIGURE 14.17

SOLVE

Step 1 Write forces in component form. Gravity is in the negative y direction, and thrust is in the positive x direction.

$$\vec{F}_g = -F_g\hat{\jmath} \quad (1)$$

$$\vec{F}_{\text{thrust}} = F_{\text{thrust}}\hat{\imath} \quad (2)$$

For each foot, the normal force is in the positive y direction, and static friction is in the negative x direction. For example, consider foot 1. The total force $\vec{F}_1$ exerted by the floor is due to the force of static friction $\vec{F}_{S1}$ and the normal force $\vec{F}_{N1}$.

$$\vec{F}_{N1} = F_{N1}\hat{\jmath}$$

$$\vec{F}_{S1} = -F_{S1}\hat{\imath}$$

$$\vec{F}_1 = -F_{S1}\hat{\imath} + F_{N1}\hat{\jmath} \quad (3)$$

Write a similar equation for $\vec{F}_2$.

$$\vec{F}_2 = -F_{S2}\hat{\imath} + F_{N2}\hat{\jmath} \quad (4)$$

Write the position vectors in component form. Gravity is applied at the center of mass, and thrust is applied at a height h above the center of the fan's X-shaped base.

$$\vec{r}_{\text{CM}} = r_{\text{CM}}\hat{\jmath} \quad (5)$$

$$\vec{r}_{\text{thrust}} = h\hat{\jmath} \quad (6)$$

Express the point of contact for the forces acting on each foot in terms of the leg length b. Refer to the coordinate system shown in Figure 14.17B.

$$\vec{r}_1 = b\sin 45°\hat{\imath} + b\cos 45°\hat{k}$$

$$\vec{r}_1 = \frac{b\sqrt{2}}{2}\hat{\imath} + \frac{b\sqrt{2}}{2}\hat{k} \quad (7)$$

$$\vec{r}_2 = \frac{b\sqrt{2}}{2}\hat{\imath} - \frac{b\sqrt{2}}{2}\hat{k} \quad (8)$$

Step 2a Apply the force balance condition in component form, which gives us two equations.

$$\sum F_x = F_{\text{thrust}} - F_{S1} - F_{S2} = 0 \quad (9)$$

$$\sum F_y = F_{N1} + F_{N2} - F_g = 0 \quad (10)$$

Step 2b Find the torques. With our coordinate system, the position vectors measured from the origin are the ones we need for calculating torque using Equation 14.7.

The torque due to gravity is zero because both $\vec{F}_g$ and $\vec{r}_{\text{CM}}$ only have y components (Eqs. 1 and 5). (Whenever two vectors are parallel or antiparallel, their cross product is zero.)

$$\vec{\tau}_g = \vec{r}_{\text{CM}} \times \vec{F}_g = 0$$

Find the torque due to the thrust using the position vector (Eq. 6) and the force (Eq. 2). All but the last term in Equation 14.7 are zero for the torque due to thrust.

$$\vec{\tau}_{\text{thrust}} = \vec{r}_{\text{thrust}} \times \vec{F}_{\text{thrust}} = -(r_yF_x)\hat{k} = -hF_{\text{thrust}}\hat{k} \quad (11)$$

Example continues on page 404 ▶

Find the torque exerted by forces on foot 1 and foot 2 using Equation 14.7. The force and position vectors for foot 1 are given by Equations (3) and (7).

$$\vec{\tau}_1 = \vec{r}_1 \times \vec{F}_1 = \left(\frac{b\sqrt{2}}{2}\hat{\imath} + \frac{b\sqrt{2}}{2}\hat{k}\right) \times (-F_{S1}\hat{\imath} + F_{N1}\hat{\jmath})$$

$$\vec{\tau}_1 = -\frac{b\sqrt{2}}{2}F_{N1}\hat{\imath} - \frac{b\sqrt{2}}{2}F_{S1}\hat{\jmath} + \frac{b\sqrt{2}}{2}F_{N1}\hat{k}$$

The force and position vectors for foot 2 are given by Equations (4) and (8). Notice that the torque exerted on each foot has x, y, and z components.

$$\vec{\tau}_2 = \frac{b\sqrt{2}}{2}F_{N2}\hat{\imath} + \frac{b\sqrt{2}}{2}F_{S2}\hat{\jmath} + \frac{b\sqrt{2}}{2}F_{N2}\hat{k}$$

Apply the torque balance condition in component form, which gives us three equations. The thrust only contributes to the z component of the torque (Eq. 11).

$$\sum \tau_x = \frac{b\sqrt{2}}{2}(-F_{N1} + F_{N2}) = 0 \quad (12)$$

$$\sum \tau_y = \frac{b\sqrt{2}}{2}(-F_{S1} + F_{S2}) = 0 \quad (13)$$

$$\sum \tau_z = \frac{b\sqrt{2}}{2}(F_{N1} + F_{N2}) - hF_{\text{thrust}} = 0 \quad (14)$$

Step 3: Algebra. We have five equations (Eqs. 9, 10, 12, 13, and 14), and we are interested in finding the maximum thrust F_{thrust}. Use Equation (10) to find an expression for the total normal force.

$$F_{N1} + F_{N2} = F_g \quad (15)$$

Substitute Equation (15) into Equation (14) and solve for thrust.

$$\frac{b\sqrt{2}}{2}(F_g) - hF_{\text{thrust}} = 0$$

$$F_{\text{thrust}} = \frac{\sqrt{2}b}{2h}F_g = \frac{\sqrt{2}(12\text{ in.})}{2(36\text{ in.})}(20\text{ lb})$$

$$F_{\text{thrust}} = 4.7\text{ lb} = 21\text{ N}$$

CHECK and THINK

We found five equations by applying the equilibrium conditions, but we only needed two of them to solve for the thrust. In solving a complicated equilibrium problem such as this one, you may write down more equations than you need. You may find those unused equations useful if you need to solve for some other quantity.

14-4 Elasticity and Fracture

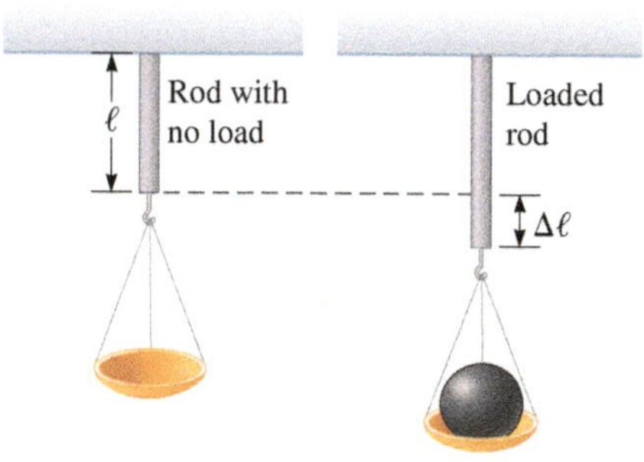

FIGURE 14.18 A load is hung from a rod of length ℓ, resulting in a new length $\ell + \Delta\ell$. The strain ε of the rod is $\varepsilon = \Delta\ell/\ell$.

Robert Hooke (1635–1703) and Isaac Newton (1642–1727) were contemporaries, but they each took a very different approach to science. Newton was interested in astronomy, cosmology, and mathematics. Hooke was interested in more practical, mundane matters such as buildings, ships, and fleas. Today we would probably consider Hooke to be primarily an *applied* physicist and Newton primarily a *theoretical* physicist.

Hooke was interested in how objects in static equilibrium responded to being pulled and pushed. That is, he wanted to know how stationary objects reacted to being *loaded.* Rather than trying to come up with a theory, his approach was to collect data first. For example, he hung loads—objects of known weight—on rods made of different materials and of different sizes (Fig. 14.18). He measured the length of the rods before and after the load was attached and found that the rods stretched when loaded and returned to their unstretched length when the load was removed. He concluded that there is no such thing as a rigid object. In other words, all objects are stretched or compressed when forces are applied, even if only slightly.

Stress

STRESS Major Concept

Augustin Cauchy (1789–1857), an engineer and academic, continued to study the behavior of loaded objects. He defined **stress** σ as the magnitude of the force applied to an object per unit area (Fig. 14.19):

$$\sigma \equiv \frac{F}{A} \quad (14.9)$$

The SI units of stress are N/m^2 and $1\ N/m^2 = 1$ Pa, where Pa stands for *pascals*. There are four types of stress: *tensile stress*, *compression stress*, *shear stress*, and *volume stress*, depending on where and how the force is applied to the object. **Tensile stress** on an object occurs when a force pulls perpendicular to a surface of the object (Fig. 14.19). Tensile stress, compression stress, and shear stress are discussed in this chapter. Volume stress is discussed in Chapter 15.

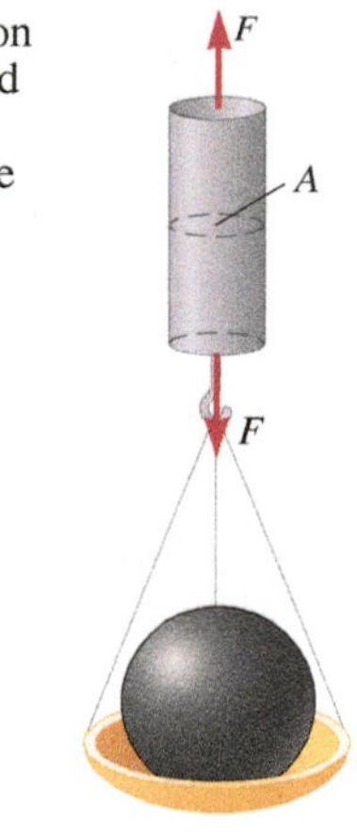

FIGURE 14.19 Stress on a rod is the force applied per unit cross-sectional area. Here is an example of *tensile* stress.

Strain

Imagine performing experiments similar to Hooke's (Fig. 14.18) in which you measure the rod's length as you load objects of known weight into the pan. From these data, you could calculate the stress σ on the rod by dividing the force applied to the rod by the rod's cross-sectional area. You could also calculate the fractional increase in the rod's length by dividing the increase $\Delta\ell$ in the rod's length by its original length ℓ. The fractional increase $\Delta\ell/\ell$ is known as the **strain**. The symbol for strain is the lowercase Greek letter epsilon, ε. Because strain is the fractional increase in the rod's length

STRAIN ✪ Major Concept

$$\varepsilon = \Delta\ell/\ell \tag{14.10}$$

it is dimensionless. The words *stress* and *strain* sound alike and are often confused. It might help to remember the phrase, "A stress on a rod causes a strain."

Tensile Deformation

Figure 14.20 shows a graph of stress as a function of tensile strain for the rod from experimental data. This graph reveals three phenomena.

First, as long as the stress σ is small, the strain is directly proportional to the stress. Over the linear response range on the graph in Figure 14.20,

$$\sigma = Y\varepsilon \tag{14.11}$$

where Y is known as **Young's modulus** and is the slope of the line. Because strain ε is dimensionless, Young's modulus must have the same dimensions as stress. Young's modulus is a measure of the rod's stiffness and depends on its composition. Table 14.1 lists Y for various substances. As long as the stress put on the rod is in the linear response range, the rod returns to its original length when the load is removed. An **elastic object** is one that returns to its original length after the load has been removed.

YOUNG'S MODULUS ✪ Major Concept

ELASTIC OBJECT ✪ Major Concept

Second, as a greater load is placed on the rod past the linear response range, the rod responds by stretching an even greater amount. Now, even when the load is removed, the rod remains somewhat stretched rather than returning to its original size.

TABLE 14.1 Approximate values for moduli and strength of materials.

Material	Young's modulus Y (10^9 N/m^2)	Shear modulus G (10^9 N/m^2)	Tensile strength σ_{ten} and compression strength σ_{cmp} (10^6 N/m^2)
Aluminum	73	30	110
Bone	21 tensile		
	18 compressive		170 (compression only)
Cement	17		40 (compression only)
Diamond	1200		
Glass	70	20	50-1000
Rubber	7×10^{-3}		
Steel	210	80	380 (tensile)
			210 (compression)
Wood (Douglas fir)	11	4.1	130 (tensile only)

Note: These values are reasonable for classroom use, but if your work requires more precision, you should do your own research or experimentation.

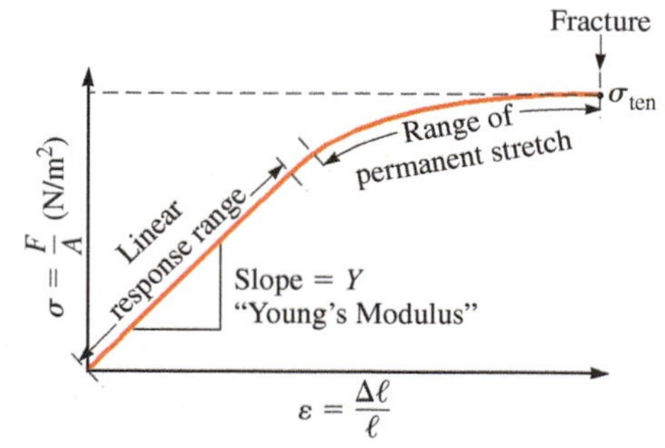

FIGURE 14.20 A graph of stress σ (vertical axis) as a function of strain ε (horizontal axis).

The **elastic limit** is the maximum stress that can be placed on an object without permanently deforming it. The elastic limit of an object is *not* necessarily equal to the maximum stress of the linear response range.

TENSILE STRENGTH

Major Concept

Third, the rod fractures at some maximum tensile stress known as the **tensile strength** σ_{ten}. Experiments show that tensile strength depends on the material of the rod. Table 14.1 includes the tensile strength of many materials. Let's take a microscopic look at what happens to the rod as it is stretched. As in Section 5-7, we model the rod's molecules as particles and the bonds between the molecules as stiff springs (Fig. 14.21). When a rod is loaded, the vertical bonds stretch, but the horizontal bonds do not. As the load is increased, the bonds obey Hooke's law and stretch in proportion to the force applied. When the load is removed, the bonds shrink back to their original size. When the load is too great, however, the molecular bonds no longer obey Hooke's law, and even if the load were to be removed, the bonds would no longer shrink to their original size. On the macroscopic level, the rod has been permanently stretched. When an even greater load is placed on the rod, the molecular bonds break, and the rod fractures.

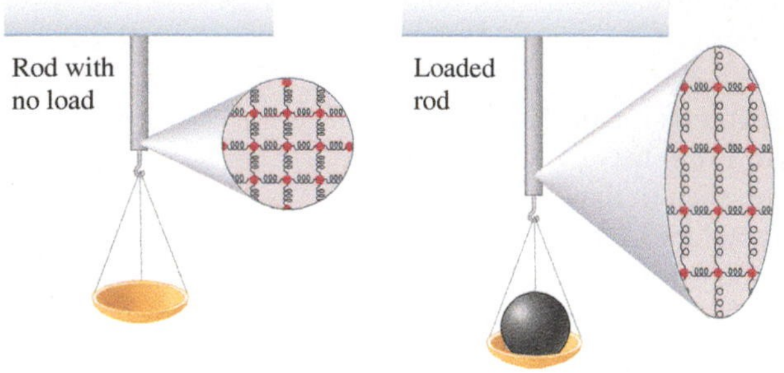

FIGURE 14.21 A microscopic model illustrates the strain on a rod as it reacts to a stress exerted on it by a hanging load. Each molecular bond stretches vertically, obeying Hooke's law.

Compressive Deformation

Engineers must also consider structures that are compressed by a load and structures that are not vertical. Whether the object is compressed or stretched, vertical or not, $\sigma = Y\varepsilon$ (Eq. 14.11) applies (Fig 14.22). If the object is compressed, however, $\Delta\ell$ is the *decrease* in the length of the object. Also, Young's *compression* modulus may not necessarily equal Young's *tensile* modulus (see, for example, the entries for bone in Table 14.1).

COMPRESSIVE STRENGTH

Major Concept

Like a tensile load, a compressive load may cause an object to fracture (Fig. 14.23). The **compressive strength** σ_{cmp} of a material—the stress at which the object fractures under compression—does not necessarily equal the tensile strength σ_{ten} of the material (Table 14.1).

Shear Deformation

Figure 14.24 shows another way a stationary object may be deformed. In this case, a *shear stress* has skewed the house. A **shear stress** τ results when a force is applied tangentially to one surface of a stationary object (Fig. 14.25). Mathematically, shear stress is given by

The Greek letter τ is used for both torque and shear stress.

$$\tau = \frac{F}{A} \tag{14.12}$$

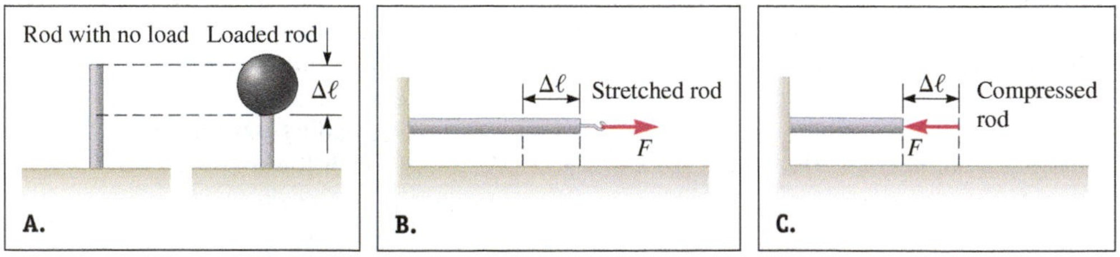

FIGURE 14.22 **A.** A rod is compressed by placing a load on top of it. **B.** A rod is stretched by a horizontal force. **C.** A rod is compressed by a horizontal force.

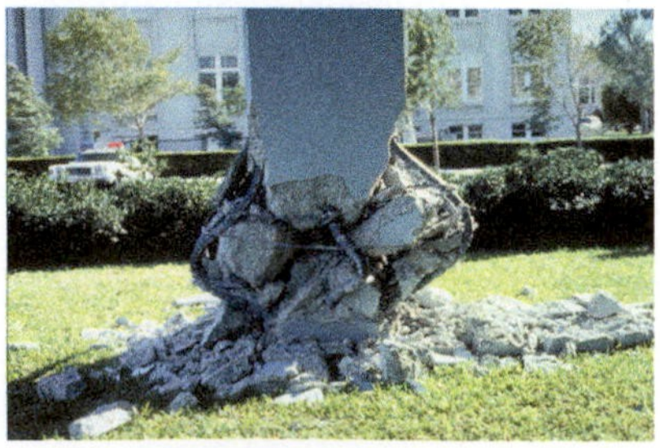

FIGURE 14.23 A column that fractured under a compressive load.

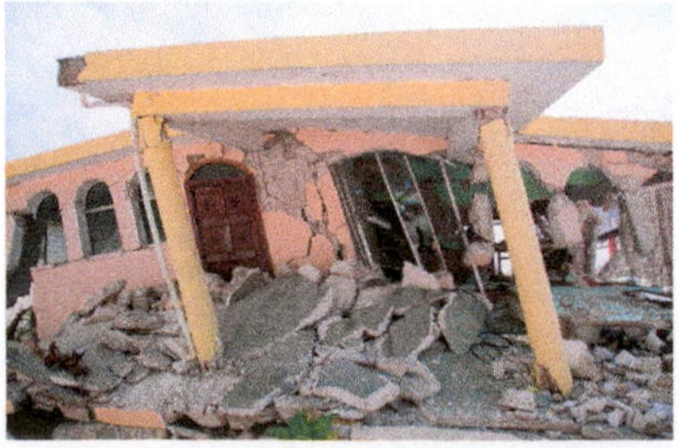

FIGURE 14.24 A house that suffered shear deformation in an earthquake.

where F is the magnitude of the force applied tangentially to the surface of area A. The **shear strain** γ is given by

$$\gamma = \frac{\Delta x}{h} \qquad (14.13)$$

where Δx is the amount the sheared face moves and h is the height of that face (Fig. 14.25). For small shear stresses τ, the shear strain γ is proportional to the stress:

$$\tau = G\gamma \qquad (14.14)$$

where G is called the **shear modulus**, also known as the torsion modulus. Like Young's modulus, G depends on the material making up the object (Table 14.1).

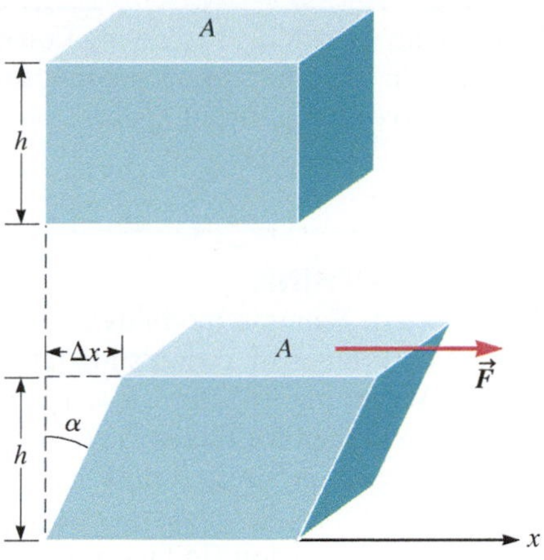

FIGURE 14.25 When a force is applied tangentially to one surface of a stationary object, a shear stress τ is exerted on the object, and a shear strain γ results.

SHEAR MODULUS

Major Concept

CONCEPT EXERCISE 14.5

Imagine two vertical rods initially of equal length. One rod is made of steel, and the other is made of rubber. Suppose you hang a load of the same mass from each rod. You are just able to eyeball the stretch in the rubber rod. Do you expect to notice any stretch in the steel rod? Explain.

EXAMPLE 14.6 Seeing Isn't Always Believing

In Figure 14.26, a car seems to be hanging from some gooey substance. Estimate Young's modulus for this unidentified material.

© Tim Hornikel

FIGURE 14.26 A car appears to hang from gooey supports.

INTERPRET and ANTICIPATE

Young's modulus can be found from $\sigma = Y\varepsilon$ (Eq. 14.11), so we need to estimate the strain ε of the substance and the stress σ exerted on it to find Y. The gooey substance appears to be very stretchy, so we expect Young's modulus to be very low.

SOLVE

Estimate the stress by assuming each gooey support experiences a force F equal to one-fourth of the car's weight F_g. A sedan typically weighs about 1.5 tons. Because this car is hanging, it seems likely that many of the heavy components (the engine, perhaps) may have been removed. So, let's assume the car weighs about 1 ton.

$$1 \text{ ton} = 2000 \text{ lb} \times \frac{1 \text{ N}}{0.2248 \text{ lb}}$$

$$1 \text{ ton} \approx 9000 \text{ N}$$

$$F = \frac{F_g}{4} = \frac{9000 \text{ N}}{4} \approx 2200 \text{ N}$$

The cross-sectional area of each gooey attachment is roughly equal to half the area of a tire's outer surface (the part that would be available to touch the road). Estimate that area by a rectangle of length $L \approx 1$ m and width $W \approx 0.3$ m.

$$A = LW \approx (1 \text{ m})(0.3 \text{ m})$$

$$A \approx 0.3 \text{ m}^2$$

Estimate the stress by dividing the force F by the cross-sectional area A (Eq. 14.9).

$$\sigma = \frac{F}{A} \qquad (14.9)$$

$$\sigma = \frac{2200 \text{ N}}{0.3 \text{ m}^2} = 7300 \text{ N/m}^2$$

To estimate the strain, imagine raising the car upward until the gooey substance is compressed into a tight rectangular solid. Perhaps the substance would form a rectangular solid with a height of 5 cm. Its stretched length is a bit longer than the height of the car; let's estimate it to be 1.5 m. Find the strain by dividing the change in length by the original length of the substance.

$$\varepsilon = \frac{\Delta \ell}{\ell}$$

$$\varepsilon = \frac{1.5 \text{ m} - 5 \times 10^{-2} \text{ m}}{5 \times 10^{-2} \text{ m}} = 29$$

Example continues on page 408 ▶

Find Young's modulus by solving Equation 14.11 for Y. Because we estimated the values from the photo to one significant figure, we report our estimate of Young's modulus to one significant figure. Don't worry if your estimate is off by a factor of a few.	$\sigma = Y\varepsilon$ (14.11) $Y = \frac{\sigma}{\varepsilon} = \frac{7300\,\text{N/m}^2}{29}$ $Y \approx 300\,\text{N/m}^2$

CHECK and THINK

As expected, Young's modulus for this substance is very low. The numbers in column 2 of Table 14.1 (values of Young's modulus) are to be multiplied by 10^9 N/m^2. Is our answer unrealistic? According to Table 14.1, Young's modulus for rubber (7×10^6 N/m^2) is about 20,000 times greater than for this mysterious substance. To increase our estimate of Young's modulus, our estimate of the stress σ must be increased or our estimate of the strain ε must be decreased. It is unlikely that the stress is much greater than our estimate because the car cannot weigh much more than 1 ton (in fact, it probably weighs much less), and, from the photo, the cross-sectional area of the gooey substance cannot be much smaller than the corresponding surface area of the tire. Could our estimate of ε be off by four orders of magnitude? From the photo, the stretched length of the substance is clearly between 1 and 2 m, so our estimate of that quantity is quite good. The original length of the substance was an assumption. Perhaps the substance is not actually that gooey; it may be carved wood or molded metal that has been artistically created to look gooey. In that case, the unstretched length may be very close to the stretched length, which would make ε smaller and Y larger. It seems likely that Figure 14.26 is a photo of a clever illusion and that the substance is not four orders of magnitude more stretchable than rubber.

EXAMPLE 14.7 CASE STUDY Margin of Safety

The yellow airplane in the case study hangs from two steel cables (Fig. 14.14, page 399). Cable 1 has a diameter of 1.2 cm, and cable 2 has a diameter of 0.3 cm. Find the maximum weight of a plane that can be hung from these two cables and compare your answer with the airplane's actual weight (9.3×10^3 N), which we found in Example 14.3.

INTERPRET and ANTICIPATE

When the tension per unit area in one of the cables equals the tensile strength of steel, that cable will break. So, we should find the maximum tension in each cable from the tensile strength of steel and the cross-sectional area of the cable. We can then use $F_{1,\,\text{max}} + F_{2,\,\text{max}} = m_{\text{max}}g$ (Eq. 1 in Example 14.3) to find the maximum weight of a plane that could be hung from these cables. We expect the maximum weight to be several times greater than the actual weight of the airplane.

SOLVE

The tensile strength is the maximum tensile stress. Use $\sigma = F/A$ (Eq. 14.9) to find the maximum tension in the cables.	$F_{1,\,\text{max}} = \sigma_{\text{ten}}A_1$ $F_{2,\,\text{max}} = \sigma_{\text{ten}}A_2$
Write the cross-sectional area in terms of each cable's diameter, $A = \pi(d/2)^2$.	$F_{1,\,\text{max}} = \sigma_{\text{ten}}\pi\left(\frac{d_1}{2}\right)^2$ $F_{2,\,\text{max}} = \sigma_{\text{ten}}\pi\left(\frac{d_2}{2}\right)^2$

Consult Table 14.1 to find the tensile strength of steel.	$F_{1,\,\text{max}} = (380 \times 10^6\,\text{N/m}^2)\pi\left(\frac{1.2 \times 10^{-2}\,\text{m}}{2}\right)^2 = 4.3 \times 10^4\,\text{N}$ $F_{2,\,\text{max}} = (380 \times 10^6\,\text{N/m}^2)\pi\left(\frac{0.3 \times 10^{-2}\,\text{m}}{2}\right)^2 = 2.7 \times 10^3\,\text{N}$

The maximum weight of the plane equals the maximum tension in the cables.	$F_{g,\,max} = F_{1,\,max} + F_{2,\,max}$ $F_{g,\,max} = (4.3 \times 10^4\ N) + (2.7 \times 10^3 N) = 4.57 \times 10^4\ N$
For the required comparison, divide the maximum weight that can be supported by the airplane's actual weight.	$\dfrac{F_{g,\,max}}{F_g} = \dfrac{4.57 \times 10^4\ N}{9.30 \times 10^3\ N} = 4.91$

CHECK and THINK

As expected, the maximum weight that the cables can support is several times the weight of the hanging airplane. It is common for designers to include this sort of safety margin in their plans.

You might be wondering if each cable is designed to support roughly five times what it actually supports. To answer that question, divide the maximum tension in each cable found in this example by the actual tension in each cable found in Example 14.3. We find that both cables have the same margin of safety.	$\dfrac{F_{1,\,max}}{F_1} = \dfrac{4.3 \times 10^4\ N}{8.7 \times 10^3\ N} = 4.9$ $\dfrac{F_{2,\,max}}{F_2} = \dfrac{2.7 \times 10^3 N}{5.50 \times 10^2 N} = 4.9$

In this chapter, we looked at a special case of motion, that is, no motion at all. At first, it may seem like a boring case when you compare static equilibrium to motion of stars as they are devoured by a giant black hole or to the accurate dive made by a hawk pursuing its prey or to the flight of the yellow plane before it was hung on display in the Udvar-Hazy Center (Fig. 14.14). When we are standing underneath a plane suspended from the ceiling by a few steel cables, however, static equilibrium is exactly what we want. Structural engineers have an important job in ensuring static equilibrium. In fact, every day we are surrounded by structures that are designed to be in static equilibrium, and if they fail to remain in static equilibrium, the consequences can be tragic. The job of a structural engineer is made more complicated because structures are never truly in static equilibrium. Structures are stretched, compressed, and sheared during the course of normal use and during extraordinary situations such earthquakes and major storms. In Chapter 19, we return to these ideas when we see how changes in temperature can further complicate an engineer's work.

SUMMARY

Underlying Principle

Equilibrium is a special case of Newton's second law, and **static equilibrium** is a special case of equilibrium. An object is in equilibrium if both its translational and angular accelerations are zero. An object in *static* equilibrium is at rest relative to an observer.

Major Concepts

1. There are three types of static equilibrium:
   a. If an object is in **stable static equilibrium** and it is displaced by a small amount, the forces acting on it will return it to its equilibrium position.
   b. If an object is in **unstable static equilibrium** and it is displaced by a small amount, the forces acting on it will move it away from its equilibrium position.
   c. The forces exerted on an object in **neutral static equilibrium** do not maintain its equilibrium, so if such an object is displaced, it is neither restored to its equilibrium position nor is it forced farther away.

2. There are two equilibrium conditions:
   a. According to the **force balance condition**, the net force acting on the object must be zero:

$$\vec{F}_{tot} = 0 \quad (14.1)$$

   b. According to the **torque balance condition**, the net torque acting on the object must be zero:

$$\vec{\tau}_{tot} = 0 \quad (14.2)$$

3. **Stress** is the magnitude of the force applied to an object per unit area.
   a. **Tensile stress** σ results when a force pulls perpendicular to a side of a stationary object.
   b. **Compression stress** σ results when a force pushes perpendicular to a side of a stationary object. For a tensile or compression stress,

$$\sigma \equiv \frac{F}{A} \quad (14.9)$$

   c. **Shear stress** τ results when a force is applied tangentially to one surface of a stationary object:

$$\tau = \frac{F}{A} \quad (14.12)$$

4. **Strain** is a dimensionless measure of the object's deformation due to shear. The fractional increase (or decrease) in an object's length $\Delta\ell/\ell$ due to a tensile (or compression) shear is known simply as the strain,

$$\varepsilon = \Delta\ell/\ell \quad (14.10)$$

The **shear strain** γ that results from a shear stress is given by

$$\gamma = \frac{\Delta x}{h} \quad (14.13)$$

where Δx is the amount that the sheared face moves and h is the height of that face.

5. As long as the tensile (or compressive) stress σ is small, the strain is directly proportional to the stress:

$$\sigma = Y\varepsilon \quad (14.11)$$

where Y is known as **Young's modulus**.

6. An **elastic** object returns to its original length after a load has been removed. The **elastic limit** is the maximum stress that can be placed on an object without permanently deforming it.
7. The **tensile strength** σ_{ten} of a material is the maximum tensile stress that may be applied to an object made of that material without fracturing it. The **compressive strength** σ_{cmp} of a material—the stress at which the object fractures under compression—does not necessarily equal the tensile strength σ_{ten}.
8. For small shear stresses τ, the shear strain γ is proportional to the stress:

$$\tau = G\gamma \quad (14.14)$$

where G is called the **shear modulus**.

Tools

Cross Products

The cross product of two vectors $\vec{R} = \vec{A} \times \vec{B}$ may be expressed in component form as

$$\vec{R} = (A_yB_z - A_zB_y)\hat{\imath} + (A_zB_x - A_xB_z)\hat{\jmath} + (A_xB_y - A_yB_x)\hat{k} \quad (14.5)$$

Using Equation 14.5 to express the torque is often helpful:

$$\vec{\tau} = \vec{r} \times \vec{F} = (r_yF_z - r_zF_y)\hat{\imath} + (r_zF_x - r_xF_z)\hat{\jmath} + (r_xF_y - r_yF_x)\hat{k} \quad (14.7)$$

PROBLEM-SOLVING STRATEGY Static Equilibrium

INTERPRET and ANTICIPATE

Draw a modified free-body diagram.

1. Represent the object as a **simple shape**.
2. Draw all the **forces** exerted on the object at each **point of application**.
3. Choose a **rotation axis**.
4. Choose a **coordinate system**.
5. Draw **position vectors**.

SOLVE

1. Write the **forces and position vectors in component form** (or list magnitudes).
2. Apply **equilibrium conditions**.
   a. Apply the force balance condition.
   b. Find the torques and **apply the torque balance condition**.
3. **Do algebra** before substituting values.

PROBLEMS AND QUESTIONS

A = algebraic **C** = conceptual **E** = estimation **G** = graphical **N** = numerical

14-1 What Is Static Equilibrium?

Problems 1–3 are grouped.

1. **C** A ball is attached to a strong, lightweight rod (Fig. P14.1). The rod is supported by a horizontal pin near the top. The ball is at rest. Is the ball in static equilibrium? If not, why not? If so, which type of equilibrium is it—stable, unstable, or neutral? *Hint*: What would happen if you displaced the ball slightly?

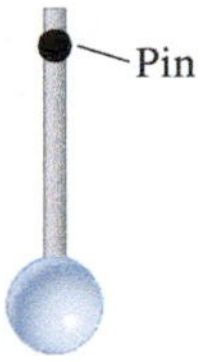

FIGURE P14.1

2. **C** A ball is attached to a strong, lightweight rod (Fig. P14.2). The rod is supported by a horizontal pin near the bottom. The ball is at rest. Is the ball in static equilibrium? If not, why not? If so, which type of equilibrium is it—stable, unstable, or neutral? *Hint*: What would happen if you displaced the ball slightly?

FIGURE P14.2

3. **C** Two identical balls are attached to a strong, lightweight rod (Fig. P14.3). The rod is supported by a horizontal pin at the system's center of mass. The balls are at rest. Is the system in static equilibrium? If not, why not? If so, which type of equilibrium is it—stable, unstable, or neutral? *Hint*: What would happen if you displaced the balls slightly?

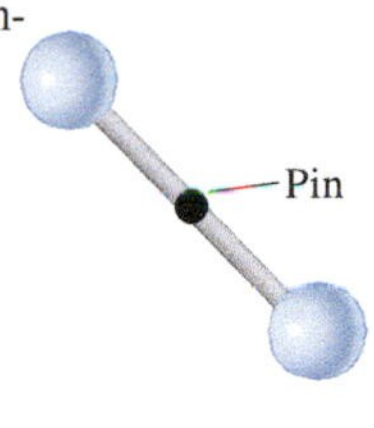

FIGURE P14.3

4. **C** While working on homework together, your friend stands her pencil on its sharpened tip. You say, "Nice trick. How did you do it?" Why do you think some trick had to be involved? Would you ask the same question if the pencil were lying on its side or if the pencil were unsharpened?
5. **C** Consider the sketch of a portion of a roller-coaster track seen in Figure P14.5. Identify places on the track that could be considered possible locations of static equilibrium for a roller-coaster car were the car to be placed at any spot on the track. Which places are candidate locations for stable, unstable, and neutral static equilibrium?

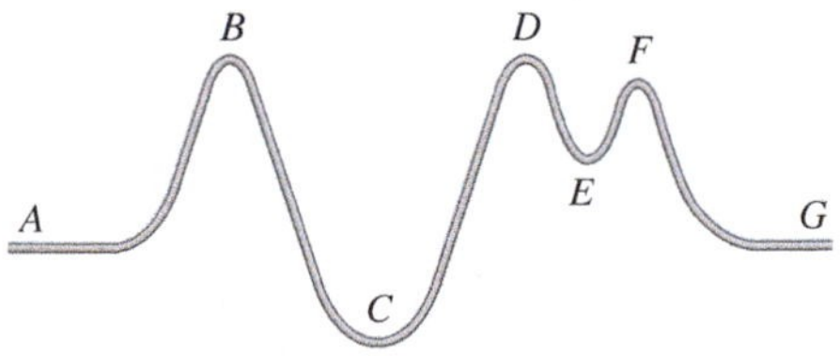

FIGURE P14.5

14-2 Conditions for Equilibrium

6. **N Review** Where is the center of mass of the L-shaped uniform metal object shown in Figure P14.6?

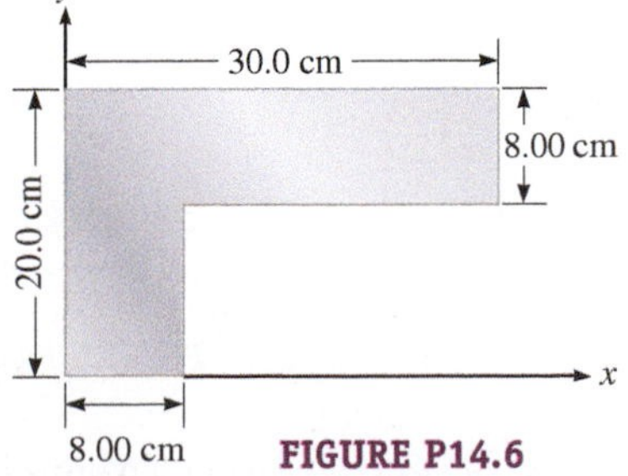

FIGURE P14.6

7. **N** A heavy iron gate is stuck closed. Opening the gate requires pushing hard along the y axis and lifting at the same time along the z axis. A force of $(920\hat{\jmath} + 110\hat{k})$ N is applied to the top right corner of the gate such that the position vector from the axis of rotation to the top right corner is given by $\vec{r} = 3.0\hat{\imath}$ m. What is the torque on the gate due to this force?
8. Suppose you have two vectors, $\vec{A} = 7.95\hat{\imath}$ and $\vec{B} = 5.13\hat{\jmath}$.
   a. **N** What is the vector cross product of these two vectors, $\vec{R} = \vec{A} \times \vec{B}$?
   b. **C** Given the direction of the resultant vector and the directions of the two original vectors, what can you say in general about the relationships between the three vectors involved in the cross product?
   c. **N** Now suppose the two vectors are $\vec{A} = 7.95\hat{\imath}$ and $\vec{B} = 3.00\hat{\imath} + 5.13\hat{\jmath}$. What is their cross product in this case?
   d. **C** Given the direction of the resultant vector and the directions of the two original vectors, can you further generalize the relationship between the three vectors involved in the cross product? How are three vectors necessarily oriented relative to one another?
9. **A** The keystone of an arch is the stone at the top (Fig. P14.9). It is supported by forces from its two neighbors, blocks A and B. Each block has mass m and approximate length L. What can you conclude about the force exerted by each block, $\vec{F}_A$ and $\vec{F}_B$, for the keystone to remain in static equilibrium? That is, show that the equilibrium conditions are satisfied by the components of the forces F_{Ax}, F_{Ay}, F_{Bx}, and F_{By}. Assume the arch is symmetric.

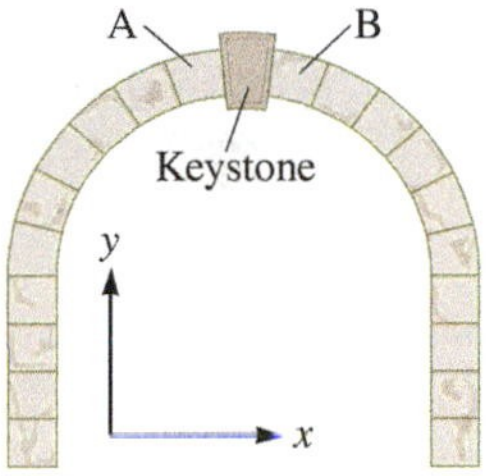

FIGURE P14.9

Problems 10 and 11 are paired.

10. **C** Stand straight and comfortably with your feet about shoulders' width apart. Your center of mass is located somewhere in your abdomen directly above the point between your feet. Now bend forward, hinging from your hips. What do you notice about the motion of your lower body? Explain why it moves. If you have trouble noticing any motion, hold a dumbbell or other heavy object in each hand. As you bend forward, allow your hands to move toward the ground.
11. **C** Stand straight and comfortably with your feet about shoulders' width apart. Now lift one of your legs so that you are balanced on only one foot. Notice how your body moves. Explain why it moves.
12. **N Review** Three uniform spheres are placed in the xy plane as follows: The center of mass of an 8.00-kg sphere is located at (2.00, 2.00) m, the center of mass of a 5.00-kg sphere is located at (0, 5.00) m, and the center of mass of a 2.00-kg sphere is located at (3.00, 0) m. Where should a 6.00-kg sphere be placed if the center of mass of the four-sphere system is to be located at the origin?

Problems 13 and 14 are paired.

13. **A** The uniform block of width W and height H in Figure P14.13 rests on an inclined plane in static equilibrium. The plane is slowly raised until the block begins to tip over. Assume the tipping occurs before the block starts to slide. Find an expression for the angle θ at which the block is on the verge of tipping.

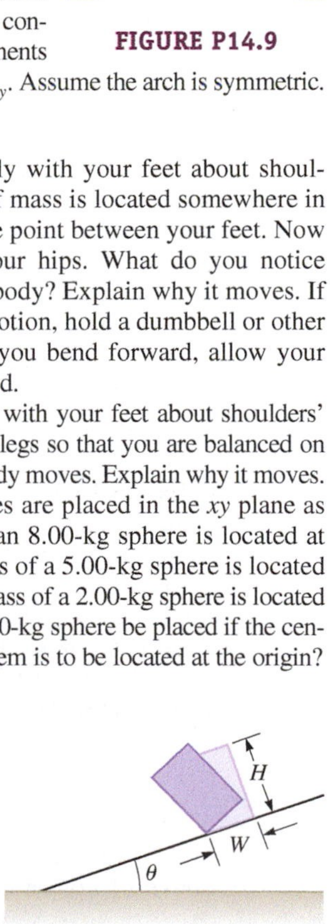

FIGURE P14.13 Problems 13 and 14.

14. **A** If the block in Problem 13 tips before it slides, determine the minimum coefficient of static friction.

15. **N** Find the cross product $\vec{R} = \vec{A} \times \vec{B}$, where

$$\vec{A} = (23.45\hat{\imath} + 106.3\hat{\jmath} - 92.81\hat{k}) \text{ and}$$
$$\vec{B} = (-5.010\hat{\imath} + 2.146\hat{\jmath} - 3.114\hat{k}).$$

Give your answer in component form and also state the magnitude of the result.

16. **A** Is the vector cross product commutative? In other words, does $\vec{A} \times \vec{B} = \vec{B} \times \vec{A}$? Expound on your answer with a mathematical proof or prove that it is not true by example.

Problems 17–19 are grouped.

17. A force $\vec{F} = (65.8\hat{\imath} + 96.3\hat{\jmath})$ N is exerted on an object at a point whose position from the axis of rotation is given by $\vec{r} = -5.80\hat{\imath}$ m.
   a. N Find the torque exerted by this force.
   b. C What happens to your answer if the force's x component increases while the y component remains constant?
   c. C What happens to your answer if the force's y component increases while the x component remains constant?

18. A force $\vec{F} = (65.8\hat{\imath} + 96.3\hat{\jmath})$ N is exerted on an object at a point whose position from the axis of rotation is given by $\vec{r} = -5.80\hat{k}$ m.
   a. N Find the torque exerted by this force.
   b. C What happens to your answer if the force's x component increases while the y component remains constant?
   c. C What happens to your answer if the force's y component increases while the x component remains constant?

19. **N** A force $\vec{F} = (65.8\hat{\imath} + 96.3\hat{k})$ N is exerted on an object, and the torque that results is given by $\vec{\tau} = (57.1\hat{\jmath})$ N · m. Find $\vec{r} = r_z\hat{k}$.

Problems 20 and 21 are paired.

20. **N** A man poses for a photo while holding a dumbbell. His arm is bent as shown in Figure P14.20. The torque exerted by the dumbbell around his elbow is 38.9 N · m. What is the mass of the dumbbell?

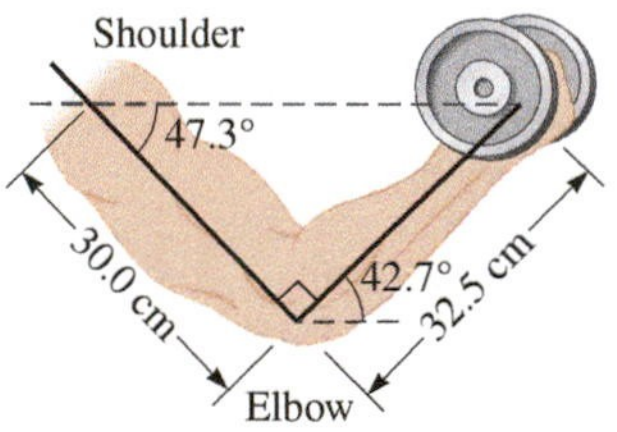

FIGURE P14.20 Problems 20 and 21.

21. **N** If the torque exerted by the dumbbell around the man's elbow in Figure P14.20 is 38.9 N · m, what is the torque exerted by the dumbbell around his shoulder?

14-3 Examples of Static Equilibrium

22. **N** The inner planets of our solar system are represented on a mobile constructed from drinking straws and light strings for a school project (Fig. P14.22). The mass of the piece representing the Earth is 25.0 g, and the mass of the straws can be ignored. (The lengths shown are not proportional to actual distances in the solar system.) What is the mass of the piece representing **a.** Mars, **b.** Venus, and **c.** Mercury?

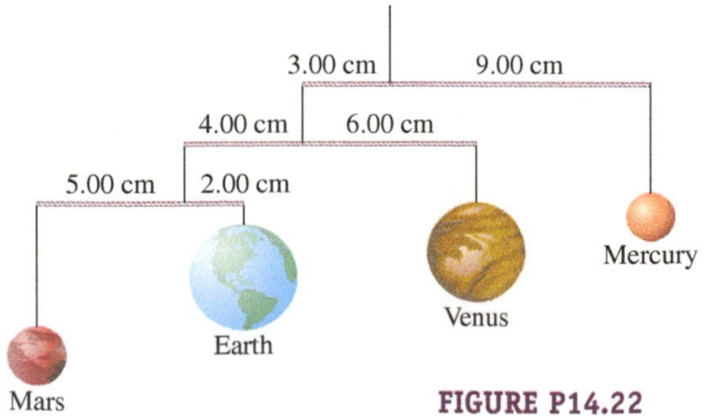

FIGURE P14.22

23. **N** Two Boy Scouts, Bobby and Jimmy, are carrying a uniform wooden tent pole 4.50 m in length and having mass $m = 23.0$ kg. Bobby is holding the pole 60.0 cm from the back, and Jimmy is holding the pole 40.0 cm from the front. If the pole is held horizontally by the two boys, what is the force Bobby and Jimmy each exert on the beam?

24. **G** Draw a modified free-body diagram for the bottle-holder system shown in Figure P14.24. Notice that the wine bottle is horizontal. *Hint*: Draw a line that passes through the system's center of mass.

FIGURE P14.24

Problems 25 and 26 are paired.

25. A painter of mass 87.8 kg is 1.45 m from the top of a 6.67-m ladder. The ladder rests against a wall. Friction between the ladder and the wall is negligible. The ladder's mass is 5.50 kg. Assume the ladder is set up according to Occupational Safety and Health Administration standards so that it makes a 75.5° angle with the floor.
   a. N What is the minimum coefficient of static friction required for the painter's safety?
   b. C Would rubber ladder tips on dry concrete be safe?

26. Consider the situation in Problem 25. Tests have shown that when people try to eyeball the proper angle for the ladder, they are often off by as much as 9°. Suppose the painter sets up the ladder at too small an angle.
   a. N At what maximum angle would rubber ladder tips on dry concrete be unsafe?
   b. C How likely is it for a person to make such a dangerous mistake?

27. **N** Children playing pirates have suspended a uniform wooden plank with mass 15.0 kg and length 2.50 m as shown in Figure P14.27. What is the tension in each of the three ropes when Sophia, with a mass of 23.0 kg, is made to "walk the plank" and is 1.50 m from reaching the end of the plank?

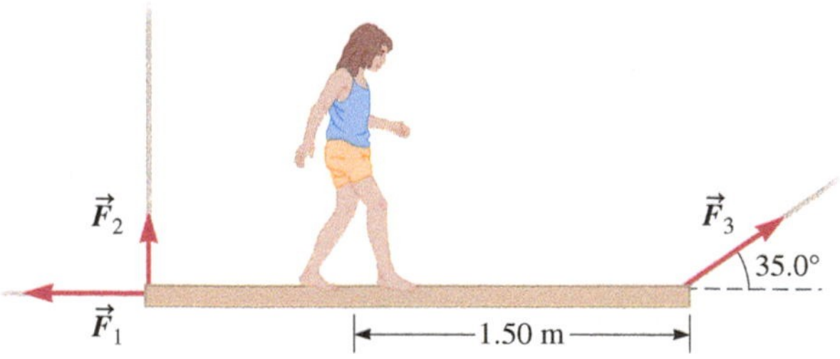

FIGURE P14.27

28. **E** CASE STUDY After having worked through the case study, you decide to visit the Udvar-Hazy Center at the National Air and Space Museum. While there, you spend more time

thinking about how the exhibits are displayed than you did on prior visits. You see the aircraft in Figure P14.28A, and you think it would be much nicer if the front supports were replaced by a single cable. You imagine that you wouldn't need to drill a hole into the top of the aircraft. Instead, you could attach one end of the cable by slipping a small yellow ring over the nose, and you could attach the other end to a vertical steel column. Your plan is shown in Figure P14.28B. Estimate the tension in the cable by considering the dimensions and weight of the plane in the case study. Keep in mind that the aircraft in Figure P14.28 is smaller than the one in the case study.

A. B.

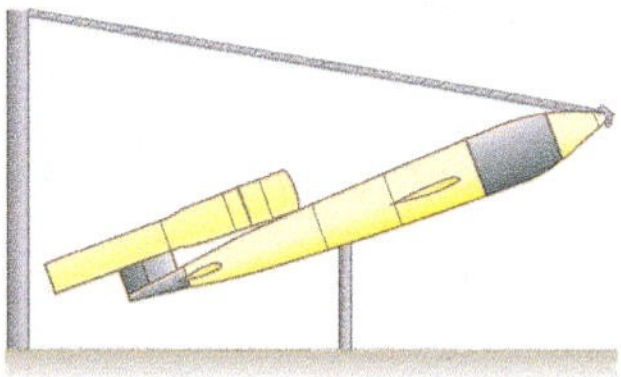

FIGURE P14.28

29. N A uniform beam rests on a cart with a portion of one end hanging out (Fig. P14.29). The overhanging portion is one-fourth of the total length L of the beam. A gradually increasing force F acts on the end of the beam at point P. The opposite side of the beam begins to rise when the magnitude of the force reaches 2.0×10^2 N. What is the beam's weight?

FIGURE P14.29

30. N A 5.45-N beam of uniform density is 1.60 m long. The beam is supported at an angle of 35.0° by a cable attached to one end. There is a pin through the other end of the beam (Fig. P14.30). Use the values given in the figure to find the tension in the cable.

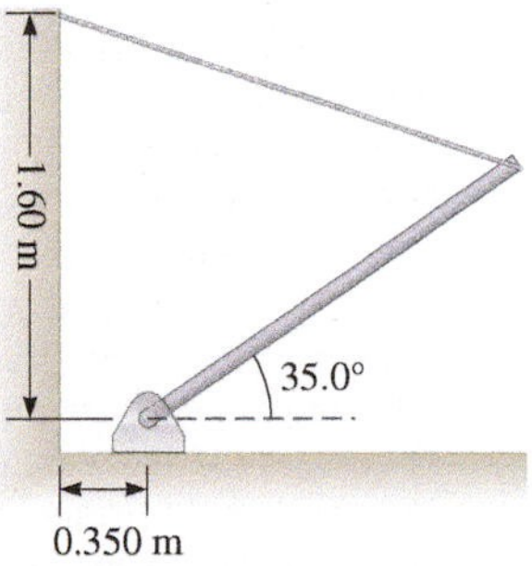

FIGURE P14.30

31. N A wooden door 2.1 m high and 0.90 m wide is hung by two hinges 1.8 m apart. The lower hinge is 15 cm above the bottom of the door. The center of mass of the door is at its geometric center, and the weight of the door is 260 N, which is supported equally by both hinges. Find the horizontal force exerted by each hinge on the door.

32. N A 215-kg robotic arm at an assembly plant is extended horizontally (Fig. P14.32). The massless support rope attached at point B makes an angle of 15.0° with the horizontal, and the center of mass of the arm is at point C. **a.** What is the tension in the support rope? **b.** What are the magnitude and direction of the force exerted by the hinge A on the robotic arm to keep the arm in the horizontal position?

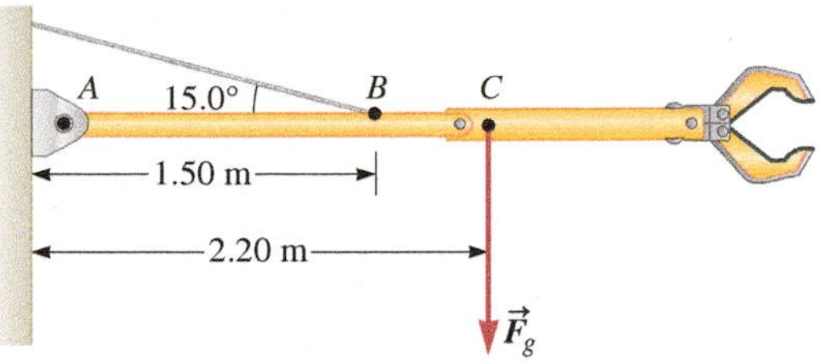

FIGURE P14.32

Problems 33 and 34 are paired.

33. N One end of a uniform beam that weighs 2.80×10^2 N is attached to a wall with a hinge pin. The other end is supported by a cable making the angles shown in Figure P14.33. Find the tension in the cable.

34. N For the uniform beam in Problem 33, find the horizontal and vertical components of the force exerted by the hinge pin on the beam.

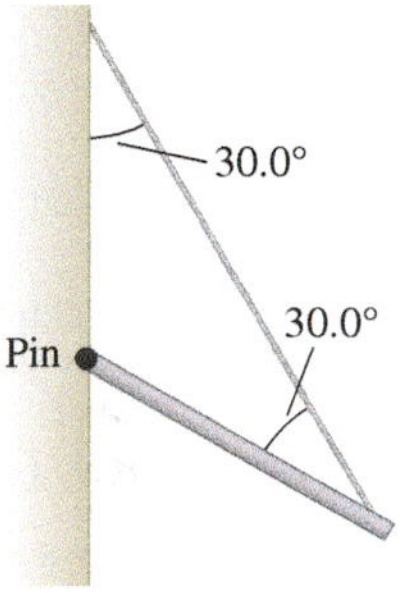

FIGURE P14.33 Problems 33 and 34.

35. N The owner of the Galaxy Café wishes to suspend a fictional spacecraft outside the front door (Fig. P14.35). The spacecraft weighs 1000 lb. The triangular structure is much lighter than the spacecraft, and its weight can be neglected. Use the values for length given in the figure and find the normal forces exerted by the pins at points A and B.

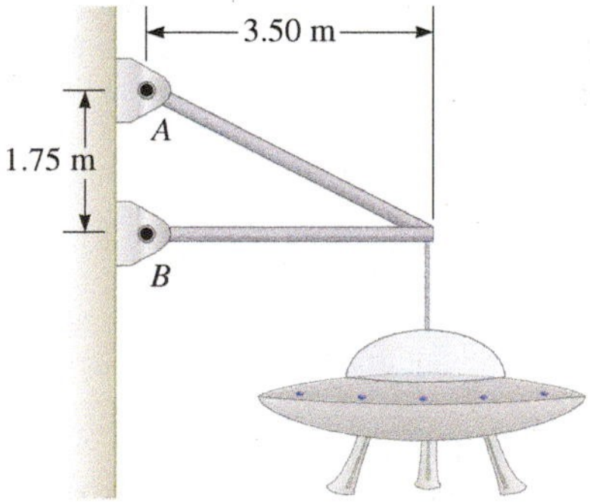

FIGURE P14.35

36. N A square plate with sides of length 4.0 m can rotate about an axle passing through its center of mass and perpendicular to the plate as shown in Figure P14.36. There are four forces acting on the plate at different points. The rotational inertia of the plate is 24 kg · m². Is the plate in equilibrium?

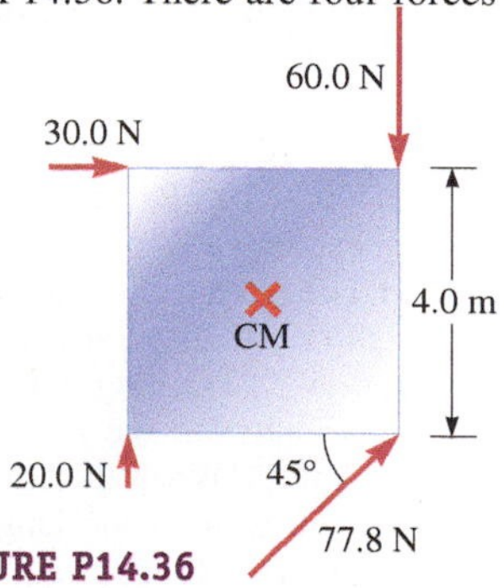

FIGURE P14.36

37. **N** The circular sign in Figure P14.37 has a mass of 16.7 kg. Use the angle shown in the figure to find the tension in the cable.

FIGURE P14.37

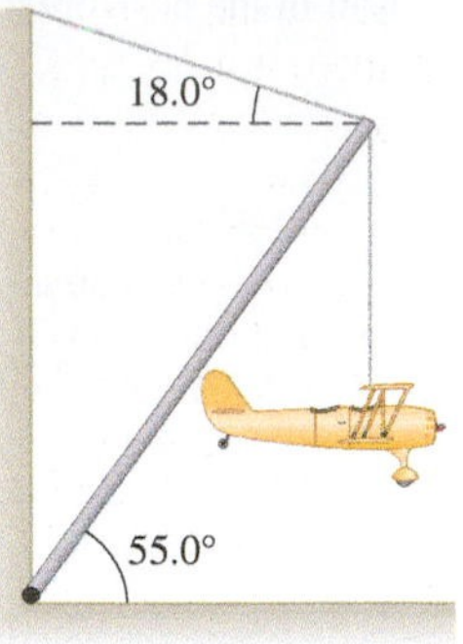

FIGURE P14.38

38. **N** At a museum, a 1300-kg model aircraft is hung from a lightweight beam of length 12.0 m that is free to pivot about its base and is supported by a massless cable (Fig. P14.38). Ignore the mass of the beam. **a.** What is the tension in the section of the cable between the beam and the wall? **b.** What are the horizontal and vertical forces that the pivot exerts on the beam?

14-4 Elasticity and Fracture

39. **N** A uniform wire ($Y = 2.0 \times 10^{11}$ N/m^2) is subjected to a longitudinal tensile stress of 4.0×10^7 N/m^2. What is the fractional change in the length of the wire?

40. **C** A brass wire and a steel wire, both of the same length, are extended by 1.0 mm under the same force. Is the cross-sectional radius of the brass wire more, less, or equal to the cross-sectional radius of the steel wire? Explain. Young's moduli for brass and steel are 1.0×10^{10} N/m^2 and 2.0×10^{11} N/m^2, respectively.

41. **N** CASE STUDY In Example 14.3, we found that one of the steel cables supporting an airplane at the Udvar-Hazy Center was under a tension of 9.30×10^3 N. Assume the cable has a diameter of 2.30 cm and an initial length of 8.00 m before the plane is suspended on the cable. How much longer is the cable when the plane is suspended on it?

Problems 42–44 are grouped.

42. **N** A carbon nanotube is a nanometer-scale cylindrical tube composed of carbon atoms. One of its interesting properties is a very large Young's modulus, measured to be more than 1.000×10^{12} N/m^2. A tensile stress of 5.3×10^{10} N/m^2 is exerted on a particular nanotube with a Young's modulus measured at 2.130×10^{12} N/m^2. Assuming the nanotube obeys Hooke's law, by what percentage does the atomic spacing increase?

43. **N** A nanotube with a Young's modulus of 1.000×10^{12} Pa is subjected to a stress of 3.14×10^{11} Pa by being pulled at its ends. Assuming the tube had an initial length of 8.12×10^{-6} m, what is the new length of the nanotube?

44. **N** Consider a nanotube with a Young's modulus of 2.130×10^{12} N/m^2 that experiences a tensile stress of 5.3×10^{10} N/m^2. Steel has a Young's modulus of about 2.000×10^{11} Pa. How much stress would cause a piece of steel to experience the same strain as the nanotube?

45. **N** The spring in Figure P14.45 has a spring constant of 7.50×10^2 N/m, and it is relaxed when the rod is vertical. The rod is 2.50 m long, with uniform density. The horizontal distance between the pin in the spring and the pin in the rod is 6.75 m. **a.** When the bar is tilted by 41.6° from the vertical as shown, what is the length of the spring? **b.** If the system in Figure P14.45 is in equilibrium, what is the weight of the bar?

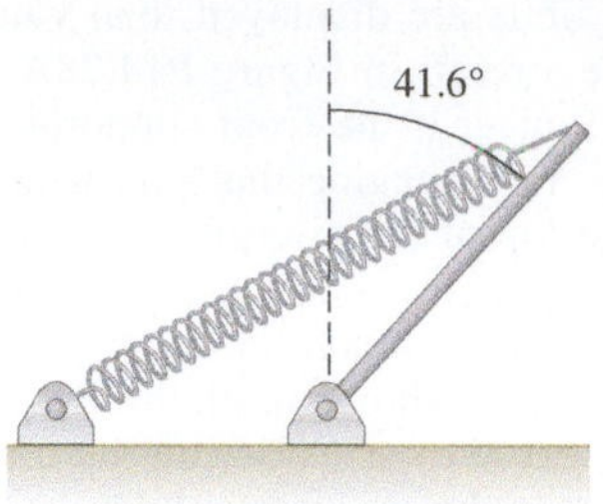

FIGURE P14.45

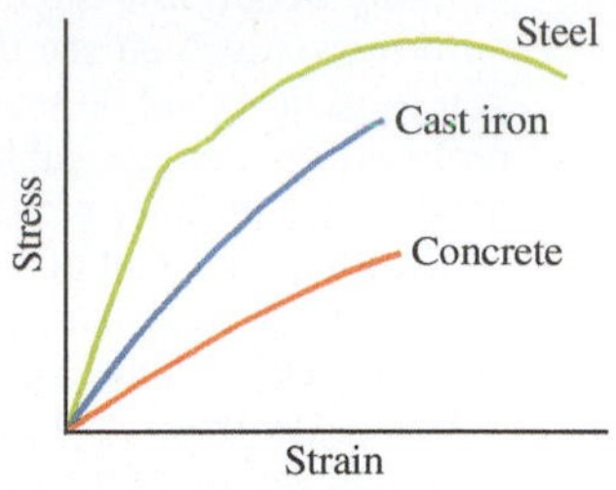

FIGURE P14.46

46. **C** Use the graph in Figure P14.46 to list the three materials from greatest Young's modulus to smallest. Explain your reasoning.

47. **N** A 5.00-m-long copper wire with a diameter of 5.00 mm and Young's modulus 11.0×10^{10} N/m^2 is used to suspend a large bucket of mass 355 kg. What is the increase in the wire's length due to this load?

48. **G** A company is testing a new material made of recycled plastic for use in construction. Figure P14.48 is a graph of the stress as a function of strain for this material. Find Young's modulus for this material and compare your result with the data for conventional building materials listed in Table 14.1.

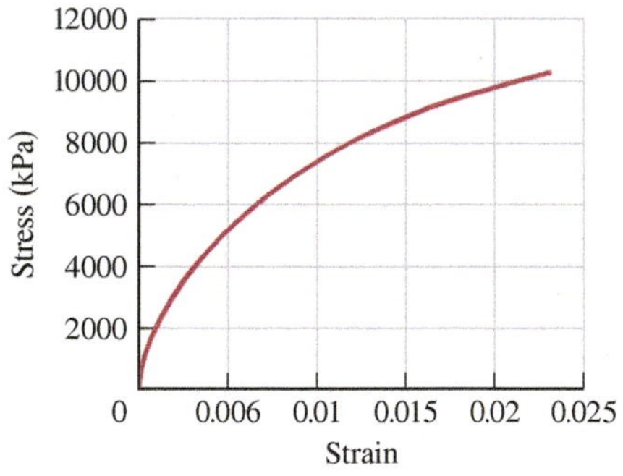

FIGURE P14.48

49. **N** A tubular steel support is 3.0 m tall and shortens 0.60 mm under a compressive force of 70 kN. If the inner radius of the tube is 0.80 times the outer radius, what is the outer radius? Young's modulus for steel is 2.0×10^{11} Pa.

Problems 50–53 are grouped.

50. **N** A steel rod has a radius of 15.0 cm and a length of 1.25 m. The rod is held firmly in place, and a 32.5-kN force is applied perpendicular to one face of the rod so that it compresses. What is the strain of the rod?

51. **N** An aluminum rod has a radius of 15.0 cm and a length of 1.25 m. The rod is held firmly in place, and a 32.5-kN force is applied perpendicular to one face of the rod so that it compresses. What is the strain of the rod? (If you worked Problem 50, compare your answers.)

52. **N** A steel rod has a radius of 15.0 cm and a length of 1.25 m. The rod is held firmly in place, and a 32.5-kN force is applied perpendicular to one face of the rod so that it stretches. What is the strain of the rod? (If you worked Problem 50, compare your answers.)

53. **N** A steel rod has a radius of 15.0 cm and a length of 1.25 m. The rod is held firmly in place, and a 32.5-kN force is applied parallel to one face of the rod so that it is sheared. What is the (shear) strain of the rod? (If you worked Problem 50, compare your answers.)

54. **N** A structural piece of material consists of a solid cylindrical rod 2.00 m in length with one half having a diameter of 1.00 cm and the other half having a diameter of 0.750 cm. It is made from an alloy with a Young's modulus of 1.30×10^{11} N/m². If the piece is put under a tensile force of 15 kN, by how much does it elongate?

55. **N** While trying to catch a badly thrown basketball, Kobe Bryant briefly slides across the basketball court. The court exerts a frictional force of 25.0 N on each foot. The footprint of his size 14 shoes has an area of 420 cm², and the rubber soles (shear modulus 6.00×10^5 N/m²) are 19.0 mm tall. What is the horizontal offset of the upper and lower surfaces of the shoe soles?

56. **A** A metal wire of length L is stretched an amount ΔL by a weight F_g. What is the fractional change in the wire's volume, $\Delta V/V$?

57. **A** A copper rod with length 1.4 m and cross-sectional area 2.0 cm² is fastened to a steel rod of length L and cross-sectional area 1.0 cm². The compound structure is pulled on each side by two forces of equal magnitude 6.00×10^4 N (Fig. P14.57). Find the length L of the steel rod if the elongations (ΔL) of the two rods are equal. Use the values $Y_{steel} = 2.0 \times 10^{11}$ Pa and $Y_{Cu} = 1.1 \times 10^{11}$ Pa.

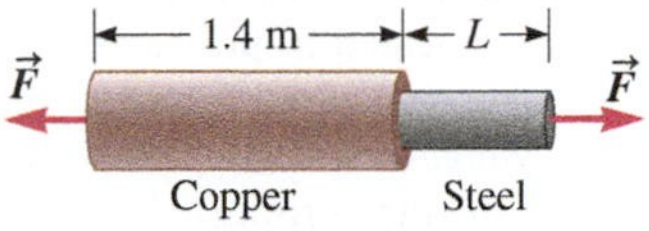

FIGURE P14.57

General Problems

58. **A, C** In Example 14.5, it was stated that at the maximum thrust the front two feet of the fan are just lifting off the floor; in this problem, you'll show that this statement is true. Return to Figure 14.17 (page 403), which shows all 10 forces exerted on the fan. Write down the force balance and the torque balance equations. Then, reason that as F_{thrust} approaches its maximum, F_{N3} and F_{N4} both approach zero.

59. **N** **Review** A uniform circle of mass $m_1 = 3.00$ kg, a uniform triangle of mass $m_2 = 1.00$ kg, and a uniform square of mass $m_3 = 6.00$ kg are placed in the xy plane as shown in Figure P14.59. What are the x and y coordinates of the center of mass of the three-object system?

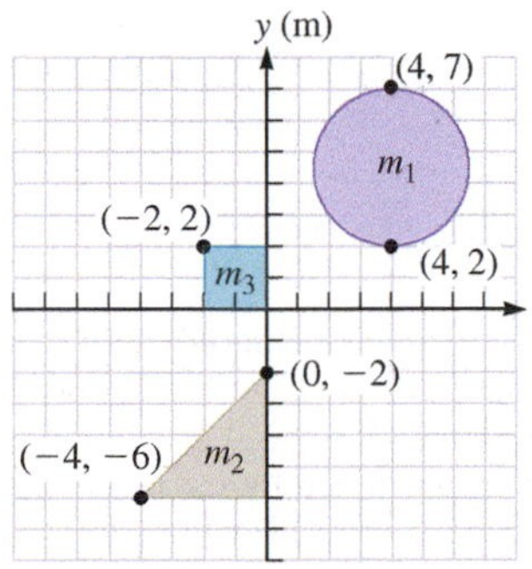

FIGURE P14.59

60. **N** Bruce Lee was famous for breaking concrete blocks with a single karate chop. From slow-motion video, the speed of his 1.50-kg hand descending on a block was estimated to be 15.0 m/s, which decreased to a speed of 0.500 m/s in the 2.50×10^{-3} s during which his hand made contact with and broke through the block. The maximum shear stress a concrete block can be subjected to before breaking is 9.50×10^5 N/m². **a.** What was the force exerted by Lee's hand on the block? **b.** If a typical concrete block broken by Lee was 3.00 cm thick and 15.2 cm wide, what is the shear stress experienced by the concrete block? **c.** Will the concrete block succumb to Lee's karate chop?

Problems 61 and 62 are paired.

61. **A** A disk of radius R and mass M is rotating and experiencing an angular acceleration $\vec{\alpha} = \alpha\hat{k}$. Define the additional torque that must be applied for the disk to be in equilibrium. Write your answer as a vector in terms of α, R, and M.

62. A disk of radius R and mass M is rotating and experiencing an angular acceleration $\vec{\alpha} = \alpha\hat{k}$. There is actually a range of possible values for the magnitude of a single force that could be applied for the disk to be in equilibrium.

a. **N** What is the minimum magnitude of a single force that, when applied, causes the disk to be in equilibrium?

b. **C** Explain why there is a range of possible values for this force and what bounds you can place on its magnitude.

63. **A** Show that when an object is suspended by a single rope, the object's center of mass must be someplace along the line that extends through the length of the rope.

64. **A** One end of a metal rod of weight F_g and length L presses against a corner between a wall and the floor (Fig. P14.64). A rope is attached to the other end of the rod. Find the magnitude of the tension in the rope if the angle β between the rod and the rope is 90°.

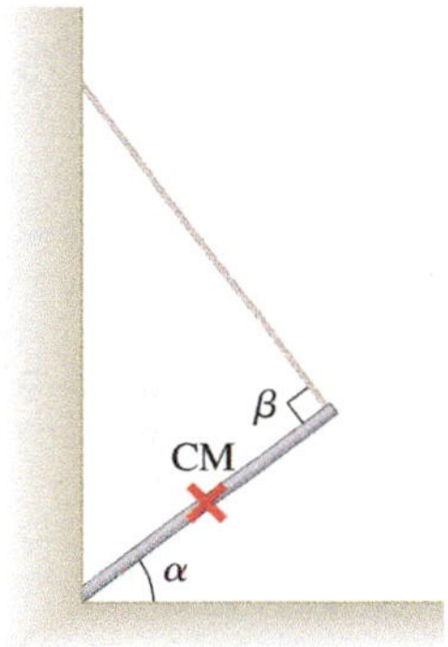

FIGURE P14.64

65. **N** A 150-kg pile driver starts from rest and impacts a steel beam 15.0 cm in radius that is 7.60 m below its original position. The pile driver rebounds with half its speed before impact after 0.065 s. What is the average strain in the beam during the impact?

66. **N** A steel cable 2.00 m in length and with cross-sectional radius 0.350 mm is used to suspend from the ceiling a 10.0-kg model aircraft that is flying in a horizontal circle with an angular speed of 6.00 rad/s. What is the strain produced in the cable?

Problems 67 and 68 are paired.

67. **N** While working out at the gym, a Marine finds that the fan (Fig. 14.16, page 402) is too low to keep her cool. She gets the idea to hang the fan from two cables. She removes its pedestal, and the fan's remaining mass is 10.3 kg. When the fan is off, the cables are vertical (Fig. P14.67A). When the fan is on, the cables make an angle θ with the vertical (Fig. P14.67B). She attaches the cables to the fan using pins that allow the fan to remain in the same orientation when it is operating. Assume the fan is exerting the maximum thrust (21.0 N). Use any of the necessary values given and derived in Example 14.5 to find θ.

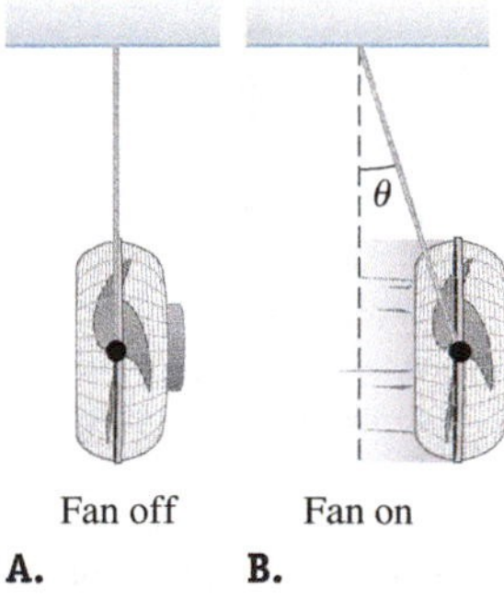

FIGURE P14.67 Problems 67 and 68.

68. **N** Return to the scenario described in Problem 14.67 and use any of the necessary values given and derived in Example 14.5 (page 402) to find the tension in each cable.

69. **A** A dresser of mass m is pushed with force F_P at height h above the ground (Fig. P14.69). The dresser has width W and total height H, and the center of mass is exactly in the middle of the dresser. If the

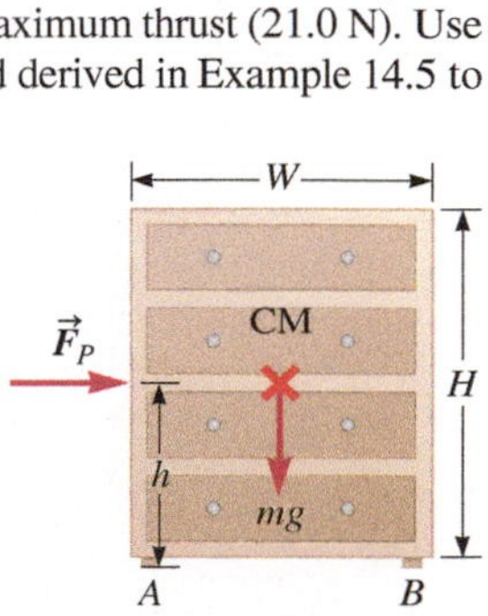

FIGURE P14.69

force is increased, the dresser can be tipped over. At the largest force we can apply before the dresser tips, the dresser is in static equilibrium, and the force between the back legs, labeled A, and the ground is zero. Using this fact, determine an equation for the force required to tip over the dresser in terms of the variables defined above.

70. **N** Two metal wires of the same material—lengths L and $3L$, respectively, and diameters $3D$ and D, respectively—are under different tensions that stretch each by the same amount ΔL. What is the ratio of the two tensions in the wires?

71. **N** Brianna is bored while at lunch and begins to play with her food. She has a block of lime gelatin dessert and pushes on the block parallel to the top surface. She notices that the strain experienced by the gelatin was about 0.150 and that the applied stress was about 100.0 Pa. What is the shear modulus of lime gelatin?

Problems 72 and 73 are paired.

72. **N** A steel rod has a radius of 15.0 cm and a length of 1.25 m. The rod is held firmly in place. What is the maximum **a.** compressive force and **b.** tensile force that can be applied to one face?

73. **N** An aluminum rod has a radius of 15.0 cm and a length of 1.25 m. The rod is held firmly in place. What is the maximum **a.** compressive force and **b.** tensile force that can be applied to one face? (If you worked Problem 72, compare your answers.)

74. **N** We know from studying friction forces that static friction increases with increasing normal force between the surfaces, which becomes important for vehicles traveling on icy or snowy roads that have coefficients of static friction much smaller than those of dry pavement. In particular, the greater the normal force on the drive wheels (those coupled to the engine), the better the traction. The horizontal position of the center of mass of a typical compact automobile is located 1.1 m toward the rear as measured from the front wheel axle. The wheelbase (distance from the front wheel axle to the rear wheel axle) is 2.7 m. Assume the car is stationary on level ground and has a weight of 12,000 N. Determine the total normal force on the two front tires and on the two rear tires. Which do you suppose are the drive wheels in this case?

75. **N** Ruby, with mass 55.0 kg, is trying to reach a box on a high shelf by standing on her tiptoes. In this position, half her weight is supported by the normal force exerted by the floor on the toes of each foot as shown in Figure P14.75A. This situation can be modeled mechanically by representing the force on Ruby's Achilles tendon with $\vec{F}_A$ and the force on her tibia as $\vec{F}_T$ as shown in Figure P14.75B. What is the value of the angle θ and the magnitudes of the forces $\vec{F}_A$ and $\vec{F}_T$?

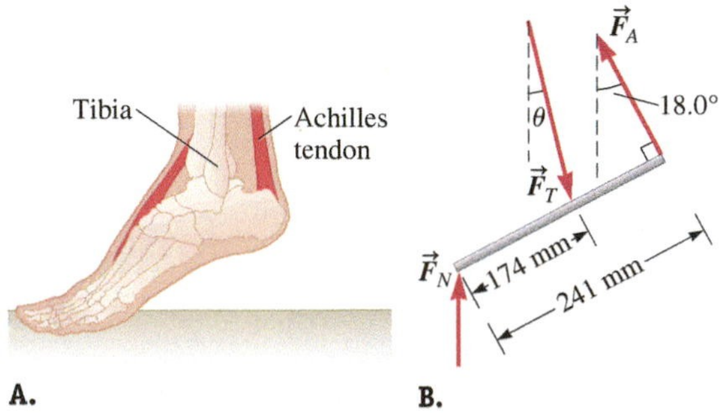

FIGURE P14.75

76. **N** An object is being weighed using an unequal-arm balance (Fig. P14.76). When the object is in the left pan, a downward force of 3.0 N must be exerted on the right pan to balance the gravitational force on the object. When the object is in the right pan, a downward force of 2.0 N must be exerted on the left pan to balance the gravitational force on the object. Determine the magnitude of the gravitational force on the object.

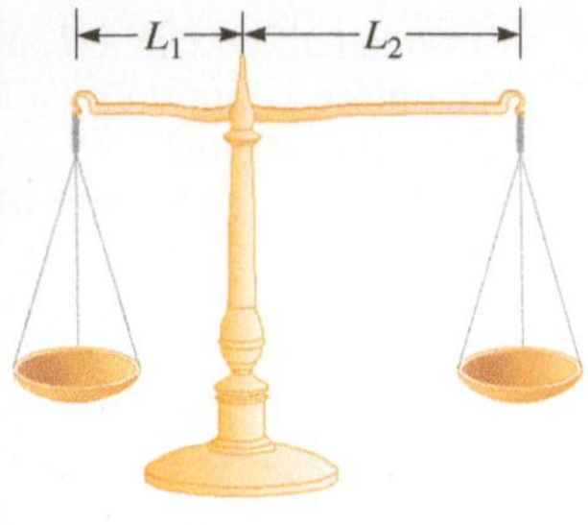

FIGURE P14.76

77. **N** The normal force exerted on an object by a surface is distributed over the area of contact between the object and the surface. The effective point of application of the normal force must then be located to produce a torque equal to the net torques of the distributed force (much like how we locate the center of mass when determining the torque due to the weight of an object). Consider a filing cabinet of height 1.3 m and width 0.40 m resting on a level, horizontal floor. The cabinet has a weight of 7.0×10^2 N and is loaded so that its center of mass is at its geometric center. Now suppose someone leans against it, pushing horizontally with a force of 50.0 N at the top edge, and the cabinet remains in static equilibrium (Fig. P14.77). Locate the effective point of application of the normal force due to the floor as measured from the left side of the cabinet.

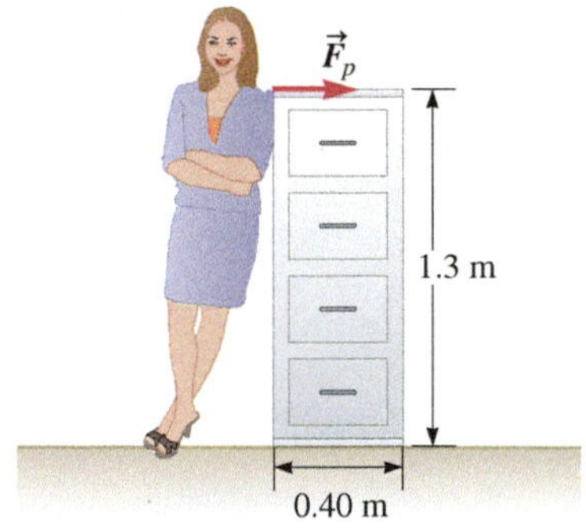

FIGURE P14.77

78. **A** A massless, horizontal beam of length L and a massless rope support a sign of mass m (Fig. P14.78). **a.** What is the tension in the rope? **b.** In terms of m, g, d, L, and θ, what are the components of the force exerted by the beam on the wall?

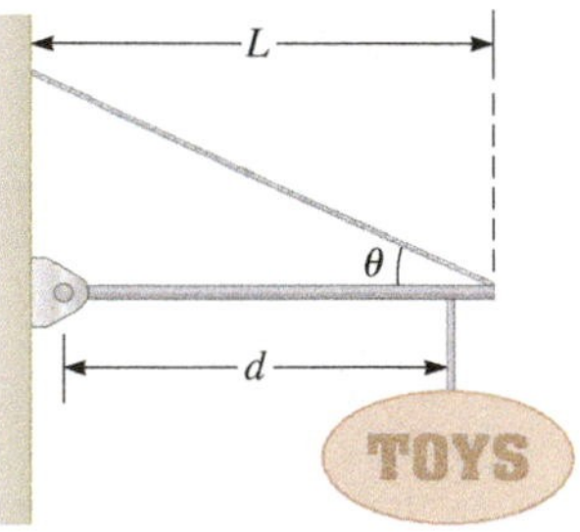

FIGURE P14.78

Problems 79 and 80 are paired.

79. **N** A rod of length 4.00 m with negligible mass is hinged to a wall. A rope attached to the end of the rod runs up to the wall at an angle of exactly 45°, helping support the rod, while a sign of weight 10.0 N is hanging by two ropes attached to the bottom of the rod. The ropes make an angle of exactly 30° with the rod as shown in Figure P14.79. Another sign with a weight of 10.0 N is attached to the top of the rod with its center of mass at the midpoint of the rod. The entire system is in equilibrium. Find the magnitude of the tension in the rope above the rod that is also attached to the wall.

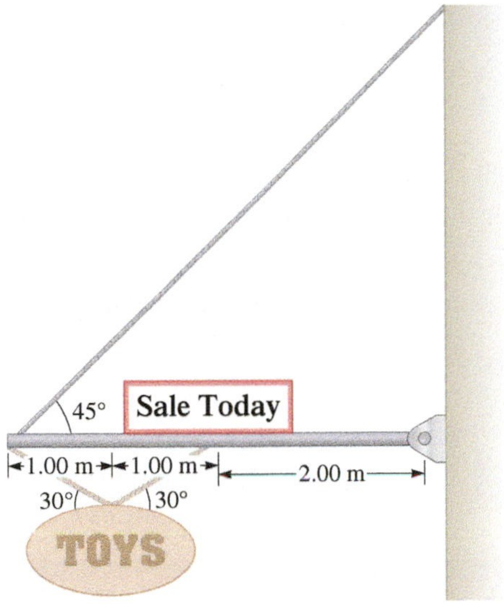

FIGURE P14.79 Problems 79 and 80.

80. **N** In Problem 79, find the magnitude of the vertical and horizontal components of the force that the hinge must exert on the rod to keep the system in equilibrium.

81. **N** A horizontal, rigid bar of negligible weight is fixed against a vertical wall at one end and supported by a vertical string at the other end. The bar has a length of 50.0 cm and is used to support a hanging block of weight 400.0 N from a point 30.0 cm from the wall as shown in Figure P14.81. The string is made from a material with a tensile strength of 1.2×10^8 N/m^2. Determine the largest diameter of the string for which it would still break.

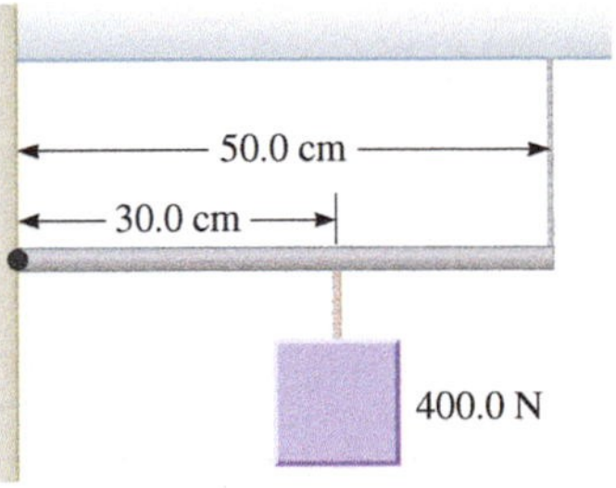

FIGURE P14.81

82. A rope is used to connect a uniform ladder with length L and mass M to a wall (Fig. P14.82). The ladder rests on a rough, horizontal surface with a coefficient of static friction μ_s, and the angle θ is such that the ladder is on the verge of slipping. Assume the ladder is in rotational and translational equilibrium.

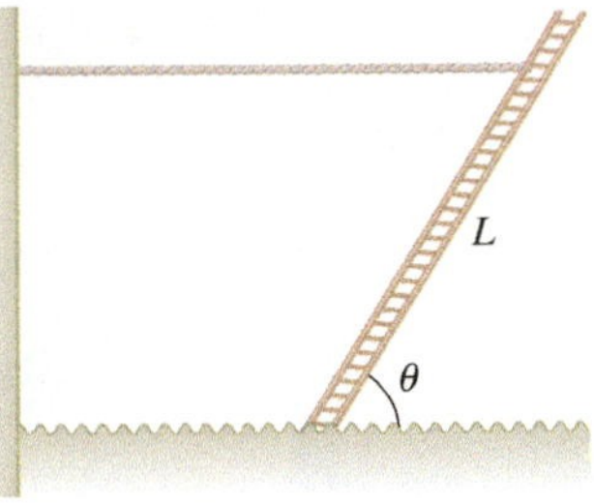

FIGURE P14.82

a. **G** Draw a free-body diagram of the forces acting on the ladder.

b. **A** If the ladder is in rotational equilibrium, what is the tension in the rope in terms of M, g, and θ?

c. **A** Obtain a second expression for the tension in the rope in terms of μ_s, M, and g by considering the ladder in translational equilibrium.

d. **A** What is the coefficient of static friction μ_s in terms of the angle θ?

e. **C** What would occur if the ladder were moved slightly so as to reduce the angle θ?

83. **N** A 185-kg uniform steel beam 8.00 m in length rests against a frictionless vertical wall, making an angle of 55.0° with the horizontal. A 75.0-kg construction worker begins walking up the beam. The coefficient of static friction between the beam and the ground is 0.750. What are the horizontal and vertical forces exerted on the beam by the ground when the worker has walked 3.00 m along the beam?

84. Two smooth planes support a uniform wooden beam of length L and mass M (Fig. P14.84).

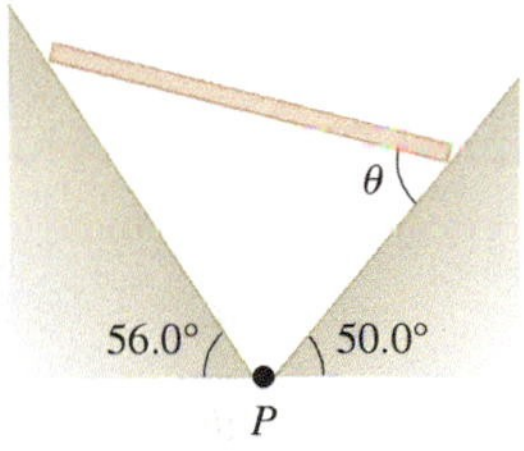

FIGURE P14.84

a. **C** Show that for the beam to be in equilibrium, its center of mass must coincide vertically with point P.

b. **N** What is the angle θ for which the beam is in equilibrium?

c. **C** Is the equilibrium found in part (b) stable or unstable? Explain.

85. **A** The length of a metal wire is L_1 when the tension in it is F_{T1}, and it is L_2 when the tension in it is F_{T2}. What is the actual length of the wire under no tension in terms of these lengths and tensions?

15 Fluids

Key Questions

How can Newton's second law and the work–energy theorem be used to predict the behavior of fluids?

When a solid interacts with a fluid, how do we describe the force exerted by the fluid?

Underlying Principles

No new *fundamental* physical principles are introduced in this chapter.

Major Concepts

1. Ideal fluid
2. Static fluid
3. Pressure
4. Archimedes's principle
5. Pascal's principle
6. Ideal fluid flow
7. Continuity equation
8. Bernoulli's equation

Special Cases

A fluid is a special case (state) of matter that can flow. Liquids and gases are fluids.

Our lives depend on fluids such as the air we breathe and the water we drink. In fact, the human body consists mostly of water. Engineers design airplanes to fly in air and boats to float in water. Meteorologists observe fluids (the Earth's atmosphere and oceans) to predict the weather. Architects must understand fluids in order to design plumbing and heating systems. Chefs know that fat floats to the top of a sauce and that the ravioli is done when it rises to the top of the water.

In this chapter, we focus our attention on fluids and how they interact with solid objects. We will introduce many new concepts and apply Newton's second law and the work–energy theorem. No new fundamental principles are necessary.

15-1 What Is a Fluid?

FLUID **Special Case**

The term **fluid** (a substance that can flow) applies to both liquids and gases, but not to solids. Previous chapters dealt with the properties and state of motion of solid objects such as hockey pucks, freight trains, and ancient monuments. How would we describe or predict the properties and motion of liquids and gases? In the following case study, three students discuss the interaction between air, a fluid, and an airplane, a solid object. Their discussion continues throughout the chapter as they attempt to build an understanding or model of fluid behavior and fluid interactions.

CASE STUDY Part 1: Flying Airplanes

Avi: I was just thinking about an air show I saw. They had these planes flying upside down (Fig. 15.1), and I can't figure out why that works. It makes me think that I don't understand how planes fly.

Shannon: Airplanes fly because air exerts a force on their wings.

Cameron: I know you have to be right about the air, but there has to be something about that plane's motion, too. I mean the plane has to keep moving, and it can't just fly any old way and stay up there. Imagine a plane pointing straight down. There will be a little air drag on the plane, but basically it is going to fall out of the sky.

Avi: I think you're kind of getting at the problem. Planes shouldn't be able to fly any old way. The net force of the air has to counteract gravity or else the plane will fall. So, how can you possibly design a plane to fly both upside down *and* right side up?

FIGURE 15.1 How can an airplane fly upside down?

CONCEPT EXERCISE 15.1

Imagine an airplane flying at constant velocity. What must be the net force on the plane? Draw a free-body diagram for the plane.

Fluid Model

To understand why some substances can flow and others cannot, let's see how a microscopic model of a solid differs from microscopic models for liquids and gases. A solid is made up of molecules that are strongly bonded together. We modeled the molecules as particles and the bonds as stiff springs (Fig. 14.21, page 406). Because of the strong bonds, the molecules in a solid have limited motion. They can move slightly as the bonds are stretched, compressed, or sheared, but they cannot slide past one another.

The molecules in a gas are not bonded to one another; the force between them is very weak. Gas molecules move freely throughout their container. The gas fills its container, and the molecules interact with one another only when they collide. We model them as very small, deformable objects similar to tiny, very springy rubber balls that collide elastically with one another and with the walls of their container (Fig. 15.2A). The gas molecules whiz past one another and barely affect the other's motion. Gas is a fluid because the gas molecules are free to flow.

Liquid is the state of matter between solid and gas. The molecules in a liquid are bound together, but the bonds are weak compared with those in a solid. Liquid molecules can easily slide past one another; so, like a gas, a liquid can freely flow. The molecules in a liquid are much closer together than are the molecules in a gas, however. We model the liquid molecules as particles that are tightly packed together and weakly connected (Fig. 15.2B).

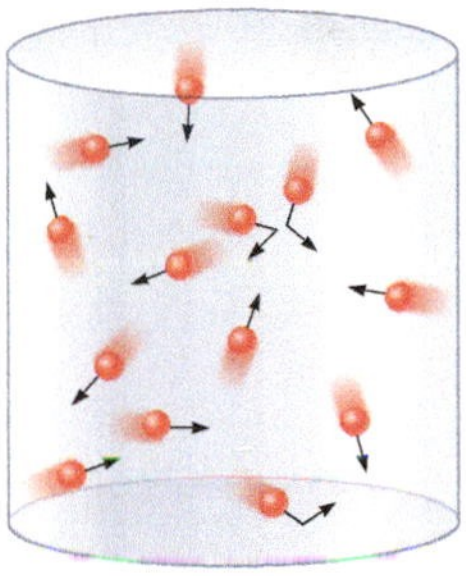

A. Gas

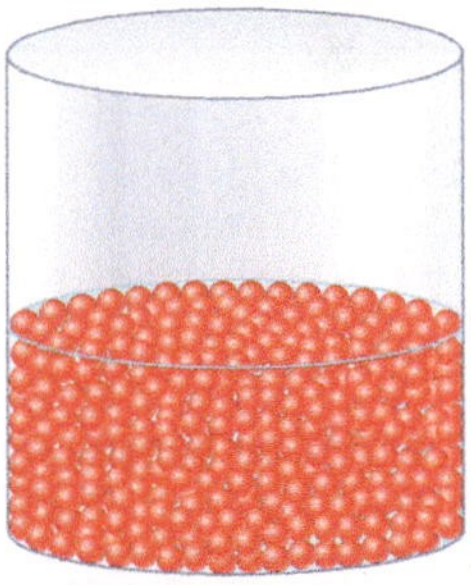

B. Liquid

FIGURE 15.2 **A.** Gas molecules are far apart. They collide elastically with one another and the walls of their container. **B.** Molecules in a liquid are closely packed, but they are free to slip past one another.

Because the particles in a liquid are much closer together than are the particles in a gas, gases can be compressed easily, whereas liquids cannot. We say that a liquid is nearly *incompressible*. **Incompressible** describes a substance whose volume *cannot* change. On the other hand, gas is *compressible*. The volume of a **compressible** substance is changeable.

In this chapter, we will limit our study to incompressible fluids. Incompressibility is one of the properties of an **ideal fluid**. Therefore, much of what we study is a very good approximation to the behavior of real liquids. Because gases are, in fact, compressible, there are many gas behaviors that we cannot discuss here. Gas compression, however, is discussed in Chapters 19 through 22.

IDEAL FLUID Major Concept

15-2 Static Fluid on the Earth

STATIC FLUID Major Concept

In this and the next three sections, we focus on *static fluids*, as in a cup of coffee, the water in a bathtub, or the (still) air in your room. A **static fluid** is able to flow but is not flowing at the time we study it. The individual particles in a static fluid are still in motion, but there is no macroscopic motion of the fluid as a whole. In Section 15-6, we turn our attention to fluids in motion.

If we limit our study to incompressible fluids, how does that limit our study of gases? We will study fluids on or near the surface of the Earth. Therefore, at least one force is exerted on the fluid—the Earth's gravity. Imagine a tall cylinder of air that stretches from sea level to the top of Mount Everest, approximate elevation 8850 m (Fig. 15.3). Gravity exerts a downward force on the air molecules, so you might expect to find all the molecules packed closely together on the Earth's surface like tennis balls in a bucket. Unlike the tennis balls in a bucket, however, air molecules are constantly in motion, colliding elastically with one another and with the walls of their container. All that motion means that the gas particles spread out to fill their container. So, there is a continuous decrease in density as a function of altitude. Recall that density is mass per unit volume (Eq. 1.1). The density of dry air at sea level is $\rho_0 = 1.2929\ \text{kg/m}^3$. The density of air at the summit of Mount Everest is 30% of the density near sea level, or $\rho_{\text{top}} \approx 0.4\,\text{kg/m}^3$. (Climbers must take bottled oxygen when they ascend tall mountains such as Everest.) This variation in density means that we cannot approximate such a large cylinder of air as an incompressible fluid. In this chapter, we are limited to studying much smaller containers of air in which the density is nearly uniform.

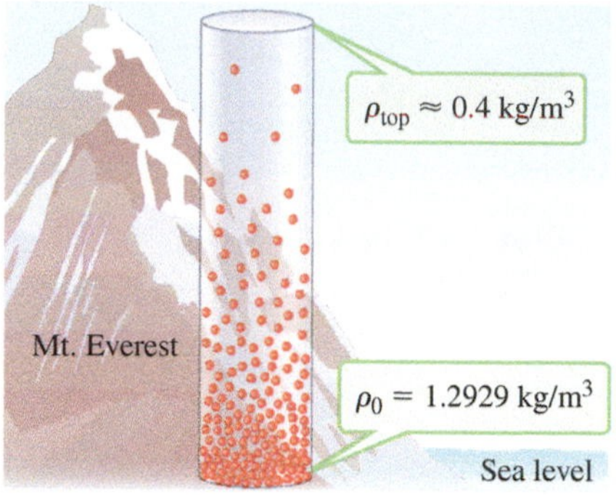

FIGURE 15.3 The density of air molecules is greater near sea level than it is near the top of Mount Everest.

Exactly what size container are we talking about? If we imagine a cylinder as tall as the Willis (Sears) Tower in Chicago (Fig. 15.4), the density at the top is $\rho_{\text{top}} \approx 0.94\rho_0 \approx 1.22\ \text{kg/m}^3$. Depending on what we wish to calculate, such a slight variation in the density may not be ignorable. In this chapter, we usually consider containers that can fit in a laboratory. Even in a container as large as a typical college building, we may safely assume air density is essentially uniform. Unless otherwise specified, we will assume the density of dry air is uniform and given by $\rho = 1.29\ \text{kg/m}^3$.

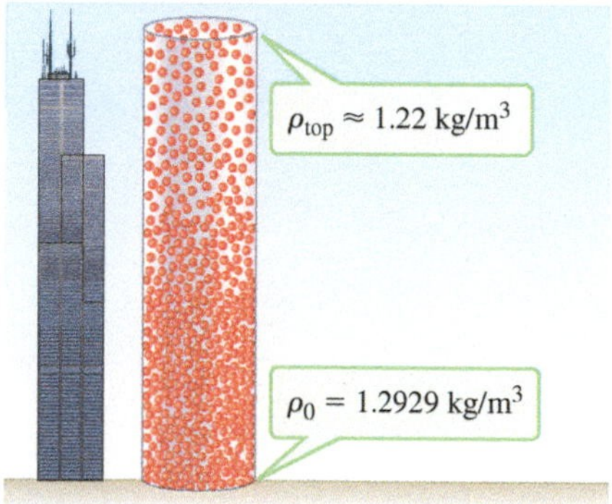

FIGURE 15.4 The air density at the top of the Sears (Willis) Tower is 94% of the density at the bottom of the tower.

Just as the Earth's gravity increases the density of gas molecules near the bottom of a container, we might expect to find an increase in the density of liquid molecules near the bottom of a similar container. Because the molecules in a liquid are much closer together than in air, however, it is difficult to compress them further. The density of seawater near the surface of the Earth is $1.025 \times 10^3\ \text{kg/m}^3$, and even 1000 m below the surface, the density of seawater increases only slightly to $1.028 \times 10^3\ \text{kg/m}^3$. In this chapter, assume liquids are incompressible with a uniform density. Some fluid density values are given in Table 15.1.

TABLE 15.1 Typical density of common fluids.

Fluid	Density ρ (kg/m³)[a]
Acetone	784.58
Air	1.29
Ethyl alcohol	785.06
Gasoline	737.22
Helium gas	0.18
Hydrogen gas	0.089
Mercury	13,600
Sunflower oil	920
Seawater	1025.18
Water (at 4°C)	1000.00
Water	998.2

[a]Densities are at room temperature (15°–25°C) except where indicated.

15-3 Pressure

Although we are usually unaware of the fluid—air—that surrounds us, a few common experiences may remind us that it is present. For example, if you quickly change elevation by flying in an airplane, driving down a mountain, or even taking an elevator in a very tall skyscraper, you may experience discomfort in your ears. Usually, when it is quiet, the eardrum is bombarded on both sides by air molecules; so, there is no net force on your eardrum, and you don't notice the air pounding on it. If your eardrum gets bombarded more often on one side, you notice a net force on the eardrum. A tube on the inside of the eardrum connects to the throat, allowing air to get to the inside of the eardrum (Fig. 15.5). Rapid elevation changes cause discomfort because the air takes a moment to equalize on both sides. The discomfort may be worse if you have a head cold that closes the tube connecting the inside of the eardrum to the throat. Many people yawn or swallow in such situations in an effort to make their ears "pop" and equalize the air.

Definition and Units of Pressure

Because eardrums and other surfaces are bombarded by a huge number of air molecules, it is not practical to measure the force exerted by each microscopic particle. Instead, we measure a macroscopic quantity called *pressure* that results from the force of the large number of microscopic particles. Consider the simple system shown in Figure 15.6, a piston in a tube with no friction between the piston and the tube's walls. On the left side of the piston, there is a gas, but there is nothing (a vacuum) on the other side. The gas molecules collide with the piston, and because there is no friction or other forces acting on the piston, the piston accelerates to the right. Now imagine holding the piston at rest by applying a force, $\vec{F}$. The force you would need to apply must be equal in magnitude to the force exerted by gas molecules on the other side of the piston. Now imagine the same experiment, but with a much larger piston. More gas molecules would collide with the larger piston, so we would need to apply a larger force to keep the piston at rest. In fact, the force we must apply is proportional to the area of the piston: $F \propto A$. The force exerted per unit area is a constant, however. The **pressure** (P) exerted by the gas on the piston equals the magnitude of the force we must exert per unit area:

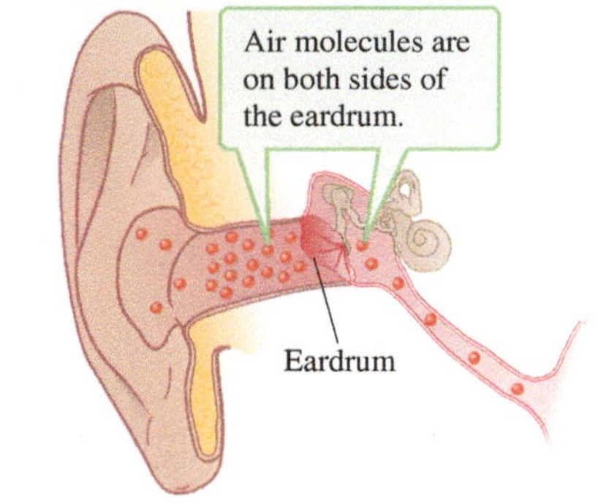

FIGURE 15.5 Usually, the air pressure on the outside of the eardrum equals the air pressure on the inside because a tube connects inside of the eardrum to the throat.

$$P = \frac{F}{A} \tag{15.1}$$

PRESSURE ✪ **Major Concept**

Pressure is a scalar quantity with the dimensions of force per area. The SI unit of pressure is the pascal (Pa):

$$1\ \text{Pa} = 1\ \text{N/m}^2$$

We use an uppercase P for pressure and a lower case $\vec{p}$ for momentum (Eq. 10.1).

There are other common units for pressure. For example, one *atmosphere* (atm) is the approximate pressure exerted by the Earth's atmosphere at sea level:

$$1\ \text{atm} = 1.01325 \times 10^5\ \text{Pa}$$

Another commonly used pressure unit is the *millimeter of mercury* (mm Hg), also called the *torricelli* (torr) in honor of the inventor of the mercury barometer, Evangelista Torricelli (Section 15-5):

$$1\ \text{Pa} = 7.501 \times 10^{-3}\ \text{mm Hg} = 7.501\ \text{mtorr}$$

The U.S. customary pressure unit that we see written on our car and bike tires is *pounds per square inch* (psi or lb/in.^{2}):

$$1\ \text{Pa} = 1.45 \times 10^{-4}\ \text{psi}$$

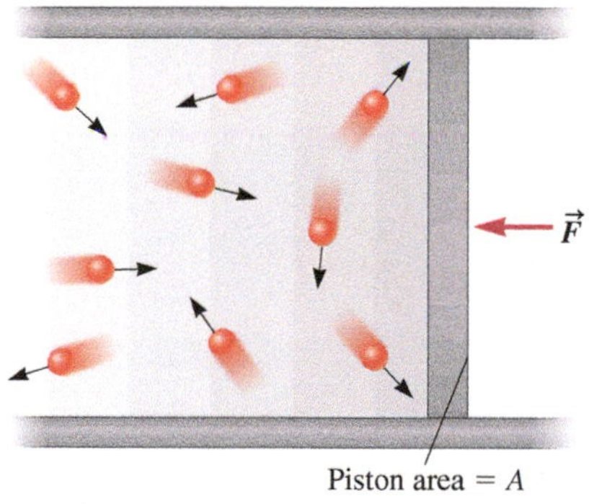

FIGURE 15.6 A piston of cross-sectional area A is in a frictionless tube. Air pressure on one side of the piston is balanced by a force on the other side of the piston. The magnitude of the force exerted must equal $F = PA$.

Pressure Variation with Depth in a Static Fluid

The Earth's gravity affects the pressure in a static fluid such as the water in a swimming pool. Let's look at the forces that act on a portion of the water. As shown in Figure 15.7A, we imagine a cylinder-shaped column of water with a cross-sectional area A. We have chosen a downward-pointing y axis so that the column of water extends from the surface to a depth y. Three vertical forces are exerted on this column of water, shown in Figure 15.7B: (1) the Earth's gravity exerts a force $\vec{F}_g$ in the positive y direction, (2) the air above the column of water also exerts a force $\vec{F}_{air}$ in

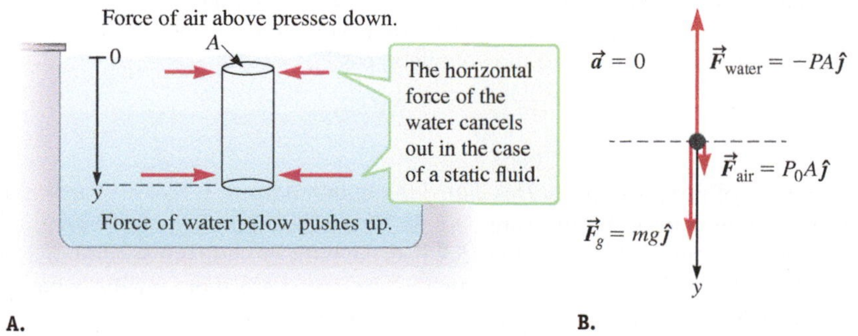

FIGURE 15.7 **A.** We imagine selecting a cylinder-shaped portion of pool water. **B.** A free-body diagram for the cylinder-shaped portion of water, showing forces acting in the vertical direction.

the positive y direction, and (3) the water below the column exerts an upward force $\vec{F}_{water}$ in the negative y direction. Because the fluid is static, we may apply Newton's second law (in component form) with the acceleration set equal to zero (Eq. 14.3):

$$\sum F_y = F_g + F_{air} - F_{water} = 0 \tag{15.2}$$

If the air pressure is P_0 and the cross-sectional area of the column is A, the force exerted by the air on the column of water is (Eq. 15.1):

$$F_{air} = P_0 A \tag{15.3}$$

Similarly, the force exerted by the water below the column is

$$F_{water} = PA \tag{15.4}$$

where P is the water pressure just below the column at depth y. The weight of the column of water is given by the usual expression $F_g = mg$, where m is the mass of the water column. Because we are modeling the water as an incompressible fluid, we assume the density ρ of water is uniform throughout the column. Therefore, we can write the mass m of the column in terms of the water density ρ and the volume of the column, $V = Ay$. The weight of the column can be expressed as

$$F_g = mg = (\rho V)g = (\rho Ay)g \tag{15.5}$$

Substitute Equations 15.3 through 15.5 into Equation 15.2 and rearrange the terms:

$$\rho Ayg + P_0 A - PA = 0$$

Pressure P of an incompressible, static fluid on the Earth as a function of depth y below the surface

$$P = \rho gy + P_0 \tag{15.6}$$

Equation 15.6 says that the pressure of a fluid increases with depth due to the weight of the column of fluid above.

Pressure and weight are both scalars. When you are swimming under water, the water pressure along the surface of your body is roughly uniform, and the force exerted by the water on you is perpendicular to your surface at every point. If you swim to the bottom of a deep pool, you experience some discomfort in your ears because you are holding your breath with air at the atmosphere's pressure on the inside of your eardrum. The water outside your eardrum has greater pressure near the bottom of the pool than it does near the surface due to the weight of all the water above you. The discomfort does not change if you turn your head because pressure is a scalar, independent of direction. All that matters is your depth. The deeper you swim, the more water is pressing down on you, and the greater your ear discomfort.

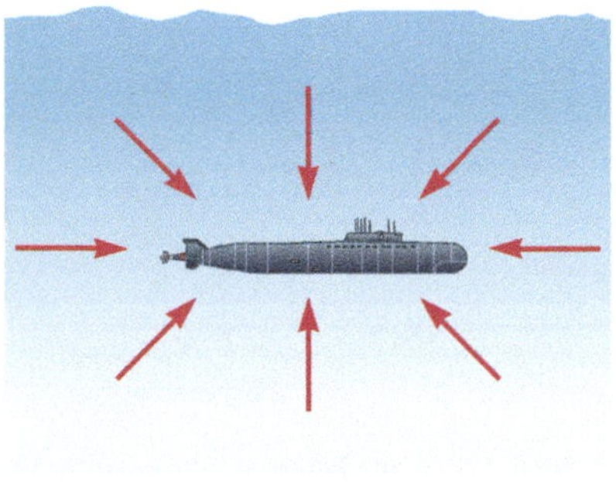

FIGURE 15.8 When an object is submerged in a fluid, the fluid presses the object equally in all directions. The result is the object is compressed, but its shape does not change.

Change in an Object's Volume with Pressure

In this chapter, we model fluids that are *un*compressed, but an object (whether a solid, liquid, or gas) immersed in the fluid may be compressed. For example, if you immerse a balloon or even a submarine in seawater, the fluid presses uniformly all over the object's surface. The result is that the object is compressed, but its shape does not change (Fig. 15.8). The total pressure exerted by the fluid on the object is called the **volume stress**. The **volume strain** is a measure of the object's compression, expressed as the change in the object's volume divided by its original volume, $\Delta V/V_i$. As discussed in Section 14-4, a stress on an object may produce a linear response so that as the stress increases the strain increases proportionally. Often, the volume strain is proportional to the volume stress:

$$P = B\frac{\Delta V}{V_i} \tag{15.7}$$

The constant of proportionality B is called the **bulk modulus**; it has the same dimensions as pressure, so its SI units are pascals. The bulk modulus is a property of the material making up the object. Table 15.2 lists the bulk modulus of several materials. Materials with a low bulk modulus (such as mercury) are more easily compressed than materials with a high bulk modulus (such as steel).

TABLE 15.2 Bulk modulus for several materials.

Material	Bulk modulus B ($\times 10^9$ Pa)
Aluminum	76
Brass	61
Mercury	27
Silicon	98
Steel	140
Tungsten	200
Water	2.2

CONCEPT EXERCISE 15.2

We often hear the word *pressure* in everyday conversations. Give examples from everyday use of the word *pressure* that are consistent with the physics definition and give examples that are inconsistent. Explain.

CONCEPT EXERCISE 15.3

Estimate the water pressure near the bottom of a swimming pool located near sea level. What is the difference in pressure between the surface and the bottom of the pool?

CONCEPT EXERCISE 15.4

Imagine a swimming pool empty of water, but full of air. Estimate the air pressure near the bottom of this swimming pool. Again, assume the pool is located near sea level. Is it reasonable to assume the air pressure is constant over a container of such volume?

15-4 Archimedes's Principle

When you think about the weight of the Earth's atmosphere pressing down on you and everything else, it may seem amazing that you are not crushed flat like a pancake against the surface of the Earth. Three physics students discuss this apparent paradox.

CASE STUDY Part 2: Airplanes and Helium Balloons

Avi: I know that somehow the air must exert an upward force on an airplane, but I can't figure out how. In fact, it seems to me that air should exert a downward force. If you have a plane on the runway, the plane has the weight of the whole atmosphere pressing it down. I can't figure out how it gets up at all.

Shannon: I think I have a way to explain it. Imagine that you have a helium balloon. You know it wants to float.

Cameron: But that is because it is lighter than air. Planes aren't lighter than air. Planes fly because they are moving.

Shannon: Yes. I agree that planes are more complicated. But I think that if we just talk about a helium balloon, it will help us understand more about atmospheric pressure. The balloon doesn't just have the weight of the atmosphere pressing it down. It also has air below pushing it up. The air pressure below the balloon must be greater than the pressure above the balloon, so the net force of the air is upward.

Shannon is correct. The weight of the atmosphere doesn't just press *down* on you. The air is all around you, and the atmospheric pressure pushes you in all directions, even upward. Normally, the air does not exert a net force on you. However, if you evacuate the air (or just reduce the amount of air) on one side of an object, the Earth's atmosphere exerts a net force on the object. For example, you may get a suction cup to hang from the underside of a horizontal surface by burping the suction cup and reducing the amount of air inside. We often use vacuum cleaners to reduce the air on one side of tiny objects, say spilled coffee grounds. Then the air on the other side forces the dirt into the vacuum cleaner's bag (Fig. 15.9).

FIGURE 15.9 The pressure inside the vacuum cleaner is lower than the pressure on the other side of the coffee grounds. The higher pressure forces the grounds into the vacuum cleaner.

Our experience with devices such as vacuum cleaners sometimes makes it more difficult to think about how fluid pressure affects objects in the fluid. For example, when we want to buy a new vacuum cleaner, the salesperson explains that brand X has more *suction* than brand Y. The word *suction* may give you the idea that the vacuum *pulls* the dirt, but air molecules don't have hooks. Instead, the air molecules push the dirt as they move from a region of high pressure to a region of lower pressure.

If the fluid pressure exerted on an object varies across its surface, the fluid exerts a net force on the object. The net force points from high pressure to low pressure. For a static fluid near the Earth's surface, $P = \rho g y + P_0$ (Eq. 15.6), so the pressure below the object is greater than the pressure above it. Therefore, the fluid exerts a net upward force on the object. Imagine carrying your friend across a room; now imagine carrying the same person across a swimming pool in which both of you are mostly submerged. It is much easier to carry your friend in the swimming pool because of the net upward force of the water. Why doesn't the net upward force of the air help when you try to carry your friend across the room? Remember that the air pressure in a container the size of a small building is nearly constant, so the net upward force of the air is very small and doesn't make a noticeable difference.

The Buoyant Force

The net upward force exerted by a fluid on an object is called the **buoyant force** $\vec{F}_B$. The magnitude of the buoyant force was discovered by Archimedes of Syracuse (c. 287–212 BCE), a Greek mathematician and scientist who reportedly was taking a bath when he was inspired.[1] According to **Archimedes's principle**, *the magnitude of the buoyant force is equal to the weight of the fluid displaced by the object.*

Archimedes's principle can be deduced by looking at the forces exerted within a fluid. Figure 15.10A shows an imaginary box enclosing part of the fluid in the container. Two forces are exerted on the box of fluid: the downward force $\vec{F}_g$ due to the Earth's gravity and the upward buoyant force $\vec{F}_B$ from the remaining fluid. The box of fluid is in static equilibrium, so there is no net force on it:

$$\sum F_y = F_B - F_g = 0$$

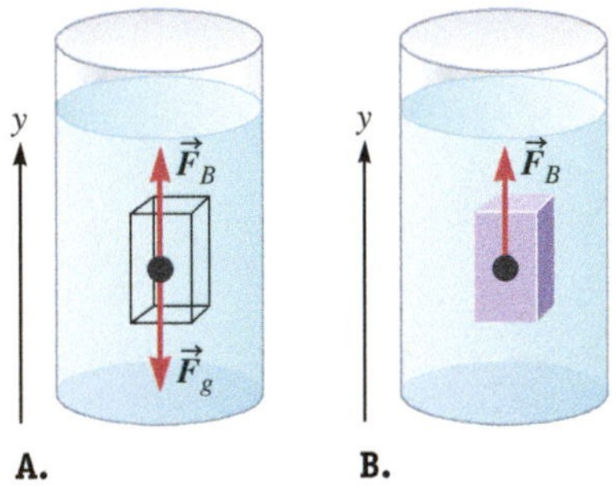

FIGURE 15.10 **A.** The magnitude of the buoyant force exerted on an imaginary "box" of fluid equals the weight of fluid in the box. **B.** According to Archimedes's principle, the magnitude of the buoyant force exerted by a fluid is equal to the weight of the fluid displaced by the object.

Therefore, the magnitude of the buoyant force equals the weight of the fluid in the box: $F_B = F_g$. What would happen to the buoyant force if we had an actual object that exactly filled the space of the imaginary box (Fig. 15.10B)? The answer is that nothing would happen; the buoyant force would not change because the buoyant force is the result of the net force exerted by the *surrounding* fluid. Replacing the box of fluid with an object of the same volume does not affect the surrounding fluid. Therefore, it doesn't matter whether we are considering a real box-shaped object or just a boxed-off portion of the fluid: The buoyant force is the same in either case. *The buoyant force equals the weight of the fluid that would fit in the volume of the submerged object.*

The argument illustrated in Figure 15.10 may leave you feeling like you've just watched a magic act as an object is swapped in to replace a boxed-off portion of fluid. In order to fit the box-shaped object into the fluid without changing the level of the fluid, a volume of fluid V_{disp} (for "*disp*laced") equal to the object's volume V_{obj} must be removed. The argument helps us arrive at Archimedes's principle, but it is outside our normal experience because usually when we place an object in a fluid, we don't first remove a volume of fluid equal to the object's volume. (We don't fill the bathtub to the top before we get in, however, because we know the water will rise when we enter!)

In Figure 15.11, we start with a container of fluid and then place an object in the fluid so that the object is totally submerged. The volume of the object V_{obj} equals the

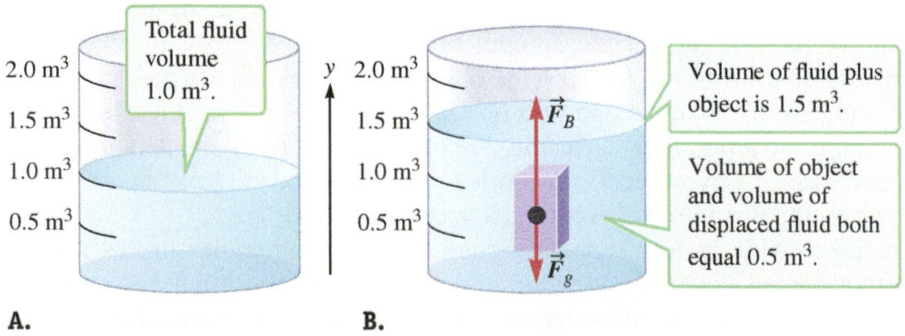

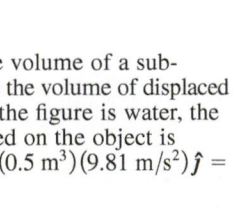

FIGURE 15.11 The volume of a submerged object equals the volume of displaced fluid. If the fluid in the figure is water, the buoyant force exerted on the object is $\vec{F}_B = (1000\ \text{kg/m}^3)(0.5\ \text{m}^3)(9.81\ \text{m/s}^2)\hat{\jmath} = 4.9 \times 10^3$ N.

[1]Archimedes's story was reported by ancient Roman writer Vitruvius in Book 9 of *De architectura*.

volume of the displaced fluid, $V_{obj} = V_{disp}$. The magnitude of the buoyant force equals the weight of this volume of displaced fluid:

$$F_B = F_g = m_f g$$

where m_f denotes the mass of *displaced fluid*. It is convenient to write the mass m_f of the displaced fluid in terms of its volume V_{disp} and density ρ_f. The buoyant force is then

$$F_B = \rho_f V_{disp} g \tag{15.8}$$

ARCHIMEDES'S PRINCIPLE

✪ **Major Concept**

EXAMPLE 15.1 A Totally Submerged Object

A totally submerged object released from rest may rise up, sink farther down, or just rest in place, depending on the relative densities of the object and the fluid. Show that after an object submerged in a fluid is released from rest, its acceleration is given by

$$a_y = \left(\frac{\rho_f - \rho_{obj}}{\rho_{obj}}\right)g \tag{15.9}$$

In the **CHECK and THINK** step, discuss the three cases of remaining at rest, sinking, and rising.

INTERPRET and ANTICIPATE

This example is much like those in Chapters 5 and 6. Use a free-body diagram (Fig. 15.11B) and apply Newton's second law to find the acceleration.

SOLVE

As shown in Figure 15.11B, two forces act on the object: gravity and the buoyant force. Apply Newton's second law in component form along the y direction.	$\sum F_y = F_B - F_g = m_{obj} a_y$ (1)
The object's mass m_{obj} can be expressed in terms of the object's volume and density.	$m_{obj} = \rho_{obj} V_{obj}$ (2)
Write the buoyant force and the gravitational force in terms of the density of the fluid and the density of the object.	$F_B = \rho_f V_{disp} g$ (3) $F_g = \rho_{obj} V_{obj} g$ (4)
Substitute Equations (2) through (4) into Equation (1).	$\rho_f V_{disp} g - \rho_{obj} V_{obj} g = \rho_{obj} V_{obj} a_y$ (5)
Because the object is totally submerged, the volume of the displaced fluid equals the object's volume, $V_{disp} = V_{obj}$. We rewrite Equation (5) in terms of V_{obj}.	$\rho_f V_{obj} g - \rho_{obj} V_{obj} g = \rho_{obj} V_{obj} a_y$
Cancel V_{obj} from both sides of the equation.	$\rho_f g - \rho_{obj} g = \rho_{obj} a_y$
Isolate the acceleration on the left side of the equation.	$a_y = \left(\frac{\rho_f - \rho_{obj}}{\rho_{obj}}\right)g$ ✓ (15.9)

CHECK and THINK

If the object's density equals the fluid's density, the buoyant force equals the object's weight. (Compare Eqs. 3 and 4.) In such a case, $\rho_{obj} = \rho_f$, and the object's acceleration is zero. Because the object was initially at rest, it remains at rest.

If the object is denser than the fluid ($\rho_{obj} > \rho_f$), the acceleration is negative (relative to the upward-pointing y axis in Fig. 15.11). In this case, the gravitational force exceeds the buoyant force.

Finally, if the object is less dense than the fluid ($\rho_{obj} < \rho_f$), the acceleration is positive. In this case, the buoyant force exceeds the gravitational force. According to Equation 15.9, if the fluid vanishes (that is, if $\rho_f \rightarrow 0$), the object is in free fall with $a_y = g$ as expected.

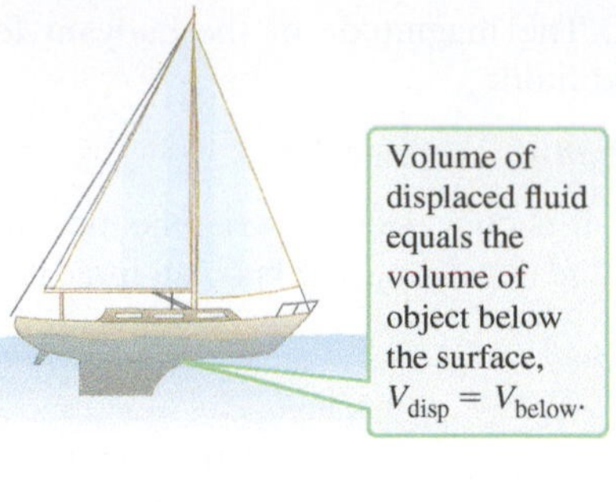
Volume of displaced fluid equals the volume of object below the surface, $V_{disp} = V_{below}$.
A.

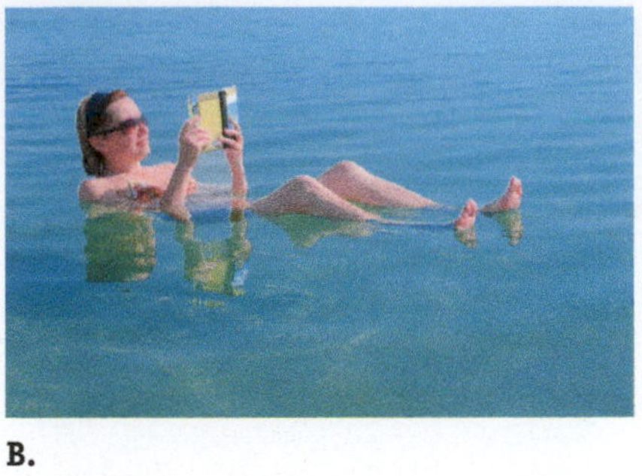
© vicspacewalker/Shutterstock.com
B.

FIGURE 15.12 **A.** For a floating, partially submerged object, the volume of the displaced fluid equals the volume of the object that is below the surface. **B.** Bathers float in salty water because their weight is balanced by the buoyant force. This woman floats easily in the Dead Sea.

Equation 15.9 only works for a totally submerged object such as a fish in the ocean because only then is the volume of the displaced fluid equal to the volume of the object. Many objects—such as a boat on the water or a person on the Dead Sea (Fig. 15.12)—float partially submerged, however. In these cases, the volume of the displaced fluid does not equal the object's total volume: $V_{disp} \neq V_{obj}$. Instead, the volume of the displaced fluid equals the volume of the portion of the object *below* the surface of the fluid: $V_{disp} = V_{below}$.

EXAMPLE 15.2 An Object Floating Near the Surface

Consider the sunbather in Figure 15.12B floating with much of her body above the surface of the water. Show that the fraction of the object below the surface of the fluid is determined by the relative densities of the object and the fluid such that

$$\frac{V_{below}}{V_{obj}} = \frac{\rho_{obj}}{\rho_f} \qquad (15.10)$$

INTERPRET and ANTICIPATE

This example is much like Example 15.1 for the submerged object. The free-body diagram is similar, with gravity pointing down and the buoyant force pointing up (Fig 15.11B).

SOLVE

Apply Newton's second law in component form along the y direction. Because the sunbather is in equilibrium, her acceleration is zero, and the two forces must be equal in magnitude.	$\sum F_y = F_B - F_g = m_{obj}a_y$ $F_B = F_g$ (1)
The sunbather displaces a volume of water V_{below}. The weight of the displaced water equals the magnitude of the buoyant force on her.	$F_B = \rho_f V_{below} g$ (2)
Write her weight in terms of her density ρ_{obj} and her volume V_{obj}.	$F_g = \rho_{obj} V_{obj} g$ (3)
Substitute Equations (2) and (3) into Equation (1). The free-fall acceleration g cancels.	$\rho_f V_{below} g = \rho_{obj} V_{obj} g$ $\rho_f V_{below} = \rho_{obj} V_{obj}$
Solve for the fraction of her body below the surface of the water.	$\frac{V_{below}}{V_{obj}} = \frac{\rho_{obj}}{\rho_f}$ ✓ (15.10)

CHECK and THINK

It makes sense that each side of the equation is dimensionless. Also, Equation 15.10 shows that the fraction of the object below the surface of the fluid is determined by the relative densities of the object and the fluid. The more dense the object (large ρ_{obj}), the more of its volume is below the surface.

CONCEPT EXERCISE 15.5

Estimate the fraction of the sunbather's body (Fig 15.12B) below the surface of the water in two ways.

a. The human body is made mostly of water; therefore, its density is close to that of water. Very lean people are somewhat denser than water, whereas overweight people are somewhat less dense. So, let's assume the density of the sunbather is about $\rho_{obj} \approx 1000 \text{ kg/m}^3$. The water in the Dead Sea is much denser than seawater: $\rho_f \approx 1250 \text{ kg/m}^3$. Use $V_{below}/V_{obj} = \rho_{obj}/\rho_f$ (Eq. 15.10) to estimate the fraction (and percent) of the sunbather's body that is below the water.

b. Next use Figure 15.12 to check your estimate.

CONCEPT EXERCISE 15.6

CASE STUDY "Lighter than Air"

Cameron says that a helium balloon rises because it is lighter than air. In what way is this statement correct? How can you make Cameron's statement more precise?

EXAMPLE 15.3 Wearable Lead

Apparent weight and *weight* are different. In Section 5-7, apparent weight was defined as the reading on a scale, and weight was defined as the magnitude of the gravitational force. (See Eq. 5.7.)

If an object is submerged in a fluid such as water, its apparent weight may be less than its weight. Sometimes a scuba (for *s*elf-*c*ontained *u*nderwater *b*reathing *a*pparatus) diver is less dense than seawater. To descend, a diver must wear 20.0 lb of lead on his weight belt (Fig. 15.13). What is the apparent weight of this much lead when it is submerged in seawater? The density of lead is $11{,}350 \text{ kg/m}^3$.

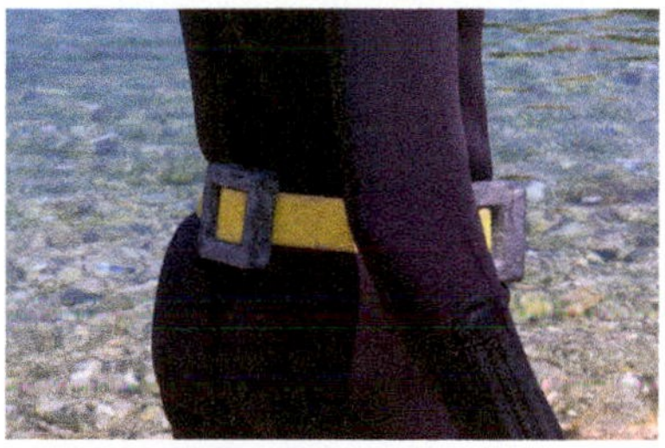

FIGURE 15.13 Lead weights on a diver's belt.

INTERPRET and ANTICIPATE

Imagine using a spring scale to weigh the lead as a single piece when it is submerged (Fig. 15.14). The three forces on the piece of lead are the spring force $\vec{F}_H$, gravity $\vec{F}_g$, and the buoyant force $\vec{F}_B$. By Newton's third law, the magnitude of the force exerted by the spring on the lead is equal to the magnitude of the force exerted by the lead on the spring. Therefore, the reading on the scale is the magnitude of $\vec{F}_H$. We expect the magnitude of the spring force to be less than the weight: $F_H < F_g$.

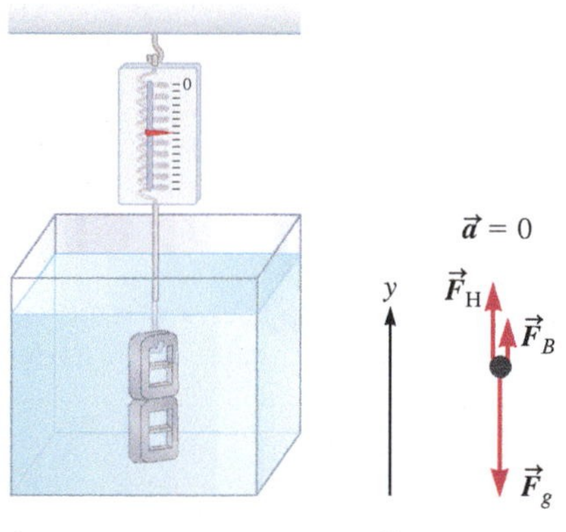

FIGURE 15.14 Finding the apparent weight of lead used by a diver.

SOLVE

Use the free-body diagram to apply Newton's second law in the *y* direction with the acceleration set equal to zero.	$\sum F_y = F_H + F_B - F_g = 0$ $F_H = F_g - F_B$
Substitute Equation 15.8 with $V_{disp} = V_{lead}$ for F_B.	$F_H = m_{lead}g - \rho_f V_{lead} g$
Express the volume of the lead V_{lead} in terms of its mass m_{lead} and density: $\rho_{lead} = m_{lead}/V_{lead}$.	$F_H = m_{lead}g - \rho_f g\left(\frac{m_{lead}}{\rho_{lead}}\right) = m_{lead}g\left(1 - \frac{\rho_f}{\rho_{lead}}\right)$
Substitute values, starting with the given 20.0 lb of lead. The density of seawater is listed in Table 15.1. (We could have, of course, converted the 20.0 lb directly into newtons.)	$m_{lead} = 20.0 \text{ lb}(1 \text{ kg}/2.20 \text{ lb}) = 9.09 \text{ kg}$ $F_H = (9.09 \text{ kg})(9.81 \text{ m/s}^2)\left(1 - \frac{1025.18 \text{ kg/m}^3}{11{,}350 \text{ kg/m}^3}\right)$ $F_H = 81.1 \text{ N}$

Example continues on page 428 ▶

CHECK and THINK

As expected, the apparent weight (about 18 lb) is less than the weight of the lead (20 lb). The apparent weight depends not just on the object's density and mass, but also on the density of the surrounding fluid. If the scuba diver were in freshwater ($\rho = 1000\ kg/m^3$) instead of seawater ($\rho = 1025\ kg/m^3$), the lead's apparent weight would be greater.

15-5 Measuring Pressure

Measuring fluid pressure is common in many engineering, scientific, and everyday applications. If you wish to go scuba diving, for instance, there are many safety checks you should perform before jumping into the ocean. One of the most important things to check before you dive is the pressure in your scuba tank. There are several types of devices for measuring pressure.

In this section, we explore two standard pressure-measuring devices: the manometer and the barometer. A manometer is used to measure the pressure of some fluid relative to atmospheric pressure. A barometer is used to measure atmospheric pressure P_0. (To measure the pressure in a scuba tank, you would use a practical device such as a Bourdon-tube pressure gauge.) Studying the operation of a manometer or barometer will give us a deeper understanding of fluid pressure (Section 15-3) and, in particular, of the variation of pressure with depth in a fluid (Eq. 15.6).

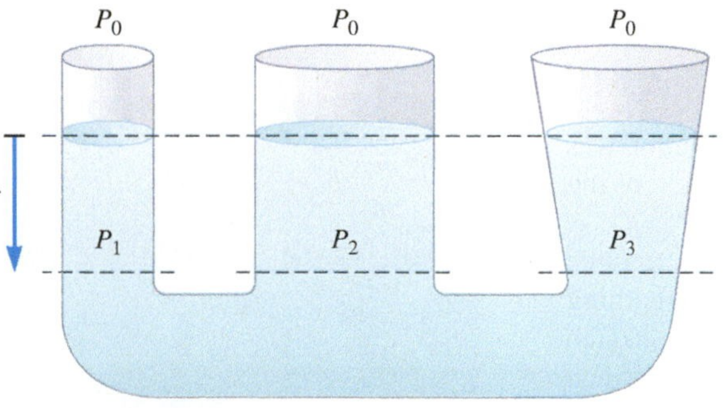

FIGURE 15.15 It may seem surprising, but the fluid comes to the same height in all columns of this complex tube, regardless of column shape or cross-sectional area. Also, the fluid pressure is the same at all points on a horizontal line through the static, incompressible fluid, so $P_1 = P_2 = P_3$.

To find pressure, manometers and barometers measure the height of a fluid. If you had a fluid in a symmetrical U-shaped tube open at both ends, your intuition would probably tell you that the fluid comes to the same height in both sections of the tube. But what about the more complex tube shown in Figure 15.15? Think about the pressure at the same depth y below the surface in each column for any such tube. If the fluid were not at the same height in all columns, the pressures would not be equal at equal depths. The fluid would be forced to move from a high-pressure region to a low-pressure region until it came to the same height in each column. We can make two general statements. First, a static, incompressible fluid comes to the same height in all open regions of a container. Second, the fluid pressure is the same at all points on a horizontal plane through the static, incompressible fluid.

Pascal's Principle

PASCAL'S PRINCIPLE

Major Concept

Next we need to know about pressure changes in a fluid. According to **Pascal's principle,** *a change in the pressure at one point in an incompressible fluid appears undiminished at all points in the fluid and on the walls of its container.* In other words, any *change* in fluid pressure has exactly the same value everywhere in the fluid. As an example, if we start with a fluid in an open container (Fig. 15.16A), the pressure at the surface of the fluid is atmospheric pressure P_0. The pressure at all points below the surface is given by $P_i = \rho gy + P_0$ (Eq. 15.6). Now suppose that we change the surface pressure by attaching this container to another container full of gas at pressure P_{gas} (Fig. 15.16B). Now the pressure at the surface of the fluid is P_{gas}, and the pressure at any point below the surface is $P_f = \rho gy + P_{gas}$. We find the change in pressure ΔP:

$$\Delta P = P_f - P_i$$

$$\Delta P = (\rho gy + P_{gas}) - (\rho gy + P_0)$$

$$\Delta P = P_{gas} - P_0 \tag{15.11}$$

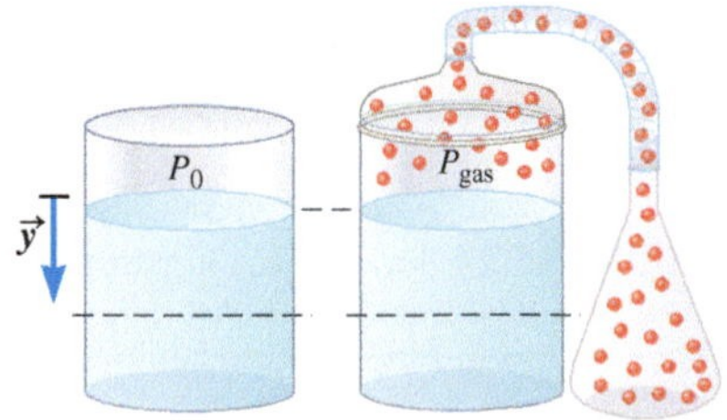

FIGURE 15.16 A. Originally, the pressure at the surface of the fluid is atmospheric pressure P_0. **B.** The pressure at the surface of the fluid is now P_{gas}. According to Pascal's principle, a change in pressure is transmitted undiminished to the entire fluid and the walls of the container.

Equation 15.11 shows that every point in the fluid experiences the same change in pressure independent of depth y.

EXAMPLE 15.4 Hydraulic Lever

One of the most useful applications of Pascal's principle is the hydraulic automobile lift. The basic design is shown in Figure 15.17. A U-shaped tube is filled with a fluid. The ends of the tube are capped off by pistons. A vehicle is placed on the larger piston of area A_L, and a force is applied to the smaller piston of area A_S to raise the vehicle for service.

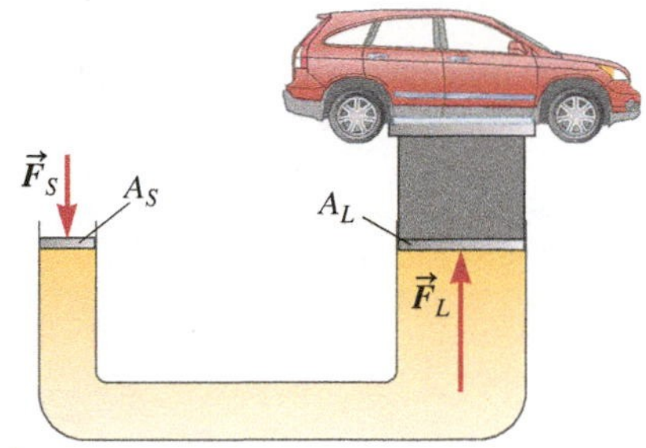

FIGURE 15.17

A A force of magnitude F_S is applied to the small piston. Find the magnitude F_L of the force on the large piston in terms of the force F_S exerted on the smaller piston.

INTERPRET and ANTICIPATE

The force on the small piston increases the pressure in the hydraulic fluid. According to Pascal's principle, that pressure is transmitted undiminished to all parts of the fluid and to the walls of the container that includes the large piston. So, the pressure change on the small piston equals the pressure change on the large piston: $\Delta P_S = \Delta P_L$. We expect to find that the force exerted on the small piston is less than the force on the larger piston ($F_S < F_L$); otherwise, no one would use a hydraulic lift.

SOLVE

Start with Pascal's principle. Use Equation 15.1 and solve for F_L.

$$\Delta P_S = \Delta P_L$$

$$\frac{F_S}{A_S} = \frac{F_L}{A_L}$$

$$F_L = \frac{A_L}{A_S} F_S \quad (1)$$

CHECK and THINK

Since $A_L > A_S$, we find that $F_S < F_L$, which is what makes devices like hydraulic lifts useful. A small force is magnified by the factor A_L/A_S.

$$\frac{A_L}{A_S} > 1$$

$$F_L > F_S$$

B A large SUV of mass $m = 2.67 \times 10^3$ kg is at rest on the car lift in Figure 15.17. Both pistons have a circular cross-sectional area. The small piston has radius $R_S = 3.81$ cm, and the large piston has radius $R_L = 20.3$ cm. Find the magnitude of the force F_S exerted on the small piston. (Don't worry that we have simplified the hydraulic lift.)

INTERPRET and ANTICIPATE

In this part, start with Equation (1). The two forces on the vehicle are the upward force due to the lift $\vec{F}_L$ and the downward gravitational force $\vec{F}_g$. Because the vehicle is at rest, the magnitude of these two forces must be equal. We expect to find that F_S is less than the weight of the vehicle.

SOLVE

Apply Equation (1) with $F_L = F_g$. Write the weight of the vehicle in terms of its mass.

$$F_L = \frac{A_L}{A_S} F_S = F_g = mg$$

Make substitutions and solve for F_S. Use the familiar formula for the area of a circle.

$$mg = \frac{A_L}{A_S} F_S$$

$$F_S = \frac{A_S}{A_L} mg = \frac{\pi R_S^2}{\pi R_L^2} mg = \left(\frac{R_S}{R_L}\right)^2 mg$$

$$F_S = \left(\frac{3.81 \text{ cm}}{20.3 \text{ cm}}\right)^2 (2.67 \times 10^3 \text{ kg})(9.81 \text{ m/s}^2)$$

$$F_S = 9.23 \times 10^2 \text{ N}$$

CHECK and THINK

As expected, F_S is less than the weight of the vehicle ($mg = 2.62 \times 10^4$ N). In fact, the force required to hold up this SUV is about the same as a typical man's weight.

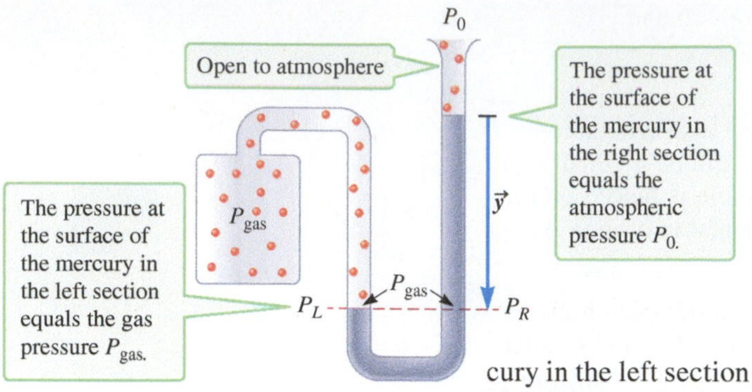

FIGURE 15.18 An open-tube manometer is used to measure gas pressure. The gauge pressure is found by measuring the height y: $P_{gauge} = \rho gy$.

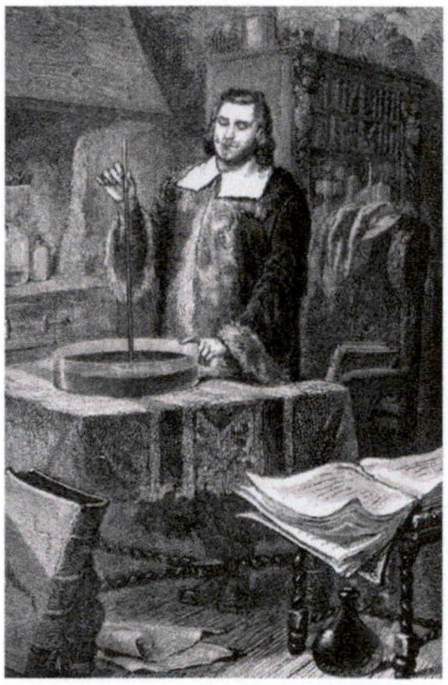

FIGURE 15.19 Evangelista Torricelli invented the mercury barometer in 1644.

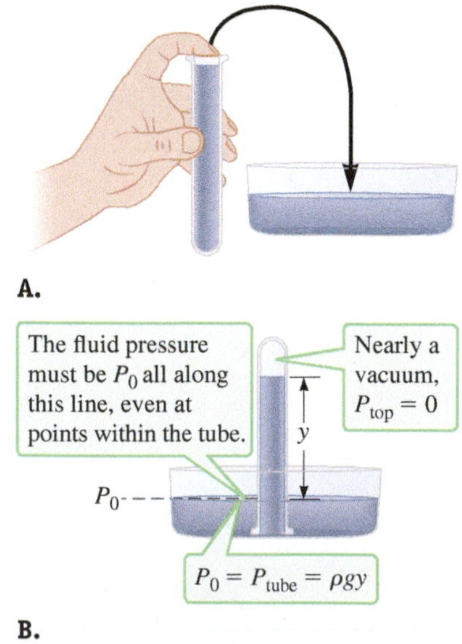

FIGURE 15.20 How a mercury barometer is made. **A.** A glass tube is filled with mercury and sealed off. **B.** The tube is inverted in a dish of mercury.

Manometers, Gauge Pressure, and Absolute Pressure

An open-tube **manometer** is a U-shaped tube with one end open to the atmosphere and the other end attached to a container full of a fluid (usually a gas) whose pressure P_{gas} you would like to measure (Fig. 15.18). The U-shaped tube contains an incompressible fluid of known density ρ such as mercury or water; just to be specific, let's assume it is mercury. As long as the gas pressure does not equal atmospheric pressure—$P_{gas} \neq P_0$—the height of the mercury in the left section does not equal the height of the mercury in the right section. If we draw a horizontal plane that passes through a point on the surface of the mercury in the left section and a point at a depth y in the right section, the pressure at those two points must be equal:

$$P_L = P_R \tag{15.12}$$

The pressure on the left equals the gas pressure, so

$$P_L = P_{gas} = P_R \tag{15.13}$$

We use $P_R = \rho gy + P_0$ (Eq. 15.6) to find the pressure on the right and write the gas pressure as

$$P_{gas} = \rho gy + P_0 \tag{15.14}$$

Equation 15.14 gives the **absolute pressure** of the gas: $P_{abs} = P_{gas} = \rho gy + P_0$. As you can see, the absolute pressure depends on the atmospheric pressure P_0. When measuring the pressure in our tires or scuba tanks, we are more interested in the **gauge pressure** P_{gauge} *relative* to atmospheric pressure:

$$P_{gauge} \equiv P_{abs} - P_0 \tag{15.15}$$

The gauge pressure just depends on the height of the mercury in the manometer:

$$P_{gauge} = P_{gas} - P_0$$
$$P_{gauge} = \rho gy + P_0 - P_0$$

$$P_{gauge} = \rho gy \tag{15.16}$$

In terms of practical measurements, if you want the gauge pressure, you only need to measure the height y of the mercury in the manometer. If you need to know the absolute pressure, you must measure y and know the atmospheric pressure P_0.

Barometers

Often it is okay to assume $P_0 = 1\text{ atm} = 1.01 \times 10^5$ Pa, but atmospheric pressure depends on altitude and weather conditions, so when a more precise value is needed, atmospheric pressure should be measured. A **barometer** is a device used to measure atmospheric pressure. Italian physicist Evangelista Torricelli (1608–1647) invented the mercury barometer in 1644 (Fig. 15.19). A simple way for Torricelli to build a mercury barometer would have been for him to pour mercury into a dish and also into a long, glass tube. The tube is permanently sealed by the glass on one end, and he may have temporarily sealed the other end with his finger (Fig. 15.20A). He would then invert the tube into a dish of mercury and remove his finger. When done, some (not all) of the mercury in the tube flows into the dish. Because the tube is initially filled with mercury and nothing is allowed to enter the tube, there is essentially nothing—that is, a vacuum—above the mercury in the tube (Fig. 15.20B). Therefore, the pressure at the top surface of the mercury in the tube is essentially zero: $P_{top} = 0$. The atmosphere presses down on the mercury in the dish, so the pressure at that surface is atmospheric pressure P_0. We find P_0 by measuring the height of the mercury in the tube. According to Equation 15.6, the pressure P_{tube} at the depth y is

$$P_{tube} = \rho gy + P_{top}$$

Because there is a vacuum above the mercury in the tube, $P_{top} = 0$ and

$$P_{tube} = \rho g y + 0 = \rho g y$$

Finally, because the pressure must be equal all along the horizontal plane, $P_{tube} = P_0$, and

$$P_0 = \rho g y \tag{15.17}$$

If we know the density of the fluid (usually mercury) in the barometer and the acceleration due to gravity g, by measuring y we can find the atmospheric pressure P_0, which is why millimeters of mercury (mm Hg) may be used as a pressure unit.

EXAMPLE 15.5 Building a Barometer

A If you wish to build your own mercury barometer, what is the minimum possible length of your tube?

INTERPRET and ANTICIPATE

Atmospheric pressure will push the mercury up in the tube. The greater the atmospheric pressure, the higher the mercury will go. If the tube is too short, the mercury will be forced to the very top, and you won't be able to measure the pressure. Assume the atmospheric pressure to be measured is roughly the pressure near sea level. Of course, if the pressure increases due to weather changes or because you take the barometer to a lower elevation such as the Dead Sea, you might find that the "sea-level" barometer is too short. From Figure 15.19, we estimate the barometer to be between 2 ft and 3 ft tall.

SOLVE

Solve Equation 15.17 for y and substitute the appropriate values. The density of mercury is given in Table 15.1.

$$P_0 = \rho g y \tag{15.17}$$

$$y = \frac{P_0}{\rho g} = \frac{1.01325 \times 10^5 \text{ Pa}}{(13{,}600 \text{ kg/m}^3)(9.81 \text{ m/s}^2)}$$

$$y = 0.760 \text{ m}$$

CHECK and THINK

The barometer's height ($y \approx 2.5$ ft) is in the expected range. You can also check the answer from the conversion 1 atm = 760 mm Hg (or torr); it says that 1 atmosphere of pressure raises mercury to a height of 760 mm, or 0.760 m. You may also have heard a weather forecaster report the pressure as 29.92 inches. That is because 29.92 in. (about 2.5 ft) is about 760 mm, so the pressure is 1 atm.

B Mercury vapor is particularly harmful to human health. To avoid the health hazards associated with handling mercury, you decide to build a water barometer. How long would your tube need to be? Why did Torricelli use mercury?

INTERPRET and ANTICIPATE

In this part, use the density of water instead of the density of mercury. We expect the water barometer to be larger than the mercury barometer.

SOLVE

Substitute the appropriate values, including the density of water from Table 15.1.

$$y = \frac{P_0}{\rho g} = \frac{1.01325 \times 10^5 \text{ Pa}}{(1000 \text{ kg/m}^3)(9.81 \text{ m/s}^2)}$$

$$y = 10.3 \text{ m}$$

CHECK and THINK

As expected, the water barometer is larger than the mercury barometer. In fact, a water barometer would need to be roughly three stories tall. It would be difficult—if not impossible—for Torricelli to have a glass tube of this size. However, as of 2013, the tallest barometer in the world (located at Portland State Universiy in Oregon) is made of glass tubes and is more than 14 m tall.

FIGURE 15.21 A. Fluid flow can be complex and turbulent as in these river rapids. **B.** A calmly flowing river is an example of ideal fluid flow.

IDEAL FLUID FLOW

✪ Major Concept

15-6 Ideal Fluid Flow

We now shift our attention from fluids at rest to moving fluids such as the water flowing in a river. When water passes through rapids (Fig. 15.21A) or runs close to the shore, fluid motion can be complicated, but the motion of water in a calm river (Fig. 15.21B) is much simpler. In this book, we study the simple motion of **incompressible** fluids, known as **ideal fluid flow.**

There are three characteristics of ideal fluid flow: It is *steady*, it is *irrotational*, and it is *nonviscous*. To understand steady flow, imagine watching cars on an amusement park ride and consider just one small portion of the track near the top of a hill (Fig. 15.22). Each car has the same velocity as it passes over that part of the track. In **steady flow** (also known as **laminar flow**), the fluid's velocity at any fixed point is constant, but a fluid in steady flow may accelerate as it moves along its path. Again imagine cars on the amusement park ride. If you consider another part of the track near the bottom of the hill, you find that each car passing over that part of the track has a greater velocity than what it had near the top. Therefore, each car accelerates as it moves along the track.

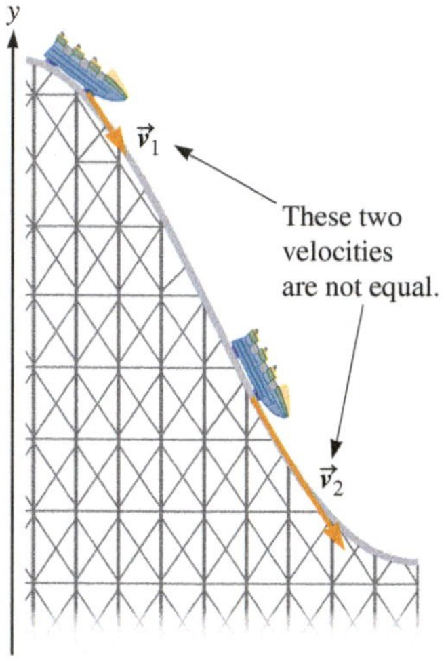

FIGURE 15.22 Cars on a roller coaster are like fluid molecules in steady flow.

An ideal flow is also irrotational; if a test object such as a leaf were placed in a fluid such as a river it would not rotate. The river may have a curved path, but the leaf would always face the same direction, just like the passengers on a Ferris wheel (Fig. 12.2A, page 332).

The third characteristic of ideal flow—the nonviscous part—is analogous to the motion of an object without resistive forces such as kinetic friction or drag. Just as kinetic friction resists motion and causes solid objects to stop, viscosity resists flow and causes fluids to come to rest. Imagine a race between a puddle of honey and puddle of water. Honey has a higher viscosity than water, so the water flows more easily than the honey. More precisely, **viscosity** is a measure of a fluid's resistance to deformation when a shear stress is applied to the fluid.

Although kinetic friction cannot be completely eliminated from any real system, we have seen that it is very useful to study ideal, frictionless systems. Likewise, viscosity cannot be completely eliminated from any fluid. Both kinetic friction and viscosity can often be ignored, however. Some fluids such as water, mercury, and alcohol have low viscosities. We say that these fluids are "thin." "Thick" fluids like honey, oil, and hot lava have higher viscosities. The viscosity of a fluid usually depends on its temperature such that viscosity is lower at higher temperatures. You probably know that the oil in your car is thicker and more viscous when the engine is cold than it is after you warm up the engine, and chocolate sauce kept in the refrigerator is thicker than when it is in a hot fondue pot. Viscosity also tends to be lower far from any walls or obstructions. If you imagine a lava flow, the viscosity is higher near the banks of the flow than near the center. Ideal fluid flow is assumed to be **nonviscous**, meaning that the viscosity is zero; in other words, we assume there is no resistance to flow.

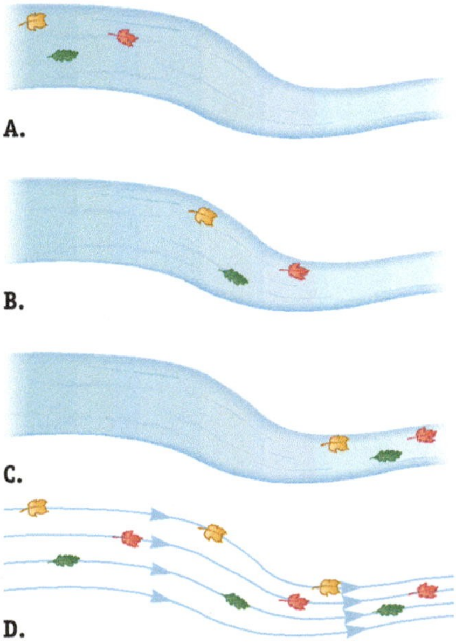

FIGURE 15.23 A–C. Three snapshots of a river show the motion of several leaves. **D.** Streamlines show the path of several leaves flowing in the river. Each leaf follows a streamline.

The study of moving fluids requires a few special tools. When studying the motion of solid objects, it is helpful to make motion diagrams. In Section 2-2, we imagined making a motion diagram of Mars by taking photos of the planet every few nights. Motion diagrams are not helpful for studying flowing fluids. Imagine taking a picture of a river at regular time intervals. You would not be able to detect any changes in the river that would help you study the water's flow. If a few leaves were floating on the water when you took your photos, however, you would be able to detect their motion (parts A–C of Fig. 15.23). Each leaf is attached to one small part of the water known as a **fluid element**. So, the motion diagram for the leaves shows the path of the fluid elements. A fluid element's path is called a **streamline**, and several streamlines are usually drawn to illustrate a flow (Fig. 15.23D). In practice, the leaves shown in Figure 15.23 are too spread out to trace streamlines adequately. Often, dyes or smoke particles (Fig. 15.24) are added to a fluid to better trace the streamlines.

Streamlines are a visualization tool, and interpreting a sketch of a fluid's streamlines is much like interpreting a sketch of a particle's path. A particle's velocity is tangent to its path. Similarly, the velocity of a fluid element is tangent to its streamline (Fig. 15.25), which leads to a few important facts about streamlines. (1) Because a fluid element must have only one velocity at any instant, streamlines cannot cross. If they did cross, at the intersection the fluid element would have two velocities simultaneously. (2) Streamlines for ideal fluid flow are constant in time. That is, the sketch of the streamlines is frozen in time, but each fluid element is in motion along its streamline. (3) The distance between streamlines is related to the speed of the fluid in that region. Where the streamlines are close together, the fluid is moving quickly. Widely separated streamlines characterize slow flow. We will discuss this idea more fully in the next section.

Insects in flight: direct visualization and flow measurements R J Bomphrey 2006 Bioinspir. Biomim. 1 S1 © 2015 IOP Publishing

FIGURE 15.24 Smoke particles in a wind tunnel show the streamlines of air as it flows past a dragonfly.

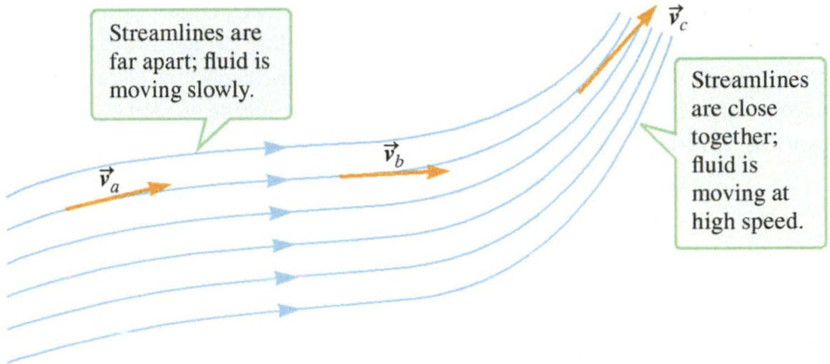
Streamlines are far apart; fluid is moving slowly.
$\vec{v}_c$
Streamlines are close together; fluid is moving at high speed.
$\vec{v}_b$
$\vec{v}_a$

FIGURE 15.25 Streamlines show the path of several fluid elements. Fluid velocity vectors are tangent to the streamlines.

15-7 The Continuity Equation

Zak creates his own shower by partially blocking the opening of the hose with his thumb (Fig. 15.26A). When he removes his thumb, the water trickles out of the end of the hose (Fig. 15.26B). These two situations differ in the size of the hole that the water passes through. Water is essentially incompressible, so all the water that enters the hose in a certain time interval—say1 second—must leave the hose in the same time interval, whether through a partially open hole or a completely open hole. Because the partially open hole is smaller than the completely open hole, the water must pass through it faster to get the same amount of water through the hose in the 1-second time interval.

To quantify this common experience, imagine a fluid flowing through a tube that is wide on one end and narrow on the other (Fig. 15.27). Let's derive an expression for the speed v of the fluid in terms of the cross-sectional area A of the tube. During a time interval Δt, a volume V of fluid enters the left region. Because the fluid is incompressible, an equal volume V of fluid must leave the right region in the

A.
B.
Photos © Dr. Jenny Spinner/St. Joseph's University

FIGURE 15.26 **A.** Zak creates a shower by reducing the cross-sectional area of the hose. When the cross-sectional area is small, the flow is fast. **B.** When the cross-sectional area is large, the flow is slow.

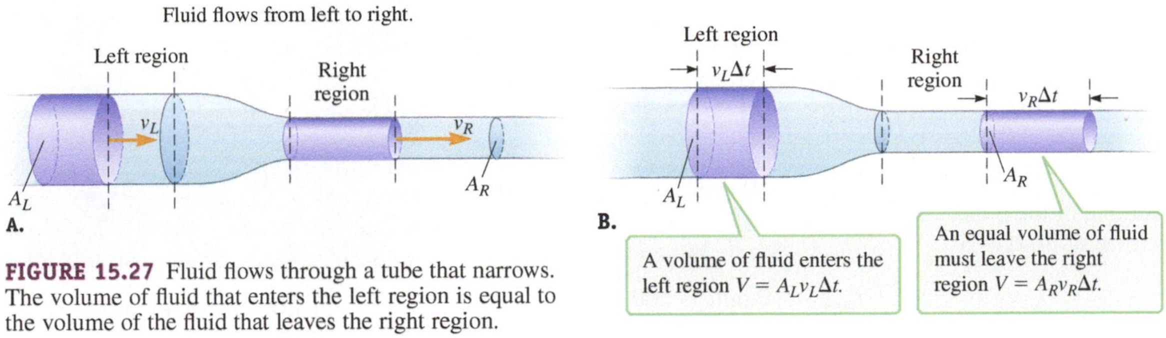
Fluid flows from left to right.
Left region
Right region
v_L
v_R
A_L
A_R
A.
Left region
$v_L\Delta t$
Right region
$v_R\Delta t$
A_L
A_R
B.
A volume of fluid enters the left region $V = A_Lv_L\Delta t$.
An equal volume of fluid must leave the right region $V = A_Rv_R\Delta t$.

FIGURE 15.27 Fluid flows through a tube that narrows. The volume of fluid that enters the left region is equal to the volume of the fluid that leaves the right region.

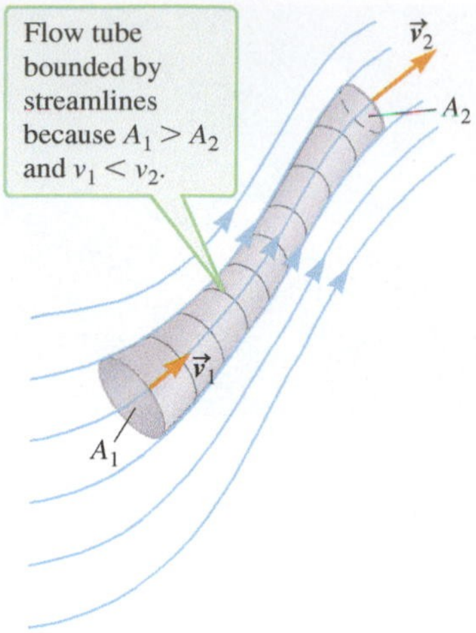
Flow tube bounded by streamlines because $A_1 > A_2$ and $v_1 < v_2$.
$\vec{v}_2$
A_2
$\vec{v}_1$
A_1

FIGURE 15.28 An imaginary tube known as a *flow tube* whose boundaries consist of streamlines.

same time Δt. We can write this volume in terms of the fluid's speed. On the left, the fluid's speed is v_L, so in the time interval Δt, the fluid's displacement is $v_L\,\Delta t$ along the axis of the tube. The volume of fluid that enters the left region is the cross-sectional area of the left side of the tube multiplied by the displacement of the fluid:

$$V = A_L v_L\,\Delta t \tag{15.18}$$

Similar reasoning can be used to find the volume of fluid exiting the right region:

$$V = A_R v_R\,\Delta t \tag{15.19}$$

Because the volume of fluid entering the right region equals the volume of fluid leaving the left region, we set Equation 15.18 equal to Equation 15.19:

$$A_R v_R\,\Delta t = A_L v_L\,\Delta t$$

$$A_R v_R = A_L v_L \tag{15.20}$$

Although Equation 15.20 had an actual tube (Fig. 15.27) in mind, Equation 15.20 also holds for an imaginary tube known as a **flow tube** whose boundaries consist of streamlines (Fig. 15.28). Because streamlines cannot cross, the fluid moves in the flow tube just as it would in a real tube. If the cross-sectional area is A_1 on one end of the flow tube and A_2 on the other end, the fluid speed at these two ends is given by Equation 15.20:

CONTINUITY EQUATION

Major Concept

$$A_1 v_1 = A_2 v_2 \tag{15.21}$$

Equation 15.21 is called the **continuity equation.**

The quantity Av is known as the **volume flow rate** $R = Av$. Volume flow rate has the dimensions of volume per time, so the SI units are cubic meters per second, or m^3/s. We can restate the continuity equation as follows: *In ideal fluid flow, the volume flow rate is uniform* (that is, it has the same value over the length of the flow tube).

One consequence of the continuity equation is that when a fluid flows from a tube of given cross-sectional area through a region with a smaller cross-sectional area, the fluid speed must increase. Another version of this statement is as follows: *In regions where the streamlines are closely crowded, the fluid flows quickly* (Fig. 15.28).

EXAMPLE 15.6 Keep Your Pets Alive

Koi are a fish known to produce a lot of waste. If you wish to keep a koi fishpond (Fig. 15.29), you will need a very good water filter. Professional-grade filters are very large and may be buried in the ground near the pond. The effectiveness of the filter is determined by the length of time the polluted water is held in it. Ideally, you would like the polluted water to remain in the filter for at least 10 min. A large pond may hold 6500 gallons, and the entire pond should be processed through the filter every 3 hours. What is the required volume flow rate (in cubic meters per second and gallons per minute)? What size filter (in cubic meters and in gallons) is required?

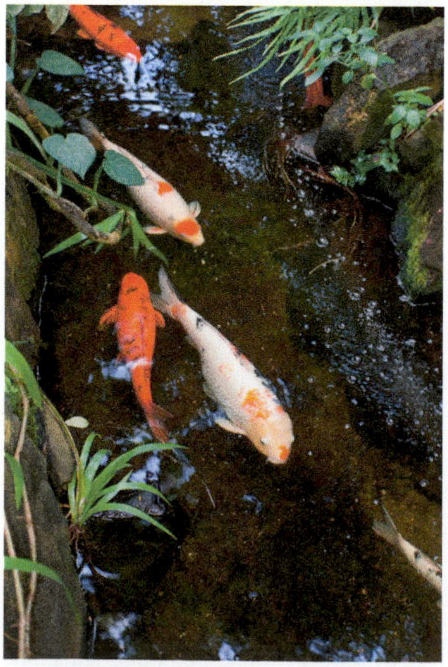

FIGURE 15.29 Koi fish.

INTERPRET and ANTICIPATE

Use the fact that the entire pond must be processed in 3 hours to find the required volume flow rate. Once that value is known, use the fact that the water should remain in the filter for 10 min to find the filter's size.

SOLVE

First, convert the volume of the pond, filter time, and processing time to SI units.

$$V_{\text{pond}} = 6500\text{ gal}\left(\frac{1\text{ m}^3}{264\text{ gal}}\right) = 24.6\text{ m}^3$$

$$t_{\text{process}} = 3\text{ h} = 180\text{ min} = 1.08 \times 10^4\text{ s}$$

$$t_{\text{filter}} = 10\text{ min} = 600\text{ s}$$

The required volume flow rate R comes from the whole volume V_{pond} having to be filtered in time $t_{process}$.	$R = \dfrac{V_{pond}}{t_{process}} = \dfrac{24.6\,\text{m}^3}{1.08 \times 10^4\,\text{s}} = 2.28 \times 10^{-3}\,\text{m}^3/\text{s}$ $R = \dfrac{V_{pond}}{t_{process}} = \dfrac{6500\,\text{gal}}{180\,\text{min}} = 36.1\,\text{gal/min}$
Use the volume flow rate and filter time to find the volume of the filter.	$V_{filter} = Rt_{filter} = (2.28 \times 10^{-3}\text{m}^3/\text{s})(600\,\text{s}) = 1.4\,\text{m}^3$ $V_{filter} = Rt_{filter} = (36.1\,\text{gal/min})(10\,\text{min}) = 360\,\text{gal}$

CHECK and THINK

Let's take a moment to appreciate the size of this filtering system. First, a typical 10-minute shower uses 30 to 50 gal of water, so the shower volume flow rate is 3 to 5 gal/min. The koi fishpond system has about 10 times the volume flow rate of a typical shower. A fairly large kitchen refrigerator has an interior volume of about 200 gal, so the required koi filter is nearly the size of two refrigerators. Such large filters are usually installed underground when the pond is constructed.

EXAMPLE 15.7 A Most Educational Demonstration

You have probably seen science demonstrations using common household items. In one such experiment, 13 Mentos mints are dropped quickly into a 2-L bottle of cola. The cola is forced rapidly out of the bottle (Fig. 15.30). Assuming it takes 0.6 s to empty the 2-L bottle, estimate the maximum height of the cola spray.

FIGURE 15.30

INTERPRET and ANTICIPATE

We have enough information (bottle volume and emptying time) to estimate the volume flow rate. From that estimate, use the equation of continuity to find the speed of the cola as it exits the bottle. (We must estimate the area of the bottle opening.) To find its maximum height, model the cola as a projectile launched straight up and assume gravity is the only force acting on the cola after it leaves the bottle.

SOLVE

Find the volume flow rate using the fact that the 2 L are emptied in 0.6 s.	$V_{bottle} = 2\,\text{L} \times \dfrac{1\,\text{m}^3}{1000\,\text{L}} = 2 \times 10^{-3}\,\text{m}^3$ $R = \dfrac{V_{bottle}}{t} = \dfrac{2 \times 10^{-3}\,\text{m}^3}{0.6\,\text{s}}$ $R = 3.3 \times 10^{-3}\,\text{m}^3/\text{s}$
Estimate the area of the bottle opening. The diameter of the opening is about 1 in., which means that the radius is about 1.3 cm.	$A_{opening} = \pi r^2 \approx \pi(1.3 \times 10^{-2}\,\text{m})^2$ $A_{opening} \approx 5.3 \times 10^{-4}\,\text{m}^2$
According to the continuity equation, the volume flow rate is a constant.	$R = A_{opening}v_{cola}$ $v_{cola} = \dfrac{R}{A_{opening}} = \dfrac{3.3 \times 10^{-3}\,\text{m}^3/\text{s}}{5.3 \times 10^{-4}\,\text{m}^2} = 6.3\,\text{m/s}$
There are many ways to find the correct expression for the maximum height. Here, we have chosen to use kinematics (Eq. 2.13). The initial speed v_{0y} is the speed v_{cola} found above, and the final speed $v_y = 0$ when $\Delta y = y_{max} - 0 = y_{max}$.	$v_y^2 = v_{0y}^2 + 2a_y\,\Delta y$ (2.13) $0 = v_{cola}^2 - 2gy_{max}$ $y_{max} = \dfrac{v_{cola}^2}{2g} = \dfrac{(6.3\,\text{m/s})^2}{2(9.81\,\text{m/s}^2)} \approx 2\,\text{m}$

CHECK and THINK

The exit speed of the cola may seem a bit high at roughly 14 mph, but the maximum height (2 m ≈ 6 ft) seems to be about right as seen in Figure 15.31.

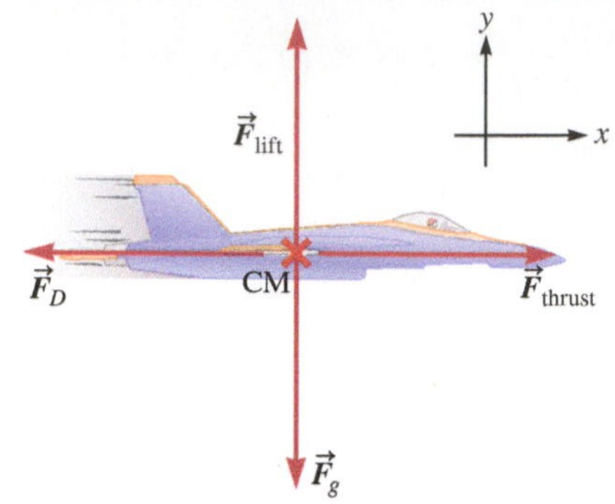

FIGURE 15.31 Free-body diagram for an airplane flying at constant velocity in the positive x direction.

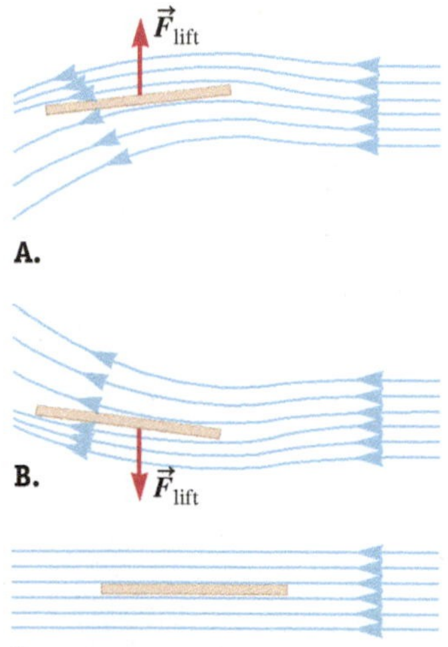

FIGURE 15.32 A flat door is in a wind tunnel. **A.** If the door's right edge is tilted upward, air moving past the wind is deflected downward, and the air exerts an upward force on the board known as *lift*, $\vec{F}_{\text{lift}}$. **B.** If the door's right edge is tilted downward, the air is deflected upward, and the lift is downward. **C.** If the door is not tilted, there is no net deflection of the air, and there is no lift exerted on the board.

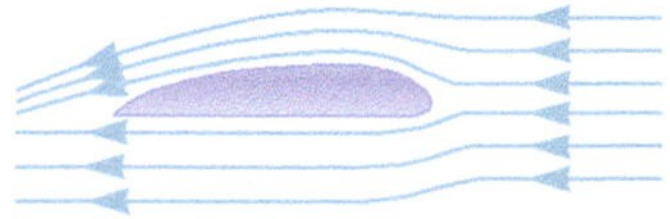

FIGURE 15.33 Cross section of a wing shows that it has a curved upper surface and flat lower surface. This design ensures that when the plane is flying level, air moving past the wing is deflected downward. Flaps on the wings allow pilots to further adjust the shape of the wing to control the lift.

15-8 Bernoulli's Equation

In this section, we return to the issues raised in our case study, namely how can an airplane fly upside down as well as right side up; indeed, how can it fly at all given that it is heavier than an equal volume of air? In Section 15-4, Cameron said that "planes fly because they are moving." It is true that there must be relative motion between the airplane wings and the surrounding air. In this section, we'll see why motion is necessary.

Let's start by drawing a free-body diagram for a jet plane flying at constant velocity in the positive x direction (Fig. 15.31). Gravity acts in the negative y direction (downward in the figure). Air drag (Eq. 6.5) and thrust (Eq. 10.16) act in opposite directions along the x axis. Because the plane is flying at constant velocity, there is no net force acting on it. Because the only forces acting in the x direction are thrust and drag, the magnitude of the thrust must equal the magnitude of the drag force: $F_{\text{thrust}} = F_D$. There must also be an upward force exerted on the plane; otherwise, the plane would accelerate downward. This upward force is known as **lift** ($\vec{F}_{\text{lift}}$ in Fig. 15.31). Lift is exerted by air primarily on the wings of the plane.

For air to exert lift on a wing, there must be relative motion between the air and the wing. As a first approximation, imagine removing a simple, flat door from its hinges to use as a wing. Figure 15.32 shows the short edge of the door (a simple board). Imagine the board tilted with its right edge upward in a wind tunnel (Fig. 15.32A). The air flows from right to left. The streamlines shown on the right are parallel and equally spaced, showing that the air is moving uniformly in a straight path toward the board. The streamlines near the board, however, are bent downward, showing that the board deflects the air downward. So, the board exerts a downward force on the air. According to Newton's third law, the air must exert an upward force on the board. This upward force is lift. If the board is tilted with its right edge downward (Fig. 15.32B), the air is deflected upward. In this case, the air must exert a downward force on the board, which we still call "lift" even though that terminology seems odd. If the board is level, the air is not deflected up or down, and there is no lift exerted on the board (Fig. 15.32C). So, although you might think that a plane, or at least its wing, must be tilted as in Figure 15.32A to have upward lift, wings are designed so that the air is deflected downward even when the plane is flying level (Fig. 15.33).

CASE STUDY Part 3: A Moving Explanation

Avi: I think I get the basic idea. The wing of the plane is designed so that when the plane moves through the air, the wing forces the air downward. That means that the air must force the plane upward. I just can't see how this is related to air pressure and helium balloons.

Cameron: Helium balloons rise because they are less dense than air. Sometimes you even see helium balloons that just hover in the middle of the room. That's because they've lost some helium and now are just as dense as the air around them.

Shannon: The helium balloon rises in the air because the air exerts a buoyant force on the balloon. And that buoyant force is greater than the balloon's weight.

Avi: That idea works for a balloon, but a plane weighs a lot. There is no way the buoyant force is greater than its weight!

Cameron: That's right. Because the plane is denser than air, the buoyant force cannot get it off the ground.

Shannon: You're both right. Look at it this way. If you blow a bubble under water, it rises to the surface. That's because the water pressure below the bubble is greater than the water pressure above the bubble. When the plane is just sitting on the runway, the air pressure above the wing is about the same as the air pressure below the wing. When the plane moves, the air pressure below the wing is greater than the air pressure above the wing.

Shannon is correct again. When the plane is stationary, the air pressure is nearly uniform over its entire surface, so the air does not exert a net force on the plane. When the air moves relative to the wings, however, the air pressure above the wings is lower than the air pressure below them, and the air exerts a net force—lift—on the wings.

Pressure in a Moving Fluid

We have just discovered that the motion of a fluid may change its pressure. To derive an expression for the pressure in a moving fluid, we need to take a conservation of energy approach (Chapter 8). Consider a fluid moving in a bent pipe (Fig. 15.34). In this scenario, a piston moves to the right, pushing fluid to the right and upward.

Our system consists of the Earth and the portion of fluid shown in orange in Figure 15.34. The piston and the rest of the fluid (shown in blue) are external to the system and may do work on it. The piston applies a constant force $\vec{F}_{\text{piston}} = F_{\text{piston}}\hat{\imath}$, displacing the point of contact between the fluid and the piston by the amount Δx_1. Therefore, the work done by the piston is positive and is given by

$$W_{\text{piston}} = F_{\text{piston}}\,\Delta x_1 \tag{15.22}$$

According to Newton's third law, the fluid in the system exerts a force on the piston equal in magnitude to F_{piston}. If the piston has an area A_1, we can write F_{piston} in terms of the fluid pressure P_1 in region 1: $F_{\text{piston}} = P_1A_1$. Equation 15.22 now becomes

$$W_{\text{piston}} = P_1A_1\,\Delta x_1 \tag{15.23}$$

As shown in Figure 15.34B, the fluid in the system (shown in orange) pushes the external fluid (shown in blue) in the positive x direction by applying a force $\vec{F}_{\text{system}} = F_{\text{system}}\hat{\imath}$. The area of the pipe in region 2 is A_2, so we can write $\vec{F}_{\text{system}}$ in terms of the pressure P_2 in region 2: $\vec{F}_{\text{system}} = P_2A_2\hat{\imath}$. According to Newton's third law, the external fluid exerts a force on the system equal in magnitude to $\vec{F}_{\text{system}}$, but in the negative x direction: $\vec{F}_{\text{external}} = -P_2A_2\hat{\imath}$. The point of contact between the system and the external fluid is displaced by amount Δx_2 in the positive x direction. Therefore, the external fluid does negative work on the system, given by

$$W_{\text{external}} = -P_2A_2\,\Delta x_2 \tag{15.24}$$

There are no other external forces exerted on the system, so the total work done on the system is just the sum of Equations 15.23 and 15.24:

$$W_{\text{tot}} = P_1A_1\,\Delta x_1 - P_2A_2\,\Delta x_2 \tag{15.25}$$

Comparing parts A and B of Figure 15.34, you see that the system fluid that was initially in region 1 moves out of region 1 and that the system fluid that was just to the left of region 2 moves into region 2. Because the fluid is incompressible, the volume of region 1 must equal the volume of region 2:

$$V_1 = V_2 \equiv V$$

$$A_1\,\Delta x_1 = A_2\,\Delta x_2 = V \tag{15.26}$$

Substitute Equation 15.26 into Equation 15.25:

$$W_{\text{tot}} = P_1V - P_2V$$

$$W_{\text{tot}} = (P_1 - P_2)V \tag{15.27}$$

We now have an expression for the work done on a system that includes fluid moving between two regions.

Next, we need to calculate the changes in the system's kinetic and potential energies. Consider the change in kinetic energy first. Because we assume ideal fluid flow, the speed of a fluid element (and therefore its kinetic energy) depends only on its location in the pipe. Because there is system fluid in the middle region at both the initial and final times, fluid in the middle region does not come into the calculation of the *change* in kinetic energy. The change in kinetic energy ΔK only depends on the fluid that moves out of region 1 and the fluid that moves into region 2. The mass m of the fluid in region 1 equals the mass m of the fluid in region 2 because these two regions have equal volumes and the fluid is incompressible. Therefore,

$$\Delta K = \tfrac{1}{2}mv_2^2 - \tfrac{1}{2}mv_1^2 \tag{15.28}$$

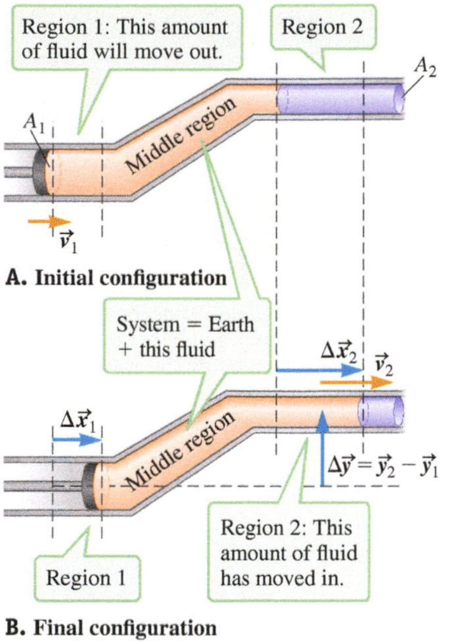

FIGURE 15.34 A piston forces fluid in a pipe to move up and to the right. The fluid in orange is part of our Earth–fluid system. The rest of the fluid, shown in blue, and the piston are outside the system and do work on the system. Fluid moves out of region 1, and fluid moves into region 2. There is no net change in the fluid in the middle region.

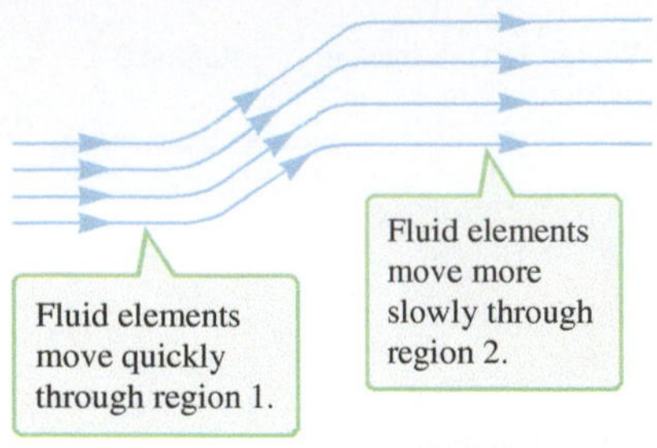
Fluid elements move more slowly through region 2.
Fluid elements move quickly through region 1.

FIGURE 15.35 Fluid elements move into and out of regions 1 and 2, but the overall fluid flow represented by the streamlines does not change in time.

The same sort of reasoning can be applied to the change in gravitational potential energy. Because the *change* in gravitational potential energy is determined by the *change* in vertical position (along the y axis), we only need to consider the fluid that moves out of region 1 and the fluid that moves into region 2:

$$\Delta U_g = mg\,\Delta y = mg(y_2 - y_1) \tag{15.29}$$

In ideal fluid flow, there are no viscous (resistive) forces, so there is no change in the fluid's internal energy. Therefore, we can express the conservation of energy using the work–mechanical energy theorem (Eq. 9.26):

$$W_{\text{tot}} = \Delta K + \Delta U$$

Substitute Equations 15.27, 15.28, and 15.29 for total work, change in kinetic energy, and change in potential energy, respectively:

$$(P_1 - P_2)V = \left(\tfrac{1}{2}mv_2^2 - \tfrac{1}{2}mv_1^2\right) + mg(y_2 - y_1)$$

Divide both sides by the volume V:

$$P_1 - P_2 = \frac{1}{V}\left(\frac{1}{2}mv_2^2 - \frac{1}{2}mv_1^2\right) + \frac{m}{V}g(y_2 - y_1)$$

Eliminate m and V by using $\rho = m/V$ (Eq. 1.1):

$$P_1 - P_2 = \left(\tfrac{1}{2}\rho v_2^2 - \tfrac{1}{2}\rho v_1^2\right) + \rho g(y_2 - y_1)$$

Finally, we group terms for region 1 on one side of the equation and terms for region 2 on the other side:

BERNOULLI'S EQUATION
Major Concept

$$P_1 + \tfrac{1}{2}\rho v_1^2 + \rho g y_1 = P_2 + \tfrac{1}{2}\rho v_2^2 + \rho g y_2 \tag{15.30}$$

Equation 15.30 is known as **Bernoulli's equation**. It is a statement of the work–energy theorem for ideal fluid flow. Bernoulli's equation looks much like the work–energy theorem (Eq. 9.31) with ΔE_{int} set to zero, but there is an important notational difference between Bernoulli's equation and the work–energy theorem. In the work–energy theorem, the subscripts i and f stand for *i*nitial and *f*inal times. Between these times, the configuration and motion of the system may change. In Bernoulli's equation, the subscripts 1 and 2 represent two different regions in an ideal fluid flow. Fluid elements move in and out of these regions, but the overall fluid flow represented by the streamlines does not change in time (Fig. 15.35).

With this difference in mind, we use Bernoulli's equation much as we use the conservation of energy principle. We used the conservation of energy principle when we had information about a system's energy at one particular time and needed information about another time. Likewise, we use Bernoulli's equation when we have information (such as pressure, speed, height) about a flowing fluid in one region and need information for another region.

PROBLEM-SOLVING STRATEGY Applying Bernoulli's Equation

:• INTERPRET and ANTICIPATE

Step 1 Draw a simple **sketch indicating your coordinate system**. The y axis is most important because vertical position coordinates are needed in Bernoulli's equation. Include streamlines if they help you visualize the problem.

Step 2 **Choose region 1 and region 2.** Usually, you know a lot of information about one region and need to know something about the other region. Label these regions on your sketch.

:• SOLVE

Step 3 Write **Bernoulli's equation.**

Step 4 **Compare the parameters** P, v, and y in the two regions. Often, one parameter is equal in the two regions, or you may consider a parameter to be approximately equal in the two regions.

Step 5 The **continuity equation** $A_1v_1 = A_2v_2$ (Eq. 15.21) is often helpful in finding or relating the speed in the two regions. If one cross-sectional area is very large compared with the other area, sometimes the fluid speed in the large area is approximately zero.

Step 6 **Do algebra** combining information in steps 3 through 5.

EXAMPLE 15.8 Doing Your Own Plumbing Job

At the restaurant where you work at your summer job, a large kitchen sink (1.20 m long, 0.46 m wide, and 0.36 m deep) is clogged and needs to be bailed out. Once there is about 5.0 cm of water left in the bottom, it becomes nearly impossible to continue bailing out the sink by scooping water into containers. Instead, you decide to siphon the remaining water with a hose of radius 1.2 cm into a cooking pot on the floor 0.79 m below the sink bottom.

A Find the speed of water emerging from the siphon hose.

B Find the volume flow rate from the hose.

C How long will it take you to empty the sink of that last 5.0 cm of water?

INTERPRET and ANTICIPATE

Apply Bernoulli's equation for part A, using the steps listed above. For part B, calculate the volume V of water in the sink from the dimensions given in the example opening. Then find the volume flow rate $R = Av$ by finding the cross-sectional area A of the hose and using the speed v of the water found in part A. The time needed to empty the sink comes from dividing the volume of water by the volume flow rate: $t = V/R$.

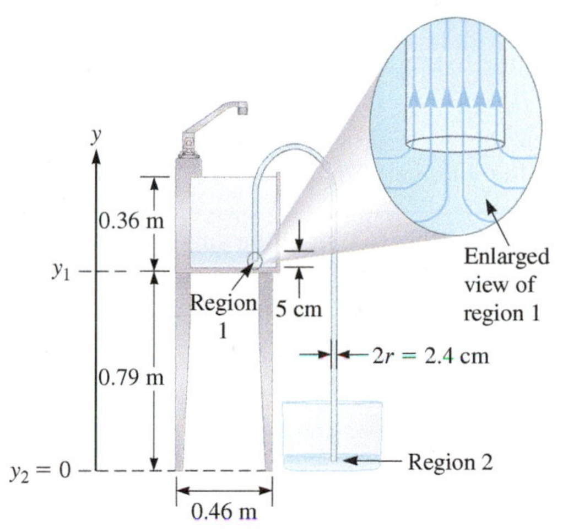

FIGURE 15.36

Steps 1 and 2 Sketch, add coordinate system, and choose regions. In our sketch (Fig. 15.36), we show the sink from the side. This perspective allows us to easily sketch the hose and pot. We indicate our coordinate system and our two regions. Region 1 is at the surface of the water near the entrance to the hose, and region 2 is at the exit. We have set $y_2 = 0$. The figure also shows an enlarged view of region 1 with streamlines in the sink spread out and then bending into the entrance of the hose and becoming closer together. The spacing of the streamlines indicates that the water moves slowly when it is in the sink, but speeds up in the hose.

A Speed of water, v_2

SOLVE

Step 3 Use **Bernoulli's equation** to find v_2 (the speed of the water as it emerges from the hose).

$$P_1 + \tfrac{1}{2}\rho v_1^2 + \rho g y_1 = P_2 + \tfrac{1}{2}\rho v_2^2 + \rho g y_2 \qquad (15.30)$$

Step 4 Compare *P*, *v*, and *y* in the two regions. The air pressure is assumed to be constant in the kitchen, so the pressure in regions 1 and 2 is roughly equal. The other parameters—speed v and vertical position y—are different in the two regions.

$$P_1 = P_2$$

Step 5 Relate speeds using the **continuity equation.** The streamlines in Figure 15.36 indicate that the water in the sink moves slowly and water in the hose moves quickly. You can confirm these speeds with the continuity equation $A_1 v_1 = A_2 v_2$ (Eq. 15.21): The area of the sink is large, and the cross-sectional area of the hose is small; so, the water in region 1 moves much slower than the water in region 2. We approximate the speed in region 1 as zero.

$$v_1 \approx 0$$

Step 6 Do algebra. Use the approximations $P_1 = P_2$ and $v_1 \approx 0$ found above to simplify Bernoulli's equation.

$$\cancel{P_1} + \tfrac{1}{2}\rho v_1^2 + \rho g y_1 = \cancel{P_2} + \tfrac{1}{2}\rho v_2^2 + \rho g y_2$$

$$\tfrac{1}{2}\rho (0)^2 + \rho g y_1 = \tfrac{1}{2}\rho v_2^2 + \rho g y_2$$

$$\rho g y_1 = \tfrac{1}{2}\rho v_2^2 + \rho g y_2$$

Solve for v_2.

$$v_2 = \sqrt{2g(y_1 - y_2)} \qquad (1)$$

Example continues on page 440 ▶

CHECK and THINK
Equation (1) is the same free-fall expression we would find for a particle dropped from rest (Section 2-10), which makes sense because gravity is the only force exerted on the water.

SOLVE, continued Substitute values into Equation (1) to find v_2.	$v_2 = \sqrt{2(9.81\,\text{m/s}^2)(0.79\,\text{m} - 0)}$ $v_2 = 3.9\,\text{m/s}$
B Volume flow rate, R **SOLVE** To find the volume flow rate, we need v_2 and the cross-sectional area of the hose.	$A = \pi r^2 = \pi(1.2 \times 10^{-2}\text{m})^2 = 4.5 \times 10^{-4}\text{m}^2$ $R = Av = (4.5 \times 10^{-4}\text{m}^2)(3.9\,\text{m/s})$ $R = 1.8 \times 10^{-3}\text{m}^3/\text{s}$
C Time to empty sink **SOLVE** When the water is in the sink, it forms a rectangular box. Find the volume of that box.	$V = lwd = (1.20\text{m})(0.46\text{m})(0.050\text{m}) = 2.8 \times 10^{-2}\text{m}^3$
Finally, the time to empty the sink is the volume divided by the volume flow rate.	$t = \dfrac{V}{R} = \dfrac{2.8 \times 10^{-2}\text{m}^3}{1.8 \times 10^{-3}\text{m}^3/\text{s}} = 1.6\text{ s}$

CHECK and THINK
Of course, this time seems short. In practice, it is difficult to get the water to flow. Some people suck on the opposite end as you would a straw (which is not a great idea if the fluid you need to siphon is toxic). There are also handheld pumps that may be used to start the fluid flowing. In practice, the more water there is in the sink, the easier it is to start the siphon and keep it going. So, instead of scooping out water, it is better to siphon the whole amount. In fact, people with large fish tanks often use a siphon to empty it.

EXAMPLE 15.9 A Venturi Tube

Figure 15.37 shows a horizontal tube in which the middle section—known as the *neck*—is narrower than the rest. This tube is known as a *Venturi* tube, and it may be used to model blood flow through capillaries (small blood vessels) or air flowing between buildings in a large city. Venturi tubes are also used to measure fluid speeds and to mix gasoline with air in automobile carburetors. The cross-sectional area of the wide part of the Venturi tube is A_1 and that of the neck is A_2, where $A_1 > A_2$.

A Compare the speed v_2 and pressure P_2 of the fluid in the neck to the speed v_1 and pressure P_1 of the fluid in the rest of the tube.

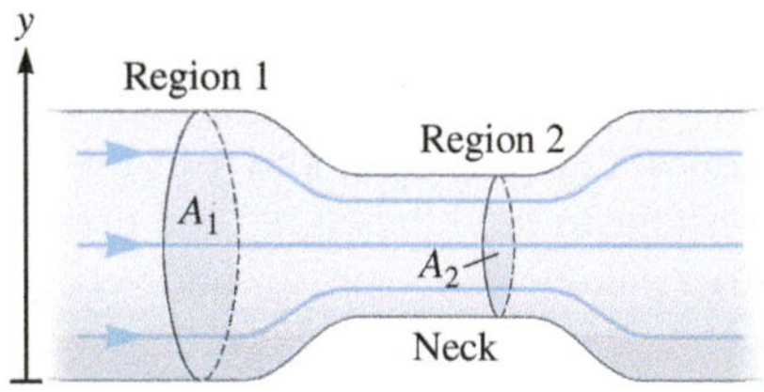

FIGURE 15.37 A Venturi tube.

INTERPRET and ANTICIPATE
Because we are asked to compare quantities in two regions, we expect to give two answers of the form $Q_1 > Q_2$, $Q_1 < Q_2$, or $Q_1 = Q_2$. We can use the continuity equation to make the comparison for speed. Use the streamlines in Figure 15.37 to anticipate the results. In region 1, the streamlines are far apart, whereas in region 2, they are closer together, so we expect $v_2 > v_1$. The comparison for pressure requires Bernoulli's equation.

SOLVE Begin with the continuity equation.	$A_1v_1 = A_2v_2$

Solve for v_2 and use $A_1 > A_2$.	$v_2 = \frac{A_1}{A_2}v_1$ (1) $v_2 > v_1$

CHECK and THINK

As expected, the fluid speed is higher in region 2 than in region 1.

SOLVE, continued Use Bernoulli's equation to find the pressure in each region. **Steps 1 (sketch)** and **2 (region selection)** are completed in Figure 15.37. So, we move on to **step 3:** Write **Bernoulli's equation**.	$P_1 + \frac{1}{2}\rho v_1^2 + \rho g y_1 = P_2 + \frac{1}{2}\rho v_2^2 + \rho g y_2$ (15.30)
In **step 4**, we **compare parameters** P, v, and y in the two regions. Figure 15.37 shows that y is the same all along the horizontal tube. We expect the pressure to be different in the two regions, and we have already found that $v_2 > v_1$.	$y_1 = y_2$
We've already applied the **continuity equation (step 5)**, and so we are on to **step 6: Do algebra.**	$P_1 + \frac{1}{2}\rho v_1^2 + \rho g y_1 = P_2 + \frac{1}{2}\rho v_2^2 + \rho g y_2$ $P_1 + \frac{1}{2}\rho v_1^2 = P_2 + \frac{1}{2}\rho v_2^2$ $P_1 - P_2 = \frac{1}{2}\rho(v_2^2 - v_1^2)$ (2)
Use the previous result, $v_2 > v_1$.	$(v_2^2 - v_1^2) > 0$ $\quad P_1 - P_2 > 0$ $P_1 > P_2$

CHECK and THINK

Our result that the pressure in region 1 is greater than the pressure in region 2 confirms the Venturi effect: In region 2, the speed is higher and the pressure is less, whereas in region 1, the speed is lower and the pressure is greater.

B Derive an expression for the speed v_2 of the fluid in the neck in terms of P_1, P_2, A_1, A_2, and ρ.

INTERPRET and ANTICIPATE

This part of the example is a continuation of part A. Pick up **step 5** (apply **continuity equation**) to eliminate v_1 from Equation (2).

SOLVE **Step 5** Substitute Equation (1), which came from the continuity equation, into Equation (2).	$v_1 = \frac{A_2}{A_1}v_2$ $P_1 - P_2 = \frac{1}{2}\rho\left[v_2^2 - \left(\frac{A_2}{A_1}v_2\right)^2\right] = \frac{1}{2}\rho v_2^2\left(1 - \frac{A_2^2}{A_1^2}\right)$
Solve for v_2.	$v_2^2 = \frac{2(P_1 - P_2)}{\rho\left(1 - \frac{A_2^2}{A_1^2}\right)} = \frac{2A_1^2(P_1 - P_2)}{\rho(A_1^2 - A_2^2)}$ $v_2 = \sqrt{\frac{2A_1^2(P_1 - P_2)}{\rho(A_1^2 - A_2^2)}}$ (3)
CHECK and THINK There are two things to check about Equation (3). First, make sure that it gives the square root of a positive number.	$\rho > 0$ $\quad A_1^2 > 0$ $\quad P_1 > P_2$ $\quad A_1^2 > A_2^2$ $\frac{2A_1^2(P_1 - P_2)}{\rho(A_1^2 - A_2^2)} > 0$

Example continues on page 442 ▶

Second, make sure we have the dimensions (L/T) of speed.	$\sqrt{\frac{[\![A_1^2]\!][\![(P_1 - P_2)]\!]}{[\![\rho]\!][\![(A_1^2 - A_2^2)]\!]}} = \sqrt{\frac{[\![(P_1 - P_2)]\!]}{[\![\rho]\!]}} = \sqrt{\frac{\text{F/L}^2}{\text{M/L}^3}}$ $\sqrt{\frac{(\text{M} \times \text{L/T}^2)/\text{L}^2}{\text{M/L}^3}} = \sqrt{\frac{\text{M} \times \text{L/T}^2}{\text{M/L}}} = \sqrt{\frac{\text{L}^2}{\text{T}^2}}$ $[\![v_2]\!] = \frac{\text{L}}{\text{T}}$

Equation (3) gives us a way to design a device—known as a *Venturi meter*—to measure the speed in a fluid of known density ρ. If you use a tube of known cross-sectional areas A_1 and A_2 and measure the pressure in regions 1 and 2 using, for example, a manometer (Section 15-5), you can solve for v_2. Of course, once you know v_2, you can use the continuity equation to solve for v_1.

EXAMPLE 15.10 CASE STUDY Flying Planes

Let's return to the case study one more time. Consider a small plane flying at constant velocity. The plane has mass $m = 1353$ kg, including its payload. The area of both wings combined is $A_{\text{wings}} = 16.3\ \text{m}^2$.

A If the air speed below the wing is 153 m/s, what is the air speed above the wing?

INTERPRET and ANTICIPATE

We know from the free-body diagram in Figure 15.31 that the magnitude of the lift exerted by air on the wings must equal the weight of the plane. Because lift is upward (the air exerts an upward force on the wings), we expect the air pressure below the wings to be greater than the pressure above the wings.

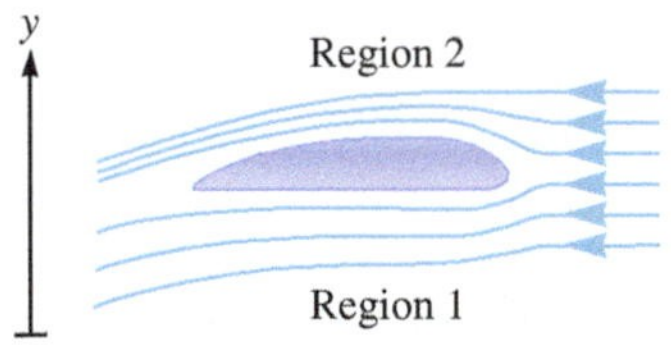

FIGURE 15.38

Relate air pressure to lift by applying Bernoulli's equation. **Steps 1 (sketch)** and **2 (choose regions)**: We have chosen an upward pointing y axis with region 2 above the wing and region 1 below the wing (Fig. 15.38).

The streamlines in region 2 are closer together than the streamlines in region 1, so air speed in region 2 is greater: $v_2 > v_1$. From Example 15.9, the pressure is lower in the region with the higher speed, so $P_2 < P_1$ as expected.

SOLVE

Step 3 Write **Bernoulli's equation.**	$P_1 + \frac{1}{2}\rho v_1^2 + \rho g y_1 = P_2 + \frac{1}{2}\rho v_2^2 + \rho g y_2$ (15.30)
Step 4 Compare parameters. The only remaining parameter to compare in the two regions is the y position. The top of the wing is only slightly above the bottom, so these regions are essentially at the same y position.	$y_1 \approx y_2$
Simplify Bernoulli's equation as in Example 15.9.	$P_1 - P_2 = \frac{1}{2}\rho(v_2^2 - v_1^2)$ (1)
We do not need **Step 5** in this part of the example (relating speeds using the continuity equation). Instead, we relate the pressure difference $P_1 - P_2$ to the lift F_{lift} on the wings. The lift is a result of the net pressure on the wings multiplied by the wing area.	$F_{\text{lift}} = (P_1 - P_2)A_{\text{wings}}$
The magnitude of the lift must equal the weight of the plane because the plane is not accelerating.	$F_g = F_{\text{lift}} = (P_1 - P_2)A_{\text{wings}}$ $P_1 - P_2 = \frac{F_g}{A_{\text{wings}}}$ (2)

Substitute Equation (1) into Equation (2).	$\frac{1}{2}\rho(v_2^2 - v_1^2) = \frac{F_g}{A_{\text{wings}}}$
Solve for v_2.	$(v_2^2 - v_1^2) = \frac{2F_g}{\rho A_{\text{wings}}}$ $v_2 = \sqrt{\frac{2F_g}{\rho A_{\text{wings}}} + v_1^2} = \sqrt{\frac{2mg}{\rho A_{\text{wings}}} + v_1^2}$
Substitute given values. The density of air is listed in Table 15.1.	$v_2 = \sqrt{\frac{2(1353\text{ kg})(9.81\text{ m/s}^2)}{(1.29\text{ kg/m}^3)(16.3\text{ m}^3)} + (153\text{ m/s})^2} = 157\text{ m/s}$

CHECK and THINK

This result is a very small difference in air speed, 4 m/s or roughly 3% of the given air speed below the wing. It might seem incredible that it is a great enough difference to create the required lift, but because the wings are large, the lift force is great enough to balance the weight of the plane.

B Answer Avi's concern about airplanes that can fly upside down. Do planes truly fly upside down?

SOLVE

For a plane to have an upward lift, its wings must force the air downward as in Figure 15.38. If you compare the streamlines on the far right of the wing with those to the far left of the wing, you can see that the ones on the left are lower than the ones on the right.

Now, imagine that plane rolled by exactly 180° around its long axis. The result would be Figure 15.39A, in which the wings would force the air upward. According to Newton's third law, the air must push the wings, and therefore the plane, downward. In practice, airplanes are not flown truly upside down. Instead, the plane is tilted slightly upward as in Figure 15.39B. For many airplanes, this upward tilt is just a few degrees, not noticeable to spectators watching an air show.

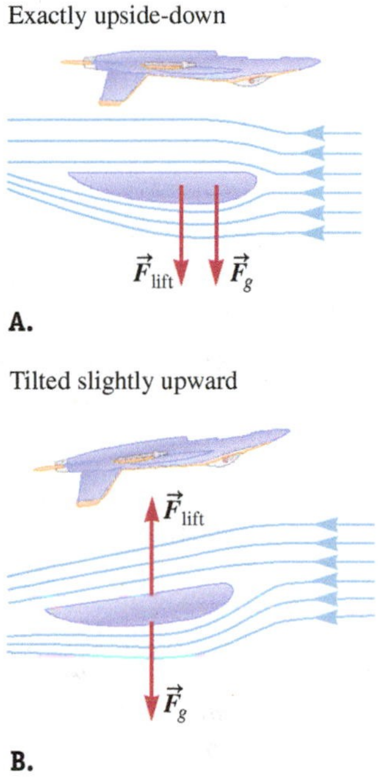

FIGURE 15.39

A Final Note

In this chapter, we restricted our study to ideal fluid flow, in which air is assumed to be incompressible, and airflow around airplane wings is assumed to be laminar (steady). This restriction allowed us to get a good idea of how planes fly and to make calculations as we did in Example 15.10. However, air is actually compressible, and the airflow around a wing is not ideal. The details of a plane's design must be examined in a wind tunnel and are beyond the scope of this book.

SUMMARY

Underlying Principles

Newton's second law is used to derive Archimedes's principle, and the work–energy theorem is used to derive Bernoulli's equation.

Major Concepts

1. An **ideal fluid** is incompressible.
2. The molecules in a **static fluid** move, but overall the fluid does not flow.
3. **Pressure** is a scalar given by

$$P = \frac{F}{A} \quad (15.1)$$

and pressure as a function of depth is

$$P = \rho g y + P_0 \quad (15.6)$$

4. The buoyant force is the net upward force exerted by a static fluid. According to **Archimedes's principle**, the magnitude of the buoyant force is equal to the weight of the fluid displaced by the object.
5. According to **Pascal's principle**, a *change* in the pressure at one point in an incompressible fluid appears undiminished at all points in the fluid and on the walls of its container.
6. The **ideal fluid flow** of an incompressible fluid is *steady* and *nonviscous*. In **steady flow**, also known as **laminar flow**, the fluid's velocity at any fixed point is a constant. In **nonviscous flow**, there is no resistance.
7. The **continuity equation** is

$$A_1 v_1 = A_2 v_2 \quad (15.21)$$

The continuity equation says that in ideal fluid flow the volume flow rate ($R = Av$) is uniform.

8. **Bernoulli's equation** is a statement of the work–energy theorem for ideal fluid flow:

$$P_1 + \tfrac{1}{2}\rho v_1^2 + \rho g y_1 = P_2 + \tfrac{1}{2}\rho v_2^2 + \rho g y_2 \quad (15.30)$$

Special Cases

A **fluid** is a substance that can flow. Therefore, liquids and gases are fluids, but solids are not.

PROBLEM-SOLVING STRATEGY Bernoulli's Equation

INTERPRET and ANTICIPATE

1. Draw a simple **sketch**, including your **coordinate system**.
2. **Pick region 1 and region 2** and indicate these regions on your sketch.

SOLVE

3. Write **Bernoulli's equation**.
4. **Compare the parameters** P, v, and y in the two regions. Make approximations if appropriate.
5. If necessary, use the **continuity equation** (Eq. 15.21). (Keep in mind that the speed in the large area is approximated as zero.)
6. **Do algebra** to solve for the required quantity.

PROBLEMS AND QUESTIONS

A = algebraic C = conceptual E = estimation G = graphical N = numerical

15-2 Static Fluid on the Earth

1. An alpha particle is the nucleus of a helium atom with two protons ($m_p = 1.673 \times 10^{-27}$ kg) and two neutrons ($m_n = 1.675 \times 10^{-27}$ kg) tightly bound together. Each of the particles has a radius of about 1.00 fm (femtometer).
   a. N What is the density of an alpha particle's nucleus?
   b. C How does this density compare with that of osmium, the densest natural element with a density of 2.26×10^4 kg/m^3? What does this comparison say about the space between nuclei in liquids and solids?
2. E What is the approximate weight of the air in your physics classroom?
3. N Dry air is primarily composed of nitrogen. In a classroom demonstration, a physics instructor pours 2.00 L of liquid nitrogen into a beaker. After the nitrogen evaporates, how much volume does it occupy if its density is equal to that of the dry air at sea level? Liquid nitrogen has a density of 808 kg/m^3.
4. C Why is the Earth's atmosphere denser near sea level than it is at a high altitude? Be sure to explain why the atmosphere's density is not uniform and why the air isn't all in contact with the Earth's surface.

15-3 Pressure

5. N Crater Lake in Oregon is the deepest lake in the United States, with a maximum depth of 655 m. The density of freshwater is 1.00×10^3 kg/m^3. **a.** If the air pressure above the lake, which has a surface altitude of 1800 m, is 8.00×10^4 Pa, what is the absolute pressure at the bottom of the lake? **b.** What is the force of the lake's water on each window of a submarine near the lake bottom if each of its circular windows has an area of 100 cm^2?
6. C At a crowded party, a woman wearing flat shoes accidentally steps on your foot. Now imagine that the same woman is wearing high-heeled shoes when she steps on your foot. Which hurts more? Why?

7. **C** A child lies on a bed of nails (Fig. P15.7). Explain why he is unharmed.

FIGURE P15.7

8. **N** One study found that the dives of emperor penguins ranged from 45 m to 265 m below the surface of the ocean. **a.** Using the density of seawater, what is the range of pressures experienced by the penguins between these two depths? **b.** At what depth below the surface is the pressure 10 times the normal atmospheric pressure?
9. Exploring the deep ocean is a challenge due to the extremely high pressures encountered. The deepest part of the ocean is at Challenger Deep in the Mariana Trench, with a depth of 10,924 m.
   - **a.** **N** What is the pressure at this depth?
   - **b.** **N** What is the volume strain of water compressed at this pressure?
   - **c.** **C** Comment on our assumption that water is incompressible.
10. **N** The dimensions of a room's floor are 6 m × 5 m, and its height is 4 m. What is the force on each wall of the room due to the air in the room?
11. **N** Suppose you are at the top of Mount Everest and you fill a water balloon. The air pressure at the top of Mount Everest is 58 kPa. **a.** What is the fractional change in the balloon's volume $\Delta V/V_i$ when you take it to sea level? **b.** If instead you take it 100 m below the surface of the ocean, what is the fractional change in its volume?
12. **E** In a Hollywood movie, a salty old sailor on a submarine wishes to frighten a newcomer. When the submarine is at sea level, the old sailor ties a rope to the walls, which are about 3.5 m apart, so the rope is taut. The rope is tied at the height of a man's shoulder. When the submarine is at depth, that rope is slack. Suppose the center of the rope may be pulled down to the sailor's waist while the ends remain at shoulder height. Assume the submarine is made of steel and estimate its depth. Comment on the validity of the old sailor's demonstration.
13. **N** Imagine a planet that has the same sea-level atmospheric pressure as the Earth, but whose gravitational acceleration is only 0.5 g. What is the pressure difference between the bottom and the top of a 2.10-m deep swimming pool of water on this planet?
14. **N** You may have had the experience of feeling your ears pop as you ascend or descend in the elevator of a tall building. Assume the density of air is constant, and find the difference in pressure from the bottom to the top of a 100-story building whose total height is roughly 333 m.

Problems 15 and 16 are paired.

15. **N** A 20.0-kg child sits on a four-legged stool. The radius of each of the stool's feet is 0.022 m. **a.** Ignoring the mass of the stool, what is the force that is exerted on the floor? **b.** What is the pressure applied to the floor by the stool?
16. A 20.0-kg child sits on a pillar with a circular base of radius 0.250 m.
   - **a.** **N** Ignoring the mass of the pillar, what is the force that is exerted on the floor?
   - **b.** **N** What is the pressure applied to the floor by the pillar?
   - **c.** **C** Evaluate the answers for all parts of this problem and Problem 15 (if you solved it) and explain any similarities or differences you find.
17. **N** The dolphin tank at an amusement park is rectangular in shape with a length of 40.0 m, a width of 15.0 m, and a depth of 7.50 m. The tank is filled to the brim to provide maximum splash during dolphin shows. What is the total amount of force exerted by the water on **a.** the bottom of the tank, **b.** the longer wall of the tank, and **c.** the shorter wall of the tank?
18. **N** In December 1995, a probe from the NASA spacecraft *Galileo* entered Jupiter's atmosphere. It traveled about 150 km into the atmosphere, collecting data for about 1 hour before it was vaporized. This problem is based on that probe's mission. At the top of its flight, the atmospheric pressure on the probe was about 10^{-2} Pa. The pressure exerted on the probe at the bottom of its flight was about 2.2×10^6 Pa. Find the average density of Jupiter's atmosphere over this depth. Compare your answer to the average density of Jupiter, which is about 1300 kg/m³. *Hint*: Jupiter's gravitational field near the top of its atmosphere, $R \approx 70{,}000$ km, is not the same as the Earth's gravitational field near its surface.

15-4 Archimedes's Principle

19. **N** A block of an unknown material floats in water with 75% of it below the surface. What is the density of the material?
20. **C** A battleship is an incredibly heavy object—much heavier than, say, a small pebble—yet when thrown into water, a pebble sinks, but a battleship can float. Explain how the battleship can float but the stone sinks.
21. **N** A block of density 1250.0 kg/m³ floats in an unknown fluid with two-thirds of its volume below the surface. What is the density of the fluid?
22. **N** A spherical submersible 2.00 m in radius, armed with multiple cameras, descends under water in a region of the Atlantic Ocean known for shipwrecks and finds its first shipwreck at a depth of 1.75×10^3 m. Seawater has density 1.03×10^3 kg/m³, and the air pressure at the ocean's surface is 1.013×10^5 Pa. **a.** What is the absolute pressure at the depth of the shipwreck? **b.** What is the buoyant force on the submersible at the depth of the shipwreck?
23. **N** What fraction of an iceberg floating in the ocean is above sea level? Assume the density of the iceberg is 917 kg/m³.
24. **A** An object in a fluid has a density $\rho_{obj} = 1755$ kg/m³. It is submerged in a fluid of density ρ_f. Its acceleration is upward with a magnitude $g/3$. What is the density of the fluid? Ignore any drag force on the object.
25. **N** A hollow copper ($\rho_{Cu} = 8.92 \times 10^3$ kg/m³) spherical shell of mass $m = 0.950$ kg floats on water with its entire volume below the surface. **a.** What is the radius of the sphere? **b.** What is the thickness of the shell wall?
26. **N** When a wooden box is placed in a pail of water, it floats with 40% of its height above the waterline. When this box is then placed in a pail full of olive oil, it floats with 30% of its height above the oil. What is the density of **a.** olive oil and **b.** the box?
27. **N** You have probably noticed that carrying a person in a pool of water is much easier than carrying a person through air. To understand why, find the buoyant force exerted by air and by water on the person. Assume the average volume of a person is 0.45 m³, and that the person is submerged in air and water respectively.
28. **C** A straw is in a glass of juice. Peter puts his finger over the top of the straw and carefully removes the straw. Juice remains in the straw (Fig. P15.28). **a.** Explain why the juice remains in the straw. **b.** What happens if Peter removes his finger?

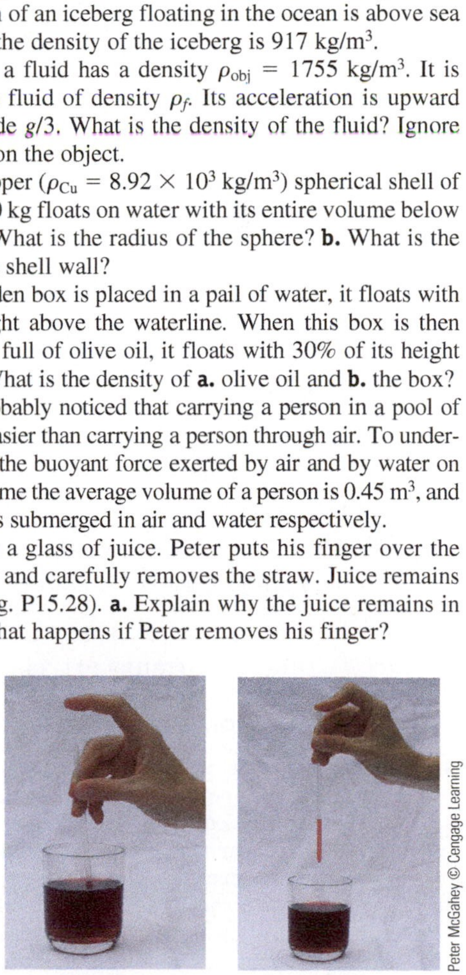

FIGURE P15.28 **A.** **B.**

29. An object of mass 3.0 kg and volume 0.0020 m³ is placed into a fluid of density 1.33×10^3 kg/m³. The fluid is at rest.
   a. **N, C** Will the object float or will it sink in the fluid?
   b. **N** If it floats, what fraction of volume is immersed in the fluid? If it sinks, find its acceleration, ignoring any drag force on the object.

30. **N** Imagine a planet that has an ocean with the same seawater density as that of the Earth, but whose gravitational acceleration is only $g/2$. A 67.5-kg person is submerged in the planet's ocean. Assume the person's volume is the same as it is on the Earth, 0.57 m³. **a.** What is the person's weight on the alien planet? **b.** What is the buoyant force exerted on the person? **c.** What is the person's acceleration, ignoring any drag forces on the person?

Problems 31 and 32 are paired.

31. **N** A ball with a volume of 0.65 m³ is floating on the surface of a pool of water. (The density of water is 1.00×10^3 kg/m³.) If 5.41% of the ball's volume is below the surface, what is the density of the ball?

32. **N** In Problem 31, what is the magnitude of the buoyant force on the ball?

33. **N** A rectangular block of Styrofoam 25.0 cm in length, 15.0 cm in width, and 12.0 cm in height is placed in a large tub of water. Assume the density of Styrofoam is 3.00×10^2 kg/m³. **a.** What volume of the block is submerged? **b.** A copper block is now placed atop the Styrofoam block so that the top of the Styrofoam block is level with the surface of the water. What is the mass of the copper block?

34. In Example 1.3 (page 15), we estimated the density of a grape and found that it had a density of 900 kg/m³. So, if we put a grape in water, we expect that it would float, but a simple kitchen experiment shows that the grape sinks. The grape will float, though, if 12.5 ml of salt is dissolved in 225 ml of water. The dry density of salt is 1233 kg/m³. The five grapes shown in Figure P15.34A displace 12.5 ml of water.
   a. **N** Find the density of the salt water.
   b. **N** From these experimental data, calculate the density and mass of a grape.
   c. **C** In Example 1.4 (page 16), we found that a raisin is denser than a grape. Does Figure P15.34B support our previous conclusion? Explain.

Peter McGahey © Cengage Learning

A. B.

FIGURE P15.34

15-5 Measuring Pressure

35. **N** Imagine a planet that has the same sea-level atmospheric pressure as on the Earth, but whose gravitational acceleration is only 0.5g. What is the minimum required height of a mercury barometer on this planet?

36. **C** A manometer is shown in Figure P15.36. Rank the pressures at the five locations indicated from highest to lowest. Indicate equal pressures, if any.

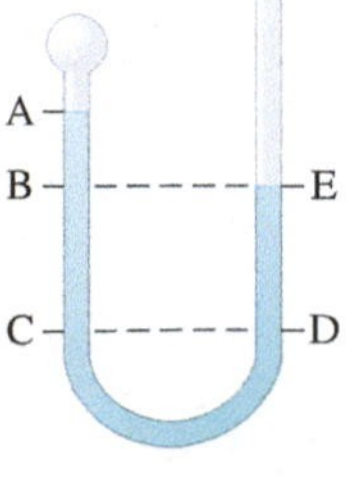

FIGURE P15.36

37. **N** The gauge pressure measured on a car's tire is 35 psi. What is the absolute pressure? Comment on your answer.

38. The gauge pressure of an *empty* scuba tank is 500 psi, whereas a full tank is at 3000 psi. The tank is connected to an open-tube manometer.
   a. **N** What is the height of the mercury if the tank is empty?
   b. **N** What is the height of the mercury if the tank is full?
   c. **C CHECK and THINK** step: Is it practical to use a manometer or a barometer on a scuba tank? Explain.

39. **N** Calculate the atmospheric pressure in pascals if the height of a mercury column in a barometer is 760 mm. The density of mercury is 13.6×10^3 kg/m³.

40. **N** To allow a car to slow down or stop, hydraulic brakes transmit forces from a master cylinder to the brake pads through a fluid. Imagine this system as a tube filled with an incompressible fluid and a piston on each end. A force of 95.0 N is applied to a piston 2.65 cm in diameter on one end of the tube. **a.** What is the magnitude of the force that is exerted on the piston 5.15 cm in diameter on the other side? **b.** If the 2.65-cm piston is displaced by 1.00 cm, by how much is the 5.15-cm piston displaced?

41. **N** A hurricane cannot develop unless the height of the column of mercury in a barometer falls below 749 mm. During Hurricane Katrina in 2005, drops in the mercury level of barometers as large as 75.0 mm from the normal level of 760 mm were recorded. If normal atmospheric pressure is 1.013×10^5 Pa, what was the lowest atmospheric pressure recorded during this hurricane?

42. **N** In a hydraulic lever such as that seen in Example 15.4, the cross-sectional areas of the pistons are 0.250 m² and 1.00 m². What is the minimum force that should be applied to the narrower piston to support a car with a mass of 1.20×10^3 kg that is resting on the larger piston?

43. **N** A schematic of a homemade hydraulic jack is shown in Figure P15.43, with cylindrical tube ends of different radii, $r_1 = 0.0500$ m and $r_2 = 0.100$ m, respectively. **a.** If a downward force of $F_1 = 15.0$ N is applied to the end with the smaller radius, how much force F_2 would the jack exert on a car if it were sitting above the other end? **b.** If the smaller piston is pushed down a distance $d_1 = 0.200$ m, through what distance d_2 would the other end move?

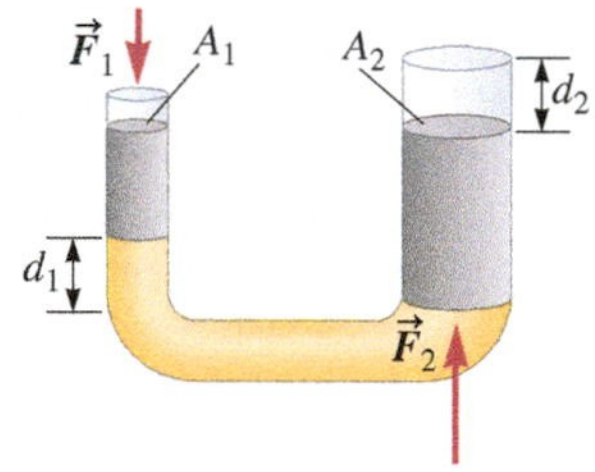

FIGURE P15.43

15-7 The Continuity Equation

44. **N** Water enters a smooth, horizontal tube with a speed of 2.0 m/s and emerges out of the tube with a speed of 8.0 m/s. Each end of the tube has a different cross-sectional radius. Find the ratio of the entrance radius to the exit radius.

Problems 45 and 46 are paired.

45. A fluid flows through a horizontal pipe (Fig. P15.45). The fluid enters the pipe with speed v_0 from the left.
   a. **G** Sketch streamlines for the fluid flow in this pipe and describe how the speed varies as it flows through the pipe.
   b. **A** The pipe has a circular cross section with a radius given by the expression $r(x) = B + Cx^2$, where the units of the constants B and C are such that r has units of centimeters. Derive an expression for the speed $v(x)$ of the fluid as a function of position for $0 < x < 1$.

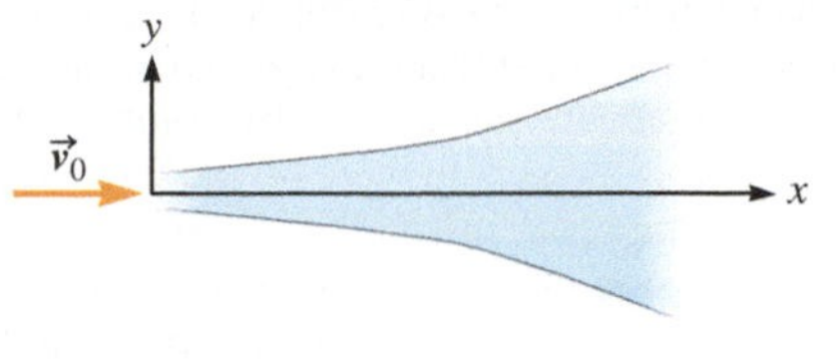

FIGURE P15.45 Problems 45 and 46.

46. Consider the pipe of variable radius described in Problem 45.

a. **G** If $v_0 = 1.35$ m/s, $B = 4.5$ cm, $C = 0.025$ cm^{-1}, and the pipe is 25 cm long, plot $v(x)$ versus x for $0 < x < 25$ cm.

b. **N** Water enters the pipe at $x = 0$ and leaves the pipe at $x = 25$ cm. Given the information in part (a), what is the volume flow rate for the water entering the pipe?

c. **N** What is the volume flow rate for water leaving the pipe?

47. **N** When connected to a standpipe, a fire hose 5.08 cm in diameter must deliver a minimum flow rate of 1.00×10^3 L/min. **a.** At this rate, with what speed does water leave the fire hose? **b.** A standard fire hose nozzle has an opening 0.950 cm in radius. What is the speed of the water leaving the fire hose with the nozzle attached?

48. A fluid flows through a horizontal pipe that widens, making a 45° angle with the y axis (Fig. P15.48). The thin part of the pipe has radius R, and the fluid's speed in the thin part of the pipe is v_0. The origin of the coordinate system is at the point where the pipe begins to widen. The pipe's cross section is circular.

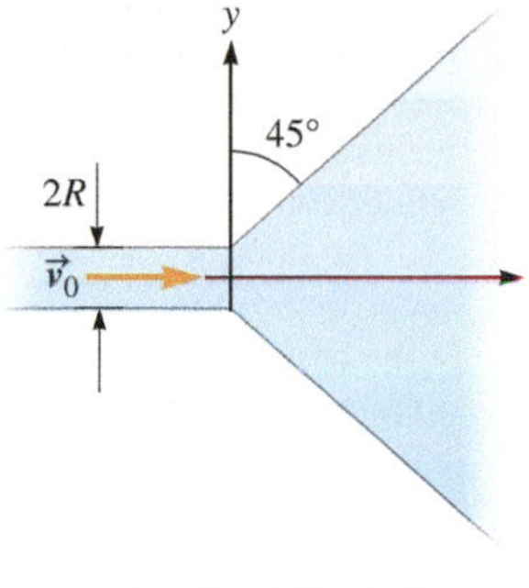

FIGURE P15.48

a. **A** Find an expression for the speed $v(x)$ of the fluid as a function of position for $x > 0$.

b. **G** Plot your result: $v(x)$ versus x.

49. **N** Water is flowing through a pipe that has a constriction opening into a region with a wider cross-sectional area. If the pipe regions are cylindrical with radii of 0.10 m and 0.35 m, respectively, and the water is moving with a speed of 1.50 m/s in the wider section, what is the speed of the water in the constricted section?

50. The stream of water flowing from a faucet is wider near the top as shown in Figure P15.50. (It is easier to observe if the faucet does not have an aerator.)

a. **C** Explain this phenomenon.

b. **A** The cross-sectional area of the stream in a region near the top is A_1, and the cross sectional area lower region is A_2. If the distance between the two regions is h, find an expression for the speed of the water in top region in term of A_1, A_2, and h.

c. **N** If $A_1 = 1.4$ cm^2, $A_2 = 0.25$ cm^2, and $h = 6.5$ cm, what is the speed at the top of the flow?

FIGURE P15.50

15-8 Bernoulli's Equation

51. **N** A water cannon 2.25 m long and with a cross-sectional area of 10.0 cm^2 is aimed straight up and ejects water out of the top at a speed of 9.5 m/s. It is fed by a pump at the bottom of the cannon, which has a cross-sectional area of 45.0 cm^2. What is the difference in pressure of the water between the pump and the length of the cannon? Assume the density of the water is 998.2 kg/m^3.

52. Figure P15.52 shows a *Venturi meter*, which may be used to measure the speed of a fluid. It consists of a Venturi tube through which the fluid moves and a manometer used to measure the pressure difference between regions 1 and 2. The fluid of density ρ_{tube} moves from left to right in the Venturi tube. Its speed in region 1 is v_1, and its speed in region 2 is v_2. The neck's cross-sectional area is A_2, and the cross-sectional area of the rest of the tube is A_1. The manometer contains a fluid of density ρ_{mano}.

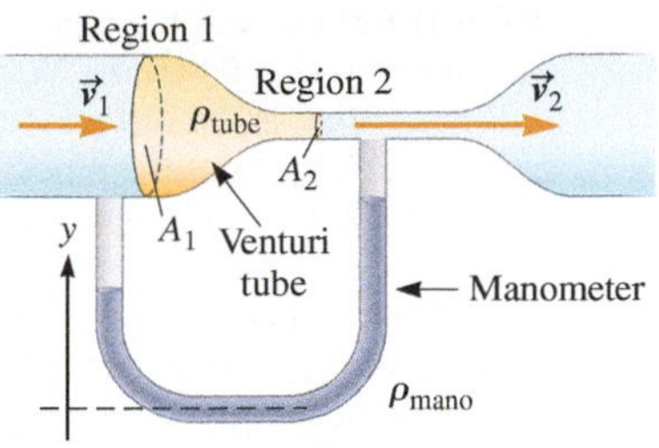

FIGURE P15.52

a. **C** Do you expect the fluid to be higher on the left side or the right side of the manometer?

b. **A** The speed v_2 of the fluid in the neck comes from measuring the difference between the heights $(y_R - y_L)$ of the fluid on the two sides of manometer. Derive an expression for v_2 in terms of $(y_R - y_L)$, A_1, A_2, ρ_{tube}, and ρ_{mano}.

53. **N** At a fraternity party, drinking straws have been joined together to make a giant straw that will be used to drink punch placed in a bowl on the ground from atop the fraternity house building. What is the maximum allowable height of the building if the partygoers are successful in drinking the punch? Assume the density of the punch is the same as the density of water.

54. **N** Liquid toxic waste with a density of 1752 kg/m^3 is flowing through a section of pipe with a radius of 0.312 m at a velocity of 1.64 m/s. **a.** What is the velocity of the waste after it goes through a constriction and enters a second section of pipe with a radius of 0.222 m? **b.** If the waste is under a pressure of 850,000 Pa in the first section of pipe, what is the pressure in the second (constricted) section of pipe?

55. **N** Water is flowing in the pipe shown in Figure P15.55, with the 8.00-cm diameter at point 1 tapering to 3.50 cm at point 2, located $y = 12.0$ cm below point 1. The water pressure at point 1 is 3.20×10^4 Pa and decreases by 50% at point 2. Assume steady, ideal flow. What is the speed of the water **a.** at point 1 and **b.** at point 2?

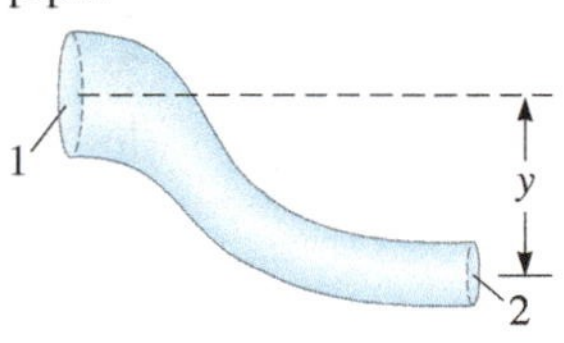

FIGURE P15.55

56. **C** A baseball is thrown through still air. In the reference frame of the ball, the air is moving around the ball. In Figure P15.56A, streamlines are shown around a ball that is *not* spinning. Figure P15.56B shows streamlines around a ball that is spinning counterclockwise as seen from above. **a.** Compare the airspeed in region L with the airspeed in region R in each case. **b.** Compare the air pressure in region L with the air pressure in region R in each case. **c.** Describe the motion of the ball in each case.

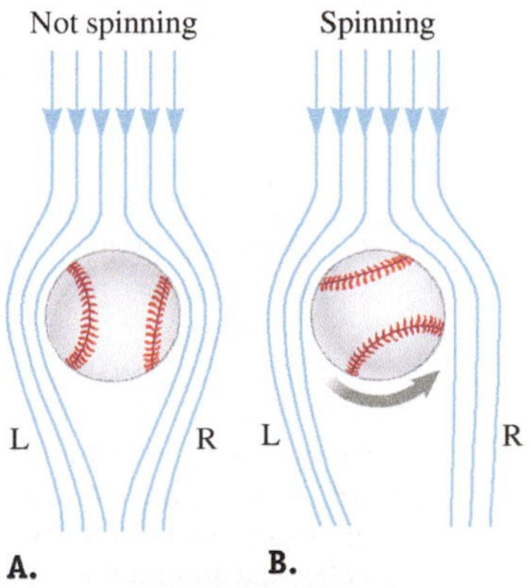

FIGURE P15.56

57. **N** Water flows through a pipe that gradually descends from a height of 6.78 m to the ground. Near the top, the cross-sectional area is 0.400 m^2, and the pipe gradually widens so that its area near the ground is 0.800 m^2. Water leaves the pipe at a speed of 16.8 m/s. What is the difference in the water pressure between the top and bottom of the pipe?

58. **N** Air flows horizontally with a speed of 108 km/h over a house that has a flat roof of area 20.0 m^2. Find the magnitude of the net force on the roof due to the air inside and outside the house. The density of air is 1.30 kg/m^3, and the thickness of the roof is negligible.

59. **N** A cylindrical tank of height 0.40 m is open at the top and has a diameter of 0.16 m. It is filled with water up to a height of 0.16 m. Find the time it takes to empty the tank through a hole of radius 5.0×10^{-3} m in its bottom.

General Problems

60. **N** A 75.0-kg man is sitting on a three-legged 15.0-kg barstool with circular legs of diameter 2.50 cm. What is the pressure exerted by each leg of the barstool on the floor?

61. **N** A rural water tank filled to capacity is subjected to subzero temperatures overnight, freezing the water inside the tank. Water expands by about 9.00% as it freezes, and the bulk modulus of ice is 2.00×10^9 N/m^2. What is the increase in pressure inside the tank?

62. **N** A spherical beach ball is 7.00 cm in radius and has a mass of 50.0 g. What is the force that must be exerted on the ball to keep it completely submerged in a swimming pool?

Problems 63–65 are grouped.

63. **C** CASE STUDY **Before airplanes, people were able to fly using hot-air balloons.** In June 1783, two brothers, Joseph and Jacques Montgolfier, demonstrated that a balloon full of hot air rises. They got the idea by observing that smoke rises and that a paper bag placed over a fire expanded and rose as well. They believed that such bags were filled with a unique gas, later called Montgolfier gas. Why do you suppose their hot-air balloon rose?

64. **E** CASE STUDY **Before airplanes, people were able to fly using hot-air balloons.** The Montgolfier brothers (Problem 63) experimented with hot-air balloons. The first flight of live animals (a sheep, a rooster, and a duck) was launched from Versailles. France, on September 19, 1783, and the first human passengers were launched about a month later, on October 15. These hot-air balloon demonstrations drew the attention of more than 100,000 spectators, including Louis XVI, Marie Antoinette, and Benjamin Franklin. The Montgolfier brothers believed that the balloons were full of a unique gas later called Montgolfier gas, but today we know that it was just hot air. Use information from the first human flight to estimate the density of Montgolfier gas. The balloon's capacity was 60,000 cubic feet. There may have been one or two adults on the ride. The balloon was tethered, and it gently rose 25 m, staying aloft for about 4 min.

65. **N** CASE STUDY **Before airplanes, people were able to fly using hot-air balloons.** The Montgolfiers' hot-air balloon (Problems 63 and 64) proved that human flight was possible, but their balloon was impractical. As the air cooled, the balloon descended, and it was dangerous to try to maintain a fire under a balloon made of cloth and paper. Jacques Alexandre Cesar Charles replaced the hot-air balloon with a silk balloon full of hydrogen gas. He launched his balloon from the Tuileries Gardens in Paris on December 1, 1783. This balloon carried a barometer and thermometer, making it the first scientific balloon flight. Assume Charles's balloon was a sphere 4.0 m in diameter full of hydrogen gas. Find the buoyant force exerted by the air on his balloon.

66. **N** Imagine a planet that has a sea with the same water density (1250 kg/m^3) as the Dead Sea on the Earth, but whose gravitational acceleration is only $g/2$. A 67.5-kg person floats in the planet's sea. What fraction of the person is above the water's surface? Think about your answer by comparing it with what you would find if the person were on the Earth.

67. **N** At a processing plant, olive oil of density 875 kg/m^3 flows in a horizontal section of hose that constricts from a diameter of 3.00 cm to a diameter of 1.00 cm. Assume steady, ideal flow. **a.** What is the volume flow rate if the change in pressure between the two sections of hose is 5.00 kPa? **b.** What is the volume flow rate if the change in pressure between the two sections of hose is 12.5 kPa?

68. **N** A device designed for undersea exploration is a solid copper sphere ($B = 120.0 \times 10^9$ Pa) with a radius of 2.50 m. Find the change in volume of the copper sphere when the device is at a depth of 100.0 m below the surface of the ocean.

69. **N** A student is designing a piece of equipment that holds a camera and remains neutrally buoyant in fresh water. That is, the device will remain at the same height at which it is placed under water. This device has a mass of 25 kg and a volume of 0.025 m^3. If it is instead brought to the ocean and released under water, what would be the acceleration experienced by the device? Would it accelerate up or down?

70. **N** A vessel contains oil of density 8.00×10^2 kg/m^3 resting on top of mercury of density 13.6×10^3 kg/m^3. A solid, homogeneous sphere floats with half its volume immersed in the mercury and the other half in the oil. What is the density of the sphere?

71. **N** The density of air in the Earth's atmosphere decreases according to the function $\rho = \rho_0 e^{-h/h_0}$, where $\rho_0 = 1.20$ kg/m^3 is the density of air at sea level and h_0 is the scale height of the atmosphere, with an average value of 7640 m. What is the maximum payload that a balloon filled with 2.50×10^3 m^3 of helium ($\rho_{He} = 0.179$ kg/m^3) can lift to an altitude of 10.0 km?

72. **N** A manometer containing water with one end connected to a container of gas has a column height difference of 0.60 m (Fig. P15.72). If the atmospheric pressure on the right column is 1.01×10^5 Pa, find the absolute pressure of the gas in the container. The density of water is 1.0×10^3 kg/m^3.

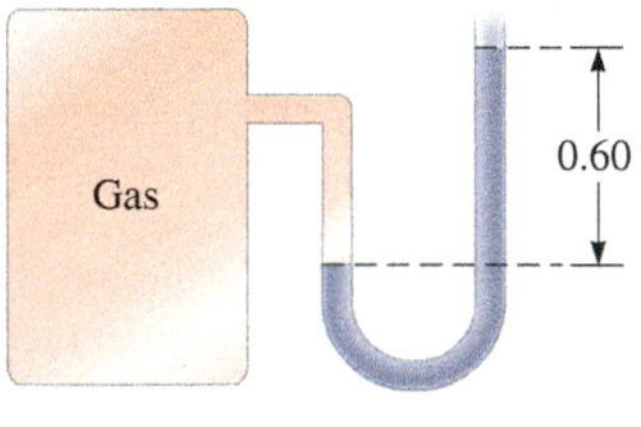

FIGURE P15.72

73. **N** In the novel *20,000 Leagues Under the Sea*, Jules Verne's *Nautilus* dove to the bottom of the sea by taking on seawater ($\rho_{seawater} = 1.03 \times 10^3$ kg/m^3) into three floodable tanks. Captain Nemo wishes to dive at a constant speed of 2.00 m/s. Model the submarine as a cylinder 70.0 m in length and 8.00 m in diameter. If the resistive force on the 1.50×10^6 kg submarine is 1.20×10^5 N in the upward direction, what is the mass of seawater the *Nautilus* must take on for this descent?

Problems 74 and 75 are paired.

74. Ancient Romans built long aqueducts to bring water from mountain springs into their cities. In imperial times, when Rome had more than one million inhabitants, the aqueducts delivered more than 1 m^3 of water per person daily.

a. **E** Place a lower limit on the volume rate of flow into Rome during this time.

b. **E** Roughly how much water do you consume each day?

c. **N** Assuming your water usage is typical, what is the volume rate of flow into your city?

75. Ancient Romans built long aqueducts to bring water from mountain springs into their cities. The longest was 59 miles. These aqueducts passed through variable terrain, and the

Romans used several different types of structures to keep the water flowing. One of the most costly and difficult structures to build was the inverted siphon (Fig. P15.75), which was used to cross valleys. It was expensive because it required lead pipes, and the lead had to be imported. The pipes were run down one side of the valley and up the other side to a slightly lower elevation. A major problem facing the ancient engineers was that it was hard to make joints strong enough to withstand the great pressure. The inverted siphon in Alatri, Italy, is 3000 m long and 101 m deep. Assume the pipe's diameter is constant.

a. C Where do you expect the maximum pressure?

b. N Calculate the change in the water's pressure between the top and bottom of the siphon. Express your answer in pascals and atmospheres.

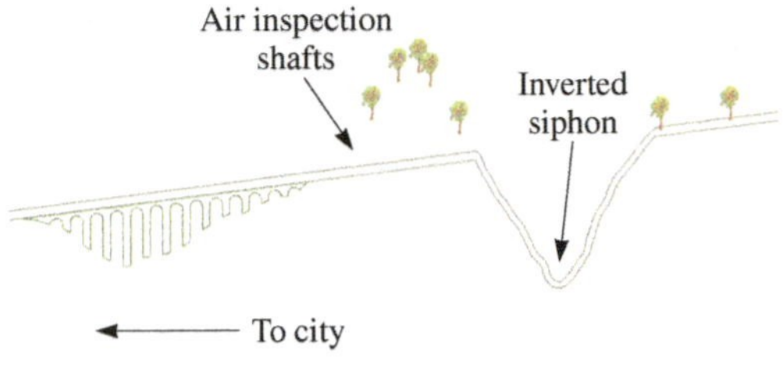

FIGURE P15.75

76. N A mother is administering cold medicine with the density of water using an oral syringe to her child. The medicine is pushed by a plunger from the barrel with a cross-sectional area of 1.25 cm^2 into the opening with a cross-sectional radius of 1.20 mm. The syringe is held horizontally, and it is a distance of 7.00 cm in the horizontal direction and 5.00 cm in the vertical direction from the child's mouth. Assume atmospheric pressure is 1.000 atm and neglect air resistance. **a.** What is the time of flight of the medicine from the syringe into the child's mouth? **b.** With what speed must the medicine leave the opening of the syringe to reach the child's mouth? **c.** What is the speed with which the mother must push the plunger for the medicine to reach the child's mouth? **d.** What is the pressure of the medicine in the opening of the syringe? **e.** What is the pressure in the barrel of the syringe?

77. N Strato-lab was a 1950s project in which large balloons were used to carry astronauts in training to the Earth's stratosphere. In one flight, a balloon 39.0 m in diameter was used to lift a total payload of 595 kg. Assume the volume of the balloon was filled with helium gas of density 0.179 kg/m^3. The air density was 1.20 kg/m^3. **a.** What was the buoyant force acting on the Strato-lab balloon? **b.** What was the net force on the Strato-lab? **c.** How much more mass could the Strato-lab have carried had it ascended at a constant velocity?

78. N CASE STUDY Shannon uses the example of a helium balloon to explain the buoyant force. Large helium "blimp" balloons are sometimes used as an advertisement (Fig. P15.78). The blimp balloon has a volume of 42.8 m^3, and the mass of the empty blimp is 13.6 kg. It is held down by either a large-link steel chain or a large-link aluminum chain. Each link of steel has a mass of 2.6 kg, and each link of aluminum has a mass of 0.87 kg. The chain rests on the ground but is not attached to it. The density of helium gas is 0.180 kg/m^3. **a.** How many links hang from the blimp if the steel chain is used? **b.** Compare your answer with the number of links that would hang if the aluminum chain were used instead.

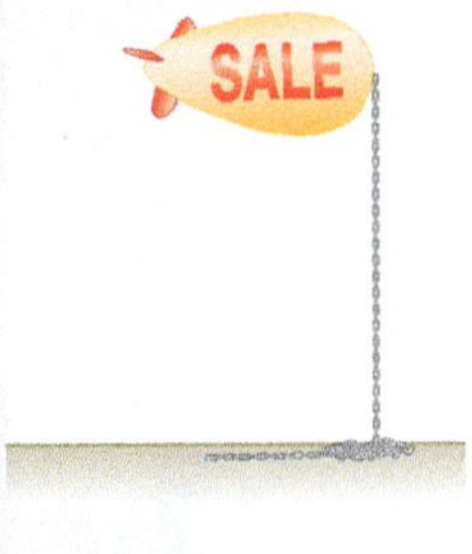

FIGURE P15.78

79. Ships crossing the Atlantic bound for Canada often cross the Laurentian Abyss, which has a maximum depth of 6.00 km, corresponding to a pressure of 6.00×10^7 N/m^2. The density of seawater is 1.03×10^3 kg/m^3.

a. N What is the change in volume of 1.00 L of seawater if a ship sinks from the surface into the deepest point of the Laurentian Abyss?

b. N What is the density of seawater at the deepest point of the Laurentian Abyss?

c. C Is the assumption that water is an incompressible fluid justified?

80. N Carolyn, living on the fourth floor of an apartment building, opens a faucet tap 1.25 cm in radius. The tap is 12.0 m above the main water supply pipe to the building. The main water pipe is 4.00 cm in radius and has a volume flow rate of 1.50 L per second. **a.** What is the speed of the water exiting Carolyn's faucet? **b.** If no other taps are open in the building, what is the gauge pressure in the main water supply pipe?

81. A A uniform wooden board of length L and mass M is hinged at the top of a vertical wall of a container partially filled with a certain liquid (Fig. P15.81). (If there were no liquid in the container, the board would hang straight down.) Three-fifths of the length of the board is submerged in the liquid when the board is in equilibrium. Find the ratio of the densities of the liquid and the board.

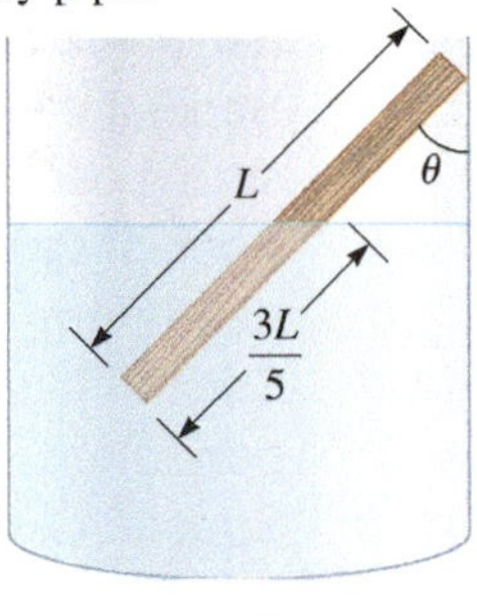

FIGURE P15.81

16 Oscillations

Key Questions

What is an oscillator?

How can we describe the kinematics, dynamics, and energy of an oscillator mathematically?

Underlying Principles

No new *fundamental* physical principles are introduced in this chapter.

Major Concepts

1. Simple harmonic motion (SHM) and simple harmonic oscillator (SHO)
2. Angular frequency
3. Amplitude
4. Phase and initial phase
5. Damped harmonic motion
6. Time constant
7. Critically damped, underdamped, and overdamped oscillators
8. Driven oscillator
9. Resonance

Special Cases

1. Object–spring oscillator
2. Simple pendulum
3. Physical pendulum
4. Torsion pendulum

We are surrounded by time-keeping devices, all of which are based on periodic motion. There are clocks on the classroom wall, next to our beds, and displayed on our phones. It doesn't matter whether you spent a lot of money on your wristwatch or found it in your box of cereal; the same physics makes it run. Something has to move around and around, back and forth, or in and out *periodically*, repeating at regular intervals.

The earliest timekeepers were based on the rotation of the Earth, but the Earth's rotational period varies slightly. (See Problem 12.66; the Earth's rotation is slowing down.) Also, you know from your own experience that we would like to measure time to the nearest minute, not day. In hopes of building a more accurate and precise device, inventors looked for objects that either oscillated back and forth or vibrated in and out at regular time intervals. Until the early part of the 20th century, the most accurate clocks were based on the oscillations of a pendulum (Fig. 16.1A). A pendulum clock can be designed so that the pendulum bob oscillates back and forth with a very regular period, but it must be vertically oriented to work. Wind-up wristwatches are based on the back-and-

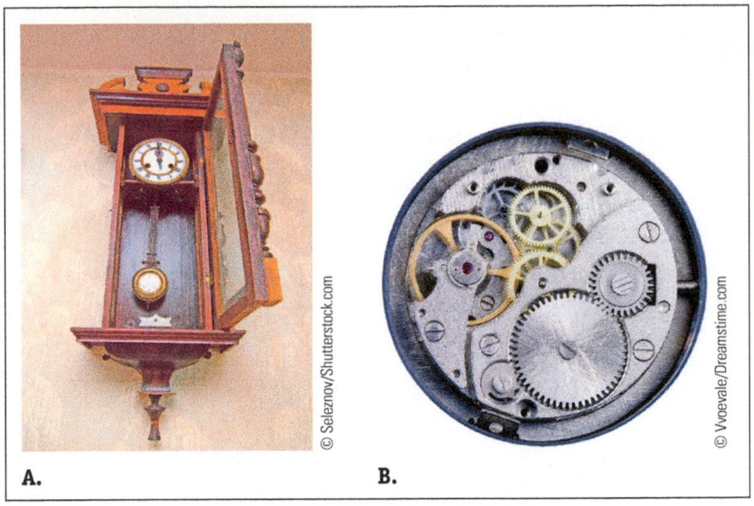

FIGURE 16.1 Timekeeping devices employ periodic motion. **A.** Pendulum clock. **B.** Wind-up watch.

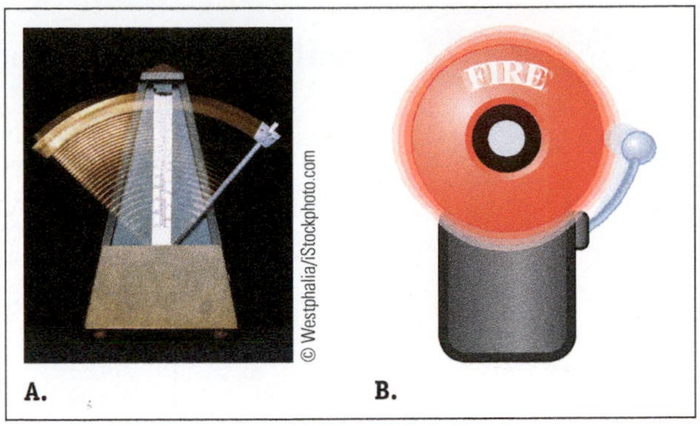

FIGURE 16.2 Oscillation versus vibration. **A.** The back-and-forth motion of a metronome is an oscillation. **B.** The ringing of a bell is a vibration.

forth rotation of a small wheel (Fig. 16.1B). Around the 1930s, quartz crystals were introduced into watches and clocks. When a quartz crystal is placed in an electrical circuit, the crystal expands and contracts with a regular period. In 1949, the first atomic clock, based on the vibration of ammonium, was built. Today the most accurate timepieces are based on the vibration of the cesium atom.

In this chapter, we study oscillations such as the motion of a pendulum bob or a particle attached to a spring. Oscillations are at the heart of wave motion (Chapters 17 and 18), common electrical circuits (Chapter 33), light (Chapters 35–38), and many aspects of modern physics (Chapters 40–43).

16-1 Picturing Harmonic Motion

Your beating heart, orbiting planets, and a child playing on a swing are all examples of objects whose motion repeats. Any motion that repeats at regular time intervals is **periodic**. We have studied two examples of periodic motion: circular motion (Chapter 4) and rotation (Chapter 12). In both of these cases, the object continues moving in just one angular direction; it does not reverse direction. We now consider motion that reverses direction periodically.

Many physicists use the words *oscillation* and *vibration* interchangeably to refer to repetitive motion. In this book, we make a slight distinction between the two terms. An **oscillation** is the back-and-forth motion of a particle or object over the same path; examples are the back-and-forth motion of a balance wheel in a watch and the motion of a metronome (Fig. 16.2A). A **vibration** is the expansion and contraction of an object or the back-and-forth motion of parts of the object. A particle such as the bob at the end of a pendulum may oscillate, but because a particle has no spatial extent, it cannot vibrate. Objects such as molded gelatin, a fire alarm bell (Fig. 16.2B), and a water molecule may change in shape or size and so can vibrate. An object's vibration may be modeled as the oscillation of particles making up the object.

The term ***harmonic motion*** describes periodic oscillations or vibrations. The word *harmonic* shows the important role periodic oscillations play in making sound and music. We've seen harmonic motion in previous chapters when studying the motion of a particle attached to a spring. In this chapter, we take a closer look at harmonic motion by considering an experiment involving a relatively simple example of a harmonic oscillator.

CASE STUDY Experiment, Part 1: Bouncy

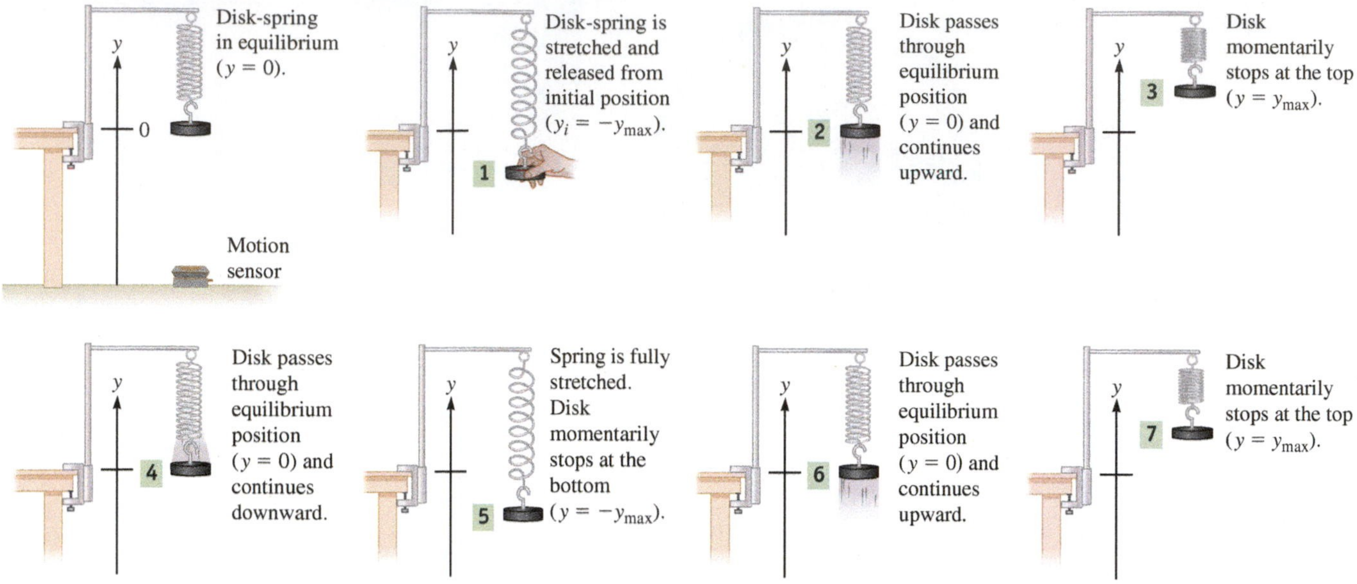
Disk-spring in equilibrium ($y = 0$).
Motion sensor
Disk-spring is stretched and released from initial position ($y_i = -y_{max}$).
Disk passes through equilibrium position ($y = 0$) and continues upward.
Disk momentarily stops at the top ($y = y_{max}$).
Disk passes through equilibrium position ($y = 0$) and continues downward.
Spring is fully stretched. Disk momentarily stops at the bottom ($y = -y_{max}$).
Disk passes through equilibrium position ($y = 0$) and continues upward.
Disk momentarily stops at the top ($y = y_{max}$).

FIGURE 16.3 Crall and Whipple's experiment consists of a disk oscillating at the end of a vertical spring. A motion sensor on the floor measures the disk's position, velocity, and acceleration.

Crall and Whipple hang a thick rubber disk of mass $m = 0.100$ kg from a vertical spring with spring constant $k = 0.987$ N/m. They choose an upward-pointing y axis with its origin set to the equilibrium position of the disk (Fig. 16.3). Whipple stretches the spring so that the disk is initially at $y_i = -0.40$ m and releases it from rest (1). The disk oscillates, first moving upward and passing through its equilibrium position (2) until it reaches its maximum height (3). It stops momentarily at this maximum height and then falls back down, again passing through its equilibrium position (4). The disk continues to fall until it reaches its initial position again (5). Now the motion repeats from the beginning: The disk moves upward, passes through its equilibrium position (6), continues to its maximum height (7), and so on.

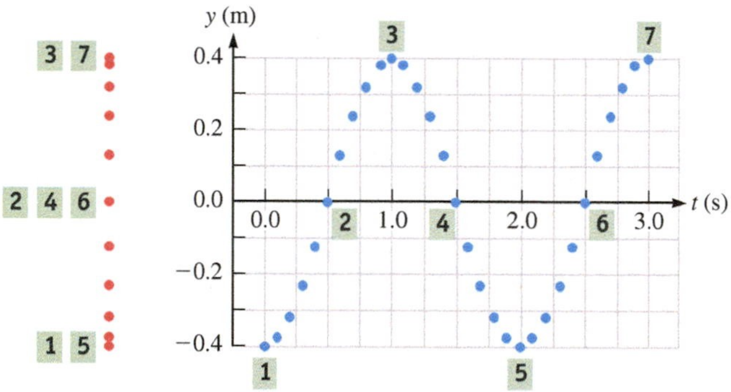
y (m)
0.4
0.2
0.0
−0.2
−0.4
0.0
1.0
2.0
3.0
t (s)

A. Motion diagram B. Position-vs.-time graph

FIGURE 16.4 Data from the experiment in Figure 16.3.

Crall and Whipple made a motion diagram for the rubber disk, labeling the points 1 through 7 as described above (Fig. 16.4A). The points near the equilibrium position (2, 4, 6) are more spread out than the points near the top (3, 7) or near the bottom (1, 5). From Section 2-2, the disk must be moving faster when it is near equilibrium than when it is near either the top or bottom of its range.

To study the disk's kinematics, Crall and Whipple made a position-versus-time graph (Fig. 16.4B) that looks like a stretched-out version of the motion diagram (Section 2-4). Compare the labeled points 1 through 7 in Figures 16.3 and 16.4. Points 2, 4, and 6 are at $y = 0$ because the disk is at the equilibrium position at those times ($t_2 = 0.5$ s, $t_4 = 1.5$ s, $t_6 = 2.5$ s). Points 1 and 5 are at the bottom of the graph because at these times ($t_1 = 0$, $t_5 = 2.0$ s) the disk is at its lowest position ($y = -0.4$ m). Points 3 and 7 are at the top of the graph because at these times ($t_3 = 1.0$ s, $t_7 = 3.0$ s) the disk is at its highest position ($y = 0.4$ m). The disk's position-versus-time graph (Fig. 16.4B) is consistent with the motion diagram (Fig. 16.4A) and the sketch (Fig. 16.3).

Crall and Whipple then connected the points in the position-versus-time graph (Fig. 16.5A) with a smooth curve and computed the slopes of lines tangent to that curve to find velocities at various times (Fig. 16.5B). Compare the four tangent lines at 2, 3, 4, and 5 in the position-versus-time graph to the velocities at these points in Figure 16.5B. At points 3 and 5, the disk is momentarily at rest ($v = 0$). At point 2, the disk is moving quickly upward. Finally, at point 4, the disk is moving quickly downward. These results are exactly what we expected from the motion diagram (Fig. 16.4A) and the sketch (Fig. 16.3).

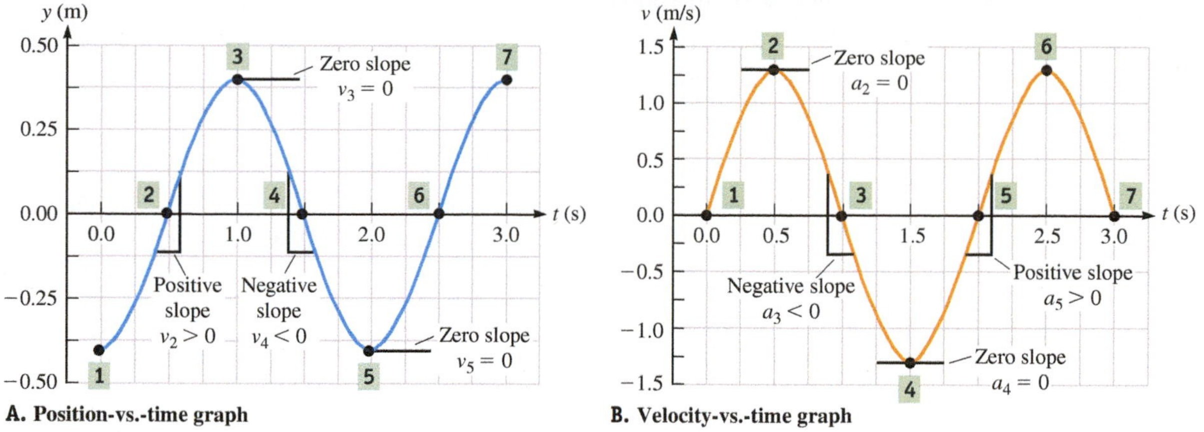

FIGURE 16.5 Graphical analysis of the motion experiment data in Figure 16.3.

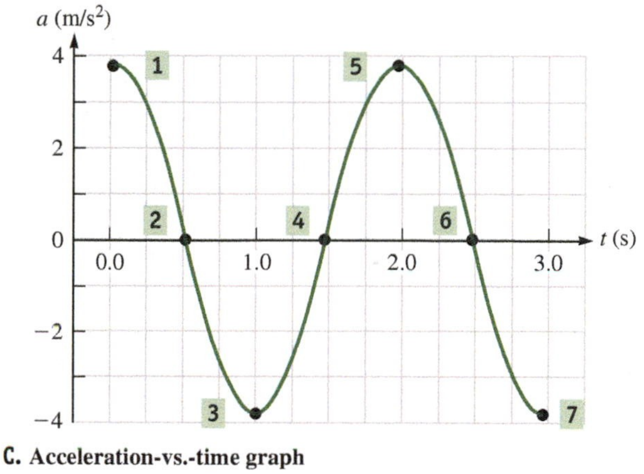

Crall and Whipple made an acceleration-versus-time graph (Fig. 16.5C) following a similar procedure. They connected the points in the velocity-versus-time graph with a smooth curve and then found the slope of that curve at various times. As before, compare the four tangent lines shown in the velocity-versus-time graph (Fig. 16.5B) with the accelerations in Figure 16.5C. At points 2 and 4, the acceleration is zero. At point 3, the acceleration is relatively great and pointing downward. At point 5, the acceleration is relatively great and pointing upward.

CONCEPT EXERCISE 16.1

CASE STUDY Disk Velocity and Acceleration

For each position listed, state the velocity and acceleration of the disk in Crall and Whipple's experiment (Figs. 16.3 through 16.5). There may be more than one possible answer for each given position.

a. $y = 0$ **b.** $y = 0.4$ m **c.** $y = -0.4$ m

16-2 Kinematic Equations of Simple Harmonic Motion

SIMPLE HARMONIC MOTION (SHM); SIMPLE HARMONIC OSCILLATOR (SHO)

Major Concepts

Crall and Whipple's rubber disk displays a relatively uncomplicated periodic oscillation known as **simple harmonic motion (SHM)**. In this section, we work through the kinematic equations that describe SHM and the mathematical signature of SHM. A **simple harmonic oscillator (SHO)** is a particle, object, or system displaying simple harmonic motion such as Crall and Whipple's disk.

Any single repetition of periodic motion is called a **cycle**. In Figure 16.3 or Figure 16.4, the disk's motion from 1 to 5, from 2 to 6, or from 3 to 7 is one cycle. Do not mistake half a cycle for a full cycle; for example, the motion from 1 to 3, from 2 to 4, or from 3 to 5 is *half* of a cycle.

The time to complete one cycle is called the **period**, symbolized by an uppercase T. From Figure 16.4B, the period of Crall and Whipple's oscillator is 2.0 s. The number of cycles per unit time is known as the **frequency**. The frequency f is the inverse of the period:

$$f = \frac{1}{T} \tag{16.1}$$

The SI unit of frequency is the *hertz* (Hz), and

$$1\,\text{Hz} = 1\,\text{cycle/s} = 1\,\text{s}^{-1}$$

In the mathematical description of harmonic motion, frequency f is often multiplied by 2π to define a new quantity known as the **angular frequency** ω such that

ANGULAR FREQUENCY

Major Concept

$$\omega \equiv 2\pi f = \frac{2\pi}{T} \tag{16.2}$$

Angular frequency is measured in radians per second. For Crall and Whipple's data, the angular frequency is

$$\omega = \frac{2\pi}{T} = \frac{2\pi}{2.0\,\text{s}} = 3.1\,\text{rad/s}$$

SHM has a single, constant angular frequency.

In Crall and Whipple's experiment, the disk's motion is one-dimensional along the y axis. Using their experimental data, we will write expressions for the y component of position, velocity, and acceleration. Similar equations could be written for oscillations along the x or z direction if needed.

Position Versus Time in Simple Harmonic Motion

The curve in the position-versus-time graph (Fig. 16.5A) is described by either a sine function or a cosine function. In fact, **the position versus time of any simple harmonic oscillator can be described by a single sine or cosine.** In this book, we have arbitrarily decided to use the cosine function for this time-varying position:

A similar equation can be written for SHM along the x or z axis.

$$y(t) = y_{\text{max}} \cos(\omega t + \varphi) \tag{16.3}$$

The maximum displacement of the oscillator *from the equilibrium position* (here, $y = 0$) is called the **amplitude** y_{max}. The amplitude is always positive, but the position may take on positive, zero, or negative values. In Figure 16.5A, the amplitude is $y_{\text{max}} = 0.40$ m. Do not mistake the maximum displacement for the amplitude. For example, the disk's displacement from 1 to 3 is 0.80 m, which is twice the amplitude. **In SHM, the amplitude is constant**, so the particle always moves between $-y_{\text{max}}$ and y_{max}.

AMPLITUDE; PHASE; INITIAL PHASE

Major Concepts

The argument $(\omega t + \varphi)$ of the cosine is called the **phase**. The **initial phase** φ is the phase when $t = 0$ as determined by the initial position $y_i \equiv y(0)$:

$$y(0) = y_{\text{max}} \cos(0 + \varphi)$$
$$y_i = y_{\text{max}} \cos\varphi$$

$$\varphi = \cos^{-1}\left(\frac{y_i}{y_{\text{max}}}\right) \tag{16.4}$$

Both the phase and the initial phase are expressed in radians. Usually, the initial phase is between $-\pi$ and π rad.

The sign of the initial phase cannot be determined from Equation 16.4 because the cosine function is symmetric about the y axis: $\cos\pi = \cos(-\pi)$. Here is a rule for choosing that sign: If the oscillator's position is increasing at $t = 0$—that is, if the oscillator is initially moving in the *positive* direction—you should choose the *negative* initial phase.

Of course, the converse is true: If the oscillator's position is decreasing at $t = 0$, you should choose a *positive* initial phase.

The initial phase for Crall and Whipple's data is

$$\varphi = \cos^{-1}\left(\frac{-0.4\,\text{m}}{0.4\,\text{m}}\right) = \cos^{-1}(-1) = \pm\pi$$

$$\varphi = -\pi$$

We chose the negative sign because the disk initially moves upward in the positive y direction. For Crall and Whipple's data, Equation 16.3 becomes

$$y(t) = (0.40\,\text{m})\cos(3.1t - \pi) \tag{16.5}$$

Velocity Versus Time in Simple Harmonic Motion

We can find an expression for the simple harmonic oscillator's velocity by taking the time derivative of its position function (Eq. 16.3):

$$v_y(t) = \frac{dy(t)}{dt} = \frac{d[y_{max}\cos(\omega t + \varphi)]}{dt} = y_{max}\frac{d\cos(\omega t + \varphi)}{dt}$$

Use the chain rule (Appendix A):

$$v_y(t) = -y_{max}\omega \sin(\omega t + \varphi) \quad (16.6)$$

A similar equation can be written for SHM along the x or z axis.

The term $y_{max}\omega$ is the amplitude of the velocity curve and also the maximum speed (Fig. 16.5B):

$$v_{max} = y_{max}\,\omega \quad (16.7)$$

We can write Equation 16.6 in terms of the maximum speed:

$$v_y(t) = -v_{max}\sin(\omega t + \varphi) \quad (16.8)$$

The maximum speed is always positive: $v_{max} > 0$. We can read v_{max} from Crall and Whipple's graph (Fig. 16.5B) or use Equation 16.7. Either way, we find

$$v_{max} = y_{max}\,\omega = (0.40\,\text{m})(3.1\,\text{rad/s}) \approx 1.3\,\text{m/s}$$

So, for Crall and Whipple's data, Equation 16.8 becomes

$$v_y(t) = -(1.3\,\text{m/s})\sin(3.1t - \pi)$$

Acceleration Versus Time in Simple Harmonic Motion

Finding an expression for the acceleration of a simple harmonic oscillator requires taking the time derivative of the velocity (Eq. 16.6):

$$a_y(t) = \frac{d[-y_{max}\omega \sin(\omega t + \varphi)]}{dt} = -y_{max}\omega\frac{d\sin(\omega t + \varphi)}{dt}$$

Again, use the chain rule:

$$a_y(t) = -y_{max}\omega^2\cos(\omega t + \varphi) \quad (16.9)$$

The term $y_{max}\omega^2$ is the amplitude of the acceleration curve (Fig. 16.5C) and also the maximum acceleration:

$$a_{max} = y_{max}\,\omega^2 \quad (16.10)$$

We can write Equation 16.9 in terms of the maximum acceleration:

$$a_y(t) = -a_{max}\cos(\omega t + \varphi) \quad (16.11)$$

A similar equation can be written for SHM along the x or z axis.

The maximum acceleration is always positive: $a_{max} > 0$.

Compare parts A and C of Figure 16.5 and notice that the acceleration curve looks like an upside-down version of the position curve. In fact, the acceleration (Eq. 16.9) can be written in terms of the position (Eq. 16.3):

$$a_y(t) = -\omega^2[y_{max}\cos(\omega t + \varphi)]$$

$$a_y(t) = -\omega^2 y(t) \quad (16.12)$$

Equation 16.12 is the chief feature or "hallmark" of simple harmonic motion: The periodic oscillator's acceleration is proportional to its position and is directed opposite to its displacement from equilibrium. In Section 16-4, we will derive this feature by considering the forces that act on the oscillator.

The acceleration of Crall and Whipple's oscillator is given by either Equation 16.9 or Equation 16.12.

$$a_y(t) = -(3.1\,\text{rad/s})^2[(0.40\,\text{m})\cos(3.1t - \pi)] = -(3.8\,\text{m/s}^2)\cos(3.1t - \pi) \quad (16.13)$$

CONCEPT EXERCISE 16.2

CASE STUDY Choosing the Starting Time

Suppose Crall sets the initial time $t = 0$ to the moment when the disk is at point 2 in Figure 16.4 and Whipple sets the initial time to the moment when the disk is at point 3.

a. Write expressions for $y(t)$, $v_y(t)$, and $a_y(t)$ for each researcher's choice of the initial time.

b. Compare your expressions and comment on the significance of their differences and similarities.

CONCEPT EXERCISE 16.3

For each expression, identify the angular frequency ω, period T, initial phase φ and amplitude y_{max} of the oscillation. All values are in SI units.

a. $y(t) = 0.75\cos(14.5t)$ **b.** $v_y(t) = -0.75\sin(14.5t + \pi/2)$
c. $a_y(t) = 14.5\cos(0.75t + \pi/2)$

16-3 Connection with Circular Motion

The equations that describe simple harmonic motion might remind you of uniform circular motion (UCM; Section 4-6). Understanding the close connection between SHM and UCM will help you digest the mathematics in the previous section.

Figure 16.6 shows a motion diagram of a ball moving in a vertical circle at a constant angular speed ω. The origin of our chosen coordinate system is at the center of the circle. The ball's initial position is at point 1, and it then moves counterclockwise. The ball completes one revolution as it passes through 1 again.

Uniform circular motion is periodic; that is, the motion repeats with a regular time interval. The ball, however, does not move back and forth along a single path. Instead, it continues to move in one direction along the same path. Therefore, the ball is not oscillating, so you probably wouldn't expect the SHM equations to describe the ball's kinematics.

Actually there are two simple harmonic oscillators hidden in the ball's circular motion. To see these two oscillators, think of lights above the circle showing the ball's shadow on the floor below and lights on the left showing the ball's shadow on the right wall. Figure 16.6 shows motion diagrams (in gray) for both the shadow on the floor and the one on the wall. The shadow on the floor moves back and forth along a line parallel to the x axis, whereas the shadow on the wall moves up and down along a line parallel to the y axis.

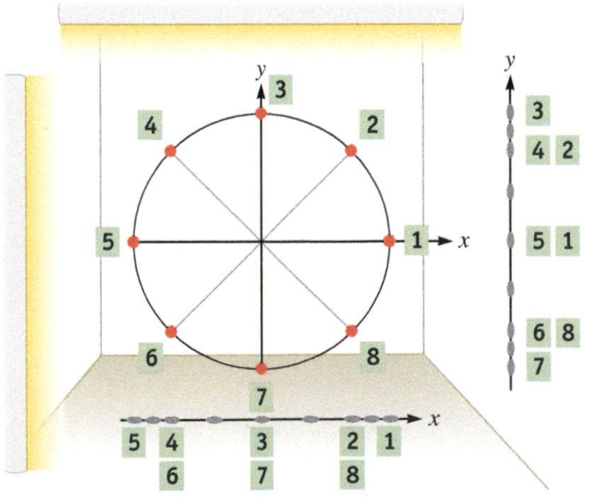

FIGURE 16.6 A ball travels in uniform circular motion. Lights on the ceiling show the ball's shadow on the floor, and lights on the left wall show the ball's shadow on the right wall.

The motion diagrams of both shadows look like Crall and Whipple's motion diagram (Fig. 16.4A). The points are close together near the ends and far apart near the middle, so the shadow moves more slowly near the ends of its path than in the middle. To confirm that the shadow's motion is simple harmonic, we must verify that the kinematic equations describe the motion we see.

Let's start with position. (You may be asked to find the velocity and acceleration for homework; see Problem 17.) Figure 16.7 shows the ball at some arbitrary time t at position $\vec{r}(t) = \vec{x}(t) + \vec{y}(t)$ or angular position $\theta(t) = \omega t$, where ω is the ball's angular speed. As usual, we can use trigonometry to find the x component of the ball's position:

$$x(t) = r\cos\omega t \tag{16.14}$$

Because the shadow on the floor is always directly below the ball, Equation 16.14 also gives this shadow's position along the x axis. The magnitude r is a constant equal to the radius of the circle. From Figure 16.7, that radius is also the maximum x or y value of:

$$r = x_{max} = y_{max} \tag{16.15}$$

We can use this result to rewrite the shadow's position (Eq. 16.14):

$$x(t) = x_{\max} \cos \omega t \tag{16.16}$$

Compare Equation 16.16 with the position of a simple harmonic oscillator, $x(t) = x_{\max} \cos(\omega t + \varphi)$ (Eq. 16.3). Equation 16.16 gives the position of a simple harmonic oscillator along the x axis with the initial phase φ set to zero.

We used the same symbol ω for both the angular speed of the ball and the angular frequency of the shadow because these two quantities are equal in magnitude; the ball and the shadow have the same period. Be careful not to confuse angular frequency and angular speed, however. The distinction between these two quantities is important when studying pendulums (Section 16-6).

Now let's find the y component of the ball's position. From Figure 16.7, we find

$$y(t) = y_{\max} \sin(\omega t) \tag{16.17}$$

Use a trigonometric identity (Appendix A) to write Equation 16.17 in terms of a cosine:

$$y(t) = y_{\max} \cos\left(\omega t - \frac{\pi}{2}\right) \tag{16.18}$$

Comparing Equation 16.18 with Equation 16.3, we have obtained the position of a simple harmonic oscillator with an initial phase $\varphi = -\pi/2$.

We just found another way to think about UCM: as the combination of two simple harmonic oscillators along perpendicular axes. In Figure 16.6, one oscillator is along the x axis and the other along the y axis. These two oscillators have equal amplitudes $r = x_{\max} = y_{\max}$ and equal angular frequencies ω, but their initial phases differ by $\pi/2$. In other words, these oscillators are $\pi/2$ *out of phase*.

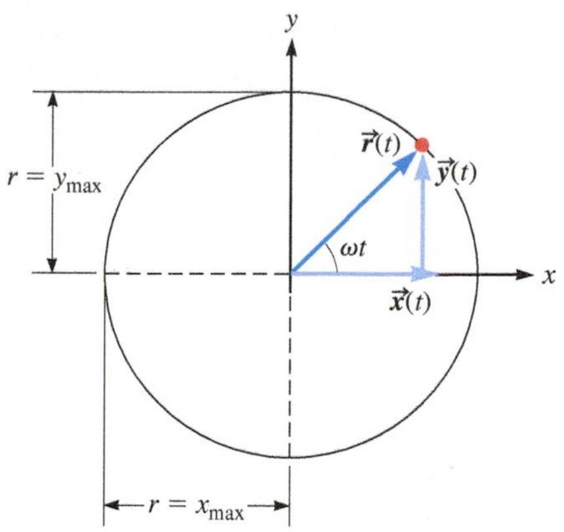

FIGURE 16.7 A ball in uniform circular motion. At an arbitrary time t, the ball's position is $\vec{r}(t)$, and its angular position is ωt.

EXAMPLE 16.1 Earth Satellite

Model the orbit of a satellite as UCM around the Earth with an orbital period of 90.0 min, ignoring the Earth's motion. Use the coordinate system shown in Figure 16.8. The Sun is at some distant position on the positive y axis, and the satellite moves clockwise as shown. Initially, the satellite is in opposition, on the opposite side of the Earth from the Sun. Find expressions for the x component of the satellite's position, velocity, and acceleration. *Hint*: Think of the shadow that would be cast by the satellite due to sunlight.

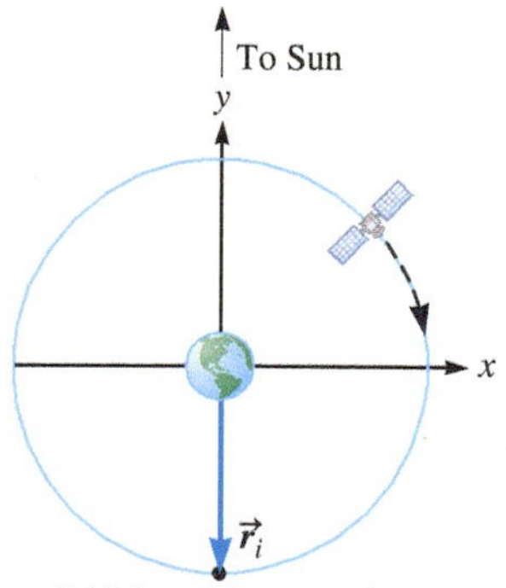

FIGURE 16.8

INTERPRET and ANTICIPATE

Once we have the position equation, it is relatively simple to find velocity and acceleration equations. The hint suggests that we compare Figure 16.8 with Figure 16.6. The satellite's solar shadow, like the ball's shadow on the floor, is along the x axis, but the initial positions and directions of motion differ in the two figures. The ball's initial position is $\vec{r}_i = x_{\max}\hat{\imath}$, and it moves counterclockwise. The satellite's initial position is $\vec{r}_i = -y_{\max}\hat{\jmath}$, and it moves clockwise. We expect our results to be like the x components of position, velocity, and acceleration for the ball, but with specific values for the satellite's amplitude $x_{\max}$, angular speed ω, and initial phase φ.

SOLVE

A To find the x component of the satellite's position, start with Equation 16.3 written in terms of x. We need to find $x_{\max}$, ω, and φ.	$x(t) = x_{\max} \cos(\omega t + \varphi)$ (16.3)
We can find the satellite's angular speed from its period using Equation 16.2. This angular speed is equal in magnitude to the angular frequency of the shadow's motion.	$T = (90 \text{ min})(60 \text{ s/min}) = 5.40 \times 10^3 \text{ s}$ $\omega = \frac{2\pi}{T} = \frac{2\pi}{5.4 \times 10^3 \text{ s}}$ $\omega = 1.16 \times 10^{-3} \text{ rad/s}$

Example continues on page 458 ▶

Use Kepler's third law (Eq. 7.5) to find x_{max}. As in the case of the ball in Figures 16.6 and 16.7, $r = x_{max}$.	$$T^2 = \left(\frac{4\pi^2}{GM_\oplus}\right)x^3_{max} \quad (7.5)$$ $$x_{max} = \left(\frac{GM_\oplus T^2}{4\pi^2}\right)^{1/3}$$ $$x_{max} = \left[\frac{(6.67 \times 10^{-11}\ N\cdot m^2/kg^2)(5.98 \times 10^{24}\ kg)(5.40 \times 10^3\ s)^2}{4\pi^2}\right]^{1/3}$$ $$x_{max} = 6.65 \times 10^6\ m$$
The initial phase comes from Equation 16.4, using $x_i = 0$. As the satellite moves clockwise from its initial position, x decreases, so we should choose the positive sign.	$$\varphi = \cos^{-1}\left(\frac{x_i}{x_{max}}\right) = \cos^{-1} 0 = \pm\frac{\pi}{2}$$ $$\varphi = \frac{\pi}{2}$$
Substitute the results for x_{max}, ω, and φ into Equation 16.3.	$$x(t) = (6.65 \times 10^6\ m)\cos\left(1.16 \times 10^{-3}t + \frac{\pi}{2}\right) \quad (1)$$

CHECK and THINK

Equation (1) has the form we expected. It is the same equation we would find for a satellite whose initial position is on the same side as the Sun, $\vec{r}_i = +y_{max}\hat{j}$, and whose orbit is counterclockwise.

INTERPRET and ANTICIPATE

There are at least two different ways to find (B) velocity and (C) acceleration. We could take the time derivative of position (Eq. 1) to find velocity and then take the time derivative of velocity to find acceleration, or we could use Equation 16.6 for velocity and Equation 16.12 for acceleration. Let's use the second approach here.

SOLVE **B** Write Equation 16.6 for velocity but in the x direction and substitute values found above.	$$v_x(t) = -x_{max}\omega \sin(\omega t + \varphi) \quad (16.6)$$ $$v_x(t) = -(6.65 \times 10^6\ m)(1.16 \times 10^{-3}\ rad/s)\sin\left(1.16 \times 10^{-3}t + \frac{\pi}{2}\right)$$ $$v_x(t) = -(7.71 \times 10^3\ m/s)\sin\left(1.16 \times 10^{-3}t + \frac{\pi}{2}\right)$$
C Write Equation (16.12) for acceleration but in the x direction and substitute values found above.	$$a_x(t) = -\omega^2 x(t)$$ $$a_x(t) = -(1.16 \times 10^{-3}\ rad/s)^2(6.65 \times 10^6\ m)\cos\left(1.16 \times 10^{-3}t + \frac{\pi}{2}\right)$$ $$a_x(t) = -(8.95\ m/s^2)\cos\left(1.16 \times 10^{-3}t + \frac{\pi}{2}\right)$$
CHECK and THINK The solutions for velocity and acceleration have the form we expected, and we can further check the acceleration equation. The maximum acceleration of the shadow along x should equal the centripetal acceleration of the satellite, $a_{max} = a_c = 8.95\ m/s^2$. In this case, the centripetal acceleration should equal the acceleration due to gravity at the satellite's position, $a_{max} = a_c = g(x_{max})$, where $g(x_{max})$ is given by Equation 7.11. Aside from a little rounding error, the numbers are in agreement.	$$g(x_{max}) = \frac{GM_\oplus}{x^2_{max}} \quad (7.11)$$ $$g(x_{max}) = \frac{(6.67 \times 10^{-11}\ N\cdot m^2/kg^2)(5.98 \times 10^{24}\ kg)}{(6.65 \times 10^6\ m)^2}$$ $$g(x_{max}) = 9.02\ m/s^2 \approx a_{max}$$

16-4 Dynamics of Simple Harmonic Motion

In Sections 16-1 and 16-2, we *described* the motion of the disk in Crall and Whipple's experiment. In this section, we examine the *cause* or dynamics of that motion. In particular, we find two ways to express the net force on the disk.

First, according to Newton's second law, all we need to do is multiply the mass ($m = 0.100$ kg, page 452) by its acceleration:

$$\sum F_y(t) = ma_y(t) = -ma_{\text{max}} \cos(\omega t + \varphi) \qquad (16.19)$$

For Crall and Whipple's data (Eq. 16.13), we find

$$\sum F_y(t) = (0.100 \text{ kg})(-3.8 \text{ m/s}^2) \cos(3.1t - \pi)$$

$$\sum F_y(t) = -(0.38 \text{ N}) \cos(3.1t - \pi) \qquad (16.20)$$

So, the net force depends on time. A graph of force versus time looks similar to the acceleration-versus-time graph with each value on the vertical axis multiplied by mass (Fig. 16.9).

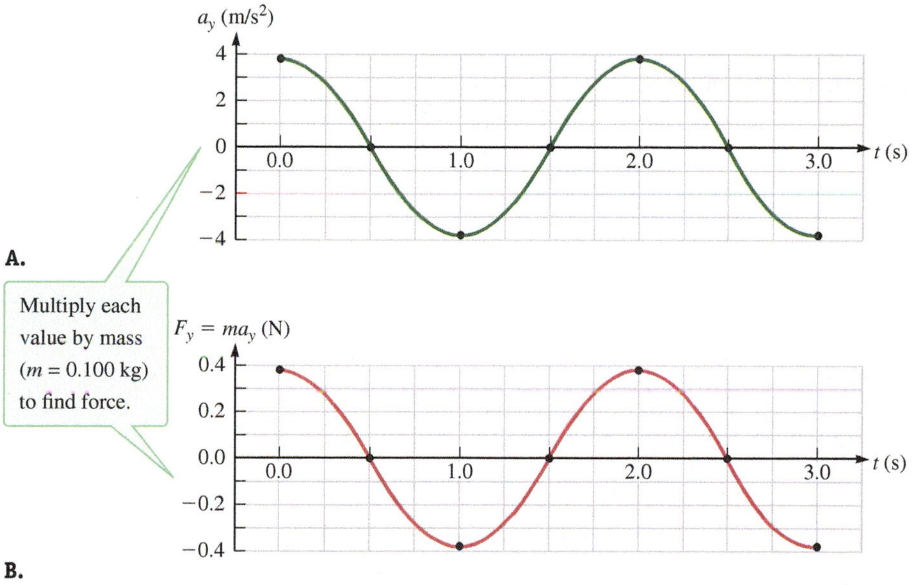

FIGURE 16.9 A. Acceleration data for Crall and Whipple's experiment, repeated from Figure 16.5C. **B.** To find force versus time, multiply by mass.

Equation 16.19 is one expression for the net force on a simple harmonic oscillator; we found it without considering the specific forces—gravity and the spring force—exerted on the disk. Let's consider these forces to find a second expression for the net force by taking a closer look at how Crall and Whipple set up their experiment. First, they hung a spring of length ℓ from its support (Fig. 16.10A). Next, they hung the disk from the spring (Fig. 16.10B). The spring then stretched to its new length $\ell + \Delta\ell$. They set the origin ($y = 0$) of the coordinate system to the disk's position. The disk was at rest and two forces—gravity and the spring force obeying Hooke's law—were exerted on it as shown in the free-body diagram in Figure 16.10B. Because the acceleration was zero when the disk was at rest, these forces were equal in magnitude at that time:

$$F_H = F_g$$

$$k\,\Delta\ell = mg \qquad (16.21)$$

Finally, Whipple set the disk into motion. Figure 16.10C shows it at some arbitrary time when the disk is at $\vec{y}(t)$, measured from the origin at $y = 0$. Two forces are exerted on the disk: the downward force of gravity and the upward force exerted by the spring. According to Hooke's law, the upward force exerted by the spring

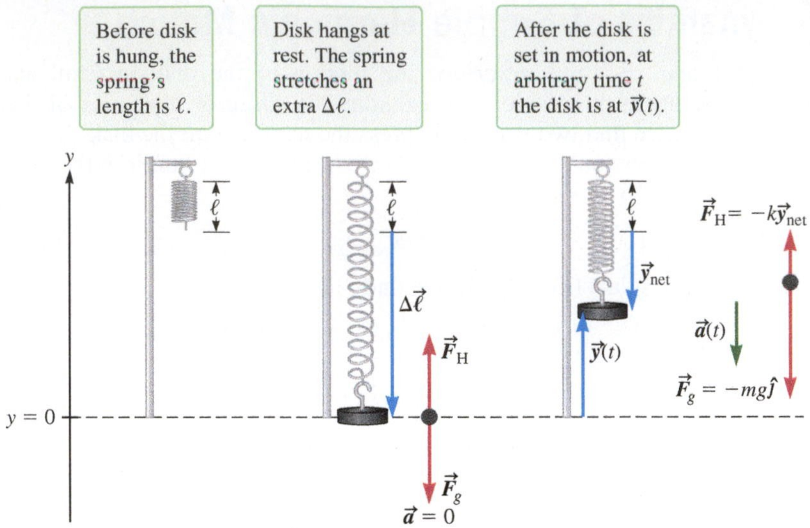

Before disk is hung, the spring's length is ℓ.
Disk hangs at rest. The spring stretches an extra Δℓ.
After the disk is set in motion, at arbitrary time t the disk is at $\vec{y}(t)$.
y
y = 0
ℓ
Δ$\vec{\ell}$
$\vec{F}_H$
$\vec{F}_g$
$\vec{a} = 0$
$\vec{y}_{net}$
$\vec{y}(t)$
$\vec{F}_H = -k\vec{y}_{net}$
$\vec{a}(t)$
$\vec{F}_g = -mg\hat{\jmath}$
A.
B.
C.

FIGURE 16.10 **A.** Vertical spring of length ℓ. **B.** Crall and Whipple then hung a disk from the spring, stretching it by an amount Δℓ. Two forces are exerted on the disk: the downward force of gravity and the upward force exerted by the spring. With the disk at rest, $F_H = F_g$. **C.** The disk oscillates up and down. At an arbitrary time t, the disk is at $y(t)$.

depends on its net stretch, which is $\vec{y}_{net}(t) = -[\Delta\ell - y(t)]\hat{\jmath}$ in the coordinate system shown. Therefore,

$$\vec{F}_H = -k\vec{y}_{net}$$

$$\vec{F}_H = k[\Delta\ell - y(t)]\hat{\jmath} \tag{16.22}$$

The total force exerted on the disk is

$$\sum\vec{F} = \vec{F}_H + \vec{F}_g = m\vec{a}$$

$$\sum\vec{F} = k[\Delta\ell - y(t)]\hat{\jmath} - mg\hat{\jmath} = ma_y(t)\hat{\jmath}$$

Regrouping terms,

$$\sum\vec{F} = (k\,\Delta\ell - mg)\hat{\jmath} - ky(t)\hat{\jmath} = ma_y(t)\hat{\jmath}$$

Simplify by substituting $k\,\Delta\ell = mg$ (Eq. 16.21):

$$\sum\vec{F} = (mg - mg)\hat{\jmath} - ky(t)\hat{\jmath} = ma_y(t)\hat{\jmath}$$

$$\sum\vec{F} = -ky(t)\hat{\jmath} = ma_y(t)\hat{\jmath} \tag{16.23}$$

So, another way to express the net force on a simple harmonic oscillator is in terms of its position $y(t)$:

$$\sum F_y = -ky(t) = -ky_{max}\cos(\omega t + \varphi) \tag{16.24}$$

where we have substituted Equation 16.3, $y(t) = y_{max}\cos(\omega t + \varphi)$.

Let's apply Equation 16.24 to Crall and Whipple's data by multiplying Equation 16.5 for the disk's position, $y(t) = (0.40\text{ m})\cos(3.1t - \pi)$, by minus one times the spring constant $k = 0.987$ N/m (case study, page 452):

$$\sum F_y = -(0.987\text{ N/m})(0.40\text{ m})\cos(3.1t - \pi)$$

$$\sum F_y = -(0.39\text{ N})\cos(3.1t - \pi)$$

which (except for a slight error) is what we found before (Eq. 16.20). So, we have just found a second way to make a force-versus-time graph: Start with a position-versus-time graph and multiply the values on the vertical axis by $-k$ (Fig. 16.11).

Moreover, by writing Equation 16.23 as

$$a_y(t) = -\frac{k}{m}y(t) \tag{16.25}$$

we gain a deeper insight into Equation 16.12, the hallmark of SHM $[a_y(t) = -\omega^2 y(t)]$. Equation 16.25 expresses the hallmark of SHM for the special case of an object–spring oscillator. The negative sign indicates that acceleration and position vectors point in opposite directions. By using Newton's laws as we did in this section, we

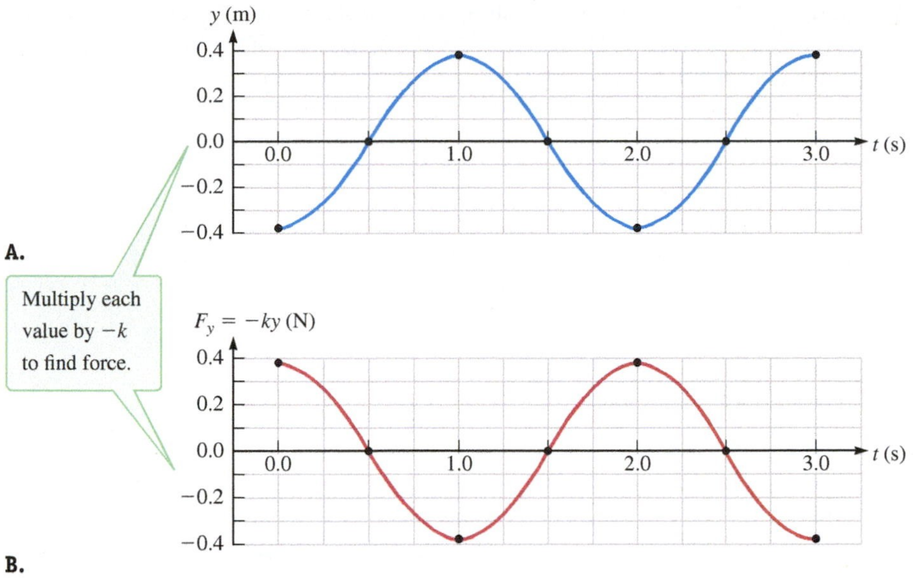

FIGURE 16.11 **A.** Crall and Whipple's position-versus-time graph. **B.** A force-versus-time graph is made by multiplying the position-versus-time graph by $-k$. Compare with Figure 16.9.

see why they are in opposite directions. The net force exerted on the oscillator is a restoring force (Eq. 16.24), which always points toward the equilibrium position ($y = 0$) in the opposite direction of the displacement.

16-5 Special Case: Object–Spring Oscillator

OBJECT–SPRING OSCILLATOR

Special Case

In this section, we continue to study the special case of a simple harmonic oscillator consisting of an object attached to a spring. We just found that for this specific oscillator, the hallmark of SHM is

$$a_y(t) = -\omega^2 y(t) = -\frac{k}{m}y(t)$$

where we have combined Equations 16.12 and 16.25. By canceling $-y(t)$, we find

$$\omega = \omega_s \equiv \sqrt{\frac{k}{m}} \tag{16.26}$$

Equation 16.26 gives the angular frequency of an oscillating object with mass m attached to a spring with spring constant k, so we have added the subscript s to denote the spring. The angular frequency is related to the frequency and period by Equation 16.2 ($\omega \equiv 2\pi f = 2\pi/T$) so once you know one of these parameters, you know all three. It might seem amazing, but **the frequency of SHM does not depend on the amplitude**. In other words, once Crall and Whipple chose their disk (mass m) and spring (spring constant k), the frequency was already set, and it did not matter how far Whipple initially displaced the disk.

The spring does not need to be vertical like the one Crall and Whipple used. Figure 16.12 shows a block attached to a horizontal spring that oscillates on a frictionless table. Three forces are exerted on the block: the normal force, gravity, and the spring force. The block does not accelerate in the y direction, so gravity is balanced by the normal force. The spring force accelerates the block in the x direction. When the block is at $x = 0$, the spring is relaxed. At some arbitrary time t, the block is at $\vec{x}(t)$. For $x(t) > 0$, the spring is stretched and exerts a force in the negative x direction, accelerating the block in that direction. From Hooke's law and Newton's second law,

$$\vec{F}_H(t) = m\vec{a}(t)$$
$$-k\vec{x}(t) = -kx(t)\hat{\imath} = ma_x(t)\hat{\imath}$$

$$a_x(t) = -\frac{k}{m}x(t) \tag{16.27}$$

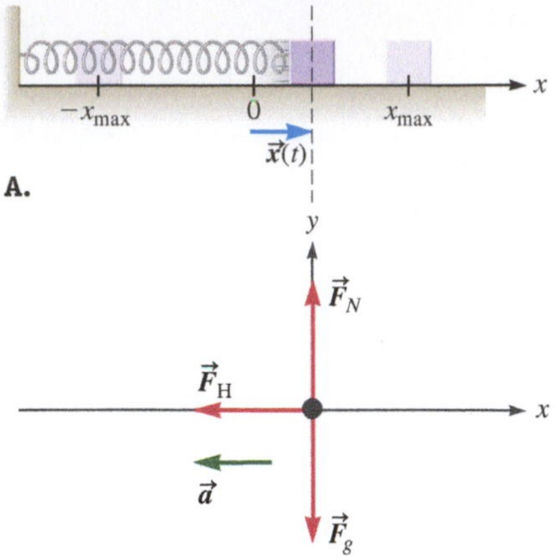

FIGURE 16.12 **A.** A block is connected to a horizontal spring. **B.** Free-body diagram for the block.

Equation 16.27 also looks like the hallmark of SHM (Eq. 16.12), so the block's angular frequency is given by Equation 16.26. As in the case of the vertical spring, the frequency does not depend on the amplitude of the oscillation.

We can use our derivation of Equation 16.26 to come up with a general procedure for finding the angular frequency of any simple harmonic oscillator. First, use Newton's laws to find an expression for the object's acceleration. Next, compare the object's acceleration with $a_y(t) = -\omega^2 y(t)$ (Eq. 16.12) to see if the acceleration obeys this hallmark of SHM. If it does, identify factors that determine ω, the angular frequency. We will use this procedure in the next three sections to find the angular frequency of several common oscillators.

CONCEPT EXERCISE 16.4

CASE STUDY Angular Frequency

Use Equation 16.26 to predict the angular frequency and period of Crall and Whipple's oscillator. How does your prediction compare with their actual data (Fig. 16.5)?

EXAMPLE 16.2 CASE STUDY A Stronger Spring

Crall and Whipple repeat their experiment using a stiffer spring of spring constant 65.3 N/m. If everything else is exactly the same, find new expressions for position, velocity, acceleration, and force.

INTERPRET and ANTICIPATE

Our results should be similar to the previous expressions for Crall and Whipple's data, but a stiffer spring means that the restoring force is greater, so we expect a greater acceleration and speed. The stiffer spring also means a higher angular frequency.

SOLVE

Equation 16.26 gives the new angular frequency.

$$\omega_s = \sqrt{\frac{k}{m}} = \sqrt{\frac{65.3\ \text{N/m}}{0.100\ \text{kg}}}$$

$$\omega_s = 25.6\ \text{rad/s}$$

The amplitude and initial phase are unchanged, so the position equation is the same as Equation 16.5 except for the new angular frequency.

$$y(t) = (0.40\ \text{m})\cos(25.6t - \pi)$$

The velocity comes from Equation 16.6.

$$v_y(t) = -y_{max}\omega \sin(\omega t + \varphi) \qquad (16.6)$$

$$v_y(t) = -(0.40\ \text{m})(25.6\ \text{rad/s})\sin(25.6t - \pi)$$

$$v_y(t) = -(10.2\ \text{m/s})\sin(25.6t - \pi)$$

The acceleration comes from Equation 16.9.

$$a_y(t) = -y_{max}\omega^2 \cos(\omega t + \varphi) \qquad (16.9)$$

$$a_y(t) = -(0.40\ \text{m})(25.6\ \text{rad/s})^2\cos(25.6t - \pi)$$

$$a_y(t) = -(262\ \text{m/s}^2)\cos(25.6t - \pi)$$

Finally, find the force by multiplying the acceleration by the mass.

$$F_y = ma_y = (0.100\ \text{kg})[-(262\ \text{m/s}^2)\cos(25.6t - \pi)]$$

$$F_y = -(26.2\ \text{N})\cos(25.6t - \pi)$$

CHECK and THINK

As expected, the maximum speed $v_{max} = 10.2$ m/s, the maximum acceleration $a_{max} = 262$ m/s^2, and the maximum force $F_{max} = 26.2$ N are all greater in the case of the stiffer spring than with the original spring. The amplitude y_{max} is unchanged, however.

EXAMPLE 16.3 Measuring Mass and Force

A block of unknown mass is attached to a horizontal spring ($k = 75$ N/m) as in Figure 16.12. The block oscillates on a frictionless table between $x = \pm 1.2$ m. The block's period is measured to be 2.7 s.

What is the mass of the block?

INTERPRET and ANTICIPATE

The mass is related to the period (and angular frequency) of the block.

SOLVE

Solve Equation 16.26 for mass.

$$\omega_s = \sqrt{\frac{k}{m}} \qquad (16.26)$$

$$m = \frac{k}{\omega_s^2}$$

Use Equation 16.2 to write mass in terms of period.

$$T = \frac{2\pi}{\omega_s} \qquad (16.2)$$

$$T^2 = \frac{4\pi^2}{\omega_s^2}$$

$$m = \frac{k}{4\pi^2}T^2 = \frac{75\ \text{N/m}}{4\pi^2}(2.7\ \text{s})^2 = 13.8\ \text{kg} = 14\ \text{kg}$$

CHECK and THINK

Verify for yourself that the results for mass have the expected dimensions. This example describes a way to find the mass of an object without using a gravitational field.

16-6 Special Case: Simple Pendulum

Today when we think of Galileo, we tend to focus on his work in astronomy. Galileo's initial fame, however, was based on a much more down-to-Earth discovery. When he was 20 years old, he used his pulse to time the period of a lamp swinging overhead in a cathedral. Galileo discovered that the period did *not* depend on the amplitude of the lamp's oscillation. The cathedral lamp is an example of a simple pendulum. A **simple pendulum** consists of a particle—known as the bob—hanging from a massless, unstretchable string or rod. The pendulum bob oscillates back and forth along a portion of a circular path. Of course, no real pendulum is such an idealized simple pendulum, but many real pendulums can be modeled in this way. Galileo's discovery had important practical significance. Because of its regular period, the pendulum was the basis of the most accurate clocks for nearly 300 years (Fig. 16.1A). Clock pendulums are usually approximated as simple pendulums.

Let's find the angular frequency of a simple pendulum. Figure 16.13 shows a simple pendulum consisting of a bob at the end of a string of length ℓ. The origin of the coordinate system is placed at the point where the string is attached to the support, so the bob's position $\vec{r}$ is measured from that point. As the bob swings along its circular arc, the magnitude of $\vec{r}$ is constant and equal to the length of the string, $r = \ell$. The x axis has been chosen to point straight down so that when the bob hangs vertically, its angular position is $\theta = 0$. When the bob is to the right of the vertical, its angular position is positive. Its angular position is negative to the left of the vertical.

At some arbitrary time t, the bob's angular position is $\theta(t)$ as shown in Figure 16.13A. Figure 16.13B shows a free-body diagram for the bob at that moment. Two forces are exerted on the bob: gravity and the tension force. In our model of a simple pendulum, the string is unstretchable. Therefore, the bob cannot move radially (along the radius of the circular arc), only tangentially. The forces in Figure 16.13B are broken into components that are either parallel (or antiparallel) or perpendicular

SIMPLE PENDULUM **Special Case**

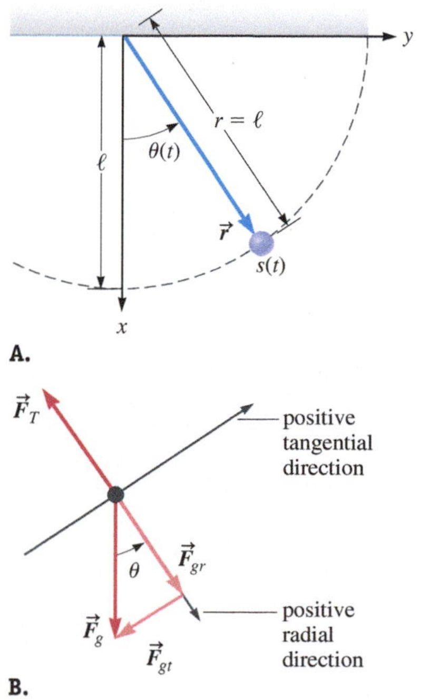

FIGURE 16.13 A. A simple pendulum consists of a particle at the end of a string of length ℓ. **B.** Free-body diagram for the pendulum bob.

to the bob's velocity. In this case, those two directions are radial and tangential to the circular arc. The tension force $\vec{F}_T$ has only a radial component, but gravity $\vec{F}_g$ has a *r*adial component $\vec{F}_{gr}$ and a *t*angential component $\vec{F}_{gt}$. Because the bob does not move radially, we only need to consider the tangential component to find expressions for the bob's SHM.

The tangential component of gravity $F_{gt} = -mg \sin\theta(t)$ provides a restoring force just like the spring provides a restoring force on Crall and Whipple's disk. In the case of a simple pendulum, gravity causes the bob to oscillate. Newton's second law gives an expression for the tangential acceleration $a_t(t)$:

$$\sum F_t = ma_t$$

$$-mg \sin\theta(t) = ma_t(t)$$

$$-g \sin\theta(t) = a_t(t)$$

We have a mixture of angular parameters $\theta(t)$ and translational parameters $a_t(t)$. Equation 12.14, $a_t = \ell\alpha$, gets rid of the translational parameter, so we can write $a_t(t)$ in terms of the bob's angular acceleration $\alpha(t)$:

$$-g \sin\theta(t) = \ell\alpha(t)$$

$$\alpha(t) = -\frac{g}{\ell}\sin\theta(t)$$

We have just found the angular acceleration of the bob as a function of its angular position, but the result does *not* fit the form of Equation 16.12 that characterizes SHM, $\alpha(t) = -\omega^2\theta(t)$. For SHM, $\alpha(t)$ must be proportional to $\theta(t)$, but we found that $\alpha(t)$ is proportional to the *sine* of $\theta(t)$. We are forced to conclude that a simple pendulum oscillates, but its motion is *not* simple harmonic.

If, however, we limit the bob's angular position $\theta(t)$ to small values—say no more than 15°—its motion is very nearly simple harmonic. From Appendix A, the sine of a small angle measured in radians approximately equals that angle:

$$\sin\theta \approx \theta \quad \text{for } \theta \ll 1 \text{ rad}$$

If the bob's angular displacement is small, we can write the angular acceleration as

$$\alpha(t) \approx -\frac{g}{\ell}\theta(t) \tag{16.28}$$

Equation 16.28 has the same mathematical form as the hallmark of SHM if the angular frequency is given by

$$\omega_{\text{smp}} = \sqrt{\frac{g}{\ell}} \tag{16.29}$$

where the subscript "smp" stands for simple pendulum. *For a small angular displacement, a simple pendulum's motion is simple harmonic with an angular frequency that depends only on the length of the string and the local gravitational acceleration.*

The kinematic equations for position (Eq. 16.3), velocity (Eq. 16.8), and acceleration (Eq. 16.11) may be written in terms of the *angular* position, *angular* speed, and *angular* acceleration of a simple pendulum by using the same analogy we used in Chapter 12:

Be careful; the symbol ω is used both for angular frequency and angular speed. You must distinguish between the two quantities from the context. For instance, when used inside the argument of a sine or cosine function, ω is the angular frequency, so ω_{smp} in Equation 16.31 represents angular frequency.

$$y \to \theta \qquad v \to \omega \qquad a \to \alpha \tag{16.30}$$

For example, Equation 16.3 becomes

$$\theta(t) = \theta_{\text{max}} \cos(\omega_{\text{smp}} t + \varphi) \tag{16.31}$$

where θ_{max} is the maximum angular position or amplitude of the bob.

CONCEPT EXERCISE 16.5

Calculate the fractional difference $|(\sin\theta - \theta)/\sin\theta|$ for $\theta = 0, \pm5°, \pm10°$, and $\pm15°$. *Hint*: Work in radians using two significant figures.

CONCEPT EXERCISE 16.6

Using the analogy in Equation 16.30,

a. Write an expression for angular acceleration from Equation 16.11, $a_y(t) = -a_{max} \cos(\omega t + \varphi)$.
b. Write an expression for the maximum angular speed from Equation 16.7, $v_{max} = y_{max}\omega$.
c. Write an expression for the maximum angular acceleration from Equation 16.10, $a_{max} = y_{max}\omega^2$.
d. What does each ω represent in your expressions?

EXAMPLE 16.4 Pendulum on the Moon

A simple pendulum in a clock is displaced to $\theta_i = 5.00°$ (8.73×10^{-2} rad) and released from rest. When the pendulum passes through its equilibrium position, the bob's speed is 0.300 m/s.

A What are the length, angular frequency, and period of the pendulum?

INTERPRET and ANTICIPATE

We must interpret two important pieces of information from the example statement. The amplitude of the oscillation is $\theta_{max} = 8.73 \times 10^{-2}$ rad, and the bob's maximum *translational* speed is $v_{max} = 0.300$ m/s. The maximum translational speed is *not* the same as the maximum *angular* speed ω_{max}.

SOLVE

Find an expression for the maximum *angular* speed from the maximum *translational* speed (Eq. 12.13). In this case, the radius of the circular arc is the length of the pendulum's rod.

$$v_{max} = \ell\omega_{max} \quad (12.13)$$

$$\omega_{max} = \frac{v_{max}}{\ell} \quad (1)$$

By analogy with Equation 16.7, the bob's maximum angular speed ω_{max} depends on the angular frequency ω_{smp} and the amplitude of the oscillation. (Also see Concept Exercise 16.6b.)

$$\omega_{max} = \theta_{max}\omega_{smp} \quad (2)$$

Substitute $\omega_{smp} = \sqrt{g/\ell}$ (Eq. 16.29) for the angular frequency ω_{smp} of a simple pendulum into Equation (2).

$$\omega_{max} = \theta_{max}\sqrt{\frac{g}{\ell}} \quad (3)$$

Substitute Equation (1) into Equation (3) and solve for ℓ.

$$\frac{v_{max}}{\ell} = \theta_{max}\sqrt{\frac{g}{\ell}} \qquad v_{max} = \ell\theta_{max}\sqrt{\frac{g}{\ell}} = \theta_{max}\sqrt{g\ell}$$

$$v_{max}^2 = \theta_{max}^2 g\ell$$

$$\ell = \frac{v_{max}^2}{g\theta_{max}^2} = \frac{(0.300 \text{ m/s})^2}{(9.81 \text{ m/s}^2)(8.73 \times 10^{-2} \text{ rad})^2}$$

$$\ell = 1.20 \text{ m}$$

Now that we know the pendulum's length, use $\omega_{smp} = \sqrt{g/\ell}$ (Eq. 16.29) to find angular frequency.

$$\omega_{smp} = \sqrt{\frac{g}{\ell}} = \sqrt{\frac{9.81 \text{ m/s}^2}{1.20 \text{ m}}}$$

$$\omega_{smp} = 2.86 \text{ rad/s}$$

Use Equation 16.2 to find the period.

$$T = \frac{2\pi}{\omega_{smp}} = \frac{2\pi}{2.86 \text{ rad/s}}$$

$$T = 2.20 \text{ s}$$

Example continues on page 466 ▶

CHECK and THINK

The dimensions are correct, and the length of the pendulum seems to fit a pendulum clock (Fig. 16.1A). Because angular frequency and angular speed are both represented by ω, one of the trickiest parts of this problem is keeping clear in your mind the distinction between the two; carefully labeled subscripts are your best aid in doing so.

B If the bob is displaced to $\theta_i = 10.0°$ instead, what are the angular frequency and period of the pendulum and the maximum speed of the bob?

INTERPRET and ANTICIPATE

The initial displacement of the bob is the amplitude. So, displacing the bob to 10.0° doubles the pendulum's amplitude. The angular frequency and period do not depend on amplitude and so remain unchanged—$\omega_{smp} = 2.87$ rad/s and $T = 2.20$ s—but the maximum speed of the bob increases.

SOLVE

The new maximum speed comes from combing Equations (1) and (2). Convert the new θ_{max} to radians and substitute values.

$$\omega_{max} = \frac{v_{max}}{\ell} = \theta_{max}\omega_{smp}$$

$$v_{max} = \ell\theta_{max}\omega_{smp}$$

$$v_{max} = (1.20 \text{ m})(2)(8.73 \times 10^{-2} \text{ rad})(2.86 \text{ rad/s}) = 0.600 \text{ m/s}$$

CHECK and THINK

Because the amplitude θ_{max} has doubled while the length and angular frequency are constant, the new maximum speed is double the previous value of 0.300 m/s.

$$(v_{max})_{new} = \theta_{max}(\ell\omega_{smp}) = \theta_{max}(v_{max})_{old}$$

$$(v_{max})_{new} = 2(0.300 \text{ m/s}) = 0.600 \text{ m/s} \checkmark$$

C If the same pendulum clock were on the Moon ($g = 1.62$ m/s^2), what would its period be?

INTERPRET and ANTICIPATE

The gravitational acceleration on the surface of the Moon is lower than that on the Earth, so the restoring force on the pendulum bob is weaker. We expect the bob to have a lower acceleration, giving the pendulum a longer period.

SOLVE

Combine Equations 16.2 and 16.29 to find the period.

$$T = \frac{2\pi}{\omega_{smp}} \quad \text{and} \quad \omega_{smp} = \sqrt{\frac{g}{\ell}}$$

$$T = 2\pi\sqrt{\frac{\ell}{g}} = 2\pi\sqrt{\frac{1.19 \text{ m}}{1.62 \text{ m/s}^2}}$$

$$T = 5.39 \text{ s}$$

CHECK and THINK

As expected, the period is longer on the Moon than on the Earth.

PHYSICAL PENDULUM

Special Case

16-7 Special Case: Physical Pendulum

A **physical pendulum** is any real pendulum that *cannot* be modeled as having its mass concentrated in a small bob at the end of a relatively massless string. Isaac in Figure 16.14A is a physical pendulum because his mass is well distributed over his entire body. If his amplitude is not far from his equilibrium position, however, his motion can still be modeled as SHM.

Isaac rotates as he swings back and forth, so we apply the concepts and mathematical tools of rotation (Chapter 12) to find an expression for his angular acceleration α.

We set the z axis along Isaac's rotation axis so that his motion is in the xy plane (Fig. 16.14A). We model Isaac as a rigid object and draw a free-body diagram by representing him as a simple shape (Fig. 16.14B). At some arbitrary time t, his center of mass is at $\vec{r}_{CM}$, which has an angular position $\theta(t)$. Two forces are exerted on him: the normal force of the bar on the back of his knees and gravity exerted on his center of mass. To find his angular acceleration, we need to find the net torque exerted on him. Because the bar runs along the rotation axis, the normal force cannot exert a torque on Isaac, so we have omitted the normal force from our free-body diagram. The only torque exerted on Isaac is due to gravity. From the definition of torque and from Newton's second law in rotational form (Eqs. 12.23 and 12.24), Isaac's angular acceleration is

$$\vec{\tau} = \vec{r} \times \vec{F} = I\vec{\alpha}$$

Substituting the parameters from Figure 16.14B and solving for α gives

$$\vec{\tau} = -r_{CM}F_g \sin\theta(t)\hat{k} = I\alpha(t)\hat{k}$$

$$-r_{CM}mg \sin\theta(t)\hat{k} = I\alpha(t)\hat{k}$$

$$\alpha(t) = -\frac{mgr_{CM}}{I}\sin\theta(t)$$

As in the case of a simple pendulum, a physical pendulum is *not* a simple harmonic oscillator because $\alpha \propto \sin\theta$ instead of $\alpha \propto \theta$. If the maximum angular displacement is small ($\theta_{max} \leq 15°$), however, the physical pendulum (like the simple pendulum) may also be modeled as a simple harmonic oscillator:

$$\alpha(t) = -\frac{mgr_{CM}}{I}\theta(t) \tag{16.32}$$

Comparing Equation 16.32 with Equation 16.12, we find that the angular frequency of a physical pendulum is

$$\omega_{phy} = \sqrt{\frac{mgr_{CM}}{I}} \tag{16.33}$$

where the subscript "phy" stands for *phy*sical pendulum.

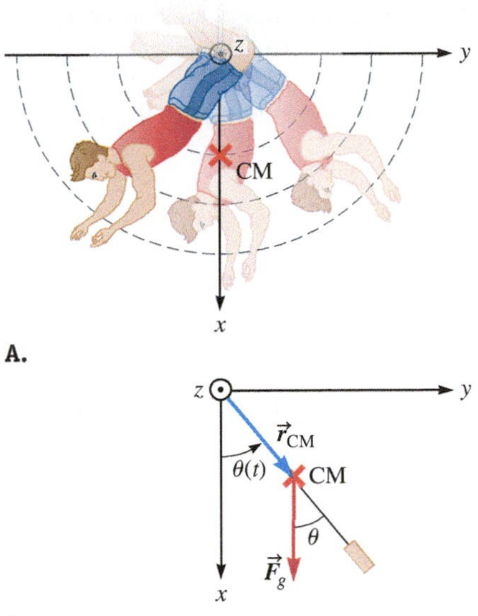

FIGURE 16.14 **A.** Isaac hangs by his knees and swings from a horizontal bar. **B.** Isaac is modeled as a physical pendulum.

EXAMPLE 16.5 A Comfortable Stroll

When you walk, your legs may be modeled as physical pendulums (Fig. 16.15). Walking is most comfortable and consumes the least amount of energy when you swing your legs at their natural angular frequency given by Equation 16.33. For the typical leg of a tall man (2 m in height), the pendulum period is 0.9 s. The length of his leg is roughly $L \approx 1$ m, and its mass is roughly one-seventh of the man's total mass, which is $m = 95$ kg. Assume his leg's center of mass is roughly $L/3$ from his hip socket and use these data to estimate the rotational inertia of a tall man's leg.

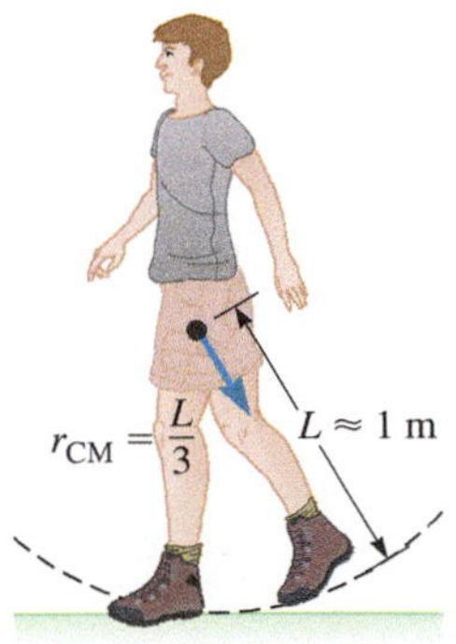

FIGURE 16.15

INTERPRET and ANTICIPATE

The man's leg is modeled as a physical pendulum, so its angular frequency depends on its rotational inertia. In Example 13.6 (page 367), we found the rotational inertia of a thin rod rotated about an axis through one end: $I = \frac{1}{3}mL^2$. If we treat the leg as a uniform rod pivoted at the hip, its rotational inertia would be $I = \frac{1}{3}[(95/7)\text{ kg}](1\text{ m})^2 = 4.5\text{ kg}\cdot\text{m}^2$. We should obtain a result less than this one with our physical pendulum model because we are taking the center of mass to be closer to the pivot (that is, center of mass at $L/3$, not at $L/2$).

SOLVE

Begin with Equations 16.2 and 16.33.

$$T = \frac{2\pi}{\omega_{phy}} \tag{16.2}$$

$$\omega_{phy} = \sqrt{\frac{mgr_{CM}}{I}} \tag{16.33}$$

Example continues on page 468 ▶

Substitute to obtain an expression for the period that includes the rotational inertia.	$T = 2\pi\sqrt{\dfrac{I}{mgr_{CM}}}$
Solve for I and substitute the given information.	$I = \dfrac{mgr_{CM}T^2}{4\pi^2} = \dfrac{[(95/7)\text{ kg}](9.81\text{ m/s}^2)(1/3\text{ m})(0.9\text{ s})^2}{4\pi^2}$ $I = 0.91\text{ kg}\cdot\text{m}^2$

CHECK and THINK

As expected, our result is lower than if we had modeled the leg as a thin, uniform rod rotated around an axis through one end.

TORSION PENDULUM

Special Case

16-8 Special Case: Torsion Pendulum

Computer scientist Danny Hillis published an essay titled "The Millennium Clock" in *Wired* magazine. In his article, Hillis said that he wanted

> to build a clock that ticks once a year. The century hand advances once every one hundred years, and the cuckoo comes out on the millennium. I want the cuckoo to come out every millennium for the next 10,000 years.[1]

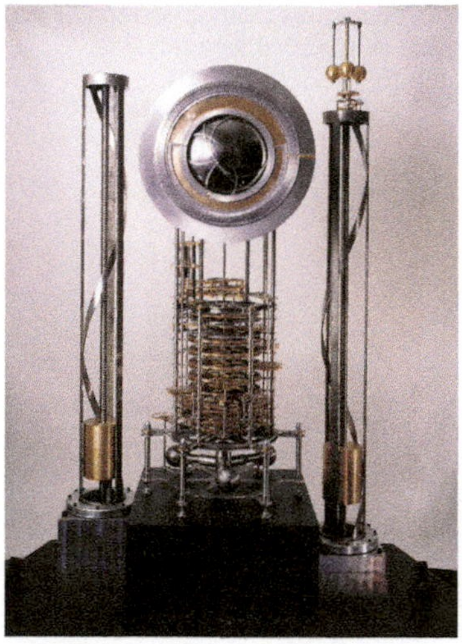

Photo by Rolfe Horn courtesy of The Long Now Foundation

FIGURE 16.16 The prototype of Hillis's 10,000-year "millennium clock" is in the Science Museum in London.

A prototype for Hillis's clock was built (Fig. 16.16) and bonged twice at midnight on January 1, 2000. Unlike the clocks in Figure 16.1A, Hillis's clock is based on a *torsion pendulum* (Fig. 16.17).

A **torsion pendulum** consists of an object such as the three-armed wheel in Figure 16.17A or the tire in Figure 16.17B connected to a *torsion spring*. A **torsion spring** resists being twisted just as a regular spring resists being stretched or compressed. For example, imagine a parent rotating and then releasing the tire (Fig. 16.17B), the torsion spring (the chains) provides a restoring force, twisting the tire and the child back toward the equilibrium position. In a torsion pendulum, the object rotates back and forth, passing the equilibrium position much like the disk moves up and down in Crall and Whipple's experiment.

In this book, we only study springs that obey Hooke's law ($\vec{F}_H = -k\vec{x}$). Similarly, we only study torsion springs that obey an analogous relation for torque:

$$\tau = -\kappa\theta \tag{16.34}$$

where θ is the object's angular position measured from equilibrium and κ (lowercase Greek letter kappa) is the **torsion spring constant**. The torsion spring constant depends on physical properties of the spring and has the SI units of newton meters per radian.

[1]http://www.longnow.org/essays/millennium-clock/.

Photo by Rolfe Horn courtesy of The Long Now Foundation

A.

Peter McGahey © Cengage Learning

B.

FIGURE 16.17 A. The interior of the millennium clock shows that a metal ribbon is the torsion spring that causes the balls to rotate back and forth. **B.** A child on a tire swing rotates back and forth due to a torsion spring (the chains).

The negative sign in Equation 16.34 means that the torque's direction always accelerates the torsion pendulum toward its equilibrium position (Fig. 16.18). Newton's second law for rotation (Eq. 12.24) leads to an expression for the angular acceleration of the object having rotational inertia I:

$$\tau = I\alpha(t) = -\kappa\theta(t)$$

$$\alpha(t) = -\frac{\kappa}{I}\theta(t) \tag{16.35}$$

Comparing Equation 16.35 with Equation 16.12 shows that the torsion pendulum is a simple harmonic oscillator with an angular frequency of

$$\omega_{\text{tor}} = \sqrt{\frac{\kappa}{I}} \tag{16.36}$$

There is no need to use the small-angle approximation as we did for the other pendulums.

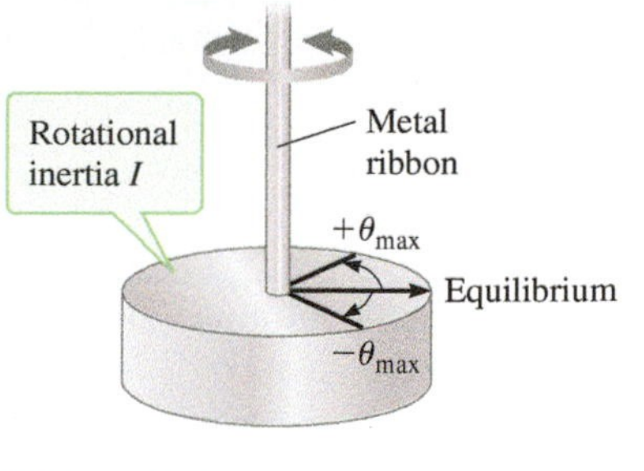

FIGURE 16.18 A basic torsion pendulum.

EXAMPLE 16.6 Hillis's Prototype Clock

The period of the three-armed torsion pendulum in Hillis's prototype clock is 60 s. Each ball (Fig. 16.17A) is made of tungsten ($\rho = 1.935 \times 10^4$ kg/m^3) and weighs about 7 lb. Estimate the torsion spring constant used in the clock. For comparison, the balance wheel in a wind-up watch has a rotational inertia $I \sim 10^{-6}\,\text{kg}\cdot\text{m}^2$ and a torsion spring constant $\kappa \sim 10^{-4}\,\text{N}\cdot\text{m/rad}$. Also, the period of the prototype clock's tick is about two orders of magnitude longer than that of a wind-up watch.

INTERPRET and ANTICIPATE

Our estimate of κ will be based on the angular frequency of a torsion pendulum (Eq. 16.36), which we can obtain from the given period. First estimate the rotational inertia of the three-armed pendulum in Figure 16.17A.

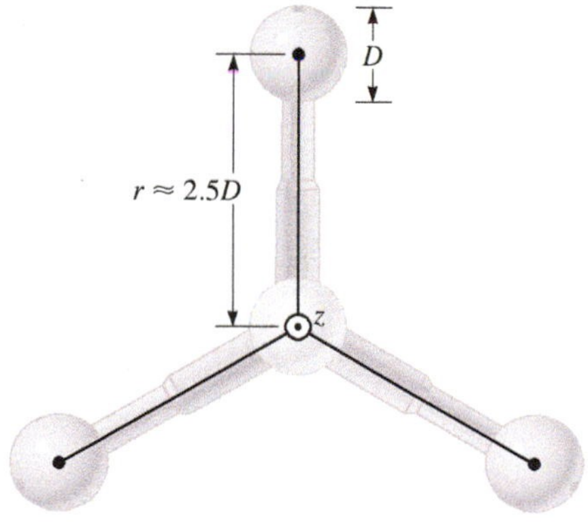

FIGURE 16.19

SOLVE

From Figure 16.17A, it looks like the arms are about twice as long as a ball's diameter D. We estimate the distance from the rotation axis labeled z in Figure 16.19 to the ball's center of mass to be about two and a half times the diameter: $r \approx 2.5D$.

Estimate the ball's diameter from its mass and density. A 7-lb ball has a mass of about 3 kg. The volume of a sphere is $V = \frac{4}{3}\pi R^3$.	$m = \rho V$ $V = \frac{4}{3}\pi R^3 = \frac{4}{3}\pi\left(\frac{D}{2}\right)^3$ $m = \rho\frac{4}{3}\pi\left(\frac{D}{2}\right)^3 = \rho\frac{\pi}{6}D^3$
Solve for D. Because this result is an estimate, keep just one significant figure.	$D = \left(\frac{6}{\pi}\frac{m}{\rho}\right)^{1/3} = \left(\frac{6}{\pi}\cdot\frac{3\text{ kg}}{1.935 \times 10^4\text{ kg/m}^3}\right)^{1/3}$ $D = 6.67 \times 10^{-2}\text{ m} \approx 7\text{ cm}$
Find the rotational inertia of the three-armed pendulum from Equation 13.4. Because $r \approx 2.5D$ and $D \approx 7$ cm, $r \approx 18$ cm. (We are modeling the three balls as three particles and ignoring the mass of the arms.)	$I = \sum_{i=1}^{n} m_i r_i^2$ (13.4) $I = 3mr^2 = 3(3\text{ kg})(18 \times 10^{-2}\text{ m})^2 = 0.29\text{ kg}\cdot\text{m}^2$ $I \approx 0.3\text{ kg}\cdot\text{m}^2$

Example continues on page 470 ▶

Finally, to estimate the torsion spring constant, use the period and $\omega_{tor} = \sqrt{\frac{\kappa}{I}}$ (Eq. 16.36).

$$\omega_{tor} = \sqrt{\frac{\kappa}{I}} = \frac{2\pi}{T} \qquad (16.36)$$

$$\frac{\kappa}{I} = \frac{4\pi^2}{T^2}$$

Solve for κ and substitute values, keeping just one significant figure in the answer.

$$\kappa = \frac{4\pi^2}{T^2}I = \frac{4\pi^2}{(60\text{ s})^2}(0.29\text{ kg}\cdot\text{m}^2)$$

$$\kappa = 3.20 \times 10^{-3}\text{ kg}\cdot\text{m}^2/\text{s}^2 \approx 0.003\text{ N}\cdot\text{m/rad}$$

CHECK and THINK

Perhaps the greatest sources of error in our answer are that we did not include the mass of the three arms in our estimate of rotational inertia and we modeled the balls as particles. So, we have underestimated I, and κ should be somewhat greater than our value.

To check our answer, let's compare the period of the prototype clock with the period of a wind-up watch. As expected, the period of the clock is about two orders of magnitude greater than for a watch.

$$\frac{T_{clock}}{T_{watch}} = \sqrt{\frac{I_{clock}}{\kappa_{clock}}\frac{\kappa_{watch}}{I_{watch}}}$$

$$\frac{T_{clock}}{T_{watch}} \sim \sqrt{\frac{0.3\text{ kg}\cdot\text{m}^2}{0.003\text{ N}\cdot\text{m/rad}}\cdot\frac{10^{-4}\text{ N}\cdot\text{m/rad}}{10^{-6}\text{ kg}\cdot\text{m}^2}}$$

$$\frac{T_{clock}}{T_{watch}} \sim 10^2 \checkmark$$

16-9 Energy in Simple Harmonic Motion

So far, we have analyzed simple harmonic motion based on the force approach. For example, in Crall and Whipple's experiment (case study, page 452), the disk is at rest in its equilibrium position before Whipple displaces it. Afterward, the restoring force causes the disk to oscillate in SHM up and down while passing above and below equilibrium.

We can gain a greater insight into SHM by taking a conservation of energy approach as in Chapter 8. First, we must define the system to include the oscillating object and the source of the restoring force. In the case of a block attached to a horizontal spring (Fig. 16.12), we would include the block and the spring in the system. If the system is isolated and there are no dissipative forces like friction, we expect the mechanical energy E of the system (Eq. 8.11) to be constant:

$$E = K + U = \text{constant}$$

Because the system is oscillating, both the kinetic and potential energies are continually changing, but the sum of these two time-varying quantities is constant.

DERIVATION Mechanical Energy of a Simple Harmonic Oscillator

We show that in the absence of dissipative forces the mechanical energy of an isolated simple harmonic oscillator is a constant given by

$$E = \tfrac{1}{2}my_{max}^2\omega^2 \qquad (16.37)$$

To show formally that mechanical energy is conserved, we use the kinematic equations in Section 16-2 to find expressions for the kinetic energy $K(t)$ and the potential energy $U(t)$. We then add these expressions together to find an expression for the mechanical energy.

To find an expression for the kinetic energy $K(t)$ of a simple harmonic oscillator as a function of time, we substitute the speed of a simple harmonic oscillator (Eq. 16.6) into the definition of kinetic energy (Eq. 8.1).

$$K = \tfrac{1}{2}mv^2 = \tfrac{1}{2}mv_y^2(t) \qquad (16.38)$$

$$K(t) = \tfrac{1}{2}m[-y_{max}\omega\sin(\omega t + \varphi)]^2$$

$$K(t) = \tfrac{1}{2}my_{max}^2\omega^2\sin^2(\omega t + \varphi) \qquad (16.39)$$

Finding an expression for the potential energy as a function of time $U(t)$ takes a few more steps. First, we must find an expression for the net force F_{tot} acting on the object by substituting the hallmark of SHM (Eq. 16.12) into Newton's second law.	$F_{tot}(y) = ma_y(t) = -m\omega^2 y(t)$ (16.40)
Second, we must integrate the force to find an expression for the potential energy as a function of position $U(y)$.	$U(y_f) - U(y_i) = -\int_{y_i}^{y_f} F_{tot}(y)dy$ (8.3)
Substitute Equation 16.40 into Equation 8.3 and integrate.	$U(y_f) - U(y_i) = -\int_{y_i}^{y_f} -m\omega^2 y\, dy$ $U(y_f) - U(y_i) = \frac{1}{2}m\omega^2(y_f^2 - y_i^2)$
The usual convention is to choose $y_i = 0$ as the equilibrium position, set $U(y_i) = 0$, and drop the subscript f.	$U(y) = \frac{1}{2}m\omega^2 y^2$ (16.41)
We now have a simple form for $U(y)$, and we know how y varies with time. To find $U(t)$, substitute the time-dependent position of a simple harmonic oscillator (Eq. 16.3) into Equation 16.41.	$U(t) = \frac{1}{2}m\omega^2[y_{max}\cos(\omega t + \varphi)]^2$ $U(t) = \frac{1}{2}my_{max}^2\omega^2\cos^2(\omega t + \varphi)$ (16.42)
Both the kinetic energy (Eq. 16.39) and the potential energy (Eq. 16.42) increase and decrease as the oscillator moves back and forth. Add them together to find the mechanical energy.	$E = K(t) + U(t)$ $E = \frac{1}{2}my_{max}^2\omega^2\sin^2(\omega t + \varphi) + \frac{1}{2}my_{max}^2\omega^2\cos^2(\omega t + \varphi)$ $E = \frac{1}{2}my_{max}^2\omega^2[\sin^2(\omega t + \varphi) + \cos^2(\omega t + \varphi)]$
The term in brackets is the trigonometric identity $\sin^2\alpha + \cos^2\alpha = 1$ (Appendix A).	$E = \frac{1}{2}my_{max}^2\omega^2$ ✓ (16.37)

COMMENTS

The mass m, the amplitude y_{max}, and the angular frequency ω are all constants for SHM, so the total mechanical energy E is indeed a constant for an isolated system with no dissipative forces. Equation 16.37 holds for any simple harmonic oscillator. Our next step is to consider the special case of an object–spring oscillator such as Crall and Whipple's system.

Energy of an Object–Spring Oscillator

For an object–spring oscillator, the angular frequency is $\omega^2 = k/m$. So, the system's potential energy (Eq. 16.41) becomes

$$U(y) = \frac{1}{2}m\omega^2 y^2 = \frac{1}{2}ky^2$$

as we found for a spring in Equation 8.10. The total mechanical energy of the object–spring oscillator is

$$E = \frac{1}{2}my_{max}^2\omega^2 = \frac{1}{2}ky_{max}^2 \quad (16.43)$$

which is the maximum value of the system's potential energy:

$$E = U_{max} = \frac{1}{2}ky_{max}^2 \quad (16.44)$$

The potential energy is a maximum when the object momentarily stops at either end of its path ($y = \pm y_{max}$).

The maximum kinetic energy (Eq. 16.38) occurs when the object moves though its equilibrium position ($y = 0$) and its speed is a maximum ($v = v_{max}$):

$$K_{max} = \frac{1}{2}mv_{max}^2 \quad (16.45)$$

When the object is at the equilibrium position, the system's potential energy is zero, so the mechanical energy equals the maximum kinetic energy:

$$E = K_{max} = \frac{1}{2}mv_{max}^2$$

Because the mechanical energy is a constant, we write

$$E = K_{max} = U_{max} \quad (16.46)$$

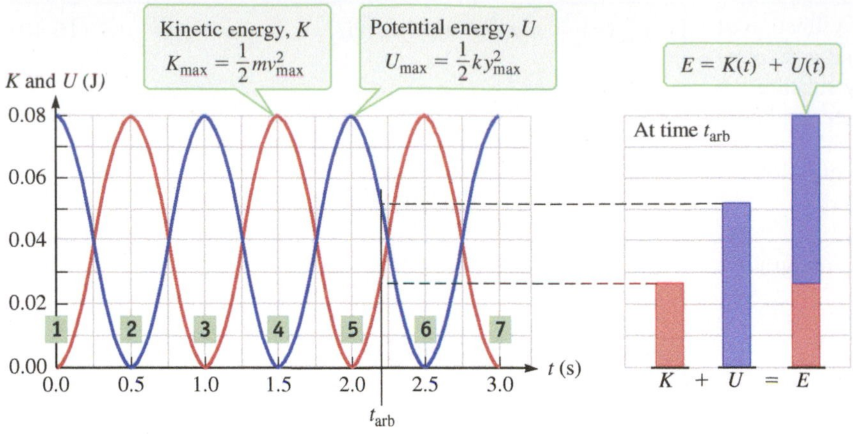

FIGURE 16.20 Kinetic energy (red) and potential energy (blue) versus time for Crall and Whipple's data.

Let's visualize a typical oscillating system's energy using Crall and Whipple's data (case study, page 452). Figure 16.20 shows the kinetic and potential energies versus time on the same graph. To make the kinetic energy graph, square the velocity data in Figure 16.5B and multiply by $\frac{1}{2}m$. To make the potential energy graph, square the position data in Figure 16.5A and multiply by $\frac{1}{2}k$. The times labeled 1 through 7 in Figure 16.20 correspond to the same configurations and times in Figures 16.3 through 16.5. For example, when Whipple first releases the disk at time 1, the system's kinetic energy is zero, and its potential energy is at a maximum ($U_{max} \approx 0.08$ J). When the disk passes through its equilibrium position at time 2, the kinetic energy is at its maximum ($K_{max} \approx 0.08$ J), and the potential energy is zero.

To show that the sum of the kinetic and potential energies in Crall and Whipple's experiment is constant, we draw energy bars in Figure 16.20. We have drawn a kinetic energy bar and a potential energy bar at an arbitrary time t_{arb}. When we stack those energy bars on top of each other, their combined height reaches the total mechanical energy ($E \approx 0.08$ J) of the system. Adding the kinetic and potential energy bars at any arbitrary time gives the same result: $E \approx 0.08$ J. Figure 16.20 shows that the mechanical energy $E \approx 0.08$ J also equals the maximum of either the kinetic or the potential energy.

The conservation of energy approach gives an additional description of simple harmonic motion that the force approach cannot supply. As the object oscillates back and forth, potential energy is converted into kinetic energy and back again repeatedly. The kinetic energy is at its maximum when the object whizzes through its equilibrium position. The potential energy is at its maximum when the object stops momentarily at either end of its path. The mechanical energy is constant. No energy is lost or gained by the system, and oscillation continues indefinitely, again provided that the system is isolated and that no dissipative forces are present.

EXAMPLE 16.7 Today's Lab: A Horizontal Spring

In laboratories all over the world, physics students observe the motion of objects attached to springs. In one such laboratory, a horizontal spring of spring constant $k = 170$ N/m is attached to an object of mass $m = 0.85$ kg. As in Figure 16.12, friction between the object and the table is negligible. A student stretches the spring so that the object's initial position is $x_i = 0.20$ m and releases it at $t = 0$. Find the maximum speed.

INTERPRET and ANTICIPATE

There are at least two ways to approach this problem. (1) We'll calculate the mechanical energy and then find the maximum speed based on the maximum kinetic energy. (2) As a check, we'll use the angular frequency and amplitude to find the maximum speed.

SOLVE

We can find the maximum potential energy from the given amplitude using $U_{max} = \frac{1}{2}ky_{max}^2$ (Eq. 16.44) expressed in terms of horizontal position x.	$U_{max} = \frac{1}{2}kx_{max}^2 = \frac{1}{2}(170\text{ N/m})(0.20\text{ m})^2$ $U_{max} = 3.4$ J
The maximum potential energy equals the total mechanical energy and also equals the maximum kinetic energy (Eq. 16.46). The maximum speed then comes from Equation 16.45.	$E = K_{max} = U_{max} = 3.4$ J $K_{max} = \frac{1}{2}mv_{max}^2 = 3.4$ J $v_{max} = \sqrt{\frac{2K_{max}}{m}} = \sqrt{\frac{2(3.4\text{ J})}{0.85\text{ kg}}} = 2.8\text{ m/s}$

CHECK and THINK

To check the result, use Equation 16.26 to calculate the angular frequency and use Equation 16.7 (in terms of x) to find v_{max}. We find the same answer using this approach.

$$\omega_s = \sqrt{\frac{k}{m}} = \sqrt{\frac{170\ \text{N/m}}{0.85\ \text{kg}}} = 14\ \text{rad/s}$$

$$v_{max} = x_{max}\,\omega_s \qquad (16.7)$$

$$v_{max} = (0.20\ \text{m})(14\ \text{rad/s}) = 2.8\ \text{m/s}\ \checkmark$$

16-10 Damped Harmonic Motion

A simple harmonic oscillator is an *ideal* oscillator that never stops. A real oscillator such as Crall and Whipple's rubber disk on a spring will not continue to move up and down forever. Many real oscillators can be well approximated as ideal simple harmonic oscillators if their motion continues for much longer than one period. For example, Crall and Whipple's oscillator took several minutes to stop on its own—much longer than the measured period ($T = 2.0$ s), so their system is well modeled as a simple harmonic oscillator. Crall and Whipple then returned to the laboratory and tried another experiment with slightly modified equipment.

DAMPED HARMONIC MOTION

✪ Major Concept

CASE STUDY Experiment, Part 2: Not Very Bouncy

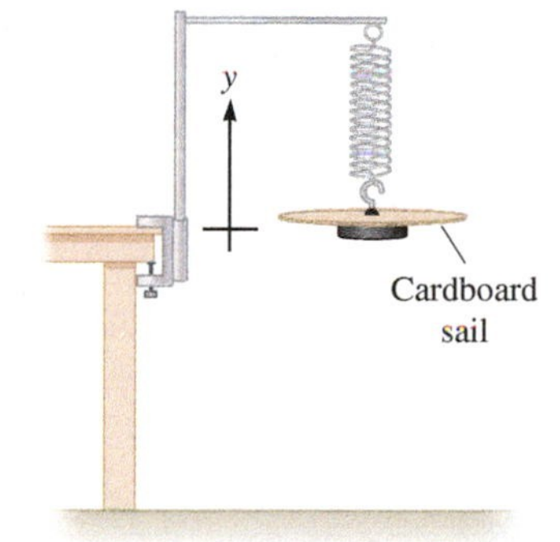

FIGURE 16.21 Crall and Whipple modified their experimental setup by adding a cardboard sail just above the disk.

This time, our experimenters added a cardboard sail just above the disk (Fig. 16.21), displaced and released the disk as in the first experiment, and collected new position and velocity data (Fig. 16.22). Because of air drag, their new oscillator stopped in a few seconds after completing just a few cycles. Their new system cannot be modeled as a simple harmonic oscillator because its position and velocity data in Figure 16.22 do not fit the cosine and sine function of Equations 16.3 and 16.6.

Crall and Whipple's sail experiment must be modeled as **damped harmonic motion**. Dissipative forces such as friction and drag cause a **damped oscillator** to lose mechanical energy and stop. We can use Crall and Whipple's position and velocity data to calculate and plot the energy of their damped oscillator at various times. Figure 16.23 shows that $K(t)$ and $U(t)$ are converted back and forth as in simple harmonic motion, but over a short time both decrease. The energies $K(t)$ and $U(t)$ still add up to the total mechanical energy, but as the mechanical energy decreases to zero, the oscillator stops.

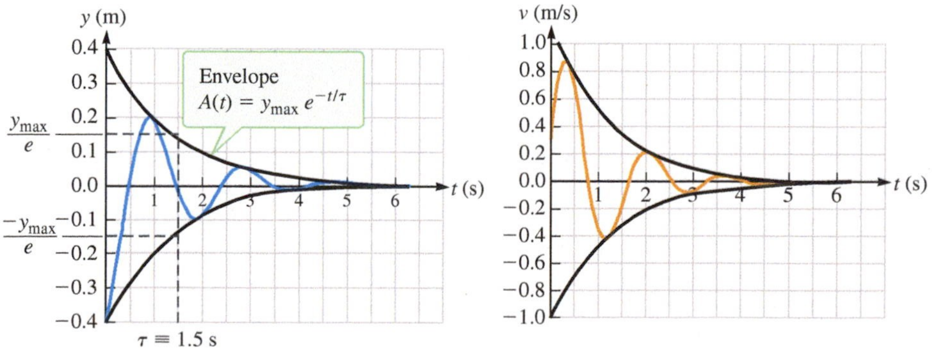

A. Position-versus-time **B. Acceleration-versus-time**

FIGURE 16.22 A. In damped harmonic motion, oscillations decrease until the oscillator comes to rest in its equilibrium configuration. **B.** The speed and energy of a damped harmonic oscillator decrease to zero.

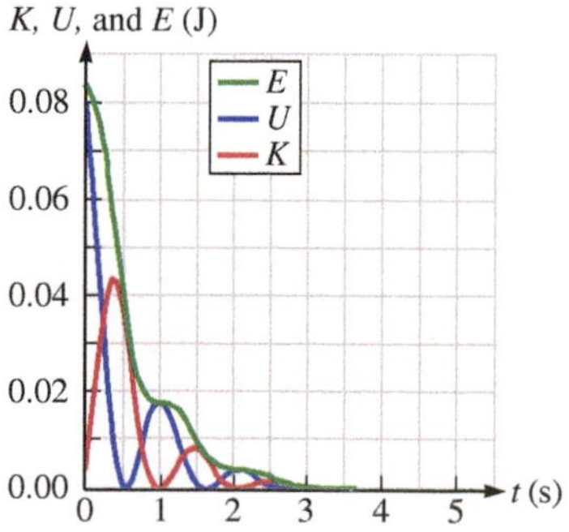

FIGURE 16.23 The mechanical energy (green), potential energy (blue), and kinetic energy (red) of a damped harmonic oscillator decrease to zero.

Describing Damped Oscillations: The Time Constant

Because the damped oscillator's motion cannot be described by the kinematic equations in Section 16-2, we must derive new equations. Using dynamics in Section 16-4 led us to Equation 16.23 for the *undamped* SHO:

$$-ky(t) = ma_y(t) \tag{16.23}$$

Substitute $a_y(t) = d^2y(t)/dt^2$ (Eq. 4.17):

$$-\frac{k}{m}y(t) = \frac{d^2y(t)}{dt^2} \tag{16.47}$$

Equation 16.47 involves derivatives of the variable $y(t)$ and is therefore a differential equation. It says that if we take two time derivatives of $y(t)$, we get back $y(t)$ multiplied by a constant $(-k/m)$. Equation 16.47 cannot be solved for $y(t)$ by direct integration; the techniques for solving such differential equations are beyond the scope of this book. One solution is Equation 16.3:

$$y(t) = y_{\max}\cos(\omega t + \varphi) \tag{16.3}$$

As an exercise, you should verify that Equation 16.3 satisfies Equation 16.47 for the undamped oscillator (Problem 52).

We can arrive at a relation for position versus time for a damped oscillator in a similar way. The sail on Crall and Whipple's oscillator exerts a drag force $\vec{F}_D = -b\vec{v}$ (Eq. 6.4) on the oscillator, so Equation 16.23 becomes

$$-ky(t) - bv_y(t) = ma_y(t) \tag{16.48}$$

By the definitions of velocity and acceleration, Equation 16.48 becomes a differential equation for $y(t)$:

$$-ky(t) - b\frac{dy(t)}{dt} = m\frac{d^2y(t)}{dt^2} \tag{16.49}$$

Equation 16.49 for damped harmonic motion looks similar to Equation 16.47 for SHM except for the extra term involving the first derivative of $y(t)$. The solution to Equation 16.49 is

$$y(t) = A(t)\cos(\omega_D t + \varphi) \tag{16.50}$$

which is similar to the solution for SHM (Eq. 16.3) except that the amplitude $A(t)$ is time-dependent, *not* constant. This amplitude is given by

$$A(t) = y_{\max}e^{-t/\tau} \tag{16.51}$$

Equation 16.51 describes a decreasing amplitude $A(t)$. The **time constant** τ is a measure of how quickly the amplitude decreases and is defined as

TIME CONSTANT ✪ **Major Concept**

$$\tau \equiv \frac{2m}{b} \tag{16.52}$$

where b is sometimes referred to as the *damping constant* or *damping coefficient*. Problem 54 asks you to verify that the time constant has dimensions of time.

Figure 16.22A illustrates the decreasing amplitude $A(t)$ and the time constant τ. The cosine part of Equation 16.50 says that Crall and Whipple's disk is still oscillating, corresponding to the peaks and valleys on the y-versus-t graph. The amplitude is decreasing (Eq. 16.51), however, so the peaks and valleys get smaller as t increases. Equation 16.51 describes an exponentially decreasing envelope, and the peaks and valleys in Figure 16.22A fit inside the envelope.

The time constant τ determines how quickly the envelope closes in, as is best seen by finding the amplitude $A(\tau)$, when $t = \tau$:

$$A(\tau) = y_{\max}e^{-\tau/\tau} = y_{\max}e^{-1}$$

$$A(\tau) = \frac{y_{\max}}{e} \approx \frac{y_{\max}}{2.718} \approx 0.3679y_{\max}$$

The time constant τ is the time it takes the amplitude envelope A to drop from y_{max} to y_{max}/e, or just more than one-third of its original value. In Figure 16.22A, $y_{max} \approx 0.4$ m, so $y_{max}/e \approx 0.15$ m. We can read the time constant off this y-versus-t graph as $\tau \approx 1.5$ s.

By substituting Equations 16.51 and 16.52 into Equation 16.50, the time-varying position of a damped oscillator can be written as

$$y(t) = y_{max}\, e^{-bt/2m} \cos(\omega_D t + \varphi) \quad (16.53)$$

Frequency of Damped Harmonic Motion

The angular frequency ω_D of damped harmonic motion is not the same as the angular frequency ω of simple harmonic motion:

$$\omega_D = \sqrt{\omega^2 - \frac{1}{\tau^2}} = \sqrt{\omega^2 - \frac{b^2}{4m^2}} \quad (16.54)$$

where ω is the angular frequency of the corresponding undamped system. If the dissipative force disappears, $b \to 0$ and $\omega_D \to \omega$. Equation 16.54 says that the angular frequency of a damped oscillator is less than the angular frequency of the simple harmonic oscillator that would result if the dissipative force were removed. Therefore, the damped oscillator has a longer period than the corresponding simple harmonic oscillator. The angular frequency of Crall and Whipple's damped oscillator is given by

$$\omega_D = \sqrt{\omega^2 - \frac{1}{\tau^2}} = \sqrt{(\pi\,\text{rad/s})^2 - \frac{1}{(1.5\text{ s})^2}} = 3.07\text{ rad/s}$$

which is less than $\omega = 3.14$ rad/s of their simple harmonic oscillator.

The strength of the dissipative force depends on b. If a strong dissipative force (large b) is exerted on an oscillator, the oscillator will have a short time constant τ and a low angular frequency ω_D. Such an oscillator will stop quickly without completing very many cycles. If the dissipative force is very strong such that $b^2/4m^2 > \omega^2$, then ω_D according to Equation 16.54 is imaginary, and the system does not oscillate.

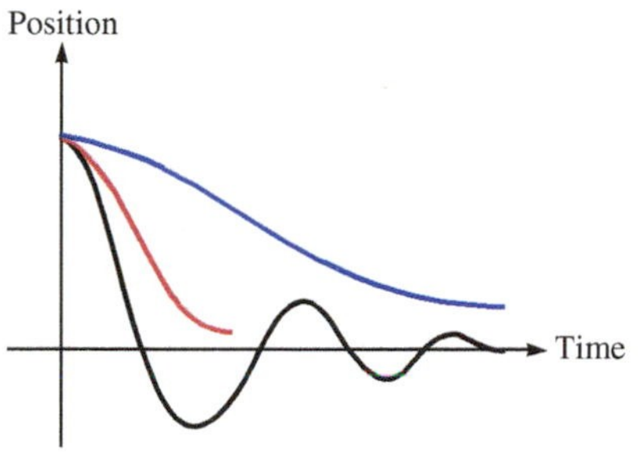

FIGURE 16.24 Position versus time for a critically damped oscillator (red), underdamped oscillator (black), and overdamped oscillator (blue).

Equation 16.54 is used to classify damped oscillators into three types, depending on the relative strength of the dissipative force (Fig. 16.24). A **critically damped** oscillator's angular frequency is zero, so

CRITICALLY DAMPED, UNDERDAMPED, AND OVERDAMPED OSCILLATORS

Major Concept

$$\omega_D = \sqrt{\omega^2 - \frac{b^2}{4m^2}} = 0$$

$$\omega = \frac{b}{2m}$$

$$b = 2m\omega$$

After a critically damped oscillator is displaced from its equilibrium position, it returns to equilibrium in the shortest possible time without passing through to the other side of equilibrium. A weaker dissipative force ($b < 2m\omega$) is exerted on an **underdamped** oscillator, which oscillates briefly before stopping at its equilibrium position. Finally, an **overdamped** oscillator results when a stronger-than-critical dissipative force ($b > 2m\omega$) is exerted. When an overdamped oscillator is displaced from its equilibrium position, it returns directly to equilibrium, but more slowly than does a critically damped oscillator.

If you are designing a clock, you should avoid damping so that the pendulum, watch spring, crystal, or whatever else is oscillating does so for as long as possible. Other systems such as the shock absorbers on your car or bicycle are deliberately damped. A shock absorber consists of a spring attached to a piston (Fig. 16.25). The piston moves through a viscous fluid that exerts a dissipative force. When you hit a pothole, a properly working shock absorber will quickly bring your vehicle back to equilibrium without oscillating. Therefore, the shock absorber should be critically damped. If the shock absorber is underdamped, your vehicle will bounce up and down several times after you hit the pothole. If the shock absorber is overdamped, your vehicle will take a longer time to return to equilibrium.

FIGURE 16.25 The shock absorber attaches the vehicle's frame to the wheel; it is designed to act as a critically damped oscillator, reducing oscillations and making the ride more comfortable.

EXAMPLE 16.8 My Cat Lives with a Physicist

A cat toy consists of a lightweight, hollow ball ($m = 2.0$ g) hanging on the end of a thin string of length $\ell = 32$ cm. A pet owner displaces the ball by 6° and releases it. The cat enthusiastically watches the ball swing back and forth, but the ball's amplitude is reduced to 1° in 24 s. Both the cat and the owner are disappointed. What is the time constant?

INTERPRET and ANTICIPATE

You might be tempted to use Equation 16.52, which defines the time constant τ, but you have not given enough information to use that equation. Instead, use Equation 16.51, noting that the amplitude drops from 6° to 1° in 24 s. We expect that because the amplitude has dropped by more than one-third, the 24-s observation interval is greater than the time constant.

SOLVE

Write Equation 16.51 for an angular amplitude envelope $A_\theta(t)$. The maximum angular amplitude is θ_{max}.

$$A_\theta(t) = \theta_{max} e^{-t/\tau}$$

To solve for τ, take the natural logarithm of both sides (Appendix A).

$$\frac{A_\theta(t)}{\theta_{max}} = e^{-t/\tau}$$

$$\ln\left(\frac{A_\theta(t)}{\theta_{max}}\right) = \ln e^{-t/\tau} = -\frac{t}{\tau}$$

$$\tau = -\frac{t}{\ln\left(A_\theta(t)/\theta_{max}\right)}$$

Substitute the values $t = 24$ s, $A_\theta(t) = 1°$, and $\theta_{max} = 6°$. There is no need to convert the angles to radians because the ratio of the angles in the logarithm is unitless.

$$\tau = -\frac{24\text{ s}}{\ln(1°/6°)} = 13\text{ s}$$

CHECK and THINK

As expected, the time constant is less than 24 s. The length of the string was extraneous information. In Problem 55, we explore ways to improve this toy.

16-11 Driven Oscillators

DRIVEN OSCILLATOR

Major Concept

In 1851, French physicist Jean Bernard Leon Foucault used a 67.5-m-long piano wire to hang a 28-kg pendulum bob from the dome of the Pantheon in Paris. His pendulum provided the first nonastronomical demonstration of the Earth's rotation. A restored version of Foucault's pendulum still oscillates in the Pantheon today (Fig. 16.26). Although Foucault's pendulum oscillates for many cycles, like all real pendulums it is damped and so would lose energy and eventually stop. There is a person whose job is to push or *drive* the pendulum's bob several times during the workday, however. The worker's push does work on the pendulum, transferring energy from the environment (the employee) to the system. The force exerted by the employee is called a **driving force**. Such an oscillator is damped and driven, so we refer it as a **damped-driven oscillator** or just a **driven oscillator**.

Although you have never been employed at the Pantheon to drive Foucault's pendulum, you have probably pushed a child on a swing, which is like driving a pendulum. You stand on one side of the pendulum and push on the bob just as it moves away from you (Fig. 16.27A). The force-versus-time graph in Figure 16.27A illustrates two important properties of the driving force. (1) The driving force is not constant; you do not push on the bob continuously. (2) The driving force is periodic with roughly the same period as the oscillator; you push on the bob every time it returns to the same position near you.

FIGURE 16.26 Foucault's pendulum inside the Pantheon in Paris, France. The circular scale gives the time.

You could be more efficient at transferring energy to the pendulum if you had a friend stand on the opposite side of the pendulum (Fig. 16.27B). He would push on the bob when the bob moves away from him (in the negative x direction), and you would continue to push on the bob as it moves away from you (in the positive x direction). The new force-versus-time graph in Figure 16.27B shows your force as the positive spikes and his force as the negative spikes. There are still gaps in the force-versus-time graph, however. A motor can be designed to exert a still more efficient driving force F_{drv}, represented by a sine or cosine function. (If you have ever seen a contemporary version of Foucault's pendulum, it may have been driven by such a motor.) We arbitrarily choose the cosine function (Fig. 16.27C):

$$F_{\text{drv}} = F_{\text{max}} \cos \omega_{\text{drv}} t \quad (16.55)$$

The driving force has an angular frequency ω_{drv}.

If an oscillator is damped and driven, (at least) three forces are exerted on it:

1. A restoring force $\vec{F}_{\text{restore}}$. In the case of Foucault's pendulum, the restoring force is exerted by gravity. In the case of Crall and Whipple's experiments, the restoring force is exerted by the spring: $\vec{F}_{\text{restore}} = -k\vec{y}$.
2. A dissipative force. The only dissipative force we study here has the form $\vec{F}_D = -b\vec{v}$, which is the result of air drag.
3. A a driving force of the form $F_{\text{drv}} = F_{\text{max}} \cos \omega_{\text{drv}} t$.

When we apply Newton's second law to a damped-driven oscillator, we have to add all three of these forces:

$$\sum \vec{F} = \vec{F}_{\text{restore}} + \vec{F}_D + \vec{F}_{\text{drv}} = m\vec{a} \quad (16.56)$$

Imagine that a driving force is applied to Crall and Whipple's damped oscillator (Fig. 16.22) and apply Equation 16.56:

$$-ky - bv_y + F_{\text{max}} \cos \omega_{\text{drv}} t = ma_y \quad (16.57)$$

Except for the additional driving term ($F_{\text{max}} \cos \omega_{\text{drv}} t$), Equation 16.57 looks like Equation 16.48, $-ky(t) - bv_y(t) = ma_y(t)$. As we did for that equation, we rewrite Equation 16.57 as a differential equation for $y(t)$ using the definitions of velocity and acceleration:

$$-ky(t) - b\frac{dy(t)}{dt} + F_{\text{max}} \cos \omega_{\text{drv}} t = m\frac{d^2y(t)}{dt^2}$$

which is traditionally written as

$$F_{\text{max}} \cos \omega_{\text{drv}} t = m\frac{d^2y(t)}{dt^2} + b\frac{dy(t)}{dt} + ky(t) \quad (16.58)$$

As before, we will provide a solution to this differential equation without carrying out the solution method. When the driving force is first exerted on the oscillator, the system's mechanical energy increases and the amplitude increases. After a while, however, the amplitude A_{drv} is constant, and the solution to Equation 16.58 is

$$y(t) = A_{\text{drv}} \cos (\omega_{\text{drv}} t + \varphi_{\text{drv}}) \quad (16.59)$$

as you may be asked to verify for homework (Problem 82). Equation 16.59 looks much like the time-varying position of a simple harmonic oscillator (Eq. 16.3). The most significant difference is that for a damped-driven oscillator, the amplitude A_{drv} depends on the driving force's angular frequency ω_{drv}:

$$A_{\text{drv}} = \frac{F_{\text{max}}}{\sqrt{m^2(\omega^2 - \omega_{\text{drv}}^2)^2 + b^2\omega_{\text{drv}}^2}} \quad (16.60)$$

The angular frequency ω with no subscript is the angular frequency that the oscillator would have in the absence of damping or driving. In other words, ω is the angular frequency of the equivalent simple harmonic oscillator and is called the **natural angular frequency**. The natural angular frequency is given by Equation 16.26, 16.29, 16.33, or 16.36, depending on the type of oscillator.

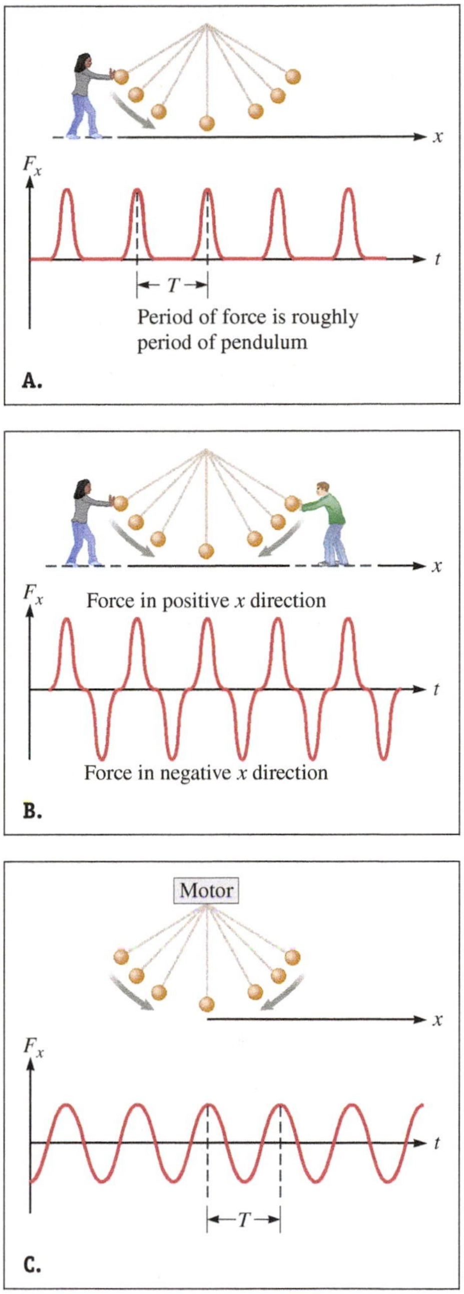

FIGURE 16.27 **A.** A pendulum is driven by a person who pushes in the positive x direction. **B.** The pendulum is driven more efficiently when a second person additionally pushes in the negative x direction. **C.** A pendulum is driven by a motor so that the driving force is $F_{\text{drv}} = F_{\text{max}} \cos \omega_{\text{drv}} t$.

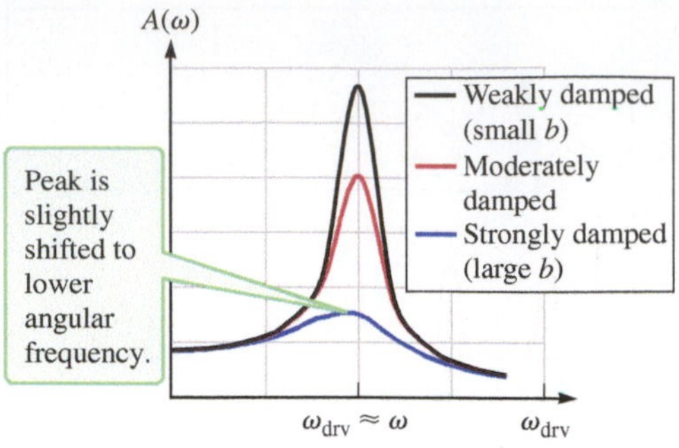

FIGURE 16.28 The amplitude A of a damped-driven oscillator peaks when the driving angular frequency equals the oscillator's natural angular frequency $\omega_{drv} = \omega$.

RESONANCE Major Concept

Because Equation 16.60 is a messy function of ω_{drv}, it is helpful to sketch a graph of A_{drv} versus ω_{drv} (Fig. 16.28). The amplitude of a damped-driven oscillator peaks when the driving angular frequency *approximately* equals the oscillator's natural angular frequency, $\omega_{drv} \approx \omega$. The driving force transfers energy to the system most effectively when its angular frequency matches the natural angular frequency of the oscillator. The term **resonance** describes the condition of an oscillator that is driven nearly at its natural angular frequency.

Often the terms *natural frequency* f and *driving frequency* f_{drv} are used instead of natural *angular* frequency and driving *angular* frequency. The natural frequency is given by Equation 16.2:

$$f = \frac{\omega}{2\pi}$$

and the driving frequency is

$$f_{drv} = \frac{\omega_{drv}}{2\pi}$$

Finally, the natural frequency is also called the **resonance frequency**.

Imagine using a motor to drive Foucault's original pendulum ($\ell = 67.5$ m). The natural angular frequency of the pendulum is given by Equation 16.29:

$$\omega = \omega_{smp} = \sqrt{\frac{g}{\ell}} = \sqrt{\frac{9.81 \text{ m/s}^2}{67.5 \text{ m}}}$$

$$\omega = 0.381 \text{ rad/s}$$

and its natural frequency is

$$f = \frac{0.381 \text{ rad/s}}{2\pi} = 6.06 \times 10^{-2} \text{ Hz}$$

Suppose you start with the pendulum at rest and the motor's frequency set at zero. You slowly increase the motor's frequency f_{drv}. As you do, you find that the pendulum does not respond until the motor's frequency is close to the natural frequency of the pendulum. The pendulum's response is most pronounced when $f_{drv} = 6.06 \times 10^{-2}$ Hz, and we say that the motor matches the pendulum's resonance. As you continue to *increase* the motor's frequency past resonance, the pendulum's amplitude *decreases*, and its motion actually dies out. The motor is still running, but it cannot transfer energy to the pendulum because its driving frequency is not close to the pendulum's resonance frequency.

EXAMPLE 16.9 CASE STUDY Driving Crall and Whipple's Oscillator

Suppose Crall and Whipple use a driving force of the form $F_{max} \cos \omega_{drv} t$ to compensate for the mechanical energy lost by their damped harmonic oscillator in Figure 16.21. They set the driving frequency equal to the natural frequency of their oscillator. Recall that the disk's mass is 0.100 kg and that the time constant of their damped oscillator is 1.5 s. If the amplitude A_{drv} of the motion is the same as that of their simple harmonic oscillator ($y_{max} = 0.40$ m), find F_{max}.

INTERPRET and ANTICIPATE

The maximum force exerted by the driving force is related to the amplitude of the driven oscillator.

SOLVE

If the driving frequency equals the natural frequency, the driving *angular* frequency equals the *angular* frequency of the equivalent simple harmonic oscillator, $\omega_{drv} = \omega = 3.1$ rad/s. Simplify Equation 16.60 by setting the driving frequency equal to the natural frequency.

$$A_{drv} = \frac{F_{max}}{\sqrt{m^2(\omega^2 - \omega_{drv}^2)^2 + b^2\omega_{drv}^2}} \quad (16.60)$$

$$A_{drv} = \frac{F_{max}}{\sqrt{0 + b^2\omega^2}} = \frac{F_{max}}{b\omega}$$

$$F_{max} = b\omega A_{drv} \quad (1)$$

Equation 16.52 relates the given time constant τ to the parameter b.	$\tau \equiv \frac{2m}{b}$ (16.52) $b = \frac{2m}{\tau} = \frac{2(0.100\text{ kg})}{1.5\text{ s}} = 0.13\text{ N}\cdot\text{s/m}$
In this case, the amplitude of the driven oscillator equals the amplitude of the undamped oscillator: $A_{drv} = y_{max}$. Substitute into Equation (1) to find F_{max}.	$F_{max} = b\omega A_{drv} = b\omega y_{max}$ $F_{max} = (0.13\text{ N}\cdot\text{s/m})(3.1\text{ rad/s})(0.40\text{m}) = 0.16\text{N}$

CHECK and THINK

Because the driving frequency equals the resonance frequency, even a small driving force is effective at producing a large amplitude. The maximum force is considerably less than the disk's weight, $w = 0.98$ N. If the driving frequency is set much higher or lower, the required driving force F_{max} must increase in order to achieve the same amplitude.

SUMMARY

Underlying Principles

We applied kinematics, dynamics (Newton's laws of motion), and conservation of energy to **oscillations**, a special case of periodic motion in which a particle moves back and forth over a path.

Major Concepts

1. **Simple harmonic motion (SHM)** is oscillatory motion that is relatively easy to describe mathematically. The hallmark of SHM is

$$a_y(t) = -\omega^2 y(t) \qquad (16.12)$$

The kinematic equations describing a **simple harmonic oscillator (SHO)** are as follows:

position: $y(t) = y_{max}\cos(\omega t + \varphi)$ (16.3)

velocity: $v_y(t) = -y_{max}\omega\sin(\omega t + \varphi)$ (16.6)

acceleration $a_y(t) = -y_{max}\omega^2\cos(\omega t + \varphi)$ (16.9)

2. **Angular frequency** is written in terms of frequency or period as

$$\omega \equiv 2\pi f = \frac{2\pi}{T} \qquad (16.2)$$

3. The maximum displacement from the origin is called the **amplitude** y_{max}.
4. The argument $(\omega t + \varphi)$ of the cosine or sine function is called the **phase**. The **initial phase** φ is determined by the initial position $y_i \equiv y(0)$ such that

$$\varphi = \cos^{-1}\left(\frac{y_i}{y_{max}}\right) \qquad (16.4)$$

5. **Damped harmonic motion** occurs when dissipative forces cause an oscillator to lose mechanical energy. The position of a damped harmonic oscillator is given by

$$y(t) = y_{max}e^{-bt/2m}\cos(\omega_D t + \varphi) \qquad (16.53)$$

Its angular frequency is given by

$$\omega_D = \sqrt{\omega^2 - \frac{1}{\tau^2}} = \sqrt{\omega^2 - \frac{b^2}{4m^2}} \qquad (16.54)$$

6. The **time constant** τ is a measure of how quickly the amplitude of a damped oscillator is decreasing:

$$\tau \equiv \frac{2m}{b} \qquad (16.52)$$

7. An oscillator is **critically damped** if $b = 2m\omega$. If $b < 2m\omega$ the oscillator is **underdamped,** and if $b > 2m\omega$ the oscillator is **overdamped.**
8. An oscillator is a **driven oscillator** if a driving force transfers energy from the environment to the oscillator system.
9. **Resonance** describes the condition of an oscillator that is driven at its natural angular frequency.

Special Cases

Angular frequency of an **object–spring oscillator**:

$$\omega = \omega_s \equiv \sqrt{\frac{k}{m}} \quad (16.26)$$

Angular frequency of a **simple pendulum**:

$$\omega_{\text{smp}} = \sqrt{\frac{g}{\ell}} \quad (16.29)$$

Angular frequency of a **physical pendulum**:

$$\omega_{\text{phy}} = \sqrt{\frac{mgr_{\text{CM}}}{I}} \quad (16.33)$$

Angular frequency of a **torsion pendulum**:

$$\omega_{\text{tor}} = \sqrt{\frac{\kappa}{I}} \quad (16.36)$$

PROBLEMS AND QUESTIONS

A = algebraic **C** = conceptual **E** = estimation **G** = graphical **N** = numerical

16-1 Picturing Harmonic Motion

1. **G** CASE STUDY For each velocity listed, state the position and acceleration of the rubber disk in Crall and Whipple's experiment (Figs. 16.3–16.5). There may be more than one possible answer for each given velocity. **a.** $v_y = 1.3$ m/s **b.** $v_y = -1.3$ m/s **c.** $v_y = 0$
2. **G** CASE STUDY For each acceleration listed, state the position and velocity of the disk in Crall and Whipple's experiment (Figs. 16.3–16.5). There may be more than one possible answer for each given acceleration. **a.** $a_y = 3.8\text{ m/s}^2$ **b.** $a_y = -3.8\text{ m/s}^2$ **c.** $a_y = 0$

16-2 Kinematic Equations of Simple Harmonic Motion

3. **N** A particle in simple harmonic motion has a period of 4.3 s. What is the particle's acceleration when it is at $\vec{r} = -1.65\hat{\jmath}$ m?
4. **N** A simple harmonic oscillator's position is given by $y(t) = (0.850\text{ m})\cos(10.4t - 5.20)$. Find the oscillator's position, velocity, and acceleration at each of the following times. **a.** $t = 0$ **b.** $t = 0.500$ s **c.** $t = 2.00$ s
5. **N** A simple harmonic oscillator's velocity is given by $v_y(t) = (0.850\text{ m/s})\sin(10.4t - 5.20)$. Find the oscillator's position, velocity, and acceleration at each of the following times. **a.** $t = 0$ **b.** $t = 0.500$ s **c.** $t = 2.00$ s
6. **N** A simple harmonic oscillator's acceleration is given by $a_y(t) = (0.850\text{ m/s}^2)\cos(10.4t - 5.20)$. Find the oscillator's position, velocity, and acceleration at each of the following times. **a.** $t = 0$ **b.** $t = 0.500$ s **c.** $t = 2.00$ s
7. **N** The equation of motion of a simple harmonic oscillator is given by $x(t) = (8.0\text{ cm})\cos(10\pi t) - (6.0\text{ cm})\sin(10\pi t)$, where t is in seconds. **a.** Find the amplitude. **b.** Determine the period. **c.** Determine the initial phase.
8. **N** The expression $x = 8.50\cos(2.40\pi t + \pi/2)$ describes the position of an object as a function of time, with x in centimeters and t in seconds. What are the **a.** frequency, **b.** period, **c.** amplitude, and **d.** initial phase of the object's motion? **e.** What is the position of the particle at $t = 1.45$ s?
9. **A** A simple harmonic oscillator has amplitude A and period T. Find the minimum time required for its position to change from $x = A$ to $x = A/2$ in terms of the period T.
10. **N** A particle is moving in simple harmonic motion with an amplitude of 4.0 cm and a maximum velocity of 10.0 cm/s. Assume the initial phase is 0. **a.** At what position is its velocity 4.0 cm/s? **b.** What is its velocity when its position is 2.0 cm?
11. A 1.50-kg mass is attached to a spring with spring constant 33.0 N/m on a frictionless, horizontal table. The spring–mass system is stretched to 4.00 cm beyond the equilibrium position of the spring and is released from rest at $t = 0$.
    a. **N** What is the maximum speed of the 1.50-kg mass?
    b. **N** What is the maximum acceleration of the 1.50-kg mass?
    c. **A** What are the position, velocity, and acceleration of the 1.50-kg mass as functions of time?

16-3 Connection with Circular Motion

12. **C** A bicycle pedal has reflective stripes on the front and back (Fig. P16.12) You are in a parked car with its headlights on, and a cyclist is riding toward you. It is very dark, and the car's headlights only illuminate the front reflective stripe. Describe the motion you see.

FIGURE P16.12

13. **C** A conical pendulum consists of a small ball of mass m attached to a string of length L. It executes uniform circular motion in a horizontal plane such that the string makes a fixed angle θ with the vertical, sweeping out the surface of a cone. The angular frequency of this motion is given by

$$\omega = \sqrt{\frac{g}{L\cos\theta}}$$

Suppose a shadow of the moving mass is projected onto a vertical screen adjacent to the pendulum by a light source in the plane of motion. Describe the motion of the shadow.

14. **C** When the Earth passes a planet such as Mars, the planet appears to move backward for a time, a phenomenon known as **retrograde motion**. Ancient astronomers believed that the Earth did not move and that the planets moved around the Earth. They also believed that uniform circular motion was perfect and that heavenly objects such as planets exhibited this perfect motion. How do you suppose ancient astronomers accounted for retrograde motion? Include a sketch with your explanation.

Problems 15 and 16 are paired.

15. **N** A point on the edge of a child's pinwheel is in uniform circular motion as the wheel spins counterclockwise with a frequency of 1.53 Hz. The point is at the location $x = 30.00$ cm and $y = 0$

when a stopwatch is started to track the motion (Fig. P16.15). **a.** What is the period of the circular motion? **b.** What is the velocity of the point at the instant described? **c.** What is the acceleration of the point at the instant described?

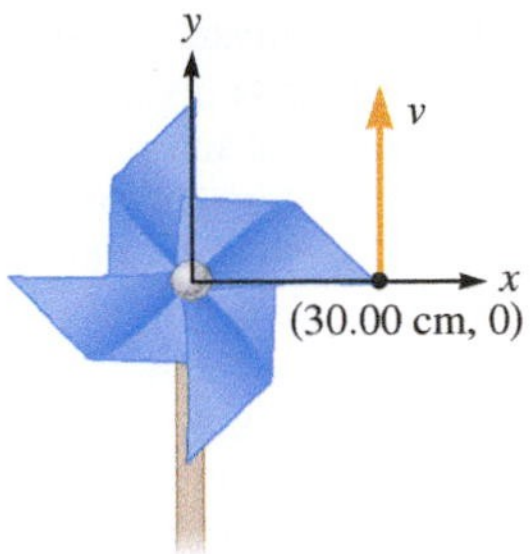

FIGURE P16.15 Problems 15 and 16.

16. Consider the scenario in Problem 15 (Fig. P16.15).

a. **C, N** How do the maximum values of the x and y components of the velocity compare? Discuss the validity of your answer.

b. **N** When is the earliest time at which the x and y components of the velocity are the same?

c. **G** Sketch the path of the point and mark the location of two points where the velocity components are equal.

17. **A** Find expressions for the y component of the ball's velocity and acceleration in Figure 16.6 (page 456).

16-4 Dynamics of Simple Harmonic Motion

18. A jack-in-the-box undergoes simple harmonic motion after it pops out of its box with a frequency of 3.4 Hz and an amplitude of 15 cm.

a. **N** What is the maximum acceleration experienced by the jack-in-the-box?

b. **N** If the mass of the jack-in-the-box is 0.210 kg, what is the maximum net force it experiences during the motion?

c. **A** If the initial phase angle is $\pi/3$ rad, write an equation that describes the net force as a function of time.

19. **C, N** A uniform plank of length L and mass M is balanced on a fixed, semicircular bowl of radius R (Fig. P16.19). If the plank is tilted slightly from its equilibrium position and released, will it execute simple harmonic motion? If so, obtain the period of its oscillation.

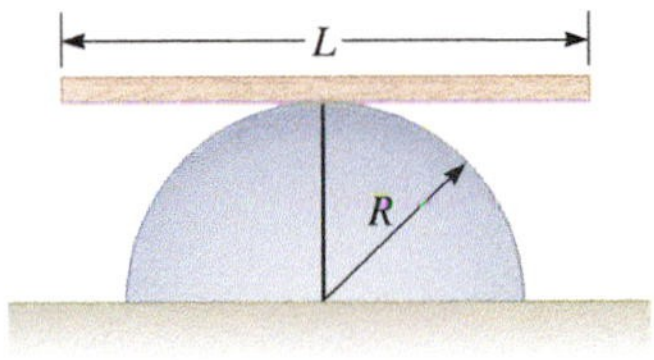

FIGURE P16.19

16-5 Special Case: Object–Spring Oscillator

20. **C** If Crall and Whipple performed their experiment (Fig. 16.3) in a laboratory on a high mountain where $g = 9.72\ \text{m/s}^2$, what changes would there be to $y(t)$, $v_y(t)$, $a_y(t)$, and $F_y(t)$?

21. **N** A block of mass $m = 5.94$ kg is attached to a spring with spring constant $k = 1592$ N/m and rests on a frictionless surface. The block is pulled, stretching the spring a distance of 0.150 m, and is held still. The block is then released and moves in simple harmonic motion about the equilibrium position. **a.** What is the frequency of this oscillation? **b.** Where is the block located 3.24 s after it is released? **c.** What is the velocity of the mass at that time?

22. **A** A block of mass m rests on a frictionless, horizontal surface and is attached to two springs with spring constants k_1 and k_2 (Fig. P16.22). It is displaced to the right and released. Find an expression for the angular frequency of oscillation of the resulting simple harmonic motion.

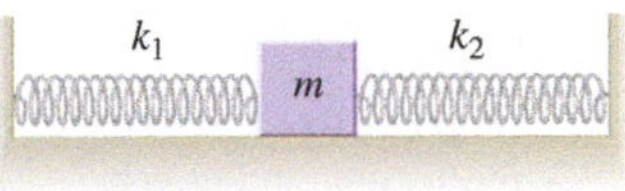

FIGURE P16.22 Problems 22 and 81.

23. **N** It is important for astronauts in space to monitor their body weight. In Earth orbit, a simple scale only reads an apparent weight of zero, so another method is needed. NASA developed the body mass measuring device (BMMD) for Skylab astronauts. The BMMD is a spring-mounted chair that oscillates in simple harmonic motion (Fig. P16.23). From the period of the motion, the mass of the astronaut can be calculated. In a typical system, the chair has a period of oscillation of 0.901 s when empty. The spring constant is 606 N/m. When a certain astronaut sits in the chair, the period of oscillation increases to 2.37 s. Determine the mass of the astronaut.

© NASA

FIGURE P16.23

24. **N** A popular item for infants consists of a seat that hangs from a spring attached to the top of a door frame. Most infants quickly learn to bounce up and down in simple harmonic motion. Suppose the total mass of an infant and the chair is 11.0 kg and the spring constant is 1.20×10^3 N/m. Find the amplitude of her oscillation such that she just loses contact with the seat when she reaches maximum height above her equilibrium position.

25. **A** A spring of mass m_s and spring constant k is attached to an object of mass M and set into simple harmonic motion on a frictionless, horizontal table. All portions of the spring are assumed to oscillate in phase, and the velocity of each segment dx of the spring with mass dm can be assumed to be proportional to the distance x of that segment from point A in Figure P16.25. **a.** What is the kinetic energy of the system at the instant the object is moving with speed v? **b.** What is the frequency of oscillation of the system?

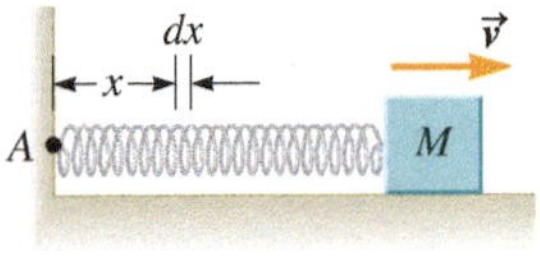

FIGURE P16.25

16-6 Special Case: Simple Pendulum

26. **N** In an undergraduate physics lab, a simple pendulum is observed to swing through 75 complete oscillations in a time period of 2.25 min. What are the **a.** period and **b.** length of the pendulum?

27. **C** A simple pendulum of length L hangs from the ceiling of an elevator. **a.** While the elevator is moving up with constant acceleration a, is the period of the pendulum affected? If so, how? **b.** Now suppose we hang a particle of mass m on a spring of spring constant k and attach it to the ceiling of the same elevator. How does an upward acceleration a affect the period of this simple harmonic oscillator?

28. **A** We do not need the analogy in Equation 16.30 to write expressions for the *translational* displacement of a pendulum bob along the circular arc $s(t)$, *translational* speed $v(t)$, and *translational* acceleration $a(t)$. Show that they are given by

$$s(t) = s_{\max} \cos(\omega_{\text{smp}} t + \varphi)$$

$$v(t) = -v_{\max} \sin(\omega_{\text{smp}} t + \varphi)$$

$$a(t) = -a_{\max} \cos(\omega_{\text{smp}} t + \varphi)$$

respectively, where $s_{\max} = \ell\theta_{\max}$ with ℓ being the length of the pendulum, $v_{\max} = s_{\max}\omega_{\text{smp}}$, and $a_{\max} = s_{\max}\omega_{\text{smp}}^2$.

29. **N** Dr. Chaos uses his watch to hypnotize unsuspecting citizens in attempting to rob them. His watch is a simple pendulum that he sets into simple harmonic motion. The length of the pendulum is 0.150 m, and the mass at the end is 3.25 kg. Use the equations described in Problem 28 to model the motion of the watch when it is pulled back to a maximum angle of 5.00° and released. Assume the initial phase is zero. What are the **a.** period of motion, **b.** maximum speed of, and **c.** maximum acceleration of the watch?

30. **N** A simple pendulum near sea level (g = 9.81 m/s^2) has a period of 1.50 s. If you observe the motion of the pendulum on a high mountain (g = 9.72 m/s^2), what is the new period?

31. A simple pendulum comprised of a bob of mass m = 122 g and a lightweight string of length 75.0 cm is released from rest from an initial angle of 23.0° from the vertical.

a. **N** Using the approach of simple harmonic motion, what is the maximum speed of the pendulum bob?

b. **N** Using the approach of simple harmonic motion, what is the maximum angular acceleration of the bob?

c. **C** How does the speed found in part (a) compare with that found by using a conservation of energy approach?

32. **N** A simple pendulum is constructed from a bob of mass m = 150 g and a lightweight string of length ℓ = 1.50 m. What are the periods of oscillation for this pendulum **a.** in a physics lab at sea level, **b.** in an elevator accelerating upward at 2.00 m/s^2, **c.** in an elevator accelerating downward at 2.00 m/s^2, and **d.** in a school bus accelerating horizontally at 2.00 m/s^2?

16-7 Special Case: Physical Pendulum

33. **C** Three scenarios are listed, each involving a physical pendulum. In each case, determine what happens to the period. Does the period increase, decrease, or stay the same? **a.** A young man is crouched on a moving tire swing and then stands up. **b.** A gymnast swings by her arms from a high bar and then swings by her knees. **c.** A trapeze artist swings alone, standing on top of a bar. Another person then grabs the bottom of the bar so that the two people swing together.

34. **A** Show that angular frequency of a physical pendulum $\omega_{\text{phy}} = \sqrt{mgr_{\text{CM}}/I}$ (Eq. 16.33) equals the angular frequency of a simple pendulum $\omega_{\text{smp}} = \sqrt{g/\ell}$ (Eq. 16.29) in the case of a particle at the end of a string of length ℓ.

35. **A** A uniform annular ring of mass m and inner and outer radii a and b, respectively, is pivoted around an axis perpendicular to the plane of the ring at point P (Fig. P16.35). Determine its period of oscillation.

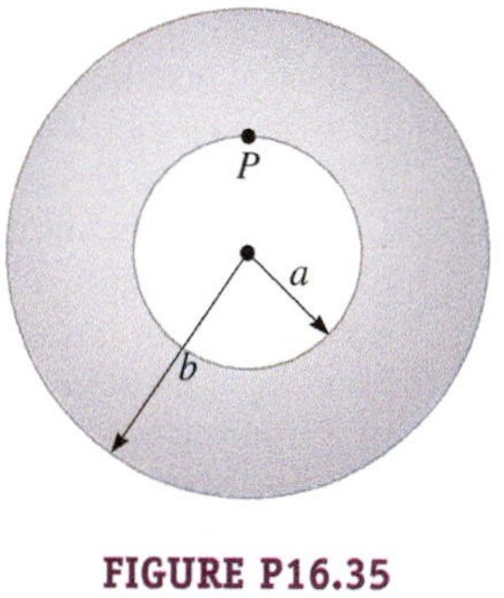

FIGURE P16.35

36. **N** A child works on a project in art class and uses an outline of her hand on a sheet of construction paper to draw a turkey (Fig. P16.36). The teacher pins the turkey to the bulletin board in the front of the classroom by using a thumbtack. The student notices that if she flicks her finger on the end of the turkey, it oscillates back and forth with a frequency of about 1.65 Hz. If the rotational inertia of the paper turkey is 1.25 × 10^{-5} kg · m^2 and its mass is 0.005 kg, what is the distance between the thumbtack and the center of mass of the turkey?

FIGURE P16.36

37. **N** Three thin sticks of equal mass, each of length 20.0 cm, are connected at the ends to form an equilateral triangle. When the triangle is pivoted around an axis perpendicular to the plane of the triangle and passing through one of the vertices, what is the period of the triangle's oscillation as a physical pendulum?

38. **N** A physical pendulum in the form of a thin rod of length 1.00 m is pivoted at point P, a distance x from the pendulum's center of mass. Find the value of x for which the period of the pendulum will be a minimum.

39. **E** In the short story *The Pit and the Pendulum* by 19th-century American horror writer Edgar Allen Poe, a man is tied to a table directly below a swinging pendulum that is slowly lowered toward him. The "bob" of the pendulum is a 1-ft steel scythe connected to a 30-ft brass rod. When the man first sees the pendulum, the pivot is roughly 1 ft above the scythe so that a 29-ft length of the brass rod oscillates above the pivot (Fig. P16.39A). The man escapes when the pivot is near the end of the brass rod (Fig. P16.39B). **a.** Model the pendulum as a particle of mass m_s = 2 kg attached to a rod of mass m_r = 160 kg. Find the pendulum's center of mass and rotational inertia around an axis through its center of mass. (Check your answers by finding the center of mass and rotational inertia of just the brass rod.) **b.** What is the initial period of the pendulum? **c.** The man saves himself by smearing food on his ropes so that rats chew through them. He does so when he has no more than 12 cycles before the pendulum will make contact with him. How much time does it take the rats to chew through the ropes?

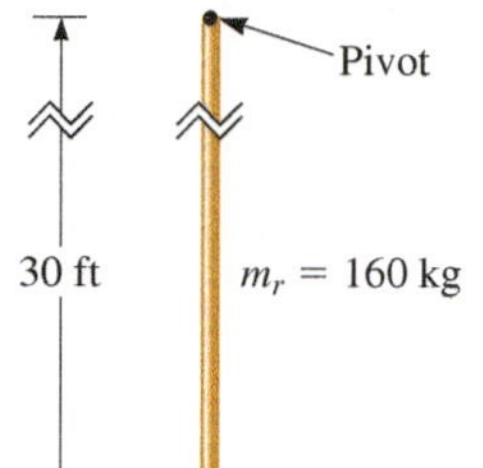

FIGURE P16.39

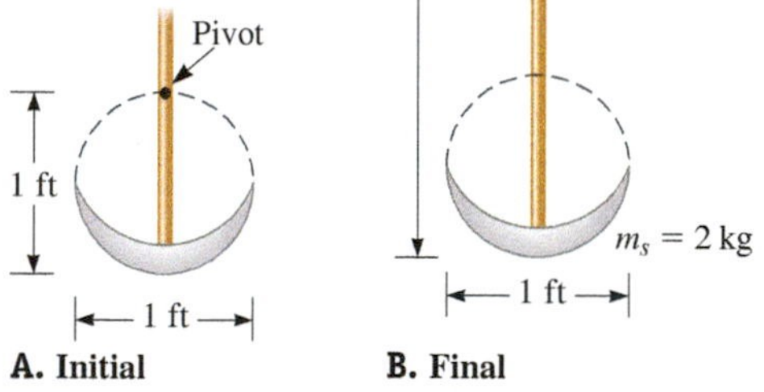

16-8 Special Case: Torsion Pendulum

40. **C** A pendulum clock that keeps time on the Earth such as that in Figure 16.1A cannot be used on the Moon because the pendulum's period would be different there. Can Hillis's millennium clock based on a torsion pendulum (Fig. 16.17, page 468) be used on the Moon to keep accurate time? Explain.

41. **N** A restaurant manager has decorated his retro diner by hanging (scratched) vinyl LP records from thin wires. The records have a mass of 180 g, a diameter of 12 in., and negligible thickness. The records oscillate as torsion pendulums. **a.** Records hung from a small hole near their rims have a period of roughly 3.5 s (Fig. P16.41A). What is the torsion spring constant of the wire? **b.** If a record is hung from its center hole using a wire of the same torsion spring constant (Fig. P16.41B), what is its period of oscillation?

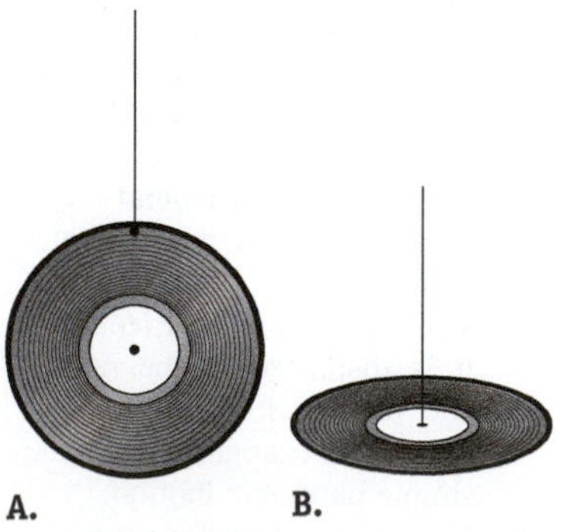

FIGURE P16.41

16-9 Energy in Simple Harmonic Motion

42. **N** A simple harmonic oscillator consists of a 695-g block attached to a lightweight spring. The total energy of the system is 8.50 J, and its period of oscillation is 0.330 s. **a.** What is the maximum speed of the block? **b.** What is the force constant of the spring? **c.** What is the amplitude of the motion of the block?

43. **N** A wooden block ($m = 0.600$ kg) is connected to a spring and undergoes simple harmonic motion with an amplitude of oscillation of 0.075 m. The frequency of the motion is 12.50 Hz. **a.** What is the spring constant? **b.** What is the maximum speed of the block? **c.** What is the speed of the block when it is 0.015 m away from the equilibrium position?

44. **N** A box of mass 0.900 kg is attached to a spring with $k = 125$ N/m and set into simple harmonic motion on a frictionless, horizontal table. The amplitude of motion is 5.00 cm. **a.** What is the total energy of the box–spring system? **b.** What is the speed of the box when the spring is compressed by 2.00 cm? **c.** What is the kinetic energy of the box at this position? **d.** What is the potential energy of the box–spring system at this position?

Problems 45 and 46 are paired.

45. **N** A block of mass $m = 1.23$ kg is attached to the end of a spring with a spring constant of 565 N/m. The block rests on a frictionless surface, is pulled to the right, and is held there. When released, the block undergoes simple harmonic motion. **a.** If the maximum speed of the block is 7.12 m/s, what is the amplitude of the motion? **b.** What is the speed of the block when it is halfway between the equilibrium point and the maximum displacement ($x = \frac{1}{2}A$)?

46. **N** In Problem 45, the block's maximum speed is 7.12 m/s. There is an instant in the motion when the potential energy of the system is equal to the kinetic energy of the block. At what position x does this situation occur?

47. **N** This problem is a follow-up to Example 16.7 (page 472). A horizontal spring of spring constant $k = 170$ N/m is attached to an object of mass $m = 0.85$ kg. Friction between the object and the table is negligible. A student stretches the spring so that the object's initial position is $x_i = 0.20$ m and releases the system at $t = 0$. Find the kinetic and potential energies at $t = 2.5$ s.

48. **N** An object of mass 0.20 kg executes simple harmonic motion along the x axis with a frequency $f = 25/\pi$ Hz. At a particular moment, the object has 0.50 J of kinetic energy and 0.40 J of potential energy. Find the amplitude of oscillation.

16-10 Damped Harmonic Motion

49. **N** A car of mass 2.00×10^3 kg is lowered by 1.50 cm when four passengers, each of mass 70.0 kg, sit down in it. **a.** Determine the damping constant b of the shock absorbers that will provide critical damping. **b.** Suppose for the same car the shock absorbers are so worn that they provide almost no damping. Find the period of up-and-down oscillation of the car after hitting a bump in the road.

50. **N** The bob of a simple pendulum is displaced by an initial angle of 22.0° and released from rest. Because of friction, the amplitude of oscillation is observed to be half the initial value after 765 s. What is the value of the time constant τ for damping for this pendulum?

51. **C** A object–spring oscillator is critically damped. The damping force exerted on the object is of the form $F_D = -bv$. If the object's mass is increased, is the oscillator still critically damped? If not, is it underdamped or overdamped? Explain.

52. **A** Show that for an object–spring oscillator, Equation 16.3

$$y(t) = y_{max} \cos(\omega t + \varphi)$$

is a solution to Equation 16.47,

$$-\frac{k}{m}y(t) = \frac{d^2y(t)}{dt^2}.$$

53. **A** Show that Equation 16.50,

$$y(t) = A(t)\cos(\omega_D t + \varphi)$$

is a solution to Equation 16.49,

$$-ky(t) - b\,dy(t)/dt = m\,d^2y(t)/dt^2$$

where $A(t) = y_{max}e^{-t/\tau}$ (Eq. 16.51), $\tau \equiv 2m/b$, and $\omega_D = [\omega^2 - (b^2/4m^2)]^{1/2}$.

54. **A** Verify that the time constant $\tau \equiv 2m/b$ has the dimensions of time.

55. **C** A cat toy consists of a lightweight, hollow ball ($m = 2.0$ g) hanging on the end of a thin string of length $\ell = 32$ cm. A pet owner displaces the ball by 6° and releases it. The cat enthusiastically watches the ball swing back and forth, but the ball's amplitude is reduced to 1° in 24 s. Both the cat and owner are disappointed. Would it help the cat if the hollow ball were filled with BBs, increasing its mass but not changing its aerodynamics? Explain your answer.

16-11 Driven Oscillators

56. **C** A university has a nonoperating Foucault pendulum. Senior students at the university decided to donate a new motor. Avi, Cameron, and Shannon have been asked to install the motor and bring the pendulum back to life. The pendulum looks much like the one in the Pantheon (Fig. 16.27, page 477) with a bob at the bottom of a steel cable of length 40 ft. It should have a 7.00-s period. The students have connected the motor and turned it on, but the pendulum is not responding. The motor has two dials, one for frequency and one for amplitude. It has one display for the frequency. Read the three students' discussion and answer the questions.

Shannon: It says that the frequency is about 0.9 Hz. That seems really low. Let's just crank up the frequency.

Avi: I calculated the frequency for a simple pendulum with a 40-ft cable, and that's about right. I think we just need to turn up the amplitude.

Shannon: It's already maxed out.

Avi: Really? Then I would say that the motor is broken. Or maybe the seniors didn't pay for a strong enough motor. Let's get our money back.

Cameron: Avi, you used the formula for a *simple* pendulum. No real pendulum is a simple pendulum. You need to model it as a physical pendulum.

Shannon: Cameron, you can calculate that if you want. I'm just going to turn up the frequency.

a. What happens as Shannon slowly turns up the frequency?
b. Do you agree with Avi's calculation?
c. Is Cameron correct that no real pendulum can be modeled as a simple pendulum? If so, calculate the correct frequency.
d. The amplitude is maxed out, and the pendulum is not oscillating. Do those problems mean that the motor is defective? In other words, does the pendulum just need a bigger push than this motor is delivering?

57. **N** To demonstrate the concept of resonance to your son and his friends, you suspend your smartphone by a lightweight string of length L and set the phone on vibrate. Using your landline phone, you call the cell phone, which vibrates with a frequency of 0.900 Hz, causing your makeshift pendulum to oscillate at a very large amplitude. What is the length L of the string you used in this experiment?

58. **N** An ideal simple harmonic oscillator comprises a 255-g ball hanging from a lightweight, vertical spring with spring constant $k = 8.50$ N/m. The system is driven by a sinusoidal force of amplitude 3.00 N. If the ball vibrates with an amplitude of 75.0 cm, what is the frequency of the driving force?

General Problems

Problems 59–65 are grouped.

59. **G** Table P16.59 gives the position of a block connected to a horizontal spring at several times. Sketch a motion diagram for the block.

TABLE P16.59

Time of measurement, t (s)	Position of block, x (m)
0	5.00
0.25	3.50
0.50	0
0.75	−3.50
1.00	−5.00
1.25	−3.50
1.50	0
1.75	3.50
2.00	5.00
2.25	3.50
2.50	0

60. **G** Use the position data for the block given in Table P16.59. Sketch a graph of the block's **a.** position versus time, **b.** velocity versus time and **c.** acceleration versus time. There is no need to label the values of velocity or acceleration on those graphs.

61. **C** Consider the position data for the block given in Table P16.59. What are the signs of the block's velocity and acceleration at the first five times listed?

62. **A** Use the data in Table P16.59. Write an expression for the magnitude of the block's **a.** position, **b.** velocity, and **c.** acceleration. Assume the maximum position observed is the amplitude.

63. Use the data in Table P16.59 for a block of mass $m = 0.250$ kg and assume friction is negligible.

a. **N** What is the spring constant k?

b. **N** What is the maximum force exerted on the block?

c. **C, N** If you replace the block with a new block of mass $m = 0.125$ kg, which parameters (x_{max}, v_{max}, a_{max}, F_{max}) change? For each parameter that changes, give its new value.

64. Use the data in Table P16.59 for a block of mass $m = 0.250$ kg and assume friction is negligible.

a. **N** Write an expression for the force F_H exerted by the spring on the block.

b. **G** Sketch F_H versus t.

65. Consider the data for a block of mass $m = 0.250$ kg given in Table P16.59. Friction is negligible.

a. **N** What is the mechanical energy of the block–spring system?

b. **N** Write expressions for the kinetic and potential energies as functions of time.

c. **G** Plot the kinetic energy, potential energy, and mechanical energy as functions of time on the same set of axes.

66. **N** A mass on a spring undergoing simple harmonic motion completes 4.00 cycles in 14.0 s. **a.** What is the period of motion for this system? **b.** What is the frequency, in hertz, of this system? **c.** What is the angular frequency of this system?

67. A particle initially located at the origin undergoes simple harmonic motion, moving first in the positive z direction, with a frequency of 3.20 Hz and an amplitude of 1.40 m. The particle oscillates between $z = 1.40$ m and $z = -1.40$ m.

a. **A** What is the equation describing the particle's position as a function of time?

b. **N** What is the maximum speed of the particle?

c. **N** What is the maximum acceleration of the particle?

d. **N** What is the total distance covered by the particle in the first 2.50 s of this motion?

68. **A** Consider the system shown in Figure P16.68 as viewed from above. A block of mass m rests on a frictionless, horizontal surface and is attached to two elastic cords, each of length L. At the equilibrium configuration, shown by the dashed line, the cords both have tension F_T. The mass is displaced a small amount as shown in the figure and released. Show that the net force on the mass is similar to the spring-restoring force and find the angular frequency of oscillation, assuming the mass behaves as a simple harmonic oscillator. You can assume the displacement is small enough to produce negligible change in the tension and length of the cords.

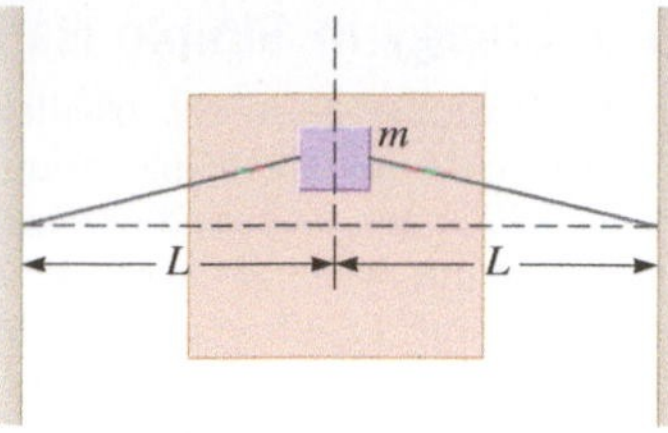

FIGURE P16.68

69. **N** A simple pendulum comprising a bob of mass $m = 2.50$ kg and a massless string of length 1.25 m is observed to have a speed of 3.24 m/s at its equilibrium position. **a.** What is the period of this pendulum? **b.** What is the total energy of the pendulum? **c.** What is the maximum angular displacement of the pendulum from its equilibrium position?

70. **C** An object–spring oscillator and a simple pendulum are each set into simple harmonic motion on the Earth and are each found to oscillate with a frequency of 4.80 Hz. If they were instead set into simple harmonic motion on the Moon, how would their frequencies compare? Explain any similarities or differences.

71. **C** A hollow, metal sphere is filled with water, and a small hole is made at the bottom. The sphere hangs by a long thread and is made to oscillate. How will the period of oscillation change over time if water is allowed to flow through the hole until the sphere is empty?

72. **N** A system exists in which a block of mass $m_1 = 0.500$ kg connected to a spring is extended to a displacement of 5.00 cm and released from rest. The block then oscillates with simple harmonic motion on a frictionless, horizontal tabletop. What is the maximum frequency of motion for which a second block, of mass $m_2 = 0.125$ kg, placed atop the first block, will not slide off? Assume the coefficient of static friction between the two blocks is $\mu_s = 0.650$.

73. **A** Determine the period of oscillation of a simple pendulum of length L suspended from the ceiling of a car that rolls down an inclined plane of angle α (Fig. P16.73). Dissipative forces between the car and the plane are negligible.

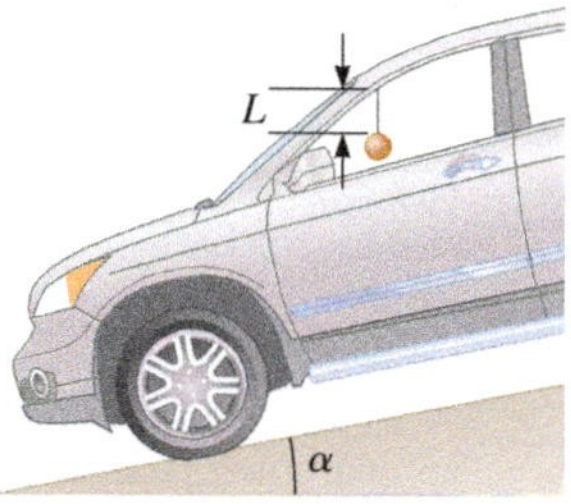

FIGURE P16.73

74. The total energy of a simple harmonic oscillator with amplitude 3.00 cm is 0.500 J.

a. **N** What is the kinetic energy of the system when the position of the oscillator is 0.750 cm?

b. **N** What is the potential energy of the system at this position?

c. **N** What is the position for which the potential energy of the system is equal to its kinetic energy?

d. **C** For a simple harmonic oscillator, what, if any, are the positions for which the kinetic energy of the system exceeds the maximum potential energy of the system? Explain your answer.

75. **A** A spherical bob of mass m and radius R is suspended from a fixed point by a rigid rod of negligible mass whose length from the point of support to the center of the bob is L (Fig. P16.75). Find the period of small oscillation.

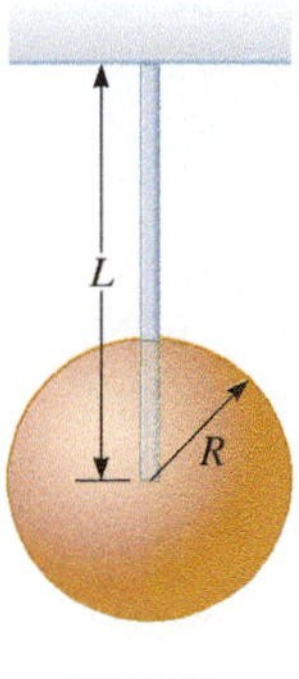

FIGURE P16.75

76. **N** The frequency of a physical pendulum comprising a nonuniform rod of mass 1.25 kg pivoted at one end is observed to be 0.667 Hz. The center of mass of the rod is 40.0 cm below the pivot point. What is the rotational inertia of the pendulum around its pivot point?

77. A lightweight spring with spring constant $k =$ 225 N/m is attached to a block of mass $m_1 =$ 4.50 kg on a frictionless, horizontal table. The block–spring system is initially in the equilibrium configuration. A second block of mass $m_2 =$ 3.00 kg is then pushed against the first block, compressing the spring by $x =$ 15.0 cm as in Figure P16.77A. When the force on the second block is removed, the spring pushes both blocks to the right. The block m_2 loses contact with the spring–block 1 system when the blocks reach the equilibrium configuration of the spring (Fig. P16.77B).

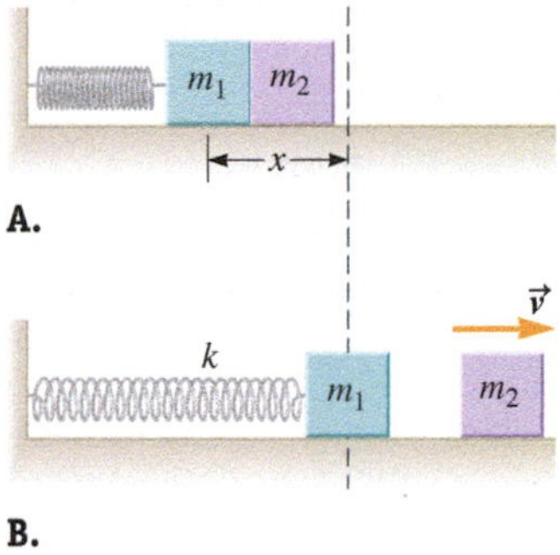

FIGURE P16.77

a. **N** What is the subsequent speed of block 2?

b. **C** Compare the speed of block 1 when it again passes through the equilibrium position with the speed of block 2 found in part (a).

78. **A** Determine the angular frequency of oscillation of a thin, uniform, vertical rod of mass m and length L pivoted at the point O and connected to two springs (Fig. P16.78). The combined spring constant of the springs is k ($k = k_1 + k_2$), and the masses of the springs are negligible. Use the small-angle approximation ($\sin\theta \approx \theta$).

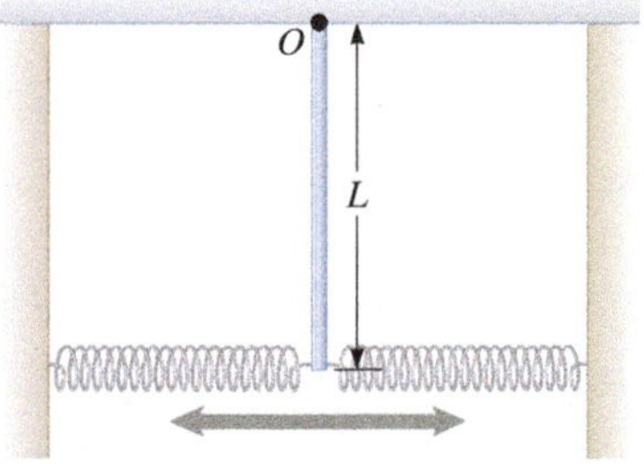

FIGURE P16.78

79. **N** Air resistance in a lab causes the motion of a 5.00-kg disk attached to a vertical spring with spring constant $k = 5.000 \times 10^3$ N/m to be damped at a rate given by a damping coefficient $b = 4.50$ N · s/m. **a.** What is the frequency of the damped oscillation of the system? **b.** What is the percentage by which the amplitude of motion decreases after each cycle?

Problems 80 and 81 are paired.

80. **A** Two springs, with spring constants k_1 and k_2, are connected to a block of mass m on a frictionless, horizontal table (Fig. P16.80). The block is extended a distance x from equilibrium and released from rest. Show that the block executes simple harmonic motion with a period given by

$$T = 2\pi\sqrt{\frac{m(k_1 + k_2)}{k_1 k_2}}$$

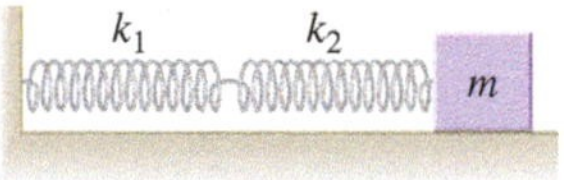

FIGURE P16.80

81. **A** Figure P16.22 shows a system in which two springs, with spring constants k_1 and k_2, are connected to a block of mass m on a frictionless, horizontal table. The block is moved a distance x to the right and released from rest. Show that the block executes simple harmonic motion with a period given by

$$T = 2\pi\sqrt{\frac{m}{(k_1 + k_2)}}$$

82. **A** Show that Equation 16.59,

$$y(t) = A_{drv}\cos(\omega_{drv}t + \varphi_{drv})$$

is a solution to Equation 16.58,

$$F_{max}\cos\omega_{drv}t = m\frac{d^2y(t)}{dt^2} + b\frac{dy(t)}{dt} + ky(t)$$

where $A_{drv} = \dfrac{F_{max}}{\sqrt{m^2(\omega^2 - \omega_{drv}^2)^2 + b^2\omega_{drv}^2}}$ (Eq. 16.60). *Hint:* Use $e^{i\alpha} = \cos\alpha + i\sin\alpha$.

17 Traveling Waves

Key Questions

How can we describe a traveling wave mathematically (that is, its kinematics)?

What is carried or transferred by a wave?

Underlying Principles

One-dimensional wave equation

Major Concepts

1. Longitudinal wave
2. Transverse wave
3. Harmonic wave
4. Wave function of transverse harmonic wave
5. Wave function of longitudinal harmonic wave
6. Intensity
7. Law of refraction
8. Diffraction
9. Doppler shift equation
10. Shock wave

Special Cases

1. Speed of transverse wave on a rope 2. Sound

Sometimes, a small group of friends in a large stadium or music hall will start a "wave" going around the audience. The friends all stand up in unison, lifting their arms above their heads and then sitting back down together (Fig. 17.1). They hope that a group of people—say to their immediate left—will imitate them soon afterward. If the people to the left of those people imitate them and so on, the wave propagates around the audience. If the stadium seating is arranged in a loop, the wave can propagate around and around the loop until people grow tired of the activity.

FIGURE 17.1 Fans in a stadium create a wave by standing and lifting their arms. The enthusiasm travels around the stadium.

A stadium wave has the features of traveling waves we study in this chapter. A disturbance—people standing and then sitting—travels around the audience. Each person rises and falls in place; no one walks around the stadium. Enthusiasm or energy, however, flows around the stadium.

17-1 Introducing Mechanical Waves

Suppose you would like to ask one of your classmates to join you for a movie. You could write a note and toss it on your (soon-to-be) friend's desk. A second—perhaps more effective—method is to simply ask. In the first case, you have used matter to transport your message; paper was displaced from your hand to your friend's desk. In the second case, no matter was transported. Your vocal cords created a disturbance in the air, and that disturbance—not the air molecules—traveled to your friend's eardrums. The disturbance is a sound wave. If you don't want to drop off a note or ask in person, you could send a text message. The text message also uses a wave in the form of an electromagnetic signal traveling from your cell phone to a relay tower and then to your friend's phone.

A **wave** is a disturbance in a medium (or field). For a sound wave, the medium is air as disturbed by a moving object such as your vocal cords. Waves are distinguished in part by the medium through which they travel. In this chapter and the next, we study **mechanical waves** (waves in a material medium such as a solid or fluid). In Chapters 34 through 38, we study light and radiation (electric and magnetic field waves requiring no material medium for propagation).

Traveling waves disturb a medium as they move from one region to another region. For example, when you ask your friend a question, your vocal cords make air molecules oscillate. The resulting disturbance in the air moves away from you. When this disturbance reaches your friend's eardrums, they oscillate. No matter what type of medium a wave travels in, it transports energy and momentum, but the medium itself is not carried along with the wave.

CASE STUDY Rolling Stones and Kidney Stones

College students discuss the warning label that came with Shannon's new earbuds.

Shannon: Check out all the warning labels on my new earbuds. Do you really think that sound can damage your hearing? Everyone is just worried about getting sued.

Avi: I think it's true. In high school health class, they made us read this article in *Rolling Stone* about how all these old rock stars were losing their hearing. Now they are trying to make people more aware of the dangers of loud music.

Shannon: But those guys didn't even wear earbuds. I just don't believe that music can damage your ears. Maybe an explosion can bust your eardrums because it blows particles right through your body.

Cameron: Look, it doesn't have to be that dramatic. When I volunteered at the hospital last summer, they broke up kidney stones with ultrasound.

Shannon: Well, isn't using *ultrasound* on a kidney stone like jackhammering a boulder? It's not like I'm jamming a pencil in my ear canal.

Avi: Yeah, but sound is a wave. It sends air molecules down your ear canal. It's like when you use a sandblaster to take the paint off a car. The air molecules coming out of an earbud just blast your eardrums.

Are any of these students correct? In this case study, we'll explore the damage that sound can do to your hearing as well as the use of sound to break up kidney stones. In the process, we'll learn more about waves, what they transmit, and how we hear.

17-2 Pulses

To see how a disturbance travels in a medium, imagine squeezing together several coils of a stretched-out horizontal spring that has a small bell dangling from the far end (Fig. 17.2). Your hands have created a disturbance in the medium (the spring). When you release the coils, the disturbance travels to the right. Beads on the coils of the spring highlight the wave. Notice that the coils never move very much, and, after the wave has passed

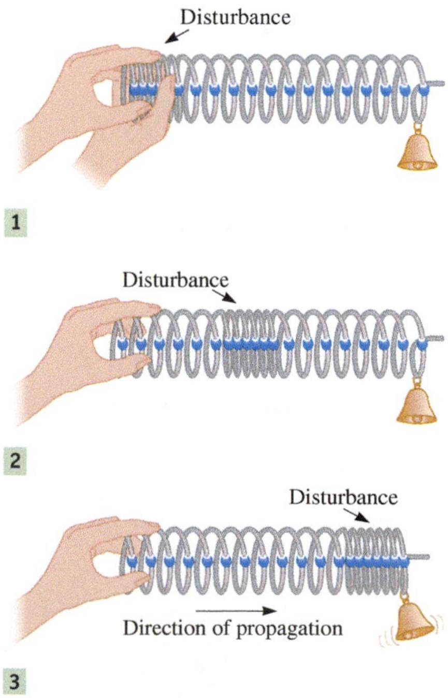

FIGURE 17.2 1 You create a longitudinal pulse by compressing a few coils together and releasing them. 2 The pulse travels to the right as the coils oscillate. 3 The pulse reaches the bell, causing it to ring.

LONGITUDINAL WAVE; TRANSVERSE WAVE

Major Concepts

a particular part of the spring, the coils return to their original positions. Contrast the coils' motion to the motion of the disturbance. The disturbance propagates the full length of the spring from your hands to the opposite end, causing the hanging bell to oscillate.

We will model a wave as the motion of a collection of particles represented by beads on the coils in Figure 17.2. As the wave propagates along the spring, each particle's motion forward and back is parallel to the spring's long axis. The wave in Figure 17.2 is an example of a **longitudinal wave**, one in which the motion of each particle is parallel to the direction of the wave's propagation. It might help to remember that in a *long*itudinal wave, the particles in the medium jiggle *along* the direction of propagation.

In a **transverse wave**, particles in the medium move perpendicular to the direction of propagation. Figure 17.3 shows how you could set up a transverse wave in the horizontal spring. By flicking your hand up and then back down, you create a disturbance that is perpendicular to the spring's long axis. Each bead moves up and then back down while the disturbance moves to the right.

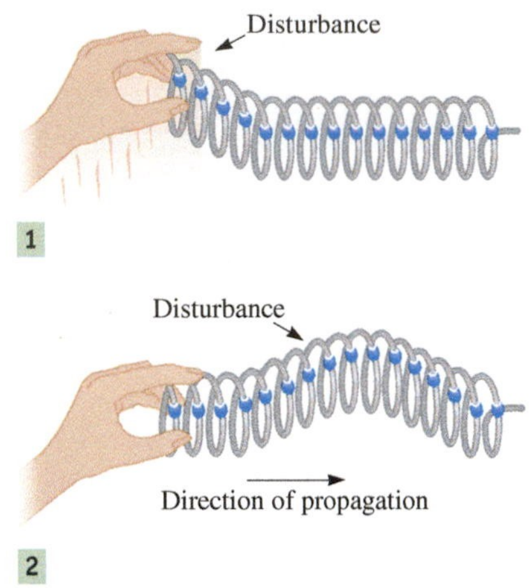

FIGURE 17.3 1 You create a transverse pulse by shaking the spring up and down. 2 The pulse travels to the right.

Wave Function for a Particular Pulse

The waves in Figures 17.2 and 17.3 are both **pulses** because they are discrete disturbances that do not repeat. Let's take a closer look at a single transverse pulse traveling in the positive x direction along a long, straight medium such as a stretched rope (Fig. 17.4). Again we model the rope as a collection of particles and use beads to help picture those particles. Our goal is to come up with a **wave function**, a mathematical description of the wave's kinematics.

Before there is a pulse, the rope is at rest; all the beads lie along the x axis (Fig. 17.4A). Mathematically, we can write $y(x) = 0$, meaning that $y = 0$ for all values of x. If no one ever plucks the rope so that there is never a pulse on it, we can write $y(x, t) = 0$, which means that $y = 0$ for all values of x at all times t. The notation (x, t) means that y (the wave function) depends on x and t.

If the rope is plucked at $t = 0$, a pulse is created. Imagine taking a photo at $t = 0$ (Fig 17.4B) so that you could easily see the shape of the pulse. To make our discussion specific, we have chosen a pulse with a simple shape. At $t = 0$, the rope's shape is mathematically described by

$$y(x) = \frac{1}{x^2 + 1} \tag{17.1}$$

where the constants are in appropriate SI units. Equation 17.1 is unique to this particular pulse shape. If the rope is plucked differently, the shape of the pulse will be different, and Equation 17.1 must be changed. The mathematical description $y(x)$ is known as the **profile** of the wave. The profile is only a function of the position x. To find an expression for the wave function $y(x, t)$ that is a function of both position x and time t, we consider the pulse at later times.

Let's start by considering one specific later time, say $t = 1$ s. At this time, the profile (shape) is unchanged, but the pulse has moved to the right by 2 m as you can see by examining the peak (Fig. 17.4C). The wave function must produce the same profile, but shifted to the right. At $t = 1$ s, the mathematical description is

$$y(x) = \frac{1}{(x - 2)^2 + 1}$$

where the 2 in $(x - 2)$ is needed because the pulse has shifted by 2 m. If you are wondering why there is a minus sign in $(x - 2)$ instead of a plus sign, notice that the peak of the pulse is $y_{peak} = 1$ m. If you substitute $x = 2$ m, you will find that the peak of the pulse is still 1 m, but if there had been a plus sign instead, as in

$$y(x) = \frac{1}{(x + 2)^2 + 1}$$

the peak would have been 1/17 m. So, only the negative sign is consistent with the profile of the wave. (We'll explore a leftward-moving pulse in part B of Example 17.1.)

We seek a way to describe the pulse mathematically at any arbitrary time t. Think of parts B through D of Figure 17.4 as a series of photos taken every second from $t = 0$ to $t = 2$ s. We see that the peak moves to the right at speed $v_x = 2$ m/s. So, the

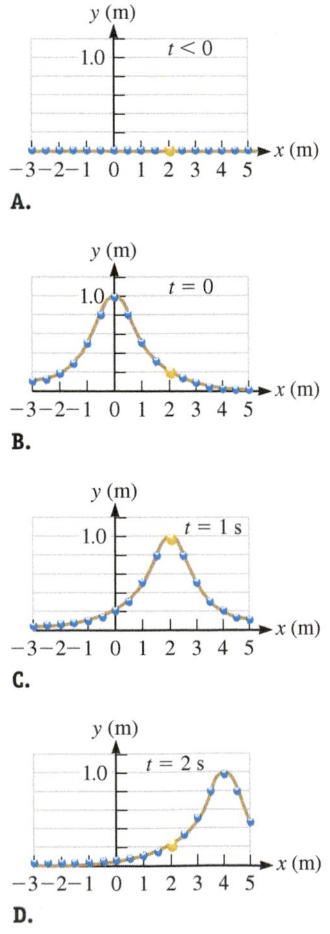

FIGURE 17.4 A transverse pulse on a rope moves to the right.

pulse shifts to the right by $v_x t$ relative to its initial position shown in Figure 17.4B. The wave function of this particular pulse is

$$y(x, t) = \frac{1}{(x - v_x t)^2 + 1} \tag{17.2}$$

where $v_x = 2$ m/s.

Equation 17.2 is not a general expression for a wave function; it is only the wave function of this particular pulse. Like all wave functions, however, this one provides a full description of the wave on the rope, giving the vertical position y of a rope particle (a bead) as a function of that particle's horizontal position x and time t. Wave functions are complicated because they depend on two variables x and t. Let's think about each variable individually.

First, if we substitute a specific time in Equation 17.2, it is like freezing the pulse in time by taking a photo. The wave function at a specific time is a mathematical description of the profile (the rope's shape).

If we instead substitute a specific position x into Equation 17.2, we then have a description of the up-and-down motion of one particle (one bead). Imagine watching the gold bead at $x = 2$ in Figure 17.4 through a narrow slit so that you cannot see the rest of the rope. The up-and-down motion you would see through the slit is best represented by a motion diagram (Fig. 17.5A). As usual, its position-versus-time graph (Fig. 17.5B) looks like a stretched-out version of its motion diagram; this graph also looks similar to the pulse's profile (Fig. 17.4), but don't confuse the two. The mathematical description of this gold bead's motion comes from substituting $x = 2$ m into Equation 17.2:

$$y(t) = \frac{1}{(2 - v_x t)^2 + 1} \tag{17.3}$$

where $v_x = 2$ m/s as before. Equation 17.3 gives the vertical position y as a function of time of only the gold bead.

Now that we've thought about the wave function at one particular time (by taking a photo) and at one particular position (by looking though a narrow slit at one particular bead), we see how the wave function describes the rightward motion of the pulse. According to the wave function, each bead only moves vertically along the y axis (as seen through the slit). No bead moves horizontally along the x axis. At any instant (as seen in a photo), the beads form a shape that is the profile of the pulse. According to the wave function, as the beads move vertically, that profile moves along the x axis.

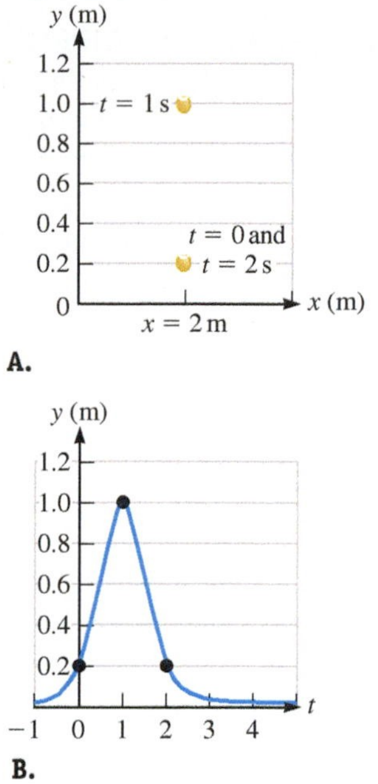

FIGURE 17.5 **A.** Motion diagram for the gold bead located at $x = 2$ m (Fig. 17.4). **B.** Position-versus-time graph for this gold bead.

CONCEPT EXERCISE 17.1

As we've seen before, terms used in physics often differ in meaning from the same terms used in everyday language. How does the use of the word *pulse* in one instance, *Be sure to check your pulse to ensure a productive workout*, differ from the physics definition of *pulse*?

CONCEPT EXERCISE 17.2

A graph of a pulse's profile and a position-versus-time graph for one particle in the medium look similar. For a transverse pulse traveling in the x direction, on both graphs the vertical axis is y. What is plotted on the horizontal axis in each graph?

EXAMPLE 17.1 Test It for Yourself

A Verify that Equation 17.3 with $v_x = 2$ m/s gives the position of the gold bead in Figures 17.4 and 17.5 by substituting $t = 0$, 1, 2, and 3 s.

Example continues on page 490 ▶

INTERPRET and ANTICIPATE
If Equation 17.3 describes the motion of the gold bead, we expect to find values for y that are consistent with those shown in Figure 17.5B.

SOLVE
Substitute $v_x = 2$ m/s and the times into Equation 17.3.

$$y(t) = \frac{1}{(2 - v_x t)^2 + 1} \quad (17.3)$$

$$y(t) = \frac{1\ \text{m}^3}{[2\ \text{m} - (2\ \text{m/s})t]^2 + 1\ \text{m}^2}$$

$$y(0) = 0.2\ \text{m} \qquad y(1) = 1\ \text{m}$$

$$y(2) = 0.2\ \text{m} \qquad y(3) = 0.06\ \text{m}$$

CHECK and THINK
Our results are consistent with the positions in Figure 17.5B.

B If the pulse in Figure 17.4 has the same profile but moves to the left with the same speed, what changes are needed in Equations 17.2 and 17.3?

INTERPRET and ANTICIPATE
If the pulse moves to the left instead of to the right, the peak of the pulse at $t = 1$ s will be at $x = -2$ m. Our revised equation should give $y(-2, 1) = 1$ m. We also expect that the gold bead at $x = 2$ m should only move slightly because the pulse moves away from that bead and to the left. We'll check that the gold bead's vertical position y is never above 0.2 m for all times. (That condition is the "mirror image" of what happens to the bead at $x = -2$ m in Fig. 17.4.)

SOLVE
Because the pulse moves to the left, replace $v_x = 2$ m/s with $v_x = -2$ m/s in Equations 17.2 and 17.3 to find the new wave function.

$$y(x, t) = \frac{1}{[x + (2\ \text{m/s})t]^2 + 1}$$

CHECK and THINK
As expected, at $t = 1$ s, the wave function gives $y(-2, 1) = 1$ m, showing that the pulse moves to the left.

$$y(-2, 1) = \frac{1\ \text{m}^3}{[(-2\ \text{m}) + (2\ \text{m/s})(1\ \text{s})]^2 + 1\ \text{m}^2}$$

$$y(-2, 1) = 1\ \text{m}$$

We can then determine the vertical position y of the gold bead by substituting $x = 2$ m.

$$y(2, t) = \frac{1\ \text{m}^3}{[2\ \text{m} + (2\ \text{m/s})t]^2 + 1\ \text{m}^2}$$

The maximum vertical position of the gold bead occurs when $t = 0$. At any later time, its vertical position is closer to zero because the denominator is larger. We have confirmed that $y(2, t) < 0.2$ m for the gold bead in the case of a leftward-moving pulse.

$$y(2, 0) = \frac{1\ \text{m}^3}{[2\ \text{m} + (2\ \text{m/s})(0)]^2 + 1\ \text{m}^2} = 0.2\ \text{m}$$

17-3 Harmonic Waves

A wave's profile is determined by how the disturbance was created. For example, the gentle up-and-down motion of the hand in Figure 17.3 creates a nearly triangular pulse. If a rope is jiggled up and down many times, the wave's profile is much longer (Fig. 17.6). Any real wave has a beginning and an end determined by the time the disturbance began and ended. A **periodic wave** is an ideal wave that repeats endlessly. Of course, you cannot produce a periodic wave by jiggling the end of a rope because that result would require you to jiggle an infinitely long rope in the same way forever.

Many real waves, however, can be modeled as periodic waves. Figure 17.7 displays a note held by an opera singer using a microphone whose signal was fed to an

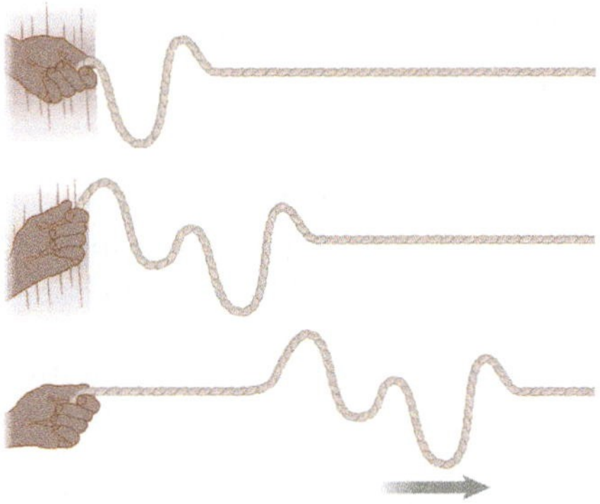

FIGURE 17.6 By jiggling the rope several times, you can generate a complicated wave profile. The resulting wave is neither periodic nor harmonic.

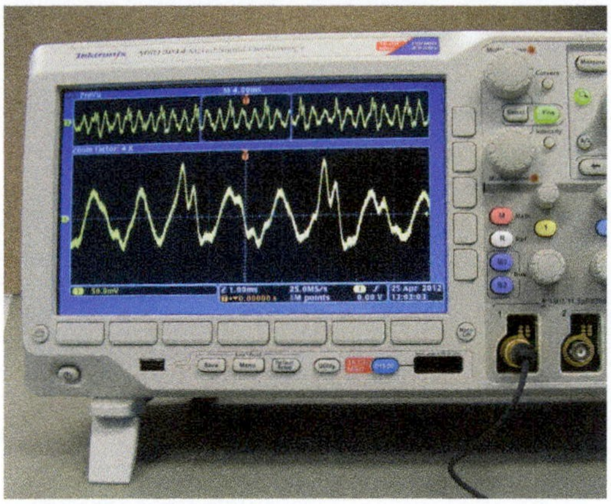

© Victor Sluiter

FIGURE 17.7 Oscilloscope trace of a single note held by opera singer.

oscilloscope. Although the oscilloscope trace has a mathematically complicated wave function, it can be modeled as a periodic wave. In this chapter, we focus on simpler periodic waves with wave functions in the form of a single sine or cosine. Such waves are called **harmonic waves**.

HARMONIC WAVE ✪ **Major Concept**

Transverse Harmonic Waves

The easiest way to generate a harmonic wave is to use a simple harmonic oscillator (SHO) to create the disturbance. In Figure 17.8A, a vertical spring is attached to one bead at the end of a rope. Think of Figure 17.8A as a snapshot that cannot show motion. The end bead oscillates up and down in simple harmonic motion (SHM) similar to the disk in Crall and Whipple's experiment (Fig. 16.3, page 452). Because the spring in Figure 17.8A oscillates in SHM consistently for a long time, the resulting wave can be modeled as a harmonic wave.

The wave moves to the right, but no bead in Figure 17.8A moves to the right. Instead, the wave causes each bead to oscillate up and down in SHM. Imagine using two thin slits to observe the motion of the gold and silver beads. Figure 17.8B shows motion diagrams for these two beads, similar to the motion diagram for Crall and Whipple's disk (Fig. 16.4, page 452). The beads move faster near the centers of their paths and slower near the ends. The main difference between the gold and silver bead's motion is that when one is moving upward, the other is moving downward. For example, as the gold bead moves downward from point 1 to point 4 (left), the silver bead moves upward from its own point 1 to point 4 (right).

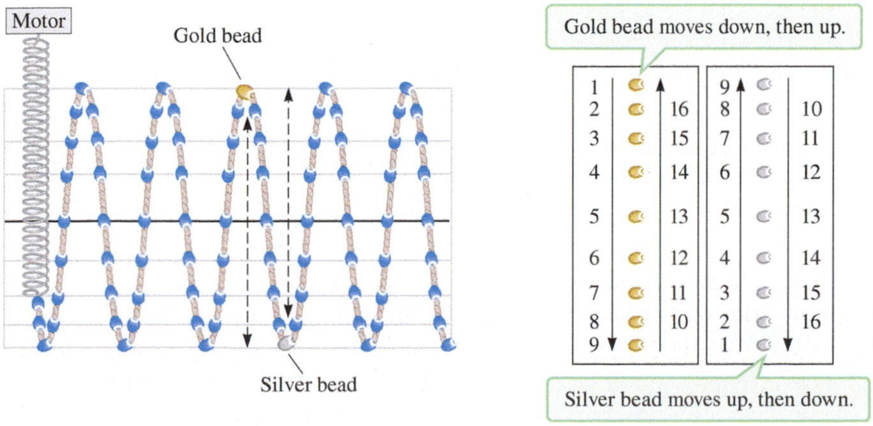

A. **B.**

FIGURE 17.8 **A.** A transverse harmonic wave is created on a beaded rope by a vertical spring driven by a motor. The resulting wave is harmonic and (nearly) periodic. **B.** Each bead oscillates up and down in simple harmonic motion.

The vertical position y of each bead can be described by the equation for SHM (Eq. 16.3), but with two different initial phases:

$$y_{gold}(t) = y_{max} \cos(\omega t + \varphi_{gold})$$
$$y_{silver}(t) = y_{max} \cos(\omega t + \varphi_{silver})$$

The angular frequency and amplitude are the same for both beads. In fact, each bead on the rope obeys Equation 16.3 with an initial phase that depends on its horizontal position x along the rope, and each bead moves with the same angular frequency ω and amplitude y_{max}.

Our observation of the gold and silver beads provides important information about the wave function $y(x, t)$ of a harmonic wave: The wave function gives the vertical position y of a series of simple harmonic oscillators whose initial phase depends on their horizontal position x. In the coordinate system in Figure 17.9, the x axis passes through the equilibrium position of the rope, and y is perpendicular to x. A sine or cosine may be used for a harmonic wave function, and we arbitrarily choose the sine:

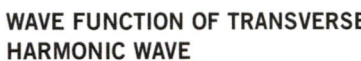

WAVE FUNCTION OF TRANSVERSE HARMONIC WAVE

Major Concept

$$y(x, t) = y_{max} \sin(kx - \omega t) \tag{17.4}$$

where k is the **angular wave number**. The angular wave number is constant for a particular wave, closely related to the **wavelength** λ of the wave:

$$k = \frac{2\pi}{\lambda} \tag{17.5}$$

The wavelength is the length of a single repetition of the wave pattern as found from a photo (with a scale) or profile of the wave (Fig. 17.9). The SI unit of wavelength is the meter, so the angular wave number is measured in radians per meter.

At a single horizontal position x, Equation 17.4 becomes the position equation for SHM (Eq. 16.3). Let's check Equation 17.4 for a particular bead in Figure 17.8A, say the one at $x = \pi/2k$:

$$y\left(\frac{\pi}{2k}, t\right) = y_{max} \sin\left[k\left(\frac{\pi}{2k}\right) - \omega t\right] = y_{max} \sin\left[\left(\frac{\pi}{2}\right) - \omega t\right]$$

Using the trigonometric identity $\sin(\pi/2 - \theta) = \cos\theta$ (Appendix A),

$$y\left(\frac{\pi}{2k}, t\right) = y_{max} \cos \omega t$$

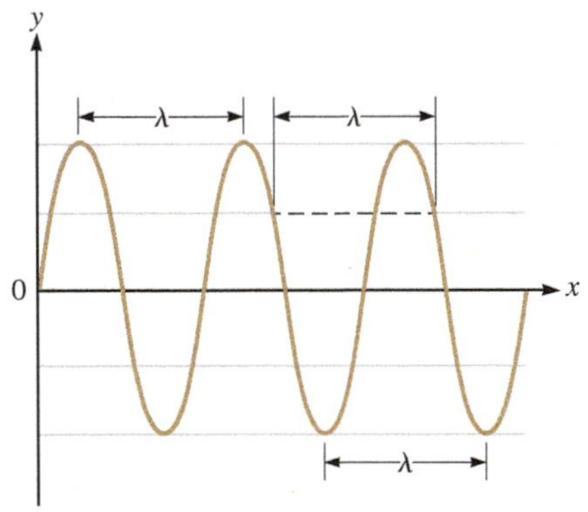

FIGURE 17.9 The profile of a harmonic wave. The wavelength may be found from the peak-to-peak distance, the valley-to-valley distance, or any other distance representing one complete wave cycle.

which is the equation for the vertical position y of an SHO (Eq. 16.3) with the initial phase equal to zero. The angular frequency ω is related to the period T and frequency f of each bead's oscillation by Equation 16.2:

$$\omega \equiv 2\pi f = \frac{2\pi}{T} \tag{16.2}$$

EXAMPLE 17.2 A Harmonic Wave Is Really a Collection of Oscillators

Show that Equation 17.4 can be rewritten in the form of Equation 16.3, $y(t) = y_{max} \cos(\omega t + \varphi)$. Find an expression for the initial phase in terms of each particle's horizontal position x.

INTERPRET and ANTICIPATE

Sine and cosine functions are connected through trigonometric identities. This example is asking us to use an identity to rewrite the sine function in terms of a cosine function. In doing so, we will discover an expression for the initial phase $\varphi(x)$. We expect φ to be in radians.

SOLVE

Equation (1) relates the sine and cosine functions (Appendix A).

$$\sin\theta = \cos\left(\frac{\pi}{2} - \theta\right) \tag{1}$$

Apply this trigonometric identity to Equation 17.4 by setting $\theta = kx - \omega t$.	$y(x, t) = y_{max} \sin (kx - \omega t) = y_{max} \cos \left[\frac{\pi}{2} - (kx - \omega t)\right]$ $y(x, t) = y_{max} \cos \left[\omega t + \left(\frac{\pi}{2} - kx\right)\right]$ (2)
Find an expression for $\varphi(x)$ by comparing Equation (2) with Equation 16.3.	$y(t) = y_{max} \cos (\omega t + \varphi)$ (16.3) $\varphi(x) = \frac{\pi}{2} - kx$ (3)

CHECK and THINK

Equation (2) is the equation of an SHO with the phase given by Equation (3). Because the angular wave number k is measured in radians per meter and the position x is in meters, the phase is in radians as expected. We can interpret our results this way: A harmonic wave is a collection of simple harmonic oscillators (Eq. 2) each of whose phase depends on its horizontal position (Eq. 3).

Longitudinal Harmonic Waves

Equation 17.4 is the wave function for a *transverse* harmonic wave. The wave moves along the x axis, and each particle in the medium (each "bead") oscillates along the y axis, perpendicular to the wave's motion. With just a slight modification, we can write Equation 17.4 for a *longitudinal* harmonic wave. First, imagine how such a wave might be established. Figure 17.10 shows a long, horizontal spring attached to a simple pendulum. Beads on the spring's coils help visualize the particle model. (Each bead represents a piece of the medium, which in this case is the spring.) The pendulum is displaced slightly so that the motion of the bob is roughly along the x axis. As the pendulum swings back and forth, the beads oscillate in SHM. All the beads oscillate with the same angular frequency and amplitude, but with differing initial phase constants φ that depend on the coil's equilibrium position.

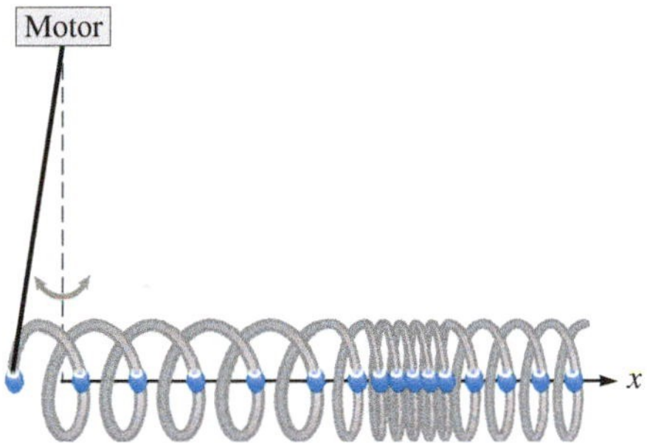

FIGURE 17.10 A motor drives a simple pendulum that is used to create a harmonic longitudinal wave on a spring.

The beads on the spring (Fig. 17.10) oscillate much like the beads on the rope (Fig. 17.8). The difference is that each bead on the spring is displaced back and forth in the $\pm x$ direction instead of in the $\pm y$ direction. We can modify Equation 17.4 to come up with the wave function of this longitudinal wave on the spring. To avoid confusion with the x that appears in Equation 17.4 and to keep the notation compact, we will write horizontal displacement as S without a Δ. The wave function for a longitudinal wave is written as

WAVE FUNCTION OF LONGITUDINAL HARMONIC WAVE

Major Concept

$$S(x, t) = S_{max} \sin (kx - \omega t) \quad (17.6)$$

where S_{max} is the amplitude or maximum displacement of any bead from its equilibrium position.

Figure 17.11 connects the longitudinal wave function (Eq. 17.6) with the motion of particles in the medium. Figure 17.11A shows the particles in their equilibrium positions, corresponding to beads on the spring before the pendulum began its motion. After the pendulum is set in motion, particles will be displaced from equilibrium. Figure 17.11B shows a graph of displacement S versus horizontal position x for time t long after the pendulum is set in motion. To find a bead's displacement, draw a line that extends from its position in Figure 17.11A to Figure 17.11B and then read the displacement off the vertical S axis. The displacement may then be used to find the bead's position at time t. Figure 17.11C shows the beads at time t. We will use Figure 17.11 in Section 17-5 to model sound waves.

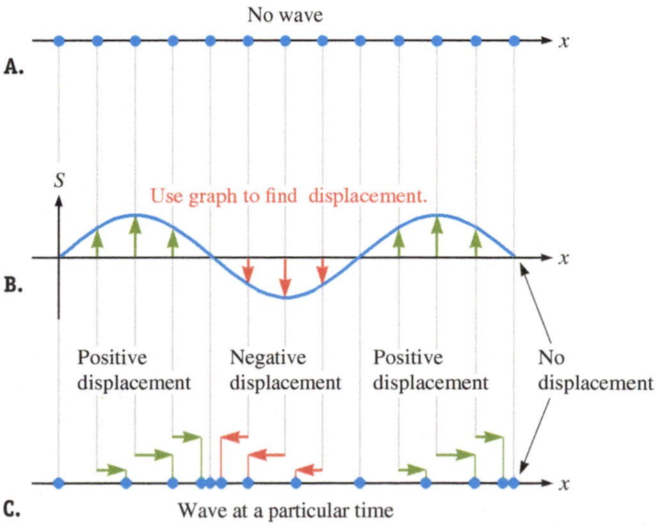

FIGURE 17.11 A. Initially, the particles are uniformly distributed. **B.** The graph of S versus x gives the displacement of each particle. **C.** The new positions of the particles are obtained by applying the displacements from part B to the particle positions from part A.

Speed of a Harmonic Wave

To find the propagation speed of a harmonic wave, imagine riding along with one of the peaks in Figure 17.8A. Because you are at a peak and traveling at the propagation speed of the wave, you always see the bead at its maximum displacement y_{max}. When we substitute $y = y_{max}$ into the wave function Equation 17.4, we find

$$y(x, t) = y_{max} = y_{max} \sin(kx - \omega t)$$
$$\sin(kx - \omega t) = 1$$

For $\sin(kx - \omega t) = 1$, the argument $kx - \omega t$ must equal a constant of the form

$$kx - \omega t = \frac{\pi}{2}, \frac{5\pi}{2}, \frac{9\pi}{2}, \ldots$$

For the purpose of finding the propagation speed, we can choose $kx - \omega t = \pi/2$. Solve for horizontal position x:

$$x = \frac{\pi}{2k} + \frac{\omega}{k}t$$

Take the time derivative to find the propagation velocity in the x direction:

$$v_x = \frac{dx}{dt} = \frac{d}{dt}\left(\frac{\pi}{2k} + \frac{\omega}{k}t\right)$$

$$v_x = \frac{\omega}{k} \tag{17.7}$$

Equation 17.7 is the propagation speed of a harmonic wave; in this case, the wave is traveling in the positive x direction. Using Equations 17.5 and 16.2, the propagation speed can also be written as

$$v_x = \frac{\lambda}{T} = \lambda f \tag{17.8}$$

If the wave is traveling in the negative x direction instead, the wave function is given by

$$y(x, t) = y_{max} \sin(kx + \omega t) \tag{17.9}$$

and (as you may show in Problem 10) the velocity is given by

$$v_x = -\frac{\omega}{k} \tag{17.10}$$

CONCEPT EXERCISE 17.3

A longitudinal wave function is given by $S(x, t) = 0.75 \sin(0.30x - 655t)$ in SI units.

a. What are the amplitude, angular wave number, and angular frequency?
b. What is the propagation velocity?

17-4 Special Case: Transverse Wave on a Rope

The profile of a wave is determined by how the disturbance was created, but the propagation speed is determined by the properties of the medium. For a transverse wave on a rope, the two properties affecting wave speed are the tension and mass per unit length. For a given rope, a higher tension means a faster wave. For a given tension, a thinner rope (smaller mass per unit length) results in a faster wave. Mathematically, the propagation speed of a wave on a rope of linear mass density μ and under tension F_T is

SPEED OF TRANSVERSE WAVE ON A ROPE

Special Case

$$v_x = \sqrt{\frac{F_T}{\mu}} \tag{17.11}$$

DERIVATION Propagation Speed of a Transverse Wave on a Rope

We will derive Equation 17.11. The profile of the wave doesn't matter, so we choose a simple pulse traveling to the right. Imagine that you are traveling along with the pulse so it appears to be stationary. From your point of view, however, the rope appears to be pulled to the left at a constant speed through a clear tube shaped like the pulse. Figure 17.12 illustrates three snapshots that you might take of the rope from your moving reference frame. The pulse appears stationary, but the colored beads move to the left.

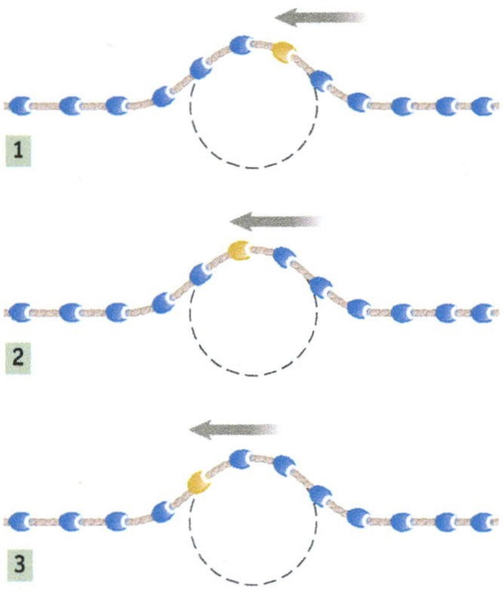

FIGURE 17.12 As seen from a reference frame that moves with the pulse, the rope moves to the left from time 1 to time 3.

The three beads near the top of the pulse momentarily move in a circle. No matter what the pulse's shape, you can always find a small enough portion of the rope to approximate its motion as circular as seen from the reference frame that moves along with the wave.

Start with a free-body diagram for a small segment of the rope marked by a single bead (Fig. 17.13). The bead momentarily moves in a circle of radius r. From the center of the circle, the bead subtends a small angle $\Delta\theta$, so the bead's diameter is $r\,\Delta\theta$. The segment of rope has mass Δm. (The bead is just there to mark that portion of the rope, so we treat it as massless.) Tension in the rope pulls the segment in two directions, both tangent to the rope and making an angle $\Delta\theta/2$ with the x axis.

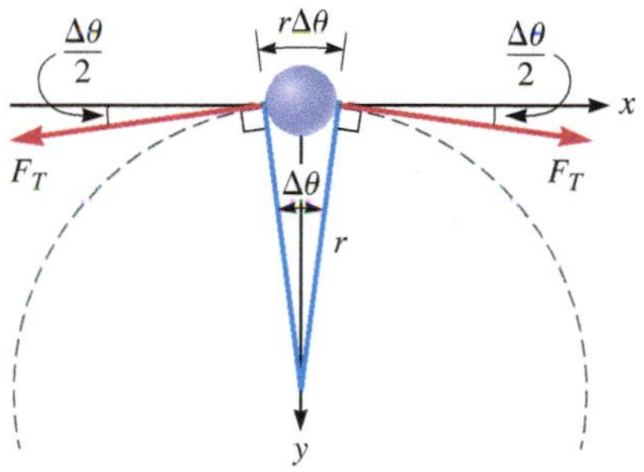

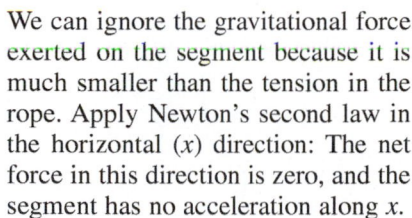

FIGURE 17.13 A rope particle momentarily moves in a circular path.

We can ignore the gravitational force exerted on the segment because it is much smaller than the tension in the rope. Apply Newton's second law in the horizontal (x) direction: The net force in this direction is zero, and the segment has no acceleration along x.

$$\sum F_x = F_T \cos\left(\frac{\Delta\theta}{2}\right) - F_T \cos\left(\frac{\Delta\theta}{2}\right) = 0$$

There is a net force in the y direction (chosen as positive downward), pointing to the center of the circle.

$$\sum F_y = F_T \sin\left(\frac{\Delta\theta}{2}\right) + F_T \sin\left(\frac{\Delta\theta}{2}\right) = \Delta m a_y$$

$$\sum F_y = 2F_T \sin\left(\frac{\Delta\theta}{2}\right) = \Delta m a_y \qquad (1)$$

This net force results in centripetal acceleration (Eq. 4.36).

$$a_y = a_c = \frac{v_x^2}{r} \qquad (2)$$

Substitute Equation (2) into Equation (1).

$$2F_T \sin\left(\frac{\Delta\theta}{2}\right) = \Delta m \frac{v_x^2}{r}$$

The segment of rope is small so that the angle $\Delta\theta$ is small; therefore, we use the small-angle approximation $\sin(\Delta\theta/2) \approx \Delta\theta/2$ and isolate the speed on one side.

$$2F_T\left(\frac{\Delta\theta}{2}\right) \approx \Delta m \frac{v_x^2}{r}$$

$$F_T \Delta\theta \approx \Delta m \frac{v_x^2}{r}$$

$$v_x^2 \approx F_T \frac{r\,\Delta\theta}{\Delta m}$$

Derivation continues on page 496 ▶

The rope's linear mass density μ is defined to be its mass per unit length. The mass of the small segment is Δm, and its length is the diameter of the bead $r\Delta\theta$. The resulting mass per unit length is $\mu = \Delta m/r\Delta\theta$.

$$v_x^2 = \frac{F_T}{\mu}$$

$$v_x = \sqrt{\frac{F_T}{\mu}} \checkmark \qquad (17.11)$$

COMMENTS

Equation 17.7 ($v_x = \omega/k$) holds for any harmonic wave, but our result (Eq. 17.11) holds only for a transverse wave on a rope. Nonetheless, our derivation leads to some insight into the speed of waves traveling in other media. When the pulse passes through some portion of the rope, it deforms the rope. If the pulse passes through quickly, the rope quickly returns to its straight configuration. How could the two factors—tension and linear mass density—in Equation 17.11 cause the pulse to move quickly? First, consider the free-body diagram. If there is a lot of tension in the rope, the bead at the top is quickly pulled down, and the rope is quickly restored to its original configuration. Second, if the mass per unit length is very great, the portion of the bead displaced has a lot of inertia. It resists being moved back to its original position. So, roughly speaking, the speed of a wave in a medium depends on the force that restores its original configuration and on the medium's ability to resist returning to equilibrium (inertia), which we can express as

$$\text{speed} = \sqrt{\frac{\text{restoring ability}}{\text{inertia resisting the return to equilibrium}}} \qquad (17.12)$$

We'll use this conceptual relationship to find the speed of waves in other media.

EXAMPLE 17.3 Piano String Factory

Piano wire must be both strong and flexible. The wire is placed under high tension and then subjected to repeated blows by the piano's hammer (Fig. 17.14), so modern piano strings are made from steel. Consider a thick piano wire of diameter $d = 4.8$ mm and linear mass density $\mu = 64.9$ g/m. If the propagation speed of a wave on this wire is 129 m/s, what is the tension in the wire? The tensile strength of steel is 3.8×10^8 N/m^2 (Table 14.1). **CHECK and THINK:** Will the wire break?

FIGURE 17.14 Inner workings of a piano.

INTERPRET and ANTICIPATE

We can find the tension in the piano wire using Equation 17.11, and we expect that the string does not break because if it did break, it wouldn't be useful in a piano.

SOLVE

Solve Equation 17.11 for tension.

$$v_x = \sqrt{\frac{F_T}{\mu}} \qquad (17.11)$$

$$v_x^2 = \frac{F_T}{\mu}$$

$$F_T = \mu v_x^2 = (64.9 \times 10^{-3}\ \text{kg/m})(129\ \text{m/s})^2$$

$$F_T = 1.08 \times 10^3\ \text{N}$$

CHECK and THINK

To determine if the wire will break, find the tensile stress $\sigma = F_T/A$ (Eq. 14.9) resulting from this force.

$$\sigma = \frac{F_T}{A} = \frac{F_T}{\pi r^2} = \frac{F_T}{\pi(d/2)^2}$$

$$\sigma = \frac{(1.08 \times 10^3\ \text{N})}{\pi(4.8 \times 10^{-3}\ \text{m}/2)^2}$$

$$\sigma = 6.0 \times 10^7\ \text{N/m}^2$$

Because the tensile strength of steel (3.8×10^8 N/m^2) is greater than the tensile stress on the wire, the piano wire does not break.

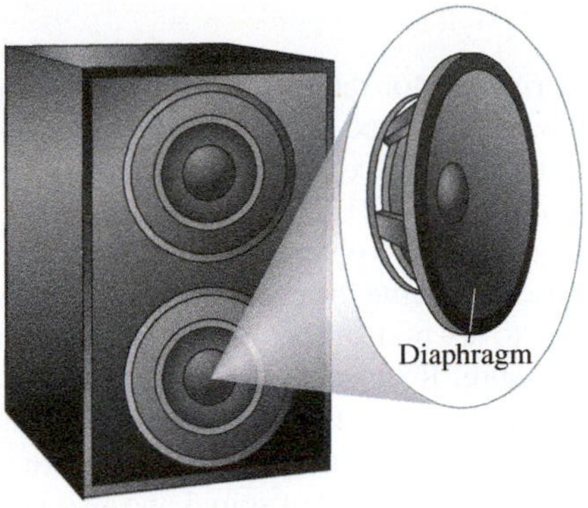

FIGURE 17.15 A stereo speaker creates longitudinal waves in air.

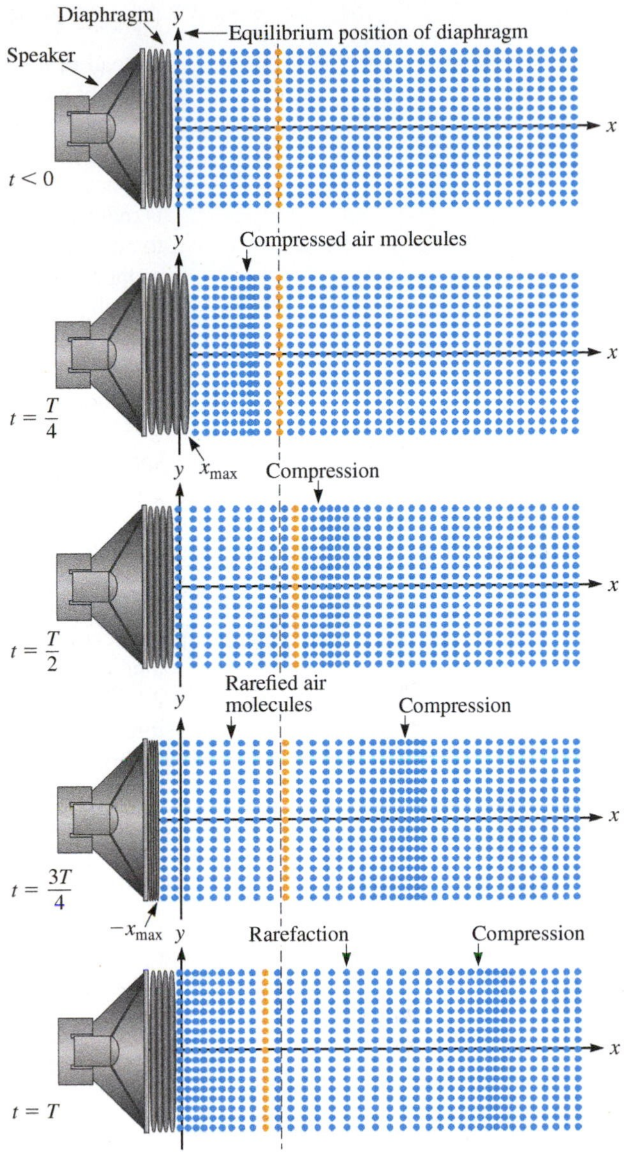

FIGURE 17.16 Longitudinal wave in a column of air near a speaker diaphragm. The gold molecules move back and forth over a short distance along the x axis. (The relative shape and size of the diaphragm are exaggerated.)

17-5 Sound: Special Case of a Traveling Longitudinal Wave

A basic speaker is a relatively simple device (Fig. 17.15). A diaphragm made of a lightweight material such as paper oscillates back and forth, creating longitudinal waves in air known as **sound waves** or **sound**.

Imagine attaching a speaker to the end of a tube filled with air. This scenario could be a model for one of Shannon's earbuds (case study, page 487) at the end of an ear canal. To keep our model simple, we'll ignore the diaphragm's curvature. Before the diaphragm begins to oscillate (that is, at $t < 0$), air molecules are uniformly distributed throughout the tube. The air is modeled as a collection of particles (Fig. 17.16).

The diaphragm begins by moving to the right in the positive x direction, so at $t = T/4$, the diaphragm is at x_{max}, and the air molecules close to the diaphragm are compressed. The air molecules farther downstream are not yet affected by the diaphragm's motion.

The air molecules oscillate back and forth along the x axis much as the coils in Figure 17.10. No molecule moves down the entire length of the tube. The pattern of compression moves in the positive x direction, but each individual molecule oscillates only around its own equilibrium position. Examine the gold molecules in Figure 17.16; they move slightly to the right and left, but never are very far from their original positions.

At $t = T/2$, the diaphragm passes through its equilibrium position as it moves in the negative x direction. At $t = 3T/4$, the diaphragm is at $-x_{max}$, and a region of low density (rarefied) air molecules has formed to the right of the diaphragm. The compression has also moved farther to the right. By $t = T$, one cycle is complete. Both a compression and a rarefaction are propagating to the right, and a new compression is just forming near the diaphragm.

If the diaphragm continues to move back and forth in SHM, the resulting sound wave is harmonic, and we would hear a single tone or note. Usually, a speaker is used to transmit spoken words or music, so the motion of a diaphragm is usually much more complicated than simple harmonic. A detailed study of such complicated motion is beyond the scope of this textbook.

CONCEPT EXERCISE 17.4

CASE STUDY **Sound from Earbuds**

In the case study (page 487), Avi models the sound from an earbud as sand from a sandblaster. Is this model a good model for sound? Explain.

Speed of Sound

Instead of a formal derivation for the speed of sound v_s, we use our derivation of the speed of a transverse wave on a rope as a guide. For simplicity, imagine a compression pulse traveling along the positive x axis through a column of fluid such as air (Fig. 17.16). Like the rope in Figure 17.12, the air molecules are distorted by the pulse. In this case, as the pulse passes, the volume occupied by a particular group of molecules is reduced. The decrease in volume means an increase in pressure. This pressure works to restore the medium to its undisturbed state. The fractional change in volume is proportional to the pressure. Recall that the bulk modulus is the constant of proportionality in Equation 15.7: $P = B(\Delta V/V_i)$. For our work here, we can think of the bulk modulus B as a measure of the medium's ability to restore its original configuration.

To find an expression for the speed of sound, we also need a measure of the fluid's "inertia." Here, that "inertia" is its density (mass per unit volume) ρ.

Using the conceptual relationship we found in Equation 17.12 when deriving the speed of a wave on rope, the speed of sound in a fluid depends on the fluid's density ρ (inertia) and bulk modulus B (restoring ability):

$$v_s = \sqrt{\frac{\text{restoring ability}}{\text{inertia resisting the return to equilibrium}}} = \sqrt{\frac{B}{\rho}} \tag{17.13}$$

According to Equation 17.13, sound travels faster in a fluid with lower density than one with higher density. The speed of sound is also higher in a less compressible fluid (one with greater B). Although water is denser than air, it is less compressible than air; as a result, the speed of sound is higher in water than in air.

For an ideal gas, the ratio B/ρ depends on the gas temperature, so the speed of sound depends on temperature. For a sound wave in room-temperature air, $v = 343$ m/s. At other air temperatures T_C (measured in Celsius), the speed of sound is given by

$$v = 331\ \text{m/s}\sqrt{1 + \frac{T_C}{273°\text{C}}} \tag{17.14}$$

When sound waves propagate in a solid rod, the longitudinal compression causes a slight transverse expansion. For a fluid, there is no such transverse motion. The speed of sound in a solid rod is given by a relation similar to Equation 17.13 except that the bulk modulus is replaced by Young's modulus (Section 14-4):

$$v_s = \sqrt{\frac{Y}{\rho}} \tag{17.15}$$

Table 17.1 provides the speed of sound in several media.

TABLE 17.1 Speed of sound in various media.

Medium	Speed of sound, v_s (m/s)
Air (0°C)	331.45
Air	343.37
Aluminum (rolled)	6420
Brass	4700
Helium	965
Human tissues at body temperature	1540
Hydrogen	1284
Glass	5640
Mercury	1450
Seawater	1531
Steel	5960
Water (0°C)	1402
Water	1496.7

Speed of sound is at room temperature except where indicated. Air is assumed to be dry.

CONCEPT EXERCISE 17.5

The bulk modulus of water is 2.2×10^9 Pa (Table 15.2). The density of water is $10^3\ \text{kg/m}^3$ (Table 15.1). Find the speed of sound in water and compare your answer with the value given in Table 17.1.

Pressure Waves

So far, we've been thinking of a harmonic sound wave in a tube (Fig. 17.16) like the longitudinal harmonic wave on the spring in Figure 17.10. When studying sound, though, it is not convenient or practical to account for the displacement of the many molecules involved. Instead, we measure changes in the fluid's density or pressure.

So, let's think of sound as a pressure wave passing through the medium. Before the wave passes through the region, the pressure equals the equilibrium pressure P_0 of the fluid. We denote a change in the pressure relative to the equilibrium pressure as $\Delta P = P - P_0$. In a sound wave, the pressure oscillates between $P_0 + \Delta P_{max}$ and $P_0 - \Delta P_{max}$.

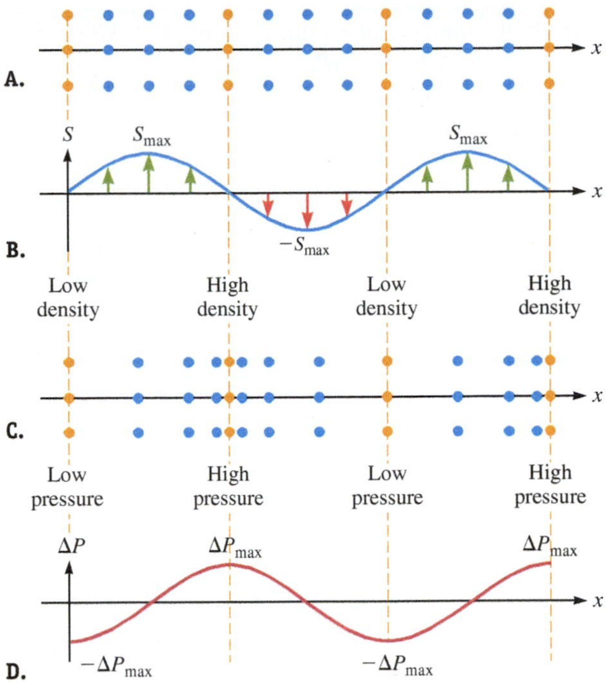

FIGURE 17.17 **A.** Initially, the particles are uniformly distributed. **B.** The graph of S versus x gives the displacement of the particles. **C.** The particles have oscillated into a new arrangement. **D.** The density and pressure are highest where $S = 0$ and are lowest when S is at a maximum or minimum.

Now, to come up with a mathematical expression for the pressure wave, let's return to the speaker causing air molecules to oscillate in a tube. Before the speaker is turned on, air molecules are evenly distributed (Fig. 17.17A). After the speaker has been on for some time, a sound wave travels in the positive x direction, and we can use Equation 17.6 to find the horizontal displacement $S(x, t)$ of the molecules:

$$S(x, t) = S_{max} \sin (kx - \omega t) \tag{17.6}$$

Figure 17.17B shows a graph of S versus x at a particular time t. Figure 17.17C shows the corresponding position of air molecules at that time. Figure 17.17 is very similar to Figure 17.11 except that instead of a single row of particles to model the spring, the air molecules are modeled by a three-dimensional grid of particles.

Figure 17.17C indicates regions of high and low density, and pressure is directly proportional to density. High-density regions are also high-pressure regions, and low-density regions are low-pressure regions.

Figure 17.17D shows ΔP versus x. The relative pressure ΔP is zero when the displacement S is a maximum or minimum. When the displacement is zero, the relative pressure is a maximum or minimum. The relative pressure is 90° or $\pi/2$ radians out of phase with the displacement, so the pressure wave corresponding to Equation 17.6 is

$$\Delta P = \Delta P_{max} \sin\left(kx - \omega t - \frac{\pi}{2}\right)$$

$$\Delta P = \Delta P_{max} \cos (kx - \omega t) \tag{17.16}$$

PRESSURE WAVE **Special Case**

where maximum relative pressure is given by

$$\Delta P_{max} = \rho \omega v_s S_{max} \tag{17.17}$$

This maximum relative pressure (also called "pressure amplitude") depends on properties of the medium (density and speed of sound) and on properties of the wave (amplitude and angular frequency). Another way to think about it is that the change in pressure—a macroscopic quantity—caused by the wave depends on the microscopic motion (amplitude and angular frequency) of the molecules.

EXAMPLE 17.4 CASE STUDY No Whispering

Hair cells inside your ear convert sound waves into electrical signals that are sent to your brain. Each hair cell has about 100 closely packed tufts of hair. Even an extremely small displacement of one of these tufts, roughly 10^{-11} m, can send a signal to the brain. Suppose you have just barely heard a whisper at $f = 1000$ Hz. Assume the displacement amplitude of the sound wave is $S_{max} = 1 \times 10^{-11}$ m, $v_s = 343$ m/s (speed of sound in air), and the density of air is 1.3 kg/m^3 (Table 15.1). Find the pressure amplitude ΔP_{max}. Compare your answer to atmospheric pressure at sea level.

INTERPRET and ANTICIPATE

We expect to find a numerical answer of the form $\Delta P_{max} =$ _____ Pa.

SOLVE

Use Equation 16.2 to find the angular frequency.	$\omega = 2\pi f = 2\pi(1000 \text{ Hz}) = 6280 \text{ rad/s}$

Example continues on page 500 ▶

Substitute numerical values into Equation 17.17 to find ΔP_{max}.	$\Delta P_{max} = \rho\omega v_s S_{max} = (1.3\ \text{kg/m}^3)(6.3 \times 10^3\ \text{rad/s})(343\ \text{m/s})(1 \times 10^{-11}\ \text{m})$ $\Delta P_{max} = 3 \times 10^{-5}\ \text{Pa}$

CHECK and THINK

Our result is in the expected form. Atmospheric pressure at sea level is 1 atm ≈ 1×10^5 Pa (Section 15-3). The quietest whisper you can hear has a pressure amplitude 10 orders of magnitude lower than the normal air pressure, making the ear a very sensitive pressure detector.

EXAMPLE 17.5 CASE STUDY No Shouting

A loud sound can permanently damage your ears. The maximum pressure amplitude ΔP_{max} the human ear can withstand is roughly 30 Pa at $f = 1000$ Hz. What is the corresponding displacement of the hair tufts? Use the same speed of sound and air density as in Example 17.4 and compare your answer with the value of S_{max} given there.

INTERPRET and ANTICIPATE

This example is similar to Example 17.4, but here we are given ΔP_{max} and need to find S_{max}. We expect S_{max} to be greater than 10^{-11} m.

SOLVE

Solve Equation 17.17 for S_{max}.	$\Delta P_{max} = \rho\omega v_s S_{max}$ (17.17) $S_{max} = \dfrac{\Delta P_{max}}{\rho\omega v_s}$
Substitute numerical values.	$S_{max} = \dfrac{30\ \text{Pa}}{(1.3\ \text{kg/m}^3)(6.3 \times 10^3\ \text{rad/s})(343\ \text{m/s})}$ $S_{max} = 1.1 \times 10^{-5}\ \text{m}$

CHECK and THINK

A displacement of 10^{-5} m is about the thickness of a red blood cell. It may be hard to believe that such a small displacement of a hair can cause permanent damage to your hearing, but this displacement is six orders of magnitude larger than the minimum displacement required for you to hear a whisper. Not only is the ear a very sensitive pressure detector, it has an amazing range. It is worth protecting, so Shannon should heed the warning labels on the earbuds.

17-6 Energy Transport in Waves

A sound wave such as one from your earbuds carries energy and momentum from the speaker, but no air molecules travel from the speaker to your ears while that happens. Waves transport energy without transporting matter, and this statement is true of all waves. For example, in Figure 17.2, each coil oscillates around its equilibrium position, but no matter is transported along the spring. The hand, though, does work on a few coils, increasing their mechanical energy. This energy is transported along the spring, causing the bell at the end to oscillate when the wave arrives. In this section, we find expressions for mechanical energy and power transmitted by a wave.

Energy and Power of a Transverse Harmonic Wave

Particles in a medium oscillate as a harmonic wave passes through. Because the particles are in motion, we can associate kinetic energy with their motion. Also, as it passes, a wave distorts the medium (the system), changing its configuration. A change in configuration is associated with a change in potential energy. So, when a wave exists in a medium, the medium (the system) has mechanical energy (kinetic energy plus potential energy).

Consider a harmonic transverse wave on rope. We can find the energy E_λ associated with a portion of the rope that is one wavelength long. Each particle of the rope behaves as an SHO, and from Equation 16.37, the mechanical of an SHO is $E = \frac{1}{2}my_{max}^2\omega^2$. It doesn't make sense to consider the mass of each particle that makes up a rope. Instead, we work with the linear mass density μ (mass per unit length). The mass in one wavelength's portion of the rope is then found by multiplying the linear mass density by the wavelength: $m = \mu\lambda$. So, the average mechanical energy in one wavelength of a transverse harmonic wave is

$$E_\lambda = \frac{1}{2}\mu\omega^2 y_{max}^2 \lambda \quad (17.18)$$

Often, we are not interested in the energy associated with a wave, but in the rate at which a wave *transfers* energy. Think of one particle (say the gold bead) as a transverse wave on a rope passes (Fig. 17.8A). The spring at the end of the rope does work on the end bead, increasing its mechanical energy. This mechanical energy travels along the rope. As one wavelength of the wave passes the gold bead, mechanical energy is transferred to that bead. At first, the bead is at rest with no kinetic energy. As the wave passes, the bead moves, gaining kinetic energy. The bead momentarily stops at the top, so its kinetic energy is momentarily zero. The bead then picks up speed and kinetic energy. After passing through equilibrium, it slows down again, losing kinetic energy. The bead completes one full cycle with the passage of each wavelength of the wave. So, E_λ is also the amount of mechanical energy transferred (to the gold bead) in one period T.

To find the average power P transferred by the wave, we divide the energy transferred in one period by one period:

$$P_{av} = \frac{\Delta E}{\Delta t} = \frac{E_\lambda}{T} = \frac{\frac{1}{2}\mu\omega^2 y_{max}^2 \lambda}{T}$$

The symbol P is used for both power and pressure. Be sure to get the proper meaning from the context.

Substitute Equation 17.8 to express the power transferred by a transverse harmonic wave in terms of its propagation speed v_x:

$$P_{av} = \frac{1}{2}\mu\omega^2 y_{max}^2 v_x \quad (17.19)$$

Energy and Power of Sound

We can use our results for the transverse harmonic wave to find the energy and power in a harmonic sound wave. Again, start with Equation 16.37 for the energy of an SHO, $E = \frac{1}{2}my_{max}^2\omega^2$. Just as the beads on the rope oscillate between $+y_{max}$ and $-y_{max}$ (Fig. 17.8A), the particles (molecules) in the tube of air oscillate between $+S_{max}$ and $-S_{max}$ (Figs 17.16 and 17.17). Clearly, we need to replace y_{max} with S_{max}.

Next, we need the mass m of a small volume of air of density ρ. Consider a sound wave in a column of cross-sectional area A (Fig. 17.18). The volume of the box of length λ is $V = A\lambda$. So, the mass in the box is

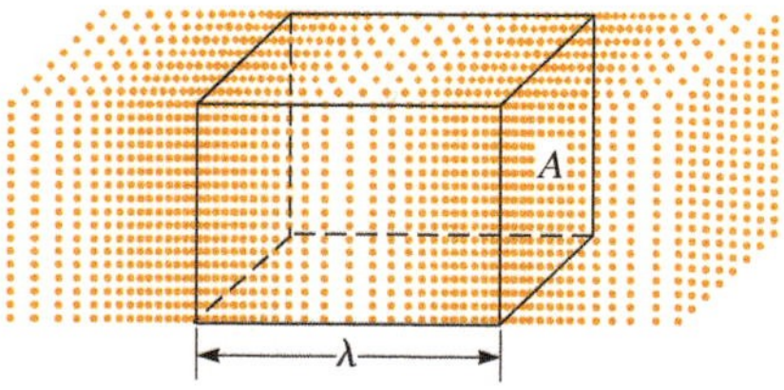

FIGURE 17.18 A sound wave in a column of air with cross-sectional area A. The volume of the box of length λ is $V = A\lambda$.

$$m = \rho V = \rho A\lambda$$

The mechanical energy in one wavelength of a harmonic sound wave is

$$E_\lambda = \frac{1}{2}\rho A\omega^2 S_{max}^2 \lambda \quad (17.20)$$

As before, we find the average rate at which energy is transferred by dividing by the period. So, the average power transmitted by sound is given by

$$P_{av} = \frac{E_\lambda}{T} = \frac{\frac{1}{2}\rho A\omega^2 S_{max}^2 \lambda}{T}$$

or

$$P_{av} = \frac{1}{2}\rho A\omega^2 S_{max}^2 v_s \quad (17.21)$$

where v_s is the speed of sound.

Comparing the average power transmitted by a harmonic wave on a rope (Eq. 17.19) with that transmitted by a harmonic sound wave (Eq. 17.21), shows that they have much in common. In both cases, the average power transmitted depends on properties of the medium (mass density and speed) as well as properties of the wave itself (amplitude and angular frequency). Perhaps it is no surprise that a dense medium with a high wave speed or a wave with a large amplitude and high frequency transmits power effectively.

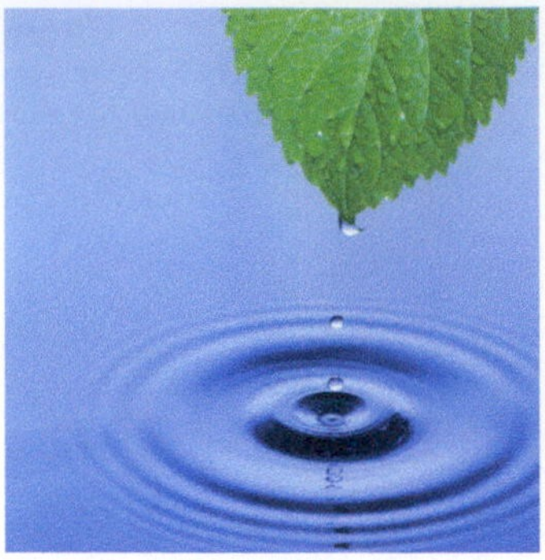

A.

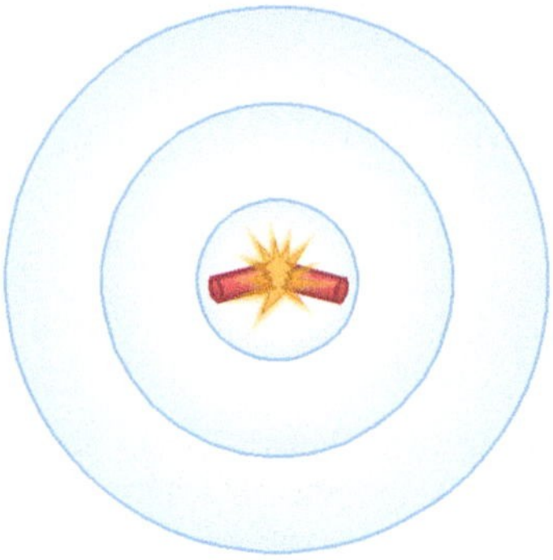

B.

FIGURE 17.19 **A.** A two-dimensional water wave on the surface of a still pond. **B.** A three-dimensional sound wave traveling outward from a source at its center.

CONCEPT EXERCISE 17.6

CASE STUDY **Sound from Earbuds Revisited**

In the case study (page 487), Avi models the sound from an earbud as sand from a sandblaster. In Concept Exercise 17.4, we argued that no air molecules are shot out of a speaker. Sound does not transmit particles, but it can damage your hearing. What does sound transmit and how can it damage your hearing? *Hint*: Examples 17.4 and 17.5 briefly explain human hearing.

17-7 Two- and Three-Dimensional Waves

So far, we have studied one-dimensional waves (those that travel along a single, straight axis). When a drop of water falls into a still pond, it creates circular ripples that travel along the two-dimensional plane of the water's surface (Fig. 17.19A). If a firecracker explodes, it creates spherical ripples of sound traveling outward in three dimensions (Fig. 17.19B).

Figure 17.20 represents the circular waves in Figure 17.19A or a cross section of the spherical waves in Figure 17.19B. The source of the disturbance is shown as a dot at the center of the wave. Such a source is known as a **point source**. Circles in Figure 17.20 represent the wave crests, also called the **wave fronts**. The distance between wave fronts is one wavelength.

Like a photo, Figure 17.20 shows the traveling wave at just one particular instant. The wave fronts move outward and expand as the wave travels. **Rays** are lines perpendicular to the wave fronts. The arrowhead on a ray indicates the direction of the wave's motion. Rays are not vectors; to distinguish them from vectors, we draw the arrowhead on the line, but not at the end.

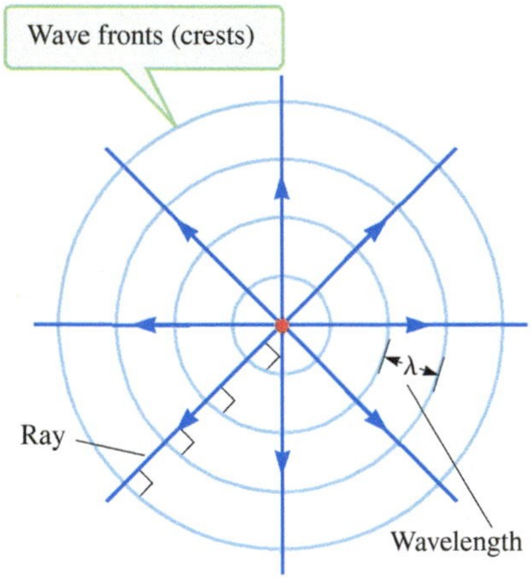

FIGURE 17.20 The wave fronts of a spherical or circular wave. The rays are perpendicular to the wave fronts.

In a small region far from the sources in Figures 17.19 and 17.20, the wave fronts are nearly parallel. **Plane waves** have straight, flat wave fronts as shown in Figure 17.21A. Waves far from a point source can be modeled as plane waves. You could also make a plane wave by using a large, straight source such as a rod oscillating up and down with its length parallel to the surface of the water as in Figure 17.21B.

Intensity and Loudness

If you hear fireworks exploding in midair or a clap of thunder, the sound is louder the closer you are to the source. A thunderclap can be modeled as a source that emits sound uniformly in all directions. The energy and power transmitted by the sound wave is uniformly distributed over any sphere centered on the source (Fig. 17.22). **Intensity** I is defined as the average power P_{av} per unit area A perpendicular to the direction of propagation:

INTENSITY ✪ **Major Concept**

$$I \equiv \frac{P_{av}}{A} \tag{17.22}$$

Intensity is a scalar with SI units of watts per square meter.

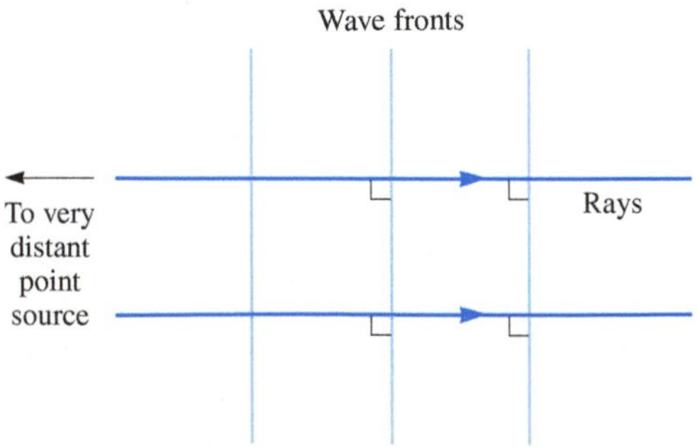

A.

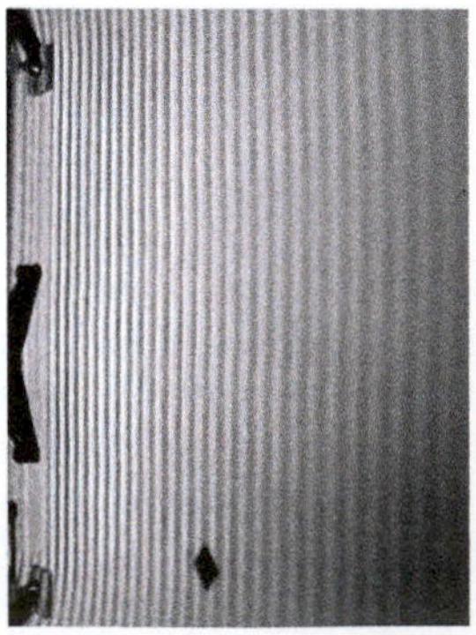

B.

FIGURE 17.21 **A.** The wave fronts of a plane wave are straight lines or planes. **B.** Plane waves can be generated on a water surface by oscillating a horizontal rod.

The same power passes through each sphere shown in Figure 17.22. The area of the sphere is greater for a more distant observer ($A_2 > A_1$), however, so the intensity of the sound wave is lower, $I_2 < I_1$. The intensity of a point source decreases as distance r from the source increases:

$$I = \frac{P_{av}}{4\pi r^2} \tag{17.23}$$

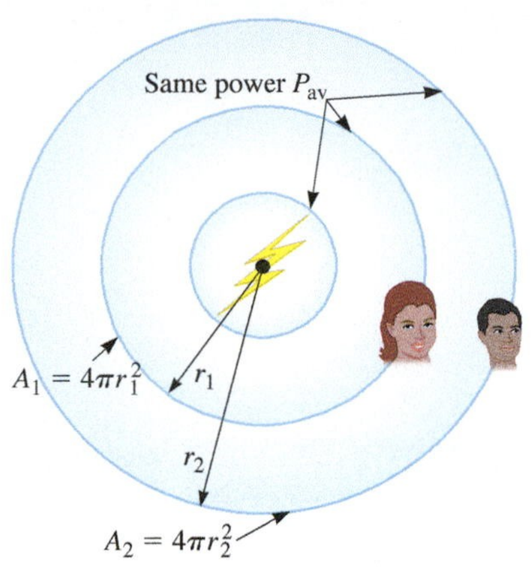

FIGURE 17.22 Intensity decreases as the distance from the source increases.

The average power of a sound wave is given by Equation 17.21, so the intensity can be written as

$$I = \frac{P_{av}}{A} = \frac{\frac{1}{2}\rho A\omega^2 S_{max}^2 v_s}{A} = \frac{1}{2}\rho\omega^2 S_{max}^2 v_s$$

We can eliminate the displacement amplitude S_{max} in favor of the pressure amplitude by using Equation 17.17:

$$I = \frac{1}{2}\rho\omega^2\left(\frac{\Delta P_{max}}{\rho\omega v_s}\right)^2 v_s$$

Simplifying, the intensity of a sound wave is given by

$$I = \frac{1}{2}\frac{\Delta P_{max}^2}{\rho v_s} \tag{17.24}$$

The intensity range of human hearing spans 12 orders of magnitude, from just barely audible ($I_0 = 10^{-12}$ W/m^2) to painful ($I = 1$ W/m^2). Human response to or perception of sound depends logarithmically on the intensity. American inventor Alexander Graham Bell defined **sound level** β in terms of base 10 logarithms as

Remember that P_{av} is average **power** and ΔP_{max} is the **pressure amplitude**.

$$\beta \equiv 10 \log\left(\frac{I}{I_0}\right) \tag{17.25}$$

where $I_0 = 10^{-12}$ W/m^2. Sound level is measured in **decibels** (dB), although the quantity defined in Equation 17.25 is actually dimensionless. The threshold of human hearing is

$$\beta = 10 \log\left(\frac{I_0}{I_0}\right) = 0 \text{ dB}$$

and sound is painful at

$$\beta = 10 \log\left(\frac{1 \text{ W/m}^2}{10^{-12} \text{ W/m}^2}\right) = 120 \text{ dB}$$

Table 17.2 provides the sound level of a number of sources.

TABLE 17.2 Typical sound levels.

Intensity I (W/m^2)	Sound level β (dB)	Examples	Time required for onset of hearing damage
10^{-12}	0	Barely audible	
10^{-9}	30	Soft whisper	
10^{-8}	40	Quiet bedroom; mosquito buzzing	
10^{-7}	50	Gentle breeze; refrigerator	
10^{-6}	60	Normal conversation	
10^{-5}	70	Noisy restaurant; vacuum cleaner	Some damage if continuous
10^{-4}	80	City traffic; factory noise	More than 8 hours
10^{-3}	90	Truck traffic; power lawn mower	Less than 8 hours
10^{-2}	100	Chain saw; subway	2 hours
1	120	Rock concert; nearby thunderclap	Immediate danger
10^2	140	Machine gun; nearby jet plane	Immediate danger
10^4	160	Rocket launchpad	Hearing loss inevitable

People can hear sounds in the frequency range from 20 Hz to 20,000 Hz. This range varies from person to person and generally decreases with increasing age. Sound with frequency higher than the upper limit of human hearing ($f > 20{,}000$ Hz) is known as **ultrasound**, and sound with a frequency below the lower limit ($f < 20$ Hz) is known as **infrasound**. Seismographs that measure earthquakes operate in the infrasound, and large animals such as elephants and whales communicate with infrasound. Ultrasound is used in medical applications to probe inside the human body and to deliver energy inside the body without surgery. Some animals such as dogs and bats can hear ultrasound.

EXAMPLE 17.6 CASE STUDY Ultrasound to the Rescue

Mineral deposits such as calcium can build up in the kidneys to form kidney stones. Although some kidney stones are as small as a grain of sand, others have been the size of golf balls. Often, the patient does not notice kidney stones until they move into the ureter (Fig. 17.23), causing a great deal of pain. Before the mid-1980s, kidney stones were surgically removed, but now physicians use lithotripsy to break up the stones into smaller pieces that easily pass through the ureter with very little pain. In lithotripsy, ultrasound pulses deliver energy to the kidney stones, as Cameron mentioned (page 487). Typically, a patient is exposed to thousands of pulses during a 45-minute procedure.

Assume the average intensity of a pulse near the lithotripter is $I_0 = 6.5 \times 10^8 \text{ W/m}^2$ and the face area of the pulse is $A = 1.9 \times 10^{-4} \text{ m}^2$. In a single medical procedure, 2000 such pulses are used to break up a kidney stone 15.5 mm in diameter that is 10 cm from the lithotripter.

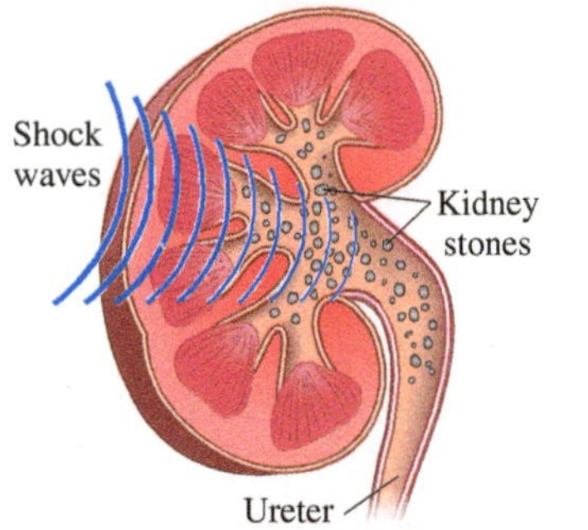

FIGURE 17.23 Kidney stones form in the kidney and pass into the ureter.

A What is the average power output of a single pulse?

INTERPRET and ANTICIPATE

Ultrasound is just high-frequency sound, so all the equations developed for sound can be applied to ultrasound. We expect a numerical result with the SI units of power (watts).

SOLVE

Rearrange Equation 17.22 to find P_{av}.

$$P_{av} = IA = (6.5 \times 10^8 \text{ W/m}^2)(1.9 \times 10^{-4} \text{ m}^2)$$

$$P_{av} = 1.2 \times 10^5 \text{ W}$$

CHECK and THINK

Our result has the expected units. For comparison, a single incandescent lightbulb in your room might put out 100 W. A single pulse of the lithotripter is approximately equivalent to 1200 such lightbulbs, or roughly the number of lightbulbs in a small public building.

B What is the intensity of the ultrasound wave near the kidney stone?

INTERPRET and ANTICIPATE

Intensity decreases with distance, so we expect a numerical value that is smaller than the intensity near the lithotripter ($I < I_0$).

SOLVE

Use Equation 17.23 and the result from part A to find the intensity of the ultrasound wave at the kidney stone.

$$I = \frac{P_{av}}{4\pi r^2} = \frac{1.2 \times 10^5 \text{ W}}{4\pi (0.10 \text{ m})^2}$$

$$I = 9.5 \times 10^5 \text{ W/m}^2$$

CHECK and THINK

As expected, $I < 6.5 \times 10^8 \text{ W/m}^2$.

C What is the average power delivered to the kidney stone in one pulse?

INTEPRET and ANTICIPATE
We need the area of the kidney stone and the intensity of the pulse near the kidney stone from part B.

SOLVE
Solve Equation 17.22 for P_{av} and use the kidney stone's diameter $d = 15.5$ mm.

$$P_{av} = IA = I(\pi d^2/4)$$

$$P_{av} = (9.5 \times 10^5 \text{ W/m}^2)\frac{\pi(15.5 \times 10^{-3} \text{ m})^2}{4}$$

$$P_{av} = 1.8 \times 10^2 \text{ W}$$

CHECK and THINK
Because the intensity drops off dramatically with increasing distance, the power delivered to the kidney stone is roughly equivalent to just two 100-W incandescent lightbulbs. The remaining power is delivered to tissues in the body. According to some studies, this power can damage healthy tissues.

D If the duration of a single pulse is 5.2 μs, what is the total energy delivered to the kidney stone by 2000 pulses?

INTERPRET and ANTICIPATE
First, find the energy delivered by a single pulse. Then, multiply by 2000 to find the energy delivered during the entire medical procedure.

SOLVE
The energy E delivered by a single pulse is the product of the power delivered to the kidney stone and the duration Δt of the pulse.

$$E = P_{av}\Delta t = (1.8 \times 10^2 \text{ W})(5.2 \times 10^{-6} \text{ s})$$

$$E = 9.4 \times 10^{-4} \text{ J}$$

Multiply by 2000 to find the total energy delivered to the kidney stone.

$$E_{tot} = 2000E = (2000)(9.4 \times 10^{-4} \text{ J}) = 1.9 \text{ J}$$

CHECK and THINK
Although 2 J may not seem like a lot of energy, laboratory experiments have shown that even 0.5 J can break up artificial kidney stones.

EXAMPLE 17.7 Buying the Right Speakers

A certain manufacturer's middle-of-the-line speakers each have an input power $P_{in} = 60$ W and reach a maximum sound level $\beta = 95$ dB at 1 m in front of each speaker. Use this information to estimate the power output P_{out} in the form of sound and the efficiency (P_{in}/P_{out}) of this manufacturer's product.

INTERPRET and ANTICIPATE
Use the sound level to find intensity (Eq. 17.25). To find the power, assume the intensity is uniform over a sphere of radius 1 m. Speakers are designed to send sound waves out primarily in the forward direction, so our assumption means that we will overestimate P_{out}. We expect that some energy is lost in producing sound, so $P_{out} < P_{in}$, and the efficiency should be less than 1.

Example continues on page 506 ▶

SOLVE

Use Equation 17.25 to relate the intensity at 1 m from the speaker to the sound level at that distance. The reference intensity I_0 is always 10^{-12} W/m². (The inverse of the log function is exponential base 10; see Appendix A.)

$$\beta = 10\log\left(\frac{I}{I_0}\right)$$

$$I = I_0 10^{\beta/10} = (10^{-12}\ \text{W/m}^2)10^{95/10}$$

$$I = 3.2 \times 10^{-3}\ \text{W/m}^2$$

Intensity is power per unit area. Assume the speaker emits sound uniformly in all directions and use Equation 17.23 to find the power output.

$$I = \frac{P_{\text{out}}}{4\pi r^2} \qquad (17.23)$$

$$P_{\text{out}} = 4\pi r^2 I = 4\pi(1\ \text{m})^2(3.2 \times 10^{-3}\ \text{W/m}^2)$$

$$P_{\text{out}} = 4.0 \times 10^{-2}\ \text{W}$$

Find the efficiency by dividing the output power by the input power.

$$\frac{P_{\text{out}}}{P_{\text{in}}} = \frac{4.0 \times 10^{-2}\ \text{W}}{60\ \text{W}} = 7 \times 10^{-4} = 0.07\%$$

CHECK and THINK

As expected, the efficiency is less than 1 (and perhaps smaller than you expected). Basic home speakers have an efficiency of around 0.5 to 4%, so this speaker is much less efficient than the average. You may wish to purchase a more efficient speaker than this one.

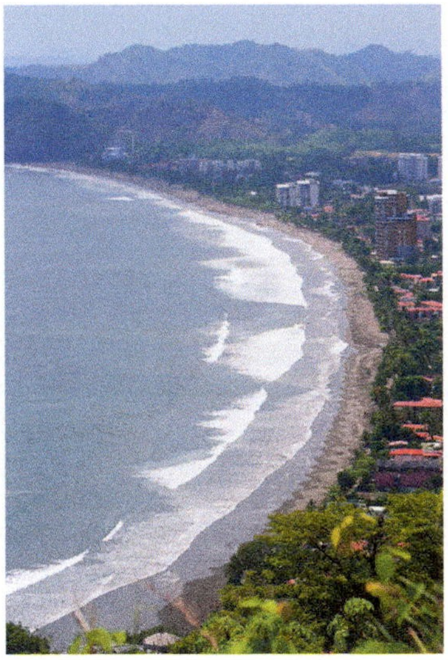

© Jodi Jacobson/iStockphoto.com

FIGURE 17.24 Ocean waves are parallel to the shoreline.

17-8 Refraction and Diffraction

Imagine that you wish to share a letter with a person who is on the other side of a heavy, closed door. The only way to get the letter to the person is through the mail slot. Now imagine that you wish to talk to the person on the other side of the door. You would probably be able to carry on a conversation even if neither of you put your ears or mouths near the mail slot because sound waves are able to bend around obstacles and through holes. Particles do not bend in the same way, so the letter must be slipped through the mail slot. In this section, we explore this unique feature, the bending of waves.

Refraction

Have you ever wondered why water waves near a beach are nearly parallel to the shore (Fig. 17.24), but waves in the middle of the sea may be seemingly random? The reason is that as the waves move through shallow water near the shore, they are bent or **refracted**. No matter what their original orientation is, when they get near the shore, they are refracted so that the wave fronts become roughly parallel to the shoreline.

A two- or three-dimensional wave is refracted when it is transmitted from one medium to another medium having a different wave propagation speed. Because the speed of water waves is lower in the shallow water near the shore than in deep water, waves are refracted near the shore. You can picture this phenomenon by imagining a marching band as it encounters a change in medium from pavement to mud (Fig. 17.25A). The band models a plane wave, each row of band members represents a wave front, and the distance between adjacent rows is a wavelength. All the musicians obediently march at the same pace, but the length of their stride is shortened by the mud. As the first few musicians step into the mud, they slow down (Fig. 17.25B). The first few rows are bent. After the entire band has entered the mud (Fig. 17.25C), all the rows are nearly parallel

FIGURE 17.25 A band marches from pavement to mud. Although this analogy is contrived, a real marching band (on a single surface) does turn by this method. The analogy illustrates that as a wave moves from one medium to another, it bends or refracts.

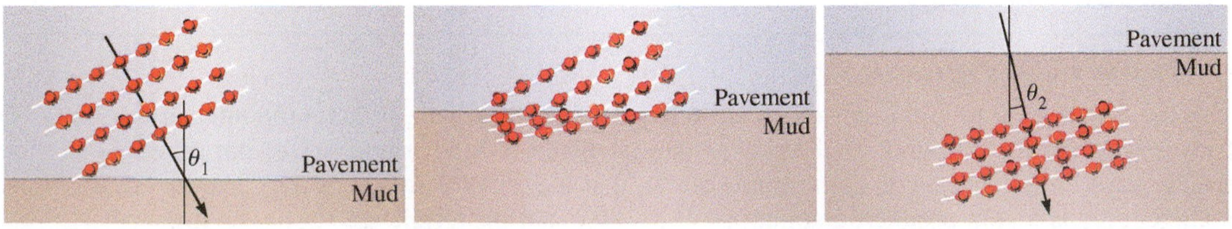

A. **B.** **C.**

to the line that separates the mud from the pavement. In a natural setting such as a seashore, the wave propagation speed drops off slowly and the refraction of the water waves is gradual, but the wave fronts still end up roughly parallel to the shore.

The ray in Figure 17.25A makes an angle θ_1 with the line perpendicular to the pavement–mud boundary. The ray in Figure 17.25C makes an angle θ_2 with the perpendicular, and $\theta_2 < \theta_1$. We can find the mathematical relationship between these two angles by taking a closer look at two of the musicians as they enter the mud (Fig. 17.26). Musician 2 is at the boundary between pavement and mud, and musician 1 must march a distance d_1 before entering the mud. Musician 1 marches on the pavement at speed v_1 parallel to ray 1. In a time interval Δt, he marches the distance $d_1 = v_1\Delta t$ to the boundary. During this same time interval, musician 2 marches in the mud on a path parallel to ray 2. Musician 2's speed in the mud is v_2, so she marches a distance $d_2 = v_2\Delta t$. Figure 17.26 shows two right triangles that share a common hypotenuse h. Using the definition of the sine function,

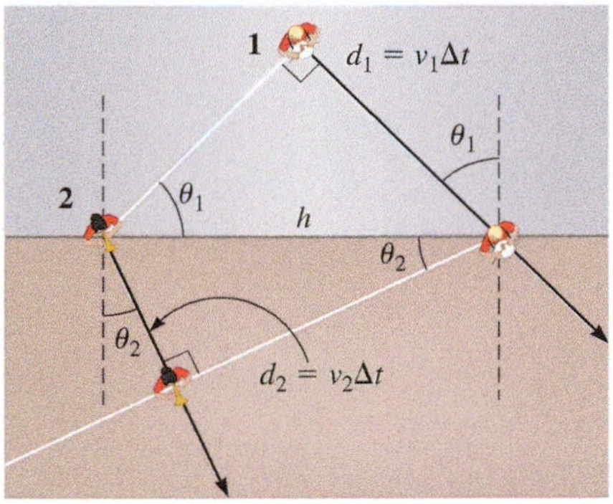

FIGURE 17.26 Two band members approach the mud.

$$h = \frac{d_1}{\sin\theta_1} = \frac{d_2}{\sin\theta_2}$$

Rewriting the distances d_1 and d_2 in terms of the speeds v_1 and v_2,

$$\frac{v_1\Delta t}{\sin\theta_1} = \frac{v_2\Delta t}{\sin\theta_2}$$

After canceling Δt and rearranging, we are left with the **law of refraction**:

LAW OF REFRACTION

Major Concept

$$\sin\theta_2 = \frac{v_2}{v_1}\sin\theta_1 \qquad (17.26)$$

where θ_1 and θ_2 are measured with respect to the **normal**, the line perpendicular to the boundary between medium 1 and medium 2. If the propagation speed in medium 2 is lower than in medium 1 (that is, if $v_2 < v_1$), the ray bends toward the perpendicular, $\theta_2 < \theta_1$. In this case, the wave fronts tend to align with the boundary (see Figs. 17.25 and 17.26). On the other hand, if the propagation speed in medium 2 is higher than in medium 1 (that is, if $v_2 > v_1$) then the ray bends away from the perpendicular, $\theta_2 > \theta_1$.

Diffraction

We just showed that waves bend at an interface between media in which they have different speeds. Waves also bend when they encounter an obstacle that partially blocks the wave fronts. The bending of waves passing through a hole or around an obstacle is known as **diffraction**. The amount of diffraction depends of the relative size of the obstacle or hole compared to the wavelength. If the hole or obstacle is much larger than the wavelength, diffraction is small. If the obstacle or hole is comparable in size to the wavelength, diffraction becomes significant.

DIFFRACTION Major Concept

Diffraction allows waves to transmit a disturbance and its energy around obstacles. Again imagine communicating with a friend on the other side of a closed door: Energy is transferred from your mouth to your friend's ears through sound waves that bend around the door and through the mail slot.

17-9 The Doppler Shift

If you have ever been to an auto race, you have probably noticed that as the cars move toward you, you hear a high-frequency hum that sounds like a swarm of bees. Just after the cars pass, you hear a low-frequency roar. You may also have heard a change in frequency if you have ever waited by a railroad track as a train passed. The pitch of the approaching train whistle sounds high, and the pitch of the receding whistle sounds low.

The change in frequency you perceive in these cases is known as the **Doppler**[1] **shift** and is due to the relative motion between you and the sound source. When the

[1]Christian Doppler, a 19th-century Austrian physicist, predicted that relative motion causes a frequency shift for both sound and light waves.

source moves toward you, you hear a higher frequency than you would were the source stationary with respect to air. When the source moves away from you, you hear a lower frequency than you would were the source stationary. In this section, we focus on sound, but there is a Doppler shift for other waves as well, including water waves and light.

The relative motion is what matters. If you were moving and the source were stationary with respect to the medium, you would still hear a Doppler shift. For example, if you are approaching a parked emergency vehicle with its siren blaring, you hear the siren at a higher frequency than what was originally emitted. After you pass the vehicle, you hear the siren at a lower frequency.

Stationary Source, Moving Observer

Figure 17.27 shows a stationary siren. Hannah is approaching the siren from the right, and Aaron is moving to the left away from the siren. The four parts of the figure show four instants from $t = 0$ to $t = 3T$. The siren creates spherical wave fronts that expand outward. You can see their motion by following the history of one wave front such as the innermost wave front shown in green in Figure 17.27A. In Figure 17.27B, the green wave front has expanded by one wavelength λ. By $t = 3T$, the green wave front has expanded to the edge of the panel in Figure 17.27D.

Initially, Hannah is at the red wave front. If she were not moving, she would have encountered three other wave fronts—orange, yellow, and green—in a time interval of $3T$. So, if Hannah were stationary, she would hear a frequency

$$f = \frac{3\,\text{cycles}}{3T} = \frac{1}{T}$$

During that time interval of $3T$, however, suppose she moves toward the siren a distance 2λ. Her speed is given by

$$v_{\text{obs}} = \frac{2\lambda}{3T}$$

The subscript "obs" stands for observer. We use $v_s = \lambda/T$ (Eq. 17.8) to write her speed in terms of the speed of sound v_s:

$$v_{\text{obs}} = \tfrac{2}{3}v_s$$

which is unrealistically fast, but that won't change our results. Because she is traveling toward the siren, she encounters five wave fronts—orange, yellow, green, blue, and violet—instead of just three. Hannah hears a frequency given by

$$f_{\text{obs}} = \frac{5\,\text{cycles}}{3T} = \frac{5}{3}\frac{1}{T} = \frac{5}{3}f$$

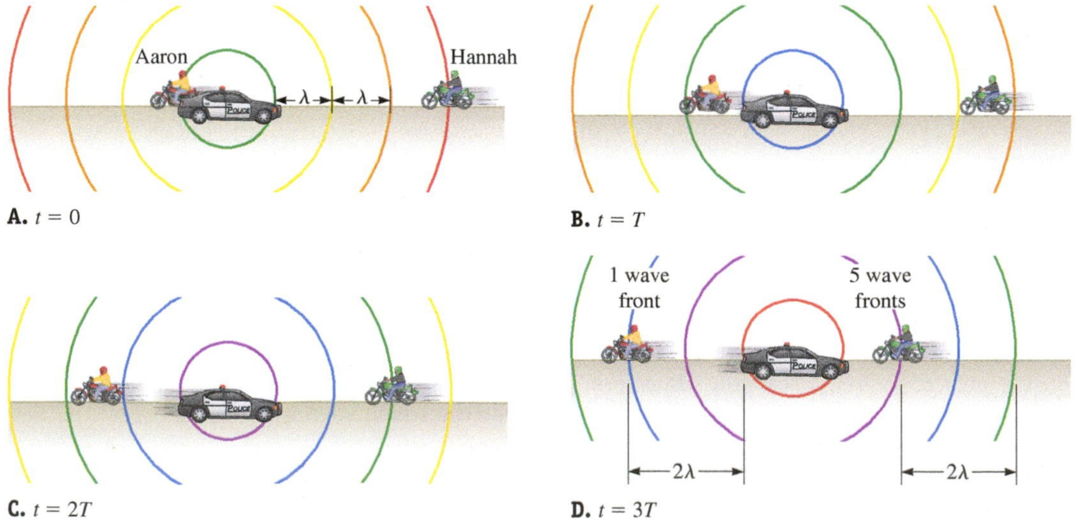

FIGURE 17.27 An observer moving toward a source hears a higher frequency than when standing still, and an observer moving away from a source hears a lower frequency.

Aaron is moving away from the source at the same speed as Hannah, $v_{\text{obs}} = \frac{2}{3}v_s$. Initially, he is at the green wave front (Fig. 17.27A). During the time interval of $3T$, he only encounters one other wave front, the blue one. Thus, the frequency Aaron hears is

$$f_{\text{obs}} = \frac{1 \text{ cycle}}{3T} = \frac{1}{3}f$$

DERIVATION Frequency Observed by Moving Observer

We'd like a general expression for the frequency heard by an observer in terms of the observer's speed and the source's speed. The frequency heard by a moving observer when the source is stationary is:

$$f_{\text{obs}} = \left(\frac{v_s \pm v_{\text{obs}}}{v_s}\right)f \tag{17.27}$$

where we choose the plus sign for an approaching observer and the minus sign for a receding observer. We will show that this expression is consistent with the frequency heard by the people in Figure 17.27.

Because Hannah moves toward the siren at v_{obs}, the relative speed between her and the sound wave is $v_s + v_{\text{obs}}$. She encounters the wave fronts more rapidly than she would were she stationary. She hears a frequency given by Equation 17.8, where v_x is replaced by her relative speed.	$v_x = \lambda f$ $f = \frac{v_x}{\lambda}$ (17.8) $f_{\text{obs}} = \frac{v_s + v_{\text{obs}}}{\lambda}$ (1)
Equation (1) can easily be modified for an observer moving away from the source. In this case, the relative speed is given by $v_s - v_{\text{obs}}$, and only the numerator is modified.	$f_{\text{obs}} = \frac{v_s - v_{\text{obs}}}{\lambda}$ (2)
Combine Equations (1) and (2) to include the possibility of an approaching observer or a receding observer.	$f_{\text{obs}} = \frac{v_s \pm v_{\text{obs}}}{\lambda}$ (3)
The source's wavelength λ is not affected by the observer's motion, so it is still given by Equation 17.8, where v_x is the speed of the sound wave v_s.	$\lambda = \frac{v_s}{f}$ (4)
Substitute Equation (4) into Equation (3) to find the frequency heard by an observer moving toward or away from a source.	$f_{\text{obs}} = \frac{v_s \pm v_{\text{obs}}}{(v_s/f)}$ $f_{\text{obs}} = \left(\frac{v_s \pm v_{\text{obs}}}{v_s}\right)f$ ✓
COMMENTS Check this result for Hannah in Figure 17.27, who is approaching the source at $v_{\text{obs}} = \frac{2}{3}v_s$.	$f_{\text{obs}} = \left(\frac{v_s + \frac{2}{3}v_s}{v_s}\right)f = \left(1 + \frac{2}{3}\right)f$ $f_{\text{obs}} = \frac{5}{3}f$ ✓
Now check this result for Aaron, who is moving away from the source at the same speed.	$f_{\text{obs}} = \left(\frac{v_s - \frac{2}{3}v_s}{v_s}\right)f = \left(1 - \frac{2}{3}\right)f$ $f_{\text{obs}} = \frac{1}{3}f$ ✓

Moving Source, Stationary Observer

If an observer is stationary with respect to the medium but the source is moving, the observer still hears a higher or lower frequency, depending on whether the source is approaching or receding. Figure 17.28 shows a source moving to the right. At time

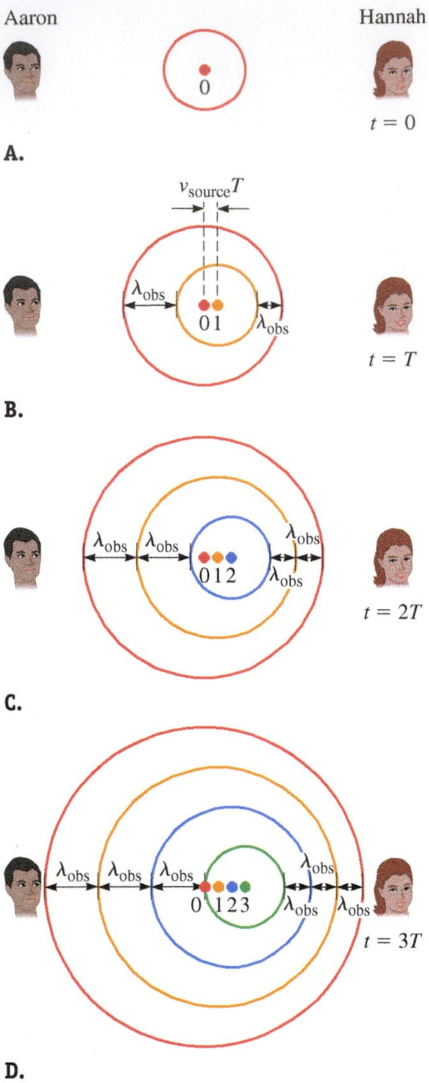

FIGURE 17.28 When the source is moving, an observer in front hears a higher frequency as the source approaches than she would if the source were stationary, and an observer in back hears a lower frequency as the source recedes.

$t = 0$ (Fig. 17.28A), the source is at position 0, and a single wave front shown in red is centered on this position. At $t = T$, the source has moved to position 1. The original red wave front has expanded, and a new wave front shown in orange and centered on position 1 has appeared. Parts C and D of Figure 17.28 show that as the source continues to move to the right, each wave front expands in a sphere centered on its own initial position. The result is that the wave fronts ahead of the source are crowded together and those behind the source are spread apart. Therefore, the wavelength is shorter in front of the source and longer behind the source.

Imagine that Hannah stands directly in front of the source. At $t = 0$, the source emits a wave front that travels outward toward her (Fig. 17.28A). One period later, at $t = T$, the source emits another wave front. If the source did not move, Hannah would find that the distance between wave fronts is λ and the frequency is $f = v_s/\lambda$. The source moves to the right, however, so the wave front emitted at $t = T$ is shifted toward her. She observes a wavelength λ_{obs} as the distance between the two wave fronts, but shortened by the distance that the source moved. The source's speed is v_{source}, so it moves a distance $v_{source}T$ in one period. The wavelength Hannah observes is then given by

$$\lambda_{obs} = \lambda - v_{source}T$$

The frequency she hears is

$$f_{obs} = \frac{v_s}{\lambda_{obs}} = \frac{v_s}{\lambda - v_{source}T}$$

Now use $f = 1/T$ and $v_x = \lambda/T$ (Eqs. 16.1 and 17.8) to write the observed frequency f_{obs} of a source moving at speed v_{source} toward a stationary observer,

$$f_{obs} = \frac{v_s}{(v_s/f) - (v_{source}/f)}$$

$$f_{obs} = \frac{v_s}{v_s - v_{source}} f \tag{17.28}$$

where v_s is the speed of sound and f is the frequency the source would emit if it were stationary.

If the source is moving away from the observer, the observed wavelength is lengthened by the amount $v_{source}T$: $\lambda_{obs} = \lambda + v_{source}T$. The observed frequency is lowered and is given by

$$f_{obs} = \frac{v_s}{v_s + v_{source}} f \tag{17.29}$$

Combining Equations 17.28 and 17.29, we have one expression for a moving source and a stationary observer:

$$f_{obs} = \frac{v_s}{v_s \mp v_{source}} f \tag{17.30}$$

where we choose the minus sign for a source moving toward the observer and the plus sign for a source moving away from the observer.

Source and Observer Both Moving

It is possible that both the source and the observer are moving with respect to the medium. For example, imagine driving along the highway and hearing a siren on a moving police car. Both you (the observer) and the source (the police siren) are moving with respect to the medium (the air). We can combine Equations 17.27 and 17.30 to cover all the possible combinations of source and observer motion. The all-purpose equation is

DOPPLER SHIFT EQUATION
Major Concept

$$f_{obs} = \left(\frac{v_s \pm v_{obs}}{v_s \mp v_{source}}\right) f \tag{17.31}$$

where f_{obs} is the observed frequency, v_s is the speed of sound, v_{obs} is the observer's speed relative to the medium, and v_{source} is the source's speed. The signs of these speeds depend on the relative direction of motion.

1. If the observer moves toward the source, choose the top sign (+) for v_{obs} in the numerator of Equation 17.31. If the source moves toward the observer, choose the top sign (−) for v_{source} in the denominator. Either choice (or both) increases the observed frequency as source and observer approach each other. (It helps to remember that both *t*oward and *t*op begin with *t*.)
2. If the observer moves away from the source, choose the bottom sign (−) in the numerator of Equation 17.31. If the source moves away from the observer, choose the bottom sign (+) in the denominator. Either choice (or both) decreases the observed frequency as source and observer recede from each other.

If the source and the observer are moving at the same speed in the same direction, there is no Doppler shift according to Equation 17.31. For example, if the source is followed by the observer and $v_{obs} = v_{source}$, then

$$f_{obs} = \left(\frac{v_s + v_{obs}}{v_s + v_{source}}\right)f = \left(\frac{v_s + v_{obs}}{v_s + v_{obs}}\right)f$$

$$f_{obs} = f$$

where we have chosen the top sign in the numerator because the observer is heading toward the source and the bottom sign in the denominator because the source is heading away from the observer. So, if the observer is moving along with the source, the observed frequency equals the frequency of a stationary source. (No matter how fast an ambulance is going, the ambulance driver always hears the same frequency from the siren.)

EXAMPLE 17.8 Where's the Fire?

A fire truck's siren has a frequency of 875 Hz. You maintain your speed at 65 mph (29.0 m/s) on a multilane highway as the fire truck comes up from behind at 80 mph (35.7 m/s). Assume the speed of sound in air is 343.4 m/s.

A What frequency do you hear as the fire truck approaches? What frequency do you hear after the truck passes you?

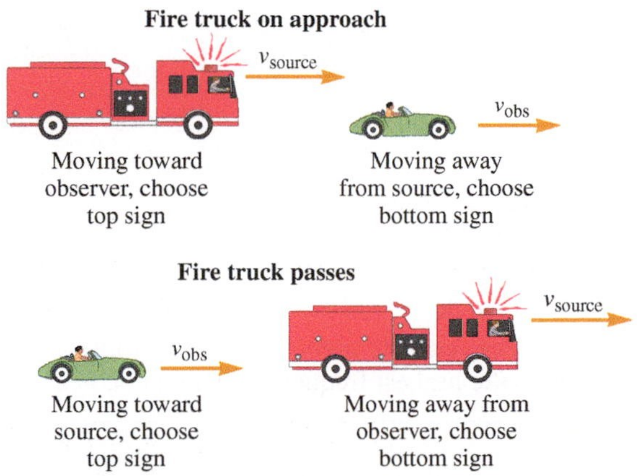

FIGURE 17.29

INTERPRET and ANTICIPATE

Use Equation 17.31 twice with the same speed values, changing signs as appropriate to each situation (fire truck approaches car and after fire truck passes car). A sketch helps in picking out the correct signs (Fig. 17.29). We expect to hear a higher frequency when the fire truck is approaching and a lower frequency after it passes.

SOLVE

Substitute speed values into Equation 17.31 and choose signs, first for when the truck is approaching the car. The source moves toward the observer and that the observer moves away from the source *at the same time*. We need the top sign in the denominator and the bottom sign in the numerator.

$$f_{obs} = \left(\frac{v_s \pm v_{obs}}{v_s \mp v_{source}}\right)f$$

$$f_{obs} = \left(\frac{v_s - v_{obs}}{v_s - v_{source}}\right)f \text{ after choosing signs}$$

$$f_{obs} = \frac{(343.4 \text{ m/s}) - (29.0 \text{ m/s})}{(343.4 \text{ m/s}) - (35.7 \text{ m/s})}(875 \text{ Hz})$$

$$f_{obs} = 894 \text{ Hz} \text{ as fire truck approaches}$$

Example continues on page 512 ▶

CHECK and THINK
As expected, the siren frequency is shifted higher when the fire truck is approaching the observer.

SOLVE
After the truck passes, only the signs change.

$$f_{obs} = \left(\frac{v_s + v_{obs}}{v_s + v_{source}}\right) f$$

$$f_{obs} = \frac{(343.4 \text{ m/s}) + (29.0 \text{ m/s})}{(343.4 \text{ m/s}) + (35.7 \text{ m/s})}(875 \text{ Hz})$$

$$f_{obs} = 860 \text{ Hz} \text{ after truck passes}$$

CHECK and THINK
As expected, the siren frequency is shifted lower when the fire truck is receding from the observer.

B If you are driving on the other side of the highway in the opposite direction as the fire truck, what frequency do you hear as the fire truck approaches? What frequency do you hear after the fire truck passes?

INTERPRET and ANTICIPATE
The solution resembles part A, but we need a new sketch to get the signs correct (Fig. 17.30). We expect an even higher frequency upon approach than in part A because the relative speed of approach is greater. Similarly, we expect an even lower frequency after the truck passes than in part A due to the greater relative speed of recession.

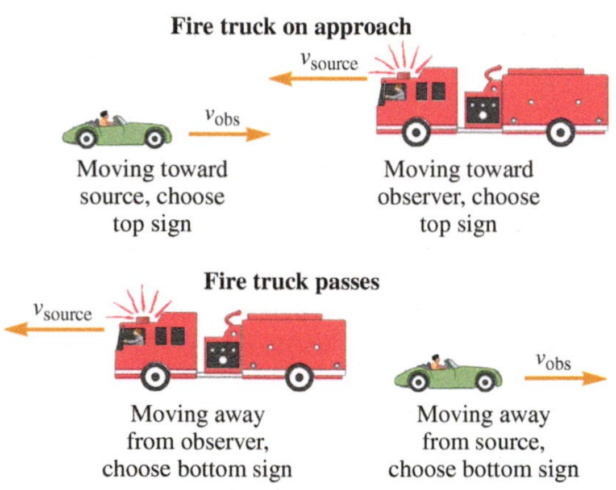

FIGURE 17.30

SOLVE
When the observer and source both move toward each other, choose the top sign for each speed in Equation 17.31.

$$f_{obs} = \left(\frac{v_s \pm v_{obs}}{v_s \mp v_{source}}\right) f$$

$$f_{obs} = \left(\frac{v_s + v_{obs}}{v_s - v_{source}}\right) f$$

$$f_{obs} = \frac{(343.4 \text{ m/s}) + (29.0 \text{ m/s})}{(343.4 \text{ m/s}) - (35.7 \text{ m/s})}(875 \text{ Hz})$$

$$f_{obs} = 1060 \text{ Hz} \text{ as fire truck approaches}$$

CHECK and THINK
As expected, this result is higher than the "approach frequency" in part A. In fact, this result corresponds to the highest frequency that can be heard from this particular siren at these speeds.

SOLVE
When the observer and source both move away from each other, choose the bottom sign for each speed.

$$f_{obs} = \left(\frac{v_s - v_{obs}}{v_s + v_{source}}\right) f$$

$$f_{obs} = \frac{(343.4 \text{ m/s}) - (29.0 \text{ m/s})}{(343.4 \text{ m/s}) + (35.7 \text{ m/s})}(875 \text{ Hz})$$

$$f_{obs} = 726 \text{ Hz} \text{ after truck passes}$$

CHECK and THINK
As expected, this result is lower than the "receding frequency" in part A and corresponds to the lowest frequency that can be heard from this siren at these speeds.

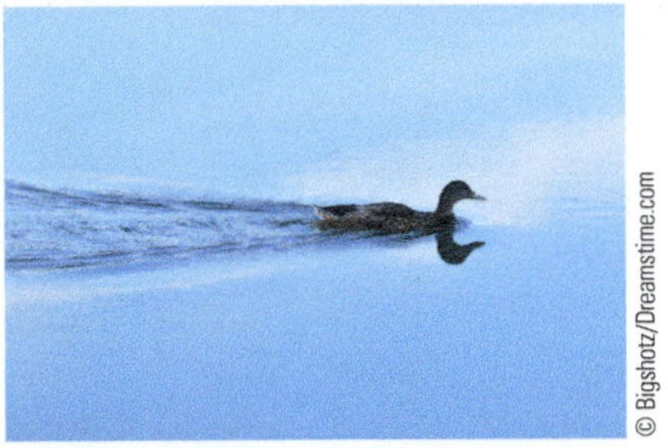

FIGURE 17.31 A swimming duck moves faster in the water than waves do and creates a wake.

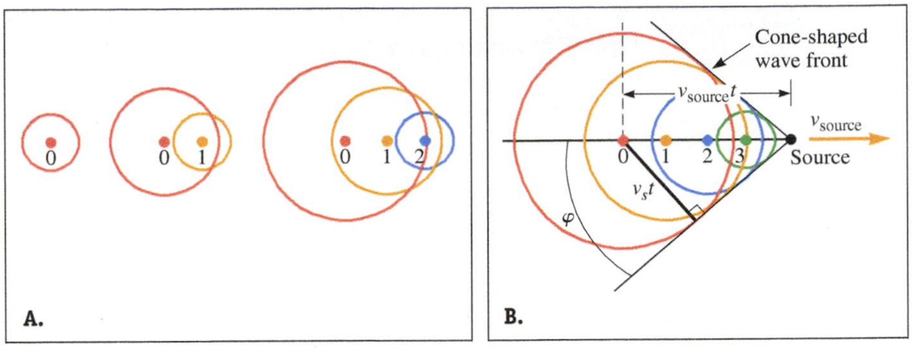

FIGURE 17.32 A shock wave forms when the source is moving faster than the speed of the wave. Compare with Figure 17.28.

Shock Waves

SHOCK WAVE ✪ Major Concept

The source in Figure 17.28 is moving at a speed that is less than the speed of the wave it produces. What happens if the source is moving faster than waves travel in that medium? Some jet airplanes and certain race cars exceed the speed of sound in air, and even the most modest boats and good swimmers can exceed the speed of the water waves they produce. When the source moves faster than the wave it emits, a cone-shaped wave front known as a **shock wave** forms (Fig. 17.31).

Figure 17.32 illustrates how the shock wave forms for a source emitting spherical sound waves in air. The source moves to the right at a speed v_{source} that is greater than the sound speed v_s. The leading edge of each subsequent spherical wave front is to the right of the previous wave front. For example, the leading edge of the blue wave front is to the right of the orange wave front in Figure 17.32A. The combination of the offset spherical waves produces a cone-shaped shock wave.

The half-angle of the cone is known as the **Mach angle** φ and can be found with trigonometry. In Figure 17.32B, the source is at the tip of the cone, having moved a distance $v_{source}t$ from position 0. During that same time interval t, the spherical sound wave that was emitted from position 0 has a radius v_st. As shown, $v_{source}t$ is the hypotenuse of a right triangle, and v_st is the leg opposite the Mach angle φ. So,

$$\sin\varphi = \frac{v_st}{v_{source}t} = \frac{v_s}{v_{source}} \tag{17.32}$$

The ratio v_{source}/v_s is known as the **Mach number** M. If an object moves faster than the speed of sound, then $M > 1$, and the object is said to be *supersonic*.

If you have ever have been near a moving supersonic jet airplane, you have probably heard a loud *sonic boom*. You hear the boom when the shock wave reaches you, some time after the airplane has passed (Fig. 17.33). Because the shock wave is the sum of many spherical waves, it has a very large amplitude and carries a lot of energy. The shock wave and sonic boom are continuously produced, but you can only hear the boom for the moment that shock wave makes contact with you. When you are in front of the cone-shaped shock wave, you cannot hear any sound from the plane because the sound waves are behind the cone. If you are inside the cone, you hear the normal Doppler-shifted sound from the plane.

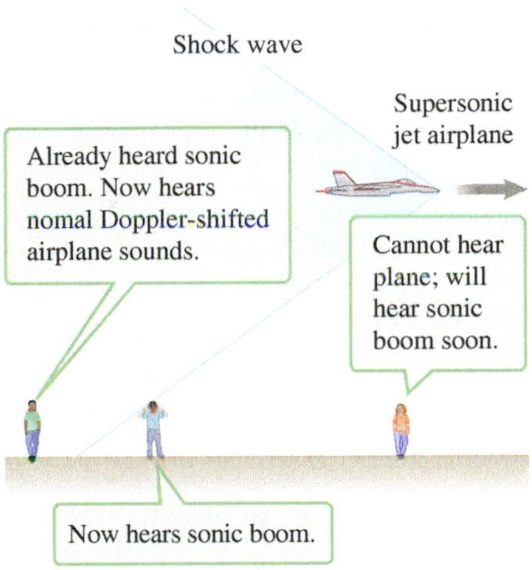

FIGURE 17.33 A sonic boom is only heard when a shock wave is in contact with the listener.

17-10 The Wave Equation

We began this chapter by thinking about fans standing and sitting back down, forming a "wave" around the stadium. Of course, people in a stadium stand up and sit back down all the time, but a wave forms only when the choreography is just right. So, when is a disturbance a wave, and when is it just a disturbance? The answers are given by a (partial) differential equation known as the **wave equation**. If a disturbance satisfies the wave equation, it is a wave. In this section, we derive a somewhat simple version of this equation.

ONE-DIMENSIONAL WAVE EQUATION

❶ Underlying Principle

DERIVATION One-Dimensional Wave Equation

Figure 17.34 shows a small segment of rope at a particular moment as a transverse wave passes. We will apply Newton's second law to that small segment and show that it leads to the wave equation:

$$\frac{\partial^2 y(x,t)}{\partial x^2} = \frac{1}{v_x^2}\frac{\partial^2 y(x,t)}{\partial t^2} \tag{17.33}$$

In this differential equation, the symbol ∂ denotes a *partial* derivative. The wave function y depends on x and t. When you take a partial derivative, you treat one of these variables as a fixed constant and take the derivative with respect to the other variable. (This derivation is much like the derivation of the propagation speed of a transverse wave on page 495.)

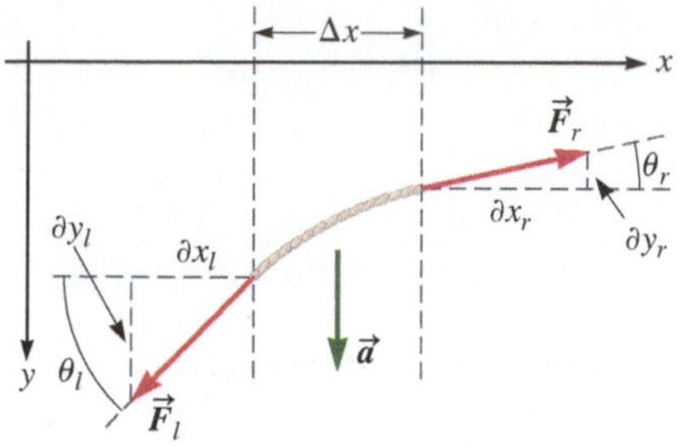

FIGURE 17.34 A small portion of a rope at one instant as a transverse wave passes. Here the notation ∂ denotes small changes. Angles are exaggerated for clarity.

There are two forces exerted on the rope segment. (We can ignore gravity because this segment is very small with very little mass.) Each force has a magnitude equal to the tension F_T in the rope. The force $\vec{F}_r$ on the right pulls up and to the right at an angle θ_r with the x axis, whereas $\vec{F}_l$ pulls down and to the left at an angle θ_l. Both angles are small. The segment is accelerated downward.

Apply Newton's second law in component form starting with the x direction. The angles are small, so $\cos\theta \approx 1$. Because the magnitude of the forces are equal, there is no net force in the x direction, confirming that there is no acceleration in the x direction.	$\sum F_x = F_r\cos\theta_r - F_l\cos\theta_l = ma_x$ $\sum F_x \approx F_r - F_l = ma_x$ $\sum F_x \approx F_T - F_T = 0$ $a_x = 0$
Now apply Newton's second law in the y direction. The magnitude of the force is the tension F_T.	$\sum F_y = F_l\sin\theta_l - F_r\sin\theta_r = ma_y$ $F_T(\sin\theta_l - \sin\theta_r) = ma_y$
Because the angles are small, $\sin\theta \approx \tan\theta$.	$F_T(\tan\theta_l - \tan\theta_r) = ma_y$
Use the right triangles in Figure 17.34 to write the tangent of each angle in terms of the differential distances ∂x and ∂y. (Although y is a function of x and t, here we are only concerned with the segment at one fixed time shown in Fig. 17.34).	$F_T\left[\left(\frac{\partial y}{\partial x}\right)_l - \left(\frac{\partial y}{\partial x}\right)_r\right] = ma_y$
As in our derivation of the propagation speed, we write the mass of the segment in terms of the rope's mass per unit length μ and the length Δx of the segment. (Don't worry about the slight upward bend of the segment; the segment's length is very close to Δx.)	$F_T\left[\left(\frac{\partial y}{\partial x}\right)_l - \left(\frac{\partial y}{\partial x}\right)_r\right] = (\mu\Delta x)a_y$
The acceleration is the second (partial) time derivative of position y.	$F_T\left[\left(\frac{\partial y}{\partial x}\right)_l - \left(\frac{\partial y}{\partial x}\right)_r\right] = (\mu\Delta x)\frac{\partial^2 y}{\partial t^2}$
Rearrange terms.	$\frac{(\partial y/\partial x)_l - (\partial y/\partial x)_r}{\Delta x} = \left(\frac{\mu}{F_T}\right)\frac{\partial^2 y}{\partial t^2}$ (1)
As the segment's length goes to zero, the term on the left becomes a (partial) derivative.	$\lim_{\Delta x\to 0}\frac{(\partial y/\partial x)_l - (\partial y/\partial x)_r}{\Delta x} = \frac{\partial}{\partial x}\left(\frac{\partial y}{\partial x}\right) = \frac{\partial^2 y}{\partial x^2}$
Rewrite Equation (1) in terms of this second (partial) derivative.	$\frac{\partial^2 y}{\partial x^2} = \left(\frac{\mu}{F_T}\right)\frac{\partial^2 y}{\partial t^2}$
Finally, use $v_x = \sqrt{F_T/\mu}$ (Eq. 17.11) to eliminate the term in parentheses.	$\frac{\partial^2 y(x,t)}{\partial x^2} = \frac{1}{v_x^2}\frac{\partial^2 y(x,t)}{\partial t^2}$ ✓ (17.33)

COMMENTS

We derived the wave equation for a transverse one-dimensional wave rather than for a more general case, but the form of the equation we derived is fundamental enough for us to appreciate its significance.

The wave equation tells us that a disturbance is a wave if the medium's curvature $\partial^2 y(x, t)/\partial x^2$ is proportional to the medium's acceleration $\partial^2 y(x, t)/\partial t^2$, where the constant of proportionality depends on the wave's propagation speed. One of the amazing things about nature is that the wave equation describes so many diverse phenomena. The wave equation applies to the mechanical waves in this chapter, including transverse and longitudinal waves, waves on water, and sound waves in air. Moreover, the wave equation describes the propagation of light through a vacuum and of the gravitational force through the Universe. According to quantum mechanics, the wave equation is a fundamental description of matter. The overarching goal of physics is to discover the fundamental laws of nature. The wave equation is a great achievement of physics, illustrating the simple beauty of nature behind its apparent diversity.

SUMMARY

Underlying Principles

One-dimensional wave equation:

$$\frac{\partial^2 y(x, t)}{\partial x^2} = \frac{1}{v_x^2}\frac{\partial^2 y(x, t)}{\partial t^2} \tag{17.33}$$

Major Concepts

1. A **longitudinal wave** is a mechanical wave in which the back-and-forth motion of each particle is parallel to the direction of wave propagation.
2. A **transverse wave** is a mechanical wave in which the particles in the medium move perpendicular to the direction of wave propagation.
3. A **harmonic wave** is a periodic wave in which each particle of the medium oscillates in simple harmonic motion (SHM).
4. The **wave function** of a **transverse** harmonic wave is

$$y(x, t) = y_{\text{max}} \sin(kx - \omega t) \tag{17.4}$$

5. The **wave function** of a **longitudinal** harmonic wave is

$$S(x, t) = S_{\text{max}} \sin(kx - \omega t) \tag{17.6}$$

In Equations 17.4 and 17.6, the angular wave number is $k = 2\pi/\lambda$ (Eq. 17.5) and the angular frequency is $\omega = 2\pi f$ (Eq. 16.2); the speed of the wave is $v = \omega/k$ (17.7).

6. **Intensity** I is defined as the average power P_{av} per unit area A perpendicular to the direction of propagation:

$$I \equiv \frac{P_{\text{av}}}{A} \tag{17.22}$$

The intensity of a point source decreases as the distance r from the source increases:

$$I = \frac{P_{\text{av}}}{4\pi r^2} \tag{17.23}$$

7. **Refraction** is the bending of a wave as it travels between media having different wave propagation speeds v_1 and v_2. The **law of refraction** states that

$$\sin\theta_2 = \frac{v_2}{v_1}\sin\theta_1 \tag{17.26}$$

where θ is the angle between the direction of wave propagation and the normal to the boundary between media.

8. **Diffraction** is the bending of a wave due to an obstacle or hole. If the obstacle or hole is comparable in size to the wavelength, diffraction becomes significant.
9. The **Doppler shift** indicates the change in frequency due to relative motion between the wave source and the observer. For sound waves, the observed frequency is

$$f_{\text{obs}} = \left(\frac{v_s \pm v_{\text{obs}}}{v_s \mp v_{\text{source}}}\right) f \tag{17.31}$$

where v_s is the speed of sound. Speeds v_{obs} and v_{source} are measured with respect to the medium (usually air). See page 511 for rules on choosing the signs.

10. A **shock wave** is a cone-shaped wave front created when a wave source moves faster than waves travel in the medium.

Special Cases

1. **Transverse waves on a rope** travel at speed

$$v_x = \sqrt{\frac{F_T}{\mu}} \quad (17.11)$$

where F_T is the tension in the rope and μ is the linear mass density.
The power transported is

$$P_{av} = \tfrac{1}{2}\mu\omega^2 y_{max}^2 v_x \quad (17.19)$$

2. **Sound** is a longitudinal wave in a medium with bulk modulus B and density ρ. The speed of sound is

$$v_s = \sqrt{\frac{B}{\rho}} \quad (17.13)$$

A harmonic sound wave can be written as a **pressure wave**:

$$\Delta P = \Delta P_{max} \cos(kx - \omega t) \quad (17.16)$$

Sound intensity:

$$I = \frac{1}{2}\frac{\Delta P_{max}^2}{\rho v_s} \quad (17.24)$$

Sound level:

$$\beta \equiv 10 \log\left(\frac{I}{I_0}\right) \quad (17.25)$$

where $I_0 = 10^{-12}$ W/m^2.

PROBLEMS AND QUESTIONS

A = algebraic C = conceptual E = estimation G = graphical N = numerical

17-1 Introducing Mechanical Waves

1. **C** A dog swims from one end of a pool to the opposite end. Is the dog's motion described as a wave? Explain.
2. **C** A boat crosses a still pond, causing a person in an inner tube floating on the pond to bob up and down. Use the terms *particle motion*, *wave*, and *oscillates* to describe this situation.

17-2 Pulses

3. **C** If the pulse in Figure 17.4 is twice as tall so that $y(0, 0) = 2$ m, what changes are needed in the wave function (Eq. 17.2)

$$y(x, t) = \frac{1}{(x - v_x t)^2 + 1}$$

4. A pulse on a rope has a profile given by

$$y(x) = \frac{1}{4x^2 + 1}$$

The pulse moves to the right at $v_x = 2$ m/s.
   a. **A** Write an expression for the wave function.
   b. **G** How does this new pulse compare with the one in Figure 17.4? *Hint*: This problem is best approached graphically.

5. **N** The wave function for a traveling pulse, similar to the one discussed in Section 17-2, is

$$y(x, t) = \frac{8}{1 + (x + 4t)^2}$$

where x and y are in meters and the time t is in seconds. In which direction and with what speed is the pulse propagating?

Problems 6 and 7 are paired.

6. **G** The wave function of a pulse is

$$y(x, t) = \frac{7.5}{(x + 3.5t)^2 + 0.5}$$

where the values are in the appropriate SI units. **a.** Consider a particle at $x = 5.0$ m and sketch a graph of its vertical position y versus time. **b.** Repeat part (a) for a particle at $x = -5.0$ m. Compare your results in the **CHECK and THINK** step.

7. The wave function of a pulse is

$$y(x, t) = \frac{7.5}{(x + 3.5t)^2 + 0.5}$$

where the values are in the appropriate SI units.
   a. **N** What are the speed and direction of the pulse? *Hint*: Sketch the wave pulse at $t = 2$, 4, and 6 s as in Figure 17.4.
   b. **A** Write the wave function for a pulse with the same profile and speed but moving in the opposite direction.

8. **G** The wave function of a transverse pulse on a rope is

$$y(x, t) = \frac{2}{(3x - 4.0t)^2 + 1}$$

where the values are in the appropriate SI units. Sketch a graph of the wave's profile.

17-3 Harmonic Waves

9. **N** A sinusoidal traveling wave is generated on a string by an oscillating source that completes 120 cycles per minute. What is the wavelength of the wave if individual crests are observed to travel 15.0 m in 1.00 min?
10. **A** If a harmonic wave is traveling in the negative x direction and its wave function is $y(x, t) = y_{max} \sin(kx + \omega t)$, show that $v_x = -\omega/k$.
11. **C** Suppose a simple harmonic oscillator creates a transverse wave on a rope as in Figure 17.8A (page 491). If the angular frequency of the oscillator is increased and the speed of the wave on the rope remains unchanged, what property or properties of the wave must change? Explain your answers.
12. The equation of a harmonic wave propagating along a stretched string is represented by $y(x, t) = 4.0 \sin(1.5x - 45t)$, where x and y are in meters and the time t is in seconds.
   a. **C** In what direction is the wave propagating?
   b–e. **N** What are the **b.** amplitude, **c.** wavelength, **d.** frequency, and **e.** propagation speed of the wave?

13. N Neonatal ultrasound machines typically operate in the 7.00-Mhz to 18.0-MHz range. What is the wavelength of 7.00-MHz ultrasound traveling through human tissue at 1497 m/s?

14. As in Figure 17.8A (page 491), a simple harmonic oscillator is attached to a rope, creating a transverse wave. The oscillator has an angular frequency of 2.34 rad/s and an amplitude of 34.6 cm.
   a. N If the propagation speed of the wave is 2.67 m/s, what is its wavelength?
   b. A Write an expression for the wave function.
   c. C, A If the oscillator's angular frequency doubles and the propagation speed remains the same, what changes in the wave function? Write the new wave function.

15. N The wave function

$$y(x, t) = (0.0500 \text{ m}) \sin\left(3\pi x + \frac{\pi}{4}t\right)$$

describes a transverse wave on a rope, with x and y in meters and t in seconds.
What are the **a.** wavelength, **b.** period, and **c.** speed of the wave? **d.** At $t = 1.00$ s, what is the transverse velocity of a rope element located at $x = 0.200$ m? **e.** At $t = 1.00$ s, what is the transverse acceleration of a rope element located at $x = 0.200$ m? Answer each part using three significant figures.

16. A harmonic transverse wave function is given by

$$y(x, t) = (0.850 \text{ m}) \sin(15.3x + 10.4t)$$

where all values are in the appropriate SI units.
   a. N What are the propagation speed and direction of the wave's travel?
   b. N What are the wave's period and wavelength?
   c. N What is the amplitude?
   d. C If the amplitude is doubled, what happens to the speed of the wave?

Problems 17 and 18 are paired.

17. A graph (profile) of a traveling wave at a moment in time is shown in Figure P17.17. The wave is traveling in the negative x direction with a propagation speed of 56.0 cm/s.
   a. N What is the amplitude of the wave?
   b. N What is the frequency of the wave?
   c. A Write the transverse harmonic wave function for this wave.

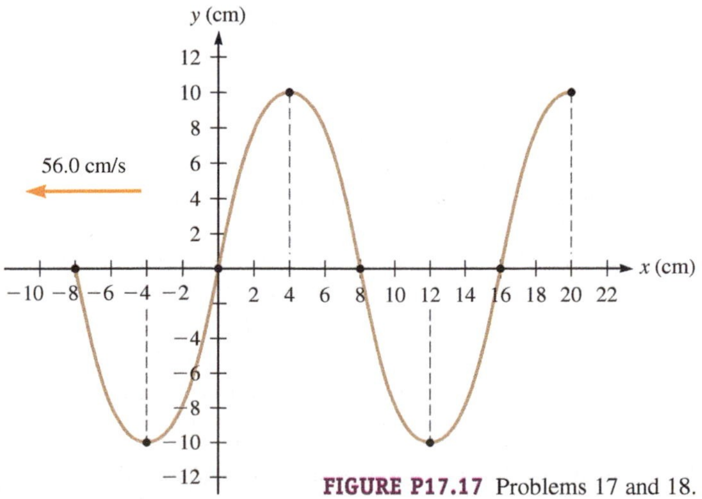

FIGURE P17.17 Problems 17 and 18.

18. In Problem 17,
   a. N What is the period of the wave?
   b. G Plot the wave over the same range of x values shown when $t = T/2$ (Assume the graph shown is for $t = 0$.)

19. N A longitudinal harmonic wave is given by

$$S(x, t) = (0.850) \sin(10.4x - 5.20t)$$

where all values are in the appropriate SI units. **a.** What are the period and wavelength of this wave? **b.** Consider a particle whose equilibrium position is $x = \lambda/2$. Find the displacement S of this particle at $t = 0, T/4, T/2, 3T/4$, and T. **c.** What is the position of the particle at these times?

17-4 Special Case: Transverse Wave on a Rope

20. C Your laboratory instructor gives your team a long, very thin thread. Your challenge is to find its mass precisely. You have a scale, several cylinders of different weights each with a hook, a meterstick, and a timer. Your lab has familiar equipment such as pulleys and vertical and horizontal supports. The thread is very light and strong, similar to spider's silk. When you place it on the scale, you find that the scale is not sensitive enough to report a meaningful mass. Come up with a better procedure to measure the thread's mass. Describe your procedure and your plan to interpret your data.

21. N A copper wire with linear mass density 6.30×10^{-2} kg/m is placed under 2.25×10^3 N of tension. What is the speed of traveling waves on this wire?

22. A A string of length L and mass M hangs freely from a fixed point on the ceiling. **a.** Determine the propagation speed of a transverse wave traveling along the string at any position y. *Hint*: The tension is not constant throughout the string. The tension at any point along the string is equal to the weight $g\,dm$ of the string below that point. **b.** Determine the time required for a transverse pulse to travel the entire length of the string.

23. N A wave on a string with linear mass density 5.00×10^{-3} kg/m travels at 105.0 m/s. What is the tension in the string?

24. N A traveling wave on a thin wire is given by the equation $y = (0.112 \text{ m}) \sin(4.35x + 46.7t)$, where all values are in the appropriate SI units. **a.** What is the propagation speed of this wave? **b.** If the linear mass density of the string is 3.20×10^{-3} kg/m, what is the tension in the string?

25. A sinusoidal wave with amplitude 1.00×10^{-2} m and frequency 325 Hz is observed to be traveling with a speed of 55.0 m/s on a wire held under 195 N of tension.
   a. A What is the wave function for this wave in SI units?
   b. N What is the mass per unit length for the wire?

26. As in Figure 17.8A, a simple harmonic oscillator is attached to a rope, creating a transverse wave of wavelength 7.47 cm. At the other end of the rope is a pulley and a hanging block of mass 6.30 kg. The oscillator has an angular frequency of 2340 rad/s and an amplitude of 34.6 cm.
   a. N What is the linear mass density of the rope?
   b. C If the angular frequency of the oscillator doubles, what properties (speed, angular wave number, amplitude) of the wave must change? Give the new values of these properties.
   c. C If the frequency remains at its original value but the mass of the hanging block is doubled, what properties of the wave must change? Give the new values of these properties.

27. N An aluminum wire with a radius of 1.00 mm is held under tension. If the speed of traveling waves on the wire is 180 m/s and the density of aluminum is 2.70 g/cm³, what is the tension in the wire?

28. C People of the Incan empire built a long series of roads and bridges in the highlands of Peru. The bridges were made of rope and had to be rebuilt yearly by nearby villagers (Fig. P17.28).

In the construction of a new bridge, two long ropes were secured across the gorge. A worker then scooted across the two ropes while attaching smaller lengths of rope to form the footpath. Suppose it is your job to scoot out over the gorge on just two ropes. You may want to check the tension in both ropes to make sure it is the same in each rope. Explain how you could use a traveling pulse to check the tension.

FIGURE P17.28

17-5 Sound: Special Case of a Traveling Longitudinal Wave

29. **N** Assume the smallest amplitude detectable by your ear is $S_{min} = 1 \times 10^{-11}$ m. If $f = 1000$ Hz, what is the smallest-pressure amplitude you can hear under the ocean? Compare your result with that in Example 17.4 (quietest whisper in air, $\Delta P_{max} = 3 \times 10^{-5}$ Pa). The speed of sound in seawater is 1531 m/s. Answer using two significant figures.

30. **N** Dolphins have been trained to understand human voice commands. Imagine a trainer standing on a platform with her head and shoulders 5.20 m above the surface of a pool, with a dolphin waiting 10.4 m directly below the trainer (Fig. P17.30). Assume the speed of sound in air is 343 m/s and in water is 1497 m/s.

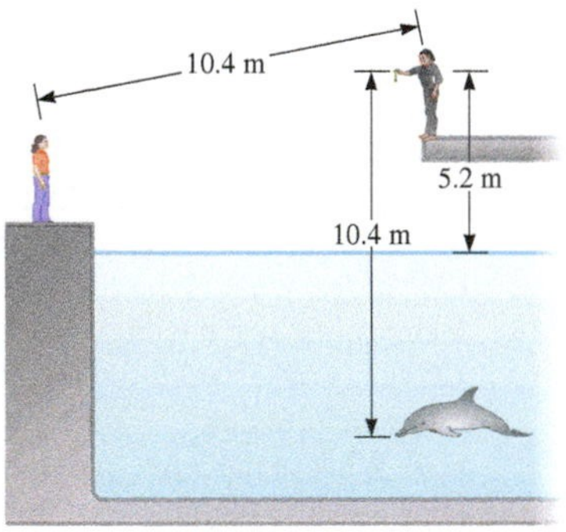

FIGURE P17.30

a. **N** If the trainer gives the command to jump, how long will it take the dolphin to hear the command?

b. **C** Would the dolphin hear the command before or after another person who is 10.4 m from the trainer? Justify your answer.

31. A whistle is often used in dog-training exercises and is an integral part of field-marking competitions, where signals and commands are relayed to the dog via whistle. Suppose the whistle produces a sound wave with a frequency of 25,100 Hz (outside the range of human hearing) and the propagation speed of sound in air is 343 m/s.

a. **N** What is the wavelength of this sound from the whistle?

b. **N** What is the wave number k corresponding to the wavelength you found?

c. **A** Write the wave function for the longitudinal sound wave described above, assuming the amplitude of the sound wave is $S_{max} = 1.57 \times 10^{-6}$ m and it is moving to the right (in the positive x direction).

Problems 32 and 33 are paired.

32. **N** Seismic waves travel outward from the epicenter of an earthquake. A single earthquake produces both longitudinal seismic waves known as P waves and transverse waves known as S waves. Both transverse and longitudinal waves can travel through solids such as rock. Longitudinal waves can travel through fluids, whereas transverse waves can only be sustained near the surface of a fluid, not inside the fluid. When seismic waves encounter a fluid medium such as the liquid outer core of the Earth, only the longitudinal P wave can propagate through. Geophysicists can model the interior of the Earth by knowing where and when S and P waves were detected by seismographs after an earthquake (Fig. P17.32). Assume the average speed of an S wave through the Earth's mantle is 5.4 km/s and the average speed of a P wave is 9.3 km/s. After an earthquake, a seismograph finds that the P wave arrives 1.5 min before the S wave. How far is the epicenter from the detector?

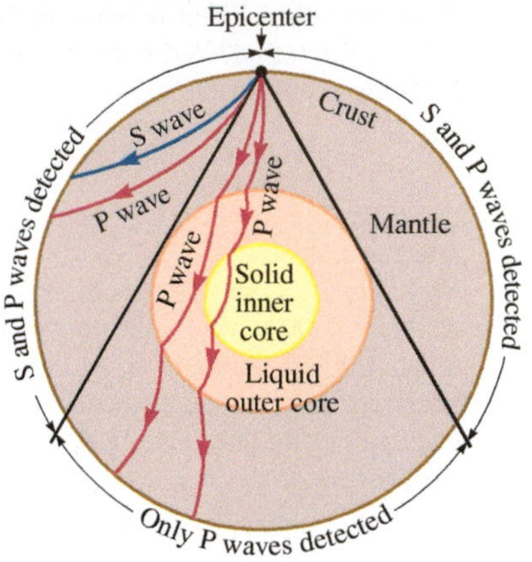

FIGURE P17.32

33. **N** On July 17, 2006, an earthquake hit just south of Java, Indonesia (between Australia and Malaysia). The time of the earthquake was 08:19:25.02 UTC (Coordinated Universal Time). The strong earthquake was detected by more than 400 seismograph stations worldwide. Table P17.33 provides the arrival time of the P wave and the distance to three such stations. Assume the P wave traveled through the crust and the crust's average density is 3300 kg/m^3. Estimate the bulk modulus of the Earth's crust. *Hint*: See Problem 32.

TABLE P17.33 Java earthquake data.

Arrival time at seismograph (UTC)	Distance from epicenter to station (km)
08:21:55.58	1190
08:22:05.09	1230
08:23:29.89	1970

17-6 Energy Transport in Waves

34. **N** A harmonic wave travels on a string with a linear mass density of 4.51×10^{-3} kg/m. It travels with a speed of 112 m/s and has a frequency of 65.0 Hz. If the wave has an amplitude of 0.118 m, what is the amount of energy in three wavelengths of the wave?

Problems 35 and 36 are paired.

35. **A** Show that the kinetic energy in one wavelength of a harmonic transverse wave on a rope (Fig. 17.8A) is given by $K_\lambda = (1/4)\mu\omega^2 y_{max}^2 \lambda$.

36. **A** Use the result of Problem 35 to show that the potential energy in one wavelength of a harmonic transverse wave on a rope (Fig. 17.8A) is given by $U_\lambda = (1/4)\mu\omega^2 y_{max}^2 \lambda$.

37. **N** A 5.00-m rope with a mass of 0.650 kg is held under tension. If 1.00 kW of power is supplied to the rope, what is the amplitude of sinusoidal waves that will be generated with a wavelength of $\pi/4$ m and speed 50.0 m/s?

38. **C** Suppose you have a transverse harmonic wave on a rope. If you wish to increase the power transferred by the wave, what should you do to the rope?

39. **N** Sound with a wavelength of 6.59 m is produced in air where it travels with a propagation speed of 343 m/s. The average power of the sound wave is 40.0 W. If the density of the air in this region of space happens to be 1.20 kg/m^3, what is the energy contained in one wavelength of the wave?

40. **N** An oscillating lever delivers a maximum power of 0.500 kW to a string with linear mass density of 1.50×10^{-2} kg/m that is under 125 N of tension. What is the maximum angular frequency of the lever that will result in the generation of sinusoidal waves with an amplitude of 3.00 cm?

17-7 Two- and Three-Dimensional Waves

41. **N** A sound produces a pressure amplitude of $\Delta P_{max} = 32$ Pa. Determine the intensity of this sound wave. Assume the density of air is 1.3 kg/m^3 and the speed of sound is 343 m/s.

42. **CASE STUDY** Besides music, other loud sounds can damage your hearing. When standing 1 m from an operating chainsaw, the sound level is 105 dB.
   a. **G** Plot the sound level versus distance from the chainsaw.
   b. **N** At what distance from the chainsaw is the sound level equivalent to half the sound level at 1 m? (Read this distance off your graph.)

Problems 43 and 44 are paired.

43. **N** Sunlight is an example of a wave that, like sound, can be thought of as being emitted in all directions with its power uniformly distributed over the surface of any sphere centered on the source (the Sun). Therefore, the intensity of sunlight will be different when measured at the locations of the various planets (Fig. P17.43). The intensity of sunlight that reaches the Earth's atmosphere is about 1400 W/m^2. The Sun-Earth distance is 1 AU (1.496×10^{11} m). **a.** What is the intensity that reaches Mars if Mars is 1.5 times farther from the Sun than the Earth? **b.** Saturn is about 10 times farther from the Sun than the Earth. What is the intensity of sunlight at Saturn?

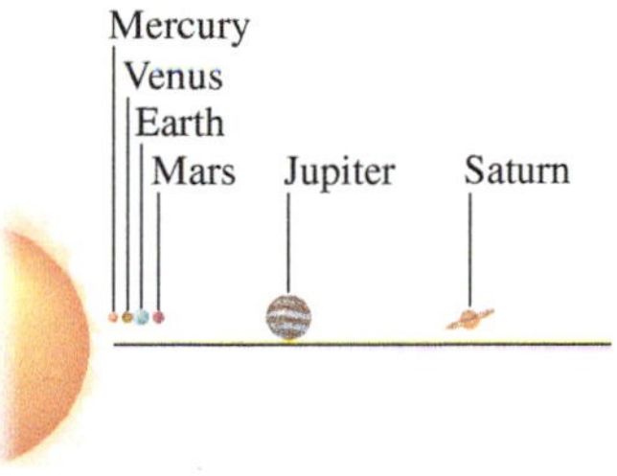

FIGURE P17.43 Problems 43 and 44.

44. Consider the discussion of light waves in Problem 43. Solar cells absorb energy from light waves and convert it to electrical energy. A certain solar cell has an area of 2.00 m^2 and converts 10% of the absorbed energy into usable electric energy. Assume normal incidence of the light on the cell.
   a. **N** What is the power output of the solar cell if placed in space near the Earth?
   b. **N** What is the power output of the solar cell if placed in space near Mars?
   c. **C** Consider your answers to parts (a) and (b). Describe an issue that must be considered when scientists and engineers plan research missions to Mars and beyond. In what practical ways can researchers address this issue? Why might we choose to use nuclear power rather than solar power when planning a mission to Mars?

45. **N** A violin is played with an initial intensity I_i increasing to a final intensity I_f. If the final intensity is 10 times the initial intensity, what is the difference in sound level between these two extremes?

46. **N** What is the sound level of a sound wave with intensity 6.40×10^{-5} W/m^2?

47. **N** A listener finds that the sound level of a flute is 3 dB higher than the sound level of a cello. How does the intensity of the flute compare with the intensity of the cello?

48. **N** The speaker system at an open-air rock concert forms a ring around the entire circular stage and delivers 50,000 W of power output. Assume the sound radiates in all directions equally as if it were generated by an isotropic point source and assume the sound energy is not absorbed by air. **a.** At what distance is the sound from the speakers barely audible? Note that your answer will be far too large since the model we are using for sound level ignores the power absorbed by the medium (air). How does your answer compare to the radius of the Earth? **b.** What is the closest distance audience members can be to the speakers if the sound is not to be painful to their ears?

17-8 Refraction and Diffraction

Problems 49 and 50 are paired.

49. **N** The Earth's mantle is denser near the core than it is near the crust, so the speed of seismic waves varies throughout the interior. Therefore, seismic waves are refracted as they travel deep within the Earth. This refraction is used to help model the density of interior layers. A seismic longitudinal P wave traveling at 8.5 km/s encounters a region in which the propagation speed is 12.5 km/s (Fig. P17.49). If the P wave strikes the boundary between regions at an angle of 65° with respect to the boundary, what is the angle of refraction θ in the dense region?

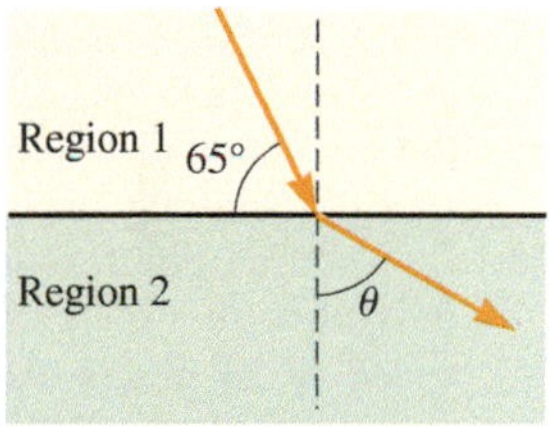

FIGURE P17.49

50. **N** A seismic P wave strikes a boundary between two types of material having the same bulk modulus. At the boundary, the density increases abruptly from 3400 kg/m^3 to 4100 kg/m^3 (Fig. P17.50). If the wave front of the P wave is planar and makes a 25° angle with the boundary, what is the angle of refraction θ?

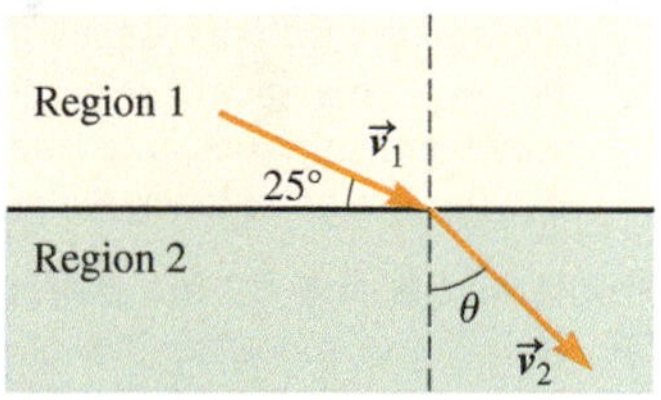

FIGURE P17.50

Problems 51 and 52 are paired.

51. **N** Light can be modeled as a wave. You can see because your eye collects the light waves that bounce off objects (or light produced by objects). When viewing a fish in the water, Wanda's eye collects light that has bounced off the fish, such as the ray indicated in Figure P17.51. The direction of wave propagation is altered due to refraction as the light wave leaves the water. The speed of light in air is 3.00×10^8 m/s and in water is 2.25×10^8 m/s. Given these speeds and the information in the figure, find the angle θ measured from the vertical.

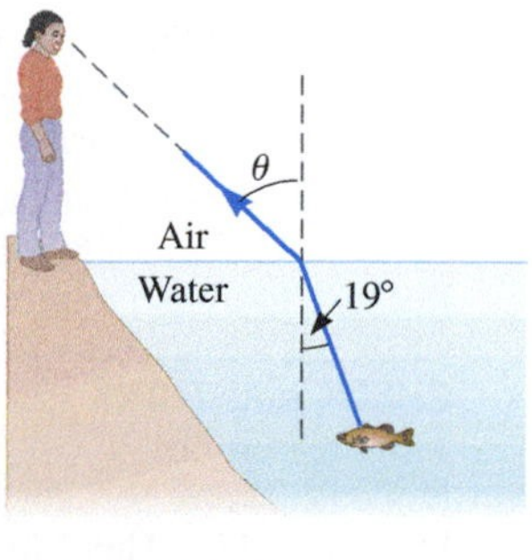

FIGURE P17.51 Problems 51 and 52.

52. **C** Spearfishing from shore or a dock is a practical means by which to catch fish and is still practiced in parts of the world. Use Figure P17.51 (Problem 51) to think about Wanda's aim when she attempts to spear a fish. Should she aim at a location farther away or closer than that at which she perceives the fish?

53. **C** Sometimes when you hear an airplane, it is difficult to locate its position in the sky immediately. Assume the density of air at the plane's altitude is less than the density of air near the Earth's surface. How can refraction make it difficult to locate the source of the sound?

54. **C** Using the concept of diffraction, discuss how the sound from a concert stage is affected if you sit behind a large pillar (greater than 1 m wide) near the middle of the concert hall. Is the answer different for treble (shorter-wavelength) or bass (longer-wavelength) sounds? Explain your answer.

17-9 The Doppler Shift

55. **N** For an ecology project, Maria is recording some ambient sounds in the woods. A hawk emits a cry with a frequency of 8.00×10^2 Hz. If the hawk is flying toward Maria's microphone at a speed of 18 m/s as it sustains this sound frequency, what is the frequency that is recorded by the microphone? Assume the speed of sound in air is 343 m/s.

56. **C** A source emits sound with a frequency of 440 Hz and is moving with a speed of 24 m/s. An observer driving a car at a speed of 24 m/s is chasing the source. Assume the speed of sound is 343 m/s. Will the frequency heard by the observer be higher than, lower than, or equal to 440 Hz? Explain your answer.

57. **N** An ambulance traveling eastbound at 140.0 km/h with sirens blaring at a frequency of 7.00×10^2 Hz passes cars traveling in both the eastbound and westbound directions at 55.0 km/h. What is the frequency observed by the eastbound drivers **a.** as the ambulance approaches from behind and **b.** after the ambulance passes them? What is the frequency observed by the westbound drivers **c.** as the ambulance approaches them and **d.** after the ambulance passes them?

Problems 58 and 59 are paired.

58. **N** Joe, a skateboarder, wants to measure his downhill speed, thinking that he may reach 45 mph. While riding down the hill, he holds a panic alarm that operates at a frequency of 1250 Hz. Rochelle, who has perfect pitch, claims that she can identify any note on a piano from 27.5 Hz to 4186.0 Hz.

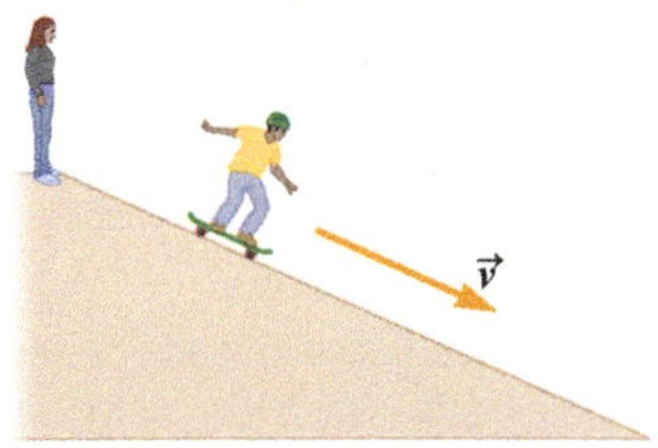

FIGURE P17.58

a. Rochelle listens from the top of the hill as Joe rolls away (Fig. P17.58). If Joe really reaches 45 mph, what frequency will Rochelle hear? **b.** From the top of the hill, Rochelle hears a C6, which has a frequency of 1046.5 Hz. What is Joe's speed? In the **CHECK and THINK** step, describe whether or not your answer seems reasonable.

59. **N** Consider the scenario in Problem 58. This time, Rochelle listens from the bottom of the hill as Joe rolls toward her (Fig. P17.59). If the skateboarder really reaches 45 mph, what frequency will Rochelle hear?

FIGURE P17.59

60. **N** The Mach number of a supersonic jet is 3.5. What is the Mach angle?

Problems 61 and 62 are paired.

61. **C** You are sitting a certain distance away from a speaker that is playing a 450-Hz sound at a constant power. Assume the speed of sound in air is 343 m/s. **a.** Describe what happens to the frequency and wavelength of the sound you hear if the speaker is moved toward your location. **b.** Describe what happens to the frequency and wavelength of the sound you hear after the speaker has moved past you and continues receding.

62. **(G, C)** In Problem 61, **a.** Sketch an image of the wave fronts being emitted by the speaker as the speaker approaches you. In what way does the sketch illustrate the change in the wavelength you observe as the speaker is moving toward your location? **b.** Sketch an image of the wave fronts being emitted by the speaker as it recedes away from you. In what way does the sketch illustrate the change in the wavelength you observe as the speaker is moving away from your location?

63. **N** A woman sees a supersonic plane directly overhead and then hears a sonic boom 3.7 s later. If the plane's altitude is 1.75×10^3 m, what is the Mach number of the plane? Assume the speed of sound is 331 m/s. Ignore any potential effects due to refraction of the sound.

General Problems

64. **A** Begin with the wave function of a one-dimensional transverse wave on a string and show that it leads to the wave equation

$$\frac{\partial^2 y(x, t)}{\partial x^2} = \frac{1}{v_x^2}\frac{\partial^2 y(x, t)}{\partial t^2}$$

In this differential equation, the symbol ∂ denotes a *partial* derivative. The wave function y depends on x and t. When you take a partial derivative, treat one of these variables as constant and take the derivative with respect to the other variable.

65. How far are you located from a lightning strike if the thunderclap arrives 7.50 s after you see the lightning flash? Assume the speed of light in air is 3.0×10^8 m/s and the speed of sound at this location is 3.40×10^2 m/s.

66. **N** A traveling wave on a thin wire is given by the equation $y = (0.405 \text{ m}) \sin[0.354x - 1150t]$, where all values are in the appropriate SI units. What are the **a.** propagation speed and **b.** period of the wave? **c.** How far horizontally does the wave travel during one period?

67. **N** A traveling sinusoidal wave is described by the wave function

$$y(x, t) = (0.600 \text{ m}) \sin\left(7.50\pi t - \pi x + \frac{\pi}{2}\right)$$

where x and y are in meters and t is in seconds. What are the **a.** wavelength, **b.** frequency, **c.** amplitude, and **d.** speed of propagation of the wave?

68. **A** A sinusoidal wave with amplitude 4.50 cm, wavelength 10.0 cm, and frequency 1.50 Hz is observed to travel in the negative direction. **a.** What is the wave function for this wave if $y(x, t) = 4.50$ cm at $x = 0$ and $t = 0$? **b.** What is the wave function for this wave if $y(x, t) = 4.50$ cm at $x = 5.00$ cm and $t = 0$?

69. **N** If the intensity of a sound wave increases by 1%, what is the change in the sound level of the wave?

70. **N** The equation of a traveling wave is given as

$$y(x, t) = 0.50 \sin^2(2.0x - 4.0t)$$

where x and y are in meters and the time t is in seconds. Find the amplitude, wavelength, and angular frequency of the wave. *Hint*: You will need to use the trigonometric identity $\cos(2\theta) = 1 - 2\sin^2\theta$.

71. **N** A block of mass $m = 5.00$ kg is suspended from a wire that passes over a pulley and is attached to a wall (Fig. P17.71). Traveling waves are observed to have a speed of 33.0 m/s on the wire. **a.** What is the mass per unit length of the wire? **b.** What would the speed of waves on the wire be if the suspended mass were decreased to 2.50 kg?

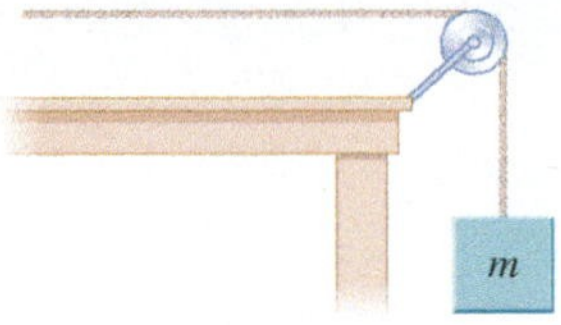

FIGURE P17.71

72. **A** The equation of a harmonic wave propagating along a stretched string is given as

$$y(x, t) = y_{\text{max}} \sin\left[2\pi\left(\frac{x}{a} - ft\right)\right]$$

where a is a constant and f is the frequency. If the magnitude of the maximum vertical speed

$$v_{y,\,\text{max}} = \left.\frac{\partial y}{\partial t}\right|_{\text{max}}$$

is equal to three times the propagation speed, what is the value of a? Express your answer in terms of y_{max}.

73. **N** A wave on a string with mass density 0.450 g/cm under tension is described by

$$y = (0.650 \text{ m}) \sin\left(4\pi x + 5\pi t - \frac{3\pi}{4}\right)$$

where x and y are in meters and t is in seconds. **a.** What is the average rate of energy transfer along the wave on the string? **b.** What is the mechanical energy content of each cycle of this wave?

74. **N** Sound waves leaving a speaker deliver a total energy of 275 J in 4.00 s to a location where the energy of the sound waves is spread over an area of 5.00 m^2. **a.** What is the intensity of the sound at this location? **b.** What is the sound level at this location? **c.** What is the intensity of the sound at a point farther away where the energy is spread over an area of 20.00 m^2? **d.** What is the sound level at this new location?

75. **N** The wave function $y(x, t) = (0.400 \text{ m}) \sin(4.00x + 25.0t)$ describes the motion of a transverse wave on a string, with x and y in meters and t in seconds. **a.** For an element of the string located at $x = 0.100$ m, what is the time interval that elapses between the first and second times that the element is located at $y = 0.300$ m? **b.** In the time interval found in part (a), what is the distance covered by the wave?

76. **N** A stationary observer receives sound waves from two tuning forks, each oscillating at a frequency of 512 Hz. One of the tuning forks is approaching the observer, and the other tuning fork is receding from the stationary observer with the same speed as the first one. The observer hears a frequency difference (beat frequency; see Chapter 18) of 4 Hz when comparing the sound from each tuning fork. Find the speed of each tuning fork. Assume the speed of sound in air is 343 m/s.

77. **N** A siren emits a sound of frequency 1.44×10^3 Hz when it is stationary with respect to an observer. The siren is moving away from a person and toward a cliff at a speed of 15 m/s. Both the cliff and the observer are at rest. Assume the speed of sound in air is 343 m/s. What is the frequency of the sound that the person will hear **a.** coming directly from the siren and **b.** reflected from the cliff?

78. **C, N** Female *Aedes aegypti* mosquitoes emit a buzz at about 4.00×10^2 Hz, whereas male *A. aegypti* mosquitoes typically emit a buzz at about 6.00×10^2 Hz. As a female mosquito is approaching a stationary male mosquito, is it possible that he mistakes the female for a male because of the Doppler shift of the sound she emits? How fast would the female have to be traveling relative to the male for him to make this mistake? Assume the speed of sound in the air is 343 m/s.

79. **N** A careless child accidentally drops a tuning fork vibrating at 450 Hz from a window of a high-rise building. How far below the window is the tuning fork when the child hears sound waves with frequency 425 Hz? Remember to account for the time required for the sound to reach the child.

80. Tsunamis, waves that displace large volumes of water, are generated by underwater disturbances, including earthquakes, volcanic eruptions, and meteorite impacts. The speed of the wave in a tsunami depends on the depth of the water d and is given by $v = \sqrt{gd}$. A meteorite 50.0 m in diameter impacts the Atlantic Ocean near Greenland, generating a tsunami that can be modeled as a series of circular waves emanating from the point of impact. Satellites measure the speed of the waves to be 300 m/s and their amplitude to be 4.00 m, a distance 1.00 km from the point of impact.

   a. **C** Why does the amplitude of the tsunami wave increase as the wave approaches the Greenland shore?

   b. **N** If the waves emanating from the point of impact are measured to have an amplitude of 0.500 m and a speed of 400 m/s, what will be their amplitude when the depth of the water decreases to 5.00 m near the Greenland coast? Assume the wave power does not diminish between the source and the coast.

81. **A** A wire with a tapered cross-sectional area is placed under a tension force F_T. The radius of the wire is given by $r = 2.50 - 0.200x$. If the density of the wire is ρ, what is the speed of a wave as a function of position along this wire?

18 Superposition and Standing Waves

Key Questions

What happens when more than one wave exists in a single medium?

How do traveling waves and standing waves differ?

Underlying Principles

1. Superposition
2. Fourier's theorem

Major Concepts

1. Boundary conditions
   a. Fixed-end reflection
   b. Free-end reflection
2. Law of reflection
3. Constructive and destructive interference
4. Standing wave
5. Beats

Special Case: Standing Waves

1. String fixed at both ends
2. Tube open at both ends
3. Tube open at one end and closed at the other end

When you talk to a friend, watch ripples on a pond, or play a guitar, you are experiencing a combination of many waves. In Chapter 17, we focused on the physics and mathematics of a single harmonic wave traveling in a medium, but we rarely encounter a single harmonic wave. More often, we find a combination of harmonic waves. In this chapter, we develop the mathematics used to describe the combination of waves, known as **superposition**. We will explore many interesting physical phenomena such as reflection, interference, standing waves, and beats.

18-1 Superposition

In Chapter 17, we studied one wave at a time traveling in some medium such as a transverse pulse on a rope (Fig. 17.4). Now let's see what happens when two (or more) waves travel in the same medium. Suppose you work with a friend to send two pulses on a rope. The rope is taut; you stand at one end of the rope, and your friend stands at the other end. You pluck the rope on your end, and your friend plucks the rope on her end; the result is two pulses traveling in opposite directions toward each other (Fig. 18.1). What happens when the two pulses meet?

Each pulse is a disturbance on the rope. Both disturbances exist on the rope simultaneously, and the result is the sum of the two disturbances (Fig. 18.2). The re-

SUPERPOSITION

Underlying Principle

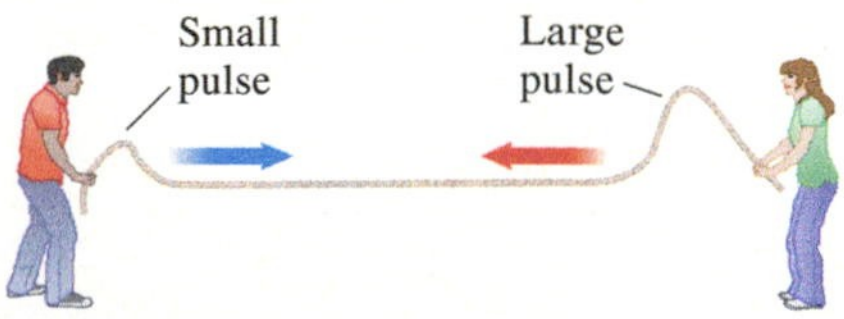

FIGURE 18.1 Two pulses on the same rope are made by people standing on opposite ends.

sulting wave function $y(x, t)$ on the rope is found by adding (**superimposing**) the wave functions $y_1(x, t)$ and $y_2(x, t)$ of the individual pulses:

$$y(x, t) = y_1(x, t) + y_2(x, t)$$

At any particular time, the wave on the rope is found by adding the contribution of the pulses at each point x. The resultant wave is the **superposition** of the individual pulses. Notice this quirk in our language: The verb is to super*im*pose, but the letters *im* are dropped from the noun: superposition.

You might have expected that the result of two waves on a rope is the superposition of the individual waves, but this result leads to quite an amazing consequence: The two pulses pass through each other (Fig. 18.2). Each individual pulse continues to travel undisturbed with its original velocity and amplitude.

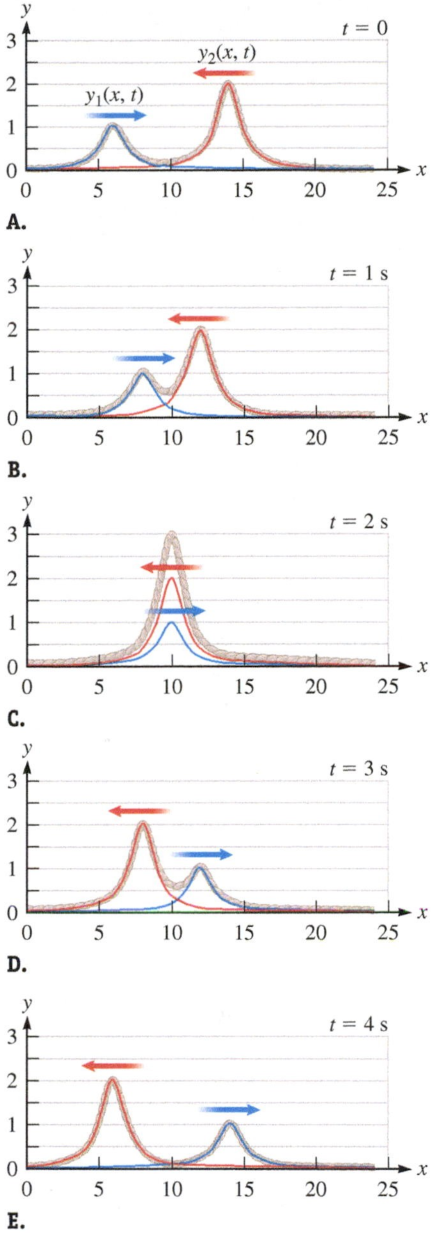

FIGURE 18.2 Two transverse pulses travel on the same rope. The individual pulse wave functions $y_1(x, t)$ in blue and $y_2(x, t)$ in red are plotted at five times. The resulting wave function $y(x, t)$ on the rope may be found by superimposing the individual pulse wave functions: $y(x, t) = y_1(x, t) + y_2(x, t)$.

CASE STUDY Part 1: A Big-Band Concert

Superposition is important whenever two or more waves exist in the same medium. In a band concert, the instruments emit sound waves through several speakers; those sound waves are reflected off the walls, the ceiling, the floor, and even the people in the room. All those sound waves travel in the same medium, the air. The music that reaches your ear is the superposition of all those waves. In addition, the sound produced by the instruments is a superposition of sound waves in the body of the instrument itself.

William Basie (1904–1984), better known as "Count" Basie, led a big-band jazz ensemble with about 20 members playing more than a dozen different instruments. Big-band music was extremely popular in the 1930s and 1940s, and there are still big bands today, including the Count Basie Orchestra. That orchestra performs concerts all over the world, including at festivals in Tokyo and Toronto, at universities, and in formal concert halls. In this case study, we'll explore how a concert hall or an outdoor venue affects the music. Of course, the richness of a big band comes from the great number of instruments, each adding its own voice to the music. So, we'll also explore how three of the big band's instruments (the flute, clarinet, and guitar) produce their unique sounds.

CONCEPT EXERCISE 18.1

As shown in Figure 18.3, two pulses traveling along the same rope toward each other have the same amplitude. Draw a sketch of the rope at the instant the pulses completely overlap.

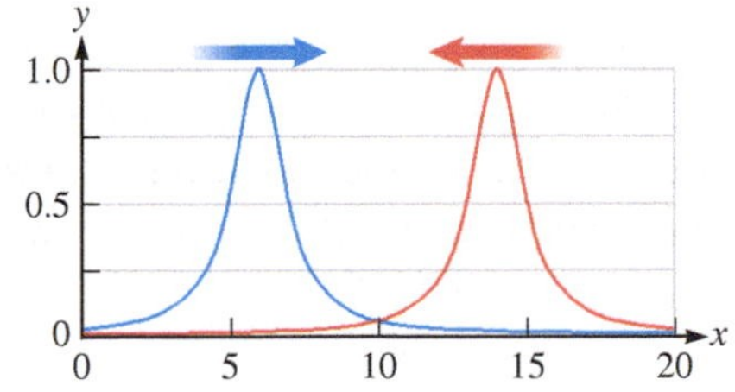

FIGURE 18.3

CONCEPT EXERCISE 18.2

As shown in Figure 18.4, two pulses traveling along the same rope toward each other have the same amplitude, but one is upside down. Draw a sketch of the rope at the instant the pulses completely overlap.

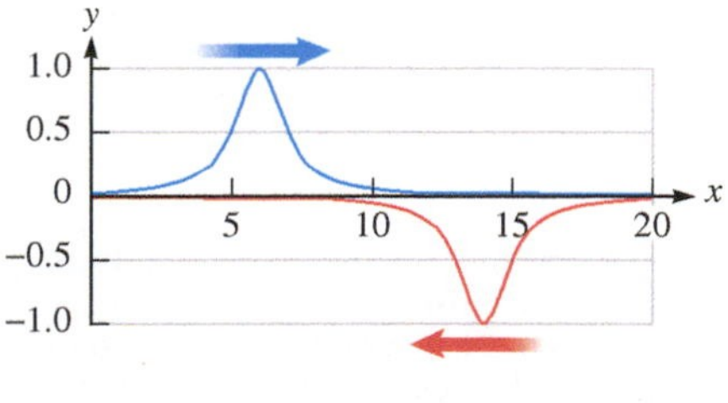

FIGURE 18.4

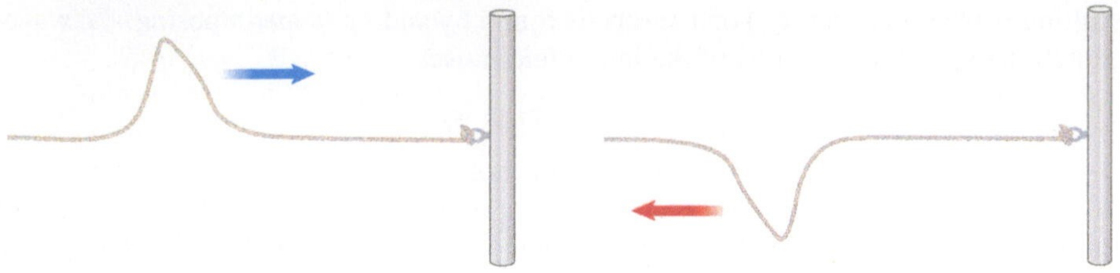

A. B.

FIGURE 18.5 When a pulse reflects from a fixed end, the reflected pulse is inverted.

18-2 Reflection

When you shout into a gorge or at a large wall, you hear your voice echo because when a wave travels to the end of a medium—in this case, the air—the wave is reflected back into the medium. The reflection requires that there be a **boundary**, or an end to the medium. The nature of the reflected wave depends on the state of the boundary known as the **boundary conditions**. Here we consider two important boundary conditions.

BOUNDARY CONDITIONS: FIXED-END REFLECTION AND FREE-END REFLECTION

Major Concept

Fixed-End Reflection

To study reflection, let's look at a pulse on a rope. A pulse shaped something like a shark's fin travels to the right along a horizontal rope (Fig. 18.5A). The rope is tied to a rigid ring on the right, so the right end is *fixed* and cannot move. As the pulse—called the *incident wave*—approaches the fixed end, it exerts a force on the rigid ring. According to Newton's third law, the ring exerts a force on the rope; as a result, there is a *reflected* pulse that looks like an inverted shark's fin traveling to the left (Fig. 18.5B).

Imagine that instead of having a rope fixed on the right end, we had two shark-fin pulses traveling toward each other on a long rope (Fig. 18.6). One pulse looks like a shark's fin traveling to the right, and the other is an inverted shark's fin traveling to the left. The pulses pass through each other as expected. Now cover up the right-hand portion of the rope shown in the gray region and labeled "imaginary part." The remaining rope on the left looks like the incident pulse and the reflected pulse in Figure 18.5. So, the formation of the reflected pulse is similar to the superposition of two pulses on a rope. To find the reflected pulse, just *imagine* a pulse traveling along a similar rope toward the boundary at the same speed as the incident pulse. The imaginary pulse must be inverted and flipped so that when the two meet at the fixed boundary, they cancel each other.

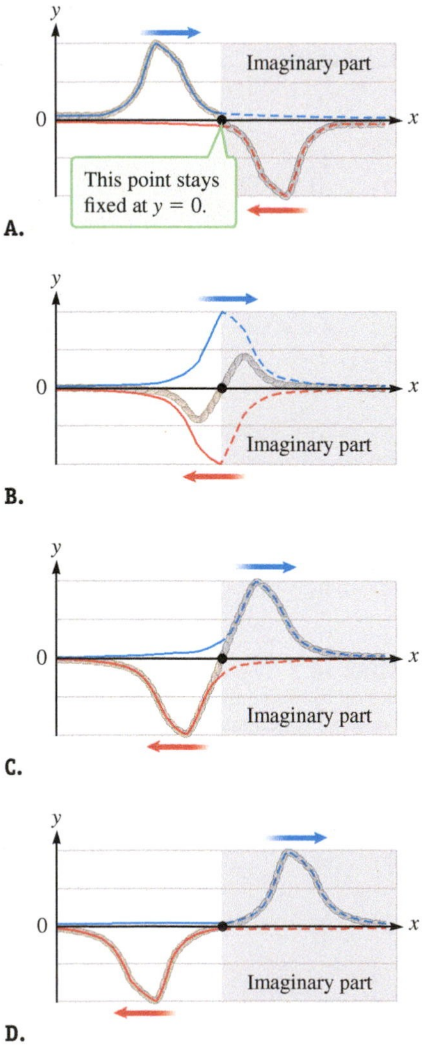

FIGURE 18.6 A pulse travels to the right along a rope that is fixed at one end. Imagine a second pulse traveling to the left. The two pulses must cancel at the fixed end of the rope, so the imaginary pulse must be inverted.

Free-End Reflection

What happens if the right-hand end of the rope is attached to a ring that is *free* to slide along a pole (Fig. 18.7)? As the pulse approaches the free end, the ring slides upward along the pole, rising to a height that is twice that of the pulse (Fig. 18.7B). The ring is pulled back down by the tension in the rope, and a reflected pulse travels to the left. In the case of a free end, the reflected pulse is upright and looks like a shark's fin traveling to the left.

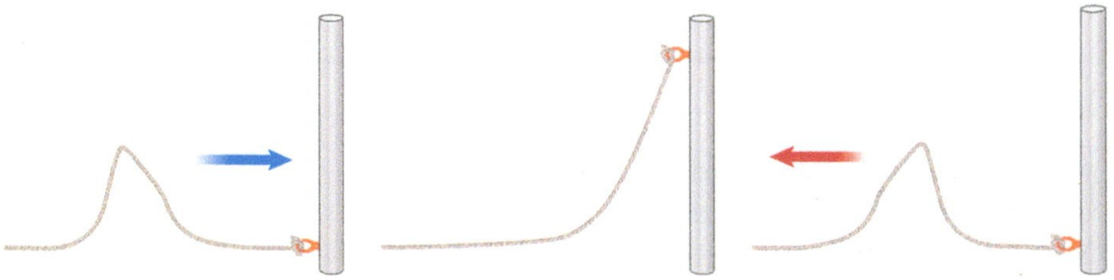

A. B. C.

FIGURE 18.7 When a pulse reflects from a free end, the reflected pulse is upright.

As in the case with a fixed right end, it is helpful to imagine free-end reflection as the superposition of two pulses along a rope. In Figure 18.8, two pulses shaped like shark's fins travel toward each other and pass through each other. Again cover up the right-hand side of the rope shown in gray and compare the remaining portion of the rope with the incident and reflected pulses in Figure 18.7. As before, the formation of the reflected pulse is similar to the superposition of two pulses on the rope. In this case, however, the imaginary pulse must be flipped but not inverted so that when the two pulses meet at the free boundary, they add together.

FIGURE 18.8 A pulse travels to the right along a rope that is free at the right-hand end, imagine a second pulse traveling to the left. The two pulses add together, so when the two pulses "meet" at the free end of the rope, the amplitude is doubled.

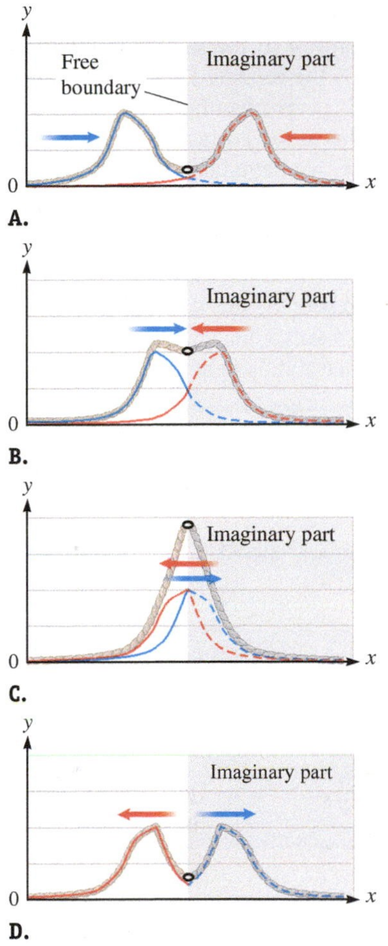

CONCEPT EXERCISE 18.3

A wave pulse travels to the left on a rope as shown in Figure 18.9. Draw a sketch of the reflected pulse if the left end is

a. fixed. **b.** free.

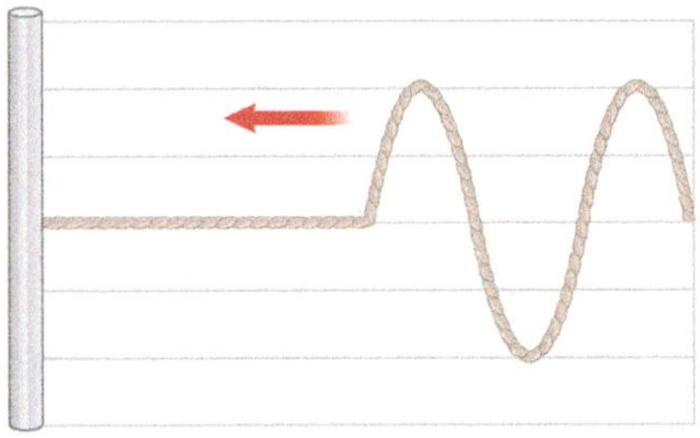

FIGURE 18.9

Law of Reflection

A wave on a rope is one-dimensional, so a reflected wave must simply travel in the opposite direction as the incident wave. A two- or three-dimensional wave does not necessarily reflect along the path of the incoming wave. Figure 18.10 shows an incident ray that reflects from a boundary. The incident ray makes an angle θ_i with the line that is perpendicular to the boundary (called the **normal**), and the reflected ray makes an angle θ_r with that perpendicular line. According to the **law of reflection**, the incident angle equals the reflected angle:

$$\theta_i = \theta_r \tag{18.1}$$

LAW OF REFLECTION

✪ **Major Concept**

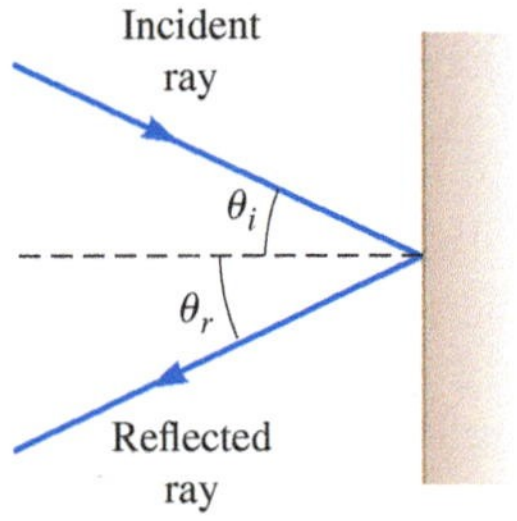

FIGURE 18.10
Law of reflection, $\theta_i = \theta_r$.

EXAMPLE 18.1 A Whisper Dish

Sophia, standing roughly 30 ft from her father, whispers. He can clearly understand her because he is standing in front of a whisper dish (Fig. 18.11A). The dish is designed so that the sound waves from a distant source such as Sophia are reflected toward a single point. The father's ears are near that point, so he hears the sum of these many reflected waves.

The sound waves near Sophia are roughly hemispherical. Far from Sophia, the waves are nearly plane waves, and the rays are a series of parallel lines (Section 17-7). Figure 18.11B shows three such rays near the surface of a spherically shaped whisper dish. Sketch the three corresponding reflected rays.

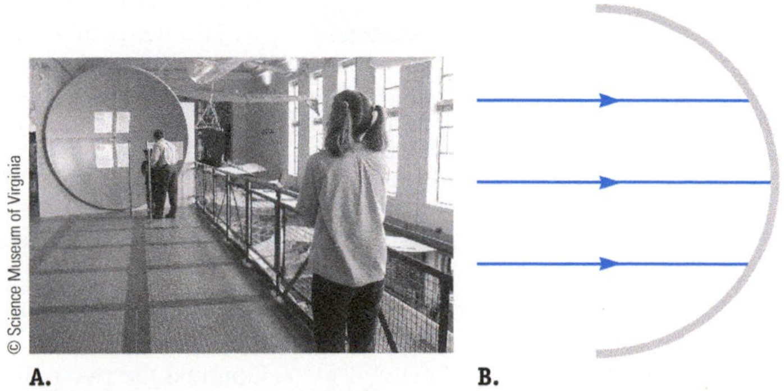

FIGURE 18.11 **A.** A whisper dish in use. **B.** Parallel rays strike the whisper dish.

Example continues on page 526 ▶

INTERPRET and ANTICIPATE

The law of reflection tells us how to draw the reflected rays. We expect from the description of the whisper dish that the reflected rays intersect.

SOLVE

According to the law of reflection (Eq. 18.1), the incident angle equals the reflected angle. Both angles are measured from the line that is perpendicular to the surface. For a sphere, radii are perpendicular to the surface (along the dashed lines in Fig. 18.12). In the figure, we have labeled the incident rays A, B, and C, and the corresponding reflected rays a, b, and c.

CHECK and THINK

As expected, the three reflected rays (a, b, and c) intersect, and even a soft whisper from a distant source can easily be heard if a listener stands near the dish close to the point of intersection.

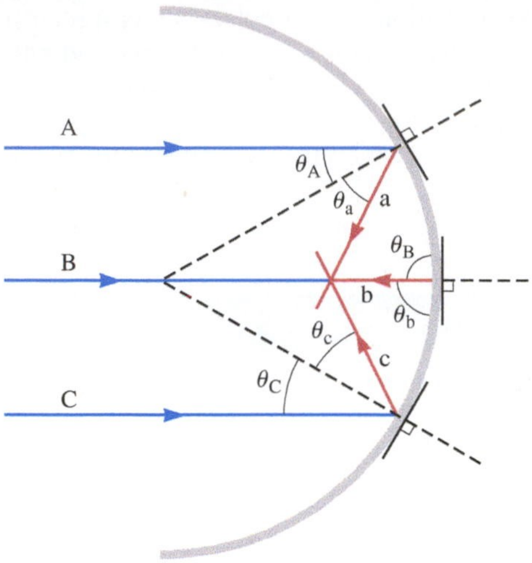

FIGURE 18.12 Reflected rays are shown in red.

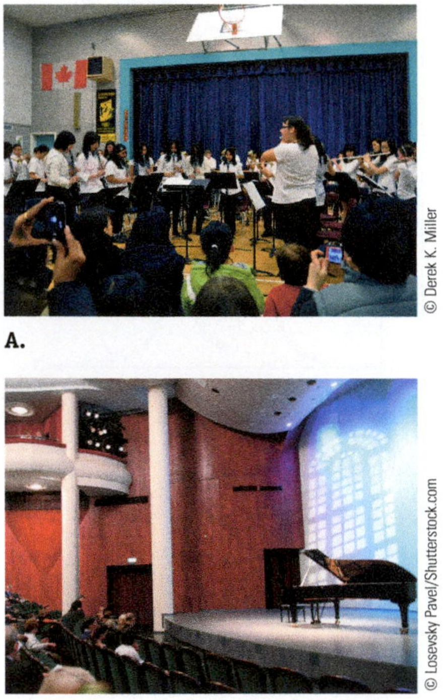

A.

B.

FIGURE 18.13 A. Music sounds dull in a school gym or cafeteria because the walls and ceiling are smooth. **B.** Walls in concert halls are made of many rough and curved panels.

CASE STUDY Part 2: Designing Concert Halls and Auditoriums

Musicians who go on tour play in many different venues, and they know that some rooms make better concert halls than others. A major factor that determines a good hall from a bad one is reflection of the sound.

When sound waves are reflected off a surface, we may hear either an echo or a reverberation. A sound affects the human brain for about 0.1 s. If the reflected wave reaches your ears in less than 0.1 s, you perceive that the sound is prolonged and hear a reverberation. Because the reflection must reach you in a short time, you generally experience reverberation in a small space (such as when you sing in the shower). If, on the other hand, the reflected wave takes longer than 0.1 s to reach your ears, the reflection is heard as a second separate sound (an echo). You probably have heard an echo by shouting at the wall of a canyon or a building.

A concert hall designer does not want the hall to have an obvious echo, but does want it to have just the right amount of reverberation. So, the hall should be designed so that reflections come to each person from many different directions. The walls and ceiling of a well-designed concert hall are not made of smooth, hard materials such as concrete, which is why school gyms and cafeterias don't make good concert halls (Fig. 18.13A). During a concert in a school's gym, a listener will hear reflections off the walls (and perhaps the ceiling). Because the walls are smooth and hard, the reflected sound heard by any individual in the audience must have come from one or just a few places. Musicians say that such a room is "dead."

The walls of a well-designed concert hall are made of a rough material and are broken into several panels oriented at various angles so that the reflected sound heard by any one individual comes from many places in the room (Fig. 18.13B). Such a room with the correct amount of reverberation is said to be "live." Because there are no people in the audience reflecting the sound waves, musicians report that even a well-designed concert hall can sound dead when they rehearse.

18-3 Interference

Contrast the two pulses in Figure 18.2 with the motion of two particles on a collision path. The waves pass through each other, but the particles don't. So, we need to think about the encounter of two (or more waves) differently than the way we think about

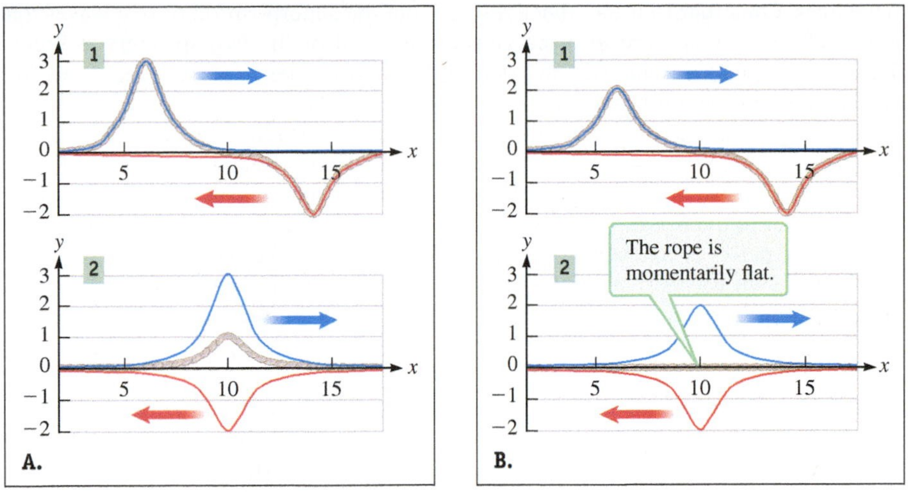

FIGURE 18.14 **A.** 1 Two pulses (red and blue) move toward each other. 2 The resulting pulse exhibits partial destructive interference. **B.** 1 Two pulses (red and blue) move toward each other. 2 The resulting pulse is completely flat, so it shows fully destructive interference.

colliding particles. When two waves exist in the same medium, they don't collide; they *interfere* with each other. Their interference creates a new wave in the medium.

Interference in Pulses and One-Dimensional Waves

The wave pattern in Figure 18.2 results from the interference of the two traveling wave pulses. When the resulting pulse's height is greater than that of both original waves, the result is *constructive interference* as in Figure 18.2C.

When the resulting pulse's height is smaller than that of each original wave as in Figure 18.14A, the result is *destructive interference*. If the two pulses have the same amplitude but one is inverted with respect to the other, the resulting wave profile is momentarily flat as in Figure 18.14B.

Now suppose there are two harmonic waves traveling in one dimension in the same medium. For example, imagine two speakers sending out harmonic sound waves in a column of air as in Figure 18.15. Each wave is represented by a single cosine function as in Equation 17.16, $\Delta P = \Delta P_{\text{max}} \cos(kx - \omega t)$. Such a sound wave wouldn't sound like music; it is just a single tone that you might have heard such as the tone broadcast by an emergency alarm system. Because the waves are traveling in the same medium, they have the same speed. In addition, suppose these speakers produce the same amplitude for each wave.

Now let's further assume the speakers are arranged so that these two waves are in phase, which means that all the peaks of one wave line up with all the peaks of the other wave (Fig. 18.15A). The superposition of two harmonic waves that are identical produces a harmonic wave that has twice the amplitude of either one of the origi-

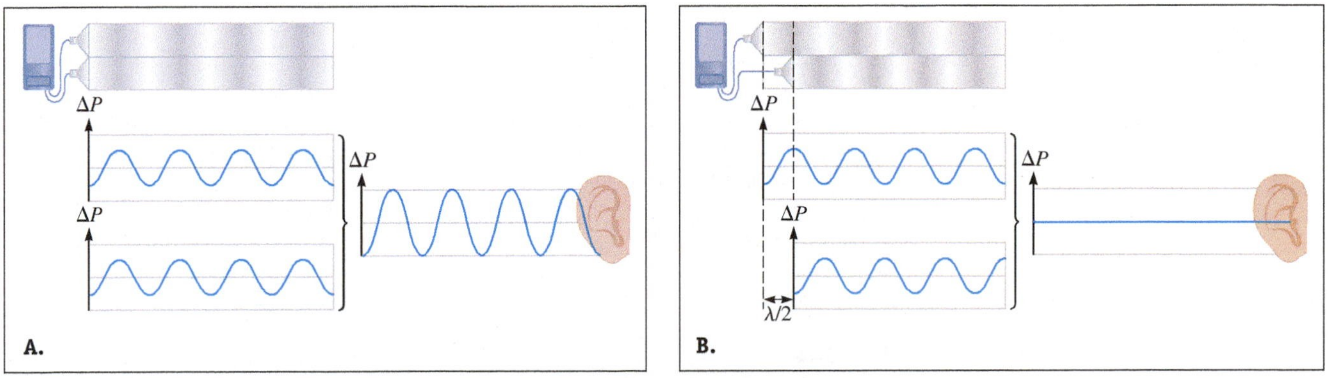

FIGURE 18.15 **A.** The two speakers create two harmonic sound waves that are in phase. The amplitude of the wave heard by the person is two times greater than the sound heard from either speaker alone. **B.** The two speakers create two harmonic sound waves that are 180° or $\lambda/2$ out of phase. The two waves cancel, and the person cannot hear a sound.

nal waves. **Constructive interference** results from the superposition of two waves that are **in phase**. If you were to place your ear in front of the two speakers in Figure 18.15A, the pressure changes on your eardrum would be twice as great with two speakers than it would be if there were only one speaker.

CONSTRUCTIVE AND DESTRUCTIVE INTERFERENCE

Major Concept

You might have guessed that two speakers always produce twice as much pressure variation on your eardrum as one speaker, but it is not true in all cases. In Figure 18.15B, the speakers have been rearranged so that one speaker is half of a wavelength closer to the person's ear. Now the two waves are out of phase by 180°, meaning that the peaks of one wave line up with the valleys of the other wave. The superposition of two waves that are 180° out of phase results in **destructive interference**. If you place your ear in front of the two speakers in Figure 18.15B, the pressure change on your eardrum would be zero; the pressure would equal atmospheric pressure, and you wouldn't hear a sound.

What would happen if the speaker were moved another $(1/2)\lambda$ closer to the person's ear so that the two speakers were separated by λ? The two waves would once again be in phase, and the result would be constructive interference. If the difference in the distance traveled by the two waves is an integer multiple of the wavelength, the result is constructive interference. Mathematically, we write the condition for constructive interference as

CONDITION FOR CONSTRUCTIVE AND DESTRUCTIVE INTERFERENCE

Major Concept

$$\Delta d = n\lambda \qquad (n = 0, 1, 2, 3, \ldots) \tag{18.2}$$

where Δd is the difference in the distance traveled by the two waves. If the difference in the distance traveled by the two waves is a half-integer number of wavelengths, the result is destructive interference:

$$\Delta d = \frac{n}{2}\lambda \qquad (n = 1, 3, 5, \ldots) \tag{18.3}$$

Two- and Three-Dimensional Interference

FIGURE 18.16 Two circular waves in the water interfere constructively and destructively.

We usually listen to sound waves that travel in three dimensions instead of one-dimensional sound waves in a column of air. Two- and three-dimensional waves also interfere constructively and destructively according to the conditions given in Equations 18.2 and 18.3. For example, circular waves on the surface of the water interfere constructively and destructively. In Figure 18.16, the interference pattern of two circular waves on the surface of water is best seen in the lower center of the photo, where you can see a grid of light and dark spots.

Figure 18.17 models the interference of two identical circular harmonic waves. The solid circles represent the wave's peaks, and the dashed circles represent the valleys; so, the distance between any two solid circles or between any two dashed circles is one wavelength λ. As shown in Figure 18.17A, constructive interference occurs at any point where two peaks or two valleys intersect. Destructive interference occurs at any point where one peak meets one valley.

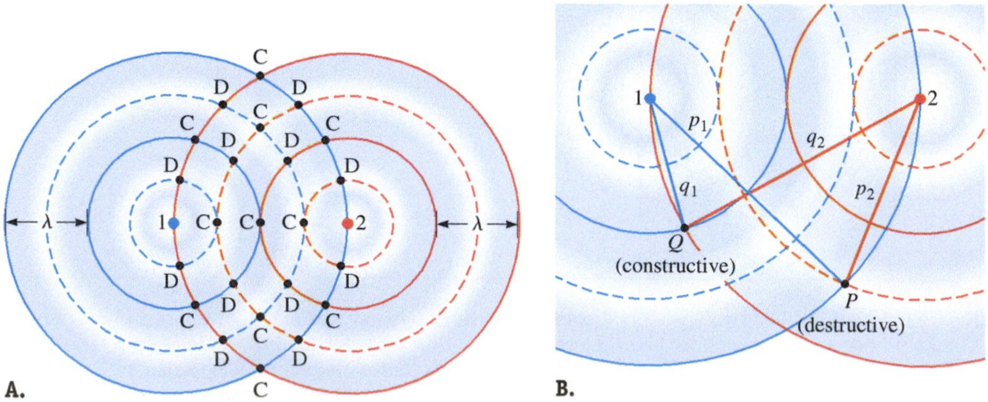

FIGURE 18.17 A. Constructive interference occurs where two valleys or two peaks meet, as indicated by "C." Destructive interference occurs where a valley and a peak meet, as indicated by "D." **B.** Close-up of part A.

At point Q in Figure 18.17B, there is constructive interference. The distance from source 1 to point Q is q_1, and the distance from source 2 to Q is q_2. From the figure, we find that

$$q_1 = \lambda$$

and

$$q_2 = 2\lambda$$

The difference in the distance to point Q traveled by the two waves is

$$\Delta d = |q_2 - q_1| = |2\lambda - \lambda| = \lambda$$

This equation fits Equation 18.2, which says that for constructive interference, Δd must be an integer multiple of λ. For point Q, that integer is $n = 1$.

At point P in Figure 18.17B, there is destructive interference. The distance from source 1 to point P is p_1, and distance from source 2 to P is p_2. From the figure, we find that

$$p_1 = 2\lambda$$

and

$$p_2 = \frac{3}{2}\lambda$$

The difference in the distance to point P traveled by the two waves is

$$\Delta d = |p_2 - p_1| = \left|\frac{3}{2}\lambda - 2\lambda\right| = \frac{1}{2}\lambda$$

This equation fits Equation 18.3 which says that for destructive interference, Δd must be a half-integer multiple of λ. For point P, the integer $n = 1$.

CONCEPT EXERCISE 18.4

Noise cancellation headphones use a microphone to pick up background noise and a speaker to produce a sound wave canceling the noise. These headphones are used on planes or in noisy dormitory rooms to reduce distracting sounds, allowing the wearer to study a physics textbook. Figure 18.18 is a graph of ΔP versus t for the noise produced by a human voice (perhaps a roommate). Sketch a corresponding graph of ΔP versus t that must be generated by the speaker so as to cancel the noise.

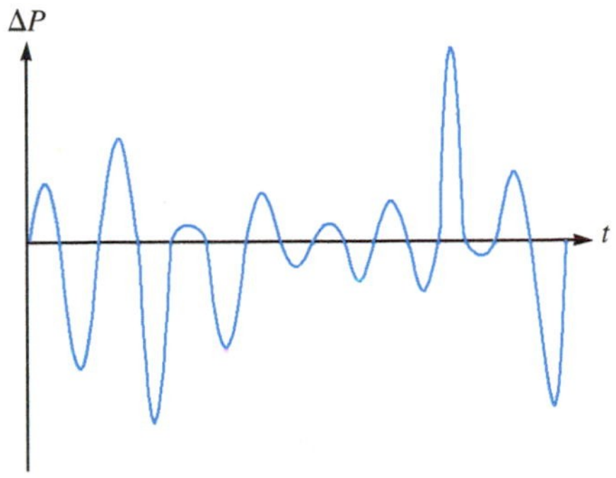

FIGURE 18.18

EXAMPLE 18.2 CASE STUDY Choosing Your Seats

When buying tickets to a concert, you are asked to choose your seats. Usually, seat prices are based on the view, but you are interested in the sound. Suppose you are offered either seats D8 and D9 or E4 and E5. (Figure 18.19 includes distances that are not usually provided on the seating chart.) You know that before a performance, instruments may be tuned with an electronic tuner that can produce a harmonic sound wave with a frequency of 440 Hz. So, you assume it is important to hear the 440-Hz note clearly. Also assume each speaker produces hemispherical waves and the speed of sound in this (hot, humid) auditorium is 350 m/s.

A Should you choose the seats in row D or E?

INTERPRET and ANTICIPATE

We'd like to hear the 440-Hz note. So, we do not want to choose seats if there is destructive interference between the speakers at this frequency.

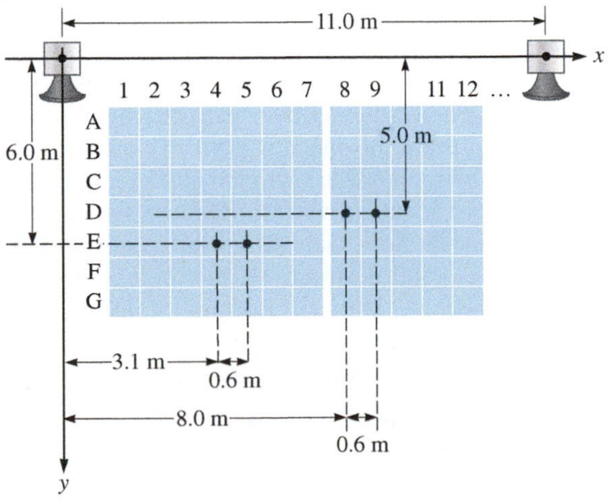

FIGURE 18.19 Not drawn to scale; for example, the aisle (between 7 and 8) is not narrower than the width of a seat.

Example continues on page 530 ▶

SOLVE

First, find the corresponding wavelength of sound in air (Eq. 17.8).	$v = \lambda f$ (17.8) $\lambda = \frac{v}{f} = \frac{350\text{ m/s}}{440\text{ Hz}} = 0.80\text{ m}$
Next, find the distance between each speaker and each seat. Let's start with speaker 1 and the seats in row D. (We keep an extra significant figure in this intermediate step.)	$d_1 = \sqrt{x^2 + y^2} = \sqrt{x^2 + (5.0\text{ m})^2}$ D8: $d_1 = \sqrt{(8.0\text{ m})^2 + (5.0\text{ m})^2} = 9.43\text{ m}$ D9: $d_1 = \sqrt{(8.6\text{ m})^2 + (5.0\text{ m})^2} = 9.95\text{ m}$
Now, do the same for speaker 2 and the seats in row D.	$d_2 = \sqrt{(11\text{ m} - x)^2 + y^2} = \sqrt{(11\text{ m} - x)^2 + (5.0\text{ m})^2}$ D8: $d_2 = \sqrt{(11\text{ m} - 8.0\text{ m})^2 + (5.0\text{ m})^2} = 5.83\text{ m}$ D9: $d_2 = \sqrt{(11\text{ m} - 8.6\text{ m})^2 + (5.0\text{ m})^2} = 5.55\text{ m}$
Next, find the distance between speaker 1 and the seats in row E.	$d_1 = \sqrt{x^2 + y^2} = \sqrt{x^2 + (6.0\text{ m})^2}$ E4: $d_1 = \sqrt{(3.1\text{ m})^2 + (6.0\text{ m})^2} = 6.75\text{ m}$ E5: $d_1 = \sqrt{(3.7\text{ m})^2 + (6.0\text{ m})^2} = 7.05\text{ m}$
Find the distance between speaker 2 and the seats in row E.	$d_2 = \sqrt{(11\text{ m} - x)^2 + y^2} = \sqrt{(11\text{ m} - x)^2 + (6.0\text{ m})^2}$ E4: $d_2 = \sqrt{(11\text{ m} - 3.1\text{ m})^2 + (6.0\text{ m})^2} = 9.92\text{ m}$ E5: $d_2 = \sqrt{(11\text{ m} - 3.7\text{ m})^2 + (6.0\text{ m})^2} = 9.45\text{ m}$
Find the absolute value of the difference in the distance Δd between a seat and each speaker.	D8: $\|d_2 - d_1\| = \|\Delta d\| = \|5.83\text{ m} - 9.43\text{ m}\| = 3.60\text{ m}$ D9: $\|\Delta d\| = 4.40\text{ m}$ E4: $\|\Delta d\| = 3.17\text{ m}$ E5: $\|\Delta d\| = 2.40\text{ m}$
Divide by 0.80 m to write each difference in terms of wavelength.	D8: $\|\Delta d\|/\lambda = 3.60\text{ m}/0.80\text{ m} = 4.5$ D8: $\|\Delta d\| = 4.5\lambda$ D9: $\|\Delta d\| = 5.5\lambda$ E4: $\|\Delta d\| = 4.0\lambda$ E5: $\|\Delta d\| = 3.0\lambda$
The best seats are the ones where there is constructive interference, which are the ones where Δd is an integer multiple of the wavelength.	Choose seats E4 and E5.

CHECK and THINK

Destructive interference at seats D8 and D9 implies that a 440-Hz note cancels at these two seats. Does it mean that there is silence there? No. The situation described in this example is *ideal* (as opposed to real). There are only two speakers, and they only emit two identical harmonic waves. In an actual auditorium, no place is perfectly quiet because (1) musical instruments produce a superposition of harmonic waves at many different frequencies (Section 18-9); (2) speakers usually do not produce identical waves; (3) there are often more than two speakers, and the sound reflects from many surfaces such as the walls, ceiling, and seats; and (4) the number and location of the constructive (and destructive) interference points is continually changing as the waves from each speaker travel outward. Because of destructive interference, music aficionados nevertheless claim that the sound is richer at some seats than at others.

B How would your answer to part A change if the tuner were to produce a 220-Hz wave instead of a 440-Hz wave?

INTERPRET and ANTICIPATE

It isn't necessary to redo all the work in part A. Just find how changing the frequency of the tone played through the speakers affects the answers.

SOLVE

Because the frequency is lower, the corresponding wavelength is longer.

$$\lambda = \frac{v}{f} = \frac{350\text{ m/s}}{220\text{ Hz}} = 1.6\text{ m}$$

Divide the absolute value of each difference Δd by 1.6 m to write each difference in terms of the new wavelength.

D8: $|\Delta d| = 2.3\lambda$ D9: $|\Delta d| = 2.8\lambda$

E4: $|\Delta d| = 2.0\lambda$ E5: $|\Delta d| = 1.5\lambda$

In this case, the only seat where the note cannot be heard is E5. Because both seats in row D can hear the note, they are the better choice.

Choose seats D8 and D9.

CHECK and THINK

Each musical instrument emits many harmonic waves. Because the interference pattern depends on the wavelength (or, equivalently, frequency), only a few of the instrument's harmonic waves interfere at any one point in the auditorium.

18-4 Standing Waves

A guitar produces a sound wave in the air that travels outward from the guitar. The wave on the guitar string is *not* a traveling wave, however. Each part of the string oscillates back and forth as it would for any wave on a string, but the wave pattern (the envelope) does not move along the string's length (Fig. 18.20). Such a wave is known as a **standing wave**.

STANDING WAVE **Major Concept**

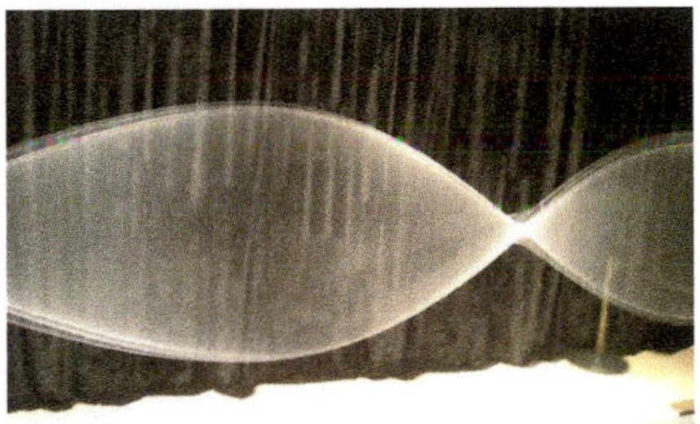

© Debora Katz

FIGURE 18.20 For a standing wave, each part of the string oscillates back and forth, but the wave pattern does not move along the string's length.

Producing a Standing Wave

You and a friend might produce a standing wave on a stretched string by repeatedly shaking opposite ends in a certain way. Suppose the string's length is L. You and your friend agree to create identical harmonic waves of wavelength $\lambda = 2L/3$. The topmost panels of Figure 18.21 show these two waves traveling on the same string at four different times. The waves have the same amplitude and wavelength. The next row of panels shows that the superposition of these two waves is a standing wave that does not move to the right or the left.

We can represent pieces of the string with beads. Any bead on the string—for example, the bead shown in black in Figure 18.21 (middle)—oscillates up and down in simple harmonic motion. The wave pattern is fixed, so each bead's amplitude is determined by its position along the string. Figure 18.22 shows the string at five different times. The black bead has a relatively small amplitude. Some beads do not oscillate at all and are said to be at the **nodes** of the standing wave. Beads with the greatest amplitude are at the **antinodes**.

Wave Function of a Standing Wave

Because a standing wave on a string may be produced by the superposition of two traveling waves, we can find the wave function for a standing wave by adding together the wave functions for two identical harmonic waves traveling in the opposite direction. A harmonic wave traveling in the positive x direction is given by Equation 17.4:

$$y_1(x, t) = y_{max} \sin(kx - \omega t) \tag{18.4}$$

and a harmonic wave traveling in the negative x direction is given by

$$y_2(x, t) = y_{max} \sin(kx + \omega t) \tag{18.5}$$

The resulting wave is the superposition of the two traveling waves:

$$y(x, t) = y_1(x, t) + y_2(x, t)$$
$$y(x, t) = y_{max} \sin(kx - \omega t) + y_{max} \sin(kx + \omega t)$$
$$y(x, t) = y_{max}[\sin(kx - \omega t) + \sin(kx + \omega t)]$$

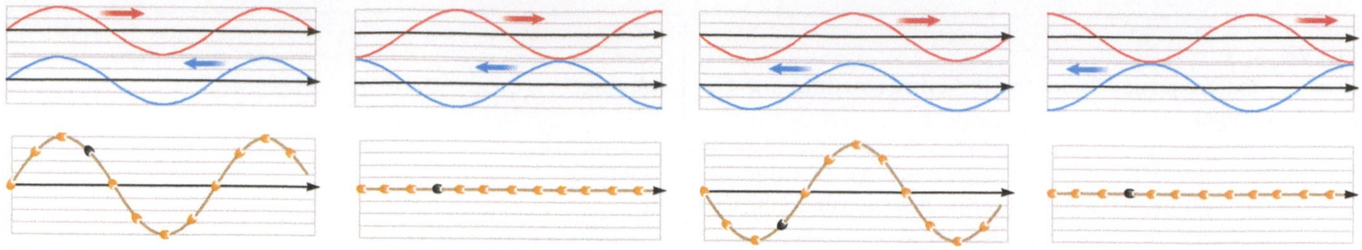

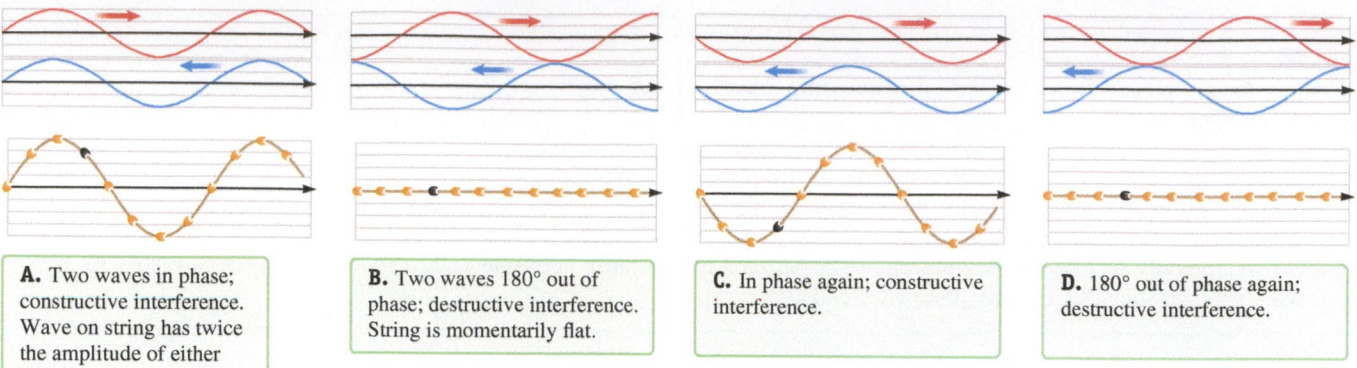

A. Two waves in phase; constructive interference. Wave on string has twice the amplitude of either traveling wave.

B. Two waves 180° out of phase; destructive interference. String is momentarily flat.

C. In phase again; constructive interference.

D. 180° out of phase again; destructive interference.

FIGURE 18.21 Two harmonic waves traveling in opposite directions on the same string. The superposition of the two waves results in a standing wave. The wave pattern does not move along the string. Each bead on the rope moves up and down in simple harmonic motion, and its amplitude is determined by its position along the rope.

We can use a trigonometric identity $\sin\alpha + \sin\beta = 2\sin\left[\frac{1}{2}(\alpha + \beta)\right]\cos\left[\frac{1}{2}(\alpha - \beta)\right]$ (Appendix A) with $\alpha = (kx - \omega t)$ and $\beta = (kx + \omega t)$ to simplify this expression:

$$y(x, t) = y_{max}\{2\sin\tfrac{1}{2}[(kx - \omega t) + (kx + \omega t)]\cos\tfrac{1}{2}[(kx - \omega t) - (kx + \omega t)]\}$$

$$y(x, t) = 2y_{max}\sin(kx)\cos(-\omega t)$$

$$y(x, t) = [2y_{max}\sin(kx)]\cos(\omega t) \qquad (18.6)$$

since $\cos(-\omega t) = \cos\omega t$.

Equation 18.6 is the wave function for a standing transverse wave, which appears to be very different mathematically from Equation 18.4 for a traveling wave. However, comparing Equation 18.6 with Equation 16.3 for a simple harmonic oscillator, $y(t) = y_{max}\cos(\omega t + \varphi)$, shows that those two equations look mathematically similar. Equation 18.6 describes a simple harmonic oscillator with a phase constant φ of zero and an amplitude given by the absolute value of the term in square brackets, $|2y_{max}\sin(kx)|$. Equation 18.6 supports what we find graphically in Figures 18.21 and 18.22: *A standing harmonic wave is made of many simple harmonic oscillators, each of whose amplitude is determined by its position x.*

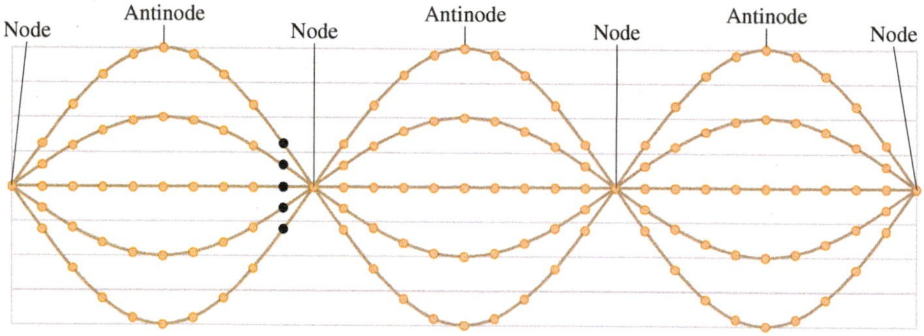

FIGURE 18.22 The amplitude at the nodes of a standing wave is zero; at the nodes, the string remains at rest. The amplitude at the antinodes is a maximum.

Position of Nodes and Antinodes

Because the nodes of a standing wave have zero amplitude, we can find the position of the nodes by setting $|2y_{max}\sin(kx)|$ to zero and solving for x:

$$|2y_{max}\sin(kx)| = 0$$

$$\sin(kx) = 0$$

$$kx = n\pi \qquad (n = 0, 1, 2, 3, \dots)$$

Substitute $k = 2\pi/\lambda$ (Eq. 17.5) for the wave number k:

$$x = \frac{n\pi}{k} = \frac{n\pi}{(2\pi/\lambda)}$$

$$x = n\frac{\lambda}{2} \qquad (n = 0, 1, 2, 3, \dots) \qquad (18.7)$$

Equation 18.7 gives the position of the nodes, showing that the distance between adjacent nodes (Fig. 18.22) is $\lambda/2$.

We can find the position of the antinodes by finding the locations of the maximum amplitude. Set $|2y_{max}\sin(kx)|$ equal to $2y_{max}$ and solve for x:

$$|2y_{max}\sin(kx)| = 2y_{max}$$
$$\sin(kx) = \pm 1$$
$$kx = \frac{1}{2}\pi, \frac{3}{2}\pi, \frac{5}{2}\pi$$
$$kx = (n + \tfrac{1}{2})\pi \qquad (n = 0, 1, 2, 3, \dots)$$

Again, substitute $k = \dfrac{2\pi}{\lambda}$ for wave number k and solve for the positions of the antinodes:

$$x = \left(n + \frac{1}{2}\right)\frac{\lambda}{2} \qquad (n = 0, 1, 2, 3, \dots) \tag{18.8}$$

The distance between adjacent antinodes is one-half of a wavelength ($\lambda/2$), and the distance between a node and an adjacent antinode is a one-fourth of a wavelength ($\lambda/4$).

Standing Waves in Musical Instruments

When a musician plays an instrument, a standing wave is established in the instrument. How can a solo musician do so when, according to Figure 18.21, we need two traveling waves to make a standing wave? Imagine that you attach one end of a string to a fixed ring as in Figure 18.5, but instead of creating a pulse shaped like a shark's fin you oscillate your end of the string in simple harmonic motion. A harmonic wave travels from you toward the ring, reflects from the ring, and creates a second wave traveling toward you. The reflected wave has the same amplitude and wavelength as the incident wave you created. The superposition of these two waves creates a standing wave on the rope just as if you had a friend at the other end helping you. The basis of musical instruments is the formation of standing waves from the superposition of reflected waves. In the next three sections, we focus on standing waves that form in three instruments in a big-band ensemble: the guitar, the flute, and the clarinet.

EXAMPLE 18.3 Making Standing Waves

In a large laboratory, two waves with the same amplitude and wavelength travel in opposite directions along the same rope. The two waves interfere, and the result is a standing wave on the rope. One traveling wave is described by

$$y_1(x, t) = 0.150\sin(2.50x - 6.35t) \tag{1}$$

where numerical values have the appropriate SI units.

A Write an expression for the standing wave and find the amplitude of the antinodes.

INTERPRET and ANTICIPATE

Equation (1) is the wave function of a wave traveling in the positive x direction. The identical wave traveling in the negative x direction must by given by $y_2(x, t) = 0.150\sin(2.50x + 6.35t)$. The sum of the two waves $y_1 + y_2$ is a standing wave.

SOLVE

Identify y_{max}, k, and ω from Equation (1) and substitute these values into Equation 18.6 to find an expression for the standing wave $y(x, t)$.

$$y_{max} = 0.150\text{ m}$$
$$k = 2.50\text{ rad/m}$$
$$\omega = 6.35\text{ rad/s}$$
$$y(x, t) = 2(0.150)\sin(2.50x)\cos(6.35t)$$
$$y(x, t) = 0.300\sin(2.50x)\cos(6.35t)$$

Example continues on page 534 ▶

The amplitude A of an antinode is the maximum amplitude of the standing wave, and it is twice the amplitude of either of the traveling waves.	$A = 2y_{max} = 0.300\text{ m}$

B What is the separation between adjacent nodes? What is the separation between a node an adjacent antinode?

SOLVE The distance between adjacent nodes is half a wavelength (Eq. 18.7). Use Equation 17.5 to find λ from k.	$\frac{\lambda}{2} = \frac{1}{2}\left(\frac{2\pi}{k}\right) = \frac{\pi}{k} = \frac{\pi}{2.50\text{ rad/m}} = 1.257\text{ m}$
The distance between a node and an adjacent antinode is one-fourth of a wavelength.	$\frac{\lambda}{4} = \frac{1}{4}\left(\frac{2\pi}{k}\right) = \frac{\pi}{2k} = \frac{\pi}{2(2.50\text{ rad/m})} = 0.6283\text{ m}$

C If there are seven nodes, including a node on one end, and if there is an antinode on the other end, find the length of the rope. (The antinode could be generated by attaching one end of the rope to a motor.)

INTERPRET and ANTICIPATE
It is helpful to sketch a standing wave with seven nodes as described (Fig. 18.23). We can then use our results from part B to find the length of the rope by adding up the distance between adjacent nodes and one node adjacent to an antinode.

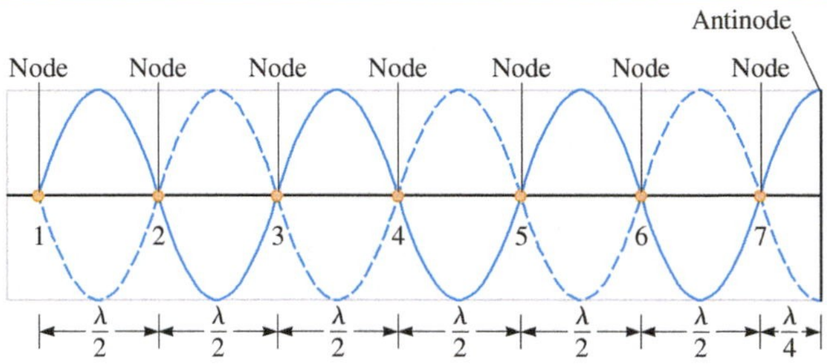

FIGURE 18.23

SOLVE The length L of the rope is the sum of the six distances between adjacent nodes $(6\lambda/2)$ and the distance between the last node and adjacent antinode $(\lambda/4)$. (Use an extra significant figure from results of the intermediate steps.)	$L = 6\frac{\lambda}{2} + \frac{\lambda}{4} = 6(1.257\text{ m}) + (0.6283\text{ m}) = 8.17\text{ m}$

CHECK and THINK
The rope may seem a bit long (nearly 30 ft). According to the problem statement, the rope is in a *large* laboratory, so 30 ft is not unreasonable.

FIGURE 18.24 A guitarist plays several different notes on a single string by holding the string down against different frets on the neck of the guitar.

18-5 Guitar: Resonance on a String Fixed at Both Ends

A guitar string is fixed at both ends. One end is held by the guitarist, and the other is attached to the guitar's bridge. A guitarist plays many different notes on a single guitar string by using one hand to hold down the end of the string at different points along the neck and strumming or plucking the string with the other hand (Fig. 18.24). Waves reflect from both ends of the string, and the resulting superposition creates a standing wave. Sound from the waving string is then enhanced or amplified in a way that depends on the type of guitar.

Because the ends of the string are fixed, they cannot oscillate; so, the standing wave that forms on the string must have a node on each end. Therefore, only certain wavelengths will fit on the string. Figure 18.25 shows the family of such waves on a string of fixed length L. The wavelength of the standing waves that fit on the string depends on the number of antinodes n such that

$$L = n\frac{\lambda_n}{2}$$

or

$$\lambda_n = 2\frac{L}{n} \qquad (n = 1, 2, 3, \ldots) \tag{18.9}$$

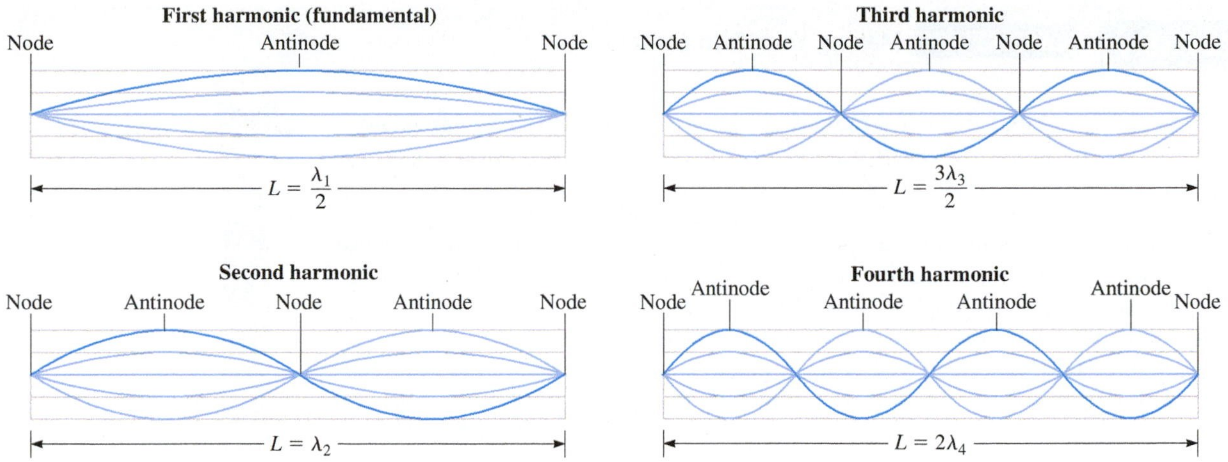

FIGURE 18.25 When a string is fixed at both ends, only standing waves that have nodes on each end can oscillate on the string. The wavelength of harmonic waves that can be sustained is given by $\lambda_n = 2L/n$, where n is the number of antinodes and L is the length of the string.

The integer subscript n is called the **harmonic number**, and each possible standing wave is referred to as the nth **harmonic**. For example, the last wave shown in Figure 18.25 has four antinodes; it is the fourth harmonic, and $\lambda_4 = L/2$.

It is often more convenient to work in terms of the wave's frequency instead of its wavelength. We find frequency by substituting Equation 18.9 into $v = \lambda f$ (Eq. 17.8):

$$f_n = \frac{v}{\lambda_n}$$

$$f_n = n\left(\frac{v}{2L}\right) \qquad (n = 1, 2, 3, \ldots) \qquad (18.10)$$

where v is the speed of the traveling waves on the string. The first harmonic ($n = 1$) is referred to as the **fundamental**, and $f_1 = v/2L$ is the **fundamental frequency**. The frequency of the higher-number harmonics is easily written in terms of the fundamental frequency:

STANDING WAVE; STRING FIXED AT BOTH ENDS

Special Case

$$f_n = nf_1 \qquad (18.11)$$

For a particular string, the distance between the two fixed ends determines the fundamental frequency (Eq. 18.10). So, when the guitarist moves his hand along the neck of the guitar, he changes the distance between the fixed ends and the note sounded by the string.

The speed of a wave on a string is given by $v = \sqrt{F_T/\mu}$ (Eq. 17.11), so the fundamental frequency can be written as

$$f_1 = \frac{1}{2L}\sqrt{\frac{F_T}{\mu}}$$

where F_T is the tension in the string and μ is the string's mass per length. Each guitar string has a different μ and F_T, and the guitarist tunes the instrument by adjusting the tension in the strings.

Another way to model a standing wave on a string such as the ones in Figure 18.25 comes from Equation 18.6, which shows that a standing wave is made of many oscillating particles. An extended object made of many oscillating particles is **vibrating**. In Section 16-11, we found that a particle—such as the bob at the end of a simple pendulum—oscillates only at a particular frequency known as the natural or resonance frequency, and when you try to drive the pendulum, the maximum amplitude occurs when the driving frequency equals the resonance frequency. The same sort of thing is true for a vibrating object except that such an object has a family of resonance frequencies. The object must be driven at one of its resonance frequencies; if not it won't vibrate very much. In the case of a string fixed at both ends, the resonance frequencies are given by Equation 18.10.

EXAMPLE 18.4 CASE STUDY Making Your Own Guitar

A guitar usually has six strings that are kept at roughly the same tension. Each string has a unique mass per unit length. In principle, a guitarist can play any note on any string by changing the length of the string that is free to vibrate. If he does not press on the string, the length of string that vibrates extends from the "0" fret to the bridge on the other end of the guitar (Fig. 18.26). When a guitarist wishes to play a higher note on the same string, he presses the string against one of the frets on the neck. Each fret marks the position for a particular musical note. Notice that the frets are not evenly spaced along the neck of the guitar.

Suppose you wish to make your own guitar. One particular string is 0.650 m long and sounds an A at $f_A = 440$ Hz when it is played open, without being pressed down. Find the location of the frets starting from "0" so that the next 12 notes, A-sharp, B, C, C-sharp, D, D-sharp, E, F, F-sharp, G, G-sharp and (octave) A, may be played. The fundamental frequency of these higher notes is given by

$$f_m = 2^{m/12} f_A \qquad (1)$$

where m is an integer that corresponds to each note such that $m = 1$ for A-sharp and $m = 12$ for (octave) A.

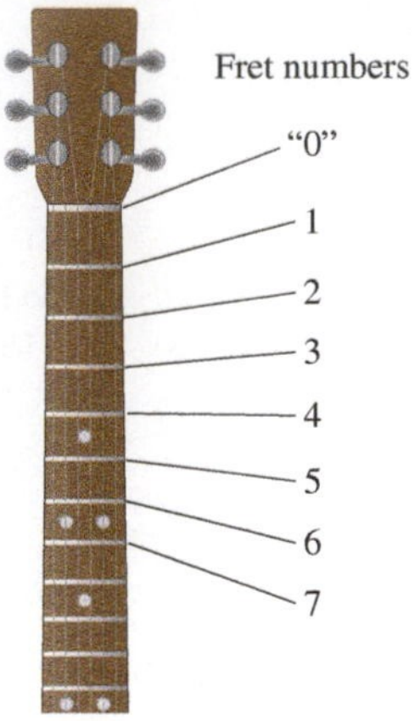

FIGURE 18.26 Frets on a guitar.

INTERPRET and ANTICIPATE

We only need the fundamental harmonic f_m to find the length L of the string for each note m, and we can derive an equation for L_m as a function of m. The position of each fret is then given by $L_A - L_m$, where $L_A = 0.650$ m. To check the results, we compare them to the location of the frets on an actual guitar (Fig. 18.26).

SOLVE

Set $n = 1$ in Equation 18.10 to write an expression for the length L_A of the string sounding an A at $f_A = 440$ Hz.

$$f_A = \frac{v}{2L_A}$$

$$L_A = \frac{v}{2f_A} \qquad (2)$$

To find L for a higher-frequency note, substitute Equation (1) for f_m into Equation 18.10 with $n = 1$.

$$f_m = \frac{v}{2L_m} \text{ from Equation 18.10}$$

$$L_m = \frac{v}{2f_m} = \frac{v}{2(2^{m/12})f_A}$$

Use Equation (2) to rewrite L_m in terms of L_A.

$$L_m = \frac{L_A}{2^{m/12}}$$

The position of each fret measured from "0" is the difference between L_A and L_m.

$$L_A - L_m = L_A - \frac{L_A}{2^{m/12}}$$

$$L_A - L_m = L_A(1 - 2^{-m/12}) \qquad (3)$$

Table 18.1 provides the frequency of each note (Eq. 1) and the position of each fret (Eq. 3).

TABLE 18.1

Note	m	f (Hz)	L_m (m)	$L_A - L_m$ (m)
A	0	440.00	0.650	0.000
A-sharp	1	466.16	0.614	0.036
B	2	493.88	0.579	0.071
C	3	523.25	0.547	0.103
C-sharp	4	554.37	0.516	0.134
D	5	587.33	0.487	0.163
D-sharp	6	622.25	0.460	0.190
E	7	659.26	0.434	0.216
F	8	698.46	0.409	0.241
F-sharp	9	739.99	0.386	0.264
G	10	783.99	0.365	0.285
G-sharp	11	830.61	0.344	0.306
Octave A	12	880.00	0.325	0.325

CHECK and THINK

If you examine the positions in Table 18.1, you will see that the frets are not evenly spaced. (The space between A and A-sharp is 0.036 m, and the space between is G-sharp and octave A is 0.019 m.) The frets get closer together as they get closer to the bridge, just as in the real guitar shown in Figure 18.26.

18-6 Flute: Resonance in a Tube Open at Both Ends

In a wind instrument such as the flute (Fig. 18.27), a standing wave forms in a tube of air. As with standing waves on a string, waves in a tube of air reflect from the ends, resulting in superposition of the reflected waves. If the tube is closed at one end, there is a node at that (closed) end. If the tube is open at one end, there is an antinode at that (open) end. The resonance frequencies depend on whether one or both ends of the tube are open.

A flute is best modeled as a tube open at both ends. In our model, we consider a straight tube of length L. Figure 18.28 shows the motion of air molecules vibrating at the fundamental frequency in a tube open at both ends. The sound wave is longitudinal, with the air molecules oscillating back and forth parallel to the tube. When the fundamental harmonic standing wave is established, there are antinodes at each end because both ends of the tube are open, and there is a single node at the tube's center. Molecules near the end of the tube are at the antinode of the wave and move with the greatest amplitude, bringing them beyond the end of the tube as well as deep inside it. Molecules near the center of the tube are at the node and do not move at all.

Figure 18.29 gives us a simpler way to represent the position of air molecules as they oscillate in a tube. In Figure 18.29A, a thin metal rod is held at its center by a clamp. A hammer strikes one end of the rod parallel to its long axis. Molecules in the rod oscillate back and forth along its length, creating a longitudinal wave. The wave reflects from the ends, resulting in a standing longitudinal wave. Molecules near the ends are free to oscillate, so they are at antinodes. Molecules near the center are fixed in place by the clamp, so there is a node at the center. The standing wave in the metal rod is similar to the standing wave in the tube of air illustrated in Figure 18.28. Rather than representing the positions of the numerous metal molecules, it is much simpler to plot their displacement $S(x, t)$ versus their position x along the rod's length. Figure 18.29B shows S versus x at two different times separated by half a period ($T/2$). The result looks much like a transverse wave. In fact, it might help to picture such a transverse wave in the rod. Instead of using the hammer to strike the rod along its axis, imagine striking the rod perpendicular to its long axis (Fig. 18.29C). Because the ends are free to oscillate, antinodes are established at the ends. Because the center is held fixed by the clamp, a node is established at the center. Figure 18.29C shows the shape of the rod at two different times separated by half a period, a shape that resembles the graph of S versus x.

Now we are ready to represent the harmonic series of standing longitudinal waves in a tube open at both ends. Figure 18.30 shows a graph of S versus x like Figure 18.29B for each harmonic. There is an antinode at each end and a node at any point that never moves. As in Figure 18.29C, we can imagine that there are "clamps" at the nodes that prevent the molecules from moving. The first harmonic has one node in the middle, and half a wavelength fits between the two ends of the tube: $L = \lambda_1/2$.

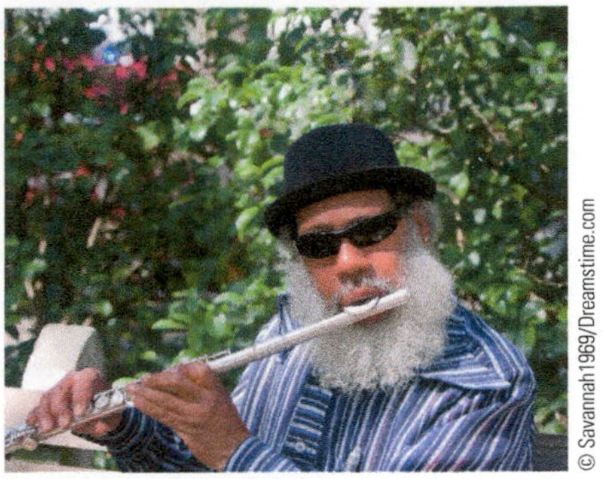

FIGURE 18.27 A musician plays a flute.

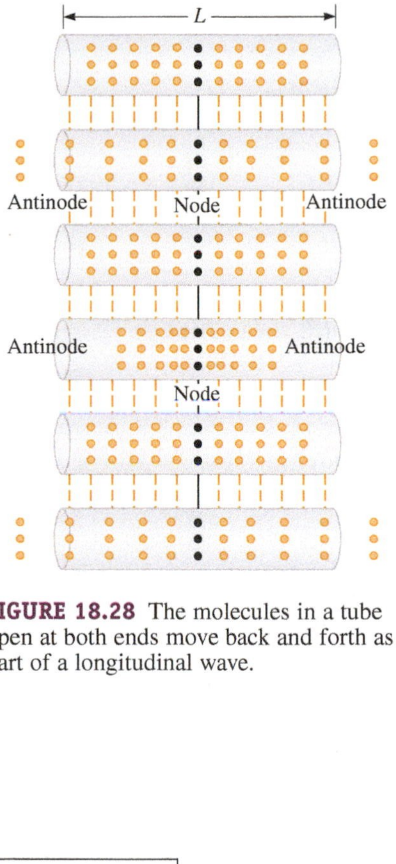

FIGURE 18.28 The molecules in a tube open at both ends move back and forth as part of a longitudinal wave.

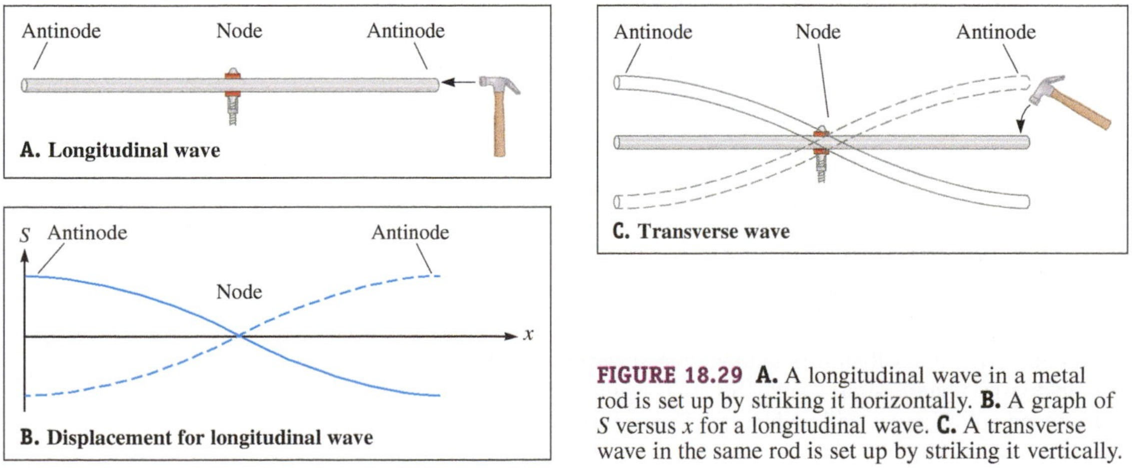

FIGURE 18.29 **A.** A longitudinal wave in a metal rod is set up by striking it horizontally. **B.** A graph of S versus x for a longitudinal wave. **C.** A transverse wave in the same rod is set up by striking it vertically.

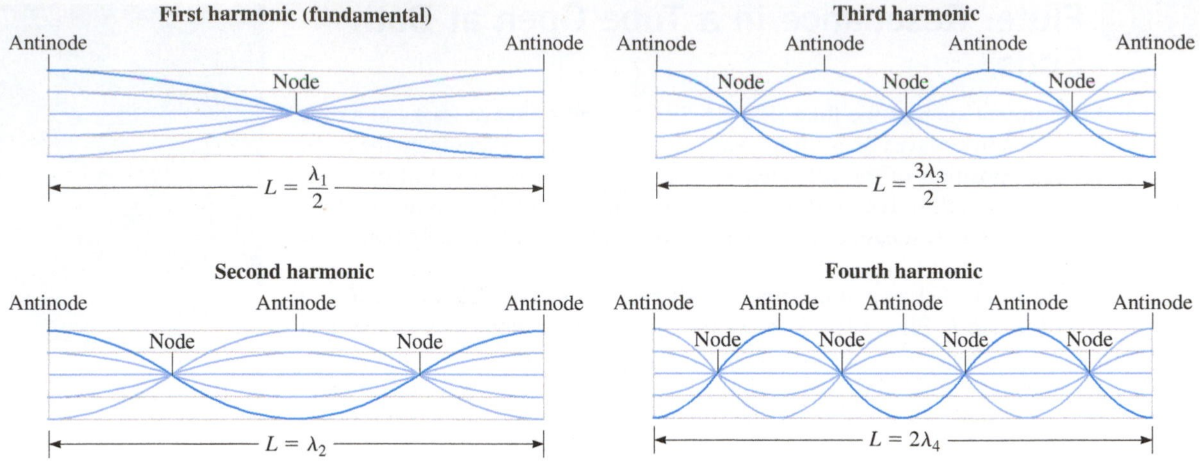

FIGURE 18.30 First four harmonics in a tube open at both ends. There must be an antinode on each end, so only those waves whose wavelength is given by $\lambda_n = 2L/n$, where n is the number of nodes, can fit in the tube. Compare with Figure 18.25.

Examining the higher harmonics in Figure 18.30 shows that the wavelength of the harmonic depends on the number of nodes n such that

$$\lambda_n = 2\frac{L}{n} \qquad (n = 1, 2, 3, \ldots) \tag{18.12}$$

The frequency of the harmonics comes from substituting Equation 18.12 into $f_n = v/\lambda_n$ (Eq. 17.8), where v is the speed of the longitudinal wave:

STANDING WAVE; TUBE OPEN AT BOTH ENDS

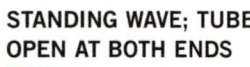

Special Case

$$f_n = n(v/2L) = nf_1 \qquad (n = 1, 2, 3, \ldots) \tag{18.13}$$

where $f_1 = \dfrac{v}{2L}$ is the fundamental frequency.

Equations for the wavelength (Eq. 18.12) and frequency (Eq. 18.13) of the harmonics in a tube open at both ends are similar to those for the wavelength (Eq. 18.9) and frequency (Eq. 18.10) of the harmonics on a string fixed at both ends. The integer n refers to the harmonic number in both cases, but in the case of a string fixed at both ends, n is the number of *antinodes*, and in the case of a tube open at both ends, n is the number of *nodes*.

EXAMPLE 18.5 CASE STUDY An Outdoor Concert

Many musicians hate to play outdoor concerts because the air temperature is not regulated and it is therefore difficult to keep their instruments in tune. Musicians often tune their instruments indoors before the performance. Suppose a flute is tuned such that the note A has a fundamental frequency of 440 Hz indoors where the temperature is $T_{in} = 24.0°C$ (75°F). Recall that the speed of sound in air as a function of temperature is given by Equation 17.14:

$$v = (331\,\text{m/s})\sqrt{1 + \frac{T_C}{273°\text{C}}}$$

A What is the fundamental frequency of the flute's A played outdoors in the summer where the temperature is $T_{out} = 35.0°C$ (95°F)? (Treat 440 Hz as three significant figures.)

INTERPRET and ANTICIPATE

Because the temperature outdoors is higher than it is indoors, the speed of sound is greater outdoors. The length of the flute doesn't change significantly. So, by Equation 18.13, the higher speed of sound means that the frequency of all notes will be higher. (The instrument will sound sharp.)

SOLVE

Use Equation 17.14 to determine v_{out}, the speed of sound outdoors at $T_{out} = 35.0°C$, and v_{in}, the speed of sound indoors at $T_{in} = 24.0°C$.

$$v_{out} = (331\text{ m/s})\sqrt{1 + \frac{35.0°C}{273°C}} = 351.6\text{ m/s}$$

$$v_{in} = (331\text{ m/s})\sqrt{1 + \frac{24.0°C}{273°C}} = 345.2\text{ m/s}$$

CHECK

As expected, the speed of sound is greater outdoors than it is indoors.

SOLVE

Use Equation 18.13 to solve for the fundamental frequency f_{out} when the flute is played outdoors in terms of the fundamental frequency f_{in} when the flute is played indoors.

$$\frac{f_{out}}{f_{in}} = \frac{v_{out}/2L}{v_{in}/2L} = \frac{v_{out}}{v_{in}}$$

$$f_{out} = \left(\frac{v_{out}}{v_{in}}\right)f_{in} = \left(\frac{351.6\text{ m/s}}{345.2\text{ m/s}}\right)(440\text{ Hz}) = 448\text{ Hz}$$

CHECK and THINK

As expected, the outdoor frequency is higher than the indoor frequency. The 8-Hz difference would certainly be noticeable to listeners.

B What is the fundamental frequency of the flute's A played outdoors in the fall where the temperature is $T_{out} = 10°C$ (50°F)? How could the flutist adjust his instrument so that it plays an A at the fundamental frequency of 440 Hz? (Round your answer to the nearest millimeter.)

INTERPRET and ANTICIPATE

In this part, the outdoor temperature is lower than that indoors, so we expect the speed of sound outdoors to be less and the flute's frequency to be lower than that indoors. According to $f_n = n(v/2L)$, the musician can correct for this difference by making his instrument shorter.

SOLVE

Following the procedure in part A, find the speed of sound and the frequency outdoors.

$$v_{out} = (331\text{ m/s})\sqrt{1 + \frac{10.0°C}{273°C}} = 337.0\text{ m/s}$$

$$f_{out} = \left(\frac{v_{out}}{v_{in}}\right)f_{in} = \left(\frac{337.0\text{ m/s}}{345.2\text{ m/s}}\right)(440\text{ Hz}) = 430\text{ Hz}$$

CHECK

As expected, the speed of sound and the frequency are both lower outdoors than they are indoors.

SOLVE

Use Equation 18.13 to find ΔL, the amount by which the flute needs to be shortened. Set the desired ("final") frequency to 440 Hz when the speed of sound outdoors is $v_{out} = 337.0$ m/s.

$$f_i = \frac{v_{out}}{2L_i} = 430\text{ Hz} \quad \text{and} \quad f_f = \frac{v_{out}}{2L_f} = 440\text{ Hz}$$

$$L_i = \frac{v_{out}}{2f_i} \quad \text{and} \quad L_f = \frac{v_{out}}{2f_f}$$

$$\Delta L = L_f - L_i = \frac{v_{out}}{2f_f} - \frac{v_{out}}{2f_i} = \frac{v_{out}}{2}\left(\frac{1}{f_f} - \frac{1}{f_i}\right)$$

$$\Delta L = \left(\frac{337.0\text{ m/s}}{2}\right)\left(\frac{1}{440\text{ Hz}} - \frac{1}{430\text{ Hz}}\right)$$

$$|\Delta L| = 8.91 \times 10^{-3}\text{ m} \approx 9\text{ mm}$$

CHECK and THINK

In principle, the flutist can compensate for the cold weather by making his flute about 9 mm shorter than its indoor length. Because there is a physical limit to how short any instrument can be made, however, musicians may adjust how they blow or may use alternate fingering. Of course, they try to keep their instruments warm, perhaps by using their own body heat. At least one flutist is known to put his flute into his pants leg between songs.

FIGURE 18.31 A musician plays the clarinet.

18-7 Clarinet: Resonance in a Tube Closed at One End and Open at the Other End

Some wind instruments such as a clarinet are best modeled by a tube of air open at one end and closed at the other end. A clarinet has an antinode at the open end as does the flute, but it differs in having a node at the closed end. The "closed" end of the clarinet has a reed that the musician holds in his mouth (Fig. 18.31), whereas the flute has a hole that the musician blows across without closing it off (Fig. 18.27).

As in the case of the flute, we represent a series of harmonic waves in the tube by a set of S versus x sketches (Fig. 18.32). The first harmonic has just one node at the closed end and a single antinode at the open end. One-fourth of a wavelength ($\lambda/4$) fits in the tube of length L. The next harmonic has an additional node and antinode in the tube, so three-fourths of a wavelength ($3\lambda/4$) fit there.

Looking at the higher harmonics in Figure 18.32, we find the wavelength of the nth harmonic is given by

STANDING WAVE; TUBE CLOSED AT ONE END Special Case

$$\lambda_n = 4\frac{L}{n} \qquad (n = 1, 3, 5, \ldots) \tag{18.14}$$

Only the odd harmonics exist in a tube that is open at just one end. We still use the number n to refer to the harmonics. For example, the last pattern shown in Figure 18.32 is referred to as the seventh harmonic (not the fourth). The harmonic number is *not* equal to either the number of nodes or antinodes. Instead, n is one less than the number of nodes plus antinodes:

$$n = \begin{pmatrix}\text{number}\\ \text{of nodes}\end{pmatrix} + \begin{pmatrix}\text{number}\\ \text{of antinodes} & -1\end{pmatrix}$$

The frequency of the harmonics is found by substituting Equation 18.14 into $f_n = v/\lambda_n$ (Eq. 17.8), where v is the speed of the longitudinal sound wave:

$$f_n = n\frac{v}{4L} = nf_1 \qquad (n = 1, 3, 5, \ldots) \tag{18.15}$$

where $f_1 = v/4L$ is the fundamental frequency.

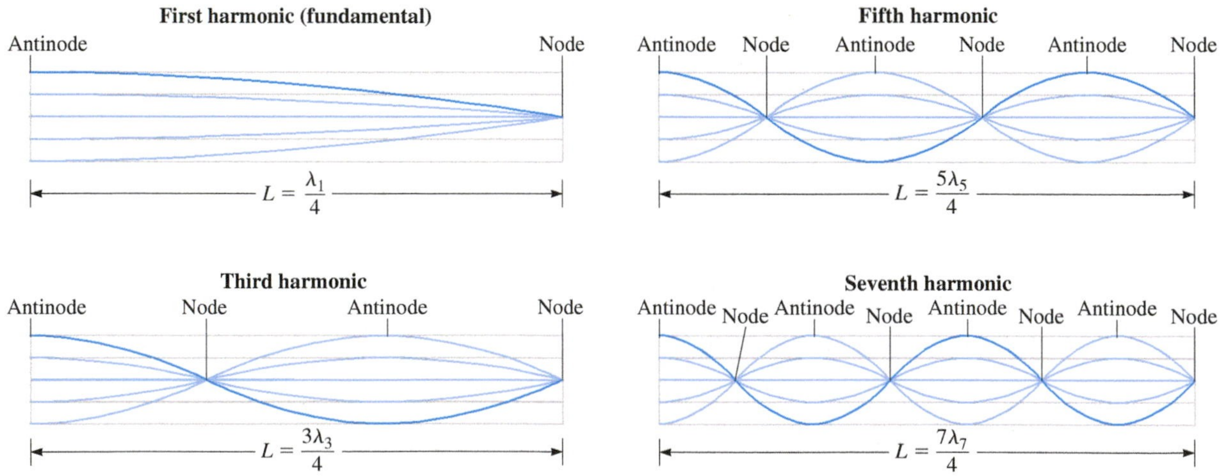

FIGURE 18.32 First four harmonics in a tube open at one end and closed at the other end. There must be an antinode at the open end and a node at the closed end, so only those waves whose wavelength is given by $\lambda_n = 4L/n$, where n is an odd integer, can fit in the tube. A tube that is only open at one end has only odd harmonics. Compare with Figures 18.25 and 18.30.

EXAMPLE 18.6 Aluminum Rod

In a physics demonstration, an aluminum rod is used to create sound waves. The rod of length $L = 1.50$ m is fixed at one end and free at the other end. By striking the free end, a standing longitudinal wave is set up in the rod. What audible frequencies are produced in the rod? (As discussed in Section 17-7, people can hear sounds in the frequency range from 20 Hz to 20,000 Hz.) The speed of sound in aluminum is 6420 m/s.

INTERPRET and ANTICIPATE

Because the rod is fixed at one end and free at the other end, it has a node at the fixed end and an antinode at the free end. The harmonics look like those in Figure 18.32. We need to find all the harmonic frequencies set up in the rod in the range from 20 Hz to 20,000 Hz.

SOLVE

Find the fundamental frequency by setting $n = 1$ in Equation 18.15.	$f_1 = \dfrac{v}{4L} = \dfrac{6420\ \text{m/s}}{4(1.50\ \text{m})} = 1070\ \text{Hz}$
Find the higher frequencies by multiplying the fundamental frequency by odd integers n beginning with 3. Most people cannot hear the $n = 19$ harmonic, so the audible frequencies are f_1 through f_{17}.	$f_3 = 3f_1 = 3210\ \text{Hz}$ $f_5 = 5f_1 = 5350\ \text{Hz}$ $f_7 = 7f_1 = 7490\ \text{Hz}$ $f_9 = 9f_1 = 9630\ \text{Hz}$ $f_{11} = 11f_1 = 11{,}770\ \text{Hz}$ $f_{13} = 13f_1 = 13{,}910\ \text{Hz}$ $f_{15} = 15f_1 = 16{,}050\ \text{Hz}$ $f_{17} = 17f_1 = 18{,}190\ \text{Hz}$ $f_{19} = 19f_1 = 20{,}330\ \text{Hz}$

CHECK and THINK

It might seem surprising that so many audible frequencies are produced at once, but the amplitude of the harmonic waves can vary greatly. Many such waves have very small amplitudes and are barely noticeable.

18-8 Beats

BEATS Major Concept

Before a performance, members of a jazz ensemble tune their instruments. If a piano is part of the ensemble, the pianist plays an A at 440 Hz. The other musicians tune their instruments to match the piano. The guitarist, for example, will sound an A on her instrument while the pianist plays an A. If the guitar is out of tune, the musicians will hear a modulated sound wave known as a *beat,* in which the sound alternates between louder and quieter (greater and smaller intensity). A **beat** is the interference of two waves that are at slightly different frequencies. The guitarist adjusts the tension of the string until the intensity of the sound is stable (in other words, until the beats disappear).

Figure 18.33 shows how beats are generated. When two nearly identical harmonic waves in a medium—such as sound waves in air—are superimposed, beats result if their frequencies are just slightly different. To see how, consider two time instants t_c and t_d. At time t_d, the two waves are nearly out of phase and nearly cancel each other at that instant (Fig. 18.33C). So, in the case of sound waves, it is momentarily very quiet at time t_d. At time t_c, the two waves are nearly in phase; the result is constructive interference and—in the case of a sound wave—a loud sound is heard at time t_c.

To better understand the beats in Figure 18.33C, let's add two harmonic waves that are identical except that one has angular frequency ω_1 and the other has a slightly different angular frequency ω_2. Because we are thinking about sound waves, we'll use Equation 17.6, $S(x, t) = S_{max} \sin(kx \pm \omega t)$, for longitudinal wave displacement. We hear the beats at a fixed position x, so we only need the time portion of this equation. The blue and red waves in Figure 18.33 are described by

$$S_1(x, t) = S_{max} \sin(\omega_1 t)$$
$$S_2(x, t) = S_{max} \sin(\omega_2 t)$$

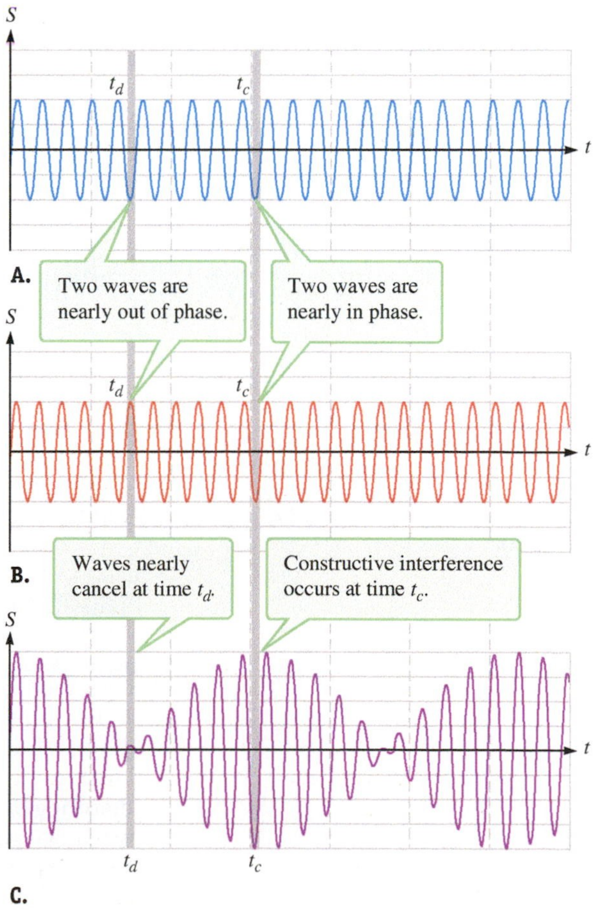

FIGURE 18.33 Beats result when two harmonic waves that are nearly identical except for a slight difference in frequency are superimposed. Adding the waves in parts **A.** and **B.** together produces the wave in part **C.**

respectively. The superposition of these two waves is the sum

$$S = S_1 + S_2 = S_{max}[\sin(\omega_1 t) + \sin(\omega_2 t)]$$

Use the trigonometric identity $\sin\alpha + \sin\beta = 2\cos[\frac{1}{2}(\alpha - \beta)]\sin[\frac{1}{2}(\alpha + \beta)]$ from Appendix A to rewrite S:

$$S = 2S_{max}\cos[\tfrac{1}{2}(\omega_1 - \omega_2)t]\sin\left[\frac{\omega_1 + \omega_2}{2}t\right] \tag{18.16}$$

Finally, we define two new angular frequencies, the average angular frequency,

$$\omega_{av} = \frac{\omega_1 + \omega_2}{2} \tag{18.17}$$

and the angular beat frequency,

$$\omega_{beat} = |\omega_1 - \omega_2| \tag{18.18}$$

(Recall that $\cos(-\theta) = \cos\theta$, and an angular frequency should be positive.) Rewriting Equation 18.16 in terms of these angular frequencies gives

$$S = 2S_{max}\cos(\tfrac{1}{2}\omega_{beat}t)\sin(\omega_{av}t) \tag{18.19}$$

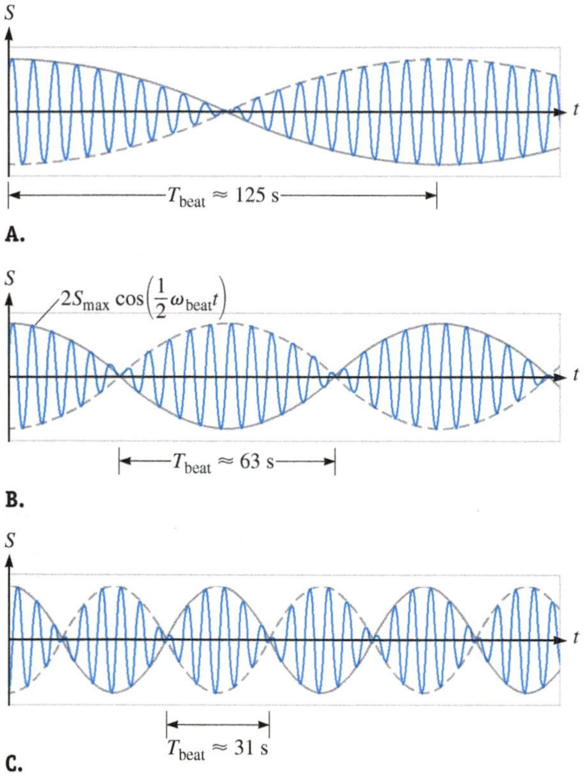

C.

FIGURE 18.34 A. Beats for $\omega_{beat} = 0.05$ rad/s ($f_{beat} \approx 8$ mHz). **B.** Beats for $\omega_{beat} = 0.1$ rad/s ($f_{beat} \approx 16$ mHz). **C.** Beats for $\omega_{beat} = 0.2$ rad/s ($f_{beat} \approx 32$ mHz).

Equation 18.19 describes the graph in Figure 18.33C. The superposition of two harmonic waves with slightly different frequencies results in a wave with angular frequency ω_{av}. The amplitude depends on time according to $S = 2S_{max}\cos(\frac{1}{2}\omega_{beat}t)$, oscillating with angular frequency $\frac{1}{2}\omega_{beat}$.

In most applications such as music, it is customary to talk about the beat frequency,

$$f_{beat} = \frac{\omega_{beat}}{2\pi} = \left|\frac{\omega_1}{2\pi} - \frac{\omega_2}{2\pi}\right| = |f_1 - f_2| \tag{18.20}$$

and the average frequency,

$$f_{av} = \frac{\omega_{av}}{2\pi} = \frac{f_1 + f_2}{2} \tag{18.21}$$

Figure 18.34 shows S versus t for three sets of beats. The beat period is the time between successive amplitude maxima (Fig. 18.34A) or successive amplitude minima (parts B and C of Fig. 18.34). The beat period is related to the beat frequency and angular frequency by

$$\omega_{beat} = 2\pi f_{beat} = \frac{2\pi}{T_{beat}} \tag{18.22}$$

Figure 18.34 also shows three different beat patterns. Figure 18.34A results from two waves with the smallest difference in angular frequency $|\omega_1 - \omega_2| = 0.05$ rad/s, so this pattern has the lowest beat frequency ($f_{beat} \approx 8$ mHz) and the longest beat period ($T_{beat} \approx 125$ s). So, the more beats per unit time (greater f_{beat}) that a musician hears, the more out of tune her instrument is.

CONCEPT EXERCISE 18.5

CASE STUDY Tuning the Guitar

Before a performance, a piano is tuned so that the middle A string emits sound at 440 Hz. The guitarist sounds an A on her instrument while that note is played simultaneously on the piano and hears beats at a frequency $f_{beat} = 2$ Hz. As she tightens the string of her guitar, the beat frequency decreases. The beats completely disappear when her guitar is in tune with the piano. What was the fundamental frequency of her guitar before it was tuned?

EXAMPLE 18.7 Speed of a Submarine

Sonar (*so*und *na*vigation *a*nd *r*anging) uses sound waves in water to detect other vessels. Two submarines are under water, each emitting identical sonar waves at 990 Hz. Submarine A is at rest, and submarine B is moving toward sub A. On sub A, beats are detected with a beat frequency of 11 Hz. What is the speed of sub B? The speed of sound waves in seawater is 1531 m/s (Table 17.1).

INTERPRET and ANTICIPATE

Sub B is the source of sonar waves that are detected by sub A. According to the Doppler effect, because the source (sub B) is moving toward the detector (sub A), the observed sonar frequency f_{obs} is higher than the emitted frequency f_{emit} = 990 Hz. Sub A produces its own sonar wave at 990 Hz that is superimposed on the sonar wave produced by sub B. Because the waves are identical except for the slight difference in frequency due to the Doppler effect, beats are created. Typical submarine speeds are between 10 and 100 km/h, so we should get an answer in this range.

SOLVE

Use Equation 18.20 to find the frequency f_{obs} of the sonar wave detected by sub A. Because sub B is moving toward the detector, assign it the higher of two possible frequencies.

$$f_{beat} = |f_1 - f_2| = |f_{obs} - f_{emit}|$$

$$11\text{ Hz} = |f_{obs} - 990\text{ Hz}|$$

$$f_{obs} = 979\text{ Hz or } 1001\text{ Hz; choose } 1001\text{ Hz}$$

The speed v_{source} of the source (sub A) can be found from the all-purpose Doppler effect relation (Eq. 17.31). In the case of a stationary observer, we have $v_{obs} = 0$. Because the source is moving toward the observer, we choose the minus sign (top sign) in the denominator. Solve for v_{source}.

$$f_{obs} = \left(\frac{v_s \pm v_{obs}}{v_s \mp v_{source}}\right) f_{emit} \quad (17.31)$$

$$f_{obs} = \left(\frac{v_s \pm 0}{v_s - v_{source}}\right) f_{emit} \text{ after choosing signs}$$

$$f_{obs}(v_s - v_{source}) = v_s f_{emit}$$

$$v_{source} = \frac{v_s(f_{obs} - f_{emit})}{f_{obs}}$$

$$v_{source} = \frac{(1531\text{ m/s})(1001\text{ Hz} - 990\text{ Hz})}{1001\text{ Hz}}$$

$$v_{source} = 16.82\text{ m/s} = 60.57\text{ km/h}$$

CHECK and THINK

Sub B's speed is in the expected range. Sonar is important on a submarine because there are no windows (and it is hard to see through water). Without sonar, a sub may not discover the location and velocity of other subs (or undersea creatures) until there is a collision. In February 2009, a British sub and a French sub collided in the Atlantic Ocean. Both subs were equipped with active and passive sonar. Active sonar sends a sound wave out and then "listens" for the echo from other objects, whereas passive sonar just listens. When on patrol, subs generally use just passive sonar because active sonar makes their presence known. Passive sonar tells you the direction (bearing) of the other sub, but not its distance. The colliding subs were probably using only their passive systems.

18-9 Fourier's Theorem

Any wind instrument can be modeled as either a tube open at both ends or closed at one end and open at the other, and strings on any instrument are fixed at both ends. It would seem at first glance, then, that various musical instruments should make very similar sounds. When two different instruments play the same note, however, it is possible to identify the type of instrument being played because an instrument does not produce a pure, single harmonic wave. In this section, we take a brief look at more complicated waves.

French physicist Jean Baptiste Joseph Fourier discovered that any wave can be mathematically represented by the superposition of harmonic (sine and cosine) wave functions; this statement is known as **Fourier's theorem**. For example, the sound wave produced by a musical instrument is a complex wave that cannot be represented by

FOURIER'S THEOREM
Underlying Principle

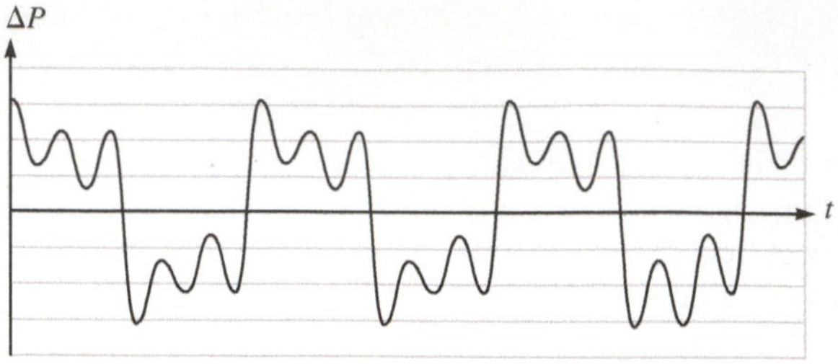

ΔP
t

FIGURE 18.35 A clarinet produces a complex pressure wave. The wave is periodic with a frequency called the *fundamental frequency* f_1.

a *single* sine or cosine function; instead, it is represented by a sum of sine and cosine functions.

Figure 18.35 represents a complex pressure wave that may be produced by an instrument such as a clarinet. This sound wave is periodic with a frequency f_1 called the **fundamental frequency**. According to Fourier's theorem, this periodic wave can be represented by a sum of harmonic waves. The first term in the sum is a wave with this same frequency, so f_1 is also called the **first harmonic frequency**. Subsequent harmonic waves in the sum have frequencies that are integer multiples of the fundamental frequency called the **second harmonic frequency** $2f_1$, the **third harmonic frequency** $3f_1$, and so on. The six harmonic waves shown in Figure 18.36 add together to produce the clarinet's sound wave in Figure 18.35. The amplitude of each harmonic wave indicates the relative importance of that harmonic frequency in the complex wave. For example, the first harmonic sine wave (Fig. 18.36A) is much stronger than the fifth harmonic sine wave (Fig. 18.36C). Even more striking, the clarinet's sound wave does not contain any even harmonics (second, fourth, sixth). Different instruments playing at the same fundamental frequency sound different from one another because of the relative strength of the harmonics that make up the complex sound waves.

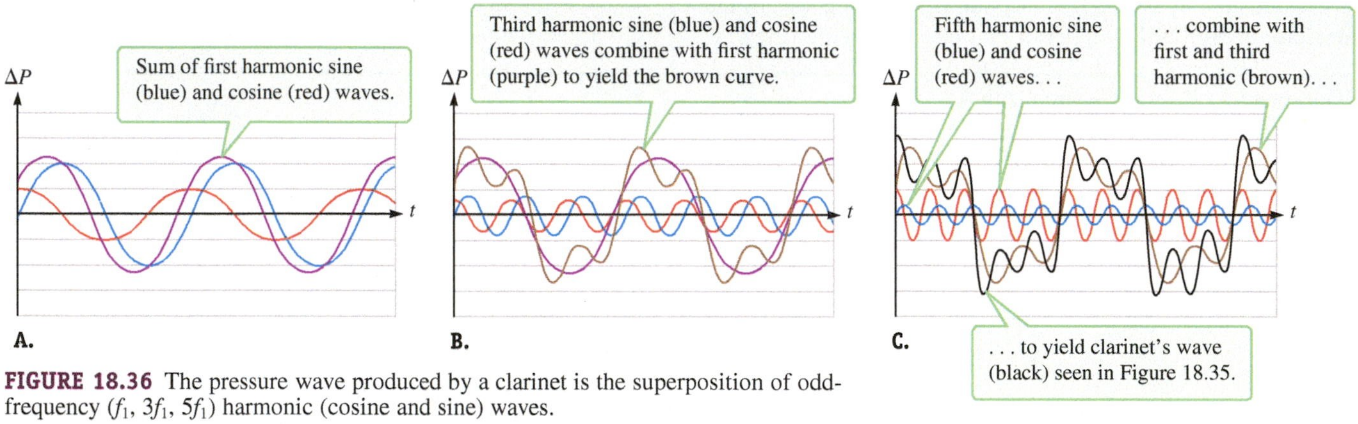

Sum of first harmonic sine (blue) and cosine (red) waves.
Third harmonic sine (blue) and cosine (red) waves combine with first harmonic (purple) to yield the brown curve.
Fifth harmonic sine (blue) and cosine (red) waves. . .
. . . combine with first and third harmonic (brown). . .
. . . to yield clarinet's wave (black) seen in Figure 18.35.
ΔP
t
A.
B.
C.

FIGURE 18.36 The pressure wave produced by a clarinet is the superposition of odd-frequency (f_1, $3f_1$, $5f_1$) harmonic (cosine and sine) waves.

CASE STUDY Part 3: Guitar String

When a guitarist strums or plucks a string, he distorts the shape of the string as shown in Figure 18.37A. According to Fourier's theorem, the pulse on the string is made of a number of harmonic waves (Fig. 18.37B). Only the harmonic waves with frequencies that resonate with the string will cause significant vibration. The resulting standing wave is the superposition of many harmonic waves. The fundamental harmonic is the strongest, but the higher harmonics give the instrument a rich sound. Tones without the higher harmonics don't sound like music, but instead sound like an emergency alarm. If the same note were to be played on a different instrument—say a violin—the same fundamental frequency would be heard, but the relative strength of the higher harmonics would be different. A characteristic set of higher harmonics gives each instrument its unique sound, and many instruments give a big band its rich sound.

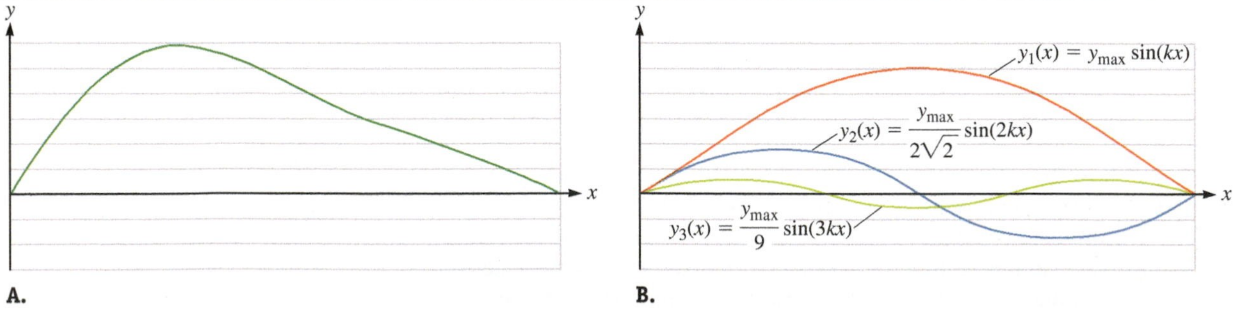

y
x
$y_1(x) = y_{max} \sin(kx)$
$y_2(x) = \frac{y_{max}}{2\sqrt{2}} \sin(2kx)$
$y_3(x) = \frac{y_{max}}{9} \sin(3kx)$
A.
B.

FIGURE 18.37 **A.** A plucked guitar string has a roughly triangular profile. **B.** The guitar string's profile is made of many harmonic waves. These three harmonic waves nearly reproduce the profile in part A.

SUMMARY

Underlying Principles

1. When more than one wave exists in a medium, the resulting wave function is found by **superimposing** (adding) the individual wave functions. For two such wave functions $y_1(x, t)$ and $y_2(x, t)$,

$$y(x, t) = y_1(x, t) + y_2(x, t)$$

2. According to **Fourier's theorem**, any wave can be mathematically represented by the superposition of harmonic (sine and cosine) wave functions.

Major Concepts

1. A **boundary condition** is the state of the medium's end.
   a. The profile of a wave reflected from a **fixed end** is inverted (Fig. 18.5).
   b. The profile of a wave reflected from a **free end** is upright (Fig. 18.7).
2. **Law of reflection:** For reflected waves, the incident angle equals the reflected angle:

$$\theta_i = \theta_r \qquad (18.1)$$

where angles are measured between the direction of propagation and the line normal (perpendicular) to the reflecting surface (Fig. 18.10).

3. The condition for **constructive interference** is

$$\Delta d = n\lambda \qquad (n = 0, 1, 2, 3, \ldots) \qquad (18.2)$$

The condition for **destructive interference** is

$$\Delta d = \frac{n}{2}\lambda \qquad (n = 1, 3, 5, \ldots) \qquad (18.3)$$

where Δd is the difference in distance traveled by the two interfering waves.

4. In a **standing wave**, each particle of the medium oscillates back and forth, but unlike in a traveling wave, the wave pattern does not propagate. The wave function for a standing harmonic wave is

$$y(x, t) = [2y_{max} \sin(kx)]\cos(\omega t) \qquad (18.6)$$

Particles that do not oscillate at all are at the **nodes** of the standing wave. Nodes are at

$$x = n\frac{\lambda}{2} \qquad (n = 0, 1, 2, 3, \ldots) \qquad (18.7)$$

Particles with the greatest amplitude are at the **antinodes**. Antinodes are at

$$x = \left(n + \frac{1}{2}\right)\frac{\lambda}{2} \qquad (n = 0, 1, 2, 3, \ldots) \qquad (18.8)$$

5. The interference of two waves that are at slightly different frequencies is a **beat**, given by

$$S = 2S_{max} \cos\left(\tfrac{1}{2}\omega_{beat}t\right)\sin(\omega_{av}t) \qquad (18.19)$$

where $\omega_{beat} = |\omega_1 - \omega_2|$ and $\omega_{av} = (\omega_1 + \omega_2)/2$. The quantities f_{beat} and f_{av} are used in music and other applications:

$$f_{beat} = \frac{1}{T_{beat}} = \frac{\omega_{beat}}{2\pi} = |f_1 - f_2| \qquad (18.20)$$

and

$$f_{av} = \frac{1}{T_{av}} = \frac{\omega_{av}}{2\pi} = \frac{f_1 + f_2}{2} \qquad (18.21)$$

Special Case: Standing Waves

1. Harmonic frequencies for a standing wave on a **string fixed at both ends** (Fig. 18.25) are given by

$$f_n = n\left(\frac{v}{2L}\right) = nf_1 \qquad (18.10 \ \& \ 18.11)$$

where $n = 1, 2, 3, \ldots$ and v is the speed of waves traveling on the string. All harmonics are present.

2. Harmonic frequencies for a standing wave in a **tube open at both ends** (Fig. 18.30) are given by

$$f_n = n\frac{v}{2L} = nf_1 \qquad (n = 1, 2, 3, \ldots) \qquad (18.13)$$

All harmonics are present.

3. Harmonic frequencies for a standing wave in a **tube open at one end and closed at the other** (Fig. 18.32) are given by

$$f_n = n\frac{v}{4L} = nf_1 \qquad (n = 1, 3, 5, \ldots) \qquad (18.15)$$

Only the odd harmonics are present.

PROBLEMS AND QUESTIONS

A = algebraic C = conceptual E = estimation G = graphical N = numerical

18-1 Superposition

1. There are two waves on a string given by $y_1(x, t) = 2\cos(3t - 10x)$ and $y_2(x, t) = 2\cos(3t + 10x)$.
   a. **C** In what ways do the two waves differ?
   b. **N** Find the wave that results on this string. *Hint*: Recall that $\cos\alpha + \cos\beta = 2\cos\left[\frac{1}{2}(\alpha + \beta)\right]\cos\left[\frac{1}{2}(\alpha - \beta)\right]$.
   c. **C** What is the resultant wave's amplitude?
2. **G** Two pulses travel in opposite directions along a string. One pulse is traveling to the right. The other pulse is inverted, with half the width and twice the amplitude of the first pulse, moving to the left as shown in Figure P18.2. Sketch the shape of the string at the moment in time when the centers of each pulse are at the same location.

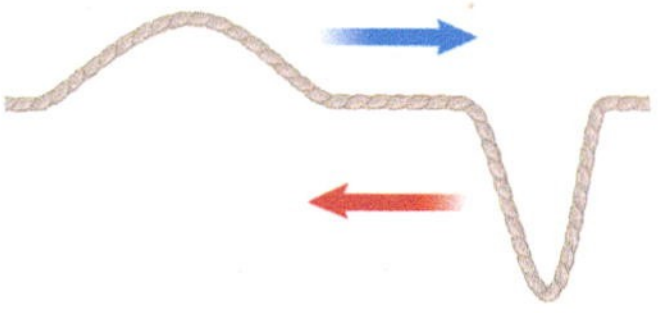
FIGURE P18.2

3. Two waves in the same medium are given by $y_1(x, t) = 2.5\sin(8x - 3.2t)$ and $y_2(x, t) = 2.5\sin(6x - 2.4t)$.
   a. **C** In what ways do the two waves differ?
   b. **N** Find the wave that results on this string. *Hint*: Recall that $\sin\alpha + \sin\beta = 2\sin\left[\frac{1}{2}(\alpha + \beta)\right]\cos\left[\frac{1}{2}(\alpha - \beta)\right]$.
   c. **C** What is the resultant wave's amplitude?
4. **G** Two individual waves exist in a medium as shown in Figure P18.4. Graphically find the superposition of these waves.

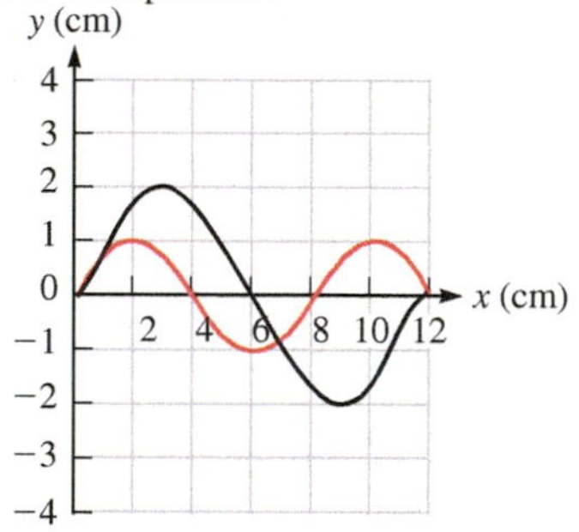

FIGURE P18.4

5. **N** The wave functions $y_1(x, t) = (0.150\text{ m})\sin(3.00x - 1.50t)$ and $y_2(x, t) = (0.250\text{ m})\cos(6.00x - 3.00t)$ describe two waves superimposed on a string, with x and y in meters and t in seconds. What is the displacement y of the resultant wave at **a.** $x = 0.100$ m and $t = 0$, **b.** $x = 1.00$ m and $t = 1.00$ s, and **c.** $x = 3.00$ m and $t = 3.00$ s?

18-2 Reflection

Problems 6 and 7 are paired.

6. **G** The wave function for a pulse on a rope is given by
$$y(x, t) = \frac{0.43}{(x - 13.6t)^2 + 1}$$
where all constants are in the appropriate SI units. Sketch the wave profile for **a.** the incident pulse, **b.** the reflected pulse if the end is free, and **c.** the reflected pulse if the end is fixed.
7. **A** The wave function for a pulse on a rope is given by
$$y(x, t) = \frac{0.43}{(x - 13.6t)^2 + 1}$$
where all constants are in the appropriate SI units. What is the wave function for the reflected pulse if the end is **a.** free or **b.** fixed?
8. **G** Jeff and Zak are sound tourists, exploring amazing sound effects around the world. At St. Paul's Basilica in Vatican City, they test out the whispering gallery inside the circular wall of the dome. If Jeff faces Zak who stands on the opposite side of the gallery, Zak cannot hear him, but if Jeff cups his mouth and whispers toward the wall while Zak puts his ears near the wall on the opposite side, Jeff's message is easily heard. Figure P18.8 shows two rays emitted by Jeff after they have reflected from the wall he faces. Draw the rays' paths as they reflect to Zak.

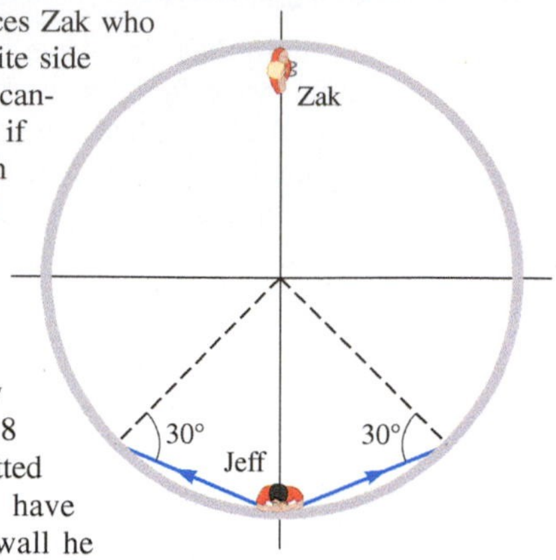

FIGURE P18.8

9. CASE STUDY Sounds that reach your ear within 0.1 s of the initial sound are perceived as *reverberation* rather than as a distinct echo. In a music hall, a little reverberation can be a good thing. Too much can be distracting. Because sound reflects off the many surfaces in the hall several times, a listener may hear many reflections of the same sound. Each reflection absorbs some of the wave's energy, and the reflected waves get quieter. So, after a short time, the reflected waves are unheard. If the surfaces don't absorb enough energy, however, the reflected waves may be perceived for too long, and an echo is heard. Often, panels are placed in the interior of a music hall to control the reflected sounds. Let's consider a simple example. (We won't worry about speakers, reflections off of people, furniture, or other objects.) A single source of sound is on stage, and a listener sits directly in front of the source (Fig. P18.9). There are six reflecting panels set inside the hall. Suppose each panel absorbs sound so that each reflected wave has 10% of the intensity of the incident wave. The range of human hearing is quite large, so let's say that if the intensity is reduced by a factor of one million, the sound is just barely audible.
   a. **N** How many reflections can the sound wave undergo before becoming inaudible?
   b. **A** Use the distances shown in Figure P18.9 to write an expression for the duration Δt of a sound heard due to its many reflections in terms of the speed of sound v_s in the room.

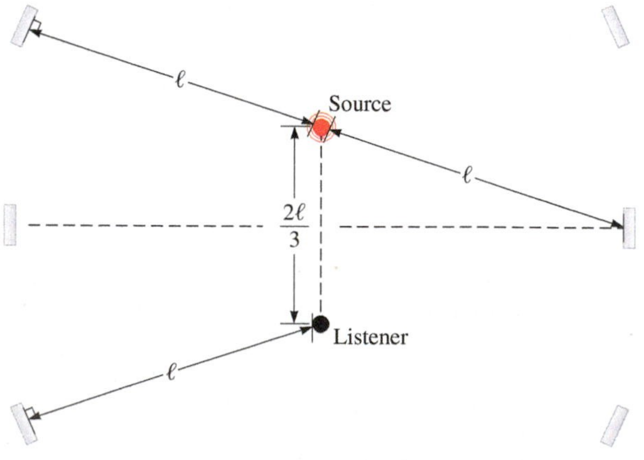

FIGURE P18.9

c. **N** The duration can be longer than 0.1 s because reflected waves are quieter than the original sound and because the music being played overwhelms the quieter reflections. So, a duration Δt of about 1 s is quite good. Find the distance ℓ that gives a duration of 1 s.

10. At many broadcast sporting events, you might notice that some crew members use microphones surrounded by a parabolic dish to record the sounds on the field (Fig. P18.10).

a. **C** What advantage does the parabolic dish provide? Why is it necessary at large events?

b. **G** Draw a sketch of a side view of a parabolic dish and show how sound would be reflected by the dish. Also indicate where the microphone should be placed for the best recording of the sound.

© Michael King/Cornell University

FIGURE P18.10

11. **C** Some animals use echolocation to navigate. Although the process may be complicated, describe in principle how emitting a sound pulse can allow a creature to locate other objects.

18-3 Interference

12. Two speakers, facing each other and separated by a distance d, each emit a pure tone of the same amplitude A with frequency f. The speed of each of the sound waves is v_s. A listener stands between the speakers, a distance x from one of the speakers.

a. **A** What frequencies would cause a dead spot (complete destructive interference) at the listener's position?

b. **N** If the speakers are separated by 5.00 m with the listener 2.00 m from one of the speakers, what is the lowest frequency for which there is a dead spot? The speed of sound in air is 343 m/s.

13. Two waves traveling on the same string are given by $y_1(x, t) = y_{max}\sin(kx - \omega t)$ and $y_2(x, t) = y_{max}\sin(kx - \omega t + \varphi)$. The two waves have the same amplitude y_{max}, angular wave number k, and angular frequency ω, but are out of phase by φ.

a. **A** Show that the resulting wave on the string is given by

$$y(x, t) = \left[2y_{max}\cos\frac{\varphi}{2}\right]\sin\left(kx - \omega t + \frac{\varphi}{2}\right)$$

where the absolute value of the term in brackets is the amplitude $A = |2y_{max}\cos(\varphi/2)|$. *Hint*: Recall that $\sin\alpha + \sin\beta = 2\cos[\frac{1}{2}(\alpha - \beta)]\sin[\frac{1}{2}(\alpha + \beta)]$.

b. **N** What is the amplitude if $\varphi = 0$? If $\varphi = \pi/2$? If $\varphi = \pi$?

14. **N** Two speakers generate harmonic sound waves that are in phase at 686 Hz. The speakers face the same direction. The speed of sound is 343 m/s. Find the phase difference between the two waves received at **a.** 4.00 m from each speaker, **b.** 3.00 m from one speaker and 3.50 m from the other, **c.** 3.00 m from one speaker and 4.00 m from the other, and **d.** 3.00 m from one speaker and 3.25 m from the other.

15. **C** It is thought by some that ancient monuments like Stonehenge were constructed with the knowledge that the sounds created by drummers drumming would become augmented or diminished at certain locations, which we now know is due to the interference of sound waves at these locations in the monument. Figure P18.15 shows an image of the Maryhill World War I Memorial Monument in Maryhill, Washington, which is a circular structure similar to Stonehenge. If a drum is struck repeatedly at the center such that the sound reflects off the outer wall, describe what will determine the locations where the sound is louder than expected and locations where the sound is diminished. Discuss these locations in terms of the radius of the circle, the speed of sound in air, and the rate at which the drum is struck.

© Michael D. Martin

FIGURE P18.15

18-4 Standing Waves

16. **N** As in Figure P18.16, a simple harmonic oscillator is attached to a rope of linear mass density 5.4×10^{-2} kg/m, creating a standing transverse wave. There is a 3.6-kg block hanging from the other end of the rope over a pulley. The oscillator has an angular frequency of 43.2 rad/s and an amplitude of 24.6 cm. **a.** What is the distance between adjacent nodes? **b.** If the angular frequency of the oscillator doubles, what happens to the distance between adjacent nodes? **c.** If the mass of the block is doubled instead, what happens to the distance between adjacent nodes? **d.** If the amplitude of the oscillator is doubled, what happens to the distance between adjacent nodes?

FIGURE P18.16

17. **N** A standing wave on a string is described by the equation $y(x, t) = 1.25\sin(0.0350x)\cos(1450t)$, where x is in centimeters, t is in seconds, and the resulting amplitude is in millimeters. **a.** What is the length of the string if this standing wave represents the first harmonic vibration of the string? **b.** What is the speed of the wave on this string?

18. **N** The resultant wave from the interference of two identical waves traveling in opposite directions is described by the wave function $y(x, t) = (2.40)\sin(0.0450x)\cos(8.00t)$, where x and y are in meters and t is in seconds. What are the **a.** frequency, **b.** wavelength, and **c.** speed of the two interfering waves?

19. **N** A standing transverse wave on a string of length 60 cm is represented by the equation $y(x, t) = 4.0\sin(\pi x/15)\cos(96\pi t)$, where x and y are in centimeters and t is in seconds. **a.** What is the maximum value of the standing wave at the point $x = 5.0$ cm? **b.** Where are the nodes located along the string for this particular standing wave? **c.** What is the vertical velocity v_y of the string at $x = 7.5$ cm when $t = 0.25$ s?

20. **C** People of the Inca Empire built a long series of roads and rope bridges in the highlands of Peru (Fig. P18.20). Imagine a couple of brave kids playing on such a bridge. By jumping up and down at a frequency of 0.55 Hz, they set up a standing wave with a node on each end and one in the middle. If the bridge is 40.7 m long, what is the speed of the wave?

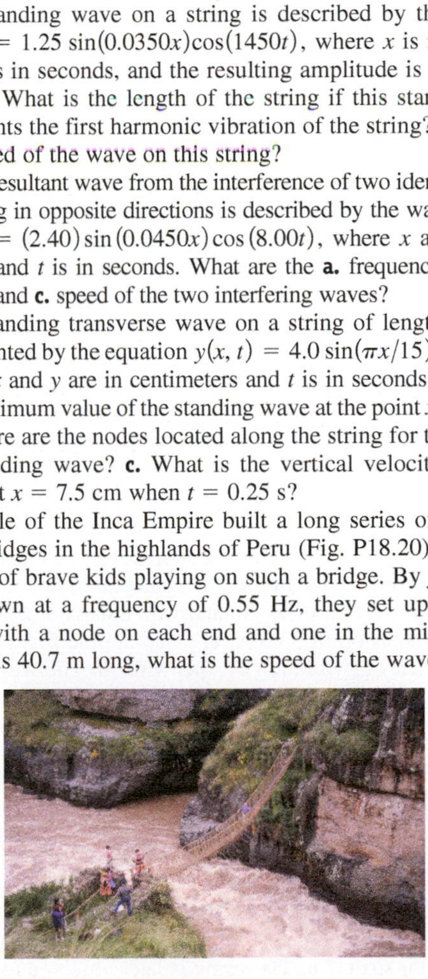
Photo by Rutahsa Adventures, Inc.

FIGURE P18.20

21. **N** A standing wave is described by $y(x, t) = [34.0 \sin(4.15x)]\cos(215t)$, where all constants are in appropriate SI units. What are the **a.** period, **b.** wavelength, and **c.** maximum y displacement of this wave?

22. **N** A standing wave is described by $y(x, t) = [34.0 \sin(4.15x)]\cos(215t)$, where all constants are in appropriate SI units. What are the wave functions for the two traveling waves that make up this standing wave?

23. **N** Two harmonic waves travel in opposite directions in the same medium. They have the same period (34.2 s), wavelength (6.23 m), and amplitude (3.45 m). What is the wave function of the resulting standing wave?

18-5 Guitar: Resonance on a String Fixed at Both Ends

24. **N** A violin string vibrates at 294 Hz when its full length is allowed to vibrate. If it is now held so that two-thirds of it is allowed to vibrate in the same harmonic, what frequency will it produce?

25. **N** Two successive harmonics on a string fixed at both ends are 66 Hz and 88 Hz. What is the fundamental frequency of the string?

26. **N** Two strings on a musical instrument are tuned to play at the fundamental frequencies 622.25 Hz (D-sharp) and 554.37 Hz (C-sharp). **a.** If the strings have the same length and are held under the same tension, what is the ratio (D-sharp to C-sharp) of their masses? (This situation describes a guitar or violin.) **b.** If the strings are held under the same tension and have the same mass per unit length, what is the ratio (D-sharp to C-sharp) of their lengths? (This situation describes a piano or harp.)

27. **N** When a string fixed at both ends resonates in its fourth harmonic, the wavelength is 0.345 m. How long is the string?

28. **N** A 50.0-cm-long copper wire with radius 0.100 cm and density 8.96 g/cm^3 is placed under 45.0 N of tension. **a.** What is the fundamental frequency of vibration for the wire? **b.** What are the next two harmonic frequencies for standing waves on this wire?

29. **N** As Toki tunes his guitar, he observes that by playing with the tension on a string he is able to change the string's fundamental frequency. He changes the fundamental frequency from 520.0 Hz to 550.0 Hz. The length of the string is 0.600 m. **a.** What is the wavelength of the fundamental harmonic in each case? **b.** What is the range of velocities of traveling waves on the string as he performs the change?

30. **A** A string fixed at both ends resonates in its fourth harmonic. The tension in the string is quadrupled. What happens to **a.** the speed of the wave, **b.** its wavelength, and **c.** its frequency?

31. **A** Two strings fixed at both ends oscillate in the fourth harmonic. String A has twice the mass per unit length of string B. The length of the strings and the tension in each are identical. Find the ratio (A to B) of their **a.** wavelengths and **b.** frequencies.

32. **N** A wire 1.50 m in length and with a mass of 25.0 g is placed under a tension of 33.0 N. **a.** What are the first two harmonic frequencies of vibration for this wire? **b.** The wire is observed to have a node at 30.0 cm from one end. What are the possible harmonic modes of vibration of the wire?

Problems 33 and 34 are paired.

33. **C** A standing wave exists on a string fixed at both ends. If you touch the string at a node, what happens to the standing wave?

34. **C** If you touch the string in Problem 33 at an antinode, what happens to the standing wave?

35. **N** A 0.530-g nylon guitar string 58.5 cm in length vibrates with a fundamental frequency of 196 Hz. **a.** What is the tension in the guitar string? **b.** The string is later observed to vibrate with two antinodes. What is the frequency of vibration?

36. **N** A sonometer is a single-stringed musical instrument on a wooden box with two movable bridges (Fig. P18.36). The string is fastened at one end, and the other end is attached to a hanging weight holder that is extended over a pulley. The distance between the end of the pulley and the fastened end define the length of the string. By adding weights to the weight holder, you can change the tension in the string. A particular sonometer wire has a length of 114 cm. Where should the two bridges be placed along this length so as to divide the wire into three segments whose fundamental frequencies are in the ratio 1:3:4? Assume the tension is the same throughout the wire.

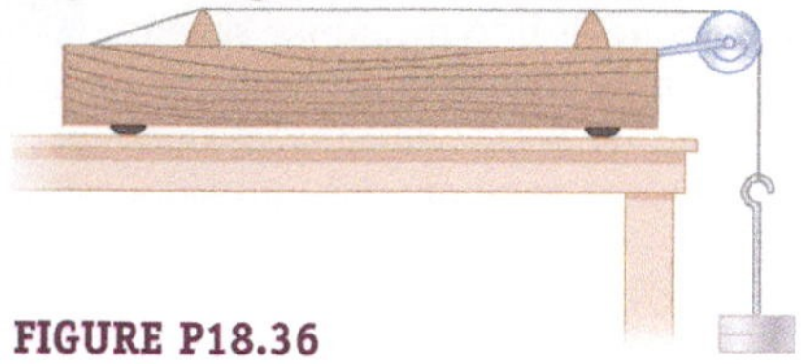

FIGURE P18.36

37. **N** A 2.20-g guitar string 60.0 cm in length is placed under a tension of 53.0 N and plucked so that it vibrates with its third harmonic frequency. **a.** Which is longer, the wavelength of the wave on the guitar string or the wavelength of the sound it emits in the room-temperature air surrounding the string? **b.** What is the ratio of the wavelength of the wave on the string to the wavelength of the sound it produces in air?

18-6 Flute: Resonance in a Tube Open at Both Ends

38. **E** A barrel organ is shown in Figure P18.38. Such organs are much smaller than traditional organs, allowing them to fit in smaller spaces and even allowing them to be portable. Use the photo to estimate the range in fundamental frequencies produced by the organ pipes in such an instrument. Assume the pipes are open at both ends. How does that range compare to a piano whose strings range in fundamental frequency from 21.7 Hz to 4186.0 Hz?

© Andrzej Barabasz (Chepry)

FIGURE P18.38

39. **N** A very large organ may have pipes that are as short as about 1 ft and as long as 16 ft. Imagine such an organ kept in a church whose temperature in the middle of winter is 55°F and in the middle of summer is 90°F. Find the seasonal change in both the shortest and the longest pipe's fundamental frequency. Assume the pipes are open at both ends.

40. **C** A standing wave exists in a pipe open at both ends. If you were to close off one end with your hand, what happens to the wave?

41. **N** The Channel Tunnel, or Chunnel, stretches 37.9 km undersea across the Strait of Dover in the English Channel between France and the United Kingdom. What are the resonant frequencies along this direction for the air in the Chunnel?

42. **N** A dog whistle emits frequencies above the range of human hearing, but within the range that dogs can hear. If a dog whistle is modeled as a tube open at both ends, how long would the whistle need to be if the fundamental frequency of the whistle is at the upper limit of human hearing, assumed to be 20,000 Hz?

Problems 43 and 44 are paired.

43. **N** A musical artist fits one piece of PVC (plastic) piping around another to create a variable-length tube in which harmonics can be excited by causing the tube to vibrate. She places a microphone on the inside of the tube to record the sound and send it to an amplifier. The tube has a minimum length of 1.2 m and a maximum length of 2.00 m, and the speed of sound in air in the tube is 343 m/s. Suppose she wants to create a sound that has a harmonic frequency of 540.0 Hz. Find each possible length of the tube such that a harmonic frequency of 540.0 Hz is created.

44. **C** In Problem 43, where should the musical artist place the microphone to best sample the sound? Consider that she would like all existing harmonic frequencies to be recorded by the microphone.

18-7 Clarinet: Resonance in a Tube Closed at One End and Open at the Other End

45. **N** If the aluminum rod in Example 18.6 were free at both ends, what audible frequencies would be heard? Compare your results with the results of Example 18.6 and explain the difference.

46. At the end of a party, Ezra finds that he can play the beginning notes (B, A, and G) to the children's song *Three Blind Mice* by blowing across the top of three identical soft-drink bottles (Fig. P18.46). One is completely empty and sounds in the fundamental harmonic, but the other two still have some liquid inside. Assume the sound is reflected from the surface of the liquid and the bottles are modeled as pipes closed at one end and open at the other end. Each bottle resonates at one of the three notes $f_B = 493.88$, $f_A = 440.00$ Hz, and $f_G = 392.00$ Hz.

FIGURE P18.46

a. **C** Are the three frequencies the fundamental frequency of each of the respective bottles? Explain your answer.

b. **C** Which bottle is empty? How do you know? (Assume the bottles are opaque so you cannot see the liquid level.)

c. **N** How tall is each bottle?

d. **N** What is the height of the liquid in the partially filled bottles?

47. **N** The third harmonic of an organ pipe that is closed at one end is in resonance with the first harmonic of an organ pipe that is open at both ends. Find the ratio of the lengths of the two pipes $(L_{\text{closed}}/L_{\text{open}})$.

48. **E** If you blow across the top of a plastic bottle, it's possible to hear an audible tone. Estimate the frequency for a 2-liter soda bottle if the frequency heard is the fundamental frequency.

Problems 49 and 50 are paired.

49. **C** A tube that is closed at one end and open at the other is filled with room-temperature air, where the speed of sound is 343 m/s. Striking the tube produces a characteristic set of harmonic wavelengths and frequencies. **a.** What happens to the harmonic wavelengths if the tube is suddenly filled with water, where the speed of sound in water is 1482 m/s? **b.** What happens to the characteristic frequencies when the water is added?

50. A tube 1.00 m long that is closed at one end and open at the other is filled with room-temperature air, where the speed of sound is 343 m/s. Then water is used to fill the tube. The speed of sound in water is 1482 m/s.

a. **N** Calculate the wavelength of the fundamental and the third harmonic both before and after the water is added.

b. **N** Calculate the fundamental and third harmonic frequencies both before and after the water is added.

c. **C** If someone were striking the tube and the sound inside were amplified through a speaker, would a listener hear a higher-pitched tone or a lower-pitched tone after the water is added? Explain your answer.

51. **N** An auto accident with a deer can be fatal for both the people and the deer. One way to prevent such an accident is with a deer whistle. Online directions tell you how to make your own deer whistle out of PVC pipes. According to the directions, you need a 16-in. length of pipe. One end is closed. The pipe is attached vertically to the front of the vehicle. Determine the fundamental frequency produced by such a pipe. Would it be audible to you?

52. **N** A cylindrical pipe of length 1.0 m is fitted with a thin, flexible diaphragm in the middle of the pipe (Fig. P18.52). Each half of the pipe is of equal length and represents a pipe that is closed at one end and open at the other. The portions A and B contain hydrogen and oxygen respectively. Each half of the pipe is set into vibration. What is the minimum harmonic number in each of the portions A and B for which they have the same frequency? Assume the speed of sound in hydrogen is 1300 m/s and in oxygen is 300 m/s.

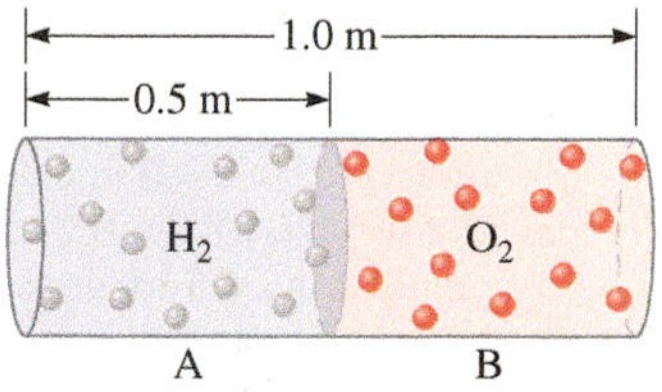

FIGURE P18.52

18-8 Beats

53. **N** CASE STUDY Before a band plays at an outdoor concert, the instruments are tuned indoors. As the concert goes on for a while, the temperature drops. One of the clarinetists was unable to keep her clarinet warm while she was required to play her second instrument, but the other clarinetist used his body heat to keep his clarinet warm. What beat frequency will be heard if the two identical clarinets attempt to play D (293.66 Hz) while one perfectly tuned clarinet is at 20.0°C and the other is at 3.00°C?

54. **C** Dog whistles operate at frequencies above the range of human hearing. Explain how two such whistles operating at slightly different frequencies may be used to make a sound audible to a person.

55. **N** A violinist and a pianist each play an A note. The piano was recently tuned to 440.0 Hz, but when the musicians play together, there are clear beats heard at a rate of about three per second. The violinist tightens the violin string, and the number of beats increases to about five beats per second. What is the frequency of the violin string after it is tightened?

56. **N** Two identical wires under tension are initially vibrating at 200.0 Hz. The tension in one string is increased slightly, resulting in 3.00 beats per second from the vibration of the two strings. What is the frequency with which the higher tension wire is vibrating?

57. **N** Find the beat frequency between the hypothetical two waves given by $y_1 = A \sin(0.250x - 420\pi t)$ and $y_2 = A \sin(0.250x - 426\pi t)$. Note that these two waves could not be in the same medium because their speeds are different.

58. **N** In a classroom demonstration, two pipes of length 1.25 m open at both ends oscillate in their fundamental harmonic. One pipe has a collar allowing it to be lengthened by 15.0 cm. As one pipe is slowly lengthened, beats are heard. **a.** Find the maximum beat frequency. **b.** What is the length of the longer pipe when the beat frequency heard is half of its maximum?

59. **N** The air in an organ pipe is set into vibration and compared with the sound from a tuning fork of frequency 256 Hz. When the air in the pipe is at a temperature of 16°C and both objects are vibrating, 24 beats are heard in 12 s. The experiment is repeated with another tuning fork that has a slightly lower frequency than the previous one, and the beat frequency is not observed. Given that information, what change in temperature of the air in the pipe is necessary to bring the pipe and the original tuning fork into resonance, where no beat frequency is observed?

18-9 Fourier's Theorem

60. At $t = 0$, three waves are given by

$$y_1(x, 0) = \frac{4}{\pi}\sin\left(\frac{\pi x}{2}\right)$$

$$y_2(x, 0) = \frac{4}{3\pi}\sin\left(\frac{3\pi x}{2}\right)$$

$$y_3(x, 0) = \frac{4}{5\pi}\sin\left(\frac{5\pi x}{2}\right)$$

a. **G** Sketch these three wave functions from $x = -4.0$ m to 4.0 m.

b. **G** Sketch $y(x, 0) = y_1(x, 0) + y_2(x, 0) + y_3(x, 0)$ in the interval $x = -4.0$ m to $x = 4.0$ m.

c. **C** Are $y_1(x, 0)$, $y_2(x, 0)$, and $y_3(x, 0)$ the harmonic components of a square wave, a triangular wave (a series of equilateral triangle shapes), or a sawtooth wave (a series of right triangular shapes)? Explain.

A spreadsheet program may be helpful.

General Problems

61. **N** What is the amplitude of the resultant wave for two sinusoidal waves with equal amplitudes of 6.50 cm traveling in the same direction on a taut string, but 90.0° out of phase?

62. **G** A pulse on a string travels to the right (Fig. P18.62). It moves one box to the right per millisecond. The right end of the string is tied to a post so that it cannot move up and down at that location. Sketch the shape of the string at a time 6 milliseconds later.

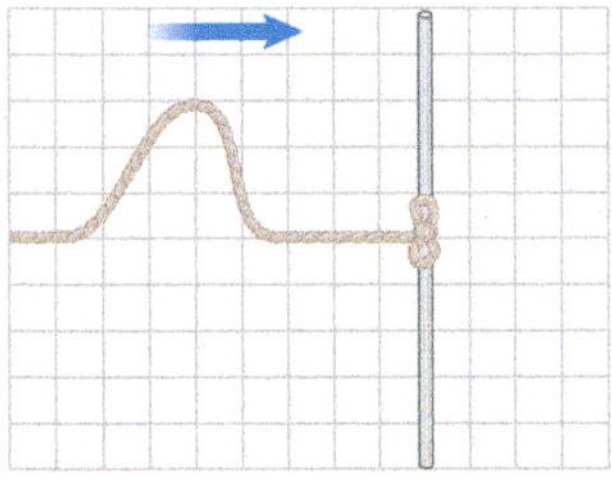

FIGURE P18.62

63. The functions

$$y_1 = \frac{2}{(2x + 5t)^2 + 4} \quad \text{and} \quad y_2 = \frac{-2}{(2x - 5t - 3)^2 + 4}$$

describe two pulses traveling along a stretched string.

a. **C** What is the direction of travel for each of the waves?

b. **N** At what position x do the two pulses always cancel each other?

c. **N** At what time t do the two pulses cancel each other everywhere?

64. **N** The low E string and the high E string on a guitar are tuned to have fundamental frequencies of 82.4 Hz and 329.6 Hz, respectively. The difference in the fundamental frequency is primarily due to the different thickness, and therefore different mass per unit length, of the two strings. We could, however, try to design a guitar using the same type of strings for both notes by varying either the length of the string or its tension.

a. If the mass per unit length and the length were kept the same for both strings, what would be the ratio of the tension in the high E string to the tension in the low E string?

b. If the mass per unit length and the tension were kept the same for both strings and we found a way to alter the length of the strings, what would be the ratio of the length of the high E string to the length of the low E string?

65. **N** The wave functions

$$y_1 = (3.00 \text{ m})\sin(2.00\pi x - 15.0\pi t)$$

$$y_2 = (3.00 \text{ m})\sin\left(2.00\pi x - 15.0\pi t + \frac{\pi}{2}\right)$$

describe two sinusoidal waves traveling on the same string, where x, y_1, and y_2 are in meters and t is in seconds. What are the **a.** amplitude, **b.** frequency, and **c.** wavelength of the resultant wave?

66. **N** The wave functions

$$y_1(x, t) = (1.20 \text{ m})\sin\left(0.350\pi x - \frac{\pi}{2}t\right)$$

$$y_2(x, t) = (1.20 \text{ m})\sin\left(0.350\pi x + \frac{\pi}{2}t\right)$$

describe two transverse sinusoidal waves traveling along a rope, where x and y are in meters and t is in seconds. What is the maximum transverse displacement of a rope element located at **a.** $x = 1.00$ m and **b.** $x = 3.00$ m? **c.** Where are the first three antinodes located along the rope?

67. **N** An air column in a variable-length pipe, which is closed at one end, is in resonance with a vibrating tuning fork of frequency 264 Hz. If the pipe has a maximum length of 1.0 m and a minimum length of 0.20 m, what is (are) the possible length(s) of the pipe? Assume the speed of sound in air is 334 m/s.

68. **N** A standing wave with three antinodes is formed on a 0.890-m-long wire under tension. **a.** Which harmonic mode of vibration does this wave represent? **b.** What is the wavelength of this wave? **c.** What is the number of nodes in this standing wave?

69. Two successive harmonic frequencies of vibration in a pipe are 532 Hz and 684 Hz.

a. **C** Is the pipe open at both ends, or is it closed at one end and open at the other?

b. **N** What is the fundamental frequency of the pipe?

70. **N** In Oshin's home theater, an oscillator operating at 425 Hz drives a pair of identical speakers placed 4.00 m apart and facing each other in phase. Where are the relative minima of sound pressure located on the line joining the two speakers?

71. **N** A block of mass $m = 2.50$ kg is suspended by a 10.0-g wire that is 2.25 m long and is draped over a massless pulley. The other end of the wire is attached to a wall (Fig. P18.71). If the distance between the wall and the pulley is 1.50 m, what is the fundamental frequency of vibration for the horizontal portion of the wire?

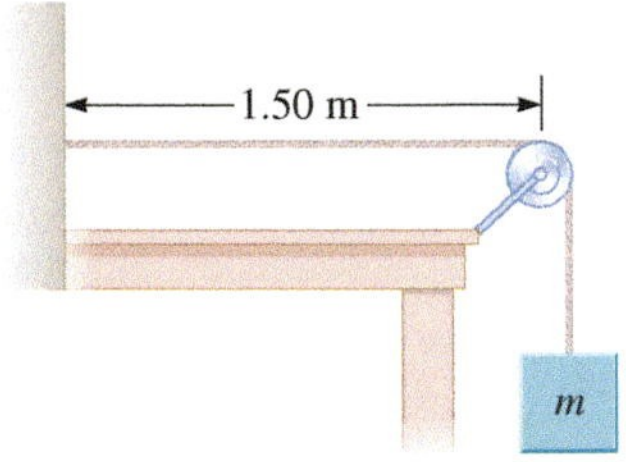

FIGURE P18.71

72. **G** The example of beats shown in Figure 18.34 (page 542) is the result of two waves with similar frequencies and equal amplitudes. If the two waves have different amplitudes instead, is there still a beat frequency? To answer, consider two waves that are incident at the same point in space such that $S_1(t) = \sin(10t)$ and $S_2(t) = 3\sin(10.5t)$. Plot the superposition of these waves versus time and comment on the effect on the beats. A spreadsheet program may be helpful.

73. **N** A pipe is observed to have a fundamental frequency of 345 Hz. Assume the pipe is filled with air ($v = 343$ m/s). What is the length of the pipe if the pipe is **a.** closed at one end and **b.** open at both ends?

74. **N** The wave function for a standing wave on a 2.00-m-long string is given by $y = 0.0500 \sin(2\pi x)\cos(76\pi t)$, where x and y are in meters and t is in seconds. **a.** How many antinodes are present? **b.** What is the fundamental frequency of this wave?

75. **N** A cylindrical glass tube is partially filled with water to create an air column with a length of 0.20 m. The air column is vibrating at its fundamental frequency. After the water column is slowly lowered, the same frequency is heard. What should be the new length of the air column so that the third harmonic has the same frequency as the previous fundamental frequency? Treat the air column as a case of resonance with one open end and one closed end.

76. **C** A common technique of tuning a guitar is to use a string that is currently in tune to create a sound with the proper frequency that should also be the fundamental frequency of the next string. When the next string is close but not exactly in tune and both are plucked at the same time, a beat frequency can be heard. If this test is performed, explain how the guitarist knows whether to increase or decrease the tension in the string being tuned. Try to construct a logic flow chart (step-by-step instructions) for tuning a string in this manner. Consider that the second string's tension might be too high or too low initially.

77. **N** A pipe open at both ends has a fundamental frequency of 145 Hz. A second pipe, which is closed at one end, vibrates at this frequency in its fourth harmonic mode. What is the length of the pipe **a.** open at both ends and **b.** closed at one end?

78. **N** After a successful business meeting, a salesman steps into his hotel room's shower stall and closes the door, creating a rectangular pipe closed at both ends with dimensions 1.20 m $\times$ 1.20 m $\times$ 2.43 m. He begins to sing, covering frequencies from 80.0 Hz to 1100.0 Hz. If nodes form at the opposite sides of the shower stall, which frequencies will be in resonance and therefore sound the richest? Assume the speed of sound in the shower stall is 360 m/s.

79. **N** In attempting to tune the G string of a guitar to 196 Hz, a musician hears 4.00 beats/s between the guitar string and a reference tone of 196 Hz, indicating that the current tuning of the string is not correct. **a.** What are the possible frequencies to which the guitar string is currently tuned? **b.** The musician now loosens the guitar string and hears 2.00 beats/s. To what frequency is the guitar's G string now tuned? **c.** By what percentage must the now current tension in the guitar string change for the string to be tuned to 196 Hz?

80. **N** An accurate electronic tuner with an adjustable frequency is used to determine the actual frequency of a tuning fork marked as 512 Hz. When the tuner is set to 514 Hz and the tuning fork is sounded, a beat frequency of 2 Hz is heard. When the tuner is set to 510 Hz and the tuning fork is sounded, the beat frequency is 6 Hz. What is the frequency of the tuning fork?

19 Temperature, Thermal Expansion, and Gas Laws

Key Questions

How is temperature measured?

What is the connection between temperature and other macroscopic properties?

Underlying Principles

1. Thermodynamics
2. Zeroth law of thermodynamics

Major Concepts

1. Temperature
2. Thermal contact
3. Thermal equilibrium
4. Thermal linear expansion
5. Thermal volume expansion
6. Equation of state
7. Equilibrium state
8. Ideal gas
9. Mole
10. Avogadro's number
11. Kelvin scale
12. Absolute zero

Special Cases: Equations of State

1. Charles's law
2. Gay-Lussac's law
3. Boyle's law
4. Avogadro's law
5. Ideal gas law

Proteales.etsy.com Photo by Samantha Walker

FIGURE 19.1 Mood rings change color depending on the temperature of the wearer.

Many fads—love beads, bell-bottom pants, pet rocks—took hold in the United States in the 1970s. Although some fads fade away forever, others, such as the mood ring (Fig. 19.1), reappear. The claim is that the "stone" of the ring changes color according to the mood of the wearer. The stone is actually a glass shell filled with liquid crystals whose molecular structure depends on their temperature. The molecular structure determines which colors of light are absorbed and which are reflected. The stone's color actually depends on the temperature of the liquid crystals, not on the mood of the wearer.

The color of mood rings is an example of a property that depends on temperature. The color of other objects also depends on temperature. Blue stars are hotter than red stars, for instance, and the heating coil on a stove top turns bright red when it is hot. Matter has other temperature-dependent properties. For example, most materials expand when they are hot and contract when they are cold. This chapter is about temperature: What is it, and how does it affect an object or system?

19-1 Thermodynamics and Temperature

When you slide your textbook across your desk, the work you do on the book–desk system goes into the internal *thermal energy* of the system. You observe that the book and the desktop get warmer. **Thermal energy** is associated with the kinetic energy of the particles that make up the objects in the system (in this case, the book and the desk).

Thermodynamics is the study of thermal energy and the transfer of energy through *heat* and work. To understand thermodynamics, we connect macroscopic observations to microscopic phenomena. For example, when the book slides on the desktop, we cannot directly observe the increase in the molecules' kinetic energy (that is, we cannot see the molecules moving faster), but we can sense the increase in thermal energy by placing our hand against each surface and feeling the increased warmth. Our experience is macroscopic, or large scale. Much of this chapter focuses on the macroscopic property of *temperature* and its connection with other macroscopic properties such as pressure, volume, and density. **Temperature** is a measure of the average kinetic energy of the particles in a system. For instance, the molecules in a cup of ice-cold tea have on average less kinetic energy than the molecules in a cup of hot tea.

THERMODYNAMICS
Underlying Principle

TEMPERATURE **Major Concept**

Units of Temperature

Temperature is one of the seven base quantities that make up the SI (Système International d'Unités) set of units (see Appendix B and Section 1-5). The SI unit for temperature is the **kelvin**, abbreviated simply as K.

We do *not* say "degrees Kelvin" or write "°K."

For 18 chapters, we have been working with three of those seven SI base units: kilogram, meter, and second. In this chapter, we add two more to our list: *kelvin* and *mole*. Here we focus on kelvin; we'll return to mole later in the chapter. Each fundamental unit must have a precise definition or standard. The SI standard for the kilogram is a specific platinum–iridium alloy cylinder kept at the International Bureau of Weights and Measures in Sèvres, France.

Some standards are given in terms of a definition or procedure that can be reproduced in a laboratory. The SI standard for temperature is a prescription or recipe for measuring temperature that can be done in any well-equipped laboratory as discussed in Section 19-7. For now, it is sufficient to know how to convert from temperature scales in common use—Fahrenheit and Celsius—to the Kelvin scale (also known as the *absolute scale*). A **temperature scale** is defined by assigning values to two well-separated reference temperatures.

Until the 1960s and 1970s, most English-speaking countries used the Fahrenheit scale, but now the only industrialized country in the world to continue using it is the United States. The Fahrenheit scale was developed by Dutch physicist Daniel Gabriel Fahrenheit (1686–1736). Today on the Fahrenheit scale, the freezing point of water is 32°F, and the boiling point of water is 212°F. Fahrenheit did not originally use water to set the reference temperatures for his scale. Instead, he set the zero of his scale to the temperature of a mixture of water, ice, and sea salt, and he set the other reference temperature to healthy human body temperature at 96°F. It is not clear why he intended to set human body temperature to 96°F instead of some round number like 100°F. On the contemporary Fahrenheit scale, healthy human body temperature is 98.6°F.

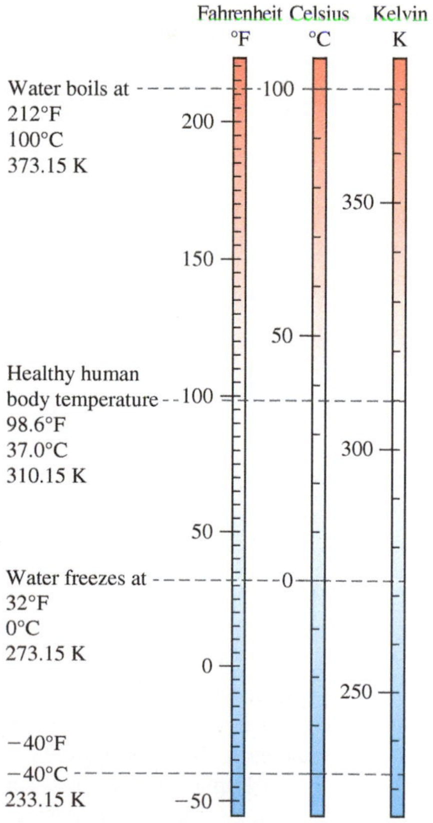

FIGURE 19.2 Three commonly used scales for measuring temperature.

The Celsius scale (sometimes known as the centigrade scale) is in common use worldwide. It was developed by Swedish astronomer Anders Celsius (1701–1744) about 30 years after the Fahrenheit scale. On the Celsius scale, the freezing point of water is 0°C, and the boiling point of water is 100°C.

In the United States, the Celsius scale is used in laboratories, and the Fahrenheit scale is used in medicine, in homes, and in weather reports. Figure 19.2 shows the Kelvin, Fahrenheit, and Celsius scales, allowing easy comparison. As shown in Figure 19.2, water freezes at

$$T_f = 273.15\,\text{K} = 0°\text{C} = 32°\text{F}$$

Water boils at

$$T_b = 373.15\,\text{K} = 100°\text{C} = 212°\text{F}$$

So, the difference in temperature between the boiling point and the freezing point of water is

$$T_b - T_f = 100\text{ K} = 100°\text{C} = 180°\text{F}$$

In other words, a 1 K increase in temperature equals a 1°C increase; the separation between increments of 1 K on the Kelvin scale equals the separation between increments of 1° on the Celsius scale. The only difference between the Kelvin scale and the Celsius scale is the zero point; zero on the Celsius scale is 273.15 K. To convert from the Celsius scale to the Kelvin scale, you only need to add 273.15:

$$T(\text{K}) = T(°\text{C}) + 273.15 \tag{19.1}$$

The relationship between the Fahrenheit scale and the Celsius scale is a little more complicated, however, because an increment of 1° on the Fahrenheit scale is smaller than an increment of 1° on the Celsius scale. Based on the difference between the boiling point and the freezing point of water, an increase of 1°C equals an increase of 1.8°F. So, the separation between degree marks on the Fahrenheit scale is 1/1.8 = 5/9 of the separation between degree marks on the Celsius scale. In addition, the zero points are not the same; zero on the Celsius scale is 32°F. The conversion from Fahrenheit to Celsius is given by

$$T(°\text{C}) = \tfrac{5}{9}[T(°\text{F}) - 32] \tag{19.2}$$

Equation 19.2 can be inverted to find the conversion from Celsius to Fahrenheit:

$$T(°\text{F}) = \tfrac{9}{5}T(°\text{C}) + 32 \tag{19.3}$$

© Ddiror/Dreamstime.com

FIGURE 19.3 Thermometers play an important role in our daily lives including in industrial applications.

CONCEPT EXERCISE 19.1

The Fahrenheit scale remains useful in part due to personal experience with temperature. For example, the average temperature range in a typical midlatitude city such as Gdansk, Poland (formerly Dansk, where Fahrenheit was born), is 25°F to 78°F over a year.

a. Find the temperature range in Gdansk, Poland, in degrees Celsius and in kelvins.
b. Many people say that 0°F is a very cold day and 100°F is an unpleasantly hot day. Convert these temperatures to the Celsius and Kelvin scales.
c. When someone has an infection, her body may respond with a fever. A high fever ($T >$ 105°F) can damage the brain. Convert this temperature to degrees Celsius.

CASE STUDY Part 1: Inventing the Thermometer

You know that water boils at 212°F, but three centuries ago scientists were studying the properties of boiling water. Without a thermometer, you wouldn't know that all pots of water boil at the same temperature, and you might speculate that some other property is more important. When Fahrenheit learned that a French scientist, Guillaume Amontons, found that water boils at a particular temperature, he was inspired to design and build his own thermometer.

Having been surrounded by thermometers and accurate temperature measurements our whole lives (Fig. 19.3), it is hard to imagine living without them, but for most of human history people didn't have thermometers. Since ancient times, however, many human endeavors—such as pottery, metallurgy, and cooking—required at least an estimate of temperature. In many cases, temperature was estimated by touch, but that is not possible if the object is very hot or very cold. The temperature of a hot metal was estimated from the color of its glow, and a kiln or oven's temperature may be estimated from its effects on other objects: Does it melt wax? Boil water? Melt lead? Accurate thermometers are necessary in thermodynamics because the laws of thermodynamics involve temperature. The case study in this chapter describes the story and physics of thermometers.

Although some types of thermometers we use today are based on recent technology, many thermometers can trace their roots to the end of the 16th century, especially to Galileo Galilei. The basic idea behind his thermometer and most of the thermometers we use today is that many macroscopic properties of matter depend on its temperature. These properties can be measured and exploited. For example, a mercury thermometer determines temperature based on expansion of the mercury as it warms.

19-2 Zeroth Law of Thermodynamics

Imagine working in a laboratory and needing to measure the temperature of some liquid in a flask. You put a mercury thermometer in the liquid and watch the mercury rise or fall (Fig. 19.4). You don't read the thermometer until the mercury level stops because the level of the mercury depends on temperature. If the level is changing, the temperature of the mercury is changing. The mercury's temperature will continue to change until it has the same temperature as the liquid in the flask. Then, by reading the temperature of the mercury from the thermometer scale, you have measured the temperature of the liquid.

This simple laboratory technique for measuring temperature illustrates the concepts of thermal contact and thermal equilibrium. When two objects are in **thermal contact**, thermal energy can pass from one object to the other. If the objects remain in thermal contact for a sufficient period of time, they will eventually reach thermal equilibrium and have the same temperature. In other words, two objects are in **thermal equilibrium** if they are in thermal contact with each other and their temperatures do not change. In the simple laboratory experiment (Fig. 19.4), the mercury thermometer is brought into thermal contact with the liquid. Thermal energy passes either from the thermometer to the liquid or from the liquid to the thermometer until they reach thermal equilibrium. When they are in thermal equilibrium, the thermometer's temperature equals the temperature of the liquid.

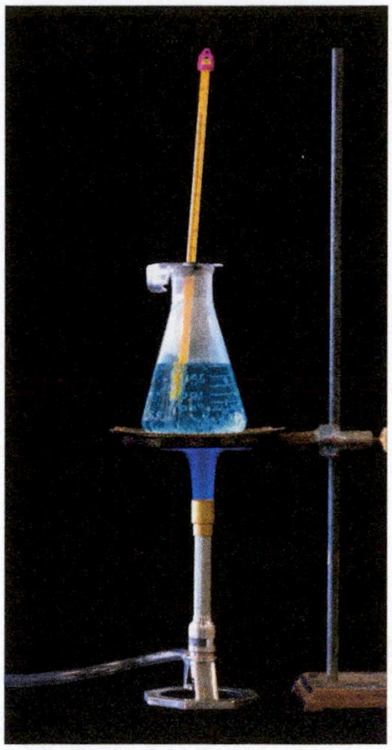

FIGURE 19.4 When using a mercury thermometer, you must wait until the mercury stops moving before reading the temperature.

THERMAL CONTACT; THERMAL EQUILIBRIUM

Major Concept

Some processes require objects to be in thermal equilibrium (in other words, to have the same temperature) before they are brought into thermal contact. Suppose you are following a recipe for homemade mayonnaise that requires you to mix egg yolks (usually stored in the refrigerator) and oil (usually stored at room temperature) only after they have reached thermal equilibrium. Because the eggs are initially colder than the oil, you should put the yolks in a small container and put the container in a warm-water bath.

Without putting them in contact with each other, how can you be sure that the egg yolks and oil are in thermal equilibrium? You could use a thermometer to measure their temperatures. Suppose you put the thermometer in the oil and, after a short time, the oil and thermometer are in thermal equilibrium. Their temperature—read off the thermometer—is 22°C. You then place the thermometer in the egg yolks, and after a while the yolks and the thermometer are in thermal equilibrium with a temperature of 22°C. You conclude that because both the oil and the yolks are in thermal equilibrium with the thermometer at 22°C, they must be in thermal equilibrium with each other. That is correct and, in fact, is an example of the **zeroth law of thermodynamics**:

ZEROTH LAW OF THERMODYNAMICS

Underlying Principle

If two objects A and B are in thermal equilibrium with a third object C, then A and B they are in thermal equilibrium with each other.

We can write the zeroth law of thermodynamics mathematically:

$$\text{If } T_A = T_C \text{ and } T_B = T_C, \text{ then } T_A = T_B.$$

There are four laws of thermodynamics. The other laws were stated before the zeroth law. You might think that it makes more sense to designate the zeroth law as the next-highest-number law, but because the zeroth law is more fundamental than the other laws, it was given a lower number. The zeroth law is considered more fundamental because it contributes an additional way to define temperature. **Temperature** is the macroscopic property of a system that determines whether that system is in thermal equilibrium with another system.

CONCEPT EXERCISE 19.2

Explain this statement: *A thermometer reports its own temperature.* How is it possible to use a thermometer to measure another object's temperature?

19-3 Thermal Expansion

FIGURE 19.5 A gap in the road prevents the concrete from cracking when the road expands or contracts.

When an engineer or architect designs a bridge, a large building, or even a sewer system, she must leave a gap between pieces (Fig. 19.5). The gap is necessary because most building materials (concrete and metals) expand as their temperature increases and contract as their temperature decreases. Without these gaps, the structures could buckle and fracture.

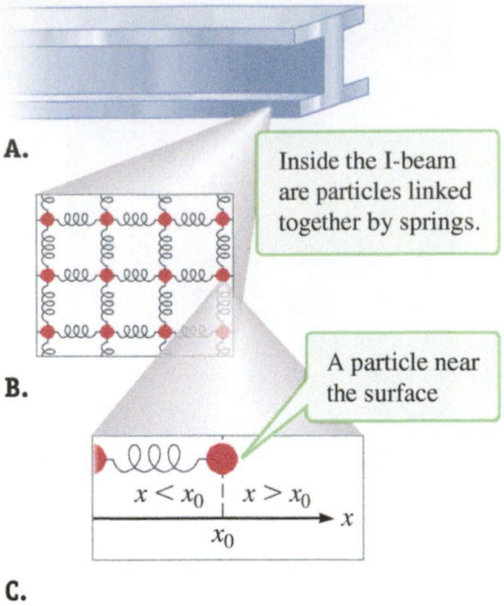

FIGURE 19.6 **A.** A steel beam. **B.** A model of the molecules that make up the steel beam. Particles are attached to one another by springs. **C.** A close-up of one particle near the surface.

Microscopic Model of Thermal Expansion

To see how temperature changes can cause a steel beam or similar object to expand, let's model the molecules in the object as particles and the molecular bonds as springs (parts A and B of Fig. 19.6) as we did in Section 14-4 when studying tensile and compressive stresses. Consider one such molecule at the surface of the beam with its horizontal bond (Fig. 19.6C). The potential energy curve for the molecule–bond system is shown in Figure 19.7. The system's potential energy is plotted on the vertical axis, and the molecule's position is plotted on the horizontal axis. When the molecule is at $x = x_0$, the molecular bond is relaxed, and the system's potential energy is at its minimum. When the molecule moves to the left ($x < x_0$), the bond is compressed, and when the molecule moves to the right ($x > x_0$), the bond is stretched.

Unlike the curve for the particle–spring system in Figure 8.32C (page 233), the potential energy curve in Figure 19.7 for a molecule–bond system is not a parabola. Instead, it is asymmetric, rising rapidly as $x \to 0$ and approaching zero as x gets very large. This curve makes sense: When the molecule is near $x = x_0$, the bond acts like a spring obeying Hooke's law, and the potential energy curve is nearly parabolic. As the molecule moves to the left, it runs up against other molecules that repel it, and the potential energy curve shoots upward. However, as the molecule moves to the right, there is nothing but the bond to keep it attached, and the molecule may overstretch and even break that bond.

As the temperature of the steel beam increases, so does its thermal energy. The molecules in a hot beam have more kinetic energy on average than do the molecules in a cool beam. According to $E = K + U$ (Eq. 8.11), the mechanical energy E of the molecules must increase as their kinetic energy K increases. Energy graphs for the molecule–bond system (near the surface) in a cool beam and a hot beam are shown in Figure 19.8. A molecule in the cool beam moves about as far to the right of its relaxed position x_0 as it moves to the left (Fig. 19.8A). On the other hand, because the asymmetry of the potential energy curve is more pronounced at higher energy, a molecule in the hot beam moves much farther to the right (outward) than it moves to the left (Fig. 19.8B). Once the surface bonds lengthen, interior bonds are able to lengthen and the beam expands.

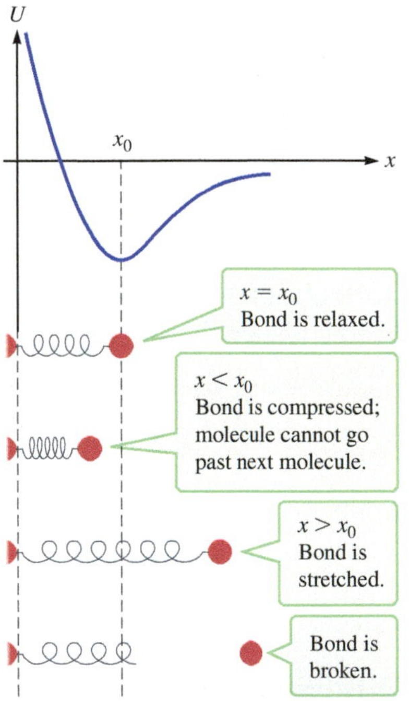

FIGURE 19.7 For a particle near the surface, the potential energy curve for a particle–bond system is not symmetric, so the particle moves farther to the right than it does to the left.

CONCEPT EXERCISE 19.3

Explain this statement: *If the potential energy curve were symmetric (as in Fig. 8.32C), hot materials would not expand.*

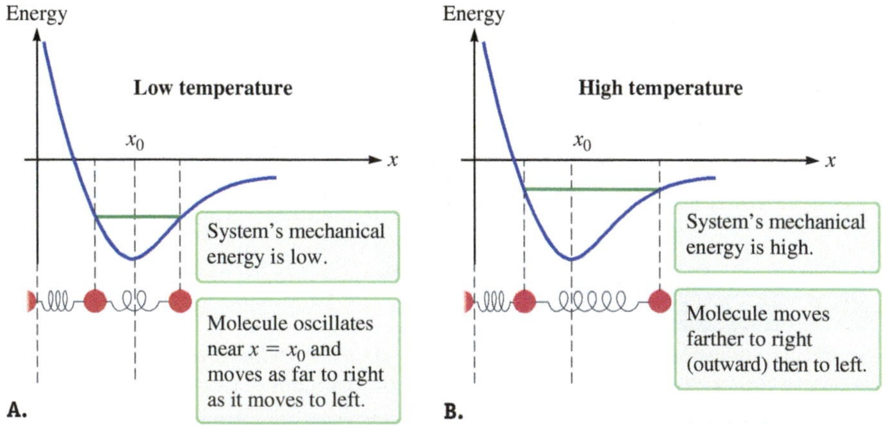

FIGURE 19.8 Due to energy conservation, mechanical energy E is represented by a straight horizontal line on an energy graph (Section 8-8). **A.** The energy graph for a cool beam shows that the mechanical energy E is low, and each surface molecule oscillates back and forth very close to its equilibrium position x_0. The asymmetry of the potential energy curve is not noticeable, and the bonds act much like springs obeying Hooke's law. **B.** The energy graph for a hot beam shows that the mechanical energy E is high, and each surface molecule's oscillation takes it far from its equilibrium position. The asymmetric nature of potential energy is noticeable, and the molecules move much farther outward than they move inward.

Macroscopic Observation of Thermal Expansion

On the macroscopic scale, when a solid rod of length L_0 experiences a temperature change ΔT, we observe a **thermal linear expansion** given by

THERMAL LINEAR EXPANSION

Major Concept

$$\Delta L = \alpha L_0 \Delta T \quad (19.4)$$

where ΔL is the change in the rod's length and α is called the **coefficient of linear expansion**. The coefficient of linear expansion has the dimensions of inverse temperature and depends on the type of material. Values of α for some common materials are given in Table 19.1.

Equation 19.4 is consistent with the microscopic model illustrated in Figure 19.8. If ΔT is positive, the molecule's mechanical energy increases, and its average distance to its neighbor increases. The initial length L_0 of the rod is proportional to its number of molecules and bonds; the more molecules and bonds in the rod, the greater its change in length. The coefficient of linear expansion α is related to the shape of the potential energy curve; a material with a large α must have a very asymmetric potential energy curve.

TABLE 19.1 Coefficients of linear expansion near room temperature.

Material	α ($\times 10^{-6}$ K^{-1} or °C^{-1})	Material	α ($\times 10^{-6}$ K^{-1} or °C^{-1})
Aluminum	22–24	Glass	4.0–9.0
Brass	18.7	Iron	10.4–12.0
Brick	5.5	Marble	12
Bronze	18.0	Mortar	7.3–13.5
Cement	10.0	Steel	13.0
Concrete	14.5	Wood (oak, parallel to grain)	4.9
Copper	16.8	Wood (oak, across grain)	5.4

These values are approximate and can be used for problems in this book if no other values are given.

DERIVATION Thermal Volume Expansion

Equation 19.4 works well for long, thin objects such as metal rods. For other shapes such as cubes or spheres, each dimension expands according to Equation 19.4, resulting in a change in the object's volume. Consider a rectangular solid whose original dimensions are L_0, W_0, and H_0 (Fig. 19.9) so that the original volume is

$$V_0 = L_0 W_0 H_0$$

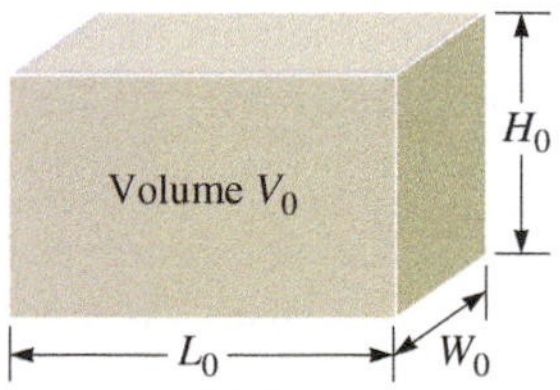

FIGURE 19.9

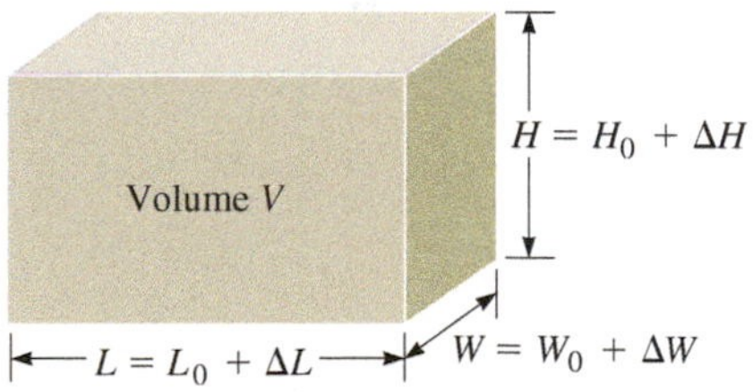

FIGURE 19.10

If the object experiences a temperature change ΔT, each dimension changes according to Equation 19.4. We will show that the change in the object's volume, known as **thermal volume expansion**, is given (approximately) by

THERMAL VOLUME EXPANSION

Major Concept

$$\Delta V \approx \beta V_0 \Delta T \quad (19.5)$$

where β is the coefficient of volume expansion: $\beta \approx 3\alpha$.

When the temperature increases, the object expands. Our sketch (Fig. 19.10) shows what happens when we apply $\Delta L = \alpha L_0 \Delta T$ (Eq. 19.4) to each dimension of the box.

Find the new length L, width W, and height H algebraically for thermal expansion.	$L = L_0 + \Delta L = L_0 + \alpha L_0 \Delta T = L_0(1 + \alpha\,\Delta T)$ $W = W_0 + \Delta W = W_0 + \alpha W_0 \Delta T = W_0(1 + \alpha\,\Delta T)$ $H = H_0 + \Delta H = H_0 + \alpha H_0 \Delta T = H_0(1 + \alpha\,\Delta T)$
The new volume V is the new length multiplied by the new height multiplied by the new width.	$V = LHW = L_0 W_0 H_0 (1 + \alpha\,\Delta T)^3 = V_0(1 + \alpha\,\Delta T)^3$

Derivation continues on page 558 ▶

Expand the cubed term.	$V = V_0[1 + 3\alpha\,\Delta T + 3(\alpha\,\Delta T)^2 + (\alpha\,\Delta T)^3]$
The increase in volume is represented as the difference between the expanded box and the original-size box (Fig. 19.11).	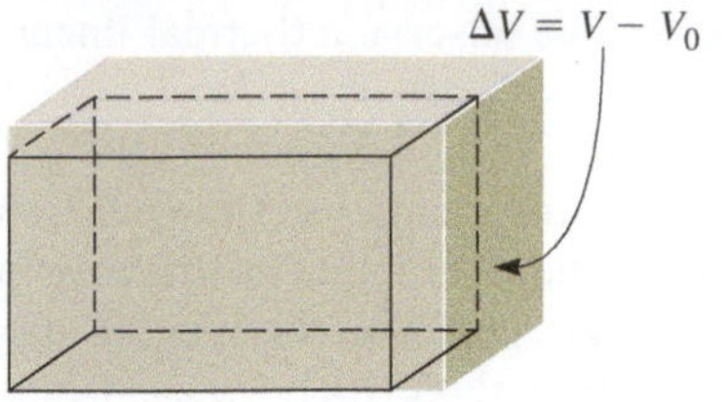 FIGURE 19.11
The change in the object's volume is the new volume V minus the original volume V_0.	$\Delta V = V - V_0 = V_0[1 + 3\alpha\,\Delta T + 3(\alpha\,\Delta T)^2 + (\alpha\,\Delta T)^3] - V_0$ $\Delta V = V_0[3\alpha\,\Delta T + 3(\alpha\,\Delta T)^2 + (\alpha\,\Delta T)^3]$
As long as the expansion is much smaller than the object's original size, $\alpha\Delta T \ll 1$, so the terms $(\alpha\,\Delta T)^2$ and $(\alpha\,\Delta T)^3$ are even smaller than $\alpha\,\Delta T$ and can be ignored.	$\Delta V \approx V_0 3\alpha\,\Delta T$
Use the definition of the coefficient of volume expansion, $\beta \approx 3\alpha$.	$\Delta V \approx \beta V_0\,\Delta T$ ✓

COMMENT

The coefficient β also depends on the type of material (and temperature); values for some common materials are given in Table 19.2.

TABLE 19.2 Coefficients of volume expansion near room temperature.

Material	β ($\times 10^{-6}$ K^{-1} or °C^{-1})
Aluminum	66–74
Brass	60
Copper	51
Ethanol (liquid)	750
Ethyl and isopropyl alcohol (liquid)	1100
Gasoline (liquid)	950
Glass	12–27
Mercury (liquid)	180
Steel	36
Water (liquid)	210

These values are approximate and can be used for problems in this book if no other values are given.

Several points need to be made about Equations 19.4 and 19.5:

1. If the object cools off, its temperature decreases ($\Delta T < 0$). According to Equations 19.4 and 19.5, the object contracts ($\Delta L < 0$ or $\Delta V < 0$).
2. The coefficients of linear expansion α and volume expansion β depend on the temperature of the object. We ignore that complication in this textbook and simply note that the values in Tables 19.1 and 19.2 are valid when the object is at approximately room temperature.
3. Because liquids and gases conform to the shape of their containers, the linear expansion coefficient has no meaning for liquids or gases. Volume expansion coefficients for several liquids are given in Table 19.2.
4. Equations 19.4 and 19.5 are only valid if ΔL or ΔV are small compared with L_0 or V_0.

The second and fourth points make it difficult to model gases with Equation 19.5 because β for gases has a strong dependence on temperature and because ΔV can be very large for a gas. In Sections 19-5 and 19-6, we will develop a better model for gases.

CONCEPT EXERCISE 19.4

No physical object is truly one-dimensional. Imagine a steel rod with a circular cross section whose temperature increases by $\Delta T = 1$ K. If the length of the rod is 1 m and its radius is 1 cm, find its change in length and radius. Is it reasonable to model such a rod as a one-dimensional object?

EXAMPLE 19.1 CASE STUDY Building an Alcohol Thermometer

In 1709, Fahrenheit invented the alcohol thermometer using "spirit of wine" (also known as ethyl alcohol, ethanol, or grain alcohol) as the expanding fluid. Consider making a simple version of his thermometer using dyed isopropyl alcohol in a container with a straw poking through the top (Fig. 19.12). Fahrenheit put his thermometer into a mixture of water, ice, and salt and marked the level of the alcohol as zero on his scale. If you wish to use the Celsius scale, you would place your thermometer in a bath of water and ice, leaving out the salt. This mixture corresponds to 0°C.

Alcohol boils at 78°C, so if you put your thermometer in boiling hot water, the alcohol will expand up the straw as its temperature rises until it reaches its boiling point. (Such a thermometer cannot operate above 78°C.)

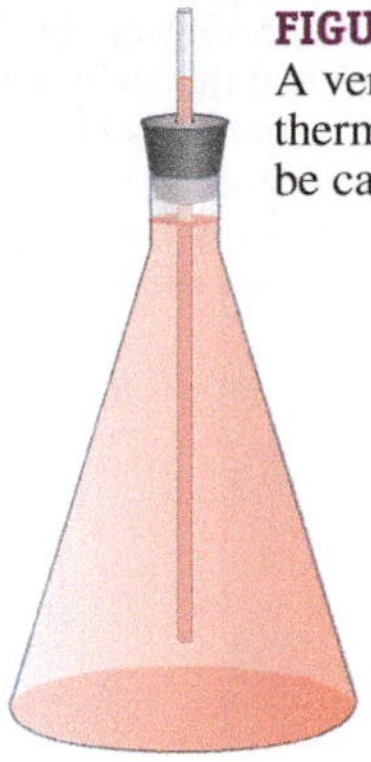

FIGURE 19.12 A very simple thermometer to be calibrated.

A If your thermometer contains an entire bottle of isopropyl alcohol (volume 473 mL at room temperature, 20°C) and the diameter of your straw is 5 mm, find the length of straw required to mark off the boiling point of alcohol. Assume only the alcohol expands appreciably.

INTERPRET and ANTICIPATE

As the alcohol's temperature increases, its volume increases accordingly. Because a liquid's shape is determined by its container, the alcohol will rise up the straw. The thinner the straw, the greater the alcohol's height.

SOLVE

The change in alcohol volume ΔV is proportional to the change in alcohol height Δh in a straw of radius r.	$\Delta V = \pi r^2 \Delta h$
Equation 19.5 relates ΔV to ΔT.	$\Delta V = \beta V_0 \Delta T = \pi r^2 \Delta h$
Solve for Δh. The coefficient of volume expansion is found in Table 19.2. The initial volume is 473 mL = 0.473 × 10^{-3} m^3. The alcohol's temperature changes from room temperature (T = 20°C) to its boiling point (T = 78°C), so ΔT = 58°C. The radius of the straw is 2.5 mm.	$\Delta h = \dfrac{\beta V_0 \Delta T}{\pi r^2}$ $\Delta h = \dfrac{(1100 \times 10^{-6}\,°\text{C}^{-1})(0.473 \times 10^{-3}\,\text{m}^3)(58°\text{C})}{\pi(2.5 \times 10^{-3}\,\text{m})^2}$ $\Delta h = 1.5\ \text{m}$

CHECK and THINK

Even if the alcohol in the straw is level with the rest of the alcohol at room temperature, the straw would need to be 1.5 m or 5 ft tall to measure the required temperature! Clearly, such a large thermometer is not practical. There are a number of ways to make the thermometer smaller, such as using less alcohol (smaller V_0) and a wider straw (larger r). Fahrenheit did not try to measure such high temperatures with his early thermometer. If the maximum temperature you wished to measure was of a healthy person (only 37°C), our simple thermometer's straw would only need to be about 0.5 m or about 1.5 ft tall.

B Why do you suppose Fahrenheit invented the mercury thermometer 5 years later in 1714? *Hint*: The boiling point of mercury is 357°C.

INTERPRET and ANTICIPATE

Think about the problems you would encounter using an alcohol thermometer: (1) It cannot measure very high temperatures, and (2) it must be very large.

SOLVE

(1) Because the boiling point of mercury is so high, a mercury thermometer can easily measure the temperature of other hot substances such as boiling water.

(2) The volume expansion coefficient for mercury is about six times smaller than that for alcohol (Table 19.2). So, if the alcohol in your thermometer (Fig. 19.12) were replaced with mercury, the straw could be six times shorter. It would only need to be about 0.25 m or about 10 in. tall to measure the required temperature.

CHECK and THINK

Today, alcohol thermometers such as the one in the back of your refrigerator are used to measure relatively cool objects over a relatively small range of temperatures.

Example continues on page 560 ▶

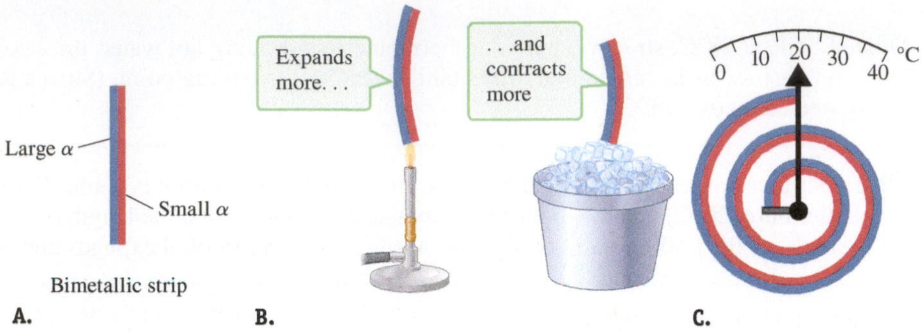

FIGURE 19.13 **A.** A bimetallic strip consists of two different metals bonded together. **B.** When the strip is heated or cooled, the strip bends because each metal expands or contracts by a different amount. **C.** Bimetallic strips are commonly used in thermometers and thermostats.

CASE STUDY Part 2: Bimetallic Strip Thermometer

In many contemporary applications, a thermometer made of glass and liquid is not practical. One alternative is the bimetallic strip thermometer. The basic physics behind this device is shown in Figure 19.13. Two strips of different metals are bonded together (Fig. 19.13A). Each metal has a different coefficient of linear expansion, so when the bimetallic strip's temperature changes, each metal expands or contracts by a different amount, causing the strip to bend (Fig. 19.13B). In a thermometer, the strip is formed into a coil and attached to a pointer (Fig. 19.13C). A scale is calibrated so that the pointer's position indicates temperature. As the temperature increases, the coil becomes tighter, and the pointer is pushed to the right. As the temperature decreases, the coil becomes looser, and the pointer is pulled to the left.

19-4 Thermal Stress

Figure 19.5 shows an example of the gaps architects and engineers put into their designs to allow structures to expand and contract. What would happen if a designer left out such a gap? Imagine, for example, a steel beam whose ends are free. If the beam's temperature rises, the beam will expand; similarly, if its temperature drops, the beam will contract. Now imagine that the ends of the steel beam are clamped and that the clamps that keep the beam from changing length apply compressive or tensile stress to the beam (Section 14-4) when its temperature changes. Because this stress is a result of the beam's temperature change, it is called **thermal stress**. If the thermal stress is large enough, the structure may crack or burst.

Thermal stress is quantified as the stress needed to keep the beam's length constant when its temperature changes. To find the thermal stress on a beam clamped at both ends, let's first use $\Delta L = \alpha L_0 \Delta T$ (Eq. 19.4) to find the fractional change ε in the beam's length if it were allowed to expand when its temperature increases by ΔT:

$$\varepsilon = \frac{\Delta L}{L_0} = \alpha \, \Delta T \tag{19.6}$$

To keep the beam's length constant, the clamps must apply a compressive stress that produces a strain equal in magnitude to ε. This compressive stress σ is given by $\sigma = Y\varepsilon$ (Eq. 14.11), where Y is Young's modulus. The thermal stress comes from substituting Equation 19.6 into Equation 14.11:

$$\sigma = Y\alpha|\Delta T| \tag{19.7}$$

We have taken the absolute value of ΔT because stress is a magnitude of force per area (Eq. 14.9), but the sign of ΔT determines whether the thermal stress is compressive or tensile. If the temperature increases ($\Delta T > 0$), the thermal stress must be compressive. In this case, if the thermal stress meets or exceeds the compressive strength σ_{cmp} of the beam, it will fracture. A decrease in temperature can also damage a structure. If the temperature decreases ($\Delta T < 0$), the thermal stress must be tensile. In this case, if the thermal stress meets or exceeds the tensile strength σ_{ten} of the beam, it will fracture. The tensile and compressive strengths of various materials are given in Table 14.1 (page 405).

EXAMPLE 19.2 Car Talk

On August 12, 2006, Gerald from Portland, Oregon, called the Tappet brothers on the National Public Radio program *Car Talk*.[1] Gerald had loaned his car to his girlfriend, who had parked it in the direct sunshine on a very hot day ($T \approx 100°F$). The next day, Gerald discovered a crack in his windshield. He told his girlfriend that she was at fault because she had turned on the air conditioner. He asked the Tappet brothers if turning on the air conditioner could have caused the crack. The Tappet brothers said that it was very unlikely for two reasons:

1. The air conditioner *gradually* cools the glass so that "it does not immediately go from 150°F to 60°F."
2. They said, "Windshields are designed to take those kinds of temperature differentials. . . . Most windshields have a bunch of dots around the perimeter or some kind of a transition zone so that where the window does touch the rest of the car there is allowance for expansion and movement of the glass so that it doesn't crack."

A third reason was provided by Gerald's girlfriend:

3. Windshields are made of tempered glass.

The Tappet brothers agreed with Gerald's girlfriend. Tempered glass has a tensile strength up to six times that of regular glass. In addition, tempered glass has a low coefficient of linear expansion α. The Tappet brothers recommended that Gerald apologize to his girlfriend and take her out to dinner.

Assume the temperature of the windshield decreased from 150°F to 60°F and the car manufacturer erred by omitting the transition zone between the glass and the rest of the car.

A Find the thermal stress on the windshield and determine if the glass will break. Use the maximum value of α listed in Table 19.1 for glass. Young's modulus for glass is 70×10^9 N/m^2 (Table 14.1).

INTERPRET and ANTICIPATE

By choosing the maximum value for α and considering the compressive strength of normal glass, we are checking to see if such a temperature change could possibly crack normal glass if the glass were improperly installed. If the thermal tensile stress on the glass due to the air conditioning is less than the tensile strength of the glass, the glass shouldn't break. If normal glass wouldn't crack, tempered glass, which is stronger and has a much lower α, shouldn't crack either.

SOLVE

We need to find the temperature change in kelvins, which is equivalent to finding the temperature change in Celsius degrees. An expression for the change in temperature is derived from Equation 19.2.

$$\Delta T(°C) = T_f(°C) - T_i(°C)$$

$$\Delta T(°C) = \tfrac{5}{9}[T_f(°F) - 32] - \tfrac{5}{9}[T_i(°F) - 32]$$

$$\Delta T(°C) = \Delta T(K) = \tfrac{5}{9}\Delta T(°F) \qquad (19.8)$$

$$\Delta T(K) = \tfrac{5}{9}(60°F - 150°F) = -50\ K$$

Calculate the thermal stress from Equation 19.7 and use the given value for Young's modulus. The maximum value of α for glass from Table 19.1 is 9.0×10^{-6} K^{-1}.

$$\sigma = Y\alpha|\Delta T|$$

$$\sigma = (70 \times 10^9\ N/m^2)(9.0 \times 10^{-6}\ K^{-1})|-50\ K|$$

$$\sigma = 3.2 \times 10^7\ N/m^2$$

Because $\Delta T < 0$, the thermal stress on the glass is tensile. Compare the thermal stress to the lowest tensile strength of normal glass found in Table 14.1 (50×10^6 N/m^2). We find that the tensile strength of glass is stronger than the thermal stress. So, the glass should survive the change in temperature.

$$\sigma_{ten} = 50 \times 10^6\ N/m^2 = 5.0 \times 10^7\ N/m^2$$

$$\sigma_{ten} \approx 1.5\sigma$$

Example continues on page 562 ▶

[1]National Public Radio *Car Talk* show number 200632 originally aired on August 12th, 2006; http://www.cartalk.com/ct/review/show.jsp?showid=200632.

CHECK and THINK

The Tappet brothers were correct in thinking that the temperature change would not cause the glass to break. Even if the glass were installed improperly, the thermal stress is small compared to the tensile strength of glass. Taking into account that tempered glass is even stronger and has a lower α, it becomes even less likely that turning on the air conditioner cracked the glass.

B Why did the Tappet brothers give their first reason that using the air conditioner did not crack the windshield (gradual cooling)?

SOLVE

If you have dropped ice cubes into a cup of hot chocolate to cool it off quickly, you have probably noticed that the ice cubes crack. Temperature differences within an object can set up thermal stresses in different parts of the object. As another example, if you pour hot tea into a cold glass, the glass may break because the thermal stress between the cold and hot parts of the glass exceeds its strength. The Tappet brothers' first reason is a statement that the windshield's temperature remains relatively uniform as it cools, so they don't expect significant thermal stress within the glass itself. In the program transcript, one of the brothers says that turning on the air conditioner is "not like someone is throwing dry ice on the windshield from the inside."

FIGURE 19.14 The water on the surface of a lake freezes and insulates the water below, so the water below remains liquid. Life is safe below the ice, except for the fish who take the bait.

Thermal Expansion of Water

Most materials contract as they are cooled. That is, most coefficients of volume expansion β in $\Delta V = \beta V_0 \Delta T$ (Eq. 19.5) are positive numbers, so if the material is cooled, ΔT is negative and so is ΔV. Water—perhaps the most importance substance for life on the Earth—does not always contract as its temperature drops, however.

When water is cooled from 4°C to its freezing point at 0°C, it expands rather than contracts. When water is between 0°C and 4°C, β is negative, so a decrease in temperature results in an increase in volume. When the water's temperature is above 4°C, β is positive, and a decrease in temperature means a decrease in volume. (Water ice contracts as its temperature decreases.) Therefore, the density of liquid water is greatest at 4°C.

This somewhat odd property of water has important implications for life (Fig. 19.14). Imagine a lake in the winter when the air temperature drops below 0°C. The water at the surface of lake is in contact with the cold air, so it cools first. When the water below the surface is above 4°C, the colder water at the surface sinks because it is denser than the water below. The warmer water from below rises up to the surface, where it in turn cools and sinks. This process continues until the entire lake is at 4°C and of uniform density. As the water at the surface continues to cool, it expands and becomes less dense. Now this colder water remains on the surface and continues to cool. When the lake's surface temperature reaches 0°C, the water at that surface turns to ice, which is less dense than water and therefore floats at the top. The ice on the surface helps insulate the water below, which remains liquid with a temperature of about 4°C. Some plants and animals survive the winter in this cold liquid water.

If water were like other substances and continually contracted as it cooled, the lake would freeze from the bottom up, and the entire lake would probably turn to ice. Plants and animals could not survive the winter in solid ice.

CONCEPT EXERCISE 19.5

Suppose you wish to make sure there is a very tight fit between a metal peg and a hole. The hole is just a little too small for the peg to fit. Should you try to heat the peg or cool it before inserting it?

EXAMPLE 19.3 Holey Sheet Metal

Three holes are drilled into a rectangular sheet of aluminum when its temperature is 95°C. The dimensions of the holes and the aluminum are given in Figure 19.15. The aluminum is cooled, and its new length is $L_f = 4.990$ m. (Assume $\alpha = 22.2 \times 10^{-6}\ \text{K}^{-1}$.)

A Find the final (cooler) temperature of the aluminum.

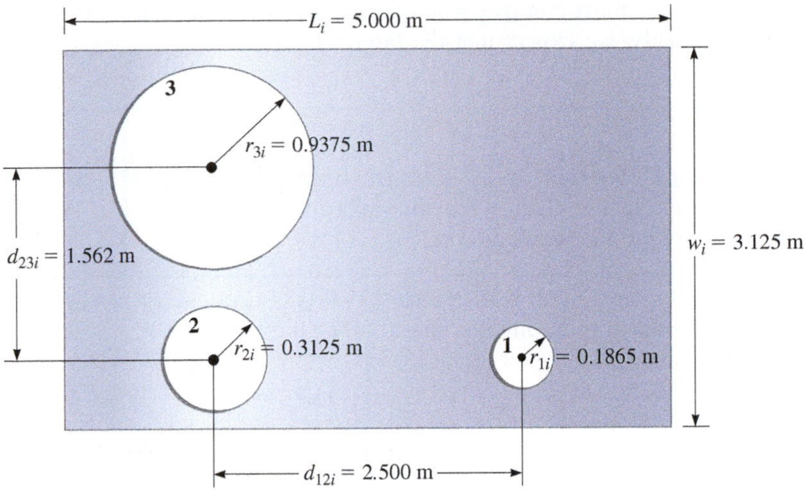

FIGURE 19.15 Three holes in a sheet of aluminum at 95°C.

INTERPRET and ANTICIPATE

The aluminum sheet contracts as it cools. Its new dimensions are determined by the coefficient of linear expansion for aluminum and the change in temperature, and we can use this information to find the new temperature.

SOLVE

Start by solving Equation 19.4 for the new temperature T_f.

$$\Delta L = \alpha L_i \Delta T$$

$$\Delta T = \frac{\Delta L}{\alpha L_i} = \frac{L_f - L_i}{\alpha L_i}$$

$$T_f = \frac{L_f - L_i}{\alpha L_i} + T_i$$

Substitute values.

$$T_f = \frac{(4.990\ \text{m}) - (5.000\ \text{m})}{(22.2 \times 10^{-6}\,°\text{C}^{-1})(5.000\ \text{m})} + 95.0°\text{C} = 4.9°\text{C}$$

CHECK and THINK

The temperature of the aluminum must drop by 90.1°C for its length to decrease by 1 cm. This temperature change is nearly the same as that required to go from boiling water to ice. Does such a great change in temperature make sense? Think about baking with an aluminum cookie sheet. Initially, the sheet is at room temperature, about 20°C. The oven temperature is about 400°F or about 200°C, so the temperature change is about 180°C. Have you ever noticed that the sheet is larger when you take it out of the oven? You probably haven't because the change in size of the sheet is proportional to its size, which is small (only about 0.5 m long), and the change is too small to notice. For the much larger aluminum sheet in this example, the temperature change is great enough to get a noticeable change in size.

B Find the change in width Δw of the rectangular sheet; the change in radius Δr_1, Δr_2, and Δr_3 of each circular hole; and the change in hole separations Δd_{12} and Δd_{23} between the warmer and cooler temperatures.

INTERPRET and ANTICIPATE

The entire aluminum sheet with its holes experiences the same change in temperature ($\Delta T = -90.1°\text{C}$). According to Equation 19.4, the change in size ΔL depends on the original length L_0, so we don't expect the width of the rectangular aluminum sheet to change by as much as the length: $|\Delta w| < |\Delta L|$. Likewise, we expect that the distance between holes 1 and 2 should change more than the distance between holes 2 and 3: $|\Delta d_{12}| > |\Delta d_{23}|$.

Perhaps the most difficult prediction to make is whether the holes contract or expand when the temperature decreases. Think about the aluminum disks that were punched out to make the holes. At the high temperature T_i, each aluminum disk fits perfectly into its respective hole. If the holes were not punched out and the aluminum sheet were cooled to T_f, those disks would still be part

Example continues on page 564 ▶

of the aluminum sheet and would still fit perfectly into their holes. Now imagine punching out the holes when the aluminum is hot at T_i and then cooling all the aluminum, including the disks, to temperature T_f. The aluminum disks must still fit perfectly inside their respective holes. Because the disks must contract when they are cooled, the holes must also contract by the same amount. Finally, we expect that the largest hole should experience the greatest change, so we predict that $|\Delta r_3| > |\Delta r_2| > |\Delta r_1|$.

SOLVE

Use Equation 19.4 repeatedly for each linear dimension. The fractional length change ε is the same for each part of the sheet, so find that value first.

$$\Delta L = \alpha L_i \Delta T \qquad \varepsilon = \frac{\Delta L}{L_i} = \alpha\,\Delta T$$

$$\varepsilon = (22.2 \times 10^{-6}{}^\circ\text{C}^{-1})(-90.1^\circ\text{C}) = -2.00 \times 10^{-3}$$

The change in any length is then just the original length multiplied by the fractional change.

$$\Delta w = w_i\varepsilon = (3.125\text{ m})(-2.00 \times 10^{-3}) = -6.25 \times 10^{-3}\text{ m} = -6.25\text{ mm}$$

$$\Delta d_{12} = d_{12,i}\varepsilon = (2.500\text{ m})(-2.00 \times 10^{-3}) = -5.00 \times 10^{-3}\text{ m} = -5.00\text{ mm}$$

$$\Delta d_{23} = d_{23,i}\varepsilon = -3.12\text{ mm}$$

$$\Delta r_1 = r_{1i}\varepsilon = (0.1865\text{ m})(-2.00 \times 10^{-3}) = -3.73 \times 10^{-4}\text{ m} = -0.373\text{ mm}$$

$$\Delta r_2 = r_{2i}\varepsilon = -0.625\text{ mm}$$

$$\Delta r_3 = r_{3i}\varepsilon = -1.88\text{ mm}$$

CHECK and THINK

Check the predictions we made for the relative changes.

$|\Delta w| < |\Delta L|$ — 0.625 mm < 1 cm ✓

$|\Delta d_{12}| > |\Delta d_{23}|$ — 5.00 mm > 3.12 mm ✓

$|\Delta r_3| > |\Delta r_2| > |\Delta r_1|$ — 1.88 mm > 0.625 mm > 0.373 mm ✓

19-5 Gas Laws

In the previous sections, we focused on liquids and solids; gases require a different approach. Usually, a gas—such as the air in your room—fills its container. Under the right conditions, a gas expands when its temperature rises and contracts when its temperature drops. If you place a balloon filled with air in liquid nitrogen, which has a temperature of 77 K (−196°C), the balloon collapses, and if you then allow the balloon to return to room temperature, it returns to its previous size (Fig. 19.16). As the air temperature changed its volume could change because its container—a balloon—is flexible. (A well-sealed, inflexible container may break when the air inside expands or contracts.)

So, $\Delta V = \beta V_0 \Delta T$ (Eq. 19.5) is not a good model for gas volume changes because it does not take into account the interrelationships between temperature, pressure,

A. B. C.

© Charles D. Winters

FIGURE 19.16 When balloons are placed in a flask of liquid nitrogen, their temperature drops and so does their volume. When the balloons are removed from the liquid nitrogen, their temperature and volume are restored.

volume, and the amount of a gas. We need to find an equation or series of equations that relate these macroscopic physical variables; such an equation is called an **equation of state**. The state of a gas is a list of its physical properties (for example, temperature, pressure, and volume) that can vary. We study gases that are in an **equilibrium state**, meaning that the physical properties are at least temporarily constant and uniform throughout the gas. The equations of state described in this section were found experimentally and are valid for gases that are not very dense and are at approximately atmospheric pressure.

EQUATION OF STATE
Major Concept

EQUILIBRIUM STATE
Major Concept

Consider again the balloons in Figure 19.16. When the air temperature in a balloon is low, the balloon's volume is small. The air pressure is constant, approximately atmospheric pressure. In the late 1700s, French scientist Jacques Charles carefully carried out experiments on gases held at constant pressure. He found that gas volume under such conditions is proportional to its temperature. **Charles's law** is written as

$$V \propto T \quad (\text{if } P \text{ is constant}) \tag{19.9}$$

or

$$\frac{V_f}{V_i} = \frac{T_f}{T_i} \quad (\text{if } P \text{ is constant}) \tag{19.10}$$

CHARLES'S LAW
Special Case

Charles never published his results. They were published by another French scientist, Joseph Louis Gay-Lussac, in the early 1900s. Gay-Lussac conducted his own experiments and found that if the gas volume is held constant using a container with rigid walls such as a scuba tank, pressure is proportional to temperature expressed in kelvins. **Gay-Lussac's law** is expressed as

$$P \propto T \quad (\text{if } V \text{ is constant}) \tag{19.11}$$

or

$$\frac{P_f}{P_i} = \frac{T_f}{T_i} \quad (\text{if } V \text{ is constant}) \tag{19.12}$$

GAY-LUSSAC'S LAW
Special Case

Charles's law is valid for a gas held at constant pressure, and Gay-Lussac's law is valid for a gas held at constant volume, but what happens if the temperature is held constant and the pressure and volume are allowed to change? To answer, imagine a squeezing a balloon full of air. As you decrease the volume, the air pressure goes up. The absolute pressure (Eq. 15.14) and volume are inverses of each other, as found by Irish chemist Robert Boyle through careful experimentation in the mid-1600s. **Boyle's law** may be written as

$$P \propto \frac{1}{V} \quad (\text{if } T \text{ is constant}) \tag{19.13}$$

or

$$P_f V_f = P_i V_i \quad (\text{if } T \text{ is constant}) \tag{19.14}$$

BOYLE'S LAW
Special Case

Equations 19.9 through 19.14 are all called *laws*, but they are really *empirical fits* to experimental data. Empirical fits model the data but do not attempt to explain the causes (as a theory does). Kepler's *laws* of orbital motion are really empirical fits as well (Chapter 7). Kepler's laws fit Tycho's celestial data, but they do not provide a theoretical explanation for the planets' motion. Such explanation comes from Newton's laws of motion and gravity. In Chapter 20, we will provide a theoretical basis for the gas laws. In the next section, we combine the empirical fits in Equations 19.9 through 19.14 into a single equation.

CONCEPT EXERCISE 19.6

Sketch the following graphs:

a. V as a function of T for Charles's law
b. P as a function of T for Gay-Lussac's law
c. P as a function of V for Boyle's law

19-6 Ideal Gas Law

Each gas law from the previous section (due to Charles, Gay-Lussac, and Boyle) was discovered by holding one of three variables (pressure, volume, or temperature) constant while allowing the other two to change. In many practical situations, none of those three variables is constant. For example, if you fill a balloon with helium gas and let it go, it will rise high into the atmosphere where it will experience lower pressure, larger volume, and colder temperatures. All three variables change, and we must use a combination of Charles's, Gay-Lussac's, and Boyle's laws to arrive at an equation of state. Mathematically, the combination is

$$PV \propto T \tag{19.15}$$

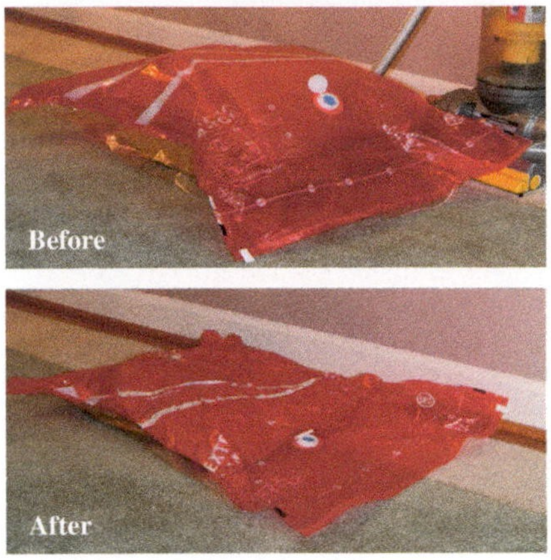

FIGURE 19.17 If you remove much of the air from this storage bag, the bag's volume decreases.

Simply combining these three gas laws doesn't describe the state of all gases because Equation 19.15 is only valid for a fixed amount of a particular gas; it doesn't tell you anything about the state of the gas if the *amount* of gas changes or what the state would be for a different *type* of gas. For instance, what happens to the state of helium in a balloon if some of it leaks out? And how does the state of a helium-filled balloon compare with the state of an oxygen-filled balloon? The goal in this section is to come up with one gas law that takes into account the amount and type of gas.

How do we measure the *amount* of gas? There are several ways, but one particularly important measure is a count of the number of particles N in the gas. By *number of particles*, we mean either the number of atoms or the number of molecules, depending on whether the gas is monatomic (like He) or polyatomic (like water vapor, H_2O).

Your experience tells you the number of particles in a gas is proportional to its volume. For example, before sealing a storage bag, you remove as much of the air as possible (Fig. 19.17). Removing air reduces the volume. This simple procedure illustrates that a gas held at constant temperature and pressure (room temperature and atmospheric pressure in this case) has a volume proportional to the amount of gas.

Now, imagine two balloons filled to the same volume at the same temperature and pressure, but one balloon is filled with oxygen and the other with helium. Will these two balloons have the same number of particles? In 1811, Amedeo Avogadro, an Italian scientist, came up with an insightful hypothesis that deals with the *type* of gas. According to **Avogadro's hypothesis** (now also known as **Avogadro's law**), *equal volumes of gas at the same temperature and pressure contain the same number of particles independent of the type of gas.* Mathematically, we can write Avogadro's law as

AVOGADRO'S LAW Special Case

$$V \propto N \quad \text{(at the same temperature and pressure)} \tag{19.16}$$

So, because the oxygen balloon and the helium balloon have the same volume (temperature and pressure), they contain equal numbers of gas particles.

We now combine Charles's, Gay-Lussac's, and Boyle's laws with Avogadro's law to arrive at the **ideal gas law**:

IDEAL GAS LAW Special Case

$$PV = Nk_BT \tag{19.17}$$

where k_B is **Boltzmann's constant**, P is absolute pressure, and T is absolute temperature. Boltzmann's constant has the same value for all gases:

$$k_B = 1.38 \times 10^{-23}\ \text{J/K}$$

IDEAL GAS Major Concept

An **ideal gas** is a hypothetical gas made up of identical particles that do not interact with one another. In practice, a real gas may be modeled as an ideal gas if its density is low, its pressure is less than about one atmosphere, and the gas temperature is not near its liquefaction point (boiling point of the liquid). The ideal gas law (Eq. 19.17) is also known as the **equation of state for an ideal gas** because it relates P, V, and T. In Section 20-6, we present slightly modified versions of Equation 19.17 that models the equation of state for a real gas under nonideal conditions. For the rest of this chapter, assume all gases can be modeled by the ideal gas law.

CONCEPT EXERCISE 19.7

The mass of an oxygen molecule is 5.3×10^{-26} kg, and that of a helium atom is 6.6×10^{-27} kg. One balloon is full of oxygen gas (O_2), and another is full of helium gas (He). If the balloons are at the same temperature and pressure and have the same volume, find the ratio of their masses and densities.

Avogadro's Number

When an administrative assistant of a large company orders office supplies, he orders pencils by the *gross*. One gross of pencils equals 12 dozen, or 144 pencils. The gross is a convenient unit for business because it is the correct size for what a company needs, and it is far more convenient than specifying some number of individual pencils and figuring out the cost of that special order.

MOLE Major Concept

Counting individual molecules in a gas is far more inconvenient than counting pencils, so it is more convenient to count in *moles*. The **mole** (abbreviated "mol") is the SI unit for the number of particles in a substance. One mole is defined as the amount of substance containing as many particles as there are atoms in 12 g of carbon-12. Exactly 12 g of carbon-12 contain **Avogadro's number** N_A of atoms, where

Avogadro never measured N_A. The quantity was measured in the middle of the twentieth century and is named to honor him.

$$N_A = 6.022 \times 10^{23}$$

AVOGADRO'S NUMBER Major Concept

A mole is like a dozen or a gross in that a dozen means 12 of something, a gross means 144 of something, and a mole means 6.022×10^{23} of something. If there is a mole of oxygen gas in a balloon, there are 6.022×10^{23} oxygen molecules in that balloon.

The ideal gas law can be written in terms of the number of moles n instead of the number of molecules N. The number of moles comes from dividing the number of molecules by Avogadro's number:

$$n = \frac{N}{N_A} \tag{19.18}$$

Solving for N and substituting into the ideal gas law (Eq. 19.17) gives

$$PV = nN_Ak_BT \tag{19.19}$$

Avogadro's number multiplied by Boltzmann's constant, N_Ak_B, is usually combined to form another constant R known as the **universal gas constant**:

$$R = N_Ak_B$$
$$R = 8.315\ \text{J/(mol} \cdot \text{K)} \tag{19.20}$$

Substituting Equation 19.20 into Equation 19.19, we can write the ideal gas law in terms of the universal gas constant and the number of moles:

$$PV = nRT \tag{19.21}$$

In practice, the number of moles n of a gas is often determined from its mass. To determine this number, you must first look up the **molar mass** M_{mol} for that type of gas. The molar mass is the mass of 1 mole of particles. Although the SI units of molar mass are kilograms per mole, we usually work in grams per mole for convenience.

The molar mass in grams per mole can be found on the Periodic Table of Elements in Appendix B. Usually, the key on a periodic table gives each element's **atomic mass**—the mass of one atom—in atomic mass units; $1\ \text{u} = 1.6605402 \times 10^{-27}$ kg. Molar mass in grams per mole is numerically equivalent to the atomic mass in atomic mass units, u.

EXAMPLE 19.4 How Much Is in the Balloon?

By weighing a balloon both before and after it is filled with molecular oxygen gas (O_2), a student determines that the mass of the oxygen is 18.3 g. According to the periodic table (Appendix B), the atomic mass of oxygen is 15.999 u. Find the number of moles n of oxygen molecules and the number of molecules N in the balloon.

Example continues on page 568 ▶

INTERPRET and ANTICIPATE

The molar mass of molecular oxygen (O_2) is twice the molar mass of atomic oxygen (O) found in the periodic table. Divide the mass of oxygen in the balloon by the molar mass of molecular oxygen to find the number of moles. The number of molecules comes from multiplying the number of moles by Avogadro's number, the number of molecules per mole.

SOLVE First, find the molar mass of molecular oxygen by multiplying the molar mass of atomic oxygen by 2.	$M_{mol} = 2(15.999 \text{ g/mol}) = 31.998 \text{ g/mol}$
Find the number of moles of oxygen gas in the balloon by dividing the mass of the gas M_{oxygen} by its molar mass M_{mol}.	$n = \frac{M_{oxygen}}{M_{mol}} = \frac{18.3\text{ g}}{31.998\text{ g/mol}} = 0.572\text{ mol}$
Use Equation 19.18 to find the number of molecules.	$n = \frac{N}{N_A} \qquad N = N_A n$ $N = (6.022 \times 10^{23}\text{ molecules/mol})(0.572\text{ mol})$ $N = 3.44 \times 10^{23}\text{ molecules}$

CHECK and THINK

Our result makes sense. The molar mass of molecular oxygen is about 32 g. Therefore, if the mass of oxygen in the balloon were 32 g, there would be 1 mole of oxygen in the balloon, and the number of molecules would equal Avogadro's number. The mass of oxygen in the balloon is just over half the molar mass of molecular oxygen, however, so the number of molecules N is just over half of Avogadro's number.

EXAMPLE 19.5 Standard Temperature and Pressure (STP)

Standard temperature and pressure (STP) is defined to be 0°C and 1 atm. Find the volume of 1 mol of an ideal gas at STP. Give your answer in cubic meters (m^3) and liters (L).

INTERPRET and ANTICIPATE

Because this example involves an ideal gas, we can use either expression of the ideal gas law (Eq. 19.17 or 19.21). The expression $PV = nRT$ is easier to use here because it is written in terms of moles, one of the parameters given in the example.

SOLVE Set $V = V_{STP}$ and solve Equation 19.21 for volume.	$PV_{STP} = nRT \qquad V_{STP} = \frac{nRT}{P}$
Convert the given pressure and temperature to SI units and find V_{STP}. Then, convert V_{STP} to liters using 1000 L = 1 m^3.	$P = 1\text{ atm} = 1.0133 \times 10^5\text{ Pa}$ $T = 0°C = 273.15\text{ K}$ $V_{STP} = \frac{(1\text{ mol})(8.315\text{ J/(mol} \cdot \text{K))}(273.15\text{ K})}{1.0133 \times 10^5\text{ Pa}}$ $V_{STP} = 2.241 \times 10^{-2}\text{ m}^3 = 22.41\text{ L}$

CHECK and THINK

Our result says that 1 mol of any gas at STP fills a volume equal to about the size of ten 2-L soda bottles. This handy result makes future problem solving easier. If a gas is at STP and you know the number of moles of that gas, you can find its volume by multiplying the number of moles n by V_{STP}. For example, if the balloon in Example 19.4 is at STP, its volume is 0.572×22.41 L = 12.8 L or about six soda bottles.

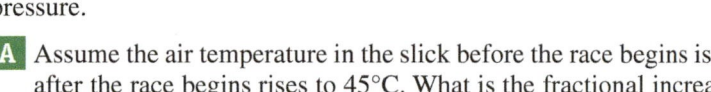

EXAMPLE 19.6 Tire Blowouts

To help prevent blowouts, manufacturers of car and bicycle tires often recommend keeping the tires at a slightly lower pressure during the hottest part of the summer. When the tire is in motion, friction between it and the road can raise the temperature of the air in the tire. The increased temperature can change the air volume and pressure.

Racing slicks are large, smooth tires used on certain types of race cars. Unlike the tires on an ordinary car, the air density in a racing slick is meant to be low. Often, wrinkles appear in the racing slick when the car is near the starting line (Fig. 19.18). Soon after the car speeds along the track, the tires appear firm and round. When the car is racing, the air in the slicks increases in temperature, volume, and pressure.

FIGURE 19.18 At the starting line, a racing slick shows wrinkles. When the tire warms up, it inflates.

A Assume the air temperature in the slick before the race begins is 15°C and soon after the race begins rises to 45°C. What is the fractional increase in the quantity PV for air in the tire?

INTEPRET AND ANTICIPATE

We model the air in the slick as an ideal gas, which is probably valid because the density is low. Because the temperature increases, we expect that the quantity PV will also increase.

SOLVE

Use the subscript f for the parameters during the race and the subscript i for the parameters before the race. Write down the ideal gas law (Eq. 19.17 or 19.21) once for each set of parameters. (We have arbitrarily used Eq. 19.17.) The amount of air in the slick does not change as long as there are no leaks. So, when the quantities during the race are divided by the quantities before the race, N cancels.

$$(PV)_f = Nk_BT_f \quad (19.17)$$
$$(PV)_i = Nk_BT_i$$
$$\frac{(PV)_f}{(PV)_i} = \frac{T_f}{T_i} \quad (1)$$

The fractional change in a quantity is the quantity's final value minus its initial value divided by its initial value.

$$\frac{(PV)_f - (PV)_i}{(PV)_i} = \frac{(PV)_f}{(PV)_i} - 1 \quad (2)$$

Substitute Equation (1) into Equation (2) and calculate the fractional change in PV. Because the ideal gas law is written in SI units, we must convert all temperatures to kelvins.

$$\frac{(PV)_f - (PV)_i}{(PV)_i} = \frac{T_f}{T_i} - 1$$
$$\frac{(PV)_f - (PV)_i}{(PV)_i} = \frac{(45 + 273.15)\text{K}}{(15 + 273.15)\text{K}} - 1 = 0.104$$

CHECK and THINK

As expected, the fractional change in PV is positive (meaning that the quantity has increased).

B For the cars that most of us drive, there is usually only a small change, if any, in the volume of air in the tires, and an increase in temperature mostly means a change in pressure. In racing slicks, the change in volume is small but not negligible. If the tire volume increases by 5% and the initial gauge pressure in the tire is 55.2 kPa (about 8 psi), what is the gauge pressure during the race?

INTERPRET and ANTICIPATE

A 5% increase in volume can be interpreted as a fractional increase of 0.05. In part A, we found the fractional increase in the quantity PV. We can use that result to find the new pressure. If the volume did not change at all, we would expect the fractional increase in pressure to equal the fractional increase in temperature, which is 0.104. Because the volume increases, we expect the fractional increase in pressure to be less than 0.104. The ideal gas law is written in terms of absolute pressure and not gauge pressure, so we must find the gauge pressure from the absolute pressure.

Example continues on page 570 ▶

SOLVE Use the result from part A to find an expression for P_f. The result will be an absolute pressure.	$\frac{(PV)_f - (PV)_i}{(PV)_i} = \frac{(PV)_f}{(PV)_i} - 1 = 0.104$ $\frac{P_f V_f}{P_i V_i} = 1.104$ $P_f = 1.104\left(\frac{V_i}{V_f}\right)P_i \qquad (3)$
Use the fractional change in V to find a value for (V_i/V_f).	$\frac{V_f - V_i}{V_i} = \frac{V_f}{V_i} - 1 = 0.05$ $\frac{V_i}{V_f} = \frac{1}{1.05}$
From Equation 15.15, absolute pressure is the sum of the gauge pressure and atmospheric pressure. Calculate the initial absolute pressure and substitute values into Equation (3).	$P_i = 55.2 \times 10^3 \text{ Pa} + 1.013 \times 10^5 \text{ Pa} = 1.565 \times 10^5 \text{ Pa}$ $P_f = (1.104)\left(\frac{1}{1.05}\right)(1.565 \times 10^5 \text{ Pa}) = 1.65 \times 10^5 \text{ Pa}$
Find the gauge pressure during the race by subtracting the atmospheric pressure from P_f.	$P_{\text{gauge}} = P_f - P_0$ from Equation 15.15 $P_{\text{gauge}} = 1.65 \times 10^5 \text{ Pa} - 1.013 \times 10^5 \text{ Pa}$ $P_{\text{gauge}} = 6.325 \times 10^4 \text{ Pa}$
CHECK and THINK To check our result, let's calculate the fractional increase in the absolute pressure.	$\frac{P_f - P_i}{P_i} = \frac{1.65 \times 10^5 \text{ Pa} - 1.565 \times 10^5 \text{ Pa}}{1.565 \times 10^5 \text{ Pa}} = 0.0543$

As expected, the fractional increase in absolute pressure is less than 0.104. The gauge pressure increases from about 8 psi to about 9 psi, an increase of roughly 13%. In the cars we normally drive, such an increase would mean that a cool gauge pressure of about 30 psi would increase to nearly 34 psi after driving. The recommended pressure given by the tire manufacturer is meant to be measured when the tire is cool, so you should not drive your car for more than about 1 mile before measuring its gauge pressure.

19-7 Temperature Standards

The Kelvin temperature scale (Section 19-1) is named for Lord Kelvin, born in 1824 in Belfast, Ireland, and originally named William Thomson. Kelvin, an impressive scientist and engineer, was given his title of nobility for his success in laying a submarine telegraph cable from England to France in 1866. (Kelvin is the name of the river that runs past the university in Glasgow, Scotland, where he worked.) In 1848, Kelvin published an article outlining the need for an absolute temperature scale.

To understand the problems Kelvin raised in his article, imagine two different laboratories, one using an alcohol thermometer and the other using a mercury thermometer. Each thermometer is calibrated according to the Celsius scale so that the freezing point of water is at 0°C and the boiling point of water is at 100°C. The two thermometers do not necessarily agree at intermediate temperatures, however. The mercury thermometer may measure a person's temperature as 36.5°C, whereas the alcohol thermometer may measure the same temperature as 37.8°C. The researchers are left wondering which temperature, if either, is correct. Kelvin wrote that researchers in different laboratories must be able to make independent temperature measurements that agree with one another and that an absolute temperature scale must not depend on the material used in the thermometer. Today, these difficulties have been overcome by (1) the absolute temperature scale in use by agreement of the international scientific community and (2) the use of constant-volume ideal gas thermometers.

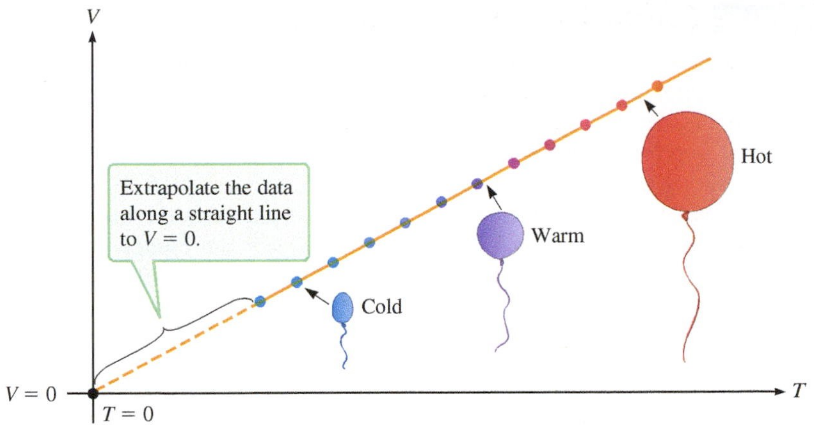

FIGURE 19.19 A graph of volume as a function of temperature shows that as the temperature of a balloon decreases, so does its volume. These data follow a straight line, and *absolute zero* is found by extrapolating the line back to zero volume.

KELVIN SCALE Major Concept

The **Kelvin scale** (also known as the **absolute scale**) in today's SI system of units is based on the principles laid out by Lord Kelvin in the mid-1800s. Kelvin used Charles's law (Eq. 19.9) to define zero on the absolute scale. According to Charles's law, as gas temperature decreases, its volume decreases. Kelvin reasoned that there must be a limit to how cold a gas could be because the gas volume could not be negative. He took zero to be the theoretical limit of any gas's volume. For a balloon filled with gas at constant pressure, a graph of the volume as a function of temperature would fit a line according to Charles's law (Fig. 19.19). Kelvin's idea was to extrapolate that line back to zero volume and set the corresponding temperature to zero on the absolute scale ($T = 0$ K). This temperature is known as **absolute zero**, equal to –273.15°C. No matter what type of gas you use, your data for volume as a function of temperature would fit a line extrapolating back to absolute zero ($T = -273.15$°C).

ABSOLUTE ZERO Major Concept

This definition of absolute zero does not require that any actual gas or object be at absolute zero. In fact, no measurement of absolute zero has been made. Three researchers—William D. Phillips, Steven Chu, and Claude Cohen-Tannoudji—won the Nobel Prize in Physics in 1997 for cooling atoms to about 900 nK, however, and in 2003, a team of researchers from the Massachusetts Institute of Technology reported that they had lowered a gas's temperature to 500 pK.[2] We will return to the topic of absolute zero in later chapters, but for now it is sufficient to know that absolute zero is the lowest temperature that is theoretically possible.

Because the lowest possible temperature is zero on the absolute temperature scale, only one reference temperature is needed to calibrate an actual thermometer. The reference temperature used for the Kelvin scale is the temperature of the **triple point of water**, the pressure and the temperature at which water exists in three phases: solid, liquid, and gas. The triple point of water is a condition that is more precisely reproducible than either the boiling point or freezing point of water. For example, because water boils at a lower temperature when it is at a lower pressure, people who live at high altitude (greater than 3000 ft) must modify the cooking instructions when they boil a pot of rice or pasta. Likewise, researchers in a laboratory located in the mountains would have a hard time calibrating a thermometer based on the boiling point of water because it would boil off before reaching 100°C.

Water can only exist in all three phases at once when the pressure and the temperature are at particular values. The triple point of water is defined to have a temperature $T = 273.16 \text{ K} = 0.01$°C and requires that the water vapor pressure be 610 Pa, roughly 6×10^{-3} atm. So, when you see water in all three phases, you know the temperature (and pressure) precisely.

The triple point requires a fairly low pressure. To maintain and establish the required temperature and pressure, laboratories use a **triple-point cell** as shown in Figure 19.20. A triple-point cell is a glass tube partially filled with a mixture of water and ice, above which is water vapor. The tube has a large depression so that a thermometer bulb may be brought into thermal contact with the cell.

A.

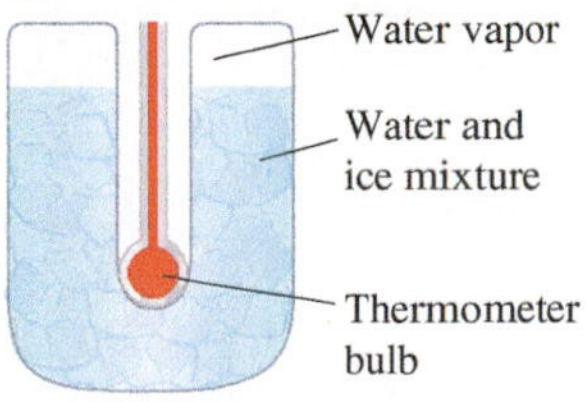

B.

FIGURE 19.20 Triple-point cells are used to keep water at its triple point: a mixture of solid, liquid, and gas.

[2]See A. E. Leanhardt et al., "Cooling Bose-Einstein Condensates Below 500 Picokelvin," *Science* 301, no. 5639 (2003): 1513–1515.

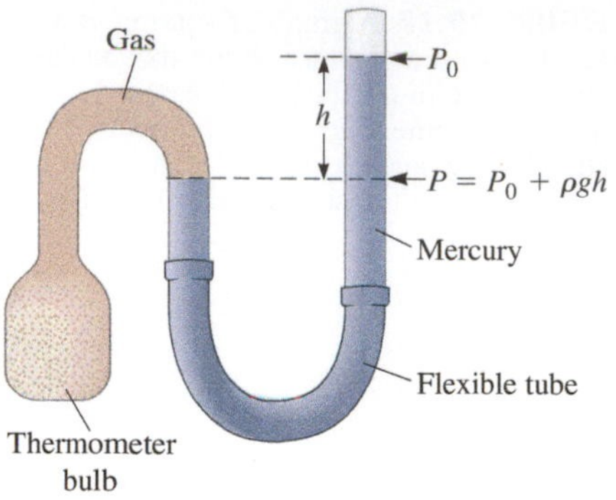

FIGURE 19.21 In a constant-volume gas thermometer, a gas is held at a constant volume so that a change in temperature means that only its pressure can change. The change in pressure is measured by a mercury manometer that may be calibrated in terms of gas temperature.

CASE STUDY Constant-Volume Gas Thermometer

In a laboratory where temperature must be measured carefully, a constant-volume ideal gas thermometer is used to calibrate the Kelvin scale. This thermometer is the final device in our case study on the construction of thermometers. A **constant-volume ideal gas thermometer** is shown in Figure 19.21. The bulb contains a low-density gas and is connected to a mercury manometer (Section 15-5). We'll assume such a gas is well modeled as an ideal gas. The bulb is placed in thermal contact with an object whose temperature is to be measured. No gas can escape from the thermometer, so the amount of gas N or n in the ideal gas law (Eqs. 19.17 and 19.21) is constant.

In general, as the gas temperature changes, both its volume and pressure also change; the constant-volume thermometer, however, is designed so that the experimenter can keep the gas volume constant. The mercury is in a U-shaped tube with the middle part of the tube made of a flexible material so that raising or lowering the right-hand tube of the mercury manometer (Fig. 19.21) keeps the gas volume at its original value, depending on the temperature of the object in thermal contact with the gas.

Figure 19.22 shows how to operate a constant-volume gas thermometer. Initially, the gas temperature is T_i, and the mercury in the right-hand tube is at height h_i above the gas–mercury boundary in the left tube (step 1). The gas temperature T_i is read by measuring the height h_i of the mercury. Now if you increase the gas temperature (step 2), the gas volume increases. To return the gas to its original volume, you must raise the right-hand tube of mercury (step 3). The volume of gas in step 3 is the same as the original volume of gas in step 1. The final temperature T_f of the gas is read by measuring the new height $h_i + \Delta h$ of the mercury.

Now let's take a mathematical look at the constant-volume thermometer. According to $P = P_0 + \rho gy$ (Eq. 15.6), the initial gas pressure P_i is given by the height of the mercury h_i:

$$P_i = P_0 + \rho gh_i \tag{19.22}$$

where P_0 is the atmospheric pressure and ρ is the density of mercury. According to the ideal gas law (Eq. 19.17), the absolute pressure P_i is proportional to the initial gas temperature T_i:

$$P_i = \frac{Nk_B}{V}T_i \tag{19.23}$$

With the volume held fixed, a change in the gas's temperature means a change in its pressure and nothing more, so N and V are unchanged and do not require a subscript to indicate initial and final values. The height of the mercury in the right-hand tube

FIGURE 19.22 1 Gas in the constant-volume gas thermometer is at T_i, and the mercury in the manometer is at height h_i above the gas–mercury boundary. 2 The thermometer is placed in thermal contact with an object, and its temperature increases. The increased temperature increases the gas volume. 3 By raising the right-hand tube of the mercury manometer, the gas volume is restored to its original size. The mercury is now at $h_i + \Delta h$ above the gas–mercury boundary. This increase in height is a result of the increase in gas temperature and pressure.

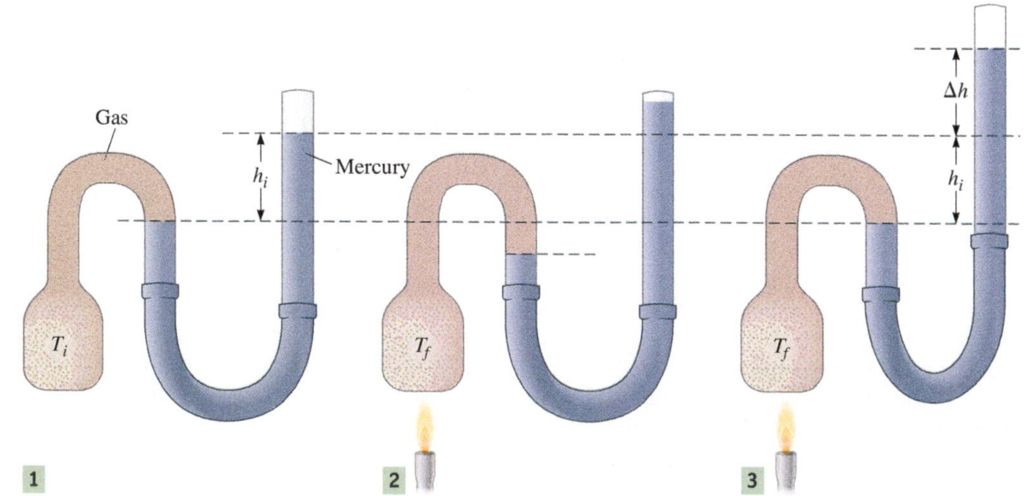

in Figure 19.21 is used to find the pressure of the gas and therefore the gas temperature according to Equation 19.23.

Similar equations hold when the gas is at its final temperature T_f as in Figure 19.22, far right (step 3). When the gas is at this higher temperature, its pressure P_f can be written as

$$P_f = P_0 + \rho g h_f$$

$$P_f = P_0 + \rho g(h_i + \Delta h) = (P_0 + \rho g h_i) + \rho g \, \Delta h$$

$$P_f = P_i + \rho g \, \Delta h \tag{19.24}$$

The final temperature T_f of the gas comes from the ideal gas law:

$$P_f = \frac{Nk_B}{V} T_f \tag{19.25}$$

For calibration, the thermometer is initially put in contact with a triple-point cell so that its initial temperature is $T_3 = 273.16$ K and the gas pressure in the thermometer (not the water vapor pressure in the triple-point cell) has some value P_3. We can find the temperature T of other objects in terms of the triple point temperature T_3 by dividing Equation 19.25 by Equation 19.23:

$$\frac{P}{P_3} = \left(\frac{Nk_B/V}{Nk_B/V}\right)\left(\frac{T}{T_3}\right)$$

$$T = \frac{P}{P_3} T_3 \tag{19.26}$$

where P is the gas pressure when the thermometer is in thermal equilibrium with an object at temperature T.

Equation 19.26 is a prescription for how to use a constant-volume gas thermometer. First, put the thermometer in thermal contact with a triple-point cell and measure the height h_i of the mercury. From this height, use $P_3 = P_0 + \rho g h_i$ (Eq. 19.22) to calculate the gas pressure P_3 in the thermometer. Now put the thermometer in thermal contact with an object whose temperature you wish to measure. As in Figure 19.22, adjust the height of the mercury until the gas volume equals its original volume. Measure the height of the mercury and use Equation 19.24 to find the gas pressure P at this new temperature T. The temperature T is found from Equation 19.26 with $T_3 = 273.16$ K. In Example 19.7, we see how to turn this prescription into tick marks made on the thermometer.

EXAMPLE 19.7 CASE STUDY Making a Constant-Volume Ideal Gas Thermometer

You are asked to design your own constant-volume gas thermometer (Fig. 19.23). Your thermometer uses alcohol ($\rho = 785.06$ kg/m^3) instead of mercury. The thermometer maintains 0.500 L of O_2 gas in its bulb. The total mass of the O_2 gas in the bulb is 0.500 g. Your job is to set the tick marks on the scale shown in Figure 19.23 so that each tick mark is 0.5 K. What is the separation in centimeters between adjacent tick marks?

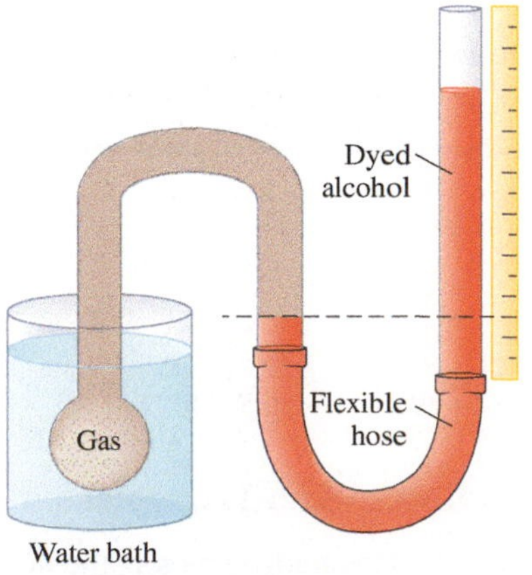

FIGURE 19.23 How far apart are the tick marks?

INTERPRET and ANTICIPATE

Figure 19.23 illustrates your task. For each change in temperature of 0.5 K, your job is to find the change Δh in the alcohol's height. The amount and volume of the O_2 gas are held constant. Only its pressure changes, resulting in a change in the height of the alcohol.

Example continues on page 574 ▶

SOLVE

Start with Equation 19.26 written in terms of an initial temperature T_i and pressure P_i and a final temperature T_f and pressure P_f.	$T_f = \frac{P_f}{P_i}T_i$
Substitute $P_f = P_i + \rho g\,\Delta h$ (Eq. 19.24) for P_f.	$T_f = \frac{P_i + \rho g\,\Delta h}{P_i}T_i$
Solve for ΔT.	$T_f = \frac{\cancel{P_i}T_i}{\cancel{P_i}} + \rho g\,\Delta h\frac{T_i}{P_i}$ $T_f - T_i = \Delta T = \rho g\,\Delta h\frac{T_i}{P_i}$
Use $P_i = (Nk_B/V)T_i$ (Eq. 19.23) to eliminate T_i/P_i.	$\Delta T = \rho g\,\Delta h\left(\frac{V}{Nk_B}\right)$
Solve for Δh. Remember that ρg has to do with the fluid in the manometer (in this case, alcohol), whereas Nk_B/V has to do with the gas in the bulb (in this case, oxygen).	$\Delta h = \frac{1}{\rho g}\left(\frac{Nk_B}{V}\right)\Delta T$ (1)
Find N for the oxygen gas in the bulb. For molecular oxygen (O_2), multiply the molar mass of oxygen found in the periodic table by 2.	number of moles $n = \frac{\text{mass of oxygen present}}{2M_{mol}}$ $n = \frac{0.500\,\text{g}}{2(15.999\text{ g/mol})} = 1.56 \times 10^{-2}$ mol number of molecules $N = nN_A$ $N = (1.56 \times 10^{-2}\text{ mol})(6.022 \times 10^{23}\text{ molecules/mol})$ $N = 9.41 \times 10^{21}$ molecules of O_2
Substitute numerical values into Equation (1). The density of alcohol is given in the opening statement. Convert the bulb volume into SI units: $V = 0.500\text{L} = 0.500 \times 10^{-3}\text{ m}^3$.	$\Delta h = \frac{(9.41 \times 10^{21})(1.38 \times 10^{-23}\text{ J/K})(0.5\text{ K})}{(785.06\text{ kg/m}^3)(9.81\text{ m/s}^2)(0.500 \times 10^{-3}\text{ m}^3)}$ $\Delta h = 1.69 \times 10^{-2}\text{ m} = 1.69\text{ cm}$

CHECK and THINK

So, the distance between each tick mark is 1.69 cm. To check our answer, imagine that we replace the alcohol with mercury. Because mercury is denser than alcohol, we would expect that it would not rise as high as the alcohol and the tick marks would therefore be closer together, due to increased ρ in Equation (1). As one further check, imagine that we put more gas in the bulb, increasing the gas pressure (if its volume is unchanged). A greater gas pressure would mean that the alcohol rises higher and the tick marks would be farther apart, due to increased N in Equation (1).

No matter what type of gas is used in constant-volume gas thermometers, the temperatures they measure agree. These thermometers are bulky and awkward to use, however. In practice, constant-volume gas thermometers are used to calibrate other thermometers that are easier to manage.

SUMMARY

Underlying Principles

1. **Thermodynamics** is the study of thermal energy and the transfer of energy through heat and work.
2. **Zeroth law of thermodynamics:** If two objects A and B are in thermal equilibrium with a third object C, they are in thermal equilibrium with each other. Mathematically,

$$\text{If } T_A = T_C \text{ and } T_B = T_C, \text{ then } T_A = T_B.$$

Major Concepts

1. **Temperature** is a measure of the average kinetic energy of the particles in a system. It is the macroscopic property of a system that determines whether that system will be in thermal equilibrium with another system.
2. When two objects are in **thermal contact**, thermal energy can pass from one object to the other.
3. Two objects are in **thermal equilibrium** if they are in thermal contact with each other and their temperatures do not change.
4. Most building materials (concrete and metals) expand as their temperature increases and contract as their temperature decreases. **Thermal linear expansion** is given by
$$\Delta L = \alpha L_0 \Delta T \quad (19.4)$$
5. **Thermal volume expansion** is approximately given by
$$\Delta V \approx \beta V_0 \Delta T \quad (19.5)$$
6. An **equation of state** is an equation or series of equations that relate the macroscopic physical variables (pressure, temperature, volume, and amount) of a gas.
7. When a gas is in an **equilibrium state**, its physical properties are constant and uniform throughout the gas.
8. An **ideal gas** is a hypothetical gas made up of identical particles that do not interact with one another. A real gas may be modeled as an ideal gas if its density is low, its pressure is less than about 1 atm, and the gas temperature is not near its liquefaction point.
9. The SI unit for the number of particles in a substance is the **mole**. One mole is the amount of substance containing as many particles as there are atoms in 12 g of carbon-12.
10. Exactly 12 g of carbon-12 contain **Avogadro's number** N_A of atoms, where $N_A = 6.022 \times 10^{23}$.
11. The **Kelvin scale (absolute scale)** is the temperature scale in the SI system.
12. On the Kelvin scale, $T = 0$ K is known as **absolute zero** and is found by extrapolating the temperature an ideal gas would have when its volume is zero. The triple point of water is defined as 273.16 K.

Special Cases: Equations of State

1. **Charles's law:**
$$V \propto T \quad \text{(if } P \text{ is constant)} \quad (19.9)$$
2. **Gay-Lussac's law:**
$$P \propto T \quad \text{(if } V \text{ is constant)} \quad (19.11)$$
3. **Boyle's law:**
$$P \propto \frac{1}{V} \quad \text{(if } T \text{ is constant)} \quad (19.13)$$
4. **Avogadro's law:** Equal volumes of gas at the same temperature and pressure contain the same number of particles independent of the type of gas. Mathematically,
$$V \propto N \text{ (at the same } T \text{ and } P\text{)} \quad (19.16)$$
5. The **ideal gas law** is a combination of Charles's, Gay-Lussac's, Boyle's, and Avogadro's laws:
$$PV = Nk_BT \quad (19.17)$$

PROBLEMS AND QUESTIONS

A = algebraic **C** = conceptual **E** = estimation **G** = graphical **N** = numerical

19-1 Thermodynamics and Temperature

1. **N** Convert the following temperatures from the Fahrenheit scale to the Celsius and Kelvin scales: **a.** The seawater temperature near Hawaii is about 76°F. **b.** McDonald's was sued for serving dangerously hot coffee. Their coffee was served between 180°F and 190°F. **c.** The coldest day recorded on Earth was −129°F in Vostok, Antarctica.
2. **C** Correct the following sentences by changing the temperature units. **a.** It is best to cook a turkey slowly at 250 K. **b.** A healthy dog's temperature is 374°C. **c.** A comfortable room temperature is 22°F. **d.** A crisp fall day in New England is about 50°C.
3. **N** Stars (like the Sun) that are actively converting hydrogen into helium in their cores are known as main sequence stars. Their surface temperatures range from about 1700 K to 53,000 K. Express this temperature range on the Fahrenheit scale.

Problems 4 and 5 are paired.

4. **C** Figure P19.4 shows a graph of the two temperature scales—Fahrenheit and Celsius—versus the Kelvin scale. Which line represents which scale? Justify your answer.

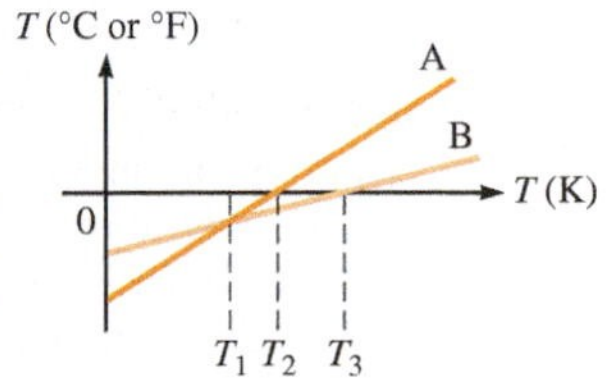

FIGURE P19.4 Problems 4 and 5.

5. **N** In Figure P19.4, what are the values of the temperatures T_1, T_2, and T_3 on all three scales?
6. **N** Hypothermia occurs when the human body temperature falls below 35.0°C, and hyperpyrexia occurs when the human body temperature exceeds 41.5°C. What are the equivalent temperatures on the **a.** Fahrenheit scale and **b.** Kelvin scale?
7. **N** The lowest temperature conceivable is absolute zero (0 K). What Fahrenheit temperature corresponds to 0 K?
8. **N** Liquid oxygen, a primary fuel component of rocket boosters, boils at a temperature of 90.2 K. What is the boiling point of liquid oxygen on the **a.** Celsius scale and **b.** Fahrenheit scale?

19-2 Zeroth Law of Thermodynamics

9. C Object A is placed in thermal contact with a very large object B of unknown temperature. Objects A and B are allowed to reach thermal equilibrium; object B's temperature does not change due to its comparative size. Object A is removed from thermal contact with B and placed in thermal contact with another object C at a temperature of 40°C. Objects A and C are of comparable size. The temperature of C is observed to be unchanged. What is the temperature of object B?

10. C A healthy boy wishes to miss school. His father says that the boy can stay home if he has a fever. The father leaves the boy with an alcohol thermometer in his mouth. (There are no lights or heaters near the boy's bed.) What is one thing the boy can do to increase the thermometer's temperature without leaving the bed?

11. C Answer each of the following questions and explain your reasoning. **a.** After you hold an ice cube for several minutes, are your hand and the ice cube in thermal equilibrium? **b.** You store ice cream and ice pops in your freezer for several days. Are they in thermal equilibrium? **c.** You swim in the ocean near Hawaii for several hours. Is your bathing suit in thermal equilibrium with the ocean? **d.** You toss your bathing suit in a sink full of cool water and let it soak overnight. Is your bathing suit in thermal equilibrium with the water?

19-3 Thermal Expansion and 19-4 Thermal Stress

12. C In the past, the coolant from a car's radiator would flow onto the ground; cars today have an overflow container to collect the excess. Why is the overflow container necessary? *Hint*: Imagine that you fill the car's radiator on a cold day.

13. C If a metal lid is stuck on a glass jar, a trick for opening it is to hold the lid and neck of the jar under hot water. **a.** Why does this method help? **b.** What would happen if the jar and lid were both made of the same type of metal? Explain.

14. N The tallest building in Chicago is the Willis Building (formerly the Sears Tower), which reaches 527 m including the antenna. The lowest and highest temperatures recorded in Chicago so far were -33.0°C (-27°F) and 44.0°C (111°F), respectively. If we assume the building is a steel structure, what is the difference in the height of the building at these two extreme temperatures? The coefficient of linear expansion for steel is $13.0 \times 10^{-6}\ °\text{C}^{-1}$.

15. N A bridge spanning a river gorge is made up of concrete sections 15.24 m in length, poured and cured at 15.0°C. What is the minimum spacing between bridge sections required to prevent buckling if the summertime temperatures at the bridge reach 43.0°C?

16. N A thin strip of copper 0.250000 m long is stored at 68.0°F. **a.** If the strip of copper is heated to 90.0°F, find the new length of the copper strip. Keep six significant figures as you work and report your answer to the same precision as the initial length. The coefficient of linear expansion for copper is $16.8 \times 10^{-6}\ \text{K}^{-1}$. **b.** What is the thermal stress on the copper?

17. N At 22.0°C, the radius of a solid aluminum sphere is 7.00 cm. **a.** At what temperature will the volume of the sphere have increased by 3.00%? **b.** What is the increase in the sphere's radius if it is heated to 250°C? Assume $\alpha = 22.2 \times 10^{-6}\ \text{K}^{-1}$ and $\beta = 66.6 \times 10^{-6}\ \text{K}^{-1}$.

18. C CASE STUDY Suppose you wish to replace a mercury thermometer, which is often used to measure a person's temperature, with a water thermometer. Estimate the dimensions of such a thermometer and comment on the pros and cons of using water as the fluid.

19. N A sphere of diameter 7.0 cm and mass 266.5 g floats in a liquid that has a temperature of 0°C. As the temperature is raised, the sphere begins to sink as the liquid reaches a temperature of 35°C. If the density of the liquid is 1.527 g/cm³ at 0°C, find the coefficient of volume expansion of the liquid. Ignore the volume expansion of the sphere.

20. An exterior wooden door fits comfortably in a steel frame on a day when the outdoor temperature is 50°F. The wood has been cut so that the grain runs vertically. The door is in a climate with a yearly temperature range from 0°F to 100°F. Keep four significant figures as you work and report final answers to one significant figure. Use values from Table 19.1.

a. C Would you expect the door to stick on a hot day or on a cold day? Explain.

b. N On a day when the temperature is 50°F, the door is 36 in. wide and 80.5 in. tall. What are its maximum and minimum width and height, given the range of temperatures?

c. N On a day when the temperature is 50°F, there is a 0.50-cm gap between the door and the frame. The steel frame has a uniform width of 2.5 in. What are the maximum and minimum gaps between the door and the frame, given the range of temperatures?

d. C Will the door ever get stuck in its frame? Explain.

21. N The distance between telephone poles is 30.50 m in a neighborhood where the temperature ranges from -35°C to 40°C. If you hang a copper cable between two adjacent poles on a day when the temperature is 22.30°C, what is the minimum length of the copper cable you must use for the cable to remain connected to the poles all year? Assume the cable is straight, and ignore the effect of gravity on the cable. Consider α to have four significant figures and report your answer to four significant figures.

22. N A copper cube with density $8.96 \times 10^3\ \text{kg/m}^3$ and mass 10.0 kg at 20.0°C is heated to 130°C. What is **a.** the density and **b.** the mass of the copper cube at this new temperature?

23. N A section of the Alaska railroad needs urgent repairs during a cold winter day with a temperature of -10.0°C, when 11.89-m-long steel beams are delivered to the repair site. **a.** What is the length of each beam in the summertime, when the temperature can reach 27.0°C? **b.** What is the thermal stress experienced by one of the beams between these two temperatures?

24. A clock with an iron pendulum is made so as to keep the correct time when at a temperature of 20°C. Assume the coefficient of linear expansion of iron is $12 \times 10^{-6}\ °\text{C}^{-1}$ and the pendulum is a simple pendulum.

a. C Will the clock be ahead or behind the correct time if the temperature rises to 30°C?

b. N What is the amount of time lost or gained each day?

25. N You notice that a concrete sidewalk is cracked and wonder if it is a result of thermal stresses. Assume the lowest temperature in the winter is -20°C, the highest temperature in summer is 40°C, and the edges of the 1-m-wide concrete sidewalk are fixed in place in the winter. Comment on whether thermal stresses might be responsible for the cracks in the sidewalk as summer temperatures are reached. Assume Young's modulus for concrete is $17 \times 10^9\ \text{N/m}^2$ and the coefficient of linear expansion is $10 \times 10^{-6}\ \text{K}^{-1}$. The compressive strength of concrete is $40 \times 10^6\ \text{N/m}^2$.

19-5 Gas Laws

26. C **Review** The gas laws are empirical, like Kepler's laws. Let's use Kepler's first law to think about the difference between an empirical model and a theory. According to Kepler's first law, planetary orbits are elliptical. Newton's laws of motion and of universal gravity provide the theoretical bases for Kepler's third law. If a planet is discovered that has a nonelliptical orbit (such as the rosette orbit of Mercury), does that discovery pose a challenge to Kepler's first law, Newton's laws, neither, or both? Explain your reasoning.

27. N A plastic bottle is closed outdoors on a cold day when the temperature is -13.0°C and is later brought inside where the temperature is 21.0°C. What is the pressure of the air in the bottle after it reaches room temperature, assuming the air in the bottle was at a pressure of 1.00 atm upon reaching thermal equilibrium with the outdoor temperature?

Problems 28 and 29 are paired.

28. **N** An 80-ft^3 scuba tank is filled with air at room temperature (22°C) so that the gauge pressure is 2987 psi. The tank is forgotten in the Sun for hours, and the tank's temperature reaches 125°F. Ignore the slight increase in the tank's size. What is the gauge pressure at this higher temperature?

29. **N** A scuba tank is filled as described in Problem 28. The tank is forgotten on the shore in winter for hours, and the tank's temperature drops to -45°F. Ignore the slight decrease in the tank's size. What is the gauge pressure at this lower temperature?

30. A 0.025-m^3 tank filled with helium had been kept in a refrigerated room at a temperature of 3.15°C and a pressure of 2.00 atm. A careless researcher moves the tank into the hallway while working in the room and forgets to put it back. The air in the hallway is at standard atmospheric pressure (1.00 atm) and room temperature (20.0 °C).

a. **C** What happens to the pressure in the tank as it remains in the hallway for a significant amount of time?

b. **A** Write an equation for the pressure of the helium as a function of its temperature.

31. **N** A balloon is filled with air outdoors on a cold day when the temperature is -13.0°C so that it occupies a volume of 3.00 L. It is then brought inside where the temperature is 21.0°C. Assuming the pressure in the balloon remains constant, what is the volume of the balloon indoors?

32. **E** Paul gets a 1.50-ft-diameter helium-filled balloon as a birthday present. He wants to preserve the balloon, so he stores it in the freezer (along with his leftover cake). Assume no helium leaks out. Estimate **a.** the volume and **b.** the diameter of the balloon when Paul finds it in the freezer the next day. Report your answers in US customary units. Consider Figure 19.16 and comment on your results.

19-6 Ideal Gas Law

33. **N** The ideal gas in a container is under a pressure of 20.0 atm at a temperature of 27.0°C. If half of the gas is released from the container and the temperature is increased by 50.0°C, what is the final pressure of the gas?

34. **N** Dexter has developed a special knockout gas that is kept in a 0.0650-m^3 cylinder at a temperature of 17.25°C. **a.** If the container is holding 10 mol of gas, what is the gas pressure? **b.** If the radius of the top of the cylinder is 0.30 m, what is the force the gas exerts on the top of the cylinder?

Problems 35 and 36 are paired.

35. **G** Figure P19.35 shows a graph of pressure P as a function of volume V for an ideal gas in a balloon with a tight seal so that no gas leaks out. Describe what is happening to the gas. What is changing, and what is constant?

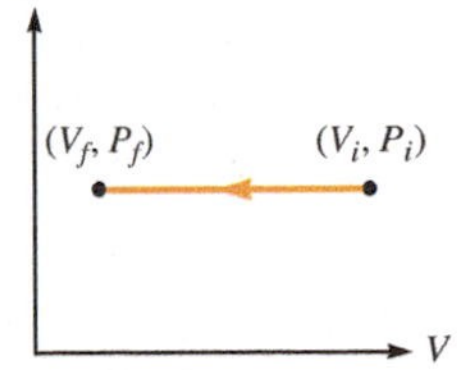

FIGURE P19.35

36. **G** Figure P19.36 shows a graph of pressure P as a function of volume V for an ideal gas in a balloon. The temperature of the balloon remains constant. Describe what is happening to the gas. What is changing and what is constant?

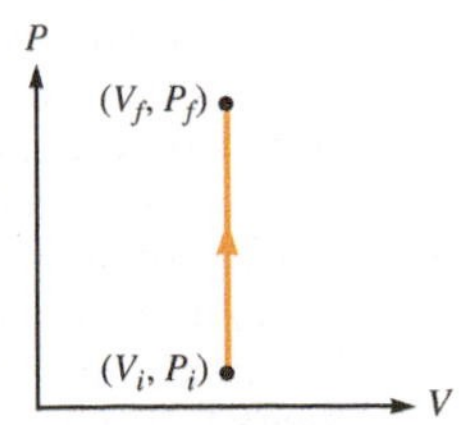

FIGURE P19.36

37. **N** An oxygen tank is filled at $T = 20.3$°C to an absolute pressure of 1.00 atm. The tank is then stored in the trunk of a car where the temperature reaches 46.8°C. What is the pressure in atmospheres?

38. **N** A very good laboratory vacuum can evacuate the gas in a chamber so that its pressure is about 10^{-12} Pa. Assuming the gas is at room temperature ($T \approx$ 20°C), estimate the number density (number per unit volume) of gas molecules in the chamber.

39. **N** What is the ratio of the number of molecules per cubic meter on top of Mount Everest at a pressure of 30.0 kPa and temperature of 2.00×10^2 K (around -100°F) compared with that at sea level (assuming standard temperature and pressure)?

40. On a hot summer day, the density of air at atmospheric pressure at 35.0°C is 1.1455 kg/m^3.

a. **N** What is the number of moles contained in 1.00 m^3 of an ideal gas at this temperature and pressure?

b. **N** Avogadro's number of air molecules has a mass of 2.85×10^{-2} kg. What is the mass of 1.00 m^3 of air?

c. **C** Does the value calculated in part (b) agree with the stated density of air at this temperature?

41. **N** A thermos is constructed such that there is an evacuated region between two walls that acts as a lining, leading to good thermal insulation. **a.** If the pressure in the lining is 1.32×10^{-6} atm at room temperature (293 K) and the volume of the lining is 30.0 cm^3, how many molecules are in the evacuated lining? **b.** How many moles are in the thermos lining?

42. **N** To decrease the pressure within a tank containing 2.00 mol of an ideal gas from 50.0 atm to 8.00 atm, a fraction of the gas is removed. If the volume and the temperature of the gas in the container remain constant, what is the number of moles of the gas that must be removed for the decrease in pressure to occur?

43. **N** Most of the Universe is made up of hydrogen, often found in huge clouds. The Orion Nebula is an example. The number density of hydrogen molecules (H_2) in such a cloud is estimated to be between 10^9 and 10^{12} molecules per cubic meter. The cloud temperature is typically between 10 K and 30 K. What is the range in such a cloud's gas pressure? Give your answer in pascals and atmospheres and comment on your results.

44. **N** The gauge pressure of a truck tire is 2.72 atm at 22.0°C. As the truck is laden with deliveries and driven around town, the temperature of its tires increases to 52.0°C. What is the gauge pressure in the tire at 52.0°C if the volume of the tires is assumed not to change?

45. **N** The 100-m^3 European JET tokamak fusion device achieves a near vacuum pressure of 5.00×10^{-6} Pa. If the temperature within the device is 22.0°C while it is not operating, what is the number of molecules in the device at this pressure?

46. A transmission electron microscope (TEM) is a very sensitive instrument that uses high-speed electrons to view materials on the nanometer-size scale. The sample you wish to observe must be kept in an ultrahigh vacuum chamber, where the pressure in the chamber is on the order of 100 nanopascals. Suppose the volume of a TEM vacuum chamber is 0.520 m^3 and there are about 3.00×10^{13} molecules/m^3 in the air inside the chamber. Treating the air as an ideal gas, what is the temperature of the air inside the chamber if the pressure is 100 nanopascals?

19-7 Temperature Standards

Problems 47 and 48 are paired.

47. **C** CASE STUDY A constant-volume ideal gas thermometer such as the one shown in Figure 19.21 (page 572) is constructed and calibrated at sea level. It is then carefully shipped to Denver, which is 1 mile above sea level. Do the researchers in the Denver laboratory need to recalibrate the thermometer? Explain.

48. **C** A triple-point cell such as the one shown in Figure 19.20 (page 571) is constructed and calibrated at sea level. It is then shipped to Denver, which is 1 mi above sea level. Can the researchers in the Denver lab use the triple-point cell just as in a laboratory at sea level, or will special adjustments be needed? Explain.

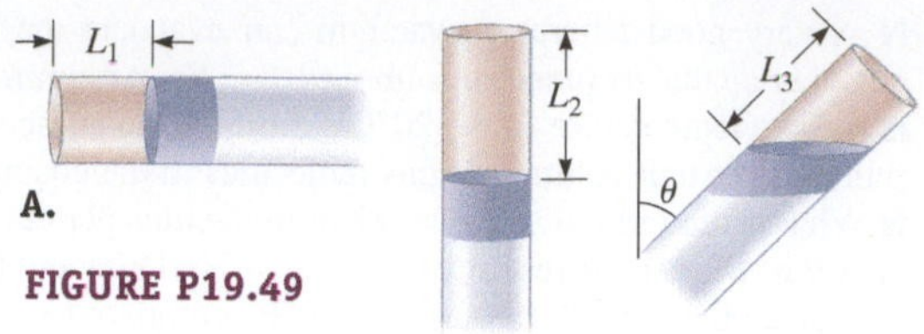

FIGURE P19.49

49. A An ideal gas is trapped inside a tube of uniform cross-sectional area sealed at one end as shown in Figure P19.49. A column of mercury separates the gas from the outside. The tube can be turned in a vertical plane. In Figure P19.49A, the column of air in the tube has length L_1, whereas in Figure P19.49B, the column of air has length L_2. Find an expression (in terms of the parameters given) for the length L_3 of the column of air in Figure P19.49C, when the tube is inclined at an angle θ with respect to the vertical.

50. C CASE STUDY When Lord Kelvin defined absolute zero, why couldn't he choose the zero point to be the temperature of some measurable phenomenon such as the temperature of liquid helium? *Hint*: Does the Kelvin scale include negative temperature?

51. C CASE STUDY A researcher misuses a constant-volume thermometer by failing to lift the right tube in Figure 19.22, when the bulb is in contact with an object that is hotter than the triple point of water. Does the researcher measure a temperature that is too high or too low? Explain.

52. C CASE STUDY When a constant-volume thermometer is in thermal contact with a substance whose temperature is lower than the triple point of water, how does the right tube in Figure 19.22 need to be moved? Explain.

53. N An air bubble starts rising from the bottom of a lake. Its diameter is 3.60 mm at the bottom and 4.00 mm at the surface. The depth of the lake is 2.50 m, and the temperature at the surface is 40.0°C. What is the temperature at the bottom of the lake? Consider the atmospheric pressure to be 1.01×10^5 Pa and the density of water to be 1.00×10^3 kg/m^3. Model the air as an ideal gas.

54. C Rather than use the triple point of water as the basis for precisely determining a temperature scale, you would like a temperature reference that is easier to produce. You're aware that the transition between freezing and melting depends on pressure, but you consider defining a specific temperature standard as the "freezing point of pure water at atmospheric pressure." What advantage does the triple point of water have over this definition?

General Problems

55. N To facilitate transport by tanker trucks, natural gas is often liquefied, which is achieved by decreasing the temperature of the gas below -162°C. What is the temperature on **a.** the Kelvin scale and **b.** the Fahrenheit scale at which natural gas liquefies?

56. N Jerome's physics lab on heat expansion involves a copper ring with a diameter of 3.000 cm and a stainless steel ball with a diameter of 3.015 cm. The linear expansion coefficient for copper is 17×10^{-6} °C^{-1} and that for stainless steel is 11×10^{-6} °C^{-1}, and both the copper ring and the steel ball are at 22.0°C. **a.** In the first phase of the experiment, Jerome warms the copper ring so that the stainless steel ball will just slip through. How hot is the copper ring when that occurs? **b.** In the second phase of the experiment, both the copper ring and the stainless steel ball are warmed together. What is the temperature at which the ball slips through the ring?

57. N A royal clock pendulum is one that has a period of precisely 1.000 s at a temperature of 20.0°C. One such clock pendulum, made of copper, is placed in a room with temperature 5.00°C. What is the change in the period of the pendulum?

58. N A university lecture hall is 12.0 m wide, 15.0 m deep, and 4.00 m high. The temperature in the hall is 22.0°C, and the pressure is 1.00 atm. What is the number of molecules of air in the lecture hall?

Problems 59 and 60 are paired.

59. N A solid aluminum sphere has a radius of 0.5700 m when it is at a temperature of 90.0°F. What is the change in volume of the sphere when it is heated to a temperature of 110.0°F?

60. C A solid aluminum sphere is heated from an initial temperature to a higher temperature. If, instead, the aluminum sphere were hollow but started with the same radius at the same initial temperature and was then heated to the same final temperature, how would the new volume of the hollow sphere compare with that of the solid aluminum sphere? Explain your reasoning.

61. N A rigid tank contains 3.00×10^{-2} m^3 of an ideal gas at a pressure of 5.15×10^5 Pa and a temperature of 300.0 K. How many **a.** moles and **b.** molecules of the gas are contained in the tank?

62. N In December 2006, two Missouri representatives filed legislation requiring that the price of motor fuel be adjusted for volume changes caused by temperature fluctuations. The industry standard assumes the temperature is 60°F. In the summer, when the temperature easily exceeds 60°F in Missouri, gasoline expands and its density decreases. Because gasoline is sold by volume, you buy fewer gasoline molecules (less energy) per gallon during the hot months. This change in volume is good for consumers in the winter months, when cold temperatures increase gasoline density and consumers get more molecules of gasoline per gallon. The industry has accepted temperature-based price adjustments in cold places like Canada, but not in warm places. In 2006, it was estimated that Missouri consumers paid $12 million extra due to the expansion of hot fuel. **a.** The density of gasoline at 60°F is 737.22 kg/m^3. What is the density of gasoline on a very hot day ($T = 95$°F)? **b.** If you buy 12.0 gal of gasoline at 60°F, what mass of gasoline have you purchased? **c.** If you buy 12.0 gal of gasoline at 95°F, what mass of gasoline have you purchased? **d.** If gasoline costs $2.50 per gallon (in 2006), how much money did a consumer lose by buying gasoline on a very hot day?

63. A A glass container of volume V_0 is completely filled with a liquid, and its temperature is then increased by ΔT. Find an expression for the volume of the liquid that will overflow. Assume the coefficient of linear expansion of glass is α_g and the coefficient of volume expansion of the liquid is β_ℓ, where $\beta_\ell > 3\alpha_g$.

64. G Some substances (such as water between 0°C and 4°C) expand when they are cooled and contract when they are heated. Sketch a potential energy curve for molecules in such a substance. Explain your sketch.

65. N A 30-quart (28.4-L) pressure cooker initially at 22.0°C is filled with 1.00 L of water and brought to a temperature of 450°C. What is the pressure within the pressure cooker?

66. An aluminum canister is filled with oxygen for use by hikers. The canister holds 20.0 L at a gauge pressure of 2200 psi at 20.0°C. The hiker empties the canister by taking 75 breaths from it. Assume each breath fills the same volume and the air is an ideal gas at constant temperature.
   a. G Plot the (absolute) pressure versus number density ($\rho = N/V$) of oxygen in the canister.
   b. C Does the hiker get the same number of oxygen molecules for each breath? If not, does the hiker get more oxygen molecules on the first breath or on the last breath?

67. N A tank contains gas at 15.0°C pressurized to 14.0 atm. The temperature of the gas is increased to 90.0°C, and half the gas is removed from the tank. What is the pressure of the remaining gas in the tank?

68. C Consider making a bimetallic strip from two metals listed in Table 19.1. Which two metals would create the most curvature (tightest circle) when the strip's temperature is changed? Explain.

69. **N** Chemical vapor deposition (CVD) is a chemical process used to manufacture several materials, including thin films and carbon nanotubes. Consider the manufacture of silicon dioxide (SiO_2) thin films (which are used in electronic devices) achieved using silane, SiH_4. Suppose the silane is drawn into a CVD chamber with a volume of 3.00 m^3 and it can be modeled as an ideal gas. The pressure of the silane gas is 0.050 atm, and it is at 825 K. **a.** Find the number of moles of silane in the chamber. **b.** What is the total mass of the silane in the chamber?

Problems 70 and 71 are paired.

70. **C** While studying a colony of bacteria, you find that the number N of bacteria grows exponentially, $N = N_0 e^{t/\tau}$, where N_0 is the initial number of bacteria and τ is the time constant. You further find that at $t = 54\tau$, the colony reaches a maximum and does not grow any further. Have you discovered an empirical fit or a theoretical model? Explain.

71. **C** You believe that a colony of bacteria should grow exponentially, following $N = N_0 e^{t/\tau}$, where N_0 is the initial number of bacteria and τ is the time constant until the colony reaches the maximum size. This is because as long as resources are available, each individual bacterium cell will divide into two new daughter cells, which will divide into two more cells and so on. Is your idea an empirical fit or a theoretical model? Explain.

72. **A** A steel plate has a circular hole drilled in its center. If the diameter of the hole varies according to thermal linear expansion, show that the area of the original circle A_0 changes with an increase in temperature ΔT, following the approximate relation $\Delta A \approx CA_0\,\Delta T$, where $C = 2\alpha$. *Hint:* $(\alpha\,\Delta T)^2 \ll \alpha\,\Delta T$.

73. **N** Before climbing a tough hill, a bicycle tire is inflated with air at 1.00 atm and 20.0°C. **a.** What is the change in the volume of the air in the tire in terms of the initial volume V_i if the pressure in the tire increases to 7.00×10^5 Pa and its temperature doubles? **b.** As the rider crests the hill, the tire temperature increases to 70.0°C, and the tire expands its volume by 3.50%. What is the absolute pressure in the bicycle tire?

74. **A** A gas is in a container of volume V_0 at pressure P_0. It is being pumped out of the container by a piston pump. Each stroke of the piston removes a volume V_s through valve A and then pushes the air out through valve B as shown in Figure P19.74. Derive an expression that relates the pressure P_n of the remaining gas to the number of strokes n that have been applied to the container.

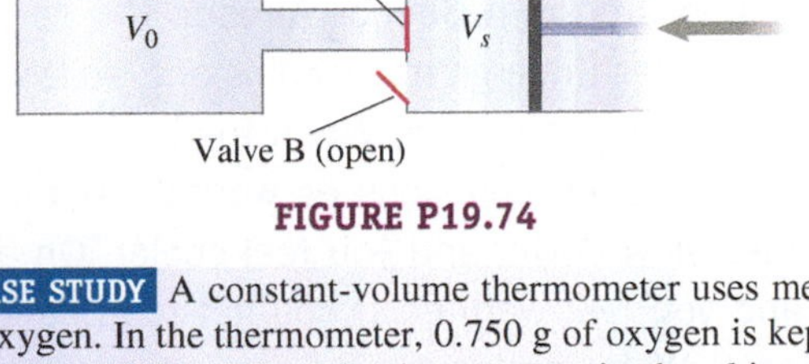

FIGURE P19.74

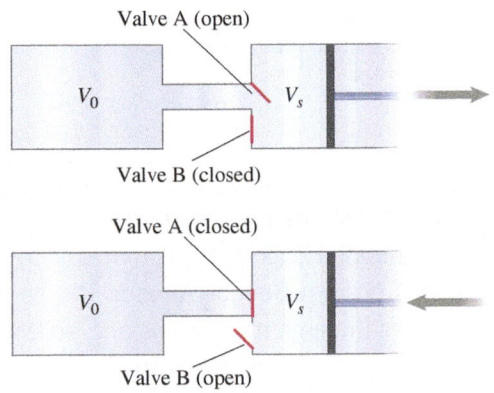

75. **N** CASE STUDY A constant-volume thermometer uses mercury and oxygen. In the thermometer, 0.750 g of oxygen is kept at a volume of 1.000 L. When the thermometer is placed in contact with a triple-point cell, the mercury rises to 1.674 cm. When the thermometer is placed in contact with an unknown substance, the mercury rises to 7.896 cm. What is the temperature (in kelvins) of the unknown substance?

76. A gas is confined to a cylindrical volume with a movable piston on one end. Initially, the volume of the container is 0.175 m^3, and the gas is at 305 K. The gas is heated, but the piston is moved such that the temperature of the gas remains constant.
   a. **A** If there are 2.00 mol of gas in the container, write an equation for the pressure of the gas as a function of the volume of the container.
   b. **G** Display your equation from part (a) on a graph, from the initial volume to five times the initial volume.

77. **N** The boiling point and the freezing point of water are marked as 80° and 10°, respectively, on an arbitrary temperature scale using a homemade thermometer. What is the Celsius temperature when this thermometer reads 59° on its arbitrary temperature scale? Assume there is a linear relationship between the Celsius scale and this arbitrary temperature scale.

78. **N** Children playing in the neighborhood park on a cold day find a section of PVC pipe 50.0 cm long and open at both ends, and they begin blowing through it to produce trumpet-like sounds. The pipe is initially at 2.00°C and warms to 25.0°C when warm air is blown through it. What is the change in the fundamental frequency of the pipe after it warms? The coefficient of thermal expansion for PVC is $54 \times 10^{-6}\,°C^{-1}$.

79. **N** The legendary Fender Stratocaster electric guitar has a scale length of 65.78 cm over which the steel strings of the guitar are stretched. One such string, 0.0254 cm in diameter, has a fundamental frequency of oscillation of 196 Hz at a temperature of 20.0°C. The density of steel in the string is 7.80 g/cm^3, its Young's modulus is 200 GPa, and its coefficient of linear expansion is $11 \times 10^{-6}\,°C^{-1}$. **a.** If the density of the steel in the string is 7.80 g/cm^3, what is the mass per unit length of the guitar string? **b.** What is the tension in the guitar string? **c.** What is the tension in the guitar string if its temperature increases to 45.0°C during an intense guitar solo? **d.** What is the fundamental frequency of the guitar string at 45.0°?

80. **G** A cylinder filled with an ideal gas is sealed by a piston on one end as shown in Figure P19.80. The piston can freely slide up and down the cylinder without any loss of gas. **a.** The cylinder is held over a burner so that the gas temperature rises (Fig. P19.80, middle). If the pressure is constant, sketch a graph of pressure as a function of volume and a graph of temperature as a function of volume. Describe what you would observe as the gas is heated. **b.** Now the temperature of the gas remains constant as you press the piston downward (Fig. P19.80, right). Sketch graphs of pressure and temperature as functions of volume. Compare your results with Charles's, Gay-Lussac's, and Boyle's laws and comment.

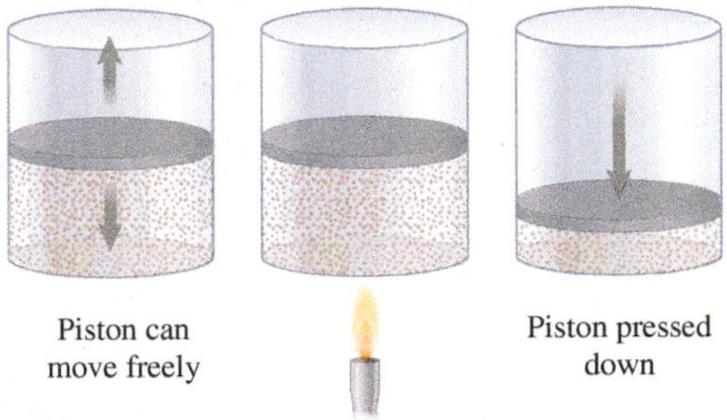

FIGURE P19.80

81. **N** Two glass bulbs of volumes 500 cm^3 and 200 cm^3 are connected by a narrow tube of negligible volume. The apparatus is filled with air and sealed. The initial pressure of the air is 1.0×10^5 Pa, and the initial temperature is 17°C. The smaller bulb is immersed in a large amount of ice at 0°C, and the larger bulb is immersed in a large pot of boiling water at a temperature of 100°C for a long period of time. Determine the final pressure of the air in the bulbs. Ignore any change in the volume of the glass. Model the air as an ideal gas.

20 Kinetic Theory of Gases

Key Question

How does the kinetic theory—a microscopic model—account for the macroscopic properties of ideal and real gases?

Underlying Principles

Kinetic theory

Major Concepts

1. Connection between average kinetic energy and temperature
2. Connection between rms velocity and temperature
3. Maxwell-Boltzmann speed distribution
4. Most probable speed
5. Average speed
6. Mean free path
7. Van der Waals equation of state
8. Critical temperature; critical point; critical pressure
9. Saturated vapor pressure
10. Relative humidity

Tools

1. Root mean square
2. Pressure–volume (*PV*) diagram
3. Phase diagram

As you read this sentence, air molecules are flying into you at an average speed of roughly 500 m/s or about 1250 mph. At that speed, you could get from Baltimore to Minneapolis or from Buenos Aires to Rio de Janeiro in about an hour, but the air molecules undergo many collisions, so it would take a molecule much longer to cross that distance. When the molecules hit your skin, you experience those high-speed collisions as warmth. If the air temperature drops, the molecules slow down, and you feel cooler. On a hot day, the molecules speed up, and you feel warmer. If you get too hot, your body produces perspiration (sweat). Water molecules on your skin evaporate, removing energy as they leave and cooling you off. In this chapter, we explore a microscopic model for gases known as the kinetic theory, which successfully explains their macroscopic properties and behavior.

20-1 What Is the Kinetic Theory?

KINETIC THEORY
Underlying Principle

The **kinetic theory** is a model for matter stating that all matter is made up of particles in motion. The particles may be either atoms or molecules. In a solid, the particles oscillate back and forth around some equilibrium position (Fig. 19.6). In a solid, there is no net motion of molecules through the body of the sample. In a liquid, however, the particles may oscillate and slide past one another as they travel throughout the sample (Fig. 15.2B). The particles in a gas move randomly and freely throughout its container (Fig. 15.2A).

The kinetic theory can be applied to an ideal gas or a real gas. Until we reach Section 20-6, we focus on gases that are well described by the ideal gas law, $PV = Nk_BT$ (Eq. 19.17). A real gas may be modeled as an ideal gas if it is at low pressure, has a low density, and is not close to its liquefaction point. In applying the kinetic theory to gases that follow the ideal gas law, we start with the following five assumptions.

1. **A gas consists of a large number of molecules or atoms, and they can be modeled as particles.** That is, the molecules or atoms are assumed to have no physical extent, so they do not vibrate or rotate. The total volume occupied by the particles is negligible compared to the volume occupied by the gas as a whole, so the particles are small compared to the average distance between them.
2. **The particles in an ideal gas do not interact with one another except when they collide.** When the particles are far apart, they do not exert forces on one another. Thus, only contact forces (not field forces) are exerted.
3. **The particles make elastic collisions with the walls, and the duration of each such collision is short.**
4. **The particles are free to move in any direction at any speed.**
5. **The gas is made up of identical particles.** Therefore, only one type of gas is present, such as pure oxygen.

In this chapter, we apply the kinetic theory to derive the ideal gas law and other properties of ideal gases. When we apply the theory to some real gases, we may need to change one or more of the five assumptions. These assumptions will be referred to by number in subsequent sections.

CASE STUDY Part 1: The Earth's Atmosphere

Life on the Earth depends on its atmosphere (Fig. 20.1). Not only do we breathe the atmosphere, it also regulates the Earth's temperature and weather. The Earth's temperature is typically between 250 K and 300 K, making it suitable for life. A different balance of gases in the atmosphere could cause major climate changes and make the planet uninhabitable.

You only need to consider our Moon to imagine what our planet might be like if it didn't have an atmosphere. The sunlit side of the Moon reaches temperatures of about 380 K (above the boiling point of water), and the dark side may be as cold as 110 K (about −170°C, or about 100°C colder than Antarctica).

Even a planet with an atmosphere can be unlivable if its atmosphere has the wrong composition. For example, Venus is the second planet from the Sun, but it is the hottest planet in the solar system. Venus's temperature is about 700 K, or about 250 K higher than it would be without an atmosphere. Venus has this high temperature because 96% of its atmosphere is carbon dioxide (CO_2). Carbon dioxide is called a greenhouse gas because it traps infrared radiation, causing the atmosphere to retain solar energy that would otherwise escape into space. The number of greenhouse gases in our atmosphere has been increasing; as a result, the temperature of the Earth has been increasing, a phenomenon known as **global warming**. As you are no doubt aware, the recent increases in carbon dioxide and other greenhouse gases in our atmosphere are most likely due to human activities. If the percentage of greenhouse gases continues to rise, the Earth may become unlivable. In this case study, we will study the Earth's atmosphere and our interaction with it.

FIGURE 20.1 The atmosphere of the Earth is contained by the Earth's gravity. If the temperature of the Earth were higher, more components of the atmosphere would escape.

20-2 Average and Root-Mean-Square Quantities

When we apply the kinetic theory to a gas, we must consider a large number of particles (assumption 1, Section 20-1). As an analogy, imagine studying the entire human population. Some insights come from focusing on one particular person, but to understand the population as a whole, we often measure properties that have been averaged over all people. Likewise, macroscopic observations of gases depend on properties that are averaged over all the particles making up the gas. In this section, we describe two ways to find the *mean* or average. The second—the root mean square—is useful in the kinetic theory of gases. To better understand the root mean square, we start with a common way to find an average.

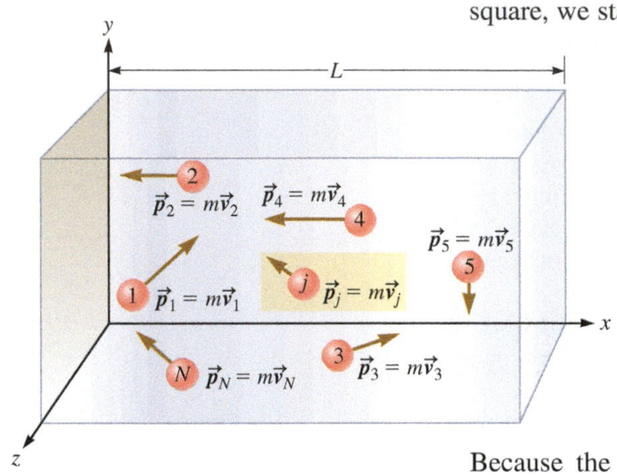

FIGURE 20.2 A gas in a box is made up of many moving, identical particles. For convenience. we number each particle from 1 to N and refer to an arbitrary particle as the jth particle.

Suppose we want to find the average kinetic energy of a set of gas particles. The gas consists of a large number N of identical particles of mass m (assumptions 1 and 5) contained in a box (Fig. 20.2). For convenience, we number each particle from 1 to N and refer to an arbitrary particle as the jth particle. The particles move randomly (assumption 4), so the jth particle has an arbitrary momentum $\vec{p}_j = m\vec{v}_j$. The particles cannot rotate or vibrate (assumption 1), so the kinetic energy of the jth particle is purely due to its translational motion and is given by $K_j = \frac{1}{2}mv_j^2$ (Eq. 8.1). The average kinetic energy is found by adding the kinetic energy of each particle and dividing by the number of particles:

$$K_{av} = \frac{1}{N}\sum_{j=1}^{N} K_j = \frac{1}{N}\sum_{j=1}^{N}\frac{1}{2}mv_j^2$$

Because the particles are identical (assumption 5), we can bring m outside the summation.

$$K_{av} = \frac{1}{2}m\left(\frac{1}{N}\sum_{j=1}^{N} v_j^2\right) = \frac{1}{2}mv_{av}^2 \tag{20.1}$$

where the average of the speed squared is

$$v_{av}^2 = \frac{1}{N}\sum_{j=1}^{N} v_j^2 \tag{20.2}$$

ROOT MEAN SQUARE Tool

Equation 20.2 leads to another kind of average known as the **root mean square (rms)** that is useful when we apply the kinetic theory to a gas. For example, the root-mean-square *speed* is found by (1) *squaring* the speed of each particle, (2) calculating the average (also known as the *mean*) of the speed squared, and (3) taking the square *root* of the average of the speed squared. Equation 20.2 includes the first two steps: The speed of each particle is squared, and the squares are averaged. To find the **rms speed**, take the square root of Equation 20.2:

$$v_{rms} = \sqrt{v_{av}^2} = \sqrt{\frac{1}{N}\sum_{j=1}^{N} v_j^2} \tag{20.3}$$

The root mean square of a quantity does not necessarily equal the mean of that quantity. For example, imagine some stretch of highway where there are about as many cars headed northbound as there are cars headed southbound. If the cars travel at roughly the same speed and if you choose a coordinate system so that north is positive and south is negative, the average velocity of all the cars traveling north and south would be near zero. The rms velocity would *not* be zero because squaring all the velocities eliminates negative values before taking the average. In fact, an rms quantity is never negative, so rms speed is the same as rms velocity.

In Section 20-3, we'll show that the temperature of a gas is closely related to the rms speed of its particles. To do so, we need to derive an expression for the particles' average momentum, and to do that, it is helpful to write the average velocity squared v_{av}^2 in terms of one scalar component v_x. First consider just the jth particle and write v_j^2 in terms of the x, y, and z components of $\vec{v}_j$ (Eq. 3.13):

$$v_j^2 = v_{jx}^2 + v_{jy}^2 + v_{jz}^2 \tag{20.4}$$

When we substitute Equation 20.4 into Equation 20.2, we find that v_{av}^2 can be written as

$$v_{av}^2 = \frac{1}{N}\sum_{j=1}^{N}(v_{jx}^2 + v_{jy}^2 + v_{jz}^2)$$

$$v_{av}^2 = \frac{1}{N}\sum_{j=1}^{N}v_{jx}^2 + \frac{1}{N}\sum_{j=1}^{N}v_{jy}^2 + \frac{1}{N}\sum_{j=1}^{N}v_{jz}^2$$

Each term on the right side of the equal sign is the average velocity squared in that particular direction:

$$v_{av}^2 = v_{av,x}^2 + v_{av,y}^2 + v_{av,z}^2 \tag{20.5}$$

In Figure 20.2, the particles' motion is random (assumption 4). So, there is no preferred direction for the motion of all the particles. Therefore, the average velocity squared should be the same in all three dimensions x, y, and z:

$$v_{av,x}^2 = v_{av,y}^2 = v_{av,z}^2$$

We arbitrarily replace $v_{av,y}^2$ and $v_{av,z}^2$ in Equation 20.5 with $v_{av,x}^2$:

$$v_{av}^2 = v_{av,x}^2 + v_{av,x}^2 + v_{av,x}^2$$

$$v_{av}^2 = 3v_{av,x}^2 \tag{20.6}$$

So, the average velocity squared is three times the average velocity in any one single dimension.

The last average property we need to derive is the average of the x component of momentum squared:

$$p_{av,x}^2 = \frac{1}{N}\sum_{j=1}^{N}p_{jx}^2 = \frac{1}{N}\sum_{j=1}^{N}m^2v_{jx}^2 = m^2\left(\frac{1}{N}\sum_{j=1}^{N}v_{jx}^2\right) \tag{20.7}$$

The term in parentheses in Equation 20.7 is $v_{av,x}^2$:

$$p_{av,x}^2 = m^2v_{av,x}^2 = \tfrac{1}{3}m^2v_{av}^2 \tag{20.8}$$

where we have used Equation 20.6.

EXAMPLE 20.1 Highway Velocities

The velocity of vehicles on a straight stretch of highway is given in Table 20.1. A coordinate system was chosen so that north is positive. Find the average velocity and the rms velocity.

TABLE 20.1 **Velocities on a straight stretch of highway.**

Number	Velocity (mph)	Number	Velocity (mph)	Number	Velocity (mph)	Number	Velocity (mph)
1 southbound	−67	10 southbound	−70	19 northbound	62	28 northbound	74
2 southbound	−92	11 southbound	−100	20 northbound	76	29 northbound	56
3 southbound	−88	12 southbound	−80	21 northbound	67	30 northbound	51
4 southbound	−82	13 southbound	−74	22 northbound	66	31 northbound	80
5 southbound	−76	14 southbound	−63	23 northbound	70	32 northbound	63
6 southbound	−78	15 southbound	−80	24 northbound	79	33 northbound	69
7 southbound	−55	16 northbound	55	25 northbound	65	34 northbound	62
8 southbound	−81	17 northbound	73	26 northbound	67	35 northbound	79
9 southbound	−77	18 northbound	65	27 northbound	82		

INTERPRET and ANTICIPATE

We expect the average and rms velocities to be very different because nearly half the velocities are negative.

Example continues on page 584 ▶

SOLVE

The sum of the velocities (v_{tot}) is fairly small because nearly half the velocities are negative.	$\vec{v}_{tot} = \sum_{j=1}^{N} \vec{v}_j = (-67 - 92 + \cdots + 79)$ mph north $\vec{v}_{tot} = 198$ mph north
Find the average velocity v_{av} by dividing by the total number of vehicles.	$\vec{v}_{av} = \frac{\vec{v}_{tot}}{N} = \frac{198\text{ mph}}{35}$ north $=$ 5.7 mph north
To find the rms velocity v_{rms}, square each velocity and add the results.	$v_{tot}^2 = \sum_{j=1}^{N} v_j^2 = (-67\text{ mph})^2 + (-92\text{ mph})^2 + \cdots + (79\text{ mph})^2$ $v_{tot}^2 = 185{,}992\,(\text{mph})^2$
Next, divide by the number of vehicles to find the average of the squared velocities.	$v_{av}^2 = \frac{v_{tot}^2}{N} = \frac{185{,}992(\text{mph})^2}{35} = 5314\,(\text{mph})^2$
Finally, take the square root to find the rms velocity.	$v_{rms} = \sqrt{v_{av}^2} = \sqrt{5314(\text{mph})^2} =$ 73 mph

CHECK and THINK

As predicted, the average velocity and the rms velocity are very different. Interpreting an average velocity of nearly zero is somewhat ambiguous. It may be that roughly equal numbers of vehicles are traveling in opposite directions at roughly the same speeds, or it may be that a few very fast vehicles are moving in one direction and many very slow vehicles are moving in the other direction. Depending on what you would like to know about traffic patterns, the average velocity may not be helpful. For example, if you want to know if drivers are exceeding the speed limit, the average velocity isn't useful. The rms velocity may be more helpful; because the velocities are squared first, the rms calculation does not depend on the sign (direction) of the velocity. The rms velocity is slightly higher than the average speed because squaring the values first means that higher speeds have a slightly greater weight when taking the average.

CONCEPT EXERCISE 20.1

In Example 20.1, we found that the rms value of a quantity was greater than the average of that quantity. Is it possible for the rms value to equal the average? If so, give an example of six numbers whose rms value equals their average. If not, explain why not.

20-3 The Kinetic Theory Applied to Gas Temperature and Pressure

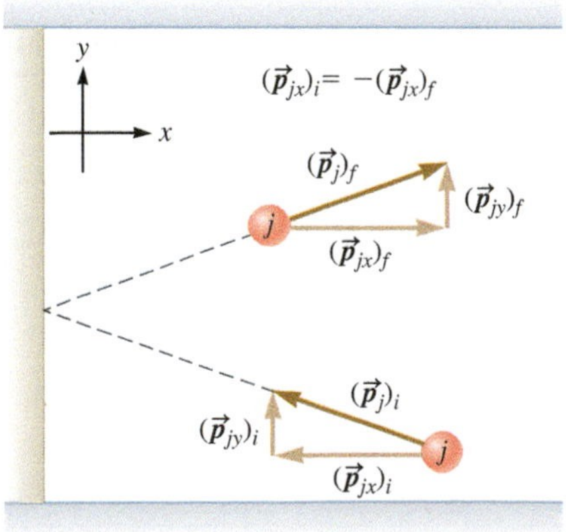

FIGURE 20.3 When the jth particle collides with the wall on the left, only the x component of its momentum is changed. The y component is identical before and after the collision.

One of the major goals of the kinetic theory is to connect the microscopic model of a gas to our macroscopic observations. In this section, we connect the particles' average kinetic energy and rms speed to the temperature of an ideal gas.

The pressure we observe in a gas is due to the total force per unit area exerted by the particles on the container walls. Once again, consider a gas inside a box. To find the total force, start by focusing on the jth particle as it collides with the shaded wall at the left in Figure 20.3. The particle collides elastically with the wall (assumption 3), and the wall exerts a force in the positive x direction on the particle. According to Newton's second law $\vec{F} = d\vec{p}/dt$ (Eq. 10.2), only the x component of the particle's momentum is subject to change as a result of the collision. Because the wall has a much greater mass than the particle, the particle reverses direction, and its velocity after the collision is $\vec{v}_{1f} \approx -\vec{v}_{1i}$ (Fig. 11.13B). So, the change in the particle's momentum is

$$\Delta p_{jx} = (p_{jx})_f - (p_{jx})_i = 2(p_{jx})_i = 2p_{jx} \qquad (20.9)$$

where we have dropped the subscript i in the last quantity for simplicity. The symbol p_{jx} now denotes the particle's initial momentum.

We use the impulse–momentum theorem $\vec{I}_{tot} = \Delta\vec{p} = \vec{F}_{av}\,\Delta t$ (Eqs. 11.3 and 11.4) to find the average force $(F_{jx})_{av}$ exerted by the left wall on the jth particle:

$$I_x = \Delta p_x = (F_{jx})_{av}\,\Delta t_{collision}$$

$$(F_{jx})_{av} = \frac{\Delta p_x}{\Delta t_{collision}} \tag{20.10}$$

The impulse Δp_x is the area under the force-versus-time curve (Fig. 20.4). We have an expression for Δp_x (Eq. 20.9), so once we find an expression for $\Delta t_{collision}$, we will have an expression for $(F_{jx})_{av}$. Finding $\Delta t_{collision}$ can be tricky because the duration of the collision is so short. In this situation, however, there is a clever solution to this problem. Imagine that at $t = 0$, the jth particle is at the *right* wall, and after some time interval Δt, the particle is back at the right wall, having collided with the left wall. (During that time interval, the particle may bounce off other walls, but because those collisions cannot change the x component of its momentum, we may ignore such collisions in our analysis.) The only force in the x direction that is exerted on the particle during this time interval is due to the left wall, so the area under the force-versus-time curve is equal to that of the thin rectangular box shown in Figure 20.4. The height of this box is the average x component of force exerted on the particle during the time Δt.

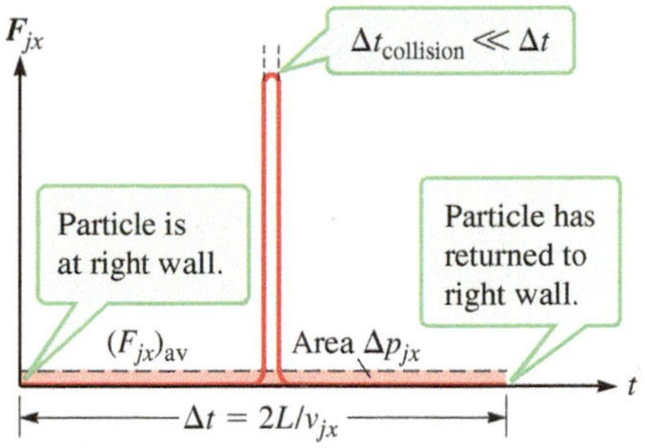

FIGURE 20.4 The change in the jth particle's momentum is the area under the force-versus-time curve (red), which also equals the area of the very thin shaded rectangular box $\Delta p_{jx} = (F_{jx})_{av}\,\Delta t$.

All we need to do is find an expression for Δt, the time the particle takes to go from the right wall to the left wall and back again. The collision with the left wall only changes the direction of the particle's momentum. So, the jth particle's velocity parallel to the x axis reverses direction, but its magnitude v_{jx} remains unchanged. The time for the particle to return to the right wall is given by the total distance traveled ($2L$ for a box of length L) divided by that magnitude v_{jx}:

$$\Delta t = \frac{2L}{v_{jx}} \tag{20.11}$$

We find an expression for the average force exerted by the left wall on the jth particle by substituting Equations 20.11 and 20.9 into Equation 20.10:

$$(F_{jx})_{av} = \frac{2p_{jx}}{2L/v_{jx}}$$

$$(F_{jx})_{av} = \frac{p_{jx}v_{jx}}{L} = \frac{p_{jx}^2}{mL} \tag{20.12}$$

where we have used $p_{jx} = mv_{jx}$.

By Newton's third law, the average force exerted by the jth particle on the left wall $(f_{jx})_{av}$ is equal in magnitude and opposite in direction to the force exerted by the wall on the particle $(F_{jx})_{av}$. So, from Equation 20.12,

$$(f_{jx})_{av} = -\frac{p_{jx}^2}{mL} \tag{20.13}$$

Equation 20.13 is the average force exerted by the jth particle on the left wall. If there were only a few particles in the box, the wall would experience a relatively strong force when it was hit by a particle and no force for some period of time when it was not being hit. According to assumption 1, however, there are many particles in the box. So, the wall is continually bombarded by particles, and the force of the particles on the wall is a constant that we can find by adding the average force exerted by each particle:

$$f_{av,x} = \sum_{j=1}^{N}(f_{jx})_{av} = -\sum_{j=1}^{N}\frac{p_{jx}^2}{mL} = -\frac{1}{mL}\sum_{j=1}^{N}p_{jx}^2$$

Multiply the right side by N/N, where N is the number of particles:

$$f_{av,x} = -\frac{N}{mL}\left(\frac{1}{N}\sum_{j=1}^{N}p_{jx}^2\right)$$

The term in parentheses is the average of the momentum's x component squared:

$$f_{av,x} = -\frac{N}{mL}p_{av,x}^2$$

Now substitute $p^2_{av,x} = \frac{1}{3}m^2v^2_{av}$ (Eq. 20.8):

$$f_{av,x} = -\frac{N}{mL}\left(\frac{1}{3}m^2v^2_{av}\right) = -\frac{N}{3L}mv^2_{av}$$

The pressure on the left wall is the absolute value of the force exerted by the particles divided by the area of the wall:

Uppercase P stands for pressure, and lowercase p stands for momentum.

$$P = \frac{|f_{av,x}|}{A} = \frac{N}{3LA}mv^2_{av} \tag{20.14}$$

The volume V of the box is LA, so

$$P = \frac{N}{3V}mv^2_{av}$$

$$PV = \frac{N}{3}mv^2_{av} \tag{20.15}$$

Uppercase V stands for volume, and lowercase v stands for velocity (speed).

Compare Equation 20.15 with the ideal gas law $PV = Nk_BT$ (Eq. 19.17) to find

$$PV = Nk_BT = \frac{N}{3}mv^2_{av}$$

$$k_BT = \tfrac{1}{3}mv^2_{av}$$

$$v^2_{av} = \frac{3k_BT}{m} \tag{20.16}$$

Substitute into Equation 20.1 for average kinetic energy:

CONNECTION BETWEEN AVERAGE KINETIC ENERGY AND TEMPERATURE

Major Concept

$$K_{av} = \tfrac{1}{2}mv^2_{av} = \tfrac{3}{2}k_BT \tag{20.17}$$

Equation 20.17 confirms what we stated in Section 19-1: **The absolute temperature is a measure of the average translational kinetic energy of (ideal) gas particles.** Specifically, temperature is directly proportional to the average translational kinetic energy.

The absolute temperature is also a measure of the rms velocity of the particles. To find the rms velocity, take the square root of each side of Equation 20.16 and use Equation 20.3:

CONNECTION BETWEEN RMS VELOCITY AND TEMPERATURE

Major Concept

$$v_{rms} = \sqrt{v^2_{av}} = \sqrt{\frac{3k_BT}{m}} \tag{20.18}$$

Equations 20.17 and 20.18 connect the microscopic motion of gas particles to macroscopic observations of the gas's absolute temperature. According to these equations, a higher temperature means that the particles have more kinetic energy and are moving, on average, faster, which gives us another way to interpret absolute zero (Section 19-7). No one has observed absolute zero, but if a system were at absolute zero, the average kinetic energy and rms velocity of its particles would be zero. So, the particles would not be moving. According to quantum mechanics (Chapters 40–42), however, this situation is not possible; particles actually have some kinetic energy even at absolute zero.

CONCEPT EXERCISE 20.2

If the temperature of a gas is doubled, what happens to the average kinetic energy and rms speed of its particles?

CONCEPT EXERCISE 20.3

Gas 1 and gas 2 are at the same temperature. Gas 1 is made up of molecules whose mass is four times that of gas 2's molecules: $m_1 = 4m_2$. Compare the average kinetic energy of the molecules in the two gases. Compare the rms speed of the molecules in the two gases.

EXAMPLE 20.2 CASE STUDY The Earth's Atmosphere

Hydrogen and helium are the most abundant elements in the Universe, together making up almost 100% of it, whereas all other elements make up just a fraction of a percent. The Sun and the four giant planets Jupiter, Saturn, Uranus, and Neptune have a lot of pure hydrogen and helium in their atmospheres. The inner planets Mercury, Venus, the Earth, and Mars have essentially no pure atmospheric hydrogen or helium. (Hydrogen compounds such as water are found in the Earth's atmosphere.)

A planet's atmosphere is not contained by the walls of a box but instead is bound by gravity. Molecules and atoms that are moving at the escape speed

$$v_{esc} = \sqrt{\frac{2GM}{R}} \tag{8.17}$$

or faster will leave the atmosphere, where M is the mass of the planet and R is its radius.

The rms speed of gas particles is determined by their temperature and mass (Eq. 20.18), so different planets have different gases in their atmosphere depending on the planet's temperature and mass. If $v_{rms} = v_{esc}$ for a given type of particle, that gas will leave the atmosphere in only a few days. The condition for retaining a certain gas in the atmosphere for several billion years (the age of the solar system) is $10v_{rms} \leq v_{esc}$. Find the mass (in kilograms and atomic mass units) of the lightest molecule or atom that can be retained in the Earth's atmosphere. Use the periodic table (Appendix B) to determine why there is no hydrogen or helium gas in the Earth's atmosphere. Assume the atmospheric temperature is a uniform 277 K.

INTERPRET and ANTICIPATE

According to $v_{rms} = \sqrt{3k_BT/m}$, lighter particles have a greater rms velocity. By setting $10v_{rms} = v_{esc}$, we can solve for the mass m_{min} of the lightest particles in the atmosphere. We expect to find that major components of the atmosphere such as nitrogen gas (N_2) and oxygen gas (O_2) are easily retained by the Earth's gravity.

SOLVE

Set $10v_{rms} = v_{esc}$. Use $v_{rms} = \sqrt{3k_BT/m}$ (Eq. 20.18) and $v_{esc} = \sqrt{2GM/R}$ (Eq. 8.17).	$10v_{rms} = v_{esc}$ $10\sqrt{\frac{3k_BT_\oplus}{m_{min}}} = \sqrt{\frac{2GM_\oplus}{R_\oplus}}$
Square each side and solve for m_{min}.	$100\left(\frac{3k_BT_\oplus}{m_{min}}\right) = \frac{2GM_\oplus}{R_\oplus}$ $m_{min} = \frac{150k_BR_\oplus T_\oplus}{GM_\oplus}$
Find m_{min} in kilograms and then convert to atomic mass units (u).	$m_{min} = \frac{150(1.38 \times 10^{-23}\ \text{J/K})(6.387 \times 10^6\ \text{m})(277\ \text{K})}{(6.67 \times 10^{-11}\ \text{N} \cdot \text{m}^2/\text{kg}^2)(5.97 \times 10^{24}\ \text{kg})}$ $m_{min} = 9.2 \times 10^{-27}\ \text{kg}$ $m_{min} = 9.2 \times 10^{-27}\left(\frac{1\ \text{u}}{1.66 \times 10^{-27}\ \text{kg}}\right) = 5.5\ \text{u}$

CHECK and THINK

Our result means that the Earth's gravity can retain gas particles whose mass is at least 5.5 u. According to the periodic table, atomic hydrogen and helium have atomic masses of 1 and 4 u, respectively. Both are too light to be retained in the atmosphere. Even molecular hydrogen gas (H_2) is too light to be retained. We expected molecular nitrogen and oxygen gases to be easily retained, however. The molecular masses of N_2 and O_2 are 28 and 32 u, respectively, both much heavier than the minimum retainable mass.

20-4 Maxwell-Boltzmann Distribution Function

Because the kinetic theory of gases has been so successful, it might be hard to imagine that there were once alternative theories. One alternative model was a *static* model, according to which gas atoms (such as those in the surrounding air) were in constant contact with one another, normally at rest and not moving around randomly as we assumed in Figure 20.2. Instead, this model attributed the gas pressure we experience to the many atoms pressed against us. Proponents of the static model believed that atoms were small, fundamental objects whose size could change. Further, if there were a net motion of the gas, such as when you breathe air into or out of your lungs, all the gas atoms would move together at the same velocity.

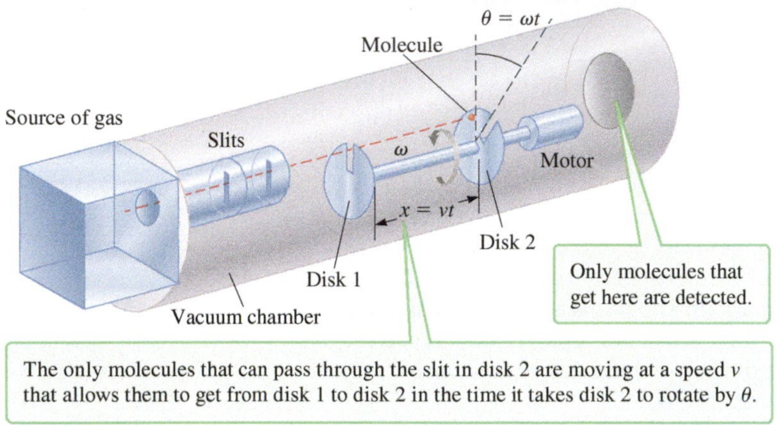

FIGURE 20.5 Molecules escape through the slits and travel in a beam toward the spinning disks. Only molecules within a certain range of speeds will make it all the way to the detector.

Figure 20.5 shows an apparatus that may be used to collect evidence supporting either the static model or the kinetic theory. On the left is a source of gas particles. The particles escape through a series of slits to the right of the source, creating a narrow beam of particles in a vacuum chamber. The beam is aimed at two subsequent disks rotating at angular speed ω. Each disk has a narrow slit allowing molecules to pass. The distance between the rotating disks is x, and the slit in disk 2 is offset by an angle θ from the slit in disk 1. The time it takes the slit in disk 2 to line up with the beam (after the beam has passed through the slit in disk 1) is $t = \theta/\omega$, so particles traveling at speed $v = x/t = x(\omega/\theta)$ will pass through the slit in disk 2, and the other particles will be blocked. Actually, because the slits have some finite width, a range of speeds can get through disk 2. The particles passing through disk 2 are counted by the detector on the right end of the vacuum chamber. By changing the angular speed ω of the disks, particles of different speed ranges are counted.

To find the number of particles in a particular speed range, ω is set for that range. Each bar in Figure 20.6 comes from counting the number of particles at a particular ω. The height of each bar is determined by dividing the number of particles in each particular range by the total number of particles detected in all the speed ranges. So, the height of each bar represents the fraction of particles that are in that particular speed range. If the static model were correct, all the particles would be in contact with one another and moving at the same speed. Therefore, the static model predicts that there should only be one bar of height 1. In Figure 20.6, however, we find a variety of ranges of possible speeds, which supports the kinetic theory.

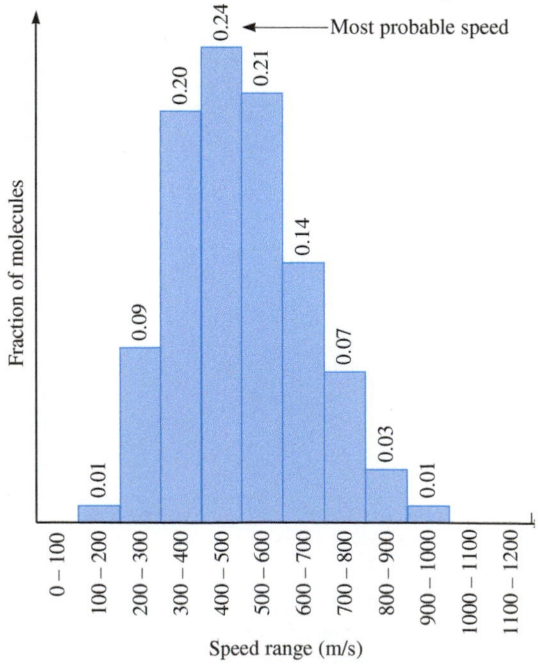

FIGURE 20.6 The height of each bar represents the fraction of particles in the sample that are in the particular speed range.

Figure 20.6 tells us more about the motion of the particles in the gas. The height of each bar is the fraction of particles in that particular speed range and also the probability that a randomly selected particle's speed is in that range. In this case, the most probable speed is somewhere in the range of 400 to 500 m/s.

We see that the most probable speed is not in the very center of the distribution, but is somewhat lower than the median (middle) speed. In other words, there are more particles with speeds greater than this most probable speed than particles with lower speeds. By adding the fractions of particles that are below the most probable range, we find that $(0.01 + 0.09 + 0.20) = 0.30$ of the particles are at speeds lower than the most probable speed. The fraction of particles with speeds higher than the most probable speed is $(0.21 + 0.14 + 0.07 + 0.03 + 0.01) = 0.46$.

If the temperature of the source of particles in Figure 20.5 increases, the most probable speed increases as shown in Figure 20.7. Compare the speed distribution for the cool gas (blue bars) with the speed distribution for the hot gas (red bars). The most probable speed for the hot gas is in the range of 900 to 1000 m/s, or about 500 m/s greater than that for the cool gas. The hot gas also has a greater overall range of speeds. The high-

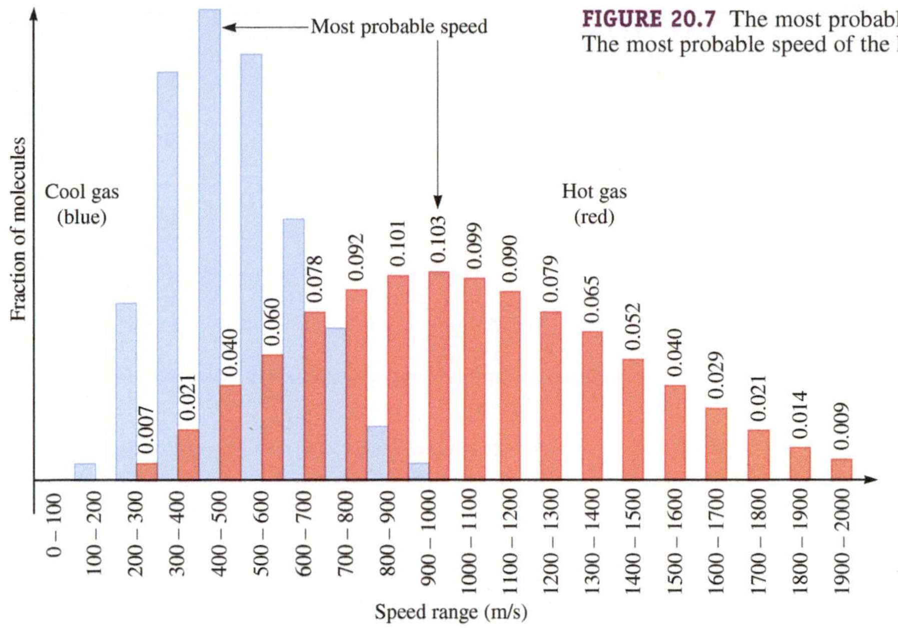

FIGURE 20.7 The most probable speed depends on the temperature of the gas. The most probable speed of the hotter gas is greater than that of cooler gases.

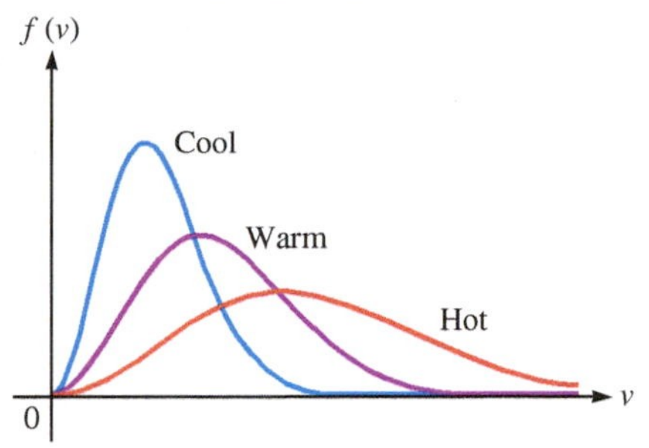

FIGURE 20.8 Maxwell-Boltzmann distribution function for a gas at three different temperatures.

est range of speeds in the hot gas is 1900 to 2000 m/s or about 1000 m/s faster than the top speeds in the cool gas.

Maxwell-Boltzmann Distribution

Imagine reducing the slit width in the rotating disks in Figure 20.5. As the widths are reduced, the range of speeds represented by each bar in Figures 20.6 and 20.7 gets smaller and smaller. As the width approaches zero, the distribution becomes continuous (Fig. 20.8) and is given by

$$f(v) = 4\pi\left(\frac{m}{2\pi k_B T}\right)^{3/2} v^2 e^{-(mv^2/2k_B T)} \tag{20.19}$$

MAXWELL-BOLTZMANN SPEED DISTRIBUTION ✪ Major Concept

Equation 20.19 is known as the **Maxwell-Boltzmann speed distribution**. The function $f(v)$ is called a distribution function, and it gives the *probability of a particle's speed per unit speed interval*. The probability that a randomly chosen particle has a speed between v and $(v + dv)$ is the probability per unit speed interval $f(v)$ multiplied by the width dv of the speed interval: $f(v)dv$. Because probability does not have dimensions, the probability per unit speed $f(v)$ must have the SI units of seconds per meter.

DERIVATION Most Probable Speed

Figure 20.8 shows the Maxwell-Boltzmann speed distribution for a hot, a warm, and a cool gas. As we saw before, the distribution is more spread out for a hot gas than it is for a cool gas. Figure 20.8 also shows that the most probable speed in a hot gas is greater than the most probable speed in a cool gas. We will show that the most probable speed is a function of temperature given by

$$v_{mp} = \sqrt{\frac{2k_B T}{m}} \tag{20.20}$$

MOST PROBABLE SPEED ✪ Major Concept

The most probable speed v_{mp} occurs at the peak of the distribution, so we can find an expression for the most probable speed by setting the derivative of Equation 20.19 equal to zero.	$\frac{df(v)}{dv} = 0$ at $v = v_{mp}$ $\frac{d}{dv}\left[4\pi\left(\frac{m}{2\pi k_B T}\right)^{3/2} v^2 e^{-(mv^2/2k_B T)}\right] = 0$
Pull the constant terms outside the derivative.	$4\pi\left(\frac{m}{2\pi k_B T}\right)^{3/2} \frac{d}{dv}[v^2 e^{-(mv^2/2k_B T)}] = 0$

Derivation continues on page 590 ▶

To complete the derivative, use the chain rule (Appendix A).	$4\pi\left(\frac{m}{2\pi k_B T}\right)^{3/2}\left[2ve^{-(mv^2/2k_BT)} - v^2\left(\frac{2mv}{2k_BT}\right)e^{-(mv^2/2k_BT)}\right] = 0$ $4\pi\left(\frac{m}{2\pi k_B T}\right)^{3/2}\left[2ve^{-(mv^2/2k_BT)}\right]\left[1 - v^2\frac{m}{2k_BT}\right] = 0$
At the most probable speed, the last term must equal zero.	$\left[1 - v_{mp}^2\frac{m}{2k_BT}\right] = 0$
Solve for the most probable speed.	$v_{mp} = \sqrt{\frac{2k_BT}{m}}$ ✓ (20.20)

DERIVATION Average Speed

To find the average speed, we add the speed of each particle and divide by the number of particles. We can extend that procedure by integrating instead of adding. So, we will integrate over a continuous range of speeds using the Maxwell-Boltzmann speed distribution to show that the average speed of the particles in a gas is given by

$$v_{av} = \sqrt{\frac{8k_BT}{\pi m}} \quad (20.21)$$

AVERAGE SPEED
Major Concept

Because there are more particles at speeds above the most probable speed v_{mp} than below (Fig. 20.6), we expect the average speed v_{av} to be greater than the most probable speed: $v_{av} > v_{mp}$.

Let's call the total number of gas particles N and the number of particles in a particular speed range dN. Write an expression for dN.	$dN(v) = \begin{pmatrix}\text{number of}\\ \text{particles}\end{pmatrix}\begin{pmatrix}\text{fraction of}\\ \text{particles per}\\ \text{unit speed range}\end{pmatrix}\begin{pmatrix}\text{width}\\ \text{of speed}\\ \text{range}\end{pmatrix} = Nf(v)dv$ (1)
Find the average speed by adding the speed of each particle (integrating in this case for a continuous distribution) and dividing by the number of particles.	$v_{av} = \frac{\int v\,dN(v)}{N}$
Substitute Equation (1) for dN. Now the integral is over dv, whose range is from 0 to ∞.	$v_{av} = \frac{\int vNf(v)dv}{N}$ $v_{av} = \int_0^\infty vf(v)dv$
Substitute the Maxwell-Boltzmann distribution (Eq. 20.19) for $f(v)$.	$v_{av} = 4\pi\left(\frac{m}{2\pi k_BT}\right)^{3/2}\int_0^\infty v^3e^{-(mv^2/2k_BT)}dv$
Integrate by parts (Appendix A) or look up the integral.	$v_{av} = 4\pi\left(\frac{m}{2\pi k_BT}\right)^{3/2}\left(\frac{2k_B^2T^2}{m^2}\right)$
Simplify this expression.	$v_{av} = \sqrt{\frac{8k_BT}{\pi m}}$ ✓ (20.21)

COMMENTS
To check our results, compare the most probable speed (Eq. 20.20) with the average speed (Eq. 20.21). As expected, the average speed is greater than the most probable speed.

$$v_{mp} = \sqrt{\frac{2k_BT}{m}} \approx 1.4\sqrt{\frac{k_BT}{m}}$$

$$v_{av} = \sqrt{\frac{8k_BT}{\pi m}} \approx 1.6\sqrt{\frac{k_BT}{m}}$$

$$v_{av} > v_{mp}$$

CASE STUDY Part 2: Smog

Smog is often seen as a brownish-yellow haze around an urban center (Fig. 20.9). Spending even a couple of hours in smog can irritate your eyes, nose, and throat. Repeated or extended exposure to smog can cause respiratory illnesses and may prematurely age the lungs. High levels of smog can damage crops and vegetation. Various government agencies issue smog warnings, and you might have noticed that smog warnings tend to be worse on hot days. In this part of the case study, we use the Maxwell-Boltzmann distribution to explain why smog depends on temperature.

The term *smog* originally referred to a mixture of *smo*ke and *fog*. Today, it refers to a noxious mixture of many air pollutants (including gases and particulate matter) suspended in the atmosphere near the surface of the Earth, and it is more precisely called *photochemical smog*, a result of sunlight (*photons*) and *chemical* reactions in the atmosphere.

FIGURE 20.9 Smog over Los Angeles harbor.

Many human activities produce hydrocarbons (molecules composed of hydrogen and carbon) and nitrogen oxides (molecules composed of nitrogen and oxygen) that react in the atmosphere to become smog. Industrial societies are generally making some effort to reduce the production of such pollutants. For example, cars produce both hydrocarbons and nitrogen oxides, but catalytic converters on cars destroy these pollutants before they become part of the atmosphere. Even with such measures in place, many cities such as Los Angeles, California, still have serious smog problems and are looking for other ways to reduce smog. One way is to reduce the city's temperature.

From your experience in the kitchen, you know that the rate at which a chemical reaction takes place depends on temperature. To preserve food, you put it in the refrigerator or freezer, and when you want to cook something (change its chemical composition), you put that food in the oven or on the stovetop.

Chemical reactions in the atmosphere also depend on temperature. A chemical reaction is a result of a collision between two or more molecules, forming one or more new molecules. For a chemical reaction to occur, the two colliding molecules must have a certain minimum amount of energy: The molecules must exceed a certain minimum relative speed when they collide. Atmospheric hydrocarbons, nitrogen oxides, hydrogen oxides, and oxygen must react with one another to make smog. These various types of molecules have Maxwell-Boltzmann speed distributions. (In the case of a mixture of gases such as our atmosphere, the Maxwell-Boltzmann distribution holds for each type of molecule separately.) When the atmosphere is at a high temperature, there are many fast particles as shown in Figure 20.8 for the hot gas, so when two molecules collide, it is likely that they will be going fast enough to undergo a chemical reaction. In a much cooler gas, very few molecules may have enough energy to undergo a chemical reaction. For this reason, more smog is produced on hot days than on cool days. In fact, smog is not a problem on days that are cooler than about 70°F.

Smog in urban centers is made worse because cities are warmer than the surrounding countryside and warmer than they used to be. For example, in the 1930s, Los Angeles was full of orchards. In 1934 (one of the hottest years on record), the highest temperature in Los Angeles was 97°F. Today, those orchards have been replaced by buildings, roads, and parking lots. The high temperature of Los Angeles has risen by more than 8°F; in 2010, the hottest temperature in Los Angeles was 113°F. Some solutions to reducing smog focus on lowering the city's temperature by making the pavement and rooftops reflective and by planting trees. Lowering the temperature means that fewer atmospheric molecules will undergo the chemical reactions that produce smog.

20-5 Mean Free Path

Imagine waking up to the sight and scent of fresh hot coffee. Aromatic molecules from the coffee travel to your nose. You probably don't notice much of a time delay between when you see the coffee and when you smell it because your nose isn't very far from the cup. In Annapolis, Maryland, a small coffee company roasts coffee beans in a residential neighborhood (Fig. 20.10A). The people in the shop are the first to smell the coffee. On a calm day without wind, residents cannot smell the coffee for several minutes or more, depending on how far they are from the roaster. If the molecules were truly point particles, those molecules would not collide with one another but would travel in straight lines out from the roaster at hundreds of meters per second. Residents who live within a few blocks of the roaster would smell the coffee in

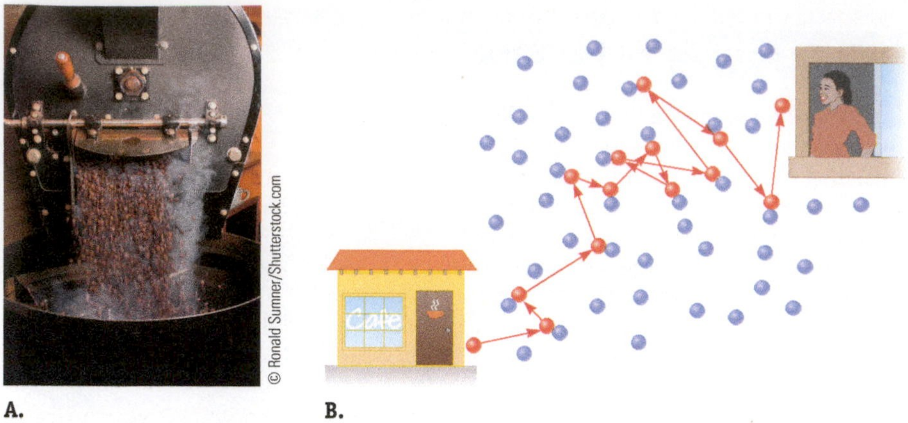

FIGURE 20.10 **A.** A coffee roaster in a small residential neighborhood. **B.** On a calm day, aromatic coffee molecules travel in a zigzag path throughout the neighborhood.

a fraction of a second. The molecules are not collisionless point particles, though. Instead, they collide as they move in the air, and with each collision their directions of motion are altered so that the molecules travel along a zigzag path (Fig. 20.10B).

Between each collision, a molecule's trajectory is a straight line. If there are many molecules or if the molecules are very large, the straight line parts of their paths will be fairly short. The **mean free path** is the average length of the straight portion of a molecule's path. When two populations of particles encounter each other, such as when two gases mix or when two galaxies of stars merge, the mean free path is a measure of how likely the particles are to collide. A long mean free path means that collisions are rare.

In this section, we relax assumption 1 (that the molecules are particles with no physical extent) to find an expression for a molecule's mean free path. Let's trace the path of one molecule shown in red in Figure 20.11A; the molecule undergoes three collisions along this short path in a time Δt. We model all the molecules as hard spheres of radius r, and for now we assume the red molecule is moving at speed v_{av} and the remaining molecules are at rest. It is somewhat easier to derive the red molecule's mean free path if we unbend its path as shown in Figure 20.11B. Now we can see that the red molecule sweeps out a cylindrical space of diameter $d = 2r$. Figure 20.11B also shows a cylinder of diameter $4r$. The red molecule will collide with any other molecule whose center is within this larger cylinder. The number of such molecules is given by the density of molecules N/V multiplied by the volume of the cylinder. The length of this large cylinder is $v_{av}\,\Delta t$, so its volume is $\pi(2r)^2 v_{av}\,\Delta t$. The number of collisions that occur in the time Δt is given by the number of molecules whose center is within this cylinder:

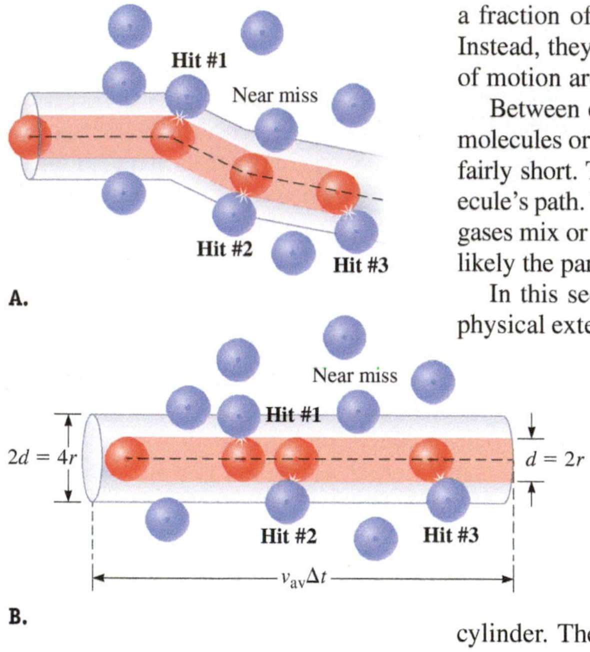

FIGURE 20.11 **A.** The red particle undergoes three collisions along this short path. **B.** For convenience, this scenario is drawn as a straight path.

$$\text{number of collisions} = \frac{N}{V}\pi(2r)^2 v_{av}\,\Delta t \tag{20.22}$$

The mean free path λ is the total distance traveled in the time Δt divided by the number of collisions in that time:

$$\lambda \approx \frac{v_{av}\,\Delta t}{(N/V)\pi(2r)^2 v_{av}\,\Delta t}$$

$$\lambda \approx \frac{1}{4\pi r^2(N/V)} \quad \text{(approximate)} \tag{20.23}$$

Equation 20.23 is approximate because it was derived under the assumption that only the red molecule moves. Of course, that assumption is not valid because the other molecules must also be moving. The number of collisions occurring in the time Δt depends on the *relative* speed of the colliding molecules; so, to find a more accurate expression, we replace v_{av} in Equation 20.22 with v_{rel}. For a Maxwell-Boltzmann distribution, the relative speed is $v_{rel} = \sqrt{2}v_{av}$, so the mean free path is given by

MEAN FREE PATH ✪ **Major Concept**

$$\lambda = \frac{1}{4\sqrt{2}\pi r^2(N/V)} \tag{20.24}$$

The average time t_{mf} between collisions is called the **mean free time**, found by dividing the mean free path by the average speed of the molecules:

$$t_{mf} = \frac{\lambda}{v_{av}} \tag{20.25}$$

Diffusion

Diffusion is a process that mixes the particles of two substances (such as the molecules of two gases) as a result of the particles' kinetic energy. For example, the aromatic molecules of coffee mix with air molecules. The zigzag path of the aromatic molecules (Fig. 20.10B) is called a **random walk**.

In Example 20.4, we will estimate the time it takes aromatic coffee molecules traveling in air to make their way throughout the neighborhood by diffusion. To do so, we would like to know how far (on average) an aromatic molecule can travel in some time Δt. To start, imagine an aromatic molecule that is constrained to move in just one dimension along the x axis. After each collision (with a particle in the air), the molecule is just as likely to move to the left as to the right, and after some time Δt, the magnitude of its displacement is Δx. Now consider N molecules that start at the origin and walk randomly along the x axis (Fig. 20.12). The average number of steps each molecule takes in a time Δt is

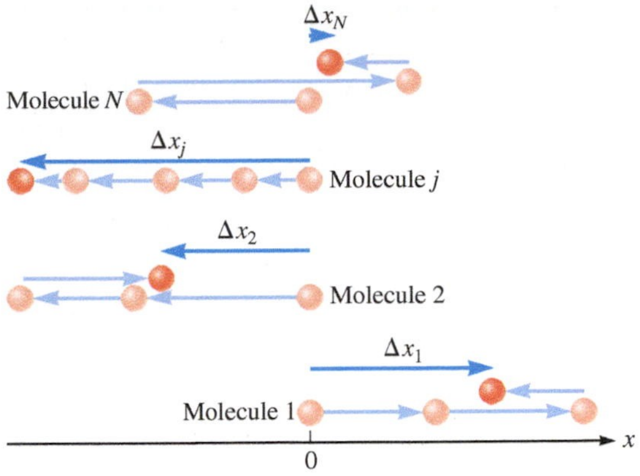

FIGURE 20.12 After each collision, a molecule is just as likely to move to the left as to the right. The average displacement of a collection of molecules depends on their mean free path and the mean free time.

$$\text{average number of steps} = \frac{\Delta t}{\text{average time between collisions}} = \frac{\Delta t}{t_{mf}}$$

Some molecules have a positive displacement and some a negative one, but we are *not* interested in their direction; we are only interested in the average distance the molecules have traveled from the origin. So, it is convenient to find an expression for the square of the average displacement Δx^2_{av}. Because the average size of a molecule's step is its mean free path λ,

$$\Delta x^2_{av} = (\lambda^2) \times (\text{average number of steps}) = \lambda^2(\Delta t/t_{mf})$$

The mean free path and the mean free time both depend on size and density of the two types of molecules. These two factors are usually combined as $D = \lambda^2/2t_{mf}$, where D is known as the **diffusion constant**. So, in one dimension, the average magnitude of the displacement is

$$\Delta x_{av} = \sqrt{2D\,\Delta t}$$

The diffusion constant has the dimensions of length squared per time; it depends on the properties of both gases, so it is reported for diffusion of one type of molecule in a particular medium at a given temperature. Table 20.2 gives D for several gases in air. Usually, the molecules may move in all three dimensions, and the average displacement is given by

$$\Delta r_{av} = \sqrt{6D\,\Delta t} \tag{20.26}$$

TABLE 20.2 Diffusion constant for gases in air.

Gas or vapor	D (m²/s)	T (°C)
Alcohol vapor	1.37×10^{-5}	40.4
Carbon dioxide	1.39×10^{-5}	0.0
Carbon disulfide	1.02×10^{-5}	19.9
Ether vapor	8.9×10^{-6}	19.9
Hydrogen, H_2	6.34×10^{-5}	0.0
Oxygen, O_2	1.78×10^{-5}	0.0
Water vapor	2.39×10^{-5}	8.0

Adapted from *The Handbook of Chemistry and Physics*, 68th ed. (Boca Raton, FL: CRC Press, 1987–1988), p. F-47.

EXAMPLE 20.3 It's in the Air

The air in your room is made up mostly of nitrogen and oxygen. Approximate these molecules as spheres of radius $r = 2.0 \times 10^{-10}$ m.

A Estimate the average distance between the air molecules in your room. Compare your answer to the size of a molecule.

:• INTERPRET and ANTICIPATE

We can find the number density (N/V) of air molecules from the ideal gas law. Number density is molecules per unit volume with the dimensions $(L)^{-3}$, so to find a distance L per molecule, we take the reciprocal of the cube root of the number density. We expect—based on assumption 1—that the average distance between molecules is greater than the size of the molecules.

Example continues on page 594 ▶

SOLVE

Use the ideal gas law (Eq. 19.17) to find N/V. Take room temperature to be 20°C (293 K) and the air pressure to be 1 atm (1.01×10^5 Pa). Keep an extra significant figure in the calculations for now.

$$PV = Nk_BT \quad (19.17)$$

$$\frac{N}{V} = \frac{P}{k_BT} = \frac{1.01 \times 10^5 \text{ Pa}}{(1.38 \times 10^{-23}\text{ J/K})(293 \text{ K})}$$

$$\frac{N}{V} = 2.5 \times 10^{25}\text{ molecules/m}^3$$

To find the average distance d_{av} between molecules, take the reciprocal of the cube root of the number density.

$$d_{av} = \left(\frac{N}{V}\right)^{-1/3} = [2.5 \times 10^{25}\text{ m}^{-3}]^{-1/3}$$

$$d_{av} = 3.4 \times 10^{-9}\text{ m}$$

Compare the average distance between molecules to the radius of a molecule by taking a ratio.

$$\frac{d_{av}}{r} = \frac{3.4 \times 10^{-9}\text{ m}}{2.0 \times 10^{-10}\text{ m}} = 17$$

$$d_{av} \approx 20r$$

CHECK and THINK

As expected, the average distance between molecules is greater than the size of an individual molecule. To help visualize such a separation, imagine sitting in a classroom auditorium. If the whole room is filled, the separation between you and the next person is perhaps half the size of your diameter. You could easily hand that person a pencil. Now imagine having about 20 times more space between you and the next person. You might share the row with just one other person, and there might be no one sitting in front of you for 10 rows. Now imagine trying to hand a pencil to someone. The molecules in Figure 20.10B are drawn too close together to represent air molecules at STP.

B Estimate the mean free path of the air molecules in your room. Compare your answer to the average distance between molecules found in part A.

INTERPRET and ANTICIPATE

Use the number density found in part A to find the mean free path. From Figures 20.10B and 20.11, we expect the mean free path to be greater than the average separation between molecules.

SOLVE

Substitute values into Equation 20.24.

$$\lambda = \frac{1}{4\sqrt{2}\pi r^2(N/V)} \quad (20.24)$$

$$\lambda = \frac{1}{4\sqrt{2}\pi(2.0 \times 10^{-10}\text{ m})^2(2.5 \times 10^{25}\text{ m}^{-3})}$$

$$\lambda = 5.6 \times 10^{-8}\text{ m}$$

Take a ratio to compare the mean free path to the average separation between molecules.

$$\frac{\lambda}{d_{av}} = \frac{5.6 \times 10^{-8}\text{ m}}{3.4 \times 10^{-9}\text{ m}} = 16$$

$$\lambda \approx 20d_{av}$$

CHECK and THINK

Because the mean free path is greater than the average separation between molecules, the molecules do not necessarily collide with their nearest neighbors as shown in Figure 20.11B.

C What is the mean free time of the air molecules, and with what frequency are they undergoing collisions? Assume air is about 70% N_2 and 30% O_2 by mass. *Hint*: Use Equation 20.21.

INTERPRET and ANTICIPATE

The mean free time depends on the mean free path and the average speed of the molecules. We can calculate the average speed from the temperature of the gas. The frequency of collisions is the number of collisions per unit time, which we can find from the reciprocal of the mean free time.

SOLVE Equation 20.21 gives the average speed of the molecules.	$v_{av} = \sqrt{\frac{8k_BT}{\pi m}}$ (20.21)
We also need the molecule's mass. The mass of N_2 is 28.0 u, and the mass of O_2 is 32.0 u. Find the average mass of air molecules from the assumption that air is 70% N_2 and 30% O_2.	$m_{av} = 0.7(28.0\,\text{u}) + 0.3(32.0\,\text{u})$ $m_{av} = 29.2\text{ u} \times \left(\frac{1.66 \times 10^{-27}\text{ kg}}{1\text{ u}}\right) = 4.85 \times 10^{-26}\text{ kg}$
Substitute numerical values. As in part A, we can assume the temperature is 20°C = 293 K.	$v_{av} = \sqrt{\frac{8(1.38 \times 10^{-23}\text{ J/K})(293\text{ K})}{\pi(4.85 \times 10^{-26}\text{ kg})}} = 4.61 \times 10^2\text{ m/s}$
Now use Equation 20.25 to find the mean free time. Substitute the mean free path λ from part B.	$t_{mf} = \frac{\lambda}{v_{av}} = \frac{5.6 \times 10^{-8}\text{ m}}{4.61 \times 10^2\text{ m/s}} = 1.2 \times 10^{-10}\text{ s}$
The frequency f of the collisions is the reciprocal of the mean free time.	$f = \frac{1}{t_{mf}} = \frac{1}{1.2 \times 10^{-10}\text{ s}} = 8.2 \times 10^9\text{ collisions/s}$

CHECK and THINK

The mean free time depends on the speed of the molecules; slower molecules have a greater mean free time. The high frequency of collisions explains why molecules take such a long time to mix by diffusion.

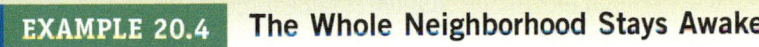

EXAMPLE 20.4 The Whole Neighborhood Stays Awake

If the aromatic molecules from the coffee roaster (Fig. 20.10A) did not undergo any collisions, they would pass through the neighborhood in just a second or so. Many molecular collisions occur (Fig. 20.10B), though, and if the air were calm, it would take years for the aroma to fill the neighborhood. Assuming the diffusion coefficient is the same as for water vapor, estimate to one significant figure the time it would take to fill a neighborhood with the scent of coffee by diffusion alone.

INTERPRET and ANTICIPATE

Our primary job is to estimate the size of the neighborhood and then find Δt, the time for the aromatic molecules to diffuse into the air in the neighborhood. From the opening statement, we expect that Δt is several years.

SOLVE

Let's assume that the coffee roaster sits in the middle of a neighborhood that is about 0.5 km in radius. The average displacement of the molecules must then be about 500 m for people in the whole neighborhood to smell the coffee.

The aromatic molecules travel in three dimensions, so we can use Equation 20.26, which we solve for Δt. Use D for water vapor from Table 20.2 and report the final answer to one significant figure. (D in Table 20.2 is for water vapor at a chilly 8°C, but the value of D is good enough for our one-significant-figure estimate.)	$\Delta r_{av} = \sqrt{6D\,\Delta t}$ (20.26) $\Delta t = \frac{\Delta r_{av}^2}{6D} = \frac{(500\text{ m})^2}{6(2.39 \times 10^{-5}\text{ m}^2/\text{s})}$ $\Delta t = 1.7 \times 10^9\text{ s} \times \left(\frac{1\text{ yr}}{3.15 \times 10^7\text{ s}}\right)$ $\Delta t = 50\text{ yr}$

CHECK and THINK

As we expected, it would take about half a century for the coffee aroma to diffuse throughout the neighborhood. Neighborhood residents, however, know that they can smell the coffee soon after the shop starts roasting it. The aroma is actually transported by currents in the air (Chapter 21).

CONCEPT EXERCISE 20.4

When people smell something unpleasant such as cigarette smoke, they sometimes fan the air. Does fanning the air help get rid of the smell? Explain your answer.

20-6 Real Gases: The Van der Waals Equation of State

The ideal gas law works (that is, it does a good job of describing a real gas) if the gas density is low enough. For many real gases, however, the ideal gas law is not a good approximation. In those cases, we must use the **Van der Waals equation of state**, which frees us from two of the assumptions of the kinetic theory. As in the last section, we can disregard assumption 1 (that the molecules are particles with no physical extent). Instead, we model gas molecules as spheres of radius r. Also, we disregard assumption 2 (that the molecules only exert contact forces on one another). The molecules actually exert an attractive field force on one another due to the electromagnetic force that we study in Parts III and IV.

Think about the motion of a single gas molecule. It cannot move through the entire volume V of the container because the other molecules take up some of that volume. Instead, the volume $V_{\text{available}}$ that the molecule can move through is reduced by the volume occupied by the other molecules:

$$V_{\text{available}} = V - V_{\text{molecules}} \tag{20.27}$$

It is common convention to write the volume occupied by the molecules in terms of b, the volume of molecules per mole. The volume occupied by the molecules is then the number of moles n multiplied by b, and Equation 20.27 is written as

$$V_{\text{available}} = V - nb \tag{20.28}$$

Because molecules take up some volume, we correct the ideal gas law by replacing V in the ideal gas law $PV = nRT$ (Eq. 19.21) by $V_{\text{available}}$ in Equation 20.28:

$$P_{\text{Clausius}}(V - nb) = nRT \tag{20.29}$$

Equation 20.29 is called the **Clausius equation of state.** It models the molecules as spheres of radius r, but it does not take into account the field force they exert on one another.

We can modify the Clausius equation of state to include that attractive field force. Imagine the motion of one molecule as it approaches one of the container's walls, where it is pulled back by the attractive force of the other molecules. The force the molecule exerts on the wall is smaller than it would be if there were no field forces. Because all the molecules experience such a force, the pressure exerted on the walls is reduced:

$$P = P_{\text{Clausius}} - P_{\text{reduction}} \tag{20.30}$$

We can find an expression for the reduction in pressure $P_{\text{reduction}}$. The force experienced by each molecule as it approaches the wall is proportional to the molar density of the molecules n/V. Further, the pressure (force per unit area) on the walls is also proportional to the density of the molecules, so the reduction in pressure is proportional to the density squared. We write the reduction in pressure as $P_{\text{reduction}} = a(n/V)^2$ and substitute this equation and Equation 20.29, $P_{\text{Clausius}} = nRT/(V - nb)$, into Equation 20.30:

$$P = \frac{nRT}{V - nb} - a\left(\frac{n}{V}\right)^2 \tag{20.31}$$

Equation 20.31 is the Van der Waals equation of state. It is often written as

VAN DER WAALS EQUATION OF STATE

Major Concept

$$\left(P + a\frac{n^2}{V^2}\right)(V - nb) = nRT \tag{20.32}$$

so that the right side follows the same form as the ideal gas law. The constants a and b depend on the type of gas and are determined experimentally (Table 20.3).

TABLE 20.3 Constants a and b in the Van der Waals equation of state.

Gas	a (Pa · m^6/mol^2)	b (m^3/mol)
Carbon dioxide, CO_2	0.359	4.27×10^{-5}
Carbon monoxide, CO	0.149	3.99×10^{-5}
Helium, He	3.46×10^{-3}	2.37×10^{-5}
Nitrogen, N_2	0.139	3.91×10^{-5}
Oxygen, O_2	0.136	3.18×10^{-5}
Water vapor	0.546	3.05×10^{-5}

Adapted from *The Handbook of Chemistry and Physics*, 68th ed. (Boca Raton, FL: CRC Press, 1987–1988), p. D-188.

Comparing Van der Waals and the Ideal Gas Equations of State

We expect that the Van der Waals equation of state should reduce to the ideal gas law when the density of the gas is very low or when the volume is very large. When the volume is very large,

$$P \gg a\frac{n^2}{V^2}$$

and

$$V \gg nb$$

so that the Van der Waals equation (Eq. 20.32) is identical to the ideal gas law in that case:

$$\left(P + a\frac{n^2}{V^2}\right)(V - nb) \approx (P + 0)(V - 0) = nRT$$

A graph of pressure as a function of volume—a ***PV* diagram** for short—is an important visualization tool. We can use a *PV* diagram to compare the ideal gas law to the Van der Waals equation of state (Fig. 20.13). Figure 20.13A is the *PV* diagram for a gas at constant temperature. Comparing the two laws for a gas at the same pressure, shows that the Van der Waals gas has a smaller volume because attractive forces between molecules tend to pull them together. (In an ideal gas, we assume there are no such long-range forces between molecules, so the volume is larger.) The two curves are nearly the same, however, when the gas volume is large, which is consistent with modeling a real gas as an ideal gas when the volume is large.

PRESSURE–VOLUME (*PV*) DIAGRAM

Tool

Figure 20.13B shows *PV* diagrams for a gas at two different temperatures, showing that a warm gas behaves much more like an ideal gas even when its volume is small. At higher temperatures, the molecules are moving faster, and the attractive force between the molecules has less influence on their motion. We will return to *PV* diagrams in the next section.

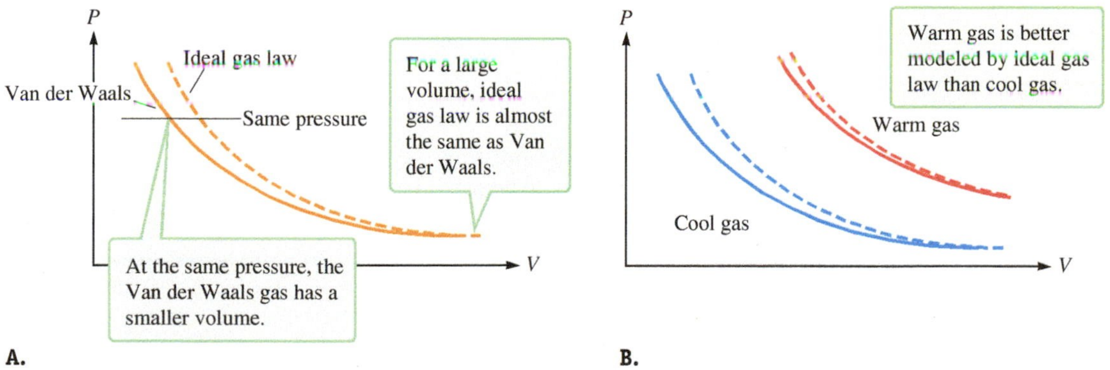

FIGURE 20.13 **A.** *PV* diagram for a gas using the ideal gas law (dashed curve) and Van der Waals equation of state (solid curve). **B.** *PV* diagrams for gases at two different temperatures using the ideal gas law (dashed curves) and Van der Waals equation of state (solid curves).

EXAMPLE 20.5 Real Gas Versus Ideal Gas

In this example, we compare the pressure found by using the ideal gas law with the pressure found by using the Van der Waals equation of state. In all cases, there are 1000 mol of oxygen gas (O_2) at a warm room temperature of 300 K. Give your answers to three significant figures.

A Use the ideal gas law to calculate the pressure exerted by the oxygen gas in a container approximately the size of a laboratory with a volume of 100 m^3.

INTERPRET and ANTICIPATE

This example is a straightforward problem that we could have tackled in Section 19-6. The density (1000 mol/100 m^3) is less than the density of air in your room at atmospheric pressure, so we expect our answer to be less than atmospheric pressure.

Example continues on page 598 ▶

SOLVE
Solve the ideal gas law (Eq. 19.21) for pressure.

$$PV = nRT$$

$$P = \frac{nRT}{V} = \frac{(1000\,\text{mol})[8.314\,\text{J/(mol}\cdot\text{K)}](300\,\text{K})}{100\,\text{m}^3}$$

$$P = 2.49 \times 10^4\,\text{Pa}$$

CHECK and THINK
As expected, the pressure is less than atmospheric pressure (1.01×10^5 Pa).

B Now use the Van der Waals equation of state to calculate the pressure exerted by the oxygen gas in a container with a volume of 100 m^3.

INTERPRET and ANTICIPATE
We need to use the Van der Waals equation for pressure with the constants a and b for oxygen given in Table 20.3. Because the density is very low, we expect that the pressure should be close to the pressure we found in part A.

SOLVE
Substitute appropriate values into Equation 20.31.

$$P = \frac{nRT}{V - nb} - a\left(\frac{n}{V}\right)^2 \tag{20.31}$$

$$P = \frac{(1000\,\text{mol})[8.314\,\text{J/(mol}\cdot\text{K)}](300\,\text{K})}{100\,\text{m}^3 - (1000\,\text{mol})(3.18 \times 10^{-5}\,\text{m}^3/\text{mol})} - \left(\frac{0.136\,\text{Pa}\cdot\text{m}^6}{\text{mol}^2}\right)\left(\frac{1000\,\text{mol}}{100\,\text{m}^3}\right)^2$$

$$P = 2.49 \times 10^4\,\text{Pa}$$

CHECK and THINK
The pressure found using the Van der Waals equation of state matches the pressure we found using the ideal gas law to three significant figures. We can see that it is okay to approximate such a low-density gas as ideal.

C Use the ideal gas law to calculate the pressure exerted by the same amount of oxygen gas at the same temperature in a container about the size of a large duffel bag (volume 1 m^3).

INTERPRET and ANTICIPATE
Use the same process as in part A. This time, the density (1000 mol/m^3) is greater than the density of the air in your room at atmospheric pressure, so we expect to find that the pressure is greater than atmospheric pressure.

SOLVE
Substitute values into the ideal gas law.

$$P = \frac{nRT}{V} = \frac{(1000\,\text{mol})[8.314\,\text{J/(mol}\cdot\text{K)}](300\,\text{K})}{1\,\text{m}^3}$$

$$P = 2.49 \times 10^6\,\text{Pa}$$

CHECK and THINK
As expected, the pressure is about 25 times greater than atmospheric pressure.

D Now use the Van der Waals equation of state to calculate the pressure exerted by that oxygen gas in a container with a volume of 1 m^3.

INTERPRET and ANTICIPATE

Repeat the process from part B. Because the density is great, we expect that the pressure should differ from the pressure we found in part C.

SOLVE

Substitute appropriate values into Van der Waals equation of state in the form of Equation 20.31.

$$P = \frac{nRT}{V - nb} - a\left(\frac{n}{V}\right)^2 \tag{20.31}$$

$$P = \frac{(1000\text{ mol})[8.314\text{ J/(mol}\cdot\text{K)}](300\text{ K})}{1\text{ m}^3 - (1000\text{ mol})(3.18\times10^{-5}\text{ m}^3\text{/mol})} - \left(\frac{0.136\text{ Pa}\cdot\text{m}^6}{\text{mol}^2}\right)\left(\frac{1000\text{ mol}}{1\text{ m}^3}\right)^2$$

$$P = 2.44 \times 10^6\text{ Pa}$$

CHECK and THINK

The pressure found using the Van der Waals equation of state is 5×10^4 Pa (about 0.5 atm) lower than the pressure we found using the ideal gas law. In this low-volume case, if we model the gas as an ideal gas, our calculations are incorrect to three significant figures. We can easily measure such a discrepancy.

20-7 Phase Changes

The molecular forces in a real gas can cause it to change phase, that is, become liquid or solid. (In an ideal gas, the particles do not exert an attractive field force on one another, so an ideal gas would not change phase.) In this section, we look at the conditions required for phase changes. Pressure–volume (*PV*) diagrams are one way to visualize these conditions.

Figure 20.13A shows that at the same pressure, a real gas has a smaller volume than an ideal gas, and we account for this smaller volume in terms of the attractive field force between molecules. At a sufficiently low temperature (or at a high pressure), these attractive forces can cause the gas to become a liquid. Follow the *PV* diagram (Fig. 20.14) for a liquefying gas from point *A* (on the right) to point *D* (on the left). As the gas liquefies, its pressure remains constant and its volume decreases; then, as the pressure of the liquid increases, its volume is nearly constant.

Under the pressure and volume conditions on the portion of the curve between points *B* and *C* in Figure 20.14, the substance exists as both gas and liquid. In fact, there is a range of pressure, temperature, and volume conditions in which the sub-

CRITICAL TEMPERATURE; CRITICAL POINT; CRITICAL PRESSURE

Major Concepts

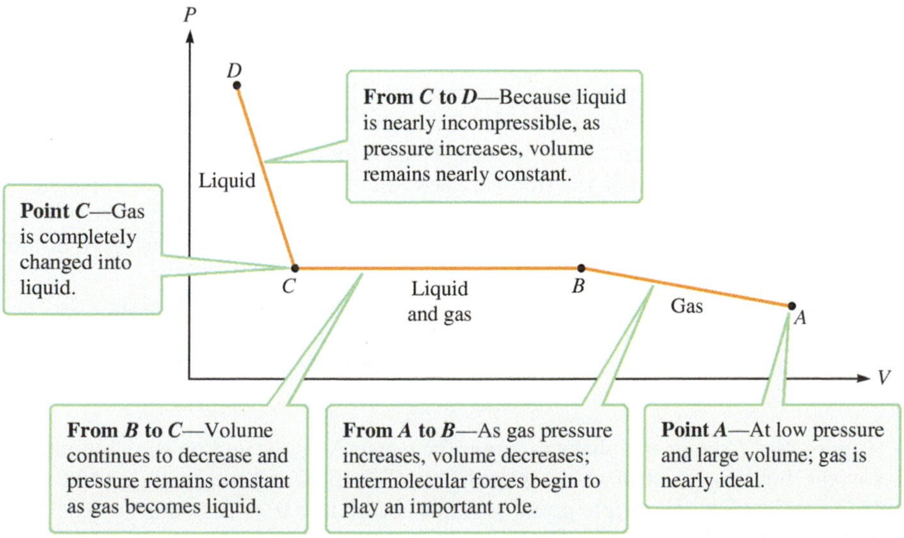
P
V
D
C
B
A
Liquid
Liquid and gas
Gas
From C to D—Because liquid is nearly incompressible, as pressure increases, volume remains nearly constant.
Point C—Gas is completely changed into liquid.
From B to C—Volume continues to decrease and pressure remains constant as gas becomes liquid.
From A to B—As gas pressure increases, volume decreases; intermolecular forces begin to play an important role.
Point A—At low pressure and large volume; gas is nearly ideal.

FIGURE 20.14 As a gas liquefies, its pressure remains constant, and its volume decreases as shown by the horizontal line between points *B* and *C*.

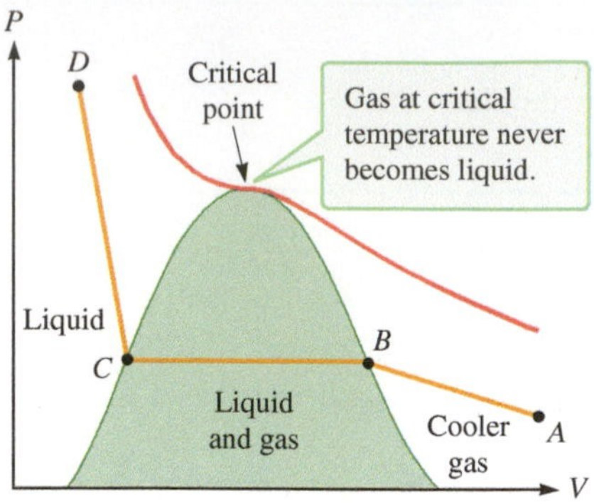

FIGURE 20.15 *PV* diagram for a hot gas (red) and cooler gas (orange). The green area represents the range where the substance exists in equilibrium between the liquid and gas phases.

stance exists in equilibrium between the liquid and gas phases. The green area in Figure 20.15 represents the range of liquid–gas equilibrium conditions.

If a gas is warmer than the **critical temperature**, it cannot become a liquid no matter how high the pressure. The red curve in Figure 20.15 shows a gas at its critical temperature. When a gas is at its critical temperature, its *PV* curve does not pass through the green region; instead, it touches the green region at the **critical point** where the curve is horizontal. The pressure at the critical point is known as the **critical pressure.** As more pressure is applied to the gas, it becomes denser. It acquires properties similar to those of a liquid, but no liquid forms. Cooler gases pass through the wide part of the green region, becoming liquid. Critical temperatures and pressures of five common gases are found in Table 20.4.

The term **vapor** describes a substance that is in the gaseous phase and below its critical temperature, and the term **gas** is reserved for a substance above its critical temperature. So, the substance represented by the orange curve in Figure 20.15 should be properly referred to as a *vapor* because its temperature is lower than the critical temperature.

PHASE DIAGRAM Tool

A **phase diagram** (another way to visualize phase changes) is a graph of pressure as a function of temperature. Curves on the phase diagram show the conditions under which a substance may change between two of the three states: solid, liquid, and gas (or vapor). The phase diagram in Figure 20.16 holds for a substance such as carbon dioxide. We often think of carbon dioxide as a gas that we exhale; it is found naturally as a liquid venting under the sea (right inset). Carbon dioxide in its solid form is known as dry ice, and its vapor is used to create an eerie atmosphere (left inset). The three curves in Figure 20.16 represent the equilibrium conditions between two phases. The orange curve represents the pressure and temperature combinations that allow the substance to exist as a solid and as a vapor. The green curve defines the equilibrium between the solid and liquid phases, and the purple curve does the same for the liquid and vapor phases.

TABLE 20.4 Critical temperatures and pressures.

Substance	Critical temperature (°C)	Critical pressure (Pa)
Carbon dioxide, CO_2	31	7.38×10^6
Hydrogen, H_2	–239.9	1.30×10^6
Nitrogen, N_2	–147	3.39×10^6
Oxygen, O_2	–118.4	5.08×10^6
Water, H_2O	374.1	2.21×10^7

Adapted from *The Handbook of Chemistry and Physics*, 68th ed. (Boca Raton, FL: CRC Press, 1987–1988), p. F-66.

A block of dry ice is a solid, so its pressure and temperature are represented by a point labeled *A* in the yellow region in Figure 20.16. If the temperature of the dry ice increases sufficiently and the pressure remains constant, the new pressure and temperature are represented by point *B*. Because point *B* is in the red region, the carbon dioxide has become a vapor. The process of changing from a solid to a vapor without becoming a liquid is called **sublimation**. Sublimation only occurs at low pressure. If the dry ice were at a higher pressure such as at point *C*, as the temperature increases the carbon dioxide first becomes a liquid (point *D*). As the temperature increases further, the substance becomes a vapor (point *E*). Of course, if the temperature increases above the critical temperature, the substance is properly referred to as a gas.

Figure 20.16 shows the following two other important points on the phase diagram for any substance.

1. The **critical point**, which also appeared on the *PV* diagram (Fig. 20.15), represents the critical temperature and pressure (Table 20.4). A gas at a pressure greater than the critical pressure does not undergo a distinct phase transition to a liquid as its temperature decreases. That is, for a substance above the critical pressure, there is no distinction between liquid and gas phases. For substances we are familiar with, the critical pressure is much greater than atmospheric pressure, so we don't usually observe the blending of the gas and liquid phases.
2. The **triple point** is a unique combination of temperature and pressure that allows the substance to exist in all three phases simultaneously. The triple point of water is used as a reference point for the absolute temperature scale (Section 19-7).

Figure 20.17 shows the phase diagram for water. As indicated on the diagram, at 1 atm the freezing point of water is at 0°C or 273.15 K and the boiling point is at 100°C or 373.15 K. The triple point of water occurs at 6.0×10^{-3} atm or 0.61 kPa and 0.01°C or 273.16 K. (The triple point of carbon dioxide occurs at 5.1 atm or 520 kPa and −56.4°C or 217 K.) The most important difference between the phase diagram for water (Fig. 20.17) and the phase diagram for a substance like carbon

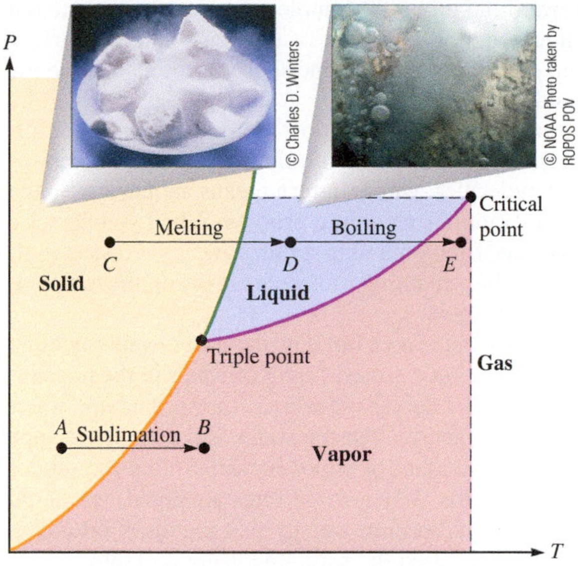

FIGURE 20.16 A phase diagram is a graph of pressure as a function of temperature. Carbon dioxide is shown in liquid form escaping from vents deep in the ocean and in solid form known as dry ice.

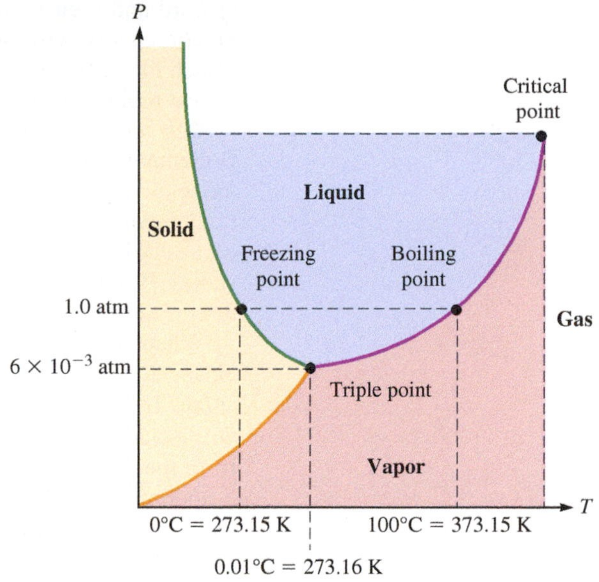

FIGURE 20.17 A phase diagram for water shows that the curve (green) between the solid and liquid phases slopes up and to the left from the triple point. Note that the temperature axis is not linear.

dioxide (Fig. 20.16) is in the shape of the curve between the solid and liquid phases (shown in green). For carbon dioxide, this curve slopes upward and to the right from the triple point, as is true for all substances that contract when freezing. For water and other substances that expand when freezing, the curve slopes upward and to the left from the triple point. For such a substance, a lower temperature is required to freeze it at higher pressure.

CONCEPT EXERCISE 20.5

a. In what state (phase) is water when it is at 100°C and its pressure is above its critical pressure?

b. In what state is water when it is at 100°C and its pressure is between 1 atm and its critical pressure?

c. In what state is water when it is at 100°C and its pressure is below 1 atm?

20-8 Evaporation

In Section 20-7, we considered phase changes between three states of matter. In this section, we look at the interplay of the liquid and gaseous states. Although much of what follows can be applied to any substance, our focus is primarily on water and its transition between a liquid and a vapor.

Evaporation is the process whereby atoms or molecules in a liquid gain enough kinetic energy to leave the liquid and enter the gaseous (vapor) phase. **Condensation** is the opposite process, in which atoms or molecules in a gas become liquid. A substance is in equilibrium when the rate of evaporation equals the rate of condensation and the relative amount of liquid and vapor remains constant. If the temperature is increased after equilibrium is reached, the evaporation rate exceeds the condensation rate, and the relative amount of liquid decreases.

We can use the kinetic theory to explain how temperature can change the rate of evaporation. The molecules in a liquid are strongly attracted to one another, but a molecule with sufficient kinetic energy near the liquid surface may leave. What happens next depends on the molecule's kinetic energy. If the kinetic energy is below a certain value, the molecule will fall back into the liquid (much like a ball thrown upward and then falling back toward the ground). A molecule with a great enough kinetic energy will escape the liquid and become part of the gas above the surface (much like an object with a speed greater than the escape speed of the Earth). So the fastest molecules escape the liquid and become a gas.

The molecules in a liquid have a speed distribution similar to the Maxwell-Boltzmann distribution (Fig. 20.8). An increase in the liquid's temperature increases the relative number of high-speed molecules, which in turn means an increase in the rate of evaporation. Therefore, evaporation is a cooling process. When you perspire on a very hot day, the water evaporates from your skin, taking away the fastest molecules. Because the average speed of the remaining molecules is lower, the temperature of the water is lower, and you feel cooler.

When you set your pet's water dish outside on a hot day, the water evaporates and disappears from the dish, but what if you set a closed bottle of water in the hot sunshine? If the bottle has been moved from a cooler place to a warmer one, the water will evaporate at a faster rate than it condenses, but the water vapor cannot escape from the sealed bottle. For a while, the amount of vapor increases until the rate of evaporation equals the rate of condensation. When those rates are equal, the liquid and vapor are then in equilibrium. At equilibrium, we say that the air is **saturated**, and the resulting pressure of the vapor is called the **saturated vapor pressure**.

SATURATED VAPOR PRESSURE

Major Concept

The saturated vapor pressure depends only on temperature and not on the volume of the container. Suppose the volume of the bottle was suddenly reduced (perhaps by squeezing the bottle). The volume of the liquid would not change because liquids are incompressible, but the volume of the vapor would decrease, so the vapor density would increase. More vapor molecules would collide with the liquid's surface each second. The condensation rate would exceed the evaporation rate until a new equilibrium was reached at the same saturated vapor pressure.

Both the evaporation rate and the saturated vapor pressure increase as temperature increases. Table 20.5 gives the saturated vapor pressure for water at several temperatures from $-50°C$ to 150°C, and Figure 20.18 is a graph of saturated vapor pressure for water in the temperature range from 20°C to 40°C.

Evaporating is not the same as boiling. A liquid **boils** when the saturation vapor pressure equals the external pressure. Consider a pot of water on a hot stove at sea level. As the water temperature increases, tiny vapor bubbles begin to form. If the vapor pressure inside the bubbles is less than the external pressure, the bubbles collapse. As the water temperature continues to increase, the vapor pressure in the bubbles increases until it equals the external pressure. Bubbles rise to the surface, and the vapor escapes. At sea level, the external pressure is 1 atm $= 1.013 \times 10^5$ Pa, and according to Table 20.5, that value corresponds to a water tempera-

TABLE 20.5 Saturated vapor pressure for water.

Temperature (°C)	Temperature (K)	Saturated vapor pressure (Pa)
Ice, –50	223	3.939
Ice, –25	248	7.011×10^1
Freezing point, 0	273	6.103×10^2
10	283	1.227×10^3
15	288	1.705×10^3
20	293	2.337×10^3
25	298	3.166×10^3
30	303	4.242×10^3
35	308	5.621×10^3
40	313	7.452×10^3
50	323	1.233×10^4
60	333	1.991×10^4
70	343	3.115×10^4
80	353	4.733×10^4
90	363	7.008×10^4
Boiling point, 100	373	1.013×10^5
125	398	2.320×10^5
150	423	4.759×10^5

Adapted from *The Handbook of Chemistry and Physics*, 68th ed. (Boca Raton, FL: CRC Press, 1987–1988), pp. D-189–D-191.

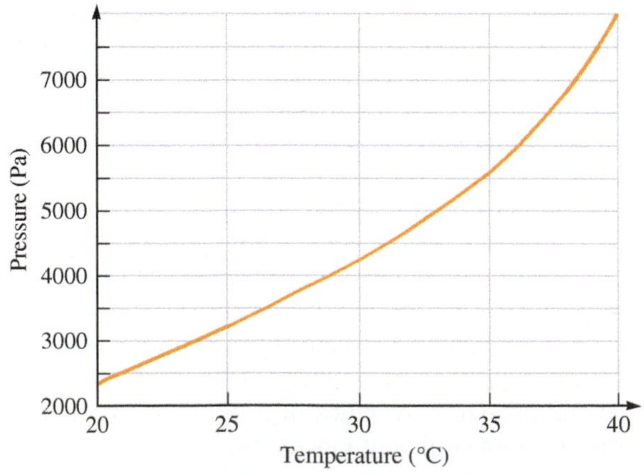

FIGURE 20.18 Saturated vapor pressure of water.

ture of 100°C, as you would expect. If the pot of water is at a higher elevation, however, the external pressure is less than 1 atm, and the water will boil at a lower temperature. Thus, cooking instructions must be modified for people who live at high altitudes.

Humidity

Air is a mixture of several gases: about 78% nitrogen by volume, 21% oxygen, and less than 1% argon and carbon dioxide. The pressure that would be exerted by each gas if the other gases were not present is called the **partial pressure** of that gas. The total air pressure is the sum of the partial pressures due to each gas. If no water vapor is in the air, the partial pressure of water is zero. The partial pressure of water cannot exceed the saturated vapor pressure at that temperature if water is to remain in the vapor phase. On a hot day, there may be more water vapor in the air than on a cooler day, and we say that it is "muggy."

Relative humidity (*RH*) is the ratio of the partial pressure of water P_{water} to the saturated vapor pressure P_{svp} at some temperature, expressed as a percentage:

$$RH = \frac{P_{water}}{P_{svp}} \times 100\% \tag{20.33}$$

RELATIVE HUMIDITY
Major Concept

The maximum partial pressure of water is the saturated vapor pressure, giving a relative humidity of 100%. In this case, we say that the air is *saturated*.

It is possible for the partial pressure of water to exceed the saturated vapor pressure temporarily, which can happen if the temperature drops as it does at night. When the temperature is high during the day, the relative humidity may get quite high. When the temperature drops, the saturated vapor pressure decreases. If the partial pressure of water is greater than the saturated vapor pressure at that lower nighttime temperature, water condenses into a liquid in the form of rain, fog, or dew, which is why we often find dew in the early morning.

In fact, dew can be used to measure the relative humidity. As the air cools, it reaches a temperature T_{DP} at which the partial pressure of water equals the saturated vapor pressure, known as the **dew point**. You can measure the dew point with a metal can and a thermometer (Fig. 20.19). Pour some water into the can and slowly add ice to lower the temperature. You need to measure the temperature of the water, so it is best to use the thermometer to mix the water as you add ice. The metal can must be in thermal contact with the air as you cool the can. Watch for dew to form on the outside of the can; the temperature at which dew forms is the dew point. Because the partial pressure of water equals the saturated vapor pressure at the temperature of the dew point, once you know the dew point you can look up the partial pressure of water in either Table 20.5 or Figure 20.18.

FIGURE 20.19 To measure the dew point, cool an object that is in thermal contact with the air and measure the temperature at which dew forms on the object.

EXAMPLE 20.6 CASE STUDY **Would You Rather Be in Phoenix?**

Some people like to live in a humid climate, and others prefer a dry one. Houston, Texas, is known for it high humidity. On a typical August day, the high temperature in Houston may reach 93°F, and the low temperature is about 73°F. The average dew point in August is close to the low temperature: T_{DP} = 72°F. Phoenix, Arizona, is known for its dry climate. In February, its relative humidity is 28% in the middle of the afternoon when the temperature reaches its high.

A Assume the dew point is constant throughout the day. What is the relative humidity in Houston when the temperature is at its typical August high point? At its low point?

INTERPRET and ANTICIPATE

At the dew point, the partial pressure of water equals the saturated vapor pressure. To find the relative humidity, find the saturated vapor pressure at the high and low temperatures. Because Houston is known for high humidity, we expect its relative humidity to be higher than that in Phoenix, and because Houston's low temperature is just above the dew point, we expect the relative humidity at that temperature to be nearly 100%.

Example continues on page 604 ▶

SOLVE First, convert all the Fahrenheit temperatures to Celsius (Eq. 19.2).	$T_{hi}(°C) = \frac{5}{9}[T_{hi}(°F) - 32]$ $T_{hi}(°C) = \frac{5}{9}[93°F - 32] = 34°C$ $T_{lo} = 73°F = 23°C$ $T_{DP} = 72°F = 22°C$
The partial pressure of water equals the saturated vapor pressure at the dew point. Use Figure 20.18 to find P_{svp} at 22°C. From the graph, P_{svp} is about 2.7×10^3 Pa.	$P_{water} = P_{svp}(22°C) \approx 2.7 \times 10^3$ Pa
Find the relative humidity from Equation 20.33. Now we need to find the saturated vapor pressure at the high and low temperatures. Again, use Figure 20.18 to read off P_{svp} at the high temperature $T_{hi} = 34°C$. We estimate P_{svp} at 34°C as 5.3×10^3 Pa.	$RH = \frac{P_{water}}{P_{svp}} \times 100\%$ (20.33) $RH_{hi} = \frac{2.7 \times 10^3 \text{Pa}}{5.3 \times 10^3 \text{Pa}} \times 100\%$ $RH_{hi} = 51\%$
At the low temperature $T_{lo} = 23°C$, P_{svp} is about 2.8×10^3 Pa.	$RH = \frac{2.7 \times 10^3 \text{Pa}}{2.8 \times 10^3 \text{Pa}} \times 100\%$ $RH_{lo} = 96\%$

CHECK and THINK

As expected, the relative humidity is greater than in Phoenix and nearly 100% at the low temperature. Dew does not form, however, because the actual temperature does not get down to the dew point.

B In Phoenix, the average high temperature in February is 70°F, and the average low temperature is 44°F. Roughly what is the typical February dew point in Phoenix? Does dew form there on a typical February day?

INTERPRET and ANTICIPATE

Use the relative humidity at the high temperature (28%, given in the problem statement) to find the partial pressure of water. The dew point occurs at the temperature where the partial pressure of water equals the saturated vapor pressure. We expect the dew point for Phoenix to be lower than the dew point for Houston. For dew to form, the temperature in Phoenix must be lower than the temperature in Houston.

SOLVE As in part A, convert all temperatures to Celsius.	$T_{hi} = 70°F = 21°C$ $T_{lo} = 44°F = 6.7°C$
The relative humidity is 28% when the temperature is at its high. Solve Equation 20.33 for the partial pressure of water, P_{water}.	$RH = \frac{P_{water}}{P_{svp}} \times 100\% = 28\%$ (20.33) $\frac{P_{water}}{P_{svp}} = 0.28$ $P_{water} = 0.28 P_{svp}$
Read the saturated vapor pressure at 21°C from Figure 20.19. We get 2.5×10^3 Pa.	$P_{water} = 0.28(2.5 \times 10^3 \text{Pa}) = 7.0 \times 10^2 \text{Pa}$

At the dew point, the saturated vapor pressure equals the partial pressure of water, so we need to find the temperature that corresponds to $P_{svp} = 7.0 \times 10^2$ Pa. That pressure is below the range in Figure 20.18, but it can be estimated from Table 20.5. According to the table, the dew point is between 0°C and 10°C. In fact, it is much closer to 0°C, so let's estimate it to be about 1°C. Because the low temperature in Phoenix on a typical February day is nearly 7°C, no dew forms.

CHECK and THINK

As expected, the dew point in Phoenix is much lower than it is in Houston. For dew to form in Phoenix, the temperature must be near freezing, whereas in Houston, it would still be balmy.

CONCEPT EXERCISE 20.6

CASE STUDY **Global Warming and Evaporation**

An increase in the amount of greenhouse gases is causing the temperature of the Earth to increase. As mentioned in Section 20-1, this change is known as global warming. Would you expect the evaporation rate to increase or decrease as a result of global warming?

CASE STUDY Part 3: Water Cycle and Global Dimming

The **water cycle** is the continuous circulation of the Earth's water stored in oceans, in other bodies of water such as lakes, and in the soil. Water evaporates from these sources and condenses in the atmosphere as clouds, and the cycle continues. Clouds transport water, and **precipitation** (such as rain, snow, and hail) occurs when the condensed water vapor in the atmosphere falls to the Earth. The water cycle is critical to life.

FIGURE 20.20 An evaporation pan is used to monitor the Earth's water cycle.

One way to monitor the Earth's water cycle is with an **evaporation pan**, an open metal container with carefully measured dimensions (Fig. 20.20). The pan is usually set on a palette, and the water level is measured daily. Measurements going back at least 50 years show an overall decrease in evaporation rate (the volume of water evaporating per day) until the 1990s. A decrease in evaporation rate indicates a slowdown in the water cycle, which could have serious consequences for all life on the Earth.

The measured decrease in the evaporation rate from the 1950s to the 1990s is believed to be an indication of **global dimming**, a long-term decrease in the amount of solar radiation that reaches the Earth's surface. One explanation for global dimming is an increase in particulate air pollution such as sulfates, nitrates, and soot. Such particles in the atmosphere block sunlight from reaching the Earth's surface, so the surface becomes cooler. The temperature decrease causes a decrease in the evaporation rate. The global dimming trend seems to have reversed since the 1990s, perhaps due to legislation such as the Clean Air Act, which has reduced the amount of pollution going into the air in the United States.

It may seem ironic that human activities produce pollutants that may cause both global warming and global dimming. Global warming mainly affects the troposphere (up to 20 km), whereas global dimming mainly affects the Earth's surface; in any event, global dimming has probably masked the effects of global warming. In fact, a strange solution to the problem of global warming proposes burning sulfur in the stratosphere to block out some sunlight. This solution was proposed several decades ago and has been recently revived by Paul Crutzen, winner of the 1995 Nobel Prize in Chemistry, as an "escape plan" because greenhouse gas reduction efforts have not been sufficient. Climatologist Mikhail Budyko—who is credited with first proposing the idea—warned, however, that such action would cause different climate changes in different regions in the world. For instance, some countries might become deserts, whereas others might become jungles. It may be better to increase our efforts to cut back on greenhouse gas emissions.

SUMMARY

❶ Underlying Principles

The **kinetic theory** states that all matter is made up of particles in motion. When the kinetic theory is applied to an ideal gas, the following five assumptions are made:

1. A gas consists of a large number of molecules or atoms that can be modeled as particles.
2. The particles in an ideal gas do not interact with one another.
3. The particles make elastic collisions with the walls, and the duration of each such collision is short.
4. The particles are free to move in any direction at any speed.
5. The gas is made up of identical particles.

Major Concepts

1. The **average kinetic energy** of gas particles is directly proportional to the gas **temperature**:

$$K_{av} = \tfrac{1}{2}mv_{av}^2 = \tfrac{3}{2}k_B T \quad (20.17)$$

2. The **root-mean-square velocity** of gas particles is proportional to the square root of the **temperature**:

$$v_{rms} = \sqrt{v_{av}^2} = \sqrt{\frac{3k_B T}{m}} \quad (20.18)$$

3. **Maxwell-Boltzmann speed distribution:** The probability of a particle's speed per unit speed interval (Fig. 20.8) is

$$f(v) = 4\pi\left(\frac{m}{2\pi k_B T}\right)^{3/2} v^2 e^{-(mv^2/2k_B T)} \quad (20.19)$$

For Maxwell-Boltzmann speed distribution,

4. the **most probable speed** is given by

$$v_{mp} = \sqrt{\frac{2k_B T}{m}} \quad (20.20)$$

5. and the **average speed** is given by

$$v_{av} = \sqrt{\frac{8k_B T}{\pi m}} \quad (20.21)$$

6. **Mean free path:** The average length of the straight portion of a molecule's zigzag path (between collisions) is

$$\lambda = \frac{1}{4\sqrt{2}\pi r^2(N/V)} \quad (20.24)$$

7. **Van der Waals equation of state:** The relation between pressure, volume, and temperature in a real gas, in which assumptions 1 and 2 for an ideal gas are not valid, is

$$\left(P + a\frac{n^2}{V^2}\right)(V - nb) = nRT \quad (20.32)$$

where a and b are experimentally determined constants for each gas.

8. **Critical temperature** is the temperature of the **critical point**. At $T > T_{crit}$, a gas cannot become a liquid no matter how high the pressure. The pressure at the critical point is known as the **critical pressure**.
9. **Saturated vapor pressure** is the pressure of a vapor when the rate of evaporation equals the rate of condensation so that the liquid and vapor are in equilibrium.
10. **Relative humidity (RH)** is the ratio of the partial pressure of water P_{water} to the saturated vapor pressure P_{svp} at some temperature, expressed as a percentage:

$$RH = \frac{P_{water}}{P_{svp}} \times 100\% \quad (20.33)$$

Tools

1. The **root mean square (rms)** of some quantity can be found as follows:
   a. Square each value.
   b. Take the average (mean) of those squared values, and
   c. Take the square root of the averaged squared values.

   The rms velocity is given by

$$v_{rms} = \sqrt{v_{av}^2} = \sqrt{\frac{1}{N}\sum_{j=1}^{N} v_j^2} \quad (20.3)$$

2. A **pressure–volume (*PV*) diagram** is a graph of pressure as a function of volume (Fig. 20.13).
3. A **phase diagram** is a graph of pressure as a function of temperature. Curves on the phase diagram show the conditions under which a substance may change between two of the three states: solid, liquid, and gas (or vapor); see Figures 20.16 and 20.17.

PROBLEMS AND QUESTIONS

A = algebraic C = conceptual E = estimation G = graphical N = numerical

20-1 What Is the Kinetic Theory?

1. **C** Use the kinetic theory to explain why a gas fills its container.
2. **C** One of the five assumptions for an ideal gas is worse for a molecular gas than for an atomic gas. Which one? Explain your answer.
3. **C** CASE STUDY Is the Earth's atmosphere contained by walls? Why doesn't the atmosphere escape? Why is the Earth's atmospheric pressure greater at sea level than it is on the tops of mountains?

20-2 Average and Root-Mean-Square Quantities

4. **A** The speeds of three molecules are v, $4v$, and $8v$, respectively. What are the average and rms speeds for the system consisting of these molecules?
5. **N** A system consists of 10 identical particles with the following speeds: two particles with speeds of 4.00 m/s, one particle with a speed of 8.00 m/s, four particles with speeds of 9.00 m/s, and three particles with speeds of 2.00 m/s. What is the **a.** average speed and **b.** rms speed of the particles in this system?

6. Five particles have velocities given by

$$\vec{v}_1 = (3.4\hat{\imath} + 5.6\hat{\jmath} + 6.7\hat{k})\ \text{m/s}$$

$$\vec{v}_2 = (2.4\hat{\imath} - 6.6\hat{\jmath} - 3.7\hat{k})\ \text{m/s}$$

$$\vec{v}_3 = (-0.4\hat{\imath} + 1.6\hat{\jmath} - 10.7\hat{k})\ \text{m/s}$$

$$\vec{v}_4 = (9.2\hat{\imath} - 9.8\hat{\jmath} + 3.6\hat{k})\ \text{m/s}$$

$$\vec{v}_5 = (0.9\hat{\imath} + 3.6\hat{\jmath} + 4.0\hat{k})\ \text{m/s}$$

N Find their **a.** average velocity, **b.** average speed, and **c.** rms speed.
d. **C** Compare your results and comment.

7. **A** For two particles with velocities $2v_0\hat{\imath}$ and $-v_0\hat{\imath}$, where $v_0 > 0$, consider the magnitude of the average velocity, the average of the two speeds, and the root-mean-square velocity. Order the magnitude of these three quantities from lowest to highest.

8. **N** **CASE STUDY** Nitrous oxide (N_2O) is a gas commonly used for its anesthetic effects in dentistry or surgery. When it interacts with oxygen, nitrous oxide gives rise to nitric oxide (NO), which will interact with ozone. Thus, nitrous oxide is a greenhouse gas that affects the ozone layer in our stratosphere. Suppose a container contains 3 mol of nitrous oxide, where the rms speed of the molecules in the container is 411.1 m/s. What is the average kinetic energy of the molecules inside the gas?

9. **N** Particles in an ideal gas of molecular oxygen (O_2) have an average momentum in the x direction of 2.726×10^{-23} kg·m/s. What is **a.** their root mean square speed and **b.** the root mean square of their x component of velocity.

10. **N** A cylinder has a nearly perfect vacuum within its interior except for 12 identical atoms. The atoms are moving with speeds 1.50 km/s, 5.00 km/s, 8.00 km/s, 1.25 km/s, 2.95 km/s, 4.40 km/s, 5.60 km/s, 3.50 km/s, 7.80 km/s, 2.70 km/s, 2.90 km/s, and 6.00 km/s. What is the **a.** average speed and **b.** rms speed of the 12 atoms in the cylinder?

11. **N** Table P20.11 gives the hourly temperature for Sudbury, Ontario, for November 1, 2006. Find the average temperature and rms temperature for that day.

TABLE P20.11

Time	Temperature (°C)	Time	Temperature (°C)
00:00	0.5	12:00	2.1
01:00	−0.7	13:00	2.2
02:00	−0.9	14:00	2.3
03:00	−1.3	15:00	1.3
04:00	−1.4	16:00	1.1
05:00	−1.4	17:00	0.4
06:00	−1.7	18:00	−0.4
07:00	−1.9	19:00	−0.5
08:00	−1.4	20:00	−0.7
09:00	−0.5	21:00	−0.8
10:00	0.5	22:00	−0.9
11:00	1.1	23:00	−1.2

20-3 The Kinetic Theory Applied to Gas Temperature and Pressure

12. **N** Find the rms speed of hydrogen molecules (H_2) at a temperature of 27°C.

13. **N** Oxygen gas (O_2) at 673 K is confined in a cylinder. What is the rms speed of oxygen molecules in the container?

14. **C** Three physics students are discussing the concept of temperature and have the following conversation.

Avi: The temperature is basically the average kinetic energy of the molecules. You have to find the average speed of the molecules and then use $k_BT = \frac{1}{2}mv^2$.

Shannon: I think you mean *velocity*. I think it's probably safer to use the average velocity of the molecules.

Cameron: I'm not sure. What was the rms velocity? Maybe we need to find that instead? Does it matter?

Who is on the right track? Explain your reasoning.

15. **N** The mass of a single hydrogen molecule is approximately 3.32×10^{-27} kg. There are 5.64×10^{23} hydrogen molecules in a box with square walls of area 49.0 cm^2. If the rms speed of the molecules is 2.72×10^3 m/s, calculate the pressure exerted by the gas.

16. **C** What is $\Delta t_{\text{collision}}$ in Equation 20.10? Why was a "trick" needed to find $\Delta t_{\text{collision}}$? Describe the trick. In particular, what is Δt in Equation 20.11?

17. **N** The noble gases neon (atomic mass 20.1797 u) and krypton (atomic mass 83.798 u) are accidentally mixed in a vessel that has a temperature of 90.0°C. What are the **a.** average kinetic energies and **b.** rms speeds of neon and krypton molecules in the vessel?

18. **C** In what way is $K_{\text{av}} = \frac{1}{2}mv_{\text{av}}^2 = \frac{3}{2}k_BT$ (Eq. 20.17) a connection between the microscopic and the macroscopic properties of an ideal gas?

19. **N** An 8.00×10^{-3} m^3 cylinder is filled with 3.00 mol of hydrogen gas. If the pressure within the cylinder is 5.00×10^5 Pa, what is the average translational kinetic energy of the hydrogen molecules in the cylinder?

20. **E** Estimate the rms speed of the oxygen molecules in your room. Estimate the rms speed of the oxygen molecules on a very cold day.

21. **N** In a cubical container with sides 4.00 cm in length, 2.00×10^{24} helium atoms, each with mass 6.64×10^{-27} kg, move with an rms speed of 1.50 km/s. What is the magnitude of the average force exerted on a wall by the helium gas in the container?

Problems 22 and 23 are paired.

22. **G** A particular roulette wheel is made up of 26 boxes, each labeled by one of the letters of the alphabet such that each letter is used. On each spin of the wheel, a ball lands in one of the 26 boxes. Plot the letters versus the frequency of the ball landing in a particular square. What letter is most probable?

23. **G** A particular roulette wheel is made up of 26 boxes, each labeled by only some of the letters of the alphabet (A–H). On each spin of the wheel, a ball lands in one of the 26 boxes. Figure P20.23 is a bar graph of the letters versus the frequency of the ball landing in a particular square. Determine the number of boxes labeled by each letter. What letter is most probable?

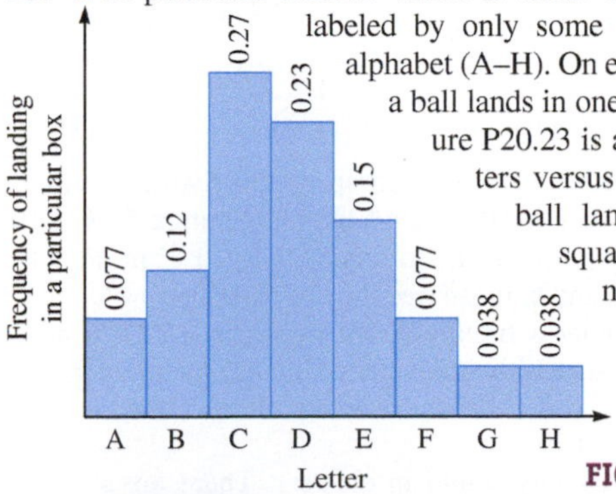

FIGURE P20.23

24. **C** **CASE STUDY** New stars form out of gaseous material (mostly molecular hydrogen) in a galaxy. Spiral galaxies like our own Milky Way have plenty of gas, so new stars are still forming. Some galaxies do not have enough gas for stars to form because the gas has escaped. A dwarf elliptical galaxy has an escape speed of 42 km/s. If the gas temperature is around 10^6 K, will the gas be retained in the galaxy? What does your answer tell you about star formation in the dwarf elliptical galaxy?

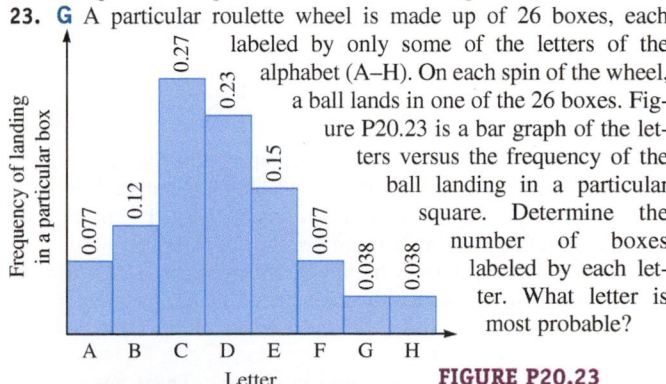

25. N CASE STUDY Because the Moon is about the same distance from the Sun as is the Earth, assume the Moon is at the same temperature as the Earth. Find the mass (in kilograms and atomic mass units) of the lightest molecule or atom that can be retained in the Moon's atmosphere. Comment on your results.

20-4 Maxwell-Boltzmann Distribution Function

26. G For the Maxwell-Boltzmann speed distribution, the most probable speed is less than the average speed. Sketch a speed distribution for which the most probable speed **a.** equals the average speed and **b.** is greater than the average speed.
27. N Consider a container of oxygen gas (O_2) at room temperature, 22°C. **a.** Find the most probable speed, average speed, and rms speed for the oxygen gas. **b.** How do you expect your results to change for nitrogen (N_2) instead of oxygen? Try it and compare your results.

Problems 28 and 29 are paired.

28. G Plot the Maxwell-Boltzmann distribution function for a gas composed of nitrogen molecules (N_2) at a temperature of 295 K. Identify the points on the curve that have a value of half the maximum value. Estimate these speeds, which represent the range of speeds most of the molecules are likely to have. The mass of a nitrogen molecule is 4.68×10^{-26} kg.
29. N Consider the Maxwell-Boltzmann distribution function plotted in Problem 28. For those parameters, determine the rms velocity and the most probable speed, as well as the values of $f(v)$ for each of these values. Compare these values with the graph in Problem 28.
30. A container of helium is at 20°C.
    a. N If you increase the temperature to 25°C, by what amount does the most probable speed increase?
    b. N If you increase the temperature to 100°C, by what amount does the most probable speed increase?
    c. C Compare your results and comment.
31. A A gas is cooled and trapped in a two-dimensional region (known as an optical lattice) by laser beams and magnets. The speed distribution function of the gas molecules obeys the two-dimensional Maxwell-Boltzmann distribution law,

$$f(v) = 2\pi\left(\frac{m}{2\pi k_B T}\right)ve^{-(mv^2/2k_BT)}$$

where m is the mass of a gas molecule, T is the temperature, and v is the speed. What is the **a.** most probable speed and **b.** average speed of the gas molecules?

20-5 Mean Free Path

32. N The Sun's surface is mainly composed of hydrogen atoms, with density $\rho = 2.1 \times 10^{-4}$ kg/m^3 and temperature 5.75×10^3 K. The radius of a hydrogen atom is 5.29×10^{-11} m. Find the **a.** mean free path and **b.** mean free time of hydrogen.
33. N A 5.00-L cylinder is filled with nitrogen gas at 22.0°C and a pressure of 75.0 atm. If the diameter of the nitrogen molecule is 1.08×10^{-10} m, what is the mean free path for nitrogen molecules in this cylinder?
34. N Galaxies are usually found in clusters. There are so many galaxies that it is common for them to pass through one another, but the stars rarely collide. A typical star has a radius of $0.63R_\odot$. The density of stars is about 0.098 star per cubic parsec (1 pc = 3.086×10^{16} m). **a.** What fraction of the galaxy's volume is occupied by stars? Use this result to help you think about your answer to part (c). **b.** Find the mean free path of a star through a galaxy. **c.** If an intruder star travels 1000 pc through a galaxy, what is the probability of a collision?
35. N Gaseous water condenses at a temperature of 100.0°C and a pressure of 1.0 atm. What is the mean free path of gaseous water molecules at this temperature? (Assume water molecules have a radius of 0.15 nm.)
36. N The mean free path of the molecules of an ideal gas under pressure P and temperature T is λ. By what factor will the mean free path increase or decrease when **a.** the pressure changes to $P/6$ while remaining at a temperature T and **b.** the temperature changes to $2T$ while the pressure changes to $2P$?
37. N A very good laboratory vacuum can evacuate the air in a chamber so that its pressure is about 10^{-12} Pa. Assume air is an ideal gas at room temperature ($T \approx 20$°C) with molecules of radius $r = 2.0 \times 10^{-10}$ m. **a.** Estimate the number density (number per unit volume) of gas molecules in the chamber. **b.** Estimate the mean free path and compare it with the mean free path at atmospheric pressure in Example 20.3.
38. E In Example 20.4, we found that it would take about 50 years to fill a neighborhood with the scent of coffee purely by diffusion. Because we know that the time is actually much shorter than 50 years, we conclude that currents help spread the aromatic molecules quickly. Estimate the time it would take for the scent of coffee to fill a home kitchen purely by diffusion. What do you conclude?
39. Monica and Kennedy are going to have a race of sorts. They measure the distance across a room as 10.0 m, declaring one end the starting line and the other the finish line. They each have a container filled with a different gas and plan to release the gases simultaneously, timing the travel of the gases across the room. (Don't worry about the details such as how they detect the gases and don't worry about chemical reactions.)
    a. C If Monica has a container of hydrogen and Kennedy has a container of oxygen, who wins the race? Explain your answer.
    b. N Assume the room's temperature is 0°C. By what factor is the one gas faster than the other?

20-6 Real Gases: The Van der Waals Equation of State

40. C You find carbon dioxide at its triple point in all three phases in a single container. Describe the phases from the bottom of the container to the top. Compare these phases to what you would find for water at its triple point.
41. E Use the value of b for oxygen from Table 20.3 (3.18×10^{-5} m^3/mol) to estimate the radius of an oxygen molecule.
42. N There are 1789 moles of carbon dioxide in a container of volume 15.75 m^3 at a temperature of 45.6°C. Can you model this situation as an ideal gas? Show your work and explain your conclusion.
43. N At very high pressures, the volume of a real gas is larger than that predicted by the ideal gas law because real molecules occupy space, which the ideal gas law neglects. At what pressure would the Van der Waals equation of state result in a volume that was 2% larger than that predicted by the ideal gas law for 1 mol of nitrogen at a temperature of 0°C? Assume $a = 0.139$ Pa·m^6/mol^2 and $b = 3.91 \times 10^{-5}$ m^3/mol.
44. C Examine the Van der Waals equation of state (Eq. 20.32):

$$\left(P + a\frac{n^2}{V^2}\right)(V - nb) = nRT$$

When the density of the confined gas is severely reduced, the Van der Waals equation is approximately equivalent to the ideal gas law ($PV = nRT$) as described in Section 20-6. Explain why that makes sense. In other words, qualitatively explain why the pressure and volume would no longer need to be modified as they are in the Van der Waals equation.

20-7 Phase Changes

45. E Figure P20.45 shows a phase diagram of carbon dioxide in terms of pressure and temperature. **a.** Use the phase diagram to explain why dry ice (solid carbon dioxide) sublimates into vapor at atmospheric pressure rather than melting into a liquid. At what temperature does the dry ice sublimate when at atmospheric pressure? **b.** Estimate what pressure would be needed to liquefy carbon dioxide at room temperature.

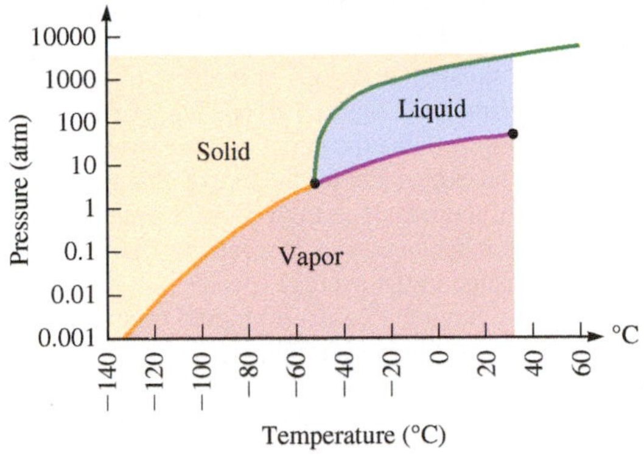

FIGURE P20.45

46. Initially, a substance has a pressure and temperature P_i and T_i.
    a. G The substance undergoes a process at constant pressure. At the end of the process, its new pressure and temperature are given by $P_f = P_i$ and $T_f > T_i$. Draw a curve on a phase diagram that represents this process.
    b. G Instead, the substance undergoes a process at constant temperature. At the end of the process, its new pressure and temperature are given by $P_f > P_i$ and $T_f = T_i$. Draw a curve on the same phase diagram that represents this process.
    c. C Come up with a general statement about how these two different processes can be recognized on a phase diagram.

Problems 47 and 48 are paired.

47. C Consider water at 1.0 atm and initially at some temperature below 0°C. Its temperature is then raised to a value above the critical temperature while the pressure is held constant. *Hint*: Consult Figure 20.17. **a.** Describe the phases that water passes through during this process. **b.** If the curve between the solid phase and the liquid phase sloped upward and to the right as it does for carbon dioxide, how would your answer to part (a) change?
48. C Consider water at 0°C and initially at some pressure just below the critical pressure. Its pressure is then lowered to a value near zero while the temperature is held constant. *Hint*: Consult Figure 20.17. **a.** Describe the phases that water passes through during this process. **b.** If the curve between the solid phase and the liquid phase sloped upward and to the right as it does for carbon dioxide, how would your answer to part (a) change?

20-8 Evaporation

49. N At high altitude, water boils at 95°C. What is the external pressure at that altitude?
50. C On a humid day, would you expect the dew point to be particularly high or particularly low? Explain.
51. N In a pressure cooker, water boils at 120°C. What is the pressure inside the cooker?
52. Minneapolis, Minnesota, is famous for its cold winters. On a typical January day, the high temperature is 21°F, and the low temperature is 3°F. The dew point is 5°F.
    a. N Calculate the high relative humidity and the low relative humidity.
    b. C Does dew form? Explain your answer.

53. N On a certain day in Orlando, Florida, the temperature of the air and dew point were 16°C and 7.6°C, respectively. Find the relative humidity of the air on that day. Assume the saturated vapor pressure of water at 7°C, 8°C, and 16°C are 7.5 mm Hg, 8.0 mm Hg, and 12.0 mm Hg, respectively.
54. N On a summer day, the temperature is 34.5°C, and the partial pressure of water vapor in air is 2.75×10^3 Pa. What is the relative humidity?

General Problems

55. N A container is filled with a gas at a temperature of 300 K such that its pressure is 2.02×10^5 Pa. If the average kinetic energy of a molecule in the gas is to be **a.** tripled and **b.** halved, by what factor must the rms speed decrease or increase? In each case, state whether the rms speed decreases or increases.

Problems 56 and 57 are paired.

56. A A box with volume V confines 18 particles with identical mass m and the following speeds: three particles with speed v, one particle with speed $2v$, four particles with speed $3v$, three particles with speed $4v$, five particles with speed $5v$, and two particles with speed $6v$. What are the **a.** most probable speed, **b.** average speed, and **c.** rms speed of the particles in this box?
57. A Consider again the box and particles with the speed distribution described in Problem 56. **a.** What is the average pressure exerted by the particles on the walls of the box? **b.** What is the average kinetic energy per particle in this box?
58. N The average translational kinetic energy and the rms speed of molecules in a container of oxygen at a particular temperature are 6.20×10^{-21} J and 434 m/s, respectively. What are the corresponding values when the temperature is doubled?
59. N The average kinetic energy of an argon atom in a spherical container filled with argon gas and with a volume of 1.00×10^{-2} m³ is 4.50×10^{-21} J. If the pressure inside the container is 2.00 atm, how many moles of argon are in the container?
60. N For the exam scores given in Table P20.60, find the average score and the rms score.

TABLE P20.60

Number	Score	Number	Score
1	67	19	72
2	92	20	86
3	88	21	77
4	82	22	76
5	76	23	80
6	78	24	89
7	22	25	75
8	81	26	77
9	77	27	92
10	70	28	84
11	100	29	66
12	89	30	61
13	74	31	90
14	63	32	73
15	94	33	79
16	55	34	72
17	83	35	89
18	75		

61. **N** The total translational kinetic energy of hydrogen molecules contained in a sealed 10.0-L container is 7.5×10^3 J. Determine the pressure of the gas molecules. *Hint*: See Equation 21.2, $NK_{av} = \frac{3}{2}Nk_BT$.

Problems 62 and 63 are paired.

62. **N** A 0.500-m^3 container is filled with 30.0 mol of carbon dioxide gas (CO_2) such that its pressure is 2.00 atm. What is the rms speed of the molecules in the gas?

63. **N** Consider the carbon dioxide gas in Problem 62. Suppose the temperature of the gas is tripled. **a.** What is the new rms speed of the molecules in the gas? **b.** If one end of the container has an area of 0.0325 m^2, what will be the average force exerted on that end by the carbon dioxide?

64. **N** A spherical balloon with a radius of 10.0 cm is filled with oxygen gas at atmospheric pressure and at a temperature of 15.0°C. What are the **a.** number of molecules, **b.** average kinetic energy, and **c.** rms speed of oxygen in the balloon?

65. **N** A lifeguard pours chlorine into a pool and assumes it will eventually diffuse throughout the pool. Assuming diffusion alone is acting and the diffusion constant of chlorine in water is 2×10^{-9} m^2/s, how long would it take for the chlorine to diffuse from one end of the pool across the 100-m length to the other side?

66. **N** At what temperature will the mean free path of the molecules of an ideal gas be twice that at 27°C, if the pressure is kept constant but the volume changes?

67. **N** Determine the rms speed of an atom in a helium balloon at standard temperature and pressure ($T = 273.15$ K and $P = 1.00 \times 10^5$ Pa). Is it lower or higher than the rms speed of a nitrogen molecule in the atmosphere?

Problems 68 and 69 are paired.

68. **C** Consider a gas filling two connected chambers that are separated by a removable barrier (Fig. P20.68). The gas molecules on the left (red) are initially at a higher temperature than the ones on the right (blue). When the barrier between the two chambers is removed, the molecules begin to mix and move from one chamber to the other. **a.** Describe what happens to the temperature in the left chamber and in the right chamber as time goes on, once the barrier is open. Discuss in terms of the mixing of the molecules from each gas. **b.** Describe what happens to the most probable speed and average speed in the left chamber and in the right chamber as time goes on, once the barrier is open. Do they increase or decrease by the same factor? Explain.

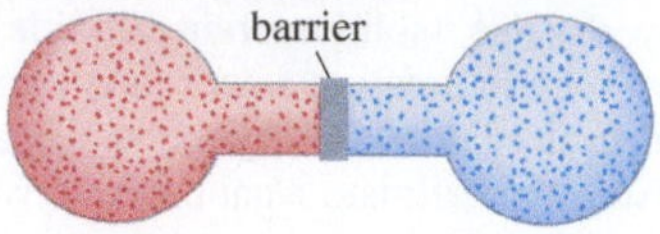

FIGURE P20.68 Problems 68 and 69.

69. Consider the situation described in Problem 68 and focus on the right chamber shown in Figure P20.68. Suppose the gas filling the chambers is O_2. Assume the initial average speed is 900.0 m/s.

a. **A** Derive an expression for the rate of change of temperature in terms of the rate of change of the average speed.

b. **N** If the temperature rises at a rate of 2.00 K/s, what is the rate of change of the average speed?

70. **N** The relative humidity of air in a closed room at 20°C is 50%. What will be the relative humidity of air if the temperature is increased to 25°C? The saturated vapor pressure of water at 20°C is 17.4 mm of mercury (Hg) and at 25°C is 23.6 mm of Hg.

Problems 71 and 72 are paired.

71. **N** A 0.500-m^3 container holding 3.00 mol of ozone (O_3) is kept at a temperature of 250 K. Assume the molecules have radius $r = 2.50 \times 10^{-10}$ m. What are the **a.** mean free path and **b.** mean free time between collisions for an ozone molecule in the container?

72. Consider the gas described in Problem 71.

a. **N** What is the average distance between molecules?

b. **C** Compare the mean free path found in Problem 71 with your answer to part (a) of this problem. Comment and explain the differences you find in comparing these quantities.

73. **N** CASE STUDY Although the nebula from which our solar system formed was comprised mostly of hydrogen and helium, the terrestrial planets including the Earth and Mars quickly lost these gases from their atmospheres. The mass of a helium atom is 6.64×10^{-27} kg. What is the temperature at which the average speed of helium atoms exceeds the escape speed from **a.** the Earth (11.2 km/s) and **b.** Mars (5.00 km/s)? **c.** If the average temperature of air at sea level is 20.0°C, how is it possible for helium to escape from the Earth's atmosphere?

Heat and the First Law of Thermodynamics

21

Underlying Principles

First law of thermodynamics

Major Concepts

1. Heat
2. Thermal energy
3. Heat capacity
4. Specific heat
5. Thermodynamic process
6. Degrees of freedom
7. Principle of equipartition of energy

Special Cases

Six specific thermodynamic processes:

1. Adiabatic
2. Isothermal
3. Constant volume
4. Constant pressure
5. Cyclic
6. Free expansion

Three specific ways for heat to flow:

1. Conduction
2. Convection
3. Radiation

Tools

Energy bar charts

Key Questions

What is heat?

How can heat be incorporated into the principle of conservation of energy?

What are some examples of processes that involve heat?

During spring break, you go to the beach with your friends. One afternoon, you take your physics textbook and a cold drink so that you can study in the Sun. You forget your cooler, though, so your cold drink doesn't stay cold very long. The drink's increased temperature is a clear indication that its thermal energy increased, but where did that energy come from? Of course, the answer is shining down on you: the Sun. How could the Sun, which is more than 93 million miles away, transfer energy to your drink? From Chapter 9, energy can be transferred to a system if the environment does work on that system. The drink was not displaced, so clearly nothing did any work on it. Therefore, the energy must have been transferred by other means. This other means of transferring energy is known as *heat*. The Sun transfers energy to the rest of the solar system through heat. If you had put your drink in the cooler, your drink would have been fairly well insulated and unable to absorb very much energy. The next time you forget your cooler, try wrapping your drink in a towel.

21-1 What Is Heat?

When you hold your cold hands in front of a nice hot campfire (Fig. 21.1), they warm up because energy is transferred from the environment (the fire) to the system (your hands). **Heat** is the energy transferred from the environment to the system (or from the system to the environment) due to their temperature difference. The term *heat* is often misused and misunderstood.

HEAT ✪ Major Concept

Yuri Arcurs/Shutterstock.com

FIGURE 21.1 Holding your hands near a campfire is one way to warm them.

Historical theories about heat contribute to the misunderstanding of that term. In the late 1700s, many scientists believed that heat was a fluid called "caloric." Caloric was believed to flow from hot objects to cooler objects, much like lemonade flowing from a pitcher into a glass. An 18th-century scientist, for example, might have said that when you put a cold potato in a hot oven, caloric flowed from the oven to the potato. The oven would have less caloric in it, and the potato would gain caloric. Today, we often use the same sort of language for heat. We might say that heat flows from the oven to the potato, but this wording gives the wrong impression because *heat is not a substance*, and it does not flow from the oven to the potato in the same way that liquids flow from one container into another.

By our contemporary definitions, neither heat nor work describes the energy contained *in* a system (or in the environment). Instead heat and work only describe energy that is transferred between a system and its environment. Work describes the transfer of energy due to forces between macroscopic objects. Heat is work on the microscopic scale. Again, imagine a cool potato surrounded by hot air molecules in the oven. Because the air in the oven is hotter than the potato, the air molecules have a greater average kinetic energy than the potato molecules. Therefore, during a collision, it is more likely that an air molecule will lose kinetic energy and a potato molecule will gain energy. So, on the whole, thermal energy is transferred from the hot air to the potato, and we say that *heat flows from the oven to the potato*. The wording is the same as above, but now the phrase means that the hot air in the oven environment did work on the potato at the microscopic level, transferring energy.

James Joule described his experiments in a paper, *On the Mechanical Equivalent of Heat*, but the Royal Society of London did not allow him to include his conclusion about heat. Joule worked on the establishment of units involved in thermodynamics, and today's SI unit for energy is named in his honor.

If the potato's temperature is equal to the temperature of the air in the oven, equal numbers of air molecules lose energy and gain energy when they collide with the potato molecules. So, when the potato is in thermal equilibrium with the oven, there is no heat flow, meaning that there is no net work done on the microscopic level.

Caloric fluid was never detected, and by the 19th century, other models for heat were pursued. James Joule (1818–1889), an English scientist and brewer, conducted experiments leading to the idea that heat is work on the microscopic level. Usually, we cannot directly observe the kinetic and potential energies of microscopic particles. Instead, we observe macroscopic properties such as temperature. If the temperature of a system rises, we can infer that energy was transferred to the system from the environment either by heat or by work.

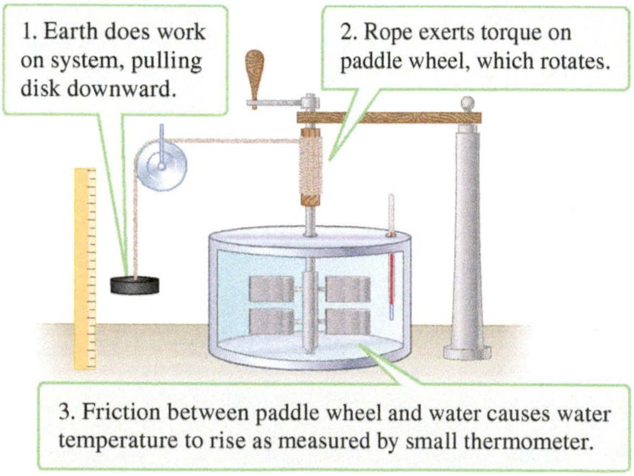
1. Earth does work on system, pulling disk downward.
2. Rope exerts torque on paddle wheel, which rotates.
3. Friction between paddle wheel and water causes water temperature to rise as measured by small thermometer.

FIGURE 21.2 Joule's apparatus. The system consists of the device, including the water in the can. The Earth—which is not in the system—does work on the system as the disk falls. The falling disk is connected by a rope to a paddle wheel. The paddle wheel rotates in the can filled with water, and the water's temperature rises.

Properties such as temperature that describe the condition or state of the system (and the environment) are called **state variables**. Energy may be transferred through work or through heat. If energy is transferred between the environment and the system, one or more of these state variables will change, and we say that "the state of the system changes."

Joule focused on changing one state variable—temperature—of a system consisting of a can of water. He found that the water temperature could be raised by two different methods. As every cook knows, if Joule placed the water on a hot stove, the water's temperature increased due to heat from the environment. What is very interesting is that he could also raise the temperature of water by allowing gravity to do macroscopic work on the system. Joule's system included a can of water and a paddle wheel attached to a hanging disk, but not the Earth (Fig. 21.2). The Earth exerts a downward force on the hanging disk, pulling it downward. The Earth does positive work on the system, transferring energy to it and ultimately causing the water temperature to rise. Joule's experiment showed that heat and work are both ways to transfer energy between a system and its environment.

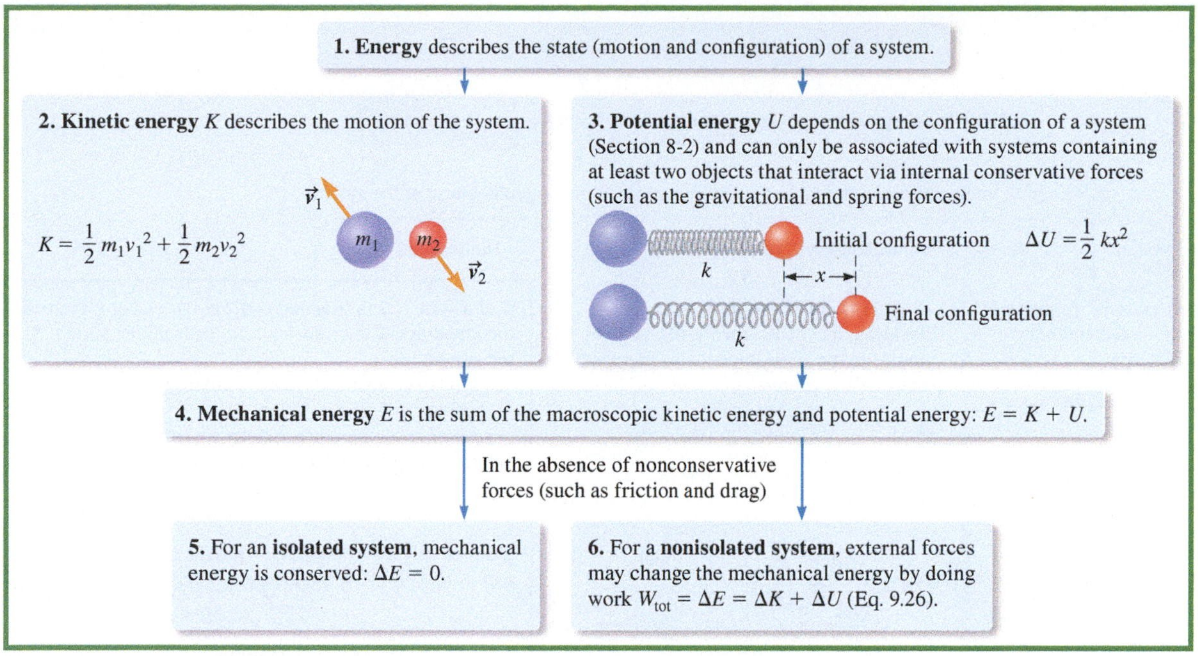

FIGURE 21.3 This concept map is a review of energy concepts from Chapters 8 and 9.

21-2 How Does Heat Fit into the Conservation of Energy?

In this chapter, we extend the concepts of energy conservation and the work–energy theorem from Chapters 8 and 9 to include heat. Review this material by studying Figure 21.3 to set heat and the first law of thermodynamics (Section 21-3) in context. Pay particular attention to when energy is transferred between a nonisolated system and its environment (panel 6 in Fig. 21.3). The total work W_{tot} done on a system by external forces in the absence of nonconservative forces (such as friction and drag) changes the system's mechanical energy by the work–mechanical energy theorem:

$$W_{tot} = \Delta E = \Delta K + \Delta U \tag{9.26}$$

Thermal Energy, Work, and Heat

The concepts of work and energy (Fig. 21.3) are convenient when studying a macroscopic system consisting of a small number of objects that are well modeled as particles (such as a tossed ball and the Earth). When the objects in the system are not well modeled as particles or when there are many particles in the system, though, we must modify and add to these concepts.

Figure 21.4 maps how we think about energy in the case of a system consisting of deformable objects or a large number of particles. The following are key points corresponding to Figure 21.4.

1. Energy still describes the state of the system.
2. Kinetic energy and potential energy are defined as in Figure 21.3, but those terms are reserved for the macroscopic state of the system. For example, in a system consisting of a hamburger at rest in a frying pan, we would say that the hamburger's kinetic energy is zero because the hamburger is at rest (in our frame). Microscopically, however, molecules within the hamburger are in motion.
3. Molecular motion is accounted for by **thermal energy** (Section 9-7), the sum of each microscopic particle's kinetic energy plus the total potential energy stored in the forces between microscopic particles. In the absence of phase changes, thermal energy depends on the temperature of the system.

THERMAL ENERGY
Major Concept

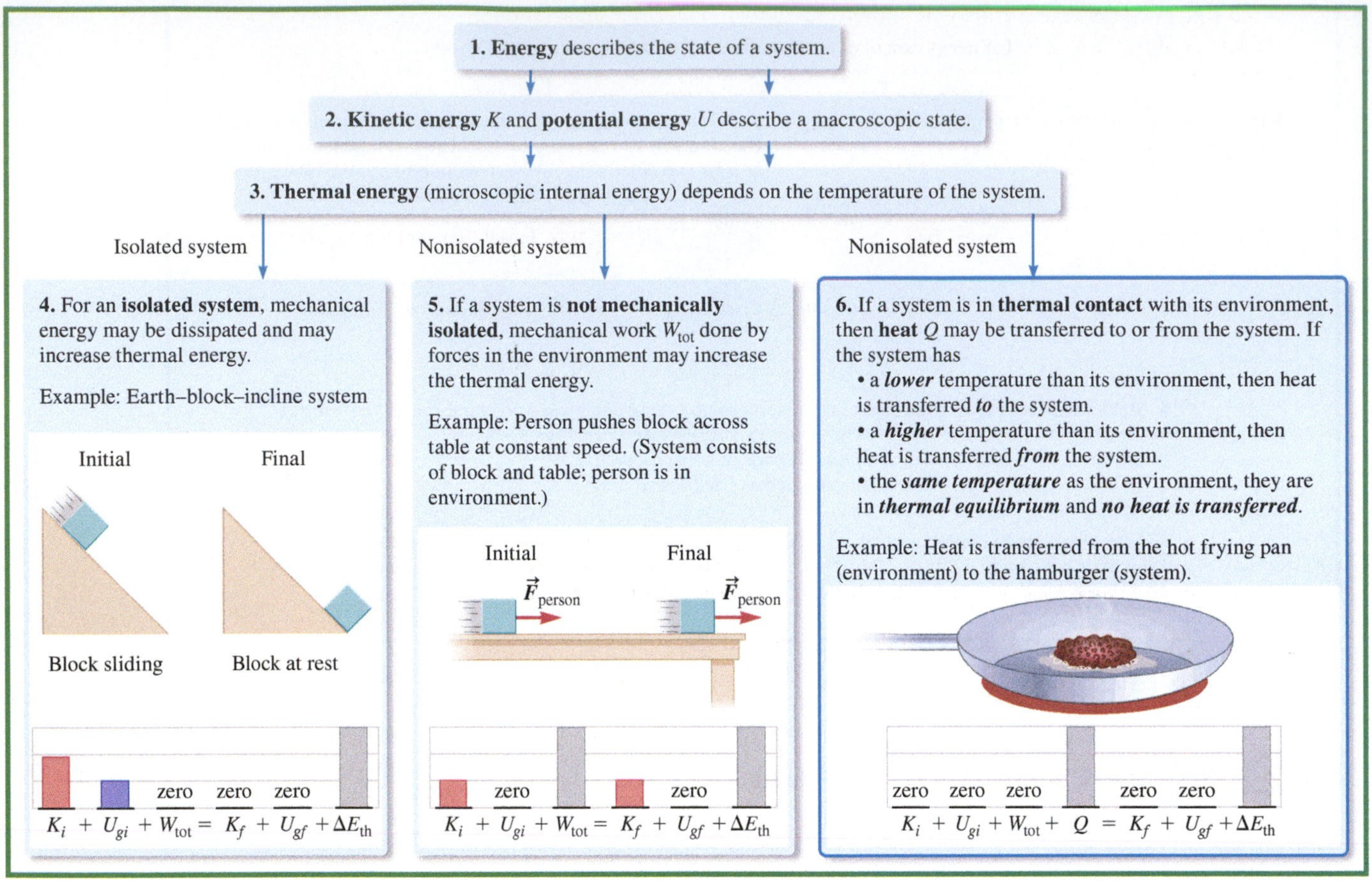
1. Energy describes the state of a system.
2. Kinetic energy K and potential energy U describe a macroscopic state.
3. Thermal energy (microscopic internal energy) depends on the temperature of the system.
Isolated system
Nonisolated system
Nonisolated system
4. For an isolated system, mechanical energy may be dissipated and may increase thermal energy.
Example: Earth–block–incline system
Initial
Final
Block sliding
Block at rest
zero zero zero
$K_i + U_{gi} + W_{tot} = K_f + U_{gf} + \Delta E_{th}$
5. If a system is not mechanically isolated, mechanical work W_{tot} done by forces in the environment may increase the thermal energy.
Example: Person pushes block across table at constant speed. (System consists of block and table; person is in environment.)
Initial
Final
$\vec{F}_{person}$
$\vec{F}_{person}$
zero zero
$K_i + U_{gi} + W_{tot} = K_f + U_{gf} + \Delta E_{th}$
6. If a system is in thermal contact with its environment, then heat Q may be transferred to or from the system. If the system has
• a lower temperature than its environment, then heat is transferred to the system.
• a higher temperature than its environment, then heat is transferred from the system.
• the same temperature as the environment, they are in thermal equilibrium and no heat is transferred.
Example: Heat is transferred from the hot frying pan (environment) to the hamburger (system).
zero zero zero zero zero
$K_i + U_{gi} + W_{tot} + Q = K_f + U_{gf} + \Delta E_{th}$

FIGURE 21.4 This concept map shows how heat (panel 6) fits into the concepts of energy.

Any changes in thermal energy depend on what contact the system has with its environment.

4. If a system is **isolated** from its environment, no energy may be exchanged. In this case, the system's mechanical energy may be transformed into thermal energy if dissipative forces such as kinetic friction are present. When a block slides down a rough, inclined plane, mechanical energy is transformed into thermal energy, and the block and incline are both warmer.
5. If the system is not mechanically isolated from its environment, the environment may do work on the system, and that work can alter the system's thermal energy. When a person (who is in the environment) pushes the block across a table at constant velocity, the positive work done by the person increases the block–table system's thermal energy. The block and the table are warmer.

Methods 4 and 5 of changing the thermal energy were discussed in Chapter 9. We now formally introduce a third way for thermal energy to change as described in the blue outlined box (panel 6) in Figure 21.4.

HEAT ✪ Major Concept

6. If the system is in **thermal contact** with its environment, energy may be transferred by heat. **Heat** Q is the amount of energy transferred between a system and its environment due to their difference in temperature. Consider a system consisting of a cool hamburger at rest in a hot frying pan. The frying pan is in the environment. The hamburger remains at rest, so its kinetic energy remains zero. The system consists only of the hamburger, so on the *macroscopic* scale there are no internal forces. No external forces do work on the hamburger. Energy (heat) is transferred from the hot frying pan to the hamburger, increasing the hamburger's thermal energy. We need to add a bar for heat Q to the bar chart (panel 6). Because heat is work on the *microscopic* level, we place Q next to W_{tot}. In the case of a hamburger in a hot frying pan, the bar for heat Q is the

same size as the bar for the change in thermal energy ΔE_{th}. In a sense, the next section is about how adding a bar for heat changes the conservation of energy principle.

CONCEPT EXERCISE 21.1

Decide which of the following statements are incorrect according to the definition of *heat* used in physics. Explain your answers and provide corrections where possible.

a. "The heat in Texas will make you tired."
b. "When you slide a book across a table, kinetic friction converts kinetic energy into heat."
c. From a science textbook: "The steam engine marked the first time in human history that heat was used to do work."

21-3 The First Law of Thermodynamics

FIRST LAW OF THERMODYNAMICS
❶ Underlying Principle

The first law of thermodynamics is both an extension and a special case of the work–energy theorem. In Equation 9.30, we wrote the work–energy theorem as

$$\begin{pmatrix}\text{initial}\\ \text{mechanical}\\ \text{energy}\end{pmatrix} + \begin{pmatrix}\text{energy}\\ \text{transferred}\\ \text{(total work)}\end{pmatrix} = \begin{pmatrix}\text{final}\\ \text{mechanical}\\ \text{energy}\end{pmatrix} + \begin{pmatrix}\text{change in}\\ \text{internal}\\ \text{energy}\end{pmatrix}$$

First, we broaden the work–energy theorem to include the idea that energy may be transferred by heat:

$$\begin{pmatrix}\text{initial}\\ \text{mechanical}\\ \text{energy}\end{pmatrix} + \begin{pmatrix}\text{energy}\\ \text{transferred}\\ \text{(total work)}\end{pmatrix} + \begin{pmatrix}\text{energy}\\ \text{transferred}\\ \text{(heat)}\end{pmatrix} = \begin{pmatrix}\text{final}\\ \text{mechanical}\\ \text{energy}\end{pmatrix} + \begin{pmatrix}\text{change in}\\ \text{internal}\\ \text{energy}\end{pmatrix}$$

This equation is a very general expression for the conservation of energy in a system because it accounts for the energy transferred *by work* and *through heat*.

Our next step is to write this general expression for a special case. Let's start by taking a close look at the last term, the change in internal energy. Of course, mechanical energy is "internal" to the system, but the term ***internal energy*** refers to *microscopic* internal energy. Internal energy is not a new form of energy; it is potential energy and kinetic energy on the microscopic level. For convenience, internal energy is further divided into other categories such as thermal energy, chemical energy, and nuclear energy. Throughout this textbook, our focus is primarily on thermal energy. So, as we did in Section 9-8, we write this last term for the special case in which the only category of internal energy we deal with is thermal energy:

$$\begin{pmatrix}\text{initial}\\ \text{mechanical}\\ \text{energy}\end{pmatrix} + \begin{pmatrix}\text{energy}\\ \text{transferred}\\ \text{(total work)}\end{pmatrix} + \begin{pmatrix}\text{energy}\\ \text{transferred}\\ \text{(heat)}\end{pmatrix} = \begin{pmatrix}\text{final}\\ \text{mechanical}\\ \text{energy}\end{pmatrix} + \begin{pmatrix}\text{change in}\\ \text{thermal}\\ \text{energy}\end{pmatrix}$$

Our final step is to write the conservation of energy for the special case in which the mechanical energy of the system is constant. In this case, the initial mechanical energy equals the final mechanical energy, and we have:

$$\begin{pmatrix}\text{energy}\\ \text{transferred}\\ \text{(total work)}\end{pmatrix} + \begin{pmatrix}\text{energy}\\ \text{transferred}\\ \text{(heat)}\end{pmatrix} = \begin{pmatrix}\text{change in}\\ \text{thermal}\\ \text{energy}\end{pmatrix}$$

Using symbols, we have the **first law of thermodynamics**:

$$W_{tot} + Q = \Delta E_{th} \tag{21.1}$$

The sign convention used throughout this book is that *if energy is transferred from the environment to the system, the sign of W_{tot} or Q is positive.*

The first law is a statement of the conservation of energy for a system that may exchange energy with its environment by work or through heat, and the only form of internal energy that may change is thermal energy. For the rest of this chapter and throughout Chapter 22, we consider systems whose energy changes are completely described by the first law of thermodynamics.

If a system is **mechanically isolated** (or simply **isolated**), energy cannot be transferred between the system and its environment by work. Likewise, if a system is **thermally insulated** (or simply **insulated**), energy cannot be transferred between the system and its environment through heat.

Blend Images/Stockphoto.com

FIGURE 21.5 This case study is about you, how you exchange energy with your environment and how you maintain your healthy body temperature. Of course, it all begins with you getting energy from the food you eat.

CASE STUDY Part 1: You

Every day, you eat food that supplies your body with chemical energy (Fig. 21.5). If you are a typical adult, you eat about 2400 Cal (1.00×10^7 J) per day. Some of this energy is transformed into macroscopic kinetic energy that allows you to walk, allows you to scan your eyes across this page, and keeps your heart beating.

The *calorie*, a commonly used unit of energy, comes from the word *caloric*. In the United States, the food energy used by the human body is reported in Calories (with an uppercase C), where 1 Calorie = 1000 calories. In European countries, the energy content of food is reported in kilocalories (kcal), where 1 kcal = 1 Calorie.

Most of the time, your body maintains its healthy temperature of 37°C. To regulate your temperature, your body must exchange heat with the environment. Throughout this chapter, we consider the human body in terms of the first law of thermodynamics. We'll see how forensic experts can use the temperature of a corpse to estimate the time of death, how much work you do just by breathing, and how you maintain your body temperature in different environments such as deserts, oceans, and space.

CONCEPT EXERCISE 21.2

In each situation listed, an object's temperature increases. For each situation, decide if energy was transferred to the object through heat or work. Explain your answers.

a. A pitcher of water is removed from the refrigerator and left on the kitchen counter.
b. A rubber ball is bounced (repeatedly) against the floor.
c. A meteor falls through the Earth's atmosphere.

21-4 Heat Capacity and Specific Heat

There is a close connection between the temperature of a system and its thermal energy. To see that connection, consider the special case of an ideal gas whose molecules may be modeled as particles that do not exert any forces on one another (except during a collision). On the microscopic level, the gas has kinetic energy but no potential energy. Therefore, the thermal energy of an ideal gas is just the sum of the kinetic energies of its particles. The average kinetic energy of the particles is given by $K_{av} = \frac{3}{2}k_B T$ (Eq. 20.17). The thermal energy of an ideal gas (total kinetic energy of the particles) is found by multiplying the average kinetic energy by the number N of particles:

$$E_{th} = NK_{av} = \frac{3}{2}Nk_B T = \frac{3}{2}nRT \tag{21.2}$$

where $Nk_B = nR$ (Eqs. 19.18 and 19.20).

The term *heat capacity* suggests that it has the same dimensions as heat and that a system has a capacity to "hold" heat. Neither of these impressions is true, so it is best to think of heat capacity as a name for the constant of proportionality in Equation 21.3.

According to Equation 21.2, we can observe a change in thermal energy by just measuring the system's temperature change. (This technique holds not just for ideal gases, but for other systems as well.) As long as a system is not undergoing a phase change, a change in the system's thermal energy is proportional to a change in the system's temperature:

$$\Delta E_{th} = \mathcal{C}\Delta T \tag{21.3}$$

HEAT CAPACITY ✪ Major Concept

The constant $\mathcal{C}$ has the dimensions of energy per temperature (as does Boltzmann's constant k_B in Eq. 21.2). Unfortunately, $\mathcal{C}$ was named when people still thought of heat as a substance, and it is called **heat capacity**.

Let's consider situations in which heat is exchanged but no work is done (such as a hamburger in a hot frying pan). We start with systems that are solids or liquids and will return to gases in Section 21-8. When no work is done on the system, the first law of thermodynamics becomes

$$\Delta E_{th} = Q$$

So, we can write Equation 21.3 as

$$Q = \mathcal{C}\Delta T \tag{21.4}$$

Heat capacity $\mathcal{C}$ depends on the *type* and the *amount* of material making up the system.

For convenience, we separate heat capacity's dependence on the *amount* of material from its dependence on the *type* of material, so we write Equation 21.4 as

$$Q = mc\Delta T \tag{21.5}$$

where $\mathcal{C} = mc$. The amount of material is expressed in terms of the system's mass m, and the constant c is known as the **mass specific heat capacity** or simply the **specific heat**. The specific heat is the heat capacity of a material per unit mass. The dimensions of specific heat are energy per mass per temperature, so its SI units are J/(kg · K).

SPECIFIC HEAT Major Concept

As in Section 19-6, it is often convenient to express the amount of a substance in terms of the number of moles, so we define the **molar specific heat capacity** or just **molar specific heat** as the heat capacity of a material per unit mole. It has the SI units J/(mol · K). We write Equation 21.4 in terms of the molar specific heat C:

$$Q = nC\Delta T \tag{21.6}$$

where the amount of substance is expressed as the number of moles n. The heat capacity $\mathcal{C} = nC$. Table 21.1 provides the specific heat and molar specific heat for a number of different solids and liquids.

In this textbook, heat capacity is denoted by an uppercase script $\mathcal{C}$, specific heat by a lowercase c, and molar specific heat by an uppercase C.

The first law of thermodynamics (Eq. 21.1) states that when energy is transferred to or from a system (by work or heat), a state variable (thermal energy) changes. Heat capacity and specific heat allow us to rewrite the first law of thermodynamics in terms of a change in temperature. First, substitute heat capacity (Eq. 21.3):

$$W_{tot} + Q = \Delta E_{th} = \mathcal{C}\Delta T$$

Then, eliminate heat capacity $\mathcal{C}$ and rewrite this equation in terms of the specific heat c ($\mathcal{C} = mc$):

$$\Delta E_{th} = W_{tot} + Q = mc\Delta T \tag{21.7}$$

Because a change in kelvin temperature is numerically the same on the Celsius scale, specific heat expressed as J/(kg · °C) is numerically equivalent to J/(kg · K).

Let's see how Equation 21.7 fits in with Joule's experiment (Fig. 21.2). In Joule's experiment, no heat was transferred to the system (a container of water), so $Q = 0$.

TABLE 21.1 Specific heat and molar specific heat for solids and liquids near room temperature and at atmospheric pressure (except where noted).

Substance	Specific heat c [J/(kg · K)]	Molar specific heat C [J/(mol · K)]	Substance	Specific heat c [J/(kg · K)]	Molar specific heat C [J/(mol · K)]
Aluminum	900	24.4	Lead	128	26.5
Beef, flank	2340		Marble	860	
Brass	380		Mercury	140	28.1
Cabbage	3940		Milk	3930	
Copper	386	24.5	Seawater	3900	
Ethyl alcohol	2430	110.4	Silver	230	25.5
Glass	840		Water (ice at −5°C)	2100	37.6
Gold	128	25.6	Water (liquid)	4190	75.4
Granite	790		Wood	1700	
Iron or steel	450	25.1			

Note: Specific heat and molar specific heat depend on temperature, but for small changes in temperature, they may be considered constant.

The Earth did work on the system, however; so, for Joule's experiment, Equation 21.7 is

$$W_{tot} = mc\Delta T \quad (21.8)$$

When Joule placed the water on a hot stove, no work was done ($W_{tot} = 0$), so $Q = mc\,\Delta T$. Joule's experiment showed that energy transferred by work (Eq. 21.8) or by heat (Eq. 21.5) raises the temperature of the water, leading to the idea that heat is work on the microscopic level.

CONCEPT EXERCISE 21.3

Come up with your own terms to replace *heat capacity*, *specific heat*, and *molar specific heat*.

CONCEPT EXERCISE 21.4

Suppose you double the number of particles in an ideal gas without changing the gas temperature. What happens to the average kinetic energy of the gas particles? What happens to the thermal energy?

EXAMPLE 21.1 Joule's Experiment, or Another Way to Make a Pot of Tea

Suppose you wish to make a liter of hot tea by using Joule's equipment (Fig. 21.2). The density of water is 1 kg/L, so 1 L of water has a mass of 1 kg. The specific heat of water is listed in Table 21.1.

A Start with a more modest goal of raising the water temperature just 1 K. If the hanging disk has a mass of 5.00 kg, how far must it fall to raise the temperature of 1.00 L of water 1.00 K? If you imagine building a tower from which to drop the disk, this distance would be the minimum required height for the tower.

INTERPRET and ANTICIPATE

The system consists of Joule's apparatus, including 1.00 kg of water in the can. The Earth is in the environment, and it does work on the system as the hanging disk falls. From the amount of work required to raise the water 1.00 K, find the distance through which the disk must fall.

SOLVE

Using the specific heat of liquid water from Table 21.1, find the work required.

$$W_{req} = mc\Delta T \quad (21.8)$$

$$W_{req} = (1.00\text{ kg})\left(4190\frac{\text{J}}{\text{kg}\cdot\text{K}}\right)(1\text{ K}) = 4190\text{J}$$

The work done by gravity on the falling disk is the weight of the disk multiplied by its displacement (Eq. 9.15). There are no other external forces doing work on the system.

$$W_{tot} = \vec{F}\cdot\Delta\vec{y} = mg\,\Delta y$$

$$\Delta y = \frac{W_{tot}}{mg}$$

If all the work done by gravity goes into heating the water, we can set the required work W_{req} equal to the work done by gravity W_{tot}. Joule's system may not be 100% efficient, and some of the work done by gravity may go into other parts of the system. (For example, kinetic friction may raise the temperature of the rope, pulley, and paddle wheel.) So, our result is a lower limit to how far the disk must fall.

$$\Delta y = \frac{W_{tot}}{mg} = \frac{4190\text{ J}}{(5.00\text{ kg})(9.81\text{ m/s}^2)}$$

$$\Delta y = 85.4\text{ m}$$

CHECK and THINK

If Joule used a 5-kg weight, he would have needed to drop the weight off a tower roughly 30 stories tall to see a rise in temperature of 1 K (= 1°C). Of course, he could reduce the height needed by increasing the mass of the disk. Perhaps by using a strong rope, he could replace the 5-kg disk with a 25-kg disk; the tower would then only need to be 17 m or about 6 stories tall.

B Replace the 5-kg disk with a 25-kg disk. If the water is initially at room temperature (22°C), how far must the disk fall to raise the water temperature to the boiling point? (Tea should be made with boiling water.)

INTERPRET and ANTICIPATE

We need only two modifications to our work in part A. First, calculate the work required to raise the water temperature from 22°C to 100°C. Then recalculate the work done by gravity on the more massive disk.

SOLVE The difference in temperature on the Celsius scale is the same as that on the Kelvin scale. Find the work required to raise the 1 kg of water to the new temperature.	$W_{req} = mc\Delta T$ (21.8) $W_{req} = (1\text{ kg})\left(4190\dfrac{\text{J}}{\text{kg}\cdot\text{K}}\right)(78\text{ K}) = 3.3 \times 10^5\text{ J}$
As before, assume all the work done by gravity goes into heating the water.	$\Delta y = \dfrac{W_{tot}}{mg} = \dfrac{3.3 \times 10^5\text{ J}}{(25\text{ kg})(9.81\text{ m/s}^2)} = 1.3 \times 10^3\text{ m}$

CHECK and THINK

To make a pot of tea using Joule's equipment, we would need a tower nearly 450 stories tall. Clearly, it is much more convenient to use a kitchen stove.

EXAMPLE 21.2 A Calorimeter

A water calorimeter is a device used to measure the specific heat of a substance (Fig. 21.6). A well-insulated container of mass m_c is filled with water of mass m_w. The specific heats of water c_w and the container c_c are known. A thermometer fits through the lid of the container, and very little (if any) heat can pass through the container. Suppose a hot sample of the substance is placed in the water. Eventually, the sample and the water are in thermal equilibrium. The mass of the sample is m_s. By measuring the change in the sample's temperature ΔT_s and the change in the water's temperature ΔT_w, the sample's specific heat c_s can be calculated. Find an expression for the sample's specific heat.

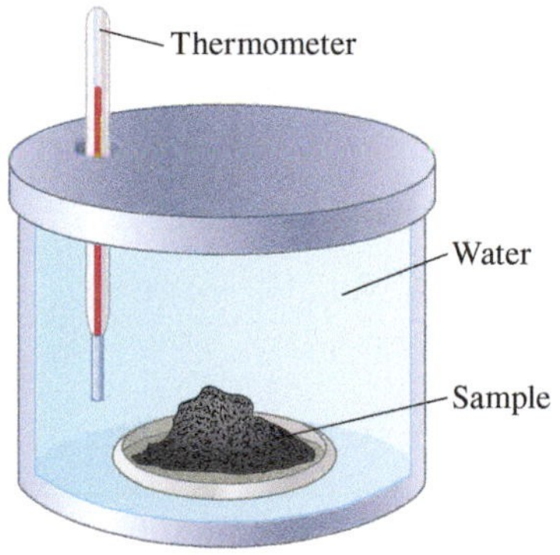

FIGURE 21.6 A water calorimeter is used to find the specific heat of a sample of some material.

INTERPRET and ANTICIPATE

Because the container is insulated, any energy lost by the sample must be gained by the water and its container. Mathematically, we write $-\Delta E_s = \Delta E_w + \Delta E_c$, where *s*, *w*, and *c* stand for *s*ample, *w*ater, and *c*alorimeter, respectively. We expect that if the sample has a relatively large change in temperature ΔT_s, it must have a low specific heat.

SOLVE Use the first law of thermodynamics (Eq. 21.7) to express changes in thermal energy in terms of changes in temperature. The water is in thermal equilibrium with its container, so they both have the same change in temperature ΔT_w.	$\Delta E_{th} = mc\,\Delta T$ (21.7) $-\Delta E_s = \Delta E_w + \Delta E_c$ $-m_s c_s\,\Delta T_s = m_w c_w\,\Delta T_w + m_c c_c\,\Delta T_w$
Solve for c_s.	$c_s = -\dfrac{(m_w c_w + m_c c_c)\,\Delta T_w}{m_s\,\Delta T_s}$ (1)

CHECK and THINK

First, because the dimensions of mass and temperature cancel, we see that our expression has the correct dimensions for specific heat. Second, our expression also has the correct sign. We expect the water temperature to increase and the sample's temperature to decrease. The minus sign in front of the expression ensures that $c_s > 0$, as expected. Finally, as predicted, if ΔT_s is large, then c_s is small.

EXAMPLE 21.3 CASE STUDY Time of Death

You have probably seen a movie where someone is found dead and the "lab report" gives the time of death. Have you ever wondered how forensic scientists can really determine the time of death? There are many methods, but one simple way is based on the core temperature of the corpse.

A living male adult consumes about 2400 Cal (1.00×10^7 J) per day. So a man extracts on average about 116 J/s from his food, which, while alive, he must lose to maintain his temperature. To estimate the time of death from a corpse's temperature, we make several assumptions: (1) at the time of death, the person's temperature was normal; (2) the corpse's temperature is above the environment's temperature; (3) the corpse continues to lose energy at a constant rate of 116 J/s; and (4) the environment's temperature has been constant.

Suppose a 65-kg male corpse with a core temperature of 35°C is found at 3:25 PM. The room's temperature has been regulated for days at 21°C. The average specific heat of the human body is $3470\ \text{J}/(\text{kg}\cdot\text{K})$. Estimate the time of death.

INTERPRET and ANTICIPATE

We assumed that the corpse loses thermal energy at a constant rate of 116 J/s. Because a change in temperature is proportional to a change in thermal energy, we can use the drop in temperature relative to normal (living) body temperature, 37°C, to find the time of death.

SOLVE

Start with the first law of thermodynamics to write an expression for ΔE_{th} in terms of ΔT.	$\Delta E_{th} = mc\Delta T$ (21.7)
Divide both sides by Δt to find an expression for the rate at which thermal energy and temperature change with time.	$\dfrac{\Delta E_{th}}{\Delta t} = mc\dfrac{\Delta T}{\Delta t}$
Solve for Δt. We were given that $\Delta E_{th}/\Delta t = 116\ \text{J/s}$. The temperature change is the same on the Kelvin scale as it is on the Celsius scale.	$\Delta t = mc\dfrac{\Delta T}{\Delta E_{th}/\Delta t}$ $\Delta t = (65\ \text{kg})\left(3470\dfrac{\text{J}}{\text{kg}\cdot\text{K}}\right)\dfrac{(2\ \text{K})}{116\ \text{J/s}}$ $\Delta t = 3.9 \times 10^3\ \text{s} \times \left(\dfrac{1\ \text{hr}}{3600\ \text{s}}\right)$ $\Delta t = 1{:}05$
This result is the elapsed time; we find the time of death t_{death} by subtracting the elapsed time from the time the body was found.	$t_{death} = 3{:}25\ \text{PM} - 1{:}05$ $t_{death} = 2{:}20\ \text{PM}$

CHECK and THINK

Our assumption that a dead body loses energy at the same rate as a living body is too simple. From your own experience, you know that the rate at which your thermal energy changes depends on your level of activity. In fact, a dead body loses energy at a somewhat lower rate than we assumed. The rule of thumb is that a dead body's temperature decreases 0.8 K per hour, so a drop in temperature of 2 K means that the time elapsed is 2:30 or the time of death was approximately 12:55 PM. (Of course, even this rule of thumb is too simple. The corpse cannot continue to lose energy at a constant rate forever.)

21-5 Latent Heat

Consider a simple experiment you could do by placing a block of ice in a hot container. The system is the ice, everything else is in the environment, and a thermometer is kept in contact with the ice. Because there are no external forces doing work on the system and as long as the ice remains below its melting point, the total heat transferred to it is directly proportional to its temperature change $Q = mc\,\Delta T$ (Eq. 21.5). A graph of temperature versus total heat transferred (Fig. 21.7) is a straight line whose slope is $1/mc$.

On the microscopic level, as heat is transferred to the ice, the water molecules oscillate more vigorously. Once it reaches its melting point (0°C at atmospheric pressure), the ice goes from its solid to its liquid phase. Although the ice melts, there is *no* change in the system's temperature (Fig. 21.7). During the phase change, the first law of thermodynamics still holds: $Q = \Delta E_{th}$, but $\Delta T = 0$. Heat is still flowing from the environment to the system, so the thermal energy of the system continues to increase. On the microscopic level, the increase in the system's thermal energy loosens the bonds between the molecules, so in the liquid phase the molecules can slip around one another.

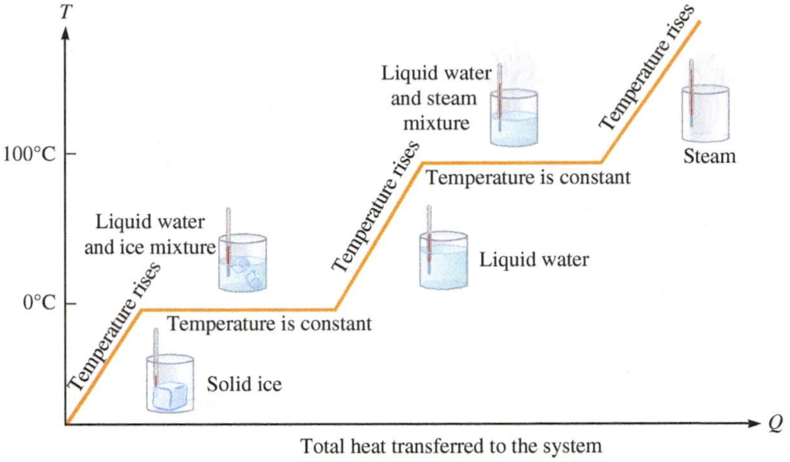

FIGURE 21.7 Temperature as a function of heat transferred to water (beginning as ice). When heat is added to the system and its temperature remains constant, the system is undergoing a phase change.

Latent heat L (also known as the **heat of transformation**) is the energy per unit mass that must be transferred to the system for the system to transform completely from one phase to the next. The heat required to melt a solid is found by multiplying the latent heat by the mass of the substance:

$$Q = mL_F \tag{21.9}$$

where L_F is the **heat of fusion**. The heat of fusion is the latent heat required to melt or freeze a substance.

Sign convention: If the substance is melting, heat must flow from the environment to the system, so Q is positive. If the substance is freezing, heat must flow from the system to the environment, so Q is negative.

Once the system has completely transformed into a liquid, a change in water temperature is directly proportional to the total heat added to the system (Fig. 21.7). On the microscopic level, the molecules move more rapidly as energy is transferred to the system.

When the water reaches its boiling point (100°C at atmospheric pressure), the liquid begins to transform into a gas. During this phase change, there is also no change in the system's temperature (Fig. 21.7). Heat is still flowing from the environment to the system, so the thermal energy of the system continues to increase. On the microscopic level, when a system changes from a liquid to a gas, the increase in the system's thermal energy breaks the bonds between molecules; in the gas phase, the molecules have little interaction with one another.

The **heat of vaporization** L_V is the latent heat required to transform a substance from a liquid to a gas or from a gas to a liquid. The heat required for such a transformation is

$$Q = mL_V \tag{21.10}$$

Again by our sign convention, if the substance is boiling, heat is transferred to the system, so Q is positive. If the substance is condensing, heat is transferred from the system, so Q is negative.

The heats of fusion and of vaporization for many substances are given in Table 21.2. The heat of vaporization is several times greater than the heat of fusion because

TABLE 21.2 Melting temperature, boiling temperature, and latent heat at 1 atm.

Substance	Melting temperature (K)	Heat of fusion L_F (J/kg)	Boiling temperature (K)	Heat of vaporization L_V (J/kg)
Ammonia	195.2	3.3×10^4	239.6	1.37×10^5
Copper	1356	2.07×10^5	2868	4.73×10^6
Ethyl alcohol	159	1.04×10^5	477	8.5×10^5
Hydrogen	14.0	5.8×10^4	20.3	4.55×10^5
Iron	2081	2.89×10^5	3296	6.34×10^6
Lead	600	2.5×10^4	2017	8.58×10^5
Mercury	234	1.14×10^4	630	2.96×10^5
Nitrogen	63	2.6×10^4	77.2	2.00×10^5
Oxygen	54.2	1.39×10^5	90	2.13×10^5
Silver	1235	1.05×10^5	2323	2.336×10^6
Tungsten	3410	1.84×10^5	6173	4.8×10^6
Water	273	3.33×10^5	373	2.256×10^6

the molecules in a liquid are loosely bound, whereas the molecules in a gas are essentially free. When a solid is melting, the latent heat required is the amount of energy per mass needed to loosen the bonds. More latent heat is required to break the bonds when a substance transforms from a liquid to a gas.

Evaporating is not the same as boiling. Recall from Section 20-8 that evaporation is a cooling process whereby the fastest particles escape from the liquid. During evaporation, the liquid's thermal energy and temperature decrease.

EXAMPLE 21.4 "Mom, How Long Should I Put This Food in the Microwave?"

Microwave ovens are very efficient at transferring heat to water. Because the major component of food is water, microwave ovens can speed up cooking so much that it is possible to take food from the freezer and eat it just minutes later. Imagine preparing a meal that has been in a cold freezer at −20°C. Model the food as 0.500 kg of water (initially ice). People generally like to eat hot food at a temperature of about 75°C. If the power transferred by the microwave oven to the food is 700 W, what time should you set on the oven timer?

INTERPRET and ANTICIPATE

This example is described by Figure 21.7. Heat Q_1 transferred to the ice will first raise its temperature. When the ice reaches its melting point, additional heat Q_2 will melt the ice without changing its temperature. After the ice completely melts, additional heat Q_3 will raise the temperature of the water. Calculate the total heat Q_{tot} that must be transferred to the ice to turn it into hot water. Then use the power rating of the microwave oven to find the cooking time.

SOLVE

Find the amount of heat Q_1 required to raise the ice temperature from −20°C to 0°C, where ΔT is the same on the Kelvin scale as it is on the Celsius scale. The specific heat of ice is given in Table 21.1.

$$Q_1 = mc\,\Delta T \qquad (21.5)$$

$$Q_1 = (0.500\text{ kg})[2.100 \times 10^3\text{ J/(kg}\cdot\text{K)}](20\text{ K})$$

$$Q_1 = 2.10 \times 10^4\text{ J}$$

Once the ice reaches 0°C, it begins to melt. Find the heat Q_2 required to melt all the ice. The latent heat of fusion is found in Table 21.2.

$$Q_2 = mL_F \qquad (21.9)$$

$$Q_2 = (0.500\text{ kg})(3.33 \times 10^5\text{ J/kg}) = 1.67 \times 10^5\text{ J}$$

After the ice has been changed into water, additional heat Q_3 increases the water's temperature. Find Q_3, this time with the specific heat of liquid water (Table 21.1) and not ice.

$$Q_3 = mc\,\Delta T \qquad (21.5)$$

$$Q_3 = (0.500\text{ kg})[4.19 \times 10^3\text{ J/(kg}\cdot\text{K)}](75\text{ K})$$

$$Q_3 = 1.57 \times 10^5\text{ J}$$

The total heat required is the sum of the heat required for each stage.

$$Q_{\text{tot}} = Q_1 + Q_2 + Q_3$$

$$Q_{\text{tot}} = (2.10 \times 10^4\text{ J}) + (1.67 \times 10^5\text{ J}) + (1.57 \times 10^5\text{ J})$$

$$Q_{\text{tot}} = 3.45 \times 10^5\text{ J}$$

Power is the rate of change of energy (Section 9-9). To find the time required, divide the total heat required by the power transferred.

$$\frac{\Delta Q_{\text{tot}}}{\Delta t} = 700\text{ W}$$

$$\Delta t = \frac{\Delta Q_{\text{tot}}}{700\text{ W}} = \frac{3.45 \times 10^5\text{ J}}{700\text{ J/s}} = 4.93 \times 10^2\text{ s} \times \left(\frac{1\text{ min}}{60\text{ s}}\right)$$

$$\Delta t = 8.22\text{ min} = 8\text{ min }13\text{ s}$$

CHECK and THINK

Our result seems about right. A cook's rule of thumb for microwave cooking is 6 min per pound. The food's mass is 0.500 kg, so it weighs slightly more than a pound, and our result is a little more than 6 min.

21-6 Work in Thermodynamic Processes

The state of a system is described by state variables such as temperature, pressure, volume, and thermal energy. Often, a *PV* diagram (Section 20-6) is used to visualize the state of a gaseous system (Fig. 21.8). A **thermodynamic process** is a way for a system to change from some initial state to some final state. On a *PV* diagram, a thermodynamic process is represented by a curve or *path* connecting the initial and final states of the system, and there are many possible paths that can connect two endpoints. The word *path* here means a curve on the *PV* diagram, not through space.

THERMODYNAMIC PROCESS
Major Concept

To record a thermodynamic process as a path, the gas must expand (or contract) slowly so that all the gas stays in equilibrium at the same pressure and temperature and so that its volume is well-defined. Such a slow process is called **quasistatic** (meaning "almost static"). Throughout this section and the next, we primarily study quasistatic thermodynamic processes of an ideal gas.

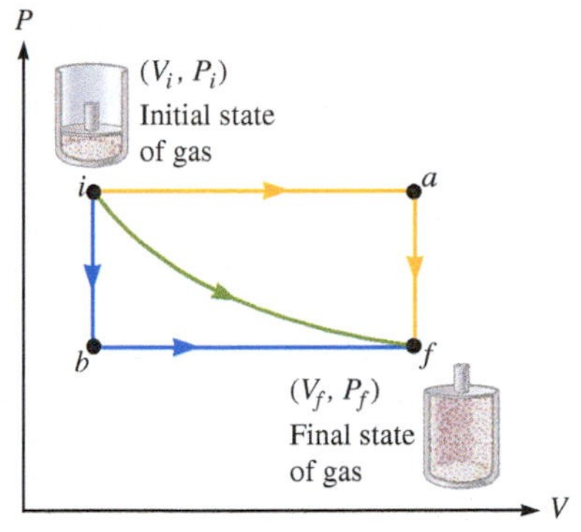

FIGURE 21.8 The state of a system is represented by a point on a *PV* diagram. Points *i* and *f* here represent an initial state and a final state of an ideal gas that expands. Three curves or paths connecting the initial state to the final state represent three different thermodynamic processes.

Consider an ideal gas in a container with a movable, frictionless piston on one end (Fig. 21.9). The gas is the system, and everything else (including the container and piston) is in the environment. When the gas does a small amount of work dW on the piston, the piston moves a short distance dy (Fig. 21.9A). The area of the piston is A, and the force exerted by the gas on the piston is PA. When the force exerted by the gas and the displacement of the piston are both upward, the work done by the gas to raise the piston by dy is positive and given by

$$dW = PA\, dy = P\, dV$$

where dV is the small change in the gas's volume. If the gas continues to do work on the piston, raising it to a new height and increasing the volume of the container (Fig. 21.9B), the total work done by the gas is

$$W = \int dW = \int_{V_i}^{V_f} P\, dV = \text{work done } by \text{ the system} \tag{21.11}$$

(energy *leaving* the system)

where V_i is the initial volume of the gas and V_f is the final volume.

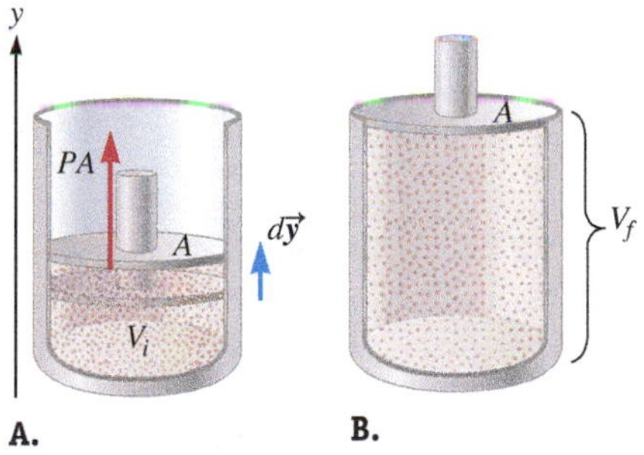

FIGURE 21.9 A. An ideal gas in a container with a movable piston. **B.** The gas raises the piston, doing positive work on the piston.

According to Equation 21.11, the work done by the gas on its environment is the area under the curve connecting the initial state of the gas to its final state on a *PV* diagram. Because there are many possible paths connecting the endpoints, however, the work done by the gas depends on the particular thermodynamic process (or path). For example, both path *A* and path *B* (Fig. 21.10A) represent **expansion processes** (final volume V_L is greater than the initial volume V_S). The area under the curve for process *A* is greater, however, so the gas does more work on its environment during process *A*.

In Figure 21.10A, $W_A > W_B$ because during process *A*, the gas pressure is higher than it is in process *B*. So, the force ($F = PA$) exerted by the gas on the piston is greater during process *A*. Because the force is greater and the piston is lifted by the

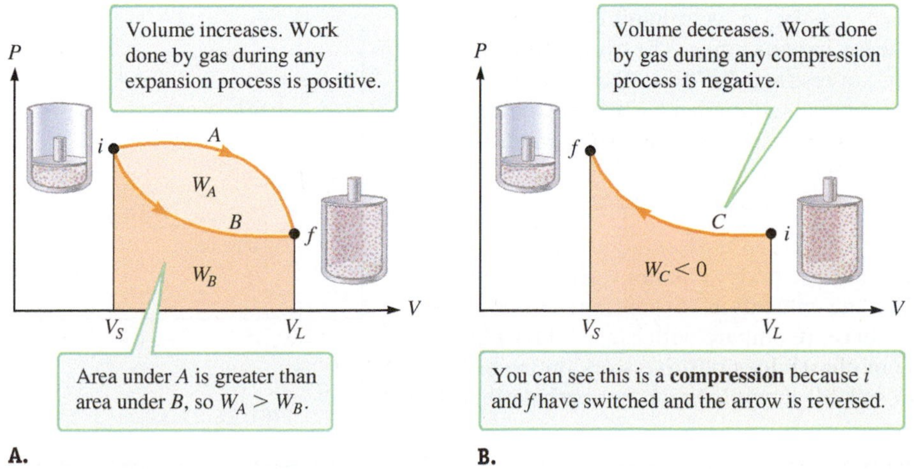

FIGURE 21.10 A. Two thermodynamic expansion processes. **B.** A compression process.

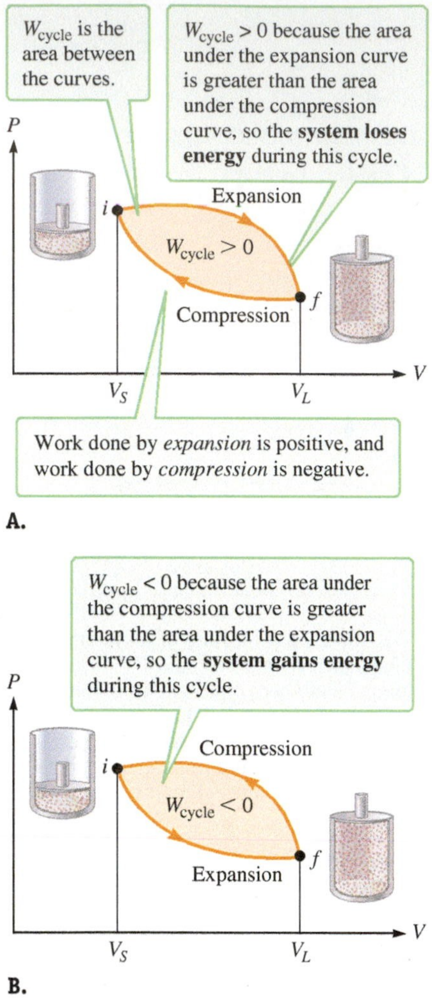

W_{cycle} is the area between the curves.
$W_{cycle} > 0$ because the area under the expansion curve is greater than the area under the compression curve, so the **system loses energy** during this cycle.
P
i
Expansion
$W_{cycle} > 0$
f
Compression
V
V_S
V_L
Work done by *expansion* is positive, and work done by *compression* is negative.
A.
$W_{cycle} < 0$ because the area under the compression curve is greater than the area under the expansion curve, so the **system gains energy** during this cycle.
P
i
Compression
$W_{cycle} < 0$
f
Expansion
V
V_S
V_L
B.

FIGURE 21.11 Two thermodynamic cycles.

same amount during either process, the gas does more work during process A than during process B (that is, more energy leaves the system during process A).

What happens if the piston (Fig. 21.10B) were lowered so that gas volume is reduced from a large volume V_L to a small volume V_S? In this **compression process**, the gas pressure would still push upward on the piston, but the piston's displacement would be downward in the opposite direction. During the compression, the work done by the gas is negative. On a *PV* diagram, the work done by the gas during compression is the negative of the area under the curve. You can interpret the negative work done by the gas during a compression as the piston doing *positive* work on the gas. So, during a compression, energy is transferred from the environment (piston) to the system (gas).

Now imagine that the gas undergoes an expansion and then a compression process that returns it to its original volume. A process that returns the system to its original state is known as a **thermodynamic cycle** (Fig. 21.11). The work done by the gas is the area under the curve, which is positive during expansion and negative during compression. The net work done by the gas during a cycle is represented by the area between the two curves.

The area under a path in a *PV* diagram is the work done *by the system* on its environment (Eq. 21.11), but that is *not* the same as the work in the first law of thermodynamics $W_{tot} + Q = \Delta E_{th}$ (Eq. 21.1). The work W_{tot} in the first law of thermodynamics is the work exerted *by the environment* on the system. By Newton's third law, the work done by the environment on the system has the opposite sign:

$$\begin{pmatrix} \text{work done} \\ \textit{by the environment} \\ \text{on the system} \end{pmatrix} = -\begin{pmatrix} \text{work done} \\ \textit{by the system} \\ \text{on the environment} \end{pmatrix} \quad (21.12)$$

To match the notation in the first law of thermodynamics, we will use the subscript "tot" for the total work W_{tot} exerted *by the environment* on the system. By inserting a negative sign into Equation 21.11, we find an expression for the work done by the environment:

$$W_{tot} = -\int_{V_i}^{V_f} P\,dV = \text{work done by the environment} \quad (21.13)$$
$$(\text{energy } \textit{entering} \text{ the system})$$

EXAMPLE 21.5 CASE STUDY Work You Do When Breathing

Since the moment you were born, you have been breathing air. During each cycle of inhalation and exhalation, your body does work on the air. With each inhalation, your chest expands your lungs, and with each exhalation, your lungs contract.

Figure 21.12 shows a *PV* diagram for the air in your lungs. The cycle shown in green is for a single breath taken while you are resting. The cycle shown in red is for a single breath taken while you are exerting yourself, perhaps while exercising. Use the figure to estimate the *work you do* on the air while you are resting and how much more work you do while you are exercising.

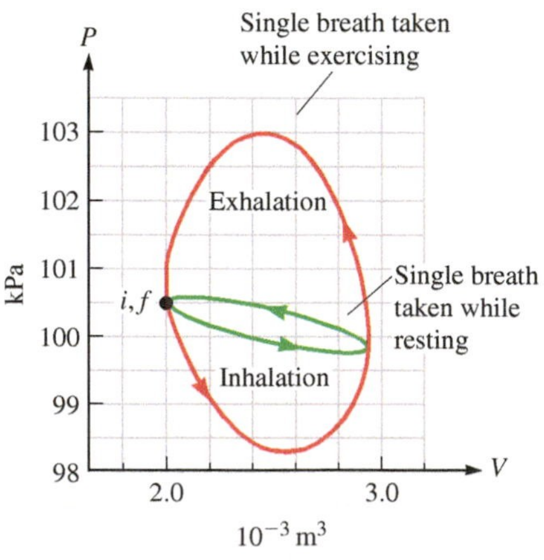

P
Single breath taken while exercising
103
102
101
100
99
98
kPa
Exhalation
i, f
Single breath taken while resting
Inhalation
2.0
3.0
V
10^{-3} m^3

FIGURE 21.12

INTERPRET and ANTICIPATE

The system is the air, and the environment is your body. The net work done by the air is negative because the area under the inhalation (expansion) curve is smaller than the area under the exhalation curve. (Compare with Fig. 21.11B.) In the case of breathing, the environment (your chest) does positive work on the air (Eq. 21.12). Find the *positive* area enclosed by the inhalation and exhalation cycle.

SOLVE

Find the area A enclosed by one rectangle on the grid. The width of one rectangle is 0.2×10^{-3} m³, and its height is 0.5×10^3 Pa. The area has the dimension of energy.

$$A = hw = (0.2 \times 10^{-3}\ \text{m}^3)(0.5 \times 10^3\ \text{Pa}) = 0.10\ \text{m}^3\ \text{Pa}$$

$$1\ \text{m}^3\ \text{Pa} = 1\ \text{m}^3\frac{\text{N}}{\text{m}^2} = 1\ \text{m}^3\frac{\text{N}}{\text{m}^2} = 1\ \text{N}\cdot\text{m} = 1\ \text{J}$$

$$A = 0.10\ \text{J}$$

Estimate the number of rectangles enclosed by the green cycle for the breaths taken while resting. You can probably estimate to the nearest one-half of a rectangle. Multiply the number of rectangles by the area of each rectangle to find the work you exert on the air when resting, W_{rest}.

3.5 rectangles

$$W_{\text{rest}} = 3.5A = 3.5(0.10\ \text{J})$$

$$W_{\text{rest}} = 0.35\ \text{J}$$

CHECK and THINK

When you are asleep, your body uses 4800 J per minute. You breathe about 10 times per minute. Our results show that you need only 3.5 J per minute for breathing when you sleep, less than 0.1% of the sleeping energy consumption. Most of the energy consumed when sleeping is used by your liver, spleen, brain, skeletal muscles, kidney, and heart.

SOLVE

Estimate the number of rectangles enclosed by the red cycle and multiply by A to find the work W_{exer} you exert to breathe while you are exercising.

34.5 rectangles

$$W_{\text{exer}} = 34.5A = 34.5(0.10\ \text{J})$$

$$W_{\text{exer}} = 3.45\ \text{J}$$

Subtract the work you do while you are at rest from the work you do while you are exercising to find the extra work required, W_{xtra}.

$$W_{\text{xtra}} = W_{\text{exer}} - W_{\text{rest}}$$

$$W_{\text{xtra}} = 3.45\ \text{J} - 0.35\ \text{J} = 3.1\ \text{J}$$

CHECK and THINK

When you exercise, your breathing rate is 150 breaths per minute, which means that you need an extra 460 J per minute when you undergo this very strenuous exercise. For some people, this difference is noticeable, and they might find such rapid breathing exhausting.

21-7 Specific Thermodynamic Processes

In Section 21-6, we used a *PV* diagram and Equation 21.13 to find the work done by the environment on a gaseous system. In this section, we apply the first law of thermodynamics to six specific thermodynamic processes for gaseous systems, illustrated using *PV* diagrams and energy bar charts.

Because the first law of thermodynamics does not involve mechanical energy, an energy bar chart (Fig. 21.13) is somewhat simplified. To see what is required, let's rewrite the first law of thermodynamics (Eq. 21.1):

$$W_{\text{tot}} + Q = \Delta E_{\text{th}} = E_{\text{th},f} - E_{\text{th},i}$$

where $E_{\text{th},i}$ is the initial thermal energy and $E_{\text{th},f}$ is the final thermal energy. Bringing the initial thermal energy to the left side of the equation, we have

$$E_{\text{th},i} + W_{\text{tot}} + Q = E_{\text{th},f} \tag{21.14}$$

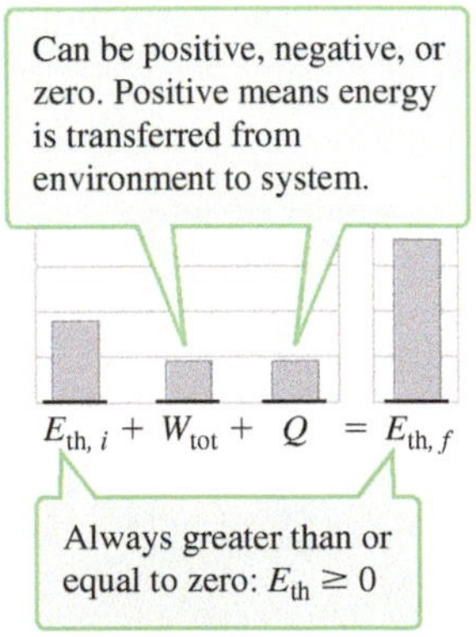
Can be positive, negative, or zero. Positive means energy is transferred from environment to system.
$E_{\text{th},i}$ + W_{tot} + Q = $E_{\text{th},f}$
Always greater than or equal to zero: $E_{\text{th}} \geq 0$

FIGURE 21.13 A bar chart helps when using the first law of thermodynamics.

ENERGY BAR CHARTS Tool

Adiabatic Process ($Q = 0$)

For the first specific thermodynamic process, consider the expansion of an ideal gas in a cylindrical container with a movable piston on top (Fig. 21.9). During an **adiabatic process**, no heat flows into or out of the system. You might guess that the system must be insulated to ensure that there is no heat flow. Although that is one possibility, if a thermodynamic process is fast enough, it is possible that little or no heat flows during the process even if a system is not insulated. For example, the rapid

ADIABATIC PROCESS Special Case

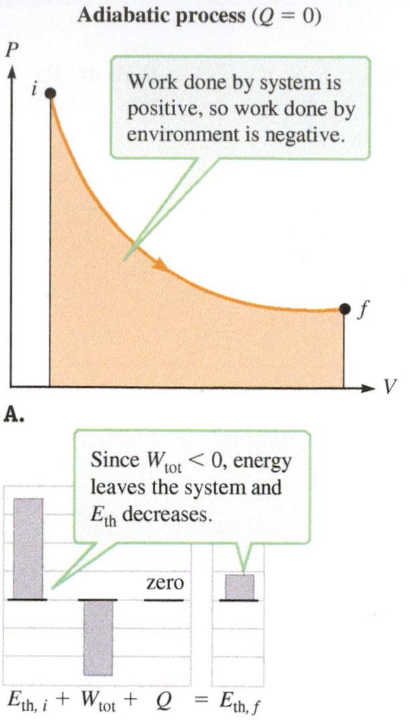

FIGURE 21.14 A. *PV* diagram for an adiabatic expansion. **B.** Bar chart for an adiabatic expansion.

expansion of the gas in your car's engine may be approximated as an adiabatic process. Such a process can be fast enough to be adiabatic, but slow enough to be considered quasistatic.

As the gas expands adiabatically, its pressure decreases as shown in the *PV* diagram (Fig. 21.14A). The work done *by the gas* in lifting the piston is the area under the curve (positive for an expanding gas). The work done *by the environment*, W_{tot}, has the opposite sign (in this case, negative). So, W_{tot} is represented by a bar below the line (Fig. 21.14B).

Initially, as long as the gas is not at absolute zero, it has some thermal energy as represented by a bar above the line (Fig. 21.14B). Because the process is adiabatic, no heat is exchanged with the environment as indicated by the "zero" written on the line for heat. According to the first law of thermodynamics (Eq. 21.14), with $Q = 0$, the final thermal energy is

$$E_{th,f} = E_{th,i} + W_{tot}$$

and because the work done by the environment is negative, the final thermal energy is less than the initial thermal energy. We return to adiabatic expansion of a gas in Section 21-9.

CONCEPT EXERCISE 21.5

Figure 21.14 is drawn for a gas that expands adiabatically. Draw a *PV* diagram and an energy bar chart for a gas that is compressed adiabatically.

Isothermal Process ($\Delta T = 0$)

ISOTHERMAL PROCESS

Special Case

Isothermal means "same temperature" in Greek, so in an **isothermal process**, the temperature of the system is unchanged: $\Delta T = 0$. Because the temperature does not change, the system's thermal energy does not change: $\Delta E_{th} = 0$.

Let's add some details to our apparatus so that we can consider the isothermal expansion of an ideal gas. The walls and piston are insulated so that no heat can flow through them (Fig. 21.15). The bottom of the container is in thermal contact with a **heat reservoir**, an object so large that when heat is exchanged with the system, the reservoir's temperature does not change significantly. In practice, a heat reservoir may be a stove top with a control that allows the researcher to set its temperature.

Because the system is modeled as an ideal gas,

$$PV = Nk_BT = \text{constant}$$

$$P = \frac{Nk_BT}{V} = \frac{\text{constant}}{V}$$

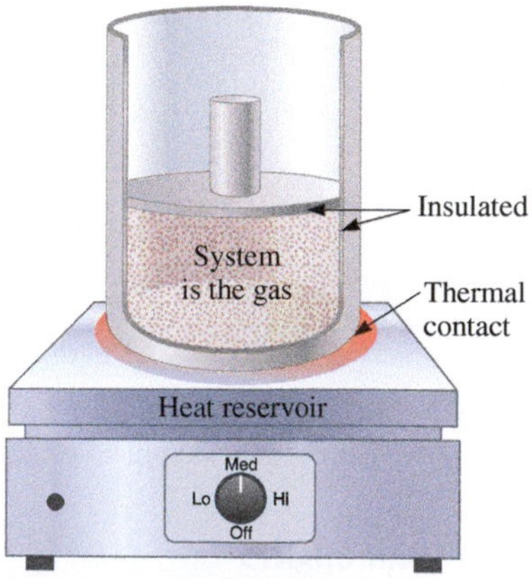

FIGURE 21.15 The container is in thermal contact with a heat reservoir. The temperature of the reservoir can be set by turning the control knob. The container's walls and piston are insulated. The piston is free to move.

which is plotted on the *PV* diagram (Fig. 21.16A) and labeled "Isothermal." Let's compare an isothermal expansion of an ideal gas to an adiabatic expansion. Suppose the gas expands from V_i to V_f. In both processes, the gas pressure drops, but the temperature drops only for the adiabatic expansion. Because temperature is proportional to pressure, the decrease in pressure is greater in the adiabatic process than in the isothermal process. So, the path in the *PV* diagram for the adiabatic process falls below the path for the isothermal process (Fig. 21.16A). Therefore, the work done *by the expanding gas* (area under the *PV* curve) in raising the piston is greater in the isothermal process.

The gas lifts a piston as it expands, which means that the environment does negative work on the system: $W_{tot} < 0$. In an isothermal process, the thermal energy remains constant. Therefore, the energy lost by the system as a result of work must be balanced by heat transferred to the system from the heat reservoir:

$$W_{tot} = -Q \text{ for an isothermal process}$$

The bar for Q is the same height as that for W_{tot} except that Q is positive (Fig. 21.16B).

We find an expression for the work done *by the environment* on an ideal gas in an isothermal process by substituting the ideal gas law into Equation 21.13:

$$W_{tot} = -\int_{V_i}^{V_f} P\,dV = -\int_{V_i}^{V_f} \frac{Nk_BT}{V}\,dV$$

For an isothermal process, T is constant, and as long as the system is closed (no gas escapes or enters), N is constant. So,

$$W_{tot} = -Nk_BT\int_{V_i}^{V_f} \frac{dV}{V}$$

Using Appendix A, we find

$$W_{tot} = -Nk_BT(\ln V_f - \ln V_i) = Nk_BT(\ln V_i - \ln V_f)$$

$$W_{tot} = Nk_BT\ln\frac{V_i}{V_f} \quad \text{for an isothermal process on an ideal gas} \qquad (21.15)$$

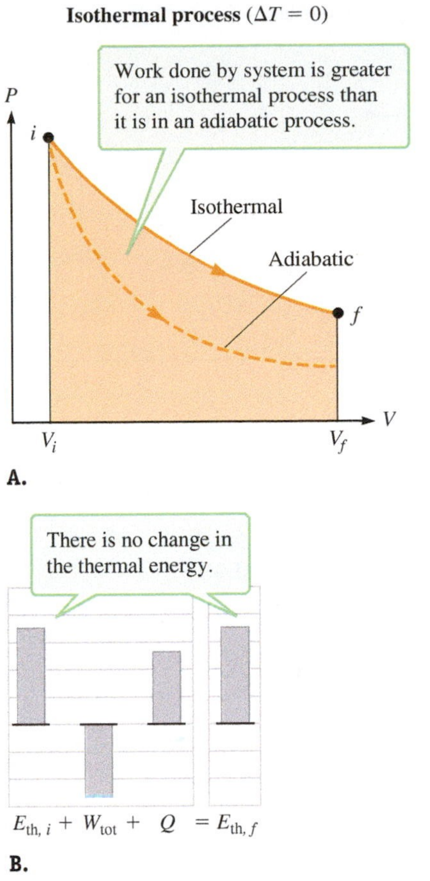

Isothermal process ($\Delta T = 0$)
Work done by system is greater for an isothermal process than it is in an adiabatic process.
P
i
Isothermal
Adiabatic
f
V
V_i
V_f
A.
There is no change in the thermal energy.
$E_{th,i}$ + W_{tot} + Q = $E_{th,f}$
B.

FIGURE 21.16 **A.** *PV* diagram for an isothermal expansion. **B.** Bar chart for an isothermal expansion.

CONCEPT EXERCISE 21.6

Figure 21.16 is drawn for a gas that expands isothermally. Draw a *PV* diagram and an energy bar chart for a gas that is compressed isothermally.

CONCEPT EXERCISE 21.7

The *PV* diagram in Figure 21.16 shows that an ideal gas must do more work on a piston in an isothermal expansion than in an adiabatic expansion. Draw a *PV* diagram for the compression of an ideal gas from V_f to V_i both isothermally and adiabatically. Compare the work done by the piston on the gas for these two compression processes.

Constant-Volume Process ($\Delta V = 0$)

CONSTANT VOLUME (ISOCHORIC) PROCESS

Special Case

In a constant-volume process, the system's volume is fixed. In practice, we must add a mechanism (such as a screw top, a clamp, or wedges) for holding the piston in place when needed.

Because the piston cannot move, it cannot do work on the gas, and the gas cannot do work on the piston:

$$W_{tot} = 0 \quad \text{for a constant-volume process}$$

Energy can still be exchanged through heat. As an example, suppose you turn down the temperature on the heat reservoir so that heat flows from the hotter system to the cooler reservoir and the gas cools. According to the ideal gas law $PV = Nk_BT$, a reduction in temperature (at constant volume) means a reduction in pressure, which is shown on the *PV* diagram by a vertical, downward path (Fig. 21.17A). The vertical path has no area under it, so no work is done by the system as expected.

Because heat flows from the gas to the reservoir, energy leaves the system as represented by a bar for Q below the zero line in the bar chart (Fig. 21.17B). According to the first law of thermodynamics, the system's thermal energy must be reduced by the amount of heat that flows out:

$$\Delta E_{th} = Q \quad \text{for a constant-volume process}$$

In this case, $Q < 0$, so the initial thermal energy bar is higher than the final thermal energy bar (Fig. 21.17B)

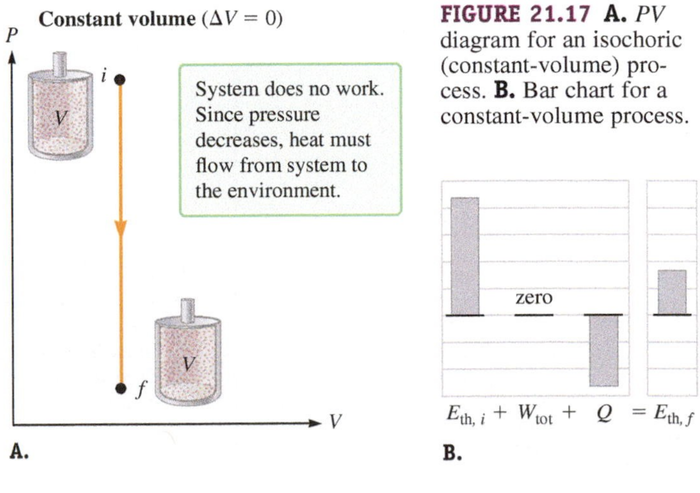

Constant volume ($\Delta V = 0$)
P
i
V
System does no work. Since pressure decreases, heat must flow from system to the environment.
f
V
A.
zero
$E_{th,i}$ + W_{tot} + Q = $E_{th,f}$
B.

FIGURE 21.17 **A.** *PV* diagram for an isochoric (constant-volume) process. **B.** Bar chart for a constant-volume process.

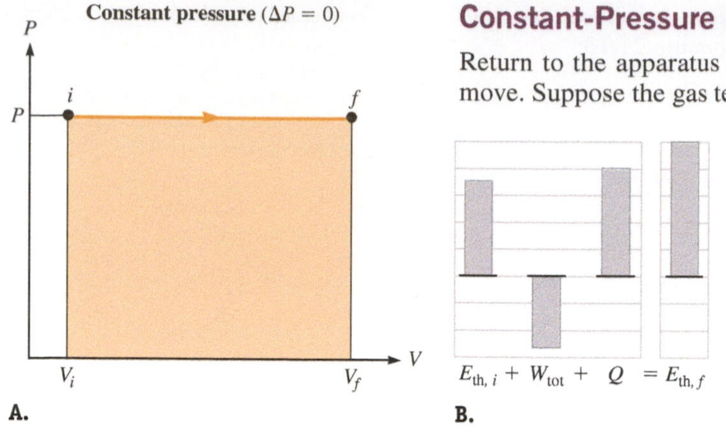
Constant pressure ($\Delta P = 0$)
P
P
i
f
V
V_i
V_f
A.
$E_{th,i}$ + W_{tot} + Q = $E_{th,f}$
B.

FIGURE 21.18 **A.** *PV* diagram for an isobaric (constant-pressure) process. **B.** Bar chart for a constant-pressure process.

Constant-Pressure Process ($\Delta P = 0$)

Return to the apparatus shown in Figure 21.15 once again with the piston free to move. Suppose the gas temperature is lower than the heat reservoir's temperature at first, so heat flows from the reservoir to the gas. Energy enters the system, and because the piston is free, the gas pushes the piston up. The volume and temperature of the gas increase, but its pressure remains constant. On a *PV* diagram, a constant-pressure process is represented by a horizontal line (Fig. 21.18A). The area under the path is easily found as the area of a rectangle of height *P* and width ΔV:

$$\text{work done by gas} = \text{area under path} = P\,\Delta V$$

The work done by the gas in this case is positive. Energy leaves the system, and W_{tot} (work done by environment) is negative as represented by the bar in Figure 21.18B:

CONSTANT-PRESSURE (ISOBARIC) PROCESS Special Case

$$W_{tot} = -P\Delta V \text{ for a constant-pressure process} \tag{21.16}$$

If the system is an ideal gas, we can substitute the ideal gas law to find the work done on the system,

$$W_{tot} = -\frac{Nk_BT_i}{V_i}(V_f - V_i) = -\frac{Nk_BT_f}{V_f}(V_f - V_i)$$

$$W_{tot} = Nk_BT_i\left(1 - \frac{V_f}{V_i}\right) = Nk_BT_f\left(\frac{V_i}{V_f} - 1\right) \tag{21.17}$$

for a constant-pressure ideal gas.

Other than the first law of thermodynamics, there is no particular constraint on the heat or change in the system's thermal energy. In Figure 21.19, we have arbitrarily shown a system that has heat flowing into the system and an increase in the system's thermal energy.

Cyclic Process ($\Delta E_{th} = 0$)

CYCLIC PROCESS Special Case

During a **cyclic process**, energy may be exchanged as a result of heat and work, but the system returns to its original state. Therefore, all the system's state variables—including its thermal energy—must return to their original values, and there is no change in the system's thermal energy: $\Delta E_{th} = 0$.

We already know that a cyclic process makes a closed path on a *PV* diagram (Fig. 21.19A) and that the area between the expansion and compression curves is the net

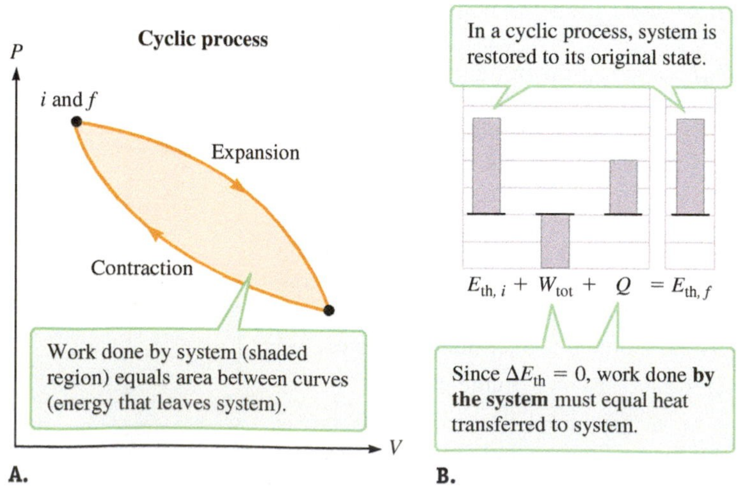
Cyclic process
P
i and f
Expansion
Contraction
Work done by system (shaded region) equals area between curves (energy that leaves system).
V
A.
In a cyclic process, system is restored to its original state.
$E_{th,i}$ + W_{tot} + Q = $E_{th,f}$
Since $\Delta E_{th} = 0$, work done **by the system** must equal heat transferred to system.
B.

FIGURE 21.19 **A.** *PV* diagram for a cyclic process. **B.** Bar chart for a cyclic process.

work exerted *by the gas*. So, the net work done *by the environment* W_{tot} is the negative of the area between the two curves (Eq. 21.11):

$$W_{tot} = -(\text{area between curves on } PV \text{ diagram})$$

In the cyclic process in Figure 21.19, W_{tot} is negative; the system loses energy through work.

Because the thermal energy returns to its original value after one cycle, the system must gain energy through heat. The amount of heat that flows into the system must equal the amount of energy the system loses through work:

$$W_{tot} = -Q \quad \text{for a cyclic process}$$

The energy transferred out of the system by work W_{tot} is shown as a bar below the line representing zero, and the heat that flows into the system is shown as a bar of equal height but above that line (Fig. 21.19B). We will look at cyclic processes in more detail in Chapter 22.

CONCEPT EXERCISE 21.8

Suppose that the cyclic process in Figure 21.19 is reversed so that the gas starts with a large volume and is compressed by the piston (upper path), and then the gas lifts the piston, returning the gas to its original volume (lower path). Draw a *PV* diagram.

a. Is the total work done by the gas on the piston (energy leaving the system) during the cycle positive, negative, or zero?
b. Is the total work done by the piston on the gas (energy entering the system) during the cycle positive, negative, or zero?
c. Is the net flow of heat from the environment to the system or from the system to the environment?

Free Expansion ($Q = W = 0$)

A gas fills the left side of a container in Figure 21.20. The container is well insulated so that heat cannot be exchanged with the environment. When the valve is opened, the gas expands to fill both sides of the container. During this expansion, the gas pressure is not the same at all points, and its volume is not clearly defined. So, this expansion process is not quasistatic and cannot be plotted on a *PV* diagram (Fig. 21.21A).

Because there is no piston for the gas to lift, the gas does no work. An expansion that involves no work or heat is called a **free expansion**. Because $W_{tot} = 0$ and $Q = 0$, there is no change in the system's thermal energy, and $\Delta E_{th} = 0$. The bar chart for a free expansion is shown in Figure 21.21B.

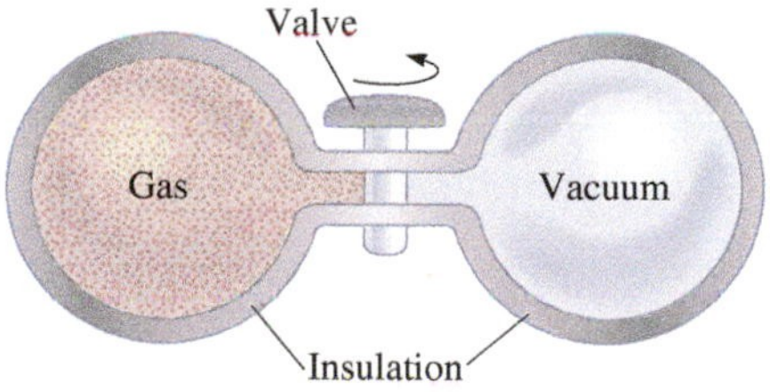

FIGURE 21.20 This apparatus allows a gas to expand freely.

FREE EXPANSION Special Case

CONCEPT EXERCISE 21.9

Which of the following quantities depend on the path taken in a *PV* diagram, and which quantities only depend on the endpoints?

a. Change in thermal energy
b. Change in temperature
c. Change in volume
d. Change in pressure
e. Heat transferred
f. Work done by the environment
g. Work done by the system

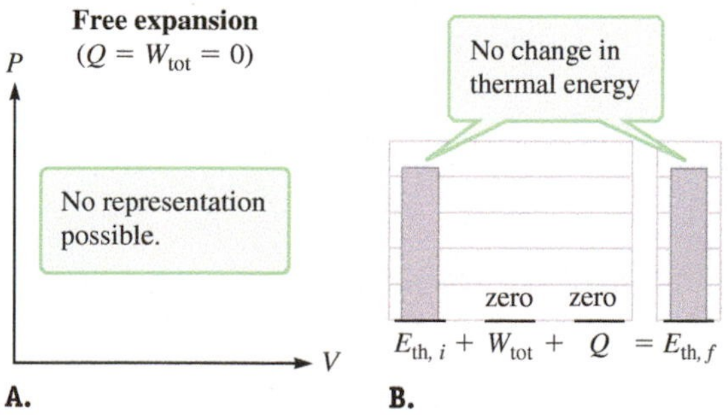

FIGURE 21.21 **A.** Free expansion cannot be plotted on a *PV* diagram. **B.** Bar chart for free expansion. A free expansion is adiabatic ($Q = 0$) and no work is done ($W = 0$).

EXAMPLE 21.6 Comparing Processes

As shown on the PV diagram in Figure 21.22, a gas may be compressed either isothermally or by a two-part process consisting of a constant-pressure stage followed by a constant-volume stage. Find the work done by the environment and the heat transferred to the system in each process. Assume it is an ideal gas. The initial position on the PV diagram is at $(19.00\ \text{m}^3, 1.29 \times 10^4\ \text{Pa})$ and the final position is at $(5.00\ \text{m}^3, 4.89 \times 10^4\ \text{Pa})$.

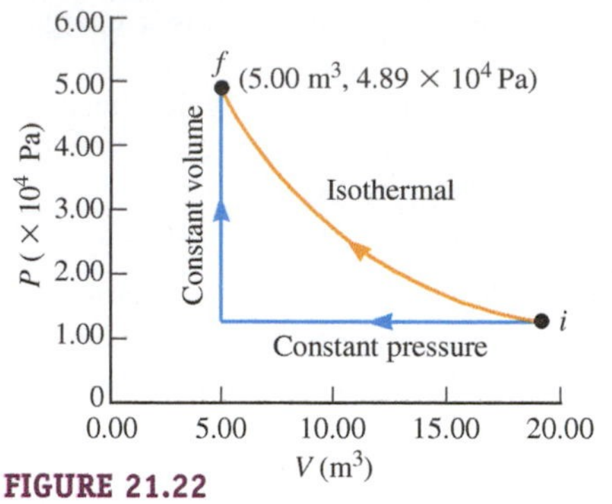

FIGURE 21.22

INTERPRET and ANTICIPATE

By inspection of the PV diagram, the area under the isothermal curve is greater than the area under the combined constant-pressure–constant-volume curve, so we expect that the environment must do more work in the isothermal process. Constructing a bar chart (Fig. 21.23) helps provide an idea of the amount of heat transferred during each process. The change in thermal energy is the same whether the gas is compressed isothermally or by the two-step process. Furthermore, $\Delta E_{\text{th}} = 0$ and $W_{\text{tot}} = -Q$ for any isothermal process. Therefore, the thermal energy bars are all identical in the two bar charts. In the isothermal process, more work is done by the environment, and more heat must flow out of the system.

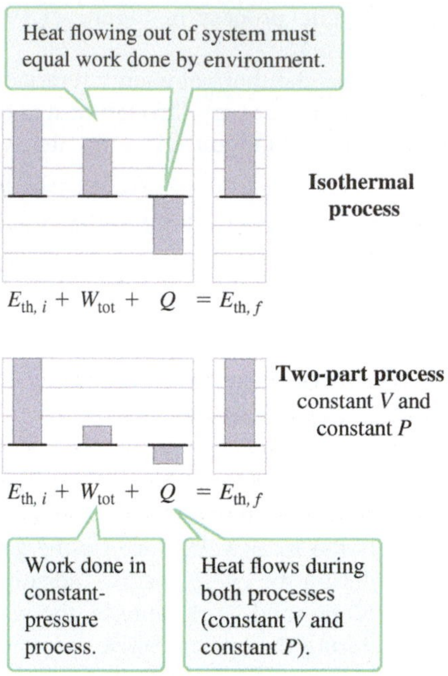

FIGURE 21.23

SOLVE

Step 1 Isothermal process. Because we know the initial pressure but not the initial temperature, we can use the ideal gas law and Equation 21.15 to find an expression for the work W_1 done by the environment in terms of P_i. (We could have also used the final pressure; our choice is arbitrary.)

$$W_1 = Nk_BT\ln\frac{V_i}{V_f} \qquad (21.15)$$

$$Nk_BT_i = P_iV_i$$

$$W_1 = P_iV_i\ln\frac{V_i}{V_f}$$

Substitute values found on the PV diagram.

$$W_1 = (1.29 \times 10^4\ \text{Pa})(19.0\ \text{m}^3)\ln\left(\frac{19.0\ \text{m}^3}{5.00\ \text{m}^3}\right)$$

$$W_1 = 3.27 \times 10^5\ \text{J (isothermal)}$$

Because there is no change in thermal energy, the heat Q_1 that flows out of the system must equal the work that the environment does on the system.

$$\Delta E_{\text{th}} = W_1 + Q_1 = 0$$

$$Q_1 = -W_1$$

$$Q_1 = -3.27 \times 10^5\ \text{J (isothermal)}$$

Step 2 Two-part process (constant-pressure process followed by **constant-volume process).** No work is done during the constant-volume process, so the only work done by the environment during this two-part process is given by Equation 21.16. During the constant-pressure part, the pressure is P_i.

$$W_2 = -P\,\Delta V = -P_i(V_f - V_i) \qquad (21.16)$$

$$W_2 = -(1.29 \times 10^4\ \text{Pa})(5.00\ \text{m}^3 - 19.0\ \text{m}^3)$$

$$W_2 = 1.81 \times 10^5\ \text{J (two-part process)}$$

The heat that leaves the system must equal the work done by the environment. Heat flows during both parts of the two-part process. Heat flows out of the system during the constant-pressure stage, and heat flows into the system during the constant-volume stage. The heat we find here is the net heat that flows out of the system.

$$\Delta E_{\text{th}} = W_2 + Q_2 = 0$$

$$Q_2 = -W_2$$

$$Q_2 = -1.81 \times 10^5\ \text{J (two-part process)}$$

CHECK and THINK

As expected, the environment does more work during the isothermal process than during the two-step process ($W_1 > W_2$). During the two-step process, compression occurs in the constant-pressure stage when the gas is at a lower pressure P_i. Therefore, the force exerted by the environment is smaller for the same change in volume than it is for the isothermal process in which pressure rises continuously.

21-8 Equipartition of Energy

Molar Specific Heat of Gases

In Section 21-4, we skipped over the specific heat of gases because the specific heat of a gas depends on the thermodynamic process. When a solid or liquid is in thermal contact with a heat reservoir, heat is exchanged, but the work done on the system is typically zero: $W_{tot} = 0$. Thus, for a solid or a liquid, $\Delta E_{th} = Q$. Because a change in temperature is directly proportional to a change in thermal energy, heat entering (or leaving) the system is directly proportional to the change in the system's temperature. The molar specific heat C of a substance in the solid or liquid phase is the constant of proportionality in

$$Q = nC\,\Delta T \tag{21.6}$$

The situation for a gas is more complicated because work is often done on (or by) the system, so $W_{tot} \neq 0$. Because the work done depends on the path between endpoints, the amount of heat required to raise the temperature of a gas by a certain amount ΔT depends on the thermodynamic process. Figure 21.24A shows the PV diagram for an ideal gas at two different temperatures. The curve in red is for the hot gas, $P = Nk_BT_{hot}/V$, and the blue curve is for the cool gas, $P = Nk_BT_{cool}/V$. Figure 21.24A also shows two possible paths (constant volume and constant pressure) for raising the gas temperature. Both processes result in the same temperature change ΔT, and because thermal energy is proportional to temperature, both processes result in the same thermal energy change ΔE_{th}. Both work and heat depend on the particular process, however. We define the amount of heat Q_V required during a constant volume process to be

$$Q_V \equiv nC_V\Delta T \tag{21.18}$$

where C_V is the molar specific heat of the gas at constant volume. We also define the amount of heat Q_P required during a constant pressure process to be

$$Q_P \equiv nC_P\Delta T \tag{21.19}$$

where C_P is the molar specific heat of the gas at constant pressure.

Values for C_V and C_P for various gases are provided in Table 21.3. From the data, we see that (1) $C_P > C_V$, and (2) their difference is approximately constant: $C_P - C_V = R$, where $R = 8.315\ \mathrm{J/(mol \cdot K)}$.

To explain these experimental results, let's take a close look at each process for raising the gas temperature. If the gas goes through a constant-volume process, no work is involved. Using Equation 21.18 and $W_{tot} = 0$, the first law of thermodynamics (Eq. 21.1) becomes

$$Q_V = \Delta E_{th} = nC_V\Delta T \tag{21.20}$$

If the gas goes through a constant-pressure process, the gas lifts the piston (energy leaves the gas), so the environment does negative work on the system (Eq. 21.16), and the amount of heat that flows from the environment is given by Equation 21.19. So, the first law of thermodynamics becomes

$$\Delta E_{th} = W_{tot} + Q_P$$

$$\Delta E_{th} = -P\Delta V + nC_P\Delta T \tag{21.21}$$

The change in thermal energy and temperature must be the same no matter which path is taken, so Equation 21.20 must equal Equation 21.21. Before looking at this

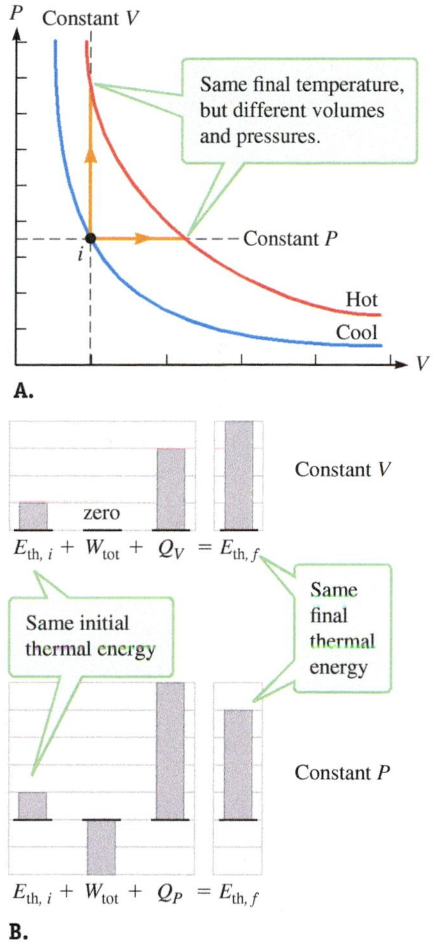

FIGURE 21.24 **A.** PV diagram for an ideal gas at two different temperatures with two possible paths (constant volume and constant pressure) for raising the gas temperature. **B.** Bar charts for constant-volume and constant-pressure processes.

TABLE 21.3 Molar specific heat for gases at room temperature except where noted.

Gas	C_P [J/(mol· K)]	C_V [J/(mol· K)]	$C_P - C_V$ [J/(mol· K)]	$\gamma = C_P/C_V$
Monatomic				
Ar	20.8	12.5	8.3	1.67
He	20.8	12.5	8.3	1.67
Ne	20.8	12.7	8.1	1.64
Diatomic				
CO	29.3	21.0	8.3	1.40
H_2	28.8	20.4	8.3	1.41
N_2	29.1	20.8	8.3	1.40
O_2	29.2	21.1	8.3	1.40
Triatomic				
CH_4	35.5	27.1	8.4	1.31
CO_2	37.0	28.5	8.5	1.30
H_2O (at 100°C)	35.4	27.0	8.4	1.30
Polyatomic				
C_2H_6	51.7	43.2	8.5	1.20

situation algebraically, consider the energy bar charts (Fig. 21.24B). The bars for $E_{\text{th},\,i}$ are the same in both processes, as are those for $E_{\text{th},\,f}$. The difference is that in the constant-pressure process, negative work is done, and in the constant-volume process, no work is done by the environment. As a result, the heat entering the gas is greater in the constant-pressure case:

$$Q_P > Q_V$$
$$nC_P\Delta T > nC_V\Delta T$$
$$C_P > C_V$$

To find exactly how much greater, let's first use the ideal gas law to find an expression for $P\,\Delta V$ in Equation 21.21:

$$\Delta(PV) = \Delta(nRT)$$

Because Equation 21.21 is for the constant-pressure process, the only quantity that changes on the left is V. Because no gas escapes or is added, n is constant, and the only quantity that changes on the right is T. We find that

$$P\Delta V = nR\Delta T \tag{21.22}$$

which we can substitute into Equation 21.21:

$$\Delta E_{\text{th}} = -nR\Delta T + nC_P\Delta T \tag{21.23}$$

Set Equation 21.23 equal to Equation 21.20:

$$-nR\Delta T + nC_P\Delta T = nC_V\Delta T$$

Solve for $C_P - C_V$:

$$C_P - C_V = R \tag{21.24}$$

This result is nearly what is found experimentally as shown in the fourth column in Table 21.3.

EXAMPLE 21.7 Raising Gas Temperature: Constant Pressure Versus Constant Volume

The initial pressure, volume, and temperature of a sample of He gas are $P_i = 2.45 \times 10^5$ Pa, $V_i = 0.75$ m^3, and $T_i = 235$ K, respectively. The final temperature of the gas is $T_f = 354$ K. The gas is well modeled by the ideal gas law.

A For a constant-volume process, what are the final pressure and volume of the gas? Also, find the work done by the environment, the heat transferred to the gas, and the change in thermal energy.

INTERPRET and ANTICIPATE

The final volume equals the initial volume. Given the initial and final temperatures, use the ideal gas law to find the final pressure. From Figure 21.24A, we expect the final pressure to be greater than the initial pressure. We can also use the ideal gas law to find the number of moles, which we need to find the heat absorbed. Because the temperature increases, we expect that the heat entering the system is positive.

SOLVE

First, use the ideal gas law to find the number of moles n. No gas escapes or is added, so n is a constant. (We'll need this result for part B.)

$$P_iV_i = nRT_i$$

$$n = \frac{P_iV_i}{RT_i} = \frac{(2.45 \times 10^5\ \text{Pa})(0.75\ \text{m}^3)}{\left(8.315\ \frac{\text{J}}{\text{mol}\cdot\text{K}}\right)(235\ \text{K})}$$

$$n = 94.0\ \text{mol}$$

The final volume equals the initial volume.

$$V_f = V_i = 0.75\ \text{m}^3$$

Because there is no change in volume, no work is done.

$$W_{\text{tot}} = 0$$

Find the final pressure from the ideal gas law.

$$V_f = V_i = 0.75\ \text{m}^3$$

$$P_f = \frac{nRT_f}{V_f} = \frac{(94.0\ \text{mol})\left(8.315\ \frac{\text{J}}{\text{mol}\cdot\text{K}}\right)(354\ \text{K})}{0.75\ \text{m}^3}$$

$$P_f = 3.69 \times 10^5\ \text{Pa}$$

CHECK and THINK

As expected, the final pressure is greater than the initial pressure.

SOLVE

For a constant-volume process, heat transferred to the gas is given by Equation 21.20. The molar specific heat C_V of helium is listed in Table 21.3.

$$Q_V = nC_V\Delta T \qquad (21.20)$$

$$Q_V = (94.0\ \text{mol})\left(12.5\frac{\text{J}}{\text{mol}\cdot\text{K}}\right)(354\ \text{K} - 235\ \text{K})$$

$$Q_V = \Delta E_{\text{th}} = 1.40 \times 10^5\ \text{J}$$

CHECK and THINK

As expected, the heat entering the gas is positive, and the thermal energy increases.

B If the process is instead at constant pressure, what are the final pressure and volume of the gas? Find the work done by the environment, the heat transferred to the gas, and the change in thermal energy.

INTERPRET and ANTICIPATE

The final pressure equals the initial pressure. The number of moles is the same as in part A, so use the ideal gas law to find the final volume. From Figure 21.24A, we expect the final volume to be greater than the initial volume. During a constant-pressure process, energy is transferred by both work and heat (Fig. 21.24B). Find the work done on the gas using $W_{\text{tot}} = -P\,\Delta V$ (Eq. 21.16) and the heat transferred to the gas using $Q_P \equiv nC_P\,\Delta T$ (Eq. 21.19). From Figure 21.24B, we expect the work done on the gas to be negative (energy leaving the system), the heat transferred to the gas to be positive, and the change in thermal energy to be the same as it is in part A. We also expect that more heat is transferred in the constant-pressure process than in the constant-volume process.

SOLVE

The final pressure equals the initial pressure.

$$P_f = P_i = 2.45 \times 10^5\ \text{Pa}$$

Example continues on page 634 ▶

Find the final volume from the ideal gas law.	$V_f = \frac{nRT_f}{P_f} = \frac{(94.0 \text{ mol})\left(8.315 \frac{\text{J}}{\text{mol}\cdot\text{K}}\right)(354 \text{ K})}{2.45 \times 10^5 \text{ Pa}}$ $V_f = 1.13 \text{ m}^3$

CHECK and THINK

As expected, the final volume is greater than the initial volume.

SOLVE For a constant-pressure process, heat transferred to the gas is given by Equation 21.19. The molar specific heat C_P of helium is listed in Table 21.3.	$Q_P = nC_P\Delta T$ (21.19) $Q_P = (94.0 \text{ mol})\left(20.8\frac{\text{J}}{\text{mol}\cdot\text{K}}\right)(354 \text{ K} - 235 \text{ K})$ $Q_P = 2.33 \times 10^5 \text{ J}$
The work done by the environment during a constant-pressure process is given by Equation 21.16.	$W_{\text{tot}} = -P\Delta V$ (21.16) $W_{\text{tot}} = -(2.45 \times 10^5 \text{ Pa})(1.13 \text{ m}^3 - 0.75 \text{ m}^3)$ $W_{\text{tot}} = -9.31 \times 10^4 \text{ J}$
The change in thermal energy is the sum of the energies entering the system through work and heat.	$\Delta E_{\text{th}} = W_{\text{tot}} + Q_P$ $\Delta E_{\text{th}} = (-9.31 \times 10^4 \text{ J}) + (2.33 \times 10^5 \text{ J}) = 1.40 \times 10^5 \text{ J}$

CHECK and THINK

As expected, energy leaves the system through work ($W_{\text{tot}} < 0$), energy enters the system through heat ($Q > 0$), and the change in thermal energy is the same in the constant-pressure process (part B) as it is in the constant-volume process (part A). The constant-pressure process requires more heat than the constant-volume process because energy is lost by the system through work. Thus, C_P is greater than C_V.

Degrees of Freedom

The values of C_V or C_P in Table 21.3 depend on the structure of the gas molecules. Monatomic molecules, those made up of just one atom, are well modeled as particles with no internal structure. Diatomic, triatomic, and polyatomic molecules are not well modeled as particles; instead, we must account for their internal structure.

To find C_V for a monatomic ideal gas, start with $E_{\text{th}} = \frac{3}{2}nRT$ (Eq. 21.2) to find an expression for ΔE_{th}.

$$\Delta E_{\text{th}} = \tfrac{3}{2}nR\Delta T \qquad (21.25)$$

Set $\Delta E_{\text{th}} = nC_V\Delta T$ (Eq. 21.20) equal to Equation 21.25:

$$\Delta E_{\text{th}} = nC_V\Delta T = \tfrac{3}{2}nR\Delta T \qquad (21.26)$$

Solve for C_V for a monatomic ideal gas:

$$C_V = \tfrac{3}{2}R \text{ monatomic} \qquad (21.27)$$

Equation 21.27 is in good agreement with experimental results: $C_V = 12.5 \text{ J/(mol}\cdot\text{K)}$ for a monatomic ideal gas, which is close to the values in Table 21.3.

More complicated molecules can rotate and vibrate, so they cannot be modeled as particles. Equation 21.25, $\Delta E_{\text{th}} = \frac{3}{2}nR\,\Delta T$, does not hold for complicated molecules. So, our next step is to find a more general expression for ΔE_{th}. We start by considering the ways molecules may move.

DEGREES OF FREEDOM
Major Concept

First, a particle (such as a monatomic molecule) can undergo translational motion in up to three independent directions—*x*, *y*, and *z*—and we say that it has three *degrees of freedom*. In thermodynamics, **degrees of freedom** are the number of independent ways a molecule can have energy. (It is possible for the particle to change its motion in one dimension without changing its motion in the other two dimensions.)

Now, consider a diatomic molecule (Fig. 21.25) modeled as two particles connected by a massless spring. The molecule can rotate around two independent axes (Fig. 21.25A), and the particles can also vibrate in and out from the center of mass (Fig. 21.25B). If the molecule vibrates, it has two forms of energy: kinetic energy due the motion of the particles and potential energy due to the compression and extension of the "spring." In addition, the entire molecule can have translational motion in three dimensions. A diatomic molecule therefore can have up to seven degrees of freedom:

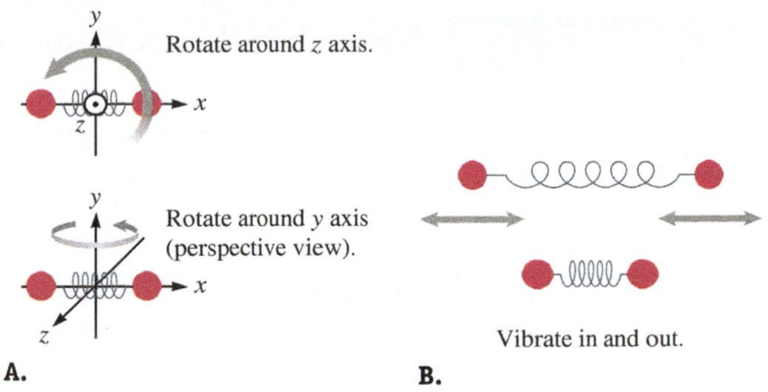

FIGURE 21.25 A diatomic molecule can **A.** rotate around two different axes and **B.** vibrate in and out.

- Three degrees of freedom for its translational kinetic energy in three dimensions
- Two degrees of freedom for its rotational kinetic energy around two different axes
- Two degrees of freedom for its kinetic energy and potential energy of vibration

Not all these seven degrees of freedom must be exhibited; the ones that are exhibited are known as the **active degrees of freedom**.

In Section 20-3, we found that the average kinetic energy of the particles in an ideal gas is $K_{avg} = \frac{3}{2}k_BT$. The "3" in the average kinetic energy is due to the individual particles' three degrees of freedom (Eq. 20.6). The **principle of equipartition of energy** states that energy is shared equally between the active degrees of freedom, each one having on average $\frac{1}{2}k_BT$ of energy.

PRINCIPLE OF EQUIPARTITION OF ENERGY ✪ Major Concept

How do the principle of equipartition of energy and a molecule's active degrees of freedom determine the molar specific heat of a gas? To answer, we need to generalize our derivation of C_V. We replace $\Delta E_{th} = \frac{3}{2}nR\Delta T$ (Eq. 21.25), which is good for monatomic molecules, with a more general expression for ΔE_{th} for polyatomic molecules:

$$\Delta E_{th} = \frac{(\text{d.o.f.})}{2}nR\Delta T \tag{21.28}$$

where "d.o.f." is the number of active *d*egrees *o*f *f*reedom. We set Equation 21.28 equal to $\Delta E_{th} = nC_V\Delta T$ (Eq. 21.20):

$$\Delta E_{th} = nC_V\Delta T = \frac{(\text{d.o.f.})}{2}nR\,\Delta T$$

and solve for C_V:

$$C_V = \frac{(\text{d.o.f.})}{2}R \tag{21.29}$$

A more complicated molecule may have more active degrees of freedom, and according to Equation 21.29, the corresponding gas has a greater molar specific heat. The molar specific heat of diatomic and more complicated molecules depends on the gas temperature because more degrees of freedom become active at higher temperatures.

At room temperature, the diatomic molecules in Table 21.3 have five active degrees of freedom: three for translational motion and two for rotational motion. According to Equation 21.29, such a diatomic gas should have a molar specific heat of

$$C_V = \frac{(\text{d.o.f.})}{2}R = \frac{5}{2}R = \frac{5}{2}[8.315\,\text{J}/(\text{mol}\cdot\text{K})] = 20.8\,\text{J}/(\text{mol}\cdot\text{K})$$

which is confirmed experimentally (Table 21.3).

Although the molar specific heats in Table 21.3 are for gases at room temperature, many of those values are, in fact, fairly good for a broad range of temperatures (150 K to 750 K). At much cooler temperatures, however, diatomic molecules tend to stop rotating, so they have only three active degrees of freedom, and their molar specific heat is $C_V = \frac{3}{2}R$. At much higher temperatures, diatomic molecules tend to rotate and vibrate, and their molar specific heat is $C_V = \frac{7}{2}R$.

EXAMPLE 21.8 Monatomic Versus Diatomic Gas

We have two containers; one holds 1 mol of a monatomic gas, and the other holds 1 mol of a diatomic gas. Both gases start with an initial temperature $T_i = 245$ K and reach a final temperature $T_f = 295$ K by a constant-volume process. Find the initial thermal energy $E_{\text{th},i}$ and heat Q_V absorbed by each gas.

INTERPRET and ANTICIPATE

There is 1 mol of each gas, but the diatomic gas has more degrees of freedom than the monatomic gas. Therefore, at the same temperature, we expect the thermal energy will be greater for the diatomic gas. Both gases experience the same change in temperature, so the thermal energy increases more for the diatomic gas than the monatomic gas. Therefore, because the process is constant-volume (no work), more heat must be required in the diatomic case.

SOLVE

Equipartition of energy allows $\frac{1}{2}k_BT$ of energy for each active degree of freedom.	$K_{\text{avg}} = (\text{d.o.f.})(\frac{1}{2}k_BT)$
For a gas, thermal energy is the number of molecules N multiplied by their average kinetic energy (Eq. 21.2). Write N in terms of n (number of moles) and N_A (Avogadro's number).	$E_{\text{th}} = NK_{\text{avg}} \qquad N = nN_A$ $E_{\text{th}} = nN_AK_{\text{avg}}$ $E_{\text{th}} = \frac{(\text{d.o.f.})}{2}nN_Ak_BT$
Use $N_Ak_B = R$ (Eq. 19.20).	$E_{\text{th}} = \frac{(\text{d.o.f.})}{2}nRT$
A monatomic gas has three degrees of freedom.	$E_{\text{th},i} = \frac{3}{2}nRT_i$ $E_{\text{th},i} = \frac{3}{2}(1\text{ mol})\left(8.315\frac{\text{J}}{\text{mol}\cdot\text{K}}\right)(245\text{ K})$ $E_{\text{th},i} = 3.06 \times 10^3\text{ J}$ (monatomic)
A diatomic gas has five active degrees of freedom in the given temperature range.	$E_{\text{th},i} = \frac{5}{2}nRT_i$ $E_{\text{th},i} = \frac{5}{2}(1\text{ mol})\left(8.315\frac{\text{J}}{\text{mol}\cdot\text{K}}\right)(245\text{ K})$ $E_{\text{th},i} = 5.09 \times 10^3\text{ J}$ (diatomic)
Because the process is constant volume (page 627), the change in thermal energy equals the heat absorbed. The change in thermal energy is given by Equation 21.28, where the change in temperature is the same (50 K) for both gases.	$\Delta E_{\text{th}} = Q_V = \frac{(\text{d.o.f.})}{2}nR\Delta T \qquad (21.28)$ $\Delta T = 50\text{ K}$
First solve for a monatomic gas.	$Q_V = \frac{3}{2}(1\text{ mol})\left(8.315\frac{\text{J}}{\text{mol}\cdot\text{K}}\right)(50\text{ K})$ $Q_V = 6.2 \times 10^2\text{ J}$ (monatomic)
Now solve for a diatomic gas.	$Q_V = \frac{5}{2}(1\text{ mol})\left(8.315\frac{\text{J}}{\text{mol}\cdot\text{K}}\right)(50\text{ K})$ $Q_V = 1.0 \times 10^3\text{ J}$ (diatomic)

CHECK and THINK

As expected, the diatomic gas has more thermal energy than the monatomic gas, and more heat is required to raise the temperature of the diatomic gas by the same amount.

EXAMPLE 21.9 Solids

The equipartition of energy principle holds for solids as well as gases. Model a solid (a crystal) as a collection of particles connected by springs (Fig. 21.26). Each particle can vibrate around its equilibrium position as if it were connected by springs to its neighbors. How many degrees of freedom are there per particle? What is the molar specific heat of a solid? Assume all degrees of freedom are active and remember that a solid's volume is nearly constant.

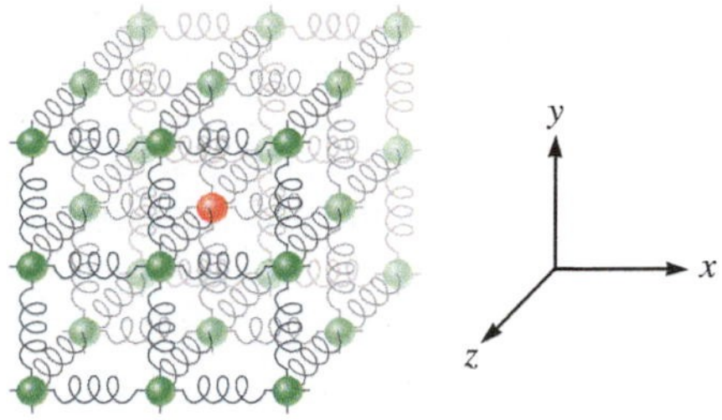

FIGURE 21.26 Particle-spring model of a solid.

INTERPRET and ANTICIPATE

Once we determine the number of active degrees of freedom, the equipartition of energy principle allows us to find the molar specific heat.

SOLVE

In Figure 21.26, the particle shown in red near the center of the solid is connected by springs to its neighbors so that it can oscillate in the x, y, and z directions independently. The particle has one degree of freedom for potential energy and kinetic energy in each of these directions.

6 degrees of freedom

The molar specific heat depends on the number of active degrees of freedom.

$$C_V = \frac{(\text{d.o.f.})}{2}R \tag{21.29}$$

$$C_V = \frac{6}{2}R = 3R = 3 \times 8.315 \frac{\text{J}}{\text{mol} \cdot \text{K}}$$

$$C_V = 24.9 \frac{\text{J}}{\text{mol} \cdot \text{K}}$$

CHECK and THINK

Our result agrees very well with the values for molar specific heat given in Table 21.1 for solids.

21-9 Adiabatic Processes Revisited

Adiabatic processes play an important role whenever a gas expands or contracts rapidly enough to experience a change in temperature without any exchange of heat. An example is the engine, which played a major role in the development of thermodynamics and is a major topic of Chapter 22. So, in this section, we take another look at the adiabatic process. Our goal is to derive equations that connect the state variables.

DERIVATION Adiabatic Path

Figure 21.27 shows a PV diagram and bar chart for an ideal gas that is initially at some cool temperature and then is adiabatically compressed (by a piston) until it reaches a higher temperature. Show that the adiabatic path on the PV diagram is given by

$$P_f V_f^{\gamma} = P_i V_i^{\gamma} = \text{constant} \tag{21.30}$$

where γ is the ratio of the specific heats:

$$\gamma \equiv \frac{C_P}{C_V} \tag{21.31}$$

(The ratio γ is given in Table 21.3.)

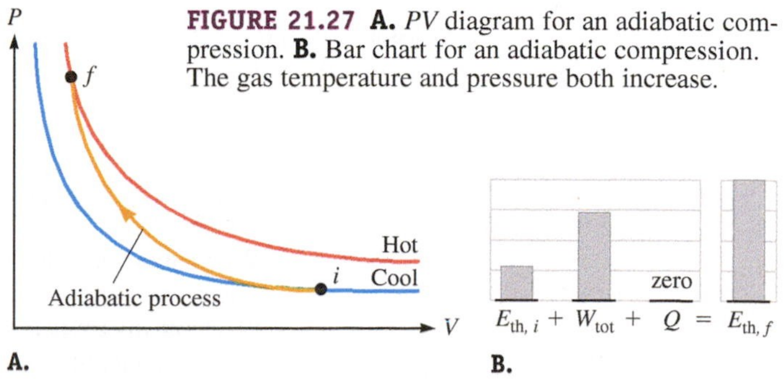

FIGURE 21.27 A. PV diagram for an adiabatic compression. **B.** Bar chart for an adiabatic compression. The gas temperature and pressure both increase.

Derivation continues on page 638 ▶

Consider a small increase in the gas's thermal energy dE_{th} as a result of a small amount of work dW done by the piston on the gas. No heat is exchanged. So, according to the first law of thermodynamics, the work done by the piston equals the change in thermal energy.	$dE_{\text{th}} = dW \quad (1)$
The piston compresses the gas, so the gas volume decreases by a small amount. The work dW done by the piston is positive, and the change in volume dV is negative.	$dW = -P\,dV$
Eliminate P using the ideal gas law $PV = nRT$.	$dW = -\dfrac{nRT}{V}dV \quad (2)$
The gas temperature rises slightly, and we can use $\Delta E_{\text{th}} = nC_V\Delta T$ (Eq. 21.26) to write an expression for dE_{th} in terms of dT.	$dE_{\text{th}} = nC_V\,dT \quad (3)$
Substitute Equations (2) and (3) into Equation (1) and simplify.	$nC_V\,dT = -\dfrac{nRT}{V}dV$ $\dfrac{dT}{T} = -\dfrac{R}{C_V}\dfrac{dV}{V} \quad (4)$
Using Equation 21.24 and $\gamma \equiv C_P/C_V$ (Eq. 21.31), we can rewrite the constant R/C_V in terms of γ.	$C_P - C_V = R \quad (21.24)$ $\dfrac{C_P}{C_V} - \dfrac{C_V}{C_V} = \dfrac{R}{C_V}$ $1 - \gamma = -\dfrac{R}{C_V} \quad (5)$
Substitute Equation (5) into Equation (4) and integrate. The limits on the left are from T_i to T_f, and the limits on the right are from V_i to V_f. The gas temperature rises ($T_i < T_f$) and the volume decreases ($V_i > V_f$).	$\dfrac{dT}{T} = (1-\gamma)\dfrac{dV}{V}$ $\displaystyle\int_{T_i}^{T_f}\frac{dT}{T} = (1-\gamma)\int_{V_i}^{V_f}\frac{dV}{V}$ $\ln T_f - \ln T_i = (1-\gamma)(\ln V_f - \ln V_i)$
Combine the natural logarithm functions (Appendix A). Equation 21.32 is important because it connects two state variables, temperature and volume.	$\ln\dfrac{T_f}{T_i} = (1-\gamma)\ln\dfrac{V_f}{V_i} = \ln\left(\dfrac{V_f}{V_i}\right)^{(1-\gamma)}$ $\dfrac{T_f}{T_i} = \left(\dfrac{V_f}{V_i}\right)^{(1-\gamma)} \quad (21.32)$
Use the ideal gas law to eliminate temperatures and simplify.	$\dfrac{P_fV_f}{P_iV_i} = \left(\dfrac{V_f}{V_i}\right)^{(1-\gamma)} = \dfrac{V_f}{V_i}\left(\dfrac{V_f}{V_i}\right)^{-\gamma}$ $\dfrac{P_f}{P_i} = \left(\dfrac{V_f}{V_i}\right)^{-\gamma} = \left(\dfrac{V_i}{V_f}\right)^{\gamma}$ $P_fV_f^{\gamma} = P_iV_i^{\gamma} = \text{constant}$ ✓ $\quad (21.30)$

COMMENTS

Equations 21.30 and 21.32 along with the ideal gas law hold whenever an ideal gas expands or contracts adiabatically. We will return to these equations when we study engines in Chapter 22.

21-10 Conduction, Convection, and Radiation

When there is a temperature difference between a system and its environment, heat flows from the hotter to the colder of these two. In this section, we consider three mechanisms for heat transfer: conduction, convection, and radiation. These mechanisms do not change the first law of thermodynamics; instead, they are three specific ways to exchange energy by heat. In a given situation, one or more of them may be operating.

Conduction

Thermal conduction (or simply **conduction**) is a process for transporting thermal energy requiring a physical connection between the system and its environment through an object called a **conductor**.

CONDUCTION **Special Case**

When you hold the dry end of a metal spoon whose other end is submerged in a hot mug of coffee, your fingers may feel warm or even burn. That is because heat is transported from the hot coffee (the environment) to your fingers (the system) by conduction through the spoon. On the microscopic level, the spoon's molecules that are in contact with the hot coffee begin to vibrate vigorously due to collisions with the hot coffee molecules. Thermal energy is transported throughout the spoon until thermal equilibrium is reached.

The power (rate at which heat is transported) by conduction depends on the temperature difference between the system and the environment. Figure 21.28 shows a cylindrical conductor connecting a hot reservoir with temperature T_{hot} to a cooler reservoir with temperature T_{cool}. The conductor is made out of a single substance of length ℓ and cross-sectional area A. The amount of heat Q that flows per time interval Δt has dimensions of power and is given by

$$\frac{Q}{\Delta t} = k\frac{A}{\ell}(T_{hot} - T_{cool}) \tag{21.33}$$

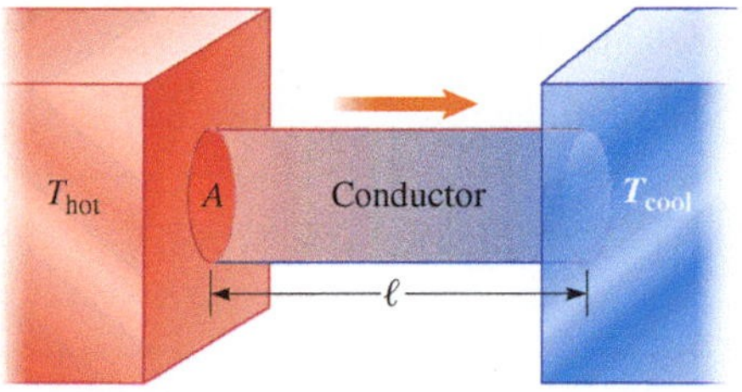

FIGURE 21.28 A conductor transports heat from the hot reservoir at T_{hot} to the cool reservoir at T_{cool}.

where k is the **thermal conductivity** of the substance. Thermal conductivity has the dimensions of power per length per temperature, with SI units W/m · K. Thermal conductivity depends on the type of material, and selected values are provided in Table 21.4. A substance such as polyurethane with a low thermal conductivity is called a good **thermal insulator** or simply an **insulator**. A substance such as a metal with a high thermal conductivity is called a good **thermal conductor** or simply a **conductor**. Many metal objects conduct heat effectively because metals have free electrons that may travel for great distances throughout the sample. These free electrons collide with one another and with the metal ion cores, speeding up the transportation of thermal energy.

TABLE 21.4 Thermal conductivity.

Substance	$k\left(\frac{W}{m \cdot K}\right)$	Conductor or insulator?
Air	0.026	Insulator
Aluminum	235	Conductor
Brick	0.84	Insulator
Concrete	0.84	Insulator
Copper	401	Conductor
Cork	0.042	Insulator
Diamond[a]	2300	Conductor
Glass	1.0	
Goose down	0.025	Insulator
Fiberglass	0.048	Insulator
Helium	0.15	Insulator
Human tissue	0.2	Insulator
Hydrogen	0.18	Insulator
Ice	2	
Iron	79.5	Conductor
Lead	35	Conductor
Polyurethane	0.024	Insulator
Silver	428	Conductor
Stainless steel	14	
Water	0.56	Insulator
Wood	0.11	Insulator
Wool	0.040	Insulator

[a]Diamond is not a metal, but it is a good thermal conductor. It is not a good electrical conductor, however.

CASE STUDY Part 2: Why You Wear Clothes

From Table 21.4, air has low conductivity and is therefore a good insulator. You probably feel most comfortable when your skin temperature is 34°C. How can you maintain such a high skin temperature when you are in a room with an air temperature of 22°C? As heat is transferred from your body to the surrounding air, the air around you would eventually reach 34°C, and heat would no longer be transferred from you to the air. This temperature might keep you comfortable even without clothing if you remained stationary with a warm bubble of surrounding air. People usually need to move around, however, and air also moves, so each of us is continually surrounded with fresh, cool air. Clothing traps air near the skin, and this trapped air moves around with the wearer so that there is much less conduction of heat than without clothing. Cold-weather clothing made of wool or down feathers contains many pockets for trapping air. Many animals stay warm by trapping air in their fur or feathers. When they are particularly cold, their fur or feathers stand up, creating large air pockets. People do the same thing when they get "goose bumps," raising the hair on their arms and legs. This hair is mostly vestigial, however, so goose bumps don't help people trap very much air.

Water is more than 21 times more conductive than air (Table 21.4), so you feel cold when you are in water at 85°F (30°C), but warm when you are in air at that temperature. If you wish to swim in much colder water, you would need to wear a wetsuit or a drysuit. Wearing a wetsuit in water is much like wearing clothes in air. You heat the water near your skin, which the wetsuit traps, so as you swim you are surrounded by a bubble of warm water. Of course, no wetsuit is perfect, and you will continue to lose energy slowly. If you want to swim in much colder water, you must use a drysuit, which keeps a layer of warm air near your skin. Because air is a better insulator than water, you lose less energy by conduction. In Example 21.10, we explore the effects of a drysuit.

EXAMPLE 21.10 CASE STUDY Your Body in Cold Water

You want to swim in a lake at 15°C while wearing a drysuit that surrounds you with a layer of air 0.75 cm thick. When you enter the water, your skin temperature is a comfortable 34°C. You can estimate your surface area A in square meters from your mass m in kilograms and your height h in meters from the following empirical relationship (Problem 1.60):

$$A = 0.202\, m^{0.425} h^{0.725}$$

For this example, assume $m = 84$ kg and $h = 1.8$ m. (You may wish to recalculate with your own specifications.)

A Find the power you lose due to conduction.

INTERPRET and ANTICIPATE

The system is your body, the conductor is the layer of air, and the environment is the cold lake.

SOLVE First, find your surface area A.	$A = 0.202 m^{0.425} h^{0.725}$ $A = (0.202)(84\,\text{kg})^{0.425}(1.8\text{ m})^{0.725} = 2.0\text{ m}^2$
Find the power you lose through conduction from Equation 21.33 using the conductivity of air listed in Table 21.4. The temperature difference can be calculated in Celsius degrees or kelvin.	$\dfrac{Q}{\Delta t} = k\dfrac{A}{\ell}(T_{\text{hot}} - T_{\text{cool}})$ $\dfrac{Q}{\Delta t} = \left(0.026\dfrac{\text{W}}{\text{m}\cdot\text{K}}\right)\left(\dfrac{2.0\,\text{m}^2}{7.5\times 10^{-3}\,\text{m}}\right)(34°\text{C} - 15°\text{C})$ $\dfrac{Q}{\Delta t} = 130\,\text{W}$

CHECK and THINK

This result is comparable to the 116 W (= 116 J/s) we know that a person loses under normal conditions (Example 21.3).

B If you stay in the water for half an hour, how much energy do you lose, and what is your skin temperature? Assume the specific heat of skin is the value given in Example 21.3 $(3470\text{ J/kg}\cdot\text{K})$.

INTERPRET and ANTICIPATE

Because we know the power lost, we can find the energy lost in half an hour by multiplying by the time. This energy is lost as heat, which is related to change in temperature. We expect that your skin temperature will be lower than 34°, but—assuming half an hour is a safe amount of time to spend in the lake—we expect that your temperature will still be much warmer than the water temperature.

SOLVE The total energy lost by conduction is the power lost multiplied by the time you are in the lake.	$\dfrac{Q}{\Delta t} = 130\text{ J/s}$ $Q = (130\text{ J/s})(0.5\times 3600\text{ s}) = 2.3\times 10^5\text{ J}$

CHECK and THINK

This result is about the amount of energy you get from eating an apple.

SOLVE Find the change in your skin temperature from Equation 21.5 with the given mass $m = 84$ kg and specific heat $c = 3470\text{ J/kg}\cdot\text{K}$. Because heat flows out of the system, we must include a negative sign.	$Q = -mc\,\Delta T$ (21.5) $\Delta T = -\dfrac{Q}{mc} = -\dfrac{2.3\times 10^5\text{ J}}{(84\text{ kg})(3470\frac{\text{J}}{\text{kg}\cdot\text{K}})} = -0.79\text{ K}$

Solve for your skin temperature when you leave the lake. The temperature difference is the same in Celsius degrees as it is in kelvin.	$T_f - T_i = -0.79°C$ $T_f = T_i - 0.79°C = 34°C - 0.79°C$ $T_f = 33.2°C$

CHECK and THINK

As expected, your skin temperature is slightly cooler than when you first went in the lake. In Problem 62, you will compare this result to what you would find if you didn't use a drysuit.

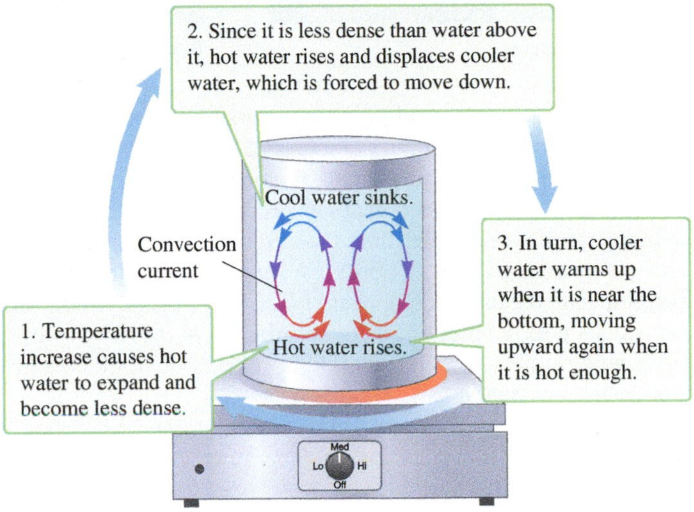

FIGURE 21.29 Convection cells bring hot water upward and cold water downward.

Convection

On a cold day, you might warm your hands by holding them near a warm fire (Fig. 21.1). The only substance between your hands and the fire is air, and according to Table 21.4, air is not a good conductor. So, how can heat get to you?

CONVECTION Special Case

In fluids (gases and liquids), the **convection** process transports heat by the large-scale motion of molecules through the fluid. If you place a pot of cool water on a hot stove, a cycle of rising warm fluid and sinking cool fluid sets up convection currents (Fig. 21.29).

An important difference between convection and conduction is that in conduction, the molecules vibrate near their equilibrium positions but do not travel throughout the conductor. During convection, molecules travel throughout the fluid, so convection transports heat and matter. For example, when you make a pot of coffee, aromatic coffee molecules are carried by convection currents in the air (Example 20.4).

Radiation

RADIATION Special Case

Life on the Earth depends on energy transferred from the Sun. There is essentially no medium between the Earth and the Sun, however, so there can be no conduction and no convection. Instead, heat is transferred by **thermal radiation**, electromagnetic waves that can travel through empty space and carry energy (Chapter 34). One of the primary examples of electromagnetic radiation is visible light. Electromagnetic waves are characterized by their wavelength. Ultraviolet (UV) radiation has a shorter wavelength than light. You cannot see UV radiation, but when your skin is exposed to it, the skin may tan or even burn. Infrared (IR) radiation has a longer wavelength than light, and you feel its effect when you stand in front of a hot fire. In fact, when you are near a fire, most of the heat transferred to you is by radiation; convection plays only a minor role.

The power emitted by an object through radiation depends on its surface area A, temperature T, and composition. The amount of heat Q radiated away per time interval Δt is given by the **Stefan-Boltzmann equation**:

$$\frac{Q}{\Delta t} = \sigma \varepsilon A T^4 = \text{power lost through radiation} \qquad (21.34)$$

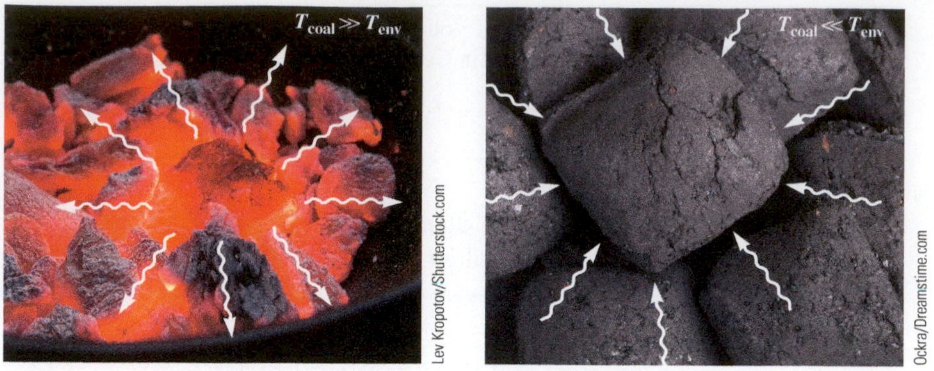

A. Glowing charcoal **B.** Black charcoal

FIGURE 21.30 A. When the temperature of an object is warmer than its surroundings, it gives off more radiation than it absorbs. **B.** When the temperature of an object is cooler than its surroundings, it absorbs more radiation than it gives off.

TABLE 21.5 Emissivity.

Substance	ε
Aluminum	0.02
Brass	0.03
Brick (red)	0.93
Carbon	0.95
Coal	0.95
Cotton cloth	0.77
Cement	0.96
Glass	0.80
Gold (polished)	0.02
Ice	0.97
Iron (dull)	0.94
Iron (polished)	0.28
Platinum	0.05
Soil	0.38
Snow	0.82–0.89
Silver (polished)	0.01
Water	0.67
Wood	0.80–0.90

where ε (Greek letter epsilon) is the **emissivity**, a number between 0 and 1 that depends on the type of material (Table 21.5). Emissivity is a measure of a substance's ability to emit or absorb radiation. A good emitter ($\varepsilon \approx 1$) is also a good absorber. For example, a very black object such as coal is usually a good emitter, with $\varepsilon = 0.95$. When coal is much hotter than its surroundings, it gives off considerable radiation and glows brightly (Fig. 21.30A). When coal is cooler than its surroundings, it absorbs much radiation and appears black (Fig. 21.30B). On the other hand, a shiny object is a poor emitter and poor absorber ($\varepsilon \approx 0$); most of the radiation is reflected off a shiny object such as when light reflects off a mirror. In SI units, the **Stefan-Boltzmann constant** σ is

$$\sigma = 5.6703 \times 10^{-8} \frac{\text{W}}{\text{m}^2 \cdot \text{K}^4} \tag{21.35}$$

If a system has a temperature T_{sys} above absolute zero and is in an environment with a temperature T_{env}, the system will absorb radiation from the environment and emit radiation at the same time. The net power absorbed is the power absorbed minus the power emitted:

$$\left.\frac{Q}{\Delta t}\right|_{net} = \left.\frac{Q}{\Delta t}\right|_{abs} - \left.\frac{Q}{\Delta t}\right|_{emit} = \sigma\varepsilon A T_{env}^4 - \sigma\varepsilon A T_{sys}^4$$

$$\left.\frac{Q}{\Delta t}\right|_{net} = \sigma\varepsilon A(T_{env}^4 - T_{sys}^4) = \text{net power absorbed} \tag{21.36}$$

Power Absorbed from Sunlight

A star such as our Sun has an emissivity of essentially 1, so for a star we write Equation 21.34 as

$$\frac{Q}{\Delta t} = 4\pi R^2 \sigma T^4 \tag{21.37}$$

where R is the radius of the star. The power radiated by the Sun is

$$\left.\frac{Q}{\Delta t}\right|_{\odot} = 4\pi R_{\odot}^2 \sigma T_{\odot}^4 = 4\pi(6.96 \times 10^8 \text{ m})^2 \left(5.67 \times 10^{-8} \frac{\text{W}}{\text{m}^2 \cdot \text{K}^4}\right)(5770 \text{ K})^4$$

$$\left.\frac{Q}{\Delta t}\right|_{\odot} = 3.83 \times 10^{26} \text{ W}$$

In the same way that a spherical sound wave's intensity I (power per unit area) decreases with distance (Section 17-7), the intensity of the Sun's radiation also decreases. The Sun's intensity at the Earth must be calculated from Equation 17.23:

$$I = \frac{1}{4\pi r^2}\left(\left.\frac{Q}{\Delta t}\right|_{\odot}\right)$$

where r is the distance between the Sun and the Earth. The intensity of the Sun's radiation at the top of the Earth's atmosphere is called the **solar constant** and is

$$I = \frac{1}{4\pi(1.50 \times 10^{11}\ \text{m})^2}(3.83 \times 10^{26}\ \text{W}) = 1350\ \text{W/m}^2$$

According to NASA, the averaged yearly solar constant measured by satellites is 1368 W/m^2. The Earth's atmosphere absorbs some of this radiation, and the intensity near the surface of the Earth on a clear day is about 1000 W/m^2.

Let's come up with an expression for the power absorbed by an object such as bowl of water lying in the Sun (Fig. 21.31). First, we need to multiply the radiation intensity by the effective area of the object. The effective area is the area perpendicular to the Sun's rays. We must also multiply by the emissivity ε of the object to account for the object's ability to absorb radiation. The amount of heat Q absorbed by an object per time interval Δt is given by

$$\frac{Q}{\Delta t} = (1000\ \text{W/m}^2)\varepsilon A \cos\theta = \text{power absorbed from Sun's radiation} \quad (21.38)$$

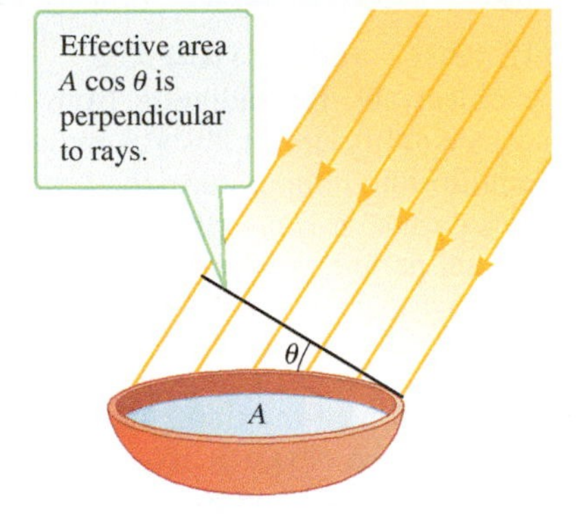

FIGURE 21.31 The power absorbed by a bowl of water exposed to sunlight depends on its cross-sectional area and the angle θ.

CASE STUDY Part 3: What Should You Wear in the Desert?

You probably like to wear white clothes in hot weather because white objects generally absorb less radiation than do dark or black objects. Bedouins who dwell in deserts such as the Sahara, Negev, and Sinai traditionally wear black robes, including a black hood (Fig. 21.32). If black robes absorb more radiation than white robes, why do Bedouins wear black? This question was answered by four researchers in an article published in *Nature* in 1980.[1] The researchers found that a black robe can be as much as 6°C warmer than a white robe, but the skin temperature of the wearer does not depend on the color of the robe. The hot black robe warms the air inside, which escapes through the porous fabric while fresh cooler air enters through the bottom of the robe. Thus, the wearer has his own convection cell with a breeze that makes the wearer more comfortable, if not cooler. Also, because the white robe reflects a lot of sunlight, looking at a white robe in the desert may be as uncomfortable to the eyes as sunlight reflected off snow, which can be blinding.

[1] A. Shkolnik, C. R. Taylor, V. Finch, and A. Borut, "Why Do Bedouins Wear Black Robes in Hot Deserts?" *Nature* 283: 373–374 (1980).

FIGURE 21.32 Bedouins wear black robes in the desert. The black robes allow for stronger convection currents, keeping the wearer cool and comfortable.

EXAMPLE 21.11 CASE STUDY What Should You Wear in Orbit?

When we think of outer space, we often think of a very cold environment and suppose a spacesuit must be designed to keep astronauts warm. It is true that objects with little exposure to the Sun's radiation get very cold. The dark side of the Moon has a low temperature of about 110 K (–163°C). If you were to come into contact with such a cold object, you would lose energy by conduction, so spacesuits are designed to keep the feet and hands warm. The interior of a spacesuit can become very hot, however, especially if the suit is worn in sunlight. For example, on the sunny side of the Moon, the temperature is 387 K (114°C).

An object such as an astronaut in space is oriented so that one side points toward the Sun (Fig. 21.33). The object maintains this orientation for a long time so that the side oriented toward the Sun comes into equilibrium. Find the equilibrium temperature of the sunny side of an object near the Earth (1 AU from the Sun).

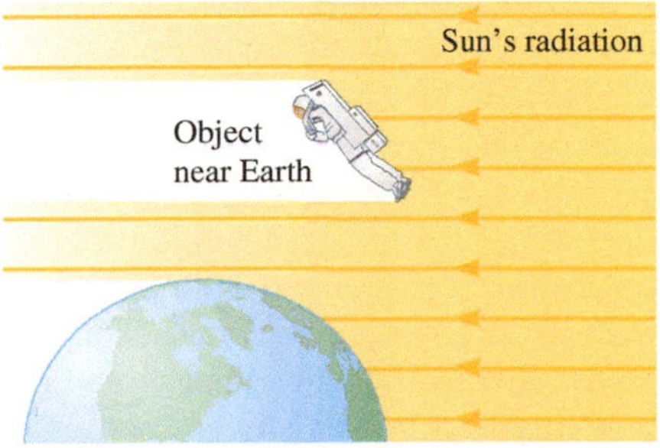

FIGURE 21.33

Example continues on page 644 ▶

INTERPRET and ANTICIPATE

The object absorbs radiation from the Sun and also emits radiation. Before the sunny side reaches equilibrium, it absorbs radiation at a higher rate than it emits. When the object reaches its equilibrium temperature, the power absorbed equals the power emitted, so the net power absorbed is zero. Because the Moon is near the Earth, has a very thin atmosphere, and has a slow rotational period, we can check our results against the high temperature of the Moon.

SOLVE

The object is near the Earth and above the atmosphere, so the intensity of sunlight intercepting the object is given by the solar constant. To find the power absorbed, modify Equation 21.38 by using the solar constant.

$$\left.\frac{Q}{\Delta t}\right|_{\text{abs}} = (1350\,\text{W/m}^2)\varepsilon A \cos\theta$$

The sunlight is direct as shown in Figure 21.33, so we set $\theta = 0$.

$$\left.\frac{Q}{\Delta t}\right|_{\text{abs}} = (1350\,\text{W/m}^2)\varepsilon A$$

In equilibrium, the net power absorbed is zero (power absorbed equals power emitted). The power emitted is given by Equation 21.34, where T_{obj} is the equilibrium temperature of the object.

$$\left.\frac{Q}{\Delta t}\right|_{\text{abs}} = \left.\frac{Q}{\Delta t}\right|_{\text{emit}}$$

$$(1350\,\text{W/m}^2)\varepsilon A = \sigma\varepsilon A T_{\text{obj}}^4$$

Solve for T_{obj}. The Stefan-Boltzmann constant is given in Equation 21.35.

$$T_{\text{obj}} = [(1350\,\text{W/m}^2)/\sigma]^{1/4} = \left[\frac{1350\,\text{W/m}^2}{5.670 \times 10^{-8}\,(\text{W/m}^2 \cdot \text{K}^4)}\right]^{1/4}$$

$$T_{\text{obj}} = 393\,\text{K} = 120°\text{C}$$

CHECK and THINK

As expected, our result is very close to the high temperature on the sunny side of the Moon and shows that objects in direct sunlight near the Earth get very hot. Spacesuits are designed to keep an astronaut at a comfortable temperature. Spacesuits are well insulated with Mylar and other materials that help prevent the absorption of radiation, but the astronaut and the electrical components inside the suit warm up just as they would on the Earth. On the Earth, people and machines give off heat through conduction, convection, and radiation. In space, there are generally no conductors or air, so heat cannot be transferred in space through conduction or convection. Radiation is too slow to keep the astronaut at a constant comfortable temperature. In fact, if the temperature inside the suit becomes warm enough, the astronaut will sweat, making the inside of the suit uncomfortably humid. The solution is a liquid cooling garment (Fig. 21.34) worn against the astronaut's skin and laced with tubes of water. Water tubes are also used to cool the electronics. Heat is transmitted to the water through conduction, and thermal energy in the water is then ejected into space either through evaporation or sublimation. Each pound of water evaporated removes about 1000 J of thermal energy.

FIGURE 21.34 A liquid cooling garment for astronauts.

One final note: The astronaut shown in Figure 21.33 is not exposed to radiation on the side facing away from the Sun. In the absence of heat transfer from the sunny side of the object, the dark side of the object would radiate energy into space until it reaches the temperature of its environment (roughly 3 K in deep space). Temperatures in the solar system are more moderate than in deep space. Objects such as planets or moons rotate so that both sides of the object have some time in the Sun. Planets and moons also have an atmosphere (if very thin sometimes), enabling heat transfer by convection. The Earth's Moon has a rotational period of roughly one month, so any particular spot on the Moon spends roughly two weeks per month in darkness and two weeks in sunshine. This rotation plus the Moon's thin atmosphere maintain its surface low temperature around 110 K, well above the temperature of deep space.

SUMMARY

Underlying Principles

The **first law of thermodynamics** is a statement of conservation of energy for a system that may exchange energy with its environment by work or through heat:

$$W_{tot} + Q = \Delta E_{th} \quad (21.1)$$

where W_{tot} and Q are positive if energy is transferred from the environment to the system.

Major Concepts

1. **Heat** is defined as energy Q transferred between a system and its environment due to their temperature difference.
2. **Thermal energy** is the sum of each microscopic particle's kinetic energy plus the total potential energy stored in the forces between microscopic particles in a system.
3. **Heat capacity** is the constant of proportionality in Equation 21.3:
$$\Delta E_{th} = \mathcal{C}\,\Delta T \quad (21.3)$$
4. A **thermodynamic process** is a way for system to change from an initial state to a final state.
5. The mass specific heat capacity or simply the **specific heat** c is related to the heat capacity by $\mathcal{C} = mc$. When no work is done on the system, the heat transferred is then given by
$$Q = mc\Delta T \quad (21.5)$$
6. **Degrees of freedom** indicate the number of independent ways a molecule can have energy. Not all degrees of freedom are active at a given temperature.
7. The **principle of equipartition of energy** states that energy is shared equally between the active degrees of freedom, each one having on average $\frac{1}{2}k_BT$ of energy.

Special Cases

Six specific thermodynamic processes:

Process	Heat entering system	Work done by environment (energy entering system)	Change in thermal energy
Adiabatic ($Q = 0$)	$Q = 0$	—	—
Isothermal ($\Delta T = 0$)	$Q = -W_{tot}$	For an ideal gas, $W_{tot} = Nk_BT \ln \frac{V_i}{V_f}$ (21.15)	$\Delta E_{th} = 0$
Cyclic ($\Delta E_{th} = 0$)		$Q = -W_{tot}$	$\Delta E_{th} = 0$
Free expansion	$Q = 0$	$W_{tot} = 0$	$\Delta E_{th} = 0$
Constant volume ($\Delta V = 0$)	$Q = nC_V\Delta T$	$W_{tot} = 0$	$\Delta E_{th} = Q$
Constant pressure ($\Delta P = 0$)	$Q = nC_P\Delta T$	$W_{tot} = -P\Delta V$ (21.16) For an ideal gas, $W_{tot} = Nk_BT_i\left(1 - \frac{V_f}{V_i}\right)$ (21.17)	—

For an **adiabatic** expansion or compression of an ideal gas, the state variables P, V and T are related by

$$\frac{T_f}{T_i} = \left(\frac{V_f}{V_i}\right)^{(1-\gamma)} \quad (21.32)$$

and

$$P_fV_f^{\gamma} = P_iV_i^{\gamma} \quad (21.30)$$

There are three ways to transfer heat.

1. **Conduction** is a process for transporting thermal energy requiring a physical connection between the system and its environment through an object (a *conductor*). The heat Q flowing per time interval Δt as a result of conduction is
$$\frac{Q}{\Delta t} = k\frac{A}{\ell}(T_{hot} - T_{cool}) \quad (21.33)$$
where k is the **thermal conductivity**.
2. **Convection** transports heat by the large-scale motion of molecules through a fluid.
3. (Thermal) **radiation** transfers heat via electromagnetic waves. The heat Q radiated away per time interval Δt is given by the **Stefan-Boltzmann equation**:
$$\frac{Q}{\Delta t} = \sigma\varepsilon\, AT^4 \quad (21.34)$$
where ε is the **emissivity** and ε is the **Stefan-Boltzmann constant.**

Tools

We modify the **energy bar charts** from Chapters 8 and 9 for the first law of thermodynamics by adding a bar for heat (Q) and removing the mechanical energy bars (K and U). See Figure 21.13, page 625.

PROBLEMS AND QUESTIONS

A = algebraic C = conceptual E = estimation G = graphical N = numerical

21-1 What Is Heat?

1. **C** Come up with a word or phrase to replace *heat*. Explain how your new term fits the concept of heat.
2. **C** To keep the contents of a house thermally isolated from the external environment, the walls and roof of the building are insulated. **a.** Is it fair to say that *insulation keeps heat in*? **b.** Is insulation only important in the cold months? Explain.

Problems 3 and 56 are paired.

3. **C** You extend an impromptu invitation to a friend for dinner. The only food you have is a couple of frozen steaks. You wish to defrost the steaks before grilling them. You defrost one by using your microwave oven, and you defrost the other by placing it in a bowl of very warm water. In each case, decide whether energy is transferred by heat or by work. Explain your answers.
4. **C** If a system is in thermal equilibrium with its environment, is it possible to transfer energy from the system to the environment through heat, work, both, or neither? Is it possible to transfer energy from the environment to the system through heat, work, both, or neither? Explain your answers.

21-2 How Does Heat Fit into the Conservation of Energy?

5. **C** The words *isolated* and *insulated* sound similar. In our everyday language, we use these words metaphorically; we may say that "Bobby is isolated" or "Bobby is insulated." What subtle difference exists between these two phrases? Explain how each word is used in physics. What is the major difference between these two terms in physics?
6. **C** A can of paint is vigorously shaken by a paint mixer. Would you expect the temperature of the paint to increase, decrease, or remain unchanged? Explain your answer.
7. **N** Late for his morning workout, a 75.0-kg man eats a 200.0-Calorie energy bar for breakfast. **a.** What is the energy equivalent of this energy bar in joules? **b.** On a stair-climbing exercise machine, each "step" can be thought of as increasing the gravitational potential energy of the man–Earth system. If the equivalent height of each step is 12.0 cm, how many steps must the man take to work off the energy of the energy bar? Assume the efficiency of the human body in converting chemical energy to mechanical energy is 20.0%.
8. **C** A wooden block slides along a horizontal aluminum plane, eventually coming to rest. Explain what happened to the kinetic energy of the block–plane system.
9. **C** After driving your car for an hour, you place your hand on the hood of the car and discover that it is very hot. Explain how the temperature increased. Would your answer change if you were driving at night? Would your answer change if you were driving on a very cold winter night?

21-3 The First Law of Thermodynamics

Problems 10 through 13 are grouped.

10. **C** Is it possible for heat to be transferred to a system but for the system's thermal energy to remain unchanged? Explain.
11. **N** A system experiences no change in thermal energy when the environment does 15.0 J of work on it. According to the first law of thermodynamics, how much heat is transferred to the system? Does energy enter or leave the system by work? Does energy enter or leave the system by heat?
12. **C** According to the first law of thermodynamics, is it possible for the thermal energy of an object to decrease if heat is entering the object from its surroundings? Justify your answer.
13. **N** A system's thermal energy increases by 15.0 J when the environment does 15.0 J of work on it. According to the first law of thermodynamics, how much heat is transferred to the system? Does energy enter or leave the system by work? Does energy enter or leave the system by heat?
14. **C** CASE STUDY To maintain your current weight, suppose you must eat 2100 Cal per day. If you only eat 1900 Cal per day, you find that you lose weight. Explain why consuming less energy per day results in a mass loss. Where did the mass go?

21-4 Heat Capacity and Specific Heat

15. **N** If 30.0 g of milk at a temperature of 4.00°C is added to a 255-g cup of coffee at a temperature 90.0°C, what is the final temperature of the mixture? Assume coffee has a specific heat of 4.19×10^3 J/(kg · K) and milk has a specific heat of 3.93×10^3 J/(kg · K).
16. **N** A block of metal of mass 0.250 kg is heated to 150.0°C and dropped in a copper calorimeter of mass 0.250 kg that contains 0.160 kg of water at 30°C. The calorimeter and its contents are insulated from the environment and have a final temperature of 40.0°C upon reaching thermal equilibrium. Find the specific heat of the metal. Assume the specific heat of water is 4.190×10^3 J/(kg · K) and the specific heat of copper is 386 J/(kg · K).

Problems 17 and 18 are paired.

17. **C** Is it possible for a cool bathtub of water to have more thermal energy than a hot cup of water? Explain.
18. **C** Suppose a bathtub of cool water has more thermal energy than a hot potato. If the hot potato is placed in the water, will heat flow from the potato to the water or from the water to the potato? Explain.
19. **N** If you wish to keep a cup of coffee warm, is it better to use a glass cup or a steel cup? Before you decide, find the final temperature of 0.125 kg water poured into **a.** a glass cup and **b.** a steel cup. In each case the water has an initial temperature of 95°C. Each cup has a mass of 0.130 kg and an initial temperature of 22°C. Comment on your results.

20. **N** From Table 21.1, the specific heat of milk is 3.93×10^3 J/(kg · K), and the specific heat of water is 4.19×10^3 J/(kg · K). Suppose you wish to make a large mug (0.500 L) of hot chocolate. Each liquid is initially at 5.00°C, and you need to raise their temperature to 80.0°C. The density of milk is about 1.03×10^3 kg/m³, and the density of water is 1.00×10^3 kg/m³. **a.** How much heat must be transferred in each case? **b.** If you use a small electric hot plate that puts out 455 W, how long would it take to heat each liquid?

21. **N** A 125-g sample of lead with specific heat $c_{Pb} = 128$ J/(kg · °C) at 85.0°C and a 320-g sample of silver with $c_{Ag} = 230.0$ J/(kg · °C) at 34.0°C are added to 0.500 kg of water at 22.0°C where $c_W = 4190$ J/(kg · °C) in an insulated container. Assuming the system is thermally isolated, what is its final equilibrium temperature?

22. **E** You bought a new electric teakettle. The manufacturer claims that the teakettle's power output is 5.2 kW. You decide to test it by heating 1 L of water. The water is initially at room temperature, and in about 3 minutes it reaches its boiling point. What do you make of the manufacturer's claim?

23. **N** An ideal gas is confined to a cylindrical container with a movable piston on one end. The 3.57 mol of gas undergo a temperature change from 300.0 K to 350.0 K. If the total work done on the gas during this process is 1.00×10^4 J, what is the energy transferred as heat during this process? Is the heat flow into or out of the system?

24. **N** Suppose 5.15 kg of helium gas is at a temperature of 30.0°C. The specific heat of helium is 5193 J/kg · K. **a.** Find the heat exchanged between this system and the surroundings when the temperature of the gas is lowered to 10.0°C. **b.** What is the heat exchanged if the helium is heated from 30.0°C to 50.0°C? **c.** What is the heat capacity of helium?

21-5 Latent Heat

25. **N** You place frozen soup ($T = -17$°C) in a microwave oven for 3.5 min. The oven transfers 650 W to the soup. Model the soup as 0.35 kg of water (initially ice). What are the temperature and state of the soup when the oven stops?

26. **N** A 25-g ice cube at 0.0°C is heated. After it first melts, the temperature increases to the boiling point of water (100.0°C), and the water then boils to form 25 g of water vapor at 100.0°C. How much energy in total is added to the ice/water? Which process (melting, increasing temperature, or boiling) requires the most energy? Water has a latent heat of vaporization of 2.256×10^6 J/kg, a latent heat of fusion of 3.33×10^5 J/kg, and specific heat of 4190 J/(kg · K).

27. **N** What is the energy required to transform an ice cube of mass $m = 45.0$ g and temperature -5.00°C to steam at 120.0°C?

Problems 28 and 29 are paired.

28. **N** A jeweler must melt 45.75 g of silver. Assume the silver starts at room temperature and reaches a temperature just above its melting point ($T = 1241$ K). Use the values found in Tables 21.1 and 21.2. How much heat is transferred **a.** in order for the whole process to take place, **b.** in order to raise the silver to its melting point and, **c.** during the phase change? In what way are your results approximations?

29. **N** A jeweler must melt 45.75 g of copper. Assume the copper starts at room temperature ($T = 22.3$°C) and reaches a temperature just above its melting point ($T = 1361$ K). Use the values found in Tables 21.1 and 21.2. How much heat is transferred **a.** in order for the whole process to take place, **b.** in order to raise the copper to its melting point and, **c.** during the phase change? In what way are your results approximations? If you solved Problem 28, compare your results and comment.

30. **N** Two 40.0-g ice cubes initially at 0°C are added to 450 g of water at 22.0°C. Assuming this system is insulated and ignoring heat transfer with the glass, what is the equilibrium temperature of the mixture?

Problems 31 and 32 are paired.

31. **N** Consider the latent heat of fusion and the latent heat of vaporization for H_2O, 3.33×10^5 J/kg and 2.256×10^6 J/kg, respectively. How much heat is needed to **a.** melt 2.00 kg of ice and **b.** vaporize 2.00 kg of water? Assume the temperatures of the ice and steam are at the melting point and vaporization point, respectively.

32. **C** Consider the latent heat of fusion and the latent heat of vaporization for H_2O, 3.33×10^5 J/kg and 2.256×10^6 J/kg, respectively. Why is the latent heat of vaporization so much larger than the latent heat of fusion? Think about what is happening to the molecules as the phase changes occur in each case. Why is more energy needed to change the same amount of material to vapor than is necessary to melt it?

21-6 Work in Thermodynamic Processes

33. **N** How much energy is transferred from a system by heat during a thermodynamic process in which the work done on the system is 4.00×10^2 kJ and the system's thermal energy decreases by 2.50×10^2 kJ?

34. **N** A thermodynamic cycle is shown in Figure P21.34 for a gas in a piston. The system changes states along the path $ABCA$. **a.** What is the total work done by the gas during this cycle? **b.** How much heat is transferred? Does heat flow into or out of the system?

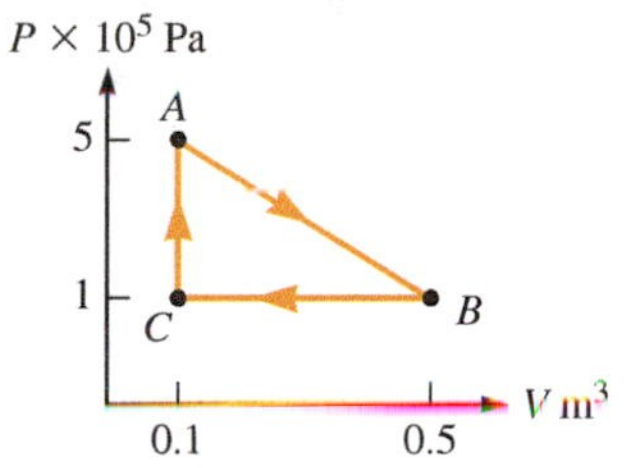

FIGURE P21.34

35. **A** During a process in which the pressure and volume of a gas are directly proportional, an ideal gas is taken from the initial state ($4P_0$, $4V_0$) to the final state (P_0, V_0). Assuming no gas is lost, what is the work done on the gas during this process?

36. **N** Figure P21.36 shows a cyclic thermodynamic process $ABCA$ for an ideal gas. **a.** What is the net energy transferred into the system by heat during each cycle? **b.** What would be the net energy transferred into the system by heat if the cycle followed the path $ACBA$ instead?

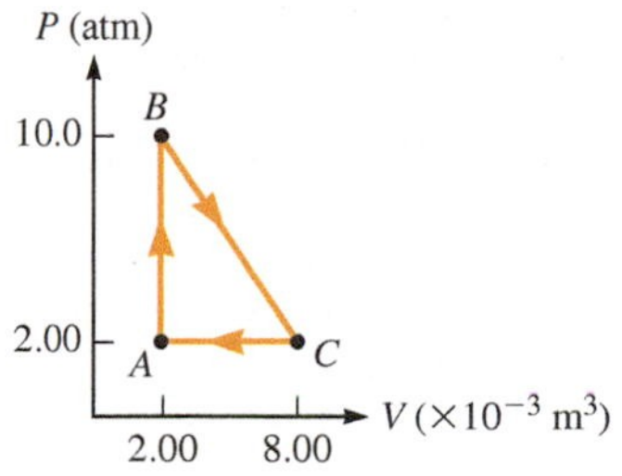

FIGURE P21.36

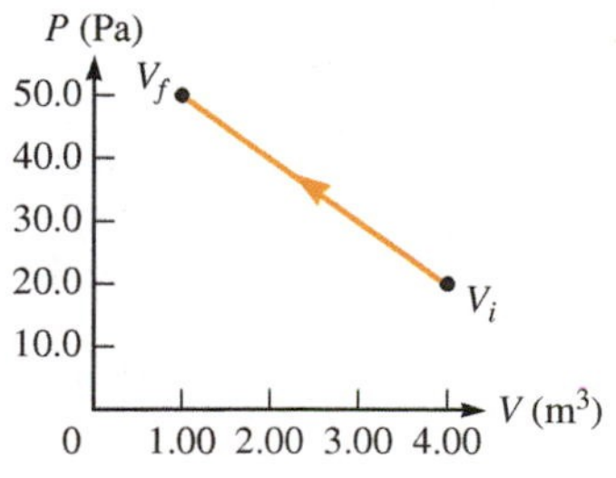

FIGURE P21.37

37. **N** Figure P21.37 shows a PV diagram for a gas that is compressed from V_i to V_f. Find the work done by the **a.** gas and **b.** environment during this process. Does energy enter the system or leave the system as a result of work?

38. **C** CASE STUDY You have injured an ankle and need to soak it in ice water. Can you model yourself as a heat reservoir if **a.** you put your foot in a bucket of ice water or **b.** you soak your whole body in a lake of ice water? Explain.

21-7 Specific Thermodynamic Processes

39. N A gas is at a constant pressure of 2.4×10^5 N/m^2. Its volume increases by 1.2×10^{-2} m^3 as 3200 J of energy is transferred to the gas in the form of heat. Find the change in thermal energy of the gas.

40. C You observe the state of a gas at two different times, measuring P, V, and T, and discover that its temperature is unchanged. How would comparing the initial and final values of the volume and pressure help you determine if the gas has undergone an isothermal expansion, a cyclic process, or some other process?

41. N During an isothermal process, 0.500 mol of an ideal gas expands to a final volume of 8.00 L and a pressure of 2.03×10^5 Pa while doing 2.50 kJ of work on its surroundings. **a.** What is the initial volume of the gas? **b.** What is the temperature of the gas?

42. An ideal gas at $P = 2.50 \times 10^5$ Pa and $T = 295$ K expands isothermally from 1.25 m^3 to 2.75 m^3. The gas then returns to its original state through a two-part process: constant-pressure followed by constant-volume.

a. G Draw a PV diagram for this gas.
b. C What is the change in thermal energy?
c. N Find the work done by the environment on the gas.
d. N Find the heat that flows into the gas.

43. N Suppose 3.67 mol of a monatomic gas is at a temperature of 300.0 K and undergoes a constant-pressure expansion from an initial volume of 0.025 m^3 to a final volume of 0.065 m^3. **a.** What is the final temperature of the gas? **b.** What is the change in thermal energy of the gas as it undergoes this process? **c.** What is the work done on the gas?

44. In Figure P21.44, an ideal gas undergoes a change in state from A to C by two different paths: ABC and AC.

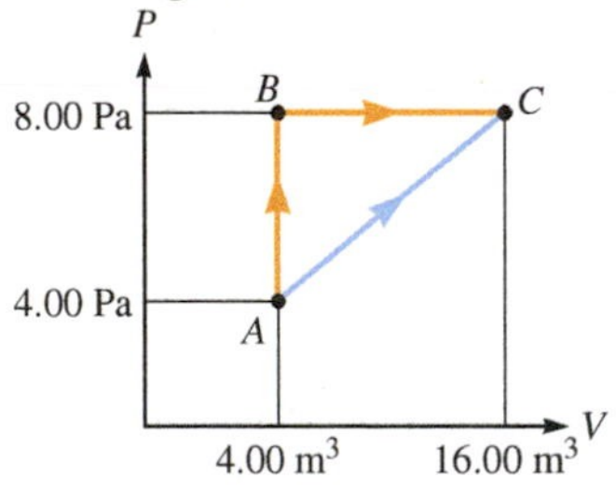

FIGURE P21.44

a. C Along which path is the least amount of work done by the gas?
b. N If the thermal energy of the gas at A is 12 J and the heat transferred to the gas along the path AC is 220 J, find the thermal energy of the gas at C.
c. N If the thermal energy of the gas at state B is 25 J, find the amount of heat added to the gas to change its state from A to B.

45. N Figure P21.45 shows a cyclic process $ABCDA$ for 1.00 mol of an ideal gas. The gas is initially at $P_i = 1.50 \times 10^5$ Pa, $V_i = 1.00 \times 10^{-3}$ m^3 (point A in Fig. P21.45). **a.** What is the net work done on the gas during the cycle? **b.** What is the net amount of energy added by heat to this gas during the cycle?

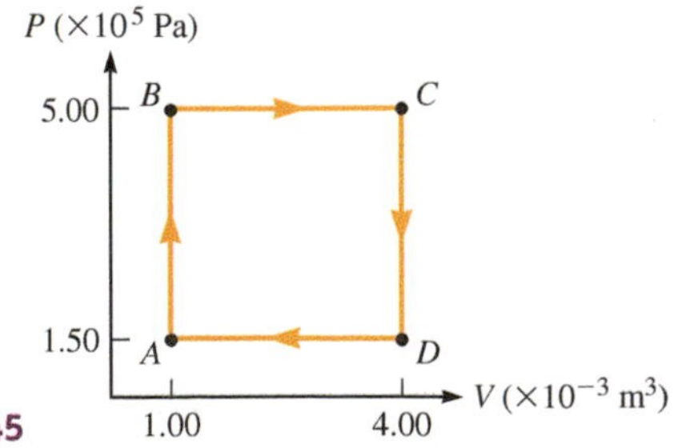

FIGURE P21.45

46. An ideal gas expands such that no heat is exchanged with the surroundings. It then undergoes a process for which no work is done to return to the original pressure. Finally, it undergoes a constant-pressure process to return to the initial state of the gas.

a. C What thermodynamic process is occurring in each of the three steps described? Explain your reasoning.
b. G Sketch this cycle on a PV diagram.

47. N A monatomic gas undergoes the set of thermodynamic processes shown in Figure P21.47, where $P_A = 2.00 \times 10^5$ Pa, $P_B = 1.00 \times 10^5$ Pa, $V_A = 0.0550$ m^3, and $V_C = 0.0750$ m^3. (The subscripts refer to the labeled points on the PV diagram.) There are 5.50 mol of gas. **a.** Find the temperature at B and C. **b.** What is the total work done on the gas during one cycle? **c.** How much heat is transferred when the gas goes from B to C?

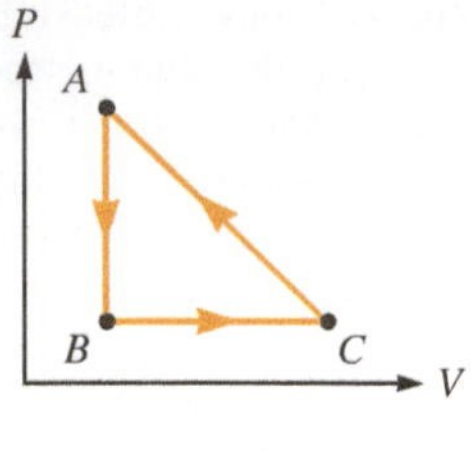

FIGURE P21.47

21-8 Equipartition of Energy

48. N Use the data in Table 21.3 to find the active number of degrees of freedom in a triatomic gas at room temperature.

49. N In a constant-pressure expansion, a diatomic gas performs 84 J of work on the environment at room temperature. Calculate the heat transferred to the gas during this process.

50. N A sample of a monatomic gas is in a container with a movable piston such that it is maintained at constant pressure. If 25.0 J of heat were transferred into the gas, the temperature would increase by 75.0°C. **a.** If, instead, the piston was initially locked in place such that the *volume* remains fixed, by how much would the temperature increase if 25.0 J of heat were transferred into the container? **b.** If, instead, the same number of moles of a *diatomic* gas were contained in the box and 25.0 J of heat were transferred in at constant pressure, by how much would the temperature increase? (Assume there are five active degrees of freedom for the diatomic gas.)

51. N A gas mixture contains 3.0 moles of oxygen (O_2) and 9.0 moles of argon (Ar) at a temperature of 300.0 K. What is the total thermal energy of the system consisting of both gases? Ignore any vibrational degrees of freedom.

52. N A diatomic gas is in a container with a fixed volume of 0.250 m^3. Assume the diatomic gas has five degrees of freedom. **a.** What is the temperature of the gas if there are 6.00 mol present and the gas is under a pressure of 1.50 atm? **b.** If the gas is to be heated by 225 K, how much heat is necessary to accomplish this task? **c.** Repeat parts (a) and (b) for a monatomic gas. Assume the monatomic gas has three degrees of freedom.

21-9 Adiabatic Processes Revisited

53. N An ideal gas is compressed adiabatically so that its volume is cut in half. **a.** If the gas is monatomic, find the ratio of its final pressure to its initial pressure, P_f/P_i. **b.** If the gas is diatomic at room temperature, find the ratio of its final pressure to its initial pressure, P_f/P_i. Explain the difference you find between the two gases in terms of the number of degrees of freedom.

54. A monatomic ideal gas is initially at a pressure of 1.00 atm in a 1.00-L cylindrical container with a piston on one side. It is compressed to a volume of 0.250 L.

a. G Plot the pressure versus volume during this process, assuming the process is either (i) adiabatic or (ii) isothermal.
b. N Determine the final pressure, assuming the process is either (i) adiabatic or (ii) isothermal.

55. A diatomic ideal gas at room temperature undergoes an adiabatic process such that its final pressure is 2.75 times its initial pressure.

a. C Did the gas expand or contract?
b. N What is the ratio of its final volume to its initial volume?

21-10 Conduction, Convection, and Radiation

56. C You extend an impromptu invitation to a friend for dinner. The only food you have is a couple of frozen steaks. You wish to defrost the steaks before grilling them. You defrost one by using your microwave oven, and you defrost the other by placing it in a bowl of very warm water. In each case, decide whether heat is transferred by conduction, convection, radiation, or some combination of these mechanisms. Explain your answers.

57. **N** Most of the energy emitted by an incandescent lightbulb is in the form of radiation. Assume all the power of a 75-W lightbulb is released from a tungsten filament at a temperature of 2800 K with an emissivity of 0.35. What is the effective area of the filament?

58. **a. E** Estimate the rate at which heat is transmitted through a typical window of a house by thermal conduction when the outside temperature is 0°C.
b. C Why do you suppose many thermally efficient windows are double-paned, with two pieces of glass separated by a small air gap?

59. **N** A lake is covered with ice that is 2.0 cm thick. The temperature of the ambient air is −20°C. Find the rate of thickening of ice. Assume the thermal conductivity of ice is 200.0 W/(m · K), the density of ice is 9.0×10^2 kg/m^3, and the latent heat of fusion is 3.33×10^5 J/kg.

60. **N** A concerned mother is dressing her child for play in the snow. The child's skin temperature is 36.0°C, and the outside air temperature is 2.00°C. If the emissivity of the child's skin is 0.790 and he has 1.10×10^{-2} m^2 of exposed skin area, what is the amount of energy transferred from his body to the surroundings in 1.00 h?

61. **N** On a hot summer day, a section of the interstate highway near Los Angeles receives 985 W/m^2 of sunlight. If the ground below the highway can be considered an insulator and the highway surface transfers energy by radiation only, what is the equilibrium temperature of the highway surface?

62. **N** CASE STUDY In Example 21.10, we calculated your skin temperature if you used a drysuit while swimming in a lake at 15°C. If you used a normal bathing suit, what would your body temperature be in half an hour? Assume your hair and the suit maintain a thin layer of water with an average thickness of 2 mm near your skin.

63. Imagine 1 kg of water in a container that does not allow heat transfer from the surrounding air to the water by conduction or convection. This container allows radiation to pass freely into it, but little radiation is allowed to escape.
a. E Estimate the time it would take to make a pot of tea if the container were left in the Sun on a clear day. Tea is properly made with boiling water.
b. C People often make *sun tea* by leaving water with tea leaves in a clear glass jar in the Sun. They claim that the tea tastes smoother because, in contradiction to part (a), the water does not boil! Explain why the water doesn't boil.

64. **C** Figure P21.64 shows a silver teapot. Explain why the handle has two small pieces of a ceramic material.

Wylius/iStockphoto.com

FIGURE P21.64

General Problems

65. **N** A water calorimeter is a device used to measure the specific heat of a substance (Fig. 21.6). A 0.125-kg sample has a temperature of 1030 K when it is placed into a container with 1.00 kg of water, both at a temperature of 280 K. The container is made of steel and has a mass of 0.250 kg. After the sample, the water, and its container reach thermal equilibrium, their common temperature is 293 K. What is the specific heat of the substance?

66. **C** An ideal gas is in the apparatus shown in Figure 21.9. Process A and process B (Fig. P21.66) represent two thermodynamic processes that expand the system from an initial state *i* to a final state *f*. **a.** If you wish to *minimize* the work done by the gas during the process, describe how to use process A or B to do so. Include a description of the apparatus during each part of the process. **b.** Repeat part (a) assuming you wish to *maximize* the work done by the gas.

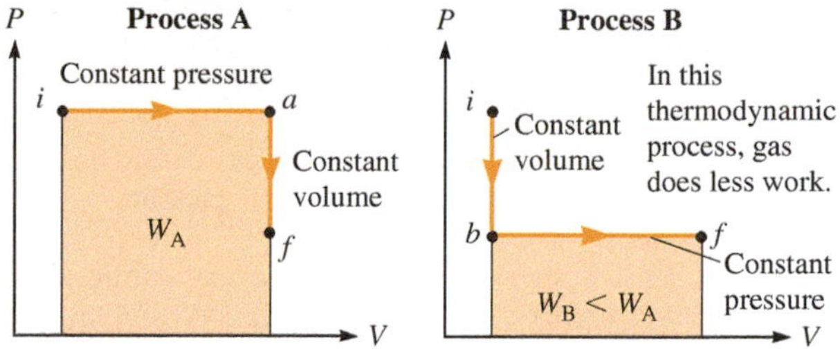

FIGURE P21.66

67. **N** Many scientists are concerned about the effects of melting icebergs and polar ice on the Earth's climate and its evolving patterns. **a.** Given a large amount of ice with a mass of 2540.0 kg at a temperature of −20.0°C, find the heat necessary to completely melt the ice so that it becomes water at 0°C. **b.** What energy flow into or out of the water is necessary to refreeze the water at 0°C?

68. A 350-g sample of an unknown (nongaseous) material experiences a 20.0°C increase in temperature after absorbing 5.53×10^3 J of energy.
a. N What is the specific heat of this material?
b. C Which substance listed in Table 21.1 best matches the specific heat of this unknown material?

69. **N** Three 100.0-g ice cubes initially at 0°C are added to 0.850 kg of water initially at 22.0°C in an insulated container. **a.** What is the equilibrium temperature of the system? **b.** What is the mass of unmelted ice, if any, when the system is at equilibrium?

70. **N** In a constant-pressure process, an ideal gas at a pressure of 800 Pa and initial temperature of 325 K increases in volume from 2.00 m^3 to 5.50 m^3. During the process, 3.00 kJ of energy is transferred into the gas by heat. **a.** What is the change in the internal energy of the gas? **b.** What is the final temperature of the gas?

71. **N** An ice cube of mass 0.200 kg at 0°C is placed in an insulated container that is at 227°C. The specific heat c of the container varies linearly with temperature as $c = A + BT$, where $A =$ 419 J/(kg · K) and $B = 8.38 \times 10^{-3}$ J/(kg · K^2). If the final temperature of the container is 27.0°C upon reaching thermal equilibrium with the ice cube, determine the mass of the container.

72. **C** In Chapter 20, we related temperature to the average kinetic energy of the molecules in an ideal gas. We wouldn't say that a baseball flying through the air is at a higher temperature than a baseball sitting on the ground, however, even though the kinetic energy of the moving ball is larger than that of the stationary ball. Explain how these two ideas are consistent with each other.

73. **N** A blacksmith forging a 950.0-g knife blade drops the steel blade, initially at 700.0°C, into a water trough containing 40.0 kg of water at 22.0°C. Assuming none of the water boils away or is lost from the trough and no energy is lost to the surrounding air, what is the final temperature of the water–knife blade system? Assume (c_{steel} = 450 J/kg · °C).

74. **A** In Figure 21.27, we saw that the gas initially has a large volume, low pressure, and low temperature and finally it has smaller volume, higher pressure, and higher temperature. Check our derivation of the adiabatic path by making sure these changes are correctly predicted by the equations $P_f V_f^\gamma = P_i V_i^\gamma$ = constant and $T_f/T_i = (V_f/V_i)^{(1-\gamma)}$.

75. **A** A container contains equal moles of two ideal gases A and B at temperature T. Gas A is monatomic, and gas B is diatomic. Find the average thermal energy per molecule at temperature T. Assume the diatomic molecule has five active degrees of freedom.

76. A vessel with a movable piston contains 1.00 mol of an ideal gas with initial pressure $P_i = 2.03 \times 10^5$ Pa, initial volume $V_i = 1.00 \times 10^{-2}$ m^3, and initial temperature $T_i = 305$ K.

a. **N** What is the work done on the gas during a constant-pressure compression, after which the final volume of the gas is 3.00 L?

b. **N** What is the work done on the gas during an isothermal compression, after which the final pressure of the gas is 5.00 atm?

c. **N** What is the work done on the gas during a constant-volume process, after which the final pressure of the gas is 5.00 atm?

d. **G** Sketch each of the processes in parts (a) through (c) on a *PV* diagram.

77. **N** The temperature of a 10.0-kg block of lead is increased from 18.0°C to 55.0°C at a constant pressure of 1.00 atm. The coefficient of linear expansion of lead is 28.0°C^{-1}, its density is 11.3×10^3 kg/m^3, and its specific heat is 128 J/(kg · °C). **a.** What is the work done on the lead block during this process? **b.** How much energy is added to the block by heat during this process? **c.** What is the change in the internal energy of the block during this process?

78. **C** All main-sequence ("living") stars convert hydrogen nuclei into helium nuclei in their hot cores. When the supply of hydrogen in the core is used up, the star dies (though aspects of the death may vary depending on the star's mass). In high-mass stars, the energy produced in the core makes its way out primarily through radiation (Fig. P21.78A). In low-mass stars, however, energy is transferred mainly through convection (Fig. P21.78B). Low-mass stars live longer than high-mass stars. The outer layers in both types of stars are primarily hydrogen. Explain how convection helps extend the lifetime of low-mass stars.

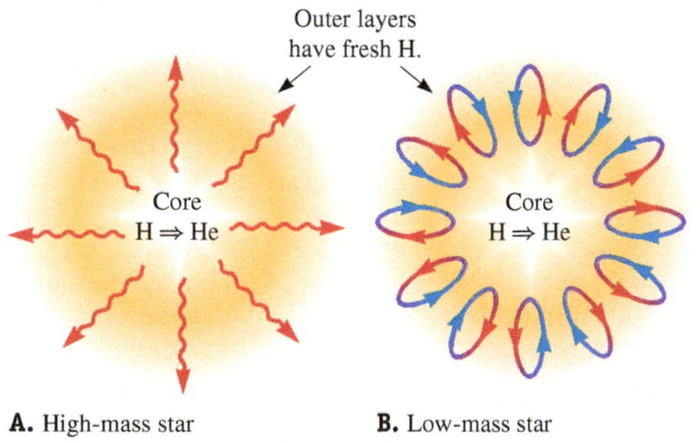

FIGURE P21.78

79. **N** How much faster does a cup of tea cool by 1°C when at 373 K than when at 303 K? Consider the tea to be a blackbody (an ideal object whose emissivity is 1) and assume the room temperature is 293 K.

80. **N** The *PV* diagram in Figure P21.80 shows a set of thermodynamic processes that make up a cycle *ABCDA* for a monatomic gas, where *AB* is an isothermal expansion occurring at a temperature of 375 K. There are 2.00 mol of gas undergoing the cycle with $P_A = 1.01 \times 10^6$ Pa, $P_B = 5.05 \times 10^5$ Pa, and $P_C = 2.02 \times 10^5$ Pa. **a.** Find the volumes V_A and V_B. **b.** Find the work done in each process of the cycle and the total work done for the whole cycle. **c.** Find the change in thermal energy during the constant-volume process *BC*.

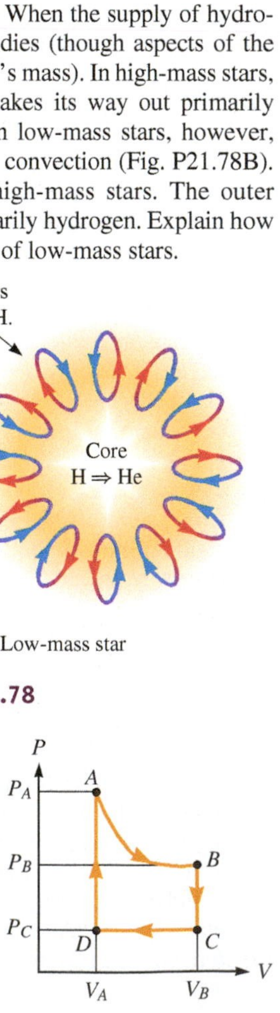

FIGURE P21.80

81. **N** An insulated canister contains 4.00×10^{-2} m^3 of liquid argon at 87.3 K, which is the boiling point for argon. A small, 7.00-W heat lamp within the canister is switched on. If the latent heat of vaporization for argon is 1.61×10^5 J/kg, what is the mass of the argon that evaporates in the first 1.00 h?

82. **A** In an adiabatic process, the relationship between pressure and volume is PV^γ = constant, where γ is the adiabatic exponent that depends on the properties of the gas. Two different adiabatic processes *AD* and *BC* intersect two isothermal processes *AB* at temperature T_1 and *DC* at temperature T_2 as shown in Figure P21.82. Is the ratio V_A/V_D greater than, less than, or equal to the ratio V_B/V_C?

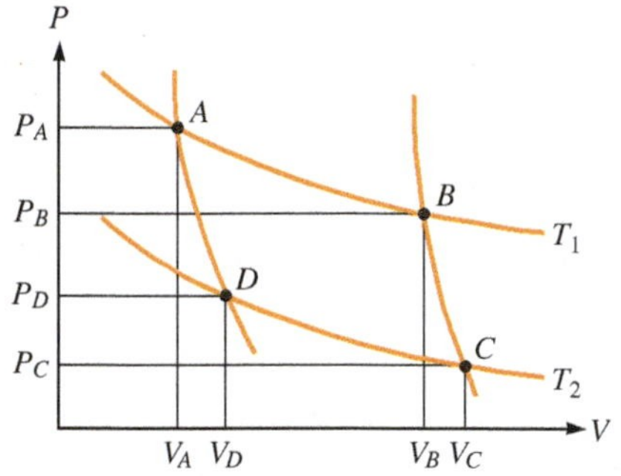

FIGURE P21.82

83. **N** Between 1793 and 1837, the U.S. penny was made entirely of copper and had a mass of 3.20 g. A coin collector accidentally drops one such penny from the balcony of his high-rise apartment, 38.0 m above the street. If 40.0% of the change in gravitational energy of the penny–Earth system is converted into internal energy of the penny, what is the increase in the temperature of the coin just before it impacts with the sidewalk?

84. **N** In the 1970s cartoon *The Super Friends*, the Wonder Twins helped Superman and others fight crime. One of the Wonder Twins was Zan, who was able to take on the form of H_2O in its various phases (ice, liquid water, water vapor). After taking a physics class, he decides to conduct a little experiment by turning into an ice igloo in the park on a hot day. **a.** If his total mass as an igloo is 65.88 kg and he starts with a temperature of −10.00°C, how much heat must flow into him to completely melt his body? **b.** He continues absorbing energy up to the point where he would start to vaporize (100.0°C). How much more heat is required to raise his temperature to the vaporization point? **c.** How much more heat is required to turn him into water vapor at 100.0°C? **d.** If we model the water vapor molecules making up his body as having only six degrees of freedom, what is the average kinetic energy of one of the water vapor molecules that make up Zan?

85. **N** On a cold winter day when the outside temperature is −5.00°C, the interior of a house is heated to a comfortable 22.0°C. The dining room window of the house is 1.50 m tall by 2.50 m long, and the glass windowpane has a thickness of 0.450 cm. **a.** What is the rate of energy transfer through the windowpane? **b.** If the interior and exterior temperatures remain constant for 6.00 h, how much energy is transferred through the windowpane during this time period?

Entropy and the Second Law of Thermodynamics

22

Underlying Principles

Second law of thermodynamics

Major Concepts

1. Efficiency
2. Reversible and irreversible processes
3. Entropy

Special Cases: Three Specific Engines

1. Carnot
2. Otto and internal combustion
3. Refrigerator

Tools

Energy transfer diagram

Key Questions

Why are some processes that are allowed by the first law of thermodynamics never seen in nature?

Why is it impossible to build a 100% efficient engine?

If you want to make a group of kids laugh, play a movie backward. For example, imagine a scene in which someone fries an egg. In the backward version, a completely cooked egg floats up from the plate into a tilted frying pan, the pan is set on the stove, and heat is transferred away from the egg as it becomes *un*cooked. When the egg is completely raw, it flies straight upward into a cracked shell, and finally the parts of the shell come together. Kids laugh at this movie because they can see that something is wrong—eggs don't spontaneously fly up into cracked shells—but why not?

Why can't an egg use the heat transferred from the stove to gain macroscopic kinetic energy and then turn that kinetic energy into potential energy by flying upward? There is nothing in the work–energy theorem or the first law of thermodynamics that would prevent such a conversion of energy. In this chapter, we introduce the second law of thermodynamics, which would be violated if an egg used heat from the stove to launch itself upward. The first law of thermodynamics is an extension and a special case of the work–energy theorem, but the second law of thermodynamics is a completely separate and independent principle, involving a new concept: entropy.

22-1 Second Law of Thermodynamics, Clausius Statement

Imagine making a potato salad. After cooking the potatoes, you place them in a very large sink full of cool water. The water is cooler than the potatoes:

$$T_{\text{water}} < T_{\text{potato}}$$

Assume the potatoes are small compared to the amount of water, so there are many more water molecules than potato molecules:

$$N_{\text{water}} \gg N_{\text{potato}}$$

Thermal energy E_{th} is proportional to the number of molecules and the temperature (Eq. 21.2), and because there are many more water molecules than potato molecules, there is more thermal energy in the cool water than in the hot potatoes:

$$N_{\text{water}}T_{\text{water}} > N_{\text{potato}}T_{\text{potato}}$$
$$E_{\text{th, water}} > E_{\text{th, potato}}$$

It might seem reasonable for heat to flow from the water to the potatoes because there is more thermal energy in the water. If that happened, the potatoes would get hotter by sitting in cool water. Experience, however, tells us that the hot potatoes cool off because heat flows from the potatoes to the cool water until they are at the same temperature.

Nothing about the first law of thermodynamics or the conservation of energy principle tells us which way heat will be transferred. Another law—the *second law of thermodynamics*—tells us which way heat flows. There are many versions of the second law; we start with the one about heat transfer. This statement is attributed to Rudolf Julius Emanuel Clausius (1822–1888), a German physicist and mathematician. The **Clausius statement** of the **second law of thermodynamics** is as follows:

SECOND LAW OF THERMODYNAMICS (CLAUSIUS STATEMENT)

Underlying Principle

Heat flows naturally from a hot object to a cooler object; heat does not naturally flow from a cold object to a hotter object.

CASE STUDY Part 1: Engines and Refrigerators

Why is the word *naturally* included in this statement of the second law of thermodynamics? *Naturally* is included because it is possible for heat to be transferred from a cooler system to a hotter environment if work is done on the system. For example, a refrigerator does work on the cool air inside it so as to transfer heat from that air to the hotter room. So, the word *naturally* is included in Clausius's wording to indicate that in the absence of work done, heat flows from a hot object to a cooler object.

The word *naturally* is also a reminder that the second law of thermodynamics was discovered because people were trying to use the principles of thermodynamics to design and build new machines, the most important of which is the *heat engine*. Examples of heat engines are the steam engines that were once used to power trains, internal combustion engines that make most of today's automobiles run, and many power plants that supply households with electricity.

A **heat engine** is a device that absorbs energy from its environment through heat and then does work on its environment. Our case study throughout this chapter involves both heat engines and refrigerators. A refrigerator is the opposite of a heat engine; it takes in energy as work and gives off energy as heat.

You might hear the phrase *a heat engine turns heat into work*. Such a phrase is actually shorthand for the idea that energy enters the engine through heat and leaves the engine through work. It is okay to use this phrase as long as you keep in mind what it really means.

22-2 Heat Engines

We have already studied a crude heat engine such as the simple device shown in Figure 21.15 and reproduced here in Figure 22.1. At first in part A, the piston is held in place, and the heat reservoir is warmer than the gas. As a result, heat Q_h is transferred to the gas, and its pressure rises. The piston is then freed in part B, and energy is transferred from the gas to the piston as the gas does positive work W_{out} in lifting the piston. This device is a heat engine because the system took in energy as heat and released energy as work. You can imagine that this device may be useful if you wish to do something like lift an elephant.

We are only interested in engines that run continuously in a thermodynamic cycle, returning repeatedly to their initial state. Figure 22.2 is a *PV* diagram for a system that undergoes an adiabatic and constant-volume cycle. As before, we consider the gas to be the system and everything else (the heat reservoir, the piston, and the Earth) to be in the environment. The cylindrical walls and piston are insulated, so the system can only exchange heat with the heat reservoir under its base.

Figure 22.2 also introduces a new visualization tool known as an **energy transfer diagram**. The system (the gas) is represented as a rectangle in the middle of the diagram. A hot reservoir (with a temperature higher than that of the system) is always drawn above the rectangle, and a cold reservoir (with a temperature lower than that of the system) is always drawn below the rectangle. Thick arrows represent energy transferred into or out of the system as either heat or work. There is no need to draw a symbol for thermal energy because we only study heat engines that operate in a cycle, and in a cycle, the thermal energy returns to its original state.

Developers of practical heat engines were interested in the net work W_{eng} done by the engine, which is the difference between the energy transferred out of the system by work and the energy transferred in:

$$W_{eng} = W_{out} - W_{in} \tag{22.1}$$

Let's find an expression for W_{eng} by applying the first law of thermodynamics (Eq. 21.1),

$$\Delta E_{th} = W_{tot} + Q$$

to one cycle of the heat engine.

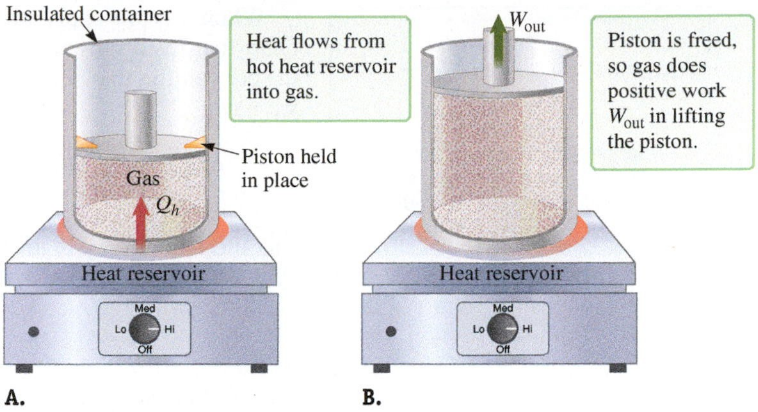

FIGURE 22.1 The basic principles of a heat engine are illustrated with this simple apparatus. **A.** The temperature of the heat reservoir may be adjusted by turning a knob. The piston may be held in place by wedges. The system is the gas inside the cylinder. Heat flows from the reservoir into the system. **B.** The system does work on the piston, raising it to a new height.

ENERGY TRANSFER DIAGRAM

Tool

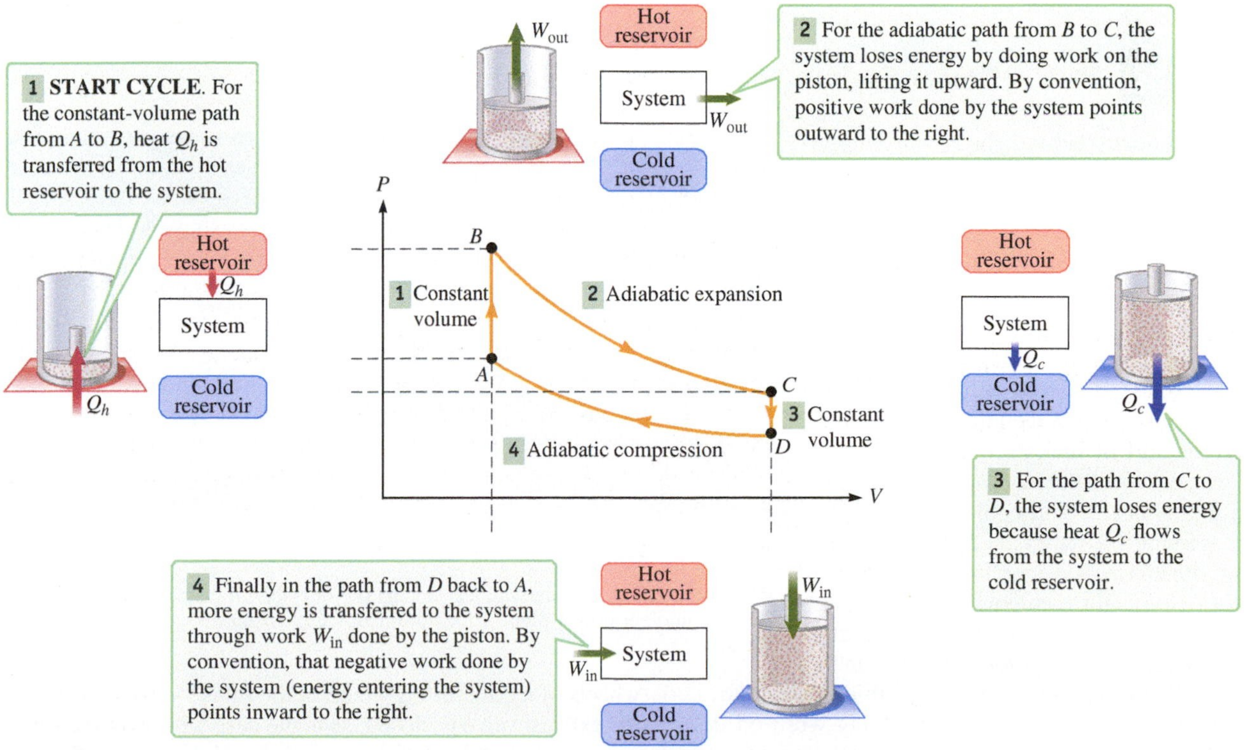

FIGURE 22.2 The system undergoes a cyclic thermodynamic process from state A to B to C to D and back to A. This figure shows the connection between a series of energy transfer diagrams and the *PV* diagram.

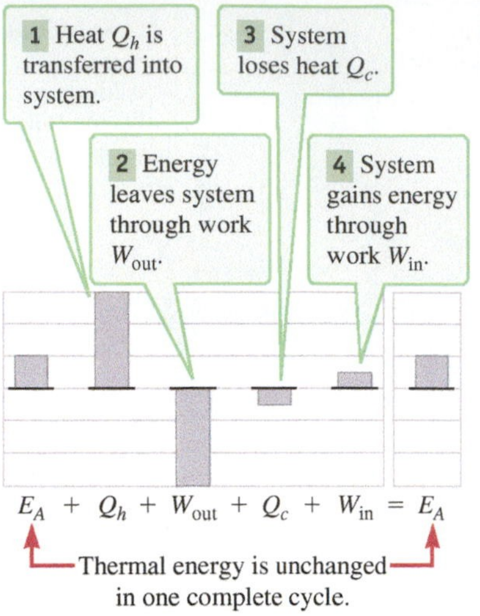
1 Heat Q_h is transferred into system.
3 System loses heat Q_c.
2 Energy leaves system through work W_{out}.
4 System gains energy through work W_{in}.
$E_A + Q_h + W_{out} + Q_c + W_{in} = E_A$
Thermal energy is unchanged in one complete cycle.

FIGURE 22.3 The bar chart for the thermodynamic cycle is made by considering the energy transferred during each of the four processes indicated in Figure 22.2. Notice that the thermal energy is unchanged.

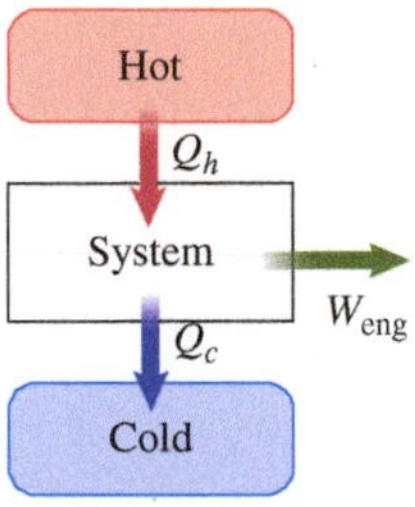
Hot
Q_h
System
W_{eng}
Q_c
Cold

FIGURE 22.4 An energy transfer diagram for a heat engine.

Whenever we apply the first law, a bar chart is helpful. Figure 22.3 is the bar chart that results from the four-part cycle in Figure 22.2. Using this bar chart and the convention that energy entering the system either by work or by heat is positive, we have

$$\Delta E_{th} = E_A - E_A = \underset{\text{(process 1)}}{(0 + Q_h)} + \underset{\text{(process 2)}}{(-W_{out} + 0)} + \underset{\text{(process 3)}}{(0 - Q_c)} + \underset{\text{(process 4)}}{(0 + W_{in})}$$

The thermal energy is unchanged at the end of the cycle. So, the net result of one cycle is $(Q_h - Q_c) - (W_{out} - W_{in}) = 0$, and

$$(Q_h - Q_c) = (W_{out} - W_{in}) \tag{22.2}$$

By substituting $W_{eng} = W_{out} - W_{in}$ (Eq. 22.1) into Equation 22.2, we find that the work done by the engine W_{eng} on its environment equals the net heat that enters the engine:

$$W_{eng} = (Q_h - |Q_c|) \tag{22.3}$$

Equation 22.3 may be represented as an energy transfer diagram as shown in Figure 22.4. The negative sign in Equation 22.3 explicitly takes care of the heat Q_c being transferred out of the system. To ensure that we don't substitute a negative number into Q_c, we use an absolute value sign.

CONCEPT EXERCISE 22.1

Imagine a perfect engine in which all the heat that enters is completely converted to work. Estimate the heat required to lift an elephant from the ground into a truck using the heat engine in Figure 22.1. Would a real engine require more heat or less heat to do the job?

CASE STUDY Part 2: Steam Engines and Internal Combustion Engines

The fundamental ideas behind the steam engine are ancient, dating to around 200 BCE in Greece, but the steam engine didn't change the world until the Industrial Revolution in the 18th and 19th centuries (Fig. 22.5A). Today we power cities with giant steam engines located inside power plants (Fig. 22.5B).

Figure 22.6 is a schematic diagram of a steam engine. Heat is transferred from a hot reservoir to water in the boiler shown on the left. The hot reservoir needs some sort of fuel, such as a fossil fuel like coal or gas, or even a nuclear fuel like uranium. The system that is heated and cooled in an engine is called the **working substance**; in a steam engine, the working substance is water in its liquid and vapor form. In a typical steam engine, the water is kept at very high pressure so that it vaporizes at a very high temperature. High-temperature steam passes through the intake valve as shown in middle of Figure 22.6 and is then in contact with a piston, doing work on it. The piston may be connected to a wheel known as a crank. The up-and-down motion of the piston rotates the crank, which may be attached to any number of devices such as the wheels of a train. In Figure 22.6, the intake valve is open and the exhaust valve is closed, as is the case whenever the steam is expanding and the piston is moving downward. When the steam is being compressed and the piston is moving upward, the intake valve is closed and the exhaust valve is open. The

Courtesy Muskegon Heritage Museum

JoLin/Shutterstock.com

A. **B.**

FIGURE 22.5 A. Steam engines such as the one shown changed the world. **B.** This nuclear power plant is located in Bay City, Texas.

exhausted steam moves through the condenser, which is in thermal contact with a cold reservoir. Heat is transferred from the hot steam to the cold reservoir, and the steam condenses back into liquid water. That water is then recycled into the boiler, and the whole process continues.

The cold reservoir may be a river or another natural body of water. Heat transferred from the hot steam to the body of water is called **thermal pollution**. Natural bodies of water are becoming warmer, which is a problem for life in the water. Unfortunately, the second law of thermodynamics tells us that all engines must give off some heat (Section 22-3).

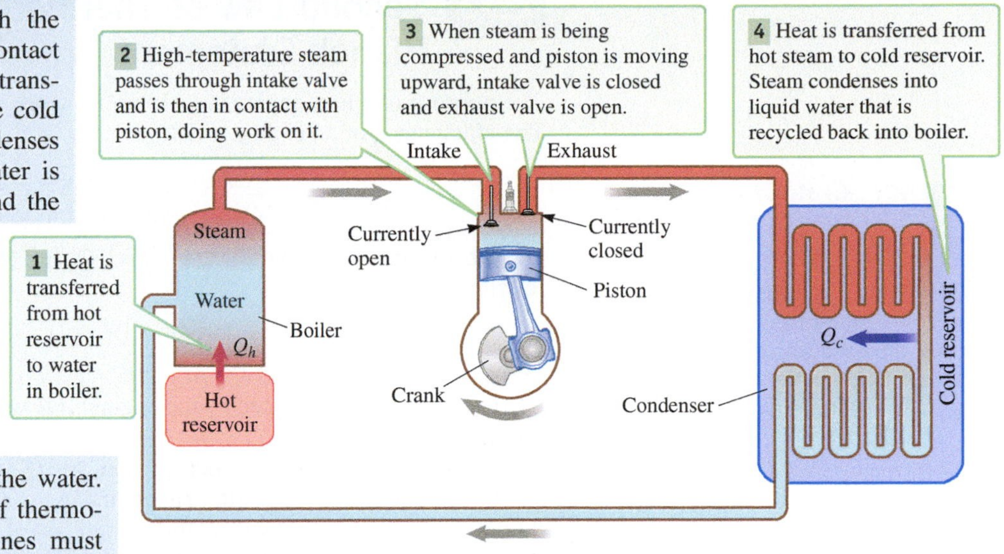

2 High-temperature steam passes through intake valve and is then in contact with piston, doing work on it.
3 When steam is being compressed and piston is moving upward, intake valve is closed and exhaust valve is open.
4 Heat is transferred from hot steam to cold reservoir. Steam condenses into liquid water that is recycled back into boiler.
1 Heat is transferred from hot reservoir to water in boiler.
Intake
Exhaust
Steam
Water
Q_h
Hot reservoir
Boiler
Currently open
Currently closed
Piston
Crank
Condenser
Q_c
Cold reservoir

FIGURE 22.6 Components of a simplified steam engine.

Steam engines used to provide electricity are similar to the one shown in Figure 22.6, but with the piston replaced by a rotating turbine as shown in Figure 22.7. High-pressure steam from the boiler pushes the paddle wheel of the turbine around, and low-pressure steam is exhausted through a pipe on the other side of the turbine. We return to electricity generation in Chapter 32 when we describe how the motion of the turbine is used to produce an electric current.

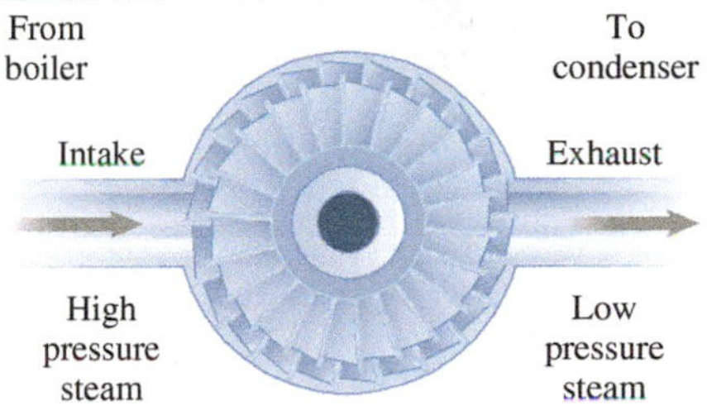

From boiler
To condenser
Intake
Exhaust
High pressure steam
Low pressure steam

FIGURE 22.7 A steam engine may be used to rotate a turbine.

The engine in most cars is an **internal combustion engine** in which the working substance is a mixture of air and gasoline vapor. The main difference between the steam engine (Fig. 22.6) and the internal combustion engine is that the high temperature in an internal combustion engine is achieved by igniting the air–gasoline mixture *in* the cylinder instead of maintaining a hot reservoir *outside* the engine. The internal combustion engine is described as having four *strokes* or movements as shown in Figure 22.8.

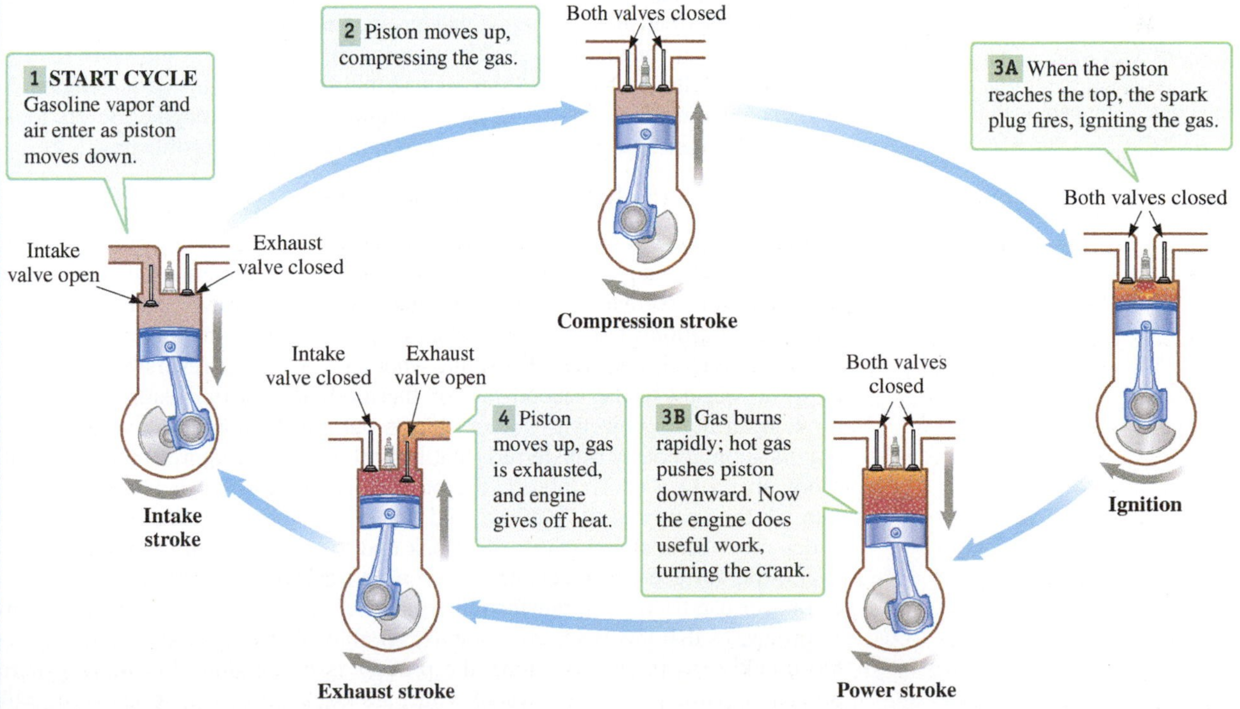

1 START CYCLE Gasoline vapor and air enter as piston moves down.
2 Piston moves up, compressing the gas.
Both valves closed
3A When the piston reaches the top, the spark plug fires, igniting the gas.
Both valves closed
Intake valve open
Exhaust valve closed
Compression stroke
Intake valve closed
Exhaust valve open
Both valves closed
4 Piston moves up, gas is exhausted, and engine gives off heat.
3B Gas burns rapidly; hot gas pushes piston downward. Now the engine does useful work, turning the crank.
Intake stroke
Ignition
Exhaust stroke
Power stroke

FIGURE 22.8 The processes of a four-stroke internal combustion engine such as a car engine. (For each cycle, the piston moves up twice and down twice for a total of four *strokes*.)

22-3 Second Law of Thermodynamics, Kelvin-Planck Statement

The purpose of a heat engine is to do useful work by taking in a net amount of heat. The **efficiency** e of an engine is defined as

EFFICIENCY ✪ Major Concept

$$e \equiv \frac{W_{\text{eng}}}{Q_h} = \frac{\text{what you get from the engine}}{\text{what you put into the engine}} \tag{22.4}$$

By substituting Equation 22.3, $W_{\text{eng}} = (Q_h - |Q_c|)$, into Equation 22.4, we can rewrite efficiency as

The negative sign takes into account that Q_c is energy that *leaves* the system, and the absolute value in Equation 22.5 makes sure we don't eliminate this negative sign.

$$e = \frac{(Q_h - |Q_c|)}{Q_h} = 1 - \frac{|Q_c|}{Q_h} \tag{22.5}$$

where Q_c is the heat expelled by the engine. It is waste, so the smaller $|Q_c|$, the more efficient the engine is.

A perfect engine would expel no heat. If $|Q_c|$ were zero, then e would equal 1, and we would have a 100% efficient engine. A 100% efficient engine cannot be built, though, because it would violate the second law of thermodynamics. In fact, another way to state the second law of thermodynamics is as follows:

SECOND LAW OF THERMODYNAMICS (KELVIN-PLANCK STATEMENT) ❶ Underlying Principle

There can be no perfect (100% efficient) heat engine.

The **Kelvin-Planck statement** of the second law of thermodynamics says the same thing:

It is impossible to construct a cyclic heat engine that takes in heat and produces only work without also giving off waste heat.

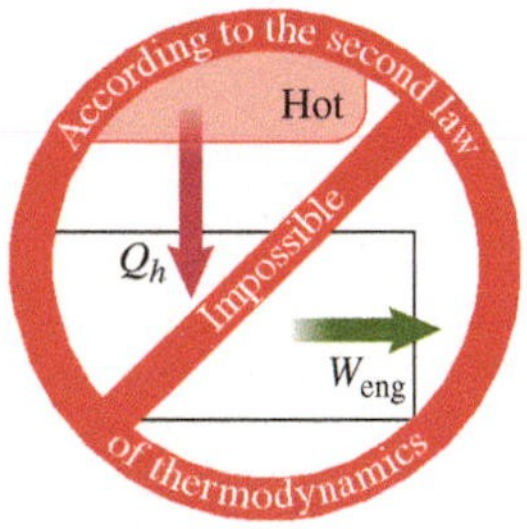

FIGURE 22.9 According to the second law of thermodynamics, all heat engines must expel heat. Compare with Figure 22.4, where waste heat and a cold reservoir must be present.

Figure 22.9 illustrates the Kelvin-Planck statement of the second law of thermodynamics. Although no perfect engine can be built, it is possible to maximize an engine's efficiency, as discussed in the next section.

22-4 The Most Efficient Engine

Sadi Carnot (1796–1832), a French physicist, was interested in maximizing the efficiency of heat engines. In 1824, he published *Reflections on the Motive Power of Fire* in which he described an ideal engine that has maximum efficiency. According to Carnot, the most efficient engine is *reversible*, meaning that the working substance returns to its original state after each cycle. Today we call this ideal engine the **Carnot engine**. No real engine can be made reversible, so no real engine is a Carnot engine. Before looking at the Carnot engine, we must first consider reversible and irreversible processes.

Reversible and Irreversible Processes

REVERSIBLE AND IRREVERSIBLE PROCESSES ✪ Major Concept

A **reversible process** is an *ideal* thermodynamic process in which a system that goes from state i to state f can go back again to state i, exchanging the same amount of heat and work so that the system and the environment have been returned to their original conditions. Other processes are known as **irreversible processes**.

Three general conditions determine whether a process is reversible or irreversible. First, according to the second law of thermodynamics (Clausius statement), heat flows from hotter objects to cooler objects but not the other way around. So, if energy is spontaneously transferred (that is, with no work required) through heat from a hot object to a cooler object, the process is irreversible. It would take work to transfer the energy back from the cold object to the hot one. Second, when a block slides down an inclined plane, the block and the plane warm up due to friction between the surfaces, so mechanical energy is transformed into thermal energy. You can never use friction to transform thermal energy into mechanical energy, however, so processes that involve friction are irreversible. Third, if the system changes states too quickly (as in an explosion), the process is irreversible. Put more generally, a process is reversible if the system can pass back through a series of equilibrium states in reverse order. (Explosions don't happen in reverse.) A reversible process is an ideal process because it must be carried out infinitely slowly so that the process can be a considered a series of equilibrium states taking the system from state i to state *f*. All real processes are irreversible, but a real process can be approximated as a reversible process if the following conditions hold:

1. The system's temperature is close to that of the environment. The environment's temperature may be changed, but this change must be slow.
2. Friction and other dissipative forces are very small so that little mechanical energy is transformed into thermal energy.
3. The process must be done very slowly—quasistatically—so that the system is always in (or very close to) an equilibrium state.

© gaspr13/iStockphoto.com

FIGURE 22.10

CONCEPT EXERCISE 22.2

You suspend a pot of water above a campfire (Fig. 22.10). Assume the pot is tightly sealed. In the evening, water boils and becomes a vapor trapped in the pot, and then the fire dies out. During the night, the water condenses back into liquid.

a. Has the water undergone an approximately reversible cycle?
b. Has the wood undergone an approximately reversible cycle?

Carnot Engine

CARNOT ENGINE Special Case

Let's imagine a Carnot engine made out of an ideal apparatus as in Figure 22.1. For this ideal apparatus, there is no friction between the piston and the walls of the container, the system temperature is very close to that of the heat reservoir, and all changes are made slowly. Furthermore, assume the working substance is an ideal gas.

The working substance in a Carnot engine goes through four reversible processes in each cycle. Any thermodynamic cycle that is made up of reversible processes is known as a **reversible cycle**. For the Carnot engine, the four reversible processes are an isothermal ($\Delta T = 0$) expansion, an adiabatic ($Q = 0$) expansion, an isothermal compression, and an adiabatic compression. Together, these four processes are known as the **Carnot cycle**. Figure 22.11 illustrates the Carnot cycle on a *PV* diagram.

FIGURE 22.11 The Carnot engine is an ideal engine; no real engine is as efficient as the Carnot engine. This engine undergoes a cyclic process made up of four parts. For each of these parts be sure you can identify the illustration, bar chart, and curve on the *PV* diagram.

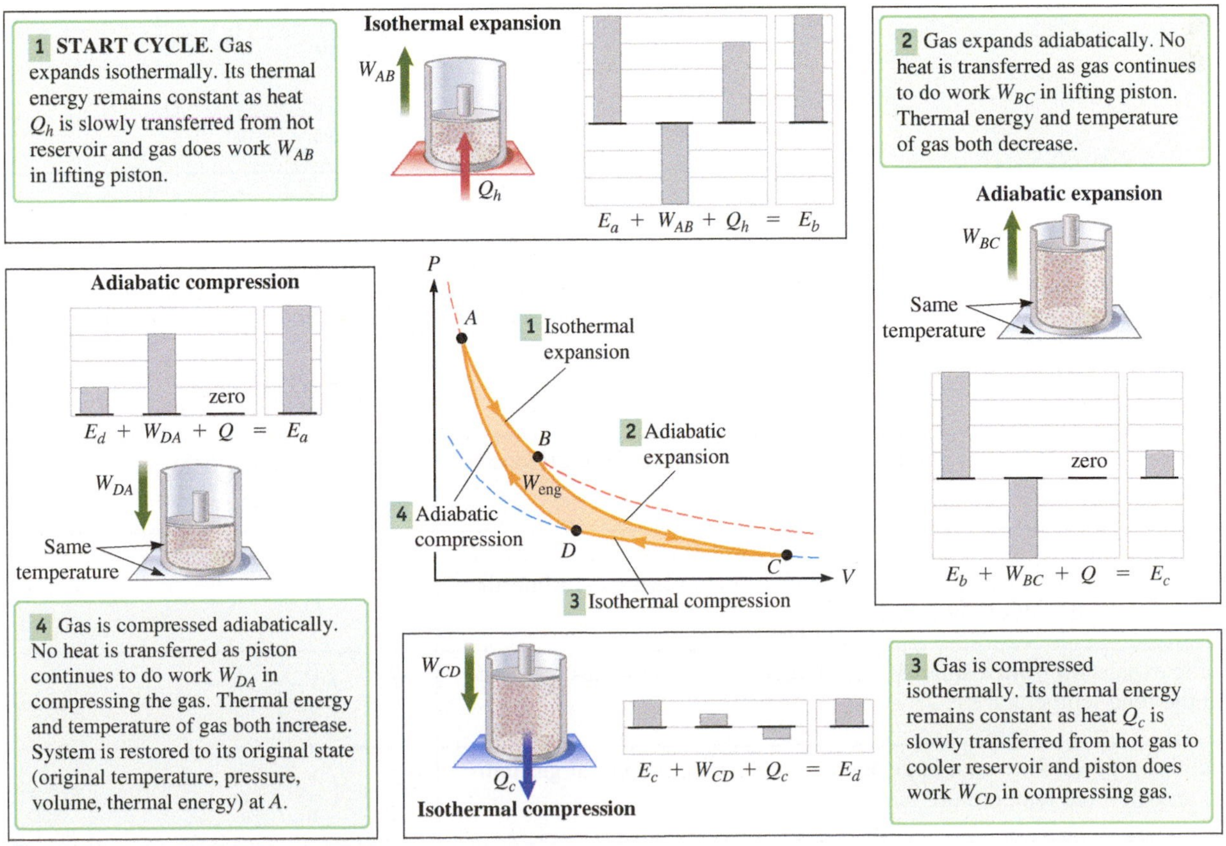
1 START CYCLE. Gas expands isothermally. Its thermal energy remains constant as heat Q_h is slowly transferred from hot reservoir and gas does work W_{AB} in lifting piston.
Isothermal expansion
W_{AB}
Q_h
$E_a + W_{AB} + Q_h = E_b$
2 Gas expands adiabatically. No heat is transferred as gas continues to do work W_{BC} in lifting piston. Thermal energy and temperature of gas both decrease.
Adiabatic expansion
W_{BC}
Same temperature
zero
$E_b + W_{BC} + Q = E_c$
Adiabatic compression
zero
$E_d + W_{DA} + Q = E_a$
W_{DA}
Same temperature
P
A
1 Isothermal expansion
B
2 Adiabatic expansion
W_{eng}
4 Adiabatic compression
D
C
V
3 Isothermal compression
4 Gas is compressed adiabatically. No heat is transferred as piston continues to do work W_{DA} in compressing the gas. Thermal energy and temperature of gas both increase. System is restored to its original state (original temperature, pressure, volume, thermal energy) at A.
W_{CD}
Q_c
Isothermal compression
$E_c + W_{CD} + Q_c = E_d$
3 Gas is compressed isothermally. Its thermal energy remains constant as heat Q_c is slowly transferred from hot gas to cooler reservoir and piston does work W_{CD} in compressing gas.

The total work done by the engine is the sum of the work done by the gas as the piston is raised in the isothermal and adiabatic expansions minus the work done by the piston as gas is compressed in the isothermal and adiabatic compressions:

$$W_{\text{eng}} = W_{AB} + W_{BC} - W_{CD} - W_{DA}$$

The work W_{eng} done by the engine can also be found from the area enclosed by the four paths in the *PV* diagram. This area is positive, so the total work done by the engine is positive, and the engine does useful work just as we would expect.

DERIVATION Carnot Efficiency

The Carnot engine is an ideal engine that has the best possible efficiency. We will derive **Carnot's theorem**, which states that the efficiency of this ideal, reversible engine is

$$e_{\text{Carnot}} = 1 - \frac{T_c}{T_h} \quad (22.6)$$

We can find an expression for its efficiency using either $e = W_{\text{eng}}/Q_h$ (Eq. 22.4) or $e = 1 - |Q_c|/Q_h$ (Eq. 22.5). Our job is somewhat simpler if we use Equation 22.5. So, we need to find the heat Q_h absorbed by the engine during the isothermal expansion and the heat Q_c lost during the isothermal compression.

From the bar charts for the two isothermal processes in Figure 22.11, the work done by the gas (negative of the work done by the environment) equals the heat absorbed by the gas.

$$W_{AB} = Q_h$$

$$W_{CD} = Q_c$$

Equation 21.15 gives the work done *by the piston* (environment) on an ideal gas that expands isothermally. To find the work done *by the gas*, multiply by -1. (The "-1" flips the numerator and denominator in the natural log.)

$$W_{\text{by piston}} = Nk_BT\ln\frac{V_i}{V_f} \quad (21.15)$$

$$W_{\text{eng}} = -Nk_BT\ln\frac{V_i}{V_f}$$

$$W_{\text{eng}} = Nk_BT\ln\frac{V_f}{V_i} \quad (22.7)$$

Use Equation 22.7 to write an expression for Q_h. The temperature of the hot reservoir is T_h.

$$Q_h = W_{AB} = Nk_BT_h\ln\frac{V_B}{V_A}$$

Now find $|Q_c|$, where the temperature of the cold reservoir is T_c. Equation 22.7 is positive when the gas expands; when the gas contracts, we need to multiply by -1 to find the absolute value of the heat transferred.

$$|Q_c| = -W_{CD} = -Nk_BT_c\ln\frac{V_D}{V_C}$$

$$|Q_c| = Nk_BT_c\ln\frac{V_C}{V_D}$$

Find the ratio of the heat lost ($|Q_c|$) to the heat gained (Q_h) by dividing these two expressions.

$$\frac{|Q_c|}{Q_h} = \frac{T_c\ln(V_C/V_D)}{T_h\ln(V_B/V_A)} \quad (22.8)$$

Adiabatic processes connect states B and C and states D and A. Use Equation 21.32 (page 638) for the adiabatic expansion or compression of an ideal gas to find the ratio of the volumes at each set of endpoints.

$$\frac{T_f}{T_i} = \left(\frac{V_f}{V_i}\right)^{(1-\gamma)} \quad (21.32)$$

For the adiabatic expansion from a small V_B to a larger V_C, the initial gas temperature is T_h, and the final temperature is T_c.

$$\frac{T_c}{T_h} = \left(\frac{V_C}{V_B}\right)^{(1-\gamma)} \quad (22.9)$$

For the adiabatic compression from a large V_D to a smaller V_A, the initial gas temperature is T_c, and the final temperature is T_h.

$$\frac{T_h}{T_c} = \left(\frac{V_A}{V_D}\right)^{(1-\gamma)} \quad (22.10)$$

By substituting Equation 22.10 into Equation 22.9, we find that the ratio of volumes are equal.	$\frac{T_c}{T_h} = \left(\frac{V_C}{V_B}\right)^{(1-\gamma)} = \left(\frac{V_D}{V_A}\right)^{(1-\gamma)}$ $\frac{V_C}{V_B} = \frac{V_D}{V_A}$ and $\frac{V_C}{V_D} = \frac{V_B}{V_A}$
The terms involving the natural logarithms cancel out of Equation 22.8.	$\frac{\lvert Q_c \rvert}{Q_h} = \frac{T_c \ln(V_C/V_D)}{T_h \ln(V_C/V_D)}$ $\frac{\lvert Q_c \rvert}{Q_h} = \frac{T_c}{T_h}$ (22.11)
Finally, substitute this result into Equation 22.5 to find the efficiency of a Carnot engine in terms of the temperatures of the hot and cold reservoirs.	$e = 1 - \frac{\lvert Q_c \rvert}{Q_h}$ (22.5) $e_{\text{Carnot}} = 1 - \frac{T_c}{T_h}$ ✓ (22.6)

COMMENTS

A Carnot engine is an ideal engine. All real, irreversible engines are less efficient than a Carnot engine. In practice, a well-designed irreversible engine may reach about 80% of the Carnot efficiency.

Third Law of Thermodynamics

The temperatures in Equation 22.6 must be expressed on the Kelvin scale. The efficiency of the Carnot engine only depends on the temperature of the two reservoirs. If the temperature of the cold reservoir is lowered or if the temperature of the hot reservoir is raised, the engine becomes more efficient. You might imagine that a perfect (100% efficient) engine could be made if the hot reservoir were infinitely hot, but it would take an infinite amount of energy to achieve an "infinite" temperature. Having dismissed the idea of an infinitely hot reservoir, you might imagine that a perfect engine could be made if the cold reservoir were at absolute zero ($T_c = 0$). Experiments show, however, that the closer the temperature gets to absolute zero, the harder it is to reduce the temperature further. In fact, the **third law of thermodynamics** says that it is not possible to lower the temperature to zero by any finite process. The third law of thermodynamics means that there are no perfect engines.

EXAMPLE 22.1 Lifting an Elephant

You wish to use a steam engine to lift an elephant into a truck as in Concept Exercise 22.1. The steam engine operates using a hot reservoir at $T_h = 250°\text{C}$, and the cold reservoir is the surrounding air at $T_c = 32°\text{C}$.

A What is the maximum efficiency of the steam engine?

INTERPRET and ANTICIPATE

The maximum efficiency is the efficiency of a Carnot engine. No real engine can operate more efficiently than this ideal engine.

SOLVE Substitute the given temperatures into Equation 22.6 after converting them to the Kelvin scale.	$e_{\text{Carnot}} = 1 - \frac{T_c}{T_h} = 1 - \frac{305\text{ K}}{523\text{ K}} = 0.42 = 42\%$

CHECK and THINK

Even an ideal engine is only 42% efficient at these operating temperatures, meaning that only 42% of the heat absorbed from the hot reservoir goes into useful work. More than half the energy absorbed is simply expelled as heat into the cold reservoir.

Example continues on page 660 ▶

B Estimate the heat required to lift a single elephant into a truck using this engine. *Hint*: See Concept Exercise 22.1.

INTEPRET and ANTICIPATE

In Concept Exercise 22.1, we estimated the amount of work (as heat) that must be done by an engine to lift the elephant into the truck, assuming a 100% efficient engine. (All the heat absorbed by such an engine goes into work done.) No engine operates at 100% efficiency. Even an ideal Carnot engine is not 100% efficient. Use the efficiency of the Carnot engine from part A to estimate the minimum heat required.

SOLVE

The work required to lift the elephant is about 6×10^4 J from Concept Exercise 22.1. We found the efficiency of the engine in part A. Use Equation 22.4 to find the heat absorbed, Q_h.

$$e = \frac{W_{eng}}{Q_h} \tag{22.4}$$

$$Q_h = \frac{W_{eng}}{e} = \frac{6 \times 10^4 \text{ J}}{0.42} \approx 1 \times 10^5 \text{ J}$$

CHECK and THINK

Of course, a real engine would require an even greater input of heat. While this isn't a lot of energy (about the same as a few grapes), it may be better to just guide the elephant up a ramp.

CASE STUDY Part 3: Otto Cycle

The internal combustion engine in your car can be modeled by the **Otto cycle** (Fig. 22.12). The Otto cycle consists of two constant-volume processes and two adiabatic processes. The Otto cycle is ideal and reversible. A real car engine (Fig. 22.8) can only be approximated as an Otto cycle because a real car engine cycle is *ir*reversible.

Compare the *PV* diagrams of the Carnot cycle (Fig. 22.11) and the Otto cycle (Fig. 22.12). In the Carnot cycle, heat is transferred into or out of the engine isothermally, but when heat is transferred during the Otto cycle, the temperature is changing. Even the ideal Otto cycle is not as efficient as the Carnot cycle.

FIGURE 22.12 The Otto engine is an ideal engine. The engine "stroke" positions have been moved relative to their positions in Figure 22.8 to help you identify where each stroke occurs on the *PV* diagram.

OTTO ENGINE

Special Case

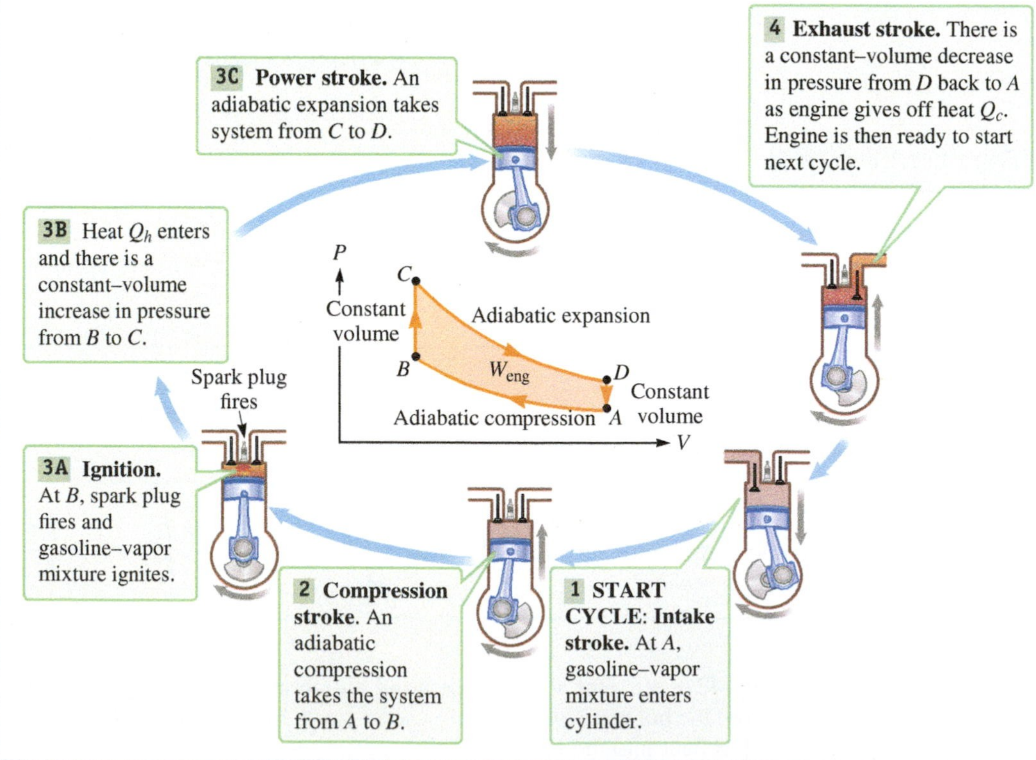
3C Power stroke. An adiabatic expansion takes system from C to D.
4 Exhaust stroke. There is a constant-volume decrease in pressure from D back to A as engine gives off heat Q_c. Engine is then ready to start next cycle.
3B Heat Q_h enters and there is a constant-volume increase in pressure from B to C.
P
C
Constant volume
Adiabatic expansion
B
W_{eng}
D
Constant volume
Adiabatic compression
A
V
Spark plug fires
3A Ignition. At B, spark plug fires and gasoline–vapor mixture ignites.
2 Compression stroke. An adiabatic compression takes the system from A to B.
1 START CYCLE: Intake stroke. At A, gasoline–vapor mixture enters cylinder.

EXAMPLE 22.2 CASE STUDY Otto Engine Efficiency

A Find an expression for the efficiency of an Otto engine in terms of the maximum volume V_{max} and minimum volume V_{min} of the system.

INTERPRET and ANTICIPATE

Follow a procedure similar to our derivation of the efficiency of a Carnot engine (Eq. 22.6). The main difference is that in the Otto engine, heat is transferred in a constant-volume process instead of in an isothermal process. From Figure 22.12, a great change in volume means a large area enclosed on the *PV* diagram. So, we expect that the greater the change in volume, the more efficient an Otto engine will be.

SOLVE

Heat is transferred in a constant-volume process from *B* to *C*, so we can use Equation 21.20 to write an expression for the heat transferred in terms of the molar specific heat at constant volume, C_V. Because $T_C > T_B$, heat Q_h is positive, meaning that energy enters the engine.

$$Q_h = nC_V \Delta T \quad (21.20)$$

$$Q_h = nC_V(T_C - T_B)$$

Heat Q_c is transferred out of the engine during the constant-volume process from *D* to *A*. Because $T_D > T_A$, heat Q_c is negative, meaning that energy leaves the engine. Equation 22.5 requires the absolute value of Q_c.

$$Q_c = nC_V \Delta T = nC_V(T_A - T_D)$$

$$|Q_c| = -nC_V(T_A - T_D) = nC_V(T_D - T_A)$$

To find the efficiency, substitute Q_h and Q_c into Equation 22.5.

$$e = 1 - \frac{|Q_c|}{Q_h} \quad (22.5)$$

$$e_{\text{Otto}} = 1 - \frac{nC_V(T_D - T_A)}{nC_V(T_C - T_B)} = 1 - \frac{(T_D - T_A)}{(T_C - T_B)} \quad (1)$$

Adiabatic processes connect states *C* and *D* and states *A* and *B*. Use Equation 21.32 for the adiabatic expansion or compression of an ideal gas to eliminate *T* in favor of *V* at each set of endpoints.

$$\frac{T_f}{T_i} = \left(\frac{V_f}{V_i}\right)^{1-\gamma} \quad (21.32)$$

$$\frac{T_D}{T_C} = \left(\frac{V_D}{V_C}\right)^{1-\gamma} \qquad \frac{T_B}{T_A} = \left(\frac{V_B}{V_A}\right)^{1-\gamma}$$

Because the processes from *D* to *A* and from *B* to *C* are constant-volume processes, we can eliminate two of the four volumes to get expressions in terms of the maximum and minimum volumes.

$$V_A = V_D = V_{max}$$

$$V_B = V_C = V_{min}$$

$$\frac{T_B}{T_A} = \left(\frac{V_B}{V_A}\right)^{1-\gamma} = \left(\frac{V_{min}}{V_{max}}\right)^{1-\gamma} \quad (22.12)$$

$$\frac{T_D}{T_C} = \left(\frac{V_D}{V_C}\right)^{1-\gamma} = \left(\frac{V_{max}}{V_{min}}\right)^{1-\gamma} \quad (22.13)$$

Find T_D from Equation 22.13.

$$T_D = T_C\left(\frac{V_{max}}{V_{min}}\right)^{1-\gamma} \quad (2)$$

Find T_A from Equation 22.12.

$$T_A = \frac{T_B}{(V_{min}/V_{max})^{1-\gamma}} = T_B\left(\frac{V_{max}}{V_{min}}\right)^{1-\gamma} \quad (3)$$

Substitute Equations (2) and (3) into Equation (1).

$$e_{\text{Otto}} = 1 - \frac{T_C(V_{max}/V_{min})^{1-\gamma} - T_B(V_{max}/V_{min})^{1-\gamma}}{T_C - T_B}$$

$$e_{\text{Otto}} = 1 - \frac{(T_C - T_B)(V_{max}/V_{max})^{1-\gamma}}{T_C - T_B}$$

$$e_{\text{Otto}} = 1 - \left(\frac{V_{max}}{V_{min}}\right)^{1-\gamma} \quad (22.14)$$

Example continues on page 662 ▶

CHECK and THINK

From Table 21.3, $\gamma > 1$, so the exponent in Equation 22.14 is negative. The greater the ratio V_{max}/V_{min}, the smaller the quantity $(V_{max}/V_{min})^{1-\gamma}$ and the greater the efficiency, as expected.

B Typically, $V_{max}/V_{min} \approx 8$ and $\gamma = 1.4$ for air. Find the typical efficiency of an Otto engine.

INTERPRET and ANTICIPATE

In this part of the example, we are given a chance to find the efficiency of an Otto engine numerically by substituting into Equation 22.14 from part A. We expect that the Otto engine is more efficient than an actual engine.

SOLVE Substitute values into Equation 22.14.	$e = 1 - \left(\frac{V_{max}}{V_{min}}\right)^{1-\gamma} = 1 - (8)^{1-1.4} = 1 - (8)^{-0.4}$ $e = 1 - 0.44 = 0.56$

CHECK and THINK

The efficiency of an Otto engine is about 56%. A typical car engine has an efficiency between 15% and 35%, so the Otto engine is more efficient than a real car engine, as expected. An Otto engine is an ideal engine whose model does not include friction; also, the gasoline vapor–air mixture is treated as an ideal gas that changes slowly through a series of equilibrium steps.

EXAMPLE 22.3 CASE STUDY Gas Mileage

When checking the specs on a car you are interested in buying, you might see that the car gets 32 miles per gallon on the highway at 65 mph and its engine is rated at about 200 hp (horsepower). In this example, we estimate the work done by the engine per second and compare our result to the car's specs. The heat of combustion for gasoline (the energy released per unit mass) is about $L_{com} = 44$ MJ/kg. The density of gasoline is about $\rho = 740$ kg/m^3.

A How much heat does the engine absorb per second on the highway? This result is the power input P_{in}. Give your answer in watts and horsepower.

INTERPRET and ANTICIPATE

Find the gasoline consumed in 1 hour of highway driving from the given specs. Then, by using the heat of combustion, find the heat absorbed by the engine in 1 hour. Dividing by the number of seconds in 1 hour will give the power input to the engine.

SOLVE If the car travels at 65 mph for 1 hour and its gas mileage is 32 mi/gal, find the volume of gasoline consumed in 1 hour. Convert that volume to SI units.	$V = 65\,\text{mi}\left(\frac{1\text{ gal}}{32\,\text{mi}}\right) = 2.03\text{ gal}$ $V = 2.03\text{ gal} \times \frac{3.79\text{ L}}{1\text{ gal}} \times \frac{10^{-3}\text{ m}^3}{1\text{ L}}$ $V = 7.69 \times 10^{-3}\,\text{m}^3$ in 1 h
The mass of gasoline consumed in 1 hour follows from its density.	$m = \rho V = (740\text{ kg/m}^3)(7.69 \times 10^{-3}\text{ m}^3)$ $m = 5.69\,\text{kg}$ in 1 h
The heat of combustion is similar to heat of transformation (Section 21-5), so we can find the heat absorbed in 1 hour as a result of combustion by multiplying the mass of gasoline consumed by the heat of combustion.	$Q_h = mL_{com} = (5.69\text{ kg})(44 \times 10^6\text{ J/kg})$ $Q_h = 2.5 \times 10^8$ J in 1 h

The power input is the heat absorbed per second, so divide Q_h by the number of seconds in 1 hour. To find power in horsepower, use the conversion factor 1 hp = 746 W.	$P_{in} = \frac{Q_h}{\Delta t} = \frac{2.5 \times 10^8 \text{ J}}{3600 \text{ s}} = 7.0 \times 10^4 \text{ W}$ $P_{in} = 93 \text{ hp}$

CHECK and THINK

This result may seem low. The car's specs say that the output power is more like 200 hp. Because the output power must be less than the input power, we might expect the input power to be several hundred or even a thousand horsepower. We will address this issue in parts B and C.

B If the engine has an efficiency of 30% (a very efficient real engine), what is the power output P_{out} of the engine? Give your answer in horsepower.

INTERPRET and ANTICIPATE

Equation 22.4 is written in terms of energy input Q_h and energy output W_{eng}. We must modify this equation so that we can work with power input and power output. We expect the output power to be less than the input power.

SOLVE Power is energy per unit time interval, so rewrite Equation 22.4 in terms of power.	$e = \frac{W_{eng}/\Delta t}{Q_h/\Delta t} = \frac{P_{out}}{P_{in}}$
Rearrange to express the power output as the efficiency multiplied by the power input.	$P_{out} = eP_{in} = (0.30)(93 \text{ hp})$ $P_{out} = 28 \text{ hp}$

CHECK and THINK

As expected, the output power is less than the input power. Our result is about one order of magnitude below the specs given by the car manufacturer because the manufacturer's reported number is not the power output during highway driving. Instead, the engine is tested in a laboratory off the road. The engine is loaded and run at different rates; the power output reported is the maximum output of the engine during the test. When you drive, you will probably never cause the engine to put out as much power.

C When the engine is being tested, its maximum power output is found to be 250 hp. At what rate is gasoline consumed by the engine if its efficiency is 20% at its maximum output? Give your answer in kilograms per second and gallons per hour.

INTERPRET and ANTICIPATE

Reverse the procedure from parts A and B to find the amount of gasoline consumed per hour. We expect a higher rate of gasoline consumption during the test than is actually consumed during highway driving.

SOLVE Reversing the process in part B, find the power input by dividing the power output by the efficiency of the engine.	$P_{in} = \frac{P_{out}}{e} = \frac{250 \text{ hp}}{0.20} = 1250 \text{ hp} \times \frac{746 \text{ W}}{1 \text{ hp}}$ $P_{in} = 9.33 \times 10^5 \text{ W}$
The power input is the heat absorbed mL_{com} per time interval Δt.	$P_{in} = \frac{mL_{com}}{\Delta t}$
Solve for the rate at which gasoline is consumed.	$\frac{m}{\Delta t} = \frac{P_{in}}{L_{com}} = \frac{9.33 \times 10^5 \text{ W}}{44 \times 10^6 \text{ J/kg}}$ $\frac{m}{\Delta t} = 2.1 \times 10^{-2} \text{ kg/s}$

Example continues on page 664 ▶

Use the density of gasoline to convert to gallons per hour.	$\frac{2.1 \times 10^{-2}\ \text{kg}}{\text{s}} \times \frac{1\ \text{m}^3}{740\ \text{kg}} \times \frac{1\ \text{L}}{10^{-3}\ \text{m}^3} \times \frac{1\ \text{gal}}{3.79\ \text{L}} \times \frac{3600\ \text{s}}{1\ \text{hr}}$ $= 27\ \text{gal/hr}$

CHECK and THINK

As expected, the rate of gasoline consumption during the test is greater—by about 13 times—than the rate during normal highway driving. At this rate of consumption, most cars would be out of gas in just about half an hour.

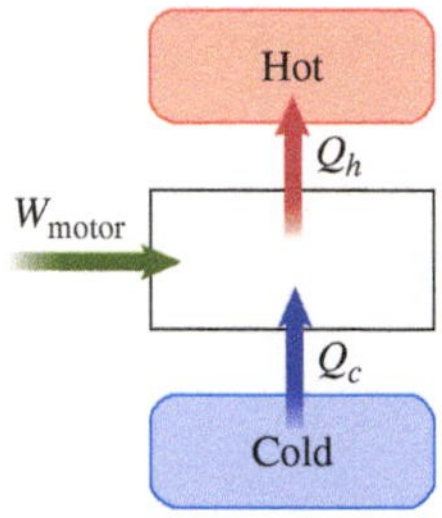

FIGURE 22.13 The energy transfer diagram for a refrigerator looks like that for a heat engine running in reverse. Compare with Figure 22.4.

22-5 Case Study: Refrigerators

A refrigerator absorbs heat from the cold food compartment and gives off heat to the warm room. It might seem like this process is a violation of the second law of thermodynamics, which says that heat flows naturally from a hot object to a colder object, not from a cold object to a hotter one (Clausius statement, Section 22-1). The key to the Clausius statement is the word *naturally* because heat can flow from a cold object to a hotter object if work is involved. A refrigerator does not violate the second law of thermodynamics. As shown in Figure 22.13, a refrigerator is like a heat engine running in reverse: The refrigerator motor does work so that heat is absorbed from the low-temperature compartment inside the fridge and expelled into the warm room. You may have felt a rush of warm air coming from behind or beneath your refrigerator.

Figure 22.14A identifies the major components of a refrigerator. A fluid known as a refrigerant is the working substance. As the refrigerant flows through a tube, it undergoes transitions between its liquid and vapor phases. The boiling point of the refrigerant is lower than the temperature you would like inside the freezer. For example, ammonia boils at −27°F, so it may be used as a refrigerant.[1]

Figure 22.14B is a schematic showing the function of each component of the refrigerator. The refrigerant on the left side of the circuit is at low temperature and low pressure; on the right side of the circuit, it is at high temperature and high pressure. The sum of the heat transferred to the refrigerant from the freezer and food compartment is Q_c. When the refrigerant enters the condenser it is a vapor, which gives off heat Q_h and becomes a liquid. The cycle then begins again as the cold, low-pressure liquid enters the freezer compartment.

A measure of the refrigerator's performance known as the **coefficient of performance** κ is given by

REFRIGERATOR **Special Case**

$$\kappa = \frac{|Q_c|}{|W_{\text{motor}}|} = \frac{\text{what you get from the refrigerator}}{\text{what you put into it}} \tag{22.15}$$

According to the second law of thermodynamics, the motor must do work to transfer heat from the cold interior of the refrigerator to the hot exterior. Because the work done by the motor cannot be zero, the coefficient of performance κ cannot be infinite.

As in the case of an engine, the work done by the motor can be written in terms of the net heat transferred:

$$|W_{\text{motor}}| = |Q_h| - |Q_c| \tag{22.16}$$

The coefficient of performance may be written as

$$\kappa = \frac{|Q_c|}{|Q_h| - |Q_c|} \tag{22.17}$$

[1]In the past, a common refrigerant was chlorofluorocarbon (CCl_2F_2), a member of the Freon family, but because it and similar substances deplete the stratospheric ozone layer, they are no longer used as refrigerants.

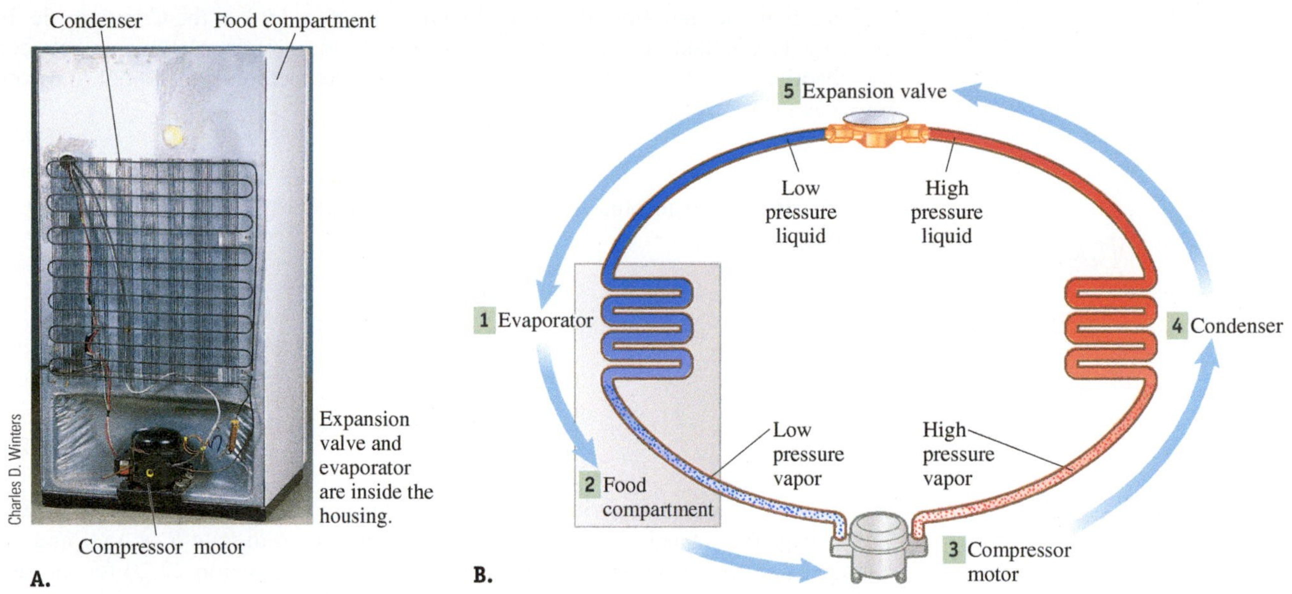

FIGURE 22.14 **A.** Components of a refrigerator. **B.** A schematic of the processes in a refrigerator shows refrigerant passing counterclockwise through a closed tube. 1 Liquid refrigerant passing through the evaporator absorbs heat from the air and becomes a vapor. 2 Vapor entering the food compartment is colder than the air in the compartment, so heat is transferred to the refrigerant. 3 Next the vapor enters the compressor, whose piston compresses it adiabatically. Positive work W_{motor} is done on the vapor and it leaves the compressor at a higher pressure and higher temperature. 4 Vapor entering the condenser gives off heat and becomes a hot, high-pressure liquid. 5 This liquid expands adiabatically at a rate controlled by the expansion valve so that the liquid is greatly cooled.

If you imagine an ideal refrigerator as a Carnot engine running in reverse, the best possible coefficient of performance may be written in terms of the cold temperature T_c inside and the hot temperature T_h outside (Problem 33):

$$\kappa = \frac{T_c}{T_h - T_c} \tag{22.18}$$

CONCEPT EXERCISE 22.3

On a hot summer day, can you cool your room by opening the door to your refrigerator?

22-6 Entropy

ENTROPY Major Concept

Neither the Clausius nor the Kelvin-Planck statement of the second law of thermodynamics explains how we know when we are watching a movie played backward. A more general and mathematical statement of the second law of thermodynamics requires another state variable known as *entropy*. By using the concept of entropy in the second law, we can explain why some processes are not reversible.

Entropy is like potential energy in that only changes in entropy are physically meaningful. A mathematical definition of the **change in entropy** ΔS for an isothermal process at temperature T is

$$\Delta S = \frac{Q}{T} \tag{22.19}$$

where ΔS is the change in the system's entropy and Q is the amount of heated added to the system. Entropy has the dimensions of energy per temperature, so its SI units are joules per kelvin.

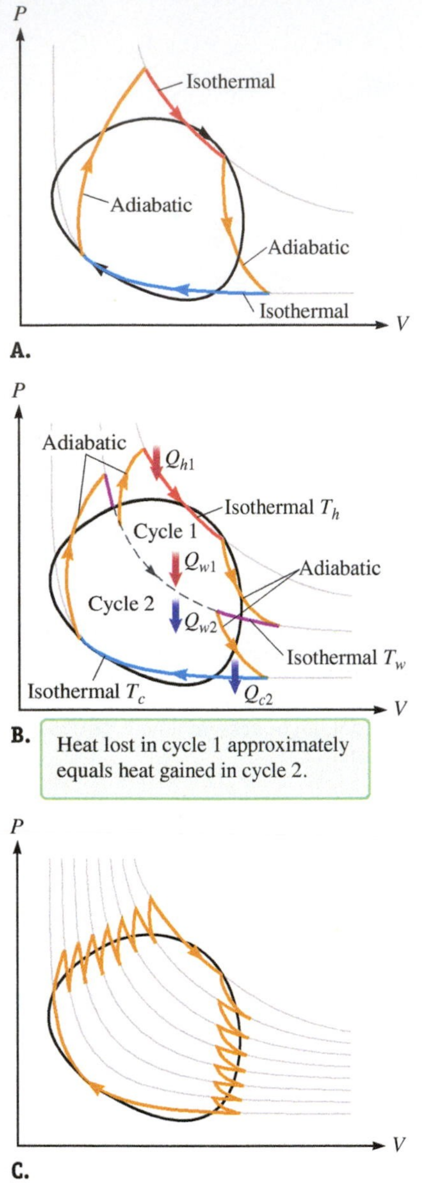

FIGURE 22.15 **A.** A reversible thermodynamic cycle is approximated by a single Carnot cycle. **B.** Two Carnot cycles approximate the thermodynamic cycle. **C.** More Carnot cycles are a better model for this arbitrary cycle.

Let's build our intuition of entropy by taking another look at the Carnot cycle. In deriving the efficiency of the Carnot engine, we found a simple relationship between the absorbed heat Q_h and emitted heat Q_c and the temperature of the hot and cold reservoirs (Eq. 22.11):

$$\frac{|Q_c|}{Q_h} = \frac{T_c}{T_h}$$

We can remove the absolute value signs if we remember that the emitted heat Q_c is negative:

$$\frac{-Q_c}{Q_h} = \frac{T_c}{T_h}$$

$$\frac{-Q_c}{T_c} = \frac{Q_h}{T_h}$$

$$\frac{Q_c}{T_c} + \frac{Q_h}{T_h} = 0 \tag{22.20}$$

Equation 22.20 can be rewritten in terms of entropy. In the *PV* diagram in Figure 22.11 (page 657), label the change in entropy from state *A* to state *B* as ΔS_{AB} and the change in entropy from state *B* to state *A* as ΔS_{BA}. So, Equation 22.20 for the two isothermal processes is

$$\Delta S_{AB} + \Delta S_{BA} = 0$$

There is no heat exchanged in the adiabatic processes, so according to Equation 22.19, there is no change in entropy during these two processes ($\Delta S_{BC} = \Delta S_{DA} = 0$). Thus the total change in entropy around the Carnot cycle is zero: $\Delta S_{\text{cycle}} = 0$.

Recall that state variables such as pressure, volume, temperature, and thermal energy depend only on the endpoints in a *PV* diagram and not on the path. If a system goes through a thermodynamic cycle, it must return to the same point in the *PV* diagram, and the state variables must also return to their original values. Thus, the total change in a state variable during a cyclic process is zero. The change in entropy around a Carnot cycle is zero, supporting the idea that entropy is a state variable.

In fact, the change in entropy is zero around any reversible cyclic path. Figure 22.15 shows an arbitrary cycle on a *PV* diagram. Any reversible cyclic process can be approximated by a series of Carnot cycles. In Figure 22.15A, the cyclic path is approximated by a single Carnot cycle. The approximation improves if more Carnot cycles are used.

In Figure 22.15B, the cycle is approximated with portions of two Carnot cycles (cycle 1 and cycle 2) indicated by the solid curves. Heat is only transferred during the isothermal processes. For the complete solid-curve cycle shown in Figure 22.15B, most of the heat is transferred during the isothermal processes at T_h and at T_c. Little heat is transferred for the portion of the cycle at the intermediate warm temperature T_w because the process along that isotherm is very short. So, the change in entropy for this complete cycle is approximately

$$\Delta S_{\text{complete}} \approx \frac{Q_{h1}}{T_h} + \frac{Q_{c2}}{T_c}$$

To see this, consider the entropy changes for Carnot cycles 1 and 2:

$$\Delta S_1 = \frac{Q_{h1}}{T_h} + \frac{Q_{w1}}{T_w} = 0 \quad \text{and} \quad \Delta S_2 = \frac{Q_{w2}}{T_w} + \frac{Q_{c2}}{T_c} = 0$$

The heat Q_{w2} absorbed during cycle 2 is approximately equal to the negative of the output heat Q_{w1} from cycle 1. When we add the changes in entropy for both cycles, we find

$$\Delta S_1 + \Delta S_2 = \frac{Q_{h1}}{T_h} + \frac{Q_{w1}}{T_w} + \frac{Q_{w2}}{T_w} + \frac{Q_{c2}}{T_c} = 0$$

$$\Delta S_1 + \Delta S_2 \approx \frac{Q_{h1}}{T_h} + \frac{Q_{w1}}{T_w} - \frac{Q_{w1}}{T_w} + \frac{Q_{c2}}{T_c} = 0$$

$$\Delta S_1 + \Delta S_2 \approx \frac{Q_{h1}}{T_h} + \frac{Q_{c2}}{T_c} = 0$$

which is the same as $\Delta S_{\text{complete}}$ for the complete solid cycle shown in Figure 22.15B. So, the change in entropy around the cycle is approximately zero: $\Delta S_{\text{complete}} \approx 0$.

Figure 22.15C shows that the approximation becomes better if more Carnot cycles are used. The same argument we made for two Carnot cycles may be used to show that the change in entropy around any reversible cyclic path is zero. In the limit of infinitely many Carnot cycles, we write the change in entropy around a reversible cycle as

$$\Delta S = \oint \frac{dQ}{T} = 0 \quad (22.21)$$

where the circle on the integral sign means that the integration is over a closed path on a PV diagram and dQ is an infinitesimal amount of heat.

The integral in Equation 22.21 can be calculated for either a clockwise or a counterclockwise path, and you can begin the calculation at any point on the cycle. For example, in Figure 22.16, you could integrate along path A from point i clockwise through point f and continue along path B clockwise back to point i:

$$\Delta S = \oint \frac{dQ}{T} = \left[\int_i^f \frac{dQ}{T}\right]_{\text{path A}} + \left[\int_f^i \frac{dQ}{T}\right]_{\text{path B}} = 0$$

By switching the limits on the integral for path B, we find

$$\Delta S = \left[\int_i^f \frac{dQ}{T}\right]_{\text{path A}} - \left[\int_i^f \frac{dQ}{T}\right]_{\text{path B}} = 0$$

$$\left[\int_i^f \frac{dQ}{T}\right]_{\text{path A}} = \left[\int_i^f \frac{dQ}{T}\right]_{\text{path B}}$$

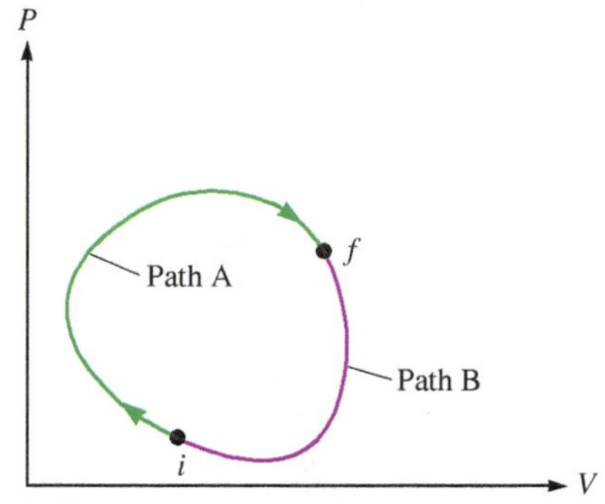

FIGURE 22.16 The change in entropy does not depend on path.

Just as you would expect for a state variable, the difference in entropy ΔS does not depend on the path:

$$\Delta S = S_f - S_i = \left[\int_i^f \frac{dQ}{T}\right]_{\text{any reversible path}} \quad (22.22)$$

Equation 22.22 is for any *reversible* process, but all real processes are *ir*reversible. To find the change in entropy ΔS for a real process, we must imagine some ideal, reversible process that would take the system from state i to state f and then calculate ΔS using Equation 22.22. Because ΔS does not depend on the path, ΔS calculated for the ideal process equals ΔS for the real process.

DERIVATION Entropy Change in an Isothermal Process

We will show that the entropy change during a reversible isothermal process is given by

$$\Delta S = \frac{Q}{T} \text{ (isothermal)} \quad (22.19)$$

In an isothermal process, temperature is constant. We will use Equation 22.22 with T constant. Our result can then be used for any isothermal process.

Because T is constant, we pull it out of the integral.	$\Delta S = \int_i^f \frac{dQ}{T}$ (22.22) $\Delta S = \frac{1}{T}\int_i^f dQ$
The integral is just the total amount of heat Q transferred during the process from state i to state f.	$\Delta S = \frac{Q}{T}$ ✓ (22.19)

COMMENTS

Equation 22.19 is valid for any real process that can be modeled as reversible and isothermal.

EXAMPLE 22.4 Put the Ice Tray Away!

Imagine leaving a tray of 14 ice cubes out of the freezer. As ice cubes melt, their temperature remains constant at 0°C. Each cube is 4.00 cm on a side. Find the entropy change when the ice melts. The density of ice is 0.917×10^3 kg/m^3, and its latent heat of fusion is 3.33×10^5 J/kg.

INTERPRET and ANTICIPATE

The ice melts slowly, so we can model the process as reversible and isothermal and thus can use Equation 22.19. To find the heat required to melt the ice, use $Q = mL_F$ (Eq. 21.9). First find the mass of the 14 ice cubes.

SOLVE

To find the mass of the ice cubes, calculate their volume and use the density of ice.	$V_{\text{cube}} = \ell^3 = (4.00 \times 10^{-2}\text{ m})^3 = 6.40 \times 10^{-5}\text{ m}^3$ $m_{\text{cube}} = \rho V_{\text{cube}} = (0.917 \times 10^3\text{ kg/m}^3)(6.40 \times 10^{-5}\text{ m}^3) = 5.87 \times 10^{-2}\text{ kg}$ $m = (14\text{ cubes})m_{\text{cube}} = 14(5.87 \times 10^{-2}\text{ kg}) = 0.822\text{ kg}$
Find the heat required to melt all the ice using Equation 21.9 and the latent heat of fusion (Table 21.2).	$Q = mL_F = (0.822\text{ kg})(3.33 \times 10^5\text{ J/kg}) = 2.74 \times 10^5\text{ J}$
Finally, find the change in entropy. The temperature must be expressed in kelvins.	$\Delta S = \frac{Q}{T} = \frac{2.74 \times 10^5\text{ J}}{273\text{ K}}$ (22.19) $\Delta S = 1.00 \times 10^3\text{ J/K}$

CHECK and THINK

Entropy is a new and abstract concept. This example helps build intuition about entropy. When a tray of ice cubes melts, the change in entropy is about 1000 J/K.

DERIVATION Entropy Change in a Constant-Pressure Process

We show that the entropy change during a reversible constant-pressure process is given by

$$\Delta S = mc_P \ln\left(\frac{T_f}{T_i}\right) \tag{22.23}$$

where c_p is the mass specific heat capacity at constant pressure.

This derivation is much like the previous one (entropy of an isothermal process) except now the process is constant pressure instead of isothermal. Because the temperature is not constant, we cannot pull T outside the integral. Instead, we find an expression for dQ when pressure is constant.

Equation 21.5 gives the heat required to change the temperature of a substance. We can modify this equation for a small change in temperature dT, including a subscript P for specific heat at constant pressure.	$Q = mc\Delta T$ (21.5) $dQ = mc_P\, dT$
Substitute dQ into Equation 22.22 and integrate.	$\Delta S = \int_i^f \frac{dQ}{T} = \int_i^f \frac{mc_P\, dT}{T} = mc_P \int_i^f \frac{dT}{T}$ $\Delta S = mc_P \ln\left(\frac{T_f}{T_i}\right)$ ✓ (22.23)

COMMENTS

Equation 22.23 is valid for any process in which the temperature changes from T_i to T_f as long as the final pressure equals the initial pressure. If the substance is cooled so that $T_f < T_i$, the change in entropy is negative (entropy decreases).

EXAMPLE 22.5 Mixing Barrels of Water

You have two barrels of water. Barrel 1 contains $m_1 = 35$ kg of water at temperature $T_1 = 30.0°C$, and barrel 2 contains $m_2 = 75$ kg of water at temperature $T_2 = 90.0°C$. If you mix the two barrels of water in an isolated tub of negligible heat capacity so that heat is only exchanged between the hot and cold water, find the temperature of the completely mixed water. Find the change in entropy of the water from barrel 1, the change in entropy of the water from barrel 2, and the change in entropy of the Universe. *Hint*: The specific heat of water is listed in Table 21.1.

INTERPRET and ANTICIPATE

Mixing the water is a constant-pressure process. Because heat is only exchanged between the water, the heat lost by the hot water must equal heat gained by the cold water. We expect the temperature of the mixed water to be between 30.0°C and 90.0°C. We also expect that the cold water's entropy increases and the hot water's entropy decreases. Because the only heat exchanged is between the water, we can add the two entropy changes to find the entropy change for the Universe.

SOLVE

Set the heat lost by the hot water equal to the heat gained by the cold water and use $Q = mc\Delta T$ (Eq. 21.5) to find the temperature of the mixture. Because hot water loses energy, we must remember Q_{lost} is negative. The final temperature of the mixture is T_f for both the hot water and the cold water. It is okay to work in Celsius for this part of the example, but we'll need this final temperature in kelvins for the next step.

$$Q_{lost} = Q_{gained}$$

$$-m_2 c\,\Delta T_{hot} = m_1 c\,\Delta T_{cold}$$

$$-m_2(T_f - T_2) = m_1(T_f - T_1)$$

$$(m_1 + m_2)T_f = m_1 T_1 + m_2 T_2$$

$$T_f = \frac{m_1 T_1 + m_2 T_2}{m_1 + m_2}$$

$$T_f = \frac{(35.0\,\text{kg})(30.0°\text{C}) + (75.0\,\text{kg})(90.0°\text{C})}{35.0\,\text{kg} + 75.0\,\text{kg}}$$

$$T_f = 70.9°\text{C} = 343.9\text{ K}$$

CHECK and THINK

As expected, the temperature of the mixture is between the two original temperatures. Because there is initially more hot water, the final temperature is higher than the average of the two original temperatures.

SOLVE

Find the entropy change for cold water from barrel 1 and for hot water from barrel 2 using Equation 22.23. The temperatures must be in kelvins. The specific heat of water from Table 21.1 is valid near room temperature and atmospheric pressure.

$$\Delta S = mc_P \ln \frac{T_f}{T_1} \tag{22.23}$$

$$\Delta S_1 = (35.0\,\text{kg})(4190\text{ J/kg}\cdot\text{K})\ln\left(\frac{343.9\text{ K}}{303\text{ K}}\right)$$

$$\Delta S_1 = 1.86 \times 10^4\text{ J/K}$$

$$\Delta S_2 = (75.0\,\text{kg})(4190\text{ J/kg}\cdot\text{K})\ln\left(\frac{343.9\text{ K}}{363\text{ K}}\right)$$

$$\Delta S_2 = -1.70 \times 10^4\text{ J/K}$$

CHECK and THINK

As expected, the entropy of the cold water from barrel 1 increased, and the entropy of the hot water from barrel 2 decreased.

SOLVE

Because heat is only exchanged between the hot water and the cold water, the only entropy changes are in the water. The entropy change in the Universe as a result of mixing the two barrels of water is the sum of ΔS_1 and ΔS_2.

$$\Delta S_{Universe} = \Delta S_1 + \Delta S_2$$

$$\Delta S_{Universe} = (1.86 \times 10^4\text{ J/K}) - (1.70 \times 10^4\text{ J/K})$$

$$\Delta S_{Universe} = 1.6 \times 10^3\text{ J/K}$$

Example continues on page 670 ▶

CHECK and THINK

The entropy of the Universe increases as a result of mixing the two barrels of water. This example is another illustration of the second law of thermodynamics. (See Section 22-7.)

EXAMPLE 22.6 An Ideal Gas Doubles in Volume

The free expansion of an ideal gas is described in Section 21-7. It can be shown (Problem 70) that the entropy change of an ideal gas during a free expansion is given by

$$\Delta S = nR\ln\left(\frac{V_f}{V_i}\right) \tag{22.24}$$

Use this expression to find the entropy change when 1 mol of an ideal gas undergoes a free expansion that doubles its volume as in Figure 21.20.

INTERPRET and ANTICIPATE

We will develop some intuition about entropy by substituting values into Equation 22.24. Because the gas expands, we expect that the entropy increases.

SOLVE

Substitute given values into Equation 22.24.

$$\Delta S = (1\text{ mol})(8.31\text{ J/mol}\cdot\text{K})\ln\left(\frac{2V_i}{V_i}\right) = 5.76\text{ J/K}$$

CHECK and THINK

As expected, the entropy increased. Compare with Example 22.4: When a tray of ice cubes melts, the entropy of the water increases by about 1000 J/K. In this example, we found that when a gas doubles in volume—picture a balloon of gas doubling in size—the change in entropy is only about 6 J/K.

22-7 Second Law of Thermodynamics, General Statements

SECOND LAW OF THERMODYNAMICS (GENERAL STATEMENTS)

Underlying Principle

The concept of entropy as a state variable allows us to make several general statements of the second law of thermodynamics. One of them is as follows:

The entropy of an isolated system never decreases: $\Delta S \geq 0$.

The equal sign holds for a reversible process, $\Delta S = 0$. Entropy always increases in the case of an irreversible process, $\Delta S > 0$.

The Universe is defined to be everything, so the ultimate isolated system is the Universe. In addition, all real processes are irreversible. Therefore, according to the second law of thermodynamics:

The entropy of the Universe must be increasing:

$$\Delta S_{\text{Universe}} > 0$$

In thermodynamics, we often choose part of the Universe as a system and then take everything else to be in the environment. Another way to state the second law of thermodynamics for the Universe is as follows:

The total entropy of a system plus its environment must increase as a result of any real process:

$$\Delta S_{\text{sys}} + \Delta S_{\text{env}} > 0$$

It is thus possible that the entropy of one part of the Universe decreases during some process, but, if so, the entropy of another part of the Universe increases so that the overall entropy of the Universe increases.

The second law of thermodynamics is very different from other underlying principles we've studied. For example, compare the second law of thermodynamics to the conservation of energy principle. On the one hand, conservation of energy states that the energy of the Universe remains constant. If energy increases in one part of the Universe, energy must decrease in another part of the Universe by the same amount. On the other hand, the second law of thermodynamics says that the entropy of the Universe is always increasing.

To understand the importance of this distinction, imagine mixing 1 liter of hot water with 1 liter of cool water. Energy is transferred in the form of heat from the hot water to the cool water, resulting in 2 liters of warm water with a temperature between the two original temperatures. Such a transfer of energy obeys the principle of energy conservation and the second law of thermodynamics. Consider the entire 2 liters of water to be an isolated system. The total energy of the water remains constant, but the entropy increases.

Now imagine that you have 2 liters of warm water, and later you find that the water has spontaneously separated itself into two parts: 1 liter is hot, and the other is cold. According to the principle of energy conservation, such a process is possible—the sum of the water's thermal energy remains constant—but according to the second law of thermodynamics, such a process is impossible because the entropy of the water would decrease. Recall Example 22.5, in which we mixed hot water and cold water and found that the entropy of the Universe increased by $\Delta S_{\text{Universe}} = 1.6 \times 10^3$ J/K. If the warm water mixture were to separate spontaneously into hot water and cold water, the entropy of the Universe would decrease by $\Delta S_{\text{Universe}} = -1.6 \times 10^3$ J/K.

The Arrow of Time

The second law of thermodynamics has broad consequences. It says that processes that decrease the entropy of the Universe are not possible. At the beginning of this chapter, we thought about how we know when we are watching a movie in reverse. We know that an egg will not spontaneously fly off a plate and into a hot frying pan, become uncooked, and then fly up into the open eggshell that closes up without a trace. How do we know that this order of events is impossible? It is impossible because the entropy of the Universe would have to decrease. When we watch a movie, we can tell whether the movie is being played forward or in reverse by observing changes in entropy, so entropy is *time's arrow* because it tells us which way time is going. In Sections 22-8 and 22-9, we develop a broader understanding of entropy and the consequences of the second law of thermodynamics.

CONCEPT EXERCISE 22.4

You have considerable intuition about whether some process increases or decreases the entropy of the Universe. Just imagine watching the process in reverse. If it doesn't make sense in reverse, entropy must have increased. This test isn't always foolproof; some process may *seem* reversible, but the entropy of the Universe still increases. Use your intuition to decide whether the entropy of the Universe increases during the following processes:

a. A car crashes into a brick wall.
b. An ice cube floating in a large cup of water melts.
c. A puddle of water freezes to ice slowly.
d. The frozen puddle of water melts slowly.

EXAMPLE 22.7 Clausius Statement

Heat is transferred from a hot reservoir at T_h to a cold reservoir at T_c. Recall that the temperature of a heat reservoir is assumed to be constant. Find an expression for the net change in the entropy of the Universe $\Delta S_{\text{Universe}}$ and then show that heat cannot flow from the cold reservoir to the hot reservoir.

Example continues on page 672 ▶

INTERPRET and ANTICIPATE

A heat reservoir is so large that when heat flows into or out of the reservoir, its temperature stays approximately constant. Therefore, we can use $\Delta S = Q/T$ (Eq. 22.19) to find the change in entropy of each heat reservoir.

SOLVE

Heat flows out of the hot reservoir, so heat is negative ($-Q$), and entropy decreases.	$\Delta S_{hot} = -\dfrac{Q}{T_h}$
Heat flows into the cold reservoir, so heat is positive (Q), and entropy increases.	$\Delta S_{cold} = \dfrac{Q}{T_c}$
Find the net entropy change of the Universe by adding the entropy changes for the hot and cold reservoirs.	$\Delta S_{Universe} = \Delta S_{hot} + \Delta S_{cold} = -\dfrac{Q}{T_h} + \dfrac{Q}{T_c}$ $\Delta S_{Universe} = Q\left(\dfrac{1}{T_c} - \dfrac{1}{T_h}\right)$ (1)
We have used the convention that when heat flows from hot to cold, Q is positive. The high temperature is greater than the low temperature, so the Universe's entropy increases.	$T_h > T_c$ $\dfrac{1}{T_h} < \dfrac{1}{T_c}$ $\left(\dfrac{1}{T_c} - \dfrac{1}{T_h}\right) > 0$ $\Delta S_{universe} > 0$
If heat flows from the cold to the hot reservoir, we must modify Equation (1) by inserting a negative sign because $Q \to -Q$, and we find that the entropy of the Universe decreases. The second law of thermodynamics does not allow the entropy of the Universe to decrease, so we conclude that heat cannot spontaneously flow from cold to hot.	$\Delta S_{Universe} < 0$

CHECK and THINK

We have just found that the general statement of the second law of thermodynamics (entropy of the Universe does not decrease) includes the Clausius statement (heat does not naturally flow from cold to hot).

22-8 Order and Disorder

ENTROPY Major Concept

As a state variable, entropy must tell us something about the condition of a system. Just as kinetic energy—another state variable—tells us about the motion in a system, **entropy** is a measure of the *disorder* in a system.

We see order and disorder in our lives everyday. A cup of black coffee and a carton of cream have more order than a blended cup of coffee with cream. When the coffee and cream are separated, all the coffee molecules are in one container, and all the cream molecules are in a different container. When you stir the cream into your coffee, disorder increases because the coffee and cream molecules are randomly distributed. As another example, a drawer full of socks rolled together by type has more order than a laundry basket full of clothes thrown in arbitrarily.

When the entropy of a system increases, the system becomes more disordered. In Example 22.4, we found that when ice melts, the entropy of the water increases. On the microscopic level, water molecules in cubes of ice are very orderly, with each molecule holding a position in a three-dimensional crystalline structure (Fig. 22.17A). When the ice melts, the bonds weaken; the structure then falls apart, becoming more disordered (Fig. 22.17B).

This statement is true in general; when heat flows into a system, the system's disorder increases because of an increase in molecular motion. From Equation 22.19, $\Delta S = Q/T$. When heat is added to a relatively cold system, the ratio Q/T is large, so there is a great change in entropy. When the system is cold, it has very little molecular motion and is relatively ordered. Adding heat causes a relatively large increase in the particles' motion and therefore a relatively large increase in the system's disorder. A system that is initially hot has considerable motion and therefore much disorder. When heat is added, it does little to increase the motion, and the increase in entropy Q/T is therefore relatively small.

Consider another example. When hot water is mixed with cold water, the entropy of the water increases (Example 22.5). Initially, the water is ordered by its temperature; afterward, that order is lost because the molecules are mixed together randomly. As molecules collide with one another, energy is redistributed among them until they become a single population of randomly moving particles at some temperature between the original temperatures.

The idea that mixing hot and cold parts of a system increases the system's entropy has an important practical implication. Suppose the hot and cold parts are the hot and cold reservoirs in a heat engine or refrigerator. For example, the hot part could be the air in the kitchen, and the cold part could be the air inside the refrigerator. Without two separate heat reservoirs, heat engines and refrigerators won't operate. Put more generally, when entropy increases, disorder increases, and the ability of the system to do useful work decreases.

In fact, in any natural process, energy is *degraded*. Of course, energy is conserved, but some of the energy goes into a more disordered, useless form. For example, when a block slides down an inclined plane, potential energy goes into kinetic energy and thermal energy. Kinetic energy is an ordered and useful form of energy that may be transformed back into potential energy, but thermal energy is disordered. That energy cannot be transformed back into mechanical energy, and we say that friction has "dissipated" some energy.

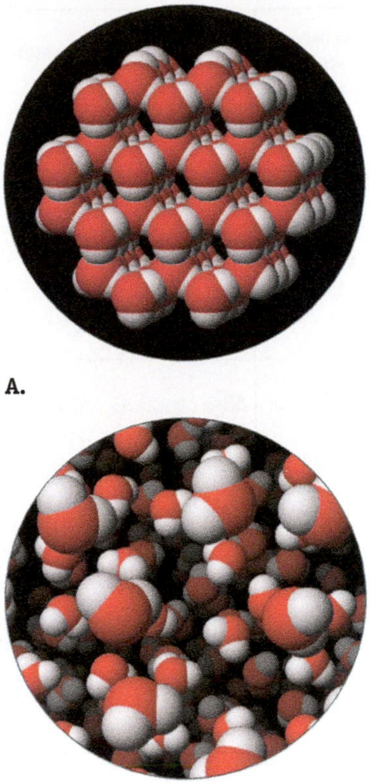

A.

B.

FIGURE 22.17 Frozen water (**A**) is more ordered than liquid water (**B**).

The idea that entropy is a measure of disorder has an important implication for the entire Universe. According to the second law of thermodynamics, the entropy of the Universe is increasing. So, another way to state the second law of thermodynamics is as follows:

The Universe is becoming more disordered.

SECOND LAW OF THERMODYNAMICS (GENERAL STATEMENT)

Underlying Principle

The idea that the Universe is becoming more disordered may not sit well with you; somehow, you came to exist, and you are clearly a highly ordered part of the Universe. In fact, evolution is a process that increases order and reduces entropy. According to the second law of thermodynamics, any process such as evolution that decreases entropy in one part of the Universe must increase entropy in another part of the Universe. Evolution—and all biological processes—produces waste molecules. These waste molecules are very disordered, and the overall entropy of the Universe increases as a result.

22-9 Entropy, Probability, and the Second Law

Entropy is a *quantitative* measure of the disorder of a system. To illustrate how to calculate the disorder in a system, let's look at a simple system consisting of four coins: a penny, a nickel, a dime, and a quarter. Each coin can show either a head or a tail. If you observe that all four coins show the same face (either four tails or four heads), you would say the system had a high degree of order. If the system shows a mixture of heads and tails, however, you would conclude the system had more disorder.

The state of the system as a whole (in this case, the tally of heads and tails) is called the **macroscopic state**. The term *macroscopic state* is equivalent to the term *state* in previous sections. State variables such as pressure, temperature, volume, and entropy describe the macroscopic state of a system. The **microscopic state** of a system is the condition of each particle that makes up the system. The microscopic state of the four-coin system specifies the condition (head or tail showing) of each coin.

Macroscopic state	Microscopic state of each coin	Number of microscopic states
4 heads		1
3 heads 1 tail		4
2 heads 2 tails		6
1 head 3 tails		4
4 tails		1

FIGURE 22.18 For this four-coin system, a microscopic state indicates the condition of each coin. The macroscopic state is the total number of heads and the total number of tails.

If the system is the air in your room, its microscopic state is the velocity and position of each air molecule. Ordinarily, we do not know the microscopic state of a thermodynamic system because there are far too many particles to have information about each one of them at every given moment. For illustration purposes, we have chosen the four-coin system because it is so simple that we can observe its microscopic state.

Imagine arranging the coins in all possible microscopic states as shown in Figure 22.18, indicating the condition of each coin. For example, in the first microscopic state in the second row, the penny is tail up and the other coins are heads up. The macroscopic state gives the total number of heads and tails, so all the microscopic states in the second row have the same macroscopic state: three heads and one tail. From Figure 22.18, there are 16 possible microscopic states that are categorized into just five macroscopic states: (1) all heads, (2) three heads and one tail, (3) two heads and two tails, (4) one head and three tails, and (5) all tails.

Entropy S is a macroscopic state variable, and it is directly related to the number $\mathcal{W}$ of possible microscopic states associated with a particular macroscopic state:

$$S = k_B \ln \mathcal{W} \tag{22.25}$$

where k_B is Boltzmann's constant. Like potential energy, only differences in entropy are physically significant. By convention associated with Equation 22.25 if $\mathcal{W} = 1$, then $S = 0$, but it is possible to add an arbitrary constant S_0 so that if $\mathcal{W} = 1$, then $S = S_0$. Throughout this book, we use the convention that $S_0 = 0$.

Taking the four-coin example as an illustration, the macroscopic state consisting of all heads or all tails has the most order and the lowest entropy. In fact, because $\mathcal{W} = 1$ for either of these macroscopic states, the entropy of either of these states is zero:

$$S = k_B \ln \mathcal{W} = k_B \ln(1) = 0$$

The macroscopic state consisting of two heads and two tails has the greatest number of microscopic states ($\mathcal{W} = 6$) and thus also has the greatest entropy.

So, the macroscopic state with the greatest number of microstates is the most disordered state. Think of it this way: The description *two heads and two tails* gives very little information about each coin. You don't know if the penny is head up or tail up, and if you guess, you have only a 50% chance of being correct. Compare that situation to the macroscopic state of *all heads up*. That statement contains all the information about each coin; you know that the penny is head up. In general, the more information provided by the macroscopic state about the microscopic state of each particle, the more ordered that state is.

This four-coin example gives us another way to think about the second law of thermodynamics. Imagine tossing the four coins on a table. What macroscopic state are you most likely to find? You wouldn't count on finding, say, all heads or all tails because there is only a 1 in 16 chance of getting one of those states. The most probable outcome is two heads and two tails because there is a 6 in 16 chance of getting that macroscopic state. So, the most probable outcome is the macroscopic state with the greatest entropy.

Another general way to state the second law of thermodynamics is as follows:

SECOND LAW OF THERMODYNAMICS (GENERAL STATEMENT)

Underlying Principle

The Universe is moving from a state of low probability to a state of higher probability.

Stating the second law in terms of probability means that it is not *impossible* for heat to flow spontaneously from cold to hot or for entropy to decrease; it just means that is very *improbable*. In terms of the four-coin example, if you toss the coins enough times, once in a while you will get all heads. That this outcome is unlikely is not the same as saying it is forbidden. We have never observed heat spontaneously flowing from cold to hot or all the air molecules in a room contracting into a corner. In both of these examples, the number of particles is very large, and it is extremely unlikely to find a large number of particles in an ordered state.

To illustrate, imagine tossing 100 coins instead of just four. The number of microstates is found using

$$\mathcal{W} = \frac{N_{\text{tot}}!}{N_h! N_t!} \qquad (22.26)$$

where N_h is the number of heads, N_t is the number of tails, and N_{tot} is the total number of coins, $N_{\text{tot}} = N_h + N_t$. (The symbol "!" stands for factorial. For example, $3! = 3 \times 2 \times 1 = 6$.) Table 22.1 lists a few macroscopic states for these 100 coins. There are a total of about 1.27×10^{30} microstates, so if you toss 100 coins, you have about a 1 in 1.3×10^{30} chance of throwing all heads. Still, 100 is a small number of objects; consider that 1 mole of particles is more than 10^{23} particles. Imagine finding 10^{23} coins, all heads up. It is very unlikely that such a great number of particles would be found in an ordered state. Put another way, you may measure state variables such as the pressure, temperature, and volume of the air in your room, but those measurements would tell you almost nothing about the microstate of a particular oxygen molecule that you inhaled half an hour ago.

TABLE 22.1 Macroscopic states and microscopic states for a collection of 100 coins.

Macroscopic state		Number of microscopic states $\mathcal{W}$
Heads	Tails	
100	0	1
99	1	100
90	10	1.73×10^{13}
80	20	5.36×10^{20}
60	40	1.37×10^{28}
55	45	6.14×10^{28}
54	46	7.35×10^{28}
53	47	8.44×10^{28}
52	48	9.32×10^{28}
51	49	9.89×10^{28}
50	50	1.01×10^{29}
49	51	9.89×10^{28}
48	52	9.32×10^{28}
47	53	8.44×10^{28}
46	54	7.35×10^{28}
45	55	6.14×10^{28}
40	60	1.37×10^{28}
20	80	5.36×10^{20}
10	90	1.73×10^{13}
1	99	100
0	100	1

CONCEPT EXERCISE 22.5

Use Table 22.1 to find the following:

a. The probability of finding 50 heads and 50 tails in a toss of 100 coins
b. The probability of finding between 45 and 55 heads

EXAMPLE 22.8 Entropy of Free Expansion Revisited

In Example 22.6, we calculated the change in entropy when 1 mole of gas undergoes a free expansion and doubles in volume (Fig. 21.20). Repeat that calculation using Equation 22.25.

INTERPRET and ANTICIPATE

We expect to find the same result, 5.76 J/K (Example 22.6). Initially, the gas occupies a volume V_i and has a number of microscopic states $\mathcal{W}_i$. When the valve is open, no work is done, so the velocity of the molecules is unchanged. The number of microscopic states increases, however, because each molecule has twice as much volume in which to move. The set of possible positions of each molecule has doubled, so the number of microstates for all N molecules has increased by 2^N. This increase in microstates is directly related to the increase in entropy.

SOLVE

The number $\mathcal{W}_f$ of microscopic states when the volume is $V_f = 2V_i$ is 2^N multiplied by the initial number of microscopic states.	$\mathcal{W}_f = 2^N \mathcal{W}_i \qquad (1)$
Use $S = k_B \ln \mathcal{W}$ (Eq. 22.25) to find an expression for the change in entropy.	$S_i = k_B \ln \mathcal{W}_i \qquad S_f = k_B \ln \mathcal{W}_f$ $\Delta S = S_f - S_i = k_B \ln \mathcal{W}_f - k_B \ln \mathcal{W}_i$ $\Delta S = k_B \ln \dfrac{\mathcal{W}_f}{\mathcal{W}_i}$

Example continues on page 676 ▶

Eliminate $\mathcal{W}_f$ with Equation (1).	$\Delta S = k_B \ln \frac{2^N \mathcal{W}_i}{\mathcal{W}_i} = k_B \ln 2^N$ $\Delta S = Nk_B \ln 2$
Substitute values. For 1 mole of molecules, $N = N_A$.	$\Delta S = N_A k_B \ln 2$ $\Delta S = (6.02 \times 10^{23})(1.38 \times 10^{-23}\ \text{J/K}) \ln 2$ $\Delta S = 5.76\ \text{J/K}$

CHECK and THINK

As expected, this result is exactly what we found in Example 22.6. This time, we did not need to think about a specific thermodynamics process and only needed to consider the increase in the number of microscopic states.

EXAMPLE 22.9 Who Stole My Air?

You and your roommate have split your room in half (Fig. 22.19). Estimate the probability of finding all the air molecules in your half of the room, leaving your roommate with no air.

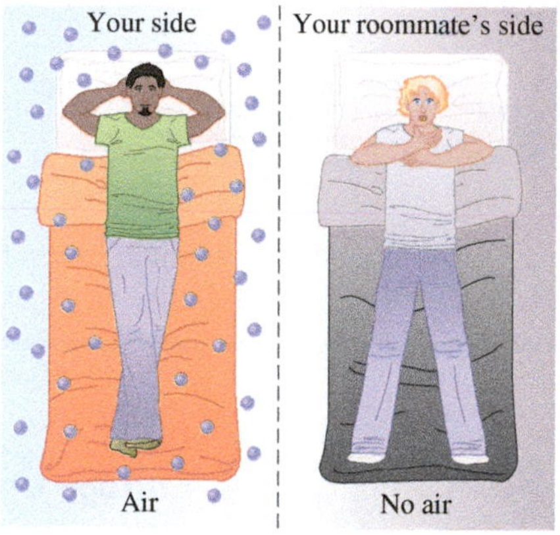

FIGURE 22.19

INTERPRET and ANTICIPATE

This example involves the opposite of a free expansion, and we know that such a process does not naturally occur. We expect the probability of finding all the air on one side of your room to be extremely low, nearly zero.

SOLVE

Suppose there are N molecules filling the entire room. The chance of finding one particular molecule on your side of the room is 50% or $\frac{1}{2}$, the same as the probability of flipping a single coin and having it land head up. The chance of finding two particular molecules on your side of the room is $(\frac{1}{2})(\frac{1}{2}) = (\frac{1}{2})^2 = \frac{1}{4}$, the same as the probability of tossing two coins and having them both land heads up. The probability of finding three particular molecules on your side of the room or of throwing three coins heads up is $(\frac{1}{2})(\frac{1}{2})(\frac{1}{2}) = (\frac{1}{2})^3 = \frac{1}{8}$.

Find a general expression for the probability $\mathcal{P}$ of finding N particular molecules on your side of the room or of throwing N coins heads up.	$\mathcal{P} = (\frac{1}{2})^N$ $\mathcal{P} = 2^{-N}$
There are probably a few moles of air in your room. For the purpose of our estimate, we can assume there is 1 mole of air so that $N \approx N_A$.	$\mathcal{P} \approx 2^{-N_A}$ $\mathcal{P} \approx 2^{-(6 \times 10^{23})}$

CHECK and THINK

This probability is so close to zero that if you try to enter it into your calculator (which probably displays 21 digits), you will find all zeros. To get a sense of how small this probability truly is, try entering 2^{-600} into your calculator; you will find that it is approximately 10^{-181}. So, even if you only had 600 molecules in your room, it is *very* unlikely that you will find them all on just your side. The probability of finding an entire mole of molecules in just half the available space is so unlikely that is essentially impossible. So, it is no surprise that the second law of thermodynamics is often stated so strongly (heat does not spontaneously flow from cold to hot) instead of in probabilistic terms (heat is unlikely to flow from cold to hot).

SUMMARY

Underlying Principles

The **second law of thermodynamics** can be expressed in many different ways.

1. **Clausius statement:**

   Heat flows naturally from a hot object to a cooler object; heat does not naturally flow from a cold object to a hotter object.

2. **Kelvin-Planck statement:**

   It is impossible to construct a cyclic heat engine that takes in heat and produces only work without also giving off waste heat.

3. The concept of entropy allows for several **general and mathematical statements** of the second law of thermodynamics:
   a. *The entropy of an isolated system never decreases:* $\Delta S \geq 0$.
   b. *The entropy of the Universe must be increasing:* $\Delta S_{\text{Universe}} > 0$.
   c. *The total entropy of a system plus its environment must increase as a result of any real process:* $\Delta S_{\text{sys}} + \Delta S_{\text{env}} > 0$.
   d. *The Universe is becoming more disordered.*
   e. *The Universe is moving from a state of low probability to a state of higher probability.*

Major Concepts

1. **Efficiency** e of an engine:

$$e \equiv \frac{W_{\text{eng}}}{Q_h} = 1 - \frac{|Q_c|}{Q_h} \quad \text{(22.4 and 22.5)}$$

2. **Reversible process** is an ideal thermodynamic process in which a system that goes from state i to state f can go back again from state f to state i, exchanging the same amount of heat and work so that the system and the environment have been returned to their original conditions. Other processes are **irreversible processes**.
3. **Entropy** is a state variable measuring a system's disorder. Only changes in entropy are physically meaningful.

a. The change in entropy ΔS for an isothermal process at temperature T is

$$\Delta S \equiv \frac{Q}{T} \quad (22.19)$$

b. The change in entropy around a reversible cycle is

$$\Delta S = \oint \frac{dQ}{T} = 0 \quad (22.21)$$

c. The difference in entropy ΔS does not depend on the path:

$$\Delta S = S_f - S_i = \left[\int_i^f \frac{dQ}{T}\right]_{\text{any reversible path}} \quad (22.22)$$

d. The entropy of a system depends on the number of microscopic states $\mathcal{W}$:

$$S = k_{\text{B}} \ln \mathcal{W} \quad (22.25)$$

Special Cases: Three Specific Engines

1. A **Carnot** engine goes through four reversible processes known as the **Carnot cycle:** (1) isothermal expansion ($\Delta T = 0$), (2) adiabatic expansion ($\Delta Q = 0$), (3) isothermal compression, and (4) adiabatic compression. The Carnot engine is an ideal engine with efficiency

$$e_{\text{Carnot}} = 1 - \frac{T_c}{T_h} \quad (22.6)$$

   No real engine can deliver this efficiency.
2. The **Otto** engine is a good model for the **internal combustion** engine. The ideal, reversible **Otto cycle** consists of four processes: two constant-volume processes and two adiabatic processes. The efficiency of the Otto cycle is

$$e_{\text{Otto}} = 1 - \left(\frac{V_{\text{max}}}{V_{\text{min}}}\right)^{1-\gamma} \quad (22.14)$$

   where $\gamma = C_P/C_V$ (Eq. 21.31).
3. A **refrigerator** is a heat engine running in reverse; its motor does work, and heat is absorbed from the low-temperature compartment inside and is expelled into the warm room. The **coefficient of performance** κ is given by

$$\kappa = \frac{|Q_c|}{|W_{\text{motor}}|} \quad (22.15)$$

Tools

Energy transfer diagrams are useful for studying heat engines. The system is represented as a rectangle in the middle of the diagram, the hot reservoir is drawn above the system, and the cold reservoir is drawn below the system. Thick arrows represent energy transferred into or out of the system.

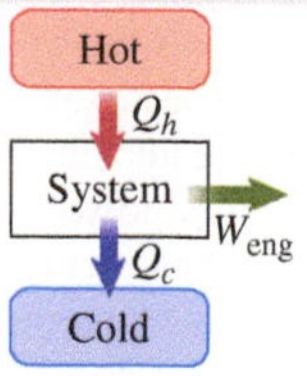

PROBLEMS AND QUESTIONS

A = algebraic C = conceptual E = estimation G = graphical N = numerical

22-1 Second Law of Thermodynamics, Clausius Statement

1. C If heat naturally flowed from cold objects to hotter ones, what would happen if you put a hot potato into a large cold body of water such as a lake? What can you say about the final temperature of the potato and of the lake?

22-2 Heat Engines

2. G Figure P22.2 shows a Carnot cycle. The system expands isothermally in path A, expands adiabatically in path B, is compressed isothermally in path C, and is compressed adiabatically in path D, returning to the initial state. Draw an energy transfer diagram, similar to Figure 22.4, for this cycle.

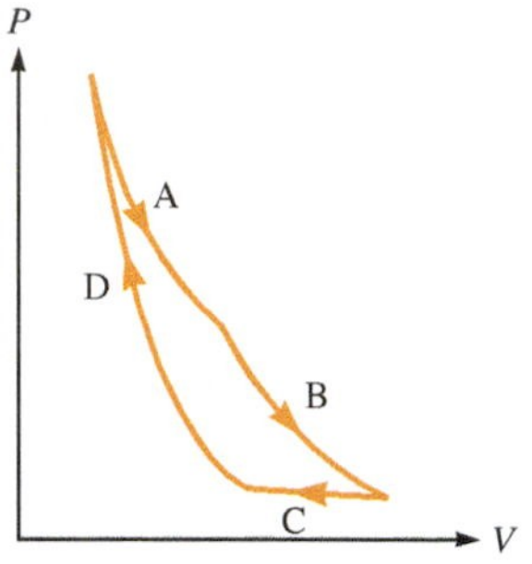

FIGURE P22.2

3. G Use a *PV* diagram such as the one in Figure 22.2 (page 653) to figure out how you could modify an engine to increase the work done.

22-3 Second Law of Thermodynamics, Kelvin-Planck Statement

4. N During each cycle, a heat engine does 50.0 J of work and absorbs 425 J of energy from a hot reservoir. **a.** What is the efficiency of this engine? **b.** What is the amount of energy that is exhausted to the cold reservoir during one cycle?
5. N An engine is able to convert 30.0% of the energy it absorbs from a reservoir into work. **a.** What is the thermal efficiency of this engine? **b.** What is the percentage of the energy absorbed that is exhausted into the cold reservoir?
6. C According to the Kelvin-Planck statement of the second law, it is impossible to construct a cyclic heat engine that takes in heat and produces only work without also giving off waste heat. Is the opposite possible? In other words, can a device operate in a cycle by having work done on it by the environment and give off heat without doing any work? If so, give an example. If not, explain why not.
7. N An engine with an efficiency of 0.36 can supply a power of 140 hp. At what rate does it exhaust heat to the environment? Express your answer in watts.
8. C An environmentally conscious driver asked Click and Clack (the Tappet Brothers), who host the National Public Radio show *Car Talk*, if it was environmentally friendly to drive a convertible with its top down and run the heater. Click and Clack assured the driver that such a practice does not consume any additional energy than driving the car with the heater off, so it is as environmentally friendly as driving with the heater off. (They did, however, warn that driving the car with the top down and the air conditioner running did consume more energy and was therefore worse for the environment.) Why did Click and Clack say that it was okay to drive with the top down and the heater on, but not okay to turn the air conditioner on with the top down?
9. N A heat engine operates in contact with a cold reservoir and a hot reservoir such that 1.620×10^4 J of heat is transferred into the cold reservoir and 7.481×10^4 J of heat is absorbed from the hot reservoir. **a.** What is the work done by the engine on its environment? **b.** What is the efficiency of the engine?
10. N A heat engine with 40.0% efficiency produces 2.50×10^4 W of power while exhausting 1.00×10^4 J to the cold reservoir during each cycle. **a.** How much energy does the engine absorb from the hot reservoir during each cycle? **b.** How long does each cycle take?

22-4 The Most Efficient Engine

11. N What is the maximum efficiency for a heat engine operating between a 22.0°C reservoir and a 535°C reservoir?
12. C What are the similarities and differences between a Carnot engine and a perfect engine?
13. N An internal combustion engine ignites a mixture of fuel and air that reaches a temperature of 1800 K and exhausts gas at 440 K. In the process, the volume of the gas expands such that its maximum volume is nine times larger than the minimum volume. The adiabatic coefficient for air is $\gamma = 1.4$. **a.** What is the maximum efficiency possible for the engine, assuming gas in the piston follows the Otto cycle? **b.** What is the ideal efficiency for a Carnot cycle operating between the same high and low temperatures?
14. C CASE STUDY If a wealthy car enthusiast says that she wants to buy an Otto engine, what would you tell her?
15. N What is the efficiency of a Carnot engine operating between a hot reservoir at 800.0 K and a cold reservoir at 400.0 K?
16. C Does the type of working substance affect the efficiency of a Carnot engine? Does the type of working substance affect the efficiency of real engines? Explain.
17. N A Carnot heat engine operates in contact with a cold reservoir and a hot reservoir such that 4.310×10^5 J of heat is transferred into the cold reservoir and 9.678×10^5 J of heat is absorbed from the hot reservoir. **a.** If the temperature of the cold reservoir is 315.0 K, what is the temperature of the hot reservoir? **b.** What ratio of volumes for an Otto cycle would cause it to have the same efficiency as the above engine? Assume $\gamma = 1.4$.
18. C Why is the size of a car engine important?
19. N A Carnot engine operating between a reservoir at 25.0°C and a reservoir at 340.0°C has a power output of 1.10×10^4 W. **a.** How much energy is extracted from the hot reservoir during each hour of operation? **b.** How much energy is exhausted to the cold reservoir during each hour of operation?
20. CASE STUDY Model a car engine as an Otto engine.
    a. N If the engine has an efficiency of 25%, what is the compression ratio $r \equiv V_{max}/V_{min}$ of the Otto engine? Assume $\gamma = 1.4$.
    b. C How does that ratio compare with a real engine's compression ratio, assuming the real engine's efficiency is 25%?
21. N A car engine has an efficiency of 22%. If it produces 1.20×10^4 J of work, how much heat does it expel?
22. In 1816, Robert Stirling, a Scottish minister, patented an engine now known as the **Stirling engine**. The Stirling engine follows a four-process cycle: two isothermal processes and two constant-volume processes as shown in Figure P22.22. The working substance used in the Stirling engine never leaves the engine. There are no exhaust valves and no internal combustion; instead, the Stirling engine uses an external hot reservoir. Thus, the Stirling engine is very quiet and is used in applications where a quiet engine is important, such as on

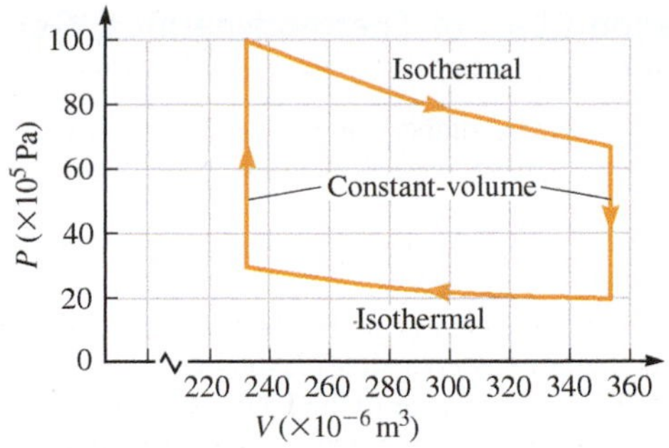

FIGURE P22.22

submarines. Today, the engine is being developed for a wider range of applications.

a. **E** Use Figure P22.22 to estimate the work done by a Stirling engine in one cycle.

b. **N** If the engine delivers 500 hp, how many cycles does it make per second?

c. **N** If the engine has an efficiency of 20%, how much input power does the engine require?

23. **A** CASE STUDY Consider the Otto engine as depicted in Figure 22.12. **a.** Show that the efficiency of an Otto engine can be expressed as

$$e_{\text{Otto}} = 1 - \frac{T_A}{T_B} = 1 - \frac{T_D}{T_C}$$

where the subscripts A through D refer to the states indicated in Figure 22.12.

b. The lowest temperature during an Otto cycle is T_A and the highest temperature is T_C. Find the difference $(e_{\text{Carnot}} - e_{\text{Otto}})$ for Carnot and Otto engines operating between these high and low temperatures. Does your answer to this part make sense? Explain.

Problems 24 and 25 are paired.

24. In the late 1800s, Rudolf Diesel invented an internal combustion engine that does not require a spark plug. Figure P22.24 is the PV diagram for an ideal **diesel engine**. Air is taken into the cylinder (at A) and compressed adiabatically to a high pressure (A to B). Then, diesel fuel (petroleum) is injected into the cylinder (at B). The fuel is ignited by the high pressure and high temperature of the air, and no spark is needed. Heat enters the system at constant pressure (B to C), and the gas then expands adiabatically (C to D). Finally, the pressure drops at constant volume from D back to A. As shown in the figure, the gas has its maximum volume V_{max} during the constant-volume process from D to A, its minimum volume V_{min} is at point B, and at point C its volume is V_{int}, an intermediate value.

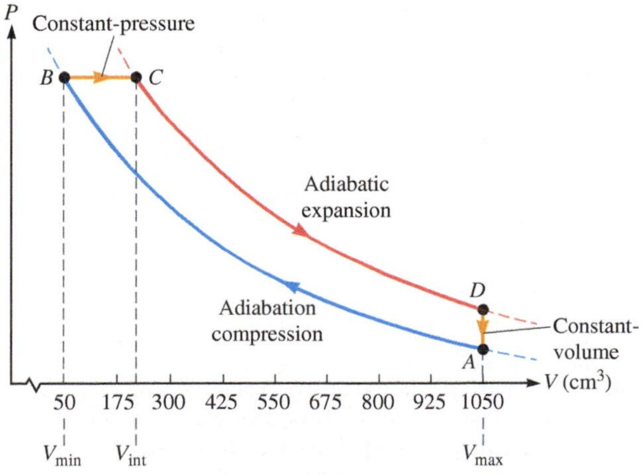

FIGURE P22.24 Problems 24 and 25.

a. **A** Show that the efficiency of a diesel engine is given by

$$e_{\text{diesel}} = 1 - \frac{1}{\gamma}\left(\frac{r^{-\gamma} - R^{-\gamma}}{r^{-1} - R^{-1}}\right)$$

where $r = V_{\text{max}}/V_{\text{int}}$, $R = V_{\text{max}}/V_{\text{min}}$, and γ is the ratio of specific heats, $\gamma = C_P/C_V$.

b. **E** Usually, air is used, so $\gamma = 1.4$. Estimate the efficiency of the diesel engine represented by Figure P22.24.

25. A diesel engine (Problem 24) has the volume ratios $r = 4.57$ and $R = 21.0$. When air is taken into the cylinder, the pressure is 1.00 atm, and the temperature is 22.5°C.

a. **N** What is the temperature of the air when the fuel is injected?

b. **C** The air–fuel mixture must be at least 483 K for ignition to take place. Is ignition achieved? What minimum intake temperature would achieve ignition?

22-5 Case Study: Refrigerators

26. **N** What is the maximum possible coefficient of performance for an ideal refrigerator operating between a high-temperature reservoir at 2.0 K and a low-temperature reservoir at 0.020 K?

27. **N** A refrigerator is a heat engine running in reverse. What is the coefficient of performance κ of a Carnot engine with an efficiency of 43.0% that is run in reverse?

28. **C** In Concept Exercise 22.3, we found that you could not cool your house by opening your refrigerator door. Any baker, however, will tell you that you can heat a bakery by turning on the oven. Why can you heat a (small) house by opening the oven, but you cannot cool it by opening the refrigerator? Be sure your explanation illustrates the fundamental difference between refrigerators and ovens.

29. **N** What is the best possible coefficient of performance for a household refrigerator in which an internal temperature of 3.0°C is desired? Assume room temperature is 20.0°C.

30. **C** An air conditioner is similar to a refrigerator. By doing work, the air conditioner takes heat Q_c from the cool building and expels heat Q_h out into the hot environment. Why is a room air conditioner placed in a window or through a hole in the wall so that part of the unit is outside the room?

31. **N** The coefficient of performance of a refrigerator that takes 210.0 J of energy from the cold reservoir during each cycle is $\kappa = 4.00$. **a.** What is the work done during each cycle? **b.** How much energy is exhausted into the hot reservoir during each cycle?

32. **C** Consider the coefficient of performance of an ideal refrigerator $\kappa = T_c/(T_h - T_c)$ as in Equation 22.18. Notice that the performance depends on the temperatures of the hot and cold reservoirs. Suppose you plug in a refrigerator and the temperatures of the hot and cold reservoirs are initially close in value. Model your refrigerator as an ideal one and describe what happens to the coefficient of performance as time goes on and you allow the refrigerator to run continuously.

33. **A** Using the derivation of the Carnot engine efficiency as a guide (Section 22-4), derive Equation 22.18, the coefficient of performance for an ideal refrigerator: $\kappa = T_c/(T_h - T_c)$.

34. A *heat pump* is a device that can heat a building in winter by taking heat Q_c from outside at low temperature and giving off heat Q_h to the warm interior by doing work. The operation of a heat pump is much like that of a refrigerator. Heat exchangers inside and outside the building resemble the coils inside the refrigerator and the condenser outside it (Fig. 22.14).

a. C Draw an energy transfer diagram for a heat pump.

b. A Because the goal of a heat pump is to deliver heat Q_h into the building, the coefficient of performance is defined as $\kappa \equiv |Q_h|/W$. Assume an ideal heat pump can be modeled as a Carnot cycle and derive an expression for the ideal heat pump's coefficient of performance in terms of the indoor and outdoor temperatures.

c. N What is the coefficient of performance for an ideal heat pump if the indoor temperature is 75.0°F and the outdoor temperature is 20.0°F?

22-6 Entropy

35. N What is the change in entropy of 100.0 g of water when heated so that its temperature changes from 20.0°C to 100.0°C? The specific heat capacity of water is 4186 J/(kg · K).

36. E Estimate the change in entropy of the Universe if you pop a balloon.

37. N An ideal gas, initially at atmospheric pressure and a temperature of 300.0 K, is compressed from a volume of 1.00 L to 0.50 L. What is the change of entropy of the gas if this process is performed **a.** adiabatically and **b.** isothermally?

38. E Estimate the change in entropy of a pond when it freezes.

39. N What is the change in entropy of a 125-g sample of liquid water at 0°C that slowly freezes at this temperature?

40. N An ideal gas undergoes a free expansion, doubling in volume. If there are 15 mol of the gas, what is the change in the gas's entropy? Use Example 22.6 to check your results.

41. N A 30.0-mol monatomic carbon gas undergoes a constant-pressure process at a pressure of 2.35 atm from an initial volume of 0.525 m^3 to a final volume of 0.340 m^3. What is **a.** the work done on the gas, **b.** the heat transferred and **c.** the change in entropy during this process?

42. N A cold reservoir at $T_c = 15.0°C$ and a hot reservoir at $T_h = 450.0°C$ are connected via a steel bar. The energy transferred from the hot reservoir to the cold reservoir in this irreversible process is 1.85×10^4 J. What is the change in the entropy of **a.** the cold reservoir and **b.** the hot reservoir during this process? **c.** Ignoring the change in entropy of the steel bar, what is the change in the entropy of the Universe during this process?

43. N A cold block of copper is placed in contact with a hot block of copper. The blocks have the same volume, and each has a mass of 0.345 kg. The original temperature of the cold copper is 6.25°C, and the original temperature of the hot copper is 45.2°C. **a.** What is the final equilibrium temperature? **b.** What is the change in entropy for each block? **c.** What is the change in entropy of the Universe?

44. N A railway car with mass $m = 5.50 \times 10^3$ kg moving at 8.00 m/s has a totally inelastic head-on collision with a second identical railway car also moving at 8.00 m/s. The air temperature at the collision site is initially 15.0°C. Consider the system to consist of the two railway cars and the surrounding air. If the kinetic energy lost in the collision goes into the air surrounding the cars, what is the change in entropy of the air at the collision site as a result of the collision?

45. N A nonporous membrane divides a vessel in the form of two identical 5.00-L chambers that are welded together (Fig. P22.45). Chamber A is filled with 1.22 mol of argon gas, and chamber B is filled with 1.22 mol of krypton gas. What is the change in entropy of the system if the membrane is removed and the gases mix together?

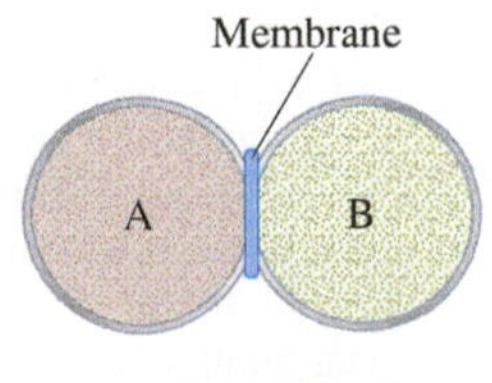

FIGURE P22.45

22-7 Second Law of Thermodynamics, General Statements

46. C If you stretch a rubber band, does the entropy of the rubber band change? If so, does it increase or decrease? How do you know?

47. A spring obeys Hooke's law and has a spring constant of 125 N/m. You stretch the spring from 1.60 m to 1.85 m at 22.2°C.

a. N What is the change in the spring's entropy?

b. C Is the process reversible? Explain.

c. N What is the change in the Universe's entropy?

48. C Two students discuss entropy on a very cold day.

Avi: So, the water in the birdbath that froze last night is more ordered now that it's ice, right? And the entropy has gone down?

Cameron: Well, I think entropy can't decrease in natural processes, so it seems like it can't decrease. Maybe it stays the same?

Avi: I thought it just had to do with heat coming out of the water to make ice. Is entropy something else?

Cameron: Entropy is unchanged for a reversible process and increases for an irreversible process.

What part of this conversation do you agree with? Which claims do you disagree with?

22-8 Order and Disorder

49. C Consider this analogy: A neatly arranged sock drawer is like a cool system, and a messy basket of laundry is like a hot system. Use this analogy to explain why adding the same amount of heat to either system causes the cool system's entropy to increase much more than the hot system's entropy.

50. C Describe three examples from your everyday experience that are representative of the second law of thermodynamics expressed as "the Universe is becoming more disordered."

51. C A physics student tells her dad that she cannot fold her laundry because it would violate the second law of thermodynamics. How does her dad (also a physicist) convince her that folding laundry does not violate any laws of physics?

52. C A watch is clearly a very ordered system but, the process of producing a watch cannot violate the second law of thermodynamics. If making a watch produces an ordered system, how can the overall entropy of the Universe increase? Name as many specific parts of the Universe as possible whose entropy must have increased in the process.

53. C The principles of biological evolution and the development of life on planet Earth seem to require a decrease in entropy to occur. Does that violate the second law of thermodynamics? Justify your answer by explaining why it is a violation or by proposing what could account for a necessary decrease in entropy, allowing life to develop.

22-9 Entropy, Probability, and the Second Law

54. E You and your roommate have split your room in two; your side of the room is one-third the volume of the whole room. Estimate the probability of finding all the air molecules in your part of the room, leaving your roommate with no air.

55. N If you roll two identical six-sided dice, what is the probability that you roll "snake eyes" (both dice show a single dot on the top surface)?

56. N If you toss a coin 100 times, what is the probability of finding between 0 and 10 heads? How does your answer compare with your answer to Concept Exercise 22.5? Use this comparison to make sense of your answer to this problem.

57. N What is the entropy of a freshly shuffled deck of 52 playing cards?

58. Feeling lucky, George is rolling a pair of dice at a craps table.
   a. N His first roll of the dice is a 2. How many microscopic states are there for this roll?
   b. N His second roll of the dice is an 8. How many microscopic states are there for this roll?
   c. C Which of the two rolls corresponds to a state with lower entropy?

Problems 59 and 60 are paired.

59. G Using Table 22.1 (page 675) and given that there are 1.27×10^{30} microstates in total, plot the probability for each outcome versus the number of heads for values listed in the table. In words, describe the shape of the curve and what it says about the likely outcome of flipping 100 coins.
60. a. N Using Equation 22.26, create a table similar to Table 22.1, but do so for the case of flipping six coins. Include the outcomes of 0, 1, 2, 3, 4, 5, and 6 heads as possible microscopic states. Calculate the number of microscopic states, the probability, and the entropy for each macroscopic state.
   b. G Plot the entropy of each macroscopic state versus number of heads.

General Problems

61. N The Carnot efficiency of a heat engine operating between two reservoirs is 54.0%. If the temperature of the hot reservoir is 355°C, what is the temperature of the cold reservoir?
62. C As you have seen in the discussion of the Carnot engine, it is impractical to think that we could have a cold reservoir at absolute zero or an infinitely energetic hot reservoir, which means that we cannot not have a 100% efficient engine. What other limitations do we have when it comes choosing two real heat reservoirs? Consider what reservoirs we might have readily available and the energy required to create a hot or cold reservoir.

Problems 63 and 64 are paired.

63. C Consider two different scenarios: (1) An ice cube melts to form water. (2) The same amount of water boils to form steam. Which of these processes would lead to a larger change in entropy? Justify your answer.
64. **a.** N What is the entropy change for 5.0 g of ice that melts at 0.0°C to form water? The heat of fusion of water is 333 J/g. **b.** N What is the entropy change for 5.0 g of water that boils at 100°C to form steam? The heat of vaporization for water is 2260 J/g.
65. N A Carnot engine operates between a cold reservoir with $T_c =$ 22.0°C and a hot reservoir with $T_h =$ 300.0°C. **a.** Consider keeping the temperature of the hot reservoir fixed at 300.0°C while the temperature of the cold reservoir is variable. What is the rate of change in efficiency of the engine per change in T_c (what is de_{Carnot}/dT_c)? **b.** Consider keeping the temperature of the cold reservoir fixed at 22.0°C while the temperature of the hot reservoir is variable. What is the rate of change in efficiency of the engine per change in T_h (what is de_{Carnot}/dT_h), when $T_h =$ 300.0°C?
66. N A 3.00-mol diatomic ideal gas undergoes a free expansion. The constant temperature during the process is 350 K, and the gas begins with a pressure of 1.5 atm and finishes with a pressure of 1.0 atm. What is the change in entropy of the gas as it undergoes this free expansion?

Problems 67 and 68 are paired.

67. N A 1.55-kg sample of liquid water at 50°C is heated and boiled away, and eventually it is converted entirely to water vapor. The resulting water vapor is then heated until it reaches a final temperature of 115°C. Assume the specific heat of the liquid water and the water vapor are 4190 J/(kg · K) and 2010 J/(kg · K), respectively, and the latent heat of vaporization is 2.256×10^6 J/kg. No work is performed on or by the water or the water vapor. **a.** What is the change in entropy of the liquid water as it is heated to the vaporization point? **b.** What is the change in entropy of the water vapor as it is heated from the vaporization point to the final temperature of 115°C? **c.** What is the change in entropy of the water as it undergoes the phase transformation?
68. C Consider heating water as described in Problem 67. **a.** Describe what is happening in the water and water vapor as they are heated that would indicate that the entropy is increasing. What might we observe happening that leads us to say that the "entropy has increased" in each case? **b.** What is happening during the phase change from liquid to gas that indicates that the entropy is increasing?
69. N One mole of an ideal gas is taken through a Carnot cycle in which the gas extracts 1.55 kJ of energy from the hot reservoir during each cycle. The isothermal expansion for this cycle occurs at 300.0°C, and the isothermal compression occurs at 85.0°C. **a.** How much energy is exhausted to the cold reservoir during each cycle? **b.** What is the net work done by the gas during each cycle?
70. A The free expansion of an ideal gas is described in Section 21-7. Show that the entropy change of an ideal gas during a free expansion (Fig. 21.20) is given by Equation 22.24,

$$\Delta S = nR \ln\left(\frac{V_f}{V_i}\right)$$

Hint: A free expansion is not a reversible process, so we cannot use Equation 22.22,

$$\Delta S = S_f - S_i = \left[\int_i^f \frac{dQ}{T}\right]_{\text{any reversible path}}$$

Instead, we must consider an equivalent reversible process that takes the gas from its initial state to its final state. Like a free expansion, a reversible isothermal process can take the ideal gas from state *i* to state *f* with no change in thermal energy. Start by applying the first law of thermodynamics to an isothermal expansion.
71. N A system consisting of 10.0 g of water at a temperature of 20.0°C is converted into ice at −10.0°C at constant atmospheric pressure. Calculate the total change in entropy of the system. Assume the specific heat of water is 4.19×10^3 J/(kg · K), the specific heat of ice is 2.10×10^3 J/(kg · K), and the latent heat of fusion is 3.33×10^5 J/kg.
72. C When a gas undergoes a free expansion, its entropy increases according to

$$\Delta S = nR \ln\left(\frac{V_f}{V_i}\right)$$

Explain the increase in entropy in terms of the change in order (or disorder).
73. N Figure P22.73 illustrates the cycle *ABCA* for a 2.00-mol sample of an ideal diatomic gas, where the process *CA* is a reversible isothermal expansion. What is **a.** the net work done by the gas during one cycle? **b.** How much energy is added to the gas by heat during one cycle? **c.** How much energy is exhausted from the gas by heat during one cycle? **d.** What is the efficiency of the cycle? **e.** What would be the efficiency of a Carnot engine operated between the temperatures at points *A* and *B* during each cycle?

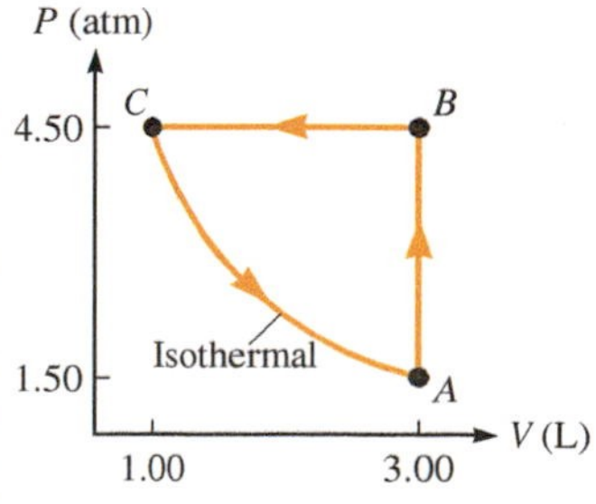

FIGURE P22.73

74. **N** Experimental measurements of heat capacities of graphite at constant pressure show that over a wide range of temperatures, the molar specific heat varies with temperature as

$$C_p = a + bT - \frac{c}{T^2}$$

where $a = 16.86$ J/K, $b = 4.77 \times 10^{-3}$ J/K^2, and $c = 8.54 \times 10^5$ J · K. Calculate the change in entropy when one mole of graphite is heated at constant pressure from 250°C to 45°C.

75. **N** A cylinder with a movable piston contains 2.00 mol of a diatomic ideal gas at atmospheric pressure and a temperature of 325.0 K. The air within the cylinder is cooled slowly to 250.0 K. **a.** If the volume of the cylinder does not change, what is the change in entropy of the gas? **b.** If the pressure inside the cylinder does not change, what is the change in entropy of the gas?

76. **C** You pop a DVD of your favorite movie into your DVD player. How might the example of watching the playback of a movie be thought of as being similar to the Clausius statement of the second law of thermodynamics? Consider the ability to watch the movie as it is playing or rewinding. In which direction do things make sense?

77. **N** During each cycle, a heat engine extracts 3.25×10^4 J from a hot reservoir at $T_h = 400$°C and does 4.00 kJ of work. The cold reservoir is at $T_c = 22.0$°C. **a.** What is the change in entropy of the Universe for each cycle? **b.** How much more work would a Carnot engine operating between these reservoirs have performed?

78. **N** The coefficient of performance of a household heater that consumes 850.0 W of power is $\kappa = 2.95$, where $\kappa = |Q_h|/W$ represents the coefficient of performance for a heater. **a.** If the heater is in operation for 6.00 h each night, how much energy does it deliver? **b.** What is the energy extracted by the heater from the outside air (the cold reservoir) during this time?

79. **N** A sample of 1.00 mol of a monatomic ideal gas is initially at atmospheric pressure and occupies a volume of 5.00 L. It then undergoes a reversible isothermal expansion followed by a reversible adiabatic contraction such that its final pressure is 1.00 atm and its final volume is 20.0 L. What is the change in entropy of the gas after the two processes?

80. **N** Consider a chemical reaction A + B → D. The three substances have molar specific heat capacities at constant pressure given by $C_A = 2\sqrt{T}$, $C_B = 4\sqrt{T}$, and $C_D = 3\sqrt{T}$ where T is in kelvin. If this reaction occurs at a temperature of 300.0 K, use Eq. 22.22 to determine how much heat is absorbed or released if 1.0 mol of substance D is produced. To determine the relative entropy of substances A, B, and D, respectively, assume the heat capacities of the reactants A and B and the product D are zero at $T = 0$ K.

81. **N** Consider a system A with two subsystems A_1 and A_2. The total number of microstates of A_1 and A_2, respectively, are 10^{10} and 2×10^{10}. **a.** What is the number of microstates available to the combined system, A? **b.** What are the entropies of the system A and the subsystems A_1 and A_2?

PART THREE
Electricity

Electric Forces

23

Underlying Principles

1. Conservation of charge
2. Electrostatic force
3. Coulomb's law

Major Concepts

1. Charge
2. Quantization of charge
3. Conductors and insulators
4. Ground

Tools

Hybrid free-body diagram

Key Questions

What are two ways to give objects a net charge?

How do we find the magnitude and direction of the electrostatic force exerted by one charged object on another?

Why do charged objects attract neutral ones?

You stick a decal on the rear window of your car. The sticker can easily be peeled off and put back on again. You decorate for a party by rubbing balloons on your head, and then they stay on the wall (Fig. 23.1). You slide out of the car and find that your skirt is clinging to your legs. After running the dryer without a dryer sheet, you find that your socks are stuck to your clothes. These are all examples of what is commonly called "static cling." Static cling is a result of the *electric force* acting between objects. In physics terms, we say that an electrostatic force attracts the decal to the car's window, the balloons to the wall, your skirt to your legs, and the socks to your clothes. The electric force is one of the fundamental forces in physics and forms the basis of what we study throughout Part III.

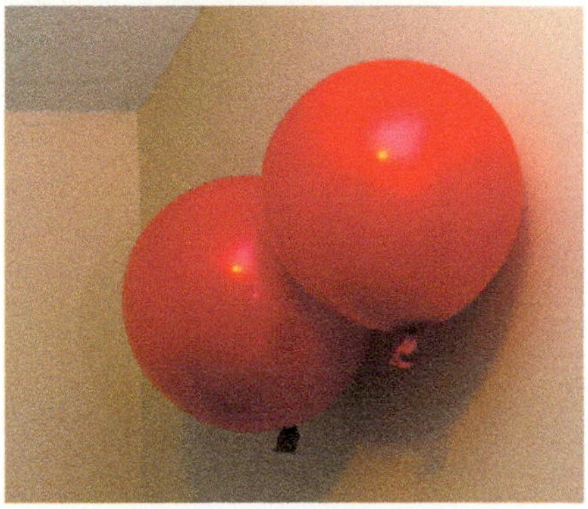

FIGURE 23.1 Visible evidence of the electric force—static cling makes the air-filled balloons stick to the wall.

23-1 Another Fundamental Force

Until the middle of the 19th century, scientists believed that electricity and magnetism were two separate forces. Then the Scottish mathematician and theoretical physicist James Clerk Maxwell discovered that the two forces were *unified*. In other words, the electric force and the magnetic force are parts of a more fundamental force known today as the **electromagnetic force**.

Today, the electromagnetic force is considered one of the four fundamental forces (Section 5-10), and Maxwell's work has inspired physicists to look for theories and experimental evidence showing that all the fundamental forces are unified into a broader, more general force (Fig. 23.2). The electromagnetic and weak nuclear forces are also combined in what is called the **electroweak force**. There are theories—known as grand unification theories (GUTs)—that support the idea that the electroweak force and the strong force may be unified. However, as yet no experimental evidence supports these GUTs. Finally, theories of how gravity may be unified with the other three forces are still being worked out. Physicists hope that such a theory will be developed and supported so that there will be a single *theory of everything*.

The theory of everything and GUTs are beyond the scope of this book. In fact, the strong and weak forces are only briefly mentioned. However, Figure 23.2 gives us perspective. In this part of the book, we study electricity—one of the fundamental branches of physics. In the next part, we tackle magnetism. By the end of Part IV, we learn how Maxwell discovered the unification of these two forces. Part V (optics) is devoted to phenomena that are governed by their unification.

Of all the forces shown in Figure 23.2, the only force we have encountered by name in earlier chapters is gravity. So you might wonder how all the other forces we have studied—friction, the normal force, tension, drag, the spring force—fit into the framework of only four fundamental forces. All of these forces (other than gravity) are macroscopic manifestations of the electromagnetic force acting on the microscopic level between atoms and molecules. In fact, much of what we study in the rest of this book involves the forces between particles on the microscopic level, which have important consequences on the macroscopic level, as illustrated by the next case study.

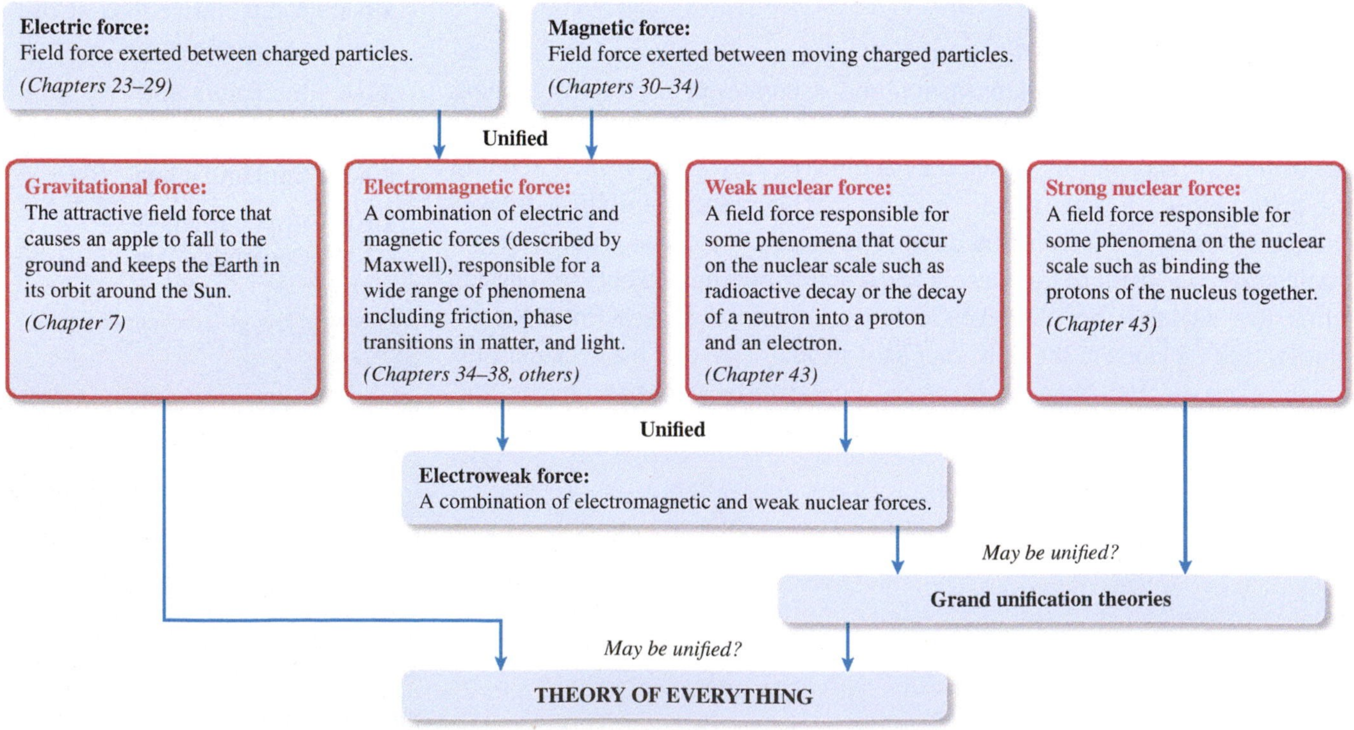

Electric force:
Field force exerted between charged particles.
(Chapters 23–29)
Magnetic force:
Field force exerted between moving charged particles.
(Chapters 30–34)
Unified
Gravitational force:
The attractive field force that causes an apple to fall to the ground and keeps the Earth in its orbit around the Sun.
(Chapter 7)
Electromagnetic force:
A combination of electric and magnetic forces (described by Maxwell), responsible for a wide range of phenomena including friction, phase transitions in matter, and light.
(Chapters 34–38, others)
Weak nuclear force:
A field force responsible for some phenomena that occur on the nuclear scale such as radioactive decay or the decay of a neutron into a proton and an electron.
(Chapter 43)
Strong nuclear force:
A field force responsible for some phenomena on the nuclear scale such as binding the protons of the nucleus together.
(Chapter 43)
Unified
Electroweak force:
A combination of electromagnetic and weak nuclear forces.
May be unified?
Grand unification theories
May be unified?
THEORY OF EVERYTHING

FIGURE 23.2 A concept map showing the relationships of the four fundamental forces (outlined in red) and their known or proposed unification.

CASE STUDY Fires at the Gas Pump

On the evening of December 12th, 2004, a young woman drove up to a gas pump. She removed the gas cap and put the pump's nozzle in her SUV's tank. (Fig. 23.3 shows a similar situation.). Then she went back inside her SUV and slid across the seat. When she got back out of the car and reached for the nozzle, flames erupted from the nozzle. Fortunately the fire was relatively small, and no one was harmed.

Such accidents can be more serious. Vehicles have been totaled; people have been seriously burned, and some have even died. From September 1999 through January 2000, 36 fires during refueling were reported to Robert N. Renkes, Executive VP and General Counsel for the Petroleum Equipment Institute (PEI). As a result, PEI asked readers of its newsletter and website to report incidents of refueling fires. Here is what PEI found:

1. Most people (94%) who reported on the type of footwear they were wearing at the time of the fire said they were wearing rubber-soled shoes. We'll assume the woman in the case study was also wearing rubber-soled shoes.
2. Many of the people reported that they had gone back inside the vehicle after they put the nozzle in their tank.
3. One person reported that she had been petting her small dog, which was in her arms. She put the dog down, and when she reached for the nozzle, a fire started.
4. Many people reported hearing a popping sound or feeling a shock just before the fire began.
5. No one reported using a cell phone while fueling. In other studies, researchers tried to ignite gasoline vapor with cell phones and were unable to do so.

Warning signs are posted at gas pumps. Typically these signs indicate that smoking and using a cell phone or other electronic devices are forbidden. What advice would you post on warning signs in order to prevent the sort of refueling fire started by the woman in our case study?

Stuart Monk/Shutterstock.com

FIGURE 23.3 A woman pumps gas into a vehicle.

23-2 Models of Electrical Phenomena

Let's begin by turning our attention back 250 years. In the mid 1700's, people performed electrostatic demonstrations for entertainment in the evenings. Figure 23.4 shows a gathering in 1746. The boy suspended in the center of the group has been *charged*. The woman on the right is about to be shocked as she touches the boy's nose.

Benjamin Franklin (1706–1790) hosted and attended such parties, and he developed many of the concepts we study in this chapter. Franklin—one of the Founding Fathers of the United States and a signer of the Declaration of Independence—was first a scientist and inventor. As we'll discuss, Franklin made major contributions to the field of electricity.

On a dry day, such as in the middle of winter, you might find that after you comb your hair, it is attracted to the comb. Such phenomena were the basis of evening entertainment in the 18th century. People rubbed objects together such as wool against amber or silk against glass (Fig. 23.5A). They found that after rubbing, the objects were attracted to each other; a silk cloth was attracted to a glass rod that it was rubbed against (Fig. 23.5B)—just as your comb is attracted to your hair. They also found that rubbed objects could be used to attract other objects. For instance, small bits of paper were attracted to a comb that had been rubbed through hair (Fig. 23.5C).

Franklin's explanation of such phenomena is at the heart of our contemporary understanding of electricity. According to Franklin, all objects are full of an *electric fluid*. When you bring two objects close together, it is possible to transfer some electric fluid from one object to the other. Afterward, one object has a surplus of

Image Asset Management Ltd./Alamy

FIGURE 23.4 During Benjamin Franklin's time, people entertained one another with static electricity demonstrations. Here Jean-Antoine Nollet is using a rubbed rod to charge a boy suspended by a silk rope. The charged boy makes bits of paper fly upward. A woman is about to discharge the boy and create a spark by touching his nose.

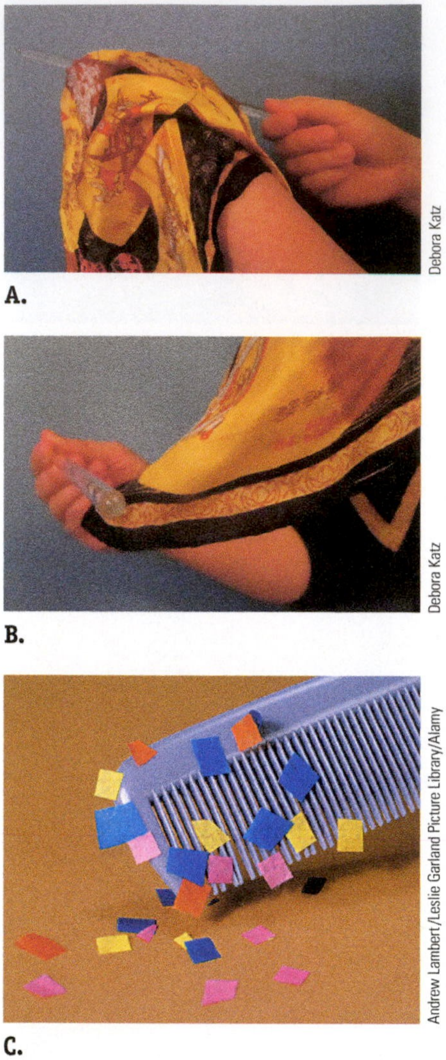

FIGURE 23.5 **A.** Silk is rubbed against glass, and afterward **B.** the silk is attracted to the glass. **C.** After running a comb through hair on a dry day, the comb attracts paper.

electric fluid and the other has a deficit of electric fluid. Franklin said an object with a surplus of electric fluid is *positive* or *plus*, and an object with a deficit of electric fluid is *negative* or *minus*. According to Franklin, both a glass rod and a silk cloth have some amount of electric fluid initially, and after they are rubbed together, some of the electric fluid is transferred from the silk to the glass (Fig. 23.6A). The silk then has a deficit of electric fluid, so Franklin said the silk is *negative*. The glass has a surplus of electric fluid, so he said it is *positive*. Franklin made an arbitrary choice in this model; he assumed the electric fluid was transferred from the silk to the glass. There was no way for Franklin to know whether that was true. Because there was no experimental evidence to determine which way the electric fluid flowed, Franklin could have imagined that electric fluid was transferred from the glass to the silk. Subsequent scientists have kept his arbitrary choice, however, and that has important implications for our contemporary model of electric charge.

Today we have experimental evidence supporting the model that all objects are made up of atoms, and atoms are made up of three types of particles—neutrons, protons, and electrons. The neutrons and protons are tightly packed into the central region of the atom known as the nucleus. The electrons move rapidly outside the nucleus, forming a cloud. According to this contemporary model, when two objects are brought near each other, it is possible for electrons that are loosely bound to the atoms in one object to move to the other object. The protons and neutrons are tightly bound in the nucleus, and so they are not transferred. Therefore, when glass is rubbed against silk, electrons are transferred from the glass to the silk. Afterward, the glass has a surplus of protons and the silk has a deficit of protons (Fig. 23.6B).

Our sign convention is based on the relative number of electrons and protons in a given object. An object that has an equal number of electrons and protons is said to be **neutral**. An object that has a deficit of electrons has a surplus *of protons*; such an object is said to be **positive**. An object that has a surplus of electrons has a *deficit of protons*; such an object is said to be **negative**.

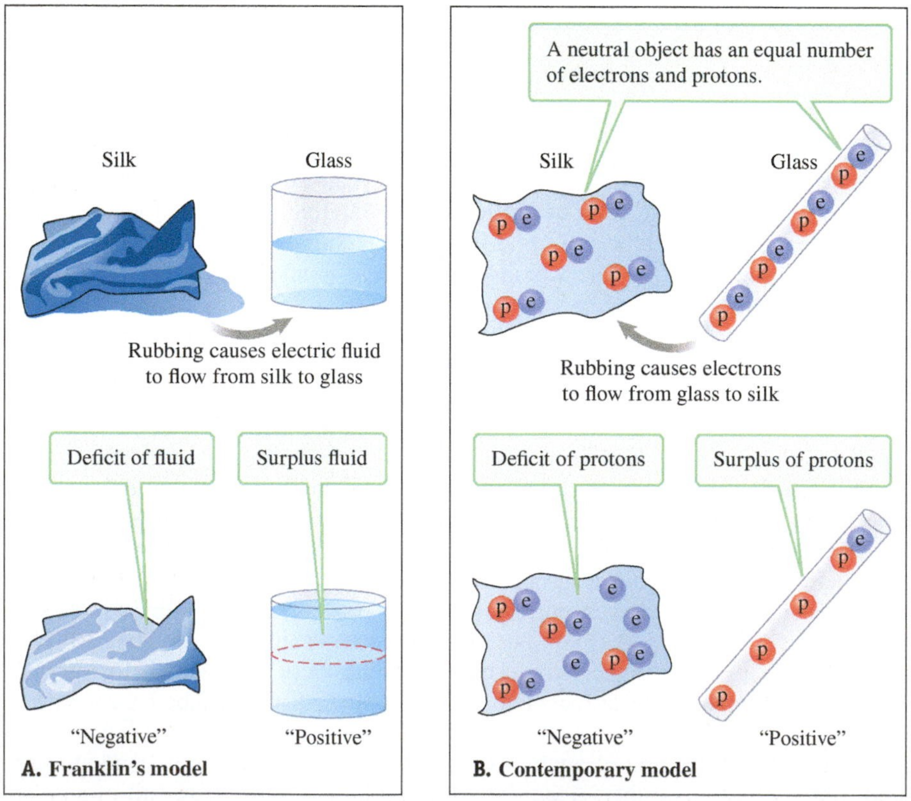

FIGURE 23.6 Comparison of Ben Franklin's model and our contemporary model of electric charge.

Charge

CHARGE ✪ Major Concept

Subatomic particles have an intrinsic property called **charge** that determines the particle's role in electrical phenomena. The charge (symbolized by q or Q) of any object or particle may be positive, negative, or zero.

The SI unit of charge is the **coulomb**, abbreviated C. Just as the mass of subatomic particles is an intrinsic property determined by experimentation, so is charge. Experiments have shown that a neutron has no charge, a proton's charge is positive, and an electron's charge is negative. Experiments have also shown that the magnitudes of the proton and electron charges are equal. This magnitude is known as the **elementary charge** e. The charge of a proton to four significant figures is

$$q_{\text{proton}} = +e = 1.602 \times 10^{-19}\text{ C}$$

and so the charge of an electron is

$$q_{\text{electron}} = -e = -1.602 \times 10^{-19}\text{ C}$$

An object is said to be **neutral** or **uncharged** if it contains an equal number of electrons and protons. An object is **positively charged** if the number of protons is greater than the number of electrons, and the object's charge q is

$$q = +Ne \tag{23.1}$$

where N is the number of excess protons. An object is **negatively charged** if the number of electrons is greater than the number of protons, and the object's charge q is

$$q = -Ne \tag{23.2}$$

where N is the number of excess electrons. Generally, the number N of excess protons or electrons is small compared to the total number of protons or electrons in the object.

QUANTIZATION OF CHARGE ✪ Major Concept

Because the charge of any object is an integer N times the elementary charge e, we say that charge is **quantized**, which means that charge is delivered in small packets or multiples of the elementary charge. So the charge of an object may be

$$q = \pm 1e, \pm 2e, \pm 3e, \ldots \pm Ne$$

An object cannot have a fractional number of excess electrons or protons—an object cannot have a charge of $\pm 1.75e$, for example.

When we study macroscopic objects such as glass rods and plastic combs, we are unaware of the quantization of charge because we cannot detect a charge as small as e in ordinary observations. For example, suppose you rub a glass rod with silk just until you can tell that the rod is charged (due to its behavior). Afterward, suppose you measure the charge of this glass rod, and you find it is $q = (53.4 \pm 0.1)$ nC (a pretty accurate measurement). The error in your measurements is $q = 0.1$ nC, which amounts to

It has been discovered that protons are not fundamental particles. Instead, they are made up of quarks. Quarks have a fractional charge, but because an object cannot have a single "excess" quark, this will not affect our discussions.

$$\delta N = \frac{\delta q}{e} = \frac{1 \times 10^{-10}\text{C}}{1.602 \times 10^{-19}\text{C}} = 6 \times 10^{8}\text{ protons}$$

You cannot come close to detecting single units of the elementary charge by conducting such an experiment. So, on the macroscopic scale, it seems like charge is continuous because each packet of elementary charge is so small.

CONSERVATION OF CHARGE ❶ Underlying Principle

In Franklin's model, electric fluid is transferred from one object to another without being created, lost, or destroyed in the process. **Conservation of charge** is also an underlying principle in our contemporary model of physics. According to this principle, the net charge in the universe is constant. If you rub glass with silk and find that the glass rod's charge has increased by, say, 15 nC, then the silk's charge must have decreased by the same amount, or –15 nC.

Sketching Charge

By now you know that a good sketch is important in solving problems and understanding physics in general. We cannot possibly draw all the atoms in an object or even the number of excess charged particles, which can be in the millions.

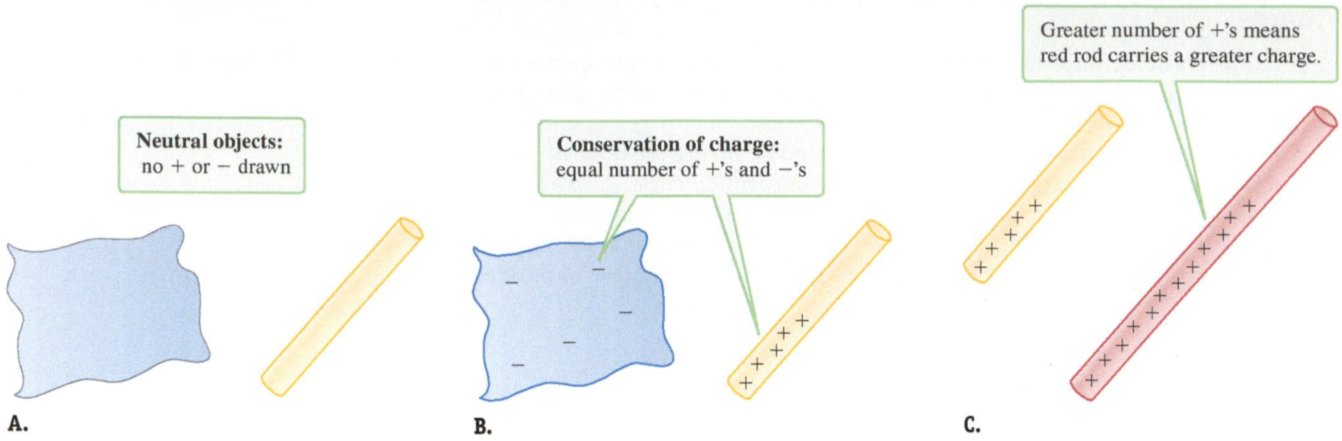

FIGURE 23.7 When you are sketching, indicate only the excess charged particles. **A.** Neutral objects are shown without any + or − signs. **B.** The rod and cloth are charged by rubbing. When charge is transferred from one neutral object to another, both become charged. The number of excess positive particles on one object must equal the number of excess negative particles on the other object. **C.** The two rods were charged independently. (The cloths used to charge these rods are not shown in the sketch.)

Instead, we represent the excess charge with plus or minus signs (Fig. 23.7). The actual number of signs is somewhat arbitrary, but usually between 1 and 12 will work. A sketch, however, should illustrate the fact that charge is conserved. So, if charge is transferred from one neutral object to another (Fig. 23.7A), the number of pluses on one object must equal the number of minuses on the other object (Fig. 23.7B). Charge conservation does not mean that the number of pluses and minuses must be the same on all sketches. For example, if one object has more positive charge than another object, you should draw more plus signs on that object (Fig. 23.7C).

CONCEPT EXERCISE 23.1

Initially a glass rod and a piece of silk are neutral. After you rub the silk against the rod, the glass rod has a surplus of 3.33×10^{11} protons. What is the charge q of the silk?

23-3 A Qualitative Look at the Electrostatic Force

Franklin learned that some rubbed objects attract one another and some repel one another. Suppose you try some of Franklin's experiments with two glass rods and two silk handkerchiefs. Initially, you find that there is no attraction or repulsion between any of these objects. Then you rub one glass rod with one of the silk handkerchiefs and the other rod with the other handkerchief (Fig. 23.8A). You find that (1) both glass rods attract the silk handkerchiefs (Fig. 23.8B); (2) the silk handkerchiefs repel each other (Fig. 23.8C); and (3) the glass rods repel each other (Fig. 23.8D). This simple experiment illustrates important properties of the **electrostatic force**, the force exerted by one charged object on another charged object. In particular, the electrostatic force depends on the charges of the objects involved:

ELECTROSTATIC FORCE

Underlying Principle

1. *Neutral objects do not attract or repel each other.* The silk and glass were initially uncharged, and there was no attraction or repulsion between them.
2. *Oppositely charged objects are attracted to each other.* The negatively charged silk is attracted to the positively charged glass (Fig. 23.8B).
3. Two negatively charged objects repel each other, such as the silk handkerchiefs (Fig. 23.8C). Two positively charged objects repel each other, such as the two glass rods (Fig. 23.8D). To summarize: Objects that have charges of the same sign are mutually repelled.

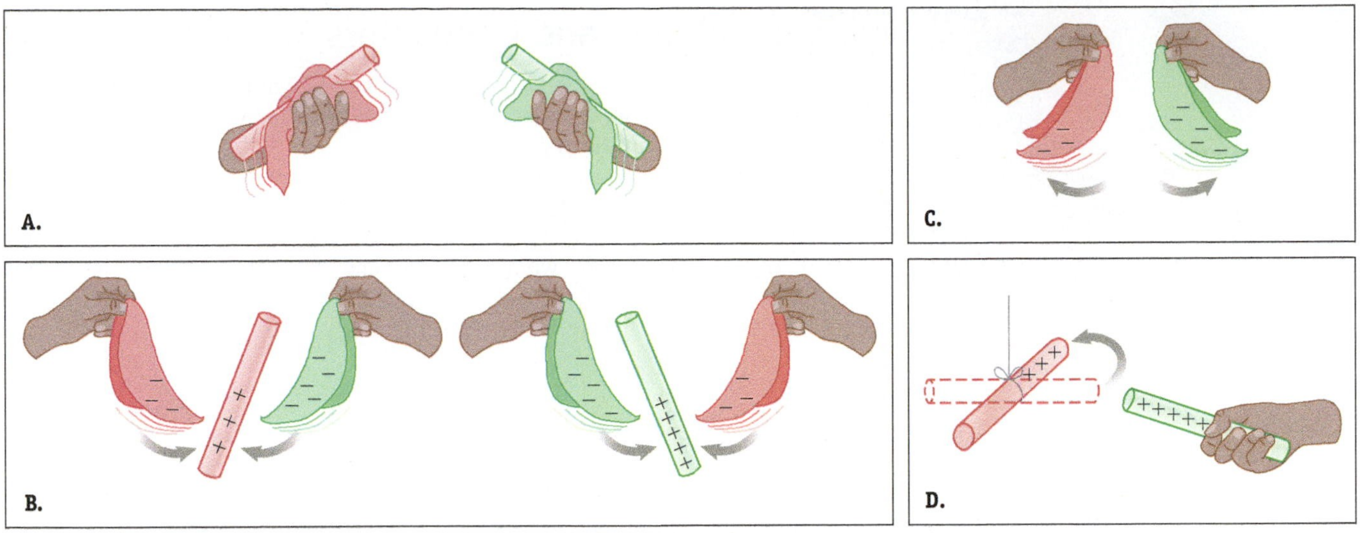

FIGURE 23.8 **A.** Two silk cloths are rubbed against two different glass rods. **B.** Each silk cloth is attracted to each rod. **C.** The two silk cloths repel each other. **D.** The two glass rods repel each other.

You may find it helpful to use the phrase *opposites attract and likes repel* to remember that the electrostatic force between objects that have charges of the opposite sign is attractive, and objects that have charges of the same sign are mutually repelled.

You may wonder why the electrostatic force isn't called simply the *electric* force. The *static* part is included because we are looking at the force between static (nonmoving) charged particles. Of course, electrons must move from one object to another in order to charge the objects, but, for now, we are not concerned about the forces exerted during that transfer. We are concerned with the force one charged object exerts on the other after the movement has stopped.

When charged particles move, we must also think about magnetism. Until Chapter 30, we will assume that magnetism is negligible even if charged objects move.

CONCEPT EXERCISE 23.2

a. In Figure 23.8, why are there three plus signs on the red rod and three minus signs on the red cloth?

b. Which object in Figure 23.8 has the greatest positive charge? How do you know?

CONCEPT EXERCISE 23.3

When wool is rubbed against amber, the wool becomes positively charged and the amber becomes negatively charged. If you rub a glass rod with a silk cloth and an amber rod with a wool cloth, are the rods attracted to or repelled by each other? Are the cloths attracted to or repelled by each other? Is the silk cloth attracted to the amber rod or repelled by the amber rod? Is the wool cloth attracted to the glass rod or repelled by the glass rod? Explain your answers.

CONCEPT EXERCISE 23.4

Three objects A, B, and C are charged. Suppose object A is repelled by object B and attracted to object C.

a. Is object B attracted to object C? Explain.

b. Is it possible to find three charged objects A, B, and C such that A and B attract each other, and A and B are both attracted to C? Explain.

K. Miri Photography/Shutterstock.com

FIGURE 23.9 Why does this technician wrap rubber over these electrical lines before repairing them?

23-4 Insulators and Conductors

Before beginning repair work, a technician covers electrical lines with rubber sheets (Fig. 23.9). He may also wear a rubber suit or rubber sheets over his arms. Why does all this rubber protect him from getting shocked?

INSULATOR Major Concept

The electrons in some materials—such as rubber—do not move freely. Such materials are known as **insulators**. When a surplus of charged particles (positive or negative) builds up on some part of an insulator, the excess remains there. So, if you hold one end of a rubber rod while the far end is being rubbed with silk (Fig. 23.10A), the far end of the rod will acquire a surplus of electrons and those electrons will remain at that end, never flowing into your hand (Fig. 23.10B).

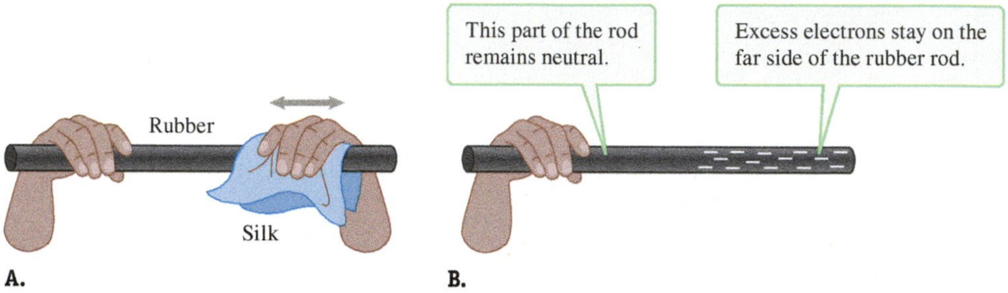
Rubber
Silk
A.
This part of the rod remains neutral.
Excess electrons stay on the far side of the rubber rod.
B.

FIGURE 23.10 **A.** When a rubber rod is rubbed with a silk cloth, **B.** the excess charge stays on the part of the rod that was in contact with the cloth.

CONDUCTOR Major Concept

If you try to charge a copper rod in the same way, you will find that you cannot build up charge on the copper. Copper is an example of a **conductor**. A conductor is a material in which the charged particles (usually electrons) can flow freely. When you hold one end of the copper rod and rub the other end with silk, electrons are transferred from the silk to the copper rod, and those excess electrons are free to flow. Because *likes repel*, the electrons move away from one another, which means they travel through the rod into your hand (Fig. 23.11). The human body is also a conductor, so charged particles move freely through your body toward the Earth. If there are no insulators between you and the ground—such as when you are barefoot—the charge will continue to flow into the Earth. As a result, you and the copper rod remain neutral despite the rod being rubbed with silk.

GROUND Major Concept

The Earth often serves as a charge reservoir known as a **ground**. A ground can accept or provide electrons freely, and it is so large that the addition or subtraction of electrons has a negligible effect on it. So, the ground remains essentially neutral at all times. The copper rod in Figure 23.11 is connected to the ground through your body. When something is connected to the ground by a conductor, we say that it is *grounded*. Every building (with a contemporary electrical system) has a wire connected between the electrical system and a copper pipe. The copper pipe is

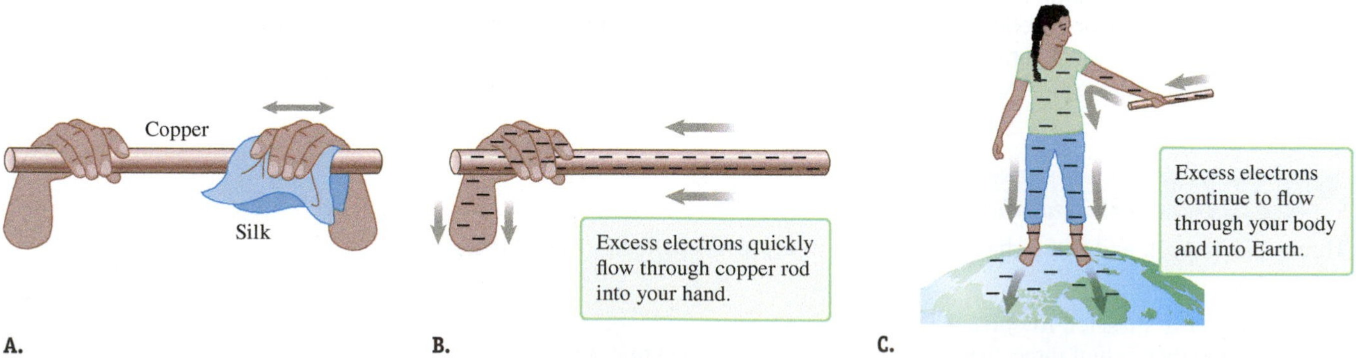
Copper
Silk
A.
Excess electrons quickly flow through copper rod into your hand.
B.
Excess electrons continue to flow through your body and into Earth.
C.

FIGURE 23.11 **A.** When you rub a copper rod with a cloth, **B.** the excess charge moves all over the rod and **C.** through your body into the Earth.

connected through other copper pipes to the Earth. This ensures that the third connection on electrical sockets is *grounded* (Fig. 23.12). The connection to the ground keeps electrical appliances from building up excess charge.

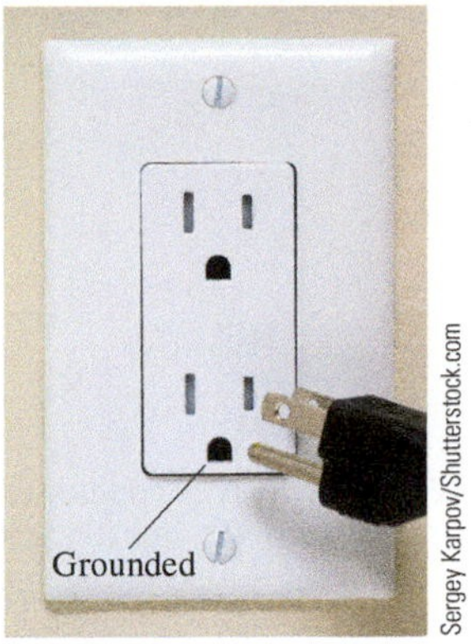

FIGURE 23.12 The third prong of a three-prong electrical plug is connected to the ground to prevent an appliance from building up dangerous excess charge.

Microscopic Model

Metals, tap water (which contains impurities), and the human body (which is similar to tap water) are all good conductors. Nonmetals such as rubber, plastic, glass, silk, wool, dry air, and pure water are good insulators. What makes some materials conductors and others insulators?

To answer this question, let's take a microscopic look at a metal conductor such as copper. Each atom of copper has 29 protons in the nucleus and 29 electrons swarming around that nucleus in a cloud. Much as the swarm of asteroids in the asteroid belt is held in place by gravity, the cloud of electrons is bound to the copper nucleus by electrical attraction. The outermost electrons, however, are more weakly bound to the nucleus than are the inner electrons. When many copper atoms are grouped together as they are in a piece of metal, the binding of each atom's outermost electron is further weakened by the presence of the other atomic nuclei. In fact, the outermost electron is no longer bound to a particular atom. Instead, the outermost electrons from all the atoms are free to move around the entire piece of metal (Fig. 23.13A). Typically, a metal has one free electron per atom, but some metals may have more than one free electron per atom.

In an insulator, the electrons are more tightly bound to their particular nucleus. When many insulator atoms are grouped together in a substance, the electrons of each particular atom stay bound to that atom (Fig. 23.13B).

Charging by Direct Contact

From Figure 23.11, we see that if we wish to build up charge on a conductor, we must insulate it from any large objects that might serve to ground the conductor. Figure 23.14A shows one possible solution—a spherical conductor is placed on a pedestal made of an insulator. As always, electrons may flow throughout the conductor, but now they do not have a conducting pathway to a ground. You can charge the conductor by touching it with a charged object such as a negatively charged plastic rod (Fig. 23.14B). When the rod is in contact with the sphere, some of the electrons are transferred from the rod to the sphere. Remember that charge cannot flow through the rod because it is an insulator, so you may need to roll the rod around the surface of the sphere in order to transfer charge from many parts of the rod. Because electrons move freely throughout the conductor and because *likes repel*, the electrons quickly redistribute themselves in the conductor, moving as far apart as possible. For a spherical conductor, "as far apart as possible" means that the electrons are uniformly distributed on the outside surface of the sphere. If you try to charge a conductor of another shape, the charge is again distributed on the outside surface, although for nonspherical shapes the charge distribution is not uniform.

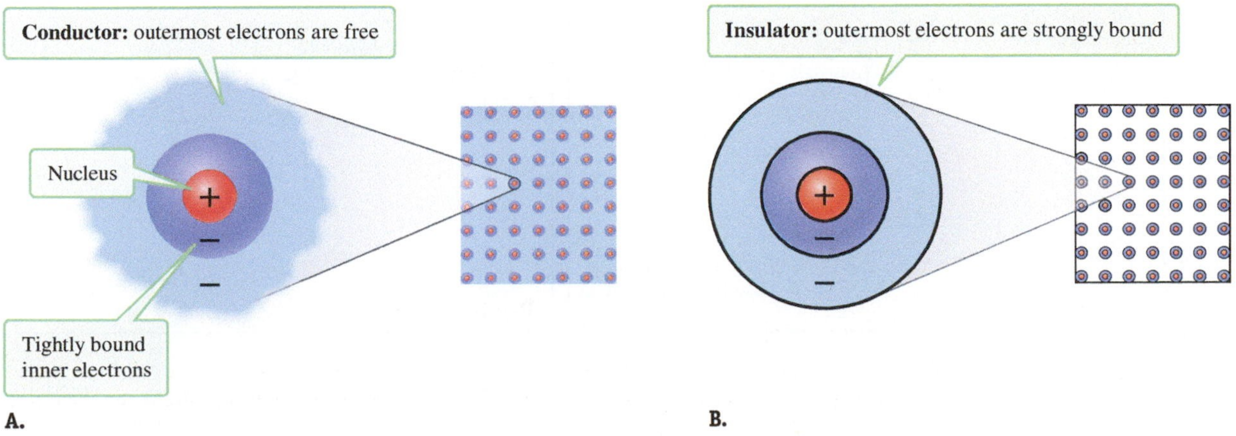

FIGURE 23.13 **A.** The outermost electrons move freely throughout a conductor. **B.** In an insulator, the electrons of each particular atom stay bound to that atom.

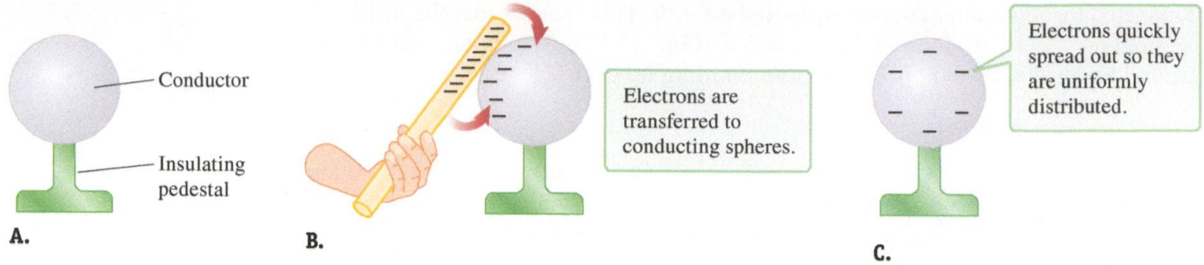

FIGURE 23.14 **A.** A conductor rests on top of an insulating pedestal. **B.** When the conductor is touched by a charged rod, some charge is transferred from the rod to the conductor. **C.** The excess charge quickly distributes over the surface of the conductor. The insulating pedestal prevents the excess charge from flowing into the Earth.

Charging by Induction

There is another way to build up charge on a conductor. Suppose you use the same equipment as in Figure 23.14. This time, however, you hold the negatively charged rod near but *not* touching the sphere (Fig. 23.15). Because the electrons in the sphere are free to move, they flow to the side opposite the rod. If you ground the side of the conducting sphere where there is a surplus of electrons, those electrons will flow to the ground. When you remove the connection to the ground, the net charge of the sphere is positive and the ground remains essentially neutral. No electrons are lost from the rod; it has the same charge throughout the process. In fact, you could reuse the rod to charge another conductor by induction without having to recharge the rod itself.

In this process—known as **induction**—the charged rod never touches the sphere. The charge that is induced on the sphere has the opposite sign as the charged rod. This is true whenever you charge an object by induction. Compare this to Figure 23.14, in which the sphere is in direct contact with the rod and the sphere ends up being negatively charged—the same as the rod. When you charge an object through direct contact with another charged object, both objects have charges of the same sign as a result of their contact.

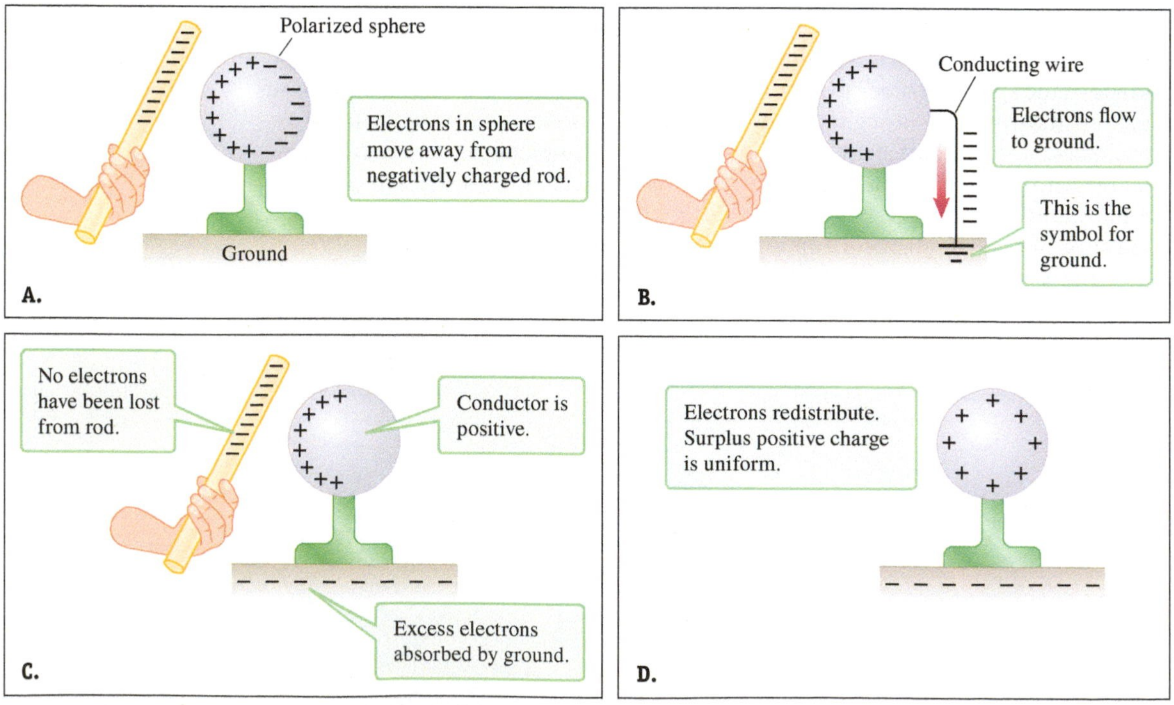

FIGURE 23.15 **A.** When a charged rod is held near a neutral conductor, the conductor becomes polarized. **B.** The far side of the conductor is connected to the ground, and some electrons flow out of the conductor into the Earth. **C.** The connection is removed. **D.** When the rod is removed, the electrons redistribute, leaving the surplus positive charge uniformly distributed over the surface of the conductor.

Polarization and Induced Dipoles

The conducting sphere in Figure 23.15A has been *polarized* by the charged rod. The term **polarized** describes an object that is neutral but has a net separation between its positive and negative charged particles. If you remove the rod, the electrons will move back and the sphere will not be polarized.

If you wish to entertain your friends at a Franklin-style party, here are instructions for a great trick. You need a glass swizzle stick, an empty metal can, and a silk handkerchief. Place the empty metal can on its side on a flat, horizontal surface. Charge the glass stick by rubbing it with your silk handkerchief. The glass stick will then be positively charged. Hold the charged glass stick near the can. It is best to align the long axis of the stick with the long axis of the can (Fig. 23.16A). The can is neutral, but because it is a conductor, the presence of the positively charged glass stick polarizes the can when electrons in the can move closer to the stick. The negatively charged particles are closer to the stick than are the positively charged particles, so the net effect is that the can is attracted to the stick and begins to roll along the table (Fig. 23.16B). With practice, you can get the can to roll quickly. Then by holding the stick on the other side of the can, you can make it slow down, stop, and reverse direction.

This party trick demonstrates why neutral insulators and charged objects are attracted to one another. For example, rubber balloons rubbed on hair are negatively charged and attracted to the ceiling, a neutral insulator. Likewise, a comb rubbed through hair is positively charged and can be used to pick up paper, another neutral insulator. The charged particles in an insulator are not free, so an insulator cannot be polarized as a whole. However, the atoms and molecules that make up an insulator can be polarized. When this happens, there is a slight separation of the positive from the negative charge within the atom or molecule (Fig. 23.17). We say that the atoms (or molecules) have become *dipoles*. A **dipole** is a neutral object in which the charged particles are distributed so that there is a net positive side and a net negative side. So, if a positively charged object (a comb) is brought near an insulator (paper), the atoms (or molecules) in the paper become dipoles such that their individual electron clouds are shifted slightly toward the charged object. The net effect of this slight shift is that the insulator is attracted by the charged object.

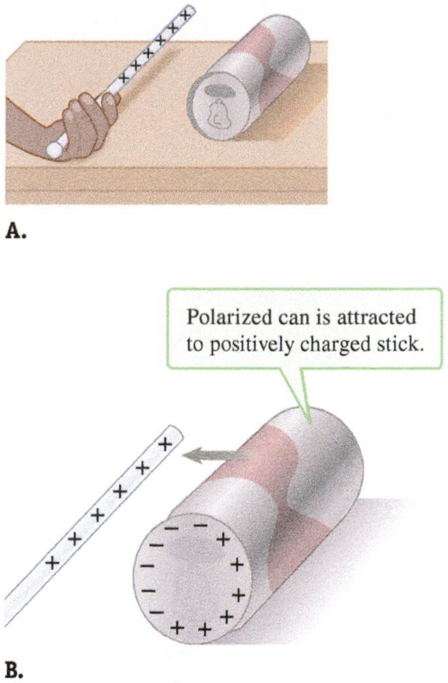
A.
Polarized can is attracted to positively charged stick.
B.

FIGURE 23.16 **A.** A charged rod is held near a neutral metal can. **B.** The can becomes polarized.

Sparks

Any insulator will allow conduction under extreme conditions. Under normal conditions, dry air is a good insulator. (Moist air, such as on a humid day or in a rainstorm, does not insulate as well as dry air.) When two oppositely charged objects are separated by air, charge is not usually transferred through air from one object to the other. However, if the charged objects are extremely large or the objects are very close together, it is possible for the air to transfer the charge momentarily. When this happens, we often see and hear a dramatic spark—as in the case of lightning.

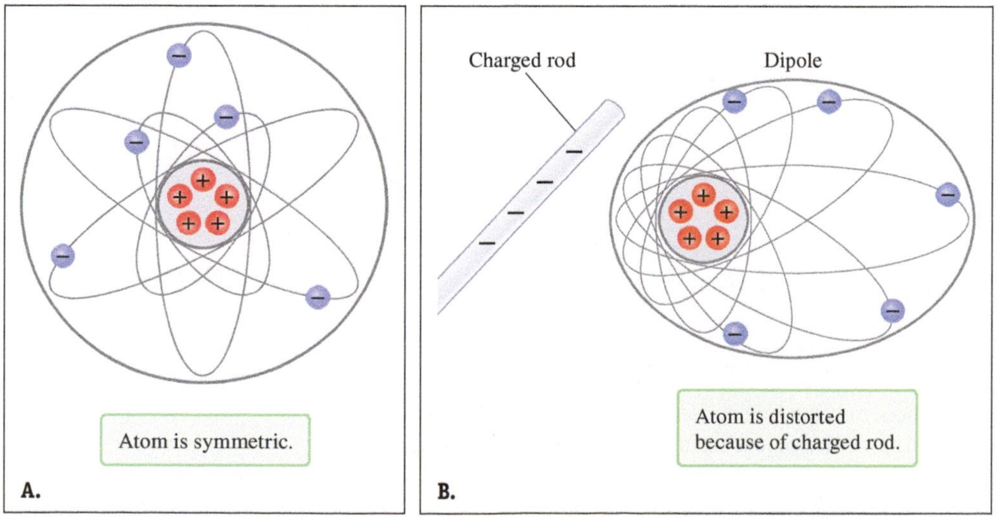
Atom is symmetric.
A.
Charged rod
Dipole
Atom is distorted because of charged rod.
B.

FIGURE 23.17 **A.** A neutral symmetrical atom. **B.** When a charged rod is held near the atom, the atom is distorted so that it is best modeled as a dipole.

You have probably experienced somewhat less dramatic cases of sparks. For example, if you pull apart clothes stuck together by static cling, you may hear a crackling sound due to charged particles moving through the air between the clothes. On a dry winter day, if you rub your socks on the carpet as you walk and then reach for the doorknob, you may see, hear, and feel a spark just before you touch the knob. Sparks were most entertaining for people in the 18th century. The woman in Figure 23.4 is about to draw a spark by touching the charged boy's nose. Some people found it amusing to charge up a woman so that when a man kissed her, he felt a shock.

CONCEPT EXERCISE 23.5

The following scenarios involve a metal ball and a charged glass rod. No other objects are involved.

a. A small metal ball is attracted to a negatively charged glass rod. Is the metal ball necessarily positively charged? Explain.
b. A small metal ball is repelled by a negatively charged glass rod. Is the metal ball necessarily negatively charged? Explain.

CONCEPT EXERCISE 23.6

a. Why is the charged boy in Figure 23.4 suspended by a silk rope? Could the rope be replaced by a metal chain?
b. Why is the boy able to attract paper?
c. Why will the woman draw a spark when she touches the boy's nose?

EXAMPLE 23.1 "Improvise!" or "You Don't Have to Dress Like James Bond to be Cool"

Suppose you wish to perform the metal can trick (Fig. 23.16), but you cannot find a glass stick and you don't wear a silk handkerchief in your lapel pocket. You can find a plastic swizzle stick, and you are wearing a wool sweater. If you rub the plastic stick on your wool sweater, the plastic stick will become negatively charged—the opposite of the glass stick rubbed with silk. Will the trick still work? Explain.

INTERPRET and ANTICIPATE

We know from Figure 23.16 that a positively charged stick attracts a neutral conductor because electrons in the conductor that are free to move are attracted to the stick. The conductor becomes polarized, so the negative electrons are closer to the positive stick and there is a net attraction. The question is, "Will a negatively charged object attract a neutral conductor?"

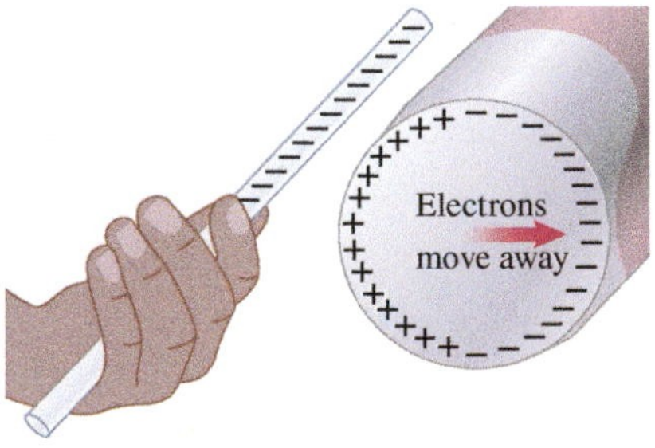

FIGURE 23.18

SOLVE

When the negatively charged stick is held near the metal can, the electrons in the can are repelled. Those electrons that are free to move go to the far side of the can (Fig. 23.18). That leaves a surplus of protons on the side near the stick. The net effect is that the can is attracted to the stick. So the can rolls toward the stick just as if you had used a positively charged glass stick.

CHECK and THINK

We can make a more general statement about conductors: A neutral conductor is always attracted to a charged object, independent of the sign of the object's charge. (A similar statement can be made about neutral insulators.)

EXAMPLE 23.2 CASE STUDY Safety Precautions at the Gas Pump

In our case study, a woman returned to her car after putting the pump nozzle in her tank. She slid across the seat. When she got out of the car and reached for the nozzle, a fire erupted.

A Why did a fire erupt?

SOLVE

This is much like the situation in Figure 23.4. When the 18th-century woman reaches for the nose of the charged boy, there is a spark as charged particles are transferred between them (Concept Exercise 23.6). When the woman at the gas pump reached for the nozzle, charge was transferred and there was a spark. This time the spark passed through a mixture of gasoline vapor and air. Just as in the engine of your car, when the spark plug ignites the gasoline vapor-air mixture, the spark caused the gas fumes around the nozzle to ignite.

B Why is it significant that the woman slid across her seat in the car? Does the type of clothing matter?

SOLVE

In order for there to be a spark between two objects, one or both of the objects must have a net charge. In Figure 23.4, the boy has a net charge. The woman at the gas pump built up a net charge on herself by sliding across the seat. The seat was probably leather or vinyl, and her clothes were probably wool, cotton, or some synthetic fabric. Just as rubbing a plastic rod with a wool cloth transfers electrons and results in charging both objects, the woman's clothes and car seat were charged by her sliding motion. Without knowing the type of fabric in her clothes or the type of material on her car seat, we cannot be sure whether her clothes were positive or negative after sliding. In order to be specific, let's assume her clothes were negatively charged. This assumption won't affect the major points of the case study, however.

C Is it significant that the woman was wearing rubber-soled shoes?

SOLVE

The human body is a good conductor, and the woman's charged clothing either transferred charged particles to her or caused her to be polarized like the can in Figure 23.16. If she had been grounded before she came close to the nozzle, then she and her clothes would have been discharged as electrons flowed from her into the Earth. However, rubber is a good insulator. So, like the pedestal in Figure 23.14, her rubber shoes prevented the woman from discharging; thus, she had a net charge when she touched the nozzle.

D What advice should be posted at the gas pump?

SOLVE

1. Don't get back into your car while your gas pump nozzle is in the tank and there is likely to be gas vapor near the nozzle. Most of the time, you need to slide across the seat to get into or out of your car. When you do this, you are probably building up a net charge.
2. If you must get back into your car, be sure to ground yourself far from the nozzle. You may want to touch the ground or the metal supports of the gas station awning with your bare hand far from your open gas tank. Notice that these warnings agree with the precautions posted at the gas pump.

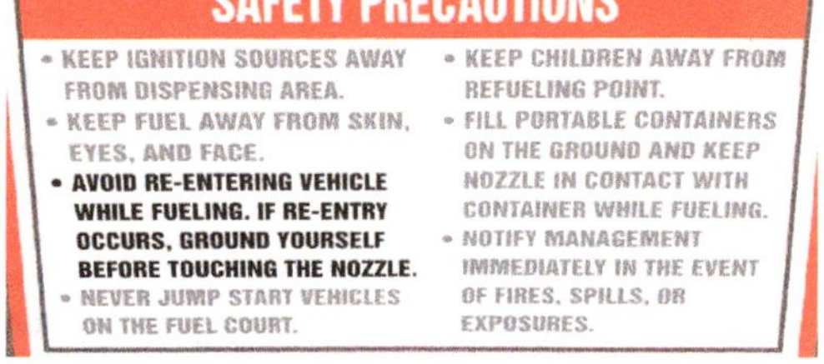
SAFETY PRECAUTIONS
• KEEP IGNITION SOURCES AWAY FROM DISPENSING AREA.
• KEEP FUEL AWAY FROM SKIN, EYES, AND FACE.
• AVOID RE-ENTERING VEHICLE WHILE FUELING. IF RE-ENTRY OCCURS, GROUND YOURSELF BEFORE TOUCHING THE NOZZLE.
• NEVER JUMP START VEHICLES ON THE FUEL COURT.
• KEEP CHILDREN AWAY FROM REFUELING POINT.
• FILL PORTABLE CONTAINERS ON THE GROUND AND KEEP NOZZLE IN CONTACT WITH CONTAINER WHILE FUELING.
• NOTIFY MANAGEMENT IMMEDIATELY IN THE EVENT OF FIRES, SPILLS, OR EXPOSURES.
Debora Katz

FIGURE 23.19

−15 nC, write "$q_1 = -15$ nC" near its dot. If no information is given, it is best to assume the particle's charge is positive.

4. Usually you are interested in finding the total electrostatic force exerted on one of the particles, called the *subject*. **Circle the subject.** Then use your knowledge of the electrostatic force to **draw vectors representing the force exerted by all the other objects on the subject.** Place the tail of each vector on the subject. Although the magnitude of the forces is not critical at this stage, it is important to indicate the direction. Draw the vector toward a particular object if that object has the opposite sign as the subject (because the force is attractive). Draw the vector away from the object if that object has the same sign as the subject (because the force is repulsive).

5. **Indicate the subject's acceleration.**
6. **Choose a coordinate system.**

:• SOLVE

The drawing of the subject—with force vectors extending from it—is treated as a free-body diagram. Then Newton's second law is applied. Often that means:

Step 1. Writing the electrostatic force (Eq. 23.3) in component form, and then

Step 2. Algebraically combining the components.

Step 3. Be careful about signs. You already took the sign of the charge into account when you determined the direction of each vector. When you substitute in values for the charge, you (probably) need only the **absolute value of the charge.**

EXAMPLE 23.5 Static Equilibrium

Two small spherical insulators are separated by 2.5 cm, which is much greater than either of their diameters. Both carry positive charge, one +60.0 μC and the other +6.66 μC. A third charged sphere remains at rest between the two spheres and along the line joining them. What is the position of this third charged sphere? In the **CHECK and THINK** step, answer this question: What can you say about the sign and magnitude of this charge?

:• INTERPRET and ANTICIPATE

Start with a hybrid free-body diagram (Fig. 23.25).

1. and 2. Each sphere is represented by a dot at its center.
3. The charges on the first two spherical insulators are labeled q_1 and q_2. Because we are given numerical information about the charges, that information is included. The charge on the third sphere is (arbitrarily) labeled q_0, and no numerical information is given about it.
4. We have circled the subject, q_0. Only two electrostatic forces are exerted on the subject: $\vec{F}_1$, the force exerted *by* q_1, and $\vec{F}_2$, the force exerted *by* q_2. We have assumed that q_0 is positive, so both objects repel the subject. Therefore, $\vec{F}_1$ and $\vec{F}_2$ are drawn *away* from their respective objects.
5. The acceleration is zero as indicated.
6. A coordinate system has been chosen.

The forces acting on the third sphere are balanced, so we expect the third charged sphere to be closer to the sphere that has the smaller charge.

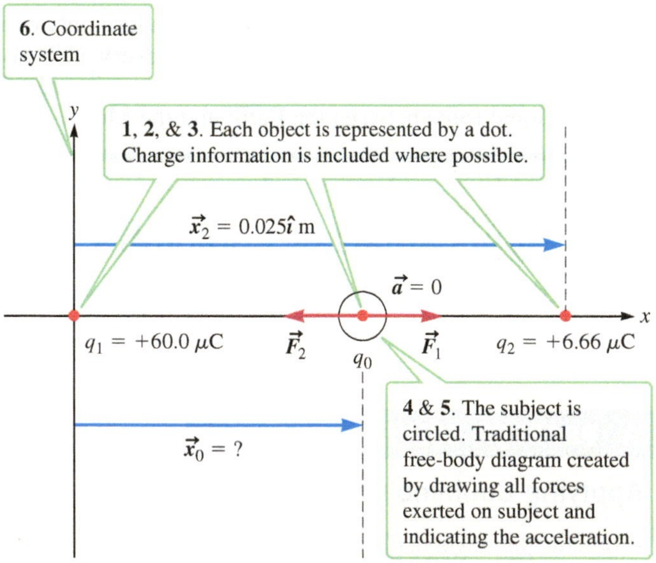

FIGURE 23.25

:• SOLVE

The region in the center of Figure 23.25 is like any traditional free-body diagram. We apply Newton's second law as usual, having chosen a coordinate system. Because the acceleration is zero, the sum of the forces is zero.	$\sum \vec{F} = \vec{F}_1 + \vec{F}_2 = 0$
These forces must have equal magnitude and point in opposite directions.	$\vec{F}_1 = -\vec{F}_2$ $F_1 = F_2$

The magnitude of the electrostatic force is given by Coulomb's law, Equation 23.3. Coulomb's constant k cancels, as does the subject's charge q_0.	$k\frac{\lvert q_1q_0\rvert}{r_1^2} = k\frac{\lvert q_2q_0\rvert}{r_2^2}$ $\frac{\lvert q_1\rvert}{r_1^2} = \frac{\lvert q_2\rvert}{r_2^2}$ (1)
The center-to-center distances r_1 and r_2 come from Figure 23.25.	$r_1 = x_0$ $r_2 = x_2 - x_0$
Substitute r_1 and r_2 into Equation (1), and solve for x_0. We choose the positive square root; the negative square root gives a value for x_0 that is not between the original two spheres.	$\frac{\lvert q_1\rvert}{(x_0)^2} = \frac{\lvert q_2\rvert}{(x_2 - x_0)^2}$ $\left(\frac{x_2 - x_0}{x_0}\right) = \pm\sqrt{\left\lvert\frac{q_2}{q_1}\right\rvert}$ $x_0 = x_2\left(\sqrt{\left\lvert\frac{q_2}{q_1}\right\rvert} + 1\right)^{-1} = (0.025\text{ m})\left(\sqrt{\left\lvert\frac{6.66\ \mu\text{C}}{60.0\ \mu\text{C}}\right\rvert} + 1\right)^{-1}$ $x_0 = 1.9 \times 10^{-2}\text{m} = 1.9\text{ cm}$

CHECK and THINK

According to our results, the subject is nearly 2 cm from sphere 1 and only about half a centimeter from sphere 2. So, as we expected, the subject is closer to the sphere that has the smaller charge. Because q_0 cancels, neither its magnitude nor its sign matters. In other words, we cannot tell whether q_0 is positive or negative, and we cannot determine its magnitude.

EXAMPLE 23.6 Right Triangle

Three small charged spheres lie on the vertices of a right isosceles triangle as shown in Figure 23.26. The sides of equal length are 0.654 m long. The right angle is at the origin of the coordinate system. The sphere at the origin has a charge of –15 mC. A sphere with a charge of +40 mC is on the x axis, and a sphere with a charge of +20 mC is on the y axis. Find the net electrostatic force exerted on the sphere at the origin. Give your answer in component form and in terms of magnitude and direction.

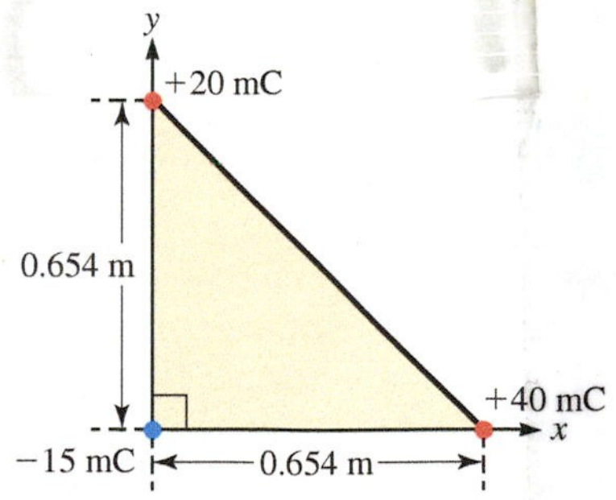

FIGURE 23.26

INTERPRET and ANTICIPATE

Figure 23.27 is the hybrid free-body diagram.

1. and 2. Each sphere is represented by a dot at its center.
3. The charge on each sphere is given in the figure. The sphere at the origin is the subject because we are asked to find the net electrostatic force exerted on it.
4. The subject is circled. Sphere 1 exerts an attractive force along the x axis, so $\vec{F}_1$ points along the x axis toward sphere 1. Sphere 2 also exerts an attractive force, so $\vec{F}_2$ points along the y axis toward sphere 2. Both spheres 1 and 2 are a distance ℓ away from sphere 0. However, sphere 1 has twice as much charge as sphere 2, so we have drawn $\vec{F}_1$ twice as long as $\vec{F}_2$. To anticipate our result, we use geometric addition to find $\vec{F}_{tot} = \vec{F}_1 + \vec{F}_2$.
5. We don't have any information about the acceleration, so it is not indicated on the diagram.

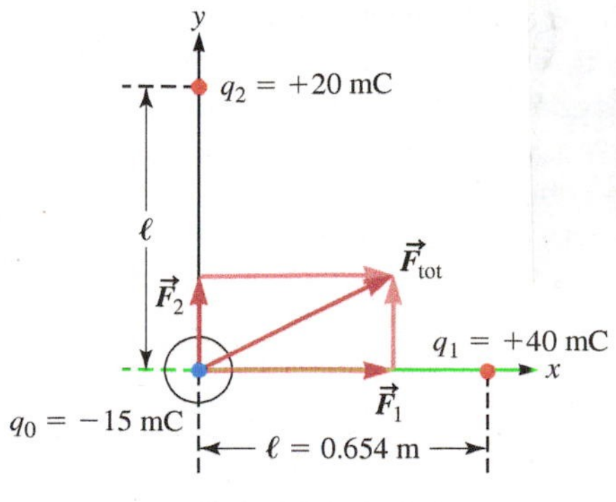

FIGURE 23.27

Example continues on page 702 ▶

SOLVE

$\vec{F}_1$ has only an x component and $\vec{F}_2$ has only a y component. Adding these components gives a general expression for the net electrostatic force $\vec{F}_{tot}$ in component form.

$$\vec{F}_{tot} = \vec{F}_1 + \vec{F}_2$$

$$\vec{F}_{tot} = F_1\hat{\imath} + F_2\hat{\jmath} \quad (1)$$

The magnitudes F_1 and F_2 come from Equation 23.3.

$$F_1 = k\frac{|q_1q_0|}{\ell^2} = (8.99 \times 10^9\ \text{N}\cdot\text{m}^2/\text{C}^2)\frac{|(40 \times 10^{-3}\,\text{C})(-15 \times 10^{-3}\,\text{C})|}{(0.654\ \text{m})^2}$$

$$F_1 = 1.26 \times 10^7\ \text{N}$$

$$F_2 = k\frac{|q_2q_0|}{\ell^2} = (8.99 \times 10^9\ \text{N}\cdot\text{m}^2/\text{C}^2)\frac{|(20 \times 10^{-3}\,\text{C})(-15 \times 10^{-3}\,\text{C})|}{(0.654\ \text{m})^2}$$

$$F_2 = 6.31 \times 10^6\ \text{N}$$

To write the net electrostatic force in component form, substitute F_1 and F_2 into Equation (1).

$$\vec{F}_{tot} = (1.26 \times 10^7\hat{\imath} + 6.31 \times 10^6\hat{\jmath})\ \text{N}$$

The magnitude and direction come from $A = \sqrt{A_x^2 + A_y^2}$ and $\theta_A = \tan^{-1}(A_y/A_x)$ (Eqs. 3.12 and 3.14, respectively).

$$F_{tot} = \sqrt{(F_{tot})_x^2 + (F_{tot})_y^2}$$

$$F_{tot} = \sqrt{(1.26 \times 10^7\,\text{N})^2 + (6.31 \times 10^6\,\text{N})^2}$$

$$F_{tot} = 1.41 \times 10^7\,\text{N}$$

$$\theta = \tan^{-1}\frac{(F_{tot})_y}{(F_{tot})_x} = \tan^{-1}\left(\frac{6.31 \times 10^6\ \text{N}}{1.26 \times 10^7\ \text{N}}\right)$$

$$\theta = 26.6°$$

CHECK and THINK

Our results fit the vector we anticipated in Figure 23.27. Both the anticipated vector and the vector we found are in quadrant I and have a magnitude greater than that of either $\vec{F}_1$ or $\vec{F}_2$ alone.

EXAMPLE 23.7 Square

Suppose a fourth small charged sphere is added to the three in Example 23.6 so that now the four spheres are at the vertices of a square. If the fourth sphere has a charge of –15 mC, what is the net electrostatic force on the sphere at the origin? Use the same coordinate system. Give your answer in component form and in terms of magnitude and direction.

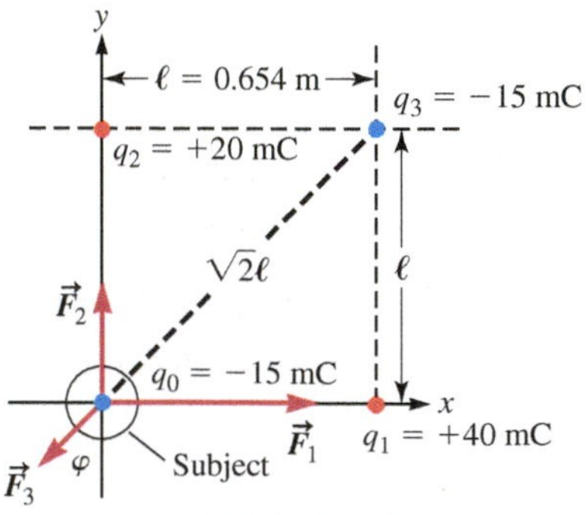

FIGURE 23.28

INTERPRET and ANTICIPATE

In this problem, we can update Figure 23.27 in order to draw our hybrid free-body diagram (Fig. 23.28). The fourth charge is represented by a dot and called sphere 3. Its charge is written on the diagram. Because sphere 3 has a charge of the same sign as sphere 0, sphere 3 repels sphere 0. Therefore, force $\vec{F}_3$ is drawn pointing away from sphere 3. The other forces are unchanged. Sphere 3 has the smallest magnitude of charge and because it is the farthest from sphere 0, we have drawn it shorter than the other forces. We anticipate that our result will be similar to that for Example 23.6.

SOLVE

Since $\vec{F}_1$ and $\vec{F}_2$ are unchanged, we only need to find an expression for $\vec{F}_3$ in component form. $\vec{F}_3$ is along the same line as the diagonal of the square, so φ is 45° ($\sin 45° = \cos 45° = \sqrt{2}/2$).

$$\vec{F}_3 = F_{3x}\hat{\imath} + F_{3y}\hat{\jmath}$$

$$\vec{F}_3 = (-F_3\sin\varphi)\hat{\imath} + (-F_3\cos\varphi)\hat{\jmath}$$

$$\vec{F}_3 = (-\tfrac{\sqrt{2}}{2}F_3)\hat{\imath} + (-\tfrac{\sqrt{2}}{2}F_3)\hat{\jmath}$$

The magnitude of F_3 comes from Equation 23.3. Remember that $q_0 = q_3 = -15$ mC.	$F_3 = k\dfrac{\lvert q_3 q_0\rvert}{(\sqrt{2}\ell)^2} = (8.99 \times 10^9\ \text{N}\cdot\text{m}^2/\text{C}^2)\dfrac{(-15 \times 10^{-3}\,\text{C})^2}{2(0.654\,\text{m})^2}$ $F_3 = 2.36 \times 10^6$ N
Write $\vec{F}_3$ in component form.	$\vec{F}_3 = (-\frac{\sqrt{2}}{2}2.36 \times 10^6\,\hat{\imath} - \frac{\sqrt{2}}{2}2.36 \times 10^6\,\hat{\jmath})$ N $\vec{F}_3 = (-1.67 \times 10^6\,\hat{\imath} - 1.67 \times 10^6\,\hat{\jmath})$ N
Add $\vec{F}_3$ to $\vec{F}_1 + \vec{F}_2$ from Example 23.6 to find the new $\vec{F}_{\text{tot}}$.	$\vec{F}_{\text{tot}} = (F_1 + F_{3x})\hat{\imath} + (F_2 + F_{3y})\hat{\jmath}$ $\vec{F}_{\text{tot}} = (1.26 \times 10^7\ \text{N} - 1.67 \times 10^6\ \text{N})\hat{\imath} + (6.31 \times 10^6\ \text{N} - 1.67 \times 10^6\ \text{N})\hat{\jmath}$ $\vec{F}_{\text{tot}} = (1.09 \times 10^7\,\hat{\imath} + 4.64 \times 10^6\,\hat{\jmath})$ N
Find the magnitude and direction as in Example 23.6.	$F_{\text{tot}} = \sqrt{(F_{\text{tot}})_x^2 + (F_{\text{tot}})_y^2} = \sqrt{(1.09 \times 10^7\,\text{N})^2 + (4.64 \times 10^6\,\text{N})^2} = 1.18 \times 10^7\,\text{N}$ $\theta = \tan^{-1}\dfrac{(F_{\text{tot}})_y}{(F_{\text{tot}})_x} = \tan^{-1}\left(\dfrac{4.64 \times 10^6\,\text{N}}{1.09 \times 10^7\,\text{N}}\right) = 23.1°$

CHECK and THINK

As anticipated, our results here are similar to the results from Example 23.6.

EXAMPLE 23.8 Equilateral Triangle

Three particles with charges q_0, q_1, and q_2 are at the vertices of an equilateral triangle of side ℓ (Fig. 23.29). Particle 1 and particle 2 lie on the x axis at $x = -\ell/2$ and $x = +\ell/2$, respectively. Particle 0 is on the y axis. Show that the net electrostatic force exerted on particle 0 is given by

$$\vec{F}_{\text{tot}} = \frac{kq_0}{2\ell^2}\left[(q_1 - q_2)\hat{\imath} + \sqrt{3}(q_1 + q_2)\hat{\jmath}\right]$$

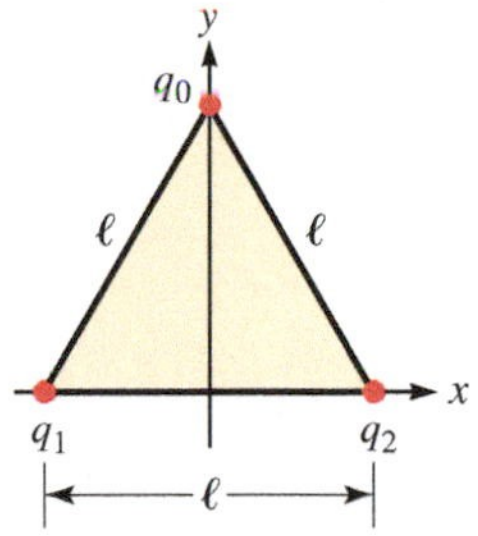

FIGURE 23.29

INTERPRET and ANTICIPATE

The three particles and the coordinate system are shown in Figure 23.30. No information is given about the charge of each particle, so only the symbols q_0, q_1, and q_2 are included on the diagram. We'll assume these charges are positive, but that does not affect the general form of the vector equation for force that we must develop. The subject is charge q_0. Two repulsive forces are exerted on the subject, but we do not know their magnitudes. (To keep the figure clean, we haven't indicated the acceleration.)

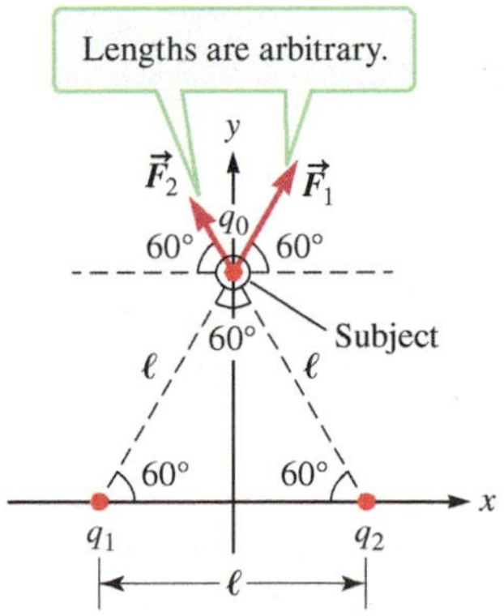

FIGURE 23.30

SOLVE

Find a general expression for $\vec{F}_{\text{tot}}$ in component form.	$\vec{F}_{\text{tot}} = \vec{F}_1 + \vec{F}_2 = (F_{1x} + F_{2x})\hat{\imath} + (F_{1y} + F_{2y})\hat{\jmath}$ $\vec{F}_{\text{tot}} = (F_1 \cos 60° - F_2 \cos 60°)\hat{\imath} + (F_1 \sin 60° + F_2 \sin 60°)\hat{\jmath}$ (1)
Use Equation 23.3 to find F_1 and F_2.	$F_1 = k\dfrac{q_1 q_0}{\ell^2}$ $F_2 = k\dfrac{q_2 q_0}{\ell^2}$

Example continues on page 704 ▶

Substitute F_1 and F_2 into Equation (1), and use $\cos 60° = \frac{1}{2}$ and $\sin 60° = \frac{\sqrt{3}}{2}$ to reduce the expression for $\vec{F}_{tot}$.	$\vec{F}_{tot} = \left(k\frac{q_1 q_0}{\ell^2}\cos 60° - k\frac{q_2 q_0}{\ell^2}\cos 60°\right)\hat{\imath} + \left(k\frac{q_1 q_0}{\ell^2}\sin 60° + k\frac{q_2 q_0}{\ell^2}\sin 60°\right)\hat{\jmath}$ $\vec{F}_{tot} = \left(k\frac{q_1 q_0}{\ell^2}\frac{1}{2} - k\frac{q_2 q_0}{\ell^2}\frac{1}{2}\right)\hat{\imath} + \left(k\frac{q_1 q_0}{\ell^2}\frac{\sqrt{3}}{2} + k\frac{q_2 q_0}{\ell^2}\frac{\sqrt{3}}{2}\right)\hat{\jmath}$ $\vec{F}_{tot} = \frac{kq_0}{2\ell^2}[(q_1 - q_2)\hat{\imath} + \sqrt{3}(q_1 + q_2)\hat{\jmath}]$ ✓ (2)

CHECK and THINK

This is exactly what we were asked to find. Notice that if q_0 were negative, there would be an overall negative sign, and the total force would be in the opposite direction. What if $q_1 = q_2$? Then we expect that (1) $F_1 = F_2$; (2) the x components of $\vec{F}_1$ and $\vec{F}_2$ cancel; and therefore (3) the total force points straight up along the y axis. That is exactly what we find with Equation (2); the first term is zero if $q_1 = q_2$, and the second term combines to give $\vec{F}_{tot} = (\sqrt{3}kq_0 q_1/\ell^2)\hat{\jmath}$.

EXAMPLE 23.9 A Show Stopper, or There Has to Be a Typo

In a physics demonstration, two identical conducting spheres are attached on top of two insulating pedestals. Each sphere, including its pedestal, has a mass of 0.58 kg. Their center-to-center distance is 2.50 m (Fig. 23.31). Initially, sphere 1 has a charge of +10.0 C and sphere 2 is neutral. A conducting wire is momentarily placed between the two spheres. The spheres remain at rest after the wire is removed.

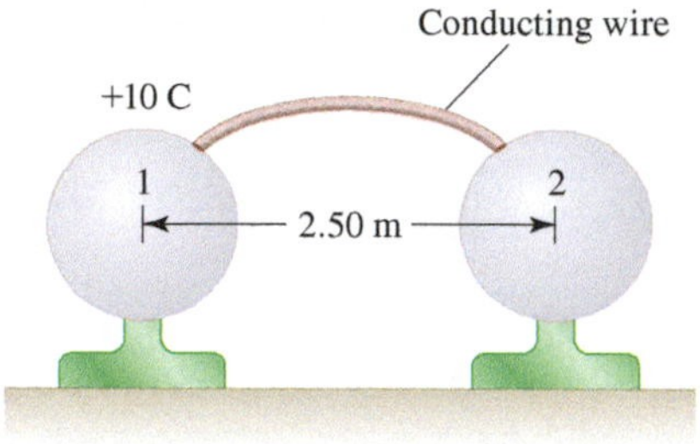

FIGURE 23.31

A What is the minimum coefficient of static friction μ_s between the pedestal and the tabletop?

INTERPRET and ANTICIPATE

Electrons in sphere 2 are attracted to sphere 1 because it is positively charged. When the spheres are joined by the conducting wire, electrons are free to flow from sphere 2 to sphere 1. Because the spheres are identical, each ends up with half of the initial charge. When the wire is removed, each sphere has a charge of +5.0 C. Because they have charges of the same sign, the spheres repel each other.

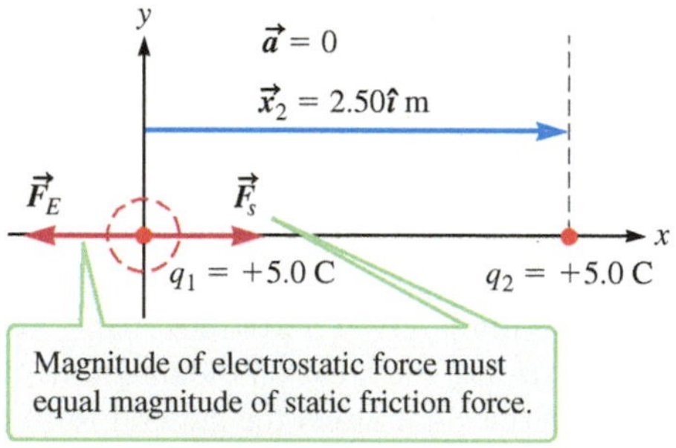

FIGURE 23.32

Let's arbitrarily choose sphere 1 as the subject and place it at the origin of the coordinate system (Fig. 23.32). Two forces are exerted on the subject—the repulsive electrostatic force exerted by sphere 2 and static friction due to the tabletop.

SOLVE

The spheres remain at rest, so the electrostatic force exerted on sphere 1 must be balanced by static friction.	$F_E = F_s$
The magnitude of the electrostatic force is very large in this case.	$F_E = k\frac{\lvert q_1 q_2\rvert}{r^2}$ $F_E = (8.99 \times 10^9\ \text{N}\cdot\text{m}^2/\text{C}^2)\frac{(5.0\,\text{C})^2}{(2.50\,\text{m})^2}$ $F_E = 3.6 \times 10^{10}$ N

Because the electrostatic force is so great, let's imagine that static friction is at its maximum. This gives us the minimum value of μ_s using $F_{s,\max} = \mu_s F_N$ (Eq. 6.1) for $F_{s,\max}$. The tabletop is level, so the magnitude of the normal force equals the sphere's weight.

$$F_E = F_{s,\max} = \mu_s F_N = \mu_s mg$$

$$\mu_s = \frac{F_E}{mg} = \frac{3.6 \times 10^{10}\ \text{N}}{(0.58\ \text{kg})(9.81\ \text{m/s}^2)}$$

$$\mu_s = 6.3 \times 10^9$$

CHECK and THINK

This answer seems **highly improbable**. According to Table 6.1, our coefficient of static friction is more than nine orders of magnitude greater than the typical value (and our answer is the *minimum* possible value). We can resolve this by recognizing that 5.0 C is a great amount of charge. If you see a result like this, you might guess that a prefix such as μ was missing so that the charge was $+5.0\ \mu\text{C}$ instead of $+5.0$ C. If that were the case, the electrostatic force would be 0.036 N. This is 12 orders of magnitude smaller than what we found, and the minimum value for μ_s would have been 0.006—a reasonable value.

B Now let's consider a more reasonable physics demonstration. After the conducting wire is removed, each conducting sphere has a charge of $+7.5\ \mu\text{C}$. The coefficient of static friction between the pedestal and the tabletop is $\mu_s = 0.006$. What is magnitude of either sphere's initial acceleration?

INTERPRET and ANTICIPATE

As in part A, we choose sphere 1 as the subject and place it at the origin of the coordinate system (Fig. 23.33). Now the acceleration may be nonzero, and we'll assume it is in the same direction as the electrostatic force.

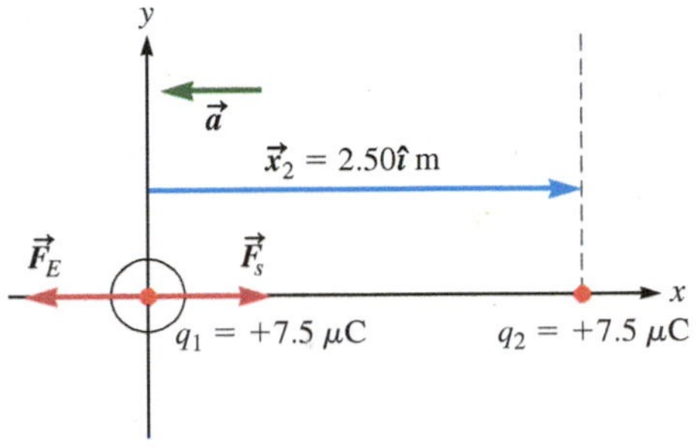

FIGURE 23.33

SOLVE

Now, when we apply Newton's second law, the sum of the force is not zero.

$$\sum F_x = F_s - F_E = -ma$$

The magnitude of the electrostatic force is not nearly as great as it was in part A.

$$F_E = k\frac{|q_1 q_2|}{r^2}$$

$$F_E = (8.99 \times 10^9\ \text{N}\cdot\text{m}^2/\text{C}^2)\frac{(7.5 \times 10^{-6}\ \text{C})^2}{(2.50\ \text{m})^2}$$

$$F_E = 8.1 \times 10^{-2}\ \text{N}$$

To know whether the sphere moves, we need to compare F_E to the maximum value of static friction $F_{s,\max} = \mu_s F_N$ (Eq. 6.1).

$$F_{s,\max} = \mu_s F_N = \mu_s mg$$

$$F_{s,\max} = (0.006)(0.58\ \text{kg})(9.81\ \text{m/s}^2)$$

$$F_{s,\max} = 3.4 \times 10^{-2}\ \text{N}$$

$$F_{s,\max} < F_E$$

Because $F_{s,\max} < F_E$, the sphere does accelerate in the direction of the electrostatic force.

$$a = \frac{F_{s,\max} - F_E}{-m} = \frac{3.4 \times 10^{-2}\ \text{N} - 8.1 \times 10^{-2}\ \text{N}}{-0.58\ \text{kg}}$$

$$a = 8.1 \times 10^{-2}\ \text{m/s}^2$$

CHECK and THINK

In this demonstration, we see that the spheres accelerate slowly away from each other. Keep in mind that this is a very low coefficient of static friction. The spheres may need to be on an air table to achieve such low friction.

EXAMPLE 23.10 The Strong Force

The helium nucleus has two neutrons and two protons. If the distance between the protons is about 2×10^{-15} m, find the electrostatic force between the protons.

INTERPRET and ANTICIPATE

You may find this situation easy to visualize without a diagram. (It is very similar to Examples 23.3 and 23.4.) However, we include a diagram (Fig. 23.34) for clarity and practice. The protons are dots labeled q_1 and q_2. Both carry the same amount of charge $+e$. We have chosen proton 1 as the subject, and we use a coordinate system so that unit vector $\hat{r}$ points from proton 1 toward proton 2. The protons are both positive, so the force on each of them is repulsive.

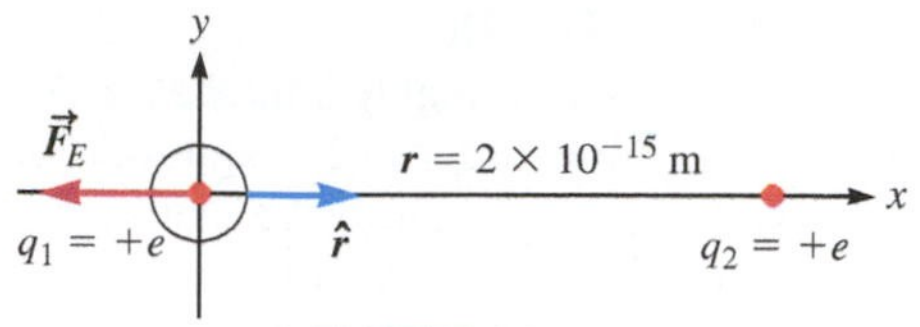

FIGURE 23.34

SOLVE

Use Equation 23.4 to find the electrostatic force on proton 1. The magnitude of the force on proton 2 is the same, but the two forces are in opposite directions.

$$\vec{F}_{[2 \text{ on } 1]} = -k\frac{q_1 q_2}{r^2}\hat{r} \qquad (23.4)$$

$$\vec{F}_E = -k\frac{e^2}{r^2}\hat{r}$$

$$\vec{F}_E = -(8.99 \times 10^9 \text{ N}\cdot\text{m}^2/\text{C}^2)\frac{(1.60 \times 10^{-19}\text{ C})^2}{(2 \times 10^{-15}\text{ m})^2}\hat{r}$$

$$\vec{F}_E = -60\,\hat{r}\text{ N}$$

CHECK and THINK

This is an enormous repulsive force acting on a particle of very low mass. If this were the only force exerted on a proton, its acceleration would be $a = F_E/m_p = (60\text{ N})/(2 \times 10^{-27}\text{ kg}) = 3 \times 10^{28}\text{ m/s}^2$. Protons subjected to such a net force would go flying out of the nucleus. Protons stay in the nucleus because another force—known as the strong nuclear force—is attractive and holds them inside.

In this chapter, we began to study a new branch of physics—electricity. When an object has a net charge, it exerts an electrostatic force on other charged objects or on neutral objects. The electrostatic force between particles is similar to gravity because both forces are inversely proportional to the square of the distance between the particles. However, unlike gravity, the electrostatic force can be either attractive or repulsive.

SUMMARY

Underlying Principles

1. **Conservation of charge.** The net charge in the Universe is constant. Charge may transfer from one object to another, but it is not created, lost, or destroyed.
2. **Electrostatic force** between charged objects is (part of) one of the fundamental forces in physics. If two objects are oppositely charged, they are mutually attracted. If two objects have charges of the same sign, they are mutually repelled. (*Opposites attract; likes repel.*)
3. **Coulomb's law.** For two charged objects that can be modeled as particles with charges q_1 and q_2, the magnitude of the electrostatic force is given by

$$F_E = k\frac{|q_1 q_2|}{r^2} \qquad (23.3)$$

where r is the distance between the particles. If polar coordinates are used so that the unit vector $\hat{r}$ points from q_1 toward q_2, Coulomb's law is written in vector form as

$$\vec{F}_{[2 \text{ on } 1]} = -k\frac{q_1 q_2}{r^2}\hat{r} \qquad (23.4)$$

Major Concepts

1. **Charge** q or Q is an intrinsic property that determines an object's role in electrical phenomena. A proton carries charge $+e$ and an electron carries charge $-e$, where e is the **elementary charge**:

$$e = (1.602\,176\,565 \times 10^{-19} \pm 000\,000\,035 \times 10^{-19})\text{C}$$

2. Charge is **quantized**.
   a. An object is **positively charged** if the number of protons is greater than the number of electrons:
   $$q = +Ne \qquad (23.1)$$
   where N is the number of excess protons.
   b. An object is **negatively charged** if the number of electrons is greater than the number of protons:
   $$q = -Ne \qquad (23.2)$$
   where N is the number of excess electrons.
3. a. A material in which electrons do not move freely is an **insulator**. In the microscopic structure of an insulator, the electrons of each particular atom stay bound to that atom.
   b. A **conductor** is a material in which the charged particles (usually electrons) can freely flow. In a metal atom, the outermost electrons are weakly bound to the nucleus. Within a sample of metal, the outermost electrons from all the atoms are free to move around in the entire sample. Typically in a metal there is one free electron per atom.
4. A **ground** is a large reservoir of charge that can accept or provide electrons freely while remaining neutral at all times. The Earth often serves as a ground. When something is connected to ground by a conductor, we say that it is **grounded**.

PROBLEM-SOLVING STRATEGY Applying Coulomb's Law

INTERPRET and ANTICIPATE

Draw a "*hybrid*" free-body diagram:

1. **Represent each object.**
2. **Represent the objects as dots.**
3. **Label information about the charges.** If no information is given, it is best to assume the particle's charge is positive.
4. **Circle the subject, and draw vectors representing the forces exerted on the subject.**
5. **Indicate the subject's acceleration.**
6. **Choose a coordinate system.**

SOLVE

The subject—with force vectors extending from it—is treated as a free-body diagram. Then Newton's second law is applied:

1. Write the electrostatic force in component form, and then
2. Algebraically combine the components.
3. When you substitute in values for the charge, you probably need only the **absolute value of the charge**.

PROBLEMS AND QUESTIONS

A = algebraic C = conceptual E = estimation G = graphical N = numerical

23-1 Another Fundamental Force

1. C What is the difference between a contact force and a field force? List all the forces presented in Chapters 1 through 22. Which are field forces? Which are contact forces? Which forces are macroscopic manifestations of the electromagnetic force? What does that tell you about contact forces? Explain your answers.

23-2 Models of Electrical Phenomena

2. C Many textbooks claim Franklin decided that moving charged particles are positive. How would you correct this claim? Think about these questions in developing your answer: Did Franklin's model include particles? What did the terms *positive* and *negative* mean to Franklin?
3. N An object has a charge of 35 nC. How many excess protons does it have?
4. E As part of a demonstration, a physics professor rubs wool against a plastic disk about the size and mass of a small dinner plate. Afterward, the disk has a charge of about -75 μC. Estimate the fractional increase in the number of electrons.
5. N A single coulomb represents a large amount of charge. A sphere has a net charge of -1.00 C. How many excess electrons does the sphere have?
6. N A sphere has a net charge of 8.05 nC, and a negatively charged rod has a charge of -6.03 nC. The sphere and rod undergo a process such that 5.00×10^9 electrons are transferred from the rod to the sphere. What are the charges of the sphere and the rod after this process?

Problems 7 and 8 are paired.

7. A glass rod is initially neutral. After it is rubbed with silk, its charge is 45.7 μC.
   a. C Has the rod gained or lost mass? Explain.
   b. N How much mass has the rod gained or lost?

8. E After an initially neutral glass rod is rubbed with an initially neutral silk scarf, the rod has a charge of 45.7 μC. Estimate the fractional increase or decrease in the scarf's mass. Would this change in mass be easily noticed?

9. N A 50.0-g piece of aluminum has a net charge of +4.20 μC. Aluminum has an atomic mass of 27.0 g/mol, and its atomic number is 13.
   a. Calculate the number of electrons that were removed from the initially neutral aluminum to produce this charge.
   b. Determine what fraction of the original number of electrons this removed number represents.
   c. By how much did the original mass of the aluminum decrease after charging?

23-3 A Qualitative Look at the Electrostatic Force

10. C You walk around a carpeted floor in your socks and get a shock when you reach for a doorknob.
    a. Use Franklin's model to explain why you feel the shock.
    b. Use the contemporary model to explain why you feel the shock. Compare your answers.

11. C A silk scarf is rubbed against glass, and a wool scarf is rubbed against plastic. (Initially, all four objects were neutral.) Afterward, it is found that the glass is attracted to the plastic. Will the silk be attracted to the wool? Explain.

23-4 Insulators and Conductors

12. C CASE STUDY A person in Franklin's time may have been able to provide the same advice we came up with in Example 23.2. However, an 18th-century person may have had a different reason for his or her advice. What part of that person's explanation would be the same as ours? What would be different?

13. C Why does the technician in Figure 23.9 cover the electrical lines with rubber and perhaps wear a rubber suit?

14. C Here is another party trick for you. Take a wooden stick, such as a chopstick. It is best if the stick has a uniform circular cross section. Suspend the stick from a string, or place it on an insulated pivot. Charge an object such as a comb or plastic stir stick by rubbing it on some material such as a sweater or your hair. Now bring the charged object near the wooden stick. What happens to the wooden stick? Explain your answer.

15. N A charge of −36.3 nC is transferred to a neutral copper ball of radius 4.35 cm. The ball is not grounded. The excess electrons spread uniformly on the surface of the ball. What is the number density (number of electrons per unit surface area) of excess electrons on the surface of the ball?

Problems 16 and 17 are paired.

16. N Two identical conductors are brought into contact. Initially, one conductor has a charge of +30.0 μC. What is the charge of each conductor afterward? Does it matter how the contact is made?

17. N Two identical insulators are brought into contact. Initially, one insulator has a charge of +30.0 μC. What is the charge of each insulator afterward? Does it matter how the contact is made?

18. C An **electrophorus** is a device developed more than 200 years ago for the purpose of charging objects. The insulator on top of a pedestal is rubbed with a cloth, such as wool (Fig. P23.18A). A conductor is placed on top of the insulator, and the conductor is connected to ground by a conducting wire (Fig. P23.18B). (The conductor has an insulating handle, so charge cannot be transferred between the person and the conductor.) The conductor is then removed (Fig. P23.18C). The conductor may then be used to transfer charge to other objects. If the insulator's charge after being rubbed with the wool is negative, what is the charge of the conductor when it is removed?

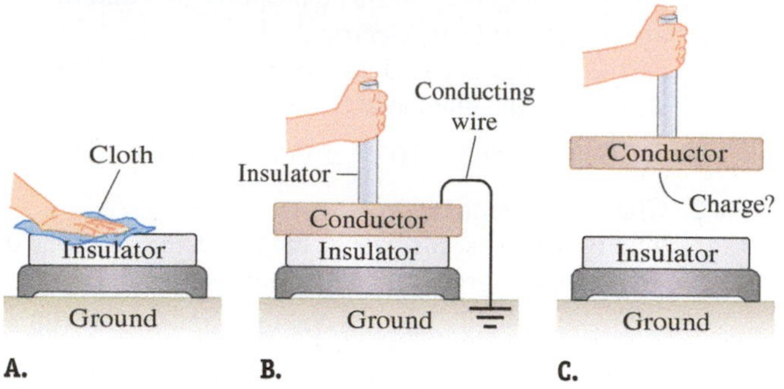

FIGURE P23.18 Problems 18 and 19.

19. C Consider the electrophorus from Problem 18. Suppose only the left half of the insulator is rubbed before the conductor is placed on top and then grounded. Then the conductor is removed. How is the conductor's charge distributed?

20. C An **electroscope** is a device used to measure the (relative) charge on an object (Fig. P23.20). The electroscope consists of two metal rods held in an insulated stand. The bent rod is fixed, and the straight rod is attached to the bent rod by a pivot. The straight rod is free to rotate. When a positively charged object is brought close to the electroscope, the straight movable rod rotates. Explain your answers to these questions:
    a. Why does the rod rotate in Figure P23.20?
    b. If the positively charged object is removed, what happens to the electroscope?
    c. If a negatively charged object replaces the positively charged object in Figure P23.20, what happens to the electroscope?
    d. If a charged object touches the top of the fixed conducting rod and is then removed, what happens to the electroscope?

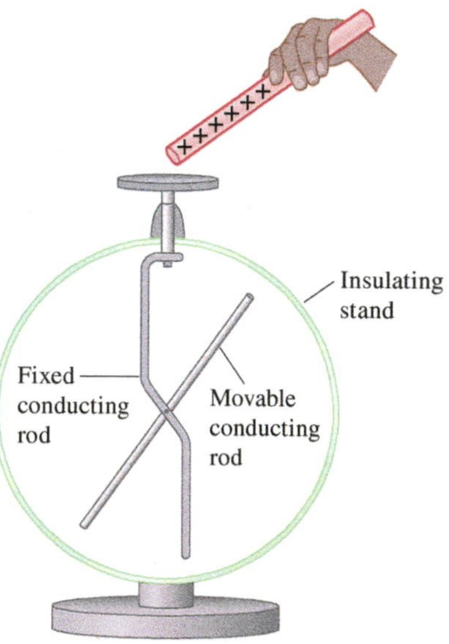

FIGURE P23.20

23-5 Coulomb's Law

21. Two particles with charges of +5.50 nC and –8.95 nC are separated by 3.00 m.
    a. N What is the magnitude of the electrostatic force between the particles?
    b. C Is this force attractive or repulsive?

22. **N** Particle A has a charge of 34.5 nC, and particle B has a charge of −54.3 nC. The attractive force between them has a magnitude of 2.70×10^{-4} N. How far apart are the particles?
23. **N** Two coins are placed on a horizontal insulating surface a distance of 1.5 m apart and given equal charges. They experience a repulsive force of 2.3 N. Calculate the magnitude of the charge on each coin.
24. **A** Show that Equation 23.5 is consistent with the rules *like charges repel* and *opposite charges attract.*
25. **A** Particle A has q_A and particle B has charge q_B. When they are separated by a distance r_i, they experience an attractive force F_i. The particles are moved without altering their charges. Now they experience an attractive force with a magnitude of $4F_i$. Find an expression for their new separation.
26. Two charged particles are placed along the y axis. The first particle, at the origin, has a charge of −25.0 μC, and the second particle, at y = 55.0 cm, has a charge of 15.0 μC.
    a. **C** Is the electric force between the two particles attractive or repulsive? Explain.
    b. **N** What is the magnitude of the electric force between the two particles?

Problems 27 and 28 are paired.

27. **N** A 1.75-nC charged particle located at the origin is separated by a distance of 0.0825 m from a 2.88-nC charged particle located farther along the positive x axis. Both particles are held at their locations by an external agent.
    a. What is the electrostatic force on the 2.88-nC particle?
    b. What is the electrostatic force on the 1.75-nC particle?
28. **N** A 1.75-nC charged particle located at the origin is separated by a distance of 0.0825 m from a 2.88-nC charged particle located farther along the positive x axis. If the 1.75-nC particle is kept fixed at the origin, where along the positive x axis should the 2.88-nC particle be located so that the magnitude of the electrostatic force it experiences is twice as great as it was in Problem 27?
29. **A** Two particles with charges q_1 and q_2 are separated by a distance d, and each exerts an electric force on the other with magnitude F_E.
    a. In terms of these quantities, what separation distance would cause the magnitude of the electric force to be halved?
    b. In terms of these quantities, what separation distance would cause the magnitude of the electric force to be doubled?
30. **A** An electron with charge $-e$ and mass m moves in a circular orbit of radius r around a nucleus of charge Ze, where Z is the atomic number of the nucleus. Ignore the gravitational force between the electron and the nucleus. Find an expression in terms of these quantities for the speed of the electron in this orbit.
31. **N** Two electrons in adjacent atomic shells are separated by a distance of 5.00×10^{-11} m.
    a. What is the magnitude of the electrostatic force between the electrons?
    b. What is the ratio of the electrostatic force to the gravitational force between the electrons?
32. Two small, identical metal balls with charges 5.0 μC and 15.0 μC are held in place 1.0 m apart. In an experiment, they are connected for a short time by a conducting wire.
    a. **C** What will be the charge on each ball after this experiment?
    b. **N** By what factor will the magnitude of the electrostatic force on either ball change after this experiment is performed?
33. **N** Two identical spheres each have a mass of 5.0 g and they are 1.0 m apart. What should be the identical charges on each of the spheres so that the magnitude of their electrostatic repulsion equals the magnitude of the gravitational force between them?
34. **N** One end of a light spring with force constant k = 125 N/m is attached to a wall, and the other end to a metal block with charge q_A = 2.00 μC on a horizontal, frictionless table (Fig. P23.34). A second block with charge q_B = −3.60 μC is brought close to the first block. The spring stretches as the blocks attract each other so that at equilibrium, the blocks are separated by a distance d = 12.0 cm. What is the displacement x of the spring?

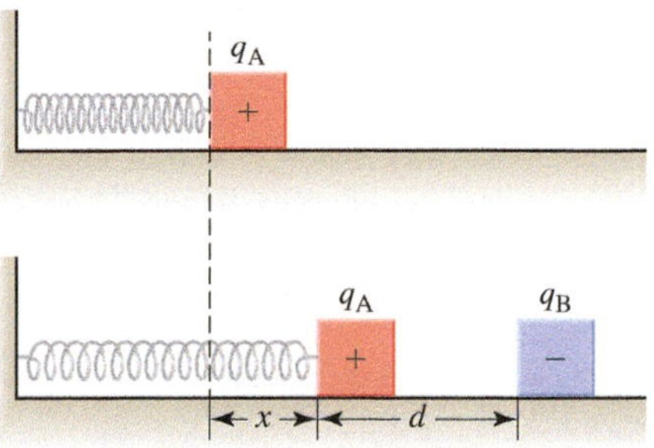

FIGURE P23.34

35. **N** Two 25.0-g copper spheres are placed 75.0 cm apart. Each copper atom has 29 electrons, and the molar mass of copper is 63.5463 g/mol. What fraction of the electrons from the first sphere must be transferred to the second sphere for the net electrostatic force between the spheres to equal 100 kN?

23-6 Applications of Coulomb's Law

36. **C** Three charged particles lie along a single line. Is it possible for one of the charges at either end to have zero net electrostatic force exerted on it? Explain.
37. **N** Given the arrangement of charged particles shown in Figure P23.37, find the net electrostatic force on the 5.00-nC charged particle located at the origin.

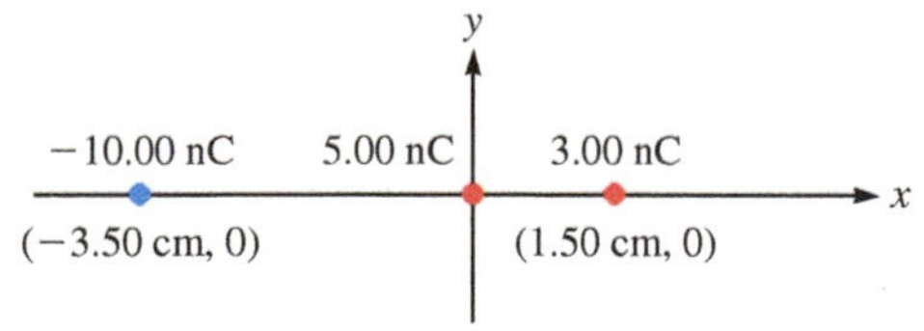

FIGURE P23.37

Problems 38 and 39 are paired.

38. **N** Given the arrangement of charged particles in Figure P23.38, find the net electrostatic force on the 5.65-μC charged particle.

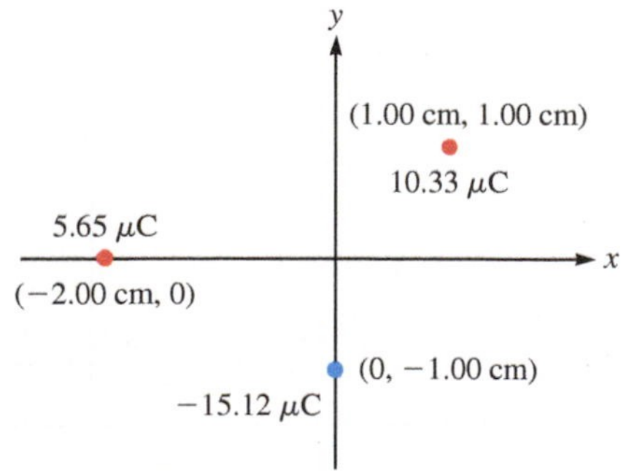

FIGURE P23.38 Problems 38 and 39.

39. **N** Given the arrangement of charged particles in Figure P23.38, find the net electrostatic force on the 10.33-μC charged particle.

40. **N** Three charged metal spheres are arrayed in the xy plane so that they form an equilateral triangle (Fig. P23.40). What is the net electrostatic force on the sphere at the origin?

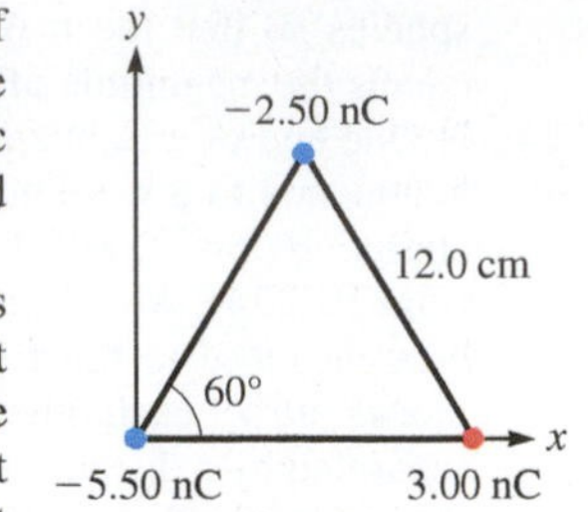

FIGURE P23.40

41. **N** Charges A, B, and C are arrayed along the y axis, with $q_A = -4.00$ nC, $q_B = 2.40$ nC, and $q_C = 5.30$ nC (Fig. P23.41). The distance between charges A and B is $y_1 = 1.00$ m, and the distance between charges B and C is $y_2 = 2.50$ m.

a. What is the net electrostatic force on charge A?

b. What is the net electrostatic force on charge B?

c. What is the net electrostatic force on charge C?

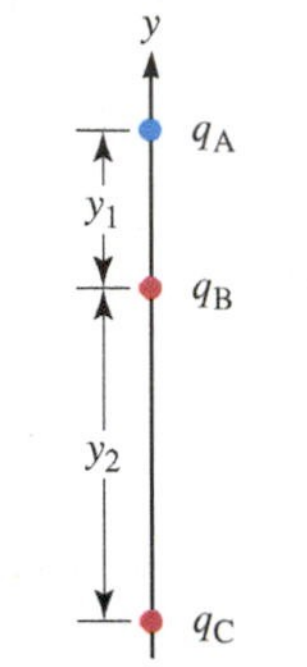

FIGURE P23.41

42. Three identical conducting spheres are fixed along a single line. The middle sphere is equidistant from the other two so that the center-to-center distance between the middle sphere and either of the other two is 0.125 m. Initially, only the middle sphere is charged, with $q_{middle} = +35.6$ nC. The middle sphere is later connected by a conducting wire to the sphere on the left. The wire is removed and then used to connect the middle sphere to the sphere on the right. The wire is again removed.

a. **C** What is the charge on each sphere?

b. **C** Which sphere experiences the greatest electrostatic force?

c. **N** What is the magnitude of that force?

43. **N** Charges A, B, and C are arranged in the xy plane with $q_A = -5.60\ \mu$C, $q_B = 4.00\ \mu$C, and $q_C = 2.30\ \mu$C (Fig. P23.43). What are the magnitude and direction of the electrostatic force on charge B?

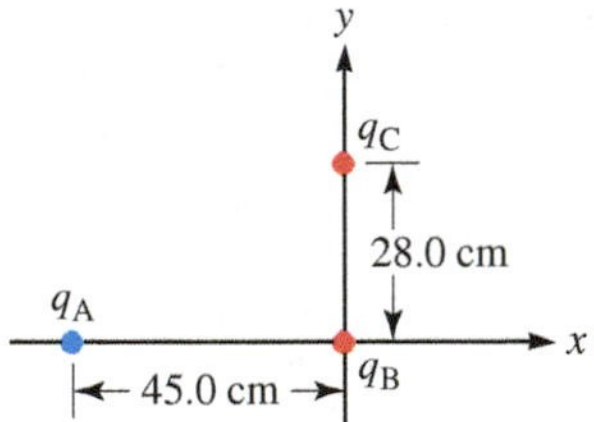

FIGURE P23.43

44. **N** In an early attempt to understand atomic structure, Niels Bohr modeled the hydrogen atom as an electron in uniform circular motion about a proton with the centripetal force caused by Coulomb attraction. He predicted the radius of the electron's orbit to be 5.29×10^{-11} m. Calculate the speed of the electron and the frequency of its circular motion.

45. **A** A particle with charge q is located at the origin, and a particle with charge $2q$ is on the positive x axis, a distance d from the origin. The particles are not free to move. In terms of q and d, at what coordinates should a third particle with charge q be placed so that it experiences no net electrostatic force?

46. **N** Figure P23.46 shows four identical conducting spheres with charge $q = +6.50$ nC placed at the corners of a rectangle of width 20.0 cm and height 80.0 cm. What are the magnitude and direction of the net electrostatic force on the charge on the lower right-hand corner?

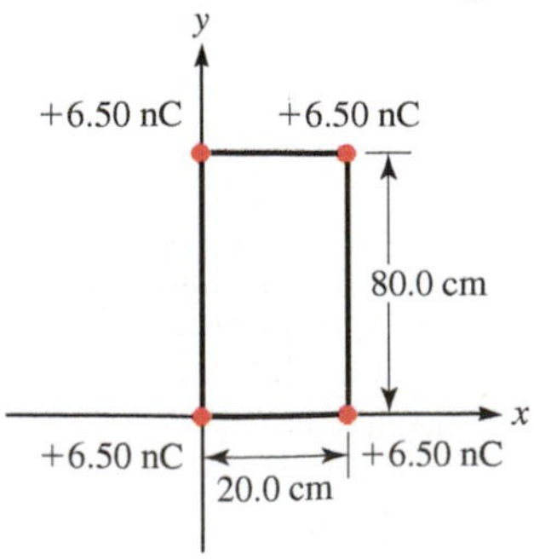

FIGURE P23.46

47. **N** A sphere of mass 5.00 g carries a positive charge of 30.0 nC and remains stationary when placed 5.00 cm directly above a second charged sphere that is fixed to a tabletop. What must be the charge of the second sphere?

48. **N** Two metal spheres of identical mass $m = 4.00$ g are suspended by light strings 0.500 m in length. The left-hand sphere carries a charge of 0.800 μC, and the right-hand sphere carries a charge of 1.50 μC. What is the equilibrium separation between the centers of the two spheres?

49. **N** Figure P23.49 shows two identical small, charged spheres. One of mass 4.0 g is hanging by an insulating thread of length 20.0 cm. The other is held in place and has charge $q_1 = -3.6\ \mu$C. The thread makes an angle of 18° with the vertical, resulting in the spheres being aligned horizontally, a distance r apart. Determine the charge q_2 on the hanging sphere.

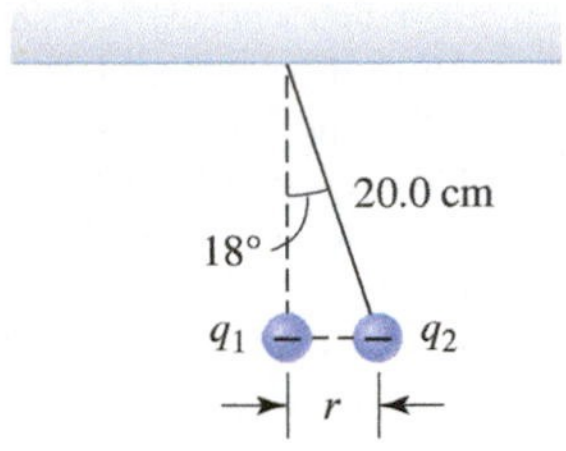

FIGURE P23.49

50. Two small spherical conductors are suspended from lightweight vertical insulating threads. The conductors are brought into contact (Fig. P23.50, left) and released. Afterward, the conductors and threads stand apart as shown at right.

a. **C** What can you say about the charge of each sphere?

b. **N** Use the data given in Figure P23.50 to find the tension in each thread.

c. **N** Find the magnitude of the charge on each sphere.

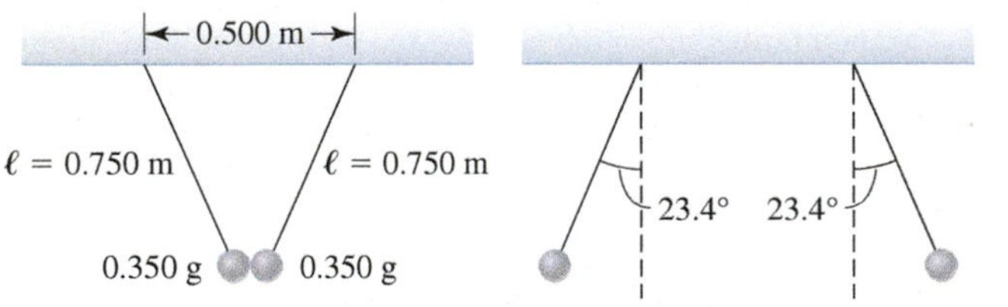

FIGURE P23.50

Problems 51 and 52 are paired.

51. **A** Four equally charged particles with charge q are placed at the corners of a square with side length L, as shown in Figure P23.51. A fifth charged particle with charge Q is placed at the center of the square so that the entire system of charges is in static equilibrium. What are the magnitude and sign of the charge Q?

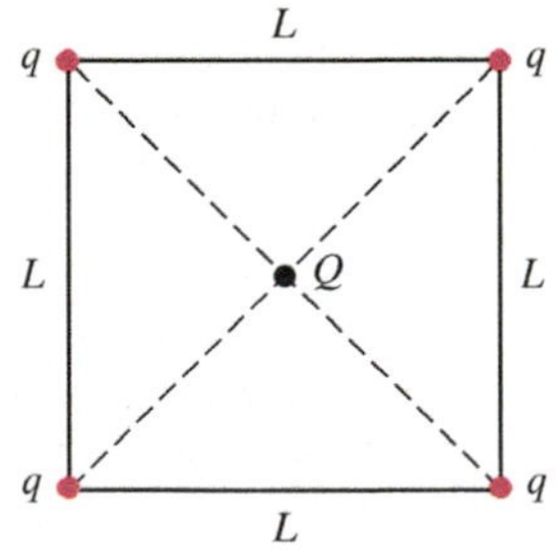

FIGURE P23.51

52. **C** Four charged particles q, $-q$, q, and $-q$ are fixed at the corners of a square with side length L as shown in Figure P23.52. If another charged particle of magnitude Q is placed at the center of the square, will it be in static equilibrium? Does the sign of the charge Q matter? Explain.

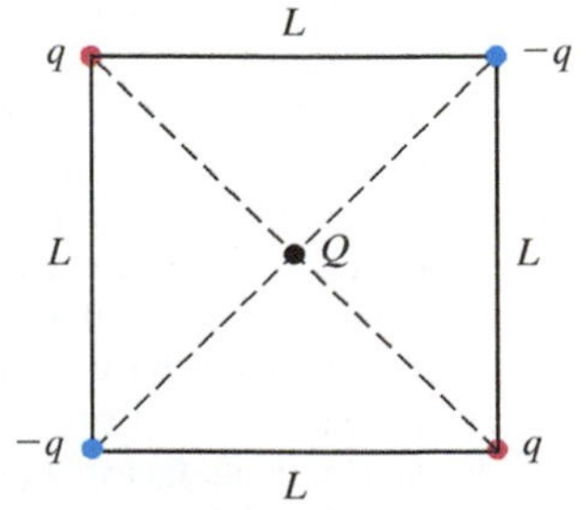

FIGURE P23.52

53. **N** A metal sphere with charge +8.00 nC is attached to the left-hand end of a nonconducting rod of length $L = 2.00$ m. A second sphere with charge +2.00 nC is fixed to the right-hand end of the rod (Fig. P23.53). At what position d along the rod can a charged bead be placed for the bead to be in equilibrium?

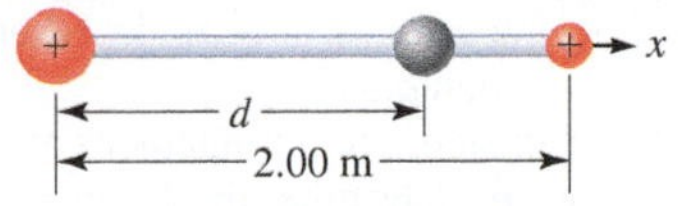

FIGURE P23.53

54. **A** A particle with charge q is located at the origin, and a particle with charge $-2q$ is on the positive x axis, a distance d away from the origin. The particles are not free to move. In terms of q and d, at what coordinate should a third particle with charge q be placed so that it experiences no net electrostatic force?

55. **A** Three small metallic spheres with identical mass m and identical charge $+q$ are suspended by light strings from the same point (Fig. P23.55). The left-hand and right-hand strings have length L and make an angle θ with the vertical. What is the value of q in terms of k, g, m, L, and θ?

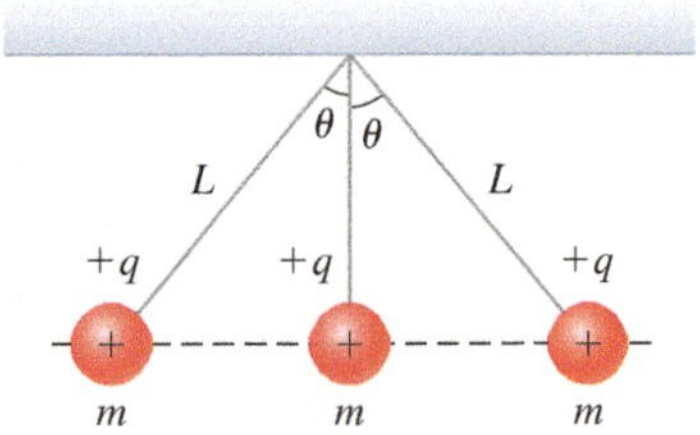

FIGURE P23.55

General Problems

56. **C** How does a negatively charged rubber balloon stick to a neutral wooden ceiling?

57. **N** How many electrons are in a 1.00-g electrically neutral steel paper clip? The molar mass of steel is approximately that of iron, or 55.845 g/mol, and a neutral iron atom has 26 electrons.

58. The Ni^{2+} and S^{2-} ions in nickel sulfide are separated by a distance of 0.680 nm.

a. **N** What is the magnitude of the electrostatic force between the ions in nickel sulfide?

b. **C** How would the magnitude of the electrostatic force found in part (a) change if the nickel ion was replaced by an iron (Fe^{2+}) ion?

59. **N** A small metal sphere with a charge of 2.25 μC is placed at the origin, and a second sphere with unknown charge q is placed at $y = 80.0$ cm. A negatively charged sphere is found to be in equilibrium when placed at $y = 62.0$ cm. What is the charge q on the second sphere?

60. **N** Two otherwise identical, small conducting spheres have charges $+5.0$ μC and -2.0 μC. When placed a distance r apart, each experiences an attractive force of 3.0 N. The spheres are then touched together and moved back to a distance r apart. Find the magnitude of the new force on each sphere.

61. **N** Three charged particles are arranged in the xy plane as shown in Figure P23.61, with $q_A = 6.40$ μC, $q_B = -2.30$ μC, and $q_C = 3.80$ μC. What is the net electrostatic force on the particle with charge q_A?

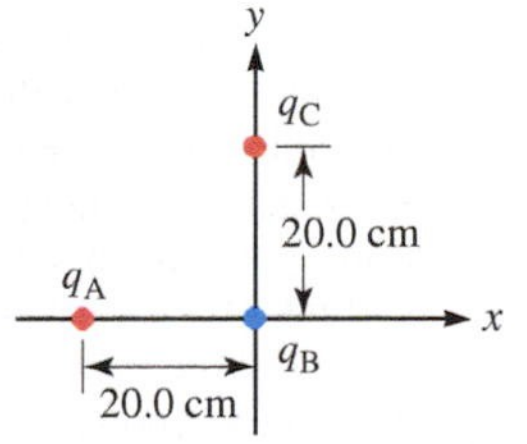

FIGURE P23.61

62. **A** We saw in Figure 23.16 that a neutral metal can was attracted to a positively charged glass rod because the rod polarized the can. Let's model this situation in a very rough sense. Suppose a charged sphere C is held near two other charged spheres A and B as shown in Figure P23.62. Let's call the magnitude of the electrostatic force exerted by sphere C on A "F_A" and that on B "F_B." Derive an expression for the ratio F_A/F_B. Comment on why a neutral conductor is attracted to a charged object.

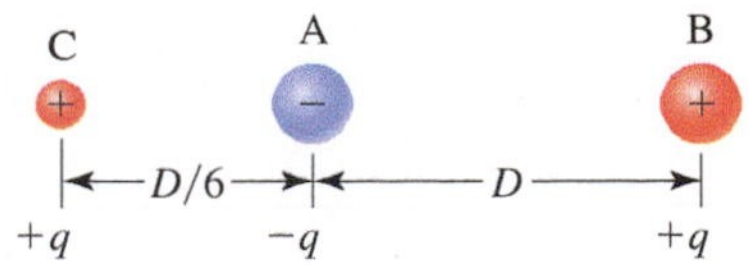

FIGURE P23.62

63. **N** A $+4.0$-μC charged particle is located at the origin of a coordinate system, and a -1.0-μC charged particle is located at $x = 2.0$ cm.

a. Calculate the net electrostatic force on a third particle with charge $+2.0$ μC located at $x = 5.0$ cm.

b. Compare your answer to part (a) with the force due to a $+3.0$-μC charged particle located at the origin on a $+2.0$-μC charged particle at $x = 5.0$ cm.

64. **E** In Figure P23.64, a boy's hair is attracted to his comb after the comb has been run through his hair. Estimate the amount of charge transferred between his raised hair and the comb.

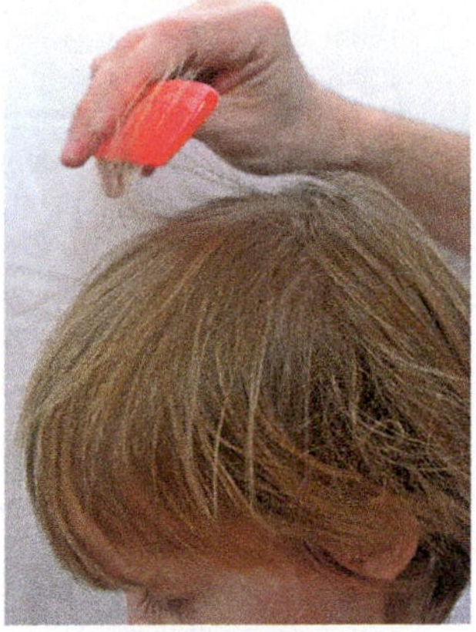

Peter McGahey/Cengage Learning

FIGURE P23.64

65. **A** Figure P23.65 shows two identical conducting spheres, each with charge q, suspended from light strings of length L. If the equilibrium angle the strings make with the vertical is θ, what is the mass m of the spheres?

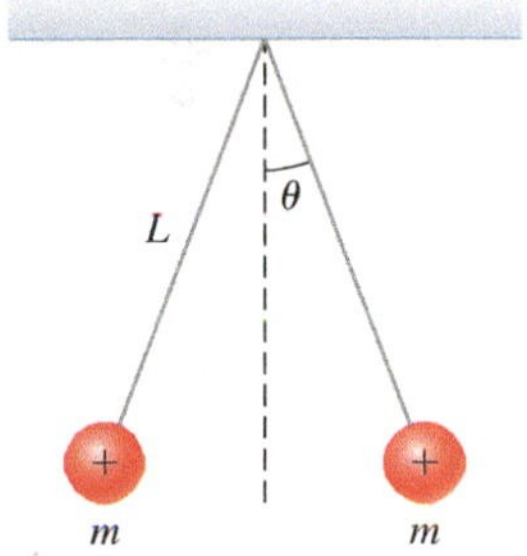

FIGURE P23.65

Problems 66 and 67 are paired.

66. **N** Two helium-filled, spherical balloons, each with charge q, are tied to a 5.00-g mass with strings of negligible mass, and the system floats in equilibrium as shown in Figure P23.66. The distance between the balloons is 60.0 cm, and the strings are 100.0 cm long. Ignore the weight of the balloon material, and assume that the density of air is 1.29 kg/m^3 and the density of helium inside the balloons is 0.200 kg/m^3.

a. Find the magnitude of the charge q on each balloon. Assume that the charge on each balloon acts as if it were concentrated at its center.

b. Find the volume of each balloon.

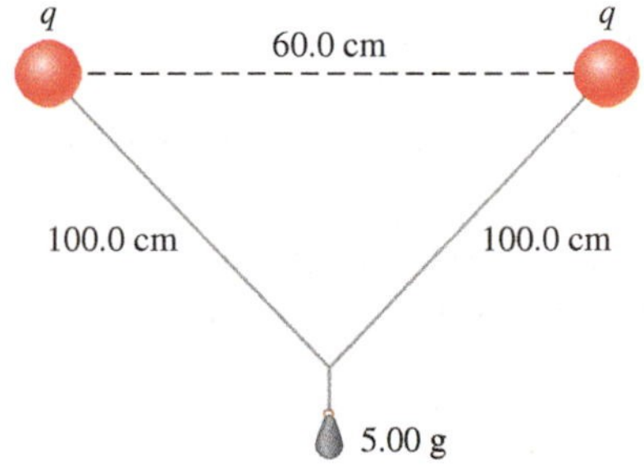

FIGURE P23.66

67. **N** Two small metallic spheres, each with a mass of 2.00 g, are suspended from a common point by two strings of negligible mass and of length 10.0 cm. When the spheres have an equal amount of charge, the two strings make an

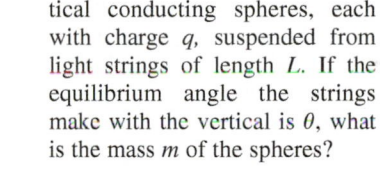

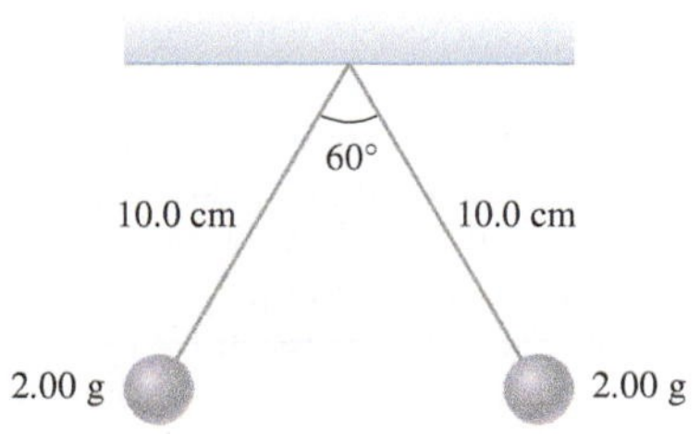

FIGURE P23.67

angle of 60° with each other as shown in Figure P23.67. Calculate the magnitude of the charge on each sphere.

68. **A** Two positively charged spheres with charges $4e$ and e are separated by a distance L and held motionless. A third charged sphere with charge Q is set between the two spheres and along the line joining them. The third sphere is in static equilibrium. What is the distance between the third charged sphere and the sphere that has charge $4e$?

69. **N** The two ends of a light spring with force constant $k = 145$ N/m are connected to identical metal blocks that are at rest on a horizontal, frictionless table. The equilibrium length of the spring is $x_0 = 34.0$ cm. Electrons are slowly stripped from both blocks, giving each an identical charge $+q$. The repulsive electric force between the blocks stretches the spring by 14.0 cm. Assuming the blocks can be modeled as particles, what is the charge q on each block?

70. **N** Three charged spheres are at rest in a plane as shown in Figure P23.70. Spheres A and B are fixed, but sphere C is attached to the ceiling by a lightweight thread. The tension in the string is 0.240 N. Spheres A and B have charges $q_A = 28.0$ nC and $q_B = -28.0$ nC. What charge is carried by sphere C?

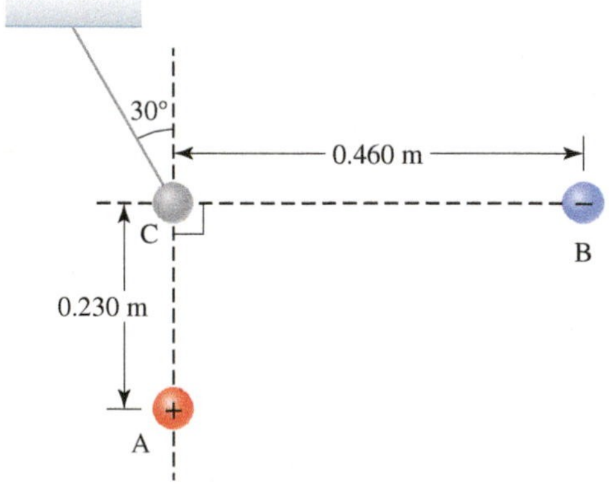

FIGURE P23.70

71. **N** A small sphere with charge $Q = 2.0\ \mu$C is held in place directly above another similar sphere with total charge $q = -1.0\ \mu$C. The lower sphere has a mass of 40 g and is suspended in place by the electrostatic force of attraction with the upper sphere. Calculate the distance r between the centers of the spheres.

72. **N** Three particles with charges of 1.0 μC, $-1.0\ \mu$C, and 0.50 μC are placed at the corners A, B, and C of an equilateral triangle with side length 0.10 m as shown in Figure P23.72. Find the net force on the charge at point C.

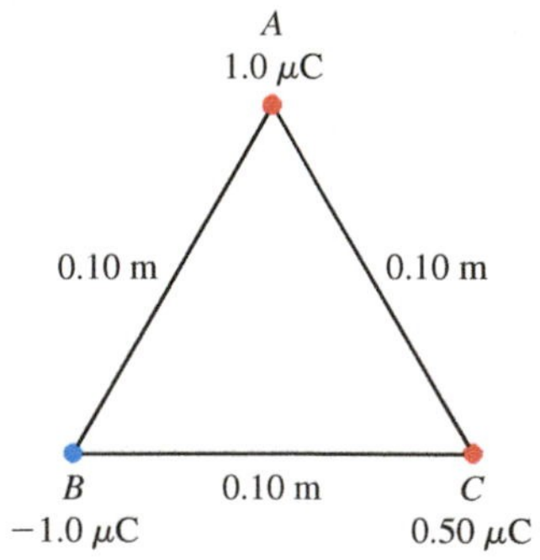

FIGURE P23.72

Problems 73 and 74 are paired.

73. **A** Two positively charged particles, each with charge Q, are held at positions $(-a, 0)$ and $(a, 0)$ as shown in Figure P23.73. A third positively charged particle with charge q is placed at $(0, h)$.

a. Find an expression for the net electric force on the third particle with charge q.

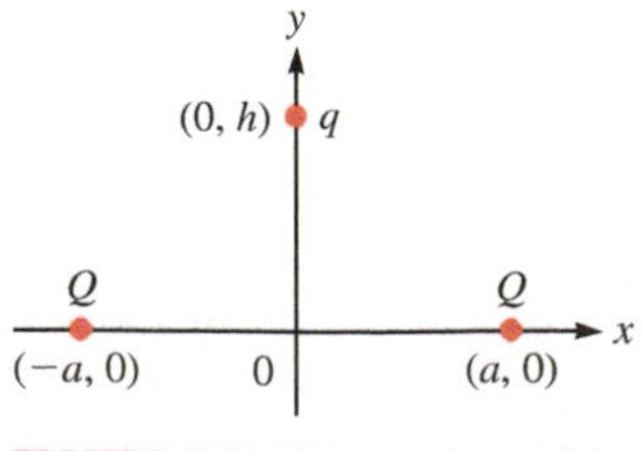

FIGURE P23.73 Problems 73 and 74.

b. Show that the two charges Q behave like a single charge $2Q$ located at the origin when the distance h is much greater than a.

74. Consider the arrangement of charges in Problem 73.

a. **A** Determine the location on the y axis where the charge q will experience the greatest force.

b. **G** Sketch the magnitude of the force on charge q as a function of y for positive values of y.

75. **N** Eight small conducting spheres with identical charge $q = -2.00\ \mu$C are placed at the corners of a cube of side $d = 0.500$ m (Fig. P23.75). What is the total force on the sphere at the origin (sphere A) due to the other seven spheres?

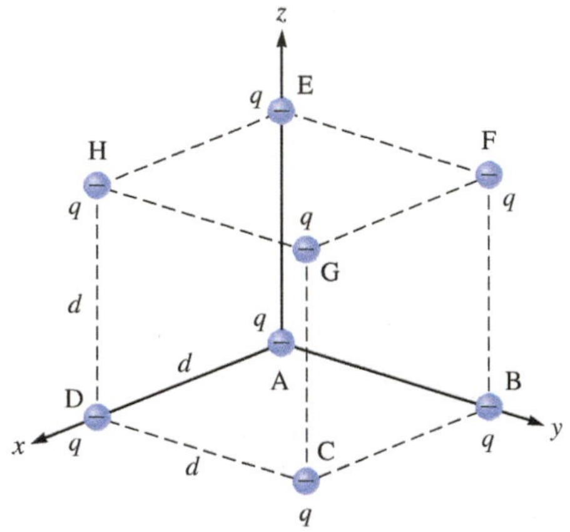

FIGURE P23.75

Problems 76, 77, and 78 are grouped.

76. **C** Two identical spheres carry identical positive charge. Static friction between the spheres and the level surface they sit on just barely keeps them at rest. Then they are placed closer together on a different level surface, and they are barely at rest due to static friction. Is the coefficient of static friction between a sphere and the second surface higher or lower than on the first? Explain.

77. **A** Two identical spheres of mass m carry identical positive charge q. Static friction between the spheres and the level surface they sit on just barely keeps them at rest. Their center-to-center separation is r. Find an expression for the coefficient of static friction between a sphere and the surface.

78. **N** Two identical spheres of mass $m = 7.5 \times 10^{-5}$ kg carry an identical positive charge $q = 89$ nC. Static friction between the spheres and the level surface they sit on just barely keeps them at rest. Their center-to-center separation is $r = 0.50$ m. Find the coefficient of static friction between a sphere and the surface.

79. **A** Two positive charges, each with charge Q, are held at positions $x = -a$ and $x = +a$ along the x axis of a coordinate system. A small sphere of mass m and positive charge q is placed at the origin, but it can move freely back and forth on the x axis. Show that if the sphere is displaced by an amount $x \ll a$ and released, it will undergo simple harmonic motion about the origin with an angular frequency given by $\omega = \sqrt{4kQq/ma^3}$.

Electric Fields

24

Underlying Principles

Electrostatic field

Major Concepts

1. Source and subject
2. Electric dipole
3. Electric dipole moment
4. Electrostatic force exerted by electrostatic field

Special Cases

1. Electric field sources
   a. Particle (or spherical object)
   b. Dipole
   c. Charged rod
   d. Charged ring
   e. Charged disk
   f. Near a charged surface
2. Dipole in an electric field

Tools

Electric field lines

Key Questions

What is an electric field?

How can we find expressions for the electric field due to sources of various geometries?

How do we find the electrostatic force exerted on subjects in an electric field?

An electric precipitator cleans the air on a submarine; a drop of ink in your printer is accelerated toward the paper; a potato cooks in a microwave oven. These objects are accelerated by electrostatic forces. The charged objects used inside many common devices are not spherical, so we cannot use Coulomb's law or the methods from Chapter 23 to calculate the forces involved. In this chapter, we learn to account for the nonspherical shapes that often play a significant role in electrostatics. To do this, we'll introduce the concept of an electrostatic field, which is analogous to the gravitational field.

24-1 What Are Fields?

The best way to study nonspherical shapes is by calculating the *field* they produce. To understand what a field is, we start by reviewing the more familiar gravitational field. Suppose you were thinking about placing several spacecraft with different masses into orbit at some distance r from the center of the Earth (Section 7-4). You would need to find the gravitational force exerted by the Earth on each spacecraft using Newton's law of universal gravity. Following the form of Equation 7.4, we write

$$\vec{F}_G(r) = -G\frac{M_\oplus m_{SC}}{r^2}\hat{r} \tag{24.1}$$

where the unit vector $\hat{r}$ points outward from the Earth toward a spacecraft, and m_{SC} is the mass of a particular spacecraft.

Because the gravitational force exerted on each spacecraft depends only on its mass m_{SC}, it is helpful to rewrite Equation 24.1 as

$$\vec{F}_G(r) = m_{SC}\left(-G\frac{M_\oplus}{r^2}\hat{r}\right) = m_{SC}\,\vec{g}(r)$$

where $\vec{g}(r)$ is known as the **gravitational field** of the Earth for points outside the Earth or on its surface, $r \geq R_\oplus$, as given by Equation 7.12 (page 197):

$$\vec{g}(r) \equiv -\frac{GM_\oplus}{r^2}\hat{r} \tag{7.12}$$

The dimensions of the gravitational field are force per mass.

The influence exerted by some physical source over a region of space is called a **field**. In this case, the field's region is the environment extending from the Earth's surface outward and the source is the Earth. Every position in a field is assigned a particular value of some quantity. For example, a temperature field is a *scalar field* and every position is associated with a temperature. If the field is a **vector field**, each position in the field is associated with a particular magnitude and direction. The Earth's gravitational field is a vector field, and Equation 7.12 assigns a magnitude and direction to every position in that field.

The word *vector* is often omitted when we refer to specific vector fields, so we use *gravitational field* rather than *gravitational vector field*. The term *field* also refers to the mathematical description of the source's influence.

Although the terms *force* and *field* may sound alike, they have different meanings. Let's use this case of spacecraft near the Earth (Fig. 24.1) to distinguish between the two:

1. The field has a **source** that influences its surroundings. We are interested in knowing the gravitational force on a variety of spacecraft, so we choose the Earth as the source of the gravitational field. At each position r, the magnitude and direction of the Earth's gravitational field $\vec{g}(r)$ **depend only on properties of the source**—in this case, the Earth's mass $M_\oplus$. In fact, even if all the spacecraft vanished, $\vec{g}(r)$ would still have the same magnitude and direction given by Equation 7.12.

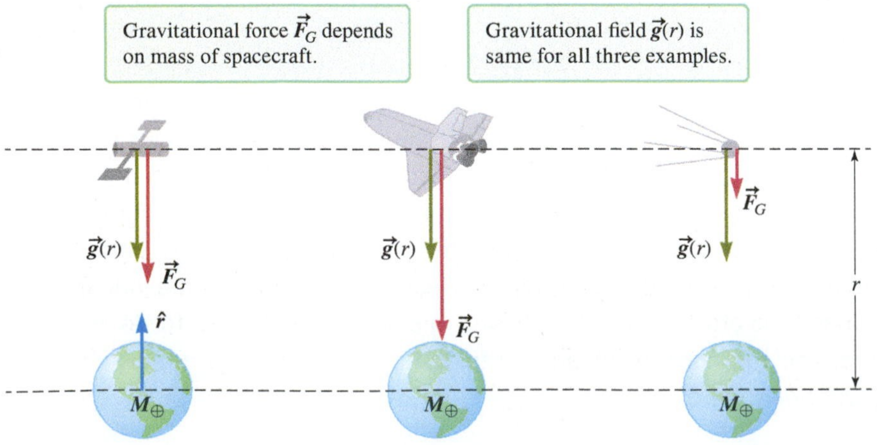
Gravitational force $\vec{F}_G$ depends on mass of spacecraft.
Gravitational field $\vec{g}(r)$ is same for all three examples.
$\vec{g}(r)$
$\vec{F}_G$
$\hat{r}$
$M_\oplus$
r

FIGURE 24.1 The gravitational field of the Earth and the gravitational force exerted on three different spacecraft. The gravitational field is the same in all three cases, but $\vec{F}_G$ is strongest for the middle spacecraft, because it has the greatest mass, and weakest for the rightmost spacecraft, because it has the smallest mass.

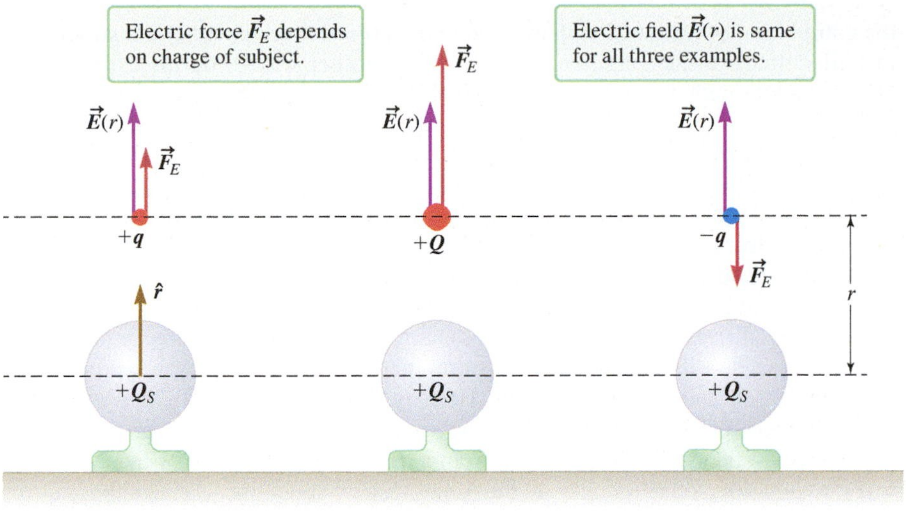
Electric force $\vec{F}_E$ depends on charge of subject.
Electric field $\vec{E}(r)$ is same for all three examples.
$\vec{E}(r)$
$\vec{F}_E$
$+q$
$+Q$
$-q$
$\hat{r}$
r
$+Q_S$

FIGURE 24.2 The electrostatic field of a charged sphere and the electrostatic force exerted on various charged particles. The electrostatic field is the same in all three cases, but the electrostatic force varies, depending on the charged particle.

2. **Subjects** are objects that may be placed in the source's field. The **force** exerted on a subject **depends on properties of the field and the subject.** In this case, the gravitational force exerted on the spacecraft depends on the mass m_{SC} of the spacecraft and the Earth's gravitational field $\vec{g}(r)$ at the location of the spacecraft.

SOURCE AND SUBJECT

Major Concept

Like gravity, the electrostatic force is also a field force. We apply the concept of a field to the electrostatic force by analogy with gravity. Figure 24.2 shows a large sphere resting on an insulated pedestal. The sphere has positive charge $+Q_S$. We consider the sphere to be the source of the field. Three subjects—a particle with charge $+q$, a sphere with a greater charge $+Q$ $(>q)$, and a particle with charge $-q$—are each placed in turn a distance r from the center of the source.

The electrostatic force exerted by the source on each subject depends on both the source's charge $+Q_S$ and the subject's charge q_{subject}. The electrostatic force $\vec{F}_E$ is strongest for the middle subject because it has the greatest charge. Also notice that $\vec{F}_E$ on the rightmost subject points toward the source because that subject has a negative charge, whereas for the other two subjects $\vec{F}_E$ points away from the source because both source and subjects are positively charged.

The **electrostatic *field*** (also known as the **electric *field***) $\vec{E}(r)$ does not depend on the subject at all. The electrostatic field $\vec{E}(r)$—like the gravitational field—is a vector field that depends only on properties of the source and position r from the source.

The concept of a field is very convenient when we want to study the influence of one particular source on a number of different subjects. The field concept is also helpful when we consider sources of different shapes. In our study of gravity and in electrostatics so far, we have considered only sources with spherical symmetry (planets, moons, spheres, shells, and particles). However, in electrostatics, we are often interested in sources that are *not* spherical. For example, in the next case study, we explore the 18th-century controversy surrounding the best shape to use for lightning rods.

CASE STUDY The Shape of Lightning Rods

In the middle of the 18th century, people loved to play with electricity (Fig. 23.4, page 685), but most people thought lightning was another phenomenon altogether—explosions of atmospheric gas, something like the explosions of gunpowder.

Benjamin Franklin thought otherwise. He believed that lightning was a colossal electrical spark just like the small sparks people found so amusing. To support his theory, he suggested that a metal rod be placed on top of a tall structure to capture "electric fluid"—what we call charged particles. While Franklin was waiting for

FIGURE 24.3 Franklin's kite experiment showed that lightning is a giant electrical spark. His son William was outside, assisting him in flying the kite.

the completion of Christ Church in Philadelphia (there were no other tall structures in Philadelphia at the time), he came up with another way to do his experiment. He used a kite with a metal wire attached (Fig. 24.3). The wire was connected to a string, which when wet would act as a conductor. At the end of the string was a metal key connected to a Leyden jar—a device used to store charge. (Leyden jars were among the props normally used in demonstrations of electricity at that time.) Franklin held a piece of dry silk, which insulated him, and then proceeded to collect charge from his flying kite. Franklin showed that a Leyden jar charged by clouds produced all the same effects as Leyden jars charged in the home. So he concluded that lightning is an electrical phenomenon, like a giant spark.

Don't try this experiment yourself. Franklin took a number of precautions—for example, he didn't fly the kite during a storm but only as the storm was approaching. He kept himself and the silk string dry. He may have just been lucky, however. The Swedish professor George William Richman was killed performing a similar experiment when a spark about a foot long jumped from a metal rod to Richman's head.

Franklin—like other people at the time—knew about the dangers posed by lightning. As people started building taller structures, those structures became more likely targets of lightning strikes that could cause fires, destroying property and lives. In 1769, lightning struck a British gunpowder magazine, killing more than a thousand people and destroying a whole town. Once Franklin understood that lightning was a giant spark, he invented a way to protect against lightning strikes—the lightning rod.

Franklin published the following recommendations for lightning rods in *Poor Richard* (1753):

1. Just outside each building an iron rod **should be planted 3 feet to 4 feet into the moist ground.**
2. The rod should extend 6 feet to 8 feet **above the tallest part of the structure.**
3. On top of the rod should be a foot of brass wire **sharpened to a fine point.**

24-2 Special Case: Electric Field of a Charged Sphere

We wish to find the electric fields generated by sources of various shapes. We start by finding the electric fields outside sources that have spherical symmetry (solid spheres, spherical shells, and particles).

Compare Figures 24.1 and 24.2. Because the charged, spherical source has the same spherical shape as the Earth, we can rely on our work on the Earth's gravitational field. The electrostatic force exerted by the source on each subject is found by using Coulomb's law:

$$\vec{F}_E = k\frac{Q_S q_{\text{subject}}}{r^2}\hat{r} \tag{23.5}$$

where the unit vector $\hat{r}$ points outward from the source (at the origin) toward the subject, and q_{subject} is the charge of a particular subject.

The electrostatic force $\vec{F}_E$ exerted on the three possible subjects in Figure 24.2 depends on each subject's charge q_{subject}. So—just as we rewrote Newton's law of gravity—we can rewrite Coulomb's law as

$$\vec{F}_E(r) = q_{\text{subject}}\left(k\frac{Q_S}{r^2}\hat{r}\right)$$

$$\vec{F}_E(r) = q_{\text{subject}}\vec{E}(r) \tag{24.2}$$

where $\vec{E}(r)$ is the electric field generated outside of any spherical source and

ELECTRIC FIELD DUE TO A CHARGED PARTICLE Special Case

$$\vec{E}(r) = k\frac{Q_S}{r^2}\hat{r} \tag{24.3}$$

Like $\vec{g}(r)$, $\vec{E}(r)$ depends only on the source and the distance r from that source. In fact, if the three subjects in Figure 24.2 vanished, $\vec{E}(r)$ would be unchanged; *the electric field* $\vec{E}(r)$ *never depends on the properties of the subject.*

Compare the electric field (Eq. 24.3) with the Earth's gravitational field (Eq. 7.12). Mathematically, these two fields are similar:

1. The magnitude of each field drops off with the distance squared.
2. The direction is radial (along $\hat{r}$). In the case of the gravitational field, the direction is given by $-\hat{r}$ (toward the source), whereas for the electrostatic field, the direction is $+\hat{r}$ (outward from the source toward the subject) if the source has a positive charge.
3. Both fields have a constant of proportionality (G in the case of gravity and k in the case of the electrostatic field).
4. Each field depends on only one property of the source (the mass of the Earth in the case of gravity and the charge of the source in the case of electrostatics).

There is one very important difference between the electrostatic field and the gravitational field. The gravitational field always points toward the center of the source because there is only one kind of mass. However, there are two kinds of charge—positive and negative. If the source is positive, the electrostatic field points away from the source (Fig. 24.4). If source is negative, the electrostatic field points toward the source.

Figure 24.4 and Equation 24.2 can be used to develop a general prescription for finding the electric field due to a source of any shape. Imagine a test particle with a small amount of positive charge $+q_{\text{test}}$. You move this particle to various locations in the field of the source (Fig. 24.4). For each position, you measure or calculate the electrostatic force exerted on the test particle. Because the test particle is positive, the electrostatic force points in the same direction as the electrostatic field. The magnitude of the electric field is found from Equation 24.2:

$$E(r) = \frac{F_E(r)}{q_{\text{test}}} \tag{24.4}$$

where the test particle is the subject and its charge is always positive: $q_{\text{test}} > 0$. So the magnitude of the electric field vectors is directly proportional to the magnitude of the electrostatic force vectors. We see from Figure 24.4C that outside a positively

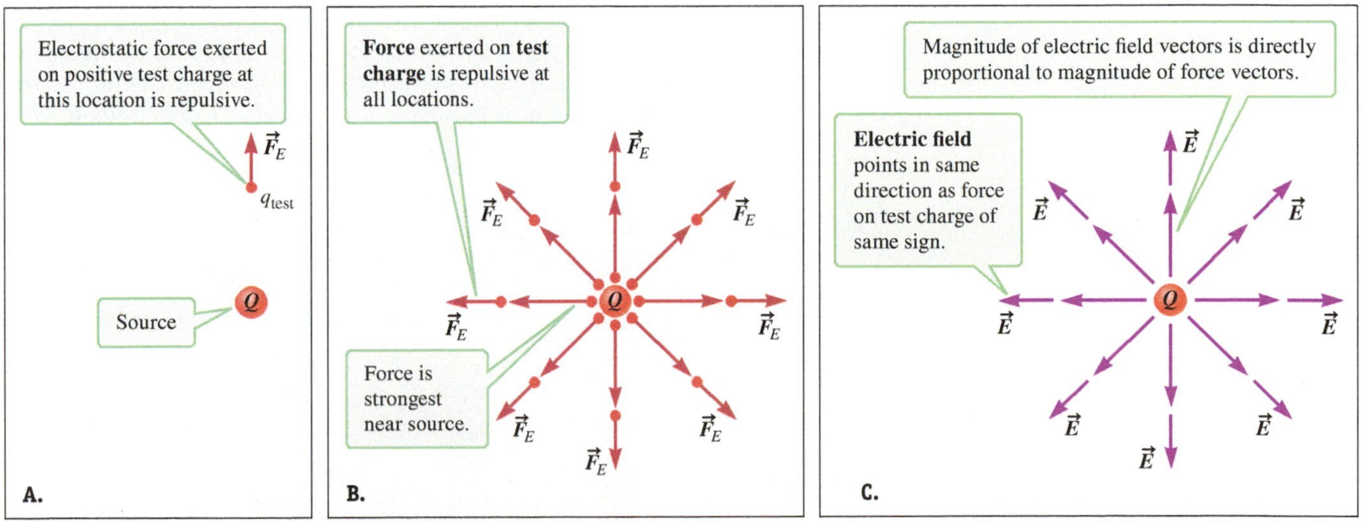

FIGURE 24.4 **A.** The electric force exerted by a positively charged sphere on a positively charged test particle. **B.** The electric force exerted on a test particle is repulsive at 16 different locations. **C.** The electric field due to the sphere at all 16 locations points outward. The magnitude of the electric field vectors is directly proportional to the magnitude of the force vectors, so the field is strongest near the source. Compare to Figure 7.19 (page 197) for the gravitational field.

charged spherical source, the electrostatic field points radially away from the source and is strongest near the source.

We see from Equation 24.4 that the dimensions of the electric field are force per charge, so the SI units of electric field are N/C. Some values of the electric fields associated with various sources are given in Table 24.1.

TABLE 24.1 Rough magnitude of the electric field near various sources.

Source	Magnitude of E (N/C)
Near surface of uranium nucleus	3×10^{21}
At orbit of ground-state electron in hydrogen atom	6×10^{11}
Near surface of charged comb or balloon rubbed on hair	10^3
Earth's lower atmosphere (fair weather)	10^2

CONCEPT EXERCISE 24.1

In a few sentences, explain how you know that $\vec{E}(r) = (kQ_S/r^2)\hat{r}$ (Eq. 24.3) is consistent with Figure 24.4C.

CONCEPT EXERCISE 24.2

What is the magnitude of the electric field due to a charged particle at its exact location ($r = 0$)?

24-3 Electric Field Lines

ELECTRIC FIELD LINES ⦿ Tool

We draw the arrowhead in the middle of the line, not at the end. This helps us avoid confusing field lines with vectors.

As with all vectors, the length of an arrow represents the vector's magnitude. In Figure 24.4C, longer vectors show that the electric field is stronger near the source, and shorter vectors show that the electric field farther from the source is weaker. Drawing vectors for each situation takes considerable time and some skill, because many must be carefully drawn at a number of locations. A much easier way to represent a vector field is by drawing *field lines* as in Figure 24.5. **Electric field lines** provide a way to visualize the magnitude and direction of the electric field. *Electric field lines must either originate at a positively charged object or terminate on a*

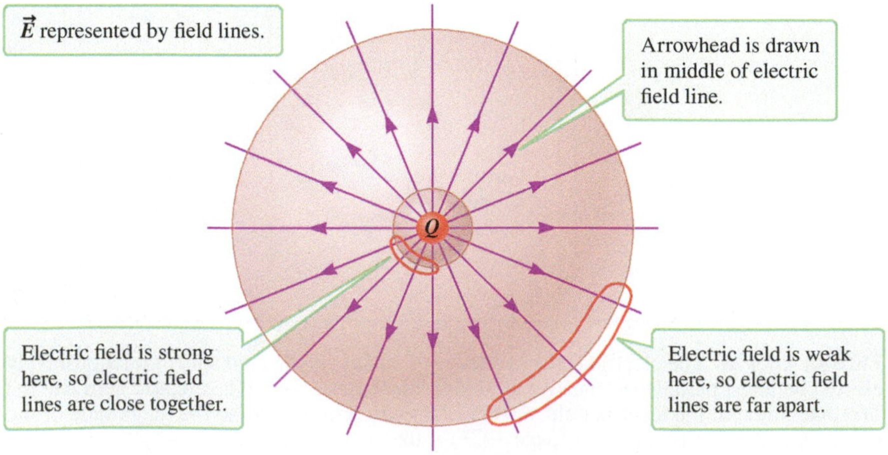
$\vec{E}$ represented by field lines.
Arrowhead is drawn in middle of electric field line.
Q
Electric field is strong here, so electric field lines are close together.
Electric field is weak here, so electric field lines are far apart.

FIGURE 24.5 The electric field of a spherical positive source represented by field lines. The surfaces of the two concentric spheres are perpendicular to the field lines. Compare this to Figure 24.4C, which is the arrow representation.

negatively charged object (or both). For example, in Figure 24.5, each electric field line originates at the positively charged source.

The magnitude of the electric field is related to the density of the electric field lines. To find the density of the electric field lines, imagine a set of surfaces perpendicular to the field lines; the more lines per unit area passing through one of these surfaces, the stronger the electric field. In Figure 24.5, the electric field lines are farther apart (less dense) in the region far from the source, showing that the electric field is weaker there.

The direction of the electric field at any position is tangent to the electric field line and in the direction indicated by the arrowhead on the field line. Figure 24.6 shows electric field lines along with vectors representing the electric fields at four locations. Notice that the electric field lines cannot cross because then there would be two possible directions for the electric field at their point of intersection.

When representing electric fields visually, we are often forced to draw two-dimensional representations of three-dimensional situations. The electric field in Figure 24.5 is three-dimensional; we must imagine the third dimension because we cannot draw it on a two-dimensional surface. Likewise, you must imagine that the vector arrows in Figure 24.4C emerge from the paper (or screen) both toward and away from you so that they point radially outward in all directions. Electric field lines from a spherical positive charge extend outward in all directions like the quills of a balled-up porcupine.

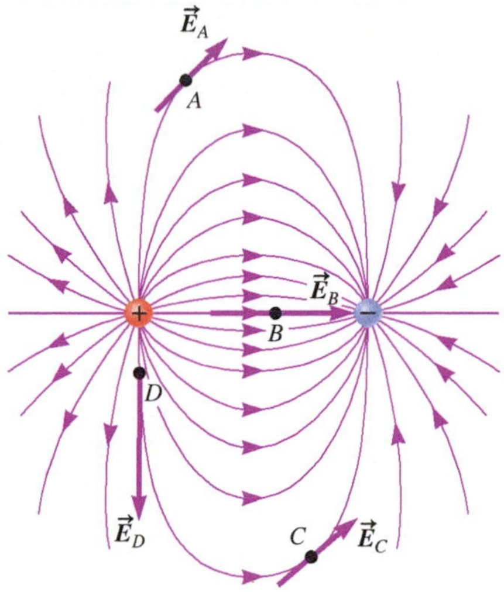

FIGURE 24.6 The electric field at any point such as *A*, *B*, *C*, or *D* is tangent to an electric field line. The direction of the field is determined by the arrow on the field line. The magnitude of the electric field is represented by the field line density. Here $E_D > E_B > E_A = E_C$.

CONCEPT EXERCISE 24.3

Which lines in Figure 24.7 *cannot* represent an electric field? Explain.

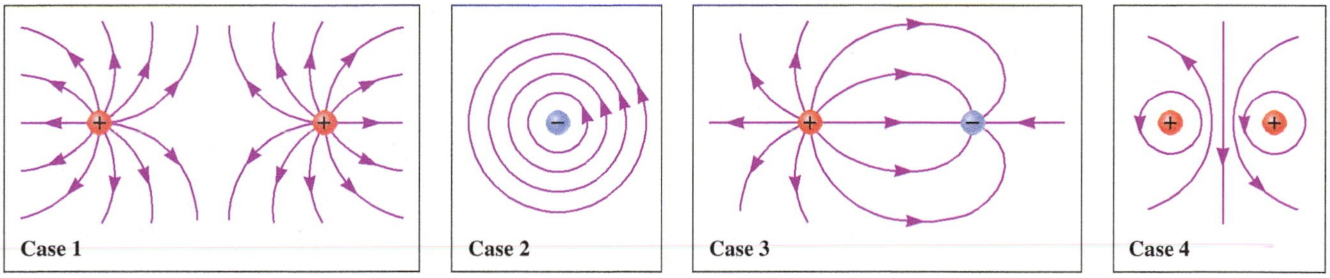

FIGURE 24.7

EXAMPLE 24.1 A Negative Spherical Source

Consider a negatively charged spherical source, and use $\vec{E}(r) = (kQ_S/r^2)\hat{r}$ (Eq. 24.3) for the electric field. Draw a diagram of the sphere's electric field using field lines, similar to Figure 24.5.

INTERPRET and ANTICIPATE

Equation 24.3 works regardless of the sign of the source charge. When Q_S has a negative value, the magnitude of the electric field is still higher near the sphere and lower farther away. The main difference from Figure 24.5 is that the negative charge changes the direction of the electric field, so now the electric field points inward—toward the source.

Example continues on page 720 ▶

SOLVE

Draw electric field lines that reflect this description (Fig. 24.8). The electric field lines are closer together near the sphere than they are farther away, and the arrowheads point inward.

CHECK and THINK

Imagine placing a *positive* test particle at several locations near a negatively charged sphere. The positive test charge is attracted to the negative sphere. This is consistent with Figure 24.8 because the electrostatic force on a test charge points in the same direction as the electric field. Both point inward in this case.

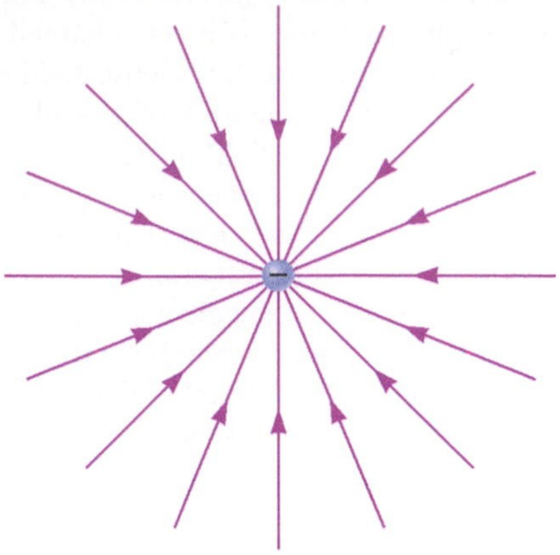

FIGURE 24.8

24-4 Electric Field of a Collection of Charged Particles

In this section, we focus on sources that are collections of charged particles. When we consider the electric field due to an unfamiliar source, it is helpful to visualize it by sketching electric field lines. For example, Figure 24.9 shows two identical positively charged particles. Here are a few tips for drawing electric field lines when the source is a collection of particles (or charged spherical objects):

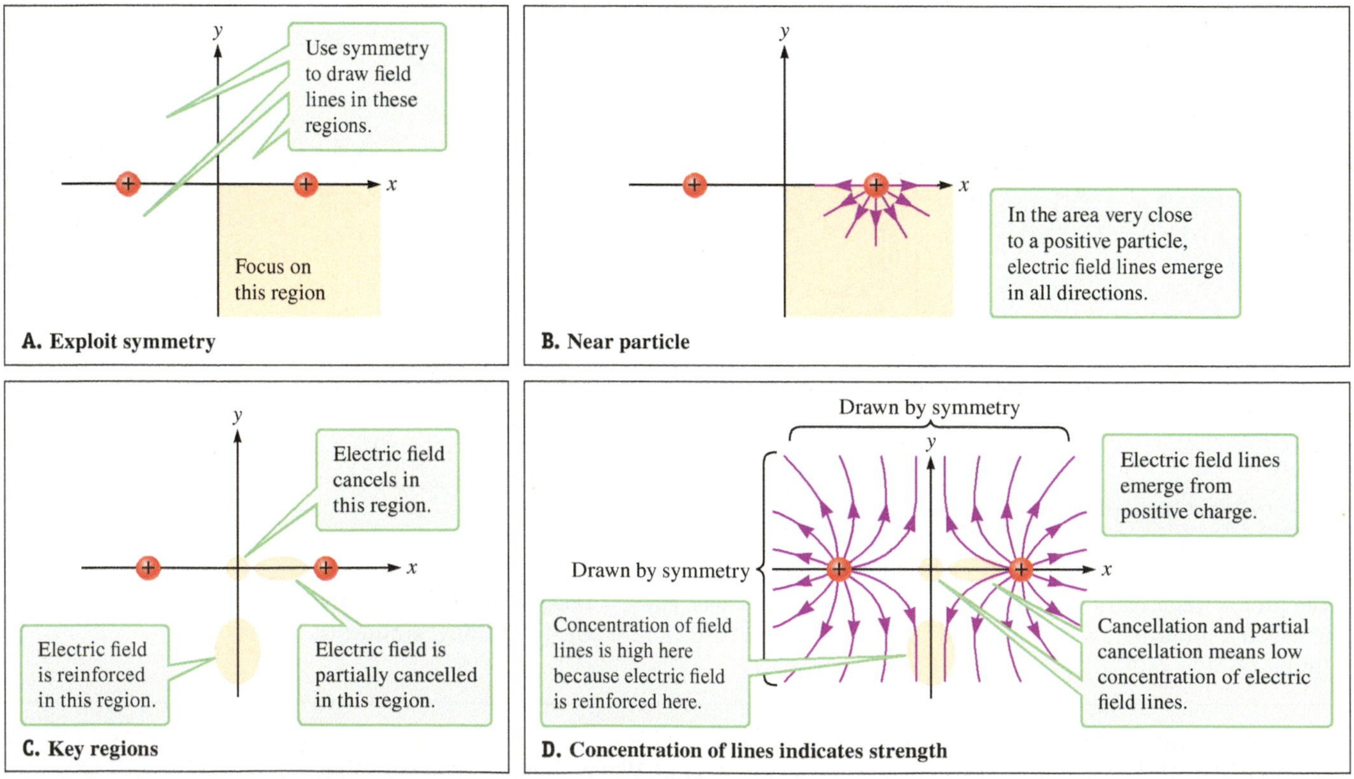

FIGURE 24.9 This source consists of two positively charged particles. Four tips for drawing electric field lines: **A.** Exploit symmetry and focus on one region. **B.** Draw field lines emerging from a positive source (or terminating on a negative source). **C.** Determine how the presence of other charged particles affects the electric field in the various regions. **D.** Make sure the concentration of field lines represents the relative strength of the field in various regions.

1. **Exploit symmetry (Fig. 24.9A).** Often the collection of particles is symmetrically arranged. You can draw the electric field lines in one region, and then exploit the symmetry to draw the electric field lines in the remaining regions. So start by focusing your attention on one region.
2. **Draw field lines near one or more particles.** Electric field lines must originate at a positive particle, terminate on a negative particle, or both (Fig. 24.9B). Because we are considering a collection of particles, the electric field lines *near* each particle must look something like the electric field lines in Figure 24.5 or 24.8.
3. **Determine the relative strength of the electric field in a few key regions.** Regions that are far from a particular particle may be strongly influenced by the presence of the other particles (Fig. 24.9C). The electric fields due to all charged particles must be added vectorially. In a region, it is possible for the electric fields due to various particles to cancel, partially cancel, or reinforce one another.
4. Sketch the field lines so that the **concentration of lines corresponds to the relative strength of the electric field** in the various regions you considered (Fig. 24.9D). Be sure the concentration of electric field lines is higher near the particles and in regions where the electric field is reinforced by the presence of other charged particles. Also be sure the concentration of electric field lines is low where the electric field is partially cancelled.

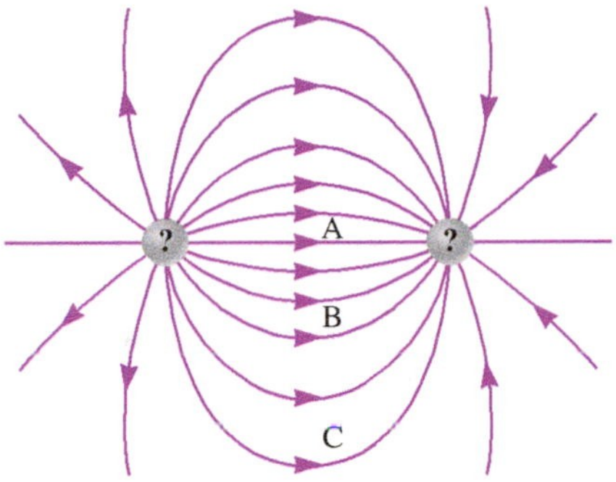

FIGURE 24.10

CONCEPT EXERCISE 24.4

Figure 24.10 shows a source that consists of two charged particles.

a. What is the sign of the charge on each particle?
b. In which region (A, B, or C) is the electric field the weakest?
c. In which region (A, B, or C) is the electric field the strongest?

PROBLEM-SOLVING STRATEGY Electric Field due to a Collection of Charged Particles

The electric field due to a single particle is given by $\vec{E}(r) = (kQ_S/r^2)\hat{r}$ (Eq. 24.3). How do we find the electric field when a source consists of more than one charged particle? To find a mathematical expression or a numerical value of the electric field at some particular position, we must add the electric fields due to each particle, taking into account the magnitude and direction of the vector fields.

INTERPRET and ANTICIPATE

Draw a diagram showing *all the electric fields* present at some particular position. There are four elements to this diagram:

1. **Labeled source particles and test particle.** Draw and label the particles that make up the source. Also draw an imaginary test particle at the location where you wish to find the electric field. The **test particle is always assumed to have a positive charge.**
2. The **electric field vectors** at the location of the test charge. These field vectors are found from the electric force exerted on the test charge by each particle in the source. The direction of the force is along the line joining the test particle to the particular source particle. To get the length of the vector, estimate the magnitude of the force by taking into account the magnitude of the source particle's charge and the distance to the test particle. Because the electrostatic force is parallel and proportional to the electric field, **label these vectors as electric fields,** using a subscript to indicate the source particle that produced that electric field vector.
3. The **net electric field vector** at the location of interest. Find this graphically in order to anticipate the result. Draw the resultant vector and label it $\vec{E}$.
4. A **coordinate system** and other **geometric details.** Your diagram is similar to a free-body diagram for the test charge, and your experience with free-body diagrams will help you choose a good coordinate system. Be sure to include details such as distances and angles.

SOLVE

Once the diagram is complete, find a mathematical expression for the electric field at the point of interest by adding the electric field vectors algebraically as you would any vectors. (We are *not* applying Newton's second law, so acceleration is not involved.)

EXAMPLE 24.2 Two Positively Charged Particles

Find an expression for the electric field at a point equidistant between two identical particles, each with charge $+Q$. The separation between the particles is $2a$, and the distance between each particle and the point is $\sqrt{2}a$.

INTERPRET and ANTICIPATE

The two charged particles exert a repulsive force on each other, but we can assume they are fixed in place, and we are not interested in calculating the force they exert on each other. We are interested in the electric field created by the combination of these two charged particles.

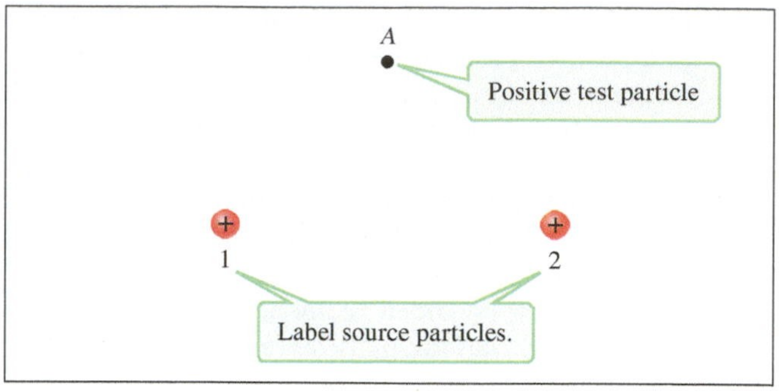

FIGURE 24.11

We draw a diagram with all four elements. **Label the source particles** 1 and 2. The **test particle** is drawn at *A*—a point equidistant from the two source particles as stated in the problem (Fig. 24.11).

There are two **electric field vectors** at *A*, the location of the (imaginary) test particle—one from particle 1 and another from particle 2. We estimate the magnitude and direction of these vectors from the force the two source particles exert on the (positively charged) test particle. The source particles carry the same charge and are equidistant from *A*, so they exert forces of equal magnitude on the test particle. These vectors are **labeled as electric fields** (Fig. 24.12).

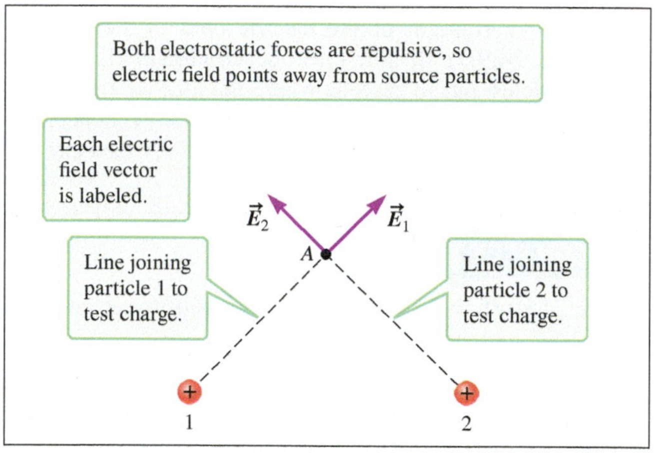

FIGURE 24.12

Graphically, add the individual electric field vectors to show the **net electric field vector** at *A* (Fig. 24.13).

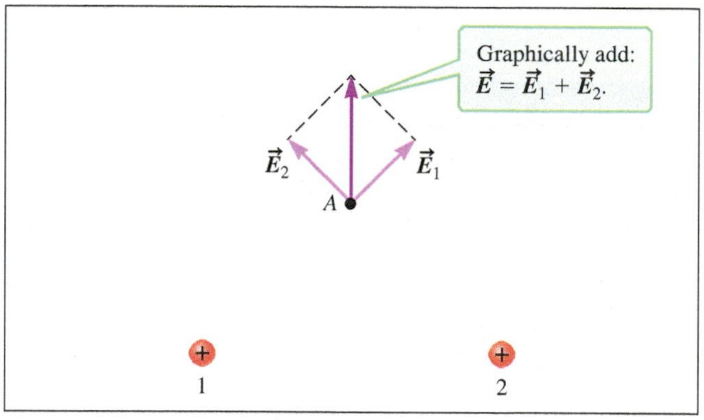

FIGURE 24.13

Add a **coordinate system** and other **geometric details**. In this case, we see that point *A* and the two source particles form a triangle, which we label in terms of the parameters given. From Figure 24.14, we expect that the *x* components of the field cancel and the *y* components add, so that the resulting electric field points in the positive *y* direction.

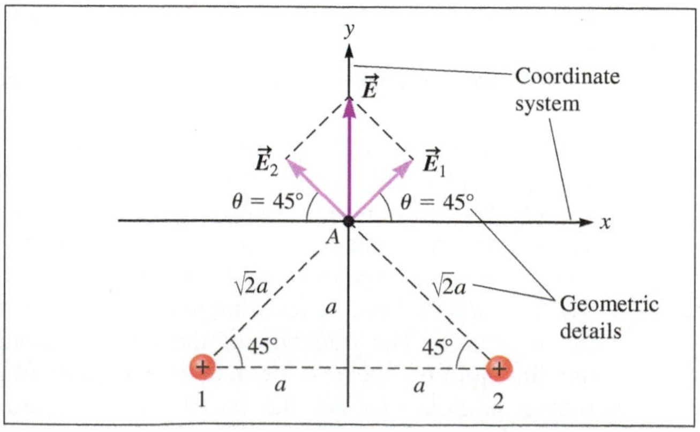

FIGURE 24.14

SOLVE Write $\vec{E}_1$ and $\vec{E}_2$ from Figure 24.14 in component form.	$\vec{E}_1 = (E_1 \cos\theta)\hat{\imath} + (E_1 \sin\theta)\hat{\jmath}$ $\vec{E}_2 = (-E_2 \cos\theta)\hat{\imath} + (E_2 \sin\theta)\hat{\jmath}$
Since $\theta = 45°$, simplify the expressions using $\sin 45° = \cos 45° = \sqrt{2}/2$.	$\vec{E}_1 = \frac{\sqrt{2}}{2}E_1(\hat{\imath} + \hat{\jmath})$ $\vec{E}_2 = \frac{\sqrt{2}}{2}E_2(-\hat{\imath} + \hat{\jmath})$
Add these vectors.	$\vec{E} = \vec{E}_1 + \vec{E}_2 = \frac{\sqrt{2}}{2}E_1(\hat{\imath} + \hat{\jmath}) + \frac{\sqrt{2}}{2}E_2(-\hat{\imath} + \hat{\jmath})$ $\vec{E} = \frac{\sqrt{2}}{2}[(E_1 - E_2)\hat{\imath} + (E_1 + E_2)\hat{\jmath}]$ (1)
The magnitudes E_1 and E_2 can be found from $E(r) = kQ_S/r^2$ (Eq. 24.3). Both source particles have positive charge $+Q$ and both are $\sqrt{2}a$ away from point A. We find that the magnitudes E_1 and E_2 are equal.	$E_1 = k\frac{Q}{(\sqrt{2}a)^2} = \frac{1}{2}\frac{kQ}{a^2}$ $E_2 = k\frac{Q}{(\sqrt{2}a)^2} = \frac{1}{2}\frac{kQ}{a^2} = E_1$
Substitute the magnitudes of E_1 and E_2 in the expression for $\vec{E}$ (Eq. 1).	$\vec{E} = \frac{\sqrt{2}}{2}[(E_1 - E_1)\hat{\imath} + (E_1 + E_1)\hat{\jmath}]$ $\vec{E} = \sqrt{2}E_1\hat{\jmath} = \frac{\sqrt{2}}{2}\frac{kQ}{a^2}\hat{\jmath}$

CHECK and THINK

As expected, the x components cancel and the y components add; the net electric field is in the positive y direction.

Electric Dipole

An **electric dipole** consists of two charged particles of magnitude Q but with opposite signs (Fig. 24.15). The particles are attracted to each other, but their separation is maintained so that the distance between them is d. The electric dipole is an important special case because in many circumstances, atoms and molecules can be modeled as dipoles. For example, in Section 23-4, we saw that the atoms or molecules in an insulator can be modeled as dipoles when the insulator is near a charged object. Other molecules are natural dipoles; for example, a water molecule is best modeled as a dipole even when it is not near any charged objects.

ELECTRIC DIPOLE ✪ **Major Concept**

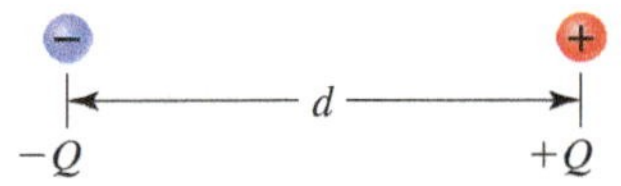

FIGURE 24.15 An electric dipole consists of two charged particles, one with charge $+Q$ and the other with charge $-Q$. The charged particles are attracted to each other, but maintain a separation d.

EXAMPLE 24.3 Special Case: Electric Field due to a Dipole

Find an expression for the electric field at a point a distance x to the right of the center of the dipole shown in Figure 24.15. (Other positions are given as homework; see Problem 18.)

INTERPRET and ANTICIPATE

Include all four elements in your sketch.

1. **Label the source particles and draw the** (positive) **test charge** (Fig. 24.16).

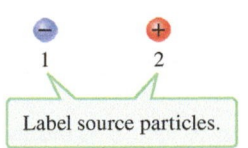

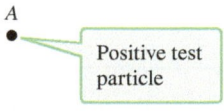

FIGURE 24.16

Example continues on page 724 ▶

2. **Sketch the electric field vectors for each source particle** (Fig. 24.17). The test charge is positive and particle 1 is negative, so particle 1 exerts an attractive force on the test charge, whereas particle 2 is positive and exerts a repulsive force on the test charge. (Remember the electric field points in the same direction as the electric force for a positive test charge.) Because particle 1 is farther away than particle 2, E_1 , E_2.

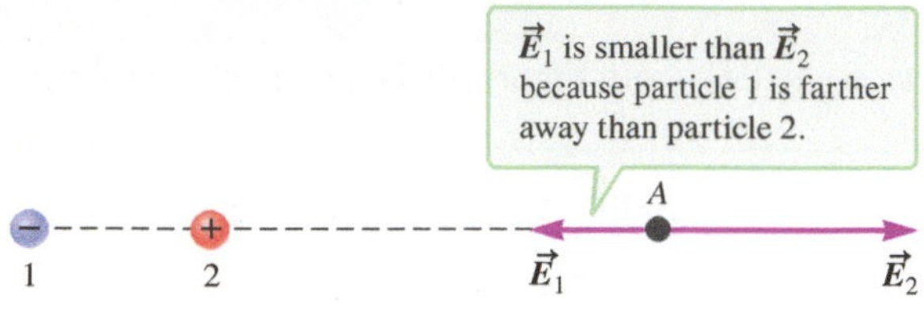

FIGURE 24.17

3. **Add the electric field vectors graphically** (Fig. 24.18). The sum of the electric fields at A is $\vec{E} = \vec{E}_1 + \vec{E}_2$.

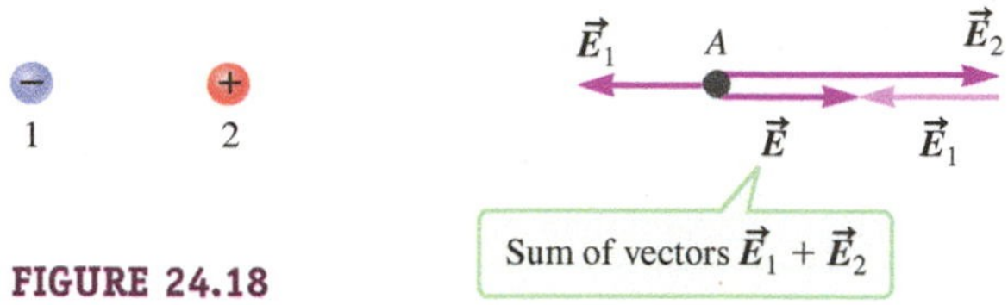

FIGURE 24.18

4. **Choose a coordinate system and add geometric details.** We have chosen the origin of coordinates to be midway between the source charges. From Figure 24.19, we expect that the electric field vector $\vec{E}$ points in the positive x direction.

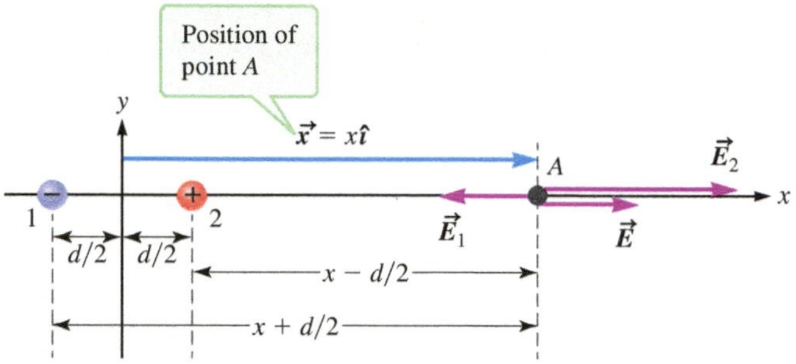

FIGURE 24.19

SOLVE

Vectors $\vec{E}_1$ and $\vec{E}_2$ have only x components; $\vec{E}_1$ is in the negative x direction and $\vec{E}_2$ is in the positive x direction.

$$\vec{E}_1 = -\vec{E}_1\hat{\imath} \qquad \vec{E}_2 = \vec{E}_2\hat{\imath}$$

Add these vectors.

$$\vec{E} = \vec{E}_1 + \vec{E}_2 = -E_1\hat{\imath} + E_2\hat{\imath} = (E_2 - E_1)\hat{\imath} \tag{1}$$

The magnitudes of E_1 and E_2 can be found from Equation 24.3. Each particle has a charge of magnitude Q. The center-to-center distances r between the test charge and each of the two sources are shown in Figure 24.19. For particle 1 this distance is $x + d/2$, and for particle 2 the distance is $x - d/2$.

$$E(r) = k\frac{|Q_S|}{r^2} \tag{24.3}$$

$$E_1 = k\frac{|-Q|}{(x + d/2)^2} = k\frac{Q}{(x + d/2)^2}$$

$$E_2 = k\frac{Q}{(x - d/2)^2}$$

Substitute E_1 and E_2 into Equation (1).

$$\vec{E} = kQ\left(\frac{1}{(x - d/2)^2} - \frac{1}{(x + d/2)^2}\right)\hat{\imath}$$

Find a common denominator and simplify this expression.

$$\vec{E} = kQ\left(\frac{(x + d/2)^2 - (x - d/2)^2}{(x - d/2)^2(x + d/2)^2}\right)\hat{\imath}$$

$$\vec{E} = kQ\frac{2xd}{(x^2 - d^2/4)^2}\hat{\imath} \tag{2}$$

CHECK and THINK

If the denominator in Equation (2) is zero, then E is infinite. Let's see under what circumstance the denominator will be zero.

$$x^2 - d^2/4 = 0$$

$$x = \pm\frac{d}{2}$$

The electric field is infinite if point A is on either of the charged particles. This makes sense because the electric field of any charged particle goes to infinity as r goes to zero (Concept Exercise 24.2). Finally, as expected, for $x > d/2$ as in Figure 24.19, the electric field $\vec{E}$ points in the positive x direction.

Far from an Electric Dipole: The Dipole Moment

ELECTRIC DIPOLE MOMENT
Major Concept

The electric field found in Example 24.3 depends on our choice of coordinates. If another person decided to use an x axis pointing in a different direction, that person would have to modify our expression to make it fit that choice of coordinate system. Because dipoles are particularly important special cases, we can avoid such complications by writing the electric field for a dipole in terms of an **electric dipole moment**, $\vec{p}$. The electric dipole moment—also called the *dipole moment*—is a vector that points from the negatively charged particle toward the positively charged particle in a dipole. So, in Figure 24.19, the dipole moment points in the positive x direction. The magnitude of the dipole moment is Qd. The result in Example 24.3 is written as

$$\vec{E} = \frac{2kx\vec{p}}{(x^2 - d^2/4)^2} \tag{24.5}$$

Equation 24.5 does not include a unit vector and so it does not depend on a particular coordinate system; it is good for finding the electric field due to a dipole at a distance x (from the midpoint of the dipole) along the line that passes through both particles.

Often we are interested in the electric field of a dipole at a position that is far compared to the separation between the two particles. We find an expression for the electric field in that case by making the approximation $x \gg d/2$. So the denominator of Equation 24.5 reduces to x^4:

$$\vec{E} \approx \frac{2kx\vec{p}}{(x^2 - 0)^2} \approx \frac{2kx\vec{p}}{x^4}$$

and the electric field is approximately

MAGNITUDE OF THE ELECTRIC FIELD DUE TO A DIPOLE FOR $r \gg d$
Special Case

$$\vec{E} \approx \frac{2k\vec{p}}{x^3} \text{ when } x \gg d/2 \tag{24.6}$$

Equation 24.6 is a good approximation for any position far from an electric dipole along the line that passes through both charges. In fact, the magnitude of Equation 24.6 is good for *any* position r in *any direction* far from an electric dipole:

$$E \approx \frac{2kp}{r^3} \tag{24.7}$$

where $r \gg d$ and d is the separation between the charges.

If a point is *very* far from the dipole—in other words, if $r \to \infty$—then the two charged particles that make up the dipole appear to cancel each other. We expect that the electric field should be zero for a position so far away, and that is exactly what we find if we take the limit of Equation 24.7:

$$\lim_{r\to\infty} E = \lim_{r\to\infty} \frac{2kp}{r^3} \to 0$$

If you model a water molecule as a dipole (Fig. 24.20), you can use Equation 24.5 to find the electric field near the molecule along the line that passes through its midpoint (the x axis shown in Fig. 24.20). You can use Equation 24.7 to find the magnitude of the electric field for any position far ($r \gg d$) from the water molecule. Finally, the electric field at a point very far from the water molecule is zero because the charged particles are at essentially the same distance from that point and so the electric field due to each particle cancels out.

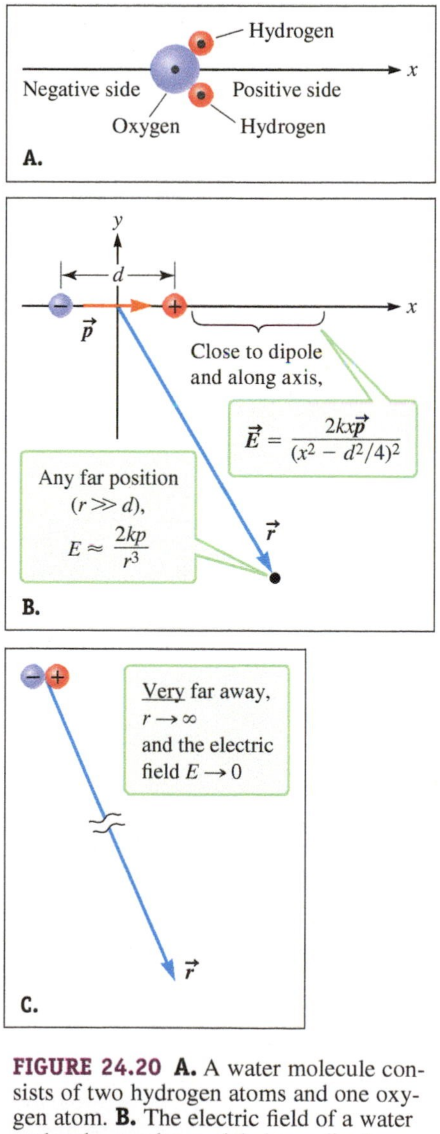

FIGURE 24.20 **A.** A water molecule consists of two hydrogen atoms and one oxygen atom. **B.** The electric field of a water molecule may be modeled as a dipole. The electric field on the x axis is given by Equation 24.5. The magnitude of the electric field for $r \gg d$ is given by Equation 24.7. **C.** The electric field very far from a water molecule is zero.

CONCEPT EXERCISE 24.5

A water molecule is made up of two hydrogen atoms and one oxygen atom, with a total of 10 electrons and 10 protons. The molecule is modeled as a dipole with an effective separation $d = 3.9 \times 10^{-12}$ m between its positive and negative charges. What is the magnitude of the water molecule's dipole moment?

24-5 Electric Field of a Continuous Charge Distribution

When you rub a plastic rod with wool or shuffle your socks across the carpet, charged particles are transferred between the objects. In fact, the charged particles are so numerous that they essentially form a continuous or smooth charge distribution. Because of this great number of particles, adding up the electric field due to each particle means solving an integral. Before we present a problem-solving strategy for finding the electric field of a continuous charge distribution, we need to discuss charge density.

Charge Density

We use the same symbol for both volume charge density and mass density (Eq. 1.1). Both represent an intrinsic property (charge or mass) per unit volume. Be careful not to confuse these two uses of the same symbol.

When applying calculus to a continuous distribution of charged particles, we use the source's *charge density* and integrate over the appropriate region of space to account for the total charge. There are three types of charge density:

1. **Volume charge density** ρ (lowercase Greek letter rho) is the amount of charge per unit volume (Fig. 24.21A). For a uniform charge distribution,

$$\rho \equiv \frac{Q}{V} \tag{24.8}$$

Volume charge density has the dimensions charge per volume, with SI units C/m^3. The amount of charge dq in a small volume dV can be written in terms of the volume charge density:

$$\begin{pmatrix}\text{charge contained}\\ \text{in small volume}\end{pmatrix} = \begin{pmatrix}\text{charge per}\\ \text{unit volume}\end{pmatrix} \times (\text{volume})$$

$$dq = \rho dV \tag{24.9}$$

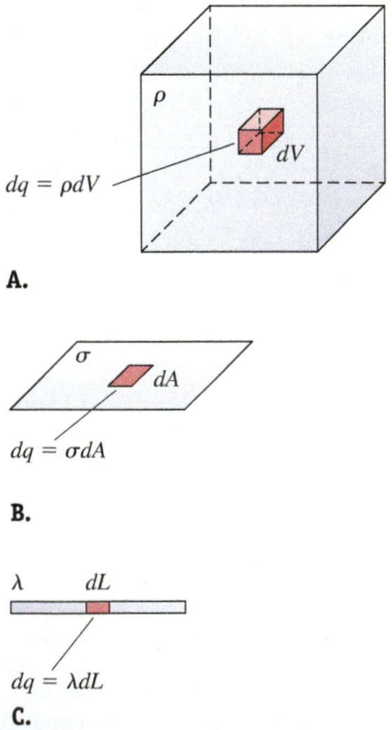

FIGURE 24.21 **A.** Volume charge density. **B.** Surface charge density. **C.** Linear charge density.

2. **Surface charge density** σ (lowercase Greek letter sigma) is the amount of charge per unit area (Fig. 24.21B). For a uniform charge distribution,

$$\sigma \equiv \frac{Q}{A} \tag{24.10}$$

The dimensions of σ are charge per area, with SI units C/m^2. The amount of charge dq in a small area dA can be written in terms of the surface charge density:

$$\begin{pmatrix}\text{charge contained}\\ \text{in small area}\end{pmatrix} = \begin{pmatrix}\text{charge per}\\ \text{unit area}\end{pmatrix} \times (\text{area})$$

$$dq = \sigma dA \tag{24.11}$$

3. **Linear charge density** λ (lowercase Greek letter lambda) is the amount of charge per unit length (Fig 24.21C). For a uniform charge distribution,

$$\lambda \equiv \frac{Q}{L} \tag{24.12}$$

The dimensions of λ are charge per length, with SI units C/m. The amount of charge dq in a small length element dL can be written in terms of the linear charge density:

$$\begin{pmatrix}\text{charge contained in} \\ \text{small length element}\end{pmatrix} = \begin{pmatrix}\text{charge per} \\ \text{unit length}\end{pmatrix} \times (\text{length})$$

$$dq = \lambda dL \tag{24.13}$$

CONCEPT EXERCISE 24.6

a. Figure 24.22A shows a rod of length L and radius R with excess positive charge Q. The excess charge is uniformly distributed over the entire outside surface of the rod. Write an expression for the surface charge density σ. Write an expression in terms of σ for the amount of charge dq contained in a small segment of the rod of length dx.

b. Figure 24.22B shows a very narrow rod of length L with excess positive charge Q. The rod is so narrow compared to its length that its radius is negligible and the rod is essentially one-dimensional. The excess charge is uniformly distributed over the length of the rod. Write an expression for the linear charge density λ. Write an expression in terms of λ for the amount of charge dq contained in a small segment of the rod of length dx. Compare your answers with those for part (a). Explain the similarities and differences.

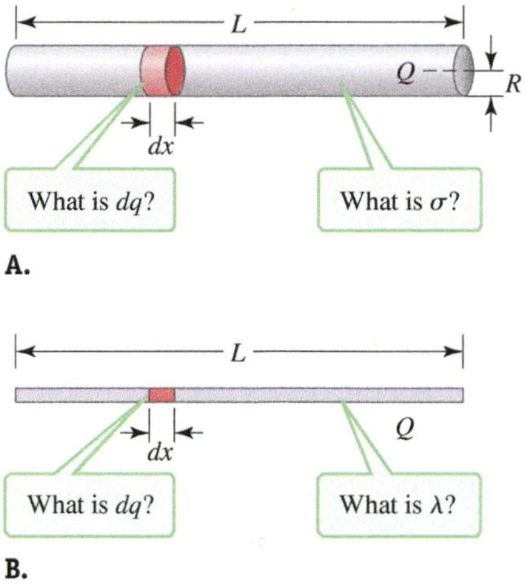

FIGURE 24.22

PROBLEM-SOLVING STRATEGY

Electric Field due to a Continuous Distribution of Charged Particles

The overall plan for finding the electric field due to a continuous distribution of charged particles is based on the procedure we followed for a collection of charged particles. The first step is to imagine slicing up the distribution into tiny pieces, with each piece so small that it can be modeled as a single charged particle. We consider just a small number of pieces—usually one or two—to get an expression for the electric field produced by those pieces. Then we integrate that expression to find the electric field produced by the entire continuous distribution.

INTERPRET and ANTICIPATE

As for a collection of charged particles, draw a diagram showing the electric field at some particular position. There are four elements to this diagram:

1. **Labeled pieces** of the source and **an imaginary test particle.** Imagine slicing up the distribution into small pieces. It is usually helpful to indicate two of these pieces on your sketch. In choosing which two pieces, exploit the symmetry of the problem. Also draw an imaginary test particle (assumed positive) at the location where you wish to find the electric field. If you are not told whether the charge distribution is positive or negative, assume it is positive.
2. The **electric field vectors** at the location of the test charge, as found from the electric force exerted on the test charge by each piece of the source you've indicated. The direction of the force is along the line joining the test particle to the particular piece of the source. To get the length of the vector, estimate the magnitude of the force by taking into account the distance to the test particle. As long as the charge distribution is uniform, each piece has the same small amount of charge dq, which makes your estimation task easy. **Label these vectors as electric fields $d\vec{E}_{\text{sub}}$**, using a subscript to indicate the piece of the source that produced each electric field vector.
3. The **net electric field vector** at the location of interest. Find this graphically in order to anticipate the result. Draw the resultant vector and label it $d\vec{E}$.
4. A **coordinate system** and other **geometric details.**

SOLVE

There are two major steps:

Step 1 Find a **mathematical expression** for the infinitesimal electric field $d\vec{E}$ in terms of an infinitesimal spatial variable (dx, dy, dz, dr, ds, or $d\theta$). The magnitude of the electric field due to an infinitesimal amount of charge dq comes from modifying $\vec{E}(r) = (kQ_S/r^2)\hat{r}$ (Eq. 24.3) by replacing Q_S with dq:

$$dE(r) = k\frac{dq}{r^2} \tag{24.14}$$

The direction will come from your diagram. In this step, you are likely to write dq in terms of the charge density, so Equations 24.8 through 24.13 are very useful.

Problem-Solving Strategy continues on page 728 ▶

Step 2 Integrate this expression over the entire charge distribution.

CHECK and THINK

Here are two things to check once you have found an expression for the electric field $\vec{E}$:

1. Make sure the **dimensions** of the electric field are force per charge.
2. If the charge distribution is finite, the **electric field** for a point far from the distribution compared to its size **should approach the electric field of a charged particle,** $\vec{E}(r) = (kQ_S/r^2)\hat{r}$.

24-6 Special Cases of Continuous Distributions

ELECTRIC FIELD DUE TO A CHARGED ROD Special Case

In this section we practice finding the electric field due to three different continuous charge distributions using the problem-solving strategy from Section 24-5. All three of these are special cases, but they are important ones that you are likely to refer to many times. In the first two cases, we start by following all the steps outlined in Section 24-5. In the third case, we use a shortcut.

EXAMPLE 24.4 Special Case: Charged Rod

In Chapter 23, we considered rods of glass or plastic that have an excess charge. Now find an expression for the electric field at point A directly above the midpoint of a very thin rod of length 2ℓ (Fig. 24.23). The excess charge Q is uniformly distributed along the length of the rod.

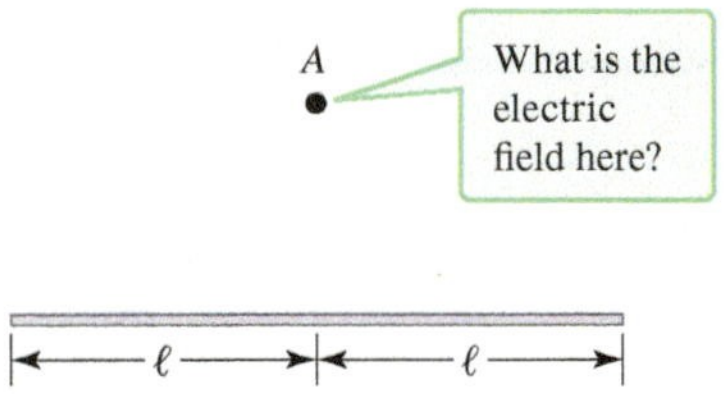

FIGURE 24.23

INTERPRET and ANTICIPATE

If you need to calculate the electric field due to any continuous charge distribution such as the charged rod in this problem, you cannot use Equation 24.3 directly because that equation is only for a single charged particle or sphere. We must derive an expression for the charged rod, using the strategy in Section 24-5. Here we start with a diagram that has all four elements. We imagine a positive **test particle** at point A and slice the rod into thin pieces (Fig. 24.24). Two pieces are **labeled** R and L for right and left. We have exploited the symmetry of the problem by choosing two pieces equidistant from point A.

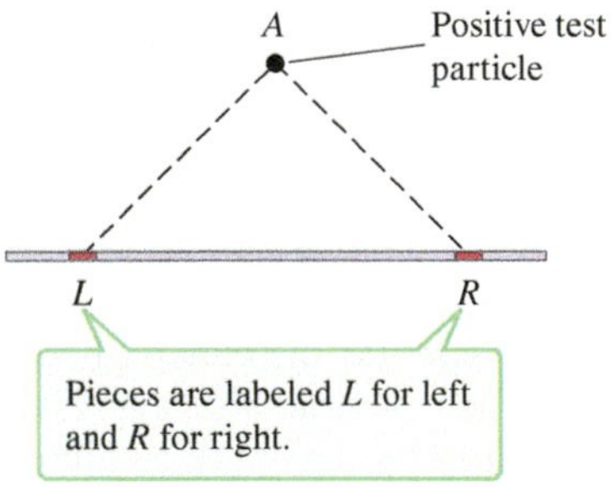

FIGURE 24.24

Draw the **electric field vectors** at the location of the test charge (Fig. 24.25). Model each piece L and R as a charged particle, and draw each electric field vector. Each piece has a positive charge dq.

Add the electric field vectors of the left and right pieces graphically, and sketch the **net electric field vector** at point A. The horizontal components cancel and the vertical components add, so we need to find only an expression for the vertical component (Fig. 24.26).

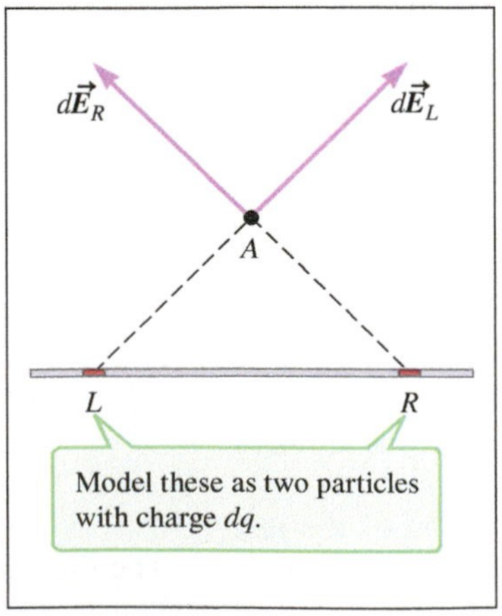

FIGURE 24.25

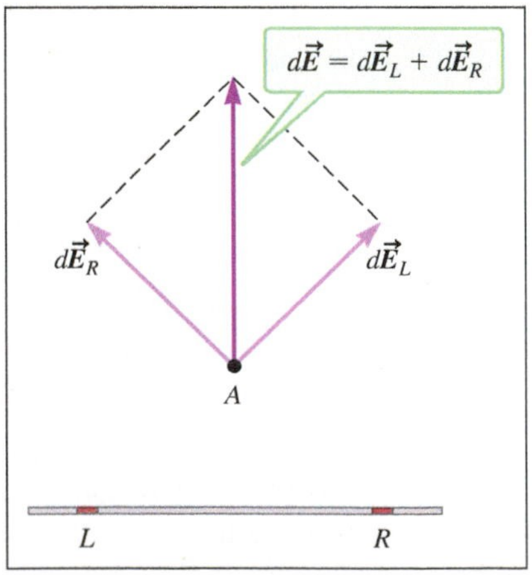

FIGURE 24.26

Add a **coordinate system** and other **geometric details.** Now compare Figure 24.27 with Figure 24.14; they are very similar. In both figures, two vectors are added such that their x components cancel. So we expect that $\vec{E}$ points in the positive y direction.

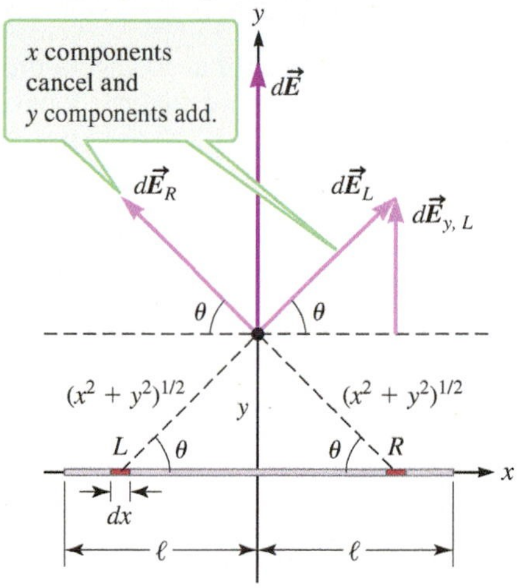

FIGURE 24.27

SOLVE

Step 1 Find a **mathematical expression** for the infinitesimal electric field $d\vec{E}$. From Figure 24.27, we reason that $d\vec{E}$ equals two times the y component of either $d\vec{E}_R$ or $d\vec{E}_L$. Let's arbitrarily work with $d\vec{E}_L$.

Write $d\vec{E}_L$ in component form.	$d\vec{E}_L = (dE_L \cos\theta)\hat{\imath} + (dE_L \sin\theta)\hat{\jmath}$
$d\vec{E}$ equals two times the y component of $d\vec{E}_L$.	$d\vec{E} = 2\,dE_L \sin\theta\hat{\jmath}$ (1)
The magnitude dE_L is found from Equation 24.14. The distance from piece L to point A is $r = \sqrt{x^2 + y^2}$ (Fig. 24.27).	$dE(r) = k\dfrac{dq}{r^2}$ (24.14) $dE_L = k\dfrac{dq}{(\sqrt{x^2 + y^2})^2} = \dfrac{kdq}{x^2 + y^2}$
Substitute the magnitude dE_L into the expression for $d\vec{E}$ (Eq. 1).	$d\vec{E} = \dfrac{2kdq}{x^2 + y^2}\sin\theta\hat{\jmath}$
The length of either piece L or R is dx. So we modify Equation 24.13 to write $dq = \lambda dx$, and replace dq.	$d\vec{E} = \dfrac{2k\lambda dx}{x^2 + y^2}\sin\theta\hat{\jmath}$ (2)
Step 2 **Integrate.** Equation (2) gives only the y component of the electric field produced by two small pieces of the charged rod as shown in Figure 24.27. We must now integrate this expression over the entire rod. We can integrate Equation (2) over only one variable—x. Because $\sin\theta$ depends on x, we must rewrite it in terms of x. Use trigonometry and the right triangle shown in Figure 24.27 to come up with an expression, and eliminate $\sin\theta$.	$\sin\theta = \dfrac{y}{(x^2 + y^2)^{1/2}}$ $d\vec{E} = \dfrac{2k\lambda dx}{x^2 + y^2}\dfrac{y}{(x^2 + y^2)^{1/2}}\hat{\jmath} = \dfrac{2k\lambda y dx}{(x^2 + y^2)^{3/2}}\hat{\jmath}$ (3)
Now we are ready to integrate over the left half of the rod from $x = -\ell$ to 0. The integral on the left of the equal sign is what we want, the electric field at point A. Pull the constants out of the integral on the right, treating y as a constant for a given point A.	$\int d\vec{E} = \int_{-\ell}^{0} \dfrac{2k\lambda y dx}{(x^2 + y^2)^{3/2}}\hat{\jmath}$ $\vec{E} = 2k\lambda y\hat{\jmath}\int_{-\ell}^{0} \dfrac{dx}{(x^2 + y^2)^{3/2}}$
This integral can be found in Appendix A.	$\vec{E} = 2k\lambda y\hat{\jmath}\left[\dfrac{1}{y^2}\dfrac{x}{\sqrt{x^2 + y^2}}\right]_{-\ell}^{0}$
Substitute the limits of integration and simplify the expression.	$\vec{E} = \dfrac{2k\lambda}{y}\dfrac{\ell}{\sqrt{\ell^2 + y^2}}\hat{\jmath}$
Finally we must write our expression in terms of the parameters given in the problem—in this case, the total charge Q of the source. The rod's length is 2ℓ and its linear charge density is λ, so $Q = 2\ell\lambda$.	$\vec{E} = \dfrac{kQ}{y}\dfrac{1}{\sqrt{\ell^2 + y^2}}\hat{\jmath}$ (24.15)

Example continues on page 730 ▶

CHECK and THINK

There are two things for us to check. First, are the **dimensions** force per charge? Our dimensional analysis looks good.

$$[\![E]\!] = \frac{(\text{F}\cdot\text{L}^2/\text{C}^2)\text{C}}{\text{L}}\frac{1}{\sqrt{\text{L}^2+\text{L}^2}}$$

$$[\![E]\!] = \frac{(\text{F}\cdot\text{L}^2/\text{C})}{\text{L}^2} = \frac{\text{F}}{\text{C}}$$

Second, make sure the electric field very far away ($y \gg \ell$) **approaches that of a charged particle** (Eq. 24.3). Indeed, we arrive at the form we expect for a charged particle at a distance y.

$$\vec{E} = \frac{kQ}{y}\frac{1}{\sqrt{\ell^2+y^2}}\hat{\jmath} \approx \frac{kQ}{y}\frac{1}{\sqrt{y^2}}\hat{\jmath} \text{ if } y \gg \ell$$

$$\vec{E} \approx \frac{kQ}{y^2}\hat{\jmath}$$ ✓

EXAMPLE 24.5 Special Case: Charged Ring

ELECTRIC FIELD DUE TO A CHARGED RING ▶ Special Case

A person wearing wool clothes uses a plastic hula hoop of radius $R = 1.20$ m. The hula hoop has negative charge $Q = -75$ nC spread uniformly on it. First find an expression for the electric field at point A, and then find a numerical result for a point 0.60 m above the center of the hoop (Fig. 24.28).

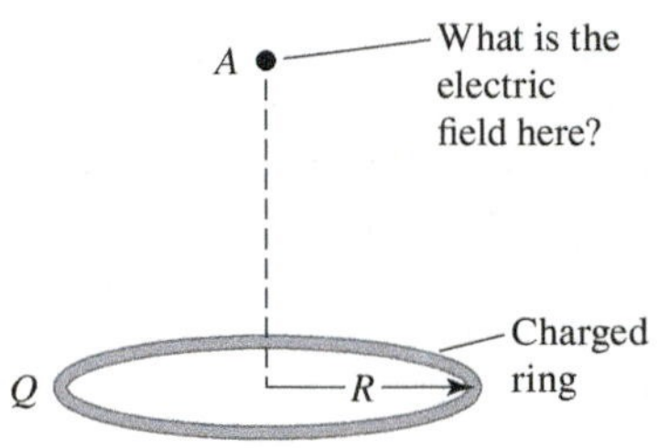

FIGURE 24.28

INTERPRET and ANTICIPATE

We must derive an expression for a charged ring (or hoop). We draw a diagram with all four elements.

Figure 24.29 shows the first two elements. We have chosen and **labeled two pieces**—L and R—positioned symmetrically on either side of point A, and drawn a positive **test charge** at point A. We model each piece as a charged particle and draw each **electric field vector**. Let's assume for now that each piece has a positive charge dq so we can come up with a general expression that is good for any charged ring. In a later substitution, we will take into account the hoop's negative charge.

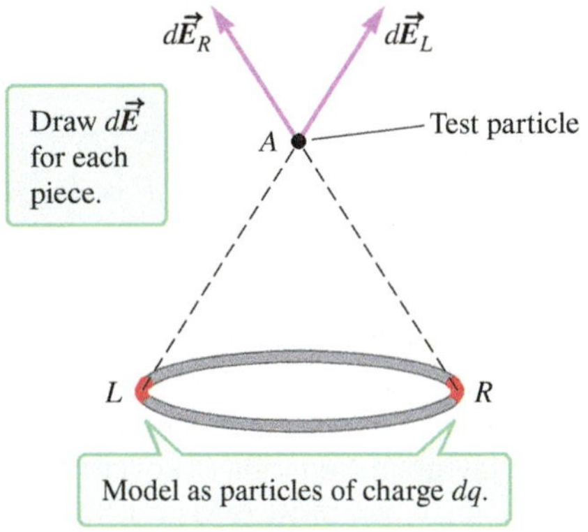

FIGURE 24.29

Figure 24.30 includes the next two elements. The electric field vectors are added to find the **net electric field vector**. The horizontal components cancel and the vertical components add, so we need to find only an expression for the vertical component. Our figure includes a **coordinate system** (y axis only) and **geometric details**.

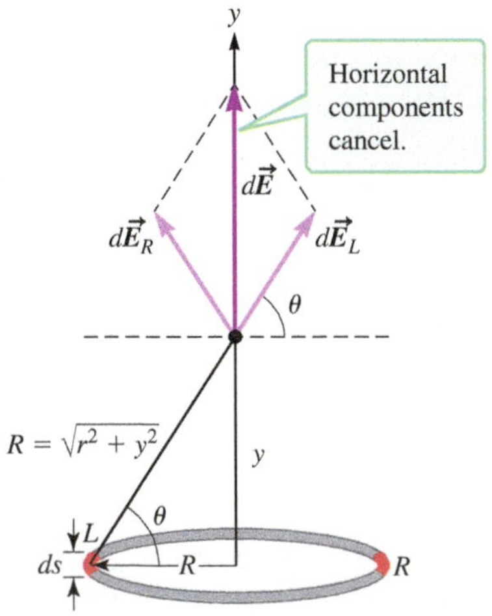

FIGURE 24.30

SOLVE **Step 1** Find an **expression** for $d\vec{E}$ in terms of a spatial variable—in this case, a small length ds of the ring (Fig. 24.30). As in the case of the charged rod, we see that $d\vec{E}$ equals two times the y component of either $d\vec{E}_R$ or $d\vec{E}_L$. Let's arbitrarily work with $d\vec{E}_L$.	$d\vec{E} = 2dE_L \sin\theta\hat{\jmath}$ (1)
The magnitude dE_L is found from Equation 24.14 for an infinitesimal amount of charge. The distance from the piece to the point of interest is given by $r = \sqrt{R^2 + y^2}$ (Fig. 24.30).	$dE(r) = k\dfrac{dq}{r^2}$ (24.14) $dE_L = k\dfrac{dq}{(R^2 + y^2)}$
Substitute dE_L into Equation (1).	$d\vec{E}_y = 2k\dfrac{dq}{(R^2 + y^2)}\sin\theta\,\hat{\jmath}$
Replace dq by modifying Equation 24.13, this time referring to a small length of the ring as ds.	$dq = \lambda ds$ $d\vec{E} = 2k\dfrac{\lambda ds}{(R^2 + y^2)}\sin\theta\,\hat{\jmath}$ (2)
Step 2 As we **integrate** around the left half the ring from $s = 0$ to $s = \pi R$, all the other parameters (R, y, and θ) in Equation (2) remain constant. Therefore, it is not necessary to replace $\sin\theta$ before integration. The left side is what we want—the electric field at point A.	$\int d\vec{E} = \int_0^{\pi R} 2k\dfrac{\lambda ds}{(R^2 + y^2)}\sin\theta\,\hat{\jmath}$ $\vec{E} = 2k\dfrac{\lambda}{(R^2 + y^2)}\sin\theta\,\hat{\jmath}\int_0^{\pi R} ds$ $\vec{E} = 2k\dfrac{\lambda}{(R^2 + y^2)}\sin\theta\hat{\jmath}[s]\Big\vert_0^{\pi R}$ $\vec{E} = \dfrac{k2\pi R\lambda}{(R^2 + y^2)}\sin\theta\hat{\jmath}$ (3)
We need to find the electric field in terms of the parameters given. The problem statement gave us Q, R, and y but not λ or θ, so now we eliminate these unknown parameters. To eliminate λ, use Equation 24.12.	$\lambda = \dfrac{Q}{L}$ (24.12) $Q = L\lambda = 2\pi R\lambda$ (4)
To eliminate θ, use trigonometry and the right triangle shown in Figure 24.30.	$\sin\theta = \dfrac{y}{r} = \dfrac{y}{(R^2 + y^2)^{1/2}}$ (5)
Finally, substitute Equations (4) and (5) into Equation (3).	$\vec{E} = \dfrac{kQy}{(R^2 + y^2)^{3/2}}\hat{\jmath}$ (24.16)
CHECK and THINK Before substituting values, let's check our expression. First, the **dimensions** are force per charge as expected (Problem 25). Second, if point A is far from the ring ($y \gg R$), the electric field **approaches that of a point charge.**	$\lim_{y \gg R}\vec{E} = \dfrac{kQy}{(R^2 + y^2)^{3/2}}\hat{\jmath} \rightarrow \dfrac{kQy}{(y^2)^{3/2}}\hat{\jmath}$ $\vec{E} \approx \dfrac{kQy}{y^3}\hat{\jmath} \approx \dfrac{kQ}{y^2}\hat{\jmath}$ ✓
SOLVE Substitute numerical values.	$\vec{E} = \dfrac{(8.99 \times 10^9\,\text{N}\cdot\text{m}^2/\text{C}^2)(-75 \times 10^{-9}\,\text{C})(0.60\,\text{m})}{((1.20\,\text{m})^2 + (0.60\,\text{m})^2)^{3/2}}\hat{\jmath}$ $\vec{E} = -1.7 \times 10^2\hat{\jmath}\,\text{N/C}$

CHECK and THINK

Because the hoop has a negative charge, the electric field points downward in the negative y direction. The magnitude of the electric field seems reasonable because it is about an order of magnitude less than the electric field near a charged comb (Table 24.1).

EXAMPLE 24.6 Special Case: Charged Disk

Use the algebraic result of Example 24.5:

$$\vec{E} = \frac{kQy}{(R^2 + y^2)^{3/2}}\hat{\jmath} \qquad (24.16)$$

to derive an expression for the electric field at a point directly above the center of a positively charged disk as shown in Figure 24.31.

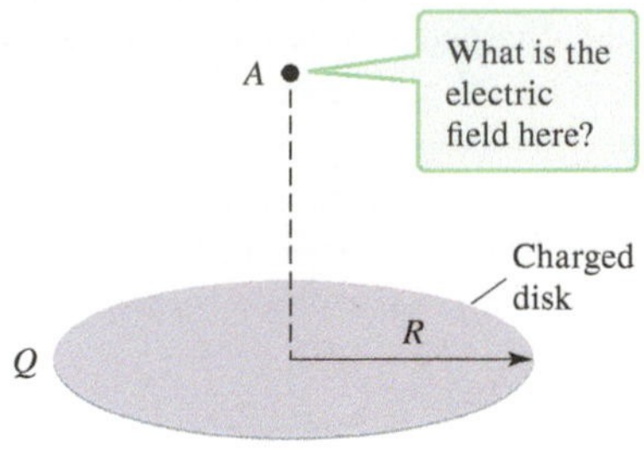

FIGURE 24.31

INTERPRET and ANTICIPATE

Using our result from Example 24.5—the electric field due to a charged ring—makes our task simpler here. Figure 24.32 combines elements 1 through 3. There is a positive **test charge** at point A. We have divided and **labeled** the disk into ring-shaped **pieces** 1 and 2. From Example 24.5, the **electric field** at point A due to each ring is straight up. **Add the electric fields** produced by these two thin rings to find $d\vec{E}$.

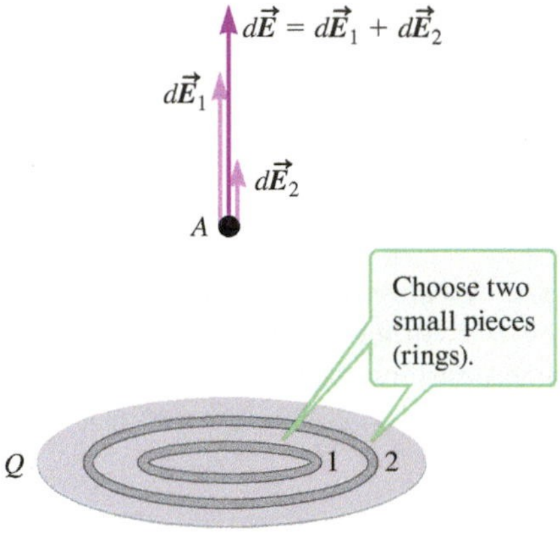

FIGURE 24.32

The electric field due to each ring adds without any component cancelling, so there is no need to consider more than one ring. Figure 24.33 shows a simplified sketch with a **coordinate system** and **geometric details**. Only one thin ring is shown, with radius r and thickness dr. If you imagine cutting the ring and uncurling it, it would be a rectangle of length $2\pi r$ and width dr, so the area of the ring is $dA = 2\pi r\, dr$.

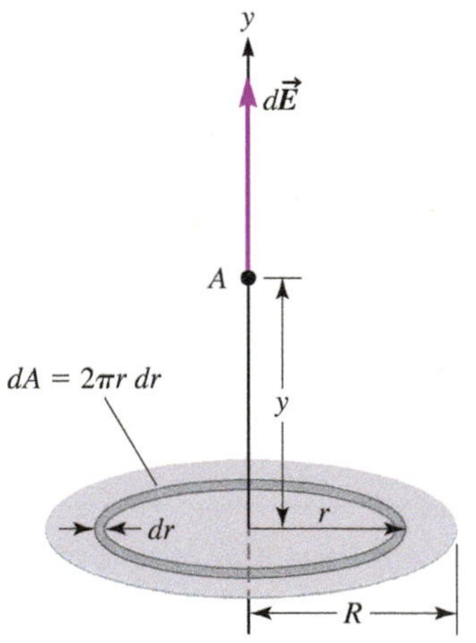

FIGURE 24.33

SOLVE

Step 1 Find an expression for $d\vec{E}$, the electric field due to that one ring of radius r. In this case we can use Equation 24.16, replacing Q with dq and R with lowercase r.

$$d\vec{E} = \frac{k(dq)y}{(r^2 + y^2)^{3/2}}\hat{\jmath} \qquad (1)$$

Step 2 Substitute $dq = \sigma(2\pi r)dr$ into Equation (1) and **integrate** over the entire disk from $r = 0$ to $r = R$ (Problem 81).

$$d\vec{E} = \frac{k\sigma 2\pi r y\, dr}{(r^2 + y^2)^{3/2}}\hat{\jmath} \qquad \int d\vec{E} = \int_0^R \frac{k\sigma 2\pi r y\, dr}{(r^2 + y^2)^{3/2}}\hat{\jmath}$$

$$\vec{E} = k\sigma 2\pi y\hat{\jmath}\int_0^R \frac{r\, dr}{(r^2 + y^2)^{3/2}}$$

$$\vec{E} = 2\pi k\sigma\left[1 - \frac{y}{\sqrt{R^2 + y^2}}\right]\hat{\jmath} \qquad (24.17)$$

ELECTRIC FIELD DUE TO A CHARGED DISK Special Case

CHECK and THINK The **dimensions** are correct—force per charge (Problem 26). To check that the electric field **approaches that of a charged particle** if point A is far away($y \gg R$), start by rearranging Equation 24.17.	$\vec{E} = 2\pi k\sigma\left[1 - \frac{y}{y\sqrt{(R/y)^2 + 1}}\right]\hat{j}$ $\vec{E} = 2\pi k\sigma\left[1 - (1 + (R/y)^2)^{-1/2}\right]\hat{j}$
Then approximate $[1 + (R/y)^2]^{-1/2}$ for the case when R/y is small (Appendix A) and simplify.	$\vec{E} \approx 2\pi k\sigma\left[1 - \left(1 - \frac{1}{2}\frac{R^2}{y^2}\right)\right]\hat{j}$ $\vec{E} \approx k\frac{\sigma\pi R^2}{y^2}\hat{j}$
Use $\sigma = Q/A$ (Eq. 24.10) to write this in terms of the total charge on the disk $Q = \sigma A = \sigma\pi R^2$. This is the electric field of a charged particle at a point y as required.	$\vec{E} \approx k\frac{Q}{y^2}\hat{j}$ ✓

24-7 Case Study: The Shape of Lightning Rods

After conducting his kite-flying experiments that supported his theory of lightning as a giant spark, Franklin came up with a plan to avoid lightning strikes. He knew that objects could be charged by rubbing. He also knew that you cannot charge a metal object—a conductor—by rubbing it if you connect the conductor to the ground. Today we know why: The Earth acts as a source and sink of electrons. If you build up a few excess electrons on a grounded conductor, these electrons quickly travel through the conductor and along the pathway to ground (Fig. 24.34A).

Franklin reasoned that during a storm, the atmosphere builds up excess charge much as a glass rod builds up excess charge when it is rubbed with silk. He knew from his experience with charged objects that a spark can make its way through the air. His indoor experiments demonstrated that if he used a pointed object (like a knitting needle) to draw charge from an object through the air, only a small spark occurred compared to the larger spark drawn to a blunt object (like his thumb). Franklin reasoned that when a small charge builds up in the atmosphere, a pointed lightning rod will draw the small charge through the air and into the Earth continuously (Fig. 24.34B). If only a small charge travels through the air, it will not be visible and there will be no giant spark of lightning.

Without a lightning rod, charge still builds up in the atmosphere. Because buildings are connected to the Earth, they attract charge of the opposite sign (Fig. 24.34C). When sufficient charge builds up, the air acts as a conductor and a large charge is transferred in a giant lightning spark. Such a violent spark causes great damage to buildings and can be very dangerous.

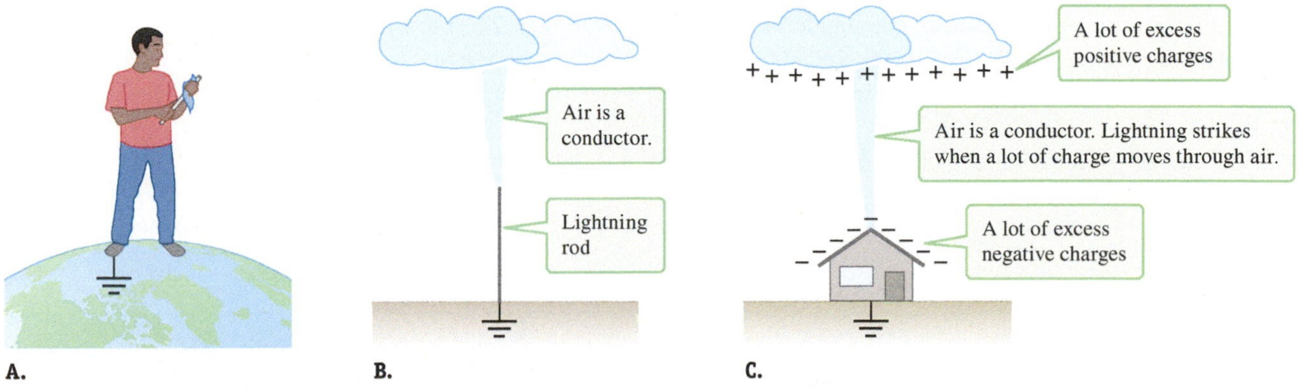

FIGURE 24.34 **A.** Franklin knew that a grounded person cannot charge a conducting rod by rubbing. **B.** According to Franklin, lightning strikes could be prevented if a town put up a lot of lightning rods. Then the atmosphere would be connected to ground, so it could not build up charge. **C.** Without a lightning rod, Franklin reasoned, the atmosphere discharges in a violent strike.

Benjamin Wilson—a contemporary of Franklin's—believed that lightning rods should be blunt. He argued that a pointed rod would draw down lightning that might have just passed harmlessly overhead. Wilson argued that pointed lightning rods were more dangerous than having no rods at all.

In this section, we explore both types of rods (Fig. 24.35). In order for air to break down and become a conductor, the electric field in the air must be 3×10^6 N/C. Let's assume that in order for a lightning rod to work, the electric field at its surface must equal that breakdown electric field. We will calculate the amount of charge on the surface of each conductor. The one with the least amount of charge is the better design because a smaller amount of charge on the surface of the conductor means a smaller amount of charge travels through the air. Calculating the charge on each rod gives us a way to compare their effectiveness. A rod that does not require much charge to have a strong electric field on its surface will not generate a big spark to discharge the atmosphere. Because both designs use conductors, we will assume the excess charge is uniformly spread over the surface of the lightning rod.

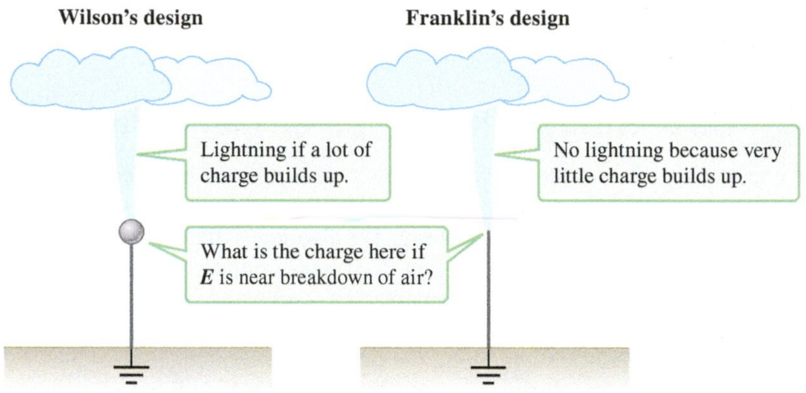

FIGURE 24.35 If there is a ball at the end of the lightning rod, a lot of charge builds up before the atmosphere discharges. If there is no ball at the end of the lightning rod, little charge builds up because the atmosphere continually discharges through the rod.

EXAMPLE 24.7 CASE STUDY Wilson's Blunt Lightning Rod

At the end of Wilson's rod was a cannonball (Fig. 24.35). Cannonballs in the middle of the 18th century varied in size, but let's assume a moderate-sized cannonball with a radius of 4 inches, or $R = 0.1$ m. Because the ball is so much larger than the thickness of the supporting rod, we will ignore that rod and model Wilson's device as a suspended sphere connected to ground. If the electric field on the surface of the ball is $E = 3 \times 10^6$ N/C, what is the charge on the ball?

INTERPRET and ANTICIPATE

The ball is a sphere, so we can use the magnitude of $\vec{E}(r) = (kQ/r^2)\hat{r}$ (Eq. 24.3) to find the charge on the ball.

SOLVE

Solve for the charge Q on the surface of the sphere, where $r = R$.

$$E(R) = k\frac{Q}{R^2}$$

$$Q = \frac{ER^2}{k} = \frac{(3 \times 10^6\ \text{N/C})(0.1\,\text{m})^2}{(8.99 \times 10^9\ \text{N}\cdot\text{m}^2/\text{C}^2)}$$

$$Q = 3.3 \times 10^{-6}\,\text{C} = 3.3\,\mu\text{C}$$

CHECK and THINK

This is not very much charge, especially compared to the amount of charge involved in a lightning strike, which is on the order of hundreds of coulombs. However, our job is not done. We need to see how Franklin's pointed rod would do under the same conditions.

EXAMPLE 24.8 CASE STUDY Franklin's Pointed Lightning Rod

Franklin's lightning rod is very thin. We don't have an expression for the electric field near the end of a very thin rod. (Equation 24.15 gives the field at a distance y from the midpoint of a charged rod but is not valid near the tip.)

A Find an expression for the electric field at the tip of Franklin's pointed lightning rod. Evaluate Franklin's design by considering a point very close to the end of the rod.

INTERPRET and ANTICIPATE

Figure 24.36 shows all four elements of our usual sketch. A positive **test charge** is at point A just above the end of the rod. The rod is divided into many pieces. **One piece** is shown as well as the **electric field** $d\vec{E}$ produced by that piece. Any piece will produce an upward-pointing electric field, so there is no cancellation. The figure includes a **coordinate system** and **geometric details.**

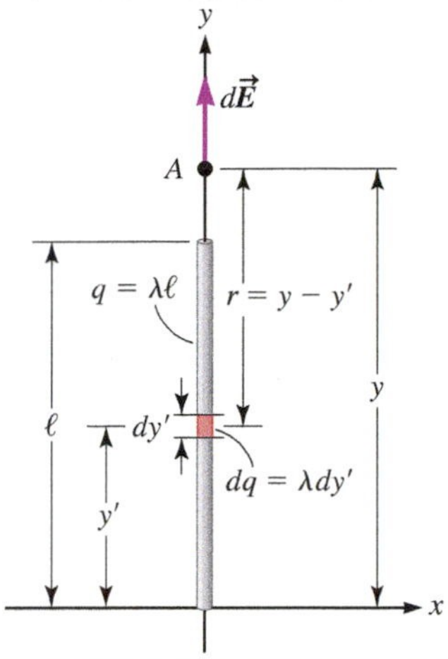

FIGURE 24.36

SOLVE

Step 1 Use Equation 24.14 to derive an **expression** for $d\vec{E}$.

$$d\vec{E} = k\frac{dq}{r^2}\hat{\jmath}$$

Use the coordinate system in Figure 24.36 to express the distance from the piece of charge to the point of interest as $r = y - y'$. The infinitesimal charge dq comes from $dq = \lambda dy'$ (Eq. 24.13).

$$d\vec{E} = k\frac{\lambda dy'}{(y - y')^2}\hat{\jmath}$$

Step 2 **Integrate** over the entire length of the rod from $y = 0$ to $y = \ell$. Complete the integration and evaluate between the limits (Problem 82).

$$\int d\vec{E} = \int_0^\ell k\frac{\lambda dy'}{(y - y')^2}\hat{\jmath}$$

$$\vec{E} = k\lambda\hat{\jmath}\int_0^\ell \frac{dy'}{(y - y')^2} = \frac{k\lambda\ell}{y(y - \ell)}\hat{\jmath}$$

The total charge is $q = \lambda\ell$. Substitute this expression for q.

$$\vec{E} = \frac{kq}{y(y - \ell)}\hat{\jmath} \tag{1}$$

CHECK and THINK

Before testing Franklin's design, first check Equation (1) for the **dimensions** of force per charge (Problem 40). Also, if point A is far from the rod $(y \gg \ell)$, the electric field **approaches the electric field due to a charged particle.**

$$\lim_{y \gg \ell} \vec{E} = \frac{kq}{y(y - \ell)}\hat{\jmath} \rightarrow \frac{kq}{y(y)}\hat{\jmath}$$

$$\vec{E} \approx \frac{kq}{y^2}\hat{\jmath} \checkmark$$

SOLVE

Now use Equation (1) to evaluate Franklin's design. As point A gets closer to the end of the rod, the electric field increases. At the tip of the rod, the electric field approaches infinity. So, in principle, even a very tiny charge on the rod would lead to a very high electric field at its end. Franklin's rod would continually draw a small amount of charge from the atmosphere and prevent a large lightning strike.

$$\lim_{y \to \ell} E = \lim_{y \to \ell}\left[\frac{kq}{y(y - \ell)}\right]$$

$$\lim_{y \to \ell} E \rightarrow \frac{1}{0} \rightarrow \infty$$

Example continues on page 736 ▶

B It seems that an ideal Franklin rod (one that is infinitely thin) would easily win over the blunt Wilson rod. As one final check, let's model the *end* of a Franklin rod as a tiny ball of radius $R = 2$ mm—something like the end of a knitting needle. If the electric field on the surface of the ball is $E = 3 \times 10^6$ N/C, what is the charge on the ball? (We'll consider a thin-rod model in Problem 42.)

SOLVE

Repeat the calculation from Example 24.7 for the Wilson rod, this time for a much smaller ball. Solve for the charge q on the surface of the sphere, where $r = R$. The charge on Franklin's rod is about 2500 times lower than on Wilson's.

$$E(R) = k\frac{q}{R^2}$$

$$q = \frac{ER^2}{k} = \frac{(3 \times 10^6\ \text{N/C})(0.002\ \text{m})^2}{(8.99 \times 10^9\ \text{N}\cdot\text{m}^2/\text{C}^2)}$$

$$q = 1.3 \times 10^{-9}\ \text{C} = 1.3\ \text{nC}$$

CHECK and THINK

Even if we model the end of Franklin's rod as a tiny ball, we find that it requires far less charge than Wilson's rod ($q \ll Q$) in order to produce a very large electric field at its tip. According to Franklin, this small amount of charge would travel across the air and keep the atmosphere discharged. Franklin actually suggested that if lightning rods were used all over a town, it would be possible to prevent the local atmosphere from building up any charge at all.

Lightning Rods on Buckingham Palace

In the 18th century, no one could carry out the calculations in Examples 24.7 and 24.8 showing that less charge travels through the air in the case of pointed rods. Nevertheless, the practical question of what shape of rods should be placed on a building needed to be answered. In 1771, the British Parliament needed to know how the arsenal at Purfleet on the Thames should be protected from lightning. The Royal Society formed a committee to investigate and make a recommendation. Both Franklin and Wilson served on the committee. The committee considered the path the lightning took when it struck an unprotected building, and it considered the effectiveness of lightning rods that had been in use for a few decades. Although Wilson objected, the committee recommended that pointed rods be installed on the arsenal. King George III must have been impressed by the committee's recommendation because he ordered that his house—now called *Buckingham Palace*—be protected by pointed lightning rods as well.

The story may have ended there, but two things happened. First, Franklin fell out of favor with British royalty when he supported America's war of independence. Second, lightning struck a building on the grounds of the Purfleet arsenal, although the protected gunpowder magazine was not hit. With Franklin out of the country, the king gave Wilson permission to test the two types of lightning rods in the London Pantheon. Wilson charged an enormous cylinder that was 16 inches in diameter and 155 feet long. The cylinder was hung from the ceiling by silk to simulate a charged-up atmosphere. Under the cylinder was a model of the Purfleet arsenal with both blunt and pointed lightning rods. The model was on a trolley so that it could be rolled under the charged cylinder, simulating a storm rolling in over a town. Both rods were found to draw sparks, but the pointed rods drew sparks from a greater distance than the blunt rods. Wilson argued that this made the pointed rods more dangerous. Franklin's supporters argued they were safer because the "damaging power" of the strikes was weaker.

In fact, the difference between a blunt and a pointed rod is irrelevant in the case of a full-scale lightning storm. The charged particles in the atmosphere are at an altitude of several kilometers. Imagine looking at the ends of a blunt rod and a pointed rod from a distance of a few kilometers. You could not distinguish one from the other, and the exact shape of the rod makes little difference. So neither Franklin nor Wilson was completely correct. Wilson was wrong to think that pointed rods were more dangerous than no rods at all, and Franklin's rod was too far from the charged particles to quietly discharge the atmosphere.

So, lightning still strikes structures equipped with lightning rods. Lightning rods are important, however, because they provide a pathway to ground that prevents damage to structures. Although some contemporary lightning protection systems are fairly complicated, many are relatively simple rods based on Franklin's original pointed design. We will continue to explore lightning protection in Chapter 25.

24-8 Charged Particle in an Electric Field

The payoff of writing an expression for the electric field produced by some source is that once you have done that, it is relatively easy to find the force exerted on some subject. In this section, we calculate the force exerted on a single charged particle in an electric field.

No matter how complicated the electric field due to some source, the force exerted by that source on a single particle with charge q is

$$\vec{F}_E = q\vec{E} \tag{24.18}$$

ELECTROSTATIC FORCE EXERTED BY ELECTROSTATIC FIELD

Major Concept

Keep in mind that the source—a charged particle, a collection of particles, or a continuous distribution of charged particles—sets up the electric field $\vec{E}$. A subject—in this case, a single particle with charge q—is in the source's electric field, and the force on the subject is given by Equation 24.18. If the subject is positively charged, the force exerted on it by the source is in the same direction as the electric field. If the subject is negatively charged, the force is in the opposite direction to the electric field. After you have calculated the force exerted on the subject, you can use all the tools of Part I to calculate the kinematics and dynamics of the particle.

EXAMPLE 24.9 Millikan's Oil Drop Experiment

Figure 24.37 shows the apparatus used by Robert Millikan (1868–1953) in the early 20th century to measure the elementary charge e. Although his measurement was off, he won the Nobel Prize as a result of his experiment. Today, students use similar equipment to reproduce Millikan's measurement. The apparatus is a cylinder divided into two chambers. Oil drops are sprayed into the upper chamber, and some of them become positively or negatively charged in the process. A few drops make it through a small hole in the disk that separates the two chambers. An experimenter can view these drops with the aid of a microscope. The disks that make up the ceiling and floor of the lower chamber are connected to a battery. When the experimenter switches on the battery, the upper disk is positively charged and the lower disk is grounded.

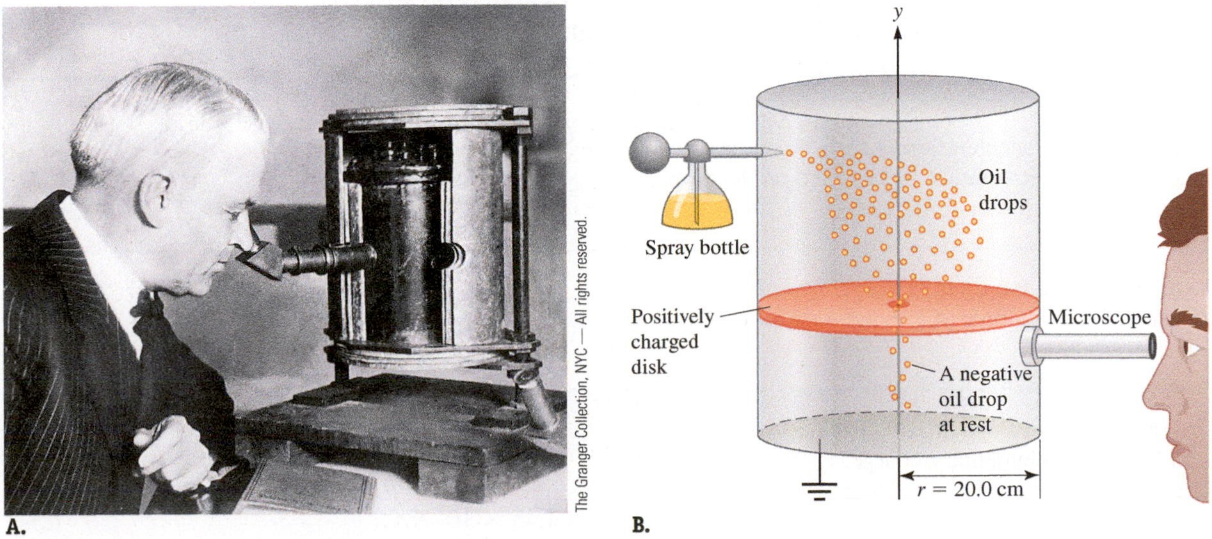

FIGURE 24.37 **A.** Millikan measured the elementary charge e using this apparatus. **B.** Diagram showing the major components of Millikan's experiment.

Example continues on page 738 ▶

In this example, we consider a single charged oil drop of mass $m = 7.26 \times 10^{-16}$ kg. The oil drop has one excess electron and is at rest in front of the microscope.

A The oil drop is relatively close to the charged disk, corresponding to $y \ll R$ in Figure 24.33. Use the result of Example 24.6 to show that the electric field near a charged surface is

$$\vec{E} \approx 2\pi k\sigma\hat{j} \tag{24.19}$$

ELECTRIC FIELD NEAR A CHARGED SURFACE Special Case

INTERPRET and ANTICIPATE

Begin with Equation 24.17, the result of Example 24.6, to find an approximate expression for the electric field due to a charged disk at point A (Fig. 24.33) close to the disk.

SOLVE

If point A is close to the disk compared to the disk's radius ($y \ll R$), then the second term in Equation 24.17 is essentially zero. The electric field is uniform—a constant value—for all points near the disk.

$$\vec{E} = 2\pi k\sigma\left[1 - \frac{y}{\sqrt{R^2 + y^2}}\right]\hat{j} \tag{24.17}$$

$$\vec{E} \approx 2\pi k\sigma\hat{j} \; ✓ \tag{24.19}$$

CHECK and THINK

Equation 24.19 has the right dimensions—force per charge. What seems odd is that the electric field is uniform. We cannot check this expression in the limit of a very distant point because it is good only for points close to the disk. However, Equation 24.19 is a good approximation for any point near a charged plane; the shape of the plane does not matter (disk or rectangle) as long as the point of interest is close to the plane and not near its edges. This is an important equation, so it appears in the summary of this chapter.

B What are the magnitude and direction of the electric field produced by the charged disk at the position of the oil drop?

INTERPRET and ANTICIPATE

We can approach this part of the problem as we would any problem involving Newton's second law. A free-body diagram for the charged oil drop (Fig. 24.38) shows two forces: the downward force of gravity and the upward force exerted by the positively charged disk above the drop.

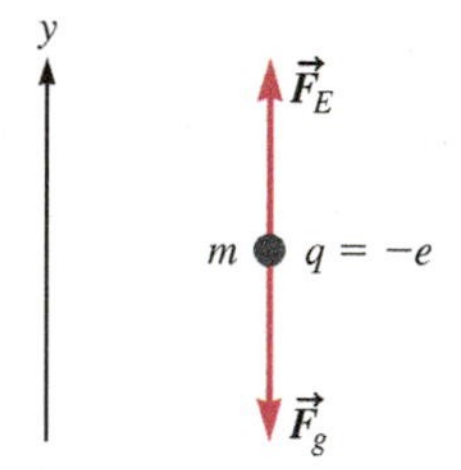

FIGURE 24.38 $\vec{a} = 0 \Rightarrow \vec{F}_E = -\vec{F}_g$

SOLVE

The oil drop is at rest, so its acceleration is zero. According to Newton's second law, the electrostatic force exerted on the drop must be balanced by the gravitational force.

$$\sum F_y = F_E - F_g = 0$$

$$F_E = F_g$$

The magnitude of the electric force can be found from $\vec{F}_E = q\vec{E}$ (Eq. 24.18). As usual, the magnitude of gravitational force is the oil drop's weight.

$$|q|E = mg$$

Solve for the magnitude of the electric field. The mass of the oil drop is given, and $|q|$ is the elementary charge.

$$E = \frac{mg}{|q|} = \frac{(7.26 \times 10^{-16}\text{ kg})(9.81\text{ m/s}^2)}{|-1.60 \times 10^{-19}\text{ C}|}$$

$$E = 4.45 \times 10^4\text{ N/C}$$

The oil drop is negatively charged, so the force exerted on it by the electric field is in the *opposite* direction to the electric field. Because the electrostatic force must be upward in the positive y direction, the electric field must be downward in the negative y direction.	$\vec{E} = -4.45 \times 10^4 \hat{\jmath}\ \text{N/C}$

C If the radius of the cylinder is $R = 20.0$ cm, estimate the amount of excess charge on the disk.

INTERPRET and ANTICIPATE

Because we know the magnitude of the electric field from part B, we can use Equation 24.19 (part A) to find the surface charge density on the disk. Then we find the area of the disk from its radius, enabling us to find its excess charge.

SOLVE

Solve for the charge density σ.

$$E \approx 2\pi k\sigma$$

$$\sigma \approx \frac{E}{2\pi k} \approx \frac{4.45 \times 10^4\ \text{N/C}}{2\pi(8.99 \times 10^9\ \text{N}\cdot\text{m}^2/\text{C}^2)}$$

$$\sigma \approx 7.88 \times 10^{-7}\ \text{C/m}^2$$

Use $\sigma = Q/A$ (Eq. 24.10) to find the total charge on the disk.

$$Q = \sigma A = \sigma(\pi R^2) \approx (7.88 \times 10^{-7}\ \text{C/m}^2)\pi(0.200\ \text{m})^2$$

$$Q \approx 9.90 \times 10^{-8}\ \text{C} \approx 99.0\ \text{nC}$$

CHECK and THINK

The excess charge must be positive in order for the electric field to point downward in the lower chamber.

EXAMPLE 24.10 Jumping Through the Hoop

A particle with charge $q = -30.0$ nC and mass $m = 50.0\ \mu$g is placed 2.00 m directly above the middle of a charged ring (Fig. 24.39). The ring has positive charge $Q = +60.0$ nC and radius $R = 0.500$ m.

A Consider the ring to be the source, and plot the (scalar component of) electric field produced by the ring from $y = -2.00$ m to $y = 2.00$ m. According to your graph, where is the magnitude of the electric field the greatest?

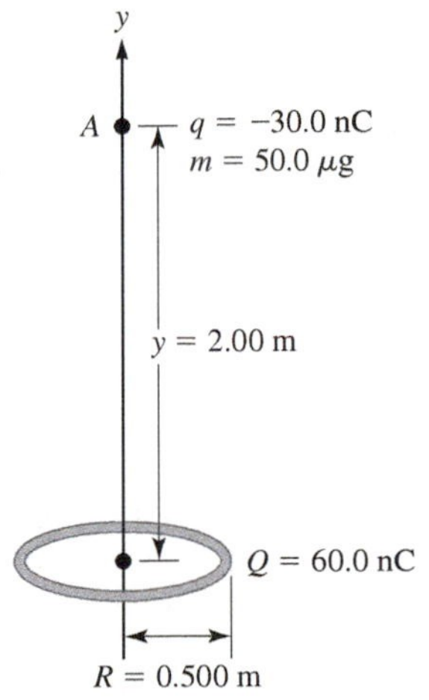

FIGURE 24.39

INTERPRET and ANTICIPATE

Our job amounts to plotting Equation 24.16—the result of Example 24.5—for the values given in the problem statement.

SOLVE

Start with the scalar component. Then substitute the numerical values of all constants and reduce the function to its simplest form before plotting.

$$E_y(y) = \frac{kQy}{(R^2 + y^2)^{3/2}}$$

$$E_y(y) = \frac{(8.99 \times 10^9\ \text{N}\cdot\text{m}^2/\text{C}^2)(60.0 \times 10^{-9}\ \text{C})y}{[(0.500\ \text{m})^2 + y^2]^{3/2}}$$

$$E_y(y) = \frac{(5.39 \times 10^2\ \text{N}\cdot\text{m}^2/\text{C})y}{[(0.250\ \text{m}^2) + y^2]^{3/2}} \quad (1)$$

Example continues on page 740 ▶

Using a spreadsheet program or working by hand, calculate E_y for values of y between $y = -2.00$ m and $y = 2.00$ m. The results are shown in Figure 24.40.

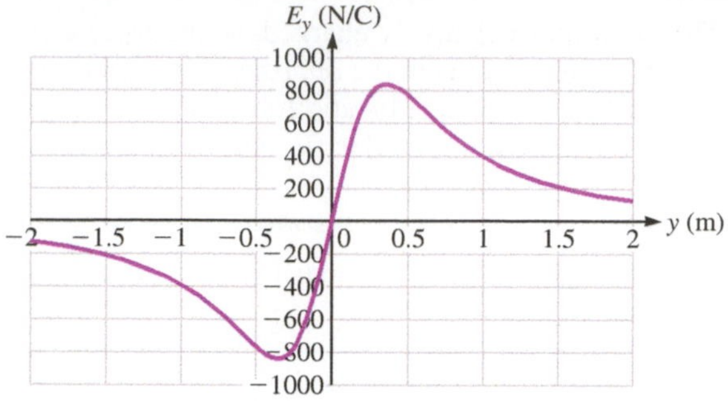

FIGURE 24.40

Inspect the graph (Fig. 24.40) to find the greatest magnitude of the electric field and read off the positions. There are two points where the electric field reaches its greatest magnitude.

The magnitude of the electric field is greatest at $y \approx \pm 0.4$ m.

CHECK and THINK

Because of the symmetry, it is not surprising that the electric field has a maximum at points equidistant from the ring.

B The charged particle is the subject in the ring's electric field. Plot the (scalar component of) force exerted by the ring on the particle from $y = -2.00$ m to $y = 2.00$ m. According to your graph, where is the magnitude of the force the greatest?

INTERPRET and ANTICIPATE

According to $\vec{F}_E = q\vec{E}$ (Eq. 24.18), to find an expression for the force, we multiply the expression for the electric field by the charge of the subject.

SOLVE

Multiply Equation 24.16 by q, the charge of the subject.

$$F_y = qE_y = q\frac{kQy}{(R^2 + y^2)^{3/2}} = \frac{kQqy}{(R^2 + y^2)^{3/2}}$$

As in part A, substitute numerical values for the constants before plotting them. We already did some of this work in Equation (1).

$$F_y(y) = \frac{(-30.0 \times 10^{-9}\,\text{C})(5.39 \times 10^2\,\text{N}\cdot\text{m}^2/\text{C})y}{[(0.250\,\text{m}^2) + y^2]^{3/2}}$$

$$F_y(y) = -\frac{(1.62 \times 10^{-5}\,\text{N}\cdot\text{m}^2)y}{[(0.250\,\text{m}^2) + y^2]^{3/2}}$$

Calculate F_y for values of y between $y = -2.00$ m and $y = 2.00$ m. The results are shown in Figure 24.41.

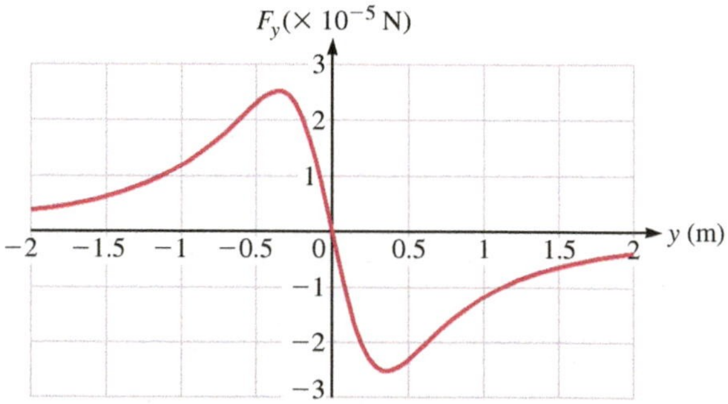

FIGURE 24.41

Inspect the graph (Fig. 24.41) to find the greatest magnitude of the force and read off the positions. There are two points where the force reaches its greatest magnitude.

The magnitude of the force is greatest at $y \approx \pm 0.4$ m.

CHECK and THINK

Our results here are consistent with part A. First, because force is proportional to the electric field, you should expect the strongest force to occur where the electric field is greatest. Second, the subject has a negative charge, so the force exerted on it is in the opposite direction to the electric field. Figure 24.41 is an inverted (upside down) version of Figure 24.40.

C Assume that the only force acting on the particle is the electrostatic force. Use your graph from part B to find the magnitude of the maximum force exerted on the particle. Compare your answer to the particle's weight. What is the magnitude of the particle's maximum acceleration?

INTERPRET and ANTICIPATE

The strongest force is found from reading the value of the peak in Figure 24.41. Once we know that force, we can find the acceleration from Newton's second law.

SOLVE

Read the value of the peak from Figure 24.41.	$F_{max} = 2.5 \times 10^{-5}\,N$
Find the weight of the particle from its mass.	$F_g = mg$ $F_g = (50.0 \times 10^{-9}\,kg)(9.81\,m/s^2)$ $F_g = 4.90 \times 10^{-7}\,N$
Compare the maximum force exerted by the ring to the weight of the particle by forming a ratio. The maximum electrostatic force is about 50 times greater than the particle's weight, so we expect the particle's acceleration to be about 50 times greater than the free-fall acceleration.	$\frac{F_{max}}{F_g} = \frac{2.5 \times 10^{-5}\,N}{4.90 \times 10^{-7}\,N} = 51$ $F_{max} = 51F_g$
Use Newton's second law to find the particle's maximum acceleration, assuming the only force exerted on the particle is the electrostatic force exerted by the charged disk.	$F_{max} = ma_{max}$ $a_{max} = \frac{F_{max}}{m} = \frac{2.5 \times 10^{-5}\,N}{50.0 \times 10^{-9}\,kg} = 500\,m/s^2$

CHECK and THINK

As expected, the particle's acceleration is nearly 50 times greater than free-fall acceleration. This means that if we were to take gravity into account, rather than assuming that the only force is the electrostatic force, we would find that the particle's acceleration is only slightly different from our value.

24-9 Special Case: Dipole in an Electric Field

When you make popcorn in a microwave oven, the oven creates an oscillating electric field. (The electric field inside the oven points first one way, then the opposite way, and so on.) Food contains water molecules, and the oscillating electric field causes the water molecules to rotate back and forth. So, the electric field increases the thermal energy of the water molecules. Those water molecules transfer thermal energy to other molecules, and the food cooks. How does the electric field cause water molecules to rotate? Remember that water is modeled as a dipole (Fig 24.20). Therefore, to answer the question, we need to consider a dipole in an electric field.

Figure 24.42 shows a dipole in a uniform electric field. (It might help to imagine that the dipole is near the surface of a charged disk.) The electric field lines are evenly spaced, which indicates an electric field with the same value everywhere. The force on the positive particle $\vec{F}_{(+)}$ points in the positive x direction, and the force on the negative particle $\vec{F}_{(-)}$ points in the negative x direction. Because the electric field has the same strength at both locations and the two charges have the same magnitude, these two forces have equal magnitudes. The result is that the net force on the electric dipole in a uniform electric field is zero, so the dipole will have no translational acceleration.

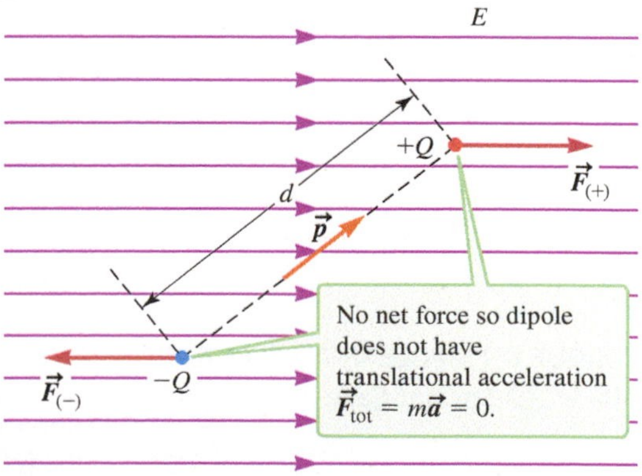

FIGURE 24.42 No net force is exerted on an electric dipole in a uniform electric field. The magnitude of the dipole moment is $p = Qd$.

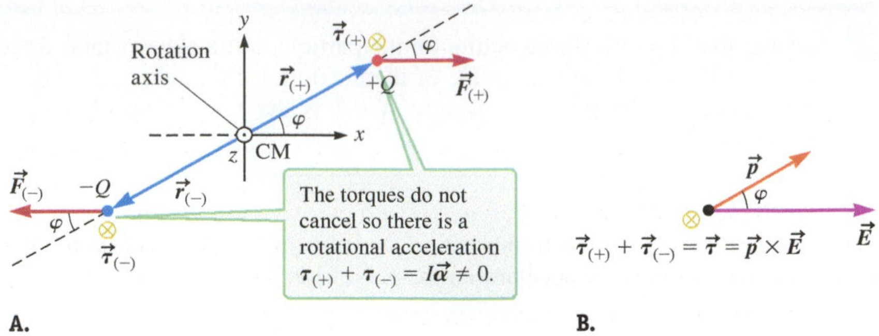

FIGURE 24.43 **A.** The net force is zero, but the net torque is not zero. **B.** The net torque may be written in terms of the dipole moment and the electric field.

However, the two forces create a net torque (Fig. 24.43A). You may imagine the dipole being like a dumbbell; if you press on the ends of the dumbbell with antiparallel forces, the dumbbell will rotate. The same thing happens here: The electric field causes the dipole to rotate around its center of mass. We can use $\tau = rF\sin\varphi$ (Eq. 12.16) and the right-hand rule to calculate the torques around the center of mass. For the torque on the positive charge,

$$\vec{\tau}_{(+)} = -r_{(+)}F_{(+)}\sin\varphi\hat{k}$$

The torque on the negative charge is

$$\vec{\tau}_{(-)} = -r_{(-)}F_{(-)}\sin\varphi\hat{k}$$

The magnitudes of the forces are equal; let's use the symbol F_E for both:

$$F_E \equiv F_{(+)} = F_{(-)}$$

The magnitudes of the position vectors both equal $d/2$:

$$r_{(+)} = r_{(-)} = \frac{d}{2}$$

Therefore, the torque exerted on the positive charge equals the torque exerted on the negative charge:

$$\vec{\tau}_{(+)} = \vec{\tau}_{(-)} = -\frac{d}{2}F_E\sin\varphi\hat{k}$$

The total torque exerted by the electric field on the dipole is the sum of the torques on both charges:

$$\vec{\tau} = \vec{\tau}_{(+)} + \vec{\tau}_{(-)}$$

$$\vec{\tau} = -\frac{d}{2}F_E\sin\varphi\hat{k} - \frac{d}{2}F_E\sin\varphi\hat{k}$$

$$\vec{\tau} = -dF_E\sin\varphi\hat{k} \tag{24.20}$$

The magnitude of the force comes from $\vec{F}_E = q\vec{E}$ (Eq. 24.18), so the torque on the dipole is

$$\vec{\tau} = -d(QE)\sin\varphi\hat{k}$$

The dipole moment is $p = Qd$, so the torque is

$$\vec{\tau} = -pE\sin\varphi\hat{k} \tag{24.21}$$

Equation 24.21 is the cross product of $\vec{p}$ and $\vec{E}$ (Fig. 24.43B):

TORQUE ON DIPOLE IN AN ELECTRIC FIELD **Special Case**

$$\vec{\tau} = \vec{p} \times \vec{E} \tag{24.22}$$

When a water molecule is in a microwave oven, the electric field reverses direction rapidly. Imagine the electric field in Figure 24.43 pointing to the right, then to the left, and so on. The torque on the water molecule switches direction, pointing in the negative z direction, then in the positive z direction, and so on. The water molecule rotates clockwise, then counterclockwise, and so on. All this rotation increases the thermal energy of the water and cooks your food. (This is something like the paddle wheels Joule used in his container of water. See Fig. 21.2, page 612.)

Potential Energy of a Dipole in an Electric Field

To derive an expression for potential energy, let's consider the dipole and the source of the electric field to be part of a system. With the source inside the system, potential energy can be stored by the electric field just as when we consider the Earth to be part of a system and then potential energy is stored by the Earth's gravitational field. Imagine an external agent exerts a torque in the positive z direction, so that the dipole in Figure 24.43 rotates counterclockwise at constant angular velocity with no change in kinetic energy or thermal energy. Then the work done by the external agent equals the change in the system's potential energy:

$$K_i + U_i + W = K_i + U_f + \Delta E_{\text{th}}$$
$$W = U_f - U_i = \Delta U$$

The work done by the external agent is given by

$$\Delta U = W = \int_{\varphi_i}^{\varphi_f} \tau d\varphi \tag{13.18}$$

As always, only changes in potential energy are significant, so potential energy is usually measured from some reference configuration. For a dipole in an electric field, the reference configuration is shown in Figure 24.44A, where the dipole moment is perpendicular to the electric field, $\varphi = 90°$. Then a counterclockwise rotation (Fig. 24.44B) to $\varphi > 90°$ represents an increase in potential energy, and a clockwise rotation (Fig. 24.44C) to $\varphi < 90°$ represents a decrease in potential energy. With this convention for the reference configuration, Equation 13.18 becomes

$$U = W = \int_{90°}^{\varphi} \tau d\varphi \tag{24.23}$$

If the dipole rotates at a constant angular velocity, the torque exerted by the external agent must be balanced by the torque exerted by the electric field. Therefore, the magnitude of the torque exerted by the external agent is $\tau = pE \sin \varphi$ (Eq. 24.21). The external agent does positive work in rotating the dipole from $\varphi_i = 90°$ to $\varphi_f = \varphi$, and Equation 24.23 becomes

$$U = W = \int_{90°}^{\varphi} pE \sin \varphi \, d\varphi \tag{24.24}$$

We integrate (Appendix A) and drop the W:

$$U = -pE(\cos \varphi - \cos 90°)$$
$$U = -pE \cos \varphi \tag{24.25}$$

Equation 24.25 is the (negative) dot product of $\vec{p}$ and $\vec{E}$:

$$U = -\vec{p} \cdot \vec{E} \tag{24.26}$$

POTENTIAL ENERGY OF DIPOLE IN AN ELECTRIC FIELD Special Case

with the reference configuration set to $\varphi_i = 90°$.

When there is no external agent, as in the case of a water molecule in a microwave oven, rotation of the dipole results in a change in the potential energy and a change in the dipole's kinetic energy, thermal energy, or both.

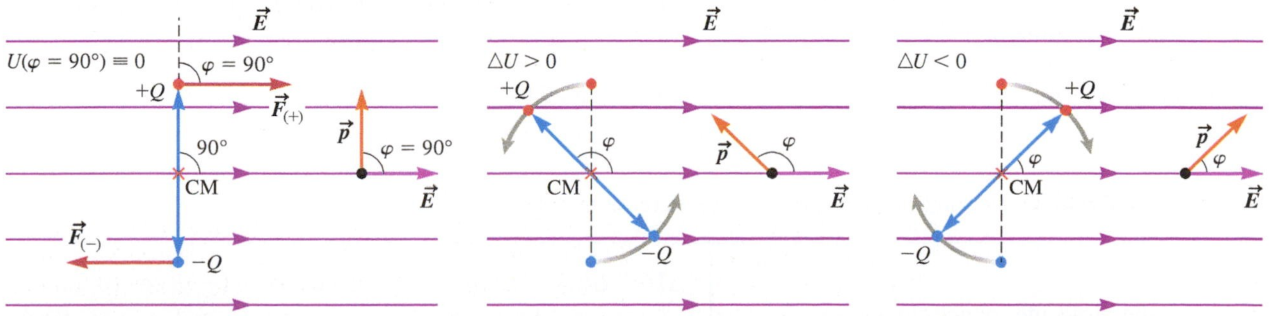

A. Reference configuration **B. Potential energy increases.** **C. Potential energy decreases.**

FIGURE 24.44 **A.** Choose the reference configuration for the angular position φ. **B.** When $\varphi > 90°$, the potential energy increases. **C.** When $\varphi < 90°$, the potential energy decreases.

EXAMPLE 24.11 Rotating Water Molecules

In Concept Exercise 24.5, we found that a water molecule is made up of two hydrogen atoms and one oxygen atom, for a total of 10 electrons and 10 protons. The molecule is modeled as a dipole with an effective separation $d = 3.9 \times 10^{-12}$ m between its positive and negative charges. The water molecule's dipole moment is $p = 6.2 \times 10^{-30}$ C·m. Consider water molecules in the lower atmosphere where the electric field has a magnitude of about 325 N/C. Assume the field is uniform.

A What is the maximum torque exerted on a water molecule?

INTERPRET and ANTICIPATE

The magnitude of the torque depends on the sine of the angle between $\vec{p}$ and $\vec{E}$. The sine function is maximum when $\varphi = 90°$, so we need to find the torque when $\varphi = 90°$.

SOLVE

Substitute numerical values for the magnitude of $\vec{\tau} = -pE\sin\varphi\hat{k}$ (Eq. 24.21).

$$\tau = pE\sin\varphi = (6.2 \times 10^{-30}\,\text{C}\cdot\text{m})(325\,\text{N/C})\sin 90°$$

$$\tau = 2.0 \times 10^{-27}\,\text{N}\cdot\text{m}$$

CHECK and THINK

This torque may seem tiny, but the rotational inertia for an object as small and light as a water molecule is also very small, so the molecule's angular acceleration is still significant.

B What is the change in potential energy if a water molecule rotates from $\varphi = 170°$ to $180°$? From $\varphi = 90°$ to $100°$? From $\varphi = 10°$ to $0°$?

INTERPRET and ANTICIPATE

In this part of the problem we use $U = -\vec{p}\cdot\vec{E}$ (Eq. 24.26) to calculate the change in potential energy due to the rotation of a water molecule. From Figure 24.44, we expect an increase in potential energy when the molecule rotates counterclockwise and a decrease when it rotates clockwise.

SOLVE

For $\varphi = 170°$ to $180°$, substitute numerical values to find ΔU. Remember that 1 N·m = 1 J.

$$\Delta U = U(180°) - U(170°)$$

$$\Delta U = (-pE\cos 180°) - (-pE\cos 170°)$$

$$\Delta U = -(6.2 \times 10^{-30}\,\text{C}\cdot\text{m})(325\,\text{N/C})(\cos 180° - \cos 170°)$$

$$\Delta U = 3.1 \times 10^{-29}\,\text{N}\cdot\text{m} = 3.1 \times 10^{-29}\,\text{J}$$

CHECK and THINK

As expected, this counterclockwise rotation results in an increase in potential energy.

SOLVE

For $\varphi = 90°$ to $100°$, substitute numerical values to find ΔU. Remember that at the reference configuration ($\varphi = 90°$), the potential energy is zero.

$$\Delta U = U(100°) - U(90°) = U(100°) - 0 = -pE\cos 100°$$

$$\Delta U = -(6.2 \times 10^{-30}\,\text{C}\cdot\text{m})(325\,\text{N/C})\cos 100°$$

$$\Delta U = 3.5 \times 10^{-28}\,\text{J}$$

CHECK and THINK

As expected, this counterclockwise rotation also results in an increase in potential energy.

SOLVE

For $\varphi = 10°$ to $0°$, the molecule rotates clockwise.

$$\Delta U = U(0) - U(10°) = (-pE\cos 0) - (-pE\cos 10°)$$

$$\Delta U = -(6.2 \times 10^{-30}\,\text{C}\cdot\text{m})(325\,\text{N/C})(\cos 0 - \cos 10°)$$

$$\Delta U = -3.1 \times 10^{-29}\,\text{J}$$

CHECK and THINK

As expected, this clockwise rotation results in a decrease in potential energy.

Notice that all the rotations go through 10°, but the change in potential energy is not the same in all three cases. To understand this, note that the torque exerted by the electric field works to align the dipole moment with that field. The system has the lowest potential energy ($-pE$) when $\vec{p}$ is parallel to $\vec{E}$. The system has the highest potential energy when $\vec{p}$ is antiparallel to $\vec{E}$. Because of the cosine function in $|U| = pE\cos\varphi$, the change in potential energy is greatest when the dipole moves between two configurations nearly perpendicular to the field ($\varphi \approx 90°$). In this example, the rotation of the molecule when it was nearly perpendicular to the field ($\varphi = 90°$ to $100°$) has an order-of-magnitude greater change in potential energy than in the other two cases.

SUMMARY

Underlying Principles

Electrostatic field or **electric field** $\vec{E}(r)$ is a vector field resulting from a charged source.

Major Concepts

1. When two objects interact through a field force, one object is the **source** and the other is the **subject**. The source produces the electric field, and the subject is influenced by that field. The electrostatic field depends only on properties of the source, whereas the electrostatic force depends on both the source and the subject.
2. An **electric dipole** consists of two charged particles of magnitude Q but with opposite signs (Fig. 24.15). The particle separation is maintained at a distance d.
3. The **electric dipole moment** $\vec{p}$ is a vector pointing from the negative charge toward the positive charge. The magnitude of the dipole moment is $p = Qd$.
4. The **electrostatic force exerted by an electric field** on a single particle with charge q is:

$$\vec{F}_E = q\vec{E} \tag{24.18}$$

Special Cases

1. **Electric field sources**

a. The electric field due to a **particle** or outside any **charged spherical object**:

$$\vec{E}(r) = k\frac{Q_S}{r^2}\hat{r} \tag{24.3}$$

b. The magnitude of the electric field at a position r far away in any direction from an electric **dipole** ($r \gg d$):

$$E \approx \frac{2kp}{r^3} \tag{24.7}$$

where p is the magnitude of the electric dipole moment.

c. The electric field at a point on the perpendicular bisector (y axis) of a **charged rod** of length 2ℓ and net charge Q (Fig. 24.23):

$$\vec{E} = \frac{kQ}{y}\frac{1}{\sqrt{\ell^2 + y^2}}\hat{j} \tag{24.15}$$

d. The electric field at a point on the y axis passing through the center of a **charged ring** of radius R and net charge Q (Fig. 24.28):

$$\vec{E} = \frac{kQy}{(R^2 + y^2)^{3/2}}\hat{j} \tag{24.16}$$

e. The electric field at a point on the y axis that passes through the center of a **charged disk** of surface charge density σ and radius R (Fig. 24.31):

$$\vec{E} = 2\pi k\sigma\left[1 - \frac{y}{\sqrt{R^2 + y^2}}\right]\hat{j} \tag{24.17}$$

f. For **points near a charged surface** (such as $y \ll R$ in Fig. 24.33), the electric field is a constant:

$$\vec{E} \approx 2\pi k\sigma\hat{j} \tag{24.19}$$

2. Dipole in an electric field
A dipole with dipole moment $\vec{p}$ in electric field $\vec{E}$ experiences a torque

$$\vec{\tau} = \vec{p} \times \vec{E} \tag{24.22}$$

and the potential energy of the dipole-source system is

$$U = -\vec{p} \cdot \vec{E} \tag{24.26}$$

with the reference configuration set to $\varphi_i = 90°$.

Tools

Electric field lines provide a way to visualize the magnitude and direction of an electric field. Electric field lines must originate or terminate (or both) on a charged object (Figs. 24.5 and 24.6). The magnitude of the electric field is shown by the density of the electric field lines: A high density of field lines indicates a region with a relatively strong electric field.

PROBLEM-SOLVING STRATEGY

Electric Field due to a Collection of Charged Particles or due to a Continuous Distribution of Charged Particles

The strategies for finding the electric field due to a collection of particles or due to a continuous distribution of particles are similar. The main difference is that for a continuous distribution you must first imagine slicing up the distribution into tiny pieces, with each piece so small that it can be modeled as a single charged particle.

INTERPRET and ANTICIPATE

Make a diagram showing *all the electric fields* at a particular location. There are four elements to this diagram:

1. **Labeled source particles (or pieces)** and the **test particle**
2. The **electric field vectors** at the location of the test charge
3. The **net electric field vector**
4. A **coordinate system** and other **geometric details**

SOLVE

For a ***collection of particles,*** find a mathematical expression for the electric field at point A by adding the electric field vectors.

For a ***continuous distribution,*** there are two steps:

1. Find a **mathematical expression** for the infinitesimal electric field $d\vec{E}$ in terms of an infinitesimal spatial variable. The magnitude of the electric field due to an infinitesimal amount of charge dq is

$$dE(r) = k\frac{dq}{r^2} \tag{24.14}$$

The direction will come from your diagram.

2. **Integrate** this expression over the entire charge distribution.

CHECK and THINK

1. Make sure the **dimensions** of the electric field are force per charge.
2. If the charge distribution is finite, the **electric field** for a point far from the distribution compared to its size **should approach the electric field of a charged particle,** $\vec{E}(r) = (kQ_S/r^2)\hat{r}$.

PROBLEMS AND QUESTIONS

A = algebraic C = conceptual E = estimation G = graphical N = numerical

24-1 What Are Fields?

1. **C** The terms electrostatic *force* and electrostatic *field* may sound alike. To help keep them straight, identify and write down the standard symbol and SI units for each one. Which one requires a source and a subject? Which requires only a source?
2. **C** Imagine you are planning to send a probe to orbit Jupiter, a planet that has many moons. Explain how you would use the concept of a gravitational field. What would you consider to be part of the source? What would be the subject?

24-2 Special Case: Electric Field of a Charged Sphere

3. **N** A sphere has a charge of −89.5 nC and a radius of 4.65 cm. What is the magnitude of its electric field 3.15 cm from its surface?
4. **G** Plot E_r (scalar component) versus r for a positively charged sphere and then for a negatively charged sphere using $\vec{E}(r) = (kQ_S/r^2)\hat{r}$ (Eq. 24.3). Compare your graphs and check for consistency.

5. **N** A sphere with a charge of -3.50 nC and a radius of 1.00 cm is located at the origin of a coordinate system.
   a. What is the electric field 1.75 cm away from the center of the sphere along the positive y axis?
   b. If a particle with a charge of 5.39 nC were placed at that location, what would be the electrostatic force on this charge?
6. **N** Is it possible for a conducting sphere of radius 0.10 m to hold a charge of 4.0 μC in air? The minimum field required to break down air and turn it into a conductor is 3.0×10^6 N/C.
7. **N** What is the net charge of the Earth if the magnitude of its electric field near the terrestrial surface is 1.30×10^2 N/C? Assume the Earth is a sphere of radius 6.40×10^6 m.

24-3 Electric Field Lines

8. **G** For each sketch of electric field lines in Figure P24.8, compare the magnitude of the electric field in region A to the magnitude of the electric field in region B.

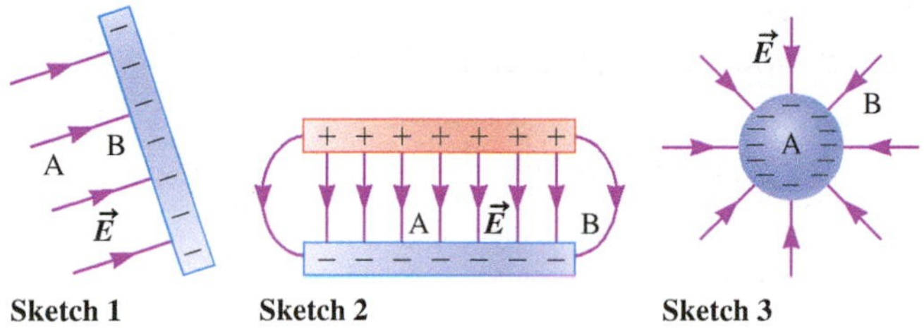

FIGURE P24.8

9. **G** Sketch the electric field lines for a source that consists of two particles carrying opposite charges. The negative particle has more excess charge than the positive one.
10. **C,G** Two large neutral metal plates, fitted tightly against each other, are placed between two particles with charges of equal magnitude but opposite sign, such that the plates are perpendicular to the line connecting the charges (Fig. P24.10). What will happen to each plate when they are released and allowed to move freely? Draw the electric field lines for the particles-plates system.

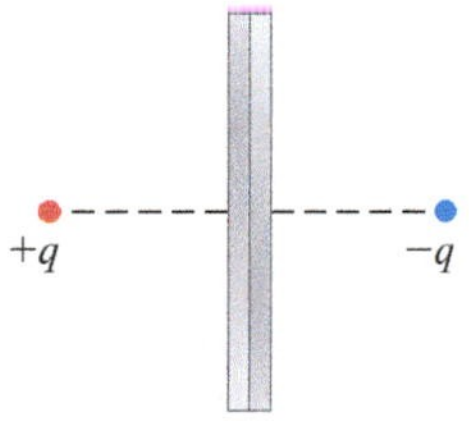

FIGURE P24.10

24-4 Electric Field of a Collection of Charged Particles

11. **N** Given the two charged particles shown in Figure P24.11, find the electric field at the origin.

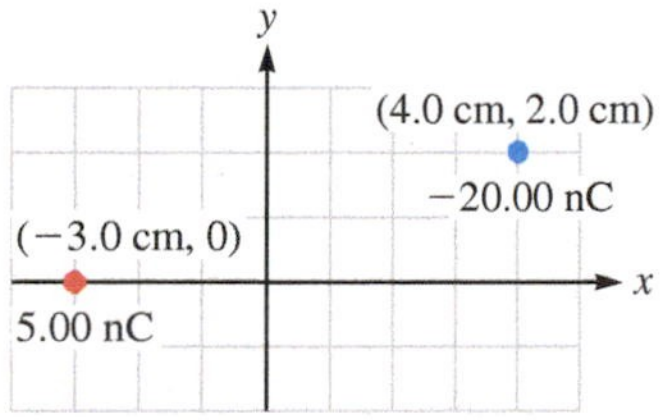

FIGURE P24.11

12. **N** A particle with charge $q_1 = +5.0$ μC is located at $x = 0$, and a second particle with charge $q_2 = -3.0$ μC is located at $x = 15$ cm. Determine the location of a third particle with charge $q_3 = +4.0$ μC such that the net electric field at $x = 25$ cm is zero.
13. **N** A small sphere with charge $q_A = +5.20$ μC is at the origin, and a second sphere with charge $q_B = -3.70$ μC is located at $x = 2.00$ m. At what finite distance from the origin is the electric field equal to zero?

Problems 14 and 15 are paired.

14. **A** A particle with charge q on the negative x axis and a second particle with charge $2q$ on the positive x axis are each a distance d from the origin. Where should a third particle with charge $3q$ be placed so that the magnitude of the electric field at the origin is zero?
15. **N** A particle with charge 2.57 μC on the negative x axis and a second particle with charge 5.14 μC on the positive x axis are each a distance 0.0569 m from the origin. Where should a third particle with charge 7.71 μC be placed so that the magnitude of the electric field at the origin is zero?
16. **N** Figure P24.16 shows three charged particles arranged in the xy plane at the coordinates shown, with $q_A = q_B = -3.30$ nC and $q_C = 4.70$ nC. What is the electric field due to these particles at the origin?

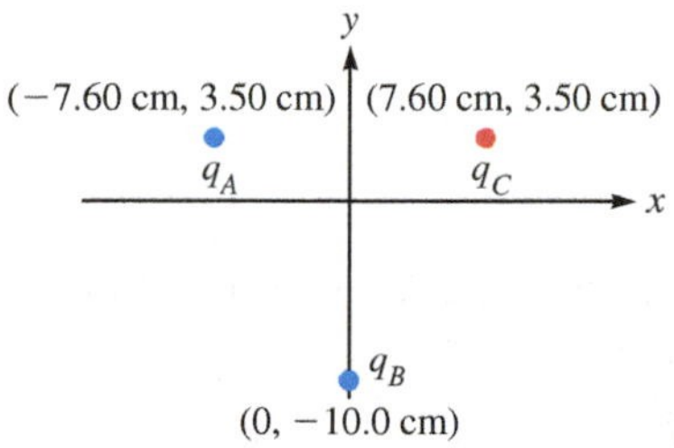

FIGURE P24.16

Problems 17, 18, and 19 are grouped.

17. **N** Figure P24.17 shows a dipole. If the positive particle has a charge of 35.7 mC and the particles are 2.56 mm apart, what is the electric field at point A located 2.00 mm above the dipole's midpoint?
18. **A** Find an expression for the electric field at point A for the dipole source shown in Figure P24.17. Show that when $y \gg d$, the electric field is given by $\vec{E} \approx -k\vec{p}/y^3$.

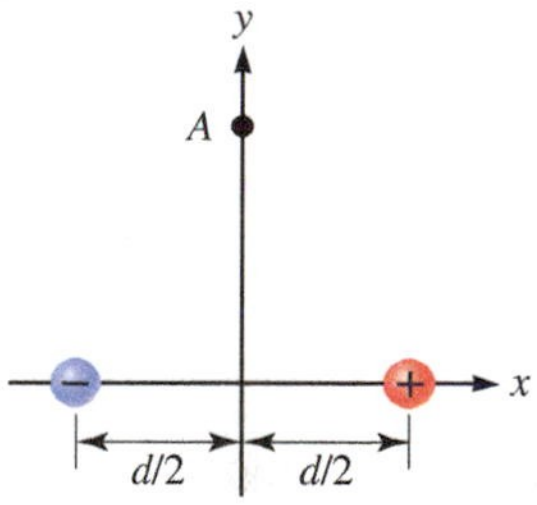

FIGURE P24.17
Problems 17, 18, and 19.

19. **N** Figure P24.17 shows a dipole (not drawn to scale). If the positive particle has a charge of 35.7 mC and the particles are 2.56 mm apart, what is the (approximate) electric field at point A located 2.00 m above the dipole's midpoint?
20. **N** Figure P24.20 shows three charged spheres arranged along the y axis.
    a. What is the electric field at $x = 0$, $y = 3.00$ m?
    b. What is the electric field at $x = 3.00$ m, $y = 0$?

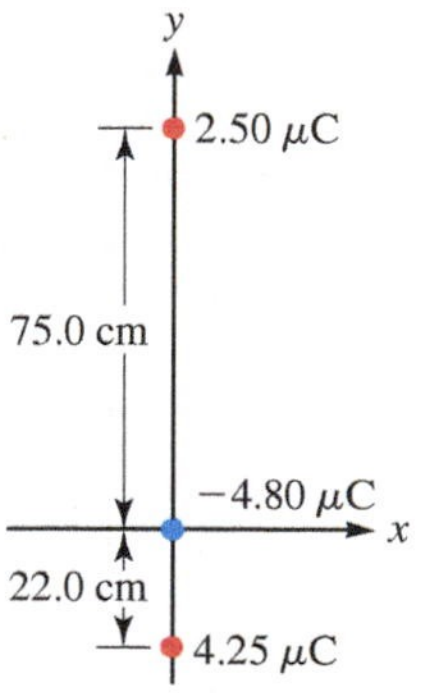

FIGURE P24.20

24-5 Electric Field of a Continuous Charge Distribution

21. **N** Often we have distributions of charge for which integrating to find the electric field may not be possible in practice. In such cases, we may be able to get a good approximate solution by dividing the distribution into small but finite particles and taking the vector sum of the contributions of each. To see how this might work, consider a very thin rod of length $L = 16$ cm with

uniform linear charge density $\lambda = 50.0$ nC/m. Estimate the magnitude of the electric field at a point P a distance $d = 8.0$ cm from the end of the rod by dividing it into n segments of equal length as illustrated in Figure P24.21 for $n = 4$. Treat each segment as a particle whose distance from point P is measured from its center. Find estimates of E_P for $n = 1, 2, 4$, and 8 segments.

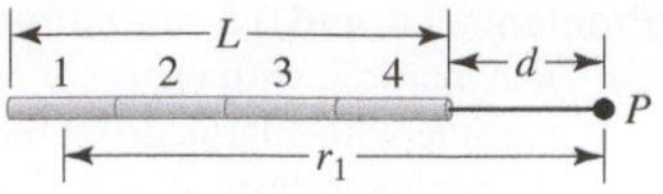

FIGURE P24.21

22. **G** Plot the magnitude of the electric field of a charged rod for points along its perpendicular bisector where Equation 24.15 holds:

$$\vec{E} = \frac{kQ}{y}\frac{1}{\sqrt{\ell^2 + y^2}}\hat{j}$$

How does your graph compare to that for a charged sphere (Problem 4)?

Problems 23 and 24 are paired.

23. **A** A positively charged rod with linear charge density λ lies along the x axis (Fig. P24.23). Find an expression for the magnitude of the electric field at the position P a distance x away from the origin, where $x > L$.

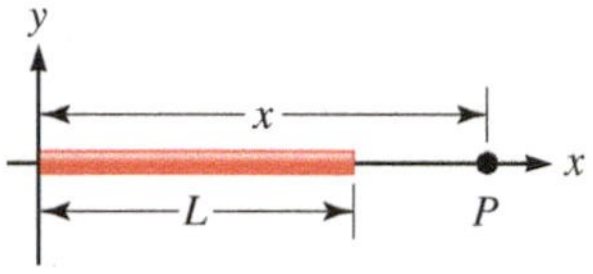

FIGURE P24.23 Problems 23 and 24.

24. **N** A positively charged rod of length $L = 0.250$ m with linear charge density $\lambda = 2.33$ mC/m lies along the x axis (Fig. P24.23). Find the electric field at the position P a distance 0.375 m away from the origin.

24-6 Special Cases of Continuous Distributions

25. **A** Check the dimensions of

$$\vec{E} = \frac{kQy}{(R^2 + y^2)^{3/2}}\hat{j}$$

from Example 24.5 (Special Case: Charged Ring, page 730).

26. **A** Check the dimensions of

$$\vec{E} = 2\pi k\sigma\left[1 - \frac{y}{\sqrt{R^2 + y^2}}\right]\hat{j}$$

from Example 24.6 (Special Case: Charged Disk, page 732).

27. **A** Find an expression for the position y (along the positive axis perpendicular to the ring and passing through its center) where the electric field due to a charged ring is a maximum. Also find an expression for the electric field at that point.

28. The electric field at a point on the perpendicular bisector of a charged rod was calculated as the first example of a continuous charge distribution, resulting in Equation 24.15:

$$\vec{E} = \frac{kQ}{y}\frac{1}{\sqrt{\ell^2 + y^2}}\hat{j}$$

a. **A** Find an expression for the electric field when the rod is infinitely long.

b. **C** An infinitely long rod with uniform linear charge density λ also contains an infinite amount of charge. Explain why this still produces an electric field near the rod that is finite.

29. **N** In Example 24.6 (page 732), we found that the electric field of a charged disk approaches that of a charged particle for distances y that are large compared to R, the radius of the disk. To see a numerical instance of this, calculate the magnitude of the electric field a distance $y = 3.0$ m from a disk of radius $R = 3.0$ cm that has a total charge of 6.0 μC using the exact formula, Equation 24.17:

$$\vec{E} = 2\pi k\sigma\left[1 - \frac{y}{\sqrt{R^2 + y^2}}\right]\hat{j}$$

Then calculate the magnitude of the electric field at $y = 3.0$ m, assuming that the disk is a particle of the same total charge at the origin, and compare your answers.

Problems 30 and 31 are paired.

30. **A** Find an expression for the magnitude of the electric field at point A midway between the two rings of radius R shown in Figure P24.30. The ring on the left has a uniform charge q_1 and the ring on the right has a uniform charge q_2. The rings are separated by distance d. Assume the positive x axis points to the right, through the center of the rings.

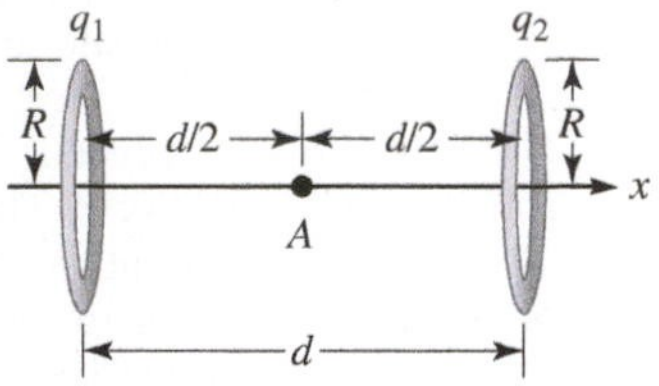

FIGURE P24.30 Problems 30 and 31.

31. **N** What is the electric field at point A in Figure P24.30 if $d = 1.40$ m, $R = 0.500$ m, $q_1 = 15.0$ nC, and $q_2 = -25.0$ nC? Assume the positive x axis points to the right, through the center of the rings.

Problems 32 and 33 are paired.

32. **A** A charged rod is curved so that it is part of a circle of radius R (Fig. P24.32). The excess positive charge Q is uniformly distributed on the rod. Find an expression for the electric field at point A in the plane of the curved rod in terms of the parameters given in the figure.

33. **N** If the curved rod in Figure P24.32 has a uniformly distributed charge $Q = 35.5$ nC, radius $R = 0.785$ m, and $\varphi = 60.0°$, what is the magnitude of the electric field at point A?

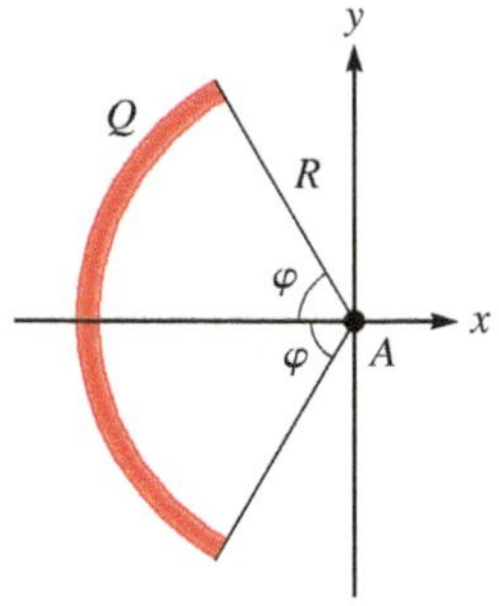

FIGURE P24.32 Problems 32 and 33.

34. **N** A plastic rod of length $\ell = 24.0$ cm is uniformly charged with a total charge of $+12.0$ μC. The rod is formed into a semicircle with its center at the origin of the xy plane (Fig. P24.34). What are the magnitude and direction of the electric field at the origin?

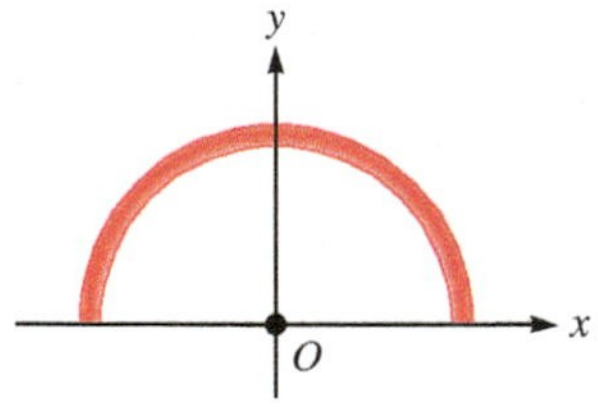

FIGURE P24.34

Problems 35 and 36 are paired.

35. **N** A positively charged disk of radius $R = 0.0366$ m and total charge 56.8 μC lies in the xz plane, centered on the y axis (Fig. P24.35). Also centered on the y axis is a charged ring with the same radius as the disk and a total charge of -34.1 μC. The ring is a distance $d = 0.0050$ m above the disk. Determine the electric field at the point P on the y axis, where P is $y = 0.0100$ m above the origin.

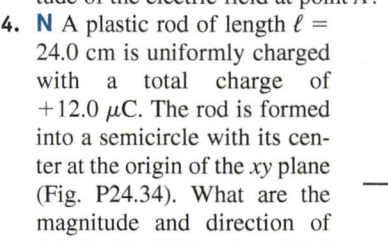

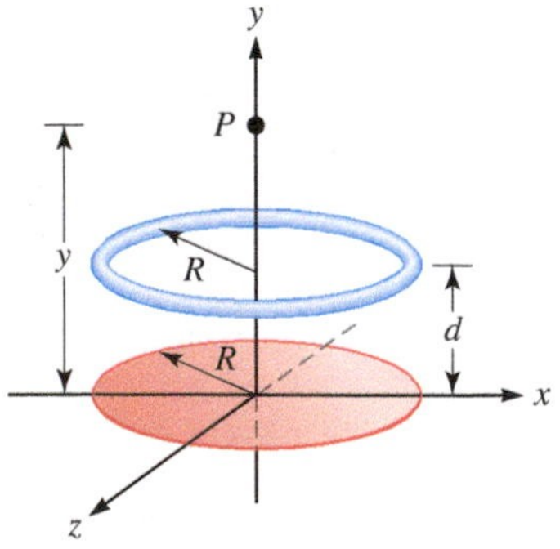

FIGURE P24.35 Problems 35 and 36.

36. **A** A positively charged disk of radius R and total charge Q_{disk} lies in the xz plane, centered on the y axis (Fig. P24.35). Also centered on the y axis is a charged ring with the same radius as the disk and total charge Q_{ring}. The ring is a distance d above the disk. Determine the electric field at the point P on the y axis, where P is above the ring a distance y from the origin.

37. **N** A uniformly charged conducting rod of length ℓ = 30.0 cm and charge per unit length $\lambda = -3.00 \times 10^{-5}$ C/m is placed horizontally at the origin (Fig. P24.37). What is the electric field at point A with coordinates (0, 0.400 m)?

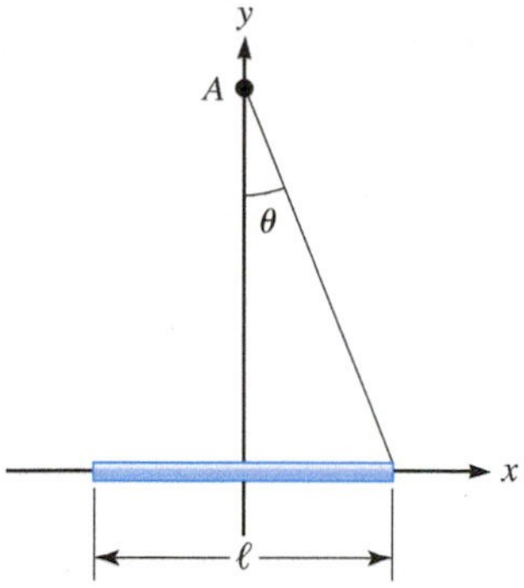

FIGURE P24.37

38. **G** Plot the scalar component of the electric field due to a charged disk for points along its axis as given by Equation 24.17:

$$\vec{E} = 2\pi k\sigma\left[1 - \frac{y}{\sqrt{R^2 + y^2}}\right]\hat{\jmath}$$

How does your graph compare to that for a charged ring (Fig. 24.40, page 740)?

24-7 Case Study: The Shape of Lightning Rods

39. **N** Assume the electric field on Franklin's pointed lightning rod is the breakdown electric field of 3×10^6 N/C, and estimate the charge on the tip. How does that compare to the charge on Wilson's rod (Example 24.7, page 734)?

40. **A** Check the dimensions of

$$\vec{E} = \frac{kq}{y(y - \ell)}\hat{\jmath}$$

from Example 24.8 (page 735).

41. **N** In Example 24.7 (page 734), we analyzed Wilson's lightning rod. The large spherical conductor was later shown to be less effective than Franklin's pointed rod. In fact, large spherical conductors are often used in high-charge applications to prevent breakdown of the surrounding air (or another insulator). Suppose a spherical conductor must be able to hold a charge of 50.0 μC in air. What must be its minimum diameter?

42. **A** In Example 24.8, Franklin's thin lightning rod has a uniform linear charge density. Consider a different situation in which the thin rod is an insulator with a nonuniform distribution of charge. Referring to the geometry in Figure 24.36 (page 735), suppose the charge density of the rod is given by $\lambda(y) = \lambda_0 y/\ell$. This represents a charge per unit length that linearly increases in magnitude from zero at the bottom to λ_0 at the top of the insulating rod. Find an expression for the electric field at point A, at a position $y > \ell$ above the end of the rod.

24-8 Charged Particle in an Electric Field

43. **N** What are the magnitude and direction of a uniform electric field perpendicular to the ground that is able to suspend a particle of mass m = 2.00 g carrying a charge of +6.00 μC in midair, assuming gravity and the electrostatic force are the only forces exerted on the particle?

44. **N** An electron is in a uniform upward-pointing electric field.

a. If the electron experiences a downward acceleration of 9.81 m/s^2, what is the magnitude of the electric field? (Ignore gravity.)

b. What is the gravitational force on this electron? Is it okay to ignore gravity? Explain.

45. **N** A proton is at rest and in equilibrium, floating above a positively charged disk. What is the surface charge density of the disk? Assume the proton is very near the surface of the charged disk.

46. **N** A uniform electric field of 725 N/C is produced within a particle accelerator. Starting from rest, what are

a. the speed of an electron placed in this field after 23.0 ns have elapsed, and

b. the speed of a proton placed in this field after 23.0 ns have elapsed?

47. **N** A very large disk lies horizontally and has surface charge density $\sigma = -2.3$ nC/m^2. An electron is released at the surface. (It begins from rest and moves vertically upward.) Ignoring gravity, find the speed of the electron when it is 1.0 mm above the disk.

48. **N** An electron is released from rest in a uniform electric field of 465 N/C near a particle detector. The electron arrives at the detector with a speed of 2.40×10^6 m/s.

a. What was the uniform acceleration of the electron?

b. How long did the electron take to reach the detector?

c. What distance was traveled by the electron?

d. What is the kinetic energy of the electron when it reaches the detector?

49. **N** In Figure P24.49, a charged particle of mass m = 4.00 g and charge q = 0.250 μC is suspended in static equilibrium at the end of an insulating thread that hangs from a very long, charged, thin rod. The thread is 12.0 cm long and makes an angle of 35.0° with the vertical. Determine the linear charge density of the rod.

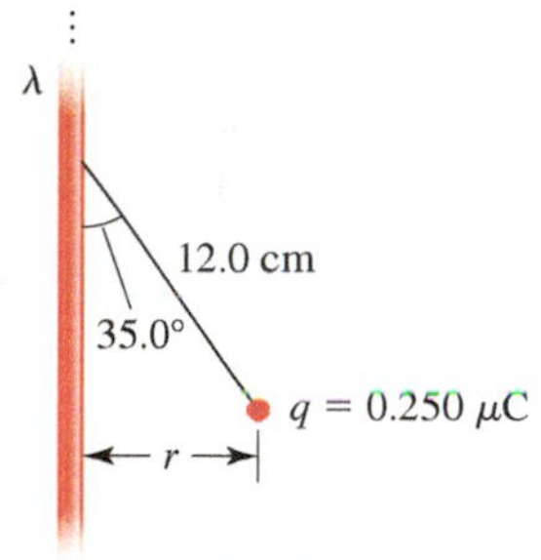

FIGURE P24.49

50. **N** Three charged spheres are suspended by nonconducting light rods of length L = 0.625 m from the point O. The rods are fixed in position so that the middle rod is vertical and the rods on the left and right make an angle of 30.0° with the vertical (Fig. P24.50).

a. What is the total electric field due to the charged spheres at point O?

b. What is the electric force on a particle with charge q = +5.00 μC placed at point O?

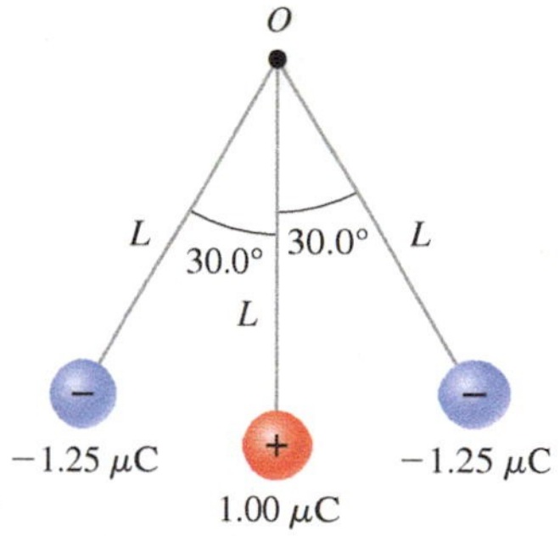

FIGURE P24.50

51. **N** Figure P24.51 shows four small charged spheres arranged at the corners of a square with side d = 25.0 cm.

a. What is the electric field at the location of the sphere with charge +2.00 nC?

b. What is the total electric force exerted on the sphere with charge +2.00 nC by the other three spheres?

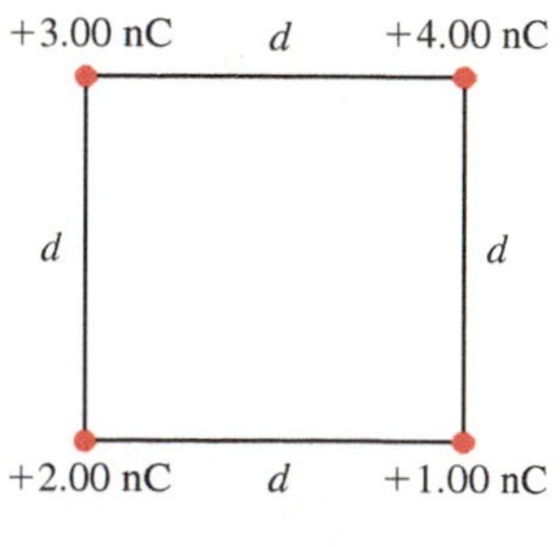

FIGURE P24.51

52. **N** A beam of protons in a particle accelerator is observed to be moving to the right with a kinetic energy per proton of 3.00×10^{-15} J. What are the magnitude and direction of the electric field that will stop the protons in this beam in 1.00 m?

53. **N** A uniform electric field given by $\vec{E} = (2.65\hat{\imath} - 5.35\hat{\jmath}) \times 10^5$ N/C permeates a region of space in which a small negatively charged sphere of mass 1.30 g is suspended by a light cord (Fig. P24.53). The sphere is found to be in equilibrium when the string makes an angle $\theta = 23.0°$.

a. What is the charge on the sphere?

b. What is the magnitude of the tension in the cord?

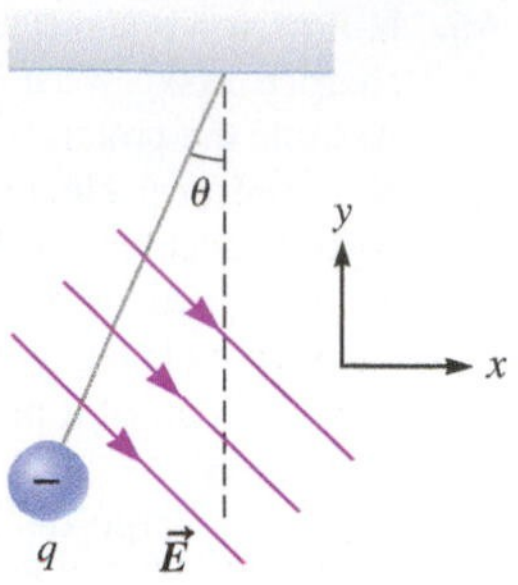

FIGURE P24.53

54. A uniformly charged ring of radius $R = 25.0$ cm carrying a total charge of -15.0 μC is placed at the origin and oriented in the *yz* plane (Fig. P24.54). A 2.00-g particle with charge $q = 1.25$ μC, initially at the origin, is nudged a small distance x along the x axis and released from rest. The particle is confined to move only in the x direction.

a. **A** Show that the particle executes simple harmonic motion about the origin.

b. **N** What is the frequency of oscillation for the particle?

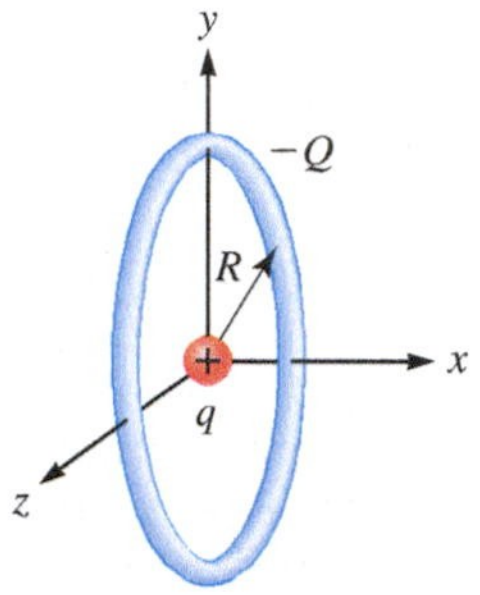

FIGURE P24.54

24-9 Special Case: Dipole in an Electric Field

55. **C** Avery runs a comb through his hair, which causes the comb to acquire a net charge. When he holds the comb near a lightly flowing stream of water coming out of a faucet, the water is attracted toward the comb (Fig. P24.55). Given that a water molecule can be modeled as an electric dipole, explain why the stream of water is deflected by the comb.

56. **C** In Chapter 23, we saw that neutral insulating objects, like bits of paper, can be attracted to charged objects because dipoles are induced in the bits of paper. Suppose instead of a comb (Fig. 23.5C, page 686), we put the bits of paper near a charged object that produces a uniform electric field, like a large, uniformly charged flat sheet. Would the paper be attracted to the sheet? Explain.

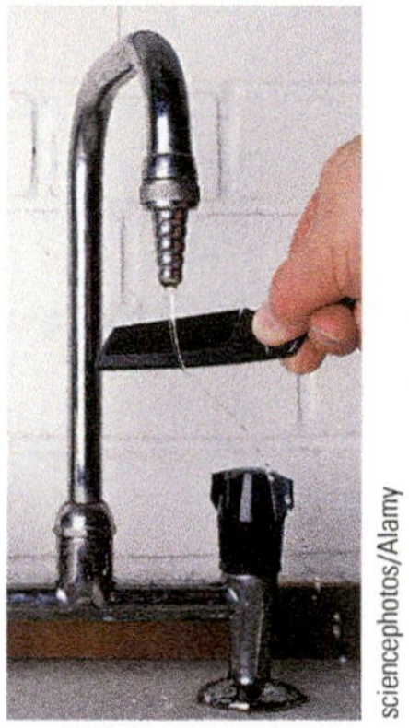

FIGURE P24.55

57. **N** A potassium chloride molecule (KCl) has a dipole moment of 8.9×10^{-30} C·m. Assume the KCl molecule is in a uniform electric field of 325 N/C. What is the change in the system's potential energy when the molecule rotates

a. from $\varphi = 170°$ to $180°$, **b.** from $\varphi = 90°$ to $100°$, and

c. from $\varphi = 10°$ to $0°$?

58. **N Review** Find the magnitude of the angular acceleration α for the water molecule in Example 24.11 (page 744).

General Problems

59. **N** Two metallic spheres of the same size with charges $+9.0$ μC and -4.0 μC are kept 0.20 m apart. Find the location where the electric field is zero if the $+9.0$-μC charge is at the origin and the -4.0-μC charge is at $x = 0.20$ m.

60. **C** Which equations in the chapter are worth memorizing? Be sure to state if the equation has certain limitations such as "good only for spherical source."

61. A total charge Q is distributed uniformly on a metal ring of radius R.

a. **N** What is the magnitude of the electric field in the center of the ring at point O (Fig. P24.61)?

b. **A** What is the magnitude of the electric field at the point A lying on the axis of the ring a distance R from the center O (same length as the radius of the ring)?

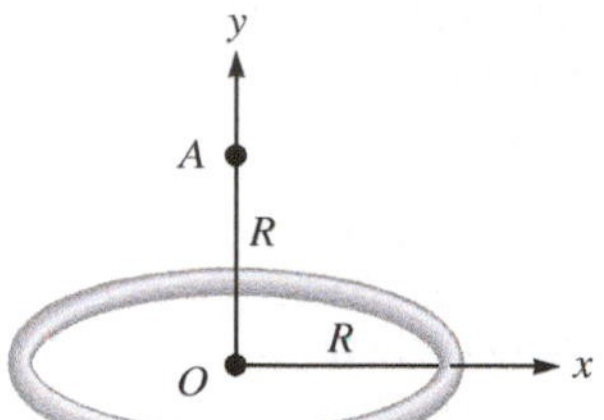

FIGURE P24.61

62. **A** A simple pendulum has a small sphere at its end with mass m and charge q. The pendulum's rod has length L and its weight is negligible. The pendulum is placed in a uniform electric field of strength E directed vertically upward. What is the period of oscillation of the sphere if the electric force is less than the gravitational force on the sphere? Assume the oscillations are small.

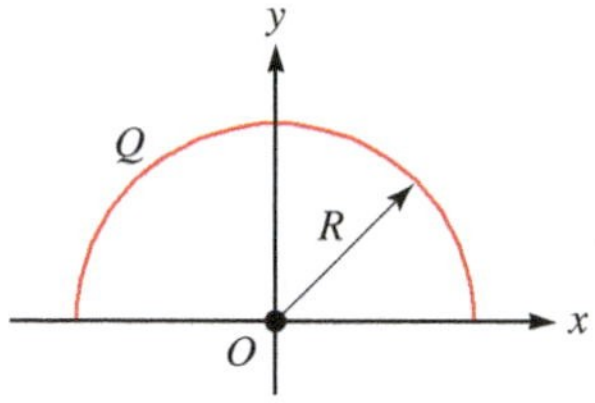

FIGURE P24.63

63. **A** A thin, semicircular wire of radius R is uniformly charged with total positive charge Q (Fig. P24.63). Determine the electric field at the midpoint O of the diameter.

Problems 64 and 65 are paired.

64. **A** An infinite number of positively charged particles, each with charge q, are placed along the x axis at $x = 1$ m, $x = 2$ m, $x = 4$ m, $x = 8$ m, ..., as shown in Figure P24.64. Determine the electric field at $x = 0$ due to this set of charges. *Hint*: Use the formula for a geometric series: $1 + r + r^2 + r^3 + r^4 + \cdots = 1/(1 - r)$.

FIGURE P24.64

65. **A** An infinite number of charged particles, each with charge $\pm q$, are placed along the x axis at $x = 1$ m, $x = 2$ m, $x = 4$ m, $x = 8$ m, and so on. What is the electric field at $x = 0$ if consecutive charges have opposite sign as shown in Figure P24.65? *Hint*: Use the formula for a geometric series: $1 + r + r^2 + r^3 + r^4 + \cdots = 1/(1 - r)$.

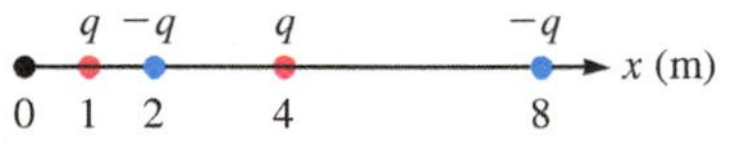

FIGURE P24.65

66. **N** Three identical cylinders made of solid plastic are 15.0 cm in length and 4.50 cm in radius. What is the total charge on

a. the first cylinder, which has a uniform charge density of 235 μC/m³ throughout the cylinder;

b. the second cylinder, which has a uniform charge density of 12.5 μC/m² on its entire surface including its two ends; and

c. the third cylinder, which has a uniform charge density of 12.5 $\mu C/m^2$ only on its curved lateral surface, not including its two ends?

67. **N** The total charge on a uniformly charged ring with diameter 25.0 cm is $-54.0\ \mu C$. What is the magnitude of the electric field along the ring's axis at a distance of
a. 2.50 cm, **b.** 12.5 cm, **c.** 25.0 cm, and
d. 2.00 m from its center?

68. **N** The charge density of a uniformly charged disk 0.500 m in diameter is $2.45 \times 10^{-2}\ C/m^2$. What is the magnitude of the electric field along the disk's axis at a distance of
a. 2.50 cm, **b.** 25.0 cm, **c.** 50.0 cm, and
d. 5.00 m from its center?

69. **N** A thin wire with linear charge density

$$\lambda = \lambda_0 y_0\left(\frac{1}{4} + \frac{1}{y}\right)$$

extends from $y_0 = 1.00$ m to infinity. If $\lambda_0 = 1.45 \times 10^{-5}$ C/m, what is the magnitude of the electric field due to this wire at the origin (y is measured in meters)?

Problems 70, 71, and 72 are grouped.

70. **A** Two positively charged spheres are shown in Figure P24.70. Sphere 1 has twice as much charge as sphere 2. Find an expression for the electrostatic field at point A in terms of the parameters shown in the figure.

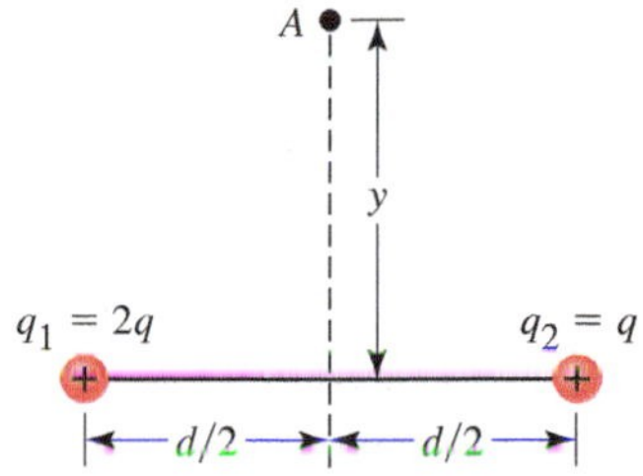

FIGURE P24.70 Problems 70, 71, and 72.

71. **N** Two positively charged spheres are shown in Figure P24.70. Sphere 1 has twice as much charge as sphere 2. If $q = 6.55$ nC, $d = 0.250$ m, and $y = 1.25$ m, what is the electric field at point A?

72. **N** Two positively charged spheres are shown in Figure P24.70. Sphere 1 has twice as much charge as sphere 2; $q = 6.55$ nC, $d = 0.250$ m, and $y = 1.25$ m. Suppose an electron is placed at point A and released.
a. What is the electron's acceleration?
b. If the electron were replaced by a proton, what is the proton's acceleration?

73. **N** The total charge on a uniformly charged, horizontal rod of length $\ell = 26.0$ cm is $+43.0\ \mu C$. What are the magnitude and direction of the electric field produced by the rod at a distance of 20.0 cm from the left end of the rod along its axis?

74. **N** Two parallel plates are placed 5.60 cm apart, creating a uniform electric field of magnitude 525 N/C between the plates. At the same instant in time, a proton is released from rest from the positive plate and an electron is released from rest from the negative plate. If we ignore the electrical attraction between the two particles, at what distance from the positive plate do the particles pass each other?

75. **N** A conducting rod carrying a total charge of $+9.00\ \mu C$ is bent into a semicircle of radius $R = 33.0$ cm, with its center of curvature at the origin (Fig.P24.75). The charge density along the rod is given by $\lambda = \lambda_0 \sin\theta$, where θ is measured clockwise from the $+x$ axis. What is the magnitude of the electric force on a 1.00-μC charged particle placed at the origin?

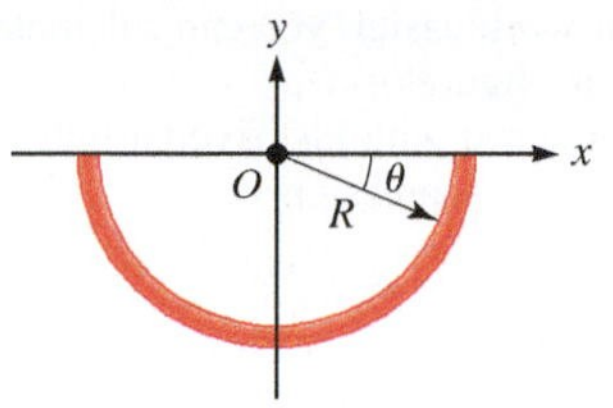

FIGURE P24.75

76. **A** A uniformly charged ring of radius d and total charge Q is oriented in the xz plane and placed at $y = 0$. Show that the maximum electric field due to this ring occurs at $y = d/\sqrt{2}$ and has magnitude $E_{max} = 2kQ/(3\sqrt{3}d^2)$.

Problems 77, 78, 79, and 80 are grouped.

77. **A** When we find the electric field due to a continuous charge distribution, we imagine slicing that source up into small pieces, finding the electric field produced by the pieces, and then integrating to find the electric field. Let's see what happens if we break a finite rod up into a small number of finite particles. Figure P24.77 shows a rod of length 2ℓ carrying a uniform charge Q modeled as two particles of charge $Q/2$. The particles are at the ends of the rod. Find an expression for the electric field at point A located a distance ℓ above the midpoint of the rod using each of two methods:
a. modeling the rod with just two particles and
b. using the exact expression

$$\vec{E} = \frac{kQ}{y}\frac{1}{\sqrt{\ell^2 + y^2}}\hat{j}$$

c. Compare your results to the exact expression for the rod by finding the ratio of the approximate expression to the exact expression.

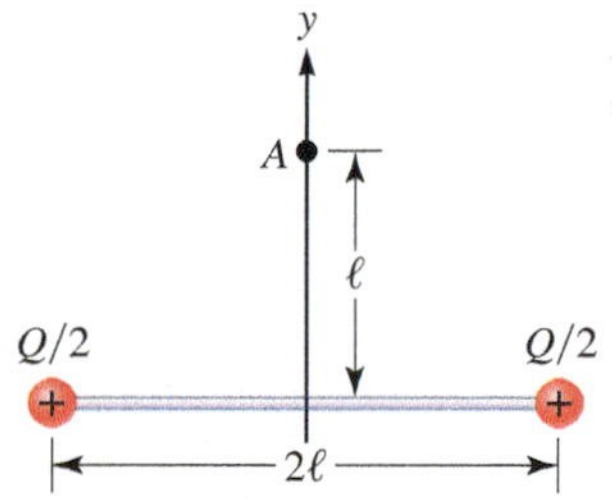

FIGURE P24.77 Problems 77 and 78.

78. **N** Consider the model discussed in Problem 77. Figure P24.77 shows a rod of length 1.00 m carrying a uniform charge $Q = 0.42$ nC modeled as two particles of charge $Q/2$. Find the electric field at point A located a distance 0.50 m above the midpoint of the rod using each of two methods:
a. modeling the rod with just two particles and
b. using the exact expression

$$\vec{E} = \frac{kQ}{y}\frac{1}{\sqrt{\ell^2 + y^2}}\hat{j}$$

c. What is the ratio of your approximate result to the electric field calculated exactly?

79. **A** Consider the model discussed in Problem 77. Figure P24.79 shows a rod of length 2ℓ carrying a uniform charge Q modeled as five particles of charge $Q/5$. Two particles are at the ends of the rod, and the rest are evenly spaced along the rod. Find an expression for the electric field at point A located a distance ℓ above the midpoint of the rod using each of two methods

(to make your work easier, you can calculate any square root to three significant figures):

a. modeling the rod with just five particles and

b. using the exact expression

$$\vec{E} = \frac{kQ}{y}\frac{1}{\sqrt{\ell^2 + y^2}}\hat{\jmath}$$

c. Compare your approximate result to your exact expression for the rod by finding the ratio of the approximate expression to the exact expression.

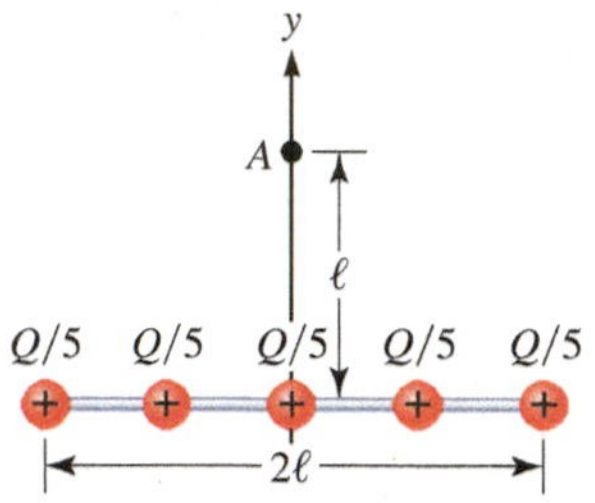

FIGURE P24.79 Problems 79 and 80.

80. **N** Consider the model discussed in Problem 77. Figure P24.79 shows a rod of length 1.00 m carrying a uniform charge $Q = 0.42$ nC modeled as five particles of charge $Q/5$. Two particles are at the ends of the rod, and the rest are evenly spaced along the rod. Find the electric field at point A located a distance 0.50 m above the midpoint of the rod using each of two methods (to make your work easier, you can calculate any square root to three significant figures):

a. modeling the rod with five particles and

b. using the exact expression

$$\vec{E} = \frac{kQ}{y}\frac{1}{\sqrt{\ell^2 + y^2}}\hat{\jmath}$$

c. What is the ratio of your approximate result to the electric field calculated exactly?

81. **A** Return to Example 24.6 (page 732) for the charged disk. By integrating, verify Equation 24.17:

$$\vec{E} = k\sigma 2\pi y\hat{\jmath}\int_0^R \frac{r\,dr}{(r^2 + y^2)^{3/2}} = 2\pi k\sigma\left[1 - \frac{y}{\sqrt{R^2 + y^2}}\right]\hat{\jmath}$$

82. **A** CASE STUDY Return to Example 24.8 (page 735) about Franklin's lightning rod. By integrating, show that

$$\vec{E} = k\lambda\hat{\jmath}\int_0^\ell \frac{dy'}{(y - y')^2} = \frac{k\lambda\ell}{y(y - \ell)}\hat{\jmath}$$

Gauss's Law

25

Underlying Principles

Gauss's law

Major Concepts

1. Gaussian surface
2. Electric flux

Special Cases

1. Electric field due to sources with
   a. Linear symmetry
   b. Spherical symmetry (outside, shell theorem, and inside solid sphere)
   c. Planar symmetry
2. Electric field inside and outside a charged conductor

Key Questions

What is electric flux?

What is Gauss's law?

How do we use Gauss's law to find the electric field due to charged sources with various symmetries?

On your birthday, you find a gift-wrapped box with your name on it hidden in the closet. You are excited because the box is the right size to hold that special gift you are hoping for. Still, you cannot just rip off the paper to see what it is. Instead, you pick it up to estimate its weight. You tap it lightly. Does it sound hollow? You might even shake it to see whether it contains loose pieces.

Now, instead of a birthday present, your physics professor hands you a sealed box. She tells you that there is a plastic object inside, and your job is to figure out what the object's net charge is—positive, negative, or zero. You ask yourself, "What would Ben Franklin do?" Then you suspend a lightweight ball from an insulating thread and give the test ball a net positive charge. You hold the test ball near the box, looking for a response on all sides. If the ball is repelled, there must be a net positive charge in the box. If the ball is attracted, the box must contain a net negative charge. If the ball does not respond at all, the box must have no net charge. In this chapter, we study Gauss's law, which relates the amount of net charge inside the box to the electric field passing through the box.

25-1 Qualitative Look at Gauss's Law

Another title for this chapter could have been "Another Look at the Electric Field" because we will learn how to find the electric fields due to sources of various shapes. In Chapter 24, we found the electric field by considering the electrostatic force the source exerts on a small positive test charge. This approach is important: It illustrates the close connection between the electric field produced by a source and the electrostatic force exerted by that source on some charged particle. In this chapter, we use another method—based on *Gauss's law*—to find the electric field.

Let's compare the method we learned in Chapter 24 to the method in this chapter (Fig. 25.1). A source—a collection of charged particles or a charged object—produces an electric field. In Chapter 24, we used Coulomb's law to come up with an expression for the electric field produced by a single charged particle (or outside a charged spherical object). With this as our only tool, we used "brute force" to calculate the electric field produced by sources with other shapes by adding up (or integrating) the electric fields produced by the many tiny charged pieces that make up the source.

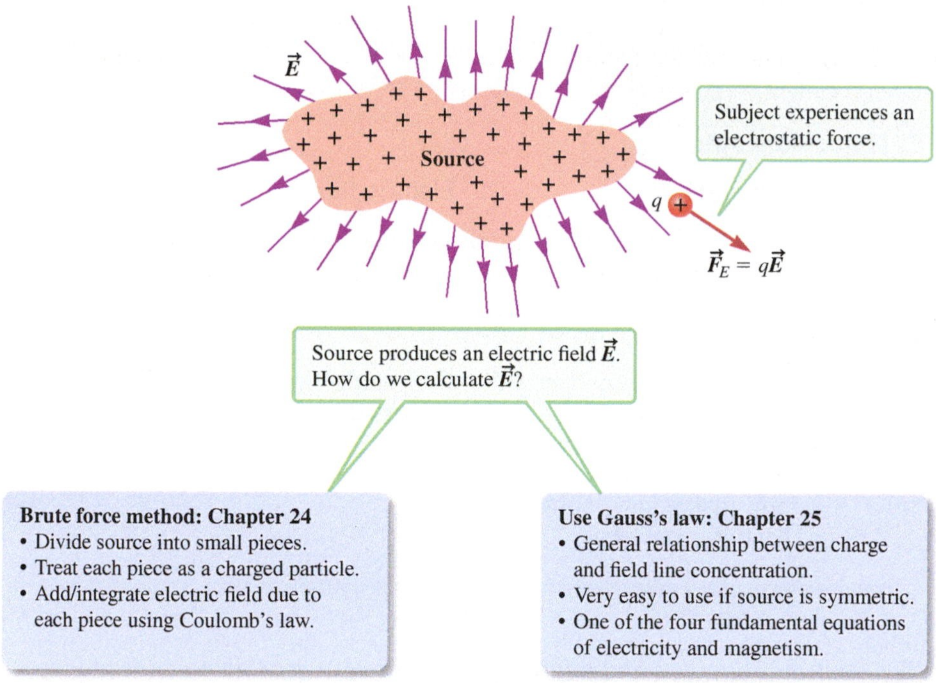

FIGURE 25.1 Concept map showing two approaches to calculating a source's electric field.

The method we learn in this chapter—based on Gauss's law—is more abstract. Instead of considering test charges and the electrostatic force, we find the electric field from the source's charge by calculating the number of electric field lines that pass through an area. Gauss's law is more general than Coulomb's law because it is good for charge distributions of any shape, whereas Coulomb's law is good only for spherically shaped charge distributions. In practice, we apply Gauss's law to sources that are symmetrical, and in those cases, Gauss's law is much easier to use than the brute force method based on Coulomb's law.

What's in the Box?

GAUSSIAN SURFACE
✪ Major Concept

Gauss's law is expressed mathematically. To get an overview, let's start with a qualitative understanding. Gauss's law makes use of an imaginary, closed, three-dimensional surface known as a **Gaussian surface**. Figure 25.2 shows three Gaussian surfaces—boxes, in this case—each with a set of electric field lines. Like a gift box, these boxes do not interfere with their contents in any way; the boxes only prevent us from seeing directly inside. Although you cannot see inside, the electric field lines pass through the walls of the boxes without being altered. In each case, we can figure out something about the contents of the box from the electric field lines that pass through it. In Figure 25.2A, the electric field lines emerge from the box, so we conclude that whatever is in the box must have a net positive charge. The electric field lines enter the box in Figure 25.2B, so we conclude that the box must contain a net negative charge. Finally, the electric field lines pass through (enter and exit) the box in Figure 25.2C, so we conclude that there is no net charge inside the box. It must either be empty or contain a neutral object.

In Figure 25.2, we imagined that each Gaussian surface was opaque, like a cardboard box you might use to hold a present. Gauss's law does not require that such a box actually exist. Instead, a Gaussian surface is always created in our imagination, so we are free to imagine a surface of any shape or size that will simplify our job of determining the electric field due to a charged source.

For example, to find the magnitude of the electric field in Fig. 25.2A, it is best to use a *spherical* Gaussian surface. You can imagine a hollow ball with a positively charged source inside (Fig. 25.3). Because Gaussian spheres exist only in our imagination, they never affect the charged source inside them or the electric field due to that source. The source inside the Gaussian sphere and the resulting electric field in Figure 25.3 are identical to the charge inside the box and the electric field in Figure 25.2A.

We can learn about Gauss's law by considering two concentric Gaussian spheres (Fig. 25.3). So far, we don't know the shape of the source inside, but we know that the source has a net positive charge. Let's assume that it is a positively charged particle (charge Q) at the common center of the two Gaussian spheres.

The electric field in Figure 25.3 is arbitrarily represented by eight lines. You might have thought that more or fewer lines would make the drawing better. But no matter how many electric field lines are drawn, we find that the same number of lines penetrate each sphere. Therefore, we suspect some physical quantity is constant for both spheres. That physical quantity is called *electric flux*, or simply *flux*, a major part of Gauss's law. Before we define flux precisely in the next section, let's continue to develop a conceptual understanding of Gauss's law by coming up with a physical interpretation for the constant number of electric field lines that penetrate each Gaussian sphere in Figure 25.3. We'll explore three different physical quantities: (1) the electric field E, (2) the surface area A, and (3) the product of the electric field and the surface area EA. One of these three quantities is constant for the two spheres.

We know from $E = kQ/r^2$ (Eq. 24.3) that the electric field decreases with increasing distance from the source. So, the electric field at positions on the larger spherical surface is weaker than the electric field at positions on the smaller spherical surface. The electric field is not constant for the two spheres.

The surface area of a sphere depends on the sphere's radius, $A(r) = 4\pi r^2$, so the surface area of the small sphere is less than the surface area of the large sphere. Area is obviously not constant between spheres of different radii.

Because the electric field is weaker on the sphere with the larger area, we can find our constant quantity by multiplying the electric field strength on the surface of one sphere by the surface area of that sphere. For the small sphere,

$$E(r)A(r) = \frac{kQ}{r^2}4\pi r^2 = 4\pi kQ \tag{25.1}$$

where $E(r) = kQ/r^2$ is the electric field on the surface of the small sphere and $A(r) = 4\pi r^2$ is its surface area. For the large sphere,

$$E(R)A(R) = \frac{kQ}{R^2}4\pi R^2 = 4\pi kQ \tag{25.2}$$

where $E(R) = kQ/R^2$ is the electric field on the surface of the large sphere and $A(R) = 4\pi R^2$ is its surface area.

From Equations 25.1 and 25.2, we see that when we multiply the electric field on the surface of a sphere by its surface area, we get the constant $EA = 4\pi kQ$. The physical interpretation of the constant number of field lines that penetrate the two Gaussian surfaces in Figure 25.3 is that the quantity EA is a constant. As we'll soon define, EA is the electric flux through either Gaussian sphere. Furthermore, from Equations 25.1 and 25.2, we see that the flux EA is a constant that depends only on

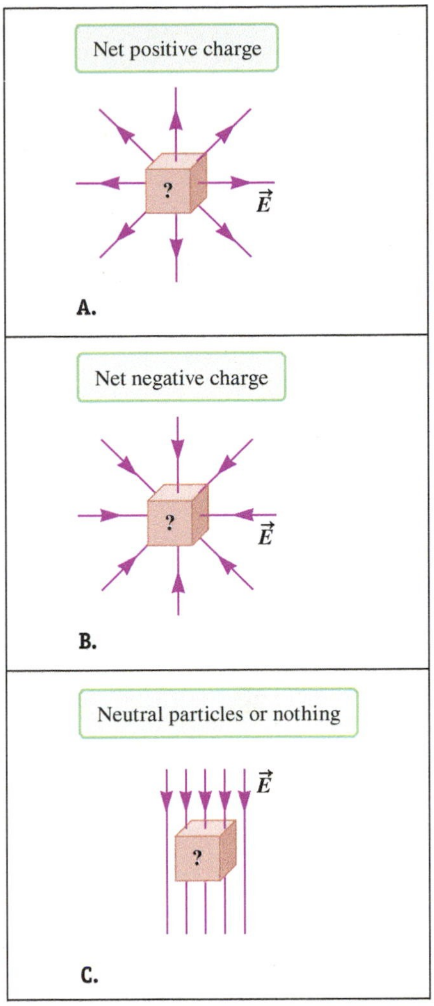

FIGURE 25.2 What is inside each box?

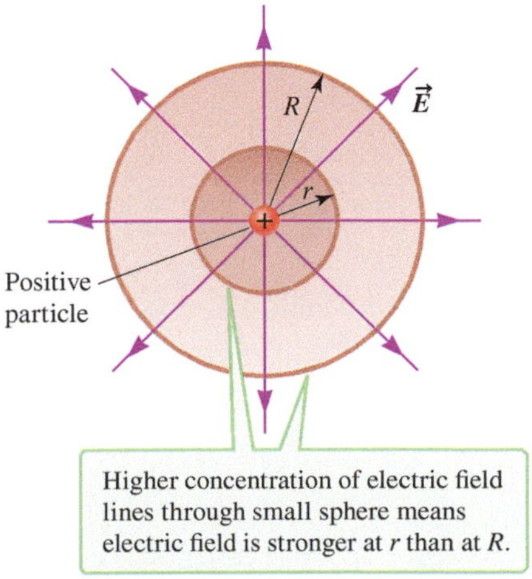

FIGURE 25.3 Gaussian spheres surround a positively charged particle. The number of electric field lines (eight in this case) that pass through each sphere is a constant proportional to EA.

the amount of charge Q inside the Gaussian sphere. Conceptually, Gauss's law relates the flux to the amount of charge enclosed by that Gaussian surface.

Symmetry

Why did we decide to use spherical Gaussian surfaces in Figure 25.3 instead of boxes as in Figure 25.2A? The reason is that spheres *exploit the symmetry* of the source. Whenever you come up with a Gaussian surface, you should try to exploit such symmetry.

"Exploiting the symmetry" means choosing a Gaussian surface that has the same symmetry as the source. There are many ways to test for symmetry, but we need to think about only one test—rotation. Consider the uppercase letter **A**. In Figure 25.4A, we have chosen a coordinate system such that **A** lies in the xy plane. Because a written letter must lie on the paper (in this case, the xy plane), we imagine rotating **A** by 180° around the x axis and then around the y axis. Both of these rotations return the letter to the xy plane. If we rotate the letter **A** around the x axis by 180°, the letter is upside down. If, instead, we rotate **A** around the y axis by 180°, the letter remains unchanged. We say that **A** is symmetrical with respect to a 180° rotation around the y axis, but not the x axis. How about the z axis? The letter lies in the xy plane, so we can test rotation around the z axis by imagining that we spin the letter around the z axis. For example, in the fourth panel of Figure 25.4A, the letter **A** has been rotated by 90° around the z axis, and the letter looks sideways. In fact, except for a 360° rotation, when we rotate **A** around the z axis, it does not look like an upright **A** again. So, **A** is not symmetrical with respect to rotations around the z axis.

Now suppose you want to find another letter that has the same symmetry as **A**. You might choose **U**, the only other vowel that has the same symmetry. To test it, imagine rotating **U** by 180° around the x axis, and you see that—just like **A**—it is upside down. Then imagine rotating **U** by 180° around the y axis; it remains unchanged. Finally, try rotating **U** by any amount (except 360°) around the z axis. You find that **U**—like **A**—is not symmetrical with respect to rotations around the z axis.

Not all letters have the same symmetry as **A**. For example, look at **Z** in Figure 25.4B. If you rotate **Z** around the x axis or the y axis by 180°, the letter looks upside down. However, if you rotate **Z** around the z axis by 180°, the letter is right side up. So, **Z** is symmetrical with respect to 180° rotations around the z axis.

The letters **A** and **Z** are equally symmetrical; both are symmetrical with respect to a 180° rotation around a single axis. Some letters have much more symmetry, however. For example, consider the letter **O**, which is unchanged by a 180° rotation around either the x axis or the y axis. You can also rotate **O** by any amount around the z axis, and it will look the same. So, **O** is symmetrical with respect to a 180° rotation around either the x axis or the y axis, and with respect to any rotation around the z axis.

In this chapter, we study three types of symmetrical charge distributions: linear, spherical, and planar. We'll use rotation to test for these types of symmetry.

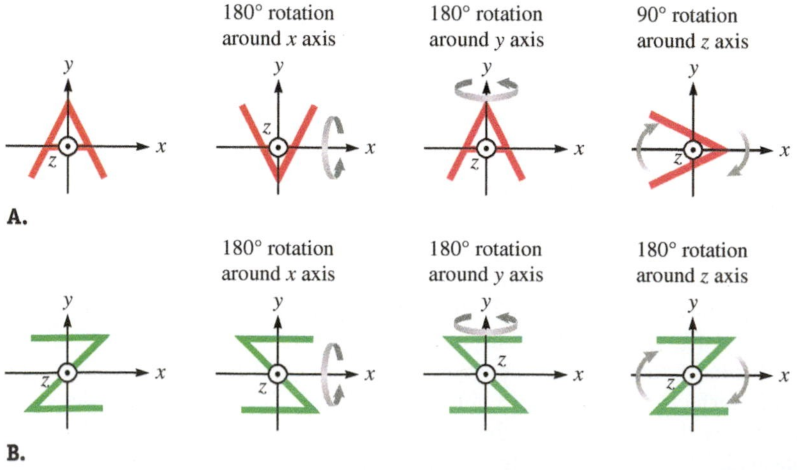

FIGURE 25.4 **A.** A symmetry. Rotation by 180° around the y axis leaves the letter unchanged. **B.** Z symmetry. Rotation by 180° around the z axis leaves the letter unchanged.

CONCEPT EXERCISE 25.1

a. List all the uppercase letters that have the same symmetry as **A**.
b. List all the uppercase letters that have the same symmetry as **Z**.
c. List all the uppercase letters that have the same symmetry as **O**.

CASE STUDY Stuck in the Rain

Avi, Cameron, and Shannon are driving along a rural road when a powerful thunderstorm rolls in. They see lightning strikes nearby.

Avi: We should pull over and find shelter.

Shannon: There's nothing around here. Just fields. Not even a farmhouse in sight.

Cameron: The safest place is in the car. Just pull over so we don't slide off the road. We'll wait it out in the car.

Avi: No way. Didn't you see how close that lightning is? We're right in the middle of this, and I don't want to get struck by lightning. If lightning hits the car, it will blow up the gas tank, and I am not interested in going up in smoke. Remember how a small spark can cause gasoline to explode (page 685)?

Cameron: Look. The gas is in a sealed metal tank. And we are in a sealed metal car. The lightning can't get to the gas or us. We are much safer in here.

Avi: I think it is safer out there. There are some trees in the fields. The trees will act like lightning rods. We'll get wet out there, but at least we won't get hurt.

Cameron: I totally disagree. Let's let Shannon decide.

Peter McGahey Cengage Learning

FIGURE 25.5 Antistatic bags protect sensitive electronics.

What would you decide? Your answer will help explain why cell phones don't work in closed elevators and how antistatic bags protect sensitive electronics (Fig. 25.5).

25-2 Flux

Gauss's law involves finding the *electric flux*. We can use a flowing fluid as a mechanical analogy. Figure 25.6A shows streamlines (Section 15-6) for water flowing through a pipe that narrows. Streamlines are like electric field lines. In both cases, the absolute number of field lines or streamlines is chosen arbitrarily,

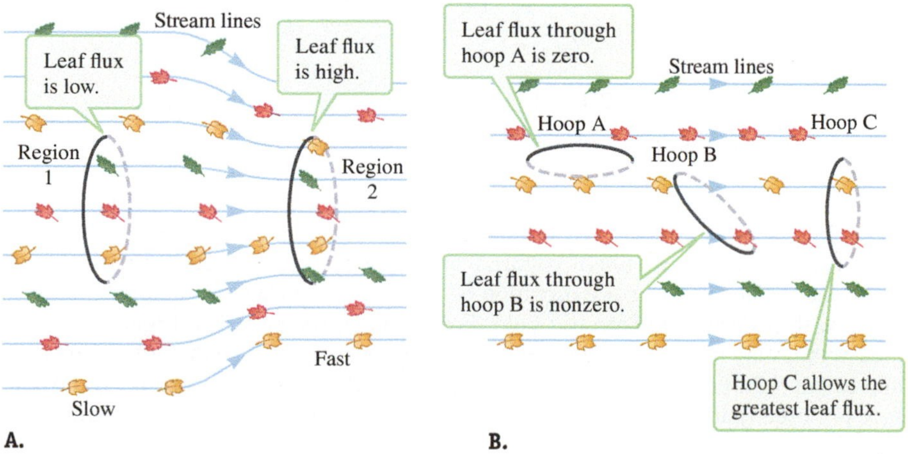

FIGURE 25.6 **A.** Streamlines for water in a pipe that narrows. The streamlines are more concentrated in region 2, indicating that the water's speed is higher in region 2. **B.** Three hoops in a stream. Hoop A is parallel to the flow, allowing no water to flow through it. Hoop B is tilted at an angle φ to the streamlines, allowing some water to flow through it. Hoop C is perpendicular to the water's flow, allowing for the maximum flow of water through it.

but once the choice has been made, the relative density of the lines in the drawing is important. The density of electric field lines indicates the relative magnitude of the electric field. The density of streamlines indicates the fluid's relative speed (Fig. 25.6A). Imagine that leaves are scattered on top of the water, making it easier to observe the fluid flow. Suppose you place a hoop of area A in the stream so that the hoop is perpendicular to the flow. If you place the hoop in region 1, relatively few leaves penetrate the hoop per second. If you place the hoop in region 2, many more leaves penetrate the hoop per second. The number of leaves that flow through the hoop (per unit time) can be defined as the "leaf flux." The word *flux* is Latin for "flow," so leaf flux is the number of leaves that flow through the hoop (per unit time).

We can learn about flux from this analogy. First, the leaf flux depends on the area of the hoop: A larger area allows more leaves to flow through. Second, the angle between the hoop and the streamlines affects the flux (Fig. 25.6B).

In Gauss's law, the magnitude of the electric field on a Gaussian surface is represented by the number of field lines that penetrate the surface and is related to the *electric flux*. The word *flux* is used here for illustration: Nothing is flowing when we study an electrostatic field. However, the electric field lines drawn on paper look like streamlines, so the analogy with fluid flow helps us visualize the calculation of electric flux. In Figure 25.7, a uniform electric field is indicated by evenly spaced field lines, with three identical circular disks in the region covered by the field lines. (These disks are not closed surfaces, so they are not Gaussian surfaces.) The number of electric field lines penetrating each disk represents the electric flux through that disk. Disk A is parallel to the field lines, so no electric field can penetrate it; the electric flux through disk A is zero. Disk B is tilted at an angle φ to the field lines, so those lines can penetrate it; the electric flux is nonzero. However, the electric flux through disk B is not as great as the electric flux through disk C. Disk C is perpendicular to the electric field lines, so it allows the greatest electric flux.

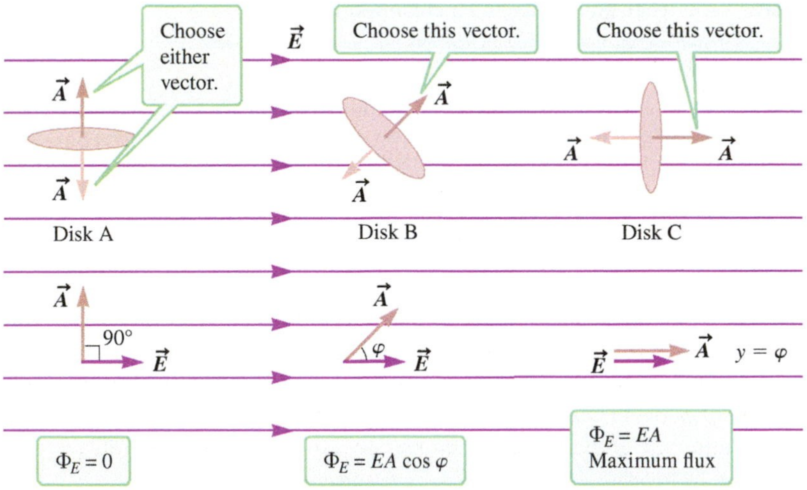

FIGURE 25.7 Electric field lines look like streamlines, but nothing is really flowing. To find the electric flux through the three disks, you must choose the direction of $\vec{A}$. Then you must maintain that direction throughout all calculations. The electric flux in each case is found from $\Phi_E = \vec{E} \cdot \vec{A} = EA\cos\varphi$. For disk A, $\varphi = 90°$, so the electric flux is zero. For disk C, $\varphi = 0$, so the electric flux is EA, which is the maximum flux. The electric flux through disk B has a value between zero and EA.

Area Vectors and Electric Flux

A mathematical description of electric flux requires us to define an **area vector** $\vec{A}$. The magnitude of the area vector is the area of the surface, with the dimensions of length squared. The direction is defined to be perpendicular to the surface. Of course, for a disk or other "open" surface that is not part of a three-dimensional volume, such as the lid taken off a tub of yogurt, there are two possible directions perpendicular to the surface. Choose the direction of $\vec{A}$ so that it makes the smallest possible angle with respect to the electric field. For example, for disk B in Figure 25.7, the area vector may point either up and to the right, or down and to the left. Choose up and to the right because that gives the smallest angle between $\vec{A}$ and $\vec{E}$. If the surface is closed (part of a surface surrounding a three-dimensional volume, such as the lid attached to a yogurt tub), choose $\vec{A}$ so that it points outward.

ELECTRIC FLUX ✪ Major Concept

The **electric flux** Φ_E is given by the dot product of the electric field and the area vector:

$$\Phi_E = \vec{E} \cdot \vec{A} \quad (25.3)$$

The electric flux has the dimensions electric field times area, so its SI units are $N \cdot m^2/C$. Like any dot product, the electric flux can be written in terms of the magnitude of each vector (E and A) and the angle φ between them:

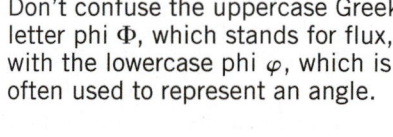

Don't confuse the uppercase Greek letter phi Φ, which stands for flux, with the lowercase phi φ, which is often used to represent an angle.

$$\Phi_E = EA \cos \varphi \quad (25.4)$$

When we apply either Equation 25.3 or Equation 25.4, it is often best to sketch $\vec{E}$ and $\vec{A}$ with their tails touching and φ between the two vectors (Fig. 25.7).

Here's another way to think about Equations 25.3 and 25.4: The electric flux Φ_E is the magnitude of the electric field $|\vec{E}|$ times the magnitude of the component of the area vector $\vec{A}_\parallel$ that is *parallel* to the electric field (Fig. 25.8):

$$\Phi_E = EA_\parallel \quad (25.5)$$

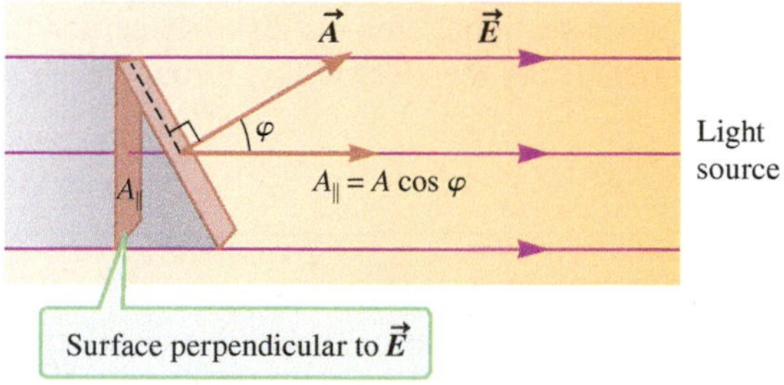

FIGURE 25.8 You can find $A_\parallel$—the component of $\vec{A}$ that is parallel to the electric field—from the shadow cast by $\vec{A}$. Then the electric flux is $\Phi_E = EA_\parallel$.

where $A_\parallel = A \cos \varphi$. If you were to shine a light parallel to the electric field, the shadow cast by a surface of area A would have an area equal to the magnitude of $\vec{A}_\parallel$ (Fig. 25.8).

A More General Expression for Electric Flux

Equations 25.3 and 25.4 work well when the angle between $\vec{A}$ and $\vec{E}$ is a constant over the entire surface, as it is for all three disks in Figure 25.7, but that is not always the case. The surface in Figure 25.9 is a hemisphere—like what you might get by blowing soap film through a ring. The electric field is uniform, but the angle between the electric field and the area is not uniform over the whole surface. We must imagine dividing the surface into many small pieces, each with an area vector $d\vec{A}$. Each piece is small enough that the angle between $d\vec{A}$ and $\vec{E}$ is constant. The flux through one small piece is $d\Phi_E = \vec{E} \cdot d\vec{A}$, and the flux through the entire surface comes from adding up—that is, integrating—the fluxes through each small piece:

$$\Phi_E = \int \vec{E} \cdot d\vec{A} \quad (25.6)$$

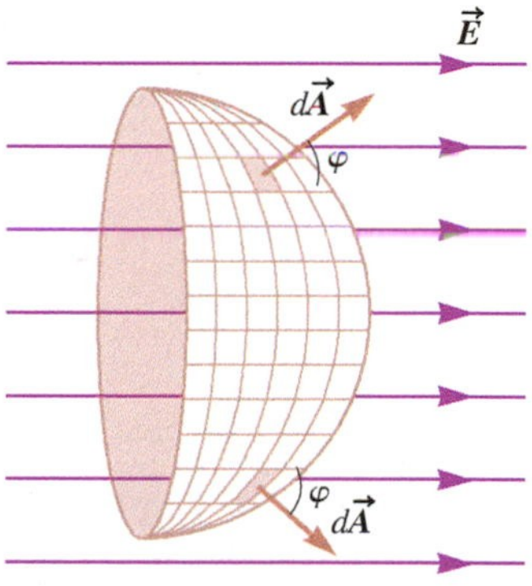

FIGURE 25.9 When we need to calculate the flux through a curved surface, we imagine breaking that surface up into many small, flat surfaces.

Equation 25.6 is a more general equation for the electric flux than Equations 25.3 and 25.4. In fact, Equation 25.6 is valid when the electric field is not uniform. In practice, calculating the integral in Equation 25.6 may be difficult or even impossible if the angle φ between $d\vec{A}$ and $\vec{E}$ is not constant, so we will choose a Gaussian surface such that the angle is a constant usually equal to 0, 90°, or 180°.

CONCEPT EXERCISE 25.2

The terms *electric force*, *electric field*, and *electric flux* sound similar and are easily confused. Write a short description of each term, including the SI units.

CONCEPT EXERCISE 25.3

An electric field is described by $\vec{E} = (15\hat{\imath} + 25\hat{\jmath})$ N/C. Find the electric flux through a surface whose area vector is $\vec{A} = (0.65\hat{\imath} + 0.35\hat{\jmath})\ m^2$.

EXAMPLE 25.1 Three Surfaces in a Uniform Electric Field

Figure 25.10 shows a uniform electric field of magnitude 15.0 N/C that has three surfaces. Surface A is a disk perpendicular to the electric field so that $\vec{A}$ is parallel to $\vec{E}$ and $\varphi = 0$. The radius of the disk is $r = 0.37$ m. Surface B is tilted with respect to the electric field so that $\varphi = 25°$ but is otherwise identical to disk A. Surface C is an open hemisphere of radius $r = 0.37$ m. The hemisphere is oriented so that its cross-sectional area—a disk—is perpendicular to the electric field. Find the electric flux through each surface.

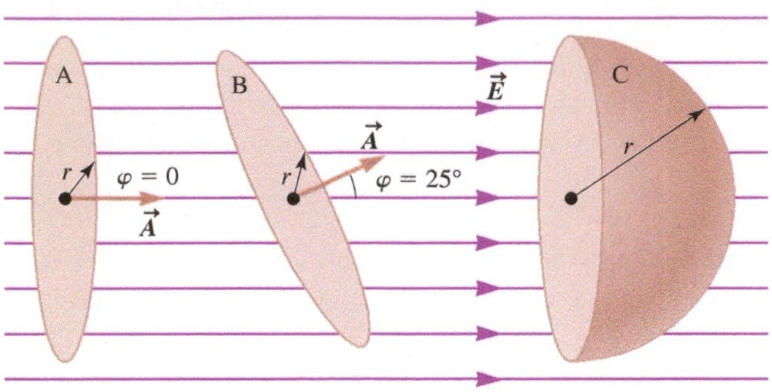

FIGURE 25.10

INTERPRET and ANTICIPATE

When calculating numerical values for the electric flux, we expect our answers to have the units $\text{N}\cdot\text{m}^2/\text{C}$. Imagine shining a light parallel to the electric field. We would find that surface B has the smallest shadow, so we expect surface B to have the smallest electric flux.

SOLVE

Disk A. Because the electric field and angle φ are both uniform over the entire surface A, we can use either Equation 25.3 or Equation 25.4 to find the flux through the disk.

$$\Phi_A = \vec{E}\cdot\vec{A} \quad (25.3)$$

$$\Phi_A = EA\cos\varphi \quad (25.4)$$

Because $\varphi = 0$, the electric flux is EA.

$$\Phi_A = EA\cos\varphi = EA\cos 0 = EA$$

The area of the disk is $A = \pi r^2$.

$$\Phi_A = EA = E(\pi r^2)$$

$$\Phi_A = (15.0\ \text{N/C})(\pi)(0.37\,\text{m})^2$$

$$\Phi_A = 6.5\ \text{N}\cdot\text{m}^2/\text{C}$$

Disk B. The process is similar to that for disk A, but in this case $\varphi = 25°$.

$$\Phi_B = EA\cos\varphi \quad (25.4)$$

$$\Phi_B = E(\pi r^2)\cos\varphi$$

$$\Phi_B = (15.0\ \text{N/C})(\pi)(0.37\ \text{m})^2\cos 25°$$

$$\Phi_B = 5.8\ \text{N}\cdot\text{m}^2/\text{C}$$

Hemisphere C. The hemisphere seems like a much more complicated calculation; you may be tempted to break up the surface into many small pieces and integrate using Equation 25.6. This would work, but there is a much simpler way. Imagine shining a light parallel to the electric field so that the shadow cast by the hemisphere is a disk of radius $r = 0.37$ m. According to Equation 25.5, we can find the electric flux from the area of the shadow, $A_{\parallel}$. In this case the area of the shadow is identical to the area of disk A, so $\Phi_C = \Phi_A$.

$$\Phi_C = EA_{\parallel} \quad (25.5)$$

$$\Phi_C = E(\pi r^2)$$

$$\Phi_C = \Phi_A = 6.5\,\text{N}\cdot\text{m}^2/\text{C}$$

CHECK and THINK

Our answers have the units we expected, and the electric flux is smallest through surface B as expected. For the hemisphere, we did not have to integrate because the dot product tells us that we need to use only the component of the area vector that is parallel to the electric field when we calculate electric flux.

25-3 Gauss's Law

According to **Gauss's law**, the net electric flux Φ_E through a closed (Gaussian) surface is proportional to the net charge q_{in} inside the surface. Mathematically, we write Gauss's law as

$$\Phi_E = \frac{q_{in}}{\varepsilon_0} \tag{25.7}$$

where $1/\varepsilon_0$ is the constant of proportionality. The quantity ε_0 is called the **permittivity of free space** or the **permittivity constant** and has the value

$$\varepsilon_0 = 8.85 \times 10^{-12} \frac{C^2}{N \cdot m^2} \tag{25.8}$$

The permittivity constant ε_0 is related to Coulomb's constant k by

$$k = \frac{1}{4\pi\varepsilon_0} = 8.99 \times 10^9 \, N \cdot m^2/C^2 \tag{25.9}$$

Figure 25.11 illustrates Gauss's law for a dipole electric source and four Gaussian surfaces. Just by looking at the field lines that penetrate each Gaussian surface, you can determine whether the net flux is positive, negative, or zero: If the net electric field lines point *outward*, the electric flux is *positive*; if they point *inward*, the electric flux is *negative*; if there are as many lines pointing inward as outward, the electric flux is *zero*. According to Gauss's law, the net flux tells the sign of the charge enclosed by the Gaussian surface. For example, you can see that the net electric flux Φ_E through surfaces A and B is zero, so according to Gauss's law, there is no net charge q_{in} inside either of these surfaces. You can see that the net flux through C is positive, so according to Gauss's law, the charge inside is positive. Finally, the flux through D is negative and the charge inside is negative.

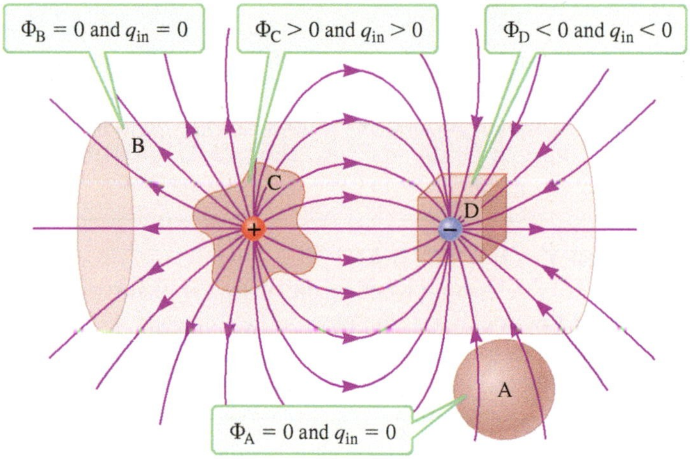

FIGURE 25.11 Dipole with four Gaussian surfaces. Surface A is an empty sphere, and all electric field lines pass through it. Surface B is a cylinder that contains equal amounts of positive and negative charge, so the net charge enclosed is zero. There are as many field lines pointing outward from surface B as pointing into it. Surface C is irregularly shaped like a potato skin. It contains a positive charge, and field lines point outward from it. Surface D is a cube that contains negative charge, and field lines point into it.

To find the electric field using Gauss's law, we calculate the net electric flux through a Gaussian surface. We use the general equation for electric flux (Eq. 25.6) to write an expression for the net flux through a Gaussian surface:

$$\Phi_E = \oint \vec{E} \cdot d\vec{A} \tag{25.10}$$

The circle on the integral symbol indicates that the integral is taken over a *closed* surface. Combining Equations 25.7 and 25.10 is a convenient way to write Gauss's law:

$$\Phi_E = \oint \vec{E} \cdot d\vec{A} = \frac{q_{in}}{\varepsilon_0} \tag{25.11}$$

GAUSS'S LAW
Underlying Principle

CONCEPT EXERCISE 25.4

Which of the following expressions are correct forms of Gauss's law?

a. $\Phi_E = \oint \vec{E} \cdot d\vec{A}$ **b.** $q_{in} = \varepsilon_0 \oint \vec{E} \cdot d\vec{A}$ **c.** $\oint \vec{E} \cdot d\vec{A} = 4\pi k q_{in}$

CONCEPT EXERCISE 25.5

Find the electric flux through the three Gaussian surfaces in Figure 25.12.

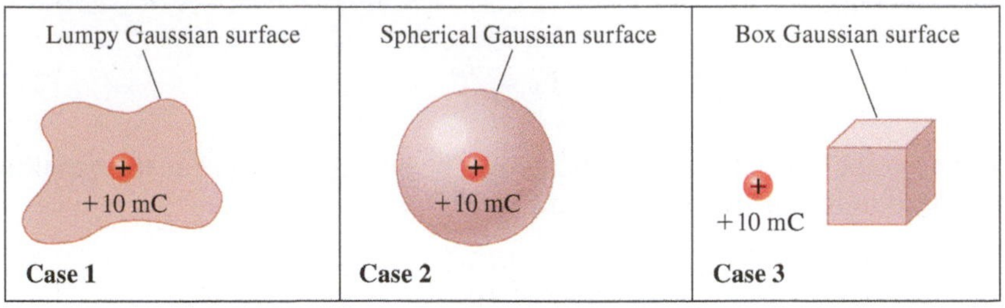

FIGURE 25.12

EXAMPLE 25.2 Comparing Gaussian Surfaces

Figure 25.13 shows two Gaussian surfaces in uniform electric fields. In part A the Gaussian surface is a square box, and in part B the Gaussian surface is a closed cylinder. In each case, integrate $\Phi_E = \oint \vec{E} \cdot d\vec{A} = q_{in}/\varepsilon_0$ (Eq. 25.11) to find the amount of charge enclosed in each Gaussian surface.

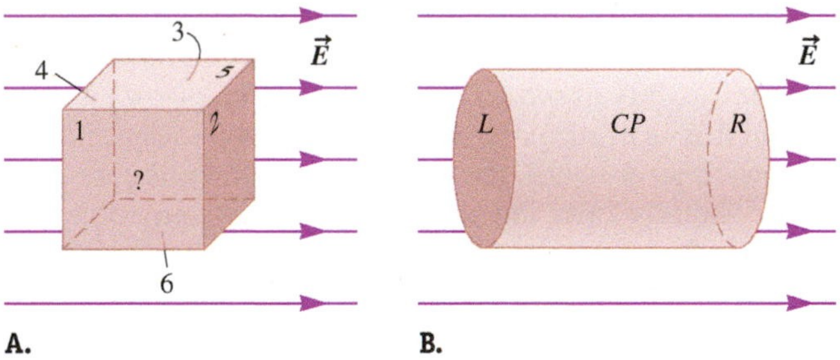

FIGURE 25.13

INTERPRET and ANTICIPATE

We can easily determine whether the charge is positive, negative, or zero by looking at the electric field lines (Fig. 25.13). Because the electric field lines pass completely through the Gaussian surfaces, the net charge inside each surface is zero.

SOLVE

Although we already know the results, we are asked to use integration in order to practice using Gauss's law with different Gaussian surfaces.

A **closed box** is made up of six sides. According to Gauss's law, we must integrate over the entire closed box, so we break up the integral into six pieces, one for each side of the box. The six subscripts on the integrals correspond to the six sides of the box (Fig. 25.13A).

$$\Phi_E = \oint \vec{E} \cdot d\vec{A}$$

$$\Phi_E = \int_1 \vec{E} \cdot d\vec{A} + \int_2 \vec{E} \cdot d\vec{A} + \int_3 \vec{E} \cdot d\vec{A} + \int_4 \vec{E} \cdot d\vec{A} + \int_5 \vec{E} \cdot d\vec{A} + \int_6 \vec{E} \cdot d\vec{A}$$

To calculate the six dot products, we need to know the angle φ between $\vec{E}$ and $d\vec{A}$ for each side. A separate drawing of each side is helpful (Fig. 25.14). The area vector $d\vec{A}$ for each side is perpendicular to that side and points outward.

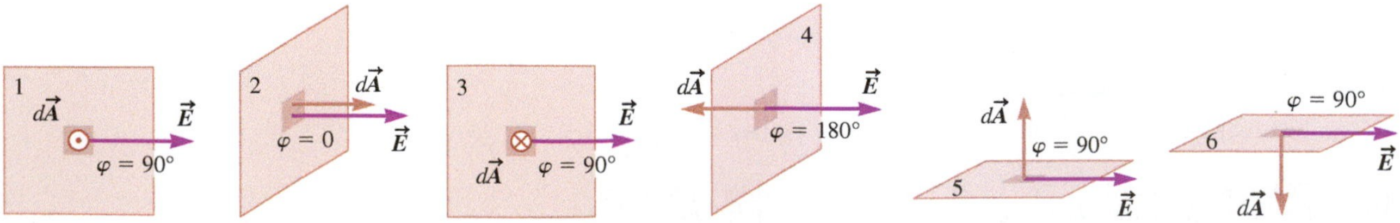

FIGURE 25.14

The dot product is zero whenever the angle φ between $\vec{E}$ and $d\vec{A}$ is 90°, so the dot product is zero for sides 1, 3, 5, and 6. That means these four integrals are zero, and we are left with just two integrals.

$$\Phi_E = \oint \vec{E} \cdot d\vec{A}$$

$$\Phi_E = 0 + \int_2 \vec{E} \cdot d\vec{A} + 0 + \int_4 \vec{E} \cdot d\vec{A} + 0 + 0$$

$$\Phi_E = \int_2 \vec{E} \cdot d\vec{A} + \int_4 \vec{E} \cdot d\vec{A}$$

Write each dot product in terms of the magnitude of the vectors and the angle between them.

$$\Phi_E = \int_2 E\,dA\cos\varphi + \int_4 E\,dA\cos\varphi$$

$$\Phi_E = \int_2 E\,dA\cos 0 + \int_4 E\,dA\cos 180°$$

$$\Phi_E = \int_2 E\,dA + \int_4 -E\,dA$$

The electric field is uniform, so as we integrate over dA, E is a constant that we can pull outside the integrals.

$$\Phi_E = E\int_2 dA - E\int_4 dA$$

The integrals are identical: Each equals the area A of one square side.

$$\Phi_E = EA - EA = 0$$

The net electric flux is zero, and according to Gauss's law, that means the charge enclosed by the Gaussian surface is zero.

$$\Phi_E = \frac{q_{in}}{\varepsilon_0} = 0$$

$$q_{in} = 0$$

CHECK and THINK

This is exactly what we predicted because the number of electric field lines entering the box is equal to the number of electric field lines leaving the box.

SOLVE

The **closed cylinder** is made up of three surfaces—the left cap, the right cap, and the curved part. These are labeled L, R, and CP (Fig. 25.13B). Break the integral up into three pieces, one for each surface in Figure 25.13B.

$$\Phi_E = \oint \vec{E} \cdot d\vec{A}$$

$$\Phi_E = \int_L \vec{E} \cdot d\vec{A} + \int_{CP} \vec{E} \cdot d\vec{A} + \int_R \vec{E} \cdot d\vec{A}$$

Figure 25.15 shows the angle φ between $\vec{E}$ and $d\vec{A}$ for each surface.

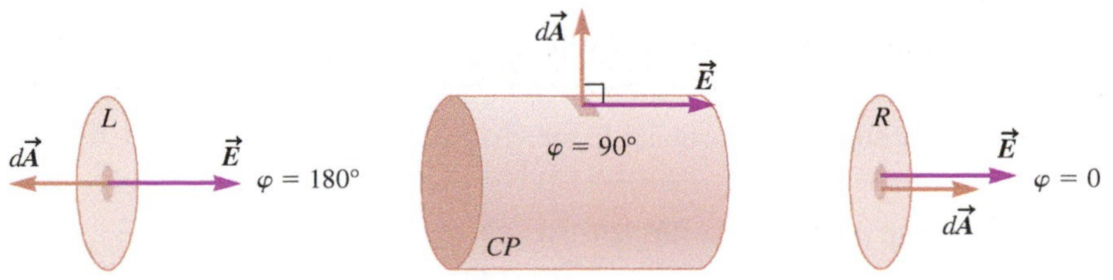

FIGURE 25.15

The area vector points outward for all small pieces of the curved part, while the electric field points to the right. Therefore, the angle $\varphi = 90°$ for the curved part and the corresponding dot product and integral are zero, leaving two integrals.

$$\Phi_E = \int_L \vec{E} \cdot d\vec{A} + \int_R \vec{E} \cdot d\vec{A}$$

Example continues on page 764 ▶

Write each dot product in terms of the magnitude of the vectors and the angle between the vectors.	$\Phi_E = \int_L E\,dA \cos 180° + \int_R E\,dA \cos 0$ $\Phi_E = -\int_L E\,dA + \int_R E\,dA \quad (25.12)$
As for the box, the electric field is constant and the integrals are identical. This time, each is the area of the end cap.	$\Phi_E = -EA + EA = 0$
The electric flux is zero, so the charge inside the Gaussian surface is zero.	$\Phi_E = \frac{q_{in}}{\varepsilon_0} = 0 \quad q_{in} = 0$

CHECK and THINK

We find the same result whether we use a Gaussian cylinder or a box. The process is similar, but there is slightly less work when we use the cylinder because the Gaussian cylinder is made up of just three surfaces instead of the six surfaces of a box. Because we are free to choose a convenient Gaussian surface, it may be helpful to choose a closed cylinder instead of a box when that is possible.

EXAMPLE 25.3 Nonuniform Electric Field

The Gaussian surface in Figure 25.16 is a closed cylinder. An electric field points to the right throughout, made up of two uniform components with magnitude $E_L = 40.9$ N/C on the left side of the cylinder and $E_R = 81.8$ N/C on the right side. Integrate $\Phi = \oint \vec{E} \cdot d\vec{A} = q_{in}/\varepsilon_0$ (Eq. 25.11) to find the amount of charge enclosed in the Gaussian surface. Express your result in terms of the surface charge density (charge per unit area) σ. (Assume σ is uniform.)

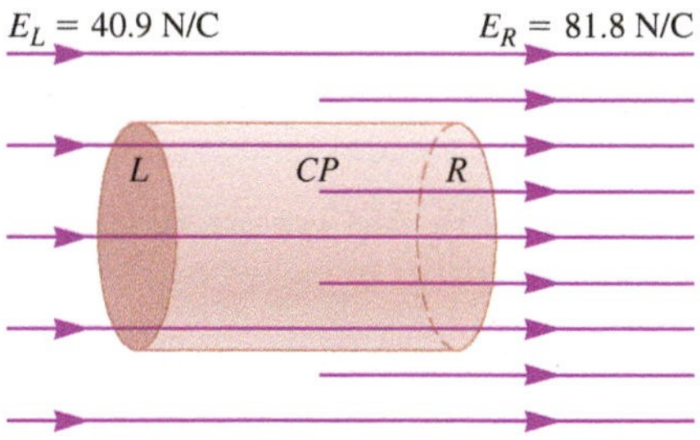

FIGURE 25.16

INTERPRET and ANTICIPATE

This is similar to Example 25.2 (Fig. 25.13B), except the electric field is not uniform. Because more electric field lines emerge from the Gaussian surface than enter it, we expect the net charge inside to be positive.

SOLVE Much of the work we need to do was already done in Example 25.2. We can start with Equation 25.12.	$\Phi_E = -\int_L E\,dA + \int_R E\,dA \quad (25.12)$
The electric field is constant over each end cap, so it can be pulled outside the integrals. The magnitude of the electric field depends on the position, and the subscripts refer to the magnitudes in Figure 25.16.	$\Phi_E = -E_L \int_L dA + E_R \int_R dA$
The integrals are identical, equaling the area A of each end cap.	$\Phi_E = -E_L A + E_R A = A(E_R - E_L)$
According to Gauss's law, the electric flux is proportional to the charge inside.	$\Phi_E = A(E_R - E_L) = \frac{q_{in}}{\varepsilon_0}$
The surface charge density σ is charge per unit area (Eq. 24.10).	$\sigma = \frac{q_{in}}{A} = \varepsilon_0(E_R - E_L) \quad (24.10)$ $\sigma = (8.85 \times 10^{-12}\,\text{C}^2/\text{N}\cdot\text{m}^2)(81.8\,\text{N/C} - 40.9\,\text{N/C})$ $\sigma = 3.62 \times 10^{-10}\,\text{C/m}^2$

CHECK and THINK

As expected, the charge inside the Gaussian surface is positive.

PROBLEM-SOLVING STRATEGY **Finding the Electric Field Using Gauss's Law**

Gauss's law says that the net electric flux Φ_E through a closed (Gaussian) surface is proportional to the net charge q_{in} inside the surface, and in the preceding examples we used Gauss's law to find that net enclosed charge. In the rest of this chapter, we focus on using Gauss's law to find the electric field for different special cases. So now we present a four-step problem-solving strategy.

INTERPRET and ANTICIPATE

Step 1 Sketch the electric field lines (Section 24-3). You may need to draw more than one perspective. Your sketch will help you choose the best Gaussian surface and anticipate your results.

Step 2 Choose a Gaussian surface. You should choose a closed surface that exploits the symmetry of the situation. In other words, you must pick a Gaussian surface that has the same symmetry as the source and its electric field—just as the letter **A** has the same symmetry as the letter **U**. In practice, you should try to pick a Gaussian surface so that each piece of the surface has an angle φ between $\vec{E}$ and $d\vec{A}$ equal to 0, 90°, or 180°. The most commonly used Gaussian surfaces are spherical shells, closed cylinders, and closed boxes. Add the Gaussian surface to your sketch.

SOLVE

Step 3 Integrate to find the flux in $\Phi_E = \oint \vec{E} \cdot d\vec{A} = q_{in}/\varepsilon_0$ (Eq. 25.11). Because your Gaussian surface exploits the symmetry of the situation, the electric field is often constant over some portion of the integral and the integral is usually very simple. It may help to add a few area vectors $d\vec{A}$ to your sketch, perpendicular to the surface and pointing outward so that you can easily see the angle between $\vec{E}$ and $d\vec{A}$.

Step 4 Determine the amount of **charge inside** the Gaussian surface. There may be charged objects inside or outside the Gaussian surface, or a portion of a continuous distribution may be inside the Gaussian surface and the rest outside. Only the net charge q_{in} inside the Gaussian surface is part of Gauss's law. This net charge may be positive, negative, or zero.

SOLVE

As in Chapter 24, there are two things to check once you have found an expression for the electric field $\vec{E}$:

1. Make sure the **dimensions** of the electric field are force per charge.
2. If the charge distribution is finite, the electric field for a point far from the distribution compared to its size should approach the electric field of a charged particle $\vec{E}(r) = (kQ_S/r^2)\hat{r}$. We can also **compare to another known electric field.** For example, after we find the electric field produced by an infinitely long charged rod, we can compare that result to the electric field of a finite rod found in Chapter 24. (See Concept Exercise 25.6.)

25-4 Special Case: Linear Symmetry

Imagine a rod such as a pencil. If you look down the end of the pencil and rotate it by any amount, the rod is symmetrical with respect to such a rotation (Fig. 25.17A). If you rotate that rod by 180° around a perpendicular axis that passes through its center, the rod is also symmetrical with respect to such a rotation (Fig. 25.17B). Now imagine an infinitely long rod. Like the finite rod (Fig. 25.17A), the infinite rod is symmetrical with respect to any rotation around its long axis. However, an infinitely long rod is more symmetrical than a finite rod because an infinitely long rod is symmetrical with respect to a 180° rotation around *any* perpendicular axis (Fig. 25.17C). An infinitely long rod is said to possess *linear symmetry*. Of course, no real rod is infinitely long, but a real rod can be modeled as infinitely long for positions close to the rod but not near its ends.

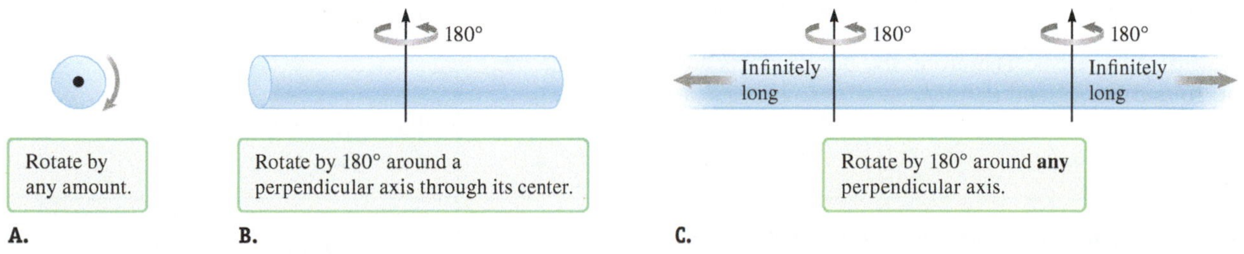

180°
180°
180°
Infinitely long
Infinitely long
Rotate by any amount.
Rotate by 180° around a perpendicular axis through its center.
Rotate by 180° around **any** perpendicular axis.
A.
B.
C.

FIGURE 25.17 Special case of linear symmetry. **A.** Rotation by any amount around the long axis. **B.** Rotation by 180° around a perpendicular axis passing through the center. **C.** Rotation by 180° around any axis perpendicular to the long axis.

EXAMPLE 25.4 $\vec{E}$ for a Charge Distribution with Linear Symmetry

It is relatively easy to find the electric field due to a source with linear symmetry using Gauss's law and the four steps listed in the problem-solving strategy (page 765). You can imagine an infinitely long rod with excess positive charge spread uniformly. Because the rod is infinitely long, the total charge is infinite, so it is not very practical to use the total charge. Instead, we work with the linear charge density λ (Eq. 24.12). Show that the electric field due to this source is

$$\vec{E} = \frac{1}{2\pi\varepsilon_0}\frac{\lambda}{r}\hat{r} \tag{25.13}$$

ELECTRIC FIELD FOR A SOURCE WITH LINEAR SYMMETRY

Special Case

where r is the perpendicular distance from the infinitely long charged rod.

INTERPRET and ANTICIPATE

Step 1 Sketch the electric field lines, which point outward and are perpendicular to the rod (Fig. 25.18). We use two views to show the electric field clearly.

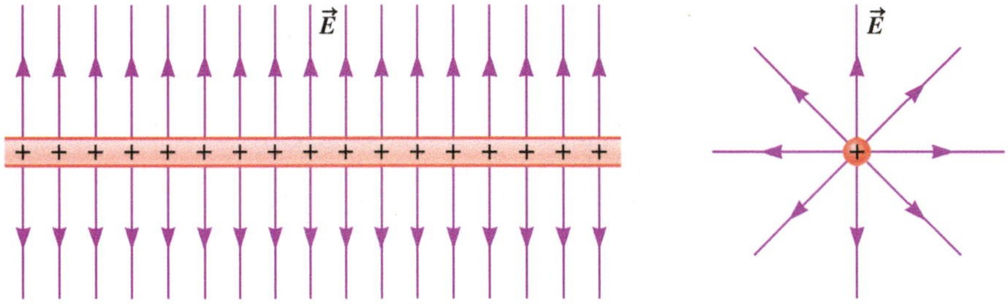

FIGURE 25.18

Step 2 Choose a Gaussian surface that exploits symmetry. A closed cylinder with radius r has the symmetry we need. We add the Gaussian surface to our sketch (Fig. 25.19). The Gaussian cylinder does not need to be infinitely long. Instead, the linear symmetry of the charge distribution means that the length of the cylinder and its position are arbitrary. To be specific, we use a cylinder of length ℓ shown in the middle of the side-view sketch.

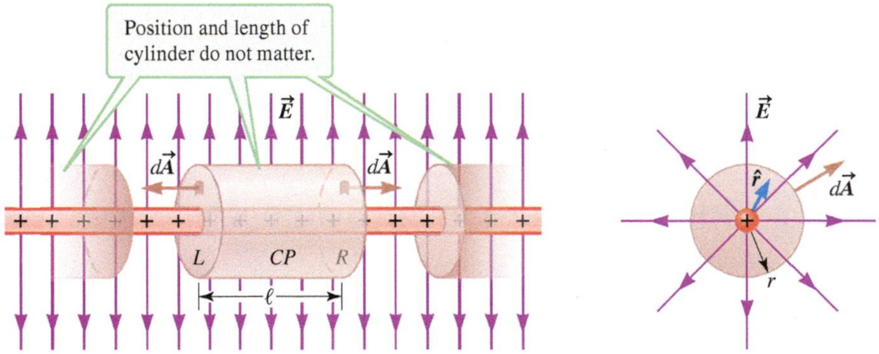

FIGURE 25.19

SOLVE

Step 3 Integrate. As we found in Example 25.2, the closed cylinder is made up of three surfaces—the left cap, the right cap, and the curved part—labeled in Figure 25.19 as L, R, and CP.

$$\Phi_E = \oint \vec{E}\cdot d\vec{A}$$

$$\Phi_E = \int_L \vec{E}\cdot d\vec{A} + \int_{CP} \vec{E}\cdot d\vec{A} + \int_R \vec{E}\cdot d\vec{A}$$

The area vector $d\vec{A}$ for surface L points to the left, while that for R points to the right. Because the electric field points outward, the angle φ between $\vec{E}$ and $d\vec{A}$ is 90° for both of these surfaces. Therefore, the dot products are zero and the integrals are zero for surfaces L and R.

$$\Phi_E = 0 + \int_{CP} \vec{E}\cdot d\vec{A} + 0$$

$$\Phi_E = \int_{CP} \vec{E}\cdot d\vec{A}$$

Only one integral remains. The area vector $d\vec{A}$ for surface CP points outward just like the electric field itself, so the angle φ is zero for this entire surface.

$$\Phi_E = \int_{CP} E\,dA\cos 0 = \int_{CP} E\,dA$$

The electric field is not uniform, but the density of the field lines is the same all over the curved surface, so $\vec{E}$ has the same magnitude over that surface and we can pull E outside the integral. The integral is then just the area of the curved part A_{CP}.	$\Phi_E = E\int_{CP} dA = EA_{CP}$
The area of the curved part is the circumference of the circle times the length of the cylinder.	$A_{CP} = 2\pi r\ell$ $\Phi_E = E2\pi r\ell$ (1)
Step 4 Find the **charge inside** the Gaussian cylinder. Because the charge per unit length is λ, multiply λ by the length of the cylinder to find the charge inside.	$q_{\text{in}} = \lambda\ell$ (2)
Finally, we substitute Equations (1) and (2) into Gauss's law (Eq. 25.7) and solve for E. As expected, the length of the cylinder does not matter (ℓ cancels out).	$\Phi_E = \frac{q_{\text{in}}}{\varepsilon_0}$ (25.7) $E2\pi r\ell = \frac{\lambda\ell}{\varepsilon_0}$ $E = \frac{1}{2\pi\varepsilon_0}\frac{\lambda}{r}$
From Figure 25.19, we see that the electric field points outward from the line of charge in the positive $\hat{r}$ direction, so we write the electric field vector in component form.	$\vec{E} = \frac{1}{2\pi\varepsilon_0}\frac{\lambda}{r}\hat{r}$ ✓ (25.13)
CHECK and THINK At a minimum, we need to check that our expression has the dimensions of electric field—force per charge. In Concept Exercise 25.6, you will perform a second check.	$[E] = \left[\frac{1}{\varepsilon_0}\right]\frac{[\lambda]}{[r]} = \frac{\text{F}\cdot\text{L}^2}{\text{C}^2}\frac{\text{C}}{\text{L}}\frac{1}{\text{L}}$ $[E] = \frac{\text{F}}{\text{C}}$

This example gave us practice using Gauss's law to find the electric field produced by a source. The result is used to model many real charge distributions, so you can expect to see it often.

CONCEPT EXERCISE 25.6

Check Equation 25.13 by showing that it is consistent with

$$\vec{E} = \frac{kQ}{y}\frac{1}{\sqrt{\ell^2 + y^2}}\hat{\jmath} \qquad (24.15)$$

for a finite rod of length 2ℓ.

25-5 Special Case: Spherical Symmetry

A spherical object is highly symmetrical. Imagine rotating a sphere by any amount through any axis that passes through its center. The sphere remains unchanged. In this section, we find the electric field due to a source with spherical symmetry, such as a charged particle, solid sphere, or spherical shell. The charge is uniformly distributed throughout the solid sphere and over the shell. Both a solid sphere and a spherical shell have interiors, but a charged particle does not (because a particle has no spatial extent). We will use Gauss's law to find the electric field both inside and outside these spherical sources. Coulomb's law is valid outside these charge distributions, so we expect our answer will be the same as what we found using Coulomb's law directly (Eq. 24.3).

EXAMPLE 25.5 $\vec{E}$ Outside a Charge Distribution with Spherical Symmetry

For a solid sphere or spherical shell, show that the electric field outside the source (at $r \geq R$, where R is the radius of the sphere or shell) is

$$\vec{E} = \frac{1}{4\pi\varepsilon_0}\frac{q}{r^2}\hat{r} \tag{25.14}$$

ELECTRIC FIELD OUTSIDE A SOURCE WITH SPHERICAL SYMMETRY

Special Case

INTERPRET and ANTICIPATE

Step 1 Sketch the electric field lines. When the source (a particle, solid sphere, or shell) is positively charged, the electric field lines point outward in all directions, like the quills of a porcupine (Fig. 25.20).

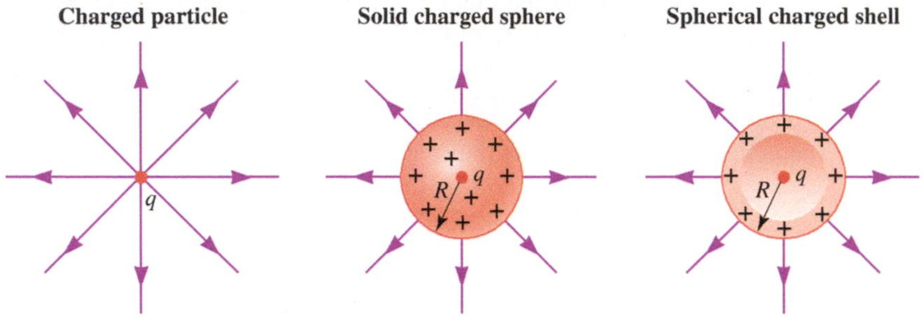

FIGURE 25.20

Step 2 Choose a Gaussian surface. A spherical Gaussian surface has the same symmetry as a spherically symmetrical source (Fig. 25.21). Because we are interested in the electric field outside the source, the radius r of the Gaussian sphere must be greater than or equal to the radius R of the source: $r \geq R$.

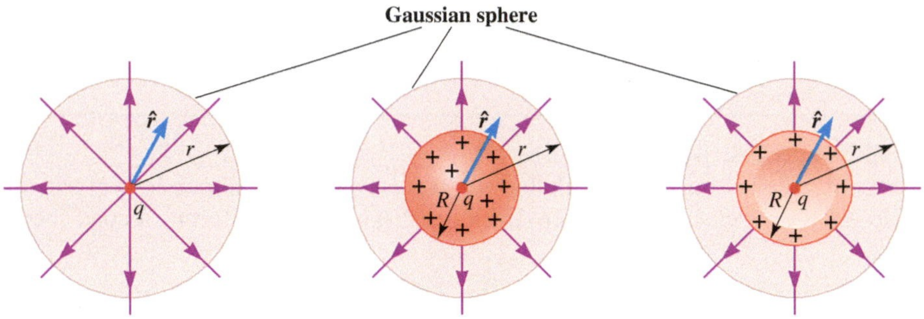

FIGURE 25.21

SOLVE

Step 3 Integrate. A sphere has only one surface, so there is no need to break up the integral into multiple parts. The area vectors point outward like the electric field vectors, and the angle φ is zero over the entire sphere.

$$\Phi_E = \oint \vec{E} \cdot d\vec{A}$$

$$\Phi_E = \oint E\, dA \cos 0 = \oint E\, dA$$

The electric field is not uniform, but it has constant magnitude over the surface of the sphere. We pull E out of the integral, which then becomes the surface area of the Gaussian sphere, $A = 4\pi r^2$.

$$\Phi_E = E\oint dA = EA$$

$$\Phi_E = E(4\pi r^2) \tag{1}$$

Step 4 Find the **charge inside**. The charge inside the Gaussian sphere in all three cases is the charge q of the source.

$$q_{\text{in}} = q \tag{2}$$

Substitute Equations (1) and (2) into Gauss's law (Eq. 25.7), and solve for E.

$$\Phi_E = \frac{q_{\text{in}}}{\varepsilon_0} \tag{25.7}$$

$$E(4\pi r^2) = \frac{q}{\varepsilon_0}$$

$$E = \frac{1}{4\pi\varepsilon_0}\frac{q}{r^2}$$

Write the electric field using the unit vector $\hat{r}$ (Fig. 25.21).

$$\vec{E} = \frac{1}{4\pi\varepsilon_0}\frac{q}{r^2}\hat{r} \;✓ \tag{25.14}$$

CHECK and THINK

With the substitution of Coulomb's constant (Eq. 25.9), we find that Equation 25.14 is the same as Equation 24.3, as expected outside a source with spherical symmetry.

$$k = \frac{1}{4\pi\varepsilon_0} \tag{25.9}$$

$$\vec{E}(r) = k\frac{q}{r^2}\hat{\boldsymbol{r}} \tag{24.3}$$

According to Equation 25.14, the electric field outside a charged sphere or shell $(r \geq R)$ is the same as the electric field due to a charged particle located at the center of the distribution. But Equation 25.14 tells us nothing about the electric field inside the sphere or shell $(r < R)$. We can use Gauss's law to find the electric field inside these sources. The case of a charged shell is a little easier, so let's start with it.

EXAMPLE 25.6 $\vec{E}$ Inside a Charged Spherical Shell

According to the **shell theorem**, the electric field inside any charged shell is zero and the electric field outside is given by Equation 25.14. Show that the electric field inside a charged spherical shell is $\vec{E} = 0$.

SHELL THEOREM Special Case

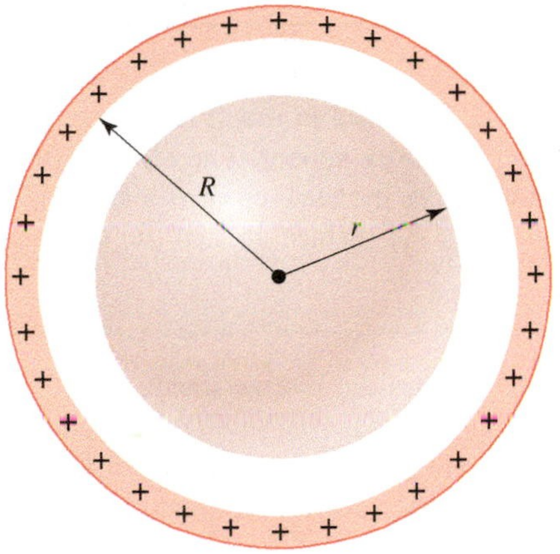

FIGURE 25.22

INTERPRET and ANTICIPATE

Finding the electric field inside the shell does not require all four steps. Figure 25.22 shows a **Gaussian sphere** inside the charged spherical shell, but no field lines are drawn. You might guess that you could draw inward-pointing field lines, but what will happen when they get to the center? They cannot terminate there because that would require a negatively charged particle in the center. They cannot cross the center and terminate on the opposite wall of the shell because the shell would have to be negative.

SOLVE

Step 4 We skip to finding the **charge inside**. There is no charge inside the Gaussian sphere.

$$q_{\text{in}} = 0$$

Because there is no charge inside the Gaussian sphere, the flux through it is zero.

$$\Phi_E = \frac{q_{\text{in}}}{\varepsilon_0} = 0$$

So, the integral in Gauss's law is zero.

$$\Phi_E = \oint \vec{E} \cdot d\vec{A} = 0$$

Because the area of the Gaussian surface cannot be zero and the angle φ is not necessarily zero, the electric field must be zero.

$$\vec{E} = 0 \checkmark$$

inside a charged shell

CHECK and THINK

We had trouble even imagining how we could draw the electric field inside the spherical shell, so it makes sense that the electric field inside the shell is zero. In a sense, the electric field lines do point inward from every segment of the wall, but they cancel in the shell so there is no net electric field inside.

EXAMPLE 25.7 $\vec{E}$ Inside a Charged Solid Sphere

Show that the electric field inside a uniformly charged solid sphere is

$$\vec{E} = \frac{1}{4\pi\varepsilon_0}\frac{q}{R^3}r\hat{r} \qquad (25.15)$$

ELECTRIC FIELD INSIDE A SOLID SPHERE

Special Case

INTERPRET and ANTICIPATE

We can combine the first two steps of the problem-solving strategy. The **electric field** points outward in all directions (Fig. 25.23). (You might be tempted to think the electric field points inward, but remember it must point outward beyond the sphere. So, unless the net electric field reverses direction at the surface, it must point outward everywhere.) Also, a **Gaussian sphere** with $r < R$ exploits the symmetry.

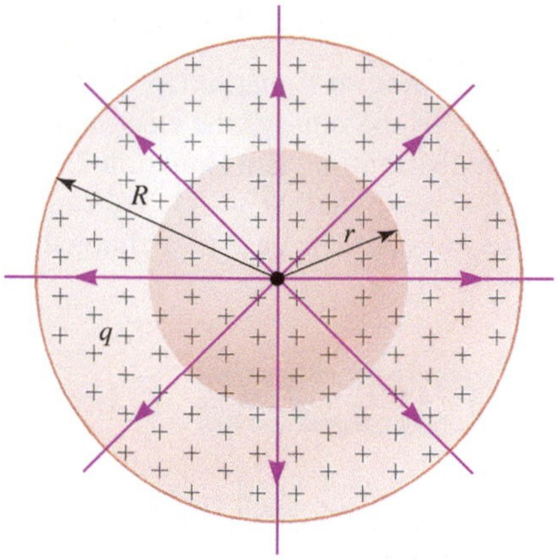

FIGURE 25.23

SOLVE

Step 3 Integrate in much the same way as in Example 25.5. The angle φ is zero over the entire sphere. The electric field is not uniform, but it has constant magnitude over the surface of the sphere. We pull E out of the integral, which then becomes the surface area of the Gaussian sphere, $A = 4\pi r^2$.

$$\Phi_E = \oint \vec{E}\cdot d\vec{A}$$

$$\Phi_E = \oint E\,dA\cos 0 = \oint E\,dA$$

$$\Phi_E = E\oint dA = EA = E(4\pi r^2) \qquad (1)$$

Step 4 Find the **charge inside**. The charged sphere has radius R and total charge q. The Gaussian surface is smaller than the charged sphere ($r < R$), so the Gaussian surface encloses some fraction (less than 1) of the sphere's charge. To find that fraction, first write an expression for the volume charge density ρ (Eq. 24.8) of the entire charged sphere.

$$\rho = \frac{q}{V} \qquad (24.8)$$

The volume V of the charged sphere can be found from its radius R.

$$V = \frac{4}{3}\pi R^3$$

Write an expression for the volume charge density.

$$\rho = \frac{3q}{4\pi R^3} \qquad (2)$$

The amount of charge inside the Gaussian surface is the volume of the Gaussian surface V_G times the volume charge density.

$$q_{in} = V_G\rho \qquad (3)$$

Find the volume enclosed by the Gaussian surface V_G from its radius r.

$$V_G = \frac{4}{3}\pi r^3 \qquad (4)$$

Substitute Equations (2) and (4) into Equation (3) to find the charge inside the Gaussian surface.

$$q_{in} = \left(\frac{4}{3}\pi r^3\right)\left(\frac{3q}{4\pi R^3}\right) = \left(\frac{r}{R}\right)^3 q \qquad (5)$$

CHECK and THINK

As expected, $r < R$, so the fraction $(r/R)^3 < 1$ and the charge inside the Gaussian surface is smaller than the total charge.

SOLVE

To find the electric field inside the sphere, substitute Equation (5) into Equation 25.7 (Gauss's law) and use Equation (1) for the electric flux.

$$\Phi_E = \frac{q_{in}}{\varepsilon_0} \qquad (25.7)$$

$$E(4\pi r^2) = \frac{1}{\varepsilon_0}\left(\frac{r}{R}\right)^3 q \qquad E = \frac{1}{4\pi\varepsilon_0}\frac{q}{R^3}r$$

Use the unit vector $\hat{r}$ to write the electric field vector.	$\vec{E} = \frac{1}{4\pi\varepsilon_0}\frac{q}{R^3}r\hat{r}$ ✓	(25.15)

CHECK and THINK

Equation 25.15 shows that the electric field gets stronger with increasing distance from the center of the sphere. It is left as a homework exercise (Problem 38) to show that at the surface of the sphere ($r = R$), Equations 25.15 and 25.14 are identical.

25-6 Special Case: Planar Symmetry

You decide to show your school pride and put a static decal on the back window of your car (Fig. 25.24). Because it has excess charge, the decal clings to the window. We can model this source as a charged, very thin sheet.

Think about the symmetry of a sheet such as a piece of printer paper. It is symmetrical with respect to 180° rotations around an axis perpendicular to the sheet and passing through its center. Now imagine an infinite sheet. An infinite sheet is symmetrical with respect to any rotation around any axis that is perpendicular to its surface. Such a sheet is said to possess **planar symmetry**. It is easier to apply Gauss's law to a source with planar symmetry than to a finite sheet. Of course, no real sheet is infinite: A real sheet may be modeled as an infinite sheet for points near its surface but not near its edges.

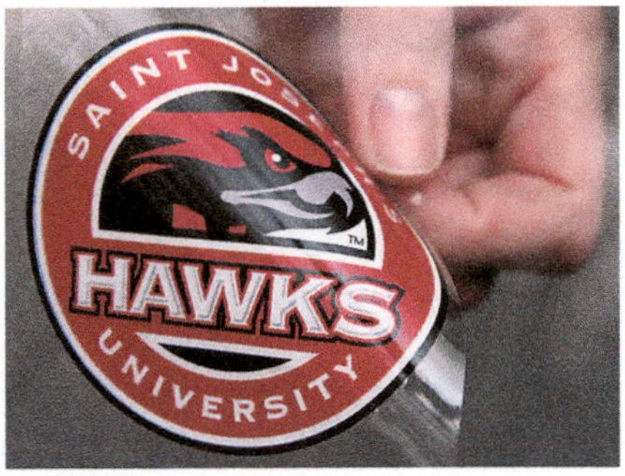

Dr. Jenny Spinner/St. Joseph's University

FIGURE 25.24 A static decal can be modeled as a charged thin sheet.

EXAMPLE 25.8 $\vec{E}$ for a Charge Distribution with Planar Symmetry

Figure 25.25 shows a continuous distribution of positively charged particles uniformly distributed to form a very thin sheet. Its surface charge density is σ. Use Gauss's law to show that the magnitude of the electric field due to such a source is

$$E = \frac{\sigma}{2\varepsilon_0} \tag{25.16}$$

ELECTRIC FIELD FOR A SOURCE WITH PLANAR SYMMETRY **Special Case**

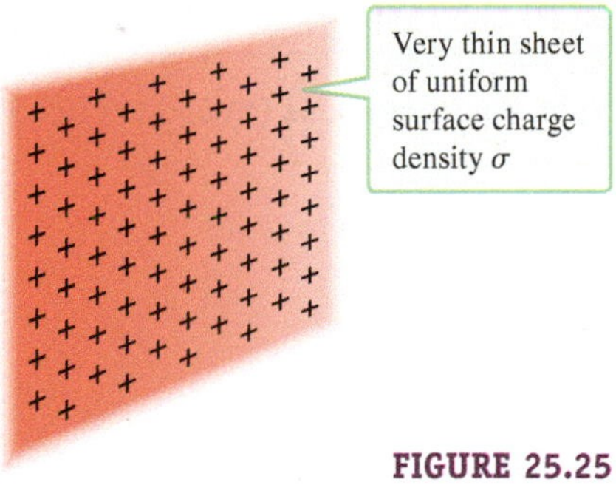

FIGURE 25.25

INTERPRET and ANTICIPATE

Step 1 Sketch the electric field lines. Because the sheet is positively charged, the electric field points outward. Because of the planar symmetry, the electric field is uniform and perpendicular to the surface, seen edge-on in Figure 25.26.

Step 2 Choose a Gaussian surface. A closed cylinder perpendicular to and intersecting the sheet has the symmetry we need (Fig. 25.27). Because of the planar symmetry, the cylinder may have any size and be located anywhere on the sheet. Because we chose the size of the Gaussian cylinder arbitrarily, we don't expect its length or area to be in our final expression.

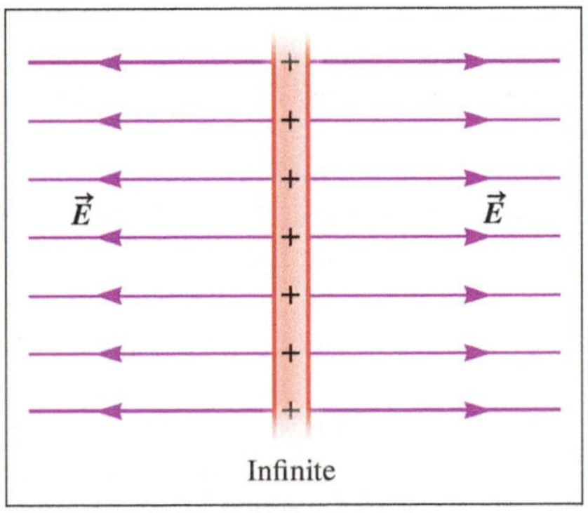

FIGURE 25.26

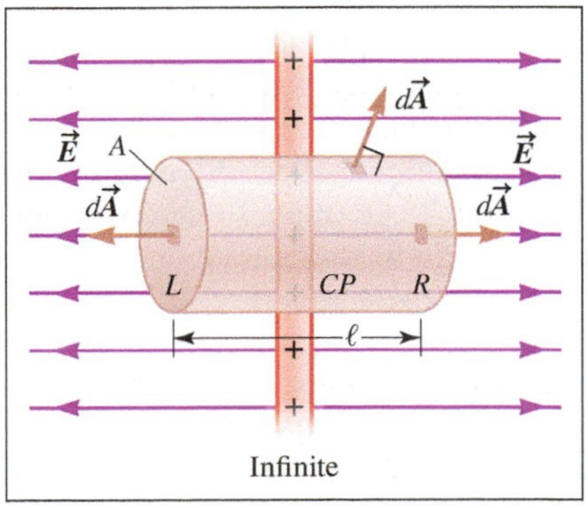

FIGURE 25.27

Example continues on page 772 ▶

SOLVE

Step 3 Integrate. The Gaussian cylinder is made up of three surfaces: the right cap R, the curved part CP, and the left cap L. The area vectors $d\vec{A}$ on the curved part are perpendicular to the electric field, so the integral over surface CP is zero.

$$\Phi_E = \oint \vec{E} \cdot d\vec{A}$$

$$\Phi_E = \int_L \vec{E} \cdot d\vec{A} + \int_{CP} \vec{E} \cdot d\vec{A} + \int_R \vec{E} \cdot d\vec{A}$$

$$\Phi_E = \int_L \vec{E} \cdot d\vec{A} + \int_R \vec{E} \cdot d\vec{A}$$

The area vectors on surface R point to the right, as does the electric field on that side of the sheet. The angle φ is thus zero for surface R. The area vectors on surface L point to the left, as does the electric field on that side of the sheet. The angle φ is thus also zero for surface L.

$$\Phi_E = \int_L E\, dA \cos 0 + \int_R E\, dA \cos 0$$

$$\Phi_E = \int_L E\, dA + \int_R E\, dA$$

The electric field is uniform, so E can be pulled outside both integrals. These integrals both equal the area A of an end cap.

$$\Phi_E = E\int_L dA + E\int_R dA$$

$$\Phi_E = EA + EA = 2EA \tag{1}$$

Step 4 Find the **charge inside** the Gaussian surface. The sheet has a uniform surface charge density σ. To find the amount of charge enclosed by the Gaussian surface, multiply the surface charge density by the area A of an end cap.

$$q_{in} = \sigma A \tag{2}$$

Substitute Equations (1) and (2) into Gauss's law (Eq. 25.7), and solve for E.

$$\Phi_E = \frac{q_{in}}{\varepsilon_0} \tag{25.7}$$

$$2EA = \frac{\sigma A}{\varepsilon_0}$$

$$E = \frac{\sigma}{2\varepsilon_0} \quad ✓ \tag{25.16}$$

CHECK and THINK

As expected, the area A of the end caps cancels out, so the electric field does not depend on the size of the Gaussian surface.

The infinite sheet of charge is used to model some real charge distributions. You will see Equation 25.16 often, so here are some things to know. First, $E = \sigma/2\varepsilon_0$ gives the magnitude of the electric field due to an infinite charged sheet. The direction of the field depends on the sign of the charge and is reversed when going from one side of the sheet to the other. If the sheet is positively charged, the field direction is away from the sheet. If the sheet is negatively charged, the field direction is toward the sheet.

Second, according to $E = \sigma/2\varepsilon_0$, the electric field due to an infinite charged sheet is uniform, with the same value at all positions. Of course, this is true for the ideal source—an infinite sheet. In practice, it is true for positions that are close to the surface of a finite sheet and not near an edge.

EXAMPLE 25.9 Two Infinite Sheets

Figure 25.28 shows a side view of two infinite sheets. The sheet on the left has excess positive charge uniformly distributed so that its surface charge density is $\sigma_+ = 48.0\,\mu\text{C/m}^2$. The sheet on the right has excess negative charge uniformly distributed so that its surface charge density is $\sigma_- = -24.0\,\mu\text{C/m}^2$. Find the net electric field at points *A*, *B*, and *C*.

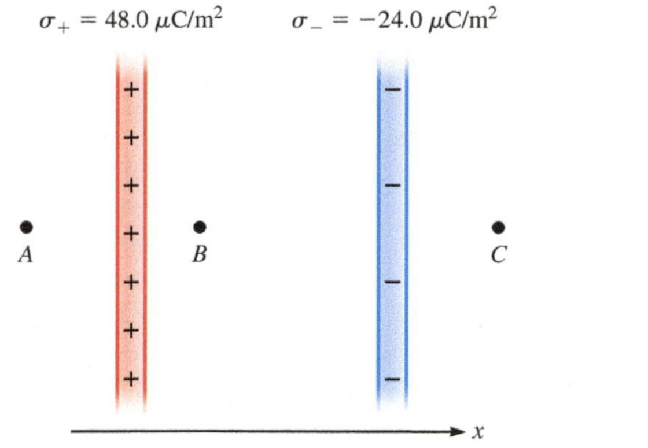

FIGURE 25.28

INTERPRET and ANTICIPATE

Because we already have found an expression for the electric field produced by an infinite sheet, there is no need to start with Gauss's law. The electric field at each point is a result of the electric field due to both sheets. For an infinite sheet, the magnitude of the electric field depends only on the magnitude of the surface charge density (Eq. 25.16). Because the positive sheet has a higher surface charge density, it produces a stronger electric field. Sketch the electric field vectors due to the positive sheet ($\vec{E}_+$) and the negative sheet ($\vec{E}_-$) at each point *A*, *B*, and *C* (Fig. 25.29). $\vec{E}_+$ points away from the positive sheet, and $\vec{E}_-$ points toward the negative sheet. Add these vectors graphically in order to anticipate the results.

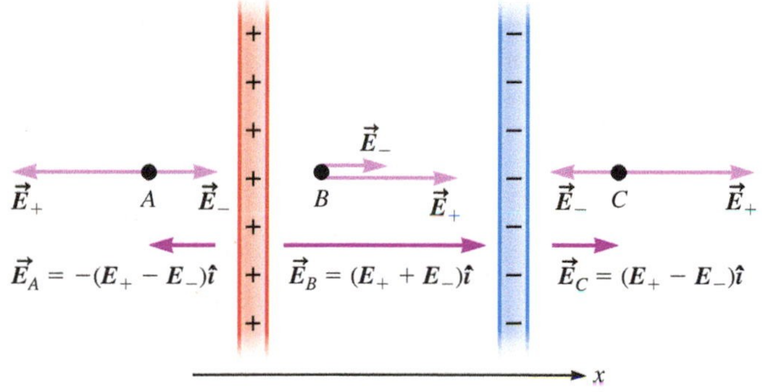

FIGURE 25.29

SOLVE

The magnitude of the electric field produced by each sheet is a constant given by $E = \sigma/2\varepsilon_0$ (Eq. 25.16).

$$E_+ = \frac{\sigma}{2\varepsilon_0} = \frac{48.0 \times 10^{-6}\ \text{C/m}^2}{2(8.85 \times 10^{-12}\ \text{C}^2/\text{N}\cdot\text{m}^2)}$$

$$E_+ = 2.71 \times 10^6\ \text{N/C}$$

$$E_- = \frac{\sigma}{2\varepsilon_0} = \frac{24.0 \times 10^{-6}\ \text{C/m}^2}{2(8.85 \times 10^{-12}\ \text{C}^2/\text{N}\cdot\text{m}^2)}$$

$$E_- = 1.36 \times 10^6\ \text{N/C}$$

Write the electric field at each point in component form, and then add the electric fields due to each sheet.

Point *A*: $\vec{E}_+$ due to the positive sheet points in the negative *x* direction, and $\vec{E}_-$ due to the negative sheet points in the positive *x* direction.

$$\vec{E}_A = E_-\hat{\imath} - E_+\hat{\imath} = (E_- - E_+)\hat{\imath}$$

$$\vec{E}_A = [(1.36 - 2.71) \times 10^6]\,\hat{\imath}\ \text{N/C}$$

$$\vec{E}_A = -1.35 \times 10^6\,\hat{\imath}\ \text{N/C}$$

Point *B*: Both $\vec{E}_+$ and $\vec{E}_-$ point in the positive *x* direction.

$$\vec{E}_B = E_+\hat{\imath} + E_-\hat{\imath} = (E_+ + E_-)\hat{\imath}$$

$$\vec{E}_B = [(2.71 + 1.36) \times 10^6]\hat{\imath}\ \text{N/C} = 4.07 \times 10^6\hat{\imath}\ \text{N/C}$$

Example continues on page 774 ▶

Point *C*: $\vec{E}_+$ points in the positive x direction, and $\vec{E}_-$ points in the negative x direction.	$\vec{E}_C = -E_-\hat{\imath} + E_+\hat{\imath} = (E_+ - E_-)\hat{\imath}$ $\vec{E}_C = [(2.71 - 1.36) \times 10^6]\hat{\imath}\text{ N/C} = 1.35 \times 10^6\hat{\imath}\text{ N/C}$

CHECK and THINK

Our results match the electric field vectors we found graphically (Fig. 25.29). At point A, the electric field points to the left (negative x direction), and at points B and C, the electric field points to the right (positive x direction). The magnitude of the electric field is greatest at point B, and the magnitude at point A equals the magnitude at point C.

As a final thought, imagine that the sheets were identical except that one was positive and the other negative. The magnitudes of the surface charge densities on the sheets would be equal, so $E_+ = E_-$ and the electric field outside the two sheets would cancel ($E_A = E_C = 0$). The electric field between the sheets would be reinforced ($E_B = 2E_+$ or $E_B = 2E_-$).

25-7 Special Case: Conductors

In this section, we consider the special case of a charged conductor. Imagine charging a solid spherical conductor such as a cannonball on an insulating pedestal (Fig. 25.30A), perhaps by touching a positively charged rod to the surface of the ball. Very quickly the excess charge on the cannonball spreads out because of the repulsive force between like charges. After the charges stop moving, the conductor is in **electrostatic equilibrium**. When a conductor is in electrostatic equilibrium, the excess charge is found only on its surface; there is no excess charge in the body of the conductor as shown in the cross-sectional view of the charged cannonball in Figure 25.30B.

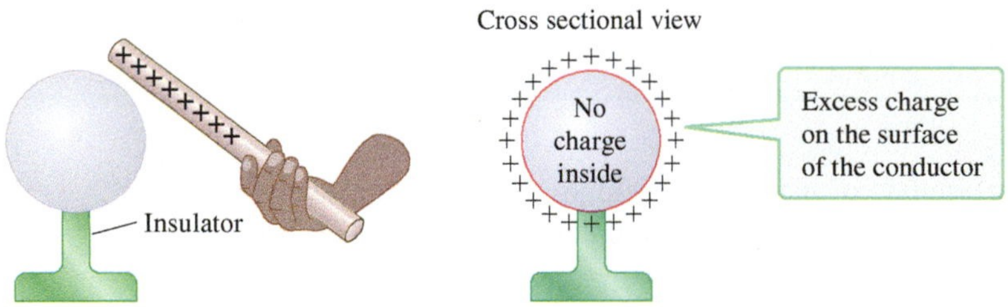

A. B.

FIGURE 25.30 **A.** Charging an isolated conductor. **B.** The excess charge rests on the conductor's outer surface.

ELECTRIC FIELD INSIDE A CONDUCTOR

Special Case

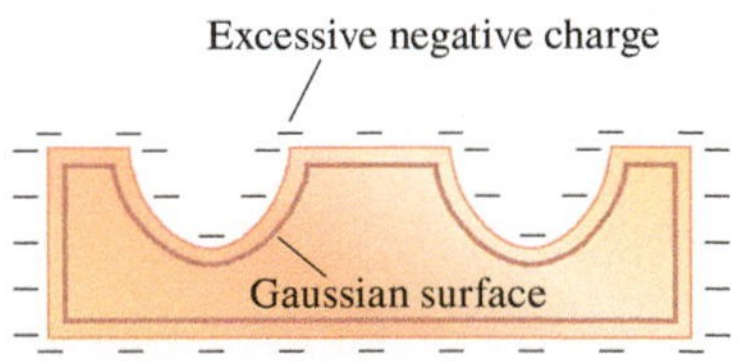

FIGURE 25.31 The excess charge is found only on the surface of this solid conductor, and the electric field inside is zero. We choose a Gaussian surface just inside the conductor.

According to Gauss's law, the electric field inside a charged conductor in electrostatic equilibrium is zero. If we choose a Gaussian spherical surface that lies just beneath the surface of the conductor, there is no charge inside it and the electric flux must be zero:

$$\Phi_E = \oint \vec{E} \cdot d\vec{A} = \frac{q_{\text{in}}}{\varepsilon_0} = 0$$

Because the area of the Gaussian sphere is *not* zero and the angle φ is not necessarily zero, the electric field inside the conductor must be zero:

$$\vec{E} = 0 \text{ inside a charged conductor}$$

The shape of the conductor and the sign of the excess charge don't change the fact that the electric field inside a conductor is zero. Figure 25.31 shows a solid isolated conductor. If you give the conductor excess negative charge, the excess charge will rest on the surface. Again, we can choose a Gaussian surface that lies just beneath the surface of the conductor. There is no charge inside that surface, so the electric field inside the conductor is zero.

Where is the excess charge if the conductor has a cavity, like a hollow locket (Fig. 25.32A)? Figure 25.32B shows a cross-sectional view of such a cavity within a conductor. In this case, the conductor has excess negative charge. To figure out how much charge is on the inner surface of the cavity, imagine embedding a Gaussian surface in the conductor just outside the cavity. There is no electric field in the body of the conductor, so the electric flux through the Gaussian surface must be zero:

$$\Phi_E = \oint \vec{E} \cdot d\vec{A} = \oint 0\, d\vec{A} = 0$$

Because the electric flux is zero, there is no charge inside the Gaussian surface:

$$\Phi_E = \frac{q_{\text{in}}}{\varepsilon_0} = 0$$

So, there is no charge inside the cavity of a charged conductor and no electric field in the cavity either.

When you charge a conductor, the excess charge quickly distributes on the conducting surface. If the conductor is spherical or an infinitely large sheet, the surface charge density σ is uniform. However, if the conductor is some other shape, such as the conductor in Figure 25.31 or the heart-shaped locket in Figure 25.32, the surface charge density is greater at the more pointed parts of the conductor (Fig. 25.33). We expect that the electric field outside those pointed places is stronger. In fact, the electric field just outside a conductor is proportional to the local surface charge density.

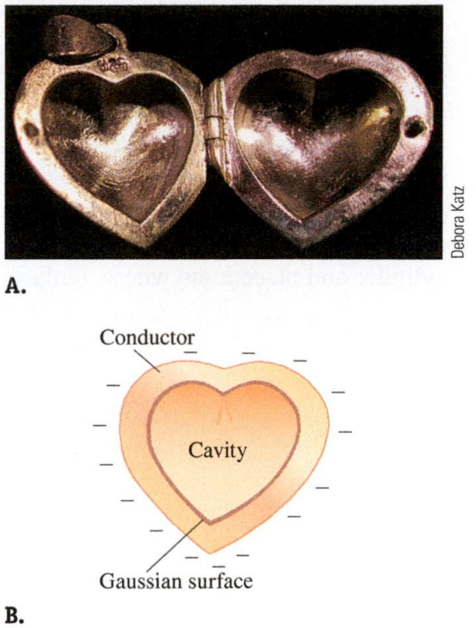

FIGURE 25.32 **A.** An open locket. When the locket is closed, it has an empty cavity. Any excess charge is found on the outside surface of the closed locket, and the electric field inside the locket is zero. **B.** Gaussian surface just outside the cavity of a heart-shaped locket.

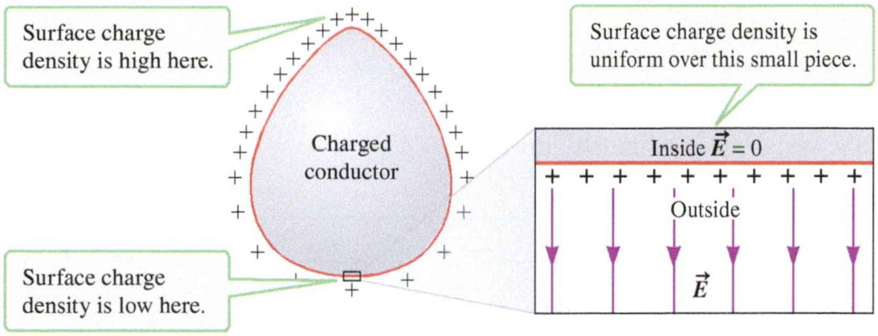

FIGURE 25.33 Excess charge is concentrated on the pointed parts of a conductor. *Inset*: Close to a charged conductor, the electric field is uniform.

EXAMPLE 25.10 $\vec{E}$ Just Outside a Charged Conductor

Use Gauss's law to show that the magnitude of the electric field just outside any charged conductor is

ELECTRIC FIELD OUTSIDE A CONDUCTOR
Special Case

$$E = \frac{\sigma}{\varepsilon_0} \quad (25.17)$$

Hint: Near the conducting surface, the surface appears to be a flat (Fig. 25.33, close-up), infinite plane, just as the surface of the ocean seems to be flat from our perspective standing on a beach. So, the surface charge density σ over a small part of the conductor is uniform.

Example continues on page 776 ▶

INTERPRET and ANTICIPATE

Step 1 Sketch the electric field lines. Figure 25.34 shows a close-up of a small portion of a charged conductor. Because the surface appears flat on this scale, the electric field must be perpendicular to the surface and directed outward. Inside the body of the conductor, the electric field is zero.

Step 2 Choose a Gaussian surface. Near the conductor, the source has planar symmetry. So, as in the case of an infinite sheet, we can choose any size Gaussian cylinder and place it anywhere. Only one such cylinder is shown here.

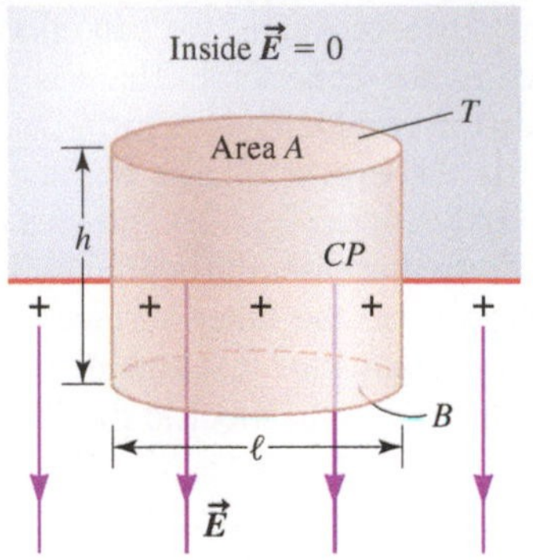

FIGURE 25.34

SOLVE

Step 3 Integrate. The Gaussian cylinder is made up of three surfaces—the top cap T, the curved part CP, and the bottom cap B.	$\Phi_E = \oint \vec{E} \cdot d\vec{A}$ $\Phi_E = \int_T \vec{E} \cdot d\vec{A} + \int_{CP} \vec{E} \cdot d\vec{A} + \int_B \vec{E} \cdot d\vec{A}$
The electric field over surface T is zero because the electric field is zero inside the conductor. The area vectors for the surface CP are perpendicular to the electric field outside the conductor, and the electric field inside the conductor is zero, so the integral over the surface CP is zero. This leaves only the integral over surface B.	$\Phi_E = 0 + 0 + \int_B \vec{E} \cdot d\vec{A}$ $\Phi_E = \int_B \vec{E} \cdot d\vec{A}$
The area vectors for surface B point downward in the same direction as $\vec{E}$, so $\varphi = 0$.	$\Phi_E = \int_B E\,dA \cos\varphi = \int_B E\,dA \cos 0$
The electric field is uniform over surface B, so E can be pulled outside the integral. The integral is the area A of the end cap.	$\Phi_E = \int_B E\,dA = E\int_B dA$ $\Phi_E = EA \quad (1)$
Step 4 Find the **charge inside** the Gaussian surface. As in the case of the infinite sheet, the charge inside the Gaussian surface is the surface charge density times the area of the end cap.	$q_{in} = \sigma A \quad (2)$
Substitute Equations (1) and (2) into Gauss's law (Eq. 25.7), and solve for E.	$\Phi_E = \dfrac{q_{in}}{\varepsilon_0} \quad (25.7)$ $EA = \dfrac{\sigma A}{\varepsilon_0}$ $E = \dfrac{\sigma}{\varepsilon_0}$ ✓ $\quad (25.17)$

CHECK and THINK

Equation 25.17 is valid for an infinite conductor with uniform charge density σ. It can also be used *just outside* a conductor of any shape, where σ is the local surface charge density. The electric field is perpendicular to the surface, pointing outward if the conductor is positively charged and inward if it is negatively charged.

A *capacitor* is a common device consisting of two large, oppositely charged conducting parallel plates. Imagine that initially the large plates are far apart and each plate has the same surface charge density σ_i (Fig. 25.35). Each plate has two surfaces, and when the plates are far apart, the excess charge is uniformly distributed on both surfaces of each plate as shown. The electric field due to the positive plate is perpendicular to the surfaces and outward, and the electric field due to the

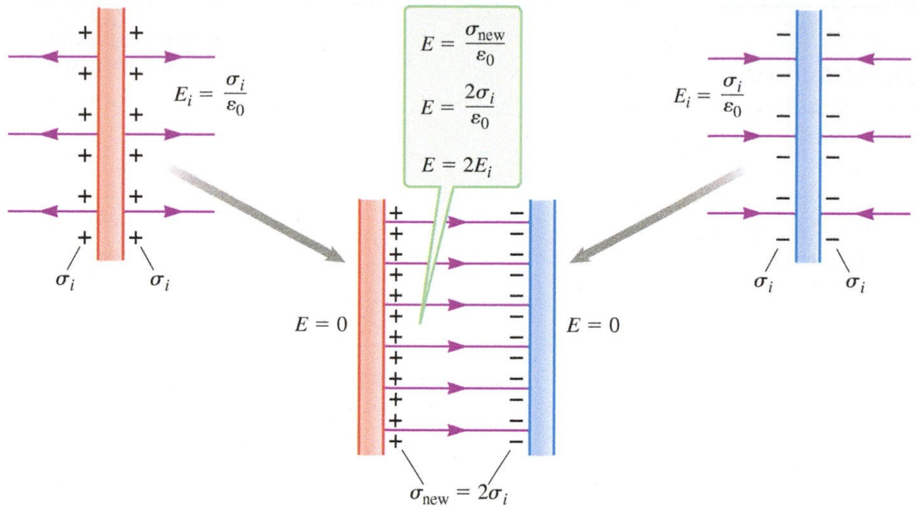

FIGURE 25.35 Two oppositely charged conducting plates, initially far apart, are brought close together (but not touching). The resulting device is a type of *capacitor*. We'll encounter these devices in Chapters 26 through 34.

negative plate is similarly perpendicular but inward. Because the plates have the same surface charge density, the magnitude of each one's electric field is the same, $E_i = \sigma_i/\varepsilon_0$ (Eq. 25.17).

Now imagine bringing these two charged plates closer together, keeping them parallel as shown in the lower middle portion of Figure 25.35. Because the plates have charges of opposite sign, the excess charge on each plate moves to the inner surfaces. This doubles the surface charge density on these inner surfaces: $\sigma_{\text{new}} = 2\sigma_i$. The electric field between the plates is still given by $E = \sigma/\varepsilon_0$ (Eq. 25.17), but the surface charge density is now σ_{new} on either plate, so the electric field has doubled:

$$E_{\text{new}} = \frac{\sigma_{\text{new}}}{\varepsilon_0} = \frac{2\sigma_i}{\varepsilon_0} \tag{25.18}$$

$$E_{\text{new}} = 2E_i$$

Because no excess charge is left on either of the outer faces, the surface charge density there is zero. So, according to Equation 25.17, the electric field outside the two plates is zero (Fig. 25.35).

CONCEPT EXERCISE 25.7

Is it possible for the charged solid sphere in Figure 25.23 to be a conductor in electrostatic equilibrium? Explain.

EXAMPLE 25.11 Hanging Inside a Conductor

Figure 25.36 shows a cross-sectional view of a thick spherical conductor. The conductor is neutral, and a small charged sphere ($q = +29.5\ \mu\text{C}$) hangs from an insulating thread. The sphere is not in the center of the conductor; instead, it is closer to the left side as shown.

A Find the charge q_{wall} on the wall of the cavity and the charge q_{out} on the outer surface of the conductor. Start by sketching the electric field for all regions—inside the cavity, inside the body of the conductor, and outside the conductor. In **CHECK and THINK**, discuss this sketch.

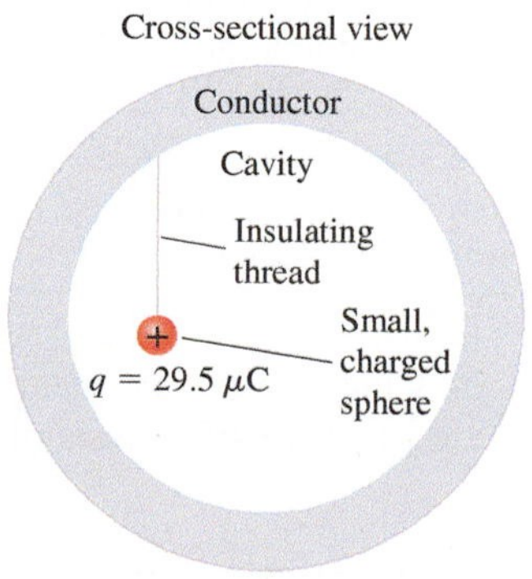

FIGURE 25.36

Example continues on page 778 ▶

INTERPRET and ANTICIPATE

The positive charge on the small sphere attracts electrons in the conductor. These electrons move close to the walls of the cavity. If the positively charged sphere were in the center of the cavity, the electrons would be uniformly distributed on the cavity wall. However, the electrons are more concentrated on the left side of the cavity because the positive sphere is closer to the left side (Fig. 25.37). The electric field in the body of any conductor in electrostatic equilibrium is zero. By choosing a Gaussian sphere that is concentric with the conductor and embedded in it, we can determine the amount of charge on the walls of the cavity.

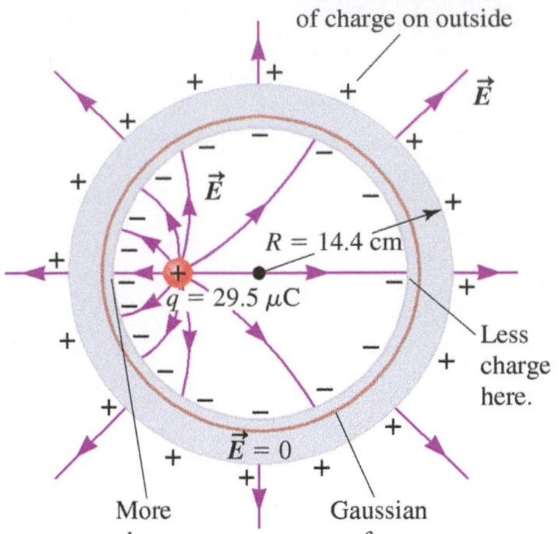

FIGURE 25.37 The Gaussian surface chosen for this problem is inside the conducting shell.

SOLVE

The electric flux through the Gaussian surface (Fig. 25.37) is zero because the electric field in the body of the conductor is zero. Therefore, the net charge inside the Gaussian sphere is zero, so the total charge on the inside wall of the cavity is negative, equal in magnitude to the charge on the small sphere inside.

$$\Phi_E = \oint \vec{E} \cdot d\vec{A} = 0 = \frac{q_{in}}{\varepsilon_0}$$

$$q_{in} = 0 = q + q_{wall}$$

$$q_{wall} = -q = -29.5\ \mu\text{C}$$

Because the conductor is neutral, the charge on its surface must be positive and equal to the charge on the cavity wall.

$$q_{out} = -q_{wall}$$

$$q_{out} = -(-29.5\ \mu\text{C}) = +29.5\ \mu\text{C}$$

The positive charge q_{out} on the outer surface of the conductor is uniformly distributed, as excess charge always is on the surface of a spherical conductor. To see why this is so, imagine that free electrons move toward the cavity wall, leaving positively charged ions in place.

CHECK and THINK

Figure 25.37 shows the electric field in all regions. Inside the cavity, the electric field is stronger on the left side where there is a higher concentration of charge. Outside the spherical conductor, the electric field looks like that of a charged particle located at the center of the entire collection of objects (and replacing them). This last sentence may seem confusing. Think of it this way. Suppose the positive sphere inside the conductor were at the center; then by symmetry the surface charge density on the outside of the conductor would be uniform. Because the electric field is zero inside the body of the conductor, moving the sphere off the center doesn't change the uniform charge distribution on the conductor's outer surface. (The particles on the outer surface having no way of "knowing" that the sphere moved.)

B If the radius of the conductor is $R = 14.4$ cm, what is the magnitude of the electric field just outside the conductor?

INTERPRET and ANTICIPATE

The electric field just outside any conductor depends only on the surface charge density, so we need to find the surface charge density so that we can find the electric field.

SOLVE

The charge is uniformly distributed on a sphere of radius R. We divide the charge q_{out} by the surface area of the sphere.

$$\sigma = \frac{q_{out}}{A} = \frac{q_{out}}{4\pi R^2} = \frac{29.5 \times 10^{-6}\text{C}}{4\pi(14.4 \times 10^{-2}\text{ m})^2}$$

$$\sigma = 1.13 \times 10^{-4}\text{C/m}^2$$

The magnitude of the electric field just outside a conductor is given by Equation 25.17.

$$E = \frac{\sigma}{\varepsilon_0} \quad (25.17)$$

$$E = \frac{1.13 \times 10^{-4}\ \text{C/m}^2}{8.85 \times 10^{-12}\ \text{C}^2/\text{N}\cdot\text{m}^2}$$

$$= 1.28 \times 10^7\ \text{N/C}$$

CHECK and THINK

From Figure 25.37, we see that the electric field outside the conductor is equivalent to the electric field produced by a charged particle located at the center of the objects. We can check our result by using the relationship for electric field derived from Coulomb's law (Eq. 24.3) to calculate the field at a distance of 14.4 cm from such a fictitious particle with charge +29.5 μC.

$$E = \frac{kq}{r^2} \quad (24.3)$$

$$E = \frac{(8.99 \times 10^9\ \text{N}\cdot\text{m}^2/\text{C}^2)(29.5 \times 10^{-6}\ \text{C})}{(14.4 \times 10^{-2}\ \text{m})^2}$$

$$E = 1.28 \times 10^7\ \text{N/C}$$

EXAMPLE 25.12 CASE STUDY What Should Shannon Do?

Avi, Cameron, and Shannon are caught in a thunderstorm. Cameron thinks they should wait out the storm in the car. Avi is worried that if lightning hits the car, the gas tank may explode. Shannon is expected to make a decision. What would you decide?

SOLVE

Cameron is right. The car is made of metal, a conductor. It is true that the car has windows made of glass (an insulator), but the car can still be modeled as a conductor with a cavity. The inside of the car—where passengers sit—is like the cavity inside the locket (Fig. 25.32). When a car is struck by lightning, the charge stays on the outside surface of the car and the electric field inside is zero. The cavity inside a conductor is shielded from the external electric field (Fig. 25.38; such a device is called a *Faraday cage*). In addition, the gasoline is in a metal tank that acts as a Faraday cage.

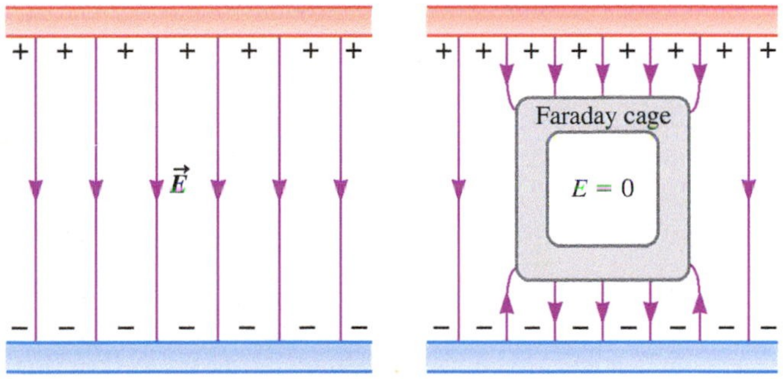

FIGURE 25.38 A Faraday cage is a hollow conductor that shields its interior from external electric fields.

We can use these ideas to explain how antistatic bags work (Fig. 25.5) and why your cell phone doesn't work in a closed elevator. The antistatic bag is coated with a conductor, and sensitive equipment is sealed inside. If the bag comes in contact with a charged object, the outside of the bag becomes charged but the inside remains uncharged, and the electric field inside the bag is zero. The antistatic bag is a Faraday cage. In order for your cell phone to work, it must be able to receive signals that consist of changing electric and magnetic fields. If your cell phone is inside a closed elevator, the elevator acts as a Faraday cage. The electric field inside the elevator is zero, so your cell phone cannot receive signals.

Avi is correct that the trees will attract lightning, and because trees are taller than people, they are more likely to get hit (Section 24-7). When a tree is struck by lightning, however, excess charge travels through the surface of the Earth. If you are in contact with the Earth nearby, considerable excess charge is likely to travel through you.

Trees and people have been known to survive a direct lightning strike. If a tree is very wet, the water (which is not pure) acts as a Faraday cage. The excess charge stays on the outside coating of water. However, the large amount of energy delivered by lightning can cause the water to vaporize and burn the tree.

SUMMARY

Underlying Principles

Gauss's law relates the electric flux that penetrates a closed surface to the charge enclosed by that surface:

$$\Phi_E = \oint \vec{E} \cdot d\vec{A} = \frac{q_{\text{in}}}{\varepsilon_0} \quad (25.11)$$

where Φ_E is the electric flux through a Gaussian surface.

Major Concepts

1. A **Gaussian surface** is an imaginary, closed, three-dimensional surface.
2. **Electric flux** is represented by the number of field lines penetrating a surface and is given by

$$\Phi_E = \int \vec{E} \cdot d\vec{A} \quad (25.6)$$

For the net flux through a Gaussian surface,

$$\Phi_E = \oint \vec{E} \cdot d\vec{A} \quad (25.10)$$

where the symbol $\oint$ indicates that the integral is taken over a closed surface.

Special Cases

1. Electric field due to sources with
   a. **Linear symmetry** (an infinitely long charged rod):

$$\vec{E} = \frac{1}{2\pi\varepsilon_0} \frac{\lambda}{r} \hat{r} \quad (25.13)$$

   b. **Spherical symmetry** (a particle, sphere, or spherical shell)
   For a particle outside a sphere or shell,

$$\vec{E} = \frac{1}{4\pi\varepsilon_0} \frac{q}{r^2} \hat{r} \quad (25.14)$$

   Inside a charged spherical shell, $\vec{E} = 0$ (the **shell theorem**).
   Inside ($r < R$) a solid sphere of radius R with uniform volume charge density,

$$\vec{E} = \frac{1}{4\pi\varepsilon_0} \frac{q}{R^3} r\hat{r} \quad (25.15)$$

   c. **Planar symmetry** (infinite charged sheet):

$$E = \frac{\sigma}{2\varepsilon_0} \quad (25.16)$$

2. Electric field inside and outside a charged conductor

   The **electric field inside a conductor** in electrostatic equilibrium is zero, $\vec{E} = 0$.

   Just **outside a charged conductor**,

$$E = \frac{\sigma}{\varepsilon_0} \quad (25.17)$$

   where σ is the local surface charge density of a conductor of any shape, or the uniform surface charge density on an infinite conducting sheet.

PROBLEM-SOLVING STRATEGY Finding the Electric Field Using Gauss's Law

INTERPRET and ANTICIPATE

1. **Sketch the electric field lines.**
2. **Choose a Gaussian surface.**

SOLVE

3. **Integrate $\Phi_E = \oint \vec{E} \cdot d\vec{A}$.**
4. Determine the amount of **charge inside** the Gaussian surface.

CHECK and THINK

There are two checks:

1. Make sure the **dimensions** of the electric field are force per charge.
2. **Compare to another known electric field.**

PROBLEMS AND QUESTIONS

A = algebraic C = conceptual E = estimation G = graphical N = numerical

25-1 Qualitative Look at Gauss's Law

1. C Which word or name has the same symmetry as the letters in the name **ZAK**? (Explain your answer.)
   a. NUT
   b. SUE
   c. CAL
   d. BIG
2. C Describe the symmetry of the number **8**.

Problems 3, 4, and 5 are grouped.

3. C A nonconducting source is hidden inside a box. Your goal is to find the sign of the source's net charge. You suspend a small, lightweight, positively charged ball just outside each of the six faces of the box. The ball is attracted to all the faces. What do you conclude about the source's net charge?
4. C A nonconducting source is hidden inside a box. Your goal is to find the sign of the source's net charge. You suspend a small, lightweight, positively charged ball just outside each of the six faces of the box. The ball shows no deflection. What do you conclude about the source's net charge?
5. C A nonconducting source is hidden inside a box. Your goal is to find the sign of the source's net charge. You suspend a small, lightweight, positively charged ball just outside each of the six faces. The ball is attracted to one face and repelled by the face on the opposite side. What do you conclude about the source's net charge?
6. C Discuss the case study on page 757 with your friends, roommates, or family. What do people think is the right decision—wait out the storm in the car or wait outside? What makes sense to you, and why?
7. C A positively charged sphere and a negatively charged sphere are in a sealed container. The only way the charged spheres can be examined is by observing the electric field outside the container.
   a. Given the depiction of the electric fields in Figure P25.7A, is the net electric flux through the container zero, positive, or negative? Explain your answer.
   b. Two different spheres are placed inside a container. Given the depiction of the electric fields in Figure P25.7B, is the net electric flux through the container zero, positive, or negative? Explain your answer.

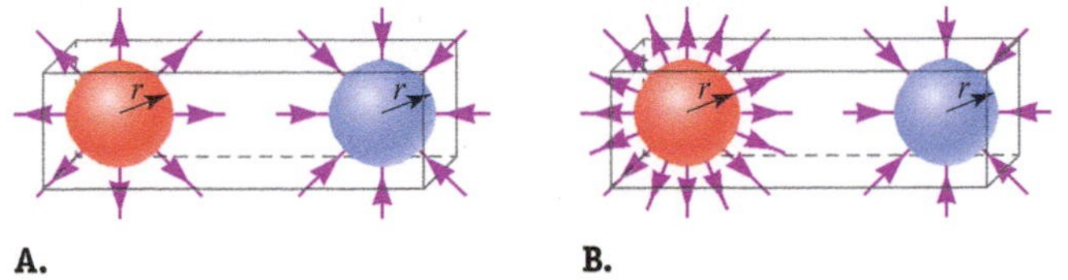

FIGURE P25.7

25-2 Flux

8. N A circular hoop of radius 0.50 m is immersed in a uniform electric field of 12.0 N/C. The electric field is at an angle of 30.0° to the plane of the hoop. Determine the electric flux through the hoop.
9. N A rectangular loop with dimensions 5.50 cm by 10.0 cm is placed in a uniform electric field and rotated to find the orientation that produces a maximum electric flux through the loop of $7.65 \times 10^4\ \text{N}\cdot\text{m}^2/\text{C}$. What is the magnitude of this electric field?
10. N If the hemisphere (surface C) in Figure 25.10 (page 760) is tilted so that its disk-shaped cross section makes a 25° angle with the electric field, what is the electric flux through the hemisphere? Use Example 25.1 to check your result.
11. N A Ping-Pong paddle with surface area $3.80 \times 10^{-2}\ \text{m}^2$ is placed in a uniform electric field of magnitude 1.10×10^6 N/C.
    a. What is the magnitude of the electric flux through the paddle when the electric field is parallel to the paddle's surface?
    b. What is the magnitude of the electric flux through the paddle when the electric field is perpendicular to the paddle's surface?
12. The electric field in some region is $\vec{E} = (35\hat{\imath} + 70\hat{\jmath})$ N/C.
    a. G Sketch an electric field vector.
    b. N, G What is the maximum electric flux through a disk of radius 15 cm? Write an expression for $\vec{A}$, and sketch the disk along with the electric field vector.
    c. N, G Write an expression for $\vec{A}$ that produces no electric flux through the disk, and sketch the disk along with the electric field vector.

Problems 13, 14, and 15 are grouped.

13. N A pyramid has a square base with an area of 4.00 m² and a height of 3.5 m. Its walls are four isosceles triangles. The pyramid is in a uniform electric field of 655 N/C pointing downward (Fig. P25.13). What is the electric flux through the square base?
14. N For the pyramid in Problem 13, what is the electric flux through all four walls combined?
15. N For the pyramid in Problem 13, what is the electric flux through one of the four walls?

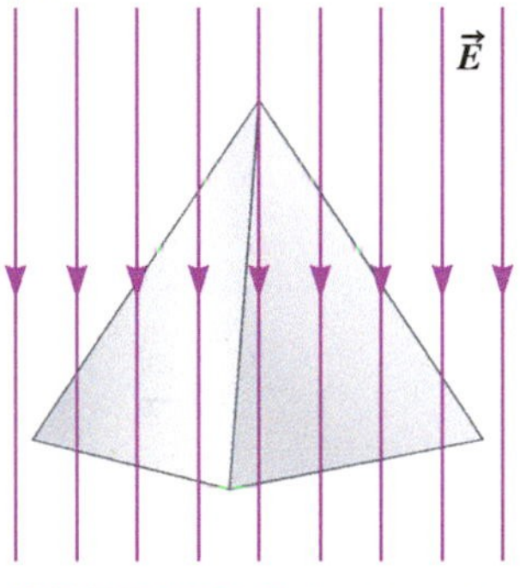

FIGURE P25.13 Problems 13, 14, and 15.

Problems 16 and 17 are paired.

16. A A circular loop with radius r is rotating with constant angular velocity ω in a uniform electric field with magnitude E. The axis of rotation is perpendicular to the electric field direction and is along the diameter of the loop. Initially, the electric flux through the loop is at its maximum value. Write an equation for the electric flux through the loop as a function of time in terms of r, E, and ω.
17. A A circular loop with radius r is rotating with constant angular velocity ω in a uniform electric field with magnitude E. The axis of rotation is perpendicular to the electric field direction and is along the diameter of the loop. Initially, the electric flux through the loop is zero. Write an equation for the electric flux through the loop as a function of time in terms of r, E, and ω.

25-3 Gauss's Law

18. N The net electric flux through a Gaussian surface is $-456\ \text{N}\cdot\text{m}^2/\text{C}$. What is the net charge of the source inside the surface?

19. **N** What is the net electric flux through each of the four surfaces shown in Figure P25.19?

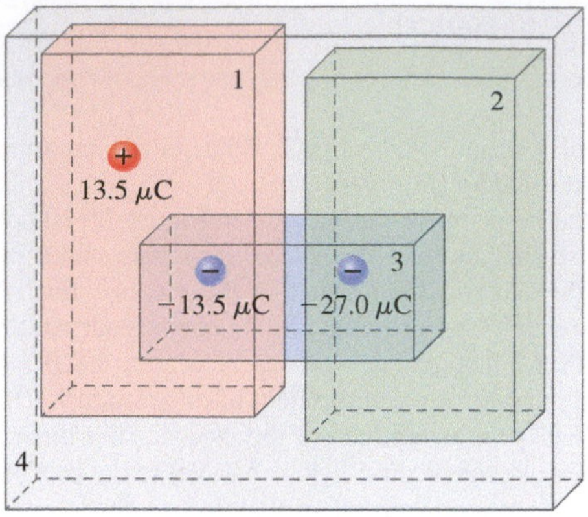

FIGURE P25.19

20. An isolated system consists of a single particle with charge 56.0 μC placed at the center of a cube of side 1.25 m.
   a. **N** What is the flux through each of the faces of the cube?
   b. **C** Would the answer to part (a) change if the particle was moved away from the center of the cube?
21. **N** The colored regions in Figure P25.21 represent four three-dimensional Gaussian surfaces A through D. The regions may also contain three charged particles, with $q_A = +5.00$ nC, $q_B = -5.00$ nC, and $q_C = +8.00$ nC, that are nearby as shown. What is the electric flux through each of the four surfaces?

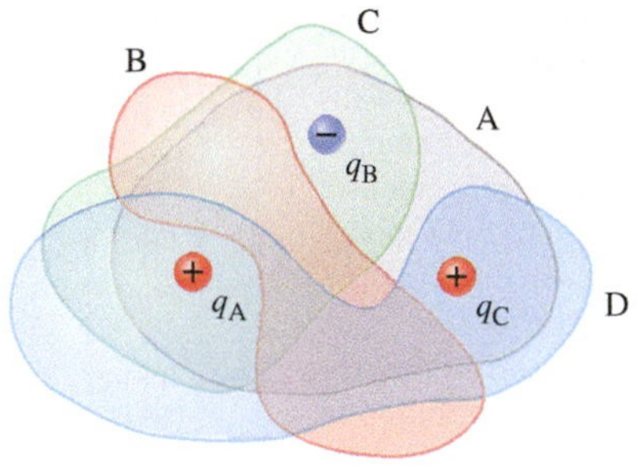

FIGURE P25.21

22. **N** A particle with a charge of 3.0 μC is placed at the center of a cubical Gaussian surface with side length 5.0 cm. Find the electric flux through one face of the cube.
23. **A** A particle with charge q is placed at a corner of a cube with side length a. Determine the net electric flux through the cube.
24. **N** Three particles and three Gaussian surfaces are shown in Figure P25.24. All the surfaces are three-dimensional. Use the net electric flux through each surface indicated on the figure to find the charge of each particle.

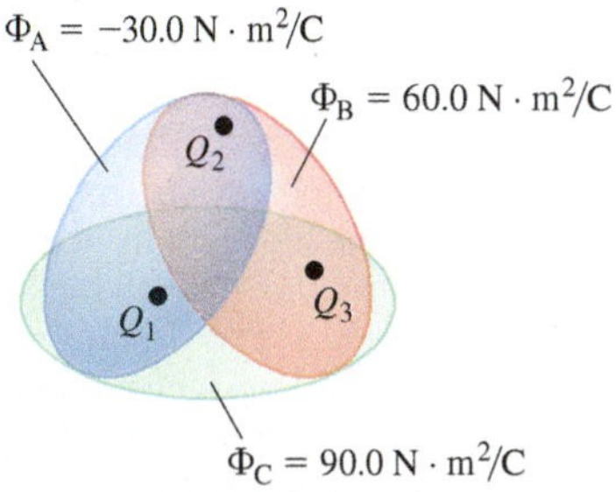

FIGURE P25.24

Problems 25 and 26 are paired.

25. **A** Using Gauss's law, find the electric flux through each of the closed Gaussian surfaces A, B, C, and D shown in Figure P25.25.

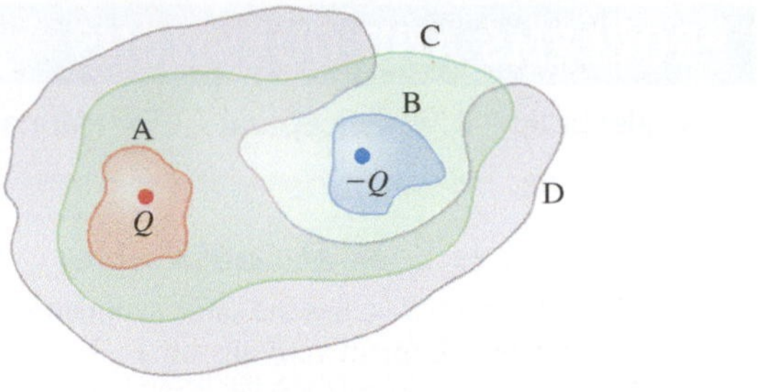

FIGURE P25.25

26. **N** Three point charges $q_1 = 2.0$ nC, $q_2 = -4.0$ nC, and $q_3 = -3.0$ nC are placed as shown in Figure P25.26. Find the electric flux through each of the closed Gaussian surfaces C_1, C_2, C_3, and C_4.

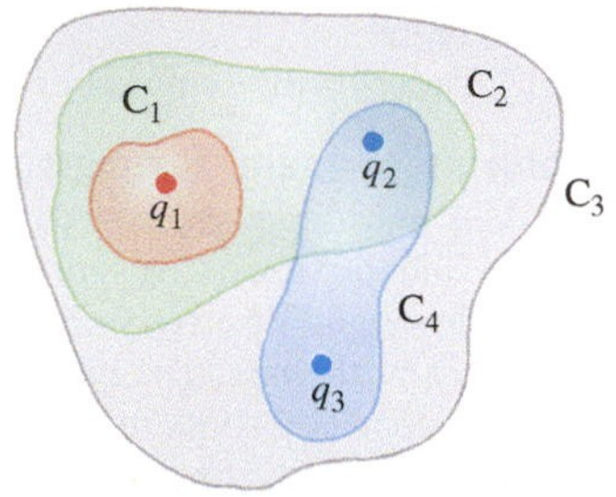

FIGURE P25.26

27. **N** A spherical shell of radius 1.00 m contains a single charged particle with $q = 78.0$ nC at its center.
   a. What is the total electric flux through the surface of the shell?
   b. What is the total electric flux through any hemispherical portion of the shell's surface?

25-4 Special Case: Linear Symmetry

28. **N** A very long, thin wire fixed along the x axis has a linear charge density of 3.2 μC/m.
   a. Determine the electric field at point P a distance of 0.50 m from the wire.
   b. If there is a test charge $q_0 = +2.0$ μC at point P, what is the magnitude of the net force on this charge? In which direction will the test charge accelerate?
29. **A** Figure P25.29 shows a very long tube of inner radius a and outer radius b that has uniform volume charge density ρ. Find an expression for the electric field between the walls of the tube—that is, for $a < r < b$.

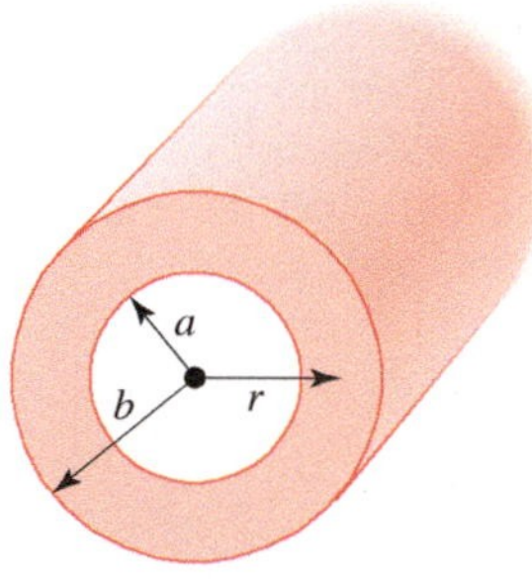

FIGURE P25.29

30. **G** Two very long, thin, charged rods lie in the same plane (Fig. P25.30). One rod is positively charged with charge per unit length $+\lambda$, and the other is negatively charged with charge per unit length $-\lambda$. The perpendicular distance between the rods is R. Using the coordinate system shown in the figure, sketch the electric field as a function of r from $-R$ to $+2R$.

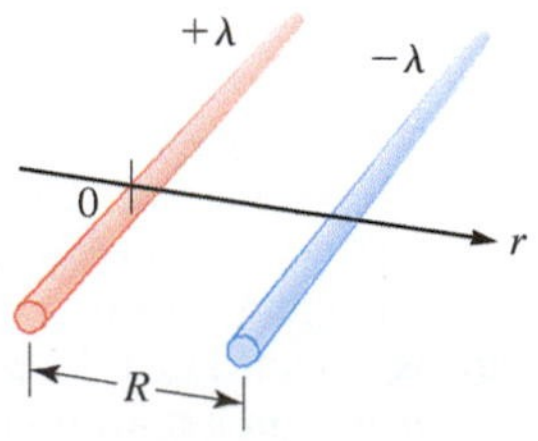

FIGURE P25.30

31. **N** A long, straight copper rod has charge per unit length $\lambda = 0.200\ \mu C/m$ and a diameter of 8.00 cm. What is the electric field at each of the following radial distances from the axis of the rod?
   a. 2.00 cm
   b. 8.00 cm
   c. 2.00 m

32. **N** Two long, thin rods each have linear charge density $\lambda = 6.0\ \mu C/m$ and lie parallel to each other, separated by 20.0 cm as shown in Figure P25.32. Determine the magnitude and direction of the net electric field at point P, a distance of 15.0 cm directly above the right rod.

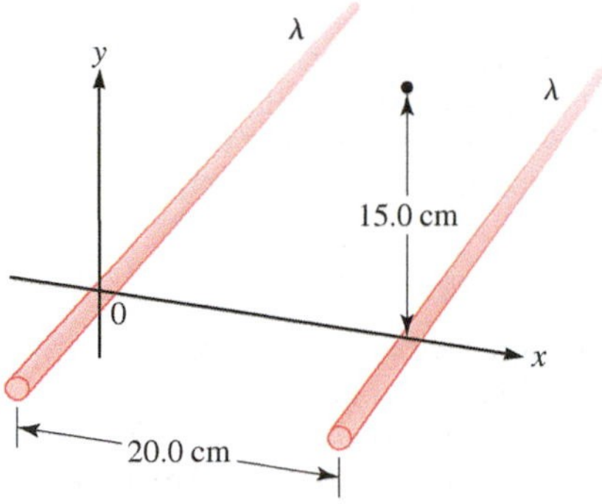

FIGURE P25.32

33. **A** Figure P25.33 shows a very long, thick rod with radius R, uniformly charged throughout. Find an expression for the electric field inside the rod ($r < R$). Use Equation 25.13,

$$\vec{E} = \frac{1}{2\pi\varepsilon_0}\frac{\lambda}{r}\hat{r}$$

to check your solution at the surface, where $r = R$.

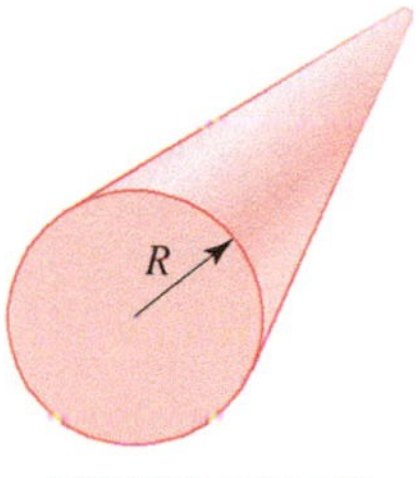

FIGURE P25.33

34. **A** A very long line of charge with a linear charge density, λ, is parallel to another very long line of charge with a linear charge density, -2λ. Both lines are parallel to the y-axis, and are the same distance r from the y-axis, where the first wire is to the left of the origin and the second is to the right. Use Gauss's law and the principle of superposition to find an expression for the magnitude of the electric field at the origin.

35. **N** Two infinitely long, parallel lines of charge with linear charge densities $3.2\ \mu C/m$ and $-3.2\ \mu C/m$ are separated by a distance of 0.50 m. What is the net electric field at points A, B, and C as shown in Figure P25.35?

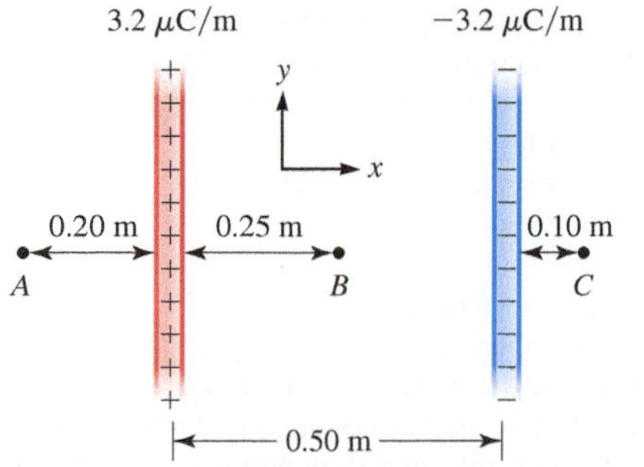

FIGURE P25.35

36. **N** An infinitely long wire with uniform linear charge density $\lambda = 2.80\ \mu C/m$ is surrounded by an insulating shell with inner radius $R_1 = 4.00$ cm and outer radius $R_2 = 6.00$ cm and uniform volume charge density $\rho = 1.05\ mC/m^3$. What is the electric field at radial distances of
   a. $r = 3.00$ cm,
   b. $r = 5.00$ cm, and
   c. $r = 10.0$ cm?

25-5 Special Case: Spherical Symmetry

37. **N** A particle with a charge of $55.0\ \mu C$ is at the center of a thin spherical shell of radius $R = 12.0$ cm that has uniform surface charge density σ. Determine the value of σ such that the net electric field outside the shell is zero.

38. Excess charged particles are uniformly distributed throughout a sphere as shown in Figure 25.23 (page 770).
   a. **A** Show that at the surface of a sphere ($r = R$),

$$\vec{E} = \frac{1}{4\pi\varepsilon_0}\frac{q}{r^2}\hat{r} \text{ and } \vec{E} = \frac{1}{4\pi\varepsilon_0}\frac{q}{R^3}r\,\hat{r}$$

   (Eqs. 25.14 and 25.15) are identical.
   b. **G** Use Equations 25.14 and 25.15 to plot the electric field for a spherical charge distribution for all values of r.

Problems 39 and 40 are paired.

39. **N** A uniform spherical charge distribution (as shown in Fig. 25.23, page 770) has a total charge of 45.3 mC and radius $R = 15.2$ cm. Find the magnitude of the electric fields at $r = 0$, 7.60 cm, 15.2 cm, and 22.8 cm.

40. **N** For the uniform spherical charge distribution in Problem 39, find the electric force exerted on an electron placed at $r = 0$, 7.60 cm, 15.2 cm, and 22.8 cm.

Problems 41 and 42 are paired.

41. **N** Two uniform spherical charge distributions (Fig. P25.41) each have a total charge of 45.3 mC and radius $R = 15.2$ cm. Their center-to-center distance is 37.50 cm. Find the magnitude of the electric field at point A midway between the two spheres.

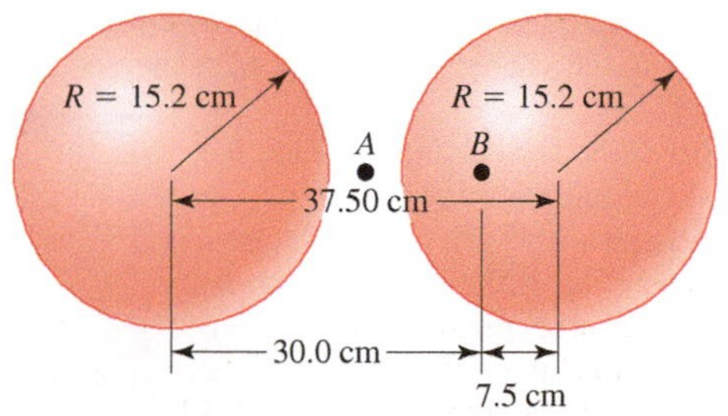

FIGURE P25.41 Problems 41 and 42.

42. **N** Two uniform spherical charge distributions (Fig. P25.41) each have a total charge of 45.3 mC and radius $R = 15.2$ cm. Their center-to-center distance is 37.50 cm. Find the magnitude of the electric field at point B, 7.50 cm from the center of one sphere and 30.0 cm from the center of the other sphere.

43. **A** The nonuniform charge density of a solid insulating sphere of radius R is given by $\rho = cr^2$ ($r < R$), where c is a positive constant and r is the radial distance from the center of the sphere. For a spherical shell of radius r and thickness dr, the volume element $dV = 4\pi r^2 dr$.
   a. What is the magnitude of the electric field outside the sphere ($r > R$)?
   b. What is the magnitude of the electric field inside the sphere ($r < R$)?

44. **N** A sphere with a radius of 0.230 m has a uniform charge density and a total charge of 70.9 mC. What is the magnitude of the electric field at each of the following locations:
 a. a distance of 0.100 m from the center,
 b. a distance of 0.230 m from the center, and
 c. a distance of 0.500 m from the center?

25-6 Special Case: Planar Symmetry

45. **N** What is the magnitude of the electric field just above the middle of a large, flat, horizontal sheet carrying a charge density of 98.0 nC/m^2?

46. **N** What is the force on a proton placed at point A in Figure 25.28 (page 773)?

47. **N** The infinite sheets in Figure P25.47 are both positively charged. The sheet on the left has a uniform surface charge density of 48.0 $\mu C/m^2$, and the one on the right has a uniform surface charge density of 24.0 $\mu C/m^2$.
 a. What are the magnitude and direction of the net electric field at points A, B, and C?
 b. What is the force exerted on an electron placed at points A, B, and C?

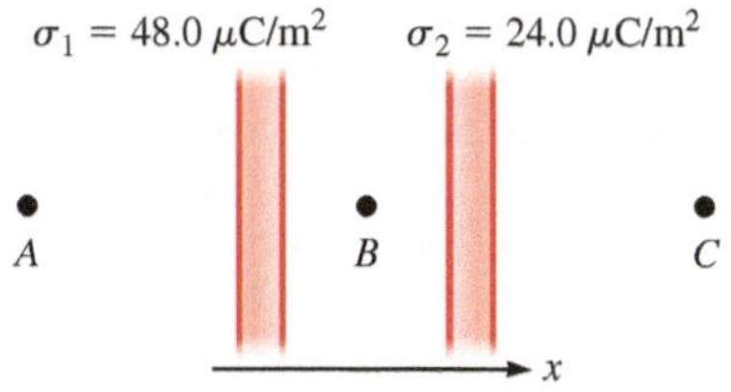

FIGURE P25.47

48. A large vertical sheet of nonconducting material has a uniform charge density of 47.0 $\mu C/m^2$.
 a. **N** What is the electric field a horizontal distance of 3.00 cm from the sheet?
 b. **C** How would the result in part (a) change if the distance from the sheet was increased by a factor of 10 or more?

49. **N** A charge-neutral, square aluminum plate with edge d = 33.0 cm is placed horizontally in a uniform electric field of 4.50×10^4 N/C directed vertically, or in the $+y$ direction.
 a. What is the charge density on each face of the aluminum plate?
 b. What is the total charge on each face of the aluminum plate?

50. **G** Two large sheets are perpendicular to each other (Fig. P25.50). One sheet has an excess charge per unit area of $-\sigma$, and the other has an excess charge per unit area of $+\sigma$. Sketch the electric field in the region bordered by the two sheets.

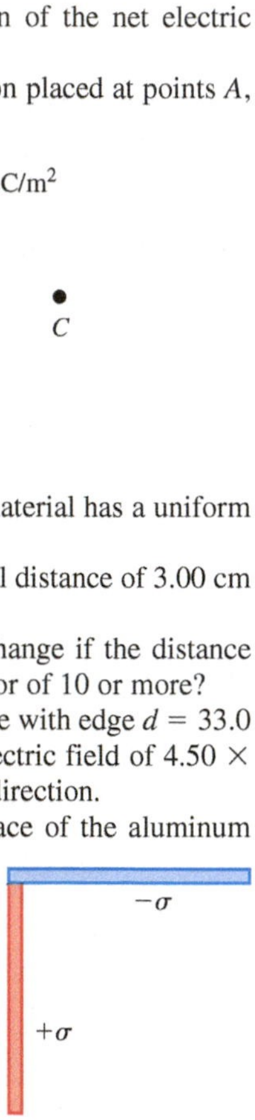

FIGURE P25.50

Problems 51 and 52 are paired.

51. A very large, flat slab has uniform volume charge density ρ and thickness $2t$. A side view of the cross section is shown in Figure P25.51.
 a. **A** Find an expression for the magnitude of the electric field inside the slab at a distance x from the center.
 b. **N** If ρ = 2.00 $\mu C/m^3$ and $2t$ = 8.00 cm, calculate the magnitude of the electric field at x = 3.00 cm.

52. **N** Find the surface charge density σ of a sheet of charge that would produce the same electric field as that of a very large flat slab of uniform charge density ρ = 2.00 $\mu C/m^3$ and thickness $2t$ = 5.00 cm (Fig. P25.51).

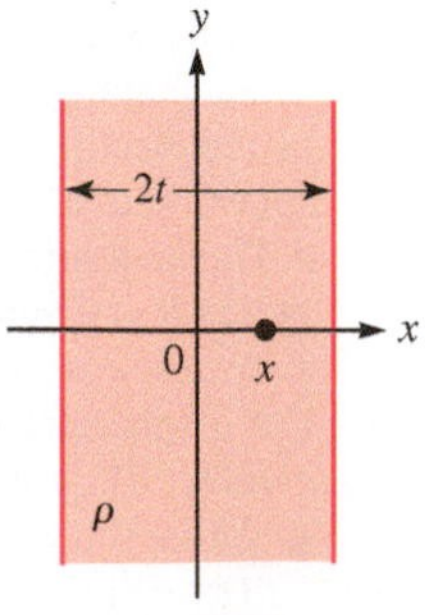

FIGURE P25.51 Problems 51 and 52.

25-7 Special Case: Conductors

53. **N** The electric field at a point 1.0 m from the center of a charged spherical conductor of radius 0.20 m is -12 N/C. Determine the number of excess electrons in the conductor.

54. **G** Model a metal can such as an empty oil barrel as having linear symmetry. If positive charge is uniformly distributed on the walls of the barrel, sketch the electric field both inside ($r < R$) and outside ($r \geq R$) the barrel as a function of r.

55. **N** If the magnitude of the surface charge density of the plates in Figure P25.55 is σ = 99.5 nC/m^2, what is the magnitude of the electric field between the plates? If an electron is placed between the plates, what is the magnitude of the electric force on it?

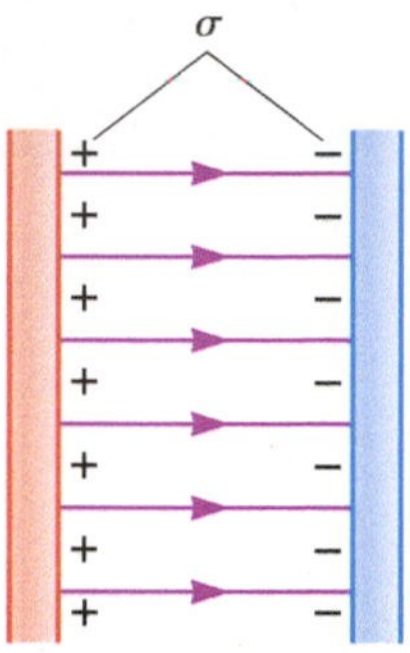

FIGURE P25.55

56. **N** A spherical conducting shell with a radius of 0.200 m has a very small charged sphere suspended in its center. The sphere has a charge of 24.6 mC, and the conducting shell has a charge of -24.6 mC. What is the magnitude of the electric field at distances of
 a. 0.100 m and
 b. 0.300 m from the center of the shell?

57. **N** A charged rod is placed in the center along the axis of a neutral metal cylinder (Fig. P25.57). The rod has a total charge of 38.3 μC uniformly distributed. What are the charges on the inner and outer surfaces of the metal cylinder? (Ignore the ends.)

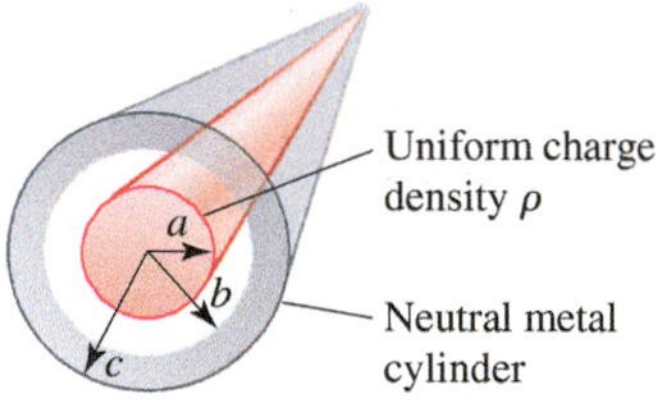

FIGURE P25.57 Problems 57 and 58.

58. A charged rod is placed in the center along the axis of a neutral metal cylinder (Fig. P25.57). The rod has positive charge uniformly distributed. (Ignore the ends.)
 a. **A** Find expressions for the electric fields in all regions: $r < a$, $a < r < b$, $b < r < c$, and $r > c$.
 b. **G** Plot your expressions on one graph. Is the electric field continuous or discontinuous? Explain.

Problems 59 and 60 are paired.

59. **N** A thick spherical conducting shell with an inner radius of 0.200 m and an outer radius of 0.250 m has a very small charged sphere suspended in its center. The sphere has a charge of 24.6 mC, and the conducting shell has a net charge of -24.6 mC. What is the net amount of charge that resides on
 a. the inner and
 b. the outer surface of the shell?
 What is the charge density on
 c. the inner and
 d. the outer surface of the shell?

60. **N** A thick spherical conducting shell with an inner radius of 0.200 m and an outer radius of 0.250 m has a very small charged sphere suspended in its center. The sphere has a charge of 24.6 mC, and the conducting shell has a net charge of -24.6 mC. What is the magnitude of the electric field at distances of
 a. 0.100 m,
 b. 0.225 m, and
 c. 0.500 m from the center of the shell?

General Problems

61. **N** A rectangular plate with sides 0.60 m and 0.40 m long is lying in the xy plane. The plate is placed in a uniform electric field $\vec{E} = (-4.0\hat{\imath} + 5.0\hat{\jmath} + 3.0\hat{k})$ N/C. Calculate the flux through the plate.

62. **N** A circular loop of radius 15.0 cm is placed in a uniform electric field given by $\vec{E} = (5.65 \times 10^5\ \text{N/C})\hat{k}$. What is the electric flux through the loop if it is oriented
 a. parallel to the xy plane and
 b. parallel to the xz plane?
 c. What is the electric flux through the loop if it is oriented at a 45.0° angle to the xy plane?

63. **N** A total charge of 78.0 μC is uniformly distributed on the surface of a thin spherical shell of radius 22.0 cm. What is the magnitude of the electric field at distances of
 a. 5.00 cm and
 b. 44.0 cm from the center of the spherical shell?

64. **N** A uniform spherical charge distribution has a total charge of 45.3 mC and a radius of 15.2 cm. It is surrounded by a thin spherical shell with a uniform charge distribution. The uniform shell's net charge is 35.5 mC. The shell's radius is 20.2 cm, and it is concentric with the solid sphere. Find the electric field at points A and B located as shown on Figure P25.64.

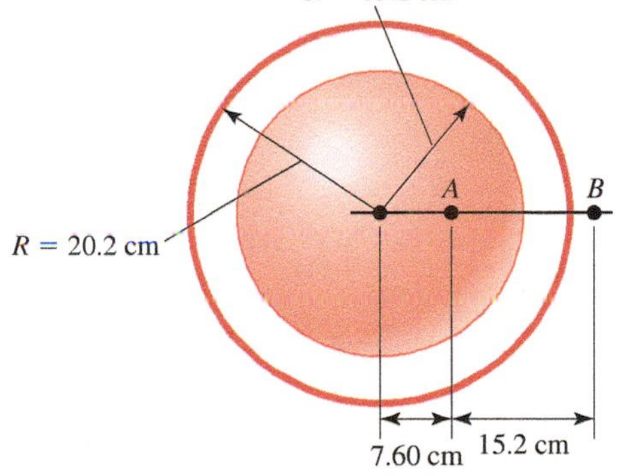

FIGURE P25.64

65. **N** A rectangular surface extends from $x = 0$ to $x = 50.0$ cm and from $y = 0$ to $y = 24.0$ cm in the xy plane in a region of space permeated by a nonuniform electric field given by $\vec{E} = (3.00z\hat{\imath} + 2.00y\hat{\jmath} - 4.00x\hat{k})$ N/C. What is the electric flux through the surface?

66. **N** A uniform electric field $\vec{E} = 1.57 \times 10^4 \hat{\imath}$ N/C passes through a closed surface with a slanted top as shown in Figure P25.66.
 a. Given the dimensions and orientation of the closed surface, what is the electric flux through the slanted top of the surface?
 b. What is the net electric flux through the entire closed surface?

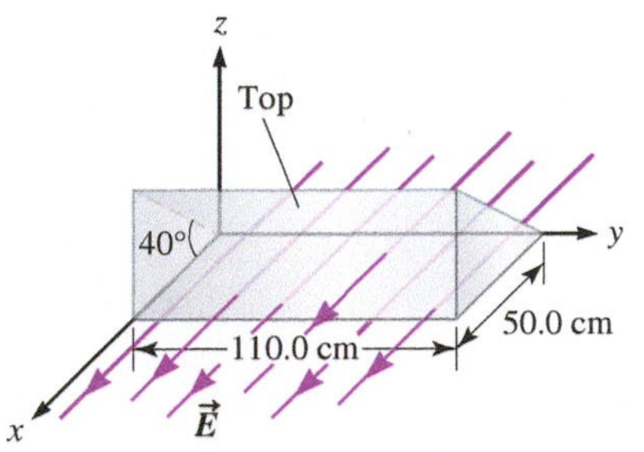

FIGURE P25.66

67. **N** A solid plastic sphere of radius $R_1 = 8.00$ cm is concentric with an aluminum spherical shell with inner radius $R_2 = 14.0$ cm and outer radius $R_3 = 17.0$ cm (Fig. P25.67). Electric field measurements are made at two points: At a radial distance of 34.0 cm from the center, the electric field has magnitude 1.70×10^3 N/C and is directed radially outward, and at a radial distance of 12.0 cm from the center, the electric field has magnitude 9.10×10^4 N/C and is directed radially inward. What are the net charges on
 a. the plastic sphere and
 b. the aluminum spherical shell?
 c. What are the charges on the inner and outer surfaces of the aluminum spherical shell?

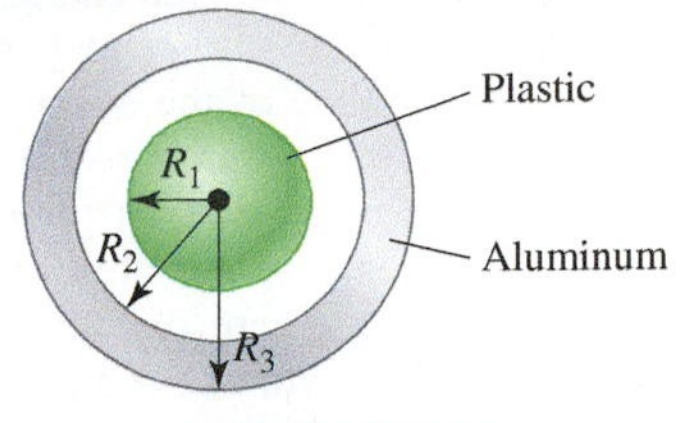

FIGURE P25.67

68. **C** Examine the summary on page 780. Why are conductors and charged sources with linear symmetry, spherical symmetry, and planar symmetry categorized as special cases rather than major concepts or underlying principles?

69. **N** A total charge of –33.0 μC is uniformly distributed throughout the volume of a solid sphere 60.0 cm in diameter.
 a. What is the magnitude of the electric field at the center of the sphere?
 What is the magnitude of the electric field
 b. 15.0 cm,
 c. 30.0 cm, and
 d. 90.0 cm from the center of the sphere?

Problems 70 and 71 are paired.

70. **N** A cube with sides of 20.0 cm is placed in a nonuniform electric field $\vec{E}(x,y,z) = -4.0x\,\hat{\imath} + 5.0y\,\hat{\jmath} + 3.0z\,\hat{k}$, where x, y, and z are measured in meters and E is in N/C. Determine the electric flux through each of the six faces of the cube, assuming the cube has a corner at the origin and sides along the x, y, and z axes.

71. **N** A cube with side length 20.0 cm is placed in a uniform electric field $\vec{E}$, directed along the z axis. The cube is oriented so that the field is perpendicular to two of its sides. What is the net flux through the cube?

72. **A** A coaxial cable is formed by a long, straight wire and a hollow conducting cylinder with axes that coincide. The wire has charge per unit length $\lambda = 2\lambda_0$, and the hollow cylinder has net charge per unit length $\lambda = 3\lambda_0$. Use Gauss's law to answer these questions: What are the charges per unit length on
 a. the inner surface and
 b. the outer surface of the hollow cylinder?
 c. What is the electric field a radial distance d from the axis of the coaxial cable?

73. **N** The electric field due to a very long wire at a point 0.50 m from the wire is 12 N/C. Determine the amount of charge in a 15-cm length of the wire.

74. **N** Charge is distributed uniformly on the curved surface of a horizontal cylindrical shell 10.0 cm in diameter and 1.50 m in length. The midpoint of the axis of the cylinder is at the origin. The magnitude of the electric field at $y = 25.0$ cm is measured to be 2.35×10^4 N/C.
 a. What is the total charge on the cylindrical shell?
 b. What is the magnitude of the electric field at $y = 5.00$ cm?

Problems 75 and 76 are paired.

75. **A** A solid sphere of radius R has a spherically symmetrical, nonuniform volume charge density given by $\rho(r) = A/r$, where

r is the radial distance from the center of the sphere in meters, and A is a constant such that the density has dimensions M/L^3.

a. Calculate the total charge in the sphere.

b. Using the answer to part (a), write an expression for the magnitude of the electric field outside the sphere—that is, for some distance $r > R$.

c. Find an expression for the magnitude of the electric field inside the sphere at position $r < R$.

76. G A solid sphere of radius R has a spherically symmetrical, nonuniform volume charge density given by $\rho(r) = A/r$, where r is the radial distance from the center of the sphere in meters, and A is a constant such that the density has dimensions of M/L^3. Sketch a graph of the magnitude of the electric field as a function of distance for $0 < r < 3R$.

77. N A very large, horizontal conducting square plate with sides of length 2.0 m and thickness 3.0 mm is given a total charge of 40.0 μC. Calculate the magnitude of the electric field near the center of the plate and just above its surface.

78. A The charge density within a spherically symmetrical charge distribution is given by $\rho = c/r^2$, where r is the radial distance from the center of the distribution and c is a positive constant. What is the electric field as a function of radial distance within this charge distribution?

79. N A particle with charge $q = 7.20$ μC is surrounded by a spherical shell of radius $R = 1.50$ m. What is the electric flux through the spherical cap with half angle $\theta = 30.0°$ (Fig. P25.79)?

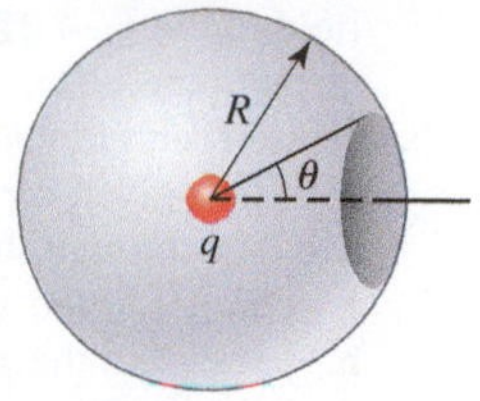

FIGURE P25.79

80. A A sphere with radius R has a charge density given by $\rho = cr^3$. Use Gauss's law to find an expression for the magnitude of the electric field at a distance r from the center of the sphere, where **a.** $r < R$ and **b.** $r > R$.

Electric Potential

26

Underlying Principles

No new principles are introduced in this chapter.

Major Concepts

1. Electric potential energy
2. Electric potential
3. Finding the electric potential V from the electric field $\vec{E}$
4. Finding the electric field $\vec{E}$ from the electric potential V

Special Cases

1. Electric potential energy for a system of two charged particles
2. Electric potential due to
   a. Source with spherical symmetry
   b. Dipole
   c. Ring
   d. Disk
   e. Source with planar symmetry

Tools

Contour map

Key Questions

What are electric potential and electric potential energy?

How are electric potential and electric potential energy related to the electric force and the electric field?

How do electric potential and electric potential energy make our study of electricity easier?

You have heard of voltage (electric potential). A car battery is 12 volts (V). The D batteries in your flashlight and the AAA batteries in your calculator are 1.5 V. In Europe the wall outlet is 240 V, and in the United States it is 120 V. On average, there is about 1 mV across your chest and 70 mV across your nerve cells. What do these numbers mean? In this chapter, we explain how these numbers are related to electric fields and forces.

This chapter is about *electric potential* (also called *voltage*), a scalar quantity. We know from our experience with forces (vectors) and energy (scalars) that using scalars allows us to analyze complicated situations. The concept of electric potential enables us to study complicated systems such as the circuits in common devices like flashlights and cameras as well as the human body.

26-1 Scalars Versus Vectors

In mechanics, we focused on forces, drew free-body diagrams, and used Newton's second law (Chapters 5 and 6). Then, in Chapter 8, we introduced energy, a scalar. For a conservative force such as gravity, we associated potential energy with the configuration of a system. The benefit of this new scalar was that we could use the conservation of energy principle to study complicated situations.

Our study of electricity will follow a similar path. We have learned to manipulate two vector quantities—electrostatic force $\vec{F}_E$ and electric field $\vec{E}$—with tools similar to those we used in mechanics. Like gravity, the electrostatic force is conservative, so now we can apply the conservation of energy principle to electricity, which allows us to study complicated and practical situations. (In Problem 62, you will show that the electrostatic force passes the tests for conservative forces.) We now introduce two scalar quantities: (1) *potential energy,* which is familiar from mechanics, and (2) *electric potential*, also known as *voltage*—an entirely new idea. Because gravity is similar to the electrostatic force both conceptually and mathematically, we use gravity throughout this chapter as an analogy.

CASE STUDY Electric Potential in the Human Body

As in a car, much of the mechanics of the human body is run by a complex electrical system. The first evidence that electricity plays a role in muscle action was discovered in 1786 by the Italian anatomist Luigi Galvani. Galvani connected the legs of a dead frog to pieces of two different metals (forming a crude battery) and found that the legs contracted. A dispute between Galvani and the Italian physicist Alessandro Volta led Volta to invent the battery.

The role that electricity plays in the human body goes beyond muscle contraction. Electricity controls organs and nerves. The brain—and therefore human thought—is controlled by electricity. The senses take in information and transmit that information via electrical signals through nerves. The brain sends electrical signals through nerves to the rest of the body. As in mechanics, a complex system like the human body is often best studied in terms of scalar quantities. In this chapter's case study, we consider parts of the body's electrical system in terms of its electric potential.

26-2 Gravity Analogy

Imagine a system consisting of the Earth and a spacecraft at some distance r from the Earth's center, with the Earth as the source and the spacecraft as the subject. We associate four quantities with this situation: (1) gravitational force $\vec{F}_G$, (2) gravitational field $\vec{g}$, (3) gravitational potential energy U_G, and (4) gravitational potential V_G. These four quantities are arranged in a pattern in Figure 26.1. The first row is for vectors: gravitational force $\vec{F}_G$ and gravitational field $\vec{g}$. The second row is for scalars: gravitational potential energy U_G and gravitational potential V_G. The first column is for quantities that depend on both the source and the subject: gravitational force $\vec{F}_G$ and gravitational potential energy U_G. The second column is for quantities that depend only on the source: gravitational field $\vec{g}$ and gravitational potential V_G.

All but the last quantity—gravitational potential V_G—are familiar from earlier chapters. The **gravitational potential** V_G, like the gravitational potential energy, is a scalar quantity. The most important difference between the gravitational potential energy and the gravitational potential is that *the gravitational potential depends only on the source*. The gravitational potential is like the gravitational field in that both depend on the mass of the source, but neither depends on the mass of the subject.

The terms *gravitational potential energy* and *gravitational potential* are very similar; you must pay close attention when reading so you do not confuse them. One way to make sure you are reading carefully is to highlight the word *energy* every time you see it.

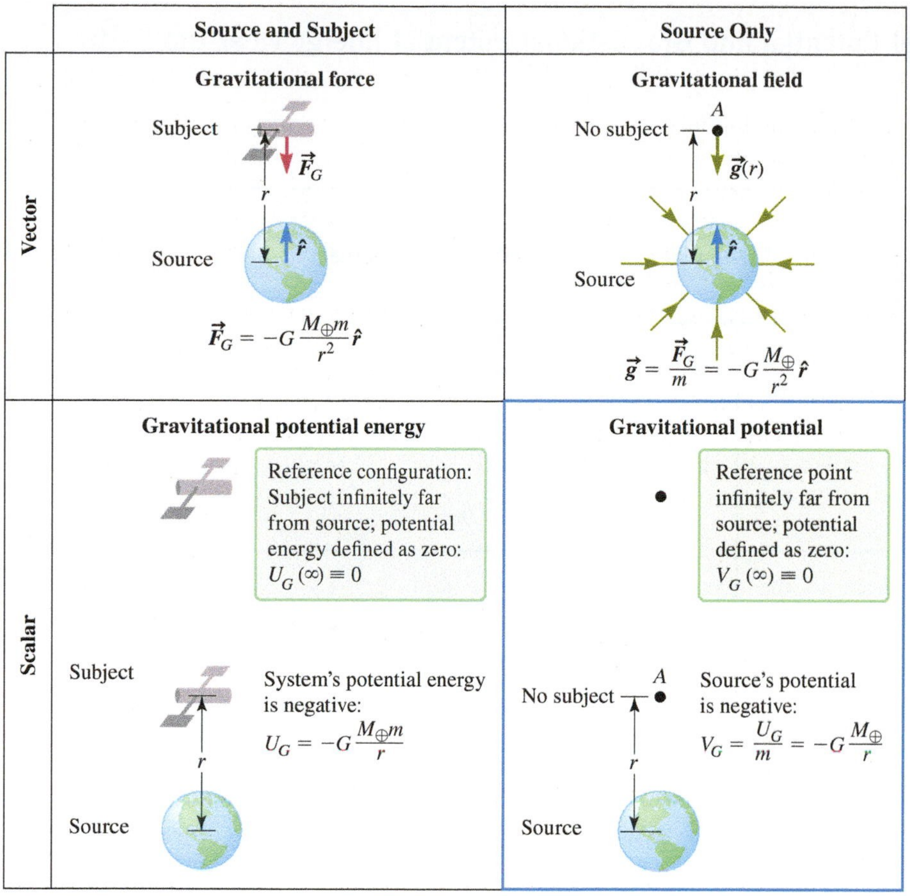

FIGURE 26.1 The gravitational influence of the Earth can be described in terms of two vector quantities and two scalar quantities.

Just as we can find the gravitational field by dividing the gravitational force by the mass m of a test particle, we can find the gravitational potential by dividing the gravitational potential energy by the mass of a test subject:

$$V_G = \frac{U_G}{m} \tag{26.1}$$

So, dividing $U_G = -GM_\oplus m/r$ (Eq. 8.7) by m, we find that the gravitational potential is given by

$$V_G = -G\frac{M_\oplus}{r} \tag{26.2}$$

The gravitational potential has the dimensions energy per mass, so it has the SI units J/kg.

Only *differences* in the potential are physically meaningful, just like only differences in potential energy are physically meaningful. When working with potential, it is usually convenient to set a reference point at which the potential is defined to be zero. In the case of universal gravity, the most convenient reference point is at infinity: $V_G(\infty) \equiv 0$. At all other points, the gravitational potential is negative and given by Equation 26.2.

CONCEPT EXERCISE 26.1

Complete the analogies by filling in the blanks, and explain your answers. *Hint*: Consult Figure 26.1.

a. Gravitational force is to gravitational field as gravitational potential energy is to ____________.

b. Gravitational force is to gravitational potential energy as gravitational field is to ____________.

EXAMPLE 26.1 Gravitational Potential and Gravitational Potential Energy Near the Earth

A Find the gravitational potential at the surface of the Earth ($r = R_\oplus$). Compare it to the gravitational potential at $r = 2R_\oplus$ and at $r = 3R_\oplus$.

INTERPRET and ANTICIPATE

Use Equation 26.2 three times to find the gravitational potential at three places near the source (the Earth).

SOLVE

Substitute values to find V_1, the gravitational potential at the surface of the Earth ($r = R_\oplus$).

$$V_G = -G\frac{M_\oplus}{r} \qquad (26.2)$$

$$V_1 = -G\frac{M_\oplus}{R_\oplus} = -\frac{(6.67 \times 10^{-11}\,\text{N}\cdot\text{m}^2/\text{kg}^2)(5.97 \times 10^{24}\,\text{kg})}{(6.38 \times 10^6\,\text{m})}$$

$$V_1 = -6.26 \times 10^7\,\text{J/kg} \text{ at the Earth's surface}$$

The term in parentheses at right equals V_1. Write V_2, the gravitational potential at $r = 2R_\oplus$, in terms of V_1 as found above.

$$V_2 = -G\frac{M_\oplus}{2R_\oplus} = \frac{1}{2}\left(-\frac{GM_\oplus}{R_\oplus}\right) = \frac{1}{2}V_1$$

$$V_2 = -3.13 \times 10^7\,\text{J/kg at } 2R_\oplus$$

Similarly, write V_3, the gravitational potential at $r = 3R_\oplus$, in terms of V_1 and substitute.

$$V_3 = -G\frac{M_\oplus}{3R_\oplus} = \frac{1}{3}\left(-\frac{GM_\oplus}{R_\oplus}\right) = \frac{1}{3}V_1$$

$$V_3 = -2.09 \times 10^7\,\text{J/kg at } 3R_\oplus$$

CHECK and THINK

Because of the convention that the gravitational potential is zero at infinity and negative at all other points, the gravitational potential is lowest at the surface of the Earth and greater (less negative) at points that are farther away.

B Now imagine a system that consists of the Earth and subject 1 with mass $m_1 = 15.0$ kg. Find the gravitational potential energy of the system when subject 1 is at the surface of the Earth ($r = R_\oplus$). Compare your result to the gravitational potential energy when subject 2 with mass $m_2 = 30.0$ kg is at $r = 2R_\oplus$ and subject 3 with mass $m_3 = 45.0$ kg is at $r = 3R_\oplus$. As part of the **CHECK and THINK** step, answer this question: If the mass of the subject were the same in each case—for instance, $m_1 = m_2 = m_3 = 15.0$ kg—what would change and what would stay the same?

INTERPRET and ANTICIPATE

Use the gravitational potential found in part A to find the gravitational potential energy in the three cases. According to Equation 26.1, multiplying the gravitational potential by the mass of the subject gives the gravitational potential energy.

SOLVE

To find the gravitational potential energy U_1 when subject 1 is at the surface, multiply V_1 by the mass m_1 of the first subject.

$$V_G = \frac{U_G}{m} \qquad (26.1)$$

$$U_G = mV_G$$

$$U_1 = m_1V_1 = (15.0\,\text{kg})(-6.26 \times 10^7\,\text{J/kg}) = -9.39 \times 10^8\,\text{J}$$

Find the gravitational potential energy U_2 when subject 2 is at $r = 2R_\oplus$ by multiplying the mass of subject 2 by the gravitational potential at this position. Write the result in terms of U_1.

$$U_2 = m_2V_2 = (30.0\,\text{kg})(-3.13 \times 10^7\,\text{J/kg}) = -9.39 \times 10^8\,\text{J}$$

$$U_2 = U_1$$

Find the gravitational potential energy U_3 when subject 3 is at $r = 3R_\oplus$ by multiplying the mass of subject 3 by the gravitational potential at this position. Write the result in terms of U_1. (If we ignore the slight rounding error, $U_3 = U_1$.)	$U_3 = m_3V_3 = (45.0\,\text{kg})(-2.09 \times 10^7\,\text{J/kg}) = -9.40 \times 10^8\,\text{J}$ $U_3 = U_1$

CHECK and THINK

The gravitational potential depends only on the Earth, and V_G increases (becomes less negative) for points that are farther from the surface of the Earth. The gravitational potential *energy* U_G, however, is the same in all three cases because that energy depends on both the gravitational potential and the mass of the subject. The masses of the three subjects were carefully chosen to compensate for the variation in the gravitational potential. If all three masses were equal ($m_1 = m_2 = m_3 = 15.0$ kg), the gravitational potential energy of the system at position 1 ($r = R_\oplus$) would be unchanged. However, the gravitational potential energy at the other positions would be $U_2 = U_1/2$ and $U_3 = U_1/3$, mirroring the relationships we found in part A for the gravitational potential.

26-3 Electric Potential Energy U_E

Now we use gravitational potential energy (lower left panel in Fig. 26.1) as an analogy for *electric potential energy*. Like gravity, the electrostatic force is conservative, so we can associate the potential energy with it. The **electric potential energy** U_E—like the gravitational potential energy U_G—is a scalar that depends on both the source and the subject. In the case of gravity, the important property of each object is its mass. In the case of electricity, the important property is the charge. Electric potential energy depends on the charge of both the source and the subject. Also, like gravity, electric potential energy depends on the configuration of the system.

ELECTRIC POTENTIAL ENERGY
✪ **Major Concept**

In this section, we find the electric potential energy of a system that consists of two objects: a spherical source with charge Q and a particle with charge q. Although this is a special case, in subsequent sections we will find the electric potential energy associated with systems that have sources of other shapes.

The SI unit for electric potential energy is the same as the SI unit for any type of energy—the joule (J).

Special Case: Electric Potential Energy Involving a Charged Spherical Source

We can find an expression for the electric potential energy of the charged particle-sphere system in two ways. First, by comparing the gravitational force exerted on a spacecraft $\vec{F}_G = -(GM_\oplus m/r^2)\hat{r}$ (Eq. 7.4) to the electrostatic force exerted on a positively charged particle $\vec{F}_E = (kQq/r^2)\hat{r}$ (Eq. 23.5), we can come up with a translation between the variables:

$$-G \rightarrow k$$
$$M_\oplus \rightarrow Q$$
$$m \rightarrow q$$

We can use this translation of variables in Equation 8.7 to find an expression for the electric potential energy of the charged particle-sphere system:

$$U_G = (-G)\,\frac{M_\oplus m}{r} \tag{8.7}$$

$$\downarrow \quad \downarrow \;\; \downarrow$$

$$U_E = k\,\frac{Q\;\;q}{r} \tag{26.3}$$

DERIVATION Electric Potential Energy for a Particle-Sphere System

Our second way of arriving at Equation 26.3 is more formal than the first. Using the same procedure we used in Section 8-4 to find the gravitational potential energy of a spacecraft-Earth system, we show that the electric potential energy for the particle-sphere system is:

$$U_E(r) = \frac{kQq}{r} \tag{26.3}$$

ELECTRIC POTENTIAL ENERGY FOR A SYSTEM OF TWO CHARGED PARTICLES Special Case

The change in potential energy is found by taking the path integral of any conservative force from an initial position r_i to a final position r_f (Eq. 8.3).	$\Delta U = -\int_{r_i}^{r_f} F_r dr \quad (8.3)$
Substitute Coulomb's law $F_E = kQq/r^2$ (Eq. 23.5) for F_r.	$\Delta U_E = -\int_{r_i}^{r_f} F_E dr \quad (26.4)$ $\Delta U_E = -\int_{r_i}^{r_f}\left(k\frac{Qq}{r^2}\right)dr = -kQq\int_{r_i}^{r_f}\left(\frac{1}{r^2}\right)dr$
Integrate, substitute limits, and reduce.	$\Delta U_E = -kQq\left[-\frac{1}{r}\right]\Bigg\vert_{r_i}^{r_f} = kQq\left[\frac{1}{r}\right]\Bigg\vert_{r_i}^{r_f}$ $\Delta U_E = \frac{kQq}{r_f} - \frac{kQq}{r_i}$
Using the usual convention that the electric potential energy is zero when the particle and sphere are infinitely far apart, we imagine that the particle started at infinity ($r_i = \infty$) and its final position is $r_f = r$.	$\Delta U_E = U_E(r) - U_E(\infty) = U_E(r) - 0$ $\Delta U_E = U_E(r)$ $U_E(r) = \frac{kQq}{r}$ ✓ $\quad (26.3)$

COMMENTS

Although we assumed that the source is a sphere with charge Q and the subject is a particle with charge q, Equation 26.3 gives the **electric potential energy** for any system that consists of two charged, spherically symmetrical objects. For convenience, we'll refer to these objects as particles.

The electric potential energy of a particle-sphere system (Eq. 26.3) is mathematically similar to the gravitational potential energy of the spacecraft-Earth system: $U_G = -GM_{\oplus}m/r$ (Eq. 8.7). Both the gravitational potential energy and the electric potential energy are zero when the particles are infinitely far apart. However, there is one major difference when $r < \infty$. In that case, the gravitational potential energy is negative, whereas the electric potential energy may be negative or positive depending on the signs of the charges. If Q and q have the same sign, the system's potential energy is positive, but if Q and q have opposite signs, the system's potential energy is negative. This difference between the gravitational potential energy and the electric potential energy arises because gravity is always an attractive force, whereas the electric force is attractive only when the source and the subject have charges of opposite sign.

CONCEPT EXERCISE 26.2

A system consisting of a charged sphere and a proton has positive electric potential energy $U_E = +15.0 \times 10^{-20}$ J (with $U_E(\infty) = 0$).

a. What is the sign of the sphere's charge?

b. If the proton were replaced by an electron, what would be the new system's electric potential energy?

Special Case: Electric Potential Energy Stored in a Collection of Charged Particles

If a system consists of more than two charged particles, you can find the system's electric potential energy by applying $U_E(r) = kQq/r$ (Eq. 26.3) to each pair and adding the results. Because electric potential energy (like any type of energy) is a scalar, there are no directions to take into account.

One way to understand the electric potential energy stored in a collection of charged particles is to imagine assembling the collection from particles that are infinitely far apart. Starting with the first particle in place and the other particles infinitely far away, imagine moving each particle into place one at a time, keeping a running tally of the system's total potential energy as shown in the next example.

EXAMPLE 26.2 System of Four Charged Particles

Calculate the electric potential energy of the system shown in Figure 26.2. Model all four objects as charged particles.

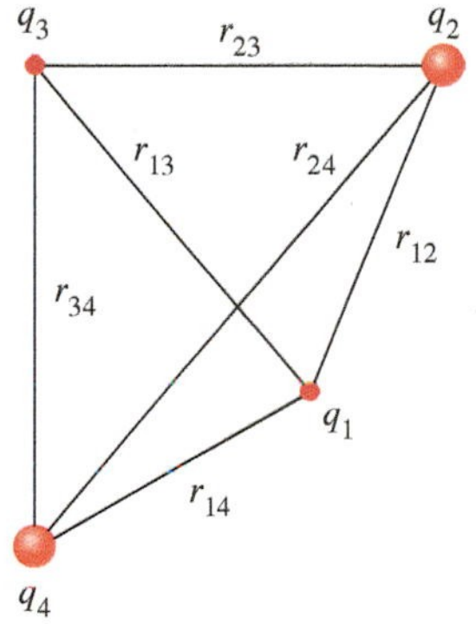

FIGURE 26.2

INTERPRET and ANTICIPATE

We expect to find a sum of terms, each with the form of Equation 26.3.

SOLVE

Start with particle 1 (Fig. 26.3). The system's potential energy is initially zero because the particles are infinitely far apart: $U_i = U(\infty) \equiv 0$.

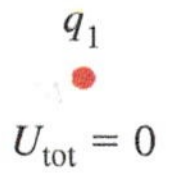

FIGURE 26.3

Next, we imagine that particle 2 moves into place so that it is at a distance r_{12} from particle 1 (Fig. 26.4).

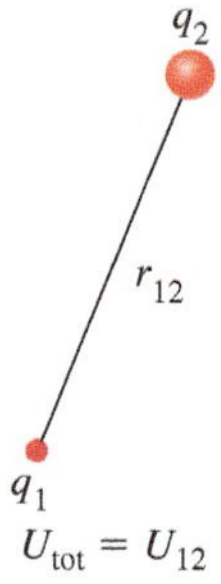

FIGURE 26.4

The system's electric potential energy U_{12} after these two particles are in place is given by Equation 26.3.

$$U_{12} = k\frac{q_1 q_2}{r_{12}}$$

Initially, the system's electric potential energy was zero because the particles were infinitely far apart.

$$U_i = U(\infty) \equiv 0$$

The system's change in potential energy is equal to the final potential energy. So far, the system's total electric potential energy is U_{12} because the other two particles (q_3 and q_4) are infinitely far away.

$$\Delta U_{\text{2nd particle}} = U_f - U_i$$

$$\Delta U_{\text{2nd particle}} = U_{12} - U(\infty) = U_{12}$$

Example continues on page 794 ▶

Next, imagine that particle 3 comes into place (Fig. 26.5)

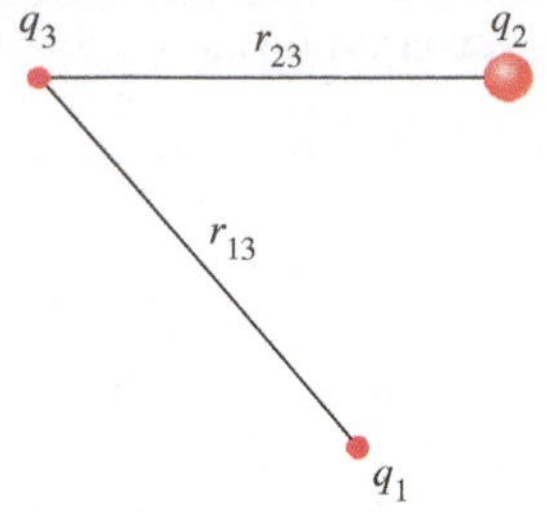

FIGURE 26.5

Once particle 3 is in position, we find the change in the system's potential energy by adding the potential energies of each pair that includes particle 3.	$\Delta U_{\text{3rd particle}} = U_{13} + U_{23}$
Before particle 3 moved into place, the system already had electric potential energy U_{12} due to the first two particles. So, the total electric potential energy of the system (so far) comes from adding the energy the system had when just q_1 and q_2 were present to the change the system undergoes when q_3 is added.	$U_{\text{tot}}\,(3 \text{ particles}) = \Delta U_{\text{3rd particle}} + U_{12}$ $U_{\text{tot}}\,(3 \text{ particles}) = U_{13} + U_{23} + U_{12}$

Finally, imagine that particle 4 comes into place (Fig. 26.6).

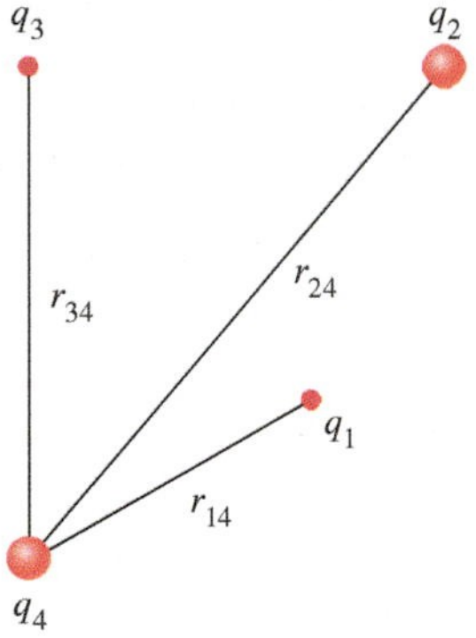

FIGURE 26.6 q_4

Once particle 4 is in position, we find the change in the system's potential energy by adding the potential energies of each pair involving particle 4.	$\Delta U_{\text{4th particle}} = U_{14} + U_{24} + U_{34}$ (26.5)
Before particle 4 moved into place, the system had the electric potential energy due to the first three particles. So, the total electric potential energy of the system comes from adding the energy the system had with only q_1, q_2, and q_3 present to the change the system undergoes when q_4 is added.	$U_{\text{tot}}\,(4 \text{ particles}) = \Delta U_{\text{4th particle}} + (U_{13} + U_{23} + U_{12})$ $U_{\text{tot}}\,(4 \text{ particles}) = (U_{14} + U_{24} + U_{34}) + (U_{13} + U_{23} + U_{12})$ $U_{\text{tot}}\,(4 \text{ particles}) = U_{12} + U_{13} + U_{14} + U_{23} + U_{24} + U_{34}$
Each term is given by $U_E(r) = kQq/r$ (Eq. 26.3), and the electric potential energy of the system shown in Figure 26.6 is given by this expression for U_{tot}.	$U_{\text{tot}} = k\frac{q_1q_2}{r_{12}} + k\frac{q_1q_3}{r_{13}} + k\frac{q_1q_4}{r_{14}} + k\frac{q_2q_3}{r_{23}} + k\frac{q_2q_4}{r_{24}} + k\frac{q_3q_4}{r_{34}}$

CHECK and THINK

Equation 26.3 is used once for each pair in the system, and the total electric potential energy is the sum of all such terms. The total electric potential energy may be positive, negative, or zero depending on the signs and magnitudes of all the charges in the collection.

Another way to think about the electric potential energy of a collection of charged particles is to imagine the amount of work W_{tot} necessary for an external force to assemble the collection. As in Example 26.2, consider a system of four particles that are initially infinitely far apart. An external force—perhaps a person—brings the four particles together one at a time. The system's kinetic energy does not change in any of the steps—the particles are initially at rest when they are infinitely far apart, and they are at rest when they are in their places in the system. Therefore, the work done by the external force equals the change in the system's potential energy, which is U_{tot} because initially the potential energy is zero, $U(\infty) = 0$:

$$W_{tot} = U_f - U_i = U_{tot} - U(\infty)$$

$$W_{tot} = U_{tot}$$

So, for the four-particle system in Figure 26.6, the work done *by an external force* is

$$W_{tot} = k\frac{q_1q_2}{r_{12}} + k\frac{q_1q_3}{r_{13}} + k\frac{q_1q_4}{r_{14}} + k\frac{q_2q_3}{r_{23}} + k\frac{q_2q_4}{r_{24}} + k\frac{q_3q_4}{r_{34}}$$

Remember that this is the work done by an *external* force, not the internal electric force each particle exerts on the other particles in the system.

The work done by the external force may be positive, negative, or zero depending on the signs and magnitudes of all the charges. If the work done is positive, energy is transferred from the environment to the system, as is the case when the charges have the same sign. Imagine bringing two positively (or two negatively) charged particles together. The particles exert a repulsive force on each other, and the external force must do positive work to bring them together. In other words, the system's electric potential energy increases as a result of bringing two like charges together.

If the work done is negative, energy is transferred out of the system and into the environment, as is the case when the particles have opposite signs. A negatively and a positively charged particle exert an attractive force on each other just as two massive objects exert an attractive gravitational force on each other. The work done by the external force in this case is negative; in other words, the system's electric potential energy decreases as a result of bringing the particles together. This is exactly like when you lower a brick off the back of a truck to the ground. If the system consists of the brick and the Earth, you do negative work in lowering the brick, and the system loses gravitational potential energy.

CONCEPT EXERCISE 26.3

A water molecule is made up of two hydrogen atoms and one oxygen atom, with a total of 10 electrons and 10 protons. The molecule is modeled as a dipole with an effective separation $d = 3.9 \times 10^{-12}$ m between its positive and negative particles. What is the electric potential energy stored in the dipole? What does the sign of your answer mean?

EXAMPLE 26.3 Particles in a Triangle

A Three negatively charged particles are held together at the vertices of an equilateral triangle with sides $r = 0.09$ m (Fig. 26.7). The magnitudes of the charges are q, $2q$, and $3q$ where $q = -360$ nC. Find the electric potential energy of the system shown in Figure 26.7.

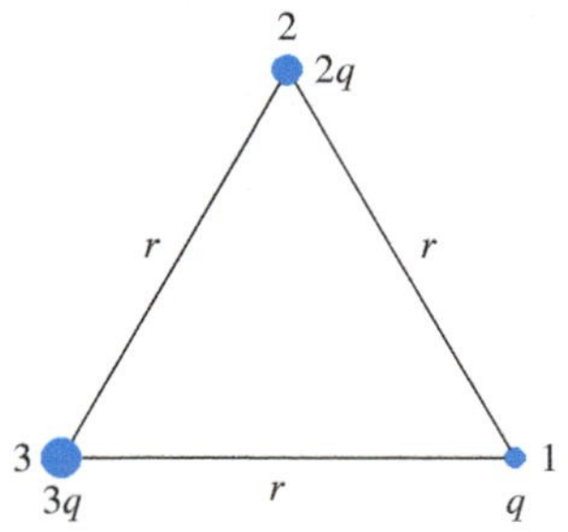

FIGURE 26.7

INTERPRET and ANTICIPATE

Conceptually, we can think of bringing the three particles together from infinity. Because all the charges have the same sign, an external force would do positive work on the system in order to put this configuration together. Because the system's electric potential energy would increase in the assembly process, we expect $U_{tot} > 0$.

Example continues on page 796 ▶

SOLVE Apply $U_E(r) = kQq/r$ (Eq. 26.3) to each pair of charges, and add.	$$U_{tot} = k\frac{q_1q_2}{r_{12}} + k\frac{q_1q_3}{r_{13}} + k\frac{q_2q_3}{r_{23}}$$
Substitute the charge of each particle and the distance between the particles.	$$U_{tot} = k\frac{q(2q)}{r} + k\frac{q(3q)}{r} + k\frac{(2q)(3q)}{r}$$ $$U_{tot} = 2\frac{kq^2}{r} + 3\frac{kq^2}{r} + 6\frac{kq^2}{r} = 11\frac{kq^2}{r}$$
Substitute values.	$$U_{tot} = 11\frac{(8.99 \times 10^9\ \mathrm{N \cdot m^2/C^2})(-360 \times 10^{-9}\ \mathrm{C})^2}{(0.09\ \mathrm{m})}$$ $$U_{tot} = 0.14\ \mathrm{J}$$

CHECK and THINK

As expected, the system has positive electric potential energy. Because there is a repulsive force between all the negatively charged particles that make up the system, an external force would do positive work in bringing them together.

B If the particles are released, what is the system's kinetic energy when the particles are infinitely far apart?

INTERPRET and ANTICIPATE

An energy bar chart (Fig. 26.8) is helpful when applying the conservation of energy principle. Initially, the particles are held in place, so their kinetic energy is zero. In part A, we found that the system's electric potential energy is positive. The particles repel one another, so when they are released they will fly apart, reducing their electric potential energy and increasing their kinetic energy. When the particles are infinitely far apart, their electric potential energy is zero. There are no external forces acting on the system and no change in the system's internal (thermal) energy.

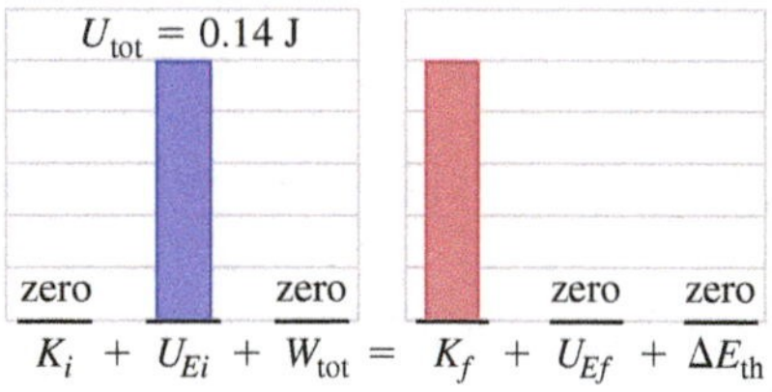

FIGURE 26.8

SOLVE Apply the work-energy theorem (Eq. 9.31).	$$K_i + U_{Ei} + W_{tot} = K_f + U_{Ef} + \Delta E_{th} \quad (9.31)$$ $$0 + U_{tot} + 0 = K_f + U(\infty) + 0$$ $$U_{tot} = K_f$$
The kinetic energy of the system when the particles are very far apart is equal to the electric potential energy of the system in its triangle configuration.	$$K_f = U_{tot} = 0.14\ \mathrm{J}$$

CHECK and THINK

Our result matches the bar chart (Fig. 26.8), which shows the final kinetic energy bar at the same height as the initial potential energy bar.

26-4 Electric Potential V

Electric potential is analogous to gravitational potential (lower right panel of Fig. 26.1). The **electric potential** V_E is a scalar that depends on the source's charge. Although the terms *electric potential* and *electric potential energy* are similar, it is important to distinguish between them. Any type of potential *energy* is associated with a system, so it must involve properties of both the source and the subject. The term *potential* refers to a property of the source alone. In our Earth-spacecraft analogy (Fig. 26.1), the gravitational potential depends only on the mass of the Earth, whereas the gravitational potential energy depends on the masses of both the Earth and the spacecraft. A less formal term for electric potential is **voltage.**

ELECTRIC POTENTIAL (VOLTAGE) Major Concept

To emphasize the distinction between electric potential and electric potential energy in the next few paragraphs, we add the term *voltage* in parentheses whenever we refer to electric potential.

In Figure 26.1, we found the gravitational potential by dividing the gravitational potential energy by the mass of a test subject. The electric potential (voltage) V_E can be found from the electric potential energy in a similar way: Divide the electric potential energy U_E by the charge q of a test subject:

$$V_E = \frac{U_E}{q} \quad (26.6)$$

The dimensions of electric potential (voltage) are energy per charge, so the SI units are joules per coulomb, or volts (V):

$$1 \text{ J/C} = 1 \text{ V}$$

The symbol for the unit—volts—is an uppercase roman (not italic) V, and the symbol for electric potential (voltage) is an uppercase italic V. When you see these symbols handwritten, you must determine their meaning from the context.

Just as only *changes* in potential energy are physically important, only *differences* in electric potential (voltage) are physically important. The electric potential (voltage) at any particular point is assigned a convenient value by convention or by definition; then you measure the potential at some other point *relative* to this assigned potential (Fig. 26.9).

Suppose a source sets up an electric potential (voltage) difference between points A and B and a subject moves from point A to point B. The source–subject system's electric potential *energy* changes as a result of the subject's motion from A to B. The change in the system's electric potential energy ΔU_E is found by multiplying the subject's charge q by the difference in the electric potential (voltage) ΔV_E:

$$\Delta U_E = q\Delta V_E \quad (26.7)$$

The SI units for ΔU_E are joules, as you would expect for energy. However, the joule is often inconvenient to use when the subject has a small number of excess protons (or electrons). In such cases, the change in energy is very small. Consider a proton as the subject, with charge $q = e$. If the proton moves from point A to point B (Fig. 26.9), the change in the system's electric potential energy is

$$\Delta U_E = q\Delta V_E = e(V_B - V_A)$$
$$\Delta U_E = (1.60 \times 10^{-19} \text{C})(4.0 \text{V})$$
$$\Delta U_E = 6.40 \times 10^{-19} \text{J}$$

To avoid such a small number, another energy unit known as the **electron volt** (eV) may be used. It is related to SI units by

$$1 \text{ eV} = 1.60 \times 10^{-19} \text{J}$$

It is convenient to think of the electron volt this way: If a single proton (or electron) moves through an electric potential difference (voltage) of 1 V, the (magnitude of the) change in the system's electric potential energy is 1 eV.

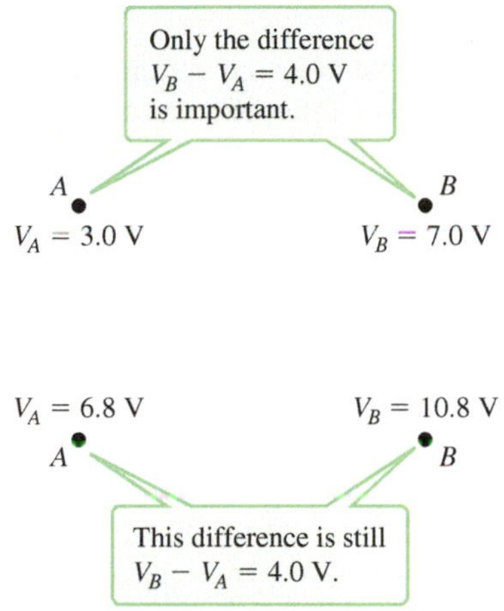

FIGURE 26.9 You are free to choose the zero point of the electric potential, but electric potential difference is independent of your choice. Here the difference in the electric potential (voltage) between points A and B is $V_B - V_A = 4.0$ V. If $V_A = 3.0$ V, then $V_B = 7.0$ V. However, if $V_B = 10.8$ V, then $V_A = 6.8$ V.

EXAMPLE 26.4 CASE STUDY Synapses

When your senses detect an event, an electrical signal travels to your brain along long, thin nerve cells known as *neurons*. Gaps between neurons (called synapses, Fig. 26.10) play an important role in regulating the signals. In order for a signal to reach the brain, it must pass through many synapses. One way for a signal to get through the synapse is by sending a chemical compound known as a neurotransmitter through the gap. The first neurotransmitter to be discovered was acetylcholine (ACh). In this problem, we model the transmission of one ACh molecule through a synapse. This first model is crude, but as we continue to study electricity in the next chapters, we will develop better models.

Consider a single ACh molecule traveling from point i to point f through a synapse. The potential difference (voltage) is $V_f - V_i = -55$ mV. The ACh molecule has one excess proton and mass $m = 2.4 \times 10^{-25}$ kg. The system consists of the ACh molecule and the two nerve cells on either side of the synapse.

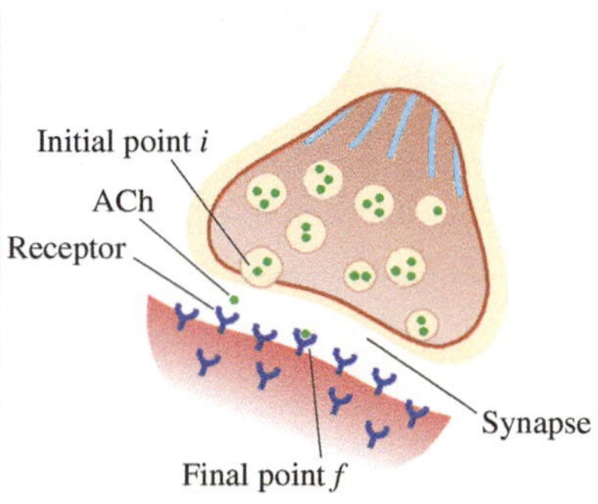

FIGURE 26.10 A synapse is a gap between nerve cells.

Example continues on page 798 ▶

A Find the change in the system's electric potential energy in electron volts and in joules.

B Assume the ACh molecule starts from rest, and find its speed when it reaches point f.

INTERPRET and ANTICIPATE

The source of the electric potential (voltage) is the nerve cells. The subject is the ACh molecule. The electric potential difference (voltage) is negative, $V_f - V_i < 0$, and the subject (ACh) is positively charged, so the system's electric potential energy will decrease: $U_f - U_i < 0$.

As shown in the bar chart (Fig 26.11), we have chosen a reference configuration so that $U_f = 0$. We have assumed there are no external forces doing work on the system ($W = 0$) and no dissipative forces ($\Delta E_{th} = 0$). These assumptions allow us to solve for the ACh molecule's final speed by setting $U_i = K_f$.

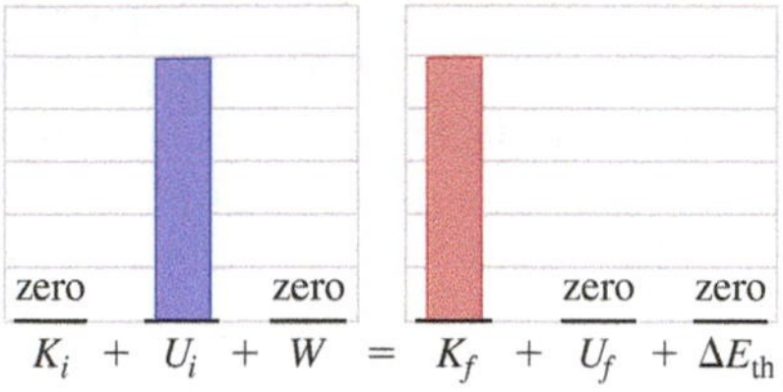

FIGURE 26.11

SOLVE

A The change in the system's electric potential energy comes from Equation 26.7. The ACh has one excess proton, so its charge is $q = e$. Working in electron volts is convenient because we do not need to substitute a value for e. As usual, the "m" in front of the unit stands for "milli."

$$\Delta U_E = q\Delta V_E \quad (26.7)$$

$$\Delta U_E = e(-55\,\text{mV}) = -55\,\text{meV}$$

Convert from electron volts to joules using $1\text{ eV} = 1.60 \times 10^{-19}\text{J}$.

$$\Delta U_E = (-55 \times 10^{-3}\,\text{eV})\left(\frac{1.6 \times 10^{-19}\,\text{J}}{1\text{ eV}}\right)$$

$$\Delta U_E = -8.8 \times 10^{-21}\,\text{J}$$

SOLVE

B Use $U_i = K_f$ to find the ACh molecule's final speed when it arrives at the other side of the synapse.

$$\Delta U_E = U_f - U_i = 0 - U_i$$

$$U_i = -\Delta U_E = K_f$$

$$-\Delta U_E = \frac{1}{2}mv_f^2$$

$$v_f = \sqrt{\frac{2(-\Delta U_E)}{m}} = \sqrt{\frac{2(8.8 \times 10^{-21}\,\text{J})}{(2.4 \times 10^{-25}\,\text{kg})}}$$

$$v_f = 2.7 \times 10^2\,\text{m/s}$$

CHECK and THINK

Our results have the correct dimensions and signs. The speed of the signal in human nerve cells is about 100 m/s. The speed of the ACh molecule we found is probably too high. Contrary to our assumptions, there are probably dissipative forces, so not all the potential energy goes into the molecule's kinetic energy. In Chapter 28, we learn to deal with such dissipative forces, which create resistance to the flow of charged particles. The speed of nerve signal transmission is important for good health. Diseases such as multiple sclerosis result when the transmission speed is atypical.

Special Case: Electric Potential (Voltage) Due to a Charged Particle

The source in Figure 26.12 has spherical symmetry, such as a charged sphere, spherical shell, or particle. For convenience, we refer to this source as a particle with charge Q.

According to $V_E = U_E/q$ (Eq. 26.6), we can find the electric potential by dividing the electric potential energy by the charge q of a test subject. Dividing $U_E(r) = kQq/r$ (Eq. 26.3) by q, we find the electric potential due to a charged particle:

ELECTRIC POTENTIAL (VOLTAGE) OUTSIDE A SPHERICAL SOURCE
Special Case

$$V_E = k\frac{Q}{r} \tag{26.8}$$

Equation 26.8 is valid outside any source with spherical symmetry that possesses excess charge Q.

As we know from our experience with gravitational potential (Fig. 26.1), it is usually convenient to set a reference point at which the potential is defined to be zero. In the case of universal gravity, the most convenient reference point is at infinity. The same is true in the case of a charged particle; the most convenient reference point is at infinity where the electric potential is defined to be zero: $V_E(\infty) \equiv 0$. At all other points, the electric potential is given by Equation 26.8.

Although the gravitational potential is zero at infinity and negative at all other points (Fig. 26.1), the electric potential due to a charged source particle is zero at infinity and may be positive or negative at other points depending on the sign of the source. If the source is positive, the electric potential is positive at points other than infinity. If the source is negative, the electric potential is negative (like the gravitational potential) at points other than infinity. Figure 26.12 summarizes the electric force, electric field, electric potential energy, and electric potential for a charged particle. Compare Figures 26.1 and 26.12, and you see that equations and concepts concerning electricity are similar to those concerning gravity.

Visualizing Electric Potential

There are two main tools for visualizing electric potential. One is a graph of electric potential as a function of position. Figure 26.13A is a graph of V_E versus r, where r is the distance from a charged particle (Eq. 26.8). The graph shows how quickly the

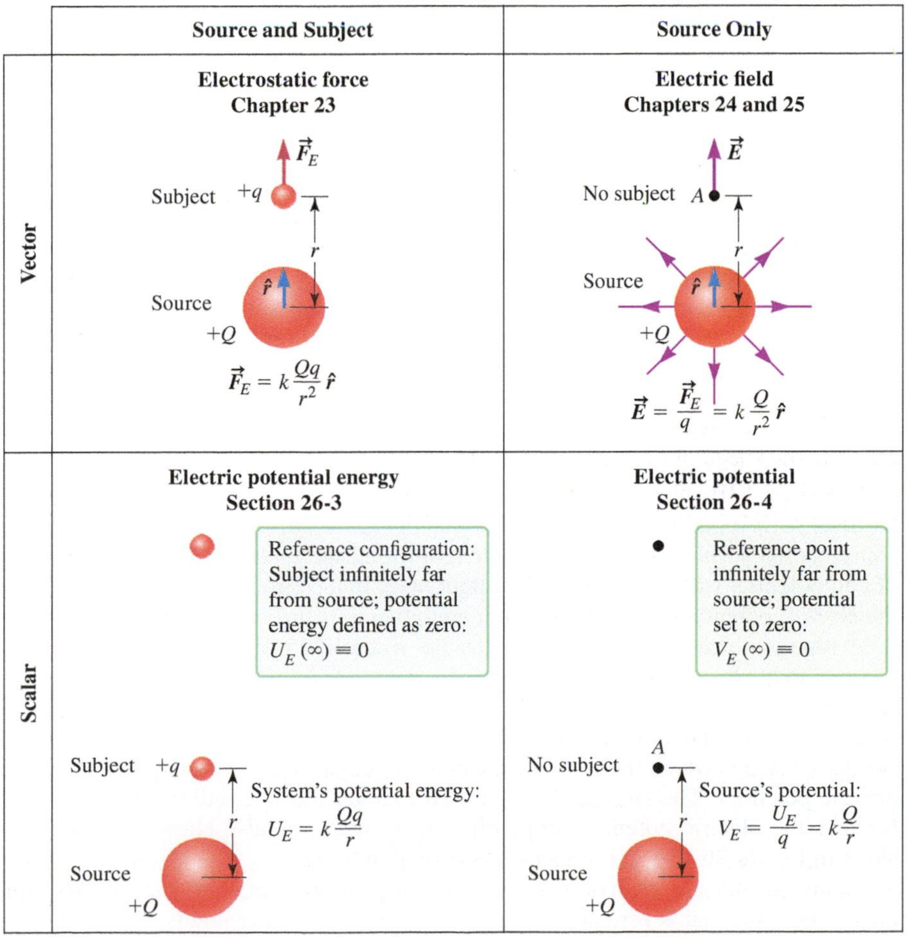

FIGURE 26.12 Compare this figure to Figure 26.1. The quantities electric force, electric field, electric potential energy, and electric potential are all analogous to the quantities for gravity. Use this analogy to help build your understanding of electricity.

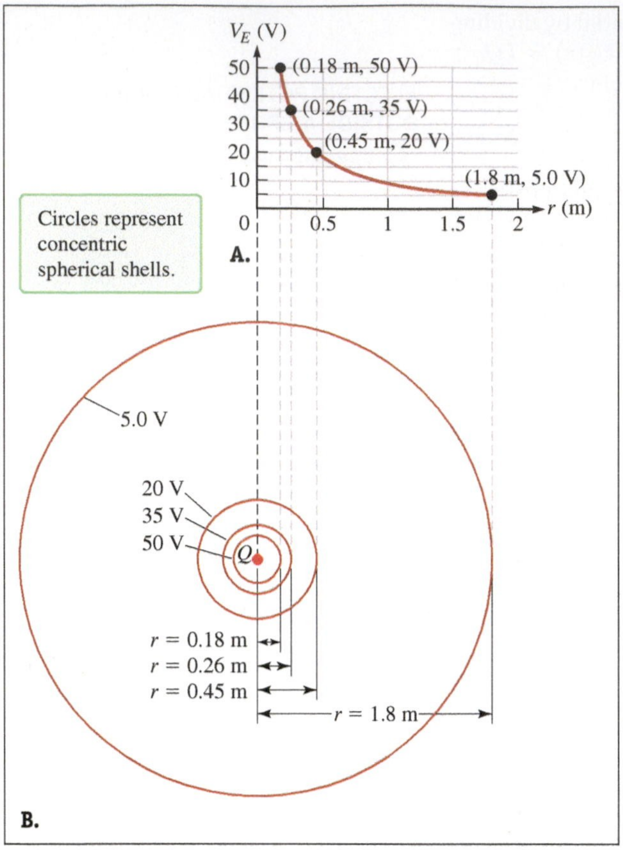

FIGURE 26.13 **A.** Graph of the electric potential near a positively charged source particle. **B.** A contour map corresponding to the graph in part A. A contour map consists of a number of equally spaced equipotential surfaces. In this case, there is 15 V between adjacent surfaces.

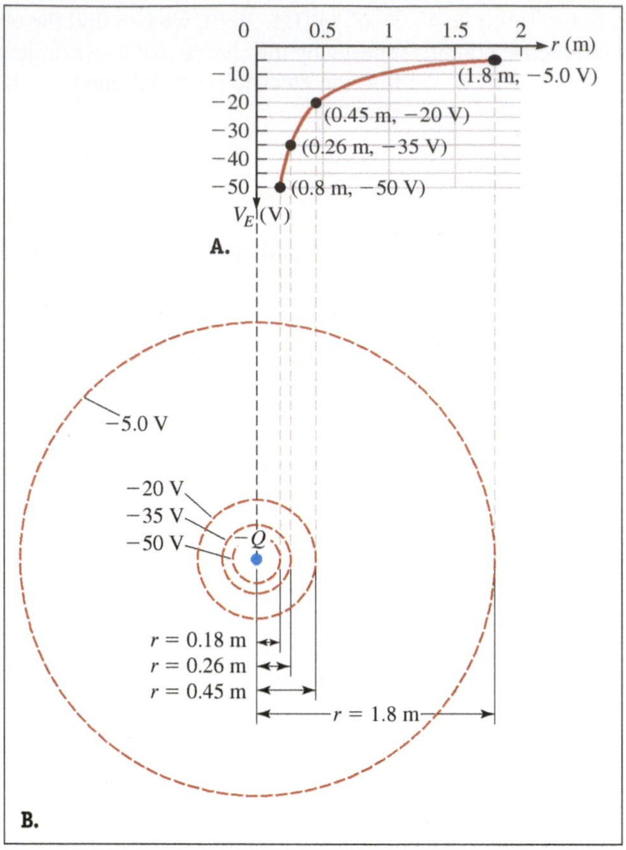

FIGURE 26.14 **A.** Graph of the electric potential near a negatively charged source particle. **B.** Corresponding contour map. Notice that we use dashed lines for negative potentials.

electric potential drops off with distance from the source. For this particle, the net charge is $Q = 1 \times 10^{-9}$ C, and the electric potential decreases from 50 V at 0.18 m from the source to 5 V at 1.8 m away, dropping off sharply near the source as can be seen by the steep slope for positions closer than about 1 m.

All points that are at the same distance r from a given charged particle have the same electric potential (Fig. 26.13). Imagine a series of concentric spherical shells centered on the particle; the electric potential is the same for all points on one shell. For example, all points on a shell of radius $r = 0.18$ m are at an electric potential of 50 V. A surface on which all the points are at the same electric potential is called an **equipotential surface**. In addition to a graph, a series of equipotential surfaces provides another way to visualize the electric potential.

Figure 26.13B shows four equipotential surfaces for a positively charged source particle, corresponding to the labeled points in Figure 26.13A. The number on each equipotential surface is the electric potential for all the points on that surface. A set of equipotential surfaces is called a **contour map**. In a contour map, the difference in the electric potentials between adjacent equipotential surfaces is constant.

CONTOUR MAP Tool

A topographical map that you might use when hiking is another example of a contour map. In a topographical map, the contours represent elevations, and the spacing of the contours corresponds to changes in the terrain. For example, closely spaced contours indicate steep terrain. Likewise, on a contour map of equipotential surfaces, regions where the contours are close together indicate a sharp change in the electric potential. The surfaces in Figure 26.13B are close together near the source because the electric potential drops off quickly there, as also shown by the steep slope in Figure 26.13A at distances less than about 1 m.

In contour maps, we represent positive equipotential surfaces by solid curves and negative equipotential surfaces by dashed curves.

Figure 26.14A is a graph of V_E versus r for a negatively charged particle. In this case, the electric potential is zero at infinity and negative at all other points. Figure 26.14A looks like an upside-down version of Figure 26.13A. Again, we use four labeled points to make a contour map (Fig. 26.14B). Compare the contour maps (Figs. 26.13B and 26.14B): The equipotential surfaces are close together near the source in both cases, indicating that the electric potential changes sharply near the source. If these were topographical hiking maps, Figure 26.13B would represent a hill that gets very steep as you climb toward the summit, whereas Figure 26.14B would represent a valley that gets very steep the closer you get to the bottom.

CONCEPT EXERCISE 26.4

Match the topographical maps in Figure 26.15 with the corresponding landscapes.

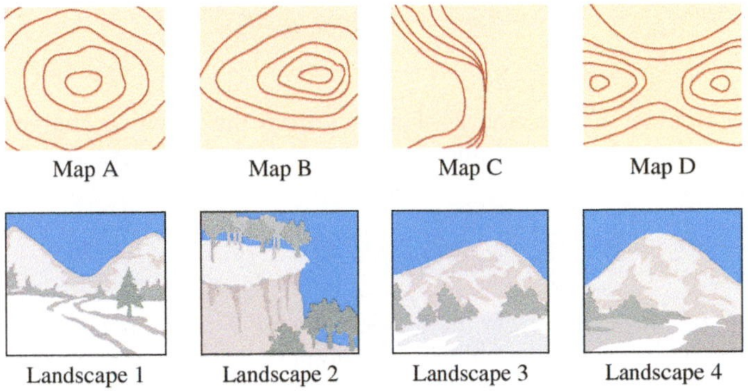

FIGURE 26.15

26-5 Special Case: Electric Potential Due to a Collection of Charged Particles

In this section, we drop the subscript *E* from the symbol V_E, so that the symbol for electric potential is simply *V*. We drop the word *electric* from the phrase *electric potential difference.*

If the source of an electric field is a collection of charged particles, the electric potential at a particular point in space is found by adding the electric potentials due to each particle. Because the potential is a scalar, there is no direction to take into account when we do this addition, although for each particle we must (1) use the same convention for the zero of electric potential, $V(\infty) = 0$, and (2) take into account the signs of the electric potentials due to each charge. The electric potential at some point in space due to a collection of *n* charged particles is given by

$$V = k\sum_{i=1}^{n} \frac{q_i}{r_i} \tag{26.9}$$

where q_i is the charge of the *i*th particle and r_i is its distance to the point at which we calculate the electric potential.

Special Case: Electric Potential Due to a Dipole

Let's find the electric potential at point *A* due to the dipole shown in Figure 26.16A. Each particle in a dipole has a charge whose absolute value is *q*. The negative particle is at a distance r_- and the positive particle is at a distance r_+ from point *A*. According to Equation 26.9, the electric potential at *A* is the sum of the electric potentials due to each charged particle:

$$V = V_+ + V_-$$

$$V = k\frac{q}{r_+} + k\frac{(-q)}{r_-}$$

Finding a common denominator and adding these terms, we find that the electric potential is

$$V = kq\left(\frac{r_- - r_+}{r_+ r_-}\right) \tag{26.10}$$

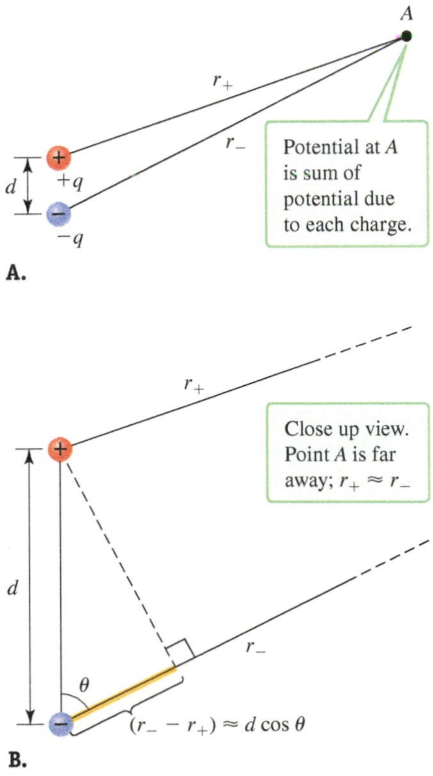

FIGURE 26.16 **A.** Finding the electric potential at point *A* near a dipole. **B.** Use trigonometry to find the extra distance from the negative particle to point *A*.

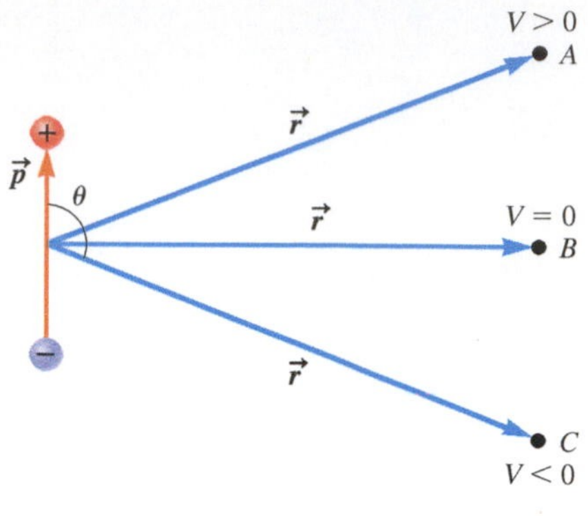

FIGURE 26.17 The dipole moment points from the negative to the positive charge. The electric potential is positive at point A, zero at point B, and negative at point C.

We are often interested in regions of space that are relatively far from the dipole compared to the charge separation d. In such cases, the distances r_- and r_+ are nearly equal, and we write

$$r \equiv r_+ \approx r_-$$

so that the denominator in Equation 26.10 becomes $r_+r_- \approx r^2$. The small difference in the numerator can be written in terms of the separation d and the angle θ (Fig. 26.16B):

$$r_- - r_+ \approx d\cos\theta$$

Substituting these approximations into Equation 26.10, we find the electric potential at a distant point from a dipole is given by

$$V \approx k\frac{qd\cos\theta}{r^2}$$

The dipole moment $\vec{p}$ (Section 24-4) points from the negative charge toward the positive charge and has magnitude qd. Write the electric potential in terms of the dipole moment p,

ELECTRIC POTENTIAL DUE TO A DIPOLE Special Case

$$V \approx k\frac{p\cos\theta}{r^2} \quad (26.11)$$

where angle θ is measured from the dipole moment $\vec{p}$ to the position vector $\vec{r}$ that extends from the center of the dipole to the point of interest. The sign of the electric potential depends on the angle θ (Fig. 26.17).

Figure 26.18 shows a contour map for the dipole in Figure 26.16. Most of the equipotential surfaces look like flattened spherical shells. Solid lines indicate positive electric potentials, and dashed lines indicate negative electric potentials. The solid black line in the center of the figure represents the zero-potential surface, a flat plane. For a dipole, the electric potential is zero both at infinity and at any point on the plane shown in black. Notice that the sharpest change in the potential is between the two charges, near the plane.

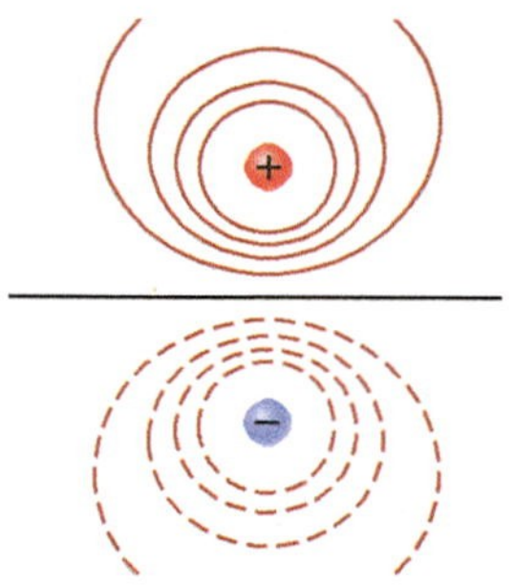

FIGURE 26.18 Contour map for an electric dipole.

CONCEPT EXERCISE 26.5

Which term or phrase is a synonym for *electric potential*?

a. gravitational potential
b. potential energy
c. voltage
d. electric field

EXAMPLE 26.5 Electric Potential Due to Three Particles

Figure 26.19 shows a source that consists of three charged objects. Objects 1 and 3 are particles with charges q_1 and q_3, respectively. Object 2 is a sphere with charge q_2. Point A is in the same plane as the source; its distance to each object is given in the figure.

A Use the numerical values in Figure 26.19 to find the electric potential at point A. (All values provided have two significant figures.)

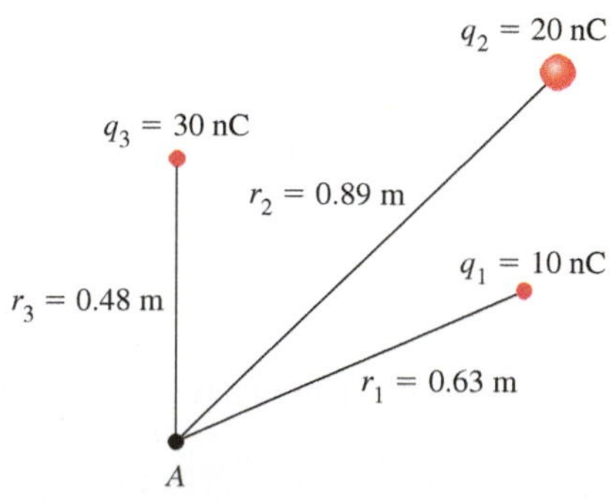

FIGURE 26.19

INTERPRET and ANTICIPATE

All objects that make up the source are positively charged, so we expect the electric potential at point A to be positive (using the usual convention that the potential at infinity is zero).

SOLVE

The electric potential outside a spherical charge distribution is the same as the electric potential due to a charged particle, so all three objects in the source may be modeled as particles. The electric potential at point A is the sum of the electric potentials due to three charged particles (Eq. 26.9).

$$V = k\sum_{i=1}^{n}\frac{q_i}{r_i} \quad (26.9)$$

$$V = k\left(\frac{q_1}{r_1} + \frac{q_2}{r_2} + \frac{q_3}{r_3}\right) \quad (1)$$

Substitute the numerical values of charges and distances given in Figure 26.19. Remember that 1 J/C = 1 V. (We'll need to keep an extra significant figure for part B: $V = 9.07 \times 10^2$ V.)

$$V = (8.99 \times 10^9\,\text{N}\cdot\text{m}^2/\text{C}^2) \times \left[\frac{10 \times 10^{-9}\,\text{C}}{0.63\,\text{m}} + \frac{20 \times 10^{-9}\,\text{C}}{0.89\,\text{m}} + \frac{30 \times 10^{-9}\,\text{C}}{0.48\,\text{m}}\right]$$

$$V = 9.1 \times 10^2\,\text{V}$$

CHECK and THINK

As expected, the electric potential is positive.

B How much work is required (by an external force) to move a 40-nC charge from very far away (from infinity) to point A?

INTERPRET and ANTICIPATE

There are at least two ways we can do this calculation. We'll use one to anticipate our results and the other to solve the problem. Of course, we expect the two answers to agree. The first method is based on an idea from Section 26-3: The work done by an external force is equal to the change in the system's electric potential energy. So, we need to calculate ΔU when the fourth charge is moved from infinitely far away to point A.

The required task is essentially what we did in Example 26.2 to find Equation 26.5 when we added the fourth particle to the system.

$$W = \Delta U_{\text{4th particle}} = U_{14} + U_{24} + U_{34} \quad (26.5)$$

The electric potential energy stored by each pair of particles is given by $U_E(r) = kQq/r$ (Eq. 26.3).

$$W = k\frac{q_1q_4}{r_1} + k\frac{q_2q_4}{r_2} + k\frac{q_3q_4}{r_3} = kq_4\left(\frac{q_1}{r_1} + \frac{q_2}{r_2} + \frac{q_3}{r_3}\right) \quad (2)$$

$$W = (8.99 \times 10^9\,\text{N}\cdot\text{m}^2/\text{C}^2)(40 \times 10^{-9}\,\text{C}) \times \left(\frac{10 \times 10^{-9}\,\text{C}}{0.63\,\text{m}} + \frac{20 \times 10^{-9}\,\text{C}}{0.89\,\text{m}} + \frac{30 \times 10^{-9}\,\text{C}}{0.48\,\text{m}}\right)$$

$$W = 3.6 \times 10^{-5}\,\text{J} \quad \text{(anticipated result)}$$

SOLVE

We can also find the change in the electric potential energy and therefore the work done by an external force by multiplying the potential difference between two points (A and ∞) by the amount of charge that moves between those two points (Eq. 26.7). The electric potential at infinity is zero, so the potential difference is equal to our results in part A.

$$W = \Delta U = q\Delta V \quad (26.7)$$

$$W = q_4[V - V(\infty)] = q_4 V$$

$$W = (40 \times 10^{-9}\,\text{C})(9.1 \times 10^2\,\text{V})$$

$$W = 3.6 \times 10^{-5}\,\text{J} \checkmark$$

CHECK and THINK

We found the same result using both methods. In fact, the two methods really aren't different. Compare Equation (2) for work W to Equation (1) for electric potential V. Equation (2) can be found by multiplying q_4 by V in Equation (1).

26-6 Electric Potential Due to a Continuous Distribution

In this section, we show how to find the electric potential V due to a continuous distribution, such as a charged rod or disk, by modeling the distribution as a large number of charged particles. The process is much like finding the electric field $\vec{E}$ for the distribution (Section 24-5), but finding the electric potential (a scalar) is somewhat easier. In Section 26-7, we show how to find the electric potential from the electric field.

PROBLEM-SOLVING STRATEGY

Electric Potential for a Continuous Charge Distribution

INTERPRET and ANTICIPATE

As in many problem-solving strategies, draw a **sketch** of the source and choose a **coordinate system**.

SOLVE

Step 1 Divide the source into small **pieces**, and write an **expression for dV** for each piece. Each small piece has charge dq and can be modeled as a charged particle. The electric potential dV for each small piece is given by a modified form of Equation 26.8:

$$dV = k\frac{dq}{r} \qquad (26.12)$$

where r is the distance between a particular piece and the point where you would like to find the electric potential. Write your expression in terms of your coordinate system, and express dq in terms of the charge density (Section 24-5).

Step 2 **Integrate** over the entire source to find V.

CHECK and THINK

In addition to checking the dimensions, there are two other checks for a finite source:

1. Verify that at **a point far away** compared to the size of the source, the electric potential has the same form as that of a **charged particle** ($V = kQ/r$).
2. Verify that at a point **infinitely far away, the electric potential is zero.**

EXAMPLE 26.6 Electric Potential due to a Charged Ring (Special Case)

Show that the electric potential at point A due to a uniformly charged ring of radius R and charge Q is

$$V = \frac{kQ}{\sqrt{R^2 + x^2}} = \frac{1}{4\pi\varepsilon_0}\frac{Q}{\sqrt{R^2 + x^2}} \qquad (26.13)$$

Point A is a distance x from the center of the ring (Fig. 26.20).

ELECTRIC POTENTIAL DUE TO A RING Special Case

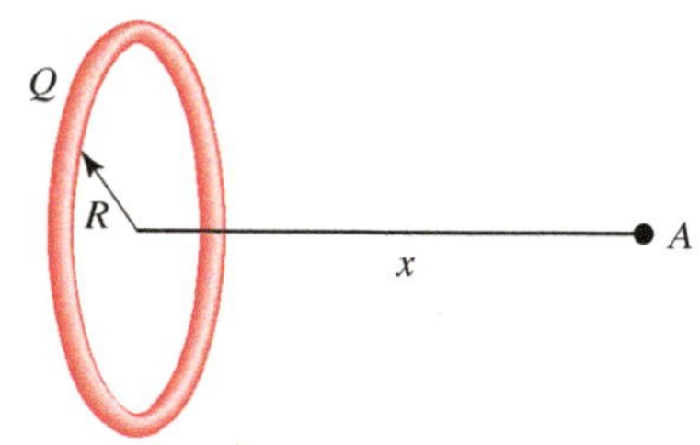

FIGURE 26.20

INTERPRET and ANTICIPATE

Sketch the source and choose a **coordinate system.** The origin of our coordinate system is at the center of the ring, and the x axis is aligned with the ring's axis so that point A is a distance x from the center of the ring (Fig. 26.21). The y axis points upward, and the z axis is coming out of the page.

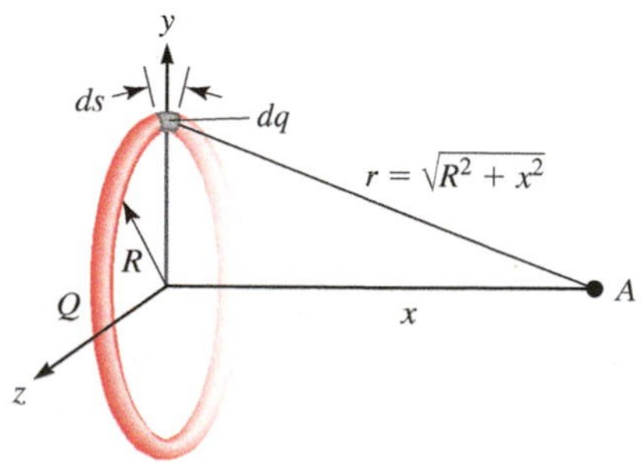

FIGURE 26.21

SOLVE **Step 1** Divide the source into **pieces**, and write an **expression for *dV*** for each piece. Figure 26.21 shows one piece at the top of the ring. The electric potential due to that one piece is given by Equation 26.12.	$dV = k\dfrac{dq}{r}$ (26.12)
The distance r to point A is the hypotenuse of a right triangle with sides x and R. Write r in terms of x and R.	$r = \sqrt{R^2 + x^2}$ (1)
The piece has charge dq and length ds. Write dq in terms of ds and the ring's linear charge density λ (Eq. 24.13).	$dq = \lambda ds$ (24.13)
The electric potential at point A due to one piece comes from substituting for r (Eq. 1) and dq (Eq. 24.13).	$dV = k\dfrac{\lambda ds}{\sqrt{R^2 + x^2}}$
Step 2 **Integrate** over the source. In this case, we must integrate from $s = 0$ to $s = 2\pi R$ (the circumference of the ring). The integral on the left is the electric potential at point A due to the entire ring. The radius R of the ring and the distance x do not depend on s, so they are pulled outside the integral along with the constants k and λ.	$\int dV = \int_0^{2\pi R} k\dfrac{\lambda ds}{\sqrt{R^2 + x^2}}$ $V = \dfrac{k\lambda}{\sqrt{R^2 + x^2}}\int_0^{2\pi R} ds$
Complete the integral and evaluate between the limits.	$V = \dfrac{k\lambda}{\sqrt{R^2 + x^2}}\, s\Big\|_0^{2\pi R} = \dfrac{k\lambda}{\sqrt{R^2 + x^2}}(2\pi R)$
The ring's charge Q is the linear charge density times the ring's circumference, $Q = (\lambda)(2\pi R)$.	$V = \dfrac{kQ}{\sqrt{R^2 + x^2}}$ ✓
It is common practice to replace Coulomb's constant k with $k = 1/(4\pi\varepsilon_0)$ (Eq. 25.9).	$V = \dfrac{1}{4\pi\varepsilon_0}\dfrac{Q}{\sqrt{R^2 + x^2}}$ ✓ (26.13)
CHECK and THINK To check our result, we imagine point A moving away from the ring ($x \gg R$). First, we see that the electric potential has the same form as that from a charged particle. Then, as we continue to imagine point A moving toward infinity ($x \to \infty$), we find that the electric potential drops off to zero.	$\lim_{x \gg R} V = \lim_{x \gg R}\dfrac{kQ}{\sqrt{R^2 + x^2}} \to \dfrac{kQ}{\sqrt{x^2}} = \dfrac{kQ}{x}$ ✓ $\lim_{x \to \infty} V = \dfrac{kQ}{x} \to 0$ ✓

EXAMPLE 26.7 Electric Potential due to a Charged Disk (Special Case)

Show that the electric potential at point A due to a uniformly charged disk of radius R and excess surface charge density σ is

$$V = 2\pi k\sigma[(\sqrt{R^2 + x^2}) - x] = \frac{\sigma}{2\varepsilon_0}[(\sqrt{R^2 + x^2}) - x] \qquad (26.14)$$

Point A is a distance x from the center of the disk (Fig. 26.22).

ELECTRIC POTENTIAL DUE TO A CHARGED DISK Special Case

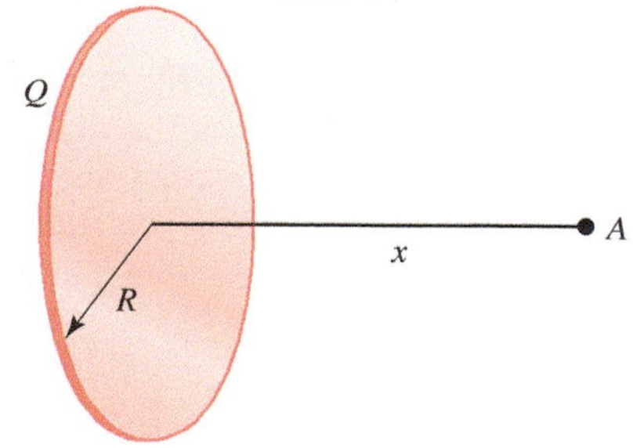

FIGURE 26.22

Example continues on page 806 ▶

INTERPRET and ANTICIPATE

Sketch the source and choose a **coordinate system.** This source is very similar to the charged ring, which we can use as a shortcut. The origin of our coordinate system is at the center of the disk, and the x axis is aligned with the disk's axis so that point A is a distance x from the center of the disk (Fig. 26.23). The y axis points upward, and the z axis points out of the page.

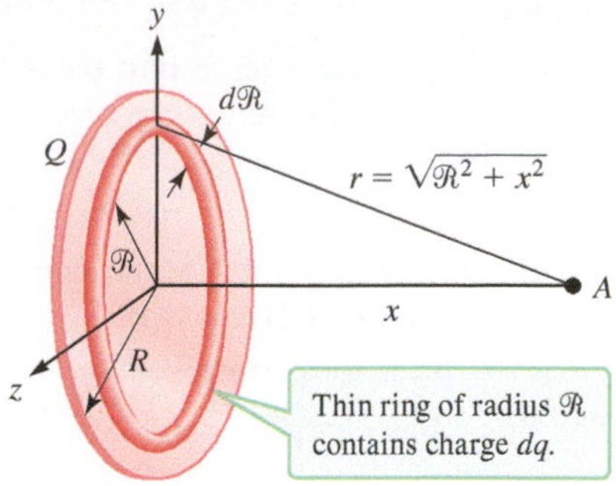

FIGURE 26.23

SOLVE

Step 1 Divide the source into **pieces**, and write an **expression for dV** for each piece. As a shortcut, we divide the disk into concentric rings. One such ring with charge dq and radius $\mathcal{R}$ is shown in Figure 26.23. (Notice that $\mathcal{R}$ is the radius of a ring, and R is the radius of the disk.) The potential dV due to a ring is given by Equation 26.13. (We could also start with Equation 26.12 and substitute $r = \sqrt{\mathcal{R}^2 + x^2}$).

$$dV = \frac{k\,dq}{\sqrt{\mathcal{R}^2 + x^2}} \qquad (26.13)$$

The area of the ring is its circumference $2\pi\mathcal{R}$ times its width $d\mathcal{R}$. Write dq in terms of the area of the ring and the surface charge density σ (Eq. 24.11).

$$dq = \sigma dA \qquad (24.11)$$

$$dq = \sigma(2\pi\mathcal{R})d\mathcal{R}$$

The electric potential at point A due to one ring comes from substituting for dq.

$$dV = k\frac{\sigma(2\pi\mathcal{R})d\mathcal{R}}{\sqrt{\mathcal{R}^2 + x^2}}$$

Step 2 Integrate over the source. In this case, we must integrate over the surface of the disk from a ring with $\mathcal{R} = 0$ to a ring with $\mathcal{R} = R$. The integral on the left is the electric potential at point A due to the entire disk. This time we cannot pull the term involving $\mathcal{R}$ outside the integral.

$$\int dV = \int_0^R k\frac{\sigma(2\pi\mathcal{R})d\mathcal{R}}{\sqrt{\mathcal{R}^2 + x^2}}$$

$$V = k2\pi\sigma\int_0^R \frac{\mathcal{R}\cdot d\mathcal{R}}{\sqrt{\mathcal{R}^2 + x^2}}$$

Complete the integral (Problem 81), and evaluate between the limits.

$$\int_0^R \frac{\mathcal{R}\cdot d\mathcal{R}}{\sqrt{\mathcal{R}^2 + x^2}} = (\sqrt{R^2 + x^2}) - x$$

Use Equation 25.9 to rewrite Coulomb's constant k in terms of the permittivity constant ε_0.

$$V = 2\pi k\sigma[(\sqrt{R^2 + x^2}) - x] \quad ✓$$

$$V = \frac{\sigma}{2\varepsilon_0}[(\sqrt{R^2 + x^2}) - x] \quad ✓ \qquad (26.14)$$

CHECK and THINK

To check the result, take the limit as the point A moves away from the disk. First, replace the surface charge density by $\sigma = Q/\pi R^2$.

$$V = 2\pi k\frac{Q}{\pi R^2}[(\sqrt{R^2 + x^2}) - x]$$

Then factor x out of the square root term.

$$V = 2k\frac{Q}{R^2}[(x\sqrt{(R/x)^2 + 1}) - x]$$

So, as A moves away from the disk ($x \gg R$), the electric potential takes on the same form as that due to a charged particle. As A continues toward infinity ($x \to \infty$), the electric potential drops off to zero.

$$\lim_{x \gg R} V \to 2k\frac{Q}{R^2}\left[x\left(\frac{1}{2}\frac{R^2}{x^2} + 1\right) - x\right]$$

$$\lim_{x \gg R} V \to 2k\frac{Q}{R^2}\left(\frac{1}{2}\frac{R^2}{x}\right) \to k\frac{Q}{x} \quad ✓$$

$$\lim_{x \to \infty} V = \lim_{x \to \infty}\left(k\frac{Q}{x}\right) \to 0 \text{ as } x \to \infty \quad ✓$$

26-7 Connection Between Electric Field $\vec{E}$ and Electric Potential V

In the preceding sections, we found an expression for the electric potential that results from a complicated charged source by breaking up the source into small pieces and adding (or integrating) the electric potentials associated with each small piece. In this section, we show another approach for finding the electric potential V based on the source's electric field $\vec{E}$.

We seek to relate the electric potential (a scalar) to the electric field (a vector); both quantities depend on only the source and not the subject. The relationship between the electric potential V and the electric field $\vec{E}$ is analogous to the relationship between the potential energy U and the force $\vec{F}$ (Fig. 26.12). To find the change in the potential energy, we take a path integral of the force:

$$\Delta U = -\int_{r_i}^{r_f} \vec{F} \cdot d\vec{r} \tag{9.24}$$

By analogy, the electric potential difference is found by taking the path integral of the electric field:

$$\Delta V = -\int_{r_i}^{r_f} \vec{E} \cdot d\vec{r} \tag{26.15}$$

FINDING THE ELECTRIC POTENTIAL V FROM THE ELECTRIC FIELD $\vec{E}$

Major Concept

We can derive Equation 26.15 formally from $\Delta U = -\int_{r_i}^{r_f} \vec{F} \cdot d\vec{r}$ (Eq. 9.24). Substitute $\vec{F}_E = q\vec{E}$ (Eq. 24.2) for the electric force:

$$\Delta U = -\int_{r_i}^{r_f} q\vec{E} \cdot d\vec{r} = -q\int_{r_i}^{r_f} \vec{E} \cdot d\vec{r}$$

where q is the subject's charge. According to $\Delta V = \Delta U/q$ (Eq. 26.7), we simply divide ΔU by q to find the electric potential difference:

$$\Delta V = -\int_{r_i}^{r_f} \vec{E} \cdot d\vec{r} \quad ✓ \tag{26.15}$$

EXAMPLE 26.8 Electric Potential due to a Charged Particle

From the electric field $\vec{E}(r) = (kQ_S/r^2)\hat{r}$ (Eq. 24.3) produced by a positively charged particle, find the (usual) expression for its electric potential.

INTERPRET and ANTICIPATE

Sketch the electric field and include a path from r_i to r_f (Fig. 26.24). We expect our result to be the same as $V = kQ/r$ (Eq. 26.8).

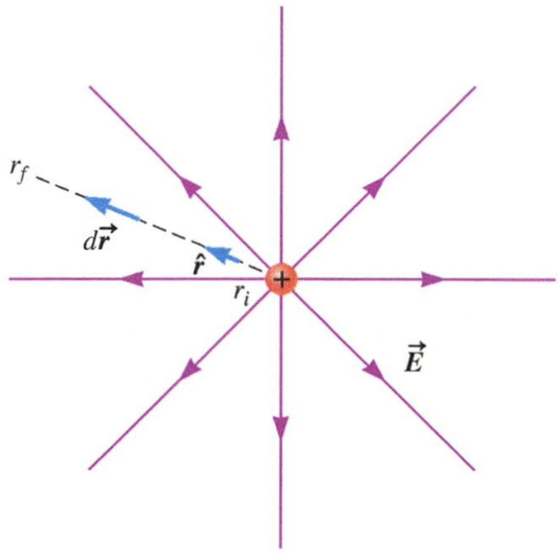

FIGURE 26.24

SOLVE

For a particle, the electric field $\vec{E}$ and $d\vec{r}$ (a small portion of the path from r_i to r_f) both point in the $\hat{r}$ direction, so their dot product is $E\,dr$. Substitute the magnitude of the electric field E for a charged particle.

$$\Delta V = -\int_{r_i}^{r_f} \vec{E} \cdot d\vec{r} \tag{26.15}$$

$$\Delta V = -\int_{r_i}^{r_f} E\,dr = -\int_{r_i}^{r_f} \left(k\frac{Q}{r^2}\right) dr$$

Example continues on page 808 ▶

Pull the constants outside the integral and integrate.	$\Delta V = -kQ\int_{r_i}^{r_f}\left(\frac{1}{r^2}\right)dr$ $\Delta V = -kQ\left[-\frac{1}{r}\right]\Big\|_{r_i}^{r_f} = kQ\left[\frac{1}{r}\right]\Big\|_{r_i}^{r_f}$
Evaluate between limits and reduce.	$\Delta V = \frac{kQ}{r_f} - \frac{kQ}{r_i}$
Using the usual convention that the electrostatic potential is zero at infinity, we find an expression for the electric potential at a distance r from the source, in agreement with Equation 26.8.	$\Delta V = V(r) - V(\infty) = V(r) - 0 = V$ $V = \frac{kQ}{r}$ ✓ (26.8)

CHECK and THINK

We started with the particle's electric field and found its electric potential as expected.

EXAMPLE 26.9 Electric Potential due to a Charged Sheet

Sources with planar symmetry (Section 25-6), such as a very large charged sheet (Fig. 26.25, edge view), are common and important. The magnitude of the electric field for such a source is uniform and given by $E = \sigma/2\varepsilon_0$ (Eq. 25.16).

A Use path 1 shown in Figure 26.25 to find the potential difference ΔV_{AB} between points A and B. (We'll consider path 2 in the next part, and path 3 in a later derivation.)

INTERPRET and ANTICIPATE

To find the potential difference between a pair of points, we must calculate the path integral $\Delta V = -\int_{r_i}^{r_f}\vec{E}\cdot d\vec{r}$. Path 1 is a vertical line between points A and B.

ELECTRIC POTENTIAL DUE TO A SOURCE WITH PLANAR SYMMETRY ▶ Special Case

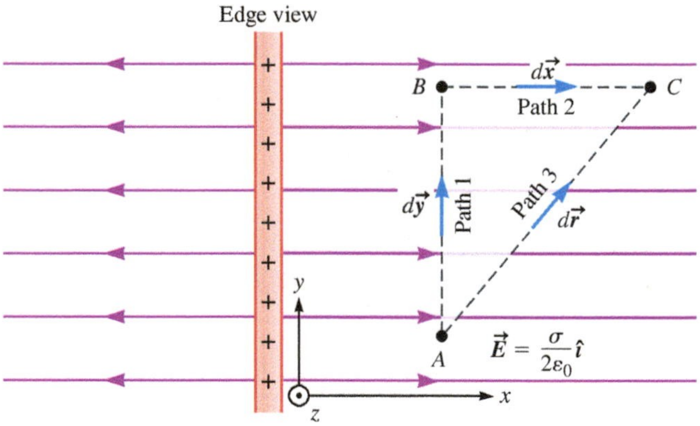

FIGURE 26.25

SOLVE

Using the coordinate system in the figure, write the electric field in component form.	$\vec{E} = \frac{\sigma}{2\varepsilon_0}\hat{\imath}$ (1)
In general, the path element $d\vec{r}$ may be written in terms of its vector components.	$d\vec{r} = d\vec{x} + d\vec{y} + d\vec{z}$
Because point B is directly above point A, path 1 is purely in the y direction. The x and z components of $d\vec{r}_1$ are zero.	$d\vec{r}_1 = 0 + d\vec{y} + 0 = d\vec{y}$ $d\vec{r}_1 = dy\hat{\jmath}$ (2)
We set up the path integral (Eq. 26.15) from A to B by substituting Equation (1) for $\vec{E}$ and Equation (2) for $d\vec{r}_1$.	$\Delta V_{AB} = -\int_{r_i}^{r_f}\vec{E}\cdot d\vec{r}_1$ (26.15) $\Delta V_{AB} = -\int_A^B\left(\frac{\sigma}{2\varepsilon_0}\hat{\imath}\right)\cdot d\vec{y} = -\int_A^B\left(\frac{\sigma}{2\varepsilon_0}dy\right)\hat{\imath}\cdot\hat{\jmath}$
The dot product of the perpendicular vectors $\hat{\imath}\cdot\hat{\jmath} = 0$, so the potential difference between points A and B is zero.	$\Delta V_{AB} = 0$

CHECK and THINK

Our result shows that $V_A = V_B$, so points A and B are on the same equipotential surface. If we choose any two other points along the same vertical line, we find that they are also at the same electric potential. Furthermore, from the symmetry of the situation, if we choose a point directly in front of or behind point A in the z direction perpendicular to the page, that point is also at the same electric potential as point A. So, the equipotential surfaces are sheets parallel to the charged source sheet (Fig. 26.25). In fact, the charged sheet itself is an equipotential surface.

B Use path 2 to show that the potential difference ΔV_{BC} between points B and C is given by

$$\Delta V = -\frac{\sigma}{2\varepsilon_0}\Delta x \qquad (26.16)$$

where $\Delta x = x_C - x_B$ is their separation.

INTERPRET and ANTICIPATE

This is similar to part A. The main difference here is that path 2 is horizontal.

SOLVE

Path 2 from B to C is in the x direction, so path element $d\vec{r}_2$ has only an x component.	$d\vec{r} = d\vec{x} + d\vec{y} + d\vec{z} = d\vec{x} + 0 + 0$ $d\vec{r}_2 = d\vec{x} = dx\hat{\imath}$ (3)
The electric field is still given by Equation (1), so we substitute Equations (1) and (3) into Equation 26.15 (the path integral).	$\Delta V_{BC} = -\int_{r_i}^{r_f} \vec{E}\cdot d\vec{r}_2$ (26.15) $\Delta V_{BC} = -\int_B^C \left(\frac{\sigma}{2\varepsilon_0}\hat{\imath}\right)\cdot d\vec{x}$ $\Delta V_{BC} = -\int_B^C \left(\frac{\sigma}{2\varepsilon_0}dx\right)\hat{\imath}\cdot\hat{\imath}$
The dot product $\hat{\imath}\cdot\hat{\imath} = 1$.	$\Delta V_{BC} = -\int_B^C \frac{\sigma}{2\varepsilon_0}dx$
Pull the constants out and integrate.	$\Delta V_{BC} = -\frac{\sigma}{2\varepsilon_0}\int_B^C dx = -\frac{\sigma}{2\varepsilon_0}x\Big\vert_B^C$ $V_C - V_B = -\frac{\sigma}{2\varepsilon_0}(x_C - x_B)$ ✓ (26.16)

CHECK and THINK

The negative sign in Equation 26.16 means that the electric potential at point B is higher than the electric potential at point C. If point B were closer to the charged sheet, it would be at an even higher electric potential. In fact, the highest electric potential is found on the charged sheet. Writing Equation 26.16 in terms of the electric field ($\Delta V = -E\Delta x$) allows you to find the potential difference due to any source that has planar symmetry. It depends only on the horizontal separation Δx between the points. There is no potential difference between points that are vertically separated.

The Relationship Between Equipotential Surfaces and Electric Field Lines

Let's continue to think about the infinite sheet in Example 26.9. We argued that the equipotential surfaces are parallel sheets and the charged sheet is also an equipotential surface. Figure 26.26 is a contour map for the infinite sheet, including its electric field lines. This figure shows two general relationships between the equipotential surfaces and the electric field lines: (1) *the electric field lines are perpendicular to the equipotential surfaces*, and (2) *the electric field lines point from high electric potential to low electric potential.*

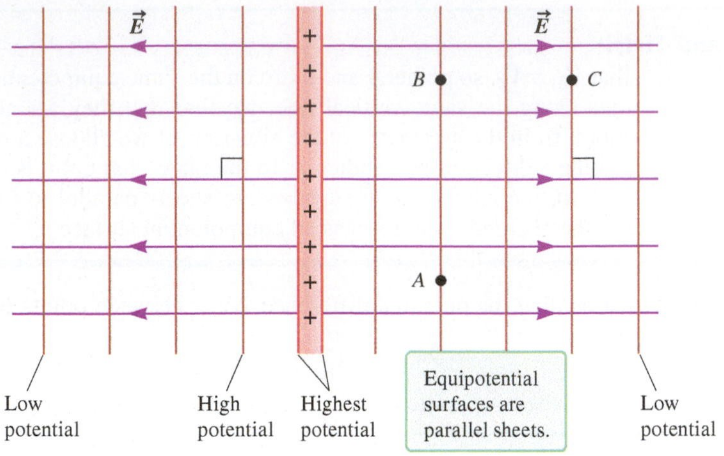

FIGURE 26.26 The equipotential surfaces for an infinite sheet. The electric potential of these parallel sheets depends on their distance from the source. The closer sheets (on both sides of the source) are at higher electric potential than the more distant sheets.

Furthermore, often the surface of the source defines an equipotential surface. (This is always the case for a charged conductor in electrostatic equilibrium.) If the source is positively charged (and an equipotential surface), it is at the highest electric potential. If the source is negatively charged (and an equipotential surface), it is at the lowest electric potential.

Figure 26.27A shows the electric field lines and equipotential surfaces outside a charged spherical source. The equipotential surfaces are concentric spheres perpendicular to the electric field lines. The field lines point outward in the direction of decreasing electric potential. The surface of the positively charged sphere has the highest electric potential. If the source were a particle instead of a sphere, it would still be at the highest electric potential, but it would consist only of a point, not a surface.

Figure 26.27B shows a dipole source that consists of two charged spheres. The equipotential surfaces are perpendicular to the electric field lines. The surfaces look like squashed spheres except for the surface directly between the two charged spheres, which is a flat sheet. The electric field lines point from high electric potential to low electric potential. The surface of the positive sphere is at the highest (positive) electric potential, and the surface of the negative sphere is at the lowest (negative) electric potential. The flat equipotential surface is at zero electric potential.

CONCEPT EXERCISE 26.6

If the contours in Figure 26.26 represent the topography of some landscape, describe that landscape in words and in a sketch.

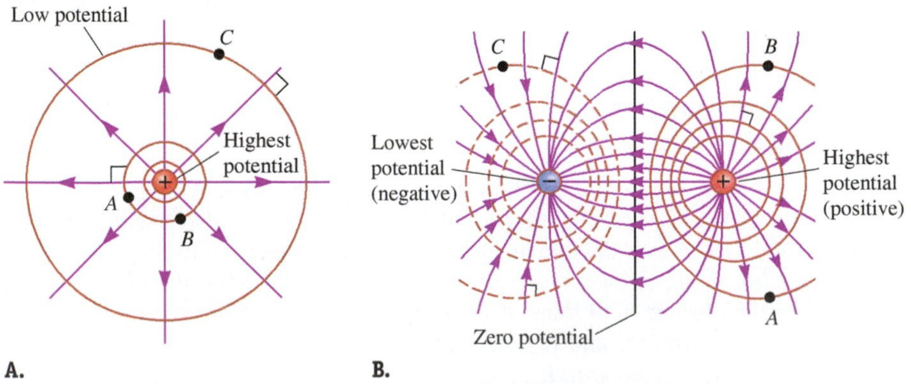

FIGURE 26.27 A. The equipotential surfaces and electric field lines for a charged sphere. **B.** The equipotential surfaces and electric field lines for a dipole. In both cases, the electric field lines are perpendicular to the surfaces and point from high electric potential to low electric potential.

DERIVATION Electric Potential Difference Is Path-Independent

For all three sources shown in Figures 26.26 and 26.27, points A and B are at the same electric potential and point C is at a lower electric potential. When we calculate the potential difference, the particular path we use does not matter; only the endpoints are important. To illustrate this, we return to the infinite charged sheet in Figure 26.25 and calculate the potential difference between points A and C using two different paths. The first path is L-shaped, made up of path 1 from A to B and path 2 from B to C. The other path is the diagonal line labeled path 3 in Figure 26.25. We will calculate ΔV_{AC} using the L-shaped path and path 3, and show that the two results are the same.

Using the L-shaped path is fairly easy because we have already calculated the potential difference along paths 1 and 2. All we need to do is add our results.	$\Delta V_{AC} = \Delta V_{AB} + \Delta V_{BC}$
We found that the potential difference between points A and B is zero ($\Delta V_{AB} = 0$), so the potential difference between A and C equals the potential difference between B and C.	$\Delta V_{AC} = 0 + \Delta V_{BC}$ $\Delta V_{AC} = \Delta V_{BC} \quad (1)$
Now we use path 3 from A to C. The path is in the xy plane, so path element $d\vec{r}_3$ has no z component.	$d\vec{r} = d\vec{x} + d\vec{y} + d\vec{z}$ $d\vec{r}_3 = d\vec{x} + d\vec{y} + 0 = dx\hat{\imath} + dy\hat{\jmath}$
As before, the electric potential difference is found by integrating Equation 26.15. The electric field is in the positive x direction, and its magnitude comes from $E = \sigma/2\varepsilon_0$ (Eq. 25.16).	$\Delta V_{AC} = -\int_{r_i}^{r_f} \vec{E} \cdot d\vec{r}_3 \quad (26.15)$ $\Delta V_{AC} = -\int_A^C \left(\frac{\sigma}{2\varepsilon_0}\hat{\imath}\right) \cdot (d\vec{x} + d\vec{y})$ $\Delta V_{AC} = -\int_A^C \left[\left(\frac{\sigma}{2\varepsilon_0}dx\right)\hat{\imath}\cdot\hat{\imath} + \left(\frac{\sigma}{2\varepsilon_0}dy\right)\hat{\imath}\cdot\hat{\jmath}\right]$
The dot products $\hat{\imath}\cdot\hat{\imath} = 1$ and $\hat{\imath}\cdot\hat{\jmath} = 0$.	$\Delta V_{AC} = -\int_B^C \frac{\sigma}{2\varepsilon_0}dx$
This is exactly the same integral we solved in going along path 2 from B to C. Because point A is directly below point B, they have the same x coordinate, $x_A = x_B$. We just found Equation (1) using path 3. Thus, the potential difference between points A and C is path-independent.	$\Delta V_{AC} = -\frac{\sigma}{2\varepsilon_0}\int_A^C dx = -\frac{\sigma}{2\varepsilon_0}x\Big\vert_A^C$ $V_C - V_A = -\frac{\sigma}{2\varepsilon_0}(x_C - x_A)$ $V_C - V_A = -\frac{\sigma}{2\varepsilon_0}(x_C - x_B)$ $V_C - V_A = V_C - V_B$ $\Delta V_{AC} = \Delta V_{BC}$ ✓

COMMENTS

We showed path independence for this particular case, but our result is generally true. The *electric potential difference between two points is path-independent*—only the location of the endpoints matters. We have seen this sort of path independence before: In Section 8-3, we showed that the change in the potential energy depends only on the final and initial configurations of the system, no matter what configurations the system takes in between.

EXAMPLE 26.10 A Sheet of Charge

Figure 26.28 shows a large charged sheet and four equipotential surfaces (A–D) above it. The equipotential surfaces are equally spaced so that the distance between adjacent surfaces, including the charged sheet, is 0.25 m. The reference point has been chosen such that the charged sheet is at +16.0 V. The electric potentials at the other surfaces are shown in the figure.

A Add electric field lines to the contour map in Figure 26.28, and then find the sign of the excess charge and the charge density of the sheet.

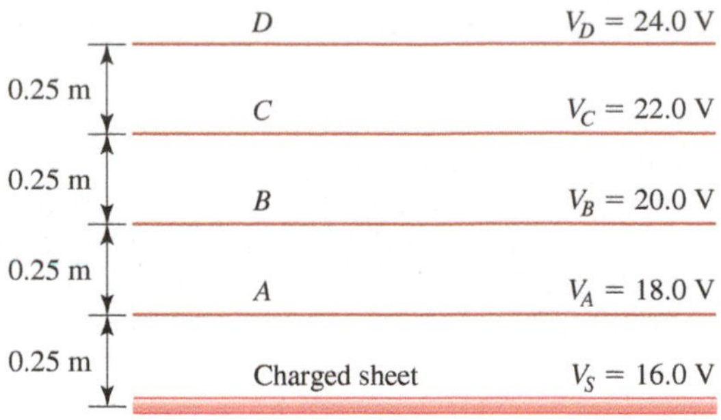

FIGURE 26.28

INTERPRET and ANTICIPATE

Electric field lines must be perpendicular to equipotential surfaces, so the field lines must be vertical. Electric field lines point from high electric potential to low electric potential. In this case, the equipotential surface at the top of the figure is at a high electric potential and the charged sheet is at the lowest electric potential, so the electric field lines must point downward (Fig. 26.29). Electric field lines terminate on negative charges, so we expect that the sheet must have an excess negative charge. In the figure, we have included an upward-pointing y axis.

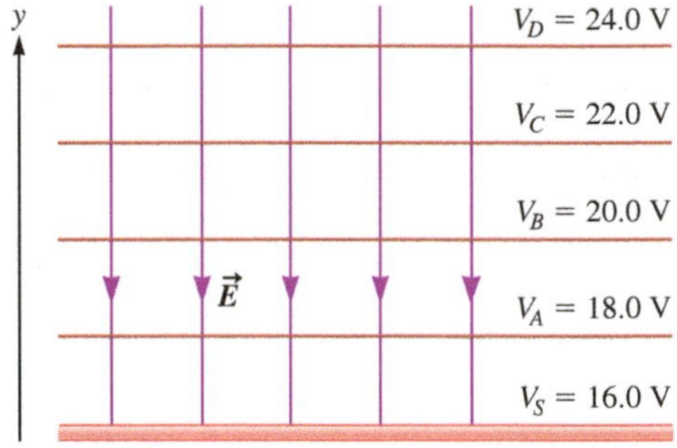

FIGURE 26.29 Electric field lines for the sheet in Figure 26.28.

SOLVE

The surface charge density is found from the potential difference between any two equipotential surfaces (Eq. 26.16). Let's work this twice, first using surfaces A and B.

$$V_B - V_A = -\frac{\sigma}{2\varepsilon_0}(y_B - y_A) \qquad (26.16)$$

$$\sigma = -2\varepsilon_0 \frac{V_B - V_A}{y_B - y_A}$$

$$\sigma = -2(8.85 \times 10^{-12}\ \text{C}^2/\text{N}\cdot\text{m}^2)\frac{(20.0\ \text{V} - 18.0\ \text{V})}{(0.50\ \text{m} - 0.25\ \text{m})}$$

$$\sigma = -1.42 \times 10^{-10}\ \text{C/m}^2 = -0.142\ \text{nC/m}^2$$

CHECK and THINK

As expected, the excess charge is negative. Now let's work the problem again using the charged sheet and the equipotential surface D to check our result. We find the same answer no matter which surfaces we choose to work with.

$$\sigma = -2\varepsilon_0 \frac{V_D - V_S}{y_D - y_S}$$

$$\sigma = -2(8.85 \times 10^{-12}\ \text{C}^2/\text{N}\cdot\text{m}^2)\frac{(24.0\ \text{V} - 16.0\ \text{V})}{(1.00\ \text{m} - 0\ \text{m})}$$

$$\sigma = -1.42 \times 10^{-10}\ \text{C/m}^2$$

$$\sigma = -0.142\ \text{nC/m}^2 \checkmark$$

B Imagine a subject with excess charge $q = +0.50$ C moving from equipotential surface A to surface C (Fig. 26.28). If the system consists of the charged sheet and the subject, what is the change in the system's electric potential energy?

INTERPRET and ANTICIPATE

The positively charged subject is attracted to the negatively charged source, so an external force must do work to lift the subject. The electric potential energy of the subject–sheet system must increase. The situation is analogous to an object near the Earth's surface being lifted against gravity; in that case, the system's gravitational potential energy increases.

SOLVE

The change in the system's electric potential energy is found by multiplying the subject's charge by the difference in the potential between the two points (Eq. 26.7).

$$\Delta U_E = q\Delta V = q(V_C - V_A) \quad (26.7)$$

$$\Delta U_E = (0.50\,\text{C})(22.0\,\text{V} - 18.0\,\text{V}) = 2.0\,\text{J}$$

CHECK and THINK

As expected, the system gains potential energy (+2.0 J) as the positively charged subject is moved away from the negatively charged sheet.

26-8 Finding $\vec{E}$ from V

In the preceding section, we integrated the electric field $\vec{E}$ of a source to find the electric potential V. In this section, we reverse that process by starting with the electric *potential* and finding the electric *field*. We expect this reverse process to involve derivatives, but before we start taking derivatives, let's consider a very simple case that doesn't require calculus.

Figure 26.30A shows a contour map that consists of equally spaced planes (Δx is the same for each pair of surfaces). The potential difference ΔV between the contours is uniform. Because the geometry of this situation is simple, it is easy to find the electric field. First, we draw the electric field lines (Fig. 26.30B) using the two rules on page 809 (1) electric field lines are perpendicular to equipotential surfaces, and (2) electric field lines point from high electric potential to low electric potential. We are interested in finding the electric field $\vec{E}$, so we combine $E = \sigma/2\varepsilon_0$ and $\Delta V = -(\sigma/2\varepsilon_0)\Delta x$ (Eqs. 25.16 and 26.16) to write E in terms of ΔV:

$$\Delta V = -E\Delta x \quad (26.17)$$

$$E = -\frac{\Delta V}{\Delta x} \quad (26.18)$$

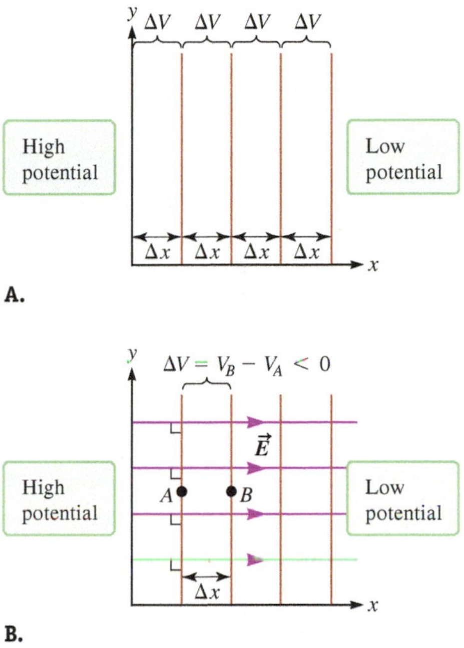

FIGURE 26.30 **A.** The equipotential surfaces are uniform. **B.** Electric field lines are perpendicular to equipotential surfaces.

The electric field is a vector. In this case, $\vec{E}$ points in the x direction, so we write it as

$$\vec{E} = -\frac{\Delta V}{\Delta x}\hat{\imath} \quad (26.19)$$

Why does Equation 26.19 have a negative sign when the electric field points in the positive x direction? Let's take a closer look. The potential difference between the equipotential surfaces is uniform, so let's arbitrarily consider contours through points A and B (Fig. 26.30B), for which the potential difference is

$$\Delta V = V_B - V_A$$

Contour A is at a higher electric potential than B ($V_A > V_B$), so their potential difference is negative:

$$\Delta V = V_B - V_A < 0$$

Because the potential difference is negative, the negative sign in Equation 26.19 ensures that $\vec{E}$ points in the correct direction—in this case, the positive x direction.

Equation 26.19 gives us another set of SI units for the electric field. Potential difference is measured in volts and distance is measured in meters, so the electric field may be measured in either volts per meter or newtons per coulomb:

$$1\,\text{N/C} = 1\,\text{V/m}$$

CASE STUDY $\vec{E}$ and V in the Human Heart

In the case of a uniform electric field—the sort of field you expect from a very large charged sheet—we can use Equation 26.19 to find the electric field from the electric potential. What do we do when the source produces a more complicated electric field, such as in the human chest?

Your heart's rhythm is controlled by electrical signals that can be monitored with an electrocardiogram (EKG), a recording of the electric potential on the chest. For an instant during each heartbeat, the equipotential surfaces look something like Figure 26.31A. We can use the electric potential surfaces on the chest to illustrate how to find the electric field.

Consider a small region near the center of the pattern (Fig. 26.31, inset) where the contours look equally spaced and parallel—like a rotated version of Figure 26.30B. We can always focus on a small enough region so that the equipotential surfaces are well modeled as parallel sheets, and we can draw the electric field lines in this tiny region so that they are (1) perpendicular to the equipotential surfaces and (2) pointing from high electric potential to low electric potential (Fig. 26.31B).

We have chosen a coordinate system such that the x axis is aligned with the electric field. The separation dx between the equipotential sheets is uniform and infinitesimal. The difference in the potential dV between adjacent contours is uniform and small. Equation 26.19 in this case becomes a derivative:

$$\vec{E} = -\frac{dV}{dx}\hat{\imath} \tag{26.20}$$

Equation 26.20 is all you need to find the electric field if the electric potential changes along only one dimension. However, Figure 26.31A shows that the electric potential over the whole chest region is more complicated. Let's use this contour map of the whole chest to see how we can extend Equation 26.20 to an expression that applies even in a more complicated situation. This time, we model the heart as a dipole (Fig. 26.32) in order to better take into account the complicated contours. Consider a region on the side near the person's left arm. The inset in Figure 26.32 shows a close-up of this region along with the same coordinate system we used in Figure 26.31B. The electric field in this region is nearly straight up, so $\vec{E}$ has both x and y components: $\vec{E} = E_x\hat{\imath} + E_y\hat{\jmath}$. The field components are found by taking the derivative of the electric potential in each direction separately:

$$E_x = -\frac{\partial V}{\partial x}, \quad E_y = -\frac{\partial V}{\partial y}, \quad E_z = -\frac{\partial V}{\partial z} \tag{26.21}$$

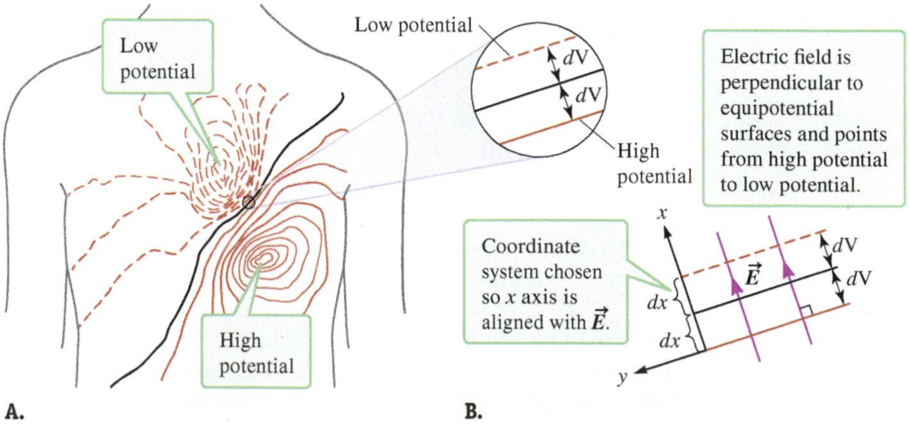

FIGURE 26.31 **A.** Equipotential surfaces on the human chest. *Inset*: Equipotential surfaces for a small region on the chest. **B.** Electric field lines for the small region in the inset.

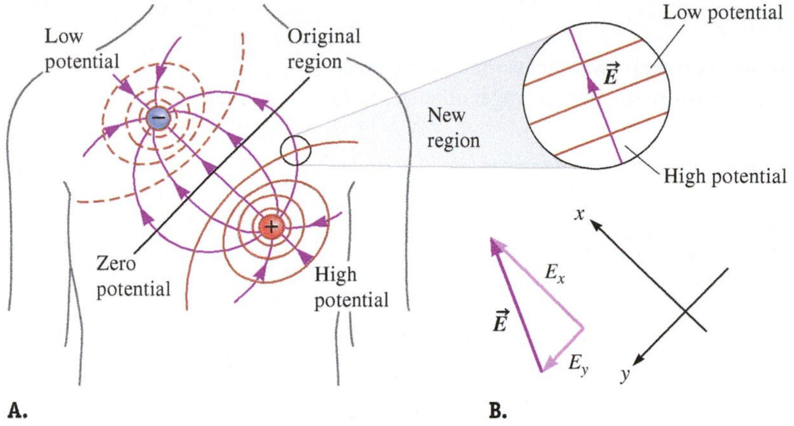

FIGURE 26.32 Modeling the human heart as a dipole electric source.

We have included the z component for completeness, although it is not shown in Figure 26.32.

The symbol ∂ indicates a **partial derivative** (Section 17-10, page 514). In a partial derivative, all variables are held fixed except the one you are differentiating with respect to. For example, if V is a function of x and y as in $V(x, y) = (3x - 6y)$ V, the electric field found by taking the partial derivatives is

$$E_x = -\frac{\partial V}{\partial x} = -\frac{\partial(3x - 6y)}{\partial x} = -3 \text{ V/m}$$

$$E_y = -\frac{\partial V}{\partial y} = -\frac{\partial(3x - 6y)}{\partial y} = 6 \text{ V/m}$$

$$\vec{E} = (-3\hat{\imath} + 6\hat{\jmath}) \text{ V/m}$$

Equations 26.21 are valid for Cartesian coordinates. When the source has spherical or linear symmetry, you may wish to use polar coordinates so that the radial component of the electric field is given by

FINDING THE ELECTRIC FIELD $\vec{E}$ FROM THE ELECTRIC POTENTIAL V

Major Concept

$$E_r = -\frac{\partial V}{\partial r} \tag{26.22}$$

(We don't need to worry about the azimuthal θ component in this book.)

EXAMPLE 26.11 CASE STUDY The Electric Field Inside You

Measurements of the electric potential at several sites on the skin allow doctors to monitor important internal activities. For example, electroencephalograms (EEGs) involve recordings from 10 to 20 electrodes placed on a person's head and can aid in diagnosing brain diseases such as epilepsy (Fig. 26.33). Use the data in Table 26.1 to estimate the magnitude of the electric field in a few body regions.

TABLE 26.1

Location and activity	Approximate potential difference	Approximate separation between electrodes
Head; resting	50 μV	10 cm
Forearm; stimulation	0.7 mV	3 cm
Center of chest; light activity	1 mV	4 cm

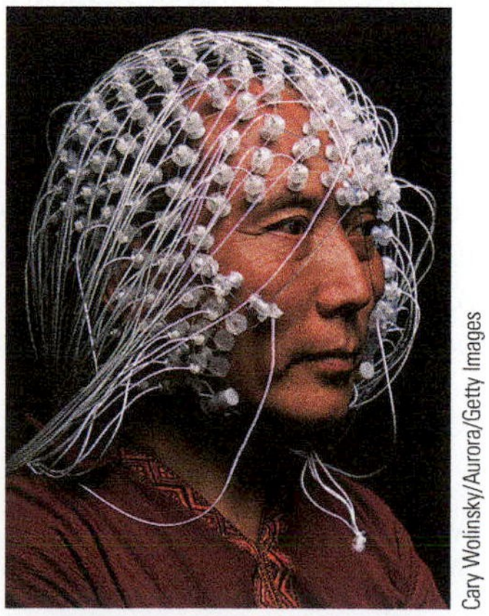

FIGURE 26.33 Electrodes placed on a person's head measure the electric potential on the scalp at each location.

Example continues on page 816 ▶

INTERPRET and ANTICIPATE

Because we are given only the potential difference between two electrodes, we cannot calculate the derivative, but we can use the data to estimate the average electric field between the electrodes in each case.

SOLVE

In this case, the absolute value of Equation 26.18 gives the average electric field.	$E_{av} = \left\| -\frac{\Delta V}{\Delta x} \right\|$ (26.18)
Head:	$E_{av} \approx \frac{50 \times 10^{-6}\,\text{V}}{0.1\,\text{m}} \approx 5 \times 10^{-4}\,\text{V/m}$ $E_{av} \approx 0.5\,\text{mV/m}$ (head)
Forearm:	$E_{av} \approx \frac{0.7 \times 10^{-3}\,\text{V}}{0.03\,\text{m}} \approx 2 \times 10^{-2}\,\text{V/m}$ $E_{av} \approx 20\,\text{mV/m}$ (forearm)
Chest:	$E_{av} \approx \frac{1 \times 10^{-3}\,\text{V}}{0.04\,\text{m}} \approx 2.5 \times 10^{-2}\,\text{V/m}$ $E_{av} \approx 30\,\text{mV/m}$ (chest)

CHECK and THINK

These electric fields are weak compared to those due to many other common sources. For example, near a charged comb, the electric field is about 100,000 times stronger. Of course, the values we calculated are for measurements made on the skin. Inside the body, the electric field is considerably stronger.

Although the electric field on the skin is very weak, it could possibly be used for networking! In spring 2005, a Japanese company developed a technology that can send data over the surface of the skin at a typical broadband connection speed.[1] The hope is to use this technology to connect headphones in your ears to a phone in your pocket with no wire or infrared connection.

EXAMPLE 26.12 Linear Symmetry

Figure 26.34 shows a positively charged, infinitely long rod of radius R and linear charge density λ. The equipotential surfaces are infinitely long, concentric cylinders. The rod itself is an equipotential surface and is chosen as the reference so that its potential is zero. The electric potential on all the other surfaces is then negative and given by

$$V(r) = \frac{\lambda}{2\pi\varepsilon_0} \ln \frac{R}{r} \tag{26.23}$$

where $r > R$. Using Equation 26.23, find an expression for the electric field due to this source.

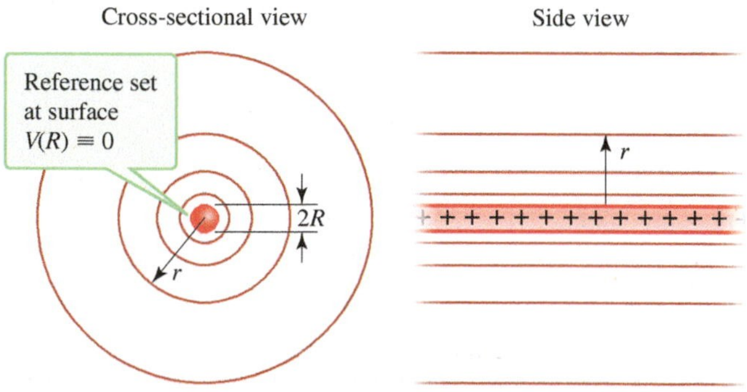

FIGURE 26.34 Two views of a long, positively charged rod.

[1]See *The Guardian*, March 20, 2005, p. 12.

INTERPRET and ANTICIPATE

We add electric field lines to Figure 26.34, perpendicular to the equipotential surfaces and pointing from high electric potential to low electric potential. In this case, the electric field lines point outward (Fig. 26.35). This situation has linear symmetry, and in Example 25.4 (page 766) we used Gauss's law to find the electric field due to such a source. So, we expect to find that the electric field is given by Equation 25.13, $\vec{E} = (\lambda/2\pi\varepsilon_0 r)\hat{r}$.

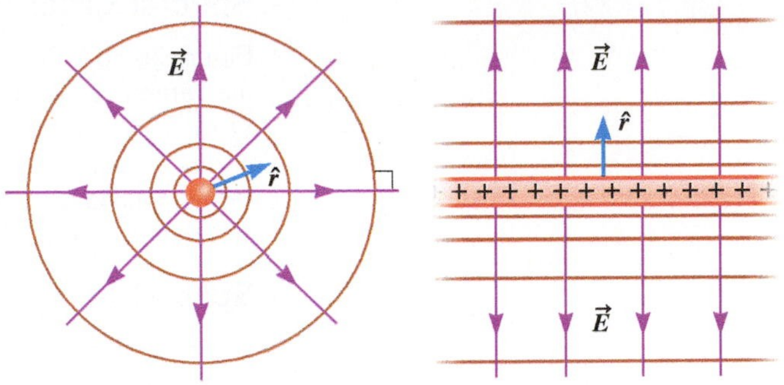

FIGURE 26.35 Electric field lines and unit vectors for the rod.

SOLVE

Because of the symmetry in this problem, we use a polar coordinate system (Fig. 26.35) with the unit vector $\hat{r}$ pointing outward from the center of the rod. Use Equation 26.22 to find the electric field, substituting Equation 26.23 for V.

$$E_r = -\frac{\partial V}{\partial r} \quad (26.22)$$

$$E_r = -\frac{\partial}{\partial r}\left(\frac{\lambda}{2\pi\varepsilon_0}\ln\frac{R}{r}\right) = -\frac{\lambda}{2\pi\varepsilon_0}\frac{\partial}{\partial r}\left(\ln\frac{R}{r}\right)$$

Write the natural logarithm of a fraction as two terms.

$$E_r = -\frac{\lambda}{2\pi\varepsilon_0}\frac{\partial}{\partial r}(\ln R - \ln r)$$

$$E_r = -\frac{\lambda}{2\pi\varepsilon_0}\left(\frac{\partial}{\partial r}\ln R - \frac{\partial}{\partial r}\ln r\right)$$

The radius R of the rod is a constant, so the first derivative is zero.

$$E_r = -\frac{\lambda}{2\pi\varepsilon_0}\left(0 - \frac{\partial}{\partial r}\ln r\right) = \frac{\lambda}{2\pi\varepsilon_0}\frac{\partial}{\partial r}\ln r$$

The derivative of the natural logarithm can be found in Appendix A.

$$E_r = \frac{\lambda}{2\pi\varepsilon_0}\frac{1}{r}$$

Finally, write the electric field in component form.

$$\vec{E} = \frac{1}{2\pi\varepsilon_0}\frac{\lambda}{r}\hat{r}$$

CHECK and THINK

As expected, this is exactly what we found using Gauss's law. Equation 26.23 is the electric potential due to a source with linear symmetry. Nerves and muscles in the body are often modeled as such sources.

26-9 Graphing E and V

We have studied sources of various geometries and compiled formulas for the electric field $\vec{E}$ and the electric potential V in numerous special cases. One way to organize all this information is by sketching graphs.

Because the electric potential is found by taking a path integral of the electric field, it is helpful to make a pair of graphs for each source to display the connection between V and E. The two graphs are V versus position and E versus position (usually r or x). Because the electric field is found by taking the (negative) spatial derivative of the electric potential, each point on the E-versus-position graph is the negative of the slope of the tangent to the V-versus-position graph. In this section, we will show such paired graphs for three charged sources: a particle, two infinite sheets, and an isolated conductor.

Special Case: Particle Source

Figure 26.36A shows a positively charged particle as the source. As usual for a particle, the reference point has been chosen at infinity where the electric potential is zero: $V(\infty) = 0$. The electric potential is high (positive) near the particle. The electric field points outward in all directions. Each point on the E-versus-r graph is -1 times the slope of the tangent line at the corresponding point on the V-versus-r plot (Fig. 26.36B), and the electric field follows an inverse-square law $(1/r^2)$ (Fig. 26.36C).

Special Case: Two Infinite Charged Sheets

Figure 26.37A shows a source that consists of two infinite parallel sheets. We have chosen a coordinate system with the x axis pointing to the right and the origin on the left sheet. The sheet on the left has excess negative charge, and the sheet on the right has excess positive charge. The charge on each sheet is uniformly distributed, and the surface charge densities are equal in magnitude. In Figure 25.35 (page 777), we found that the electric field is zero outside the sheets and uniform between them, pointing from the positive sheet to the negative sheet.

The electric field points from high to low electric potential, so the negative sheet is at a lower electric potential than the positive sheet. It is convenient to choose our reference for the electric potential such that the surface of the negative sheet is at $V = 0$. The electric potential V increases uniformly from the negative sheet to the positive sheet (Fig. 26.37B). For all positions to the left of the sheets, the electric potential is zero, and for all positions to the right of the sheets, the electric potential is a constant equal to the electric potential of the positive sheet.

Each point on the graph of E_x versus x is the negative of the slope of the V-versus-x graph (Fig. 26.37C). For the regions outside the sheets, $E = 0$ whenever V is constant, whether $V = 0$ or some other constant.

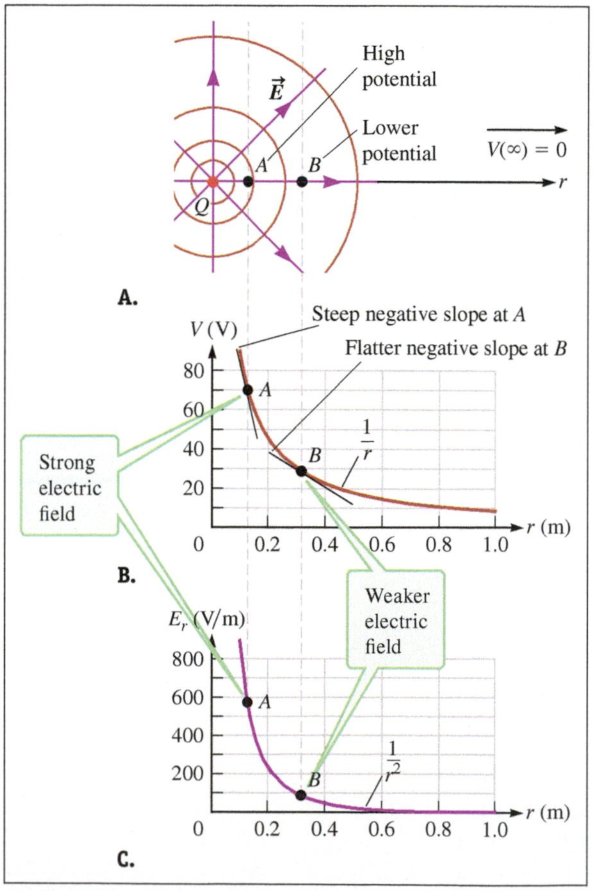

FIGURE 26.36 Special case of point particle source. **A.** Its electric field lines and equipotential surfaces. **B.** V versus r and **C.** E versus r for this source found from $E_r = -\partial V/\partial r$.

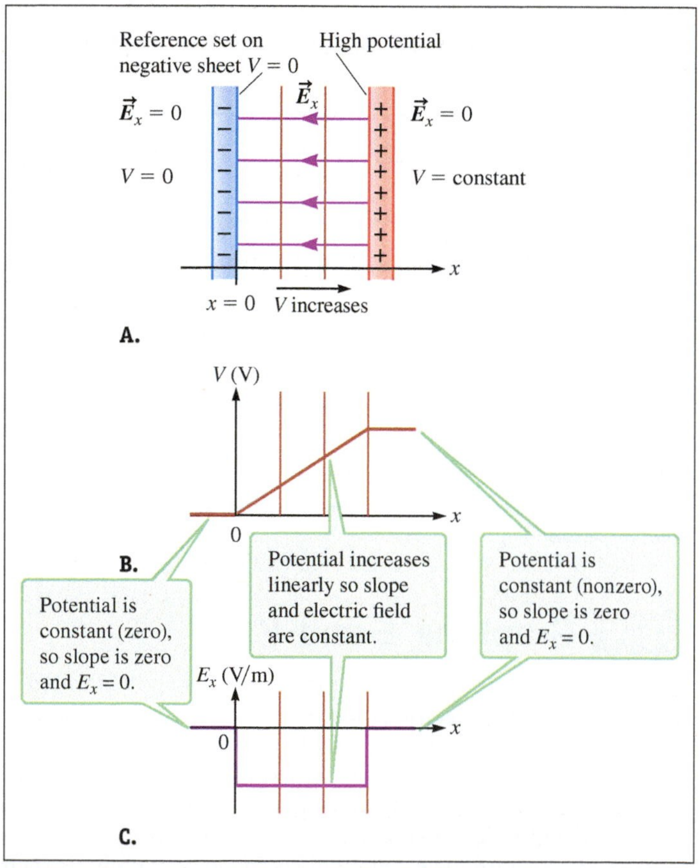

FIGURE 26.37 Special case of two infinite sheets. **A.** The negative sheet is on the left and the positive sheet is on the right, so the electric field points to the left. The electric potential on the right is higher than that on the left. **B.** V versus x and **C.** E versus x for this source. E is negative because it points in the negative x direction.

Special Case: Isolated Charged Conductor

Conductors play important practical roles in our use of electricity. So our last special case about an arbitrarily shaped charged conductor will help us in the next three chapters when we study the practical uses of electricity.

For now, we are concerned only with a conductor in electrostatic equilibrium. Imagine charging a conductor by touching it with a charged rod. At first, the conductor is not in equilibrium. The charged particles are free to travel throughout the conductor, and they do until (1) the excess charge rests only on the surface, (2) the electric field inside the body of the conductor is zero, and (3) the electric potential is uniform over the surface. Because the electric field inside the conductor is zero, the electric potential inside the body of the conductor is constant, although not necessarily zero. In fact, the electric potential in the body of the conductor is equal to the electric potential V_S at its surface. Think of it this way: If the electric field was not zero inside the conductor, or if the electric potential V_S was not the same all over the conductor, charged particles would move and the conductor would no longer be in electrostatic equilibrium.

Figure 26.38A shows an irregularly shaped, positively charged conductor with an r axis that runs from the interior of the conductor to the right. The electric field is zero in the body of the conductor. Just outside the conductor, the electric field is uniform ($E = \sigma/\varepsilon_0$, Eq. 25.17). Far from the conductor, the electric field looks like the electric field due to a positively charged particle ($E = kQ/r^2$).

Because the electric field far from a charged conductor of any shape looks like the electric field due to a charged particle, we typically set the reference at infinity where the electric potential is zero. With this convention, the electric potential V_S at the surface of the conductor is positive if the conductor is positively charged and negative if the conductor is negatively charged.

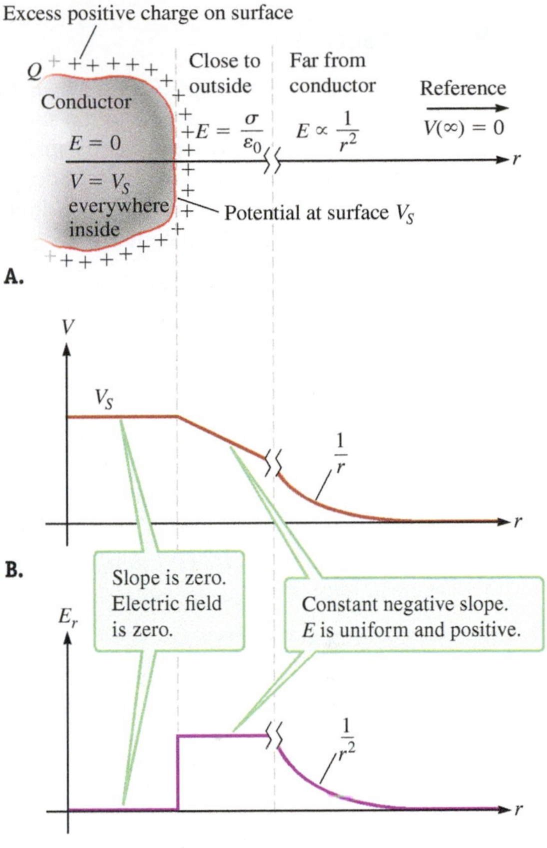

FIGURE 26.38 Special case of an isolated conductor. **A.** The conductor is positively charged and irregularly shaped. **B.** V versus r and **C.** E versus r for this source.

Now look at the electric potential as we work our way from $r = 0$ outward (Fig. 26.38B). Inside the body of the conductor, the electric potential is a constant, V_S. Outside the conductor but close to its surface, the electric potential drops off linearly. Finally, far from the conductor, the electric potential drops off as $1/r$ as it does when the source is a particle.

As usual, the graph of E_r versus r is the negative of the slope of the V-versus-r graph. Note that this graph (Fig. 26.38C) is consistent with the sketch (Fig. 26.38A).

CONCEPT EXERCISE 26.7

CASE STUDY EKG

One way for doctors to monitor a patient's heart is with an EKG. A potential difference is measured across two (or more) points on the patient's chest. The EKG is a graph of electric potential versus time (Fig 26.39). Can you use an EKG to plot electric field versus position? If so, explain how to find that graph. If not, explain why not.

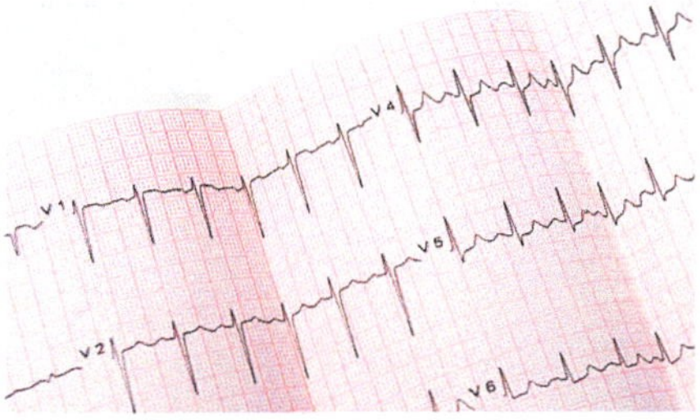

Lisovskaya Natalia/Shutterstock.com

FIGURE 26.39 In an EKG, electric potential as a function of time is recorded at several different points.

SUMMARY

Underlying Principles

No new principles are introduced in this chapter.

Major Concepts

1. **Electric potential energy U_E** is a scalar that depends on both the source's charge and the subject's charge. Only changes in U_E are physically important.
2. **Electric potential V_E** is a scalar that depends only on the source's charge. Find V_E by dividing the electric potential energy U_E by the charge q of a test subject:

$$V_E = \frac{U_E}{q} \tag{26.6}$$

Only differences in V_E are physically important. Electric potential differences depend only on the endpoints, not on the particular path between them.

3. **Finding the electric potential from the electric field.** The electric potential difference ΔV between two points can be found by taking the path integral of the electric field $\vec{E}$ between those points:

$$\Delta V = -\int_{r_i}^{r_f} \vec{E} \cdot d\vec{r} \tag{26.15}$$

4. **Finding the electric field from the electric potential.** The electric field is found by taking the partial derivative of V. In Cartesian coordinates,

$$E_x = -\frac{\partial V}{\partial x}, \quad E_y = -\frac{\partial V}{\partial y}, \quad E_z = -\frac{\partial V}{\partial z} \tag{26.21}$$

In polar coordinates,

$$E_r = -\frac{\partial V}{\partial r} \tag{26.22}$$

Each point on an E-versus-position graph is the negative of the slope of the tangent on the corresponding V-versus-position graph.

Special Cases

1. **Electric potential energy U_E in systems with a spherically symmetrical source** (a particle, outside a sphere or spherical shell where r is greater than the radius of the sphere or spherical shell):

$$U_E(r) = \frac{kQq}{r} \tag{26.3}$$

By the usual convention, $U_E = 0$ when the source and subject are infinitely far apart.

If a system has more than two charged particles, the system's electric potential energy is found by applying Equation 26.3 to each pair and then adding the results.

2. Electric potential V due to:
   a. A source with **spherical symmetry**. For a particle source with excess charge Q,

$$V = k\frac{Q}{r} \tag{26.8}$$

The electric potential due to a collection of n charged particles is

$$V = k\sum_{i=1}^{n}\frac{q_i}{r_i} \tag{26.9}$$

where q_i is the charge of the ith particle and r_i is its distance to the point at which the electric potential is calculated.

   b. **Dipole** source: $V \approx k\dfrac{p\cos\theta}{r^2}$ (26.11)

   c. **Ring** source:

$$V = \frac{kQ}{\sqrt{R^2 + x^2}} = \frac{1}{4\pi\varepsilon_0}\frac{Q}{\sqrt{R^2 + x^2}} \tag{26.13}$$

   d. **Disk** source:

$$V = 2\pi k\sigma[(\sqrt{R^2 + x^2}) - x] = \frac{\sigma}{2\varepsilon_0}[(\sqrt{R^2 + x^2}) - x] \tag{26.14}$$

   e. A source with **planar symmetry:**

$$\Delta V = -E\Delta x \tag{26.17}$$

Tools

A surface on which all the points are at the same electric potential (voltage) is called an **equipotential surface**. A set of equipotential surfaces separated by equal steps in electric potential is known as a **contour map**.

For any source:

1. Electric field lines are perpendicular to equipotential surfaces.
2. Electric field lines point from high electric potential to low electric potential.
3. If the source is a continuous distribution, its surface defines an equipotential surface; if the source is positively charged, the electric potential at the source is highest; if the source is negatively charged, the source is at the lowest electric potential.

PROBLEM-SOLVING STRATEGY Electric Potential for a Continuous Charge Distribution

INTERPRET and ANTICIPATE

Draw a **sketch** of the source and choose a **coordinate system**.

SOLVE

1. Divide the source into small **pieces** and write an **expression for** dV for each piece.
2. **Integrate** over the entire source to find V.

CHECK and THINK

For a finite source:

1. Verify that **at a point far away** compared to the size of the source, the electric potential has the same form as that of a **charged particle**.
2. Verify that **at a point infinitely far away, the electric potential is zero.**

PROBLEMS AND QUESTIONS

A = algebraic C = conceptual E = estimation G = graphical N = numerical

26-1 Scalars Versus Vectors

1. **C** What does it mean when a force is negative? What does it mean when the potential energy is negative?
2. **C Review** Return to Chapter 8 and the potential energy associated with both gravity near the surface of the Earth and universal gravity.
   - **a.** What does the term *reference configuration* mean in the context of gravity?
   - **b.** Suppose a system consists of an apple and the Earth, with the reference configuration set so that the system's gravitational potential energy is zero when the apple is on your desk. If the apple is above the desk, is the potential energy positive, negative, or zero?
   - **c.** What is the conventional reference configuration for universal gravity? Why is that a convenient reference configuration? If a system consists of an apple and the Earth, does the system's gravitational potential energy increase, decrease, or stay the same as you raise the apple upward?

26-2 Gravity Analogy

3. **C Review** A system consists of a planet and a star, with the planet in an elliptical orbit. As the planet orbits, which of these quantities change because they depend on the mass of the planet? Explain your answers.
   - **a.** Gravitational potential energy
   - **b.** Gravitational potential
   - **c.** Kinetic energy
   - **d.** Gravitational force
   - **e.** Gravitational field
4. **C** Try to complete Table P26.4 from memory. If you must look back in this chapter or other chapters for information, note the page number, figure number, or equation number that helped you.

TABLE P26.4

Term	Mathematical symbol	Vector or scalar?	Source and subject or source only?	SI units
Gravitational force				
Gravitational field				
Gravitational potential energy				
Gravitational potential				

5. **C** Try to complete Table P26.5 from memory. If you must look back in the chapter for information, note the page number, figure number, or equation number that helped you.

TABLE P26.5

Term	Mathematical symbol	Vector or scalar?	Source and subject or source only?	SI units
Electrostatic force				
Electric field				
Electric potential energy				
Electric potential				

26-3 Electric Potential Energy U_E

6. **C** Can you associate electric potential energy with an isolated charged particle? Explain.

Problems 7 and 28 are paired.

7. **N** Consider the final arrangement of charged particles shown in Figure P26.7. What is the work necessary to build such an arrangement of particles, assuming they were originally very far from one another?

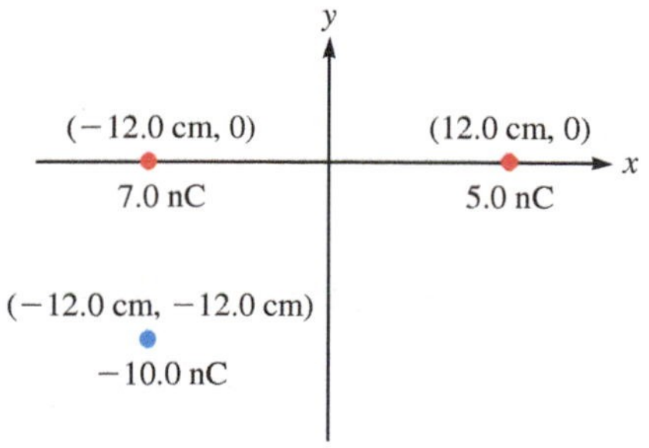

FIGURE P26.7 Problems 7 and 28.

Problems 8 and 9 are paired.

8. **C** Using the usual convention that the electric potential energy is zero when charged particles are infinitely far apart, rank the electric potential energy from least to greatest for the systems shown in Figure P26.8. Explain your answers.

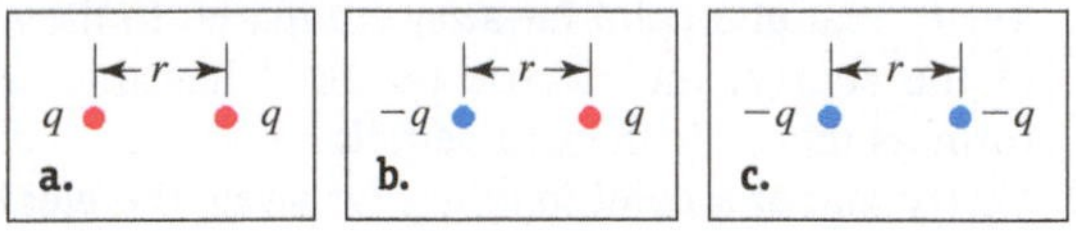

FIGURE P26.8

9. **A** Find an expression for the electric potential energy associated with each system in Figure P26.8 in terms of the quantities provided on the figure.

10. A hydrogen atom consists of an electron and a proton. Model the hydrogen atom as a dipole with separation $d = 10^{-10}$ m.

a. **E** Estimate the electric potential energy of the hydrogen atom.

b. **N** How much work does an external force do in liberating the electron from the atom?

c. **C** If the external force does more than the work you found in part (b), what can you say about the electron's motion when it is very far from the proton?

11. **N** What is the work that a generator must do to move 1.80×10^{10} protons from a location with electric potential 12.4 V to a location with electric potential 43.0 V?

12. **N** How far should a $+3.0$-μC charged particle be from a -5.5-μC charged particle so that the electric potential energy of the pair of particles is -0.90 J?

13. **N** A proton is fired from very far away directly at a fixed particle with charge $q = 1.28 \times 10^{-18}$ C. If the initial speed of the proton is 2.4×10^5 m/s, what is its distance of closest approach to the fixed particle? The mass of a proton is 1.67×10^{-27} kg.

Problems 14, 15, and 16 are grouped.

14. Four charged particles are at rest at the corners of a square (Fig. P26.14). The net charges are $q_1 = q_2 = 2.65$ μC and $q_3 = q_4 = 5.15$ μC. The distance between particle 1 and particle 3 is $r_{13} = 1.75$ cm.

a. **N** What is the electric potential energy of the four-particle system?

b. **C** If the particles are released from rest, what will happen to the system? In particular, what will happen to the system's kinetic energy as their separations become infinite?

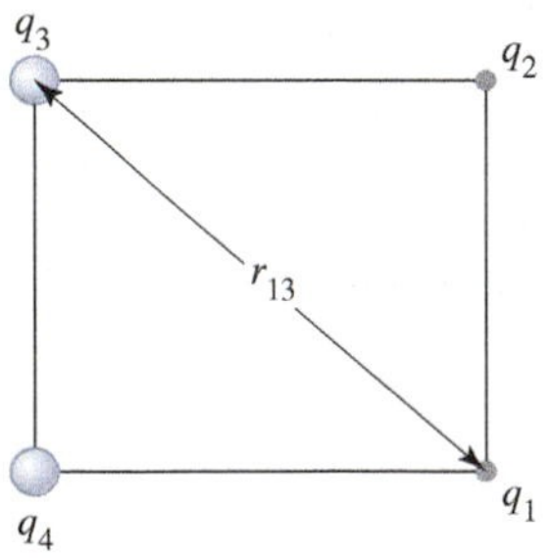

FIGURE P26.14 Problems 14, 15, and 16.

15. Four charged particles are at rest at the corners of a square (Fig. P26.14). The net charges are $q_1 = q_2 = -2.65$ μC and $q_3 = q_4 = -5.15$ μC. The distance between particle 1 and particle 3 is $r_{13} = 1.75$ cm.

a. **N** What is the electric potential energy of the four-particle system?

b. **C** If the particles are released from rest, what will happen to the system? In particular, what will happen to the system's kinetic energy as their separations become infinite?

16. Four charged particles are at rest at the corners of a square (Fig. P26.14). The net charges are $q_1 = q_2 = +2.65$ μC and $q_3 = q_4 = -5.15$ μC. The distance between particle 1 and particle 3 is $r_{13} = 1.75$ cm.

a. **N** What is the electric potential energy of the four-particle system?

b. **C** If the particles are released from rest, what will happen to the system? In particular, what will happen to the system's kinetic energy?

17. **N** Eight identical charged particles with $q = 1.00$ nC are to be arrayed at the vertices of a cube of side $d = 1.00$ cm. What is the work required to assemble the particles into this arrangement?

26-4 Electric Potential *V*

18. **N** A conducting sphere with a radius of 0.25 m has a total charge of 6.00 mC. A particle with a charge of -2.00 mC is initially 0.35 m from the sphere's center and is moved to a final position 0.50 m from the sphere's center.

a. What is the difference in electric potential between the particle's final and initial positions, $\Delta V = V_f - V_i$?

b. What is the change in the system's electric potential energy?

19. **N** The speed of an electron moving along the y axis increases from 4.40×10^6 m/s at $y = 10.0$ cm to 7.00×10^6 m/s at $y = 2.00$ cm.

a. What is the electric potential difference between these two points?

b. Which of the two points is at a higher electric potential?

20. **G** Figure P26.20 is a topographic map.

a. Rank A, B, and C by elevation from the lowest point to the highest point.

b. Rank A, B, and C by slope from the steepest slope to the flattest slope.

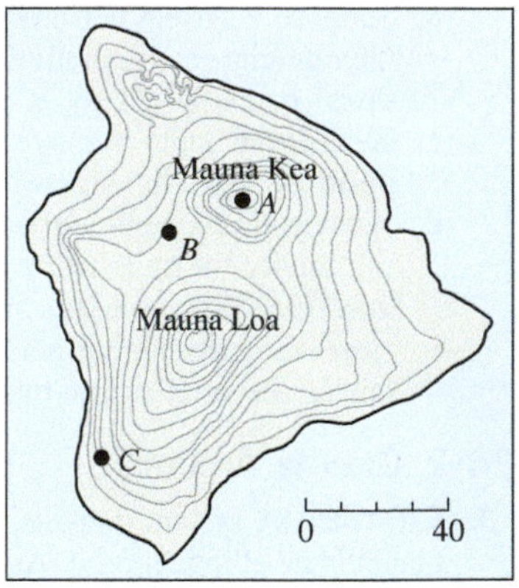

FIGURE P26.20

21. **N** At a point in space, the electric potential due to a charged particle is measured to be 2.55×10^3 V, and the magnitude of the electric field is measured to be 875 V/m.

a. How far is the charged particle from this point?

b. What is the magnitude of the charge on the particle?

22. **C** Explain the difference between $U_E(r) = kQq/r$ and $V_E = kQ/r$ (Eqs. 26.3 and 26.8). Which depends on both the source and the subject? Describe the source in each case. When do you use each equation? Can you use either equation if the source has linear symmetry? Explain.

23. **N** Suppose a single electron moves through an electric potential difference of 1.5 V (the potential difference between the terminals of an AAA battery). What is the change in the system's electric potential energy? Give your answer in eV and in J.

24. Two point charges, $q_1 = -2.0$ μC and $q_2 = 2.0$ μC, are placed on the x axis at $x = 1.0$ m and $x = -1.0$ m, respectively (Fig. P26.24).

a. **N** What are the electric potentials at the points P (0, 1.0 m) and R (2.0 m, 0)?

b. **N** Find the work done in moving a 1.0-μC charge from P to R along a straight line joining the two points.

c. **C** Is there any path along which the work done in moving the charge from P to R is less than the value from part (b)? Explain.

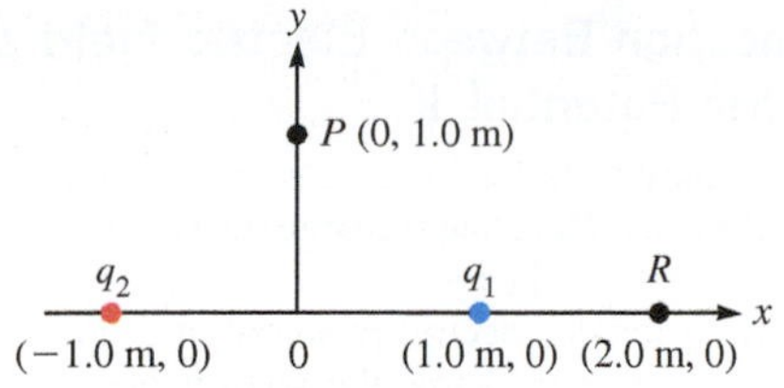

FIGURE P26.24

25. **N** Separating the electron from the proton in a hydrogen atom takes 2.21×10^{-18} J of work. Through what electric potential difference does the electron move?
26. **C** Can a contour map help you visualize the electric potential energy? Explain.

26-5 Special Case: Electric Potential Due to a Collection of Charged Particles

27. **N** A particle with charge $q_A = -6.75\ \mu C$ is located at (0, 3.25 cm), and a second particle with charge $q_B = 3.20\ \mu C$ is located at (0, −2.75 cm). What is the electric potential due to the two charges
    a. at the origin and
    b. at (3.00 cm, 0)?
28. **N** Find the electric potential at the origin given the arrangement of charged particles shown in Figure P26.7.
29. **N** Twenty-seven identical spherical drops of mercury are charged simultaneously to the same electric potential of 10.0 V. What will the electric potential be if all the charged drops are combined to form one large spherical drop?
30. Figure P26.30 shows a source that consists of two negatively charged particles, one with charge $-q$ and the other with charge $-2q$.
    a. C Assuming the usual convention, where is the electric potential zero?
    b. C Is the electric potential higher at point A or at point B? Explain.
    c. N Find the electric potential at point A and at point B.

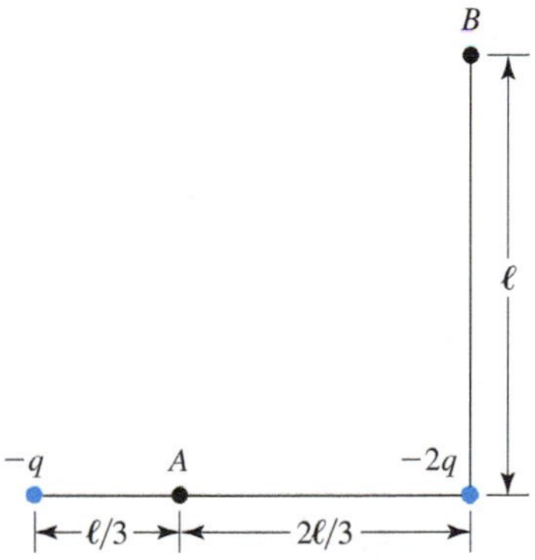

FIGURE P26.30

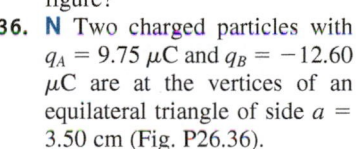

31. **N** A spherical conductor of radius 15.0 cm has net charge $Q = 0.72\ \mu C$. Calculate the electric potential at the following distances from its center, assuming the electric potential goes to zero at an infinite distance from the sphere:
    a. 30.0 cm,
    b. 15.0 cm, and
    c. 4.0 cm.

Problems 32 and 33 are paired.

32. **A** A source consists of three charged particles located at the vertices of a square (Fig. P26.32). Find an expression for the electric potential at point A located at the fourth vertex.
33. **N** A source consists of three charged particles located at the vertices of a square (Fig. P26.32), where the square has sides of length 0.243 m. The charges are $q_1 = 35.0$ nC, $q_2 = -65.0$ nC, and $q_3 = 56.5$ nC. Find the electric potential at point A located at the fourth vertex.

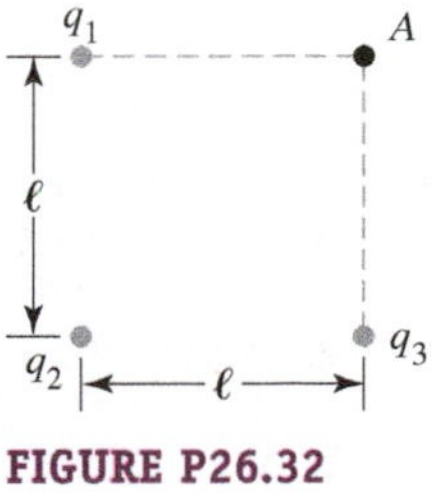

FIGURE P26.32
Problems 32 and 33.

34. **N** Two identical metal balls of radii 2.50 cm are at a center-to-center distance of 1.00 m from each other (Fig. P26.34). Each ball is charged so that a point at the surface of the first ball has an electric potential of $+1.20 \times 10^3$ V and a point at the surface of the other ball has an electric potential of -1.20×10^3 V. What is the total charge on each ball?

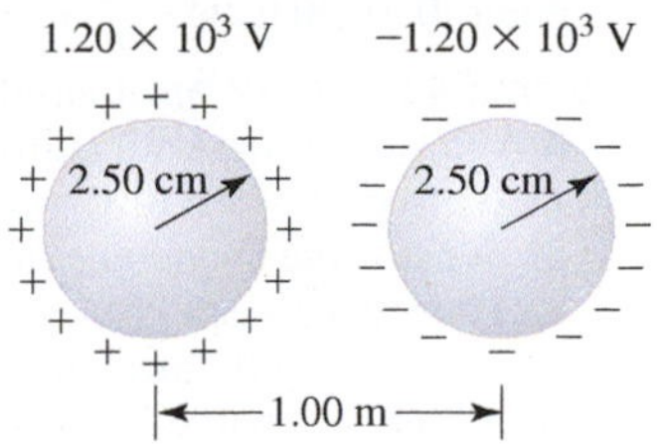

FIGURE P26.34

35. **N** Figure P26.35 shows four particles with identical charges of $+5.75\ \mu C$ arrayed at the vertices of a rectangle of width 25.0 cm and height 55.0 cm. What is the change in the electric potential energy of this system if particles A, B, and C are held in place and particle D is brought from infinity to the position shown in the figure?

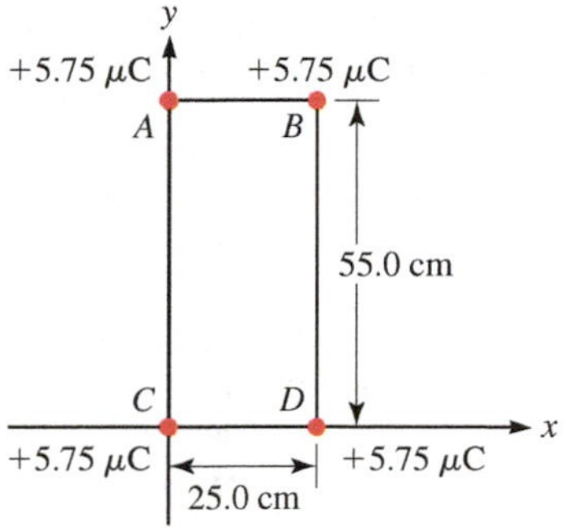

FIGURE P26.35

36. **N** Two charged particles with $q_A = 9.75\ \mu C$ and $q_B = -12.60\ \mu C$ are at the vertices of an equilateral triangle of side $a = 3.50$ cm (Fig. P26.36).
    a. What is the electric potential due to these charges at point A, halfway between the charges?
    b. What is the electric potential due to these charges at point P, the apex of the triangle?

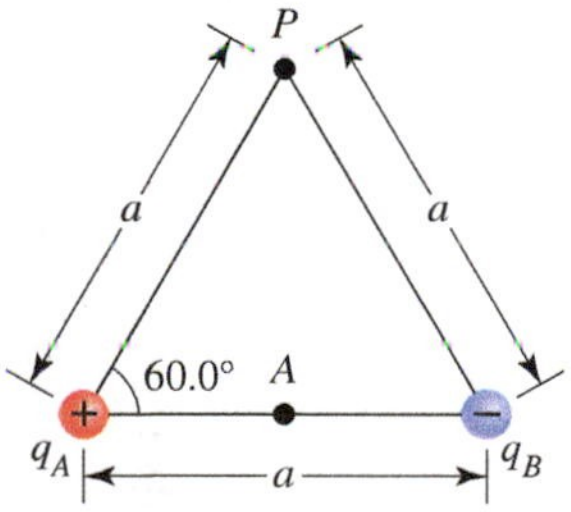

FIGURE P26.36

37. **N** Two charged particles with $q_1 = 5.00\ \mu C$ and $q_2 = -3.00\ \mu C$ are placed at two vertices of an equilateral tetrahedron whose edges all have length $s = 4.20$ m (Fig. P26.37). Determine what charge q_3 should be placed at the third vertex so that the total electric potential at the fourth vertex is 2.00 kV.

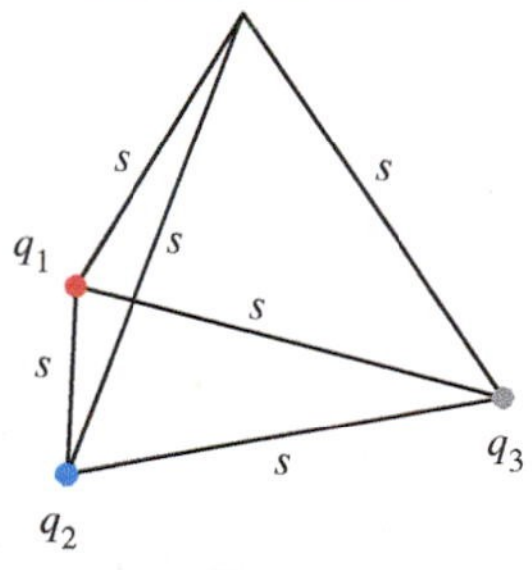

FIGURE P26.37

26-6 Electric Potential Due to a Continuous Distribution

38. Consider the charged ring in Figure 26.20 (page 804).
    a. G Sketch V versus x.
    b. C Where is the electric potential greatest?
    c. C Compare a sketch of V versus x for a small ring to that for a large ring. What differences are there in the graphs? Does the point at which the electric potential maximum is found depend on the size of the ring? Explain.
39. **N** A ring of radius 0.75 m has an excess charge of $-892\ \mu C$, uniformly distributed. Find the electric potential at the center of the ring.

40. **N** A uniformly charged ring with total charge $q = 3.00\ \mu C$ and radius $R = 10.0$ cm is placed with its center at the origin and oriented in the xy plane. What is the difference between the electric potential at the origin and the electric potential at the point (0, 0, 30.0 cm)?

Problems 41 and 42 are paired.

41. **A** A line of charge with uniform charge density λ lies along the x axis from $x = -a$ to $x = a$.
   a. What is the magnitude of the electric potential at $(0, y)$?
   b. How much work is necessary to move a particle with charge q from very far away to $(0, y)$?

42. **N** A line of charge with uniform charge density $\lambda = 2.00 \times 10^{-3}$ C/m lies along the x axis from $x = -0.250$ m to $x = 0.250$ m.
   a. What is the magnitude of the electric potential at (0, 1.000 m)?
   b. How much work is necessary to move a particle with a charge of −5.00 nC from very far away to (0, 1.000 m)?

Problems 43 and 54 are paired.

43. **A** Consider a thin rod of total charge Q and length L (Fig. P26.43). Show that the electric potential at point P, a distance x from the end of the rod, is given by

$$V(x) = \frac{kQ}{L}\ln\left(\frac{x+L}{x}\right)$$

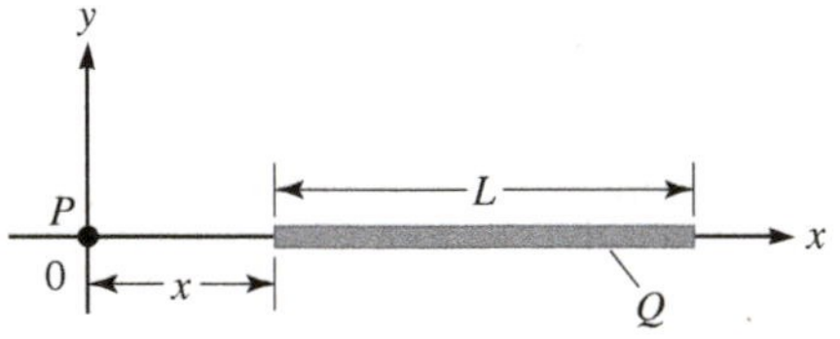

FIGURE P26.43 Problems 43 and 54.

44. **N** Figure P26.44 shows a rod of length $\ell = 1.00$ m aligned with the y axis and oriented so that its lower end is at the origin. The charge density on the rod is given by $\lambda = a + by$, with $a = 2.00\ \mu C/m$ and $b = -1.00\ \mu C/m^2$. What is the electric potential at point P with coordinates (0, 25.0 cm)? A table of integrals will aid you in solving this problem.

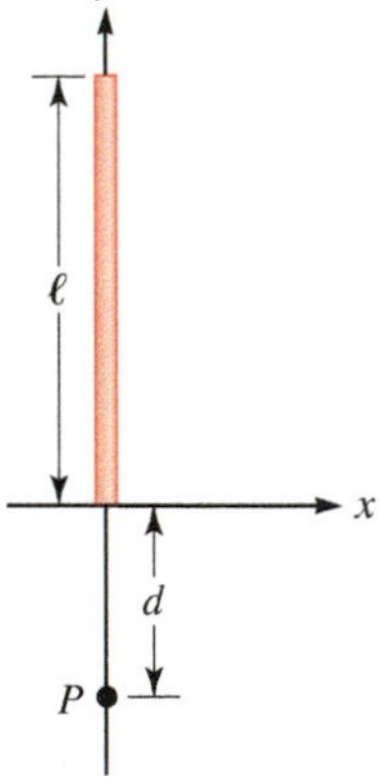

FIGURE P26.44

45. **N** The charge density on a disk of radius $R = 12.0$ cm is given by $\sigma = ar$, with $a = 1.40\ \mu C/m^3$ and r measured radially outward from the origin (Fig. P26.45). What is the electric potential at point A, a distance of 40.0 cm above the disk? *Hint*: You will need to integrate the nonuniform charge density to find the electric potential. You will find a table of integrals helpful for performing the integration.

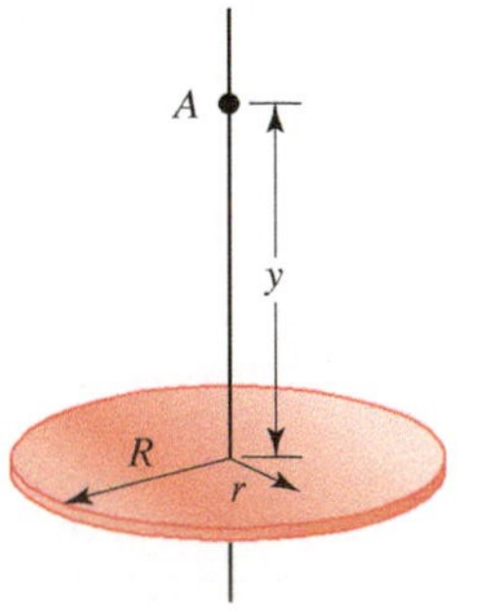

FIGURE P26.45

26-7 Connection Between Electric Field $\vec{E}$ and Electric Potential V

46. **A** Start with the expression for the electric field due to an infinitely long rod of radius R and linear charge density λ at a perpendicular distance r away from the rod: $\vec{E} = (\lambda/2\pi\varepsilon_0 r)\hat{r}$ (Eq. 25.13). Find an expression for the electric potential at a point r outside the rod. *Hint*: You already know the answer from Example 26.12 (page 816), and now you must show how V is derived from $\vec{E}$.

47. **N** In some region of space, the electric field is given by $\vec{E} = Ax\hat{\imath} + By^2\hat{\jmath}$. Find the electric potential difference between points whose positions are $(x_i, y_i) = (a, 0)$ and $(x_f, y_f) = (0, b)$. The constants A, B, a, and b have the appropriate SI units.

48. **N** A particle with charge 1.60×10^{-19} C enters midway between two charged plates, one positive and the other negative. The initial velocity of the particle is parallel to the plates and along the midline between them (Fig. P26.48). A potential difference of 300.0 V is maintained between the two charged plates. If the lengths of the plates are 10.0 cm and they are separated by 2.00 cm, find the greatest initial velocity for which the particle will not be able to exit the region between the plates. The mass of the particle is 12.0×10^{-24} kg.

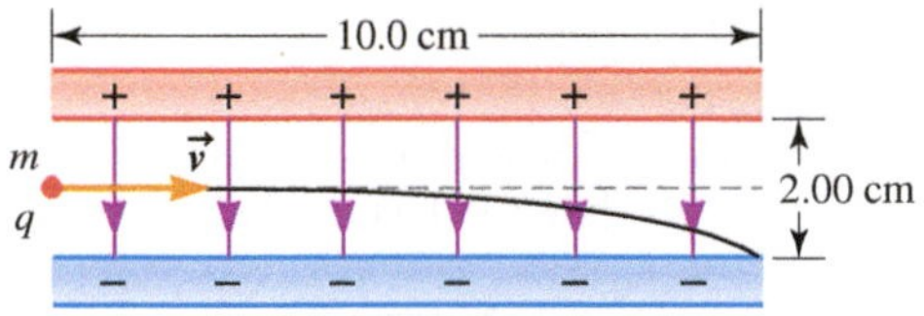

FIGURE P26.48

49. A conducting sphere with radius R has total charge Q.
   a. **A** Find the relationship between the magnitude of the electric field and the electric potential on the surface of the conducting sphere.
   b. **N** For a sphere of radius 16 cm, calculate the maximum surface electric potential at which the surrounding air begins to break down. Take the dielectric strength of (maximum sustainable electric field in) air to be 3.0×10^6 V/m.

50. **A** Show that the interior of a conductor in electrostatic equilibrium is always an equipotential volume.

26-8 Finding $\vec{E}$ from V

51. **A** The electric potential for a system is given by $V(r) = V_0 e^{-ar}/r$, where V_0 and a ($a > 0$) are constants. Determine the magnitude of the electric field.

52. **E** CASE STUDY The resting electric potential across the surface or membrane (perpendicular to the long axis) of a nerve cell is −70 mV (Fig. P26.52). The thickness of the membrane that separates the inside from the outside of the nerve cell is about 6×10^{-9} m. Estimate the magnitude of the electric field. Which way does the electric field point?

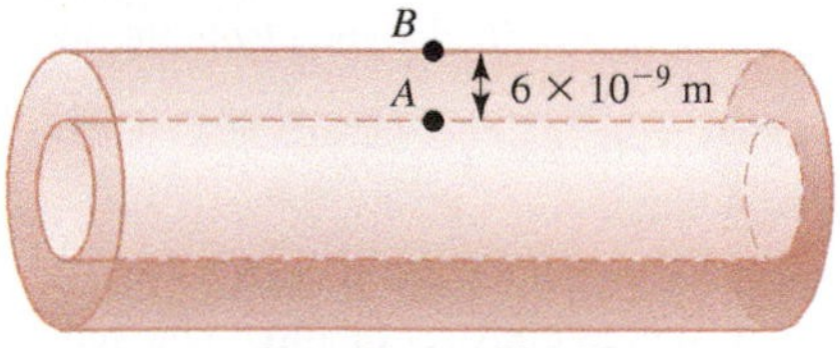

FIGURE P26.52

53. **N** The function $V = b + cy$ describes the electric potential in the region between $y = -4.00$ m and $y = 4.00$ m, with $b = 12.5$ V and $c = 4.50$ V/m.
   a. What is the electric potential at $y = -4.00$ m, at $y = 0$, and at $y = 4.00$ m?
   b. What are the magnitude and direction of the electric field at $y = -4.00$ m, at $y = 0$, and at $y = 4.00$ m?

54. **A** According to Problem 43, the electric potential at point P, a distance x from the end of the rod, is given by

$$V(x) = \frac{kQ}{L}\ln\left(\frac{x+L}{x}\right)$$

Use this equation to find the electric field at a distance x from the end of the charged rod.

55. The electric potential is given by $V = 4x^2z + 2xy^2 - 8yz^2$ in a region of space, with x, y, and z in meters and V in volts.

a. **A** What are the x, y, and z components of the electric field in this region?

b. **N** What is the magnitude of the electric field at the coordinates (2.00 m, −2.00 m, 1.00 m)?

56. **N** The electric potential $V(x, y, z)$ in a region of space is given by $V(x, y, z) = V_0(2x^2 - 3y^2 - z^2)$, where $V_0 = 12.0$ V and x, y, and z are measured in meters. Find the electric field at the point (1.00 m, 1.00 m, 0).

26-9 Graphing *E* and *V*

57. Two thin, rectangular plates have charges of equal magnitude and opposite sign as shown in Figure P26.57. The electric field between the plates is constant and equal to E.

a. **G** Sketch the electric field lines and the equipotential lines between the plates.

b. **N** Determine the work required to move a particle with charge q in the closed rectangular loop $ABCD$ shown in the figure.

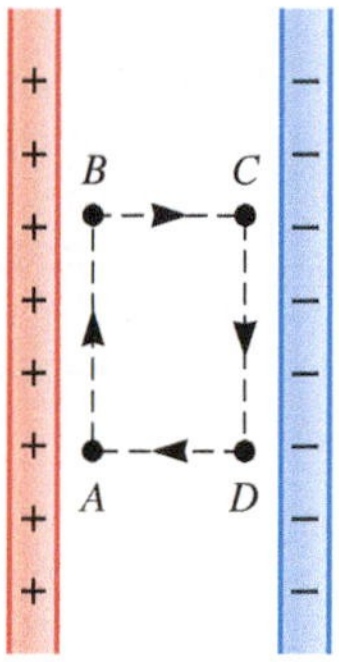

FIGURE P26.57

58. In three regions of space, the electric potential is given by

$$V(r) = 0 \text{ for } r < R$$
$$V(r) = \frac{V_0}{4R^2}r^2 \text{ for } R \le r < 2R$$
$$V(r) = V_0 \text{ for } r \ge 2R$$

a. **G** Plot V as a function of r.

b. **A** Find expressions for the electric field in all three regions.

c. **G** Plot E versus r in all three regions.

General Problems

59. **N** An aluminum sphere 22.0 cm in radius carries a total charge of 845 nC. What are the magnitudes of the electric field and the electric potential at these radial distances from the center of the sphere:

a. 5.00 cm,

b. 22.0 cm, and

c. 50.0 cm?

60. **N** A particle with charge $q = 2.7$ nC and mass $m = 0.57$ g is in a circular orbit of radius $r = 1.3$ mm around a fixed particle with charge $Q = -8.6\ \mu$C.

a. Find the speed of the orbiting particle.

b. Calculate the electric potential energy of the system.

c. What is the total energy of the system?

61. **N** The distance between two small charged spheres with charges $q_A = -8.35\ \mu$C and $q_B = +4.90\ \mu$C is 48.0 cm.

a. What is the electric potential energy due to the two spheres?

b. What is the electric potential halfway between the two spheres along the line connecting them?

62. **A** From Section 9-5, there are two ways to test a force to see whether it is path-independent—a necessary property of a conservative force. The first test involves calculating the work done by the force on a subject along several different paths. If the work done is not the same along all the paths, the force is nonconservative. The alternative test involves calculating the work done by the force on a particle along a closed path. If the work done along a closed path does not equal zero, the force is nonconservative. Show that the electrostatic force does not fail the two tests for a conservative force. *Hint*: Use Examples 9.5 and 9.6 (pages 257–259) as a guide.

63. **N** A glass sphere with radius 4.00 mm, mass 85.0 g, and total charge 4.00 μC is separated by 150.0 cm from a second glass sphere 2.00 mm in radius, with mass 300.0 g and total charge −5.00 μC. The charge distribution on both spheres is uniform. If the spheres are released from rest, what is the speed of each sphere the instant before they collide?

64. **N** A charged particle with $q_A = 1.00$ nC is located at the origin, and a second particle with charge $q_B = -3.00$ nC is located at $y = 1.50$.

a. What are the finite values of y for which the electric field is zero?

b. What are the finite values of y for which the electric potential is zero?

Problems 65, 66, and 67 are grouped.

65. Two 5.00-nC charged particles are in a uniform electric field with a magnitude of 625 N/C. Each of the particles is moved from point A to point B along two different paths, labeled in Figure P26.65.

a. **N** Given the dimensions in the figure, what is the change in the electric potential experienced by the particle that is moved along path 1 (black)?

b. **N** What is the change in the electric potential experienced by the particle that is moved along path 2 (red)?

c. **C** Is there a path between the points A and B for which the change in the electric potential is different from your answers to parts (a) and (b)? Explain.

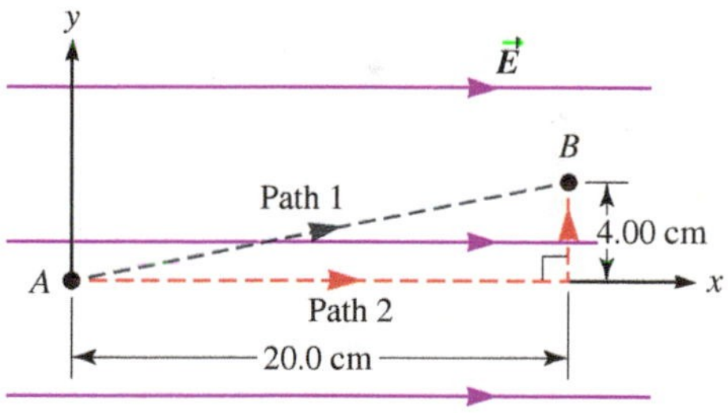

FIGURE P26.65 Problems 65, 66, and 67.

66. **N** A 5.00-nC charged particle is at point B in a uniform electric field with a magnitude of 625 N/C (Fig. P26.65). What is the change in electric potential experienced by the charge if it is moved from B to A along

a. path 1 and

b. path 2?

67. **C** A charged particle is moved in a uniform electric field between two points, A and B, as depicted in Figure P26.65. Does the change in the electric potential or the change in the electric potential energy of the particle depend on the sign of the charged particle? Consider the movement of the particle from A to B, and vice versa, and determine the signs of the electric potential and the electric potential energy in each possible scenario.

68. **N** Figure P26.68 shows three small spheres with identical charges of −3.00 nC placed at the vertices of an equilateral triangle with side $d = 2.50$ cm.

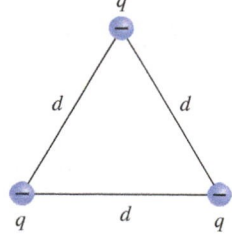

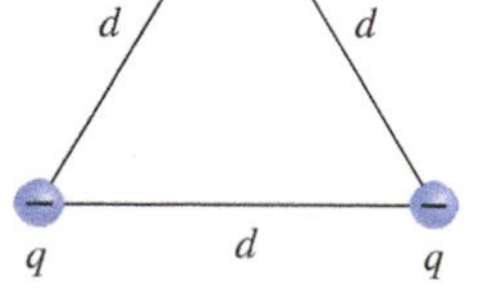

FIGURE P26.68

a. Is the electric potential due to the three spheres zero anywhere in the plane that contains the triangle, other than at infinity?

b. What is the electric potential at the location of each sphere due to the other two spheres?

69. **N** What is the work required to charge a spherical shell of radius $r = 43.0$ cm to a total charge of 67.0 μC with charges that are brought to the shell from infinity?

70. **G** For a system consisting of two identical negatively charged particles, sketch U_E versus r and V_E versus r. Comment on your graphs.

71. **N** Figure P26.71 shows three charged particles arranged at the vertices of an isosceles triangle with base $b = 1.00$ m. What is the electric potential due to the particles at point P, which is at the midpoint of the base?

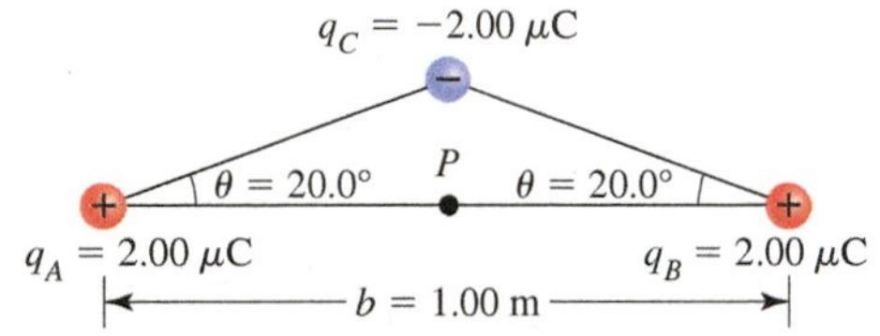

FIGURE P26.71

Problems 72, 73, and 74 are grouped.

72. **A** Figure P26.72 shows a source consisting of two identical parallel disks of radius R. The x axis runs through the center of each disk. Each disk carries an excess charge uniformly distributed on its surface. The disk on the left has a total positive charge Q, and the disk on the right has a total negative charge $-Q$. The distance between the disks is $3R$, and point A is $2R$ from the positively charged disk. Find an expression for the electric potential at point A between the disks on the x axis. Approximate any square roots to three significant figures.

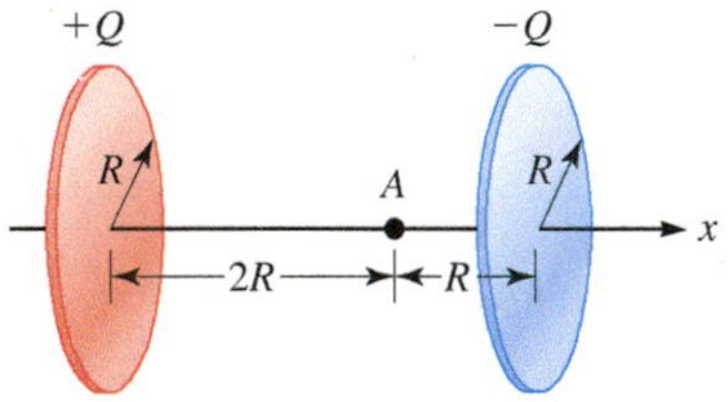

FIGURE P26.72 Problems 72, 73, and 74.

73. **A** Start with $V = 2\pi k\sigma\left[\left(\sqrt{R^2 + x^2}\right) - x\right]$ for the electric potential of a disk of radius R and excess surface charge density σ at a position x from the center of a disk on its axis, and derive an expression for the electric field at this position. *Hint*: See Example 24.6 (page 732) to check your answer.

74. **A** **Review** Consider the charged disks in Problem 72 (Fig. P26.72). Find an expression for the electric field at point A between the disks on the x axis. Approximate any square roots to three significant figures.

75. A long thin wire is used in laser printers to charge the photoreceptor before exposure to light. This is done by applying a large potential difference between the wire and the photoreceptor.

a. **A** Use Equation 26.23,

$$V(r) = \frac{\lambda}{2\pi\varepsilon_0}\ln\frac{R}{r}$$

to determine a relationship between the electric potential V and the magnitude of the electric field E at a distance r from the center of the wire of radius R ($r > R$).

b. **N** Determine the electric potential at a distance of 2.0 mm from the surface of a wire of radius $R = 0.80$ mm that will produce an electric field of 1.8×10^6 V/m at that point.

76. An electric potential exists in a region of space such that $V = 8x^4 - 2y^2 + 9z^3$ and V is in units of volts, when x, y, and z are in meters.

a. **A** Find an expression for the electric field as a function of position.

b. **N** What is the electric field at (2.0 m, −4.5 m, −2.0 m)?

77. **A** A disk with a nonuniform charge density $\sigma = ar^2$ has total charge Q and radius R. Derive an expression for the electric potential at a point along the axis of the disk a distance x from the center of the disk. A table of integrals will aid you in solving this problem.

Problems 78 and 79 are paired.

78. **N** An infinite number of charges with $q = 2.0\ \mu$C are placed along the x axis at $x = 1.0$ m, $x = 2.0$ m, $x = 4.0$ m, $x = 8.0$ m, and so on, as shown in Figure P26.78. Determine the electric potential at the point $x = 0$ due to this set of charges. *Hint*: Use the mathematical formula for a geometric series,

$$1 + r + r^2 + r^3 + r^4 + \cdots = \frac{1}{1 - r}$$

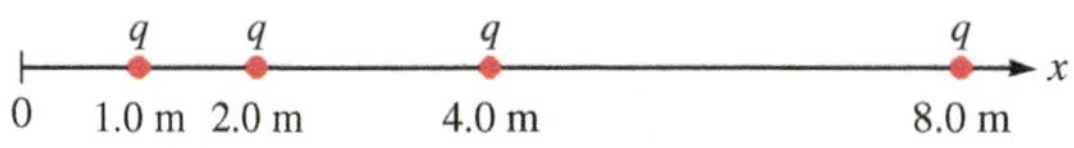

FIGURE P26.78

79. **N** An infinite number of charges with $|q| = 2.0\ \mu$C are placed along the x axis at $x = 1.0$ m, $x = 2.0$ m, $x = 4.0$ m, $x = 8.0$ m, and so on, as shown in Figure P26.79. What will be the electric potential at $x = 0$ if the consecutive charges have alternating signs as shown in Figure P26.79? *Hint*: Use the mathematical formula for a geometric series,

$$1 + r + r^2 + r^3 + r^4 + \cdots = \frac{1}{1 - r}$$

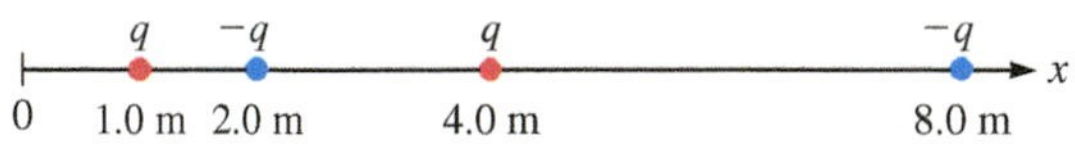

FIGURE P26.79

80. **N** Figure P26.80 shows a wire with uniform charge per unit length $\lambda = 2.25$ nC/m comprised of two straight sections of length $d = 75.0$ cm and a semicircle with radius $r = 25.0$ cm. What is the electric potential at point P, the center of the semicircular portion of the wire?

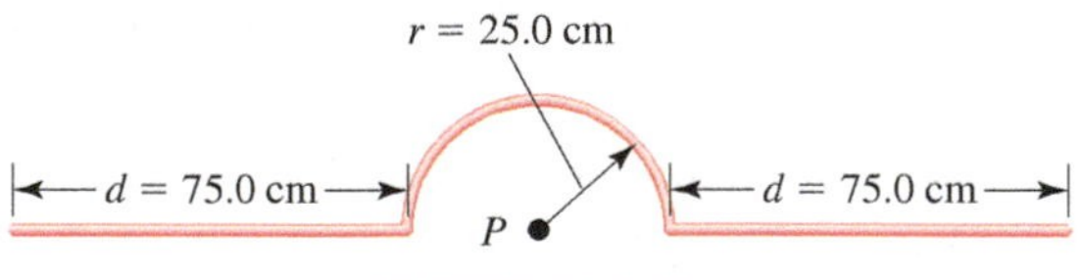

FIGURE P26.80

81. **A** Integrate to show

$$\int_0^R \frac{\mathcal{R}\cdot d\mathcal{R}}{\sqrt{\mathcal{R}^2 + x^2}} = \left(\sqrt{R^2 + x^2}\right) - x$$

which we needed in Example 26.7 (page 805).

Capacitors and Batteries

27

Underlying Principles

General form of Gauss's law

Major Concepts

1. Capacitor
2. Capacitance
3. Energy stored by a capacitor
4. Ideal battery
5. Equivalent capacitance of capacitors in series
6. Equivalent capacitance of capacitors in parallel
7. Dielectric; capacitance in the presence of a dielectric
8. Energy density stored in an electric field

Special Cases

1. Parallel-plate capacitor
2. Cylindrical capacitor

Tools

Schematic diagrams (circuit diagrams)

Key Questions

How do we apply the theoretical principles and concepts of electricity to practical devices?

Can an electric field be used to store energy?

How do we make Gauss's law more general?

Ben Franklin, like most scientists, believed the most important goal of science is to understand nature. Scientists try to answer fundamental questions about nature, such as "How does the Moon orbit the Earth?" and "What is lightning?"

Like many inventors, Franklin also believed scientific discoveries should be used to help people. Franklin invented many useful things, including bifocals, the Franklin stove, and lightning rods. This chapter is about practical ways to create an electric field and to store energy as electric potential energy. We will study two types of devices: (1) those that maintain a potential difference, such as batteries and hand-operated electric generators, and (2) those that store electric potential energy. The original device for storing electric potential energy—the Leyden jar—comes from Franklin's time.

The Library Company of Philadelphia, Gift of Benjamin Franklin Bache, 1792.

FIGURE 27.1 In the late 1700s, generators enabled experimenters to build up charge efficiently. By turning the crank, the experimenter rubbed two materials together, causing the transfer of charge from one material to the other.

27-1 The Leyden Jar

In order to perform his experiments, Franklin needed a way to generate and store charged particles. He could charge a glass rod by rubbing it with a piece of silk, but this soon grew tiring and was not efficient for building up a large amount of charge. So, 18th-century experimenters used generators like the ones shown in Figure 27.1. A large glass sphere replaces the glass rod. A pad covered in a piece of cloth such as silk is held against the sphere, which is attached to a crank. When the experimenter turns the crank, the sphere rotates and charge is transferred between the sphere and the cloth.

Using our contemporary understanding of work and energy, consider the silk cloth and the glass sphere to be the system. The experimenter does work on the system by turning the crank, which transfers energy from the environment (the experimenter) to the silk-glass system. Although some energy may be dissipated by friction, much of the energy goes into transferring charged particles. Because the glass sphere ends up positively charged and the silk is negatively charged as a result of the work done, electric potential energy is stored in the system. Furthermore, if we consider the glass sphere as the source, we can say that the work done by the experimenter sets up the sphere's electric potential.

Experimenters in the 18th century found that after turning the crank for a few minutes, they were unable to build up any more charge on the sphere. After it acquires some positive charge, the sphere attracts negatively charged particles, so it becomes difficult to transfer more electrons to the silk. Experimenters needed a way around this limitation. In the 18th century, people thought of electricity as a fluid, so they stored the "fluid" in a **Leyden jar,** which consisted of two conductors separated by an insulator. To store even more charge, they used several Leyden jars.

Franklin built his own Leyden jars and described their construction in a 1758 letter. He lined the inside and outside of a glass jar with tinfoil. You can easily make your own Leyden jar; the key is to have two conductors (the foil) separated by an insulator (glass). The jar is charged by means of a metal conductor in contact with the glass sphere of a generator. Charged particles move from the rotating sphere to the conductor and through a thin wire to the tinfoil on the inside of the Leyden jar (Fig. 27.2). If the experimenter holds the tinfoil on the outside of the jar and is in

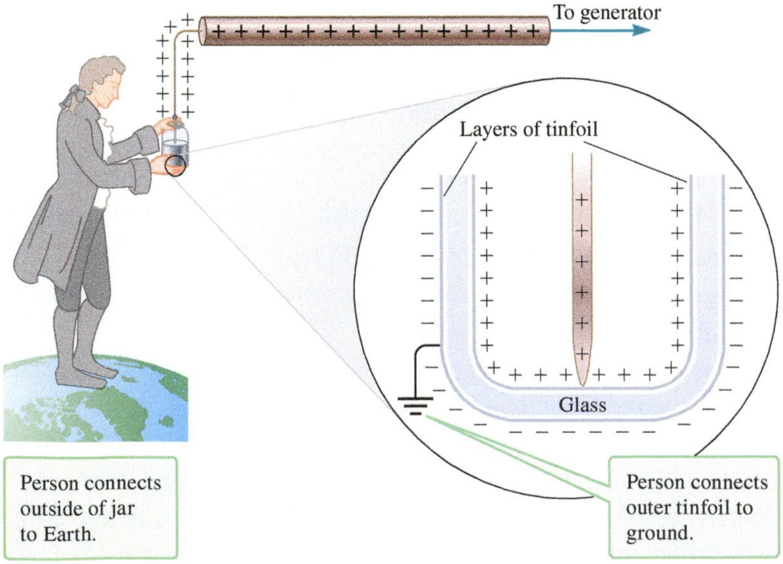
To generator
Layers of tinfoil
Glass
Person connects outside of jar to Earth.
Person connects outer tinfoil to ground.

FIGURE 27.2 Positive particles are transferred to the inner surface of the Leyden jar. The outer surface is connected to ground through the researcher's body, so it picks up negative particles.

contact with the ground, an equal magnitude of charge of opposite sign builds up on the outside of the jar (Fig. 27.2, close-up).

With our contemporary understanding of work and energy, we know that the work done by the experimenter in turning the crank is stored as electric potential energy in the Leyden jar, so the 18th-century experimenter had a way to store energy. The experimenter could later use that energy as needed by connecting the inside of the jar to the outside with a conductor (Fig. 27.3). A Leyden jar is like a simple spring-loaded toy dart gun. Compressing the spring does work on the dart-spring system, storing potential energy. Energy can be released from a loaded dart gun or a charged Leyden jar as in this chapter's case study.

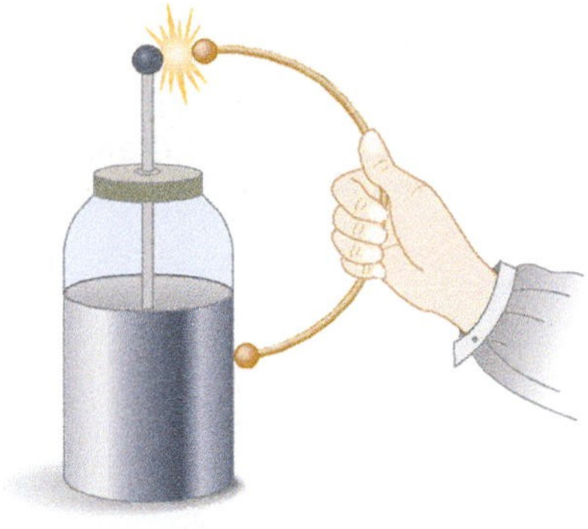

FIGURE 27.3 Discharging a Leyden jar.

CASE STUDY **Thompson Coil**

If you search on the Internet for *Thompson coils*, you are likely to find many sites describing how enthusiastic experimenters build such a device. Potential energy stored in an electric field may be used to launch a metal ring (Fig. 27.4). No springs are used, only electromagnetism. In Chapter 33, we will study the mechanism for converting the stored electric potential energy into kinetic energy. In this chapter, we will explore claims made on some websites in terms of the conservation of energy principle. You may see a demonstration of a Thompson coil in your class. Typically, a ring reaches a height of a few feet (1 m to 5 m); however, our case study is about one group of experimenters who claimed to have launched a 90-g aluminum ring about 120 m straight up!

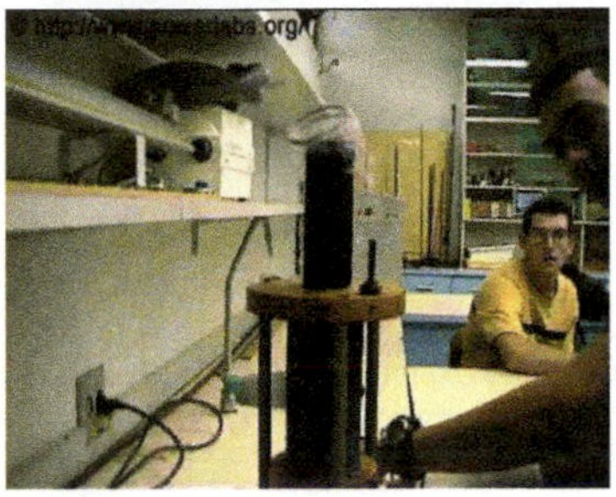

Courtesy of Sam Barros PowerLabs

FIGURE 27.4 A Thompson coil in use.

CONCEPT EXERCISE 27.1

CASE STUDY **How Big a Spring?**

Imagine the ring in the case study is launched using a spring-loaded gun instead of a Thompson coil. If the gun has a stiff spring with a spring constant of 1000 N/m, how much does the spring need to be compressed in order to propel a 90-g object 120 m up? Comment on the size of the spring. (Ignore air resistance.)

27-2 Capacitors

A Leyden jar is an example of a **capacitor**, a device that consists of two conductors known as **plates** separated by an insulator or vacuum (Fig. 27.5). The plates don't have to be flat. For Leyden jars, the plates are the metal foil lining the inside and outside of the jar, and the insulator is the glass. The purpose of the capacitor is to store electric potential energy. When the capacitor is charged, one of the plates is positive and the other has an equal magnitude of negative charge.

CAPACITOR Major Concept

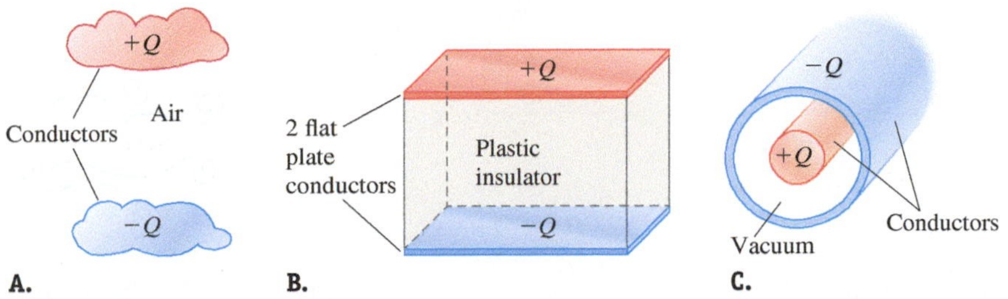

FIGURE 27.5 A capacitor consists of two conducting plates separated by an insulator such as **A.** air, **B.** plastic, or **C.** a vacuum.

The phrases "the capacitor has a charge Q" and "charge Q is stored on the capacitor" mean that the charge on the positive plate is $+Q$ and the charge on the negative plate is $-Q$.

The net charge of the entire capacitor is zero (positive charge $+Q$ on one plate and negative charge $-Q$ on the other). However, we usually refer to the charge on the positive plate rather than to the net charge of the entire capacitor. If a capacitor is charged, there is a potential difference between the two plates. Because the potential is uniform on the surface of any conductor, no matter how oddly shaped it is, the potential difference between any point on one plate and any point on the other plate is the same. Now, imagine that each plate is neutral, so the potential difference between the plates is zero. If one plate acquires a small amount of positive charge and the other a small amount of negative charge, you would expect a small potential difference between the plates. As the amount of charge on each plate increases, so does the potential difference between the plates. The potential difference ΔV between the plates is directly proportional to the magnitude of charge Q on each plate:

$$Q \propto \Delta V$$

CAPACITANCE ✪ **Major Concept**

The constant of proportionality is called the **capacitance** C:

$$Q = C\Delta V \tag{27.1}$$

The SI unit for capacitance is the **farad** (F). Using Equation 27.1, we can express farads in terms of coulombs and volts:

$$1\,\text{F} = 1\,\text{C/V}$$

You must distinguish between the symbol for capacitance (uppercase italic C) and the abbreviation for the SI unit coulomb (uppercase nonitalic C).

The capacitance depends only on geometric factors and the type of insulator between the plates. For two capacitors with the same potential difference between their plates, the one with greater capacitance has a greater charge on its plates and stores more electric potential energy.

General Expression for Energy Stored by a Capacitor

Let's use gravity as an analogy to find a general expression for the electric potential energy stored by a capacitor. Figure 27.6 shows a tray full of marbles on a table. Above the marbles is another tray, which is initially empty. The system consists of the Earth and the marbles, with the reference configuration chosen so that the gravitational potential energy is zero when all the marbles are in the lower tray. In Figure 27.6A, Rochelle (outside the system) raises the marbles one at a time to the upper tray at height h above the lower tray. If each marble has mass m, the work w she does to raise one marble to the upper tray is

$$w = mgh$$

If Rochelle lifts all N marbles, the total work W she does on the system is the sum of the work she does in lifting each marble:

$$W = \sum_{i=1}^{N} m_i gh = Nmgh = Mgh$$

where M is the total mass of the marbles. Because there is no change in the system's kinetic energy ($\Delta K = 0$) and no dissipative forces are present ($\Delta E_{\text{th}} = 0$), according

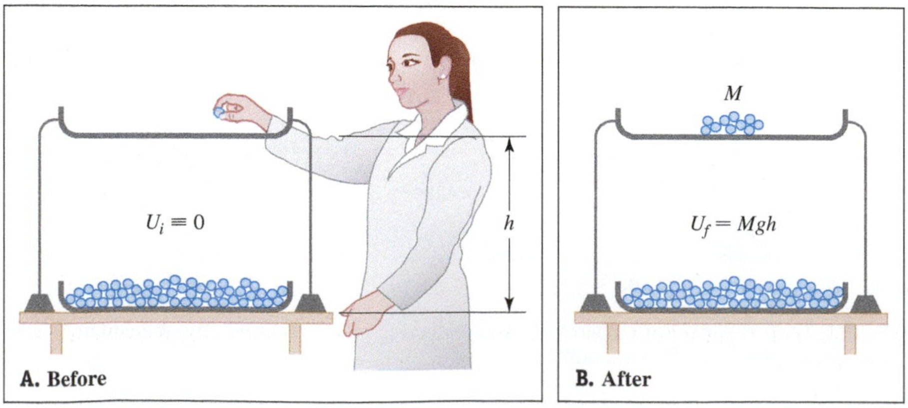

FIGURE 27.6 A person does work in lifting marbles from a lower tray to an upper tray.

to the work-energy theorem ($W = \Delta K + \Delta U + \Delta E_{th}$), the total work done by Rochelle equals the change in the system's potential energy: $W = \Delta U$. The system started with zero potential energy, so the system's potential energy after Rochelle lifted the marbles is equal to the work she did: $U_f = Mgh$ (Fig. 26.7B). In the next derivation, we use similar reasoning to find the potential energy stored by a capacitor.

DERIVATION Potential Energy Stored by a Capacitor

Figure 27.7A shows two neutral conductors separated by a vacuum. Each plate has an equal number of positive and negative particles, but for simplicity we have omitted the charges on the upper plate. As in Figure 27.6, we imagine an external force moves positive particles from the lower plate to the upper plate. (This is equivalent to moving negative particles in the opposite direction.) We will show that the potential energy stored by a capacitor (Fig. 27.7B) is given by

$$U_E = \frac{1}{2}\frac{Q^2}{C} \quad (27.2)$$

CAPACITOR'S ELECTRIC POTENTIAL ENERGY

Major Concept

and

$$U_E = \frac{1}{2}C(\Delta V)^2 \quad (27.3)$$

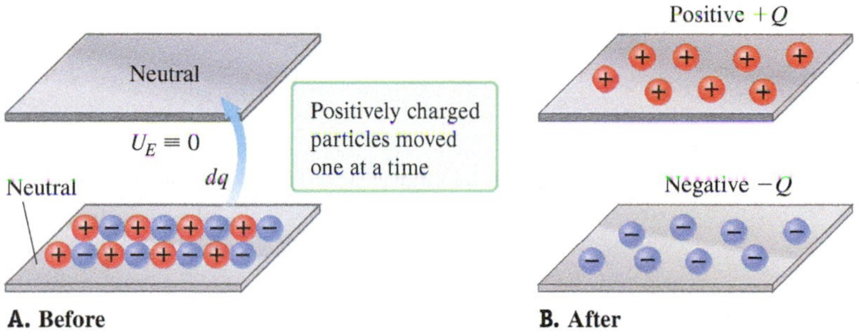

FIGURE 27.7

The system consists of the two plates. When the plates are neutral, the system is in the reference configuration with zero electric potential energy (Fig. 27.7A). The work done by the external force equals the change in the system's potential energy: $W = \Delta U_E$.

Each particle has a small amount of positive charge dq. Equation 26.7 gives the change in the system's electric potential energy dU when a single charged particle is moved to the upper plate.	$dU_E = (dq)\Delta V \quad (26.7)$
After N particles are moved, the total charge on the upper plate is $+Q$ and the total charge on the lower plate is $-Q$ (Fig. 27.7B). The total change in the system's potential energy is found by adding (or integrating) the changes in potential energy due to the relocation of all the charged particles.	$\int_0^{U_E} dU_E = \int_0^Q (dq)\Delta V$
The potential difference ΔV depends on the amount of charge already on the plates. Substitute $\Delta V = q/C$ (Eq. 27.1) before integrating.	$\int_0^{U_E} dU_E = \int_0^Q (dq)\left(\frac{q}{C}\right) = \frac{1}{C}\int_0^Q q\,dq$ $U_E = \frac{1}{C}\frac{q^2}{2}\Big\vert_0^Q$
Evaluate the integral between limits to arrive at the energy stored by a capacitor with charge Q.	$U_E = \frac{1}{2}\frac{Q^2}{C}$ ✓ $\quad (27.2)$

Derivation continues on page 832 ▶

It is sometimes convenient to write the potential energy stored by a capacitor in terms of the potential difference ΔV between its plates instead of its charge. Substitute $Q = C\Delta V$ (Eq. 27.1) for Q in Equation 27.2.

$$U_E = \frac{1}{2} C(\Delta V)^2 \quad ✓ \qquad (27.3)$$

COMMENTS

The amount of energy stored by a capacitor depends on three factors. According to $U_E = Q^2/2C$ (Eq. 27.2), more stored charge Q means more stored energy U_E, and according to $U_E = \frac{1}{2}C(\Delta V)^2$ (Eq. 27.3), a greater potential difference means more energy stored. However, the third factor is the capacitance, which is a property of the geometry and the insulator between the plates. The energy stored is inversely proportional to C in $U_E = Q^2/2C$ and directly proportional to C in $U_E = \frac{1}{2}C(\Delta V)^2$. You will explore this apparent contradiction further in Problem 6.

CONCEPT EXERCISE 27.2

Consider two different capacitors, A and B. Figure 27.8 shows a graph of the potential difference ΔV between the two plates of each capacitor versus the charge Q on the plates. Use these graphs to find the capacitance of each capacitor.

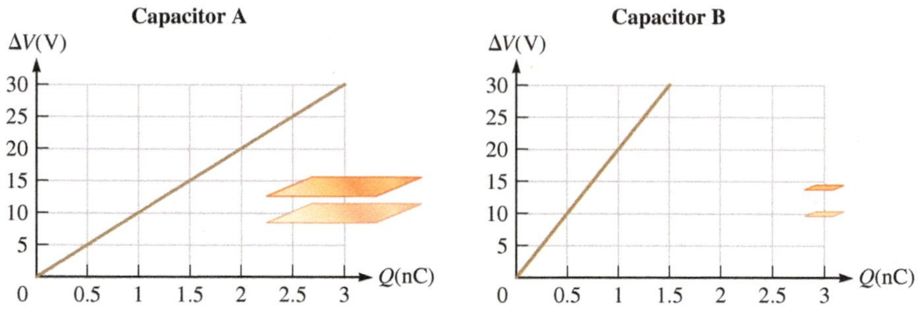

FIGURE 27.8

CONCEPT EXERCISE 27.3

a. If capacitor B in Figure 27.8 has a charge of 0.5 nC, what is the potential difference between its plates and how much potential energy does it store?

b. If capacitor A in Figure 27.8 stores the same amount of potential energy found in part (a), what is the magnitude of the excess charge on its plates? What is the potential difference between its plates? Use the capacitance C_A found in Concept Exercise 27.2.

EXAMPLE 27.1 CASE STUDY Careful Around Those Plates!

The Thompson coil experimenters in the case study on page 829 said they launched the ring by charging a bank of capacitors something like a collection of Leyden jars. We can model this bank of capacitors as a single device known as an **equivalent** capacitor (Section 27-4). The experimenters reported a potential difference of 1500 V between the plates of an equivalent capacitor. Model their collection of capacitors as a single capacitor, and find its (equivalent) capacitance. How much charge is stored by this capacitor when its potential difference is 1500 V? Ignore air resistance, and assume that the ring undergoes only translational motion. Recall that the mass of the ring is 90 g and its maximum height is 120 m. (Report your answers to two significant figures.)

INTERPRET and ANTICIPATE

Consider the capacitor, the ring, and the Earth to be the system. The ring begins at rest, and it is momentarily at rest at the top of its flight. No external force does work on the system, and there is no change in the thermal energy. A bar chart (Fig. 27.9) shows that the electric potential energy stored by the bank of capacitors is equal to the gravitational potential energy of the system when the ring is at its maximum height. Because we can find the gravitational potential energy, we can also find the electric potential energy. We expect the capacitance and stored charge to be fairly large.

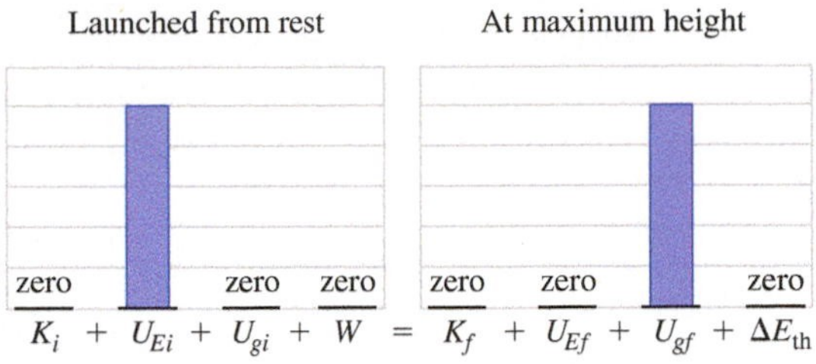

FIGURE 27.9

SOLVE

Set the potential energy stored by the capacitor equal to the maximum gravitational potential energy. U_E is related to the capacitance by $U_E = \frac{1}{2}C(\Delta V)^2$ (Eq. 27.3).

$$U_{Ei} = U_{gf}$$

$$\frac{1}{2}C(\Delta V)^2 = mgy_{\text{max}}$$

$$C = \frac{2mgy_{\text{max}}}{(\Delta V)^2} = \frac{2(0.09\,\text{kg})(9.81\,\text{m/s}^2)(120\,\text{m})}{(1500\,\text{V})^2}$$

$$C = 9.4 \times 10^{-5}\,\text{F} = 94\,\mu\text{F}$$

Use Equation 27.1 to find the charge stored by the capacitor.

$$Q = C\Delta V \qquad (27.1)$$

$$Q = (9.4 \times 10^{-5}\,\text{F})(1500\,\text{V}) = 0.14\,\text{C}$$

CHECK and THINK

The capacitance of a typical off-the-shelf capacitor ranges from a few picofarads to a few microfarads, and as expected, the capacitance here is rather large. We'll continue to explore how the experimenters were able to get such a large capacitance. Also, notice that the charge stored by the capacitor is very large. (The charge that builds up when two objects are rubbed together is on the order of a few microcoulombs.) As discussed on their website, the experimenters had to be careful not to touch both plates of the capacitor at the same time because if the capacitor had discharged through a person, the result would have probably been fatal.

27-3 Batteries

The invention of the battery was the byproduct of a heated disagreement between 18th-century scientists. The Italian anatomy professor Luigi Galvani was dissecting a frog near an electric generator. When he touched the scalpel to the frog's nerves, its legs twitched and a spark was drawn from the generator. Galvani discovered that the frog's legs also twitched when he touched the animal with two different metals (for instance, copper on the frog's spine and iron on its feet). Galvani was convinced that the frog's body produced electricity ("animal electricity").

Another Italian scientist, Alessandro Giuseppe Antonio Volta, successfully reproduced Galvani's experiment using two different metals in contact with a frog. Eventually, Volta found that if he used the same metal at both contact places, the frog's legs would *not* twitch. Volta reasoned that connecting dissimilar metals somehow produced electricity and there was no "animal electricity." Many scientists joined the argument, which went on for decades beyond the lifetimes of both Galvani and Volta. Although Galvani was correct that electricity plays a key role in muscle function, Volta's skepticism led him to invent the battery by "piling" (stacking) dissimilar types of metal.

The SI unit of electric potential—the volt—is named in Volta's honor.

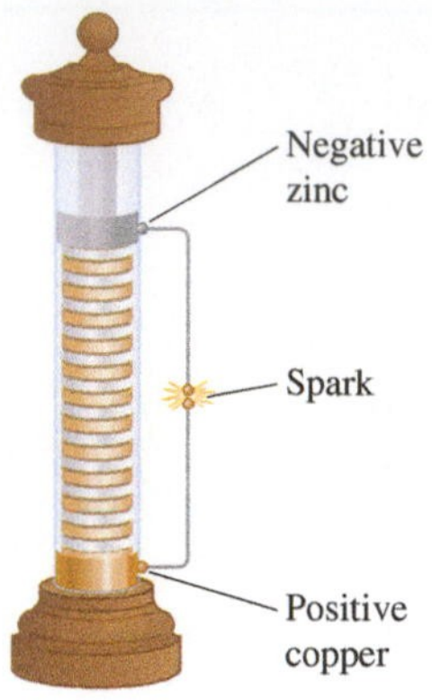

FIGURE 27.10 Volta's pile: a metal rod attached to the zinc plate at the top and another rod attached to the copper plate at the bottom. At the other end of each rod is a small metal ball with a short gap between the balls. Volta could demonstrate the electric nature of his pile by narrowing the gap between the balls so that a spark would jump that gap.

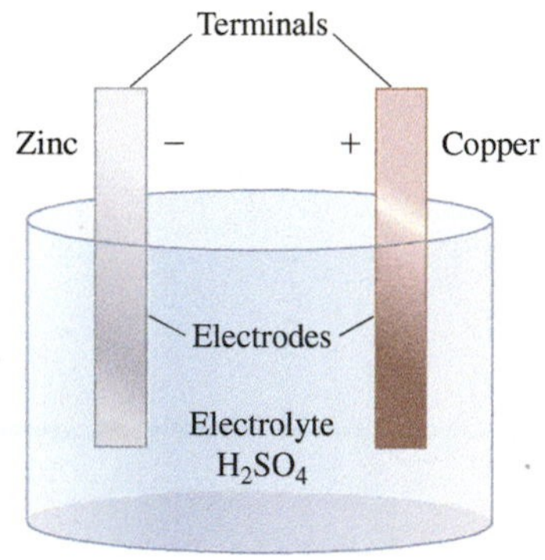

FIGURE 27.11 A wet cell consists of electrodes in a liquid electrolyte. The part of an electrode outside the electrolyte is known as a terminal. The cell maintains a potential difference between its terminals through chemical reactions.

Figure 27.10 shows one of Volta's piles consisting of many cells. Each cell is made up of three layers: two types of metal (here, copper and zinc) separated by a layer of moist paper or cloth. There is a potential difference between the top and bottom of each cell. Volta achieved a greater potential difference by stacking many cells on top of one another. A pile of such cells is called a **battery**. Today, we use the term *battery* to mean any device—cell or collection of cells—that maintains a potential difference through chemical reactions.

Many students think batteries are like capacitors, but batteries and capacitors have different purposes. A capacitor's purpose is to store charge and therefore electric potential energy, whereas a battery's purpose is to maintain a potential difference through chemical reactions. A battery is more like an 18th-century electric generator (Fig. 27.1). In both a generator and a battery, work is done in order to separate charged particles, creating an electric potential difference. In a generator, this work comes from a person rotating the crank. In a battery, the energy needed to separate charges comes from chemical reactions.

Volta found that he could also build a cell by placing strips of different metals in a cup of liquid such as salt water, forming a **wet cell**. Figure 27.11 shows a wet cell with a strip of zinc and a strip of copper in a sulfuric acid (H_2SO_4) solution. The metal strips are called **electrodes**, and the liquid solution is called an **electrolyte**. The parts of the electrodes that are above the solution are called the **terminals**.

Chemical reactions generate a potential difference between the zinc and copper terminals. Sulfuric acid (H_2SO_4) molecules break up when they are in solution into two positively charged H^+ ions and one negatively charged SO_4^{2-} ion. The SO_4^{2-} ion reacts with zinc, pulling positively charged zinc ions off that electrode. The zinc electrode begins to dissolve and becomes negatively charged. At the same time, positively charged H^+ ions are attracted to copper and remove electrons from the copper electrode. This reaction causes the copper to become positively charged. The result is that the zinc terminal becomes negative and the copper terminal becomes positive. So, there is a potential difference between the terminals; this is called the **terminal potential** or **terminal voltage**.

When the electrodes in Figure 27.11 are first put into the sulfuric acid (H_2SO_4) solution, they are neutral and the terminal potential is zero. As the chemical reactions proceed, the zinc electrode becomes negative and the copper becomes positive, so the terminal potential increases. After a period of time, the terminal potential reaches a maximum because as the zinc electrode becomes more negative, positive ions cannot easily escape; they are attracted to that electrode and are quickly pulled back. The same sort of argument can be made about electrons pulled off the positive copper electrode. The maximum terminal potential is reached when no more positive ions can successfully escape from the zinc and no more electrons can successfully escape from the copper. The terminal potential depends on the specific chemical reaction, so different electrodes in different electrolytes produce different terminal potentials. Many batteries that are commonly used in portable devices such as cell phones, portable media players, and flashlights are known as **dry cells** because the electrolyte is a paste instead of a messy liquid. Table 27.1 lists some commonly used batteries by their electrodes and gives their typical terminal potentials.

TABLE 27.1 Common types of batteries.

Electrodes	Terminal potential (V)	Uses
Zinc and carbon	1.5	AA, AAA, C, and D dry cells used in portable devices
Lead and lead dioxide	12 (6 cells)	Cars
Nickel-hydroxide and cadmium	1.2	Rechargeable
Lithium and carbon	3.7	High-end rechargeable batteries used in computers and cell phones
Lithium-iodide and lead-iodide	2.2	Cameras
Zinc and air	1.4	Hearing aids

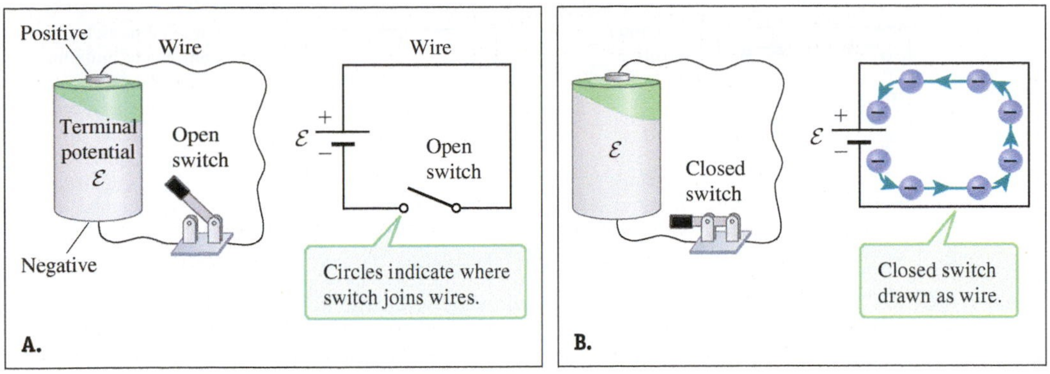

FIGURE 27.12 **A.** The switch is open, so charged particles cannot move through the wire. **B.** The switch is closed, and electrons move away from the negative terminal through the wire to the positive terminal. The battery maintains the potential difference between its terminals.

Figure 27.12 shows a typical battery, some wire, and a switch connected to form an **electrical network**. The wire and switch are conductors. If the switch is open, air in the gap prevents electrons from moving through the network (Fig. 27.12A). When the switch is closed, electrons can flow from one part of the electrical network to the other (Fig. 27.12B).

SCHEMATIC DIAGRAMS Tool

Simple sketches used to illustrate an electrical network are called **schematic diagrams**. The schematic diagrams on the right of each part of Figure 27.12 represent the electrical networks shown on the left. A wire is represented by lines that may be bent or straight as shown. A switch is represented by a short, straight line with a circle on the end. Another circle is drawn on the part of the wire across the gap. When the switch is closed, it is not necessary to draw these circles (Fig. 27.12B). A battery is represented by two parallel lines: a long line representing the positive terminal and a short line representing the negative terminal. The terminal potential $\mathcal{E}$, which looks like a curly Greek letter epsilon, is written next to the battery's symbol.

Suppose the terminals of a zinc-copper battery are connected by a wire (Fig. 27.12B). Electrons flow from the negative electrode (zinc) through the external circuit to the positive electrode (copper). As the electrons leave the zinc electrode, the zinc can once again react with the electrolyte, and more positive zinc ions move from the electrode into the electrolyte. As a result, the zinc is able to maintain its negative charge. The same sort of process maintains the copper's positive charge. Because each electrode maintains its net charge, the terminal potential remains constant. Throughout this book, we will use the term **ideal battery** to describe a device that maintains its terminal potential through chemical reactions. The symbol $\mathcal{E}$ represents the terminal potential of an ideal battery, which is actually the potential *difference* between its terminals. All the batteries in this chapter may be modeled as ideal.

IDEAL BATTERY Major Concept

Charging a Capacitor

In the 18th century, experimenters charged Leyden jars with electric generators. Today, capacitors are charged with batteries. Figure 27.13A shows an electrical network that may be used to charge a capacitor that consists of two parallel plates separated by air. The positive terminal of the battery is connected to the top plate of the capacitor, and the negative terminal is connected through a switch to the bottom plate. Figure 27.13B shows a schematic diagram for this network. All capacitors,

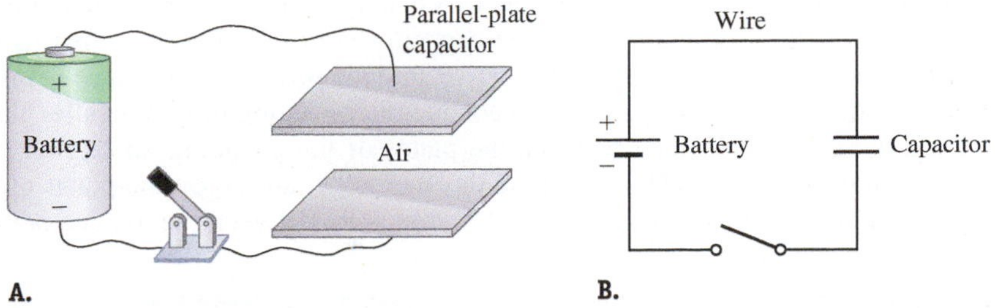

FIGURE 27.13 **A.** A battery, parallel-plate capacitor, and switch form a network. **B.** Schematic diagram for the network shown in part A. The symbol for a capacitor is similar to the symbol for a battery. However, for a capacitor, the parallel lines are the same length, and often the capacitance C is written next to the capacitor symbol.

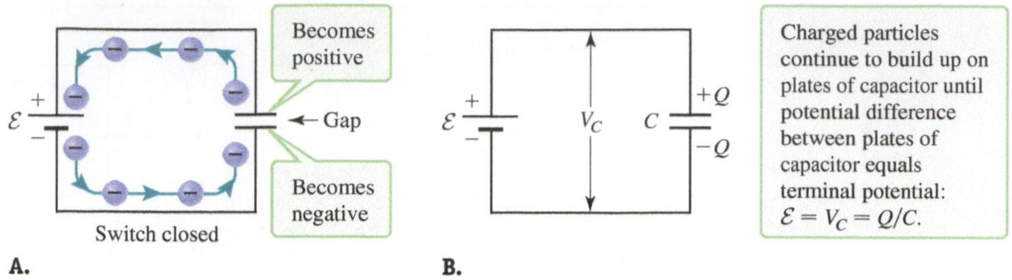

FIGURE 27.14 **A.** When the switch is closed, negative charge flows from the negative terminal of the battery toward the positive terminal. Charge cannot pass through the capacitor, so it builds up on the plates. **B.** The fully charged capacitor has the same potential difference across its plates as the battery's terminal potential.

regardless of shape, are represented in circuit diagrams by two short parallel lines of equal length.

When the switch is closed, electrons from the negative terminal begin to flow toward the positive terminal (Fig. 27.14A). This is similar to Figure 27.12B, except in that case there is no capacitor, so the particles are free to flow through the wire from the negative terminal to the positive terminal. The particles in Figure 27.14A cannot flow through the gap between the plates of the capacitor; instead, electrons build up on the lower plate. The lower plate becomes negatively charged, and then the electrons on the upper plate are repelled by those on the lower plate. Electrons flow away from the upper plate toward the battery's positive terminal. As a result, the upper plate acquires an excess positive charge exactly equal in magnitude to the excess negative charge on the lower plate.

For a while, electrons continue to build up on the lower plate and the upper plate continues to lose an equal number of electrons, which increases the potential difference between the plates of the capacitor. When that potential difference equals the battery's terminal potential, electrons stop flowing and the capacitor is said to be *fully charged* (Fig. 27.14B). It is common practice to omit the symbol Δ and represent the potential difference between the plates of a capacitor as V_C. The capacitor is fully charged when

$$V_C = \mathcal{E} \tag{27.4}$$

The amount of charge on a fully charged capacitor is found by substituting Equation 27.4 into $Q = C\Delta V$ (Eq. 27.1):

$$Q = C\mathcal{E} \tag{27.5}$$

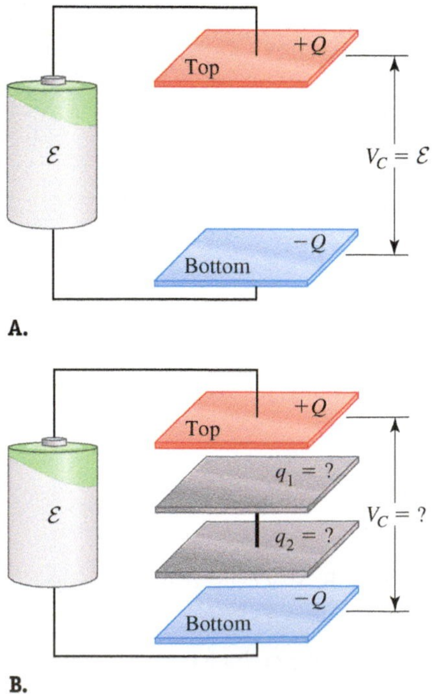

FIGURE 27.15

CONCEPT EXERCISE 27.4

Explain why electrons stop flowing when the potential difference between the plates of a capacitor equals the battery's terminal potential.

CONCEPT EXERCISE 27.5

A large parallel-plate capacitor is attached to a battery that has terminal potential $\mathcal{E}$ (Fig. 27.15A). After a period of time, the capacitor stores charge Q so that its top plate is positive and its bottom plate is negative, and the potential difference between the plates is $V_C = \mathcal{E}$. An **I**-shaped neutral conductor consisting of two parallel plates connected by a wire is slipped between the plates of the capacitor so that all four plates are parallel (Fig. 27.15B). What are the charges q_1 and q_2 on the plates of the **I**-shaped conductor? What is the potential difference V_C between the top and bottom plates of the capacitor?

EXAMPLE 27.2 Comparing Capacitors

A battery with a terminal potential of 11.7 V fully charges two different capacitors in turn. First, a capacitor with capacitance $C_1 = 31.9\ \mu\text{F}$ is connected to the battery. It is then removed and a second capacitor with double the capacitance is connected to the battery. Find and compare the charges on each capacitor and the energy they store after they have been fully charged. In the **CHECK and THINK** step, answer this question: What will happen if the second capacitor is fully charged by a battery whose terminal potential is $\frac{1}{2}(11.7)\ \text{V} = 5.85\ \text{V}$?

INTERPRET and ANTICIPATE

The same battery is used to charge both capacitors, so the terminal potential $\mathcal{E}$ is the same for both capacitors. Therefore, the capacitor with the greater capacitance will store a greater charge and more electric potential energy.

SOLVE

Substitute values into Equation 27.5 for the first capacitor. Remember that 1 F = 1 C/V.	$Q_1 = C_1\mathcal{E}$ (27.5) $Q_1 = (31.9 \times 10^{-6}\,\text{F})(11.7\,\text{V})$ $Q_1 = 3.73 \times 10^{-4}\,\text{F}\cdot\text{V} = 373\,\mu\text{C}$
Now find the charge on the second capacitor, where $C_2 = 2C_1$.	$Q_2 = C_2\mathcal{E} = 2C_1\mathcal{E}$ $Q_2 = 2(31.9 \times 10^{-6}\,\text{F})(11.7\,\text{V})$ $Q_2 = 746\,\mu\text{C}$
The second capacitor stores twice as much charge as the first capacitor, as expected.	$Q_2 = 2Q_1$

We can use either $U_E = Q^2/2C$ or $U_E = \frac{1}{2}C(\Delta V)^2$ (Eqs. 27.2 or 27.3) to find the electric potential energy stored by each capacitor. Let's arbitrarily use Equation 27.2.

Substitute values for the first capacitor.	$U_1 = \frac{1}{2}\frac{Q_1^2}{C_1} = \frac{1}{2}\frac{(3.73 \times 10^{-4}\,\text{C})^2}{(31.9 \times 10^{-6}\,\text{F})}$ $U_1 = 2.18\,\text{mJ}$
Use $Q_2 = 2Q_1$ and $C_2 = 2C_1$, and then substitute values for the second capacitor.	$U_2 = \frac{1}{2}\frac{Q_2^2}{C_2} = \frac{1}{2}\frac{(2Q_1)^2}{2C_1} = \frac{1}{2}\frac{4Q_1^2}{2C_1}$ $U_2 = \frac{Q_1^2}{C_1} = \frac{(3.73 \times 10^{-4}\,\text{C})^2}{(31.9 \times 10^{-6}\,\text{F})} = 4.36\,\text{mJ}$
As expected, the second capacitor stores twice as much energy as the first capacitor.	$U_2 = 2U_1$

CHECK and THINK

If the second capacitor were charged with a battery that has a terminal potential of 5.85 V, it would store the same amount of charge $Q_1 = Q_2$ and *half* as much energy as the first capacitor: $U_2 = \frac{1}{2}U_1$.

27-4 Capacitors in Parallel and Series

Imagine the task of designing and building an electronic network for a particular device such as a camera or travel alarm clock. The device should be powered by batteries that may be easily purchased, so your design must work with one of a small number of terminal potentials (Table 27.1). In addition, commercially available capacitors are manufactured with a limited variety of capacitances. In your design, you are

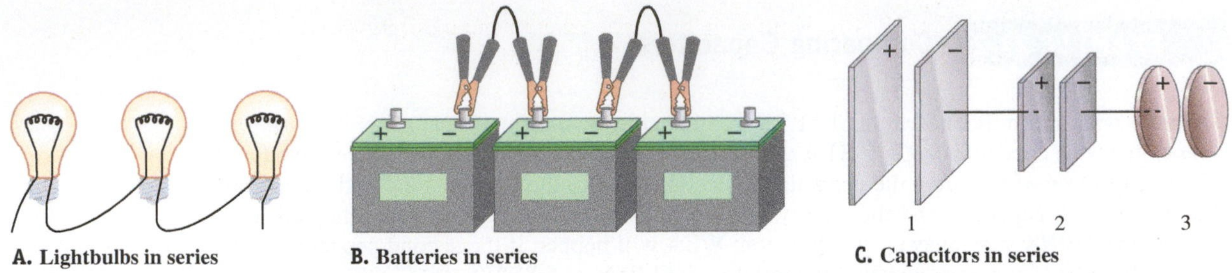

FIGURE 27.16 When elements are connected in series, one wire connects each element to the next.

likely to find you need a capacitance that cannot be purchased from any manufacturer. The way around this problem is to combine available capacitors to come up with the equivalent capacitance that you need. In this section, we learn how to find the equivalent capacitance of a combination of capacitors.

Series Capacitors

In this book, we study two ways to combine elements such as lightbulbs, batteries, or capacitors to form a network. Let's first consider a **series** network, in which one element is connected to the next with a single wire between elements. The elements in each network in Figure 27.16 are connected in series.

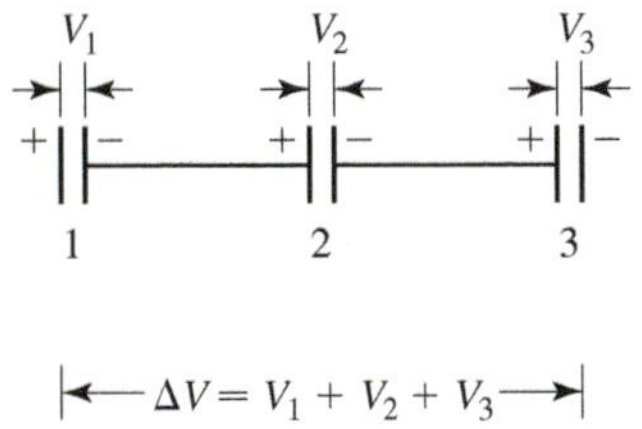

FIGURE 27.17 A schematic diagram of three capacitors connected in series.

You have probably loaded several batteries into a device. These batteries are connected in series to achieve a greater potential difference. For example, if you put two AA batteries into a flashlight, the potential difference from the free positive terminal of one battery to the free negative terminal of the other battery is 1.5 V + 1.5 V = 3.0 V. So, from your experience, you know that the potential difference across the two batteries is the sum of the terminal potentials. Likewise, the potential difference across elements connected in series is the sum of their individual potential differences. For example, the potential difference ΔV from the positive plate of capacitor 1 to the negative plate of capacitor 3 in Figure 27.17 is

$$\Delta V = V_1 + V_2 + V_3 \tag{27.6}$$

Now that we know how to find the potential difference across a series of capacitors, we can derive their equivalent capacitance.

DERIVATION Capacitors in Series

Suppose you design a network that uses a battery with terminal potential $\mathcal{E}$ to charge a capacitor, which for convenience we call the **equivalent capacitor** with capacitance C_{eq} (Fig. 27.18A). Because this equivalent capacitor is not available, you connect in series a number of capacitors that are equivalent to this one capacitor (Fig. 27.18B). We will show that the equivalent capacitance of N capacitors in series is

$$\frac{1}{C_{eq}} = \sum_{i=1}^{N} \frac{1}{C_i} = \frac{1}{C_1} + \frac{1}{C_2} + \cdots + \frac{1}{C_N} \tag{27.7}$$

To make the derivation more manageable, we'll start with $N = 3$ (Fig. 27.18) and use ideas from Concept Exercise 27.5.

All the capacitors in Figure 27.18 are fully charged, and in order for the two networks to be equivalent, the batteries must be identical.

CAPACITORS IN SERIES ✪ **Major Concept**

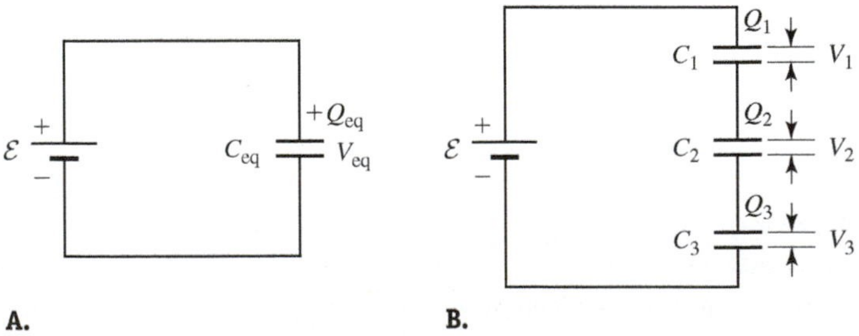

FIGURE 27.18 The capacitor in part A can be replaced by a number of capacitors connected in series as in part B.

The equivalent capacitor (Fig. 27.18A) is fully charged when the potential difference V_{eq} across its plates equals the battery's terminal potential (Eq. 27.4).

$$V_{eq} = \mathcal{E} \quad (1)$$

Similarly, when a series of capacitors is connected to the battery (Fig. 27.18B), the electrons stop flowing when the potential difference ΔV from the positive plate of capacitor 1 to the negative plate of capacitor 3 equals the battery's terminal potential $\mathcal{E}$.

$$\Delta V = \mathcal{E} \quad (2)$$

Substitute Equations 27.6 and (1) into Equation (2). Equation (3) shows that in order for the two networks to be equivalent, the voltage across the equivalent capacitor must equal the voltage across the three capacitors in series.

$$\Delta V = V_1 + V_2 + V_3 \quad (27.6)$$

$$V_1 + V_2 + V_3 = V_{eq} = \mathcal{E} \quad (3)$$

Next, we find the charge stored by the two networks. The amount of charge stored on the equivalent (single) capacitor when it is fully charged is $Q_{eq} = C_{eq}V_{eq}$ (Eq. 27.1). Then solve for V_{eq}.

For the single equivalent capacitor,

$$V_{eq} = \mathcal{E} = \frac{Q_{eq}}{C_{eq}} \quad (4)$$

To find the charge stored by the series of capacitors (Fig. 27.18B), imagine the process of charging them. Electrons flow from the battery's negative terminal and collect on the lower plate of capacitor 3, building an excess negative charge $-Q$ (Fig. 27.19). The highlighted portion in the close-up of capacitors 2 and 3 looks like an **I**-shaped conductor. Because the lower plate of capacitor 3 has a negative charge, electrons in the bottom of the **I**-shaped conductor move upward. The result is that the bottom part of the **I**-shaped conductor acquires an excess positive charge $+Q$ and the upper part acquires an excess negative charge $-Q$. Thus, capacitor 3's upper plate develops charge $+Q$ and capacitor 2's lower plate develops charge $-Q$.

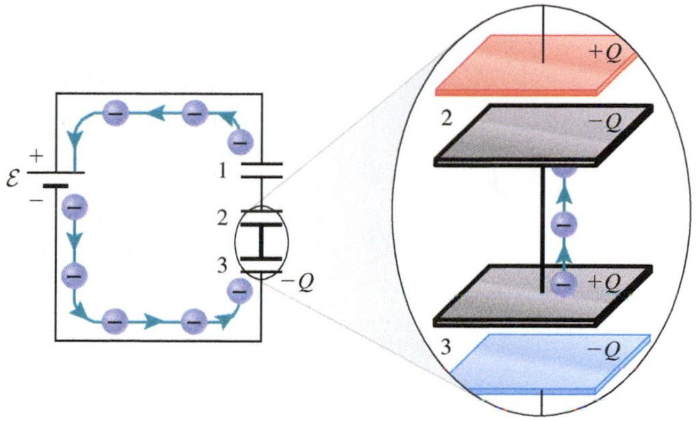

FIGURE 27.19

A similar situation occurs for capacitors 2 and 1, with each plate acquiring excess charge of either $+Q$ or $-Q$. The result is that each capacitor stores the same amount of charge.

$$Q_1 = Q_2 = Q_3 \equiv Q$$

In order for the series of capacitors (Fig. 27.18B) to function just like the equivalent capacitor in Fig. 27.18A, the charge stored on the equivalent capacitor must equal the charged stored by each individual capacitor in the series.

$$Q_1 = Q_2 = Q_3 = Q = Q_{eq} \quad (5)$$

The potential difference across each individual capacitor V_i depends on its capacitance and the amount of charge Q_{eq} it stores. So, we can rewrite Equation (3) in terms of charge and capacitance by substituting $V = Q/C$ (Eq. 27.1) for each individual capacitor.

$$V_1 + V_2 + V_3 = \mathcal{E}$$

$$\frac{Q_1}{C_1} + \frac{Q_2}{C_2} + \frac{Q_3}{C_3} = \mathcal{E}$$

Use Equation (5) to write this in terms of Q.

$$\frac{Q_{eq}}{C_1} + \frac{Q_{eq}}{C_2} + \frac{Q_{eq}}{C_3} = \mathcal{E}$$

For a series of three capacitors,

$$Q\left(\frac{1}{C_1} + \frac{1}{C_2} + \frac{1}{C_3}\right) = \mathcal{E} \quad (6)$$

We find the relationship between the capacitances of the equivalent capacitor and the series capacitors by comparing Equation (4) for the equivalent capacitor to Equation (6) for the series of three capacitors.

$$\frac{1}{C_{eq}} = \frac{1}{C_1} + \frac{1}{C_2} + \frac{1}{C_3}$$

We considered a series of three capacitors, but it could have been N capacitors, each with capacitance C_i. The charge stored by each capacitor equals the charge Q

$$\frac{1}{C_{eq}} = \sum_{i=1}^{N} \frac{1}{C_i} \ ✓ \quad (27.7)$$

Derivation continues on page 840 ▶

stored by the equivalent capacitor, and the potential differences between the plates of each capacitor add up to the potential difference between the plates of the equivalent capacitor: $V_{eq} = \sum V_i$. If we follow the same procedure in this more general case, we arrive at the general expression for the capacitance of a series of capacitors.

COMMENTS

According to Equation 27.7, the equivalent capacitance C_{eq} is smaller than the capacitance of the individual capacitors: $C_{eq} < C_1$, $C_{eq} < C_2$, ..., $C_{eq} < C_N$.

Often the quantity you need to solve for is in the denominator of Equation 27.7, so be sure to invert both sides of the equation. For example, let's find an expression for the equivalent capacitance in the case of a network consisting of two capacitors in series.

$$\frac{1}{C_{eq}} = \sum_{i=1}^{2} \frac{1}{C_i} = \frac{1}{C_1} + \frac{1}{C_2}$$

$$C_{eq} = \left(\frac{C_2 + C_1}{C_1 C_2}\right)^{-1} = \frac{C_1 C_2}{C_2 + C_1}$$

Parallel Capacitors

When you connect capacitors in series, their equivalent capacitance is *smaller* than that of the individual capacitors. When your design requires a *greater* capacitance than is available from individual capacitors, you must connect the capacitors in **parallel**. In order to connect elements in parallel, you need two wires joining each element to the next. The circuit elements in Figure 27.20 are connected in parallel.

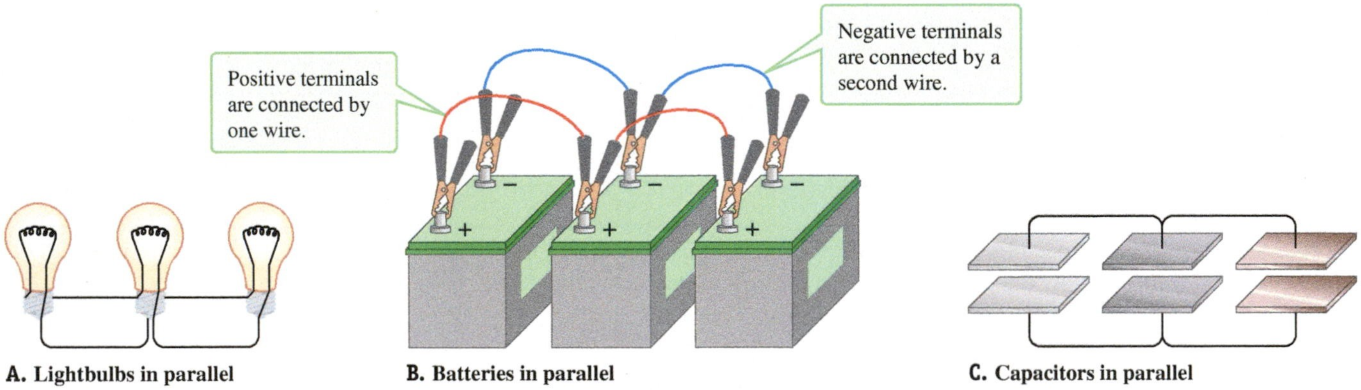
Positive terminals are connected by one wire.
Negative terminals are connected by a second wire.
A. Lightbulbs in parallel
B. Batteries in parallel
C. Capacitors in parallel

FIGURE 27.20 When circuit elements are connected in parallel, two wires connect each element to the next.

DERIVATION Capacitors in Parallel

We will show that N parallel capacitors, each with capacitance C_i, have an equivalent capacitance given by

$$C_{eq} = \sum_{i=1}^{N} C_i \qquad (27.8)$$

CAPACITORS IN PARALLEL

Major Concept

To make this derivation more manageable, we start by deriving an expression in the case of just $N = 2$ capacitors in parallel.

To develop a conceptual understanding, imagine that the two capacitors are constructed from one large one as follows: Start with a capacitor consisting of two large parallel plates separated by air. The plates have been charged so that the top plate has a net positive charge $+Q$ and the bottom plate has a net negative charge $-Q$ (Fig. 27.21A). The potential difference between the plates is V_C. Two wires have been used to connect a point on the left side of each plate to a point on the right side. The capacitance of this original capacitor is $C_{eq} = Q/V_C$.

Now imagine cutting each plate into two unequal pieces so that one plate is 1/3 the size of the original plate and the other is 2/3 (Fig. 27.21B). The smaller plate has 1/3 of the original charge, and the larger plate has 2/3 of the original charge. So, the charges on the top plates are $+Q/3$ and $+2Q/3$, while the charges on the bottom plates are $-Q/3$ and $-2Q/3$. The potential is uniform over the surface of a conductor, so the top two plates plus the connecting wire are at the same potential and the bottom two plates plus the connecting wire are at a different, lower potential. Cutting the plates does not change the potential difference between the top plates and the bottom plates; the potential difference across those plates still equals V_C.

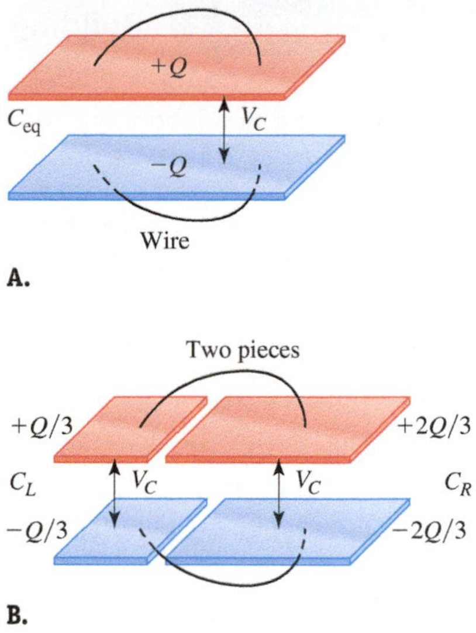

FIGURE 27.21 Constructing capacitors connected in parallel by cutting capacitor plates that are connected by thin wires.

Cutting the plates makes two new parallel-plate capacitors. Because capacitance depends on geometry, these two newly made capacitors do not have the same capacitance as the original device. Label them C_L and C_R for *l*eft and *r*ight capacitors, and use $C = q/V_C$, where q is the charge stored by each.	$C_L = \frac{Q/3}{V_C} = \frac{1}{3}\frac{Q}{V_C}$ (1) $C_R = \frac{2Q/3}{V_C} = \frac{2}{3}\frac{Q}{V_C}$ (2)
Substitute $C_{eq} = Q/V_C$ into Equations (1) and (2).	$C_L = \frac{1}{3}C_{eq}$ (3) $C_R = \frac{2}{3}C_{eq}$ (4)
The sum of C_L and C_R is equal to the capacitance of the original large capacitor.	$C_L + C_R = \frac{1}{3}C_{eq} + \frac{2}{3}C_{eq} = C_{eq}$
We imagined cutting the original capacitor into two pieces, but we could have cut it into any number of pieces N. The potential difference between the plates of each capacitor would still be the same as the potential difference V_C between the plates of the equivalent capacitor, and the total charge stored on the capacitors in parallel would equal the charge stored on the equivalent capacitor: $Q = \sum Q_i$. When there are N capacitors in parallel, the equivalent capacitance C_{eq} is found by adding the individual capacitances.	$C_{eq} = \sum_{i=1}^{N} C_i$ ✓ (27.8)

COMMENTS

Equations (3) and (4) show that the smallest capacitor (the left one) has the smallest capacitance.

CONCEPT EXERCISE 27.6

CASE STUDY **Capacitors for a Thompson coil**

The Thompson coil experimenters (page 829) claimed they used six identical capacitors connected in parallel. In Example 27.1, we found that the equivalent capacitance is 94 μF. What is the capacitance of each capacitor?

EXAMPLE 27.3 Building a Network of Capacitors

You are working in a laboratory to build a network from a design (Fig. 27.22). You have a dozen batteries with terminal potential $\mathcal{E}_0 = 1.5$ V and a dozen capacitors with capacitance $C_0 = 9.0\ \mu$F. Draw a schematic diagram showing how you would build this network from the components in the laboratory.

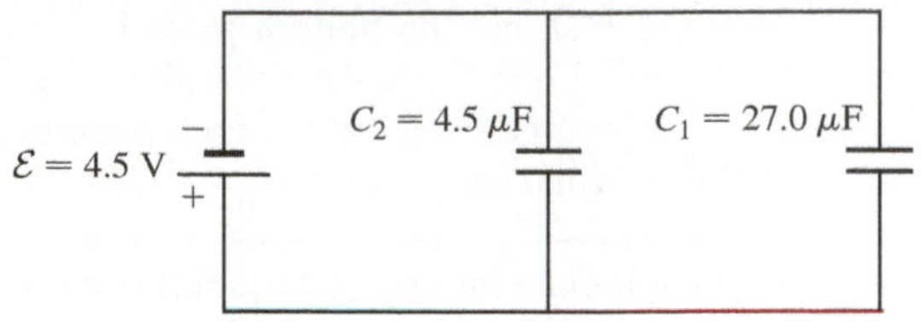

FIGURE 27.22 How can we build this network from 1.5-V batteries and 9.0-μF capacitors?

INTERPRET and ANTICIPATE

The design requires a greater terminal potential than that of an individual battery. When batteries are connected in series, their terminal potentials add, so we will use several batteries in series. The design requires C_1 to be greater than the capacitance of the laboratory's capacitors. When capacitors are connected in parallel, the resulting capacitance is the sum of the individual capacitances. So, to get C_1, we need to connect some of the laboratory's capacitors in parallel. The design also requires C_2 to be smaller than the capacitance of the laboratory's capacitors. When capacitors are connected in series, the resulting capacitance is smaller than the individual capacitances. So, to get C_2, we need to connect some of the laboratory's capacitors in series.

SOLVE

All the laboratory's batteries have the same terminal potential $\mathcal{E}_0 = 1.5$ V, so connecting $n = 3$ of these in series gives the required $\mathcal{E} = 4.5$ V.

$$\mathcal{E} = n\mathcal{E}_0$$

$$n = \frac{\mathcal{E}}{\mathcal{E}_0} = \frac{4.5\ \text{V}}{1.5\ V} = 3$$

All the laboratory capacitors have the same capacitance $C_0 = 9.0\ \mu$F, so connecting $n_1 = 3$ of these in parallel gives the required C_1 (Eq. 27.8).

$$C_{eq} = \sum_{i=1}^{N} C_i \tag{27.8}$$

$$C_1 = n_1 C_0$$

$$n_1 = \frac{C_1}{C_0} = \frac{27.0\ \mu\text{F}}{9.0\ \mu\text{F}} = 3$$

We need to connect $n_2 = 2$ of the 9.0-μF capacitors in series to achieve the required C_2 (Eq. 27.7).

$$\frac{1}{C_{eq}} = \sum_{i=1}^{N} \frac{1}{C_i} \tag{27.7}$$

$$\frac{1}{C_2} = n_2 \frac{1}{C_0}$$

$$n_2 = \frac{C_0}{C_2} = \frac{9.0\ \mu\text{F}}{4.5\ \mu\text{F}} = 2$$

As shown in Figure 27.23, we need three batteries connected in series, three capacitors connected in parallel, and two more capacitors connected in series.

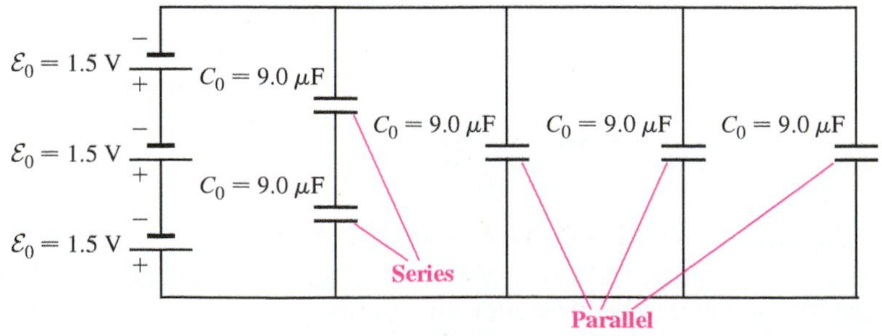

FIGURE 27.23

CHECK and THINK

One way to check our network is to think about other equivalent networks. For example, the original design (Fig. 27.22) shows two capacitors in parallel. We can find the equivalent capacitance of these two capacitors and compare it to the equivalent capacitance of the five capacitors in Figure 27.23. First, use Equation 27.8 to find the equivalent capacitance of C_1 and C_2.

For the original design (Fig. 27.22),

$$C_{eq} = C_1 + C_2 = 27.0\,\mu\text{F} + 4.5\,\mu\text{F}$$

$$C_{eq} = 31.5\,\mu\text{F}$$

For a combination of capacitors in series and parallel, as in Figure 27.23, we must use both Equations 27.7 (series) and 27.8 (parallel). As required, the equivalent capacitances are equal.

For the actual network (Fig. 27.23),

$$C_{eq} = C_{\text{parallel}} + C_{\text{series}}$$

$$C_{eq} = 3C_0 + \left(\frac{2}{C_0}\right)^{-1} = 3C_0 + \frac{C_0}{2}$$

$$C_{eq} = \frac{7}{2}C_0 = \frac{7}{2}(9\,\mu\text{F}) = 31.5\,\mu\text{F}$$

27-5 Capacitance: Special Cases

The capacitance of a capacitor depends only on its geometry and on the insulator between the plates. In this section, we develop a problem-solving strategy for finding the capacitances for different geometries, assuming a vacuum between the plates. For most practical purposes, air between the plates can be modeled as a vacuum. We apply this strategy to find the capacitances for two special cases: a parallel-plate capacitor and a cylindrical capacitor. The special case of a spherical capacitor is left as homework (Problem 32).

PROBLEM-SOLVING STRATEGY

Finding Capacitance

SOLVE

Step 1 Assume the capacitor plates are charged so that one has positive charge $+q$ and the other has negative charge $-q$. Then use Gauss's law to find an expression for the **electric field** between the plates. Refer to the problem-solving strategy "Finding the Electric Field Using Gauss's Law" on page 765.

Step 2 After the electric field is found, you need to complete a **path integral** to find the electric potential difference V_C between the plates. Choose a path from the positive plate to the negative plate, so the limits of integration are $r_i = r_+$ and $r_f = r_-$. With this path, the integral you need to solve is

$$V_C = \int_{r_+}^{r_-} E\,dr \qquad (27.9)$$

Because the positive plate is at a higher potential than the negative plate, this potential difference is positive ($V_C > 0$).

Step 3 When V_C is written in terms of Q (the charge stored by the capacitor), use $C = Q/V_C$ (Eq. 27.1) to find the capacitance.

CHECK and THINK

In addition to checking the **dimensions**, you should make sure the expression you find for capacitance **depends only on geometric properties.**

EXAMPLE 27.4 Special Case: Parallel-plate Capacitor

Two large conducting plates make up a parallel-plate capacitor (Fig. 27.24). The left plate has positive charge $+Q$, and the right plate has negative charge $-Q$. The area of each plate is A, and the plates are separated by a distance d. Show that the capacitance is

$$C = \frac{\varepsilon_0 A}{d} \quad (27.10)$$

CAPACITANCE OF PARALLEL-PLATE CAPACITOR Special Case

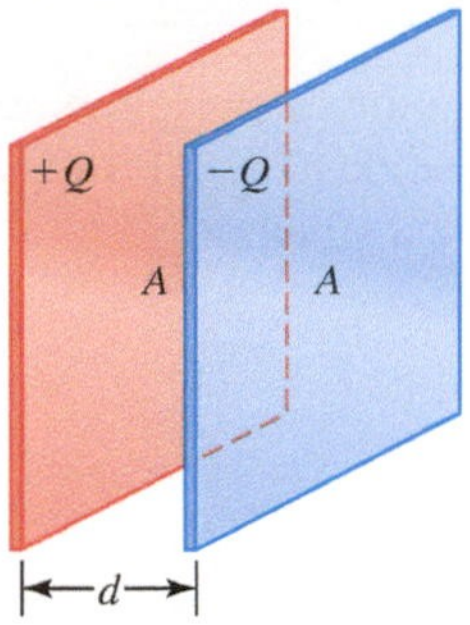

FIGURE 27.24 A parallel-plate capacitor.

INTERPRET and ANTICIPATE

The area of the plates is large compared to their separation, and because we are interested in the region between the plates, we can model each one as an infinitely large plane. Thus, we can ignore the slight bend in the electric field near the edge of the plates and model them as sources with planar symmetry (an important step for using Gauss's law). Use the three problem-solving steps to find the capacitance, assuming either a vacuum or air between the plates.

SOLVE

Step 1 Use Gauss's law to find $\vec{E}$. Finding the **electric field** between the plates of a parallel-plate capacitor is similar to finding the electric field just outside any charged conductor (Section 25-7). The electric field points from the positive plate to the negative plate (Fig. 27.25). Due to planar symmetry, the electric field is uniform and the electric field lines are evenly spaced and parallel. A Gaussian cylinder exploits this symmetry.

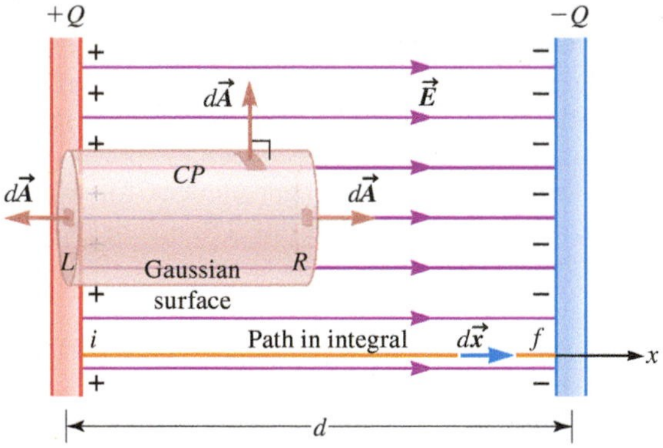

FIGURE 27.25

Integrate to find the flux through the closed Gaussian surface. Divide the integral into three parts: one for the left cap L, one for the curved part CP, and one for the right cap R. The integral over L is zero because that cap is inside the conductor, and the electric field is zero within the body of a conductor. The integral over CP is zero because $\vec{E}$ is perpendicular to $d\vec{A}$ for the curved part. The integral over R is positive because $\vec{E}$ is parallel to $d\vec{A}$. Because E is constant, pull it outside the integral, and then the integral is just the area of either cap, A_{cap}.	$\Phi_E = \oint \vec{E}\cdot d\vec{A}$ $\Phi_E = \int_L \vec{E}\cdot d\vec{A} + \int_{CP} \vec{E}\cdot d\vec{A} + \int_R \vec{E}\cdot d\vec{A}$ $\Phi_E = 0 + 0 + \int_R E\,dA = EA_{cap}$
The surface charge density σ is the total charge Q on the plate's surface divided by the area A of the plate (Eq. 24.10). The charge enclosed by the Gaussian surface is the surface charge density multiplied by the area A_{cap} of either cap.	$\sigma = \frac{Q}{A}$ (24.10) $q_{in} = \sigma A_{cap} = QA_{cap}/A$
Find the electric field by substituting the flux Φ_E and charge q_{in} into Gauss's law (Eq. 25.7).	$\Phi_E = \frac{q_{in}}{\varepsilon_0}$ (25.7) $EA_{cap} = \frac{1}{\varepsilon_0}\frac{Q}{A}A_{cap}$ $E = \frac{1}{\varepsilon_0}\frac{Q}{A}$ (27.11)

Step 2 Complete the **path integral** to find V_C. When using Equation 27.9, choose a path parallel to an electric field line and pointing from the positive plate to the negative plate. In this case, the path is in the positive x direction (Fig. 27.25). Both $d\vec{x}$ and $\vec{E}$ are in the positive x direction. Because we already took these directions into account, all we need are the magnitudes.

$$V_C = \int_{r_+}^{r_-} E\,dr \tag{27.9}$$

$$V_C = \int_{x_+}^{x_-} \left(\frac{1}{\varepsilon_0}\frac{Q}{A}\,dx\right)$$

$$V_C = \left(\frac{1}{\varepsilon_0}\frac{Q}{A}\right)\int_{x_+}^{x_-} dx$$

The integral is the length of the path from the positive to the negative plate. We have labeled the separation between the plates d (Fig. 27.25).

$$V_C = \left(\frac{1}{\varepsilon_0}\frac{Q}{A}\right)(x_- - x_+)$$

$$V_C = \frac{1}{\varepsilon_0}\frac{Q}{A}d \tag{27.12}$$

Step 3 Use $C = Q/V_C$ (Eq. 27.1) to find the capacitance. Solve Equation 27.12 for Q/V_C.

$$C = \frac{\varepsilon_0 A}{d} \; ✓ \tag{27.10}$$

CHECK and THINK

1. The permittivity constant ε_0 has the dimensions $C^2/(F \cdot L^2)$, where C is charge (Eq. 25.8). So, the **dimensions** of Equation 27.10 are charge per voltage as expected.
2. Because ε_0 is a constant, the capacitance **depends only on geometric properties** (A and d) as expected. The capacitance does not depend on the charge stored or the potential difference between the plates.

$$[\![C]\!] = \left[\!\!\left[\frac{\varepsilon_0 A}{d}\right]\!\!\right]$$

$$[\![C]\!] = \frac{C^2}{F \cdot L^2} \cdot \frac{L^2}{L} = \frac{C^2}{F \cdot L} = C \cdot \frac{C}{F \cdot L}$$

$$[\![C]\!] = \frac{C}{V}$$

Before we consider the next special case, let's express the permittivity constant $\varepsilon_0 = 8.85 \times 10^{-12}\,C^2/(N \cdot m^2)$ (Eq. 25.8) in more convenient SI units. According to Equation 27.10, the dimensions of the permittivity constant are capacitance (CAP) per length (L): $[\![\varepsilon_0]\!] = CAP/L$. So, we can use $1\,F = 1\,C/V$ to write $\varepsilon_0 = 8.85 \times 10^{-12}\,F/m = 8.85\,pF/m$. These units for ε_0 are convenient when we work with capacitance.

EXAMPLE 27.5 Special Case: Cylindrical Capacitor

A cylindrical capacitor of length L is made up of an inner plate that is a long solid cylinder with radius r_{in} and an outer plate that is a long hollow cylinder with inner radius r_{out} (Fig. 27.26). The inner plate has positive charge $+Q$, and the outer plate has negative charge $-Q$. Show that the capacitance is

$$C = \frac{2\pi\varepsilon_0 L}{\ln(r_{out}/r_{in})} \tag{27.13}$$

INTERPRET and ANTICIPATE

The plates are very long compared to the radius r_{out}, so we can ignore the slight bend in the electric field near the edge of the plates and model them as sources with linear symmetry. We find the capacitance of a cylindrical capacitor assuming either a vacuum or air between the plates.

CAPACITANCE OF CYLINDRICAL CAPACITOR

Special Case

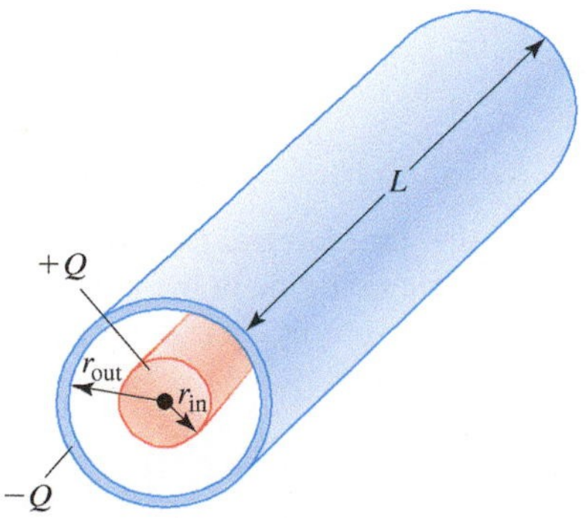

FIGURE 27.26 A cylindrical capacitor.

Example continues on page 846 ▶

SOLVE

Step 1 Use Gauss's law to find $\vec{E}$. Finding the **electric field** between the plates of a cylindrical capacitor is similar to finding the electric field due to a charged rod (Section 25-4). The electric field points from the positive plate to the negative plate (Fig. 27.27). A Gaussian cylinder of radius r and length ℓ aligned with the axes of the cylindrical plates exploits this symmetry.

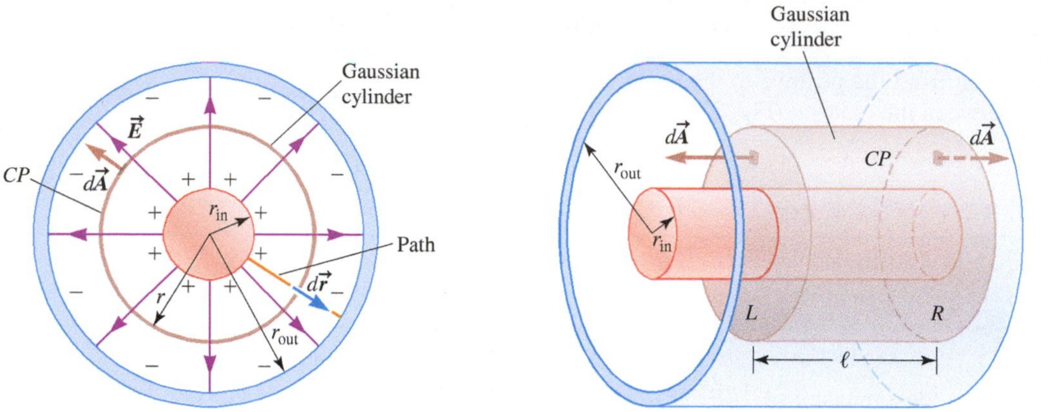

FIGURE 27.27 Gaussian cylinder aligned with the capacitor's axis.

Integrate to find the flux through the closed Gaussian surface. Divide the integral into three parts: one for the left cap L, one for the curved part CP, and one for the right cap R. The integrals over L and R are zero because $\vec{E}$ is perpendicular to $d\vec{A}$ for both caps. The integral over CP is positive because $\vec{E}$ is parallel to $d\vec{A}$ for every point on the curved part. Bring E outside the integral because it is constant. Write the area A_{CP} of the curved part in terms of the circumference $2\pi r$ of the Gaussian cylinder and its length ℓ.

$$\Phi_E = \oint \vec{E} \cdot d\vec{A}$$

$$\Phi_E = \int_L \vec{E} \cdot d\vec{A} + \int_{CP} \vec{E} \cdot d\vec{A} + \int_R \vec{E} \cdot d\vec{A}$$

$$\Phi_E = 0 + \int_{CP} E\,dA + 0$$

$$\Phi_E = EA_{CP} = E2\pi r\ell$$

The linear charge density λ is the total charge Q on the plate's surface divided by the length L of the plate (Eq. 24.12). The charge enclosed by the Gaussian surface is the linear charge density multiplied by the length ℓ of the Gaussian cylinder.

$$\lambda = \frac{Q}{L} \quad (24.12)$$

$$q_{in} = \lambda\ell = \frac{Q}{L}\ell$$

Find the electric field by substituting the flux Φ_E and charge q_{in} into Gauss's law (Eq. 25.7).

$$\Phi_E = \frac{q_{in}}{\varepsilon_0} \quad (25.7)$$

$$E2\pi r\ell = \frac{1}{\varepsilon_0}\frac{Q}{L}\ell$$

$$E = \frac{1}{2\pi\varepsilon_0}\frac{Q}{L}\frac{1}{r}$$

Step 2 Complete the **path integral** to find V_C. As shown in Figure 27.27, the path points along an electric field line in the positive r direction. Both $d\vec{r}$ and $\vec{E}$ are in the positive r direction, so only their magnitudes are needed. The antiderivative is the natural logarithm (Appendix A).

$$V_C = \int_{r_{in}}^{r_{out}} E\,dr \quad (27.9)$$

$$V_C = \int_{r_{in}}^{r_{out}} \left(\frac{1}{2\pi\varepsilon_0}\frac{Q}{L}\frac{1}{r}\right) dr$$

$$V_C = \frac{1}{2\pi\varepsilon_0}\frac{Q}{L}(\ln r_{out} - \ln r_{in})$$

$$V_C = \frac{1}{2\pi\varepsilon_0}\frac{Q}{L}\ln\frac{r_{out}}{r_{in}} \quad (1)$$

Step 3 Use $C = Q/V_C$ (Eq. 27.1) to find the capacitance. Solve Equation (1) for Q/V_C.

$$C = \frac{2\pi\varepsilon_0 L}{\ln(r_{out}/r_{in})} \quad \checkmark \qquad (27.13)$$

CHECK and THINK

1. The denominator has no **dimensions**, so the dimensions come from the numerator. We find the dimensions are capacitance.
2. As expected, the capacitance **depends only on geometric properties**—the radii of the cylinders and their length.

$$[\![C]\!] = \left[\!\!\left[\frac{2\pi\varepsilon_0 L}{\ln(r_{out}/r_{in})}\right]\!\!\right] = [\![\varepsilon_0]\!] \cdot [\![L]\!]$$

$$[\![C]\!] = \frac{\text{CAP}}{\text{L}} \cdot \text{L} = \text{CAP}$$

EXAMPLE 27.6 Making a 1-F Capacitor

The capacitance of most commonly used capacitors is small ($\sim 1\,\text{pF} < C < \sim 1\,\mu\text{F}$). A capacitance of 1 F is very large. In this problem, we estimate the size of a 1-F capacitor using cylindrical geometry. (In Problem 41, you will consider parallel-plate geometry.) If the radius of the outer plate of a 1-F cylindrical capacitor is $r_{out} = 1.5$ mm (about the size of sewing needle) and the radius of the inner plate is $r_{in} = 50\ \mu$m, how long must the capacitor be?

INTERPRET and ANTICIPATE

The capacitance of a cylindrical capacitor is directly proportional to the length of its plates, so we expect the capacitor to be very long.

SOLVE

Solve Equation 27.13 for L.

$$L = \frac{C \ln(r_{out}/r_{in})}{2\pi\varepsilon_0} = \frac{(1\,\text{F}) \ln\left[(1.5 \times 10^{-3}\,\text{m})/(50 \times 10^{-6}\,\text{m})\right]}{2\pi(8.85 \times 10^{-12}\,\text{F/m})}$$

$$L \approx 6 \times 10^{10}\,\text{m}$$

CHECK and THINK

As expected, this is a very long capacitor. In fact, it would stretch to about 40% of the distance to the Sun! Figure 27.28 shows an inexpensive 1-F capacitor that is about the size of a piece of candy. How is it possible to build such a small capacitor with such a large capacitance? The answer is that the space between the plates must be filled with an insulator (other than air).

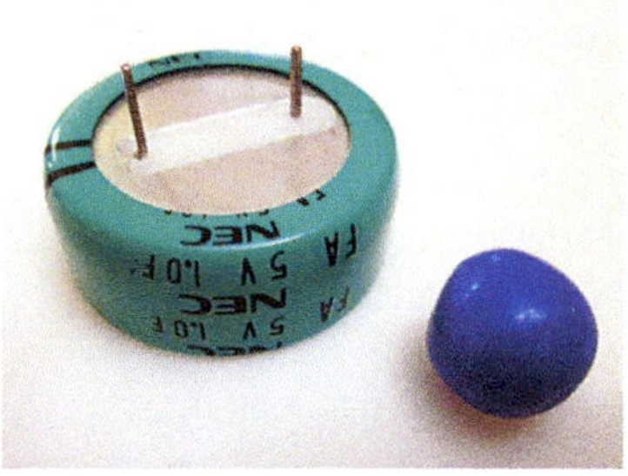

FIGURE 27.28 A 1-F capacitor, just a little larger than a piece of candy.

Debora Katz

27-6 Dielectrics

So far, we have been considering capacitors with either a vacuum or air between the plates. Capacitance can be increased by filling the space between the plates with an insulator such as plastic, paper, or pure water. When the capacitor is charged, an electric field is induced within the insulator. The term **dielectric** describes an insulator that can support an induced electric field. In this section, we modify the problem-solving strategy on page 843 to find capacitance in the presence of a dielectric. (Later, we'll develop a shortcut for finding the capacitance.)

DIELECTRIC ✪ Major Concept

With a dielectric between the plates, the **electric field** found in **Step 1** must be modified to include the field induced in the dielectric. Figure 27.29 shows how this electric field is induced for the special case of a charged parallel-plate capacitor. Start by imagining the familiar vacuum between the plates; then the capacitor's

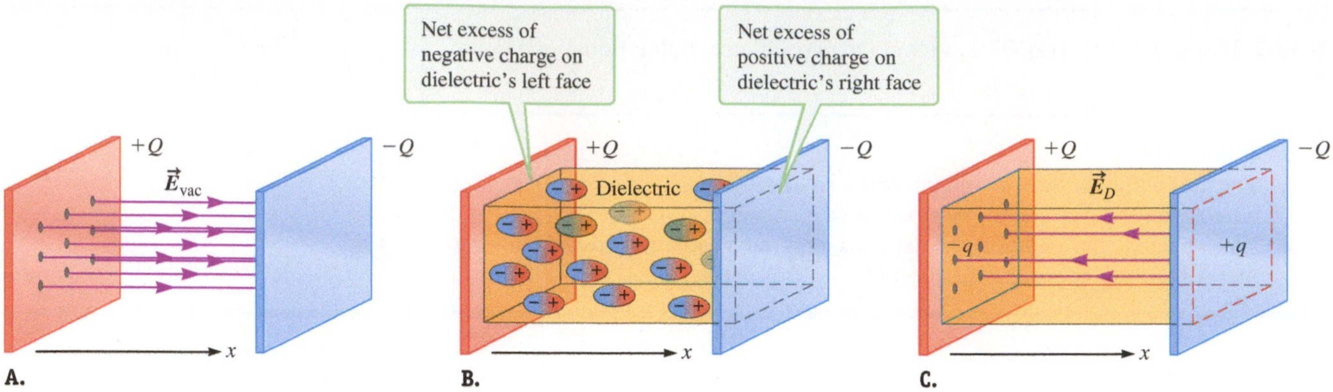

FIGURE 27.29 A. The electric field produced by the plates of the capacitor points in the positive x direction. **B.** The dielectric's molecules are polarized. **C.** The dielectric produces an electric field in the negative x direction.

electric field $\vec{E}_{vac}$ points to the right in the positive x direction (Fig. 27.29A). This electric field polarizes the molecules inside the dielectric so that each molecule's electrons are shifted slightly to the left (Fig. 27.29B). (If the dielectric is made up of dipole molecules such as water molecules, the capacitor's electric field causes the dipole molecules to line up.) The result is that the dielectric creates a second electric field $\vec{E}_D$ pointing to the left in the negative x direction (Fig. 27.29C). The net electric field between the capacitor plates is

$$\vec{E}_C = \vec{E}_{vac} + \vec{E}_D = (E_{vac} - E_D)\hat{\imath} \tag{27.14}$$

As represented by the relative number of field lines in Figure 27.29, the electric field produced by the dielectric is weaker than the vacuum-filled capacitor's electric field: $E_D < E_{vac}$. So, the net electric field points in the positive x direction (the same direction as the vacuum-filled capacitor's electric field) and is weaker than the empty capacitor's electric field. Mathematically, we write the total electric field in terms of the vacuum-filled capacitor's electric field:

$$\vec{E}_C = \frac{1}{\kappa}\vec{E}_{vac} \tag{27.15}$$

where κ is called the **dielectric constant**. So, in Step 1, we modify the electric field we found for a vacuum-filled capacitor by dividing it by the dielectric constant.

The dielectric constant is a positive number that depends only on the type of material; κ is unitless and its value is always greater than or equal to 1. If no dielectric is present (if there is a vacuum between the plates), then $\kappa = 1$. Table 27.2 lists the dielectric constants for a number of materials. Because the dielectric constant of air is very close to 1, we are able to model an air-filled capacitor as a vacuum-filled capacitor.

In **Step 2**, we need to replace the vacuum-filled electric field with E_C (Eq. 27.15). Then the potential difference between the plates is given by a modified **path integral**:

$$V_C = \int_{r_+}^{r_-} E_C\,dr = \int_{r_+}^{r_-} \frac{E_{vac}}{\kappa}\,dr = \frac{1}{\kappa}\int_{r_+}^{r_-} E_{vac}\,dr$$

where E_{vac} is the electric field magnitude if there were a vacuum between the plates. So, we can write the potential difference between the plates filled with a dielectric in terms of the vacuum potential difference:

$$V_C = \frac{1}{\kappa}V_{vac} \tag{27.16}$$

Because the dielectric constant $\kappa \geq 1$, the potential difference when a dielectric is present is *less* than when there is a vacuum.

TABLE 27.2 Dielectrics at room temperature.

Material	Dielectric constant κ	Dielectric strength ($\times 10^6$ V/m)
Air (at 1 atm)	1.00054	3
Bakelite	4.9	24
Mica	7	150
Mylar	3.2	7
Neoprene rubber	6.7	12
Nylon	3.4	14
Paper	3.5	16
Paraffin	2.2	10
Polycarbonate	2.8	30
Polyester	3.3	60
Polystyrene	2.6	24
Polyvinyl chloride	3.4	40
Porcelain	6.5	12
Pyrex	4.7	14
Quartz	4.3	8
Silicone oil	2.5	15
Strontium titanate	310	8
Teflon	2.1	60
Titania ceramic	130	
Vacuum	1	n/a
Water (liquid)	80	

Finally, in **Step 3**, we divide the charge Q stored by the potential difference V_C between the plates (Eq. 27.16):

$$C = \frac{Q}{V_C} = \frac{Q}{V_{vac}/\kappa} = \kappa \frac{Q}{V_{vac}}$$

CAPACITANCE IN THE PRESENCE OF A DIELECTRIC ✪ Major Concept

$$C = \kappa C_{vac} \tag{27.17}$$

where C_{vac} is the capacitance the capacitor would have if there were a vacuum between the plates. In practice, once you have an expression for the capacitance of a vacuum-filled capacitor, you only need to multiply by the dielectric constant to find the capacitance when a dielectric is present. (You may not need the entire problem-solving strategy.) Because the dielectric constant $\kappa \geq 1$, the capacitance when a dielectric is present is *greater* than when there is a vacuum.

Besides increasing the capacitance, a dielectric has another practical function. It provides a practical way to bring the plates of a capacitor close together without touching, which also increases the capacitance. For example, the capacitance of a parallel-plate capacitor with a dielectric is

$$C = \kappa C_{vac} = \kappa \frac{\varepsilon_0 A}{d} \tag{27.18}$$

where C_{vac} is given by Equation 27.10. If the dielectric is sandwiched between the parallel plates, the separation between the plates d equals the thickness of the dielectric. Thus, a very thin dielectric increases the capacitance.

However, there are practical limits to how thin a dielectric can be and how much charge can be stored on a capacitor's plates. As the charge increases, the electric field inside the dielectric becomes stronger. At some point, the electric field is strong enough that the dielectric *breaks down* and behaves as a conductor, allowing charge to flow through. This is not a new idea to this chapter; we considered lightning in Chapter 24. Air is normally a good insulator, but when a lot of charge builds up in a storm, air breaks down and becomes a conductor, and lightning results. The **dielectric strength** of an insulator is the maximum electric field it can tolerate without breaking down and becoming a conductor. The dielectric strengths of many materials are listed in Table 27.2. The electric field and dielectric strength both depend on the potential difference between the plates and the plate separation. In practice, the plate separation is fixed, and off-the-shelf capacitors are rated in terms of their maximum operating potential. In addition, dielectric strength depends on temperature, so capacitors are rated in terms of their operating temperatures also.

EXAMPLE 27.7 CASE STUDY Size of the Plates

We found in Concept Exercise 27.6 that each of the six capacitors used by the experimenters in the case study had a fairly large capacitance ($C = 16\ \mu F$). Such capacitors may be made by layering the plates of the capacitor with two dielectric sheets (Fig. 27.30). The layers are then rolled together so that the capacitor fits inside a cylindrical case something like a soda can. Imagine uncoiling one of these capacitors so that it resembles a large parallel-plate capacitor with a polyvinyl chloride dielectric.

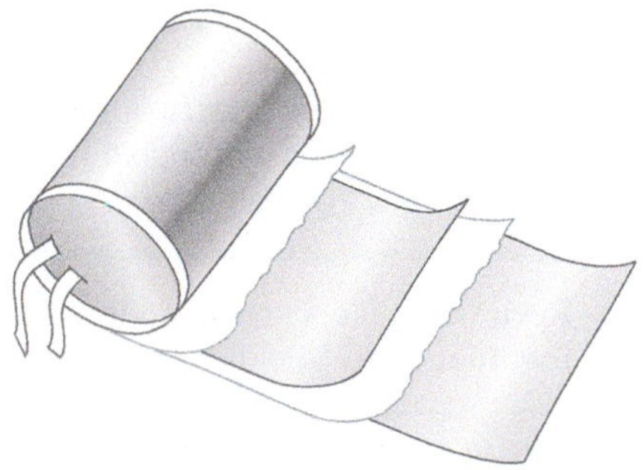

FIGURE 27.30 Layers of dielectric and metal are rolled together to form a compact capacitor with a large capacitance.

A The experimenters charged the capacitors until the potential difference between the plates was 1500 V. If this produces an electric field that is about half the dielectric strength, what is the thickness of the polyvinyl chloride dielectric?

Example continues on page 850 ▶

INTERPRET and ANTICIPATE

We must first relate the electric field E_C produced by a parallel-plate capacitor to the potential difference and the distance between the plates. Then we set E_C equal to half the dielectric strength and solve for the thickness of the polyvinyl chloride.

SOLVE

Divide $V_C = V_{vac}/\kappa$ by $E_C = E_{vac}/\kappa$ and the dielectric constant cancels out.

$$\frac{V_C}{E_C} = \frac{V_{vac}}{E_{vac}} \quad (1)$$

Substitute Equations 27.11 and 27.12 for the vacuum-filled parallel-plate capacitor's electric field and the potential difference into Equation (1). Solve for the plate separation d. [Equation 2 is used frequently; also notice that it is similar to Equation 26.16, $\Delta V = -(\sigma/2\varepsilon_0)\Delta x$.]

$$\frac{V_C}{E_C} = \left(\frac{1}{\varepsilon_0}\frac{Q}{A}d\right)\Big/\left(\frac{1}{\varepsilon_0}\frac{Q}{A}\right)$$

$$d = \frac{V_C}{E_C} \quad (2)$$

The potential between the plates is 1500 V. For the electric field, use half the dielectric strength of polyvinyl chloride found in Table 27.2.

$$d = \frac{1500\text{ V}}{\frac{1}{2}(40 \times 10^6\text{ V/m})} = 7.5 \times 10^{-5}\text{ m}$$

CHECK and THINK

This is a little thinner than a lightweight sheet of paper.

B Assuming there is no space between the plates and the dielectric, find the area of the plates. If the plates and the dielectric are rolled into a cylinder 0.18 m tall, what is the length of one plate?

INTERPRET and ANTICIPATE

We estimated the thickness of the dielectric in part A. We can use this thickness to find the area of the plates because we already know the capacitance.

SOLVE

The capacitance of a dielectric-filled parallel-plate capacitor is given by Equation 27.18. Use Table 27.2 to find κ for polyvinyl chloride.

$$C = \kappa\frac{\varepsilon_0 A}{d} \quad (27.18)$$

$$A = \frac{Cd}{\kappa\varepsilon_0} = \frac{(1.6 \times 10^{-5}\text{ F})(7.5 \times 10^{-5}\text{ m})}{(3.4)(8.85 \times 10^{-12}\text{ F/m})}$$

$$A = 39.9\text{ m}^2 \approx 40\text{ m}^2$$

The plates are long, skinny rectangles of height $h = 0.18$ m.

$$A = h\ell$$

$$\ell = \frac{A}{h} = \frac{40\text{ m}^2}{0.18\text{ m}} = 2.2 \times 10^2\text{ m}$$

$$\ell = 220\text{ m}$$

CHECK and THINK

The plates are longer than two football fields. It may be difficult to imagine rolling plates this long into a device of a reasonable size. However, the paper on a single roll of toilet paper is about the length of one football field.

C Review the facts presented by the experimenters. Does their Thompson coil apparatus seem feasible, or do you think their description is a hoax?

The apparatus seems feasible. The energy required to launch the ring could come from a large potential difference (1500 V) across six capacitors, each with a large capacitance (16 μF). Such capacitors are possible to construct and available from many manufacturers, although high-voltage capacitors can be expensive. The experimenters claim that they worked hard to find affordable capacitors. Other experimenters claim to have built their own high-voltage capacitors in order to save money, which increases already serious safety concerns.

The amount of energy stored by the capacitor bank is great enough to be fatal if a person touched one of the positive plates and one of the negative plates. In fact, the energy present is sufficient to vaporize metal. The experimenters found this to be a problem and often damaged parts of their setup.

EXAMPLE 27.8 Cylindrical Nerves

The nerve cells in the human body and in other animals are modeled as very long cylindrical capacitors. Portions of some nerves are covered with a layer of fat, known as myelin, functioning as the dielectric ($\kappa = 7$) between two plates in the cylindrical capacitor model. The immune system in some people attacks this layer of fat, changing their nerve cells' capacitance and resulting in multiple sclerosis (MS). Normally, the inner radius of the myelin layer is 2×10^{-6} m and its thickness is 5×10^{-6} m. Find the capacitance per unit length for a nerve cell. (Give your answer to two significant figures.)

INTERPRET and ANTICIPATE

The capacitance of a long, cylindrical capacitor with a vacuum between the plates (Section 27-5) is proportional to the length of the cylinder. In this problem, we need just the capacitance per unit length.

SOLVE

The expression for the capacitance of a long, cylindrical, vacuum-filled capacitor (Eq. 27.13) must be multiplied by κ because in our model of a nerve cell, the plates are filled with a dielectric.

$$C = \frac{2\pi\varepsilon_0 L}{\ln(r_{out}/r_{in})} \tag{27.13}$$

$$\frac{C}{L} = \kappa\left[\frac{2\pi\varepsilon_0}{\ln(r_{out}/r_{in})}\right] \tag{1}$$

The inner radius r_{in} of the myelin is given, and the outer radius r_{out} must be found from the thickness.

$$r_{out} - r_{in} = 5 \times 10^{-6}\text{ m}$$

$$r_{out} = r_{in} + (5 \times 10^{-6}\text{ m}) = (2 \times 10^{-6}\text{ m}) + (5 \times 10^{-6}\text{ m})$$

$$r_{out} = 7 \times 10^{-6}\text{ m}$$

Substitute values into Equation (1).

$$\frac{C}{L} = \frac{(7)2\pi(8.85 \times 10^{-12}\text{ F/m})}{\ln[(7 \times 10^{-6}\text{ m})/(2 \times 10^{-6}\text{ m})]}$$

$$\frac{C}{L} = 3.1 \times 10^{-10}\text{ F/m} = 0.31\text{ nF/m}$$

CHECK and THINK

Our result has the expected dimensions.

27-7 Energy Stored by a Capacitor with a Dielectric

Does the presence of a dielectric increase or decrease the amount of energy stored by a capacitor? The short answer is that *it depends* on whether or not the charged capacitor is connected to a battery when the dielectric is inserted. In this section, we consider a few situations in order to see which factors are important.

First, imagine there are two identical batteries, each connected to a capacitor. The capacitors are identical except that one is vacuum-filled and has capacitance C_{vac}, and the other contains a dielectric (dielectric constant κ) and has capacitance κC_{vac}. Each battery has terminal potential $\mathcal{E}$, so when either capacitor is fully charged, the potential difference across the plates is $V_C = \mathcal{E}$. The energy stored by the vacuum-filled capacitor is $U_{vac} = \frac{1}{2}C_{vac}\mathcal{E}^2$ (Eq. 27.3), and the energy stored by the dielectric-filled capacitor is $U = \frac{1}{2}\kappa C_{vac}\mathcal{E}^2$. So, the dielectric-filled capacitor stores more energy than the vacuum-filled capacitor:

$$U = \kappa U_{vac} \tag{27.19}$$

Why didn't we use $U = Q^2/2C$ (Eq. 27.2) to find the energy stored by each capacitor? Because we would first need to find the charge stored by each capacitor, and each stores a different amount. The charge stored by the vacuum-filled capacitor is $Q_{vac} = C_{vac}\mathcal{E}$ (Eq. 27.1), and the charge stored in the dielectric-filled capacitor is $Q = \kappa C_{vac}\mathcal{E}$. So, the dielectric-filled capacitor stores more charge:

$$Q = \kappa Q_{vac} \tag{27.20}$$

Now when we use $U = Q^2/2C$, we find the same result as when we used $U = \frac{1}{2}CV_C^2$ (Eq. 27.3):

$$U = \frac{1}{2}\frac{Q^2}{C} = \frac{1}{2}\frac{(\kappa Q_{vac})^2}{\kappa C_{vac}} = \kappa\frac{1}{2}\frac{Q_{vac}^2}{C_{vac}}$$

$$U = \kappa U_{vac} \tag{27.19}$$

So, we see it is better to use $U = \frac{1}{2}CV_C^2$ (Eq. 27.3) when the potential V_C is constant and $U = Q^2/2C$ (Eq. 27.2) when the charge Q is constant.

To further make this point, consider a parallel-plate capacitor is connected to a battery (Fig. 27.31A). The battery maintains a constant potential ($V_C = \mathcal{E}$) across the capacitor's plates as a dielectric is inserted between them. Inserting the dielectric *increases* (1) the capacitance $C = \kappa C_{vac}$ (Eq. 27.17), (2) the energy stored by the capacitor $U = \kappa U_{vac}$ (Eq. 27.19), and (3) the amount of charge stored by the capacitor $Q = \kappa Q_{vac}$ (Eq. 27.20).

The situation in Figure 27.31B is somewhat different; a parallel-plate capacitor is connected to a battery. The battery is then disconnected, so the potential V_C is no longer held constant. The charge Q stored by the capacitor is constant because neither plate is connected to a conductor. A dielectric is then inserted between the plates. As before, inserting the dielectric increases the capacitance, $C = \kappa C_{vac}$ (Eq. 27.17).

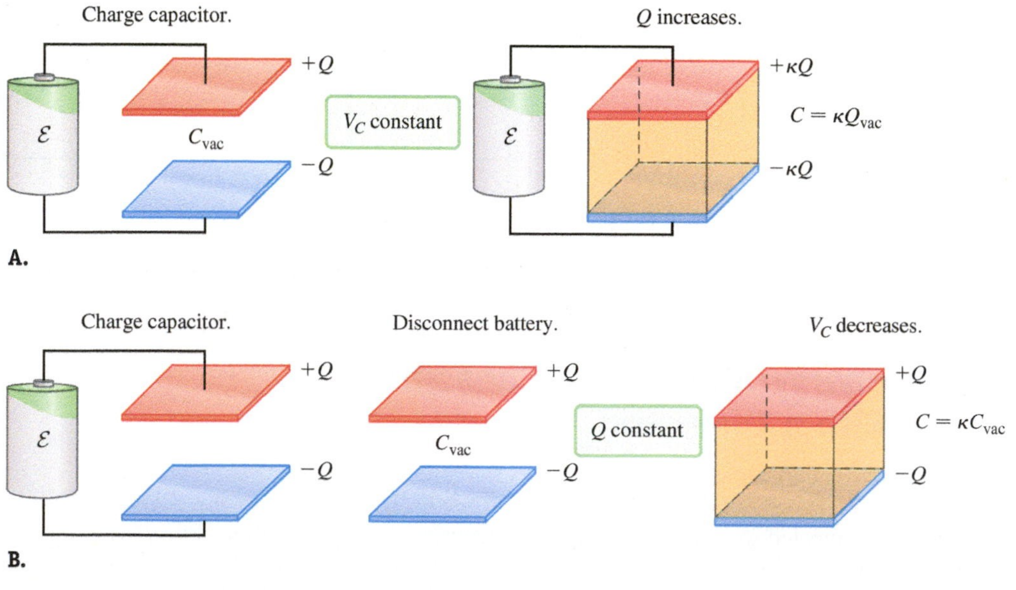

FIGURE 27.31 **A.** The battery connected to the capacitor ensures that the potential is held constant when a dielectric is slipped between the plates. As a result, the charge on the plates and the energy stored by the capacitor both increase. **B.** If the capacitor is charged and the battery is disconnected, when the dielectric is slipped between the plates, the potential decreases while the charge remains constant, and the energy stored by the capacitor decreases.

However, because the charge is constant, we find using Equation 27.2 that the energy stored by the capacitor decreases:

$$U = \frac{1}{2}\frac{Q^2}{\kappa C_{vac}} = \frac{1}{\kappa}U_{vac}$$

and the potential between the plates also decreases (Eq. 27.1):

$$V_C = \frac{Q}{\kappa C_{vac}} = \frac{1}{\kappa}\mathcal{E}$$

The difference between the two situations in Figure 27.31 can be explained as follows: Inserting a dielectric partially cancels the electric field between the plates because the electric field $\vec{E}_D$ created by the dielectric is in the opposite direction as the vacuum-filled capacitor's electric field $\vec{E}_{vac}$ (Fig. 27.29C). When no battery is connected to the capacitor, this reduced electric field means a reduced potential difference V_C and less energy stored by the capacitor. However, when the capacitor is connected to a battery and a dielectric is inserted, the battery supplies more charge to each plate in order to maintain a constant potential difference. Because the capacitor now stores more charge, it also stores more energy.

Energy Density

The electric potential energy stored by a capacitor can also be considered energy stored in the electric field between the capacitor plates. In fact, any electric field stores energy, no matter what the source of that field. We can derive an expression for the energy stored per unit volume by an electric field by thinking about the special case of a parallel-plate capacitor.

Consider a dielectric-filled parallel-plate capacitor (Fig. 27.29). The net electric field between the plates is $\vec{E}_C$, and the potential between the plates is

$$V_C = E_C d \tag{27.21}$$

The capacitance is given by $C = \kappa\varepsilon_0 A/d$ (Eq. 27.18), so the energy stored by the electric field $\vec{E}_C$ is (Eq. 27.3):

$$U_E = \frac{1}{2}CV_C^2 = \frac{1}{2}\frac{\kappa\varepsilon_0 A}{d}(E_C d)^2$$

$$U_E = \frac{1}{2}\kappa\varepsilon_0 E_C^2(Ad)$$

The dielectric is a rectangular box with volume Ad, so the energy density (energy per unit volume) of the field is

$$u_E = \frac{U_E}{Ad} = \frac{1}{2}\kappa\varepsilon_0 E^2 \tag{27.22}$$

ENERGY DENSITY STORED IN AN ELECTRIC FIELD ✪ **Major Concept**

where we have dropped the subscript from the electric field, and the symbol for energy density is a lowercase u.

Equation 27.22 was derived for the special case of a dielectric-filled parallel-plate capacitor, but it is a general expression for the energy density stored in any electric field. If the field is in a vacuum, then $\kappa = 1$. Because the dielectric constant of any insulator is greater than 1, more energy is stored in a region that contains an insulator than in a vacuum, for electric fields of equal magnitude.

CONCEPT EXERCISE 27.7

An X-ray tube at a dentist's office produces X-rays by striking a metal plate with a beam of electrons in an evacuated tube. The electric field generated in an X-ray tube is approximately 5×10^6 V/m. What is the energy density stored in this electric field?

EXAMPLE 27.9 Removing a Dielectric

A large parallel-plate capacitor consists of two metal disks of radius 9.45 cm. A Teflon dielectric has the same radius and is 0.55 cm thick. The dielectric just barely fits between the capacitor's plates, and the capacitor is connected to a battery that has a terminal potential of 3.2 V. The capacitor is fully charged, and then the battery is disconnected. Next, a person carefully removes the dielectric without touching the plates. How much work must the person do to remove the dielectric? (Assume the capacitor is horizontal so the person removes the dielectric horizontally and there is no change in the microscopic internal energy.) In the **CHECK and THINK** step, compare your result to the amount of work needed to lift the dielectric by a distance equal to its diameter. The density of Teflon is 2200 kg/m³.

INTERPRET and ANTICIPATE

Consider a system consisting of the capacitor and the dielectric. The bar chart in Figure 27.32 shows that the initial electrostatic potential energy plus the work done by the person equals the final potential energy. We expect that the dielectric is attracted to the plates of the capacitor, so the person must exert positive work to remove it. This might seem odd because removing the dielectric increases the energy stored in the capacitor in this case.

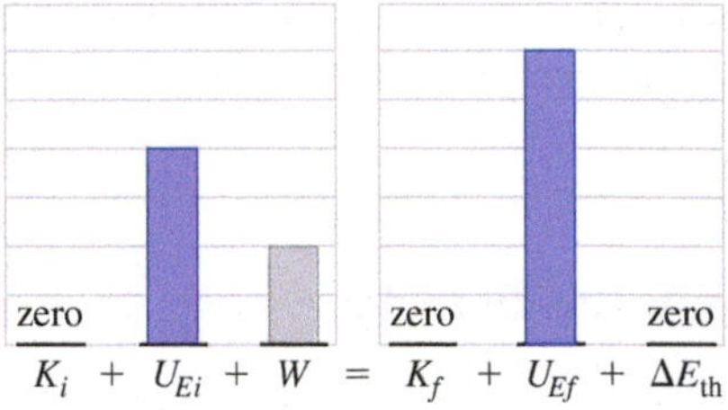

FIGURE 27.32

SOLVE

The work done by the person is the final potential energy minus the initial potential energy.	$W = U_{Ef} - U_{Ei}$
Initially, the capacitor contains a dielectric and is fully charged, so the potential difference across its plates is $\mathcal{E}$. The energy stored by the capacitor is given by Equation 27.3. Substitute the capacitance given by $C = \kappa C_{vac}$ (Eq. 27.17).	$U_{Ei} = \frac{1}{2} C V_C^2$ (27.3) $U_{Ei} = \frac{1}{2} \kappa C_{vac} \mathcal{E}^2$ (1)
It is convenient (though not necessary) to find a value for the initial potential energy at this stage. The capacitance C_{vac} of a parallel-plate capacitor is given by Equation 27.10. The area of a disk is πr^2. The dielectric constant comes from Table 27.2.	$U_{Ei} = \frac{1}{2} \kappa \left(\frac{\varepsilon_0 A}{d}\right) \mathcal{E}^2 = \frac{1}{2} \kappa \left(\frac{\varepsilon_0 \pi r^2}{d}\right) \mathcal{E}^2$ $U_{Ei} = \frac{1}{2}(2.1)\left(\frac{(8.85 \times 10^{-12}\,\text{F/m})\pi(9.45 \times 10^{-2}\,\text{m})^2}{0.55 \times 10^{-2}\,\text{m}}\right)(3.2\,\text{V})^2$ $U_{Ei} = 4.9 \times 10^{-10}\,\text{J}$
The battery is disconnected, so the charge on the plates is constant and the potential difference changes when the dielectric is removed. The charge is found from $Q = CV_C$ (Eq. 27.1).	$Q = CV_C = \kappa C_{vac} \mathcal{E}$ (2)
Because the charge is constant, it is best to find the final potential energy using Equation 27.2. The charge comes from Equation (2), and the capacitance is C_{vac} because the dielectric has been removed.	$U_{Ef} = \frac{1}{2} \frac{Q^2}{C}$ (27.2) $U_{Ef} = \frac{1}{2} \frac{(\kappa C_{vac} \mathcal{E})^2}{C_{vac}} = \frac{1}{2} \kappa^2 C_{vac} \mathcal{E}^2$
The final potential energy may be written in terms of the initial potential energy (Eq. 1).	$U_{Ef} = \kappa U_{Ei}$
The work done by the person is the final potential energy minus the initial potential energy.	$W = \kappa U_{Ei} - U_{Ei} = (\kappa - 1) U_{Ei} = (2.1 - 1)(4.9 \times 10^{-10}\,\text{J})$ $W = 5.4 \times 10^{-10}\,\text{J}$

CHECK and THINK

As expected, the person does positive work to remove the dielectric. This amount of work is tiny compared to the work needed to lift the dielectric up by $2r$. The mass of the dielectric can be found from its volume and density; then use $W = mg(2r)$.

$$m = \rho V = \rho \pi r^2 d$$

$$m = (2200\ \text{kg/m}^3)\pi(9.45 \times 10^{-2}\ \text{m})^2(0.55 \times 10^{-2}\ \text{m})$$

$$m = 0.34\ \text{kg}$$

$$W = mg(2r) = (0.34\ \text{kg})(9.81\ \text{m/s}^2)2(9.45 \times 10^{-2}\ \text{m})$$

$$W = 0.63\ \text{J} \gg 5.3 \times 10^{-10}\ \text{J}$$

27-8 Gauss's Law in a Dielectric

When we studied Gauss's law in Chapter 25, we assumed the electric field existed in a vacuum. In that case, the electric flux through a Gaussian surface is proportional to the charge inside that surface:

$$\Phi = \oint \vec{E} \cdot d\vec{A} = \frac{q_{\text{in}}}{\varepsilon_0} \tag{25.11}$$

How can we modify Gauss's law to include the presence of an insulator such as a dielectric between the plates of a capacitor?

To answer that question, let's consider a parallel-plate capacitor. First, remember that we have used Gauss's law to find the electric field between the plates of a vacuum-filled parallel-plate capacitor: $E_{\text{vac}} = Q_{\text{free}}/\varepsilon_0 A$ (Eq. 27.11); the reason for adding the subscript "free" is explained below. Next, imagine a dielectric consisting of dipole molecules between the plates of an uncharged capacitor (Fig. 27.33A). Because there is no electric field, the dipole molecules are randomly oriented. Any Gaussian surface you could imagine would contain no net charge because every molecule is neutral. (Even if the Gaussian surface happened to pass right through the middle of a few molecules, it would include positive charge from some molecules and negative charge from other molecules, with no net charge overall.)

Now, imagine charging the plates of the capacitor so that there is an electric field inside the dielectric. The electric field causes the dipole molecules to line up such that the negative side of each molecule is closer to the capacitor's positive plate (Fig. 27.33B). Now our Gaussian surface encloses the excess positive charge Q_{free} on the capacitor's plate. Also, our Gaussian surface no doubt passes through the middle of some molecules, preferentially enclosing an excess negative charge because the negative sides of the molecules all face the same way. So, the total charge q_{in} inside the Gaussian cylinder is the charge $+Q_{\text{free}}$ on the surface of the capacitor's plate plus the excess negative charge $-q_D$ from the dielectric:

$$q_{\text{in}} = Q_{\text{free}} - q_D \tag{27.23}$$

We substitute Equation 27.23 into Gauss's law and solve for the electric field inside the dielectric:

$$E_{\text{tot}} = \frac{Q_{\text{free}} - q_D}{\varepsilon_0 A} \tag{27.24}$$

This is the *total* electric field that results from the charged capacitor and the dielectric itself. The dielectric's electric field is in the opposite direction as the electric field produced by the vacuum-filled capacitor $\vec{E}_{\text{vac}}$ (Fig. 27.29), so the total electric field is weaker by a factor of κ (Eq. 27.15):

$$E_{\text{tot}} = \frac{E_{\text{vac}}}{\kappa} \tag{27.25}$$

To find an expression for the charge q_{in} inside the Gaussian surface, substitute Equations 27.24 for E_{tot} and 27.11 for E_{vac} into Equation 27.25:

$$\frac{Q_{\text{free}} - q_D}{\varepsilon_0 A} = \frac{1}{\kappa \varepsilon_0} \frac{Q_{\text{free}}}{A}$$

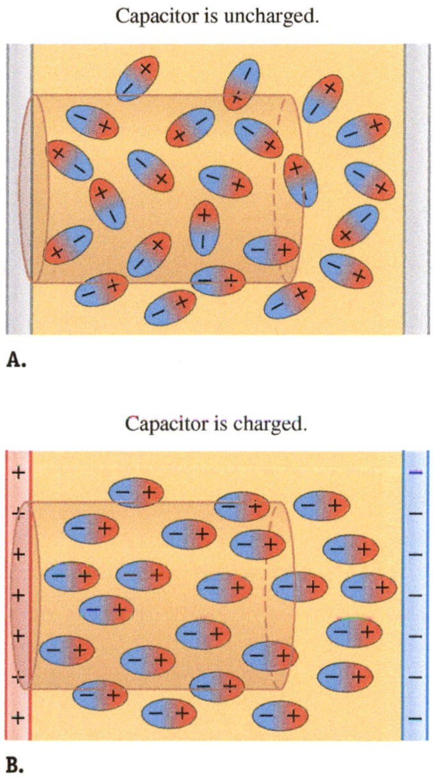
Capacitor is uncharged.
A.
Capacitor is charged.
B.

FIGURE 27.33 **A.** The dipole molecules of a dielectric are randomly oriented in the absence of an electric field. The net charge enclosed by the Gaussian surface is zero. **B.** When an electric field is present, the molecules of the dielectric line up as shown. The net charge enclosed by the Gaussian surface is $q_{\text{in}} = Q_{\text{free}} - q_D$.

$$q_{in} = Q_{free} - q_D = \frac{Q_{free}}{\kappa} \quad (27.26)$$

So we modify Gauss's law (Eq. 25.11) by replacing q_{in} with Equation 27.26:

GENERAL FORM OF GAUSS'S LAW

Underlying Principle

$$\oint \kappa\vec{E} \cdot d\vec{A} = \frac{Q_{free}}{\varepsilon_0} \quad (27.27)$$

Equation 27.27 was derived for a dielectric-filled capacitor, but it is valid in any situation. It is a more general expression of Gauss's law than Equation 25.11. (If the charged source is in a vacuum, then $\kappa = 1$ and the two expressions are identical.)

There are two major differences between the general expression for Gauss's law (Eq. 27.27) and the vacuum expression (Eq. 25.11). First, the flux integral in the general expression involves the dielectric constant κ, and κ is kept inside the integral because the dielectric constant may not have the same value throughout the entire Gaussian surface. Second, the right side of the general expression for Gauss's law involves Q_{free} instead of q_{in}. The subscript "free" is a reminder that the charged particles on the plates of the capacitor are free to move throughout the conductor and would move if—for example—we changed the potential across the plates. The free charge Q_{free} (on the plates) is not the same as the charge enclosed by the Gaussian surface because it does not include the excess charge q_D embedded in the dielectric. We already took q_D into account when we introduced κ in Gauss's law.

Let's find a way to summarize how quantities are affected by the presence of a dielectric. For example, compare the equations for the capacitance of a vacuum-filled parallel-plate capacitor $C_{vac} = \varepsilon_0 A/d$ (Eq. 27.10) to that of dielectric-filled one $C = \kappa\varepsilon_0 A/d$ (27.18). Mathematically we can alter the equation for a vacuum-filled capacitor by replacing ε_0 with $\kappa\varepsilon_0$ to arrive at the equation for a dielectric-filled capacitor. We extend this comparison to come up with a general rule: *the electrostatic equations involving the permittivity constant ε_0 are altered by replacing ε_0 with $\kappa\varepsilon_0$ to take into account the effects of the presence of a dielectric material with dielectric constant κ.*

The general form of Gauss's law does not play a major role in this textbook. In practice, you may not need to use this general form, but it is the underlying principle. All the principles of electricity and magnetism can be expressed in four equations, and Gauss's law is one of them.

SUMMARY

Underlying Principle

General form of Gauss's law:

$$\oint \kappa\vec{E} \cdot d\vec{A} = \frac{Q_{free}}{\varepsilon_0} \quad (27.27)$$

where κ is the dielectric constant and Q_{free} is the free charge enclosed in the Gaussian surface. In a vacuum, $\kappa = 1$, $q_{in} = Q_{free}$, and Equation 27.27 is the same as Equation 25.11.

Major Concepts

1. A **capacitor** is a device that consists of two conductors (**plates**) separated by an insulator or vacuum. When charged, the plates carry charges of equal magnitude and opposite sign. A capacitor stores electric potential energy.
2. **Capacitance** C is the constant of proportionality in
$$Q = CV_C \quad (27.1)$$
where $V_C = \Delta V$, the potential difference across the capacitor's plates. The SI unit for capacitance is the **farad**: 1 F = 1 C/V.

Capacitance depends only on geometry and the type of insulator between the plates.

3. The **energy stored by a capacitor** with charge Q is

$$U_E = \frac{1}{2}\frac{Q^2}{C} \quad (27.2)$$

Another convenient expression for this potential energy is

$$U_E = \frac{1}{2}CV_C^2 \quad (27.3)$$

4. An **ideal battery** is a device that maintains its terminal potential $\mathcal{E}$ through chemical reactions.

5. The **equivalent capacitance of N capacitors in series** is

$$\frac{1}{C_{eq}} = \sum_{i=1}^{N}\frac{1}{C_i} \quad (27.7)$$

6. The **equivalent capacitance of N capacitors in parallel** is

$$C_{eq} = \sum_{i=1}^{N} C_i \quad (27.8)$$

7. A **dielectric** is an insulator that can support an induced electric field. A dielectric placed between the plates of a capacitor increases its capacitance by a factor κ, the **dielectric constant**:

$$C = \kappa C_{vac} \quad (27.17)$$

8. The **energy density** u_E (energy per unit volume) **stored in an electric field** $\vec{E}$ is:

$$u_E = \frac{1}{2}\kappa\varepsilon_0 E^2 \quad (27.22)$$

Special Cases

1. **Parallel-plate capacitor** of plate area A and separation d:

$$C = \frac{\varepsilon_0 A}{d} \quad (27.10)$$

2. **Cylindrical capacitor** of inner radius r_{in} and outer radius r_{out}:

$$C = \frac{2\pi\varepsilon_0 L}{\ln(r_{out}/r_{in})} \quad (27.13)$$

PROBLEM-SOLVING STRATEGY Finding Capacitance

SOLVE

1. Use Gauss's law to find an expression for the **electric field** between the plates. Refer to the problem-solving strategy "Finding the Electric Field Using Gauss's Law" on page 765.
2. Complete a **path integral** to find the electric potential difference V_C:

$$V_C = \int_{r_+}^{r_-} E\,dr \quad (27.9)$$

3. Use $C = Q/V_C$ (Eq. 27.1) to find the capacitance.

CHECK and THINK

In addition to checking the **dimensions**, you should make sure that the expression you found for capacitance **depends only on geometric properties**.

PROBLEMS AND QUESTIONS

A = algebraic **C** = conceptual **E** = estimation **G** = graphical **N** = numerical

27-1 The Leyden Jar

1. **C** CASE STUDY In Concept Exercise 27.1 (page 829), we ignored air resistance acting on the ring launched by the experimenters in the case study. How would including air resistance change your answer to the question about the size of the spring? Does air resistance require that more, less, or the same amount of electric potential energy be stored in the spring? Explain your answers.

2. **C** Imagine turning the crank on a generator like the one shown in Figure 27.1 (page 828). Describe the experience of turning the crank for a long time. Would it get easier or harder to turn? Explain your reasoning.

3. **C** In Franklin's time, a device for storing electric potential energy was called a **Leyden jar**. Today, we call that device a **capacitor**. Another term that is sometimes used is **condenser**. What ideas do these three terms bring to mind? What are the advantages and disadvantages of each?

27-2 Capacitors

4. **E** The first Leyden jar was probably discovered by a German clerk named E. Georg von Kleist. Because von Kleist was not a scientist and did not keep good records, the credit for the discovery of the Leyden jar usually goes to physicist Pieter Musschenbroek from Leyden, Holland. Musschenbroek accidentally discovered the Leyden jar when he tried to charge a jar of water and shocked himself by touching the wire on the inside of the jar while holding the jar on the outside. He said that the shock was no ordinary shock and his body shook violently as though he had been hit by lightning. The energy from the jar that passed through his body was probably around 1 J, and his jar probably had a capacitance of about 1 nF.
   a. Estimate the charge that passed through Musschenbroek's body.
   b. What was the potential difference between the inside and outside of the Leyden jar before Musschenbroek discharged it?
5. **N** A parallel-plate capacitor has a capacitance of 6.0 μF. If the potential difference across the plates of the capacitor is 48 V, what must be the charge on the plates of the capacitor?
6. **C** According to $U_E = \frac{1}{2}C(\Delta V)^2$ (Eq. 27.3), a greater capacitance means more energy is stored by the capacitor, but according to $U_E = Q^2/2C$ (Eq. 27.2), a greater capacitance means less energy is stored. How can both of these equations be correct?
7. **N** In Figure P27.7, capacitor 1 (C_1 = 20.0 μF) initially has a potential difference of 50.0 V and capacitor 2 (C_2 = 5.00 μF) has none. The switches are then closed simultaneously. **a.** Find the final charge on each capacitor after a long time has passed. **b.** Calculate the percentage of the initial stored energy that was lost when the switches were closed.

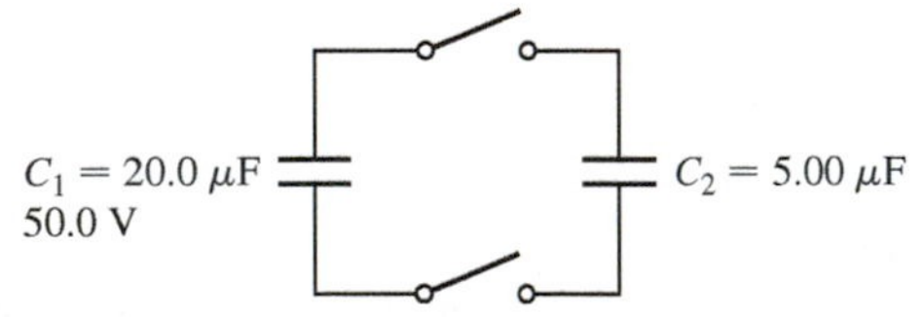

FIGURE P27.7

8. **N** A capacitor has a charge of 3.587 nC and a potential difference of 5.00 V. **a.** How much energy is stored in the capacitor? **b.** What is its capacitance?
9. **N** A 4.50-μF capacitor is connected to a battery for a long time. **a.** If the voltage of the battery is 9.00 V, how much energy is stored in the capacitor? **b.** If the voltage of the battery is increased to 24.0 V, how much energy is now stored in the capacitor?
10. **G** **CASE STUDY** Students working in a physics lab use a Thompson coil to launch a light ring (m = 25.0 g) in much the same way as the experimenters in the case study. The students vary the potential difference across their capacitor bank and measure the maximum height of the ring. Their data are shown in the table to the right. Plot these data to find the capacitance of their capacitor bank.

V (V)	h (cm)
0	0.00
10	0.32
20	2.7
30	5.8
40	9.9
50	16.2
60	21.2
70	29.3
80	39.1
90	49.6
100	61.1

Problems 11 and 12 are paired.

11. **N** A capacitor stores 37.5 mJ when the potential difference between its plates is 16.0 V. **a.** What is the charge stored by this capacitor? **b.** What is its capacitance?
12. **N** A capacitor stores 37.5 mJ when the potential difference between its plates is 16.0 V. Then the potential difference is decreased to 8.00 V. **a.** After the potential difference is reduced, what is the charge stored by the capacitor? **b.** What is its final capacitance?

27-3 Batteries

13. **N** A 1.50-V battery is used to charge a capacitor. When the capacitor is fully charged, it stores 37.5 mJ. If the capacitor were fully charged by a 3.00-V battery instead, how much energy would it store?
14. **C** When a Leyden jar is charged by a hand generator (Fig. 27.1, page 828), the work done by the person turning the crank is stored as electric potential energy in the jar. When a capacitor is charged by a battery, where does the electric potential energy come from?
15. **N** A battery is used to fully charge a capacitor of capacitance 85.25 μF. When the capacitor is fully charged, it stores 3.454 mJ. What is the terminal potential of the battery?
16. **N** A 6.50-μF capacitor is connected to a battery. What is the charge on each plate of the capacitor if the voltage of the battery is **a.** 10.0 V and **b.** 2.00 V?

27-4 Capacitors in Parallel and Series

17. **N** A pair of capacitors with capacitances C_A = 3.70 μF and C_B = 6.40 μF are connected in a network. What is the equivalent capacitance of the pair of capacitors if they are connected **a.** in parallel and **b.** in series?
18. **C** Two 1.5-V batteries are required in a flashlight. **a.** If the batteries are connected as shown in configuration 1 in Figure P27.18, what is the potential difference between points A and B? **b.** If, instead, the batteries are connected as shown in configuration 2, what is the potential difference between points A and B? **c.** Use your answers to figure out why a flashlight with two good batteries may not light up.

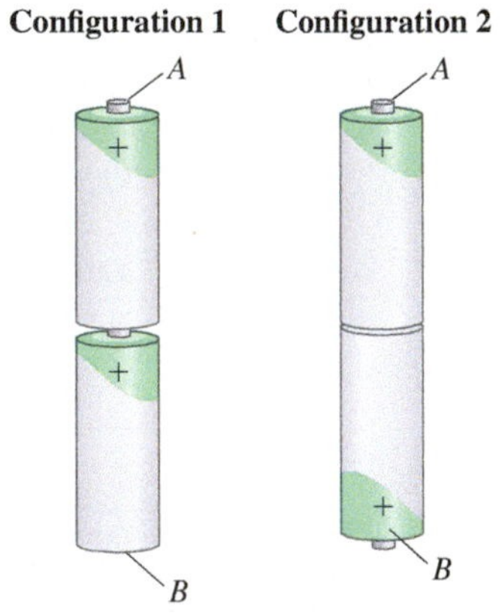

FIGURE P27.18

19. **N** Two capacitors have capacitances of 6.0 μF and 3.0 μF. Initially, the first one has a potential difference of 18.0 V and the second one 9.0 V. If the capacitors are now connected in parallel, what is their common potential difference?
20. **C** When 18th-century experimenters such as Ben Franklin and Abbé Nollet needed a large amount of electric potential energy (or a lot of charged particles), they used several Leyden jars. Did they connect the jars in series or in parallel? Draw a sketch and explain your answer.
21. **N** Calculate the equivalent capacitance between points a and b for each of the two networks shown in Figure P27.21. Each capacitor has a capacitance of 1.00 μF.

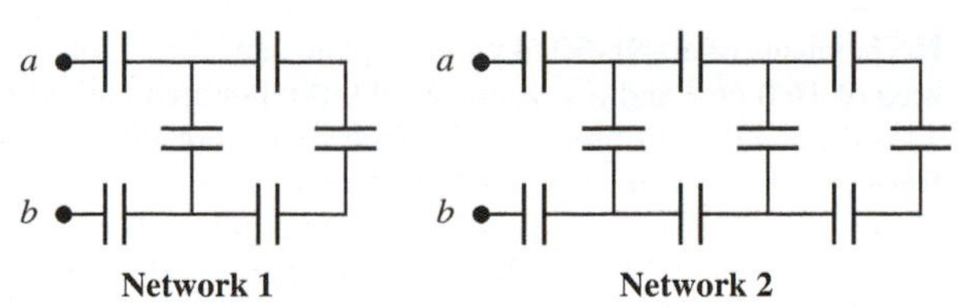

FIGURE P27.21

22. **N** A 12.0-V battery is connected to a pair of capacitors in parallel with capacitances $C_A = 9.00\ \mu F$ and $C_B = 14.0\ \mu F$. **a.** What is the equivalent capacitance of the pair of capacitors? **b.** What charge is stored by each of the capacitors? **c.** What is the potential difference across each of the capacitors?

Problems 23 and 24 are paired.

23. **A** Given the arrangement of capacitors in Figure P27.23, find an expression for the equivalent capacitance between points a and b.

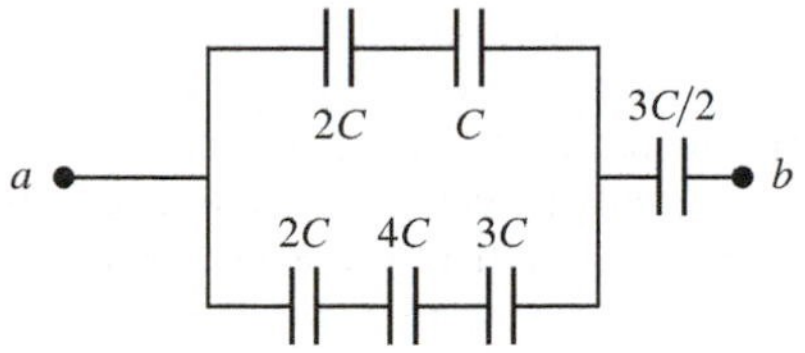

FIGURE P27.23 Problems 23 and 24.

24. **N** An arrangement of capacitors is shown in Figure P27.23. **a.** If $C = 9.70 \times 10^{-5}$ F, what is the equivalent capacitance between points a and b? **b.** A battery with a potential difference of 12.00 V is connected to a capacitor with the equivalent capacitance. What is the energy stored by this capacitor?

25. **N** You find three capacitors that have capacitances 20.00 μF, 33.00 μF, and 47.00 μF. What are the smallest and greatest equivalent capacitances that you can make with these capacitors?

Problems 26 and 27 are paired.

26. **A** Find the equivalent capacitance for the network shown in Figure P27.26.

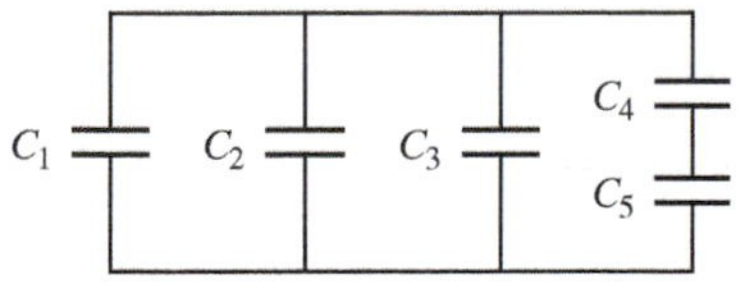

FIGURE P27.26 Problems 26 and 27.

27. **N** Find the equivalent capacitance for the network shown in Figure P27.26 if $C_1 = 1.00\ \mu F$, $C_2 = 2.00\ \mu F$, $C_3 = 3.00\ \mu F$, $C_4 = 4.00\ \mu F$, and $C_5 = 5.00\ \mu F$.

28. **N** Figure P27.28 shows three capacitors with capacitances $C_A = 1.00\ \mu F$, $C_B = 2.00\ \mu F$, and $C_C = 4.00\ \mu F$ connected to a 9.00-V battery. **a.** What is the equivalent capacitance of the three capacitors? **b.** What charge is stored in each of the capacitors? **c.** What is the potential difference across each of the capacitors?

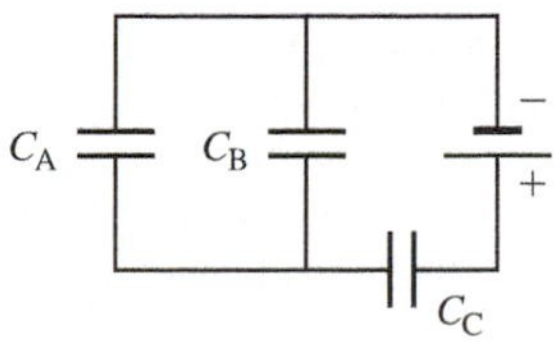

FIGURE P27.28

29. **N** The capacitances of three capacitors are in the ratio 1:2:3. Their equivalent capacitance when all three are in parallel is 120.0 pF greater than when all three are in series. Determine the capacitance of each capacitor.

30. **N** For the four capacitors in the circuit shown in Figure P27.30, $C_A = 1.00\ \mu F$, $C_B = 4.00\ \mu F$, $C_C = 2.00\ \mu F$, and $C_D = 3.00\ \mu F$. What is the equivalent capacitance between points a and b?

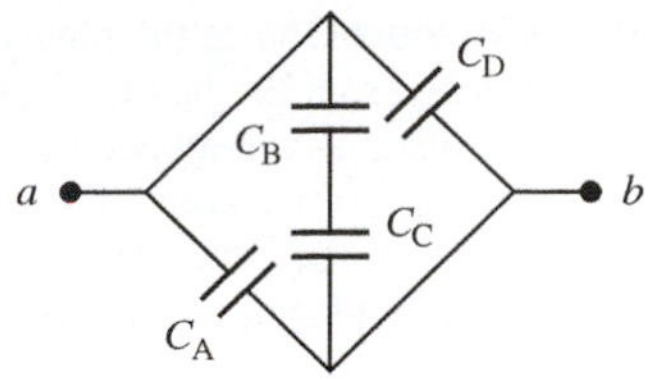

FIGURE P27.30

27-5 Capacitance: Special Cases

31. **N** The separation between the 4.40-cm^2 plates of an air-filled parallel-plate capacitor is 0.230 cm. **a.** What is the capacitance of this capacitor? If the capacitor is connected to a 9.00-V battery, find **b.** the charge stored by the capacitor and **c.** the magnitude of the electric field between its plates.

Problems 32, 33, and 34 are grouped.

32. **A** A spherical capacitor is made up of two concentric spherical conductors. The inner sphere has positive charge $+Q$, and the outer sphere has negative charge $-Q$. The radius of the inner sphere is r_{in}, and the radius of the outer sphere is r_{out}. Show that its capacitance is $C = 4\pi\varepsilon_0[r_{in}r_{out}/(r_{out} - r_{in})]$.

33. **A** Derive an expression for the capacitance of an isolated sphere of radius r by taking the limit of $C = 4\pi\varepsilon_0[r_{in}r_{out}/(r_{out} - r_{in})]$ (see Problem 32) as r_{out} approaches infinity.

34. **E** Use the expression you derived in Problem 33 to estimate the capacitance of **a.** the Earth and **b.** a cannonball placed at the top of a lightning rod (Fig. 24.35, page 734).

35. **N** The plates of an air-filled parallel-plate capacitor carry a surface charge density of 56.0 μC/m^2 when the potential difference across the plates is 225 V. What is the separation between the plates of this capacitor?

36. **N** A variable capacitor like the one shown in Figure P27.36 is often used as a tuning element in a high-frequency radio circuit. Suppose the air gap between a pair of plates is 0.300 mm and the pair behaves like a parallel-plate capacitor. **a.** Determine the effective area of the plates when they are rotated to be fully meshed (overlapped), where the capacitance of the pair has a maximum value of 100.0 pF. **b.** Determine the effective area of the plates when they are rotated to the point where the pair of plates has the minimum capacitance of 2.00 pF.

FIGURE P27.36

Problems 37 and 38 are paired.

37. **C** A charged parallel-plate capacitor is connected to a battery. The plates of the capacitor are pulled apart so that their separation doubles. What happens to **a.** the charge and **b.** the amount of energy stored by the capacitor?

38. **C** A charged parallel-plate capacitor is isolated (not connected to a battery). The plates of the capacitor are pulled apart so that their separation doubles. What happens to **a.** the charge and **b.** the amount of energy stored by the capacitor?

39. **N Review** One of the plates of a parallel-plate capacitor is suspended from the beam of a balance as shown in Figure P27.39. The distance d between the capacitor plates is 5.00 mm, and the cross-sectional area of the plates is 625 cm^2. Determine the potential difference between the capacitor plates if a mass of 4.00 g is placed on the other pan of the balance to obtain static equilibrium.

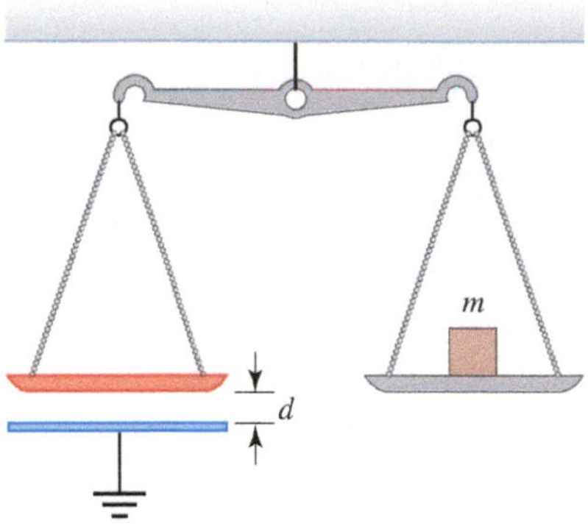

FIGURE P27.39

40. **A** Two fully charged cylindrical capacitors are connected to two identical batteries. The capacitors are identical except that the radius of the outer plate of capacitor A is twice the radius of the outer plate of capacitor B. Both inner plates have a radius that is half the outer radius of capacitor B. **a.** What is the relative capacitance of each capacitor? Express your answer as a ratio C_A/C_B. **b.** What is the relative energy stored by each capacitor? Express your answer as a ratio U_A/U_B. **c.** What is the relative charge stored by each capacitor? Express your answer as a ratio Q_A/Q_B.

41. **N** The capacitance of most commonly used capacitors is small $(\sim 1\text{ pF} < C < \sim 1\ \mu\text{F})$. A capacitance of 1 F is very large. In this problem, we estimate the size of a 1-F capacitor using parallel-plate geometry. If the plates of a 1.0-F, air-filled, parallel-plate capacitor are separated by 50.0 μm (about the diameter of a human hair), what is the area of each plate?

Problems 42 and 43 are paired.

42. **N** A 56.90-pF cylindrical capacitor carries a charge of 1.540 μC. The capacitor has a length of 1.000×10^{-3} m. **a.** What is the potential difference across the capacitor? **b.** If the radial separation between the two cylinders is 6.520×10^{-4} m, what are the inner and outer radii of the cylindrical conductors?

43. **N** A 5.69-pF spherical capacitor carries a charge of 1.54 μC. **a.** What is the potential difference across the capacitor? **b.** If the radial separation between the two spherical shells is 6.52×10^{-3} m, what are the inner and outer radii of the spherical conductors? *Hint*: See Problem 32.

27-6 Dielectrics

44. **C** If a parallel-plate capacitor is to have its capacitance doubled, describe three possible ways this can be accomplished.

45. **N** How much charge is stored by a parallel-plate capacitor with plate area 2.25×10^{-2} m^2 and plate separation 5.00×10^{-3} m that is filled with a dielectric with $\kappa = 5$ and connected to a battery with a potential difference of 12.0 V?

46. **E** Model a Leyden jar as a cylindrical capacitor with a Pyrex dielectric. Assume the jar is about the size of a large coffee cup or large soup can. **a.** Estimate the capacitance of this model Leyden jar. **b.** If you charged the jar so that the electric field was just less than the dielectric strength of Pyrex, how much energy could you store? Check your answer by using the information in Problem 4.

47. **N** The plates of an air-filled parallel-plate capacitor with a plate area of 16.0 cm^2 and a separation of 9.00 mm are charged to a 145-V potential difference. After the plates are disconnected from the source, a porcelain dielectric with $\kappa = 6.5$ is inserted between the plates of the capacitor. **a.** What is the charge on the capacitor before and after the dielectric is inserted? **b.** What is the capacitance of the capacitor after the dielectric is inserted? **c.** What is the potential difference between the plates of the capacitor after the dielectric is inserted? **d.** What is the magnitude of the change in the energy stored in the capacitor after the dielectric is inserted?

48. **E** You can build a capacitor very cheaply by using two rolls of aluminum foil and an equal amount of paper towels. Imagine layering the paper and foil together and rolling the layers as in Figure 27.30 (page 849). **a.** Estimate the capacitance of this homemade capacitor. Explain your work, including any assumptions you make. **b.** What is the maximum potential you can have across this capacitor without dielectric breakdown? **c.** What steps could you take to increase the capacitance without purchasing more equipment?

49. **N** The robot Johnny-Five is undergoing some repairs and needs to replace one of his parallel-plate capacitors. The original capacitor consisted of plates with an area of 0.0500 m^2 separated by 0.0300 m, and there was a vacuum between the plates. He finds a replacement that has the same plate separation, but the plate area is only 0.00143 m^2. In order to achieve the same capacitance as the original part, he can insert a dielectric between the plates of the new capacitor. What dielectric constant does he require?

50. **N** A vacuum-filled, cylindrical capacitor connected to a battery stores 6.75 J when it is fully charged. Paraffin is inserted between the plates without removing the capacitor from the battery. How much energy is stored in the paraffin-filled capacitor?

51. **N** A parallel-plate capacitor with a Pyrex dielectric ($\kappa = 4.70$) and a capacitance of 3.40 μF is charged to a potential difference of 230.0 V and disconnected from the source. **a.** How much work is required to withdraw the Pyrex dielectric from the capacitor? (Ignore any change in the gravitational potential energy.) **b.** What is the change in the potential difference of the capacitor after the Pyrex is removed?

27-7 Energy Stored by a Capacitor with a Dielectric

52. **C** Imagine charging two capacitors with a hand generator (Fig. 27.1). The capacitors are identical except that one has air between its plates and the other has a dielectric. **a.** If you wished to store the same amount of charge on each capacitor, which one would be more difficult to charge? **b.** If you wished to have the same potential difference between the plates of each capacitor, which one would be more difficult to charge? Explain your answers in terms of the amount of work needed in each case.

Problems 53 and 54 are paired.

53. **A** A parallel-plate capacitor with an air gap has capacitance C_0. It is connected to a battery with potential V_0 that gives it charge Q_0 and stored energy U_0. While the capacitor is still connected to the battery, a dielectric with constant $\kappa = 3$ is inserted into the air gap, completely filling it. In terms of the initial values, find the new capacitance C, charge Q, potential V, and stored energy U.

54. **A** A parallel-plate capacitor with an air gap has capacitance C_0. It is connected to a battery with potential V_0 that gives it charge Q_0 and stored energy U_0. After the capacitor is disconnected from the battery, a dielectric with constant $\kappa = 3$ is inserted into the air gap, completely filling it. In terms of the initial values, find the new capacitance C, charge Q, potential V, and stored energy U.

55. **N** A parallel-plate capacitor with plates of area $A = 0.100\ \text{m}^2$ separated by distance $d = 2.25 \times 10^{-3}$ m is connected to a battery with a potential difference of 9.00 V for a very long time. Initially, a vacuum exists between the plates. **a.** How much energy is stored in the capacitor? **b.** What is the magnitude of the electric field between the plates? **c.** The capacitor is disconnected from the battery and a dielectric with $\kappa = 4.23$ is inserted, filling the space between the plates. What is the magnitude of the electric field in the region between the plates? **d.** What is the magnitude of the electric field produced by the dielectric? **e.** How much energy is stored in the capacitor after the dielectric has been inserted?

56. Model a charged rubber balloon as a spherical source with charge Q and radius R.
    - **a.** **A** Find an expression for the energy density of the electric field just outside the balloon. If the balloon is deflated without losing charge, does the energy density of the electric field increase, decrease, or stay the same? Explain your answer.
    - **b.** **C** How does your answer to part (a) change if the balloon is made of Mylar instead of rubber?

57. **N** Five hundred 8.00-μF capacitors are connected in parallel and then charged to a potential of 25.0 kV. For how long will the stored energy light a 100.0-W bulb until no energy remains in the capacitors?

58. **E** Another simple way to make a homemade capacitor requires only a soft-lead pencil and paper. You can make two conducting parallel plates by covering the paper with graphite on both sides. The finish should be dark and glossy when you are done. It may be helpful to use thick, strong paper so that you don't puncture it when you are coloring. If you use a standard letter-sized sheet of paper, estimate the maximum amount of electric potential energy you could store with such a capacitor. Explain your work, including any assumptions you make.

59. **N** A parallel-plate capacitor with a gap of 0.200 mm is filled with a dielectric with constant $\kappa = 4.60$. **a.** When the capacitor's potential difference is 50.0 V, what is the energy density? **b.** Find the average energy density in an AA battery. A typical new alkaline AA battery stores 9.00×10^3 J of electrical energy and has a volume of 8.30 cm^3.

27-8 Gauss's Law in a Dielectric

60. A rubber balloon with a radius of 18.4 cm has a charge of 30 nC spread uniformly over its surface.
    - **a.** **N** Suppose the balloon is underwater. What is the electric field at a distance of 36.8 cm from the center of the balloon?
    - **b.** **C** How does your answer to part (a) change if the balloon is surrounded by oil instead of water?
    - **c.** **C** How does your answer to part (a) change if the balloon is made of Mylar instead of rubber?

61. **A** Find an expression for the electric field between the two conducting disks in Figure P27.61. Make sure your expression is general enough to include the possibility of a dielectric between the disks. Check your answer using the information given in Section 27-8.

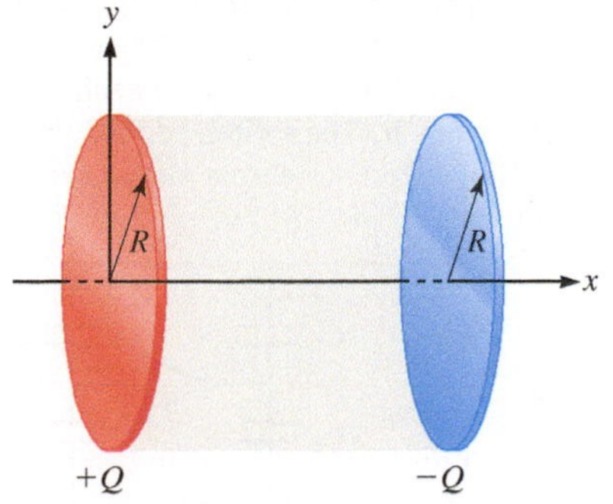

FIGURE P27.61

62. **A** An air-filled parallel-plate capacitor is charged to a certain potential difference. A dielectric is then inserted in the capacitor to completely fill the space between the plates. Then the charge on the plates is increased by a factor of three to restore the original potential difference. Determine the dielectric constant.

63. **N** Two Leyden jars are similar in size and shape, but one has glass as the dielectric and the other ebonite. The glass jar is charged, but when the charge is shared between the two jars (connected in parallel), the electric potential drops by 40% of its initial value. If the dielectric constant of glass is 3.0, find the dielectric constant of ebonite.

Problems 64, 65, and 66 are grouped.

64. **A** Nerve cells in the human body and in other animals are modeled as very long cylindrical capacitors. Portions of some nerves are covered in a layer of fat known as myelin, which functions as the dielectric between two plates in the cylindrical capacitor model. Find an expression for the electric field in the myelin as a function of the distance r from the cell's central axis (assuming a cylindrical cell with $r = 0$ at the axis).

65. **N** Nerve cells in the human body and in other animals are modeled as very long cylindrical capacitors. Portions of some nerves are covered with a layer of fat known as myelin, which functions as the dielectric ($\kappa = 7$) between two plates in the cylindrical capacitor model. The potential difference between the inner and outer walls of myelin in resting nerve cells is roughly $V_{\text{inner}} - V_{\text{outer}} = -70$ mV. Find the linear charge density on the inner (positive) plate. *Hint*: Use the result of Example 27.8.

66. **G** Nerve cells in the human body and in other animals are modeled as very long cylindrical capacitors. Portions of some nerves are covered with a layer of fat known as myelin, which functions as the dielectric ($\kappa = 7$) between two plates in the cylindrical capacitor model. The potential difference between the inner and outer walls of myelin in resting nerve cells is roughly $V_{\text{inner}} - V_{\text{outer}} = -70$ mV. Normally, the inner radius of the myelin layer is 2×10^{-6} m and its thickness is 5×10^{-6} m. Plot the magnitude of the electric field as a function of position r for a nerve cell.

General Problems

67. **N** A 1.50-nF capacitor has a charge of 6.00 nC. **a.** How much energy is stored by the capacitor? **b.** What is the potential difference across the capacitor?

68. **C** If you carefully open a 9-V battery, you discover it is made up of six small batteries (Fig. P27.68). How are these batteries connected to one another, and what is the terminal potential across each of them?

Zoltan Fabian/Shutterstock.com

FIGURE P27.68

69. **N** A 9.00-V battery is connected across two capacitors, $C_A = 12.0\ \mu F$ and $C_B = 3.0\ \mu F$, connected in series. **a.** What is the equivalent capacitance of the two capacitors? **b.** How much energy is stored by such an equivalent capacitor? **c.** What is the charge on each of the capacitors? **d.** How much energy is stored in each of the capacitors in this circuit?

70. **N** Three capacitors with capacitances $2.00 \times 10^{-3}\ \mu F$, $4.00 \times 10^{-3}\ \mu F$, and $6.00 \times 10^{-3}\ \mu F$ are connected in series. Is it possible to apply a potential difference of 11.00×10^3 V across the set if the breakdown voltage of each capacitor is 4.00×10^3 V?

71. **N** What is the maximum charge that can be stored on the 8.00-cm^2 plates of an air-filled parallel-plate capacitor before breakdown occurs? The dielectric strength of air is 3.00 MV/m.

Problems 72 and 73 are paired.

72. **N, C** In a laboratory, you find a 9.00-V battery and a 12.0-V battery. You also find a 30.0-μF capacitor and a 45.0-μF capacitor. Your challenge is to use only one battery and one capacitor to store the maximum possible energy. Which battery and which capacitor do you choose, and how much energy will the capacitor store?

73. **N, C** In a laboratory, you find a 9.00-V battery and a 12.0-V battery. You also find a 30.0-μF capacitor and a 45.0-μF capacitor. Your challenge is to store the maximum possible energy. You may use as much of this equipment as you wish. Describe your solution, and draw a schematic diagram of your network. How much energy is stored by the capacitor(s)?

74. **N** A spherical capacitor has an inner-shell radius of 4.00 cm and an outer-shell radius of 8.00 cm. (See Problem 32.) **a.** What is the capacitance of this capacitor? **b.** When connected to a battery, the capacitor carries a charge of 7.45 μC. What is the voltage of the battery?

75. **N** Figure P27.75 shows four capacitors with $C_A = 4.00\ \mu F$, $C_B = 8.00\ \mu F$, $C_C = 6.00\ \mu F$, and $C_D = 5.00\ \mu F$ connected across points a and b, which have potential difference $\Delta V_{ab} = 12.0$ V. **a.** What is the equivalent capacitance of the four capacitors? **b.** What is the charge on each of the four capacitors?

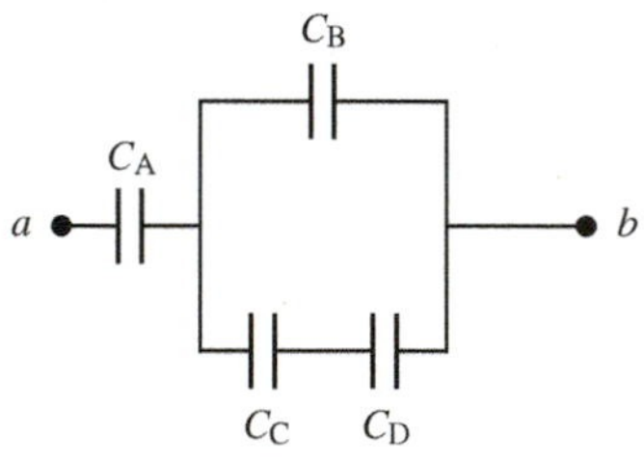

FIGURE P27.75

76. **N, C** As in Example 27.3 (page 842), you are working in a laboratory to build a network from a design (Fig. P27.76). You have a dozen batteries with terminal potential $\mathcal{E}_0 = 1.5$ V and a dozen capacitors with capacitance $C_0 = 9.0\ \mu F$. Draw a schematic diagram showing how you can build this network from the components in the laboratory.

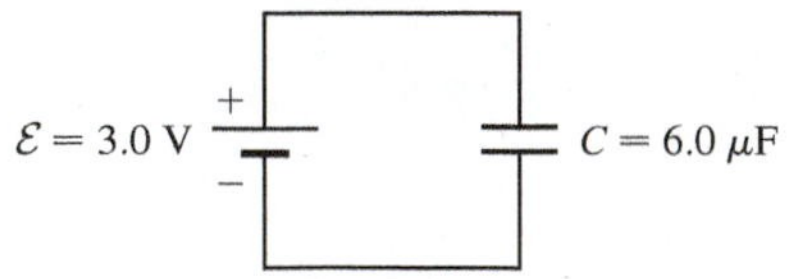

FIGURE P27.76

77. **N** The plates of an air-filled parallel-plate capacitor each have an area of 5.75 cm^2 and are separated by 0.145 cm. A 12.0-V battery is connected to the capacitor. **a.** What is the capacitance of the capacitor? **b.** How much charge is stored by the capacitor? **c.** What is the magnitude of the electric field between the plates of the capacitor? **d.** What is the surface charge density on the positive plate of the capacitor?

78. **A** A parallel-plate capacitor with plates of area A and spacing d is filled with a dielectric with constant κ. The dielectric is then pulled halfway out of the gap. Find an expression for the capacitance of the resulting configuration.

79. **N** When connected in series, two capacitors have an equivalent capacitance of 3.00 μF. The same two capacitors have an equivalent capacitance of 13.0 μF when connected in parallel. What is the capacitance of each of the capacitors?

80. **A** Show that $V_C = \int_{r_+}^{r_-} E\,dr$ (Eq. 27.9) is derived from Equation 26.15, $\Delta V = V_f - V_i = -\int_{r_i}^{r_f} \vec{E} \cdot d\vec{r}$ when a path is chosen that is parallel to an electric field line and $\Delta V = V_f - V_i = V_- - V_+ = -V_C$.

81. **N** A 90.0-V battery is connected to a capacitor with capacitance C_A. The capacitor is charged and then disconnected from the battery. Capacitor C_A is next connected to a second, uncharged capacitor with capacitance $C_B = 22.0\ \mu F$. If the voltage across the capacitors in parallel is measured to be 55.0 V, what is the capacitance C_A?

82. **N** Consider an infinitely long network with identical capacitors arranged as shown in Figure P27.82. Determine the equivalent capacitance of such a network. Each capacitor has a capacitance of 1.00 μF.

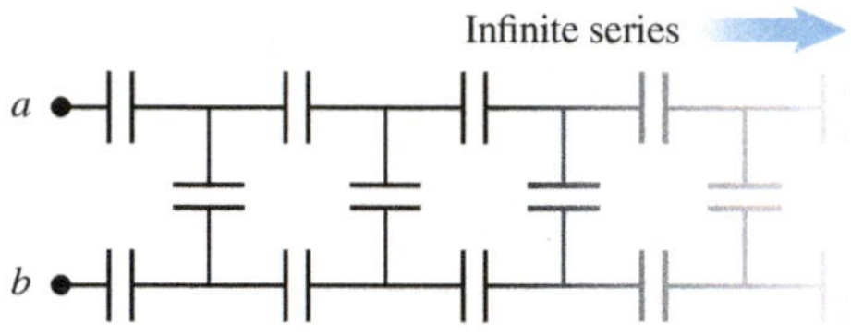

FIGURE P27.82

83. **A** A capacitor C_1 has potential difference V and stores energy U_1. It is then connected without loss of charge in parallel with a second capacitor C_2 that was initially uncharged. **a.** Find an expression for the energy U_2 stored in C_2 in terms of C_1, C_2, and U_1. **b.** Show that U_2 is a maximum when $C_1 = C_2$. *Hint*: Consider the derivative of U_2 as a function of the ratio $x = C_1/C_2$.

84. **N** What is the equivalent capacitance of the five capacitors shown in Figure P27.84? *Hint*: Note the symmetry of the circuit.

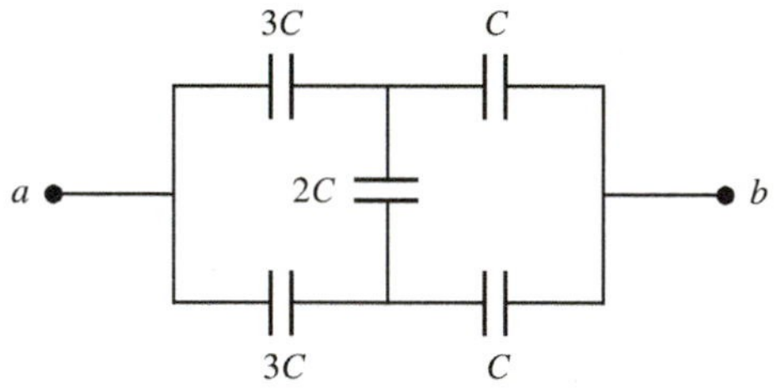

FIGURE P27.84

85. **N** The circuit in Figure P27.85 shows four capacitors connected to a battery. The switch S is initially open, and all capacitors have reached their final charge. The capacitances are $C_1 = 6.00\ \mu F$, $C_2 = 12.00\ \mu F$, $C_3 = 8.00\ \mu F$, and $C_4 = 4.00\ \mu F$. **a.** Find the potential difference across each capacitor and the charge stored in each. **b.** The switch is now closed. What is the new final potential difference across each capacitor and the new charge stored in each?

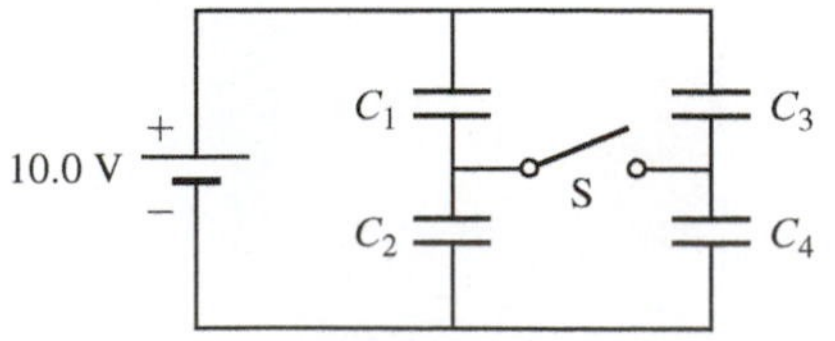

FIGURE P27.85

86. **A** The 1.00-cm gap between the horizontal plates of an air-filled, 44.0-μF, parallel-plate capacitor is slowly filled with peanut oil ($\kappa = 3.0$). How does the capacitance of the capacitor change as a function of the height h of the oil that fills the gap between its plates?

87. **A** Pairs of parallel wires or coaxial cables are two conductors separated by an insulator, so they have a capacitance. For a given cable, the capacitance is independent of the length if the cable is very long. A typical circuit model of a cable is shown in Figure P27.87. It is called a lumped-parameter model and represents how a unit length of the cable behaves. Find the equivalent capacitance of **a.** one unit length (Fig. P27.87A), **b.** two unit lengths (Fig. P27.87B), and **c.** an infinite number of unit lengths (Fig. P27.87C). *Hint*: For the infinite number of units, adding one more unit at the beginning does not change the equivalent capacitance.

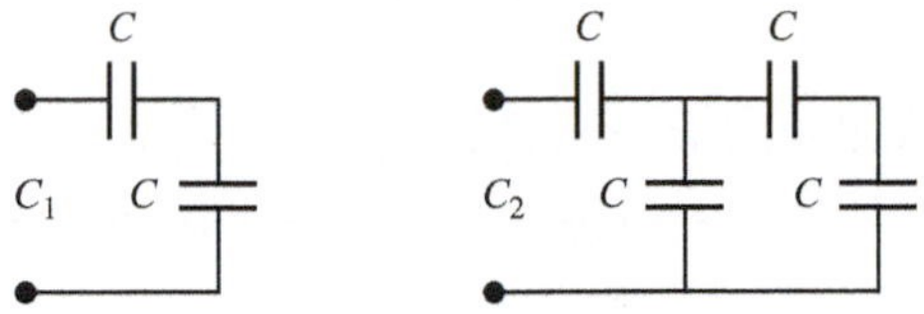

A. One unit length **B. Two unit lengths**

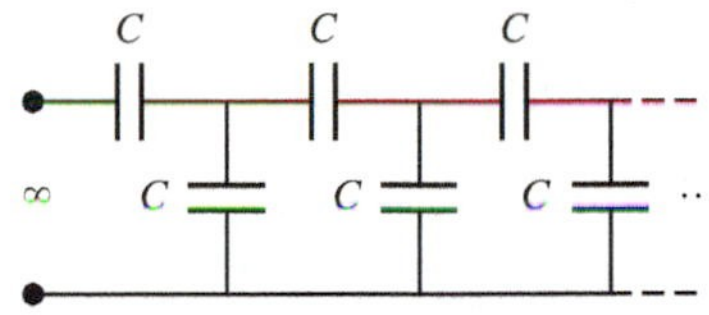

C. Infinite number of unit lengths

FIGURE P27.87

88. **N** A parallel-plate capacitor has square plates of side $s = 2.50$ cm and plate separation $d = 2.50$ mm. The capacitor is charged by a battery to a charge $Q = 4.00\ \mu$C, after which the battery is disconnected. A porcelain dielectric ($\kappa = 6.5$) is then inserted a distance $y = 1.00$ cm into the capacitor (Fig. P27.88). *Hint*: Consider the system as two capacitors connected in parallel. **a.** What is the effective capacitance of this capacitor? **b.** How much energy is stored in the capacitor? **c.** What are the magnitude and direction of the force exerted on the dielectric by the plates of the capacitor?

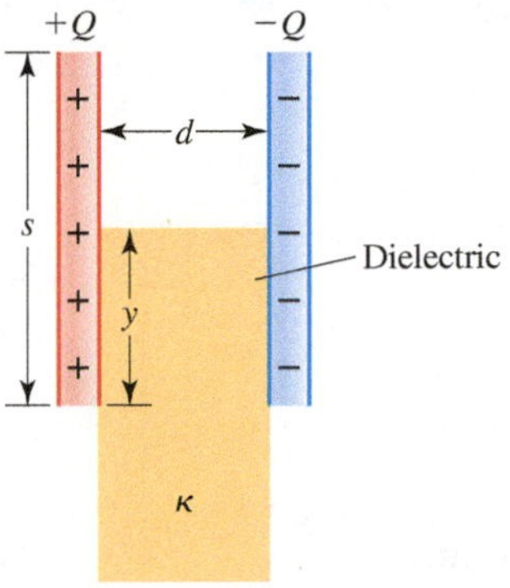

FIGURE P27.88

28 Current and Resistance

Key Questions

How does the microscopic model for the motion of electrons in a conductor relate to our macroscopic observations of current and electric potential?

What does a battery supply in a circuit?

Underlying Principles

No new principles are introduced in this chapter.

Major Concepts

1. Circuits and circuit elements
2. Current
3. Current density
4. Conductivity, resistivity, and their temperature dependence
5. Resistance and resistors
6. Ohm's law
7. Power supplied to or used by a circuit element

You dry your hair with an electric blow dryer. Your car's headlights illuminate the road. You send e-mail messages to your professor and call your friends and family on your cell phone. These are examples of the ways in which your daily routine depends on the motion of charged particles.

The past five chapters dealt with *electrostatics.* Although an electric field causes charged particles to rearrange themselves, we ignored their motion and imagined waiting until they reached equilibrium. Now we take a close look at the motion of charged particles. In metallic conductors, those charged particles are electrons whose motion is described in terms of a *current.* Electrons that carry a current do not flow smoothly like toothpaste out of a tube. Moving electrons collide with ion cores in the conducting metal and zigzag through the conductor, so that the conductor *resists* the current. In this chapter, we will connect a microscopic model of the motion of electrons to macroscopic observations of current and resistance in conductors. What do quantities we can measure (macroscopic properties) tell us about the motion of charged particles inside a conductor (the microscopic model)?

28-1 Microscopic Model of Charge Flow

This is the second of three chapters devoted to practical applications of the principles of electricity. Chapter 27 focused on capacitors (devices for storing electric potential energy) and batteries (devices that use chemical reactions to maintain a potential difference). This chapter is about the motion of charged particles within conductors. We develop a microscopic model and then connect that model to macroscopic observations. The microscopic model for the motion of charged particles underlies many practical applications, as in this chapter's case study.

CASE STUDY Dead Phone[1]

Read the following student discussion. As always, some parts of the explanations are incorrect, and other parts are on the right track. For now, think about what makes sense to you. We'll return to this case study after we learn more about current and resistance.

Shannon: Avi, you look upset. What's going on?

Avi: I went to Electronics Hut to get a car adapter for my new smartphone, and the guy wrecked my phone. He grabbed some adapter off the rack, and when he plugged it into my phone, the words kind of melted off the screen. I think I even smelled smoke. We unplugged it, but it wouldn't go back on. We tried everything—recharging the battery off the wall socket, rebooting. Nothing worked. Then I had to get the manager to say they would buy me a new phone. I have to go back later to deal with that.

Shannon: That's awful. I'm so sorry.

Cameron: Sounds like the guy fried your phone. I bet he used a high-voltage adapter. It's like the phone's equivalent of touching a high-voltage power line.

Shannon: I think you mean the current was too high. The adapter is like a battery. It's supposed to maintain a constant voltage. Besides, you normally charge your phone off the wall socket, which is about 120 V, and you would use that adapter in the car, where the battery is only 12 V. There's no way the adapter voltage was higher than 120 V.

Cameron: Look, my friend made his own adapter for his MP3 player, and he had to make sure the output voltage of the adapter was around 5 V.

Avi: The thing I really don't get is how the metal in my phone could melt. It's not like there was a big spark like lightning hitting a tree. Why did my phone go up in smoke?

Basic Model

In electronic devices such as Avi's phone, charged particles move through conductors. To understand how these devices work, we must model the motion of charged particles in a conductor. Consider copper. Each neutral copper atom has 29 protons in its nucleus surrounded by 29 swarming electrons. Electrical attraction keeps the electrons bonded to the nucleus, but the outer electrons are only weakly attracted. When many copper atoms are close together, as they are in a metal wire, the outer electrons are also attracted to the nuclei of other nearby atoms. In fact, the outermost electron is no longer bound to any one particular atom and is free to move throughout the entire conductor. These free electrons are known as **conduction electrons**. Typically, a metal conductor has one free electron per atom. We think of the metal as having two components: (1) conduction electrons free to move throughout the conductor (Fig. 28.1) and (2) ions essentially fixed in place in a lattice.

Perhaps we can model the motion of the conduction electrons in the same way we modeled the motion of gas molecules in a container. In fact, our conduction model is based on kinetic theory (Chapter 20), and the five assumptions we made about the

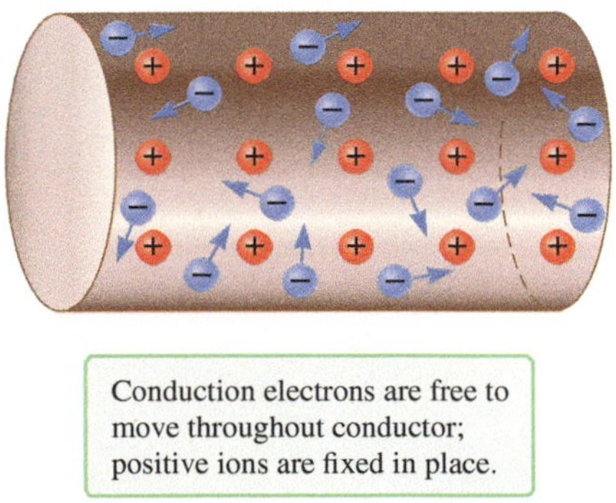
Conduction electrons are free to move throughout conductor; positive ions are fixed in place.

FIGURE 28.1 A model for the motion of conduction electrons in a conductor.

[1]This case study is based on the real experience of the textbook author.

molecules in an ideal gas are a good place to start. Table 28.1 lists the assumptions we make about the conduction electrons in a metal and, for comparison, the assumptions we made about molecules in an ideal gas.

TABLE 28.1

Ideal Gas	Conduction Electrons
1. A gas consists of a large number of molecules or atoms that can be modeled as particles.	**1.** A metal contains a large number of conduction electrons that can be modeled as particles. (Conduction electrons are small compared to the average distance between fixed ions.)
2. Particles in an ideal gas do not interact with one another.	**2.** Conduction electrons do not interact with one another. Electrons repel one another, but because there are many free electrons, we assume the net force exerted on any one electron by all the other electrons is zero.
3. Particles make elastic collisions with the walls. The duration of each collision is short.	**3.** Conduction electrons collide with the positively charged ions, which are assumed to be (essentially) at rest. The duration of a collision is short.
4. Particles are free to move in any direction at any speed.	**4.** Between collisions, conduction electrons are free to move in any direction at relatively high speeds.
5. The gas is made up of identical particles.	**5.** Conduction electrons are identical. Each has the same mass m_e and charge $-e$.

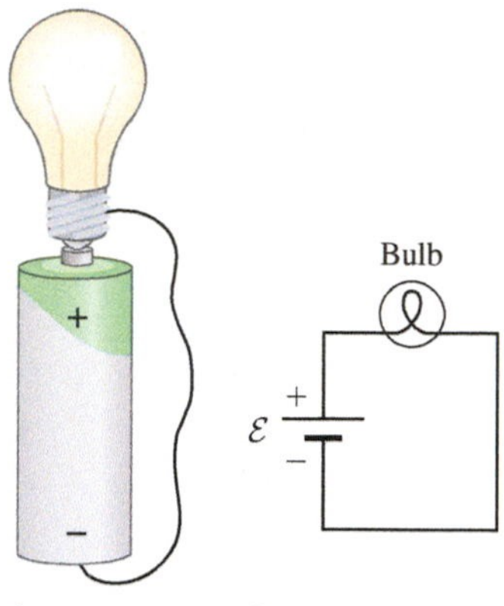

A. **B.**

FIGURE 28.2 **A.** A lightbulb's side is connected to the negative terminal of a battery, and the bulb's base is connected to the battery's positive terminal. **B.** Schematic diagram for the circuit in part A.

Throughout this chapter, we will apply this microscopic model to a simple practical device—a battery with terminal potential $\mathcal{E}$ connected to a lightbulb (Fig. 28.2A). There are several symbols for a lightbulb; in this textbook, we use the symbol shown in the schematic diagram in Figure 28.2B. The smooth lines drawn between the battery and the bulb represent wires. The network in Figure 28.2 is a closed pathway in which charged particles can flow. Such a closed network is called a **circuit**. The various devices that may be in the circuit—such as the battery and the bulb—are called the **circuit elements**. In this chapter, we consider simple circuits with only one pathway for charge to flow through. In the next chapter, we will consider more complicated circuits that may contain many branching pathways.

CIRCUITS AND CIRCUIT ELEMENTS

Major Concepts

CONCEPT EXERCISE 28.1

One of the assumptions we make about the conduction electrons in a metal is that the net force exerted on any one electron by all the other conduction electrons is zero (Table 28.1). Justify that assumption.

28-2 Current

Picture the sea of activity inside a copper penny in your pocket, where presumably there is no electric field. Conduction electrons in the penny whiz around, colliding with ions. Imagine watching just one conduction electron for a few moments (Fig. 28.3A). Because the electron's starting position at A is directly above its ending position at G, its displacement along x is zero during this time. The electron's displacement depends on the time interval during which it is observed. For example, if we had observed the electron for a slightly shorter time, it would have been at position F, and its displacement would have had a component in the negative x direction. However, if we observe a large number of conduction electrons over any length of time, we find there is no net motion of the electrons in any particular direction. (This is also true for ideal gas particles inside a container: Their motions are also random, so there is no net displacement of the gas particles.)

What happens if you put your penny in an electric field—for example, by placing it between the plates of a charged capacitor? The electric field causes the charged

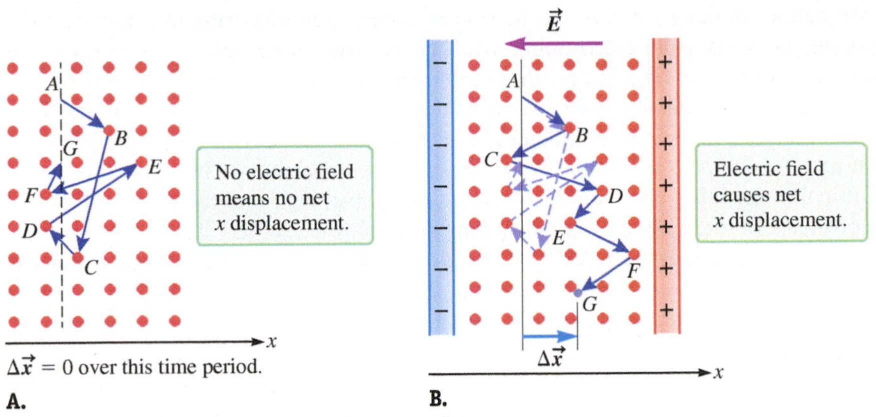

No electric field means no net x displacement.
$\Delta\vec{x} = 0$ over this time period.
A.
$\vec{E}$
Electric field causes net x displacement.
$\Delta\vec{x}$
B.

FIGURE 28.3 **A.** Path of a conduction electron in a metal with $\vec{E} = 0$. **B.** Path of a conduction electron in a metal with $\vec{E} = E_y\hat{\jmath}$. In each a case, the electron starts at point A and ends at point G. In between, it collides with five different ions at points B–F, changing direction at each of these points.

particles in the penny to rearrange themselves to reach equilibrium with no electric field inside the penny. During the short time interval *before* the particles reach equilibrium, however, the electric field inside the penny is not zero. Suppose the electric field is in the negative x direction, $\vec{E} = -E_x\hat{\imath}$, and consider the path of a single conduction electron (Fig. 28.3B). The electric field exerts a force on that electron in the positive x direction:

$$\vec{F}_E = q\vec{E} = (-e)(-E_x\hat{\imath})$$

$$\vec{F}_E = eE_x\hat{\imath}$$

When the electron is moving between collisions in this field, it is pulled in the positive x direction. As a result, the electron drifts to the right. If we observe the same conduction electron over the same period of time as in Figure 28.3A, we find the ending position at G is to the right of the starting position at A, so the electron's displacement has a positive x component (Fig. 28.3B).

Figure 28.3B shows the path of the electron both with and without an electric field in the penny for comparison. The electric field does not reduce the average number of collisions the conduction electron makes with the ions. Instead, the electron's path is shifted to the right. If we looked at all the conduction electrons in the penny, we would find (1) they undergo many collisions with the ions; (2) their motion between collisions is very fast, with speeds around 10^6 m/s (Problem 4); and (3) they tend to move toward the right at a slow speed known as the **drift velocity** in the direction opposite to the electric field. The typical drift speed (magnitude of drift velocity) v_{drift} is between 10^{-5} m/s and 10^{-4} m/s.

Let's see how this microscopic model of electron flow is connected to our macroscopic observations. You might use a voltmeter to measure the potential difference ΔV between two points R and L on the penny, where $\Delta V = V_R - V_L$. In Figure 28.4A, there is no electric field, so there is no net displacement of conduction electrons and no excess charge anywhere in the penny. The result is that the voltmeter's reading is $\Delta V = V_R - V_L = 0$, which indicates no potential difference across the penny.

In Figure 28.4B, the penny is placed in a leftward-pointing electric field so that the conduction electrons drift toward the right. If we wait until the charged particles in the penny reach equilibrium, the penny is still neutral overall but the right side of the penny has excess negative charge and the left side has excess positive charge. The voltmeter's reading is $\Delta V = V_R - V_L < 0$, indicating that the right side of the penny is now at a lower potential than the left side.

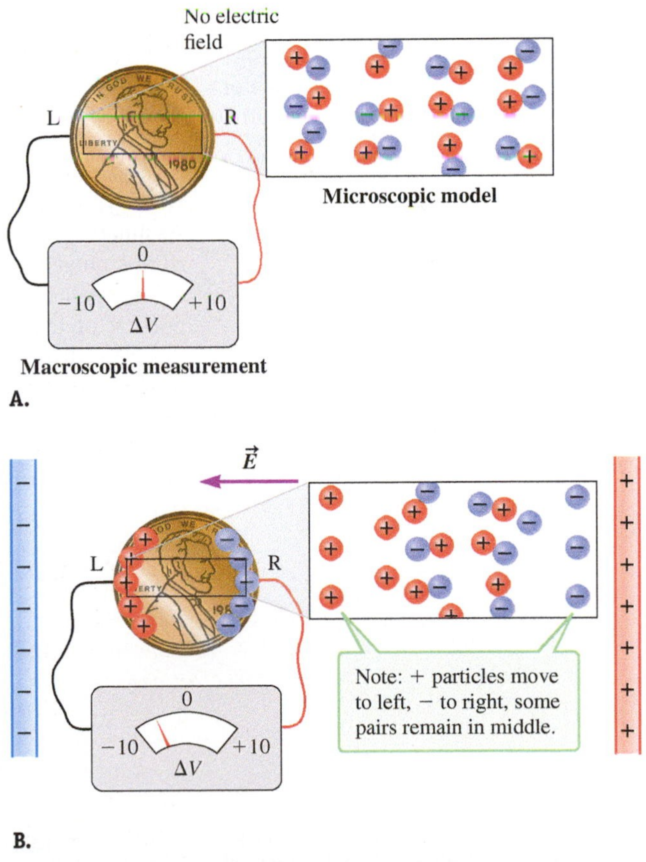

No electric field
L
R
Microscopic model
0
−10
+10
ΔV
Macroscopic measurement
A.
$\vec{E}$
L
R
Note: + particles move to left, − to right, some pairs remain in middle.
0
−10
+10
ΔV
B.

FIGURE 28.4 **A.** When $\vec{E} = 0$, $V_R = V_L$. **B.** When the penny is in an electric field $\vec{E} = -E_x\hat{\imath}$, the conduction electrons drift so that the left side of the penny becomes positive and the right side becomes negative, and $V_R - V_L < 0$.

We cannot directly observe the motion of conduction electrons in a metal, and a voltmeter is no help in determining whether the free negatively charged electrons really move or perhaps the free positively charged particles move. Without better information, 18th- and 19th-century experimenters assumed the positive particles were mobile. Suppose the *positive* particles move toward the left as a result of the leftward-pointing electric field. The bound negative particles would remain in place, and the net result is that the left side would have excess positive charge and the right side excess negative charge—exactly the way the penny appears (Fig. 28.4B). The voltmeter measures the potential difference $\Delta V = V_R - V_L < 0$ exactly as before. The case study in Chapter 30 is about an experiment showing that in metals, conduction electrons move and positive ions stay relatively motionless, but today we still find it convenient to think about positively charged particles moving in the opposite direction as electrons.

Current is the apparent motion of positively charged particles. In a metal conductor, the current is a result of the conduction electrons' motion in the opposite direction. So, in a metal conductor, the *current* is in the same direction as the electric field. Current is not always the result of electrons flowing. In ionic solutions or in semiconductors, current may be the result of the motion of positive particles.

If a penny is placed between the plates of a charged capacitor, conduction electrons flow in the opposite direction as the electric field, and very soon the particles are in equilibrium again. Such a current does not last very long. In order to study current, we must set up a steady current in some conductor, such as in a lightbulb connected to a battery. An incandescent bulb consists of a conducting wire known as a *filament* encased in a glass globe. The filament may be made of tungsten and may be coiled (Fig. 28.5). One end of the filament is connected to the metal screw threads at the base of the globe. The other end of the filament is connected to the metal foot in the center of the base. Between this foot and the metal threads is an insulator. If you want to light the bulb with a battery, you connect one end of the battery to the metal foot and the other end to the metal threads (Fig. 28.2A). Chemical reactions inside the battery maintain its terminal potential, so electrons continue to flow away from the negative terminal toward the positive terminal. In terms of current, we say that the current is directed away from the positive terminal through the filament and toward the negative terminal.

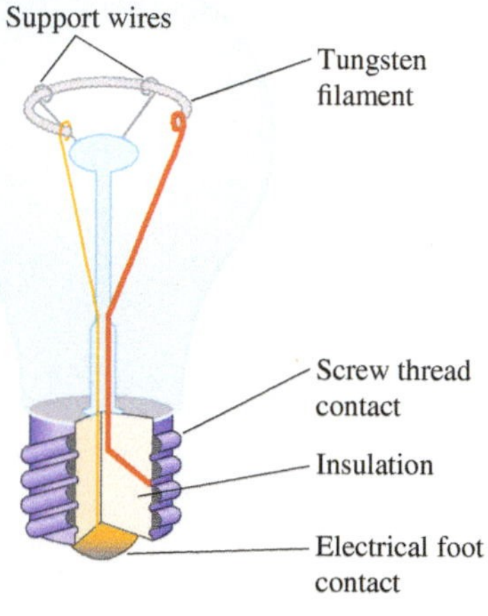

FIGURE 28.5 An incandescent lightbulb. Although compact fluorescent bulbs are in wide use, incandescent bulbs are still used in some applications. Compact fluorescent bulbs operate on a different principle and are not discussed in this chapter.

The amount of current in the filament depends on temperature. The bulb is cool when you first turn it on, and it can get very hot after it has been on for a while. Assume the bulb has been on for a few moments and its temperature is in equilibrium so that the current is steady. Figure 28.6A shows a close-up of a small portion of the filament, small enough to look like a simple cylinder. Because in this small portion of the filament the electric field points from left to right, the conduction electrons drift from right to left. This flow of electrons is modeled as a flow of imaginary positive particles from left to right (Fig. 28.6B).

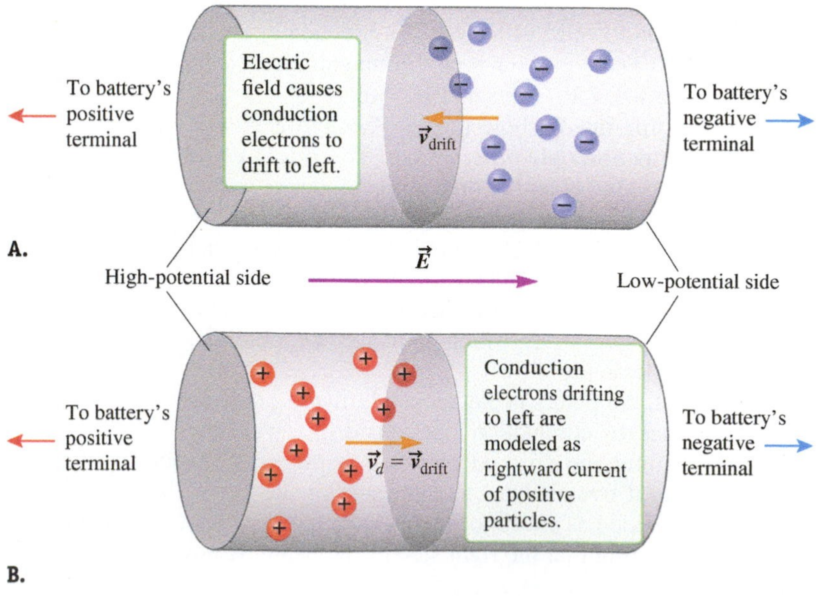

FIGURE 28.6 A. A small portion of a filament connected to a battery. Because the electric field points from high potential to low potential, the field inside the filament points from left to right. So, the conduction electrons in the filament drift to the left. **B.** We model each conduction electron as an imaginary positive particle traveling in the opposite direction.

CURRENT **Major Concept**

Current is defined as the rate at which these imaginary positive particles pass through a cross section of the filament:

$$I = \frac{dq}{dt} \tag{28.1}$$

The SI unit for current is the **ampere**, abbreviated with an upper case A. An ampere is one coulomb per second:

$$1\,\text{A} = 1\,\text{C/s}$$

The term *amp* is used informally for ampere, and *amperage* is used for current.

The ampere is one of the seven fundamental SI units. In fact, a coulomb is defined in terms of the ampere: *If there is a steady current of one ampere* (1 A), *then one coulomb* (1 C) *is the amount of charge that passes a particular cross section in one second* (1 s). The amount of charge q that passes a particular cross section in some amount of time t is found by integration:

$$q = \int dq = \int_0^t I\,dt \tag{28.2}$$

Current has magnitude given by Equation 28.1 and is directed from high potential to low potential. However, current is a *scalar*. The current follows the geometry of the wire. In the filament in Figure 28.5, the current follows the curling wire from high potential toward low potential. We often draw an arrow near a wire to indicate the direction of the current, as in Figure 28.7, but these arrows are not vectors. Because current is a scalar, vector algebra is not needed when we deal with current combinations. For example, in Figure 28.7, a current-carrying conductor is split into two pieces. The relationship between the currents in each branch is found by simple addition: $I_0 = I_1 + I_2$.

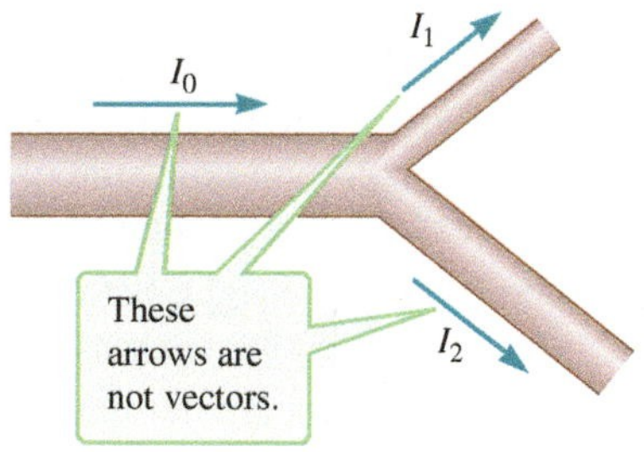

FIGURE 28.7 Current is a scalar, and no vector algebra is needed. Here, $I_0 = I_1 + I_2$.

CONCEPT EXERCISE 28.2

Typical currents in three common devices are listed in the table. Assume the currents are constant. Find the amount of charge that passes through a cross section of wire carrying these currents in 1 s.

Device	Current	Charge in 1 s?
a. Ordinary flashlight	0.5 A	
b. Wires in car starter motor	200 A	
c. Laptop computer circuit	1 pA	

EXAMPLE 28.1 CASE STUDY The Battery's Job

In an effort to understand what happened to Avi's phone, the three students discuss a very simple circuit made up of just one battery whose terminals are connected by a simple wire (Fig. 28.8). After reading their discussion, decide whom you agree with and why. You may agree with more than one student.

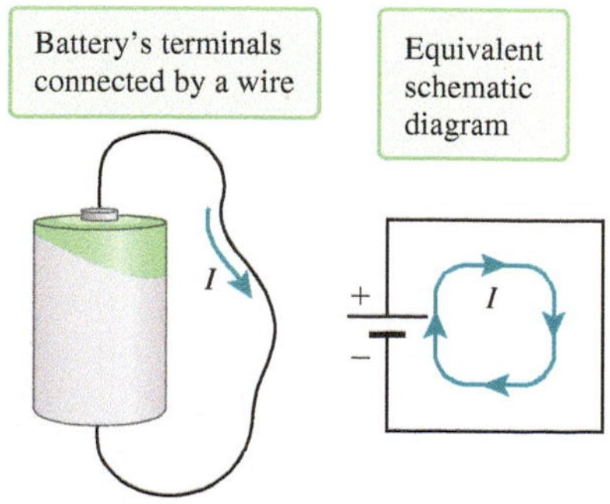

FIGURE 28.8 A wire is connected to the terminals of a simple battery.

Example continues on page 870 ▶

Avi: The battery is a source of heat. If you touch the wire, it is going to get very hot, just like a lightbulb.

Shannon: No. The battery is a source of current. The bulb will not light up unless charged particles flow through the filament. The chemical reactions in the battery generate charges, like there is a fountain inside the battery. Positive charges bubble out of the positive end. Most of the particles are converted to light by the bulb. Any particles that are left over get sucked into the negative end of the battery and reused.

Cameron: OK, current is from the positive side of the battery toward the negative side, but the battery's job is to maintain a potential difference between its terminals. It does this through chemical reactions. That's why you buy a battery by its voltage, not by its amperage.

Shannon: If the battery didn't make the current, there would be no current in the wire. It would be like a dry riverbed. It has the potential for current, but it is empty.

Cameron: The wire is not like a dry riverbed! It is more like a flat pipe full of water. The water just sits there until you pick up one end so that the water can flow down. The battery is more like the thing that picks up the end of the pipe. The battery makes sure there is a high-potential side of the wire and a low-potential side of the wire. Then the current just goes from the high side to the low side.

Shannon: If you were right, batteries would last forever. Batteries die because they cannot make any more charged particles. They just run out of juice.

Avi: I think you are both missing the bigger picture. Heat is what killed my phone. Just as the wire or a lightbulb gets hot when you connect it to a battery, my phone got *very* hot. The battery has to be a source of heat. The chemical reactions give off heat.

SOLVE

Some parts of Avi's and Cameron's statements are correct, but Shannon is simply wrong. The battery is not a source of current. Chemical reactions in the battery maintain a potential difference between its terminals. When a wire is connected to those terminals, there is a current in the wire because conduction electrons drift toward the positive end of the battery, carrying energy throughout the wire and causing the wire to get hot. Thus, the battery must be a source of *energy*. Avi is misusing the word *heat*. **Heat** is energy transferred from the environment to a system or from a system to the environment due to their temperature difference (Chapter 21).

28-3 Current Density

If all we need is a macroscopic description of a circuit, current as a scalar parameter is usually sufficient. However, our goal is to connect the macroscopic description to a microscopic model, so we need a vector quantity—**current density**. The magnitude of the current density is the current I per unit cross-sectional area A,

$$J \equiv \frac{I}{A} \tag{28.3}$$

where I is uniform over the area A. Current density is a vector that points in the same direction as the electric field,

$$\vec{J} \propto \vec{E} \tag{28.4}$$

and is a macroscopic property.

To relate current density to the microscopic motion inside a wire, consider Figure 28.6. When a battery's terminals are connected by a wire, the potential difference causes conduction electrons in the wire to drift from the negative terminal toward the positive terminal (Fig. 28.6A), equivalent to the motion of positive particles in the opposite direction. The "moving" positive particles are not the metal ions, but rather imaginary particles that replace actual conduction electrons. Each imaginary particle has the mass of an electron m_e and carries a positive charge $+e$. We imagine that (1) these positive particles undergo many collisions with fixed ions,

(2) between collisions their speed is very high, and (3) their direction of motion after each collision is random. If there is an electric field in the wire, the positive particles drift in the direction of the electric field (Fig. 28.6B). Their drift velocity $\vec{v}_d$ has the same magnitude as the drift velocity $\vec{v}_{\text{drift}}$ of the conduction electrons, but it is in the opposite direction: $\vec{v}_d = -\vec{v}_{\text{drift}}$.

DERIVATION Current Density in a Wire

CURRENT DENSITY
Major Concept

We show that the current density $\vec{J}$ in a wire is directly proportional to the drift velocity $\vec{v}_d$ of the imaginary positive particles:

$$\vec{J} = ne\vec{v}_d \tag{28.5}$$

where n is the number density of atoms in the conductor.

Figure 28.9 shows a straight wire of cross-sectional area A. An electric field in the positive x direction causes conduction electrons to drift in the negative x direction, modeled as positive particles drifting in the positive x direction. These equivalent positive particles are shown at two instants: just as they are about to pass through the cross section labeled 1 and just as they are about to pass through another section labeled 2.

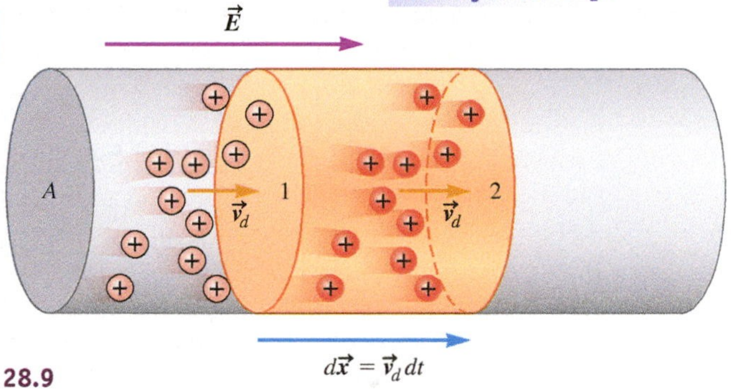

FIGURE 28.9

The positive particles travel with velocity $\vec{v}_d$, so in a time interval of dt, their displacement is $d\vec{x}$.	$d\vec{x} = \vec{v}_d\,dt$	(1)
The amount of charge dq passing through 1 in the time interval dt depends on the number of particles N passing through and the amount of charge e that each one carries.	$dq = Ne$	(2)
We find N from the number density of the actual conduction electrons. Most conductors have one conduction electron per atom, so the number density of the imaginary particles is equal to the number density n of atoms in the metal, and N is the number density of the particles n times the volume dV of that portion of the wire.	$N = n\,dV$	(3)
The volume of the orange portion of the wire is its cross-sectional area A times the length of that portion dx. Substitute Equation (1) for dx.	$dV = A\,dx = Av_d\,dt$	(4)
Find the number of particles that pass through cross section 1 in the time interval dt by substituting Equation (4) into Equation (3).	$N = nAv_d\,dt$	(5)
Find the amount of charge dq that passes through cross section 1 in the time interval dt by substituting Equation (5) into Equation (2).	$dq = neAv_d\,dt$	(6)
Substitute Equation (6) into $I = dq/dt$ (Eq. 28.1). The time interval dt cancels out.	$I = \dfrac{dq}{dt} = nev_dA$	(28.6)
Find the magnitude of the current density by dividing by the cross-sectional area (Eq. 28.3).	$J = \dfrac{I}{A} = nev_d$	
The current density points in the same direction as the electric field. From Figure 28.9, the electric field is in the same direction as the drift velocity of the imaginary positive particles. So, the current density is in the same direction as the drift velocity of the imaginary positive particles.	$\vec{J} = ne\vec{v}_d$ ✓	(28.5)

Derivation continues on page 872 ▶

COMMENTS

Equation 28.5 achieves our goal of finding a connection between the macroscopic observation of current density and the microscopic model of particle motion. On the left side of the equation is the current density—a quantity we can observe on the macroscopic level. The right side involves properties (charge and drift velocity) of the microscopic particles. The current density is also proportional to the number density n. Table 28.2 lists values of n for various metals.

TABLE 28.2 Number density, conductivity, resistivity, and temperature coefficient of resistivity for various materials near room temperature (20°C).

Substance	Number density n ($\times 10^{28}$ m^{-3})	Conductivity σ ($\Omega^{-1} \cdot$ m^{-1})	Resistivity ρ ($\Omega \cdot$ m)	Temperature coefficient of resistivity α (°C^{-1})
		Conductors		
Aluminum	6.03	3.767×10^7	2.655×10^{-8}	0.00429
Constantan		2.0×10^6	4.9×10^{-7}	0.00001
Copper	8.42	5.959×10^7	1.678×10^{-8}	0.00393
Gold	5.90	4.46×10^7	2.24×10^{-8}	0.0083
Lead	3.31	4.843×10^6	2.065×10^{-7}	0.00336
Manganin		2.3×10^6	4.4×10^{-7}	0.00001
Mercury	4.07	1.02×10^6	9.84×10^{-7}	0.00089
Nichrome		1.00×10^6	1.00×10^{-6}	0.0004
Silver	5.86	6.302×10^7	1.586×10^{-8}	0.0061
Tungsten	6.22	1.77×10^7	5.65×10^{-8}	0.0045
		Semiconductors[a]		
Carbon (graphite)		1.7 to 29×10^3	3.5 to 60×10^{-5}	-0.0005
Germanium		0.001 to 2	1 to 500×10^{-3}	-0.05
Silicon		0.02 to 10	0.1 to 60	-0.07
		Insulators		
Amber		2×10^{-15}	5×10^{14}	
Glass[b]		2×10^{-4}	5×10^3	
Rubber (hard)[b]		2×10^{-15}	5×10^{14}	
Quartz		1.3×10^{-18}	7.5×10^{17}	

[a] The conductivity and resistivity of a semiconductor depend on the presence of impurities in the material, and the ranges here reflect ranges in the amount of impurities present.

[b] There is a wide range of values for glass and hard rubber. The values here are in the midrange.

CONCEPT EXERCISE 28.3

We found $\vec{J} = ne\vec{v}_d$ (Eq. 28.5) by considering the motion of the imaginary positive particles (Fig. 28.9). If we consider the actual conduction electrons moving in the negative x direction and carrying charge $-e$, will the current density still point in the positive x direction?

EXAMPLE 28.2 That's Slow!

Assume the wire between the battery and the bulb in Figure 28.2 is made of copper and has a diameter of 1.022 mm. If the current in the wire is 1.33 A, find the magnitude of the current density and the drift speed in the wire.

INTERPRET and ANTICIPATE

Because the current is given, we only need to divide by the cross-sectional area of the wire to find the magnitude of the current density in A/m^2. Once we know the current density, we use the number density n for copper from Table 28.2 to find the drift speed. We expect to find a speed between 10^{-5} m/s and 10^{-4} m/s as stated on page 867.

SOLVE

Find the cross-sectional area of the wire, assuming it is circular.

$$A = \pi r^2 = \pi\left(\frac{d}{2}\right)^2$$

$$A = \pi\left(\frac{1.022 \times 10^{-3}\,\text{m}}{2}\right)^2$$

$$A = 8.203 \times 10^{-7}\,\text{m}^2$$

Divide the current by the cross-sectional area (Eq. 28.3).

$$J = \frac{I}{A} = \frac{1.33\,\text{A}}{8.203 \times 10^{-7}\,\text{m}^2}$$

$$J = 1.62 \times 10^6\,\text{A/m}^2$$

To calculate the drift speed, use $\vec{J} = ne\vec{v}_d$ (Eq. 28.5) and look up the number density for copper in Table 28.2. Use the definition 1 A = 1 C/s.

$$J = nev_d$$

$$v_d = \frac{J}{ne}$$

$$v_d = \frac{1.62 \times 10^6\,\text{A/m}^2}{(8.42 \times 10^{28}\,\text{m}^{-3})(1.60 \times 10^{-19}\,\text{C})}$$

$$v_d = 1.20 \times 10^{-4}\,\text{m/s}$$

CHECK and THINK

The current density has the correct units, and the drift speed is in the range we expected (very slow). At this speed, it would take an electron more than 2 hours to travel through a wire the length of your arm. The current density is a very large number because there are many charge carriers per unit volume (more than 10^{28} per cubic meter) spread throughout the wire. If you leave a lightbulb on for a short time, many of the charge carriers from the wire will never make it into the bulb. Instead, the bulb is lit by the motion of charge carriers that were already in the filament.

EXAMPLE 28.3 That's Slow Too!

The filament in an incandescent bulb is often very thin compared to the wires that connect the bulb to the battery. Suppose the filament in Figure 28.5 is made of tungsten and has a diameter of 0.045 mm. Use the information in Example 28.2 to find the current density and drift speed in the filament.

Example continues on page 874 ▶

INTERPRET and ANTICIPATE

This problem is essentially like Example 28.2. The same current must exist in both the filament and the wire. If not, charged particles would "pile up." Because the current is the same in the filament but its cross-sectional area is smaller, we expect the current density to be higher in the filament than in the wire. If the filament has the same number density of charge carriers as the copper wire, we expect the increase in the current density to mean an increase in the drift speed. However, the number density of tungsten is a little lower than that of copper (Table 28.2).

SOLVE

Find the cross-sectional area of the filament, assuming it is circular.	$A = \pi r^2 = \pi\left(\frac{d}{2}\right)^2 = \pi\left(\frac{0.045 \times 10^{-3}\,\text{m}}{2}\right)^2$ $A = 1.59 \times 10^{-9}\,\text{m}^2$
Divide the current by the cross-sectional area (Eq. 28.3).	$J = \frac{I}{A} = \frac{1.33\,\text{A}}{1.59 \times 10^{-9}\,\text{m}^2}$ $J = 8.36 \times 10^8\,\text{A/m}^2$
To calculate the drift speed, use $\vec{J} = ne\vec{v}_d$ (Eq. 28.5) and look up the number density for tungsten in Table 28.2.	$v_d = \frac{J}{ne}$ $v_d = \frac{8.36 \times 10^8\,\text{A/m}^2}{(6.22 \times 10^{28}\,\text{m}^{-3})(1.60 \times 10^{-19}\,\text{C})}$ $v_d = 8.40 \times 10^{-2}\,\text{m/s}$

CHECK and THINK

As predicted, the current density is higher in the filament than in the wire that connects the bulb to the battery. Because the number density of tungsten is not much lower than that of copper, the drift speed in the filament is about 700 times higher than the drift speed in the wire.

How can the current I be the same in both the wire and the filament, but the drift speed v_d is higher in the filament? It might help to think of an analogy. Imagine beads placed four abreast on a conveyor belt leading into a paint box (Fig. 28.10A). The speed v_A of this conveyor belt is set so that four beads pass into the paint box per second. Now imagine a narrower conveyor belt on which the beads must be placed single file (Fig. 28.10B). If these beads are to pass into the paint box at the same rate (four beads per second), this narrow conveyor belt must move at a speed four times higher: $v_B = 4v_A$. In this analogy, the rate at which beads enter the paint box is like the current, four beads per second for both conveyor belts. The width of the conveyor belt is like the cross-sectional area of each conductor (wide for the wire, narrow for the filament). The conveyor belt's speed is analogous to the drift speed. The narrow conveyor belt must have a higher speed in order for the same number of beads to enter the paint box per second, just like charge carriers in the filament must have a higher drift speed in order to maintain the same current as in the thicker wire.

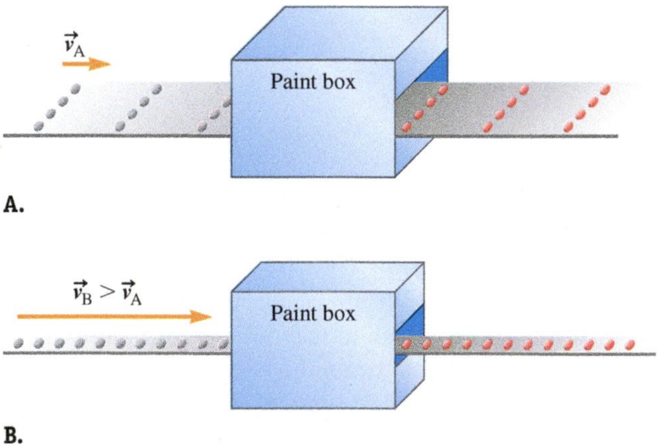

FIGURE 28.10 Paint box analogy for thick versus thin conductors.

28-4 Resistivity and Conductivity

We have modeled a conductor as a material in which charged particles (usually electrons) are free to flow, but Figure 28.3 shows that this model is not exactly correct. Because moving conduction electrons in a metal collide frequently with positive ions, a better description is: *Conduction electrons are free to walk randomly (or bump) around the entire conductor*. The frequency of the collisions determines how well the material conducts electricity. If the frequency of collisions is low, the material is a good conductor. **Electrical conductivity** (or simply **conductivity**) is a measure of a material's ability to conduct current, so it is a measure of how freely charged particles are able to flow in a given material.

Gravitational Analogy

To develop an expression for the conductivity of a material, we use a gravitational analogy. Imagine an array of pegs sandwiched between two vertical boards and organized into an array (Fig. 28.11A). The vertical boards (not shown) are parallel to the page, with one board behind the page and the other in front. Now a small rubber ball is released with some initial velocity between the vertical pegboards. The Earth's gravity accelerates the ball downward. As the ball falls, it is likely to collide with a peg. It bounces off the peg and travels in a parabolic path, just as all projectiles do, and then collides with another peg. Between collisions, the ball is a projectile accelerated downward by gravity and traveling on a parabolic path. Of course, the ball eventually makes it to the bottom.

We have chosen an upward-pointing y axis, so the acceleration of the ball between collisions is

$$\vec{a} = -g\hat{j} \tag{28.7}$$

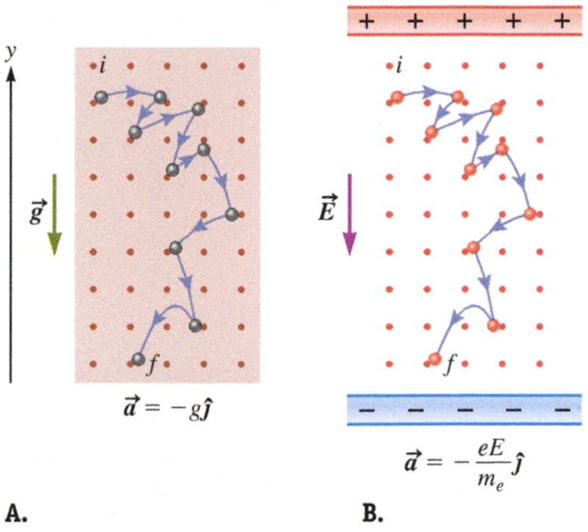

FIGURE 28.11 **A.** A rubber ball falls through a pegboard. **B.** An imaginary positively charged particle falls through fixed ions. In both cases, the path of the ball or particle between collisions is a parabola.

Because the acceleration is constant, the y component of the ball's velocity between two collisions is given by Equation 2.9:

$$v_y = v_{0y} + a_y t$$

$$v_y = v_{0y} - gt \tag{28.8}$$

The initial time in this case is the instant the ball bounces off of some peg.

We need to think about a large number of balls falling through the pegs because a large number of conduction electrons move through a conductor. If you took a snapshot of the balls at some arbitrary time, you would find the balls that had just collided with a peg moving in random directions. Thus, the average y component of the balls' velocities just after colliding with a peg is zero:

$$(v_{0y})_{\text{av}} = 0 \tag{28.9}$$

This is the average initial speed along y.

The average y component of all the balls' velocities at any arbitrary time depends on the average time between collisions. The average time between collisions is called the *mean free time*, t_{mf} (Section 20-5). To find the average y component of velocity, replace t in Equation 28.8 with the mean free time t_{mf} and use Equation 28.9 for the initial speed:

$$v_{\text{av},y} = (v_{0y})_{\text{av}} - gt_{mf} = 0 - gt_{mf}$$

$$v_{\text{av},y} = -gt_{mf} \tag{28.10}$$

In Figure 28.11B, a conductor is between the plates of a charged capacitor, with a downward-pointing electric field:

$$\vec{E} = -E\hat{j} \tag{28.11}$$

The ions are shown in a lattice similar to the pegs, and the downward-pointing electric field is analogous to the gravitational field in Figure 28.11A and causes conduction electrons to move upward along a zigzag path. However, following the usual convention, we show an imaginary positive particle of mass m_e and charge e moving in the opposite direction—heading downward along a zigzag path. Like the ball in Figure 28.11A, the particle makes many collisions and its path between collisions is a parabola. Between collisions, each particle has an acceleration given by

$$\vec{a} = -\frac{eE}{m_e}\hat{j} \tag{28.12}$$

The average downward velocity of many such particles is their drift velocity $\vec{v}_d$, which can be found by analogy with Equation 28.10. The quantity $v_{av,y}$ in Equation 28.10 is analogous to the drift speed v_d in a conductor. Replace the acceleration due to gravity with the acceleration due to the electric field (Eq. 28.12):

$$\vec{v}_d = -\left(\frac{eE}{m_e}\right)t_{mf}\hat{\jmath} \tag{28.13}$$

According to Equation 28.13, the drift velocity is higher when there is a longer average time between collisions. A higher drift velocity (longer average time between collisions) means a higher current density. Substitute Equation 28.13 into $\vec{J} = ne\vec{v}_d$ (Eq. 28.5):

$$\vec{J} = ne\left[-\left(\frac{eE}{m_e}\right)t_{mf}\hat{\jmath}\right] = \frac{ne^2}{m_e}t_{mf}(-E\hat{\jmath})$$

Then substitute $\vec{E} = -E\hat{\jmath}$ (Eq. 28.11):

$$\vec{J} = \left(\frac{ne^2}{m_e}t_{mf}\right)\vec{E} \tag{28.14}$$

The quantity (ne^2t_{mf}/m_e) is the constant of proportionality missing from $\vec{J} \propto \vec{E}$ (Eq. 28.4).

Now, imagine putting two different conducting materials between the plates of a capacitor. The electric field in each conductor is the same, but the current density in each conductor depends on the proportionality constant (ne^2t_{mf}/m_e). The conductor with the longer average time between collisions and the greater density of conduction electrons will be the better conductor and have the higher current density. This constant of proportionality is the **conductivity** σ of the material:

CONDUCTIVITY ✪ Major Concept

$$\sigma \equiv \frac{ne^2}{m_e}t_{mf} \tag{28.15}$$

Rewriting Equation 28.14 in terms of conductivity, we have

$$\vec{J} = \sigma\vec{E} \tag{28.16}$$

The symbol for conductivity is σ (lowercase Greek letter sigma), the same symbol used for surface charge density and the Stefan–Boltzmann constant. You must use context to interpret this symbol.

Conductivity is a scalar, so Equation 28.16 shows that the current density points in the same direction as the electric field. Conductivity depends only on the type of material, with values for various substances given in Table 28.2. The SI units for conductivity can be found from Equation 28.16:

$$[\![\sigma]\!] = \frac{\text{A/m}^2}{\text{V/m}} = \frac{\text{A/V}}{\text{m}}$$

In the SI system, the combination of volts per ampere (V/A) is called an **ohm**:

$$1\,\text{V/A} = 1\,\Omega \tag{28.17}$$

where Ω is the uppercase Greek letter omega. The SI units for conductivity are usually given in ohms and meters $(\Omega^{-1} \cdot \text{m}^{-1})$.

Resistivity

In many practical applications, it is more convenient to work with the reciprocal of the conductivity, known as the resistivity. **Resistivity** ρ is a measure of a material's ability to resist conducting electricity:

RESISTIVITY ✪ Major Concept

$$\rho \equiv \frac{1}{\sigma} = \frac{m_e}{ne^2t_{mf}} \tag{28.18}$$

The SI units for resistivity are $\Omega \cdot \text{m}$. The symbol for resistivity is ρ (lowercase Greek letter rho), the same symbol used for volume charge density and mass density. You must use context to interpret this symbol.

Resistivity is a scalar that depends only on the type of substance, with values for various substances given in Table 28.2.

Conductivity in Equation 28.16 may be replaced by resistivity: $\vec{J} = (1/\rho)\vec{E}$, and then

$$\vec{E} = \rho\vec{J} \tag{28.19}$$

In Table 28.2, the substances categorized as *conductors* have high conductivities and low resistivities. The substances categorized as *insulators* have low conductivities and high resistivities. The substances between conductors and insulators are known as **semiconductors**; their conductivities and resistivities depend on the amount of impurities in a given sample. Semiconductors play an important role in devices such as computers.

Temperature Dependence

The conductivity and therefore the resistivity of a metal depend on (1) the lattice structure of the ions that make up the metal, (2) impurities in the metal, and (3) temperature. Circuit elements such as a lightbulb can get very hot when there is current in the circuit, so it is important to know how temperature affects resistivity.

When a conductor is hot, the ions in the lattice vibrate vigorously. Think back to the analogy of a ball dropping through an array of pegs (Fig. 28.11A), and imagine each peg is vibrating. Each ball will then make more collisions with the pegs because effectively the pegs take up more space. Thus, the mean free path for the balls is reduced.

In a conductor, a higher temperature means more vigorously vibrating ions and more collisions for conduction electrons. An increase in the collision frequency means a decrease in the mean free time between collisions. So, as the temperature of a conductor increases, its conductivity goes down and its resistivity goes up. Mathematically, we express the resistivity at some temperature T as

$$\rho(T) = \rho_0[1 + \alpha(T - T_0)] \tag{28.20}$$

where ρ_0 is the resistivity at temperature T_0. Usually T_0 is set to room temperature. The resistivities listed in Table 28.2 are for substances at $T_0 = 20°\text{C}$. The **temperature coefficient of resistivity** α depends on the type of material. It has the dimensions of 1/temperature; for convenience, the values listed in Table 28.2 are given in terms of the Celsius scale (non-SI units). For most conductors, $\alpha > 0$, so resistivity increases at higher temperatures (Eq. 28.20).

The semiconductors in Table 28.2 have midrange conductivities. A semiconductor is similar to an insulator in that the electrons are bound to particular atoms even when many atoms are pressed together as they are in a solid. However, under the right conditions, it is possible to free an outer electron in each atom. One way to do this is to increase the temperature of the semiconductor. The increase in thermal energy causes the atoms to vibrate more vigorously, and the outermost electrons may shake free. Thus, in semiconductors, resistivity decreases as temperature increases. Mathematically, this means semiconductors have a negative temperature coefficient of resistivity: $\alpha < 0$.

This is a classical physics description; a better description involves quantum physics.

EXAMPLE 28.4 Conductors, Semiconductors, and Superconductors

In this example, we compare the temperature dependence of resistivity in a conductor and a semiconductor.

A Make a graph of resistivity as a function of temperature $\rho(T) = \rho_0[1 + \alpha(T - T_0)]$ (Eq. 28.20) for tungsten between 0 and 40°C.

B Make a similar graph for carbon in the same temperature range. For carbon, assume $\rho_0 = 3.5 \times 10^{-5}\ \Omega \cdot \text{m}$.

INTERPRET and ANTICIPATE

Equation 28.20 is the equation of a line. The sign of the slope (positive or negative) is the same as the sign of the temperature coefficient α. Because α for tungsten is positive (Table 28.2), we expect the graph for part A will be a line tilted up toward the right. Because α for carbon is negative, we expect the graph for part B will be a line tilted down toward the right.

Example continues on page 878 ▶

SOLVE

Substitute values of ρ_0 and α from Table 28.2 for each material into Equation 28.20, with $T_0 = 20°C$.

For tungsten,

$$\rho(T) = [5.65 \times 10^{-8}\,\Omega\cdot\text{m}][1 + 0.0045°\text{C}^{-1}(T - 20°\text{C})]$$

For carbon,

$$\rho(T) = [3.5 \times 10^{-5}\,\Omega\cdot\text{m}][1 - 0.0005°\text{C}^{-1}(T - 20°\text{C})]$$

Calculate $\rho(T)$ for T between 0 and 40°C. Plot these values for each material (Fig. 28.12).

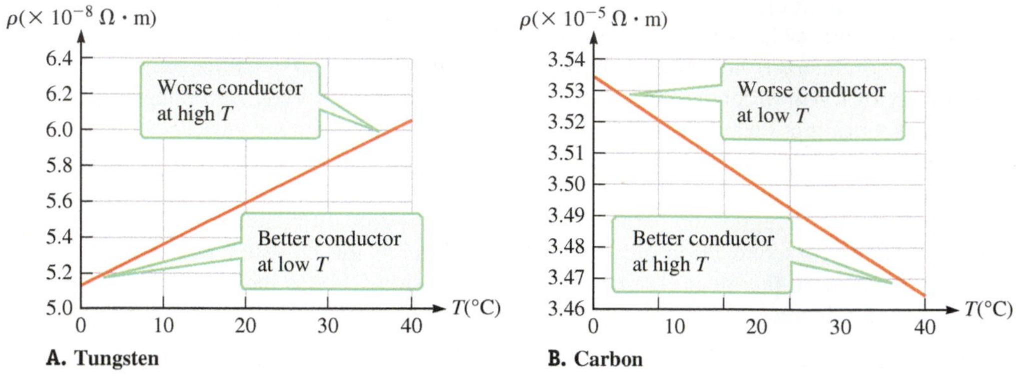

FIGURE 28.12

CHECK and THINK

The graphs are as we expected. Figure 28.12A shows that when tungsten is below room temperature, its resistivity is lower and it is a better conductor. The graph for carbon in Figure 28.12B shows just the opposite. When carbon is below room temperature, its resistivity is higher and it is a worse conductor.

The temperature dependence of resistivity is dramatic in certain materials known as **superconductors**. In 1911, Dutch physicist Heike Kamerlingh discovered that at very low temperatures (about 4 K), mercury's resistivity drops to zero. Liquid helium is used to cool the mercury to such a low temperature, but liquid helium is expensive and impractical to use. Since Kamerlingh's discovery, other materials have been found to be superconductors at somewhat higher temperatures that can be reached with liquid nitrogen, a much cheaper coolant. Superconductors are used in some magnetic-levitation (Maglev) trains.

28-5 Resistance and Resistors

In the two preceding sections, we used a microscopic model to find the current density $\vec{J}$ when there is an electric field $\vec{E}$. The macroscopic vector quantities $\vec{J}$ and $\vec{E}$ are not very useful when we are building circuits. Instead, it is more convenient to know two scalar quantities: the electric potential difference ΔV and current I.

In order to write $\vec{E} = \rho\vec{J}$ (Eq. 28.19) in terms of electric potential ΔV and current I, consider a small portion of the filament in a lightbulb as shown in Figure 28.13. The left side of the filament is at a higher potential than the right side. Let's call the slight potential difference in that small portion of the filament δV and the length of this small portion $\delta\ell$. The magnitude of the electric field E is uniform throughout the filament, so the potential difference is proportional to the electric field (Eq. 26.17):

$$\delta V = E\,\delta\ell \tag{28.21}$$

(We are concerned only with the magnitude of the electric field, so we don't need the negative sign in Equation 26.17.) Because the magnitude of the electric field is uniform, we can add up δV for each segment to arrive at the potential difference ΔV between the ends of the filament:

$$\Delta V = E\ell \tag{28.22}$$

where ℓ is the length of the entire filament.

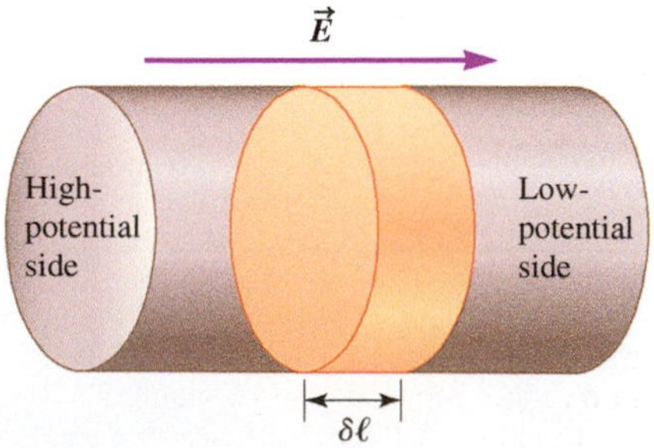

FIGURE 28.13 A small portion of a wire such as a lightbulb filament.

Working with magnitudes only, we substitute $E = \Delta V/\ell$ and $J \equiv I/A$ (Eq. 28.3) into $E = \rho J$ (Eq. 28.19):

$$\frac{\Delta V}{\ell} = \rho \frac{I}{A}$$

Solve for the potential difference ΔV between the ends of the filament:

$$\Delta V = I\left(\rho \frac{\ell}{A}\right) \tag{28.23}$$

The term in parentheses depends only on properties of the filament: (1) its resistivity ρ, (2) its length ℓ, and (3) its cross-sectional area A. The combination of these properties is called the **resistance** of the filament:

RESISTANCE ✪ Major Concept

$$R = \rho \frac{\ell}{A} \tag{28.24}$$

Equation 28.23 is usually expressed in terms of resistance:

$$\Delta V = IR \tag{28.25}$$

RESISTORS ✪ Major Concept

Resistance R is another measure of an object's ability to resist an electric current. It is different from resistivity ρ, which depends on the type of material and temperature. Resistance depends on the object's resistivity, so resistance also depends on the type of material and temperature. However, resistance R also depends on the geometry of the object. If two conductors are made from the same material and have the same temperature, they have the same resistivity ρ. Now imagine that one of the conductors is a long, thin wire like the filament in a lightbulb, and the other is a short, thick one. According to $R = \rho\ell/A$ (Eq. 28.24), the resistance R of the long, thin wire is higher than the resistance of the short, thick one.

The difference between resistivity ρ and resistance R can be explained using the analogy of balls falling through pegs (Fig. 28.14). Resistivity is represented by the spacing of the pegs. Current is represented by the number of balls that emerge from the bottom per unit time. If the arrangement of the pegs is long and narrow, very few balls emerge from the bottom per unit time, representing a small current in a long, thin wire.

The SI units of resistance are ohms (Ω). Usually the resistance of the wires in a circuit is fairly low. For example, 100 m of wire in a typical household circuit has a resistance of about 1 Ω. The resistance of an incandescent lightbulb's filament when it is operating is about 150 Ω.

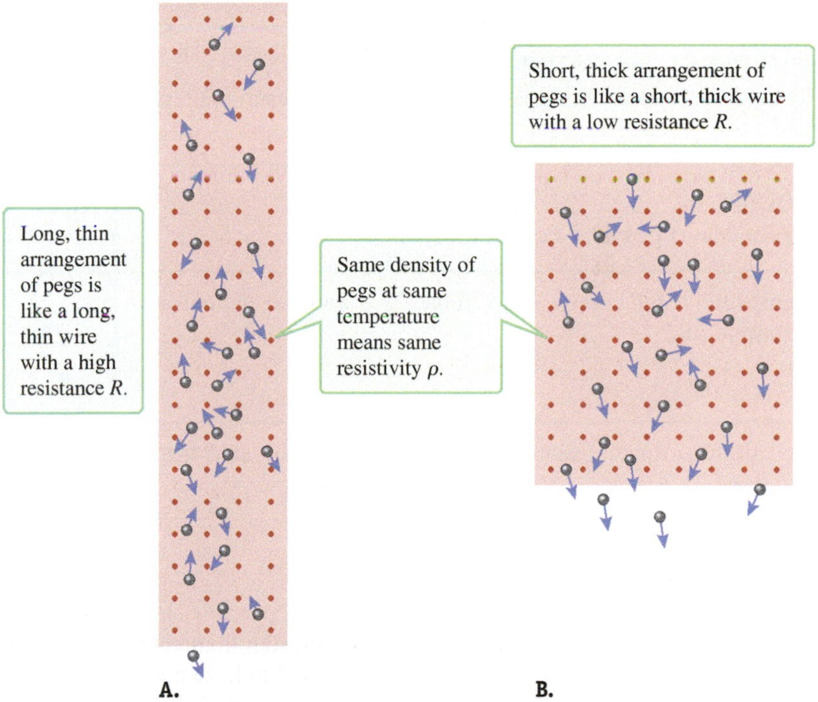
Long, thin arrangement of pegs is like a long, thin wire with a high resistance R.
Same density of pegs at same temperature means same resistivity ρ.
Short, thick arrangement of pegs is like a short, thick wire with a low resistance R.
A.
B.

FIGURE 28.14 **A.** A gravitational analogy for a long, narrow resistor. When the arrangement of the pegs is long and narrow, very few balls emerge from the bottom per unit time. **B.** A model for a short, thick resistor. When the arrangement of the pegs is short and wide, many balls emerge per unit time. In both cases, balls falling through the pegs are analogous to the motion of charged particles in a resistor. So, the current in a long, thin wire is less than in a short, thick wire of the same resistivity.

A **resistor** is a circuit element designed to have a particular constant resistance (Fig. 28.15). According to $\Delta V = IR$ (Eq. 28.25), for a given potential difference ΔV, the current in a conductor depends on the conductor's resistance. Resistors are made with resistances ranging from 10^{-2} Ω to 10^{7} Ω and are used in circuits to control the amount of current. The colored bands on a resistor indicate its nominal resistance. As shown in Table 28.3, the first two bands give the first two digits of the resistance, and the third band is a power-of-10 multiplier. If a resistor's first three bands are green, blue, and brown, its resistance is $56 \times 10^{1} = 560$ Ω. The fourth band, if present, is the precision of the resistance. If no band is present, the precision is $\pm 20\%$.

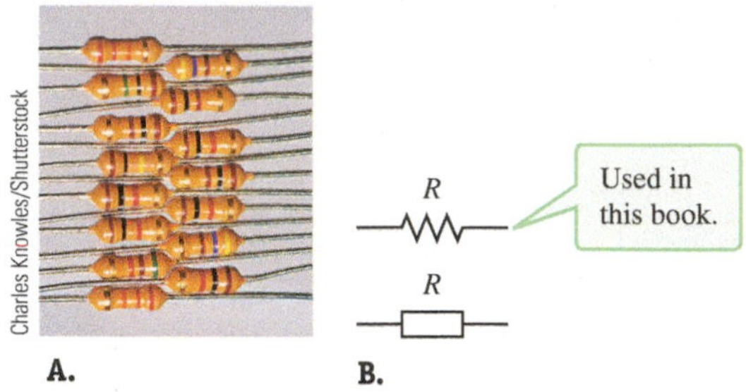

FIGURE 28.15 **A.** Some resistors. **B.** Schematic symbols for a resistor. The top symbol is used in North America and Japan; the bottom one is used in Europe.

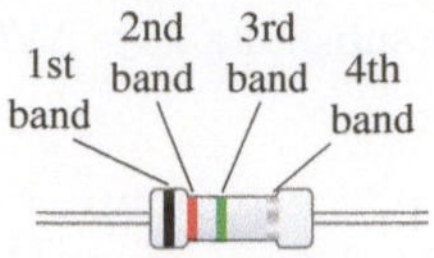

Color	1st band (1st figure)	2nd band (2nd figure)	3rd band (multiplier)	4th band (tolerance)
Black	0	0	10^0	
Brown	1	1	10^1	
Red	2	2	10^2	±2%
Orange	3	3	10^3	
Yellow	4	4	10^4	
Green	5	5	10^5	
Blue	6	6	10^6	
Violet	7	7	10^7	
Gray	8	8	10^8	
White	9	9	10^9	
Gold			10^{-1}	±5%
Silver			10^{-2}	±10%

TABLE 28.3 Color codes for resistance 4-band resistors.

EXAMPLE 28.5 Wires and Filaments

Suppose the total length of copper wire used to connect a lightbulb to a battery (Fig. 28.2) is 35 cm and the length of the tungsten filament is 9.0 cm. Use the cross-sectional areas found in Examples 28.2 and 28.3: $8.203 \times 10^{-7}\ \text{m}^2$ for the wire and $1.59 \times 10^{-9}\ \text{m}^2$ for the filament.

A Calculate the resistances of the copper wire and the tungsten filament at room temperature.

INTERPRET and ANTICIPATE

To find the resistances, look up the resistivity of each material in Table 28.2, and use the lengths and areas given. The results should be in Ω.

SOLVE

Resistance is given by Equation 28.24.

$$R = \rho\frac{\ell}{A} \tag{28.24}$$

Substitute ρ for copper from Table 28.2 and the area from Example 28.2.

$$R_{\text{wire}} = (1.678 \times 10^{-8}\ \Omega\cdot\text{m})\left(\frac{0.35\ \text{m}}{8.203 \times 10^{-7}\,\text{m}^2}\right)$$

$$R_{\text{wire}} = 7.2 \times 10^{-3}\,\Omega$$

Substitute ρ for tungsten from Table 28.2 and the area from Example 28.3.

$$R_{\text{filament}} = (5.65 \times 10^{-8}\ \Omega\cdot\text{m})\left(\frac{0.09\ \text{m}}{1.59 \times 10^{-9}\,\text{m}^2}\right)$$

$$R_{\text{filament}} = 3.2\,\Omega$$

CHECK and THINK

The results have the expected SI units. The tungsten filament has a much higher resistance than the wires. In most circuits, the wires are designed to have a very low resistance compared to the circuit elements. Usually we can ignore the resistance of the wires.

B Calculate the potential difference between the ends of the wire (one end adjacent to the battery and the other end adjacent to the bulb) if the current is 1.33 A (as in Example 28.2). Then calculate the potential difference between the ends of the filament.

INTERPRET and ANTICIPATE

The currents in the wire and in the filament are the same but the resistance of the filament is much higher, so we expect the potential difference across the ends of the filament to be much greater than that across the wire.

SOLVE Use the resistance found in part A for the wire. Use $1\ \text{V/A} = 1\ \Omega$ (Eq. 28.17) to write the potential difference in volts.	$\Delta V_{\text{wire}} = IR_{\text{wire}}$ (28.25) $\Delta V_{\text{wire}} = (1.33\ \text{A})(7.2 \times 10^{-3}\ \Omega)$ $\Delta V_{\text{wire}} = 9.6 \times 10^{-3}\ \text{V}$
Repeat this calculation for the filament.	$\Delta V_{\text{filament}} = IR_{\text{filament}} = (1.33\ \text{A})(3.2\ \Omega)$ $\Delta V_{\text{filament}} = 4.3\ \text{V}$

CHECK and THINK

As expected, the potential difference between the ends of the wire is much less than the potential difference across the filament. Again, this is part of the design used in most circuits. There is essentially no potential difference along the wire, and we usually assume $\Delta V = 0$ for any wire. In the simple circuit in Figure 28.2, the potential difference across the bulb is equal to the terminal potential of the battery. If the battery were built from the typical 1.5-V D-cell batteries in a flashlight, there would be three D cells in this circuit. (The actual terminal potential doesn't always equal the nominal value given by the manufacturer.)

CONCEPT EXERCISE 28.4

When a lightbulb burns out, its filament breaks so that there is a gap between the two sides of the filament. What happens to the current in and the resistance of the lightbulb when it burns out? Explain your answers.

28-6 Ohm's Law

In the decades that followed Franklin's work on electricity, scientists continued to experiment and develop a microscopic model of electrical conduction. At that time, it was difficult to make precise macroscopic observations in order to test models. A major breakthrough was made by German mathematics and physics teacher Georg S. Ohm, who published an article entitled *Mathematical Theory of the Galvanic Circuit* in 1827. Although many details of his microscopic model differ from our current model, his article is the basis of what we call *Ohm's law.* The ideas in his article were initially so scorned that Ohm was forced to resign from his teaching position at the Jesuit school in Cologne. Later, Ohm's law was accepted and he won the Copley Prize in 1841.

The best way to state Ohm's law is still somewhat controversial today. For example, Ohm's law is often stated mathematically as $\Delta V = IR$ (Eq. 28.25), but this statement is misleading. In this book, we will present three statements of Ohm's law and show why Equation 28.25 alone is not sufficient.

Ohm's law is not a *law* in the same sense as Newton's laws of motion. Ohm's law is more like Hooke's law for springs. Both Hooke and Ohm came up with their "laws" by empirically fitting their data to a mathematical function. Hooke hung objects of various masses from a spring or rod and measured the amount the spring or rod stretched (Chapter 14). And, just as Hooke's law does not hold for all stretched objects, Ohm's law does not hold for all conductors. Hooke's law and Ohm's law are valid in many practical situations, however, and both apply successfully in a wide variety of circumstances.

Hooke's law and Ohm's law are also mathematically similar. Hooke found that the magnitude of the force applied to a spring is directly proportional to the amount it stretches (Eq. 5.8). Ohm found that *the current I in a conductor is directly*

1ST STATEMENT OF OHM'S LAW
Major Concept

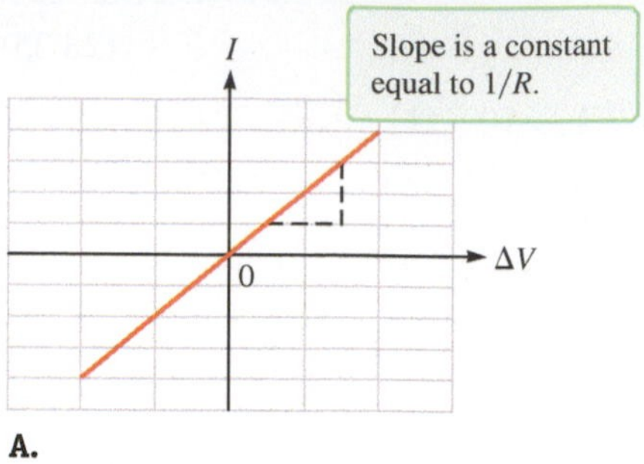

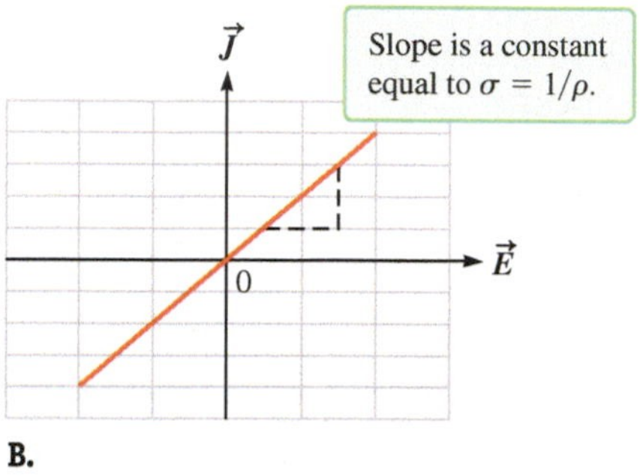

FIGURE 28.16 Alternative expressions for Ohm's law: **A.** $I \propto \Delta V$, **B.** $\vec{J} \propto \vec{E}$.

proportional to the potential difference applied across it. This is our first statement of **Ohm's law**, usually expressed mathematically as

$$\Delta V \propto I \tag{28.26}$$

This expression of Ohm's law is shown graphically in Figure 28.16A.

If you compare this proportionality relationship $\Delta V \propto I$ (Eq. 28.26) with $\Delta V = IR$ (Eq. 28.25), you can see why Equation 28.25 is called Ohm's law. If the resistance of the conductor is constant—independent of both ΔV and I—then $\Delta V = IR$ is a statement of Ohm's law, where R is the constant of proportionality. Because we have plotted I on the vertical axis and ΔV on the horizontal axis, the slope of the line in Figure 28.16A is $1/R$. For a circuit element that obeys Ohm's law, that slope is constant. The quantity $1/R$ is called the **conductance**, so the slope of the curve on a graph of I versus ΔV is the conductance.

2ND AND 3RD STATEMENTS OF OHM'S LAW ✪ Major Concept

A second statement of Ohm's law is in terms of the electric field within a conductor and the current density: *The electric field $\vec{E}$ in a conductor is directly proportional to the current density $\vec{J}$ in that conductor*:

$$\vec{E} \propto \vec{J} \tag{28.27}$$

This statement of Ohm's law is shown graphically in Figure 28.16B. According to $\vec{J} = \sigma\vec{E}$ (Eq. 28.16), the slope of the line is the conductivity σ. Because the conductivity is the reciprocal of the resistivity ($\sigma = 1/\rho$), the slope is also $1/\rho$. For a device that obeys Ohm's law, the slope in Figure 28.16B is constant, so the conductivity and resistivity are constants.

A third statement of Ohm's law is that the resistance R, conductance $1/R$, resistivity ρ, and conductivity σ are all constants. If so, the resistance and conductance of a circuit element do not depend on the potential difference across the element or the current in it, and the resistivity and conductivity do not depend on the electric field or current density within the element. No actual circuit element obeys Ohm's law under all conditions, but a circuit that obeys Ohm's law over a wide range of potential differences across it is described as **ohmic**. Resistors are examples of ohmic circuit elements.

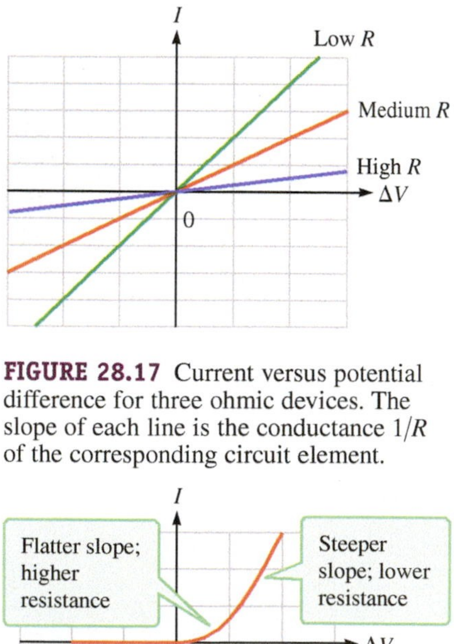

FIGURE 28.17 Current versus potential difference for three ohmic devices. The slope of each line is the conductance $1/R$ of the corresponding circuit element.

Figure 28.17 shows graphs of I versus ΔV for three different ohmic circuit elements (such as three different resistors). The curve for each circuit element is a straight line with constant slope equal to the conductance $1/R$. The line with the steepest slope represents the circuit element with the lowest resistance, and the line with the flattest slope represents the element with the highest resistance.

A circuit element that does not obey Ohm's law over any significant range of ΔV is described as **nonohmic**. Figure 28.18 is a graph of I versus ΔV for a diode, a nonohmic device. The curve for the diode is a flat line for $\Delta V < 0$; for $\Delta V > 0$, the curve is approximately given by $I \propto \Delta V^{3/2}$.

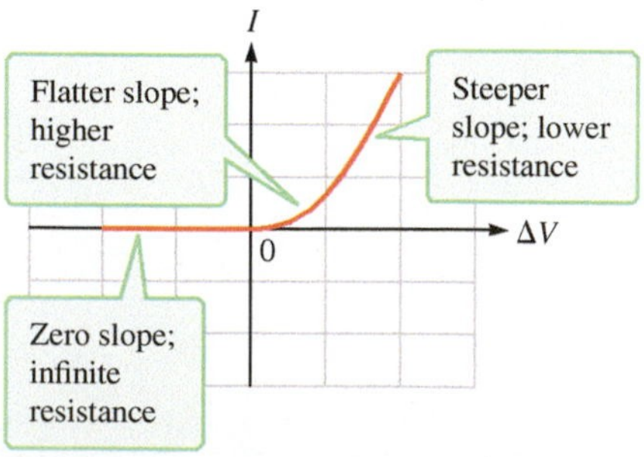

FIGURE 28.18 Current as a function of potential difference for a nonohmic device. The resistance is *not* constant.

Although $\Delta V = IR$ (Eq. 28.25) holds for a nonohmic device, the resistance is not constant. Let's use the diode as an example. Although a diode does not obey Ohm's law, any point on the curve in Figure 28.18 is described by $\Delta V = IR$. When $\Delta V < 0$, the current is very nearly zero and we'll assume $I = 0$. According to $\Delta V = IR$, the diode has (nearly) infinite resistance ($R \to \infty$) when $\Delta V < 0$. The resistance is finite but not constant when $\Delta V > 0$. Because $I \propto \Delta V^{3/2}$, the curve is flatter (has lower slope) for small values of ΔV than for large values, so the resistance is higher for small values of ΔV than for large values and is given by

$$R = \frac{\Delta V}{I} = \frac{\Delta V}{\Delta V^{3/2}} = \Delta V^{-1/2}$$

CONCEPT EXERCISE 28.5

A battery with terminal potential $\mathcal{E}$ is connected to a lightbulb (Fig. 28.2). After a long time, the bulb stops working because the filament snaps. After the filament breaks, what is the potential difference across the filament? Explain your answer.

EXAMPLE 28.6 Tungsten and Carbon Filaments: Which Is Which?

Students in a physics laboratory are given two incandescent lightbulbs, one with a carbon filament and the other with a tungsten filament. Their assignment is to figure out which is which based on electrical measurements. The students connect each bulb to a variable power supply, which is like a battery except that the terminal voltage may be adjusted by simply turning a dial. The students vary the terminal potential from roughly 10 V to 100 V and measure the current through each filament. As usual, the resistance in the wires is very low compared to the resistance in the bulb, so the terminal potential of the power supply is about equal to the potential difference between the ends of the filament. The current and voltage for each bulb are given in Table 28.4. Calculate the resistance of each bulb for each current-voltage measurement, and then plot resistance as a function of potential difference to determine which bulb is carbon and which is tungsten.

TABLE 28.4 Data taken to determine the type of filament used in each bulb. The ending zero in each case is a significant figure.

Bulb A		Bulb B	
ΔV (V)	I (A)	ΔV (V)	I (A)
10	0.10	10	0.21
21	0.27	21	0.30
30	0.41	30	0.38
41	0.59	41	0.46
50	0.75	50	0.51
60	0.97	60	0.56
70	1.17	70	0.63
80	1.36	80	0.67
90	1.55	91	0.71
		101	0.77
		110	0.79

INTERPRET and ANTICIPATE

As the students increase the terminal potential, the bulbs get brighter and hotter. We expect each bulb's temperature to increase as the students increase the terminal potential of the power supply. Carbon is a semiconductor with a negative temperature coefficient of resistivity α, whereas tungsten is a conductor with a positive α (Table 28.2). So, by plotting R as a function of V, the students can determine whether the filament is carbon or tungsten. Carbon's resistance should decrease as the terminal potential increases, and tungsten's resistance should increase.

SOLVE

Use $\Delta V = IR$ (Eq. 28.25) to find the resistance of each bulb for each measurement. Although lightbulbs are nonohmic devices, Equation 28.25 still holds (but R is not constant).

TABLE 28.5 Tabulation of R (shaded columns) for each measurement, where $R = (\Delta V)/I$. Values of ΔV and I are from Table 28.4.

Bulb A			Bulb B		
Δ (V)	I (A)	R (Ω)	Δ (V)	I (A)	R (Ω)
10	0.10	100	10	0.21	48
21	0.27	78	21	0.30	70
30	0.41	73	30	0.38	79
41	0.59	69	41	0.46	89
50	0.75	67	50	0.51	98
60	0.97	62	60	0.56	107
70	1.17	60	70	0.63	111
80	1.36	59	80	0.67	119
90	1.55	58	91	0.71	128
			101	0.77	131
			110	0.79	139

Plot R on the vertical axis and ΔV on the horizontal axis for each bulb (Fig. 28.19). Bulb A must be carbon because its resistance decreases as the terminal potential increases. Bulb B must be tungsten because its resistance increases as the terminal potential increases.

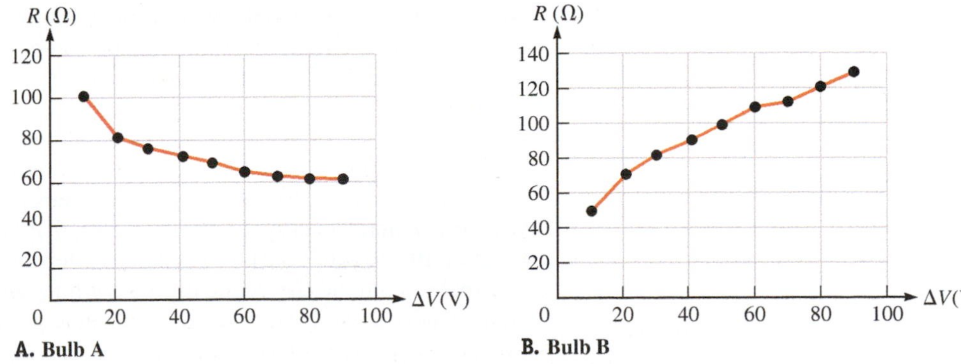

FIGURE 28.19

Example continues on page 884 ▶

CHECK and THINK

We could have plotted R as a function of I because an increase in current also means an increase in the bulb's temperature. In the next section, we will show that the power lost (through heat) by a lightbulb is proportional to both $(\Delta V)^2$ and I^2.

28-7 Power in a Circuit

In Example 28.1, Shannon argues that a battery is a source of charged particles and that devices such as lightbulbs use up those charged particles. Shannon is mistaken. The function of the battery is to maintain a specific potential difference between its terminals through chemical reactions (Section 27-3). When a device such as a lightbulb is connected to the terminals of a battery (Fig. 28.2), a current is set up in the filament because charged particles are forced to drift due to the potential difference between the filament's ends. Those drifting charged particles were already in the filament; the battery did not supply them. Shannon's intuition is only slightly misguided, however. The battery does supply something that is converted to light by the lightbulb. That "something" is energy.

Recall the basic physics of a wet cell (battery) as shown in Figure 27.11 (page 834). Initially a neutral zinc terminal and a neutral copper terminal are placed in a solution of sulfuric acid. After a period of time, the zinc terminal is negatively charged and the copper terminal is positively charged. Because positively charged particles are attracted to negatively charged particles, we conclude that the chemical reactions must have supplied energy to separate the charged particles. Once charged particles have been separated, the battery has stored electric potential energy that we can tap by connecting a device such as a lightbulb to the battery terminals.

Gravitational Analogy Revisited

The analogy with rubber balls falling through a pegboard (Fig. 28.11) can help us apply the conservation of energy principle to circuits. In order for the balls to drop through the pegboard, a person must first raise the balls to the top, doing positive work and increasing the system's gravitational potential energy. The person is analogous to a battery. The battery does work in separating the charged particles, depositing positive particles on the positive terminal and negative particles on the negative terminal. We can carry this analogy one step further. In order for a person to do the work necessary in lifting the balls, she must eat. As she digests food, chemical reactions in her body give her the energy to do the work. Chemical reactions in the battery enable the battery to do work.

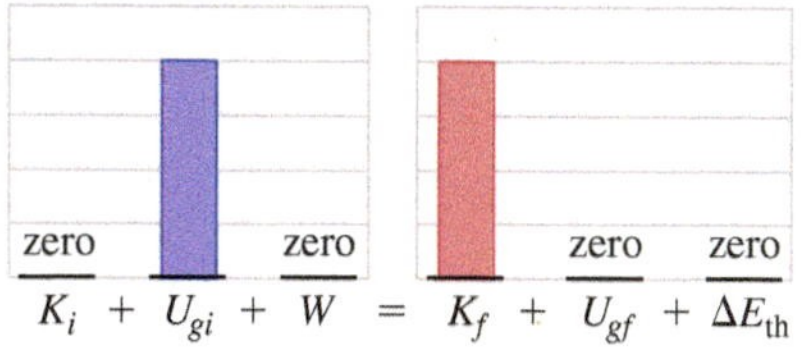

FIGURE 28.20 A bar chart for balls falling through free space. In the absence of drag forces, all the initial gravitational potential energy of the system is converted to kinetic energy.

For the moment, suppose the person drops the balls through free space so they do not collide with any pegs. The gravitational potential energy of the system is converted to kinetic energy as the balls fall back to the ground. If the balls fall through a vacuum so that there is no drag force, all the gravitational potential energy is converted to kinetic energy (Fig. 28.20).

Now suppose the balls are dropped through the array of pegs (Fig. 28.11A). Although the gravitational potential energy of the system decreases as before, not all of this potential energy goes into kinetic energy of the balls. A peg is at rest when a ball collides with it. Much of the ball's kinetic energy is transferred to the peg as a result of the collision, and the peg begins to vibrate. Let's include the pegs, the balls, and the Earth in the system so that we can take the kinetic energy associated with peg vibration into account as a change in internal energy. Consider this vibrational kinetic energy as an increase in thermal energy ΔE_{th}. (**Thermal energy** is associated with the kinetic energy of the particles that make up the objects in a system; Chapter 19.) Figure 28.21A is a bar chart for this system.

Finally, imagine the pegboard is very long and each ball undergoes many collisions. Under these conditions, the balls emerge from the bottom of the pegboard at the rate the person drops them into the top. Effectively, the balls drift downward at a constant drift speed, which they reach soon after they are dropped. The initial time

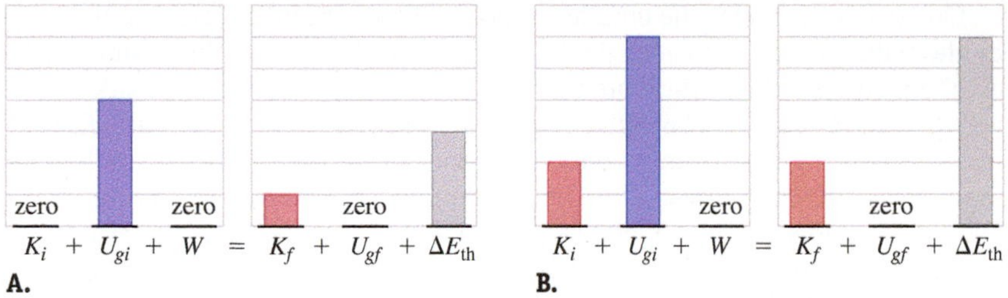

FIGURE 28.21 Bar charts for balls falling through pegboard. **A.** The balls are initially at rest, and the gravitational potential energy is converted to the kinetic energy of the balls plus the thermal energy associated with the pegs' vibration. **B.** The balls are moving at the drift speed, and their kinetic energy is unchanged.

is shortly after the balls have been dropped into the pegboard, so that the balls have already reached their constant drift speed. In this case, there is no change in kinetic energy and all the potential energy is converted to thermal energy (Fig. 28.21B).

This gravitational analogy helps explain what happens on the microscopic level when a conductor such as a lightbulb filament is connected to the terminals of a battery. The charge carriers collide with "fixed" ions just as the balls collide with pegs, causing the ions to increase their vibration so that the system's thermal energy increases.

As in the case where the balls undergo many collisions, soon after the filament is connected to the battery, the average kinetic energy of the charge carriers reaches equilibrium because collisions with the fixed ions keep the charge carriers drifting at a constant average drift speed v_d. The bar chart in Figure 28.22 is similar to Figure 28.21B, showing that the electric potential energy stored by the battery's electric field is converted to thermal energy of the ions in the conductor lattice.

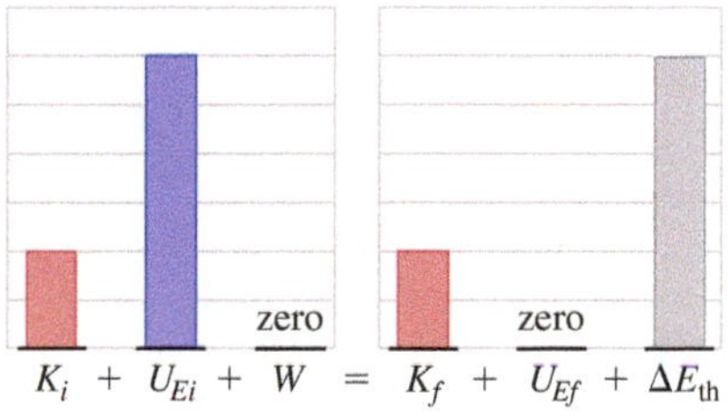

FIGURE 28.22 The bar chart for charged carriers moving in a conductor is similar to the bar chart for balls falling through a pegboard (Fig. 28.21B). The kinetic energy of the charged particles does not change. The electric potential energy stored in the battery goes into the thermal energy of the filament.

Collisions between balls and pegs in the pegboard system cause the system's internal energy to increase. This energy may leave the system in many ways. If you were in a room with such an apparatus, you would probably find it very noisy. The pegs vibrations cause air molecules to vibrate, so energy leaves the system as sound waves.

The same sort of thing happens with current in a conductor. The increase in the conductor's thermal energy leaves the system through heat by conduction, convection, and radiation. (**Heat** describes the energy transferred between a system and its environment due to their temperature difference; Chapter 21.) In the case of a lightbulb's filament, radiation takes the form of light. So, there was something right about Shannon's intuition. The battery does work on the charge carriers, increasing the system's potential energy, and the potential energy is converted to thermal energy, some of which is radiated away in the form of light.

Calculating Power

In practice, the total amount of energy supplied by a battery is not as important as the rate at which the energy is delivered. The rate of energy transfer is known as **power** (Section 9-9). To find the power delivered by a battery, consider a battery whose terminals are connected by a wire (Fig. 28.8). Let's simplify our microscopic conduction model somewhat by imagining that small, positively charged bundles move through the wire from the positive terminal to the negative terminal at the drift speed v_d. In Figure 28.23, each bundle of charge looks like a bead on a string moving clockwise.

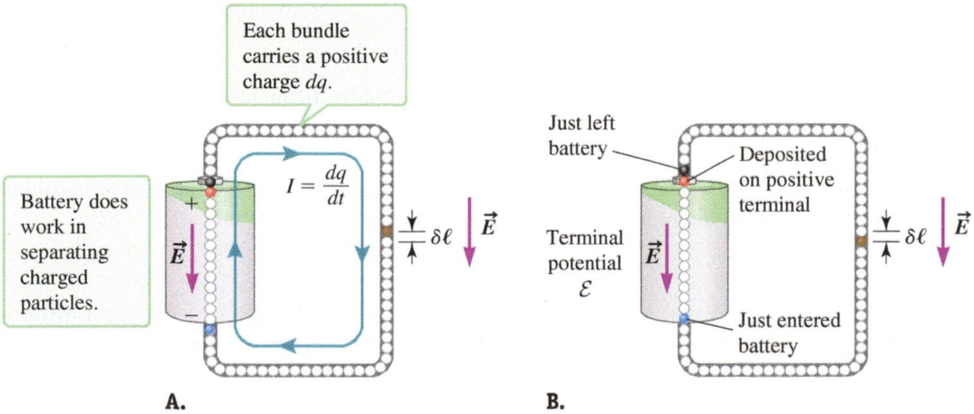

FIGURE 28.23 **A.** A bundle of charge dq shown in orange is about to be deposited on the positive terminal, and at the same time a bundle shown in blue is about to enter the battery at the negative terminal. **B.** A short interval of time dt later, the orange bundle has been deposited and the blue bundle has entered the battery. And, in this short time interval, each bundle (such as the brown ones) between the blue bundle and the orange one has advanced one step clockwise toward the positive terminal.

Our system includes the bundles of positive charge and the electric field produced by the battery. We'll consider the battery itself to be outside the system. As a result of chemical reactions, the battery moves each positive bundle inside it from the negative terminal to the positive terminal. Because there is an attraction between like charges, the battery must put energy into the system through *positive* work, so the electric potential energy of the bundles must increase. In Figure 28.23, each bundle of charge dq moves one step clockwise in a short time interval dt. The net result of the motion of all these bundles in the battery is equivalent to one bundle having moved from the negative terminal to the positive terminal through a potential difference equal to the terminal potential $\mathcal{E}$ of the battery. The movement of one bundle of charge dq from the negative to the positive terminal of the battery increases the system's electric potential energy by a small amount dU (Eq. 26.6):

$$dU = \mathcal{E}dq \tag{28.28}$$

The rate at which bundles of charge are deposited on the positive terminal is equal to the rate at which bundles of charge move around the circuit, which is the current $I = dq/dt$:

$$dq = Idt \tag{28.29}$$

Substitute Equation 28.29 into Equation 28.28:

$$dU = \mathcal{E}Idt$$

and divide by dt:

$$\frac{dU}{dt} = I\mathcal{E} \tag{28.30}$$

Equation 28.30 is the rate at which the potential energy of the system increases. Because the battery is responsible for the increase in potential energy, Equation 28.30 is the power supplied by the battery:

POWER SUPPLIED BY A BATTERY
Major Concept

$$P_{\text{bat}} = I\mathcal{E} \tag{28.31}$$

The SI unit for power is the watt, and from Equation 28.31,

$$1\text{ W} = 1\text{ A}\cdot\text{V}$$

Now let's find an expression for the rate at which energy leaves the system. The battery moves positively charged bundles inside it from the negative terminal toward the positive terminal in a direction opposite to the electric field. The battery must do work to make that happen, increasing the system's potential energy. On the other hand, a charge bundle in the wire, such as the brown one in Figure 28.23, moves from the positive terminal toward the negative terminal in the same direction as the electric field. The motion of bundles in the wire decreases the system's potential energy. The brown bundle moves a distance $\delta\ell$ in the short time interval dt. The electric field in the wire is uniform, and the brown bundle moves through a small potential difference given by $\delta V = E\delta\ell$ (Eq. 28.21). The bundle carries a positive charge dq, so the system's potential energy decreases by a small amount δU:

$$\delta U = dqE\delta\ell$$

This is the decrease in potential energy that results from the motion of just one bundle. However, every bundle in the wire must move a distance $\delta\ell$ in the time interval dt.

Therefore, the total decrease dU in potential energy in the time interval dt is found by adding up the decreases in potential energy due to each bundle:

$$dU = \sum \delta U = \sum dqE\delta\ell$$

Each bundle carries the same amount of charge and the electric field is uniform throughout the wire, so we can pull those constants outside the summation:

$$dU = dqE\sum \delta\ell = dqE\ell$$

where ℓ is the entire length of the wire. Substitute $dq = Idt$ (Eq. 28.29) for dq and $\Delta V = E\ell$ (Eq. 28.22) for $E\ell$ to find $dU = Idt\Delta V$. Then divide by dt:

$$\frac{dU}{dt} = I\Delta V \quad (28.32)$$

Equation 28.32 is the rate at which the system's potential energy decreases as a result of the current in a wire. It is a general expression for the power P used by any circuit element that carries current I and has potential difference ΔV between its ends:

POWER USED BY A CIRCUIT ELEMENT

Major Concept

$$P = I\Delta V \quad (28.33)$$

Other convenient expressions for the power used by a circuit element with resistance R can be found by substituting $\Delta V = IR$ (Eq. 28.25) into Equation 28.33, eliminating either I or ΔV:

$$P = I^2R = \frac{(\Delta V)^2}{R} \quad (28.34)$$

For a wire connected to a battery (Fig. 28.8), the potential difference between the ends of the wire equals the terminal potential $\mathcal{E}$ of the battery. In this case, Equation 28.33 becomes

$$P = I\mathcal{E} = P_{\text{bat}}$$

so the power supplied by the battery is used by the wire.

What does the wire do with that power? In the case of wire or a resistor, the power goes into vibrating the ions in the lattice (due to collisions between conduction electrons and ions); thus, the power goes into the system's thermal energy. The wire or resistor becomes hotter than its environment (the circuit board on which it is mounted and the surrounding air). As a result, energy leaves the system through heat. You have experienced this energy transfer when you hold your phone, computer, or other electronic device and feel it warm up your hand. In a poorly designed or dysfunctional device, the heat transfer rate may not be sufficient to cool the circuit. When this happens, the device may smoke, melt, or even catch fire (Fig. 28.24).

mikeledray/Shutterstock.com

FIGURE 28.24 This computer's circuit was unable to transfer energy to its environment fast enough to avoid catching on fire.

Kilowatt-Hours

Of course, electric circuits would not be particularly useful if the only thing they could do is get warmer. The power that goes into a circuit can do many useful things, such as move the speakers in your car stereo system, rotate the tub in your washing machine, and keep your food cold. In your home or dormitory, power is supplied by a utility company. Typically, households and businesses pay for the total amount of energy they consume in a month.

The SI unit for power is the watt, where

$$1\text{ W} = 1\text{ J/s}$$

You might expect that the utility company's bill would indicate your energy consumption in joules, but most companies use **kilowatt-hours** (kW · h) to measure energy. One kilowatt-hour is the amount of energy consumed by a 1000-W device such as a small blow dryer in 1 hour.

CONCEPT EXERCISE 28.6

A battery of terminal potential $\mathcal{E}$ is connected to a lightbulb (Fig. 28.2). After a long time, the bulb stops working because the filament snaps. After the filament breaks, how much power is dissipated by the filament? Explain your answer.

EXAMPLE 28.7 Record-Breaking Lightbulb

One of the earliest household devices to use electricity was the lightbulb. In 1901, Dennis Bernal donated a carbon-filament lightbulb to the fire department in Livermore, California. The bulb has been lit nearly continuously since that year. It went without power for very short time periods when the fire department moved. (It was also off for about 10 hours in 2013 when the power to the bulb was cut.) The book *Guinness World Records* has recognized this "Centennial Bulb" as the longest-burning in the world (Fig. 28.25). The bulb's power output is 4 W.

Caters News/ZUMA Press/Newscom

FIGURE 28.25 The Centennial Bulb holds the world record for the longest-illuminated lightbulb.

A Estimate the total amount of energy W consumed by the lightbulb since it was first turned on. Give your answer in kilowatt-hours and in joules.

B Today we pay about 10¢ per kW · h. How much does the fire station have to pay each month to keep the bulb lit?

C How much money would it cost to keep this bulb lit at this price for 100 years? (Assume the power radiated has been constant during this period.)

INTERPRET and ANTICIPATE

We can find the total energy radiated by the bulb during the past 100 years from its power output. The power radiated must equal the power supplied. Because this is an estimate, we will assume the bulb has been lit continuously for 100 years.

SOLVE

A The total energy consumed in 100 years is the power times 100 years. (One year is about $\pi \times 10^7$ s—good enough for an estimate.)

$$100\ \text{yr}\left(\frac{\pi \times 10^7\ \text{s}}{1\ \text{yr}}\right) = \pi \times 10^9\ \text{s} \approx 3 \times 10^9\ \text{s}$$

$$W \approx P\Delta t \approx (4\,\text{W})(3 \times 10^9\ \text{s}) \approx 1.2 \times 10^{10}\ \text{J}$$

Convert to kilowatt-hours.

$$W = (1.2 \times 10^{10}\ \text{J})\left(\frac{1\ \text{kW}\cdot\text{h}}{3.6 \times 10^6\ \text{J}}\right)$$

$$W \approx 3.3 \times 10^3\ \text{kW}\cdot\text{h}$$

B The energy used in 1 month is found in a similar way, but we need the number of hours in a month.

$$\Delta t = (30\ \text{days})\left(\frac{24\ \text{h}}{1\ \text{day}}\right) = 720\ \text{h}$$

$$W = P\Delta t = (4\ \text{W})(720\ \text{h}) = 2880\ \text{W}\cdot\text{h}$$

$$W \approx 2.9\ \text{kW}\cdot\text{h}$$

To find the cost, multiply by the price per kilowatt-hour.

$$(2.9\ \text{kW}\cdot\text{h})\left(\frac{\$0.10}{\text{kW}\cdot\text{h}}\right) = \$0.29\ \text{per month}$$

C There are 1200 months in 100 years, so multiply by 1200 to find the cost for 100 years at this rate.

$$\$0.29 \times 1200 = \$\,348.00$$

CHECK and THINK

Because this is a low-power bulb, it hasn't consumed (or given off) much energy during 100 years. The typical American home uses about 10,000 kW · h per year at an annual cost of about \$1000. During the 100+ years that the Centennial Bulb has been glowing, it has used about one-third the energy a typical American household consumes in 1 year.

EXAMPLE 28.8 CASE STUDY **Where There's Smoke...**

Answer Avi's question, "Why did my phone go up in smoke?"

SOLVE
The thermal energy of the phone increased at a rate faster than its circuits could remove energy through heat. Electronic devices usually have some mechanism to promote heat flow from the device. For example, computers often have fans that remove thermal energy from the circuitry by forced convection. Some electrical components are equipped with fins that increase their surface area and allow them to lose energy to the surroundings more efficiently.

Avi's phone was designed to lose thermal energy at some particular rate, which was sufficient before the adapter was connected to it. The adapter probably supplied the phone with too much power. (The adapter provided a voltage V that was too high, and because $P = IV$, the power was too high.) The phone was unable to radiate fast enough, so thermal energy built up and destroyed the phone.

SUMMARY

Underlying Principles

No new principles are introduced in this chapter.

Major Concepts

1. A **circuit** is a closed network of circuit elements, including metal wires, that form a pathway in which charged particles can flow. A **circuit element** is a component or device found in a circuit, such as a battery, capacitor, resistor, or lightbulb.
2. **Current** I is the (apparent) motion of positively charged particles:
$$I = \frac{dq}{dt} \tag{28.1}$$
In a metal conductor, current is a result of the motion of conduction electrons. For a particular location in the conductor, current is in the direction opposite to the electrons' drift velocity and is in the same direction as the electric field.
3. **Current density** $\vec{J}$ is a vector whose magnitude is the current I per unit cross-sectional area A:
$$J \equiv \frac{I}{A} \tag{28.3}$$
and whose direction is the same as that of the electric field, $\vec{J} \propto \vec{E}$. Current density is a macroscopic observation connected to the microscopic motion of charged particles in the conductor:
$$\vec{J} = ne\vec{v}_d \tag{28.5}$$
where $\vec{v}_d$ is the drift velocity of (imaginary) positive charge carriers and n is the number density of charge carriers in the conductor.
4. **Conductivity, resistivity, and their temperature dependence**
   a. **Conductivity** σ is a measure of a material's ability to conduct current:
$$\sigma \equiv \frac{ne^2}{m_e} t_{mf} \tag{28.15}$$
where e is the electron's charge, m_e is the electron's mass, and t_{mf} is the mean free time between collisions of conduction electrons and lattice ions.
The current density in a conductor is determined by the electric field in that conductor and its conductivity:
$$\vec{J} = \sigma\vec{E} \tag{28.16}$$
   b. **Resistivity** ρ is a measure of a material's ability to resist conducting electricity. Resistivity is the reciprocal of conductivity:
$$\rho \equiv \frac{1}{\sigma} = \frac{m_e}{ne^2 t_{mf}} \tag{28.18}$$
and
$$\vec{E} = \rho\vec{J} \tag{28.19}$$
   c. Conductivity and resistivity depend on the material's temperature. The **temperature dependence** of resistivity is given by
$$\rho(T) = \rho_0[1 + \alpha(T - T_0)] \tag{28.20}$$
where α is the temperature coefficient of resistivity and ρ_0 is the resistivity at temperature T_0, usually 20°C.

Major Concepts cont'd

5. **Resistance and resistors**
   a. **Resistance** R is a measure of an object's ability to resist an electric current. Resistance depends on the object's resistivity ρ and the geometry of the object:
   $$R = \rho\frac{\ell}{A} \quad (28.24)$$
   where ℓ is the object's length and A is its cross-sectional area.
   The current in a conductor depends on the potential difference between the ends of the conductor and its resistance:
   $$\Delta V = IR \quad (28.25)$$
   b. A **resistor** is a circuit element designed to have a particular constant resistance.
6. **Ohm's law** is an empirically derived rule that holds for some circuit elements such as resistors. Ohm's law may be stated in three ways:
   a. The potential difference ΔV between the ends of a conductor is directly proportional to the current I in that conductor:
   $$\Delta V \propto I \quad (28.26)$$
   b. The electric field $\vec{E}$ in a conductor is directly proportional to the current density $\vec{J}$ in that conductor:
   $$\vec{E} \propto \vec{J} \quad (28.27)$$
   c. The resistance R, conductance $1/R$, resistivity ρ, and conductivity σ are all constants.
   A circuit element that obeys Ohm's law over a wide range of potential differences across it is described as **ohmic.**
7. The **power supplied to or used by a circuit element** depends on the current in the circuit element and the potential difference across it:
   $$P = I\Delta V \quad (28.33)$$
   The power supplied by a battery with terminal potential $\mathcal{E}$ is
   $$P_{bat} = I\mathcal{E} \quad (28.31)$$
   The power used by a circuit element with resistance R is
   $$P = I^2R = \frac{(\Delta V)^2}{R} \quad (28.34)$$

PROBLEMS AND QUESTIONS

A = algebraic C = conceptual E = estimation G = graphical N = numerical

28-1 Microscopic Model of Charge Flow

1. **C** You have two different metal conductors. Each has the same number of conduction electrons per unit volume, but the conduction electrons in metal 1 undergo fewer collisions per unit time than those in metal 2. Which is the better conductor: metal 1 or metal 2? Explain.
2. **C** In most metals, there is one conduction electron per atom. Suppose you find a metal that on average has more than one conduction electron per atom. Do you expect it to be a better conductor than most other metals? Explain.

Problems 3, 4, and 5 are grouped.

3. **C Review** You have two antique copper pennies. One is much cooler than the other, but both have the same density and both have one conduction electron per atom. Which penny is a better conductor? Explain.
4. **E Review** Suppose you have an antique copper penny in your pocket. Modeling the conduction electrons in the penny as an ideal gas, use results from Chapter 20 to estimate their kinetic energy and root-mean-square (rms) velocity.
5. **E Review** Suppose that a heated gas comprised of electrons glows bright orange when it is in use. Estimate the average speed of the electrons in the gas. Model this gas as an ideal gas.

28-2 Current

6. **N** In a low-energy transmission electron microscope (TEM), the typical average beam current is about $-1.00\ \mu$A. If a material sample is exposed to this electron current for 10.0 min, how many electrons impact the material during that time?
7. **N** Imagine a horizontal wire in which a steady stream of 3.0×10^{18} electrons per second flow to the right. What is the current in the wire? What is the direction of the current?
8. **N** In Concept Exercise 28.2 (page 869), we found that the current in a computer circuit is about 1 pA and the current in a car starter motor is about 200 A. How long do you need to wait to have the same amount of charge pass through the wires of a computer circuit that passes through a starter motor in 1 s?

Problems 9, 10, and 11 are grouped.

9. **A** Positively charged ions move along a single axis, forming a beam. The amount of charge that passes through a cross section of the beam is given by $q = q_0 \cos \omega t$, where q_0 is the amount of charge that passes through at $t = 0$ and ω is the angular frequency. Find an expression for the current as a function of time.
10. **C, A** As described in Problem 9, positively charged ions move along a single axis to form a beam. **a.** Describe what is happening to the total charge that has passed by as time goes on. **b.** Are there moments in time when the total charge that has passed by is zero? If so, find an expression for when this occurs and explain your result.
11. **C, A** As described in Problem 9, positively charged ions move along a single axis to form a beam. **a.** Describe what is happening to the current as time goes on. **b.** Are there moments in time when the current is zero? If so, find an expression for when this occurs and explain your result.

12. **C** CASE STUDY Figure P28.12 shows a series of six identical lightbulbs connected to a battery. The bulbs could be a short strand of old-style holiday lights. Assume all the bulbs are operational. Reread the discussion among Avi, Shannon, and Cameron in the case study (Section 28-1 and Example 28.1). Then decide how Shannon and Cameron might answer these questions: Which bulb has the greatest current? Which bulb is the brightest? Which bulb is the hottest? **a.** How do you think Shannon might answer the questions? Explain. **b.** How do you think Cameron might answer the questions? Explain. **c.** What are your answers to the questions? Explain.

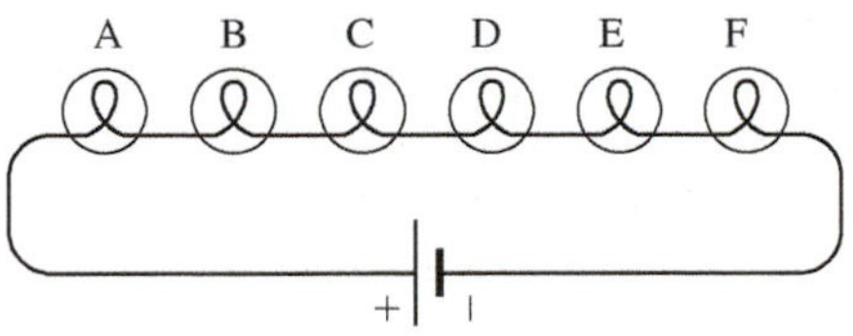

FIGURE P28.12

13. **N** The current in a metal wire of cross-sectional area $1.0 \times 10^{-6}\ \text{m}^2$ and length 1.0 m is 16 A. If the metal contains 6.0×10^{28} free electrons per cubic meter, find the drift speed in the wire.
14. **E, C** Consider the length of wire that connects the batteries to the lightbulb in a typical flashlight. **a.** Estimate the time it would take an imaginary positive charge to make its way through this length of wire. Explain how you found any values you use. **b.** How does your answer compare to the amount of time it takes before the light appears? (Use your experience from switching on a flashlight.) Comment on your results.
15. **N** The current in a wire varies with time (measured in seconds) as $I = 24\ \text{A} - (0.12\ \text{A/s}^2)t^2$. Determine the amount of charge that flows through a cross-sectional area of the wire between $t = 0$ and $t = 12$ s.

28-3 Current Density

16. **C Review** Many people say that the electrons in a metal wire travel at nearly the speed of light. Think of how you would respond to that statement. How does the drift speed compare to the speed of light? How about the rms speed? *Hint*: Consider the results from Example 28.2 and Problem 4.
17. **N** The amount of charge that flows through a copper wire 1.00 cm in radius as a function of time is given by $q = 2t^3 + 8t^2 + 2$, where q is in coulombs and t is in seconds. **a.** What is the current through the copper wire at $t = 2.00$ s? **b.** What is the current density in the copper wire at $t = 2.00$ s?

Problems 18 and 19 are paired.

18. **N** A copper wire that has a cross-sectional diameter of 3.500 mm has a current of 1.241 A. **a.** What is the current density? **b.** What is the total charge that passes by a certain location along the wire in 2.000 s?
19. **N** Consider the current described in Problem 18. How long would it take for an electron to travel a 1.00-m length of the wire?
20. **C** Imagine that you can change the thickness of a wire while the electric field in the wire remains constant. Suppose you double the cross-sectional area. What happens to **a.** the current density, **b.** the current, **c.** the number density of charge carriers, and **d.** the drift speed in the wire?

Problems 21 and 22 are paired.

21. **N** A current-carrying conductor made of aluminum gradually narrows as shown in Figure P28.21. The cross-sectional area of region 1 is twice that of region 2. In region 1, the current is 6.25 mA and the current density is $2.00\ \text{kA/m}^2$. Find the current and current density in region 2, assuming a steady-state current has already been achieved.

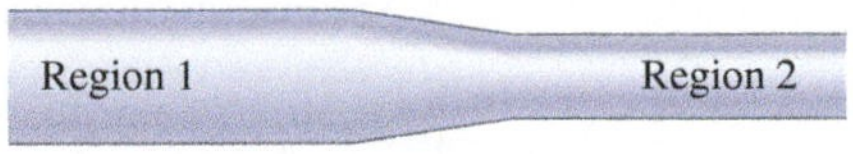

FIGURE P28.21 Problems 21 and 22.

22. **N** Consider the situation described in Problem 21. Find the drift speed in both regions of the conductor.
23. **N** A copper wire that is 2.00 mm in radius with density $8.94\ \text{g/cm}^3$ has a current of 8.00 A. The molar mass of copper is 63.5463, and each copper atom contributes one free electron. What is the drift speed of the electrons in the copper wire?
24. **C** Many people believe that when you turn on the lamp in your room, electrons from the wall outlet are immediately delivered to the lightbulb filament and then flow back to the wall via the cord running between the wall and the lamp. Explain why this could not be the case given your knowledge of the behavior of the electrons in the wire. Why, then, does the lightbulb light immediately?

28-4 Resistivity and Conductivity

Problems 25 and 26 are paired.

25. **N** The resistivity of a metallic, single-walled carbon nanotube is $3.40 \times 10^{-8}\ \Omega \cdot \text{m}$. What is the conductivity of this nanotube?
26. **N** Consider the nanotube described in Problem 25. The electron number density is $6.57 \times 10^{28}\ \text{m}^{-3}$. What is the mean free time for the electrons flowing in a current along the carbon nanotube?
27. **N** What is the electric field in an aluminum wire if the drift speed of the free electrons in the wire is measured to be 4.50×10^{-5} m/s? The resistivity of aluminum is $2.655 \times 10^{-8}\ \Omega \cdot \text{m}$.
28. **E** Fifty pennies are stacked together to make one roll. From 1793 to 1857, the American penny was made of copper. After 1857, the penny was made of copper alloys, in which copper was the predominant but not the only metal. Today's penny is copper-plated zinc. **a.** Estimate the resistivity and conductivity of a roll of pure copper pennies. **b.** Estimate the resistance of a roll of pure copper pennies. **c.** In the mid-1970s, the price of copper became so high that the U.S. Mint struck a batch of pennies from aluminum as part of a search for a replacement metal. Those pennies were never released into circulation. One of them was donated to the Smithsonian Institution, where it is now part of the coin collection in the National Museum of American History. A few aluminum pennies may be in the hands of collectors, but such pennies are illegal. Estimate the resistance of a roll of aluminum pennies.

Problems 29 and 30 are paired.

29. **C** You increase the temperature of a metal conductor by a few degrees. What happens to the **a.** mean free time t_{mf}, **b.** frequency of collisions between the charge carriers and lattice ions, **c.** density of the ions, **d.** density of the conduction electrons, and **e.** conductivity? Explain your answers.
30. **C** You increase the temperature of a semiconductor by a few degrees. What happens to the **a.** density of the ions, **b.** density of the conduction electrons, and **c.** conductivity? Explain your answers. If appropriate, compare your answers to those in Problem 29.
31. **N** Two long concentric cylinders of radii 4.0 cm and 8.0 cm are separated by aluminum. The inner cylinder has a charge per unit length of λ at any time. When the two cylinders are maintained at a constant potential difference of 2.0 V via an external source, calculate the current from one cylinder to the other if the cylinders are 100.0 cm long.

32. **G** Plot resistivity versus temperature for the data in Table P28.32. Use your graph to find the temperature coefficient of resistivity. Then use Table 28.2 to identify the material.

TABLE P28.32

T (°C)	ρ ($\Omega \cdot$ m)
0	1.87×10^{-8}
10	2.05×10^{-8}
20	2.24×10^{-8}
30	2.42×10^{-8}
40	2.61×10^{-8}
50	2.80×10^{-8}
60	2.99×10^{-8}
70	3.17×10^{-8}
80	3.36×10^{-8}
90	3.54×10^{-8}
100	3.27×10^{-8}

33. **N** Two concentric, metal spherical shells of radii $a = 4.0$ cm and $b = 8.0$ cm are separated by aluminum as shown in Figure P28.33. The inner sphere has a total charge Q at any time. If the two spheres are maintained at a potential difference of 2.0 V via an external source, calculate the current from one sphere to the other.

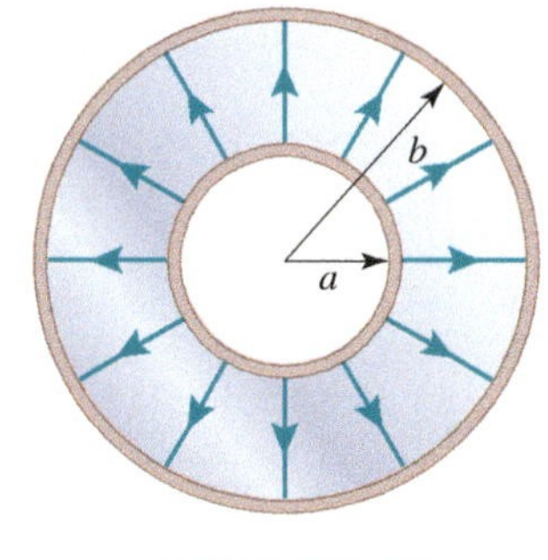

FIGURE P28.33

34. **C** A battery is connected across a conductor that has an irregular cross-sectional area as shown in Figure P28.34.
    a. Is the rate of flow of electrons per unit area at point A less than, equal to, or greater than the rate of flow of electrons per unit area at point B?
    b. Is the magnitude of the electric field at A less than, equal to, or greater than that at B?

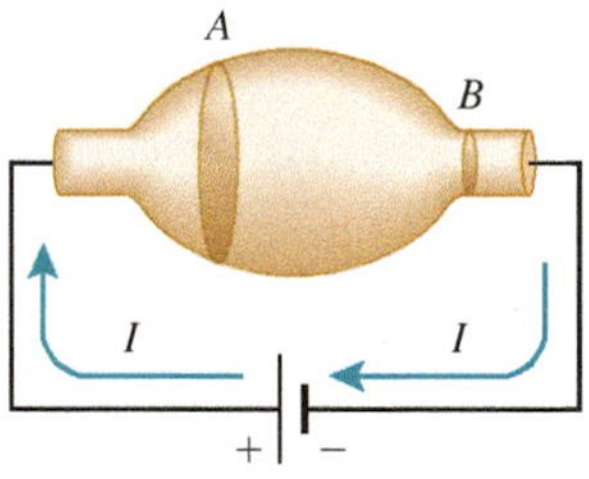

FIGURE P28.34

28-5 Resistance and Resistors

35. **N** When 110.0 V of potential difference is applied across a lightbulb, its filament develops a resistance of 195 Ω. What is the current through the bulb?
36. **A** Show that the dependence of resistance on temperature is given by $R(T) = R_0[1 + \alpha(T - T_0)]$, where α is the temperature coefficient of resistivity and R_0 is the resistance at temperature T_0.
37. **N** A copper wire has a diameter of 0.400 mm. What length of the wire has a resistance of 2.00 Ω?

Problems 38, 50, and 52 are grouped.

38. A lightbulb is connected to a variable power supply. As the potential across the bulb is varied, the resulting current and the filament's temperature are measured. The data are listed in Table P28.38.
    a. **G** Find R for each entry in Table P28.38, and then plot R as a function of T.
    b. **N** Assume that room temperature is at 293 K. Find R_0 (resistance at room temperature). Comment on your result.

TABLE P28.38

I (A)	V (V)	T (K)
0.268	7.603	2209
0.269	7.661	2216
0.266	7.682	2241
0.271	7.744	2222
0.272	7.815	2231
0.273	7.894	2243
0.274	7.947	2249
0.277	8.075	2258
0.279	8.213	2276
0.281	8.347	2292
0.284	8.457	2297
0.287	8.656	2321
0.291	8.842	2335
0.293	8.954	2346
0.295	9.059	2355
0.297	9.163	2364
0.298	9.227	2371
0.300	9.335	2380
0.302	9.445	2385
0.301	9.449	2390

39. **N** A 10.0-g sample of copper with density 8.94×10^3 kg/m^3 is made into a wire 1.00×10^3 m long. If the resistivity of copper is 1.678×10^{-8} $\Omega \cdot$ m, what is the resistance of this wire?
40. **C** For each of the following quantities, determine its fundamental dimensions (length, mass, time, current, and temperature): **a.** conductivity, **b.** resistivity, **c.** conductance, and **d.** resistance. Use your results to write the SI unit ohm (Ω) in terms of the fundamental units (meter, kilogram, second, ampere, and kelvin).
41. **C** Which of the following quantities depends only on the microscopic properties of a conductor? Which depend on both microscopic and macroscopic properties? Explain. **a.** conductivity **b.** resistivity **c.** conductance **d.** resistance
42. **N** A long, thin cylindrical conductor is made of silver. Its resistance is 78.5 Ω, and its mass is 0.0250 kg. Find the length and radius of the conductor.
43. **N** An aluminum wire is heated from 20.0°C to 90.0°C. By what percentage does the resistance of the wire change? The temperature coefficient for aluminum is $\alpha = 4.29 \times 10^{-3}$ °C^{-1}.

Problems 44 and 45 are paired.

44. **A** Two wires with different resistivities, ρ_1 and ρ_2, are supposed to have the same resistance. If the radii of the wires are r_1 and r_2, find a function for the length of the second wire in terms of the length of the first wire.
45. **N** A copper and a gold wire are supposed to have the same resistance. The copper wire has a radius of 2.50 mm, and the gold wire has a radius of 3.25 mm. If the copper wire is 1.20 m long, what must be the length of the gold wire?
46. **N** Gold bricks are formed with the dimensions $7 \times 3\frac{5}{8} \times 1\frac{3}{4}$ inches. What is the maximum resistance of a gold brick? What is the minimum resistance? Include sketches to explain how each of these resistances is achieved.
47. **N** A 75.0-cm length of a cylindrical silver wire with a radius of 0.150 mm is extended horizontally between two leads. The potential at the left end of the wire is 3.20 V, and the potential at the right end is zero. The resistivity of silver is 1.586×10^{-8} $\Omega \cdot$ m.

a. What are the magnitude and direction of the electric field in the wire? **b.** What is the resistance of the wire? **c.** What are the magnitude and direction of the current in the wire? **d.** What is the current density in the wire?

28-6 Ohm's Law

48. G In 2001, a group of scientists measured the conductivity of a single molecule. Such work is key to developing molecular-based electronics, the next step in further reducing the size of electronic devices and possibly leading to nanoscale computers. Imagine a computer no larger than a single bacterium. Computers that small could be programmed to enter the body and cure diseases on the cellular level. Figure P28.48 is based on the paper published by X. D. Cui and collaborators in 2001. It shows current (in nanoamperes) as a function of potential difference across a single molecule. **a.** Does the molecule obey Ohm's law over the entire range from −1 V to 1 V? **b.** Use the inset in Figure P28.48 to find the conductance and resistance of the molecule over the range from −0.1 V to 0.1 V.

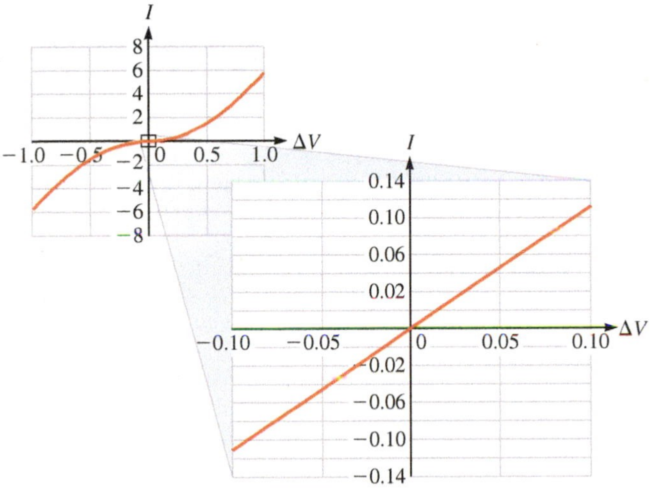

FIGURE P28.48

49. C Juanita is hard at work in her lab measuring the resistance of a new material. She observes a current of 0.10 A when applying a potential difference of 9.00 V and a current of 0.20 A when applying a potential difference of 12.00 V. Is the material ohmic? Explain your answer.

50. G A conductor is connected to a variable power supply. As the potential across the conductor is varied, the resulting current is measured. The data are listed in Table P28.38. (Ignore the temperature data.) Plot current versus potential and determine whether this conductor obeys Ohm's law. Explain your reasoning.

51. C The word *law* in physics can be misleading. In the text, we said that Ohm's law and Hooke's law are not laws in the same sense as Newton's three laws of motion. Perhaps they should be renamed "Ohm's rule" and "Hooke's rule." Explain why these may be better terms by comparing Ohm's law and Hooke's law to Newton's laws of motion.

28-7 Power in a Circuit

52. **Review** A lightbulb is connected to a variable power supply. As the potential across the bulb is varied, the resulting current and the filament's temperature are measured. The data are listed in Table P28.38.

a. N Find the power for each entry in Table P28.38.
b. N Find T^4 for each entry.
c. G Plot P as a function of T^4. Comment on your graph.

53. N A 100.0-W lightbulb is connected to a 220.0-V source. **a.** What is the current through the bulb? **b.** What is the internal resistance of the bulb?

54. C A current I exists in a loop of superconducting wire. How much power is dissipated by the wire?

55. N A two-slice bread toaster consumes 850.0 W of power when plugged into a 120.0-V source. **a.** What is the current in the toaster? **b.** What is the resistance of the coils in the toaster?

56. N, E CASE STUDY According to information on the back of Avi's phone, it draws a current of 3.0 A. Assume it uses a lithium–ion battery that has a terminal potential of 3.7 V. **a.** What is the phone's power requirement? **b.** Avi's phone was destroyed in seconds after it was plugged into the wrong adapter. Estimate how much energy was delivered by the battery to the phone in that amount of time. **c.** If the battery were replaced by a power supply with a terminal voltage of 120 V, estimate how much energy would be delivered by the power supply to Avi's phone in the amount of time it took to destroy the phone.

57. N Household water heaters use a 240.0-V rather than a 120.0-V source. What is the resistance of a water heater's heating element if it heats 40.0 gallons (151 kg) of water from 15.0°C to 60.0°C in 15.0 min?

58. **Review** Assume the filaments in the bulbs in Example 28.6 (page 883) have the same dimensions, 0.045 mm in diameter and 580 mm in length. Assume all the power supplied is radiated away by the bulb.

a. N Find the power delivered by the power supply for each measurement in Table 28.4.
b. N Assuming the emissivity of the filaments is 1, find the temperature of each filament for each entry in the table.
c. G Plot P as a function of T for each bulb on the same graph.

59. A A resistor is connected to a variable power supply. Initially, the power supply's terminal potential is V_0 and the current through the resistor is I_0. The terminal potential is then doubled. Find an expression for the new **a.** current through the resistor, **b.** power delivered by the power supply, and **c.** power used by the resistor in terms of the initial parameters.

60. C Two separate resistors, R_1 and R_2, are each connected across identical batteries with terminal potential $\mathcal{E}$. In each case, the power used by each resistor is the same. Explain how this is possible. What must be true about the resistors?

61. N **Review** The current through a 1.50-L electric kettle operating at 120.0 V is 12.5 A. If the energy delivered to the heating element of the kettle is absorbed entirely by the water, how much time does it take to get 1.50 L of water initially at 20.0°C to start boiling?

62. N High-voltage transmission lines carrying electricity from a generating station to a local switching station 125 km away carry a current of 850 A. How much power is lost because of resistance in each wire during transmission from the generating station to the local station if the wire's resistance is 0.240 Ω/km?

General Problems

63. N The USB charger for a smartphone delivers 500.0 mA of current at 5.00 V while charging the phone. What is the power consumption of the smartphone while it is charging?

64. You measure the amount of charge that passes a certain location as a function of time for several seconds in three separate trials. Your measurements in each of three trials are represented by the functions (i) $q_1 = 3t^2 + 6t$, (ii) $q_2 = 4t^2 + 2t$, and (iii) $q_3 = 2t^2 + 10t$.

a. A Find the function that represents the current in each of the three trials.

b. G Plot each of the currents you found on the same graph for $t = 0$ to $t = 5$ s (the first 5 s of each trial).

c. C In which case is the current greatest after 10 s? In which case does the most charge pass by the location in 10 s?

65. N The current through a copper wire as a function of time is given by $I(t) = 65.0 \cos(33\pi t)$, with I in amperes and t in seconds. How much charge, in coulombs, flows through the wire during the first 0.0050 s after the current is switched on?

66. N A beam of electrons incident on a target carries -0.325 mA of current. How many electrons strike the target each minute?

67. N A variable resistance is connected to a battery that has a potential difference of 24.0 V. What is the resistance if a current of **a.** 1.00 A and **b.** 1.00×10^{-2} A is observed?

Problems 68 and 69 are paired.

68. G The resistivity ρ of a cylindrical conductor with uniform cross-sectional area A increases linearly from the left end to the right end as $\rho = \rho_0 + \alpha x$, where ρ_0 and α are constants. If the current I is constant through it, obtain an expression for the electric field as a function of the distance x from the left end, and plot the electric field as a function of the distance.

69. A If the electric field in a uniform conductor is E and v_d is the corresponding drift velocity of the free electrons in the conductor, obtain an expression for the drift speed as a function of the electric field E. Assume the cross-sectional area of the conductor is A, the number density of electrons is n, and ρ is the resistivity of the conductor.

70. N A 60.0-V battery is connected across a 175-Ω resistor. What is the **a.** power delivered by the battery and **b.** power used by the resistor?

71. N, C Two separate 60.0-V batteries are each connected across separate resistors. One is connected across a 250.0-Ω resistor, and the other is connected across a 350.0-Ω resistor. **a.** How much power is delivered by the battery in each case? **b.** Which resistor consumes 1000 J of energy in the shortest time?

72. C Terrance is hard at work fixing his robot, Victor, but finds that a resistor made of a coiled-up gold wire has gone bad. In looking for a replacement part, Terrance finds he has several lead bullets he can melt down and form into a new wire. How should Terrance choose the dimensions of the new lead wire? Describe the reasons for your choices and how each dimension compares to the original gold wire's dimensions.

73. N What is the ratio of the cross-sectional areas of a silver wire and a gold wire of equal length that have an equal amount of resistance? The resistivity of silver is $1.586 \times 10^{-8}\ \Omega \cdot \text{m}$, and the resistivity of gold is $2.24 \times 10^{-8}\ \Omega \cdot \text{m}$.

74. N The resistance of a copper wire is measured to be 3.50 Ω at 22.0°C. What is the resistance of this wire at 43.0°C? The temperature coefficient for copper is $\alpha = 3.9 \times 10^{-3}\ \text{C}^{-1}$.

75. N **Review** When a metal rod is heated, its resistance changes both because of a change in resistivity and because of a change in the length of the rod. If a silver rod has a resistance of 2.00 Ω at 22.0°C, what is its resistance when it is heated to 200.0°C? The temperature coefficient for silver is $\alpha = 6.1 \times 10^{-3}\ {}^\circ\text{C}^{-1}$, and its coefficient of linear expansion is $18 \times 10^{-6}\ \text{C}^{-1}$. Assume that the rod expands in all three dimensions.

Problems 76, 77, and 78 are grouped.

76. A A 1.50-m-long wire of aluminum is constructed such that it has inner and outer radii as shown in Figure P28.76. A potential difference is maintained on either end of the wire so that current I is from one end of the wire to the other. Find an expression for the current density as a function of I and the two radii r_a and r_b.

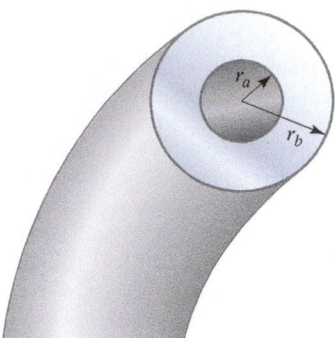

FIGURE P28.76 Problems 76, 77, and 78.

77. A Consider the situation described in Problem 76. Find an expression for the drift velocity of the electrons flowing in the wire as a function of I, the number density n, and the two radii r_a and r_b.

78. C Consider the situation described in Problem 76. Describe what happens to the current density and drift velocity as the radius r_a is decreased while the radius r_b remains constant. Why, in principle, do these quantities change or remain constant as the inner radius is decreased?

79. N Assuming that its coefficient of resistivity remains constant at $3.93 \times 10^{-3}\ {}^\circ\text{C}^{-1}$, at what temperature is the resistance of a copper wire 50% higher than its resistance at 22.0°C?

80. A Two long concentric cylinders with radii a and b are separated by a material with conductivity σ that varies as $\sigma(r) = k/r\ (r > a)$, where r is the radial distance from the axis of the two cylinders and k is a constant. A steady current I flows radially between the cylinders. Find the resistance between the cylinders. Assume the length of the cylinders is L.

81. N A generating station and a switching station 140.0 km away are connected by a high-voltage aluminum transmission line that has a current of 850.0 A. Each wire of the transmission line has a cross-sectional area of 750.0 mm^2, and the density of free electrons in aluminum is 6.022×10^{28} electrons/m^3. How much time does it take for one electron to travel from the generating station to the switching station in this wire?

82. A A conducting material with resistivity ρ is shaped into a wire of length ℓ with a tapered cross section that decreases from radius r_1 on the left end to r_2 on its right end (Fig. P28.82). If the current density in the wire is a constant as a function of the horizontal distance x along the wire, what is the resistance of this tapered wire?

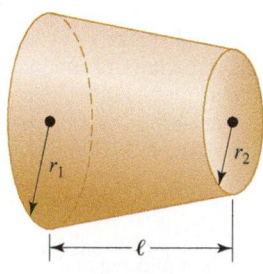

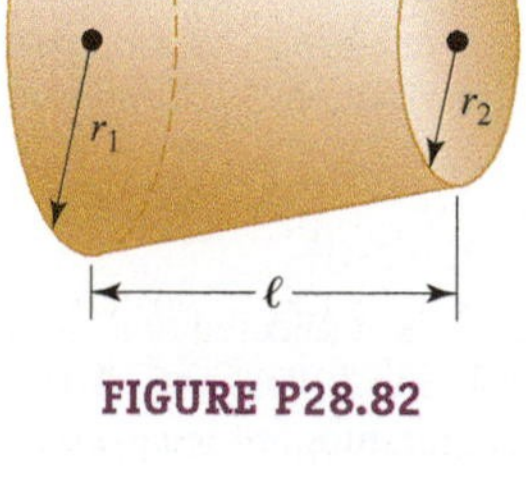

FIGURE P28.82

Direct Current (DC) Circuits

29

Underlying Principles

No new principles are introduced in this chapter.

Major Concepts

1. Kirchhoff's loop rule
2. Kirchhoff's junction rule
3. Resistors in series
4. Resistors in parallel

Special Cases

1. Emf devices (ideal and real)
2. Voltage rules in a circuit
   a. Wire rule
   b. Resistor rule
   c. Switch rule
   d. Emf rule (ideal and real)
3. *RC* circuits
   a. Charging
   b. Discharging

Key Questions

How do the concepts of electricity apply to practical devices?

How do we analyze existing circuits and design new ones?

An astronomer finds that her telescope won't move into position. Is the telescope's motor broken? The astronomer calls in an electrician, who uses a multimeter to measure the electric potential difference, current, and resistance in the circuit that contains the motor.

Scientists are not the only people who rely on electrical devices. Your home and car are full of devices such as lights, hot water heaters, refrigerators, clothes washers, and radios. When a device like a telescope motor fails, the fault may be mechanical or electrical. To find out whether something is wrong with an electric circuit, you must first know what to expect. What is the potential difference supposed to be across each circuit element? What is the expected current? What is the nominal resistance of each circuit element? Calculating these values is known as *circuit analysis*, the focus of this chapter. Once the expected values are known, a multimeter can be used to find any discrepancies and determine the reason for the failure.

29-1 Measuring Potential Differences Between Two Points

This chapter continues our practical application of the principles of electricity. Because this chapter focuses on practical applications, we often use the terminology heard in an electronics laboratory (Table 29.1). You need to be comfortable using both the formal and informal terms. Our goal is to analyze circuits, which means calculating the potential difference across, the current in, and the resistance of the various circuit elements. Circuit analysis can be used to design a new device or troubleshoot a malfunctioning one.

TABLE 29.1 Electrical terminology.

Informal term(s)	Formal term(s)	Definition	Usage
Voltage, voltage across, or voltage drop	Electric potential difference or potential difference (ΔV)	Difference in electric potential energy per unit charge (Section 26-4)	"The voltage drop across the resistor is 6.0 V." "The electric potential difference between the two ends of the resistor is −6.0 V."
Amperage or juice	Current (I)	*Apparent* motion of positively charged particles per unit time	"The amperage or current in those speakers should be 1.2 A."
DC power supply	DC emf device	Device that does work in separating charge, creating a constant electric potential difference between its terminals	"Connect a DC power supply or DC emf device to your lightbulb."
Input voltage	Terminal potential (of emf device)	Electric potential difference between the terminals of an emf device	"The input voltage must be 5.0 V." "The terminal potential must be 5.0 V."

EMF DEVICE Special Case

Of course, when an electrical device doesn't work at all, one of the first questions we ask is about the power source: "Does my phone need a new battery?" "Is the TV plugged in?" So far, the only electrical power source we have focused on is a chemical battery. There are many other types of power supplies, however, such as the electric generators cranked by a person in Ben Franklin's time or contemporary generators, which run on fossil fuels, nuclear material, wind, falling water, or sunlight. Another term for a power supply is an **emf device**. The initials "emf" come from the outdated term *electromotive force*. That term is misleading because it implies that the job of an emf device is to exert a force on the charged particles and that we can measure that force in newtons. Instead, an emf device is meant to maintain a potential difference between its terminals, which we measure in volts. So, it is common practice to refer to a power supply as an emf device without using the outdated term *electromotive force*.

Emf devices come in two major types—direct current (DC) and alternating current (AC). Again, these terms are somewhat misleading; they imply that the function of the emf device is to supply current, which is not correct. Charged particles are not made inside the emf device. Instead, the current in a circuit depends on the potential difference of the emf device and the resistance in the circuit. In a **direct current (DC)** circuit, the direction of the current does not change. In an **alternating current (AC)** circuit, the direction of the current switches back and forth, first pointing in one direction and then in the opposite direction. A DC power supply, such as a battery, maintains a constant potential difference between its terminals. In this chapter, we study only DC circuits.

If a circuit has a working power supply but is still malfunctioning, one of the next things to check is the potential difference between various points in the circuit. Let's illustrate the procedure with a simple circuit consisting of a lightbulb, an emf device, and a switch (Fig. 29.1). The schematic diagram (Fig. 29.1B) is easier to draw and makes it easier to see how the circuit elements are connected. So, from now on, we'll mostly use schematic diagrams. Learn to draw the symbols in Table 29.2 for the circuit elements.

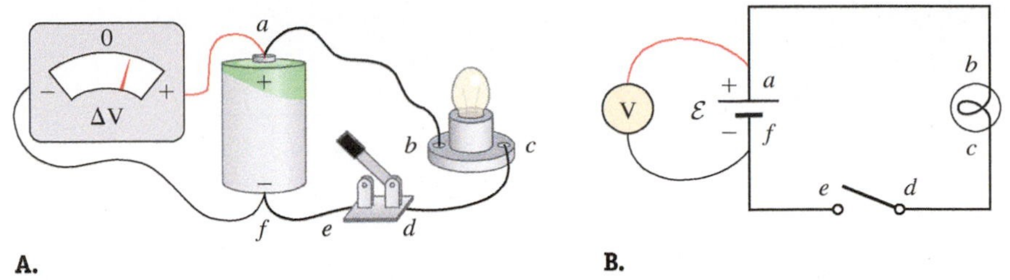

FIGURE 29.1 **A.** A simple circuit consisting of a lightbulb, a switch, wires, and an emf device. The voltmeter measures the potential difference across the battery. **B.** Schematic diagram for the circuit shown in part A.

TABLE 29.2 **Circuit elements and their symbols.**

Name	Symbol	Picture
DC emf device (battery, DC power supply)		
Resistor		
Capacitor	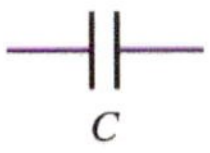	
Wire		
Lightbulb		
Switches		
	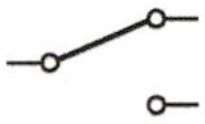	
Ground		
Voltmeter		
Ammeter		
Ohmmeter		

Battery: pryzmat/Shutterstock.com
Resistor: TheSnake19/Shutterstock.com
Capacitor: Stu49/Shutterstock.com
Wire: Ingvar Bjork/Shutterstock.com
Lightbulb: ThomasLenne /Shutterstock.com
Two-way switch: Hemera Technologies/Getty Images
Ground connection: Ntdanai/Shutterstock.com
Multimeter: StefanoT/Shutterstock.com

Using a Voltmeter

A **multimeter** is a device that has various settings for measuring voltage, current, and resistance (the last row in Table 29.2). We'll refer to a multimeter as a **voltmeter, ammeter**, or **ohmmeter** when it is set to measure potential difference, current, or resistance, respectively.

The voltmeter is connected *in parallel* to a particular circuit element using two wires known as leads. Traditionally, one lead is red and the other is black. The voltmeter reading is the potential difference between the two leads: $\Delta V = V_{\text{red}} - V_{\text{black}}$. Although there is nothing special about these colors, for convenience throughout this book, we will assume that the voltmeter has two colored leads—a red one and a black one—and that these leads are connected to the voltmeter so that it always gives the difference $\Delta V = V_{\text{red}} - V_{\text{black}}$. For example, if you wish to measure the potential difference $\Delta V = V_a - V_f$ across the battery in Figure 29.1, connect the voltmeter's red lead to point a and the black lead to point f ($V_a - V_f = V_{\text{red}} - V_{\text{black}}$).

Expected Voltages

To diagnose a circuit using a voltmeter, you first need to know what the potential difference should be across each circuit element. We use $\Delta V = IR$ (Eq. 28.25) to come up with five rules for finding the expected voltage across the circuit elements in Figure 29.1:

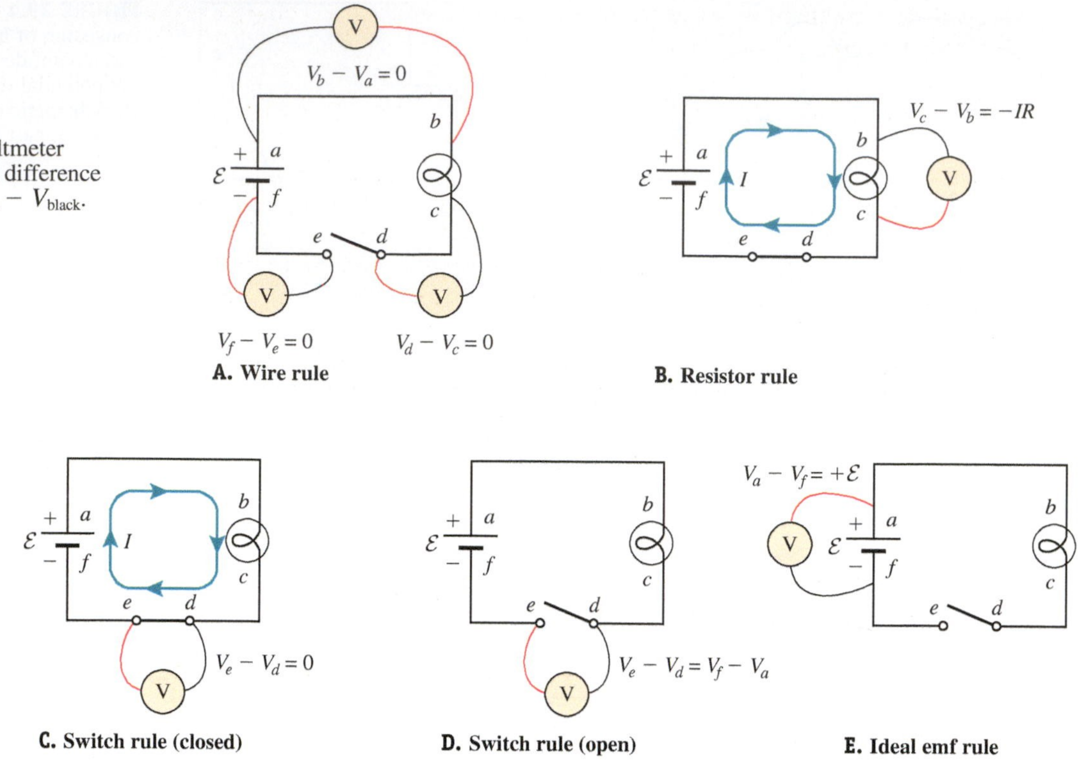

FIGURE 29.2 The voltmeter measures the potential difference between its leads: $V_{\text{red}} - V_{\text{black}}$.

1. **Wire rule.** Wires are designed to have very low resistance compared to other circuit elements. We will assume that the resistance of a wire is zero. So, the voltage across a wire is zero whether there is current in the wire or not (Fig. 29.2A):

WIRE RULE ▶ Special Case

$$\Delta V_{\text{wire}} = 0$$

2. **Resistor rule.** The current in a resistor or lightbulb is from high potential to low potential. In Figure 29.2B, the potential at point b is higher than the potential at point c, and the current through the bulb is from b to c when the switch is closed. If you connect a voltmeter in parallel across a resistor or lightbulb with the *red lead on the low-potential side* and the *black lead on the high-potential side* (Fig. 29.2B), the potential difference is negative:

$$V_{\text{red}} - V_{\text{black}} < 0$$

So, the reading on the voltmeter is negative. Informally, we say that there is a "voltage drop" across the resistor or bulb. The magnitude of that drop is given by $\Delta V = IR$ (Eq. 28.25).

Of course, if you reverse the leads so that the red lead is at point b and the black at c, the voltage reading is positive. So, to come up with a rule for the voltage across a resistor, we must first state how the voltmeter is connected. It is most convenient to write the rule in the case shown in Figure 29.2B. Then the resistor rule is: *When the current is from the black to the red lead, the voltage measured across a resistor is negative and given by*

RESISTOR RULE ▶ Special Case

$$\Delta V_R = V_{\text{red}} - V_{\text{black}} = -IR \quad (29.1)$$

(If the current is from the red to the black lead, then $\Delta V_R = +IR$.)

A resistor obeys Ohm's law (Section 28-6), so its resistance is constant and thus the current is directly proportional to the voltage drop across the resistor. Circuits that contain lightbulbs are more complicated because lightbulbs do not obey Ohm's law. When there is current in an incandescent bulb, the filament gets hot, increasing its resistance. Equation 29.1 still holds, but R is not constant.

3. **Switch rule.** If a switch is closed, it acts like a wire. A closed switch has very low resistance, and the voltage drop across it should be very small (Fig. 29.2C).

We will assume that the resistance of a closed switch and the voltage drop across it are zero:

$$\Delta V_{\text{closed switch}} = 0$$

SWITCH RULE **Special Case**

If the switch is open, there is air between the two sides of the switch. Because air is normally a good insulator, the resistance of an open switch is very high. There is no current through an open switch, but there may be a potential difference across it. For example, in Figure 29.2D, the voltage across the open switch equals the terminal voltage of the power supply.

4. **(Ideal) emf rule.** In DC power supplies, the high-potential terminal and the low-potential terminal of the emf device remain fixed and do not alternate. The plus and minus signs next to the emf device in schematic diagrams (Table 29.2) indicate the positive and negative terminals. So, the emf rule is: *If the red lead is connected to the positive terminal and the black lead to the negative terminal, the voltage measured is positive and equals the terminal potential* (Fig. 29.2E):

$$\Delta V = V_{\text{red}} - V_{\text{black}} = \Delta V_{\text{terminal}} = \mathcal{E} \qquad (29.2)$$

IDEAL EMF RULE **Special Case**

5. **Circuit elements in series.** When you connect a voltmeter so that there are two or more circuit elements between its leads, the voltage reading is the sum of the voltages across each element.

Real Versus Ideal Emf Devices

So far, we have considered only *ideal* emf devices. An **ideal DC emf device**, such as an ideal battery, maintains a constant terminal potential $\mathcal{E}$ whether there is current in the emf device or not. No real emf device can maintain its terminal potential when there is current in the device. If you close the switch while measuring the voltage across the emf device (Fig. 29.3), the terminal voltage decreases slightly. If you open the switch again, the terminal voltage returns to its earlier higher value. We can explain this observation: When you close the switch, there is current in the entire circuit, including the emf device. The current in a real emf device—like the current anywhere else in the circuit—is hindered by resistance, even inside the emf itself.

The **real emf device** in Figure 29.3 is modeled as two circuit elements: (1) an ideal emf device with no internal resistance and (2) a resistor with resistance r. Of course, we cannot put our voltmeter inside the emf device. Instead, we connect our voltmeter to the emf terminals labeled H and L (Fig. 29.3) and measure the potential difference between them. In Figure 29.3A, the emf is connected to an open switch, so there is no current in the leads or the battery, and therefore no voltage drop across the battery's internal resistor: $\Delta V_r = -Ir = 0$. The terminal potential measured by the voltmeter is

$$\Delta V_{\text{terminal}} = V_{\text{red}} - V_{\text{black}} = \mathcal{E}$$

Thus, when there is no current in a real emf device, its terminal potential is the same as that of an ideal emf device.

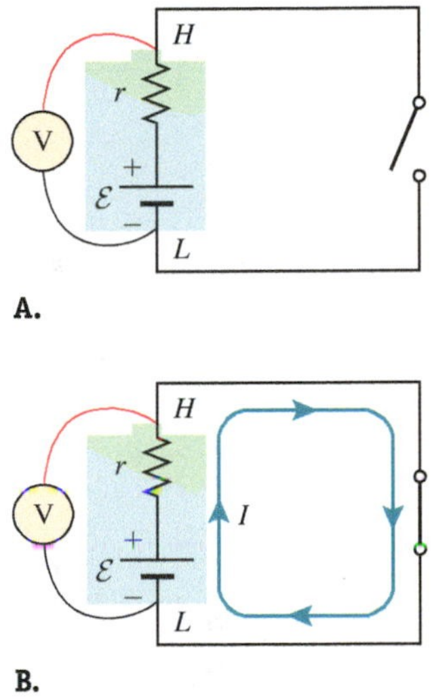

FIGURE 29.3 A real emf device is modeled as an ideal emf in series with a resistor. **A.** The terminal voltage across a real emf device is higher when there is no current in the device. **B.** The terminal voltage is lower across a real emf device when a current is present.

When the switch is closed (Fig 29.3B), current I is present in the circuit and therefore in the battery. Now the voltage drop across the internal resistor is given by $\Delta V_r = -Ir$. The terminal potential is the sum of the voltage across the ideal emf device and the voltage drop across the internal resistor. Therefore, the emf rule for a real emf device is

$$\Delta V_{\text{real emf}} = \Delta V_{\text{terminal}} = \mathcal{E} - Ir \qquad (29.3)$$

REAL EMF RULE **Special Case**

Unless otherwise specified, all emf devices in this textbook are assumed to be *ideal*, obeying $\Delta V_{\text{terminal}} = \mathcal{E}$ (Eq. 29.2).

Grounded Circuits

Only *differences* in potential are physically meaningful (Section 26-2). The rules given above for finding expected voltages are used to find the potential *difference* across a circuit element. Some circuits are **grounded**, meaning that a wire is connected from some point in the circuit to the Earth (or another large object such as the frame of a car). A point in a circuit connected to ground is at zero potential. The electric potential of all other points in the circuit is then measured with respect to the grounded point. Table 29.2 includes the ground symbol.

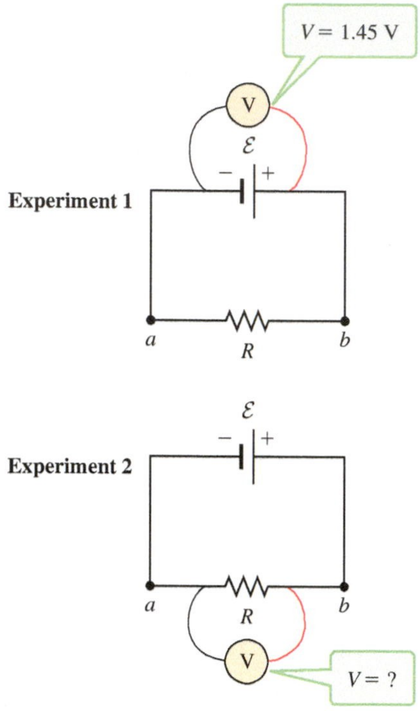

FIGURE 29.4

Because only potential differences are physically important, a grounded circuit behaves in the same way as an ungrounded circuit. Circuits are grounded for safety reasons. The circuit in most household appliances is enclosed in some sort of insulating case. Under normal conditions, the case remains neutral, but if the circuit malfunctions, it is possible for the case of an ungrounded appliance to become charged. Then, if a person touches the charged case, he or she provides the excess charge with a pathway to the ground and the person could be shocked or even electrocuted. If the circuit is grounded, the case remains neutral.

CONCEPT EXERCISE 29.1

What are the SI units of $\mathcal{E}$?

CONCEPT EXERCISE 29.2

An emf device is connected to a resistor as shown in Figure 29.4. In experiment 1, a student measures the terminal potential of the emf device and finds that it is 1.45 V. In experiment 2, the student moves the voltmeter to points *a* and *b*. What will the voltmeter read? Will the reading be positive or negative? Explain.

CASE STUDY Helping Around the House

Avi, Cameron, and Shannon are staying at Cameron's house for spring break. Cameron's mom (Carol) wants them to help fix a couple of things: (1) The hot water heater is not working properly. The hot water isn't very warm, and the supply quickly runs out when anyone tries to take a shower or bath. (2) Carol recently replaced a headlight in her antique car. Now both headlights work, but the old headlight is dimmer than the new one.

EXAMPLE 29.1 CASE STUDY What's Wrong with the Water Heater?

Carol's water heater has two large resistors. When the circuit is operating correctly, the thermal energy of these resistors heats the water primarily by conduction. The emf device is the wall outlet (an AC device), but we can treat it as a DC emf device with a terminal potential of roughly 240 V. In trying to diagnose the water heater's problems, Avi, Cameron, and Shannon discuss a simpler circuit—a single resistor connected to a 240-V battery.

Avi: If you measure the voltage across the battery, it's 240 V. If the resistor is working, you should find that the voltage drop across it is −240 V.

Cameron: Yes, but I think that's also what you would find if the resistor is cracked. So, measuring the voltage across the resistor won't help us find out whether it's dead.

Shannon: If the resistor is cracked, there's no current in it. No current means no voltage.

Decide which student's statements are true and which are false. Explain your answer.

SOLVE

Avi is correct. If the resistor is intact, the voltage drop across the resistor should equal the (negative) terminal potential of the emf device.

Cameron is correct. A crack in the resistor is like an open switch. In this case, the voltage drop across the broken resistor is 240 V. (The absolute value of the voltage is the same whether the resistor is intact or broken.)

Shannon is incorrect. No current does not necessarily mean no voltage. Consider voltage across the open switch in Figure 29.2D. As long as the emf is working, the voltage across the switch is nonzero.

EXAMPLE 29.2 Ideal and Real Batteries

A An ideal 12.0-V battery is connected to a 135-Ω resistor (Fig. 29.5). Find the current in the circuit, the power delivered by the battery, and the power consumed by the resistor.

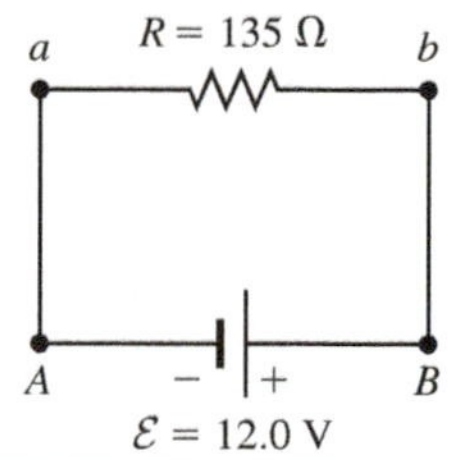

FIGURE 29.5

INTERPRET and ANTICIPATE

We expect to find that the power supplied by the battery equals the power consumed by the resistor (Section 28-7). Always indicate the direction of the current on the schematic diagram; the current in this circuit is counterclockwise (from the emf's positive terminal through the circuit to its negative terminal; Fig. 29.6).

To find the required quantities, we first need to know the voltage across the resistor and across the emf device. So, imagine connecting the red and black leads of the voltmeter so that the voltmeter measures $\Delta V = V_{\text{red}} - V_{\text{black}}$.

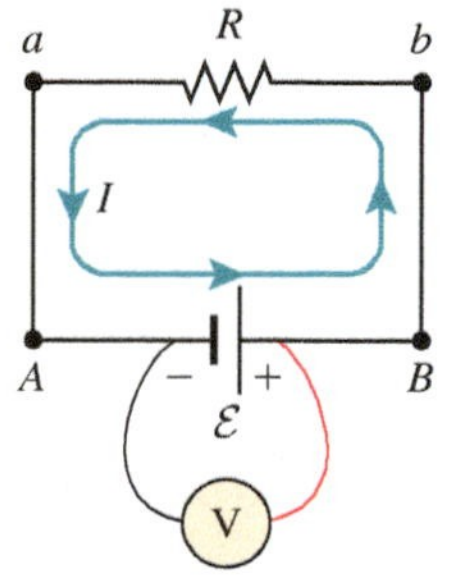

FIGURE 29.6

SOLVE

Only a wire connects point a to A, and another wire connects point b to B. Because the voltage across a wire is zero, point a is at the same potential as point A, and point b is at the same potential as point B. If you connect the red lead of the voltmeter to point B and the black lead to point A, you get the same reading as if you had connected the red lead of the voltmeter to point b and the black lead to point a.

The terminal voltage is equal to the voltage across the resistor.	$\Delta V_{\text{terminal}} = \Delta V_R$ (1)
Because this is an ideal battery, the terminal voltage is equal to the emf $\mathcal{E}$.	$\Delta V_{\text{terminal}} = \mathcal{E}$ (2)
The current is from the red lead to the black lead. So, according to the resistor rule, the voltage across the resistor is $+IR$.	$\Delta V_R = IR$ (3)
Substitute Equations (2) and (3) into Equation (1), and solve for I.	$\mathcal{E} = IR$ $I = \frac{\mathcal{E}}{R} = \frac{12.0\text{ V}}{135\ \Omega} = 8.89 \times 10^{-2}\text{ A} = 88.9\text{ mA}$
The power delivered by the battery is given by Equation 28.31.	$P_{\text{bat}} = I\mathcal{E}$ (28.31) $P_{\text{bat}} = (8.89 \times 10^{-2}\text{ A})(12.0\text{ V}) = 1.07\text{ W}$
The power used by the resistor is given by Equation 28.34.	$P_{\text{res}} = I^2R$ (28.34) $P_{\text{res}} = (8.89 \times 10^{-2}\text{ A})^2(135\ \Omega) = 1.07\text{ W}$

CHECK and THINK

As expected, the power supplied by the battery is equal to the power consumed by the resistor. In fact, this is a result of the conservation of energy. When a particle passes through the battery from the negative to the positive terminal, the electric potential energy of the system increases. When the particle moves through the resistor, it collides with the molecules, increasing the system's thermal energy. The increase in electric potential energy is equal to the increase in thermal energy. In the next section, we present a convenient rule—known as *Kirchhoff's loop rule*—to express the conservation of energy principle for circuits.

Example continues on page 902 ▶

B Now, a real battery is connected to a 135-Ω resistor (Fig. 29.7). The real battery can be modeled as an ideal emf device with $\mathcal{E} = 12.0$ V in series with a resistor with $r = 1.0\ \Omega$. Find the current in the circuit, the terminal potential of the real battery, the power consumed by the 135-Ω resistor, and the power consumed by the internal resistance of the battery.

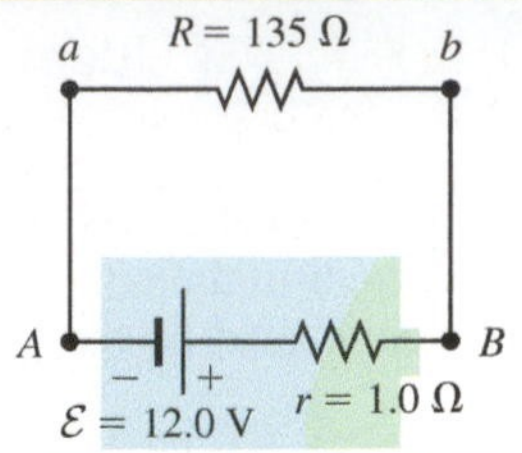

FIGURE 29.7

INTERPRET and ANTICIPATE

Follow a procedure similar to that in part A. Because this circuit has a higher resistance than the ideal emf circuit (given the added internal resistance of the emf), we expect that the current is smaller and the terminal potential is lower than 12.0 V. We also expect that the power consumed by the 135-Ω resistor plus the power consumed by the battery's internal resistance equals the power supplied by the ideal battery in part A. Figure 29.8 shows the current's direction.

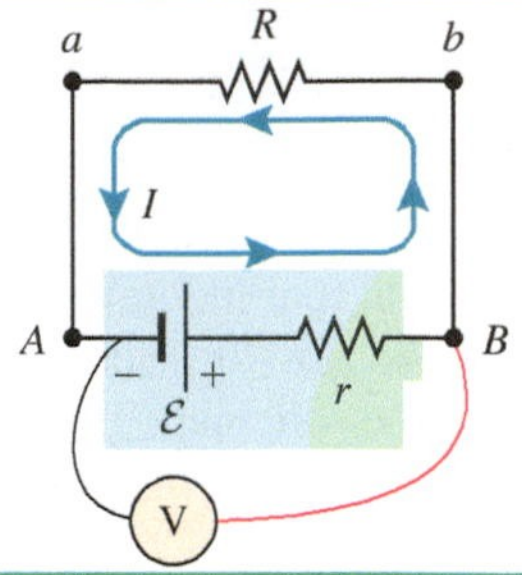

FIGURE 29.8

SOLVE

As in part A, the red lead is attached to point b or B, and the black lead is attached to point a or A. The potential difference across the resistor is $+IR$ because the current points from the red lead to the black lead.

$$\Delta V_R = IR \tag{4}$$

To find the terminal potential of the real emf, add the potential $\mathcal{E}$ across the ideal emf to the voltage drop across its internal resistance r. According to the resistor rule, the voltage across the internal resistor is negative because here the current is from the black lead to the red lead. (Notice that our result is the same as Eq. 29.3.)

$$\Delta V_{\text{terminal}} = \mathcal{E} - Ir \tag{29.3}$$

As in part A, the terminal voltage equals the voltage across the 135-Ω resistor. Substitute Equations (4) and 29.3 into $\Delta V_{\text{terminal}} = \Delta V_R$ (Eq. 1).

$$\mathcal{E} - Ir = IR$$

$$\mathcal{E} = I(R + r)$$

$$I = \frac{\mathcal{E}}{R + r} = \frac{12.0\ \text{V}}{135\ \Omega + 1.0\ \Omega} = 8.82 \times 10^{-2}\ \text{A}$$

$$I = 88.2\ \text{mA}$$

CHECK and THINK

As expected, the current is smaller in the circuit that has a real battery than in the earlier circuit with an ideal battery.

SOLVE

To find the terminal potential, substitute values into Equation 29.3.

$$\Delta V_{\text{terminal}} = \mathcal{E} - Ir$$

$$\Delta V_{\text{terminal}} = 12.0\ \text{V} - (8.82 \times 10^{-2}\ \text{A})(1.0\ \Omega)$$

$$\Delta V_{\text{terminal}} = 11.9\ \text{V}$$

CHECK and THINK

The terminal potential is only slightly lower than that of an ideal emf device. For this reason, we usually treat all emf devices as ideal.

SOLVE

The power consumed by the 135-Ω resistor and by the internal resistance is given by Equation 28.34.

For the 135-Ω resistor,

$$P_{\text{res}} = I^2R \tag{28.34}$$

$$P_{\text{res}} = (8.82 \times 10^{-2}\ \text{A})^2(135\ \Omega) = 1.05\ \text{W}$$

For the internal resistance r,

$$P_{\text{int}} = I^2r \tag{28.34}$$

$$P_{\text{int}} = (8.82 \times 10^{-2}\ \text{A})^2(1.0\ \Omega)$$

$$P_{\text{int}} = 7.8 \times 10^{-3}\ \text{W} = 7.8\ \text{mW}$$

CHECK and THINK
As expected, the power consumed by the 135-Ω resistor plus the power consumed by the emf device's internal resistance (≈ 1.06 W) equals the power supplied by the ideal battery in part A: 1.07 W. (The slight discrepancy is due to rounding error.) Another reason we usually model a real emf device as ideal is that the power consumed by the internal resistance of the emf device is much lower than the power consumed by the external 135-Ω resistor.

29-2 Kirchhoff's Loop Rule

When the German physicist Gustav Robert Kirchhoff was still a student, he discovered two rules that are essential for circuit analysis. The first is **Kirchhoff's loop rule**: *The total change in potential around any closed loop in a circuit is always zero.* Kirchhoff's loop rule is based on the principle of conservation of energy; the rule is expressed in terms of potential energy per charge (electric potential), and so it is convenient to use in circuit analysis.

KIRCHHOFF'S LOOP RULE
Major Concept

Kirchhoff's loop rule may not hold when a changing magnetic field is present. (See Section 34-4.)

PROBLEM-SOLVING STRATEGY

Using Kirchhoff's Loop Rule

The loop rule is used to write an equation involving the potential difference across each circuit element in a closed loop.

INTERPRET and ANTICIPATE
Start with a schematic diagram to which you add two elements: (1) the **current** and (2) the **starting point**.

Element 1. Indicate the **current** as a curve or line drawn near the circuit elements, with an arrow indicating the direction.

a. If there is an open switch in the circuit, there will be no current.

b. If there is a current, choose a direction for it. If there is one emf device in the circuit, the current is away from the positive terminal through the loop and toward the negative terminal. The current continues through the emf from the negative terminal toward the positive one. If there is more than one emf device, the current is in the direction of the stronger emf.

c. Sometimes you will not be able to determine the direction of the current. In those cases, make your best guess. If you are wrong, when you solve for current, you will get a negative value.

Element 2. Choose a convenient **starting point** and indicate it with a dot. The key to using the loop rule is to imagine measuring the voltage across each circuit element in the loop. By choosing and labeling a starting point, you know when you have considered the complete loop.

SOLVE
Step 1 Write expressions for the expected **voltage across each circuit element.** Imagine using a voltmeter to measure the voltage across each circuit element, always keeping the voltmeter's leads oriented the same way as you imagine moving it around the loop. The resistor rule and the emf rule in Section 29-1 can be summarized: (1) Measuring in the direction of the current results in a voltage drop across the resistor $-IR$. (2) Measuring from the negative to the positive terminal of the emf results in a positive voltage $+\mathcal{E}$. (Of course, measurements made in the opposite direction result in the opposite signs.)

Step 2 **Set the sum of the potential differences to zero.** Suppose the first voltage measurement is taken with the black lead at the starting point. When the red lead is at the starting point, you have measured the potential differences around the closed loop, and according to Kirchhoff's loop rule, the sum of the measurements must be zero.

EXAMPLE 29.3 **Two Emf Devices, a Resistor, and a Bulb**

The circuit shown in Figure 29.9 consists of two emf devices, a resistor, and a bulb. Use Kirchhoff's loop rule to show that

$$\mathcal{E}_1 - IR_1 - \mathcal{E}_2 - IR_2 = 0$$

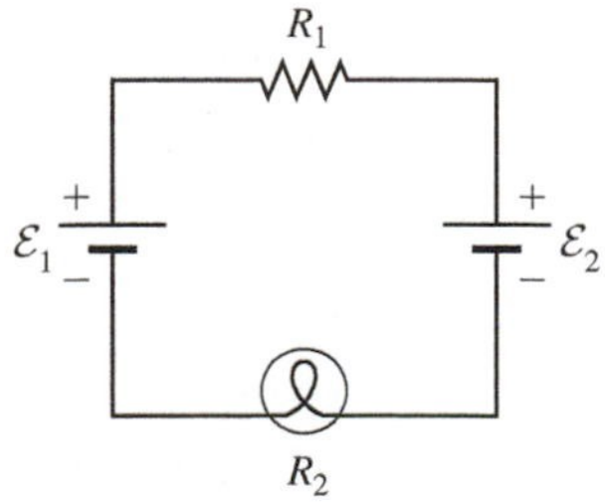

FIGURE 29.9

Example continues on page 904 ▶

INTERPRET and ANTICIPATE

Start with a schematic diagram (Fig. 29.10). Because there are two emf devices and we don't know their relative strengths, we don't know the direction of the **current**. We arbitrarily choose clockwise for the current and the bottom left corner as the **starting point**.

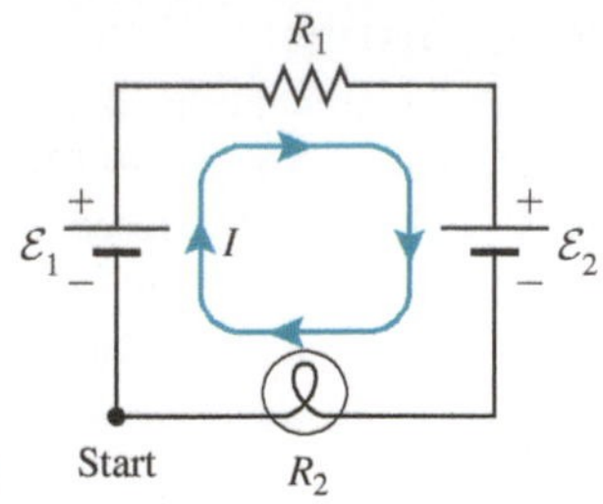

FIGURE 29.10

SOLVE

Step 1 Write expressions for the **voltage across each circuit element.** Imagine that a voltmeter is moved clockwise around the loop, with the red lead in front of the black lead at all times. (In practice, you do not need to redraw the circuit diagram as we have done here. Instead, use your finger or a pencil to represent the voltmeter as you imagine measuring the voltage around the closed loop.)

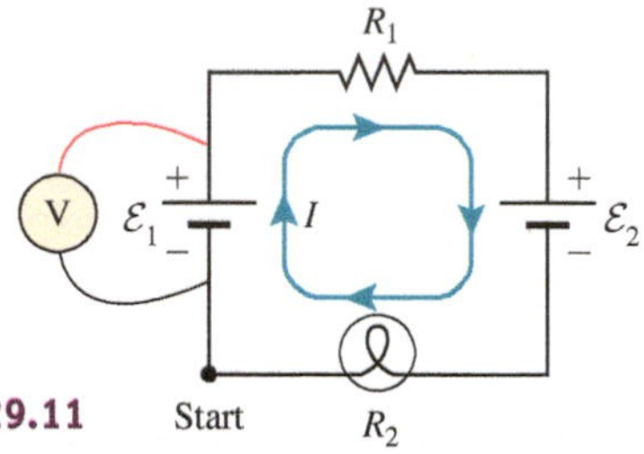

FIGURE 29.11

The emf rule across emf device 1 gives a positive potential difference because the red lead is on the positive terminal and the black lead is on the negative terminal (Fig. 29.11). We are measuring from the negative to the positive terminal, and we find

$$\Delta V_{\mathcal{E}1} = V_{\text{red}} - V_{\text{black}} = +\mathcal{E}_1 \tag{1}$$

The resistor rule for R_1 gives a negative potential difference because the current is from the black to the red lead (Fig. 29.12). Measuring in the direction of the current, we have

$$\Delta V_{R1} = V_{\text{red}} - V_{\text{black}} = -IR_1 \tag{2}$$

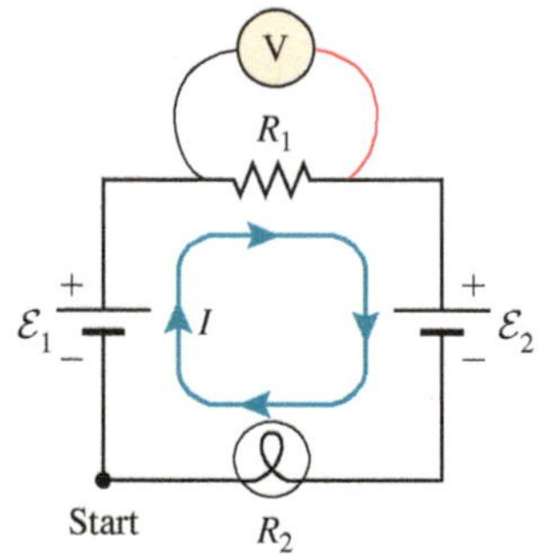

FIGURE 29.12

The voltage across emf device 2 is negative because now we are measuring from the positive to the negative terminal (Fig. 29.13):

$$\Delta V_{\mathcal{E}2} = V_{\text{red}} - V_{\text{black}} = -\mathcal{E}_2 \tag{3}$$

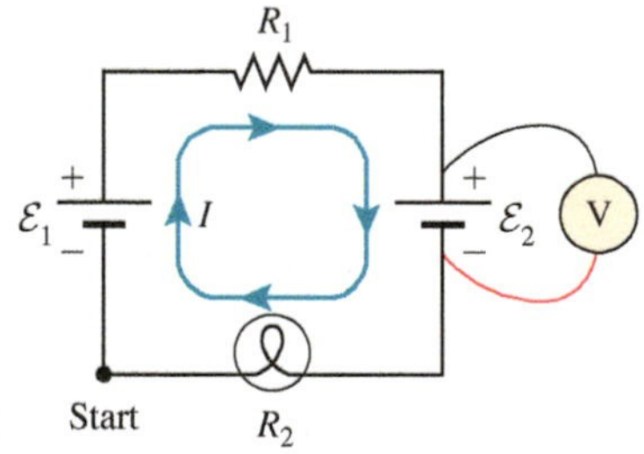

FIGURE 29.13

Finally, the resistor rule gives a negative voltage across lightbulb R_2 because we are measuring in the direction of the current (Fig. 29.14):

$$\Delta V_{R2} = V_{\text{red}} - V_{\text{black}} = -IR_2 \tag{4}$$

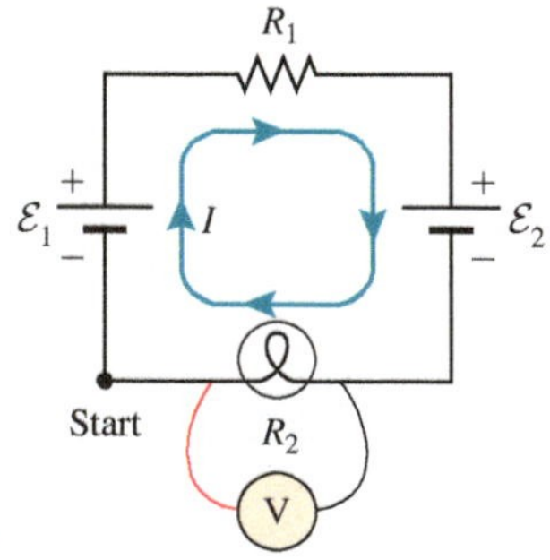

FIGURE 29.14

Step 2 Sum of potential differences is zero. According to Kirchhoff's loop rule, the sum of the four measurements (Eqs. 1–4) must be zero.

$$\Delta V_{\mathcal{E}1} + \Delta V_{R1} + \Delta V_{\mathcal{E}2} + \Delta V_{R2} = 0$$

$$\mathcal{E}_1 - IR_1 - \mathcal{E}_2 - IR_2 = 0 \checkmark$$

CHECK and THINK

This is exactly what we expected to find.

CONCEPT EXERCISE 29.3

You may wonder why anyone would ever build the circuit in Figure 29.9 with two emfs in opposite directions. One possibility is that the circuit is designed to recharge a battery. Suppose emf device 2 is a rechargeable battery and emf device 1 is another battery such as your car's battery. Resistor R_1 is in the circuit to regulate the amount of current. So, why is there a lightbulb? Lightbulbs are often used in circuits as indicators. (Does your phone have an indicator light to tell you whether it is plugged in or powered by its internal battery?) If a bulb is off, there is no current. A dim bulb indicates a small current; a bright bulb indicates a large current.

a. Suppose battery 2 starts off with a very low terminal potential and then its terminal potential gradually increases. Describe the brightness of the bulb during this process.

b. The term *recharge* is misleading. Why? What is a better term?

EXAMPLE 29.4 CASE STUDY Headlights

When Carol replaces a burnt-out headlight in her antique car, the remaining old headlight is dimmer than the new one. Avi and Shannon have learned that the old headlight has a higher resistance than the new headlight. According to their research, the old headlight has operating resistance $R_{\text{old}} = 3.2\ \Omega$, and the new headlight has operating resistance $R_{\text{new}} = 1.6\ \Omega$. Avi believes both bulbs are connected in series with the car's 12.0-V battery (Fig. 29.15).

A Find the current in each bulb and the power consumed by each bulb when the switch is closed.

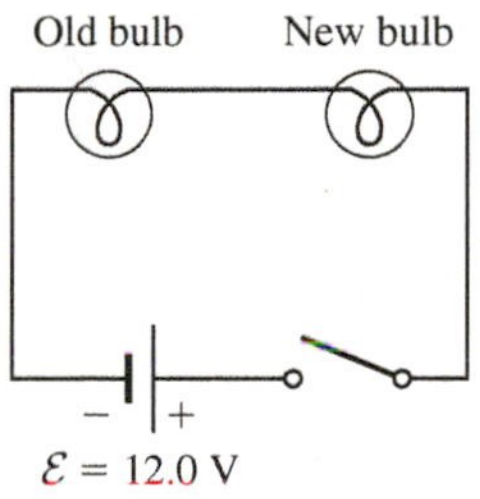

FIGURE 29.15 Avi's proposed headlight circuit.

INTERPRET and ANTICIPATE

The circuit consists of only one loop, so we expect the current to be the same through each circuit element. The power consumed by each bulb depends on the current and its resistance. We expect that the bulb with the higher resistance (the old bulb) will consume more power. To find the current, we will use Kirchhoff's loop rule.

Sketch the circuit when the switch is closed (Fig. 29.16). There is only one emf device, so it is easy to find the **current's** direction—counterclockwise as indicated. The bottom left corner is our (arbitrary) **starting point**.

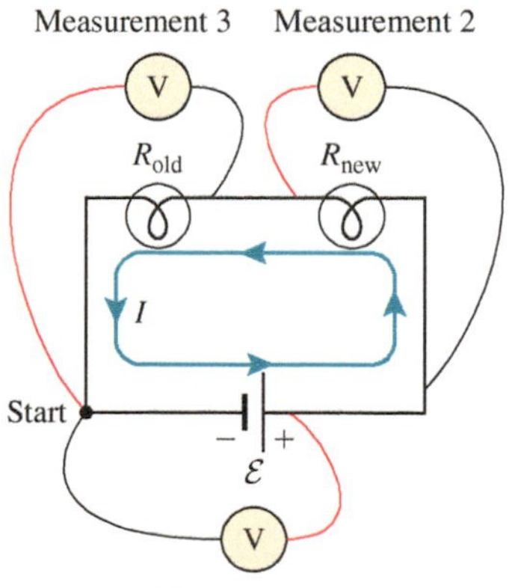

FIGURE 29.16

SOLVE

Step 1 Write expressions for the **voltage across each circuit element.** We arbitrarily decided to keep the red lead behind the black one, so the current is from black to red. For the three circuit elements, we imagine making three voltage measurements as indicated in Figure 29.16. (Drawing the voltmeter is not necessary.)

Measurement 1. The voltage across the emf is positive.	$\Delta V_{\mathcal{E}} = +\mathcal{E}$
Measurements 2 and 3. The voltage across each bulb is negative because the measurement is in the direction of the current (from black to red).	$\Delta V_{\text{new}} = -IR_{\text{new}}$ $\Delta V_{\text{old}} = -IR_{\text{old}}$

Example continues on page 906 ▶

Step 2 Sum of potential differences is zero. In the third measurement, the red lead is at the starting point, so we have measured all the potential differences. Applying Kirchhoff's loop rule, we find the sum of the potential differences is zero.	$\Delta V_{\mathcal{E}} + \Delta V_{\text{new}} + \Delta V_{\text{old}} = 0$ $\mathcal{E} - IR_{\text{new}} - IR_{\text{old}} = 0 \quad (1)$
Solve Equation (1) for current.	$\mathcal{E} = IR_{\text{new}} + IR_{\text{old}}$ $I = \frac{\mathcal{E}}{R_{\text{new}} + R_{\text{old}}} = \frac{12.0\text{ V}}{1.6\,\Omega + 3.2\,\Omega}$ $I = 2.5\text{ A}$
The power consumed by each bulb is given by Equation 28.34.	For the new bulb, $P_{\text{new}} = I^2 R_{\text{new}} \quad (28.34)$ $P_{\text{new}} = (2.5\text{ A})^2(1.6\,\Omega) = 10\text{ W}$ For the old bulb, $P_{\text{old}} = I^2 R_{\text{old}} \quad (28.34)$ $P_{\text{old}} = (2.5\text{ A})^2(3.2\,\Omega) = 20\text{ W}$

:• CHECK and THINK

As expected, the current is the same in all the circuit elements and the power consumed by the old bulb is greater than the power consumed by the new bulb because the old bulb has a higher resistance. In fact, the old bulb consumes about twice as much power as the new bulb. That doesn't fit Carol's observation that the old bulb is dimmer than the new bulb. Because the old bulb consumes more power, we would expect it to give off more light than the new bulb. There is likely a problem with Avi's circuit diagram. In the next part, we will evaluate this diagram.

B If Avi is correct about the bulbs being connected in series, what would happen to the good bulb when one bulb burns out?

:• SOLVE

When we say "a bulb burns out," what we really mean is that the filament breaks. A broken filament is like an open switch; there can be no current through either. We would expect both bulbs to be off; neither headlight would light up when one of them burns out. Of course, one headlight often burns out on a car while the other bulb remains lit. Therefore, the bulbs must not be connected to the battery in series as Avi thought.

29-3 Resistors in Series

The function of a resistor in a circuit is to regulate the amount of current. Manufacturers make a wide range of resistances, but often you must combine a number of standard resistors to come up with the resistance required. As we know from our work with capacitors, we may need to connect a number of resistors in series or parallel to come up with the desired resistance. For example, the three resistors in series in Figure 29.17A are equivalent to the one resistor in Figure 29.17B, and these two circuits are said to be *equivalent* because when the same emf device is used, both have the same current.

We use Kirchhoff's loop rule (and the problem-solving strategy on page 903) for each circuit in Figure 29.17 to come up with an expression for the equivalent resistance R_{eq} in terms of the three resistances R_1, R_2, and R_3. In both circuits, there is only one emf device, and the current is clockwise in both. Start in the bottom left corner of each circuit, and imagine measuring the potential across each circuit element clockwise around the loop. Apply the emf rule and the resistor rule.

After completing the loop, set the sum of the potential differences to zero. For the circuit in Figure 29.17A,

$$\mathcal{E} - IR_1 - IR_2 - IR_3 = 0$$

$$\mathcal{E} = I(R_1 + R_2 + R_3) \quad (29.4)$$

For the circuit in Figure 29.17B,

$$\mathcal{E} - IR_{eq} = 0$$

$$\mathcal{E} = IR_{eq} \quad (29.5)$$

Comparing Equation 29.5 to Equation 29.4, we find that

$$R_{eq} = R_1 + R_2 + R_3$$

We generalize our results for a circuit with any number of resistors in series. The equivalent resistance R_{eq} of N resistors connected in series is

RESISTORS IN SERIES
Major Concept

$$R_{eq} = R_1 + R_2 + R_3 + \cdots + R_N = \sum_{i=1}^{N} R_i \quad (29.6)$$

Compare Equation 29.6 to $C_{eq} = \sum_{i=1}^{N} C_i$ (Eq. 27.8) for capacitors in parallel; they are mathematically the same. *Resistors in series* combine like *capacitors in parallel.*

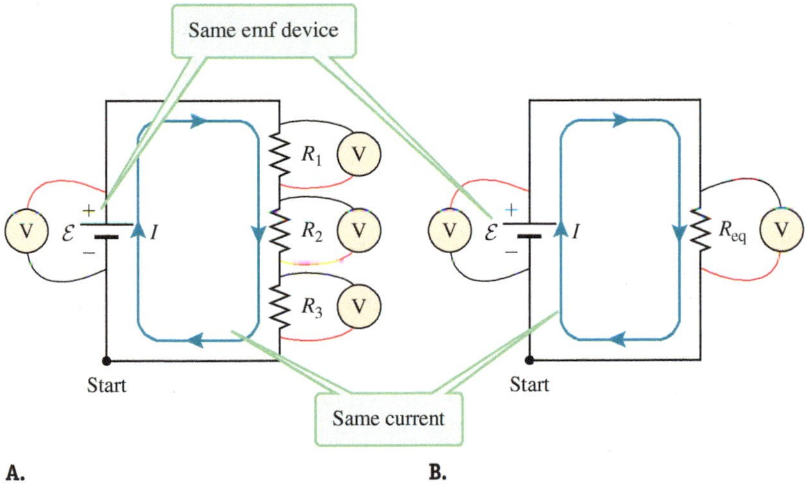

FIGURE 29.17 **A.** Three resistors connected in series. **B.** An equivalent circuit.

CONCEPT EXERCISE 29.4

Use the gravitational analogy from Section 28-4 (page 875) to explain why Equation 29.6 makes sense.

EXAMPLE 29.5 CASE STUDY Another Look at Avi's Circuit

When you analyze a circuit that involves resistors in series, it may help if you imagine the equivalent circuit. Find the equivalent resistance in Avi's circuit diagram (Fig. 29.15) for the car's headlights, and then find the current. In the **CHECK and THINK** step, compare your result to the current found in Example 29.4.

INTERPRET and ANTICIPATE

The total resistance should be higher than the resistance of either bulb, and the current we find here should equal the current we found earlier.

Example continues on page 908 ▶

SOLVE

Find the equivalent resistance by adding the resistance of each bulb (Eq. 29.6).

$$R_{eq} = \sum_{i=1}^{N} R_i \qquad (29.6)$$

$$R_{eq} = R_{old} + R_{new} = 3.2\ \Omega + 1.6\ \Omega$$

$$R_{eq} = 4.8\ \Omega$$

The current is given by Equation 29.5.

$$\mathcal{E} = IR_{eq} \qquad (29.5)$$

$$I = \frac{\mathcal{E}}{R_{eq}} = \frac{12.0\ \text{V}}{4.8\ \Omega} = 2.5\ \text{A}$$

CHECK and THINK

As expected, the equivalent resistance is higher than that of either bulb, and the current is exactly what we found in Example 29.4. When a circuit involves a large number of resistors in series, it may simplify the problem to draw a schematic diagram for the equivalent circuit and then apply Kirchhoff's loop rule to the equivalent circuit.

29-4 Kirchhoff's Junction Rule

Consider a charged particle moving in a circuit that consists of a single loop (Fig. 29.17). This charged particle must pass through each circuit element because there is nowhere else for it to go. So, the number of charged particles that pass through each circuit element per unit time is constant, and the same current exists in each circuit element.

Now consider the circuit in Figure 29.18 made up of three branches and two junctions. A **junction** is a place where three or more wires meet. A **branch** is a part of a circuit in between junctions. Starting at point a, a charged particle moves through "branch 0" to point b, which is a junction. Just as when a driver encounters a junction in a roadway, the particle can go into either branch 1 or branch 2.

In this circuit, the current is different in each branch. Because charged particles are not created or destroyed by any circuit element, and because no particle leaks out of the circuit or flows into it, all the current going *into* a junction must equal all the current going *out*. This is a loose statement of **Kirchhoff's junction rule**: *At any junction, the sum of the all the currents entering the junction equals the sum of all the currents exiting the junction.*

KIRCHHOFF'S JUNCTION RULE

Major Concept

Consider the junction at point b (Fig. 29.18). The current I_0 from branch 0 enters the junction. The sum of the currents exiting that junction is the current I_1 in branch 1 plus the current I_2 in branch 2. According to Kirchhoff's junction rule,

$$I_0 = I_1 + I_2$$

If we consider the junction at a instead of the junction at b, we find the same equation. The sum of the currents entering the junction at a is the current I_1 in branch 1 plus the current I_2 in branch 2, and the current leaving the junction is the current in branch 0:

$$I_1 + I_2 = I_0$$

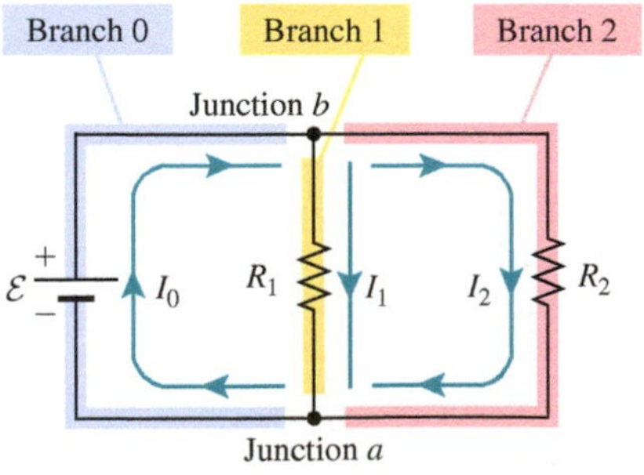

FIGURE 29.18 Kirchhoff's junction rule gives the same result when applied to the junction at a or the junction at b.

CONCEPT EXERCISE 29.5

Imagine connecting a voltmeter to points a and b in Figure 29.18 so that the voltmeter reads the potential difference $\Delta V = V_b - V_a$. *Hint*: The red lead is at point b.

a. Use the emf rule to write an expression for ΔV in terms of the emf $\mathcal{E}$.
b. Use the resistor rule to write an expression for ΔV in terms of the resistances R_1 and R_2.

CONCEPT EXERCISE 29.6

If $I_0 = 3.0$ A and $R_1 = R_2$ in Figure 29.18, what are I_1 and I_2? *Hint*: Use your answer to Concept Exercise 29.5, part (b).

29-5 Resistors in Parallel

If you try to drink a milkshake with only one straw, you find it difficult to get a satisfying amount into your mouth with each sip. The best way to drink a milkshake is with two straws. Think of it this way: Each straw resists the motion of the milkshake, but two straws used in parallel decrease the overall resistance to the milkshake flow. What if you use the two straws in series—by forcing the end of one straw into the end of the other? You would find it even more difficult to drink the milkshake with two straws in series than if you used only one straw, because resistances in series add (Eq. 29.6).

Drinking a milkshake with several straws is analogous to charged particles flowing through several resistors. If the resistors are in series (Fig. 29.17), the equivalent resistance R_{eq} is higher than the resistance of any one resistor. If the resistors are in parallel (Fig. 29.19), the equivalent resistance R_{eq} is lower than the resistance of any one of the resistors.

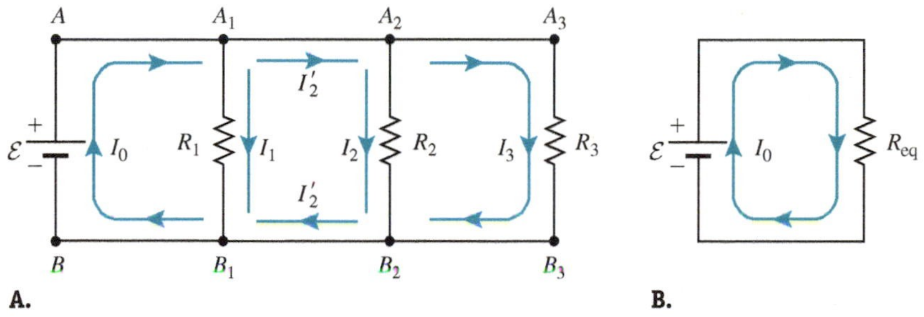

FIGURE 29.19 **A.** Three resistors connected in parallel. **B.** An equivalent circuit.

DERIVATION Resistors In Parallel

We will show that the equivalent resistance for resistors in parallel is

$$\frac{1}{R_{eq}} = \sum_{i=1}^{N} \frac{1}{R_i} = \frac{1}{R_1} + \frac{1}{R_2} + \frac{1}{R_3} + \cdots + \frac{1}{R_N} \quad (29.7)$$

RESISTORS IN PARALLEL

Major Concept

To make this derivation concrete, consider three parallel resistors and find an expression for the equivalent resistor (Fig. 29.19). Both circuits have the same emf $\mathcal{E}$ and the same amount of current I_0 in the emf device.

The circuit in Figure 29.19A is made up of three loops that involve the emf device. Let's use the loop rule for each.

For the clockwise loop from B to A to A_1 to B_1 and back to B:	$\mathcal{E} - I_1R_1 = 0$ $\mathcal{E} = I_1R_1$ (1)
For the clockwise loop from B to A to A_2 to B_2 and back to B:	$\mathcal{E} - I_2R_2 = 0$ $\mathcal{E} = I_2R_2$ (2)
For the clockwise loop from B to A to A_3 to B_3 and back to B:	$\mathcal{E} - I_3R_3 = 0$ $\mathcal{E} = I_3R_3$ (3)

Derivation continues on page 910 ▶

The current splits twice—once at junction A_1 and once at junction A_2. Let's apply the junction rule at these two points.

The current entering junction A_1 is I_0. The current leaving this junction is I_1 plus I'_2. (We get the same result if we apply the junction rule at junction B_1.)	$I_0 = I_1 + I'_2$	(4)
The current entering junction A_2 is I'_2. The current leaving this junction is I_2 plus I_3. (We get the same result if we apply the junction rule at junction B_2.)	$I'_2 = I_2 + I_3$	(5)
Combine these two applications of the junction rule (Eqs. 4 and 5).	$I_0 = I_1 + I_2 + I_3$	(6)
Substitute Equations (1) through (3) into Equation (6).	$I_0 = \frac{\mathcal{E}}{R_1} + \frac{\mathcal{E}}{R_2} + \frac{\mathcal{E}}{R_3}$ $I_0 = \mathcal{E}\left(\frac{1}{R_1} + \frac{1}{R_2} + \frac{1}{R_3}\right)$	(7)
Now apply the loop rule to the equivalent circuit in Figure 29.19B.	$\mathcal{E} - I_0 R_{eq} = 0$ $\mathcal{E} = I_0 R_{eq}$	
Solve for I_0.	$I_0 = \frac{\mathcal{E}}{R_{eq}}$	(8)
Compare Equations (8) and (7) to find an expression for the equivalent resistance R_{eq} in terms of the individual resistances R_1, R_2, and R_3.	$\frac{1}{R_{eq}} = \frac{1}{R_1} + \frac{1}{R_2} + \frac{1}{R_3}$	(9)
We derived Equation (9) using three resistors in parallel, and we can extend that equation for any number N of resistors in parallel.	$\frac{1}{R_{eq}} = \sum_{i=1}^{N} \frac{1}{R_i}$ ✓	(29.7)

COMMENTS

As expected from our analogy with drinking straws, Equation 29.7 shows that the equivalent resistance R_{eq} is lower than the resistance of the individual resistors: $R_{eq} < R_1$ and $R_{eq} < R_2$ and … and $R_{eq} < R_N$. Compare Equation 29.7 for resistors in parallel to

$$\frac{1}{C_{eq}} = \sum_{i=1}^{N} \frac{1}{C_i} \tag{27.7}$$

for capacitors in series; these equations are mathematically identical. *Resistors in parallel* combine like *capacitors in series.*

CONCEPT EXERCISE 29.7

When you solve for R_{eq} using Equation 29.7, the quantity you need is in the denominator, so you must be sure to invert both sides of the equation. It is common to forget this step. To help avoid this mistake, show that the equivalent resistance of a network consisting of two resistors in parallel is

$$R_{eq} = \frac{R_1 R_2}{R_1 + R_2} \tag{29.8}$$

EXAMPLE 29.6 CASE STUDY Current Through a Car's Battery

Avi, Cameron, and Shannon have decided that a car's headlights must be connected in parallel (Fig. 29.20).

A Find the equivalent resistance. From Example 29.4, $R_{old} = 3.2\ \Omega$ and $R_{new} = 1.6\ \Omega$.

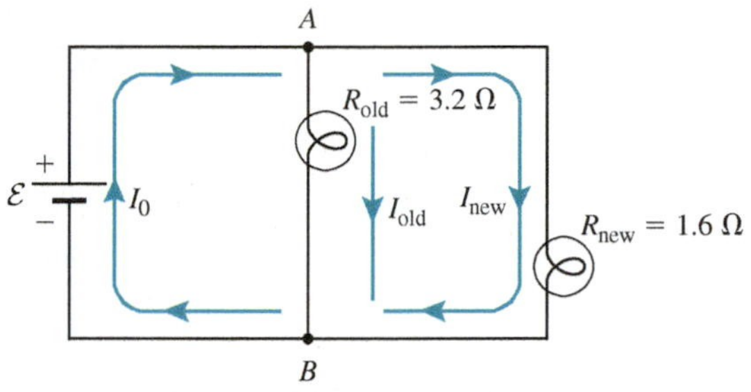

FIGURE 29.20 Parallel circuit for car headlights.

INTERPRET and ANTICIPATE

Sketch the equivalent circuit (Fig. 29.21). Because the bulbs are in parallel, we expect the equivalent resistance to be lower than the resistance of either bulb: $R_{eq} < R_{old}$ and $R_{eq} < R_{new}$.

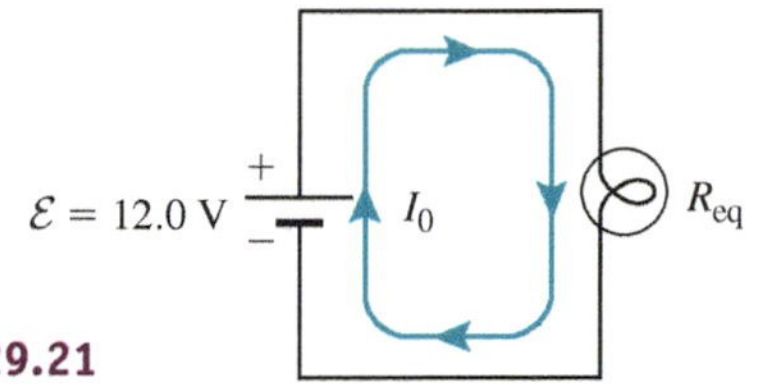

FIGURE 29.21

SOLVE

Because there are only two bulbs, apply Equation 29.8 to find the equivalent resistance.

$$R_{eq} = \frac{R_{old}R_{new}}{R_{old} + R_{new}} \tag{29.8}$$

$$R_{eq} = \frac{(3.2\ \Omega)(1.6\ \Omega)}{3.2\ \Omega + 1.6\ \Omega} = 1.07\ \Omega$$

$$R_{eq} = 1.1\ \Omega$$

CHECK and THINK

As expected, the equivalent resistance is lower than that of either bulb.

B Use your result from part A to find the current I_0 through the car's 12-V battery. (In the **CHECK and THINK** step, consider this fact: In a car's headlight circuit, there is a 15-A fuse. If the current reaches 15 A, the fuse breaks. Just like an open switch, current cannot pass through a broken fuse.)

INTERPRET and ANTICIPATE

Now that we have the equivalent resistance, we can consider the simple equivalent circuit (Fig. 29.21).

SOLVE

Use the loop rule.

$$\mathcal{E} - I_0R_{eq} = 0 \qquad I_0 = \frac{\mathcal{E}}{R_{eq}}$$

Substitute values.

$$I_0 = \frac{12.0\ \text{V}}{1.07\ \Omega} = 11.2\ \text{A} \approx 11\ \text{A}$$

CHECK and THINK

This seems like a reasonable value for the current because it is somewhat smaller than the maximum current (15 A) that can pass through the fuse. Fuses protect circuit elements from overheating. Recall from Section 28-7 that when there is current in a circuit element with resistance, that circuit element gets hot. For this reason, circuit designers include mechanisms for transferring heat to the environment. For example, your computer has a fan to help transfer heat through forced convection. Many circuits include a fuse to further protect the circuit. Fuses are rated by the maximum current they can endure. If the current reaches that value, the fuse breaks and acts as an open switch, cutting off current.

29-6 Circuit Analysis

At the beginning of this chapter, we imagined using a multimeter to troubleshoot a failed electric circuit. The multimeter can measure potential difference, current, resistance, and capacitance. Circuit analysis is the process of determining the expected values that you can measure with a multimeter. The first five sections of this chapter as well as Section 27-4 introduced rules for circuit analysis. The rules fall into three broad categories:

1. **Voltage rules** for finding the potential difference across various circuit elements (Section 29-1)
2. **Kirchhoff's rules** for writing equations (Sections 29-2 and 29-4)
3. **Capacitance and resistance combination rules** for simplifying a circuit that contains more than one capacitor, resistor, or lightbulb:

$$\frac{1}{C_{eq}} = \sum_{i=1}^{N} \frac{1}{C_i} \text{ (Eq. 27.7) and } R_{eq} = \sum_{i=1}^{N} R_i \text{ (Eq. 29.6) for } \textbf{series}$$

$$C_{eq} = \sum_{i=1}^{N} C_i \text{ (Eq. 27.8) and } \frac{1}{R_{eq}} = \sum_{i=1}^{N} \frac{1}{R_i} \text{ (Eq. 29.7) for } \textbf{parallel}$$

Now we bring these rules together in a problem-solving strategy for circuit analysis in the case of complicated circuits involving multiple branches.

PROBLEM-SOLVING STRATEGY

Circuit Analysis

INTERPRET and ANTICIPATE

As in the problem-solving strategy "Using Kirchhoff's Loop Rule" (page 903), start with a schematic diagram (Fig. 29.22) that includes these elements: (1) arrows and labels for the **currents**, (2) labels for the **junctions** and the **circuit elements** (such as resistors, capacitors, and emf devices), and (3) a dot for the **starting point** (if you plan to use Kirchhoff's loop rule).

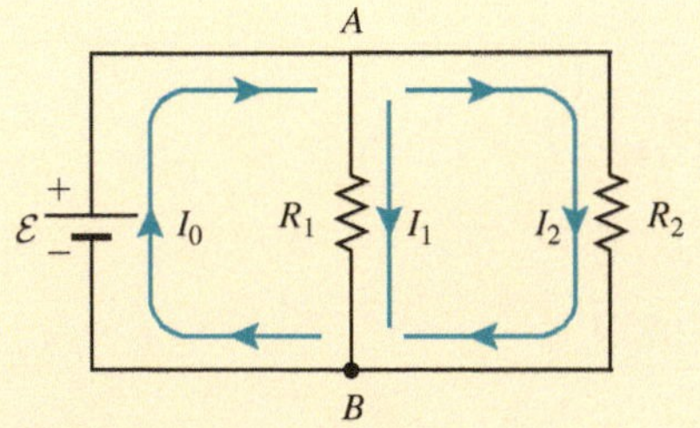

FIGURE 29.22 This schematic diagram has all three elements: (1) The three current directions are indicated and labeled I_0, I_1, and I_2. (2) The two junctions are labeled A and B, the two resistors are labeled R_1 and R_2, and the emf device is labeled $\mathcal{E}$. (3) The starting point is indicated by a dot at junction B.

SOLVE

There are three possible tactics for you to try depending on the situation. In many cases, you will not use all three tactics.

Tactic 1. Find an **equivalent circuit** and draw its schematic diagram. When a circuit involves more than one type of circuit element, finding an equivalent circuit is sometimes helpful and is particularly useful when you need to find the current in one branch of the circuit. One of the hardest tasks in finding the equivalent circuit is determining which elements are in series and which are in parallel. Use these facts to make your decision.

Elements in series (Fig. 27.16, page 838)

1. are connected by a single wire and
2. have the same current in each element.

Elements in parallel (Fig. 27.20, page 840)

1. are connected by two wires and
2. have the same voltage across each element.

Tactic 2. Apply **Kirchhoff's junction rule.** If the circuit has more than one loop, the junction rule may be used to relate the current in the various branches. Often you need to apply the junction rule to only half the junctions. For example, applying the junction rule to either junction A or B in Figure 29.22 gives the same relationship.

Tactic 3. Apply **Kirchhoff's loop rule** using the problem-solving strategy on page 903. You do not need to apply Kirchhoff's loop rule to every closed loop because the equations are mathematically dependent.

EXAMPLE 29.7 Resistors in Series and in Parallel

When there is current in the circuit in Figure 29.23, all the resistors have $R = 144\ \Omega$ and the two emf devices have the same terminal potential $\mathcal{E} = 1.48$ V. Find the current in either emf device.

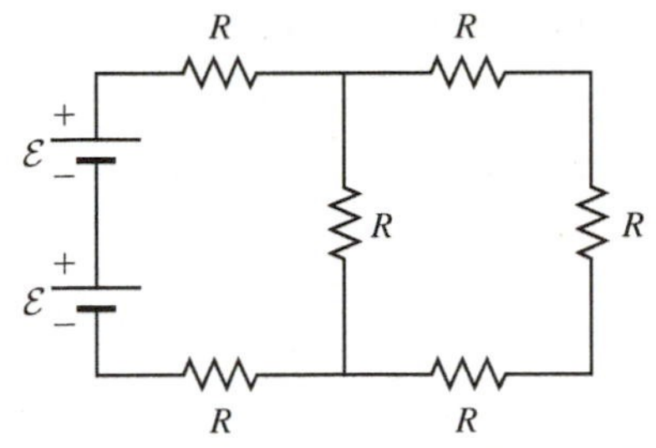

FIGURE 29.23

INTERPRET and ANTICIPATE

Draw a schematic diagram (Fig. 29.24), indicating the **currents** and labeling the **junctions** and **circuit elements**. At this time, there is no need to indicate a **starting point** because we don't plan to apply Kirchhoff's loop rule (yet).

The emf devices are in series with the two resistors adjacent to them, so the current is the same (I_0) in all four of those circuit elements. At junction A, the current splits into two branches.

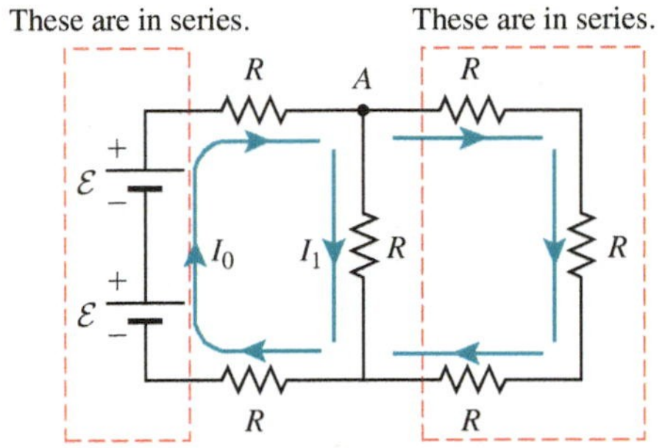

FIGURE 29.24

SOLVE

Tactic 1. Find an **equivalent circuit** and draw its schematic diagram (Fig. 29.25). It is difficult to find the equivalent circuit in just one step because of the large number of resistors and emf devices. In such a case, consider a small number of resistors at a time and find several intermediate equivalent circuits.

Because the emf devices are in series (Fig. 29.24), we can replace them with an equivalent emf device that has an emf of $2\mathcal{E}$. The three resistors in the branch farthest from the emf devices are connected by a single wire between each pair and have the same current (Fig. 29.24). Therefore, these resistors are in series and have an equivalent resistance of $3R$ (Eq. 29.6).

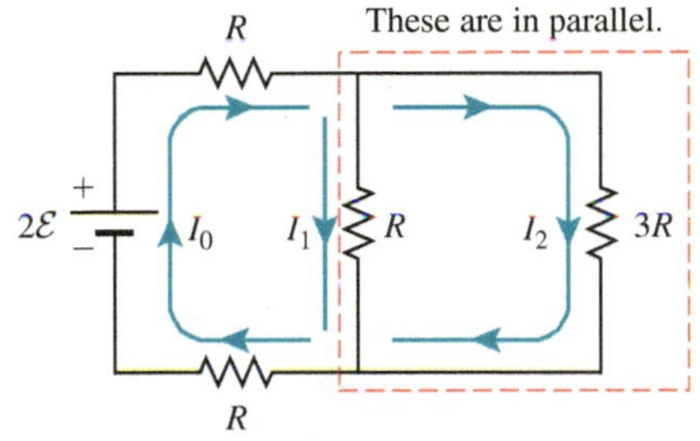

FIGURE 29.25

The two resistors in Figure 29.25 that are farthest from the equivalent emf device are connected by two wires and have the same potential difference across them. Therefore, these resistors are in parallel. Use Equation 29.8 to find their equivalent resistance, $3R/4$, and draw the new equivalent circuit (Fig. 29.26).

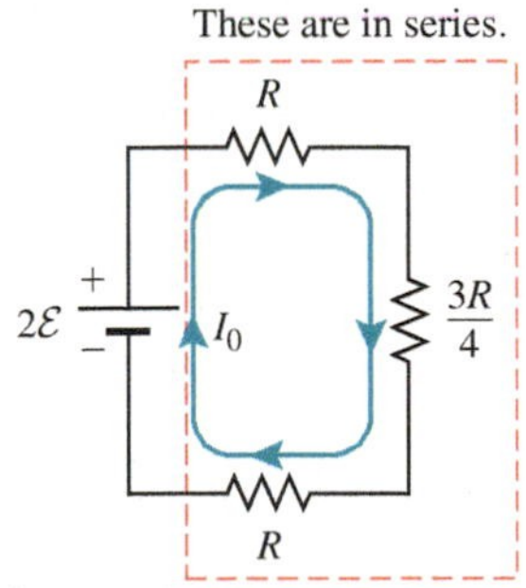

FIGURE 29.26

CHECK and THINK

Check the dimensions at intermediate steps in a complicated circuit analysis, especially if the circuit involves resistors in parallel. The equivalent resistance we found is a number times R (correct). Forgetting to invert the result in the earlier step leaves R in the denominator (wrong dimensions).

SOLVE

The three resistors in Figure 29.26 are connected by a single wire between each pair, and they have the same current. Therefore, these three resistors are in series and have an equivalent resistance of $11R/4$ (Eq. 29.6). The circuit in Figure 29.27 is equivalent to the one in Figure 29.23.

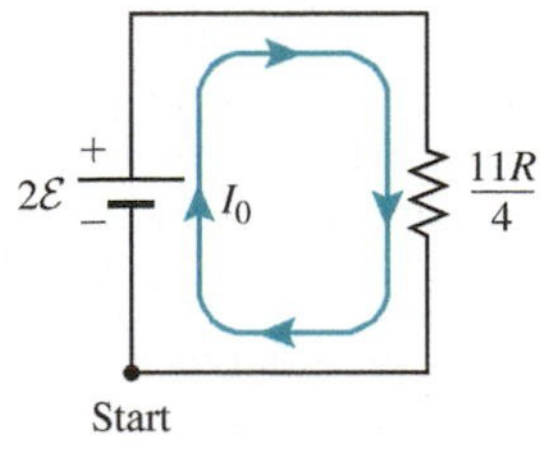

FIGURE 29.27

Example continues on page 914 ▶

Tactic 3. Apply **Kirchhoff's loop rule** to the equivalent circuit. Use the starting point indicated (Fig. 29.27), and imagine taking voltage measurements in the direction of the current.

Loop rule:

$$2\mathcal{E} - I_0\left(\frac{11R}{4}\right) = 0$$

Solve for I_0.

$$I_0 = 2\mathcal{E}\left(\frac{4}{11R}\right)$$

$$I_0 = 2(1.48\,\text{V})\left[\frac{4}{11(144\,\Omega)}\right]$$

$$I_0 = 7.47 \times 10^{-3}\,\text{A}$$

CHECK and THINK

As one final dimensional check, the SI units are correct because $1\ \text{V}/\Omega = 1\ \text{A}$.

EXAMPLE 29.8 Can't Be Simplified

Four ideal emf devices and three resistors are connected as in Figure 29.28. The emf devices are identical (terminal potential $\mathcal{E} = 3.22$ V). The resistors have resistances $R_1 = R$, $R_2 = 2R$, and $R_3 = 3R$, where $R = 57.4\ \Omega$. Find the current through each resistor.

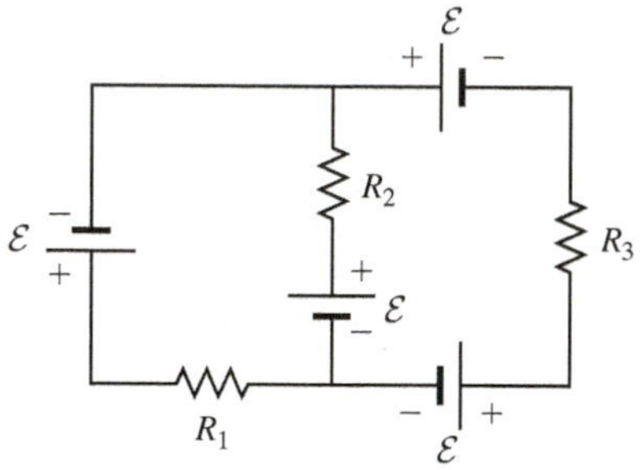

FIGURE 29.28

INTERPRET and ANTICIPATE

None of the resistors in this circuit are in series or in parallel, so we can't use Tactic 1 (page 912). We must apply Kirchhoff's rules and the voltage rules to find the current in each resistor. We'll be able to check at the end that the current in each resistor is different and the voltage across each resistor is different.

Draw a schematic diagram (Fig. 29.29), indicating the **currents** and labeling the **junction** and **circuit elements**. At this time, there is no need to indicate a **starting point**. With this many emf devices, the direction of the current in each branch is not obvious. Make arbitrary choices; if a choice is incorrect, you will end up with a negative value when you solve for that current. (There is no need to label both junctions because we only need to apply the junction rule to one.)

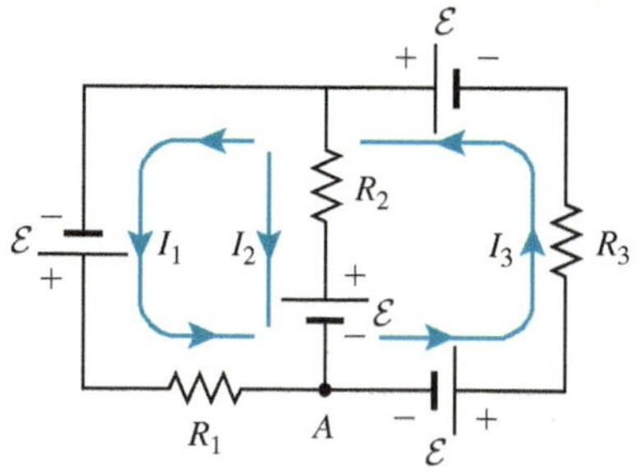

FIGURE 29.29

SOLVE

Tactic 2. Apply **Kirchhoff's junction rule** to junction A.

$$I_1 + I_2 = I_3 \quad (1)$$

Tactic 3. Apply **Kirchhoff's loop rule**. The circuit has three loops (left, right, and outer). Apply the loop rule to any two of the three loops. (If you apply the loop rule to all three loops, you will not gain extra information.) We arbitrarily choose the left and right loops.

Left loop. Start in the bottom left corner and measure counterclockwise around the left loop (Fig. 29.30).

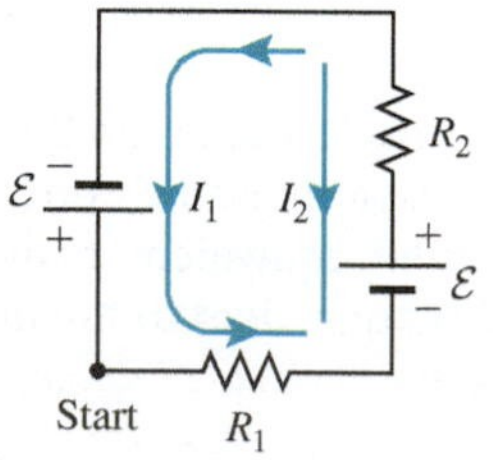

FIGURE 29.30

The voltage across R_1 is negative (a voltage drop), and the voltage across R_2 is positive. The voltage across each emf device is positive.	$-I_1R_1 + \mathcal{E} + I_2R_2 + \mathcal{E} = 0$	(2)

Right loop. Start in the bottom right corner and measure clockwise around the right loop, applying the voltage rules (Fig. 29.31).

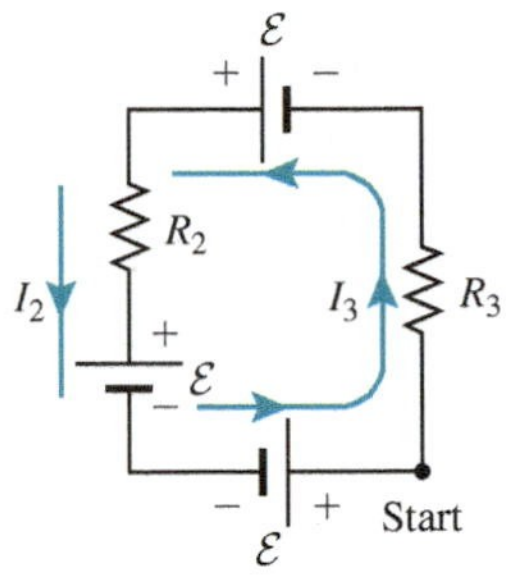

FIGURE 29.31

The voltage across both resistors is positive. The voltage across one emf device is positive and across the other two is negative.	$-\mathcal{E} + \mathcal{E} + I_2R_2 - \mathcal{E} + I_3R_3 = 0$	(3)

Algebra is all that is left. There are three unknowns—the currents I_1, I_2, and I_3—and three independent equations (1, 2, and 3). There are several ways to solve three equations simultaneously. We'll use the substitution method, showing only a few key steps.

Use Equation (1) to eliminate I_1 from Equation (2), and simplify using $R_1 = R$ and $R_2 = 2R$.	$-(I_3 - I_2)R_1 + I_2R_2 = -2\mathcal{E}$ $3I_2R - I_3R = -2\mathcal{E}$	(4)
Solve Equation (3) for I_3 and simplify using $R_2 = 2R$ and $R_3 = 3R$.	$I_2R_2 + I_3R_3 = \mathcal{E}$ $I_3 = \dfrac{\mathcal{E} - I_2R_2}{R_3} = \dfrac{\mathcal{E} - 2I_2R}{3R}$	(5)
Substitute Equation (5) for I_3 into Equation (4), solving for I_2.	$3I_2R - \left(\dfrac{\mathcal{E} - 2I_2R}{3R}\right)R = -2\mathcal{E}$ $I_2\left(\dfrac{11R}{3}\right) = -\dfrac{5\mathcal{E}}{3}$ $I_2 = -\dfrac{5\mathcal{E}}{11R} = -\dfrac{5(3.22\text{ V})}{11(57.4\ \Omega)} = -2.54 \times 10^{-2}\text{ A}$ $I_2 = -25.4\text{ mA}$	
Now that we have a value for I_2, substitute it into Equation (5) to find I_3.	$I_3 = \dfrac{3.22\text{ V} - (-2.54 \times 10^{-2}\text{A})(2)(57.4\ \Omega)}{(3)(57.4\ \Omega)}$ $I_3 = 3.56 \times 10^{-2}\text{A} = 35.6\text{ mA}$	
To find I_1, substitute I_2 and I_3 into Equation (1).	$I_1 = I_3 - I_2 = 35.6\text{ mA} - (-25.4\text{ mA})$ $I_1 = 61.0\text{ mA}$	

CHECK and THINK

The values found for I_1 and I_3 are both positive, so the directions we chose for those currents are correct. However, the value for I_2 is negative, so the direction we picked for that current is incorrect. Instead of pointing downward, I_2 points upward. The currents I_1, I_2, and I_3 through the three resistors are different, which confirms that these resistors are *not* in series.

Now let's check that the resistors are not in parallel by finding the potential difference across each one. For the moment, we are not concerned about the sign of the voltage, so we will just use $\Delta V = |IR|$ for each resistor.

Example continues on page 916 ▶

Potential difference across R_1:	$\Delta V_1 = \lvert I_1 R_1 \rvert = (61.0 \times 10^{-3}\,\text{A})(57.4\,\Omega)$ $\Delta V_1 = 3.50\,\text{V}$
Potential difference across R_2:	$\Delta V_2 = \lvert I_2 R_2 \rvert = \lvert I_2(2R) \rvert$ $\Delta V_2 = \lvert(-25.4 \times 10^{-3}\,\text{A})(2)(57.4\,\Omega)\rvert$ $\Delta V_2 = 2.92\,\text{V}$
Potential difference across R_3:	$\Delta V_3 = \lvert I_3 R_3 \rvert = \lvert I_3(3R) \rvert$ $\Delta V_3 = (35.6 \times 10^{-3}\,\text{A})(3)(57.4\,\Omega) = 6.31\,\text{V}$

The potential differences across the resistors are not equal, so they cannot be in parallel.

EXAMPLE 29.9 CASE STUDY Why Is the Old Headlight So Dim?

Avi, Cameron, and Shannon now know that car headlights are connected in parallel. In Example 29.6, we found the equivalent circuit (Fig. 29.21) and the current in the car's battery ($I_0 = 11.2$ A). In the original circuit (Fig. 29.20), find the current in each bulb and the power each consumes.

INTERPRET and ANTICIPATE

There are two unknown currents, I_{old} and I_{new}, so we need two independent equations to solve for them. More than one solution method exists; here we use Kirchhoff's junction rule once and Kirchhoff's loop rule once. Because the new headlight is brighter, we expect it to consume more power than the old bulb. We can use Figure 29.20 as our complete schematic diagram.

SOLVE **Tactic 2.** Apply **Kirchhoff's junction rule**. There are two junctions (*A* and *B*) in this circuit. Apply the junction rule to either.	$I_0 = I_{old} + I_{new}$	(1)
Tactic 3. Apply **Kirchhoff's loop rule**. Start at *B* and measure clockwise around the right loop to come up with one equation that involves both currents I_{old} and I_{new}.	$I_{old}R_{old} - I_{new}R_{new} = 0$	(2)
CHECK and THINK Rewrite Equation (2). According to Equation (3), the voltage drop across the old bulb equals the voltage drop across the new bulb, as should be true for two bulbs in parallel.	$I_{old}R_{old} = I_{new}R_{new}$	(3)
SOLVE Solve Equation (3) for I_{old} and substitute into Equation (1).	$I_{old} = \dfrac{I_{new}R_{new}}{R_{old}}$ $I_0 = \dfrac{I_{new}R_{new}}{R_{old}} + I_{new} = I_{new}\left(\dfrac{R_{new} + R_{old}}{R_{old}}\right)$	
Solve for I_{new}.	$I_{new} = I_0\left(\dfrac{R_{old}}{R_{new} + R_{old}}\right)$ $I_{new} = (11.2\,\text{A})\left(\dfrac{3.2\,\Omega}{1.6\,\Omega + 3.2\,\Omega}\right) = 7.47\,\text{A}$	(4)

Find I_{old} from Equation (1).	$I_{old} = I_0 - I_{new} = 11.2\text{ A} - 7.47\text{ A} = 3.73\text{ A}$
Find the power consumed by each bulb from Equation 28.34.	For the new bulb, $P_{new} = I_{new}^2 R_{new} = (7.47\text{ A})^2(1.6\ \Omega) = 90\text{ W}$ For the old bulb, $P_{old} = I_{old}^2 R_{old} = (3.73\text{ A})^2(3.2\ \Omega) = 45\text{ W}$

CHECK and THINK

This fits Carol's observation. The old bulb appears dimmer because it consumes less power than the new bulb. To fix the problem, use two bulbs with the same operating resistance.

We can apply our experience with this circuit to similar circuits. This circuit (Fig. 29.20) is sometimes called a current splitter because at junction A the current splits into two branches. If the two bulbs have the same resistance, the current through each bulb is half the current passing through the battery: $I_{new} = I_{old} = I_0/2$. If the bulbs have unequal resistances, as they do in this case, the branch with the lower resistance has greater current. I_{new} is twice as much as I_{old} because the resistance of the new bulb is half that of the old bulb.

29-7 DC Multimeters

This chapter is about the practical application of the principles of electricity. In Section 29-6, we studied circuit analysis—a way to predict the measurements made by a DC multimeter. In this section, we learn how a multimeter is used and designed to measure current, voltage, and resistance. Learning about multimeters not only will help you build your laboratory skills; it will also help you better understand circuits. Many laboratories have a digital multimeter like the one in Figure 29.32A. We will describe an analog multimeter (Fig. 29.32B), however, because it is partially mechanical and so its principles of operation are simpler.

The essential element of an analog meter is a **galvanometer**, which consists of an armature (a coil of wire wrapped around an iron core) that is free to rotate between the poles of a permanent magnet (Fig. 29.33). When there is current in the coil of wire, the coil becomes an electromagnet (Chapter 30). The electromagnet inside the galvanometer acts like a compass needle, a small magnet that aligns itself with the Earth's magnetic field so that the needle points toward the North Pole. The electromagnet produced by the coil in the galvanometer aligns itself with the galvanometer's permanent magnet, causing the needle attached to the armature to be deflected. The greater the current in the coil, the stronger the electromagnet and the greater the

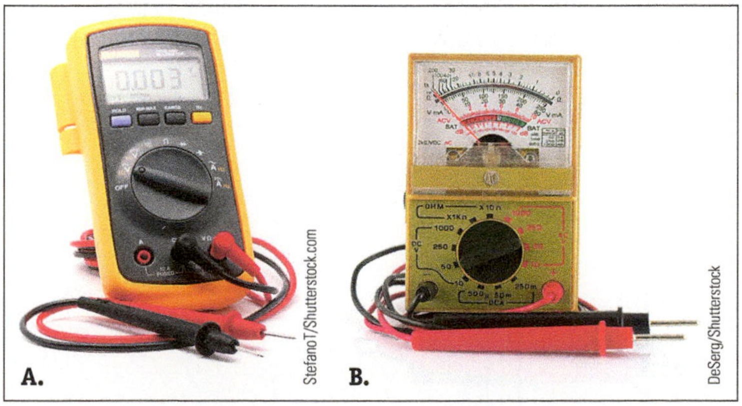

FIGURE 29.32 A multimeter is an ammeter, voltmeter, ohmmeter, and capacitance meter all in one. **A.** Digital multimeters are commonly found in student laboratories. **B.** Analog multimeters are easier to understand.

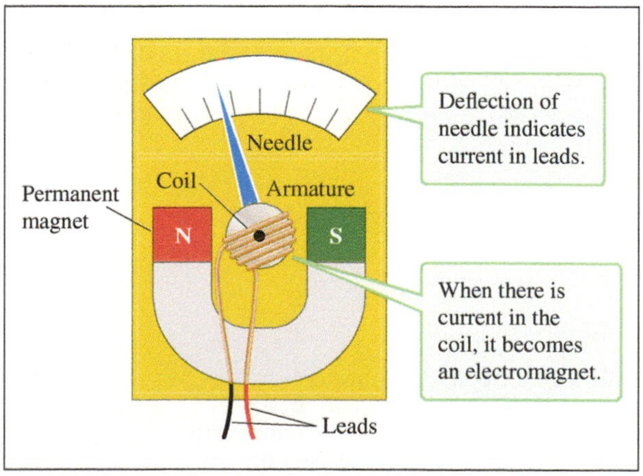

FIGURE 29.33 Current in the coil of the galvanometer causes it to become magnetic, and the coil rotates to align with the permanent magnet.

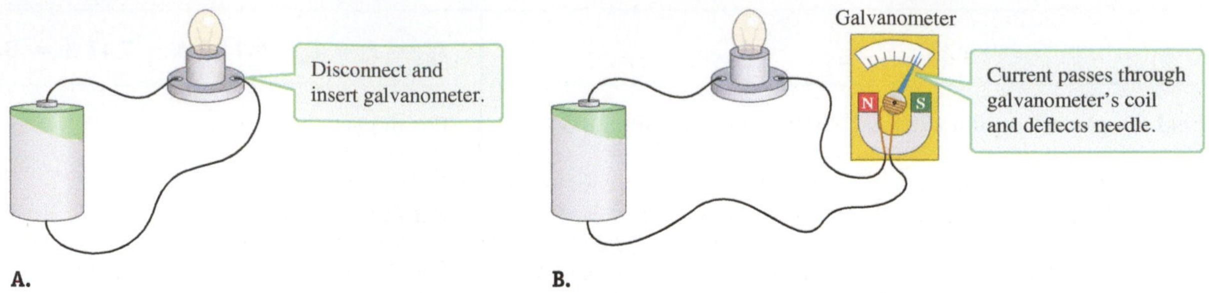

FIGURE 29.34 A. A circuit with a battery and a lightbulb. **B.** The circuit is broken so that a galvanometer can be inserted.

needle's deflection. The direction of deflection depends on the direction of the current in the coil. Thus, the deflection of the needle indicates the direction and amount of current in the coil. A spring attached to the armature (not shown in the figure) restores the needle to its original position when there is no current.

When you want to measure current in a circuit, you must break the circuit and insert the galvanometer in series (Fig. 29.34). In this textbook, the symbol for an ideal galvanometer is a circle with a **G** inside (Fig. 29.35A). An ideal galvanometer has no resistance, so it does not change the amount of current in a circuit when it is inserted. The coil of a real galvanometer contains many turns of wire, so it has an internal resistance *r.* We represent a real galvanometer by an ideal galvanometer in series with a resistor of resistance *r* (Fig. 29.35B).

In practice, a galvanometer can measure small currents up to several microamperes. If you try to use the galvanometer to measure a somewhat greater current, the needle will be deflected to its maximum position. Suppose the galvanometer you are using is designed to operate at currents up to 50 μA and you are trying to measure a current of 60 μA. The needle will deflect so that the meter reads 50 μA. You then know that the current is at least 50 μA, but it could be greater.

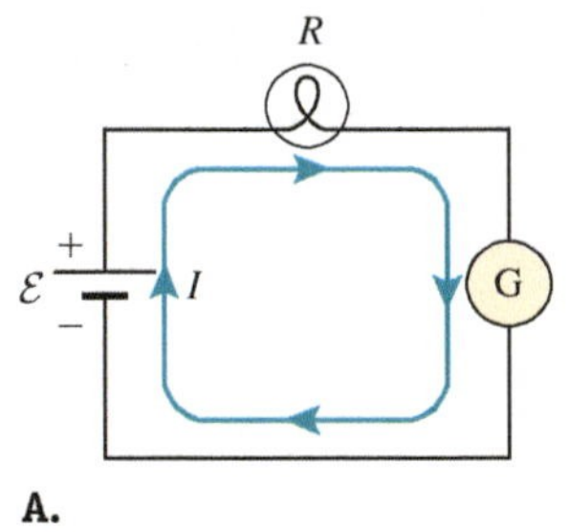

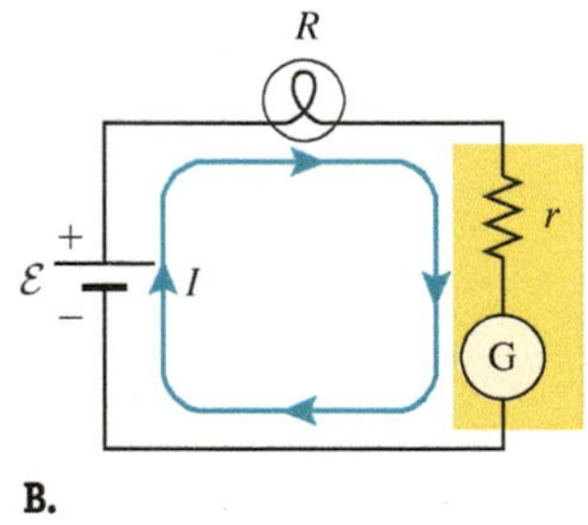

FIGURE 29.35 A. An ideal galvanometer measures current without altering the circuit. **B.** A real galvanometer has some internal resistance, so its presence alters the current. Figure 29.35B is a schematic diagram for the circuit shown in Figure 29.34B.

Ammeter Design

An **ammeter** is designed to measure a range of currents by placing a galvanometer in parallel with a **shunt resistor**. (*Shunt* is another word for "parallel.") The shunt resistor is inside the case of the multimeter, so you cannot easily see it. In Figure 29.36A, an ammeter is shown in a circuit that consists of a lightbulb and a battery. Figure 29.36B shows the same circuit with schematic detail of the inside of the ammeter. The shunt resistor has resistance R_{sh}. The galvanometer's coil has resistance r and is shown as a resistor in series with an ideal galvanometer. The shunt resistance is generally low compared to the internal resistance of the galvanometer, so only a small current I_G passes through the galvanometer and the needle is not deflected to its maximum position. The current through the shunt resistor is I_{sh}, and the current through the lightbulb and battery is $I = I_{sh} + I_G$. The designer knows the internal galvanometer resistance, and the user chooses the shunt resistance, thus determining what fraction of the total current I goes through the galvanometer so that the appropriate scale can be displayed on the meter's face.

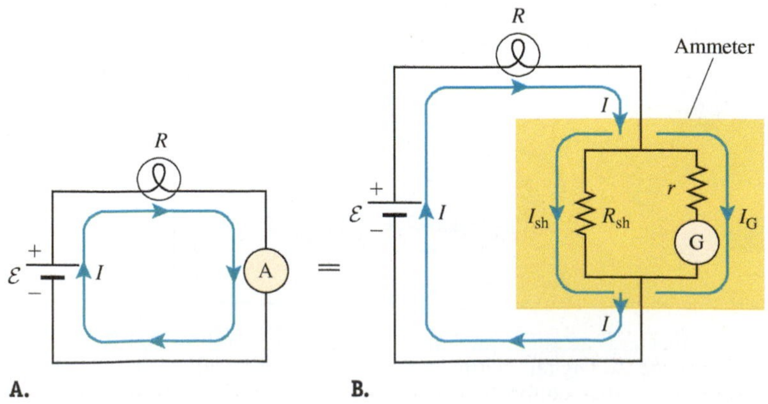

FIGURE 29.36 A. An ammeter is connected in series in a circuit. **B.** An ammeter is a galvanometer in parallel with a shunt resistor. The shunt resistor divides the current. When you use an ammeter, you select the value of the shunt resistor, which determines the range of measurable current.

Voltmeter Design

A galvanometer is also the basis for an analog **voltmeter**. Measuring voltage is often more practical than measuring current because using a voltmeter does not require you to break the circuit, as you must when using an ammeter. A voltmeter is placed in parallel with a circuit element in order to measure the potential difference across it (Fig. 29.37A). Because you do not want to disturb the circuit by deflecting much current through the voltmeter, the branch that contains the voltmeter should have a high resistance. A voltmeter is constructed by placing a galvanometer in series with a large resistor R_{ser} (Fig. 29.37B). The galvanometer measures the current I through R_{ser} and r. The voltage drop across them, $\Delta V = I(R_{ser} + r)$, equals the voltage across the circuit element—in this case, the lightbulb (and emf). The scale on the face of the meter is calibrated in volts, with that calculation "built in."

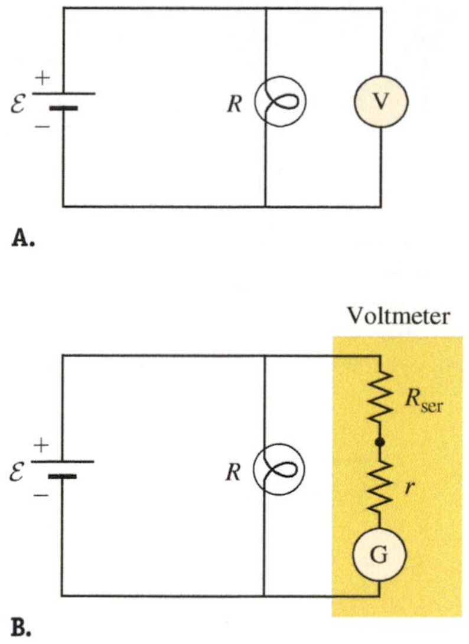

FIGURE 29.37 **A.** A voltmeter is connected in parallel with a circuit element. **B.** A voltmeter is a galvanometer in series with a large resistor.

Ohmmeter Design

An **ohmmeter** measures the resistance of a circuit element. To measure the resistance of an element such as the lightbulb in Figure 29.38A, connect the ohmmeter across the element (which should not be in a working circuit). The ohmmeter is made up of an emf device and an ammeter (Fig. 29.38B). The emf device inside the ohmmeter ensures that there is current through the circuit element. The ammeter measures that current. The current is inversely proportional to the resistance. The scale on the ohmmeter is calibrated in ohms, with that calculation "built in."

29-8 *RC* Circuits

So far, our focus has been on DC circuits in which the current does not change over time. There are many circumstances where a time-varying current is required, however. A camera's flashbulb requires a short burst of current. A car's turn signal, a patient's electronic pacemaker, and a car's intermittent windshield wipers require a current that repeatedly increases and decreases.

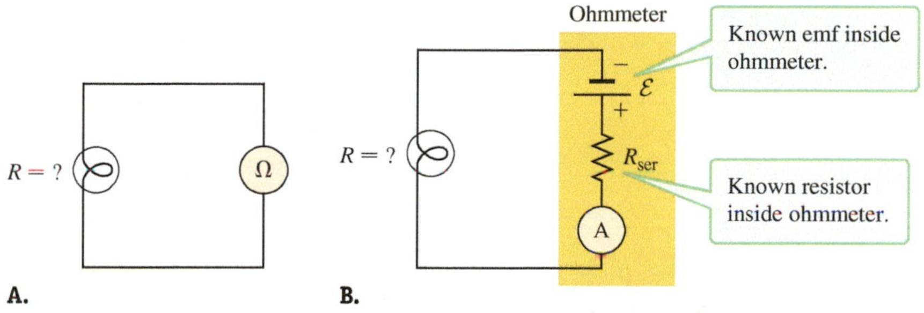

FIGURE 29.38 **A.** A lightbulb of unknown resistance is connected to an ohmmeter. **B.** An ohmmeter is an ammeter in series with an emf device and a resistor.

A circuit consisting of a resistor, a capacitor, and an emf (Fig. 29.39) can produce such time-varying currents and is called an ***RC* circuit**, where R stands for resistance and C for capacitance. The switch S has three possible positions; in Figure 29.39 it is shown open. When the switch is closed at point A, an emf device with terminal potential $\mathcal{E}$ is in the circuit. In this case, the capacitor is *charging*. When the switch is closed at point B, the emf device is not in the circuit and the capacitor is *discharging*.

We will use circuit analysis to find expressions for the current $I(t)$ in the circuit and the charge $q(t)$ stored by the capacitor, both of which are functions of time. We can imagine monitoring the current in the circuit by placing a voltmeter across the resistor. The current through the resistor is then

$$I(t) = \frac{\Delta V_R(t)}{R} \tag{28.25}$$

where $\Delta V_R(t)$ is the potential difference across the resistor as a function of time. We can monitor the charge stored by the capacitor with a second voltmeter placed across the capacitor. The charge stored by the capacitor is given by

$$q(t) = C\Delta V_C(t) \tag{27.1}$$

where $\Delta V_C(t)$ is the potential difference across the capacitor as a function of time.

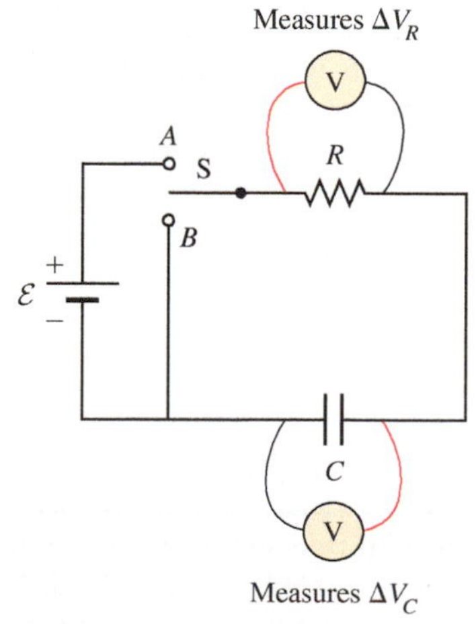

FIGURE 29.39 An *RC* circuit.

Charging a Capacitor

Initially, the capacitor is uncharged and the switch is open (Fig. 29.39). At time $t = 0$, the switch is closed at point A so that the circuit consists of a resistor, a capacitor, and the emf device (Fig. 29.40; the extraneous parts of the network are not

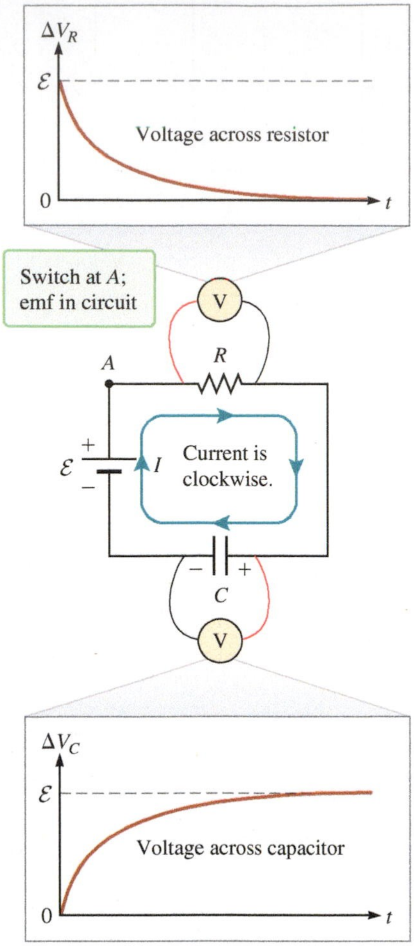

shown). The current in the circuit is initially clockwise as shown. Positive charge builds up on the right plate of the capacitor, and negative charge builds up on the left plate. Charge continues to flow until the voltage across the capacitor $\Delta V_C(t)$ equals the terminal potential $\mathcal{E}$ across the emf device.

A graph of the capacitor's voltage $\Delta V_C(t)$ versus t (lower portion of Fig. 29.40) shows: (1) Initially, the capacitor is uncharged, so the voltage is zero: $\Delta V_C(0) = 0$. (2) As the charge builds up, the measured potential difference across the capacitor is positive and increases. (3) As time approaches infinity, the capacitor reaches its maximum potential difference, the terminal potential $\mathcal{E}$ of the emf device. In practice, this maximum potential difference is reached after a long time.

Now consider the graph of the resistor's voltage $\Delta V_R(t)$ versus t (upper portion of Fig. 29.40). The voltage across the emf device must equal the sum of the voltage across the resistor and the voltage across the capacitor:

$$\mathcal{E} = \Delta V_R(t) + \Delta V_C(t) \tag{29.9}$$

The initial capacitor voltage $\Delta V_C(0) = 0$, so the voltage across the resistor equals the voltage across the emf device: $\Delta V_R(0) = \mathcal{E}$. As the voltage across the capacitor increases, the voltage across the resistor must decrease. When the voltage across the capacitor equals the voltage of the emf device ($\mathcal{E}$), the voltage across the resistor must be zero.

The graphs in Figure 29.40 give us a qualitative understanding of what happens when the capacitor is charging. The voltage across the capacitor increases as it stores more charge, and the voltage across the resistor drops as the current drops. Our goal now is to derive expressions for the charge stored by the capacitor and for the current in the circuit as functions of time.

FIGURE 29.40 A charging *RC* circuit. With the voltmeter's red lead connected to the capacitor's positive plate and the black lead connected to the negative plate, the measured potential difference across the capacitor is positive. A graph of the voltage across the capacitor as a function of time rises from zero to $\mathcal{E}$. The current is from the red lead toward the black lead, so the potential difference across the resistor is positive. The voltage across the resistor as a function of time drops from $\mathcal{E}$ to zero.

DERIVATION $q(t)$ for a Charging Capacitor

We show that the charge $q(t)$ stored by the charging capacitor (Fig. 29.40) as a function of time is given by

$$q(t) = C\mathcal{E}(1 - e^{-t/RC}) \tag{29.10}$$

CHARGE ON A CHARGING CAPACITOR
Special Case

Because the voltage across the capacitor is proportional to the charge it stores, we expect a graph of $q(t)$ versus t to look similar to the graph of $\Delta V_C(t)$ versus t.

We begin when the switch has been closed at point A for some time but the capacitor has not yet reached its maximum voltage. Use Kirchhoff's loop rule, starting in the lower left corner of the circuit and measuring clockwise in the direction of the current (Fig. 29.41). The voltmeter leads here are in the opposite orientation as those in Figure 29.40.

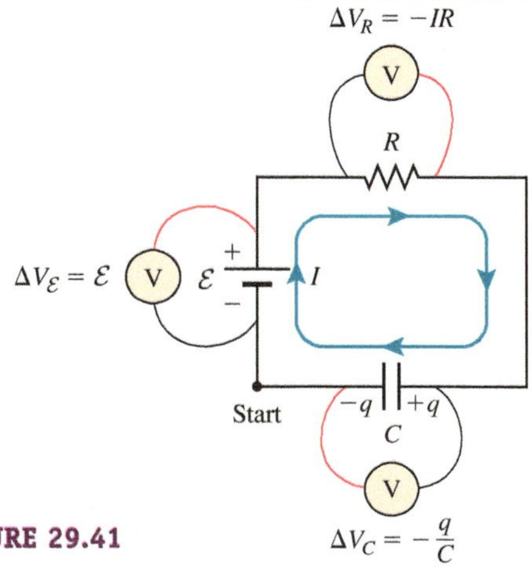

FIGURE 29.41

Use the emf rule and the resistor rule. The voltage across the capacitor is given by $q = C\Delta V_C$ (Eq. 27.1), and with the red lead connected to the negative plate of the capacitor, the measured potential difference is negative.

$$\Delta V_\mathcal{E} + \Delta V_R + \Delta V_C = 0$$

$$\mathcal{E} - IR - \frac{q}{C} = 0 \tag{29.11}$$

After substituting $I = dq/dt$ (Eq. 28.1) for the current in Equation 29.11, we are left with a differential equation.	$\mathcal{E} - \dfrac{dq}{dt}R - \dfrac{q}{C} = 0$ (2) $\dfrac{C\mathcal{E} - q}{RC} = \dfrac{dq}{dt}$ (3)
Isolate q on one side of the equation.	$\dfrac{dt}{RC} = \dfrac{dq}{C\mathcal{E} - q}$
Integrate the left side from $t = 0$ (when the switch was closed at point A) to some arbitrary time t later. Integrate the right side from $q = 0$ (because the capacitor is initially uncharged) to q (the charge stored at some arbitrary time t).	$\displaystyle\int_0^t \frac{dt}{RC} = \int_0^q \frac{dq}{C\mathcal{E} - q}$ $\displaystyle\frac{1}{RC}\int_0^t dt = \int_0^q \frac{dq}{C\mathcal{E} - q}$
The integral on the left is straightforward. The integral on the right leads to a natural logarithm (Problem 82).	$-\dfrac{t}{RC} = \ln\left(\dfrac{C\mathcal{E} - q}{C\mathcal{E}}\right)$
To eliminate the natural logarithm, take the exponential of both sides.	$e^{-t/RC} = \left(\dfrac{C\mathcal{E} - q}{C\mathcal{E}}\right)$
Solve for q.	$q(t) = C\mathcal{E}(1 - e^{-t/RC})$ ✓ (29.10)

COMMENTS

Figure 29.42 is a graph of q as a function of t. This graph looks similar to the capacitor's voltage graph in Figure 29.40, as expected. When the switch is closed at $t = 0$, the charge is zero:

$$q(0) = C\mathcal{E}(1 - e^{-0/RC}) = 0$$

as it should be. After a long time, $t \to \infty$ and the charge stored by the capacitor is at its maximum: $q_{max} = C\mathcal{E}(1 - e^{-\infty}) = C\mathcal{E}$, also as it should be.

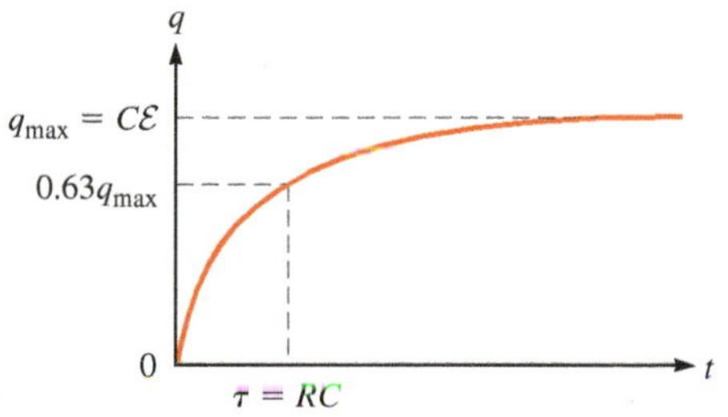

FIGURE 29.42 Charge versus time for a charging capacitor.

DERIVATION *I*(*t*) for a Charging Capacitor

We show that the current $I(t)$ in the circuit (Fig. 29.40) as a function of time is

$$I(t) = \left(\frac{\mathcal{E}}{R}\right)e^{-t/RC} \quad (29.12)$$

when the capacitor is charging.

CURRENT IN A CHARGING *RC* CIRCUIT Special Case

Because the current through the resistor is proportional to the voltage across it, we expect a graph of $I(t)$ versus t to look similar to the graph of $\Delta V_R(t)$ versus t.

Take the time derivative of Equation 29.10.	$I(t) = \dfrac{dq}{dt} = \dfrac{d}{dt}[C\mathcal{E}(1 - e^{-t/RC})]$ $I(t) = C\mathcal{E}\left[0 - \left(-\dfrac{1}{RC}\right)e^{-t/RC}\right]$ $I(t) = C\mathcal{E}\left[\left(\dfrac{1}{RC}\right)e^{-t/RC}\right]$ $I(t) = \left(\dfrac{\mathcal{E}}{R}\right)e^{-t/RC}$ ✓ (29.12)

Derivation continues on page 922 ▶

COMMENTS

As expected, a graph of I as a function of t (Fig. 29.43) looks similar to the resistor's voltage graph in Figure 29.40. When the switch is first closed at $t = 0$, the current is initially

$$I_0 = \left(\frac{\mathcal{E}}{R}\right)e^{-0/RC} = \frac{\mathcal{E}}{R}$$

as expected. After the switch has been closed for a long time, $t \to \infty$ and the current goes to zero:

$$I(t \to \infty) = \left(\frac{\mathcal{E}}{R}\right)e^{-\infty/RC} = 0$$

as we expect for a fully charged capacitor.

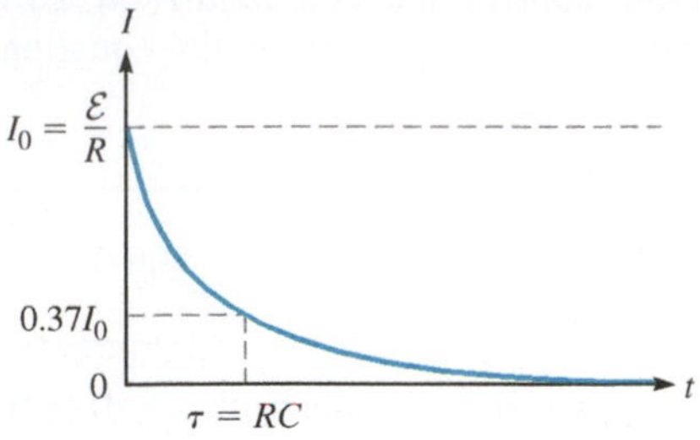

FIGURE 29.43 Current versus time in a charging RC circuit.

Time Constant

Equations 29.10 for $q(t)$ and 29.12 for $I(t)$ both depend on the quantity RC, which has dimensions of time (see Problem 83). This quantity is called the **time constant** τ:

$$\tau \equiv RC \tag{29.13}$$

The time constant is completely determined by the resistance and capacitance in the circuit. The emf does not affect the time constant.

We make our expressions for the charge (Eq. 29.10) and current (Eq. 29.12) more intuitive and convenient to use when we rewrite them in terms of the time constant τ. The charge is

$$q(t) = C\mathcal{E}(1 - e^{-t/\tau}) = q_{\max}(1 - e^{-t/\tau}) \tag{29.14}$$

where $q_{\max} = C\mathcal{E}$, and the current is

$$I(t) = \left(\frac{\mathcal{E}}{R}\right)e^{-t/\tau} = I_0e^{-t/\tau} \tag{29.15}$$

where $I_0 = \mathcal{E}/R$.

The time constant is a measure of how quickly the capacitor charges up. By substituting $t = \tau$ into Equation 29.14, we have

$$q(\tau) = q_{\max}(1 - e^{-\tau/\tau}) = q_{\max}(1 - e^{-1}) \approx 0.63q_{\max}$$

We find that the time constant τ is the time it takes the capacitor to reach about 63% of its maximum charge (Fig. 29.42). The time constant is also a measure of how quickly the current in the circuit drops. By substituting $t = \tau$ into Equation 29.15, we have

$$I(\tau) = I_0e^{-\tau/\tau} = I_0e^{-1} \approx 0.37I_0$$

We see that the time constant τ is the time the current takes to drop to 37% of its initial value (Fig. 29.43).

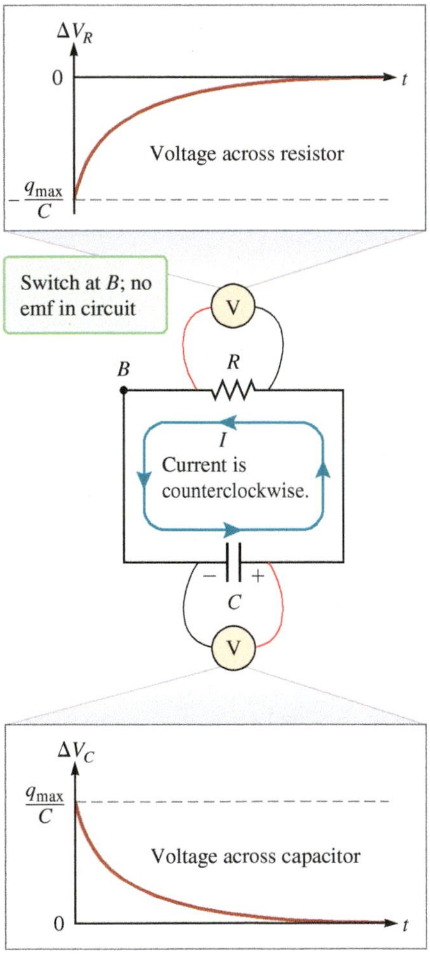

FIGURE 29.44 A discharging RC circuit. With the voltmeter's red lead connected to the capacitor's positive plate and the black lead connected to the negative plate, the measured potential difference across the capacitor is positive. A graph of the voltage across the capacitor as a function of time falls to zero. The current is from the black lead toward the red lead, so the potential difference across the resistor is negative. The voltage across the resistor as a function of time goes to zero.

Discharging a Capacitor

When the switch in Figure 29.39 has been closed at point A for a long time (at least several time constants), the capacitor is fully charged, which means $q = q_{\max}$. There is no current in the circuit. To discharge the capacitor, the switch is thrown to point B so that there is no emf device in the circuit (Fig. 29.44). For convenience, we reset the time to $t = 0$ at the moment the switch is thrown to point B. (This is like resetting a stopwatch.)

Just before the switch is thrown to B, the capacitor is fully charged and its voltage must equal the terminal potential of the emf device, so now the initial voltage across the capacitor must be $\Delta V_C(0) = q_{\max}/C = \mathcal{E}$. After the switch is thrown to B, the capacitor discharges through a counterclockwise current in the circuit. When the capacitor is completely discharged, the voltage across it must be zero. So, the voltage across the capacitor goes from $\mathcal{E}$ to zero (Fig. 29.44). Similarly, we expect the charge stored by the capacitor to decay from its maximum value.

In Problem 93, you will show that for a discharging *RC* circuit, the charge $q(t)$ on the capacitor as a function of time is

$$q(t) = q_{max} e^{-t/\tau} \tag{29.16}$$

CHARGE IN A DISCHARGING *RC* CIRCUIT Special Case

As expected, a graph of q as a function of t (Fig. 29.45) looks similar to the capacitor's voltage graph (Fig. 29.44). When the switch is closed at point B, $t = 0$ and

$$q(0) = q_{max} e^{-0/\tau} = q_{max}$$

so the charge is at its maximum. After a long time, $t \to \infty$ and $q(t \to \infty) = q_{max}(e^{-\infty}) \to 0$, so the charge stored by the capacitor falls to zero. Finally, the time constant is the time it takes the capacitor's stored charge to drop to about 37% of its maximum: $q(\tau) = q_{max}(e^{-1}) \approx 0.37 q_{max}$.

For the charging capacitor, the current is clockwise, which we set to positive (Eq. 29.15). In the discharging *RC* circuit, the current is counterclockwise (Fig. 29.44), so, as you will find in Problem 95, the current is negative and given by

$$I(t) = -I_{max} e^{-t/\tau} \tag{29.17}$$

CURRENT IN A DISCHARGING *RC* CIRCUIT Special Case

A graph of I as a function of t (Fig. 29.46) looks similar to the resistor's voltage graph in Figure 29.44. (If the capacitor was fully charged, then $I_{max} = I_0 = \mathcal{E}/R$.) After the switch has been closed for a long time, $t \to \infty$ and the current goes to zero:

$$I(t \to \infty) = -I_{max} e^{-\infty/\tau} \to 0$$

We see that the time constant is also the time it takes the current in a discharging *RC* circuit to drop to 37% of its initial value: $I(\tau) \approx 0.37 I_0$.

In practice, "a long time" is about 3 to 5 times the time constant.

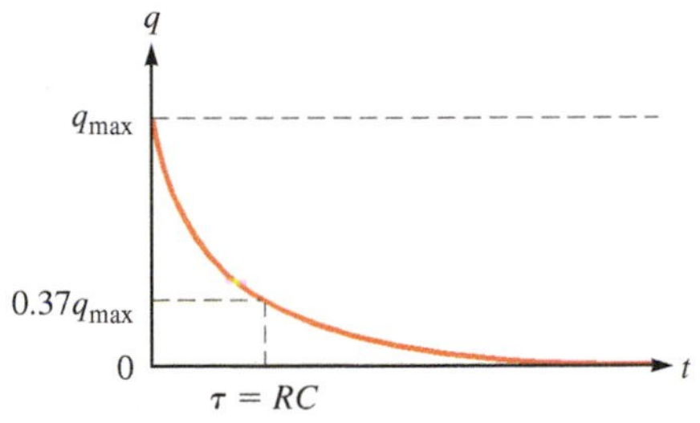

FIGURE 29.45 Charge versus time for a discharging *RC* circuit.

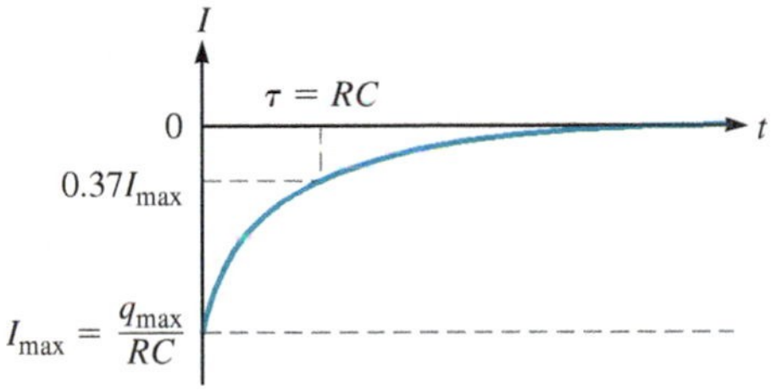

FIGURE 29.46 Current versus time for a discharging *RC* circuit.

EXAMPLE 29.10 Photography in a Cave

Your camera probably has a flash in the form of a discharge tube like a small version of the fluorescent tubes that light up your classroom. You may have seen an old-style flashbulb (Fig. 29.47) used by a newspaper photographer in a movie set in the 1950s or a disposable flashbulb from the 1970s. These old flashbulbs were similar to incandescent lightbulbs. When the photographer pressed a button, current passed through the flashbulb's filament. The flashbulb gave off considerable energy (light and thermal energy) in a short amount of time, breaking the filament and sometimes the glass housing. The photographer waited a moment for the flashbulb to cool, then tossed it out and replaced it with a new bulb. You might think flashbulbs are a thing of the past, but many photographers still use large flashbulbs to take photos in very dark conditions, such as in a cave.

The flashbulb is part of an *RC* circuit that contains two parallel switches. When switch *A* is closed (Fig. 29.48A), the capacitor is charging. When switch *B* is closed (Fig. 29.48B), the capacitor discharges through the flash. The charging circuit includes an indicator light. Assume most of the resistance in the charging circuit is due to the resistance of the indicator light, $R = 4.0\ \text{k}\Omega$. The terminal potential of the battery is 22.5 V, and the capacitance $C = 150\ \mu\text{F}$.

James Steidl/Shutterstock.com

FIGURE 29.47 An old-fashioned camera with a flashbulb.

Example continues on page 924 ▶

A Find the time constant of the charging circuit. In the **CHECK and THINK** step, answer this question: What happens to the indicator light when the capacitor is fully charged?

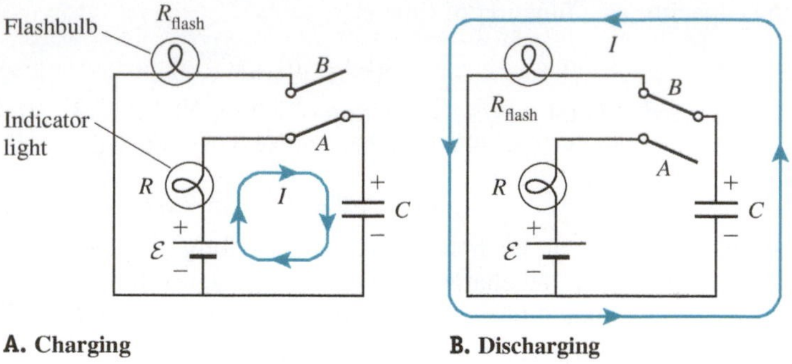

FIGURE 29.48 **A.** Charging flashbulb *RC* circuit. **B.** Discharging circuit.

INTERPRET and ANTICIPATE

The time constant depends only on the resistance and the capacitance, both of which are given. From our experience with modern flash tubes or from watching old movies, we know it takes a second or so before the flash is ready. So we expect the time constant to be roughly 1 second.

SOLVE

Substitute values into Equation 29.13.

$$\tau = RC \quad (29.13)$$

$$\tau = (4.0 \times 10^3\ \Omega)(150 \times 10^{-6}\ \text{F}) = 0.60\ \text{s}$$

CHECK and THINK

The time constant is just a little less than 1 second, which makes sense because τ is the time it takes the capacitor to reach about 63% of its maximum charge. We should wait somewhat longer than one time constant to have a fully charged capacitor.

The indicator light will be illuminated when there is current in it. When switch *A* is first closed, the current in the circuit is at its maximum ($I_0 = \mathcal{E}/R$) and the indicator light is bright. As the capacitor charges up, the current decreases (Fig. 29.43), and the indicator light grows dimmer. When the capacitor is fully charged, there is no current in the circuit, so the indicator light goes off.

B Flashbulbs are used to illuminate a scene from 4 ms to 2 s. Photographers typically carry several different flashbulbs depending on the amount of time they need to illuminate the scene. Estimate the range in the resistance of these flashbulbs.

INTERPRET and ANTICIPATE

The time the flashbulb illuminates the scene is somewhat longer than the time constant of the discharging capacitor. Let's assume the illumination time is about three times as long as the time constant. We expect the range in the resistance to correspond to the range in time.

SOLVE

Find the range in time constants using the assumption that the illumination time is 3τ.

$$\frac{4\ \text{ms}}{3} \leq \tau \leq \frac{2\ \text{s}}{3}$$

$$1.3 \times 10^{-3}\ \text{s} \leq \tau \leq 0.67\ \text{s}$$

Use Equation 29.13 to find a range of resistances. The capacitance is the same as in the charging circuit.

$$\tau = RC \quad (29.13)$$

$$R = \frac{\tau}{C}$$

$$\frac{1.3 \times 10^{-3}\ \text{s}}{150 \times 10^{-6}\ \text{F}} \leq R \leq \frac{0.67\ \text{s}}{150 \times 10^{-6}\ \text{F}}$$

$$8.7\ \Omega \leq R \leq 4.47\ \text{k}\Omega$$

$$9\ \Omega \leq R \leq 4\ \text{k}\Omega$$

CHECK and THINK

As expected, the range in the resistance corresponds to the range in illumination time (and is about a factor of 500 in each case). If you compare flashbulbs with low resistance to those with high resistance, you will see that the ones with high resistance have many more coils of filament inside.

The terminal potential of the battery was irrelevant information for this problem. Why?

SUMMARY

Underlying Principles

No new principles are introduced in this chapter.

Major Concepts

1. **Kirchhoff's loop rule:** The total change in electric potential around any closed loop in a circuit is always zero.
2. **Kirchhoff's junction rule:** At any junction point in a circuit, the sum of all the currents entering the junction equals the sum of all the currents exiting the junction.
3. The equivalent resistance R_{eq} of N **resistors connected in series** is

$$R_{eq} = \sum_{i=1}^{N} R_i \quad (29.6)$$

4. The equivalent resistance R_{eq} of N **resistors connected in parallel** is

$$\frac{1}{R_{eq}} = \sum_{i=1}^{N} \frac{1}{R_i} \quad (29.7)$$

Special Cases

1. An **emf device** is another term for a power supply. The function of an emf device is to maintain a potential difference $\mathcal{E}$ between its terminals.
   a. An **ideal DC emf device**, such as an ideal battery, maintains a constant terminal potential $\mathcal{E}$ whether there is current in the emf device or not.
   b. A **real emf device** is modeled as an ideal emf device in series with an internal resistor of resistance r.
2. Potential differences across circuit elements follow these **voltage rules**:
   a. **Wire rule:** The voltage across a wire is zero whether there is current in the wire or not:

$$\Delta V_{wire} = 0$$

   b. **Resistor rule:** The voltage measured across a resistor when the current is from the black lead to the red lead is

$$\Delta V_R = V_{red} - V_{black} = -IR \quad (29.1)$$

   c. **Switch rule:** The voltage drop across a closed switch is zero:

$$\Delta V_{closed\ switch} = 0$$

   If the switch is open, it has a very high resistance. There is no current through an open switch, but there may be a potential difference across it.

   d. **Emf rules:** (The red lead is assumed to be connected to the positive terminal in either case.)
   For an ideal emf device, the emf rule is

$$\Delta V_{ideal\ emf} = V_{red} - V_{black} = \mathcal{E}.$$

   For a real emf device, the emf rule is

$$\Delta V_{real\ emf} = V_{red} - V_{black} = \mathcal{E} - Ir.$$

3. An **RC circuit** consists of an emf device, a resistor, and a capacitor. The time constant τ is defined as

$$\tau \equiv RC \quad (29.13)$$

   a. For a **charging RC circuit**, the charge stored by the capacitor depends on time:

$$q(t) = C\mathcal{E}(1 - e^{-t/\tau}) = q_{max}(1 - e^{-t/\tau}) \quad (29.14)$$

   The current in the resistor is also time-dependent and given by

$$I(t) = \left(\frac{\mathcal{E}}{R}\right)e^{-t/\tau} = I_0 e^{-t/\tau} \quad (29.15)$$

   b. For a **discharging RC circuit**, the charge stored by the capacitor depends on time:

$$q(t) = q_{max} e^{-t/\tau} \quad (29.16)$$

   The time-dependent current in the resistor is given by

$$I(t) = -I_0 e^{-t/\tau} \quad (29.17)$$

PROBLEM-SOLVING STRATEGY Using Kirchhoff's Loop Rule

INTERPRET and ANTICIPATE
Start with a schematic diagram to which you add two elements: (1) an arrow for the **current** and (2) a dot for the **starting point**.

SOLVE
1. Write expressions for the expected **voltage across each circuit element.**
2. **Set the sum of the expected voltages to zero.**

PROBLEM-SOLVING STRATEGY Circuit Analysis

INTERPRET and ANTICIPATE
As in the problem-solving strategy "Using Kirchhoff's Loop Rule," start with a schematic diagram that includes these elements: (1) arrows and labels for the **currents**, (2) labels for the **junctions** and the **circuit elements**, and (3) a dot for the **starting point** (if you plan to use Kirchhoff's loop rule).

SOLVE
There are three possible tactics for you to try depending on the situation.
Tactic 1. Find an **equivalent circuit** and draw its schematic diagram.
Tactic 2. Apply **Kirchhoff's junction rule.**
Tactic 3. Apply **Kirchhoff's loop rule.**

PROBLEMS AND QUESTIONS

A = algebraic C = conceptual E = estimation G = graphical N = numerical

29-1 Measuring Potential Differences Between Two Points

1. **C** Study the symbols in Table 29.2. Then, without looking at the table, draw the symbols for these circuit elements:
   a. a wire,
   b. a switch,
   c. a resistor,
   d. an emf device, and
   e. a lightbulb.
2. **C** Study the terms in Table 29.1. Then, without looking at the table, provide formal synonyms for these terms:
   a. voltage,
   b. amperage,
   c. DC power supply, and
   D. input voltage.
3. **C** *True or false*: The terminal potential of a DC power supply is *not* required to be constant. Explain your answer.

Problems 4, 5, and 6 are grouped.

4. **C** Suppose you need to measure the potential difference between the points in Figure P29.4. Assume the voltmeter reading is the potential difference between the two leads: $\Delta V = V_{red} - V_{black}$. For each of the following measurements, determine at which point you would connect the red lead and at which point you would connect the black lead:
   a. $V_b - V_a$,
   b. $V_c - V_b$,
   c. $V_d - V_c$,
   d. $V_a - V_d$.

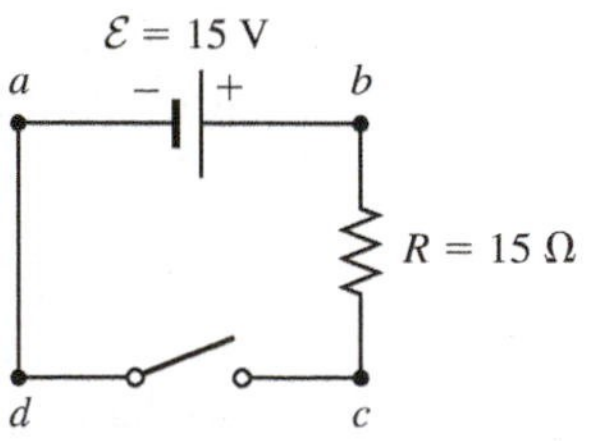

FIGURE P29.4 Problems 4, 5, and 6.

5. **N** Figure P29.4 shows a circuit with an open switch, an emf device, and a resistor. If we assume the switch remains open, use the values given in the figure to find the potential difference between the points
   a. $V_b - V_a$,
   b. $V_c - V_b$,
   c. $V_d - V_c$, and
   d. $V_a - V_d$.
6. **N** If the switch in Figure P29.4 is closed, use the values given in the figure to find the potential difference between the points
   a. $V_b - V_a$,
   b. $V_c - V_b$,
   c. $V_d - V_c$, and
   d. $V_a - V_d$.
   If you worked Problem 5, compare your answers in each case.
7. **C** A real battery (modeled as an ideal emf device in series with an internal resistor) is connected to a lightbulb. As the battery ages, its internal resistance increases while the internal ideal emf device remains unchanged. As the battery ages, what happens to
   a. the terminal potential of the real battery,
   b. the current in the circuit, and
   c. the light emitted?

Problems 8 and 9 are paired.

8. **A** Two circuits made up of identical ideal emf devices and resistors are shown in Figure P29.8. What is the potential difference $V_b - V_a$
   a. for circuit 1 and
   b. for circuit 2?
   c. Find expressions for both the current in the resistor in circuit 1 and the current in the resistor in circuit 2, and compare them.

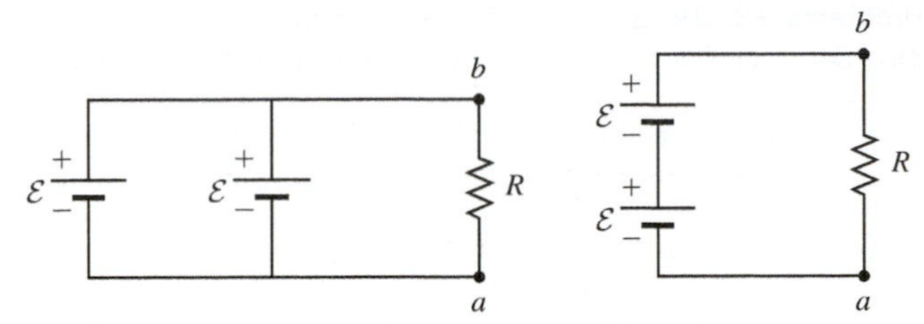

Circuit 1 Circuit 2

FIGURE P29.8 Problems 8 and 9.

9. **N** Two circuits made up of identical ideal emf devices ($\mathcal{E} = 1.67$ V) and resistors ($R = 35.9\ \Omega$) are shown in Figure P29.8. What is the potential difference $V_b - V_a$
   a. for circuit 1 and
   b. for circuit 2?

   What is the current in the resistor
   c. in circuit 1 and
   d. in circuit 2?
10. A lightbulb is connected to an ideal emf device that has an emf of 1.5 V.
    a. **N** What is the magnitude of the potential difference across the lightbulb?
    b. **C** If the bulb burns out, what is the potential difference across the lightbulb? What is the current in the circuit?
    c. **C** If the bulb is unscrewed from its socket, what is the potential difference across the empty socket?
11. **N** The terminal voltage of a real battery that delivers 15.0 W of power to a load resistor is 13.4 V, and its emf is 16.0 V. What is the resistance of
    a. the load resistor in this circuit and
    b. the internal resistor of the battery?
12. **C** Two circuit elements are connected directly end to end. Can the potential difference across the first circuit element ever be different from the potential difference across the second circuit element? Explain your answer.
13. **N** Eight real batteries, each with an emf of 5.00 V and an internal resistance of 0.200 Ω, are connected end to end in a loop as in Figure P29.13. What is the terminal voltage across one of the batteries between points a and b?

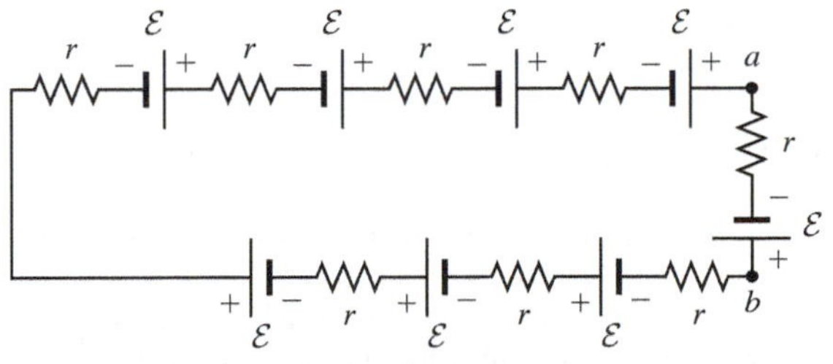

FIGURE P29.13

14. **C** Two circuit elements are connected directly end to end. Can the current through the first circuit element ever be different from the current through the second circuit element? Explain your answer.

29-2 Kirchhoff's Loop Rule

Problems 15 and 16 are paired.

15. **N** In Figure P29.15, three resistors are connected to an ideal emf device, where $\mathcal{E} = 14.8$ V, $R_1 = 13.4\ \Omega$, $R_2 = 20.5\ \Omega$, and $R_3 = 9.80\ \Omega$.
    a. What is the current through each resistor?
    b. What is the voltage across each resistor?
    c. How much power is consumed by each resistor and by the emf device?
    d. If R_3 is replaced by a new resistor that has twice the resistance, answer parts (a) through (c) for this new circuit. Check your new answers against your old answers.

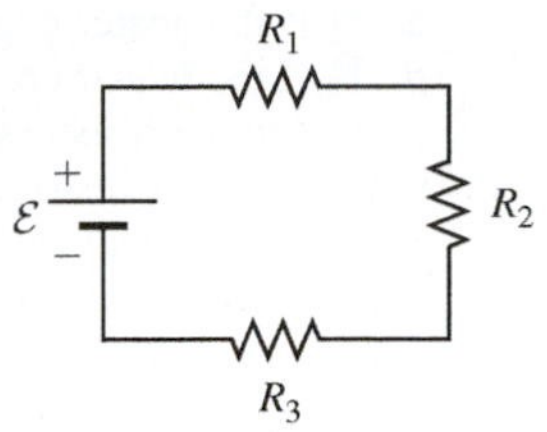

FIGURE P29.15 Problems 15 and 16.

16. **N** In Figure P29.15, three resistors are connected to an ideal emf device. The resistances are $R_1 = 13.4\ \Omega$, $R_2 = 20.5\ \Omega$, and $R_3 = 9.8\ \Omega$. The current through the last resistor is 7.55 mA.
    a. What is the current through the other two resistors?
    b. What is the terminal potential of the emf device?
17. **C** Some instructors use the following analogy to describe how the electric potential changes throughout a circuit: The flow of charge from higher to lower electric potential is like the flow of water down an incline from a greater height to a lower level. Consider an ideal emf device connected to a single resistor.
    a. Using the analogy, what happens to the water level as it crosses the emf device from the lower electric potential to the higher electric potential?
    b. What happens to the water level as it moves along the wire that connects the emf device and the resistor?
    c. Describe what happens to the water level as the resistor is crossed in the direction of the current flow.
    d. What is the "height" of the water level after it crosses the resistor compared to that at the end of the emf device with the lower electric potential?
18. **E** Students at one university have purchased large "handheld" calculators capable of doing many calculations. The professors at the same university use an older-style calculator with many fewer functions. The students' calculators need new batteries about twice per semester. The professors' calculators need new batteries about once every other year. Both types of calculators are powered by four AAA 1.5-V batteries connected in series. Assume students and professors use their calculators about the same amount of time each week. Model the calculators as simple circuits consisting of a resistor connected to the four batteries. Estimate these ratios:
    a. the resistance in a professor's calculator to the resistance in a student's calculator, and
    b. the current in a professor's calculator batteries to the current in a student's calculator batteries.

    Explain your assumptions.

Problems 19 and 20 are paired.

19. **N** An ideal emf device with $\mathcal{E} = 9.00$ V is connected to two resistors in series. One of the resistors has a resistance of 145 Ω, and the other has unknown resistance R. If the current through the emf device is 0.0155 A, what is the resistance R?
20. **A** An ideal emf device with emf $\mathcal{E}$ is connected to two resistors in series. One of the resistors has resistance R_1 and the other has unknown resistance R. If the current through the emf device is I, find an expression for the unknown resistance R in terms of the other quantities.

29-3 Resistors in Series

21. **CASE STUDY** Having fixed Carol's hot water heater problem, the students decide to work on a home space heater. They model a home space heater as a couple of resistors connected to a 120-V (DC) power supply. (Of course, the power supply is actually AC, but that does not affect the analysis in this problem.) The heater has two settings—one for 1300 W and the other for 1500 W.

a. **N** If the heater consumes 1500 W, what is its resistance?
b. **N** If the heater consumes 1300 W, what is its resistance?
c. **C** Are the resistors connected in series? Explain.

22. **C** Three series resistors (22.5 Ω, 45.0 Ω, and 90.0 Ω) are connected to an ideal emf source, so that a current I flows through the 22.5-Ω resistor first. The resistors are reordered so that the current flows through the 90.0-Ω resistor first, followed by the 22.5-Ω and the 45.0-Ω resistors.
a. What is the effect of the new order on the current I?
b. What is the effect on the voltage drop across each resistor? Explain your answers.

Problems 23 and 31 are paired.

23. **A** Six resistors with resistances $7R$, $6R$, $2R$, R, $R/2$, and $R/4$ are connected in series. What is the equivalent resistance of this combination?

Problems 24 and 25 are paired.

24. The emf devices and lightbulbs in Figure P29.24 are identical.
a. **A** Find an expression for the current in each bulb.
b. **C** List the bulbs in order from brightest to dimmest. Explain your answer.

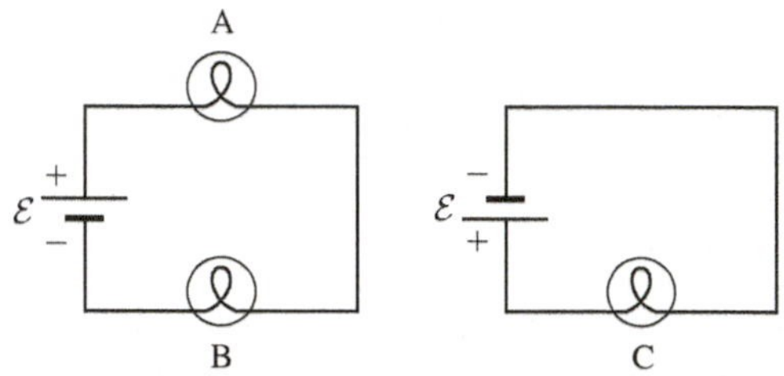

FIGURE P29.24 Problems 24 and 25.

25. **N** The emf devices and lightbulbs in Figure P29.24 are identical. The terminal potential of the emf device is 4.50 V. The current through bulb A is 9.00 mA.
a. Find the current through bulb B.
b. What is the current through bulb C?

26. **N** When three resistors with resistances in the ratio 1:2:3 are connected in series with a 12.0-V ideal battery, there is 6.00 A of current in the circuit. Determine
a. the resistance of and
b. the voltage drop across each of the three resistors.

29-4 Kirchhoff's Junction Rule

27. **N** Determine the currents through the resistors R_2, R_5, R_6, and R_7 in the set of junctions and branches shown in Figure P29.27. *Hint*: Use Kirchhoff's junction rule, be sure to consider the branches where a current is shown, and assume the branches that appear disconnected are connected to other parts of the circuit.

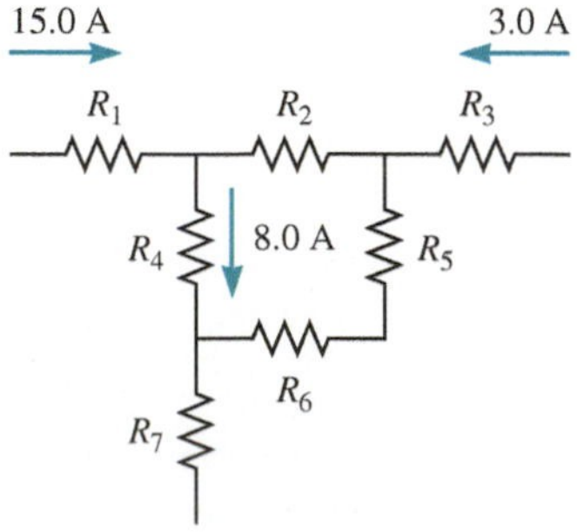

FIGURE P29.27

Problems 28, 29, 32, and 39 are grouped.

28. **C** The emf devices in the circuits shown in Figure P29.28 are identical.
a. Redraw circuit 1 in Figure P29.28, including (and labeling) the current in each branch of the circuit. Apply the junction rule to write an equation in terms of the currents you have labeled.
b. Repeat part (a) for circuit 2 in Figure P29.28.
c. Simplify your equations if necessary, and compare your answers for each circuit. What can you say about the two circuits?

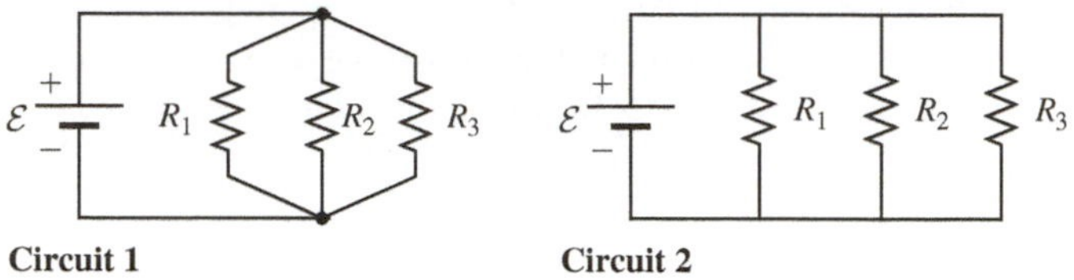

FIGURE P29.28 Problems 28, 29, 32, and 39.

29. **A** The emf devices in the circuits shown in Figure P29.28 are identical. The resistances are $R_1 = R$, $R_2 = 2R$, and $R_3 = 3R$.
a. If the current through each emf device is I_0, find an expression for the current through each resistor in circuit 1.
b. Repeat part (a) for circuit 2.

30. **C** Some people might describe Kirchhoff's junction rule as a statement that "the charge in the circuit is conserved." Describe what people are referring to when they make this statement. In what way is the charge conserved in the junction rule?

29-5 Resistors in Parallel

31. **A** Six resistors with resistances $7R$, $6R$, $2R$, R, $R/2$, and $R/4$ are connected in parallel. What is the equivalent resistance of this combination?

32. The emf in circuit 2 in Figure P29.28 is 22.5 V. The resistors have resistances $R_1 = 47.6\ \Omega$, $R_2 = 98.3\ \Omega$, and $R_3 = 50.0\ \Omega$.
a. **N** What is the current in the emf device?
b. **C** If a fourth resistor is added in parallel with the first three resistors, will the current in the emf device increase, decrease, or stay the same? Explain.
c. **C** If, instead, a fourth resistor is added in series with the emf device, will the current in the emf device increase, decrease, or stay the same? Explain.

33. **C** The French natural philosopher Abbé Nollet (1700–1770) was interested in studying the strength of an electric shock as it traveled a great distance. He assembled a large group of monks into a circle reported to be a mile in circumference. Each monk was in contact with the next monk by a short piece of wire held in their hands. Between one pair of monks was a Leyden jar that they could touch and discharge, causing a (hopefully mild) shock. Imagine that, instead, Nollet connected the two "end monks" to a Volta pile (a battery) as shown in Figure P29.33A. Alternatively, Nollet could have made a mile-long chain of the same monks using an identical Volta pile as shown in Figure P29.33B. In which circuit is the current through the Volta pile greater? In which circuit is the current through a particular monk greater? Explain your answers.

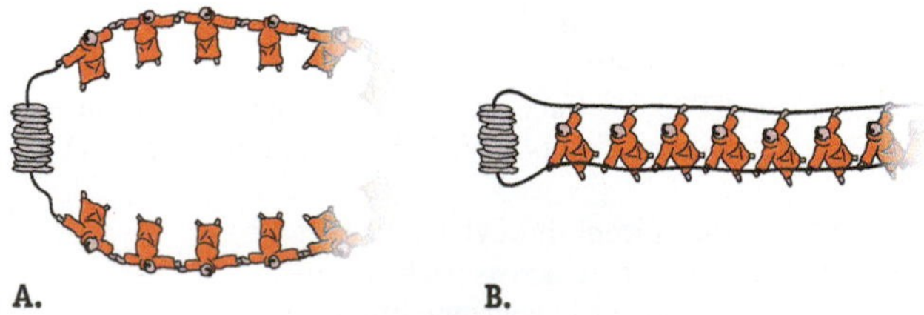

FIGURE P29.33

34. You are building a device that requires a 37.5-Ω resistor. You have only a box of a dozen 50.0-Ω resistors.

a. **C** Draw the network of resistors that best meets your design requirement.

b. **N** What is the resistance of your network?

35. **A** Figure P29.35 shows a combination of six resistors with identical resistance R. What is the equivalent resistance between points a and b?

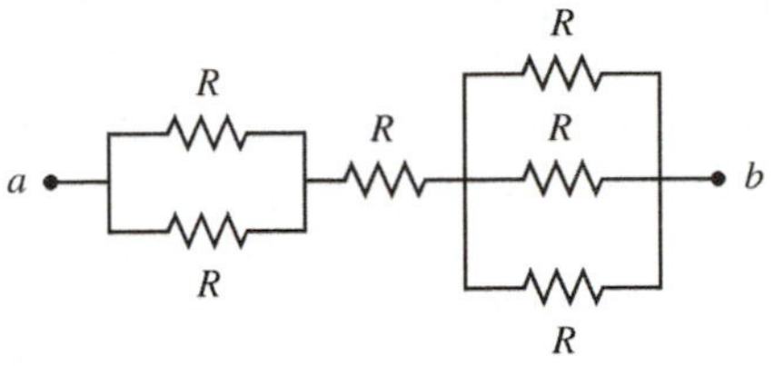

FIGURE P29.35

Problems 36 and 37 are paired.

36. **A** Each resistor shown in Figure P29.36 has resistance R. An ideal emf device ($\mathcal{E}$) is connected to points a and b via two leads (not shown in the figure). Find an expression for the current through the emf device.

37. **N** Each resistor shown in Figure P29.36 has a resistance of 100.0 Ω. An ideal emf device (120.0 V) is connected to points a and b via two leads (not shown in the figure). Find the current that flows through the emf device.

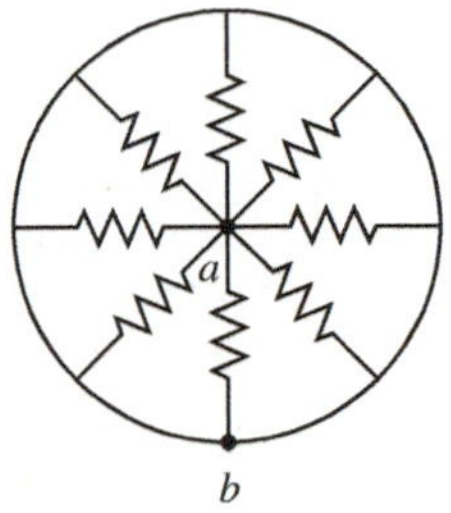

FIGURE P29.36 Problems 36 and 37.

38. **C** You have access to any number of 5.00-Ω and 75.0-Ω resistors. How can an effective resistance of 30.0 Ω be obtained by using these resistors?

29-6 Circuit Analysis

39. **N** The emf devices in the circuits shown in Figure P29.28 are identical and have a terminal potential of 7.50 V. The resistances are $R_1 = R$, $R_2 = 2R$, and $R_3 = 3R$, with $R = 15.0$ Ω.

a. Find the current through the emf device and each resistor in circuit 1.

b. Repeat part (a) for circuit 2.

40. **N** The emf in Figure P29.40 is 4.54 V. The resistances are $R_1 = 13.0$ Ω, $R_2 = 26.0$ Ω, and $R_3 = 39.0$ Ω. Find

a. the current in each resistor,

b. the power consumed by each resistor, and

c. the power supplied by the emf device.

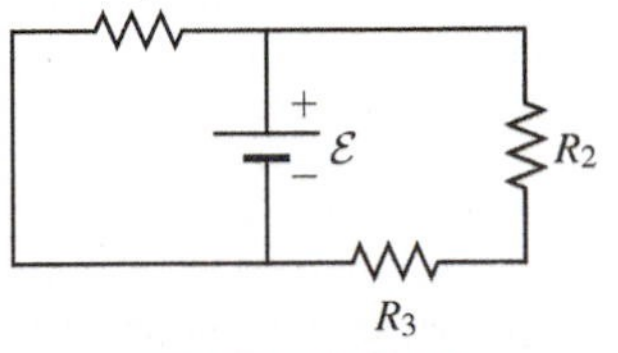

FIGURE P29.40

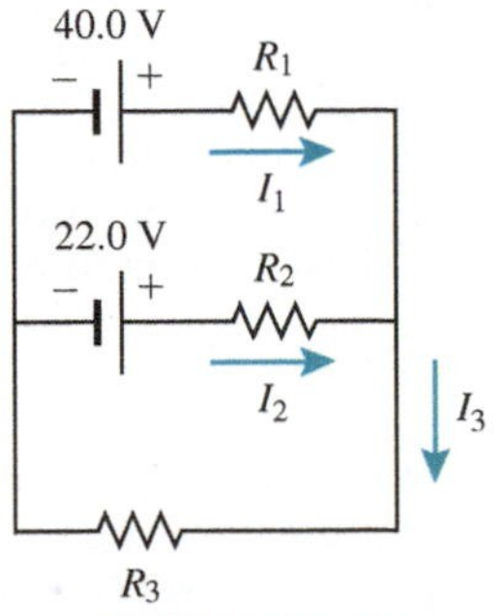

FIGURE P29.41

41. **N** Figure P29.41 shows three resistors ($R_1 = 14.0$ Ω, $R_2 = 8.00$ Ω, and $R_3 = 10.0$ Ω) and two batteries connected in a circuit.

a. What is the current in each of the resistors?

b. How much power is delivered to each of the resistors?

42. **N** Figure P29.42 shows five resistors and two batteries connected in a circuit. What are the currents I_1, I_2, and I_3?

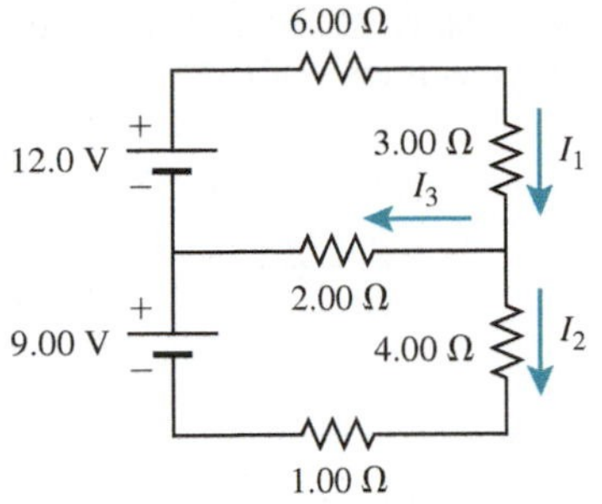

FIGURE P29.42

Problems 43 and 44 are paired.

43. **N** The emfs in Figure P29.43 are $\mathcal{E}_1 = 6.00$ V and $\mathcal{E}_2 = 12.0$ V. The resistances are $R_1 = 15.0$ Ω, $R_2 = 30.0$ Ω, $R_3 = 45.0$ Ω, and $R_4 = 60.0$ Ω. Find the current in each resistor when the switch is

a. open and

b. closed.

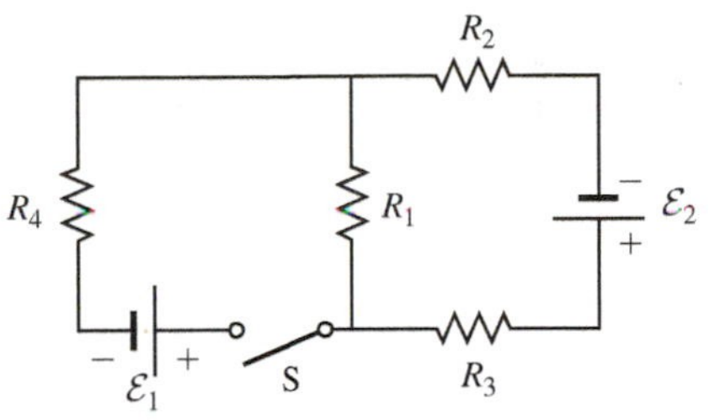

FIGURE P29.43 Problems 43 and 44.

44. **N** The emfs in Figure P29.43 are $\mathcal{E}_1 = 6.00$ V and $\mathcal{E}_2 = 12.0$ V. The resistances are $R_1 = 15.0$ Ω, $R_2 = 30.0$ Ω, $R_3 = 45.0$ Ω, and $R_4 = 60.0$ Ω. Find the power consumed by each resistor when the switch is

a. open and

b. closed.

45. **N** Figure P29.45 shows five resistors connected between terminals a and b.

a. What is the equivalent resistance of this combination of resistors?

b. What is the current through each resistor if a 24.0-V battery is connected across the terminals?

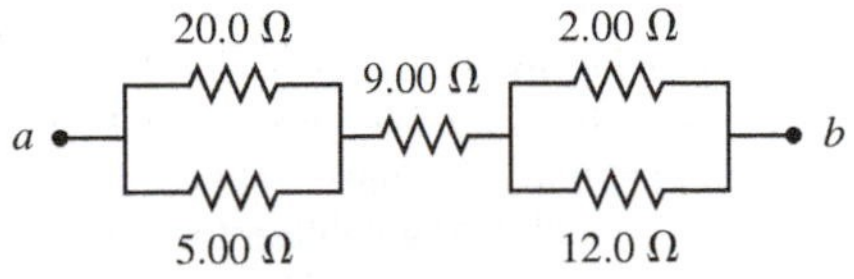

FIGURE P29.45

46. **N** Figure P29.46 shows a circuit with a 12.0-V battery connected to four resistors. How much power is delivered to each resistor?

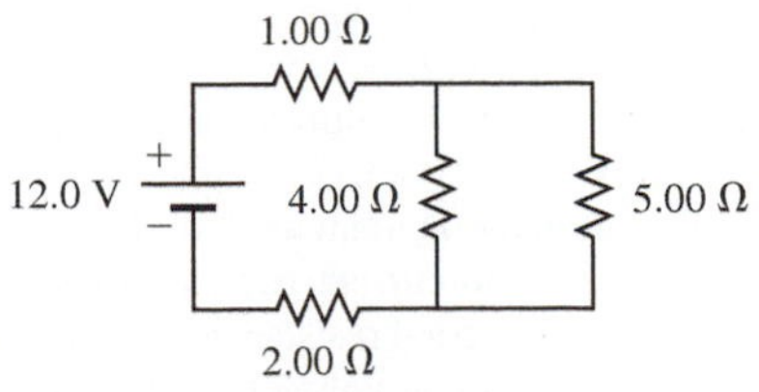

FIGURE P29.46

Problems 47 and 48 are paired.

47. **N** Two ideal emf devices are connected to a set of resistors as shown in Figure P29.47. If $\mathcal{E}_1 = 6.00$ V, $R_1 = 10.00\ \Omega$, $R_2 = 5.00\ \Omega$, $R_3 = 15.00\ \Omega$, $R_4 = 20.00\ \Omega$, and the current through R_4 is 0.250 A, what is the emf $\mathcal{E}_2$?

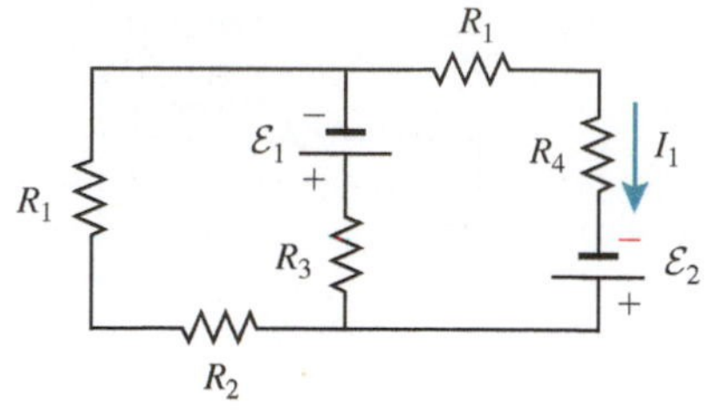

FIGURE P29.47 Problems 47 and 48.

48. **A** Two ideal emf devices are connected to a set of resistors as shown in Figure P29.47. Find an expression for the emf $\mathcal{E}_2$ in terms of $\mathcal{E}_1$, R_1, R_2, R_3, R_4, and the current through R_4, labeled I_1.

49. **N** Three resistors with resistances $R_1 = R/2$ and $R_2 = R_3 = R$ are connected as shown, and a potential difference of 225 V is applied across terminals *a* and *b* (Fig. P29.49).

a. If the resistor R_1 dissipates 75.0 W of power, what is the value of R?

b. What is the total power supplied to the circuit by the emf?

c. What is the potential difference across each of the three resistors?

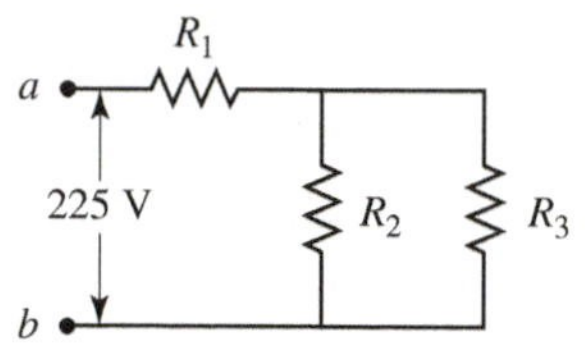

FIGURE P29.49

29-7 DC Multimeters

50. **E** An *E-meter* (Fig. P29.50) is an electric device patented by L. Ron Hubbard of the Church of Scientology (U.S. patent 3,290,589 on December 6, 1966). Originally the church used the device for medical purposes. This use was challenged in court, which ruled that the device is not medically or scientifically capable of improving the health or bodily functions of anyone. The church now states that the E-meter is strictly for the guidance of ministers of the church in confessionals and pastoral counseling. The E-meter, much like a lie detector, measures changes in the electrical resistance of human skin and is thus an ohmmeter. The resistivity of human skin is about $5 \times 10^5\ \Omega \cdot$m. Assume the E-meter uses a 6.0-V DC emf device. Estimate the current passing through the person shown in the figure holding the metal cylinders. What do you imagine the person might feel during the test?

FIGURE P29.50 An E-meter in use.

51. **N** Consider a simple circuit consisting of a 15.0-V power supply connected to a 30.0-Ω resistor.

a. What would an ideal voltmeter measure for the voltage across the resistor?

b. An ideal voltmeter is imagined to have infinite internal resistance so that no current passes through it. A real voltmeter has a high internal resistance of 50,000 Ω. What is the current through this real voltmeter? How does it compare to the current through the 30.0-Ω resistor?

52. **N** Figure P29.52 shows a circuit in which the ideal ammeter has a reading of 2.00×10^{-3} A and the ideal voltmeter has a reading of 9.00 V. What are

a. the unknown resistance R,

b. the emf $\mathcal{E}$ of the battery, and

c. the potential difference across the 1.65-kΩ resistor?

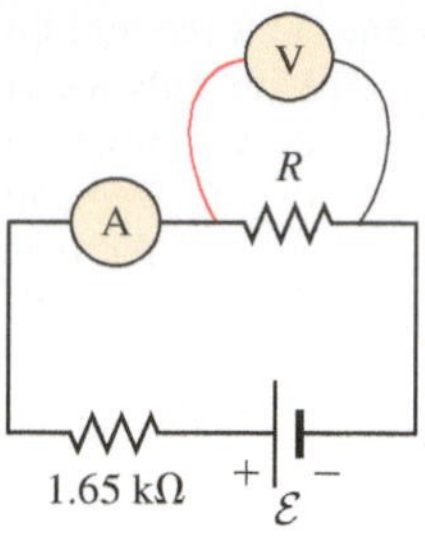

FIGURE P29.52

29-8 *RC* Circuits

53. **N** In Example 27.8 (page 851), we modeled the human nerve cell as a capacitor. Each nerve cell is a collection of nodes, and signals must be transmitted across each node. Each node may be modeled as a discharging *RC* circuit with resistance $R = 41 \times 10^6\ \Omega$ and capacitance $C = 1.5 \times 10^{-12}$ F. Find the time constant for discharging a node. How does your answer compare to the measured firing times of about 50 μs?

54. **G** A charging *RC* circuit consists of a 12.0-V emf device, a 60.0-Ω resistor, and a 150-μF capacitor.

a. Plot the charge stored by the capacitor as a function of time.

b. On your graph, indicate the times when $t = \tau$, 2τ, 3τ, and 4τ. Record the charge stored by the capacitor at each of these times. Check your results algebraically.

55. **N** A 650.0-Ω resistor is connected across the terminals of a 12.0-nF capacitor that carries an initial charge of 7.40 μC.

a. What is the magnitude of the maximum current in the resistor?

b. What is the current in the resistor 5.00 μs after the circuit is completed and the capacitor begins to discharge through the resistor?

c. How much charge remains in the capacitor 5.00 μs after the circuit is completed?

56. **N** At time $t = 0$, an *RC* circuit consists of a 12.0-V emf device, a 60.0-Ω resistor, and a 150.0-μF capacitor that is fully charged. The switch is thrown so that the capacitor begins to discharge.

a. What is the time constant τ of this circuit?

b. How much charge is stored by the capacitor at $t = 0.5\tau$, 2τ, and 4τ?

57. **N** A 210.0-Ω resistor and an initially uncharged 6.00-μF capacitor are connected in series to a 12.0-V emf source. A switch is closed to complete the circuit at $t = 0$.

a. What is the time constant of this circuit?

b. What is the maximum charge on the capacitor?

c. What is the charge on the capacitor at $t = 3\tau$?

Problems 58 and 59 are paired.

58. **A** A real battery with internal resistance r and emf $\mathcal{E}$ is used to charge a capacitor with capacitance C. A resistor R is put in series with the battery and the capacitor when charging.

a. Write an expression for the time constant of this circuit.

b. Find an expression for the time when the capacitor has reached half its maximum charge.

c. Find an expression for the current through the capacitor at this time.

59. **N** A real battery with internal resistance 0.500 Ω and emf 9.00 V is used to charge a 50.0-μF capacitor. A 10.0-Ω resistor is put in series with the battery and the capacitor when charging.

a. What is the time constant for this circuit?

b. What is the time when the capacitor has reached half its maximum charge?

c. What is the current through the capacitor at this time?

60. **N** Figure P29.60 shows a simple *RC* circuit with a 2.50-μF capacitor, a 3.50-MΩ resistor, a 9.00-V emf, and a switch.

What are
a. the charge on the capacitor,
b. the current in the resistor,
c. the rate at which the capacitor is storing energy, and
d. the rate at which the battery is delivering energy exactly 7.50 s after the switch is closed?

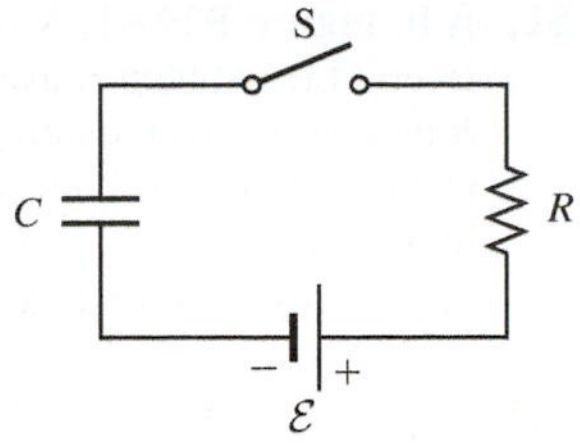

FIGURE P29.60

General Problems

61. **N** When connected in parallel, two resistors have an equivalent resistance of 125 Ω. The same two resistors have an equivalent resistance of 820 Ω when connected in series. What is the resistance of each of the resistors?

62. **A** Two lightbulbs have powers P_1 and P_2 when separately connected across an ideal battery with potential difference ΔV. The bulbs are then connected in series across the same battery. Determine the total power consumption of the bulbs in this configuration.

63. **N** A toy robot requires four D batteries rated at 1.50 V, and each delivers a charge of 1.20×10^3 milliampere-hour (mAh) or 4.32×10^3 C. If the internal resistance of the robot is 45.0 Ω, how long will one set of four batteries last?

64. **A** Ralph has three resistors, R_1, R_2, and R_3, connected in series. When connected to an ideal emf source $\mathcal{E}_1$, current I_1 flows through the resistors.
a. If the resistors are instead connected to a second source with $\mathcal{E}_2 = 2\mathcal{E}_1$, what is the new current through the resistors in terms of the first current?
b. Show that, if each resistance is doubled and the resistors are connected in series to the second emf source, the current through the resistors is equal to I_1.

65. **N** The reading on the ammeter in Figure P29.65 is 3.00 A, and the current runs from right to left through the ammeter. What is the value of current
a. I_1 and
b. I_2?
c. What is the emf $\mathcal{E}$?

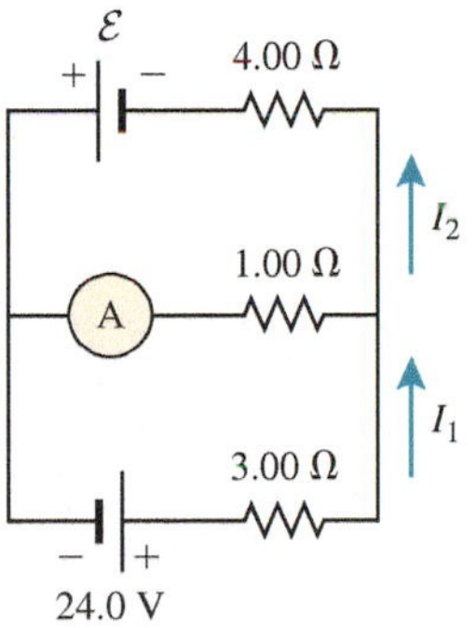

FIGURE P29.65

Problems 66, 67, and 68 are grouped.

66. **A** An ideal emf device is connected to a set of resistors as shown in Figure P29.66. Find an expression for the current through the resistor R_3 in terms of the emf and the resistances.

67. **N** An ideal emf device (24.0 V) is connected to a set of resistors as shown in Figure P29.66. If $R_1 = 22.5$ Ω, $R_2 = 52.5$ Ω, $R_3 = 125$ Ω, and $R_4 = 75.0$ Ω, find the current through resistor R_3.

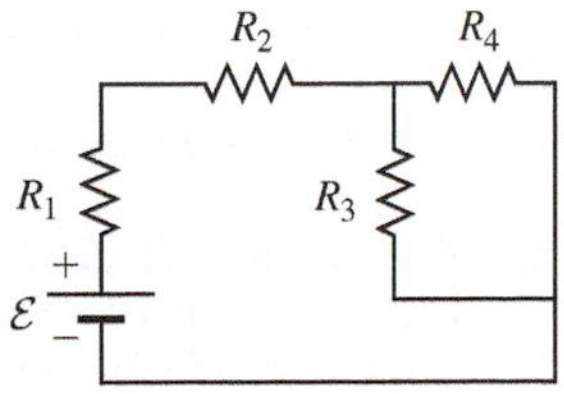

FIGURE P29.66 Problems 66, 67, and 68.

68. **N** An ideal emf device (24.0 V) is connected to a set of resistors as shown in Figure P29.66. If $R_1 = 22.5$ Ω, $R_2 = 52.5$ Ω, $R_3 = 125$ Ω, and $R_4 = 75.0$ Ω, what is the voltage drop across each resistor?

69. **N** A real battery with emf 12.0 V and internal resistance 2.00 Ω is connected to a resistance of 4.00 Ω.
a. What is the current through the battery?
b. What is the terminal voltage of the battery?

70. **N** What is the equivalent resistance between points a and b of the six resistors shown in Figure P29.70?

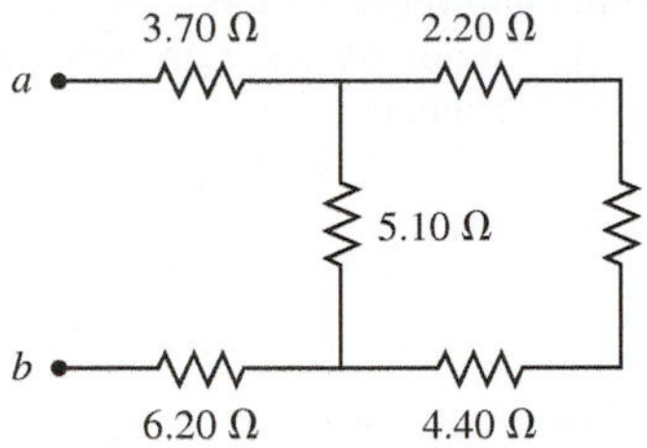

FIGURE P29.70

71. **N** Three batteries and four resistors are connected in a circuit as shown in Figure P29.71.
a. What is the current in each of the resistors?
b. What is the potential difference across the 175-Ω resistor?

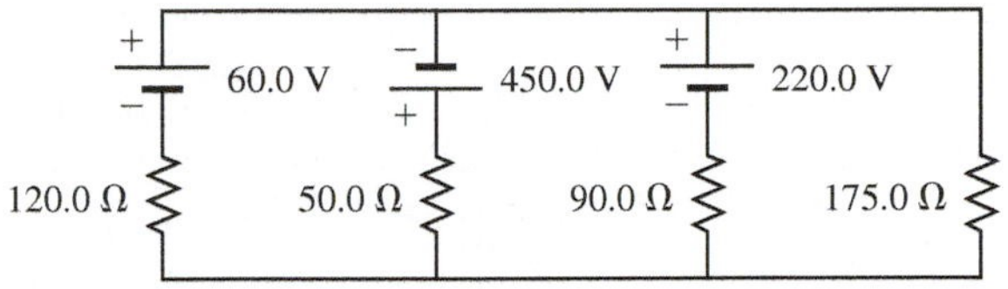

FIGURE P29.71

72. **A** A capacitor with initial charge Q_0 is connected across a resistor R at time $t = 0$. The separation between the plates of the capacitor changes as $d = d_0/(1 + t)$ for $0 \leq t < 1$ s. Find an expression for the voltage drop across the capacitor as a function of time.

73. **N** Batman is attempting to diffuse a bomb created by The Riddler and opens it to find a set of resistors connected as shown in Figure P29.73. He can tell from the coloring on several of the resistors that some are equivalent to each other, and he determines that $R_1 = 50.00$ Ω, $R_2 = 100.0$ Ω, $R_3 = 150.0$ Ω, and $R_4 = 300.0$ Ω. However, he needs to know the resistance of R_5 before he can begin diffusing the bomb. Using his Bat-Ohm-Meter, he is able to determine that the equivalent resistance between points a and b is 200.0 Ω. What is the resistance of the unknown resistor?

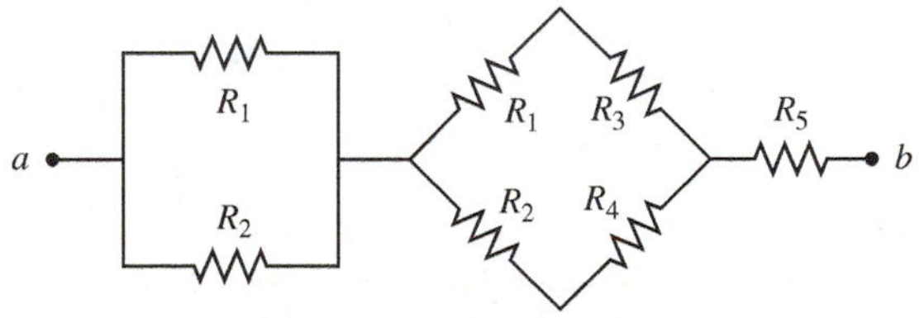

FIGURE P29.73

Problems 74 and 75 are paired.

74. **A** A capacitor with capacitance C is charged to a total charge Q. The capacitor is then connected in parallel to two resistors (R_1 and R_2) that are also connected in parallel.
a. Write an expression for the time constant of this circuit.
b. Find an expression for the time when the capacitor has lost half of its charge.
c. Find an expression for the current through the capacitor at this time.

75. **N** A capacitor with capacitance 50.0 μF is charged to a total charge of 200.0 μC. The capacitor is then connected in parallel to two resistors (10.0 Ω and 30.0 Ω) that are also connected in parallel.
a. What is the time constant for this circuit?
b. What is the time when the capacitor has lost half of its charge?
c. What is the current through the capacitor at that time?

76. **N** Figure P29.76 shows three identical resistors with $R = 125\ \Omega$ connected between terminals a and b. Each resistor can safely withstand a maximum power of 17.5 W.

a. What is the upper limit on the safe voltage that can be applied across the terminals?

b. How much power is delivered to each of the three resistors at the maximum voltage found in part (a)?

c. How much total power is delivered to the combination of three resistors?

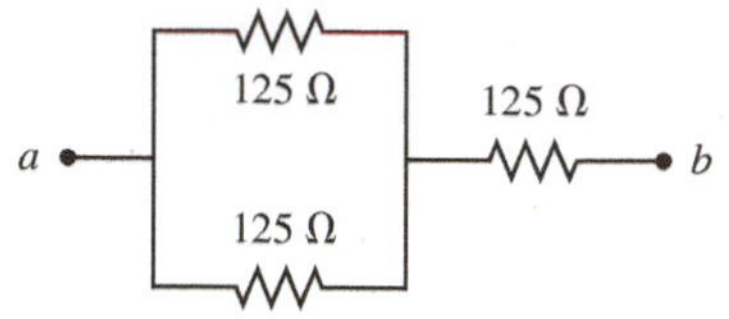

FIGURE P29.76

77. **N** Figure P29.77 shows a circuit with two batteries and three resistors.

a. How much current flows through the 2.00-Ω resistor?

b. What is the potential difference between points a and b in the circuit?

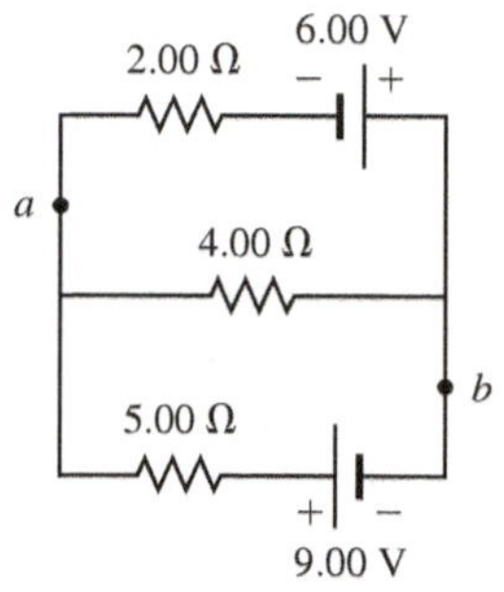

FIGURE P29.77

78. **A** In the RC circuit shown in Figure P29.78, an ideal battery with emf $\mathcal{E}$ and internal resistance r is connected to capacitor C. The switch S is initially open and the capacitor is uncharged. At $t = 0$, the switch is closed.

a. Determine the charge q on the capacitor at time t.

b. Find the current in the branch $b-e$ at time t. What is the current as t goes to infinity?

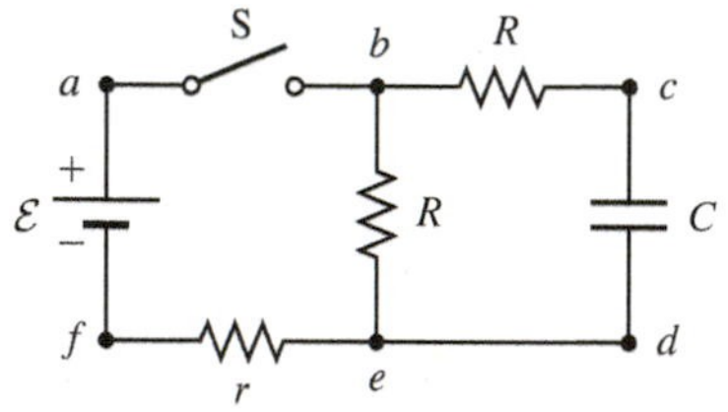

FIGURE P29.78

79. **N** A 12.0-V battery is used to charge a 4.00-μF capacitor in a simple RC circuit. After 5.00 s have elapsed, there is a potential difference of 6.60 V across the capacitor. What is the resistance of the resistor in this circuit?

80. **A** Calculate the equivalent resistance between points P and Q of the electrical network shown in Figure P29.80.

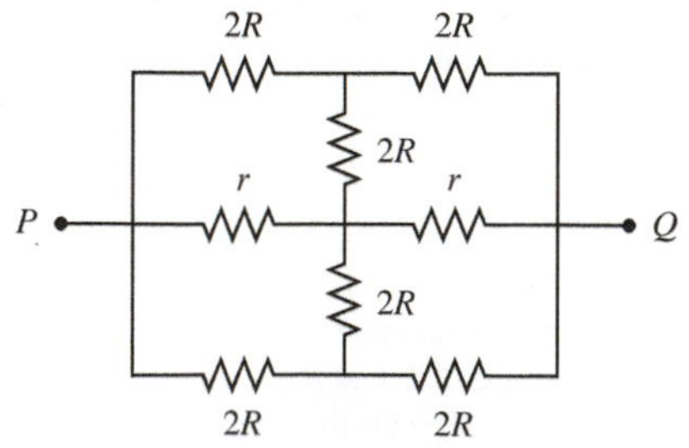

FIGURE P29.80

81. **A** In Figure P29.81, N real batteries, each with an emf $\mathcal{E}$ and internal resistance r, are connected in a closed ring. A resistor R can be connected across any two points of this ring, causing there to be n real batteries in one branch and $N - n$ resistors in the other branch. Find an expression for the current through the resistor R in this case.

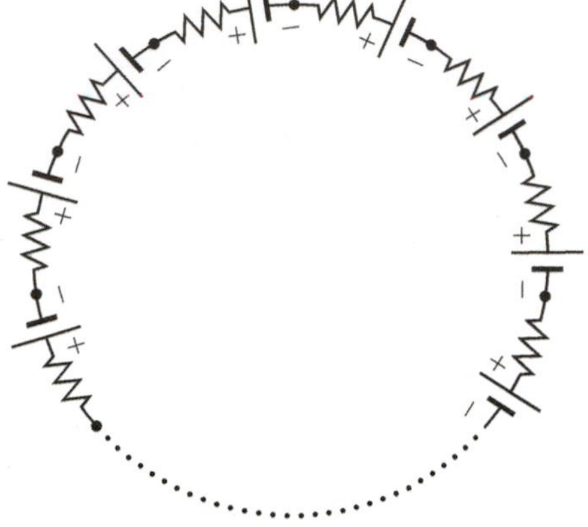

FIGURE P29.81

82. **A** By integrating, show that

$$\frac{1}{RC}\int_0^t dt = \int_0^q \frac{dq}{C\mathcal{E} - q}$$

leads to

$$-\frac{t}{RC} = \ln\left(\frac{C\mathcal{E} - q}{C\mathcal{E}}\right)$$

83. **A** Show that the time constant RC has the SI units of seconds.

Problems 84, 85, 86, and 87 are grouped.

84. **A** Figure P29.84 shows a circuit that consists of two identical emf devices. If $R_1 = R_2 = R$ and the switch is open, find an expression (in terms of R and $\mathcal{E}$) for the current I that is in the branch from point a to b.

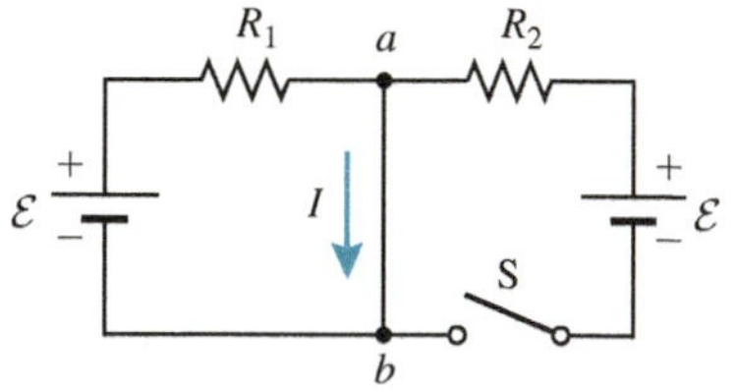

FIGURE P29.84 Problems 84–87.

85. **A** Figure P29.84 shows a circuit that consists of two identical emf devices. If $R_1 = R_2 = R$ and the switch is closed, find an expression (in terms of R and $\mathcal{E}$) for the current I that is in the branch from point a to b.

86. **A** Figure P29.84 shows a circuit that consists of two identical emf devices. If $R_1 = R$ and $R_2 = 2R$ and the switch is open, find an expression (in terms of R and $\mathcal{E}$) for the current I that is in the branch from point a to b.

87. **A** Figure P29.84 shows a circuit that consists of two identical emf devices. If $R_1 = R$ and $R_2 = 2R$ and the switch is closed, find an expression (in terms of R and $\mathcal{E}$) for the current I that is in the branch from point a to b.

Problems 88, 89, 90, 91, and 92 are grouped.

88. **A** Figure P29.88 shows a circuit that consists of two identical emf devices. If $R_1 = R_2 = R$ and the switch is open, find an expression (in terms of R and $\mathcal{E}$) for the current I_1 that is in resistor 1.

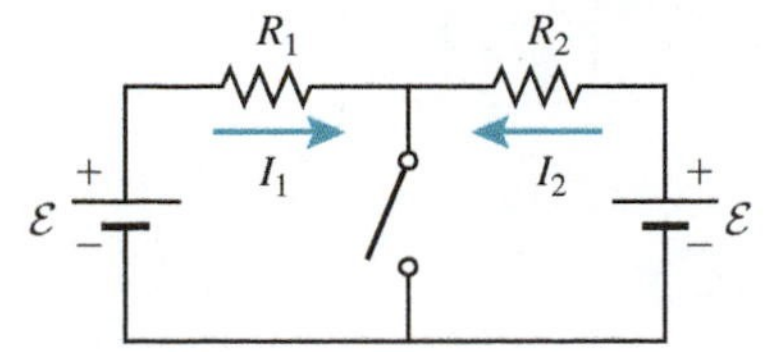

FIGURE P29.88 Problems 88–92.

89. **A** Figure P29.88 shows a circuit that consists of two identical emf devices. If $R_1 = R_2 = R$ and the switch is closed, find an expression (in terms of R and $\mathcal{E}$) for the current I_1 that is in resistor 1.

90. **A** Figure P29.88 shows a circuit that consists of two identical emf devices. If $R_1 = R$ and $R_2 = 2R$ and the switch is open, find an expression (in terms of R and $\mathcal{E}$) for the current I_1 that is in resistor 1.

91. **A** Figure P29.88 shows a circuit that consists of two identical emf devices. If $R_1 = R$ and $R_2 = 2R$ and the switch is closed, find an expression (in terms of R and $\mathcal{E}$) for the current I_1 that is in resistor 1.

92. **A** Figure P29.88 shows a circuit that consists of two identical emf devices. If $R_1 = R$ and $R_2 = 2R$ and the switch is open, find an expression (in terms of R and $\mathcal{E}$) for the current I_2 that is in resistor 2.

93. **A** Show that for a discharging RC circuit (Fig. 29.44) the charge $q(t)$ on the capacitor as a function of time is given by $q(t) = q_{\max} e^{-t/\tau}$ (Eq. 29.16). *Hint*: What term in Equation 29.14 is zero for the discharging RC circuit?

94. **C** CASE STUDY Avi, Cameron, and Shannon turn their attention to the water heater problem. The symptom is that the hot water is merely warm and the supply runs out too quickly. From their research, the students know that the hot water heater has two heating coils that act as large resistors. Current in the coils increases their thermal energy, some of which flows into the water through conduction and convection. The power supply is AC, but the students have a digital multimeter that measures the rms voltage across any circuit element (rms = root mean square; Section 20-2). For the rest of the analysis, the circuit may be treated as a DC circuit with an emf equal to the rms voltage. The students use their multimeter to make two measurements: (1) The rms voltage across each coil is found to be 247 V. (2) They turn off power to the water heater and then measure the resistance across each coil. One coil's resistance is 13 Ω. When they try to measure the other resistance, the ohmmeter gives an error message. The students want to know if the coils are connected in parallel or in series and what the error message means.

Shannon: I think the coils are in parallel like the headlights in the car (Fig. 29.20).

Cameron: No. I think they are in series. If the coils are in series, the error message means the resistance of one has dropped to nearly zero, and that would explain the problem. The one working coil is just not enough to heat the whole tank of water.

Shannon: I agree that one coil must not be working, and it has to be the one that gives us the error message. I don't think the coils can be in series. It's like the headlights in a car. If one blows, both lights go out when they are in series.

Cameron: That's because when a lightbulb burns out, the filament breaks, so there is no current passing through it. I don't think the coil is broken like that. It's just lost its resistance.

Avi: There's an easy way to tell if they are in series or in parallel—from the voltage. If the coils have the same voltage, they're in parallel.

Cameron: If they're in parallel, and the resistance of one of them is essentially zero, then all the current would go through that one. The other coil would get no current. The water would never heat up because the good coil never gets hot.

Shannon: I think the error message might mean that the coil has *infinite* resistance. That makes more sense. If the resistance is very low, as it is for a wire, the meter would give you a zero or a really small number. I don't think it would give an error message.

Figure P29.94 shows the circuits described by Shannon and Cameron. Which circuit correctly fits the hot water heater's symptoms? Explain your answer.

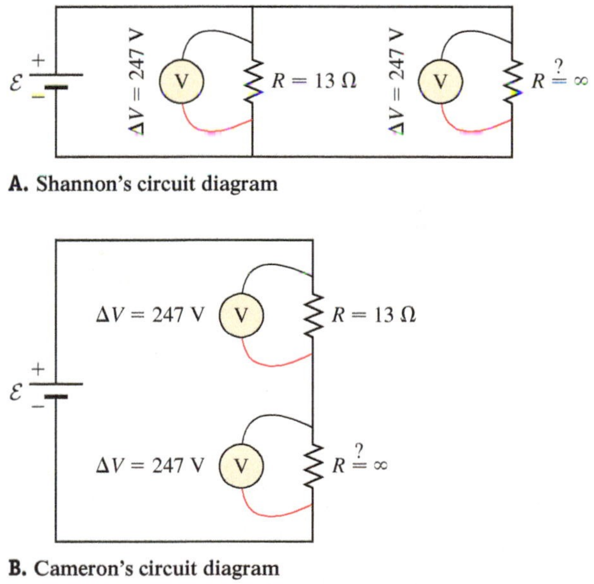

FIGURE P29.94

95. **A** Show that for a discharging RC circuit, the current $I(t)$ as a function of time is $I(t) = -I_0 e^{-t/\tau}$ (Eq. 29.17). Explain the leading negative sign.

30 Magnetic Fields and Forces

Key Questions

What are the sources of magnetic fields?

What properties of matter make it subject to a magnetic force?

How do we calculate magnetic fields and magnetic forces?

Underlying Principles

Magnetic fields and magnetic forces

Major Concepts

1. Magnetic dipole and magnetic monopole
2. Biot-Savart law
3. Magnetic dipole moment
4. Ferromagnetic materials and magnetic domains
5. Magnetic force on a moving charged particle
6. Lorentz force
7. Magnetic force on a current-carrying wire

Special Cases

1. Magnetic field due to
   a. Long, straight wire
   b. Circular wire (or current loop)
   c. Magnetic dipole
2. Magnetic force between two parallel current-carrying wires
3. Torque on a magnetic dipole
4. Magnetic potential energy stored in a dipole-magnetic field system

Tools

1. Magnetic field lines
2. Simple right-hand rule
3. Right-hand rule for direction of magnetic dipole moment

In the fall of 2007, Vikki Ortiz was on her way to a Milwaukee restaurant when her keys fell into a sewer grate. Vikki recounted her story in a blog read by Joselyn McKinley and her coworker, Dave Dulek. Joselyn and Dave's job is to rescue items from sewers. They used a large permanent magnet on a rope to retrieve the keys (Fig. 30.1).

The word *magnet* comes from the name Magnesia—a city in ancient Greece where naturally occurring magnets known as *lodestones* were found. Many ancient peoples believed lodestones were magical and even dangerous because they could attract bits of iron and attract or repel other lodestones. By the 12th century, the Chinese used lodestones as compasses. In the 1600s, magnets were studied by the scientific community. For 200 years, scientists tried to explain why magnets exert forces on bits of iron, push or pull on each other, and rotate to point toward the Earth's North Pole. We begin this part of the textbook by describing properties of magnets, magnetic fields, and magnetic forces.

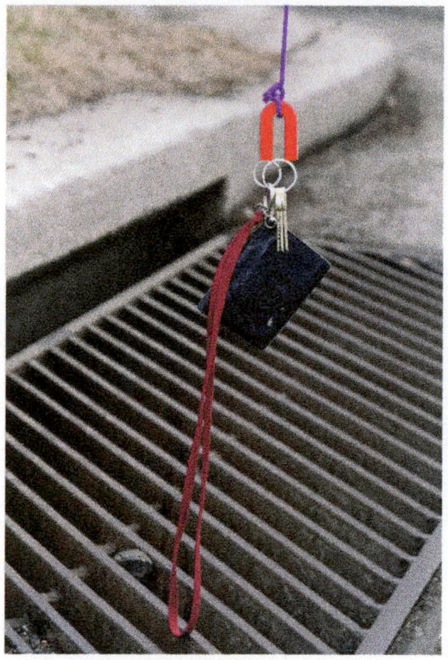

FIGURE 30.1 Using a magnet to retrieve keys from a city sewer.

30-1 Another Fundamental Force

If you play with permanent magnets, you will find that a magnet attracts certain metal objects, such as paper clips and Canadian coins. You will also learn that not all metal objects are attracted by a magnet. For example, a magnet will not pick up American coins. And, if you have two magnets, certain arrangements of the two magnets result in their mutual attraction while other arrangements result in their mutual repulsion. This repulsion can produce a stunning effect as shown in Figure 30.2; the upper magnets hover.

One of the amazing things we can see in Figure 30.2 is that, although none of the magnets are in contact with each other, they clearly exert a force on each other because the upper magnets are hovering. Magnet A experiences a downward force due to gravity, balanced by an upward **magnetic force** exerted by magnet B. Because the magnets are not in contact, we conclude that the magnetic force is a *field force*, like gravity. Physicists have identified four fundamental forces—the electromagnetic force, the gravitational force, the strong force, and the weak force (Fig. 23.2, page 684). The electromagnetic force is a combination of the electrostatic force, which we studied in Part III, and the magnetic force. All the fundamental forces are field forces.

MAGNETIC FIELDS AND MAGNETIC FORCES **Underlying Principle**

The source of a gravitational field is an object that has mass, and the source of an electric field is an object that has an excess of either positively or negatively charged particles. The source of a **magnetic field** is an object that has a net motion of charged particles—or, in the case of permanent magnets, the primary source is the "motion" of the electrons bound to their atoms. Like the gravitational and electric fields, the magnetic field is a vector field, having both a magnitude and a direction at each point in space.

FIGURE 30.2 Each magnet consists of a blue side and a yellow side. The blue sides are mutually repulsive and the yellow sides are mutually repulsive. So magnet A is repelled by and hovers above magnet B, and magnet B is repelled by and hovers above magnet C. The blue side of one magnet attracts the yellow side of another magnet (not shown).

The words *force* and *field* are similar. Read carefully so you don't confuse them.

CASE STUDY The Hall Effect

In this case study, we tie up a major loose end that first appeared in our study of electricity (Chapters 23 and 24). Ben Franklin came up with his own model for electrostatics, in which all objects are full of an "electric fluid." When Franklin rubbed one object against another, such as silk against glass, he thought electric fluid was transferred from one object to the other. Franklin called the object with excess electric fluid *positive*, and the object with a deficit of fluid *negative*. However, Franklin could not perform an experiment to tell him which way the fluid flowed, so he made an arbitrary choice. He said some of the electric fluid is transferred from the silk to the glass after rubbing. The silk then has a deficit of electric fluid and is *negative*. The glass has an excess of electric fluid and is *positive*. Subsequent scientists followed Franklin's arbitrary choice, but today we know that when glass is rubbed against silk, electrons (not electric fluid) are transferred *from* the glass *to* the silk. Because of Franklin's arbitrary choice, we say that electrons are negative and protons are positive.

Choosing electrons to be negative leads to a complication, and a bit of an inconvenience, when we study current in a conductor, such as a wire. If we think of

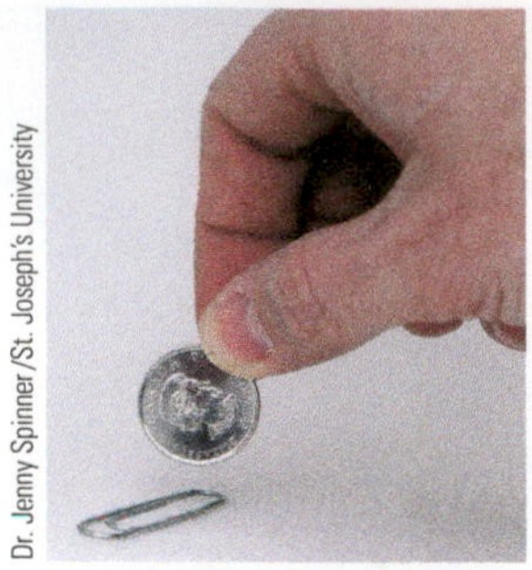

A.

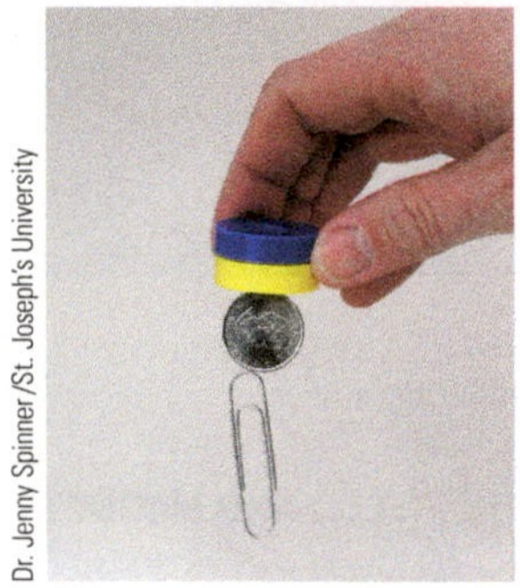

B.

FIGURE 30.3 **A.** Normally a Canadian coin is not a magnet. **B.** When the coin is near the toy magnet, the coin becomes magnetic and can pick up a paper clip.

electrons as positive charge carriers, the current in a wire is in the same direction as the electron flow. However, because we think of electrons as negative charge carriers, we use the convention that current is the result of *imaginary* positively charged particles flowing in the opposite direction. By contrast, 18th- and 19th-century experimenters assumed that current is the result of the motion of *real* positive particles. Their assumption would not be challenged by any of the circuit analyses we performed earlier in this book.

So, how do we know that, in fact, the negative particles—the electrons—are the particles that flow in a conductor? The answer comes from observing a current-carrying conductor in a magnetic field. This experiment was performed by the American physicist Edwin Hall in 1879. His observation, now called the *Hall effect*, showed that the current in a conductor results from the motion of negative charge carriers, not positive ones. To understand the Hall effect, we must first learn about magnetic fields and then about the magnetic force exerted on moving charged particles.

30-2 Revealing Magnetic Fields

We know that Canadian coins and paper clips are not normally magnetic; Canadian coins cannot be used to pick up paper clips under normal circumstances (Fig. 30.3A). When a Canadian coin is near a magnet, however, it acts like a magnet and is able to pick up a paper clip (Fig. 30.3B). Just as when we hold an electric dipole near a conductor and the conductor becomes polarized like the dipole, we infer that when a Canadian coin is near a magnet, the magnet causes changes inside the coin that make it act like a magnet. We will learn more about how objects (like the Canadian coin) are magnetized in Section 30-7.

From our experience with magnets, we know that some objects (paper clips and Canadian coins) are attracted by magnets and others (American coins) are not. This difference is due to their composition. Because iron, for example, is easy to magnetize, it responds to the presence of a magnetic field. A *bar* magnet is an example of a permanent magnet made of a magnetic material, such as iron. The iron may be bent into other shapes, such as a horseshoe, the letter C, or a donut. The geometry of these magnets is more complicated, but the basic physics is the same—so we focus much of our attention on bar magnets.

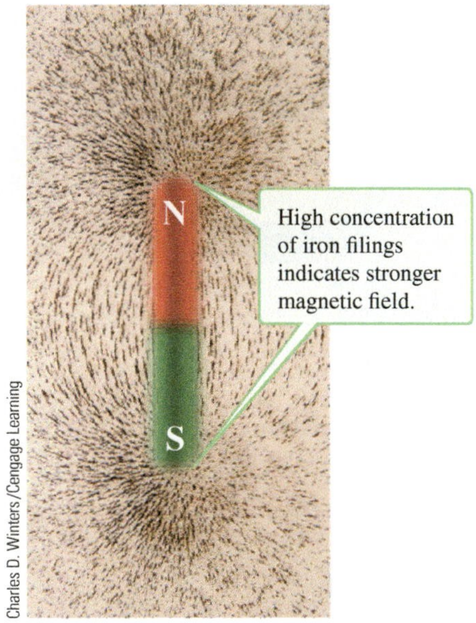

FIGURE 30.4 Iron filings trace the magnetic field near a bar magnet. (This magnet is painted red and green instead of blue and yellow. The colors are chosen for aesthetic reasons only.)

Small, needle-shaped shavings of iron known as *iron filings* line up with a magnetic field to reveal the **magnetic field lines**. Figure 30.4 shows iron filings near a bar magnet that form long arcs running from one end of the magnet to the other. The iron filings also indicate relative magnetic field strengths. Regions with a high concentration of iron filings have a high concentration of magnetic field lines, and correspond to places where the magnetic field is particularly strong. For a bar magnet, the field is particularly strong near the ends of the bar, known as **poles**.

Iron filings are helpful in indicating the relative strength of the magnetic field, but to find its direction we need to use a compass. Imagine making a compass needle from a bar magnet. To make the magnet into a useful compass, it should be marked in terms of north and south. Therefore, you need to know which way is north. When the magnet is allowed to rotate, one end of the bar will always point north—that is, in the general direction of the Earth's North Pole. This end is labeled N for *north pole*, and the other end is labeled S for *south pole*. In making two compass needles, you would find the north pole of one compass repels the other's north pole. (Their south poles also repel each other.) You would also see that the north pole of one compass attracts the other magnet's south pole. In general, *like magnetic poles repel and opposite poles attract.*

Just as iron filings align with the magnetic field, compass needles align with magnetic field lines (Fig. 30.5A). By design, the north pole of a compass needle

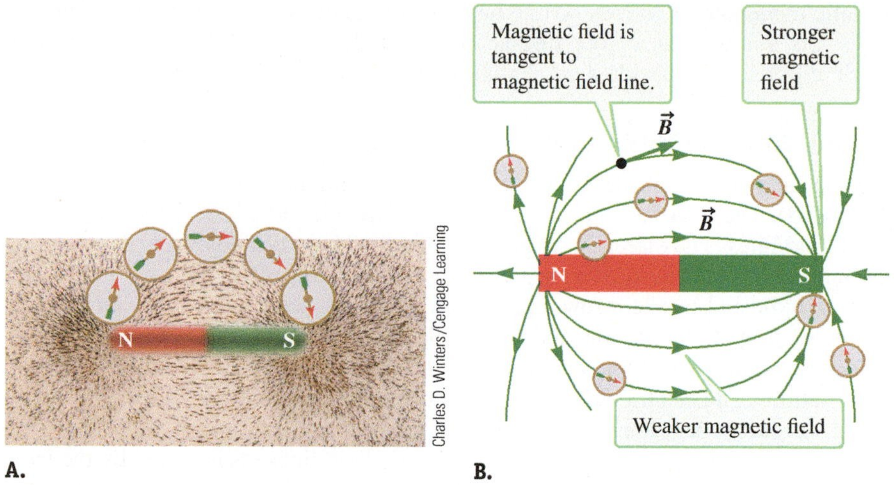

FIGURE 30.5 **A.** Compasses indicate the direction of the magnetic field. **B.** Magnetic field lines around a bar magnet.

points in the direction of the magnetic field. So, according to the compasses in Figure 30.5A, the magnetic field of a bar magnet points along the arc from its north pole to its south pole (Fig. 30.5B).

The symbol for the magnetic field is an uppercase $\vec{B}$.

Magnetic Monopoles and Dipoles

Compare Figure 30.5B to the *electric* field lines around an *electric* dipole in Figure 30.6A. The electric field around an electric dipole looks like the magnetic field around a bar magnet. The bar magnet is an example of a **magnetic dipole**. Although an electric dipole can be used as an analogy for a bar magnet, electric and magnetic dipoles are fundamentally different. The electric dipole's field is a result of stationary charged particles, whereas the bar magnet's field is a result of moving charged particles.

MAGNETIC DIPOLE **Major Concept**

This fundamental difference between an electric dipole and a bar magnet is revealed if you imagine breaking each one. First, imagine an electric dipole made up of two small spheres, one with charge $+q$ and the other with charge–q, held in place by a thin insulating rod (Fig. 30.6A). If you cut the rod in two and remove the negatively charged sphere, the electric field lines will point radially outward in all directions, as expected for a single positively charged sphere (Fig. 30.6B).

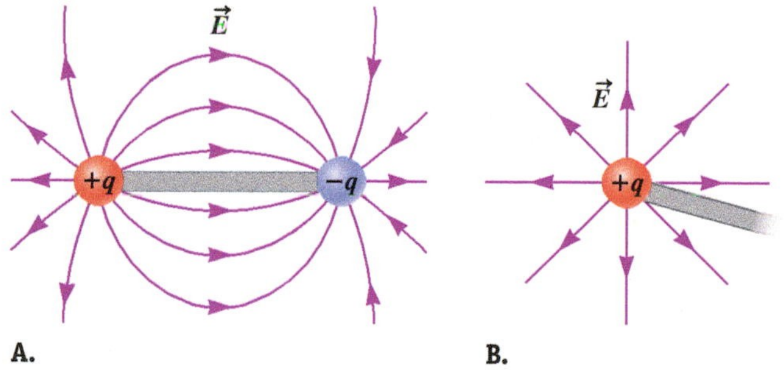

FIGURE 30.6 **A.** The electric field of an electric dipole is similar to the magnetic field of a magnetic dipole. **B.** The electric field of a broken dipole is simply the electric field of a single charged particle.

Now imagine cutting a bar magnet in two. You might expect that if you cut the bar into two equal pieces, the magnetic field lines of each piece would point radially outward (or inward). However, that is not what you find. Figure 30.7 shows iron filings around the two pieces of a broken bar magnet. Each piece has the same magnetic field pattern as the whole bar. In each piece, the iron filings form arcs that run from one end of the bar fragment to the other. A bar magnet is called a magnetic *di*pole because it has *two* poles. Even after breaking the bar magnet, we still observe two poles *in each fragment*, with magnetic field lines running from one pole to the other. Furthermore, both pieces are magnetic dipoles, no matter where you cut the bar magnet. Even if you slice off just a small portion of one end, the resulting pieces will both have magnetic field lines running from one end to the other. Both pieces are magnetic dipoles.

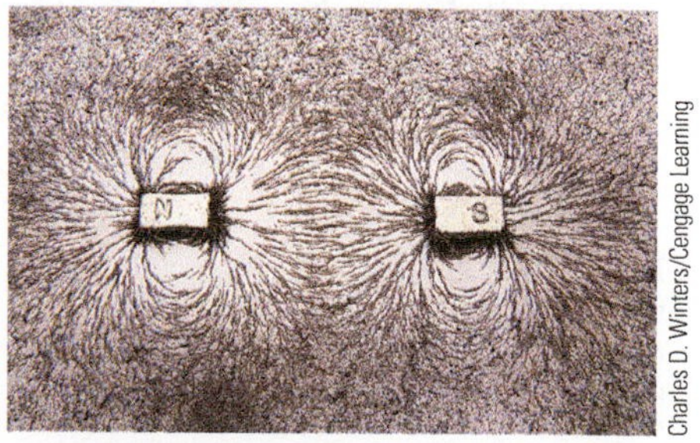

FIGURE 30.7 The magnetic field of a broken bar magnet looks like the magnetic fields of two smaller bar magnets.

A magnetic field is fundamentally different from an electric field: The source of a magnetic field never produces field lines that simply point outward or inward. Instead, the magnetic field lines always loop around, as they do for a magnetic dipole. According to physics theories, a source with only one magnetic pole and field lines pointing either outward or inward is possible; such a source is known as a

MAGNETIC MONOPOLE
Major Concept

MAGNETIC FIELD LINES Tool

Recall: Gravitational field lines originate from an object that has mass. Electric field lines originate from objects with excess positive charge and terminate on objects with excess negative charge.

magnetic monopole. However, no one has ever found evidence of a magnetic monopole.[1,2] So, for the rest of our discussion, we will assume they don't exist, and electric dipoles will provide a good analogy for many magnetic sources.

Internal and External Magnetic Field Lines

Magnetic field lines indicate both the strength and direction of the magnetic field (Fig. 30.5B). The relative strength of a magnetic field is indicated by the concentration of field lines. The direction of the magnetic field is indicated by an arrow on the field line. The magnetic field at a particular point is tangent to the line and in the direction indicated by the arrow. Outside of a bar magnet, the magnetic field lines point *away from* the north pole and *toward* the south pole.

Because no magnetic monopole is known to exist, magnetic field lines, in contrast to gravitational and electric field lines, never originate from or terminate on a source. Instead, magnetic field lines always form closed loops.

In Figure 30.5A, it may look like the magnetic field lines (as revealed by the iron filings) start at the north pole and end at the south pole. However, these are just the *external* portions of the actual magnetic field lines. The magnetic field lines continue through the body of the bar magnet from the south pole to the north pole (Fig. 30.8). So, for a bar magnet, the *external* magnetic field lines point from the north pole toward the south pole, and the *internal* magnetic field lines point from the south pole toward the north pole. Notice that the magnetic field lines inside the bar magnet are densely packed, indicating that the internal magnetic field is very strong.

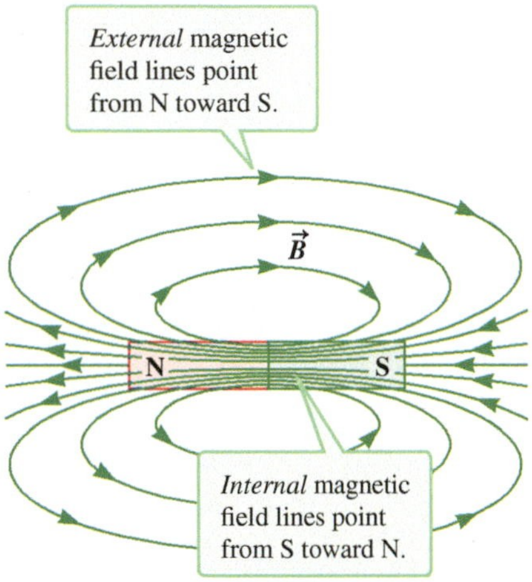

External magnetic field lines point from N toward S.
$\vec{B}$
N
S
Internal magnetic field lines point from S toward N.

FIGURE 30.8 Magnetic field lines in and around a bar magnet. (The magnetic field lines exist in three dimensions, but only two dimensions are shown.)

The Earth's Magnetic Field

The Earth's core is made up of molten iron and nickel. The swirling, convective motion of these molten metals gives rise to the Earth's magnetic field, which forms a magnetic dipole that may be modeled as a bar magnet (Fig. 30.9).

In Figure 30.5A, the north poles of the compass needles point toward the south pole of the bar magnet. Therefore, a compass's "north" points to the *south* magnetic pole of the Earth. In other words, the Earth's geographic North Pole is a *magnetic south* pole. If this seems confusing, remember how we imagined constructing a compass out of a bar magnet: We allowed the magnet to rotate so that it aligned with the Earth's magnetic field, and then we labeled as N the pole that pointed toward the Earth's geographic North Pole. Remember also that, in general, opposite magnetic poles attract.

There are a couple other complications. First, the Earth's geographic North Pole and magnetic south pole are not in exactly the same position. The geographic North Pole lies along the Earth's axis of rotation, as found by observing the North Star (Polaris). The North Star does not appear to move during the course of the night, so the Earth's rotation axis passes through the North Star. The magnetic south pole is found by using a compass and is about 11° from the geographic North Pole. Because most of us live in the midlatitudes, this slight discrepancy doesn't matter when we use a compass to navigate.

The other complication is that the Earth's magnetic poles are not fixed. The poles switch roughly every 300,000 years, so the north magnetic pole will one day be very close to the geographic North Pole, and the south magnetic pole will be very close to the geographic South Pole. The magnetic poles also wander several tens of kilometers daily.

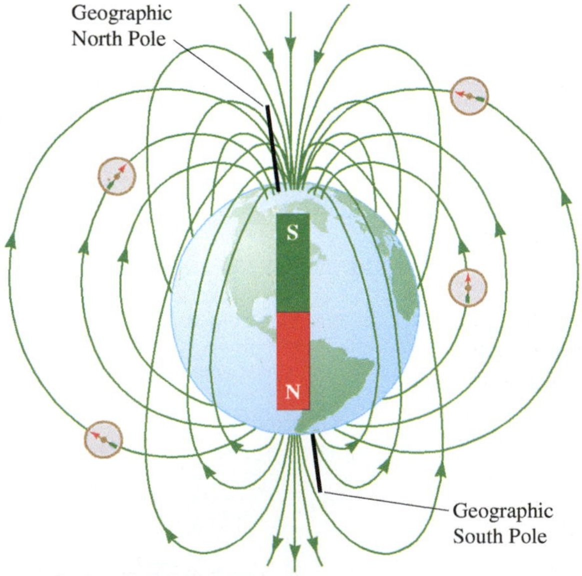

Geographic North Pole
S
N
Geographic South Pole

FIGURE 30.9 The Earth's magnetic field may be modeled as a bar magnet. The south magnetic pole is about 11° from the geographic North Pole. So a compass needle points toward the south magnetic pole, which is often good enough for navigation.

[1]On Valentine's Day (February 14, 1982), Blas Cabrera reported finding a magnetic monopole (*Science* 216: 1082–1088, June 1982). Today this is known as the Valentine's Day Monopole. No one was able to confirm the discovery.

[2]In 2009, two independent research groups reported that they had created magnetic monopoles in an artificial substance called a spin ice. The monopoles they made were not like elementary particles; instead, they were an unbound pair of north and south poles in the spin ice.

CONCEPT EXERCISE 30.1

What would the magnetic field lines of an isolated north magnetic pole look like?

30-3 Ørsted's Discovery

Although magnets such as lodestones were studied and used for many centuries, the mechanism for generating a magnetic field was not known until 1820. While Hans Christian Ørsted (Oersted), a science professor at Copenhagen University, was giving an electricity and magnetism demonstration, he placed a compass near a current-carrying wire. Ørsted (and his audience) observed that the compass needle was deflected. Months later, Ørsted conducted the experiment more carefully and got similar results. His experiment connects the theories of electricity and magnetism by leading to the idea that *moving charged particles can be the source of a magnetic field.*

You may see a version of Ørsted's demonstration in your class (Fig. 30.10). Within a simple circuit, consisting of a wire connected to a switch and an emf device, a portion of the wire is arranged to be stiff and vertical. A small horizontal platform surrounds this portion of the wire, and several small compasses rest on the platform. When the switch is open so that there is no current in the wire, the compass needles point in the general direction of the Earth's geographic North Pole, as usual (Fig. 30.10A). When the switch is closed, so that there is an upward current in the vertical portion of the wire, the compass needles are deflected, pointing counterclockwise around a circle, as seen from above the platform (Fig. 30.10B). If the emf device is reversed so that the current in the vertical wire is downward, the compass needles are deflected so that they point clockwise around the circle (Fig. 30.10C).

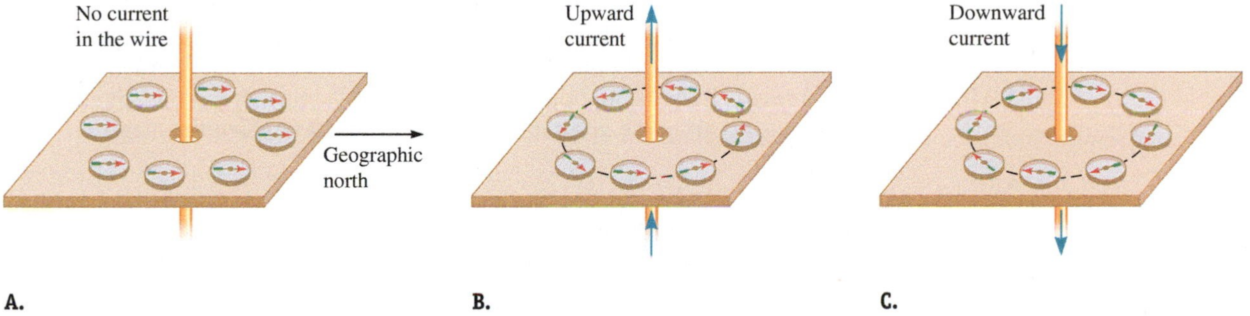

FIGURE 30.10 **A.** When there is no current in the wire, the compasses point in the general direction of the Earth's geographic North Pole (the Earth's magnetic south pole). **B.** When there is an upward current in the wire, the compass needles are deflected so that they point around a circle. **C.** When there is a downward current in the wire, the compass needles are deflected in the opposite direction from their deflection in part B.

Consider again the upward current in Figure 30.10B. If we move the platform up or down, we find that the compass needles still point in a counterclockwise loop around the wire. If we move the compasses closer to or farther from the wire, we also find that they point in a counterclockwise loop. So, the magnetic field around a long, straight, current-carrying wire wraps around the wire in cylindrical sheets (Fig. 30.11).

There is a **simple right-hand rule (SRHR)** for finding the direction of the magnetic field produced by a current-carrying wire. Imagine grabbing the wire with your right hand and pointing your right thumb in the direction of the current. The current's direction is the direction in which real (or fictitious) positive particles move. Your fingers wrap around the wire in the direction of the magnetic field (Fig. 30.11). The SRHR works even if the wire is curved. Just imagine grabbing a small portion of wire, and your fingers will wrap in the direction of the magnetic field produced by that small portion.

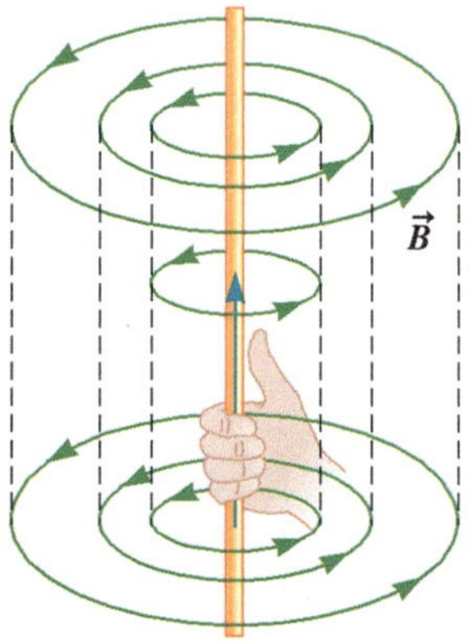

FIGURE 30.11 Using the SRHR: Place the thumb of your right hand in the direction of the current. Your fingers will wrap around the wire in the direction of the magnetic field $\vec{B}$. Here, $\vec{B}$ is counterclockwise as viewed from above.

SIMPLE RIGHT-HAND RULE (SRHR)

Tool

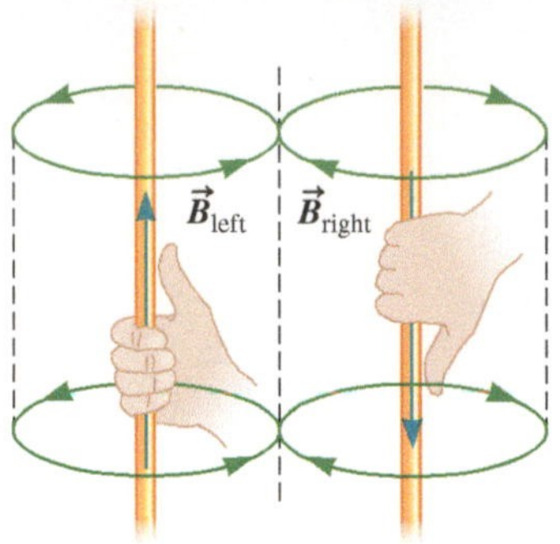

FIGURE 30.12 A wire is bent so that it forms a long, thin rectangle. The top and bottom of the rectangle are not shown in the figure. Between the left portion and the right portion of this long rectangular wire, the magnetic fields add and point into the page.

The magnetic field at any point near a curved current-carrying wire is the vector sum of the magnetic fields produced by each small segment. For example, imagine a wire bent into a long, thin rectangle. Figure 30.12 shows the left and right sides of the rectangle. The left side carries an upward current, and the right side carries a downward current. Use the SRHR twice—once for the left side and again for the right—to find the magnetic field midway between the two sides of the wire. According to the SRHR, the magnetic field due to each side of the wire points into the page, so the total magnetic field midway between the two portions points into the page (Fig. 30.12).

Compare the magnetic field produced by a bar magnet (Fig. 30.8) to the magnetic field produced by a long, straight, current-carrying wire (Fig. 30.11). Both sources produce closed magnetic field lines. The magnetic field lines associated with the bar magnet form squished loops that bunch together inside the bar, passing through the ends of the magnet at its poles. The magnetic field lines associated with the long, straight wire form concentric circles (along concentric cylinders). One major difference between a bar magnet and a long, straight wire is that the magnetic field lines of the bar magnet run into and out of the ends of the bar magnet. The places where the lines pass through the magnet are the poles. Because the loops around the wire do not pass through the wire, the wire does not have poles. Only when the field lines pass through a source, as in the case of the bar magnet, do we find it convenient to describe those places as poles.

The long, straight, current-carrying wire is an example of an **electromagnet**. The magnetic field produced by an electromagnet results from a gross motion of charged particles, such as the current in a wire. Electromagnets are often formed by coiling wire into a long cylinder, like a Slinky. By contrast, a permanent magnet is the source of a magnetic field even if the magnet carries no current. A permanent magnet's field results from the motion (and quantum-mechanical properties) of electrons within atoms, not from the gross motion of electrons throughout the magnet.

CONCEPT EXERCISE 30.2

In a wire made of copper or aluminum, the charged particles that move are really electrons. However, the conventional current is said to be due to fictitious positive particles that move in the opposite direction from the actual electrons. The simple right-hand rule is based on this convention. How does the SRHR have to be modified if we use the average motion of actual electrons instead of the usual convention for current?

30-4 The Biot-Savart Law

Biot-Savart rhymes with "Leo Guitar."

So far, our study of the magnetic field has been qualitative. Eventually, we want to calculate the magnetic *force* exerted on various *subjects*, but first we need to learn to calculate the magnetic *field* produced by various *sources.*

Soon after Ørsted's discovery, two French physics professors—Jean-Baptiste Biot and Felix Savart—experimented with current-carrying wires. By using a compass to measure the magnetic field around these wires, they discovered an empirical law known as the **Biot-Savart law** (Problem 94), which we can use to calculate the magnetic field.

The Magnetic Field Due to a Moving Charged Particle

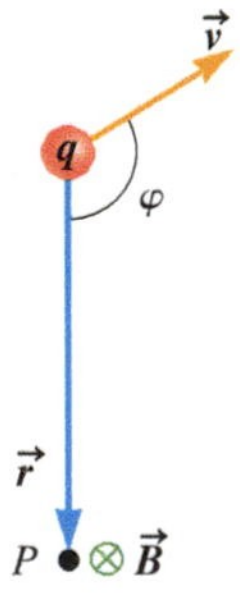

FIGURE 30.13 A charged particle moves at velocity $\vec{v}$, creating a magnetic field.

Consider first the magnetic field that results from the motion of a single charged particle, where we are interested in finding the magnetic field at point P (Fig. 30.13). The vector $\vec{r}$ points from the charged particle to point P. According to the Biot-Savart law, the magnetic field $\vec{B}$ produced by a single particle with charge q moving at velocity $\vec{v}$ is given by

BIOT-SAVART LAW FOR A SINGLE CHARGED PARTICLE Major Concept

$$\vec{B} = \left(\frac{\mu_0}{4\pi}\right) q \frac{\vec{v} \times \vec{r}}{r^3} \tag{30.1}$$

The SI unit for the magnitude of the magnetic field (also called the field strength) is the **tesla** (T), where

$$1\,\text{T} \equiv 1\frac{\text{N}\cdot\text{s}}{\text{C}\cdot\text{m}} = 1\frac{\text{N}}{\text{A}\cdot\text{m}}$$

To help develop your intuition about magnetic field strengths, the field strengths for several sources are listed in Table 30.1. The constant μ_0 is called the **permeability of free space**, with the exact value

$$\mu_0 = 4\pi \times 10^{-7}\frac{\text{T}\cdot\text{m}}{\text{A}} \tag{30.2}$$

Notice that when we substitute μ_0 into the Biot-Savart law (Eq. 30.1), the constant in parentheses is $10^{-7}\,\text{T}\cdot\text{m/A}$.

TABLE 30.1 Typical magnetic field strengths.

Source	*B* (T)
Interstellar clouds in the Milky Way galaxy	2×10^{-10}
Magnetic field produced by the human body	3×10^{-10}
Earth's magnetic field near its surface	5×10^{-5}
Sun's magnetic field near its surface	2×10^{-4}
Refrigerator magnet	5×10^{-3}
Magnet used in MRI[a]	2
World's strongest magnet[b]	45
Magnetic field near the surface of a neutron star	10^8 to 10^{10}

[a]MRI stands for *m*agnetic *r*esonance *i*maging, also known as NMR (*n*uclear *m*agnetic *r*esonance).

[b]As of 2009, this record is held by the hybrid magnet at the National High Magnetic Field Laboratory (Florida State University, Los Alamos National Laboratory, and University of Florida).

The direction of the magnetic field is determined by the sign of the charge and by the cross product in Equation 30.1. Point the fingers of your right hand in the direction of the first vector $\vec{v}$; then close your hand so that you "push" $\vec{v}$ into $\vec{r}$ (Section 12-6). For a positive charge, your thumb then points in the direction of $\vec{B}$. The magnetic field at point P is directed into the page (Fig. 30.13).

The magnitude of the cross product is given by Equation 12.22:

$$|\vec{v} \times \vec{r}| = vr\sin\varphi \tag{30.3}$$

where φ is the angle between $\vec{v}$ and $\vec{r}$. We can find the magnitude of the magnetic field B at point P by substituting Equation 30.3 into Equation 30.1 and canceling a power of r from the numerator and denominator:

$$B = \left(\frac{\mu_0}{4\pi}\right)q\frac{v\sin\varphi}{r^2} \tag{30.4}$$

Equation 30.4 shows that the strength of the magnetic field due to a single moving charged particle obeys an inverse-square law, just like the electric field due to a single stationary charged particle, $\vec{E}(r) = kQ_S\hat{r}/r^2$ (Eq. 24.3), or the gravitational field outside a massive spherical object, $\vec{g}(r) = -GM\hat{r}/r^2$ (Eq. 7.13).

The Magnetic Field Due to a Current

Now let's find the magnetic field due to a segment of current-carrying wire. In the arbitrarily shaped wire in Figure 30.14, the current is roughly from left to right. Imagine marking off segments of the wire small enough to be considered straight lines of length $d\ell$. Suppose we want to find the magnetic field at a particular point, such as P. The Biot-Savart law for the magnetic field $d\vec{B}$ produced by one such small segment of wire is a slightly modified version of Equation 30.1:

BIOT-SAVART LAW FOR A CURRENT

Major Concept

$$d\vec{B} = \left(\frac{\mu_0}{4\pi}\right) I \frac{d\vec{\ell} \times \vec{r}}{r^3} \tag{30.5}$$

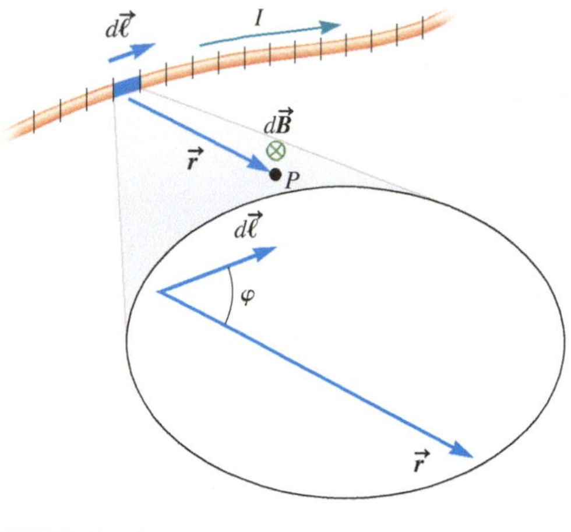

FIGURE 30.14 A current-carrying wire is broken into many small segments. The Biot-Savart law gives the magnetic field produced by each small segment at point P. According to the cross product $d\vec{\ell} \times \vec{r}$ the magnetic field $d\vec{B}$ produced by the blue segment at point P points into the page. Check this using the SRHR.

Let's apply the Biot-Savart law to the segment shown in blue in Figure 30.14. The length of the segment is described by the vector $d\vec{\ell}$, whose magnitude $d\ell$ is the linear size of the segment, measured in meters. The direction of $d\vec{\ell}$ is given by the current, which is upward and to the right for this segment. The vector $\vec{r}$ extends from the segment to the point P, where we would like to know the magnetic field. The cross product in the Biot-Savart law (Eq. 30.5) has magnitude

$$|d\vec{\ell} \times \vec{r}| = (d\ell)(r) \sin \varphi \tag{30.6}$$

where φ is the angle between $d\vec{\ell}$ and $\vec{r}$. The magnitude dB of the magnetic field produced by the blue segment at point P is found by substituting Equation 30.6 into Equation 30.5:

$$dB = \left(\frac{\mu_0}{4\pi}\right) I \frac{\sin \varphi}{r^2} d\ell \tag{30.7}$$

Equation 30.7 shows again that the magnetic field strength due to a *small* segment of a current-carrying wire is inversely proportional to r^2.

The cross product in the Biot-Savart law gives the direction of $d\vec{B}$. According to the right-hand rule for cross products (Section 12-6), the direction of $(d\vec{\ell} \times \vec{r})$ in Figure 30.14 (and therefore of the resulting vector $d\vec{B}$) is into the page. This direction must be consistent with the SRHR from Section 30-3. To check, point your right thumb in the direction of the current (upward and to the right). Then your fingers wrap around the wire in the direction of the magnetic field. Your fingers go into the paper for points like P below the wire. So the SRHR is consistent with the right-hand rule for cross products. Because the cross product and the SRHR give the same results, you can use either one to find the magnetic field direction. Then use Equation 30.7 to find the magnitude dB. When using the Biot-Savart law to find the magnetic field for a current-carrying wire, we must add the contributions of all the wire segments. In the next section, we will learn some specific steps for doing this.

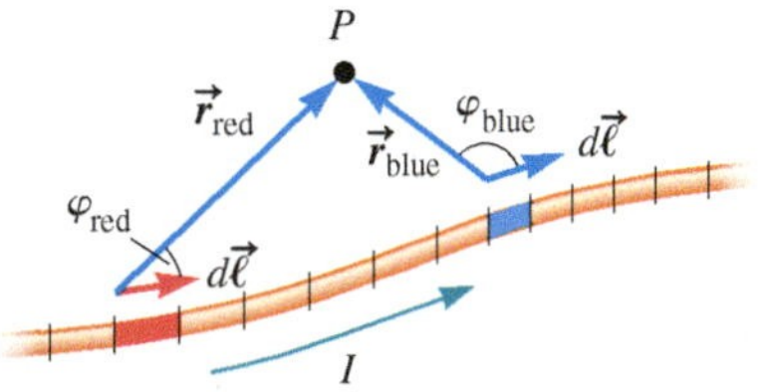

FIGURE 30.15 Which segment (red or blue) produces the stronger magnetic field at point P? *Hint*: $\varphi_{red} + \varphi_{blue} = 180°$

CONCEPT EXERCISE 30.3

In the next section, we'll learn how to calculate the total magnetic field at point P in Figure 30.15. For now, consider two segments—the blue one and the red one. Each segment has the same length $d\ell$. The red segment is farther from P, so $r_{red} > r_{blue}$; also $\varphi_{red} + \varphi_{blue} = 180°$. Use Equation 30.7 to state which segment produces the stronger magnetic field at point P.

30-5 Using the Biot-Savart Law

Using the Biot-Savart law to find the magnetic field involves a procedure similar to the one we used to find the electric field (Section 24-5). In both, we (1) imagine dividing the source into a number of small pieces, (2) find the field produced by a small piece, and (3) integrate over the entire source. In this section, we introduce a detailed problem-solving strategy and then use it to find the magnetic field produced by sources with different shapes.

PROBLEM-SOLVING STRATEGY

Using the Biot-Savart Law

INTERPRET and ANTICIPATE

We start with a diagram of the source—a current-carrying wire—and the point at which you want to find the magnetic field. There are four elements to this diagram:

1. A **coordinate system** and other **geometric details.**
2. **Labeled segments** of the source. Imagine slicing the current-carrying wire into a number of small segments. It usually helps to draw one or two of these segments, exploiting the symmetry of the problem when possible. Draw $d\vec{\ell}$, $\vec{r}$, and φ for each segment.
3. The **magnetic field vectors** produced by the segments at the point of interest. You may use either the SRHR or the cross product $(d\vec{\ell} \times \vec{r})$ from the Biot-Savart law to determine the direction of $d\vec{B}$ for each segment.
4. The **net magnetic field vector** at the point of interest. Find this graphically in order to determine whether any components cancel and to anticipate your result.

SOLVE

Step 1. Use the Biot-Savart law to find a **mathematical expression** for $d\vec{B}$, the small magnetic field produced by a single current segment at the point of interest. You may need only a component of $d\vec{B}$.

Step 2. Integrate $d\vec{B}$ to find $\vec{B}$ at the point of interest.

EXAMPLE 30.1 A Straight Wire

Figure 30.16 shows a straight wire of length L connected to a battery. The current in the wire is toward the right. Find an expression for the magnetic field at point P, directly above the midpoint of the wire.

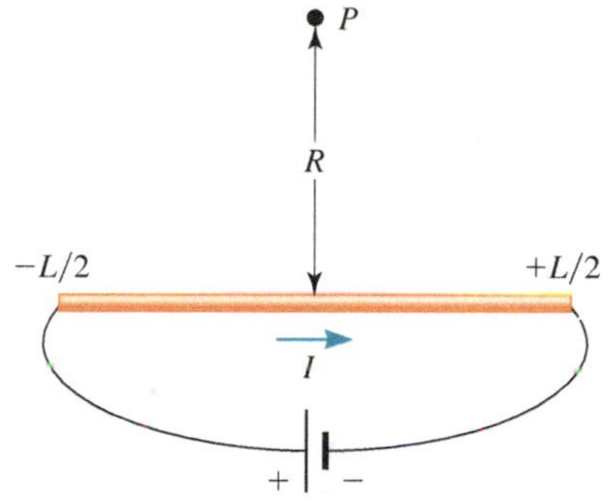

FIGURE 30.16

INTERPRET and ANTICIPATE

Start with a sketch. In this problem, we deal only with the straight wire, so we don't include the curved wire or the battery in our sketch (Fig. 30.17). We have included a **coordinate system** and some **geometric details.** Imagine slicing the wire into segments, exploit symmetry, and choose one segment from each side. We have **labeled segments** G and H. Draw $d\vec{\ell}$, $\vec{r}$, and φ for each segment. Because the wire lies along the x axis, we have labeled the length $d\vec{\ell}$ of each segment as $d\vec{x}$. Because $d\vec{x}$ is the same for both segments, we do not need a subscript. The **magnetic field vectors** produced by segments G and H both point out of the page in the positive z direction. So their **net magnetic field vector** $d\vec{B}$ also points out of the page, and we expect $\vec{B}$ to point in the positive z direction (Fig. 30.17).

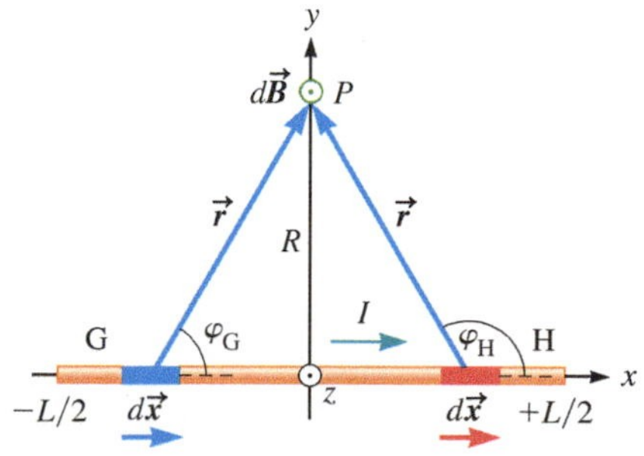

FIGURE 30.17

SOLVE

Step 1 Use the Biot-Savart law to find a **mathematical expression** for $d\vec{B}$. By symmetry, both segments G and H produce the same magnetic field $d\vec{B}_G = d\vec{B}_H$, so we can consider either one.

From Figure 30.17, $d\vec{B}$ is in the positive z direction, and the magnitude of the magnetic field is given by Equation 30.7. Substitute into Equation (1), replacing $d\ell$ with dx to match Figure 30.17.	$$d\vec{B} = (dB)\hat{k} \quad (1)$$ $$d\vec{B} = \left(\frac{\mu_0}{4\pi} I \frac{\sin\varphi}{r^2} dx\right)\hat{k} \quad (2)$$

Example continues on page 944 ▶

Step 2 Integrate Equation (2) over the entire length of the wire from $x = -L/2$ to $x = +L/2$.	$\int d\vec{B} = \int_{-L/2}^{L/2} \left(\frac{\mu_0}{4\pi} I \frac{\sin\varphi}{r^2} dx \right) \hat{k}$
In the many small segments that make up the wire, the vector $\vec{r}$ that runs from each segment to the point P varies in both magnitude and direction. So we cannot pull r or the angle φ outside the integral. We can pull out only the current and the constants.	$\vec{B} = \frac{\mu_0}{4\pi} I \left[\int_{-L/2}^{L/2} \left(\frac{\sin\varphi}{r^2} dx \right) \right] \hat{k}$ (3)

The variables x, r, and φ are interconnected. To find the connection, consider segment H and the triangle shown in Figure 30.18.

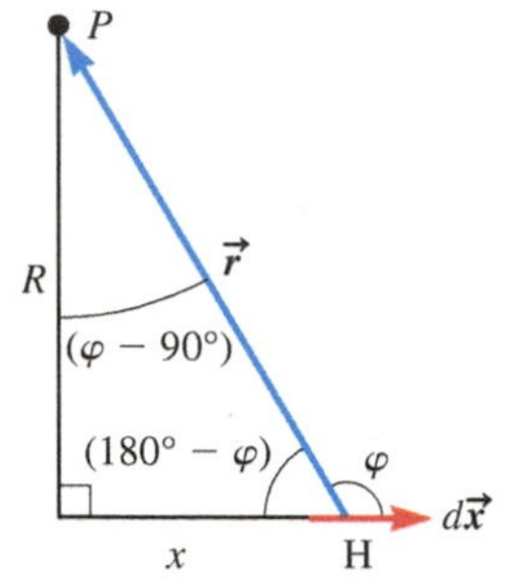

FIGURE 30.18

Use the Pythagorean theorem to express r in terms of x and the constant R. Next, use the trigonometric identity $\sin\alpha = \cos(\alpha - 90°)$. (See Appendix A.) Use this identity and Equation (4) to write $\sin\varphi$ in terms of x and R.	$r = (x^2 + R^2)^{1/2}$ (4) $\sin\varphi = \cos(\varphi - 90°) = \frac{R}{r} = \frac{R}{(x^2 + R^2)^{1/2}}$ (5)
Substitute Equations (4) and (5) into Equation (3).	$\vec{B} = \frac{\mu_0}{4\pi} I \left[\int_{-L/2}^{L/2} \frac{R}{(x^2 + R^2)^{1/2}} \frac{dx}{(x^2 + R^2)} \right] \hat{k} = \frac{\mu_0}{4\pi} I \left[\int_{-L/2}^{L/2} \frac{R\,dx}{(x^2 + R^2)^{3/2}} \right] \hat{k}$
The distance R between point P and the middle of the wire is constant, so we can pull it outside the integral.	$\vec{B} = \frac{\mu_0}{4\pi} IR \left[\int_{-L/2}^{L/2} \frac{dx}{(x^2 + R^2)^{3/2}} \right] \hat{k}$
The remaining integral can be found in Appendix A.	$\vec{B} = \frac{\mu_0}{4\pi} IR \left[\frac{x}{R^2(x^2 + R^2)^{1/2}} \right]_{-L/2}^{L/2} \hat{k}$
Evaluate between the limits and simplify.	$\vec{B} = \frac{\mu_0}{4\pi} IR \left[\frac{L/2}{R^2(L^2/4 + R^2)^{1/2}} - \frac{-L/2}{R^2(L^2/4 + R^2)^{1/2}} \right] \hat{k}$ $\vec{B} = \frac{\mu_0 I}{2\pi R} \left[\frac{L}{(L^2 + 4R^2)^{1/2}} \right] \hat{k}$ (30.8)

CHECK and THINK

As expected, the magnetic field at point P is in the positive z direction (out of the page).

A Very Long, Straight Wire

Equation 30.8 is an exact expression for the magnetic field at point P due to a straight current-carrying wire of length L (Fig. 30.16). We can model a long, straight wire as *infinitely* long when point P is very close to the wire. The limit of Equation 30.8 when L is very long compared to R is $\vec{B} \to (\mu_0 I/2\pi R)\hat{k}$. The direction of the magnetic field at point P in Figure 30.16 is out of the page in the positive z direction. (See Figure 30.17 for the coordinates.) According to the SRHR, the magnetic field wraps in concentric cylinders around an infinitely long, straight wire (Fig. 30.11). In other words, because the magnetic field does not point in the z direction for all points around the wire, it is often best to use the SRHR to find the field direction, and to remember that the magnitude of the magnetic field around an infinitely long wire is

MAGNETIC FIELD DUE TO A LONG, STRAIGHT WIRE ▶ Special Case

$$B = \frac{\mu_0 I}{2\pi r} \tag{30.9}$$

where, to be consistent with the equations for gravitational and electric fields, we use a lowercase "r" and r is the shortest distance from the wire to the point where the field is measured.

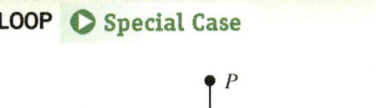

Notice that the magnetic field due to a long, straight wire has an inverse r dependence, not an inverse r^2 dependence.

EXAMPLE 30.2 Special Case: A Current Loop

MAGNETIC FIELD DUE TO A CURRENT LOOP ▶ Special Case

In addition to a long, straight, current-carrying wire, another important special case is the magnetic field produced by current in a circular loop of wire. Figure 30.19 shows a loop of wire connected to a battery. Model the loop as a circle, and show that the magnetic field at point P on the axis that runs through the center of the circle is given by

$$\vec{B} = \frac{\mu_0 I R^2}{2(R^2 + y^2)^{3/2}}\hat{\jmath} \tag{30.10}$$

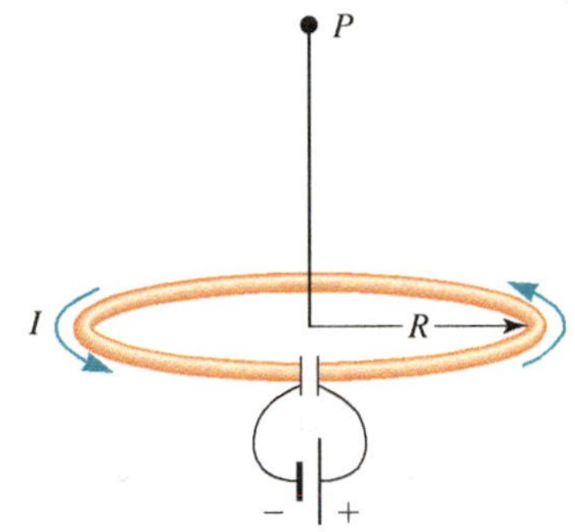

FIGURE 30.19

INTERPRET and ANTICIPATE

Sketch the current distribution (Fig. 30.20), choose a **coordinate system**, and add **geometric details**. In our coordinate system, the circular wire lies in the xz plane. Point P is on the y axis at a height y above the center of the circle. Imagine a cone with its vertex at point P and its base on the circle. The angle between the vertical and the wall of the cone is α. Now imagine dividing the loop into segments, with each segment so small that it is approximately straight. Two such **segments labeled** 1 and 2 are shown. The length element $d\vec{\ell}_1$ for segment 1 is in the positive z direction, and $d\vec{\ell}_2$ is in the negative z direction. Vectors $\vec{r}_1$ and $\vec{r}_2$ are in the xy plane, perpendicular to $d\vec{\ell}_1$ and $d\vec{\ell}_2$, respectively. Thus, $\varphi_1 = \varphi_2 = 90°$ (not shown on the figure). Sketch the **magnetic field vectors** produced by segments 1 and 2. Then add these to find the **net magnetic field vector**. It may be easiest to find the direction of $d\vec{B}$ using the cross product $(d\vec{\ell} \times \vec{r})$ in the Biot-Savart law. Consider segment 2, with $d\vec{\ell}_2$ in the negative z direction. Point your right fingers into the page; then close your hand so that you push $d\vec{\ell}_2$ into $\vec{r}_2$. Your right thumb should be parallel to the page, pointing upward and to the right (perpendicular to $\vec{r}_2$). This is the direction of $d\vec{B}_2$. The magnetic field $d\vec{B}_1$ due to segment 1 is also in the plane of the page, pointing upward and to the left (perpendicular to $\vec{r}_1$). Their magnitudes are equal: $dB_1 = dB_2$. They lie in the plane of the page, each having both an x and a y component. When these two vectors are added, their x components cancel and their resultant, $d\vec{B}_1 + d\vec{B}_2$, points in the positive y direction.

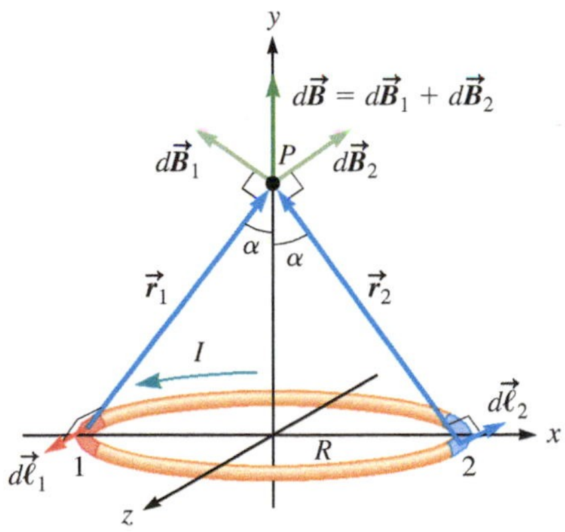

FIGURE 30.20

SOLVE

Step 1 Use the Biot-Savart law to find a **mathematical expression** for $d\vec{B}$. Any current segment will do, so we no longer need the subscripts. Remember that the horizontal components will cancel.

For all segments, $\varphi = 90°$ (the angle between $d\vec{\ell}$ and $\vec{r}$ in the Biot-Savart law; Eq. 30.5). The magnitude dB comes from Equation 30.7.	$dB = \left(\frac{\mu_0}{4\pi}\right) I \frac{\sin\varphi}{r^2} d\ell$ (30.7) $dB = \left(\frac{\mu_0}{4\pi}\right) I \left(\frac{1}{r^2}\right) d\ell$ (1)

Example continues on page 946 ▶

Let's arbitrarily choose segment 2 (Fig. 30.21). We need to integrate only the y component of $d\vec{B}$. The horizontal components (x and z) cancel.

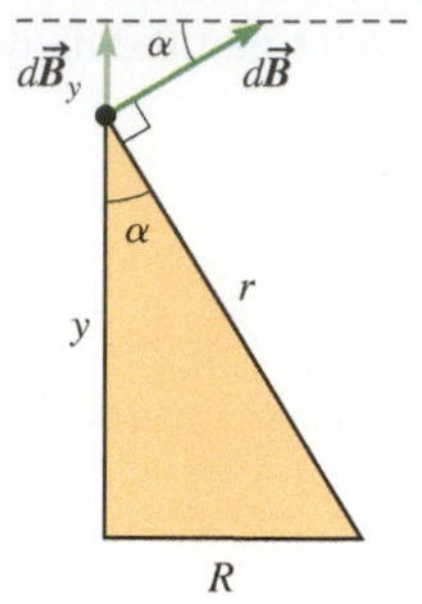

FIGURE 30.21

Apply trigonometry to the upper right triangle in Figure 30.21 to find the y component of $d\vec{B}$.	$dB_y = dB \sin\alpha$ (2)
Substitute Equation (1) into Equation (2).	$dB_y = \left(\dfrac{\mu_0}{4\pi}\right) I \dfrac{\sin\alpha}{r^2} d\ell$ (3)
Use the lower triangle highlighted in yellow in Figure 30.21 to write $\sin\alpha$ in terms of R and y.	$\sin\alpha = \dfrac{R}{r}$ $r = \sqrt{R^2 + y^2}$ (4) $\sin\alpha = \dfrac{R}{\sqrt{R^2 + y^2}}$ (5)
Substitute Equations (4) and (5) into Equation (3).	$dB_y = \left(\dfrac{\mu_0 I}{4\pi}\right)\left(\dfrac{1}{R^2 + y^2}\right)\left(\dfrac{R}{\sqrt{R^2 + y^2}}\right) d\ell$ $dB_y = \left(\dfrac{\mu_0 I}{4\pi}\right)\dfrac{R}{(R^2 + y^2)^{3/2}} d\ell$
Step 2 Integrate. To sum all the segments that make up the circular wire, we must integrate over the entire circumference. Thus, our limits of integration run from $\ell = 0$ to $\ell = 2\pi R$.	$\int dB_y = \int_0^{2\pi R} \left(\dfrac{\mu_0 I}{4\pi}\right)\dfrac{R}{(R^2 + y^2)^{3/2}} d\ell$
The parameters R and y are constant over the entire circle (Fig. 30.20), so we can pull them outside the integral along with the current and other constants.	$\int dB_y = \left(\dfrac{\mu_0 I}{4\pi}\right)\dfrac{R}{(R^2 + y^2)^{3/2}} \int_0^{2\pi R} d\ell$
Complete the integral, and evaluate between the limits.	$B_y = \left(\dfrac{\mu_0 I}{4\pi}\right)\dfrac{R}{(R^2 + y^2)^{3/2}} \ell \Big\vert_0^{2\pi R} = \left(\dfrac{\mu_0 I}{2}\right)\dfrac{R^2}{(R^2 + y^2)^{3/2}}$
Finally, write the magnetic field at point P in component form.	$\vec{B} = B_y \hat{\jmath}$ $\vec{B} = \dfrac{\mu_0 I R^2}{2(R^2 + y^2)^{3/2}} \hat{\jmath}$ ✓ (30.10)
CHECK and THINK The magnetic field at point P is in the positive y direction. A check of the SI units shows that they are teslas (T), as expected. In the next section, we will use Equation 30.10 to develop a model for bar magnets.	$[\vec{B}]_{\text{SI units}} = \dfrac{\text{T}\cdot\text{m}}{\text{A}} \dfrac{\text{A}\cdot\text{m}^2}{\text{m}^3}$ $[\vec{B}]_{\text{SI units}} = \dfrac{\text{T}\cdot\cancel{\text{m}}}{\cancel{\text{A}}} \dfrac{\cancel{\text{A}}\cdot\cancel{\text{m}^2}}{\cancel{\text{m}^3}} = \text{T}$

EXAMPLE 30.3 A Bent Wire

Figure 30.22 shows a bent wire connected to an emf device. The orange segment of the wire consists of two straight portions and a portion that is a quarter of a circle of radius R. Point P is at the center of the complete circle. Find an expression for the magnetic field at point P. Assume the portions of the wire shown in black are far from point P and contribute a negligible magnetic field.

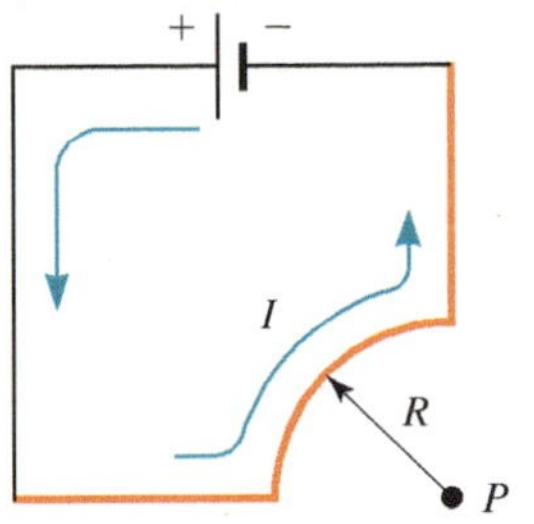

FIGURE 30.22

INTERPRET and ANTICIPATE

Sketch the current distribution (Fig. 30.23), choose a **coordinate system**, and add **geometric details**. The origin of the coordinate system is at point P (not labeled). The z axis points out of the page. Choose three **segments**: one for the horizontal portion, one for the curved portion, and one for the vertical portion—**labeled** 1, 2, and 3, respectively. The $\vec{r}$ vectors point from the length elements to the origin. For each portion of the wire, the angle φ between $d\vec{\ell}$ and $\vec{r}$ is constant. For the horizontal portion (segment 1), both $d\vec{\ell}_1$ and $\vec{r}_1$ are in the positive x direction, so $\varphi_1 = 0$. For the vertical portion (segment 3), $d\vec{\ell}_3$ points in the positive y direction and $\vec{r}_3$ points in the negative y direction, so $\varphi_3 = 180°$. The curved portion (segment 2) is part of a circle to which $d\vec{\ell}_2$ is tangent. Because $\vec{r}_2$ points toward the center of that circle, $\varphi_2 = 90°$. Use the cross product $(d\vec{\ell} \times \vec{r})$ in the Biot-Savart law to find the direction of $d\vec{B}$ for each segment. Because the cross product is zero for vectors that are parallel or antiparallel, segments 1 and 3 do not contribute to the magnetic field at point P. This leaves only segment 2. Place your right fingers in the direction of $d\vec{\ell}_2$ and close your hand in the direction of $\vec{r}_2$; your thumb points into the page. Thus, $d\vec{B}_2$ is in the negative z direction.

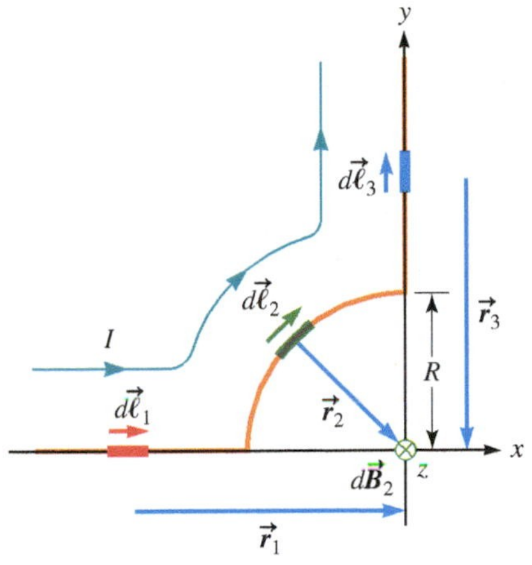

FIGURE 30.23

Only the curved portion of the wire contributes to the field at point P. To check our answer, we can use the results for a complete circular wire from Example 30.2.

SOLVE

Step 1 Use the Biot-Savart law (Eq. 30.7) to find a **mathematical expression** for dB. For every segment that makes up the curved portion, $\varphi = 90°$ and $r = R$.

$$dB = \left(\frac{\mu_0}{4\pi}\right) I \frac{\sin\varphi}{r^2} d\ell \quad (30.7)$$

$$dB = \left(\frac{\mu_0}{4\pi}\right) I \frac{\sin 90°}{R^2} d\ell = \left(\frac{\mu_0}{4\pi}\right) \frac{I}{R^2} d\ell$$

Step 2 **Integrate** over the curved portion, whose length is a quarter of the circumference of a circle: $2\pi R/4$. The integration limits are from $\ell = 0$ to $\ell = \pi R/2$.

$$\int dB = \int_0^{\pi R/2} \left(\frac{\mu_0}{4\pi}\right) \frac{I}{R^2} d\ell$$

Both I and R are constant for the entire curved portion, so we can pull them, as well as the other constants, outside the integral. Complete the integration.

$$\int dB = \left(\frac{\mu_0}{4\pi}\right) \frac{I}{R^2} \int_0^{\pi R/2} d\ell$$

$$B = \left(\frac{\mu_0}{4\pi}\right) \frac{I}{R^2} \ell \Big|_0^{\pi R/2} = \frac{\mu_0 I}{8R}$$

Taking direction into account, write an expression for the magnetic field at point P.

$$\vec{B} = -\frac{\mu_0 I}{8R} \hat{k} \quad (1)$$

CHECK and THINK

To check the magnitude of this answer, use the magnetic field of the complete circular wire found in Example 30.2. First, set $y = 0$ in Equation 30.10 to find an expression for the magnetic field at the center of the complete circle.

$$B(y) = \frac{\mu_0 I R^2}{2(R^2 + y^2)^{3/2}} \quad (30.10)$$

$$B(0) = \frac{\mu_0 I R^2}{2(R^2)^{3/2}} = \frac{\mu_0 I R^2}{2R^3} = \frac{\mu_0 I}{2R} \quad (2)$$

Example continues on page 948 ▶

Equation (2) is the magnetic field at the center of a complete circle of current, so we should divide by 4 to find the magnetic field due to a quarter of a circle. The result agrees with the magnitude of Equation (1).	$B_{\text{quarter}} = \dfrac{1}{4}\left(\dfrac{\mu_0 I}{2R}\right) = \dfrac{\mu_0 I}{8R}$ ✓

30-6 The Magnetic Dipole Moment and Modeling Atoms

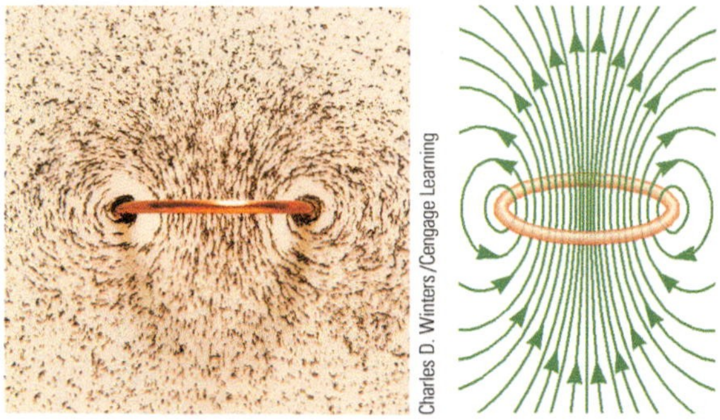

A. B.

FIGURE 30.24 **A.** Iron filings around a circular wire show the magnetic field produced by current in the wire. **B.** The magnetic field lines associated with a circular current loop look similar to the magnetic field lines of a bar magnet.

The magnetic field due to a current in a circular wire is such an important special case (Example 30.2) that it needs further exploration. We can find the magnetic field for any point on the axis perpendicular to the plane of the circle from

$$\vec{B} = \frac{\mu_0 I R^2}{2(R^2 + y^2)^{3/2}}\hat{\jmath} \tag{30.10}$$

Finding the magnetic field at other points requires numerical integration. However, we can find the magnetic field lines experimentally by sprinkling iron filings around a current-carrying circle of wire (Fig. 30.24A). Compare these field lines in Figure 30.24B to the magnetic field lines for the bar magnet in Figure 30.8. The magnetic field lines look similar, and both the circular current loop and the bar magnet are magnetic dipoles.

In Section 24-4, we wrote an expression for the electric field at a distance x from an electric dipole compared to its size d:

$$\vec{E} \approx \frac{2k\vec{p}}{x^3} \tag{24.6}$$

where $\vec{p}$ is the electric dipole moment. By analogy, we can write an expression for the magnetic field at a distance y from a magnetic dipole compared to its size R. Start with Equation 30.10 and take the limit in the case that $y \gg R$:

$$\vec{B} \approx \frac{\mu_0 I R^2}{2y^3}\hat{\jmath} \tag{30.11}$$

MAGNETIC DIPOLE MOMENT
✪ Major Concept

Also, by analogy, it is useful to define a **magnetic dipole moment** $\vec{\mu}$, often referred to as the **magnetic moment**. The magnetic dipole moment—like the electric dipole moment—is a vector. For a current loop, its magnitude is given by

$$\mu \equiv IA \tag{30.12}$$

where A is the area of the loop (of any shape). The SI units of the magnetic moment are $A \cdot m^2$. (The symbol for the permeability of free space μ_0 is similar to the symbol for the magnetic dipole moment μ. By convention, the subscript "0" denotes the permeability of free space.)

For a circular loop (Fig. 30.25), the area is $A = \pi R^2$, and the magnitude of the magnetic dipole moment is

$$\mu = I\pi R^2 \tag{30.13}$$

The direction of the magnetic moment is perpendicular to the plane of the loop, in the same direction as the magnetic field along the loop's axis. For the circular loop in Figure 30.25, the magnetic field and magnetic moment are in the positive y direction:

$$\vec{\mu} = (I\pi R^2)\hat{\jmath} \tag{30.14}$$

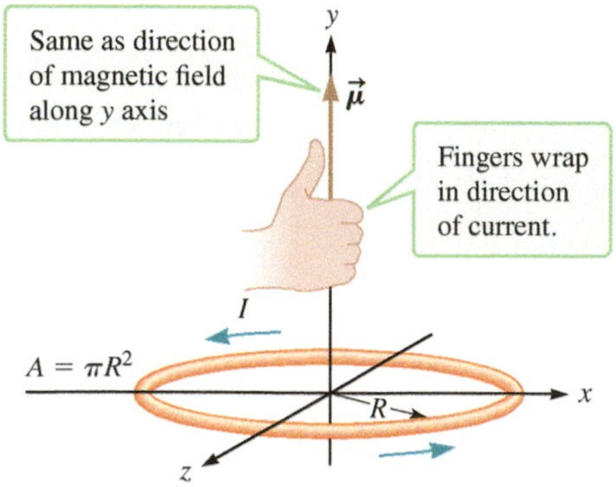

FIGURE 30.25 Find the direction of the magnetic dipole moment by using your right hand.

RIGHT-HAND RULE FOR DIRECTION OF $\vec{\mu}$ ⦿ Tool

There is a right-hand rule for finding the direction of the magnetic moment and therefore the direction of the magnetic field along the loop's axis. Wrap the fingers of your right hand in the direction of the current, and your thumb points in the direction of the magnetic moment (Fig. 30.25).

We can write an expression for the magnetic field along the axis of a distant magnetic dipole by substituting $\vec{\mu}/\pi = IR^2\hat{j}$ (from Eq. 30.14) into Equation 30.11:

$$\vec{B} \approx \left(\frac{\mu_0}{2\pi}\right)\frac{\vec{\mu}}{y^3} \tag{30.15}$$

MAGNETIC FIELD DUE TO A MAGNETIC DIPOLE Special Case

Compare Equation 30.15 for the magnetic field of a magnetic dipole to $\vec{E} \approx 2k\vec{p}/x^3$ (Eq. 24.6), for the electric field of an electric dipole. Mathematically, these expressions are similar: Both show that the field far from a dipole decreases as the cube of the distance. Equation 30.15 holds for loops of wire of any shape, as long as y is large compared to the size of the loop.

Modeling Atoms

Ørsted's discovery that a current-carrying wire deflects a compass needle showed that moving charges are the source of a magnetic field. However, a bar magnet (Fig. 30.8) is also a source of a magnetic field, yet there is no current. How does such a permanent magnet produce a field?

In a permanent magnet, the magnetic field is caused by electrons that are bound to their particular atoms, so those electrons do not generate a current. A full explanation of how these electrons produce a magnetic field requires knowledge of both relativity and quantum mechanics at a level beyond the scope of this textbook. Here we will develop a model much like the microscopic models we used to describe friction (Section 6-2), solid matter (Section 14-4), fluids (Section 15-1), gases (Chapter 20), and—most recently—current in a conductor (Section 28-1). Although our model cannot be completely accurate, it will provide a reasonable explanation as well as a mental image.

In a solid, such as a permanent magnet, each atom is fixed in a lattice structure while its electrons move around the atom's nucleus. Imagine the motion of electrons around the nucleus as the motion of planets around the Sun. Each planet orbits the Sun and spins (rotates) on its own axis. By analogy, imagine each electron orbiting the nucleus of its atom while spinning on its own axis (Fig. 30.26).

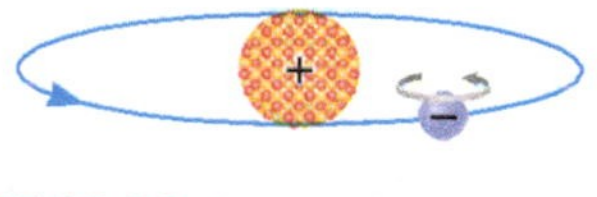

FIGURE 30.26 A mental image of an electron orbiting an atomic nucleus while spinning on its axis. (Not to scale.)

It is important to remember that this is just a mental image. In classical mechanics, an electron is considered a particle, and because particles have no spatial extent, they cannot spin (Section 2-1). In fact, the term *spin* as used in quantum mechanics does not really refer to the rotation of an electron. Instead, spin is a quantum property of an electron, just as color is a property of a shirt. With this warning in mind, we use the mental image of an electron spinning on its axis while orbiting its atomic nucleus to explain how permanent magnets work.

The Magnetic Moment of Electrons and Atoms

An electron in a circular orbit is much like the current in a circular wire (Fig. 30.25). So the orbiting electron is much like a dipole, and the magnetic field far from any particular atom is given by Equation 30.15:

$$\vec{B}_{\text{dipole}} \approx \left(\frac{\mu_0}{2\pi}\right)\frac{\vec{\mu}}{y^3}$$

The subscript reminds us that this is the field of a magnetic dipole. Because the electron carries a negative charge, when you use your right hand to find the direction of the orbiting electron's magnetic dipole moment, you must reverse the result. Wrap the fingers of your right hand in the direction of the electron's orbit in Figure 30.27A. Your thumb then points upward. But because the electron's charge is negative, you must flip your hand over, so now your thumb points downward—in the direction of the magnetic dipole moment. From Equation 30.15, we know that $\vec{B}_{\text{dipole}}$ (on the dipole's axis) is in the same direction as $\vec{\mu}$.

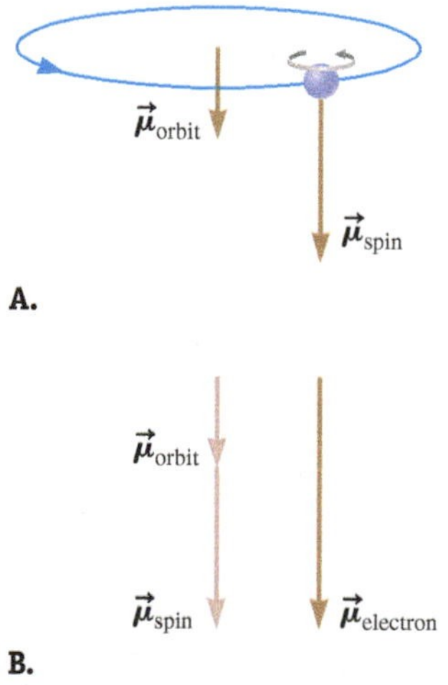

FIGURE 30.27 **A.** The magnetic moments resulting from the orbital motion and the spin of an electron. **B.** The net magnetic moment. In this special case, the orbital and spin magnetic moments are parallel.

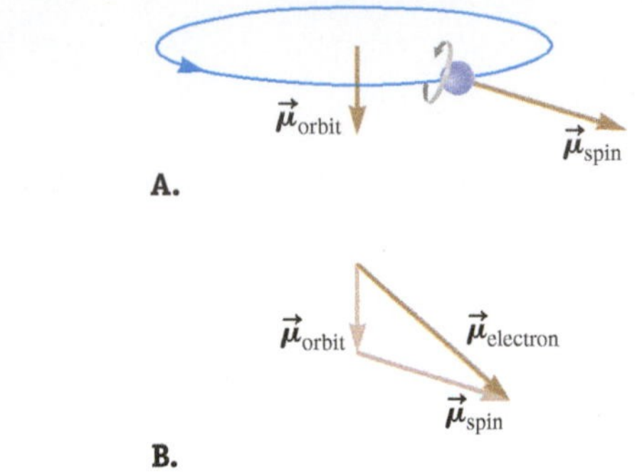

FIGURE 30.28 **A.** The electron's spin axis is often tilted with respect to its orbital axis. **B.** The net magnetic moment in this general case.

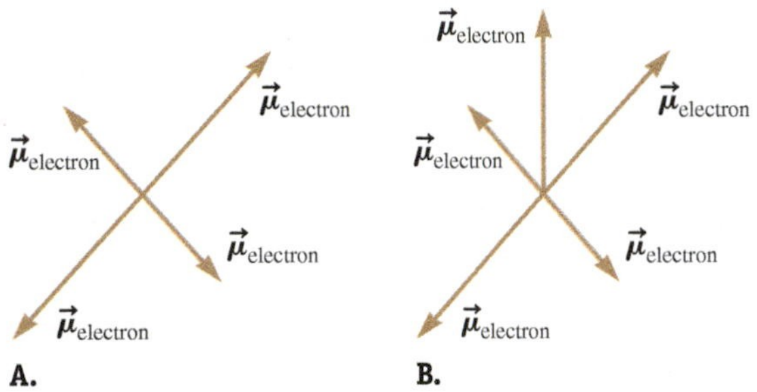

FIGURE 30.29 In these schematic representations of atoms, only the electrons' magnetic moments are shown. **A.** An atom with no net magnetic moment. **B.** An atom with a net upward magnetic moment.

Figure 30.27A shows two magnetic dipole moments for the electron. One of these ($\vec{\mu}_{orbit}$) is the result of the electron's orbital motion around the atomic nucleus. The other ($\vec{\mu}_{spin}$) is the result of the electron's spin around its axis. The net magnetic moment for an electron is the vector sum: $\vec{\mu}_{electron} = \vec{\mu}_{orbit} + \vec{\mu}_{spin}$. In this case, $\vec{\mu}_{electron}$ is in the same direction as $\vec{\mu}_{orbit}$ and $\vec{\mu}_{spin}$.

Figure 30.27 shows a special case in which the electron spins and orbits around parallel axes, so that $\vec{\mu}_{spin}$ is parallel to $\vec{\mu}_{orbit}$. Just as it is rare to find a planet whose orbital axis of rotation is parallel to its spin axis of rotation, this situation is also rare for electrons. Instead, the spin axis of the electron is often tilted with respect to the electron orbit, so that $\vec{\mu}_{spin}$ is tilted with respect to $\vec{\mu}_{orbit}$ (Fig. 30.28). The electron's net magnetic moment, the vector sum $\vec{\mu}_{electron} = \vec{\mu}_{orbit} + \vec{\mu}_{spin}$, points downward and to the right in this case. Its magnitude is slightly smaller than in the case where $\vec{\mu}_{orbit}$ and $\vec{\mu}_{spin}$ are parallel (Fig. 30.27).

Except for hydrogen, atoms are made up of many electrons, each orbiting and spinning in different directions. The net magnetic moment of the atom is the vector sum of all the electron magnetic moments: $\vec{\mu}_{atom} = \Sigma\vec{\mu}_{electron}$. In Figure 30.29A, an atom with four electrons has no net magnetic moment because the magnetic moments of the four electrons cancel: $\vec{\mu}_{atom} = \Sigma\vec{\mu}_{electron} = 0$. When an atom has an even number of electrons, their magnetic moments often cancel. In Figure 30.29B, an atom with an odd number of electrons has a net magnetic moment that is nonzero and upward.

An object is made up of many atoms; the net magnetic moment of the whole object is the vector sum of the atomic magnetic moments. In a nonmagnetic material, the vector sum of the atomic magnetic moments is zero. (The protons in an atom also have a dipole magnetic moment, but usually the net magnetic moment of the protons is small compared to the net magnetic moment of the electrons. So, the protons' magnetic moments can usually be ignored.)

EXAMPLE 30.4 The Orbital Magnetic Moment of an Electron

Assume an electron's orbit is a circle of radius $R \approx 10^{-10}$ m—about the size of an atom. Estimate the magnitude of the orbital magnetic dipole moment μ_{orbit} for an electron.

INTERPRET and ANTICIPATE

Figure 30.28A models our task. Before using $\mu_{orbit} = IA$ (Eq. 30.12), we need to find the area and current associated with the electron's orbit. We assume the electron moves in uniform circular motion, with the centripetal force provided by the electrostatic force (Coulomb's law).

SOLVE

The orbital period is the circumference divided by the electron's speed.	$$T = \frac{2\pi R}{v} \quad (1)$$
Find an expression for the current in terms of the electron's speed and orbital radius. Substitute Equation (1) into Equation (2).	$$I = \frac{\Delta q}{\Delta t} = \frac{e}{T} \quad (2)$$ $$I = \frac{ev}{2\pi R} \quad (3)$$
Substitute Equation (3) into Equation 30.13 to get an expression for the magnitude of the orbital dipole moment.	$$\mu_{orbit} = I\pi R^2 \quad (30.13)$$ $$\mu_{orbit} = \left(\frac{ev}{2\pi R}\right)\pi R^2 = \frac{evR}{2} \quad (4)$$

To use Equation (4), we need to find the electron's speed v from its centripetal acceleration. The centripetal force is equal to the electrostatic force between the electron and the other charged particles in the atom.	$F_c = m_e a_c$ $F_E = \dfrac{kq_1q_2}{R^2} = m_e a_c$

The centripetal force on the electron is the net force exerted by the other charged particles. In a neutral atom, the number of orbiting electrons equals the number of protons in the nucleus. The outermost electron feels the force from all the inner electrons as well as from the protons. (For instance, the outermost electron in iron feels the pull from 26 protons and the push from 25 electrons. Those electrons effectively shield the pull from 25 out of 26 protons. The net result is that the outermost electron feels the pull from approximately one proton.)

Estimate the magnitude of the centripetal force by using $q_1 = \lvert -e \rvert$ for the electron and $q_2 = e$ for the net charge of the protons and the other electrons.	$\dfrac{k(e)(e)}{R^2} = \dfrac{ke^2}{R^2} = m_e a_c$
Find the electron's speed from its centripetal acceleration, using $a_c = v^2/R$ (Eq. 4.36).	$a_c = \dfrac{ke^2}{m_e R^2} = \dfrac{v^2}{R}$ $v = \left(\dfrac{ke^2}{m_e R}\right)^{1/2}$ (30.16)
Substitute numbers and find the speed. Compare your answer to the speed of light (3×10^8 m/s). The electron is roughly 200 times slower than light (a reassuring answer).	$v = \left[\dfrac{(8.99 \times 10^9\,\text{N}\cdot\text{m}^2/\text{C}^2)(1.6 \times 10^{-19}\,\text{C})^2}{(9.11 \times 10^{-31}\,\text{kg})(10^{-10}\,\text{m})}\right]^{1/2}$ $v = 1.6 \times 10^6\,\text{m/s}$
Finally, substitute values into Equation (4). Remember that this is an order-of-magnitude estimate.	$\mu_{\text{orbit}} = \dfrac{(1.6 \times 10^{-19}\,\text{C})(1.6 \times 10^6\,\text{m/s})(10^{-10}\,\text{m})}{2}$ $\mu_{\text{orbit}} = 1.3 \times 10^{-23}\,\text{A}\cdot\text{m}^2$ $\sim 10^{-23}\,\text{A}\cdot\text{m}^2$

CHECK and THINK

Our order-of-magnitude estimate has the correct SI units. How can such a small magnetic moment produce a magnetic field that we can detect on the macroscopic scale? Remember that a large number of electrons must have their orbital dipole magnetic moments aligned to produce a significant macroscopic magnetic field.

30-7 Ferromagnetic Materials

FERROMAGNETIC MATERIALS AND MAGNETIC DOMAINS

Major Concepts

Permanent magnets—such as refrigerator magnets, bar magnets, and lodestones—are made from **ferromagnetic materials**, such as iron, iron alloys, cobalt, and nickel, and are known as **ferromagnets**. Each atom in a ferromagnetic material has a net nonzero magnetic moment (Fig. 30.29B). Of course, the entire object would have a net magnetic moment of zero if the atoms' magnetic moments canceled. But, in ferromagnetic materials, atoms interact strongly with their neighbors, such that the magnetic moments of nearby atoms line up. The interactions among atoms decrease with distance, so the magnetic moments are aligned over a small region known as a **magnetic domain**, or simply **domain**. A domain has between 10^{17} and 10^{21} atoms. In an unmagnetized ferromagnetic object, the magnetic moments $\vec{\mu}_{\text{domain}}$ of the domains are randomly oriented so that the net magnetic moment of the entire object is zero: $\vec{\mu}_{\text{net}} = \Sigma\vec{\mu}_{\text{domain}} = 0$, as shown in Figure 30.30.

Canadian coins are made from ferromagnetic materials. Normally a coin is unmagnetized, so Figure 30.30 is a good model for the coin's domains. What happens when the coin is in a magnetic field? Imagine each domain as a small bar magnet. In an unmagnetized coin, these atomic-sized bar magnets are randomly oriented (Fig. 30.31A). When an exterior magnet's north pole is held near the coin, the

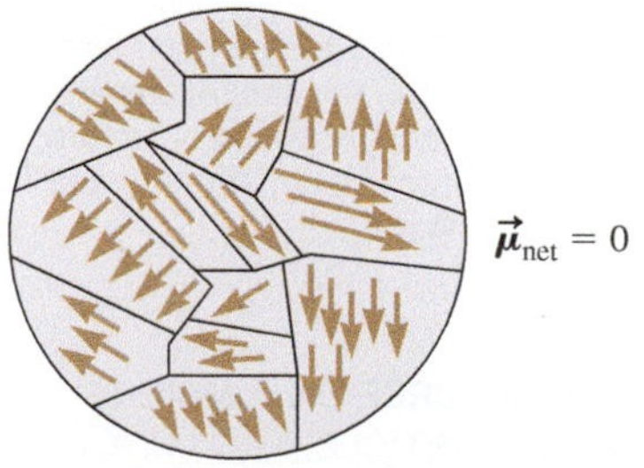

FIGURE 30.30 In an unmagnetized ferromagnetic object, the domains' magnetic moments are randomly oriented, so that the material has no net magnetic moment: $\vec{\mu}_{\text{net}} = 0$.

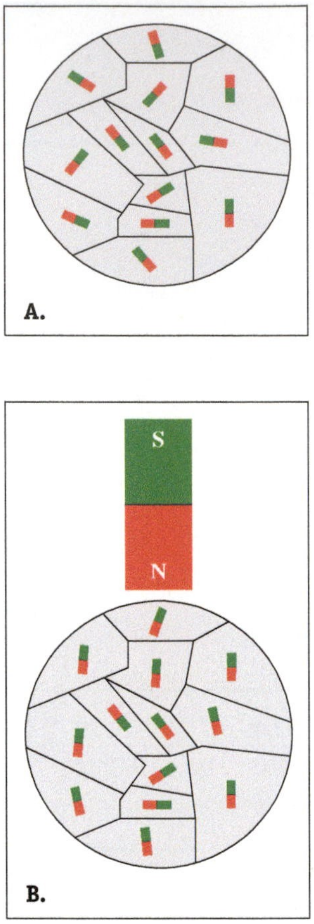

FIGURE 30.31 **A.** In an unmagnetized ferromagnetic material, each domain is modeled as a randomly oriented bar magnet. **B.** A Canadian coin is in the magnetic field of a bar magnet. In our model, the small bar magnets associated with the atoms inside the coin twist.

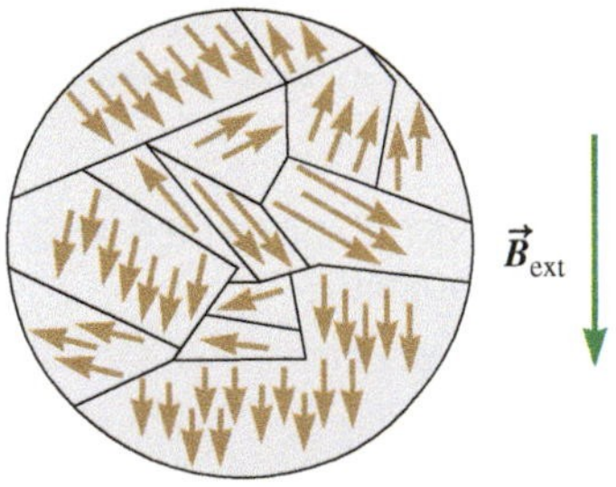

FIGURE 30.32 In a magnetized ferromagnetic material, the sum of the atomic magnetic moments is nonzero. In this case, the net dipole magnetic moment is roughly downward.

atomic-sized bar magnets inside the coin twist so that their south poles point toward the exterior magnet's north pole (Fig. 30.31B). As a result, the coin is now essentially a magnet, with its south pole near the exterior magnet's north pole; thus, the two objects are mutually attracted.

Additionally, when a ferromagnetic material is in a magnetic field, domains that are aligned or nearly aligned with the external magnetic field grow, while the other domains shrink. Figure 30.32 shows the domains in a ferromagnet that is in a downward external magnetic field $\vec{B}_{ext}$. Compare Figure 30.32 to Figure 30.30, which has no external magnetic field. The domains that have a downward-pointing magnetic moment are larger in Figure 30.32 than in Figure 30.30, resulting in a net magnetic moment (roughly) downward in the same direction as the external magnetic field. The ferromagnet is now a "permanent" magnet and will respond as such.

A "permanent" magnet will not necessarily be magnetized forever. Permanent magnets maintain their magnetization for some period of time after an external magnetic field is removed. In the case of a Canadian coin, this effect does not last long because the domains are easily agitated and quickly become randomly oriented again, returning $\vec{\mu}_{net}$ to zero. There are three ways to demagnetize a ferromagnet:

1. Strike the object, for example by hitting the ferromagnet with a hammer or dropping it onto a hard surface.
2. Raise the object's temperature. Above a threshold temperature known as the Curie temperature, a ferromagnet becomes demagnetized.

Methods 1 and 2 both increase the ferromagnet's thermal energy, which also increases the random motion of atoms so that the domains become randomly oriented, thus demagnetizing the object.

3. Place the ferromagnet in a magnetic field $\vec{B}_{ext}$ (with a component) pointing in the direction opposite its $\vec{\mu}_{net}$. This method may create a "permanent" magnet with $\vec{\mu}_{net}$ reversed with respect to the original direction.

30-8 Magnetic Force on a Charged Particle

For two objects to interact through a particular force, they must have a particular property in common (mass for gravity, net charge for electric force). Because both objects have this property in common, in principle either one can be thought of as the source or as the subject. In practice, there is usually a good reason to choose one object over the other to be the source. For two objects to interact through the magnetic force, each must have a net motion of charged particles or a net magnetic moment. The magnetic force exerted between permanent magnets, each with a net magnetic moment, is beyond the scope of this textbook. Instead, we focus on the magnetic force exerted on subjects that have a net motion of charged particles. The source may be a permanent magnet or another object with a net motion of charged particles.

MAGNETIC FORCE ON A MOVING CHARGED PARTICLE

✪ Major Concept

Start by considering a subject that is a single charged particle moving at velocity $\vec{v}$ (Fig. 30.33A). An unknown source creates a magnetic field vector $\vec{B}$ at the location of the subject. The magnetic force $\vec{F}_B$ on a particle with charge q is

$$\vec{F}_B = q(\vec{v} \times \vec{B}) \quad (30.17)$$

Because Equation 30.17 involves a cross product, we can find the magnitude of the magnetic force using Equation 12.22:

$$F_B = |q|vB\sin\varphi \quad (30.18)$$

where φ is the angle between $\vec{v}$ and $\vec{B}$ (Fig. 30.33A). From our experience with cross products (Section 12-6), we can rewrite the magnitude of the force F_B in terms of $B_\perp$ (the component of the magnetic field that is perpendicular to the subject's velocity; Fig. 30.33B). Substitute $B_\perp = B\sin\varphi$ into Equation 30.18:

$$F_B = |q|vB_\perp \quad (30.19)$$

Further, we can rewrite F_B in terms of $v_\perp$ (the component of the subject's velocity that is perpendicular to the magnetic field; Fig. 30.33C):

$$F_B = |q|v_\perp B \tag{30.20}$$

From Equations 30.18 through 30.20, we see that for a particular particle moving through a particular magnetic field, the magnetic force F_B is at a maximum when the particle is moving perpendicular ($\varphi = 90°$) to the magnetic field.

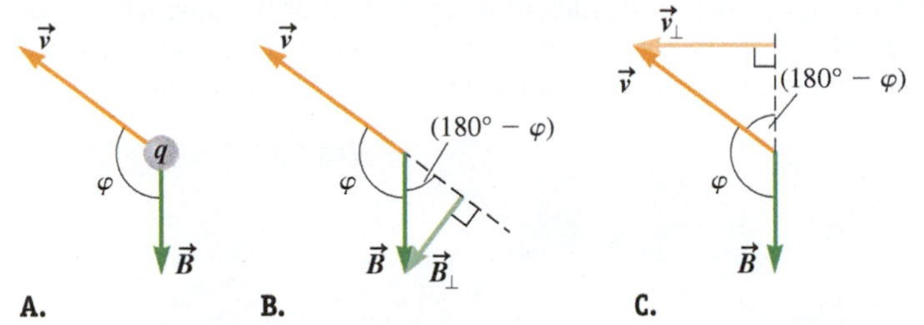

FIGURE 30.33 **A.** A charged particle is the subject of a magnetic field. We draw the subject's velocity vector $\vec{v}$ because it is important in determining the magnetic force. The cross product $(\vec{v} \times \vec{B})$ is out of the page. If q is positive, the magnetic force $q(\vec{v} \times \vec{B})$ is out of the page. If q is negative, the magnetic force $q(\vec{v} \times \vec{B})$ is into the page. **B.** The magnetic force is given by $F_B = |q|vB_\perp$. **C.** The magnetic force is also given by $F_B = |q|v_\perp B$.

There are four conditions that result in no magnetic force on the subject: (1) Perhaps it is obvious, but there is no magnetic force if there is no magnetic field; if $\vec{B} = 0$, then $\vec{F}_B = 0$. (2) There is no magnetic force exerted on a neutral particle; if $q = 0$, then $\vec{F}_B = 0$. (3) There is no magnetic force if the particle is at rest; if $\vec{v} = 0$, then $\vec{F}_B = 0$. (4) There is no magnetic force if the particle is moving parallel or antiparallel to the magnetic field; if $\varphi = 0$ or $\varphi = 180°$, then $\vec{F}_B = 0$.

The least intuitive aspect of the magnetic force is its direction. The magnetic *force* is *not* parallel or antiparallel to the magnetic *field*. The cross product in Equation 30.17, $\vec{F}_B = q(\vec{v} \times \vec{B})$, means the magnetic force $\vec{F}_B$ is perpendicular to both $\vec{B}$ and $\vec{v}$. The direction of the cross product $\vec{v} \times \vec{B}$ is given by the usual right-hand rule for cross products (Section 12-6), but the direction of $\vec{F}_B$ also depends on the sign of the particle's net charge. If the particle is positively charged, $\vec{F}_B$ is in the same direction as $\vec{v} \times \vec{B}$. In that case, when you use the right-hand rule to find the direction of the cross product, your thumb points in the direction of the magnetic force. If the particle is negatively charged, $\vec{F}_B$ is in the opposite direction from $\vec{v} \times \vec{B}$, and when you use the right-hand rule to find the direction of the cross product, you must flip your thumb over by 180° to find the direction of the magnetic force.

If you forgot how to find the direction of a cross product, return to Fig. 12.23 on page 346.

The Lorentz Force

We can summarize the main ideas of electricity (Part III) with a simple statement: *Charged particles interact through an electric force.* In addition, if the source and subject are both *moving* charged particles, they exert a magnetic force on each other. So we can make another simple statement: *Moving charged particles interact through both a magnetic force and an electric force.*

Of course, the motion of a particle depends on the observer's reference frame. If a charged source is moving in the observer's frame, there is both a magnetic field $\vec{B}$ and an electric field $\vec{E}$, and the subject (of charge q) may experience both an electric force $\vec{F}_E = q\vec{E}$ (Eq. 24.2) and a magnetic force $\vec{F}_B = q(\vec{v} \times \vec{B})$ (Eq. 30.17). We can combine these two equations to express the total electromagnetic force experienced by a particle with charge q moving with velocity $\vec{v}$:

$$\vec{F}_L = \vec{F}_E + \vec{F}_B = q(\vec{E} + \vec{v} \times \vec{B}) \tag{30.21}$$

LORENTZ FORCE **Major Concept**

To use Equation 30.21, the field vectors $\vec{E}$ and $\vec{B}$, and the subject's velocity $\vec{v}$, must all be observed in the same reference frame. It is possible that in this frame, the velocity of the subject may be zero—in which case the Lorentz force is due solely to the electric force, even if a nonzero magnetic field is present.

Equation 30.21 is called the *Lorentz force* in honor of the Dutch physicist Hendrik Anton Lorentz (1853–1928).

CONCEPT EXERCISE 30.4

Cosmic rays are high-energy charged particles produced by astronomical objects. Many of the cosmic rays that make their way to the Earth are trapped by the Earth's magnetic field and never reach the surface. These trapped cosmic rays are found in the Van Allen belts—donut-shaped zones over the Earth's equator (Fig. 30.34). These cosmic rays are mostly protons with energies of about 30 MeV. The inset in the figure shows a cosmic ray proton as it is about to enter the Earth's magnetic field. The cosmic ray's velocity is initially perpendicular to the field. Three students discuss what happens to the incoming cosmic ray. Decide which student or students are correct.

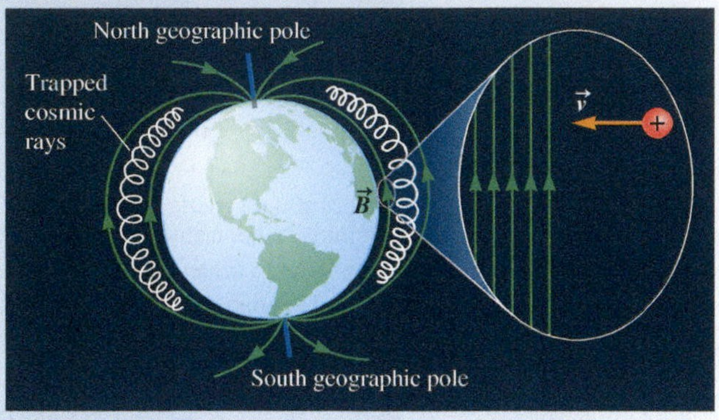

FIGURE 30.34 The Van Allen belts are donut-shaped zones of trapped cosmic rays above the Earth's surface. *Inset*: What happens to this cosmic ray as it enters the Earth's magnetic field?

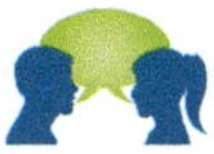

Shannon: The velocity is perpendicular to the magnetic field, so the cosmic ray just passes through the field and hits the Earth's atmosphere.

Avi: What you are saying is that the magnetic field exerts no force on the cosmic ray. Actually, it exerts a huge force because the velocity is perpendicular to the magnetic field. The force will be into the page.

Cameron: Avi is right. The cosmic ray proton is going to feel a huge magnetic force. Because it is positively charged, it will be pushed upward along the magnetic field lines.

Shannon: I never said the force was zero. There is a force, but the force is perpendicular to the magnetic field lines. In this case, that's to the left—toward the Earth.

Avi: The force is perpendicular to the magnetic field, but it also has to be perpendicular to the velocity. Because $\vec{B}$ and $\vec{v}$ are both in the plane of the page, the force must be perpendicular to the page.

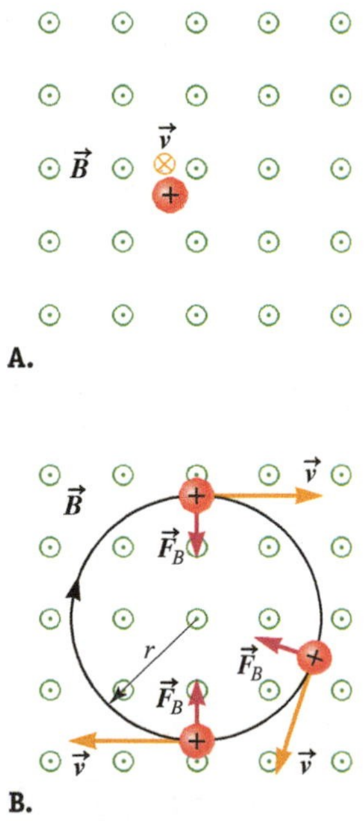

FIGURE 30.35 A. A particle moves at constant velocity antiparallel to a magnetic field. **B.** A particle moves in uniform circular motion in a magnetic field. The magnetic field does not change the particle's speed in either case.

30-9 Motion of Charged Particles in a Magnetic Field

In this section, we explore the motion of charged particles in a magnetic field. When a particle is moving either parallel or antiparallel to the magnetic field (Fig. 30.35A), $v_\perp = 0$ and there is no magnetic force acting on the particle. In this case, the magnetic field does not accelerate the particle, and the particle continues to move at constant speed in a straight line (assuming no other forces are acting).

When a particle is moving perpendicular to a uniform magnetic field (Fig. 30.35B), the particle experiences a magnetic force that is always perpendicular to the particle's velocity (Eq. 30.17). In this case (assuming no other forces are acting), the particle moves with uniform speed in a circle of radius r because the particle's acceleration is perpendicular to its velocity (Section 4-6).

The radius of the circle depends on the strength of the magnetic field. The required centripetal force F_c is supplied by the magnetic force F_B: $F_c = F_B$. The centripetal force is given by $F_c = mv^2/r$ (Eq. 6.7), and the magnetic force comes from $F_B = |q|vB \sin \varphi$ (Eq. 30.18) with $\varphi = 90°$:

$$m\frac{v^2}{r} = |q|vB \tag{30.22}$$

Solve Equation 30.22 for r:

$$r = \frac{mv}{|q|B} \tag{30.23}$$

What if a particle's initial velocity makes some arbitrary angle with respect to a uniform magnetic field? In that case, it is best to break up the velocity into two components. One component $\vec{v}_\perp$ is perpendicular to the magnetic field, and the other component $\vec{v}_\parallel$ is parallel to the magnetic field, so the total velocity is $\vec{v} = \vec{v}_\perp + \vec{v}_\parallel$. This expression is convenient because there is no magnetic force parallel to the magnetic field, so there is no acceleration parallel to the field, $\vec{a}_\parallel = 0$, and the parallel velocity $\vec{v}_\parallel$ is constant. However, the particle experiences a centripetal acceleration, which is constant in magnitude, and the particle's path is a helix (Fig. 30.36). The cross section of the helical path is a circle of radius r. The helix is a combination of linear motion and uniform circular motion. The radius of the helix is given by $r = mv_\perp/|q|B$ (Eq. 30.23), where v has been replaced by $v_\perp$.

The path of a particle in a nonuniform magnetic field can be more complicated. For example, the magnetic field shown in Figure 30.37 is weaker in the central region than it is near the top or bottom. The particle's path spirals outward (has a larger radius) as the particle moves toward the central region, and inward (with smaller radius) as it approaches the top or bottom. The particle spirals back and forth between the top and bottom, and it is effectively trapped. This magnetic field configuration is called a *magnetic bottle*. Magnetic bottles are used to contain very hot plasmas.

It is important to notice that whether the magnetic field is uniform or nonuniform, only the *direction* of the velocity can be changed by the magnetic field. The particle's *kinetic energy remains constant*. Put another way, the magnetic field can never do work on the particle.

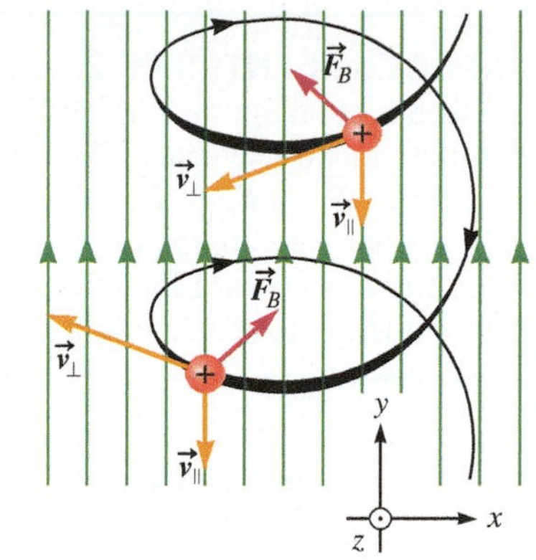

FIGURE 30.36 A positively charged particle moves in a helical path in a region with a uniform magnetic field. The particle's speed is constant.

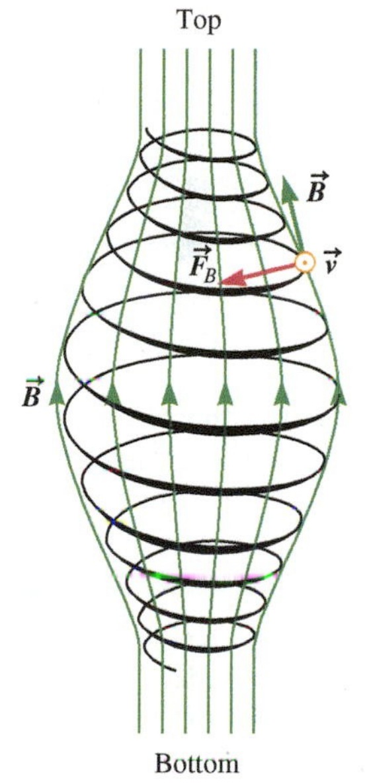

FIGURE 30.37 A positively charged particle spirals in a region with a nonuniform magnetic field. Use the right-hand rule to show that the magnetic force near the bottom points somewhat upward, and the magnetic force near the top points somewhat downward. The particle is trapped. When the particle reaches the top, it will spiral back down, and when it reaches the bottom, it will spiral back up.

CONCEPT EXERCISE 30.5

The Earth's Van Allen belts (Fig. 30.34) are a natural magnetic bottle, trapping cosmic rays that spiral back and forth between the North and South Poles. Use the shape of a cosmic ray's path to describe the variation in the magnetic field strength.

EXAMPLE 30.5 A Velocity Selector

Many laboratory experiments require a beam of charged particles, all moving with the same velocity. A *velocity selector* uses electric and magnetic fields to filter the particles by velocity. Figure 30.38 shows the basic parts of the device. The source on the left supplies a beam of charged particles with a range of velocities. The beam then passes through a region of space with a uniform electric field and a uniform magnetic field. The electric field for this velocity selector points downward, and the magnetic field points into the page. Show that only positively charged particles with speed

$$v = \frac{E}{B} \tag{1}$$

can pass straight through the velocity selector and emerge from the opening on the right.

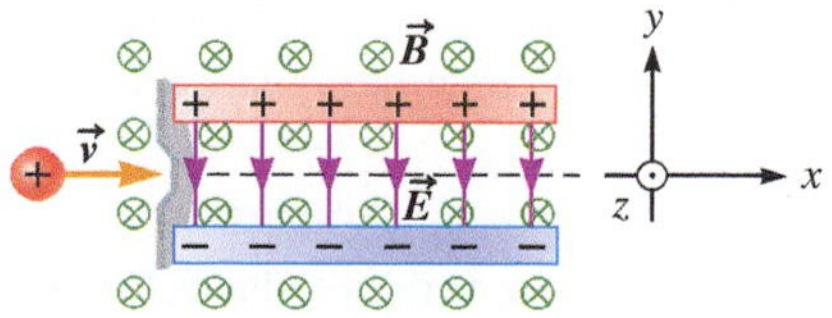

FIGURE 30.38 A velocity selector.

Example continues on page 956 ▶

INTERPRET and ANTICIPATE

Draw a free-body diagram (Fig. 30.39) for a particle that travels at constant velocity through the velocity selector. There is an electric field and a magnetic field, so the Lorentz force $\vec{F}_L = \vec{F}_E + \vec{F}_B = q(\vec{E} + \vec{v} \times \vec{B})$ (Eq. 30.21) applies. The electric force on a positively charged particle is downward in the negative y direction. The magnetic force is upward in the positive y direction. Because the particle travels at constant velocity, its acceleration is zero.

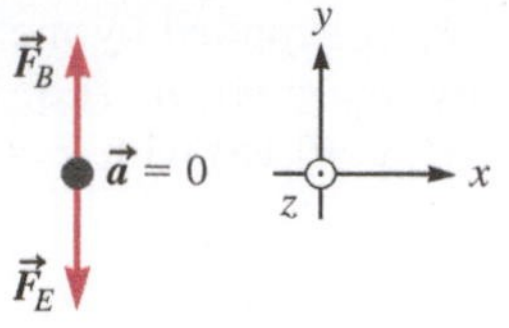

FIGURE 30.39

SOLVE

Because the particle's acceleration is zero, the (total) Lorentz force must be zero.

$$\vec{F}_L = \vec{F}_E + \vec{F}_B = m\vec{a}$$

$$\vec{F}_E + \vec{F}_B = 0 = -F_E\hat{j} + F_B\hat{j}$$

$$F_B = F_E \tag{2}$$

The velocity of the particle is perpendicular to the magnetic field, so $\varphi = 90°$ in $F_B = |q|vB \sin\varphi$ (Eq. 30.18). The magnitude of the electric force is given by Equation 24.2.

$$F_B = |q|vB \sin\varphi = |q|vB$$

$$F_E = |q|E \tag{24.2}$$

Substitute F_B and F_E into Equation (2), and solve for v.

$$|q|vB = |q|E$$

$$v = \frac{E}{B} \checkmark$$

CHECK and THINK

This is exactly what we set out to derive. Our answer does not depend on the particle's charge. What if the particle had a negative charge? Then the magnetic force would be in the negative y direction and the electric force would be in the positive y direction. The two forces would cancel for a particle traveling at the speed given by Equation (1). In practice, the experimenter varies the ratio E/B in order to select a desired particle speed. In the next example, we show how a velocity selector may be used to determine a particle's mass.

EXAMPLE 30.6 A Mass Spectrometer

A *mass spectrometer* (Fig. 30.40) is a device used to measure the mass of charged particles, such as ions. Knowing the mass often allows the researcher to identify the type of ion. A mass spectrometer uses a velocity selector, shown on the left in the figure (region 1). The particles leave the velocity selector with speed $v = E/B_1$ and enter a region with another uniform magnetic field (region 2). In region 2, the magnetic field has magnitude B_2 directed into the page. There is no electric field in this region. The particles' velocities are perpendicular to the magnetic field, so the particles travel along a circular arc and strike a photographic plate (or other detector) at point P. The radius r of the semicircular trajectory is easily measured. The particle's mass is determined from the radius measurement.

Assume a beam of ions has been singly ionized so that the particles are missing just one electron. The velocity selector's electric field is $E = 3.54 \times 10^3$ N/C, and the magnetic field in region 1 is $B_1 = 0.035$ T. The magnetic field in region 2 is $B_2 = 0.075$ T, and the radius of the semicircular path is 39.1 cm. What is the particle mass (in atomic mass units)?

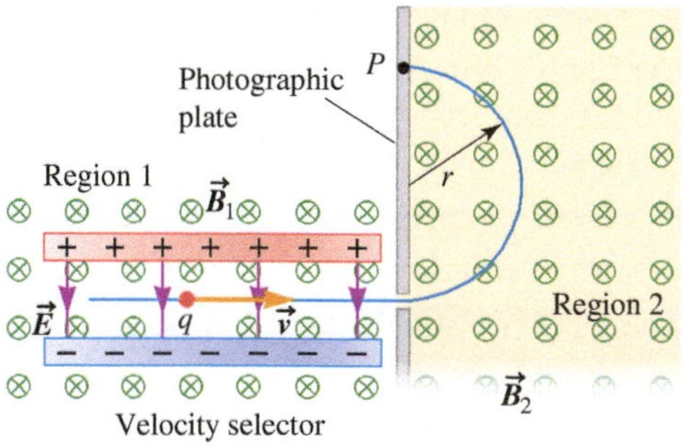

FIGURE 30.40 A mass spectrometer.

INTERPRET and ANTICIPATE

This problem combines the velocity selector from Example 30.5 with the circular motion of a charged particle in a uniform magnetic field. Once we know the particle's velocity when it enters region 2, we can find its mass. Because the required centripetal force is proportional to the particle's mass, we expect a more massive particle to move in a larger circular path.

SOLVE

First find the particle's velocity. There are two magnetic fields; here we need the magnetic field in the velocity selector (region 1).	$v = \frac{E}{B_1} = \frac{3.54 \times 10^3\ \text{N/C}}{0.035\ \text{T}} = 1.01 \times 10^5\ \text{m/s}$
Now solve Equation 30.23 for mass. Here we need the magnetic field in region 2.	$r = \frac{mv}{\lvert q \rvert B_2}$ (30.23) $m = \frac{\lvert q \rvert B_2 r}{v}$
Substitute numerical values. The ion is missing an electron, so it has excess positive charge e.	$m = \frac{\lvert 1.6 \times 10^{-19}\ \text{C} \rvert (0.075\ \text{T})(0.391\ \text{m})}{1.01 \times 10^5\ \text{m/s}} = 4.6 \times 10^{-26}\ \text{kg}$
Because this mass is so small, it is convenient to convert to atomic mass units (Section 19-6).	$m = (4.6 \times 10^{-26}\ \text{kg})\left(\frac{1\ \text{u}}{1.66 \times 10^{-27}\ \text{kg}}\right) = 28\ \text{u}$

CHECK and THINK

Let's see if we can identify this ion. It has all of its protons and neutrons, and it is missing only one electron, so it has most of its mass. According to the periodic table (Appendix B), silicon (Si) has an atomic mass of about 28 u. It seems likely that this ion is silicon, Si^+.

30-10 Case Study: The Hall Effect

Throughout Part III, we assumed a current in a metal conductor was due to the drifting of negatively charged particles (electrons) in the direction opposite to the current. If we put the conductor in a magnetic field, we can show that our assumption is correct. This experiment was performed by the American physicist Edwin Hall in 1879 and is called the **Hall effect.**

To show the Hall effect, a rectangular conductor is connected to a battery in a region with a uniform magnetic field (Fig. 30.41). The downward current through the conductor could be the result of either positively charged particles drifting downward or negatively charged particles drifting upward. An ammeter that measures the current cannot tell us the sign of the charges or the direction of the drifting particles. However, because the conductor is in a magnetic field, the magnetic force on the drifting particles will reveal which particles are really in motion. Try the next two concept exercises.

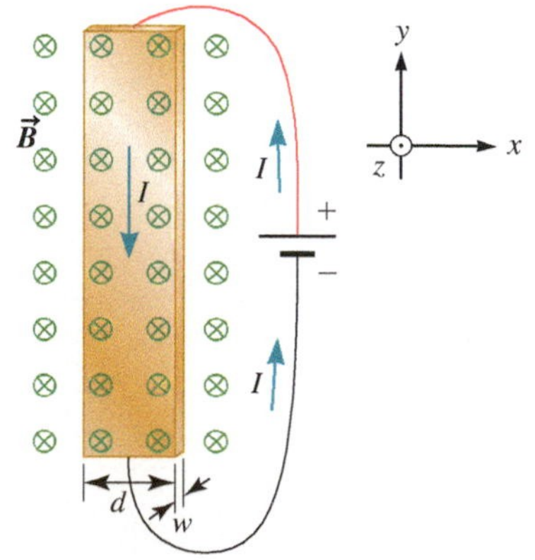

FIGURE 30.41 A rectangular conductor of width d and thickness w is attached to a battery. The current in the circuit is counterclockwise, so the current through the conductor is downward. The conductor is in a region with a uniform magnetic field pointing into the page.

CONCEPT EXERCISE 30.6

CASE STUDY Hall Effect; Positive Charge Carriers

If the current in the rectangular conductor (Fig. 30.41) is the result of positively charged particles drifting downward in the negative y direction, what is the direction of the magnetic force on these particles?

CONCEPT EXERCISE 30.7

CASE STUDY Hall Effect; Negative Charge Carriers

If the current in the rectangular conductor (Fig. 30.41) is the result of negatively charged particles drifting upward in the positive y direction, what is the direction of the magnetic force on these particles?

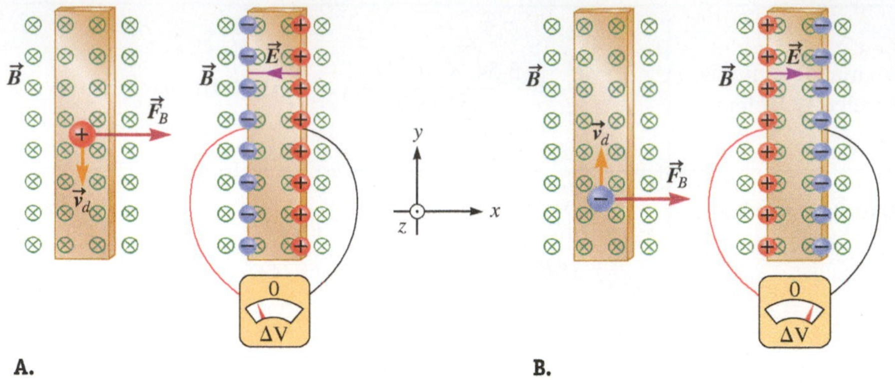

FIGURE 30.42 The Hall effect may be used to show that electrons flow in a conductor. **A.** If positively charged particles flow downward in the *negative y* direction, there is a buildup of positively charged particles on the right wall of the conductor. **B.** If negatively charged particles flow upward in the *positive y* direction, there is a buildup of negatively charged particles on the right wall of the conductor.

Figure 30.42 shows the two possible ways to have a downward current in the rectangular conductor. The current can be the result of positively charged particles drifting downward (Fig. 30.42A) or of negatively charged particles drifting upward (Fig. 30.42B). A positive particle drifting downward experiences a magnetic force to the right. After some time, excess positive charge builds up on the right wall of the conductor, leaving excess negative charge on the left wall. In contrast, a negative particle drifting upward also experiences a magnetic force to the right, and after some time there is a buildup of negative charge on the right wall and of positive charge on the left wall.

We can use a voltmeter to distinguish between these situations and thus to determine the sign of the flowing particles. The potential difference between the two walls of the conductor is called the **Hall voltage**, ΔV_{Hall}. If you attach the red lead to the left side of the conductor and the black lead to the right side, you find

$$\Delta V_{\text{Hall}} = V_{\text{red}} - V_{\text{black}} < 0$$

in the case of positive flowing charges (Fig. 30.42A). In the case of negative flowing charges (Fig. 30.42B), you find

$$\Delta V_{\text{Hall}} = V_{\text{red}} - V_{\text{black}} > 0$$

The Hall effect shows that negative charges (electrons) flow in a conductor because the Hall voltage is observed to be positive.

The Hall effect can also be used to find the number density of conduction electrons as well as their drift speed v_d. Once positive charge builds up on the left wall and negative charge on the right wall, an electric field in the conductor points from left to right in the positive x direction (Fig. 30.42B, right). An electron drifting upward is now subject to a magnetic force in the positive x direction and also to an electric force in the negative x direction. This situation is similar to that in the velocity selector (Example 30.5). When the electric and magnetic forces have the same magnitude, there is no net acceleration in the x direction and $F_B = F_E$. Using $F_B = |q|vB \sin\varphi$ (Eq. 30.18), with $\varphi = 90°$ because the drift velocity is perpendicular to the magnetic field, and $F_E = |q|E$ (Eq. 24.2), we find

$$|q|v_d B = |q|E \tag{30.24}$$

The charge cancels out:

$$v_d B = E \tag{30.25}$$

The magnitude of the electric field E comes from the Hall voltage (and Eq. 26.17):

$$\Delta V_{\text{Hall}} = Ed$$

$$E = \frac{\Delta V_{\text{Hall}}}{d} \tag{30.26}$$

where d is the width of the conductor (Fig. 30.41). Substitute Equation 30.26 into Equation 30.25:

$$v_d B = \frac{\Delta V_{\text{Hall}}}{d} \tag{30.27}$$

The drift speed may be written in terms of the current:

$$I = nev_d A \tag{28.6}$$

where n is the number density of the conduction electrons and A is the cross-sectional area of the conductor. For a rectangular conductor (Fig. 30.41), the cross-sectional area is $A = dw$, and Equation 28.6 becomes

$$I = nev_d dw \tag{30.28}$$

Solve for v_d:

$$v_d = \frac{I}{nedw} \tag{30.29}$$

Substitute Equation 30.29 into Equation 30.27:

$$\left(\frac{I}{nedw}\right)B = \frac{\Delta V_{\text{Hall}}}{d}$$

$$n = \frac{IB}{ew\Delta V_{\text{Hall}}} \tag{30.30}$$

Equation 30.30 gives the number density n of conduction electrons. The drift speed comes from Equation 30.27:

$$v_d = \frac{\Delta V_{\text{Hall}}}{Bd} \tag{30.31}$$

Because the drift speed is very low (between 10^{-5} m/s and 10^{-4} m/s), the Hall voltage is typically small and hard to measure accurately. Using Equation 30.31 directly can lead to inaccuracies in the drift speed measurement. To improve accuracy, experimenters slide the entire conductor in the direction of the current—opposite the direction of the electron's drift velocity. If the conductor in Figure 30.42B slides downward at exactly the drift speed, the conduction electrons have no net velocity with respect to the magnetic field. So the magnetic field exerts no force on the electrons, no charge builds up on the walls of the conductor, and the Hall voltage is zero. By sliding the conductor at increasing speeds until the Hall voltage reaches zero, the drift speed can be determined accurately.

EXAMPLE 30.7 CASE STUDY Hall Voltage

Suppose the rectangular conductor in Figure 30.41 is made of copper, with width $d = 2.54$ cm and thickness $w = 0.500$ cm. The number density of conduction electrons in copper is $8.42 \times 10^{28}\ \text{m}^{-3}$ (Table 28.2, page 872). The magnetic field is that of the Earth, $B = 0.50 \times 10^{-4}$ T, in a direction perpendicular to the conductor. The current is measured to be 205 A.

A Find the expected Hall voltage. Answer in **CHECK and THINK**: If your voltmeter's error is $\pm 1\ \mu$V, can you measure the Hall voltage?

INTERPRET and ANTICIPATE

First find the drift speed in copper and then, using this information, find the Hall voltage. We expect a small voltage.

SOLVE

Find the drift speed from Equation 30.29. Substitute numerical values.

$$v_d = \frac{I}{nedw} \tag{30.29}$$

$$v_d = \frac{205\ \text{A}}{(8.42 \times 10^{28}\ \text{m}^{-3})(1.60 \times 10^{-19}\ \text{C})(0.0254\ \text{m})(0.00500\ \text{m})}$$

$$v_d = 1.20 \times 10^{-4}\ \text{m/s}$$

Example continues on page 960 ▶

Solve Equation 30.31 for the Hall voltage, and substitute numerical values.

$$v_d = \frac{\Delta V_{Hall}}{Bd} \quad (30.31)$$

$$\Delta V_{Hall} = Bdv_d = (0.50 \times 10^{-4}\,\text{T})(0.0254\,\text{m})(1.20 \times 10^{-4}\,\text{m/s})$$

$$\Delta V_{Hall} = 1.5 \times 10^{-10}\,\text{V} = 1.5 \times 10^{-4}\,\mu\text{V}$$

CHECK and THINK

As expected, the Hall voltage is very small. A voltmeter with an error of $\pm 1\,\mu$V cannot distinguish this small potential difference from zero, and you cannot measure potential difference with it. In part B, we imagine using an electromagnet to increase the Hall effect.

B Instead of using the Earth's magnetic field in Figure 30.41, imagine we use a large electromagnet, producing a field with magnitude $B = 2.5$ T. What is the Hall voltage in this case?

INTERPRET and ANTICIPATE

The only change from part A is the stronger magnetic field, so we can start with $v_d = \Delta V_{Hall}/Bd$ (Eq. 30.31).

SOLVE

Solve for the Hall voltage, and substitute numerical values.

$$\Delta V_{Hall} = Bdv_d = (2.5\,\text{T})(0.0254\,\text{m})(1.20 \times 10^{-4}\,\text{m/s})$$

$$\Delta V_{Hall} = 7.6 \times 10^{-6}\,\text{V} = 7.6\,\mu\text{V}$$

CHECK and THINK

A voltmeter with an error of $\pm 1\,\mu$V can distinguish this small potential difference from zero. It measures perhaps $7\,\mu$V or $8\,\mu$V. The Hall voltage is still very small, requiring a very large current (205 A) for this measurement. The experiment could be further improved by using a larger sample of copper. The width of the sample in this problem is only 1 in. The Hall voltage ΔV_{Hall} increases linearly as the width d increases.

30-11 Magnetic Force on a Current-Carrying Wire

MAGNETIC FORCE ON A CURRENT-CARRYING CONDUCTOR

Major Concept

Equation 30.17, $\vec{F}_B = q(\vec{v} \times \vec{B})$, is a convenient way to find the magnetic force on a single charged particle. In many practical circumstances, however, the subject is not a single moving particle, but instead a current in a conductor. So, in this section, we derive an expression for the magnetic force exerted on a current-carrying wire.

DERIVATION Magnetic Force on a Current-Carrying Conductor

Figure 30.43 shows a segment of a current-carrying wire in a region with a uniform magnetic field $\vec{B}$. We define a length vector $\vec{\ell}$ equal in magnitude to the length of the wire segment and pointing in the direction of the current. We show that the magnetic force on the straight current-carrying conductor is

$$\vec{F}_B = I(\vec{\ell} \times \vec{B}) \quad (30.32)$$

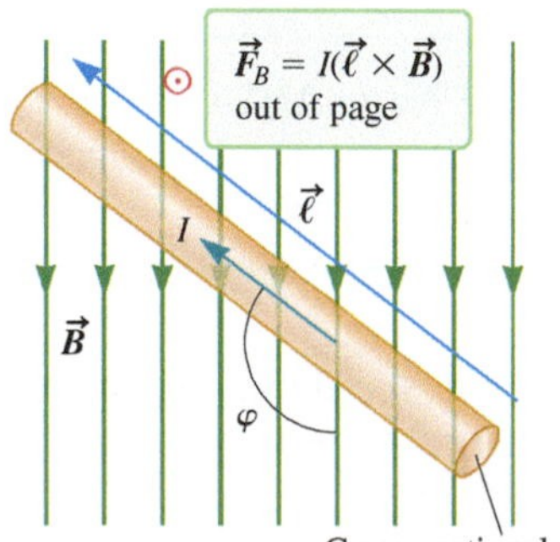

FIGURE 30.43 A straight segment of a current-carrying wire in a uniform magnetic field.

To derive an expression for the magnetic force exerted on the current-carrying wire, we first find the magnetic force on a single charged particle moving in the wire. We could consider the motion of an actual conduction electron in the direction opposite the current, but it is slightly easier to deal with the equivalent positive charge e moving at the drift velocity $\vec{v}_d$ in the direction of the current (Fig. 30.44).

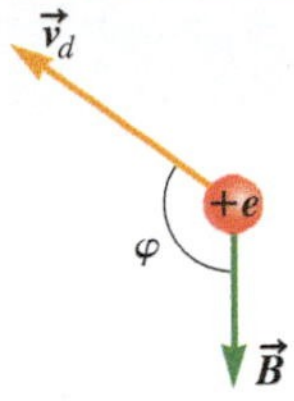

FIGURE 30.44 One equivalent positive charge moving in the direction of the current.

The magnetic force on this single particle is given by Equation 30.17.

$$\vec{f}_B = e(\vec{v}_d \times \vec{B}) \qquad (30.17)$$

The magnetic force on the entire segment of wire is the sum of the magnetic forces on all the equivalent positively charged particles. So we need to know the number of conduction electrons. The number density of conduction electrons is n and the volume of the segment is $A\ell$, so the number of conduction electrons is $nA\ell$.

The total magnetic force on the segment is the number of conduction electrons times the force on a single particle.

$$\vec{F}_B = (nA\ell)\vec{f}_B = nA\ell e(\vec{v}_d \times \vec{B})$$

$$\vec{F}_B = A\ell(ne\vec{v}_d \times \vec{B}) \qquad (1)$$

We can write Equation (1) in terms of the current density: $\vec{J} = ne\vec{v}_d$ (Eq. 28.5).

$$\vec{F}_B = A\ell(\vec{J} \times \vec{B}) \qquad (2)$$

The current density has magnitude $J = I/A$, and in this case its direction is the same as that of $\vec{\ell}$. The term in parentheses is a unit vector pointing in the same direction as $\vec{\ell}$.

$$\vec{J} = \frac{I}{A}\left(\frac{\vec{\ell}}{\ell}\right) \qquad (3)$$

Substitute Equation (3) into Equation (2). The cross-sectional area A and the magnitude ℓ of the length vector cancel.

$$\vec{F}_B = A\ell\left(\frac{I\vec{\ell}}{A\ell} \times \vec{B}\right) = I(\vec{\ell} \times \vec{B}) \ ✓ \qquad (30.32)$$

COMMENTS

1. The magnitude of the force is given by

$$F_B = I\ell B \sin\varphi \qquad (30.33)$$

where φ is the angle between $\vec{\ell}$ and $\vec{B}$. The direction of the force is given by the right-hand rule for cross products.

2. If the conductor is not straight, we find the magnetic force by imagining the conductor broken into many small segments that can each be considered a straight line with a length vector $d\vec{\ell}$ (Fig. 30.45). The magnetic force $d\vec{F}_B$ on any one of these segments is

$$d\vec{F}_B = I(d\vec{\ell} \times \vec{B}) \qquad (30.34)$$

The total force comes from integrating $d\vec{F}_B$ over the entire length of the conductor.

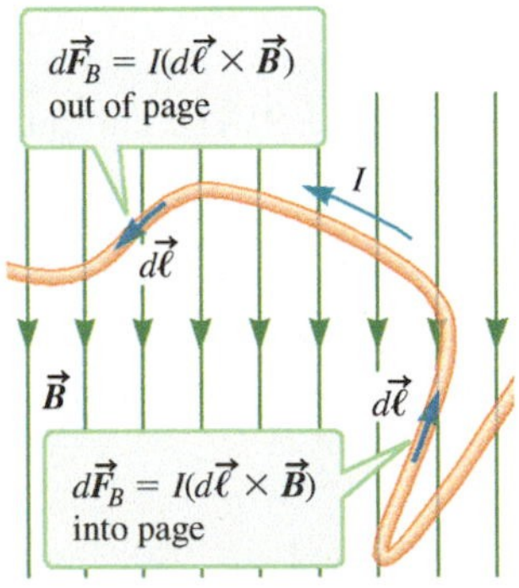

FIGURE 30.45 Imagine breaking up a curved current-carrying wire into small straight segments, like the two shown here. Then integrate the magnetic force on each segment.

EXAMPLE 30.8 Stereo Speakers and the Direction of Magnetic Force

Figure 30.46A shows the basic components of a stereo speaker. A permanent magnet shaped like a donut with a cylinder in the hole creates a radial magnetic field. In the space between the magnet's cylinder and its donut is a coil of wire called the *voice coil*. The signal from the stereo's amplifier controls the amount and direction of current in the voice coil. The permanent magnet exerts a force on the coil, causing it to move in and out. The motion of the voice coil results in the motion of the diaphragm, causing a sound wave in the surrounding air.

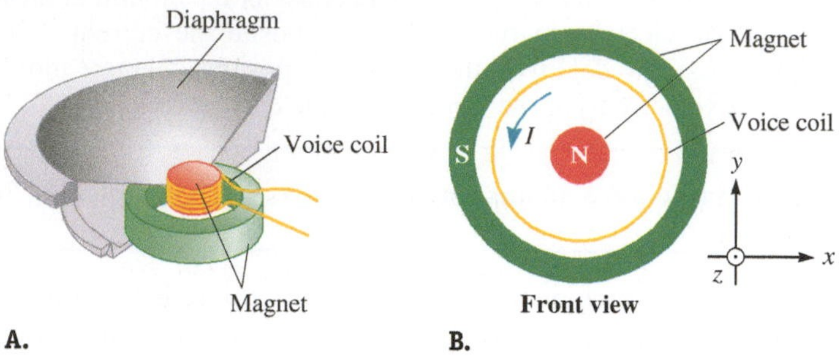

FIGURE 30.46 A. Parts of a stereo speaker. **B.** Magnet and voice coil of a stereo speaker. Find the direction of the magnetic force on the voice coil.

The magnitude and direction of the magnetic force on the voice coil depend on the current in it. Figure 30.46B shows a front view of these components. From this perspective, the voice coil looks like a single circular current loop. At the instant shown, the current is 1.5 A counterclockwise, and the magnitude of the magnetic field is 0.50 T (at the location of the coil). The radius of the voice coil is $R = 2.0$ cm. Find the magnetic force exerted on one loop of the voice coil for the instant shown in Figure 30.46B.

INTERPRET and ANTICIPATE

Because the wire loop of the coil is curved, break it up into small, straight segments. Find the magnitude of the force on one segment, and then integrate over the entire circular loop. Draw the magnetic field lines $\vec{B}$ due to the permanent magnet, pointing from the north pole to the south pole (Fig. 30.47). Next, find the magnetic force $d\vec{F}_B$ on a small, straight segment. We arbitrarily choose a segment of the coil at the bottom and draw $d\vec{\ell}$ for that segment, to the right in the direction of the current. Pointing the fingers of your right hand toward the right in the direction of $d\vec{\ell}$, imagine pushing $d\vec{\ell}$ into $\vec{B}$, which is downward near this segment. Your thumb should point into the page, so $d\vec{F}_B$ for this segment is into the page (in the negative z direction). For any other segment, $d\vec{F}_B$ is also in the negative z direction. So, the total magnetic force $\vec{F}_B$ on the coil is in the negative z direction.

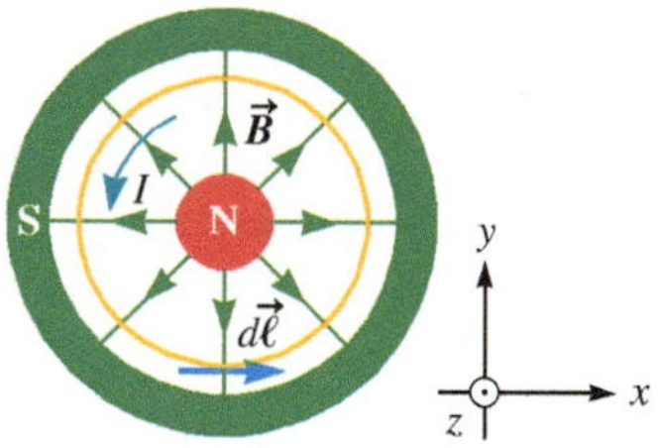

FIGURE 30.47

SOLVE

Start with Equation 30.34 for the segment shown in Figure 30.47. The magnetic field $\vec{B}$ is perpendicular to $d\vec{\ell}$, so the angle φ between $\vec{B}$ and $d\vec{\ell}$ is 90°.

$$d\vec{F}_B = I(d\vec{\ell} \times \vec{B}) \quad (30.34)$$

$$dF_B = I(d\ell)(B)\sin\varphi = I(d\ell)(B)$$

Find the total magnetic force by integrating over the entire circumference of the circular loop, from 0 to $2\pi R$. The current and the magnetic field strength are constant over the integral.

$$\int dF_B = \int I(d\ell)(B)$$

$$F_B = IB\int_0^{2\pi R} d\ell = IB2\pi R$$

Substitute numerical values. Remember that the direction is into the page.

$$F_B = 2\pi(1.5\,\text{A})(0.50\,\text{T})(2.0 \times 10^{-2}\,\text{m})$$

$$\vec{F}_B = -9.4 \times 10^{-2}\hat{k}\ \text{N}$$

CHECK and THINK

We found the magnetic force at one instant. When the current reverses direction, the force will be outward, and so on. So the speaker vibrates in and out to create a sound wave.

30-12 Force Between Two Long, Straight, Parallel Wires

In Section 30-11, we studied the magnetic force exerted on a current-carrying wire, regardless of the source of the magnetic field. In this section, we consider the special case in which the source of the magnetic field is another long, straight, current-carrying wire. Picture two wires hanging vertically, with no current in either wire (Fig. 30.48A). When the wires are connected to a battery, so that they both have current, they exert an attractive magnetic force on each other (Fig. 30.48B).

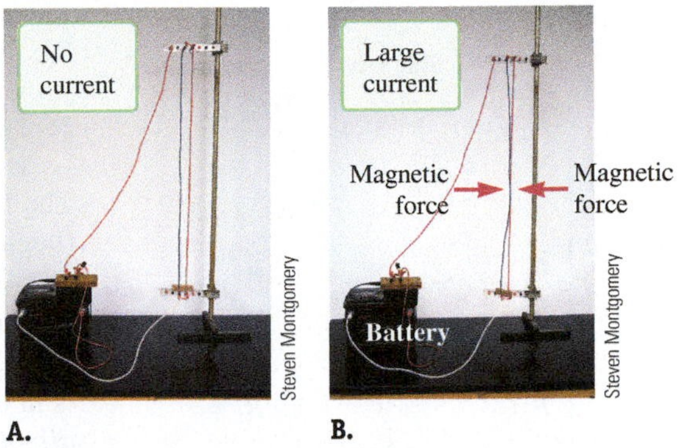

FIGURE 30.48 **A.** When there is no current in the wires, they hang vertically. **B.** When there is current in the wires, they exert an attractive magnetic force on each another. Both are bowed inward.

Figure 30.49 shows the force between two straight, parallel wires carrying current I_1 in wire 1 and I_2 in wire 2. Each wire produces a magnetic field that exerts a force on the other wire. We can arbitrarily consider wire 1 to be the source of the magnetic field and wire 2 to be the subject. Using the simple right-hand rule, we find that the direction of the magnetic field produced by wire 1 is clockwise (Fig. 30.49A). The magnetic field $\vec{B}_1$ *at the location of wire 2* is tangent to that circle in the negative y direction. From the side view (Fig. 30.49B), the magnetic field $\vec{B}_1$ at the location of wire 2 is directed into the page (in the negative y direction).

Now use Equation 30.32, $\vec{F}_B = I(\vec{\ell} \times \vec{B})$, and the right-hand rule for cross products to find the direction of the magnetic force exerted on wire 2 by the magnetic field $\vec{B}_1$. The magnetic force on wire 2 due to $\vec{B}_1$ is to the left (negative x direction), so wire 2 is attracted to wire 1. Next, let's find the magnitude of that attractive force using $F_B = I\ell B \sin\varphi$ (Eq. 30.33). The current we need is the subject's current (in this case I_2), and the magnetic field we need is the source's magnetic field B_1. The angle φ between the length vector $\vec{\ell}$ and the magnetic field $\vec{B}_1$ is 90°, so the magnitude of the force exerted by wire 1 on wire 2 is

$$F_{[1\text{ on }2]} = I_2 \ell B_1 \sin\varphi = I_2 \ell B_1 \sin 90°$$

$$F_{[1\text{ on }2]} = I_2 \ell B_1 \tag{30.35}$$

Using the notation from Section 5-9, we denote the magnetic force exerted by wire 1 on wire 2 by $\vec{F}_{[1\text{ on }2]}$.

Wire 1 is long and straight, so substitute the expression $B_1 = (\mu_0 I_1)/(2\pi r)$ (Eq. 30.9) for a long, straight wire into Equation 30.35 in order to write the force exerted by wire 1 on wire 2 in terms of their currents:

$$F_{[1\text{ on }2]} = \left(\frac{\mu_0}{2\pi}\right)\frac{I_1 I_2 \ell}{r} \tag{30.36}$$

Equation 30.36 is the force exerted by wire 1 on wire 2. By Newton's third law, the force exerted by wire 2 on wire 1 must have the same magnitude but the opposite direction:

$$\vec{F}_{[2\text{ on }1]} = -\vec{F}_{[1\text{ on }2]}$$

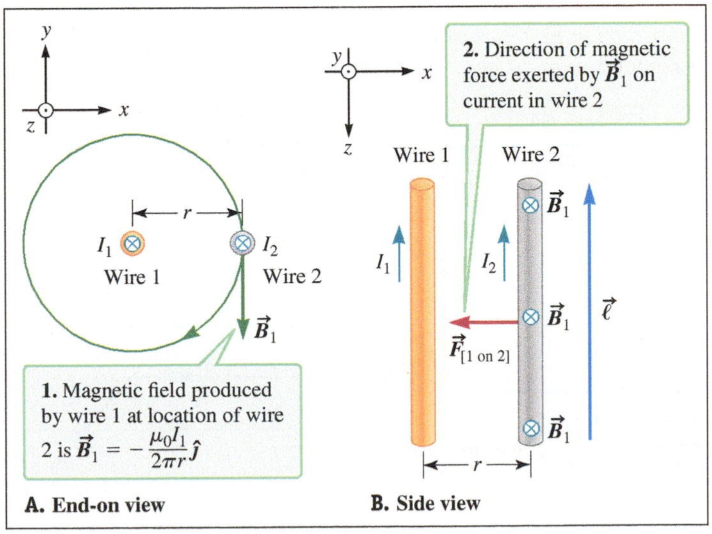

FIGURE 30.49 Two parallel, straight, current-carrying wires seen **A.** end-on and **B.** from the side. To find the magnetic field produced by wire 1, point your right thumb into the page in the direction of I_1 (in the negative z direction). Your fingers wrap clockwise as shown by the circular field line in part A. The magnetic force exerted by wire 1 on wire 2 is to the left (negative x direction) by $\vec{F}_{[1\text{ on }2]} = I(\vec{\ell} \times \vec{B})$.

Thus, Equation 30.36 gives the magnitude of the force exerted on either wire 1 or wire 2. Because the wires are very long, it is often convenient to calculate the magnitude of the force per unit length f on either wire:

$$f = \frac{F_{[1\text{ on }2]}}{\ell} = \frac{F_{[2\text{ on }1]}}{\ell}$$

MAGNETIC FORCE (PER UNIT LENGTH) BETWEEN TWO PARALLEL CURRENT-CARRYING WIRES Special Case

$$f = \left(\frac{\mu_0}{2\pi}\right)\frac{I_1 I_2}{r} \tag{30.37}$$

We have shown that when two long, straight, parallel wires *carry currents in the same direction*, they are *attracted* to each other, and the magnitude of the force per unit length is given by Equation 30.37.

What if two parallel wires carry oppositely directed currents, as in Figure 30.50? The magnetic field produced by wire 1 at the location of wire 2 is into the page (negative y), just as it was in Figure 30.49B. In this case, however, the current in wire 2 is downward, so $\vec{\ell}_2$ is in the positive z direction. Applying the right-hand rule for the cross product in $\vec{F}_B = I(\vec{\ell} \times \vec{B})$ shows that the force $\vec{F}_{[1\text{ on }2]}$ exerted by $\vec{B}_1$ on wire 2 is in the positive x direction. By Newton's third law, $\vec{F}_{[2\text{ on }1]}$ is in the negative x direction. Thus, when parallel wires *carry currents in opposite directions*, the wires *repel* each other.

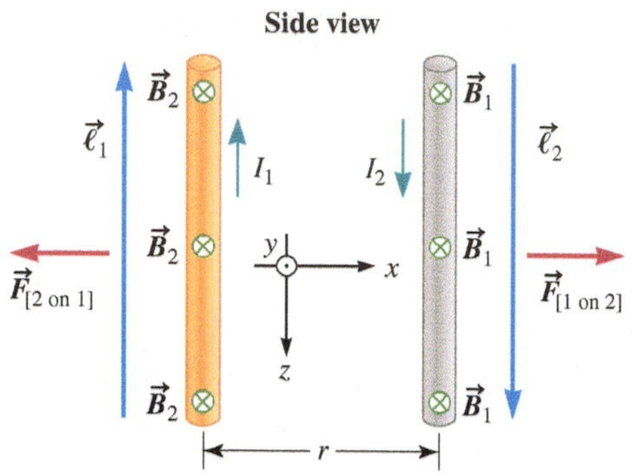

FIGURE 30.50 When they carry currents in opposite directions, long, straight, parallel wires repel each other.

CONCEPT EXERCISE 30.8

When we think about two charged particles, it is helpful to remember the phrase *opposites attract and likes repel.* Come up with a similar phrase for the magnetic force between two parallel current-carrying wires.

30-13 Current Loop in a Uniform Magnetic Field

We have studied the magnetic force exerted on a single charged particle and on a straight current-carrying wire. In this section, we turn our attention to a current-carrying loop of wire. Current loops play an important role in practical devices such as galvanometers and electric motors, and they are also important in modeling permanent magnets (Section 30-7).

To keep things simple for now, consider a source that produces a uniform magnetic field, and a rectangular current loop carrying a current I as the subject (Fig. 30.51A). Our goal is to find the net force and the net torque on the loop. The top and bottom sides of the loop have length ℓ, and the left and right sides have length w. The loop is tilted with respect to the magnetic field, which is best seen from the side (Fig. 30.51B).

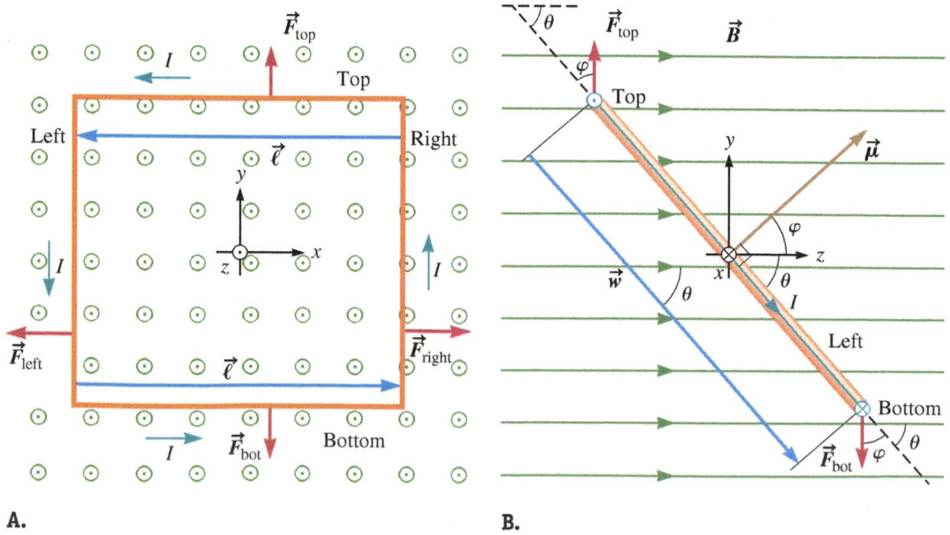

FIGURE 30.51 A current loop in a uniform magnetic field. **A.** Front view showing forces on all four sides. The net force on the loop is zero. **B.** Side view. The side labeled *right* is behind the left side and is not visible from this view. The forces exerted on the right and left sides are not shown in this view. There is a net torque on the loop.

The direction of the magnetic force comes from Equation 30.32, $\vec{F}_B = I(\vec{\ell} \times \vec{B})$, and the right-hand rule for cross products. The magnetic field is perpendicular to the length vectors $\vec{\ell}$ for the top and bottom portions of the loop, so $\varphi = 90°$ and the magnitude of the magnetic force on the top and bottom is

$$F_{top} = F_{bot} = I\ell B \tag{30.38}$$

For the top, $\vec{\ell}$ is in the negative x direction, so the magnetic force is upward in the positive y direction:

$$\vec{F}_{top} = I\ell B\hat{\jmath} \tag{30.39}$$

For the bottom, $\vec{\ell}$ is in the positive x direction, so the magnetic force is downward in the negative y direction:

$$\vec{F}_{bot} = -I\ell B\hat{\jmath} \tag{30.40}$$

The magnetic field makes an angle θ with the length vectors $\vec{w}$ for the left and right portions of the loop (Fig. 30.51B). The magnitude of the magnetic force on the right and left is $F_{right} = F_{left} = IwB \sin\theta$. After finding the direction of these forces, we have

$$\vec{F}_{left} = -IwB \sin\theta\,\hat{\imath} \tag{30.41}$$

and

$$\vec{F}_{right} = IwB \sin\theta\,\hat{\imath} \tag{30.42}$$

The total magnetic force $\vec{F}_B$ exerted on the loop is the sum of the magnetic forces exerted on each of the four sides:

$$\vec{F}_B = \vec{F}_{top} + \vec{F}_{bot} + \vec{F}_{right} + \vec{F}_{left}$$

Substitute Equations 30.39 through 30.42:

$$\vec{F}_B = (I\ell B\hat{\jmath} - I\ell B\hat{\jmath}) + (IwB \sin\theta\,\hat{\imath} - IwB \sin\theta\,\hat{\imath})$$

$$\vec{F}_B = 0$$

We considered a rectangular loop, but our result holds for a loop of any shape: *A uniform magnetic field exerts no net magnetic force on a loop of current.*

The forces $\vec{F}_{top}$ and $\vec{F}_{bot}$ exert a net torque, which rotates the loop clockwise as seen from the side (Fig. 30.51B). Let's calculate the net torque around the loop's center of mass (the origin of our coordinate system). The magnitude of the torque exerted by a force F is given by

$$\tau = rF \sin\varphi \tag{12.16}$$

where $\vec{r}$ is the position vector from the rotation axis to the point at which $\vec{F}$ is applied and φ is the angle between $\vec{F}$ and $\vec{r}$. The distance from the axis of rotation to either the top or bottom of the loop is $r = w/2$. The magnitude of each force is given by Equation 30.38, so the magnitude of the torque exerted by $\vec{F}_{top}$ or $\vec{F}_{bot}$ is

$$\tau_{top} = \tau_{bot} = \left(\frac{w}{2}\right)(I\ell B) \sin\varphi \tag{30.43}$$

The direction of the torque is given by $\vec{\tau} = \vec{r} \times \vec{F}$ (Eq. 12.23) and the right-hand rule for cross products. For both $\vec{F}_{top}$ and $\vec{F}_{bot}$, the torque is into the page in the positive x direction (Fig. 30.51B). The total torque exerted on the loop is

$$\tau = 2\tau_{top} = 2\left(\frac{w}{2}\right)(I\ell B) \sin\varphi$$

$$\vec{\tau} = [I(\ell w)B \sin\varphi]\hat{\imath} \tag{30.44}$$

The term in parentheses is the area A of the loop:

$$\vec{\tau} = [(IA)B \sin\varphi]\hat{\imath} \tag{30.45}$$

The total torque on the loop is more conveniently expressed in terms of the loop's magnetic dipole moment of magnitude $\mu \equiv IA$ (Eq. 30.12). The torque becomes

$$\vec{\tau} = [\mu B \sin \varphi]\hat{\imath} \tag{30.46}$$

The direction of the magnetic dipole moment $\vec{\mu}$ is found by wrapping the fingers of your right hand in the direction of the current; your thumb then points in the direction of the magnetic moment (Fig. 30.51B). The angle between $\vec{\mu}$ and $\vec{B}$ is φ, so the term in brackets in Equation 30.46 is the magnitude of the cross product of $\vec{\mu}$ and $\vec{B}$. Thus, the total torque is written as

TORQUE ON A MAGNETIC DIPOLE

Special Case

$$\vec{\tau} = \vec{\mu} \times \vec{B} \tag{30.47}$$

Equation 30.47 is the torque exerted on a current loop (a magnetic dipole) by a uniform magnetic field. From Section 12-6, the direction of the torque is perpendicular to the plane in which the object rotates. The torque in Figure 30.51B is in the positive x direction, and the loop rotates clockwise in the yz plane. It is helpful to remember: *The torque attempts to align* $\vec{\mu}$ *with* $\vec{B}$.

Magnetic Potential Energy

The torque on a magnetic dipole causes it to rotate back and forth. Just as we described the back-and-forth motion of a pendulum in terms of changing kinetic and potential energies, we can associate kinetic and potential energies with the rotation of a current loop. To arrive at an expression for the magnetic potential energy, we can use our work in Section 24-9 on the electric dipole. A magnetic dipole rotating in a uniform magnetic field $\vec{B}$ is analogous to an electric dipole rotating in a uniform electric field $\vec{E}$. The torque on such an electric dipole is given by

$$\vec{\tau} = \vec{p} \times \vec{E} \tag{24.22}$$

where $\vec{p}$ is the electric dipole moment. Comparing Equation 30.47 to Equation 24.22, we find

$$\vec{p} \rightarrow \vec{\mu} \quad \text{and} \quad \vec{E} \rightarrow \vec{B}$$

From Chapter 24, the electric potential energy stored in a dipole–electric field system is

$$U = -\vec{p} \cdot \vec{E} = -pE \cos \varphi \tag{24.26, 24.25}$$

where the reference configuration is set so that when the electric dipole is perpendicular to the electric field ($\varphi = 90°$), the potential energy is zero. By analogy, the magnetic potential energy of a dipole–field system is

MAGNETIC POTENTIAL ENERGY

Special Case

$$U = -\vec{\mu} \cdot \vec{B} = -\mu B \cos \varphi \tag{30.48}$$

where the reference configuration is set so that when the magnetic dipole moment is perpendicular to the magnetic field ($\varphi = 90°$), the potential energy is zero. The lowest potential energy is $U_{\text{min}} = -\mu B$, when $\varphi = 0$ and the dipole moment is aligned with the magnetic field. The highest potential energy is $U_{\text{max}} = +\mu B$, when $\varphi = 180°$ and the dipole moment is opposite to the magnetic field.

Figure 30.52 shows a current loop as it rotates clockwise in a uniform magnetic field. 1 Initially, the magnetic dipole is at rest, the kinetic energy is zero, and the potential energy is negative. Because the torque is into the page, the magnetic dipole rotates clockwise. 2 As the angle φ decreases, the potential energy decreases and the kinetic energy increases. 3 When the magnetic dipole moment is aligned with the magnetic field, the potential energy is at its minimum, the kinetic energy is at its maximum, and there is no torque. 4 Because the magnetic dipole has angular momentum, it continues to rotate in the same clockwise direction. As the angle φ increases, the potential energy increases (becomes less negative), the kinetic energy decreases, and the torque is now out of the page. 5 When the angle φ reaches its maximum, the magnetic dipole stops momentarily, the potential energy is at its maximum, the kinetic energy is zero, and the torque is still out of the page. So, the dipole begins to rotate in the other direction. In the absence of dissipative forces, the magnetic dipole would rotate back and forth as energy is converted between its potential and kinetic forms.

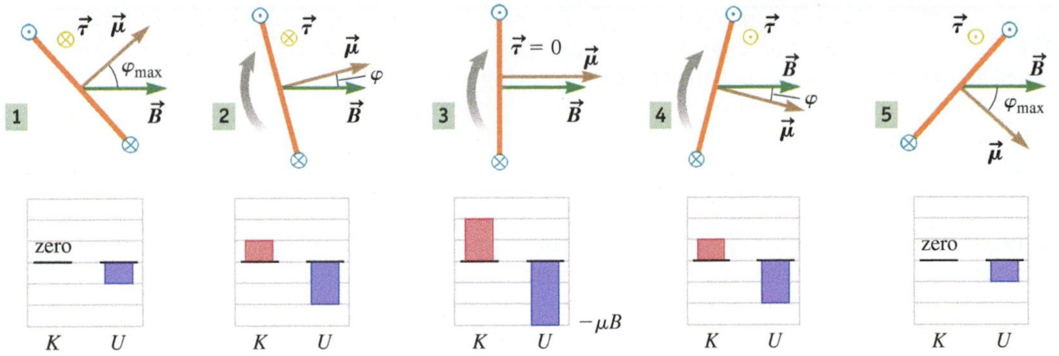

FIGURE 30.52 A current loop rotates clockwise, starting from rest.

We have considered a single current loop. If a wire is curled into a coil with N turns, the magnitude of the magnetic dipole moment is

$$\mu = NIA \quad (30.49)$$

The torque in a uniform magnetic field is still given by $\vec{\tau} = \vec{\mu} \times \vec{B}$ (Eq. 30.47), and the potential energy of the coil-field system is given by $U = -\vec{\mu} \cdot \vec{B}$ (Eq. 30.48).

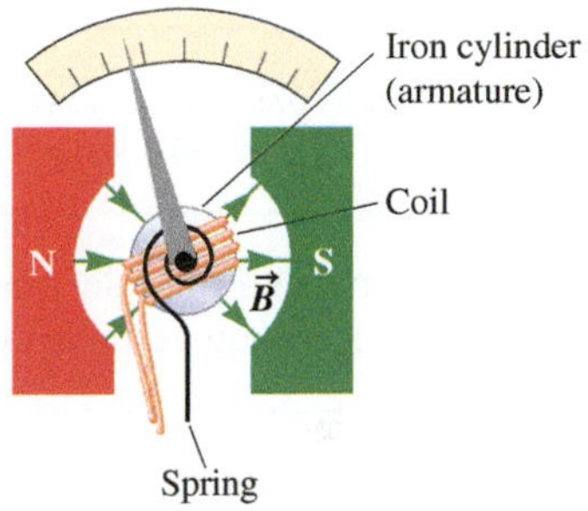

FIGURE 30.53 Basic parts of a d'Arsonval galvanometer.

Galvanometers

The essential element of an analog multimeter is a d'Arsonval **galvanometer** (Section 29-8), which consists of a rectangular coil of wire wrapped around an armature free to rotate between the poles of a permanent magnet (Fig. 30.53). The permanent magnet is bent as shown, and the armature is usually an iron cylinder. The result is that the magnetic field is nearly radial. The field lines point from the north to the south poles, crossing in the center. The field exerts a torque $\vec{\tau}_B$ on the coil when it carries a current. A spring then exerts a restoring torque $\vec{\tau}_S$ on the armature, so that the armature–coil assembly rotates until the two torques cancel. A needle attached to the armature rotates with it, and a scale printed on the galvanometer indicates the current in the coil.

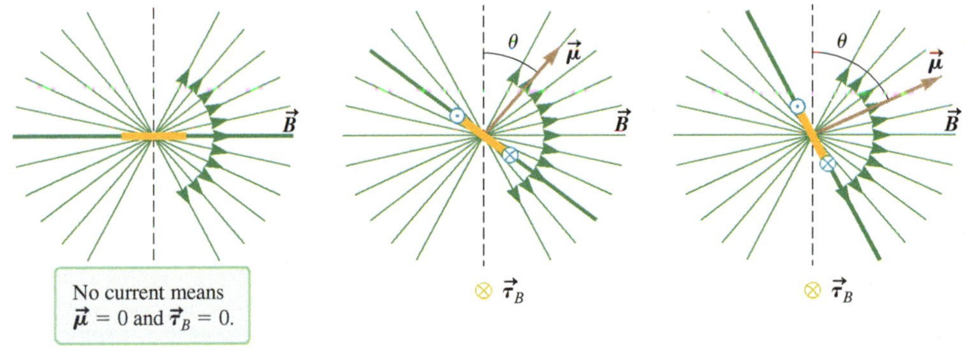

A. B. C.

FIGURE 30.54 **A.** No current in the coil. The magnetic dipole moment is zero and the magnetic torque is zero. **B.** Small current in the coil results in a small magnetic dipole moment and a magnetic torque directed into the page. **C.** Larger current means a large magnetic dipole moment and a greater magnetic torque. In all cases, $\tau_B = \tau_S$ and the coil is at rest.

Figure 30.54 shows just the coil and the radial magnetic field in a cross-sectional view like the one in Figure 30.51B. In the galvanometer, however, the magnetic field is radial, so the magnetic field that penetrates the metal of the coil (yellow in Fig. 30.54) is parallel to the plane of the coil. When there is no current in the coil, the magnetic dipole moment $\vec{\mu} = 0$ and the magnetic torque $\vec{\tau}_B = 0$ (Fig. 30.54A). This is also the relaxed position of the spring (not shown in this figure), so $\vec{\tau}_S = 0$. The current in the coil causes the armature to rotate until the magnetic torque is balanced by the spring's torque: $\tau_B = \tau_S$. The magnetic field passing through the metal of the coil is perpendicular to the coil's magnetic dipole moment $\vec{\mu}$.

EXAMPLE 30.9 Building a Galvanometer

Suppose you want to make a galvanometer using a square coil with sides of length 2.00 cm. The magnitude of the magnetic field is 0.250 T. The maximum current to be read is 1.00×10^{-4} A when the deflection of the needle is 45.0°. If the spring has torsion spring constant $\kappa = 3.00 \times 10^{-6}$ N · m/rad, how many turns does the coil need to have?

Example continues on page 968 ▶

INTERPRET and ANTICIPATE

The armature stops rotating when the magnetic torque is balanced by the spring's torque. We need to find the torque exerted by both the magnetic and spring forces when the armature has rotated $\theta = 45.0°$. Figure 30.54B provides a rough sketch.

SOLVE

The magnitude of the spring's torque comes from Equation 16.34. The spring is relaxed when the needle points straight up (indicating zero current), so θ is measured from the vertical position.

$$\tau_S = \kappa\theta \qquad (16.34)$$

The magnitude of the magnetic torque comes from Equation 30.46.

$$\vec{\tau} = [\mu B \sin\varphi]\hat{\imath} \qquad (30.46)$$

$$\tau_B = \mu B \sin\varphi$$

The magnetic field passing through the metal of the coil is perpendicular to the coil's dipole moment (Fig. 30.54), so $\varphi = 90°$.

$$\tau_B = \mu B \sin 90° = \mu B \qquad (1)$$

Substitute the magnetic dipole moment for a coil (Eq. 30.49) into Equation (1). Write the area of the square in terms of the length ℓ of one of its sides.

$$\mu = NIA \qquad (30.49)$$

$$\tau_B = NIAB = NI\ell^2 B \qquad (2)$$

The armature stops rotating when $\tau_B = \tau_S$, so set Equation (2) equal to Equation 16.34 and solve for N. Substitute numerical values, including $\theta = 45° = \pi/4$ rad.

$$\tau_B = \tau_S$$

$$NI\ell^2 B = \kappa\theta$$

$$N = \frac{\kappa\theta}{I\ell^2 B} = \frac{(3.00 \times 10^{-6}\,\text{N}\cdot\text{m/rad})(\pi/4\,\text{rad})}{(100 \times 10^{-6}\,\text{A})(2.00 \times 10^{-2}\,\text{m})^2(0.250\,\text{T})}$$

$$N = 236$$

CHECK and THINK

Dimensional analysis shows that the result is a pure number, as it should be. The galvanometer is cleverly designed. If the magnetic field were not (nearly) radial, then as the armature rotated, the magnitude of the magnetic field's torque would vary depending on $\sin\varphi$. In other words, the amount of current in the coil would no longer be directly proportional to the angle θ of the coil's deflection. In Problem 71, you can see how a galvanometer would work if the magnetic field were uniform instead of radial.

DC Motors

The direct current (DC) motor is a practical application of the magnetic torque exerted on a current loop. A motor's job is to convert electrical energy into mechanical energy. A current loop—known as a *rotor*—is rotated by the torque exerted by a permanent magnet (Fig. 30.55). The rotor is connected to a DC power supply, such as battery, through a *commutator*. The commutator is a conductor—usually shaped like a ring or disk—that is split by an insulator, such as air. At the moment shown in Figure 30.55A, the commutator is in contact with the brushes (also conductors): The current in the yellow part of the rotor is in the negative z direction, and the current in the orange part is in the positive z direction. Further, the magnetic field is in the positive x direction, $\vec{B} = B\hat{\imath}$, and the magnetic dipole moment is in the negative y direction, $\vec{\mu} = -\mu\hat{\jmath}$. By $\vec{\tau} = \vec{\mu} \times \vec{B}$ (Eq. 30.47), the torque is

$$\vec{\tau} = -\mu B(\hat{\jmath} \times \hat{\imath}) = -\mu B(-\hat{k}) = \mu B\hat{k}$$

The z axis coincides with the rotor's rotation axis, and the torque is in the positive z direction. So, if you viewed the rotor from the commutator in the negative z direction, you would see the rotor turn counterclockwise.

When the rotor has turned 90°, so that the orange portion is directly above the yellow portion (Fig. 30.55B), the split in the commutator is centered on the two brushes. Each brush is in contact with both halves of the commutator. The commutator now provides a low-resistance path for the current, so the current passes through the commutator instead of through the rotor. In this position, the rotor has no current and no magnetic moment, so no torque is exerted on it. However, the rotor does have angular momentum and rotational inertia, so it continues to rotate despite the small amount of friction between the brushes and the commutator.

Soon, each brush is back in contact with only one half of the commutator, so the current must pass through the rotor again. In Figure 30.55C, the current in the yellow part of the rotor is in the positive z direction, and the current in the orange part is in the negative z direction, because the rotor–commutator assembly is oriented 180° from its position in Figure 30.55A. However, the magnetic moment is still in the negative y direction, and the torque is again in the positive z direction. In other words, in this position, the rotor continues to rotate in the same counterclockwise direction, as seen from the left in the figure. The split-ring commutator is designed to keep the current flowing clockwise as seen from above, even as the rotor flips over. The resulting current ensures that the magnetic moment always points in the same direction (negative y).

A real DC motor is more complicated. Its rotor actually contains a coil with many turns. This construction increases its magnetic dipole moment and the torque exerted on the rotor. To increase the torque further, a strong magnet is used—often an electromagnet instead of a permanent magnet. Finally, the magnitude of the torque exerted on the rotor depends on the angle between $\vec{\mu}$ and $\vec{B}$, which increases and decreases as the rotor rotates. To maintain a more constant torque, a real rotor consists of several coils oriented at different angles.

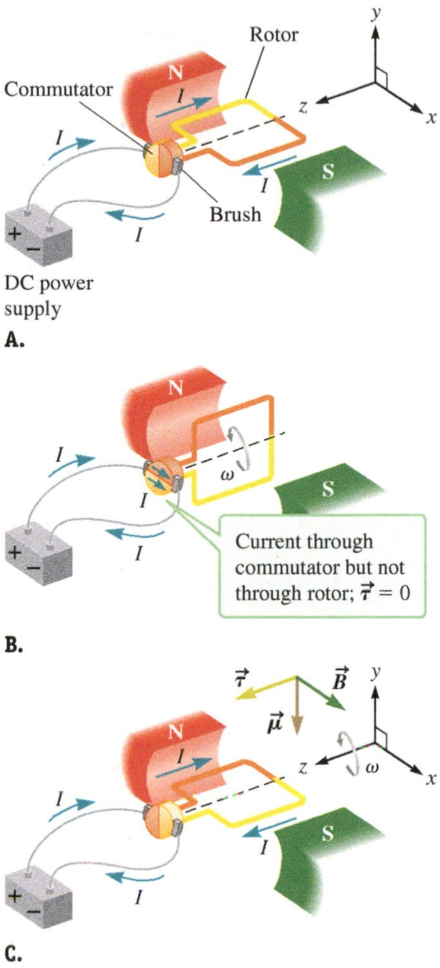

FIGURE 30.55 Components of a simple DC motor. **A.** The current is clockwise as seen from above. The magnetic moment is in the negative y direction, and the torque is in the positive z direction. **B.** When the rotor is vertical, there is no current in it. Instead, the current goes directly through the commutator. Because the rotor has angular momentum, it continues to rotate counterclockwise as seen from the left. **C.** The rotor has flipped over but the current is still clockwise as in A.

SUMMARY

Underlying Principles

Magnetic fields are produced by moving charged particles and in magnetized materials by the orbital motion and "spin" (a quantum property) of subatomic particles, such as electrons. No matter what the source, magnetic fields may exert **magnetic forces** on subjects that are moving charged particles or magnetized materials.

Major Concepts

1. A **magnetic dipole** has both a north and a south pole. A **magnetic monopole** has either a north or a south pole, but not both. So far, no magnetic monopole has been observed.
2. **Biot-Savart law:** The magnetic field $\vec{B}$ produced by a single particle with charge q moving at velocity $\vec{v}$:

$$\vec{B} = \left(\frac{\mu_0}{4\pi}\right) q \frac{\vec{v} \times \vec{r}}{r^3} \qquad (30.1)$$

The Biot-Savart law for the magnetic field $d\vec{B}$ produced by a small segment of a wire:

$$d\vec{B} = \left(\frac{\mu_0}{4\pi}\right) I \frac{d\vec{\ell} \times \vec{r}}{r^3} \qquad (30.5)$$

3. The **magnetic dipole moment** $\vec{\mu}$ (also called the **magnetic moment**) is a vector of magnitude

$$\mu \equiv IA \qquad (30.12)$$

for a loop of current I and area A. The vector direction is given by a right-hand rule (tool 3, next page). The magnetic moment direction is the same as that of the magnetic field at the center of the current loop.

4. Each atom in a **ferromagnetic material** has a net nonzero magnetic moment and interacts strongly with its neighbors, causing the magnetic moments of nearby atoms to line up. A region in which the atoms' magnetic moments are aligned is called a **magnetic domain**.
5. **Magnetic force on a moving charged particle:**

$$\vec{F}_B = q(\vec{v} \times \vec{B}) \quad (30.17)$$

6. The **Lorentz force** is the total electromagnetic force experienced by a charged particle:

$$\vec{F}_L = \vec{F}_E + \vec{F}_B = q(\vec{E} + \vec{v} \times \vec{B}) \quad (30.21)$$

7. **Magnetic force on a current-carrying wire:**

$$\vec{F}_B = I(\vec{\ell} \times \vec{B}) \quad (30.32)$$

Special Cases

1. **Magnetic field due to**
   a. an **infinitely long, straight, current-carrying wire:**

$$B = \frac{\mu_0 I}{2\pi r} \quad (30.9)$$

   b. a **circular current loop** of radius R (on axis):

$$\vec{B} = \frac{\mu_0 I R^2}{2(R^2 + y^2)^{3/2}}\hat{\jmath} \quad (30.10)$$

   where $\hat{\jmath}$ points along the loop axis.

   c. a **magnetic dipole** for distant points on the dipole axis:

$$\vec{B} \approx \left(\frac{\mu_0}{2\pi}\right)\frac{\vec{\mu}}{y^3} \quad (30.15)$$

2. **Magnetic force (per unit length) between two parallel current-carrying wires:**

$$f = \left(\frac{\mu_0}{2\pi}\right)\frac{I_1 I_2}{r} \quad (30.37)$$

   The force is attractive if the currents are in the same direction and repulsive if the currents are in the opposite directions.
3. **Torque on a magnetic dipole** in a uniform magnetic field:

$$\vec{\tau} = \vec{\mu} \times \vec{B} \quad (30.47)$$

4. **Magnetic potential energy stored in a dipole–magnetic field system:**

$$U = -\vec{\mu} \cdot \vec{B} = -\mu B \cos\varphi \quad (30.48)$$

Tools

1. **Magnetic field lines** indicate both the strength and direction of the magnetic field (Fig. 30.5B). Magnetic field lines always form closed loops. Outside a bar magnet, field lines point *away from* the north pole and *toward* the south pole. Inside a bar magnet, field lines point from the south pole toward the north pole (Fig. 30.8). Field lines are densely packed in regions where the magnetic field is relatively strong. The field direction at a particular point is tangent to the field line.
2. The **simple right-hand rule** (SRHR) gives the direction of the magnetic field produced by a current-carrying wire. Point your right thumb in the direction of the current; then your fingers wrap around the wire in the direction of the magnetic field (Fig. 30.11).
3. To find the direction of the magnetic moment $\vec{\mu}$ and of the magnetic field $\vec{B}$ along the axis of a current loop, use the **right-hand rule for the direction of the magnetic dipole moment**: Wrap the fingers of your right hand in the direction of the current; then your thumb points in the direction of $\vec{\mu}$ and $\vec{B}$ (Fig. 30.25).

PROBLEM-SOLVING STRATEGY Using the Biot-Savart Law $d\vec{B} = \left(\frac{\mu_0}{4\pi}\right)I\frac{d\vec{\ell} \times \vec{r}}{r^3}$

INTERPRET and ANTICIPATE

Draw a diagram with these four elements:

1. A **coordinate system** and other **geometric details.**
2. **Labeled segments** of the source.
3. The **magnetic field vectors** produced by the segments at the point of interest.
4. The **net magnetic field vector** at the point of interest.

SOLVE

1. Use the Biot-Savart law to find a **mathematical expression** for $d\vec{B}$, the small magnetic field produced by a single current segment at the point of interest.
2. **Integrate** $d\vec{B}$ to find $\vec{B}$ at the point of interest.

PROBLEMS AND QUESTIONS

A = algebraic **C** = conceptual **E** = estimation **G** = graphical **N** = numerical

30-1 Another Fundamental Force

1. **C** A yoga teacher tells her students to imagine their hands are magnets pulling on each other. What are the problems with this metaphor? What is a better metaphor?
2. **C** One end of a bar magnet is brought close to the end of a metal bar. This metal bar might be attracted, be repelled, or experience no magnetic force. In which of these three cases can you determine that the second metal bar is also a magnet? In which cases can you determine that the second bar is definitely not a magnet? Explain your reasoning.
3. **C** We have studied three field forces so far—gravity, the electric force, and the magnetic force. Which of these forces is associated with each of the following sources according to a stationary observer: **a.** a neutron at rest, **b.** a proton at rest, and **c.** a proton moving at velocity $\vec{v}$?

30-2 Revealing Magnetic Fields

4. **C** Why are there no gravitational dipoles?
5. **C** Because there are no magnetic monopoles, magnetic field lines must form closed loops. Consider the magnetic field lines for a bar magnet in Figure 30.8. **a.** Do all the lines form closed loops? **b.** Does the line that passes through the axis of the magnet form a closed loop? Explain.
6. **G** Copy Figure P30.6 and sketch the magnetic field lines that result from the bar magnets shown there.

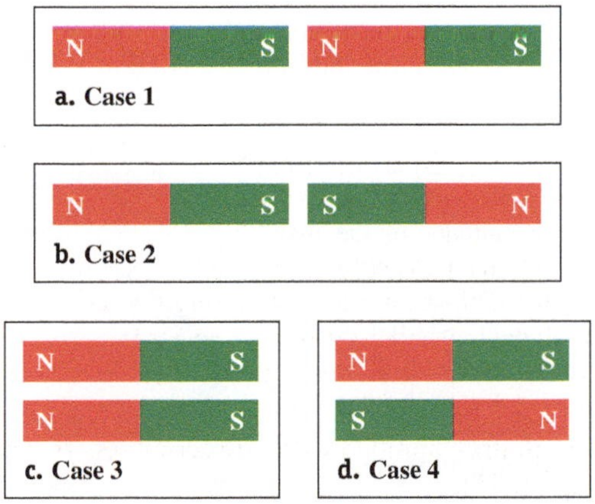

FIGURE P30.6

7. **G** A student attempts to draw the magnetic field lines near one end of a bar magnet that is held near a coin. The student produces Figure P30.7. Identify which aspects of the drawing are not consistent with the rules for correctly drawing magnetic field lines.

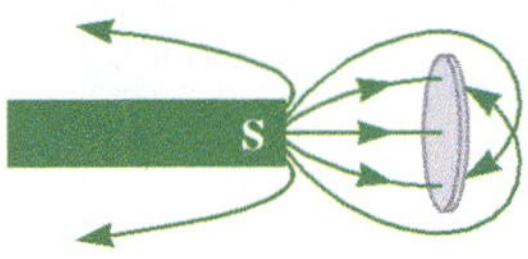

FIGURE P30.7

30-3 Ørsted's Discovery

8. **C** In your lab, you have a vertical wire that carries a current from the floor to the ceiling. A compass is moved in a plane perpendicular to the wire. When there is no current in the wire, the compass needle points parallel to the north-south direction. Assume that when there is a current in the wire, the magnetic field created by this current is much greater than the Earth's magnetic field. **a.** Describe where the compass should be placed relative to the wire so that the compass needle is undeflected when the current is turned on. **b.** Describe where the compass should be placed relative to the wire so that the compass needle is deflected by 90° when the current is turned on.
9. **C** Figure P30.9 shows very long current-carrying wires. Using the coordinate system indicated (with the z axis out of the page), state the direction of the magnetic field at point P in each case.

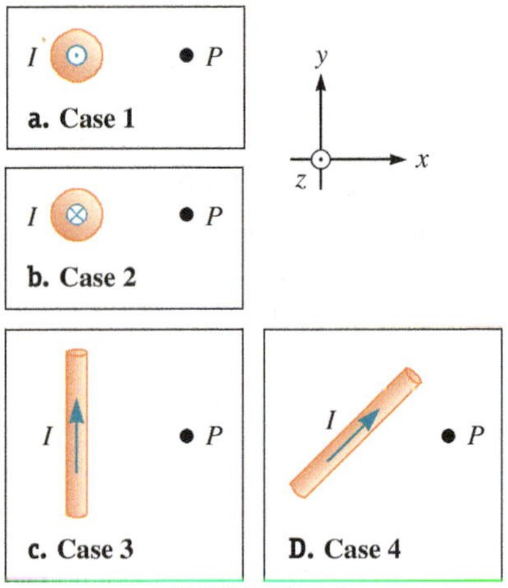

FIGURE P30.9

10. **C** Figure P30.10 shows a circular current-carrying wire. Using the coordinate system indicated (with the z axis out of the page), state the direction of the magnetic field at points A and B.

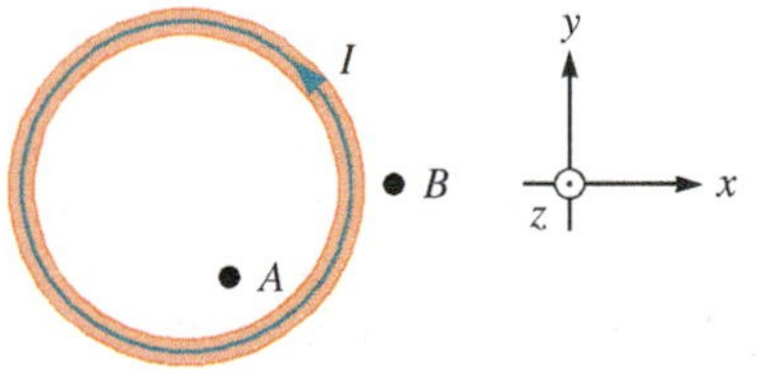

FIGURE P30.10

11. **C** Figure P30.11 shows three configurations of wires and the resultant magnetic fields due to current in the wires. What is the direction of the current that gives the resultant magnetic field shown in each case?

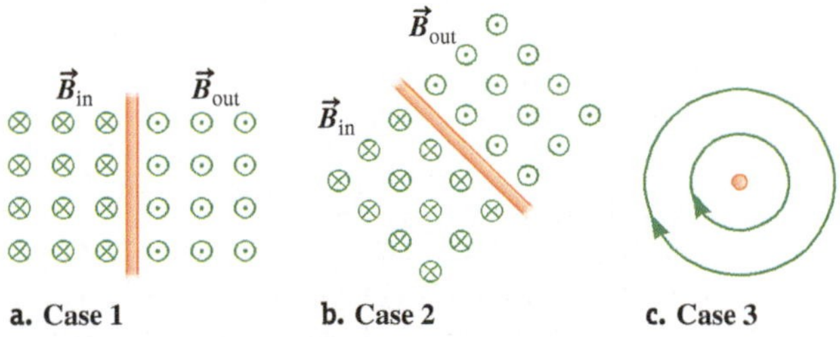

FIGURE P30.11

30-4 The Biot-Savart Law

12. N Review A proton is accelerated from rest through a 5.00-V potential difference. **a.** What is the proton's speed after it has been accelerated? **b.** What is the maximum magnetic field that this proton produces at a point that is 1.00 m from the proton?

13. A An electron moves in a circle of radius r at uniform speed around a single stationary proton. Find an expression for the magnitude of the magnetic field at the center of the circle in terms of μ_0, e, the speed v, and r.

Problems 14 and 15 are paired.

14. N One common type of cosmic ray is a proton traveling at close to the speed of light. If the proton is traveling downward, as shown in Figure P30.14, at a speed of 1.00×10^7 m/s, what are the magnitude and direction of the magnetic field at point B?

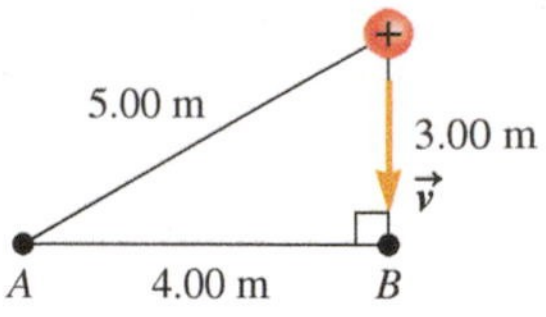

FIGURE P30.14 Problems 14 and 15.

15. N For the proton in Problem 14, what are the magnitude and direction of the magnetic field at point A?

30-5 Using the Biot-Savart Law

16. C Consider two very long, parallel wires that carry the same current but in opposite directions. Is the resulting magnetic field 0 everywhere? Does your answer depend on the distance between the wires?

17. N A straight wire carries a current of 5.00 A. What is the magnitude of the magnetic field it produces at a perpendicular distance of 50.0 cm from the wire?

Problems 18, 19, and 20 are grouped.

18. A Two long, straight, parallel wires are shown in Figure P30.18. The current in the wire on the left is double the current in the wire on the right. Find an expression for the magnetic field at points A and B. Use the indicated coordinate system to write your answer in component form.

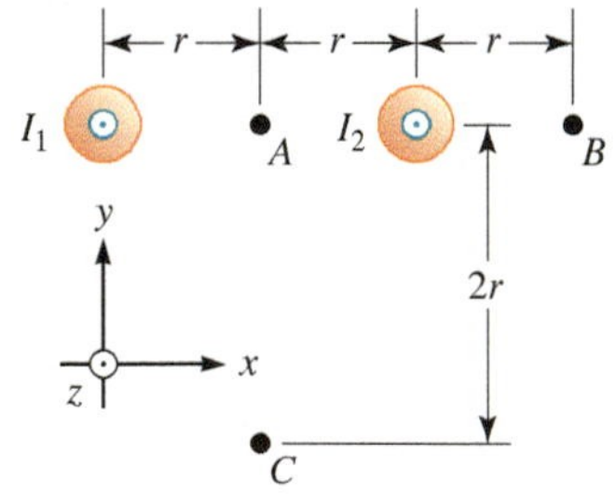

FIGURE P30.18
Problems 18–20.

19. A Two long, straight, parallel wires carry current as shown in Figure P30.18. If the currents are equal, find an expression for the magnetic field at points A and B. Use the indicated coordinate system to write your answer in component form.

20. A Two long, straight, parallel wires carry current as shown in Figure P30.18. If the currents are equal, find an expression for the magnetic field at point C. Use the indicated coordinate system to write your answer in component form.

21. N A counterclockwise current of 7.00 A flows in a square metal loop of side $d = 15.0$ cm. What is the magnitude of the magnetic field at the center of the loop?

22. A Two long, straight wires carry the same current as shown in Figure P30.22. One wire is parallel to the z axis and the other wire is parallel to the x axis as shown. Find an expression for the magnetic field at the origin.

23. N Figure P30.23 shows two infinitely long, straight, current-carrying wires located at $x = -20.0$ cm and at $x = +30.0$ cm. The left-hand wire carries current $I_1 = 2.00$ A in the negative z direction (into the page), and the current I_2 in the right-hand wire is unknown. If the magnitude of the total magnetic field due to the two wires at the origin is 4.00 μT, what are the two possible values for the magnitude and direction of the current in the right-hand wire?

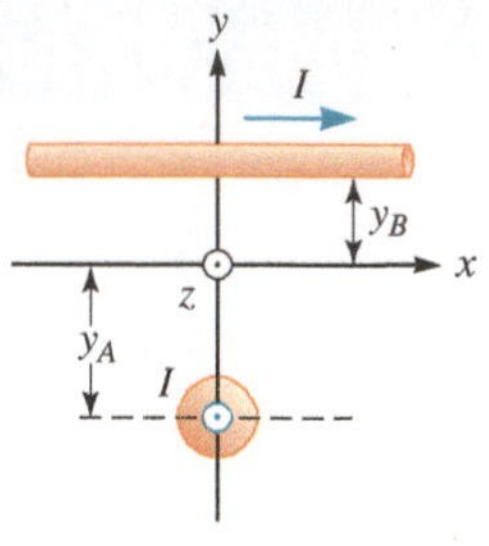

FIGURE P30.22

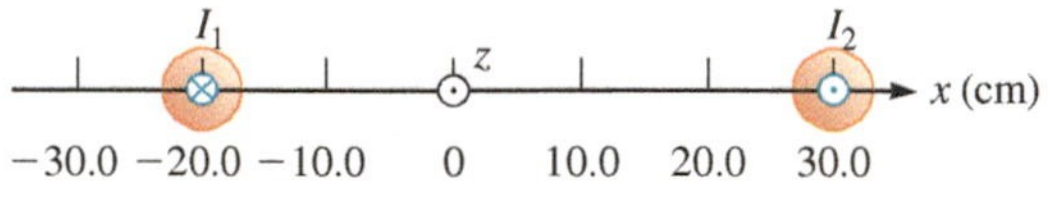

FIGURE P30.23

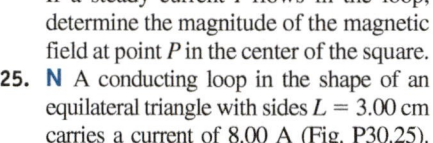

24. A A wire is bent in the form of a square loop with sides of length L (Fig. P30.24). If a steady current I flows in the loop, determine the magnitude of the magnetic field at point P in the center of the square.

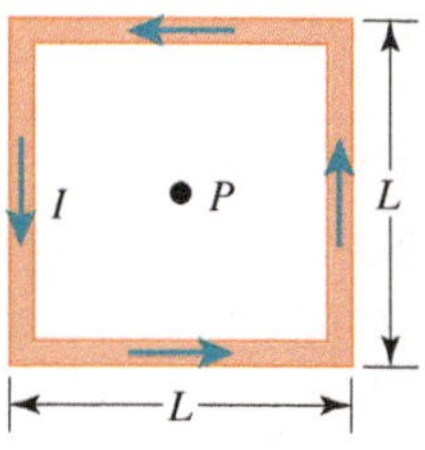

FIGURE P30.24

25. N A conducting loop in the shape of an equilateral triangle with sides $L = 3.00$ cm carries a current of 8.00 A (Fig. P30.25). What is the magnitude of the magnetic field at the center O of the loop?

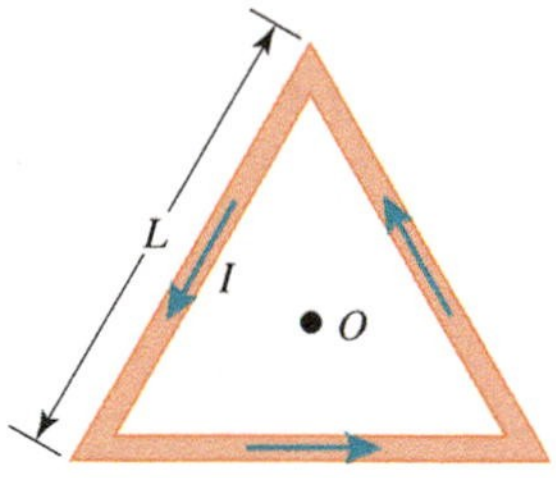

FIGURE P30.25

26. A Derive an expression for the magnetic field produced at point P due to the current-carrying wire shown in Figure P30.26. The curved parts of the wire are pieces of concentric circles. Point P is at their center.

27. N Two circular coils of radii 10.0 cm and 40.0 cm are connected in series to a battery. Find the ratio of the magnitudes of the magnetic fields at their centers.

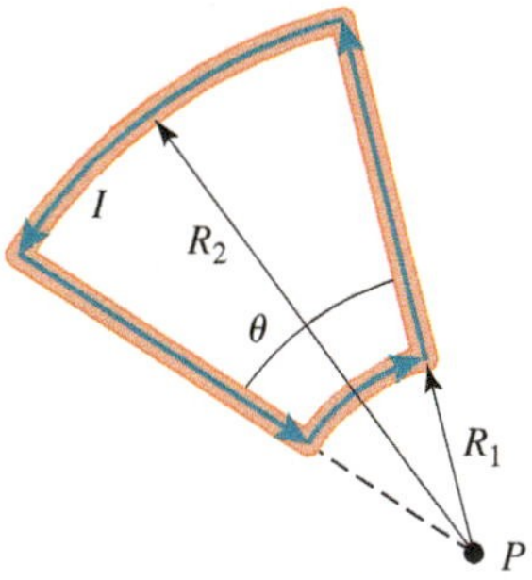

FIGURE P30.26

28. N Figure P30.28 shows three long, straight, parallel current-carrying wires with $I_1 = I_2 = I_3 = 5.00$ A into the page (the negative $\hat{k}$ direction) at the positions shown in the xy plane, with $d = 10.0$ cm. What are the magnitude and direction of the magnetic field at **a.** point A, **b.** point B, and **c.** point C?

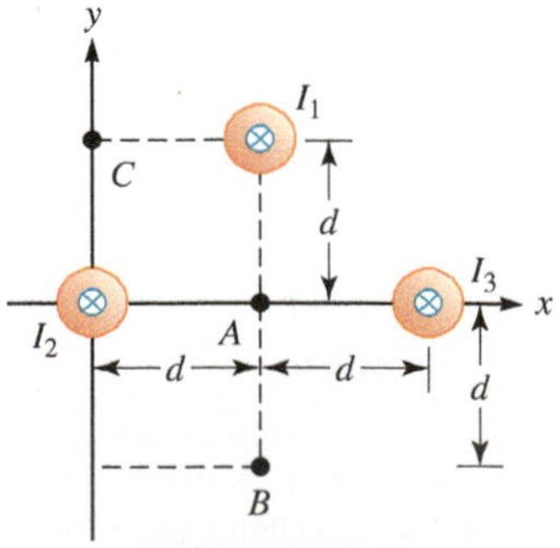

FIGURE P30.28

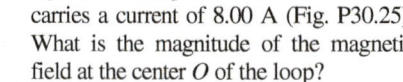

29. **N** Figure P30.29 shows two current-carrying loops with $I_1 = 4.00$ A clockwise and $I_2 = 9.00$ A counterclockwise, placed with their centers at the origin of the xy plane. **a.** If $r_1 = 10.0$ cm and $r_2 = 16.0$ cm, what are the magnitude and direction of the magnetic field due to the two loops at the origin? **b.** If r_1 is held constant at 10.0 cm, what would r_2 have to be for the magnetic field at the origin to be 0?

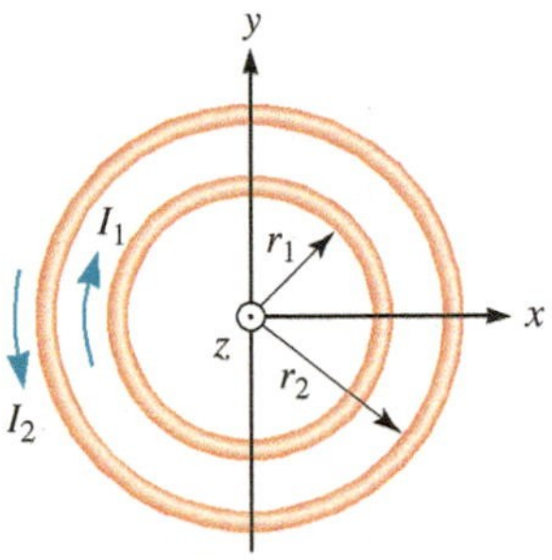

FIGURE P30.29

30-6 The Magnetic Dipole Moment and Modeling Atoms

30. **E** Model the Earth's magnetic field as a dipole produced by the motion of the molten iron and nickel core (1200 km $< R <$ 3500 km). Near the surface, the field strength is 0.5×10^{-4} T. Estimate the Earth's magnetic moment.

31. **N** A magnetic field of magnitude 0.450 T is directed parallel to the plane of a circular loop of radius 75.0 cm. A current of 9.00 mA is maintained in the loop. What is the magnetic moment of the loop?

32. **E** The world's strongest magnet according to Table 30.1 has a field strength of about 45 T. Imagine using a loop of wire and a battery to create a similar magnetic field at the center of the loop. Estimate the current and the size of the loop that you need to create this magnitude field. Can this be done in your lab room?

33. **N** A square loop of wire with side length 0.205 m carries a current of 1.50 A. What is the magnitude of the magnetic field along the axis of the square loop, 2.50 m from the center?

30-7 Ferromagnetic Materials

34. **C** The magnetic moments within several domains of a ferromagnetic material at different times are shown in Figure P30.34. What could have happened in the time between figures A and B to cause the change?

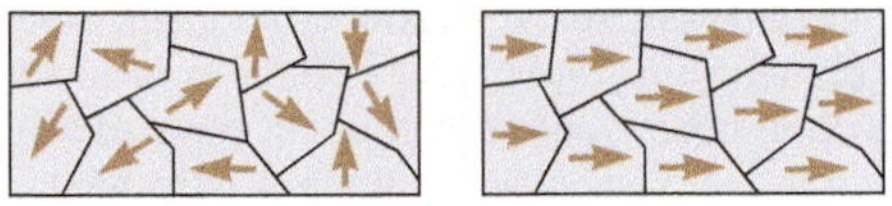

A. **B.**

FIGURE P30.34

35. **C** Normally a refrigerator is not magnetized. If you tried to stick a bar magnet to a refrigerator, would you expect both poles to be attracted to the refrigerator, neither pole to be attracted, or just one pole to be attracted? Explain.

30-8 Magnetic Force on a Charged Particle

36. **C** For each case, determine the fundamental force or forces through which each pair of particles interacts: **a.** a stationary proton and a stationary neutron, **b.** a moving proton and a stationary neutron, **c.** a stationary proton and a stationary electron, and **d.** a moving proton and a moving electron. Explain your answers, including anything special about each situation.

37. **N** At the instant shown in both sketches in Figure P30.37, a positively charged particle ($q = 15.5$ mC) travels with speed $v = 356$ m/s in a magnetic field with $B = 0.650$ T. The angle θ is 32.0°. **a.** Find the magnetic force (magnitude and direction) exerted on the particle in both cases. **b.** If the positively charged particle is replaced by a negatively charged particle ($q = -15.5$ mC), how do your results change?

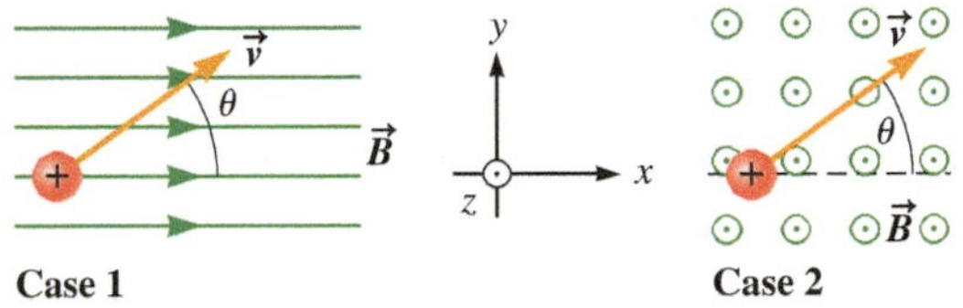

FIGURE P30.37

Problems 38 and 39 are paired.

38. **A** The magnetic field in a region is given by $\vec{B} = B_x\hat{\imath} + B_y\hat{\jmath}$. At some instant, a charged particle's velocity is given by $\vec{v} = v_x\hat{\imath} + v_y\hat{\jmath} + v_z\hat{k}$. Find an expression for the magnetic force exerted on the particle at that instant. In thinking about your result, find what happens to this force if **a.** $v_x = v_y = 0$ or **b.** $v_z = 0$.

39. **N** The magnetic field in a region is given by $\vec{B} = (0.750\,\hat{\imath} + 0.250\,\hat{\jmath})$T. At some instant, a particle with charge $q = 10.0$ mC has velocity $\vec{v} = (35.6\hat{\imath} + 107.3\hat{\jmath} + 43.5\hat{k})$ m/s. What is the magnetic force exerted on the particle at that instant? (Give your answer in component form.)

40. **C** A positively charged particle starts at rest in a region of space that has both a uniform electric field and a uniform magnetic field. Describe the resulting motion of the particle if **a.** both the electric field and the magnetic field point to the right $(+\hat{\imath})$ in this region of space, and **b.** the electric field points to the right $(+\hat{\imath})$ and the magnetic field points out of the page $(+\hat{k})$.

41. **N** A charged particle enters a region of space with a uniform magnetic field $\vec{B} = 1.88\hat{\imath}$ T. At a particular instant in time, it has velocity $\vec{v} = (1.44 \times 10^6\hat{\imath} + 2.50 \times 10^6\hat{\jmath})$ m/s. Based on the observed acceleration, you determine that the force acting on the particle at this instant is $\vec{F} = 1.50\,\hat{k}$ N. What are **a.** the sign and **b.** the magnitude of the charged particle?

42. **N** The velocity vector of a singly charged helium ion ($m_{He} = 6.64 \times 10^{-27}$ kg) is given by $\vec{v} = 4.50 \times 10^5\,\hat{\imath}$ m/s. The acceleration of the ion in a region of space with a uniform magnetic field is 8.50×10^{12} m/s^2 in the positive y direction. The velocity is perpendicular to the field direction. What are the magnitude and direction of the magnetic field in this region?

43. **N** The magnetic field in a region of space is given by $\vec{B} = 0.540\hat{\jmath}$ T. The velocity vector of an electron moving at 7.55×10^6 m/s makes an angle of 45.0° with this field. What are the magnitudes of **a.** the magnetic force on the electron and **b.** the electron's acceleration?

30-9 Motion of Charged Particles in a Magnetic Field

44. **C** Can you use a mass spectrometer to measure the mass of a proton? Can you use a mass spectrometer to measure the mass of a neutron?

45. **N** In a laboratory experiment, a beam of electrons is accelerated from rest through a 154-V potential difference. The beam then

enters a uniform magnetic field and follows a circular path of radius $r = 19.3$ cm in the field region. **a.** What is the angle between the magnetic field and the electrons' velocity? **b.** What is the magnitude of the magnetic field?

46. **A** Two particles A and B with equal charges accelerated through potential differences V and $3V$, respectively, enter a region with a uniform magnetic field. The particles move in circular paths of radii R and $2R$, respectively. Determine the ratio of the masses of particles A and B.

47. **N** A proton is launched with a speed of 3.00×10^6 m/s perpendicular to a uniform magnetic field of 0.300 T in the positive z direction. **a.** What is the radius of the circular orbit of the proton? **b.** What is the frequency of the circular movement of the proton in this field?

48. **C** A proton moves in a helical path of constant radius in a magnetic field. **a.** What can you conclude about the magnitude and direction of the magnetic field? **b.** If, instead, the radius of the helical path decreases, what can you conclude about the magnetic field in that region?

49. **A** A proton and a helium nucleus (consisting of two protons and two neutrons) pass through a velocity selector and into a mass spectrometer. The radius of the proton's circular path is r_p. Find an expression for the radius r of the helium nucleus's path in terms of r_p. (You may assume the mass of a proton is roughly equal to the mass of a neutron, and the helium nucleus has the same speed as the proton.)

50. **N** Two ions are accelerated from rest in a mass spectrometer operating with potential difference ΔV. The first ion, with mass m_1, is singly ionized and is deflected into a semicircle of radius R_1 by the uniform magnetic field in the mass spectrometer. A second, doubly-ionized ion with mass m_2 is deflected into a semicircle with twice the radius of the first ion. What is the ratio m_2/m_1?

51. **N** A uniform electric field of magnitude 124 kV/m is directed upward in a region of space. A uniform magnetic field of magnitude 0.60 T perpendicular to the electric field also exists in this region. A beam of positively charged particles travels into the region. Determine the speed of the particles at which they will not be deflected by the crossed electric and magnetic fields.

30-10 Case Study: The Hall Effect

52. **C** What are the major differences between Ben Franklin's experiments and the Hall effect? Why can't we use the potential difference measured across a circuit element, such as a resistor, to find the sign of the charge carriers?

53. **N** A rectangular silver strip is 2.50 cm wide and 0.050 cm thick. It is in a magnetic field perpendicular to its surface (Fig. 30.41, page 957). The magnetic field is uniform, with a magnitude of 1.75 T. The strip carries a current of 6.45 A. According to Table 28.2, the number density of charge carriers in silver is 5.86×10^{28} m^{-3}. Find the Hall voltage for this strip.

Problems 54 and 55 are paired.

54. **C** An archeologist finds a painted rectangular metal artifact. She believes it was once part of a necklace and it may be made of lead, but she must be careful not to damage the paint in her examination. (She cannot dip the artifact in a fluid or take a sample of the metal.) Explain how she can use the Hall effect to find the number density of charge carriers and Table 28.2 to determine whether the artifact is made of lead.

55. **N** The artifact in Problem 54 is flat, with a thickness of 1.01 cm and a width of 1.139 cm. The archeologist places it in a strong magnetic field of 5.00 T that is perpendicular to the artifact (Fig. 30.41, page 957). The accuracy of her voltage measurement is ± 50.0 nV. To minimize the possibility of damaging the artifact, she wishes to put the minimum current through it. **a.** If the artifact is made of lead, estimate the current required. **b.** Starting at the minimum, the researcher slowly turns up the current. When the current reaches six times the minimum, she measures a Hall voltage of 160.0 nV. What is the number density of charge carriers? Is the artifact made of lead? If not, use Table 28.2 to find its most likely (pure) composition.

30-11 Magnetic Force on a Current-Carrying Wire

56. **N** For both sketches in Figure P30.56, there is a 3.54-A current, a magnetic field strength $B = 0.650$ T, and the angle θ is 32.0°. Find the magnetic force per unit length (magnitude and direction) exerted on the current-carrying conductor in both cases.

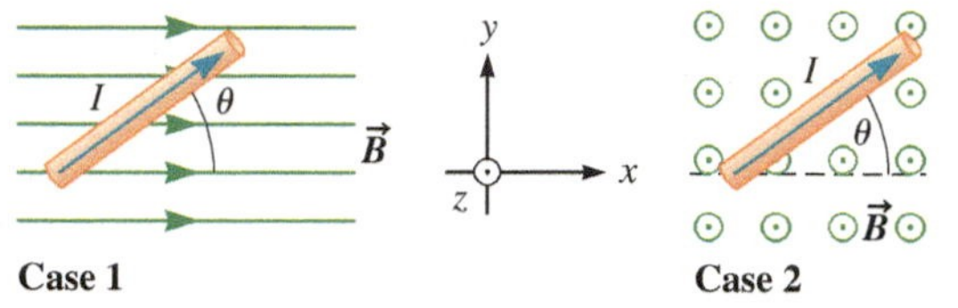

FIGURE P30.56

57. **N** A 1.40-m section of a straight wire oriented along the z axis carries 11.0 A of current in the positive z direction. The wire is placed in a region with a uniform magnetic field of 4.00 T in the positive y direction. What are the magnitude and direction of the magnetic force on the wire?

58. **E** Professor Edward Ney was the founder of infrared astronomy at the University of Minnesota. In his later years, he wore an artificial pacemaker. Always an experimentalist, Ney often held a strong laboratory magnet near his chest to see what effect it had on his pacemaker. Perhaps he was using the magnet to throw switches that control different modes of operation. An admiring student (without an artificial pacemaker) thought it would be fun to imitate this great man by holding a strong magnet to his own chest. The natural pacemaker of the heart (known as the sinoatrial node) carries a current of about 0.5 mA. Estimate the magnetic force exerted on a natural pacemaker by a strong magnet held to the chest. How do you think the student might have felt during the experiment? Explain your geometric assumptions. *Hints*: See Table 30.1 (page 941) to estimate the magnetic field, and assume the field is roughly uniform. Use Figure P30.58 to estimate the size of the sinoatrial node; your heart is about the size of your fist.

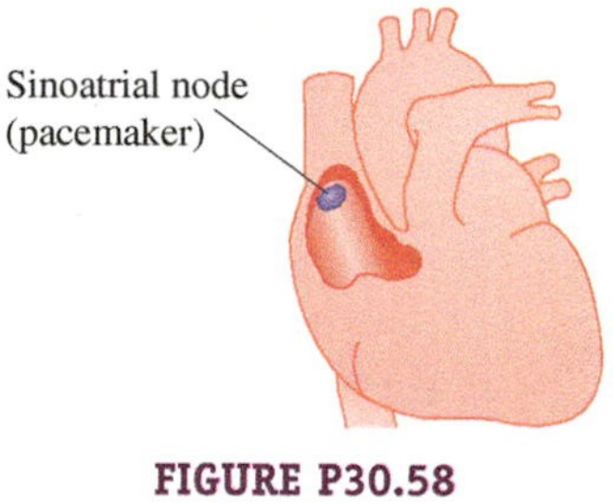

FIGURE P30.58

59. **N** A current of 5.64 A flows along a wire with $\vec{\ell} = (1.382\hat{\imath} - 2.095\hat{\jmath})$ m. The wire resides in a uniform magnetic field $\vec{B} = (-0.300\hat{\jmath} + 0.750\hat{k})$ T. What is the magnetic force acting on the wire?

60. **N** A wire with a current of $I = 8.00$ A directed along the positive y axis is embedded in a uniform magnetic field perpendicular to the current in the wire. If the wire experiences a magnetic force of 50.0 mN/m in the positive z direction, what are the magnitude and direction of the magnetic field in this region?

61. **N** A 3.60-m section of a straight wire carrying a current of 2.40 A is embedded in a 1.30-T uniform magnetic field. What is the magnitude of the magnetic force on the wire if the angle between the current in the wire and the direction of the magnetic field is **a.** 45.0°, **b.** 135.0°, and **c.** 90.0°?

62. **N** The triangular loop of wire shown in Figure P30.62 carries a current of 0.125 A, and a uniform magnetic field of 0.250 T points toward the right. Determine the force on each segment of the wire (indicate magnitude and direction) and the net force on the triangular loop.

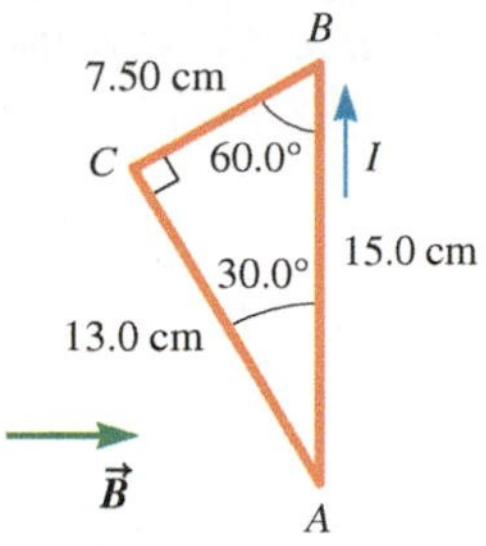

FIGURE P30.62

30-12 Force Between Two Long, Straight, Parallel Wires

Problems 63, 64, and 65 are grouped.

63. Three long, straight wires are seen end-on in Figure P30.63. The distance between the wires is $r = 0.256$ m. Each carries current $I_A = I_B = I_C = 1.63$ A. Assume the only forces exerted on wire A are due to wires B and C.

a. **G** Draw a free-body diagram for wire A.

b. **N** Find the magnetic force per unit length exerted on wire A.

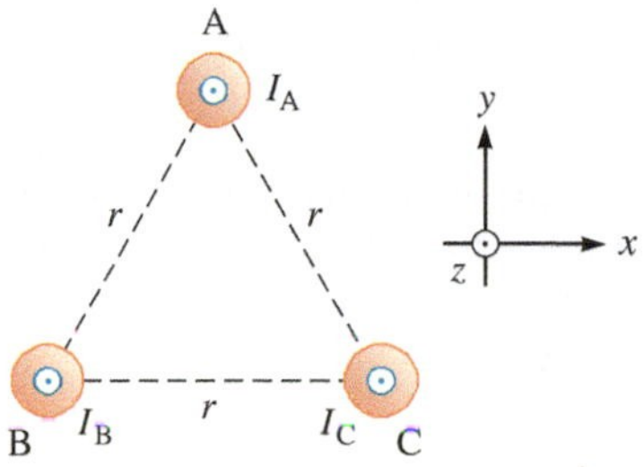

FIGURE P30.63 Problems 63–65.

64. **N** Consider the wires described in Problem 63. Find the magnetic force per unit length exerted on wire B.

65. **N** Consider the wires described in Problem 63, except now wire C carries current $I_C = 3.26$ A. Find the force per unit length exerted on **a.** wire A and **b.** wire B.

66. Two long, straight, parallel wires carry identical currents of 6.30 A in the positive x direction. The separation between the wires is 8.50 cm.

a. **C** Is the force between the wires attractive or repulsive?

b. **N** What is the magnitude of the force per unit length that each wire exerts on the other?

Problems 67, 68, and 84 are grouped.

67. **A** Three parallel current-carrying wires are shown in Figure P30.67. Find an expression for the net magnetic force per unit length on the middle wire due to the other wires.

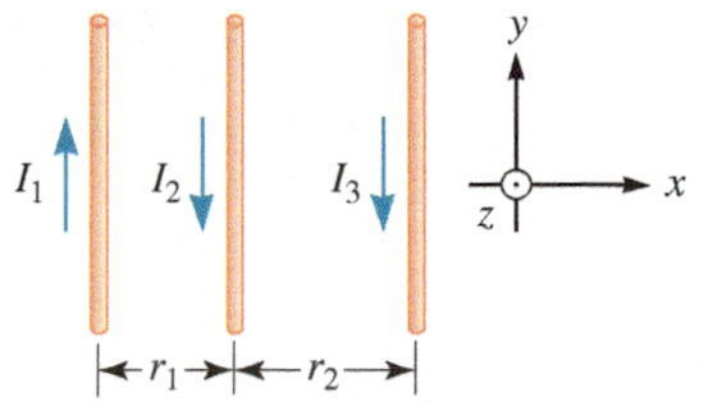

FIGURE P30.67 Problems 67, 68, and 84.

68. **N** Three parallel current-carrying wires are shown in Figure P30.67. The currents in each wire are $I_1 = 0.500$ A, $I_2 = 0.750$ A, and $I_3 = 0.250$ A. If the wires are separated by distances $r_1 = 0.185$ m and $r_2 = 0.290$ m, find the net magnetic force per unit length on the middle wire due to the other wires.

30-13 Current Loop in a Uniform Magnetic Field

69. **N** A current loop freely rotates in a uniform magnetic field ($B = 0.250$ T). The maximum torque on the loop is $0.540\ \text{N}\cdot\text{m}$. What is the magnetic dipole moment of the loop?

70. **E** A student wishes to make a compass out of a wire and a 9-V battery. She wraps copper wire around a plastic water bottle (Fig. P30.70). The bottle hangs from a thread so that it is horizontal and free to rotate. Estimate the maximum torque exerted on her homemade compass due to the Earth's magnetic field.

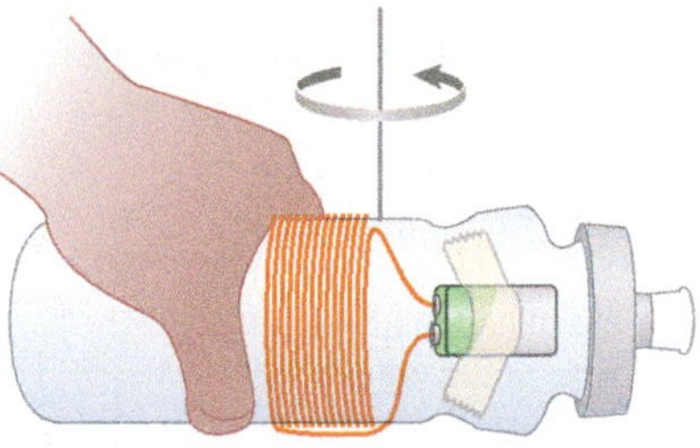

FIGURE P30.70

71. **C** In Figure 30.54 (page 967), we saw that the magnetic field inside a galvanometer is radial. Why is this an important part of its design? What would happen if the magnetic field were uniform instead? In particular, what could you say about the torque on the coil if the field were uniform?

72. **N** A straight conductor between points O and P has a mass of 0.50 kg and a length of 1.0 m and carries a current of 12 A (Fig. P30.72). It is hinged at O and is placed in a plane perpendicular to a magnetic field of 2.0 T. If the conductor begins from rest, determine the angular acceleration of the conductor due to the magnetic force. Ignore the effect of the gravitational force on the conductor.

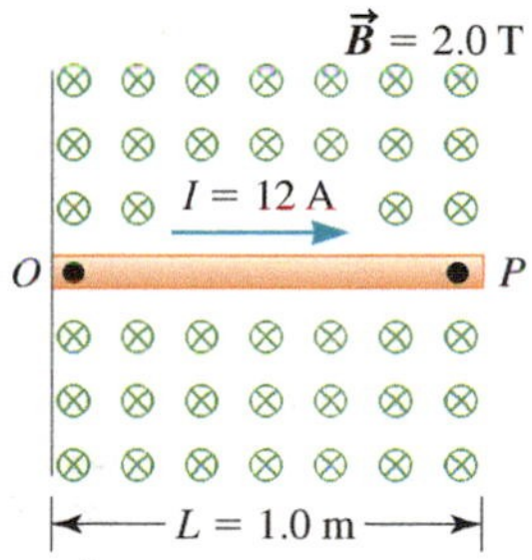

FIGURE P30.72

73. **N** A circular coil 15.0 cm in radius and composed of 145 tightly wound turns carries a current of 2.50 A in the counterclockwise direction, where the plane of the coil makes an angle of 15.0° with the y axis (Fig. P30.73). The coil is free to rotate about the z axis and is placed in a region with a uniform magnetic field given by $\vec{B} = 1.35\hat{\jmath}$ T. **a.** What is the magnitude of the magnetic torque on the coil? **b.** In what direction will the coil rotate?

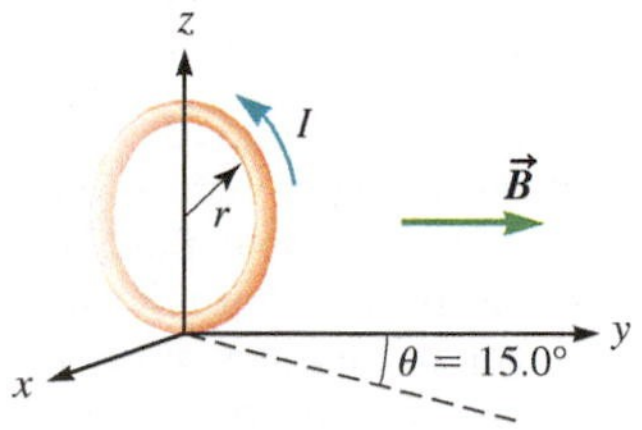

FIGURE P30.73

74. **N** A circular conductor with radius $r = 25.0$ cm, carrying a current of 3.40 A, is embedded in a 0.650-T uniform magnetic field. **a.** What is the magnitude of the maximum torque on the conductor? **b.** What are the minimum and maximum values of the potential energy of the conductor–magnetic field system?

General Problems

75. **G, E** The formula for the magnetic field along the axis of a magnetic dipole moment given by Equation 30.11,

$$\vec{B} \approx \frac{\mu_0}{2} \frac{IR^2}{y^3} \hat{\jmath}$$

is valid far from the loop ($y \gg R$), but how far is far enough? Plot the magnetic field versus distance y for this expression, along with the exact expression

$$\vec{B} = \frac{\mu_0}{2} \frac{IR^2}{(R^2 + y^2)^{3/2}} \hat{\jmath}$$

from Equation 30.10 for a loop that has a radius of 2.5 cm and a current of 0.5 A. Use your graph to estimate the distance at which the approximate value equals the exact value. Express your estimate in terms of loop's diameter.

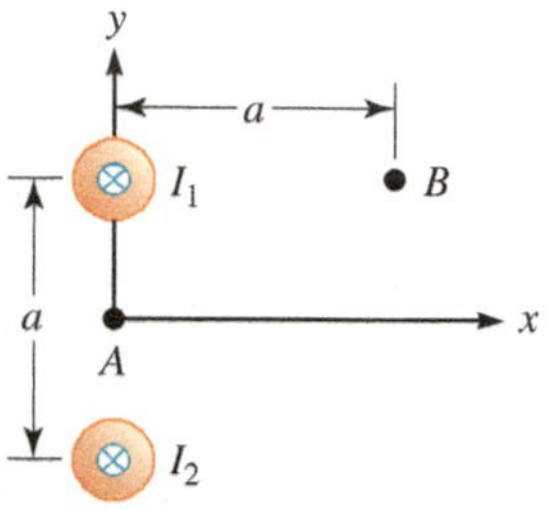

FIGURE P30.76

76. **N** Figure P30.76 shows two long, straight wires carrying currents $I_1 = 2.00$ A and $I_2 = 8.00$ A into the page. The wires are separated by a distance $a = 1.00$ m. **a.** What are the magnitude and direction of the magnetic field at point A, halfway between the two wires? **b.** What are the magnitude and direction of the magnetic field at point B, at a distance $a = 1.00$ m to the right of the top wire?

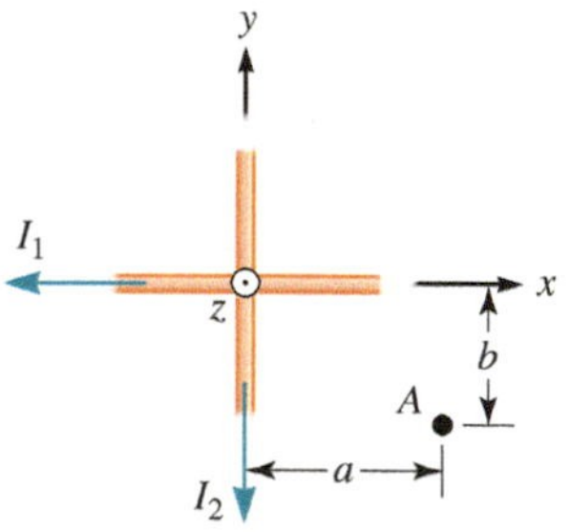

FIGURE P30.77

77. **N** Figure P30.77 shows two long current-carrying wires in the xy plane with $I_1 = 2.00$ A to the left and $I_2 = 7.00$ A downward. The wires are not in contact with each other. What are the magnitude and direction of the magnetic field due to the two wires **a.** at point A, a distance $a = 60.0$ cm and $b = -50.0$ cm in the x and y directions from the intersection of the wires, and **b.** at $z = -50.0$ cm behind the intersection of the wires?

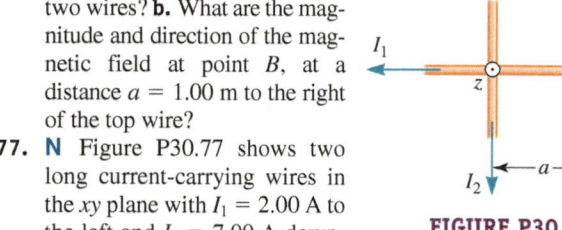

78. **N** Two long, straight, current-carrying wires run parallel to the y axis. The first wire is located at ($x = 0$, $z = 0$) with current $I_1 = 12.0$ A in the positive y direction, and the second wire, located at ($x = 95.0$ cm, $z = 0$), carries a current of 24.0 A in the negative y direction. At what location in the xy plane is the magnetic field due to the two wires equal to 0?

Problems 79, 80, and 81 are grouped.

79. **A** Consider a regular polygon with N sides carrying a steady current I. Determine the magnetic field at the center of such a polygon. Assume R is the perpendicular distance from the center of the polygon to the sides of the polygon. Assume the positive z axis points out of the page.

80. **A** In Problem 79, assume the regular polygon has a very large number of sides. Mathematically, this corresponds to taking the limit as $N \rightarrow \infty$. Obtain the analytical expression for the magnetic field at the center of such a polygon. Compare your result with the magnetic field at the center of a circular loop. Assume the positive z axis points out of the page.

81. **N** A hexagonal loop of side 6.0 cm carries a current of 2.0 A (Fig. P30.81). Determine the magnetic field at the center of the loop. Assume the positive z axis points out of the page. *Hint*: Consider the results of Problems 79 and 80.

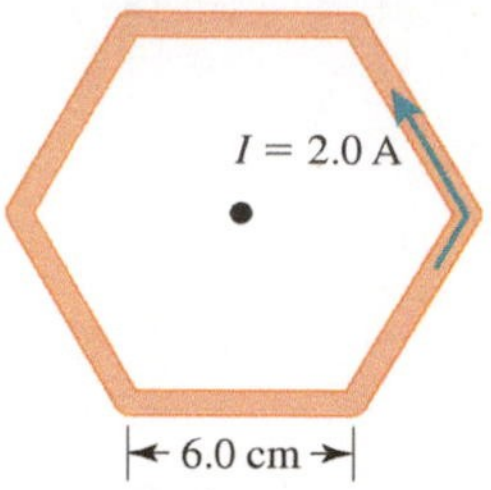

FIGURE P30.81

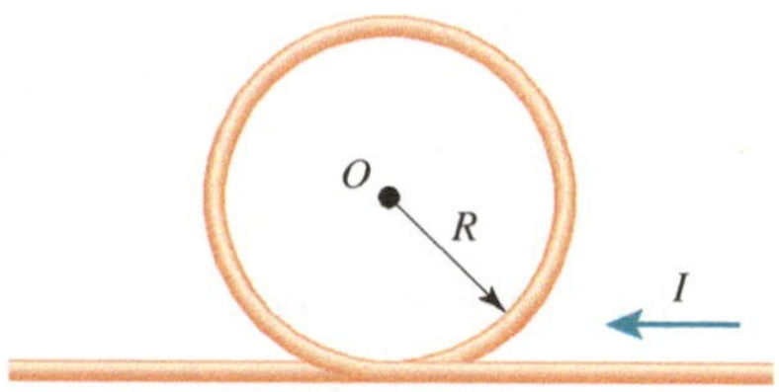

FIGURE P30.82

82. **N** A section of a long, straight wire is twisted into a coplanar loop of radius $R = 45.0$ cm (Fig. P30.82). If the current in the wire is $I = 5.00$ A, what are the magnitude and direction of the magnetic field at point O, the center of the loop? (The wire is sheathed in insulation.)

83. Two infinitely long current-carrying wires run parallel in the xy plane and are each a distance $d = 11.0$ cm from the y axis (Fig. P30.83). The current in both wires is $I = 5.00$ A in the negative y direction.

a. **G** Draw a sketch of the magnetic field pattern in the xz plane due to the two wires. What is the magnitude of the magnetic field due to the two wires

b. **N** at the origin and

c. **A** as a function of z along the z axis, at $x = y = 0$?

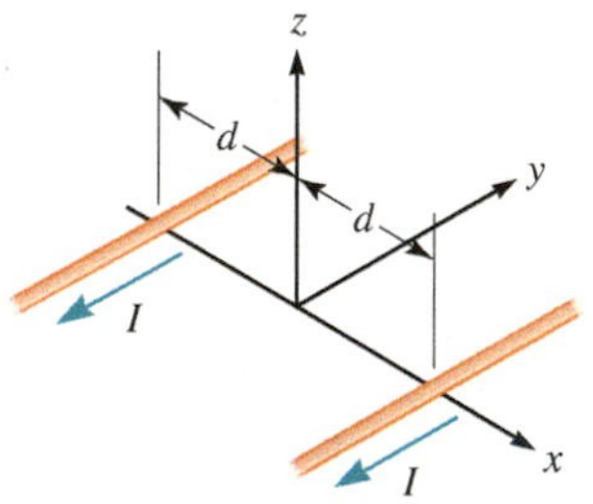

FIGURE P30.83

84. **C** Consider the current-carrying wires depicted in Figure P30.67. If the only sources of magnetic fields are the three wires, is it possible that the net magnetic force per unit length on the middle wire is 0? If yes, explain how this is possible. If no, explain what might be changed to make it possible for the net magnetic force per unit length to be 0.

85. **C** Each part of Figure P30.85 shows the velocity vector of a positively charged ion and the magnetic force it is experiencing at an instant in time. What is the direction of the magnetic field that causes the force on the ion in each of the cases shown in the figure?

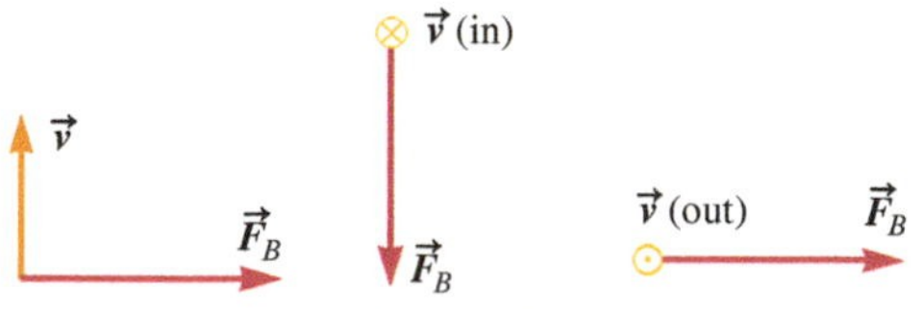

a. Case 1 **b.** Case 2 **c.** Case 3

FIGURE P30.85

Problems 86 and 87 are paired.

86. **N** An electron with velocity $\vec{v} = (3.14 \times 10^5 \hat{\imath} - 5.78 \times 10^5 \hat{\jmath})$ m/s is moving into a region of space with a uniform magnetic field $\vec{B} = (-0.500\hat{\imath} + 0.250\hat{\jmath})$ T. Find the resultant magnetic force on the electron the instant it enters the field.

87. **A** A charged particle with charge q and velocity $\vec{v} = v_x\hat{\imath} + v_y\hat{\jmath}$ is moving into a region of space with a uniform magnetic field $\vec{B} = B_x\hat{\imath} + B_y\hat{\jmath}$. Find the resultant magnetic force on the particle the instant it enters the field.

88. **N** A magnetic field of magnitude 0.450 T is directed parallel to the plane of a circular loop of radius 75.0 cm. A current of 9.00 mA is maintained in the loop. What is the magnitude of the torque the magnetic field exerts on the loop?

89. **N** A potential difference of 5.10×10^3 V is used to accelerate a proton initially at rest in a region of space with a 0.500-T uniform magnetic field. What are the **a.** minimum and **b.** maximum magnetic forces the proton can experience in this region?

90. **N** A mass spectrometer (Fig. 30.40, page 956) operates with a uniform magnetic field of 20.0 mT and an electric field of 4.00×10^3 V/m in the velocity selector. What is the radius of the semicircular path of a doubly ionized alpha particle ($m_a = 6.64 \times 10^{-27}$ kg)?

91. **A** Three long, current-carrying wires are parallel to one another and separated by a distance d. The magnitudes and directions of the currents are shown in Figure P30.91. Wires 1 and 3 are fixed, but wire 2 is free to move. Wire 2 is displaced to the right by a small distance x. Determine the net force (per unit length) acting on wire 2 and the angular frequency of the resulting oscillation. Assume the mass per unit length of wire 2 is λ and $x \ll d$.

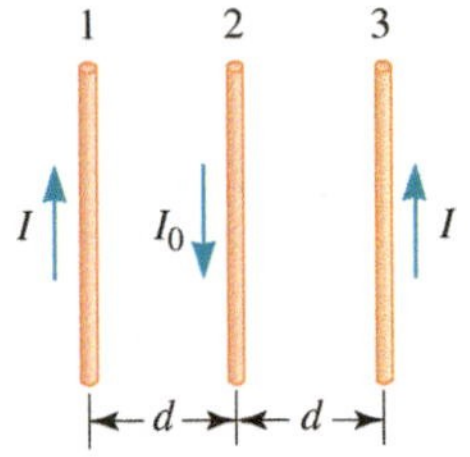

FIGURE P30.91

92. **N** A long, straight wire carrying a current of 2.0 A is placed in the plane of a conducting strip of width 8.0 cm (Fig. P30.92). The strip carries a current of 4.0 A. The distance from the wire to the near edge of the strip is 4.0 cm. Calculate the attractive force per unit length between the wire and the strip.

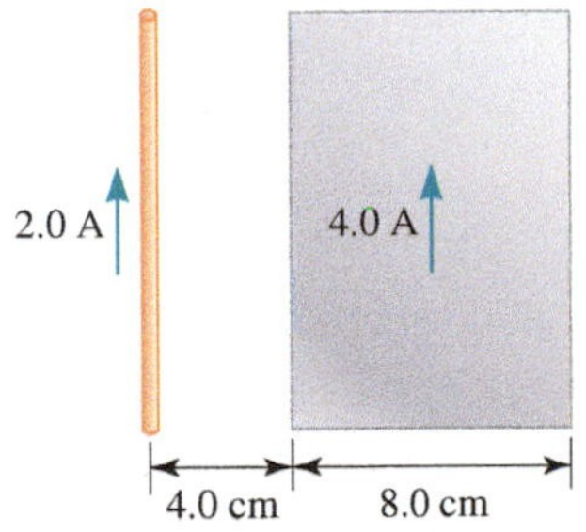

FIGURE P30.92

93. **A** A current-carrying conductor PQ of mass m and length L is placed on an inclined plane with angle of inclination θ (Fig. P30.93). A uniform magnetic field B is directed upward as shown. Assume friction is negligible. **a.** Determine the magnitude and direction of the current in the conductor so that it remains in equilibrium. **b.** If the direction of the current is reversed, will the conductor still be in equilibrium? If not, find the magnitude of the initial acceleration of the conductor.

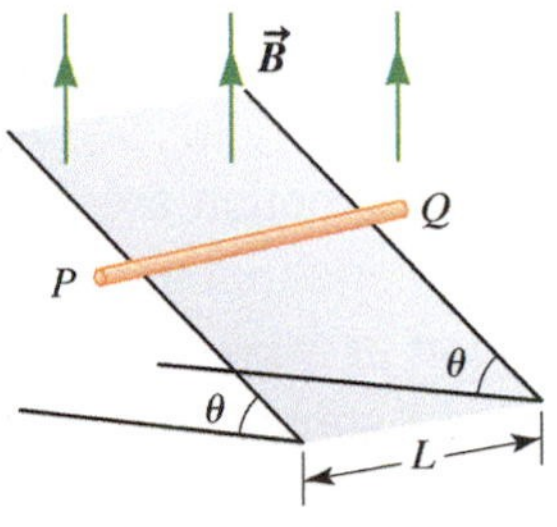

FIGURE P30.93

94. This problem is designed to show how the Biot-Savart empirical law may be found using simple laboratory equipment.[1] A wire is bent into a square U-shape and a current is set up in the wire (Fig. P30.94). The short segment of the wire has been aligned with the Earth's magnetic field $\vec{B}_{\oplus}$, so that when the current is turned off, the compass needle points in the x direction. The long segments of the wire produce magnetic fields that are nearly vertical at the location of the compass, so the needle is not deflected by these magnetic fields. When the current is on, the short segment produces a magnetic field $\vec{B}$ in the positive y direction (at the compass's location). The needle is deflected so that it points along the total magnetic field $\vec{B}_{tot}$, making an angle θ with the x axis. The compass rests on thin wooden sheets of known thickness. By varying the number of sheets, you can measure the angle θ as a function of distance r from the short segment of wire. (r comes from multiplying the thickness of a sheet by the number of sheets.)

a. **A** Assume the segment's magnetic field obeys the inverse-square Biot-Savart law, and derive an expression for $\tan\theta$ as a function of r.

b. **G** Plot $\tan\theta$ as a function of $1/r^2$ for the data in Table P30.94. Assume the error in $\tan\theta$ is ± 0.01. On this graph, sketch the expression you found in part (a). Assume the expression agrees with the data at $r = 8$ cm. (You may use centimeters on your graph.)

c. **C** Explain the discrepancy between the data and the expression predicted by the Biot-Savart law. How can you improve the experiment? *Hint*: How long is the segment described by the Biot-Savart law?

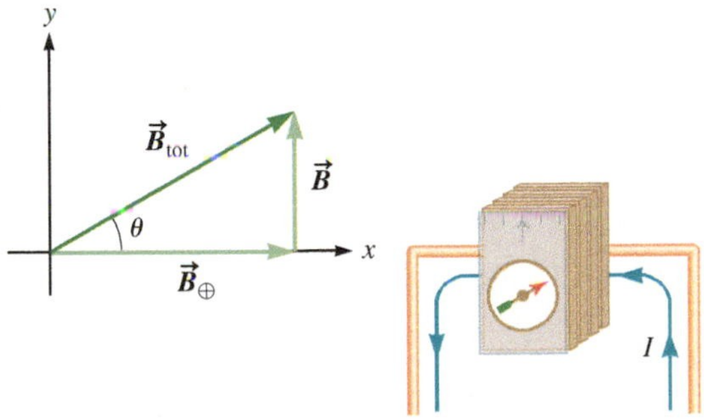

FIGURE P30.94

TABLE P30.94

θ (°)	r (cm)
22.9	2
15.1	2.67
11.1	3.33
7.88	4
5.71	4.67
3.96	5.33
2.86	6
2.86	6.67
1.76	7.33
1.76	8

95. **N** A proton enters a region with a uniform electric field $\vec{E} = 5.0\,\hat{k}$ V/m and a uniform magnetic field $\vec{B} = 5.0 \times 10^{-4}\,\hat{k}$T. The proton has initial velocity $\vec{v}_0 = 2.5 \times 10^5\hat{\imath}$ m/s. How far along the z axis does the proton travel after it undergoes three complete revolutions?

[1] Based on *The Biot-Savart Law: From Infinitesimal to Infinite* by Jeffrey A. Phillips and Jeff Sanny (*The Physics Teacher* 46: 44–47, Jan. 2008).

31 Gauss's Law for Magnetism and Ampère's Law

Key Questions

How do we state formally that no magnetic monopoles have been observed?

How do we use Ampère's law and Maxwell–Ampère's law to find the magnetic fields produced by different sources?

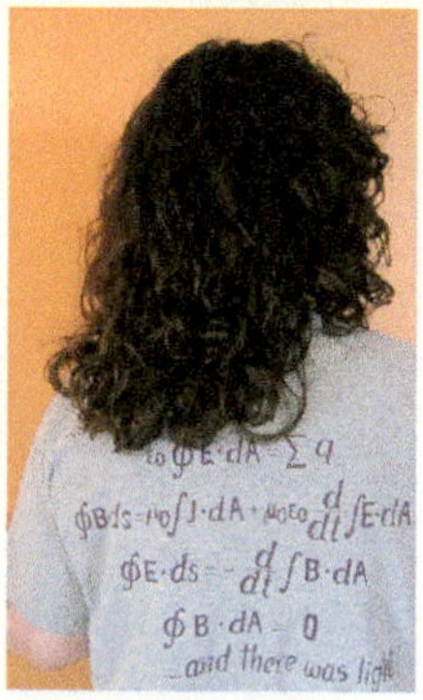

Jeff Black

FIGURE 31.1 Because Maxwell's four simple equations describe diverse phenomena concerning electricity, magnetism, and light, physicists believe they are perfect for T-shirts.

Underlying Principles

1. Gauss's law for magnetism
2. Ampère's law (Ampère–Maxwell's law)

These laws are two of the four Maxwell equations.

Major Concepts

Displacement current

Special Cases

1. Sources with linear symmetry
2. Ideal solenoid
3. Toroid

After doing hundreds of physics homework problems, you might be surprised to learn that physicists strive to make their theories as simple and eloquent as possible. An eloquent theory is an underlying principle usually expressed by a single equation or sentence that holds for a wide range of phenomena. Physicists believe there is beauty in a simple theory. Newton's laws of motion are considered beautiful because his three simple laws apply to so many situations.

Just as Newton's laws of motion are the principles that underlie classical mechanics, there are four laws that underlie electricity, magnetism, and light. These four laws are known as **Maxwell's equations**; they are considered so beautiful that they appear on T-shirts (Fig 31.1).

Newton's discovery relied on the careful observations and experiments of Galileo. Maxwell's work also depended on the careful observations and experiments of other scientists. In fact, Maxwell's equations were actually discovered by scientists who preceded him; each equation is known by the name of the scientist who originally discovered it. One such equation is Gauss's law for electricity (Chapter 25). In this chapter, we study two more of Maxwell's equations—Gauss's law for magnetism and Ampère's law.

Chapter 32 will introduce the last of Maxwell's equations—Faraday's law. Maxwell's amazing achievement was recognizing that these four laws combine to form the underlying principle for light (Chapter 34). Maxwell's equations form one of the most eloquent theories of physics.

31-1 Measuring the Magnetic Field

In 1820, Hans Christian Ørsted discovered that a compass needle is deflected by the current in a wire. Within a few months, French prodigy André Marie Ampère (1775–1836) conducted his own experiments and concluded that moving charged particles are the source of any magnetic field, including fields due to permanent magnets. Ampère's law is one of the four underlying principles of electricity and magnetism.

Today, we know that the magnetic field of a permanent magnet is largely due to the spin of electrons—an intrinsic property.

CASE STUDY Measuring the Magnitude of the Magnetic Field

In Chapter 30, we used the Biot-Savart law to find expressions for the magnetic field. Ampère's law as introduced in this chapter also allows us to find expressions for the magnetic field produced by various highly symmetrical sources. Much of this chapter focuses on finding mathematical expressions for the *magnitude* of the magnetic field for several different sources using Ampère's law. How do we know these expressions are correct? Of course, they must be verified by experiments. The goal of this case study is to confirm the expressions derived from Ampère's law using a compass and the simple techniques used to discover Biot-Savart's empirical law (similar to Problem 30.94, page 977).

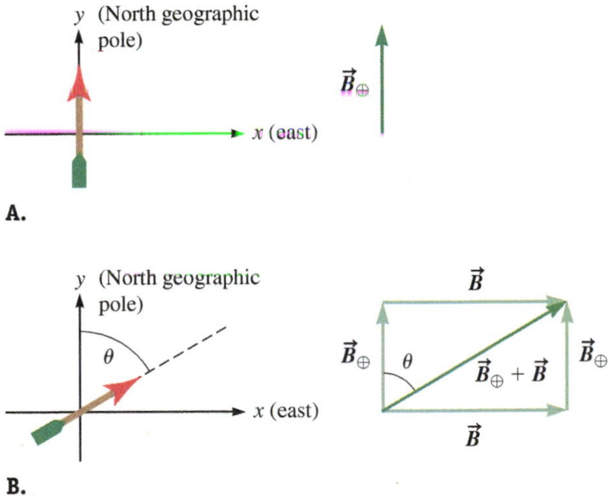

FIGURE 31.2 **A.** The Earth's magnetic field $\vec{B}_\oplus$ provides a reference position for a compass needle. The compass needle points toward the Earth's magnetic south pole (North geographic pole) when no other magnetic fields are present. **B.** When there is another magnetic field $\vec{B}$, the compass needle points in the direction of the total magnetic field $\vec{B}_\oplus + \vec{B}$.

To find the magnitude of the magnetic field, we measure the compass needle's deflection from a reference position. The Earth's magnetic field $\vec{B}_\oplus$ provides a natural reference position. When no other magnetic field is present, a compass needle points toward the Earth's south magnetic pole (the North geographic pole); let's call this the positive y direction (Fig. 31.2A). Now, imagine there is a second magnetic field $\vec{B}$ directed along the positive x axis (Fig. 31.2B). We can find the magnitude B of this magnetic field by measuring the deflection θ of the compass needle:

$$B = B_\oplus \tan \theta \tag{31.1}$$

where $B_\oplus$ is the magnitude of the Earth's magnetic field, about 0.5×10^{-4} T. Equation 31.1 holds only when the magnetic field $\vec{B}$ is perpendicular to the Earth's magnetic field $\vec{B}_\oplus$. (If $\vec{B}$ is not perpendicular to $\vec{B}_\oplus$ or some other known magnetic field, finding the magnitude B is more complicated, but we will not consider such cases.) The goals of this case study are to connect theory (Ampère's law) with observation, and to visualize the magnetic field through its effect on compass needles and graphs of its magnitude versus position.

CONCEPT EXERCISE 31.1

CASE STUDY Measuring the Magnetic Field Near a Bar Magnet

Suppose you have a bar magnet (Fig. 30.4, page 936) and you want to find the magnetic field strength along the magnet's axis as a function of distance from one of the poles. Explain how you could use a small compass to achieve your goal.

EXAMPLE 31.1 CASE STUDY Playing with Magnets

From the information in Table 30.1 (page 941), calculate the deflection of a compass needle that results **A** from being near a refrigerator magnet and **B** from being near an MRI magnet. Assume the magnetic field in each case is perpendicular to the Earth's magnetic field. Report your results to three significant figures.

INTERPRET and ANTICIPATE

The magnetic field in both cases is perpendicular to the Earth's magnetic field (Fig. 31.2B).

SOLVE

Find the deflection θ from Equation 31.1.

$$B = B_\oplus \tan\theta \tag{31.1}$$

$$\theta = \tan^{-1}\left(\frac{B}{B_\oplus}\right)$$

A According to Table 30.1, the magnetic field near a typical refrigerator magnet is $B = 5 \times 10^{-3}$ T.

For a refrigerator magnet,

$$\theta = \tan^{-1}\left(\frac{5 \times 10^{-3}\ \text{T}}{5 \times 10^{-5}\ \text{T}}\right) = 89.4°$$

B Near a typical MRI magnet, the magnetic field is $B = 2$ T.

For an MRI magnet,

$$\theta = \tan^{-1}\left(\frac{2\ \text{T}}{5 \times 10^{-5}\ \text{T}}\right) = 90.0°$$

CHECK and THINK

This example illustrates a problem with using the Earth's magnetic field to set a reference position. The magnetic field of a refrigerator magnet is much weaker than that of an MRI magnet, yet both produce almost the same deflection of the compass needle. It would be difficult to measure the difference between the two deflections. One improvement is to use a stronger magnetic field than the Earth's as the reference. Ideally, it is best to choose a reference magnetic field that is comparable to the field you want to measure.

31-2 Gauss's Law for Magnetism

Gauss's law for magnetism and Ampère's law are both similar to Gauss's law for electricity, so let's review a few important ideas about Gauss's law from Chapter 25.

Quick Review of Gauss's Law for Electricity

We begin by imagining an electric field source hidden inside a box. We can get some idea of what is inside the box (a Gaussian surface) by suspending a small, positively charged sphere from a thin thread at various places around the box (Fig. 31.3). According to Gauss's law for electricity, the net charge contained in the surface is proportional to the electric flux Φ_E through the closed surface:

$$\Phi_E = \oint \vec{E} \cdot d\vec{A} = \frac{q_{\text{in}}}{\varepsilon_0} \tag{25.11}$$

We see that the box in Figure 31.3A contains a positively charged source, so the electric flux through the Gaussian surface is positive and the electric field lines emanate from the box. The box in Figure 31.3B contains a negatively charged source, so the electric flux through the Gaussian surface is negative and the electric field lines terminate in the box. Finally, if the positively charged sphere is not deflected no matter where it is placed outside the box, we conclude that the net charge inside is zero ($q_{\text{in}} = 0$), the electric field is zero ($\vec{E} = 0$) everywhere, and the electric flux through the Gaussian surface is zero ($\Phi_E = 0$).

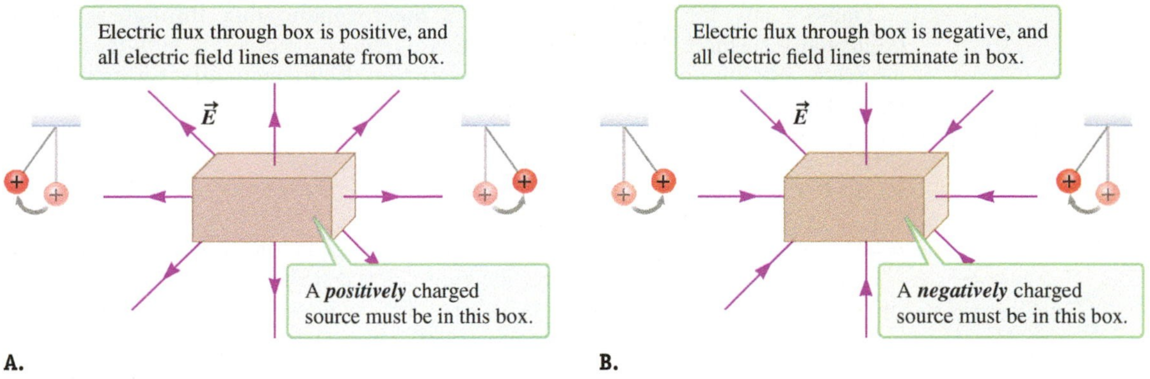

Electric flux through box is positive, and all electric field lines emanate from box.
$\vec{E}$
A *positively* charged source must be in this box.
A.
Electric flux through box is negative, and all electric field lines terminate in box.
$\vec{E}$
A *negatively* charged source must be in this box.
B.

FIGURE 31.3 A small, positively charged sphere is used to gather information about the contents of a box. According to Gauss's law for electricity, the net charge contained in the box is proportional to the electric flux through the box. **A.** The sphere is repelled by the source, so the box contains a positively charged source and the net electric flux is positive. **B.** The sphere is attracted by the source, so the box contains a negatively charged source and the net electric flux is negative.

What can we conclude if the sphere is both attracted to and repelled by the contents of the box (Fig. 31.4)? First, because the sphere is deflected, we conclude that the electric field is nonzero, pointing leftward at the two positions shown. Second, the contents of the box cannot be either positive or negative because the sphere is both attracted to and repelled by the box, so the contents must be neutral. If the contents of the box are neutral, how can the positively charged sphere be deflected? One possible answer (Fig. 31.5A) is that the box is empty and there is a positively charged source located on the right, deflecting the positively charged ball toward the left. Another possible answer (Fig. 31.5B) is that the box contains an electric dipole. When the positively charged sphere is on the right side of the box, it is closer to the negative side of the dipole, so it is attracted to the box. When the positively charged sphere is on the left side of the box, it is closer to the positive side of the dipole, so it is repelled. In both cases, there are just as many electric field lines entering the box as leaving the box, and so the electric flux through the box is zero.

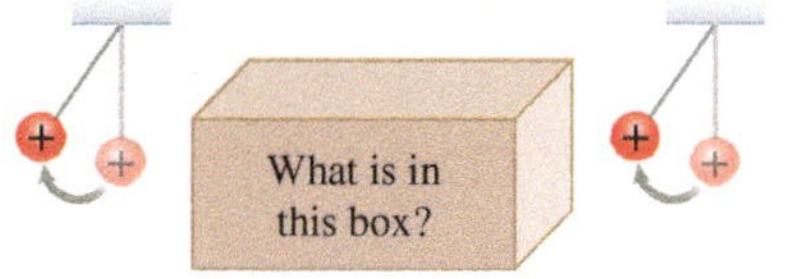

What is in this box?

FIGURE 31.4 The positively charged sphere is deflected toward the box at some points and away from the box at other points.

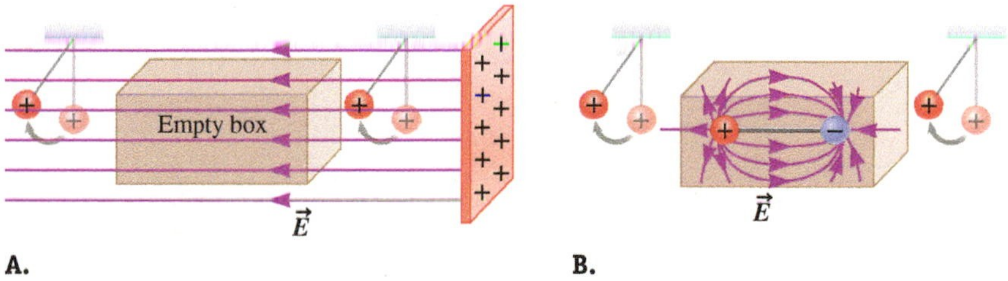

Empty box
$\vec{E}$
A.
$\vec{E}$
B.

FIGURE 31.5 The net electric flux is zero in both cases.

Magnetic Monopoles Revisited

A magnetic monopole is an idealized magnetic source that is either a north pole or a south pole, analogous to an electric source that has either a net positive or a net negative charge. There is no conclusive evidence for the existence of magnetic monopoles (Section 30-2). Magnets always have both a north and a south pole. The simplest combination of a north pole and a south pole is a magnetic dipole, analogous to an electric dipole.

Gauss's law for magnetism is a formal mathematical statement that there are no known magnetic monopoles. Just as Gauss's law for electricity involves electric flux Φ_E, Gauss's law for magnetism involves magnetic flux Φ_B. The mathematical expression for magnetic flux is similar to the expression for electric flux (Eq. 25.6). The magnetic flux through a surface is given by

$$\Phi_B = \int \vec{B} \cdot d\vec{A} \tag{31.2}$$

The SI unit for magnetic flux is the weber (Wb), named for Wilhelm Weber (1804–1891), a colleague of Gauss:

$$1 \text{ Wb} = 1 \text{ T} \cdot \text{m}^2 \tag{31.3}$$

Gauss's law for magnetism involves the magnetic flux through a closed (Gaussian) surface. *Because there are no (known) magnetic monopoles, the magnetic flux through a Gaussian surface is zero.* Mathematically, **Gauss's law for magnetism** is

GAUSS'S LAW FOR MAGNETISM

Underlying Principle

$$\Phi_B = \oint \vec{B} \cdot d\vec{A} = 0 \tag{31.4}$$

The circle on the integral symbol in Equation 31.4 indicates that the integral is taken over the entire closed surface.

To understand Gauss's law for magnetism, imagine using a compass to probe the contents of a box. The compass's north pole is attracted to south poles and repelled by other north poles. Because there are no magnetic monopoles, we will never find a situation analogous to Figure 31.3 in which the positively charged sphere is always either repelled by or attracted to the source in the box.

The compass's north pole is attracted to the box when it is placed at some points and repelled by the box at other points, analogous to Figure 31.4. Figure 31.6A shows that when the compass is to the right of the box, its north pole is attracted to the box, and when the compass is to the left of the box, the compass's north pole is repelled. One possibility for what causes the deflection of the compass needle in this case is that there is nothing in the box; the deflection is caused by an external magnetic source such as a loop of current (Fig. 31.6B). Figure 31.6C shows a second possibility; there is a dipole in the box—in this case, a bar magnet. The north poles of the compass needles are attracted to the south pole of the bar magnet on the right and repelled by the north pole on the left. In both cases (current loop outside the box and bar magnet inside the box), the number of magnetic field lines entering the box is equal to the number of field lines leaving the box. If we calculate the magnetic flux through the entire closed Gaussian surface, we find there is no net magnetic flux through the box (Eq. 31.4).

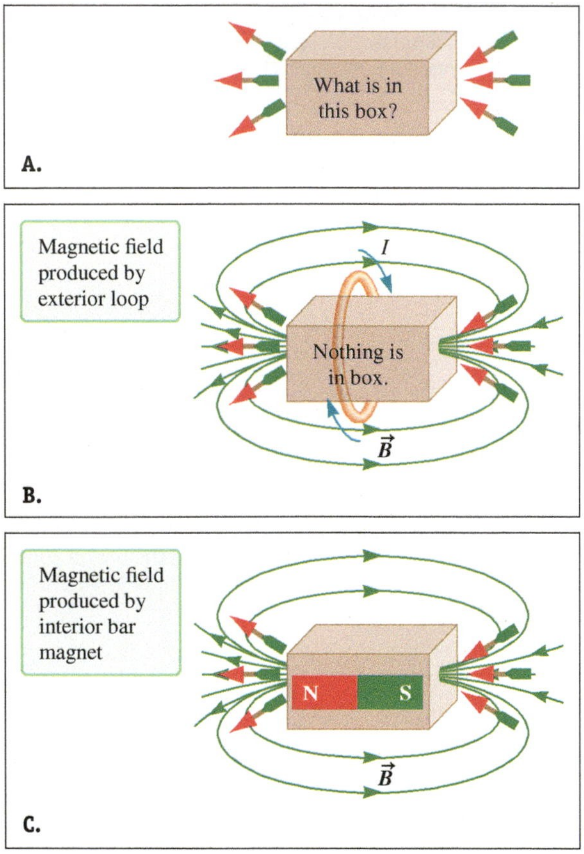

FIGURE 31.6 **A.** Analogous to Figure 31.4, we can use a compass needle to probe the magnetic contents of a box. **B.** One possible explanation for the deflection of these compass needles: an external source. This is analogous to Figure 31.5A. **C.** Another possibility: a bar magnet inside the box. The compass needles' north poles are attracted to the south pole of the bar magnet and repelled by the bar magnet's north pole. This is analogous to Figure 31.5B.

Comparing Gauss's law for magnetism to Gauss's law for electricity shows that the electric flux through a Gaussian surface may be positive, negative, or zero depending on the net charge of the source enclosed by the surface, but the magnetic flux through a Gaussian surface is always zero. This difference arises because electric monopoles exist (sources with a net charge), whereas magnetic monopoles are not known to exist. If magnetic monopoles are discovered, Gauss's law for magnetism will have to be modified; the right side of Equation 31.4 would then be proportional to the net magnetic "charge" enclosed in the Gaussian surface.

It is possible to choose a Gaussian surface near an electric dipole so that the electric flux through that surface is not zero. For example, imagine drawing a small Gaussian surface that encloses only the positive charge of the dipole. All the electric field lines pass outward through this Gaussian surface, so the net electric flux through it is positive. There is no analogy in the case of a magnetic dipole. Suppose you draw a small Gaussian surface around the north pole of the bar magnet in Figure 31.6C. Just as many magnetic field lines enter the Gaussian surface from the right as leave the surface on the left. The net magnetic flux is zero no matter how small a portion of the bar magnet you place in your Gaussian surface. This is the same as saying that if you break a bar magnet, you have two new complete bar magnets, each with a north and a south pole (Fig. 30.7, page 937).

CONCEPT EXERCISE 31.2

Does Gauss's law for magnetism state that the magnetic flux $\Phi_B = \int \vec{B} \cdot d\vec{A}$ (Eq. 31.2) is always zero? Explain.

EXAMPLE 31.2 Magnetic Flux

Find the magnetic flux through the rectangular loop lying near the long, straight wire carrying current $I = 2.50$ A (Fig. 31.7). According to the simple right-hand rule (Fig. 30.11, page 939), this current produces a magnetic field in the negative z direction on the right side of the wire. The magnetic field is stronger near the wire and gets weaker farther away according to Equation 30.9, $B = (\mu_0 I)/(2\pi r)$.

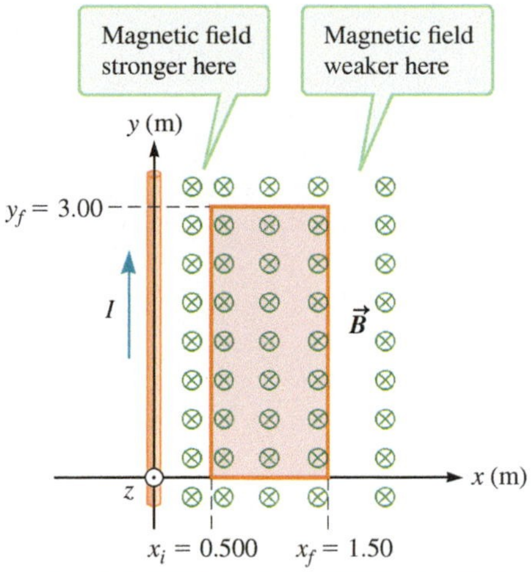

FIGURE 31.7

INTERPRET and ANTICIPATE

The current-carrying wire and the rectangular loop lie in the xy plane. The current is in the positive y direction.

SOLVE

Because the magnetic field is not uniform over the rectangular loop, we cannot pull $\vec{B}$ outside the integral in Equation 31.2. Instead, we must imagine cutting the rectangle into small vertical slices (Fig. 31.8). The direction of $d\vec{A}$ is perpendicular to the surface, so $d\vec{A}$ is in either the positive or negative z direction. Choose the negative z direction so that $d\vec{A}$ makes the smallest angle with the magnetic field.

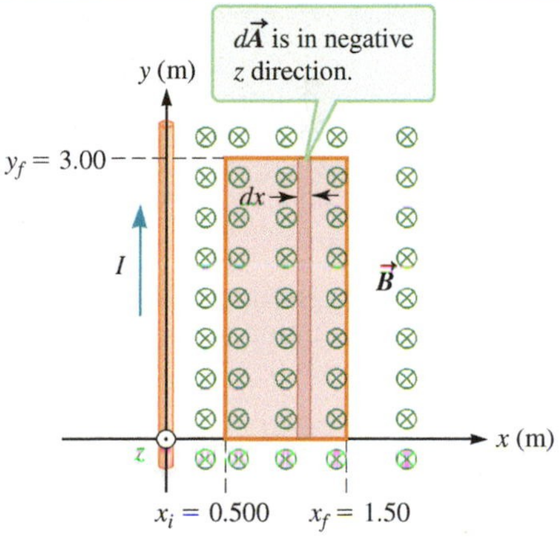

FIGURE 31.8

Because $d\vec{A}$ is parallel to $\vec{B}$, the dot product in the integral is simplified.	$\Phi_B = \int \vec{B} \cdot d\vec{A} = \int B\,dA$
The width of each slice is dx and the height is y_f, so the area of a slice is $dA = y_f\,dx$. Integrate from the left side of the rectangle ($x_i = 0.500$ m) to the right side ($x_f = 1.50$ m).	$\Phi_B = \int_{x_i}^{x_f} B y_f\,dx$
The magnetic field for a long, straight wire is given by $B = \mu_0 I/2\pi r$ (Eq. 30.9; replace r with x).	$\Phi_B = \int_{x_i}^{x_f} \left(\frac{\mu_0 I}{2\pi x}\right) y_f\,dx = \frac{\mu_0 I y_f}{2\pi} \int_{x_i}^{x_f} \left(\frac{dx}{x}\right)$
Integrate and substitute values.	$\Phi_B = \frac{\mu_0 I y_f}{2\pi} \ln\left(\frac{x_f}{x_i}\right) = (2 \times 10^{-7}\ \text{T}\cdot\text{m/A})(2.50\ \text{A})(3.00\ \text{m}) \ln\left(\frac{1.50\ \text{m}}{0.500\ \text{m}}\right)$ $\Phi_B = 1.65 \times 10^{-6}$ Wb

CHECK and THINK

The SI units work out correctly ($1\ \text{T}\cdot\text{m}^2 = 1$ Wb). The mathematical process for calculating magnetic flux is the same as the process for calculating electric flux (Section 25-2). Whereas the magnetic flux through any Gaussian surface is zero, the magnetic flux through an open surface is not necessarily zero. Magnetic flux through an open surface is very important in the next chapter, which will focus on the fourth of Maxwell's equations—Faraday's law.

31-3 Ampère's Law

Gauss's law for electricity gave us an alternative to calculating the electric field using the "brute force" method of applying Coulomb's law (Chapter 24). In practice, we used Gauss's law to find expressions for the electric field created by highly symmetrical charged sources such as an infinite line, sphere, or infinitely long cylinder (Chapter 25).

Using Coulomb's law to calculate the electric field is a lot like using the Biot-Savart law to calculate the magnetic field (Chapter 30). In both cases, we imagine slicing the source into a number of small pieces, calculating the electric or magnetic field due to each small piece, and then integrating over the entire source. What alternative do we have to the Biot-Savart law? Gauss's law for magnetism cannot be used to calculate the magnetic field because the right side of Gauss's law for magnetism equals zero. So, we get no information about the field.

However, Ampère came up with a law—like Gauss's law for electricity—that can be used to calculate the magnetic field for sources that have a high degree of symmetry. **Ampère's law** is expressed mathematically as

AMPÈRE'S LAW

Underlying Principle

$$\oint \vec{B} \cdot d\vec{\ell} = \mu_0 I_{\text{thru}} \tag{31.5}$$

Path integrals were introduced in Section 9-4 on work and then used again to calculate potential energy and potential.

The symbol $d\vec{\ell}$ is used in Biot-Savart's law to represent a piece of a wire.

The integral in Equation 31.5 is a **path integral**, and $d\vec{\ell}$ is a small piece of that path. In Ampère's law, the path integral is around a closed path known as an **Ampèrian loop** as indicated by the circle on the integral symbol. Like a Gaussian surface, an Ampèrian loop is imaginary, so we are free to choose a loop of any shape or size. We are also free to choose the direction over which we integrate. The path integral of a vector over a closed path is sometimes called the **circulation integral** or, simply, the **circulation**.

EXAMPLE 31.3 A Rectangular Loop

Consider a uniform magnetic field $\vec{B}$ and complete the circulation integral $\oint \vec{B} \cdot d\vec{\ell}$. Integrate counterclockwise over the rectangular Ampèrian loop as indicated by the direction of the $d\vec{\ell}$ vectors in Figure 31.9.

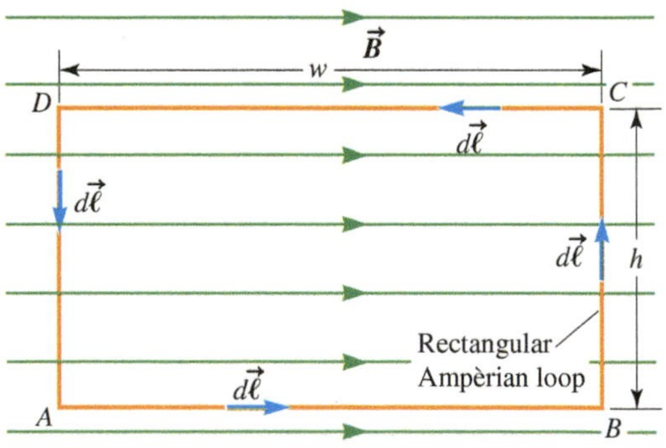

FIGURE 31.9

INTERPRET and ANTICIPATE

Because the rectangular loop consists of four straight paths, we can break up the circulation integral into four parts: $\vec{B} \cdot d\vec{\ell}$ is constant over each path, which makes the integration easy. Now we complete the four separate integrals along the four straight paths.

$$\oint \vec{B} \cdot d\vec{\ell} = \int_A^B \vec{B} \cdot d\vec{\ell} + \int_B^C \vec{B} \cdot d\vec{\ell} + \int_C^D \vec{B} \cdot d\vec{\ell} + \int_D^A \vec{B} \cdot d\vec{\ell}$$

SOLVE

From *A* to *B*: Both the magnetic field $\vec{B}$ and the direction of the path $d\vec{\ell}$ point toward the right, so the dot product is simply the product $B d\ell$. The magnetic field is constant over this path, so we can pull B outside the integral and complete the integral. The length ℓ from A to B is the width w of the rectangle (Fig. 31.9).

$$\int_A^B \vec{B} \cdot d\vec{\ell} = \int_A^B B\, d\ell = B \int_A^B d\ell = B\ell \Big|_A^B$$

$$\int_A^B \vec{B} \cdot d\vec{\ell} = Bw \tag{1}$$

From C to D: This integral is similar to the first integral, so we solve it next. The magnetic field $\vec{B}$ points toward the right and the path $d\vec{\ell}$ points toward the left. Because $\vec{B}$ and $d\vec{\ell}$ are in opposite directions, the dot product is negative. As before, we can pull B outside the integral.

$$\int_C^D \vec{B}\cdot d\vec{\ell} = -\int_C^D B d\ell = -B\int_C^D d\ell = -B\ell\Big|_C^D$$

$$\int_C^D \vec{B}\cdot d\vec{\ell} = -Bw \qquad (2)$$

From B to C and from D to A: For the path from B to C, $d\vec{\ell}$ is upward and $\vec{B}$ is toward the right, and for the last path, $d\vec{\ell}$ is downward and $\vec{B}$ is toward the right. Because $\vec{B}$ is perpendicular to $d\vec{\ell}$ for both these paths, the dot products and therefore the integrals are zero.

$$\int_B^C \vec{B}\cdot d\vec{\ell} = 0 \qquad (3)$$

$$\int_D^A \vec{B}\cdot d\vec{\ell} = 0 \qquad (4)$$

The integral over the Ampèrian loop is found by adding the results for the four paths (Eqs. 1–4).

$$\oint \vec{B}\cdot d\vec{\ell} = Bw + 0 + (-Bw) + 0$$

$$\oint \vec{B}\cdot d\vec{\ell} = 0$$

CHECK and THINK

For this rectangular loop, the circulation is zero. This result is consistent with Ampère's law, $\oint \vec{B}\cdot d\vec{\ell} = \mu_0 I_{thru}$ (Eq. 31.5), which says the circulation is proportional to the current I_{thru} through the loop. There is no current through this Ampèrian loop, so the circulation is zero.

Finding the Current Through a Loop

To find the current I_{thru}, imagine stretching a soap film over the loop (see Fig. 36.12, page 1162). A wire that goes through the film contributes to the current I_{thru} through the loop. The current may be negative or positive depending on the direction of the current relative to the direction of the Ampèrian path. The shape of the loop and the path direction are chosen for convenience. Once the path direction in the Ampèrian loop has been chosen, use your right hand to find the direction of positive current through that loop. Wrap the fingers of your right hand in the direction of the path, and then your thumb points in the direction of positive current. For example, in Figure 31.10A, your thumb points in the positive x direction, so I_1 is positive and I_2 is negative. The current through this Ampèrian loop is $I_{thru} = I_1 - I_2$. The third wire carries current I_3. Because this wire does not go through the "soap film," it does not contribute to I_{thru}.

In Figure 31.10A, the soap film is in the same plane as the Ampèrian loop. However, the right-hand rule holds even if the imaginary film is distended. For example, the Ampèrian loop in Figure 31.10B is a circle lying in the yz plane.

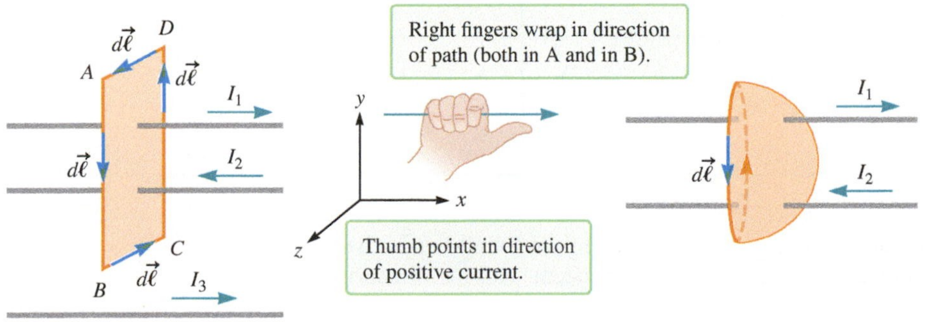
D
A
C
B
dℓ
I1
I2
I3
y
x
z
Right fingers wrap in direction of path (both in A and in B).
Thumb points in direction of positive current.

A. B.

FIGURE 31.10 A. The net current through the rectangular Ampèrian loop is $I_{thru} = I_1 - I_2$. **B.** Imagine a soap film over this circular loop is stretched into a hemisphere. (Dashed portion of the path is *behind* the plane of the page.) The net current through the circular Ampèrian loop is still $I_{thru} = I_1 - I_2$.

Imagine the soap film bubbles outward as shown, with two wires penetrating it. The direction of the Ampèrian path is similar to the direction in Figure 31.10A as indicated by the $d\vec{\ell}$ vector, so your thumb points in the positive x direction as before. The current through this Ampèrian loop is $I_{\text{thru}} = I_1 - I_2$, the same as in Figure 31.10A.

CONCEPT EXERCISE 31.3

Two circular Ampèrian loops lie in the xz plane. Curved wires carrying current I penetrate the imaginary soap films as shown in Figure 31.11. Find the current through each Ampèrian loop.

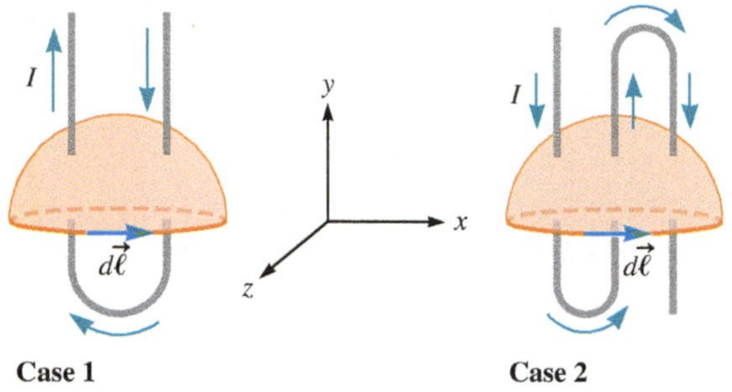

FIGURE 31.11

EXAMPLE 31.4 Finding Current

The circulation integral around the Ampèrian loop shown in case 2 of Figure 31.11 is $\oint \vec{B} \cdot d\vec{\ell} = -4.40 \times 10^{-6}\ \text{T}\cdot\text{m}$. What is the current I in the bent wire?

INTERPRET and ANTICIPATE

Ampère's law quickly leads to the answer, independent of the shape or size of the loop. All we need to know is that the circulation is proportional to the current through the film, $I_{\text{thru}} = I - 2I = -I$.

SOLVE

Start with Ampère's law (Eq. 31.5), and then divide the circulation by μ_0.

$$\oint \vec{B} \cdot d\vec{\ell} = \mu_0 I_{\text{thru}} \qquad (31.5)$$

$$I_{\text{thru}} = \frac{1}{\mu_0}\oint \vec{B} \cdot d\vec{\ell} = \frac{-4.40 \times 10^{-6}\ \text{T}\cdot\text{m}}{(4\pi \times 10^{-7}\ \text{T}\cdot\text{m/A})}$$

$$I_{\text{thru}} = -3.50\ \text{A}$$

By the right-hand rule for finding the current through a loop (Fig. 31.10), we have $I_{\text{thru}} = -I$.

$$I_{\text{thru}} = -I = -3.50\ \text{A}$$

$$I = 3.50\ \text{A}$$

CHECK and THINK

It may seem odd that the circulation integral is negative, but the current we found is positive. Remember that we found the sign of the current through the loop consistent with the direction of the path using the right-hand rule.

PROBLEM-SOLVING STRATEGY

Finding the Magnetic Field Using Ampère's Law

We can use Ampère's law to find the magnetic field for a number of different sources, much like using Gauss's law for electricity to find the electric field. The steps here are similar to those we used in Section 25-3.

INTERPRET and ANTICIPATE

Step 1 Sketch the magnetic field lines. You may need to draw more than one perspective. Your sketch will help you choose the Ampèrian loop and anticipate your results.

Step 2 Choose an Ampèrian loop. You must choose a closed path that exploits the symmetry of the situation and also choose the direction of the path. If possible, choose the direction so that $\vec{B}$ and $d\vec{\ell}$ are parallel. In practice, pick an Ampèrian loop so that each piece of the path has an angle φ equal to 0, 90°, or 180° between $\vec{B}$ and $d\vec{\ell}$. The most commonly used Ampèrian loops are circles and rectangles.

SOLVE

Step 3 Do the **circulation integral** in $\oint \vec{B} \cdot d\vec{\ell} = \mu_0 I_{\text{thru}}$ (Eq. 31.5). Because the Ampèrian loop exploits the symmetry of the situation, the magnetic field is often a constant over each portion of the integral, and so the integral is usually simple. It may help to sketch a few length vectors $d\vec{\ell}$.

Step 4 Determine the amount of **current I_{thru}** through the Ampèrian loop. Imagine a soap film stretched across the loop. In most cases, you may assume the film is taut and in the same plane as the loop. Wrap the fingers of your right hand in the direction of the path you have chosen; then your thumb points in the direction of positive current. Add the currents through the film to find I_{thru}.

31-4 Special Case: Linear Symmetry

In the next three sections, we use Ampére's law to find expressions for the magnetic field strength as a function of position for sources with commonly found shapes. We begin with a long, straight, current-carrying wire.

EXAMPLE 31.5 $\vec{B}$ Due to a Long, Straight Wire Using Ampère's Law

In Section 30-5, we found the magnetic field due to a long, straight wire carrying current I by applying the Biot-Savart law:

$$B = \frac{\mu_0 I}{2\pi r} \tag{30.9}$$

where r is the shortest distance from the wire to the point where the field is measured. The magnetic field wraps around the wire as shown in Figure 31.12. Use Ampère's law to re-derive Equation 30.9 for a long, straight wire.

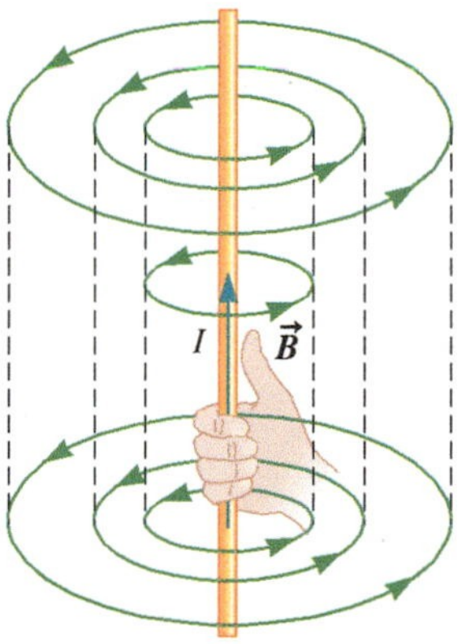

FIGURE 31.12

INTERPRET and ANTICIPATE

Step 1 Sketch the magnetic field lines. In this case, it is helpful to draw an end-on sketch of the wire coming out of the page and the magnetic field lines as concentric circles around the wire (Fig. 31.13).

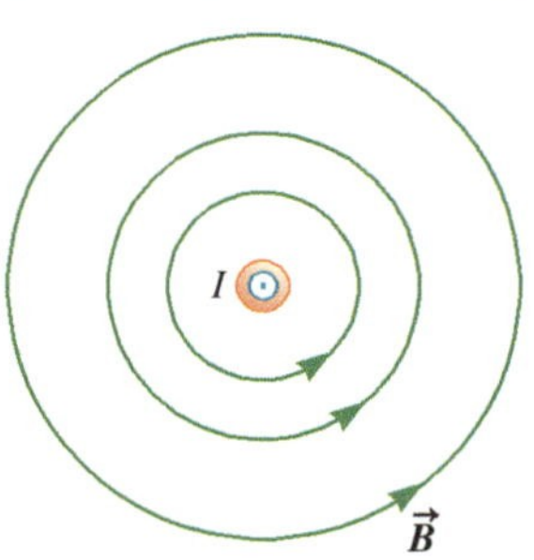

FIGURE 31.13

Example continues on page 988 ▶

Step 2 Choose an Ampèrian loop. A circular Ampèrian loop exploits the symmetry of the source (Fig. 31.14). We choose a path in the same direction as the magnetic field, so the angle φ between $\vec{B}$ and $d\vec{\ell}$ is zero all around the path. The radius of the circle is r.

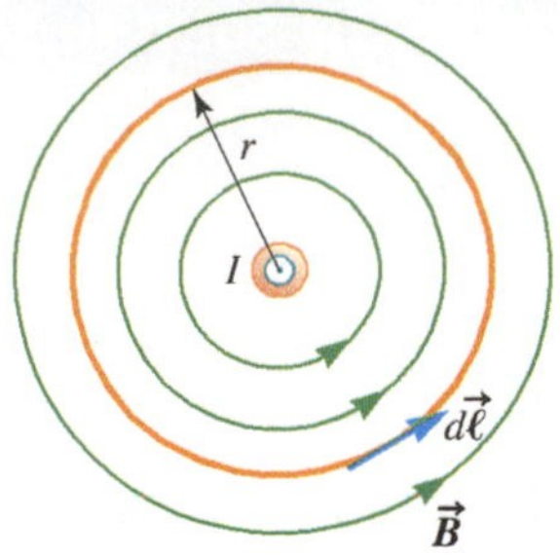

FIGURE 31.14

SOLVE

Step 3 Do the **circulation integral**. Because $\vec{B}$ and $d\vec{\ell}$ are parallel all around the circular loop, their dot product is $\vec{B}\cdot d\vec{\ell} = B\,d\ell$. The magnetic field is constant over the entire path.	$\oint \vec{B}\cdot d\vec{\ell} = \oint B\,d\ell = B\oint d\ell$
The integral $\oint d\ell$ covers the entire length of the path, which is the circumference $2\pi r$ of the circle.	$\oint \vec{B}\cdot d\vec{\ell} = B(2\pi r)$
Step 4 Find **current I_{thru}**. Imagine a soap film stretched taut across the circular Ampèrian loop (Fig. 31.14). The only current penetrating the film is the current I in the wire. Wrap the fingers of your right hand in the counterclockwise direction of the path; then your thumb points out of the page and the current through the Ampèrian loop is $I_{thru} = +I$. Substitute $I_{thru} = I$ into Equation (1).	$\oint \vec{B}\cdot d\vec{\ell} = B(2\pi r) = \mu_0 I_{thru}$ $B = \dfrac{\mu_0 I_{thru}}{2\pi r}$ (1) $B = \dfrac{\mu_0 I}{2\pi r}$ ✓ (30.9)

CHECK and THINK

This result is exactly what we found using the Biot-Savart law, but using Ampère's law is much simpler. Why would we ever use the Biot-Savart law? The Biot-Savart law can be applied to a source of any shape, but Ampère's law is best used to find the magnetic field produced by a source with a high degree of symmetry.

CONCEPT EXERCISE 31.4

CASE STUDY Magnetic Field Due to a Long, Straight Wire

In a laboratory, you measure the magnitude of the magnetic field generated by a long, straight wire, and you plot your results—B as a function of position r. Which of the graphs in Figure 31.15 best represents the magnetic field due to a long, straight wire?

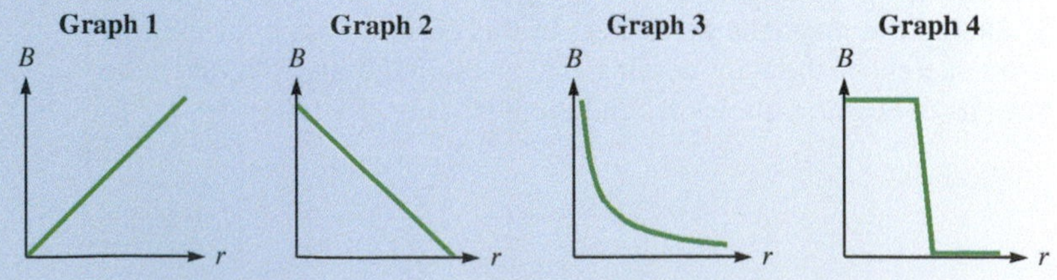

FIGURE 31.15

EXAMPLE 31.6 CASE STUDY Designing an Experiment for a Long, Straight Wire

When designing an experiment, you need to start with your expected results. Suppose you want to confirm $B = (\mu_0 I)/(2\pi r)$ (Eq. 30.9) for a long, straight wire. You decide your work will be easy if the current in the wire is exactly 1 A. Find the magnetic field strength at $r = 0.001$ m, 0.01 m, 0.1 m, 1.0 m, and 10 m from the wire. Imagine using a compass to measure the magnetic field strength at these positions. Assume the magnetic field at each position of the compass is perpendicular to the Earth's magnetic field, and find the deflection of the compass needle at each position. In **CHECK and THINK**: Sketch the deflection of the needle for the first three positions, and comment on your experimental design.

INTERPRET and ANTICIPATE

The magnetic field is stronger near the wire than it is farther away, so the compass needle will be deflected more when it is close to the wire. The question is: Will we be able to discern the deflection both near the wire and far from it?

SOLVE

To tabulate values of B, start by substituting $I = 1$ A into Equation 30.9.

$$B = \frac{\mu_0 I}{2\pi r} = \frac{(4\pi \times 10^{-7}\ \text{T}\cdot\text{m/A})(1\ \text{A})}{2\pi r}$$

$$B = \frac{2 \times 10^{-7}\ \text{T}\cdot\text{m}}{r} \tag{1}$$

We must also tabulate the angles θ (compass needle deflections), so solve Equation 31.1 for θ. We plan to use the Earth's magnetic field as the reference.

$$B = B_\oplus \tan\theta \tag{31.1}$$

$$\theta = \tan^{-1}\left(\frac{B}{B_\oplus}\right) = \tan^{-1}\left(\frac{B}{0.5 \times 10^{-4}\ \text{T}}\right) \tag{2}$$

Now substitute each value of r into Equation (1) to calculate B. Once B is calculated, substitute into Equation (2) to find θ for each distance r.

r (m)	B (T)	θ (°)
0.001	2.00×10^{-4}	76
0.01	2.00×10^{-5}	22
0.1	2.00×10^{-6}	2.3
1.0	2.00×10^{-7}	0.23
10	2.00×10^{-8}	0.023

CHECK and THINK

The sketch (Fig. 31.16) comes from the first three entries in the table. The Earth's magnetic field is in the positive y direction (straight up in this figure), and the current points out of the page. The top point (at 1 mm) is so close to the wire that the compass needle looks like it is touching the wire in the sketch. So, one problem is that we probably cannot get a compass needle that close to the wire, and we'll have to start at about 1 cm away from the wire. Another problem is that by 10 cm away, the compass needle deflection is nearly indistinguishable from zero. We will have difficulty measuring this slight deflection and will certainly not be able to measure the deflection at 1.0 m or 10 m. So we might hope to make about five reasonable measurements between about 1 cm and 10 cm. In Problem 24, you will be asked to make other improvements to this experiment.

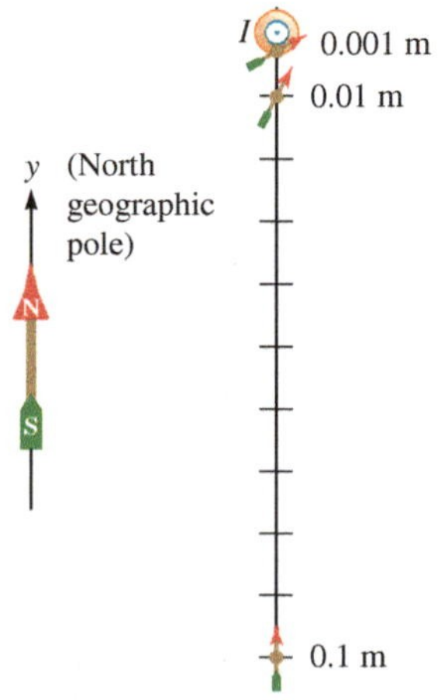

FIGURE 31.16

EXAMPLE 31.7 A Layered Cylindrical Wire

A very long, cylindrical wire has an insulating core surrounded by a cylindrical conductor carrying current I (Fig. 31.17). The current is uniformly distributed in the conducting cylinder. The radius of the insulating core is r_1, and the outer radius of the conducting cylinder is r_2. Find expressions for the magnetic field strength **A** outside the conducting cylinder ($r > r_2$), **B** inside the conducting cylinder but outside the insulator ($r_2 > r > r_1$), and **C** inside the insulator ($r < r_1$).

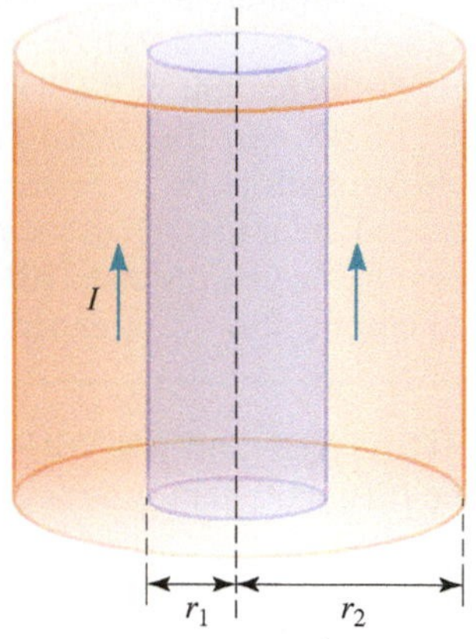

FIGURE 31.17 Current is only in the conducting cylindrical shell, not in the insulating core.

INTERPRET and ANTICIPATE

We can use Ampère's law and the problem-solving strategy to find expressions for B. The geometry resembles that of a long, straight wire, so start with an end-on view showing all three regions. We need three different Ampèrian loops—one for each region in which we must find B.

Step 1 Sketch magnetic field lines. Figure 31.18 is an end-on sketch showing the geometry of the situation. By the symmetry of the source, the magnetic field lines are circular in all regions. Because the conductor carries current I out of the page, the magnetic field lines are counterclockwise by the simple right-hand rule. (For now, don't worry if we find that there is no magnetic field in one or more regions.)

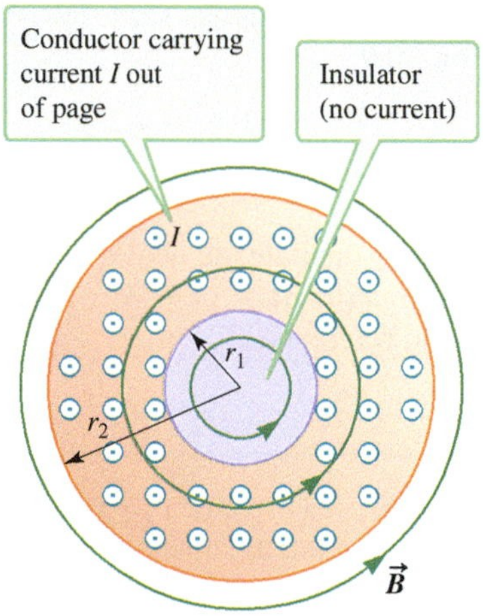

FIGURE 31.18

A Outside the hollow conducting cylinder.

Step 2 Choose an Ampèrian loop in the region outside the conductor ($r > r_2$). Because this region is much like the region outside a long, straight wire, a circular path in the direction of the magnetic field works well. The loop has radius $r > r_2$ (Fig. 31.19).

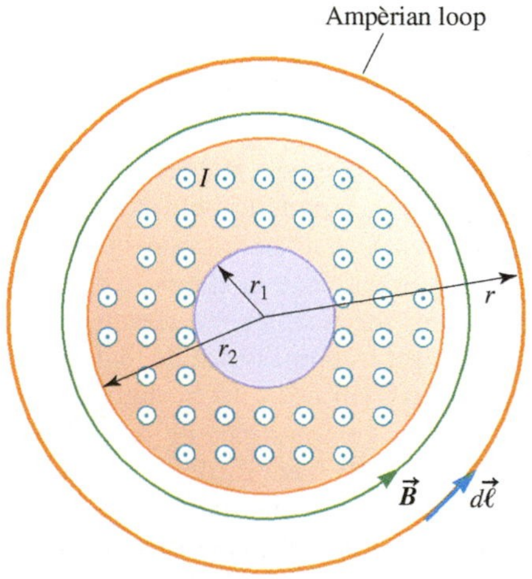

FIGURE 31.19

SOLVE

Step 3 Do the **circulation integral**. This calculation is exactly the same as for a long, straight wire.

$$\oint \vec{B} \cdot d\vec{\ell} = \oint B d\ell = B \oint d\ell = B(2\pi r)$$

Step 4 Find **current I_{thru}**. Just as for a long, straight wire, the current I penetrates an imaginary soap film stretched taut across the circular loop. Wrap the fingers of your right hand in the counterclockwise direction of the path; then your thumb points out of the page. Because I is out of the page, the current through the Ampèrian loop is $I_{thru} = +I$. Substitute $I_{thru} = I$ into Equation (1).	$\oint \vec{B} \cdot d\vec{\ell} = B(2\pi r) = \mu_0 I_{thru}$ $B = \dfrac{\mu_0 I_{thru}}{2\pi r}$ (1) $B = \dfrac{\mu_0 I}{2\pi r}$ for $r > r_2$

CHECK and THINK

The magnetic field outside the conductor is exactly what we found for the magnetic field outside a long, straight, current-carrying wire.

B Inside the conducting cylinder but outside the insulator.

INTERPRET and ANTICIPATE

Step 2 **Choose an Ampèrian loop** in the region inside the conductor ($r_2 > r > r_1$). Again, a circular path in the direction of the field works well (Fig. 31.20). The loop has radius r, where $r_2 > r > r_1$.

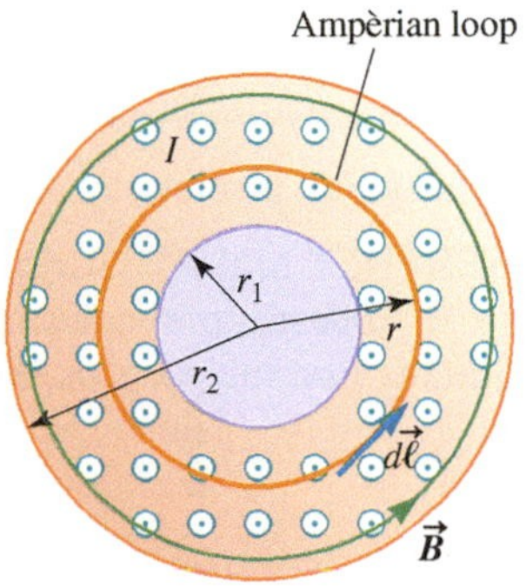

FIGURE 31.20

SOLVE **Step 3** Do the **circulation integral**. This calculation is the same as for part A.	$\oint \vec{B} \cdot d\vec{\ell} = \oint B d\ell = B \oint d\ell$ $\oint \vec{B} \cdot d\vec{\ell} = B(2\pi r)$ (2)
Step 4 Find **current I_{thru}**. Now we find a departure from the case of a long, straight wire. The current through the Ampèrian loop is *not* I; only a fraction of I penetrates an imaginary film covering the loop (Fig. 31.20). To find that fraction, begin with the current density (Eq. 28.3).	$J = \dfrac{I}{A}$ (28.3)
We need the cross-sectional area A of the conductor: a disk with a hole in it. The required area is that of the disk (radius r_2) minus that of the hole (radius r_1).	$A = \pi r_2^2 - \pi r_1^2 = \pi(r_2^2 - r_1^2)$
Substitute A into $J = I/A$, Equation 28.3.	$J = \dfrac{I}{\pi(r_2^2 - r_1^2)}$ (3)
There is no current through the insulator. The "film" area $A_{current}$ that is penetrated by the current is the area inside the Ampèrian loop (radius r) minus the area of the hole (radius r_1). (Keep in mind that $A_{current}$ does not equal the area A of the conductor.)	$A_{current} = \pi r^2 - \pi r_1^2$ $A_{current} = \pi(r^2 - r_1^2)$ (4)
The current I_{thru} penetrating that film is the current density J times $A_{current}$. Substitute Equation (3) for J and Equation (4) for $A_{current}$. The numerator and denominator do not cancel unless the Ampèrian loop is the size of the cylinder.	$I_{thru} = JA_{current}$ $I_{thru} = \left[\dfrac{I}{\pi(r_2^2 - r_1^2)}\right][\pi(r^2 - r_1^2)]$ $I_{thru} = I\dfrac{(r^2 - r_1^2)}{(r_2^2 - r_1^2)}$ (5)

Example continues on page 992 ▶

Substitute Equations (2) and (5) into Ampère's law, and solve for B.	$\oint \vec{B} \cdot d\vec{\ell} = \mu_0 I_{\text{thru}}$ $B(2\pi r) = \mu_0 I \dfrac{(r^2 - r_1^2)}{(r_2^2 - r_1^2)}$ $B = \dfrac{\mu_0 I}{2\pi} \dfrac{(r^2 - r_1^2)}{(r_2^2 - r_1^2)} \dfrac{1}{r}$ for $r_2 > r > r_1$ (6)
CHECK and THINK Equation (6) must give the same answer as Equation 30.9 at the outer surface of the conductor ($r = r_2$). Substitute $r = r_2$ into both equations; the results are the same.	From Equation 30.9, $B = \dfrac{\mu_0 I}{2\pi r_2}$ From Equation (6), $B = \dfrac{\mu_0 I}{2\pi} \dfrac{(r_2^2 - r_1^2)}{(r_2^2 - r_1^2)} \dfrac{1}{r_2} = \dfrac{\mu_0 I}{2\pi} \dfrac{1}{r_2}$

C Inside the insulator

INTERPRET and ANTICIPATE

Step 2 Choose an Ampèrian loop. Now the Ampèrian loop is a circle inside the insulator ($r < r_1$; Fig. 31.21).

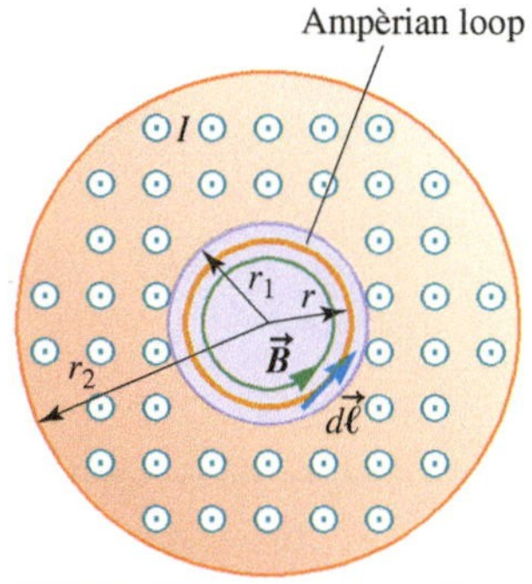

FIGURE 31.21

SOLVE **Step 3** Do the **circulation integral.** This calculation is the same as in parts A and B.	$\oint \vec{B} \cdot d\vec{\ell} = B(2\pi r)$
Step 4 Find **current I_{thru}.** Again imagine a film stretched over the Ampèrian loop; the conducting cylinder does not penetrate the film. Therefore, $I_{\text{thru}} = 0$, and by Ampère's law, the magnetic field inside the insulator is zero. (Now we know we didn't need the inner magnetic field line in Fig. 31.18 or Fig. 31.21.)	$\oint \vec{B} \cdot d\vec{\ell} = \mu_0 I_{\text{thru}}$ $B(2\pi r) = \mu_0(0)$ $B = 0$ for $r < r_1$
CHECK and THINK To check our result, we can calculate the magnetic field at the interface between the insulator and the conductor ($r = r_1$) using Equation (6). As expected, the magnetic field at $r = r_1$ is zero. In Problem 26, you will put all three parts of this example together to graph the magnetic field in all three regions.	$B = \dfrac{\mu_0 I}{2\pi} \dfrac{(r_1^2 - r_1^2)}{(r_2^2 - r_1^2)} \dfrac{1}{r_1^2} = 0$

31-5 Special Case: Solenoids

Today, we know that the magnetic field of a permanent magnet is largely due to the spin of electrons—an intrinsic property.

Just one week after Ampère theorized that moving charged particles are the source of all magnetic fields, the French physicist Dominique François Arago (1786–1853) presented a practical demonstration. Arago showed that a current-carrying coil of wire acts like a bar magnet by attracting bits of iron filings and that when the current is cut off, the coil's magnetic properties disappear. This demonstration led Ampère to try to magnetize an iron rod by placing it inside the coil. Ampère found that when the iron rod is removed from the coil, it is a permanent magnet.

A **solenoid** is a coil that consists of many windings or loops of wire. Solenoids are found in the circuits of many common electrical devices. Before we use Ampère's law to find the magnetic field strength due to a current-carrying solenoid, let's try to visualize the solenoid's magnetic field.

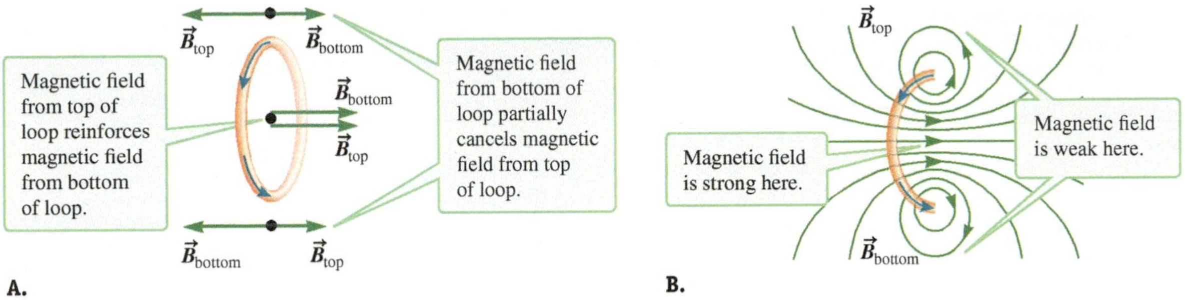

FIGURE 31.22 **A.** The current in the top portion of the loop points out of the page, and the current in the bottom portion points into the page. The direction of the magnetic field at the three points (above the loop, in the middle, and below the loop) comes from the simple right-hand rule. **B.** The lines are close together in the center of the loop because the magnetic field is strong there. Outside the loop, where the magnetic field is weaker, the lines are more widely spaced.

Each winding of a solenoid is similar to a single current loop. Figure 31.22A shows a perspective drawing of a loop that is perpendicular to the page. From this perspective, there is a current coming out of the page at the top of the loop and an equal current going into the page at the bottom. Consider a point directly above the loop; the top portion of the loop produces a magnetic field toward the left and the bottom portion produces a magnetic field toward the right. The point above the loop is closer to the top portion, so the magnetic field produced by the top portion is stronger than the bottom portion's magnetic field. The two vectors partially cancel out, resulting in a magnetic field pointing leftward for this point above the loop. A similar argument about the magnetic fields partially cancelling out can be made for the point directly below the loop, again resulting in a magnetic field that points toward the left. However, in the center of the loop, both the top portion and the bottom portion produce magnetic fields that point toward the right. The result is a strong magnetic field in the center of the loop and a weaker magnetic field outside (Fig. 31.22B).

A solenoid consists of many windings. First, consider the magnetic field produced by two loops placed near each other (Fig. 31.23). Between two top or two bottom portions, the curved magnetic field lines point in roughly opposite directions, indicating that the magnetic fields nearly cancel in these regions. The nearly straight field lines near the center of the loops are roughly in the same direction, so the magnetic fields add together and the result is that the field is strong in this region. The net result is that the magnetic field lines near the center of the two loops are tightly packed, nearly uniformly spaced, and nearly horizontal, pointing toward the right, whereas outside the loops, the magnetic field lines are widely spaced, indicating that the magnetic field is weak.

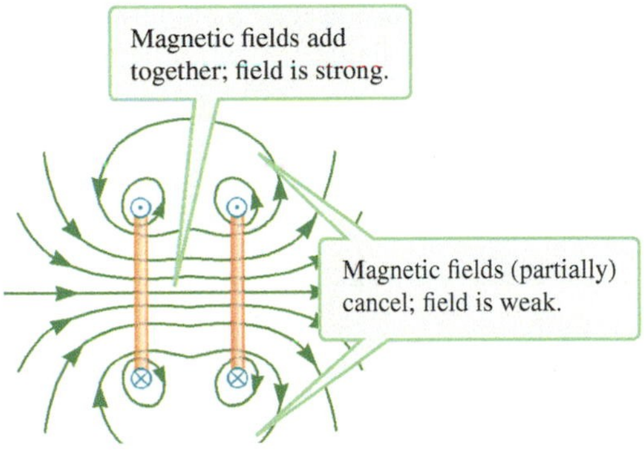

FIGURE 31.23 The magnetic fields of two loops must be added together. The combined magnetic field of two loops is strong near the center and weak outside.

Now let's look at a loosely wound solenoid (Fig. 31.24). The four turns of this solenoid produce a weak magnetic field on the outside, but inside the solenoid the magnetic field is strong and uniform. Compare the magnetic field of two loops (Fig. 31.23) to that of the loosely wound solenoid (Fig. 31.24). The solenoid has a more uniform interior magnetic field and a weaker exterior field.

In general, as the number of turns increases and as the turns are more closely packed, the interior magnetic field becomes stronger and more uniform while the exterior field becomes weaker. An **ideal solenoid** is infinitely long with tightly packed turns, a uniform interior magnetic field, and no exterior magnetic field (Fig. 31.25). In practice, a long solenoid with tightly packed turns can be approximated as an ideal solenoid in the region far from either end. Because the magnetic field inside the solenoid is uniform, the ideal solenoid is a practical device that we will encounter over and over again in electromagnetism. Throughout this textbook, you may assume all solenoids are ideal (even if they have a finite length). The direction of the magnetic field inside the solenoid is found by wrapping the fingers of your right hand in the direction of the current; then your thumb points in the field's direction.

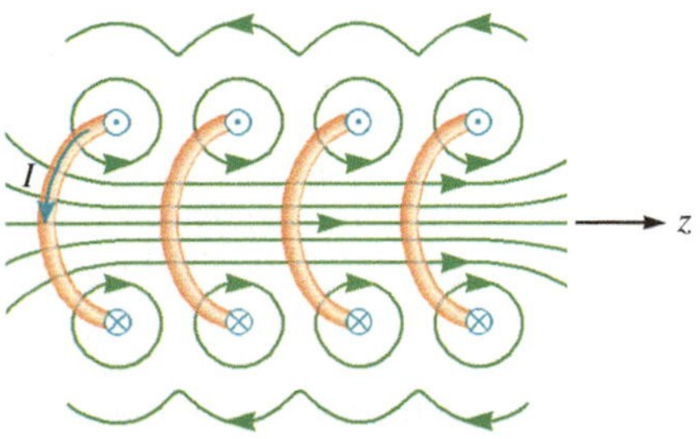

FIGURE 31.24 This solenoid has four loosely wound turns. The magnetic field in the inside is strong and nearly uniform in the z direction. The magnetic field outside is weak.

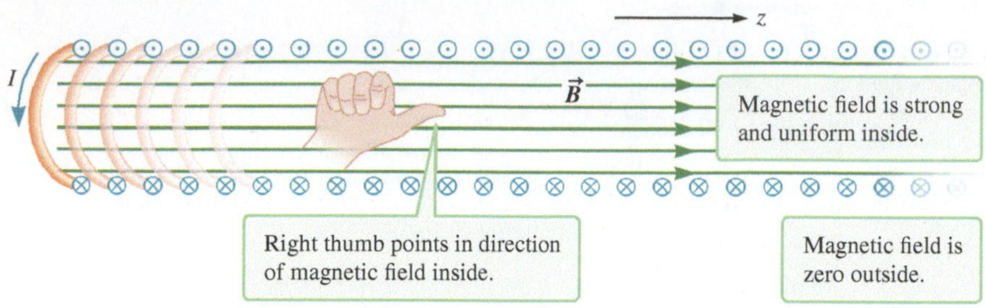

z
I
$\vec{B}$
Magnetic field is strong and uniform inside.
Right thumb points in direction of magnetic field inside.
Magnetic field is zero outside.

FIGURE 31.25 An ideal solenoid consists of tightly packed turns and is infinitely long. Only a few loops of wire are shown here. In many figures the loops are not drawn and the current is only indicated by the circled dots and circled crosses.

EXAMPLE 31.8 $\vec{B}$ for a Solenoid

We have reasoned that the magnetic field outside an ideal solenoid is zero. Now use Ampère's law to show that the magnetic field strength inside an ideal solenoid with n turns per unit length is given by

$$B = \mu_0 nI \tag{31.6}$$

INTERIOR MAGNETIC FIELD OF IDEAL SOLENOID Special Case

INTERPRET and ANTICIPATE

Steps 1 and 2 Sketch the magnetic field lines, and **choose an Ampèrian loop**. Figure 31.26 shows a slice through the solenoid. Because the interior magnetic field forms horizontal parallel lines, a rectangular Ampèrian loop exploits the symmetry. The width of the rectangle is w. We have chosen a counterclockwise path so that the part of the loop inside the solenoid is lined up with the interior magnetic field.

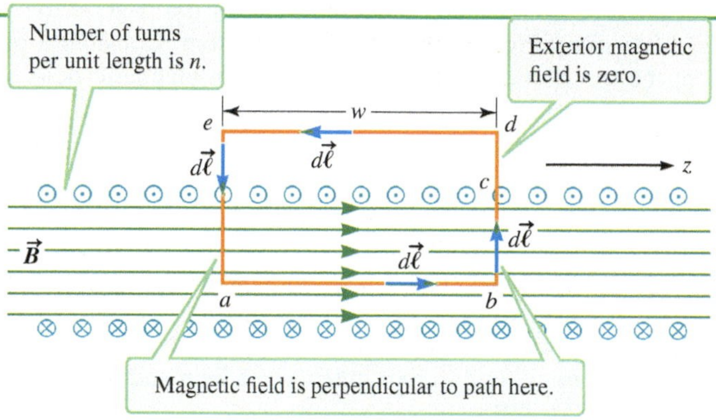

Number of turns per unit length is n.
Exterior magnetic field is zero.
w
e
d
c
a
b
z
$\vec{B}$
$d\vec{\ell}$
Magnetic field is perpendicular to path here.

FIGURE 31.26

SOLVE

Step 3 Do the **circulation integral**. Because the loop is a rectangle, break up the integral into four pieces, one for each side of the rectangle.

$$\oint \vec{B}\cdot d\vec{\ell} = \int_a^b \vec{B}\cdot d\vec{\ell} + \int_b^d \vec{B}\cdot d\vec{\ell} + \int_d^e \vec{B}\cdot d\vec{\ell} + \int_e^a \vec{B}\cdot d\vec{\ell} \tag{1}$$

The path from a to b is parallel to the magnetic field, and the magnetic field is uniform along that part of the path. The dot product is $Bd\ell$. The path length is the width of the rectangle.

$$\int_a^b \vec{B}\cdot d\vec{\ell} = \int_a^b B d\ell = B\int_a^b d\ell$$

$$\int_a^b \vec{B}\cdot d\vec{\ell} = Bw \tag{2}$$

The magnetic field is not constant over the path from b to d. Inside the solenoid from b to c, the magnetic field has some uniform value B, and outside the solenoid from c to d, the magnetic field is zero. We break up this straight path into two pieces. The integral over the path outside the solenoid (from c to d) is zero because the magnetic field is zero. The integral over the path inside the solenoid is also zero because the magnetic field is perpendicular to the path. So, the entire integral from b to d is zero.

$$\int_b^d \vec{B}\cdot d\vec{\ell} = \int_b^c \vec{B}\cdot d\vec{\ell} + \int_c^d \vec{B}\cdot d\vec{\ell}$$

$$\int_b^d \vec{B}\cdot d\vec{\ell} = \int_b^c B\,d\ell\cos 90° + \int_c^d 0$$

$$\int_b^d \vec{B}\cdot d\vec{\ell} = 0 \tag{3}$$

The integral from d to e is zero because the exterior magnetic field is zero.

$$\int_d^e \vec{B}\cdot d\vec{\ell} = 0 \tag{4}$$

Finally, the integral from e back to a is similar to the integral from b to d. Outside the solenoid, the magnetic field is zero, and inside the solenoid, the magnetic field is perpendicular to the path. So, the integral from e to a is zero.	$\int_e^a \vec{B} \cdot d\vec{\ell} = 0$ (5)
Substitute Equations (2) through (5) into the circulation integral (Eq. 1).	$\oint \vec{B} \cdot d\vec{\ell} = Bw + 0 + 0 + 0 = Bw$ (6)
Step 4 Find **current I_{thru}.** Imagine a soap film stretched taut across the rectangular Ampèrian loop in Figure 31.26. If N turns penetrate the film and each turn carries current I, the total current that penetrates the film is NI. Because an ideal solenoid is infinitely long, it has an infinite number of turns. We should characterize an ideal solenoid in terms of the number of turns per unit length (n, measured in m^{-1}) instead of the total number of turns. The number N of turns that penetrate the film is then the width w of the rectangle times n. Finally, we must determine whether the current through the loop is positive or negative. Wrap the fingers of your right hand in the direction of the path from a around to e; then your thumb points out of the page. Because the top of each turn carries current out of the page, the current through the film is positive.	$I_{thru} = NI$ $I_{thru} = nwI$ (7)
Substitute Equations (6) and (7) into Ampère's law, and solve for B.	$\oint \vec{B} \cdot d\vec{\ell} = \mu_0 I_{thru}$ $Bw = \mu_0 nwI$ $B = \mu_0 nI$ ✓ (31.6)

CHECK and THINK

At the minimum, you need to confirm that our expression has the dimensions of magnetic field. Notice that Equation 31.6 does not have any spatial dependence. (It does not involve r, x, y, or z.) This result confirms that inside an ideal solenoid, the magnetic field is uniform. Throughout this textbook, you may use Equation 31.6 to find the interior magnetic field of any solenoid, whether or not it is described as an *ideal* solenoid, and assume the exterior magnetic field is zero.

CONCEPT EXERCISE 31.5

CASE STUDY Designing an Experiment for an Ideal Solenoid

Suppose you want to confirm $B = 0$ outside and $B = \mu_0 nI$ (Eq. 31.6) inside an ideal solenoid using a compass. Your ideal solenoid has exactly 5000 turns per meter. You align your solenoid so that its magnetic field is perpendicular to the Earth's magnetic field. The compass is easiest to read when its deflection is between 10° and 80°. What range of current will work best for your experiment?

EXAMPLE 31.9 Homemade Solenoid

Suppose you want to make your own solenoid using a 9-V battery and copper wire. You wrap the wire in one neat layer around a portion of a pencil. Estimate the magnetic field strength inside your solenoid. Assume the internal resistance of the battery is about 2 Ω.

INTERPRET and ANTICIPATE

Once we know the current in the wire and the number of turns per unit length, we can find the magnetic field inside the solenoid. The number of turns per unit length comes from estimating the parameters—the radius of the pencil, the length of the solenoid, and the thickness of the wire. The current depends on these geometric factors and also on the resistivity of copper. Start by estimating the geometric factors based on your experience with pencils and wires.

Example continues on page 996 ▶

SOLVE

Using a ruler shows that a pencil is about 1 cm in diameter, so a pencil's radius is $r = 0.5$ cm. Now imagine a wire about as thick as a paper clip, which is slightly thinner than 1 mm. Finally, imagine wrapping the wire around the pencil. You will probably leave a little room at the ends for convenience, so the solenoid is about 10 cm long.

Pencil's radius $r = 0.005$ m

Wire's thickness $t = 0.0008$ m

Solenoid's length $L = 0.1$ m

The total number of turns N is found by dividing the length of the solenoid by the thickness of one turn (the thickness of the wire). If you had the solenoid in your hands, you could count the number of turns.

$$N = \frac{L}{t} = \frac{0.1\,\text{m}}{0.0008\,\text{m}}$$

$$N = 125 \text{ turns}$$

Find the number of turns per unit length by dividing the number of turns N by the length of the solenoid.

$$n = \frac{N}{L} = \frac{125}{0.1\,\text{m}} = 1250\,\text{m}^{-1} \quad (1)$$

To find the current in the wire, we first need to find the wire's resistance, so we need the length ℓ of the wire. We can estimate that length as the circumference C of one turn multiplied by the total number of turns N. The circumference of one turn is roughly the circumference of the pencil.

$$\ell = NC = N2\pi r$$

$$\ell = (125)2\pi(0.005\text{ m}) = 4\,\text{m}$$

We also need the cross-sectional area of the wire, which we find from its thickness.

$$A = \pi(t/2)^2 = \pi(0.0008\,\text{m}/2)^2 = 5 \times 10^{-7}\,\text{m}^2$$

The resistivity of copper is listed in Table 28.2 (page 872; $\rho = 1.68 \times 10^{-8}\ \Omega \cdot \text{m}$). Use Equation 28.24 to find the wire's resistance.

$$R_{\text{wire}} = \rho\frac{\ell}{A} \quad (28.24)$$

$$R_{\text{wire}} = (1.68 \times 10^{-8}\,\Omega\cdot\text{m})\left(\frac{4\,\text{m}}{5 \times 10^{-7}\,\text{m}^2}\right) = 0.13\,\Omega$$

The circuit's total resistance is the sum of the battery's internal resistance and the wire's resistance. We can see that the wire's resistance does not contribute very much.

$$R = R_{\text{internal}} + R_{\text{wire}} = (2 + 0.13)\Omega$$

$$R \approx 2\,\Omega$$

Find the current in the wire from Equation 29.1; the sign does not matter in this problem.

$$\Delta V_R = -IR \quad (29.1)$$

$$I = \frac{\Delta V_R}{R} = \frac{9\,\text{V}}{2\,\Omega} = 4.5\text{ A} \approx 4\text{ A} \quad (2)$$

Finally, to find the magnetic field inside the homemade solenoid, substitute values for n (Eq. 1) and I (Eq. 2) into Equation 31.6.

$$B = \mu_0 nI \quad (31.6)$$

$$B = (4\pi \times 10^{-7}\,\text{T}\cdot\text{m/A})(1250\,\text{m}^{-1})(4\,\text{A})$$

$$B = 6 \times 10^{-3}\,\text{T}$$

CHECK and THINK

The magnetic field produced inside a crude homemade solenoid is about two orders of magnitude stronger than the Earth's magnetic field. If such a strong magnetic field were perpendicular to the Earth's magnetic field, it would deflect a compass needle by (essentially) 90°.

FIGURE 31.27 A toroid is a solenoid bent into a donut shape.

31-6 Special Case: Toroids

Another important source of a magnetic field is a toroid (Fig. 31.27). You can make a toroid by bending a solenoid into a donut. Toroids—like solenoids—are found in the circuits of many common devices. In this section, we use Ampère's law to find an expression for the magnetic field produced by an ideal toroid.

Figure 31.28 is a cross-sectional sketch of an ideal toroid; each turn is set neatly around the donut shape and there are no wires leading in or out. Take a look at the exterior regions—inside the donut hole and outside the toroid. Because there is no magnetic field outside a solenoid, we expect there is no magnetic field in either of these exterior regions. Let's use Ampère's law to confirm our expectations. First, consider an Ampèrian loop inside the donut hole. Imagine a film stretched across this loop. No wire penetrates the film, so there is no current through the Ampèrian loop: $I_{\text{thru}} = 0$. According to Ampère's law, $\oint \vec{B} \cdot d\vec{\ell} = \mu_0 I_{\text{thru}}$ (Eq. 31.5), the circulation must be zero no matter what loop shape we choose. The only way for the circulation integral to be zero independent of the path shape is for the magnetic field to be zero. We conclude that the magnetic field in the donut hole is zero.

Next, consider an Ampèrian loop outside the toroid. Again, imagine a film stretched across the loop. In this case, each turn of the toroid penetrates the film twice—once coming out of the page and once going into the page. The net result is that the total current through the loop is zero: $I_{\text{thru}} = 0$. As in the case of the Ampèrian loop in the donut hole, the circulation integral must be zero no matter what loop shape we choose. We conclude that the magnetic field outside the toroid must also be zero.

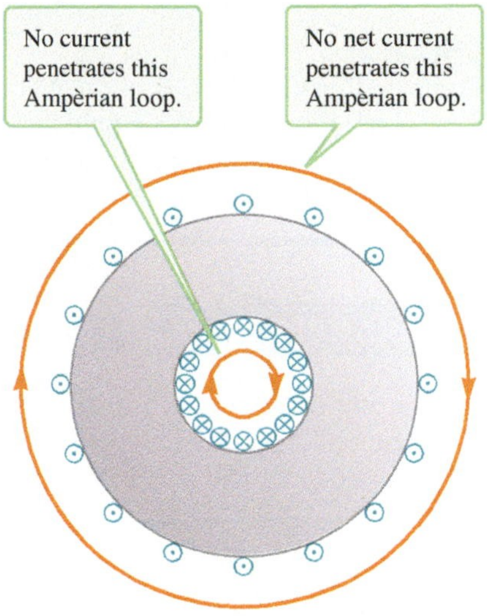
No current penetrates this Ampèrian loop.
No net current penetrates this Ampèrian loop.

FIGURE 31.28 Circular Ampèrian loops inside the donut hole and outside the toroid.

EXAMPLE 31.10 $\vec{B}$ Inside an Ideal Toroid

INTERIOR MAGNETIC FIELD OF IDEAL TOROID Special Case

Now we turn our attention to the magnetic field inside a toroid. Consider an ideal toroid that has N turns and a current I in the wire. Use Ampère's law to show that the magnetic field strength inside an ideal toroid is given by

$$B = \frac{\mu_0 NI}{2\pi r} \tag{31.7}$$

Steps 1 and 2 Sketch the magnetic field lines, and **choose an Ampèrian loop**. The direction of the magnetic field is found using the same right-hand rule as for a solenoid. For any single turn, wrap the fingers of your right hand in the direction of the current, and then your thumb points in the direction of the magnetic field. The magnetic field lines bend into concentric circles on the inside, so a circular Ampèrian loop of radius r exploits the symmetry. We have chosen a clockwise path lined up with the magnetic field (Fig. 31.29).

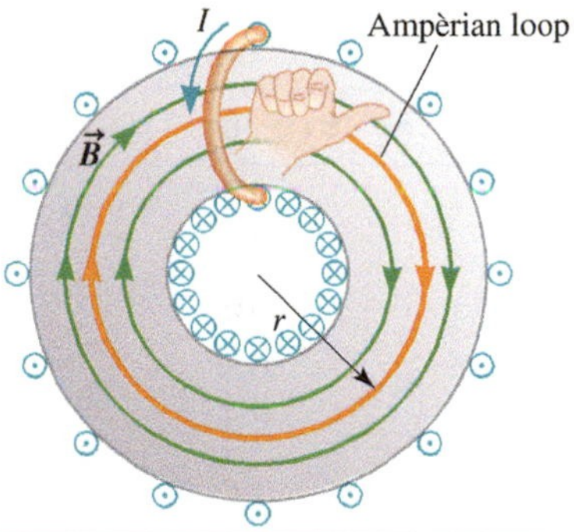
I
Ampèrian loop
$\vec{B}$
r

FIGURE 31.29

SOLVE

Step 3 Do the **circulation integral**. Because the magnetic field is parallel to and constant over the whole path, we can pull B outside the integral. The path integral is just the circumference of the Ampèrian loop.

$$\oint \vec{B} \cdot d\vec{\ell} = \oint B d\ell = B \oint d\ell$$

$$\oint \vec{B} \cdot d\vec{\ell} = B(2\pi r) \tag{1}$$

Step 4 Find **current I_{thru}**. Imagine a soap film stretched taut across the circular Ampèrian loop in Figure 31.29. All the turns penetrate the film. Each turn carries a current I, so the total current that penetrates the film is NI. Next, determine whether the current through the loop is positive or negative. Wrap the fingers of your right hand clockwise in the direction of the path, and your thumb points into the page. Because each turn carries current into the page, the current through the film is positive.

$$I_{\text{thru}} = NI \tag{2}$$

Substitute Equations (1) and (2) into Ampère's law, and solve for B.

$$\oint \vec{B} \cdot d\vec{\ell} = \mu_0 I_{\text{thru}}$$

$$B(2\pi r) = \mu_0 NI$$

$$B = \frac{\mu_0 NI}{2\pi r} \; ✓ \tag{31.7}$$

Example continues on page 998 ▶

CHECK and THINK

Compare Equation 31.7, $B = (\mu_0 NI)/(2\pi r)$, for the magnetic field inside a toroid to Equation 31.6, $B = \mu_0 nI$, for the magnetic field inside a solenoid. The magnetic field inside a toroid depends on the position and is strongest near the donut hole but gets weaker at farther positions. By contrast, the magnetic field inside a solenoid is constant throughout the interior region.

EXAMPLE 31.11 CASE STUDY Plotting *B* for a Toroid

The current in a toroid is 1.0 A, and it consists of 66 turns. Suppose the radius of the donut hole is 1.2 cm and the outer radius of the toroid is 3.7 cm. Plot the toroid's magnetic field strength as a function of r from $r = 0$ to $r = 5.0$ cm.

INTERPRET and ANTICIPATE

This is an application of the equations we found for the magnetic field produced by a toroid. In the regions outside the toroid, the magnetic field strength is zero, and inside the toroid, the magnetic field depends on the position r.

SOLVE

Calculate and tabulate several values (Table 31.1). We need Equation 31.7 only for points inside the toroid—that is, for points between $r = 1.2$ cm and 3.7 cm. So, substitute all values except r and remember to work in SI units.

$$B = \frac{\mu_0 NI}{2\pi r} \quad (31.7)$$

$$B = \frac{(4\pi \times 10^{-7}\,\mathrm{T\cdot m/A})(66)(1.0\,\mathrm{A})}{2\pi r}$$

$$B = \frac{1.32 \times 10^{-5}}{r}\,\mathrm{T}$$

TABLE 31.1 Field strength vs. *r* for toroid.

$r (\times 10^{-2}$ m)	$B (\times 10^{-4}$ T)	$r (\times 10^{-2}$ m)	$B (\times 10^{-4}$ T)	$r (\times 10^{-2}$ m)	$B (\times 10^{-4}$ T)
1.2	11.0	2.0	6.6	2.8	4.7
1.3	10.2	2.1	6.3	2.9	4.6
1.4	9.4	2.2	6.0	3.0	4.4
1.5	8.8	2.3	5.7	3.1	4.3
1.6	8.3	2.4	5.5	3.2	4.1
1.7	7.8	2.5	5.3	3.3	4.0
1.8	7.3	2.6	5.1	3.4	3.9
1.9	6.9	2.7	4.9	3.5	3.8

Plot the values from Table 31.1 as shown in Figure 31.30. Include $B = 0$ for $r < 1.2$ cm and for $r > 3.7$ cm on the graph.

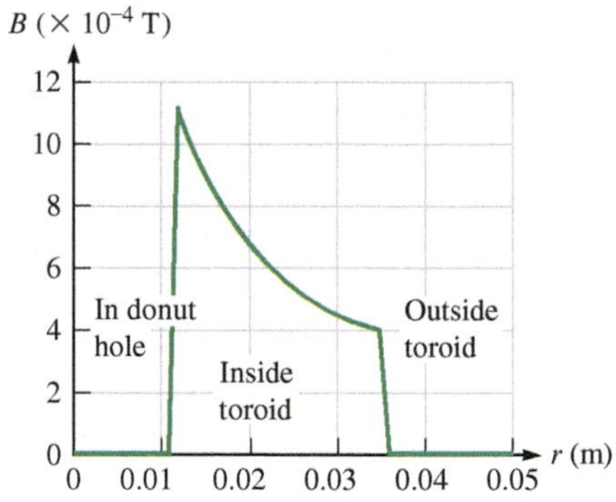

FIGURE 31.30

CHECK and THINK

Figure 31.30 clearly shows that the magnetic field inside the toroid is strongest near the donut hole. Another way to visualize this is with magnetic field lines (Fig. 31.31). The field lines inside the toroid are not evenly spaced; the concentric circles are more concentrated near the donut hole and get farther apart with increasing distance.

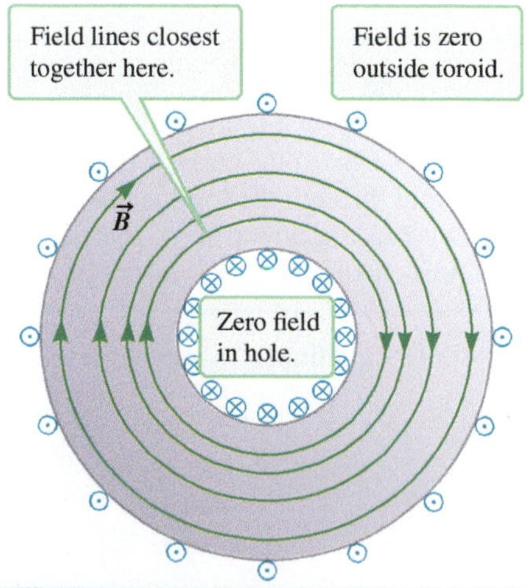

FIGURE 31.31

31-7 General Form of Ampère's Law

Maxwell's greatest insight was to combine four seemingly separate equations to describe electricity, magnetism, and light (Fig. 31.1). Each of these equations had been discovered by another researcher, and each carries the name of its discoverer. We have studied two and a half of Maxwell's equations so far: Gauss's law for electricity and Gauss's law for magnetism. The "half" is Ampère's law, which is incomplete in the form $\oint \vec{B} \cdot d\vec{\ell} = \mu_0 I_{\text{thru}}$ (Eq. 31.5). Maxwell discovered that it is missing a term, without which he would not have been able to discover that light is intimately connected to electricity and magnetism. In this section, we present Maxwell's addition to Ampère's law.

The need for Maxwell's addition to Ampère's law is best understood by considering a simple circuit that consists of an emf and a capacitor (Fig. 31.32). Imagine the switch S has been closed and the capacitor is charging. The top plate of the capacitor becomes positively charged and the bottom plate becomes negatively charged, so there is a downward-pointing electric field between the plates. As more charge builds up on the plates, the electric field grows stronger.

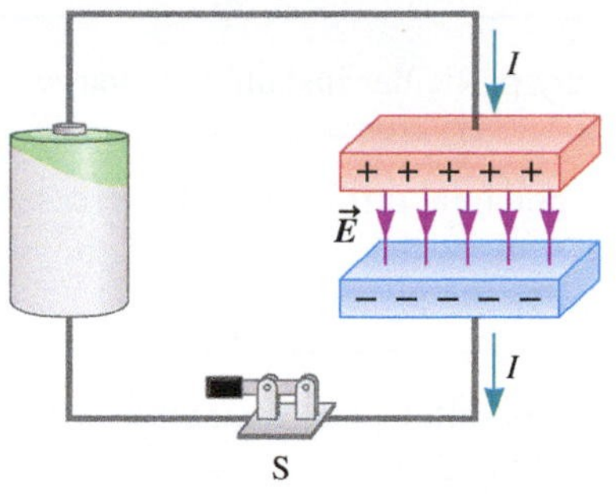

I
+ + + + +
$\vec{E}$
– – – – –
I
S

FIGURE 31.32 The switch is closed, there is current in the circuit, and the capacitor is charging. While a capacitor is charging, an electric field between the plates grows in strength.

Suppose you wish to use Ampère's law to calculate the magnetic field at a point P; you need to use an Ampèrian loop that passes through that point (Fig. 31.33). Let's try to determine the current through the loop. If you imagine a film stretched taut across the loop (Fig. 31.33A), you find that the current through the loop is just the current in the wire: $I_{\text{thru}} = I$. However, if you imagine the film is bowed outward like a soap film so that the film encompasses the capacitor's top plate (Fig. 31.33B), there is no current through the loop.

This ambiguity over the current through the Ampèrian loop leads to a contradiction in the magnetic field at point P. If the film is taut (Fig. 31.33A), the current through the Ampèrian loop is nonzero and the magnetic field at point P is also nonzero. But if the film is bowed out (Fig. 31.33B), the current through the loop is zero and, according to Ampère's law $\oint \vec{B} \cdot d\vec{\ell} = \mu_0 I_{\text{thru}}$ (Eq. 31.5), the magnetic field must be zero. Because the magnetic field cannot be both nonzero and zero at the same time, something must be wrong with Ampère's law. Maxwell realized he could correct Ampère's law by adding a new term to the right side:

$$\oint \vec{B} \cdot d\vec{\ell} = \mu_0 I_{\text{thru}} + (\text{new term})$$

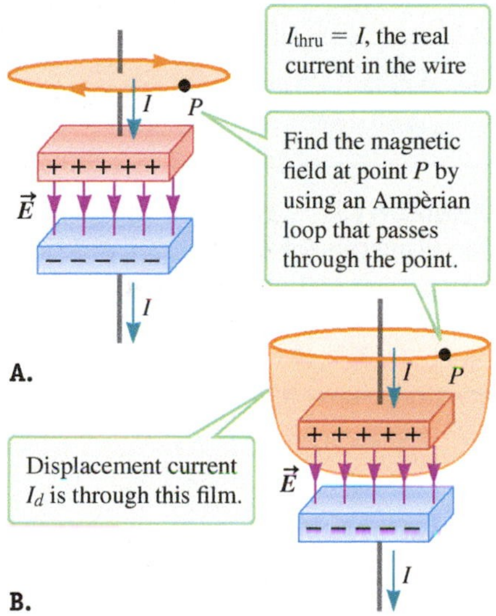

$I_{\text{thru}} = I$, the real current in the wire
I
P
+ + + + +
$\vec{E}$
– – – – –
I
Find the magnetic field at point P by using an Ampèrian loop that passes through the point.
A.
Displacement current I_d is through this film.
I
P
+ + + + +
$\vec{E}$
– – – – –
I
B.

FIGURE 31.33 **A.** An Ampèrian loop with a film stretched taut across the loop. **B.** An Ampèrian loop with the film bowed outward like a soap bubble.

Even if $I_{\text{thru}} = 0$, as it does in the case of the bowed-out film, the new term would still give a nonzero value and then the magnetic field would be nonzero.

To see what the new term involves, take another look at Figure 31.33B. Although there is no current through the film, something else penetrates it: the electric field. The new term involves the *electric flux* (Eq. 25.3, p. 759) through the film. The general form of Ampère's law is known as Ampère–Maxwell's law:

AMPÈRE–MAXWELL'S LAW
Underlying Principle

$$\oint \vec{B} \cdot d\vec{\ell} = \mu_0 I_{\text{thru}} + \mu_0 \varepsilon_0 \frac{d\Phi_E}{dt} \quad (31.8)$$

DISPLACEMENT CURRENT
Major Concept

For reasons that will become clear after its derivation, the new term (excluding the μ_0) is known as the **displacement current**. The displacement current involves the *rate of change* of the electric flux. If the electric flux is not changing, the displacement current is zero. Of course, when a capacitor is charging, the electric field is increasing, so the electric flux is changing.

DERIVATION Displacement Current

We can gain a greater insight into the displacement current I_d by deriving it. We show that for the case of a parallel-plate capacitor with no dielectric between its plates,

$$I_d = \varepsilon_0 \frac{d\Phi_E}{dt} \quad (31.9)$$

Derivation continues on page 1000 ▶

At any particular instant, the charge stored by a capacitor is given by Equation 27.1.	$Q = C\Delta V$	(27.1)
The potential difference ΔV increases as charge continues to build up on the plates, and so does the magnitude of the electric field. (Remember that d is the distance between the plates.)	$\Delta V = Ed$	(26.17)
For a parallel-plate capacitor without a dielectric, the capacitance comes from Equation 27.10.	$C = \dfrac{\varepsilon_0 A}{d}$	(27.10)
Substitute Equations 27.10 and 26.17 into Equation 27.1.	$Q = \left(\dfrac{\varepsilon_0 A}{d}\right)(Ed) = \varepsilon_0 EA$	
Substitute the electric flux $\Phi_E = EA$. This is the flux through the bowed-out film (Fig. 31.33B), and A is the area perpendicular to $\vec{E}$.	$Q = \varepsilon_0 \Phi_E$	(1)
The charge on the capacitor is increasing. The rate of charge increase is the current $I = dQ/dt$.	$I_d = \varepsilon_0 \dfrac{d\Phi_E}{dt}$ ✓	(31.9)

COMMENTS

The **displacement current** I_d is a fictitious current through the bowed-out film in Figure 31.33B. There is no real current between the plates of a capacitor. The displacement current is really a changing electric field between the capacitor's plates, which provides us with a helpful way to think about how the changing electric field between the plates can create a magnetic field at point P.

Ampère–Maxwell's law can be written in terms of two currents—the real current I_{thru} and the displacement current I_d through the loop:

AMPÈRE–MAXWELL'S LAW

Underlying Principle

$$\oint \vec{B} \cdot d\vec{\ell} = \mu_0(I_{\text{thru}} + I_d) = \mu_0 I_{\text{tot}} \tag{31.10}$$

where $I_{\text{tot}} = I_{\text{thru}} + I_d$ is the total (real plus displacement) current through the Ampèrian loop. The value of the displacement current is the same as the real current in the wire: $I_d = I$. So, whether we imagine a film stretched taut (Fig. 31.33A) or bowed out (Fig. 31.33B), we find the same value for the magnetic field.

For the taut film (Fig. 31.33A), the general form of Ampère's law is

$$\oint \vec{B} \cdot d\vec{\ell} = \mu_0 I + 0 \tag{31.11}$$

where $I_{\text{thru}} = I$ and there is (almost) no electric flux through the taut film. For the bowed-out film (Fig. 31.33B), there is no real current through the film, so $I_{\text{thru}} = 0$. Maxwell–Ampère's law becomes

$$\oint \vec{B} \cdot d\vec{\ell} = 0 + \mu_0 \varepsilon_0 \frac{d\Phi_E}{dt}$$

The second term on the right includes the displacement current (Eq. 31.9):

$$\oint \vec{B} \cdot d\vec{\ell} = 0 + \mu_0 I_d$$

and the displacement current equals the real current in the wire:

$$\oint \vec{B} \cdot d\vec{\ell} = 0 + \mu_0 I \tag{31.12}$$

This is exactly what we found (Eq. 31.11) for the taut film (Fig. 31.33A).

What about the direction of the magnetic field produced by the displacement current? The displacement current points in the same direction as the electric field if the electric flux is increasing. If the electric flux is decreasing, the displacement current is in the opposite direction as the electric field. To find the direction of the magnetic field, use the simple right-hand rule as you would for a real current. Point your right thumb in the direction of the displacement current, and your fingers naturally wrap in the direction of the magnetic field.

Finally, it is sometimes helpful to work with the **displacement current density**. Like the real current density, the displacement current density J_d is the displacement current divided by the cross-sectional area A:

$$J_d = \frac{I_d}{A} = \frac{1}{A}\left(\varepsilon_0 \frac{d\Phi_E}{dt}\right) = \frac{1}{A}\left[\varepsilon_0 \frac{d(EA)}{dt}\right]$$

$$J_d = \varepsilon_0 \frac{dE}{dt} \tag{31.13}$$

Newton and Maxwell both made major contributions in physics. Newton's laws are the basis of classical mechanics, and Maxwell's equations are the basis of electricity, magnetism, and light. Both scientists used their imagination to discover fundamental laws of nature. Newton imagined a world without friction, which enabled him to discover the law of inertia. Maxwell imagined a displacement current between the plates of a capacitor, allowing him to generalize Ampère's law and take a necessary step toward his greatest discovery that light is an electromagnetic phenomenon (Chapter 34).

EXAMPLE 31.12 CASE STUDY Magnetic Field Inside a Capacitor

A parallel-plate capacitor consists of two circular disks of radius R (Fig. 31.34). The capacitor is charging, and there is no dielectric between the plates.

A Find an expression for the magnetic field at point P between the plates of the capacitor. Point P is at a distance r from the axis, where $r < R$. The current in the wire is I.

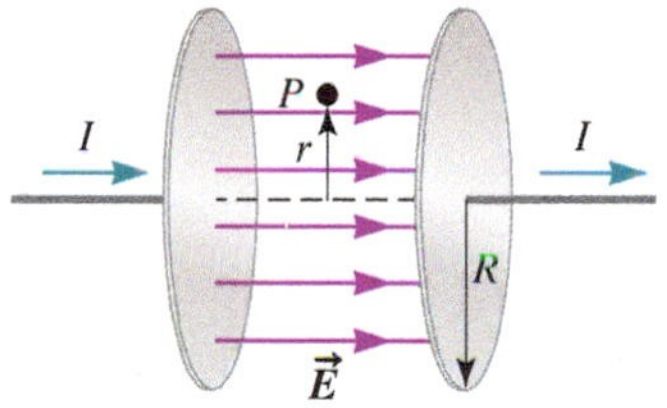

FIGURE 31.34

INTERPRET and ANTICIPATE

The same four steps we used to find the magnetic field with Ampère's law can be applied when using Ampère–Maxwell's law, provided we find the total (real and displacement) current through the Ampèrian loop.

Steps 1 and 2 Sketch the magnetic field lines, and **choose an Ampèrian loop.**
An end-on view is convenient for showing the magnetic fields. Imagine rotating Figure 31.34 so that the electric field is directed out of the page (Fig. 31.35). Because the electric flux increases while the capacitor is charging, the displacement current is in the same direction as the electric field. Use the simple right-hand rule to find the direction of the magnetic field. Point your thumb in the direction of the displacement current (out of the page); then your fingers wrap counterclockwise. To keep the figure clean, only one magnetic field line is shown in Figure 31.35. A circular Ampèrian loop exploits the symmetry of the problem. The path is in the same direction as the magnetic field.

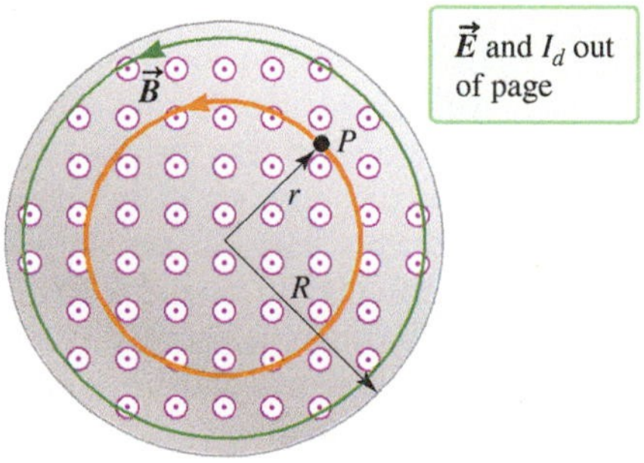

FIGURE 31.35

SOLVE

Step 3 Do the **circulation integral**. The magnetic field is parallel to the path and constant over the whole path. The path integral is just the circumference of the Ampèrian loop.

$$\oint \vec{B} \cdot d\vec{\ell} = \oint B\,d\ell = B\oint d\ell$$

$$\oint \vec{B} \cdot d\vec{\ell} = B(2\pi r) \tag{1}$$

Example continues on page 1002 ▶

Step 4 Find **current I_{tot}**. Imagine a soap film stretched taut across the circular Ampèrian loop in Figure 31.35. No wire penetrates the film, so there is no real current through the loop. However, electric field lines penetrate the loop. Because the electric field is increasing, there is a displacement current through the loop.	$I_{tot} = I_{thru} + I_d = 0 + I_d$ $I_{tot} = I_d$ (2)
The displacement current density J_d is the displacement current through the whole capacitor $(I_d)_{cap}$ divided by the area of the capacitor.	$J_d = \dfrac{(I_d)_{cap}}{\pi R^2}$
The displacement current through the capacitor equals the current in the wire in Figure 31.34. Write the displacement current density in terms of I.	$(I_d)_{cap} = I$ $J_d = \dfrac{I}{\pi R^2}$ (3)
The displacement current I_d *through the Ampèrian loop* is less than the displacement current through the whole capacitor $(I_d)_{cap}$. To find the displacement current through the Ampèrian loop, multiply the displacement current density J_d times the area A of the loop. Substitute Equations (3) and (2).	$I_d = J_d A = J_d \pi r^2 = \dfrac{I}{\pi R^2}\pi r^2$ $I_{tot} = I_d = I\dfrac{r^2}{R^2}$ (4)
Substitute Equations (1) and (4) into Ampère–Maxwell's law (Eq. 31.10), and solve for B.	$\oint \vec{B} \cdot d\vec{\ell} = \mu_0(I_{thru} + I_d) = \mu_0 I_{tot}$ $B(2\pi r) = \mu_0 I \dfrac{r^2}{R^2}$ $B = \dfrac{\mu_0 I}{2\pi}\dfrac{r}{R^2}$ (31.14)

CHECK and THINK

Equation 31.14 predicts that inside a charging capacitor, the magnetic field increases linearly with distance r from the axis. Measurements of such magnetic fields confirm Maxwell's modification to Ampère's law.

B Suppose you wish to confirm the displacement current term that Maxwell added to Ampère's law. You could use the deflection of compass needles to measure the magnetic field at several points between the plates of a large capacitor to see that it is given by Equation 31.14. The capacitor consists of two large circular plates of radius $R = 0.30$ m. You make your measurements using a very large current of 100 A. You have four compasses located at $r = 0$, 0.10 m, 0.20 m, and 0.30 m from the axis of the capacitor. Assume the magnetic field at these locations is perpendicular to the Earth's magnetic field. Find the deflection of the compass needles. Compare your results to the deflection of compass needles at the same distances due to a long, straight wire leading to the capacitor.

INTERPRET and ANTICIPATE

The goal of this part of the problem is to compare Equation 30.9, $B = (\mu_0 I)/(2\pi r)$, for the long, straight wire to Equation 31.14 for the region inside the capacitor. The latter must be derived with Maxwell's modification of Ampère's law. Ampère's law in its original form would conclude that the magnetic field inside the capacitor is zero.

SOLVE

The deflection angle comes from Equation 31.1. Substitute expressions for the magnetic fields produced by a long, straight wire (Equation 30.9) and by a capacitor, $B = (\mu_0 Ir)/(2\pi R^2)$. Substitute numerical values to complete Table 31.2.

r (m)	B (T), capacitor	θ (°), capacitor	B (T), wire	θ (°), wire
0	0	0	$\to\infty$	90
0.10	2.2×10^{-5}	24	2.0×10^{-4}	76
0.20	4.4×10^{-5}	42	1.0×10^{-4}	63
0.30	6.7×10^{-5}	53	6.7×10^{-5}	53

TABLE 31.2

$$\theta = \tan^{-1}\left(\frac{B}{B_\oplus}\right)$$

For a straight wire,

$$\theta = \tan^{-1}\left(\frac{\mu_0 I}{2\pi r B_\oplus}\right) = \tan^{-1}\left(\frac{0.4}{r}\right)$$

For a capacitor,

$$\theta = \tan^{-1}\left(\frac{\mu_0 I}{2\pi R^2 B_\oplus} r\right)$$

$$\theta = \tan^{-1}(4.44r)$$

CHECK and THINK

The last entry in the table shows that at the edge of the capacitor ($r = R$), the magnetic field produced by the displacement current equals the magnetic field produced by the real current. This is exactly what we would expect.

Compare the deflection angles. For the region between the plates of a charging (or discharging) capacitor, the deflection angle grows as we look at increasing distance because the magnetic field is strongest near the edge of the capacitor. The opposite is true for the long, straight wire because the magnetic field is strongest near the wire.

It might seem that Maxwell invented the displacement current just to fix a contradiction in Ampère's law and that his additional term does not correspond to any real phenomenon. However, experiments such as the one in Example 31.12 show that the displacement current term is a necessary part of Ampère's law. Maxwell's term predicts that there is nonuniform, nonzero magnetic field in the empty space between the capacitor's plates. The specific deflection angles predicted at each position in Example 31.12, part B, give us a way to test Maxwell's addition to Ampère's law. If Maxwell is correct about the displacement current, the magnetic field and the deflection angle should get larger as the compass is moved outward. Experiments have shown that Maxwell's addition of the displacement current to Ampère's law is correct. In fact, the displacement current is not only necessary but also fundamental because it led Maxwell to make a more important discovery about light (Chapter 34).

SUMMARY

Underlying Principles

Two of Maxwell's equations:

1. According to **Gauss's law for magnetism**, because there are no (known) magnetic monopoles, the magnetic flux through a Gaussian surface is zero. Mathematically, Gauss's law for magnetism is

$$\Phi_B = \oint \vec{B} \cdot d\vec{A} = 0 \quad (31.4)$$

The circle on the integral symbol in Equation 31.4 indicates that the integral is taken over the entire closed surface.

2. **Ampère's law** is expressed mathematically as

$$\oint \vec{B} \cdot d\vec{\ell} = \mu_0 I_{\text{thru}} \quad (31.5)$$

The general form of Ampere's law, known as **Ampère–Maxwell's law**, is

$$\oint \vec{B} \cdot d\vec{\ell} = \mu_0 I_{\text{thru}} + \mu_0 \varepsilon_0 \frac{d\Phi_E}{dt} = \mu_0 (I_{\text{thru}} + I_d)$$

(31.8) and (31.10)

Major Concepts

The **displacement current** is given by

$$I_d = \varepsilon_0 \frac{d\Phi_E}{dt} \quad (31.9)$$

The displacement current is a fictitious current; it represents not a real current but rather a changing electric field. The displacement current is a helpful way to think about how a changing electric field can create a magnetic field.

Special Cases

1. A **long, straight wire** exhibits linear symmetry. Ampère's law confirms that the magnetic field strength due to a long, straight, current-carrying wire is given by

$$B = \frac{\mu_0 I}{2\pi r} \quad (30.9)$$

2. The magnetic field inside an **ideal solenoid** is uniform; its strength is given by

$$B = \mu_0 n I \quad (31.6)$$

Outside the solenoid, the magnetic field is zero.

3. The magnetic field inside a **toroid** is given by

$$B = \frac{\mu_0 N I}{2\pi r} \quad (31.7)$$

Outside the toroid and inside the donut hole, the magnetic field is zero.

PROBLEM-SOLVING STRATEGY Finding the Magnetic Field Using Ampère's Law

INTERPRET and ANTICIPATE

1. **Sketch the magnetic field lines.**
2. **Choose an Ampèrian loop.**

SOLVE

3. Do the **circulation integral** in $\oint \vec{B} \cdot d\vec{\ell} = \mu_0 I_{\text{thru}}$ (Eq. 31.5).
4. Determine the amount of **current I_{thru}** through the Ampèrian loop.

PROBLEMS AND QUESTIONS

A = algebraic C = conceptual E = estimation G = graphical N = numerical

31-1 Measuring the Magnetic Field

1. N CASE STUDY Suppose you want to use a compass needle to measure the magnitude of a particular magnetic field. You plan to orient the field so it is perpendicular to the Earth's magnetic field. You have made a slight mistake, however, and instead the Earth's magnetic field makes an angle of 85° with respect to the field you want to measure. What is the percentage error in your observation?

Problems 2, 3, and 4 are grouped.

2. C **Review** Suppose you want to use a small, positively charged ball suspended by a light thread to map out the electric field in the space around a charged source. Describe the reaction of the ball as you place it at the two locations in front of the infinitely large, positively charged sheet shown in Figure P31.2A. Be sure to relate your description to what you know about the sheet's electric field.

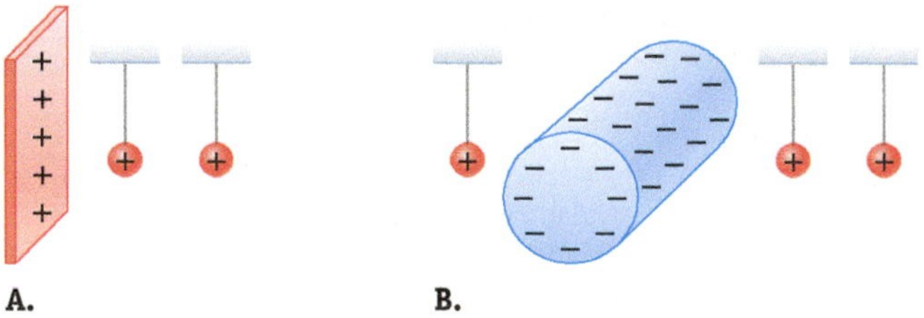

FIGURE P31.2 Problems 2 and 3.

3. C **Review** Suppose you want to use a small, positively charged ball suspended by a light thread to map out the electric field in the space around a charged source. Describe the reaction of the ball as you place it at the three locations in the space around the infinitely long, negatively charged rod shown in Figure P31.2B. Be sure to relate your description to what you know about the rod's electric field.

4. **C Review** Suppose you want to use a small, positively charged ball suspended by a light thread to map out the electric field in the space around a charged source. The deflection of the ball is used to find both the magnitude and direction of the electric field. What important role does the Earth's gravity play in your experiment?

Problems 5 and 6 are paired.

5. **N** In a laboratory exercise, you place a compass 2.00 cm from a wire as shown in Figure P31.5. Using a variable power supply, you slowly increase the current through the wire starting with $I = 0$. Assume the magnitude of the Earth's magnetic field is 5.0×10^{-5} T and that it points to the right in the figure. **a.** If you can detect a deflection of only 2°, what is the minimum current you can detect? **b.** Very large currents cause the compass needle to essentially align with the field due to the wire. Assuming you can't distinguish currents that lead to more than 88° of deflection, what is the largest current you can detect?

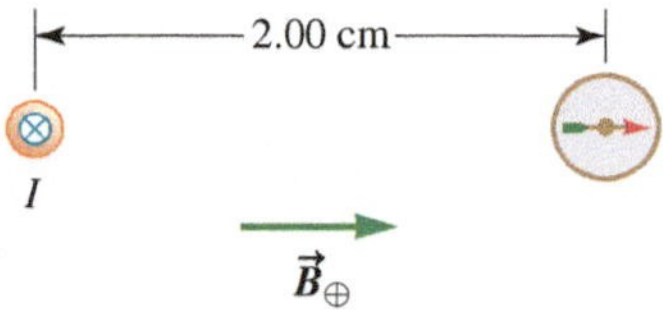

FIGURE P31.5 Problems 5 and 6.

6. **G** Plot the deflection angle of the compass needle in Problem 5 versus the current through the wire from 0 to 200 A. Is the compass needle a more sensitive measure for small currents or for large currents? Explain your answer.

31-2 Gauss's Law for Magnetism

7. **A** The magnetic field in some region is given by $\vec{B} = B_x\hat{\imath} + B_y\hat{\jmath}$. The areas A_x, A_y, A_z, B_x, and B_y are constants. What is the magnetic flux through each of these areas: **a.** $\vec{A} = A_x\hat{\imath}$, **b.** $\vec{A} = A_y\hat{\jmath}$, **c.** $\vec{A} = A_z\hat{k}$?

8. **C** Suppose the magnetic monopole is discovered. What change would be needed in Gauss's law for magnetism, if any? Write the new law, explaining your changes, if any.

9. **N** A loop of area 0.800 m² is in a uniform magnetic field $B = 0.652$ T. The magnetic flux through the loop is 0.240 Wb. What is the angle between the normal to the plane of the loop and the field?

10. **E** What is the Earth's magnetic flux through **a.** a basketball, **b.** a hula hoop standing up perpendicularly on its rim at the North Pole, and **c.** a hula hoop lying on the ground at the North Pole?

11. **N** A uniform magnetic field $\vec{B} = (0.0030\hat{\imath} + 0.0090\hat{\jmath})$ T passes through a cube with side length $s = 10.0$ cm as shown in Figure P31.11. The edges of the cube align with the x, y, and z axes. Calculate the flux through each of the six surfaces of the cube, and show that Gauss's law for magnetism (Eq. 31.4) holds.

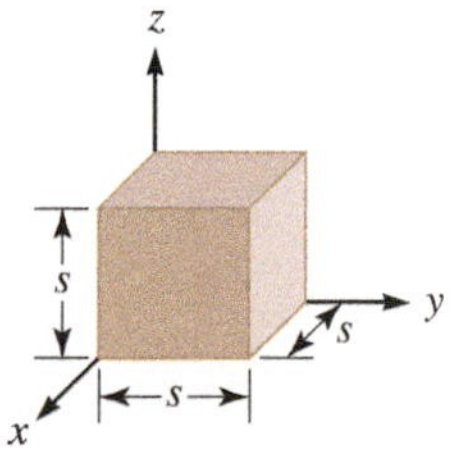

FIGURE P31.11

12. **N** A uniform magnetic field in a certain region of space has a magnitude of 0.25 T and points in a direction 30.0° from the positive x axis, 60.0° from the negative y axis, and 90.0° from the positive z axis. In each of the following cases, sketch the surface and determine the magnetic flux through the surface: **a.** $\vec{A} = 0.010\hat{\jmath}$ m² and **b.** $\vec{A} = 1.0\hat{k}$ m².

31-3 Ampère's Law

Problems 13, 14, 15, and 16 are grouped.

13. **G** Figure P31.13 shows a uniform magnetic field. **a.** Can you find an Ampèrian loop that gives a circulation integral of zero? If so, draw the loop and the field. If not, explain why not. **b.** Can you find an Ampèrian loop that gives a nonzero circulation integral? If so, draw the loop and the field. If not, explain why not.

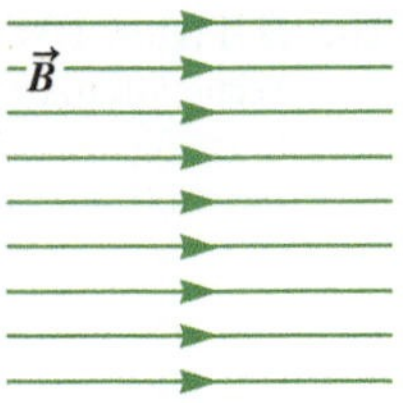

FIGURE P31.13 Problems 13 and 15.

14. **G** Figure P31.14 shows a magnetic field that is strongest near the center of the figure and weaker farther out. **a.** Can you find an Ampèrian loop that gives a circulation integral of zero? If so, draw the loop and the field. If not, explain why not. **b.** Can you find an Ampèrian loop that gives a nonzero circulation integral? If so, draw the loop and the field. If not, explain why not.

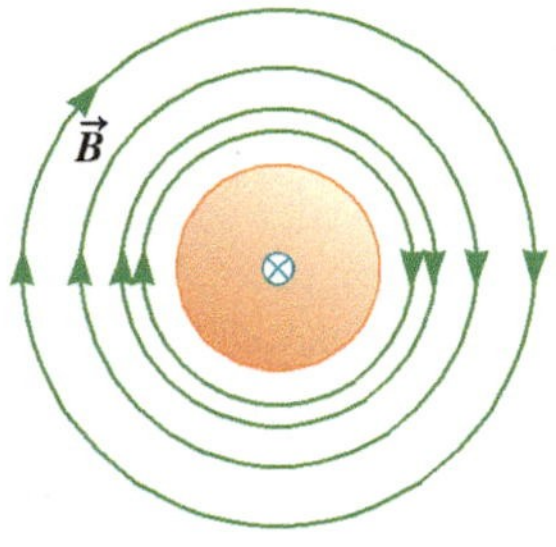

FIGURE P31.14 Problems 14 and 16.

15. **G** Figure P31.13 shows a uniform magnetic field. **a.** Can you find a (nonzero area) loop through which the magnetic flux is zero? If so, draw the loop and the field. If not, explain why not. **b.** Can you find a loop through which the magnetic flux is nonzero? If so, draw the loop and the field. If not, explain why not.

16. **G** Figure P31.14 shows a magnetic field that is strongest near the center of the figure and weaker farther out. **a.** Can you find a (nonzero area) loop through which the magnetic flux is zero? If so, draw the loop and the field. If not, explain why not. **b.** Can you find a loop through which the magnetic flux is nonzero? If so, draw the loop and the field. If not, explain why not.

17. **N** A 2.50 cm × 2.50 cm square Ampèrian loop exists in the xy plane in a region of space with a uniform magnetic field $\vec{B} = (1.25\hat{\imath} + 1.75\hat{\jmath})$ T. Two sides of the loop are parallel to the x axis, and two sides are parallel to the y axis. The integration path is such that side 1 is traversed in the positive x direction, side 2 in the negative y direction, side 3 in the negative x direction, and side 4 in the positive y direction. Calculate the contribution to the circulation integral due to each segment of the loop, and determine the net current through the loop that must be present.

Problems 18 and 19 are paired.

18. **N** Suppose the current in each wire shown in Figure P31.18 is $I = 0.50$ A. What is the result of the circulation integral around the Ampèrian loop in parts A and B of the figure?

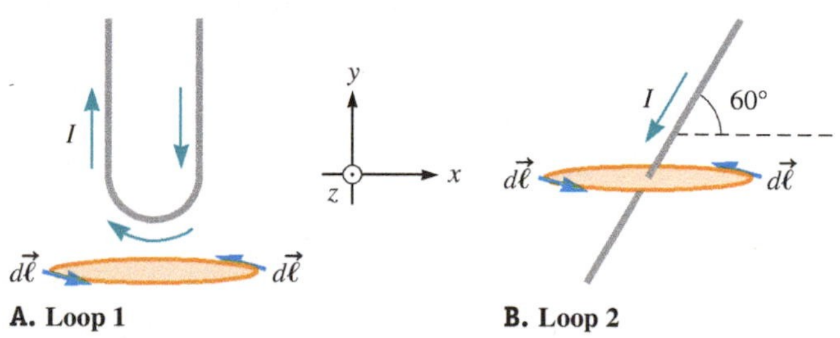

FIGURE P31.18 Problems 18 and 19.

19. **N** Suppose the circulation integral around Ampèrian loop 2 in Figure P31.18B is 3.142×10^{-7} T·m. What is the current through the loop?

20. **N** Figure P31.20 shows four current-carrying wires that are perpendicular to the page, where $I_1 = 1.0$ A, $I_2 = 2.0$ A, $I_3 = 3.0$ A, and $I_4 = 4.0$ A. Four Ampèrian loops are also shown. Find the current through each loop.

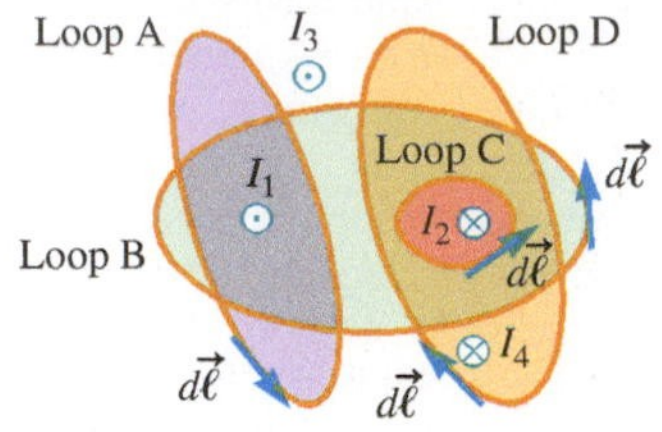

FIGURE P31.20

21. **A** A thin, infinite conducting sheet lies in the xy plane. It carries current per unit length J in the positive y direction. Find an expression for the magnitude of the magnetic field near the infinite sheet. (*Hint*: First, find the current through an Ampèrian loop perpendicular to the plane of the sheet using $I_{\text{thru}} = \int \vec{J} \cdot d\vec{\ell}$.)

Problems 22 and 23 are paired.

22. **A** A steady current I flows through a wire of radius a. The current density in the wire varies with r as $J = kr$, where k is a constant and r is the distance from the axis of the wire. Find expressions for the magnitudes of the magnetic field inside and outside the wire as a function of r. (*Hint*: Find the current through an Ampèrian loop of radius r using $I_{\text{thru}} = \int \vec{J} \cdot d\vec{A}$.)

23. **A** A steady current I flows through a wire of radius a. The current density in a wire varies with r as $J = kr^2$, where k is a constant and r is the distance from the axis of the wire. Find expressions for the magnitudes of the magnetic field inside and outside the wire as a function of r. (*Hint*: Find the current through an Ampèrian loop of radius r using $I_{\text{thru}} = \int \vec{J} \cdot d\vec{A}$.)

31-4 Special Case: Linear Symmetry

24. **C** CASE STUDY In Example 31.6, we discovered several problems in our experiment designed to measure the magnetic field due to a long, straight wire. How can this design be improved by more effectively using the same equipment?

25. **N** A magnetic field of 4.00 μT is measured at a distance of 25.0 cm from a long, straight wire with a current of 5.00 A. What is the distance from the wire at which a field of 0.500 μT will be measured?

26. **G** CASE STUDY A very long cylindrical wire has an insulating core as shown in Figure 31.17 (page 990). The core is surrounded by a cylindrical conducting shell with an inner and outer radius, carrying a current $I = 1.00$ A. The current is uniformly distributed in the conducting shell. Suppose the radius of the insulator and the inner radius of the shell is $r_1 = 0.00500$ m and the outer radius of the conductor is $r_2 = 0.0500$ m. Plot B versus r for this source. Compare your graph to that for a long, straight wire. *Hint*: Refer to Example 31.7 (page 990) and Concept Exercise 31.4 (page 988).

Problems 27, 28, and 29 are grouped.

27. A coaxial cable (sometimes called a "coax") is a long, layered cylindrical wire. You might find a coax attached to the back of your TV. Figure P31.27 is a cross-sectional view of a coax. The innermost layer is a long, straight conductor carrying current I into the page. The next layer is an insulator surrounded by another conductor carrying current I out of the page. The last layer is another insulator. The radius of each layer is labeled in the figure, and the current density is constant in each separate layer.

a. **A** Find expressions for the magnitudes of the magnetic field in each of the four layers.

b. **C** Compare your results to the magnetic field produced by a long, straight wire. Explain the advantage of using a coax.

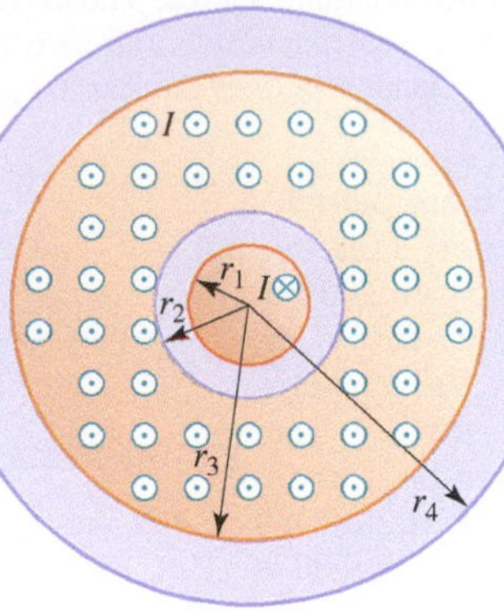

FIGURE P31.27 Problems 27, 28, and 29.

28. **G** Sketch a plot of the magnitude of the magnetic field as a function of position r for a coax (Fig. P31.27).

29. **N** A coaxial cable (Fig. P31.27) carries a current of 3.25 A. The radii are $r_1 = 0.500$ cm, $r_2 = 1.00$ cm, $r_3 = 1.50$ cm, and $r_4 = 2.00$ cm.

a. Where is the magnetic field strongest?

b. What is the magnitude of the magnetic field at this point?

30. The flow of ionic fluids can be viewed as a current, similar to electrons through a wire.

a. **A** If a fluid with charge density ρ (in C/m^3) is pumped through a cylindrical tube with radius R (in m) at speed v (in m/s), write an expression for the magnetic field versus the distance r from the center of the tube, both inside and outside the tube.

b. **G** Sketch the magnitude of the magnetic field versus distance from the center of the cylindrical tube (B vs. r), both inside and outside the tube.

Problems 31 and 32 are paired.

31. **N** Figure P31.31 shows a cylindrical conducting shell carrying current I to the right. The nonuniform current density in the conductor is given by $J = c(r - R_1)$ for the region $R_1 < r < R_2$, where c is a constant. What is the magnitude of the magnetic field at a radial distance r_A within the conductor? (*Hint*: Find the current through an Ampèrian loop of radius r using $I_{\text{thru}} = \int \vec{J} \cdot d\vec{A}$.)

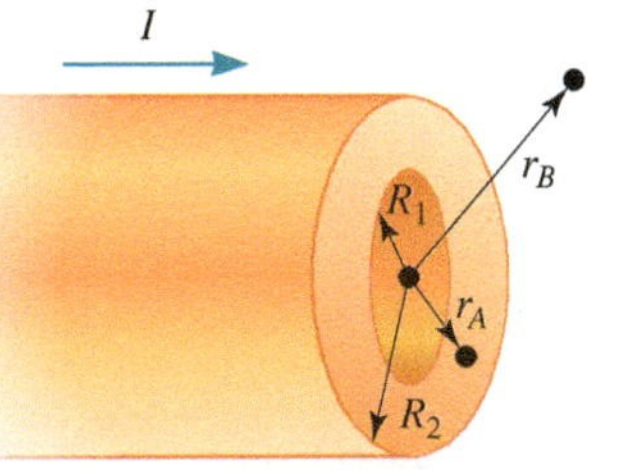

FIGURE P31.31 Problems 31 and 32.

32. **N** What is the magnitude of the magnetic field at a radial distance r_B from the center of a cylindrical conducting shell carrying current I (Fig. P31.31) in which the current density is given by $J = c(r - R_1)$ for the region $R_1 < r < R_2$, where c is a constant? (*Hint*: Find the current through an Ampèrian loop of radius r using $I_{\text{thru}} = \int \vec{J} \cdot d\vec{A}$.)

31-5 Special Case: Solenoids

33. **N** The magnitude of the magnetic field inside a very long solenoid is 1.0 T. If the current in the solenoid is 1.0 A, what is the number of turns in one meter?

34. **N** A solenoid of length 2.00 m and radius 1.00 cm carries a current of 0.100 A. Determine the magnitude of the magnetic field inside if the solenoid consists of 2000 turns of wire.

35. **N** The magnitude of the magnetic field at the center of a 25.0-cm-long solenoid with 650.0 turns is 2.00 mT. What is the current in the windings of the solenoid?

36. **N** **CASE STUDY** You want to measure magnetic fields in the range between 10 mT and 100 mT using a compass. You cannot use the Earth's magnetic field as a reference because it is too weak. Instead, you will use a solenoid to create a uniform magnetic field. If you want to keep the current smaller than 2.0 A, what is a reasonable value for the number of turns per unit length?

37. **N** A solenoid in your car is 33 cm long and has a diameter of 7.5 cm. It is made of copper wire with a radius of 0.51 mm. The turns are closely packed and form a single layer. Assume the resistivity of copper is $1.68 \times 10^{-8}\,\Omega\cdot\text{m}$. **a.** When the solenoid is connected to your battery with $\mathcal{E} = 11.6$ V, what is the magnetic field inside the solenoid? **b.** If, instead, the turns form two layers, how does your answer change?

Problems 38 and 39 are paired.

38. **N** The magnetic field inside a solenoid is 0.235 T. If the current is 4.53 A, how many turns are there per unit length?

39. **N** The magnetic field inside a solenoid 23.7 cm long is 0.235 T. If the current is 4.53 A and the diameter of the solenoid is 2.37 cm, what are the total number of turns and the length of the wire?

40. **N** A square conducting loop with side length $a = 1.25$ cm is placed at the center of a solenoid 40.0 cm long with 300 turns and aligned so that the plane of the loop is perpendicular to the long axis of the solenoid. According to an observer, the current in the single turn of the loop is 0.800 A in the counterclockwise direction, and the current in the windings of the solenoid is 8.00 A in the counterclockwise direction. **a.** What is the magnetic force on each side of the square loop? **b.** What is the net torque acting on the square loop?

41. **N** Aluminum wire 1.00 mm in radius is to be used to construct a solenoid 15.0 mm in radius so that the magnitude of the magnetic field at its center is 50.0 mT and the total resistance of the solenoid is 3.00 Ω when it carries a current of 1.00 A. How long does this solenoid need to be? The resistivity of aluminum is $2.655 \times 10^{-8}\,\Omega\cdot\text{m}$.

Problems 42 and 43 are paired.

42. **C** A straight current-carrying wire with current I_w resides inside a solenoid with n turns per unit length. A current I_s flows through the coils of the solenoid. **a.** Under what orientations will the wire experience no magnetic force? Explain your answer. **b.** Under what orientations will the wire experience a magnetic force? Explain your answer.

43. **A** A straight current-carrying wire with current I_w resides inside a solenoid with n turns per unit length. A current I_s flows through the coils of the solenoid. Find an expression for the magnitude of the maximum magnetic force per unit length the wire could experience inside the solenoid (the straight wire can be threaded through the coils of the solenoid, such that any angle is possible between the straight wire and the axis of the solenoid).

Problems 44 and 46 are paired.

44. **N** You have a spool of copper wire 5.00 mm in diameter and a power supply. You decide to wrap the wire tightly around a soda can that is 12.0 cm long and has a diameter of 6.50 cm, forming a coil, and then slide the wire coil off the can to form a solenoid. If the power supply can produce a maximum current of 125 A in the coil, what is the maximum magnetic field you would expect to produce in this solenoid? Assume the resistivity of copper is $1.68 \times 10^{-8}\,\Omega\cdot\text{m}$.

31-6 Special Case: Toroids

45. **N** The magnitude of the magnetic field inside a toroid with 1200 turns at a distance of 24 cm from the center is measured to be 0.020 T. Determine the current in the toroid.

46. **N** The solenoid in Problem 44 is wrapped into a donut to form a toroid. The solenoid (12.0 cm long) is wrapped as tightly as possible, so that the inner circumference of the toroid is the same as the length of the solenoid. What is the range of magnetic fields inside the toroid when the maximum current of 125 A is sent through the wire?

47. **N** A toroid with an inner radius of 50.0 cm and outer radius of 75.0 cm is wound with 760.0 turns of wire carrying a current of 955 A. What is the magnitude of the magnetic field along **a.** the inner radius of the toroid and **b.** the outer radius of the toroid?

48. **E** Today, nuclear power plants are based on fission reactions—reactions that split atomic nuclei. Another type of nuclear reaction is fusion, in which atomic nuclei are fused together. Fusion reactions power the Sun and may hold the key to Earth's power needs. JET (Joint European Torus) is a large device designed to study fusion reactions (Fig. P31.48). Inside JET's large donut-shaped apparatus, hydrogen plasma (a gas of free electrons and protons) is accelerated, reaching temperatures of about 100×10^6 K. Such a hot plasma is contained using a magnetic field. JET operates in short pulses, each several tens of seconds in duration. During a single pulse, JET uses 500 MW, about half of which goes to the toroid. The toroid gets very hot and must be water-cooled. The toroidal magnetic field keeps the plasma moving in its donut-shaped path. The maximum field strength is 4.5 T. If the magnetic field is produced by a 768-turn toroid, estimate the current in the toroid.

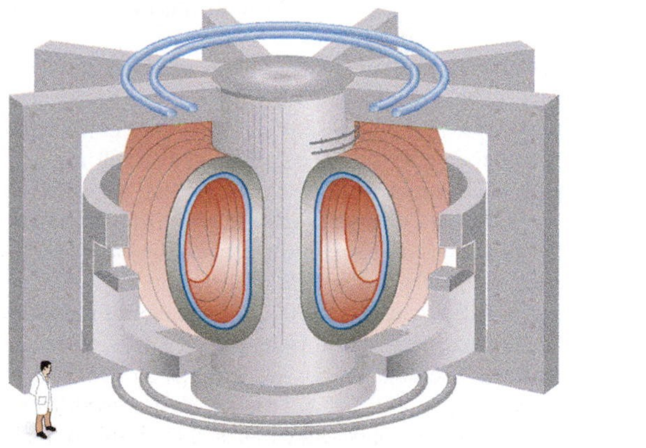

FIGURE P31.48 The Joint European Torus.

49. **N** A toroid with 150 turns of wire and inner and outer radii of 0.115 m and 0.295 m, respectively, carries a current of 1.46 A. At what location within the toroid is the magnitude of the magnetic field equal to half the value at the inner wall (the inner radius)?

Problems 50 and 51 are paired.

50. **N** The inner and outer radii of a toroid are 1.25 cm and 2.75 cm. The toroid has 64 turns and carries a current of 2.15 A. What are the strongest and weakest magnetic fields produced by this toroid?

51. **N** A toroid has an outer radius of 5.75 cm. The strongest magnetic field produced by the torus is 0.632 T and the weakest magnetic field is 0.316 T. What is the toroid's inner radius?

31-7 General Form of Ampère's Law

52. **C** In what way is the term *displacement current* misleading? In what way is the term useful?

53. **N** A parallel-plate capacitor consists of two rectangular plates, each of area $A = 36.7\ \text{cm}^2$. There is a vacuum between the plates, which are separated by 2.14 mm. When a current of 0.0935 A is charging the capacitor, find the rate at which the magnitude of the electric field is changing.

54. **N** A parallel-plate capacitor is made using circular plates of radius 10.0 cm separated by a distance of 0.50 cm. It is connected to a battery, and the voltage between the capacitor plates increases at a rate of 125,000 V/s. What is the displacement current between the capacitor plates?

Problems 55, 56, and 57 are grouped.

55. **C Review** A capacitor with circular plates of radius a separated by distance d is being charged by a battery with emf $\mathcal{E}$. A vacuum exists between the plates of the capacitor. The constructed circuit has resistance R in series with the capacitor, and the capacitor is initially uncharged when a switch is closed, completing the circuit and causing a current to flow (an RC series circuit). What happens to the magnitude of the magnetic field at a location between the plates at a distance r from the axis through the center of the plates, as time goes on while the capacitor is charging? Explain.

56. **A Review** A capacitor with circular plates of radius a separated by distance d is being charged by a battery with emf $\mathcal{E}$. A vacuum exists between the plates of the capacitor. The constructed circuit has resistance R in series with the capacitor, and the capacitor is initially uncharged when a switch is closed, completing the circuit and causing a current to flow (an RC series circuit). Find an expression for the magnitude of the magnetic field between the plates at a distance r from the axis through the center of the plates. Assume $r < a$.

57. **N Review** A capacitor with circular plates of radius 0.0534 m separated by a distance of 5.00×10^{-4} m is being charged by a battery with an emf of 12.0 V. A vacuum exists between the plates of the capacitor. The constructed circuit has a resistance of 325 Ω in series with the capacitor, and the capacitor is initially uncharged when a switch is closed, completing the circuit and causing a current to flow (an RC series circuit). Find an expression for the magnitude of the magnetic field between the plates at a distance of 0.0250 m from the axis through the center of the plates, when one time constant has passed since the switch was closed.

General Problems

58. **N** A 2.50 cm × 2.50 cm square Ampèrian loop exists in the xy plane in a region of space with a magnetic field. The magnetic field has a y component that varies along the x direction and an x component that varies along the y direction as described by $\vec{B} = (0.125y\hat{\imath} - 0.125x\hat{\jmath})\,\text{T}$. Calculate the contribution to the circulation integral due to each segment of the loop, and determine the net current through the loop that must be present.

59. **N** A uniform magnetic field $\vec{B} = 5.44 \times 10^4\hat{\imath}\ \text{T}$ passes through a closed surface with a slanted top as shown in Figure P31.59. **a.** Given the dimensions and orientation of the closed surface shown, what is the magnetic flux through the slanted top of the surface? **b.** What is the net magnetic flux through the entire closed surface?

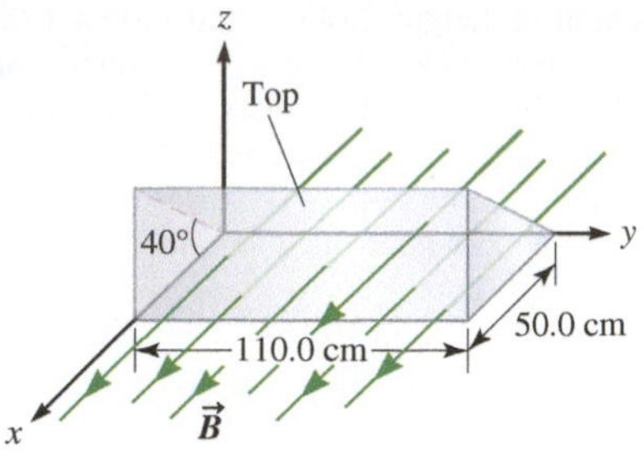

FIGURE P31.59

60. **N Review** Figure P31.60 shows four parallel wires arrayed at the indicated coordinates in the xy plane carrying identical currents of 8.00 A. The currents in wires A and B are in the positive z direction (out of the page), and the currents in wires C and D are in the negative z direction (into the page). What are the magnitude and direction of the magnetic field at the origin?

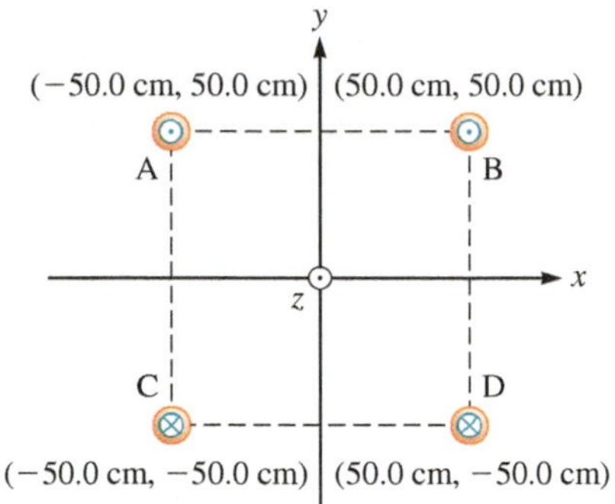

FIGURE P31.60

61. **N** A solenoid 1.25 m long with a current of 5.00 A in its windings produces a magnetic field at its center of 2.00 T. How many windings are in this solenoid?

62. **C** Review the new equations presented in this chapter. Use your own judgment to order the equations worth remembering according to their importance. Explain your reasoning.

63. **N** A current of 5.00 A flows through a 10.0-cm-long solenoid that has 55 turns. What is the magnitude of the field at the center of the solenoid?

64. **N** Figure P31.64 shows a cube with side length $d = 10.0$ cm with its center at the origin. The cube is subject to a uniform magnetic field $\vec{B} = (3\hat{\imath} - 5\hat{\jmath} - 2\hat{k})\,\text{T}$. **a.** What is the magnetic flux through the top face of the cube? **b.** What is the total magnetic flux through all six faces of the cube?

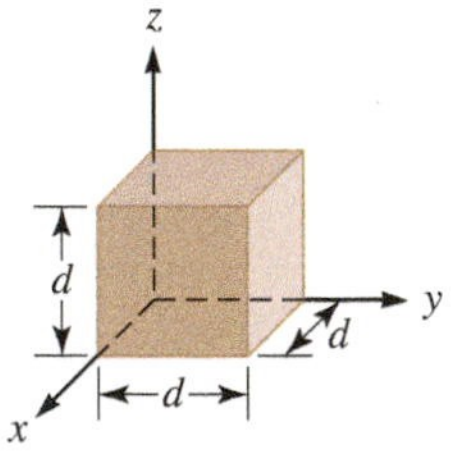

FIGURE P31.64

65. **N** A coaxial cable is constructed from a central cylindrical conductor of radius $r_A = 1.00$ cm carrying current $I_A = 8.00$ A in the positive x direction and a concentric conducting cylindrical shell with inner radius $r_B = 14.0$ cm and outer radius $r_C = 15.0$ cm with a current of 20.0 A in the negative x direction (Fig. P31.65). What are the magnitude and direction of the magnetic field **a.** at point O, a distance of 10.0 cm from the center of the coaxial cable along the y axis, and **b.** at point P, a distance of 20.0 cm from the center of the coaxial cable along the y axis?

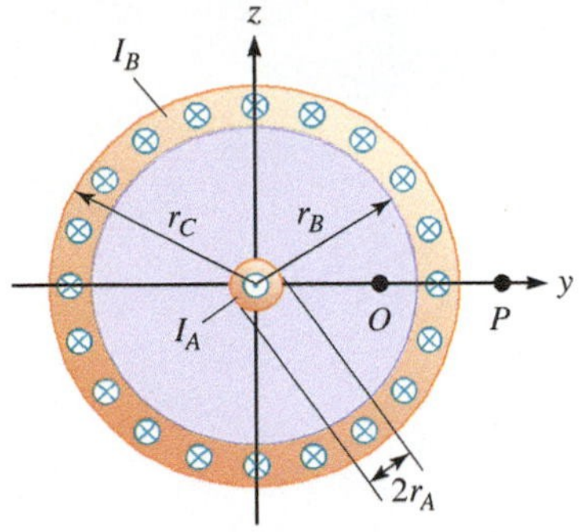

FIGURE P31.65

66. **N Review** In a laboratory vacuum chamber, a long, straight, horizontal wire with a current of 55.0 μA in the positive x direction is used to "levitate" a doubly ionized helium nucleus moving parallel to the wire with a speed of 5.10×10^5 m/s in the negative x direction. What is the distance above the wire where the magnetic force of the wire balances the gravitational force on the helium nucleus?

67. **N** Cross-sectional views of three long conductors carrying currents $I_1 = 2.0$ A, $I_2 = 1.0$ A, and $I_3 = 3.0$ A are shown in Figure P31.67. Determine the closed line integrals $\oint \vec{B} \cdot d\vec{\ell}$ around the five Ampèrian loops a through e as shown in the figure.

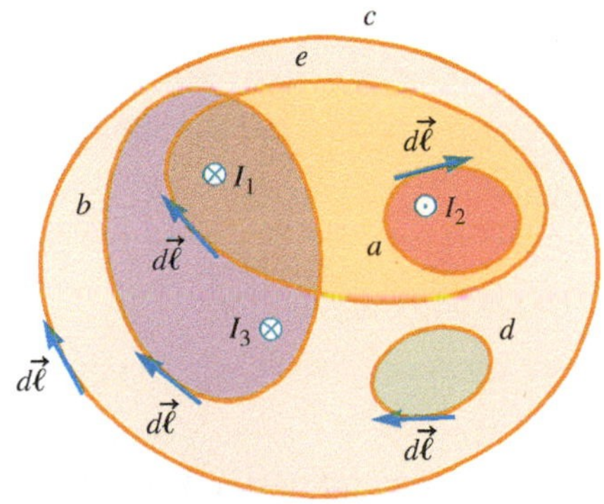

FIGURE P31.67

68. **N** A closed loop encloses several current-carrying wires with currents in different directions. The line integral $\oint \vec{B} \cdot d\vec{\ell}$ around the curve is 2.0×10^{-3} T·m. Determine the net current through the loop.

69. **N** Figure P31.69 shows a cross-sectional view of a 510-turn solenoid 50.0 cm in length and with radius $r = 2.00$ cm. The solenoid carries a current of 4.00 A. What is the magnetic flux through **a.** the circular area of radius $R = 3.50$ cm, and **b.** the gray-shaded annulus with $r_A = 5.00$ mm inner radius and $r_B = 15.0$ mm outer radius?

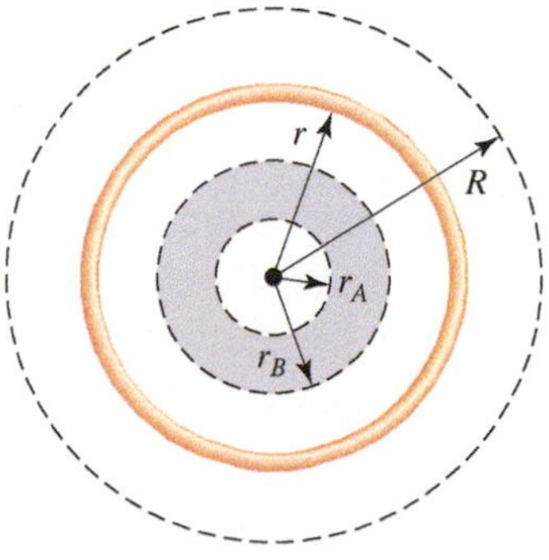

FIGURE P31.69

Problems 70 and 71 are paired.

70. **C** A square loop of wire with side length s carries current I_1 and resides inside a solenoid with n turns per unit length. The solenoid carries current I_2. Under what orientation does the square loop experience the maximum possible torque? Explain your answer.

71. **A** A square loop of wire with side length s carries current I_1 and resides inside a solenoid with n turns per unit length. The solenoid carries current I_2. Find an expression for the magnitude of the maximum possible torque experienced by the square loop of wire.

72. A straight wire carrying current I_1 resides inside a toroid, oriented along the radial direction (running from the inner wall to the outer wall). The toroid has N turns and carries current I_2. The inner and outer radii of the toroid are r_i and r_o, respectively.
    a. **A** Find an expression for the magnitude of the magnetic force per unit length on the straight wire at r_i.
    b. **A** Find an expression for the magnitude of the magnetic force per unit length on the straight wire at r_o.
    c. **C** Given your answers to parts (a) and (b), what will happen to the wire if it is subject to only these forces initially?

73. **A** An infinite slab with thickness extending from $z = -L$ to $z = L$ carries a uniform current density $\vec{J} = J\hat{\imath}$ as shown in Figure P31.73. Find an expression for the magnetic field as a function of z inside, below, and above the slab. (*Hint*: First, find the current through an Ampèrian loop perpendicular to the plane of the sheet using $I_{\text{thru}} = \int \vec{J} \cdot d\vec{A}$.)

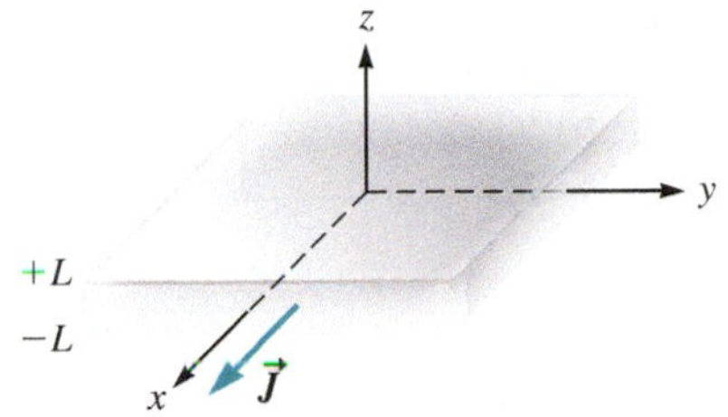

FIGURE P31.73

74. **A** Two long coaxial solenoids each carry current I but in opposite directions as shown in Figure P31.74. The inner solenoid of radius R_1 has n_1 turns per unit length, and the outer solenoid with radius R_2 has n_2 turns per unit length. Find expressions for the magnetic fields inside the inner solenoid, between the two solenoids, and outside both solenoids.

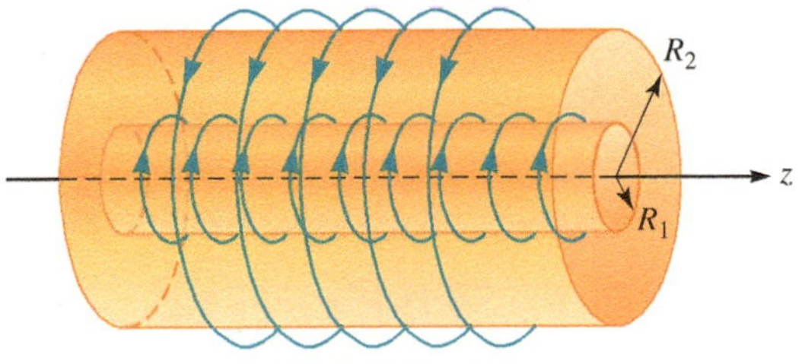

FIGURE P31.74

75. **A** A time-dependent current $I = I_0 \cos(\omega t)$ induces a time-dependent electric field

$$\vec{E}(s,t) = \frac{\mu_0 I_0 \omega}{2\pi} \sin(\omega t) \ln(R/s)\,\hat{k}$$

in a region of space. Obtain the total displacement current through the end of a cylinder with radius R that has its axis aligned with the z axis.

32 Faraday's Law of Induction

Key Questions

How can a magnetic field become a source of electric current?

How does Faraday's law lead to a practical method of generating electrical power?

Underlying Principles

1. Faraday's law, one of Maxwell's equations
2. Lenz's law

Major Concepts

Eddy current

Special Cases

1. Slide generator
2. AC generator
3. DC generator
4. Transformer

Michael Faraday was a great experimenter. After Hans Christian Ørsted discovered that electric currents are the source of magnetic fields, Faraday suspected that magnetic fields might be the source of electric currents because he believed the laws of nature are symmetrical. Many scientists look for symmetry in nature and are often rewarded with major new discoveries. Faraday's careful laboratory observations led him to discover a connection between magnetic fields and electric currents. His discovery is now known as **Faraday's law**.

Faraday saw great practical applications of his law and built the first electric generator. Today, the operation of all electric generators and electric motors is based on Faraday's law.

But that is not all. Faraday's law is our last of the four Maxwell's equations. Maxwell started with Faraday's law and Ampère's law and derived an equation showing that light is an electromagnetic wave. Faraday's law was a key component of Maxwell's major theoretical discovery that light is a combination of electricity and magnetism (Chapter 34). In this chapter, we focus on Faraday's law and its immediate practical application in our everyday lives.

32-1 Another Kind of Emf

There are two different kinds of electric field sources. We studied the first type, including sources with excess charge, such as the spheres and rods used by Ben Franklin and his contemporaries. We also studied the practical application of charged sources such as Leyden jars and batteries, which have excess positive charge on one terminal and excess negative charge on the other. It is convenient to characterize these sources in terms of the electric potential difference between their terminals, known as their *emf* (Section 29-1). As you know, a current results when the terminals of such an emf are connected by a conductor (Fig. 32.1).

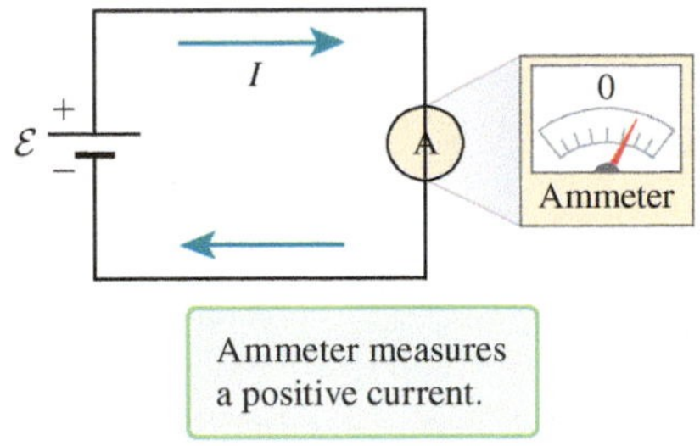
I
+
$\mathcal{E}$
−
A
0
Ammeter
Ammeter measures
a positive current.

FIGURE 32.1 A battery represents one type of source of an electric field. The top terminal has excess positive charge, and the bottom terminal has excess negative charge. The circuit includes an ammeter, and the ammeter shows there is positive (clockwise) current.

About four decades after Franklin died, the British physicist Michael Faraday (1791–1867) discovered another kind of electric field source. This source is a *changing magnetic flux* Φ_B. Like a battery, a changing magnetic flux creates current in a conductor. We usually use the phrases "the battery *sets up* a current" and "the changing magnetic flux *induces* a current."

Because Faraday's law is based on his laboratory observations, let's begin by considering a few experiments. Figure 32.2 shows a wire connected to an ammeter, but—unlike the wire in Figure 32.1—there is no battery connected to the wire. A solenoid passes through the rectangular loop with its long axis perpendicular to the page. The solenoid is connected to an external circuit (not shown) so that it creates a uniform magnetic field $\vec{B}$ pointing out of the page. Recall that outside an ideal solenoid, the magnetic field is zero. The magnetic flux through the rectangular loop is given by Equation 31.2

$$\Phi_B = \int \vec{B} \cdot d\vec{A} \quad (31.2)$$

$$\Phi_B = BA \quad (32.1)$$

where A is the cross-sectional area of the solenoid.

In Figure 32.2A, the current in the solenoid is constant, so the solenoid's magnetic field is steady. The magnetic flux through the rectangular loop (Eq. 32.1) is therefore constant. We observe no current in the rectangular loop; the ammeter needle remains stationary at 0. As you may expect, the magnetic field of the solenoid has no effect on the rectangular wire.

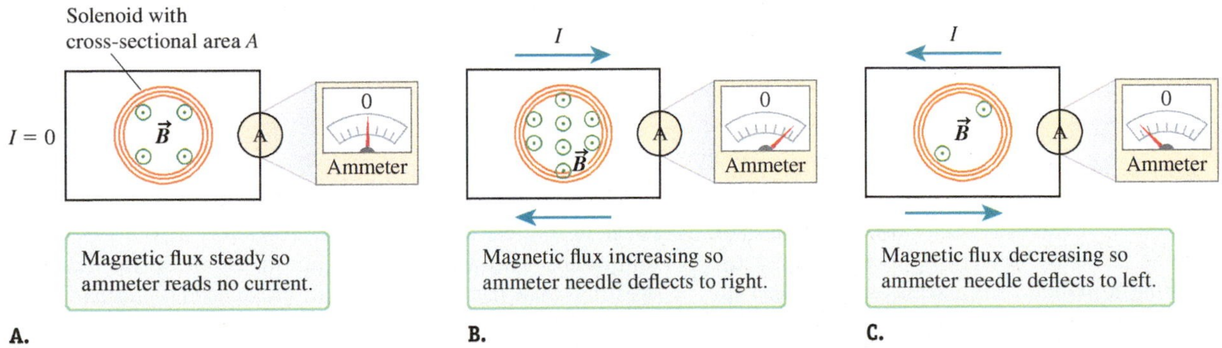
Solenoid with
cross-sectional area A
I = 0
$\vec{B}$
A
0
Ammeter
Magnetic flux steady so
ammeter reads no current.
A.
I
$\vec{B}$
A
0
Ammeter
Magnetic flux increasing so
ammeter needle deflects to right.
B.
I
$\vec{B}$
A
0
Ammeter
Magnetic flux decreasing so
ammeter needle deflects to left.
C.

FIGURE 32.2 A. A solenoid generates a magnetic field pointing out of the page. The magnetic field is steady, and there is no current in the wire. **B.** The magnetic field increases, changing the magnetic flux through the rectangular loop. A current appears in the rectangular loop, and the ammeter's needle deflects to the right as it did with a battery in the circuit (Fig. 32.1). **C.** The magnetic field decreases, again changing the magnetic flux through the rectangular loop. A current appears in the rectangular loop, but this time the ammeter's needle deflects to the left, as if you had reversed the battery in Figure 32.1.

The situation is very different when the solenoid's magnetic field is changing. If the external current in the solenoid increases, so that its magnetic field $\vec{B}$ increases, then according to $\Phi_B = BA$ (Eq. 32.1), the magnetic flux Φ_B also increases. In this case, a clockwise current appears in the rectangular loop and the ammeter's needle is deflected to the right (Fig. 32.2B). This is exactly what we found when a battery was in the circuit (Fig. 32.1). To the ammeter, there is no difference between having a battery in the circuit and having a changing magnetic flux through the circuit. This is true of any device you might put in the circuit. For example, if you replaced the ammeter with a lightbulb, the bulb would light up whether you used a battery (Fig. 32.1) or there was a changing magnetic flux (Fig. 32.2B).

What happens if the current in the solenoid decreases rather than increases? In that case, the magnetic field $\vec{B}$ still points out of the page, but it decreases (grows weaker) and the magnetic flux Φ_B also decreases. The decreasing magnetic flux causes a counterclockwise current in the loop, and so this time, the ammeter needle deflects to the left (Fig. 32.2C). This is exactly what would happen if we reversed the battery in Figure 32.1.

From the experiments in Figure 32.2, we may guess that a changing magnetic flux is required to produce a current in the loop. So far, we have imagined that the changing magnetic flux is produced by changing the magnetic field of a solenoid. A permanent magnet can also be used to create a changing magnetic flux. If you hold a bar magnet stationary near a wire loop, there is no current in the loop because the magnetic flux Φ_B through the loop does not change. However, if you move the bar magnet toward or away from the loop, the magnetic flux through the loop changes and a current appears. The direction of the current depends on which pole of the magnet is closest to the loop and whether you move the bar magnet toward or away from the loop. For example, if you point the north pole at the loop and move the magnet toward it, the ammeter needle in Figure 32.3A deflects to the left. If you move the magnet away from the loop as in Figure 32.3B, the ammeter needle deflects to the right.

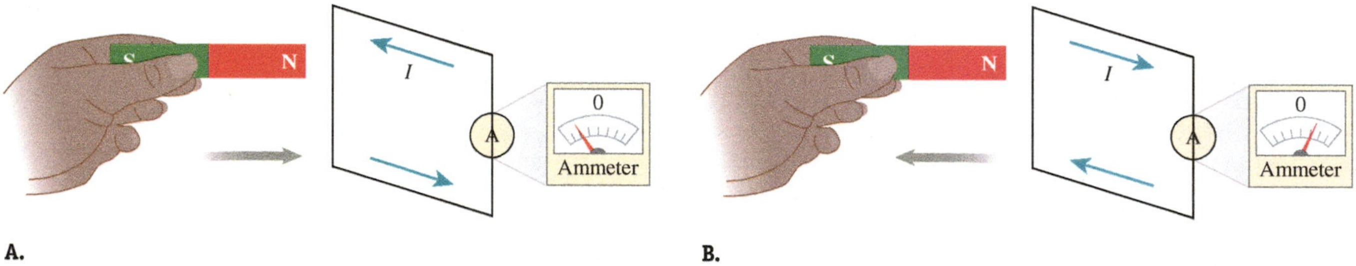

FIGURE 32.3 **A.** A bar magnet is pushed toward the loop as shown, and the ammeter needle deflects to the left. **B.** The bar magnet is pulled away from the loop, and the ammeter needle deflects to the right.

CASE STUDY AC/DC: The War of Currents

A changing magnetic flux (and Faraday's law) is the basis of the AC (alternating current) power supplies that you use every time you plug a device into a wall outlet. When the electrical infrastructure was being developed in the late 1800s, however, AC power was controversial. Some people supported developing a DC (direct current) infrastructure, while others recommended using AC. The debate was so controversial that it is described as a *war* over current. In this case study, we look at the roles played by three great inventors—Thomas Edison, Nikola Tesla, and George Westinghouse—in this war.

Edison—probably best known for inventing the incandescent bulb—believed DC was superior because it was safer. He built many steam-powered DC generators. Tesla and Westinghouse favored AC, but neither one was as famous in the field of electricity as Edison. Tesla was a young Serbian immigrant who worked as a junior engineer for Edison. Westinghouse had invented the air brake used on trains. Tesla believed DC generators and motors failed more often than their AC counterparts.

Another argument involved the location of generators. Edison's DC generators had to be located in metropolitan areas, so they had to be powered by steam engines. Both Tesla and Westinghouse recognized that AC generators could be located far from cities and so could be powered by waterfalls such as Niagara Falls (Fig. 32.4).

By the late 1880s, Edison's DC generators were used in major cities to light homes and businesses and to run motors. Westinghouse had built many AC generators, but at that time only DC motors were commercially viable, so his AC

generators were used only for lighting. In 1888, Edison wrote an 84-page book entitled *WARNING,* in which he cautioned the public not to invest in AC power. He said AC power was not as useful as DC power because there were no AC motors and AC power was deadly.

The public was moved by safety concerns. Harold Brown, an independent New York engineer, proposed a law limiting AC power to 300 V. If such a law had been passed, Westinghouse would have had to close all his operations. In this chapter, we explore the technological developments that led to the AC infrastructure we use today.

FIGURE 32.4 George Westinghouse used Niagara Falls to generate AC power. The power was used in Buffalo, New York, more than 20 miles away. This photo shows the current dam built in the 1960s.

CONCEPT EXERCISE 32.1

To calculate the magnetic flux through the rectangular loop in Figure 32.2, we used the cross-sectional area A of the solenoid in $\Phi_B = BA$ (Eq. 32.1). Why didn't we use the area of the rectangular loop?

32-2 Faraday's Law

The experiments described in Section 32-1 show that a changing magnetic flux may induce a current in a conductor, like the current set up by a battery. Both are measured by an ammeter, and both can be used to light a bulb. Just as it is convenient to characterize a battery in terms of the emf between its terminals, it is convenient to characterize a changing magnetic flux in terms of the emf it produces. One statement of **Faraday's law** is that *a changing magnetic flux through a conductor induces an emf in that conductor.* For a single conducting loop, Faraday's law is expressed mathematically as

FARADAY'S LAW

Underlying Principle

$$|\mathcal{E}| = \left|\frac{d\Phi_B}{dt}\right| \tag{32.2}$$

where Φ_B is the magnetic flux through the loop (Eq. 31.2), and $|\mathcal{E}|$ is the absolute value of the emf induced in the conductor. (In the next section, we'll learn how to find the sign of the emf, which allows us to remove the absolute-value signs from Equation 32.2.)

Let's consider a simple example of a uniform magnetic field and a single rectangular loop (Fig. 32.5). To calculate the magnetic flux, we need to find the direction of the loop's area vector. Recall that the area vector is perpendicular to the loop, so it can point either up and to the right or down and to the left (Section 25-2). Although either choice is acceptable, choosing the direction that is closest to the magnetic field is convenient and the one we make in this book—in this case, up and to the right (Fig. 32.5). The magnetic field $\vec{B}$ and the angle φ between $\vec{A}$ and $\vec{B}$ are constants. So Equation 31.2 for the magnetic flux becomes

$$\Phi_B = \int (B)(dA)\cos\varphi = B\cos\varphi \int dA$$

The integral is just the area A of the loop:

$$\Phi_B = BA\cos\varphi \tag{32.3}$$

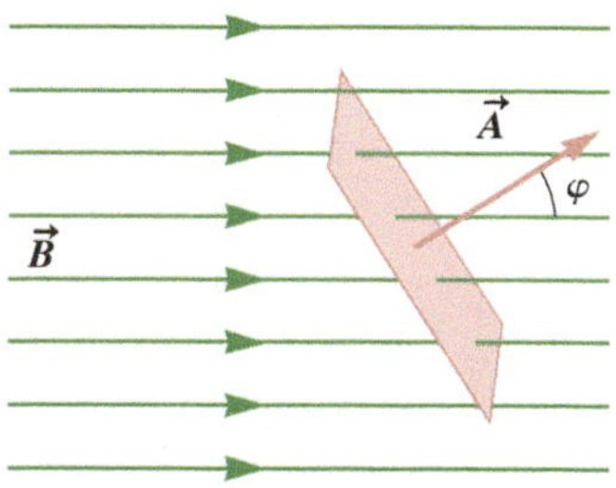

FIGURE 32.5 A rectangular conducting loop is perpendicular to the page and tilted with respect to the uniform magnetic field.

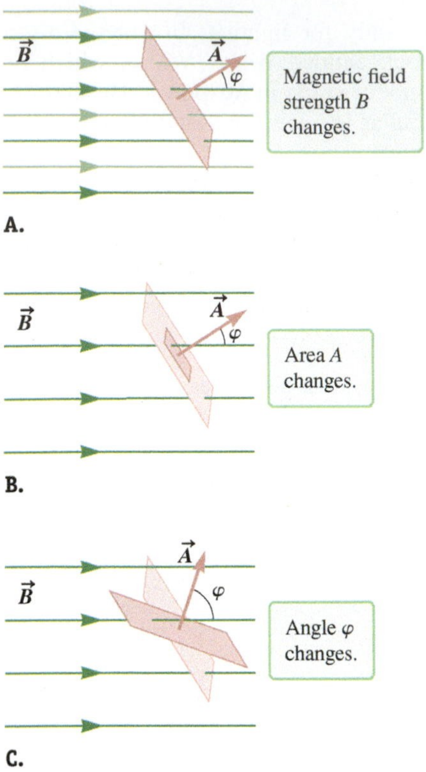

FIGURE 32.6 Three ways for the magnetic flux through the loop to change: **A.** The magnetic field gets weaker. **B.** The loop (made of a flexible material) shrinks. **C.** The loop rotates counterclockwise.

To find the absolute value of the emf induced in the loop, substitute Equation 32.3 for Φ_B into Faraday's law (Eq. 32.2):

$$|\mathcal{E}| = \left|\frac{d\Phi_B}{dt}\right| = \left|\frac{d(BA\cos\varphi)}{dt}\right|$$

Because of the time derivative in this expression, if B, A, and φ are all constants in time, no emf is induced in the loop. At least one of these parameters must change in order to induce an emf (Fig. 32.6). According to Faraday's law, during any one of these changes, an emf is induced in the loop. If the changes stop happening, the magnetic flux becomes constant and there is no longer an emf.

The changes in Figure 32.6 all show a reduction in the magnetic flux. Perhaps it is obvious from the linear dependence of B and A in $\Phi_B = BA\cos\varphi$ (Eq. 32.3) that a weaker magnetic field B and a smaller loop area A both reduce Φ_B. Because of the cosine dependence on φ, the magnetic flux Φ_B is greatest when the area vector is lined up with the magnetic field so that $\varphi = 0$. When the loop rotates counterclockwise, thus increasing φ, the magnetic flux is reduced (Fig. 32.6C). If the loop keeps rotating counterclockwise, the magnetic flux will become negative for the angle $90° < \varphi < 270°$, and the flux will be positive and increase again as the area vector moves between $\varphi = 270°$ and $360°$. Changing the angle between the area vector and the magnetic field is an important way of generating AC and DC power, so it is important to the case study.

In Figure 32.6, we imagined an emf induced in a single loop. If we replace the single loop with a coil of N turns, each turn has an induced emf. The turns are in series with one another, so the total induced emf is the sum of the individual emfs in each loop, or N times the emf in each loop. Faraday's law for a coil of N turns comes from multiplying Faraday's law for a single loop by N:

$$|\mathcal{E}| = N\left|\frac{d\Phi_B}{dt}\right| \quad (32.4)$$

where $|\mathcal{E}|$ is the (absolute value of the) net emf induced in the coil.

It is sometimes convenient to use the term **flux linkage** for the combination $N\Phi_B$. Then we can state Faraday's law (Eq. 32.4) in terms of the flux linkage: *The emf induced in a coil is equal to the time rate of change of the flux linkage.*

CONCEPT EXERCISE 32.2

Figure 32.7 shows three cases involving a single, stationary circular loop of wire. For each case, decide whether an emf is induced in the circular loop. Explain your reasoning.

Case 1: A bar magnet twists back and forth near the loop.

Case 2: A bar magnet swings toward and away from the loop but never penetrates it.

Case 3: A bar magnet is at rest so that one end penetrates the loop.

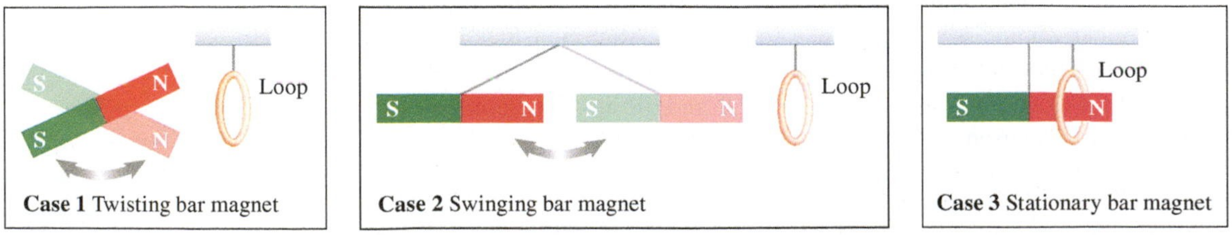

FIGURE 32.7

EXAMPLE 32.1 Lighting a Flashlight the Hard Way

A small coil has area $A = 2.83 \times 10^{-3}\ \text{m}^2$ and $N = 2550$ turns. A small bulb is attached to the coil, and the coil rests inside a long solenoid with $n = 3.65 \times 10^5$ turns per meter. The axis of the coil is aligned with the solenoid's axis. The current in the solenoid increases according to $I = \frac{1}{2}t$, where I is in amperes and t is in seconds. Find the absolute value of the emf induced in the coil, and describe the appearance of the bulb.

INTERPRET and ANTICIPATE

Sketch the problem as in Figure 32.8. In this case, the area vector is aligned with the magnetic field, so $\varphi = 0$. To use Faraday's law, first find an expression for the magnetic flux Φ_B through the loop (Eq. 31.2). Often this can be done by inspection. After finding Φ_B, substitute it into Faraday's law (Eq. 32.4) to find the induced emf. We expect to find a numerical value in volts.

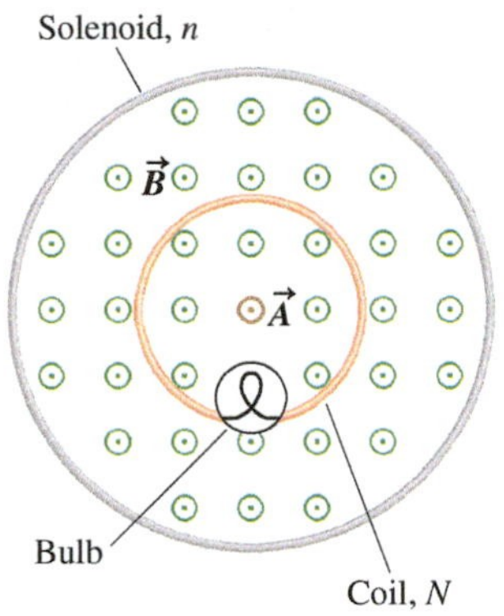

FIGURE 32.8 Cross-sectional view of the solenoid and coil.

SOLVE

Use Equation 31.2 with $\varphi = 0$. The coil is in the uniform magnetic field of the solenoid. The result looks familiar (Eq. 32.1). We find this result whenever a loop is aligned with a uniform magnetic field.

$$\Phi_B = \int \vec{B} \cdot d\vec{A} = \int B(\cos 0)dA \qquad (31.2)$$

$$\Phi_B = B\int dA = BA \qquad (32.1)$$

Because the current in the solenoid is changing, write the magnetic flux in terms of that changing current. The solenoid's magnetic field is given by Equation 31.6, and the current is $I = \frac{1}{2}t$. Substitute both of these expressions in the flux equation (Eq. 32.1).

$$B = \mu_0 nI = \mu_0 n\left(\tfrac{1}{2}t\right) \qquad (31.6)$$

$$\Phi_B = \left[\mu_0 n\left(\tfrac{1}{2}t\right)\right]A \qquad (1)$$

Substitute the magnetic flux (Eq. 1) in Faraday's law (Eq. 32.4).

$$|\mathcal{E}| = N\left|\frac{d\Phi_B}{dt}\right| = N\left|\frac{d\left[\mu_0 n\left(\frac{1}{2}t\right)\right]A}{dt}\right| \qquad (32.4)$$

Pull the constants outside the derivative and rearrange terms. In this case, the derivative is 1, so the emf is constant.

$$|\mathcal{E}| = \tfrac{1}{2}\mu_0 nNA\left|\frac{dt}{dt}\right| = \tfrac{1}{2}\mu_0 nNA$$

Substitute numerical values. This result means the bulb is lit with a constant brightness much as it is when connected to the batteries in your flashlight.

$$|\mathcal{E}| = \left(\tfrac{1}{2}\,\text{A/s}\right)(4\pi \times 10^{-7}\ \text{T}\cdot\text{m/A})(2550)(3.65 \times 10^5\ \text{m}^{-1})(2.83 \times 10^{-3}\ \text{m}^2)$$

$$|\mathcal{E}| = 1.66\,\frac{\text{T}\cdot\text{m}^2}{\text{s}}$$

CHECK and THINK

One way to check our results is to make sure the answer has the expected SI units—volts. The tesla is defined in Section 30-4 as $1\ \text{T} \equiv 1\ (\text{N}\cdot\text{s})/(\text{C}\cdot\text{m})$. A joule is a newton-meter, and a volt is a joule per coulomb (Section 26-4). So, our result is in volts as expected.

To help with future applications of Faraday's law, recall from Equation 31.3 that the SI unit of magnetic flux is the weber (Wb). So, a weber per second is a volt.

$$1\,\frac{\text{T}\cdot\text{m}^2}{\text{s}} = 1\left(\frac{\text{N}\cdot\text{s}}{\text{C}\cdot\text{m}}\right)\left(\frac{\text{m}^2}{\text{s}}\right) = 1\,\frac{\text{N}\cdot\text{m}}{\text{C}}$$

$$1\,\frac{\text{T}\cdot\text{m}^2}{\text{s}} = 1\,\frac{\text{J}}{\text{C}} = 1\ \text{V}$$

$$1\,\frac{\text{Wb}}{\text{s}} = 1\ \text{V} \qquad (32.5)$$

32-3 Lenz's Law

The experiment in Figure 32.2 shows that a changing magnetic flux induces an emf in the loop, which we observe by measuring the current. If the conductor has resistance R, the current is given by

$$I = \frac{\mathcal{E}}{R}$$

Because the induced current and the induced emf are in the same direction, it is common to use the phrases "direction of the induced current" and "direction of the induced emf" interchangeably.

where the emf is found from Faraday's law. Although current and electric potential are both scalars, a direction is associated with them. Current is the flow of positively charged particles from high potential to low potential. We think of an emf as pointing in the same direction as the current. For example, the current in Figure 32.2B is clockwise, so we say the emf induced in the loop is also clockwise. Likewise, the current in Figure 32.2C is counterclockwise, so the emf in the loop is also counterclockwise. The Russian physicist Heinrich Friedrich Emil Lenz (1804–1865) discovered a rule—known as **Lenz's law**—for finding the direction of the induced emf.

Observed Current Directions

Let's see what can be deduced from the four simple experiments in Figure 32.9. The first two experiments (parts A and B of Fig. 32.9) are the same as those in Figure 32.2, with the magnetic field out of the page. The second two experiments (parts C and D of Fig. 32.9) are similar, but with the solenoid's magnetic field into the page. We have listed three facts about each experiment—the direction of the magnetic field $\vec{B}$, the direction of the change in the magnetic field $d\vec{B}/dt$, and the direction of the observed current.

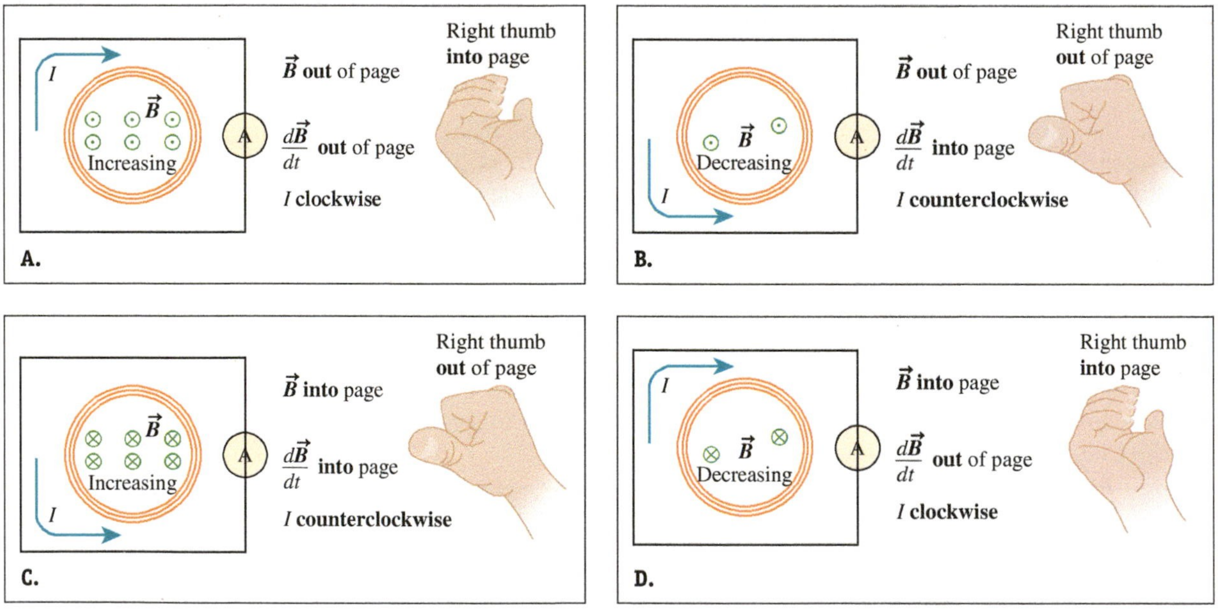

FIGURE 32.9 Lenz's law experiments. Your thumb always points in the direction opposite $d\vec{B}/dt$. An ammeter in each circuit indicates the direction of the current we observe. Notice that there is no correlation between your thumb's direction and that of the magnetic field $\vec{B}$.

Why is the direction of $d\vec{B}/dt$ important? From Faraday's law, the change in the magnetic field is what induces the current in these cases. The direction of $d\vec{B}/dt$ depends on the direction of the magnetic field $\vec{B}$ and whether that field is increasing or decreasing. When the magnetic field is increasing, $d\vec{B}/dt$ is in the same direction as $\vec{B}$. When the magnetic field is decreasing, $d\vec{B}/dt$ is in the direction opposite $\vec{B}$. You should verify this for the four experiments in Figure 32.9.

Another Right-Hand Rule

The observed current is in the plane of the page circulating either clockwise or counterclockwise, whereas $\vec{B}$ and $d\vec{B}/dt$ are perpendicular to the page. Again, a right-hand rule helps associate the direction of the current with a direction either into or out of the page. We can begin with the same right-hand rule we used to find the magnetic moment of a loop or the magnetic field produced by a loop near its center (Fig. 30.25, page 948): Wrap the fingers of your right hand in the direction of the current; then your thumb points in the direction of the loop's magnetic moment and the magnetic field that the loop produces along an axis passing through its center.

In Figure 32.9, the fingers of the right hand shown in each part wrap in the direction of the observed induced current. There is a correlation between the thumb directions and the change in the magnetic field's direction $d\vec{B}/dt$. The thumbs always point in the direction opposite $d\vec{B}/dt$. (Check the experiments in Figure 32.9 for yourself.) So, by studying the relationships in the figure, we can come up with a rule for finding the direction of the induced current. Point your right thumb in the direction *opposite* $d\vec{B}/dt$, and your fingers wrap in the direction of the induced current (and the induced emf). This is very close to what Lenz deduced, although his law is stated in more general terms.

LENZ'S LAW

Underlying Principle

According to **Lenz's law**, *the induced current produces a magnetic field that always acts to oppose the change in the magnetic flux that created the induced current.* Let's see how the law fits a couple of the experiments in Figure 32.9. In Figure 32.9A, the solenoid's magnetic field $\vec{B}_{sol}$ points out of the page through the loop and is increasing. Because the magnetic field is changing, a current is induced in the rectangular loop. The induced current is the source of a second magnetic field $\vec{B}_{loop}$. Near the center of the loop, the second magnetic field $\vec{B}_{loop}$ points either out of the page or into the page depending on the direction of the induced current. According to Lenz's law, $\vec{B}_{loop}$ *opposes the change in* $\vec{B}_{sol}$. Because $d\vec{B}_{sol}/dt$ points out of the page, $\vec{B}_{loop}$ must point into the page. When you point your thumb in the direction *opposite* $d\vec{B}/dt$, you are also pointing your thumb in the direction of $\vec{B}_{loop}$—the magnetic field created by the induced current. That's why your fingers correctly indicate the direction of the induced current.

As a second example, consider the experiment in Figure 32.9D. The solenoid's magnetic field $\vec{B}_{sol}$ points into the page and is decreasing. To oppose this change in $\vec{B}_{sol}$, the loop's magnetic field $\vec{B}_{loop}$ near the center of the loop must also point into the page. Point your right thumb in the direction of $\vec{B}_{loop}$ (into the page). Your fingers then wrap clockwise in the direction of the induced current. It may help to remember this: **If $\vec{B}_{sol}$ is decreasing, then $\vec{B}_{loop}$ is in the same direction as $\vec{B}_{sol}$, but if $\vec{B}_{sol}$ is increasing, then $\vec{B}_{loop}$ is in the direction opposite $\vec{B}_{sol}$.**

In our discussion of Lenz's law so far, we have focused on a changing magnetic field. There are two other ways for the magnetic flux Φ_B to change: The area A can change or the angle φ can change. How should we apply Lenz's law in these two cases?

First, consider a loop with a changing area in a uniform and constant magnetic field $\vec{B}_{sol}$. The key to finding the direction of the induced current is to first figure out the direction of the loop's magnetic field $\vec{B}_{loop}$. If the loop area **is shrinking,** the magnetic flux Φ_B is decreasing, so $\vec{B}_{loop}$ must be in the **same direction** as the original magnetic field $\vec{B}_{sol}$ (Fig. 32.10). If the area is **growing** instead, the magnetic flux Φ_B is increasing, so the induced current creates a magnetic field $\vec{B}_{loop}$ in the **opposite direction** as $\vec{B}_{sol}$. Once you know the direction of $\vec{B}_{loop}$, point your right thumb in that direction and your fingers wrap or point in the direction of the induced current (and induced emf).

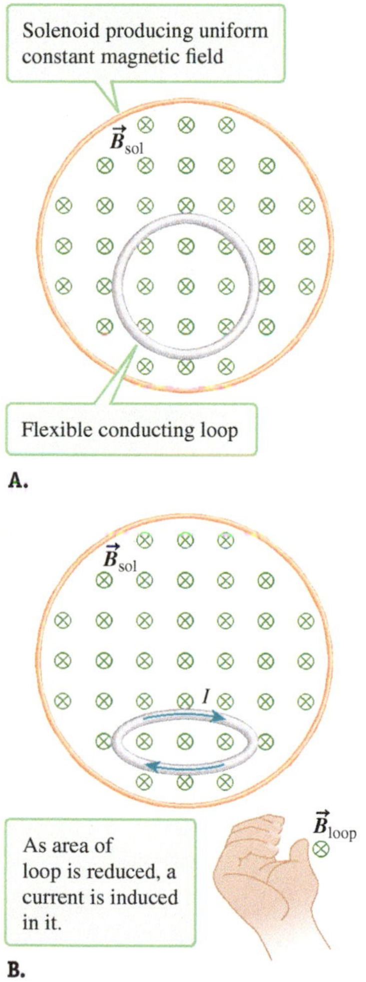
Solenoid producing uniform constant magnetic field
$\vec{B}_{sol}$
Flexible conducting loop
A.
$\vec{B}_{sol}$
I
As area of loop is reduced, a current is induced in it.
$\vec{B}_{loop}$
B.

FIGURE 32.10 **A.** A thin, flexible loop is in the plane of the page, and a uniform and constant magnetic field $\vec{B}_{sol}$ points into the page. **B.** If you reduce the area of the loop, the magnetic flux Φ_B through the loop decreases. According to Lenz's law, the induced current creates a magnetic field $\vec{B}_{loop}$ that opposes the change. In this case, the change is a reduction in magnetic flux, so to oppose the change, $\vec{B}_{loop}$ must point into the page in the same direction as $\vec{B}_{sol}$. Point your right thumb in the direction of $\vec{B}_{loop}$ (into the page), and your fingers wrap clockwise. So the induced current and emf in the loop are clockwise.

For the case of a changing angle, consider a loop wrapped around a ball that rolls in a uniform and constant magnetic field $\vec{B}_{sol}$ (Fig. 32.11). Because the angle between the area vector of the loop and the magnetic field changes, a current is induced in the loop. We can apply Lenz's law at two separate instants to find the direction of the induced current. In the first instant (Fig. 32.11A), the loop's area vector $\vec{A}$ is initially aligned with the magnetic field vector $\vec{B}_{sol}$. So, initially the magnetic flux through the loop is at its maximum. After the ball rolls to the right, the area vector points downward and to the right, so the magnetic flux through the loop has decreased. To visualize this, imagine looking along the magnetic field from the left side of the page. You would see the solenoid's magnetic field $\vec{B}_{sol}$ pointed away from you. Initially the loop would look like a large circle, but a moment later it would look like a squashed oval, exactly as it looks in Figure 32.10B. Because the magnetic flux decreases, the loop's magnetic field must point in the same direction as the original magnetic field—away from you if you are looking from the left in the figure. To a reader looking straight at the page, the field points to the right. The only way to be consistent with the idea that the loop's magnetic field must have a component to

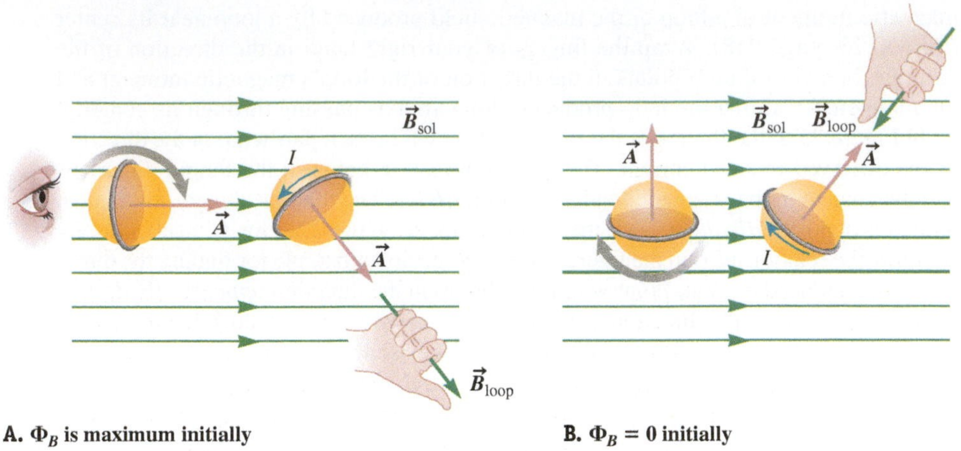

FIGURE 32.11 **A.** The magnetic flux through the loop is reduced from a maximum, so the loop's magnetic field points in roughly the same direction as the original magnetic field $\vec{B}_{sol}$. Point your right thumb in the direction of $\vec{B}_{loop}$, and your fingers wrap in the direction of the induced current. **B.** The magnetic flux increases from a minimum, so the loop's magnetic field points in roughly the opposite direction as the original magnetic field $\vec{B}_{sol}$. As before, if your right thumb points in the direction of $\vec{B}_{loop}$, your fingers wrap in the direction of the induced current.

the right is for $\vec{B}_{loop}$ to point downward and to the right as shown in Figure 32.11A. Therefore, $\vec{B}_{loop}$ has a component in the same direction as $\vec{B}_{sol}$ when rotation of the loop causes a decrease in the magnetic flux Φ_B.

In the second instant (Fig. 32.11B), the area vector $\vec{A}$ is initially perpendicular to the magnetic field vector $\vec{B}_{sol}$. So, initially the magnetic flux $\Phi_B = 0$ through the loop. The ball continues to roll to the right so that the loop's area vector points upward and to the right. The magnetic flux through the loop has increased. From your perspective at the left side of the page, initially the loop would look like a straight line, and a moment later it would look like a squashed oval, as if someone had stretched it open. Because the magnetic flux through the loop increases, the loop's magnetic field must point in the direction opposite the original magnetic field—toward you. For readers looking at the page, this is to the left. The only way to be consistent with the idea that the loop's magnetic field must have a component to the left is for $\vec{B}_{loop}$ to point downward and to the left as shown in Figure 32.11B. So, $\vec{B}_{loop}$ has a component in the direction opposite $\vec{B}_{sol}$ when rotation of the loop causes an increase in the magnetic flux Φ_B.

Combining Faraday and Lenz

We have been using a verbal statement of Lenz's law, but his law can also be expressed mathematically. The idea that an induced emf *opposes* a change in the magnetic flux means Lenz's law can be expressed as a negative sign in Faraday's law:

FARADAY-LENZ'S LAW

Underlying Principle

$$\mathcal{E} = -N\frac{d\Phi_B}{dt} \tag{32.6}$$

Although Lenz's law is the negative sign in Equation 32.6, his name is often dropped and this equation is usually referred to as **Faraday's law**.

CONCEPT EXERCISE 32.3

Find and draw the direction of $\vec{B}_{loop}$ for each case in Figure 32.12. The original magnetic field $\vec{B}_{sol}$ in each case is produced by a solenoid (not shown) and is *decreasing*. All the loops are circular and seen in perspective.

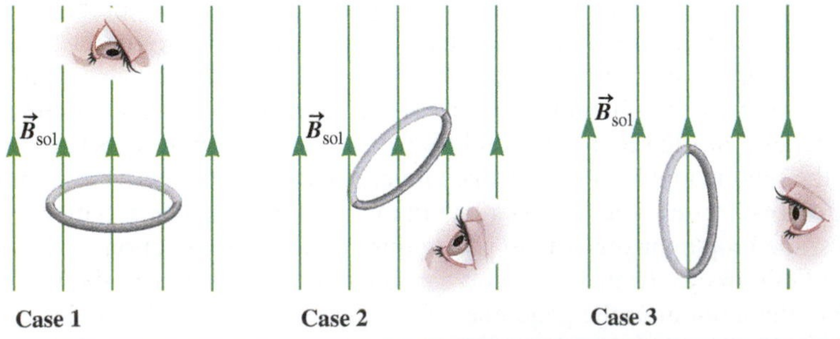

FIGURE 32.12

CONCEPT EXERCISE 32.4

For each case in Figure 32.12, indicate the direction of the induced current as seen by the observer represented by the eye.

32-4 Lenz's Law and Conservation of Energy

Both Faraday's law and Lenz's law were discovered through careful observations made in a laboratory. Although neither law was derived from previously accepted laws, these laws are consistent with existing laws. In this section, we show that Lenz's law is really a statement of the conservation of energy.

First, imagine pushing a bar magnet in a frictionless environment. If the magnet is the system and you are outside the system, you do positive work on the bar magnet. If no other forces act on the system, the positive work you do must cause the bar magnet to speed up. Its kinetic energy increases:

$$W = \Delta K$$

Now, imagine pushing the bar magnet along a rough surface so that kinetic friction acts on the magnet. If the magnet and the surface are in the system, the positive work you do goes into increasing the magnet's kinetic energy as before, but it also goes into the thermal energy of the magnet and surface:

$$W = \Delta K + \Delta E_{\text{th}}$$

In order for energy to be conserved, kinetic friction must oppose the force you exert on the bar magnet. Suppose for a moment that weren't true. Then, when you give the bar magnet a small push, doing a small amount of positive work, kinetic friction would act in the same direction as your push. The magnet would continue to speed up and gain kinetic energy, and the system's thermal energy would also continue to increase. In this imaginary scenario, you would not need to put gasoline into your car. Instead, you could get someone to give your car a small push, and it would continue to speed up throughout your journey. In fact, stopping would be a big problem.

Finally, imagine pushing the bar magnet in a frictionless environment but toward a coil as shown in Figure 32.13A. Here, the magnet's north pole points toward the coil. The magnetic flux through the coil changes, inducing a current. According to Lenz's law, the coil's induced magnetic field opposes the change. A convenient way to visualize this is to remember that a coil's magnetic field looks similar to that of a bar magnet. So, you can imagine the coil as a bar magnet with its north pole directed toward the actual bar magnet's north pole (Fig. 32.13B). The coil's north pole repels the bar magnet's north pole, much like the situation in which you push the bar magnet along a rough surface; in that case, kinetic friction opposes your force. The work you do goes into the bar magnet's kinetic energy and the system's thermal energy.

If we consider both magnets to be part of the system, the work you do on the system goes into the bar magnet's kinetic energy and the electrical energy that induces the current in the coil and creates its magnetic field. Like friction, the coil's magnetic field must resist your force on the bar magnet, so that field must be oriented as shown in Figure 32.13B, with its north pole pointing toward the bar magnet. Suppose Lenz's law were reversed, so that the coil's south pole was pointing toward the bar magnet's north pole. Now, when you give the bar magnet a small push, doing a small amount of positive work, the coil's magnetic field would act in the same direction as your push. The bar magnet would continue to speed up, gaining kinetic energy. More energy would also go into the coil's current, creating an even stronger magnet. So the work you did on the system would be much smaller than the increase in the magnet's kinetic energy and the coil's electrical energy. In this imaginary scenario, you would not need to worry about powering electrical circuits or propelling objects. If you wished to light up a bulb, you could connect the bulb to a coil like the one in Figure 32.13. Then simply give a bar

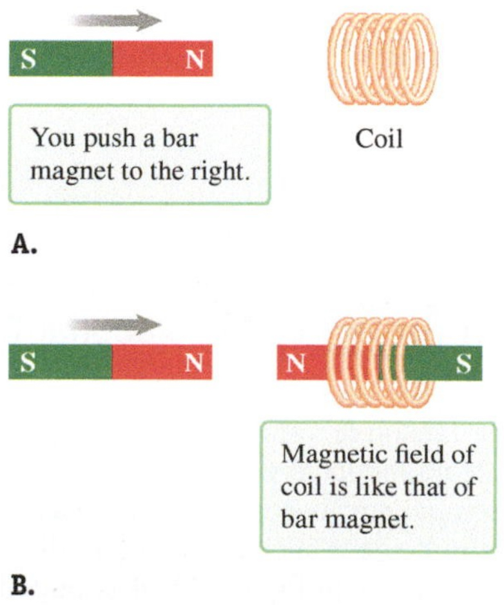

FIGURE 32.13 **A.** When you push a bar magnet toward a coil, **B.** the coil creates a second magnetic field that pushes back.

magnet a slight push toward the coil. A current would be induced in the coil, lighting the bulb. The coil would make a magnetic field that would continue to propel the bar magnet toward it, which would keep a current induced in the coil. A much bigger problem would be not to burn out the lightbulb.

Stating that kinetic friction always opposes relative motion and that induced current always opposes a change in the magnetic flux (Lenz's law) is consistent with the principle of conservation of energy. If either one of these facts were not true, it would be possible to accelerate an object or light a bulb indefinitely by putting in just a small amount of energy.

CONCEPT EXERCISE 32.5

Suppose you have two bar magnets, with one fixed to a tabletop. You briefly push the other magnet toward the fixed magnet. There is negligible friction between the sliding magnet and the tabletop. You orient the sliding magnet so that its north end points toward the fixed magnet's south pole. Does the sliding magnet move at constant speed, speed up, or slow down? How is your answer consistent with the principle of conservation of energy?

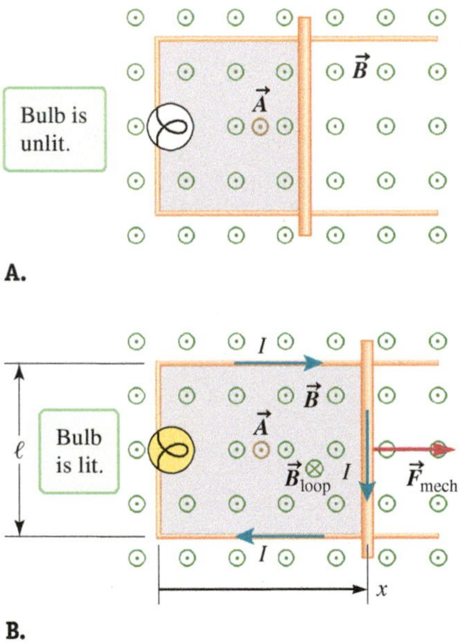

FIGURE 32.14 **A.** A slide generator consists of a U-shaped conductor and a conducting bar in a uniform and constant magnetic field. **B.** When the bar slides to the right, there is a clockwise current in the conductors and the bulb lights up. (The mechanical force is balanced by a magnetic force; Example 32.2.)

SLIDE GENERATOR Special Case

32-5 Case Study: Slide Generator

Soon after Faraday discovered his law, he realized its enormous practical applications. In 1831, he invented an electric generator, a device that converts mechanical power to electrical power. We saw a simple example of a generator in Figure 32.13; the mechanical power you put in to move the bar magnet is converted to electrical power. Of course, having people move bar magnets around is not practical. Our case study focuses on the war over AC and DC power, and in this section we consider a simple device that highlights the basic features of any generator.

Figure 32.14 shows a **slide generator**, which consists of two conductors in a uniform magnetic field. One conductor looks like a sideways U. To visualize the current in the circuit, we have included a lightbulb in this conductor. The other conductor is a bar of length ℓ that is free to slide along the tracks of the U-shaped conductor without losing electrical contact. We'll assume the kinetic friction between the two conductors is negligible.

The bar and the U-shaped conductor form a rectangle of area $A = \ell x$. As the bar slides to the right, (1) the position x of the bar increases, (2) the area of the rectangle (shown in gray) grows, and (3) the magnetic flux through the rectangle increases. So, an emf is induced in the rectangular loop. According to Lenz's law, the loop's magnetic field must be directed into the page, and the current in the loop is clockwise (Fig. 32.14B). Our next steps are to derive the emf and current induced in, and the power required and produced by, a slide generator.

DERIVATION Motional Emf

The (absolute value of the) emf induced in the rectangular loop in Figure 32.14 is sometimes called a **motional emf** because the bar must be in motion ($v \neq 0$). Like the emf of a battery, the motional emf is constant as long as the speed v is constant, and the resulting current can operate useful devices such as the lightbulb.

The bar in Figure 32.14 is pulled to the right by a mechanical force $\vec{F}_{mech}$, and it moves to the right at constant velocity $\vec{v} = d\vec{x}/dt$. We will use Faraday's law to show the motional emf is

$$\mathcal{E} = B\ell v \tag{32.7}$$

and the current in the conductor is

$$I = \frac{B\ell v}{R} \tag{32.8}$$

We are not concerned about direction here, so there is no need to include Lenz's law.

As in Example 32.1, the magnetic field is uniform over the entire area of the loop, and the area vector $\vec{A}$ is parallel to the magnetic field $\vec{B}$.	$\Phi_B = \int \vec{B} \cdot d\vec{A}$ $\Phi_B = BA$ (32.1)
Express the area A in terms of the length of the bar ℓ and its position x. As the bar slides to the right, x increases.	$\Phi_B = B(\ell x)$
Substitute the expression for the magnetic flux Φ_B into Faraday's law (Eq. 32.4).	$\mathcal{E} = N\frac{d\Phi_B}{dt} = N\frac{d(B\ell x)}{dt}$
There is only one loop, so $N = 1$. The magnetic field B and the length ℓ of the bar are constant.	$\mathcal{E} = \frac{d(B\ell x)}{dt} = B\ell\frac{dx}{dt}$
The derivative dx/dt is the constant speed v of the bar.	$\mathcal{E} = B\ell v$ ✓ (32.7)
The current in the conductor is found by dividing the motional emf by the conductor's total resistance R.	$I = \frac{B\ell v}{R}$ ✓ (32.8)

COMMENTS

Finding R in Equation 32.8 can be complicated. The bar, the legs of the U-shaped conductor, and the lightbulb are all connected in series (Fig. 32.14), so R is found by adding up the resistances of each of these elements. However, the resistance of the legs of the U-shaped conductor depends on the position x of the bar. When the bar is far along the tracks, the current must pass through a long portion of the U-shaped conductor's legs and the resistance is higher because $R = \rho L/A$ (Eq. 28.24), where L is the total length of the conductor through which current is flowing and A is its cross-sectional area.

EXAMPLE 32.2 Mechanical Power Supplied to a Slide Generator

Find an expression for the mechanical power put into the slide generator when the bar slides at constant speed in Figure 32.14.

INTERPRET and ANTICIPATE

A free-body diagram for the bar is helpful (Fig. 32.15). Two forces are exerted on the bar: the mechanical force $\vec{F}_{\text{mech}}$ to the right (as might be exerted by a person pulling the bar) and the magnetic force. The current and $\vec{\ell}$ are downward, and the magnetic field is out of the page. According to $\vec{F}_B = I(\vec{\ell} \times \vec{B})$ (Eq. 30.32) and the right-hand rule for cross products, the magnetic force on the bar is to the left.

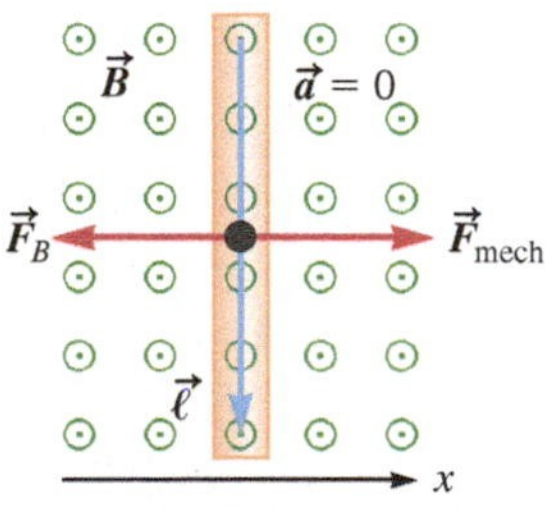

FIGURE 32.15

SOLVE

Find the magnitude of the magnetic force. The angle between $\vec{\ell}$ and $\vec{B}$ is 90°.	$F_B = I\ell B \sin\varphi = I\ell B \sin 90°$ $F_B = I\ell B$

Example continues on page 1022 ▶

Because the bar is moving at constant velocity, its acceleration is zero and the mechanical force must be balanced by the magnetic force.	$F_{\text{mech}} = F_B$ $F_{\text{mech}} = I\ell B$ (1)
The power P_{mech} put in by the mechanical force is given by $P_{\text{mech}} = \vec{F}_{\text{mech}} \cdot \vec{v}$ (Eq. 9.38). Both the mechanical force $\vec{F}_{\text{mech}}$ and the bar's velocity $\vec{v}$ are to the right in the positive x direction, so the angle between them is zero.	$P_{\text{mech}} = F_{\text{mech}} v \cos 0 = F_{\text{mech}} v$
Substitute Equation (1) for the mechanical force.	$P_{\text{mech}} = I\ell B v$
Substitute Equation 32.8 for the current.	$P_{\text{mech}} = \left(\frac{B\ell v}{R}\right)\ell B v$ $P_{\text{mech}} = \frac{(B\ell v)^2}{R}$ (32.9)

CHECK and THINK

As always, you should verify that our expression has the correct dimensions (energy per time). In the next example, we'll find the power supplied by the generator and think about both results together.

EXAMPLE 32.3 Electrical Power Supplied by a Slide Generator

Equation 32.9 is the power supplied *by the mechanical force*. Electrical power is, in turn, supplied *by the generator* and consumed by the conductors and the bulb. Find an expression for the power consumed by the conductors and the bulb. In **CHECK and THINK**, compare the power supplied by the mechanical force to the power P_{res} consumed.

INTERPRET and ANTICIPATE

Because the conductors and the bulb have resistance, they must dissipate power. Our job is to find an expression for that power in terms of B, ℓ, v, and R so that we can compare P_{mech} to P_{res}.

SOLVE

The conductors and the bulb have total resistance R, and the power they use is found from Equation 28.34.	$P_{\text{res}} = I^2 R$ (28.34)
Substitute $I = B\ell v/R$ (Eq. 32.8) for the current.	$P_{\text{res}} = \left(\frac{B\ell v}{R}\right)^2 R = \frac{(B\ell v)^2}{R}$ (32.10)

CHECK and THINK

Equation 32.10 is the power used by the conductors and bulb. It is exactly what we found for the power supplied by the mechanical force: $P_{\text{mech}} = P_{\text{res}}$. Having the power supplied by the mechanical force equal to the power used by the conducting loop is consistent with the principle of conservation of energy. In the absence of dissipative forces such as kinetic friction, the mechanical work per unit time supplied to the generator equals the energy per unit time that is used by the electric circuit. So, if you want to light a 100-W bulb, you must supply 100 W of mechanical power.

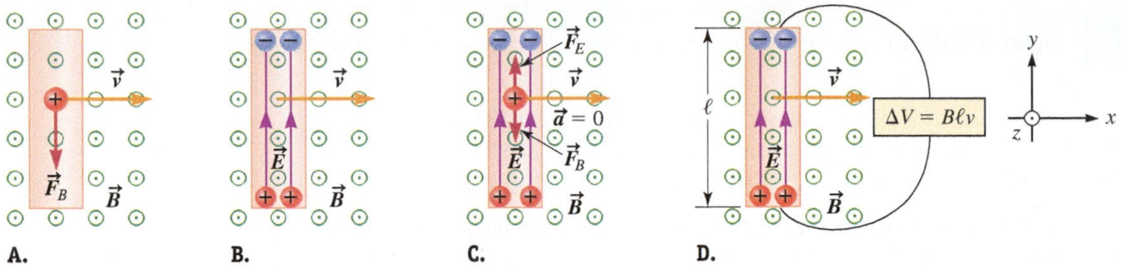

FIGURE 32.16 **A.** A conducting bar moves at constant velocity in a uniform magnetic field, leading to a downward magnetic force on the positive (imaginary) particles in the conductor. **B.** Positive charge builds up on the bottom of the bar and negative charge builds up on the top, resulting in an upward-pointing electric field. **C.** Now, the upward electric force on a positive particle is balanced by the downward magnetic force. **D.** When the particles are in equilibrium, the electric potential between the bottom and top is equal to the emf found using Faraday's law for a slide generator.

Another Look at the Motional Emf

The motional emf that is induced in the slide generator (Fig. 32.14) does *not* require the U-shaped conductor or Faraday's law. All that is necessary is the movement of the conducting bar at constant velocity through the uniform magnetic field (Fig. 32.16). Because the bar is moving, conducting electrons are moving with respect to the magnetic field, so a magnetic force is exerted on them. For convenience, we'll think about the magnetic force on the equivalent imaginary positive particles.

The bar moves at constant velocity in the positive x direction, so the imaginary positive particle's velocity $\vec{v}$ is in the positive x direction (Fig. 32.16A). The magnetic field $\vec{B}$ is out of the page in the positive z direction. According to $\vec{F}_B = q(\vec{v} \times \vec{B})$ (Eq. 30.17) and the right-hand rule for cross products, the magnetic force $\vec{F}_B$ on this imaginary positive particle is downward in the negative y direction. As a result of the magnetic force, positive charge builds up on the lower end of the conductor and negative charge on the upper end (Fig. 32.16B). This charge separation results in an electric field $\vec{E}$ in the bar pointing in the positive y direction—from the positive particles toward the negative ones. Once the electric field in the moving bar is established, an imaginary positive particle experiences two forces (Fig. 32.16C): a magnetic force $\vec{F}_B$ in the negative y direction and an electric force $\vec{F}_E$ in the positive y direction. The electric field reaches its maximum strength when these two forces are in balance. Once they are equal in magnitude, there is no net force on the imaginary charged particle, so it experiences no acceleration:

$$F_E = F_B$$

Equations 24.18 and 30.18 give the magnitudes of the electric and magnetic forces, respectively:

$$qE = qvB$$

The charge q cancels, so the maximum electric field at equilibrium is

$$E = vB \tag{32.11}$$

The electric field in the bar is uniform, so the electric potential difference between the bottom and the top of the bar (Fig. 32.16D) comes from modifying Equation 26.17:

$$\Delta V = V_{\text{bottom}} - V_{\text{top}} = E\ell$$

Substitute Equation 32.11 for the electric field:

$$\Delta V = Bv\ell = B\ell v \tag{32.12}$$

This is exactly what we found for the motional emf (Eq. 32.7) induced in the rectangular loop of the slide generator using Faraday's law. You could imagine using wires to connect the top and bottom of the bar (like the positive and negative terminals of a battery) to a lightbulb; then the bulb would be lit just as it is in Figure 32.14B.

EXAMPLE 32.4 Space Tether

Suppose you want to use a slide generator like the one in Figure 32.14 to light a small flashlight bulb. Normally, two 1.5-V batteries are used to power the bulb. The flashlight case is about 20 cm long, so let's assume the bar in Figure 32.14 is $\ell = 20$ cm long.

A If you use a uniform magnetic field of 0.15 T, at what speed must the bar be moved to produce the usual emf across the bulb?

INTERPRET and ANTICIPATE

This is a direct application of the slide generator. We expect to find a numerical result in m/s.

SOLVE

Solve Equation 32.7 for speed v, and then substitute values. The emf across the bulb when it is connected to two batteries is the sum of the emfs of both batteries (3.0 V).

$$\mathcal{E} = B\ell v \quad (32.7)$$

$$v = \frac{\mathcal{E}}{B\ell} = \frac{3.0\text{ V}}{(0.15\text{ T})(0.20\text{ m})}$$

$$v = 100\ \frac{\text{V}}{\text{T}\cdot\text{m}} \quad (1)$$

We need to obtain the expected SI units. Use Equation 32.5 to convert.

$$1\ \frac{\text{T}\cdot\text{m}^2}{\text{s}} = 1\ \frac{\text{Wb}}{\text{s}} = 1\text{ V} \quad (32.5)$$

$$\text{so } 1\ \frac{\text{V}}{\text{T}\cdot\text{m}} = 1\ \frac{\text{Wb/s}}{\text{T}\cdot\text{m}} = 1\ \frac{\text{T}\cdot\text{m}^2}{\text{T}\cdot\text{m}\cdot\text{s}} = 1\text{ m/s}$$

We have arrived at the correct units for speed. Substitute them into Equation (1).

$$v = 100\text{ m/s}$$

CHECK and THINK

We just found that the bar must move at 100 m/s or approximately 224 mph, which is not practical. You might wonder whether slide generators have any use, but NASA engineers think such generators might be used in satellites. Part B is based on an experiment conducted by NASA.

B In 1996, NASA tested a large-scale slide generator. A space shuttle deployed a tether—a long conducting wire—with a spherical satellite at the end. The tether was moved through the Earth's magnetic field at a speed of 7.7 km/s, inducing an emf in the wire. The tether broke when it was 19.6 km long. Although the magnetic field is not uniform over the length of the wire, assume it is uniform with a magnitude of 2.5×10^{-5} T. Estimate the maximum motional emf between the ends of the tether.

INTERPRET and ANTICIPATE

This is another application of motional emf. In this case, we can think of the tether as the bar in Figure 32.16.

SOLVE

Equations 32.12 and 32.7 are equivalent. Use either to find the motional emf $\mathcal{E}$. Remember to convert the speed and length to SI units. Use Equation 32.5 to express the emf in volts.

$$\mathcal{E} = B\ell v \quad (32.12)$$

$$\mathcal{E} = (2.5 \times 10^{-5}\text{ T})(19.6 \times 10^3\text{ m})(7.7 \times 10^3\text{ m/s})$$

$$\mathcal{E} = 3.8 \times 10^3\ \frac{\text{T}\cdot\text{m}^2}{\text{s}} = 3.8 \times 10^3\ \frac{\text{Wb}}{\text{s}}$$

$$\mathcal{E} = 3.8 \times 10^3\text{ V}$$

CHECK and THINK

This particular slide generator could be a very practical device. A space shuttle or satellite can produce an enormous motional emf that can then be used to produce a current in any sort of useful device. In fact, the tether actually broke because of the current. The tether was made of many layers, with the innermost layer—a conducting wire—surrounded by an insulator full of small air pockets. Normally air is a good insulator, but due to the large motional emf, the air became a conductor and allowed for a complete circuit. The current in the circuit melted the tether.

Eddy Currents

Just as kinetic friction can slow down moving objects, so can currents induced as described by Faraday's law. Figure 32.17 shows a wire conducting loop and the poles of a C-shaped magnet. Between the poles of the magnet, the magnetic field is nearly uniform. The loop is made from a nonmagnetic material. When the loop is at rest, the magnet has no effect on the loop; there is no induced current and no magnetic force. Now imagine pulling the loop to the right (Fig. 32.17A). At first there is still no induced current and no magnetic force on the loop because the magnetic flux through it is constant. However, once the left edge of the loop begins to move through the magnetic field, the magnetic flux through the loop decreases, causing an induced current in the loop (Fig. 32.17B). According to Lenz's law, the loop's magnetic field $\vec{B}_{\text{loop}}$ must point downward near its center—in the same direction as the magnet's field. Point your right thumb downward, and your fingers wrap clockwise as seen from above the loop.

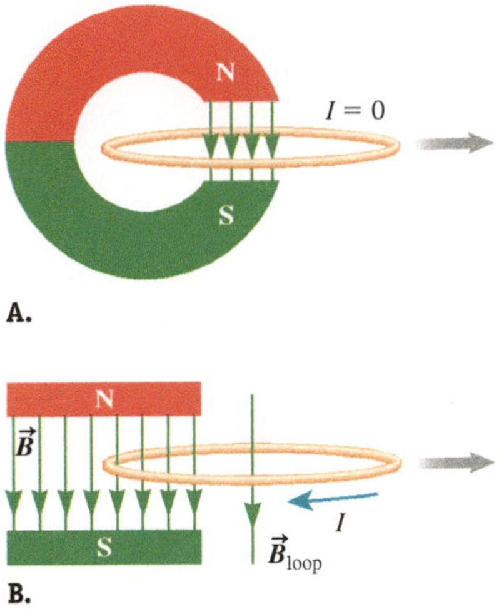

FIGURE 32.17 **A.** A loop is pulled to the right between the poles of a C-shaped magnet. There is no change in magnetic flux, so there is no current through the loop. **B.** Once the loop's left edge crosses the magnetic field lines, the magnetic flux begins to decrease and a current is induced in the loop.

Now imagine pulling two loops to the right between the magnet's poles (Fig. 32.18). Both loops experience a change in the magnetic flux. For the loop on the right, the magnetic flux is decreasing. By Lenz's law, as argued above, $\vec{B}_{\text{right}}$ must point downward and the induced current in the right loop must be clockwise as seen from above. For the loop on the left, however, the magnetic flux is increasing. This time, according to Lenz's law, its magnetic field $\vec{B}_{\text{left}}$ must point in the opposite direction—upward—as the magnet's field. From the right-hand rule, the induced current in the left loop must be counterclockwise as seen from above (Fig. 32.18).

Finally, imagine pulling a conducting metal sheet to the right through the magnet's poles (Fig. 32.19). When the sheet is at rest, no magnetic force is exerted on it. However, when the sheet is moving with respect to the magnet, the conduction electrons experience a magnetic force like the force experienced by the charged particles in the bar in Figure 32.16. As a result, **eddy currents** develop in the metal sheet. The eddy currents on the right side swirl around clockwise, and the eddy currents on the left side swirl around counterclockwise, as we found for the two loops in Figure 32.18. Eddy currents—like the currents in the two loops—obey Lenz's law by opposing the change that created them. So, if you are pulling the sheet to the right, the magnetic force on the sheet pulls the sheet back toward the left.

EDDY CURRENT ✪ **Major Concept**

A practical application of eddy currents is **magnetic braking**, which is used to damp the motion of delicate instruments and to stop some vehicles such as trains. You may have encountered a magnetic brake in your laboratory course. Typically,

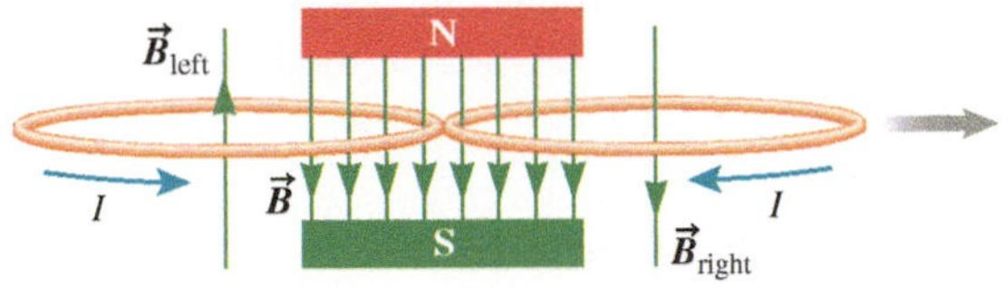

FIGURE 32.18 Two loops are pulled to the right between the poles of a C-shaped magnet. The magnetic flux through the right loop decreases, so it produces a magnetic field that points in the same direction as that of the magnet. The induced current in the right loop is clockwise as seen from above. The magnetic flux through the left loop increases, so it produces a magnetic field that points in the opposite direction as that of the magnet. The induced current in the left loop is counterclockwise as seen from above. (The loops are insulated from one another.)

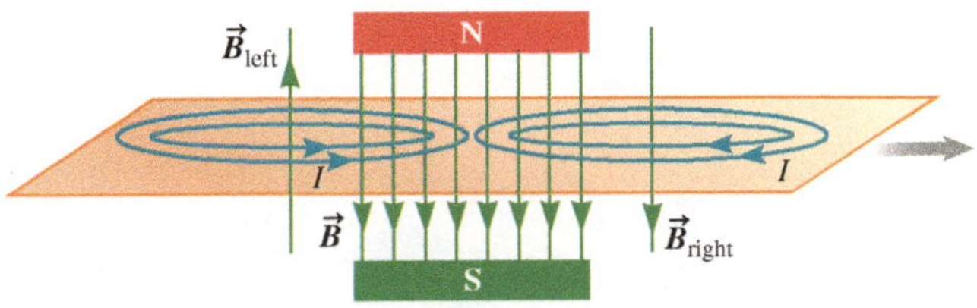

FIGURE 32.19 A conducting metal sheet is pulled to the right through the poles of a C-shaped magnet. Eddy currents are induced in the sheet. The eddy currents on the right are clockwise as seen from above, and the eddy currents on the left are counterclockwise as seen from above. Compare the eddy currents to the currents in the two loops in Figure 32.18.

the magnetic brake looks like a pendulum with a flat metal bob that swings between the poles of a magnet. Magnetic braking causes the pendulum's motion to damp, and the pendulum stops quickly. A train's magnetic brakes are electromagnets that are attached to the train and straddle the track. Normally there is no current or field in the electromagnets. When the train conductor wants to slow down or stop the train, current in the electromagnets produces a magnetic field, inducing eddy currents in the train track. Magnetic force produced by eddy currents in the track slows the train down. There is no kinetic friction between the brakes and the track, so the magnetic brakes do not wear out as quickly as normal brakes.

However, eddy currents heat up the train tracks. In terms of energy conservation, the kinetic energy of the train's motion must be transformed into some other form of energy when the train stops. If kinetic friction between the wheels and the track stops the train, both the wheels and the track gain thermal energy. If magnetic brakes are used, the track's thermal energy increases because the current in the conducting tracks encounters resistance. The power that heats the conductor is $P = I^2R$. Our case study focuses on electric generators. Even the most efficient generators lose energy due to eddy currents heating the conductors.

CONCEPT EXERCISE 32.6

A horizontal metal disk is balanced on a pivot. A strong bar magnet is moved in circles just above the surface of the disk. The disk is made of a nonmagnetic, conducting material, yet it begins to spin, following the magnet's circles. Explain how this is possible.

32-6 Case Study: AC Generators

The slide generator (Fig. 32.14) has two features that are common to all the generators we consider in this chapter: (1) The magnetic field is uniform and constant and may be due to a solenoid or a permanent magnet. Because a permanent magnet does not require a current to maintain its magnetic field, permanent magnets are used in most generators. (2) A change in the conductor induces the emf. Two features of the conductor can change—the area A and the angle φ. In the slide generator, the area A changes (grows or shrinks). We found, however, that the slide generator is usually not practical because it requires long tracks and a bar moving at very high speed. Another way to improve the slide generator is to increase the number of turns.

A more practical approach is to design a generator in which φ changes. In this section and the next, we look at two generators—AC and DC—based on a changing φ. The current in the AC (alternating current) generator increases, decreases, and changes direction. The current in the DC (direct current) generator is in a constant direction and is nearly constant in magnitude. Today, we use AC in all wall outlets. However, the decision to go with AC instead of DC was at the heart of a debate between Edison and his rivals, Westinghouse and Tesla.

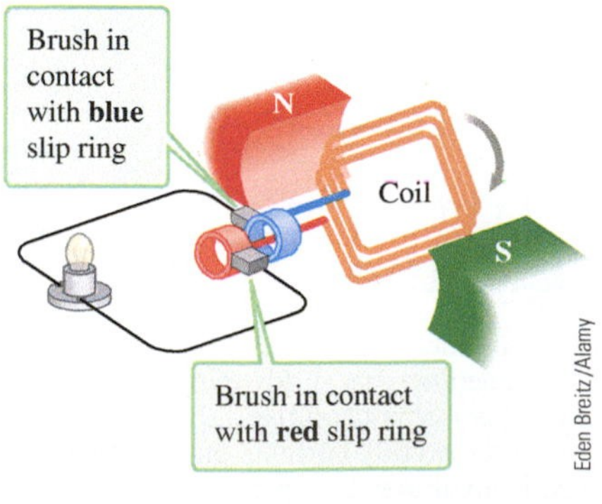

FIGURE 32.20 A basic AC generator.

Features and Operation of an AC Generator

Figure 32.20 shows the basic features of an AC generator. A coil rotates between the poles of one or more permanent magnets. Two wires protrude from the coil. The wire shown in blue is connected to the blue slip ring. The wire shown in red passes through the blue slip ring without making contact and is attached to the red slip ring. The slip rings rotate with the coil, and each ring is in contact with a stationary conducting brush. A wire is attached to the brushes to complete the circuit. Of course, the circuit can involve a number of useful devices. For simplicity, we imagine there is a lightbulb in the circuit.

Figure 32.21 shows how the coil rotation leads to a current. For simplicity, the coil is shown as a loop with only one turn. The loop is between the poles of a magnet (not shown); this permanent magnet's field lines are assumed to be horizontal and point from left to right. From our perspective (near the bulb), the loop rotates clockwise. In Figure 32.21A, the loop is not yet vertical. In Figure 32.21B, the loop is vertical so that the magnetic flux Φ_B through the loop is maximum. The magnetic flux increases as the loop rotates from time A to B and according to Lenz's law, the loop's magnetic field $\vec{B}_{\text{loop}}$ in part A must have a component in the direction opposite the permanent magnet's field $\vec{B}$. To find the direction of the current in the loop, point your right thumb in the direction of $\vec{B}_{\text{loop}}$ and trace the current through the bulb. (Follow it out through the red slip ring.) In part A, the current is from right to left through the bulb, the magnetic flux is increasing, and the emf is negative.

For a moment in Figure 32.21B, the magnetic flux does not change; its time derivative is zero, and so the induced emf is zero. Of course, the loop continues to rotate, and the magnetic flux decreases. By the time shown in Figure 32.21C, the loop's magnetic field $\vec{B}_{\text{loop}}$ must have a component in the same direction as the magnet's field $\vec{B}$. Point your right thumb in the direction of $\vec{B}_{\text{loop}}$, and follow the current out through the blue slip ring. In part C, the current through the bulb is from left to right, the magnetic flux is decreasing, and the emf is positive.

Now we are ready to look at one complete cycle. To keep the drawing simple, we show only the loop, the permanent magnet's field $\vec{B}$, and the loop's area vector $\vec{A}$. Figure 32.22 shows the loop end-on as it rotates clockwise from (1) vertical, going upside down and back to (9) vertical. At each moment, the magnetic flux is given by $\Phi_B = BA\cos\varphi$. As the loop rotates from (1) to (9), the magnetic flux through the loop decreases and then increases. The sign of the induced emf comes from Lenz's law and so is the opposite of $d\Phi_B/dt$. At times 1, 5, and 9, the flux is momentarily constant, and according to Faraday's law, the induced emf is zero at those times.

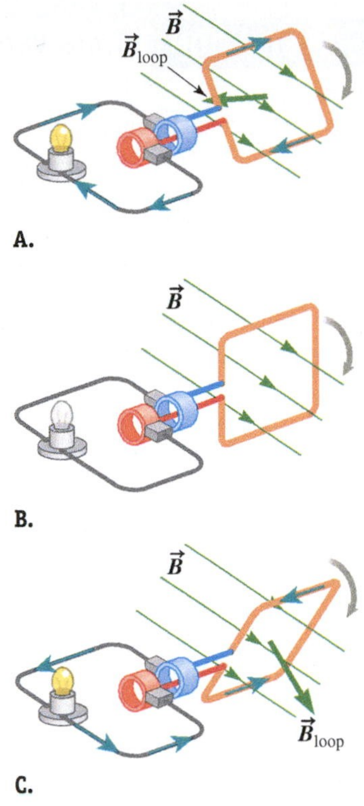

FIGURE 32.21 Imagine observing an AC generator in motion from a position near the bulb. **A.** The magnetic flux Φ_B is increasing, so $\vec{B}_{\text{loop}}$ points roughly opposite to $\vec{B}$ and the current through the bulb is to the *left*. **B.** Φ_B is at its maximum and momentarily constant, so $\vec{B}_{\text{loop}} = 0$, and there is no current. **C.** Φ_B is decreasing, so $\vec{B}_{\text{loop}}$ points roughly in the direction of $\vec{B}$ and the current through the bulb is to the *right*.

	Configuration	Magnetic flux Φ_B	Change in flux $\frac{d\Phi_B}{dt}$	Induced emf $\mathcal{E}$
1	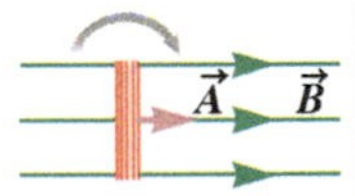	Positive, maximum	Momentarily zero (constant flux)	Zero
2	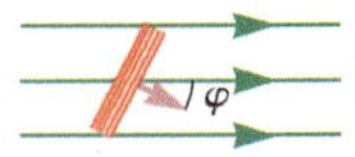	Positive	Decreasing (negative)	Positive
3	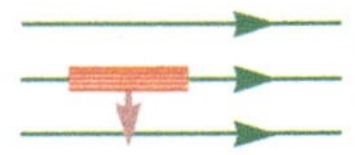	Zero	Decreasing (negative)	Positive
4	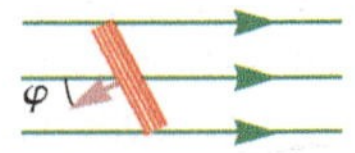	Negative	Decreasing (negative)	Positive
5		Negative, maximum	Momentarily zero (constant flux)	Zero
6	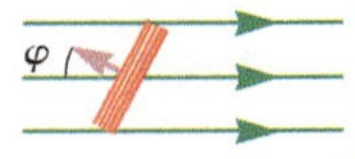	Negative	Increasing (positive)	Negative
7	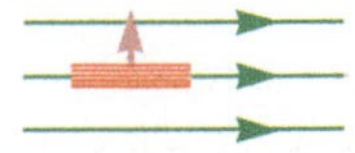	Zero	Increasing (positive)	Negative
8		Positive	Increasing (positive)	Negative
9	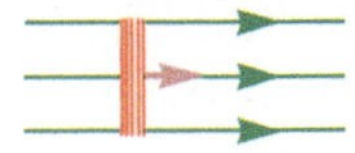	Return to positive, maximum	Momentarily zero (constant flux)	Zero

FIGURE 32.22 An AC generator loop shown schematically during one complete cycle.

DERIVATION Time-Varying AC Emf and Current

We show here that the induced emf in an AC generator (Fig. 32.20) is

$$\mathcal{E} = NBA\omega \sin \omega t = \mathcal{E}_{\max} \sin \omega t \tag{32.13}$$

AC GENERATOR Special Case

and the current is

$$I = \frac{NBA\omega}{R} \sin \omega t = I_{\max} \sin \omega t \tag{32.14}$$

Then we use Figure 32.22 to connect these results to the motion in the generator.

First, we need an expression for the magnetic flux through the loop or coil. Because the magnetic field is uniform over the loop, the magnetic flux comes from Equation 32.3.	$\Phi_B = BA \cos \varphi$ (32.3) $\Phi_B = (\Phi_B)_{\max} \cos \varphi$ (1) where $(\Phi_B)_{\max} = BA$
The coil in the generator rotates at constant angular speed ω, so we write φ in terms of ω and substitute $\varphi = \omega t$ into Equation (1).	$\Phi_B = (\Phi_B)_{\max} \cos \omega t$ (32.15)
Now substitute Equation 32.15 for magnetic flux in Faraday–Lenz's law (Eq. 32.6).	$\mathcal{E} = -N\frac{d\Phi_B}{dt}$ (32.6) $\mathcal{E} = -N\frac{d(BA \cos \omega t)}{dt}$
The magnetic field B and area A are constants, so we pull them outside the derivative.	$\mathcal{E} = -NBA\frac{d(\cos \omega t)}{dt}$ $\mathcal{E} = NBA\omega \sin \omega t = \mathcal{E}_{\max} \sin \omega t$ ✓ (32.13) where $\mathcal{E}_{\max} = NBA\omega$ (32.16)
The current is found by dividing the emf by the total resistance R in the circuit.	$I = \frac{NBA\omega}{R} \sin \omega t = I_{\max} \sin \omega t$ ✓ (32.14) where $I_{\max} = \frac{\mathcal{E}_{\max}}{R} = \frac{NBA\omega}{R}$ (32.17)

COMMENTS

The presence of a sine function in Equations 32.13 and 32.14 means the emf and current increase, decrease, and change sign periodically. Because the current *alternates*, such a generator is called an **alternating current** generator.

The coil's angular speed ω is numerically equivalent to the angular frequency ω of the emf and current oscillation that appear in Equations 32.13 and 32.14. Although *angular* frequency appears in these equations, *frequency* f is commonly used. Recall the relationship among angular frequency, frequency, and period from Equation 16.2: $\omega \equiv 2\pi f = 2\pi/T$. Standard AC power in the United States and Canada oscillates at $f = 60$ Hz. In many other countries, the standard frequency is $f = 50$ Hz.

EXAMPLE 32.5 Plotting an AC Generator's Emf

Whenever a complicated function is involved, it is helpful to make a graph. Sketch the magnetic flux Φ_B (Eq. 32.15) and induced emf $\mathcal{E}$ (Eq. 32.13) as functions of time t for one cycle. Then connect these functions to the motion of the generator by including the numbers representing the time sequence of events in Figure 32.22, and check for consistency.

INTERPRET and ANTICIPATE

It is straightforward to sketch $\Phi_B = (\Phi_B)_{max} \cos \omega t$ and $\mathcal{E} = \mathcal{E}_{max} \sin \omega t$ because they are the familiar cosine and sine graphs. Our real work is checking these graphs for consistency with the motion of the AC generator.

SOLVE

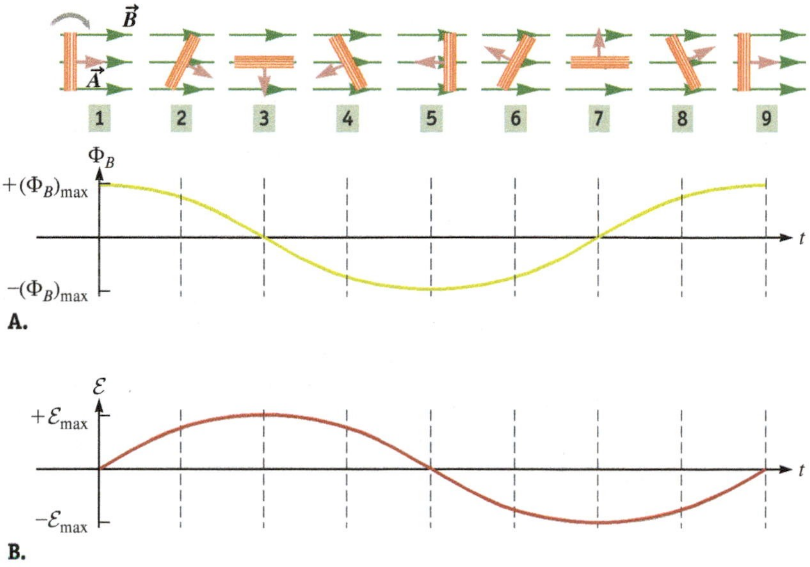

FIGURE 32.23 A. Magnetic flux though the loop and **B.** emf versus time for an AC generator. The configuration of the coil at selected instants in time is shown above the graphs.

CHECK and THINK

First, compare the sketch of the generator at the nine times to the graph of magnetic flux (Fig. 32.23A). We see that the flux starts at its maximum at 1, decreases to its minimum at 5, and increases to its maximum again at 9. Now, let's check that the two graphs are consistent. At times 1, 5, and 9, the flux is at an extreme value, so the time derivative is zero. We should find that the induced emf $\mathcal{E}$ is zero at these three times, and this is exactly what is shown (Fig. 32.23B). Further, between times 1 and 5, the flux is decreasing, and because the slope of the graph is negative (Fig. 32.23A), according to Lenz's law, we expect the induced emf to be positive during this period. Again, this is exactly what we see in Figure 32.23B. Finally, consider the magnetic flux between times 5 and 9. The flux is increasing, and the slope of the graph is positive. According to Lenz's law, we expect the induced emf to be negative during this period. Once again, this result is consistent with the graph of emf versus time.

CONCEPT EXERCISE 32.7

Suppose the lightbulb in Figure 32.20 is fluorescent. Is it always illuminated while the generator is running? Explain.

EXAMPLE 32.6 CASE STUDY AC Generator

The circular coil of an AC generator has 225 turns and an area of 2.00×10^{-2} m^2 (radius about 8 cm). The generator's magnetic field is 0.10 T. The maximum induced emf is $\mathcal{E}_{max} = 170$ V—typical of the AC power in American houses. What is the angular speed of the coil?

INTERPRET and ANTICIPATE

We expect to find a numerical answer in rad/s that is consistent with the frequency ($f = 60$ Hz) of the current's oscillation in a typical American house.

Example continues on page 1030 ▶

SOLVE

Find $\mathcal{E}_{max}$ from Equation 32.16. Solve for ω and substitute values.

$$\mathcal{E}_{max} = NBA\omega \quad (32.16)$$

$$\omega = \frac{\mathcal{E}_{max}}{NBA} = \frac{170\text{ V}}{(225)(0.10\text{ T})(2.00 \times 10^{-2}\text{ m}^2)}$$

$$\omega = 378\frac{\text{V}}{\text{T}\cdot\text{m}^2}$$

CHECK and THINK

We must make sure the units are correct and the value is consistent with what we expect for a typical American home. Equation 32.5 helps when reducing the units. Be careful; radians are dimensionless. *Angular* frequency and *angular* velocity are both in rad/s. Convert to frequency for the final answer. Our result is consistent with the frequency of current oscillation in a typical American house.

$$1\frac{\text{V}}{\text{T}\cdot\text{m}^2} = 1\frac{\text{V}}{\text{Wb}} = 1\frac{\text{Wb/s}}{\text{Wb}} = 1\text{ s}^{-1}$$

$$\omega = 378\text{ rad/s}$$

$$f = \frac{\omega}{2\pi} = \frac{378\text{ rad/s}}{2\pi\text{ rad}} = 60\text{ Hz}$$

Compare these results to those for the slide generator in Example 32.4. We found an impractical result in the case of a slide generator, but the AC generator in this example has an achievable magnetic field, coil size, and number of turns. Now imagine turning the coils at 378 rad/s (60 rps). Is that also achievable? Of course, that is much too fast to do by hand, but another source of mechanical power such as a steam engine or a waterfall could do the trick. Edison's DC generators had to be located in cities, where it is unlikely to find a waterfall, so they were powered by steam engines.

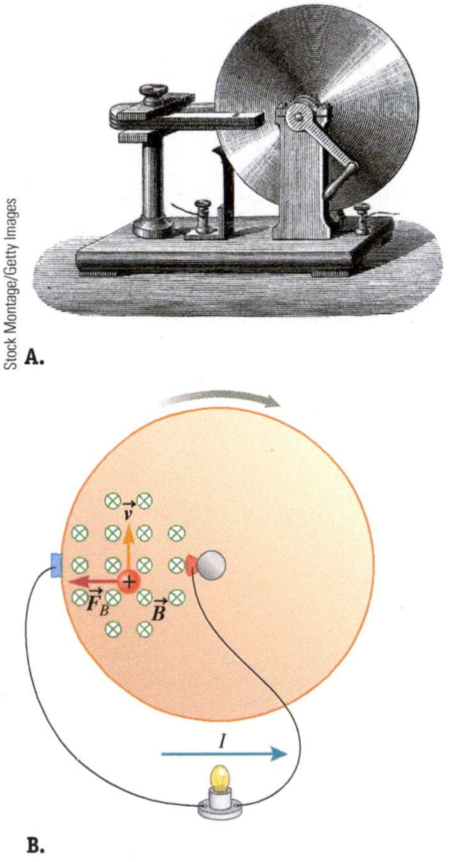

FIGURE 32.24 **A.** Faraday's electric generator (1831). **B.** As the copper disk is rotated, the magnetic field forces imaginary positive charges outward, creating a radial current from the shaft toward the edge of the disk.

The AC Motor

The AC motor is an AC generator running in reverse. In the motor, an AC current causes the coil to rotate. That mechanical rotation can be used to do any number of things, such as run an electric blender, vacuum cleaner, or power saw. Today many of our devices are based on AC motors running off of AC generators, but in the late 1880s DC motors (Fig. 30.55, page 969) were the standard. Edison worked to improve the DC generator and believed DC generators should become the standard. The DC generator is based on Faraday's generator (Section 32-7).

32-7 Case Study: Faraday's Generator and Other DC Generators

When presenting a new discovery, Faraday liked to quote Franklin: "Endeavor to make it useful." Within months of conducting the experiments that led to Faraday's law, he built a very useful device—the first electromagnetic generator as shown in Figure 32.24A.

Faraday's generator consists of a copper disk and a permanent magnet. A portion of the disk slips between the poles of the magnet as a person turns the hand crank. In Figure 32.24B, the disk is shown rotating clockwise and the permanent magnet's field points into the page. Conduction electrons move perpendicular to the magnetic field, so they experience a magnetic force. Consider an equivalent positive particle in Figure 32.24B. At the moment shown, the particle moves upward, so the magnetic force is radial toward the edge of the disk. The result is a radial current directed away from the shaft and toward the edge of the disk. One conducting brush is in contact with the shaft and another with the edge of the disk. Trace the current in the circuit through the bulb from left to right. As long as the crank is turned in the same direction, the direction of the current does not change. So, Faraday's generator is a DC (direct current) generator.

Faraday's generator was a technological breakthrough, but his actual design was impractical because it did not generate much power. In the 1870s, the Belgian engineer Zénobe-Théophile Gramme invented a practical DC generator and at the same time a DC motor (Fig. 30.55, page 969). A DC generator is a DC motor running in reverse: Both consist of a coil or loop that rotates in a magnetic field. In a DC motor, a DC power supply such as a battery produces a current in the coil. A permanent magnet exerts a torque on the coil, so electrical power is converted to mechanical power.

In a DC generator, a coil is mechanically rotated (Fig. 32.25). Because the angle φ between the coil's area and the magnetic field changes, current is induced in the coil. So, mechanical power is converted to electrical power.

Compare the AC generator in Figure 32.21 to the DC generator in Figure 32.25. The main difference is that in the AC generator the coil is connected to two separate rings, and in the DC generator the coil is connected to a single **split ring commutator**. The split ring commutator ensures that the direction of the current is constant. Because the commutator is an external device, Tesla argued, all DC generators and motors are really AC. He said the commutator is a complicated device that often causes trouble. In fact, the commutator often sparked and wore out easily. Because the commutator is fragile and must be replaced often, Tesla wanted to eliminate it from the system, leaving an AC generator.

Nevertheless, it is important to understand how DC generators operate. To see how the commutator works, consider a few moments of the coil's rotation. The coil in Figure 32.25 is represented by a single loop rotating clockwise (as seen from the bulb). The loop is between the poles of a permanent magnet so that the magnetic field is horizontal and points roughly from left to right (from a perspective near the bulb). Two conducting brushes maintain contact with the split ring commutator as the loop rotates.

In Figure 32.25A, the loop is horizontal. There is no magnetic flux through the loop, but the flux increases as the coil rotates. The loop's magnetic field $\vec{B}_{\text{loop}}$ must have a component in the direction opposite the permanent magnet's field $\vec{B}$, and the current through the bulb is from right to left. In Figure 32.25B, the loop is vertical and the magnetic flux is at its maximum. For the moment, there is no change in flux and no induced current. Finally, in part C, the loop is once again horizontal. Compare part A to part C, paying close attention to the color of the loop and the commutator. In part A, the yellow part is near the magnet's north pole; in part C, the yellow part is near the magnet's south pole and the loop has flipped over. However, the situation in both parts is similar. There is no magnetic flux through the loop, but the flux increases as the coil rotates. So the loop's magnetic field $\vec{B}_{\text{loop}}$ in part C must have a component in the direction opposite the permanent magnet's field $\vec{B}$, and the current through the bulb is from right to left exactly as in part A.

In an AC generator as shown in Figure 32.22, at times 3 and 7, when the loop flips, the induced current reverses direction. Figure 32.25 shows that the split ring commutator in the DC generator keeps the current in the same direction when the loop flips.

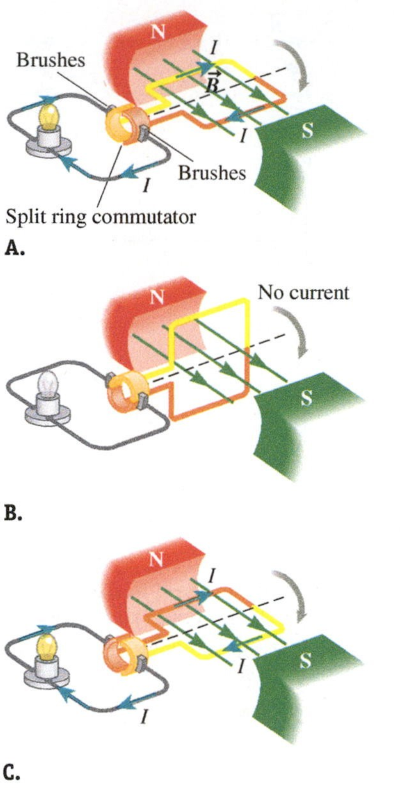

FIGURE 32.25 Compare this DC generator to the AC generator in Figure 32.21. **A.** The magnetic flux Φ_B is increasing, so $\vec{B}_{\text{loop}}$ points roughly opposite to $\vec{B}$ and the current through the bulb is to the *left*. **B.** Φ_B is at its maximum and momentarily constant, so $\vec{B}_{\text{loop}} = 0$, and there is no current. **C.** Φ_B is increasing, so $\vec{B}_{\text{loop}}$ points roughly opposite to $\vec{B}$ and the current through the bulb is to the *left* (same as in part A).

Time-Varying DC Emf and Current

A DC generator is not the same as a battery. In a battery, the current or emf is constant in both direction and magnitude. In a DC generator, the direction of the current or emf is constant, but the magnitude is not constant. In Figure 32.25A and C, there is nonzero current, but in part B, the current is zero.

The procedure for finding the induced emf in the DC generator is the same as the procedure for finding the induced emf in the AC generator with one exception. The emf in the DC generator does not reverse direction; it is always positive, so we take the absolute value of the emf for the AC generator in $\mathcal{E} = NBA\omega \sin \omega t = \mathcal{E}_{\max} \sin \omega t$ (Eq. 32.13). Because $NBA\omega$ is positive, take the absolute value of the sine function:

$$\mathcal{E} = NBA\omega |\sin \omega t| = \mathcal{E}_{\max} |\sin \omega t| \tag{32.18}$$

DC GENERATOR **Special Case**

The current is found by dividing the emf by the total resistance in the circuit:

$$I = \frac{NBA\omega}{R} |\sin \omega t| = I_{\max} |\sin \omega t| \tag{32.19}$$

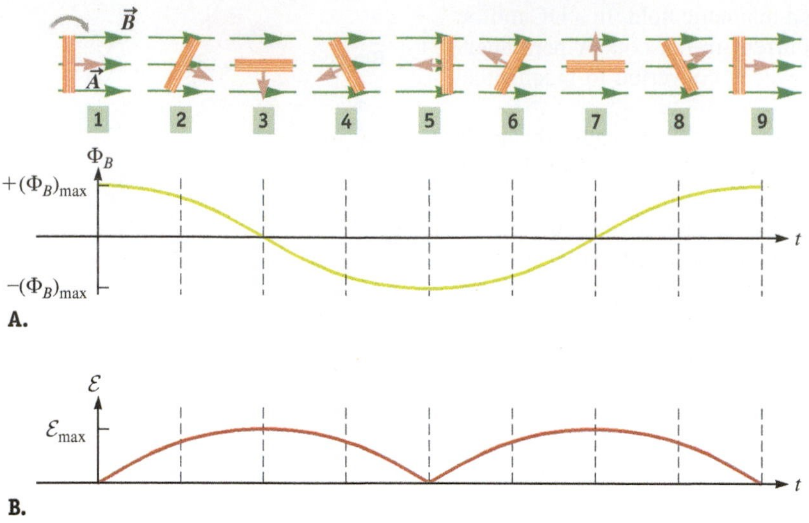

FIGURE 32.26 Graphs of **A.** the magnetic flux through the loop and **B.** the emf induced in a DC generator during one cycle. To make the comparison with the same graphs for an AC generator (Fig. 32.23) easy, the same time sequence 1 through 9 is shown.

Figure 32.26 shows the magnetic flux $\Phi_B = (\Phi_B)_{\max} \cos \omega t$ (Eq. 32.15) and the induced emf $\mathcal{E}$ (Eq. 32.18) plotted as functions of time t for one cycle. Consider Φ_B at times 1, 5, and 9. At these times, the flux is at an extreme value, so the time derivative is zero (Fig. 32.26A). We find that the induced emf $\mathcal{E}$ is zero at these three times, just as in Figure 32.23 for the AC generator. Between times 1 and 5, the flux is decreasing. The slope of the graph is negative and the induced emf is positive during this period, exactly as for the AC generator. Finally, consider the magnetic flux between times 5 and 9. The flux is increasing and the slope of the graph is positive, but because of the split ring commutator, the emf remains positive. Compare this sequence, 5 through 9, to the same time interval in Figure 32.23; the result is the opposite of what we found for the AC generator.

A battery produces an emf that is constant in magnitude, but the simple DC generator in Figure 32.25 produces an emf that varies between zero and $\mathcal{E}_{\max}$ (Fig. 32.26B). Many people—including Edison—made improvements to the DC generator. For example, including a loop perpendicular to the loop in Figure 32.25 results in an emf that varies less. By the late 1800s, the DC generator was efficient and widely used to light bulbs and run DC motors.

EXAMPLE 32.7 Gramme's DC Generator

Gramme's original DC generator was cranked by hand. It consisted of a coil with thirty turns and a diameter of roughly 51 cm. Assume the field of the permanent magnet was $B = 0.15$ T, and estimate the rotational speed needed to achieve an average emf of 110 V—the same as Edison required of his generator.

INTERPRET and ANTICIPATE

A key to solving this problem is finding an expression for the *average* emf from an equation that gives the *instantaneous* emf $\mathcal{E}$ as a function of time (Eq. 32.18).

SOLVE

Finding $\mathcal{E}_{\text{avg}}$ from $\mathcal{E} = \mathcal{E}_{\max}|\sin \omega t|$ (Eq. 32.18) means finding the average of the function $|\sin \omega t|$, which is the same as the average of $\sin \omega t$ over half a period. (The sine function is positive from 0 to $T/2$.) We can integrate $\sin \omega t$ over half a period and then divide by that time interval, $T/2$.

$$|\sin \omega t|_{\text{avg}} = \frac{1}{T/2}\int_0^{T/2} \sin \omega t \, dt = \frac{2}{T}\int_0^{T/2} \sin \omega t \, dt$$

Solve the integral by substitution (Appendix A).

$$|\sin \omega t|_{\text{avg}} = -\frac{2}{T}\left[\frac{1}{\omega}\cos \omega t\right]_0^{T/2}$$

Eliminate ω in favor of T.

$$\omega = \frac{2\pi}{T}$$

$$|\sin \omega t|_{\text{avg}} = -\frac{2}{T}\left[\frac{T}{2\pi}\cos \frac{2\pi}{T}t\right]_0^{T/2} = -\frac{1}{\pi}\left[\cos \frac{2\pi}{T}t\right]_0^{T/2}$$

Evaluate between limits.

$$|\sin \omega t|_{\text{avg}} = -\frac{1}{\pi}\left[\cos \frac{2\pi}{T}\frac{T}{2} - \cos 0\right] = -\frac{1}{\pi}[\cos \pi - \cos 0] = \frac{2}{\pi} \quad (1)$$

Start with Equation 32.18 and use Equation (1) to find an expression for the average emf.	$\mathcal{E}_{avg} = NBA\omega \lvert \sin \omega t \rvert_{avg} = NBA\omega \left(\frac{2}{\pi}\right)$
Solve for ω and substitute values. Assume the loop is circular, so from the given diameter, its area is $A = \pi r^2 = 0.20\,\text{m}^2$.	$\omega = \frac{\pi \mathcal{E}_{avg}}{2NBA} = \frac{\pi(110\,\text{V})}{2(30)(0.15\,\text{T})(0.20\,\text{m}^2)}$ $\omega \approx 190\,\text{rad/s} \approx 1800\,\text{rev/min}$

CHECK and THINK

Turning the crank 1800 times per minute (about 30 rps) is difficult for a person to do by hand for any substantial amount of time. DC generators are not usually hand-operated. Edison used a 500-hp steam engine to run the DC generator he used to light up hundreds of bulbs in the fall of 1882 in Lower Manhattan. This was one of the first steps in replacing gas-burning lamps with incandescent bulbs.

32-8 Case Study: Power Transmission and Transformers

Figure 32.27 illustrates how electrical power is transferred from a power plant to your home. When was the last time you saw a power plant? Probably on some long drive you took out of town. In the 1880s, however, Edison's DC generators were located in cities and towns because it wasn't practical to transmit electrical power more than about half a mile. Tesla pointed out to Edison that AC power plants could be located far outside the city, but Edison replied that he was not interested in high-voltage AC power because it was deadly. A person who touched any part of Edison's low-voltage DC system—from the generators to an incandescent bulb—would experience only a minor shock. Ultimately, Tesla won the argument; AC power lines run across the world. Yet Edison was right; high-voltage AC power lines are deadly. So, why is AC the standard used today? The answer has to do with the physics of power transmission as discussed in this section.

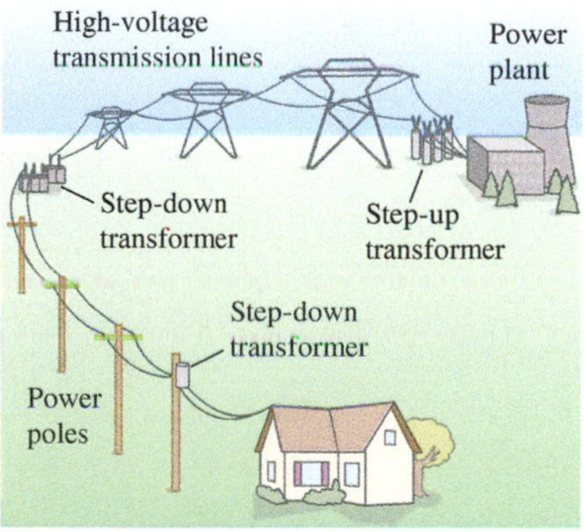

FIGURE 32.27 Electrical power is transmitted from the power plant to your home.

Root Mean Square Emf, Current, and Power

The instantaneous emf produced by an AC generator is given by $\mathcal{E} = NBA\omega \sin \omega t$ (Eq. 32.13), and the peak or maximum emf is given by $\mathcal{E}_{max} = NBA\omega$ (Eq. 32.16). The emf oscillates between $\mathcal{E}_{max}$ and $-\mathcal{E}_{max}$, so the average emf (over a whole number of periods) is zero: $\mathcal{E}_{avg} = 0$. An average emf of zero is not helpful in characterizing AC power sources. From Section 20-2, we know that the *root-mean-square* (rms) value is more useful in such circumstances. To find the root-mean-square emf $\mathcal{E}_{rms}$, we follow three steps:

1. *Square* the emf (Eq. 32.13):
$$\mathcal{E}^2 = (NBA\omega)^2 \sin^2 \omega t$$
2. Calculate the average over a whole number of periods, also known as the *mean*, of the emf squared:
$$(\mathcal{E}^2)_{avg} = [(NBA\omega)^2]_{avg}[\sin^2 \omega t]_{avg}$$
The average of the constant $[(NBA\omega)^2]_{avg}$ is $(NBA\omega)^2$. The average of the sine squared term is $\frac{1}{2}$ (Problem 55). So, the mean of the emf squared is
$$(\mathcal{E}^2)_{avg} = \frac{1}{2}(NBA\omega)^2$$
3. Take the *square root* of the average of the emf squared:
$$\mathcal{E}_{rms} = \sqrt{(\mathcal{E}^2)_{avg}} = \sqrt{\frac{1}{2}(NBA\omega)^2}$$

So the root-mean-square emf is

$$\mathcal{E}_{rms} = \frac{1}{\sqrt{2}}NBA\omega = \frac{\mathcal{E}_{max}}{\sqrt{2}} \tag{32.20}$$

In the United States and Canada, the standard rms voltage is $\mathcal{E}_{rms} = 120$ V. So the peak emf is

$$\mathcal{E}_{max} = \sqrt{2}\mathcal{E}_{rms} = \sqrt{2}(120\,\text{V}) = 170\,\text{V}$$

Now, let's consider the AC current given by $I = I_{max}\sin\omega t$ (Eq. 32.14), where the peak current is $I_{max} = \mathcal{E}_{max}/R = NBA\omega/R$ (Eq. 32.17). The average current (over a whole number of periods) is zero: $I_{avg} = 0$. The rms current is found by dividing the root-mean-square emf by the resistance:

$$I_{rms} = \frac{\mathcal{E}_{rms}}{R} = \frac{1}{\sqrt{2}}\frac{NBA\omega}{R}$$

$$I_{rms} = \frac{1}{\sqrt{2}}\frac{\mathcal{E}_{max}}{R}$$

Substitute Equation 32.17 to find a simple expression for the rms current:

$$I_{rms} = \frac{I_{max}}{\sqrt{2}} \tag{32.21}$$

Finally, we need expressions for the power supplied by an AC generator. The power supplied by any emf device is given by $P = I\mathcal{E}$ (Eq. 28.31), so the instantaneous power supplied by an AC generator comes from substituting Equations 32.14 and 32.13:

$$P = I_{max}\mathcal{E}_{max}\sin^2\omega t \tag{32.22}$$

Equation 32.22 says that the power's oscillation depends on a sine *squared*, so the average power is not zero. The average power is

$$P_{avg} = [I_{max}\mathcal{E}_{max}]_{avg}[\sin^2\omega t]_{avg}$$

$$P_{avg} = \frac{1}{2}I_{max}\mathcal{E}_{max} \tag{32.23}$$

By substituting Equations 32.20 and 32.21, we can write the average power in terms of the rms current and rms voltage:

$$P_{avg} = \frac{1}{2}(\sqrt{2}I_{rms})(\sqrt{2}\mathcal{E}_{rms})$$

$$P_{avg} = I_{rms}\mathcal{E}_{rms} \tag{32.24}$$

CONCEPT EXERCISE 32.8

The standard rms voltage in European countries is $\mathcal{E}_{rms} = 240$ V. What is the peak voltage in these countries?

Power Transmission in Wires

Edison's DC generators had a half-mile limit. Imagine the inconvenience of filling a city such as Manhattan with DC generators every mile or so. Why is the location of commercial DC generators so severely limited?

The problem with DC generators is that power is lost due to the resistance in the transmission wires. Normally, we approximate wires as having zero resistance, so we find that no power is lost. However, resistance is directly proportional to the length of a resistor or wire ($R = \rho\ell/A$; Eq. 28.24), so the resistance in long transmission lines cannot be ignored.

As a concrete example, suppose a DC generator supplies $P_{gen} = 150$ kW of power to a town just 2 or 3 miles away. The transmission lines have resistance $R = 0.25\ \Omega$.

Figure 32.28A is a simple model for the DC generator and the transmission lines. If the DC generator's emf $\mathcal{E} = 240\text{ V}$, the current in the transmission lines is given by $I = P_{\text{gen}}/V$ (Eq. 28.33):

$$I = \frac{150 \times 10^3\text{ W}}{240\text{ V}} = 625\text{ A}$$

The power lost due to resistance in the transmission lines is found from the current in the wires and their resistance, $P_{\text{loss}} = I^2R$ (Eq. 28.34):

$$P_{\text{loss}} = (625\text{ A})^2(0.25\ \Omega) \approx 98\text{ kW}$$

More than 65% of the power generated is lost in the transmission wires.

Edison's transmission lines were limited to about half a mile in order to reduce the transmission wire's resistance and reduce the power lost. He could, of course, also have reduced the resistance by making the transmission wires thicker, but this requires more copper (or whatever metal is used) and increases the cost.

In our example, the resistance is not very high (0.25 Ω), and reducing it much won't significantly reduce the power lost. In $P_{\text{loss}} = I^2R$ (Eq. 28.34), the power lost depends linearly on the wire's resistance but on the *square* of the current. The best way to reduce the power lost is to reduce the current in the wire, but this means transmitting the power at a higher voltage, exactly as Tesla proposed to do with his AC generator.

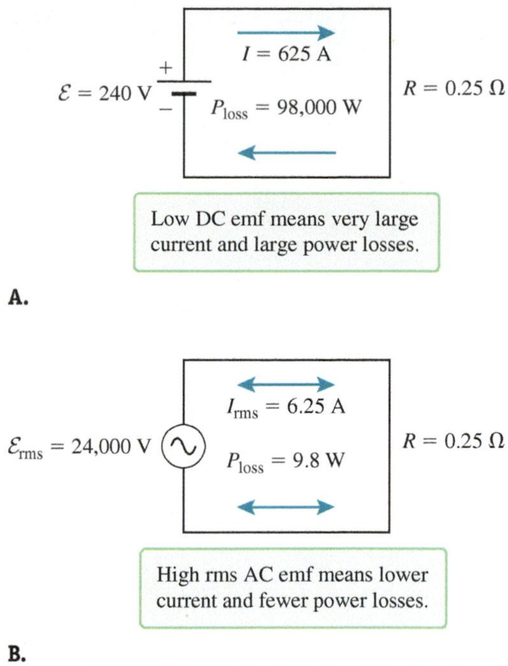

FIGURE 32.28 **A.** A DC emf device (such as a DC generator) powers a town and loses a lot of power. **B.** An AC generator powers the same town with less power lost in the wires. (These circuits only represent the resistance in the transmission wires and do not include the resistance of the devices powered in the town.)

What happens if we replace the DC generator with an AC generator in this example? As before, the AC generator supplies average power $P_{\text{gen}} = 150\text{ kW}$ to a town just 2 or 3 miles away. Figure 32.28B is a simple model for the AC generator and the transmission lines; the symbol for an AC generator is a wavy line inside a circle —(∿)—. The transmission lines have the same resistance $R = 0.25\ \Omega$. If the AC generator's rms emf is $\mathcal{E}_{\text{rms}} = 24{,}000\text{ V}$ (100 times greater than the DC generator's emf), the rms current in the transmission lines is given by Equation 32.24:

$$I_{\text{rms}} = \frac{P_{\text{gen}}}{\mathcal{E}_{\text{rms}}} = \frac{150 \times 10^3\text{ W}}{24{,}000\text{ V}} = 6.25\text{ A}$$

The average power lost in a resistor carrying an AC current is

$$P_{\text{loss}} = I_{\text{rms}}^2 R \qquad (32.25)$$

So the power lost in the transmission wires is

$$P_{\text{loss}} = (6.25\text{ A})^2(0.25\ \Omega) \approx 9.8\text{ W}$$

You may be wondering why an AC generator is necessary. As long as the voltage is high, the current in the transmission wires will be small and the power loss will be small. Why can't we just use a very high voltage DC generator? Part of the answer has to do with Edison's objection—high voltage is deadly. We don't really want to have 24,000 V between the terminals of every outlet in our house.

Transformers

TRANSFORMER Special Case

Westinghouse—Edison's rival—understood the real benefit of AC power after he learned about a European invention known today as a transformer. A **transformer** is a device that changes the voltage of alternating current; it does not work on direct current. Westinghouse's great insight was to realize that the transformer would allow AC generators to be located far from metropolitan areas so they could be powered by waterfalls. Additionally, one generator could serve the needs of a large number of people. In Figure 32.27, three transformers are shown. The **step-up transformer** near the power plant increases the voltage so the power loss in transmission is minimal. The two **step-down transformers** lower the voltage for transmission in town and into your home.

A transformer consists of two coils—the input or primary coil and the output or secondary coil. Westinghouse and his employees designed the modern transformer so that most of the magnetic flux produced by the primary coil passes through the secondary coil. Figure 32.29 shows two designs. In Figure 32.29A, the primary coil is inside the secondary coil, and in Figure 32.29B, the coils are both wrapped around an iron core.

The primary coil in Figure 32.29B is connected to an AC generator. Because the current in the coil oscillates, the magnetic field created by the coil oscillates. The coil's magnetic field creates an oscillating magnetic field in the iron core. Because the secondary coil is wrapped around that core, the magnetic flux through the secondary coil oscillates. According to Faraday's law (Eq. 32.6), the changing magnetic flux in the secondary coil induces an emf and current in the secondary coil.

The relative voltage V across the two coils depends on the relative number of turns:

$$\frac{V_S}{V_P} = \frac{N_S}{N_P} \tag{32.26}$$

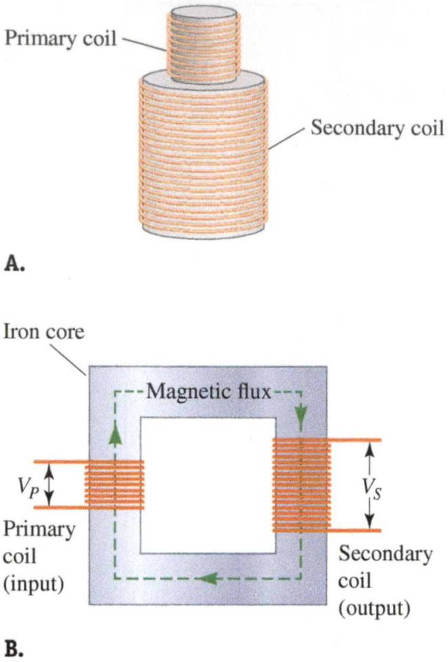

FIGURE 32.29 Two designs for transformers. **A.** The primary coil is inside the secondary coil. **B.** Both coils are wound around an iron core. The primary coil has N_P turns and carries a current I_P; the secondary coil has N_S turns and carries a current I_S.

where the subscript P stands for primary and the subscript S for secondary. In a step-up transformer, the voltage across the secondary coil is higher than the voltage across the primary coil, so the secondary coil has more turns than the primary coil: $N_S > N_P$. In a step-down transformer, the voltage across the secondary coil is lower than the voltage across the primary coil, so the secondary coil has fewer turns than the primary coil: $N_S < N_P$.

In an ideal transformer, there is no power loss from the primary to the secondary coil. In well-designed real transformers, only about 1% of the power is lost. So, if we assume the power is constant, we can find the relative current in the secondary coil:

$$P_{\text{in}} = P_{\text{out}}$$
$$I_P V_P = I_S V_S$$

Rearrange and use Equation 32.26:

$$\frac{I_S}{I_P} = \frac{V_P}{V_S} = \frac{N_P}{N_S} \tag{32.27}$$

Usually the voltages in Equation 32.27 are rms voltages.

By Equation 32.27, a step-up transformer increases the voltage $V_S > V_P$ and decreases the current $I_S < I_P$. Step-up transformers are used just before the current is sent through long power lines to reduce the power loss in those wires (Fig. 32.30A). Step-down transformers are used when a lower voltage is required. You can find step-down transformers in substations in towns, on utility poles near buildings, and in household devices (Fig. 32.30B and C).

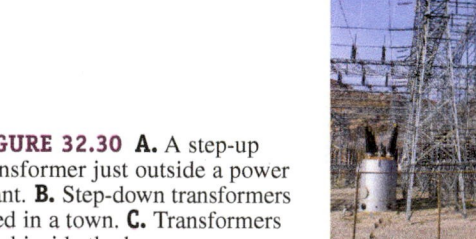

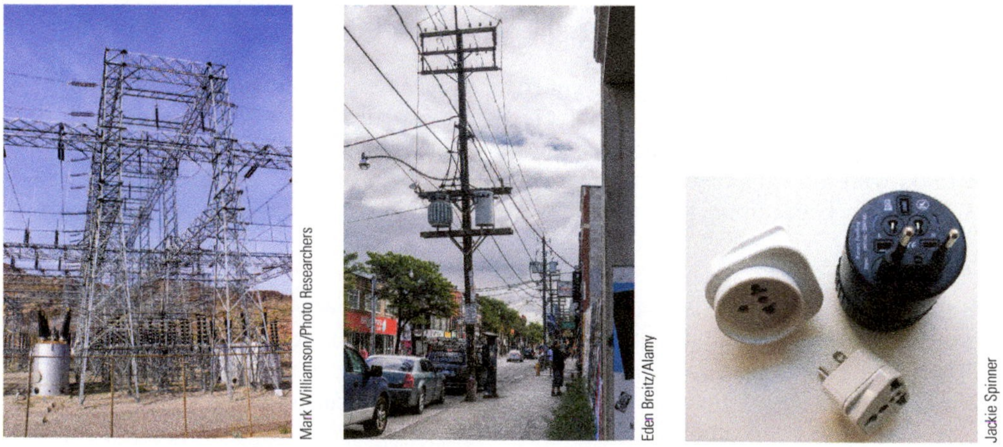

A. B. C.

FIGURE 32.30 **A.** A step-up transformer just outside a power plant. **B.** Step-down transformers used in a town. **C.** Transformers used inside the home.

CONCEPT EXERCISE 32.9

Why can't you use a transformer to change the voltage of a direct current produced by **a.** a battery and **b.** a DC generator (Figs. 32.25 and 32.26)?

EXAMPLE 32.8 Big Metal Boxes in Town

You have probably seen substations (step-down transformers) like the one in Figure 32.27 in your town. The input voltage at a substation is rather high—assume 650 kV. The output voltage is typically lower than 10,000 V—assume 6.5 kV. The current in high-voltage transmission lines is typically 2 kA. For a simple ideal transformer, find the current that leaves the substation.

INTERPRET and ANTICIPATE

This is a straightforward application of Equation 32.27, giving us a chance to build our intuition about the magnitude of the voltage and current in a typical power distribution grid.

SOLVE

The primary coil is connected to the high-voltage lines, and the secondary coil is the output of the substation. Solve for the secondary current I_S.

$$\frac{I_S}{I_P} = \frac{V_P}{V_S} \quad (32.27)$$

$$I_S = \frac{V_P}{V_S} I_P = \left(\frac{650\,\text{kV}}{6.5\,\text{kV}}\right)(2 \times 10^3\,\text{A})$$

$$I_S = 2 \times 10^5\,\text{A} = 200\,\text{kA}$$

CHECK and THINK

This may seem like a huge current. In fact, it is. Besides stepping down the voltage, the substation splits the current so it can be directed along different distribution wires to serve many regions in the town.

Who Won the War?

A turning point in the war over current was the 1893 World's Fair in Chicago. Westinghouse and Edison both submitted bids to supply electricity for the fair. Westinghouse was awarded the contract, and the fair was run on AC power. By this time, Tesla's AC motor was working. Millions of visitors saw how electric motors could be used to do tasks that were being done by people or work animals. Westinghouse also demonstrated the flexibility of AC power. His original contract required him to light 92,000 bulbs. However, by the time the fair was running, about 180,000 bulbs were needed, and his AC generator was easily able to supply the extra power. Finally, in 1893, Westinghouse won the contract to build the first great hydroelectric power plant using Niagara Falls as the source of mechanical power. The AC power generated was then transmitted to Buffalo, demonstrating that Tesla's AC system had been perfected.

SUMMARY

Underlying Principles

1. **Faraday's law**: A changing magnetic flux through a conductor induces an emf in that conductor.
2. **Lenz's law** follows from the principle of conservation of energy: The induced current produces a magnetic field that always acts to oppose the change that created the induced current. Lenz's law can be expressed mathematically as a minus sign in Faraday's law. The combination is known as **Faraday–Lenz's law**:

$$\mathcal{E} = -N\frac{d\Phi_B}{dt} \quad (32.6)$$

Major Concepts

When a conductor such as a metal sheet is moving with respect to a magnetic field, the conduction electrons in the conductor experience a magnetic force. As a result, **eddy currents** develop in the conductor.

Special Cases

1. The motional emf generated by a metal bar in a **slide generator** is
$$\mathcal{E} = B\ell v \quad (32.7)$$
2. The emf produced by an **AC generator** is
$$\mathcal{E} = NBA\omega \sin \omega t = \mathcal{E}_{\max} \sin \omega t \quad (32.13)$$
3. The emf produced by a **DC generator** is
$$\mathcal{E} = NBA\omega |\sin \omega t| = \mathcal{E}_{\max} |\sin \omega t| \quad (32.18)$$
4. A **transformer** is a device that changes the voltage of alternating current. It consists of two coils—the input or primary coil and the output or secondary coil. The relative current and voltage depend on the relative number of turns:
$$\frac{I_S}{I_P} = \frac{V_P}{V_S} = \frac{N_P}{N_S} \quad (32.27)$$

PROBLEMS AND QUESTIONS

A = algebraic **C** = conceptual **E** = estimation **G** = graphical **N** = numerical

32-1 Another Kind of Emf

1. A constant magnetic field of 0.275 T points through a circular loop of wire with radius 3.50 cm as shown in Figure P32.1.
   a. **N** What is the magnetic flux through the loop?
   b. **C** Is a current induced in the loop? Explain.

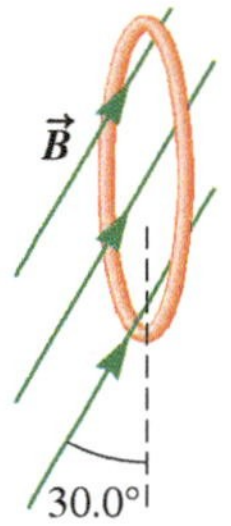

FIGURE P32.1

2. **C** In each of the following cases, a conducting loop is coaxial with a solenoid. Determine whether an emf is induced in the loop when the current in the solenoid is **a.** constant, **b.** increasing, and **c.** decreasing.
3. **A** A solenoid's long axis is perpendicular to the page and to the rectangular loop shown in Figure P32.3. The loop is completely inside the solenoid. The cross-sectional area of the solenoid is A_{sol}, and the area of the loop is A_{loop}. The solenoid's magnetic field $\vec{B}$ points out of the page. Find the magnetic flux Φ_B through the loop.

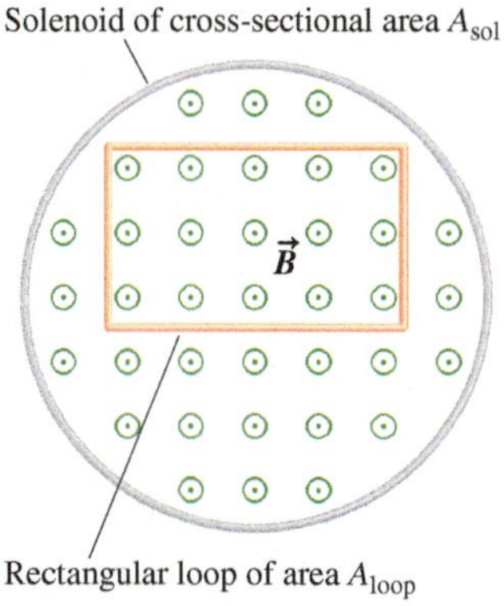

FIGURE P32.3

4. **C** Suppose magnetic monopoles exist. Would a current of monopoles be the source of an electric field, a magnetic field, or both? Explain.

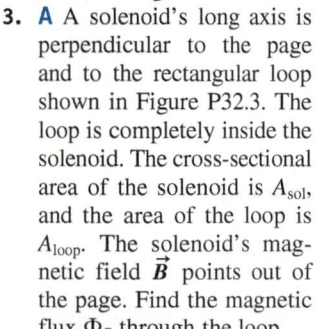

32-2 Faraday's Law

5. **N** The intensity of the Earth's magnetic field near the equator is 35.0 μT. A circular coil with 40 turns and a radius of 75.0 cm is placed so its axis points along the direction of the Earth's magnetic field. The coil is then rotated through an angle of 225° in 50.0 ms. What is the magnitude of the average emf generated in the circular coil?
6. **C** Figure P32.6 shows three situations involving a single circular loop of wire. For each case, decide whether an emf is induced in the circular loop. Explain your reasoning.
   Case 1: The loop lies in the plane of the page near a solenoid with its long axis perpendicular to the page. The solenoid's magnetic field is increasing.
   Case 2: A solenoid carries a constant current. The loop falls straight down inside the solenoid.
   Case 3: A solenoid carries a constant current. The loop is wrapped around a ball. The ball and loop roll along the inside of the solenoid.

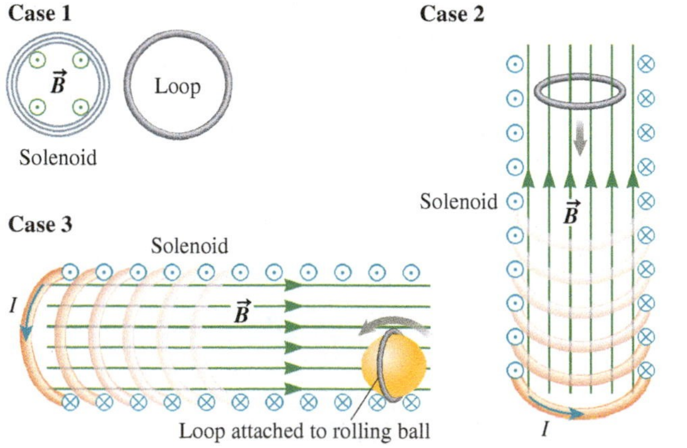

FIGURE P32.6

7. **A** A rectangular loop of length L and width W is placed in a uniform magnetic field $\vec{B}$ with its plane perpendicular to the field (Fig. P32.7). Determine the time-averaged induced emf if the loop rotates with constant angular velocity ω through an angle of 180° around an axis passing through the loop's center **a.** perpendicular to the loop and **b.** parallel to its width.

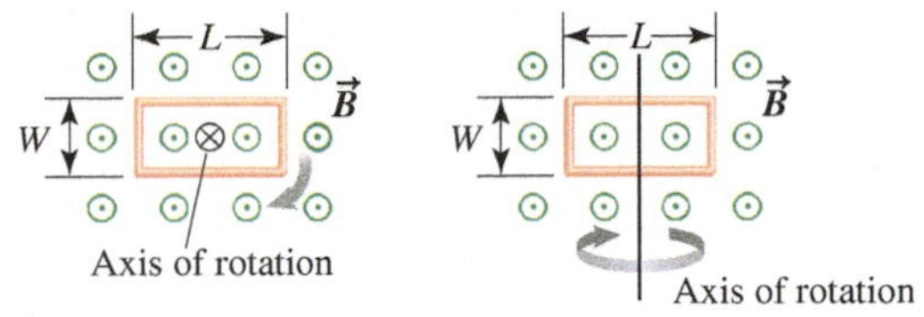

FIGURE P32.7

8. **G** The magnetic field through a square loop of wire with sides of length 3.00 cm changes with time as shown in Figure P32.8, where the sign indicates the direction of the field relative to the axis of the loop. Plot the emf induced in the loop versus time.

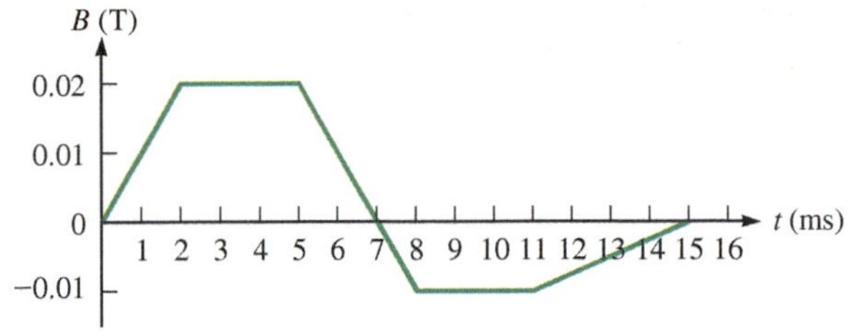

FIGURE P32.8

9. **N** A light conducting string is formed into a circular hoop with initial radius 20.0 cm. The hoop is placed in a 2.00-T uniform magnetic field so that its plane is perpendicular to the field, and the string is drawn to collapse the hoop to zero area in 125 ms. What is the average emf generated in the hoop during its collapse?
10. **N** A wire is formed into a square loop with sides d = 18.0 cm and positioned in a spatially uniform magnetic field with its plane perpendicular to the direction of the field. What is the magnitude of the average emf induced in the loop if the magnitude of the magnetic field is increased by 30.0 mT per second?

Problems 11 and 12 are paired.

11. Suppose a uniform magnetic field is perpendicular to the $8\frac{1}{2} \times$ 11-in. page of your homework and a rectangular metal loop lies on the page. The loop's sides line up with the edges of the page. The magnetic field is changing with time as described by $B = 3.75 \times 10^{-3}\,t$, where B is in teslas and t is in seconds.
    a. **C** Is the magnetic field increasing or decreasing?
    b. **N** Find the magnitude of the emf induced in the loop.
12. Suppose a uniform magnetic field is perpendicular to the 8½ × 11-in. page of your homework and a rectangular metal loop is perpendicular to the page such that one of its sides bisects the page into two long strips. The loop has the same dimensions as the page. The magnetic field is changing with time as described by $B = 3.75 \times 10^{-3}\,t^{-2}$, where B is in teslas and t is in seconds.
    a. **C** Is the magnetic field increasing or decreasing?
    b. **N** Find the emf induced in the loop.
13. **N** A square conducting loop with side length a = 1.25 cm is placed at the center of a solenoid 40.0 cm long with a current of 4.30 A flowing through its 420 turns, and it is aligned so that the plane of the loop is perpendicular to the long axis of the solenoid. The radius of the solenoid is 5.00 cm. **a.** What is the magnetic flux through the loop? **b.** What is the magnitude of the average emf induced in the loop if the current in the solenoid is increased from 4.30 A to 10.0 A in 1.75 s?

Problems 14 and 15 are paired.

14. **A** The magnetic field in a region of space is given by $\vec{B} = B_x\hat{\imath} + B_y\hat{\jmath}$. A coil of N turns is oriented so that its cross-sectional area is in the x direction: $\vec{A} = A_x\hat{\imath}$. A small bulb is connected across the ends of the coil. The total resistance of the coil and the bulb is R. Find an expression for the current through the bulb if $B_x(t) = B_0(t/t_0)^2$ and $B_y = B_0$, where B_0 and t_0 are constants.
15. **A** The magnetic field in a region of space is given by $\vec{B} = B_x\hat{\imath} + B_y\hat{\jmath}$. A coil of N turns is oriented so that its cross-sectional area is in the x direction: $\vec{A} = A_x\hat{\imath}$. A small bulb is connected across the ends of the coil. The total resistance of the coil and the bulb is R. Find an expression for the current through the bulb if $B_x(t) = B_0$ and $B_y = B_0(t/t_0)^2$, where B_0 and t_0 are constants.
16. **N** A coil of 255 turns and area 0.425 m^2 rotates in a uniform magnetic field B = 0.325 T such that the angle between the magnetic field and the area vector is given by $\varphi(t) = 6.35t$ rad. If the coil starts from rest at t = 0, what is the magnitude of the emf induced in the coil at t = 25.8 s?

32-3 Lenz's Law

17. **C** In Figure P32.17, a bar magnet oscillates back and forth near a stationary loop. Consider a half-cycle starting with the bar magnet at its farthest point from the loop. Describe the current induced in the loop from the perspective of an observer on the opposite side from the magnet (looking from the right in the figure).

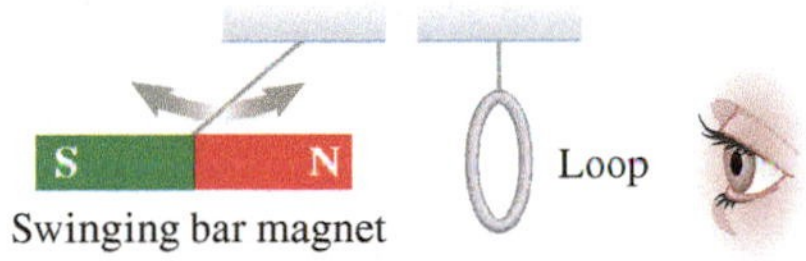

FIGURE P32.17

18. **G** Imagine dropping a bar magnet through a horizontal conducting loop so that the north pole falls through the loop first. Assume a clockwise current as seen from above the loop is positive. Sketch the current as a function of time from this perspective. Remember that gravity accelerates the bar magnet. Explain the major features of your graph.
19. **N** A square loop with side length 5.00 cm is on a tabletop in a uniform magnetic field with the field pointing perpendicular to the loop, downward into the table. The field decreases from 0.500 T to 0.100 T in 0.15 s, and the loop has a resistance of 18.0 Ω. What are the induced current and its direction, as viewed from above?
20. **A** A thin copper rod of length L rotates with constant angular velocity ω about a point O, in a plane perpendicular to a uniform magnetic field $\vec{B}$ as shown in Figure P32.20. Determine the induced emf across its ends. Consider that the emf produced between the point O and a small segment of the rod, $d\vec{\ell}$, is given by $d\mathcal{E} = Bv\,d\ell$.

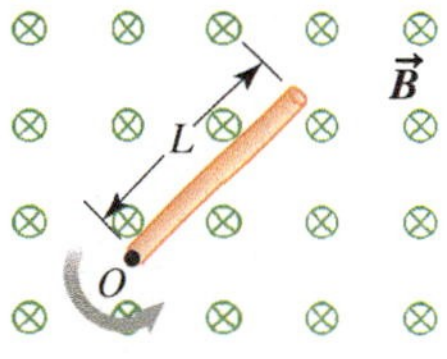

FIGURE P32.20

21. **N** Figure P32.21 shows a circular conducting loop with a 5.00-cm radius and a total resistance of 1.30 Ω placed within a uniform magnetic field pointing into the page. **a.** What is the rate at which the magnetic field is changing if a counterclockwise current $I = 4.60 \times 10^{-2}$ A is induced in the loop? **b.** Is the induced current caused by an increase or a decrease in the magnetic field with time?

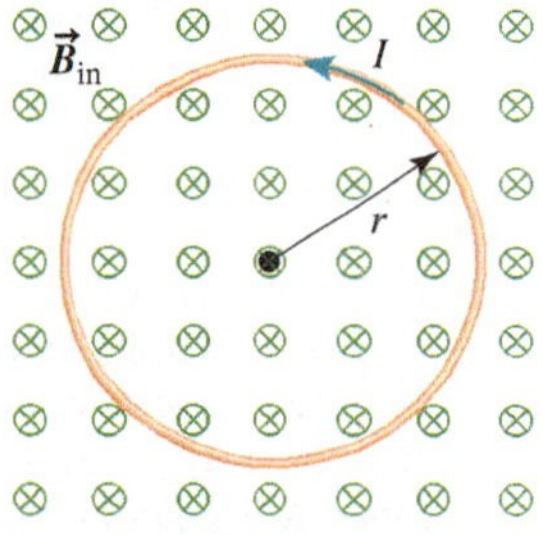

FIGURE P32.21

22. **G** Two solenoids are placed next to each other as shown in Figure P32.22. The solenoid on the left is connected to a battery and a switch. The switch is initially positioned such that the battery is not connected. It is then switched to include the battery, held in place for a minute, and then returned to the initial position. Sketch the current through the second solenoid versus time. Indicate the direction of current flow in the second solenoid as viewed from the first solenoid. Assume the solenoids have a nonzero resistance.

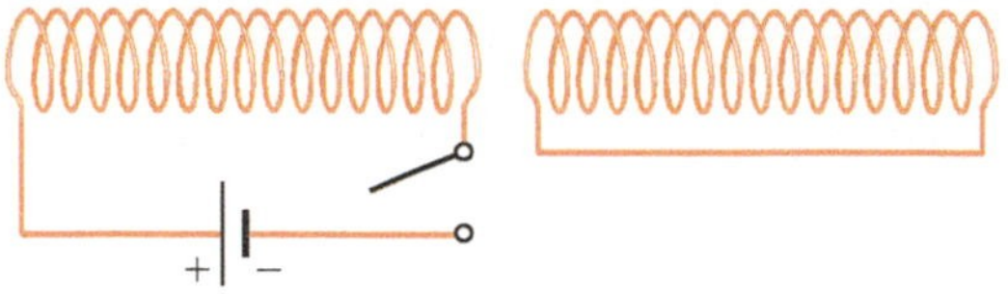

FIGURE P32.22

23. **A** A square loop with side length L, mass M, and resistance R lies in the xy plane. A magnetic field $\vec{B} = B_0(y/L)\hat{k}$ is present in the region of the space near the loop. Determine the magnitude and direction of the induced current in the loop as the loop starts moving at velocity $\vec{v} = v_y\hat{j}$.

24. **C** A small loop is arranged coaxially with a larger loop. The larger loop has a battery in series with a variable resistor such that a current is flowing clockwise in the larger loop as viewed from your vantage point. In what direction will the induced current flow in the small loop when the resistance is decreased?

25. **N** A coil with cross-sectional area 1.50×10^{-2} m^2 and 225 turns is positioned in a uniform magnetic field so that the plane of the coil is perpendicular to the direction of the field. The time-varying magnitude of the magnetic field is given by $B = 4.00 - 0.0200t - 0.00500t^2$, with B in teslas and t in seconds. What is the emf induced in the coil when $t = 2.00$ s?

26. **G** A rectangular loop lies in the plane of the page. In a small region of space, there is a uniform magnetic field that is perpendicular to the page as shown in Figure P32.26. **a.** The loop moves at a constant velocity from region L through region C to region R. Sketch a plot of the current in the loop versus time. Assume clockwise current is positive from your perspective. **b.** How does your answer change if the loop moves at a constant velocity from region R through C to L?

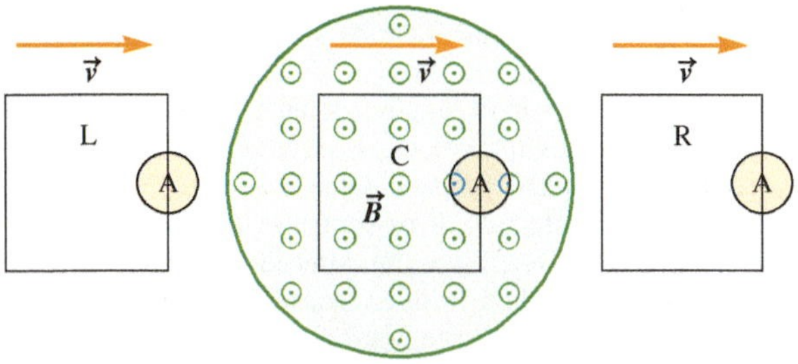

FIGURE P32.26

27. **N** A vibrating copper wire is magnetized so that it produces a variable magnetic field given by $B = 25.0 - 1.55\cos(67.0\pi t)$ perpendicular to the plane of a 14-turn coil with a cross-sectional area of 0.500 cm^2 placed nearby. How does the emf induced in the coil vary with time?

28. **A** A solenoid of area A_{sol} produces a uniform magnetic field (Fig. P32.28; shown in cross section). The solenoid's magnetic field points out of the page and is decreasing according to $B = B_0(t_0/t)^2$. A single conducting loop of area A_{loop} and resistance R is coaxial with the solenoid, with $A_{loop} < A_{sol}$. Find an expression for the current in the loop. What does the sign of your answer mean?

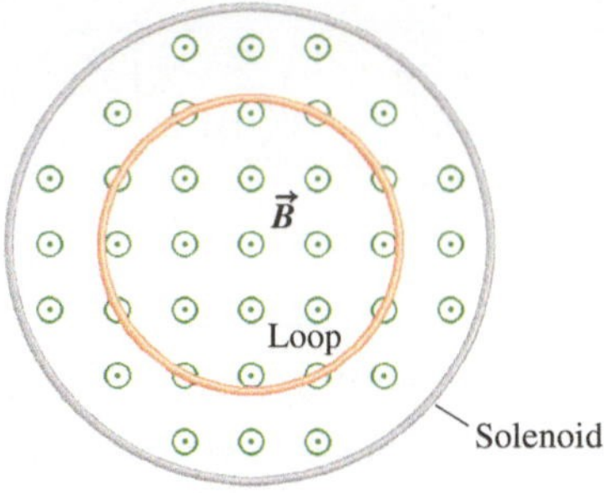

FIGURE P32.28

29. **C** Two circular conductors are perpendicular to each other as shown in Figure P32.29. Suppose conductor B carries a current. Will a current be induced in conductor A if there is a change in the current in conductor B? (The loops are insulated from one another.)

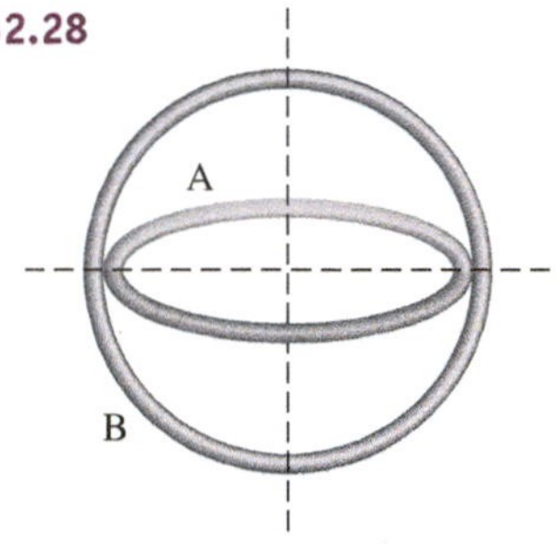

FIGURE P32.29

32-4 Lenz's Law and Conservation of Energy

Problems 30 and 31 are paired.

30. **C** Two circular conducting loops labeled A and B are close together and their axes are parallel. Loop A is connected to a power supply, and the current in A is increasing linearly with time. Loop B is not connected to a power supply, but the current in B runs through a bulb. **a.** Describe the current induced in loop B and the brightness of the bulb as a function of time. **b.** Explain how your answer is consistent with the principle of conservation of energy. **c.** If the current in loop A is decreasing instead of increasing, how do your answers change?

31. **C** Two circular conducting loops labeled A and B are close together and their axes are parallel. Loop A is connected to a power supply, and the current in A is increasing. Loop B is not connected to a power supply. Loop B is free to move toward or away from A. **a.** Does loop B move toward A, away from A, or not at all? Explain. **b.** If the current in loop A is decreasing instead of increasing, how does your answer change?

Problems 32 and 33 are paired.

32. **C** You are attending a magic show at a local theater, and the magician has the stage cleared except for a cylindrical copper pipe 2.0 m long and a small wooden block. The magician drops the wooden block through the top of the pipe, and it falls out the bottom about 0.7 s after it is released. You think this is what would have happened if the magician had simply dropped the wooden block from the same height. The magician then taps his wand three times against the pipe, while uttering an incantation, and drops a similar-looking block through the pipe. This time, however, the block takes significantly longer to fall out the bottom. You suspect the magician used a different block. What might have been different about this block? Explain how this trick could be done with a magnet and a copper pipe.

33. **C** Joanna first drops a magnet through a plastic pipe and then through a copper pipe. The pipes are the same length, but the magnet takes a much longer time to fall through the copper pipe. This means the final velocity of the magnet is much lower in one case than the other, yet the magnet began with the same total mechanical energy in each case. No nonconservative forces are exerted on the magnet–pipe system in either case. What happened to the supposed "missing energy" in the case where the magnet fell through the copper pipe?

32-5 Case Study: Slide Generator

34. **E** Estimate the magnitude of the emf induced across the ends of a commercial jet airplane's wings as a result of its motion through the Earth's magnetic field. Would it be practical to use this induced emf to power devices on the plane? Explain.

35. A slide generator has a movable bar with a length of 0.355 m on a U-shaped conductor and is in a magnetic field with a magnitude of 1.50 T, perpendicular to the plane of the generator.

a. **N** How fast must the bar move to create an induced emf with a magnitude of 12.0 V?

b. **C** Does it matter whether the bar is moving in one direction or the other along the U-shaped conductor? What is different when the bar moves in one direction versus the other?

Problems 36 and 37 are paired.

36. **A** Find an expression for the current in the slide generator in Figure 32.14 (page 1020) as a function of x. The U-shaped conductor and the bar have cross-sectional area A and resistivity ρ. The bar's length is ℓ, and the magnetic field is B. The bar is pulled at constant speed v. *Hint*: Your expression does not involve R.

37. **N** The slide generator in Figure 32.14 (page 1020) is in a uniform magnetic field of magnitude 0.0500 T. The bar of length 0.365 m is pulled at a constant speed of 0.500 m/s. The U-shaped conductor and the bar have a resistivity of $2.75 \times 10^{-8}\ \Omega \cdot \text{m}$ and a cross-sectional area of $8.75 \times 10^{-4}\ \text{m}^2$. Find the current in the generator when $x = 0.650$ m.

38. **C** CASE STUDY In 1887, Tesla was in need of funding to work on his AC system. He was introduced to an investor. The investor said he wanted Tesla to impress him, but the investor refused to watch a demonstration of the AC system. Tesla remembered that Christopher Columbus was able to make a great impression and gain an audience with Queen Isabella by balancing an egg on its end. Columbus did this by cracking the shell slightly. Isabella—of course—funded Columbus's expedition. So, Tesla said he would balance an egg on end without cracking its shell. Tesla installed a rotating magnetic field below a wooden tabletop. His hard-boiled egg was coated in copper. Not only was the egg standing on end, but it was rapidly rotating. The investor provided Tesla with the funds needed to start his own company—the Tesla Electric Company. Explain Tesla's demonstration.

39. A thin conducting bar (60.0 cm long) aligned in the positive y direction is moving with velocity $\vec{v} = (1.25\ \text{m/s})\hat{\imath}$ in a region with a spatially uniform 0.400-T magnetic field directed at an angle of 36.0° above the xy plane.

a. **N** What is the magnitude of the emf induced along the length of the moving bar?

b. **C** Which end of the bar is positively charged?

40. **A** A stiff spring with a spring constant of 1200.0 N/m is connected to a bar on a slide generator as shown in Figure P32.40. Assume the bar has length $\ell = 60.0$ cm and mass $m = 0.75$ kg, and it slides without friction. The bar connects to a U-shaped wire to form a loop that has width $w = 40.0$ cm and total resistance 25 Ω and that sits in a uniform magnetic field $B = 0.35$ T. The bar is initially pulled 5.0 cm to the left and released so that it begins to oscillate. What is the induced current in the loop as a function of time, $I(t)$? (Ignore any effects due to the magnetic force on the oscillating bar.)

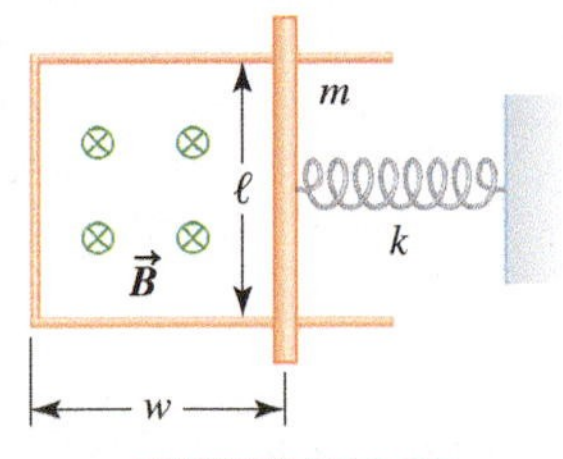

FIGURE P32.40

32-6 Case Study: AC Generators

41. **N** A generator spinning at a rate of 1.20×10^3 rev/min produces a maximum emf of 45.0 V. At what angular speed does this generator produce a maximum emf of 64.0 V?

42. **C** Suppose you have a simple homemade AC generator like the one in Figure 32.20 (page 1026) whose emf is about half of what you need to power a particular motor. How can you easily modify the generator to suit your needs? Explain your modifications specifically.

43. **N** A simple AC generator has a coil of exactly 250 turns (Fig. 32.20, page 1026). It rotates at 60.0 Hz in a magnetic field of 0.215 T. If the generator must produce a maximum emf of 170.0 V, what is the area of the coil? If the coil is square, what is the length of each side? Compare the size of the coil to some easy-to-visualize object.

44. Tesla's AC generator had three phases. To understand the basic idea behind his generators, imagine adding two loops to the AC generator in Figure 32.20 (page 1026). The three loops are set so that each is 120° from the other two.

a. **G** For each loop, make a graph of $\mathcal{E}$ versus t similar to the one in Figure 32.23B (page 1029).

b. **C** Imagine powering a motor with the three-phase generator. In what way is a three-phase generator superior to a one-phase generator?

45. **N** The magnetic flux through each turn of a 140-turn coil is given by $\Phi_B = 8.75 \times 10^{-3} \sin \omega t$, where ω is the angular speed of the coil and Φ_B is in webers. At one instant, the coil is observed to be rotating at a rate of 8.90×10^2 rev/min. **a.** What is the induced emf in the coil as a function of time for this angular speed? **b.** What is the maximum induced emf in the coil for this angular speed?

46. **C** Given a spool of wire, discuss the feasibility of using a rotating loop of wire in the Earth's magnetic field (about 10^{-4} T) to create a hand-powered AC generator to produce household voltage (170-V peak voltage with a frequency of 60 Hz).

47. **N** A square coil with a side length of 12.0 cm and 34 turns is positioned in a region with a horizontally directed, spatially uniform magnetic field of 82.0 mT and set to rotate about a vertical axis with an angular speed of 1.20×10^2 rev/min. **a.** What is the maximum emf induced in the spinning coil by this field? **b.** What is the angle between the plane of the coil and the direction of the field when the maximum induced emf occurs?

48. **N** A 44-turn rectangular coil with length $\ell = 15.0$ cm and width $w = 8.50$ cm is in a region with its axis initially aligned to a horizontally directed uniform magnetic field of 745 mT and set to rotate about a vertical axis with an angular speed of 64.0 rad/s. **a.** What is the maximum induced emf in the rotating coil? **b.** What is the induced emf in the rotating coil at $t = 1.00$ s? **c.** What is the maximum rate of change of the magnetic flux through the rotating coil?

32-7 Case Study: Faraday's Generator and Other DC Generators

49. **C** Explain Tesla's statement that all generators are really AC generators.

50. **C** Suppose you light a bulb with a battery, then Faraday's generator, then a DC generator, and finally an AC generator. Describe the brightness of the bulb as a function of time in each case. (Does the light flicker?) Do the lights in your room flicker? Explain.

51. **N, C** A typical flashlight uses two 1.5-V batteries. Could you crank Gramme's DC generator by hand to light such a bulb? Your answer should involve a short calculation similar to Example 32.7 (page 1032).

52. **C** An AC generator, a DC generator, and a DC battery with the same peak voltage are each connected to a lightbulb. Sketch the current through the lightbulb when it is connected to each generator separately, and rank them from the largest average current to the smallest average current.

32-8 Case Study: Power Transmission and Transformers

53. **N** The maximum value of the emf in the primary coil ($N_P = 1000$) of a transformer is 175 V. **a.** What is the maximum induced emf in the secondary coil ($N_S = 250$)? **b.** What is the ratio of the current in the primary coil to the current in the secondary coil?

Problems 54 and 55 are paired.

54. **A** Find the average of $f(x) = \sin x$ between $x = 0$ and 2π. What does your answer imply about the emf produced by an AC generator, averaged over one period?

55. **A** Find the average of $f(x) = \sin^2 x$ between $x = 0$ and 2π. What does your answer imply about the power supplied by an AC generator, averaged over one period?

56. **N** You have a 750-W hair dryer that is designed to work on the American power grid ($\mathcal{E}_{rms} = 120.0$ V, 60.0 Hz). **a.** What is the rms current in the dryer? **b.** If you connected the dryer to the European power grid ($\mathcal{E}_{rms} = 240.0$ V, 50.0 Hz), what would be the current in the dryer? What would happen to the dryer and why? **c.** If you used a transformer to connect the dryer to the European grid, what ratio of N_S/N_P would be required?

57. **N** An electric toothbrush charger is one example of a small transformer used in your home. The secondary coil in the toothbrush gets an induced emf from the primary coil in the charger when they are aligned. (The toothbrush coil is placed within the primary coil when it is plugged in.) **a.** If the maximum value of the emf in the primary coil ($N_P = 1000$) is 112 V, what is the maximum induced emf in the secondary coil ($N_S = 150$)? **b.** A maximum current of 0.050 A flows through the primary coil. What is the average power delivered to the toothbrush?

Problems 58 and 59 are paired.

58. **N** A step-down transformer has 65 turns in its primary coil and 10 turns in its secondary coil. The primary coil is connected to standard household voltage ($\mathcal{E}_{rms} = 120.0$ V, 60.0 Hz). **a.** What is the rms voltage in the secondary coil? **b.** If, instead, the transformer is connected to a 6-V DC battery, what is the rms voltage in the secondary coil after the first few milliseconds?

59. **N** Suppose the secondary coil in Problem 58(a) is connected to a 12.0-Ω resistor. **a.** What is the maximum current in the resistor? **b.** What is the average power dissipated by the resistor?

General Problems

60. **C** A conducting loop is placed over a lit incandescent bulb so that the filament is surrounded. Determine whether an emf is induced in the loop **a.** if the bulb is in a flashlight powered by a battery and **b.** if the bulb is in a lamp plugged into a wall outlet. Explain your answers.

Problems 61 and 62 are paired.

61. **N** A circular loop with a radius of 0.25 m is rotated by 90.0° over 0.200 s in a uniform magnetic field with $B = 1.50$ T. The plane of the loop is initially perpendicular to the field and is parallel to the field after the rotation. **a.** What is the average induced emf in the loop? **b.** If the rotation is then reversed, what is the average induced emf in the loop?

62. **C** A circular loop with a radius of 0.25 m is rotated by 90.0° over 0.200 s in a uniform magnetic field with $B = 1.50$ T. The plane of the loop is initially perpendicular to the field and is parallel to the field after the rotation. When the rotation is complete, the process is reversed and the loop is rotated back in the other direction, again in 0.200 s. Considering the induced emf and current, what is different about the two rotations? Are they indistinguishable?

63. **N** A 75-turn square coil constructed from aluminum wire is placed in a uniform 1.20-T magnetic field with the plane of the coil making an angle of 45.0° with the field direction. During the next 45.0 ms, the magnitude of the field is reduced to 0.300 T, resulting in an induced emf in the coil with magnitude 1.45 V. What is the total length of aluminum wire used to construct the coil?

64. **C** A bar magnet is dropped through a loop of wire as shown in Figure P32.64. **a.** What is the direction of the induced current as the magnet is approaching the loop, as viewed from above where the magnet begins? **b.** What is the direction of the induced current after the magnet falls through and is receding from the loop, as viewed from above where the magnet began?

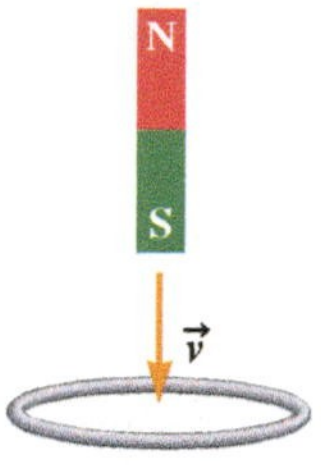

FIGURE P32.64

65. **N** An airplane with wings 10.0 m long on either side flies south, parallel to the Earth's surface, at 250.0 m/s. The horizontal component of the Earth's magnetic field is 2.0×10^{-5} Wb/m^2, and the dip angle of the Earth's magnetic field is 60.0°. Calculate the induced emf between the wing tips. (The dip angle is the angle the field makes with the horizontal plane in which the airplane resides.)

66. **N** A helicopter has blades 4.0 m long (beginning at the rotor and extending outward) rotating at 300.0 rpm (revolutions per minute). Determine the maximum voltage that might develop between the two ends of the blade while they are rotating in the Earth's magnetic field (approximately 10^{-4} T). Assume the speed of the blade is determined by the linear speed of its center of mass and its mass is uniformly distributed.

67. **N** A circular coil with 75 turns and radius 12.0 cm is placed around an electromagnet that produces a uniform magnetic field through the coil and perpendicular to the plane of the coil. As the electromagnet powers up, the field it produces increases linearly from 0 to a maximum of 3.50 T in 0.110 s. If the total resistance of the coil is 5.00 Ω, what is the magnitude of the average current induced in the circular coil as the electromagnet powers up?

68. **C** Each of the three situations in Figure P32.68 shows a resistor in a circuit in which currents are induced. Using Lenz's law, determine whether the current in each situation is from a to b or from b to a. **a.** If the current I in the wire in Figure P32.68A is increased from zero to I, what is the direction of the current induced across the resistor R? **b.** The switch in Figure P32.68B is initially closed and is thrown open at $t = 0$. What is the direction of the current induced across the resistor R immediately afterward? **c.** A bar magnet is brought close to the circuit shown in Figure P32.68C. What is the direction of the current induced across the resistor R?

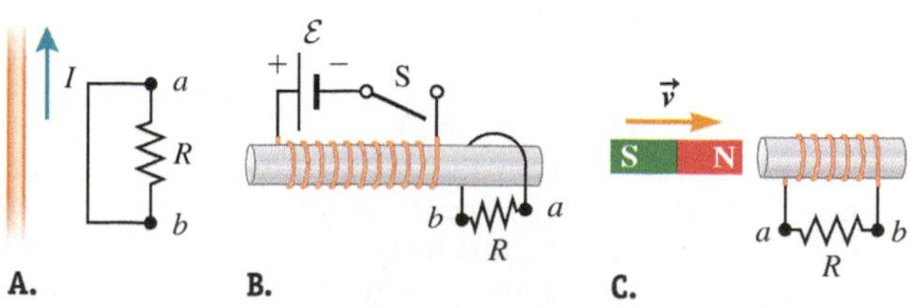

FIGURE P32.68

69. **N** A square loop with sides 1.0 m in length is placed in a magnetic field perpendicular to the plane of the loop. Half the area of the loop lies outside the magnetic field. The magnetic field varies with time as $B(t) = (0.010 - 2.00t)$ T, where t is in seconds. The loop also has a battery of emf 12.0 V as shown in Figure P32.69. Determine the resultant emf of the circuit.

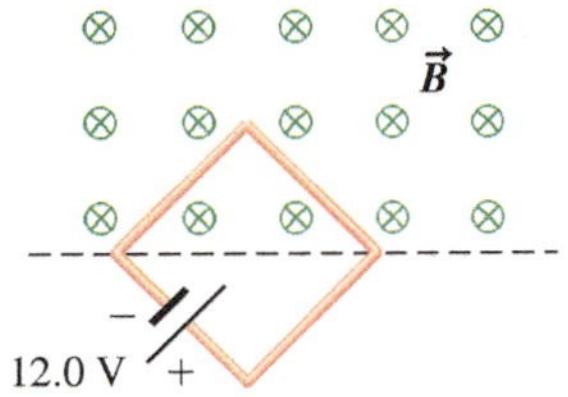

FIGURE P32.69

70. **C** You are working on an AC generator that consists of a coil rotating in a uniform magnetic field, and you discover that it is no longer creating the desired amount of power after being dropped on the floor. The coil has resistance R with N turns and cross-sectional area A, and the magnitude of the magnetic field is B. As you consider the cause of the power loss, your friend suggests that the coil may have been knocked slightly off-axis so that the axis of the coil is no longer pointing in the same direction as the static magnetic field, as desired. Could the coil being slightly off-axis explain the power loss? If no, what else might explain the loss of power? If yes, what can you do to compensate for the power loss due to the off-axis coil, assuming you are not able to adjust the coil or the magnetic field? Justify your answers.

71. **N** Two frictionless conducting rails separated by $\ell = 55.0$ cm are connected through a 2.00-Ω resistor, and the circuit is completed by a bar that is free to slide on the rails (Fig. P32.71). A uniform magnetic field of 5.00 T directed out of the page permeates the region. **a.** What is the magnitude of the force $\vec{F}_p$ that must be applied so that the bar moves with a constant speed of 1.25 m/s to the right? **b.** What is the rate at which energy is dissipated through the 2.00-Ω resistor in the circuit?

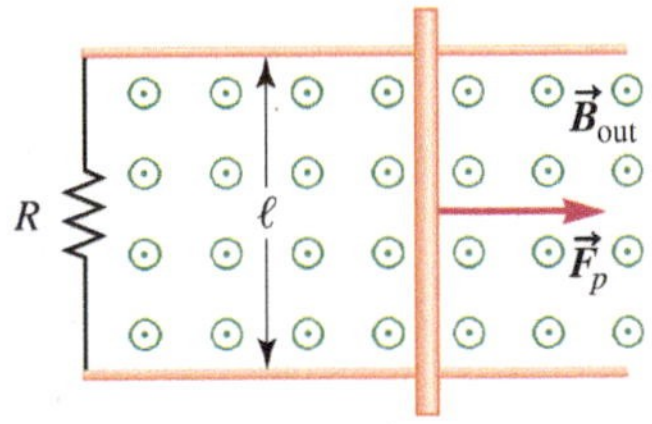

FIGURE P32.71

72. **C** **CASE STUDY** Imagine a glorious day after you've finished school; you are working as a scientist or engineer in a large research laboratory. Most likely you won't always agree with all the people who work on your team. What would you do if you disagreed with your boss, and your boss was the well-known American hero Thomas Edison? Think of Tesla. In 1882, Edison was known worldwide as the great inventor of the telegraph, phonograph, and incandescent bulb, while Tesla was a 26-year-old immigrant to the United States. Tesla originally worked for Edison, but they didn't get along and Tesla left in 1885. One of the major disagreements between the two was over the motor: Edison favored the DC motor and Tesla the AC version. In 1888, Tesla gave a presentation to the American Institute of Electrical Engineers, arguing for AC motors and generators. Present a case in favor of the AC motor and generator over their DC counterparts. Be fair in your presentation, listing the pros and cons of both AC and DC.

73. **N** The two free ends of a 320-turn circular coil with a radius of 9.00 cm are connected to a 2.40-Ω resistor. The coil is placed in a region with a uniform 3.00-T field that is initially in the upward direction so that its axis is parallel to the field. The coil then rotates about an axis perpendicular to the field by 180.0° during a time interval Δt. How much charge enters the resistor during the time interval Δt?

74. **A** Figure P32.74 shows an N-turn rectangular coil of length a and width b entering a region of uniform magnetic field of magnitude B_{out} directed out of the page. The velocity of the coil is constant and is upward in the figure. The total resistance of the coil is R. What are the magnitude and direction of the magnetic force on the coil **a.** when only a portion of the coil has entered the region with the field, **b.** when the coil is completely embedded in the field, and **c.** as the coil begins to exit the region with the field?

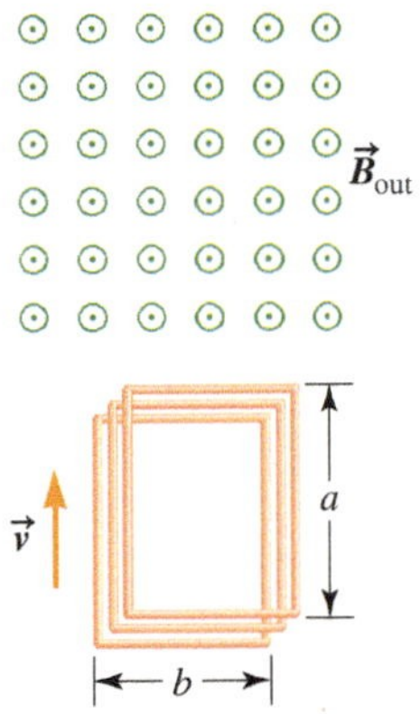

FIGURE P32.74

75. **N** A rectangular conducting loop with dimensions $w = 32.0$ cm and $h = 78.0$ cm is placed a distance $a = 5.00$ cm from a long, straight wire carrying current $I = 7.00$ A in the downward direction (Fig. P32.75). **a.** What is the magnitude of the magnetic flux through the loop? **b.** If the current in the wire is increased linearly from 7.00 A to 15.0 A in 0.230 s, what is the magnitude of the induced emf in the loop? **c.** What is the direction of the current that is induced in the loop during this time interval?

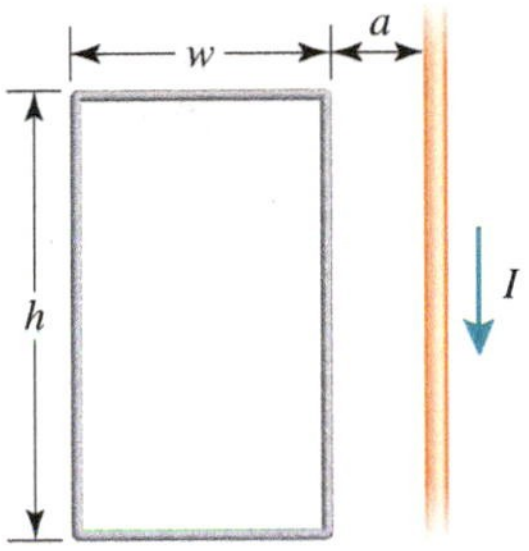

FIGURE P32.75

76. **C** Sadly, in an effort to retain the value of his patents involving the distribution of power with DC current, Edison electrocuted many animals using AC current in a series of public appearances attempting to discredit the use of AC power generation. The crowds at the time were unaware of the governing relationships among emf, current, and power, and these demonstrations caused many people to fear AC power. Given what you know about emf, current, and power, explain the faulty logic used to condemn AC power based on these demonstrations. Should people fear DC power as well? Or, should people fear neither?

77. **N** A conducting rod is pulled with constant speed v on a smooth conducting rail as shown in Figure P32.77. A constant magnetic field $\vec{B}$ is directed into the page. If the speed of the bar is doubled, by what factor does the rate of heat dissipation change?

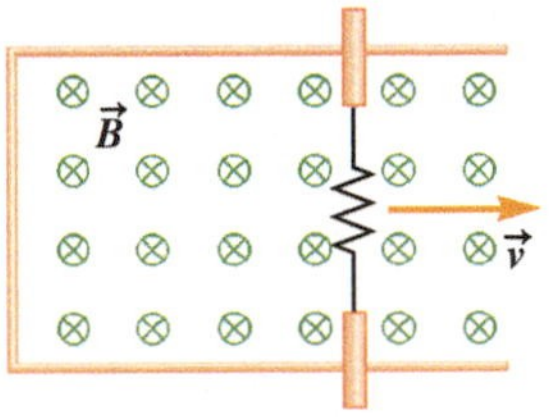

FIGURE P32.77

78. **A** A circular loop of radius a and resistance R is placed in a changing magnetic field so that the field is perpendicular to the plane of the loop. The magnetic field varies with time as $B(t) = B_0 e^{-t}$, where B_0 is a constant. Determine the electrical power in the circuit when $t = 0$.

79. **N** A conducting single-turn circular loop with a total resistance of 5.00 Ω is placed in a time-varying magnetic field that produces a magnetic flux through the loop given by $\Phi_B = a + bt^2 - ct^3$, where $a = 4.00$ Wb, $b = 11.0$ Wb/s^{-2}, and $c = 6.00$ Wb/s^{-3}. Φ_B is in webers, and t is in seconds. What is the maximum current induced in the loop during the time interval $t = 0$ to $t = 3.50$ s?

80. **A** A metal rod of mass M and length L is pivoted about a hinge at point O as shown in Figure P32.80. The axis of rotation passes through O into the page. A constant magnetic field $\vec{B}$ is applied into the page. Find the ratio of the maximum electric field inside the rod to the applied magnetic field when the rod is rotated with angular speed ω. Assume the speed of the rod is determined by the linear speed of its center of mass, and its mass is uniformly distributed.

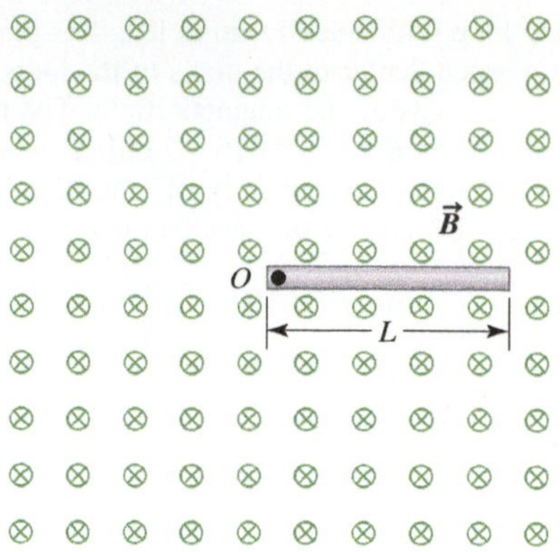

FIGURE P32.80

Inductors and AC Circuits

33

Underlying Principles

There are no new principles in this chapter; instead, we apply:

1. Faraday's law, one of Maxwell's equations
2. Lenz's law

Major Concepts

1. Inductor
2. Inductance
3. Energy density stored by a magnetic field
4. Capacitive reactance
5. Inductive reactance
6. Impedance
7. Resonance

Special Cases

1. Inductance of an ideal solenoid
2. Inductor rule
3. *RL* circuit
4. *LC* circuit
5. AC circuit with resistance
6. AC circuit with capacitance
7. AC circuit with inductance
8. Filter circuits
9. *RLC* (AC) circuit

Tools

Phasor diagrams

Key Questions

What practical devices exploit Faraday's law?

How is a pendulum like a circuit?

How do we analyze AC circuits?

Once an infrastructure has been built, creative people develop technologies that make use of that infrastructure. When cell phone towers were new, cell phones were large, clunky devices used only to make (expensive) calls. Now cell phones are compact, and we use them to send text messages and pictures to our friends. We use them to check our e-mail and get directions to a new restaurant.

The same sort of development took place after the AC power grid was established. At first, there were only a few AC-powered devices, but today we use AC circuits all the time. The AC electricity grid powers dishwashers, refrigerators, and vacuum cleaners. The development of AC circuits led to the development of communication devices such as radios, TVs, and cell phones. In this chapter, we study basic AC circuits.

INDUCTOR **Major Concept**

33-1 Inductors and Inductance

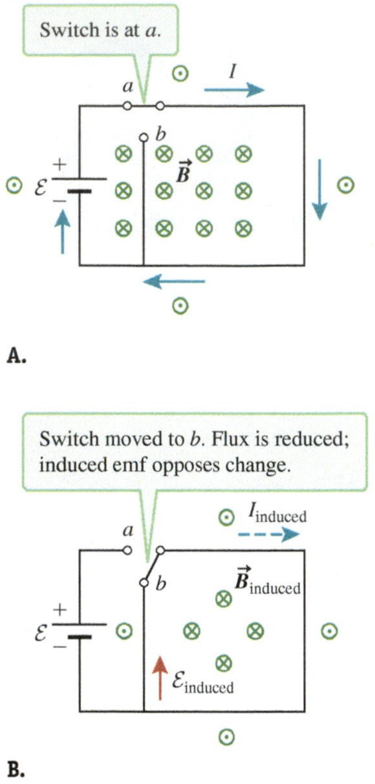

FIGURE 33.1 **A.** When the switch is closed at a, there is a magnetic flux through the loop. **B.** When the switch is moved from a to b, the magnetic flux is reduced, so an emf is induced in the loop in the same direction as the battery's emf. The energy to induce this emf was stored in the magnetic field. We use an arrow (→) to indicate the direction of the *induced* emf. The arrow points in the direction the induced emf would tend to drive the current.

Every time you open or close a switch, you turn on or shut off a magnetic field. Consider the circuit in Figure 33.1A, in which the terminals of a battery are connected by wires through a closed switch at point a. As long as the switch is at point a, the current produces a nonzero magnetic flux through the circuit. When you close the switch at point b, you disconnect the battery from the circuit so that the circuit is simply a loop (Fig. 33.1B), and you turn off a magnetic field, so the magnetic flux through the loop decreases to zero. According to Lenz's law, an emf is induced that opposes the change. So, the induced emf must point in the same direction as the battery's emf (Fig. 33.1). This emf is not very strong and the induced current does not last very long, but this circuit illustrates an important point: Whenever you flip a switch, you induce an emf. Any emf—whether it is due to a battery or is induced—supplies energy in a circuit and moves charged particles. In a battery, the energy ultimately comes from chemical reactions inside the battery cell. So, where does the induced emf get its energy? The answer is *from the magnetic field*.

The circuit in Figure 33.1 is not well designed to take advantage of the energy stored in the magnetic field because it is the change in the magnetic flux that induces the emf, and a single loop cannot "hold" much magnetic flux. An **inductor** is a circuit element that is designed to hold more magnetic flux and store magnetic field energy. One type of inductor is a solenoid. Their simplicity makes solenoids easier to study than more complicated inductors. The solenoid is such an important type of inductor that the symbol for any inductor is a curly line that looks like a solenoid (—ꝏ—).

We encountered a similar circuit element—the capacitor—in our study of DC circuits. A capacitor "holds" charge on its plates and stores potential energy in an electric field between its plates. A parallel-plate capacitor is analogous to a solenoid inductor (Table 33.1).

An ideal solenoid's magnetic field is uniform in its interior (Section 31-5), analogous to the uniform electric field between the plates of a parallel-plate capacitor. The electric field E between the plates of a particular parallel-plate capacitor is proportional to the potential difference V_C between its plates: $E = V_C/d$ (Eq. 27.21). Likewise, the magnetic field B inside a particular solenoid is proportional to the current I in the solenoid's wire: $B = \mu_0 nI$ (Eq. 31.6). Compare Equations 27.21 and 31.6: The magnetic field inside the ideal solenoid is analogous to the electric field between the plates of a capacitor, $E \rightarrow B$, and the current in the solenoid is analogous to the voltage across the capacitor, $V_C \rightarrow I$.

To find an expression for the energy stored by an inductor, let's review the energy stored by a capacitor. Imagine two parallel-plate capacitors, each connected to identical batteries and fully charged so that each capacitor has the same potential difference V_C across its plates. The capacitor with the greater capacitance stores more

TABLE 33.1 Analogy between capacitors and inductors. Refer to this table often as you read the text.

Capacitor			Inductor		
—||—			—ꝏ—		
$E = \dfrac{V_C}{d}$	(parallel-plate capacitor)	(27.21)	$B = \mu_0 nI$	(solenoid)	(31.6)
Electric field E between plates			Magnetic field B in coils		
Voltage V_C across plates			Current I in wire		
Charge Q stored by capacitor			Magnetic flux Φ_B through inductor		
Capacitance C			Inductance L		
$Q = CV_C$		(27.1)	$\Phi_B = LI$		(33.1)
$C = \dfrac{\varepsilon_0}{d}A$	(parallel-plate capacitor)	(27.10)	$L = \mu_0 n^2 \ell A = \dfrac{\mu_0 N^2}{\ell}A$	(solenoid)	(33.5)
$U_E = \dfrac{1}{2}CV_C^2$		(27.3)	$U_B = \dfrac{1}{2}LI^2$		(33.3)

charge on its plates, $Q = CV_C$ (Eq. 27.1), and more energy in its electric field, $U_E = \frac{1}{2}CV_C^2$ (Eq. 27.3).

Now, let's continue to use the capacitor as an analogy to come up with an expression for the energy stored in the magnetic field of an inductor. A capacitor holds charge on its plates, and an inductor holds magnetic flux through its coils. The charge stored Q on the plates of the capacitor is analogous to the total magnetic flux (or *flux linkage*) Φ_{tot} through the inductor, $Q \rightarrow \Phi_{tot}$. So, just as Q is proportional to V_C (Eq. 27.1), the magnetic flux through the inductor is proportional to the current:

$$\Phi_{tot} = LI \tag{33.1}$$

INDUCTANCE **Major Concept**

where the constant of proportionality L is the **inductance**. The inductance in Equation 33.1 is sometimes called the **self-inductance** because the flux Φ_{tot} through the inductor is the result of the current I in the same ("self") inductor. Comparing $\Phi_{tot} = LI$ (Eq. 33.1) to $Q = CV_C$ (Eq. 27.1) shows that inductance L is analogous to capacitance, $C \rightarrow L$. The SI unit for inductance is named the *henry* in honor of the American physicist Joseph Henry. The henry is abbreviated by an uppercase H, and

$$1\,\text{H} = 1\frac{\text{Wb}}{\text{A}} = 1\frac{\text{T}\cdot\text{m}^2}{\text{A}} \tag{33.2}$$

Applying this analogy ($V_C \rightarrow I$ and $C \rightarrow L$) to $U_E = \frac{1}{2}CV_C^2$ (Eq. 27.3), we see that the energy stored in the magnetic field of an inductor is

$$U_B = \frac{1}{2}LI^2 \tag{33.3}$$

Now, let's review capacitance so we can use the analogy to understand the inductance L. The capacitance of a capacitor depends only on geometry. For an air-filled parallel-plate capacitor, $C = \varepsilon_0 A/d$ (Eq. 27.10), where A is the plate's area and d is the separation between the plates. Again, for two parallel-plate capacitors connected to identical batteries and with equal separation d between their plates, the capacitor with the larger plate area A has a greater capacitance. Because the capacitors are connected to identical batteries, the capacitor with the larger plates stores more charge.

Like capacitance, inductance depends only on the geometry of the inductor. For a solenoid, the inductance is proportional to its cross-sectional area A:

$$L \propto A \tag{33.4}$$

Imagine two solenoids, each with the same current I through their wires. The solenoid with the larger area A has a greater inductance L and, according to $\Phi_{tot} = LI$ (Eq. 33.1), has a greater magnetic flux through its loops. Also, because $U_B = \frac{1}{2}LI^2$ (Eq. 33.3), the inductor with the larger area stores more energy.

CONCEPT EXERCISE 33.1

Suppose the switch in Figure 33.1 is closed at point b for a long time so that there is no current in the circuit. Then the switch is closed at point a. Is there an induced emf? If not, explain why not. If so, what is the direction of the induced emf and why?

EXAMPLE 33.1 Inductance of an Ideal Solenoid

INDUCTANCE OF AN IDEAL SOLENOID **Special Case**

An ideal solenoid has n turns per unit length and cross-sectional area A. Show that its inductance is given by

$$L = \mu_0 n^2 \ell A = \frac{\mu_0 N^2}{\ell}A \tag{33.5}$$

where ℓ is the length of the solenoid and N is the number of turns.

Example continues on page 1048 ▶

INTERPRET and ANTICIPATE

To solve this problem, you must first find an expression for the total magnetic flux Φ_{tot} when current I is in the solenoid.

SOLVE

Figure 31.25 (page 994) shows an ideal solenoid. When the solenoid carries a current, the resulting magnetic field is uniform and perpendicular to its cross section. The magnetic flux through each loop is BA. The total magnetic flux or flux linkage is the flux through each loop times the number of loops N.	$\Phi_{tot} = NBA$
Write the number of loops in terms of the length ℓ of the solenoid and its number of turns per unit length: $N = n\ell$.	$\Phi_{tot} = (n\ell)BA \quad (1)$
When the current in the solenoid is I, the magnetic field inside the solenoid is given by Equation 31.6.	$B = \mu_0 nI \quad (31.6)$
Substitute Equation 31.6 into Equation (1). Equation (2) is the flux linkage.	$\Phi_{tot} = (n\ell)(\mu_0 nI)A = \mu_0 n^2 \ell AI \quad (2)$
Solve Equation 33.1 for L.	$\Phi_{tot} = LI \quad (33.1)$ $L = \frac{\Phi_{tot}}{I}$
Substitute Equation (2) for Φ_{tot}. Also use $N = n\ell$ to express L in terms of N.	$L = \frac{\mu_0 n^2 \ell AI}{I}$ $L = \mu_0 n^2 \ell A = \frac{\mu_0 N^2}{\ell} A$ ✓ $\quad (33.5)$

CHECK and THINK

Equation 33.5 is the inductance of a solenoid. As expected, and analogous to the capacitor, the inductance depends only on geometry—the number of turns per unit length n, the length ℓ, and the cross-sectional area A. We will often use Equation 33.5, but because an ideal solenoid is very long, it is sometimes more convenient to have an expression for the inductance per unit length $\mathscr{L}$ (Eq. 33.6).	$\mathscr{L} \equiv \frac{L}{\ell} = \mu_0 n^2 A \quad (33.6)$

Barry Z Levine/Contributor/Getty Images

FIGURE 33.2 Jimi Hendrix in concert.

CASE STUDY Listening to Hendrix

In the late 1960s, Jimi Hendrix (1942–1970) was extremely popular (Fig. 33.2). He was known for his unique electric guitar sound. Imagine a college student in the 1960s listening to Hendrix while studying physics. This case study is about three of the devices that made this activity possible. The first device is Hendrix's electric guitar, the second is the student's radio (the type used then), and the third is the radio's speaker. All three devices depend on inductors.

The strings of an electric guitar are made of a magnetic material such as steel. Below each string are pickups that respond to the motion of the string with an induced current (Fig. 33.3A). The pickup closest to the guitar's neck is normally more sensitive to the string's low-frequency vibrations, and the pickup farthest from the neck is normally more sensitive to high frequencies. As shown in Figure 33.3B, a pickup is a small inductor consisting of a coil wrapped around a permanent magnet. The permanent magnet causes magnetic moments in the steel string to line up so that it becomes locally magnetized. When the string vibrates, the magnetic flux through the pickup coil oscillates. According to Faraday's law, the changing magnetic flux induces a current in the coil that oscillates at the same frequency as the string. The current can be amplified and put through speakers so we can hear the music.

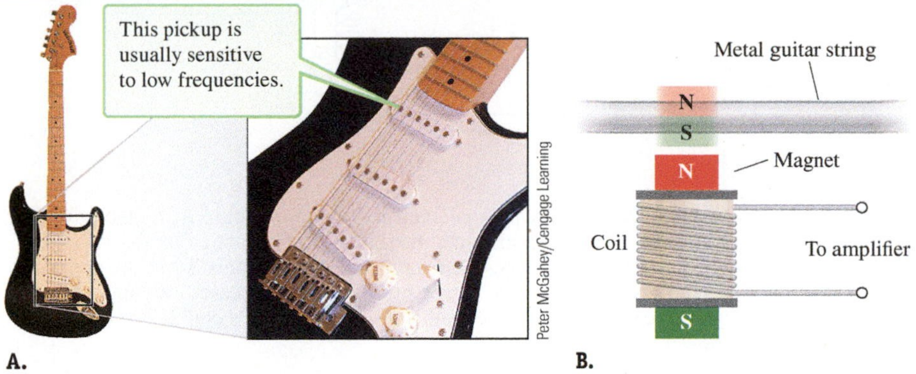

FIGURE 33.3 **A.** Some electric guitars have three sets of pickups. **B.** Each pickup is a coil of wire wrapped around a permanent magnet. The magnet causes the magnetic dipole moments in the steel string to line up so that a portion of the string is also magnetized.

CONCEPT EXERCISE 33.2

CASE STUDY Hendrix's Guitar

Jimi Hendrix—like other great guitar players—wanted to create his own unique sound. So, he rewrapped the pickup coils on his guitar, changing the number of turns. This changes the pickups' relative sensitivity to the string vibrations. Today, enthusiastic musicians make their own custom pickups. They might decrease the number of turns on the low-frequency pickup and increase the number on the high-frequency pickup. Model the pickup coil as a solenoid. Suppose you increase the number of turns by 10% on one particular pickup. Assume the cross-sectional area and length of the solenoid are essentially constant. By what amount does the inductance change?

33-2 Back Emf

To analyze practical circuits that involve inductors, we must understand how inductors behave when they carry a current. To do that, we must first introduce a DC power supply (Fig. 33.4). A DC power supply is much like a battery in that the potential difference between its terminals does not change sign. The advantage of a DC power supply over a battery is that you can adjust the terminal voltage by just turning a knob. The DC power supply's symbol looks like the symbol for a battery except an arrow is drawn across it (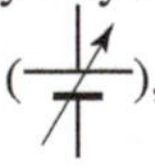), indicating that the terminal voltage is variable.

Imagine connecting an inductor to a DC power supply (Fig. 33.5). The power supply sets up a clockwise current in the circuit. If that current is constant, the inductor acts like a piece of curly wire. A solenoid inductor may, for example, have a downward-pointing magnetic field in its interior (not shown in Fig. 33.5). Because the current is constant, the magnetic flux through the inductor is constant.

If the current is changing, however, the magnetic flux through the inductor changes and an emf $\mathcal{E}_L$ is induced. In Figure 33.5A, the current is increasing as you increase the voltage on the power supply. According to Lenz's law, the induced emf $\mathcal{E}_L$ must oppose the change—in this case, an increasing clockwise current—and the induced emf's direction is counterclockwise (as indicated by the upward arrow next to the inductor in Fig. 33.5). Put simply, the inductor "wants" to create a counterclockwise current to oppose the increasing clockwise current, so the

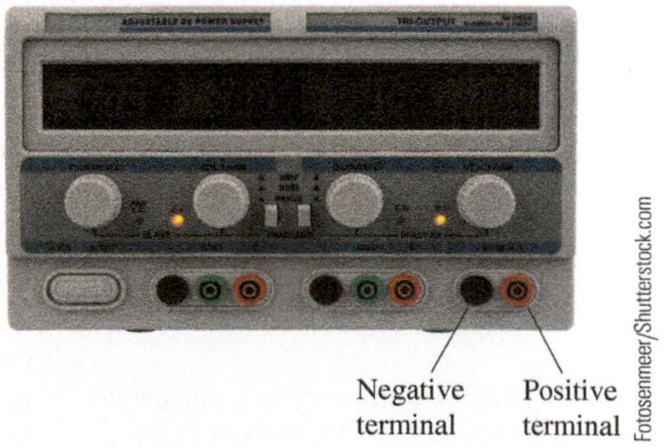

FIGURE 33.4 A variable DC power supply allows you to adjust the voltage across its terminals.

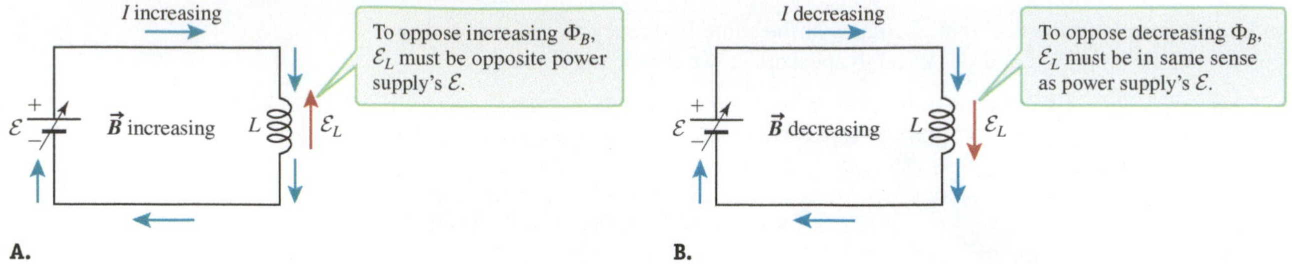
I increasing
To oppose increasing Φ_B, $\mathcal{E}_L$ must be opposite power supply's $\mathcal{E}$.
$\mathcal{E}$
$\vec{B}$ increasing
L
$\mathcal{E}_L$
A.
I decreasing
To oppose decreasing Φ_B, $\mathcal{E}_L$ must be in same sense as power supply's $\mathcal{E}$.
$\mathcal{E}$
$\vec{B}$ decreasing
L
$\mathcal{E}_L$
B.

FIGURE 33.5 A variable DC power supply is connected to an inductor. **A.** As the power supply's emf increases, the current, the magnetic field, and the magnetic flux through the inductor all increase. To oppose that change, the inductor's emf $\mathcal{E}_L$ must be in the opposite sense as the power supply's emf. **B.** As the power supply's emf decreases, the magnetic flux through the inductor decreases. To oppose that change, the inductor's emf $\mathcal{E}_L$ must be in the same sense as the power supply's emf.

inductor acts like a second power supply working against the actual power supply. Because of the induced emf, the current does not increase as quickly as it would if the inductor were replaced by a simple straight wire.

If we decrease the power supply's voltage, the current is still clockwise but decreasing (Fig. 33.5B). Now the magnetic flux through the inductor decreases. In order to oppose the change, the induced emf must work with the power supply's emf. In other words, the inductor "wants" to maintain the clockwise current. So, the induced emf $\mathcal{E}_L$ is clockwise. Because of the induced emf, the current does not decrease as quickly as it would if the inductor were replaced by a simple straight wire.

We have found the direction of the induced emf; we can find the magnitude from Faraday's law. Equation 32.4 expresses Faraday's law in terms of the flux linkage (total magnetic flux):

$$\mathcal{E}_L = \frac{d(N\Phi_B)}{dt} = \frac{d\Phi_{\text{tot}}}{dt} \tag{32.4}$$

The flux linkage is replaced using $\Phi_{\text{tot}} = LI$ (Eq. 33.1):

$$\mathcal{E}_L = \frac{d(LI)}{dt}$$

The inductance L depends only on geometry, so it is constant:

$$\mathcal{E}_L = L\frac{dI}{dt} \tag{33.7}$$

Any real inductor has some resistance because the wire it is made from has resistance. Our focus is on **ideal inductors**, however, which have no resistance. A real inductor can be modeled as an ideal inductor in series with a resistor.

When analyzing DC circuits, we listed rules for finding the potential difference across various circuit elements (Section 29-1). These rules are helpful when we apply Kirchhoff's loop rule using the steps in Section 29-2. A similar rule for ideal inductors combines the directions we found with Lenz's law (Fig. 33.5) with the magnitude we found with Faraday's law (Eq. 33.7). Imagine connecting a voltmeter across an inductor so that we measure the potential difference in the direction of the current. The red lead is downstream, and the black lead is upstream (Fig. 33.6). If the current is decreasing (Fig. 33.5B), the induced emf $\mathcal{E}_L$ is downward from the black lead to the red lead. Compare this situation to that shown in Figure 29.2E (page 898), in which a voltmeter measures a *positive* potential across an emf device. In the case of the inductor, when we measure the voltage in the direction of the current and the current is decreasing ($dI/dt < 0$), the voltage is positive. Its magnitude is given by $\mathcal{E}_L = L(dI/dt)$, Equation 33.7. Combining the direction with the magnitude, we write an expression for the potential difference:

$$\Delta V_L = V_{\text{red}} - V_{\text{black}} = -L\frac{dI}{dt} \tag{33.8}$$

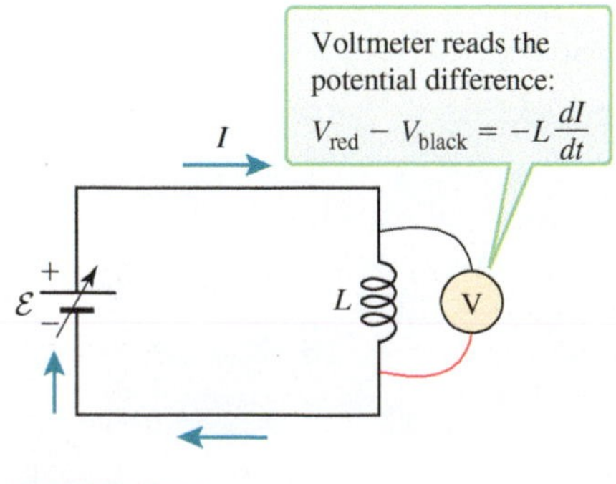
Voltmeter reads the potential difference:
$V_{\text{red}} - V_{\text{black}} = -L\frac{dI}{dt}$
I
$\mathcal{E}$
L
V

FIGURE 33.6 A voltmeter is connected across the inductor. If the current is decreasing, the voltmeter reports a positive potential difference. If the current is increasing, the voltmeter reports a negative potential difference.

where the red lead is downstream and the black lead is upstream so that the measurement is made in the direction of the current. Because $dI/dt < 0$ when the current is decreasing, the potential difference ($V_{red} - V_{black}$) is positive.

Equation 33.8 is the **inductor rule**; it holds even when the current is increasing. If the current is increasing (Fig. 33.5A), the induced emf points upward from the red lead to the black lead. In that case, our work with emf devices (Chapter 29) tells us the voltmeter measures a negative potential difference. This result is consistent with Equation 33.8 for an increasing current, $dI/dt > 0$. There are three things to keep in mind when using the inductor rule:

INDUCTOR RULE Special Case

1. The potential difference across an inductor is zero if the current is constant ($dI/dt = 0$).
2. The negative sign in $\Delta V_L = -L(dI/dt)$ (Eq. 33.8) implies that we are measuring the potential difference in the direction of the current.
3. The negative sign also indicates that the voltage across the inductor always opposes the change in the current. The inductor voltage is referred to as the **back emf** because it is fighting *back* against the changing current. Thinking of the back emf is often the easiest way to reason out the sign of the potential difference across the inductor.

(Strictly speaking, Kirchhoff's loop rule holds only in the case of conservative fields. The inductor rule is a way to preserve the use of Kirchhoff's loop rule; what follows in this chapter gives the right equations but misses on conceptual details. See Section 34-2 and Problems 13, 14, and 68 at the end of Chapter 34.)

Transformers Revisited

After electrical power is transported over great distances at high voltages and low current, transformers (Fig. 32.30, page 1036) step down the power to lower voltages for use in your home. Ultimately, the development of transformers led to the success of AC over DC power supplies (Case Study, Chapter 32). In Section 32-8, we stated the transformer equation

$$\frac{V_S}{V_P} = \frac{N_S}{N_P} \tag{32.26}$$

without deriving it. Now we can derive this equation.

Start by analyzing the primary circuit, where the primary circuit consists of an AC generator and an inductor (Fig. 33.7A). Ideally, the resistance in the primary circuit is negligible. As in any circuit analysis, use Kirchhoff's loop rule to find an expression for the potential differences around the closed loop. This is an AC circuit, so the current's direction oscillates. To apply Kirchhoff's loop rule, imagine an instant when the current is clockwise. Starting in the bottom left corner and working our way around in the direction of the current, we first encounter the AC generator, whose emf is momentarily pointing upward (the negative terminal is below the positive one momentarily). From the emf rule, the potential difference is positive. Although the emf increases and decreases periodically, we can use the rms value $\mathcal{E}_{rms}$. Because there is no (or very little) resistance, the only other circuit element is the inductor. According to the inductor rule, the potential difference measured in the direction of the current is given by $\Delta V_L = -L(dI/dt)$ (Eq. 33.8). Kirchhoff's loop rule gives

$$\mathcal{E}_{rms} - L\frac{dI}{dt} = 0$$

To be consistent with the notation in Section 32-8, we write $\mathcal{E}_{rms} = V_P$, where P stands for primary, and solve for V_P:

$$V_P = L\frac{dI}{dt}$$

Eliminate the inductance L using Equation 33.7, $\mathcal{E}_L = L(dI/dt)$, and Equation 32.4, Faraday's law, $\mathcal{E} = N(d\Phi_B/dt)$:

$$V_P = L\frac{dI}{dt} = N_P\frac{d\Phi_B}{dt} \tag{33.9}$$

where N_P is the primary solenoid's number of turns.

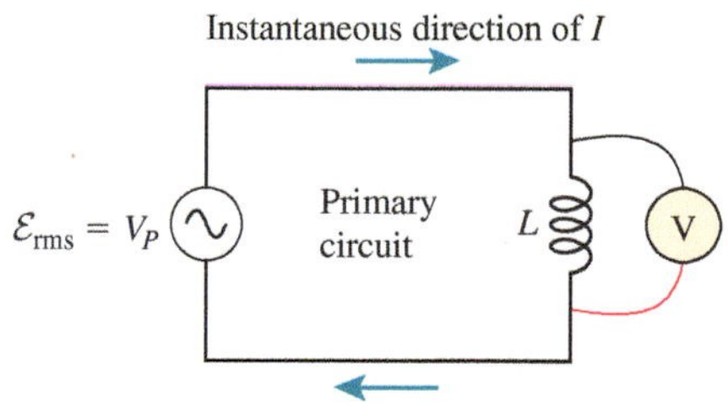

A.

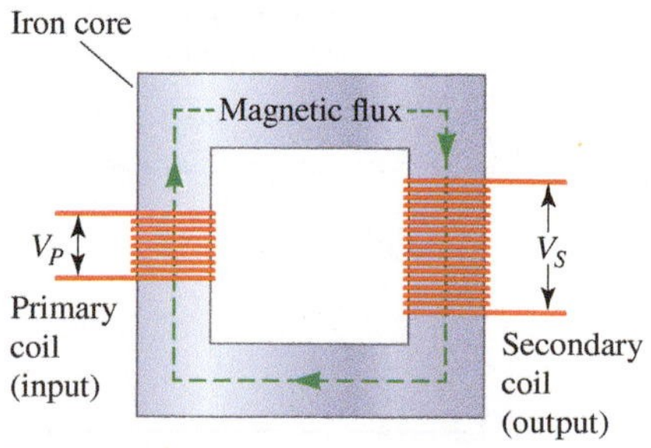

B.

FIGURE 33.7 A transformer has two circuits. **A.** The primary circuit consists of an inductor and an AC power supply. **B.** The iron core and both solenoids are shown.

Both solenoids (primary and secondary) are wound around an iron core (Fig. 33.7B). In an ideal transformer, the iron core ensures that the time-varying magnetic flux is the same through both the primary and secondary solenoids. The flux induces an emf in the secondary circuit given by Faraday's law:

$$V_S = N_S \frac{d\Phi_B}{dt} \tag{33.10}$$

where V_S is the rms induced emf in the secondary circuit and N_S is the number of turns in the secondary solenoid.

Because the time-varying magnetic flux $d\Phi_B/dt$ is the same through both solenoids, it can be eliminated from Equations 33.9 and 33.10 to arrive at the transformer equation:

$$\frac{d\Phi_B}{dt} = \frac{V_S}{N_S} = \frac{V_P}{N_P}$$

$$\frac{V_S}{V_P} = \frac{N_S}{N_P} \tag{32.26}$$

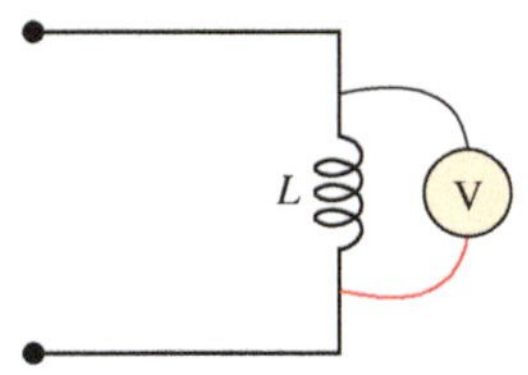

FIGURE 33.8

CONCEPT EXERCISE 33.3

Figure 33.8 shows an inductor that is part of a larger circuit.

a. If the voltmeter reads zero, what do you know about the current in the inductor?
b. If the voltmeter reads a positive value, what do you know about the current in the inductor?

RL CIRCUIT Special Case

33-3 Special Case: Resistor–Inductor (*RL*) Circuit

All real wires have resistance, so any real inductor has resistance, and circuits that involve resistance and inductance (*RL* circuits) are common. Figure 33.9A shows a circuit that consists of an emf device such as a battery, a switch, an inductor, and a resistor. The resistor may represent the inductor's own resistance, or it may be an actual resistor. In either case, the inductor alone is modeled as an ideal inductor with no other resistance.

Initially the switch is open, so there is no current in the circuit. The switch is then closed at position *a* (Fig. 33.9A). Because the current increases from zero, there is a back emf $\mathcal{E}_L$ across the inductor as shown. To find an expression for the increasing current, apply Kirchhoff's loop rule starting in the bottom left corner and working clockwise in the direction of the current. The voltage across the battery is positive going from the negative to the positive terminal; the voltage across the inductor is given by Equation 33.8, $\Delta V_L = -L(dI/dt)$, because we are measuring in the direction of the current, and the voltage across the resistor is negative because we are measuring in the direction of the current. According to Kirchhoff's loop rule,

$$\mathcal{E} - L\frac{dI}{dt} - IR = 0 \tag{33.11}$$

Equation 33.11 is a differential equation that is mathematically equivalent to the differential equation we found for the resistor–capacitor (*RC*) circuit (Section 29-8):

$$\mathcal{E} - R\frac{dq}{dt} - \frac{q}{C} = 0$$

We found the solution to this equation:

$$q(t) = C\mathcal{E}(1 - e^{-t/\tau}) \tag{29.10}$$

where the capacitive time constant is given by

$$\tau \equiv RC \tag{29.13}$$

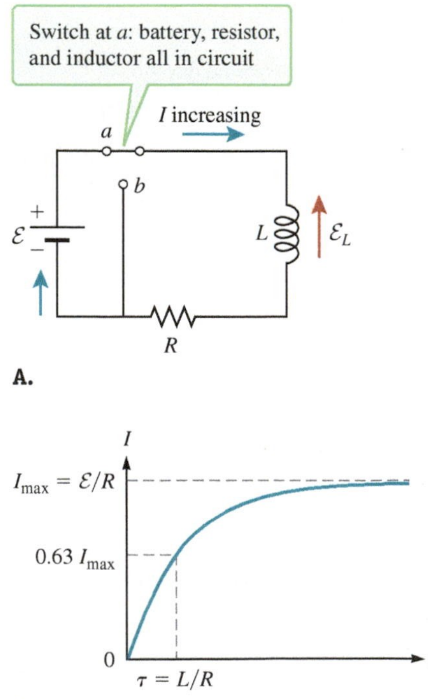

FIGURE 33.9 A. When the switch has just closed at position *a*, the current is increasing. **B.** Current versus time when the switch is at *a*.

We want to solve Equation 33.11 for I, and because we already solved the equivalent equation, we can just translate variables by comparing the two differential equations:

$$RC \text{ circuit} \rightarrow RL \text{ circuit}$$
$$\mathcal{E} \rightarrow \mathcal{E}$$
$$R \rightarrow L$$
$$q \rightarrow I$$
$$C \rightarrow \frac{1}{R}$$

Using this translation of variables, we can solve Equation 33.11:

$$I(t) = \frac{\mathcal{E}}{R}(1 - e^{-t/\tau}) \tag{33.12}$$

where the inductive time constant is

$$\tau \equiv \frac{L}{R} \tag{33.13}$$

If L is in henrys and R is in ohms, the inductive time constant τ is in seconds.

Figure 33.9B is a plot of Equation 33.12. Contrary to what you might think, when you close the switch at a, the current does not immediately reach its maximum value $I_{max} = \mathcal{E}/R$. Because the back emf opposes the increasing current, the current builds up to its maximum value over some time interval. At time $t = 0$, when the switch is first closed, the current is zero. The current gradually increases, and at $t = \tau$, the current has reached 63% of its maximum value. According to Equation 33.12, the current approaches its maximum value as $t \rightarrow \infty$. In practice, the current reaches its maximum value after a long time compared to τ, and then the back emf vanishes.

After the switch has been in position a for a while, it is switched to position b (Fig. 33.10A), so that only the resistor and inductor are in the circuit. The current does not immediately drop to zero because the back emf opposes the decrease in current. Kirchhoff's loop rule gives an expression for the decreasing current. Because the current is in the same direction as in Figure 33.9A, we start with $\mathcal{E} - L(dI/dt) - IR = 0$ (Eq. 33.11). This time, we must set $\mathcal{E} = 0$ because there is no battery in the circuit (Fig. 33.10A):

$$0 - L\frac{dI}{dt} - IR = 0$$

$$L\frac{dI}{dt} + IR = 0 \tag{33.14}$$

Equation 33.14 is the same differential equation we found for a discharging capacitor in an RC circuit (Problem 29.93):

$$R\frac{dq}{dt} + \frac{q}{C} = 0$$

whose solution is

$$q(t) = q_{max}e^{-t/\tau}$$

As before, the capacitive time constant is given by $\tau \equiv RC$ (Eq. 29.13). We can use the same translation of variables to write the solution to Equation 33.14 in the case of an RL circuit:

$$I(t) = I_{max}e^{-t/\tau} \tag{33.15}$$

where the inductive time constant is given by $\tau \equiv L/R$ (Eq. 33.13).

Figure 33.10B is a plot of Equation 33.15. When you first move the switch to position b, the current is at its maximum value. If the switch has been at position a for a long enough time, that maximum current is $I_{max} = \mathcal{E}/R$. If the switch has not

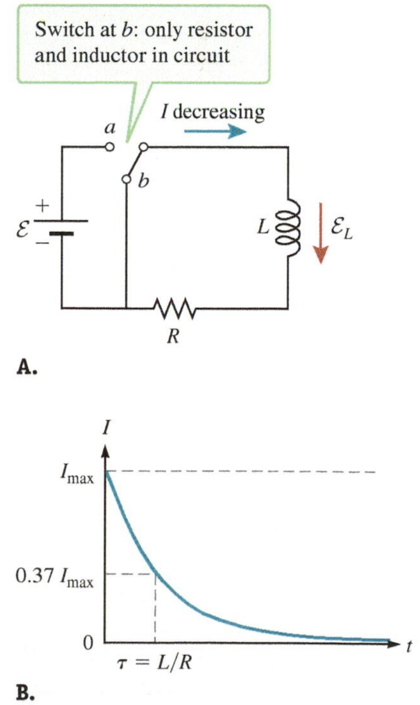

FIGURE 33.10 **A.** When the switch is moved to position b, the current decreases. **B.** Current versus time when the switch is at b.

been at position a long enough, the maximum current in Equation 33.15 is smaller: $I_{max} < \mathcal{E}/R$. Once again, after the battery is cut out of the circuit (switch at b), the current does not immediately drop to zero. Instead, the current gradually decreases because the back emf opposes the change. At time $t = \tau$, the current has dropped to 37% of its maximum value. Eventually the current reaches zero, at which time the back emf is also zero.

CONCEPT EXERCISE 33.4

In what ways is the RL circuit like an RC circuit (Section 29-8)?

EXAMPLE 33.2 Half the Maximum

It might seem odd that at $t = \tau$, the current in Figure 33.9B is at 63% of its maximum, but the current is at only 37% of its maximum in Figure 33.10B.

A At what time does the current reach half of its maximum when the switch is at a in Figure 33.9A? Give your answer in terms of τ.

INTERPRET and ANTICIPATE

We need an expression for t when the current has risen to half its maximum value. From the graph in Figure 33.9B, our answer should be less than one time constant: $t < \tau$.

SOLVE To find t when $I = \frac{1}{2}I_{max}$, substitute this value of the current into Equation 33.12.	$I(t) = \frac{\mathcal{E}}{R}(1 - e^{-t/\tau})$ (33.12) $\frac{1}{2}I_{max} = \frac{\mathcal{E}}{R}(1 - e^{-t/\tau})$
Because $I_{max} = \mathcal{E}/R$, we can eliminate $\mathcal{E}$ and R from the equation.	$\frac{1}{2}I_{max} = I_{max}(1 - e^{-t/\tau})$ $\frac{1}{2} = (1 - e^{-t/\tau})$
Solving for t involves taking a natural logarithm.	$e^{-t/\tau} = 1 - \frac{1}{2} = \frac{1}{2}$ $\ln e^{-t/\tau} = \ln \frac{1}{2}$ $-t/\tau = -0.693$ $t = 0.693\tau$

CHECK and THINK

As expected, the time for the current to reach half its maximum value is less than one time constant: $t < \tau$.

B At what time does the current reach half its maximum when the switch is at b in Figure 33.10A? Give your answer in terms of τ.

INTERPRET and ANTICIPATE

This is similar to part A, except now we need to look at the plot in Figure 33.10B to anticipate the result. Our answer here should also be less than one time constant: $t < \tau$.

SOLVE Again, find t when $I = \frac{1}{2}I_{max}$. Substitute this value of the current into Equation 33.15, remembering that I_{max} does not necessarily equal $\mathcal{E}/R$.	$I(t) = I_{max}e^{-t/\tau}$ (33.15) $\frac{1}{2}I_{max} = I_{max}e^{-t/\tau}$

As before, solving for t involves taking a natural logarithm.	$e^{-t/\tau} = \frac{1}{2}$ $\ln e^{-t/\tau} = \ln \frac{1}{2}$ $-t/\tau = -0.693$ $t = 0.693\tau$

CHECK and THINK

This is exactly what we found in part A. The time for the current to drop to half its original value is sometimes called the **half-life**. The term *half-life* is used in any situation that involves exponential decay. For example, you might have heard of carbon dating for determining the age of fossils and archeological specimens. Carbon dating works because the amount of carbon-14 in a substance decreases when the organism dies and stops metabolizing carbon. The half-life of carbon-14 is about 5700 years.

33-4 Energy Stored in a Magnetic Field

Using the analogy between capacitors and inductors (Table 33.1), we found an expression for the energy stored by an inductor: $U_B = \frac{1}{2}LI^2$ (Eq. 33.3). In this section, we use our work on *RL* circuits to derive this expression, and then we use this expression to derive a more general expression for the energy stored by a magnetic field.

DERIVATION Energy Density Stored in a Magnetic Field

Consider the *RL* circuit in Figure 33.9A when the switch has just been thrown to position *a*. First, we derive $U_B = \frac{1}{2}LI^2$ (Eq. 33.3), and then we will show that in the case of a solenoid of length ℓ and cross-sectional area A,

$$U_B = \frac{1}{2}\frac{B^2}{\mu_0}(\ell A) \tag{33.16}$$

Finally, we will show that the energy density (energy per unit volume) stored in a magnetic field is

$$u_B = \frac{U_B}{V} = \frac{1}{2}\frac{B^2}{\mu_0} \tag{33.17}$$

ENERGY DENSITY STORED BY A MAGNETIC FIELD ✪ **Major Concept**

Start with Kirchhoff's loop rule (Eq 33.11) applied to this circuit (Fig. 33.9A).	$\mathcal{E} - L\frac{dI}{dt} - IR = 0$ (33.11) $\mathcal{E} = IR + L\frac{dI}{dt}$ (1)
Multiply each side of Equation (1) by I.	$\mathcal{E}I = I^2R + LI\frac{dI}{dt}$ (2)
The left side of Equation (2) is the power supplied by the battery. The first term on the right is the power dissipated by the resistor. The last term is the power stored in the inductor's magnetic field. Thus, part of the power supplied by the battery is dissipated by the resistor, and the rest is stored in the inductor's magnetic field.	(power supplied by battery) = (power dissipated by resistor) + (power stored by inductor) $P_{\mathcal{E}} = P_R + P_L$
Consider the power stored by the inductor (last term in Eq. 2). This is the rate at which the energy is stored in the inductor's magnetic field.	$P_L = \frac{dU_B}{dt} = LI\frac{dI}{dt}$ (3)

Derivation continues on page 1056 ▶

Cancel dt in Equation (3).	$dU_B = LIdI$	
Integrate the left side from 0 to U_B (the stored energy when the current is I), and integrate the right side from 0 to the current I.	$\int_0^{U_B} dU_B = \int_0^I LI\,dI$	
The inductance L depends only on geometry, not current, so pull L outside the integral. Complete the integral and substitute limits to find Equation 33.3 as before.	$U_B = L\int_0^I I\,dI = \frac{1}{2}LI^2$	(33.3)
Equation 33.3 is the energy stored in the magnetic field of an inductor of any shape. To find the energy stored in the special case of a solenoid, substitute the solenoid's inductance as given by $L = \mu_0 n^2 \ell A$ (Eq. 33.5).	$U_B = \frac{1}{2}(\mu_0 n^2 \ell A)I^2$	
Use $I = B/(\mu_0 n)$ from Equation 31.6 to eliminate I, and simplify to find the energy stored in the magnetic field of a solenoid.	$U_B = \frac{1}{2}(\mu_0 n^2 \ell A)\left(\frac{B}{\mu_0 n}\right)^2$ $U_B = \frac{1}{2}\frac{B^2}{\mu_0}(\ell A)$ ✓	(33.16)
The energy density u_B stored in the solenoid's magnetic field is found by dividing by its volume: $V = \ell A$.	$u_B = \frac{U_B}{V} = \frac{1}{2}\frac{B^2}{\mu_0}$ ✓	(33.17)

:• **COMMENTS**

Equation 33.17 was derived for the special case of a solenoid but, because the expression is independent of geometry, it is true in general; it gives the magnetic energy density in any region that has a magnetic field.

CONCEPT EXERCISE 33.5

Skim through Chapter 27 to find an expression analogous to Equation 33.17 for the energy density stored in an electric field. Report your answer and explain.

EXAMPLE 33.3 The Earth's Magnetic Field

Estimate the energy density of the Earth's magnetic field near its surface.

:• **INTERPRET and ANTICIPATE**

We expect a (scalar) numerical answer with the SI units of joules per cubic meter.

:• **SOLVE** This is a straightforward application of Equation 33.17. The magnetic field near the Earth's surface is about 0.5×10^{-4} T.	$u_B = \frac{1}{2}\frac{B^2}{\mu_0}$ $u_B = \frac{1}{2}\frac{(0.5 \times 10^{-4}\text{ T})^2}{(4\pi \times 10^{-7}\text{ T}\cdot\text{m/A})}$ $u_B = 9.9 \times 10^{-4}\text{ T}\cdot\text{A/m}$	(33.17)
:• **CHECK and THINK** Make sure these units are right. It helps to think about the magnetic force exerted on a current-carrying conductor $\vec{F}_B = I(\vec{\ell} \times \vec{B})$ (Eq. 30.32) to write the SI units.	$1\text{ N} = 1\text{ T}\cdot\text{m}\cdot\text{A}$	

Then use 1 N · m = 1 J to find an expression for the SI units of energy density. The SI units we found are correct.	$1\ \text{N}\cdot\text{m} = 1\ \text{J} = 1\ \text{T}\cdot\text{m}^2\cdot\text{A}$ $1\ \text{J/m}^3 = 1\ \text{T}\cdot\text{m}^2\cdot\text{A/m}^3 = 1\ \text{T}\cdot\text{A/m}$ $u_B = 9.9 \times 10^{-4}\ \text{J/m}^3$

This problem helps build intuition about the energy stored in a magnetic field. Near the surface of the Earth, the magnetic energy density is about 10^{-3} J/m^3. Compare that value to the energy density in gasoline (about 10^{10} J/m^3). In Problem 18, you will calculate the energy density stored in the magnetic field near a neutron star.

33-5 Special Case: Inductor–Capacitor (*LC*) Circuit

***LC* CIRCUIT** **Special Case**

So far we have taken a close look at two circuits—the *RC* circuit and the *RL* circuit—in which the current changes by either increasing or decreasing and finally reaching a steady value, but the current does not alternate in these circuits. What would happen if we built a circuit that has an inductor and a capacitor but no resistor? The answer is that energy would be conserved and transferred back and forth between the inductor and the capacitor, and the current would alternate.

Figure 33.11 shows an ideal *LC* circuit consisting of an inductor and a capacitor. Ideally, the circuit has no resistance, but of course in any real circuit there is resistance in the wires. We consider that complication in Section 33-9. The capacitor is fully charged, and the switch is open. The capacitor stores energy in its electric field. There is no magnetic field because there is no current. When the switch is closed, the capacitor discharges, resulting in a counterclockwise current from the positive plate toward the negative plate. Once there is a current through the inductor, it stores energy in its magnetic field. Eventually, the capacitor is completely discharged and all the energy is stored in the inductor's magnetic field. A current remains in the circuit, so the capacitor charges up again and the cycle repeats. Energy is transferred back and forth between the capacitor's electric field and the inductor's magnetic field, much like the energy that is transformed from gravitational potential energy into kinetic energy in the case of an oscillating pendulum. We say the circuit *oscillates*.

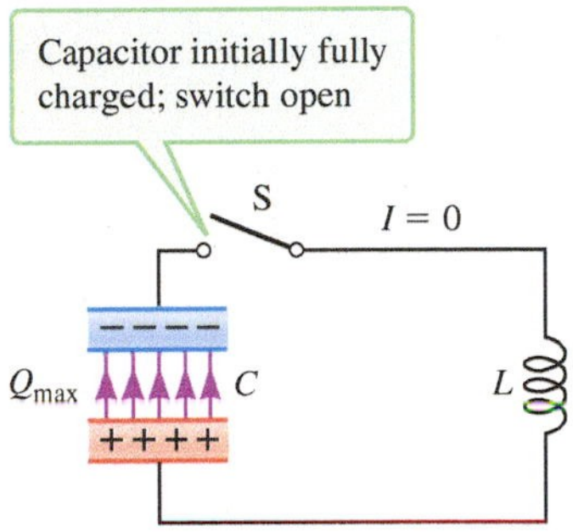

FIGURE 33.11 The capacitor has been charged by a battery that is not shown. The charged capacitor is then connected to an inductor through a switch.

Figure 33.12 shows nine key instants in one of the circuit's cycles. For each instant, an energy bar graph shows the relative amounts of energy stored in the capacitor's electric field and the inductor's magnetic field. When the capacitor is fully charged (times 1, 5, and 9), the total energy is stored in the capacitor's electric field and given by $E_{\text{tot}} = Q^2_{\text{max}}/2C$ (Eq. 27.2). When the capacitor is momentarily discharged, the current has reached it maximum value I_{max} and the magnetic field is at its maximum strength (times 3 and 7). Then the total energy is stored in the inductor's magnetic field and given by $E_{\text{tot}} = \frac{1}{2}LI^2_{\text{max}}$ (Eq. 33.3). When the capacitor is partially charged and there is current in the circuit (times 2, 4, 6, and 8), energy is stored in both the capacitor's electric field and the inductor's magnetic field. The total energy is the sum of the energy stored in the two devices: $E_{\text{tot}} = (Q^2/2C) + \frac{1}{2}LI^2$. As the circuit oscillates, energy is transferred back and forth but not lost. Thus, the total energy E_{tot} is a constant. Further, the total energy equals the maximum energy stored by the capacitor's electric field:

$$E_{\text{tot}} = \frac{1}{2}\frac{Q^2}{C} + \frac{1}{2}LI^2 = \frac{1}{2}\frac{Q^2_{\text{max}}}{C} = (U_E)_{\text{max}} \tag{33.18}$$

The total energy also equals the maximum energy stored by the inductor's magnetic field:

$$E_{\text{tot}} = \frac{1}{2}\frac{Q^2}{C} + \frac{1}{2}LI^2 = \frac{1}{2}LI^2_{\text{max}} = (U_B)_{\text{max}} \tag{33.19}$$

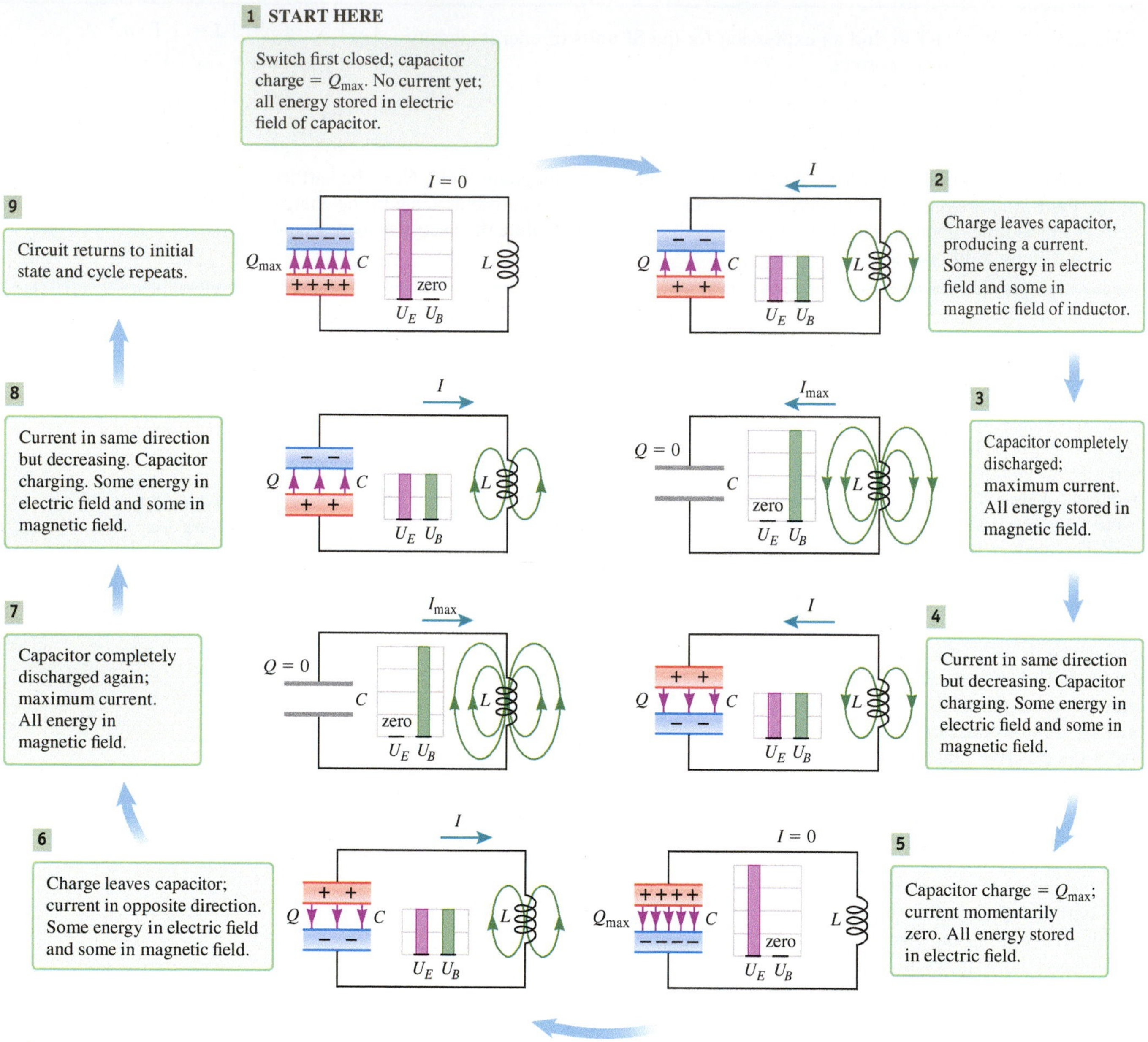

FIGURE 33.12 One complete cycle of an LC circuit. The cycle begins at time **1** with the capacitor fully charged, so its lower plate is positive. At time **5**, the circuit is halfway through its cycle; the capacitor is fully charged, but now the top plate is positive. By time **9**, the circuit has returned to its original state, after which the cycle repeats.

DERIVATION *Q* and *I* for the *LC* Circuit

We will show that the charge stored by a capacitor in an LC circuit oscillates according to

$$Q(t) = Q_{max} \cos(\omega t + \varphi) \tag{33.20}$$

$$\text{where } \omega = \sqrt{\frac{1}{LC}} \tag{33.21}$$

Also, we'll show that the current in the circuit is given by

$$I(t) = -I_{max} \sin(\omega t + \varphi) \tag{33.22}$$

$$\text{where } I_{max} = \omega Q_{max} \tag{33.23}$$

Start with the expression for constant total energy.	$E_{tot} = \frac{1}{2}\frac{Q^2}{C} + \frac{1}{2}LI^2$
The time derivative of a constant is zero. Taking the time derivative of both sides results in a differential equation set equal to zero.	$\frac{dE_{tot}}{dt} = \frac{d}{dt}\left(\frac{1}{2}\frac{Q^2}{C} + \frac{1}{2}LI^2\right) = 0$
Pull out the constants.	$\frac{1}{2}\left(\frac{1}{C}\frac{dQ^2}{dt} + L\frac{dI^2}{dt}\right) = 0$ (33.24)
Apply the chain rule (Appendix A) to each derivative.	$\frac{dQ^2}{dt} = \frac{dQ^2}{dQ}\frac{dQ}{dt} = 2Q\frac{dQ}{dt}$ and similarly for current I
Substitute these results into Equation 33.24.	$\frac{1}{2}\left[\frac{1}{C}\left(2Q\frac{dQ}{dt}\right) + L\left(2I\frac{dI}{dt}\right)\right] = 0$ $\frac{Q}{C}\left(\frac{dQ}{dt}\right) + LI\frac{dI}{dt} = 0$
Use $I = dq/dt$ (Eq. 28.1) to cancel the current I from both terms.	$\frac{Q}{C}(\cancel{I}) + L(\cancel{I})\frac{dI}{dt} = 0$
Use Equation 28.1 again to write a differential equation for Q.	$\frac{Q}{C} + L\frac{d}{dt}\left(\frac{dQ}{dt}\right) = \frac{Q}{C} + L\frac{d^2Q}{dt^2} = 0$
Rearranging this differential equation shows that it is mathematically equivalent to Equation 16.25—the equation for a simple harmonic oscillator that consists of a particle of mass m attached to a spring of spring constant k. The particle's position is $y(t)$.	$-\frac{Q(t)}{LC} = \frac{d^2Q(t)}{dt^2}$ (33.25) $-\frac{k}{m}y(t) = \frac{d^2y(t)}{dt^2}$ (16.25)
We can use the solution to Equation 16.25 here. Compare Equation 33.25 to Equation 16.25 to come up with a translation of variables.	$y(t) \rightarrow Q(t)$ $\frac{k}{m} \rightarrow \frac{1}{LC}$
We found that the particle's position oscillates sinusoidally (Eq. 16.3) and the angular frequency ω depends on the mass and the spring constant. The initial phase (the phase constant) is the phase $(\omega t + \varphi)$ when $t = 0$. See Section 16-2.	$y(t) = y_{max}\cos(\omega t + \varphi)$ (16.3) $\omega = \sqrt{\frac{k}{m}}$ (16.26)
To solve the LC circuit's differential equation, substitute the translation of variables into Equations 16.3 and 16.26.	$Q(t) = Q_{max}\cos(\omega t + \varphi)$ ✓ (33.20) $\omega = \sqrt{\frac{1}{LC}}$ ✓ (33.21)
Take the time derivative to find the current in the circuit.	$I(t) = \frac{dQ(t)}{dt} = \frac{d}{dt}[Q_{max}\cos(\omega t + \varphi)]$ $I(t) = -\omega Q_{max}\sin(\omega t + \varphi)$ $I(t) = -I_{max}\sin(\omega t + \varphi)$ ✓ (33.22) where $I_{max} = \omega Q_{max}$ ✓ (33.23)

Derivation continues on page 1060 ▶

COMMENTS

While deriving expressions for the charge stored by the capacitor (Eq. 33.20) and the current (Eq. 33.22), we found that the *LC* circuit's differential equation is mathematically equivalent to the particle–spring oscillator's equation. The charge stored by the capacitor and the position of an oscillating particle are conceptually connected.

Figure 33.13 shows graphs of Q and I versus t. Compare this to a particle connected to a spring and oscillating back and forth from $+y_{max}$ to $-y_{max}$ as in Figure 16.5A (page 453). In a similar fashion, the charge on the bottom plate in Figure 33.12 is initially $+Q_{max}$. That charge is reduced to zero, and then the charge switches sign and increases in magnitude until it reaches $-Q_{max}$. The charge is reduced in magnitude again, reaches zero, and then increases until it reaches $+Q_{max}$ again. Like the particle attached to the spring, the charge on the capacitor oscillates back and forth from $+Q_{max}$ to $-Q_{max}$ as shown in Figure 33.13A, and current is analogous to the particle's velocity (Fig. 33.13B).

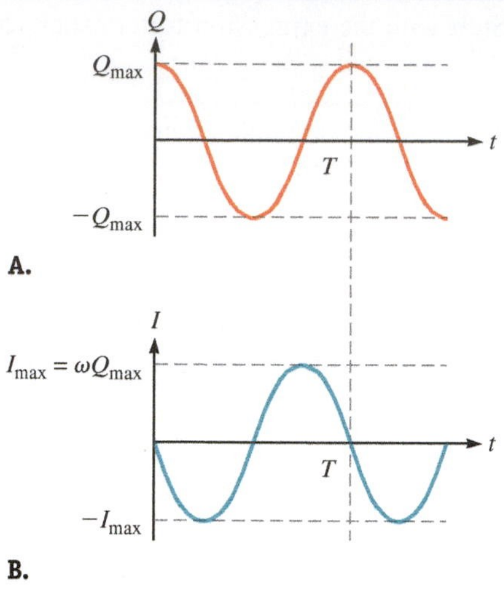

FIGURE 33.13 Charge and current versus time for an *LC* circuit.

CONCEPT EXERCISE 33.6

Take another look at the simple harmonic oscillator (Section 16-2, page 453). How is $I(t)$ analogous to $v(t)$?

Graphs of U_E and U_B for the *LC* Circuit

In Figure 33.12, we see that the energy U_E stored by the capacitor's electric field and the energy U_B stored by the inductor's magnetic field both oscillate. We can also express this fact mathematically. We found that the charge stored by the capacitor oscillates according to $Q(t) = Q_{max}\cos(\omega t + \varphi)$ (Eq. 33.20), and we know that the energy stored in the electric field depends on the stored charge, $U_E = Q^2/2C$ (Eq. 27.2). So, as you will show in Problem 27, the energy stored in the electric field oscillates according to

$$U_E(t) = \frac{Q_{max}^2}{2C}\cos^2(\omega t + \varphi) \tag{33.26}$$

The same sort of mathematical reasoning applies to the energy stored in the magnetic field. As you will show in Problem 28, the energy stored in the magnetic field is given by

$$U_B(t) = \frac{Q_{max}^2}{2C}\sin^2(\omega t + \varphi) \tag{33.27}$$

At any moment, the total energy stored by the electric and magnetic fields is the sum of Equations 33.26 and 33.27:

$$E_{tot} = U_E + U_B = \frac{Q_{max}^2}{2C}\cos^2(\omega t + \varphi) + \frac{Q_{max}^2}{2C}\sin^2(\omega t + \varphi)$$

The total energy is a constant given by

$$E_{tot} = \frac{Q_{max}^2}{2C} = \frac{1}{2}LI_{max}^2 \tag{33.28}$$

as you will show in Problem 29.

The transformation of energy between the electric and magnetic fields is illustrated in the graphs of Equations 33.26 and 33.27 shown in Figure 33.14. Both energies are always positive, and at any instant the sum of the two is a constant given by Equation 33.28.

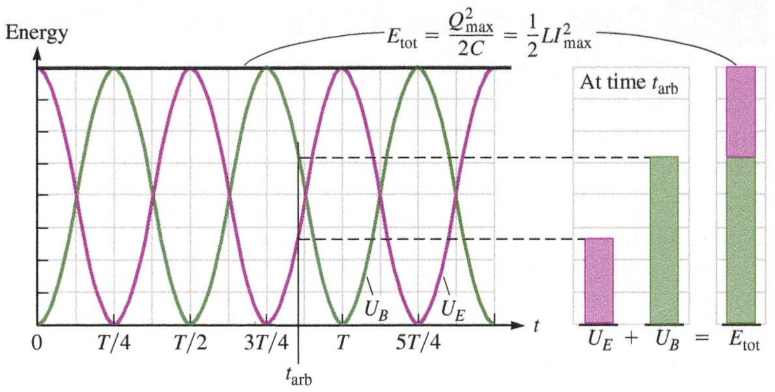

FIGURE 33.14 Energy stored in the electric field U_E and in the magnetic field U_B for the *LC* circuit. At any arbitrary time t_{arb}: $E_{tot} = U_E(t_{arb}) + U_B(t_{arb})$.

EXAMPLE 33.4 Energy Evenly Split Between the Fields

Consider the circuit in Figure 33.11 in which the capacitor is initially fully charged. The switch is closed at time $t = 0$. What is the first time after the switch is closed that half the energy is in the electric field of the capacitor and half is in the magnetic field of the inductor? Express your answer in terms of the period T of the oscillation.

INTERPRET and ANTICIPATE

The three graphs in Figures 33.13 and 33.14 show one and a half periods. Figure 33.14 shows that energy is equally split in less than a quarter of the period. We might guess that the split is halfway between 0 and $T/4$.

SOLVE

Set either U_E or U_B equal to half the total energy, and solve for t. We arbitrarily choose U_E (Eq. 33.26).	$U_E(t) = \frac{Q_{max}^2}{2C}\cos^2(\omega t + \varphi) = \frac{1}{2}E_{tot}$
Eliminate Q_{max} with $E_{tot} = Q_{max}^2/2C$ (Eq. 33.28).	$E_{tot}\cos^2(\omega t + \varphi) = \frac{1}{2}E_{tot}$
Cancel E_{tot}. The capacitor's charge is initially Q_{max}, so the phase constant $\varphi = 0$ (Fig. 33.13A).	$\cos^2(\omega t) = \frac{1}{2}$
Solve for ωt.	$\cos(\omega t) = \pm\frac{1}{\sqrt{2}}$ $\omega t = \cos^{-1}\left(\pm\frac{1}{\sqrt{2}}\right)$ $\omega t = \frac{\pi}{4}, \frac{3\pi}{4}, \frac{5\pi}{4}, \frac{7\pi}{4}, \frac{9\pi}{4}, \ldots$
We need only the first time after the switch is closed.	$\omega t = \pi/4$
Now use $\omega = 2\pi/T$ (Eq. 16.2) to find t in terms of the period.	$t = \frac{\pi}{4}\frac{1}{\omega} = \frac{\pi}{4}\frac{T}{2\pi} = T/8$

CHECK and THINK

As expected, the time for the energy to be evenly split between the capacitor and the inductor is less than a quarter of a period.

33-6 Special Case: AC Circuit with Resistance

AC CIRCUIT WITH RESISTANCE

Special Case

In Chapter 29, we applied the principles of electricity to practical direct current (DC) circuits. In this chapter, we apply these principles to alternating current (AC) circuits. In the remainder of this chapter, we will consider circuits that involve an AC power source such as an AC generator (Fig. 32.20, page 1026). The symbol for an AC power source is (∼). We start by considering circuits that contain an AC power source and just one other circuit element—a resistor, a capacitor, or an inductor. In the final section, we will consider an AC circuit that has all three of these elements. Our primary goal in each case is to find a connection between the voltage across the circuit element and the current in the circuit.

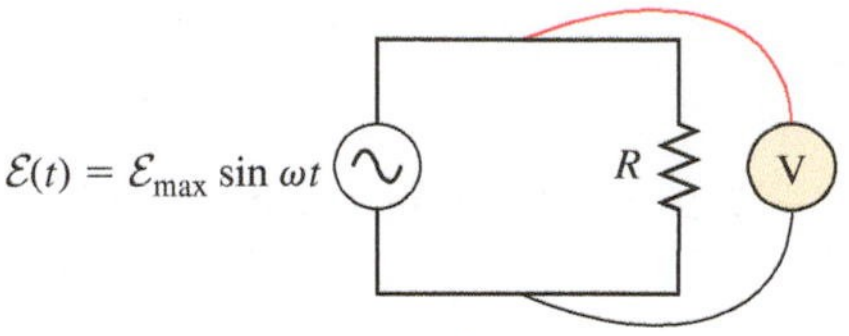

FIGURE 33.15 An AC generator and a resistor. A voltmeter measures the voltage across the AC generator and across the resistor. The voltage across the resistor equals the voltage across the generator.

Figure 33.15 shows an AC power supply connected to a resistor. The AC power supply's emf oscillates according to $\mathcal{E}(t) = \mathcal{E}_{max} \sin \omega t$ (Eq. 32.13). The voltage across the generator equals the voltage across the resistor:

$$V_R(t) = \mathcal{E}(t) = \mathcal{E}_{max} \sin \omega t \tag{33.29}$$

where V_R is the potential difference across the resistor.

To find the current in the circuit, apply Kirchhoff's loop rule at one particular instant in which the current is clockwise (not shown on the figure). Applying the loop rule and the resistance rule clockwise around the circuit in the direction of the current, we have

$$\mathcal{E}(t) - IR = 0 \tag{33.30}$$

The current as a function of time is given by

$$I(t) = \frac{\mathcal{E}(t)}{R} = \frac{\mathcal{E}_{max} \sin \omega t}{R}$$

The maximum current is

$$I_{max} = \frac{\mathcal{E}_{max}}{R} \tag{33.31}$$

and the current as a function of time may be written in terms of I_{max}:

$$I(t) = I_{max} \sin \omega t \tag{33.32}$$

Figure 33.16A is a graph of the voltage across the resistor as a function of time, and Figure 33.16B is a graph of the current as a function of time. The current and the voltage across the resistor are **in phase**, which means the current and voltage rise and fall together at the same time.

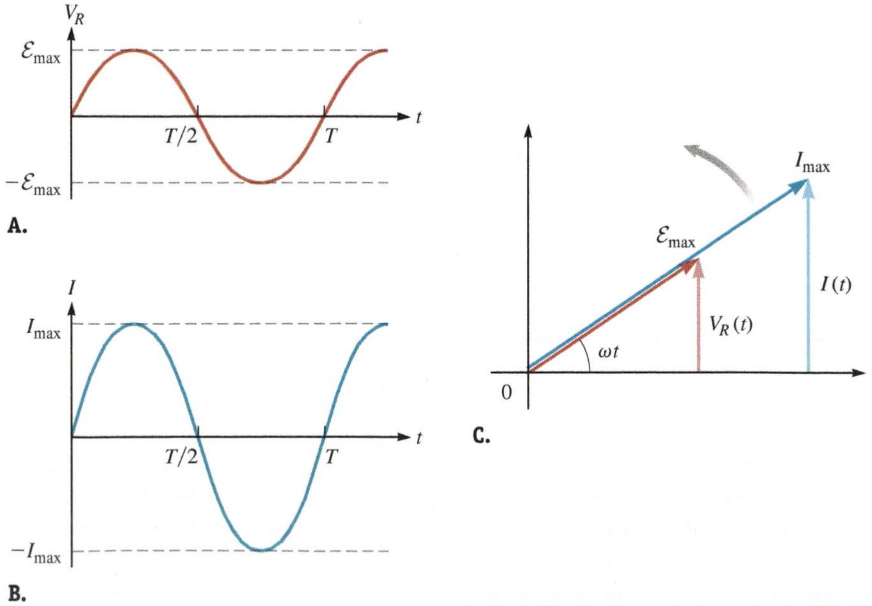

FIGURE 33.16 **A.** Voltage across the resistor in Figure 33.15. **B.** Current in the circuit. **C.** A phasor diagram shows that I and V_R are in phase because their phasors lie on top of one another as they rotate at angular speed ω around the origin.

Phasor Diagrams

In our study of AC circuits, we compare the time dependence of the current to that of the voltage across various circuit elements. We can make this comparison visually by plotting current and voltage as functions of time as in parts A and B of Figure 33.16.

A **phasor diagram** is another visual tool for making such a comparison. In a phasor diagram, a quantity such as current or voltage is represented by a vector known as a phasor. The length of the phasor represents the quantity's maximum value. To represent the time dependence of the quantity, the phasor rotates around the origin at an angular speed ω. So, at any instant t, the phasor makes an angle of ωt with the horizontal axis. The instantaneous value of the quantity is projected onto either the horizontal axis or the vertical axis. If you choose to represent the time variation using the cosine function, the projection is onto the horizontal axis. In this textbook, we have chosen to use the sine function, so the phasor's projection onto the vertical axis gives the quantity's instantaneous value. Figure 33.16C is the phasor diagram representing the current and voltage across the resistor for the AC circuit with resistance (Fig. 33.15).

PHASOR DIAGRAMS Tool

Although current and voltage may be represented by phasors, they are not vectors.

EXAMPLE 33.5 Power in an AC Circuit with Resistance

The AC generator in Figure 33.15 supplies energy, and the resistor dissipates energy through heat. Find an expression for the (instantaneous) power P dissipated by the resistor as a function of time. Plot P versus t. Indicate the average power on the plot.

INTERPRET and ANTICIPATE

The current through the resistor and the voltage across the resistor oscillate with time. The power supplied to or consumed by any circuit element at any moment is given by $P = I\Delta V$ (Eq. 28.33), so multiply the expression for current by the expression for voltage. We expect to find that the power also oscillates.

SOLVE

Substitute $I(t) = I_{max} \sin \omega t$ and $V_R(t) = \mathcal{E}_{max} \sin \omega t$ (Eqs. 33.32 and 33.29) into Equation 28.33.

$$P(t) = I(t)V_R(t) \quad (28.33)$$

$$P(t) = (I_{max} \sin \omega t)(\mathcal{E}_{max} \sin \omega t)$$

$$P(t) = I_{max}\mathcal{E}_{max} \sin^2 \omega t$$

The factor $I_{max}\mathcal{E}_{max}$ is the maximum power, and it is convenient to write the expression in terms of P_{max}.

$$P(t) = P_{max} \sin^2 \omega t \quad (33.33)$$

where $P_{max} = I_{max}\mathcal{E}_{max}$

The graph of $P(t)$ versus t (Eq. 33.33) is always positive in Figure 33.17 because the sine squared is always positive.

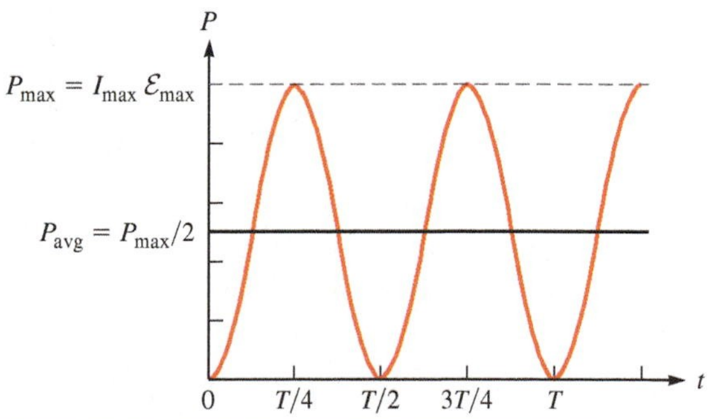

FIGURE 33.17 Power dissipated by the resistor in Figure 33.15.

The average power can be found from the rms current and rms voltage (Eq. 32.24). The rms voltage V_{rms} across the resistor equals the rms voltage $\mathcal{E}_{rms}$ across the power supply.

$$P_{avg} = I_{rms}V_{rms} \quad (32.24)$$

$$P_{avg} = I_{rms}\mathcal{E}_{rms} \quad (1)$$

Example continues on page 1064 ▶

Express the rms current and voltage in terms of their maximum values (Eqs. 32.21 and 32.20).	$I_{rms} = \frac{I_{max}}{\sqrt{2}}$ (32.21) $\mathcal{E}_{rms} = \frac{\mathcal{E}_{max}}{\sqrt{2}}$ (32.20)
Substitute these expressions for I_{rms} and V_{rms} into Equation (1).	$P_{avg} = \left(\frac{I_{max}}{\sqrt{2}}\right)\left(\frac{\mathcal{E}_{max}}{\sqrt{2}}\right) = \frac{I_{max}\mathcal{E}_{max}}{2}$
The numerator is the maximum power dissipated by the resistor, so the average power equals half the maximum power. A horizontal line on Figure 33.17 shows the average power dissipated by the resistor.	$P_{avg} = \frac{P_{max}}{2}$

CHECK and THINK

Figure 33.17 shows that the power oscillates as we expected. Suppose we were asked to find the power *supplied by the generator*. We would again use Equation 28.33 ($P = I\Delta V$). Because the voltage across the power supply equals the voltage across the resistor and the current through the power supply equals the current through the resistor, we would again find Equation 33.33 for the power supplied by the generator. Therefore, *all the power supplied by the generator is dissipated by the resistor.*

AC CIRCUIT WITH CAPACITANCE

Special Case

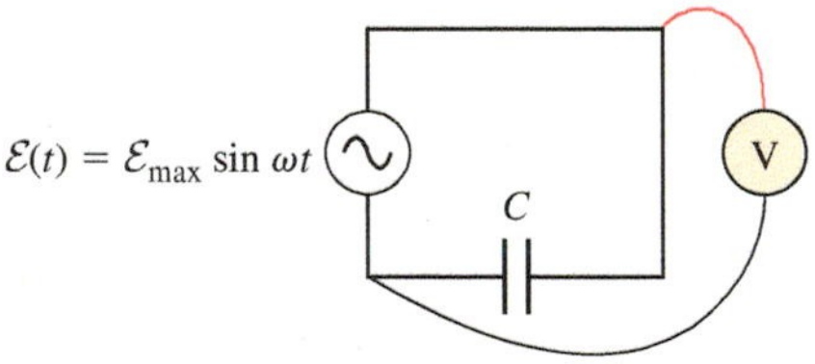

FIGURE 33.18 An AC generator and a capacitor. A voltmeter measures the voltage across the AC generator and across the capacitor. The voltage across the capacitor equals the voltage across the generator.

33-7 Special Case: AC Circuit with Capacitance

A purely capacitive AC circuit consists of an AC power source as before, but the resistor is replaced by a capacitor (Fig. 33.18). Again, imagine measuring the voltage across the power supply, which equals the voltage across the capacitor. As in the circuit described in Section 33-6,

$$V_C(t) = \mathcal{E}(t) = \mathcal{E}_{max} \sin \omega t \quad (33.34)$$

where V_C is the potential difference across the capacitor.

The voltage across a capacitor is proportional to the charge stored on its plates, $Q = CV_C$ (Eq. 27.1). So, the charge stored by the capacitor oscillates because V_C oscillates. To find an expression for $Q(t)$, substitute Equation 33.34 into Equation 27.1:

$$Q(t) = C\mathcal{E}_{max} \sin \omega t \quad (33.35)$$

We also need the current as a function of time, which we find by taking the time derivative of Equation 33.35:

$$I(t) = C\omega\mathcal{E}_{max} \cos \omega t \quad (33.36)$$

Equation 33.36 meets our goal of finding an expression for the current. In contrast to the purely resistive circuit in Section 33-6, the capacitive circuit has current and voltage that are *not* in phase: If the voltage is described by a sine function, the current is described by a cosine. In order to compare the phases, we should express both current and voltage in terms of either the sine function or the cosine function. We arbitrarily choose sine and, using the trigonometric identity $\cos \theta = \sin (\theta + 90°)$ from Appendix A, we rewrite Equation 33.36 as

Angular frequency ω is in rad/s, so we often need to express the phase constant in radians: $90° = \pi/2$ rad.

$$I(t) = C\omega\mathcal{E}_{max} \sin (\omega t + 90°) \quad (33.37)$$

The maximum current is the amplitude in front of the sine function:

$$I_{max} = C\omega\mathcal{E}_{max} \quad (33.38)$$

In Equation 33.37, the 90° is called the **phase constant**.

One more simplification is typically made. Compare Equation 33.38 to the current in the resistor circuit, $I_{max} = \mathcal{E}_{max}/R$ (Eq. 33.31). In both circuits, the maximum current is proportional to the generator's maximum emf $\mathcal{E}_{max}$. To make the equations look more similar and to understand the role of $C\omega$, a new quantity called the **capacitive reactance X_C** is defined as

CAPACITIVE REACTANCE

Major Concept

$$X_C \equiv \frac{1}{\omega C} \tag{33.39}$$

So, the current in the circuit shown in Figure 33.18 is given by

$$I(t) = \frac{\mathcal{E}_{\max}}{X_C} \sin(\omega t + 90°) \tag{33.40}$$

where the maximum current is given by

$$I_{\max} = \frac{\mathcal{E}_{\max}}{X_C} \tag{33.41}$$

Capacitive reactance has the SI units of ohms, and it is mathematically analogous to resistance. Conceptually, the reactance impedes the current. A large reactance means a small maximum current (Eq. 33.41).

Whereas resistance depends only on a particular resistor's geometry and composition, capacitive reactance depends on both the capacitor's capacitance *and the angular frequency of the generator*. When the angular frequency is low, the capacitive reactance is large (Fig. 33.19). So, at low angular frequencies, the current is strongly impeded and the maximum current is relatively low. This idea should make sense because if the AC generator were replaced by a DC generator, the angular frequency would be zero, and in this case, the capacitor prevents charge from flowing once it has been fully charged. When the angular frequency is high, the capacitive reactance is small. So, at high angular frequencies, the current is only weakly impeded and the maximum current is relatively high. As the angular frequency approaches infinity, the capacitive reactance approaches zero and the capacitor acts like an ideal wire.

Figure 33.20 shows the voltage across the capacitor and the current in the circuit as functions of time. The two sine functions are out of phase. The current reaches its peak a quarter of a cycle, or 90°, before the voltage reaches its peak. So we say *the current leads the voltage by 90°*, or *the voltage across a capacitor lags the current by 90°*.

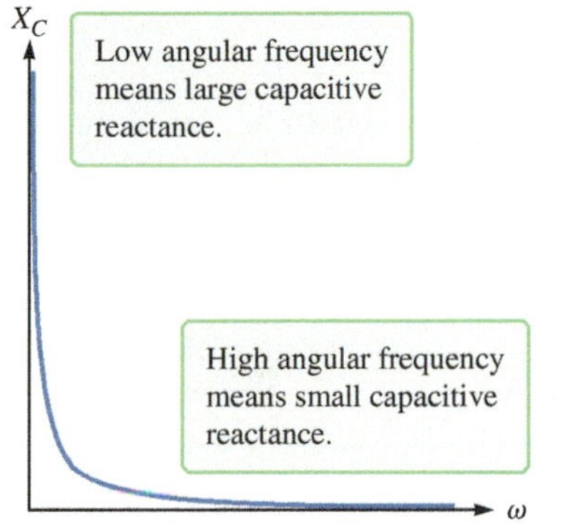

FIGURE 33.19 Capacitive reactance as a function of angular frequency.

CONCEPT EXERCISE 33.7

Read the dialogue and decide which student is correct. Give your reasons.

Shannon: You can only use the capacitive reactance to find the maximum current.

Cameron: No, you can use capacitive reactance exactly as you use resistance. So, to get the current you simply divide the voltage by X_C. It doesn't have to be the maximum current.

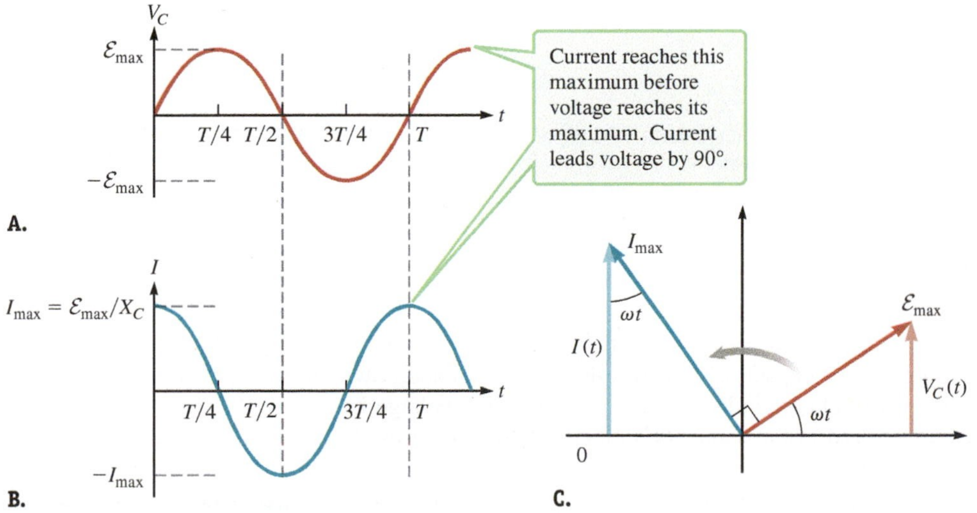

FIGURE 33.20 A. Voltage across the capacitor in Figure 33.18. **B.** Current in the circuit. **C.** Phasor diagram shows that I leads V_C by 90° as the two phasors rotate counterclockwise.

EXAMPLE 33.6 Power in an AC Circuit with Capacitance

The capacitor in Figure 33.18 stores energy in its electric field and releases that energy periodically. Find an expression for the (instantaneous) power P transferred to and from the capacitor as a function of time. Copy the graph of V_C as a function of time (Fig. 33.20A). Then plot P versus t, indicating the average power, and comment on the relationship between the two graphs.

INTERPRET and ANTICIPATE

The process for finding the power associated with this circuit is much like finding the power in the purely resistive circuit (Example 33.5). Again, we expect to find that the power transferred to and from the capacitor oscillates.

SOLVE

Substitute $I(t) = (\mathcal{E}_{max}/X_c)\sin(\omega t + 90°)$ and $V_C(t) = \mathcal{E}_{max}\sin\omega t$ (Eqs. 33.37 and 33.34) into Equation 28.33.

$$P(t) = I(t)V_C(t) \quad (28.33)$$

$$P(t) = \left[\frac{\mathcal{E}_{max}}{X_C}\sin(\omega t + 90°)\right](\mathcal{E}_{max}\sin\omega t)$$

$$P(t) = \frac{\mathcal{E}_{max}^2}{X_C}\sin(\omega t + 90°)\sin\omega t$$

A few trigonometric identities help to simplify this expression. First, use $\sin(\omega t + 90°) = \cos\omega t$.

$$P(t) = \frac{\mathcal{E}_{max}^2}{X_C}\cos\omega t\sin\omega t$$

Second, use $\sin 2\theta = 2\sin\theta\cos\theta$ (Appendix A).

$$P(t) = \frac{1}{2}\frac{\mathcal{E}_{max}^2}{X_C}\sin 2\omega t$$

The factor in front of the sine function is the maximum power, and it is convenient to write the expression in terms of P_{max}.

$$P(t) = P_{max}\sin 2\omega t \quad (33.42)$$

where

$$P_{max} = \frac{1}{2}\frac{\mathcal{E}_{max}^2}{X_C} \quad (33.43)$$

The graph of $P(t)$ versus t (Fig. 33.21B) shows that the power can be either positive or negative. The power is positive when energy is being transferred *to* the capacitor and negative when energy is being transferred *from* the capacitor. The average power (over one cycle) is zero as shown in the figure.

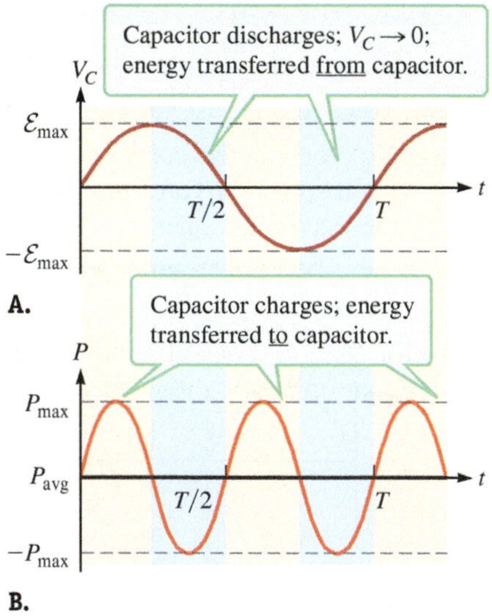

FIGURE 33.21 **A.** Voltage across the capacitor in Figure 33.18. **B.** Power transferred to and from the capacitor. In both parts, the yellow regions correspond to charging and the blue regions to discharging.

CHECK and THINK

Figure 33.21 shows that the power oscillates as we expected. When charge is building up on the capacitor, $|V_C|$ increases. The energy stored in the capacitor's electric field increases and power is transferred *to* the capacitor (positive). When charge is leaving the capacitor, the voltage between its plates approaches 0. The energy stored in the capacitor's electric field decreases and power is transferred *from* the capacitor (negative).

Special Case: *RC* Filters

The dependence of capacitive reactance on the generator's angular frequency has important practical applications, one of which is frequency filter circuits. A **filter circuit** produces negligible voltage for certain frequencies and nonnegligible voltage for other frequencies. To understand the basic idea of a filter, imagine the AC generator in Figure 33.18 is replaced by a power supply that produces an emf oscillating at many frequencies at once. For instance, the power might be supplied by a person singing into a microphone. The sound wave is made up of many frequencies (Chapter 18), so the resulting emf is made up of many frequencies. Many performers like to adjust their singing electronically by filtering out some frequencies that their voice produces. A filter circuit is inserted between the microphone and the speaker.

Let's use mathematics to see how a filter works. Because any real circuit has some resistance, consider a filter that consists of a resistor and a capacitor (Fig. 33.22A). We seek an expression for the maximum voltage across the capacitor as a function of angular frequency.

The current oscillates:

$$I(t) = I_{\max} \sin \omega t \tag{33.44}$$

The voltage across the resistor is in phase with the current:

$$V_R(t) = RI_{\max} \sin \omega t \tag{33.45}$$

The voltage across the capacitor lags the current by 90°:

$$V_C(t) = X_C I_{\max} \sin\left(\omega t - \frac{\pi}{2}\right) \tag{33.46}$$

where we have written the phase constant in radians. The AC power supply's emf is not necessarily in phase with the current or with the voltage across the other circuit elements. We express its phase with an arbitrary constant φ:

$$\mathcal{E}(t) = \mathcal{E}_{\max} \sin(\omega t + \varphi) \tag{33.47}$$

According to Kirchhoff's loop rule, at any instant t, the voltage across the AC power supply must equal the voltage across the resistor plus the voltage across the capacitor:

$$\mathcal{E}(t) = V_R(t) + V_C(t) \tag{33.48}$$

Substitute Equations 33.45, 33.46, and 33.47 into Equation 33.48:

$$\mathcal{E}_{\max} \sin(\omega t + \varphi) = RI_{\max} \sin \omega t + X_C I_{\max} \sin\left(\omega t - \frac{\pi}{2}\right) \tag{33.49}$$

To solve for $I_{\max}$, we must eliminate the phase constant φ. One way to do that is to evaluate Equation 33.49 at two special times. Let one such time be $t = 0$:

$$\mathcal{E}_{\max} \sin(0 + \varphi) = RI_{\max} \sin 0 + X_C I_{\max} \sin\left(0 - \frac{\pi}{2}\right)$$

$$\mathcal{E}_{\max} \sin \varphi = -X_C I_{\max} \tag{33.50}$$

Let the other time be when $\omega t = \pi/2$:

$$\mathcal{E}_{\max} \sin\left(\frac{\pi}{2} + \varphi\right) = RI_{\max} \sin \frac{\pi}{2} + X_C I_{\max} \sin(0)$$

$$\mathcal{E}_{\max} \cos \varphi = RI_{\max} \tag{33.51}$$

where we used the trigonometric identity $\sin(\pi/2 + \varphi) = \cos \varphi$. Now, square Equations 33.50 and 33.51 and add the results:

$$\mathcal{E}^2_{\max} \sin^2\varphi + \mathcal{E}^2_{\max} \cos^2\varphi = (-X_C I_{\max})^2 + (RI_{\max})^2$$

RC **FILTERS** **Special Case**

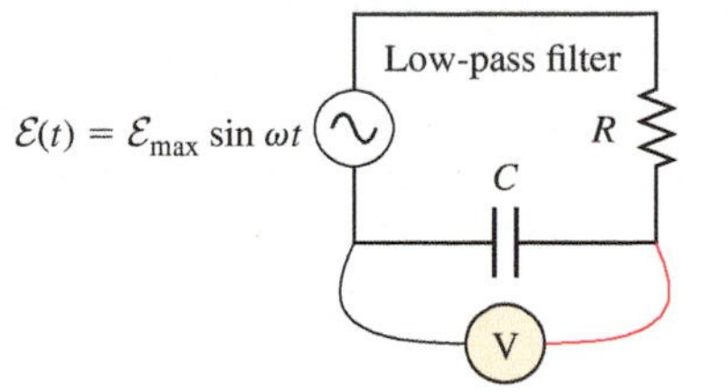

A.

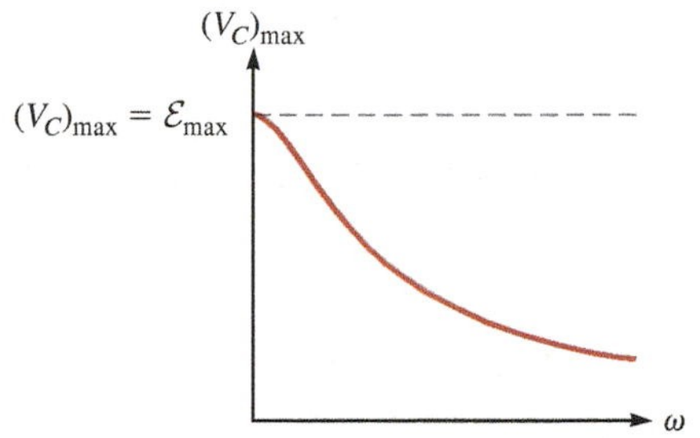

B.

FIGURE 33.22 **A.** A low-pass *RC* filter allows low frequencies to pass through the capacitor. **B.** Maximum voltage $(V_C)_{\max}$ across the capacitor in an *RC* filter is frequency-dependent.

Use the trigonometric identity $\sin^2\varphi + \cos^2\varphi = 1$ to solve for $I_{\max}$:

$$\mathcal{E}_{\max}^2 = (X_C^2 + R^2)I_{\max}^2$$

$$I_{\max} = \frac{\mathcal{E}_{\max}}{\sqrt{X_C^2 + R^2}} \tag{33.52}$$

Now that we have the maximum current, multiply by the capacitive reactance to find an expression for the maximum voltage across the capacitor:

$$(V_C)_{\max} = X_C I_{\max} = X_C \frac{\mathcal{E}_{\max}}{\sqrt{X_C^2 + R^2}} \tag{33.53}$$

Finally, we substitute $X_C \equiv 1/\omega C$ (Eq. 33.39) for capacitive reactance so that our expression is in terms of ω:

$$(V_C)_{\max} = \frac{1}{\omega C} \frac{\mathcal{E}_{\max}}{\sqrt{(1/\omega^2 C^2) + R^2}} \tag{33.54}$$

Equation 33.54 meets our goal and is best understood from a graph of $(V_C)_{\max}$ versus angular frequency ω (Fig. 33.22B). The voltage across the capacitor is very low at high frequencies. At low frequencies, the voltage across the capacitor nearly equals the voltage across the AC power supply. If the AC power supply is replaced by a microphone and a singer, the signal from the microphone has both high and low frequencies. Only the low-frequency signal produces a significant voltage across the capacitor. So, if the voltmeter in Figure 33.22A is replaced by a speaker, only the low-frequency signal passes to the speaker. If the singer makes some high squeaky notes, those are filtered out of the sound you hear. Because the low frequencies are passed through the filter, it is called a **low-pass filter**.

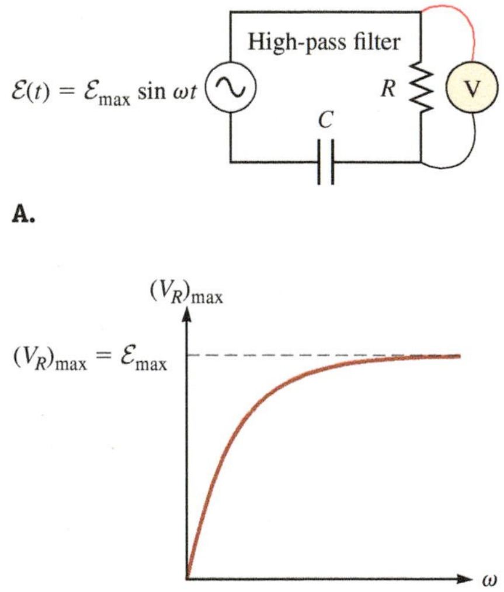

FIGURE 33.23 **A.** A high-pass RC filter allows high frequencies to pass through the capacitor. **B.** Maximum voltage $(V_R)_{\max}$ across the resistor in an RC filter is frequency-dependent.

This same circuit can also be used as a **high-pass filter** (Fig. 33.23A), allowing the high-frequency signal to pass through. According to $\mathcal{E}(t) = V_R(t) + V_C(t)$ (Eq. 33.48), when the voltage across the capacitor is low (when the frequency is high), the voltage across the resistor must be high. The maximum voltage across the resistor is found by multiplying the maximum current by the resistance:

$$(V_R)_{\max} = RI_{\max} = R\frac{\mathcal{E}_{\max}}{\sqrt{X_C^2 + R^2}} \tag{33.55}$$

Again we substitute $X_C \equiv 1/\omega C$ (Eq. 33.39) for capacitive reactance so that our expression is in terms of ω:

$$(V_R)_{\max} = \frac{\mathcal{E}_{\max} R}{\sqrt{(1/\omega^2 C^2) + R^2}} \tag{33.56}$$

Figure 33.23B shows a graph of $(V_R)_{\max}$ versus angular frequency ω. The voltage across the resistor is very low at low frequencies, but at high frequencies, the voltage across the resistor nearly equals the voltage across the AC power supply. Again, imagine the AC power supply is replaced by a microphone and a person singing. Only the high-frequency signal produces a significant voltage across the resistor. So, if the voltmeter in Figure 33.23A is replaced by a speaker, only the high-frequency signal passes to the speaker. A high-pass filter forms the basic circuit for the *tweeter* in a stereo speaker.

33-8 Special Case: AC Circuit with Inductance

AC CIRCUIT WITH INDUCTANCE

Special Case

To form a purely inductive circuit, we replace the capacitor (Fig. 33.18) with an inductor (Fig. 33.24). The voltage across the power supply equals the voltage across the inductor:

$$V_L(t) = \mathcal{E}(t) = \mathcal{E}_{\max} \sin \omega t \tag{33.57}$$

where V_L is the potential difference across the inductor. For all three circuits so far (purely resistive, capacitive, and inductive), the voltage across the circuit element is equal to the voltage across the power supply.

Consider an instant when the current in the circuit is clockwise so that the red lead of the voltmeter is on the upstream side of the inductor. We are measuring the voltage in the direction *opposite* the current, so according to the inductor rule (Eq. 33.8),

$$V_L(t) = V_{red} - V_{black} = L\frac{dI}{dt} \tag{33.58}$$

The minus sign is dropped because the voltmeter leads have been switched and $V_{red} - V_{black} = -(V_{black} - V_{red})$.

To find the current in the circuit, set Equation 33.57 equal to Equation 33.58:

$$L\frac{dI}{dt} = \mathcal{E}_{max}\sin\omega t$$

Isolate the current on one side of the equation:

$$dI = \frac{\mathcal{E}_{max}}{L}(\sin\omega t)dt$$

Then integrate:

$$\int dI = \int\frac{\mathcal{E}_{max}}{L}(\sin\omega t)dt$$

$$I = \frac{\mathcal{E}_{max}}{L}\int(\sin\omega t)dt$$

The integral of the sine function can be found in Appendix A:

$$I = -\frac{\mathcal{E}_{max}}{\omega L}\cos\omega t + \text{constant}$$

The constant of integration represents the current when the first term is zero. In other words, it is the DC current. Because there is no DC current in this circuit, we set the constant of integration to zero:

$$I = -\frac{\mathcal{E}_{max}}{\omega L}\cos\omega t$$

To compare the phase of the current to the phase of the voltage, we should express both current and voltage in terms of either the sine function or the cosine function. As in the case of the capacitor circuit, we have arbitrarily chosen the sine. Use a trigonometric identity (Appendix A) to rewrite the current as

$$I = \frac{\mathcal{E}_{max}}{\omega L}\sin(\omega t - 90°) \tag{33.59}$$

The maximum current is the amplitude in front of the sine function:

$$I_{max} = \frac{\mathcal{E}_{max}}{\omega L} \tag{33.60}$$

Analogous to the capacitive reactance X_C, the **inductive reactance** X_L is defined as

$$X_L \equiv \omega L \tag{33.61}$$

The current in the circuit shown in Figure 33.24 is thus

$$I(t) = \frac{\mathcal{E}_{max}}{X_L}\sin(\omega t - 90°) \tag{33.62}$$

where the maximum current is

$$I_{max} = \frac{\mathcal{E}_{max}}{X_L} \tag{33.63}$$

Inductive reactance has the SI units of ohms and is mathematically analogous to resistance and capacitive reactance. Like capacitive reactance, inductive reactance impedes the current; a large reactance means a small maximum current.

Also like capacitive reactance, inductive reactance depends on both the inductor's geometry *and the angular frequency of the generator*. Figure 33.25 shows a plot of the inductive reactance X_L versus the generator's angular frequency ω. Inductive reactance depends linearly on angular frequency, so when the angular frequency is

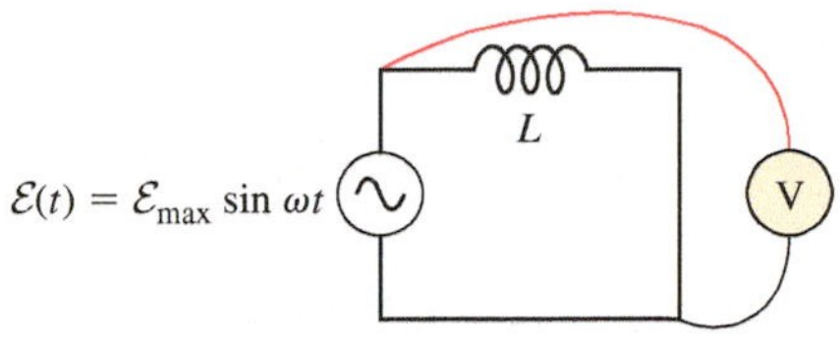
L
ℰ(t) = ℰ_max sin ωt
V

FIGURE 33.24 An AC generator and an inductor. A voltmeter measures the voltage across the AC generator and across the inductor. The voltage across the inductor equals the voltage across the generator.

INDUCTIVE REACTANCE

Major Concept

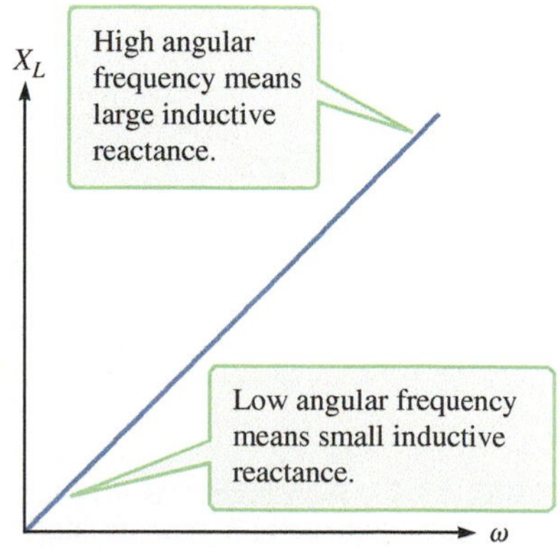
X_L
High angular frequency means large inductive reactance.
Low angular frequency means small inductive reactance.
ω

FIGURE 33.25 Inductive reactance as a function of angular frequency. Compare with Figure 33.19.

low, the inductive reactance is also low. Thus, at low angular frequencies, the current is only weakly impeded and the maximum current is large. This idea makes sense because if the AC generator were replaced by a DC generator, the angular frequency would be zero, and in that case, the inductor would not produce a back emf. Instead, the inductor would be modeled as an ideal wire, allowing current to flow freely. When the angular frequency is high, the inductive reactance is large; current is strongly impeded and the maximum current is low.

Figure 33.26 shows the voltage across the inductor and the current in the circuit as functions of time. The two sine functions are out of phase. The current reaches its peak a quarter of a cycle, or 90°, *after* the voltage reaches its peak. So, we say *the current lags the voltage by 90°*, or *the voltage across an inductor leads the current by 90°*.

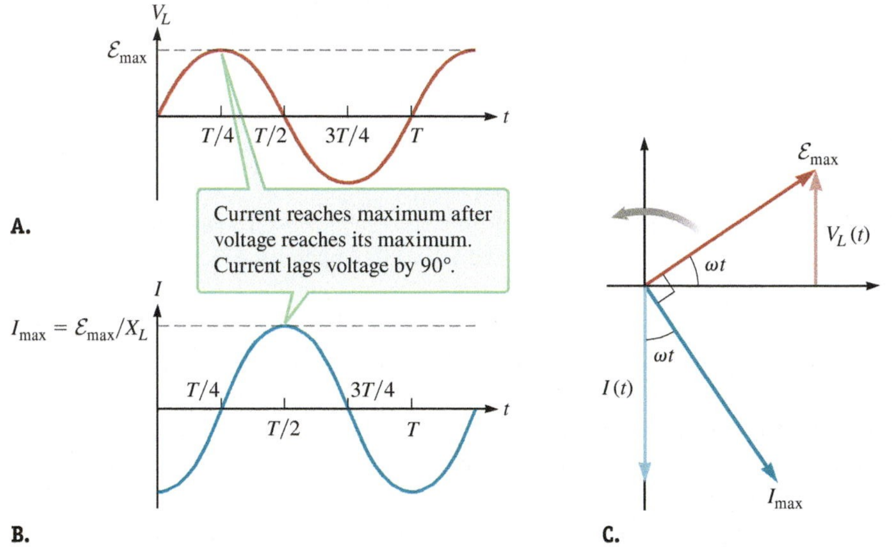

FIGURE 33.26 **A.** Voltage across the inductor in Figure 33.24. **B.** Current in the circuit. Compare with Figure 33.20. **C.** A phasor diagram shows that V_L leads I by 90°. Compare with Figure 33.20 for a capacitive circuit. One way to remember the difference between capacitive and inductive circuits is with the mnemonic CIVIL—in a ***C***apacitive circuit, ***I*** (current) leads ***V*** (voltage); ***V*** leads ***I*** in an inductive (***L***) circuit.

***RL* FILTERS** Special Case

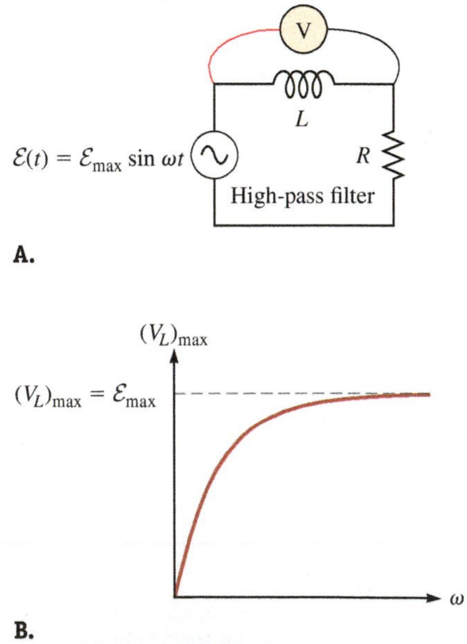

FIGURE 33.27 **A.** A high-pass *RL* filter allows high frequencies to pass through the inductor. **B.** Maximum voltage $(V_L)_{max}$ across the inductor of an *RL* filter is frequency-dependent. Compare with Figure 33.23 for a high-pass *RC* filter.

Special Case: *RL* Filters

Because inductive reactance depends on the generator's angular frequency, an inductive circuit can be used as a filter. Figure 33.27A shows a basic *RL* filter consisting of a resistor, an inductor, and an AC power supply with an adjustable frequency. We can find an expression for the maximum voltage across the inductor as a function of the angular frequency by using a process much like the one we used in Section 33-7 for the *RC* filter. Start by finding an expression for I_{max} (Problem 46):

$$I_{max} = \frac{\mathcal{E}_{max}}{\sqrt{X_L^2 + R^2}} \tag{33.64}$$

Multiply the maximum current by the inductive reactance to find an expression for the maximum voltage across the inductor:

$$(V_L)_{max} = X_L I_{max} = X_L \frac{\mathcal{E}_{max}}{\sqrt{X_L^2 + R^2}} \tag{33.65}$$

Substitute $X_L \equiv \omega L$ (Eq. 33.61) for inductive reactance so that our expression is in terms of ω:

$$(V_L)_{max} = L\omega \frac{\mathcal{E}_{max}}{\sqrt{L^2\omega^2 + R^2}} \tag{33.66}$$

Figure 33.27B is a graph of the maximum voltage across the inductor (Eq. 33.66) as a function of ω. The voltage across the inductor is very low at low frequencies. But at high frequencies, the voltage across the inductor is high, nearly equal to the voltage across the AC power supply. Because the high frequency is passed through the inductor, the circuit in Figure 33.27A is a **high-pass filter**.

This same circuit can also be used as a **low-pass filter** (Fig. 33.28A), allowing the high-frequency signal to pass through. The maximum voltage across the resistor is found by multiplying the maximum current by the resistance:

$$(V_R)_{\max} = R\frac{\mathcal{E}_{\max}}{\sqrt{X_L^2 + R^2}} \tag{33.67}$$

Substitute $X_L \equiv \omega L$ (Eq. 33.61) for inductive reactance:

$$(V_R)_{\max} = \frac{\mathcal{E}_{\max} R}{\sqrt{L^2\omega^2 + R^2}} \tag{33.68}$$

Figure 33.28B is a graph of $(V_R)_{\max}$ as a function of ω, showing that the voltage across the resistor is very low at high frequencies. At low frequencies, the voltage across the resistor nearly equals the voltage across the AC power supply.

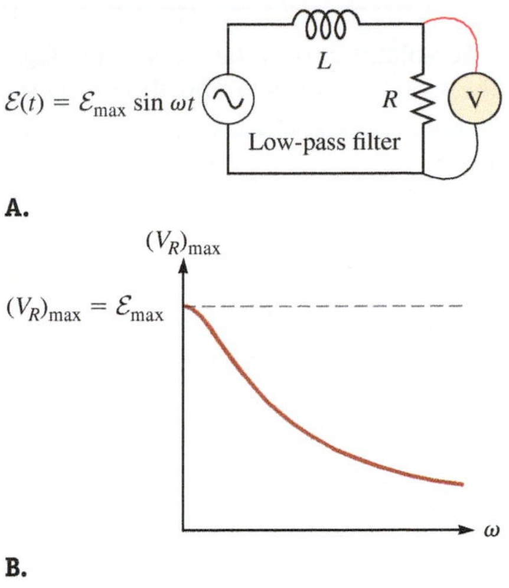

FIGURE 33.28 A. A low-pass *RL* filter allows low frequencies to pass through the resistor. **B.** Maximum voltage $(V_R)_{\max}$ across the resistor of an *RL* filter is frequency-dependent. Compare with Figure 33.22 for a low-pass *RC* filter.

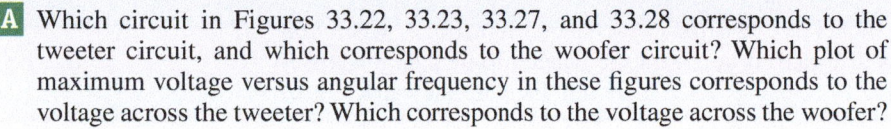

EXAMPLE 33.7 CASE STUDY A Stereo Speaker

In the 1960s, students listened to Jimi Hendrix play the guitar through stereo speakers at live concerts, broadcast on the radio, or from recordings. In any case, the input to the speakers corresponded to the notes Hendrix played on his guitar (made up of many frequencies). To help the listener hear the full range of frequencies, a good speaker is made up of two individual speakers. The smaller speaker—known as the **tweeter**—is good at producing high-frequency sound waves. The larger speaker—the **woofer**—is good at producing low-frequency sound waves. Ideally, a high-frequency voltage should be passed to the tweeter and a low-frequency voltage should be passed to the woofer. This is done with a high-pass *RC* filter and a low-pass *RL* filter (Fig. 33.29).

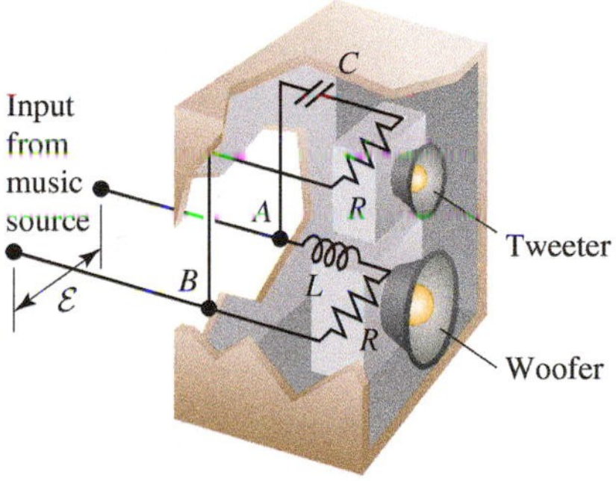

FIGURE 33.29 A stereo speaker, simplified.

A Which circuit in Figures 33.22, 33.23, 33.27, and 33.28 corresponds to the tweeter circuit, and which corresponds to the woofer circuit? Which plot of maximum voltage versus angular frequency in these figures corresponds to the voltage across the tweeter? Which corresponds to the voltage across the woofer?

INTERPRET and ANTICIPATE

Your job is to match the simple circuit diagrams in the figures listed to the drawing in Figure 33.29 and then identify the correct voltage-versus-angular frequency graphs.

SOLVE

Tweeter. The tweeter circuit in Figure 33.29 consists of a capacitor and a resistor as in Figures 33.22 and 33.23. The tweeter speaker is represented by a resistor. So, the tweeter circuit corresponds to Figure 33.23A, which has the voltmeter across the resistor (a high-pass filter).

Because the tweeter is a high-pass *RC* circuit and its voltage corresponds to the voltage across the resistor, the graph of $(V_R)_{\max}$ versus ω in Figure 33.23B is the one we want. As this graph shows, the voltage across the tweeter is high at high frequencies.

Woofer. Like the circuits in Figures 33.27 and 33.28, the woofer circuit consists of an inductor and a resistor. The woofer speaker is represented by the resistor. So, the woofer circuit corresponds to Figure 33.28A, which has the voltmeter across the resistor.

The voltage across the resistor $(V_R)_{max}$ shown in Figure 33.28B corresponds to the voltage across the woofer. As this graph shows, the voltage across the woofer is high at low frequencies.

B Assume the tweeter's resistance equals the woofer's resistance. The **angular crossover frequency** is the angular frequency at which the maximum voltage across the tweeter equals the maximum voltage across the woofer. (This is sometimes defined in terms of the current through each element, but if the resistance is the same in each, the definitions are equivalent.) Roughly speaking, the tweeter produces sounds higher than the crossover frequency and the woofer produces sounds lower than the crossover frequency. Find an expression for the crossover frequency in terms of L and C. In the **CHECK and THINK** step answer this: How can you change the crossover frequency so that the tweeter picks up lower frequencies?

INTERPRET and ANTICIPATE

In part A, we identified the graphs of maximum voltage across the tweeter and the woofer. We can plot these curves on the same axes (Fig. 33.30). To find an expression for the crossover frequency, set the maximum voltage across the resistor in the high-pass *RC* filter equal to the voltage across the resistor in the low-pass *RL* filter.

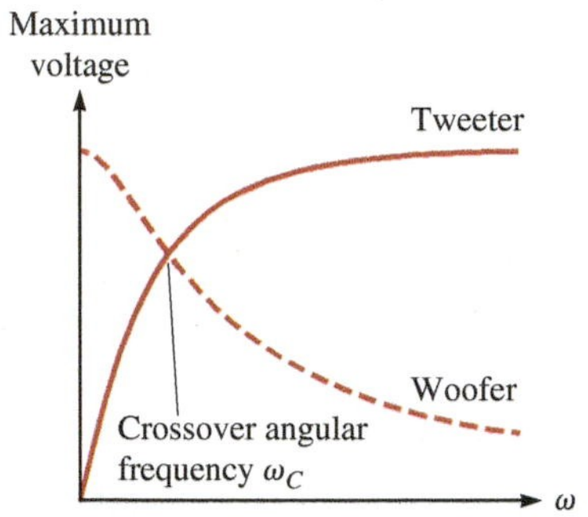

FIGURE 33.30

SOLVE

Set $(V_R)_{max}$ from Equation 33.56 equal to $(V_R)_{max}$ from Equation 33.68, and solve for the angular crossover frequency ω_c.

$$\frac{\mathcal{E}_{max} R}{\sqrt{(1/\omega_c^2 C^2) + R^2}} = \frac{\mathcal{E}_{max} R}{\sqrt{L^2\omega_c^2 + R^2}}$$

$$(1/\omega_c^2 C^2) + R^2 = L^2\omega_c^2 + R^2$$

$$1/\omega_c^2 C^2 = L^2\omega_c^2 \qquad \omega_c^4 = \frac{1}{L^2C^2}$$

$$\omega_c = \frac{1}{\sqrt{LC}}$$

CHECK and THINK

The crossover frequency depends on the inductance and the capacitance. Sometimes, when you listen to music through speakers, it sounds distorted. If you want the tweeter to pick up lower frequencies, you need the crossover frequency to decrease, so you should increase the capacitance. (Problem 54 asks you to think about what happens to the voltage across the woofer as a result.)

***RLC* (AC) CIRCUIT** Special Case

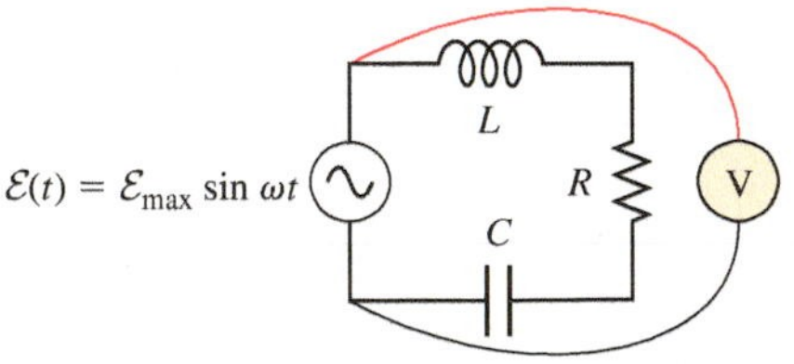

FIGURE 33.31 An AC generator, a resistor, an inductor, and a capacitor. A voltmeter measures the voltage across the AC generator and across the three other circuit elements. The voltage across the generator equals the sum of the voltages across the three other circuit elements.

33-9 Special Case: AC Circuit with Resistance, Inductance, and Capacitance

Our last circuit (Fig. 33.31) consists of an AC power supply, an inductor, a resistor, and a capacitor and is usually referred to as an *RLC* circuit. *RLC* circuits play an important role in communications equipment such as radios, televisions, and telephones.

Current and Voltage in an *RLC* Circuit

We seek expressions for the current and the AC power supply's emf. The current varies sinusoidally:

$$I(t) = I_{max} \sin \omega t \qquad (33.69)$$

We want to find I_{max} in terms of $\mathcal{E}_{max}$, R, X_C, and X_L. Equation 33.69 for $I(t)$ is the same expression we had in the resistor circuit, but it is different from the expression we had for the capacitor and inductor circuits. Figure 33.32A is a graph of Equation 33.69.

To find I_{max}, we need the circuit voltages. As for the three circuits we have presented earlier, imagine measuring the voltage across the AC generator. In this case, the voltage across the AC generator is the sum of the voltages across the three other circuit elements:

$$\mathcal{E}(t) = V_L(t) + V_R(t) + V_C(t) \tag{33.70}$$

We need expressions for each of these terms. From our earlier work, we know how each voltage is related to the current. First, the voltage across the resistor is in phase with the current (Fig. 33.32B), so we can write an expression for V_R:

$$V_R(t) = RI_{max} \sin \omega t \tag{33.71}$$

where the maximum voltage across the resistor is RI_{max}. Second, the voltage across the capacitor lags the current by 90° (or $\pi/2$), so we can write an expression for V_C:

$$V_C(t) = X_C I_{max} \sin\left(\omega t - \frac{\pi}{2}\right) = -X_C I_{max} \sin\left(\omega t + \frac{\pi}{2}\right) \tag{33.72}$$

where we have used $\sin(\alpha + \pi) = -\sin\alpha$ and the maximum voltage across the capacitor is $X_C I_{max}$. (Eq. 33.72 is plotted in Fig. 33.32C.) Third, the voltage across the inductor leads the current by 90° (or $\pi/2$), so we can write an expression for V_L:

$$V_L(t) = X_L I_{max} \sin\left(\omega t + \frac{\pi}{2}\right) \tag{33.73}$$

where the maximum voltage across the capacitor is $X_L I_{max}$. (Eq. 33.73 is plotted in Fig. 33.32D.)

Our next step is to add the voltages (Eqs. 33.71, 33.72, and 33.73) across the three circuit elements to find the voltage across the AC generator:

$$\mathcal{E}(t) = RI_{max} \sin \omega t - X_C I_{max} \sin\left(\omega t + \frac{\pi}{2}\right) + X_L I_{max} \sin\left(\omega t + \frac{\pi}{2}\right)$$

Combine like terms:

$$\mathcal{E}(t) = RI_{max} \sin \omega t + (X_L - X_C) I_{max} \sin\left(\omega t + \frac{\pi}{2}\right) \tag{33.74}$$

We need an expression for $\mathcal{E}(t)$, the AC generator's emf, in terms of $\mathcal{E}_{max}$. Figure 33.33 helps organize the information graphically. Figure 33.33A is a graph of

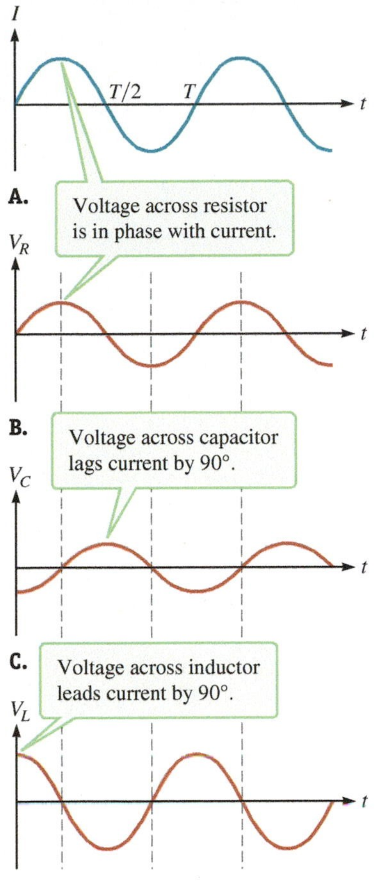

FIGURE 33.32 **A.** Current in an *RLC* circuit (Fig. 33.31). **B.** Voltage across the resistor. **C.** Voltage across the capacitor. **D.** Voltage across the inductor.

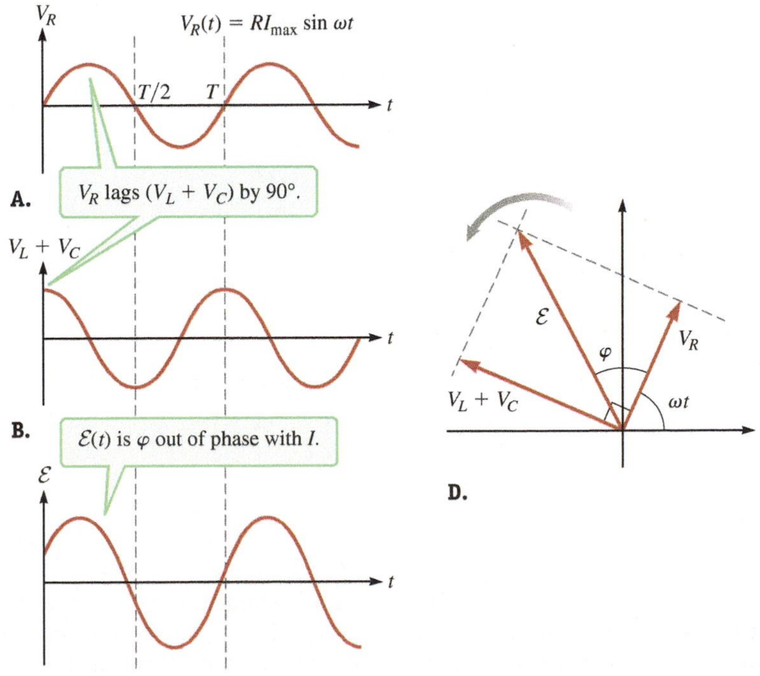

FIGURE 33.33 **A.** Voltage across the resistor from Figure 33.32B. **B.** Voltage across the inductor plus voltage across the capacitor: $V_L + V_C = (X_L - X_C)I_{max} \sin(\omega t + 90°)$. **C.** Voltage across the power supply. **D.** The corresponding phasor diagram. The current I (not shown) is in phase with V_R and so lies on its phasor. The voltage across the power supply is the sum of the three other voltages: $\mathcal{E}(t) = V_L(t) + V_R(t) + V_C(t)$.

$V_R(t)$, and part B is a graph of $V_L(t) + V_C(t)$. These graphs show that the voltage across the resistor lags the sum $V_L(t) + V_C(t)$ by 90°. When we add these values to find $\mathcal{E}(t)$, the result is out of phase with the current by some amount φ. To see this, compare Figure 33.33C for $\mathcal{E}(t)$ with the graph for the current in Figure 33.32A. We express the AC generator's emf as

$$\mathcal{E}(t) = \mathcal{E}_{\max} \sin(\omega t + \varphi) \tag{33.75}$$

Substitute Equation 33.75 into Equation 33.74:

$$\mathcal{E}_{\max} \sin(\omega t + \varphi) = RI_{\max} \sin \omega t + (X_L - X_C)I_{\max} \sin\left(\omega t + \frac{\pi}{2}\right) \tag{33.76}$$

To find $I_{\max}$, evaluate Equation 33.76 at two different times in order to eliminate φ. First, at $t = 0$:

$$\mathcal{E}_{\max} \sin \varphi = (X_L - X_C)I_{\max} \tag{33.77}$$

Then, when $\omega t = \pi/2$:

$$\mathcal{E}_{\max} \sin\left(\frac{\pi}{2} + \varphi\right) = RI_{\max} \sin\frac{\pi}{2} + (X_L - X_C)I_{\max} \sin\left(\frac{\pi}{2} + \frac{\pi}{2}\right)$$

$$\mathcal{E}_{\max} \cos\varphi = RI_{\max} + (X_L - X_C)I_{\max} \sin\pi$$

$$\mathcal{E}_{\max} \cos \varphi = RI_{\max} \tag{33.78}$$

There are two unknowns ($I_{\max}$ and φ) and two equations (33.77 and 33.78). To solve for $I_{\max}$, square these equations and add them:

$$(\mathcal{E}_{\max} \sin \varphi)^2 + (\mathcal{E}_{\max} \cos \varphi)^2 = [(X_L - X_C)I_{\max}]^2 + (RI_{\max})^2$$

$$\mathcal{E}^2_{\max} = [(X_L - X_C)^2 + R^2]I^2_{\max}$$

$$I_{\max} = \frac{\mathcal{E}_{\max}}{\sqrt{(X_L - X_C)^2 + R^2}} \tag{33.79}$$

To find φ, divide Equation 33.77 by Equation 33.78:

$$\frac{\mathcal{E}_{\max} \sin \varphi}{\mathcal{E}_{\max} \cos \varphi} = \frac{(X_L - X_C)I_{\max}}{RI_{\max}}$$

$$\frac{\sin \varphi}{\cos \varphi} = \tan \varphi = \frac{(X_L - X_C)}{R}$$

$$\varphi = \tan^{-1}\left(\frac{X_L - X_C}{R}\right) \tag{33.80}$$

Our goal of finding an expression for the current is met by $I(t) = I_{\max} \sin \omega t$ (Eq. 33.69) along with Equation 33.79. Likewise, $\mathcal{E}(t) = \mathcal{E}_{\max} \sin(\omega t + \varphi)$ (Eq. 33.75) along with Equation 33.80 meets the second goal of finding the AC generator's emf.

Impedance

The denominator in Equation 33.79 is defined to be the **impedance *Z***:

IMPEDANCE ✪ **Major Concept**

$$Z \equiv \sqrt{(X_L - X_C)^2 + R^2} \tag{33.81}$$

So, the maximum current is

$$I_{\max} = \frac{\mathcal{E}_{\max}}{Z} \tag{33.82}$$

The term *impedance* is descriptive. If a circuit has a large impedance, the maximum current is small. Like resistance, impedance has SI units of ohms. The impedance depends on the inductive reactance, the capacitive reactance, and the resistance, so *the impedance also depends on the AC generator's angular frequency*. Using the expressions for reactance $X_C \equiv 1/\omega C$ and $X_L \equiv \omega L$ (Eqs. 33.39 and 33.61), we can write impedance explicitly in terms of ω:

$$Z = \sqrt{\left(\omega L - \frac{1}{\omega C}\right)^2 + R^2} \tag{33.83}$$

The inductance L, capacitance C, and resistance R depend on the geometry (size, shape, and composition) of each circuit element.

Resonance in an *RLC* Circuit

The dependence of impedance on angular frequency has important consequences. Consider a mechanical analog: a pendulum swinging back and forth. If there were no air resistance or friction, the pendulum would swing back and forth forever, repeatedly transforming gravitational potential energy into kinetic energy, and vice versa. This is analogous to the *LC* circuit (Fig. 33.12). Energy is repeatedly transferred from the electric field to the magnetic field, and vice versa, forever, in the absence of resistance. Of course, friction and air resistance cannot be completely eliminated, so any real pendulum loses mechanical energy and eventually stops. Friction and air resistance are analogous to resistance in a circuit. Because resistance can never be eliminated, the *LC* circuit loses energy and eventually the current stops. If you want to keep a pendulum from stopping, you can drive the pendulum by pushing it periodically. From Section 16-11, the pendulum responds best (that is, with maximum amplitude) if you drive it at its natural frequency, known as the **resonance frequency**. The same is true for a circuit. If you want to keep the current in an *RLC* circuit from fading to 0, you can use an AC generator to drive the current. The current will display the greatest response if the AC generator's driving frequency is set to the circuit's resonance frequency.

The maximum current I_{max} depends on the emf's driving frequency (according to Eqs. 33.82 and 33.83). The highest value of I_{max} occurs when the driving frequency equals the resonance angular frequency ω_0, the value of ω that minimizes the impedance Z. The minimum value of the impedance is $Z_{min} = R$. When $Z = R$, the first two terms in Equation 33.83 cancel:

$$\omega_0 L = \frac{1}{\omega_0 C}$$

Solve for ω_0:

$$\omega_0 = \frac{1}{\sqrt{LC}} \tag{33.84}$$

Equation 33.84 is the resonance or natural angular frequency of the circuit. When the AC power supply's driving frequency equals the resonance frequency, we say the circuit is in **resonance** and the maximum value of the current is

$$I_{max}(\omega_0) = \frac{\mathcal{E}_{max}}{Z_{min}} = \frac{\mathcal{E}_{max}}{R} \tag{33.85}$$

RESONANCE **Major Concept**

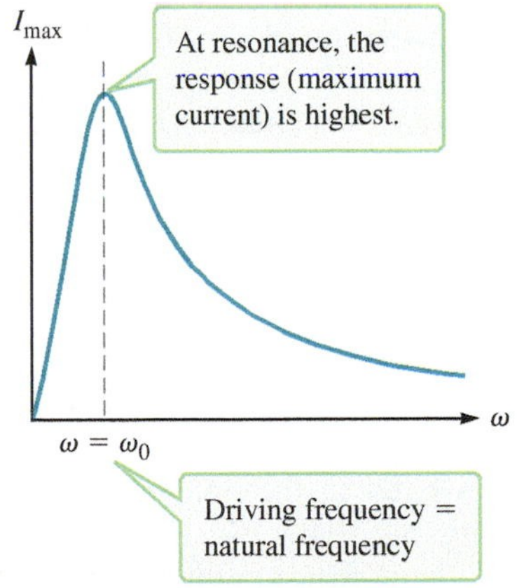
I_{max}
At resonance, the response (maximum current) is highest.
ω
$\omega = \omega_0$
Driving frequency = natural frequency

FIGURE 33.34 I_{max} as a function of angular frequency ω for an *RLC* circuit.

Figure 33.34 is a graph of I_{max} as a function of ω (Eqs. 33.82 and 33.83). The peak of the graph occurs at the resonance frequency ω_0. Imagine the AC generator in Figure 33.31 has a knob that allows you to adjust the angular frequency (the rate at which the coil in the generator rotates). As you slowly turn the knob up from $\omega = 0$, the maximum current in the circuit increases. When $\omega = \omega_0$, the maximum current has reached its highest value. As you continue to increase ω, the maximum current drops in value.

EXAMPLE 33.8 Power Factor

An ideal *LC* circuit does not dissipate energy. Of course, any real circuit has some resistance, so energy is actually dissipated. Suppose you have an *RLC* circuit with an AC power supply that adds energy at the same rate that the resistor dissipates energy. Show that the average power dissipated by the resistor is

$$P_{avg} = I_{rms}\mathcal{E}_{rms} \cos \varphi \tag{33.86}$$

(Problem 58 asks you to show that this is equal to the average power supplied.)

INTERPRET and ANTICIPATE

Start with a general expression for the average power dissipated by a resistor, and then make the appropriate substitutions for the specific case of the *RLC* circuit in Figure 33.31.

SOLVE

Start with Equation 32.25 for the average power dissipated by a resistor.	$P_{avg} = I_{rms}^2 R$ (32.25)
Solve Equation 33.78 for R.	$\mathcal{E}_{max} \cos \varphi = RI_{max}$ (33.78) $R = \dfrac{\mathcal{E}_{max}}{I_{max}} \cos \varphi$
Use $\mathcal{E}_{rms} = \mathcal{E}_{max}/\sqrt{2}$ and $I_{rms} = I_{max}/\sqrt{2}$ (Eqs. 32.20 and 32.21) to write the maximum emf and maximum current in terms of their rms values.	$R = \dfrac{\sqrt{2}\mathcal{E}_{rms}}{\sqrt{2}I_{rms}} \cos \varphi$ $R = \dfrac{\mathcal{E}_{rms}}{I_{rms}} \cos \varphi$ (1)
Substitute Equation (1) into Equation 33.25.	$P_{avg} = I_{rms}^2 \left(\dfrac{\mathcal{E}_{rms}}{I_{rms}} \cos \varphi \right)$ $P_{avg} = I_{rms}\mathcal{E}_{rms} \cos \varphi$ ✓ (33.86)

CHECK and THINK

This is what we set out to show. The factor $\cos \varphi$ is called the **power factor**; it is needed because the current in an *RLC* circuit is *not* in phase with the AC power supply's emf. Equation 33.86 shows that the average power dissipated (and supplied) depends on the phase.

Let's look at three special cases: (1) If the inductance and the capacitance are zero (if the inductor and the capacitor are replaced by wires), the circuit in Figure 33.31 consists of only a power supply and a resistor and the phase is zero. When $\varphi = 0$, the average power dissipated is at its maximum: $P_{avg} = I_{rms}\mathcal{E}_{rms}$. (2) If the *RLC* circuit is in resonance, $X_L = X_C$ and again the phase is zero. The average power dissipated (and supplied) is $P_{avg} = I_{rms}\mathcal{E}_{rms}$. (3) If the resistance is zero (an ideal situation), $\varphi = 90°$ and no power is dissipated: $P_{avg} = I_{rms}\mathcal{E}_{rms} \cos 90° = 0$.

EXAMPLE 33.9 CASE STUDY Broadcast Radio

If you lived in the San Francisco area in the 1960s, you may have found a Hendrix song playing on 104.5 FM (KFOG). You tuned your radio to 104.5 MHz. A radio receiver is basically an *RLC* circuit except that the AC generator in Figure 33.31 is replaced by the signal picked up by the antenna. Of course, many radio stations are sending out signals at many different frequencies. Ideally, you want your radio to be sensitive to only one signal at a time. When you tune your radio to a certain station, you are adjusting the *RLC* circuit's resonance frequency so that it equals the broadcast frequency of that particular station. The resonance frequency depends on the inductance and the capacitance (Eq. 33.84), so altering either one allows you to tune your radio. In old radios and in radios built at home, typically the inductance is fixed and the capacitance is varied. Figure P27.36 (page 859) shows a variable capacitor consisting of several parallel plates that can be rotated by turning a knob. The position of the plates determines the geometry of the capacitor and its capacitance. Assume a simple radio can be modeled as an *RLC* circuit with an inductance of 1.64×10^{-7} H, a resistance of 0.0346 Ω, and a variable capacitor.

A What capacitance is needed so that your radio is tuned to 104.5 MHz (KFOG)?

INTERPRET and ANTICIPATE

Your radio receiver's resonance frequency f_0 must be set equal to the broadcast frequency f of the radio station. All the equations we have worked with so far involve angular frequency ω, but the radio station's broadcast frequency f is given, so we must convert as needed.

SOLVE

Find an expression for the resonance frequency f_0 from $\omega_0 = 1/\sqrt{LC}$ (Eq. 33.84) and $\omega = 2\pi f$ (Eq. 16.2).

$$2\pi f_0 = \omega_0 = \frac{1}{\sqrt{LC}}$$

Solve for capacitance.

$$4\pi^2 f_0^2 = \frac{1}{LC}$$

$$C = \frac{1}{4\pi^2 f_0^2 L} = \frac{1}{4\pi^2(104.5 \times 10^6\,\text{Hz})^2(1.64 \times 10^{-7}\,\text{H})}$$

$$C = 1.41 \times 10^{-11}\,\text{F} = 14.1\,\text{pF}$$

CHECK and THINK

This is the capacitance when you are listening to KFOG. When you want to listen to another station, you tune that station in by changing the capacitance.

B If the radio signal produces a maximum emf of 17.5 mV, what is the maximum current in the circuit?

INTERPRET and ANTICIPATE

The radio is tuned so that it is in resonance with the broadcaster's signal. You can model this as an *RLC* circuit in which the angular frequency of the AC power supply equals the resonance angular frequency: $\omega = \omega_0$. In that case, the capacitive reactance equals the inductive reactance, $X_C = X_L$, and the impedance equals the resistance, $Z = R$. Then the maximum current depends only on the resistance and the power supply's maximum emf.

SOLVE

When an *RLC* circuit is in resonance, its maximum current is given by Equation 33.85.

$$I_{\max}(\omega_0) = \frac{\mathcal{E}_{\max}}{R} \qquad (33.85)$$

$$I_{\max}(\omega_0) = \frac{17.5 \times 10^{-3}\,\text{V}}{0.0346\,\Omega} = 5.06 \times 10^{-1}\,\text{A}$$

$$I_{\max}(\omega_0) = 506\,\text{mA}$$

CHECK and THINK

As shown in Figure 33.34, the maximum current is largest when the circuit is in resonance. If you tune your radio to another station, you change the resonance frequency of the circuit. KFOG still broadcasts its signal, but the current it produces in your radio receiver is much smaller. As long as that current is small compared to the current produced when you are tuned to the station you want to listen to, you won't get interference.

C There is another radio station in the San Francisco area at 104.9 FM (KCNL). Suppose this station produces a stronger signal so that the maximum emf in your radio is 35.0 mV (two times stronger than the station at KFOG). When your radio is tuned to KFOG (104.5 MHz), what is the maximum current produced by the KCNL signal? Do you expect KCNL to cause much interference?

INTERPRET and ANTICIPATE

Your radio is tuned so it is in resonance with the KFOG signal, not the KCNL signal. Because the radio is not in resonance with KCNL, the impedance depends on the resistance, the capacitive reactance, and the inductive reactance. The maximum current then depends on the maximum emf and the impedance.

Example continues on page 1078 ▶

SOLVE Find KCNL's broadcast angular frequency.	$\omega = 2\pi f = 2\pi(104.9 \times 10^6\,\text{Hz}) = 6.591 \times 10^8\,\text{Hz}$
The capacitive reactance is given by Equation 33.39. Use the capacitance from part A (because the radio is still tuned to KFOG), but use the angular frequency of KCNL's broadcast. We retain an extra significant figure to avoid rounding error.	$X_C = \dfrac{1}{\omega C}$ (33.39) $X_C = \dfrac{1}{(6.591 \times 10^8\,\text{Hz})(1.41 \times 10^{-11}\,\text{F})} = 107.6\,\Omega$
The inductive reactance is given by Equation 33.61. The inductance is constant for this circuit. Again, use the angular frequency of KCNL's broadcast.	$X_L = \omega L$ (33.61) $X_L = (6.591 \times 10^8\,\text{Hz})(1.64 \times 10^{-7}\,\text{H}) = 108.1\,\Omega$
Because the circuit is not in resonance with KCNL, we must use Equation 33.81 to find the circuit's impedance for KCNL's frequency.	$Z = \sqrt{(X_L - X_C)^2 + R^2}$ (33.81) $Z = \sqrt{(108.1\,\Omega - 107.6\,\Omega)^2 + (0.0346\,\Omega)^2} = 0.5\,\Omega$
The maximum current for KCNL's broadcast is given by $I_{max} = \mathcal{E}_{max}/Z$ (Eq. 33.82).	$I_{max} = \dfrac{35.0 \times 10^{-3}\,\text{V}}{0.5\,\Omega} = 70 \times 10^{-3}\,\text{A} = 70\,\text{mA}$

CHECK and THINK

KCNL's maximum emf is two times greater than KFOG's, but when your radio is tuned to KFOG (104.5 MHz), the current produced by KCNL (104.9 MHz) is about seven times smaller than the current produced by KFOG. You are not likely to hear any interference due to KCNL. You could have enjoyed listening to Hendrix on KFOG without worry.

SUMMARY

Underlying Principles

There are no new principles in this chapter.

Major Concepts

1. An **inductor** is a circuit element designed to store energy in a magnetic field.
2. **Inductance** L (sometimes called **self-inductance**) is the constant of proportionality in
$$\Phi_{tot} = LI \quad (33.1)$$
The total flux Φ_{tot} through the inductor is the result of the current I in the wire of the same (self) inductor. The SI unit for inductance is the **henry**:
$$1\,\text{H} = 1\frac{\text{Wb}}{\text{A}} = 1\frac{\text{T}\cdot\text{m}^2}{\text{A}} \quad (33.2)$$
3. **Energy stored by an inductor:** $U_B = \frac{1}{2}LI^2$ (33.3)
Energy density stored by a magnetic field:
$$u_B = \frac{U_B}{V} = \frac{1}{2}\frac{B^2}{\mu_0} \quad (33.17)$$
4. **Capacitive reactance:** $X_C = 1/\omega C$ (33.39)
5. **Inductive reactance:** $X_L = \omega L$ (33.61)
6. **Impedance:** $Z \equiv \sqrt{(X_L - X_C)^2 + R^2}$ (33.81)
7. **Resonance** occurs when the AC power supply's driving frequency equals the resonance or natural (angular) frequency. The resonance angular frequency is given by
$$\omega_0 = \frac{1}{\sqrt{LC}} \quad (33.84)$$
At resonance, the maximum current is given by
$$I_{max}(\omega_0) = \frac{\mathcal{E}_{max}}{Z_{min}} = \frac{\mathcal{E}_{max}}{R} \quad (33.85)$$

Special Cases

1. **Inductance of an ideal solenoid:**

$$L = \mu_0 n^2 \ell A = \frac{\mu_0 N^2}{\ell} A \quad (33.5)$$

2. **Inductor rule:** $\Delta V_L = V_{\text{red}} - V_{\text{black}} = -L\frac{dI}{dt}$ (33.8)

   where the red voltmeter lead is downstream and the black lead is upstream so that the measurement is made in the direction of the current.

3. An ***RL* circuit** consists of an inductor, a resistor, and a DC power supply that can be switched into or out of the circuit (Fig. 33.9). When the DC power supply is in the circuit, the current is

$$I(t) = \frac{\mathcal{E}}{R}(1 - e^{-t/\tau}) \quad (33.12)$$

   When the DC power supply is switched out of the circuit, the current decays:

$$I(t) = I_{\text{max}} e^{-t/\tau} \quad (33.15)$$

   where τ is the inductive time constant:

$$\tau \equiv \frac{L}{R} \quad (33.13)$$

4. An ***LC* circuit** is an ideal circuit that consists of an inductor and a capacitor (no resistance). Energy is transferred back and forth between the magnetic field of the inductor and the electric field of the capacitor (Fig. 33.12). The electric field energy is

$$U_E(t) = \frac{Q^2_{\text{max}}}{2C}\cos^2(\omega t + \varphi) \quad (33.26)$$

   The magnetic field energy is

$$U_B(t) = \frac{Q^2_{\text{max}}}{2C}\sin^2(\omega t + \varphi) \quad (33.27)$$

   The current in the circuit and the charge on the capacitor both oscillate:

$$I(t) = -I_{\text{max}} \sin(\omega t + \varphi) \quad (33.22)$$

$$Q(t) = Q_{\text{max}} \cos(\omega t + \varphi) \quad (33.20)$$

   where the angular frequency is

$$\omega = \sqrt{\frac{1}{LC}} \quad (33.21)$$

5. An **AC circuit with resistance** consists of an AC power supply and a resistor (Fig. 33.15). The voltage across the generator equals the voltage across the resistor:

$$V_R(t) = \mathcal{E}(t) = \mathcal{E}_{\text{max}} \sin \omega t \quad (33.29)$$

   The current in the circuit and the voltage across the resistor are **in phase**:

$$I(t) = I_{\text{max}} \sin \omega t \quad (33.32)$$

   where the maximum current is

$$I_{\text{max}} = \frac{\mathcal{E}_{\text{max}}}{R} \quad (33.31)$$

6. An **AC circuit with capacitance** consists of an AC power supply and a capacitor (Fig. 33.18). The voltage across the power supply equals the voltage across the capacitor:

$$V_C(t) = \mathcal{E}(t) = \mathcal{E}_{\text{max}} \sin \omega t \quad (33.34)$$

   The current in the circuit is

$$I(t) = \frac{\mathcal{E}_{\text{max}}}{X_C}\sin(\omega t + 90°) \quad (33.40)$$

   where the maximum current is

$$I_{\text{max}} = \frac{\mathcal{E}_{\text{max}}}{X_C} \quad (33.41)$$

   The current leads the voltage by 90°, or the voltage across a capacitor lags the current by 90°.

7. An **AC circuit with inductance** consists of an AC power supply and an inductor (Fig. 33.24). The voltage across the power supply equals the voltage across the inductor:

$$V_L(t) = \mathcal{E}(t) = \mathcal{E}_{\text{max}} \sin \omega t \quad (33.57)$$

   The current in the circuit is

$$I(t) = \frac{\mathcal{E}_{\text{max}}}{X_L}\sin(\omega t - 90°) \quad (33.62)$$

   where the maximum current is

$$I_{\text{max}} = \frac{\mathcal{E}_{\text{max}}}{X_L} \quad (33.63)$$

   The current lags the voltage by 90°, or the voltage across an inductor leads the current by 90°.

8. **a.** An ***RC* filter circuit** consists of a resistor and a capacitor (Fig. 33.22A). Low frequencies are passed through the capacitor (low-pass filter):

$$(V_C)_{\text{max}} = X_C \frac{\mathcal{E}_{\text{max}}}{\sqrt{X_C^2 + R^2}} \quad (33.53)$$

   High frequencies are passed through the resistor (high-pass filter):

$$(V_R)_{\text{max}} = R\frac{\mathcal{E}_{\text{max}}}{\sqrt{X_C^2 + R^2}} \quad (33.55)$$

   b. An ***RL* filter circuit** consists of a resistor and an inductor (Fig. 33.27A). High frequencies are passed through the inductor (high-pass filter):

$$(V_L)_{\text{max}} = X_L \frac{\mathcal{E}_{\text{max}}}{\sqrt{X_L^2 + R^2}} \quad (33.65)$$

   Low frequencies are passed through the resistor (low-pass filter):

$$(V_R)_{\text{max}} = R\frac{\mathcal{E}_{\text{max}}}{\sqrt{X_L^2 + R^2}} \quad (33.67)$$

9. An ***RLC* circuit** consists of an inductor, a resistor, a capacitor, and an AC power supply (Fig. 33.31). The current is not in phase with the AC power supply's emf. If the current is given by

$$I(t) = I_{\text{max}} \sin \omega t \quad (33.69)$$

the AC generator's emf is

$$\mathcal{E}(t) = \mathcal{E}_{max} \sin(\omega t + \varphi) \quad (33.75)$$

where φ is the phase constant:

$$\varphi = \tan^{-1}\left(\frac{X_L - X_C}{R}\right) \quad (33.80)$$

The maximum current is

$$I_{max} = \frac{\mathcal{E}_{max}}{Z} \quad (33.82)$$

The **resonance angular frequency** is

$$\omega_0 = \frac{1}{\sqrt{LC}} \quad (33.84)$$

When the AC power supply's driving frequency equals the resonance frequency, we say the circuit is in **resonance**, and the maximum value of the current is

$$I_{max}(\omega_0) = \frac{\mathcal{E}_{max}}{Z_{min}} = \frac{\mathcal{E}_{max}}{R} \quad (33.85)$$

Tool

In a **phasor diagram**, a quantity such as current or voltage is represented by a vector known as a phasor. (For examples, see Figures 33.16C, 33.20C, and 33.26C.) The length of the phasor represents the quantity's maximum value. To represent the time dependence of the quantity, the phasor rotates around the origin at an angular speed ω. At any instant t, the phasor makes an angle ωt with the horizontal axis. In this textbook, we have chosen to use the sine function for AC currents and voltages, so the phasor's projection onto the vertical axis gives the quantity's instantaneous value.

PROBLEMS AND QUESTIONS

A = algebraic C = conceptual E = estimation G = graphical N = numerical

33-1 Inductors and Inductance

1. C Two solenoids have the same number of turns per unit length, but one is short and wide and the other is long and narrow. **a.** Which inductor has the greater inductance per unit length L/ℓ? Explain. **b.** How is it possible for the two to have the same inductance L?
2. E Estimate the inductance of a typical (metal) Slinky.
3. C A solenoid is constructed by wrapping a wire of a fixed length d to form N loops of radius R such that the total length of the solenoid created is ℓ. A battery is connected to drive a current I through the solenoid. Which of the following would increase the inductance of the solenoid: **a.** squeezing the solenoid so that the loops have the same radius but are closer together, so that the total length of the solenoid is shorter; **b.** unwrapping the solenoid and using the same total length of wire d to form a new solenoid of the same length ℓ but with the coil wrapped less tightly so that the radius is larger and the number of loops is smaller; or **c.** increasing the voltage of the battery so that the current is three times larger? Justify your answer.

33-2 Back Emf

4. C The back emf in a circuit can sometimes be much greater than the emf supplied by the battery. **a.** Would you expect the inductance of such a circuit to be very high or very low? Explain. **b.** Suppose the inductance in some circuit is fixed. There is a steady current in the circuit, and you wish to cut that current. Explain how you could avoid inducing a large emf in the circuit.
5. A 15.0-mH inductor is connected to a DC power supply. The power supply's emf is fixed at 5.00 V.
   a. N If you turn down the power supply's current at a constant rate, $dI/dt = -50.0$ A/s, what is the magnitude of the back emf?
   b. N At what rate would you need to turn down the current so that the back emf is 5.00 V?
   c. C If you had a lightbulb in this circuit, how would it glow in the two cases? (Ignore the resistance of the bulb.)
6. G A voltage source is connected directly to an 80.0-mH inductor, and the current through the inductor is a triangular wave as shown in Figure P33.6. Plot the voltage across the inductor during the time interval from 0 to 16 s.

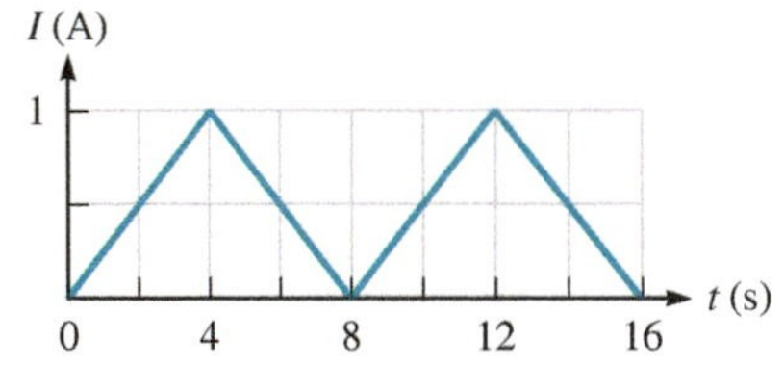

FIGURE P33.6

7. N What is the inductance of a coil in which the average emf induced is 23.0 mV when the current in the coil is increased from 4.00 A to 7.00 A in 0.330 s?
8. N A 45.0-cm-long solenoid is 8.00 cm in diameter and has 690 turns. **a.** What is the inductance of the solenoid? **b.** What is the rate of change of the current dI/dt required to produce an emf of 44.0 μV in the solenoid?

33-3 Special Case: Resistor–Inductor (RL) Circuit

9. N A series circuit contains a 4.00-H inductor, a 5.00-Ω resistor, and a 9.00-V battery. The current is initially zero when the circuit is connected at $t = 0$. At what time will the current reach **a.** 33.3% and **b.** 95.0% of its final value?
10. C When you close the switch on a flashlight, does the bulb instantaneously light up? When you open the switch on a flashlight, does the bulb instantaneously shut off? Explain your answers.

11. A battery with emf $\mathcal{E}$ is connected in series with an inductance L and a resistance R.
   a. **A** Assuming the current has reached steady state when it is at 99% of its maximum value, how long does it take to reach steady state, assuming the initial current is zero?
   b. **N** If an emergency power circuit needs to reach steady state within 1.0 ms of turning on and the circuit has a total resistance of 75 Ω, what values of the total inductance of the circuit are needed to satisfy the requirement?

12. **N** At one instant, a current of 6.0 A flows through part of a circuit as shown in Figure P33.12. Determine the instantaneous potential difference between points A and B if the current starts to decrease at a constant rate of 1.0×10^2 A/s.

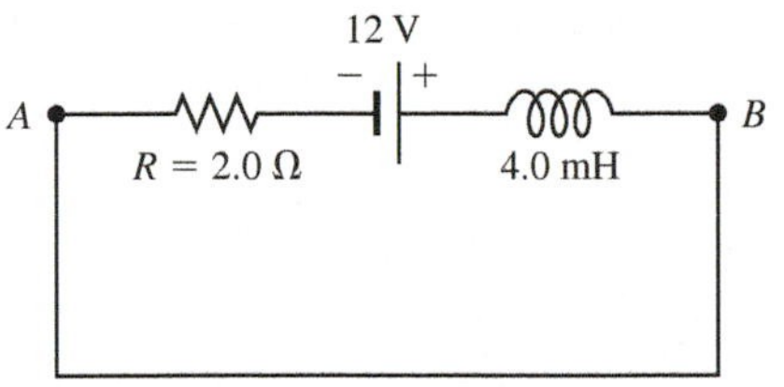

FIGURE P33.12

13. **N** The time constants for a series RC circuit with a capacitance of 5.00 μF and a series RL circuit with an inductance of 2.00 H are identical. **a.** What is the resistance R in the two circuits? **b.** What is the common time constant for the two circuits?

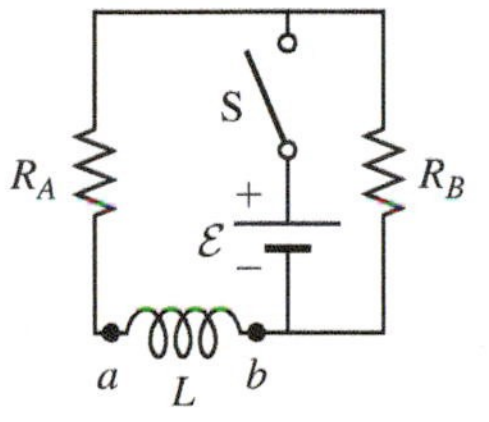

FIGURE P33.14

14. **N** After being closed for a long time, the switch S in the circuit shown in Figure P33.14 is thrown open at $t = 0$. In the circuit, $\mathcal{E} = 24.0$ V, $R_A = 4.00$ kΩ, $R_B = 7.00$ kΩ, and $L = 589$ mH. **a.** What is the emf across the inductor immediately after the switch is opened? **b.** When does the current in the resistor R_B have a magnitude of 1.00 mA?

Problems 15, 16, and 17 are grouped.

15. In Figure 33.9A (page 1052), the switch is closed at a at $t = 0$.
   a. **A** Find an expression for the total charge that passes through the resistor in one time constant.
   b. **A** After the switch is left at a for many time constants, it is switched to b (Fig. 33.10A, page 1053). Find an expression for the total charge that passes through the resistor in one time constant.
   c. **C** Compare your results (ignoring any sign difference) and comment.

33-4 Energy Stored in a Magnetic Field

16. **G** In Figure 33.9A (page 1052), the switch is closed at a at $t = 0$. Find an expression for the power dissipated by the resistor as a function of time, and sketch your result. Is the power lost greater as soon as the switch is closed or a long time after it has been closed? Does your answer make sense?

17. In Figure 33.9A (page 1052), the switch is closed at a at $t = 0$.
   a. **A** Find an expression for the total energy dissipated by the resistor in one time constant.
   b. **A** After the switch is left at a for many time constants, it is switched to b. Find an expression for the total energy dissipated by the resistor in one time constant.
   c. **C** Compare your results and comment.

18. **E** Use Table 30.1 (page 941) to estimate the energy density stored in the magnetic field near a neutron star.

19. **N** Find the energy stored in a 4.0-mH inductor when the current is 4.0 A.

20. **N** If a high-voltage power line 25 m above the ground carries a current of 1.00×10^3 A, estimate the energy density of the magnetic field near the ground and compare it to the energy density of the Earth's magnetic field.

33-5 Special Case: Inductor–Capacitor (*LC*) Circuit

21. **N** What is the inductance of an LC circuit with $C = 4.50$ μF oscillating at 76.0 Hz?

22. **N** Figure 33.12 (page 1058) shows an LC circuit whose capacitor is initially ($t = 0$) fully charged. Find the phase constant for this common situation, and write an expression for $Q(t)$.

23. **A** In the LC circuit in Figure 33.11, the inductance is $L = 19.8$ mH and the capacitance is $C = 19.6$ mF. At some moment, $U_B = U_E = 17.5$ mJ. **a.** What is the maximum charge stored by the capacitor? **b.** What is the maximum current in the circuit? **c.** At $t = 0$, the capacitor is fully charged. Write an expression for the charge stored by the capacitor as a function of time. **d.** Write an expression for the current as a function of time.

24. **G** An LC circuit is an ideal circuit. Any real circuit has resistance, so energy is dissipated. This is analogous to the damped oscillator in Section 16-10. Use this analogy and Figure 16.22 (page 473) to sketch the charge stored by the capacitor and the current in a real LC circuit with resistance as functions of time.

25. **N** A 2.0-μF capacitor is charged to a potential difference of 12.0 V and then connected across a 0.40-mH inductor. What is the current in the circuit when the potential difference across the capacitor is 6.0 V?

26. **N** Figure P33.26 shows a circuit with $\mathcal{E} = 9.00$ V, $R = 6.00$ Ω, $L = 75.0$ mH, and $C = 2.55$ μF. After a long time interval at the position a shown in the figure, the switch S is thrown to position b at time $t = 0$. What is the maximum **a.** charge on the capacitor and **b.** current in the inductor for $t > 0$? **c.** What is the frequency of oscillation of the resulting LC circuit for $t > 0$?

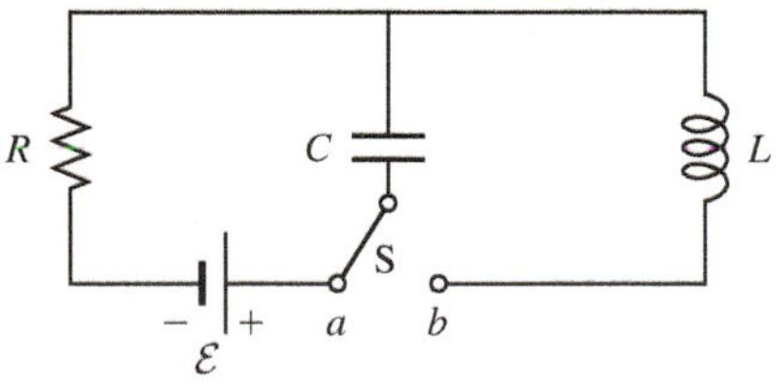

FIGURE P33.26

Problems 27, 28, and 29 are grouped.

27. **A** For an LC circuit, show that the energy stored in the electric field oscillates according to Equation 33.26, $U_E(t) = (Q_{max}^2/2C)\cos^2(\omega t + \varphi)$.

28. **A** For an LC circuit, show that the energy stored in the magnetic field oscillates according to Equation 33.27, $U_B(t) = (Q_{max}^2/2C)\sin^2(\omega t + \varphi)$.

29. **A** For an LC circuit, show that the total energy stored in the electric and magnetic fields is a constant given by Equation 33.28, $E_{tot} = Q_{max}^2/2C = \frac{1}{2}LI_{max}^2$.

30. **A** In Example 33.4 (page 1061), we found that the energy in an LC circuit is evenly split between the two fields at $t = T/8$. **a.** Find the charge and current at that time. **b.** Find the energy stored in each circuit element to confirm that the energy is evenly split at this time.

31. **N** A fully charged capacitor and a 0.20-H inductor are connected to form a complete circuit. If the circuit oscillates with a frequency of 1.2×10^3 Hz, determine the capacitance of the capacitor.

33-6 Special Case: AC Circuit with Resistance

32. **N** A 60-W lightbulb is used in an American desk lamp. **a.** What is the average power dissipated by the filament? **b.** What is the rms current in the filament? **c.** What is the maximum current in the filament? **d.** What is the maximum power dissipated by the filament? (Give your answers to two significant figures.)

33. **N** The rms current in a 29.0-Ω resistor connected to an AC source is 3.00 A. **a.** What is the rms voltage across this resistor? **b.** What is the peak voltage of the AC source? **c.** What is the average power dissipated by the resistor? **d.** What is the maximum current across this resistor?

34. **C** Suppose you connect a small lightbulb across a DC power supply that has an emf of 5.0 V. You then connect the bulb across a 60.0-Hz AC power supply that has an rms voltage of 5.0 V. Does the bulb dissipate more power on average when it is connected to the DC power supply or the AC power supply? Is the bulb brighter when it is connected to the DC power supply or the AC power supply? Explain.

Problems 35 and 36 are paired.

35. An AC generator delivers an alternating current $I(t) = (2.0\text{ A})\sin[(120\pi\text{ rad/s})t]$ to a single resistor in series with the generator.
- **a.** **N** What is the rms value of the current in the circuit?
- **b.** **N** If the resistor has a resistance of 100.0 Ω, what is the rms value of the source emf?
- **c.** **N** What is the maximum value of the source emf?
- **d.** **A** Write a function that describes the source emf as a function of time.

36. **G** An AC generator delivers an alternating current $I(t) = (2.0\text{ A})\sin[(120\pi\text{ rad/s})t]$ to a single resistor in series with the generator. Given a resistance of 100.0 Ω, draw a phasor diagram for this circuit, including the current, the potential difference across the resistor, and the source emf. Draw your diagram with the current phasor pointing toward the right along the horizontal axis.

37. **N** A simple AC circuit contains an AC source with output voltage $V_{max}\cos\omega t$ and a resistor with resistance $R = 150.0\ \Omega$. **a.** What is the angular frequency of the AC source if the first time $V_R = 0.400V_{max}$ is at $t = 1.00 \times 10^{-3}$ s? **b.** When is $V_R = 0.750V_{max}$ for the first time?

33-7 Special Case: AC Circuit with Capacitance

38. **C** Avi and Cameron are working on homework problems. They need to find the current in an AC circuit with capacitance at a particular time t. Decide who is right and why. If neither is right, make your own correct statement.

Avi: Capacitive reactance is like resistance. To find the current at some particular time, you just divide the emf by the capacitive reactance X_C.

Cameron: That works only if you want to find the maximum current. You need to divide the emf by the resistance R. Then multiply it by $\sin(\omega t + \pi/2)$.

39. **N** An 8.00-mF capacitor is connected across an AC source with an rms voltage of 68.0 V oscillating with a frequency of 50.0 Hz. What are the **a.** capacitive reactance of, **b.** rms current in, and **c.** maximum current in this circuit?

40. **C** Suppose you have an AC circuit with capacitance (Fig. 33.18, page 1064) and you wish to increase the maximum current. The AC power supply is adjustable. Without changing the capacitor, how can you increase the maximum current? (There is more than one answer.)

Problems 41, 42, and 43 are grouped.

41. An AC generator delivers an rms current of 3.50 A to a 6.25-μF capacitor connected in series. The frequency of the source emf is 60.0 Hz.
- **a.** **N** What is the capacitive reactance of the circuit?
- **b.** **N** What is the rms emf of the generator?
- **c.** **N** What are the maximum values of the current and the source emf?
- **d.** **A** Write a function for the potential difference across the capacitor as a function of time.

42. **G** An AC generator delivers an rms current of 3.50 A to a 6.25-μF capacitor connected in series. The frequency of the source emf is 60.0 Hz. Draw a phasor diagram for this circuit, including the current, the potential difference across the capacitor, and the source emf. Draw your diagram with the current phasor pointing toward the right along the horizontal axis.

43. **C** An AC generator delivers an rms current of 3.50 A to a 6.25-μF capacitor connected in series. The frequency of the source emf is 60.0 Hz. When the potential difference across the capacitor is at its maximum positive value, what is the value of the current? How does the potential difference compare to the source emf at this time? Justify and explain your answer.

44. **C** In an *ideal* AC circuit with capacitance, there is no resistance. Is any energy dissipated? How about in a *real* AC circuit with capacitance?

45. A radio telescope is designed to pick up natural signals at 1420 MHz. Unfortunately, it also picks up artificial signals at 1475 MHz. You want to use an *RC* filter to remove the signal above 1450 MHz.
- **a.** **C** Explain how an *RC* filter can be used to accomplish your task.
- **b.** **N** If the capacitance is 245 nF and the resistance is 34.7 Ω, what fraction of the artificial signal's maximum voltage $\mathcal{E}_{max}$ will the radio telescope pick up? Does your result tell you that the telescope won't suffer from artificial interference?

33-8 Special Case: AC Circuit with Inductance

46. **A** Follow the reasoning used for the *RC* filter to show that in the case of an *RL* filter, the maximum current is given by Equation 33.64:

$$I_{max} = \frac{\mathcal{E}_{max}}{\sqrt{X_L^2 + R^2}}$$

47. **N** A 68.0-mH inductor is connected across an AC source that has an rms voltage of 120.0 V oscillating at a frequency of 110.0 Hz. What are the **a.** inductive reactance of, **b.** rms current in, and **c.** maximum current in this circuit?

48. **C** Avi and Cameron are discussing another problem. Decide who is right and why.

Avi: Inductive reactance is just like resistance. To find the current, you just need to divide by the inductive reactance. So the current at $t = 0$ is zero because the emf is zero.

Cameron: That is only good for finding the maximum current.

Problems 49, 50, and 51 are grouped.

49. An AC generator with an rms emf of 15.0 V is connected in series with a 0.54-H inductor. The frequency of the source emf is 70.0 Hz.
- **a.** **N** What is the inductive reactance of the circuit?
- **b.** **N** What is the rms current in the circuit?
- **c.** **N** What are the maximum values of the current and the source emf?
- **d.** **A** Write an expression for the potential difference across the inductor as a function of time.

50. **G** An AC generator with an rms emf of 15.0 V is connected in series with a 0.54-H inductor. The frequency of the source emf is 70.0 Hz. Draw a phasor diagram for this circuit, including the current, the potential difference across the inductor, and the source emf. Draw your diagram with the current phasor pointing toward the right along the horizontal axis.

51. **C** An AC generator with an rms emf of 15.0 V is connected in series with a 0.54-H inductor. The frequency of the source emf is 70.0 Hz. When the current through the inductor is a maximum, what is the potential difference across the inductor? How does this compare to the source emf at this time? Justify and explain your answer.

52. **N** When connected to a 90.0-Hz AC source, an inductor has an inductive reactance of 33.0 Ω. What is the maximum current in this inductor if it is connected to an AC source that has an rms voltage of 120.0 V and a frequency of 110.0 Hz?

53. **N** An inductor ($L = 37.8$ mH) is connected to an AC power supply that has a maximum emf of 16.4 V and an angular frequency of 62.8 rad/s. At time $t = 0$, the emf is zero and increasing. **a.** What is the current at $t = 0$? **b.** What is the current across the inductor at $t = 25.0$ s? **c.** What is the maximum current in the circuit?

54. **C** **CASE STUDY** What happens to the woofer's voltage in Example 33.7 (page 1071) when the crossover frequency decreases?

55. **N** Design a high-pass filter using a 1.5-kΩ resistor and a 0.75-mF capacitor, and determine the approximate frequency above which voltages are able to pass through the circuit. Assume this is the frequency at which the output voltage is at least half the input voltage.

33-9 Special Case: AC Circuit with Resistance, Inductance, and Capacitance

56. **N** A radio tuner circuit is created with a 1.50-Ω resistor and a 2.50-μH inductor in series with a variable capacitor. **a.** What capacitance is needed for this circuit to have a resonance frequency of 88.7 MHz, the frequency of a local radio station? **b.** How much interference is there from the next lowest FM station at 88.1 MHz? That is, assuming the circuit is tuned to 88.7 MHz as in part (a), how much smaller is the current due to a signal at 88.1 MHz, assuming the signal strength is the same for both stations?

57. **N** A series *RLC* circuit with a resistance of 120.0 Ω has a resonant angular frequency of 4.0×10^5 rad/s. At resonance, the voltages across the resistor and inductor are 60.0 V and 40.0 V, respectively. **a.** Determine the values of L and C. **b.** At what frequency does the current in the circuit lag the voltage by 45°?

58. **A** Start with Equations 33.69 and 33.75, $I(t) = I_{max} \sin \omega t$ and $\mathcal{E}(t) = \mathcal{E}_{max} \sin(\omega t + \varphi)$, and show that the average power supplied is given by Equation 33.86, $P_{avg} = I_{rms}\mathcal{E}_{rms} \cos \varphi$.

59. **N** An AC source with $V_{rms} = 110.0$ V and $I_{rms} = 12.0$ A is connected to a series *RLC* circuit in which the current leads the voltage by 23.5°. What are the **a.** total resistance R and **b.** net reactance $(X_L - X_C)$ of this circuit?

60. **N** An AC source of angular frequency ω is connected to a resistor R and a capacitor C in series. The maximum current measured is I_{max}. While the same maximum emf is maintained, the angular frequency is changed to $\omega/3$. The measured current is now $I/2$. Determine the ratio of the capacitive reactance to the resistance at the initial frequency ω.

Problems 61 and 62 are paired.

61. An *RLC* series circuit is constructed with $R = 100.0$ Ω, $C = 6.25$ μF, and $L = 0.54$ H. The circuit is connected to an AC generator with a frequency of 60.0 Hz that delivers a maximum current of 2.00 A to the circuit.
    a. **N** What is the impedance of this circuit?
    b. **N** What are the maximum potential differences across each of the three circuit elements (R, L, and C)?
    c. **N** What is the phase angle between the source emf and the current?
    d. **A** Write expressions for the source emf and the current as functions of time.

62. **G** An *RLC* series circuit is constructed with $R = 100.0$ Ω, $C = 6.25$ μF, and $L = 0.54$ H. The circuit is connected to an AC generator with a frequency of 60.0 Hz that delivers a maximum current of 2.00 A to the circuit. Draw a phasor diagram for this circuit, including the current, the potential difference across each of the circuit elements, and the source emf. Draw your diagram with the current phasor pointing toward the right along the horizontal axis.

63. **N** A series *RLC* circuit driven by a source with an amplitude of 120.0 V and a frequency of 50.0 Hz has an inductance of 787 mH, a resistance of 267 Ω, and a capacitance of 45.7 μF. **a.** What are the maximum current and the phase angle between the current and the source emf in this circuit? **b.** What are the maximum potential difference across the inductor and the phase angle between this potential difference and the current in the circuit? **c.** What are the maximum potential difference across the resistor and the phase angle between this potential difference and the current in this circuit? **d.** What are the maximum potential difference across the capacitor and the phase angle between this potential difference and the current in this circuit?

Problems 64, 65, and 66 are grouped.

64. **N** In an *RLC* circuit (Fig. 33.31, page 1072), the resistance is 325 Ω, the inductance is 126 mH, and the capacitance is 13.7 μF (13.7×10^{-6} F). The angular frequency is 377 rad/s and $\mathcal{E}_{max} = 18.8$ V. **a.** What is the impedance? **b.** What is the maximum current? **c.** What is the phase constant φ of the power supply's emf with respect to the current?

65. **N** In an *RLC* circuit (Fig. 33.31, page 1072), the resistance is 325 Ω, the inductance is 126 mH, and the capacitance is 13.7 mF (13.7×10^{-3} F). The angular frequency is 377 rad/s and $\mathcal{E}_{max} = 18.8$ V. **a.** What is the impedance? **b.** What is the maximum current? **c.** What is the phase constant φ of the power supply's emf with respect to the current?

66. **G** In an *RLC* circuit (Fig. 33.31, page 1072), the resistance is 325 Ω, the inductance is 126 mH, the angular frequency is 377 rad/s, and $\mathcal{E}_{max} = 18.8$ V. Draw a phasor diagram if the capacitance is **a.** 13.7 μF and **b.** 13.7 mF.

General Problems

67. **N** How much energy is stored in the magnetic field of an 18.0-cm-long solenoid that has 154 turns and a radius of 0.900 cm and carries a current of 1.65 A?

68. **C** In circuits with nonzero inductance, interrupting the circuit by throwing a switch open can lead to a large spark at the switch. Why is a large spark possible, even though we generally assume a broken circuit always has zero current?

69. **N** An inductor ($L = 0.35$ H) and two resistors ($R_1 = R_2 = 2.0$ Ω) are connected to a battery with an emf of 12.0 V as shown in Figure P33.69. **a.** If the switch S is closed at time $t = 0$, determine the potential drop across the inductor at time $t = 0.040$ s. **b.** After a steady state is reached,

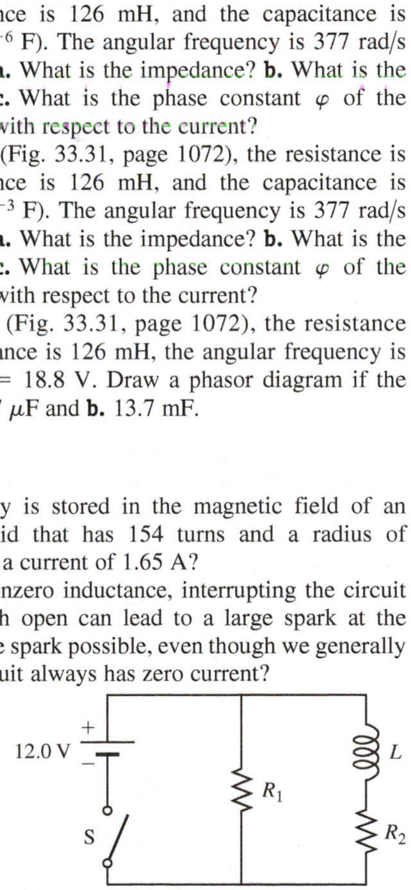

FIGURE P33.69

the switch is opened. What are the direction and magnitude of the current through R_1 at a time 0.040 s after the switch is opened?

70. **N** An ideal solenoid with 855 turns is 0.100 m long and has a cross-sectional area of 3.00×10^{-3} m^2. **a.** What is the inductance of this solenoid? **b.** If we connect this solenoid in series with a 12.0-V battery and a 185-Ω resistor, what is the maximum value of the current? **c.** Assuming the current is zero when $t = 0$ as the circuit elements are connected, at what time will the current reach 50% of its maximum value?

Problems 71 and 72 are paired.

71. **N** Figure P33.71 shows a series *RLC* circuit with a 25.0-Ω resistor, a 430.0-mH inductor, and a 24.0-μF capacitor connected to an AC source with $V_{max} = 60.0$ V operating at 60.0 Hz. What is the maximum voltage across the **a.** resistor, **b.** inductor, and **c.** capacitor in the circuit?

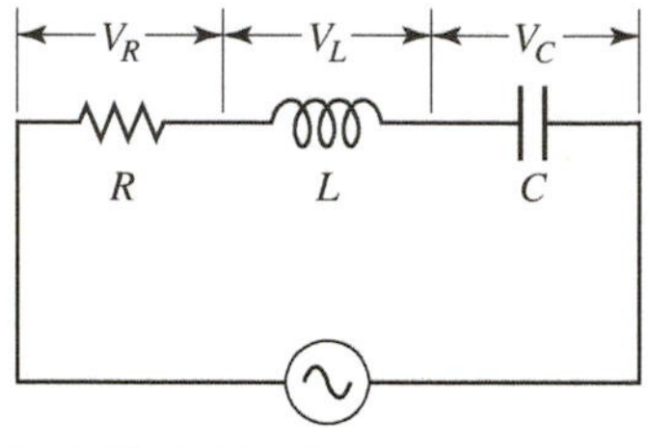

FIGURE P33.71 Problems 71 and 72.

72. **G** Figure P33.71 shows a series *RLC* circuit with a 25.0-Ω resistor, a 430.0-mH inductor, and a 24.0-μF capacitor connected to an AC source with $V_{max} = 60.0$ V operating at 60.0 Hz. Draw a phasor diagram for this circuit, including the current, the potential difference across each of the circuit elements, and the source emf. Draw your diagram with the current phasor pointing toward the right along the horizontal axis.

73. **N** The time variation of the current in a 37.0-mH inductor is given by $I = 3.00 + 4.00t - 2.00t^2$, with I in amperes and t in seconds. What is the magnitude of the emf induced in the inductor at **a.** $t = 2.00$ s and **b.** $t = 5.00$ s? **c.** For what value of t is the emf induced in the inductor 0?

74. **N** A 22.5-μF capacitor is charged by a 6.00-V battery and then connected to a 75.0-mH inductor. At what frequency does the current oscillate, and what is the maximum current?

Problems 75 and 76 are paired.

75. **N** In a series *RLC* circuit with a maximum current of 0.250 A, an AC source with $V_{max} = 115$ V operating at 60.0 Hz is connected to a 325-mH inductor, a 7.50-μF capacitor, and a resistor with unknown resistance *R*. What are the **a.** inductive reactance, **b.** capacitive reactance, and **c.** impedance of this circuit? **d.** What is the resistance *R* of the resistor in this circuit? **e.** What is the phase angle between the current and the source voltage?

76. **G** In a series *RLC* circuit with a maximum current of 0.250 A, an AC source with $V_{max} = 115$ V operating at 60.0 Hz is connected to a 325-mH inductor, a 7.50-μF capacitor, and a resistor with unknown resistance *R*. Draw a phasor diagram for this circuit, including the current, the potential difference across each of the circuit elements, and the source emf. Draw your diagram with the current phasor pointing upward along the vertical axis.

77. **N** A 45.0-V battery with an internal resistance of 13.0 Ω is connected to a 7.40-H inductor. The current is zero at $t = 0$. At what rate is the current in the inductor increasing at **a.** $t = 0$ and **b.** $t = 2.00$ s?

78. **A** Two coaxial cables of length ℓ with radii *a* and *b* are carrying currents in opposite directions as shown in Figure P33.78. Determine the inductance of the system. *Hint*: Use Ampère's law to write an expression for the magnetic field in the region between the cables, a distance *r* from the axis of the cables. Then calculate the magnetic flux through a narrow rectangular region between the cables such that the field is perpendicular to the area everywhere.

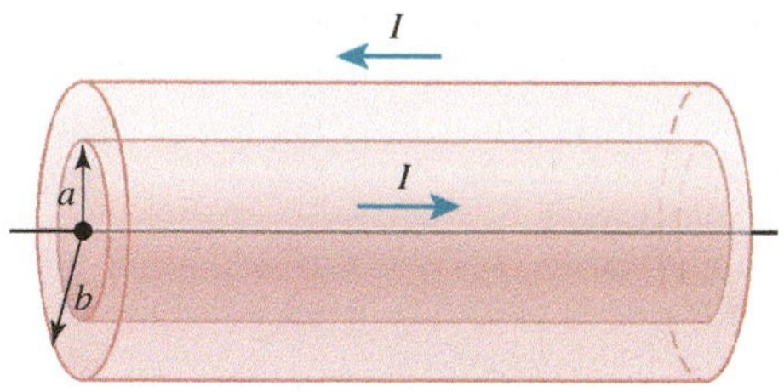

FIGURE P33.78

79. **N** An 11.0-μF capacitor is connected across an AC source with $V_{max} = 90.0$ V oscillating at 60.0 Hz. What is the maximum current in the capacitor?

Problems 80 and 81 are paired.

80. **G** A 20.0-Ω resistor, a 200.0-mH inductor, and a 200.0-μF capacitor are connected to a variable-frequency AC voltage source one at a time. The voltage source has a frequency *f* that ranges from 5.0 Hz to 100.0 Hz with a peak voltage of 10.0 V. Plot the peak current versus frequency for each circuit element.

81. **N** A 20.0-Ω resistor, a 200.0-mH inductor, and a 200.0-μF capacitor are each connected to a 120.0-V rms, 60.0-Hz household voltage one at a time. What is the peak current through each circuit element?

82. **A** A conducting bar of mass *M* and length ℓ is given an initial speed v_0 on a smooth horizontal conducting rail as shown in Figure P33.82. Assume the system has a constant inductance *L*, and note that the applied magnetic field $\vec{B}$ is into the page as shown in the figure. Find the maximum distance the bar travels before it stops. *Hint*: Use the relationship $B\ell(dx/dt) = L(dI/dt)$.

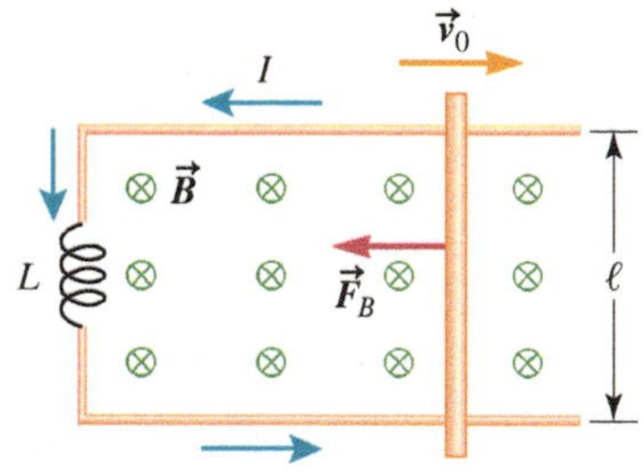

FIGURE P33.82

83. **N** An electronic device is composed of an *RLC* circuit with frequency $f = 60.0$ Hz. The capacitor in the circuit goes bad and must be replaced, but the engineer does not know the capacitance. If she wants the impedance to be 1000.0 Ω but knows only that $R = 500.0$ Ω and $L = 2.50$ H, find the two values of capacitance that will solve her dilemma. (There are two answers to this problem!)

84. **N** An inductor of inductance 2.0 mH is connected across a charged capacitor ($C = 4.0$ μF). The maximum value of the charge Q_{max} on the capacitor is 2.0×10^2 μC. **a.** When the charge on the capacitor is 1.0×10^2 μC, what is the value of $|dI/dt|$? **b.** When the charge on the capacitor is 2.0×10^2 μC, what is the value of the current?

Maxwell's Equations and Electromagnetic Waves

34

Underlying Principles

1. General form of Faraday's law
2. Maxwell's equations
3. Lorentz force
4. Wave equation for an electromagnetic wave

Major Concepts

1. Transverse electric and magnetic waves
2. Energy transferred by electromagnetic waves
3. Intensity
4. Momentum transferred by electromagnetic waves
5. Pressure exerted by electromagnetic waves
6. Polarization of electromagnetic waves and Malus's law

Key Questions

How can we express Faraday's law in a more general form?

What does Faraday's law tell us about an *induced* electric field?

How do Maxwell's equations describe electromagnetic waves?

How do scientists discover new laws of nature? There are many answers to this question, but one is based on the idea that nature is made up of symmetrical or parallel principles.

Maxwell reached his great insight by recognizing the symmetry in nature. Faraday formulated his law about 30 years before Maxwell modified Ampère's law. (We studied the two laws out of chronological order.) So Faraday's law was well known by the time Maxwell was just a young boy. Maxwell used Faraday's law and the idea that nature is based on parallel principles to expand Ampère's law. In doing so, he discovered the fundamental nature of light.

In this chapter, we gain a deeper understanding of electricity and magnetism by looking at all four of Maxwell's equations together. We also combine two of his equations to come up with a wave equation for electromagnetic waves.

34-1 Light: One Last Classical Topic

We are near the end of our study of classical physics. In fact, we have only one phenomenon left to discuss—electromagnetic waves, including light. Light is a very important topic, and the whole next part of this text is devoted to it.

All the classical physics in the preceding 33 chapters can be summed up in a few fundamental principles:

1. Newton's three laws of motion
   a. The law of inertia
   b. $\vec{F}_{\text{tot}} = d\vec{p}/dt$
   c. $\vec{F}_{[\text{B on A}]} = -\vec{F}_{[\text{A on B}]}$
2. Newton's law of universal gravity (one of the four fundamental forces)
3. The first law of thermodynamics (the principle of conservation of energy)
4. The second law of thermodynamics (the entropy of an isolated system never decreases; $\Delta S \geq 0$)
5. Maxwell's four equations and the Lorentz force (the second of the four fundamental forces)

The crowning achievement of classical physics is to describe in these few fundamental principles the enormous range of phenomena that we have studied so far and encounter so often in our daily lives. These phenomena include train crashes, the flight of airplanes, music, refrigerators, cosmic rays, electric generators, electric guitars, and nerve signals in the human body. The Universe would be a very different place if each of these phenomena was governed by different principles. Imagine designing a spacecraft if the physical principles that are true on the Earth were not valid in space.

Missing from this impressive list of phenomena, however, is light. By the early 1800s, scientists modeled light as a wave. That model led to more questions: What is waving? What kind of wave is it—longitudinal or transverse? In the mid-1800s, Maxwell showed that light is a transverse electromagnetic wave, which means electric and magnetic fields oscillate perpendicular to the direction of the wave's propagation (Section 34-4). Two other important properties of light were discovered in the 1900s: (1) Light waves require no medium; they can travel in a vacuum. By contrast, sound requires a medium such as air; there is no sound in a vacuum. (2) Light is also modeled as a particle known as a **photon**. Today, we say that light is both a wave and a particle: Sometimes we model light as a wave and sometimes as a particle, but photons are a modern concept, so we will not study them until Chapter 39.

CASE STUDY Part 1: The Power of Light

Sunlight enables life to exist on the Earth, but you probably never thought sunlight could save the Earth from destruction by a wandering comet or asteroid. The solar system contains a large number of these objects and sometimes one of them hits a planet, as in 1994 when Comet Shoemaker-Levy 9 collided with Jupiter. The comet had broken into many pieces, so there were multiple collisions; Figure 34.1 shows four impact sites along a diagonal line running upward and to the right. After the collision, Jupiter had scars that took about six months to heal. Many of the scars were several times the size of the Earth.

The Earth is not immune to such impacts. It is likely that the dinosaurs were wiped out by an impact 65 million years ago. Astronomers search for near-Earth objects such as comets and asteroids that might hit the Earth and do serious damage. If they find such an object, we could try to deflect it. If the object were predicted to strike in a few decades, we might

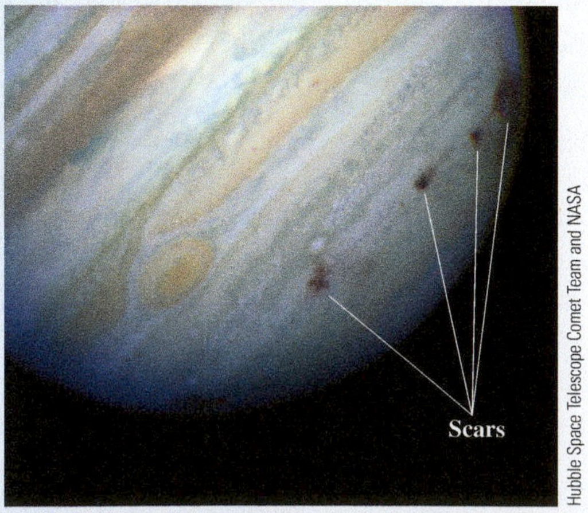

FIGURE 34.1 Jupiter was hit by Comet Shoemaker-Levy 9. Four impact sites are seen in this Hubble Space Telescope image.

consider detonating a nuclear weapon near the object. If we had even more warning, we might try to use the Sun's light to push the object off course. We could attach a solar sail to the object to harness the pressure of the Sun's radiation.

It might seem incredible that light could push a giant rock off course. This is possible, however, and it is just such an effect that limits the amount of light a star can give off without blowing itself apart. In this case study, we will look at (1) saving the Earth from a collision by using sunlight and (2) the maximum luminosity of stars.

CASE STUDY Part 2: Meteor Shower Revisited

In the middle of August, three physics professors—Black, Noir, and Kuro—watched the Perseid meteor shower. On their drive home, Black asked whether either of the other two professors knew the size of a piece of the debris. Neither of them knew the answer, but among the three professors, they knew enough information to estimate an answer. They worked in the dark without a calculator and came up with a good estimate using two different approaches. We described the first approach based on conservation of energy in Chapter 9. Here we consider their second approach, based on the light from the meteor.

The Perseids peak on August 12. The shooting stars appear to come from the constellation Perseus.

CONCEPT EXERCISE 34.1

One way to think about the fundamental principles of physics is to imagine an extraterrestrial alien society. If the alien society could discover the principle, it must be a fundamental principle of the Universe. If the aliens could discover the principle only by coincidence (if they happened to live on a planet with properties identical to those of the Earth), the principle is not fundamental. Avi and Cameron discuss the fundamental nature of the constants g and G. With whom do you agree and why?

Avi: Little g is gravitational acceleration near the Earth. It isn't even good on the Moon, so it isn't fundamental. Big G is the constant in Newton's law of universal gravity. So it must be universal.

Cameron: Both constants depend on the SI system. They aren't the same even if we use U.S. customary units. Neither one is fundamental.

34-2 Generalized Form of Faraday's Law

In order to study light, we need to revisit Faraday's law. Faraday's law in the form presented in Chapters 32 and 33 is very useful in practical situations such as finding the current induced in a conductor. For example, in Figure 34.2A, a single conducting loop is in the plane of the page and a magnetic field (uniform over the loop's area) points into the page. (The magnetic field may be generated by a large solenoid.) If the magnetic field gets stronger (because the current in the solenoid increases), the magnetic flux through the loop increases. According to Faraday's law, the changing magnetic flux induces a current in the loop like the current that would be set up with a battery that has emf $\mathcal{E}$. So, a practical way to write Faraday's law for a single loop is in terms of the emf *induced* by the changing magnetic flux: $\mathcal{E} = -d\Phi_B/dt$ (Eq. 32.6 with $N = 1$).

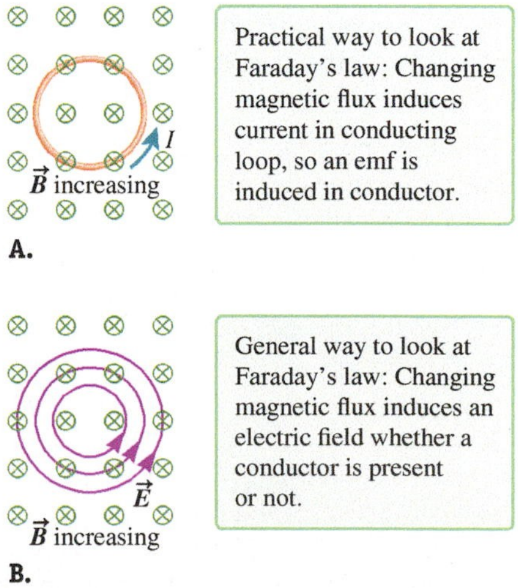

I
$\vec{B}$ increasing
A.
Practical way to look at Faraday's law: Changing magnetic flux induces current in conducting loop, so an emf is induced in conductor.
$\vec{E}$
$\vec{B}$ increasing
B.
General way to look at Faraday's law: Changing magnetic flux induces an electric field whether a conductor is present or not.

FIGURE 34.2 **A.** The changing magnetic flux induces an emf in the conducting loop. **B.** The changing magnetic flux induces an electric field. This kind of electric field is nonconservative.

To study light, however, we need a more general form of Faraday's law that involves the electric field instead of an induced emf. According to Faraday's experiments, a changing magnetic flux through a loop induces a current in the loop. You can think of that current as being set up by an emf, but you can also think of it as being set up by an electric field. To get positive charge to move in a counterclockwise circle (Fig. 34.2A), the electric field must form a counterclockwise circle (Fig. 34.2B). In fact, the conducting loop is *not* necessary, and the electric field is

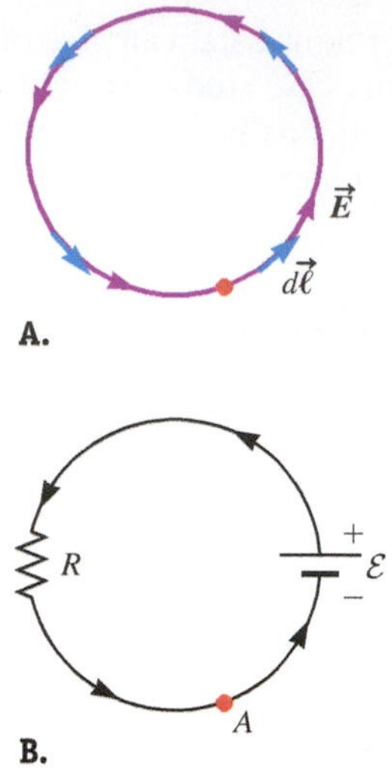

FIGURE 34.3 **A.** A particle moves in a circular path along a single electric field line. Each segment of the path is parallel to the electric field. **B.** A positively charged particle starts at point *A* and moves counterclockwise around the circuit. The particle gains energy when it goes through the battery and loses the same amount of energy when it passes through the resistor.

found throughout the indicated region of space. The electric field lines are represented by a number of circles.

This should seem odd for electric field lines, which we said must originate from a positively charged particle and/or terminate on a negatively charged particle (Sections 24-3 and 24-4). The electric field lines in Figure 34.2B neither originate nor terminate. Instead, they wrap around on themselves. And, because the conductor is unnecessary, there are no charged particles in Figure 34.2B.

What is going on here? The rules in Chapter 24 were established for electrostatic fields—that is, for electric fields whose source is stationary, electrically-charged particles. The electric field in Figure 34.2B is generated by the changing magnetic field. (The magnetic field is the result of moving charges, so the ultimate source of this electric field is also those moving charged particles.) There are two types of electric fields that result from two types of electric sources: (1) Stationary charged particles produce an electrostatic field, and (2) a changing magnetic field produces an *induced* electric field.

The closed loops of the induced electric field lines in Figure 34.2 mean that the induced electric field is *nonconservative.* To see this, imagine a positively charged particle moves in a circular path following one of the field lines (Fig. 34.3A). The work done by the electric field on this positively charged particle is given by $W = \oint \vec{F}_E \cdot d\vec{\ell}$ (similar to Eq. 9.21), where the circle on the path integral means it is taken over a closed path of which $d\vec{\ell}$ is a segment. (The original form of Eq. 9.21 contains $d\vec{r}$. We have replaced it with $d\vec{\ell}$ so that we use the same symbols as in Ampère's law, Eq. 31.5.) The particle's charge is q and the electric field around the path is $\vec{E}$, so the force exerted by the electric field is given by $\vec{F}_E = q\vec{E}$ (Eq. 24.2). Substitute Equation 24.2 into Equation 9.21:

$$W = q\oint \vec{E} \cdot d\vec{\ell} \tag{34.1}$$

where the charge q is constant as the particle moves around the closed path. At any point along the path, the electric field is parallel to the path segment (Fig. 34.3A), so at every point the dot product $\vec{E} \cdot d\vec{\ell}$ is positive. Therefore, the entire integral and thus the work around the closed path are positive:

$$W = q\oint \vec{E} \cdot d\vec{\ell} > 0 \tag{34.2}$$

An *induced* electric field is nonconservative, so its path integral depends on the particular path.

When the positive particle in Figure 34.3A completes one trip around its closed path, positive work has been done on it and its energy is greater than when it started. This result shows that the induced electric field fails the (second) test for a conservative force (Section 9-5). According to this test, the work done by a conservative force around a closed path is zero.

Compare the round trip of a particle in Figure 34.3A to that of a particle moving around the simple circuit in Figure 34.3B. Suppose a positive particle starts at point *A* in the circuit and moves in a counterclockwise direction (Fig. 34.3B). The particle passes through the battery from the negative to the positive terminal. Chemical reactions inside the battery do work on the particle, increasing its energy by $q\mathcal{E}$. Then the particle passes through the resistor, losing an amount of energy qV_R equal to the amount of work done on it by the battery: $q\mathcal{E} - qV_R = 0$. So, when the particle returns to point *A*, it has exactly the same energy it started with. This is really a statement of Kirchhoff's loop rule: The sum of the voltages around a closed loop is zero: $\mathcal{E} - V_R = 0$.

Kirchhoff's rule doesn't hold when an emf is induced in a conducting loop (Fig. 34.2A). From Faraday's law, the voltage around this loop is not zero; it is given by $\mathcal{E} = -d\Phi_B/dt$ (Eq. 32.6 with $N = 1$). When a positively charged particle moves once around the loop, positive net work W is done on it by the electric field:

$$W = q\mathcal{E} \tag{34.3}$$

Equation 34.3 must equal Equation 34.1, so

$$W = q\mathcal{E} = q\oint \vec{E} \cdot d\vec{\ell}$$

$$\mathcal{E} = \oint \vec{E} \cdot d\vec{\ell} \tag{34.4}$$

Because the potential difference around the closed loop is *not* zero when a changing magnetic field is present, we cannot use Kirchhoff's loop rule. (See Problems 13, 14, and 68.) Faraday's law still holds, however.

Now, to find a general expression for Faraday's law, substitute Equation 34.4 into $\mathcal{E} = -d\Phi_B/dt$ (Eq. 32.6):

$$\oint \vec{E} \cdot d\vec{\ell} = -\frac{d\Phi_B}{dt} \tag{34.5}$$

We can rewrite Equation 34.5 by replacing the magnetic flux with $\Phi_B = \int \vec{B} \cdot d\vec{A}$ (Eq. 31.2):

GENERAL FORM OF FARADAY'S LAW

❶ Underlying Principle

$$\oint \vec{E} \cdot d\vec{\ell} = -\frac{d}{dt}\int \vec{B} \cdot d\vec{A} \tag{34.6}$$

Equation 34.6 says that a changing magnetic flux produces an electric field. It does not matter whether there is a conductor present, or free charges, or simply a vacuum.

The negative sign in Equations 34.5 and 34.6 is Lenz's law. Conceptually, Lenz's law says the induced electric field is in the direction that opposes the change in magnetic flux. In practice, Lenz's law tells us how to find the direction of the induced electric field. The left side $\left(\oint \vec{E} \cdot d\vec{\ell}\right)$ is a circulation integral (a path integral around a closed loop), for which you must choose the shape and direction of the path. Once you have picked the direction, wrap the fingers of your right hand in that direction. Your thumb then points in the direction of the positive area vector $\vec{A}$. If the magnetic field points in the same direction as $\vec{A}$, the magnetic flux $\left(\int \vec{B} \cdot d\vec{A}\right)$ is positive. If the magnetic flux *increases*, the right side of Equation 34.6 is negative (due to Lenz's law). In order for the left side to be negative also, the electric field must point *opposite* the path's direction you chose so that $\vec{E} \cdot d\vec{\ell} < 0$. In Concept Exercise 34.2, you are asked to work out the direction of the electric field for other situations.

As a simple rule of thumb, you can imagine the path you have chosen is made of a conducting loop. Use your usual Lenz's-law method to find the direction of the current induced in that (imaginary) conducting loop. The electric field points in the same direction as that (imaginary) induced current.

PROBLEM-SOLVING STRATEGY

Finding the Electric Field Using Faraday's Law

Because Ampère's law,

$$\oint \vec{B} \cdot d\vec{\ell} = \mu_0 \varepsilon_0 \frac{d}{dt}\int \vec{E} \cdot d\vec{A}$$

(see Eqs. 25.6 and 31.8 with $I_{\text{thru}} = 0$) is similar to Faraday's law,

$$\oint \vec{E} \cdot d\vec{\ell} = -\frac{d}{dt}\int \vec{B} \cdot d\vec{A} \tag{34.6}$$

we can adapt the four steps from Chapter 31 (page 987) to come up with four steps for solving problems with Faraday's law. These steps are useful when we need to find the electric field induced by a changing magnetic flux.

INTERPRET and ANTICIPATE

Step 1 Sketch electric field lines, and apply Lenz's law to find the direction of the electric field, if possible.

Step 2 Choose a closed path. You must choose a shape that exploits the symmetry of the situation and also choose the direction of the path. If possible, choose the direction so that $\vec{E}$ and $d\vec{\ell}$ are parallel.

SOLVE

Step 3 Do the circulation integral $\oint \vec{E} \cdot d\vec{\ell}$. The electric field is often a constant over each portion of the integral, so the integral is usually very simple.

Step 4 Find the change in magnetic flux $\frac{d}{dt}\int \vec{B} \cdot d\vec{A}$, **and substitute into Faraday's law** (Eq. 34.6).

CONCEPT EXERCISE 34.2

a. Suppose you choose a direction for the path in Equation 34.6,

$$\oint \vec{E} \cdot d\vec{\ell} = -\frac{d}{dt}\int \vec{B} \cdot d\vec{A}$$

and you find that the magnetic field points in the same direction as the positive area vector. If the magnetic flux is *decreasing*, is the electric field in the same direction as the path or in the opposite direction?

b. Now you choose a new direction for the path, and the magnetic field points in the direction opposite the positive area vector. If the magnetic flux is *increasing*, is the electric field in the same direction as the path or in the opposite direction?

c. Again you choose a path such that the magnetic field points in the direction opposite the positive area vector. If the magnetic flux is *decreasing*, is the electric field in the same direction as the path or in the opposite direction?

EXAMPLE 34.1 A Circular Region

A magnetic field points out of the page and perpendicular to it. The field is uniform in space over a circular region of radius R and zero everywhere else. The field increases at a rate dB/dt.

A Find an expression for the electric field $E(r)$ at a distance r from the center of the circular region, where $r < R$.

INTERPRET and ANTICIPATE

Use the four steps of the problem-solving strategy. We expect the electric field to depend on the rate at which the magnetic field changes. If the magnetic field changes rapidly, a strong electric field should be generated.

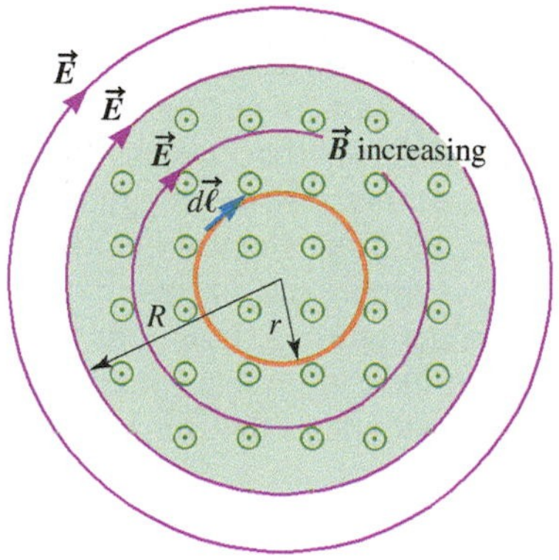

FIGURE 34.4

Step 1 Sketch electric field lines. The magnetic field points out of the page and is uniform in space over a circular region of radius R and zero everywhere else (Fig. 34.4). From the symmetry of the problem, the electric field must form concentric circles as shown. (For now, don't worry about the spacing of the electric field lines.) To find the direction of the induced electric field, imagine one of the electric field lines is a circular conductor. The current in the imaginary conductor must be clockwise to oppose the changing magnetic flux, so the induced electric field is clockwise.

Step 2 Choose a closed path. To exploit the symmetry, we choose a circular path of radius r that runs along an electric field line. We have decided to integrate clockwise around the path in the direction of the electric field. Our path is shown in Figure 34.4.

SOLVE

Step 3 Do the circulation integral. This is the integral on the left side of Faraday's law (Eq. 34.6). Because we choose to integrate clockwise in the direction of the electric field, the dot product is $E\,d\ell$. The electric field is constant over the path; that is why a circular path is a good choice. The path integral $\oint d\ell$ is the circumference of the circle.

$$\oint \vec{E} \cdot d\vec{\ell} = \oint E\,d\ell = E\oint d\ell$$

$$\oint \vec{E} \cdot d\vec{\ell} = E(2\pi r) \tag{1}$$

Step 4 Find the change in magnetic flux. First use Equation 31.2 to find the magnetic flux through the circular path. To find the direction of the area vector, wrap the fingers of your right hand clockwise in the direction of the path. Your thumb then points into the page, so the area vector points into the page while the magnetic field points

$$\Phi_B = \int \vec{B} \cdot d\vec{A} \tag{31.2}$$

out of the page. The dot product is therefore negative. The magnetic field is uniform in space through the circular path of radius $r < R$. The integral $\int dA$ is then the area of the circular path.	$\Phi_B = -\int B\,dA = -B\int dA = -BA$ $\Phi_B = -B\pi r^2$ (2)
Now **substitute** Equations (1) and (2) **into Faraday's law** (Eq. 34.6). The negative signs cancel on the right.	$\oint \vec{E}\cdot d\vec{\ell} = -\frac{d}{dt}\int \vec{B}\cdot d\vec{A}$ (34.6) $E(2\pi r) = -\frac{d}{dt}(-B\pi r^2) = \pi r^2\frac{dB}{dt}$
Solve for E as a function of r.	$E(r) = \frac{r}{2}\frac{dB}{dt}$

CHECK and THINK

As expected, the strength of the electric field depends on the rate at which the magnetic field changes. If the magnetic field changes rapidly, the electric field is strong. The electric field is zero at the center of the region ($r = 0$) and gets stronger as r increases. At the outer border of the region ($r = R$), the electric field is $E(R) = (R/2)(dB/dt)$. We'll use this value to check our result in part B.

B Find an expression for the electric field $E(r)$ at a distance r from the center of the circular region, where $r > R$.

INTERPRET and ANTICIPATE

This part is similar to part A, so we can use our previous work to take a few shortcuts. At $r = R$, both expressions should give the same answer.

Step 1 Sketch electric field lines. The electric field lines are the same as in part A (Fig. 34.5). (Now we know the electric field increases linearly as a function of r, as we roughly represent in our spacing of the field lines.)

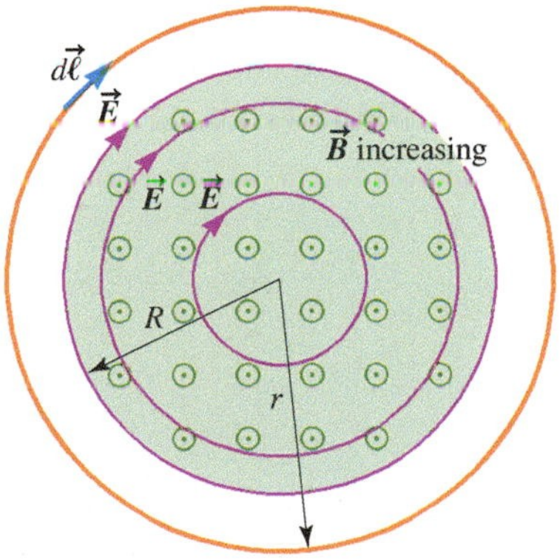

FIGURE 34.5

Step 2 Choose a closed path. The circular path must lie in the region we are interested in, so this time our circular path has radius $r > R$.

SOLVE

Step 3 Do the circulation integral. The circulation integral is identical to the one in part A.	$\oint \vec{E}\cdot d\vec{\ell} = E(2\pi r)$ (1)
Step 4 Find the change in magnetic flux. This step is different from the step in part A because the magnetic field is not uniform in space throughout the circular path. It is uniform out to radius R but zero in the donut-shaped region between R and r. So we break up the flux integral (Eq. 31.2) into two pieces covering (1) the inner region over which the magnetic field is spatially uniform and (2) the donut-shaped region over which the magnetic field is zero.	$\Phi_B = \int \vec{B}\cdot d\vec{A}$ (31.2) $\Phi_B = \left(\int \vec{B}\cdot d\vec{A}\right)_{\text{inner}} + \left(\int \vec{B}\cdot d\vec{A}\right)_{\text{donut}}$
The magnetic flux over the donut-shaped region is zero because the magnetic field is zero.	$\Phi_B = \left(\int \vec{B}\cdot d\vec{A}\right)_{\text{inner}} + 0 = \left(\int \vec{B}\cdot d\vec{A}\right)_{\text{inner}}$

Example continues on page 1092 ▶

Finding the magnetic flux through the inner region is like finding the flux through the circular path in part A. The area vector points into the page and the magnetic field points out of the page, so the dot product is negative. The magnetic field through the inner region is uniform. The remaining integral is then just the area of the inner region of radius *R*.	$\Phi_B = \int \vec{B} \cdot d\vec{A} = -\int B\,dA = -B\int dA$ $\Phi_B = -BA = -B\pi R^2$ (3)
Now **substitute** Equations (1) and (3) **into Faraday's law** (Eq. 34.6). Solve for E as a function of r.	$\oint \vec{E} \cdot d\vec{\ell} = -\frac{d}{dt}\int \vec{B} \cdot d\vec{A}$ (34.6) $E(2\pi r) = -\frac{d}{dt}(-B\pi R^2) = \pi R^2 \frac{dB}{dt}$ $E(r) = \frac{R^2}{2r}\frac{dB}{dt}$
CHECK and THINK To check the result, find E at the border of the region, where $r = R$. The electric fields from parts A and B are the same at that distance, as required.	$E(R) = \frac{R^2}{2R}\frac{dB}{dt} = \frac{R}{2}\frac{dB}{dt}$ ✓

34-3 Five Equations of Electromagnetism

MAXWELL'S EQUATIONS

Underlying Principle

The entire theory of electricity and magnetism boils down to just five equations, four of which are Maxwell's equations. James Clerk Maxwell (1831–1879) was born in Edinburgh, Scotland. Like Newton about 200 years earlier, Maxwell went to Cambridge University and worked on many of the problems of his time. He correctly described the nature of Saturn's rings, and he is one of the founders of the kinetic theory of gases (Chapter 20), but his greatest work was on electricity, magnetism, and light. Maxwell believed electricity and magnetism were part of a single unified principle. In 1864, he published *A Dynamical Theory of the Electromagnetic Field*, in which he wrote his four now-famous equations in mathematical form. To physicists, Maxwell's equations are beautiful (Fig. 31.1, page 978) because four relatively simple equations describe a wide range of phenomena. In this section, we review Maxwell's four equations and consider the achievements of Maxwell and the body of classical physics. We state Maxwell's four equations next (in the absence of a dielectric or magnetic material).

Gauss's Law for Electricity

The electric flux through a closed surface is proportional to the charge inside that surface:

$$\Phi_E = \oint \vec{E} \cdot d\vec{A} = \frac{q_{\text{in}}}{\varepsilon_0} \tag{25.11}$$

Conceptually, Gauss's law for electricity says that particles with excess charge are the source of electrostatic fields, and electric field lines emerge from positive particles and terminate on negative ones.

Gauss's Law for Magnetism

The magnetic flux through a closed surface is zero:

$$\Phi_B = \oint \vec{B} \cdot d\vec{A} = 0 \tag{31.4}$$

Gauss's law for magnetism says that so far, the existence of magnetic monopoles has not been confirmed. If magnetic monopoles are discovered, the 0 on the right side of the equation will have to be replaced so that the net magnetic flux through a closed surface is proportional to the net magnetic "charge" enclosed. Conceptually, Gauss's law for magnetism says that magnetic fields always form closed loops such as those created by magnetic dipoles (Fig. 30.8 on page 938).

Faraday's Law

The circulation integral of the electric field is proportional to the changing magnetic flux through the closed path:

$$\oint \vec{E} \cdot d\vec{\ell} = -\frac{d\Phi_B}{dt} \tag{34.5}$$

where the negative sign is Lenz's law. Conceptually, Faraday's law says a changing magnetic flux is another source of an electric field. We have referred to this as the *induced* electric field to distinguish it from the electrostatic field. Lenz's law states that the direction of the induced electric field is such that it opposes the change that created it. According to Faraday's law, we can simply say a changing magnetic field is the source of an electric field.

Ampère–Maxwell's Law

Ampère's law says that the circulation integral of the magnetic field is proportional to the current that passes through the closed path. Maxwell's addition to Ampère's law adds a second term so that the circulation integral also depends on the changing electric flux through the path:

$$\oint \vec{B} \cdot d\vec{\ell} = \mu_0 I_{\text{thru}} + \mu_0 \varepsilon_0 \frac{d\Phi_E}{dt} \tag{31.8}$$

Conceptually, Ampère–Maxwell's law says that a current and a changing electric *flux* are the sources of a magnetic field. More generally, we can say a changing electric *field* is the source of a magnetic field. Because magnetic monopoles have not been confirmed, this is the only source of a magnetic field.

The magnetic field of a permanent magnet is largely due to the spin of the electrons. (See Section 30-6.)

When Gauss, Faraday, and Ampère developed their laws, they did not express them in the mathematical form we use today. Not only was Maxwell the first to think of these four laws as being part of the same physical principle, but he was also the first to write them down mathematically in a form close to how we have given them here. This mathematical description is important in seeing how these laws come together to describe light (Section 34-4).

The Fifth Equation

An additional equation gives the Lorentz force:

$$\vec{F}_{\text{L}} = \vec{F}_E + \vec{F}_B = q(\vec{E} + \vec{v} \times \vec{B}) \tag{30.21}$$

LORENTZ FORCE

Underlying Principle

The Lorentz force is the vector sum of the force exerted by an electric field plus the force exerted by a magnetic field on a charged particle. The electric force is either parallel (in the case of positive particles) or antiparallel (in the case of negative particles) to the electric field. The magnetic force is exerted only on particles that have velocity $\vec{v}$ that is not parallel (or antiparallel) to the magnetic field. The magnetic force is perpendicular to both $\vec{v}$ and $\vec{B}$.

These five equations are the fundamental equations of electricity and magnetism. All the new equations introduced in Chapters 23 through 34 are based on these five equations and other fundamental principles such as conservation of energy. Even light can be understood in terms of these few fundamental equations.

34-4 Electromagnetic Waves

In this section, we use Maxwell's equations to derive a wave equation for electromagnetic waves, including light. Maxwell presented his derivation in the late 1870s, but the experimental evidence confirming his prediction of electromagnetic waves came about 10 years later. Let's consider some experimental evidence first.

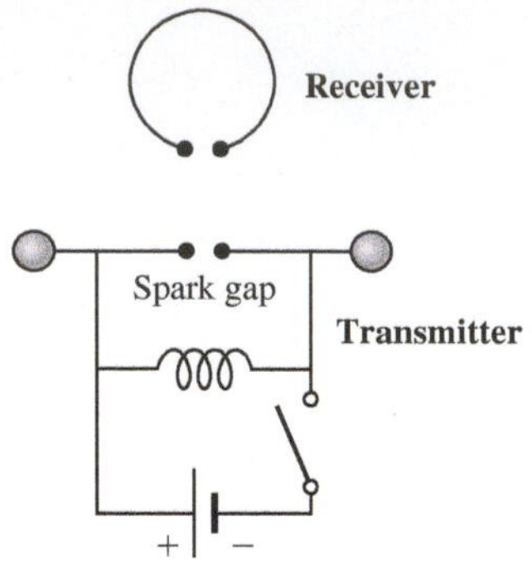

FIGURE 34.6 Hertz built the first radio transmitter and receiver. Both are essentially *LC* circuits.

Hertz's Experiments

In 1887, the German physicist Heinrich Rudolf Hertz (1857–1894) developed and performed an experiment that confirmed Maxwell's prediction. Figure 34.6 shows a very crude sketch of Hertz's basic apparatus, which consists of two parts. The *transmitter* is a circuit that produces an electromagnetic oscillation. The oscillation causes electromagnetic waves to travel outward, and the *receiver* detects those waves. The transmitter and receiver (both essentially *LC* circuits) are not in contact. The main parts of the transmitter are an induction coil and a capacitor that looks like a rod with a small gap. Hertz needed a very high potential difference between the two sides of the gap so that the air in between would become a conductor and a spark could jump across. So, Hertz used two large spheres to increase the capacitance of the circuit, allowing for a high voltage across the gap. A power supply (consisting of three batteries) was required to charge up the capacitor. The receiver is a simple loop of wire with a small gap. The loop has some inductance, and the gap is like the plates of a capacitor. The receiver does not have a separate power supply.

When the transmitter is turned on, a strong electric field builds up in its gap. The electric field ionizes the air, and acceleration of the resulting free electrons causes more ionization until the air conducts a spark. The plates of the capacitor charge and discharge periodically, while the sparks in the gap oscillate at the natural frequency of the *LC* circuit. (Recall that in an *LC* circuit, the electric field between the capacitor plates oscillates back and forth at the natural frequency; see Fig. 33.12, page 1058.) The electric field oscillation creates an electromagnetic wave. This is similar to the oscillation of a stereo speaker, which is analogous to the transmitter. The speaker's back-and-forth motion creates a sound wave in the surrounding air. Your eardrum is the receiver; the sound wave causes your eardrum to move back and forth at the same frequency as the speaker's frequency.

Hertz found that sparks in the transmitter's gap sometimes set off sparks in the receiver's gap. The receiver has its own natural frequency. Sparks in the receiver were present only when the natural frequency of the transmitter matched the natural frequency of the receiver.

Hertz also found that he could reflect the wave (the spark "signal" from the transmitter) off a metal plate just as a sound wave is reflected off a cliff. Recall that a standing wave can be formed by using the reflection of a traveling wave. Hertz was able to set up a standing electromagnetic wave similar to a sound wave in a pipe. By moving the plate from which the wave reflected, Hertz could find the nodes of the standing wave, which allowed him to determine n—the harmonic number and wavelength. Because the frequency of the wave was the natural frequency of the *LC* circuit, Hertz could find the speed of the wave using Equation 17.8 $\left(v = \lambda f\right)$. His result was a speed very close to that of light $\left(c = 3.00 \times 10^8 \text{ m/s}\right)$.

Hertz's confirmation of Maxwell's theory came about eight years after Maxwell died. Not only was it an important confirmation of the existence of electromagnetic waves, but it also led to important new technologies. Hertz's setup is essentially a radio transmitter and receiver. Unfortunately, Hertz died in 1894 and did not see the important communication devices that resulted from his work. However, his experiment provided the insight needed by the Italian electrical engineer Guglielmo Marconi (1874–1937), who invented the first wireless communication device (a wireless telegraph).

Electromagnetic Wave Transmission

The panels in Figure 34.7 show a transmitter on the left and a receiver on the right, each represented by a simple capacitor. No other parts of the circuit are shown. We wish to follow the electric field from the transmitter to the receiver. One way to visualize an electric field is to imagine the electric force exerted on a positive test particle, so imagine a whole line of test particles along the x axis from the transmitter to the receiver. Suppose the transmitter is charged up so the bottom plate is positive and the top plate is negative. Then an upward-pointing electric field pulse moves from

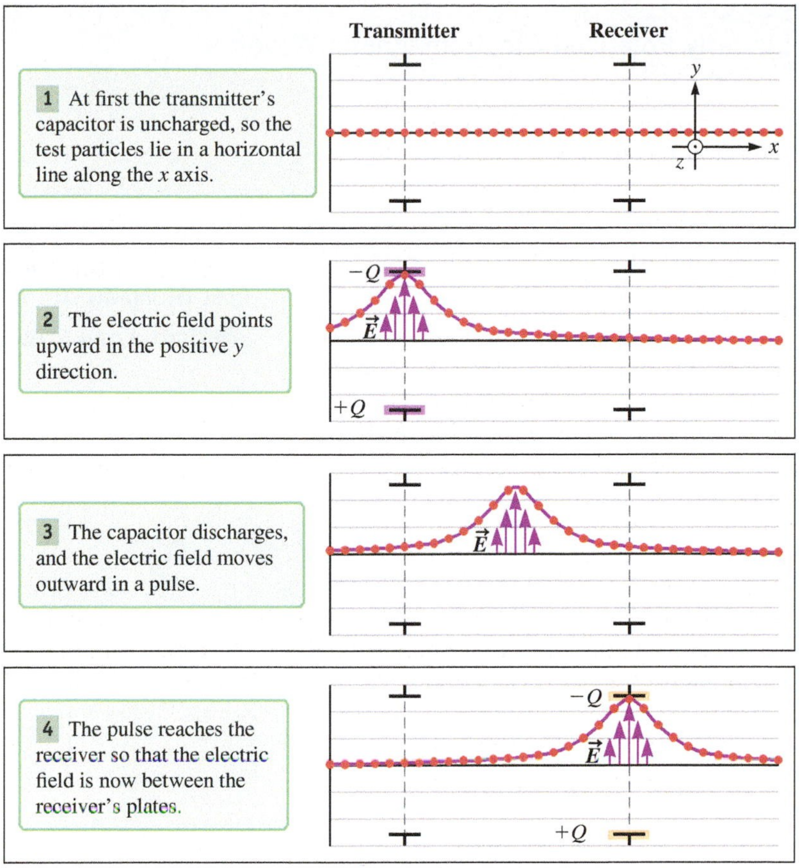
Transmitter
Receiver
1 At first the transmitter's capacitor is uncharged, so the test particles lie in a horizontal line along the x axis.
y
x
z
2 The electric field points upward in the positive y direction.
−Q
+Q
E
3 The capacitor discharges, and the electric field moves outward in a pulse.
E
4 The pulse reaches the receiver so that the electric field is now between the receiver's plates.
−Q
+Q
E

FIGURE 34.7 An electric pulse moves from the transmitter to the receiver.

the transmitter to the receiver. The magnetic field is not shown in Figure 34.7, but we can picture it. According to Ampère's law, the changing electric field causes a magnetic field. To find the direction of the magnetic field, point your right thumb in the direction of the displacement current (the y direction). Then your fingers wrap around in the direction of the magnetic field. So the magnetic field lies in the xz plane and is perpendicular to the electric field. Both fields move outward from the transmitter.

The displacement current points in the same direction as the electric field if the electric flux is increasing.

The Electromagnetic Wave Equation

To derive a wave equation for any possible electromagnetic wave is beyond the scope of this textbook. Instead, we will derive the wave equation for a plane wave traveling in the positive x direction (Fig. 34.8). The electric field oscillates parallel to the y axis, while the magnetic field oscillates parallel to the z axis, perpendicular to the electric field. The wave in Figure 34.8 is **linearly polarized**, which means the electric field is always parallel to a single axis (and therefore the magnetic field is always perpendicular to that chosen axis).

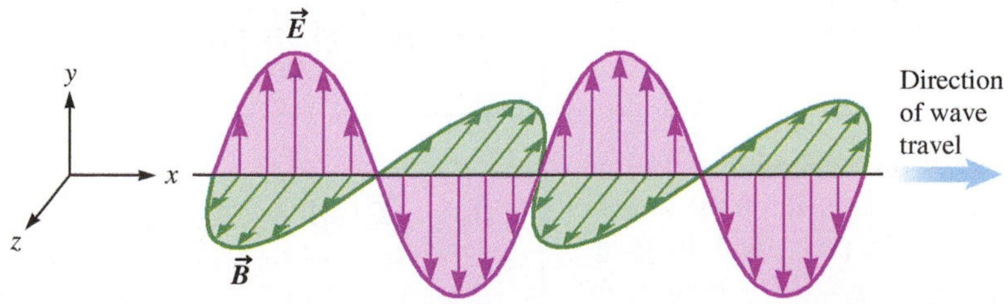
E
y
x
z
B
Direction of wave travel

FIGURE 34.8 A linearly polarized electromagnetic wave (pulse) travels in the positive x direction. For all electromagnetic waves, *the electric and magnetic fields are perpendicular to each other and to the direction of the wave's propagation.* (However, $\vec{E}$ along y and $\vec{B}$ along z were chosen arbitrarily here.)

DERIVATION Wave Equation for Linearly Polarized Electromagnetic Waves

We will use the last two of Maxwell's equations—Faraday's law (Eq. 34.5) and Ampère–Maxwell's law (Eq. 31.8)—to derive the wave equation for the electric field of a linearly polarized wave in a vacuum:

$$\frac{\partial^2 E}{\partial x^2} = \mu_0 \varepsilon_0 \frac{\partial^2 E}{\partial t^2} \tag{34.7}$$

WAVE EQUATION FOR AN ELECTROMAGNETIC WAVE

Underlying Principle

For homework (Problem 26), you will show that

$$\frac{\partial^2 B}{\partial x^2} = \mu_0 \varepsilon_0 \frac{\partial^2 B}{\partial t^2} \tag{34.8}$$

In a vacuum, you can assume there is no charge and therefore no current: $I_{thru} = 0$. Also assume the space between the transmitter and receiver is empty.

Apply Faraday's law (Eq. 34.5) to a linearly polarized wave (Fig. 34.8), using the four steps from this chapter's problem-solving strategy (page 1089) as a guide.

$$\oint \vec{E} \cdot d\vec{\ell} = -\frac{d\Phi_B}{dt} \tag{34.5}$$

Sketch electric field lines. The electric and magnetic field lines are shown in Figure 34.8. Consider a small portion of this wave. Figure 34.9 shows the electric field at one instant in a small, narrow rectangle that lies in the *xy* plane. The rectangle has height *h* and width *dx*. For this particular region at this particular instant, the electric field is weaker on the left, $\vec{E}(x)$, than it is on the right, $\vec{E}(x + dx)$.

Choose a closed path. A rectangular path exploits the symmetry of the problem. The electric field is zero along the horizontal portions of the path, and if we choose to integrate counterclockwise, the electric field is parallel to the path on the right side of the rectangle and antiparallel on the left side.

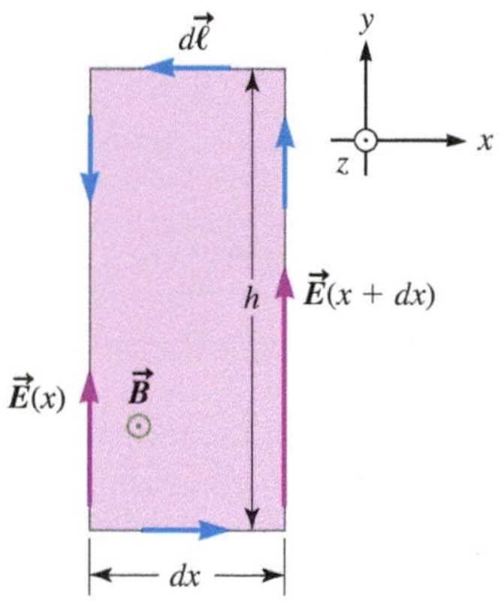

FIGURE 34.9 A small portion of the wave in Figure 34.8, showing the plane of the electric field.

Do the circulation integral. The horizontal portions of the path do not contribute to the circulation integral. The dot product over the right side of the path is positive and over the left side is negative. We take the electric field to be constant over each individual path segment. The integrals $\int d\ell$ are both equal to the height *h* of the rectangle.

$$\oint \vec{E} \cdot d\vec{\ell} = \left(\int \vec{E} \cdot d\vec{\ell}\right)_{\text{right}} + \left(\int \vec{E} \cdot d\vec{\ell}\right)_{\text{left}}$$

$$\oint \vec{E} \cdot d\vec{\ell} = E_{\text{right}} \int d\ell - E_{\text{left}} \int d\ell$$

$$\oint \vec{E} \cdot d\vec{\ell} = [E(x + dx)]h - [E(x)]h$$

The term in square brackets is the derivative of the electric field with respect to *x* times the differential *dx*. We use the partial derivative because the electric field is a function of time and position, and our derivative is taken only over position *x* while the time *t* is held constant.

$$\oint \vec{E} \cdot d\vec{\ell} = h[E(x + dx) - E(x)]_{t\,\text{constant}}$$

$$\oint \vec{E} \cdot d\vec{\ell} = h\left(\frac{\partial E}{\partial x}\right)dx \tag{34.9}$$

Find the change in magnetic flux. Figure 34.9 shows one magnetic field line perpendicular to the electric field and pointing out of the page. Wrap the fingers of your right hand counterclockwise in the direction of the path; then the magnetic field is in the same direction as the (positive) area vector. If we assume the magnetic field is uniform over this very small rectangle, the magnetic flux is easy to find. The area *A* is the area of the rectangle.

$$\Phi_B = \int \vec{B} \cdot d\vec{A} = BA$$

$$\Phi_B = B(h\,dx) \tag{34.10}$$

Substitute Equations 34.9 and 34.10 **into Faraday's law.** The height h and width dx of the rectangle are constants. The magnetic field is a function of position and time. Because we need to take the derivative with respect to time only, we again use a partial derivative, this time holding x constant. Equation 34.11 is the first of two partial differential equations we need.

$$\oint \vec{E}\cdot d\vec{\ell} = -\frac{d\Phi_B}{dt}$$

$$h\left(\frac{\partial E}{\partial x}\right)dx = -\frac{d}{dt}B(h\,dx) = -h\left(\frac{\partial B}{\partial t}\right)dx$$

$$\left(\frac{\partial E}{\partial x}\right) = -\left(\frac{\partial B}{\partial t}\right) \tag{34.11}$$

We are about halfway through our derivation. The next step is to apply Ampère–Maxwell's law (Eq. 31.8), setting $I_{\text{thru}} = 0$ because there is no current in a vacuum. Equation 34.12 is mathematically similar to Faraday's law, so the procedure is much like what we did above and follows the same four steps.

$$\oint \vec{B}\cdot d\vec{\ell} = \mu_0 I_{\text{thru}} + \mu_0\varepsilon_0\frac{d\Phi_E}{dt} \tag{31.8}$$

$$\oint \vec{B}\cdot d\vec{\ell} = \mu_0\varepsilon_0\frac{d\Phi_E}{dt} \tag{34.12}$$

Figure 34.10 is similar to Figure 34.9 and combines the first two steps. **Sketch magnetic field lines**, and **choose a closed path**. The rectangle lies in the xz plane; its length is dx and its width is w. For this particular region at this particular instant, the magnetic field is weaker on the left, $\vec{B}(x)$, than it is on the right, $\vec{B}(x + dx)$. Our path is a rectangle as shown. Wrap the fingers of your right hand in the direction of the path; the electric field then points in the same direction as the (positive) area vector.

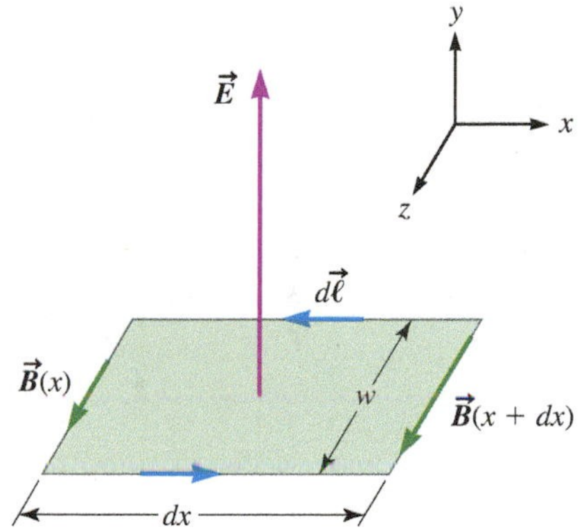

FIGURE 34.10 A small portion of the wave in Figure 34.8, showing the plane of the magnetic field.

The circulation integral is similar to the circulation integral over the electric field. Only the left and right portions of the path contribute to the integral. The dot product on the path's left is positive and on the right is negative. The magnetic field is constant over an individual path segment. The integrals $\int d\ell$ are both equal to the width w of the rectangle.

$$\oint \vec{B}\cdot d\vec{\ell} = \left(\int \vec{B}\cdot d\vec{\ell}\right)_{\text{right}} + \left(\int \vec{B}\cdot d\vec{\ell}\right)_{\text{left}}$$

$$\oint \vec{B}\cdot d\vec{\ell} = -B_{\text{right}}\int d\ell + B_{\text{left}}\int d\ell$$

$$\oint \vec{B}\cdot d\vec{\ell} = [-B(x + dx)]w + [B(x)]w$$

The term in square brackets is the derivative of the magnetic field with respect to x times the differential dx. As before, this is a partial derivative.

$$\oint \vec{B}\cdot d\vec{\ell} = -w[B(x + dx) - B(x)]_{t\,\text{constant}}$$

$$\oint \vec{B}\cdot d\vec{\ell} = -w\left(\frac{\partial B}{\partial x}\right)dx \tag{34.13}$$

Find the change in electric flux. The electric field is perpendicular to the rectangle and parallel to the area vector. If we assume the electric field is uniform over this very small rectangle, the electric flux is easy to find.

$$\Phi_E = \int \vec{E}\cdot d\vec{A} = EA$$

$$\Phi_E = E(w\,dx) \tag{34.14}$$

Substitute Equations 34.13 and 34.14 into Ampère–Maxwell's law as expressed in Equation 34.12. The length dx and width w of the rectangle are constants. The electric field is a function of position and time. Because we need to take the derivative with respect to time only, we again use a partial derivative, holding x constant. Equation 34.15 is the second of the two partial differential equations we need.

$$\oint \vec{B}\cdot d\vec{\ell} = \mu_0\varepsilon_0\frac{d\Phi_E}{dt} \tag{34.12}$$

$$-w\left(\frac{\partial B}{\partial x}\right)dx = \mu_0\varepsilon_0\frac{d}{dt}E(w\,dx)$$

$$-w\left(\frac{\partial B}{\partial x}\right)dx = \mu_0\varepsilon_0 w\left(\frac{\partial E}{\partial t}\right)dx$$

$$\left(\frac{\partial B}{\partial x}\right) = -\mu_0\varepsilon_0\left(\frac{\partial E}{\partial t}\right) \tag{34.15}$$

Derivation continues on page 1098 ▶

We are not done because both equations (Eqs. 34.11 and 34.15) involve B and E in the same expression. We need one equation for E. There is a bit of a trick to finding it.

Take the partial derivative of Equation 34.11 with respect to x. On the left side, we find the second partial derivative of E with respect to x. On the right side, we switch the order of differentiation.	$\frac{\partial}{\partial x}\left(\frac{\partial E}{\partial x}\right) = -\frac{\partial}{\partial x}\left(\frac{\partial B}{\partial t}\right)$ $\frac{\partial^2 E}{\partial x^2} = -\frac{\partial}{\partial t}\left(\frac{\partial B}{\partial x}\right)$
Substitute Equation 34.15 for $\partial B/\partial x$. This eliminates B and gives us an equation that involves only E.	$\frac{\partial^2 E}{\partial x^2} = -\frac{\partial}{\partial t}\left[-\mu_0\varepsilon_0\left(\frac{\partial E}{\partial t}\right)\right]$ $\frac{\partial^2 E}{\partial x^2} = \mu_0\varepsilon_0\frac{\partial^2 E}{\partial t^2}$ ✓ (34.7)
We can find a similar wave equation for the magnetic field using a similar trick (Problem 26).	$\frac{\partial^2 B}{\partial x^2} = \mu_0\varepsilon_0\frac{\partial^2 B}{\partial t^2}$ (34.8)

COMMENTS

Equation 34.7 is the **wave equation** for the electric field, a differential equation in which the second spatial derivative is proportional to the second time derivative. Likewise, Equation 34.8 is the wave equation for the magnetic field.

A general wave equation is given by Equation 17.33, where v is the propagation speed of the wave.	$\frac{\partial^2 y}{\partial x^2} = \frac{1}{v^2}\frac{\partial^2 y}{\partial t^2}$ (17.33)

Properties of Electromagnetic Waves

We have just shown that if we combine Ampère's law and Faraday's law, we come up with an electromagnetic wave—an electric field and a magnetic field that oscillate periodically as depicted in Figure 34.8. By comparing Equation 17.33 to either Equation 34.7 or 34.8, we find that the speed of the wave comes from the coefficient on the right side of the wave equation:

$$\frac{1}{v^2} = \mu_0\varepsilon_0$$

$$v = \frac{1}{\sqrt{\mu_0\varepsilon_0}} \tag{34.16}$$

The entire wave moves along the x axis (in a direction perpendicular to both fields) at a speed v given by Equation 34.16. Let's calculate a numerical value for the speed by substituting for the constants. We leave it as a homework problem (Problem 18) to show that the units work out to m/s.

$$v = \frac{1}{\sqrt{\mu_0\varepsilon_0}} = \frac{1}{\sqrt{\left(4\pi \times 10^{-7}\frac{\text{T}\cdot\text{m}}{\text{A}}\right)\left(8.854 \times 10^{-12}\frac{\text{C}^2}{\text{N}\cdot\text{m}^2}\right)}} = 2.998 \times 10^8\,\text{m/s}$$

$$v = \frac{1}{\sqrt{\mu_0\varepsilon_0}} = c \tag{34.17}$$

The value is known as the **speed of light** c.

Because the electromagnetic wave equations (Eqs. 34.7 and 34.8) are mathematically identical to the corresponding equation for a wave on a string (Eq. 17.33), the solutions to electromagnetic wave equations are the same as the solution to the transverse mechanical wave equation $y(x, t) = y_{max} \sin(kx - \omega t)$ (Eq. 17.4) with a simple translation of variables:

$$E(x, t) = E_{max} \sin(kx - \omega t) \quad (34.18)$$

$$B(x, t) = B_{max} \sin(kx - \omega t) \quad (34.19)$$

As with the oscillation of beads on a string, both fields ($\vec{E}$ and $\vec{B}$) are perpendicular to the direction of propagation, so this is a **transverse wave.** But now, instead of beads on a string oscillating up and down, electric and magnetic fields oscillate in both magnitude and direction (Fig. 34.8). The electric field is parallel to the y axis. It grows and shrinks periodically just as a bead on a string moves up and down along the y axis. The magnetic field is parallel to the z axis and also grows and shrinks periodically. Compare Equations 34.18 and 34.19; the speed v, wave number k, and angular frequency ω are the same for the electric and magnetic field oscillations, so they have the same frequency f and period T. Like a mechanical wave, the speed, frequency, angular frequency, angular wave number, and wavelength are connected by

TRANSVERSE ELECTRIC AND MAGNETIC WAVE ✪ Major Concept

$$v = \frac{\omega}{k} = \lambda f = c \quad (34.20)$$

Table 34.1 summarizes what we know about electromagnetic waves by comparing them to transverse waves on a string.

TABLE 34.1 Comparison between transverse wave on a string and electromagnetic wave.

	Transverse wave on a string	Linearly polarized electromagnetic wave in a vacuum
Snapshot—a picture of the wave taken at one instant	Figure 17.8A, page 491	Figure 34.8, page 1095
Wave equation(s)	$\frac{\partial^2 y}{\partial x^2} = \frac{1}{v^2}\frac{\partial^2 y}{\partial t^2}$ (17.33)	$\frac{\partial^2 E}{\partial x^2} = \mu_0\varepsilon_0 \frac{\partial^2 E}{\partial t^2}$ (34.7) and $\frac{\partial^2 B}{\partial x^2} = \mu_0\varepsilon_0 \frac{\partial^2 B}{\partial t^2}$ (34.8)
Solution to wave equation(s)	$y(x, t) = y_{max} \sin(kx - \omega t)$ (17.4) The string oscillates along the y direction, while the wave moves in the positive x direction.	$\vec{E}(x, t) = [E_{max} \sin(kx - \omega t)]\hat{\jmath}$ (34.18) $\vec{B}(x, t) = [B_{max} \sin(kx - \omega t)]\hat{k}$ (34.19) The electric field oscillates along the y direction and the magnetic field oscillates along the z direction, while the wave moves in the positive x direction.
Propagation speed—speed of the wave in the x direction	$v = \frac{\omega}{k} = \lambda f$ (17.7) and (17.8)	$v = \frac{\omega}{k} = \lambda f = c$ (34.20)

The Connection Between $\vec{E}$ and $\vec{B}$

Consider the wave on the string in Figure 17.8A (page 491). In that case, the string is connected to an oscillator. The amplitude y_{max} of the wave depends on the amplitude of the oscillator. Similarly, the amplitude E_{max} of the electric field depends on the maximum electric field produced by the transmitter. That same transmitter creates both an electric and a magnetic field, so there is a connection between the amplitude of the electric part of the wave (E_{max}) and the magnetic part (B_{max}).

Here's how we can find that connection. Start by taking the partial derivative of E (Eq. 34.18) with respect to x while holding t constant:

$$\frac{\partial E(x, t)}{\partial x} = \frac{\partial}{\partial x} E_{max} \sin(kx - \omega t) = kE_{max} \cos(kx - \omega t) \quad (34.21)$$

Next, take the partial derivative of B (Eq 34.19) with respect to t while holding x constant:

$$\frac{\partial B(x, t)}{\partial t} = \frac{\partial}{\partial t} B_{\max} \sin(kx - \omega t) = -\omega B_{\max} \cos(kx - \omega t) \quad (34.22)$$

Now substitute Equations 34.21 and 34.22 into $(\partial E/\partial x) = -(\partial B/\partial t)$ (Eq. 34.11):

$$kE_{\max} \cos(kx - \omega t) = -[-\omega B_{\max} \cos(kx - \omega t)] = \omega B_{\max} \cos(kx - \omega t)$$

Solve for $E_{\max}$:

$$kE_{\max} = \omega B_{\max}$$

$$E_{\max} = \frac{\omega}{k} B_{\max}$$

Use $\omega/k = c$ (Eq. 34.20) to eliminate the angular frequency and angular wave number:

$$E_{\max} = cB_{\max} \quad (34.23)$$

Equation 34.23 is the connection between the amplitude of the electric field and that of the magnetic field. Inserting Equation 34.23 into Equation 34.18 gives

$$E(x, t) = c[B_{\max} \sin(kx - \omega t)]$$

The term in square brackets is the magnetic field (Eq. 34.19), so

$$E(x, t) = cB(x, t) \quad (34.24)$$

At every moment, the magnitude of the electric field is c times the magnitude of the magnetic field. Although c is a very large number, the electric and magnetic fields are measured in different units, so you cannot conclude that the electric field is stronger than the magnetic field. Both fields contribute equally to the energy of the electromagnetic wave.

Let's look at the energy density stored in each field to show that they are equal. The energy density stored in an electric field in a vacuum ($\kappa = 1$) is $u_E = \frac{1}{2}\varepsilon_0 E^2$ (Eq. 27.22). Substitute $E = cB$ (Eq. 34.24):

$$u_E = \frac{1}{2}\varepsilon_0 c^2 B^2 \quad (34.25)$$

Now, we use the propagation speed $c = 1/\sqrt{\mu_0 \varepsilon_0}$ (Eq. 34.17):

$$u_E = \frac{1}{2}\varepsilon_0 \frac{1}{\mu_0 \varepsilon_0} B^2 = \frac{1}{2}\frac{B^2}{\mu_0}$$

This is the same as the energy density stored in the magnetic field, $u_B = B^2/2\mu_0$ (Eq. 33.17):

$$u_E = u_B \quad (34.26)$$

So, the electric and magnetic fields contribute an equal amount of energy to an electromagnetic wave.

CONCEPT EXERCISE 34.3

The electric part of an electromagnetic wave is given by $E(x, t) = 0.75 \sin(0.30x - \omega t)$ V/m in SI units.

a. What are the amplitudes $E_{\max}$ and $B_{\max}$?
b. What are the angular wave number and the wavelength?
c. What is the propagation velocity?
d. What are the angular frequency, frequency, and period?

34-5 The Electromagnetic Spectrum

In 1864, in front of the Royal Society of London, Maxwell declared that light and other forms of radiation are electromagnetic disturbances in the form of waves that propagate according to the laws of electricity and magnetism. In Section 34-4, we started with Maxwell's equations for electricity and magnetism and derived an equation for an electromagnetic wave. We showed that the wave moves at the speed of light.

The terms *electromagnetic wave* and *electromagnetic radiation* are interchangeable, but from now on we save the term *visible light* for electromagnetic waves that healthy human eyes can sense.

The frequency and wavelength of any electromagnetic wave in a vacuum are given by $\lambda f = c$ (Eq. 34.20). The term **electromagnetic spectrum** refers to the continuum of electromagnetic waves arranged in order by frequency (and wavelength). In practice, the electromagnetic spectrum is divided into a few frequency bands. These bands blend together; there is no strict boundary between bands, and the terms given to the bands often vary from one branch of science and technology to the next. The electromagnetic spectrum and bands shown in Figure 34.11 and listed in Table 34.2 are commonly used in introductory physics. (You may find slight discrepancies when you look at other sources.)

Unless we specify otherwise, assume the electromagnetic wave is in a vacuum with propagation speed *c*.

All electromagnetic waves are produced by accelerating charged particles. Hertz's experiment is just one example in which electrons oscillating in an *LC* circuit create an electromagnetic wave. The bands in the electromagnetic spectrum are convenient because they correspond to how the charged particles are accelerated and to methods for detecting those waves. We will briefly describe the bands listed in Table 34.2.

TABLE 34.2 The electromagnetic spectrum broken into convenient bands. Values are approximate.

Name of band	Wavelength λ (m)	Frequency f (Hz)
Radio	$> 10^{-2}$	$< 10^{11}$
Microwave	10^{-4}–1	10^{9}–10^{13}
Infrared (IR)	10^{-6}–10^{-4}	10^{12}–10^{14}
Visible light	10^{-7}–10^{-6}	10^{14}–10^{15}
Ultraviolet (UV)	10^{-9}–10^{-7}	10^{15}–10^{18}
X-rays	10^{-12}–10^{-9}	10^{17}–10^{20}
Gamma rays	$< 10^{-10}$	$> 10^{19}$

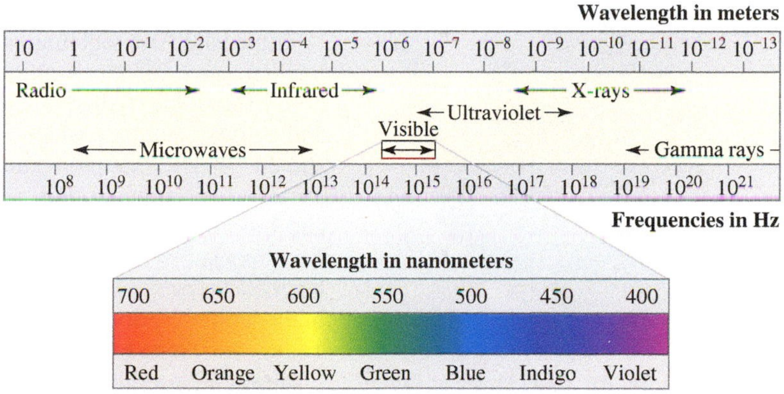

FIGURE 34.11 The electromagnetic spectrum is separated into bands for convenience. The bands overlap and do not have strict boundaries. It may help to memorize the visible colors in order from long to short wavelength by thinking of the name "Roy G. Biv," which stands for Red, Orange, Yellow, Green, Blue, Indigo, and Violet. (Values here may differ slightly from those in Table 34.2.)

Radio waves are generated by the acceleration of electrons in a circuit and are used for communication. In practice, the electrons oscillate in the antenna of a transmitter (Fig. 34.12). Radio waves are also produced by astronomical sources. For example, neutron stars produce radio waves when high-speed electrons are accelerated by very strong magnetic fields (Table 30.1). The resulting radiation covers many parts of the electromagnetic spectrum, but it is often easiest to detect in the radio band. A radio telescope receiver is much like the circuits that detect artificial radio waves.

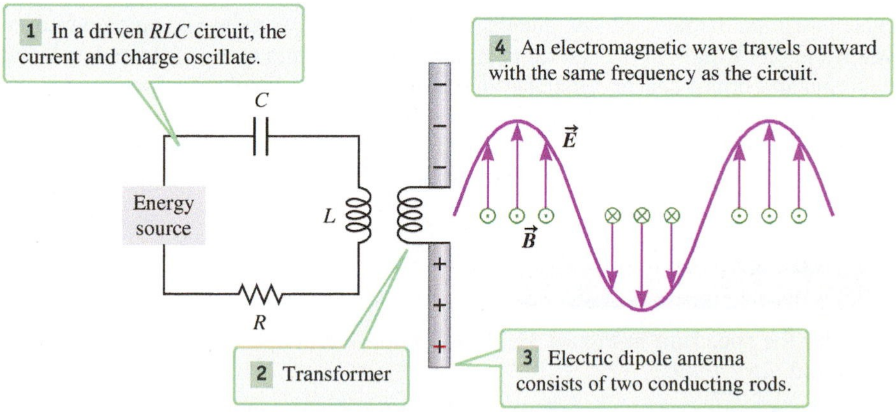

FIGURE 34.12 The basic design of a radio transmitter. **1** An *LC* circuit is driven to compensate for energy loss due to internal resistance and radiation. **2** The circuit is coupled to an antenna by a transformer, causing charge to oscillate in the antenna. **3** The current in the antenna oscillates. Effectively, the antenna has a dipole moment that oscillates in magnitude and direction. The dipole's electric field also oscillates. The changing current creates an oscillating magnetic field. **4** The changing electric and magnetic fields travel out into space as an electromagnetic wave.

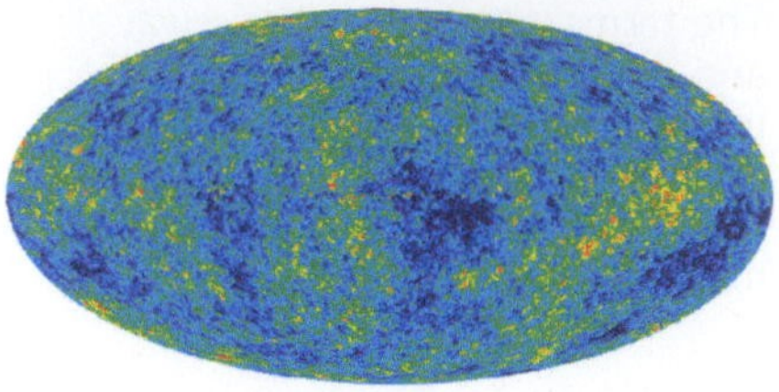

NASA

FIGURE 34.13 Microwave radiation detected from the early Universe. The image is a map of the whole sky. The wavelengths of the microwaves are used to determine temperature. The red regions are slightly hotter than the blue regions.

Microwaves are also generated by electronic devices and by natural sources. Their production and detection are similar to those of radio waves, and both wave bands are used for communication. Microwaves are better for certain applications such as **radar** (*ra*dio *d*etection *a*nd *r*anging), which was developed during World War II to detect aircraft and ships. Microwaves are transmitted, reflected by objects such as aircraft, and then detected by the radar receiver. In another common application, microwaves are used to cook food because they are absorbed by water and fat. One of the most important sources of microwaves is the Universe (Fig. 34.13). Early in the formation of the Universe, light was generated, and we observe that radiation today as microwaves that have a wavelength of about 1 mm coming from all directions. By studying this radiation, cosmologists can measure the age of the Universe and predict its fate.

Infrared radiation is produced by the motion within molecules when they vibrate and rotate. We often associate infrared radiation with a warm object such as a stovetop. The temperature of the object is a measure of its thermal energy. When you hold your hand near the stovetop, infrared radiation causes the molecules in your hand to move more quickly (Section 21-10), so your hand feels warmer.

Visible light is produced by the motion of electrons in atoms and molecules. Electrons in atoms orbit the nucleus at different distances. When an electron moves from a high orbit to a lower orbit, visible light is given off. Human eyes have two types of detectors—known as cones and rods—to detect visible light. Most of the Sun's electromagnetic radiation is in the form of visible light, and our eyes are sensitive to this radiation band because humans evolved on the Earth's surface. White light is made up of all the colors of the rainbow. Red light has the longest wavelength; violet has the shortest.

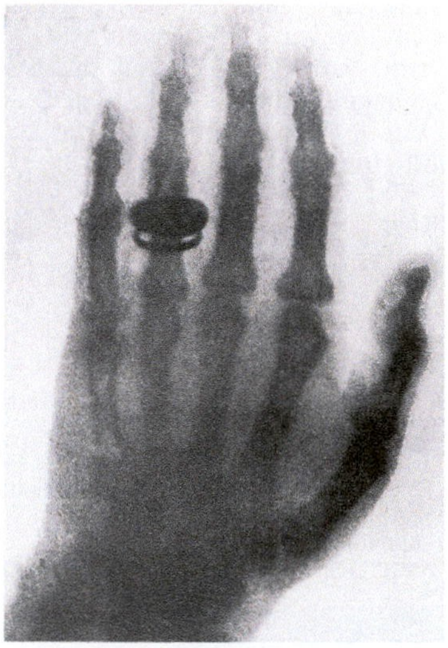

Hulton Archive/Getty Images

FIGURE 34.14 An X-ray image of Bertha Röntgen's hand.

Ultraviolet (UV) radiation is produced in much the same way as visible light. Very hot stars give off much ultraviolet radiation (as does the Sun, but not as much as hotter stars). UV radiation causes human skin to tan or burn, promotes skin cancer, and causes cataracts, which cloud our vision. Clothing, sunscreen, and sunglasses should be used for protection. In addition, ozone (O_3) in the Earth's atmosphere absorbs UV radiation, but ozone is destroyed by chemical reactions with chlorine and bromine. Chlorofluorocarbons (CFCs), formerly present in refrigerants and in many consumer products, are a reservoir for chlorine. In the mid-1980s, the ozone layer was greatly depleted and a hole was discovered over Antarctica. Every country in the United Nations agreed to phase out the use of CFCs to protect the ozone layer. Today, the ozone layer has been greatly restored.

X-rays are produced when high-energy electrons bombard metal so that the metal stops the electrons. Because the electrons decelerate, electromagnetic radiation is created. X-rays can penetrate soft tissues in the body but not bone. The German physicist Wilhelm Conrad Röntgen won the first Nobel Prize in physics in 1901 for his discovery of X-rays. One of the first X-ray photos he took was of his wife Bertha's hand (Fig. 34.14). Her bones and ring are clearly visible. Today, X-rays are used both as a medical diagnostic tool and as a therapy for cancer.

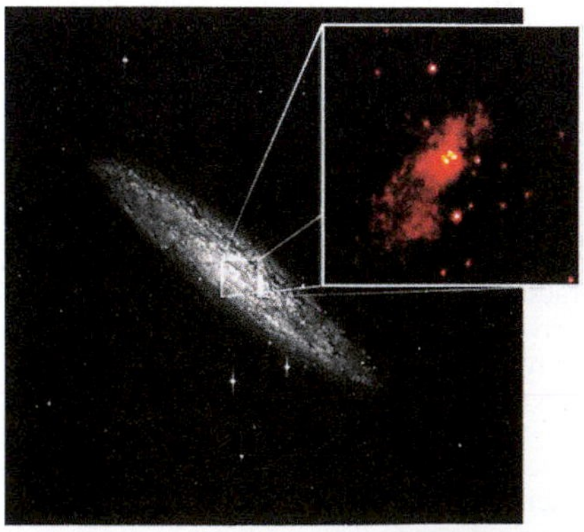

AFP/Newscom

FIGURE 34.15 Chandra, a space-based telescope, took this X-ray image (*inset*) of the central region of a starburst galaxy (NGC 253). The X-ray image may be evidence of intermediate-mass black holes in the galaxy.

Gamma rays are given off during certain nuclear reactions, such as fusion reactions in the Sun's core and fission reactions on the Earth. Gamma rays can cause significant biological damage, so—like X-rays—gamma rays can be used to destroy cancerous cells. However, gamma rays can also be very dangerous. Exposure to gamma rays can cause intestinal damage, cancer, and death. In the 1930s, women who painted radium watch and clock dials were exposed to gamma radiation, and many of them developed leukemia and breast cancer. Many survivors of the atomic bombs dropped over Hiroshima and Nagasaki also developed cancer. Both X-rays and gamma rays are produced by astronomical sources (Fig. 34.15). This radiation does not penetrate the Earth's atmosphere, so X-ray and gamma-ray detectors are put into orbit.

CONCEPT EXERCISE 34.4

Consider the spectrum in Figure 34.11 and propose an explanation of the terms in*fra*red and ultra*violet*.

CONCEPT EXERCISE 34.5

The most abundant element in the Universe is hydrogen (the subject of the case study in Chapter 42). Neutral hydrogen gives off electromagnetic waves with a wavelength of 21 cm. What is the frequency of this radiation? In what part of the electromagnetic spectrum is it found?

34-6 Energy and Intensity

Like all waves, electromagnetic waves transport energy along the direction of propagation. In 1884, the British physicist John Henry Poynting (1852–1914) published a paper entitled *On the Transfer of Energy in the Electromagnetic Field.* In this paper, he showed that the flow of energy at a point can be expressed in terms of the electric and magnetic fields at that point. Mathematically, the energy transferred per unit time is given by the magnitude of the **Poynting vector** $\vec{S}$:

ENERGY TRANSFERRED BY ELECTROMAGNETIC WAVES

Major Concept

$$\vec{S} \equiv \frac{1}{\mu_0}(\vec{E} \times \vec{B}) \tag{34.27}$$

The Poynting vector "points" in the direction of energy transport.

The magnitude S of the Poynting vector is the **power flux**, the energy transported per unit time per unit area (which is the same as the power per unit area) of a surface perpendicular to the direction of the Poynting vector. The SI units for power flux are W/m^2. The power flux depends on the position x and time t because the electric and magnetic fields depend on x and t.

To make this idea more concrete, let's use Equations 34.18 and 34.19 for a plane wave (Fig. 34.8) to find an expression for the Poynting vector. First, calculate the cross product:

$$(\vec{E} \times \vec{B}) = [E_{max} \sin(kx - \omega t)]\hat{\jmath} \times [B_{max} \sin(kx - \omega t)]\hat{k}$$

$$(\vec{E} \times \vec{B}) = [E_{max} B_{max} \sin^2(kx - \omega t)]\hat{\jmath} \times \hat{k}$$

Because $\hat{\jmath} \times \hat{k} = \hat{\imath}$, the Poynting vector is

$$\vec{S}(x, t) = \frac{1}{\mu_0}[E_{max} B_{max} \sin^2(kx - \omega t)]\hat{\imath} \tag{34.28}$$

The magnitude S of the Poynting vector at one instant is plotted in Figure 34.16. Compare the power flux S to the electric and magnetic fields in the wave at the same instant. At points on the x axis where the electric and magnetic fields are at their maxima, the power flux is at its maximum. At points where the electric and magnetic fields are zero, the power flux is also zero. The power flux is never negative, and when the electric and magnetic fields are pointing in the negative direction, the power flux is still positive, so the Poynting *vector* still points in the positive x direction.

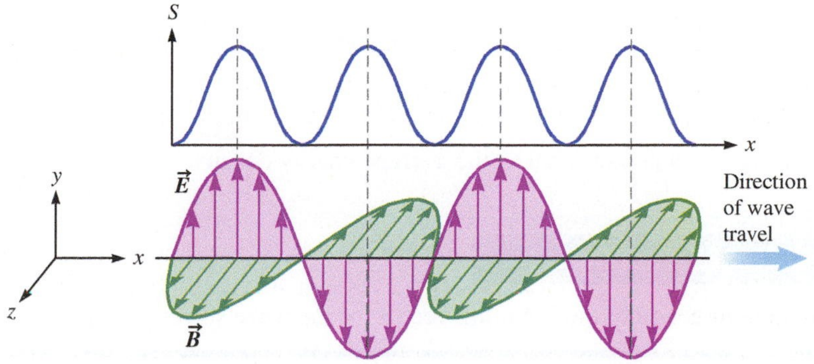

FIGURE 34.16 The power flux depends on time and position. This is the power flux at one particular time. The power flux is greatest when the fields are at their maxima. The direction of the fields does not affect the magnitude of the power flux.

The Poynting vector tells us about the power flux as a function of time and position. However, we are usually more interested in the intensity (power per unit area) of the radiation. From Section 17-7, the intensity I is the time-averaged power per unit area (flux):

$$I \equiv S_{av}$$

The average of $\sin^2(kx - \omega t)$ over a period is $\frac{1}{2}$ (Problem 76), so the intensity I is

$$I = \frac{1}{2\mu_0} E_{max} B_{max} \tag{34.29}$$

It is often convenient to express the intensity in terms of either E_{max} or B_{max} by using $E = cB$ (Eq. 34.24):

INTENSITY ✪ Major Concept

$$I = \frac{E_{max}^2}{2\mu_0 c} = \frac{c}{2\mu_0} B_{max}^2 \tag{34.30}$$

Recall that the intensity from a point source such as a distant lightbulb, radio antenna, or galaxy decreases as the distance r from the source increases:

$$I = \frac{P_{av}}{4\pi r^2} \tag{17.23}$$

where P_{av} is the average power emitted by the source.

Intensity is also related to the average energy density stored by the electric and magnetic fields, as we can show by starting with an expression for the total energy density:

$$u = u_E + u_B$$

The energy density stored by the electric field is equal to the energy density stored by the magnetic field, $u_E = u_B$ (Eq. 34.26), so the total energy density is

$$u = 2u_E = 2u_B$$

We can write the energy density in terms of the electric field or the magnetic field. Arbitrarily choosing the magnetic field, we substitute $u_B = B^2/2\mu_0$ (Eq. 33.17):

$$u = 2\left(\frac{B^2}{2\mu_0}\right) = \frac{B^2}{\mu_0}$$

Next, we substitute $B(x, t) = B_{max} \sin(kx - \omega t)$ (Eq. 34.19):

$$u = \frac{1}{\mu_0}\left[B_{max}^2 \sin^2(kx - \omega t)\right]$$

The energy density u depends on time. We need the energy density averaged over one or more periods and, as before, the average of $\sin^2(kx - \omega t)$ is $\frac{1}{2}$. So the average energy density is

$$u_{av} = \frac{1}{2\mu_0} B_{max}^2 \tag{34.31}$$

If we had worked in terms of the electric field, we would have found

$$u_{av} = \frac{1}{2}\varepsilon_0 E_{max}^2 = \frac{1}{2\mu_0 c^2} E_{max}^2 \tag{34.32}$$

Substitute Equation 34.31 or 34.32 into Equation 34.30, and the result is

$$I = cu_{av} \tag{34.33}$$

So the intensity is proportional to the average energy density.

CONCEPT EXERCISE 34.6

What is the direction of the Poynting vector for the wave shown in Figure 34.8?

EXAMPLE 34.2 Buy More Sunscreen

A salesperson once suggested that the author of this book should wear sunscreen on her face every day whether she planned to be out in the Sun or not. The salesperson argued that indoor lighting can cause as much damage as the Sun to a person with fair skin. Let's analyze this claim.

The author does have fair skin, and on a sunny day she will burn in about 20 minutes without protection. The author's face has an area of roughly $2 \times 10^{-2}\ \text{m}^2$. The Sun's power output is 3.85×10^{26} W. Estimate the amount of energy from the Sun that the author's face is exposed to in 20 minutes. While writing this textbook, the author sits about 2 m from a 150-W lightbulb. Model the lightbulb as a point source, and estimate the time the author would need to sit near the bulb in order to absorb as much energy from the bulb as she does from the Sun in 20 minutes.

INTERPRET and ANTICIPATE

Our first step is to find the intensity of sunlight on Earth. Once we know that, we can find the power delivered to the author's face by multiplying the intensity by the area of the face. Finally, find the energy delivered in the 20-minute period by multiplying the power by the time of exposure. This will tell us how much energy her face must be exposed to in order to burn. Then find the intensity of the lightbulb's radiation at her face and the power her face is exposed to from that bulb. To find the time required in the problem, divide the energy her face is exposed to in 20 minutes of sunlight by the power delivered by the bulb. We expect the answer will be greater than 20 minutes.

SOLVE

Find the intensity of sunlight at the position of the Earth using Equation 17.23. The distance between the Earth and the Sun is $1\ \text{AU} = 1.50 \times 10^{11}$ m.

$$I = \frac{P_{av}}{4\pi r^2} \quad (17.23)$$

$$I = \frac{(3.85 \times 10^{26}\,\text{W})}{4\pi(1.50 \times 10^{11}\ \text{m})^2}$$

$$I = 1360\ \text{W/m}^2$$

The power P delivered to the author's face is the intensity times the area of her face. We retain an extra significant figure for now.

$$P = IA \quad (34.34)$$

$$P = (1360\ \text{W/m}^2)(2 \times 10^{-2}\text{m}^2)$$

$$P = 27\ \text{W}$$

The energy (heat) Q delivered to her face in 20 minutes (1200 s) is the power times the time of exposure. (Q stands for energy here, not charge.)

$$Q = (27\ \text{W})(1200\ \text{s})$$

$$Q = 3 \times 10^4\ \text{J} \quad (1)$$

Now we turn our attention to the 150-W lightbulb. Find its intensity at a distance of 2 m as given by Equation 17.23.

$$I = \frac{P_{av}}{4\pi r^2} \quad (17.23)$$

$$I = \frac{(150\,\text{W})}{4\pi(2\ \text{m})^2} = 3\ \text{W/m}^2$$

Again, the power P delivered to the author's face is the intensity times her face's area.

$$P = IA = (3\ \text{W/m}^2)(2 \times 10^{-2}\text{m}^2)$$

$$P = 0.06\ \text{W} \quad (2)$$

The time for her face to be exposed to as much energy from the lightbulb as it is in a 20-minute sunbath is the energy Q (Eq. 1) divided by the power P (Eq. 2).

$$\Delta t = \frac{Q}{P} = \frac{3 \times 10^4\ \text{J}}{0.06\ \text{J/s}}$$

$$\Delta t = 5 \times 10^5\ \text{s} = 140\ \text{h}$$

CHECK and THINK

It takes about $3\frac{1}{2}$ workweeks to be exposed to as much energy from the 150-W bulb as in just 20 minutes in sunlight. Of course, during those workweeks, a person spends about 8 hours per day in total darkness. Clearly, exposure to sunlight is much more hazardous than exposure to normal indoor lighting. Nevertheless, over the course of a lifetime, a person's face is exposed to a great deal of electromagnetic energy.

EXAMPLE 34.3 CASE STUDY Meteor Estimate Revisited

In Chapter 9's case study, three physics professors estimated the size of a meteor. They worked in the dark without a calculator, no steps were written down, and their estimate was made to one significant figure. Using the conservation of energy approach (Example 9.12, page 272), they found a radius $R_{\text{mtr}} = 0.3$ mm. Now consider another approach based on light from the meteor. All three professors agree that the meteor looks about as bright as an ordinary faint star. Professor Black knows that if the Sun were about 30 light-years from the Earth, it would look like an ordinary star. Professor Noir read that meteors are visible when they are about 100 km above us. They all know that the Sun's radius is about 110 times the Earth's radius, or about $R_{\odot} \approx 6.6 \times 10^5$ km. Finally, they know that a meteor gets hot when it travels through the Earth's atmosphere. They assume it gets about as hot as the Sun's surface. They model both the Sun and the meteor as black bodies. Use their values and observations to estimate the radius of a meteor.

INTERPRET and ANTICIPATE

This problem brings together concepts from many chapters, including some ideas from Chapter 21. We expect our estimate to be consistent with the estimate from Chapter 9 based on conservation of energy.

SOLVE

The meteor is the same temperature as the Sun's surface. From Equation 21.34, the power P of radiation given off by an object is proportional to the surface area of the object and its temperature to the fourth power. Because the meteor and the Sun are assumed to be the same temperature, the ratio of their emitted powers depends only on their radii. (Because they are modeled as good emitters, both objects have $\varepsilon \approx 1$; see Section 21-10.)

$$P = \frac{Q}{\Delta t} = 4\pi R^2 \varepsilon \sigma T^4 \tag{21.34}$$

$$P_{\odot} = 4\pi R_{\odot}^2 \varepsilon \sigma T_{\odot}^4$$

$$P_{\text{mtr}} = 4\pi R_{\text{mtr}}^2 \varepsilon \sigma T_{\odot}^4$$

$$\frac{P_{\text{mtr}}}{P_{\odot}} = \frac{R_{\text{mtr}}^2}{R_{\odot}^2} \tag{1}$$

The professors observed that the meteors look as bright as ordinary stars. Professor Black knows that the Sun would look like an ordinary star if it were 30 ly from us. So the intensity of the meteor at altitude $r_{\text{mtr}} = 100$ km equals the intensity the Sun would have if it were $r_{\odot} = 30$ ly away. Use $I = P_{\text{av}}/4\pi r^2$ (Eq. 17.23) with $r_{\odot} = 30$ ly and $r_{\text{mtr}} = 100$ km.

$$\frac{P_{\odot}}{4\pi r_{\odot}^2} = \frac{P_{\text{mtr}}}{4\pi r_{\text{mtr}}^2} \tag{2}$$

Solve Equations (1) and (2) for the radius of the meteor, R_{mtr}.

$$\frac{P_{\text{mtr}}}{P_{\odot}} = \frac{4\pi r_{\text{mtr}}^2}{4\pi r_{\odot}^2} \qquad \frac{R_{\text{mtr}}^2}{R_{\odot}^2} = \frac{r_{\text{mtr}}^2}{r_{\odot}^2}$$

$$R_{\text{mtr}} = \frac{r_{\text{mtr}}}{r_{\odot}} R_{\odot} \tag{3}$$

To convert 30 ly to kilometers, remember that $c = 3 \times 10^5$ km/s and there are roughly 3×10^7 s in a year. Substitute numerical values into Equation (3), including $r_{\odot}$ in kilometers.

$$r_{\odot} = 30 \text{ ly} \approx 30 \text{ yr} \times \frac{3 \times 10^5 \text{ km}}{\text{s}} \times \frac{3 \times 10^7 \text{ s}}{\text{yr}}$$

$$r_{\odot} \approx 3 \times 10^{14} \text{ km}$$

$$R_{\text{mtr}} = \left(\frac{100 \text{ km}}{3 \times 10^{14} \text{ km}}\right)(6.6 \times 10^5 \text{ km})$$

$$R_{\text{mtr}} = 2.2 \times 10^{-7} \text{ km} \approx 0.2 \text{ mm}$$

CHECK and THINK

This result is very close to the estimate from Chapter 9. This example illustrates how physicists model a complicated problem to make an estimate. It also shows the importance of knowing a couple of numerical facts. Whenever you must make a detailed calculation, start with an estimate as a way to check your answer. This is particularly important if you need to calculate a value that is completely unknown.

34-7 Momentum and Radiation Pressure

In Chapter 17, we mentioned that waves transport momentum and we developed equations for wave pressure. Electromagnetic waves also transport momentum and exert pressure. We are familiar with the pressure exerted by a sound wave on the eardrum, but the pressure exerted by light on our skin may not be so obvious. In this section, we estimate the (very small) pressure exerted by sunlight on skin. The pressure exerted on a large solar sail attached to a deadly asteroid may be high enough to save humanity.

First, let's derive an expression for the momentum an electromagnetic wave imparts to a single particle; our results can be extended to any object.

DERIVATION Momentum Delivered by an Electromagnetic Wave

Consider a simple situation in which a single positively charged particle (charge $+q$) is at rest when an electromagnetic wave strikes it, delivering energy Q. The wave is linearly polarized as shown in Figure 34.17. We will show that the momentum p delivered to the particle is proportional to the energy Q:

$$p = Q/c \qquad (34.35)$$

MOMENTUM TRANSFERRED BY (ABSORBED) ELECTROMAGNETIC WAVE ✪ **Major Concept**

Don't confuse q for charge with Q for energy.

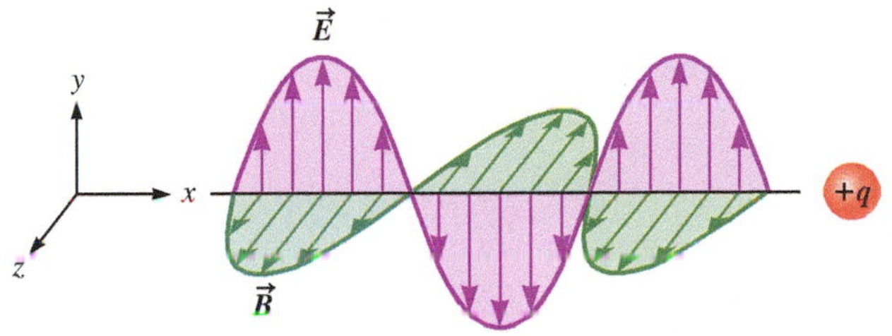

FIGURE 34.17 An electromagnetic wave traveling to the right encounters a single positively charged particle.

Consider one instant when the electric field $\vec{E}$ at the particle's location is in the positive y direction and the magnetic field $\vec{B}$ is in the positive z direction. The electric force on the particle is then in the positive y direction as given by Equation 24.2.	$\vec{F}_E = qE\hat{\jmath}$ (24.2)
This force causes the particle of mass m to accelerate (from rest) in the positive y direction.	$a_y = \dfrac{qE}{m}$
To find the particle's speed, assume it has been in contact with the wave for a short time t, and during that time, the electric force on it is constant.	$v_y = a_y t = \dfrac{qE}{m}t$ (34.36)
Because the particle is moving in the positive y direction, the magnetic field exerts a force on it in the positive x direction according to the right-hand rule for cross products (Eq. 30.17). The velocity is perpendicular to the magnetic field and its magnitude is given by Equation 34.36.	$\vec{F}_B = q(\vec{v} \times \vec{B})$ (30.17) $\vec{F}_B = qv_yB\hat{\imath} = \dfrac{q^2E}{m}tB\hat{\imath}$
The magnetic force is in the direction of the wave's propagation, and we can think of the wave as delivering momentum in the x direction. To find the momentum p_x, combine Equations 11.3 and 11.4 (page 309). The particle is initially at rest, so its change in momentum is just its final momentum in the x direction, p_x.	$\Delta\vec{p} = \int_0^t \vec{F}_B\,dt$ $(p_f - p_i)_x = \dfrac{q^2EB}{m}\int_0^t t\,dt = \dfrac{q^2EB}{2m}t^2$
Use $E = cB$ (Eq. 34.24) to eliminate B.	$p_x = \dfrac{q^2E^2}{2mc}t^2$

Derivation continues on page 1108 ▶

Substitute Equation 34.36, which is v_y. The particle started from rest, so the term $\frac{1}{2}mv_y^2$ is the increase in its kinetic energy.	$p_x = \frac{m}{2c}\left(\frac{q^2E^2}{m^2}t^2\right) = \frac{1}{c}\left(\frac{1}{2}mv_y^2\right)$
So, the momentum p delivered to the particle is proportional to the energy Q delivered to the particle by the electromagnetic radiation.	$p = \frac{Q}{c}$ ✓ (34.35)

COMMENTS

Although we considered a very simple object—a single charged particle—Equation 34.35 is still valid for more complicated, extended objects.

Reflected Versus Absorbed Waves

Equation 34.35 applies when the electromagnetic wave is *totally absorbed* by an object. It is analogous to a sticky ball thrown at a vertical wall; the ball's momentum is transferred to the wall. Now imagine the electromagnetic radiation is reflected from the surface. This case is similar to a ball that bounces off a wall so that its final momentum has the same magnitude as its initial momentum: $\vec{p}_f = -\vec{p}_i$. In that case, the change in the ball's momentum is

$$\Delta\vec{p} = \vec{p}_f - \vec{p}_i$$

$$\Delta\vec{p} = -\vec{p}_i - \vec{p}_i = -2\vec{p}_i$$

So the momentum delivered to the wall is $2\vec{p}_i$. If an electromagnetic wave is completely reflected from an object, the momentum delivered to the object is

MOMENTUM TRANSFERRED BY (REFLECTED) ELECTROMAGNETIC WAVE Major Concept

$$p = 2\frac{Q}{c} \quad (34.37)$$

If the electromagnetic wave is partially absorbed by the object, the momentum delivered is somewhere between $p = Q/c$ and $p = 2Q/c$.

Radiation Pressure

To find the pressure exerted by electromagnetic radiation on an object, start with Newton's second law written in terms of momentum $\sum\vec{F} = d\vec{p}/dt$ (Eq. 10.2). In this case, the only force we are interested in is the force in the x direction as exerted by the electromagnetic wave shown in Figure 34.17. We start with a wave that has been totally absorbed, so its momentum is given by $p = Q/c$ (Eq. 34.35) and the force is

$$F_x = \frac{d}{dt}\left(\frac{Q}{c}\right) = \frac{1}{c}\frac{dQ}{dt}$$

The rate at which the energy Q is delivered is the power, which is the intensity times the cross-sectional area A (Eq. 34.34):

$$F_x = \frac{1}{c}\frac{dQ}{dt} = \frac{\text{power}}{c} = \frac{IA}{c}$$

The force divided by the area is the pressure P:

In Equation 34.38, uppercase P stands for pressure, not power; lowercase p still stands for momentum.

$$\frac{F_x}{A} = P = \frac{I}{c} \quad (34.38)$$

Equation 34.38 is the pressure exerted by electromagnetic radiation when it is absorbed by an object. If the radiation is completely reflected, the pressure is twice as great:

PRESSURE EXERTED BY ELECTROMAGNETIC WAVES Major Concept

$$P = 2\frac{I}{c} \quad (34.39)$$

If the radiation is partially absorbed by the object, the pressure exerted on the object is somewhere between $P = I/c$ and $P = 2I/c$.

EXAMPLE 34.4 That Doesn't Hurt a Bit

On a day at the beach, you orient your face so it points straight up toward the Sun. Assume the intensity of sunlight is 1000 W/m^2 and the area of your face is $A = 2 \times 10^{-2}$ m^2. Estimate the force exerted by sunlight on your face.

INTERPRET and ANTICIPATE

Because you have pointed your face toward the Sun, the force exerted by sunlight is perpendicular to your face. Find that force by multiplying the radiation pressure by the area of your face. We expect a small number because we don't normally feel this force.

SOLVE

Sunlight is partially reflected from your skin, so the pressure exerted by sunlight on your face is somewhere between $P = I/c$ and $P = 2I/c$. Because we are only making an estimate, let's assume the sunlight is totally reflected. This assumption gives the greatest possible force.

$$P = 2\frac{I}{c} \quad (34.39)$$

The force is the pressure times the area of your face. (We have retained an extra significant figure because the leading number is 1.)

$$F = PA = \left(2\frac{I}{c}\right)A = 2\left(\frac{1000 \text{ W/m}^2}{3.00 \times 10^8 \text{ m/s}}\right)(2 \times 10^{-2} \text{ m}^2)$$

$$F = 1.3 \times 10^{-7} \text{ N}$$

CHECK and THINK

As expected, this is a very small force—about equal to the weight of a couple of hairs on your head. So it is not surprising that we don't normally think about the force exerted by sunlight. In the next examples, we will show how this force can have an enormous effect.

EXAMPLE 34.5 CASE STUDY A Stellar Weight Loss Plan

Deep inside a star, nuclear fusion reactions generate huge amounts of power. The energy makes its way to the star's surface through convection and radiation, but in this problem, we consider only radiation. Radiation exerts outward pressure on the outer layers of a star, which are also pulled inward by the gravitational attraction of the inner layers. Normally, a star is in equilibrium, so the gravitational attraction is balanced by the outward pressure. However, there is a limit to the power a star can give off without blowing off its outer layers. More massive stars generate more power and run a greater risk of shedding material. Figure 34.18 shows the most massive star known in our Milky Way galaxy, Eta Carinae. Notice the bright spot near the middle of the image and the two enormous lobes of material that the star has blown off.

The maximum amount of power that an object can give off in the form of radiation without losing mass is called the **Eddington luminosity**. (*Luminosity* is another term for power.) The outer layer of stars is made up of mostly ionized hydrogen (protons). We'll model this hydrogen as particles that have an effective cross-sectional area given by $\sigma_T = 6.65 \times 10^{-29}$ m^2 and the mass of a hydrogen atom. (The light interacts strongly with the electrons, pushing them outward; the electrons attract the protons, dragging them outward as well.) Assume the radiation from lower layers is totally absorbed.

A Derive an expression for the Eddington luminosity L_{edd} for a star of mass M.

B Once you have that expression, calculate L_{edd} for the Sun and for a star with 100 times the mass of the Sun.

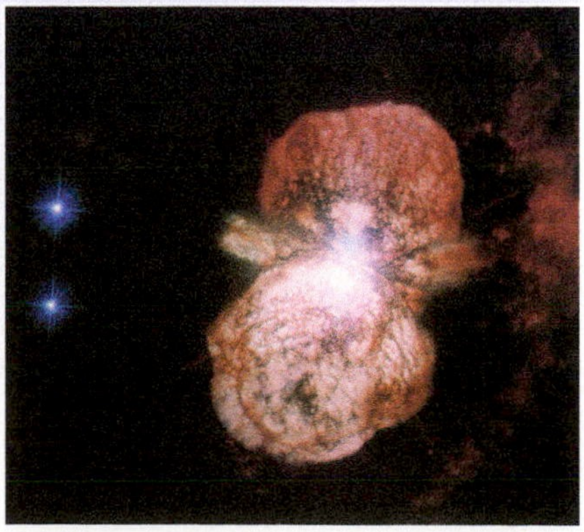

ESA/NASA Images

FIGURE 34.18 Eta Carinae may be the most massive star in our Milky Way galaxy. It gives off so much radiation that it is blowing off its outer layers.

Example continues on page 1110 ▶

INTERPRET and ANTICIPATE

To make this problem manageable, think about the forces exerted on a single particle in the outer layer of a star. Figure 34.19 shows a free-body diagram for that single particle. It is pulled toward the center of the star by gravity ($\vec{F}_G$) and pushed outward by radiation ($\vec{F}_{\text{rad}}$). At the Eddington limit, the particle does not accelerate and these two forces are balanced.

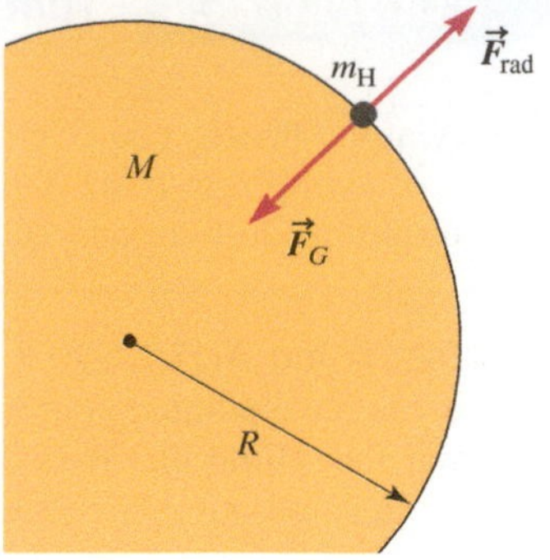

FIGURE 34.19

A SOLVE

The magnitude of the gravitational force on the particle equals the force due to radiation pressure.

$$F_G = F_{\text{rad}}$$

The gravitational force is given by Newton's law of universal gravity (Eq. 7.4). The force exerted by radiation is the pressure times the particle's cross-sectional area σ_T (Eq. 34.38). Here, the center-to-center distance in the law of universal gravity is the radius R of the star.

$$F_G = G\frac{Mm_H}{R^2} \tag{7.4}$$

$$G\frac{Mm_H}{R^2} = \left(\frac{I}{c}\right)\sigma_T \tag{1}$$

The intensity is the power divided by the surface area of the star. In this case, that power is the Eddington luminosity L_{edd}.

$$I = \frac{L_{\text{edd}}}{4\pi R^2}$$

Substitute I into Equation (1) and solve for L_{edd}.

$$G\frac{Mm_H}{R^2} = \frac{1}{c}\left(\frac{L_{\text{edd}}}{4\pi R^2}\right)\sigma_T$$

$$L_{\text{edd}} = \left(\frac{4\pi cGm_H}{\sigma_T}\right)M \tag{2}$$

CHECK and THINK

The term in parentheses in Equation (2) is a constant; only M changes as we consider different stars. Finding a value for that constant will be convenient for the calculations in part B and allows us to check the units. We find the constant term has dimensions of power per mass, so the expression for the Eddington luminosity has dimensions of power as expected.

$$\left(\frac{4\pi cGm_H}{\sigma_T}\right) = \frac{4\pi(3.00\times10^8\text{ m/s})(6.67\times10^{-11}\text{ N}\cdot\text{m}^2/\text{kg}^2)(1.67\times10^{-27}\text{ kg})}{6.65\times10^{-29}\text{ m}^2}$$

$$\left(\frac{4\pi cGm_H}{\sigma_T}\right) = 6.31\frac{\text{N}\cdot\text{m/s}}{\text{kg}} = 6.31\frac{\text{W}}{\text{kg}} \tag{3}$$

B SOLVE

For our calculations, it is convenient to convert from kilograms to solar mass units, $M_\odot$. Do this by multiplying Equation (3) by the mass of the Sun ($M_\odot = 1.98\times10^{30}$ kg).

$$L_{\text{edd}} = 1.25\times10^{31}M$$

(in watts when M is in solar mass units)

Now we are ready to calculate the Eddington luminosity for the Sun, $M = 1$.

$$L_{\text{edd}} = 1.25\times10^{31}\text{ W} \quad \text{(Sun)}$$

For the 100-solar-mass star, $M = 100$.

$$L_{\text{edd}} = 1.25\times10^{33}\text{ W}$$

(100-solar-mass star)

CHECK and THINK

The Eddington luminosity we found for the Sun is about 30,000 times greater than its actual luminosity ($L_\odot = 3.84 \times 10^{26}$ W), so the Sun's radiation is not blowing off its outer layers. Our Sun is stable (for now). However, the Eddington luminosity we found for the 100-solar-mass star is very close to the actual luminosity observed for such stars. The light given off by these stars is actually destroying them. They are blowing off their outer layers, as in the case of Eta Carinae (Fig. 34.18). Because the amount of radiation given off is greater for stars of greater mass, there is a limit to the ultimate mass of a star. Stellar masses are not greater than about 120 times the mass of the Sun because if a more massive star existed, its radiation would blow it apart. This limit was first theorized by Roberta Humphreys and Kris Davidson and is known as the *Humphreys-Davidson limit.*

EXAMPLE 34.6 CASE STUDY Solar Sailing Saves the Earth?

In 1950, an asteroid (1950DA) was discovered. It was observed again on December 31, 2000, and appears to be on a collision course with the Earth. The asteroid has about a 1-in-300 chance of colliding with us in about 800 years. Don't panic; we'll have a lot of time to figure out what we should do. One idea is to use a solar sail, harnessing the Sun's light to push the asteroid off course. Ideally, the sail would add little mass to the asteroid but would increase its area and the force due to the Sun's radiation. The mass of 1950DA is unknown, but its diameter is about 1 km. Suppose its density is 2500 kg/m^3, the same as the density of Eros (another near-Earth asteroid). Assume the only forces acting on 1950DA are the gravitational attraction of the Sun and the force from radiation on its solar sail. What is the minimum area of the sail needed to accelerate the asteroid away from the Sun? Because of uncertainties in the mass, estimate to just one significant figure.

INTERPRET and ANTICIPATE

This problem is similar to Example 34.5. The free-body diagram is the same; there is a gravitational force toward the Sun and a force due to radiation pressure away from the Sun. We want the acceleration to be away from the Sun, so the force due to radiation pressure must be greater than the gravitational force. If the acceleration is zero, the two forces have the same magnitude. Use this condition to set a lower limit to the size of the sail. If the sail is larger, the force due to radiation will be greater in magnitude than that due to gravity.

SOLVE

First, estimate the mass of 1950DA. Its exact shape does not matter, and we can say that the volume is roughly the diameter cubed. (Retain a couple extra significant figures to avoid rounding errors.)

$$V \sim d^3 \sim (1000\,\text{m})^3 = 10^9\,\text{m}^3$$

$$m = \rho V \sim (2500\,\text{kg/m}^3)(10^9\,\text{m}^3)$$

$$m \sim 2.5 \times 10^{12}\,\text{kg}$$

Consider the critical case in which the gravitational force is balanced by the radiation force. As in Example 34-5, the gravitational force is given by Equation 7.4 and the force due to radiation is the pressure times the area. In this case, the pressure is given by $P = 2I/c$ (Eq. 34.39) for a perfectly reflecting sail.

$$F_G = F_{\text{rad}}$$

$$G\frac{M_\odot m}{r^2} = 2\frac{I}{c}A \quad (1)$$

Find the intensity from $I = P_{\text{av}}/4\pi r^2$ (Eq. 17.23), where the average power is the Sun's luminosity.

$$I = \frac{L_\odot}{4\pi r^2} \quad (2)$$

Substitute Equation (2) into Equation (1) and solve for A, the area of the sail. Notice that the distance r cancels out.

$$G\frac{M_\odot m}{r^2} = \frac{2}{c}\left(\frac{L_\odot}{4\pi r^2}\right)A \qquad A = \frac{2\pi c G M_\odot m}{L_\odot}$$

$$A = \frac{2\pi(3.0 \times 10^8\,\text{m/s})(6.7 \times 10^{-11}\,\text{N}\cdot\text{m}^2/\text{kg}^2)(2 \times 10^{30}\,\text{kg})(2.5 \times 10^{12}\,\text{kg})}{4.0 \times 10^{26}\,\text{W}}$$

$$A = 2 \times 10^{15}\,\text{m}^2$$

Example continues on page 1112 ▶

CHECK and THINK

The area we just found is enormous (about four times the Earth's surface area), but our goal need not be to eject the asteroid completely from the solar system. In other words, we don't have to create a force that exceeds the Sun's gravitational force. Like other objects in the solar system, the asteroid is in orbit around the Sun; we only need to change that orbit so it doesn't cross the Earth's orbit. A more realistic plan is to send an impactor propelled by a solar sail on a collision course with the asteroid, change its orbit, and avoid a collision with the Earth (Problem 75). Of course, any real solution must take into account the gravitational force exerted by other objects such as the Earth, the Moon, and other planets. Fortunately, we have 35 generations to come up with a plan.

34-8 Polarization

Due to the reflections off the surface of the water, you cannot see into the pot in Figure 34.20A. However, if you hold a filter between your eye and the pot, you can see there is a stone under the water (Fig. 34.20B). Why does this work? In this section, we study another property of electromagnetic radiation—polarization—and find out how filters can be constructed based on this property.

A. No filter. **B.** Notice the hand holding a filter.

FIGURE 34.20 With the right filter, you can see into the water.

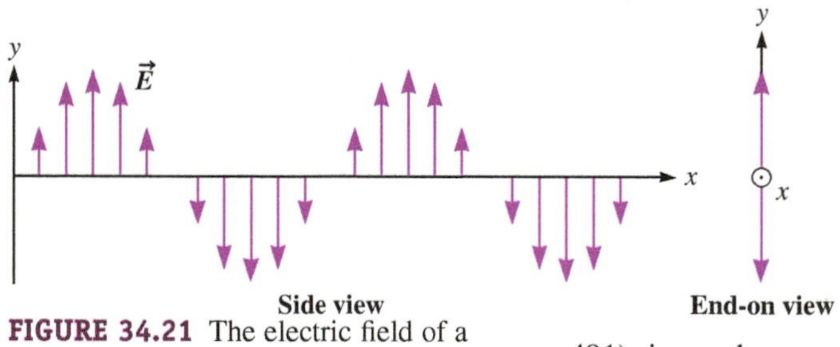

FIGURE 34.21 The electric field of a linearly polarized electromagnetic wave oscillates in one plane. In the end-on view, you see the electric field oscillating up and down along the y axis.

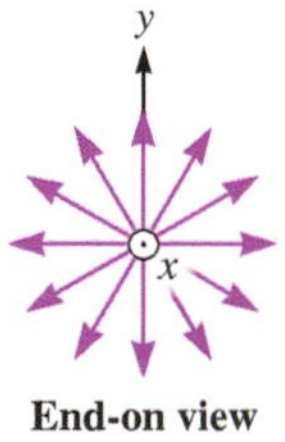

FIGURE 34.22 The electric field of unpolarized radiation oscillates in all planes perpendicular to the direction of propagation. In this end-on view, you see the electric field oscillating up and down along all radial directions.

Polarized Versus Unpolarized Radiation

First consider the electromagnetic wave in Figure 34.8. The electric field oscillates in the xy plane, and the magnetic field oscillates in the xz plane. For the rest of our discussion, we will focus on only the electric field; the magnetic field is always perpendicular to the electric field. Because the electric field oscillates in only one plane, we say the wave is **plane polarized.** A transverse wave on a string, such as in Figure 17.8A (page 491), is another example of a plane-polarized wave. The string oscillates in a single plane. Imagine watching one of the beads in Figure 17.8A from a point along the direction of propagation (an end-on view). You would see the bead oscillating up and down along a single line as in Figure 17.8B. So, we also say the wave is **linearly polarized.** The electromagnetic wave in Figure 34.21 is also linearly polarized because if you view the wave end-on, you see the electric field oscillating along a single line—the y axis.

The wave on the string in Figure 17.8A is linearly polarized because the oscillator that generated the wave oscillates up and down along a single line. In an ordinary radiation source such as a lightbulb, light is emitted by the motion of many electrons oscillating randomly. The net effect is that the electric field oscillates in randomly and rapidly changing planes. Such radiation is said to be **unpolarized** because it does not have a single plane of polarization. When we look at unpolarized radiation end-on, we see electric field vectors oscillating along all the lines perpendicular to the x axis (Fig. 34.22).

The oscillation of any one of these electric field vectors can be divided into two perpendicular components—the horizontal z component and the vertical y component. Consider one such vector $\vec{E}$ that oscillates along a line pointing from the lower left to the upper right. We divide this vector into its components at several times (Fig. 34.23). The components oscillate as a result of the electric field vector's oscillation. So, one way to think about unpolarized light is as the sum of light polarized in two perpendicular planes, each of which carries half of the whole wave's intensity.

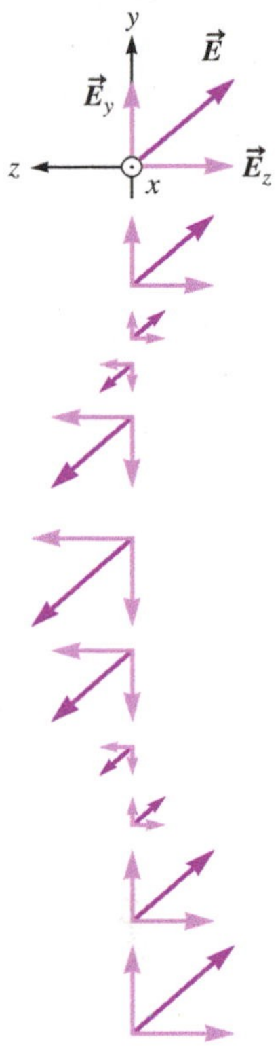

FIGURE 34.23 Any oscillating electric field vector can be broken into two vectors that oscillate perpendicular to each other. In this case, an electromagnetic wave travels in the x direction—out of the page. The electric field oscillates along a line that runs from the lower left to the upper right. The electric field is broken into two perpendicular components along y and z, and is shown at 11 different times. So, we can think of unpolarized light as equivalent to two waves with the same amplitude and perpendicular polarizations, varying independently.

Polarizers

A **polarizer** is a filter that allows only one plane of polarization to pass through. The polarized lenses in your good sunglasses are polarizers. Figure 34.24 shows unpolarized light that passes through a polarizer. Initially this light can be modeled as the sum of the light polarized in two perpendicular directions. It is helpful to imagine that one of these directions is parallel to the polarization that the filter allows through, and the other is perpendicular. The polarizer in Figure 34.24 allows the light's vertically polarized component to pass through but absorbs its horizontal component. Because half the intensity is carried in the horizontal component, only half the intensity gets through.

Let's see how a polarizer works. In the late 1920s, the American inventor Edwin Herbert Land (1909–1991) invented the sheet polarizer, which is the basis of such devices as polarized lenses in sunglasses. The polarizer consists of long chains of hydrocarbon molecules arranged in parallel rows, as represented by the horizontal parallel lines shown on the polarizer in Figure 34.24. The spacing between the rows must be smaller than the wavelength of the light to be filtered (a few hundred nanometers for visible light). The hydrocarbon molecules are coated in iodine. The conduction electrons in the coated molecules are able to move along the entire chain, so each long row of molecules acts like a thin

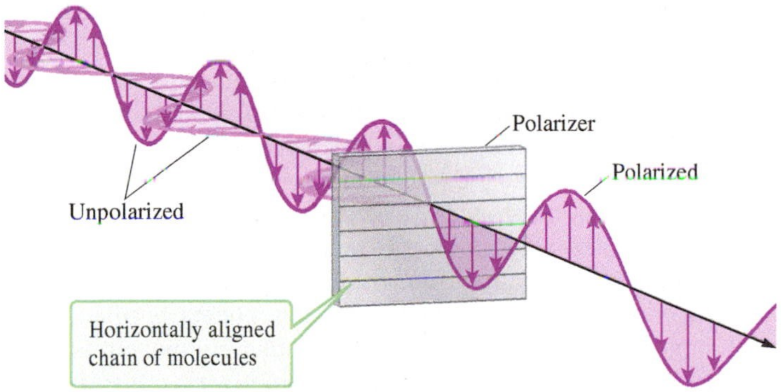

FIGURE 34.24 Only the electric field is represented here. A polarizing filter has all of its molecules aligned in the same direction. The electric field with the same orientation as the filter is absorbed by the molecules. The radiation that passes through the polarizing filter is polarized perpendicular to the chain of molecules and is reduced in intensity.

conducting wire or antenna. When unpolarized light meets the polarizer in Figure 34.24, the horizontal component of this light causes the conduction electrons to oscillate horizontally along the chain of molecules. Their oscillation results in a linearly polarized wave 180° out of phase with the horizontally polarized component of the incident light. Because they are 180° out of phase, the two waves cancel each other. In effect, the horizontal component is absorbed and only the vertical component passes through the polarizer.

Often polarizers are described in terms of their **transmission axis**, which is parallel to the component that passes through the filter. For the polarizer in Figure 34.24, the transmission axis is vertical. The transmission axis is always perpendicular to the rows (long chains) of molecules.

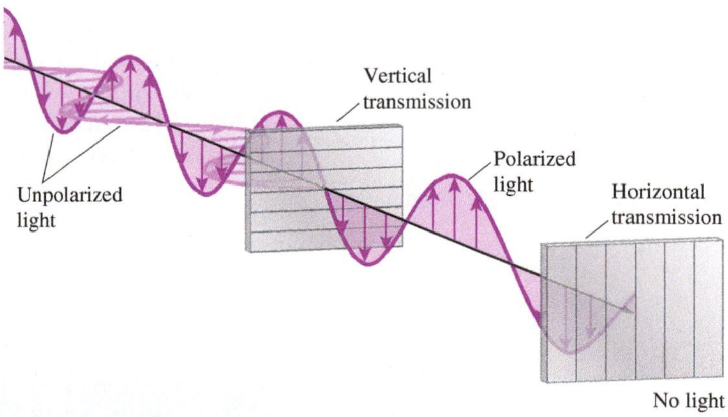

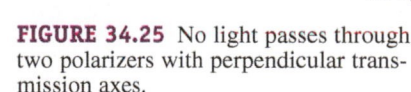

FIGURE 34.25 No light passes through two polarizers with perpendicular transmission axes.

What happens when light passes through more than one polarizer? In Figure 34.25, unpolarized light first passes through a polarizer with a vertical transmission axis. After doing so, it is vertically polarized and has half the intensity of the original

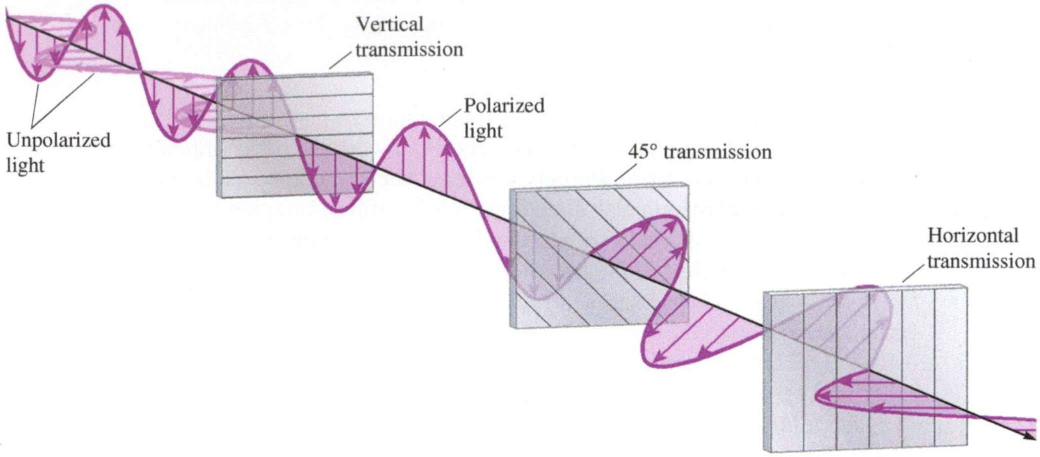

FIGURE 34.26 Some light passes through all three polarizers, each rotated 45° from the previous one. (Magnitudes are not to scale.)

unpolarized light. Now, if that vertically polarized light passes through a second polarizer with a horizontal transmission axis, no light gets through the second polarizer. In other words, the transmitted light has zero intensity.

What happens if you insert a third polarizer at some angle between vertical and horizontal—if, for example, the transmission axis is 45° from vertical? The answer depends on where you put the third polarizer. If you put it at the end, so the light encounters the polarizers in order from vertical transmission to horizontal transmission and then to 45° transmission, there is no change because no light reaches the last polarizer. What is really amazing is that if you put the 45° transmission polarizer between the vertical and horizontal transmission polarizers, some light passes through all three polarizers (Fig. 34.26). How can this happen? The first polarizer allows vertically polarized radiation through. But this vertically polarized radiation may also be broken into two components. One component is tilted by 45° and is parallel to the molecules in the second filter. The other component is perpendicular to these molecules, so it is parallel to the second polarizer's transmission axis. The second polarizer—tilted by 45°—allows radiation tilted by 135° (= 45° + 90°) relative to the original horizontal to pass through. This is the light that passes between the second and third polarizers. This radiation may in turn be broken into two components, one vertical and the other horizontal. The final polarizer allows the horizontally polarized radiation to pass. This is the radiation that passes through all three filters.

This experiment with three polarizers shows that the intensity of the radiation that passes through a filter depends on the angle between the radiation's plane of polarization and the polarizer transmission axis. This relationship was described by a French military engineer who worked for Napoleon, Etienne-Louis Malus (1775–1812). According to Malus's law, the intensity I_f of the linearly polarized radiation that passes through a polarizer is given by

MALUS'S LAW ✪ Major Concept

$$I_f = I_i \cos^2\varphi \tag{34.40}$$

where I_i is the intensity of the polarized radiation incident on the polarizer and φ is the angle between the light's plane of polarization and the polarizer's transmission axis.

CONCEPT EXERCISE 34.7

Explain why the term *all polarized* might be better than *unpolarized*.

EXAMPLE 34.7 Three Polarizers

What percentage of the incident light's intensity in Figure 34.26 passes through all three polarizers?

INTERPRET and ANTICIPATE

There are two keys to solving this problem: (1) Half the unpolarized light's intensity passes through the first polarizer, and (2) we can use Malus's law for the other two polarizers because the light that passes through them is linearly polarized.

SOLVE

Let I_v be the intensity of unpolarized light that passes through the vertical polarizer. This light has half the intensity I_0 of the unpolarized light.

$$I_v = \frac{1}{2} I_0 \quad (1)$$

The vertically polarized light encounters the second polarizer, whose transmission axis is tilted 45° with respect to the vertical (or 135° from the horizontal). Either angle may be used because in Malus's law the cosine function is squared. The intensity I_t of the light that passes through the tilted polarizer is given by Equation 34.40, with Equation (1) substituted for the light incident on this polarizer.

$$I_t = I_v \cos^2\varphi \quad (34.40)$$

$$I_t = \frac{1}{2} I_0 \cos^2 45° = \frac{1}{2} I_0 \left(\frac{1}{2}\right)$$

$$I_t = \frac{1}{4} I_0 \quad (2)$$

Use Malus's law one more time to find the intensity I_f of the light that passes through the final polarizer. The light emerging from the second polarizer is tilted 45° with respect to the horizontal, but the last polarizer's transmission axis is horizontal. Substitute Equation (2) for the intensity incident on this polarizer.

$$I_f = I_t \cos^2\varphi \quad (34.40)$$

$$I_f = \frac{1}{4} I_0 \cos^2 45° = \frac{1}{4} I_0 \left(\frac{1}{2}\right) = \frac{1}{8} I_0$$

Take the ratio I_f/I_0. We are asked for a percentage, so multiply the result by 100.

$$\frac{I_f}{I_0} = 0.125 = 12.5\%$$

CHECK and THINK

To think about these results, use Malus's law for two perpendicular polarizers (Fig. 34.25). In this case, half the intensity gets through the first polarizer, but no light gets through the second polarizer because $\cos 90° = 0$. If we put a third polarizer between these two perpendicular polarizers, some light (12.5%) passes through all three polarizers. These results are amazing, but the order of the polarizers matters. If the tilted polarizer is placed third, after the horizontal polarizer, no light is incident on the tilted polarizer and so no light gets through.

SUMMARY

Underlying Principles

1. **General form of Faraday's law:**

$$\oint \vec{E} \cdot d\vec{\ell} = -\frac{d\Phi_B}{dt} = -\frac{d}{dt}\int \vec{B} \cdot d\vec{A} \quad (34.5) \text{ and } (34.6)$$

The induced electric field is nonconservative, so $\mathcal{E} = \oint \vec{E} \cdot d\vec{\ell} \neq 0$.

2. **Maxwell's equations,** which are named for the scientists who first discovered them, describe all of electricity and magnetism. When combined, they describe electromagnetic radiation. The four equations are:

a. Gauss's law for electricity:

$$\Phi_E = \oint \vec{E} \cdot d\vec{A} = \frac{q_{in}}{\varepsilon_0} \quad (25.11)$$

b. Gauss's law for magnetism:

$$\Phi_B = \oint \vec{B} \cdot d\vec{A} = 0 \quad (31.4)$$

c. Faraday's law:

$$\oint \vec{E} \cdot d\vec{\ell} = -\frac{d\Phi_B}{dt} \qquad (34.5)$$

d. Ampère–Maxwell's law:

$$\oint \vec{B} \cdot d\vec{\ell} = \mu_0 I_{\text{thru}} + \mu_0 \varepsilon_0 \frac{d\Phi_E}{dt} \qquad (31.8)$$

3. **Lorentz force** exerted on a charged particle by electric and magnetic fields:

$$\vec{F}_L = \vec{F}_E + \vec{F}_B = q(\vec{E} + \vec{v} \times \vec{B}) \qquad (30.21)$$

4. Electromagnetic waves satisfy a **general wave equation** of the form

$$\frac{\partial^2 y}{\partial x^2} = \frac{1}{v^2} \frac{\partial^2 y}{\partial t^2} \qquad (17.33)$$

The wave equation for the electric field is

$$\frac{\partial^2 E}{\partial x^2} = \mu_0 \varepsilon_0 \frac{\partial^2 E}{\partial t^2} \qquad (34.7)$$

and the wave equation for the magnetic field is

$$\frac{\partial^2 B}{\partial x^2} = \mu_0 \varepsilon_0 \frac{\partial^2 B}{\partial t^2} \qquad (34.8)$$

Major Concepts

1. Electromagnetic waves are **transverse**. If a wave propagates along the positive x direction, its electric and magnetic fields are given by

$$\vec{E}(x, t) = [E_{\text{max}} \sin(kx - \omega t)]\hat{\jmath} \qquad (34.18)$$

$$\vec{B}(x, t) = [B_{\text{max}} \sin(kx - \omega t)]\hat{k} \qquad (34.19)$$

where $E(x, t) = cB(x, t)$ (34.24)

2. Mathematically, the **energy transferred** by an electromagnetic wave per unit time is given by (the magnitude of) the Poynting vector $\vec{S}$:

$$\vec{S} \equiv \frac{1}{\mu_0}(\vec{E} \times \vec{B}) \qquad (34.27)$$

3. The **intensity** of an electromagnetic wave is the time average of S: $I \equiv S_{\text{av}}$. In terms of the maximum electric and magnetic fields, the intensity I is

$$I = \frac{1}{2\mu_0} E_{\text{max}} B_{\text{max}} \qquad (34.29)$$

The intensity may also be expressed in terms of either E_{max} or B_{max} alone:

$$I = \frac{E_{\text{max}}^2}{2\mu_0 c} = \frac{c}{2\mu_0} B_{\text{max}}^2 \qquad (34.30)$$

For a distant point source of electromagnetic radiation, the intensity decreases as the distance r from the source increases:

$$I = \frac{P_{\text{av}}}{4\pi r^2} \qquad (17.23)$$

where P_{av} is the average power emitted by the source. The intensity is also proportional to the average field energy density:

$$I = cu_{\text{av}} \qquad (34.33)$$

4. The **momentum** p delivered by an electromagnetic wave is proportional to the energy Q delivered. When the electromagnetic wave is totally absorbed by an object, the momentum is given by

$$p = \frac{Q}{c} \qquad (34.35)$$

When the electromagnetic wave is completely reflected from the object, the momentum delivered to the object is given by

$$p = 2\frac{Q}{c} \qquad (34.37)$$

5. The **pressure** exerted by an electromagnetic wave that is totally absorbed by an object is

$$P = \frac{I}{c} \qquad (34.38)$$

If the radiation is completely reflected, the pressure is

$$P = 2\frac{I}{c} \qquad (34.39)$$

6. **Malus's law:** The intensity I_f of the linearly polarized radiation that passes through a polarizer is given by

$$I_f = I_i \cos^2\varphi \qquad (34.40)$$

where I_i is the intensity of the polarized radiation incident on the polarizer and φ is the angle between the light's plane of polarization and the polarizer's transmission axis.

PROBLEMS AND QUESTIONS

A = algebraic **C** = conceptual **E** = estimation **G** = graphical **N** = numerical

34-1 Light: One Last Classical Topic

1. **C** What do we mean when we say that light is sometimes modeled as a particle and sometimes modeled as a wave?

34-2 Generalized Form of Faraday's Law

2. **C** Suppose a positive particle moves once around the circuit in Figure 34.3B (page 1088). The amount of energy it gains going through the battery equals the amount of energy it loses going through the resistor, so when it returns to its initial position, its energy is unchanged. Now think about a positive particle that goes around the loop in Figure 34.3A. When this particle moves once around, it has gained energy. **a.** Where does that energy come from? **b.** Is this a source of *free* energy? Explain.
3. **N** A circular coil of radius 0.50 m is placed in a time-varying magnetic field $B(t) = (5.80 \times 10^{-4})\sin[(12.6 \times 10^2\,\text{rad/s})t]$ where B is in teslas. The magnetic field is perpendicular to the plane of the coil. Find the magnitude of the induced electric field in the coil at $t = 0.001$ s and $t = 0.01$ s.
4. **C** An end-on view of a coil is shown in Figure P34.4. A current in the coil is increasing toward a steady-state value. Indicate the direction (into or out of the page) of the magnetic field near the center of the coil. Indicate the direction (clockwise or counterclockwise) of the electric field inside the coil.

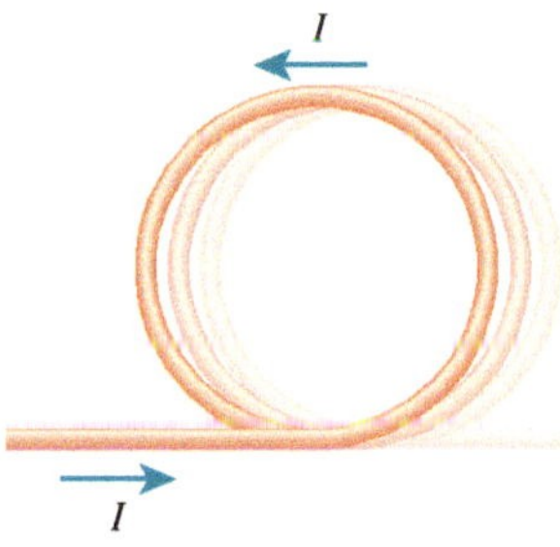

FIGURE P34.4

5. A solenoid with n turns per unit length has radius a. It is connected to a power supply that drives a current increasing linearly with time as $I(t) = Ct$, where C is a constant. Assume the magnetic field is uniform inside the solenoid.
   - **a.** **A** Find an expression for the magnetic field inside the solenoid as a function of time, and determine the magnitude of the induced electric field just inside the solenoid.
   - **b.** **N** What is the magnitude of the electric field induced just inside a solenoid that has exactly 10 turns/cm and a radius of 1.5 cm if the current increases at a rate of 0.50 A/s?

Problems 6 and 7 are paired.

6. **G** Suppose the magnetic field in Figure 34.4 (page 1090) increases at a constant rate dB/dt. Sketch $E(r)$ from $r = 0$ to $r = 2R$.
7. **A** If $B(t) = B_0 e^{-t/\tau}$ in Figure 34.4 (page 1090), find an expression for the electric field at $r = R$. Does the electric field depend on time?

34-3 Five Equations of Electromagnetism

8. **C** If you discover magnetic monopoles, what changes (if any) do you need to make to Maxwell's equations?
9. **N** A capacitor with square plates, each with an area of 36.0 cm^2 and plate separation $d = 2.54$ mm, is being charged by a 265-mA current. **a.** What is the change in the electric flux between the plates as a function of time? **b.** What is the magnitude of the displacement current between the capacitor's plates?
10. **C** Three students are discussing Maxwell's equations.

    Avi: If you don't have any electric charges at all, Maxwell's equations still tell you how electric and magnetic fields behave, right?

    Cameron: That doesn't make sense to me. Electric fields are made by charges, and magnetic fields are made by *moving* charges. So we have to have charges to describe electromagnetism. I don't see how Maxwell's equations can be used without electric charges.

    Shannon: Well, Maxwell's equations are still true, but the right sides of all the equations are zero anyway, so the electric and magnetic fields have to be zero.

    With which student do you agree? Justify your answer.
11. **N** An electric field with initial magnitude 140 V/m directed into the page and increasing at a rate of 12.0 V/(m · s) is confined to the area of radius $R = 17.0$ cm in Figure P34.11. If $r = 42.0$ cm, what are the magnitude and direction of the magnetic field at point A?

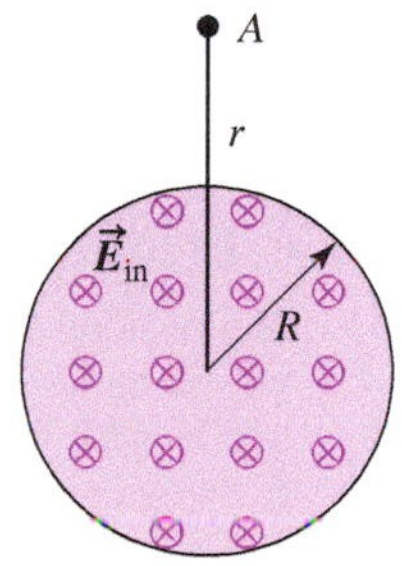

FIGURE P34.11

12. **C** Suppose a positive particle is initially at rest in Figure 34.4 (page 1090). Then the magnitude of the magnetic field begins to change. **a.** Will the particle be accelerated? **b.** If so, will the particle remain in the plane of the page? Explain your answers.

Problems 13 and 14 are paired.

13. **A** A circular conductor (Fig. P34.13) encloses a uniform magnetic field that is perpendicular to the page (not shown), pointing outward and increasing. The changing magnetic flux induces an emf $\mathcal{E}$ in the conductor and a clockwise current. The magnetic field exists only in the circular region enclosed by the conductor. If we measure the potential difference between points A and B using the two voltmeters as shown, we find $V_1 \neq V_2$. To see how that is possible, set up the path integral from A to B along path 1 to find V_1, and then set up the path integral from A to B along path 2 to find V_2. (These two paths pass through their respective voltmeters.) Finally, by combining these integrals, show that $V_2 - V_1 = \mathcal{E}$.

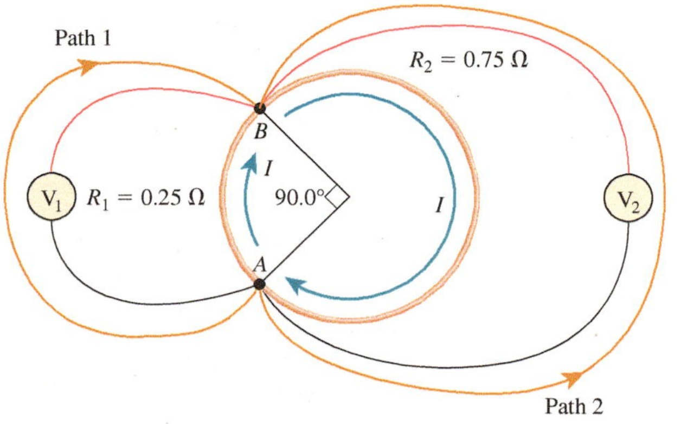

FIGURE P34.13 Problems 13 and 14.

14. **N** A circular conductor encloses a uniform magnetic field that is perpendicular to the page (not shown), pointing inward and increasing. The magnetic field outside the circular loop is zero. Suppose the changing magnetic flux induces a 1.0-V emf in the conductor and the conductor's resistance is 1.0 Ω. Imagine connecting two voltmeters to the conductor (Fig. P34.13). The clockwise distance from A to B is one-fourth the circumference. Because the resistance of a wire is proportional to its length, this portion of the loop has resistance $R_1 = 0.25\ \Omega$. Likewise, the clockwise distance from B to A is three-fourths the circumference, so this portion's resistance is $R_2 = 0.75\ \Omega$. It may seem incredible, but what a voltmeter measures depends on its placement; in Problem 13, we showed that $V_2 - V_1 = \mathcal{E}$. Find the voltage measured by both meters. *Hint*: Kirchhoff's loop rule (and other rules from Chapter 29) may be applied to a loop that does not enclose the changing magnetic flux.
15. **N** What is the acceleration of a proton moving with velocity $\vec{v} = 15.0\hat{\jmath}$ m/s through a region that has magnetic field $\vec{B} = 1.20\hat{\imath}$ T and electric field $\vec{E} = (3.00\hat{\jmath} + 2.00\hat{k})$ V/m?
16. **C** Write Maxwell's equations for the case in which there are no charges present, and explain in words the meaning of each equation.

34-4 Electromagnetic Waves

17. **N** Consider electromagnetic waves in free space. What is the wavelength of a wave that has a frequency of **a.** 2.00×10^{11} Hz and **b.** 8.00×10^{16} Hz?
18. **A** Show that the SI units of $1/\sqrt{\mu_0 \varepsilon_0}$ are m/s.
19. **N** The human eye can see light that has a maximum frequency of 7.69×10^{14} Hz. What is the corresponding wavelength of the light?
20. **N** An electromagnetic wave is given in SI units by $E(x, t) = 3.75 \sin (0.60x - \omega t)$ V/m. **a.** What is the angular frequency? **b.** What is the magnetic field at $x = 2.0$ m and $t = 3.0$ s?
21. **N** Ultraviolet (UV) radiation is a part of the electromagnetic spectrum that reaches the Earth from the Sun. It has wavelengths shorter than those of visible light, making it invisible to the naked eye. These wavelengths are classified as UVA, UVB, or UVC, with UVA the longest of the three at 320 nm to 400 nm. Both the U.S. Department of Health and Human Services and the World Health Organization have identified UV as a proven human carcinogen. Many experts believe that, especially for fair-skinned people, UV radiation frequently plays a key role in melanoma, the deadliest form of skin cancer, which kills more than 8000 Americans each year. UVB has a wavelength between 280 nm and 320 nm. Determine the frequency ranges of UVA and UVB.
22. **G** An electromagnetic wave is given in SI units by $E(x, t) = 3.75\sin (kx - 0.094t)$ V/m. Plot the magnetic energy density and the electrical energy density associated with this wave as a function of time t.
23. **N** What is the frequency of the blue-violet light of wavelength 405 nm emitted by the laser-reading heads of Blu-ray disc players?
24. **A** Write equations for both the electric and magnetic fields for an electromagnetic wave in the red part of the visible spectrum that has a wavelength of 710 nm and a peak electric field magnitude of 2.5 V/m.
25. **N** The amplitude of the electric field of an electromagnetic wave traveling in a vacuum is measured to be 4.3×10^2 V/m. What is the amplitude of the magnetic field in this wave?
26. **A** Start with Equation 34.11,
$$\left(\frac{\partial E}{\partial x}\right) = -\left(\frac{\partial B}{\partial t}\right)$$
and Equation 34.15,
$$\left(\frac{\partial B}{\partial x}\right) = -\mu_0\varepsilon_0\left(\frac{\partial E}{\partial t}\right)$$
and show that the wave equation for a linearly polarized magnetic wave in a vacuum is given by Equation 34.8,
$$\frac{\partial^2 B}{\partial x^2} = \mu_0\varepsilon_0\frac{\partial^2 B}{\partial t^2}$$
27. **N** WGVU-AM is a radio station that serves the Grand Rapids, Michigan, area. The main broadcast frequency is 1480 kHz. At a certain distance from the radio station transmitter, the magnitude of the magnetic field of the electromagnetic wave is 3.0×10^{-11} T. **a.** Calculate the wavelength. **b.** What is the angular frequency? **c.** Find the wave number of the wave. **d.** What is the amplitude of the electric field at this distance from the transmitter?
28. **N** Suppose the magnetic field of an electromagnetic wave is given by $B = (1.5 \times 10^{-10}) \sin (kx - \omega t)$ T. **a.** What is the maximum energy density of the magnetic field of this wave? **b.** What is maximum energy density of the electric field?
29. **N** The magnetic field of a plane electromagnetic wave is given by $B_z = (68.0) \sin (kx - 2.60 \times 10^6 t)$ nT. **a.** What is the amplitude of the electric field of this wave? **b.** What is the frequency f? **c.** What is the wavelength λ?
30. **A** Write equations for both the electric and magnetic fields for an electromagnetic wave (an X-ray) that has an angular frequency of 7.5×10^{18} Hz and a peak magnetic field magnitude of 10^{-10} T.
31. **N** A cell phone sends and receives electromagnetic waves. The quality of the phone's reception depends on the strength of the electric field. The stronger the electric field around a cellular telephone, the better the reception. In this scenario, however, there is a higher chance that the user's body will absorb the electric field signal, slowly leading to possible harmful effects like cancer. A guideline has been established limiting the electric field strength near a cell phone to protect the user. The "maximum permissible exposure" in this context is considered to be 100 V/m. If the amplitude of the electric field near a cell phone is 41 V/m, what is the amplitude of the magnetic field? How does it compare to the magnitude of the Earth's magnetic field near the surface, which is 5.0×10^{-5} T?
32. **N** An electromagnetic wave is traveling in the positive x direction with the electric field oscillating along the z direction. If the wavelength is 555 nm, $E_{max} = 0.050$ V/m, and the wave is in a vacuum, write equations describing the electric and magnetic fields that make up the electromagnetic wave as functions of x and t.

Problems 33 and 34 are paired.

33. **A** By substitution, show that $E(x, t) = E_{max} \sin (kx - \omega t)$ is a solution to Equation 34.7:
$$\frac{\partial^2 E}{\partial x^2} = \mu_0\varepsilon_0\frac{\partial^2 E}{\partial t^2}$$
where
$$c = \frac{1}{\sqrt{\mu_0\varepsilon_0}} = \frac{\omega}{k}$$
34. **A** By substitution, show that $B(x, t) = B_{max} \sin (kx - \omega t)$ is a solution to Equation 34.8:
$$\frac{\partial^2 B}{\partial x^2} = \mu_0\varepsilon_0\frac{\partial^2 B}{\partial t^2}$$

34-5 The Electromagnetic Spectrum

35. **C** Can you hear radio waves? Explain.
36. **C** When electromagnetic radiation shines through openings with a size that is comparable to the wavelength of the radiation,

diffraction becomes important, leading to variations in intensity due to interference of the waves. What type of electromagnetic radiation would lead to diffraction when shined on a lattice of atoms separated by 0.2 nm? What type of radiation would diffract through a row of skyscrapers separated by a couple hundred meters? Explain your answers.

37. **C** Which waves travel faster—radio waves or gamma rays?

Problems 38 and 39 are paired.

38. **C** Astronomers often speak of light from a distant source as being *red-shifted*, where the measured wavelength is made longer than that emitted by the source by some interfering effect (typically a Doppler effect). Consider the electromagnetic spectrum, and explain why astronomers might use this term to describe these wavelength measurements.

39. **C** Astronomers often speak of light from a distant source as being *red-shifted*, where the measured wavelength is made longer than that emitted by the source by some interfering effect (typically a Doppler effect). If someone described a wavelength measurement as being *blue-shifted* instead, what can you say about the original wavelength and frequency of the wave before the blue-shifting occurred?

34-6 Energy and Intensity

40. **E** Estimate the amount of energy from sunlight that your face absorbs over your entire lifetime. *Hints*: See Example 34.2. The intensity of sunlight at the Earth's surface is roughly 1000 W/m^2. (This intensity is lower than the 1360 W/m^2 used in Example 34.2 because of atmospheric absorption, scattering, and other factors.)

41. **N** Find the intensity of the electromagnetic wave described in each case: An electromagnetic wave with **a.** a wavelength of 710 nm and a peak electric field magnitude of 2.5 V/m and **b.** an angular frequency of 7.5×10^{18} rad/s and a peak magnetic field magnitude of 10^{-10} T.

42. An electric field $\vec{E} = (23.0\hat{\imath} - 55.0\hat{\jmath} + 17.0\hat{k})$ V/m and a magnetic field $\vec{B} = (5.25\hat{\imath} + 4.05\hat{\jmath} + 6.00\hat{k})$ nT are measured in a location in free space.
   a. **N** What is the Poynting vector at this location?
   b. **A** Show that the electric and magnetic fields are perpendicular to each other at this location.

43. **N** You wish to send a probe to the Moon. The probe has a radio transmitter that you test in the laboratory. When the probe is 10 m from your receiver, the intensity is I_0. When the probe is on the Moon, it sends radio waves to you. If you want to receive radio waves of the same intensity, how much stronger must the probe's electric field be when it is on the Moon?

44. **A** The intensity in Equation 34.29, $I = (1/2\mu_0)E_{max}B_{max}$, and in Equation 34.30, $I = E^2_{max}/(2\mu_0 c) = (cB^2_{max})/(2\mu_0)$, is written in terms of the maximum electric and magnetic fields. Rewrite these equations in terms of the rms field values.

45. **N** The state-of-the-art digital air surveillance radar used in civil aviation emits its signal equally in all directions (isotropically) with an average power of 22.0 kW. What is the average intensity near an aircraft on final approach to an airport, 2.00 km from the radar tower?

46. **N** At an instant in time, the electric and magnetic fields of an electromagnetic wave are given by $\vec{E} = -4.00 \times 10^{-3}\hat{k}$ V/m and $\vec{B} = -1.33 \times 10^{-11}\hat{\imath}$ T. Find the Poynting vector for this wave.

47. **N** The electric field of an electromagnetic wave traveling in the vacuum of space is described by $E = (5.00 \times 10^{-3})(kx - \omega t)$ V/m. **a.** What is the maximum value of the associated magnetic field for this electromagnetic wave? **b.** What is the average energy density of the wave?

48. **E** You may have heard that our telecommunication signals are traveling out in space and that some alien society may be watching old TV episodes or listening to your phone calls. A powerful cell phone puts out about 3 W. The closest star to our solar system is about 4 light-years away. Think about a call you made four years ago. The electromagnetic wave from that call is just reaching our nearest neighbor star. How strong is the intensity at that distance? Compare your answer to the intensity of the signal picked up by the nearest cell phone tower on the Earth.

49. **C** Determine the direction of energy flow for the following four cases, where $\vec{E}$ and $\vec{B}$ are the electric and magnetic fields, respectively: **a.** $\vec{E} = E\hat{\jmath}, \vec{B} = B\hat{k}$; **b.** $\vec{E} = -E\hat{\imath}, \vec{B} = B\hat{k}$; **c.** $\vec{E} = -E\hat{\imath}, \vec{B} = -B\hat{\jmath}$; **d.** $\vec{E} = -E\hat{\imath}, \vec{B} = -B\hat{k}$.

50. **N** A circular mirror 78.0 cm in radius is used to reflect sunlight onto a small plate with a diameter of 3.40 cm that completely absorbs the light. The intensity of sunlight at the Earth's surface is 980 W/m^2. **a.** What is the intensity of sunlight incident on the absorbing plate? **b.** What is the amplitude of the electric field at the absorbing plate? **c.** What is the amplitude of the magnetic field at the absorbing plate?

Problems 51 and 52 are paired.

51. **A** A current of magnitude I flows through a coiled wire. The coil forms a cylinder of length L and radius R. The potential difference between the two ends of the wire is V. Determine the magnitude of the Poynting vector.

52. **A** A current of magnitude I flows through a coiled wire. The coil forms a cylinder of length L and radius R. The potential difference between the two ends of the wire is V. Integrate the Poynting vector over the cross-sectional area of the wire to obtain the energy per unit time (E/t) passing through the surface of the wire $\left(E/t = \int \vec{S} \cdot d\vec{A}\right)$.

34-7 Momentum and Radiation Pressure

53. **N** Optical tweezers use light from a laser to move single atoms and molecules around. Suppose the intensity of light from the tweezers is 1.00×10^3 W/m^2, the same as the intensity of sunlight at the surface of the Earth. **a.** What is the pressure on an atom if light from the tweezers is totally absorbed? **b.** If this pressure were exerted on a hydrogen atom, what would be its acceleration? Assume the cross-sectional area is 6.65×10^{-29} m^2.

54. **N** The intensity of sunlight on the Earth is about 1.00×10^3 W/m^2. The average orbital radius of Mercury is about 40% of the Earth's orbital radius. Using this information, determine the approximate intensity of the sunlight and the radiation pressure on Mercury.

55. **N** What is the radiation pressure on a perfectly reflecting mirror due to a 0.600-W laser beam of radius 1.50 mm that is normally incident on the mirror?

56. Enrique claims he can push a toy cart across a frozen pond using a flashlight and a sail. He says that if he spreads the sail out to its maximum area of 64 cm^2 and attaches it to the 75-g toy cart, the light from the flashlight alone will push the cart. Angelique objects to Enrique's claim, noting that the surrounding sunlight is more intense than the light from the flashlight.
   a. **C** Evaluate Angelique's claim. Does the surrounding sunlight matter when the flashlight shines on the sail? In other words, does the sunlight cause a resistive force when Enrique tries to move the cart? Explain.
   b. **E, N** Estimate or research the intensity of light from a typical household flashlight when it is held about 0.01 m away from the sail. If we assume the light is perfectly reflected by the sail, what is the magnitude of the force exerted on the cart by the flashlight?
   c. **C** Given that there is a small amount of friction between the cart and the frozen pond, will Enrique observe the cart moving if the experiment is performed? Explain your answer.

57. **N** The accepted value of the intensity of sunlight at the Earth's surface is 1.36 kW/m^2. **a.** Determine the pressure exerted by sunlight if it strikes a perfect absorber. **b.** Determine the pressure exerted by sunlight if it strikes a perfect reflector.

58. **C** The expressions for the momentum transferred by radiation (Eqs. 34.35 and 34.37) and for the pressure exerted (Eqs. 34.38 and 34.39) may seem odd. To become more comfortable with them, check their dimensions.

59. **N** A perfectly reflecting circular mirror 15.0 cm in radius is placed in the path of a plane electromagnetic wave with an intensity of 3.65 W/m^2 traveling in the positive z direction. **a.** How much momentum does the wave impart to the mirror per second? **b.** What is the force exerted on the mirror by the wave?

60. **N** A budding magician holds a 5.00-mW laser pointer, wondering whether he could use it to keep an object floating in the air with the radiation pressure. This might be an idea for a new trick! Assuming the laser pointer has a circular beam 3.00 mm in diameter and the magician rigs up a totally reflecting sail on which to shine the laser, what is the maximum weight the magician could suspend with this technique?

34-8 Polarization

61. **N** Some unpolarized light has an intensity of 1365 W/m^2 before passing through three polarizing filters. The transmission axis of the first filter is vertical. The second filter's transmission axis is 30.0° from vertical. The third filter's transmission axis is 40.0° from vertical. What is the intensity of the light that emerges from the three filters?

Problems 62 and 63 are paired.

62. **N** Suppose you have a polarized light source in your lab that has an intensity of 1409 W/m^2, where the electric field oscillates in the z direction. The light travels toward a workbench where you can send it through either one of two polarizing filters. **a.** If the light passes through a filter with its transmission axis at an angle of 45.0° from the z direction, what is the intensity of the transmitted light? **b.** If, instead, the light passes through a filter with its transmission axis at an angle of 90.0° from the z direction, what is the intensity of the transmitted light?

63. **N** The light in Problem 62 travels toward a workbench and encounters two polarizing filters. **a.** If the light passes first through a filter with its transmission axis at an angle of 45.0° from the z direction, and then through a filter with its transmission axis at an angle of 90.0° from the z direction, what is the final intensity of the transmitted light? **b.** If, instead, the light passes first through a filter with its transmission axis at an angle of 90.0° from the z direction, and then through a filter with its transmission axis at an angle of 45.0° from the z direction, what is the final intensity of the transmitted light?

64. **C** You have a set of four polarizing filters, and you are free to orient their transmission axes however you prefer. If you begin with an unpolarized light wave, is it possible to halve the intensity of the wave each time it passes through a polarizer, so that the final intensity is 1/16 the original intensity? How could this be done?

65. **N** Unpolarized light passes through three polarizing filters. The first filter has its transmission axis parallel to the z direction, the second has its transmission axis at an angle of 30.0° from the z direction, and the third has its transmission axis at an angle of 60.0° from the z direction. If the light that emerges from the third filter has an intensity of 250.0 W/m^2, what is the original intensity of the light?

General Problems

66. **N** The average Earth–Sun distance is 1.00 astronomical unit (AU). At how many AUs from the Sun is the intensity of sunlight 1/25 the intensity at the Earth?

67. The magnetic field of an electromagnetic wave is given by $B(x, t) = (4.0 \times 10^{-8}) \sin[(1.4 \times 10^4 \text{ rad/m})x + \omega t]$ T.
    a. **C** In which direction is the wave traveling?
    b. **N** Determine the wave number, wavelength, and frequency of the wave.
    c. **A** Write an equation for the electric field as a function of x and t.

68. **A** As mentioned in Section 34-2, Kirchhoff's loop rule holds only for conservative fields, and the inductor rule is a way to preserve using the loop rule. To see that we get the same mathematical relationship $\mathcal{E} - L(dI/dt) - IR = 0$ as we did in Section 33-3 when we (mis-)applied Kirchhoff's loop rule to a circuit consisting of a battery, a resistor, and an inductor, correctly apply Faraday's law (Equation 34.6)

$$\oint \vec{E} \cdot d\vec{\ell} = -\frac{d}{dt}\int \vec{B} \cdot d\vec{A}$$

to the circuit shown in Figure P34.68. This circuit is essentially a resistor, a battery, and a one-loop inductor. There is negligible resistance in the wires, and the total inductance is L. When the switch closes at $t = 0$, the current is counterclockwise, while the magnetic field points out of the page and increases. *Hints*: Choose an outward-pointing $d\vec{A}$, and choose $d\vec{\ell}$ in the same direction as the current. In the **CHECK and THINK** step, describe the conceptual difference between (mis-)applying Kirchhoff's loop rule and applying Faraday's law.

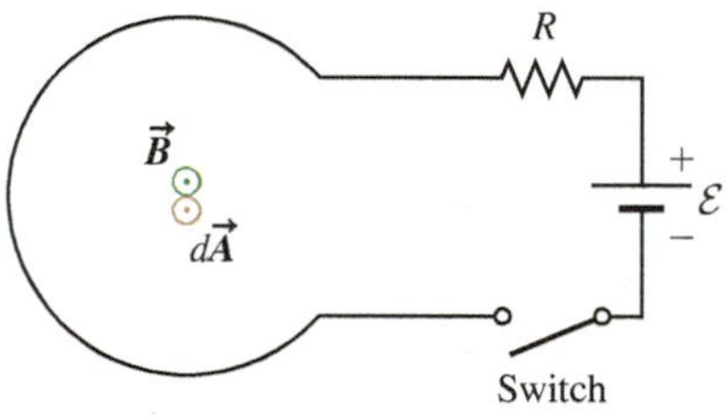

FIGURE P34.68

69. **N** What is the amplitude of the magnetic field a distance of 3.50 km away from a radio station that is broadcasting isotropically (in all directions) with a power of 43.0 kW?

70. **N** The magnetic field of an electromagnetic wave is given by $B = 1.5 \times 10^{-10} \sin(kx - \omega t)$ T. **a.** If the wavelength is 752 nm, what are the frequency, angular frequency, and wave number of this wave? **b.** What is the maximum total energy density of this wave?

71. **N** Household solar panels typically have an efficiency of 15.0%. If the intensity of sunlight is assumed to be a constant 9.80×10^2 W/m^2, what should be the total area of rooftop solar panels on a house in order to supply an average of 33.0 kW each day? Efficiency denotes the percentage of incident solar energy that is converted to electricity by the solar panel.

72. **N** A plane electromagnetic sinusoidal wave with a wavelength of 34.0 mm and a magnetic field oscillating in the yz plane with an amplitude of 96.0 nT is traveling in the positive z direction (Fig. P34.72). **a.** What is the frequency of this wave? **b.** What are the magnitude and direction of the electric field when $\vec{B} = B_{max}\hat{\jmath}$? **c.** Expressing the magnitude of the electric field as $E(z, t) = E_{max} \sin(kz - \omega t)$, obtain an expression for E that includes numerical values for E_{max}, k, and ω.

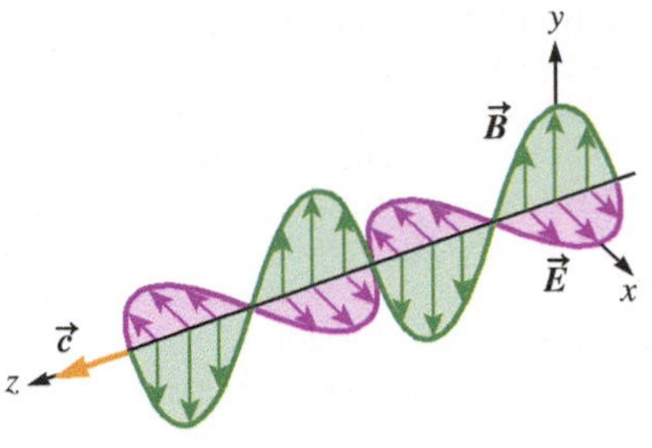

FIGURE P34.72

73. The electric field of an electromagnetic wave traveling in a vacuum is described by the equation $E = (5.00 \times 10^{-3}) \sin(kz - \omega t)$ V/m, where the electric field oscillates along the y direction.

a. C In what direction is this wave traveling?

b. N If the wavelength is 454 nm, what are the frequency, angular frequency, and wave number for this wave?

c. C Along which direction does the associated magnetic field oscillate?

d. N Write an equation describing the associated magnetic field as a function of z and t.

74. E CASE STUDY The region around the Earth is filling up with space junk such as old satellites. One idea for cleaning up space involves using sails that create drag (Fig. P34.74). Perhaps one day satellites will be equipped with sails that are deployed at the end of their missions. The NASA mission NanoSail-D was launched in 2010 to test this idea. This problem compares the drag on a solar sail due to the Earth's upper atmosphere with the force exerted on the sail by the Sun's radiation and with the gravitational force exerted on the satellite by the Earth. NanoSail-D orbited at an altitude of 6.5×10^5 m, and its solar sail had an area of 10 m^2. Assume the satellite's total mass was about 4 kg and it was in a circular orbit. Also assume the Earth's atmosphere at that altitude has a density of about 5×10^{-14} kg/m^3 and a drag coefficient $C \approx 1$, and the Earth's gravitational force provided the centripetal acceleration. Finally, assume the sail perfectly reflected the sunlight. Estimate the magnitude of the three forces exerted on the satellite.

FIGURE P34.74 An artist's conception of a solar sail.

75. E CASE STUDY In Example 34.6 (page 1111), we imagined equipping 1950DA, an asteroid on a collision course with the Earth, with a solar sail in hopes of ejecting it from the solar system. We found that the enormous size required for the solar sail makes the plan impossible at this time. Of course, there is no need to eject such an object from the solar system; we only need to change the orbit. A much more pressing problem is Apophis, a 300-m asteroid that may be on a collision course with the Earth and is due to come by on April 13, 2029. It is unlikely to hit the Earth on that pass, but it will return again in 2036. If Apophis passes through a 600-m keyhole on its 2029 pass, it is expected to hit the Earth in 2036, causing great damage. There are plans to deflect Apophis when it comes by in 2029. For example, we could hit it with a 10- to 150-kg impactor accelerated by a solar sail. The impactor is launched from the Earth to start orbiting the Sun in the same direction as the Earth and Apophis. The idea is to use a solar sail to accelerate the impactor so that it reverses direction and collides head-on with Apophis at 80–90 km/s and thereby keeps Apophis out of the keyhole. Consider the momentum in the impactor's orbit (Fig. P34.75) when the solar sail makes an angle of $\theta = 60°$ with the tangent to its orbit. Current solar sails may be about 40 m on a side, but the hope is to construct some that are about 160 m on a side. Estimate the impactor's tangential acceleration when it is about 1 AU from the Sun. Keep in mind that the sail is neither a perfect absorber nor a perfect reflector, and a heavier impactor would presumably be equipped with a larger sail. Don't be surprised by what may seem like a very small acceleration.

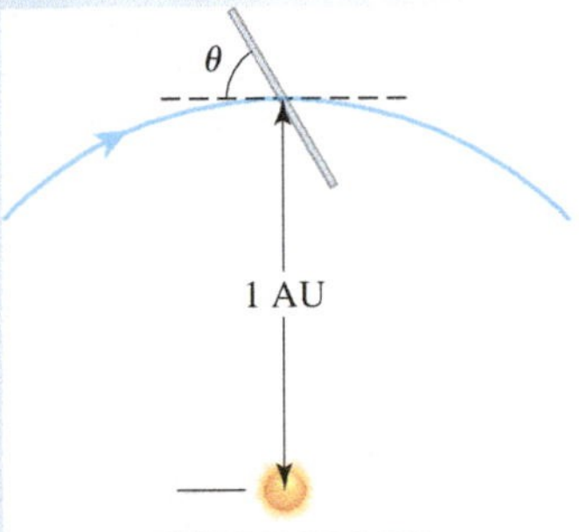

FIGURE P34.75

76. A Show that over one full cycle, the average value of $\sin^2\theta$ is $\frac{1}{2}$.

77. N A plane sinusoidal electromagnetic wave traveling in the positive z direction has a wavelength of 625 nm and a magnetic field amplitude of 48.0 μT. Obtain expressions for the electric and magnetic fields of this wave, assuming the electric field oscillates along the y direction.

78. C We know from Chapter 27 that insulating materials have a different electric permittivity (ε) than does a vacuum (ε_0). This difference can be accounted for by defining a *dielectric constant* κ such that $\varepsilon = \kappa\varepsilon_0$. How do you suspect the speed of an electromagnetic wave is affected when it enters an insulating material? Does the wave speed up, slow down, or remain at the same speed it had while traveling in a vacuum? Justify your answer.

79. N In 2010, the Japanese IKAROS satellite became the first to demonstrate the viability of using solar sails as a means of propulsion. The 315-kg spacecraft used a perfectly reflecting polyimide sail of area 2.00×10^2 m^2 and successfully reached the planet Venus. The intensity of the solar radiation near Venus is 2.62×10^3 W/m^2. Ignore the gravitational effects from the Sun and other bodies. **a.** What is the magnitude of the force exerted on the sail by the solar radiation near Venus? **b.** What is the acceleration of the spacecraft near Venus?

Problems 80, 81, and 82 are grouped.

80. A Consider two circular regions of the same cross-sectional area A with uniform magnetic fields, one with the field pointing into the page and the other pointing out of the page (Fig. P34.80). Initially, both fields have magnitude B_0, and the magnitude of each is increasing at the same rate dB/dt. Calculate $\oint \vec{E} \cdot d\vec{\ell}$ using **a.** circular path 1 and **b.** circular path 2.

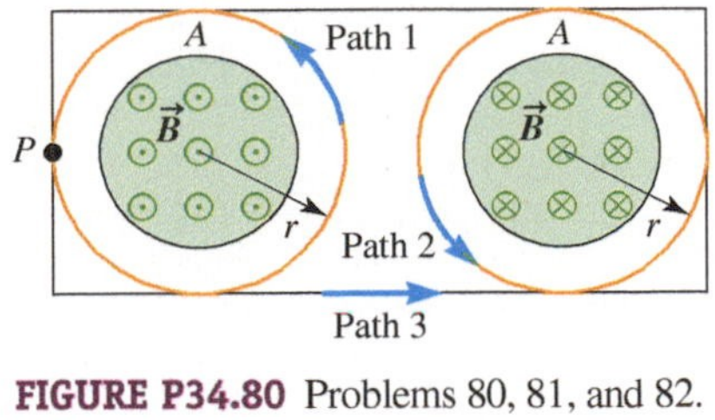

FIGURE P34.80 Problems 80, 81, and 82.

81. A Consider two circular regions of the same cross-sectional area A with uniform magnetic fields, one with the field pointing into the page and the other pointing out of the page (Fig. P34.80). Initially, both fields have magnitude B_0, and the magnitude of each is decreasing at the same rate dB/dt. Calculate $\oint \vec{E} \cdot d\vec{\ell}$ using rectangular path 3.

82. Consider two circular regions of the same cross-sectional area with uniform magnetic fields, one with the field pointing into the page and the other pointing out of the page (Fig. P34.80). Initially, both fields have magnitude B_0, and the magnitude of each is decreasing at the same rate dB/dt, so that the magnitude of either at time t is B.

a. A What is the magnetic flux through circular path 1?

b. A What is the magnetic flux through rectangular path 3?

c. C Explain why it is difficult to use either path 1 or path 3 to calculate the electric field at point P. If you could overcome that difficulty (perhaps by using a computer to integrate), would your result depend on which path you used? Explain.

83. N A linearly polarized sinusoidal electromagnetic wave moving in the positive x direction has a wavelength of 5.80 mm. The magnetic field of the wave oscillates in the xz plane and has an amplitude of 8.70 μT. The magnitude of the electric field vector for this wave can be written as $E = E_{max} \sin (kx - \omega t)$. What are **a.** the value of E_{max}, **b.** the wave number k, **c.** the angular frequency ω, **d.** the plane in which the electric field oscillates, **e.** the average Poynting vector, **f.** the radiation pressure exerted by this wave on a perfectly reflecting lightweight solar sail, and **g.** the acceleration of the solar sail if its dimensions are 5.00 m $\times$ 8.00 m and its mass is 34.5 g?

84. C In Section 34-1, we summarized classical mechanics. You may have noticed that conservation of momentum (and angular momentum) is not on that list. How is this principle already included in the list?

PART FIVE
Light

35 Diffraction and Interference

Underlying Principles

Huygens's principle

Major Concepts

1. Diffraction
2. Interference
3. Conditions for constructive and destructive interference
4. Coherence

Special Cases

1. Young's double-slit experiment (position of fringes; intensity)
2. Single-slit diffraction (position of dark fringes; intensity)
3. Double-slit diffraction (intensity)

Tools

Visual representations of light

Key Questions

What evidence supports the wave model for light?

How do we find the location and intensity of the fringes in interference and diffraction patterns?

Many people think vision is the most important human sense. It is certainly a highly developed sense. About 25% of the human brain is used for vision, so people have a preference for *visible* electromagnetic radiation. This entire part of the book is dedicated to the physics of visible light.

A doctoral graduate student in physics must pass a qualifying exam before being allowed to pursue her dissertation. As part of her qualifying exam, this book's author had to pass an oral exam during which she was asked, "How do you know that light is a wave?" How would you answer? You might be tempted to derive the wave equation from Maxwell's equations (Section 34-4). But such a derivation is based on a *theory*, whereas the question asks for *evidence*.

Before James Clerk Maxwell was born, the British physicist Thomas Young (1773–1829) considered this question. By the time Young was 19, he was highly proficient in Latin, which enabled him to master important scientific

iStockphoto.com/Meinzahn

FIGURE 35.1 Sunlight through windows produces crisp shadows that look like stars and hexagons.

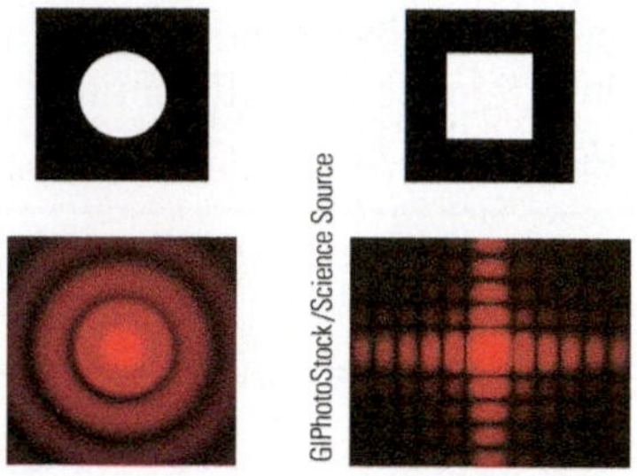

GIPhotoStock/Science Source

GIPhotoStock/Science Source

A. **B.**

FIGURE 35.2 The top row shows two apertures (not to scale). The bottom row shows the resulting patterns made by the light on a screen. When laser light passes through a small aperture, the pattern of light on a screen does not simply have the same shape as the aperture. **A.** The aperture is circular, but the light pattern shows concentric circles around a central disk. **B.** The aperture is a square, but the light pattern is a cross pattern of rectangles.

DIFFRACTION ✪ **Major Concept**

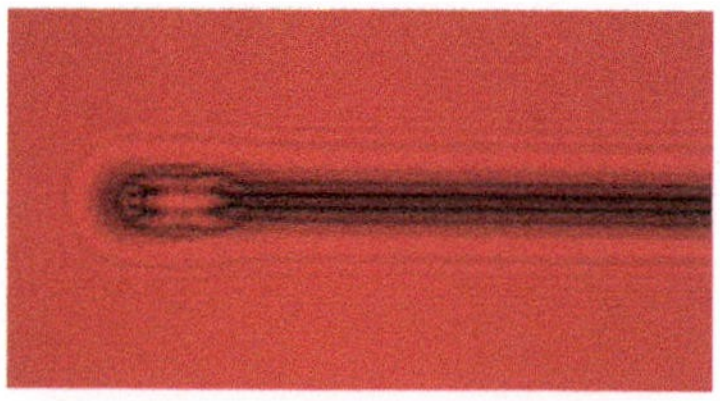

GIPhotoStock/Science Source

FIGURE 35.3 A needle illuminated by a bright red laser. Notice the diffraction pattern both inside the eye and around the outside of the needle in its shadow.

works, including Newton's *Principia* and *Opticks.* Young is best known for his 1803 experiment that provided strong evidence that light is a wave. Whereas a theory can be disproved by a single experiment, one experiment cannot prove a theory. A single experiment can only confirm or be consistent with a theory. If a number of experiments confirm a theory, the theory is accepted as a good model of nature. The wave model of light did not become widely accepted until another experiment, performed by Francois Arago in 1819 (10 years before Young died), confirmed Young's results. This chapter presents evidence that supports the wave model for light.

35-1 Light Is a Wave

Contemporary scientists model light both as a photon (particle) and as a wave, choosing the model that works best in any particular situation (Part VI). Several hundred years ago, scientists believed that light must be either a collection of particles or a wave, but not both. On one side of the debate were the French philosopher and mathematician René Descartes (1596–1650), the English physicist Robert Hooke (1635–1703), and the Dutch mathematician and astronomer Christiaan Huygens (1629–1695), who all believed that light is a wave. On the other side was Isaac Newton (1642–1727), who believed that light is made up of particles.

Figure 35.1 shows sunlight passing through an open window. The sharp shadows cast by the edges of the (glassless) window suggest that Newton was right: Light is made up of particles. Imagine that, instead of sunlight passing through the window, there was a steady stream of particles, such as paint droplets from a spray can, all moving in the same direction. The paint droplets that made it though the holes of the windows would paint the wall in the shape of the window. So, we expect particles passing through an aperture to trace the shape of the aperture. You might think observations of light passing through an aperture, such as those in Figure 35.1, would settle the debate in favor of the particle model.

If we look at laser light passing through a small aperture, however, we find evidence that does not support the particle model. In the top row of Figure 35.2 are two apertures with different shapes. In the bottom row are the resulting patterns of laser light projected onto a screen. These complicated patterns, known as **diffraction patterns**, do *not* simply trace the shapes of the apertures and cannot be explained by the particle model. Instead, diffraction patterns are a result of the interference of light *waves*.

Diffraction occurs when light passes through a narrow aperture or near the edge of an obstacle; the diffracted light bends and forms a diffraction pattern of bright and dark fringes. You may not notice diffraction because it is prominent only when the size of the aperture (or obstacle) is small enough compared to the wavelength of light. Because visible light has a wavelength of a few hundred nanometers, the aperture must be fairly small for us to observe diffraction. (For many of our applications in this chapter, we consider apertures and obstacles that are roughly a millimeter in size.)

Diffraction is not unique to light waves; we encountered the phenomenon when we studied mechanical waves. For example, the diffraction of sound waves allows you to hear around obstacles. If you sit in the back of a room so crowded that you cannot see the speaker standing at the front, you can still hear his voice. Because sound waves have wavelengths of roughly a meter, you can hear around obstacles, such as people, that are roughly a meter across. Because the wavelength of light is short compared to the width of a person, light essentially does not diffract around people. As a result, you cannot see around the people sitting in front of you.

Similarly, a diffraction pattern of bright and dark bands (or fringes) may be seen around the edges of an obstacle such as a sewing needle (Fig. 35.3). The pattern around the needle looks like thin fringes of light and darkness outlining the needle, and some of the bright fringes appear in the region that we normally think of as the needle's shadow.

Creating diffraction patterns does not require special equipment. You can make your own diffraction pattern by holding your hand in front of a bright light source, such as an open window or lightbulb. Extend two fingers so that they are parallel to each other with a gap in between. As you bring your fingers close together, look at the space between them, and in it you will see one or more dark lines running parallel to your fingers.

Both diffraction and **interference** are wave phenomena that result from the superposition of waves. Because both phenomena produce a characteristic pattern of bright and dark fringes, the distinction between interference and diffraction is arbitrary and not consistently made. However, a common distinction between the two phenomena is that diffraction involves only a single aperture (or source), whereas interference involves two or more such apertures.

INTERFERENCE ✪ Major Concept

Throughout this chapter, we'll consider the details of interference and diffraction experiments that produce such patterns of bright and dark fringes and thus confirm the wave model. Before we get into details, however, we need to consider some convenient ways to visually represent light as a wave.

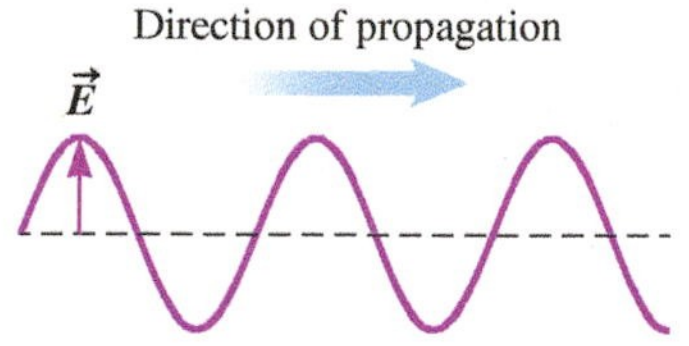

FIGURE 35.4 This representation of light looks like a wave on a string. You can imagine it is the oscillating electric field, so that the amplitude of the "string wave" represents the maximum electric field magnitude. The magnetic field oscillates in the plane that is perpendicular to both the electric field and the direction of propagation.

CONCEPT EXERCISE 35.1

Perhaps Newton never observed a diffraction pattern. If he had, what might he have concluded about light? Why do you suppose he was unable to see a diffraction pattern?

Representing Light as a Wave

From Chapter 34, we know that light is a transverse electromagnetic wave with an electric field oscillating perpendicular to a magnetic field and both fields oscillating perpendicular to the propagation direction (Fig. 34.8, page 1095). When we use this wave model, it is inconvenient and unnecessary to sketch both fields and the direction of propagation. Instead, we sometimes draw a single sinusoidal curve (Fig. 35.4), which closely resembles a transverse traveling wave on a string (Fig. 17.9, page 492). You can think of such a sketch as representing the oscillating electric field. Then you can imagine that the magnetic field oscillates perpendicular to the electric field—but there is no need to draw it.

VISUAL REPRESENTATIONS OF LIGHT ◉ Tool

However, a sketch of a single curve fails to represent a wave that may be two- or three-dimensional. To represent two-dimensional waves, such as waves on the surface of water, we imagine looking at the wave from above and draw solid lines to represent the wave crests (also called the **wave fronts**) and dashed lines to represent the wave troughs or valleys. Figure 35.5A shows a **plane parallel wave**. Each wave crest appears as a straight line parallel to all other crests. If you imagine viewing Figure 35.5A from the side, it would look like Figure 35.4, so a plane parallel wave is really a one-dimensional wave. Figure 35.5B shows a **circular wave** in which the wave crests form concentric circles. Circular waves travel in two dimensions. Because there is no good way to represent a three-dimensional wave on a two-dimensional surface, we must draw a two-dimensional slice through the wave. Figure 35.5B is exactly what we would draw to represent a three-dimensional **spherical wave** in which the wave crests form concentric spheres. Spherical waves are given off by sources such as the Sun and lightbulbs.

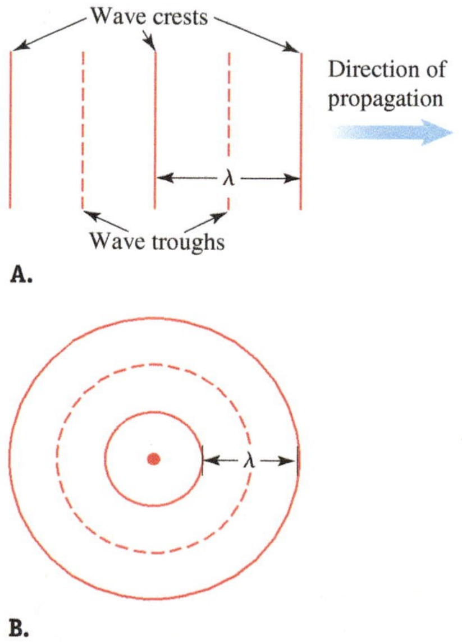

FIGURE 35.5 **A.** Representation of a plane parallel wave. **B.** Representation of a two-dimensional circular wave or a three-dimensional spherical wave.

Propagation of Light

Long before Maxwell discovered that light is an electromagnetic wave, researchers had studied the propagation of light beams. Huygens came up with a way to model the propagation of light. To understand Huygens's model, we use an analogy with waves on the surface of water. First, imagine a small ball bobbing up and down, creating circular waves on the water as in Figure 17.19A (page 502). Then imagine a rod bobbing up and down, creating parallel waves on the water as in Figure 17.21B. Huygens's idea is that if you replace the rod with a large number of small bobbing balls, the circular waves produced by the balls will add together to produce the plane parallel wave. Huygens extended this idea beyond such bobbing, or oscillation, to

HUYGENS'S PRINCIPLE

Underlying Principle

describe the forward motion of any wave front. Today we call this **Huygens's principle**: *Each point on a primary wave front serves as the source of a spherical wave, called a wavelet, that advances with a speed and frequency equal to those of the primary wave. The primary wave front at some later time is the envelope of these wavelets* (Fig. 35.6).

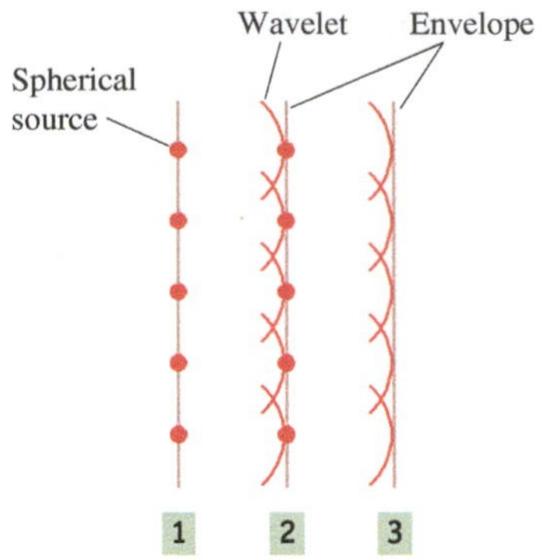

FIGURE 35.6 Huygens's principle applied to a plane parallel wave. The wave front starts at **1** and is made up of point sources creating spherical wavelets. The wave front at time **2** is the envelope of all these wavelets. This process is repeated to find the wave front at time **3** and at later times not shown.

CASE STUDY Part 1: Ghosts

A group of college students discusses a photograph.

Shannon: Check this out. I took a picture of ghosts (Fig. 35.7).

Avi: I've seen this kind of picture before. I think people call these things orbs. They're caused by dust, or something like that.

Shannon: No, I thought of that. I cleaned the lens and took another picture, and there were more orbs.

Avi: The dust I mean isn't in the camera. It's in the air.

Shannon: Are you kidding? Dust in the air would be too small, and it wouldn't have rings. Take a close look at the orb. It has a light spot in the middle and dark rings like this (Fig. 35.7, inset).

Cameron: I've seen that kind of pattern before. It kind of looks like the fringes around this sewing needle (Fig. 35.3).

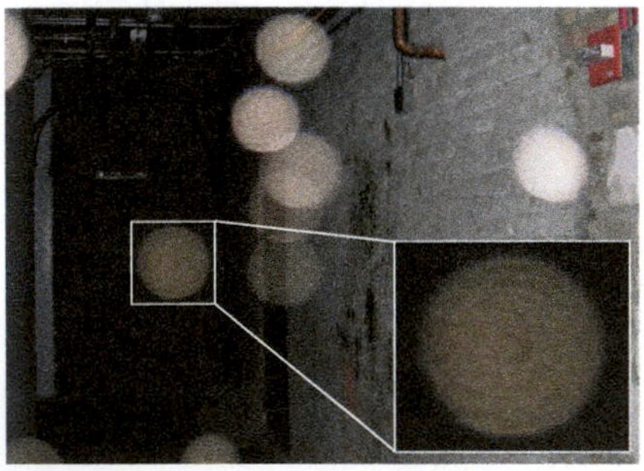

Adam Geremia

FIGURE 35.7 Some people believe orbs are ghosts. They are really diffractions of light due to dust or other small particles near the camera. *Inset:* A close-up of an orb reveals rings.

CASE STUDY Part 2: Science Competition

In 1819, the French Academy of Sciences held an essay competition to explain properties of light. At the time, most members of the Academy strongly supported the particle model for light, and they expected the competition to disprove the wave model. A physicist and military engineer named Augustin-Jean Fresnel submitted a theory based on the wave model. The French mathematician Siméon-Denis Poisson (1781–1840) was one of the judges and a supporter of the particle model. He read Fresnel's essay and thought he had found what he needed to disprove the wave model. Poisson showed that Fresnel's work predicted the existence of a bright spot in the center of the shadow cast by a round object, such as a disk or sphere. Poisson argued that, because it is absurd to expect to see a bright spot in the middle of a shadow, light must be made up of particles, not waves. In this chapter's two-part case study, we'll learn why the orb in Figure 35.7 has rings and find out whether Fresnel or Poisson was right.

CONCEPT EXERCISE 35.2

CASE STUDY Hiding?

You are kneeling behind a large sandbag in a game of paintball. On the other side of this obstacle, a friend fires paint in your direction. Do you get hit? Now imagine floating behind a rock in the ocean, with water waves moving toward the rock. Do you oscillate up and down when the wave arrives at your location? Explain your answers. Later in the case study, we'll use these situations as analogies for light that encounters an obstacle.

35-2 Sound Wave Interference Revisited

Finding the connection between two seemingly unrelated phenomena helps us learn something new and gain a deeper insight into both phenomena. In this section, we put aside our study of light for the moment in order to review sound wave interference.

Then we'll be more prepared to study the interference of light waves. Sound and light are similar, but they differ in two major ways: First, sound waves are longitudinal waves, whereas light is a transverse wave. Second, sound needs a medium, such as air, whereas light needs no medium.

We are interested in the interference pattern produced by two identical harmonic waves. Harmonic waves are mathematically simple, represented by a single sine or cosine function. In particular, we'll review sound waves emitted by speakers at a single tone (or frequency). Such a sound is not like music or speech; rather, it is something like an emergency alarm.

Section 18-3 presented three important properties of waves: (1) When two or more waves are present, the resulting wave is the **superposition** (addition) of those individual waves. (2) Constructive interference results when two identical harmonic waves are in phase; the resulting wave has twice the amplitude of either of the original waves (Fig. 18.15A, page 527). (3) Destructive interference results when two identical harmonic waves are 180° out of phase (Fig. 18.15B).

Consider the ideal situation in which two speakers produce identical harmonic sound waves in a large room as in Example 18.2. The waves that emerge from each speaker are three-dimensional hemispheres (Fig. 35.8). There are points of constructive and destructive interference throughout the room. In this ideal situation, places of constructive interference are loud and places of destructive interference are silent.

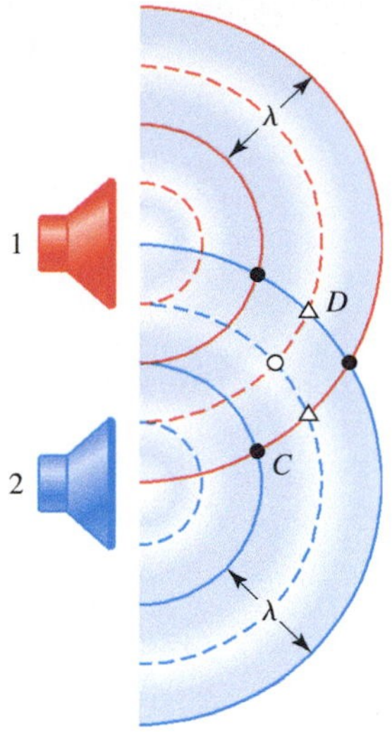

FIGURE 35.8 The filled circles (dots) are places where two peaks meet. The open circle is a place where two valleys meet. Both are places of constructive interference. The triangles are places where a valley meets a peak. These are places of destructive interference.

Whether there is constructive or destructive interference at a particular point depends on how far the point is from each speaker. If the difference in the distance traveled by two waves is an integer multiple of the wavelength, the result is constructive interference. The condition for constructive interference is

$$\Delta d = n\lambda \qquad (n = 0, 1, 2, 3, \ldots) \qquad (18.2)$$

where Δd is the difference in the distances traveled by the two waves. This condition is met by point C in Figure 35.8. To see this, use the wavelengths indicated to find the distance. C's distance from speaker 1 is $d_1 = 2\lambda$, and C's distance from speaker 2 is $d_2 = \lambda$. The difference $\Delta d = \lambda$, so $n = 1$ and there is constructive interference at point C.

CONDITIONS FOR CONSTRUCTIVE AND DESTRUCTIVE INTERFERENCE

Major Concepts

If one wave travels an extra $\lambda/2$, the two waves are 180° out of phase and interfere destructively. The distance between point D and speaker 1 is $3\lambda/2$, and D's distance from speaker 2 is 2λ (Fig. 35.8). So, point D is $\lambda/2$ farther from speaker 2 than it is from speaker 1. As a result, waves arrive at point D 180° out of phase, and point D is a place of destructive interference. More generally, if the difference in the distances traveled by two waves is a half-integer number of wavelengths, the result is destructive interference:

$$\Delta d = \left(n + \frac{1}{2}\right)\lambda \qquad (n = 0, 1, 2, 3, \ldots) \qquad (18.3)$$

Notice that we modified Equation 18.3, but the two expressions are equivalent.

CONCEPT EXERCISE 35.3

In Figure 35.8, the speakers are 2λ apart. What would happen at points C and D if speaker 2 was moved so that the speakers' separation increased to $5\lambda/2$?

Young's Double-Slit Experiment

Young's experiment, shown in Figure 35.9, looks much like Figure 35.8 with the two speakers replaced by two slits. Each slit is a source of identical light waves. As in Figure 35.8, these light waves interfere constructively or destructively depending on location. We experience constructive interference of sound waves as a loud sound and destructive interference as silence. In Young's experiment with light waves, we see places of constructive interference as bright fringes and places of destructive interference as dark fringes on a screen. Particles do not undergo interference, but waves do. So, Young's experiment supports the wave model. Young's work was not well received in England because Newton was a figure of national pride and opposition to his theories was difficult to accept.

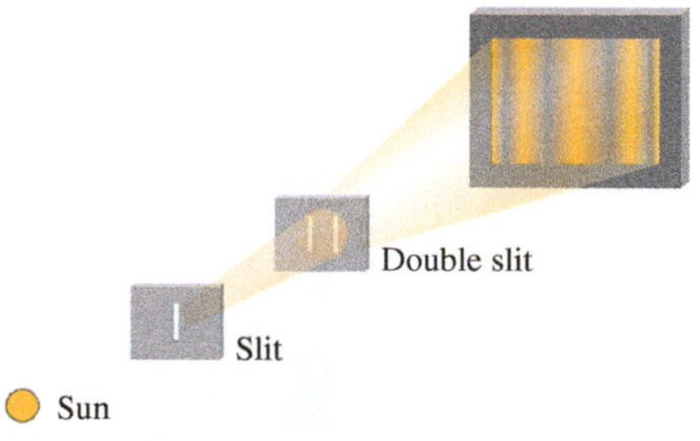

FIGURE 35.9 Young's experiment shows that light makes an interference pattern. Light passes through two small apertures. Usually the apertures are narrow slits, so Young's experiment is sometimes called the *double-slit experiment.* Young needed to use the single slit in the foreground to make sunlight that passes through the double-slit coherent. In contemporary experiments using laser light, this single slit is not necessary. (This diagram is a schematic showing only a crude representation of Young's experiment.)

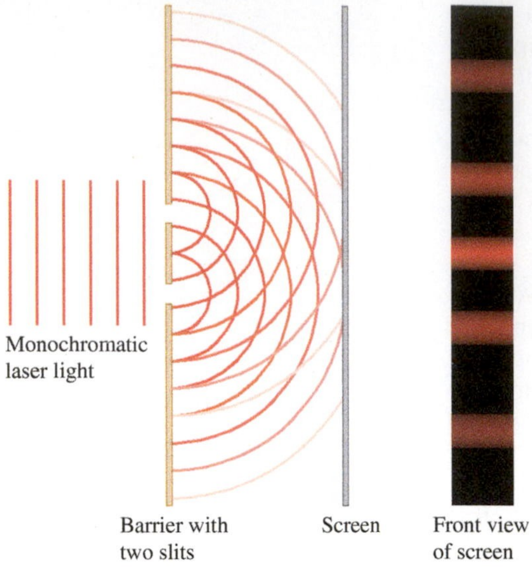

FIGURE 35.10 Schematic diagram of Young's experiment, as seen from above. The pattern on the screen cannot be seen from this perspective, but it is shown to the right. A front view of the screen is placed at the location of the screen in subsequent figures.

Figure 35.10 is a schematic drawing of Young's experiment, shown from above the apparatus. Bright light encounters a barrier that has two slits. The resulting interference pattern is projected onto the screen on the right. From this perspective, you can't see the pattern, so we have included a front view of the pattern on the screen. The brightest fringe is in the center of the pattern; the farther fringes are somewhat dimmer. The bright fringes gradually fade into the dark fringes with no sharp cutoff.

The diagram also shows the wave fronts. Before the light encounters the slits, it is a plane wave. According to Huygens's principle, we can think of the wave front as comprised of spherical wavelets. When the wave front encounters the barrier, only a couple of wavelets emerge. These hemispherical waves are like the sound waves that emerge from a speaker. As a result, the wave pattern from the two slits (Fig. 35.10) is like the pattern from the speakers (Fig. 35.8). So, the conditions for constructive and destructive interference in Young's experiment are also given by Equations 18.2 and 18.3.

COHERENCE ✪ **Major Concept**

In our daily lives we are often exposed to two light sources, so why don't we see interference patterns all the time? The answer is that four conditions must be met for us to see a fringe pattern like the one in Figure 35.10. First, the waves must be in the same medium, so they have the same propagation speed. Second, they must have the same frequency or, equivalently, the same wavelength. Third, they may differ by a (nonzero) phase constant φ, but that difference must remain constant in both space and time. The combination of the second and third conditions is called **coherence.** Two waves are said to be **coherent** if they are monochromatic sources of the same frequency and there is a constant phase difference between them. Finally, if two waves that are 180° out of phase are to cancel, the two waves must also have the same amplitude.

We don't normally see interference patterns because two ordinary light sources can maintain a constant phase difference for only about 10^{-8} s or less. Any interference pattern they produce is shifted around rapidly and randomly, so, on average, you cannot detect any interference. Also, different parts of the same light source, such as two pieces of the same lightbulb filament, do not produce coherent light waves. The atoms in any one source normally emit light of many different frequencies and amplitudes. Even if the source produces monochromatic (single frequency) light, the light waves from different parts of the source do not maintain a constant phase difference. The incoherence of light normally destroys the interference pattern. For the same reason, we normally don't notice interference when listening to music in an auditorium: The speakers in such a case do not emit coherent sound waves.

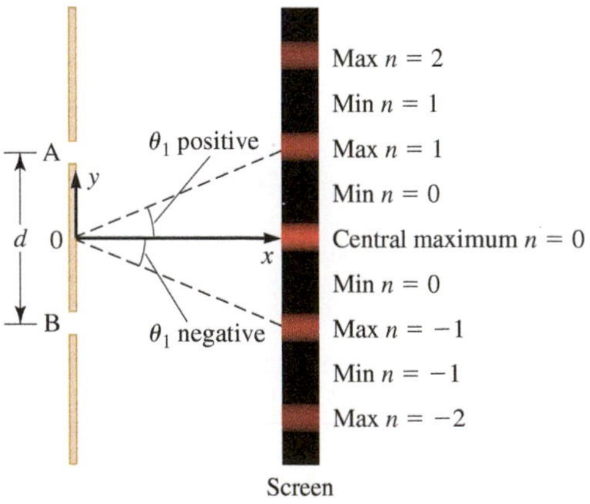

FIGURE 35.11 Places of constructive interference are bright and are known as maxima. Dark regions are places of destructive interference and are known as minima. Angles measured counterclockwise from the x axis are positive, and angles measured clockwise are negative.

In your laboratory or classroom, you will likely see a demonstration of Young's experiment using a laser. A laser produces monochromatic, coherent light. Young conducted his experiment in 1803, long before the laser was invented. He needed a source of bright light, though, so he used the Sun, but sunlight is neither monochromatic nor coherent. To create a nearly coherent source, Young placed another small aperture between the Sun and the double slits (Fig. 35.9). The light that makes it through this first slit is essentially a single source of light, which Young then split into two coherent sources. In other words, the light that emerges from the double slits is coherent. Because Young used sunlight, the interference pattern he saw showed all the colors arranged in bright and dark fringes. If he had used a monochromatic light source instead of the Sun, he would have seen the familiar monochromatic bright and dark pattern (Fig. 35.10). For the rest of our discussion of Young's experiment, we will assume that the light source is monochromatic and coherent; you can imagine it is a laser.

35-3 Young's Experiment: Position of the Fringes

In this section, we find expressions for the positions of the bright and dark fringes in Young's double-slit experiment. Figure 35.11 shows the coordinate system and the conventions we use to label these fringes. The origin of the coordinate system is the point on the barrier directly between the two slits. A positive x axis points from the origin to the screen. At the point where the x axis touches the screen ($y = 0$), there is a bright fringe. The bright fringes are also called **maxima**, and the bright fringe at $y = 0$ is called the **central maximum**. Each maximum is assigned an integer n starting with the central maximum, for which $n = 0$. Maxima on the positive y axis are assigned positive integers, and maxima on the negative y axis are assigned negative integers.

The dark fringes are called **minima** and are also assigned integers n, so you must determine from the context whether n refers to a maximum or a minimum. The labeling of dark fringes corresponds to that of bright fringes. Each dark fringe is surrounded by two bright fringes. The label n of the dark fringe is the same as the lower absolute magnitude of the labels of the two adjacent bright fringes. Consider the dark fringe just above the central maximum; it is surrounded by two bright fringes ($n = 0$ and $n = 1$). It is labeled $n = 0$, the same as the neighboring bright fringe with the lower absolute magnitude.

The position of each fringe is indicated by the angle θ measured from the x axis, normally with a subscript corresponding to the integer n. So, the position of the $n = 1$ maximum is given by the positive angle θ_1, and the position of the $n = -1$ minimum is given by the negative angle θ_1.

The central maximum is at $\theta_0 = 0$. The distance d_A from slit A to the central maximum is equal to the distance d_B from slit B to the central maximum: $d_A = d_B$ (Fig. 35.12). So, the waves that emerge from these slits arrive in phase at the position of the central maximum. Other fringes are not equidistant from both slits. Whether there is a bright fringe or a dark fringe at some location depends on the difference between the distances traveled by the light waves that emerge from each slit.

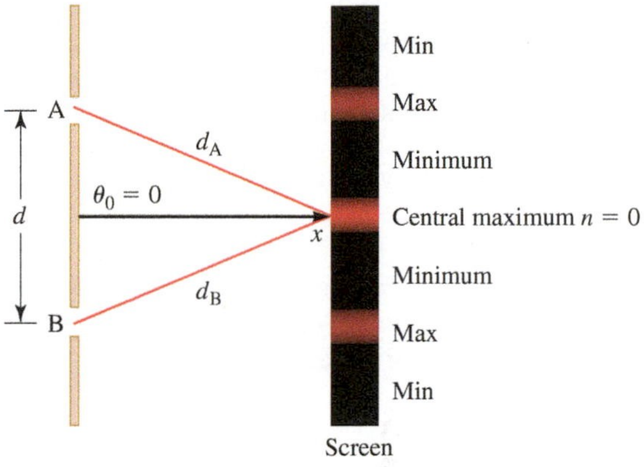

FIGURE 35.12 The distance from slit A to the central maximum equals the distance from slit B to the central maximum: $d_A = d_B$.

DERIVATION Bright and Dark Fringe Positions

YOUNG'S EXPERIMENT—POSITION OF FRINGES

Special Case

We will show that the positions of the maxima in Young's double-slit experiment are given by

$$d \sin \theta_n = n\lambda \quad (n = 0, \pm 1, \pm 2, \pm 3, \ldots) \tag{35.1}$$

and the positions of the minima are given by

$$d \sin \theta_n = \left(n + \frac{1}{2}\right)\lambda \quad (n = 0, \pm 1, \pm 2, \pm 3, \ldots) \tag{35.2}$$

where d is the separation between the slits and λ is the wavelength of the light.

In Figure 35.13, the distance from the barrier to the screen is x. We assume that the distance x is much greater than the separation d between the slits. Let's start by arbitrarily choosing a maximum at a positive position above the central maximum. The distance from slit A to this maximum is shorter than the distance from slit B to this maximum: $d_B > d_A$. This extra distance, the path-length difference $\Delta d = d_B - d_A$, determines whether the waves arrive in phase or out of phase.

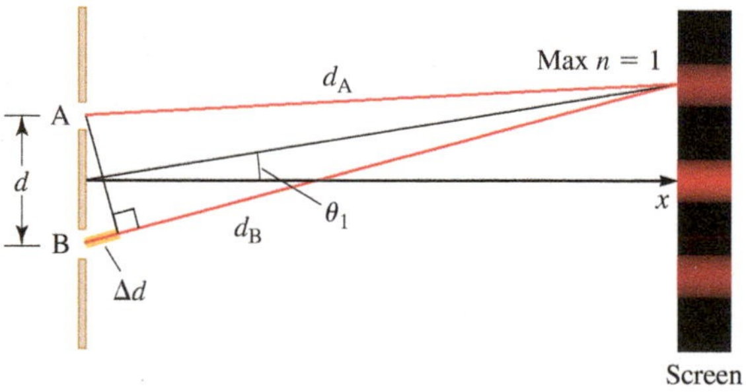

FIGURE 35.13

We must relate Δd to the angle θ using an approximation. As long as the distance to the screen is greater than the distance between the slits, we can approximate the two paths (one from each slit) to any point on the screen as nearly parallel (Fig. 35.14).

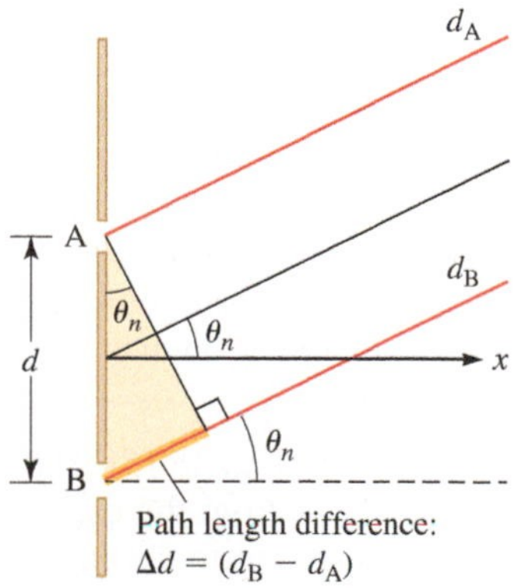

FIGURE 35.14

Derivation continues on page 1130 ▶

Angle θ_n to the nth bright fringe is measured from the x axis and also equals the angle at vertex A of the highlighted right triangle in Figure 35.14. Use this triangle to relate the angle θ to the path-length difference Δd.	$\sin\theta_n = \frac{\Delta d}{d}$ $\Delta d = d\sin\theta_n$ (35.3)
Setting the condition for constructive interference (Eq. 18.2) equal to Equation 35.3, we find an expression for the position of the positive maxima. To include the fringes below the central maximum at negative positions, we add ± to indicate that n may be positive or negative. Choose the positive sign for fringes at positive y positions (above the central maximum) and the negative sign for fringes at negative y positions (below the central maximum).	$\Delta d = n\lambda \quad (n = 0, 1, 2, 3, \dots)$ (18.2) $\Delta d = d\sin\theta_n = n\lambda \quad (n = 0, 1, 2, 3, \dots)$ $d\sin\theta_n = n\lambda \quad (n = 0, \pm1, \pm2, \pm3, \dots)$ ✓ (35.1)
From here, it is easy to find an expression for the positions of the minima. The path-length difference is still given by Equation 35.3, but there is destructive interference at the minima, so we set Equation 35.3 equal to Equation 18.3 for destructive interference. Again, we include ± to indicate that n may be positive or negative.	$\Delta d = \left(n + \frac{1}{2}\right)\lambda \quad (n = 0, 1, 2, 3, \dots)$ (18.3) $\Delta d = d\sin\theta_n = \left(n + \frac{1}{2}\right)\lambda$ $d\sin\theta_n = \left(n + \frac{1}{2}\right)\lambda \quad (n = 0, \pm1, \pm2, \pm3, \dots)$ ✓ (35.2)

COMMENTS

In Equations 35.1 and 35.2, d (the distance between slits) and λ should be in the same length units. However, it is not necessary or convenient to use meters; we typically use nanometers instead.

CONCEPT EXERCISE 35.4

When Young performed his double-slit experiment, he used sunlight. Explain why the interference pattern he saw showed a rainbow of colors.

EXAMPLE 35.1 A Birthday Present

You are given a red-light laser pointer for your birthday, but you don't know what wavelength of light it emits. You take the laser into your physics laboratory to perform a double-slit experiment. The slits are 0.350 mm apart. The distance from the barrier to the screen is 2.50 m. You measure the distance from the central maximum to the $n = 2$ maximum and find that it is 9.04 mm. What is the wavelength of your laser pointer's light?

INTERPRET and ANTICIPATE

The location of the bright fringes depends on the wavelength of the light, so if we know the position of one of the bright fringes (with respect to the central maximum), we can find the wavelength. Because the laser light is red, we expect to find an answer between 620 nm and 780 nm.

Sketch the situation (Fig. 35.15). The linear position y_2 of the $n = 2$ maximum is known from the experiment and is indicated on the figure. This corresponds to the angular position θ_2.

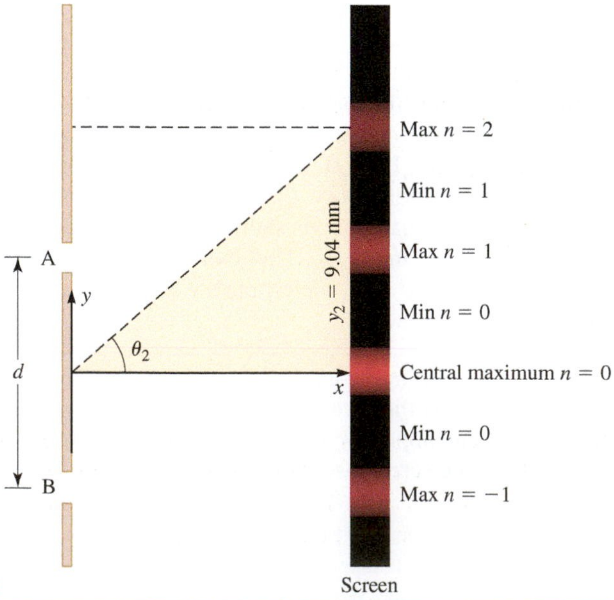

FIGURE 35.15

SOLVE
Find the angular position from the linear position using the highlighted right triangle (Fig. 35.15). Both y_2 and x must be in the same length units; we use millimeters here. (We keep an extra significant figure in this intermediate step.)

$$\tan \theta_2 = \frac{y_2}{x}$$

$$\theta_2 = \tan^{-1}\left(\frac{y_2}{x}\right) = \tan^{-1}\left(\frac{9.04 \text{ mm}}{2500 \text{ mm}}\right) = 0.2072°$$

Solve Equation 35.1 for wavelength, setting $n = 2$.

$$d \sin \theta_n = n\lambda = 2\lambda \qquad (35.1)$$

$$\lambda = \frac{d \sin \theta_n}{2} = \frac{(0.350 \text{ mm}) \sin 0.2072°}{2} = 6.33 \times 10^{-4} \text{mm}$$

$$\lambda = 633 \text{ nm}$$

CHECK and THINK
As expected, the wavelength is in the red range. Now let's think about improving this experiment. The distance was measured between the central maximum and the $n = 2$ maximum. This distance is fairly small and would have been even smaller if the $n = 1$ maximum were used instead. To make the distance easier to measure, the screen could be moved farther away. Also, it may be easier to measure the distance between extreme maxima, such as between $n = -3$ and $n = 3$.

EXAMPLE 35.2 Small Angles

Often the angular position of a fringe is a very small angle. Find an approximate expression for the y position in this case. Then repeat Example 35.1 using this approximate position equation and compare your results in the **CHECK and THINK** step.

INTERPRET and ANTICIPATE
Figure 35.15 shows the geometry of the situation, but because we are considering small angles, θ_2 is greatly exaggerated. The angle in Example 35.1 was very small ($\theta_2 = 0.207° = 3.16 \times 10^3$ rad), so a small-angle approximation should produce a result very close to the exact wavelength we found before.

SOLVE
From the highlighted triangle in Figure 35.15, we found an exact expression involving θ_n. When this angle is small, the tangent of the angle approximately equals the sine of the angle.

$$\tan \theta_n = \frac{y_n}{x}$$

$$\sin \theta_n \approx \frac{y_n}{x} \qquad (1)$$

Substitute Equation (1) into Equation 35.1 for a bright fringe.

$$d \sin \theta_n = n\lambda \quad (n = 0, \pm 1, \pm 2, \pm 3, \ldots) \qquad (35.1)$$

$$d\left(\frac{y_n}{x}\right) \approx n\lambda$$

Solve for y_n.

$$y_n \approx \frac{n x \lambda}{d} \quad (n = 0, \pm 1, \pm 2, \pm 3, \ldots) \qquad (35.4)$$

CHECK and THINK
To see whether Equation 35.4 leads to a different wavelength than in Example 35.1, solve Equation 35.4 for λ and substitute values, including $n = 2$. It makes sense that our approximate answer matches our exact answer because in this case angle θ_2 is very small.

$$\lambda \approx \frac{y_n d}{n x} \approx \frac{(9.04 \text{ mm})(0.350 \text{mm})}{2(2500 \text{mm})} \approx 6.33 \times 10^{-4} \text{ mm}$$

$$\lambda \approx 633 \text{ mm}$$

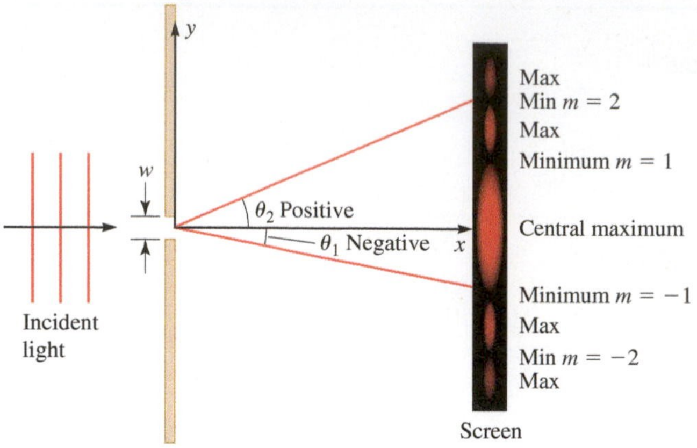

FIGURE 35.16 Coherent light incident from the left encounters a single slit of width w. A diffraction pattern is seen on a screen on the right. The origin of the coordinate system is at the center of the slit. The dark fringes are labeled by positive and negative integers.

35-4 Single-Slit Diffraction

As stated earlier, the distinction between interference and diffraction is arbitrary and not consistently made. However, a common distinction between the two phenomena is that diffraction involves only a single wave front, whereas interference involves a small number of coherent sources. So, diffraction is observed when coherent light passes through a single slit, whereas interference is observed when coherent light passes through two slits. The pattern observed in Young's double-slit experiment is called an interference pattern. However, we'll see that it is a combination of interference and diffraction. First, we need to take a close look at diffraction. To keep things relatively simple, we focus on the diffraction pattern produced by a single slit. In this section, we find a mathematical expression for the positions of the dark fringes produced by the slit.

Figure 35.16 is a schematic drawing of a single-slit diffraction experiment with a slit of width w. The coordinate system and labeling are similar to those for Young's double-slit experiment (Fig. 35.11). At the point ($y = 0$) where the x axis touches the screen, there is a bright fringe, the **central maximum**. The other bright fringes are sometimes referred to as **secondary maxima** or **side lobes**.

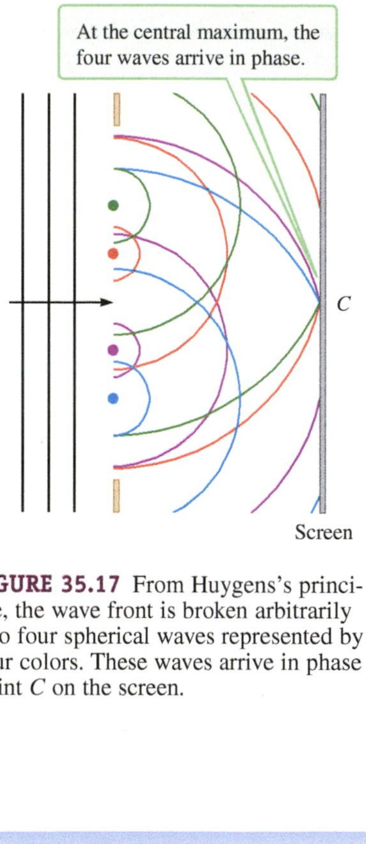

FIGURE 35.17 From Huygens's principle, the wave front is broken arbitrarily into four spherical waves represented by four colors. These waves arrive in phase at point C on the screen.

We can derive the position of the dark fringes (the minima) by following a procedure similar to that for the double-slit experiment. Here we label each minimum with an integer m. Minima at positive y positions have positive-integer labels, and those at negative y positions have negative-integer labels. This labeling convention is different from that in Figure 35.11 for two slits; in that case, there was an $n = 0$, but here there is no $m = 0$.

The position of each dark fringe is indicated by the angle θ measured from the x axis (positive if measured counterclockwise; negative if measured clockwise). Normally we use a subscript on θ that corresponds to the integer m.

Before we derive the position of the dark fringes, let's see why there is a central maximum. The key is to use Huygens's principle: Each point on a primary wave front serves as the source of a spherical wavelet that advances with a speed and frequency equal to those of the primary wave. The primary wave front at some later time is the envelope of these spherical waves (Fig. 35.6). We can model the wave front that enters the slit as the source of four spherical wavelets (Fig. 35.17). The four waves arrive in phase at point C on the screen, and the superposition of these four wavelets produces constructive interference. The choice to break up the wave front into four spherical waves is arbitrary. We could have chosen any number and found the same result: Constructive interference at C produces a central maximum. (The mathematics leading to the position of the other maxima is left for a more advanced course.)

DERIVATION Dark Fringe Positions in a Single-Slit Diffraction Pattern

We will show that the angular positions θ_m of the dark fringes in a single-slit diffraction pattern are given by

SINGLE-SLIT DIFFRACTION—POSITION OF DARK FRINGES Special Case

$$w \sin \theta_m = m\lambda \quad (m = \pm 1, \pm 2, \pm 3, \ldots) \tag{35.5}$$

We begin by finding expressions for the first and second dark fringes, and then extrapolate from these to find an expression for the position of any dark fringe.

Finding the positions of the dark fringes produced by a single slit is much like finding the positions of the fringes produced by a double slit. Break the slit into two zones A and B, each of width $w/2$. Use Huygens's principle to break up the wave front into as many spherical wavelets as we have zones. The source of one wavelet is at the top of zone A, and the source of the other is at the top of zone B. Figure 35.18 does not show the sources or the wavelets, but instead shows the distance from each source to the position of the first dark fringe on the screen. The distance d_B from source B is greater than the distance d_A from source A.

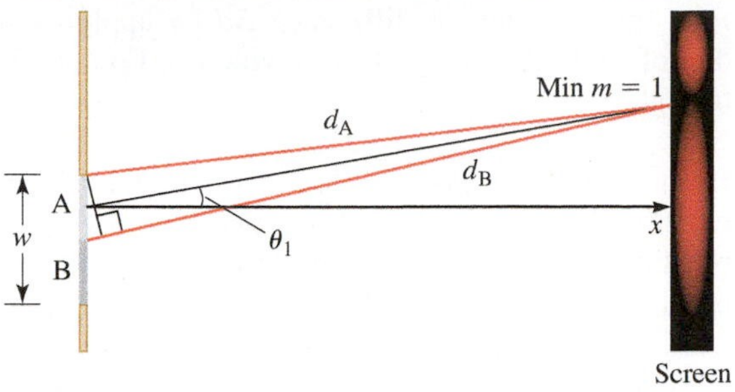

FIGURE 35.18

As in the case of Young's double-slit experiment, the path-length difference $d_B - d_A$ determines the position of the dark fringes. We again make the approximation that the distance x from the slit to the screen is much greater than the width w of the slit: $x \gg w$. In this approximation, the paths from each zone to any point on the screen are nearly parallel (Fig. 35.19).

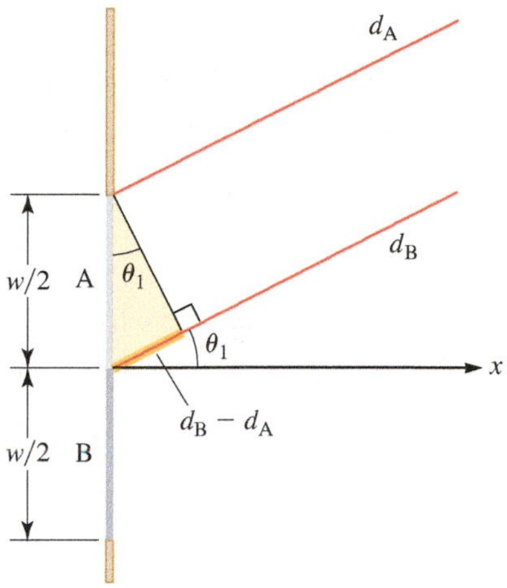

FIGURE 35.19

Use the highlighted right triangle in Figure 35.19 to find an expression for the path-length difference Δd.	$\Delta d \equiv d_B - d_A = \frac{w}{2}\sin\theta_1$
In order for there to be a dark fringe at the corresponding screen position, the two waves must arrive 180° out of phase. This occurs when the path-length difference equals half a wavelength. Equation (1) may be used to find the angular position θ_1 of the first dark fringe above or below the central maximum. (The angle below the central maximum is negative.)	$\Delta d = \frac{\lambda}{2}$ $\frac{w}{2}\sin\theta_1 = \frac{\lambda}{2}$ $w\sin\theta_1 = \lambda$ (1)

Find the position of the second dark fringe by splitting the slit into four zones, each with width $w/4$ and labeled A through D. We imagine four spherical wave sources, one at the top of each zone. The path length from each of these sources to the second dark fringe is labeled d with a subscript corresponding to the zone. As before, if $x \gg w$, the paths are approximately parallel. The path-length difference between the paths from adjacent zones is Δd.

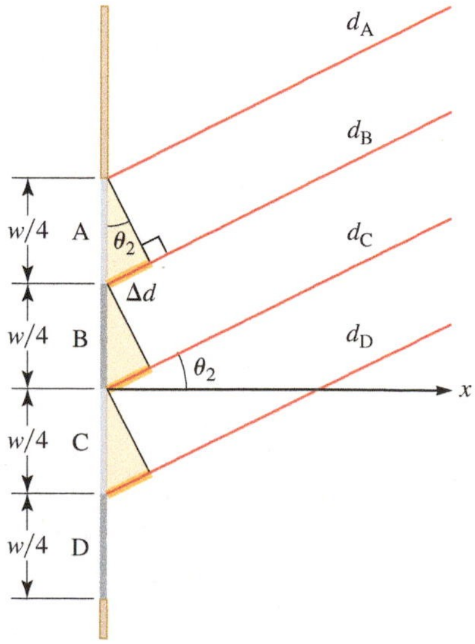

FIGURE 35.20

Derivation continues on page 1134 ▶

Find this path-length difference Δd by applying trigonometry to any one of the right triangles highlighted in Figure 35.20, such as the top triangle.	$\Delta d \equiv d_B - d_A = \frac{w}{4} \sin\theta_2$
Because θ_2 is the position of a dark fringe, the path-length difference must be half a wavelength. Equation (2) may be used to find the angular position θ_2 of the second dark fringe above or below the central maximum.	$\Delta d = \frac{\lambda}{2} = \frac{w}{4} \sin\theta_2$ $w \sin\theta_2 = 2\lambda$ (2)
We can generalize this procedure to find the angular position θ_m of any dark fringe by dividing the slit into an even number of zones n_{zones}. The path-length difference between the paths from adjacent zones is found by applying trigonometry to a right triangle.	$\Delta d = \frac{w}{n_{zones}} \sin\theta_m$
From our procedure for finding the positions of the first and second dark fringes, we see that the number of zones n_{zones} is two times the label m for the fringe.	$n_{zones} = 2m$ $\Delta d = \frac{w}{2m} \sin\theta_m$
At the position of any dark fringe, the path-length difference must be half a wavelength.	$\Delta d = \frac{\lambda}{2} = \frac{w}{2m} \sin\theta_m$ (3)
Rearrange Equation (3) and note that m is an integer other than 0 because there must be at least two zones.	$w \sin\theta_m = m\lambda \quad (m = \pm1, \pm2, \pm3, \ldots)$ ✓ (35.5)

COMMENTS

Equation 35.5 closely resembles

$$d \sin\theta_n = n\lambda \quad (n = 0, \pm1, \pm2, \pm3, \ldots) \tag{35.1}$$

which gives the angular position of the *bright* fringes produced in the double-slit interference pattern. So, you need to find a way to tell them apart. Here are two differences: (1) Equation 35.5 is used to find the angular position of the *dark* fringes, not the *bright* ones; and (2) in Equation 35.5, the first dark fringe is labeled $m = \pm1$, whereas the first bright fringe in Equation 35.1 is labeled $n = 0$.

Effect of Slit Width on the Diffraction Pattern

The diffraction pattern and, in particular, the width of the central maximum depend on the slit width. Let's say that the width Δy of the central maximum is the distance between the two first dark fringes (Fig. 35.16); that is,

$$\Delta y = 2y_1 = 2x \tan\theta_1 \tag{35.6}$$

where θ_1 is positive. For a slit illuminated by monochromatic light of wavelength λ, let's see what happens if the slit width equals the wavelength: $w = \lambda$. In that case, Equation 35.5 (with $m = \pm1$) becomes

$$\sin\theta_1 = \pm\frac{\lambda}{w} = \pm\frac{w}{w} = \pm1 \tag{35.7}$$

and so

$$\theta_1 = \pm 90°$$

When we substitute 90° into Equation 35.6, the central maximum's width is infinite: $\Delta y = 2x \tan 90° \rightarrow \infty$. In other words, if the slit width equals the light's wavelength, the first dark fringes are infinitely far from the central maximum, which completely covers the viewing screen. You will not see a diffraction *pattern*; instead, you will see that the center of the screen is very bright and the brightness fades away from the center.

By combining $\Delta y = 2x \tan \theta_1$ (Eq. 35.6) and $\sin \theta_1 = \pm\lambda/w$, we can plot Δy as a function of λ/w as in Figure 35.21. For monochromatic light, the horizontal axis is a measure of the slit width. The widest slit is on the far left, and the narrowest slit $(w = \lambda)$ for which a pattern may be observed on a flat screen is on the far right. This graph shows that for a wide slit, the central maximum is narrow, so the first dark fringes are very close together. By contrast, for a narrow slit, the central maximum is wide and the first dark fringes are far apart. And, as we already know, for the narrowest slit $(w = \lambda)$, the central maximum is infinite.

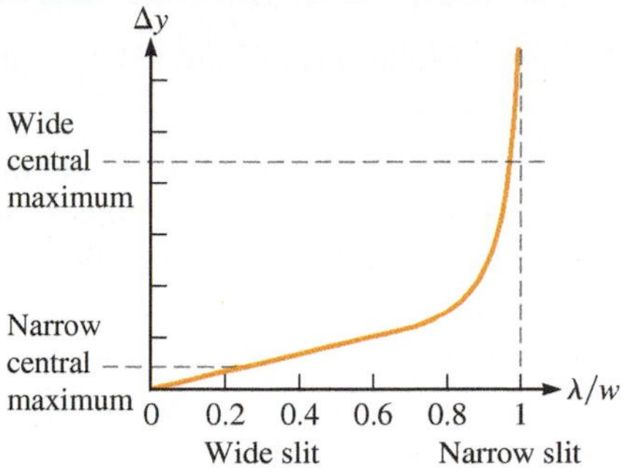

FIGURE 35.21 The position of the first dark fringe as a function of λ/w. A wide slit is on the left of the graph, and the slit width gets narrower toward the right. If λ is constant, the horizontal axis depends only on the slit width.

CONCEPT EXERCISE 35.5

When we studied Young's double-slit experiment, we mostly ignored the dark fringe pattern produced by diffraction. Use Figure 35.21 to describe situations in which that omission makes sense. Think especially about the single slit used in front of the double slit in Young's experiment (Fig. 35.9).

EXAMPLE 35.3 A Wide Slit

We just found that with a narrow slit, diffraction spreads out light into a wide central maximum, so we don't usually notice the dark fringe pattern produced. We also don't notice the diffraction pattern when light passes through a slit that is wide compared to the wavelength. Now consider a narrow beam of coherent, monochromatic light of wavelength $\lambda = 550$ nm passing through a single slit of width $w = 1.00$ cm. (This slit is nearly 20,000 times wider than the wavelength.) The light illuminates a screen at $x = 1.00$ m from the slit. On the screen, we see a bright central spot (Fig. 35.22, top). There are actually 20 dark fringes on either side of the central maximum, but you cannot see these fringes because they are too close together. The inset in Figure 35.22 is a magnified sketch of the diffraction pattern, showing the 20 fringes on either side of the central bright spot. (Even after magnification, the pattern is blurred.) Estimate the magnification required to see these fringes at the scale indicated in the figure. Proceed by finding the y positions of the $m = \pm 1, \pm 2, \pm 3, \pm 10$, and ± 20 dark fringes, and then find the magnification needed to display the images on this scale.

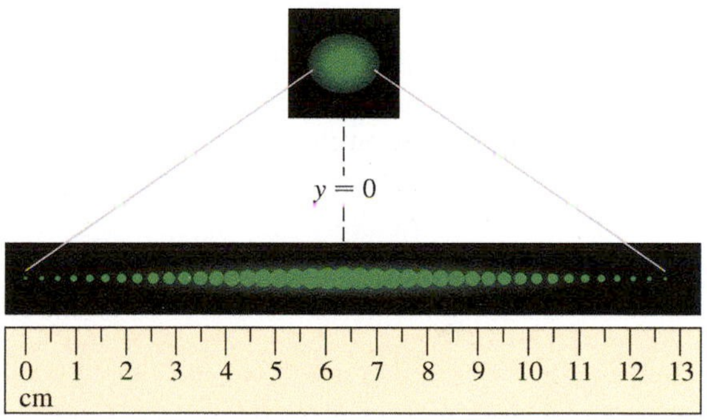

FIGURE 35.22 A bright central spot from a green laser and a magnified view of its diffraction pattern. The distance from the slit to the screen is x (out of the page, not labeled in this figure).

INTERPRET and ANTICIPATE

Because we normally don't see the dark fringes when light passes through a wide slit, we expect that all 20 fringes are close to the central maximum. The angular position should be very small, so we can use the small-angle approximation: $\sin \theta \approx \tan \theta \approx \theta$ for θ in radians.

SOLVE

Use trigonometry to write an expression for the y position of the mth dark fringe.	$y_m = \pm x \tan \theta_m$ (1)
Equation 35.5 gives the angular position of the mth dark fringe.	$w \sin \theta_m = m\lambda \quad (m = \pm 1, \pm 2, \pm 3, \ldots)$ (35.5)
Apply the small-angle approximation to Equations (1) and 35.5.	$y_m \approx \pm x\theta_m$ (2) $w\theta_m \approx m\lambda$ (3)

Example continues on page 1136 ▶

Combine Equations (2) and (3) by eliminating θ_m.

$$y_m \approx \pm mx\frac{\lambda}{w}$$

The wavelength λ, slit width w, and distance to the screen x have the same values for all dark fringes. Substitute these values. It is convenient to work in millimeters.

$$y_m \approx \pm m(1000\text{ mm})\left(\frac{5.50 \times 10^{-4}\text{ mm}}{10.0\text{ mm}}\right)$$

$$y_m \approx \pm m(5.50 \times 10^{-2}\text{ mm})$$

Substitute the 10 values of m. We show five examples here.

m	y_m (mm)
±1	$\pm5.50 \times 10^{-2}$
±2	±0.110
±3	±0.170
±10	±0.550
±20	±1.10

According to the values we just found, there are 40 dark fringes in a region only 2.20 mm across. The scale in Figure 35.22 shows fringes spread over about 12.7 cm when magnified. Find the magnification M by dividing the distance seen in the figure by the actual distance.

$$M = \frac{127\text{ mm}}{2.20\text{ mm}} = 57.7$$

$$M \approx 60\times$$

CHECK and THINK

As expected, the 40 dark fringes are all close to (less than 1.10 mm from) the central maximum. To see the fringes spread out as they are in the inset of Figure 35.22, you must magnify the diffraction pattern by about 60 times, which requires a microscope. This explains why we normally don't notice the diffraction pattern when light passes through a very wide slit (compared to its wavelength): The dark fringes are so close together that they blur into a seemingly dark region around the central maximum.

EXAMPLE 35.4 CASE STUDY What Is an Orb?

Shannon: Of course, I know I didn't take a picture of a ghost, but I still don't think I took a picture of dust. Dust should just look like a small spot.

Cameron: Yeah, the orb is big and has a very regular ring pattern.

Avi: I see that too. But what is causing the pattern? How exactly did you take the picture?

Shannon: I used my digital camera. I wiped the lens clean, so I'm sure there was no dust on the lens. I had the flash on because it was dark in the basement.

Assume the orb is due to dust in the air. Come up with a simple, conceptual reason why the orb has a ring pattern.

SOLVE

Orbs are best seen when a flash photo is taken by a camera whose flash is close to the lens, as it often is on contemporary digital cameras. The illuminated dust particle reflects light. The small dust is effectively the source of (nearly) coherent light—something like the sunlight that passed through the small slit in Young's experiment (Fig. 35.9). That light must pass through the camera's circular aperture. Notice that the diffraction pattern for a circular aperture (Fig. 35.2A) looks much like the close-up of the orb in Shannon's photo (Fig. 35.7, inset). So, it is likely that the ring pattern is caused by diffraction.

CHECK and THINK

There is usually dust in the air, yet we rarely see orbs in photographs. However, orbs are becoming more common as people use digital photography. Orbs are best seen in digital photography because orbs are most visible in the UV spectrum. Film is not very sensitive to UV, but the detectors used in digital cameras are. Also, orbs are more likely to be noticed if the dust is relatively close to the camera's lens because then the lens acts as a magnifier and the orb may appear very large on the photo.

35-5 Young's Experiment: Intensity

In Section 35-4, we derived expressions for the positions of the bright and dark fringes. In this section, we derive an expression for the intensity of the pattern. Examine the pattern in Figure 35.10; the bright fringes are *not* uniformly bright and the dark fringes are not uniformly dark. Instead, the intensity gradually changes from bright to dark and back again. So, we expect the mathematical description of the intensity to be a function that gradually varies from high intensity to zero intensity and back again.

DERIVATION Intensity of the Double-Slit Interference Pattern

Again, consider Young's double-slit experiment in which the slits are separated by distance d, the light has wavelength λ, and θ is the angular position shown in Figure 35.11. We show that the intensity of the interference pattern is given by

$$I = I_{max} \cos^2 \frac{\varphi}{2} = \frac{2E_0^2}{\mu_0 c} \cos^2 \frac{\varphi}{2} \qquad (35.8)$$

YOUNG'S EXPERIMENT—INTENSITY Special Case

where

$$\varphi = \frac{2\pi}{\lambda} \Delta d = \frac{2\pi}{\lambda} d \sin \theta \qquad (35.9)$$

For this derivation, we need only consider the electric field part of the electromagnetic waves. Let the field from slit A have subscript A and the field from slit B have subscript B. The light that passes through the two slits is coherent, so those two sets of waves have the same amplitude E_0 and angular frequency ω. The only difference between the two waves that hit any particular point on the screen is a constant phase difference φ due to the path-length difference.	$E_A = E_0 \sin \omega t$ $E_B = E_0 \sin (\omega t + \varphi)$
According to the principle of superposition, the electric field at the location of the screen is the sum of the individual waves.	$E_{tot} = E_A + E_B = E_0 [\sin \omega t + \sin (\omega t + \varphi)]$
Apply the trigonometric identity $\sin \alpha \pm \sin \beta = 2 \sin \frac{1}{2}(\alpha \pm \beta) \cos \frac{1}{2}(\alpha \mp \beta)$ (Appendix A) to simplify the expression for E_{tot}.	$E_{tot} = E_0 [2 \sin \frac{1}{2}(2\omega t + \varphi) \cos \frac{1}{2}(\varphi)]$ $E_{tot} = 2E_0 \cos \frac{\varphi}{2} \sin \left(\omega t + \frac{\varphi}{2}\right) \qquad (1)$
Equation (1) describes an electric wave whose maximum value E_{max} depends on the relative phase φ between the original two waves.	$E_{max} \equiv 2E_0 \cos \frac{\varphi}{2} \qquad (2)$

Derivation continues on page 1138 ▶

Using our definition for E_{max} (Eq. 2), we rewrite Equation (1) in a more familiar form.	$E_{tot} = E_{max} \sin\left(\omega t + \frac{\varphi}{2}\right)$
The intensity of an electromagnetic wave is given by Equation 34.30. Substitute for E_{max} from Equation (2).	$I = \frac{E_{max}^2}{2\mu_0 c}$ (34.30) $I = \frac{1}{2\mu_0 c}\left(2E_0 \cos\frac{\varphi}{2}\right)^2 = \frac{1}{2\mu_0 c}\left(4E_0^2 \cos^2\frac{\varphi}{2}\right)$ $I = \frac{2E_0^2}{\mu_0 c}\cos^2\frac{\varphi}{2}$ (3)
We define the maximum intensity I_{max} and write Equation (3) in terms of I_{max}.	$I_{max} \equiv \frac{2E_0^2}{\mu_0 c}$ $I = I_{max}\cos^2\frac{\varphi}{2} = \frac{2E_0^2}{\mu_0 c}\cos^2\frac{\varphi}{2}$ ✓ (35.8)
We still need to show that $\varphi = (2\pi/\lambda)\Delta d$ (Eq. 35.9). Let's look at a few specific cases to see if we can find a general trend. When there is no path-length difference, the two waves are in phase. When the path-length difference is half a wavelength, the two waves are 180° (π rad) out of phase. When the path-length difference is a whole wavelength, the two waves are back in phase, with a phase difference of 360° (2π).	see table below
Reviewing this short table, we find that, in general, the phase difference (in radians) is $2\pi/\lambda$ times the path-length difference, which is $\Delta d = d \sin\theta$ (Eq. 35.3).	$\varphi = \frac{2\pi}{\lambda}\Delta d = \frac{2\pi}{\lambda} d \sin\theta$ ✓ (35.9)

Δd	φ (radians)
0	0
$\lambda/2$	π
λ	2π

COMMENTS

The expression for intensity in Equation 35.8 involves a cosine function, so as expected, the intensity varies from zero to a maximum value I_{max}. Note that the intensity cannot be negative. Because the cosine function is squared, the intensity varies from zero to I_{max} as a function of φ (or, of θ).

Finding the Maxima and Minima from the Intensity

The intensity expression $I = I_{max}\cos^2(\varphi/2)$ is consistent with the expressions we found for the locations of the bright and dark fringes. In the center of a bright fringe, the intensity must be at its maximum. This occurs when $\cos(\varphi/2) = \pm 1$. So there is constructive interference when

$$\frac{\varphi}{2} = 0, \pm\pi, \pm 2\pi, \pm 3\pi, \ldots$$
$$\varphi = 0, \pm 2\pi, \pm 4\pi, \pm 6\pi, \ldots$$

Now let's use $\varphi = (2\pi/\lambda)\Delta d = (2\pi/\lambda) d \sin\theta$ (Eq. 35.9) to write this condition for constructive interference as

$$d \sin\theta = \Delta d = \frac{\lambda}{2\pi}\varphi$$
$$d \sin\theta = \Delta d = 0, \pm\lambda, \pm 2\lambda, \pm 3\lambda, \ldots$$

This is exactly the constructive interference condition we found earlier: $d \sin \theta_n = n\lambda$ (Eq. 35.1).

Likewise, in the center of a dark fringe, the intensity must be zero, which occurs when $\cos(\varphi/2) = 0$. So there is destructive interference when

$$\frac{\varphi}{2} = \pm\frac{\pi}{2}, \pm\frac{3\pi}{2}, \pm\frac{5\pi}{2}, \pm\frac{7\pi}{2}, \ldots$$

$$\varphi = \pm\pi, \pm 3\pi, \pm 5\pi, \pm 7\pi, \ldots$$

Again, when we use $\varphi = (2\pi/\lambda)\Delta d = (2\pi/\lambda) d \sin \theta$ (Eq. 35.9), this condition for destructive interference is

$$d \sin \theta = \Delta d = \frac{\lambda}{2\pi}\varphi$$

$$d \sin \theta = \Delta d = \pm\frac{\lambda}{2}, \pm\frac{3\lambda}{2}, \pm\frac{5\lambda}{2}, \pm\frac{7\lambda}{2}, \ldots$$

This is the same destructive interference condition we found earlier: $d \sin \theta_n = (n + \frac{1}{2})\lambda$ (Eq. 35.2).

Plotting Intensity

It is helpful to plot the intensity as a function of either the phase difference φ or the path-length difference $\Delta d = d \sin \theta_n$. Figure 35.23 is a graph of $I = I_{max} \cos^2(\varphi/2)$ (Eq. 35.8). The horizontal axis has been labeled in terms of both the phase difference and the path-length difference using $\varphi = (2\pi/\lambda)\Delta d$ (Eq. 35.9).

Also included on this graph is the intensity of a single source $I_A = E_0^2/(2\mu_0 c)$ (Eq. 34.30). The single source's intensity is a horizontal line on this graph because it does not depend on φ. The maximum intensity of the double-slit pattern is four times greater than the intensity of the single source. You might have expected the maximum intensity through two slits to be *twice* as great as the intensity through a single slit. The electric field's amplitude is twice as great, but the intensity depends on the electric field squared, so it is four times greater.

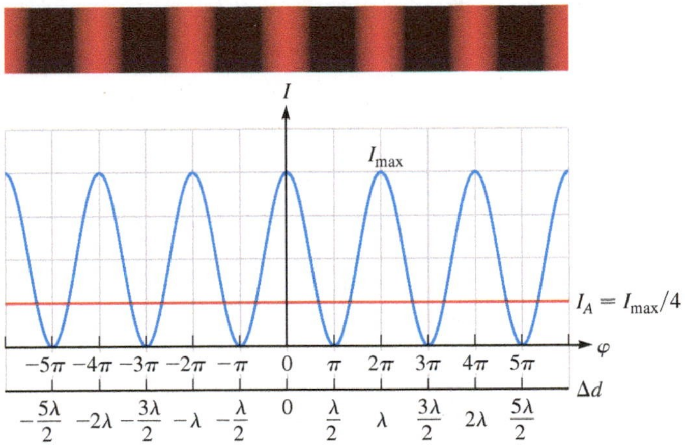

FIGURE 35.23 A graph of intensity for the double-slit interference pattern (blue) and for a single source (red).

You may have noticed something odd about the expression for intensity derived here. According to $I = I_{max} \cos^2(\varphi/2)$ or Figure 35.23, the intensity of every fringe is exactly the same. So all the bright fringes should be equally bright no matter how far they are from the central maximum. But this is not what we observe. Instead, the central maximum is the brightest fringe, and the fringes fade as a function of their distance from the central maximum (Fig. 35.10). Our derivation did not agree with this observation because we ignored the diffraction of light through the individual slits. We will consider double-slit diffraction in Section 35-7. We can ignore diffraction and use $I = I_{max} \cos^2(\varphi/2)$ (Eq. 35.8) as a good approximation as long as we consider only the fringes near the central maximum.

EXAMPLE 35.5 **Half the Maximum**

For Young's double-slit experiment, find the first three values of phase difference φ and path-length difference Δd for which the intensity is half its maximum. Include positive and negative values of each quantity.

Example continues on page 1140 ▶

INTERPRET and ANTICIPATE

The interference pattern's intensity varies from zero at a dark fringe to I_{max} at a bright fringe. We have already found the phase differences and path-length differences that produce dark and bright fringes, as shown on the graph in Figure 35.23. We can find the phase differences and path-length differences where the intensity is half its maximum by using either algebra or the graph of intensity. We use the graph in Figure 35.23 to anticipate our algebraic solution.

Imagine a horizontal line at $I = I_{max}/2$ in Figure 35.23, and read off the first three values of φ and Δd that intersect this line. Next, we use algebra to check our results against the values we found from the graph.	$\varphi = \pm\frac{\pi}{2}, \pm\frac{3\pi}{2}, \text{ and } \pm\frac{5\pi}{2}$ $\Delta d = \pm\frac{\lambda}{4}, \pm\frac{3\lambda}{4}, \text{ and } \pm\frac{5\lambda}{4}$
SOLVE Set the intensity in Equation 35.8 equal to the intensity ($I = I_{max}/2$) we're interested in.	$I = I_{max}\cos^2\frac{\varphi}{2}$ (35.8) $\frac{I_{max}}{2} = I_{max}\cos^2\frac{\varphi}{2}$
The maximum intensity cancels. The cosine function is squared, so taking its square root gives two possible solutions.	$\frac{1}{2} = \cos^2\frac{\varphi}{2}$ $\cos\frac{\varphi}{2} = \pm\frac{1}{\sqrt{2}}$
Write down the first three solutions in radians.	$\frac{\varphi}{2} = \pm\frac{\pi}{4}, \pm\frac{3\pi}{4}, \text{ and } \pm\frac{5\pi}{4}$ $\varphi = \pm\frac{\pi}{2}, \pm\frac{3\pi}{2}, \text{ and } \pm\frac{5\pi}{2}$
Use these solutions for φ and Equation 35.9 to find the first three solutions for Δd.	$\varphi = \frac{2\pi}{\lambda}\Delta d$ (35.9) $\Delta d = \frac{\lambda}{2\pi}\varphi = \pm\frac{\lambda}{4}, \pm\frac{3\lambda}{4}, \text{ and } \pm\frac{5\lambda}{4}$

CHECK and THINK

The solution we found algebraically is exactly the same as the values we read off the graph in Figure 35.23. Remember that the intensity of the double-slit interference pattern varies smoothly from zero to I_{max} and back again. Sometimes it is easy to forget this, and instead to think that the intensity is either I_{max} at a bright fringe or zero at a dark fringe, with no transitions in between.

35-6 Single-Slit Diffraction Intensity

In Section 35-5, we found the intensity of the interference pattern produced by Young's double-slit experiment. Now let's find the intensity of the single-slit diffraction pattern.

Figure 35.24 shows a single-slit pattern with four points, *c*, *p*, *d*, and *q*. Within the diffraction pattern, light intensity depends on position. The maximum intensity I_{max} is in the center of the pattern at $\theta = 0$. The intensity drops off for points that are farther from the center, such as point *p*. The intensity is zero at the first dark fringes on either side of the central maximum. Then the intensity increases, but never again reaches I_{max}. So, the intensity at point *q* is lower than I_{max}. Mathematically, the **intensity of the single-slit diffraction pattern** as a function of angular position θ is given by

SINGLE-SLIT DIFFRACTION—INTENSITY

Special Case

$$I = I_{max}\left(\frac{\sin\alpha}{\alpha}\right)^2 \quad (35.10)$$

where α is defined as

$$\alpha \equiv \frac{1}{2}\varphi = \frac{\pi w}{\lambda}\sin\theta \quad (35.11)$$

where *w* is the slit width and φ is described below.

In this section, we do not formally derive the intensity expression (Eqs. 35.10 and 35.11), but we explore the concepts behind such a derivation. We also explain how the intensity expression is related to the expression for the position of the dark fringes. Finally, we plot the intensity for different slit widths.

Where the Intensity Expression Comes From

First, we show how the intensities at the four labeled points in Figure 35.24 are determined. As when we found the position of the dark fringes, we use Huygens's principle and divide the slit into a number of zones. Consider the four zones shown in Figure 35.20. The electric (and magnetic) field at any point on the screen is the result of the superposition of the four spherical wavelets. The relative phase of those four wavelets depends on the distance each wave must travel to get to that particular point on the screen. So the total electric field E_{tot} at each point is found by adding up the electric fields of the four waves. To get the intensity, we must find E_{tot}^2 because the intensity is proportional to the square of the electric field's amplitude $I \propto E_{max}^2$ (Eq. 34.30).

Table 35.1 illustrates how the relative phase of the four wavelets determines the intensity at each of the four points. For each point, we graph the electric fields of the four wavelets, their total electric field, and the square of the total electric field. Table 35.1 also includes three parameters for each point. The first is the sine of the angular position θ because it enters into Equation 35.11. The second parameter is the phase difference φ between the wavelet from zone A and the one from zone D (Fig. 35.20). The third parameter is α as given by Equation 35.11.

At the central point c in Figure 35.24, the angular position is $\theta = 0$. The four spherical waves arrive at c in phase and interfere constructively, so the total electric field is very strong. The intensity at the central point is the maximum intensity in the pattern, I_{max}.

Now let's skip to the first dark fringe—point d, at the positive angular position given by $\sin\theta = \lambda/w$. According to Equation 35.11, α is given by

$$\alpha = \frac{\pi w}{\lambda}\left(\frac{\lambda}{w}\right) = \pi$$

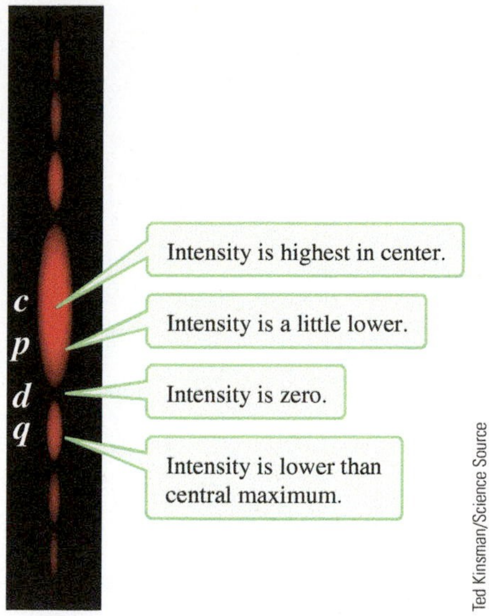

FIGURE 35.24 A diffraction pattern produced on a screen by laser light that passes through a single narrow slit. The label c stands for center and d for dark.

TABLE 35.1 Finding the intensity of points *c*, *p*, *d*, and *q* in Figure 35.24. (For point *c*, the four waves overlap and cannot be seen separately.)

Point c	Point p	Point d	Point q
$\sin\theta = 0$	$\sin\theta = 3\lambda/8w$	$\sin\theta = \lambda/w$	$\sin\theta = 3\lambda/2w$
$\varphi = 0$	$\varphi = 3\pi/4$	$\varphi = 2\pi$	$\varphi = 3\pi$
$\alpha = 0$	$\alpha = 3\pi/8$	$\alpha = \pi$	$\alpha = 3\pi/2$

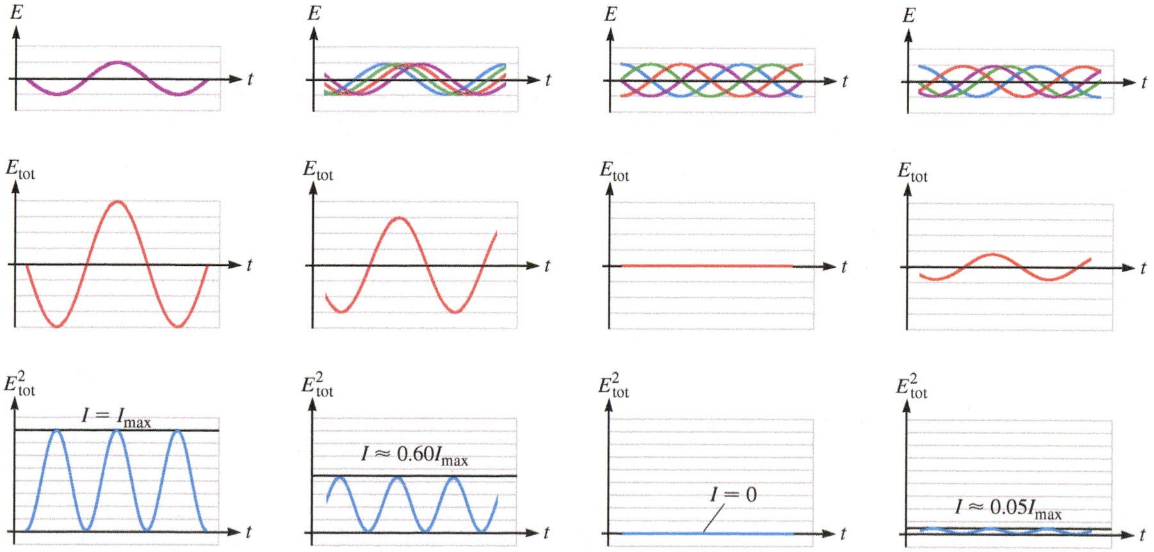

Also by Equation 35.11, the total phase difference φ between the wavelet from zone A and the one from zone D is $\varphi = 2\alpha = 2\pi$.

Let's think about what this total phase difference of $\varphi = 2\pi$ means for the four wavelets that arrive at d. Suppose the time dependence of the wave from A is given by

$$E_A(t) = E_{max} \sin(\omega t)$$

If the total phase difference across the slit is $\varphi = 2\pi$, the phase difference between each of the four adjacent zones must be $2\pi/4 = \pi/2$. So the time dependence of the waves from the other three zones must be

$$E_B(t) = E_{max} \sin\left(\omega t + \frac{\pi}{2}\right)$$

$$E_C(t) = E_{max} \sin(\omega t + \pi)$$

$$E_D(t) = E_{max} \sin\left(\omega t + \frac{3\pi}{2}\right)$$

These four electric fields plotted in Table 35.1 for point d interfere destructively and cancel, so that their superposition E_{tot} is zero. Thus, E_{tot}^2 is also zero and because the intensity is proportional to E_{max}^2, the intensity is zero, exactly as you would expect for a dark fringe.

Now let's consider point p in Figure 35.24, between the center c and the first dark fringe d. Point p falls within the central maximum, but it is not at the center of this maximum. We have chosen point p at an angular position given by $\sin\theta = 3\lambda/8w$. According to Equation 35.11, the total phase difference is $\varphi = 3\pi/4$, and $\alpha = 3\pi/8$. The wavelets do not interfere completely constructively or completely destructively. Compare the column for point p in Table 35.1 to the graphs of E_{tot}^2 for points c, d, and q, and you'll see that the intensity for p is between zero and I_{max} (about 60% of I_{max}).

Finally, consider point q in the center of the second maximum. It is halfway between the first ($\sin\theta_1 = \lambda/w$) and the second ($\sin\theta_2 = 2\lambda/w$) dark fringes, so its angular position is given by $\sin\theta = 3\lambda/2w$. According to Equation 35.11, the total phase difference is $\varphi = 3\pi$, and $\alpha = 3\pi/2$. The four wavelets that arrive at point q are not in phase. Furthermore, the intensity at point q is lower than it is at either c or p (about 5% of I_{max}).

We have explored the intensity at four points (c, d, p, and q) in the diffraction pattern by dividing the slit into four zones (A through D) of width $w/4$. In a thorough derivation of Equations 35.10 and 35.11, the slit is divided into infinitely many infinitesimal zones. The total electric field at an arbitrary point is then found from the superposition of the electric fields from these infinitesimal zones. Even without doing such a formal derivation, however, we can see how the relative intensity in the diffraction pattern is determined.

Dark Fringes

Our next goal is to connect the expression for intensity with the position of the dark fringes. To do this, we first find the values of α that correspond to zero intensity by setting $I = 0$ in Equation 35.10:

$$I = I_{max}\left(\frac{\sin\alpha}{\alpha}\right)^2 = 0$$

This expression is zero when $\sin\alpha = 0$, so

$$\alpha_m = m\pi \quad (m = \pm1, \pm2, \pm3, \ldots) \tag{35.12}$$

Substitute Equation 35.12 into Equation 35.11 to find an expression for the angular positions θ where the intensity is zero:

$$\alpha_m = \frac{\pi w}{\lambda}\sin\theta_m = m\pi \quad (m = \pm1, \pm2, \pm3, \ldots)$$

$$w\sin\theta_m = m\lambda \quad (m = \pm1, \pm2, \pm3, \ldots) \tag{35.5}$$

This is exactly what we found for the position of the dark fringes (Eq. 35.5).

Why didn't we start with $m = 0$ in Equation 35.12? Because if $\alpha = 0$, both the numerator and the denominator in Equation 35.10 are zero. To find the value of the fraction in this case, take the limit using L'Hôspital's rule:

$$\lim_{\alpha \to 0}\left(\frac{\sin \alpha}{\alpha}\right) = \lim_{\alpha \to 0}\left(\frac{\cos \alpha}{1}\right) = 1$$

So, at $\alpha = 0$ (corresponding to $\theta = 0$), the intensity equals its maximum $I(0) = I_{max}$, which obviously is not the location of a *dark* fringe. Instead, the most intense spot is in the center, at $\theta = 0$ (Fig. 35.24).

Visualizing Intensity

To visualize the intensity of the diffraction pattern produced by a variety of slit widths, imagine shining laser light of wavelength λ through three different slits. Figure 35.25 shows graphs of I versus θ for these three situations. The vertical axis is the relative intensity I/I_{max}, so its maximum value is 1. Each part of the figure extends only 30° to either side of the central point. The vertical and horizontal scales are thus the same in each case.

The diffraction pattern in Figure 35.25A comes from a slit whose width equals the wavelength of the light ($w = \lambda$). We found in this case that the first dark fringe is at $\theta = 90°$, so the screen is completely covered by the central maximum. At the edges of this small portion of the screen that extends to $\theta = \pm 30°$, the intensity drops to 40% of its maximum value.

The pattern in Figure 35.25B comes from the same laser, using a slit that is three times wider ($w = 3\lambda$). The central maximum in this case extends only to about $\theta = \pm 20°$, and at these points we see the first dark fringes. We can also see about half of a side lobe on either side of the central maximum. Notice that the side lobe is much dimmer than the central maximum.

Finally, Figure 35.25C shows the pattern produced by a slit that is six times wider than the first slit ($w = 6\lambda$). The central maximum is now much thinner, extending to only roughly $\theta = \pm 10°$. We now see two side lobes on either side of the central maximum, narrower than those in Figure 35.25B.

Comparing the three graphs in Figure 35.25 leads to three general statements. First, a wide slit produces a narrow central maximum. Second, a wide slit produces more side lobes (and dark fringes) on the screen. Third, a wide slit produces narrow side lobes. In fact, at the limit of a very wide slit, the pattern we see is like the pattern due to diffraction around an edge, as around the outside of the sewing needle shown in Figure 35.3.

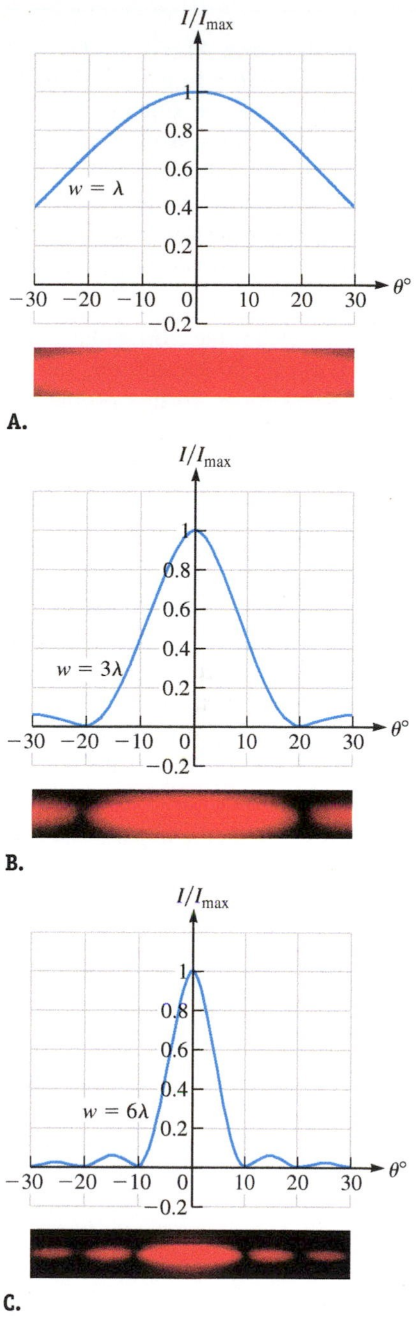

FIGURE 35.25 In all three cases, a graph of intensity versus angular position is shown above the pattern that is actually observed on the screen. These patterns were made using a slit whose width **A.** equals the wavelength of the light; **B.** is three times the wavelength of the light; and **C.** is six times the wavelength of the light.

CONCEPT EXERCISE 35.6

In Figure 35.25, we imagined using a single light source with a fixed wavelength and three different slits of widths λ, 3λ, and 6λ. We could have equally imagined using a single slit of width w and varying the wavelength. In this case, order the graphs from shortest wavelength to longest. If light of all three wavelengths passed simultaneously through a single slit, which wavelength would produce the broadest central maximum?

EXAMPLE 35.6 Half as Intense

Red light of wavelength $\lambda = 625$ nm passes through a single slit of width $w = 1875$ nm. Estimate to three significant figures the (positive) angular position θ where light in the diffraction pattern equals half of its maximum.

Example continues on page 1144 ▶

INTERPRET and ANTICIPATE

If we could set I in Equation 35.10 equal to $I = I_{max}/2$ and then solve for α, we could use Equation 35.11 to find the angular position θ. However, we cannot isolate α in Equation (1). Instead of solving for α, we must estimate its value by making a good guess at what the answer might be. We can use Figure 35.25 as a basis for our initial guess. This also sets our expectation for the final estimate.

$$I = I_{max}\left(\frac{\sin\alpha}{\alpha}\right)^2 \qquad (35.10)$$

$$\frac{I}{I_{max}} = \frac{1}{2} = \left(\frac{\sin\alpha}{\alpha}\right)^2 \qquad (1)$$

SOLVE

In Figure 35.25, the horizontal axis is angular position θ. Once we read off our initial guess for θ, we can use Equation 35.11 to make an initial guess for α. In this problem, the slit width is three times the wavelength. From Figure 35.25B, read off θ for $I/I_{max} = 0.5$.

$$\theta \approx 8° = 0.14 \text{ rad}$$

$$\alpha = \frac{\pi w}{\lambda}\sin\theta \qquad (35.11)$$

$$\alpha = \frac{\pi(1875 \text{ nm})}{(625 \text{ nm})}(\sin 0.14) = 1.3 \text{ rad}$$

Our initial guess corresponds to $\alpha = 1.3$ rad. To find a better estimate, calculate $[(\sin\alpha)/\alpha]^2$ for values near the initial guess. The goal is to find the value of α that comes closest to $[(\sin\alpha)/\alpha]^2 = 1/2$ as required in Equation (1). Make a table to help organize the work. Our initial guess gives a result that is a little greater than 1/2. When we try a smaller α, we find that the result is greater than 1/2. So we try a larger α, and we find that the result is also greater than 1/2 but getting closer. An even larger value for α shows that we have gone too far because the result is smaller than 1/2.

α (rad)	$\left(\frac{\sin\alpha}{\alpha}\right)^2$
1.3 initial guess	0.549 too high
1.25 low guess	0.576 worse
1.35 high guess	0.522 better
1.40 high guess	0.495 too low

Now we know that α is between 1.35 and 1.40. Make a new table of guesses in this range. We were asked to find an estimate to three significant figures. We check our answer to four significant figures to confirm that we don't need to round up to 1.40. So our best estimate is $\alpha = 1.39$ rad.

α (rad)	$\left(\frac{\sin\alpha}{\alpha}\right)^2$
1.36	0.517 better
1.38	0.506 better
1.39	0.500 great!
1.392	0.4997 too low

Use Equation 35.11 to find θ from our estimate of α.

$$\alpha = \frac{\pi w}{\lambda}\sin\theta \qquad (35.11)$$

$$\theta = \sin^{-1}\left(\frac{\alpha\lambda}{\pi w}\right) = \sin^{-1}\left[\frac{(1.39 \text{ rad})(625 \text{ nm})}{\pi(1875 \text{ nm})}\right]$$

$$\theta = 0.148 \text{ rad} = 8.48°$$

CHECK and THINK

As expected, the intensity drops to about half its maximum at $\theta \approx 8°$. The entire central maximum is about 40° wide (Fig. 35.25B). Because the intensity drops by 50% at roughly $\theta \approx 8°$ from the center, the inner 16° of the maximum is fairly bright compared to the outer 24°. (As a final note, some problem-solvers may use a computer-based approach instead of the method used here.)

35-7 Double-Slit Diffraction

In Section 35-3, we imagined that the slits in Young's double-slit experiment were very narrow compared to the wavelength of light ($w \ll \lambda$). In that case, the diffraction pattern from each slit was wide enough to cover the entire screen, so the interference pattern of light from the two slits dominated what we saw. In particular, the intensity of the bright fringes was nearly constant. In effect, we ignored the diffraction of light through the individual slits.

In this section, we examine the pattern produced by two slits that are wide enough that we cannot ignore diffraction. The resulting pattern is the combination of diffraction through the individual slits and interference between the two slits. This sort of pattern is often referred to as **double-slit diffraction**, a shorter way of describing the pattern produced by the interference of light from two slits and the diffraction due to each slit. We limit our study to the intensity produced by two slits of equal width.

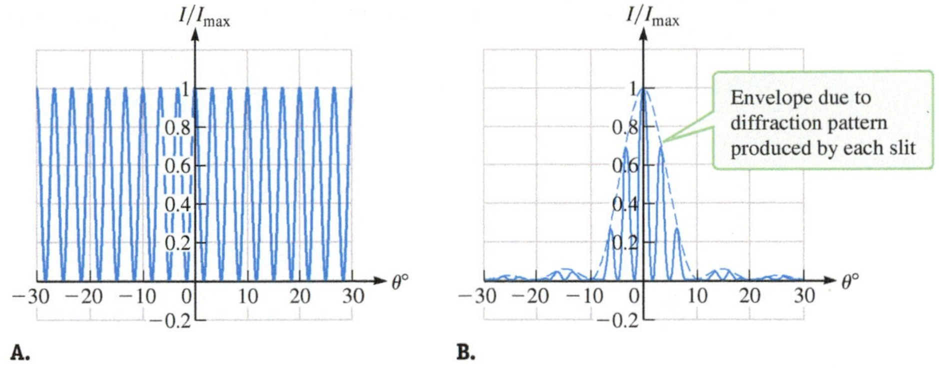

FIGURE 35.26 A. A double-slit interference pattern produced by very narrow slits. **B.** A double-slit interference pattern produced by slits of width $w = 6\lambda$.

Interference from light passing through two very narrow slits ($w \ll \lambda$) produces bright fringes with equal intensities (Fig. 35.26A). Now imagine repeating this experiment with wider slits, six times the wavelength. In this case, diffraction through each slit is important. The intensity due to diffraction through either slit *alone* would look like Figure 35.25C. But the intensity produced by double-slit diffraction looks like a *combination* of double-slit interference (Fig. 35.26A) and single-slit diffraction (Fig. 35.25C). Figure 35.26B shows that single-slit diffraction defines an envelope for the intensity we see through both slits; the fringes are in the same location as the fringes in Figure 35.26A, but their intensities are limited by the diffraction pattern in Figure 35.25C.

For easy comparison, Figure 35.27 shows both a graph of intensity and the pattern you would see on the screen. The thin fringes are due to interference of light from the two slits. Even if the slits were very narrow, the thin fringes would still be there. The fringes are also grouped in a series of broad bands due to the diffraction pattern produced by each slit. If you covered up one of the slits, the thin fringes would disappear, but the broad bands would still be there.

Mathematically, the intensity of double-slit diffraction is given by

DOUBLE-SLIT DIFFRACTION—INTENSITY
Special Case

$$I = I_{max}\left(\frac{\sin\alpha}{\alpha}\right)^2 \cos^2\beta \tag{35.13}$$

where $\alpha = (\pi w/\lambda)\sin\theta$ (Eq. 35.11) and β is given by

$$\beta = \frac{\varphi}{2} = \frac{\pi d}{\lambda}\sin\theta \tag{35.14}$$

where d is the center-to-center separation between the slits. We expect from Figure 35.26 that the intensity of double-slit diffraction is a combination of the intensities of single-slit diffraction and double-slit interference. To check, compare Equation 35.13 to Equation 35.10 for single-slit diffraction. Then compare Equation 35.13 to $I = I_{max}\cos^2\beta$ (Eq. 35.8 with $\beta = \varphi/2$) for double-slit interference. You can see that the $[(\sin\alpha)/\alpha]^2$ term comes from single-slit diffraction, while the $\cos^2\beta$ term comes from double-slit interference.

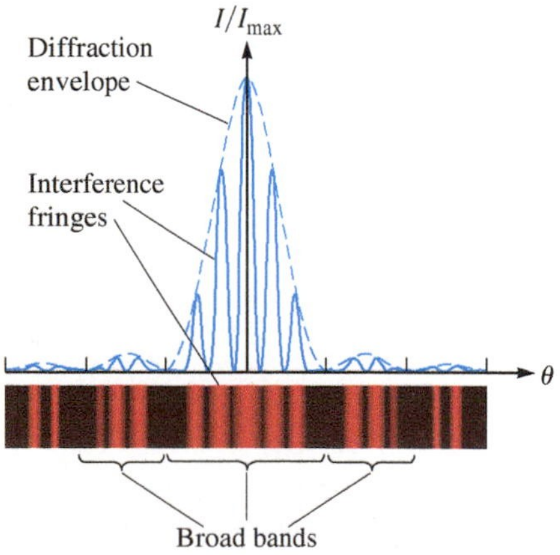

FIGURE 35.27 The relationship between the graph of intensity and the pattern you see on the screen.

Equation 35.13 actually describes the intensity of either the double-slit interference pattern or the single-slit diffraction pattern, so you don't have to remember three separate equations. To see this, first consider double-slit interference. We must imagine that the slits are very narrow ($w \ll \lambda$), so that $\alpha \to 0$ (Eq. 35.11). From L'Hôspital's rule, when $\alpha \to 0$,

$$\frac{\sin\alpha}{\alpha} \to \cos\alpha \to 1$$

So, when the slits are very narrow, Equation 35.13 reduces to

$$I = I_{max}\cos^2\beta$$

which is exactly the intensity of the double-slit interference pattern (Eq. 35.8). To find the single-slit diffraction pattern, we imagine that the two slits overlap, which means $d \to 0$ and $\beta \to 0$ (Eq. 35.14). Substitute $\beta = 0$ into Equation 35.13 to find

$$I = I_{max}\left(\frac{\sin\alpha}{\alpha}\right)^2 \cos^2 0 = I_{max}\left(\frac{\sin\alpha}{\alpha}\right)^2$$

This is the same as Equation 35.10 for the single-slit diffraction pattern.

EXAMPLE 35.7 Double-Slit Interference Versus Diffraction

You are working in a laboratory with a single laser of wavelength λ. The laser is used to illuminate one or two slits, and interference or diffraction patterns are observed as a result. Give your results to three significant figures.

A If there are two very thin slits ($w \ll \lambda$) separated by $d = 18\lambda$, at what (positive) angular position is the $m = 1$ bright fringe? Give your answer in radians and degrees. What is the relative intensity I/I_{max} of this fringe?

INTERPRET and ANTICIPATE

Because the slits are so narrow, we can ignore diffraction. This situation is exactly like double-slit interference (Section 35-3).

SOLVE

Use Equation 35.1 to find the positive position of the $m = 1$ bright fringe.

$$d \sin\theta_m = m\lambda \tag{35.1}$$

$$d \sin\theta_1 = \lambda$$

$$\theta_1 = \sin^{-1}\left(\frac{\lambda}{d}\right) \tag{1}$$

$$\theta_1 = \sin^{-1}\left(\frac{1}{18}\right)$$

$$\theta_1 = 5.56 \times 10^{-2}\,\text{rad} = 3.18°$$

To find the relative intensity, substitute Equation (1) into Equation 35.14 to find β when $m = 1$.

$$\beta = \frac{\pi d}{\lambda}\sin\theta_1 \tag{35.14}$$

$$\beta = \pi$$

Now substitute $\beta = \varphi/2 = \pi$ into Equation 35.8.

$$I = I_{max}\cos^2\beta$$

$$\frac{I}{I_{max}} = \cos^2\beta = \cos^2\pi$$

$$\frac{I}{I_{max}} = 1$$

CHECK and THINK

It makes sense that $\beta = \pi$ because $\beta = \varphi/2$ and the phase difference must be $\varphi = 2\pi$ in order for there to be constructive interference at the first bright fringe. It also makes sense that the relative intensity of the bright fringe is 1 because the intensity of all fringes in a double-slit interference pattern is the same (Fig. 35.26A).

B Now the laser is used to illuminate a single slit of width $w = 6\lambda$. What is the relative intensity at the angular position you found in part A?

INTERPRET and ANTICIPATE

The key to solving this problem is to find α from θ_1, found in the preceding part. The relative intensity is only 1 at the central maximum ($\theta = 0$), so we expect that the relative intensity at θ_1 should be less than 1.

SOLVE

Use Equation 35.11 to find α.

$$\alpha = \frac{\pi w}{\lambda} \sin\theta \quad (35.11)$$

$$\alpha = 6\pi \sin 3.18° = 1.046 \text{ rad}$$

Substitute α into Equation 35.10 to find the relative intensity for a single slit.

$$I = I_{max}\left(\frac{\sin\alpha}{\alpha}\right)^2 \quad (35.10)$$

$$\frac{I}{I_{max}} = \left(\frac{\sin 1.046 \text{ rad}}{1.046 \text{ rad}}\right)^2 = 0.684$$

CHECK and THINK

The intensity at this particular location is about 68% of the maximum, consistent with our expectations.

C The laser is used to illuminate two slits of width $w = 6\lambda$ and separation $d = 18\lambda$. What is the relative intensity at the angular position you found in part A?

INTERPRET and ANTICIPATE

Both interference and diffraction play important roles in this case. Because we are considering the same angular position, we can use the values we found for α and β in parts A and B.

SOLVE

Substitute α and β in Equation 35.13 to find the relative intensity for double-slit diffraction.

$$I = I_{max}\left(\frac{\sin\alpha}{\alpha}\right)^2 \cos^2\beta \quad (35.13)$$

$$\frac{I}{I_{max}} = \left(\frac{\sin 1.046}{1.046}\right)^2 \cos^2\pi = 0.684$$

CHECK and THINK

Our results are pretty amazing. With the use of wider slits, the intensity at this particular spot ($\theta \approx 3°$) is actually decreased to about 68% of what it was when we used two narrow slits (part A).

EXAMPLE 35.8 CASE STUDY Poisson's Spot

Recall that Poisson was an advocate for the particle model of light. When he analyzed Fresnel's competition essay, he thought he found a fatal flaw in Fresnel's use of the wave model. Fresnel's work predicted a bright spot at the center of the shadow cast by an object with a circular cross section. François Arago—another judge—tested Poisson's prediction experimentally. This experiment has been repeated many times; you may even do it in your laboratory (Fig. 35.28). At the center of the shadow is a bright spot as predicted. Although this phenomenon is often called Poisson's spot, Poisson probably was not happy to have seen it because it supported the wave model of light. The spot is sometimes called Fresnel's spot because it is a direct consequence of his work, and sometimes Arago's spot because Arago devised the experiment that confirmed its existence.

Come up with a simple, conceptual reason for the light in the center of the shadow in Figure 35.28. *Hint*: See Concept Exercise 35.2.

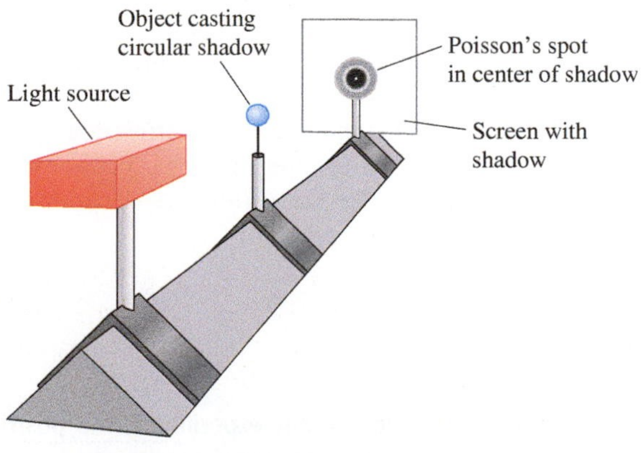

FIGURE 35.28 A bright spot is clearly seen in the center of the shadow of a circular object.

Example continues on page 1148 ▶

INTERPRET and ANTICIPATE

The key to explaining the bright spot lies in modeling light as a wave. If light obeyed only the particle model, like paint pellets hitting a sandbag, the light would be stopped and there would be no light in the shadow. But when you float at a point behind a rock in the ocean and waves arrive, you oscillate up and down because the water waves diffract around the rock and reach you.

FIGURE 35.29 Diffraction around an object causes a bright spot in its shadow.

SOLVE

Figure 35.29 shows the diffraction of a wave (such as light or water) around an object with a circular cross section. The waves that emerge around the edges of the object interfere constructively in the center. If these were light waves, you would see a bright spot in the center of the object's shadow.

CHECK and THINK

The spot in the center of the shadow in Arago's experiment and the interference pattern observed in Young's experiment are convincing evidence that light should be modeled as a wave.

SUMMARY

Underlying Principles

Huygens's principle: Each point on a primary wave front serves as the source of a spherical wave, called a wavelet, that advances with a speed and frequency equal to those of the primary wave. The primary wave front at some later time is the envelope of these wavelets (Fig. 35.6).

Major Concepts

1. **Diffraction** occurs when waves pass through a narrow opening or near the edge of an obstacle. A diffraction pattern consists of bright and dark regions.
2. Like diffraction, **interference** is a wave phenomenon that results from the superposition of waves, producing a characteristic pattern of bright and dark fringes. A common distinction between the two phenomena is that diffraction involves only a single aperture (or source), whereas interference involves two or more such apertures.
3. **Conditions for constructive interference:**
   For two waves to interfere constructively, they must be in phase, so the difference between their path lengths must be given by

$$\Delta d = n\lambda \quad (n = 0, 1, 2, 3, \ldots) \qquad (18.2)$$

Conditions for destructive interference:
For two waves to interfere destructively and cancel, they must be 180° out of phase, so the difference between their path lengths must be given by

$$\Delta d = \left(n + \frac{1}{2}\right)\lambda \quad (n = 0, 1, 2, 3, \ldots) \qquad (18.3)$$

4. Two light waves are said to be **coherent** if they are monochromatic sources of the same frequency with a constant phase difference φ between them; φ must be a constant in both space and time. In order to see an interference pattern, the light involved must be coherent.

Special Cases

1. In **Young's double-slit experiment**, the **position θ_n of the bright fringes (maxima)** is given by

$$d \sin\theta_n = n\lambda \; (n = 0, \pm1, \pm2, \pm3, \ldots) \qquad (35.1)$$

where d is the distance between the slits and λ is the wavelength of the monochromatic light source. The **position θ_n of the dark fringes (minima)** is given by

Special Cases cont'd

$$d \sin \theta_n = \left(n + \frac{1}{2}\right)\lambda \ (n = 0, \pm 1, \pm 2, \pm 3, \ldots) \quad (35.2)$$

The **intensity of the fringes in Young's double-slit experiment** is given by

$$I = I_{max} \cos^2 \frac{\varphi}{2} = \frac{2E_0^2}{\mu_0 c} \cos^2 \frac{\varphi}{2} \quad (35.8)$$

where

$$\varphi = \frac{2\pi}{\lambda} \Delta d = \frac{2\pi}{\lambda} d \sin \theta \quad (35.9)$$

2. In **single-slit diffraction**, the angular **position** θ_m **of the dark fringes** is given by

$$w \sin \theta_m = m\lambda \ (m = \pm 1, \pm 2, \pm 3, \ldots) \quad (35.5)$$

and the **intensity** is given by

$$I = I_{max} \left(\frac{\sin \alpha}{\alpha}\right)^2 \quad (35.10)$$

where $\alpha \equiv \varphi/2 = (\pi w/\lambda) \sin \theta$ (Eq. 35.11).

3. The **intensity of double-slit diffraction** is given by

$$I = I_{max} \left(\frac{\sin \alpha}{\alpha}\right)^2 \cos^2 \beta \quad (35.13)$$

where $\alpha = (\pi w/\lambda) \sin \theta$ (Eq. 35.11) and $\beta = \varphi/2 = (\pi d/\lambda) \sin \theta$ (Eq. 35.14).

Tools

Visual representations of light

1. Draw a single sinusoidal curve to represent the oscillating electric field (Fig. 35.4).
2. Draw solid lines to represent the wave crests and dashed lines to represent the wave troughs or valleys (Fig. 35.5).

PROBLEMS AND QUESTIONS

A = algebraic C = conceptual E = estimation G = graphical N = numerical

35-1 Light Is a Wave

1. **C** As shown in Figure P35.1, spray paint can be used with a stencil to produce an image that has sharp edges. Should you model the paint as waves or as particles? Explain.

Mark Phillips/Alamy

FIGURE P35.1

35-2 Sound Wave Interference Revisited

2. **G** Draw two harmonic waves that are 90° out of phase. Then draw the superposition of these two waves.
3. **C** Two speakers produce identical harmonic waves as in Figure P35.3. If the two waves are 90° out of phase when they arrive at your ear, what is the minimum distance between the speakers in terms of the wavelength?

FIGURE P35.3

4. You are seated on a couch equidistant between two speakers. The speakers are 2.0 m apart and you are seated 4.0 m away from the point between the speakers as shown in Figure P35.4. Both speakers play an A note (a constant frequency of 440 Hz) in phase and at the same volume. As you move left and right along the couch, you notice that the volume alternates between minimum and maximum due to interference effects. Assume the speed of sound is 343 m/s.

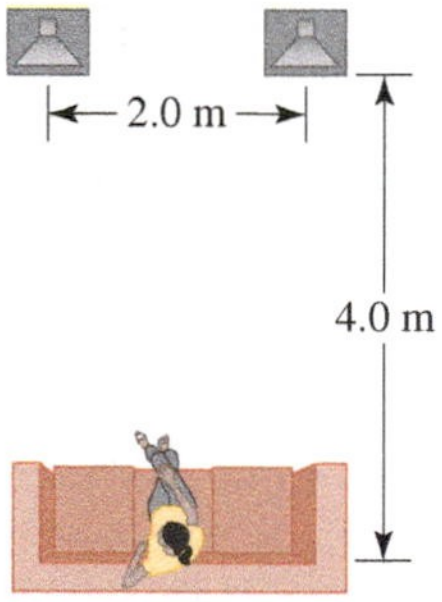

FIGURE P35.4

   a. **C** When you are seated at the center of the couch, is the sound a maximum or minimum in volume?
   b. **N** If you slide along the couch to the left or right, approximately how far do you need to move until you reach the next location that has the same volume? (*Hint*: You may wish to obtain an exact expression and then use a spreadsheet or trial and error to find an approximate solution.)

35-3 Young's Experiment: Position of the Fringes

5. **N** The first bright fringe of an interference pattern occurs at an angle of 11.5° from the central fringe when a double slit is illuminated by a 415-nm blue laser. What is the spacing of the slits?
6. **C** In Young's double-slit experiment, the positions of the bright fringes are given by $d \sin \theta_n = n\lambda$, where $n = 0, \pm 1, \pm 2, \pm 3, \ldots$ (Eq. 35.1). Mathematically, there are an infinite number of bright fringes. However, there is a physical

limit to the number of bright fringes that appear on the screen. Why is there a physical limit?

7. **N** A student shines a red laser pointer with a wavelength of 675 nm through a double-slit apparatus in which the two slits are separated by 75.0 μm. He observes the diffraction pattern on the wall 1.50 m away. What is the distance between the central bright fringe and either of the neighboring bright fringes on the wall?

8. **N** Monochromatic light is incident on a pair of slits that are separated by 0.200 mm. The screen is 2.50 m away from the slits.
 a. If the distance between the central bright fringe and either of the adjacent bright fringes is 1.67 cm, find the wavelength of the incident light.
 b. At what angle does the next set of bright fringes appear?

9. **N** In a double-slit experiment, the wavelength of monochromatic light used is 489.7 nm and the distance between the slits is 7.500 μm. How many bright fringes are created by the light passing through the slits?

10. **N** In a Young's double-slit experiment with microwaves of wavelength 3.00 cm, the distance between the slits is 5.00 cm and the distance between the slits and the screen is 100.0 cm. Determine the number of bright fringes on the screen and their distances from the central bright fringe on the screen. Assume the screen is long enough to at least show the first order bright fringes.

11. **N** A beam from a helium-neon laser with wavelength 635 nm strikes two slits separated by 0.240 mm. What is the distance between the first and third dark fringes on a screen located 4.20 m from the slits?

Problems 12, 13, and 14 are grouped.

12. **N** The wavelength of light emitted by a particular laser is 633 nm. This laser light illuminates two slits that are 50.0 μm apart.
 a. What is the angular separation between the $n = 2$ and $n = -2$ maxima?
 b. If a screen is placed 1.50 m from the slits, what is the distance between these two maxima?
 c. If you want these two maxima to be 1.00 cm apart, how far from the slits do you need to place the screen?

13. The wavelength of light emitted by a particular laser is 633 nm. This laser light illuminates two slits that are 50.0 μm apart, and a screen is 1.50 m from the slits.
 a. **N** What is the (linear) distance on the screen between the central maximum and the $n = 2$ maximum?
 b. **N** What is the (linear) distance on the screen between the central maximum and the $n = 20$ maximum?
 c. **C** Are the maxima evenly spaced? Explain.

14. The wavelength of light emitted by a particular laser is 633 nm. This laser light illuminates two slits that are 50.0 μm apart, and a screen is 1.50 m from the slits.
 a. **N** What is the (linear) distance on the screen between the central maximum and the $n = 2$ minimum?
 b. **N** What is the (linear) distance on the screen between the central maximum and the $n = 20$ minimum?
 c. **C** Are the minima evenly spaced? Explain.

15. **N** Light from a sodium vapor lamp ($\lambda = 589$ nm) forms an interference pattern on a screen 0.80 m from a pair of slits in a double-slit experiment. The bright fringes near the center of the pattern are 0.35 cm apart. Determine the separation between the slits. Assume the small-angle approximation is valid here.

16. **A** In a Young's double-slit experiment with two sources of different wavelengths, the eighth maximum of wavelength λ_1 is at a distance y_1 from the central maximum and the sixth maximum of wavelength λ_2 is at a distance y_2 from the central maximum. Find an expression for the ratio y_1/y_2.

17. **N** When a Young's double-slit experiment is carried out with 620.0-nm light, first-order bright fringes near the center of the pattern appear on a screen placed 2.50 m away each with a spacing of 1.75 cm from the central bright fringe. What is the distance between the slits?

18. **C** After shining a red laser pointer through a double-slit apparatus and observing the interference pattern on the wall, three students discuss what would happen if they used light of a shorter wavelength, such as that produced by a green laser pointer.

Avi: It's still coherent, monochromatic light, so I expect the pattern to look the same.

Cameron: But the wavelength is shorter, so the pattern will be smaller. I mean the peaks will not be spread out as far from the center.

Shannon: I think these interference effects work the other way, though—an inverse relationship. So I think the interference pattern will actually be more spread out on the wall.

Which student do you agree with, and why?

19. **N** Susan designs a double-slit experiment that uses coherent light with a wavelength of 408.0 nm and a slit separation of 0.125 mm. Peter sets up a second double-slit experiment with a slit separation of 0.250 mm. If Peter wants to produce an interference pattern that matches Susan's, what wavelength of coherent light must he use?

20. **N** In a Young's double-slit experiment, 586-nm-wavelength light is sent through the slits. A screen is held at a distance of 1.50 m from the slits. The second-order maxima appear at an angle of 2.50° from the central bright fringe. How far apart do the first-order ($m = 1$) and second-order ($m = 2$) maxima appear on the screen?

21. **N** Two slits separated by 0.130 mm are illuminated by visible light, forming an interference pattern on a screen 3.50 m away. What is the wavelength of the light if the first bright fringe of the interference pattern is 0.235° from the central fringe?

22. **N** Red light with a wavelength of 715.5 nm and violet light with a wavelength of 412.5 nm are used simultaneously in a double-slit experiment. The first maximum of the violet light is 1.975 mm from the central maximum. What is the distance between the central maximum and the first maximum of the red light?

35-4 Single-Slit Diffraction

23. **N** What is the maximum width of a single slit for which 454.6-nm light from an argon laser does not produce any diffraction minima?

24. **C** Figure P35.24 shows the diffraction patterns produced by a slit of varying width. What is the relative width of the slit in each case, from narrowest to widest?

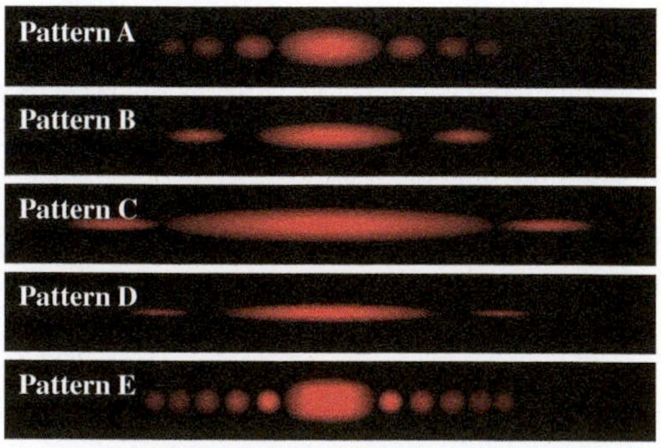

FIGURE P35.24 Problems 24 and 32.

25. **N** Monochromatic light of wavelength 529 nm is incident on a single slit. The second-order diffraction minimum is at an angle of 7.50×10^{-3} rad. What is the width of the slit?

26. **C** One day you are running late to class and realize as you hurry down the hall that you can hear your instructor even though you are not yet in front of the doorway. However, you also notice that your instructor's voice sounds lower and more muffled than normal. Why do you hear the lower frequencies (longer wavelengths) from your instructor's voice, but not the higher frequencies (shorter wavelengths)? How does the width of the doorway affect what you hear?

27. **N** A thread must have a uniform thickness of 0.525 mm. To check the thickness of the thread, you can illuminate it with a laser of wavelength 625.8 nm. A diffraction pattern like the one produced by a single slit forms on a screen.

a. If the screen is 3.00 m from the thread, how far apart are the fifth-order minima from one another?

b. If the thread's thickness increases by 20%, how far apart will the fifth-order minima be?

28. **N** Microwaves with a wavelength of 3.56 mm are passed through a single slit. If only five dark fringes appear on either side of the central maximum, what is the minimum width of the slit?

29. **N** What is the width of the central maximum in the diffraction pattern formed on a viewing screen placed 1.60 m away from a single slit of width 0.185 mm illuminated by a 410.0-nm blue laser?

30. **N** A radio wave of wavelength 21.5 cm passes through a window of width 96.3 cm.

a. What is the angular separation of the first-order minima?

b. How does your answer change for a radio wave of twice the wavelength?

31. **N** For single-slit diffraction, what ratio λ/w produces a first dark fringe at exactly $\theta_1 = 30°$?

32. **C** Suppose the diffraction patterns produced in Figure P35.24 are made by a single slit of fixed width. The slit is illuminated in turn by monochromatic light of varying wavelengths. What are the relative wavelengths of the light, from shortest to longest?

33. **N** A single slit is illuminated by light consisting of two wavelengths. One wavelength is 540.0 nm. If the first minimum of one color is located at the second minimum of the other color, what are the possible wavelengths of the other color? Is the light visible?

34. **N** A long, narrow slit 3.0 μm wide is illuminated by light of wavelength 1.20×10^3 nm. Determine the angle corresponding to the first minimum of intensity in the diffraction pattern.

35. **N** A single slit 0.450 mm wide is illuminated with monochromatic light, forming a diffraction pattern on a screen placed 2.15 m away. What is the wavelength λ of the incident light if the width of the central maximum is 5.55 mm?

36. **N** A diffraction pattern is produced by passing He-Ne laser light of wavelength 632.8 nm through a single slit. The pattern shown is viewed on a screen 2.0 m behind the slit, where the second-order minima are separated by 15.2 cm. Find the width of the slit.

35-5 Young's Experiment: Intensity

37. **N** Two identical coherent waves with different intensities interfere with each other. The intensity of the second wave is double the intensity of the first wave. Determine the ratio of the maximum intensity to the minimum intensity when the two waves interfere.

38. **C** If one slit in the double-slit experiment is blocked so that no light passes through it, what happens to the maximum intensity observed?

Problems 39 and 40 are paired.

39. **N** A double-slit experiment is conducted, and at some point on the screen the intensity is found to be 75.0% of its maximum. What is the minimum phase difference (in radians and degrees) that produces this result?

40. **N** A double-slit experiment is conducted with light of wavelength 550.0 nm, and at some point on the screen the intensity is found to be 75.0% of its maximum. The slits are 75.0×10^3 nm apart, and the screen is 1.75 m from the slits. What is the minimum separation between this point and the central maximum?

41. **N** A 520.0-nm light source illuminates two slits with a separation of 6.00×10^{-4} m, forming an interference pattern on a screen placed 4.20 m away from the slits. At a point a distance of 3.75 mm from the central maximum, **a.** what is the phase difference between the two waves originating at the slits, and **b.** what is the intensity compared to that of the central maximum?

Problems 42 and 43 are paired.

42. **A** There isn't a sharp edge between bright and dark fringes, so how do we define the width of a fringe? Assume the angular width of a bright fringe corresponds to the angular distance between points that are two-thirds of the maximum intensity. Find an algebraic expression in terms of d and λ for the angular width $\Delta\theta$ of a bright fringe. Use the small-angle approximation.

43. **N** Assume the angular width of a bright fringe corresponds to the angular distance between points that are two-thirds of the maximum intensity. A certain double-slit experiment uses green light at 555.5 nm with slit separation d_{green}. You want to design another double-slit experiment using red light at 725.0 nm. If the red fringes in your experiment are to have the same angular width as the fringes in the experiment that used green light, what slit separation must you use? Express your answer in terms of d_{green}.

44. **N** Light of wavelength 455 nm is incident on two slits separated by $d = 3.75$ mm, forming an interference pattern on a screen placed 1.85 m away. At what distance y from the central maximum is the intensity exactly half that of the maximum?

45. **N** Light of wavelength 589 nm is used to illuminate two slits with a separation of 0.270 mm. What is the percentage of the maximum intensity a distance of 0.950 cm away from the central maximum if the screen on which the interference pattern forms is 1.25 m away from the slits?

Problems 46 and 47 are paired.

46. **A** In a Young's double-slit experiment, light of wavelength λ is sent through the slits. The intensity I at angle θ_0 from the central bright fringe is lower than the maximum intensity I_{max} on the screen. Find an expression for the spacing between the slits in terms of λ, θ_0, I, and I_{max}.

47. **N** In a Young's double-slit experiment, 586-nm-wavelength light is sent through the slits. The intensity at an angle of 2.50° from the central bright fringe is 80.0% of the maximum intensity on the screen. What is the spacing between the slits?

48. **G** Red light with a wavelength of 715.5 nm and violet light with a wavelength of 412.5 nm are used simultaneously in a double-slit experiment. The slits are 651.5 nm apart. On the same graph, sketch the ratio I/I_{max} as a function of θ for both colors over the range $-50° < \theta < 50°$. Comment on the relative position and width of the maxima.

35-6 Single-Slit Diffraction Intensity

Problems 49 and 50 are paired.

49. **C** Figure P35.49 shows the intensity of the diffraction patterns produced by a slit of varying width. Rank the relative widths of the slit in each case, from narrowest to widest.

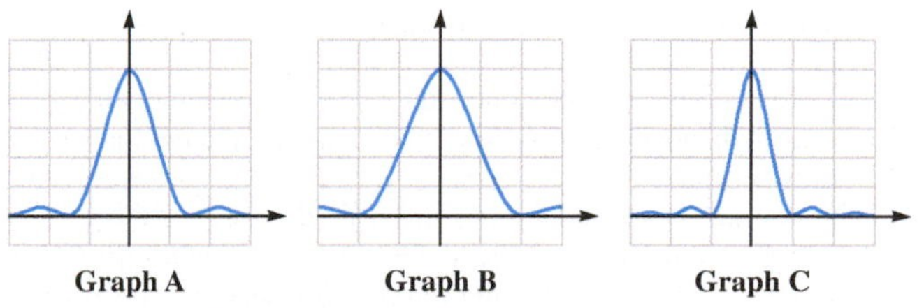

FIGURE P35.49 Problems 49 and 50.

50. **C** Figure P35.49 shows the intensity of the diffraction patterns produced by a slit of fixed width. The slit is illuminated in turn by monochromatic light of varying wavelengths. Rank the relative wavelengths of the light in each case, from shortest to longest.

51. **N** Assume the maxima produced by a single-slit diffraction pattern have angular locations α that are halfway between the adjacent minima of angular location α. Find the intensity of the first three maxima as a fraction of the maximum intensity.

52. **N** Light with a wavelength of 426 nm is incident on a single slit with a width of 4.5 μm. If the intensity of the central bright fringe is 655 W/m^2, find the intensity at these angular positions from the fringe: **a.** 20.0°, **b.** 40.0°, and **c.** 60.0°.

53. **N** Light of wavelength 750.0 nm passes through a single slit of width 2.00 μm, and a diffraction pattern is observed on a screen 25.0 cm away. Determine the relative intensity of light I/I_{max} at 15.0 cm away from the central maximum.

54. **N** Monochromatic light of wavelength 414 nm is incident on a single slit of width 32.0 μm. The distance from the slit to the screen is 2.60 m. Consider a point at y = 16.5 mm from the center of the central maximum. What is the ratio of the intensity at that point to the maximum intensity?

55. **N** For the following questions, assume the noncentral maxima in a single-slit diffraction pattern are exactly halfway between the minima immediately adjacent to them. What percentage of the central maximum's intensity is **a.** the first-order maximum's intensity, and **b.** the second-order maximum's intensity?

56. **N** Light with a wavelength of 585 nm is incident on a single slit with a width of 3.21 μm.
    a. What is the angular position of the second-order minimum?
    b. What is the ratio of the intensity at the angular position of the second-order minimum and the maximum intensity (I/I_{max})?

35-7 Double-Slit Diffraction

Problems 57 and 58 are paired.

57. **N** Light of wavelength 515 nm is incident on two slits of width 1.545 μm. The slits are 6.18 μm apart. What is the relative intensity at θ = 15.0°?

58. **N** Light of wavelength 515 nm is incident on two slits of width 1.545 μm. The slits are 6.18 μm apart. How many bright fringes are within the central maximum of the diffraction pattern?

59. **A** Two slits are separated by distance d and each has width w. If $d = 2w$, how many bright fringes are within the central maximum of the diffraction pattern?

60. **N, C** Suppose you want exactly 11 bright fringes inside the central maximum of a double-slit diffraction pattern. What must be true of the slit width and separation?

General Problems

61. **C** Two very small identical lightbulbs are connected to the same light fixture. Explain why no interference pattern is seen.

62. **C** If you spray paint through two slits, what pattern results? How does it compare to the pattern seen in Young's experiment?

63. **N** Light of wavelength 570.0 nm incident on a single slit forms a diffraction pattern on a screen placed 1.10 m from the slit. What is the width of the slit if the first and fifth dark fringes are separated by 5.30 mm?

64. **C** Much of this chapter is devoted to Young's double-slit experiment, which we will revisit in later chapters. Why is this experiment so important?

65. **N** A screen is located 2.0 m from a single slit of width 4.0 μm. If light from a helium-neon laser of wavelength 632.8 nm shines on the slit, determine the position of the first minimum on the screen.

Problems 66 and 67 are paired.

66. **N** The two slits of a double-slit apparatus each have a width of 0.120 mm and their centers are separated by 0.720 mm. What orders are missing in the diffraction pattern?

67. **N** The two slits of a double-slit apparatus each have a width of 0.120 mm and their centers are separated by 0.120 mm. What orders are missing in the diffraction pattern?

68. **N** A single slit with width 0.314 mm is illuminated with light of wavelength 490.0 nm, and a diffraction pattern with a central maximum of width 6.60 mm forms on the viewing screen.
    a. What is the distance from the slit to the viewing screen?
    b. What is the width of the first side maximum in this diffraction pattern?

69. **N** An exterior sliding door of a building 1.20 m in width opens to let in coherent microwaves with a wavelength of 28.0 cm, creating a diffraction pattern on the opposite wall of the lobby located 11.2 m from the door. If we assume the building is opaque to this wavelength, what is the distance between the second-order minimum and the central maximum on the lobby wall?

70. A single slit of width w is illuminated by light of wavelength λ so that the first-order minimum occurs at an angle θ_1.
    a. **A** Find an expression for the angular position of the first-order minimum in terms of θ_1 if the width of the slit is halved.
    b. **A, C** How does your answer change if the small-angle approximation is valid in this case?

71. **N** Monochromatic light incident on a slit of width 0.635 mm produces a diffraction pattern on a screen 1.55 m away from the slit. If the third-order dark fringe of the pattern is 3.04 mm from the center of the pattern, what is the wavelength of the light incident on the slit?

72. **C** When you arrive at a concert, you are dismayed to find that your seats are directly behind a pillar 1 m wide. (That's why the tickets were so cheap.) Your friend says not to worry and assures you that your enjoyment of the concert will not be affected. She says, "The sound will bend around the pillar, and you will still be able to hear everything." Another concertgoer suggests that you will hear the concert but everything will sound "bassier," or that you will hear primarily the lower frequencies (longer wavelengths). Explain how the pillar will affect what you hear and why.

73. **N** A narrow slit is illuminated with light of wavelength 750.0 nm, 1.0 m in front of a screen. The first minima on either side of the central maximum of the diffraction pattern observed are separated by 3.0 mm. Determine the width of the slit.

74. **N** Sound with a wavelength of 2.29 m is incident on an opening in a wall that acts as a single slit with a width of 4.59 m. The maximum intensity of the sound after passing through the opening is 1.00×10^{-6} W/m^2. The locations of absolute silence would be found in a way similar to finding the angular positions of dark fringes for light passing through a single slit. What is the angular position for the first-order (m = 1) location of silence in this example?

75. **N** When a double slit is illuminated with 622-nm light, the eighth interference fringe is observed a distance of 5.92 mm from the central maximum on a screen 3.60 m away. What is the separation of the two slits?

76. **N** Light with a wavelength of 550.0 nm is incident on a pair of slits with a separation of 0.350 mm.
    a. Find the angles corresponding to the locations of the first three orders of fringes away from the central bright fringe.
    b. If the screen is 2.00 m away from the slits, what is the distance between the first-order (n = 1) and second-order (n = 2) bright fringes?

77. **N** Consider two monochromatic sources A and B with wavelength λ such that A is initially ahead of B in phase by 66°. The waves interfere at a certain point after having traveled different paths, where B travels $\lambda/4$ farther than A. Determine the phase difference between the waves when they interfere at this point.

78. **C** Another way to construct a double-slit experiment is to use a Lloyd's mirror (Fig. P35.78). Light from the single slit strikes the screen and interferes with the light that has reflected from the mirror. Explain why at the center of the fringes there is a dark fringe instead of a bright fringe.

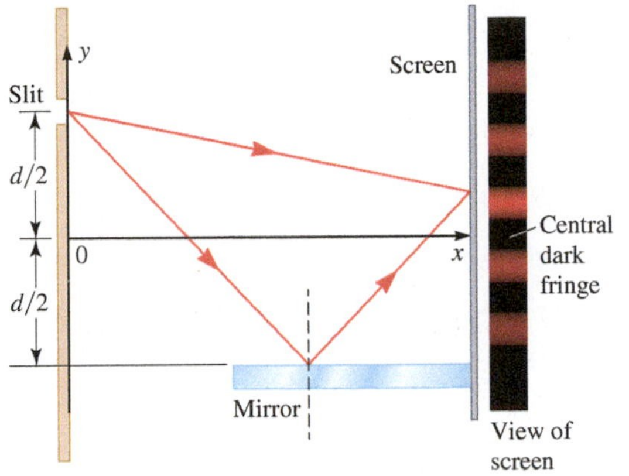

FIGURE P35.78

79. **N** Sound waves with frequency 1850 Hz are incident on a pair of slits placed 38.0 cm apart. Use 343 m/s for the speed of sound.

a. What is the angular separation between the central maximum and an adjacent maximum?

b. What would the slit separation have to be for the interference pattern of microwaves with wavelength 2.65 cm to have the same angular separation between the central maximum and an adjacent maximum?

Problems 80 and 81 are paired.

80. **G** Table P35.80 presents data gathered by students performing a double-slit experiment. The distance between the slits is 0.0700 mm, and the distance to the screen is 2.50 m. The distance y from the central maximum to other maxima is given. Plot the data and find the wavelength of the laser's light.

81. **N** Table P35.80 presents data gathered by students performing a double-slit experiment. The distance between the slits is 0.0700 mm, and the distance to the screen is 2.50 m. The intensity of the central maximum is 6.50×10^{-6} W/m². What is the intensity at $y = 0.500$ cm?

TABLE P35.80

n	y (cm)
−4	−7.52
−3	−5.11
−2	−3.52
−1	−1.70
0	0.00
1	1.63
2	3.40
3	4.80
4	6.98

82. **N** A pair of slits separated by 0.340 mm and placed 2.10 m away from a viewing screen is illuminated by a source that produces both 580-nm and 620-nm light, resulting in overlapping interference patterns. What is the shortest distance from the central maximum for which bright fringes for the two wavelengths coincide?

83. **N** Monochromatic light waves of wavelengths 400.0 nm and 560.0 nm are incident simultaneously and normally on a Young's double-slit apparatus. The separation between the slits is 0.100 mm, and the distance between the slits and the screen is 1.00 m. Determine the distance between the first two adjacent dark fringes on the screen.

84. **N** In a Young's double-slit experiment, the intensity at a point on the screen is measured to be 55.0% of the maximum.

a. What is the minimum phase difference between the two sources, measured in radians, that would give this intensity?

b. If the path difference between the light from the two slits is 106 nm, what is the wavelength of light being used in the experiment?

85. **N** Two transmitters for rival radio stations both transmitting at 20.0 MHz are located 45.0 m apart on a hilltop. Due to a tuning error, the second transmitter transmits 180° out of phase with the first. If a technician walks from the first transmitter to the second along the line joining them, what is the shortest distance he must walk to reach a point where the transmissions from the two towers are in phase?

86. **G** Three waves, which can each be represented as $y(t) = A \sin(\omega t + \varphi)$, reach the same point in space. For times $0 < t < 10.0$ s, plot the superposition of the three waves for the following conditions and comment on the qualitative differences.

a. The waves are coherent: $A = 2.0$ mm and $f = 3.0$ Hz for each wave, but they have phases of $\varphi = 0$, 0.30 rad, and 0.80 rad.

b. The waves are incoherent: The parameters of the three waves are (1) $A = 2.0$ mm, $f = 3.0$ Hz, and $\varphi = 0$; (2) $A = 1.0$ mm, $f = 4.0$ Hz, and $\varphi = 0.30$ rad; and (3) $A = 1.5$ mm, $f = 5.0$ Hz, and $\varphi = 0.80$ rad.

87. **N** A source of 485-nm light is used to perform Young's double-slit experiment. The separation of the slits is 0.185 mm, and the screen is placed 2.30 m away from the slits.

a. What is the distance between the central maximum and an adjacent maximum in the interference pattern on the screen?

b. What is the distance between the first and third dark bands in the interference pattern on the screen?

Problems 88 and 89 are paired.

88. **A** Show that the resultant intensity from light waves with amplitudes E_1 and E_2 coming through two slits of a Young's double-slit apparatus is given by

$$I = \left(\frac{1}{2\mu_0 c}\right)\left[E_1^2 + E_2^2 + 2E_1E_2\cos\varphi\right]$$

where φ is the phase difference between the two interfering waves.

89. **A** One of the slits in a Young's double-slit apparatus is wider than the other, so that the amplitude of the light that reaches the central point of the screen from one slit alone is twice that from the other slit alone. Determine the resultant intensity as a function of the direction θ on the screen, the wavelength λ of the incident light, the incident intensity I_0, and the slit separation d.

90. A laser shines through a double-slit apparatus, creating bright and dark fringes.

a. **N** If we define the bright fringes as the regions between locations where the intensity falls below 20%, what is the ratio of the width of the bright fringes to the width of the dark fringes?

b. **G** Plot the intensity of the interference pattern for a few peaks around $\varphi = 0°$ and indicate the positions for which the intensity is higher than 20% of the maximum value to confirm your answer to part (a).

36 Applications of the Wave Model

Key Questions

How do thin films generate interference patterns?

How did Michelson and Morley use interference to show that light does not require a medium?

What instruments exploit the wave nature of light?

Dean Pennala/Shutterstock

FIGURE 36.1 Oil on water creates beautiful colors.

Underlying Principles

No new principles are introduced in this chapter.

Major Concepts

1. Rayleigh's criterion
2. Dispersion
3. Resolving power

Special Cases

1. Circular aperture resolution
2. Thin-film interference
3. Diffraction grating (position and half-width of lines)

Rainbows are commonly observed in the skies of Hawaii because brief rain showers are often followed by sunny skies. The author of this book spent much of her early childhood in the Midwest, however, where rain showers are followed by dreary skies. She rarely saw rainbows in the midwestern skies. But, after a rainstorm, she often saw rainbows in parking lots because many vehicles at that time left films of gasoline or oil floating in shallow puddles. Of course, this is harmful to the environment and today's vehicles are much cleaner. Nevertheless, the rainbows observed in a thin film of oil are beautiful (Fig. 36.1). The wonderful colors we see in such polluted water are best explained by the wave model of light. Chapter 35 presented evidence that supports the wave model, and this chapter is about applying that model to various phenomena and using it to design instruments to study the Universe.

36-1 Implications of the Wave Model

Because the nature of light was hotly debated, performing experiments that supported the wave model was important in the process of scientific discovery. Such experiments often lead to more discoveries and inventions. For example, after Ben Franklin discovered that lightning is a giant electrical spark, he went on to invent the lightning rod (case study in Chapter 24). Applying the wave model allows us to explain phenomena and build new instruments. The case study in this chapter involves both the invention of a new scientific instrument and the important discovery made with that instrument.

CASE STUDY The Michelson-Morley Experiment

This case study involves the most important *failed* experiment in history. In the 19th century, most scientists thought light is a wave that propagates in a medium they called the **luminiferous ether** (usually referred to simply as the **ether**). Today we know that no such medium exists and that light travels in a vacuum. But 19th-century scientists reasoned that because light travels through the open space between the Sun and the Earth, as well as through substances such as water and glass, the ether permeates everything. These scientists wanted to characterize and measure its properties.

Luminiferous means "light-bearing" in Latin, and the Greek term *ether* designates an unknown medium.

The speed of a wave through a medium depends on the properties of that medium. For example, the speed of sound depends on the medium's bulk modulus B (or compressibility): $v_s = \sqrt{B/\rho}$ (Eq. 17.13). The speed of sound is greater in a less compressible fluid (one with a greater bulk modulus). Because the speed of light is very high, scientists reasoned that the ether was nearly incompressible (had a very great bulk modulus).

If the ether was also everywhere, then everything—you, racehorses, and the Earth—moved through the ether. If the ether created a drag or frictional force on moving objects, we would notice the resulting deceleration. For example, the Earth has been orbiting the Sun for 4.5 billion years. If the ether exerted a dissipative force on the Earth, the Earth would have spiraled in toward the Sun. Because no such dissipative force was observed, 19th-century scientists concluded that the ether was frictionless.

The ether was also supposedly at rest, and thus it should have been possible to measure the speed of objects moving with respect to the ether. In other words, the ether defined an absolute reference frame against which to measure all motion. So, in the late 1800s, Albert Michelson and Edward Morley set out to measure the speed of the Earth through the ether. Like Thomas Young's experiment, Michelson and Morley's experiment depended on observing the fringes produced by interference.

Much to their surprise, Michelson and Morley were unable to measure the speed of the Earth with respect to the ether. Instead, they found *no* evidence that the ether existed. Michelson was bothered by this failed experiment for the rest of his life, but today we understand that this is the most important experimental failure in history. The simplest explanation for the failure is that the ether does not exist and light waves do not require a medium to propagate through.

36-2 Circular Aperture Diffraction

In many practical situations, light must pass through a circular or nearly circular aperture, such as the shutter on your camera or the aperture in a telescope. Diffraction occurs whenever light passes through an aperture or near the edge of an object. In the case study in Chapter 35, we considered diffraction of light through a circular aperture and around an object with a circular cross section. In this section, we take a more mathematical approach.

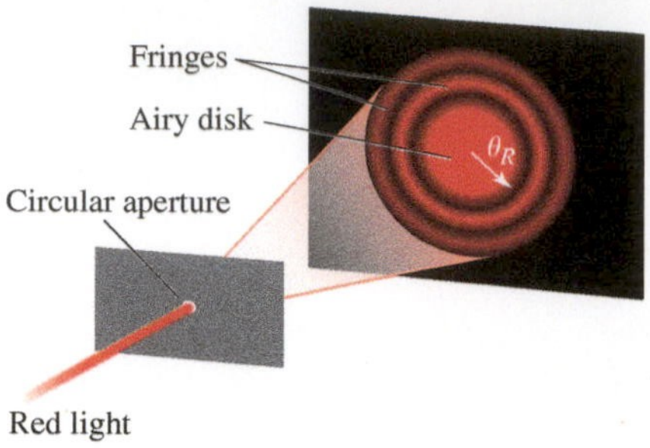

FIGURE 36.2 When light passes through a circular aperture, the diffraction pattern has a central maximum, known as an Airy disk, and fringes that form concentric rings.

In Figure 36.2, light from a single source passes through a circular aperture. The diffraction pattern that results has a circular central maximum called an **Airy disk**. The angular radius θ_R of the Airy disk is given by

$$\sin\theta_R = 1.22\frac{\lambda}{d} \quad (36.1)$$

where λ is the wavelength of the light and d is the diameter of the aperture. Both the wavelength and the diameter must be measured in the same length units. Often the angular radius is small, so it is approximately given by

$$\theta_R \approx 1.22\frac{\lambda}{d} \quad (36.2)$$

where θ_R is in radians. The central maximum is surrounded by bright and dark fringes that form concentric circles, often unnoticed because they are much fainter than the central maximum. Figure 36.3 shows a graph of the light intensity across the diffraction pattern. The central maximum is many times more intense than the bright fringes.

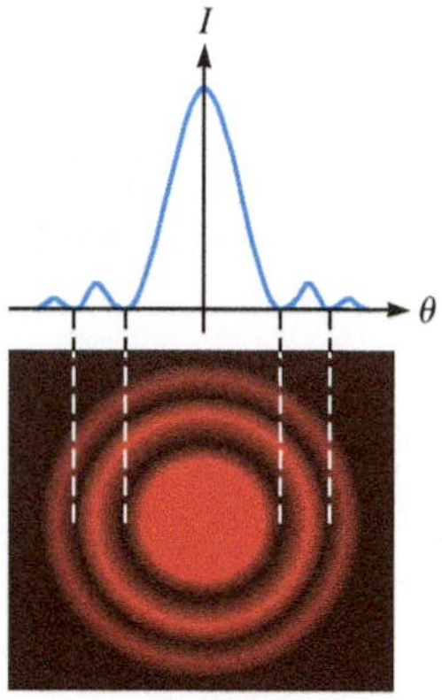

FIGURE 36.3 Light intensity as a function of angular position for diffraction through a circular aperture.

If the fringes are so faint compared to the central maximum, why do we care about diffraction through a circular aperture? Diffraction is important when the patterns of two or more sources overlap. Figure 36.4 shows graphs of intensity and the diffraction patterns that result when two point sources emit light through a circular aperture. In Figure 36.4A, the point sources are close together and their diffraction patterns overlap. The total intensity has a single central hump, so it looks similar to the intensity from a single source (Fig. 36.3). If these two sources were two very distant stars, the resulting image through a telescope would appear to be a single star (perhaps slightly elongated). We would say that the images of the two stars are **not spatially resolved**. **Resolution** is the ability of an aperture to separate the diffraction patterns produced by two sources. Resolution depends on the wavelength of the light passing through the aperture and on the aperture's size.

Figure 36.4B shows the diffraction patterns of two sources that are just barely resolved. On the intensity graph, the total intensity has a slight dip in the center, so it does not look like the intensity from a single source; instead, it hints at two sources. The image has two lobes, again hinting that there are two sources.

RAYLEIGH'S CRITERION

✪ Major Concept

According to **Rayleigh's criterion**, two sources are barely resolved when the central maximum of one diffraction pattern is centered on the first minimum of the other diffraction pattern (Fig. 36.4B). That is, the two images are barely resolved when they are separated by the angular radius of the Airy disk. So, we can use Equation 36.1 to write an expression for the minimum angular separation θ_{min} between two images that are barely resolved:

CIRCULAR APERTURE RESOLUTION

Special Case

$$\sin\theta_{min} = 1.22\frac{\lambda}{d} \quad (36.3)$$

Often θ_{min} is called the **diffraction limited resolution**, or simply the **resolution**, of an instrument with a circular aperture. Usually θ_{min} is small, and we use the small-angle approximation to write

$$\theta_{min} \approx 1.22\frac{\lambda}{d} \quad (36.4)$$

where θ_{min} is in radians. Often the angle is *very* small and so is more conveniently expressed in arc seconds, in which case Equation 36.4 becomes

$$\theta_{min} \approx 251643\frac{\lambda}{d}\ \text{arcsec} \quad (36.5)$$

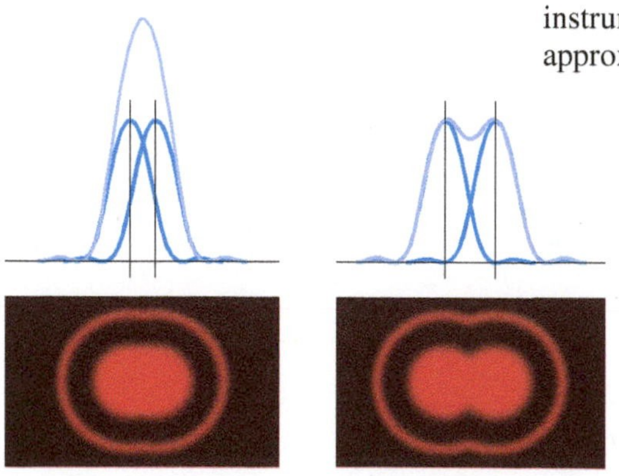

A. B.

FIGURE 36.4 **A.** These two sources appear to be a single extended source. **B.** These two sources fit Rayleigh's criterion; they are barely resolved.

Why Is There a Factor of 1.22?

We cannot derive Equation 36.3 or 36.4 without using special mathematical functions (known as Bessel functions) that are beyond the scope of this textbook. However, we can give a plausible argument using a couple of approximations and come close to the exact answer.

Let's start by finding an expression for the resolution θ_{min} of a (rectangular) slit of width w, assuming its height is much greater than its width. According to Rayleigh's criterion, two sources whose light passes through the single slit will be resolved if the central maximum of one falls no closer to the central maximum of the other than the first dark fringe of the other's diffraction pattern. So, their minimum separation is equal to the position of the first dark fringe (Eq. 35.5 with $m = 1$):

$$\sin\theta_{min} = \frac{\lambda}{w}$$

For the rest of this discussion, we will assume that the resolution θ_{min} is small and use the small-angle approximation:

$$\theta_{min} \approx \frac{\lambda}{w} \text{ (radians)} \tag{36.6}$$

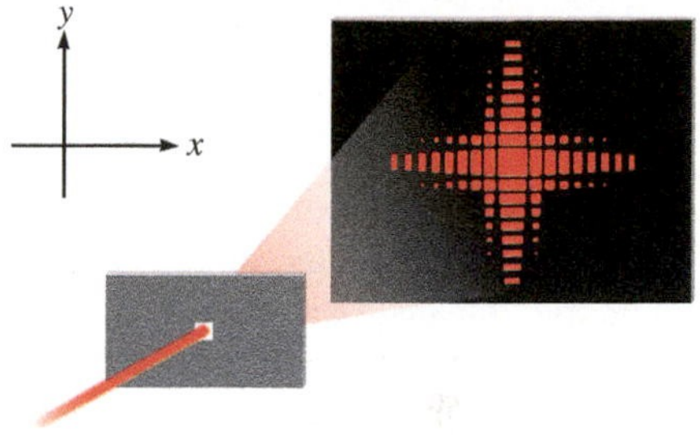

FIGURE 36.5 The diffraction pattern produced by a square aperture.

For a long, narrow slit, the diffraction pattern is perpendicular to the long axis of the slit (Fig. 35.16). Circles are not well modeled by a single narrow slit, and we improve our approximation by using a square aperture. Figure 36.5 shows the diffraction pattern that is formed by a square aperture: The pattern is stretched out along both the x and y directions. The resolution in each of these directions is given by Equation 36.6:

$$(\theta_{min})_x = (\theta_{min})_y \approx \frac{\lambda}{w}$$

Let's model a circular aperture as a square whose width equals the diameter of the circle: $w = d$ (Fig. 36.6A). The resolution (in any direction) is given by Equation 36.6, with $w = d$:

$$\theta_{min} \approx \frac{\lambda}{d}$$

From this simple model, we see why θ_{min} in Equation 36.4 depends directly on λ and inversely on d. However, Figure 36.6A is not a good model for a circular aperture because the area of the square is greater than the area of the circle. Let's try a second model, using a square whose area equals the area of the circle (Fig. 36.6B). To find the resolution, we first need to find an expression for the width of this second, smaller square. Because its area is equal to the area of the circle,

$$w^2 = \pi\left(\frac{d}{2}\right)^2$$

so the width of the square in Figure 36.6B is

$$w = \frac{\sqrt{\pi}}{2}d \tag{36.7}$$

To find the resolution, substitute Equation 36.7 into $\theta_{min} \approx \lambda/w$ (Eq. 36.6):

$$\theta_{min} \approx \frac{2}{\sqrt{\pi}}\frac{\lambda}{d} \approx 1.13\frac{\lambda}{d}$$

Now we have found a slightly closer result; instead of the factor of 1.22, our factor is 1.13. Although we did not get the exact result (Eq. 36.4), we have a way to understand where the equation comes from. The circle is modeled as a square whose width is smaller than the diameter of the circle in order to ensure that their areas are equal. We don't get the exact factor of 1.22 because the square is not quite the right shape. Its corners extend beyond the circle, and its sides fall within the circle. A better model is beyond the mathematics of this textbook.

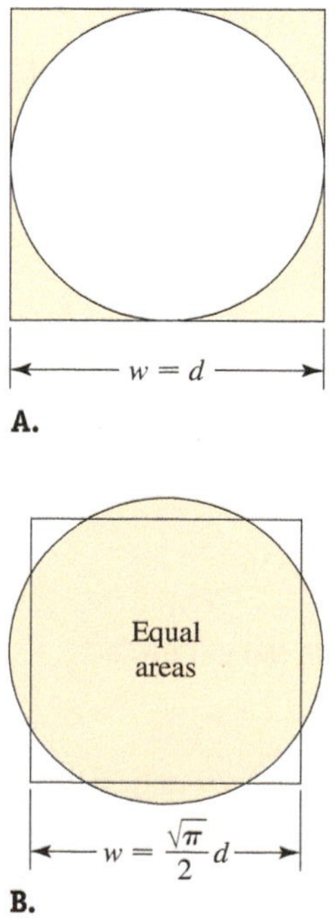

FIGURE 36.6 Model a circular aperture as a square **A.** whose width equals the diameter of the circle, and **B.** whose area equals the area of the circle.

CONCEPT EXERCISE 36.1

When it comes to telescopes, bigger is better. By *bigger*, we mean a larger-diameter primary lens or mirror. How does the diameter of a telescope affect its resolution?

EXAMPLE 36.1 Your Eye Versus Your Telescope

For your birthday, your very nice aunt gives you a 4-in. refracting telescope. The diameter of a fully open pupil is about 8 mm. How does the resolution of your telescope compare to the resolution of your fully open pupil? Assume the minimum angle is small.

INTERPRET and ANTICIPATE

Resolution depends on the diameter of the aperture and the wavelength of the light. For a telescope, the diameter of the objective lens is all that matters. The phrase "4-in. refracting telescope" means that the objective lens has a 4-in. diameter. Because the diameter of the telescope is so much larger than the diameter of your pupil (while the wavelength of the light remains the same), we expect the resolution of the telescope to be much better than that of your eye. It is convenient to work in millimeters (4 in. = 102 mm).

SOLVE

Because the angle is small, we can use Equation 36.4 for the resolution. We need only the ratio of resolutions, so wavelength cancels out.

$$\theta_{min} \approx 1.22\frac{\lambda}{d} \quad (36.4)$$

$$\frac{(\theta_{min})_{tele}}{(\theta_{min})_{eye}} = \left(1.22\frac{\lambda}{d}\right)_{tele} \bigg/ \left(1.22\frac{\lambda}{d}\right)_{eye}$$

$$\frac{(\theta_{min})_{tele}}{(\theta_{min})_{eye}} = \frac{d_{eye}}{d_{tele}} = \frac{8 \text{ mm}}{102 \text{ mm}} = 0.08$$

CHECK and THINK

The resolution of the telescope is 0.08 times the resolution of your eye. That means the telescope can produce separate images of sources that are 0.08 times closer together than the sources resolved by your eye. Let's look at a few specific values. Your eye can just barely resolve two sources that are separated by about 16 arcsec—the width of a human hair held at arm's length. The telescope can resolve two sources that are about 1.3 arcsec apart, the width of a human hair held at 13 arm's lengths.

36-3 Thin-Film Interference

THIN-FILM INTERFERENCE
Special Case

You have probably seen the rainbow of colors in the thin film of a soap bubble or an oil slick (Fig. 36.1). These colors are a result of wave interference. When light is incident on a thin film, such as oil floating on water, it undergoes two (or more) reflections before it enters your eye, and these reflected waves interfere with one another. To understand how a thin film can cause interference, we first need to take a closer look at the speed of light and what happens when light encounters the boundary between two media such as air and oil.

Speed of Light

All electromagnetic waves propagate at the speed of light c in a vacuum. If an electromagnetic wave is traveling through a transparent medium such as glass or water, its speed is lower than c because the electric and magnetic fields interact with the atoms and molecules in the medium. To see this mathematically, first remember that when we derived the wave equation for an electromagnetic wave in a vacuum, we found that the speed is given by $v_{\text{vac}} = 1/\sqrt{\mu_0\varepsilon_0}$ (Eq. 34.16). Next, remember that we take the effect of a dielectric into account by replacing the permittivity constant ε_0 with $\kappa\varepsilon_0$ in an equation where κ is the dielectric constant (Section 27-8). Similarly, we replace the permeability constant μ_0 with $\kappa_m\mu_0$. Let's model a transparent medium as a dielectric. Then the speed of light in a medium is found by replacing ε_0 with $\kappa\varepsilon_0$ and μ_0 with $\kappa_m\mu_0$ in Equation 34.16:

$$v = \frac{1}{\sqrt{\kappa_m\mu_0\kappa\varepsilon_0}} = \frac{v_{\text{vac}}}{\sqrt{\kappa_m\kappa}} = \frac{c}{\sqrt{\kappa_m\kappa}}$$

For most dielectrics, $\kappa_m \approx 1$ and $\kappa > 1$, so the speed of light in a medium is lower than the speed of light in a vacuum.

The **index of refraction** n is the (dimensionless) ratio of the speed of light in a vacuum to the speed of light in the medium:

$$n = \frac{c}{v} \qquad (36.8)$$

The indices of refraction for various media are listed in Table 36.1. We'll take a closer look at the index of refraction in Section 38-1. For now, the index of refraction for a vacuum is exactly 1 and for air is nearly 1. So the speed of light in air is nearly c. All other media have a higher index of refraction, meaning that light propagates more slowly in a medium than it does in air. (Remember that higher n means lower v.) When light is transmitted from a medium with a low index of refraction to one with a higher index of refraction, the light slows down.

Because the speed of light depends on its wavelength and frequency according to $v = \lambda f$ (Eq. 17.8), if there is a change in speed, there must be a change in either frequency or wavelength, or both. When light is transmitted from one medium to another, the frequency does not change, however, so the wavelength in the medium must change. If the speed of light in a vacuum is $c = \lambda_0 f$, then by substituting $v = c/n$ into $v = \lambda f$, we find

$$\frac{c}{n} = \frac{\lambda_0 f}{n} = \lambda f$$

where λ_0 is the wavelength in a vacuum and λ is the wavelength in the medium. So the wavelength in the medium is

$$\lambda = \frac{\lambda_0}{n} \qquad (36.9)$$

Equation 36.9 shows that the wavelength in the medium is shorter than in a vacuum because the index of refraction for all media is greater than 1.

TABLE 36.1 Index of refraction.

Medium	n
Air	1.0002926
Cubic zirconia	2.14
Diamond	2.417
Fused quartz	1.458
Heavy flint glass	1.890
Ice	1.3049
Quartz	1.54
Water	1.333
Zinc crown glass	1.517

Notes:

1. The index of refraction depends on wavelength. The values in the table are good for yellow light at $\lambda = 589$ nm. You may use these values in this chapter unless you are told otherwise.
2. You may notice that $n \approx \sqrt{\kappa}$, but don't try to find the index of refraction by using the values for the dielectric constants in Table 27.2. Those values were measured using constant electric fields, and κ is usually much lower when the fields oscillate rapidly.

Reflection and Phase Changes

The reflection of light has some similarities to the reflection of a mechanical wave on a rope. Figure 36.7 shows a pulse on a rope that is fixed to the pole on the right. With respect to the incoming pulse in Figure 36.7A, the reflected pulse in Figure 36.7B is

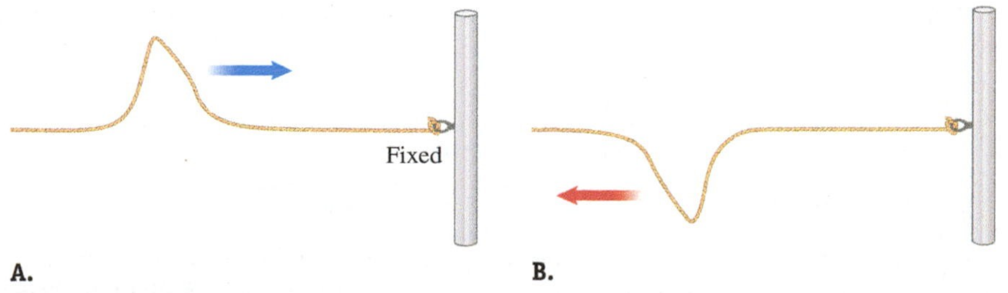

FIGURE 36.7 When a pulse reflects from a fixed end, the reflected pulse is 180° out of phase with the incoming pulse.

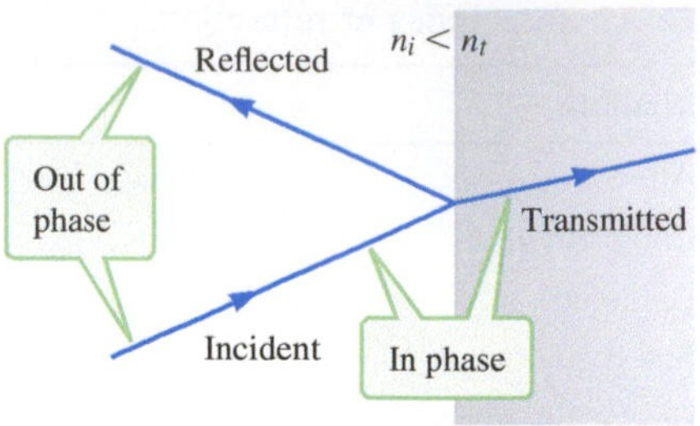

FIGURE 36.8 When light is reflected from a boundary and $n_i < n_t$, the reflected wave is 180° out of phase with the incident wave. We have sketched light as a ray here (see Chapter 37). We are still using the wave model because we are interested in the relative phase of the light, but it is more convenient in this figure to draw rays.

both inverted and flipped from left to right. This difference between incoming and reflected pulses can be described in terms of their relative phase. Imagine two such pulses approaching each other on the same rope. Where they overlap completely, the pulses cancel and the rope remains flat (Fig. 18.6, page 524). Because the reflected wave and the incident wave interfere destructively, they are 180° out of phase.

This situation is analogous to light incident in a medium with index of refraction n_i and reflected from a medium with index of refraction n_t, where $n_i < n_t$, such as when light in air is reflected from glass. *An electromagnetic wave reflected from the boundary of a medium with a higher index of refraction than that of the medium in which it is incident is 180° out of phase with the incident wave* (Fig. 36.8).

Now consider a pulse on a rope whose end is free (Fig. 36.9). As before, the incident pulse moves to the right, but now the reflected pulse is flipped from left to right while remaining upright. To find the relative phase of the two pulses, imagine that they approach each other on the same rope. Where they overlap completely, the pulses add together and the amplitude doubles (Fig. 18.8, page 525). Because the reflected wave and the incident wave interfere constructively, they are in phase.

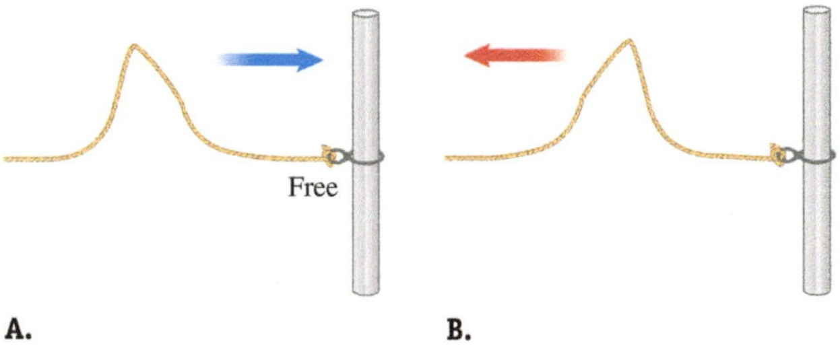

FIGURE 36.9 When a pulse reflects from a free end, the reflected pulse is upright.

This situation is analogous to light incident in a medium with index of refraction n_i and reflected from a medium with index of refraction n_t, where $n_i > n_t$, such as when light in glass is reflected at an air-glass boundary. *An electromagnetic wave reflected from the boundary of a medium with a lower index of refraction than that of the medium in which it is incident is in phase with the incident wave* (Fig. 36.10).

To keep the two possibilities straight, remember the phrase *low to high is out*, which corresponds to Figure 36.8. By contrast, in both situations ($n_i < n_t$ and $n_i > n_t$), the transmitted wave is in phase with the incident wave.

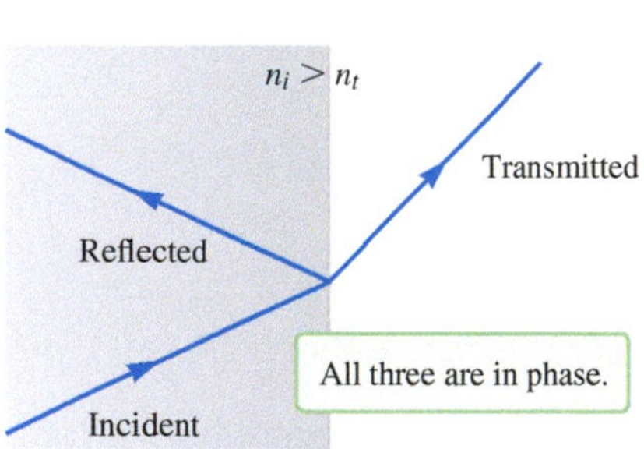

FIGURE 36.10 When light is reflected from a boundary and $n_i > n_t$, the reflected wave is in phase with the incident wave.

Thin Films: Conditions for Constructive and Destructive Interference

Now we are ready to analyze the interference produced by a thin film and come up with mathematical expressions for the conditions of constructive and destructive interference. The film has index of refraction n_f and width w (Fig. 36.11), and it is surrounded by a medium with index of refraction n_i. The film's index of refraction is higher than that of the surrounding medium: $n_f > n_i$. This film could be soap and the surrounding medium air, which is a special case of thin-film interference. There are many other possibilities: For example, the film could be surrounded by a material that has a higher index of refraction, or there could be one medium on one side of the film and another medium on the other side. Some of these other special cases are explored in the examples and homework problems.

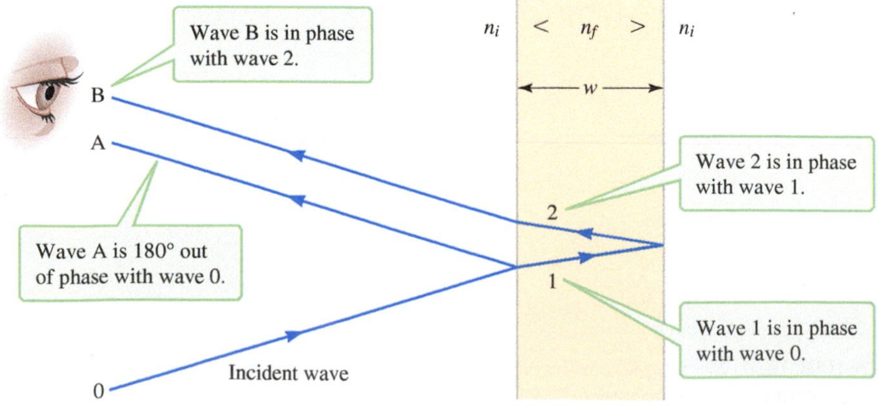

FIGURE 36.11 Wave A is out of phase with the incident wave labeled as 0. Wave 1 is in phase with wave 0, and wave 2 is in phase with wave 1. If the film is very thin, so that we can ignore the extra path length, we find that wave B is in phase with wave 0 and out of phase with wave A. In general, though, the relative phase of waves A and B depends on the thickness w of the film.

Consider one wave incident in air (Fig. 36.11). We'll trace its progress as it encounters first the front side of the film and then the back side. When the wave encounters the front side of the film, it is both reflected and refracted (transmitted). The reflected wave is labeled A. Because the wave is incident in the medium with the lower index of refraction ($n_i < n_f$), the reflected wave is 180° out of phase with the incident wave (Fig. 36.8).

The transmitted wave is labeled 1 and is in phase with the incident wave. When wave 1 encounters the back side of the thin film, it is both transmitted into the surrounding medium and reflected back into the film. We'll ignore the light transmitted into the surrounding medium and instead follow the reflected wave, labeled 2. Wave 2 is in phase with wave 1 because wave 2 is reflected from the back side of the film. For this boundary, the incident medium is the film and the transmitted medium is the air; because the incident medium has a greater index of refraction, the reflected wave does not experience a phase shift. Next, wave 2 encounters the front side of the film and is both reflected and transmitted. This time we ignore the reflection and follow the transmitted wave, labeled B. Like all transmitted waves, wave B is in phase with its incident wave (wave 2 in this case).

When your eye intercepts waves A and B, you see a bright fringe if the two waves are in phase and a dark fringe if they are 180° out of phase. The relative phase of A and B depends on two factors: (1) Waves A and B result from two different reflections. Reflection from the front of the film introduces a 180° phase shift, but reflection from the back side does not. (2) Wave B in effect travels farther than wave A because B includes the paths of waves 1 and 2 through the film. Let's separate these two factors for the moment by imagining that the film is very thin, so that the extra path length is negligible. In this case, waves A and B must be 180° out of phase because A is 180° out of phase with the incident wave (labeled 0) while B is in phase with the incident wave (labeled 0).

Now let's take into account the extra path length traveled by B. If the waves are nearly perpendicular to the film, the distance traveled either by wave 1 or by wave 2 is equal to the width of the film. So the extra path length is twice the width w of the film:

$$\Delta d = 2w \tag{36.10}$$

We just stated that if the film is very thin so that there is no path-length difference, then A and B are 180° out of phase. But if the extra path length is a half-wavelength, this path-length difference puts the two waves back in phase. In fact, as long as the extra path-length difference is an odd number of half-wavelengths, A and B are in phase. We can express this condition for constructive interference mathematically as

$$\Delta d = \left(m + \frac{1}{2}\right)\lambda_f \quad (m = 0, 1, 2, 3, \ldots) \tag{36.11}$$

In some equations, we represent a whole number with the letter m instead of n to avoid confusion with the index of refraction.

Recall that the wavelength λ_f in the film is given by $\lambda_f = \lambda_0/n_f$ (Eq. 36.9), where λ_0 is the wavelength in a vacuum (or in air). It is most convenient to write the condition for constructive interference in terms of the width of the film and the wavelength of light in a vacuum. Substituting Equations 36.9 and 36.10 into Equation 36.11 gives

$$2w = \left(m + \frac{1}{2}\right)\frac{\lambda_0}{n_f} \quad (m = 0, 1, 2, 3, \ldots) \tag{36.12}$$

Now we find the condition for destructive interference. In this case, the path-length difference must be a whole number of wavelengths. Mathematically, we have

$$\Delta d = m\lambda_f \quad (m = 0, 1, 2, 3, \ldots) \tag{36.13}$$

As before, we write the condition for destructive interference in terms of the width of the film and the wavelength of light in a vacuum:

$$2w = m\frac{\lambda_0}{n_f} \quad (m = 0, 1, 2, 3, \ldots) \tag{36.14}$$

Equations 36.12 and 36.14 may be applied to situations other than the one depicted in Figure 36.11, in which a thin film is surrounded by a medium with a lower index of refraction. These equations apply whenever the reflection at one boundary produces a 180° phase shift and the reflection at the other boundary produces no phase shift.

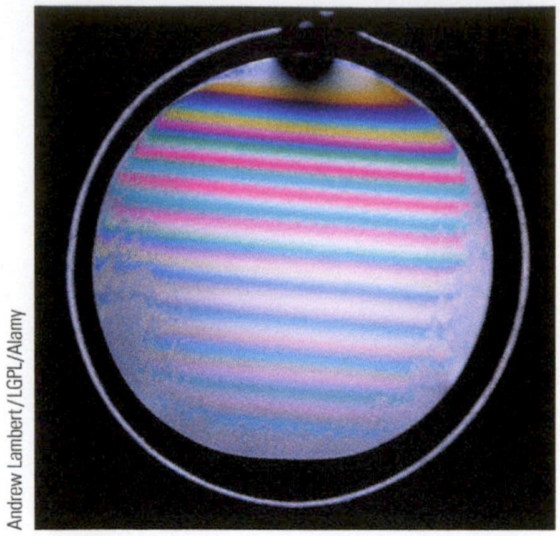

FIGURE 36.12 A thin-film interference pattern.

In other situations, the equations may need to be switched so that Equation 36.14 is for constructive and Equation 36.12 is for destructive interference.

We derived Equations 36.12 and 36.14 for the reflected waves A and B, ignoring the waves that are transmitted all the way through the soap film. The transmitted waves also produce an interference pattern, however, so you can see these patterns from either side of the film. The derivation of the conditions for constructive and destructive interference for the transmitted waves is similar to the derivation done here (Problem 12).

CONCEPT EXERCISE 36.2

We derived the conditions for constructive and destructive interference for a thin film such as the soap bubble in Figure 36.12. Use these conditions to explain why a colorful pattern is seen.

EXAMPLE 36.2 Newton's Rings

Thin-film interference may be used to test the shape of a lens. Consider a plano-convex glass lens, which has one plane surface and one convex surface as shown in Figure 36.13A. To test its shape, the lens is placed on top of a flat slab of glass so that there is a thin film of air between the lens and the slab. Monochromatic light is incident from above the lens, producing an interference pattern as shown from above in Figure 36.13B. If the lens is perfect, the pattern is a series of concentric rings called Newton's rings. (If a different pattern is seen, the lens has a defect.) Find an approximate expression for the radius r_m of the mth dark ring in terms of the lens's radius of curvature R and the wavelength of the incident light. Assume the index of refraction for air is 1 and the radius of a fringe is much smaller than the radius of curvature of the lens ($r_m \ll R$). In the **CHECK and THINK** step, explain why there is a dark spot at the center of the pattern.

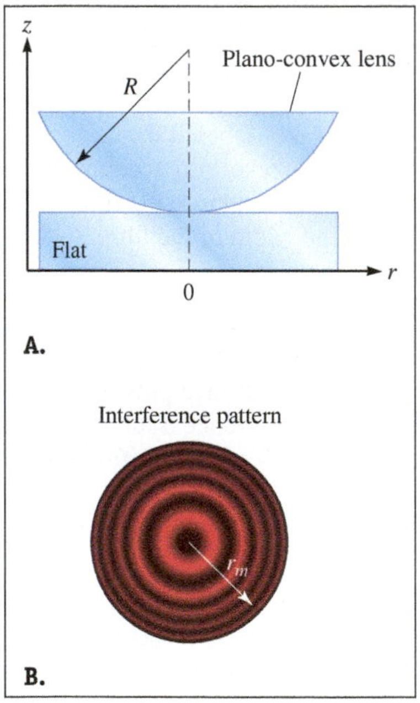

FIGURE 36.13 Apparatus for testing the shape of a lens.

INTERPRET and ANTICIPATE

When solving problems that involve thin-film interference, start by deciding whether the conditions for constructive (Eq. 36.12) or destructive (Eq. 36.14) interference apply to the situation. These equations apply whenever the reflection at one boundary produces a 180° phase shift and the reflection at the other boundary produces no phase shift. In our current problem, the film is air surrounded by glass, so we can use Figure 36.11, but now $n_i > n_f$. When the incident wave is reflected from the first boundary, there is no phase shift because the light is incident in a medium (glass) with a higher index of refraction. Wave 1 travels through the film of air. When wave 1 is reflected from the next boundary, there is a 180° phase shift because the light is incident in a medium with a lower index of refraction. So wave 2 is 180° out of phase with wave 1. There is no phase shift as wave B is transmitted in the lens. So, as in the case of the soap film in air (Fig. 36.11), for the air film in glass, one reflection has no phase shift and the other has a 180° phase shift. Thus, Equations 36.12 and 36.14 are the conditions for constructive and destructive interference, respectively. Because we are interested in one of the dark rings, which are the result of destructive interference, we should apply the condition for destructive interference to find its radius.

SOLVE

Draw a sketch showing the relevant geometry (Fig. 36.14). The lens's radius of curvature is R, and the radius of an arbitrary dark ring is r_m. The width of the air film is w at the location of that ring.

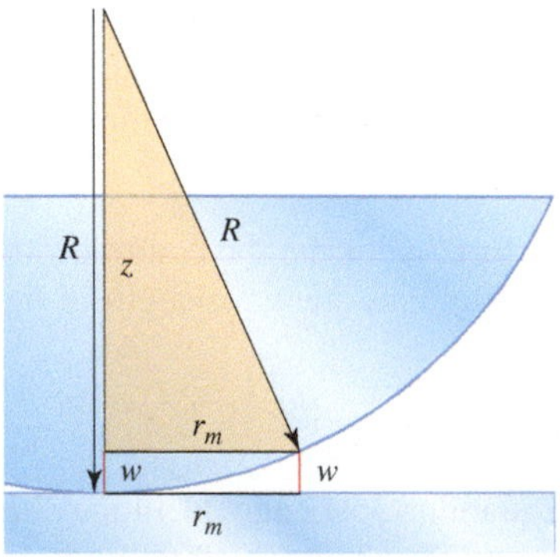

FIGURE 36.14

In Figure 36.14, we have defined z such that $z + w$ is the radius of curvature.	$z + w = R$ (1)
Square Equation (1).	$(z + w)^2 = R^2$ $z^2 + 2zw + w^2 = R^2$ (2)
Apply the Pythagorean theorem to the highlighted right triangle with sides z, r_m, and R.	$z^2 + r_m^2 = R^2$ (3)
Eliminate R^2 from Equations (2) and (3); then solve for r_m^2.	$r_m^2 = R^2 - z^2 = z^2 + 2zw + w^2 - z^2$ $r_m^2 = 2zw + w^2$ (4)
We are trying to derive an approximate expression for the situation where $r_m \ll R$. In Figure 36.14, when $r_m \ll R$, $z \approx R$ and $w \ll z$. We can approximate Equation (4) by replacing z with R and ignoring the last term w^2 because this term is much smaller than $2zw$.	$r_m^2 \approx 2Rw$ (5)
Now apply the condition for destructive interference (Eq. 36.14). In this case, the film is air with index of refraction $n_f \approx 1$. We drop the zero subscript because the wavelength of light is roughly the same in air as it is in a vacuum.	$2w = m\dfrac{\lambda_0}{n_f} \quad m = 0, 1, 2, 3, \ldots$ (36.14) $2w = m\lambda$ $w = \dfrac{m\lambda}{2}$ (6)
Substitute Equation (6) into Equation (5), and solve for r_m.	$r_m^2 \approx 2Rw \approx 2R\dfrac{m\lambda}{2}$ $r_m \approx \sqrt{m\lambda R}$

CHECK and THINK

This is the equation we were asked to derive. It gives the radius of each dark ring as a function of the wavelength and the lens's radius of curvature. By measuring r_m, you can make sure the lens has the expected radius of curvature R.

There is a dark spot in the center of the interference pattern because the lens is in contact with the flat glass slab. Light incident at the center undergoes two reflections: one involving a 180° phase shift and the other involving no phase shift. But, because no extra distance is traveled, the two reflected waves are out of phase and undergo destructive interference.

EXAMPLE 36.3 Like Oil and Water

A thin film with of an index of refraction of 1.20 lies on top of water with an index of refraction of 1.33. (See Fig. 36.1 for an example.)

A Explain why the conditions for constructive and destructive interference (Eqs. 36.12 and 36.14) are not applicable here. How can these equations be modified to fit this situation?

INTERPRET and ANTICIPATE

The key to determining whether Equations 36.12 and 36.14 apply to a situation is to see whether one reflection causes a 180° phase shift while the other causes no phase shift. (Consult Figure 36.11.) In this case, the incident light is initially in air. At the boundary between the air and the oily film, the reflected wave is 180° out of phase with the incident wave because the oil has a higher index of refraction than the air. At the next boundary between the oily film and the water, there is also a 180° phase shift because the water has a higher index of refraction than the oil. Because there are *two* 180° phase shifts, Equations 36.12 and 36.14 do **not** represent constructive and destructive interference, respectively.

Example continues on page 1164 ▶

SOLVE

Now our task is to modify these equations to fit this situation. First imagine the film is so thin that there is no extra path-length difference. In this case, the two reflected waves are in phase because both are 180° out of phase with the incident wave. So, if the path-length difference is zero or a whole number of wavelengths, the two reflected waves will remain in phase and interfere constructively. If the extra path-length difference is a half-integer number of wavelengths, the two reflected waves will be out of phase and interfere destructively. All the conditions that held when we derived Equations 36.12 and 36.14 have been switched, so all we need to do is switch the two equations.

Equation 36.12 is now the condition for destructive interference in the case of an oily film that has air on one side and water on the other.	Destructive condition in this case: $$2w = \left(m + \frac{1}{2}\right)\frac{\lambda_0}{n_f} \quad (m = 0, 1, 2, 3, \ldots) \quad (36.12)$$
Equation 36.14 is the condition for constructive interference in this case.	Constructive condition in this case: $$2w = m\frac{\lambda_0}{n_f} \quad (m = 0, 1, 2, 3, \ldots) \quad (36.14)$$

CHECK and THINK

One of the most important lessons of this problem is that you must not use Equations 36.12 and 36.14 blindly. Always consider first whether or not each reflection produces a phase shift. Note that the conditions involve the index of refraction of the film. You don't need to know the index of refraction for the surrounding media; you just need to know whether the surrounding media have higher or lower indices of refraction than the film.

B If the film is 465 nm thick and you are viewing it from directly above, for what wavelengths of visible light is the reflection brightest because of constructive interference? (Assume the Sun is directly above the film.)

INTERPRET and ANTICIPATE

In this part of the problem, we apply the condition for constructive interference that we found in part A.

SOLVE Solve the condition for constructive interference for wavelength. Then substitute numerical values for all variables except m. Keep an extra significant figure to avoid rounding errors.	$$2w = m\frac{\lambda_0}{n_f} \quad (m = 0, 1, 2, 3, \ldots) \quad (36.14)$$ $$\lambda_0 = (2n_f w)\frac{1}{m} = 2(1.20)(465\text{ nm})\frac{1}{m}$$ $$\lambda_0 = (1116\text{ nm})\left(\frac{1}{m}\right)$$
Now, to find the wavelengths of visible light, start with $m = 1$ to find all the wavelengths between 400 nm and 780 nm. (Do not include $m = 0$ because that gives an infinite wavelength.)	$$\lambda_0(1) = (1116\text{ nm})\left(\frac{1}{1}\right) = 1116\text{ nm}$$ $$\lambda_0(2) = (1116\text{ nm})\left(\frac{1}{2}\right) = 558\text{ nm}$$ $$\lambda_0(3) = (1116\text{ nm})\left(\frac{1}{3}\right) = 372\text{ nm}$$
The only wavelength that is in the visible range is green light.	$\lambda_0(2) = 558\text{ nm}$

CHECK and THINK

When you look from directly above, you see the film as a bright green patch. If you look at other patches (that are not directly below you), you see other colors because the path length for light through the film is longer.

C Now imagine viewing the film from directly below, as a scuba diver might do. For what wavelengths of visible light is the transmitted light brightest because of constructive interference?

INTERPRET and ANTICIPATE

Once again, we need to decide whether Equations 36.12 and 36.14 apply to this situation. This is our first problem involving transmitted light, so we start with the new sketch in Figure 36.15, keeping the vertical film orientation for easy comparison to Figure 36.11. Let the index of refraction of air be n_i (incident), the index of refraction of the film n_f (film), and the index of refraction of water n_t (transmitted). When the incident light in air encounters the film, it is reflected and transmitted, but we are not interested in the reflection that occurs at this boundary. The light transmitted into the film has no phase shift. Next, that light encounters the water and is again transmitted and reflected. The transmitted wave is labeled A, and the reflected wave is labeled 1. Wave 1 undergoes a 180° phase shift because the water has a higher index of refraction than the film. Next, wave 1 encounters the boundary between the film and air. Wave 1 is reflected and transmitted, but again we are not interested in the transmitted wave. Instead, follow the reflected wave, labeled 2. Wave 2 does not undergo a phase shift because the air has a lower index of refraction than the film. Finally, wave 2 encounters the film-water boundary, where once again it is reflected and transmitted. Now we are interested in the transmitted wave, labeled B. To sum up: Wave A has undergone no phase shifts due to reflection because it has undergone no reflection. Wave B has undergone one 180° phase shift due to reflection, although it has undergone two reflections. So Equations 36.12 and 36.14 may be used in this case, exactly as they appear.

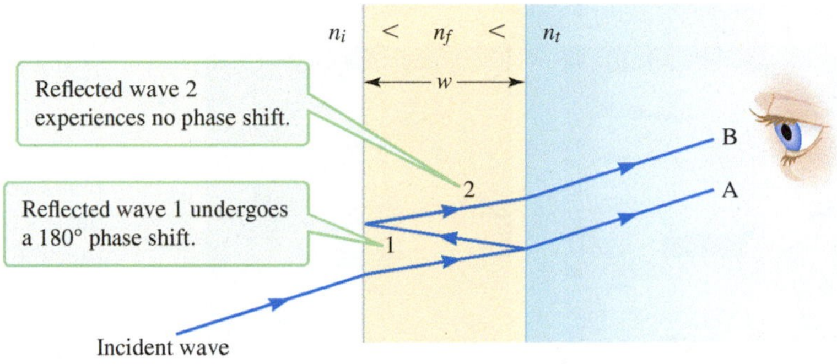

n_i < n_f < n_t
w
Reflected wave 2 experiences no phase shift.
Reflected wave 1 undergoes a 180° phase shift.
Incident wave
1
2
A
B

FIGURE 36.15

SOLVE

Now our work is much like that in part B, except we use Equation 36.12 for constructive interference. Solve for wavelength.	$2w = \left(m + \frac{1}{2}\right)\frac{\lambda_0}{n_f} \quad (m = 0, 1, 2, 3, \ldots) \quad (36.12)$ $\lambda_0 = (2n_f w)\left(m + \frac{1}{2}\right)^{-1}$ $\lambda_0 = (1116\text{ nm})\left(m + \frac{1}{2}\right)^{-1}$
To find the wavelengths of visible light, start with $m = 0$ and calculate all the wavelengths between 400 nm and 780 nm.	$\lambda_0(0) = (1116\text{ nm})(\frac{1}{2})^{-1} = 2232\text{ nm}$ $\lambda_0(1) = (1116\text{ nm})(\frac{3}{2})^{-1} = 744\text{ nm}$ $\lambda_0(2) = (1116\text{ nm})(\frac{5}{2})^{-1} = 446\text{ nm}$ $\lambda_0(3) = (1116\text{ nm})(\frac{7}{2})^{-1} = 319\text{ nm}$
In this case, two colors—red and indigo—are visible.	$\lambda_0(1) = 744\text{ nm}$ $\lambda_0(2) = 446\text{ nm}$

CHECK and THINK

An observer above the film sees green, while the diver below the surface sees burgundy (a blend of red and indigo).

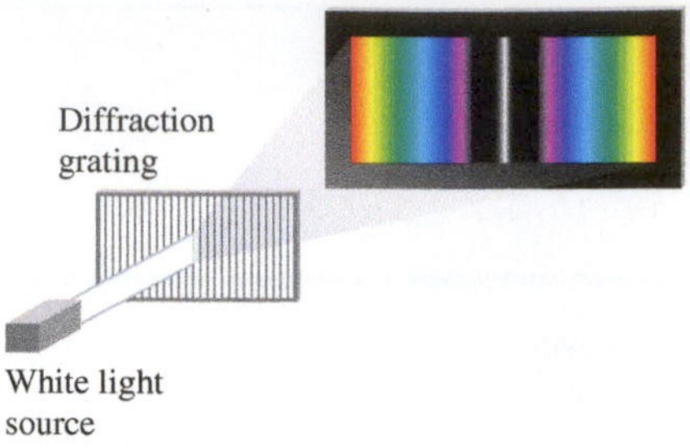

FIGURE 36.16 A diffraction grating produces a spectrum.

36-4 Diffraction Gratings

A **diffraction grating**, also known simply as a **grating**, is a device that has a large number of slits called **rulings**. Often diffraction gratings have thousands of rulings per millimeter. As in Young's double-slit experiment, light from each ruling arrives at points on a screen and interferes with light from the other rulings. Well-separated bright fringes, also known as **lines**, appear on the screen at places of constructive interference. Because the location of the lines depends on wavelength, the grating produces a "rainbow" (Fig. 36.16). This rainbow is called a **visible light spectrum**, a **color spectrum**, or just a **spectrum**.

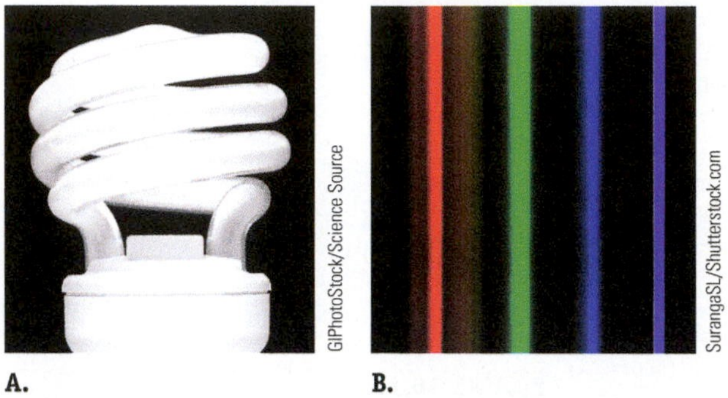

FIGURE 36.17 A. A compact fluorescent bulb. **B.** A compact fluorescent bulb's spectrum. The fluorescent bulb's spectrum is missing many colors.

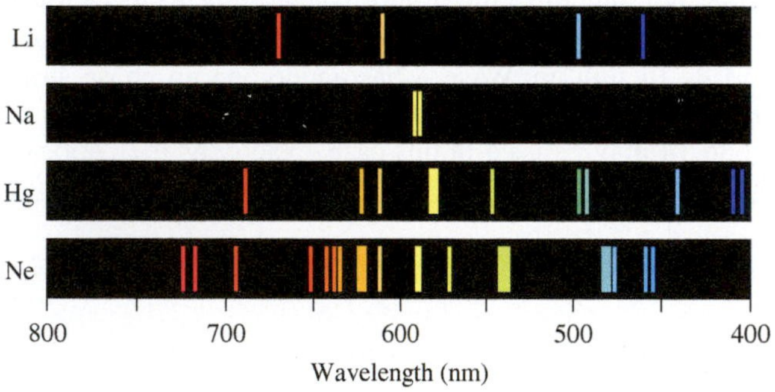

FIGURE 36.18 Each atom's spectrum is a unique set of colors.

The Hydrogen Spectrum

One of the most important uses of a diffraction grating is to spread polychromatic light out into a color spectrum. A complete full-color spectrum is generated when light from a **black body**, an *ideal* object that absorbs all incident radiation, passes through a diffraction grating. When a black body is in thermal equilibrium, it is an ideal emitter (with emissivity $\varepsilon = 1$; Section 21-10) that produces an *ideal* spectrum called a **black-body spectrum**. No real object can produce a perfect black-body spectrum, but when white light—for example, from a glowing tungsten filament—passes through a diffraction grating, a nearly full-color spectrum from red to violet appears on the screen (Fig. 36.16). Other light sources are not well modeled as black bodies. For example, a compact fluorescent bulb's spectrum is missing many colors (Fig. 36.17).

The spectrum of each type of atom or molecule is unique, showing lines of only certain colors. Figure 36.18 shows the spectra produced by several different elements. For example, the spectrum of lithium (Li) has about four well-spaced lines, whereas Na has two lines near the center of the range of visible wavelengths. One practical application of a diffraction grating is to examine the spectrum of a light source in order to determine the source's composition. This is particularly important in astronomy because the great distances involved mean that nearly all of the information we have comes from the light emitted by the objects. Observations of spectra have determined that roughly 75% of the universe is hydrogen and most of the rest is helium. All the other elements are present in only trace amounts. The hydrogen spectrum is discussed in the case study in Chapter 42.

A.

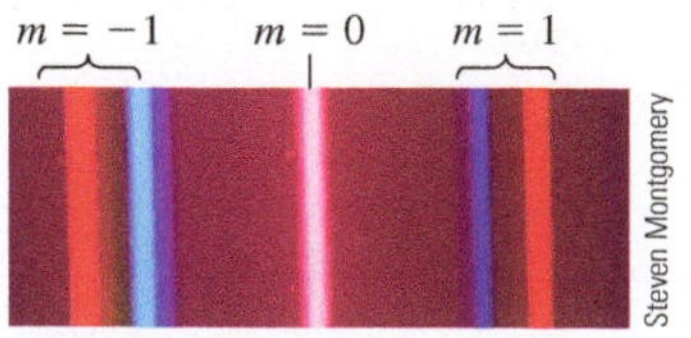

B.

FIGURE 36.19 A. Hydrogen gives off four colored lines in the visible part of the electromagnetic spectrum. The wavelength of each line in nanometers is provided on the figure. **B.** The spectrum of hydrogen observed through a poor quality diffraction grating.

Figure 36.19A shows that hydrogen ordinarily produces four strong lines in the visible part of the electromagnetic spectrum: Hα (red), Hβ (aqua), Hγ (indigo), and Hδ (violet). You may also see another faint shorter wavelength line; we'll ignore this line in our discussion. Figure 36.19B shows the spectrum of hydrogen observed in a laboratory using a diffraction grating. There are several features to notice in this spectrum. First, we see two sets of bright fringes labeled with the appropriate value of m. Second, the four expected lines are blended together so we only see two separate lines on either side of the center. Third, the central maximum ($m = 0$) is pale pink because the different colored lines overlap in the center. Fourth, the short-wavelength lines are closer to the center than are the long-wavelength lines, so compared to the $m = 1$ lines, the $m = -1$ lines are in the reverse order. The general process of using a diffraction grating to observe the spectrum of a light source may be applied to any element or compound. Our next step is to derive an expression for the position of observed lines.

DERIVATION Position of the Maxima (Lines) in a Diffraction Grating

We show that the angular position θ of the lines produced by a diffraction grating is given by

$$d \sin\theta = m\lambda \quad (m = 0, \pm1, \pm2, \pm3, \ldots) \qquad (36.15)$$

where d is the separation of adjacent rulings and λ is the wavelength of the light. The procedure for finding an expression for the position of the lines produced by a diffraction grating is very similar to that for Young's double-slit experiment (page 1129).

DIFFRACTION GRATING—POSITION OF LINES Special Case

Consider a viewing screen far from the grating, so that the paths from the rulings to any point on the screen are approximately parallel (Fig. 36.20). A point on the screen has the angular position θ measured from the horizontal axis.

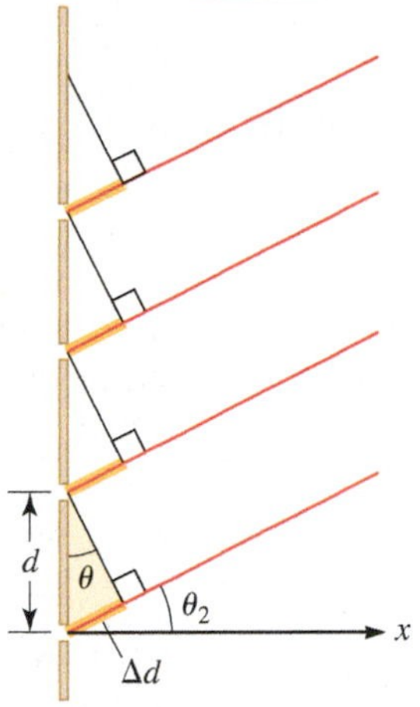

FIGURE 36.20

Consider the bottom two rulings and the paths from them to the screen in Figure 36.20. The path from the bottom ruling to the point on the screen is longer. Find the path-length difference Δd by applying trigonometry to the highlighted right triangle.	$\Delta d = d \sin\theta$ ✓ (1)
In order for a maximum to appear on the screen at the angular position θ, the waves must arrive in phase, so the path-length difference must be an integer number of wavelengths. Using this condition and Equation (1), we find an expression for the position of the lines.	$\Delta d = m\lambda \quad (m = 0, \pm1, \pm2, \pm3, \ldots)$ $d \sin\theta = m\lambda \quad (m = 0, \pm1, \pm2, \pm3, \ldots)$ ✓ (36.15)

COMMENTS

Each integer m corresponds to a different line, as labeled in Figure 36.21. The integers m are called the **order numbers**. We refer to the zeroth-order line, the first-order line, and so on.

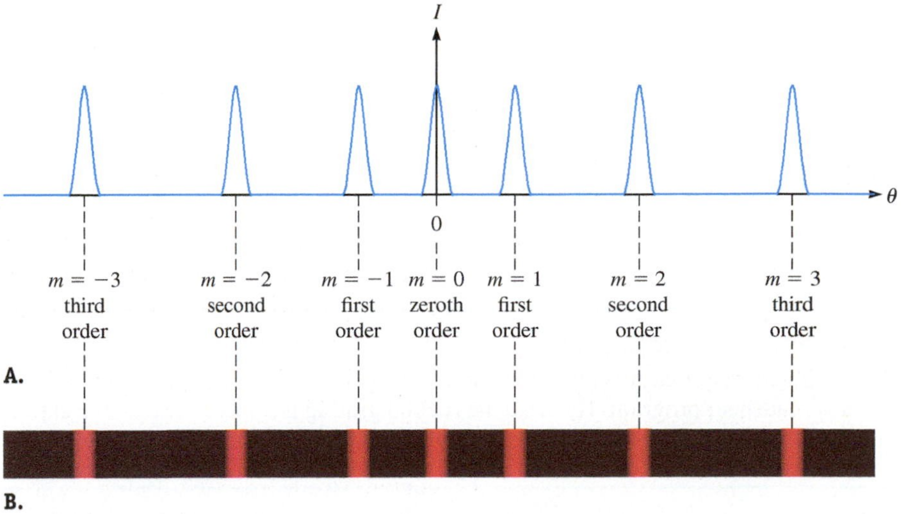

FIGURE 36.21 **A.** A graph of intensity versus angular position θ for monochromatic light passing through a diffraction grating. **B.** The corresponding pattern of lines produced by this diffraction grating. The lines are referred to by the order number m.

Width of the Maxima in a Diffraction Grating

Now suppose you don't know which element produced the observed spectrum in Figure 36.19B, so you compare this observed spectrum to the theoretical spectra shown in Figures 36.18 and 36.19A. It is hard to identify the element from its observed spectrum because the lines overlap and blend together. As a result, the diffraction grating used to produce the observed spectrum in Figure 36.19B is not very effective because the lines are not well separated. To characterize the line separation, we need an expression for the width of the lines (described in this section). In the next section, we show how to characterize the efficiency of diffraction gratings.

We want to quantify how effective a grating is at producing a spectrum. A good diffraction grating produces thin lines, which is important so that the colors are well separated. It is common to characterize the width of a line in terms of its **half-width** $\Delta\theta_{hw}$, which is measured from the line's center to the darkness found on either side (Fig. 36.22). The half-width of any line is given by

DIFFRACTION GRATING—HALF-WIDTH OF LINES **Special Case**

$$\Delta\theta_{hw} = \frac{\lambda}{Nd\cos\theta} \qquad (36.16)$$

where N is the total number of rulings in the grating (Problem 65).

Diffraction grating manufacturers often specify the total number N of rulings. We can find the ruling separation d in terms of N and the length ℓ of the grating:

$$d = \frac{\ell}{N} \qquad (36.17)$$

If the manufacturer specifies instead the linear density of the rulings n (number of rulings per unit length), we use $d = 1/n$.

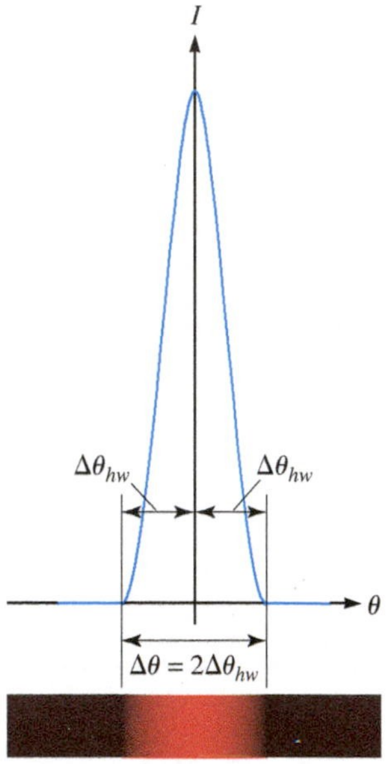

FIGURE 36.22 The width of a line is quantified in terms of the half-width from the center to the darkness on either side.

CONCEPT EXERCISE 36.3

Explain why it would be better to call a *diffraction* grating an *interference* grating.

EXAMPLE 36.4 **A Tale of Two Gratings**

Compare the hydrogen spectra produced by two different gratings by completing the specified tasks. Each grating has the same separation $d = 2000$ nm between rulings, but the first grating has 5 rulings and the second grating has 50 rulings. For each grating, find the $m = 1$ and $m = 2$ angular positions of the four hydrogen lines (Hα, Hβ, Hγ, and Hδ). Then find the half-widths of these (eight) lines. Sketch the lines, taking into account their positions and thicknesses.

INTERPRET and ANTICIPATE

This problem requires repeating the same sort of calculation several times, so you may wish to use a spreadsheet program. Because the second grating has more rulings, we expect it to produce spectral lines that are narrower than the lines produced by the first grating.

SOLVE

Equation 36.15 gives the angular position of the lines. Because we must calculate the angular positions for four different wavelengths, we substitute all the other values. Start with the first-order lines ($m = 1$).

$$d\sin\theta = m\lambda \qquad (36.15)$$

$$\theta = \sin^{-1}\left(\frac{m\lambda}{d}\right)$$

$$\theta_1 = \sin^{-1}\left(\frac{\lambda}{2.00\times10^{-6}\,\text{m}}\right)$$

We also need the second-order lines ($m = 2$).

$$\theta_2 = \sin^{-1}\left(\frac{2\lambda}{2.00 \times 10^{-6}\,\text{m}}\right)$$

Notice that the angular positions do not depend on the number of rulings, so the values of θ_1 and θ_2 can be used for both gratings. Organize these results in a table.

TABLE 36.2 Angular positions of four hydrogen lines.

Line	λ ($\times 10^{-7}$ m)	θ_1 (rad)	θ_2 (rad)
Hα	6.56	0.334	0.716
Hβ	4.86	0.246	0.508
Hγ	4.34	0.219	0.449
Hδ	4.10	0.207	0.423

The next step is to find the half-widths of these eight lines from Equation 36.16. Because the half-width depends on the number of rulings, we treat each grating separately.

$$\Delta\theta_{hw} = \frac{\lambda}{Nd\cos\theta} \quad (36.16)$$

Substitute $N = 5$ and the value of d to find an expression for the half-widths of the lines produced by the first grating.

First grating, $N = 5$:

$$\Delta\theta_{hw} = \frac{\lambda}{(5)(2.00 \times 10^{-6}\,\text{m})\cos\theta}$$

$$\Delta\theta_{hw} = \frac{\lambda}{(1.00 \times 10^{-5}\,\text{m})\cos\theta}$$

Then use the wavelengths and the angular positions from Table 36.2 to find the half-width of each of the eight lines produced by the first grating. Again, display the results in a table. Because the half-width depends on θ, we must calculate $\Delta\theta_{hw}$ separately for θ_1 and θ_2.

TABLE 36.3 Half-widths for first grating, $N = 5$.

Line	$(\Delta\theta_{hw})_1$ (rad)	$(\Delta\theta_{hw})_2$ (rad)
Hα	0.0695	0.0870
Hβ	0.0501	0.0556
Hγ	0.0444	0.0482
Hδ	0.0419	0.0450

Sketch the lines produced by the first grating (Fig. 36.23). The positions of the eight lines are given in Table 36.2. The thicknesses of the lines come from Table 36.3. Notice that the lines from the two orders overlap.

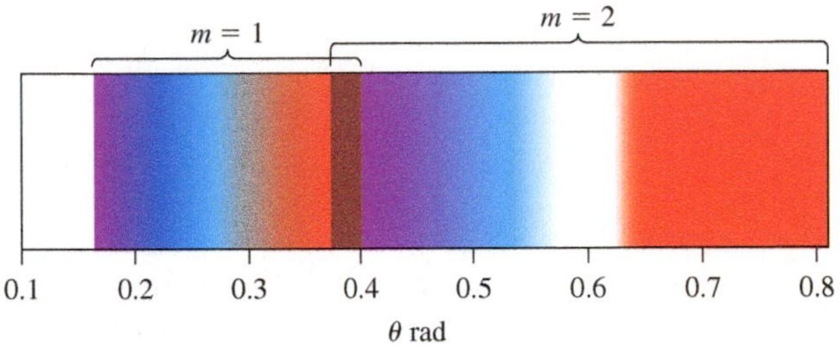

FIGURE 36.23 In a diffraction grating with only five rulings, the hydrogen lines overlap.

Because the second grating has 50 rulings, we must repeat the process of finding the half-widths of the lines it produces.

$$\Delta\theta_{hw} = \frac{\lambda}{(50)(2.00 \times 10^{-6}\,\text{m})\cos\theta}$$

$$\Delta\theta_{hw} = \frac{\lambda}{(1.00 \times 10^{-4}\,\text{m})\cos\theta}$$

Table 36.4 gives the half-widths of the eight lines produced by the second grating. Notice that these lines are thinner than the lines produced by the first grating.

TABLE 36.4 Half-widths for second grating, $N = 50$.

Line	$(\Delta\theta_{hw})_1$ (rad)	$(\Delta\theta_{hw})_2$ (rad)
Hα	0.00695	0.00870
Hβ	0.00501	0.00556
Hγ	0.00444	0.00482
Hδ	0.00419	0.00450

Example continues on page 1170 ▶

Sketch the lines produced by the second grating (Fig. 36.24). As for the first grating, the positions of the eight lines are given in Table 36.2. The thicknesses of the lines are listed in Table 36.4.

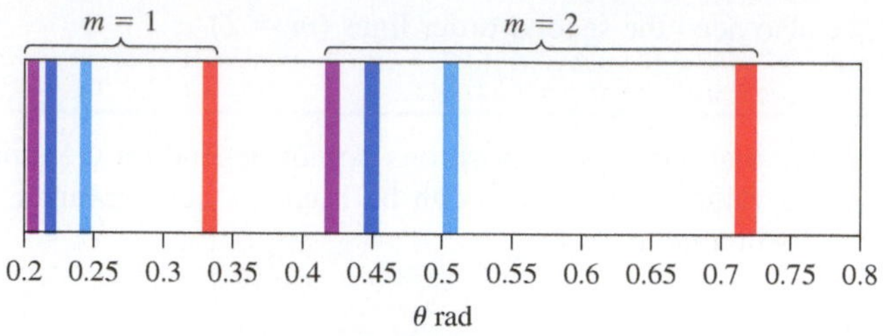

FIGURE 36.24 In a diffraction grating with 50 rulings, the hydrogen lines are well separated. Both the $m = 1$ and the $m = 2$ lines are shown.

CHECK and THINK

Compare Figures 36.23 and 36.24. The second grating has 10 times more rulings than the first grating and produces a spectrum of eight lines that are well separated. If you want to identify the element that produced the spectrum, you may compare Figure 36.24 to the theoretical spectrum in Figure 36.19A. It would be difficult to do the same with Figure 36.23 because the lines are blended together.

36-5 Dispersion and Resolving Power of Gratings

Figures 36.23 and 36.24 show how different diffraction gratings can produce spectra of different qualities. The effectiveness of a diffraction grating is quantified in terms of two parameters—*dispersion* and *resolving power*. **Dispersion** is a measure of the angular separation between the lines of a spectrum. **Resolving power** is a measure of the thickness of these lines. The gratings that produced the spectra in Figures 36.23 and 36.24 have the same dispersion, but the resolving power of the grating is greater in Figure 36.24. (The term *resolution* describes the spatial separation between the diffraction patterns produced by two different sources through a single aperture; *resolving power* describes the narrowness of the lines produced by a single source through a diffraction grating.)

Dispersion

The mathematical definition of dispersion D is

DISPERSION Major Concept

$$D \equiv \frac{\Delta\theta}{\Delta\lambda} \tag{36.18}$$

where $\Delta\theta$ is the angular separation between two lines whose wavelength difference is $\Delta\lambda$. The SI units for dispersion are radians per meter, although sometimes it is more convenient to work in degrees per meter, arc seconds per meter, or arc minutes per meter.

Don't confuse $\Delta\theta$ with $\Delta\theta_{hw}$: $\Delta\theta$ is the separation between two different lines, and $\Delta\theta_{hw}$ is the half-width (thickness) of a single line.

If the dispersion of a grating is high, its lines have a high degree of separation. When designing a grating, you can increase the dispersion by using closer rulings. In Problem 66, you will show that the dispersion of a grating is given by

$$D \equiv \frac{\Delta\theta}{\Delta\lambda} = \frac{m}{d\cos\theta} \tag{36.19}$$

where m is the order number, θ is the angular position of the corresponding line, and d is the separation between the rulings. Equation 36.19 shows that the dispersion of a grating is greater (better) if the rulings are closely spaced (d is small) and if the order number m is large. The factor $\cos\theta$ in the denominator means that the dispersion also depends on the angular position of the lines. Lines that have a large (absolute) value for θ have a greater dispersion D.

CONCEPT EXERCISE 36.4

Find the dispersion for the ($m = 1$) Hα and Hβ lines produced by the first grating in Example 36.4. How does it compare to the dispersion for the second grating?

Resolving Power

The lines in Figures 36.23 and 36.24 have the same dispersion, but they are better separated in Figure 36.24 because the lines are narrower. The resolving power R is a measure of the lines' narrowness. The narrower the lines, the higher the resolving power R, which is defined as

$$R \equiv \frac{\lambda_{av}}{\Delta\lambda} \quad (36.20)$$

RESOLVING POWER ✪ **Major Concept**

where λ_{av} is the average wavelength of two lines that are just barely separated and $\Delta\lambda$ is the difference in their wavelengths.

DERIVATION Resolving Power of a Diffraction Grating

Next we show that the resolving power of a diffraction grating is given by

$$R \equiv \frac{\lambda_{av}}{\Delta\lambda} = Nm \quad (36.21)$$

where N is the number of rulings and m is the order number.

Solve the dispersion equation (Eq. 36.19) for $\Delta\theta$.	$D \equiv \frac{\Delta\theta}{\Delta\lambda} = \frac{m}{d\cos\theta}$ (36.19) $\Delta\theta = \left(\frac{m}{d\cos\theta}\right)\Delta\lambda$
If the angular separation $\Delta\theta$ between two lines is just barely visible, the minimum of one line falls on the maximum of the other line, and their separation is their (average) half-width $\Delta\theta_{hw}$ (Eq. 36.16).	$\Delta\theta_{hw} = \Delta\theta$ $\frac{\lambda_{av}}{Nd\cos\theta} = \left(\frac{m}{d\cos\theta}\right)\Delta\lambda$
Solve for $\lambda_{av}/\Delta\lambda$.	$\frac{\lambda_{av}}{\Delta\lambda} = Nm$
Substitute the definition of the resolving power (Eq. 36.20).	$R \equiv \frac{\lambda_{av}}{\Delta\lambda} = Nm$ ✓ (36.21)

COMMENTS

According to Equation 36.21, the resolving power is greater (better) if the diffraction has more rulings (if N is large) and if the order number m is also large.

CONCEPT EXERCISE 36.5

Why isn't the spectrum produced by Young's double-slit experiment effective at producing well-separated spectral lines? (For simplicity, think about the first-order separation of a red line and a violet line.)

EXAMPLE 36.5 A Tale of Two Gratings, Revisited

Let's apply the idea of resolving power to the two diffraction gratings in Example 36.4. What resolving power is required to just barely resolve the $m = 1$ Hγ (indigo) and Hδ (violet) lines? Find the resolving power of each grating. In the **CHECK and THINK** step, compare the resolving power of each grating to the resolving power required to just barely separate the lines.

Example continues on page 1172 ▶

INTERPRET and ANTICIPATE
From Figures 36.23 and 36.24, the first grating cannot resolve the lines but the second one can. Now we show this numerically.

SOLVE
Find the average wavelength of the Hγ and Hδ lines and the difference between their wavelengths.

$$\lambda_{av} = \frac{(410 + 434)\,\text{nm}}{2} = 422\ \text{nm}$$

$$\Delta\lambda = (434 - 410)\,\text{nm} = 24\ \text{nm}$$

The resolving power required to just separate the lines is $R = \lambda_{avg}/\Delta\lambda$ (Eq. 36.20).

$$\frac{\lambda_{av}}{\Delta\lambda} = \frac{422\,\text{nm}}{24\,\text{nm}} = 17.5$$

Now, to find the first-order ($m = 1$) resolving power of the two gratings in Example 36.4, use Equation 36.21. The first grating has 5 rulings and the second grating has 50 rulings.

$$R = Nm \quad (36.21)$$

First grating:

$$R_1 = (5)(1) = 5$$

Second grating:

$$R_2 = (50)(1) = 50$$

CHECK and THINK
The resolving power R_1 of the first grating is lower than the required value of 17.5, so the two lines are blended together as shown in Figure 36.23. However, the resolving power R_2 of the second grating is several times higher than the required resolving power, so the lines are well separated as seen in Figure 36.24.

EXAMPLE 36.6 A Third Grating

Let's compare the two diffraction gratings from Example 36.4 to a third grating that has 50 rulings and a distance between rulings $d = 3000$ nm. For the first order ($m = 1$), find the angular positions and half-widths of the four hydrogen lines. Sketch this first-order spectrum and compare it to the spectrum found for the second grating in Example 36.4 (Fig. 36.24).

INTERPRET and ANTICIPATE
Use the same procedure as in Example 36.4. We expect this third grating to have the same resolution as the second grating because both have the same number of rulings ($N = 50$). However, we expect this grating to have a lower dispersion because the separation between the rulings is greater.

SOLVE
The angular positions of the lines come from Equation 36.15. Because we must calculate the angular positions for four different wavelengths, it is helpful to substitute all the other values, including $m = 1$.

$$d\sin\theta = m\lambda \quad (36.15)$$

$$\theta = \sin^{-1}\left(\frac{m\lambda}{d}\right) = \sin^{-1}\left(\frac{\lambda}{3.00 \times 10^{-6}\,\text{m}}\right)$$

Organize the results in a table.

TABLE 36.5 Angular positions of four hydrogen lines.

Line	λ ($\times 10^{-7}$ m)	θ_1 (rad)
Hα	6.56	0.221
Hβ	4.86	0.163
Hγ	4.34	0.145
Hδ	4.10	0.137

The next step is to find the widths of these four lines. The half-width comes from Equation 36.16. It depends on the separation d between rulings and the angular position of the lines, so we don't expect to find the same half-widths that we found in Example 36.4.

$$\Delta\theta_{hw} = \frac{\lambda}{Nd\cos\theta} \qquad (36.16)$$

$$\Delta\theta_{hw} = \frac{\lambda}{(50)(3.00 \times 10^{-6}\,\text{m})\cos\theta} = \frac{\lambda}{(1.50 \times 10^{-4}\,\text{m})\cos\theta}$$

Table 36.6 lists the half-widths of the four lines produced by the third grating.

TABLE 36.6 Half-widths for new grating.

Line	$(\Delta\theta_{hw})_1$ (rad)
Hα	0.00448
Hβ	0.00328
Hγ	0.00293
Hδ	0.00277

Sketch the lines produced by the third grating (Fig. 36.25). The positions of the lines are given in Table 36.5, and the thicknesses of the lines come from Table 36.6. We have included the $m = 1$ portion of the second grating's spectrum for easy comparison. (To do this, we had to rescale the horizontal axis to match the scale of the third grating's spectrum.)

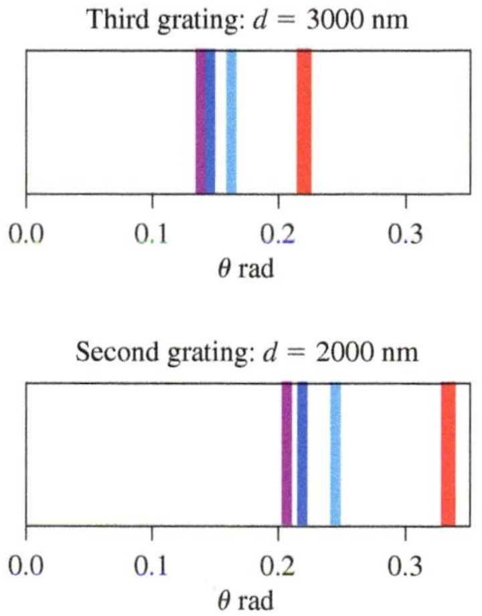

FIGURE 36.25 The $m = 1$ hydrogen spectra for the third grating (d = 3000 nm) and the second grating (d = 2000 nm). Both gratings have 50 rulings.

CHECK and THINK

A lower dispersion for the third grating means that the lines are closer to the central maximum at $\theta = 0$ and closer to one another. So the second grating is the most effective of the three. It has the greatest dispersion and the highest resolution.

X-ray Diffraction

So far, we have imagined using an artificially constructed diffraction grating to study light from a natural source such as the Sun. In such a case, we know the separation of adjacent rulings in the grating and we may wish to measure the wavelengths of light produced by the Sun. In this subsection, we see how artificially generated X-rays may be used to find the spacing of atoms in a solid, which acts as a natural three-dimensional diffraction grating.

X-rays are electromagnetic waves with wavelengths between about 1 pm and 1 nm (Section 34-5).

To get an idea of how this works, imagine finding a diffraction grating with no indication of the separation d between rulings. You want to use the diffraction grating, but first you must find d. So, you use a laser of known wavelength to create a diffraction pattern. By measuring the angular positions of the different order lines, you can deduce the ruling separation using $d\sin\theta = m\lambda$ (Eq. 36.15).

Finding the separation between atoms in a solid is somewhat more complicated. First, because the separation between atoms is on the order of 10^{-10} m, visible light has a wavelength about three orders of magnitude too long to create a diffraction pattern. So X-rays are used instead. Second, the solid is three-dimensional and the diffraction pattern is a result of the X-rays reflected from parallel planes (Fig. 36.26).

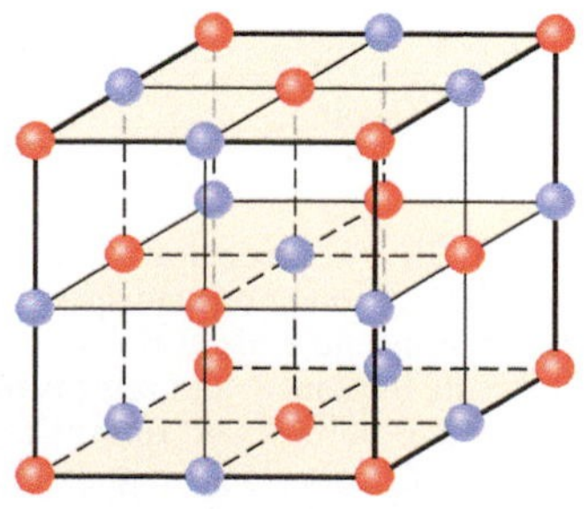

FIGURE 36.26 Atoms in a solid are arranged in a three-dimensional pattern.

Consider an X-ray beam reflected from two such parallel planes (Fig. 36.27). The beam makes an angle θ with respect to the planes. The X-ray reflected from the lower plane travels farther than the X-ray reflected from the upper plane. We can find the path-length difference by examining the small highlighted right triangle in

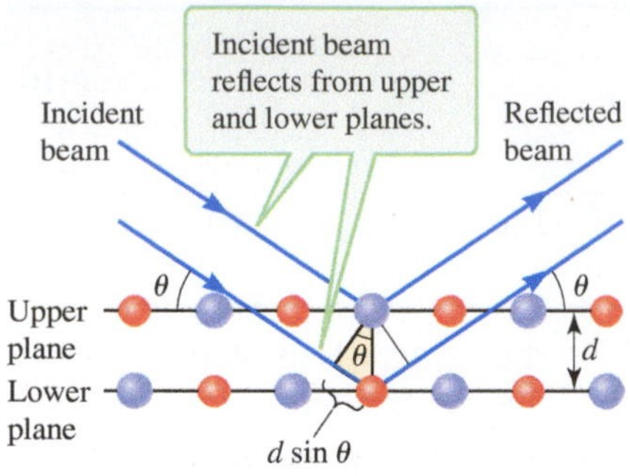

FIGURE 36.27 X-rays are reflected from two parallel planes separated by a distance *d*.

Figure 36.27. The hypotenuse of this triangle is *d*, and the extra path length traveled by the incident beam is $d \sin \theta$. Of course, this is also the extra path length traveled by the beam reflected from the lower plane, so the total path-length difference is $2d \sin \theta$. If this extra path length is an integer multiple of the X-ray wavelength, the two reflected beams interfere constructively. We've shown only two adjacent planes, but as long as the planes are parallel and evenly spaced, the same condition produces constructive interference for all planes. The condition for constructive interference is known as **Bragg's law** and is given by

$$2d \sin \theta = m\lambda \quad (m = 1, 2, 3, \ldots) \tag{36.22}$$

The spacing *d* between atomic planes is found by using X-rays of known wavelength and measuring the angular positions θ for constructive interference spots observed at order *m*. To get the three-dimensional structure, the solid must be rotated to find the spacing between other parallel planes.

Bragg's law is named for William Lawrence Bragg (1890–1971), who shared the Nobel Prize in physics with his father William Henry Bragg (1862–1942) in 1915 for formulating the condition for constructive interference (Eq. 36.22). The two men continued to work together to find the structure of other substances, including sodium chloride—table salt.

36-6 Case Study: Michelson's Interferometer

The first American to win a Nobel Prize in science was Albert Abraham Michelson (1852–1931), who was born in a part of Prussia that is now in Poland and moved to the United States at age 2. Michelson graduated in 1873 from the United States Naval Academy and subsequently became a teacher there, where he began the experiments that led to his prize-winning work.

By the time Michelson was at the academy, Young had already demonstrated that light is a wave and Maxwell had derived the electromagnetic wave equation. So, like you, Michelson was taught that light is a wave. In Michelson's time, however, scientists thought that all waves must propagate in a medium. They called the medium in which light propagates the **ether**, reasoned to be weightless, (nearly) incompressible, frictionless, transparent, and omnipresent—permeating all space and matter. Physicists in the 1800s also reasoned that there was no chemical way to detect the ether, meaning it couldn't be captured in a bottle and normal chemical experiments could not determine its properties. One of the major differences between science and religion is that in science, something exists only if it is detectable. So, for ether to exist, there must be some way to detect it and determine its physical characteristics. Michelson designed an experiment to measure the speed of the Earth with respect to the ether. What he found instead is that the ether does not exist.

Swimming Race Analogy

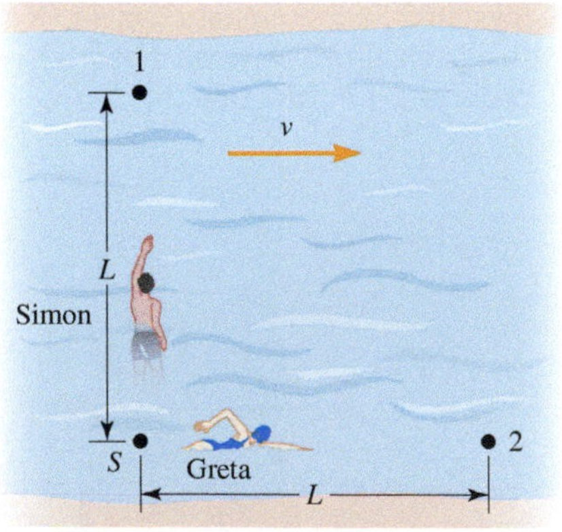

FIGURE 36.28 The river flows to the right at a uniform and constant speed *v*. The race begins and ends at point *S*. Simon swims to point 1 and returns to *S*. Greta swims to point 2 and returns to *S*. The distance *L* from *S* to 1 equals the distance from *S* to 2. Each swimmer's speed with respect to the water is *c*. Simon always wins as long as $0 < v < c$.

Michelson designed an instrument that we now call a **Michelson interferometer**, intended to allow him and his collaborator Edward Morley to measure the speed of the Earth through the ether. Michelson thought the kinematics of relative motion (Sections 4-7 and 4-8) could be applied to the situation of the Earth moving relative to the ether. He explained his experiment using an analogy with swimmers racing in a river, so we'll use this analogy too.

Figure 36.28 shows a river flowing at constant uniform speed *v* to the right with respect to the river's banks. In this river, Simon and Greta are about to have a swimming race. Both start in the river at point *S* near one bank. Simon swims to point 1 on the opposite side of the river, while Greta swims to point 2 downstream. Then both swimmers reverse course and return to their starting position at *S*. The distance *L* from the starting point *S* to point 1 equals the distance from *S* to point 2.

Now, here is the unusual thing about this race: Both swimmers swim at the same speed *c* with respect to the river. (For the moment, *c* is not the speed of light, just the speed of the swimmers.) So, if the river were not flowing (if $v = 0$), the race would be a tie; both swimmers would arrive back at *S* at the same moment, and each swimmer's time would be $\Delta t = 2L/c$.

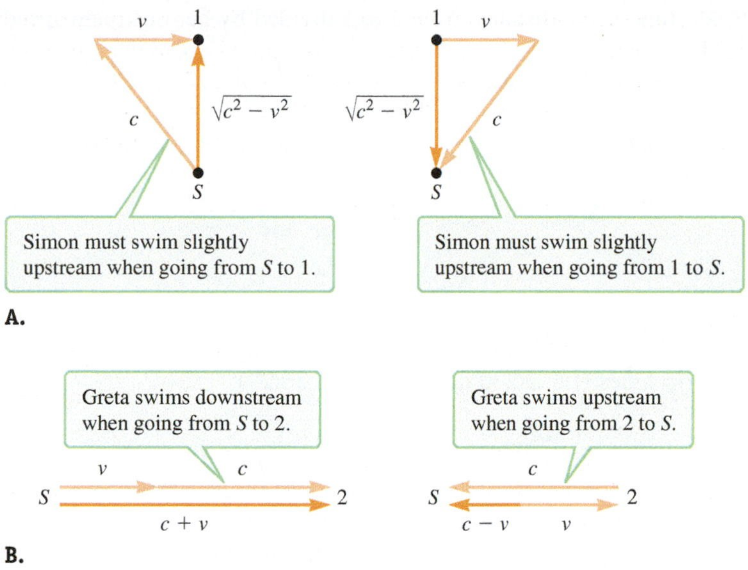

FIGURE 36.29 Each swimmer's speed with respect to the water is c. (The magnitude of each vector is written next to each vector.) **A.** Simon's velocity with respect to the river's banks must point directly toward 1 on his way out and directly toward S on his way back. So, Simon must swim slightly upstream, whether he is swimming to or from the starting position S. As a result, his speed relative to the banks is same whether he is swimming toward or away from S. **B.** Greta swims with the current when she goes from S to 2, so her speed relative to the banks equals her speed relative to the water plus the speed of the water. When she swims from 2 back to S, she swims against the current, so her speed relative to the banks equals her speed relative to the water minus the speed of the water.

However, if the river is moving at speed v, Simon always wins the race. To see why this is true, let's derive each swimmer's travel time. Simon must swim slightly upstream in order to compensate for the river's motion to the right, whether he is swimming toward or away from the starting position at S (Fig. 36.29A). We can use the Pythagorean theorem to determine Simon's speed relative to the banks:

$$v_{\text{Simon}} = \sqrt{c^2 - v^2} \tag{36.23}$$

where v is the speed of the water with respect to the banks.

By contrast, Greta's speed relative to the banks depends on whether she is swimming away from or toward the starting position. Her speed relative to the water is c in either case (Fig. 36.29B). But when she swims with the current away from S, her speed relative to the banks is higher because she is helped by the current. When she swims back toward the starting position, her speed relative to the banks is lower. So, when Greta swims from S to 2 (downstream), her speed with respect to the banks is given by

$$(v_{\text{Greta}})_{\text{down}} = c + v \tag{36.24}$$

When she swims from 2 toward S (upstream), her speed with respect to the banks is given by

$$(v_{\text{Greta}})_{\text{up}} = c - v \tag{36.25}$$

The swimmer who wins the race has the shortest round-trip time, which we need to find. Simon's speed relative to the banks is the same whether he is swimming toward or away from the starting point. The round-trip distance he travels (relative to the banks) is $2L$. So, his round-trip time is his round-trip distance divided by his speed v_{Simon} relative to the banks:

$$\Delta t_{\text{Simon}} = \frac{2L}{v_{\text{Simon}}} \tag{36.26}$$

Substitute Equation 36.23 into Equation 36.26:

$$\Delta t_{\text{Simon}} = \frac{2L}{\sqrt{c^2 - v^2}} \tag{36.27}$$

Greta's speed relative to the banks is faster when she is moving downstream, so her time to go from S to 2 is shorter than her time on the return trip. We must take this into account when we calculate her round-trip time. Her downstream time is the distance from S to 2 divided by her downstream speed:

$$\Delta t_{\text{down}} = \frac{L}{(v_{\text{Greta}})_{\text{down}}} = \frac{L}{c + v} \tag{36.28}$$

Her upstream time is the distance from 2 to S divided by her upstream speed:

$$\Delta t_{\text{up}} = \frac{L}{(v_{\text{Greta}})_{\text{up}}} = \frac{L}{c - v} \tag{36.29}$$

So, her total round-trip time is the sum of Equations 36.28 and 36.29:

$$\Delta t_{\text{Greta}} = \Delta t_{\text{down}} + \Delta t_{\text{up}} = \frac{L}{c + v} + \frac{L}{c - v}$$

$$\Delta t_{\text{Greta}} = \frac{2Lc}{c^2 - v^2} \tag{36.30}$$

One way to show that Simon always wins is to find an expression for the ratio of their round-trip times (Eq. 36.30 divided by Eq. 36.27):

$$\frac{\Delta t_{\text{Greta}}}{\Delta t_{\text{Simon}}} = \left(\frac{2Lc}{c^2 - v^2}\right)\left(\frac{\sqrt{c^2 - v^2}}{2L}\right)$$

$$\frac{\Delta t_{\text{Greta}}}{\Delta t_{\text{Simon}}} = \frac{1}{\sqrt{1 - (v/c)^2}} \tag{36.31}$$

where v is the speed of the river with respect to the banks and c is the speed of each swimmer with respect to the water. Now examine Equation 36.31 to see who wins. First, if $v = c$ (the swimmers' speed equals the river's speed), the ratio (Eq. 36.31) goes to infinity. Second, if $v > c$ (the swimmers are slower than the river), the ratio is an imaginary number. So we conclude that the river must be slower than the swimmers: $v < c$. Third, if the river does not move with respect to the banks ($v = 0$), the ratio is 1 and there is a tie as we already predicted. Finally, as long as $0 < v < c$, then $\Delta t_{\text{Greta}} > \Delta t_{\text{Simon}}$, and Greta's round-trip time is greater than Simon's so that Simon always wins.

This swimming analogy is meant to help us understand the light measured by Michelson's interferometer. There is one more useful idea we need from the analogy: Both swimmers have the same speed c with respect to the water, so if they were to swim the same distance in still water, their race would always be a tie. If Simon were to win in still water, you would conclude that he had a shorter distance to swim. So, another way to think about the race in moving water (Fig. 36.28) is to say that the moving water is *equivalent* to Simon (the perpendicular swimmer) having a shorter distance to swim than Greta (the parallel swimmer).

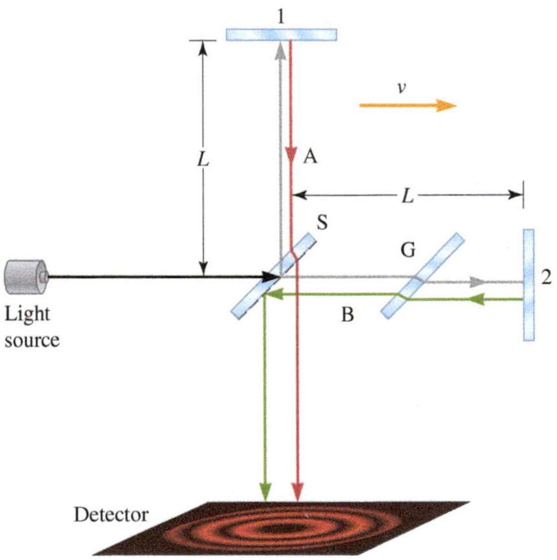

FIGURE 36.30 The ether flows to the right at a uniform and constant speed v. A half-silvered mirror S splits the light, so that half of the light travels along path A and the other half travels along path B. The light on path A travels to the mirror at 1, back through S, and to the detector. The light on path B travels to the mirror at 2. On its way both to and from 2, it passes through the glass slab G. When it returns to S, it is reflected toward the detector. The distance L from S to 1 equals the distance from S to 2. The speed of light with respect to the ether is c, and the light that follows A always gets to the detector before the light that travels along B. The result is an interference pattern produced at the detector. (The colors used are only for clarity; the light does not change color. Rays are slightly offset to avoid overlap.)

The Michelson–Morley Experiment

With the swimming analogy in mind, we return to light, the ether, and Michelson's interferometer. Michelson—like other scientists at the time—believed that the ether permeated all space, so that objects such as the Earth moved through the ether. His goal was to measure the speed of the Earth relative to the ether, which was thought to be motionless. If the Earth moved through the ether, we and everything else on the Earth would experience the ether as a river moving in the direction opposite to the Earth's motion. In our analogy, the ether is like the river flowing at constant speed v. To design his interferometer, Michelson needed a rough idea of the Earth's speed through the ether. He estimated that the speed should be similar to the Earth's orbital velocity around the Sun, or $v \approx 30$ km/s.

Figure 36.30 shows the basic design of Michelson's interferometer. We arbitrarily assume that the ether moves to the right. Light enters the interferometer from the source on the left. The light first encounters a glass plate with a half-silvered back, labeled S. This plate is called a **beam splitter** because it allows about half the light to pass through but reflects the other half. The splitter is tilted at 45° with respect to the incoming beam. The reflected light travels to the mirror at point 1, while the transmitted light travels to the mirror at point 2. In the swimming analogy, the splitter is at the starting position. The light is like the two swimmers racing from point *S*. The mirrors at points 1 and 2 are like the places where the swimmers turn around and head back toward *S*. The two paths labeled A and B are perpendicular to each other.

Let's follow the light along path A. After it is reflected from the mirror at point 1, it returns to S. Half of the light is reflected by the splitter, and half passes straight through. We are not interested in the reflected light in this case, so it is not shown. Instead, we follow the light that passes straight through S and goes to the detector at the bottom of the figure.

Next, follow the light along path B. After it is reflected from the mirror at 2, it returns to S. Half of that light passes straight through the splitter and back toward the light source. We are not interested in this light, so it is not shown. The other half of the light is reflected from the splitter and goes to the detector.

A glass slab labeled G is located along path B to make sure that the light passes through the same total thickness of glass along both paths. Before the incoming light is split, it must travel through the glass slab of the splitter in order to encounter the half-silvered surface on the back side. The light that travels along path A must pass back through that glass twice on its way to 1, and it must also pass back through the glass once on its way to the detector. However, the light that travels along path B passes through the splitter's glass only once. To make up for path A's two extra passes through the splitter's glass, a second glass slab G is inserted in path B with the same thickness and composition as the glass in the splitter. So, like the light that travels along path A, the light along path B travels through glass three times. (Slab G ensures that the race is fair!)

As in the case of the two swimmers, Michelson predicted that the light traveling on the perpendicular path (path A) always would arrive at the detector first. We can apply Equation 36.31 to Michelson's experiment by replacing Δt_{Simon} with Δt_{A} and Δt_{Greta} with Δt_{B}:

$$\frac{\Delta t_{\text{B}}}{\Delta t_{\text{A}}} = \frac{1}{\sqrt{1 - (v/c)^2}} \tag{36.32}$$

Now *c* is the speed of light, which Michelson had carefully measured in earlier experiments (at the Naval Academy). If he could also measure the ratio of the times $\Delta t_{\text{B}}/\Delta t_{\text{A}}$ that light took to travel over the two paths, he could solve for *v*, the speed of the Earth relative to the ether. Of course, the speed of light is very high compared to the anticipated speed of the Earth through the ether (30 km/s), so Michelson expected the ratio to be very close to 1:

$$\frac{\Delta t_{\text{B}}}{\Delta t_{\text{A}}} = \frac{1}{\sqrt{1 - \left(\dfrac{3 \times 10^4 \text{ m/s}}{3 \times 10^8 \text{ m/s}}\right)^2}} = 1.000000005 \tag{36.33}$$

The brilliance of Michelson's experimental design is that he did not need to measure the time delay directly. Another way to think about the time delay is in terms of a path-length difference. A shorter travel time for light along path A is equivalent to a shorter *effective* path length. As in Young's experiment, this path-length difference should result in a phase difference between the two beams of light that arrive at the detector, resulting in an interference pattern. If the interference pattern resembles a bull's eye and the path-length difference is $\lambda/2$, there should be destructive interference at the center of the pattern as shown by the dark spot in Figure 36.30.

We arbitrarily decided that the ether moves to the right in Figure 36.30, parallel to path B. That arbitrary decision led to the conclusion that path A is effectively shorter than path B. In the Michelson–Morley experiment, the entire interferometer was placed on a large piece of sandstone that floated in mercury (Fig. 36.31). This setup reduced errors introduced by the vibrations of moving vehicles and other nearby machinery and also allowed the experimenters to rotate the interferometer, thus changing the effective path lengths. For example, imagine that the two paths are turned so that they are both at 45° with respect to the ether's velocity. Then the two paths would have the same effective length and both beams would arrive at the detector at the same instant. As a result, there would be a bright spot at the center of the interference pattern due to the constructive interference of the two beams, instead of the dark spot shown in Figure 36.30.

By measuring the shift in the fringes of the interference pattern as the interferometer was rotated, Michelson hoped to measure the speed of the Earth through the ether. Let's see what he expected to find. The maximum path-length difference Δd occurs when one path is parallel to the ether's velocity as in Figure 36.30. Then

$$\Delta d = d_B - d_A \tag{36.34}$$

It is helpful to rewrite Equation 36.34 as

$$\Delta d = d_A\left(\frac{d_B}{d_A} - 1\right) \tag{36.35}$$

The ratio of the effective path lengths is the same as the ratio of the times:

$$\frac{d_B}{d_A} = \frac{c\Delta t_B}{c\Delta t_A} = \frac{\Delta t_B}{\Delta t_A} \tag{36.36}$$

Substitute Equation 36.36 into Equation 36.35:

$$\Delta d = d_A\left(\frac{\Delta t_B}{\Delta t_A} - 1\right)$$

In the Michelson–Morley experiment, $d_A \approx 22$ m and $\Delta t_B/\Delta t_A$ is given by Equation 36.33, so the effective path-length difference is

$$\Delta d = (22\text{ m})(1.000000005 - 1)$$

$$\Delta d = 1.1 \times 10^{-7}\text{ m} = 110\text{ nm} \tag{36.37}$$

As mentioned previously, Michelson and Morley rotated their interferometer. Suppose path A is originally perpendicular to the ether's velocity and then the interferometer is rotated by 90° so that path B is perpendicular. In effect, this is equivalent to lengthening path A and shortening path B by the same amount. By rotating the interferometer, the experimenters could double the difference between the two path lengths so that the pattern would shift by

$$m = \frac{2\Delta d}{\lambda} \tag{36.38}$$

fringes. Because the Michelson–Morley experiment used visible light with a wavelength of 550 nm, they expected the pattern to shift by

$$m = \frac{2(110\text{ nm})}{(550\text{ nm})} = 0.4\text{ fringe}$$

How can an interference pattern be observed to shift by a fraction of a fringe? Michelson's detector included a telescope (Fig. 36.31) that could detect shifts of as little as 0.01 fringe, so the experimenters should have easily been able to detect the shift they expected. Michelson and Morley ran their experiment at several different times of day and night, and several different times of year. The experiment was even tried at different locations on the Earth. But they never detected any shift in the pattern.

Huntington Library/SuperStock

FIGURE 36.31 In the Michelson–Morley experiment, the interferometer was placed on a stone slab floated on mercury so that it could be rotated. The apparatus of the actual interferometer includes a telescope for viewing the interference pattern.

Michelson and Morley failed to detect the Earth's motion with respect to the ether. This is considered the most important failed experiment in history because it led to a major change in our understanding. Because the experiment found no shift in the interference pattern, it demonstrated that the Earth does not move with respect to the ether. In Chapter 39, we will argue that this experiment also demonstrated that the ether does not exist. Instead, light propagates in a vacuum. Although Albert Einstein claimed that he had not heard of Michelson and Morley's experiment before he developed his theory of special relativity, their experiment provides an important physical basis for Einstein's work.

Einstein and Michelson eventually became friends, but Michelson never felt comfortable with his failure. He never warmed to Einstein's theory, and when he died in 1931, he still believed that the ether existed.

SUMMARY

Underlying Principles

No new principles are introduced in this chapter.

Major Concepts

1. According to **Rayleigh's criterion**, two sources are barely resolved when the central maximum of one diffraction pattern is centered on the first minimum of the other diffraction pattern (Fig. 36.4).
2. For a grating, **dispersion** is a measure of the angular separation between the lines (maxima):
$$D \equiv \frac{\Delta\theta}{\Delta\lambda} = \frac{m}{d\cos\theta} \quad (36.18 \text{ and } 36.19)$$
where m is the order number, θ is the angular position of the corresponding line, and d is the separation between the rulings.
3. For a grating, the **resolving power** is a measure of the thickness of the lines:
$$R \equiv \frac{\lambda_{av}}{\Delta\lambda} = Nm \quad (36.21)$$
where λ_{av} is the average wavelength of two lines that are just barely separated, $\Delta\lambda$ is the difference in their wavelengths, N is the number of rulings, and m is the order number.

Special Cases

1. The **resolution** of an instrument with a **circular aperture** is given by
$$\sin\theta_{min} = 1.22\frac{\lambda}{d} \quad (36.3)$$
2. When a **thin film** of thickness w is surrounded by a medium with a lower index of refraction ($n_f > n_i$), the condition for **constructive interference** is
$$2w = \left(m + \frac{1}{2}\right)\frac{\lambda_0}{n_f} \quad (m = 0, 1, 2, 3, \ldots) \quad (36.12)$$
and the condition for **destructive interference** is
$$2w = m\frac{\lambda_0}{n_f} \quad (m = 0, 1, 2, 3, \ldots) \quad (36.14)$$
3. For a **diffraction grating**, the angular **position** θ of the lines is given by
$$d\sin\theta = m\lambda \quad (m = 0, \pm1, \pm2, \pm3, \ldots) \quad (36.15)$$
and each line's **half-width** is given by
$$\Delta\theta_{hw} = \frac{\lambda}{Nd\cos\theta} \quad (36.16)$$

PROBLEMS AND QUESTIONS

A = algebraic **C** = conceptual **E** = estimation **G** = graphical

36-2 Circular Aperture Diffraction

1. **C** Many circular apertures are adjustable, such as the pupil of your eye or the shutter of a camera. Describe the change in the diffraction pattern as such an aperture decreases in size.
2. **C** Many of the images we regularly look at are digitized; that is, they are made up of many small individual dots. For example, a color laser printer produces images by printing many dots of various colors. If the printer is of high quality, we do not see the individual dots. Why not?
3. **N** The "hydrogen line" at 1420.4 MHz corresponds to the natural frequency of neutral hydrogen atoms and plays an important role in radio astronomy. What size dish is required so that a radio telescope receives this frequency with an angular resolution of 0.0500°?
4. **E** Post-Impressionist Georges Seurat is famous for a painting technique known as pointillism. His paintings are made up of different-colored dots, and each dot is roughly 2 mm in diameter. When viewed from far enough away, the dots blend together and only large-scale images are seen. Assuming that the human pupil is about 2.5 mm in diameter, determine an approximate minimum distance you need to stand from one of Seurat's paintings to see the dots blended together.

5. **E** Estimate the diffraction-limited resolution of the radio telescope in Arecibo, Puerto Rico. The radio telescope's diameter is 305 m, and it operates at a frequency of 300.0 MHz.

Problems 6 and 9 are paired.

6. **N** The resolution of a telescope on the Earth is not limited by diffraction. Instead, the Earth's atmosphere causes blurriness such that the best resolution on Earth is about 0.50 arcsec. The Hubble Space Telescope was placed into orbit to avoid such blurriness. Its resolution is determined by diffraction. The Hubble's mirror is 2.4 m in diameter. Estimate its diffraction-limited resolution. *Hint*: Use an estimate of 6.0×10^{-7} m (or 600 nm) for the wavelength of light in this calculation.
7. **N** A microscope with an objective lens of diameter 7.50 mm is used to view samples in light of wavelength 500.0 nm.
   a. What is the limiting resolution of the microscope with this objective lens?
   b. If the lights in the laboratory could be tuned to the range of visible wavelengths, which wavelength would produce the smallest angle of resolution?
   c. What would the angular resolution of the microscope be at the wavelength found in part (b)?
   d. What would the angular resolution of the microscope be at the wavelength found in part (b) if the microscope was immersed in water?
8. **C** In a particular microscope, you are just able to resolve two blood cells using a certain aperture. How is your ability to resolve the two cells altered if the diameter of the aperture is increased? What if it is decreased?
9. **N** Use the results from Problem 6 to find the distance between two objects on the Moon that can just barely be resolved by the Hubble Space Telescope.
10. **N** Later in this textbook, you will explore the intriguing idea that matter can exhibit wavelike behavior when moving at velocities near the speed of light. One example where this is used is in the imaging of materials in an electron microscope. A common wavelength for the electrons in a transmission electron microscope is 2.5×10^{-12} m. The diameter of the apertures is 5.0 μm. What is the diffraction-limited resolution of this microscope?
11. **N** Suppose that in a particular microscope, a diffraction-limited resolution of 0.25° is necessary to resolve two plant cells that are part of a specimen. If the microscope uses light with a wavelength of 525 nm, what is the minimum aperture diameter so that the cells can be resolved?

36-3 Thin-Film Interference

12. **A** For a thin film such as a soap bubble surrounded by a medium with a smaller index of refraction such as air, find the conditions for constructive and destructive interference produced by the transmitted light.
13. **N** The nonreflective coating on a camera lens with an index of refraction of 1.27 is designed to minimize the reflection of 589-nm light. If the lens glass has an index of refraction of 1.54, what is the minimum thickness of the coating that will accomplish this task?
14. **N** When you spread oil ($n_{oil} = 1.50$) on water ($n_{water} = 1.33$) and have white light incident from above, you might observe different reflected colors that depend on the thickness of the oil on the water at each point. Consider the wavelength of red light (750.0 nm) and the wavelength of violet light (380.0 nm), covering the visible spectrum. What is the minimum possible thickness of oil that would give rise to seeing **a.** the red light reflected and **b.** the violet light reflected?
15. Sometimes a nonreflective coating is applied to a lens, such as a camera lens. The coating has an index of refraction between the index of air and the index of the lens. The coating cancels the reflections of one particular wavelength of the incident light. Usually, it cancels green-yellow light ($\lambda = 550.0$ nm) in the middle of the visible spectrum.
    a. **A** Assuming the light is incident perpendicular to the lens surface, what is the minimum thickness of the coating in terms of the wavelength of light in that coating?
    b. **N** If the coating's index of refraction is 1.38, what should be the minimum thickness of the coating?
16. **C** Just outside of Esther's bedroom window is an illuminated sign. In red (716 nm) and blue (408 nm) letters it proclaims "Danger Ahead." Tired of light from this sign, Esther coats her glass ($n = 1.50$) window on the inside with a thin film ($n_f = 1.39$). Now, when she looks straight through her window, the sign reads "anger head." What color was emitted by the missing letters? Assume the film is the minimum thickness required to prevent the colored light transmission.
17. **N** An oil slick on water displays a variety of iridescent colors due to interference effects. Assuming the film has the minimum thickness to produce the colors observed, what thickness of the oil film will appear green ($\lambda = 550.0$ nm) or red ($\lambda = 700.0$ nm)? The index of refraction of the oil is 1.47.

Problems 18 and 19 are paired.

18. **N** A thin film ($n_f = 1.29$) coats a glass ($n = 1.56$) window on the inside to prevent the transmission of red light at 735 nm straight through the window. What is the minimum thickness of the film?
19. **N** A thin film ($n_f = 1.29$) coats a glass ($n = 1.56$) camera lens on the outside to prevent the reflection of red light at 735 nm from rays that are perpendicular to the surface. What is the minimum thickness of the film?

Problems 20 and 21 are paired.

20. **C** When you spread oil ($n_{oil} = 1.50$) on water ($n_{water} = 1.33$) and have white light incident from above, you might observe different reflected colors that depend on the thickness of the oil on the water at each point. Suppose we find a location where we see green light reflected from the thin film of oil. If we slowly increase the thickness of oil at that location, what happens to the observed reflected color?
21. **C** In Problem 20, what happens to the observed reflected color if we slowly decrease (rather than increase) the thickness of the oil at that location?
22. **E** Police measure the speed of moving vehicles with radar guns. A radar gun sends out radio waves with a typical frequency of 10 GHz. The radio waves reflect from vehicles. By measuring the Doppler shift of the reflected radio waves (Section 17-9), the radar detector measures the speed of the vehicle. A physics student wants to paint his car with a nonreflective coating that will prevent radar detectors from measuring his speed. Estimate the thickness of the coating, and describe your assumptions.
23. **N** An oil slick of thickness 325 nm and index of refraction $n_{oil} = 1.50$ forms on the surface of a still pond ($n_{water} = 1.33$). When it is viewed straight from above, what is the wavelength of visible light that is **a.** most strongly reflected and **b.** most strongly transmitted?
24. **N** An oil tanker bound for the Suez Canal runs aground in the Red Sea, spilling sweet light crude oil that forms a uniform thin film on the surface of the sea. After ten hours of weathering, the index of refraction of the crude oil has increased to 1.28, and the index of refraction of seawater at the tanker's location is 1.341. What is the minimum thickness of the oil film that will result in the strong reflection of normally incident 450-nm blue light?

36-4 Diffraction Gratings

25. **N** Light of wavelength 566 nm is incident on a grating. Its third-order maximum is at 15.7°. What is the density of rulings on this grating?
26. **N** A diffraction grating with a ruling separation of 3.00 μm is illuminated with light of wavelength 525 nm. What is the highest-order maximum visible?
27. **N** A diffraction grating is illuminated with a krypton laser operating at 416 nm. What is the spacing between rulings on the grating if the first-order maximum is observed at 18.8°?
28. **N** A grating has 3330 rulings per centimeter. When monochromatic light is incident on this grating, a fourth-order line is seen at 30.0°. What is the wavelength of the light?
29. **N** A diffraction grating is illuminated with green light of wavelength 530.0 nm, and a second-order maximum is observed at 27.5°. **a.** How many slits per centimeter does the diffraction grating have? **b.** What is the number of maxima that can be seen in the diffraction pattern?
30. **N** A diffraction grating is 5.00 cm wide and has 515 rulings per millimeter. If light of **a.** 450.0 nm and **b.** 650.0 nm is incident on the grating, what is the half-width of the central maximum in each case?
31. **N** A light source emits a mixture of wavelengths from 450.0 nm to 600.0 nm. When the light passes through a diffraction grating, two adjacent spectra barely overlap at an angle of 30.0°. How many rulings per meter are on the grating?
32. **A** A diffraction grating of length ℓ has N rulings. When the grating is illuminated with light of wavelength λ, the $m = 3$ maximum is diffracted at angle $\theta = 90.0°$. Find an expression for the wavelength of the light in terms of ℓ and N.
33. **N** A diffraction grating with 1.60×10^3 rulings per centimeter spreads white light into its spectral components. What is the angle at which the second-order maximum appears for green light with a wavelength of 530.0 nm?
34. **N** Light is incident on a grating. The third-order line is observed at 65°. At what angular position is the first-order line?
35. **N** A diffraction grating with 2.65×10^3 rulings per centimeter is illuminated by white light. **a.** How many orders of the entire visible light spectrum, from wavelength 400.0 nm to 780.0 nm, are visible? **b.** How many orders of violet light of wavelength 400.0 nm are visible?
36. **N** A diffraction grating with 3.65×10^3 rulings per centimeter is illuminated by light of wavelength 632.8 nm from a helium-neon laser. **a.** At what angle does the third-order maximum occur? **b.** At what angle would the third-order maximum occur if the experiment was carried out underwater? **c.** What is the relationship between the diffracted rays in parts (a) and (b)?

36-5 Dispersion and Resolving Power of Gratings

Problems 37 and 38 are paired.

37. **N** A grating has 3.200×10^4 rulings and is 5.00 cm wide. Light of wavelength 555 nm is incident on the grating. What is its dispersion for the first- and second-order lines?
38. **N** A grating has 3.200×10^4 rulings and is 5.00 cm wide. What is its resolving power for the first- and second-order lines?

Problems 39 and 40 are paired.

39. **N** A diffraction grating is exposed to light from a source that consists of two different wavelengths: 575 nm and 585 nm. What is the minimum number of rulings necessary to be able to resolve the **a.** $m = 1$ lines, **b.** $m = 2$ lines, and **c.** $m = 3$ lines?
40. **C** A diffraction grating is exposed to light from a source that consists of two different wavelengths: 575 nm and 585 nm. Explain why the resolving power of the grating must be higher to resolve the $m = 1$ lines versus the $m = 3$ lines.
41. **N** A diffraction grating with 3.52×10^3 rulings/cm and length 2.50 cm is illuminated by light with a wavelength of 425 nm. What is the dispersion of the first-order maximum?

Problems 42 and 43 are paired.

42. **A** A diffraction grating with n rulings per unit length and total length ℓ is illuminated by light with several similar wavelengths. Find an expression for the resolving power of this diffraction grating for the first-order maxima of two similar wavelengths of light in terms of n and ℓ.
43. **N** A diffraction grating with 2.00×10^3 rulings/cm and total length 3.00 cm is illuminated by light with several similar wavelengths. Find the resolving power of this diffraction grating for the first-order maxima of two similar wavelengths of light.
44. **N** A diffraction grating with 1.54×10^3 rulings/cm and length 3.10 cm is illuminated by light with a wavelength of 310.0 nm. What is the dispersion of the first three maxima that fall on the screen?

36-6 Case Study: Michelson's Interferometer

45. **N** CASE STUDY Michelson's interferometer played an important role in improving our understanding of light, and it has many practical uses today. For example, it may be used to measure distances precisely. Suppose the mirror labeled 1 in Figure 36.30 (page 1176) is movable. If the laser light has a wavelength of 632.5 nm, how many fringes will pass across the detector if mirror 1 is moved just 1.000 mm? If you can easily detect the passage of just one fringe, how accurately can you measure the displacement of the mirror?
46. **N** CASE STUDY Michelson's interferometer played an important role in improving our understanding of light, and it has many practical uses today. For example, it may be used to measure precisely the indices of refraction of various media. Suppose an initially evacuated tube of length 6.50 cm is placed in one of the paths in Michelson's interferometer. The tube is then filled slowly with a gas of an unknown index of refraction, and as a result 42 fringes shift. The laser light has a wavelength of 608.5 nm. You will find that the index of refraction is very close to 1. Give $n - 1$ to three significant figures.
47. **N** CASE STUDY The interferometer uses interference effects to produce an observable change with a small change in the optical path length of one path relative to the other. Michelson and Morley also used a telescope to zoom in to observe phase shifts of only 1% of the width of a fringe. **a.** In Figure 36.30, how far would mirror 2 need to move to shift the interference pattern by 1.0% of the width of a fringe, when used with light of wavelength 500.0 nm? **b.** If Michelson and Morley tried to measure the difference in travel times for light traveling along paths A and B, what time resolution would they need? That is, how much extra time is needed for the light to cover the distance found in part (a)? Assume the speed of light is 3.0×10^8 m/s.
48. **N** CASE STUDY The Michelson–Morley interferometer can also be used to determine the wavelength of light. Given light with an unknown wavelength, the position of one of the mirrors is shifted by 1.000 mm and the interference pattern shifts by exactly 4500 fringes. What is the wavelength of the light?

General Problems

Problems 49 and 50 are paired.

49. **C** Optical flats are flat pieces of glass used to determine the flatness of other optical components. They are placed at an angle above the component as shown in Figure P36.49A, and

monochromatic light is incident and observed from above, leading to interference fringes. Parts B and C of Figure P36.49 show the results of tests on two optical components. Which of the two is more flat? Explain.

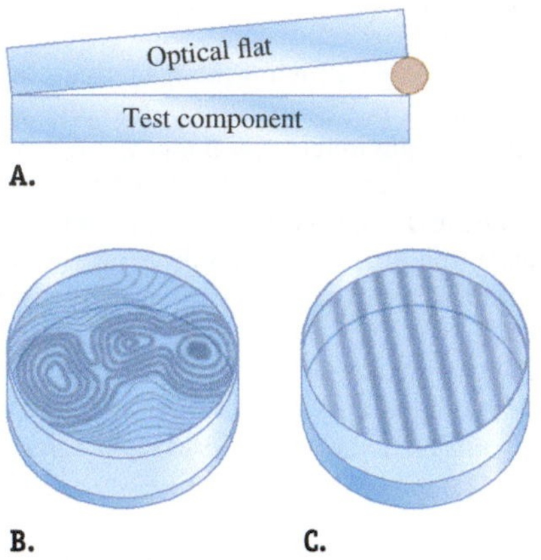

FIGURE P36.49 Problems 49 and 50.

50. **E** Optical flats are flat pieces of glass used to determine the flatness of other optical components. They are placed at an angle above the component as shown in Figure P36.49A, and monochromatic light is incident and observed from above, leading to interference fringes. Figure P36.49C shows the results of one of these tests. What is the approximate difference in the gap thickness between the left and right sides of the optical flat and the component? Is it possible to determine from this figure alone which side has the greater gap thickness (left or right)?

Problems 51 and 52 are paired.

51. **N** A thin film of material is coated on a substrate to protect the substrate while it is immersed in oil (Fig. P36.51). Light incident in the oil with a wavelength of 485 nm is reflected from the thin film. What is the minimum thickness of the thin film that results in the reflection, given the incident wavelength? Assume $n_{oil} = 1.70$, $n_{film} = 1.40$, and $n_{substrate} = 1.25$.

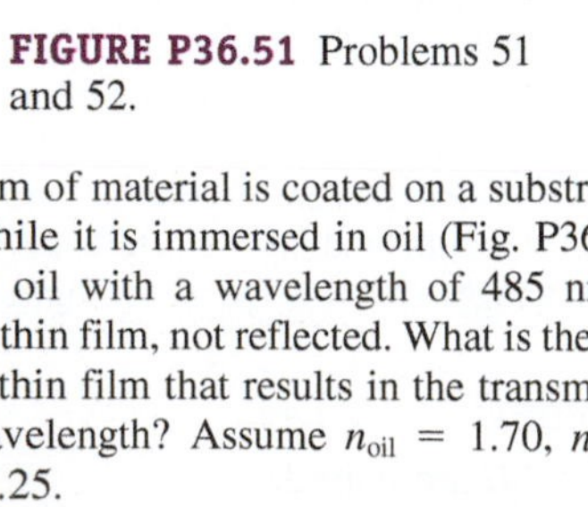

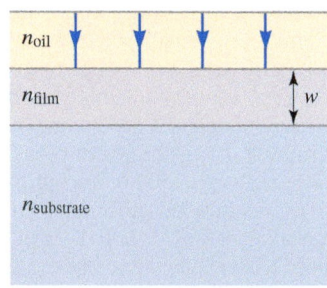

FIGURE P36.51 Problems 51 and 52.

52. **N** A thin film of material is coated on a substrate to protect the substrate while it is immersed in oil (Fig. P36.51). Light incident in the oil with a wavelength of 485 nm is transmitted through the thin film, not reflected. What is the minimum thickness of the thin film that results in the transmission, given the incident wavelength? Assume $n_{oil} = 1.70$, $n_{film} = 1.40$, and $n_{substrate} = 1.25$.

53. **N** Figure P36.53 shows two thin glass plates separated by a wire with a square cross section of side length w, forming an air wedge between the plates. What is the edge length w of the wire if 42 dark fringes are observed from above when 589-nm light strikes the wedge at normal incidence?

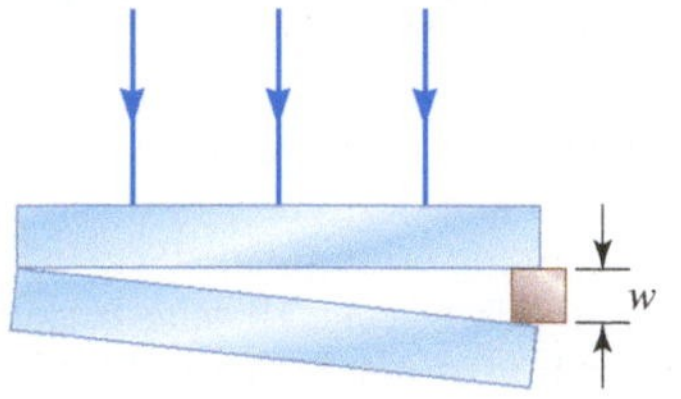

FIGURE P36.53

54. **N** Viewed from above, a thin film of motor oil with index of refraction $n = 1.28$ on a wet asphalt surface (water, $n = 1.33$) strongly reflects light of 595 nm wavelength and does not reflect any light at 476 nm. What is the minimum thickness of the film of motor oil?

55. **N** Newton's rings, discovered by Isaac Newton, are an interference pattern of dark and bright rings formed because of the air gap of increasing thickness w between a spherical surface and an adjoining flat surface (Example 36.2). Figure P36.55 shows a Newton's rings apparatus with a plano-convex lens with radius of curvature R atop a flat glass slab, both with index of refraction n_{glass}. In the configuration shown, the seventh bright fringe has a radius of 1.10 cm. After the air gap in the apparatus is filled with an unknown liquid, the radius of the seventh fringe decreases to 0.968 cm. What is the index of refraction of the unknown liquid?

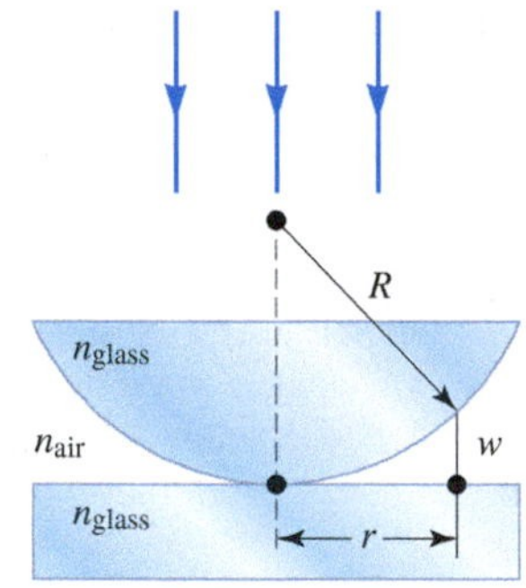

FIGURE P36.55

56. **C** A diffraction grating of length ℓ has N rulings. When the grating is illuminated with light of wavelength λ, only a certain number of maxima appear on the screen. How will the number of visible maxima change if the number of rulings is decreased or increased, while the length of the grating remains the same? Explain.

57. **N** What is the radius of the beam of an argon laser with wavelength 454.6 nm when viewed 50.0 km away from the laser if the laser's aperture has a radius of 3.00 mm?

58. **N** A diffraction grating with a length of 5.65 cm consists of 185 rulings and is illuminated with light that consists of several similar wavelengths. One of the lines corresponds to light with a wavelength of 489 nm. If this line is a first-order maximum, what are the wavelengths of the nearest first-order maxima that could be resolved, assuming they are present in the light?

59. **N** A diffraction grating with 428 rulings per millimeter is illuminated by light from a xenon ion laser. What is the wavelength of the laser light being used if the separation between the central and second-order maxima on a screen placed 2.10 m away from the grating is 67.5 cm?

60. **N** How many rulings must a diffraction grating have if it is just to resolve the sodium doublet (589.592 nm and 588.995 nm) in the second-order spectrum?

61. **N** Freeway exit signs are typically constructed with letters 25.0 cm in height and placed a similar distance apart. Assuming the diameter of the pupil is 3.00 mm in daylight and the sign is observed with light of wavelength 589 nm, at what distance can a driver distinguish the individual letters on a freeway sign?

62. **N** White light is incident on a diffraction grating that has 2.85×10^3 rulings per centimeter, producing first-order maxima at 6.55°, 8.75°, and 11.2°.
   a. What are the wavelengths responsible for these three first-order maxima?
   b. At which angles do these wavelengths produce second-order maxima?
63. **N** X-rays incident on a crystal with planes of atoms located 0.378 nm apart produce a diffraction pattern in which a first-order maximum is observed at an angle of 14.2°. **a.** What is the wavelength of the X-rays incident on the crystal? **b.** How many orders are visible in the diffraction pattern?
64. **N** Later in this book, you will explore the intriguing idea that matter can exhibit wavelike behavior when moving at velocities near the speed of light. One example where this is used is in the imaging of materials in an electron microscope. A common wavelength for the electrons in a transmission electron microscope is 2.5×10^{-12} m. Assume the diameter of the aperture is 500.0 μm. What is the required distance between the aperture and the specimen if we want to resolve two atoms separated by approximately 1.0×10^{-9} m, assuming the resolution is not dependent on other factors?
65. **A** Confirm Equation 36.16,

$$\Delta\theta_{hw} = \frac{\lambda}{Nd\cos\theta}$$

by showing that the half-width of the zeroth-order line produced by a diffraction grating is given by $\Delta\theta_{hw} = \lambda/Nd$. Do not begin with Equation 36.16, but instead use our work finding the first dark minimum produced by single-slit diffraction as a guide (Section 35-4). The angular position of the first dark minimum produced by a single slit is like the angular position of the dark fringes on either side of the zeroth-order line from a grating.
66. **A** Show that the dispersion of a diffraction grating is given by Equation 36.19,

$$D \equiv \frac{\Delta\theta}{\Delta\lambda} = \frac{m}{d\cos\theta}$$

where m is the order number, θ is the angular position of the corresponding line, and d is the separation between the rulings.

Problems 67 and 68 are paired.

67. **N** The fringe width β is defined as the distance between two consecutive maxima (or consecutive minima). In a Young's double-slit experiment, the fringe width obtained from a source of wavelength 500.0 nm is 3.90 mm. If the apparatus is immersed in a liquid that has index of refraction $n = 1.30$, what is the new fringe width? Assume the apparatus was originally in air and that the small-angle approximation applies to this situation.
68. **N** The fringe width β is defined as the distance between two consecutive maxima (or consecutive minima). When a Young's double-slit apparatus is completely immersed in a liquid, the fringe width decreases by 20%. Determine the liquid's index of refraction. Assume the apparatus was originally in air.
69. **N** A pair of closely spaced slits is illuminated with 600.0-nm light in a Young's double-slit experiment. During the experiment, one of the two slits is covered by an ultrathin Lucite plate with index of refraction $n = 1.485$. What is the minimum thickness of the Lucite plate that produces a dark fringe at the center of the viewing screen?

37 Reflection and Images Formed by Reflection

Key Questions

How do we find the location and magnification of an image formed by a mirror?

Underlying Principles

Geometric optics

Major Concepts

1. Law of reflection
2. Magnification
3. Virtual and real images
4. Mirror equation

Special Cases

1. Camera obscura
2. Plane mirrors
3. Convex spherical mirrors
4. Concave spherical mirrors

Tools

1. Ray diagrams
2. Primary rays

Imagine this. You are having dinner with the President. It feels like some food is stuck to your front teeth, but you can't excuse yourself from the table. What can you do? You can try to catch a glimpse of your image in your knife or spoon. Your knife is a lot like your bathroom mirror. You are familiar with the image of yourself that you see in a mirror. You know that if you see food on your image's left front tooth, it is really on your right tooth. If you try using the bowl of your spoon as a mirror, your image is large but upside down. If you look at the back of your spoon, your image is upright but small. In this chapter, you will see how the shape of a mirror affects the image you see.

37-1 Geometric Optics

In earlier chapters, we modeled light as a wave and applied Huygens's principle. In this chapter and the next, we will model light as a ray and explore the images formed when light reflects or refracts. This chapter is devoted to reflection and the images formed by mirrors. You probably look into your bathroom mirror every morning. Mirrors have many other everyday uses, such as in the car, in a store as part of its security system, and in the dentist's office. Mirrors are also used in major telescopes, one of which inspires the case study for this chapter (Section 37-2).

GEOMETRIC OPTICS
Underlying Principle

Ray Model

In many circumstances, the wave nature of light is not relevant to the problem at hand; only its direction of propagation is relevant. In those cases, we crudely model light as a **ray**, drawn as a line perpendicular to the wave fronts with an arrowhead indicating the direction of the wave's motion. Rays are not vectors; to distinguish them from vectors, we draw the arrow for a ray on the line, not at its end. The electric and magnetic fields oscillate perpendicular to the ray, but they are not drawn. It is easy to represent a plane parallel wave with a single ray or several parallel rays (Fig. 37.1A). A spherical wave is represented by many rays (Fig. 37.1B).

The ray model ignores the wave qualities of light. It is a good model when light encounters only obstacles and apertures that are large compared to its wavelength. For practical purposes, the ray model works for obstacles and apertures that are at least 1 mm in diameter. In this chapter, we assume all mirrors and apertures are larger than 1 mm in diameter or width. Because rays reflected from such mirrors travel in straight lines, much of our analysis involves geometry. Situations that can be studied using the ray model fall into the realm of **geometric optics** (Chapters 37 and 38). Situations involving the wave qualities of light require *physical optics*. We used physical optics when we studied such phenomena as polarization, interference, and diffraction (Chapters 34, 35, and 36).

A narrow beam of light, such as from a laser, may be represented by a single ray or a small number of parallel rays (Fig. 37.1A). However, light reflected from a complicated object such as a flower must be represented by rays pointing in many different directions. When you view such an object, you are aware only of the rays that enter your eyes. To see an entire flower, you move your eyes and head to see more of the rays. Your brain uses the ray model to interpret the object's distance and size by assuming any ray that enters the eye must have traveled along a single straight path. The straight-ray assumption made by the human brain is sometimes exploited to create optical illusions (Fig. 37.2). In this chapter and the next, we will use the straight-ray assumption to find the images generated by mirrors and lenses.

VISUAL REPRESENTATIONS OF LIGHT Tool

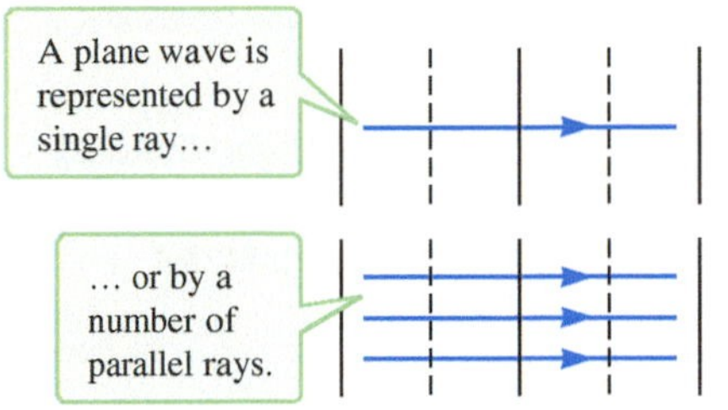
A plane wave is represented by a single ray...
... or by a number of parallel rays.

A.

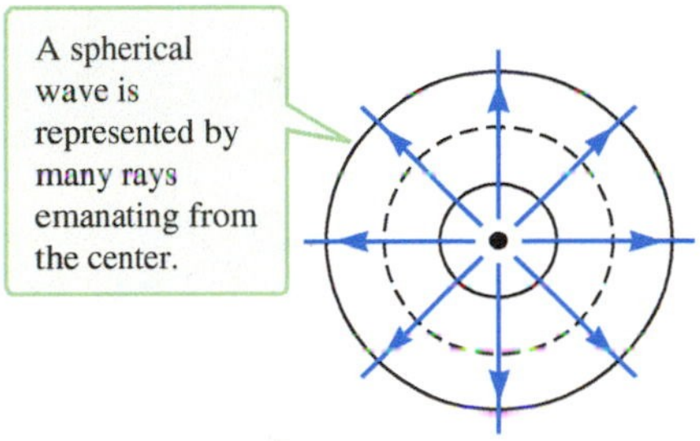
A spherical wave is represented by many rays emanating from the center.

B.

FIGURE 37.1 **A.** Representation of a plane parallel wave. **B.** Representation of a two-dimensional circular wave or a three-dimensional spherical wave. Compare to Figure 35.5.

Camera Obscura

To see how geometric optics works, let's consider a specific example—the camera obscura or pinhole camera. The term *camera obscura* is Latin for "dark chamber." This device dates back to antiquity. A camera obscura is a dark room with a small hole (the **aperture**) that allows light to enter from the **object** (Fig. 37.3). In optics, the object is anything from which light rays are emitted. It can be a source of light such as the Sun, but more often it is something that reflects light. People in the camera obscura look at an **image** (the apparent reproduction of the object) projected on a screen or wall on the opposite side of the room from the hole. In Figure 37.3, the image is formed by rays that pass through the aperture. A pinhole camera is a miniature camera obscura in which the room is replaced by a small box with a pinhole aperture. A larger hole cut into the box allows people to look at an image projected on its back wall.

When using geometric optics to analyze a situation, we draw a simple sketch called a **ray diagram** in which the object is usually represented by one or two arrows for simplicity. In Figure 37.3, we have overlaid the object with two perpendicular arrows. This practice enables us to see if the image is upright or upside down, and if it is flipped from left to right. Many rays are emitted by the object, but we need to draw only a few. Figure 37.3 shows three rays—one from the point of each arrow and the third from the bottom of the vertical arrow. Extending these rays all the way to the back wall shows that the image is both upside down and flipped (left to right).

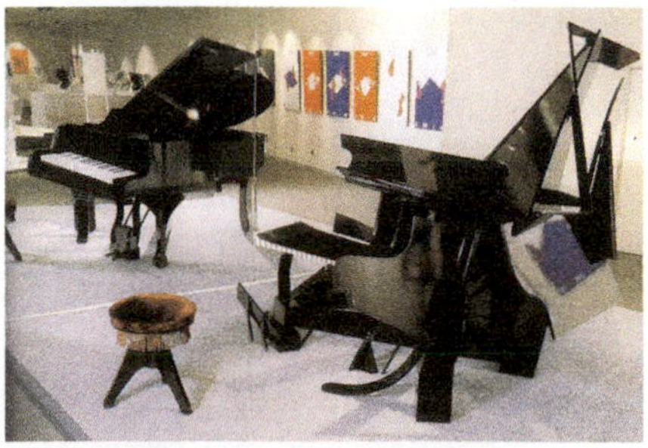
Shigeo Fukuda

FIGURE 37.2 Is there a piano in the room? Optical illusions are created because the human brain assumes all the rays that enter the eye have traveled along a straight line.

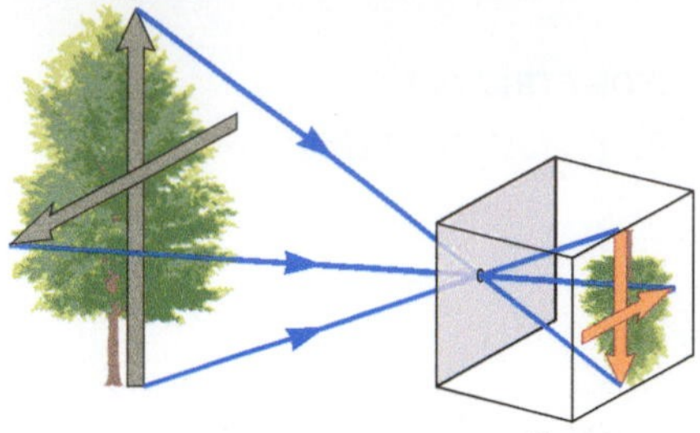

FIGURE 37.3 The camera obscura was a great form of entertainment. The image produced is upside down and reversed.

The image in a pinhole camera is usually blurry because every point on the object emits many rays that make it through the hole and to the back wall. These rays all come through at slightly different angles, so the image of any part of the object is slightly spread out over the whole image. You can improve a pinhole camera by making the hole smaller, but that also reduces the amount of light, which makes the image dimmer.

We are often interested in the **linear magnification** of an image—that is, its size relative to the object's size. Often the term *linear* is dropped. If the *o*bject's height is h_o and the *i*mage's height is h_i, the magnification M is

$$M = \frac{h_i}{h_o} \tag{37.1}$$

MAGNIFICATION ✪ Major Concept

Magnification depends on geometric factors, such as distance. To find an expression for the magnification created by a camera obscura in terms of the distance d_o between the hole and the object and the distance d_i between the hole and the image, we start with the ray diagram in Figure 37.4. This diagram shows the object, the hole, the image, and two rays—one from the top of the arrow and the other from its base. We can use geometry to find the magnification. If the dashed line bisects the object and the image, we have

$$\tan\theta = \frac{h_o/2}{d_o} = \frac{h_i/2}{d_i}$$

MAGNIFICATION OF A CAMERA OBSCURA ▶ Special Case

For a pinhole camera, both the object distance and the image distance are positive.

Solving for the magnification h_i/h_o gives

$$M = \frac{h_i}{h_o} = -\frac{d_i}{d_o} \tag{37.2}$$

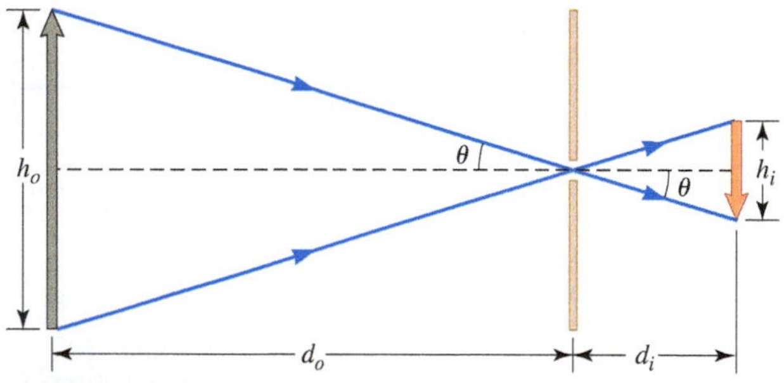

FIGURE 37.4 A ray diagram allows us to use geometry to find the magnification of a pinhole camera.

where we have inserted a negative sign by convention, the first of three we'll encounter in this chapter. The convention we use in this textbook is that if the image is inverted (upside down), its height is negative. Because the image's height is negative, the magnification is also negative. The negative sign in Equation 37.2 means the image in a pinhole camera or camera obscura is inverted.

For this pinhole camera, the image distance is less than the object distance, $d_i < d_o$, so the absolute value of the magnification is less than 1, $|M| < 1$. When $|M| < 1$, the image is smaller than the object. In everyday language, *magnification* means "making larger." But, in optics, magnification describes the relative size of the image, which may actually be smaller than the object.

EXAMPLE 37.1 Transit of Venus

On June 8, 2004, and then again on June 5, 2012, Venus crossed in front of the Sun. This event is called the transit of Venus. Venus had gone 122 years without such an event, and it will now be more than a century before the next one. Figure 37.5A shows a slightly distorted image of the Sun and Venus taken in 2012, and Figure 37.5B shows a sketch of Venus's path across the Sun. Estimate the magnification of the image in Figure 37.5A. If this image was made with a camera obscura, what was the distance from the hole to the image? The Sun's radius is 6.955×10^8 m.

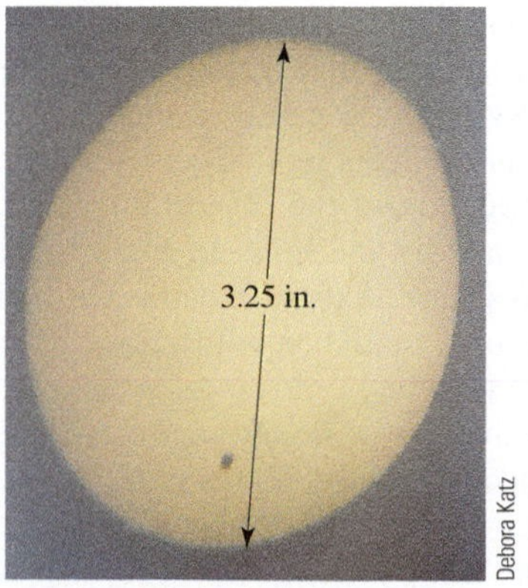

A.

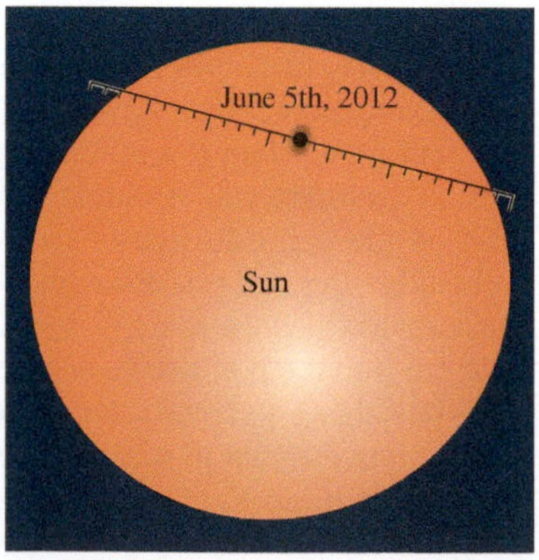

B.

FIGURE 37.5 A. Image of Venus in transit across the Sun. The image is distorted because the photographer had to stand to the side of the image in order not to block the sunlight. **B.** Sketch of Venus's path.

INTERPRET and ANTICIPATE

The photo shows that the image of the Sun has a diameter of 3.25 in., so we can find the image height h_i. The object is the Sun itself, whose radius is given inside the cover of this text. From these values, we can estimate the magnification. We expect the absolute value of the magnification to be much less than 1 because the Sun is very large and the image is very small.

SOLVE Convert the image diameter to meters. This is the image height.	$h_i = 3.25 \text{ in.} = 8.255 \times 10^{-2} \text{ m}$
The diameter of the Sun is the object height.	$h_o = 2(6.955 \times 10^{8} \text{ m}) = 1.391 \times 10^{9} \text{ m}$
Compare the sketch to the image. The sketch shows Venus crossing the top half of the Sun (Fig. 37.5B). In the image, Venus is on the bottom (Fig. 37.5A). So the image is inverted, and the magnification is negative.	$M = \frac{h_i}{h_o} = \frac{-8.255 \times 10^{-2} \text{ m}}{1.391 \times 10^{9} \text{ m}}$ $M = -5.93 \times 10^{-11}$
Now use Equation 37.2 to find the distance between the image and the hole. The object distance is the distance between the Earth and the Sun (1 AU). See Appendix B.	$M = \frac{h_i}{h_o} = -\frac{d_i}{d_o}$ (37.2) $d_i = -Md_o = -(-5.93 \times 10^{-11})(1.496 \times 10^{11} \text{ m})$ $d_i = 8.87 \text{ m}$

CHECK and THINK

The Sun is much larger than its image in Figure 37.5, and as expected the absolute value of the magnification is much less than 1. If a camera obscura was used, the hole was about 9 m (or roughly 30 ft) from the screen. This is somewhat larger than the giant camera obscura in San Francisco, but much smaller than the world's largest camera obscura ($d_i \approx 55$ ft); that camera obscura was made from an F-18 fighter plane's hangar. Most likely this image was not made with a camera obscura, but with another sort of instrument such as a telescope.

37-2 Law of Reflection

Consider a typical room full of objects illuminated by any number of lightbulbs. You can see those objects because the light from the bulbs reflects from the surface of the objects. When light reflects from a nonflat object—such as a flower, a piano, or the many walls of a room—the many reflected rays travel in many different directions. However, let's consider a simple situation in which light reflects from a flat surface, such as laser light reflected from a polished flat mirror.

Figure 37.6A shows a light beam of width w making an angle γ_i with the surface of a medium. Because of the angle γ_i, the beam first encounters the medium point at A. To see what happens, we apply Huygens's principle (Section 35-1) and consider a wavelet centered on point A. Just as the wavelet at A is beginning to emerge, the other edge of incident beam is still a distance r from medium at point B. Of course, both the wavelet centered on A and the beam move at the speed of light. So, by the time beam has reached point B, the radius of the wavelet must equal the distance r. We've only considered two points A and B on the surface, and if we applied Huygens's principle to a great number of wavelets originating from points between A and B, we would find a plane parallel wave traveling away from the glass at an angle γ_r.

With a little trigonometry, we can find the relationship between the angle γ_i that the *i*ncident light makes with the surface and the angle γ_r that the *r*eflected beam makes. Figure 37.6B shows a simpler sketch, showing the geometry we need. The distance between points A and B is h. There are two right triangles with the same hypotenuse in common. Because each one has a leg of length r, the remaining leg

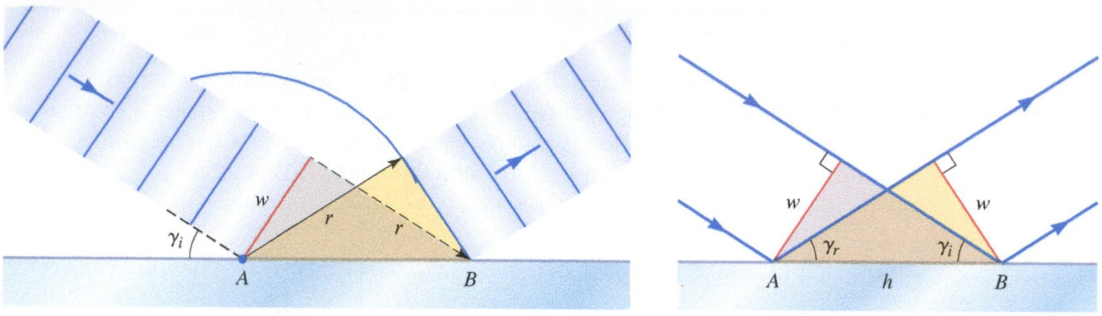

FIGURE 37.6 **A.** A beam of light reflects from a flat surface. **B.** Geometry of this reflection.

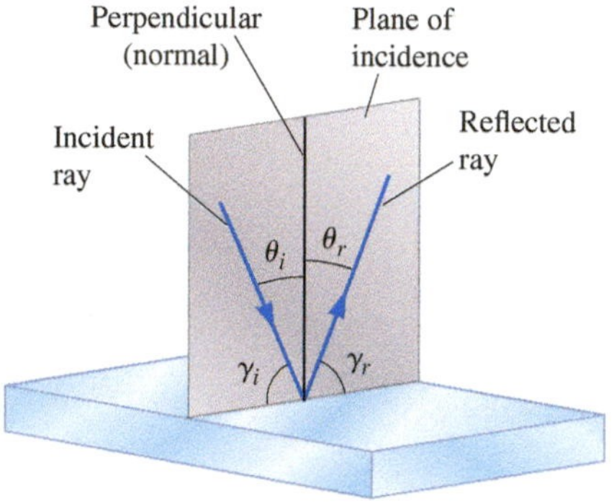

FIGURE 37.7 The law of reflection. Compare this figure to Figure 18.10.

on each triangle must also be equal. So the width of the reflected beam must be w. Use these triangles to find the sine of the two angles γ_i and γ_r:

$$\sin \gamma_i = \sin \gamma_r = \frac{w}{h}$$

Because $\sin \gamma_i = \sin \gamma_r$, the angles must be equal: $\gamma_i = \gamma_r$.

The angles γ_i and γ_r are measured between the beams and the surface. By custom, however, the angle used to express the *law of reflection* is measured between the beam and a line perpendicular to the surface. Figure 37.7 shows the angles θ_i and θ_r measured with respect to the perpendicular (called the *normal*). These angles θ_i and θ_r are referred to as the **angle of incidence** and the **angle of reflection**, respectively.

The **law of reflection** has two parts. The first part is that the angle of incidence equals the angle of reflection:

LAW OF REFLECTION

✪ **Major Concept**

$$\theta_i = \theta_r \tag{37.3}$$

The second part is that the incident ray, the reflected ray, and the perpendicular (normal) line all lie in a single plane called the **plane of incidence** (Fig. 37.7). The law of reflection applies to curved surfaces as long as you consider a portion that is small enough to be considered flat.

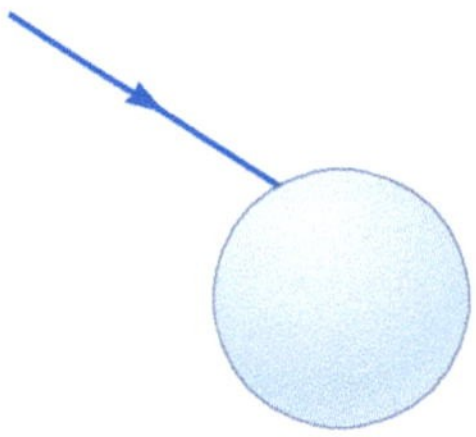

FIGURE 37.8

CONCEPT EXERCISE 37.1

A beam in air strikes a glass ball as shown in Figure 37.8. Draw the reflected ray.

EXAMPLE 37.2 The Ultimate in Security

Figure 37.9 shows the plan for a small apartment. A light source is in the left part of the apartment. A hole is in the wall on the right. Two mirrors are fixed on the long wall at the bottom of the plan. You would like to stand outside the apartment, look through the hole, and see light from the source. Where along the short wall should you place the third mirror? As part of your thinking, ask yourself: If you stood at the position of the light source, could you see out through the hole (assuming there was light outside the apartment)?

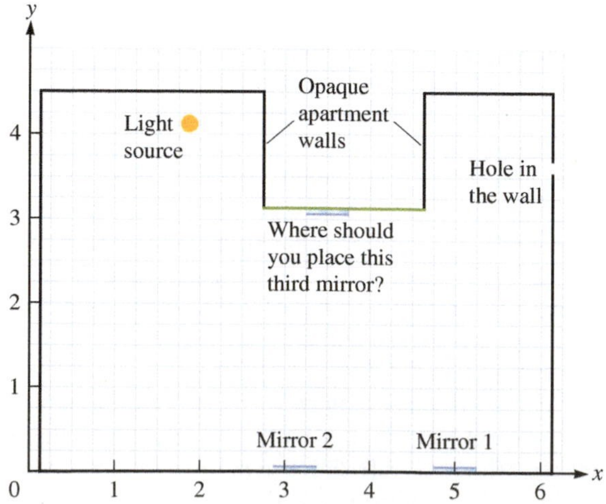

FIGURE 37.9 Apartment floor plan with three mirrors.

INTERPRET and ANTICIPATE

The physics behind this problem is the law of reflection, $\theta_i = \theta_r$. We can solve this problem by neatly drawing a few rays and then just reading the position of the third mirror off the diagram.

SOLVE

Start by drawing the line normal to each mirror, as shown in the middle of each mirror in Figure 37.10. When we find the position of the third mirror, we are really finding the position of the normal line we have drawn.

Also, draw the ray (labeled 1) from the last mirror at (5.00, 0). This is the ray you can see from that point on the mirror.

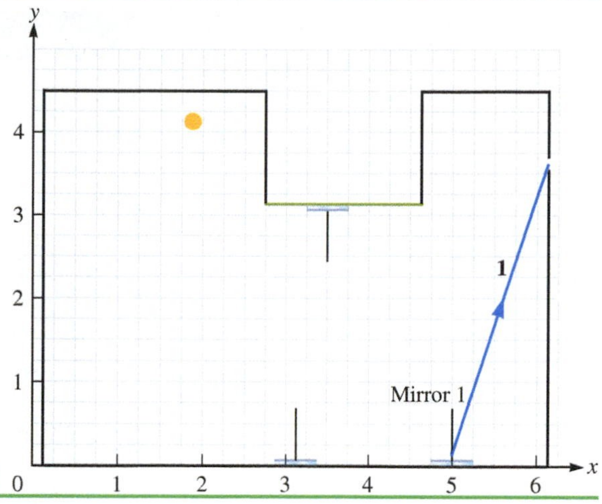

FIGURE 37.10

Apply the law of reflection ($\theta_i = \theta_r$) to this mirror and ray in order to draw the incident ray, labeled 2. The incident ray must come from the third mirror (on the green wall). By sliding the third mirror so that it can produce ray 2, we find the position of that mirror (Fig. 37.11).

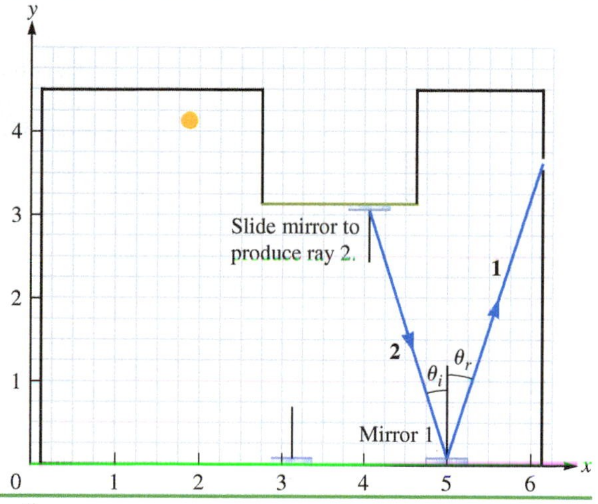

FIGURE 37.11

We are not done because it is possible that the light from the source cannot get to this third mirror. Repeating the process of using the law of reflection to find rays 3 and 4, we find that the light can make it from the source to all three mirrors and the hole (Fig. 37.12).

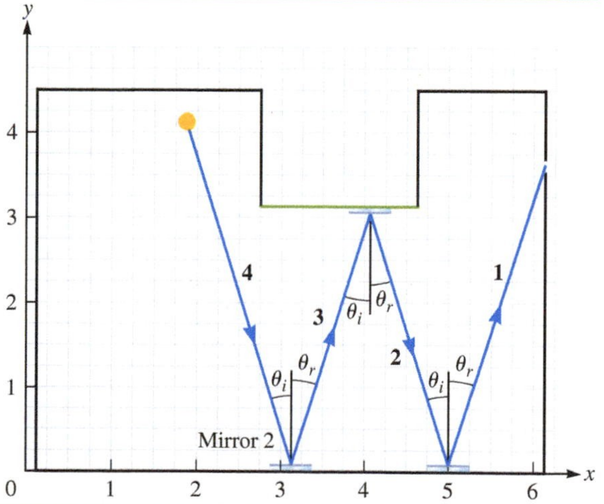

FIGURE 37.12

Examine Figure 37.12 to find the position of the mirror. Read off the position of the normal line with respect to the x axis.	The third mirror is at $x = 4.1$ m.

CHECK and THINK

If you tried to set up these mirrors on the scale shown, your placement would not need to be very precise because each mirror is about 0.5 m wide. So, if the center of the third mirror is not exactly at $x = 4.1$ m, this will still probably work. Also, there is nothing special that distinguishes reflected rays from incident rays. Suppose you stand in the room at the position of the light source. If the light source is off but there is light coming from the hole, you will be able to see that light. The "incident rays" shown in Figure 37.12 would be reflected rays, and the "reflected rays" would be incident rays. The W-shaped path would look the same, except the arrowheads would be reversed.

CASE STUDY The Hubble Space Telescope's Mirror

Smart people learn from their mistakes. This case study is about a mistake and how it was fixed. In April 1990, the Hubble Space Telescope (HST) was launched. About a month later, scientists discovered that the HST's images were blurry (Fig. 37.13A).

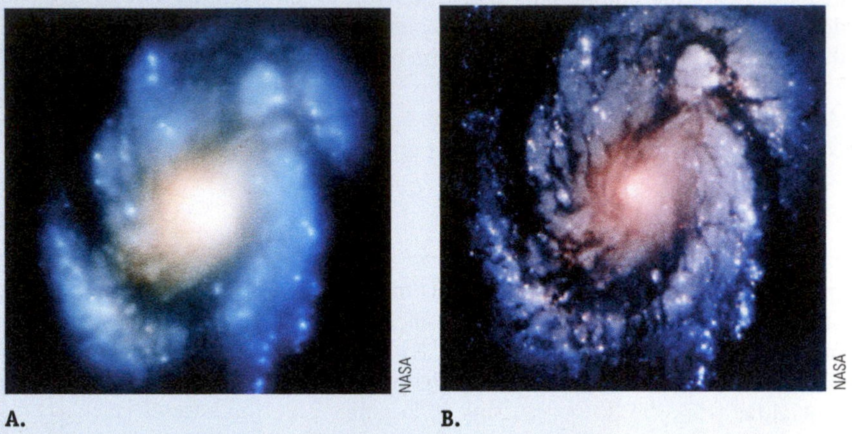

FIGURE 37.13 **A.** A picture taken by the Hubble Space Telescope before lenses were used to correct the shape of its mirror. **B.** A picture of the same astronomical object taken after the correction.

The largest telescopes are reflecting telescopes, which means they use mirrors, not lenses, to create images of astronomical objects. The HST—like large ground-based telescopes—uses an enormous dish-shaped mirror (the primary mirror) to focus light from distant objects. The HST's initial images were blurry because its primary mirror was incorrectly shaped. Three years later, this problem was corrected, and the images taken by the HST became much sharper (Fig. 37.13B). In this case study, we explore reflecting telescopes and how a misshaped primary mirror causes a blurry image.

37-3 Images Formed by Plane Mirrors

A mirror is a device that reflects more than 90% of the light incident on it, and the reflection is specular rather than diffuse. (We ignore the glass that usually covers the shiny reflective surface of a real mirror.) When you look at *diffuse* reflection from a surface such as the wall of a room, you can see reflected light no matter where you look. Even if you shine a laser at the wall, you can see the reflection (the spot) from many positions (Fig. 37.14A). When you look at a *specular* reflection, however, your eyes must be in the right place (determined by the law of reflection) in order to see the reflected light (Fig. 37.14B). The result is that walls do not form

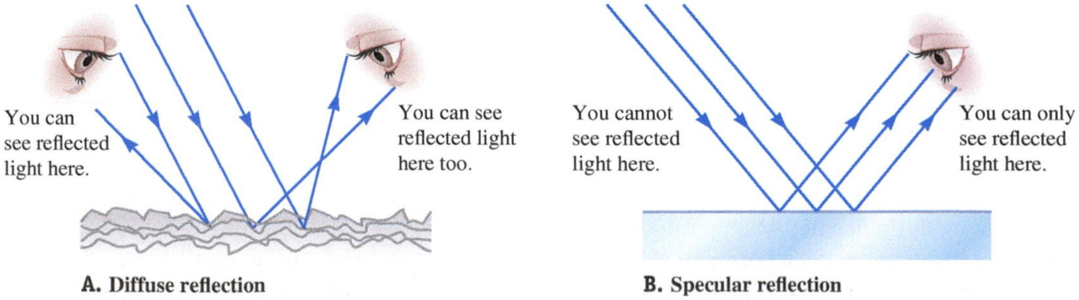

FIGURE 37.14 **A.** In diffuse reflection, you can see reflected light from many vantage points. **B.** In specular reflection, your eyes must be in the right place to see a reflection.

images, but mirrors do. For the rest of this chapter, we will study the images formed by mirrors of various shapes. In each case, our goal is to come up with expressions for the position and magnification of the image formed by the mirror. We begin with the familiar example of a **plane mirror**; such a mirror is flat and lies in a single plane.

PLANE MIRRORS Special Case

Ray Diagrams

RAY DIAGRAMS Tool

We used a crude ray diagram when we studied the camera obscura. Now we describe the elements needed to sketch a ray diagram that can be used to find the image formed by a plane mirror (Fig. 37.15):

1. A thick line representing the **mirror** seen edge-on. Light totally reflects from the thin layer on the front of the mirror. Assume an observer and an object are in front of the mirror. (Usually, the front side is shown on the left, but that is not necessary and is sometimes inconvenient.) It is not normally necessary to represent the observer.
2. An arrow representing the **object**.
3. A small number of **emitted rays** (usually four or fewer) coming from key points on the object, such as the tip and base of the arrow.
4. **Reflected rays.** For each emitted ray, use the law of reflection to draw the reflected ray.
5. An arrow representing the **image.** Consider two rays emitted from a single point, such as from the tip or the base of the arrow, and find the place where their reflected rays cross or would appear to cross. Because these rays were emitted from a single point, the image of that point is formed where the reflected rays cross (or seem to cross). Use this crossing point to sketch the image at that location. (In Figures 37.3 and 37.4, we considered rays emitted from two separate points of the object, so they do not cross to form a point on the image.)

Virtual and Real Images

In Figure 37.15, the two rays emitted from the tip of the object arrow are reflected by the mirror. These reflected rays enter the observer's eye without crossing, and the human brain assumes the rays are not bent. So, in effect, the brain traces these rays back along two straight lines that meet at an imaginary point *behind* the mirror, but there is no light behind the mirror. (Your bathroom mirror is probably hanging on a wall.) Nevertheless, your brain interprets the point where these rays appear to meet as the location of the image. We define a **real image** as one in which light rays from every point of an object **converge** or come together to create every point of the image. This convergence of light rays makes it possible to project a real image on a detector like film, a screen, or your retina. A movie camera lens and a camera obscura both produce real images. By contrast, we call the image produced by your bathroom mirror a **virtual image**. Light rays appear to **diverge** or spread from every point of a virtual image, just as they actually diverge from every point of an object. This divergence of light rays means it is not possible to project a virtual image on a screen, just as an object by itself does not project an image. A screen behind your bathroom mirror does not capture your image because there is no light there. A screen facing a bathroom mirror does not capture your image either. Nevertheless, because your eye has a lens, you see a virtual image of yourself in a plane mirror. Some curved mirrors produce real images (Section 37-4), but plane mirrors always produce virtual images.

VIRTUAL AND REAL IMAGES Major Concept

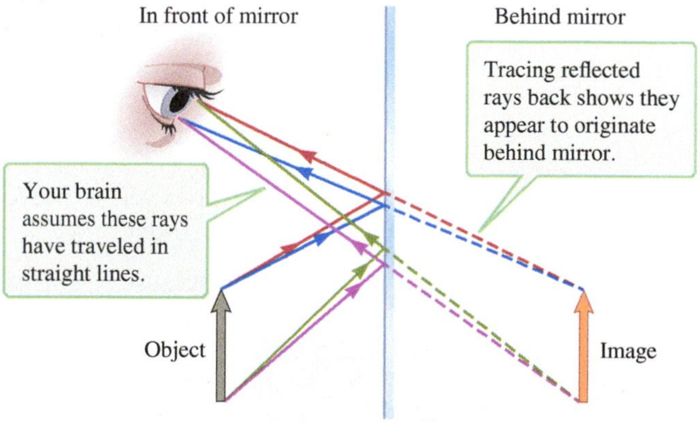
In front of mirror
Behind mirror
Tracing reflected rays back shows they appear to originate behind mirror.
Your brain assumes these rays have traveled in straight lines.
Object
Image

FIGURE 37.15 We have drawn two rays emitted from the tip and two from the base of the object's arrow. For all four emitted rays here, we have used the law of reflection to draw the reflected rays. The image you see in a plane mirror is behind the mirror, but there is no light back there. Such an image is called a *virtual image*.

Finding the Image Formed by a Plane Mirror

Your experience looking at your image in a flat mirror provides you with a set of expectations. First, imagine sharing the mirror with a roommate who is standing behind you. Your roommate's image is behind your image, so we expect that the image distance d_i is proportional to the object distance d_o. Second, the image is

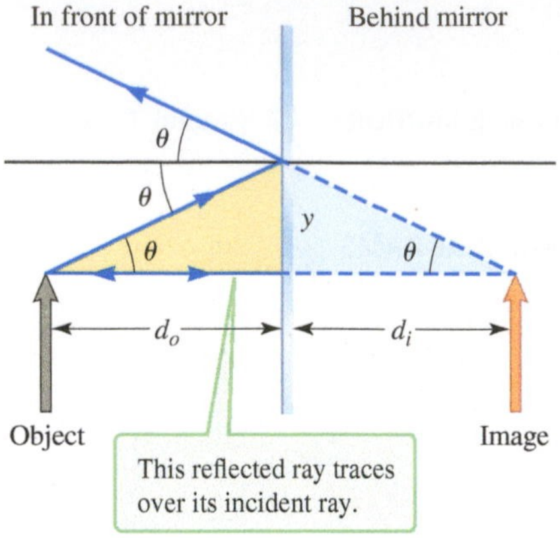

FIGURE 37.16 A simple ray diagram for a plane mirror. All the angles labeled θ are equal, so the highlighted triangles show that the image distance equals the object distance: $d_i = d_o$.

upright, so the image height h_i is positive, and we expect the magnification M also to be positive. Third, when you look in a bathroom mirror, you don't look smaller or larger than you really are, so we expect the magnification to be $M = 1$.

The simplified ray diagram in Figure 37.16 shows two rays emitted from the tip of an object (represented by the large upright arrow in front of the mirror). One ray is perpendicular to the mirror, so according to the law of reflection, its reflected ray is also perpendicular. The other ray is also emitted from the tip, but drawn in some arbitrary direction toward the mirror. The law of reflection determines its reflected ray. The two reflected rays do not cross in front of the mirror. To find the image, trace these two rays back to the point behind the mirror where they seem to meet.

The two highlighted right triangles in Figure 37.16 share a common side of length y. The base of the triangle on the left is the object distance d_o, and the base of the triangle on the right is the image distance d_i. The tangent of θ for both of these triangles is

$$\tan \theta = \frac{y}{d_o} = \frac{y}{d_i}$$

So the image distance equals the object distance:

$$d_i = -d_o \tag{37.4}$$

using a second sign convention that when the image is located behind the mirror, its distance d_i is negative. Equation 37.4 meets our expectation that the image distance d_i is proportional to the object distance d_o.

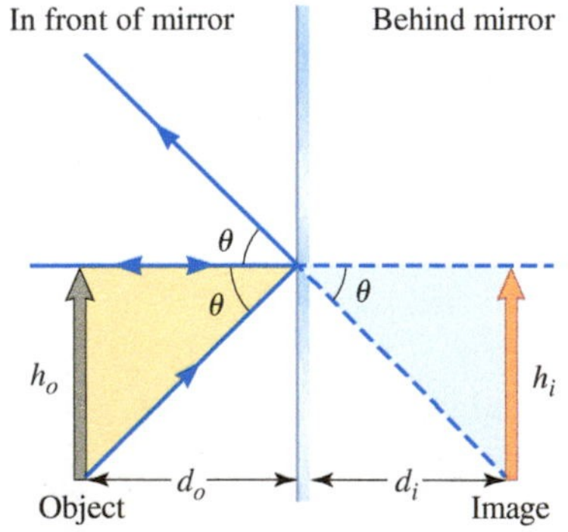

FIGURE 37.17 Another simple ray diagram for a plane mirror. All the angles labeled θ are equal, so the highlighted triangles show that the image height equals the object height: $h_i = h_o$.

Another simple ray diagram (Fig. 37.17) is helpful in finding the magnification. As before, we draw one ray perpendicularly from the tip of the object arrow to the mirror, so that ray's reflection traces back over the incident ray. The other ray is drawn from the base of the arrow, and it strikes the mirror at the same place as the first incident ray. It obeys the law of reflection, and the two reflected rays are traced back to points behind the mirror.

Take the tangent of θ for both highlighted triangles:

$$\tan \theta = \frac{h_o}{d_o} = \frac{h_i}{d_i}$$

According to Equation 37.4, the magnitudes of the object and image distances are equal, so the bases of the triangles are equal:

$$\frac{h_o}{d_o} = \frac{h_i}{d_o}$$

The image height thus equals the object height:

$$h_o = h_i$$

The magnification (Eq. 37.2) of a plane mirror is given by

$$M = \frac{h_i}{h_o} = -\frac{d_i}{d_o} = 1 \tag{37.5}$$

Because d_i is negative, Equation 37.5 meets our expectations that the magnification is positive and equal to 1.

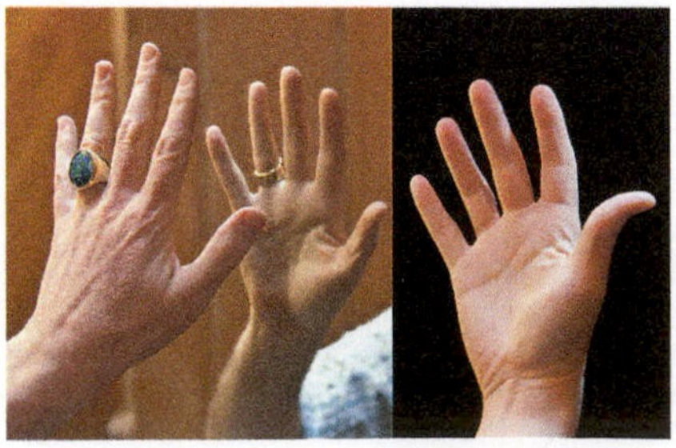

Charles D. Winters/Cengage Learning

FIGURE 37.18 The reflection of the left hand looks like the palm of the right hand.

Left Hand or Right Hand?

We just showed that your image in a bathroom mirror is upright, has a magnification of 1, is as far behind the mirror as you are in front of the mirror, and is virtual. However, the image you see every morning also appears to be reversed from left to right. The camera obscura causes a left-right reversal, and it also causes the image to be inverted (Fig. 37.3).

To see how a plane mirror can cause the left-right reversal without inverting the image, let's consider the reflection of a person's left hand. In Figure 37.18, the left hand has a ring on it, so you can keep track of which hand is which. The reflected image of the left hand in the mirror looks like the right hand, with its fingers pointing up, its palm facing out, and its thumb pointing toward the right. The only substantial difference is that the image of the left hand has a ring. So, you might say the mirror has made the left hand look like the right hand. This is the left-right reversal you see every time you look in the bathroom mirror.

However, describing what you see as a simple left-right reversal is not quite correct. Examine the actual left hand. The left fingers point up and the thumb points toward the right, as in the image of the left hand. If the person flipped her left hand so that the palm faced forward, the fingers would be up but the thumb would point toward the left. So the mirror doesn't really create a left-right reversal. The real difference is that the image of the left hand in the mirror shows the *palm* of the left hand, and the view of the actual left hand is of the back. In a sense, the mirror creates a *depth reversal*, not a left-right reversal. To make the left hand look like its image, each particle of the left hand would need to move straight through to the other side of the hand, so that the back of the hand would be behind the palm.

A depth reversal may sound complicated, but when we consider a moving object and its mirror image, the concept becomes clearer. Consider Charlie Chaplin's encounter with his twin in Figure 37.19A. When Chaplin moves his left hand toward his twin in the positive x direction, the twin creates a mirror image by moving his right hand toward Chaplin in the *negative* x direction. Now imagine the twin is replaced by a plane mirror (Fig. 37.19B). When Chaplin moves his left hand in the positive x direction, his mirror image moves its right hand in the negative x direction. To Chaplin, the twin in Figure 37.19A appears to be a mirror image because the twin imitates the depth reversal of the mirror image in Figure 37.19B.

CONCEPT EXERCISE 37.2

The terms *virtual image* and *real image* are not particularly descriptive or intuitive. Come up with your own terms, and explain your reasons for suggesting them.

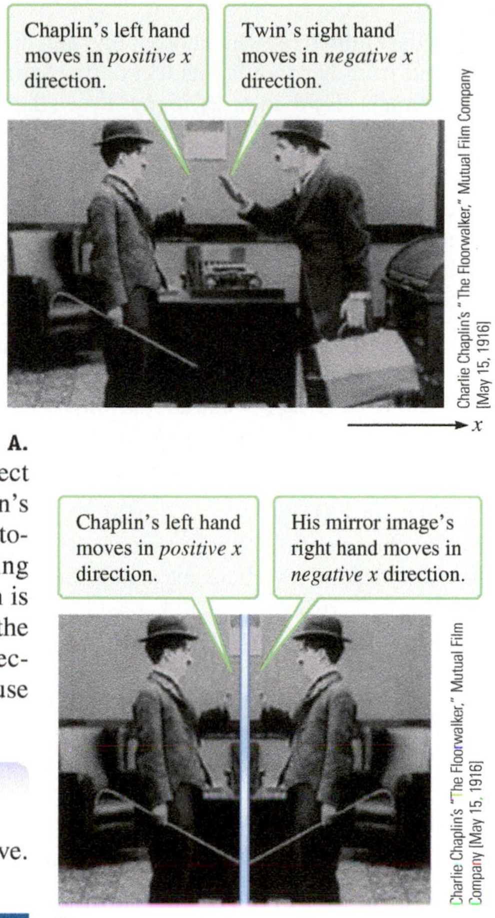
Chaplin's left hand moves in *positive x* direction.
Twin's right hand moves in *negative x* direction.
Charlie Chaplin's "The Floorwalker," Mutual Film Company [May 15, 1916]
x
A.
Chaplin's left hand moves in *positive x* direction.
His mirror image's right hand moves in *negative x* direction.
Charlie Chaplin's "The Floorwalker," Mutual Film Company [May 15, 1916]
B.

FIGURE 37.19 Charlie Chaplin (1889–1977) was a British comedian. **A.** Charlie Chaplin and his twin create a mirror image. **B.** Charlie Chaplin and his mirror image move like Chaplin and his twin in part A.

EXAMPLE 37.3 Full-Length Mirror

People often purchase a full-length mirror that is at least as tall as they are, but such a large mirror is not necessary. Suppose a woman of height h wishes to see her entire image in a mirror. What is the minimum height of the plane mirror she needs?

INTERPRET and ANTICIPATE

When solving problems in geometric optics, start with a ray diagram. Because the object—the woman—is also the observer, we have sketched her instead of a simple arrow (Fig. 37.20). The key to solving this problem is to draw one ray emitted from the top of her head and another ray from her foot. These rays come from the extreme parts of her body. If she can see their reflections, she will be able to see the rays reflected from points between these extremes.

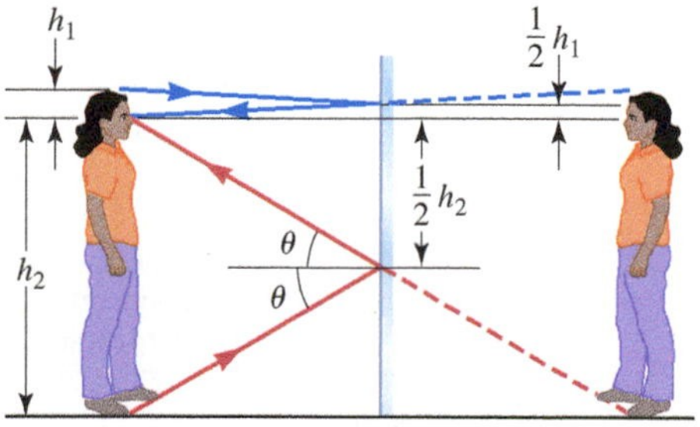
h_1
$\frac{1}{2}h_1$
$\frac{1}{2}h_2$
θ
θ
h_2

FIGURE 37.20 A ray diagram of a woman seeing her entire image in a mirror.

SOLVE

Label a few key distances. Figure 37.20 shows a horizontal line through the woman's eye and the image of her eye. The top of her head is a distance h_1 above her eye, and her foot is a distance h_2 below it, so her total height is the sum of these two distances.

$$h = h_1 + h_2 \quad (1)$$

Example continues on page 1194 ▶

After the emitted rays are chosen, find and draw the reflected rays. The ray emitted from her foot is incident on the mirror at angle θ, so its angle of reflection is also θ. In order for her to see the reflected ray, the incident ray must strike the mirror at a point halfway between her eye and her foot.	The bottom of the mirror is $h_2/2$ below the horizontal line that passes through her eye.
The same sort of geometry applies to the ray emitted from the top of her head. That ray must strike the mirror halfway between her eye and the top of her head.	The top of the mirror is $h_1/2$ above the horizontal line that passes through her eye.
Let h_{mirror} be the total height of the mirror that just enables the woman to see her entire reflection. This total height is $h_1/2$ (top of mirror to woman's eye level) plus $h_2/2$ (bottom of mirror to woman's eye level). Using Equation (1), we find the mirror must be half the height of the woman.	$h_{\text{mirror}} = \frac{h_1}{2} + \frac{h_2}{2} = \frac{h_1 + h_2}{2}$ $h_{\text{mirror}} = \frac{h}{2}$

CHECK and THINK

Notice that we did not need to consider how far the woman is from the mirror, so our answer does not depend on where she stands. She sees her entire body whether she is far from the mirror or close to it. However, we did consider where the mirror hangs on the wall. The top of the mirror must be almost level with the top of the woman's head. Why do people buy such large full-length mirrors (about their own height)? There are a number of reasons. If two people of different heights want to use a mirror of the size we calculated, it needs to be half the height of the taller person, and it needs to hang almost level with the top of that person's head. A second, much shorter person wouldn't be able to see his whole body. In the extreme case that one person is a tall adult and the other is a short child, the child's head may be below the bottom of the mirror, so he may not be able to see himself in the mirror at all.

EXAMPLE 37.4 True Mirror

Figure 37.21A shows the image of a message and a clock created by a plane mirror. The image in Figure 37.21B is not reversed because it is the image created by a second plane mirror perpendicular to the first. Two perpendicular mirrors are called a *true mirror* because the final image is not reversed. Use a ray diagram to show that a true mirror produces an image that is not reversed.

Charles D. Winters/Cengage Learning

Charles D. Winters/Cengage Learning

FIGURE 37.21 **A.** This is the image in the first mirror. **B.** A second mirror creates an image that is not reversed.

INTERPRET and ANTICIPATE

Start by sketching a side view of the two perpendicular mirrors and the object, which is represented by an arrow (Fig. 37.22). We can break down the next part of the job into steps. First draw a few emitted rays and their reflection from one of the two mirrors. In the second step, trace these reflected rays to the second mirror. Find the rays that reflect from the second mirror, and then trace these reflected rays back to the image formed by the second mirror.

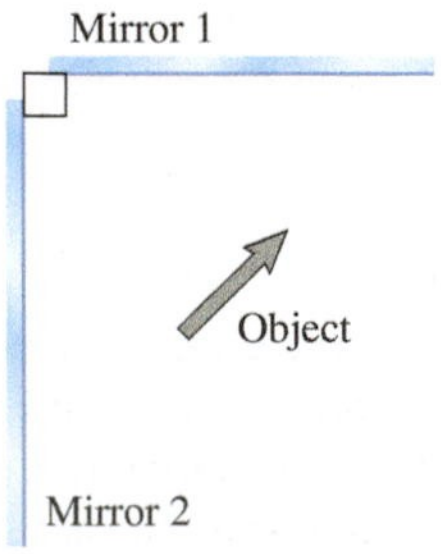

FIGURE 37.22

SOLVE

We arbitrarily choose the mirror shown at the top of Figures 37.22 and 37.23 as the first mirror. (In Problem 20, you show that the same result occurs if you choose the other mirror as the first mirror.) Consider three emitted rays; two are emitted from the tip of the arrow and the third from the base. Each ray is traced to mirror 1, and its reflection from mirror 1 is drawn. Then trace these three rays to points behind mirror 1 to show the image formed by mirror 1. We find that image 1 is reversed.

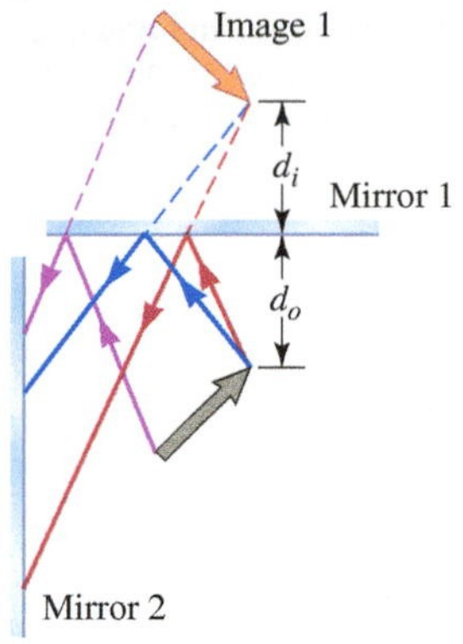

FIGURE 37.23 Image 1 is reversed.

In the second step, trace the rays reflected from mirror 1 to mirror 2 (Fig. 37.24). In doing this, we are treating image 1 as the object for mirror 2. For each of the three rays incident on mirror 2, we find the ray that reflects from that mirror. Tracing these reflected rays back to points behind mirror 2 shows the image formed by mirror 2; image 2 is the image of image 1.

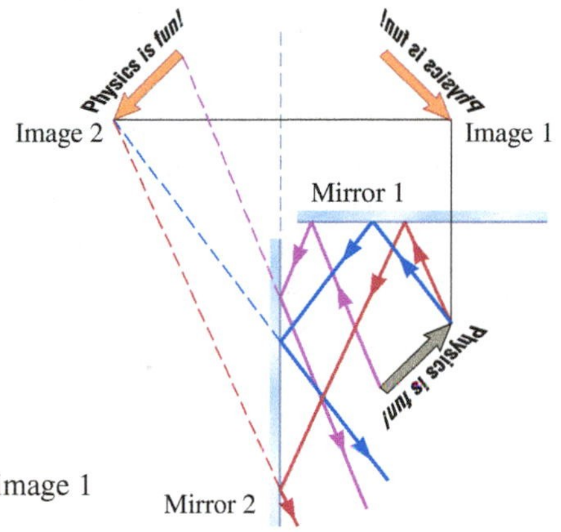

FIGURE 37.24 The image of image 1 is not reversed.

CHECK and THINK

Compare the final ray diagram (Fig. 37.24) to the photos in Figure 37.21. The words in the first image in mirror 1 are backward (reversed), which matches what we see in Figure 37.21A. The words in the second mirror (image 2) are not reversed, as in Figure 37.21B. So our ray diagram seems to match the images in Figure 37.21.

37-4 Spherical Mirrors

Mirrors can be bent into complicated shapes. When the mirror's geometry is complicated, however, it is difficult to calculate the position and magnification of the image formed by the mirror. To keep these tasks manageable, we will focus on *spherical mirrors*. A **spherical mirror's** surface forms either a sphere or part of a sphere. The reflective surface may be on the outside or the inside of the sphere. In Figure 37.25A, an artist sees his reflection in a mirror that forms a complete sphere, with the reflective surface on the outside. In Figure 37.25B, a photographer reaches for her reflection in a spherical mirror. The surface of her mirror does not form a complete sphere, and the reflective surface is on the inside of the sphere. The perimeter of a spherical mirror does not need to be a circle. Figure 37.26 shows a spherical mirror with a roughly rectangular perimeter. The surface of the mirror is part of a sphere, and you can imagine peeling it off a large sphere whose inside surface is reflective.

A plane mirror forms an upright virtual image that is the same size as the object and located as far behind the mirror as the object is in front of the mirror (Section 37-3). By contrast, the image formed by a spherical mirror may be upright or inverted; it may be larger or smaller than the object; and it may be virtual and behind the mirror, or real and in front of the mirror.

In this section, we use ray diagrams to help visualize the images produced by spherical mirrors and to derive algebraic expressions for the positions and magnifications of

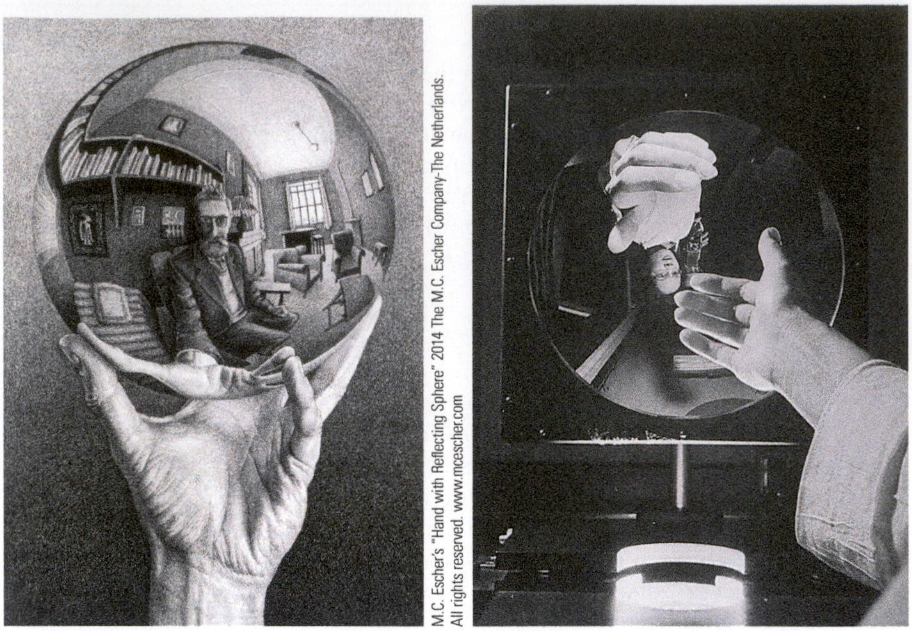

FIGURE 37.25 **A.** The image of the artist in this convex spherical mirror is upright. **B.** The image of the photographer in this concave spherical mirror is inverted.

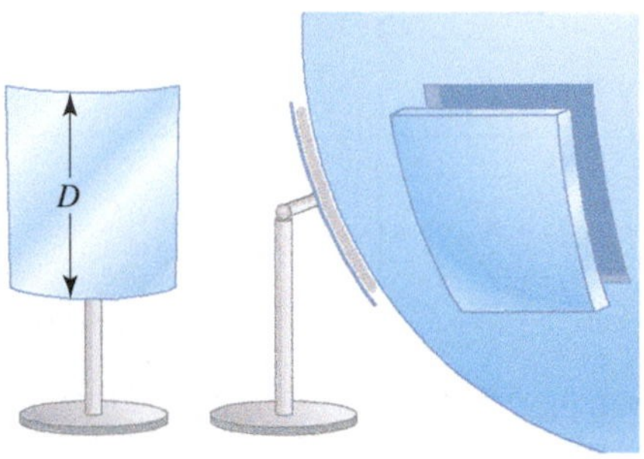

FIGURE 37.26 A spherical mirror does not have to be circular. This rectangular mirror is a spherical mirror. Its surface is a patch of a spherical surface. The inside surface of the sphere is reflective.

CONVEX AND CONCAVE SPHERICAL MIRRORS Special Case

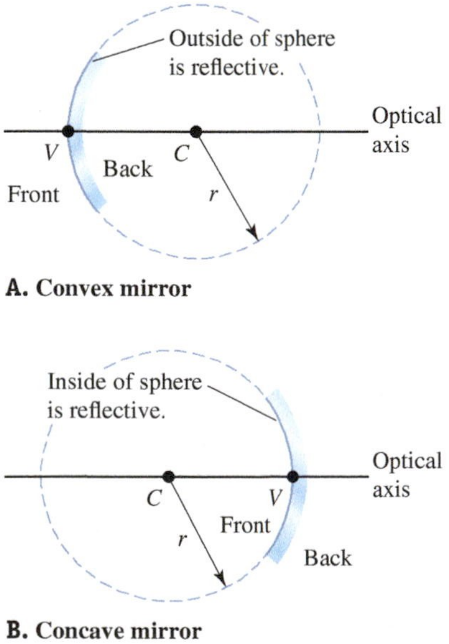

FIGURE 37.27 **A.** For a *convex* spherical mirror, the outside surface is reflective. **B.** For a *concave* spherical mirror, the inside surface is reflective. To keep the two terms straight, remember that from the front a concave mirror looks like a *cave*. The dashed circles represent the complete spheres to which these mirrors correspond and are not normally included in a ray diagram.

the images. When representing a spherical mirror on a ray diagram, we draw an edge-on view that is either a circle or a circular arc (Fig. 37.27). The radius of curvature r is the radius of the circle. The center of the circle—labeled C—is called the mirror's **center of curvature.**

In Figure 37.27A, the reflective surface of the mirror is the outside the sphere, as with the mirror in Figure 37.25A. Such a mirror is called a **convex** mirror. When you stand in front of a convex mirror, the surface bulges out toward you. Convex mirrors are used in public places such as parking garages and stores because these mirrors have a wide field of view, allowing you to look for traffic around the corner of a garage or for shoplifters in a store. In Figure 37.27B, the reflective surface of the mirror is the inside of the sphere, as with the mirror in Figure 37.25B. Such a mirror is called a **concave** mirror. When you stand in front of a concave mirror, the surface looks like a bowl or *cave.* The image formed by a concave mirror may be larger than the object, so concave mirrors are used as shaving or makeup mirrors. Concave mirrors are also used in telescopes (Section 37-7).

It is important to distinguish between the radius of curvature and the size of the mirror. The radius of curvature is the radius r of the entire sphere (Fig. 37.27). It is a measure of how curved the mirror is. A small radius of curvature means a tightly curved mirror. A flat mirror has an infinite radius of curvature. The size of the mirror is its largest linear dimension; for example, D is the height of the rectangular mirror in Figure 37.26. If the mirror happens to have a circular perimeter, D is the diameter of that circle.

It is also important to distinguish between the center of curvature and the *center of the mirror*. The **center of the mirror** lies on its surface and is also called the **vertex.** The center of curvature does not lie on the mirror. In Figure 37.27, the center of curvature is labeled C and the vertex is labeled V. The **optical axis** is the line that runs through the center of curvature C and the vertex V; it is perpendicular to the mirror at V. Sometimes the optical axis is referred to as the **principal axis** or the **central axis.**

Finding the Focal Point

Consider a simple situation in which the object is very far from the mirror. For example, you can imagine the Sun is the object and the mirror is on the Earth. Because the Sun is very far away, any rays that strike the mirror must be traveling in the same direction and are therefore parallel. To keep things simple for now, consider only mirrors that are small compared to their radii of curvature: $D \ll r$. (We'll consider larger mirrors in Section 37-7.)

Now imagine orienting a convex mirror so that its optical axis is parallel to the Sun's rays (Fig. 37.28A). The law of reflection determines each ray's path after it strikes the mirror. We find that the reflected rays **diverge** (do not cross). No real image of the Sun forms in front of the mirror. However, if we trace each of these reflected rays back along a straight path, we see that they meet at a single point behind the mirror. If these reflected rays enter your eye, you see a virtual image of the Sun. It looks like the rays emerged from the point labeled F. The point where parallel rays form a virtual image is known as the **virtual focal point**. Only the reflections of parallel rays, or rays from a very distant object, appear to emerge from the virtual focal point.

Now consider what happens when parallel rays encounter a concave mirror. As before, these rays must be parallel to the optical axis. By the law of reflection, this time we find that the reflected rays meet or **converge** at a point in front of the mirror. This point—labeled F—is called the **real focal point**. A real image of the Sun forms at the real focal point; it is like the real image formed by a camera obscura. If you put a small piece of paper (or film) at the focal point so that it doesn't block much of the Sun's light from reaching the mirror, you see an image projected on the paper. (You may even be able to light the paper on fire.)

Because the law of reflection is symmetrical (the angle of incidence equals the angle of reflection), we can reverse the arrowheads on all the rays in Figure 37.28. So, if an incoming ray is aimed at the virtual focal point of a convex mirror, the reflected ray is parallel to the optical axis (Fig. 37.28A). Further, if an incident ray passes through the real focal point of a concave mirror, its reflection is parallel to the optical axis (by ray reversal of Fig. 37.28B). Flashlight and headlight manufacturers make use of this fact. If you examine a flashlight or headlight closely, you see that the bulb is placed at or near the focal point of a concave mirror.

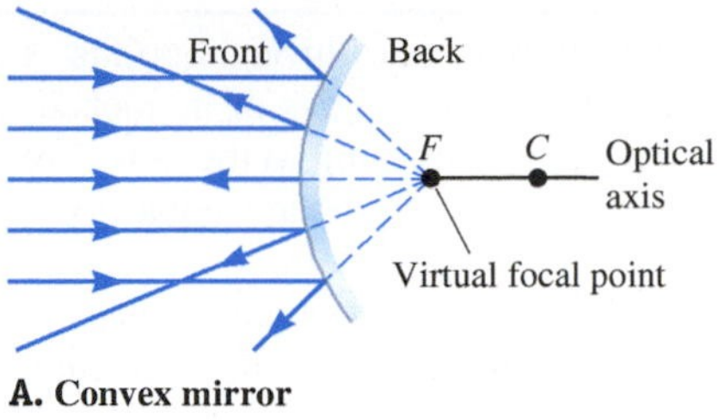

FIGURE 37.28 These mirrors are small compared to their radii of curvature. **A.** Parallel rays reflected by the convex mirror appear to come from points behind the mirror. As in the case of a flat mirror, we see a virtual image located behind the mirror. **B.** Parallel rays incident on a concave mirror are focused in front of the mirror, forming a real image.

Primary Rays

There are four rays that are not only easy to draw but also particularly helpful in ray diagrams involving spherical mirrors. These four rays are known as the **primary rays** and are described in Table 37.1.

A ray diagram does not necessarily need to show all four primary rays. Often two rays are sufficient. Then you may wish to draw a third ray to check your work. Choosing which primary rays to draw sometimes depends on geometry. Some choices may make your work easier than others, but very often all rays are equally good choices. As with drawing free-body diagrams, practice will help you draw good ray diagrams.

PRIMARY RAYS Tool

DERIVATION Focal Length f

The **focal length** f of a mirror is the distance between the focal point and the vertex of the mirror (Fig. 37.29). We show that for a small convex mirror ($D \ll r$), the focal length is half the radius of curvature:

$$f = \frac{r}{2} \tag{37.6}$$

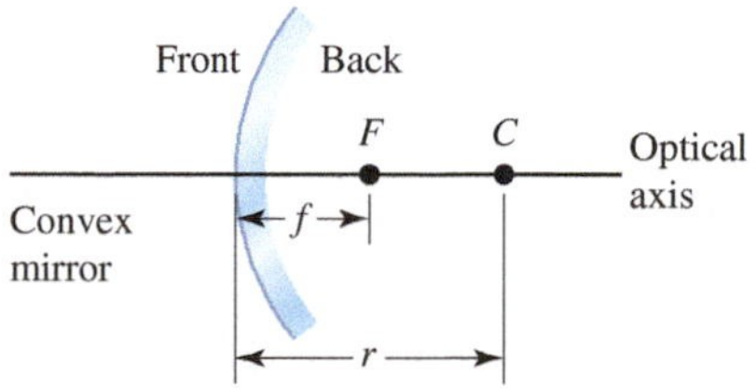

FIGURE 37.29 The focal length f is the distance between the focal point and the vertex.

In Problem 27, you are asked to repeat this derivation for a concave mirror.

We can derive the focal length by drawing just one primary ray and using geometry. Rays 3 and 4 work equally well, but we have arbitrarily decided to use ray 4. Ray 4 is aimed at the virtual focus (Fig 37.30) and reflects at point A. Its reflected ray is parallel to the optical axis.

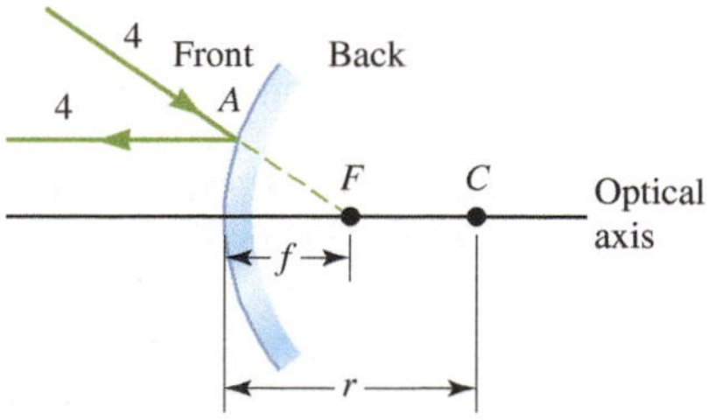

FIGURE 37.30

Derivation continues on page 1198 ▶

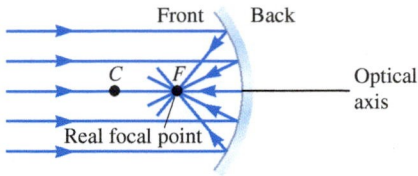

The line that runs through points A and C is perpendicular to the mirror (Fig. 37.31). The angles of incidence and reflection are measured with respect to that line. According to the law of reflection, these two angles θ are equal. In addition, these angles are equal to the two angles θ on the back side of the mirror. By the rules of geometry, the angle ACF must also be equal to θ. The triangle ACF has two equal angles, so it is an isosceles triangle. The two short sides must be equal. We have labeled their lengths ℓ.

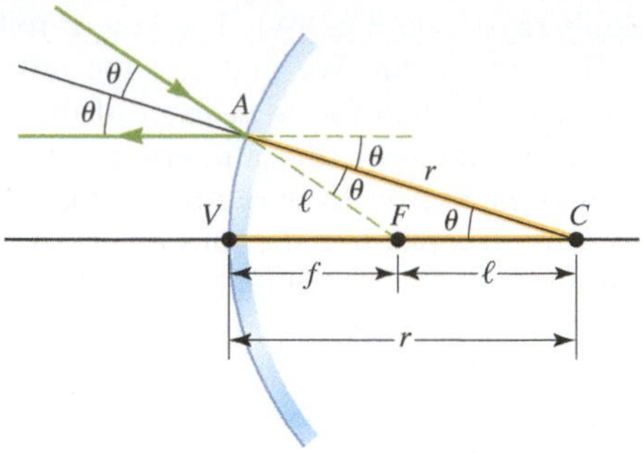

FIGURE 37.31 The triangle ACF is isosceles, so its two short sides have the same length ℓ.

The radius of curvature r is the distance ℓ from C to F plus the distance f from F to V.

$$r = \ell + f \tag{1}$$

Although Figure 37.31 is exaggerated, the mirror is small compared to its center of curvature ($r \gg D$), so the distance from the focal point F to any point on the mirror is roughly the same. Because the distance from F to V is the focal length f, the distance from F to any point on the mirror is roughly f. So the distance ℓ from F to A is roughly f.

$$\ell \approx f \tag{2}$$

Substitute Equation (2) into Equation (1), and solve for f.

$$r \approx f + f = 2f$$

$$f = \frac{r}{2} \quad ✓ \tag{37.6}$$

COMMENTS

We dropped the approximately equal sign in the result, but keep in mind that Equation 37.6 is true only when the spherical mirror is small compared to its radius of curvature. This derivation shows the value of considering primary rays. In the next two sections, we will use primary rays to find the images formed by both convex and concave mirrors for objects that are at some finite distance from the mirror.

TABLE 37.1 Primary rays for spherical mirrors. Colors shown are arbitrary and do not relate to actual color of light.

Ray	Convex Mirror	Concave Mirror
Ray 1 strikes the mirror at its vertex. Because the optical axis is perpendicular to the mirror at its vertex, the incident and reflected rays are symmetrical with respect to the optical axis.	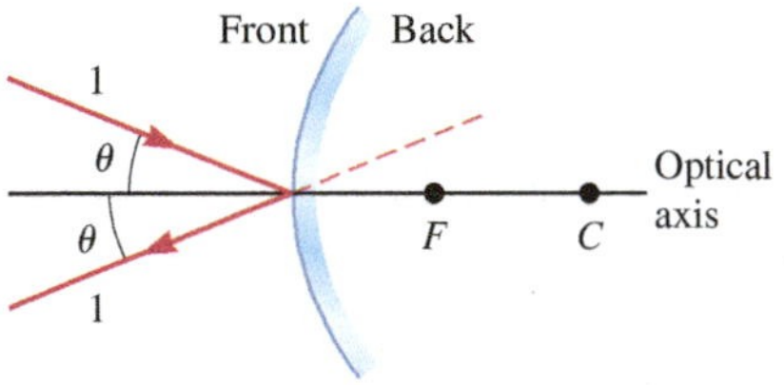	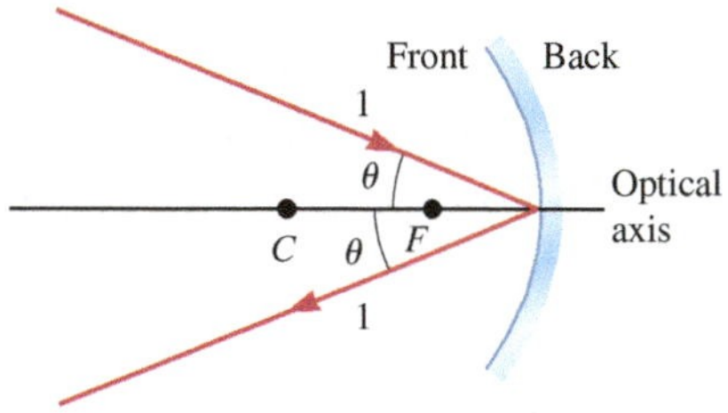
For a convex mirror, **ray 2** is aimed at the center of curvature C. For a concave mirror, ray 2 passes through the center of curvature. For either type of mirror, ray 2 is perpendicular to the mirror surface and the reflected ray is back along ray 2.	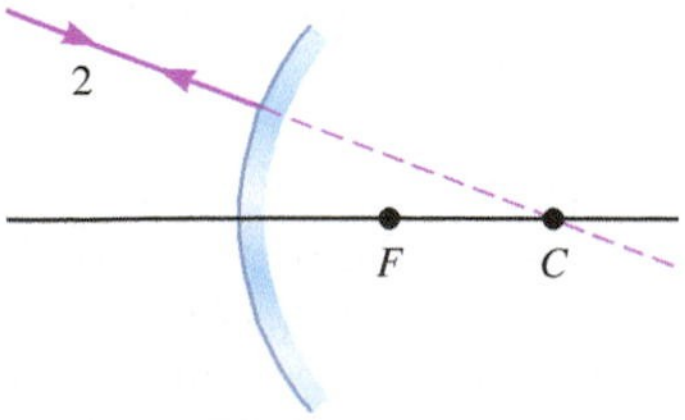	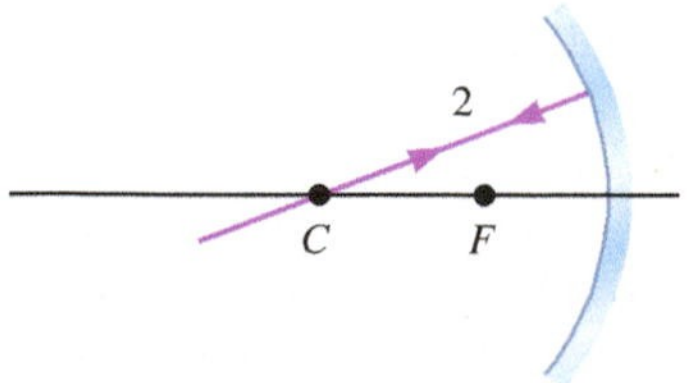

Ray	Convex Mirror	Concave Mirror
Ray 3 is parallel to the optical axis. When it reflects from a convex mirror, the reflected ray extends through the virtual focal point. When ray 3 reflects from a concave mirror, the reflected ray passes through the real focal point.	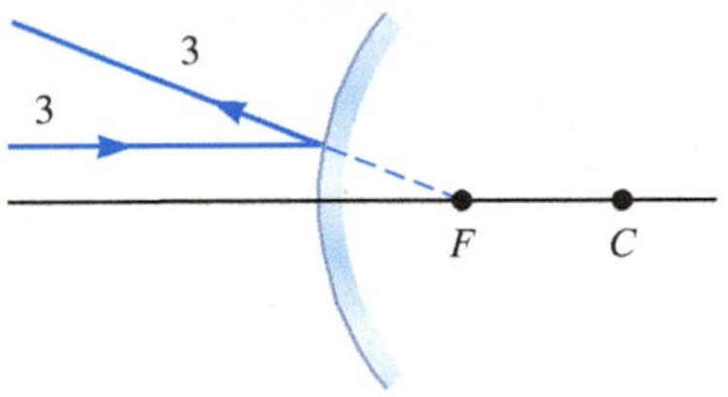	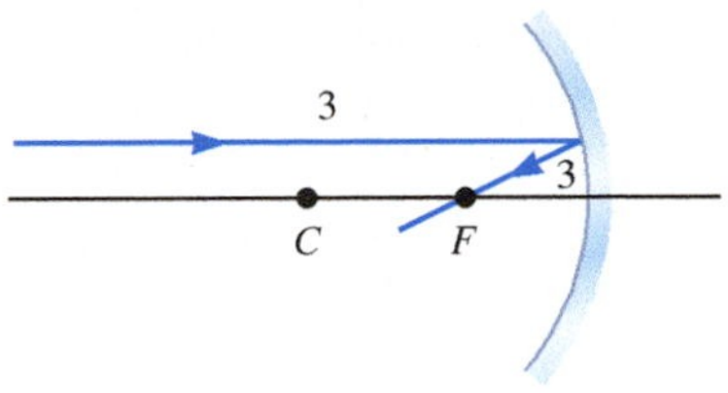
For a convex mirror, **ray 4** is aimed at the focal point. For a concave mirror, **ray 4** passes through the focal point. For either type of mirror, ray 4's reflection is parallel to the optical axis.	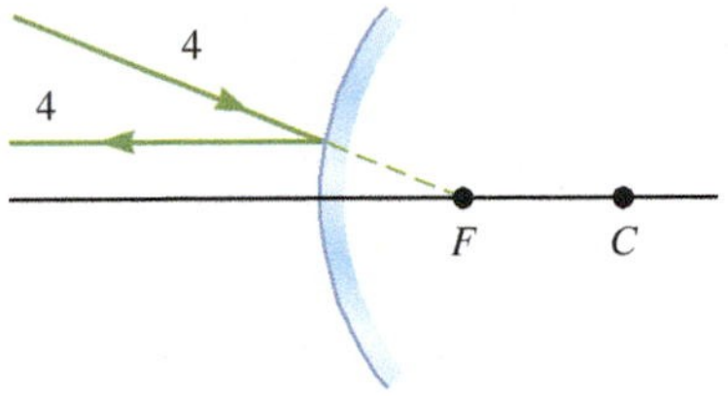	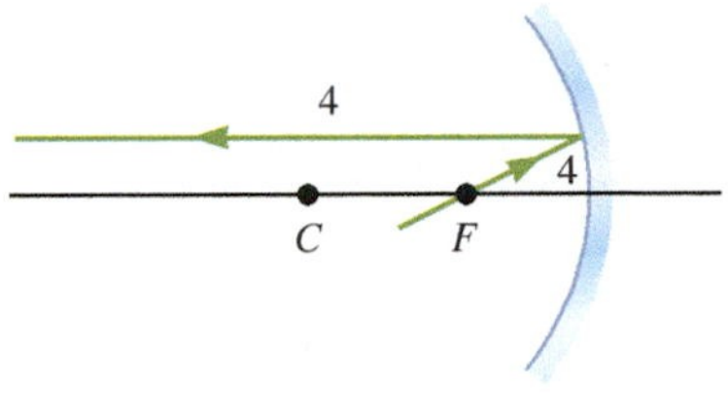

Sign Conventions

The third and final sign convention is for the focal length f and radius of curvature r. For a convex mirror, the center of curvature C and the focal point F are both behind the mirror (Figs. 37.28A and 37.29). In this case, the focal point is virtual, and by convention the focal length and radius of curvature are both negative. For a concave mirror, the center of curvature C and the focal point F are both in front of the mirror (Fig. 37.28B). In this case, the focal point is real, and by convention the focal length and radius of curvature are both positive. The three sign conventions for mirrors are summarized in Table 37.2.

TABLE 37.2 Sign conventions for mirrors.

Quantity	Positive	Negative
1. Image height h_i and magnification M	If image is **upright**	If image is **inverted**
2. Object distance d_o and image distance d_i	If object or image is **real**	If object or image is **virtual**
3. Radius of curvature r and focal length f	If mirror is **concave**	If mirror is **convex**

CONCEPT EXERCISE 37.3

Estimate the focal length of the mirror shown in Figure 37.25A.

37-5 Images Formed by Convex Mirrors

Figure 37.32 shows a convex mirror and an object whose distance from the mirror is d_o. Only an object that is very far from the mirror, so that its rays are parallel to the optical axis, forms a virtual image at the virtual focus. Thus, the object in Figure 37.32 does not form a virtual image at F. To get a rough idea of the location and size of the image, we make a ray diagram using primary rays 1 and 2. Both rays are drawn from a single point on the object (the tip of the object arrow). The two reflected rays do not intersect, but when we trace these rays back behind the mirror, their extensions intersect. Because these rays come from the tip of the object arrow,

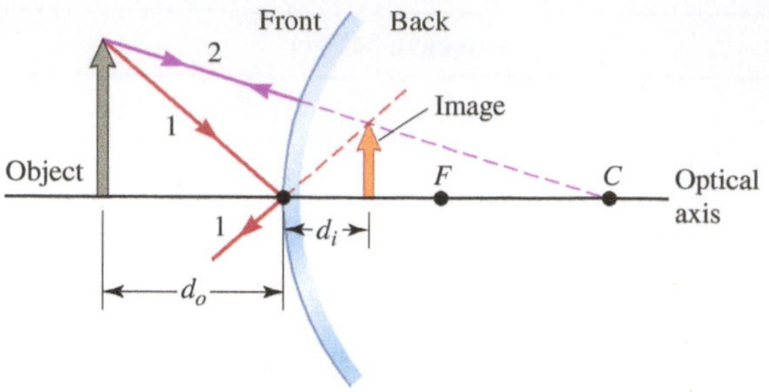

FIGURE 37.32 To find the location and size of the image created by this convex mirror, we draw two of the four primary rays. We arbitrarily choose rays 1 and 2. Ray 1 strikes the vertex of the mirror; ray 1 and its reflection are symmetrical with respect to the optical axis. Ray 2 is aimed at the center of curvature C, and its reflection is back along itself.

the image of the tip must be at this intersection. As with a plane mirror, there is no light behind the convex mirror, so this is a virtual image located a distance d_i behind the mirror.

Figure 37.32 also shows that the image points upward, just like the object. Because the image is *upright* (rather than inverted), the magnification is positive (Table 37.2). Additionally, the image is smaller than the object, so the absolute value of the magnification is less than 1 ($0 < M < 1$). Compare the image formed by the convex mirror (Fig. 37.32) to the image formed by a plane mirror (Fig. 37.17). Both images are formed behind the mirror, so both are virtual. The images formed in both mirrors are upright, so $M > 0$. However, the image and the object for a plane mirror have the same size, so $M = 1$, whereas the image is smaller than the object in a convex mirror, so $0 < M < 1$.

DERIVATION Magnification and the Mirror Equation for Spherical Mirrors

Ray diagrams provide a rough idea of the location and magnification of an image. We can get more precise results if we use geometry to derive algebraic expressions. Here we show that the magnification for a spherical mirror is given by

$$M = \frac{h_i}{h_o} = -\frac{d_i}{d_o} \tag{37.7}$$

and we also derive the **mirror equation**, which relates the image distance d_i to the object distance d_o and to the focal length f of the mirror:

$$\frac{1}{d_o} + \frac{1}{d_i} = \frac{1}{f} \tag{37.8}$$

We will derive these equations for a convex mirror, but they also apply to a concave mirror (Problems 36 and 40).

Start with a simple ray diagram (Fig. 37.33) showing only rays 1 and 3. We'll apply the sign convention later; for now, we will use absolute values of all lengths.

MAGNIFICATION OF SPHERICAL MIRROR Major Concept

MIRROR EQUATION Major Concept

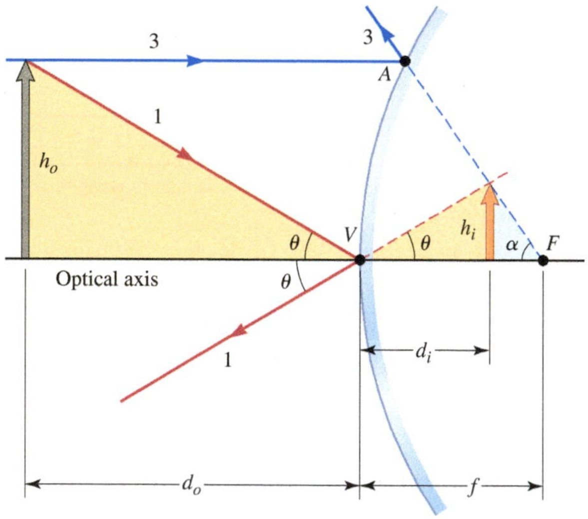

FIGURE 37.33 Geometry for deriving the mirror equation.

Use the two right triangles highlighted in yellow in Figure 37.33 to write expressions for $\tan\theta$.	$\tan\theta = \frac{\lvert h_o\rvert}{\lvert d_o\rvert} = \frac{\lvert h_i\rvert}{\lvert d_i\rvert}$	(1)
Rearrange Equation (1) to come up with an expression for the absolute value of the magnification: $\lvert M\rvert = \lvert h_i\rvert/\lvert h_o\rvert$.	$\lvert M\rvert = \frac{\lvert h_i\rvert}{\lvert h_o\rvert} = \frac{\lvert d_i\rvert}{\lvert d_o\rvert}$	(2)
The **magnification of a spherical mirror** is found by applying the first two sign conventions (Table 37.2) to Equation (1). The image height and object height are both positive because both object and image are upright, so the magnification must be positive. The image is virtual, so the image distance is negative. We must insert a negative sign in Equation (1) to be consistent with this convention.	$M = \frac{h_i}{h_o} = -\frac{d_i}{d_o}$ ✓	(37.7)

Now consider the right triangle highlighted in blue. The horizontal leg of this triangle has length $|f| - |d_i|$. Use this to find an expression for $\tan\alpha$.

$$\tan\alpha = \frac{|h_i|}{|f| - |d_i|} \qquad (3)$$

For a small spherical mirror (compared to the radius of curvature), point A where ray 3 strikes the mirror is almost directly above the vertex V (Fig. 37.34). Thus, AVF is approximately a right triangle.

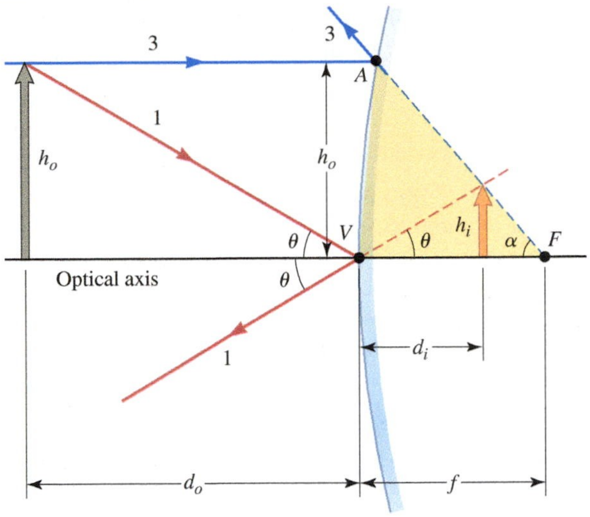

FIGURE 37.34 For a small spherical mirror, the distance from A to V is approximately the height of the object h_o.

Right triangle AVF highlighted in yellow yields another expression for $\tan\alpha$.

$$\tan\alpha = \frac{|h_o|}{|f|} \qquad (4)$$

Set Equation (3) equal to Equation (4) to find another expression for the absolute value of the magnification, $|M|$.

$$\tan\alpha = \frac{|h_i|}{|f| - |d_i|} = \frac{|h_o|}{|f|}$$

$$|M| = \frac{|h_i|}{|h_o|} = \frac{|f| - |d_i|}{|f|}$$

Use Equation (2) to eliminate the object and image heights.

$$\frac{|d_i|}{|d_o|} = \frac{|h_i|}{|h_o|} = \frac{|f| - |d_i|}{|f|} = 1 - \frac{|d_i|}{|f|}$$

Divide both sides by the image distance.

$$\frac{1}{|d_o|} = \frac{1}{|d_i|} - \frac{1}{|f|}$$

$$\frac{1}{|d_o|} - \frac{1}{|d_i|} = -\frac{1}{|f|} \qquad (5)$$

Now apply the second and third sign conventions (Table 37.2) to Equation (5). The object distance is positive because the object is real (in front of the mirror), but the image distance is negative because the image is virtual (behind the mirror). The focal length is also negative because the mirror is convex.

$$\frac{1}{d_o} + \frac{1}{d_i} = \frac{1}{f} \qquad (37.8)$$

COMMENTS

Equations 37.7 and 37.8, along with the sign conventions in Table 37.2, hold for all mirrors—even plane ones as shown in the next example. In applying geometric optics equations such as these, we can use any convenient length units because these equations do not involve fundamental constants.

CONCEPT EXERCISE 37.4

Show that all four primary rays produce the same results as those in Figures 37.32 and 37.33.

EXAMPLE 37.5 A Flat Mirror Is a Spherical Mirror

One way to think of a flat mirror is as a special case of a spherical mirror in which the radius of curvature is infinite. Start with the mirror equation and show that, in the limit where $r \to \infty$, the mirror equation is equal to $d_i = -d_o$ (Eq. 37.4) and the magnification is given by Equation 37.5:

$$M = \frac{h_i}{h_o} = -\frac{d_i}{d_o} = 1$$

INTERPRET and ANTICIPATE

Take a moment to see whether the problem makes sense. Imagine a mirror as a patch on a very large sphere (Fig. 37.26), and imagine that sphere growing in size while the size of the patch remains the same. As the sphere grows larger, the patch gets flatter. So, as $r \to \infty$ for a spherical mirror, it becomes a plane mirror.

SOLVE

Because a flat mirror does not have an inside and an outside, as a spherical mirror does, we can start with either a convex or a concave mirror. We'll choose the concave mirror arbitrarily. By Equation 37.6, as the radius of curvature approaches infinity, so does the focal length.

$$f = \frac{r}{2} \quad (37.6)$$

$$\lim_{r\to\infty}\left(f = \frac{r}{2}\right) \to \infty$$

$$f \to \infty$$

Take the limit of the mirror equation (37.8) as $f \to \infty$. The term $1/f$ goes to 0, and the image distance is the negative of the object distance (Eq. 37.4), as for a flat mirror.

$$\lim_{r\to\infty}\left(\frac{1}{d_o} + \frac{1}{d_i} = \frac{1}{f}\right) \to 0$$

$$\frac{1}{d_o} + \frac{1}{d_i} = 0, \text{ so } \frac{1}{d_i} = -\frac{1}{d_o}$$

$$d_i = -d_o \checkmark \quad (37.4)$$

The magnification follows from substituting $d_i = -d_o$ into Equation 37.2. As expected, the magnification is 1.

$$M = \frac{h_i}{h_o} = -\frac{d_i}{d_o} \quad (37.2)$$

$$M = -\frac{-d_o}{d_o} = 1 \checkmark$$

CHECK and THINK

We have shown that a flat mirror is a special case of a spherical mirror, with infinite radius of curvature and focal length. Here's a follow-up question: If an object is infinitely far from a flat mirror, where does its image form? Because the object is very far from the flat mirror, its rays are parallel. The reflected rays cross at the focal point, which is located at infinity. So, the image of an infinitely far object is located at infinity.

EXAMPLE 37.6 Find the Car's Image and Magnification

A car is 3.00 m from a convex mirror in a parking garage. The mirror's radius of curvature is 8.00 m.

A Find the image distance and the magnification. Is the image upright or inverted? Is the image real or virtual?

INTERPRET and ANTICIPATE

To anticipate our results, consider either Figure 37.33 or Figure 37.34. According to these figures, we expect the image to be behind the mirror, upright, and virtual. We'll use the mirror equation and the magnification equation to get exact results, and then confirm our expectations by using the sign conventions.

SOLVE

Find the focal length of this mirror from the radius of curvature (Eq. 37.6). Because the mirror is convex, the radius of curvature and the focal length are negative (Table 37.2).

$$f = \frac{r}{2} \tag{37.6}$$

$$f = \frac{-8.00\text{ m}}{2} = -4.00\text{ m}$$

Use the mirror equation (Eq. 37.8) to find the image distance. Be careful; the answer we want is often in the denominator. Solve for $1/d_i$ and then take the inverse (reciprocal). From the sign convention in Table 37.2, the object distance d_o is positive because the object is real. We get a negative image distance (image is virtual).

$$\frac{1}{d_o} + \frac{1}{d_i} = \frac{1}{f} \tag{37.8}$$

$$\frac{1}{d_i} = \frac{1}{f} - \frac{1}{d_o} = -\frac{1}{4.00\text{ m}} - \frac{1}{3.00\text{ m}} = -\frac{7.00}{12.0\text{ m}}$$

$$d_i = -\frac{12.0\text{ m}}{7.00} = -1.71\text{ m}$$

To find the magnification, use Equation 37.7. The positive result tells us the image is upright.

$$M = -\frac{d_i}{d_o} \tag{37.7}$$

$$M = -\frac{-1.71\text{m}}{3.00\text{m}} = 0.571$$

CHECK and THINK

Our calculations are consistent with Figures 37.33 and 37.34. The image is smaller than the object (about 60% of the object's size). This result matches the common experience of looking at images in convex mirrors (Fig. 37.35). The image of the car is smaller than the arrow sign below the mirror.

Maciej Figiel/Alamy

FIGURE 37.35 A convex mirror allows you to see traffic around the corner.

B How do your answers change if the car is 5.00 m from the mirror?

INTERPRET and ANTICIPATE

The only difference from part A is that the car is farther from the mirror. So, our expectations are the same: negative image distance and positive magnification.

SOLVE

This is the same mirror with the same focal length as in part A; simply substitute the new object distance.

$$\frac{1}{d_i} = \frac{1}{f} - \frac{1}{d_o} = -\frac{1}{4.00\text{ m}} - \frac{1}{5.00\text{ m}} = -\frac{9.00}{20.0\text{ m}}$$

$$d_i = -\frac{20.0\text{ m}}{9.00} = -2.22\text{ m}$$

Find the magnification from Equation 37.7 using the new object and image distances.

$$M = -\frac{d_i}{d_o} \tag{37.7}$$

$$M = -\frac{-2.22\text{m}}{5.00\text{m}} = 0.444$$

Example continues on page 1204 ▶

CHECK and THINK

As expected, the image is still behind the mirror ($d_i < 0$) and upright ($M > 0$). In part A, the car's distance to the mirror (3.00 m) is less than the focal length (4.00 m). In this part, the car's distance to the mirror (5.00 m) is greater than the focal length. When the car is closer, its image is closer and its magnification is greater.

C How do your answers change if the car is 4.00 m from the mirror?

INTERPRET and ANTICIPATE

In this part, the object distance is equal to the focal length of the mirror. This problem allows us to see whether this special case produces special results.

SOLVE

Substitute the new value of the object distance.

$$\frac{1}{d_i} = \frac{1}{f} - \frac{1}{d_o} = -\frac{1}{4.00\text{ m}} - \frac{1}{4.00\text{ m}} = -\frac{2.00}{4.00\text{ m}}$$

$$d_i = -\frac{4.00\text{ m}}{2.00} = -2.00\text{ m}$$

Use Equation 37.7 again to find the magnification.

$$M = -\frac{d_i}{d_o} \quad (37.7)$$

$$M = -\frac{-2.00\text{ m}}{4.00\text{ m}} = 0.500$$

CHECK and THINK

These results are interesting. When the object distance equals the focal length ($d_o = f$), the image distance is half the object distance ($d_i = 0.5d_o$) and the magnification $M = 0.5$. Combining these results with those of parts A and B, we can say that when the object distance is less than the focal length ($d_o < f$), the image distance is more than half the object distance ($d_i > 0.5d_o$) and $M > 0.5$. When the object distance is greater than the focal length ($d_o > f$), the image distance is less than half the object distance ($d_i < 0.5d_o$) and $M < 0.5$. You can add these observations to your list of expectations when solving problems involving a convex mirror.

37-6 Images Formed by Concave Mirrors

We found that images formed by plane and convex mirrors are behind the mirror, virtual, and upright. Concave mirrors are more complicated because the image may be behind the mirror, virtual, and upright, or it may be in front of the mirror, real, and inverted. The type of image depends on the position of the object. We must consider three possible positions for the object relative to the mirror: (1) closer than the focal point, (2) farther than the focal point, and (3) at the focal point.

The four primary rays for convex mirrors also apply to concave mirrors. Also, three equations

$$f = \frac{r}{2} \quad (37.6)$$

$$M = \frac{h_i}{h_o} = -\frac{d_i}{d_o} \quad (37.7)$$

and the mirror equation

$$\frac{1}{d_o} + \frac{1}{d_i} = \frac{1}{f} \quad (37.8)$$

hold for concave mirrors. The center of curvature C and the focal point F are both in front of the mirror (Fig. 37.28B). So, according to the third sign convention (Table 37.2), the radius of curvature and the focal length are both positive: $r > 0$ and $f > 0$.

Concave Mirror: Object Closer than the Focal Point

In Figure 37.36, the object distance is less than the mirror's focal length: $d_o < f$. We can use any two of the four primary rays to find the image formed by the mirror. We arbitrarily choose rays 1 and 3. Their reflected rays do not intersect in front of the mirror, but if we extend them backward, they intersect behind the mirror. The place where they intersect locates the tip of the image. The image is behind the mirror, virtual, and upright. However, unlike what happens with plane or convex mirrors, the image is larger than the object, so $M > 1$. Also, the image distance is greater than the object distance, $d_i > d_o$.

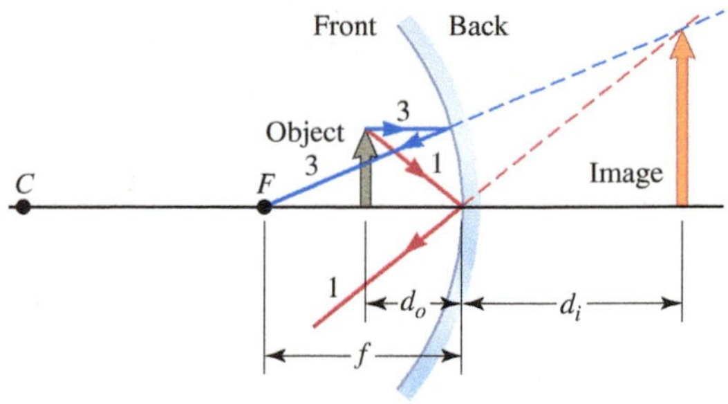

FIGURE 37.36 Ray 1 and ray 3 are used to find the image produced by a concave mirror when the object is closer than the focal point F. Ray 1 strikes the vertex; ray 1 and its reflection are symmetrical around the optical axis. Ray 3 is parallel to the optical axis. Its reflection passes through the focal point F. The reflected rays are traced backward, locating a virtual and upright image behind the mirror.

EXAMPLE 37.7 Comparing Concave and Convex Mirrors, Part 1

Find the image distance and the magnification for a concave mirror (Fig. 37.36) with focal length $f = 4.00$ m. The object distance $d_o = 3.00$ m. These are the same values from Example 37.6, part A, so compare your results here to the results from that example.

INTERPRET and ANTICIPATE

As in Example 37.6, apply the mirror equation and the magnification equation in conjunction with the sign conventions.

SOLVE

Use the mirror equation (Eq. 37.8) to find the image distance. The focal length is positive because the mirror is concave, and the object distance is positive because the object is real (Table 37.2).

$$\frac{1}{d_o} + \frac{1}{d_i} = \frac{1}{f} \tag{37.8}$$

$$\frac{1}{d_i} = \frac{1}{f} - \frac{1}{d_o} = \frac{1}{4.00 \text{ m}} - \frac{1}{3.00 \text{ m}} = -\frac{1.00}{12.0 \text{ m}}$$

$$d_i = -12.0 \text{ m}$$

Because the image distance is negative, the image is virtual and behind the mirror. The magnitude of the image distance is greater than that of the object distance. Compare this result to those for plane and convex mirrors. For a plane mirror, the image distance and the object distance have the same magnitude. For a convex mirror, the magnitude of the image distance is smaller than that of the object distance. In Example 37.6A, we found $d_i = -1.71$ m, so the image was close to that convex mirror. By contrast, the image is far from the concave mirror in this example.

Find the magnification using $M = -d_i/d_o$ (Eq. 37.7).

$$M = -\frac{d_i}{d_o} = -\frac{-12.0 \text{ m}}{3.00 \text{ m}} = 4.00$$

CHECK and THINK

The magnification is positive, so the image is upright. The magnification is greater than 1, so the image is larger than the object. Contrast this with Example 37.6, part A, in which $M = 0.571$. The convex mirror produced a small image, but this concave mirror produces an image four times larger than the object. This magnification is why a concave mirror is useful as a shaving mirror, so the shaver can see a larger image of himself.

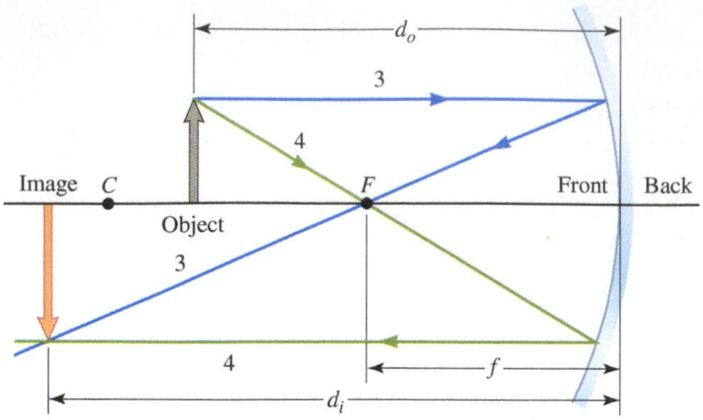

FIGURE 37.37 Ray 3 and ray 4 are used to find the image produced by a concave mirror when the object distance is greater than the focal length. Ray 3 is parallel to the optical axis. Its reflection passes through the focal point F. Ray 4 passes through the focal point on its way to the mirror. Its reflection is parallel to the optical axis. These reflected rays cross in front of the mirror, creating an image that is in front of the mirror, real, and inverted.

Concave Mirror: Object Farther than the Focal Point

In Figure 37.37, the object distance is greater than the mirror's focal length: $d_o > f$. Again, we can use any two of the four primary rays to find the image formed by the mirror. We have arbitrarily chosen rays 3 and 4. Unlike the rays reflected by a flat mirror, these reflected rays *do* meet at a point in front of the mirror. Because the rays converge in front of the mirror, it is possible to project the image on a screen, so the image is *real*. The place where the rays intersect locates the tip of the image. This point is below the optical axis, so the image is *inverted*. These results contrast with those for plane and convex mirrors.

Also, the image in Figure 37.37 is larger than the object, $|M| > 1$, but this is not always the case for real images formed by concave mirrors. Whether $|M|$ is greater or less than 1 depends on the location of the object relative to the center of curvature. When the object is between the center of curvature C and the focal point F (Fig. 37.37), its image is larger than the object ($|M| > 1$). When the object is farther than the center of curvature, however, the image is smaller than the object ($|M| < 1$). We can see this from Figure 37.37 by imagining that the arrow labeled *Image* is actually the object. Then just reverse the arrowheads on the rays to find the location of the image. The image will be smaller than the object ($|M| < 1$).

EXAMPLE 37.8 Comparing Concave and Convex Mirrors, Part 2

Find the image distance and the magnification for a concave mirror (Fig. 37.37) with focal length $f = 4.00$ m and radius of curvature $r = 8.00$ m. The object distance is $d_o = 5.00$ m, so the object is between the focal point and the center of curvature. These are the same values used in Example 37.6, part B, so compare your results here to those results.

INTERPRET and ANTICIPATE

As in Example 37.7, apply the mirror equation and the magnification equation in conjunction with the sign conventions, and compare your results to those in Example 37.6.

SOLVE

Use the mirror equation (Eq. 37.8) to find the image distance. The focal length is positive because the mirror is concave, and the object distance is positive because the object is real (Table 37.2).

$$\frac{1}{d_o} + \frac{1}{d_i} = \frac{1}{f} \quad (37.8)$$

$$\frac{1}{d_i} = \frac{1}{f} - \frac{1}{d_o} = \frac{1}{4.00\text{ m}} - \frac{1}{5.00\text{ m}} = \frac{1.00}{20.0\text{ m}}$$

$$d_i = 20.0\text{ m}$$

CHECK and THINK

Because the image distance is positive, the image is real and in front of the mirror. Also, the image distance is greater than the object distance. In Example 37.6, part B, we found $d_i = -2.22$ m, so the image was close to and behind that convex mirror. In this example, where the object is between the focal point and the center of curvature of a concave mirror, the image is far from and in front of that mirror.

SOLVE

Now, let's find the magnification using $M = -d_i/d_o$ (Eq. 37.7).

$$M = -\frac{d_i}{d_o} = -\frac{20.0\text{ m}}{5.00\text{ m}} = -4.00$$

CHECK and THINK

The magnification is negative, so the image is inverted. The absolute value of the magnification is greater than 1, so the image is larger than the object. Contrast this with Example 37.7, in which we found $M = 4.00$ also for a concave mirror. In that case, however, the object was closer to the mirror than the focal point. In this case, the object is farther from the mirror than the focal point, and although the image is four times larger than the object, this mirror would not be particularly good for shaving because the image is inverted.

EXAMPLE 37.9 Comparing Concave and Convex Mirrors, Part 3

Find the image distance and the magnification for a concave mirror with focal length $f = 4.00$ m and radius of curvature $r = 8.00$ m. The object distance is $d_o = 9.00$ m, so the object is to the left of the center of curvature in Figure 37.37.

INTERPRET and ANTICIPATE

As in Example 37.8, apply the mirror equation and the magnification equation in conjunction with the sign conventions. But now imagine the rays in Figure 37.37 are reversed, so the image should be real, in front of the mirror, and inverted. The image will be smaller than the object.

SOLVE

Use the mirror equation (Eq. 37.8) to find the image distance. The focal length is positive because the mirror is concave, and the object distance is positive because the object is real (Table 37.2).

$$\frac{1}{d_o} + \frac{1}{d_i} = \frac{1}{f} \tag{37.8}$$

$$\frac{1}{d_i} = \frac{1}{f} - \frac{1}{d_o} = \frac{1}{4.00\text{ m}} - \frac{1}{9.00\text{ m}} = \frac{5.00}{36.0\text{ m}}$$

$$d_i = 7.20\text{ m}$$

CHECK and THINK

As expected, the image distance is positive, so the image is real and in front of the mirror. In this case, the image distance is less than the object distance.

SOLVE

Find the magnification using $M = -d_i/d_o$ (Eq. 37.7).

$$M = -\frac{d_i}{d_o} = -\frac{7.20\text{ m}}{9.00\text{ m}} = -0.800$$

CHECK and THINK

The magnification is negative, so as expected, the image is inverted. The absolute value of the magnification is less than 1. So, when the object is farther away than the center of curvature of a concave mirror, the image is smaller than the object.

Concave Mirror: Object at the Focal Point

Finally, what can we say about the image when the object is exactly at the focal point of a concave mirror? To answer this question, start with the mirror equation, with $d_o = f$:

$$\frac{1}{d_i} = \frac{1}{f} - \frac{1}{d_o} = \frac{1}{f} - \frac{1}{f} = 0$$

When the object is at the focal point of the concave mirror, the image distance is infinite:

$$d_i \to \infty$$

To understand what this result means for the image, let's review how images are formed. Each point of an object emits rays in all directions. A real image is formed when those rays from each point converge on a detector such as a screen, so that each point of the object is represented by a point in the image. Likewise, a virtual image is composed of points from which rays appear to diverge, and which correspond to points in the object. For an image to form, the reflected rays from a single point must either meet or appear to meet, and their intersection is located at d_i. So, when the object is located at the focal point of a concave mirror, its image is located at infinity, $d_i \to \infty$, which means the reflected rays from each point meet at infinity. But rays that "meet at infinity" are really parallel rays; they don't meet at a finite distance, and so no image forms.

CONCEPT EXERCISE 37.5

Fill in the table by inserting a greater than (>), less than (<), or equals (=) sign in each blank.

Plane mirror	$\|d_i\|$ ___ $\|d_o\|$	M ___ 0	$\|M\|$ ___ 1
Convex mirror with $d_o < \|f\|$	$\|d_i\|$ ___ $\|d_o\|$	M ___ 0	$\|M\|$ ___ 1
Convex mirror with $d_o > \|f\|$	$\|d_i\|$ ___ $\|d_o\|$	M ___ 0	$\|M\|$ ___ 1
Concave mirror with $d_o < \|f\|$	$\|d_i\|$ ___ $\|d_o\|$	M ___ 0	$\|M\|$ ___ 1
Concave mirror with $\|r\| > d_o > \|f\|$	$\|d_i\|$ ___ $\|d_o\|$	M ___ 0	$\|M\|$ ___ 1
Concave mirror with $d_o > \|r\|$	$\|d_i\|$ ___ $\|d_o\|$	M ___ 0	$\|M\|$ ___ 1

CONCEPT EXERCISE 37.6

CASE STUDY Why Use a Secondary Mirror?

As shown in Figure 37.38, reflecting telescopes usually have two mirrors—the primary and the secondary. Light from an astronomical object strikes the primary mirror. The reflected rays then strike the secondary mirror. In the case of a telescope like the HST, rays from the secondary mirror pass through a hole in the primary mirror to an instrument such as a digital camera. The focal length of the HST's primary mirror is 57.6 m. If the secondary mirror were removed, where would the images of distant stars form? The length of the HST's tube is 13.2 m. Explain the statement "A secondary mirror allows the telescope to be compact."

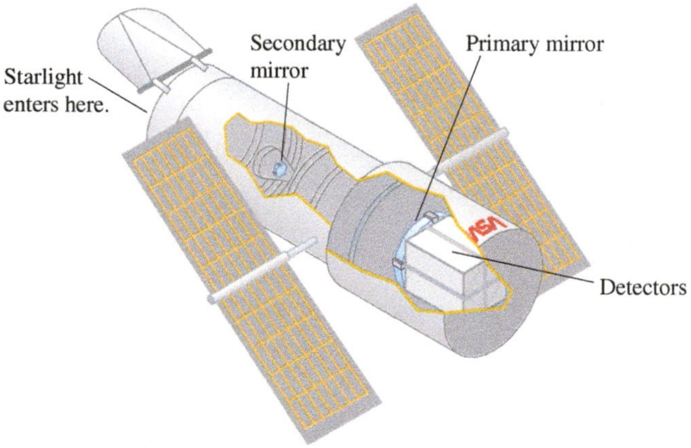

FIGURE 37.38 Diagram of the Hubble Space Telescope.

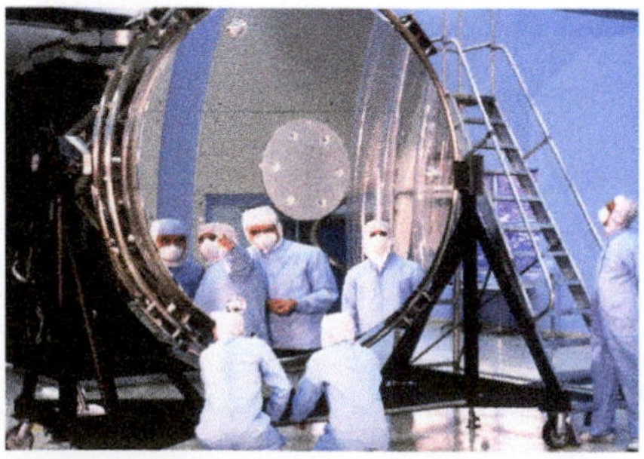

NASA/STSCI (Space Telescope Science Institute)

FIGURE 37.39 The HST's primary mirror has a hole in the center. It has been covered in this photo.

CONCEPT EXERCISE 37.7

CASE STUDY Where Are Those People?

Figure 37.39 shows the HST's primary mirror in a clean room before the telescope was launched. You can see an image of four people in the mirror, but you cannot see them in the room. Roughly how far are they from the mirror? Is their image virtual or real? In this exercise, you may model the HST's mirror as concave and spherical.

EXAMPLE 37.10 CASE STUDY Hubble's Image of the Moon

In April 1999, the HST took a picture of the Moon (Fig. 37.40). The large impact crater in the photo is called Copernicus. Use the information in Concept Exercise 37.6 to find the image distance and magnification, given also the Earth-Moon distance 3.82×10^8 m. The diameter of the Copernicus crater is about 93 km. What is the diameter of its image? In this problem, model the HST as a concave spherical mirror. To check your results, answer these questions: Is the image upright or inverted? Is it real or virtual?

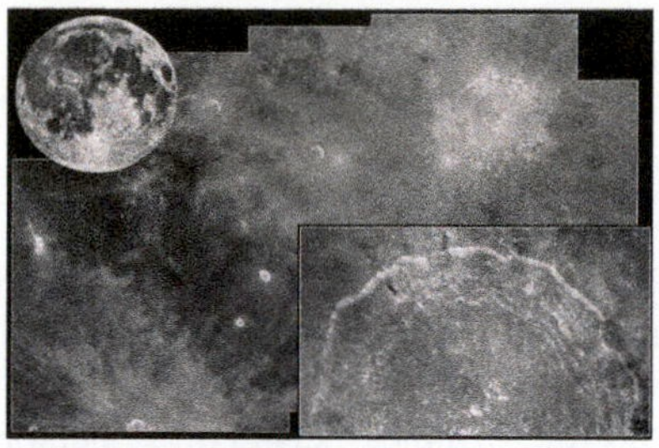

FIGURE 37.40 An HST image of the Moon. The large crater is called Copernicus. The photo of the whole Moon was taken by a ground-based telescope. The lower right image is a close-up of Copernicus.

INTERPRET and ANTICIPATE

Figure 37.37 is a good ray diagram for this situation because the Moon is very far from the focal point. This diagram shows that the image is inverted and real. So we expect that the magnification is negative and the image distance is positive.

SOLVE

The object distance is roughly the distance between the Earth and the Moon. The object distance is positive because the object is real. The focal length is 57.6 m (Concept Exercise 37.6). The focal length is positive because the mirror is concave.

$$d_o = 3.82 \times 10^8 \text{ m}$$

$$f = 57.6 \text{ m}$$

When using the mirror equation (Eq. 37.8) for a distant object, we find the image distance is roughly equal to the focal length. This makes sense because the rays from the Moon are roughly parallel, so they focus at the focal point.

$$\frac{1}{d_o} + \frac{1}{d_i} = \frac{1}{f} \quad (37.8)$$

$$\frac{1}{d_i} = \frac{1}{f} - \frac{1}{d_o} = \frac{1}{57.6 \text{ m}} - \frac{1}{3.82 \times 10^8 \text{ m}} \approx \frac{1}{57.6 \text{ m}}$$

$$d_i = 57.6 \text{ m}$$

CHECK and THINK

As expected, the image distance is positive, so the image is real and in front of the mirror. A telescope requires a real image in order to record that image as a photograph. The image must be projected onto a film or digital detector.

SOLVE

Find the magnification from Equation 37.7.

$$M = -\frac{d_i}{d_o} \quad (37.7)$$

$$M = -\frac{57.6 \text{ m}}{3.82 \times 10^8 \text{ m}} = -1.51 \times 10^{-7}$$

CHECK and THINK

As expected, the magnification is negative, so the image is inverted.

SOLVE

To find the size of the crater's image, use Equation 37.7 again, but as the ratio of image height to object height.

$$M = \frac{h_i}{h_o} \quad (37.7)$$

$$h_i = h_o M = (93 \text{ km})(-1.51 \times 10^{-7})$$

$$h_i = -1.4 \times 10^{-5} \text{ km} = -1.4 \text{ cm}$$

CHECK and THINK

The negative image height means the image is inverted. And, of course, it makes sense that the crater's image is much smaller than the crater.

EXAMPLE 37.11 Cosmetic Mirror

A mirror manufacturer reports that its cosmetic mirror has a magnification of 5×. Estimate the radius of curvature and the size of the image of a person's eye.

INTERPRET and ANTICIPATE

The mirror must be concave because neither a plane mirror nor a convex mirror produces an image that is larger than the object. For the image to be useful, it must be upright. So the person's face must be closer to the mirror than the focal point (Fig. 37.36). The key to this problem is estimating the distance between the person's face and the mirror. How close is your face to a shaving or cosmetic mirror? Perhaps 5–10 in. Let's say about 7 in. (18 cm).

SOLVE

We have estimated the object distance $d_o = 18$ cm. Use the magnification equation (Eq. 37.7) to find the image distance.

$$M = -\frac{d_i}{d_o} \quad (37.7)$$

$$d_i = -Md_o = -(5)(18 \text{ cm}) = -90 \text{ cm}$$

CHECK and THINK

The image distance is negative, as expected for a virtual image. The image is farther from the mirror than the object, consistent with Figure 37.36.

SOLVE

Because we know the object distance and the image distance, we can find the focal length of the mirror from the mirror equation (Eq. 37.8).

$$\frac{1}{d_o} + \frac{1}{d_i} = \frac{1}{f} \quad (37.8)$$

$$\frac{1}{f} = \frac{1}{18 \text{ cm}} + \frac{1}{-90 \text{ cm}}$$

$$f = 22.5 \text{ cm}$$

The radius of curvature is twice the focal length (Eq. 37.6). As with all estimates, we report our result to one significant figure.

$$f = \frac{r}{2} \quad (37.6)$$

$$r = 2f = 2(22.5 \text{ cm}) = 50 \text{ cm}$$

To estimate the size of the image of an eye, start by estimating the size of your eye. Perhaps your eye is about an inch in diameter, or about 2 cm. Use Equation 37.7 (image height to object height) to find the size of the eye's image.

$$M = \frac{h_i}{h_o} \quad (37.7)$$

$$h_i = Mh_o = (5)(2 \text{ cm}) = 10 \text{ cm}$$

CHECK and THINK

Typically a cosmetic mirror is about 20 cm tall, so if it has a radius of curvature of 50 cm, the mirror has the shape of a very shallow bowl. Such a mirror would make an image of your eye that is 10 cm in diameter, or about half the size of your hand. An image that size is useful for applying makeup. The image of your eye in such a mirror is larger than the 1.4-cm image of the Copernicus crater produced by the HST. Unlike a cosmetic mirror, a telescope is designed to gather light, not to produce an enlarged image.

37-7 Spherical Aberration

In the past three sections, we considered only spherical mirrors that are small compared to their radii of curvature: $D \ll r$. We even modeled the HST's primary mirror as a small spherical mirror. In this section, we show that a large spherical mirror produces a blurry image and that the HST's mirror cannot always be modeled as a small spherical mirror.

Figure 37.41 shows two spherical mirrors—one concave and the other convex. Each is larger than its radius of curvature ($D > r$). Consider an object such as a star that is very far away so that its rays are parallel. The rays close to the optical axis come together in front of the concave mirror at the focal point (Fig. 37.41A). If the mirror were small ($D \ll r$), all the parallel rays that struck the mirror would be reflected through the focal point. However, this mirror is large, so reflections of the outer rays do not cross at the focal point, but instead cross at a point slightly closer to the mirror. Rays don't end at the point where they cross. If you place a small screen at the focal point to capture the image of a distant star, you see the light from the inner rays that cross at a single point, and you also see light from the outer rays above and below this point. Such a screen is in the **focal plane**, perpendicular to the optical axis and passing through the focal point. A distant star is a point source of light, and ideally its image on the focal plane should be a dot. But the image produced by a large concave mirror is blurry because it includes light from the outer rays.

Large convex mirrors also produce blurry images. Again, the rays from a distant object are parallel, and the inner rays seem to diverge from the focal point behind the mirror. But the outer rays seem to diverge from a point slightly closer to the mirror (Fig. 37.41B). If you look at the virtual image in this mirror, it also looks blurry.

A defect in an optical system that results in a poor image is called **aberration. Spherical aberration** results from rays far from the optical axis, or rays that make a large angle with the optical axis of a spherical mirror. Such rays strike points on the mirror that are far from its vertex. **Paraxial rays** are those that are nearly parallel and close to the optical axis, striking near the vertex of the mirror. We considered only small mirrors in earlier sections because any rays that strike a small mirror ($D \ll r$) must be paraxial.

For many practical applications, spherical aberration is avoided by making spherical mirrors small compared to their radii of curvature, such as a cosmetic mirror (Example 37.11). However, small spherical mirrors still suffer from some spherical aberration. In the case of a cosmetic or parking garage mirror, the aberration does not diminish the mirror's usefulness. But even a little spherical aberration can be a major problem in some applications, such as in astronomy. Astronomy requires the best images possible, so telescopes are designed to avoid all aberration, including spherical aberration.

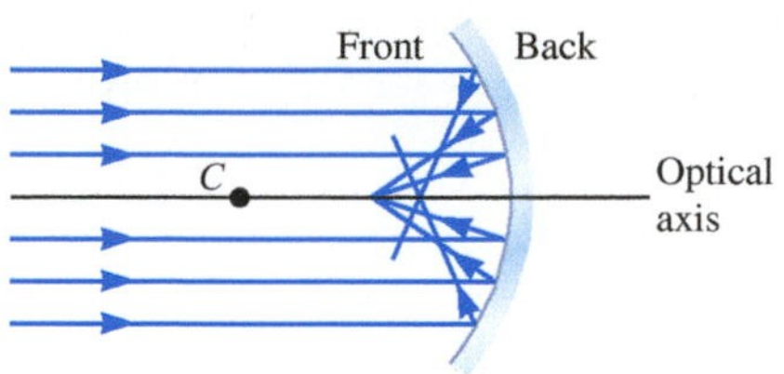

A. Concave mirror

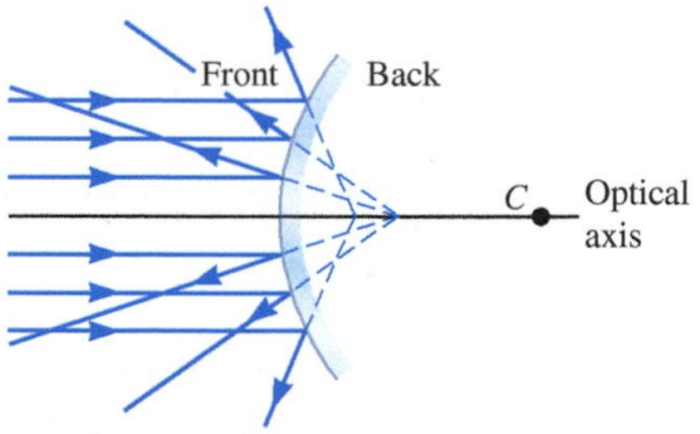

B. Convex mirror

FIGURE 37.41 **A.** Parallel rays incident on a concave mirror are focused in front of the mirror. The inner rays focus at a single point, but the outer rays focus at another point closer to the mirror. **B.** The rays reflected by a convex mirror appear to come from points behind the mirror. The inner rays appear to originate from one point, but the outer rays appear to originate from another point, closer to the mirror.

CASE STUDY Hubble's Primary Mirror

A reflecting telescope consists of two mirrors—the primary and the secondary (Fig. 37.38). The front of the primary mirror faces the sky and plays the most important role in gathering light from distant objects. For the moment, consider a special type of telescope design called a *prime focus*, which uses only a single (primary) mirror. Light from a distant object forms an image at the focal point of the mirror. The astronomer captures that image either on photographic film or on a sensor known as a charge-coupled device (CCD), like the sensor in a digital camera. Because the image must be projected on a film or sensor, the image must be real, so the primary mirror must be concave. If the primary mirror is spherical, the parallel rays do not form a single image in the focal plane.

However, if the primary mirror has a parabolic or hyperbolic shape, even the rays far from the optical axis come together at the focal point (Fig. 37.42). (A slice through a parabolic mirror is a parabola.) Astronomical telescopes often have parabolic or hyperbolic primary and secondary mirrors to avoid spherical aberration. But there are drawbacks to these mirrors. If the rays from a distant object are not parallel to the optical axis, they do not form a clear image at the focal point. When a star is at the center of a telescope's field, it forms a clear image on the focal plane. However, stars that are elsewhere than at the center of the field form distorted images. These images look like comets' tails and are known as *comas*, so this sort of aberration is called **coma.** Parabolic and hyperbolic mirrors are also difficult to make, which led to the problem with the HST's primary mirror.

The HST was designed to have two hyperbolic mirrors. The primary mirror was incorrectly shaped due to a testing procedure used during the manufacturing process. Its incorrect shape led to spherical aberration: Parallel rays did not come together at a single focal point, and the HST's images were blurry.

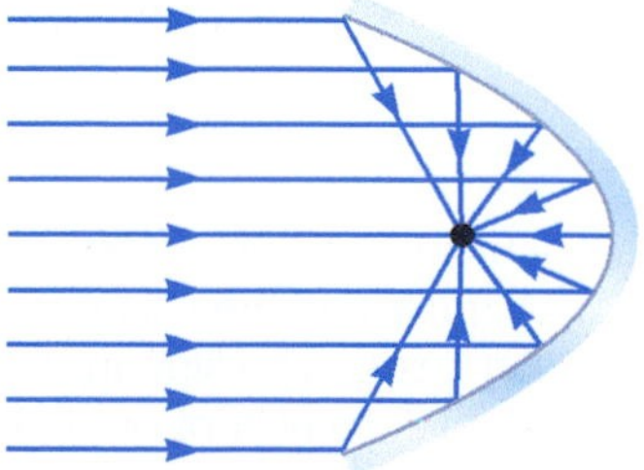

FIGURE 37.42 All rays focus at a single point for a parabolic mirror.

The problem was corrected by the addition of other optical systems. Once the actual shape of the primary mirror was measured, engineers designed a system of mirrors called COSTAR (*c*orrective *o*ptics *s*pace *t*elescope *a*xial *r*eplacement) to correct the problem. This system is like the corrective lenses given to a nearsighted person. COSTAR was installed in December 1993. Figure 37.13 shows that COSTAR was successful. There is an important lesson in this story: People make mistakes, but they can also correct those mistakes. Since 1993, the HST has contributed to major discoveries in our understanding of the Universe.

SUMMARY

Underlying Principles

Geometric optics: When light encounters only obstacles and apertures that are at least 1 mm in diameter, the ray model is valid. According to this model, light travels in straight lines represented by rays. The ray model ignores the wave properties of light.

Major Concepts

1. According to the **law of reflection**, the angle of incidence equals the angle of reflection:
$$\theta_i = \theta_r \tag{37.3}$$
The incident ray, the reflected ray, and the perpendicular (normal) line all lie in a single plane, called the plane of incidence (Fig. 37.7).
2. **Magnification** (also known as linear magnification) is given by
$$M = \frac{h_i}{h_o} = -\frac{d_i}{d_o} \tag{37.7}$$
where the sign conventions in Table 37.2 (page 1199) must be used.
3. Light rays from each point of an object converge to create each point of a **real image**, whereas light rays appear to diverge from each point of a **virtual image**. A real image can appear on a screen, but a virtual one cannot.
4. The **mirror equation** relates the image distance d_i to the object distance d_o and the focal length f of the mirror:
$$\frac{1}{d_o} + \frac{1}{d_i} = \frac{1}{f} \tag{37.8}$$
where the sign conventions in Table 37.2 must be used.

Special Cases

1. A **plane mirror** is a flat mirror that has an infinite radius of curvature and produces an upright and virtual image with $M = 1$ (Fig. 37.17). The image distance is $d_i = -d_o$ (Eq. 37.4).
2. A **convex spherical mirror** has a negative (finite) radius of curvature and produces an upright and virtual image (Fig. 37.32).
3. A **concave spherical mirror** has a positive (finite) radius of curvature. It produces either an upright, virtual image (Fig. 37.36) or an inverted, real image (Fig. 37.37).

Tools

The elements of a **ray diagram** (Fig. 37.15) are:
1. A thick line representing the **mirror** seen edge-on
2. An arrow representing the **object**
3. A small number of **emitted rays** coming from key points on the object, such as the tip or base of the arrow
4. **Reflected rays** for each emitted ray
5. An arrow representing the **image**

When drawing a ray diagram, you may find two or more of the four **primary rays** described in Table 37.1 to be particularly helpful.

PROBLEMS AND QUESTIONS

A = algebraic C = conceptual E = estimation G = graphical N = numerical

37-1 Geometric Optics

1. **N** A camera obscura is used to form an image of a distant object. If the object is 10.0 m away from the aperture and the magnification of the image is −0.15, what is the image distance?
2. **C** Because you should never stare directly into the Sun, pinhole cameras are very handy for observing solar eclipses. Come up with a simple, inexpensive design for a pinhole camera that you could use to observe an eclipse. Describe your design and how you would use the camera. A sketch may be helpful.
3. An image is formed from an object using a camera obscura. The image is one-fifth the height of the object and is oriented upside-down compared to the orientation of the object.
   a. **N** What is the magnification of the image? Answer with two significant figures.
   b. **A** Write an expression for the image distance in terms of the object distance. Answer with two significant figures.
4. **E** The image of a person seen through a camera obscura seems to be 1.5 ft tall. If the image distance is roughly 7.5 ft, estimate the distance to the person.
5. **N** We have all enjoyed sitting in the shade of a leafy tree. If the leaves of the shade tree are not too densely packed, light from the Sun is able to filter through the tiny openings created by overlapping leaves. These openings essentially form tiny pinholes that project potentially hundreds of circular images of the Sun onto the forest floor. If one of the projected images of the Sun you observe on the forest floor is 4.1 cm in diameter, how high above the ground in the tree canopy is the pinhole that projected this image? *Hint*: You may need to look up the diameter of the Sun and the distance from the Sun to the Earth.
6. **C** Figure P37.6 shows the image of a person seen through a camera obscura. Sketch the person as if viewed from the camera's location.

FIGURE P37.6

37-2 Law of Reflection

7. **C** In this chapter, we study the reflection of visible light. Have we ignored other parts of the electromagnetic spectrum because radiation at other frequencies does not reflect? If so, explain why other wavelength radiation cannot reflect. If not, give an example of other wavelength radiation reflecting from some surface.

Problems 8 and 9 are paired.

8. **A** Two mirrors are perpendicular as shown in Figure P37.8. A narrow beam of light is incident on one mirror. The angle of incidence is θ_i. Find an expression for the angle of reflection θ_r from the second mirror.
9. **N** Two mirrors are perpendicular as shown in Figure P37.8. A narrow beam of light is incident on one mirror. The angle of incidence is 22.3°. Find the angle of reflection from the second mirror.

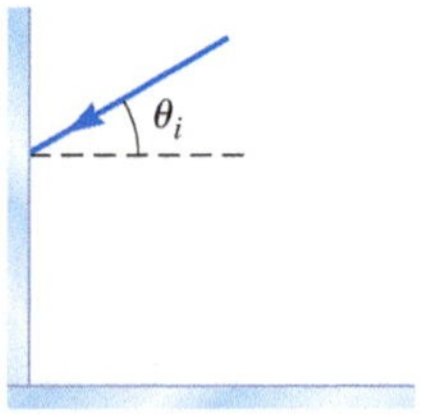

FIGURE P37.8 Problems 8 and 9.

10. **N** A light ray enters a region between two parallel mirrors at the angle shown in Figure P37.10. The mirrors are separated by 1.00 m and are 2.00 m long. How many times will the light ray be reflected before it exits the region between the mirrors?

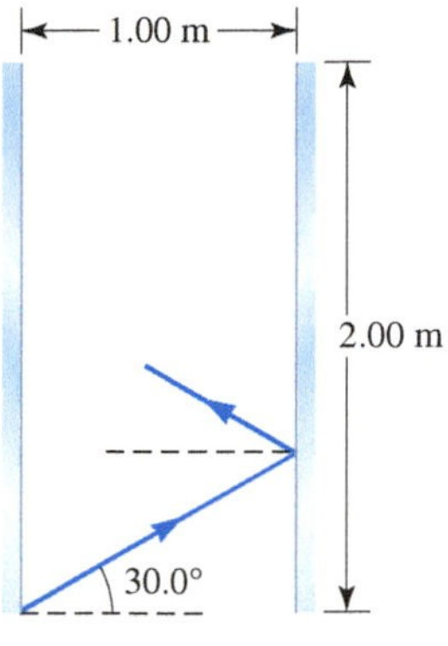

FIGURE P37.10

Problems 11 and 12 are paired.

11. **A** Two mirrors make a 45.0° angle as shown in Figure P37.11. A narrow beam of light is incident on one mirror. The angle of incidence is θ_i. Find an expression for the angle of reflection θ_r from the second mirror.

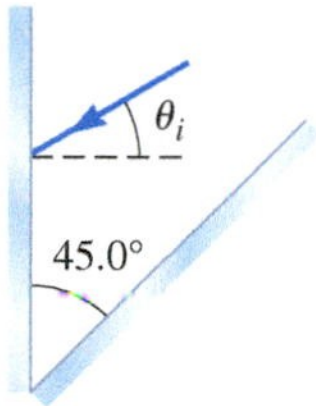

FIGURE P37.11 Problems 11 and 12.

12. **N** Two mirrors make a 45.0° angle as shown in Figure P37.11. A narrow beam of light is incident on one mirror. The angle of incidence is 22.3°. Find the angle of reflection from the second mirror.
13. **N** Figure P37.13 shows a beam of light incident on one of two parallel mirrors at an angle of 12.0°. How many total reflections will the beam undergo in the mirrors?

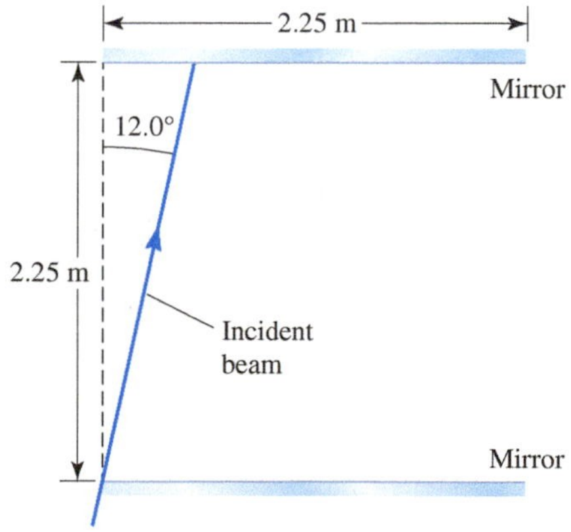

FIGURE P37.13

14. **N** One way to train a dog to stay when you leave the room is to use a mirror so that you can see the dog from outside the room. Many doors have shiny metal plates at the bottom that are convenient for this purpose (Fig. P37.14). At what angle should the door be placed with respect to the threshold in order for the trainer to see the dog? Use the dimensions given in the figure.

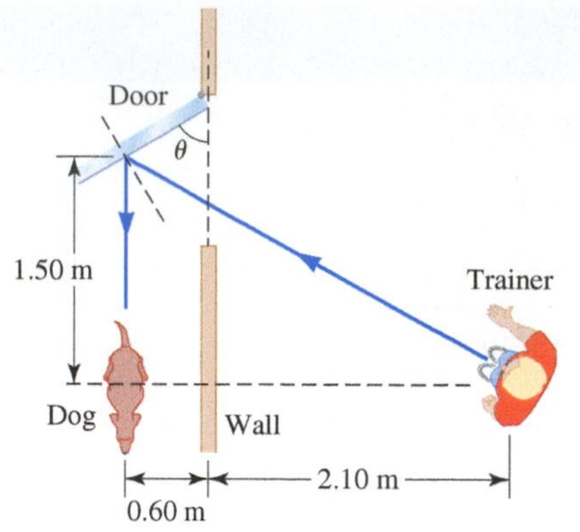

FIGURE P37.14

Problems 15 and 16 are paired.

15. **N** Light rays strike a plane mirror at an angle of 45.0° as shown in Figure P37.15. At what angle should a second mirror be placed so that the reflected rays are parallel to the first mirror?

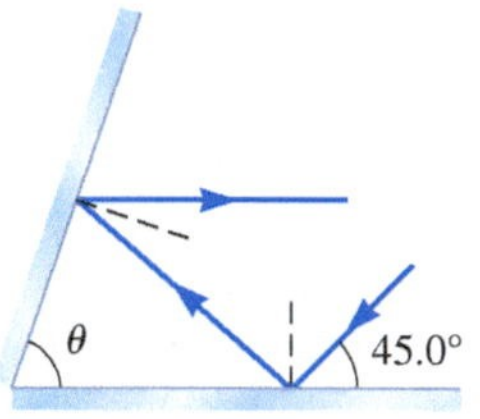

FIGURE P37.15

16. **N** Determine the angle between two plane mirrors such that a ray of light incident on the first mirror and parallel to the second mirror is reflected from the second mirror and emerges parallel to the first mirror as shown in Figure P37.16.

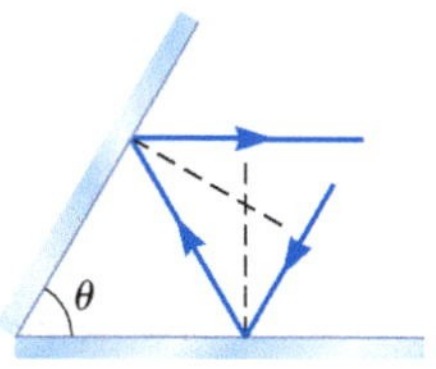

FIGURE P37.16

17. **N** A monochromatic beam of light enters a square enclosure with mirrored interior surfaces at an angle of incidence θ_i (Fig. P37.17). If each of the mirrored walls (other than the one with the opening) reflects the beam only once, what is the value of θ_i for which the beam will exit the enclosure through the same hole?

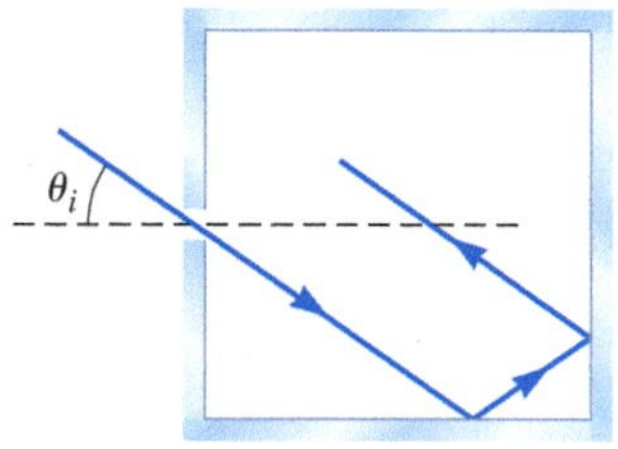

FIGURE P37.17

37-3 Images Formed by Plane Mirrors

18. **C** In Figure P37.18, you can see the painting reflected in the mirror. Why can't we see an image of the painting on the floor?

19. **N** You and your roommate share a plane mirror. She stands behind you so that her face is 18 in. behind your face. Your face is 3.5 ft from the mirror. How far does the image of your roommate's face appear to be from you? Give your answer in feet.

Dr. Jenny Spinner/St. Joseph's University

FIGURE P37.18

20. **G** Two perpendicular mirrors are used to produce an image that is not reversed (Fig. 37.21B). In Example 37.4 (page 1194), we used a ray diagram to show how this is possible. Repeat the example, but this time choose mirror 2 as the first mirror and mirror 1 as the second mirror.

21. **N** At a department store, you adjust the mirrors in the dressing room so that they are parallel and 5.0 ft apart. You stand 2.0 ft from one mirror and face it. You see an infinite number of reflections of your front and back.

a. How far from you is the first "front" image?

b. How far from you is the first "back" image?

22. **A** An object moves toward a plane mirror with speed v_p at an angle θ with respect to the normal to the interface as shown in Figure P37.22. Determine the relative velocity between the object and the image.

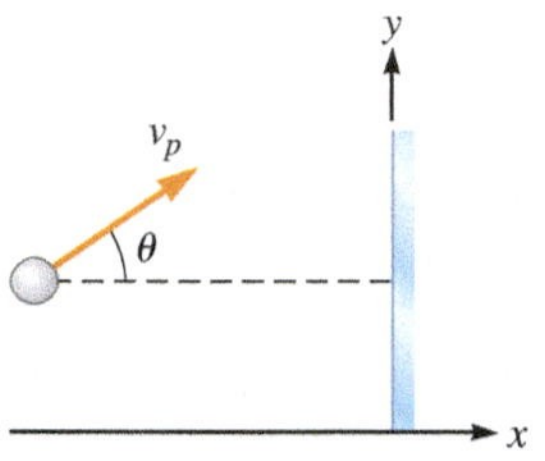

FIGURE P37.22

23. **N** A rectangular room used for ballet practice is 12.0 m long and 6.20 m wide. The ballet teacher has mounted a 1.45-m-wide mirror on one of the 12.0-m-long walls of the room so that she can at all times observe the students lined at the handrail along the opposite wall. At one instant, the teacher is facing away from the students, a distance of 1.10 m from the mirror. What is the extent of the line of students that the teacher can see while looking forward?

24. **A** Two perpendicular mirrors are used to produce a true image (Fig. P37.24). The distance from the tip of the arrow to mirror 1 is d_1, and the distance from the tip of the arrow to mirror 2 is d_2. Find an expression for the distance r between the tip of the arrow and its true image.

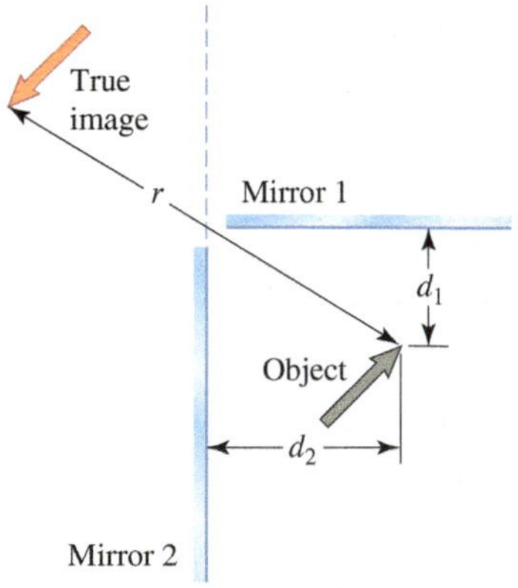

FIGURE P37.24

25. **N** Diane wants to buy a flat wall mirror in which she can see her full image. If she is 165 cm tall, what is the minimum height of the mirror she should purchase?

26. **N** A statue on a pedestal is placed in a room with flat mirrors on the parallel northern and southern walls, producing an infinite number of images in each mirror. If the walls are 7.00 m apart and the statue is 1.00 m from the northern wall, what are the distances from the statue to the first four images formed in the northern mirror?

37-4 Spherical Mirrors

27. **A** Derive $f = r/2$ (Eq. 37.6) for a concave spherical mirror.

37-5 Images Formed by Convex Mirrors

28. **N** An object is placed 30.0 cm in front of a convex mirror. If the focal length of the mirror is −20.0 cm, find the distance between the object and the final image.

29. **N** A convex mirror with a radius of curvature of −25.0 cm is used to form an image of an arrow that is 10.0 cm away from the mirror. If the arrow is 2.00 cm tall and inverted (pointing below the optical axis), what is the height of the arrow's image?

30. **N** The magnitude of the radius of curvature of a convex spherical mirror is 22.0 cm.

a. What are the location and magnification of the image of an object placed 16.0 cm in front of the mirror? Is this image upright or inverted?

b. What are the location and magnification of the image of an object placed 44.0 cm in front of the mirror? Is this image upright or inverted?

31. **N** When an object is placed 60.0 cm from a convex mirror, the image formed is half the height of the object. Where should the object be placed so that the height of the image becomes one-third the height of the object?

32. The image formed by a convex spherical mirror with a focal length of magnitude 12.0 cm is located one-fourth of the object–mirror distance from the mirror.

a. **N** What is the distance of the object from the mirror?

b. **C** Is the image formed by the mirror upright or inverted?

c. **N** What is the magnification of this image?

33. **N** An object is placed 25.0 cm from the surface of a convex mirror and an image is formed. The same object is placed 20.0 cm in front of a plane mirror and, again, an image is formed. Suppose in each case the object is located at $x = 0$ and the mirrors lie along the positive x axis. If the images formed in the two mirrors lie at the same x coordinate, determine the radius of curvature of the convex mirror.

34. At an ice cream shop, a cylindrical mirror is used to give customers the impression that they are taller and thinner than they really are.

a. **C** How should the cylinder be oriented? If the cylinder were tipped by 90°, how would the customers look?

b. **C** Is the image upright or inverted? Real or virtual?

c. **E** If the image of the customer's width is about 75% of the actual width, estimate the mirror's radius of curvature.

35. **N** A convex mirror magnifies an upright arrow so that the image of the arrow is one-third the height of the arrow. If the arrow is 35.0 cm away from the mirror, what is the focal length of the mirror?

37-6 Images Formed by Concave Mirrors

36. **A** Derive the mirror equation $1/d_o + 1/d_i = 1/f$ (Eq. 37.8) for a concave mirror.

37. **N** A dental hygienist uses a small concave mirror to look at the back of a patient's tooth. If the mirror is 1.33 cm from the tooth and the magnification is 2.00, what is the mirror's focal length?

38. **C** Come up with a procedure to find the focal length of a concave mirror, such as a spoon.

39. **N** The magnitude of the radius of curvature of a concave spherical mirror is 34.0 cm.

a. What are the location and magnification of the image of an object placed 60.0 cm in front of the mirror? Is this image real or virtual? Is it upright or inverted?

b. What are the location and magnification of the image of an object placed 34.0 cm in front of the mirror? Is this image real or virtual? Is it upright or inverted?

40. **A** Derive $M = h_i/h_o = -d_i/d_o$ (Eq. 37.7) for a concave mirror.

41. **C** A researcher wishes to capture the image of an object from a mirror using a CCD camera. Why is a concave mirror better suited to this task than a convex mirror?

Problems 42 and 43 are paired.

42. **N** A concave mirror with a radius of curvature of 25.0 cm is used to form an image of an arrow that is 10.0 cm away from the mirror. If the arrow is 2.00 cm long and inverted (pointing below the optical axis), what is the height of the arrow's image?

43. **N** A concave mirror with a radius of curvature of 25.0 cm is used to form an image of an arrow that is 30.0 cm away from the mirror. If the arrow is 2.00 cm tall and inverted (pointing below the optical axis), what is the height of the arrow's image?

44. **C** In your physics class, you may have seen a *mirage* demonstration (Fig. P37.44). One such demonstration uses a device that consists of two concave mirrors with the same radius of curvature. One mirror has a small hole near its vertex. An object, such as a strawberry, is placed at the vertex of the other mirror. The mirror with the hole is placed on top of the first mirror, so that their shiny surfaces point toward each other. The image of the strawberry hovers at the hole. Use a ray diagram to show how this works. *Hint*: The focal length of each mirror is about equal to the distance between their vertices.

Michael Levin/Opti-Gone International

FIGURE P37.44

45. **N** A concave spherical mirror with a focal length of magnitude 2.10 m is used to form an image of the Sun. The angular size of the Sun's disk on the Earth is 0.533°.

a. What is the position of the solar image formed by the mirror?

b. What is the radius of the solar image formed by the mirror?

46. The upright image formed by a concave spherical mirror with a focal length of 14.0 cm is 2.50 times larger than the object.

a. **N** What is the distance of the object from the mirror?

b. **C** Is the image formed by the mirror real or virtual?

c. **G** Draw a ray diagram showing the locations of the object and the image by tracing at least three rays.

47. **N** A concave mirror magnifies an upright arrow so that the image of the arrow is three times taller and inverted. If the arrow is 35.0 cm from the mirror, what is the focal length of the mirror?

48. An object is located 12.0 cm in front of a concave spherical mirror that has a radius of curvature of 32.0 cm.

a. **N** What are the location and magnification of the image?

b. **G** Draw a ray diagram showing the locations of the object and the image by tracing at least three rays.

49. **N** The focal length of a concave mirror is 30.0 cm. Find the two positions of the object in front of the mirror so that the image height is three times greater in magnitude than the object height.

Problems 50 and 51 are paired.

50. **N** An object in front of a concave mirror has a real image that is 11.0 cm from the mirror. The mirror's radius of curvature is 20.0 cm.
a. What is the object distance?
b. What is the magnification?

51. **N** An object in front of a concave mirror has a virtual image that is 11.0 cm from the mirror. The mirror's radius of curvature is 20.0 cm.
a. What is the object distance?
b. What is the magnification?

52. **C** CASE STUDY The United States Naval Academy owns the primary mirror to a defunct telescope. The academy staff would like to put the mirror into a new telescope, and to do so they must construct a telescope tube with the correct dimensions. The mirror is 20 in. in diameter. The length of the tube depends in part on the focal length of the primary mirror. Of course, the secondary mirror will help to fold the optics, but no further planning can be done without knowing the focal length of the primary mirror. Come up with an experiment to find the focal length. The mirror's surface must remain clean and free of scratches.

53. **N** What must be the radius of curvature of a concave mirror to form an image of the Sun 2.0 cm in diameter?

37-7 Spherical Aberration

54. **C** How does the radius of curvature compare to the size of the convex mirror in Figure 37.25A (page 1196)? Explain why the image of the artist's face seems more realistic than the image of the ceiling.

General Problems

55. An object 3.50 cm high is located 18.0 cm in front of a convex spherical mirror that has a radius of curvature of magnitude 8.00 cm.
a. **N** What is the location of the image?
b. **N** What is the magnification of the image?
c. **C** Is the image upright or inverted?
d. **N** What is the height of the image?

56. A screen is located 6.75 m from a spherical mirror. An image with $h_i = -4.00h_o$ is to be formed on the screen.
a. **C** Should a concave or convex mirror be selected for this task?
b. **N** What is the focal length of the mirror that will accomplish this task?
c. **N** What should the object distance be in this case?

Problems 57, 58, 59, and 60 are grouped.

57. **G** You see the image of a sign through a camera obscura. The sign reads FOX AND SOX. Sketch the image.

58. **G** You see the image of a sign in a plane mirror. The sign reads FOX AND SOX. Sketch the image.

59. **G** You see the image of a sign produced by a convex mirror. The sign reads FOX AND SOX. Sketch the image.

60. **G** You see the image of a sign produced by a concave mirror. The sign reads FOX AND SOX. Sketch the image. There are two possible images; sketch them both and explain under what circumstances each one is possible.

61. **A** An object is placed midway between two concave spherical mirrors as shown in Figure P37.61. The distance between the mirrors is D, and they have the same focal length. Determine the value(s) of D in terms of the focal length f for which only one image is formed in each mirror.

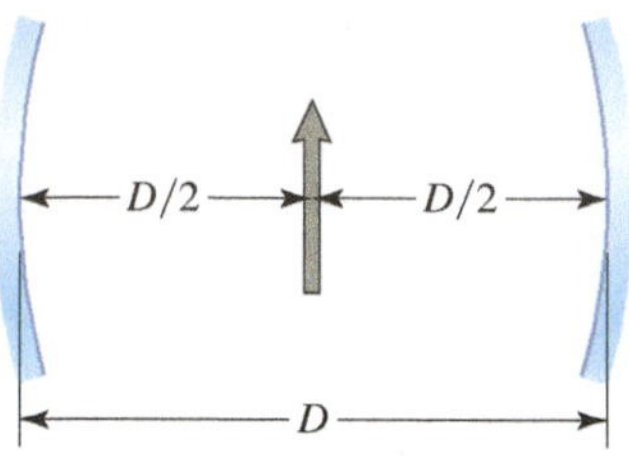

FIGURE P37.61

62. The upright image from a spherical mirror is 30.0% of the object's size and located 33.0 cm away from the object.
a. **C** Is this a convex or concave mirror?
b. **N** What is the distance between the object and the mirror?
c. **N** What is the focal length of the mirror?

63. **N** A critical characteristic of light rays that are reflected diffusely from an object such as your friend's nose is the extent to which adjacent rays diverge from each other. Imagine holding a gumball 2.0 cm in diameter in front of your friend's nose as shown in Figure P37.63. By what angle do the two tangential rays shown in the diagram diverge from each other when the center of the gumball is **a.** $D = 5.00$ cm, **b.** $D = 20.0$ cm, **c.** $D = 100.0$ cm, and **d.** $D = 100.0$ km from the tip of your friend's nose?

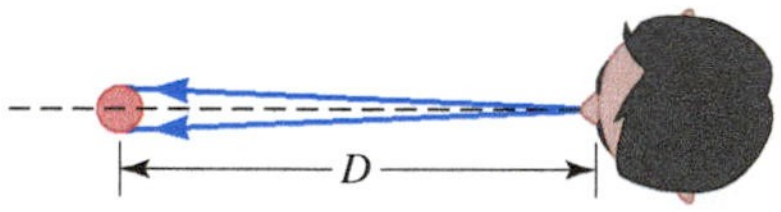

FIGURE P37.63

64. **C** An arrow points above the optical axis, and a spherical mirror forms an image of that arrow. Is it possible to determine initially whether or not the mirror forming the image is convex or concave? Why or why not? Does your answer change if you are told that the image is also upright, pointing above the optical axis?

Problems 65 and 66 are paired.

65. **N** The height of an inverted image formed by a concave spherical mirror is 5.00 times the height of the object. The object and the image are separated by 48.0 cm. What is the focal length of the mirror?

66. **N** The height of an image formed by a convex spherical mirror is 40.0% of the object's height. The object and the image are separated by 48.0 cm. What is the focal length of the mirror?

Problems 67 and 68 are paired.

67. **E** Observe your reflection in the back of a spoon. From that observation, estimate the radius of curvature of the spoon. *Hint*: Model the spoon as a spherical mirror.

68. **E** In Problem 67, you used the back of a spoon to see your face. Now imagine you flip the spoon over to see your face in its bowl. Use your estimate from Problem 67 to find the diameter of your face's image in the bowl of the spoon.

69. **N** A small convex mirror and a large concave mirror are separated by 1.00 m, and an object is placed 1.40 m to the left of the concave mirror (Fig. P37.69). The concave mirror forms an image of this object at distance $d_i = 25.0$ cm. This image is then reflected in the convex mirror, which forms an image a distance of 8.00 cm behind the convex mirror. What is the focal length of the small convex mirror?

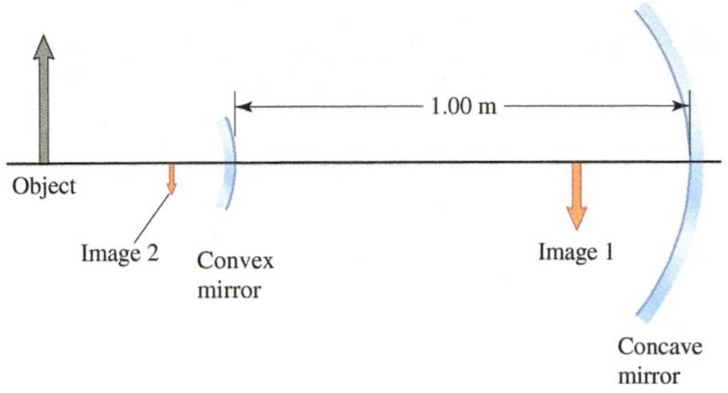

FIGURE P37.69

Problems 70 and 71 are paired.

70. Two plane mirrors are facing each other, placed on opposite walls of a room. The mirrors are separated by a distance D. A vase is placed at the midpoint between the mirrors. An infinite number of images of the vase are created in each mirror. (Try this at home!) The first image created in each mirror is of the vase itself, and those images are separated by a total distance $2D$.

a. **G** Sketch the scenario and the first image created in each mirror to verify the distance between the images.

b. **A** The first image in each mirror creates a second image in each opposite mirror. How far apart is this second set of images?

71. **A** For the scenario in Problem 70, derive an expression in terms of D for the distance between any corresponding pair of images in both mirrors. You should use a variable like n to indicate the pair of corresponding images to which you are referring. For example, the first set described here would be $n = 1$. The second set ($n = 2$) would be the set referred to in part (b) of Problem 70.

72. **C** A light ray is traveling along the positive y axis toward the origin. A rotating plane mirror is located at the origin and can be oriented to reflect the incoming ray. If you want the reflected ray to travel along the negative x axis, how should the mirror be oriented? How does the answer change if you want the reflected ray to travel along the positive x axis?

73. **A** **Fermat's principle of least time for reflection.** A ray of light traveling in a medium with speed v leaves point A and strikes a reflecting surface at point O, a horizontal distance x from point A as shown in Figure P37.73. The reflected ray reaches point B, where the horizontal distance between A and B is L.

a. Derive an expression for the time t required for the light to travel from A to B in terms of the parameters labeled in the figure.

b. Now take the derivative of t with respect to x. What is the condition for which the ray of light will take the shortest time to travel from A to B?

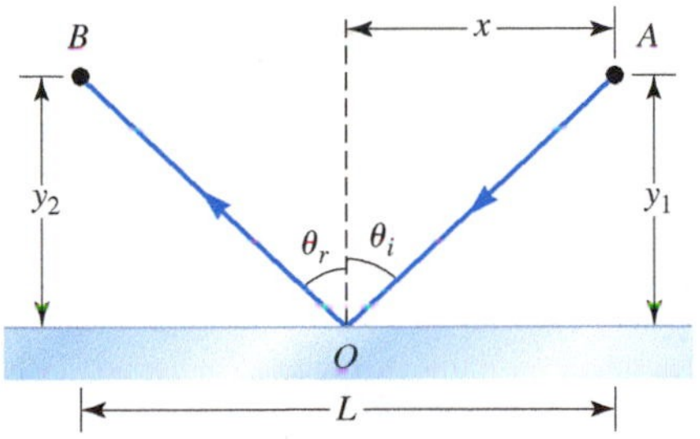

FIGURE P37.73

38 Refraction and Images Formed by Refraction

Key Questions

How do we use the law of refraction to find the location and magnification of images formed by refraction?

In what ways are thin lenses like spherical mirrors?

Underlying Principles

No new principles are introduced in this chapter.

Major Concepts

1. Law of refraction
2. Total internal reflection
3. Dispersion
4. Thin-lens equation
5. Magnification equation for thin lenses
6. Lens maker's equation
7. Angular magnification

Special Cases

1. Diverging lens
2. Converging lens

Tools

Ray diagrams and primary rays for thin lenses

Your brain is working a lot harder than you think. When you look at your dog, the light reflected from your dog enters your eye. The lens in your eye bends or refracts this light to form an image on your retina. But the image is upside down! Your dog's feet are at the top of the image, and his head is at the bottom. This upside-down image is then carried to your brain for interpretation. The world appears right-side-up to you because your brain flips the image. Newborn babies see upside-down images, and their brains learn to flip the images in a few days. If you wore lenses that inverted the image projected onto your retina, your brain would take a few days to learn how *not* to flip the image.

This chapter is about refraction and images formed by refraction. Many real images formed by lenses are inverted.

38-1 Law of Refraction

LAW OF REFRACTION
Major Concept

Chapter 37 was about mirrors and practical applications of reflection; this chapter is about lenses and practical applications of refraction. Any image you see is the result of refraction in your eye. If your vision is impaired, you may need corrective lenses. Our study of images formed by mirrors is helpful here because we use much of the same language and mathematics to study the images formed by refraction. Before we study the images formed by lenses, however, we need to consider what happens to a light ray when it passes from one transparent medium into another.

Recall that the speed of light in a medium depends on the medium's index of refraction n such that $v = c/n$ (Eq. 36.8). The index of refraction depends on the wavelength of the light and the type of medium (Table 38.1). So, the speed at which light propagates changes as it passes from one medium into another. Recall that when a wave's propagation speed changes, the wave's path bends or *refracts* (Section 17-8). So, light bends as it is transmitted from one medium into another because the speed of light depends on the medium.

TABLE 38.1 Index of refraction. (Unless otherwise specified, use n for yellow light.)

Medium	n, λ = 589 nm (yellow)	n, λ = 486 nm (blue)	n, λ = 656 nm (red)
Air	1.0002926		
Cubic zirconia	2.14		
Diamond	2.417		
Fused quartz	1.458		
Heavy flint glass	1.890	1.919	1.879
Ice	1.3049		
Quartz	1.54		
Water	1.333	1.337	1.331
Zinc crown glass	1.517	1.523	1.514

The term *refraction* is used to describe the bending of light as it propagates from an incident medium into a transmitted medium. The **law of refraction** has two parts. The first part states that the refracted ray, the incident ray, and the normal all lie in a single plane—the plane of incidence. Combining this law with the law of reflection, we can say that all three rays (incident, reflected, and refracted) lie in the plane of incidence (Fig. 38.1). The second part of the law of refraction is a mathematical relationship known as **Snell's law**, in honor of the Dutch mathematician and physicist Willebrord van Roijen Snell. You can derive Snell's law for the refraction of light by starting with $\sin\theta_2 = (v_2/v_1)\sin\theta_1$ (Eq. 17.26) and then substituting $v = c/n$ (Eq. 36.8) for the speed of light in each medium. In optics, it is convenient to use the subscripts t and i, and we write Snell's law as

$$n_t \sin\theta_t = n_i \sin\theta_i \quad (38.1)$$

where θ_t and θ_i are the *t*ransmitted and *i*ncident angles (Fig. 38.1), and n_t and n_i are the indices of refraction for the transmitted and incident media, respectively. According to Snell's law, when $n_t > n_i$, the light is bent toward the normal: $\theta_t < \theta_i$.

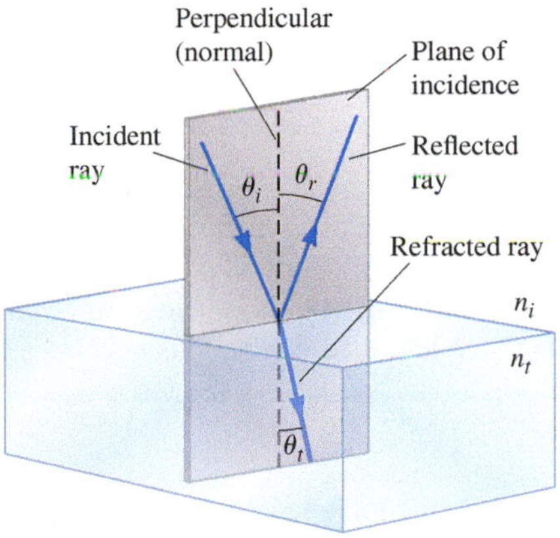

FIGURE 38.1 The angle of incidence θ_i and the angle of refraction θ_t are measured with respect to the line normal to the surface. The incident ray, the reflected ray, and the refracted ray all lie in the plane of incidence.

CONCEPT EXERCISE 38.1

Light travels from air into glass. Which sketch in Figure 38.2 correctly shows the incident, reflected, and refracted beams? *Hint*: Consider the law of reflection (Section 37-2).

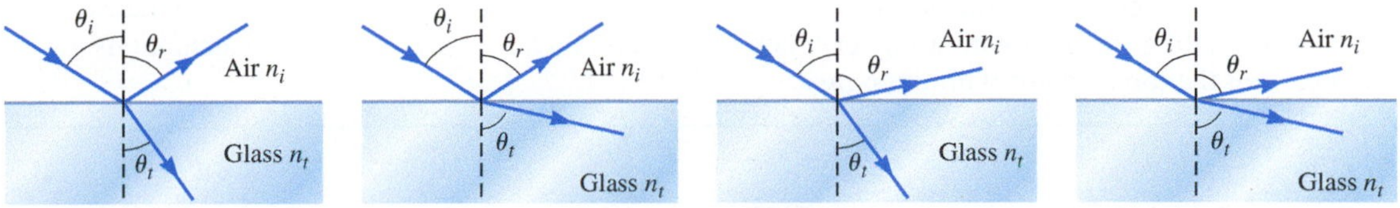

FIGURE 38.2

EXAMPLE 38.1 Glass or Diamond?

Suppose you don't know whether an object is made of glass, cubic zirconia, or diamond. You test the material by shining a laser at its surface so that the incident beam makes a 15.0° angle with the normal. You find that the refracted beam makes a 6.15° angle with the normal. What is the object made of? What is the speed of light in that medium?

INTERPRET and ANTICIPATE

Use Snell's law to find the index of refraction of the object. Then use Table 38.1 to identify the material. Once you know the index of refraction, find the speed by dividing c by that index of refraction.

SOLVE

The light travels from air with an index of refraction n_i. We know the angle of incidence and the angle of refraction, so we use Snell's law (Eq. 38.1) to solve for n_t.

$$n_t \sin\theta_t = n_i \sin\theta_i \tag{38.1}$$

$$n_t = \frac{n_i \sin\theta_i}{\sin\theta_t}$$

The incident light is in air, so we find the index of refraction for air in Table 38.1. The angles are given in the problem statement.

$$n_t = \frac{(1.00029)\sin 15.0°}{\sin 6.15°} = 2.42$$

According to Table 38.1, this is the index of refraction for diamond.

To find the speed of light in diamond, use Equation 36.8.

$$n = \frac{c}{v} \tag{36.8}$$

$$v = \frac{c}{n_t} = \frac{3.00 \times 10^8\,\text{m/s}}{2.42} = 1.24 \times 10^8\,\text{m/s}$$

CHECK and THINK

It makes sense that when light passes from air into diamond, the beam is bent toward the normal because the index of refraction of diamond is higher than the index of refraction of air. When light passes from diamond into air, the light is bent away from the normal. It also makes sense that the speed of light in diamond is much less than it is in a vacuum. In a medium with a high index of refraction, light travels slowly.

EXAMPLE 38.2 From Water into Glass

Light travels from water into flint glass as shown in Figure 38.3. The angle of incidence is 25.7°. Find the angle of reflection and the angle of refraction. Draw a sketch showing the incident, reflected, and refracted beams.

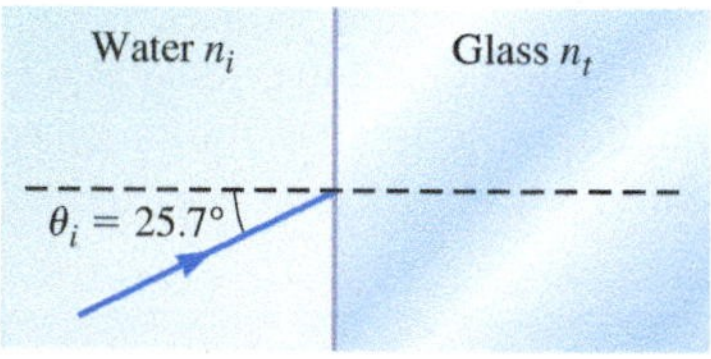

FIGURE 38.3

INTERPRET and ANTICIPATE

We need to use the law of reflection and the law of refraction to find the angles. We expect that the refracted beam is bent toward the normal because the light travels from a medium with a low index of refraction into a medium with a higher index of refraction. Once we've made our calculations, we represent the three beams with three rays.

SOLVE

Use Equation 37.3 to find the angle of reflection.

$$\theta_i = \theta_r \tag{37.3}$$

$$\theta_r = 25.7°$$

Use Snell's law (Eq. 38.1) to find the angle of refraction. The indices of refraction are listed in Table 38.1.	$n_t \sin\theta_t = n_i \sin\theta_i$ (38.1) $\sin\theta_t = \frac{n_i}{n_t}\sin\theta_i$ $\theta_t = \sin^{-1}\left(\frac{n_i}{n_t}\sin\theta_i\right) = \sin^{-1}\left(\frac{1.333}{1.890}\sin 25.7°\right)$ $\theta_t = 17.8°$

Use the angles to sketch the three beams (Fig. 38.4).

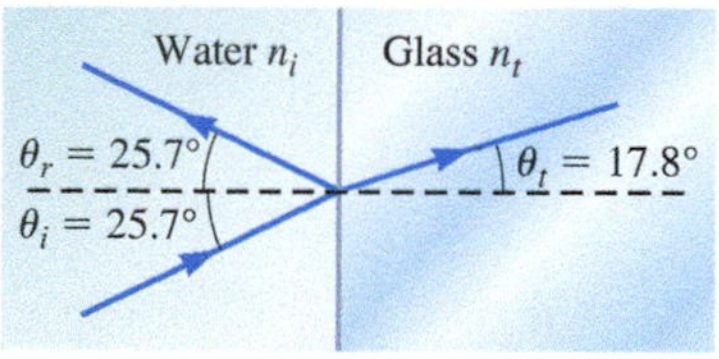

FIGURE 38.4

CHECK and THINK

As expected, the refracted beam is closer to the normal than the incident beam.

CASE STUDY Devices You Use to See

In this chapter's case study, we explore instruments with lenses that you can use to correct or improve your vision. If you are nearsighted or farsighted, you may wear glasses or contacts. When an object is so small you cannot see its details with your unaided eyes, you may use a lens called a *magnifying glass* to produce an enlarged image. If a magnifying glass cannot produce a large enough image, you may use a microscope. If you wish to see the faint light from a distant star, you may use a refracting telescope. All these instruments rely on refraction.

38-2 Total Internal Reflection

This chapter is about *refraction*, so it might seem odd that this section is about *reflection*. But you will soon see how refraction can lead to a particular kind of reflection.

Figure 38.5A shows light traveling from a medium with a high index of refraction into a medium with a lower index of refraction: $n_i > n_t$. According to Snell's law, $\sin\theta_t = (n_i/n_t)\sin\theta_i$ (Eq. 38.1), the refracted ray is bent away from the normal, so $\theta_t > \theta_i$ as shown in Figure 38.5A.

Also according to Snell's law, as the angle of incidence gets larger, the angle of refraction also gets larger. At the **critical incident angle**, the angle of refraction is 90° and the refracted ray is parallel to the surface (Fig. 38.5B). At any incident angle

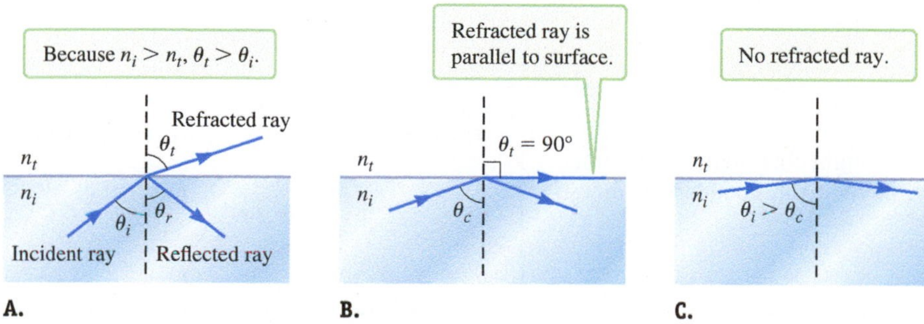

FIGURE 38.5 **A.** When the external transmitting index of refraction n_t is less than the internal incident index of refraction n_i, the refracted ray is bent away from the normal. **B.** When the angle of incidence equals the critical angle, the refracted ray is parallel to the surface. **C.** If the incident angle is greater than the critical angle, there is no refracted ray, only a reflected one.

larger than the critical angle, there is no refracted ray and no light is transmitted into the second medium (Fig. 38.5C).

Reflected rays are present in all three cases (Fig. 38.5), obeying the law of reflection $\theta_i = \theta_r$ (Eq. 37.3). If the angle of incidence is larger than the critical angle, no light is transmitted; instead, all light is reflected. This phenomenon is called **total internal reflection** and is possible only if light is incident in a medium with a higher index of refraction than that of the medium beyond the boundary. From Figure 38.5B, at the critical angle, the angle of refraction is 90°. According to Snell's law (Eq. 38.1), the critical angle is

TOTAL INTERNAL REFLECTION
Major Concept

$$n_t \sin 90° = n_i \sin \theta_c$$

$$\sin \theta_c = \frac{n_t}{n_i} \quad (38.2)$$

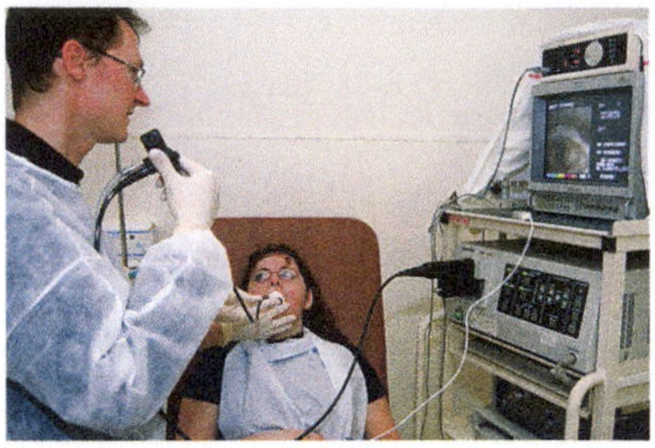
Caxhia/AGE Fotostock

FIGURE 38.6 A doctor looks inside a patient with fiber optics.

Optical fibers are a practical application of total internal reflection. An optical fiber is made of a material, such as plastic or glass, that has a relatively high index of refraction. Many fibers are often bundled together in a device. Figure 38.6 shows a medical application of an optical fiber bundle that enables a doctor to look inside a patient. Other uses are in communication systems, such as telephones, televisions, and the Internet.

CONCEPT EXERCISE 38.2

Imagine looking at your own reflection in a pond. If your reflection is clear, is that a result of total internal reflection? Explain.

EXAMPLE 38.3 A Simple Optical Fiber

A laser is directed at an optical fiber (Fig. 38.7). The fiber has an index of refraction of 1.27 and is surrounded by air. For what range of the angle θ_i is the light totally internally reflected inside the fiber?

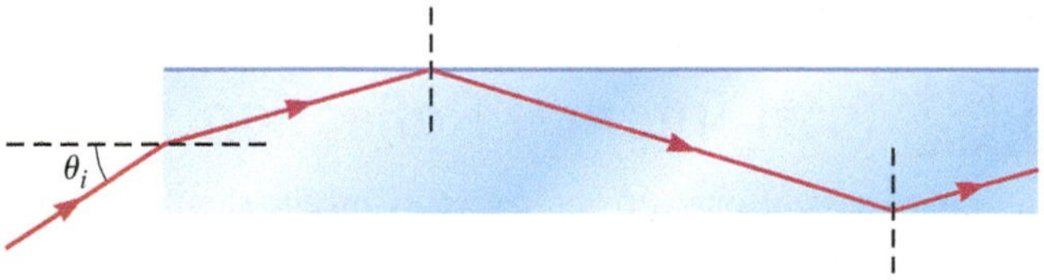

FIGURE 38.7

INTERPRET and ANTICIPATE
Begin with a sketch in which the light already inside the fiber is incident on the air–fiber boundary at the critical angle (Fig. 38.8). The light is refracted as it goes from the air into the fiber, but when the light gets to the second boundary, the refracted ray is parallel to the surface. We can use geometry and Snell's law to find θ_i for this critical case. Then any incident angle less than θ_i results in total internal reflection inside the fiber.

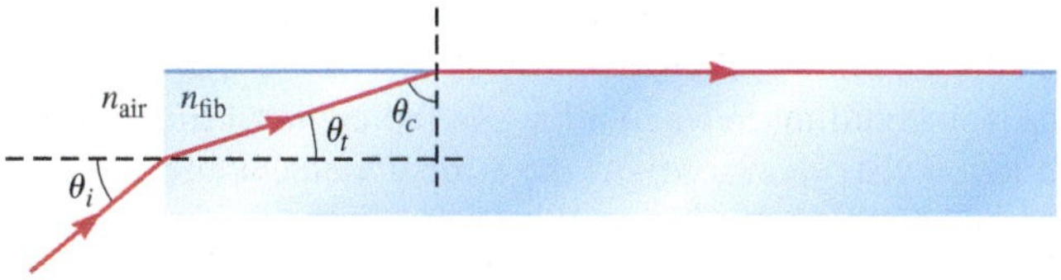

FIGURE 38.8

SOLVE
First, find the critical angle (Eq. 38.2) for the light that starts in the fiber. In that case, $n_i = n_{fib} = 1.27$; the medium beyond the boundary is air, so $n_t = n_{air} = 1.00029$ (Table 38.1).

$$\sin \theta_c = \frac{n_t}{n_i} \quad (38.2)$$

$$\theta_c = \sin^{-1}\left(\frac{n_t}{n_i}\right) = \sin^{-1}\left(\frac{1.00029}{1.27}\right)$$

$$\theta_c = 52.0°$$

Because the triangle in Figure 38.8 is a right triangle, we know the two acute angles add up to 90°.	$\theta_c + \theta_t = 90°$ $\theta_t = 90° - \theta_c$ $\theta_t = 90° - 52.0° = 38.0°$
The incident angle θ_i is related to the refracted angle θ_t by Snell's law (Eq. 38.1). At the left boundary in Figure 38.8, light is incident in air, so $n_i = n_{\text{air}} = 1.00029$ and $n_t = n_{\text{fib}} = 1.27$.	$n_t \sin\theta_t = n_i \sin\theta_i$ (38.1) $\theta_i = \sin^{-1}\left(\frac{n_t \sin\theta_t}{n_i}\right)$ $\theta_i = \sin^{-1}\left(\frac{1.27 \sin 38.0°}{1.00029}\right) = 51.4°$
If the laser is aimed at any angle less than 51.4°, its light will not leak out the sides of the fiber. In the limiting case where $\theta_i = 0$, so that the incident beam is perpendicular to the end of the fiber, the laser light is not refracted. Instead, it travels in a straight line down the fiber's long axis.	$\theta_i \leq 51.4°$

CHECK and THINK

Our answer also shows that any ray whose incident angle is greater than about 52° is transmitted out the sides of the fiber. So optical fibers do not perfectly transmit light along their axes; they leak. To help reduce leakage, the fibers should curve gently. A fiber with a sharp right angle would leak a great deal of light.

38-3 Dispersion

In the mid-1600s, people didn't know that white light is a mixture of all colors. Some people believed color was a mixture of light and darkness. But Isaac Newton performed an experiment in 1666 showing that white light can be spread out to reveal all the colors and that all those colors can be recombined to form white light. In this section, we'll see how such an experiment is possible. (In Problem 27, you'll consider the details.)

The speed of an electromagnetic wave in a medium depends on the medium's index of refraction n (Eq. 36.8). The index of refraction depends in turn on the frequency, wavelength, or color of the light (Table 38.1). For many media, the index of refraction is highest for violet light and lowest for red light. So, when white light propagates from a medium with a low index of refraction, like air, into a medium with a high index of refraction, like glass, the violet light slows down more than the red light. In addition, light bends or refracts when it is transferred from one medium into another because the speed of light is different in the two media. In general, violet light is refracted more than red light (Fig. 38.9A). The result is that white light is spread out into a broader beam that is separated by color. This broad beam looks something like a rainbow and is called a **visible light spectrum**, a **color spectrum**, or simply a **spectrum** (Fig. 38.9B). The spreading out of light by color due to differences in the index of refraction is called **dispersion**.

DISPERSION Major Concept

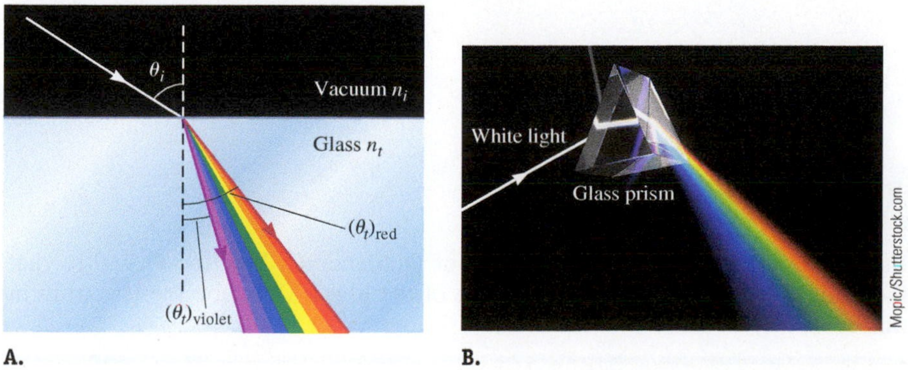

FIGURE 38.9 A. When white light (composed of red, orange, yellow, green, blue, indigo, and violet) goes from a vacuum into glass, the violet light is refracted more than the red light. The other colors are refracted in order by wavelength from red to violet. The spreading out of light by color is known as dispersion. **B.** A prism is used to spread white light into its color spectrum.

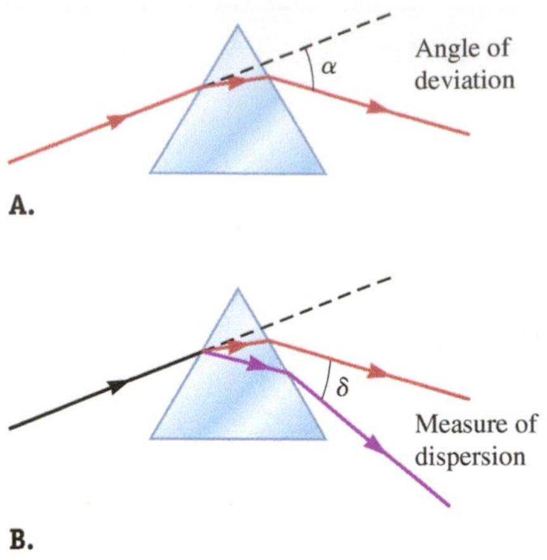

FIGURE 38.10 **A.** The angle between the original path of the light and the path of the light that emerges from the prism is the angle of deviation α. **B.** The angle between the red ray and the violet ray that emerge from the prism is the measure of dispersion δ.

A device known as a *prism* makes use of dispersion to create a color spectrum (Fig. 38.9B). Two angles—the **angle of deviation** α and the **measure of dispersion** δ—characterize the dispersion of light emerging from a prism (Fig. 38.10). The measure of dispersion (difference between the deviations for violet and red) depends on the difference between n for violet light and n for red light: The greater this difference, the greater the measure of dispersion.

Rainbows

Artificial devices such as prisms are not the only causes of dispersion. Dispersion also occurs naturally, creating beautiful phenomena such as rainbows (Fig. 38.11A). For you to see a rainbow, there must be water droplets in the air and the Sun must be behind you. Sunlight is refracted and reflected by the water droplets and into your eyes. These droplets are spread out in the air, and the position of each droplet determines which color enters your eyes (Fig. 38.11B). The highest droplets allow red light to enter your eyes, and the lowest ones allow violet light to do so. Drops in between allow other colors to enter your eyes.

Sometimes it is possible to see a second rainbow above the first. The second rainbow's colors are reversed so that violet is on top and red is on the bottom. The reversal occurs because the sunlight that reaches the secondary rainbow's droplets undergoes two reflections from the back side of each droplet (Fig. 38.11C). The net result is that the violet ray emerges below the red ray.

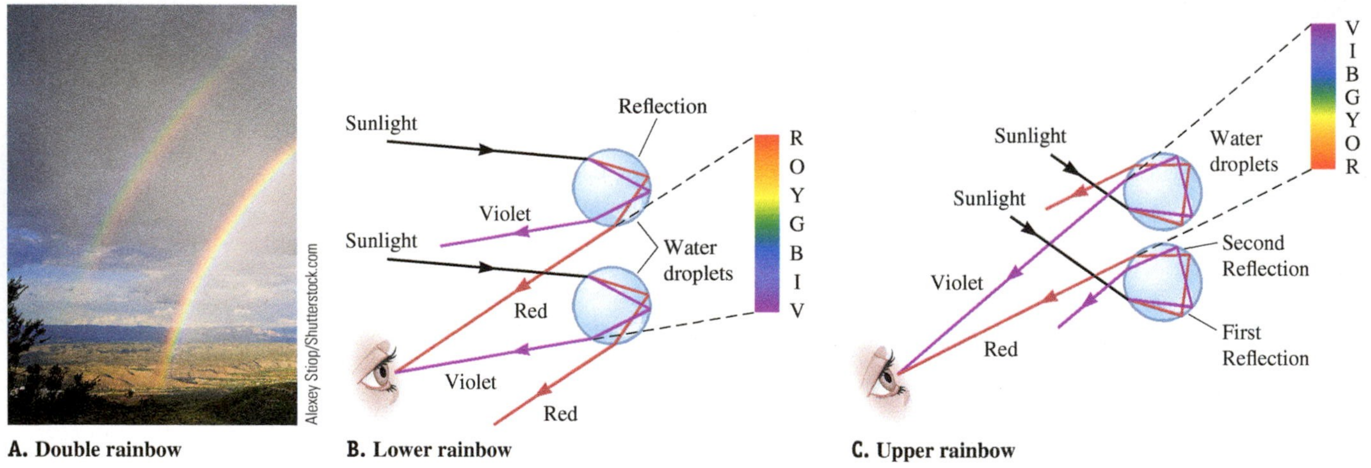

FIGURE 38.11 **A.** You will see a rainbow only when there are water droplets in the air and the Sun is behind you. Sometimes, you may see a second rainbow above the first with the colors in reverse order. **B.** Because the index of refraction depends on the color of the light, red light is not bent as much as violet light. The red ray strikes the back of each droplet at a point slightly above where the violet ray strikes it. When the rays reach the near side of the droplet, they are refracted again as they leave the water and enter the air. **C.** The second rainbow's colors are reversed because two reflections take place inside each droplet. So, violet light is seen from the highest droplets and red light from the lowest droplets.

CONCEPT EXERCISE 38.3

In Figure 38.9A, white light has angle of incidence θ_i. Imagine $\theta_i = 0$ so that the light is perpendicular to the surface of the glass. Describe the spectrum in this case.

EXAMPLE 38.4 Dispersion in a Prism

Figure 38.12 shows a beam of white light that was originally in air and is incident on a (high-dispersion) prism. The prism's index of refraction for violet light is $n_{violet} = 1.511$ and for red light is $n_{red} = 1.487$. Assume the index of refraction is the same for all colors of light in air: $n_{air} = 1.000293$. The prism's cross section is an equilateral triangle. The beam is parallel to the base of the triangle. Find the measure of dispersion δ.

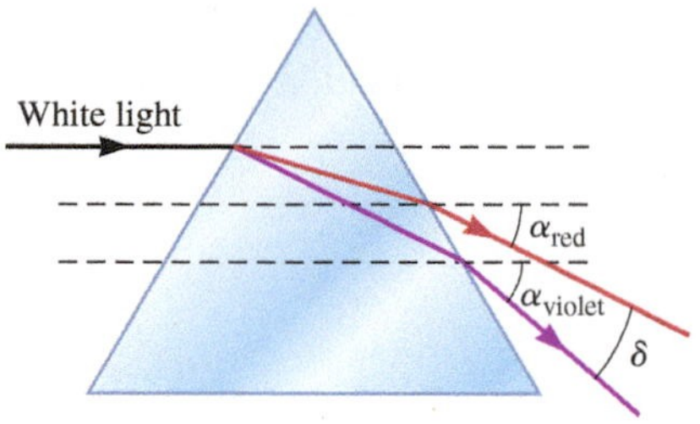

FIGURE 38.12

INTERPRET and ANTICIPATE

The measure of dispersion is the deviation of violet light minus the deviation of red light: $\delta = \alpha_{violet} - \alpha_{red}$ (Fig. 38.10B). We must find both angles of deviation before we can find the measure of dispersion. Because this is a *high-dispersion* prism, we expect the measure of dispersion to be fairly large so that the spectrum is easily seen.

Start by making a sketch of a single beam that enters the prism at L and exits at R (Fig. 38.13). The angles of incidence and refraction are labeled θ_1 through θ_4, and α is the angle of deviation between the original horizontal path of the light and the path of the ray after it has passed through the prism. From basic geometry, the internal angles of an equilateral triangle are each 60°. The only physics we need is Snell's law, but we also need to apply some geometry.

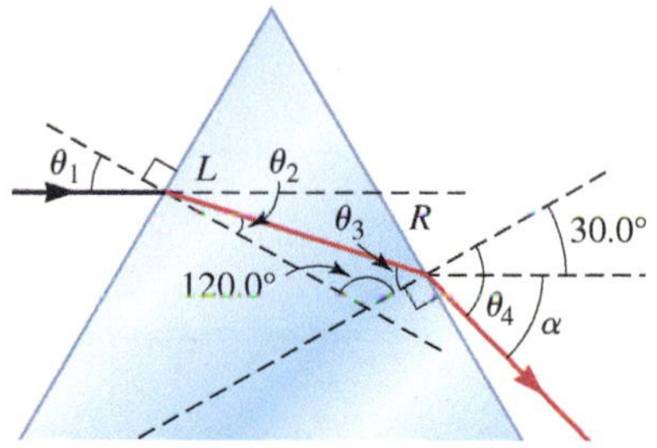

FIGURE 38.13

SOLVE

Apply Snell's law (Eq. 38.1) at the air-to-glass boundary at point L.

$$n_{air} \sin \theta_1 = n_{glass} \sin \theta_2 \qquad (1)$$

Now apply Snell's law at the glass-to-air boundary at point R.

$$n_{glass} \sin \theta_3 = n_{air} \sin \theta_4 \qquad (2)$$

Use geometry to find expressions involving $\theta_1, \theta_2, \theta_3, \theta_4$, and α (Problem 33). The angles here are derived from geometry, so they are exact.

$$\theta_2 + \theta_3 = 60° \qquad (3)$$

$$\alpha = \theta_4 - 30° \qquad (4)$$

$$\theta_1 = 30° \qquad (5)$$

Normally, we do all the algebra before substituting values. However, in this problem, it helps to substitute values now and check our preliminary results for both violet light and red light. Rearrange Equation (1), and then substitute the indices of refraction and Equation (5) into Equation (1) to find θ_2.

$$\theta_2 = \sin^{-1}\left(\frac{n_{air} \sin \theta_1}{n_{glass}}\right)$$

For violet light,

$$\theta_2 = \sin^{-1}\left(\frac{1.000293 \sin 30°}{1.511}\right) = 19.330°$$

For red light,

$$\theta_2 = \sin^{-1}\left(\frac{1.000293 \sin 30°}{1.487}\right) = 19.654°$$

CHECK and THINK

It makes sense that the angle of refraction is greater for red light because this means red light is refracted farther from the normal (closer to the original path of the light). So, as expected, red light is bent less than violet light.

Example continues on page 1226 ▶

Use $\theta_2 + \theta_3 = 60°$ (Eq. 3) to find θ_3.	For violet light, $\theta_3 = 60° - 19.330° = 40.670°$ For red light, $\theta_3 = 60° - 19.654° = 40.346°$
Next, rearrange Equation (2) to find θ_4.	$\theta_4 = \sin^{-1}\left(\frac{n_{glass} \sin \theta_3}{n_{air}}\right)$ For violet light, $\theta_4 = \sin^{-1}\left(\frac{1.511 \sin 40.670°}{1.000293}\right) = 79.877°$ For red light, $\theta_4 = \sin^{-1}\left(\frac{1.487 \sin 40.346°}{1.000293}\right) = 74.239°$
Use $\alpha = \theta_4 - 30°$ (Eq. 4) to find α. Round to the correct number of significant figures.	For violet light, $\alpha_{violet} = 79.877° - 30° = 49.88°$ For red light, $\alpha_{red} = 74.239° - 30° = 44.24°$
The difference between the angles of deviation for violet light and red light is the measure of dispersion.	$\delta = \alpha_{violet} - \alpha_{red} = 49.88° - 44.24°$ $\delta = 5.64°$

CHECK and THINK

The measure of dispersion is about 6°. As expected, this angle would be easy to see.

38-4 Refraction at Spherical Surfaces

In Section 38-3, we saw that refraction may be used to create a color spectrum. In the rest of this chapter, we'll study how refraction forms images. We start with a simple example: observing a fish's image in a still pond. Rays from a point on the fish—located a distance d_o below the water—are refracted at the boundary between the water and the air (Fig. 38.14). These refracted rays do not cross in the air. However, the diverging rays enter your eye, and as in the case of a plane mirror, your visual system in effect traces these rays back to a point a distance d_i below the water. Because the refracted rays are diverging, the resulting image cannot be projected onto a screen. Thus, the image you see is virtual; the refracted rays appear to diverge from the points that make up the image.

FIGURE 38.15 An image of a person is formed by this spherical globe.

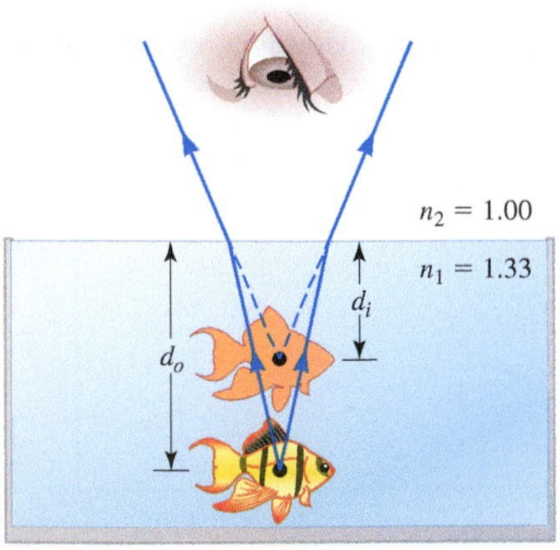

FIGURE 38.14 The virtual image of the fish is formed by refraction. This image appears above the actual fish. The eye is not drawn to scale.

Paraxial Assumption

The surface of the water in Figure 38.14 is flat. From Example 37.5, a plane mirror is a special case of a spherical mirror for which the radius of curvature is infinite. Likewise, the flat surface of the water is a special case of a spherical surface with an infinite radius of curvature. Let's look at the more general case of a spherical surface with a *finite* radius of curvature, which might form one surface of a lens.

In the image formed by the glass globe in Figure 38.15, the person's face near the center is very clear. However, around the globe's edge, you can see a very distorted image of the person's throat, hair, and background wall, an aberration we wish to avoid. For mirrors, we can ignore spherical aberration if we consider only small spherical mirrors so that the rays are paraxial. Similarly, in this chapter, you can assume the rays are paraxial unless otherwise indicated. In practice, this means our derivations hold for images formed near the optical axis.

DERIVATION Position of Image Formed by Refraction at a Spherical Surface

Light from an object travels in a medium with index of refraction n_i; it is then refracted in a medium with a higher index n_t (Fig. 38.16). The boundary between the media is spherical, and the object is small compared to the radius of curvature r, so the paraxial ray assumption holds. Call the region with the object the *front side*, so that the center of curvature C is on the back side. We will show that the object distance d_o and image distance d_i are related by

$$\frac{n_i}{d_o} + \frac{n_t}{d_i} = \frac{(n_t - n_i)}{r} \quad (38.3)$$

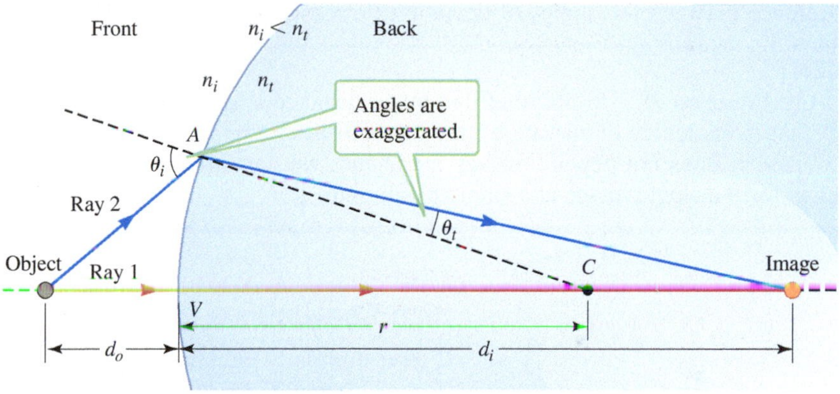

FIGURE 38.16 An image forms due to refraction at a spherical surface.

Consider two rays emitted by a single point on the object (Fig. 38.16). Ray 1 is emitted along the optical axis. It strikes the surface at the vertex V, and because it is perpendicular to the surface, it is not bent. Ray 2 strikes the surface at point A and is refracted toward the normal. The only physics we need for this derivation is Snell's law; the rest comes from geometry.

Apply Snell's law (Eq. 38.1) to ray 2 at point A.	$n_t \sin\theta_t = n_i \sin\theta_i$	(38.1)
For paraxial rays, the angles θ_t and θ_i are small. Use the small-angle approximation for the sine function.	$\sin\theta_t \approx \theta_t$ $\sin\theta_i \approx \theta_i$	
Write Snell's law in terms of this approximation.	$n_t\theta_t \approx n_i\theta_i$ radians	(1)

Our next steps involve geometry as shown by the two highlighted triangles and angles α, β, and γ in Figure 38.17.

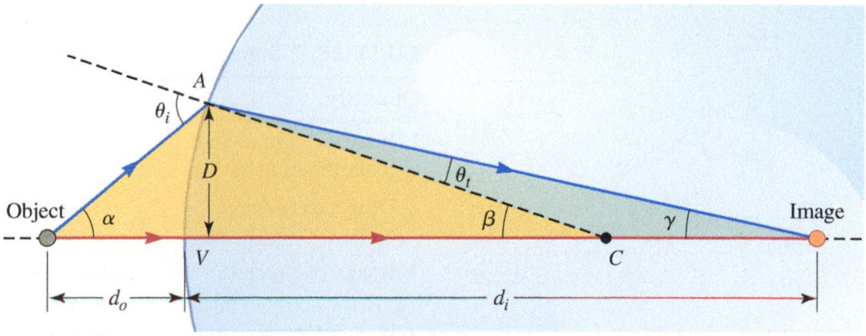

FIGURE 38.17

Derivation continues on page 1228 ▶

In Problem 36, you'll find two expressions relating α, β, and γ to the angles of incidence and reflection.	$\theta_i = \alpha + \beta$ (2) $\theta_t = \beta - \gamma$ (3)
Substitute Equations (2) and (3) into Equation (1).	$n_t(\beta - \gamma) = n_i(\alpha + \beta)$ $n_i\alpha + n_t\gamma = (n_t - n_i)\beta$ (4)
For paraxial rays, point A is almost directly above the vertex V. Call this vertical distance D (Fig. 38.17). Three right triangles share side D and can be used to find the tangents of α, β, and γ. For paraxial rays, these three angles must be small, so we use the small-angle approximation for the tangent function.	$\tan\alpha = \frac{D}{d_o} \approx \alpha$ $\tan\beta = \frac{D}{r} \approx \beta$ $\tan\gamma = \frac{D}{d_i} \approx \gamma$
Use these three approximate expressions to eliminate the angles α, β, and γ from Equation (4).	$n_i\frac{D}{d_o} + n_t\frac{D}{d_i} = (n_t - n_i)\frac{D}{r}$
The vertical distance D cancels.	$\frac{n_i}{d_o} + \frac{n_t}{d_i} = \frac{(n_t - n_i)}{r}$ ✓ (38.3)

COMMENTS

We can use Equation 38.3 to find that the image distance d_i depends only on the shape of the surface (that is, its radius of curvature), the index of refraction of both media, and the position of the object. It does not depend on any angle. So, we conclude that all paraxial rays come together to form a single image at a single position d_i.

Sign Conventions for Spherical Refractors

If you compare Equation 38.3 to the mirror equation $(1/d_o) + (1/d_i) = (2/r)$ (Eqs. 37.6 and 37.8), you find three similarities. First, in both cases, the object distance and the image distance are in the denominator of two separate terms. Second, in both cases, the radius of curvature plays the same role in determining the position of the image. Finally, both equations require a sign convention. The sign convention for a spherical refractor is similar to that for a spherical mirror.

The front of the refractor is the side from which the light originates. If the center of curvature is in the back, as in Figure 38.16, the surface is convex and the radius of curvature $r > 0$. If the center of curvature is in the front, the surface is concave and $r < 0$. If the object is in the front, $d_o > 0$ and the object is real. This is often the case, but it is possible for multiple refractions that the object is in the back—in which case $d_o < 0$ and the object is virtual. If the image is in the back, $d_i > 0$ and the image is real. As for spherical mirrors, we use the convention that an inverted image has a negative height and its magnification is negative. These sign conventions are summarized in Table 38.2.

TABLE 38.2 Sign conventions for spherical refracting surfaces.

Quantity	Positive	Negative
1. Image height h_i and magnification M	If image is **upright**	If image is **inverted**
2. Object distance d_o	If object is **real** (in front of surface)	If object is **virtual** (behind surface)
3. Image distance d_i	If image is **real** (behind surface)	If image is **virtual** (in front of surface)
4. Radius of curvature r	If surface is **convex**	If surface is **concave**

DERIVATION Magnification by Refraction at a Spherical Surface

Next, we show that the magnification of the image created by refraction at a spherical surface is given by

$$M = \frac{h_i}{h_o} = -\frac{n_i}{n_t}\frac{d_i}{d_o} \tag{38.4}$$

We use the same two media and surfaces as in the preceding derivation, but now we must consider an extended object represented by an arrow (Fig. 38.18). As before, we consider only paraxial rays.

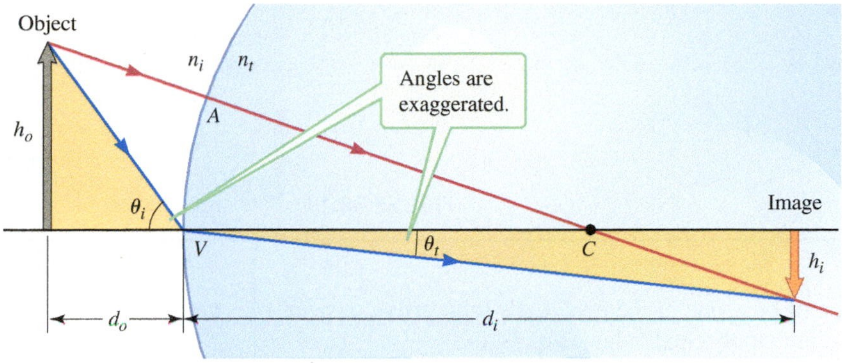

FIGURE 38.18

We draw two rays from the tip of the image. The first ray passes through the center of curvature, is perpendicular to the surface at point A, and is therefore not refracted. The other ray strikes the surface at the vertex V and is refracted toward the normal. The two rays cross to form a real inverted image in the second medium. We will apply the sign convention at the end of the derivation.

Use Snell's law (Eq. 38.1) for the ray at V.	$n_t \sin\theta_t = n_i \sin\theta_i$	(1)
Use the two right triangles highlighted in Figure 38.18 to write expressions for the tangents of θ_i and θ_t. These angles are small, so we can use the approximation that the tangent of a small angle equals the sine of the angle.	$\tan\theta_i = \frac{h_o}{d_o} \approx \sin\theta_i$	(2)
	$\tan\theta_t = \frac{h_i}{d_i} \approx \sin\theta_t$	(3)
Substitute Equations (2) and (3) into Equation (1).	$n_t \tan\theta_t \approx n_i \tan\theta_i$ $n_t\frac{h_i}{d_i} = n_i\frac{h_o}{d_o}$	
The magnification is the image height divided by the object height (Eq. 37.2). By the first sign convention in Table 38.2, we insert a negative sign for an inverted image.	$M = \frac{h_i}{h_o} = -\frac{n_i}{n_t}\frac{d_i}{d_o}$ ✓	(38.4)

COMMENTS

Compare the magnification in this case to the magnification of a spherical mirror, $M = -d_i/d_o$ (Eq. 37.7), and you notice a similarity. Both depend on $-d_i/d_o$, but in this case of refraction by a spherical surface, the magnification also depends on the indices of refraction. Throughout this chapter, many equations are similar to those we learned in Chapter 37 on reflection, except here the equations involve the index of refraction. Making comparisons like this will help you to keep the equations straight in your mind.

EXAMPLE 38.5 Looking at a Fish

In this problem, we find an expression for the image distance of the fish in Figure 38.14 in two different ways.

A Use the rays shown in Figure 38.14 to find an expression for d_i. Include the appropriate sign convention from Table 38.2 in your final expression.

INTERPRET and ANTICIPATE

Figure 38.19 is a simple ray diagram for one point on the fish. The angles must be small because the person is directly above the fish. Only slightly bent rays will make it into the person's eye.

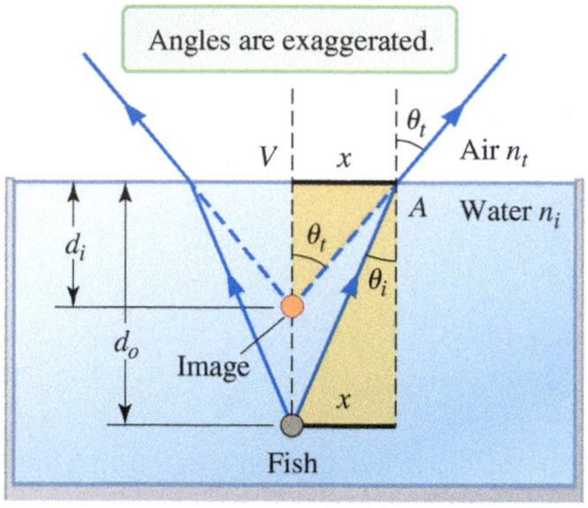

FIGURE 38.19

SOLVE

We need to consider the refraction of only one of the rays. Let's arbitrarily consider the ray that is refracted at point A and apply Snell's law (Eq. 38.1).

$$n_t \sin\theta_t = n_i \sin\theta_i \qquad (38.1)$$

Use the highlighted right triangles to find expressions for $\tan\theta_i$ and $\tan\theta_t$. The angles are small, so the tangent of each angle is roughly equal to the sine of that angle.

$$\tan\theta_t = \frac{x}{d_i} \approx \sin\theta_t$$

$$\tan\theta_i = \frac{x}{d_o} \approx \sin\theta_i$$

Substitute these approximate expressions into Snell's law.

$$n_t \frac{x}{d_i} = n_i \frac{x}{d_o}$$

Solve for d_i.

$$d_i = \frac{n_t}{n_i} d_o$$

The fish is the object, so the front side of the boundary is below the surface of the water. Because the image is virtual (in front of the surface), the image distance is negative according to the sign convention (Table 38.2).

$$d_i = -\frac{n_t}{n_i} d_o$$

B Use Equation 38.3 and consider the flat water surface as a special case of a spherical surface in which the radius of curvature is infinite to derive an expression for d_i. Check your results by comparing parts A and B.

INTERPRET and ANTICIPATE

Figure 38.19 still applies to this problem, but now we think of the flat surface as a special case of a spherical surface.

SOLVE

Start with Equation 38.3 and take the limit as the radius of curvature goes to infinity. The right side goes to 0, so the left side must also approach 0.

$$\frac{n_i}{d_o} + \frac{n_t}{d_i} = \frac{(n_t - n_i)}{r} \qquad (38.3)$$

$$\lim_{r\to\infty}\left[\frac{(n_t - n_i)}{r}\right] \to 0$$

$$\frac{n_i}{d_o} + \frac{n_t}{d_i} \to 0$$

Solve for d_i.	$$\frac{n_i}{d_o} = -\frac{n_t}{d_i}$$ $$d_i = -\frac{n_t}{n_i} d_o$$

CHECK and THINK

As expected, this is the same expression we found in part A. This time we didn't have to insert a negative sign because the sign convention is built into Equation 38.3.

EXAMPLE 38.6 A Thick Lens

Suppose you have a nearly perfect spherical vase full of water and have discovered you can use the vase as a lens to magnify the writing on a sheet of paper. When you place a message on one side of the vase and look through the opposite side, you see an upright, slightly magnified image of the message (Fig. 38.20). Model the vase as a spherical refractor (a thick lens) with a radius of 6.0 cm and an index of refraction of 1.3. Assume the air has an index of refraction of 1.0. If you hold the message 3.0 cm from the vase, find the position of the final image and its magnification. (As shown in Figure 38.20, you can also use the thick lens to project a real image onto a screen; in Problem 39, you will find the distance and magnification for this image.)

Charles D. Winters/Cengage Learning

FIGURE 38.20 An upright image of the message "Physics is Fun!" as seen through the vase. The image is greatly distorted, so the whole message cannot be read. An inverted image of a lamp is projected through the vase and onto a screen.

INTERPRET and ANTICIPATE

Both the vase of water and the globe in Figure 38.15 are thick lenses. Unlike in Example 38.5, where the light is refracted only after it passes from water into air, in this example the light is refracted twice through the thick lens. The refraction from one surface produces an image that becomes the object for the refraction from the second surface. We will therefore work this problem in two parts. We'll first find the image produced by the first surface. Then, using that image as an object, we'll find the final image produced by the second surface. Throughout this problem, we consider only paraxial rays and we model the vase of water as a single spherical refractor, ignoring the thin layer of glass.

SOLVE

In Figure 38.21, the sphere is divided into two surfaces labeled 1 and 2. The object (the message) is to the left, in front of both surfaces; your eye (not shown) is to the right, behind them. So, for the rays emitted by the object, surface 1 is convex and surface 2 is concave.

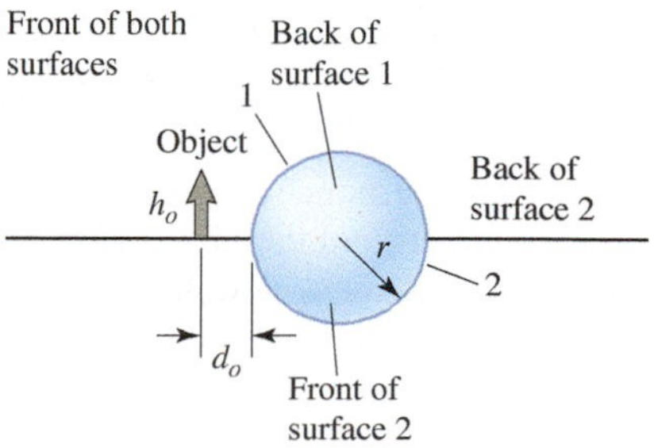

FIGURE 38.21

Solve Equation 38.3 for $1/d_i$.	$$\frac{n_i}{d_o} + \frac{n_t}{d_i} = \frac{(n_t - n_i)}{r} \quad (38.3)$$ $$\frac{1}{d_i} = \frac{1}{n_t}\left[\frac{(n_t - n_i)}{r} - \frac{n_i}{d_o}\right] \quad (1)$$

Example continues on page 1232 ▶

Substitute values, taking into account the sign convention, to find the image distance produced by surface 1. Surface 1 is convex, so r is positive. The object is real and in front of surface 1, so d_o is positive.	$\frac{1}{d_i} = \frac{1}{1.3}\left[\frac{(1.3 - 1.0)}{6.0\text{ cm}} - \frac{1.0}{3.0\text{ cm}}\right]$ $d_i = -4.6\text{ cm}$

We find that the image produced by the first surface has a negative image distance, so this image is in front of surface 1 (Fig. 38.22). We can think of it as image 1, but we must also think of it as object 2 (the object for surface 2).

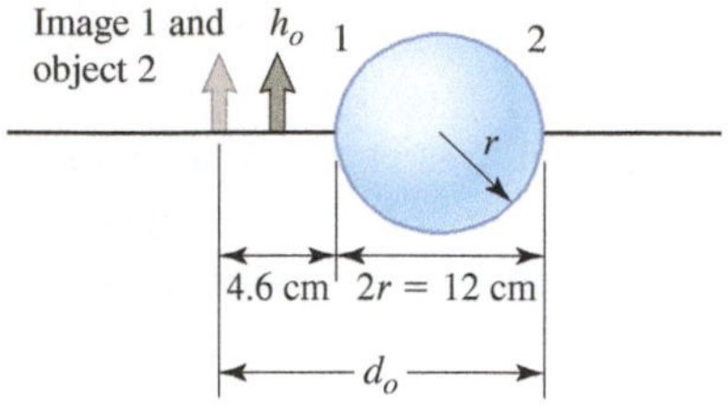

FIGURE 38.22

Before finding the position of the final image, let's find the magnification of image 1 using Equation 38.4. The result is positive, so the image is upright.	$M_1 = -\frac{n_i\, d_i}{n_t\, d_o}$ (38.4) $M_1 = -\frac{1.0\ (-4.6\text{ cm})}{1.3\ \ (3.0\text{ cm})} = 1.2$
Now consider image 1 as the object for surface 2. Find the image distance with Equation (1). The object (image 1) is in front of surface 2 (Fig. 38.22), so the object distance is positive. This distance is 4.6 cm plus the diameter (12 cm) of the spherical lens. The radius of curvature is negative because surface 2 is concave. Be careful; now the incident ray is in the lens ($n_i = 1.3$) and the transmitted ray is in the air ($n_t = 1.0$).	$\frac{1}{d_i} = \frac{1}{n_t}\left[\frac{(n_t - n_i)}{r} - \frac{n_i}{d_o}\right]$ $\frac{1}{d_i} = \frac{1}{1.0}\left[\frac{(1.0 - 1.3)}{-6.0\text{ cm}} - \frac{1.3}{16.6\text{ cm}}\right]$ $d_i = -35\text{ cm}$

CHECK and THINK

The final image you see is about 35 cm in front of the second surface, or about 23 cm in front of the first surface. So the image is in front of the object (the message). But, from the perspective of the eye at the back of these surfaces, the image is behind the vase and virtual. You cannot project this image onto a screen. Instead, you see the image by looking through the vase (Fig. 38.20).

The magnification of image 2 is also given by Equation 38.4. It is the ratio of image 2's height to object 2's height.	$M_2 = -\frac{n_i\, d_i}{n_t\, d_o} = -\frac{1.3\ (-35\text{ cm})}{1.0\ (16.6\text{ cm})} = 2.7$
To find the total magnification—the ratio of image 2's height to object 1's (the message's) height—multiply the magnifications caused by each surface.	$M = M_1 M_2 = (1.2)(2.7)$ $M = 3.2$

CHECK and THINK

The vase acts as a magnifying glass, producing an enlarged image. However, because the lens is thick, there is a great deal of aberration and the image is blurred around the edges. The image also suffers from *chromatic aberration* because the index of refraction depends on color, and different colors produce images at different places. Don't worry if you find it difficult to see the halo of colors around the image (Fig. 38.20); the photo is very small.

38-5 Thin Lenses

Unless otherwise specified, assume all lenses are thin.

A lens is an optical system with two refracting surfaces. For example, the vase of water in Figure 38.20 is a thick lens. Optical instruments such as cameras, microscopes, and the human eye are best modeled as thin lenses. The thickness of a **thin lens** is small compared to its radius of curvature. In the rest of this chapter, we study thin lenses. Our goal in this section is to find expressions for the image distance and the magnification produced by thin lenses. Our study of spherical mirrors provides a major shortcut in reaching these goals because the mirror equation (Eq. 37.8),

THIN-LENS EQUATION

Major Concept

$$\frac{1}{d_o} + \frac{1}{d_i} = \frac{1}{f} \tag{38.5}$$

is the same as the **thin-lens equation**. Also (as you show in Problem 44), the magnification equation for a thin lens is the same as that for a spherical mirror (Eq. 37.7):

LINEAR MAGNIFICATION OF LENSES

Major Concept

$$M = \frac{h_i}{h_o} = -\frac{d_i}{d_o} \tag{38.6}$$

However, thin lenses require the slightly different sign conventions given in Table 38.3. Also, the focal length of a lens depends on both its shape and the lens's index of refraction. The focal length of a thin lens is given by the *lens maker's equation*, which we derive next.

TABLE 38.3 Sign conventions for thin spherical lenses.

Quantity	Positive	Negative
1. Image height h_i and magnification M	If image is **upright**	If image is **inverted**
2. Object distance d_o	If object is **real** (in front)	If object is **virtual** (behind)
3. Image distance d_i	If image is **real** (behind)	If image is **virtual** (in front)
4. Radius of curvature r	If surface is **convex**	If surface is **concave**
5. Focal length f	If lens is **converging**	If lens is **diverging**

DERIVATION The Lens Maker's Equation

We derive the lens maker's equation, an expression for the focal length f of a thin lens:

LENS MAKER'S EQUATION

Major Concept

$$\frac{1}{f} = (n-1)\left[\frac{1}{r_1} - \frac{1}{r_2}\right] \tag{38.7}$$

where n is the lens's index of refraction. Although the lens maker's equation holds for thin lenses, we begin by first considering a thick lens and then we move to a thin lens mathematically. The lens's surfaces are both spherical, but they do not necessarily have the same radius of curvature, so the focal length depends on r_1 and r_2 for each surface (Fig. 38.23). Assume the lens is in a vacuum with an index of refraction equal to 1. (Because the index of refraction of air is 1.0003, this assumption holds well for a lens in air.)

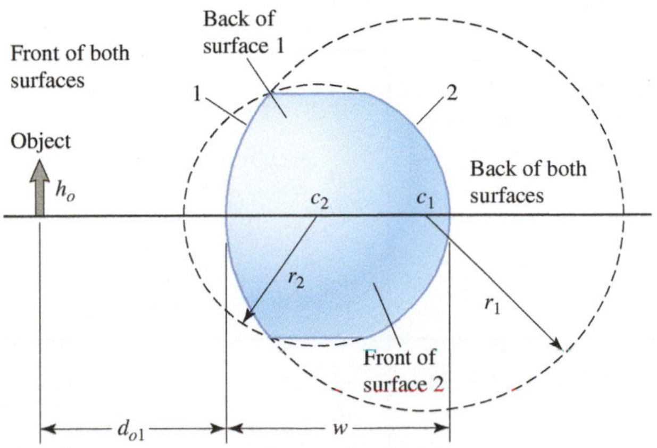

FIGURE 38.23

Derivation continues on page 1234 ▶

Our approach here is similar to that in Example 38.6 with the vase of water. An object is located in front of surface 1 (Fig. 38.23). Surface 1 produces an image that becomes the object for surface 2. The image produced by surface 2 is the final image produced by the lens.

Start by applying Equation 38.3 to surface 1. The subscript 1 is a reminder that we are working with surface 1. Light from the object passes first through the vacuum and then into the lens, so $n_i = 1$ and $n_t = n$ for the lens.

$$\frac{n_i}{d_o} + \frac{n_t}{d_i} = \frac{(n_t - n_i)}{r} \tag{38.3}$$

$$\frac{1}{d_{o1}} + \frac{n}{d_{i1}} = \frac{(n - 1)}{r_1} \tag{1}$$

Surface 1 produces one of three possible images. If the image is virtual and in front of both surfaces (Fig. 38.24), d_{i1} is negative. Because this image is also the object of surface 2, the object distance d_{o2} is positive and the object is real.

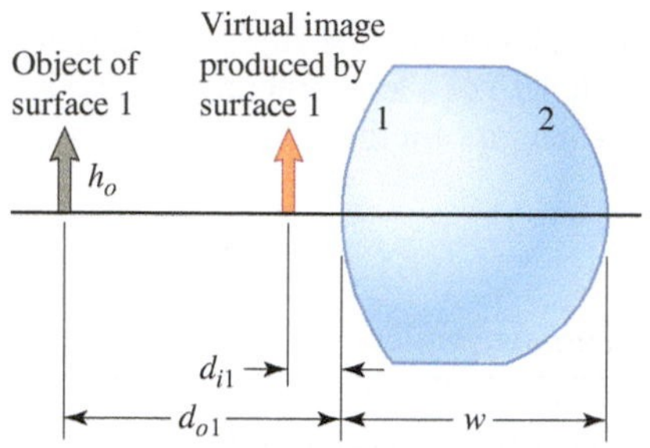

FIGURE 38.24

If surface 1 produces a real image inside the lens itself (Fig. 38.25), d_{i1} is positive. Because the image is in front of surface 2, as in the preceding case, the object for surface 2 is real.

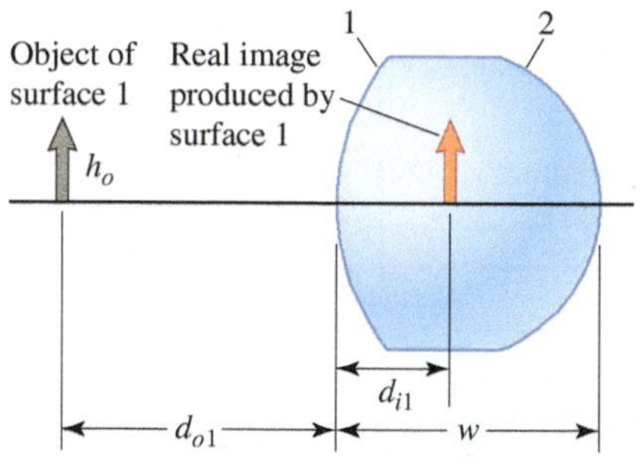

FIGURE 38.25

Finally, the image produced by surface 1 may be real and behind surface 2 (Fig. 38.26). Because this image is the object for surface 2 and is behind surface 2, the object is virtual and its distance is negative.

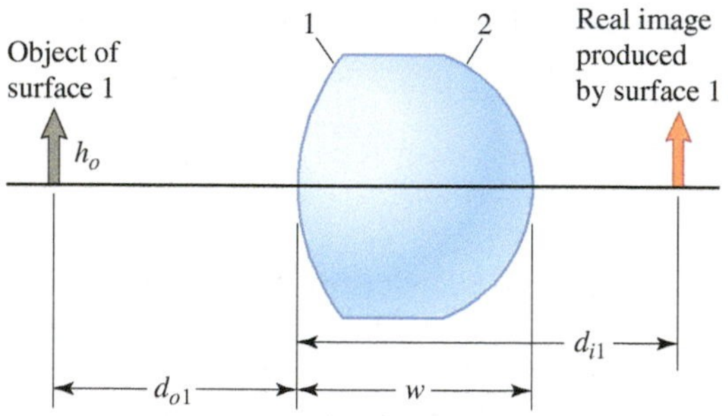

FIGURE 38.26

In all three cases, the object distance for surface 2 is given by Equation (2). (Test this for yourself in Problem 112.)

$$d_{o2} = w - d_{i1} \tag{2}$$

Now we're ready to apply Equation 38.3 to surface 2. This time the light is incident in the lens and then refracted into the vacuum. So, $n_i = n$ for the lens, and $n_t = 1$. The object distance d_{o2} is given by Equation (2).

$$\frac{n_i}{d_o} + \frac{n_t}{d_i} = \frac{(n_t - n_i)}{r} \tag{38.3}$$

$$\frac{n}{d_{o2}} + \frac{1}{d_{i2}} = \frac{(1 - n)}{r_2}$$

$$\frac{n}{w - d_{i1}} + \frac{1}{d_{i2}} = \frac{(1 - n)}{r_2} \tag{3}$$

To find the lens maker's equation, add Equations (1) and (3). For a thin lens, $w \to 0$.

$$\frac{1}{d_{o1}} + \frac{n}{d_{i1}} + \frac{n}{w - d_{i1}} + \frac{1}{d_{i2}} = \frac{(n - 1)}{r_1} + \frac{(1 - n)}{r_2}$$

The object distance d_{o1} is the distance of the original real object, while d_{i2} is the image distance of the final image. So the double subscripts are no longer necessary.

$$\frac{1}{d_{o1}} + \frac{n}{d_{i1}} - \frac{n}{d_{i1}} + \frac{1}{d_{i2}} = \frac{(n-1)}{r_1} + \frac{(1-n)}{r_2}$$

$$\frac{1}{d_{o1}} + \frac{1}{d_{i2}} = (n-1)\left[\frac{1}{r_1} - \frac{1}{r_2}\right]$$

$$\frac{1}{d_o} + \frac{1}{d_i} = (n-1)\left[\frac{1}{r_1} - \frac{1}{r_2}\right] \quad (4)$$

Compare Equation (4) to the thin-lens equation (Eq. 38.5). The left sides are equal; therefore, the right sides are equal. We have reached the lens maker's equation.

$$\frac{1}{d_o} + \frac{1}{d_i} = \frac{1}{f} \quad (38.5)$$

$$\frac{1}{f} = (n-1)\left[\frac{1}{r_1} - \frac{1}{r_2}\right] \checkmark \quad (38.7)$$

COMMENTS

Compare the focal length of a thin lens (Eq. 38.8) to the focal length for a spherical mirror $f = r/2$ (Eq. 37.6). Both depend on the radius of curvature, but the focal length for a thin lens is more complicated. First, a thin lens is made up of two curved surfaces and so depends on two radii of curvature. Also, because the light is refracted by the thin lens, its focal length depends on its index of refraction. Why does the focal length of a thin lens have the "1" in the $(n-1)$ factor?

Sign Conventions for Thin Lenses

To use the thin-lens equation (38.5), the lens maker's equation (38.7), and the magnification equation (38.6), we need the sign conventions from Table 38.3. Because all these lens equations follow from refraction at spherical surfaces, the sign conventions for thin lenses are similar to those for spherical surfaces. First determine the front and back of the lens using the same convention as for spherical surfaces. The front of the lens is the side from which light originates; the other side is the back. Because the lens is thin, we don't concern ourselves with the inside.

Comparing Table 38.3 (sign conventions for thin lenses) to Table 38.2 for spherical refracting surfaces shows they have much in common. However, a refracting surface does not have a focal length.

The sign of the lens's focal length comes from the signs of the radii of curvature. Figure 38.27 shows two thin lenses. In both cases, the object is on the left, the left side is the front, and surface 1 is the surface that the object's light encounters first. In Figure 38.27A, surface 1 bulges out toward the object. Its center of curvature is behind the lens, so surface 1 is convex. According to the fourth sign convention in Table 38.3, this surface's radius of curvature is positive: $r_1 = |r_1|$. Surface 2 bulges away from the object. Its center of curvature is in front of the lens, so surface 2 is concave. According to the sign convention, its radius of curvature is negative: $r_2 = -|r_2|$.

Now let's find the sign of the focal length using the lens maker's equation (Eq. 38.7). The term $(n - 1)$ is positive because the index of refraction n for the lens must be greater than 1. For the lens in Figure 38.27A, the term in square brackets is

$$\left[\frac{1}{|r_1|} - \frac{1}{-|r_2|}\right] = \left[\frac{1}{|r_1|} + \frac{1}{|r_2|}\right] > 0$$

so the focal length is positive. A lens with a positive focal length is called a **converging lens**, whereas a lens with a negative focal length is called a **diverging lens**.

Next, let's show that the lens in Figure 38.27B has a negative focal length and is a diverging lens. Because c_1 is in front of the lens, surface 1 is concave and its

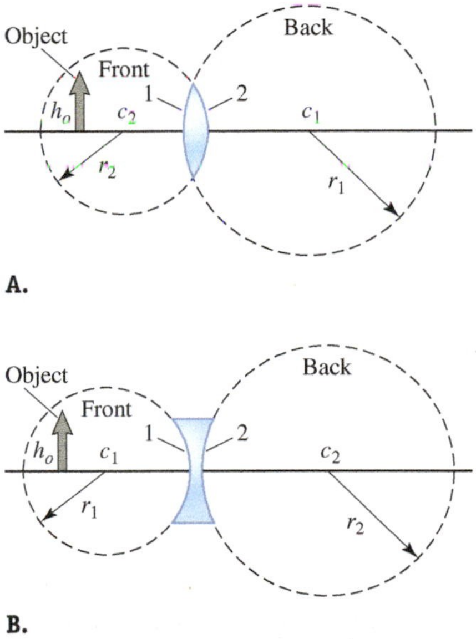

FIGURE 38.27 A thin spherical lens has two surfaces. The center of curvature for surface 1 is c_1. The center of curvature for surface 2 is c_2. **A.** Surface 1 is convex and surface 2 is concave. This is a converging lens, and so it has a positive focal length. **B.** Surface 1 is concave and surface 2 is convex. This is a diverging lens, so it has a negative focal length.

CONVERGING AND DIVERGING LENSES

Special Case

radius of curvature is negative: $r_1 = -|r_1|$. Because c_2 is behind the lens, surface 2 is convex and its radius of curvature is positive: $r_2 = |r_2|$. So we have

$$\left[\frac{1}{-|r_1|} - \frac{1}{|r_2|}\right] = -\left[\frac{1}{|r_1|} + \frac{1}{|r_2|}\right] < 0$$

which means the focal length of the lens in Figure 38.27B is negative.

Converging Versus Diverging Lenses

Figure 38.28 illustrates the difference between converging and diverging lenses. In each case, parallel rays from a very distant object are incident on the lens. The converging lens (Fig. 38.28A) bends the rays so that they come together or *converge* at a point behind the lens. The point where they come together is the *focal point* F, and a real image of the distant object is formed there. The distance from the lens to the focal point is the focal length f. If you place a screen at F, you see the real image projected onto the screen.

When parallel rays encounter a diverging lens (Fig. 38.28B), the refracted rays separate or *diverge*. Diverging rays never cross, so no real image forms. However, the diverging rays appear to originate from a point in front of the lens. This point is the focal point F of the lens. If you look through the back of the lens, you see a virtual image at the focal point. In the next two sections, we'll explore the images produced by both types of lenses when the objects are *not* infinitely far away.

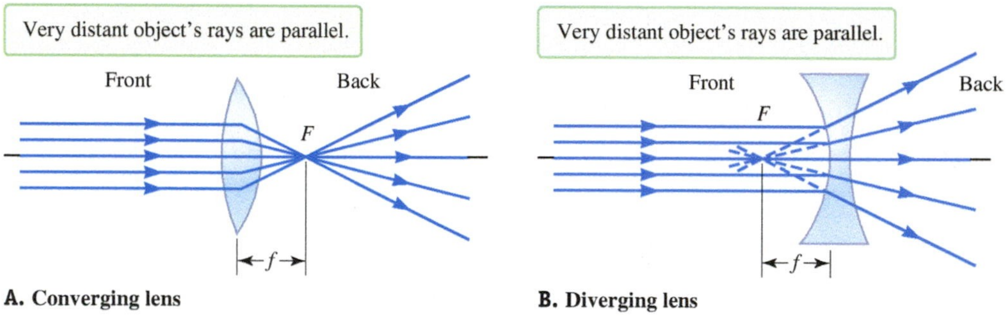

FIGURE 38.28 **A.** When parallel rays are incident on a converging lens, they are bent toward one another and a real image forms at the focal point, sometimes called the *real focal point.* **B.** When parallel rays are incident on a diverging lens, they are bent away from one another and a virtual image forms at the focal point, sometimes called the *virtual focal point.*

EXAMPLE 38.7 Convex Meniscus Lens

A Figure 38.29 shows a convex meniscus lens and an object. If $|r_1| = 6.50\,\text{cm}$ and $|r_2| = 8.50\,\text{cm}$, find the focal length and determine whether the lens is converging or diverging. The lens is made of glass with index of refraction $n = 1.55$.

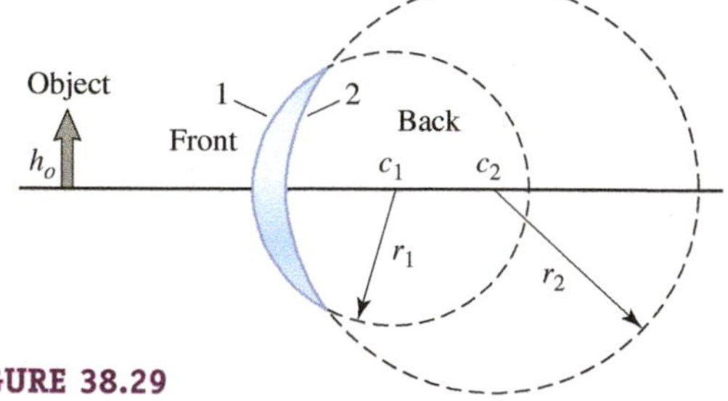

FIGURE 38.29

INTERPRET and ANTICIPATE

The key to solving this problem is to apply the sign conventions correctly (Table 38.3). When we find the focal length using the lens maker's equation, we can use the sign conventions again to determine whether the lens is converging or diverging.

SOLVE

The object is on the left, so the left side of the lens is the front. For both surfaces, the center of curvature is in the back, so both surfaces are convex and both radii of curvature are positive.

$$r_1 = 6.50\,\text{cm}$$
$$r_2 = 8.50\,\text{cm}$$

Substitute values into the lens maker's equation (Eq. 38.7). Remember to invert the equation to find f. The focal length is positive, so according to the sign convention, the lens is converging.

$$\frac{1}{f} = (n-1)\left[\frac{1}{r_1} - \frac{1}{r_2}\right] \quad (38.7)$$

$$\frac{1}{f} = (1.55-1)\left[\frac{1}{6.50\text{ cm}} - \frac{1}{8.50\text{ cm}}\right] \quad (1)$$

$$f = 50.2\text{ cm, converging}$$

CHECK and THINK

Both surfaces are convex, so both have a positive radius of curvature. If the first surface had the larger radius of curvature, the lens would have had a negative focal length (diverging).

B Suppose you flip the lens around, back to front (Fig. 38.30). Find the focal length and determine whether the lens is still converging.

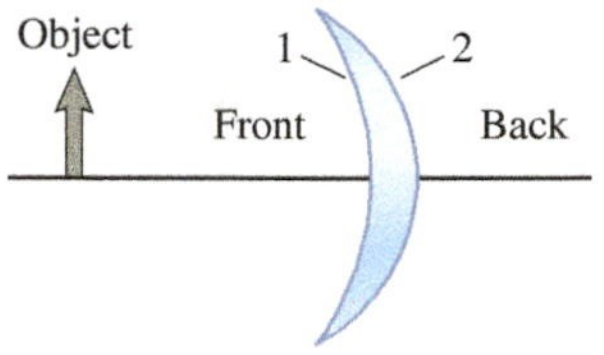

FIGURE 38.30

INTERPRET and ANTICIPATE

This is the same lens, so all the parameters are the same. However, surface 1 and surface 2 are switched, so $|r_1| = 8.50\,\text{cm}$ and $|r_2| = 6.50\,\text{cm}$.

SOLVE

The object is on the left, so the left side is the front. For both surfaces, the center of curvature is in the front, so both surfaces are concave. According to the fourth sign convention in Table 38.3, both radii of curvature must be negative.

$$r_1 = -8.50\,\text{cm}$$
$$r_2 = -6.50\,\text{cm}$$

Substitute values into the lens maker's equation (Eq. 38.7). Notice that Equation (2) is the same as Equation (1) in part A. The focal length is the same as before the lens was flipped.

$$\frac{1}{f} = (n-1)\left[\frac{1}{r_1} - \frac{1}{r_2}\right] \quad (38.7)$$

$$\frac{1}{f} = (1.55-1)\left[\frac{1}{-8.50\text{ cm}} - \frac{1}{-6.50\text{ cm}}\right]$$

$$\frac{1}{f} = (1.55-1)\left[\frac{1}{6.50\text{ cm}} - \frac{1}{8.50\text{ cm}}\right] \quad (2)$$

$$f = 50.2\text{ cm, still converging}$$

CHECK and THINK

We found that the orientation of the lens does not determine whether the lens is converging or diverging.

Ray Diagrams for Thin Lenses

In the next two sections, we study the images that result when an object is placed at a finite distance from a thin lens. The process of finding the image distance and magnification is much like the process we used with spherical mirrors. Again, we use

RAY DIAGRAMS Tool

ray diagrams to illustrate the problem. A **ray diagram** for a thin lens usually includes these five elements (Figs. 38.31 and 38.32; pages 1239 and 1240):

1. The **lens**, including a vertical line that runs through its middle. Sketch the rays as though they refract at this midline.
2. The **optical axis** is perpendicular to the line through the lens and passes through the lens's center.
3. The **focal points** F drawn on both sides of the lens.
4. An **object** represented as a simple shape, such as an arrow. The placement of the object determines the front of the lens. Because we read from left to right, we usually draw the object on the left side of the lens.
5. Two or three **primary rays** (described below) emerging from one point of the object, usually the tip of the arrow. Only two rays are needed to find the position, orientation, and magnification of the image. Sometimes a third ray is helpful to check your drawing.

Primary Rays for Thin Lenses

PRIMARY RAYS FOR THIN LENSES Tool

In Section 37-4, we listed four primary rays that are helpful when working with spherical mirrors. Likewise, there are three primary rays that are helpful when you must find the position, magnification, and orientation of an image produced by a thin lens. These primary rays are described in Table 38.4.

TABLE 38.4 Primary rays for thin lenses.

Ray	Converging lens	Diverging lens
Ray 1 passes through the center of the lens. It is not deflected because any ray that passes through the center of the lens goes through two nearly parallel surfaces.	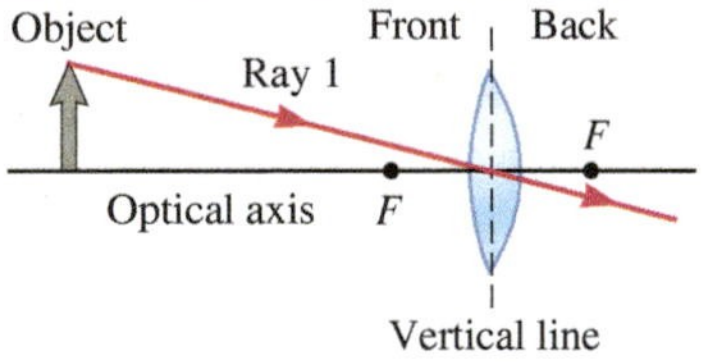	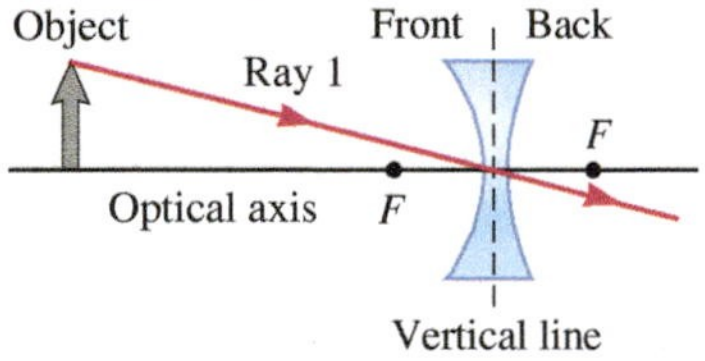
Ray 2 is parallel to the optical axis. For a converging lens, ray 2's refraction passes through the focal point on the back of the lens. For a diverging lens, ray 2's refraction is bent away from the optical axis. When you extend the refracted ray backward, it passes through the focal point on the front of the lens. Notice that you draw the bend in the ray only where it strikes the vertical line through the middle of the lens.	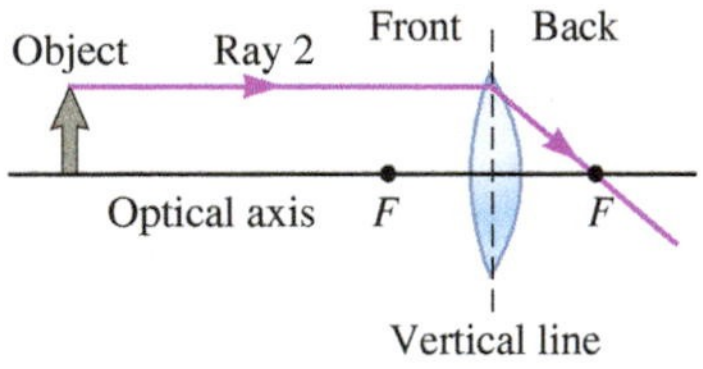	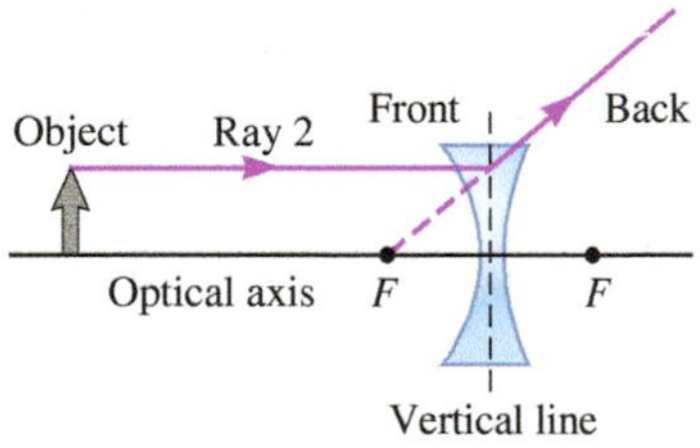
For a converging lens, **ray 3** passes through the focal point on the front of the lens. For a diverging lens, ray 3 is aimed at the focal point on the back of the lens. For both lenses, the refracted ray is parallel to the optical axis.	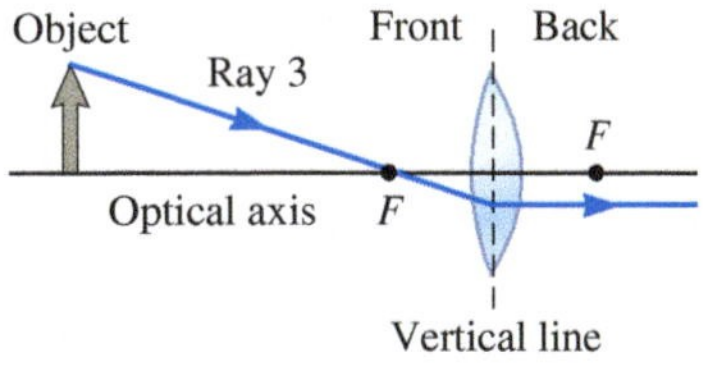	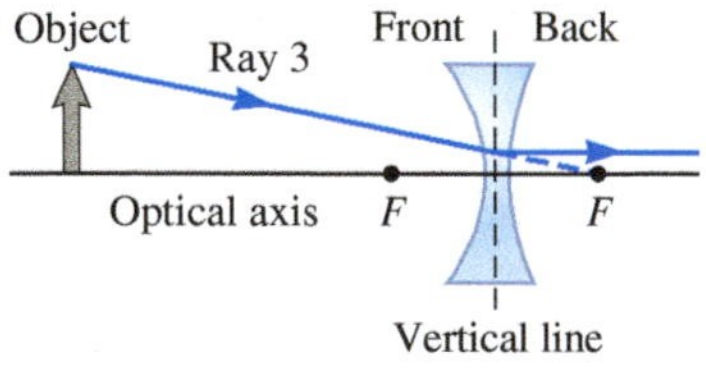

38-6 Images Formed by Diverging Lenses

We can use ray diagrams to find the position, orientation, and magnification of a diverging lens when an object is at some finite distance from it. The location of an object relative to the focal point of a diverging lens does not matter. Figure 38.31 shows an object at some arbitrary distance in front of a diverging lens along with the three primary rays. The three refracted rays that result do not cross behind the lens, so no real image is formed. However, if you look through the lens from the back, the three rays entering your eye can be traced to a point in front of the lens. The resulting image is upright, in front of the lens, virtual, and smaller than the object.

Let's connect our ray diagram (Fig. 38.31) to the thin-lens equation (Eq. 38.5). The focal length is negative because the lens is diverging, and the object distance is positive because the object is real and in front of the lens. The thin-lens equation (Eq. 38.5),

$$\frac{1}{d_i} = \frac{1}{-|f|} - \frac{1}{|d_o|}$$

shows that the image distance must be negative, which means the image is virtual and in front of the lens (Table 38.3) as shown in the figure.

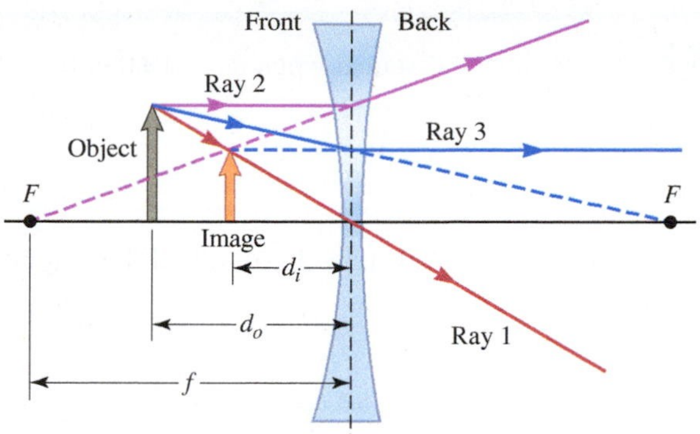

FIGURE 38.31 The three primary rays show that for a diverging lens, the image is virtual, upright, and small.

CONCEPT EXERCISE 38.4

The image produced by a diverging lens is upright and virtual no matter where the object is in front of the lens. We'll show in Section 38-7, however, that for a converging lens, the placement of the object relative to the focal length determines whether the image is virtual or real, upright or inverted, and larger or smaller than the object. Examine the convex and concave mirrors in Sections 37-5 and 37-6. Is a diverging lens more like a convex mirror or a concave mirror? Explain.

EXAMPLE 38.8 All Possible Regions for a Diverging Lens

A diverging lens has focal length $|f| = 10.0$ cm. Find the image distance and magnification if the object is

A 15.0 cm,

B 10.0 cm, and

C 5.00 cm in front of the lens.

INTERPRET and ANTICIPATE

For a diverging lens, the image is upright, in front of the lens, and smaller than the object, no matter where the real object is placed. So we expect to find $0 < M < 1$ and $d_i < 0$ in all three cases.

SOLVE

Isolate the image distance on one side of the thin-lens equation (Eq. 38.5). Substitute the focal length, which is negative because this is a diverging lens.

$$\frac{1}{d_i} = \frac{1}{f} - \frac{1}{d_o} = -\frac{1}{10.0\,\text{cm}} - \frac{1}{d_o}$$

A Find the image distance when $d_o = 15.0$ cm.

$$\frac{1}{d_i} = -\frac{1}{10.0\,\text{cm}} - \frac{1}{15.0\,\text{cm}} = -\frac{5}{30.0\,\text{cm}}$$

$$d_i = -6.00\text{ cm}$$

Use $M = -d_i/d_o$ (Eq. 38.6) to find the magnification.

$$M = -\frac{d_i}{d_o} = -\frac{-6.00\text{ cm}}{15.0\text{ cm}}$$

$$M = +0.400$$

Example continues on page 1240 ▶

B Find the image distance when $d_o = 10.0$ cm.	$\frac{1}{d_i} = -\frac{1}{10.0\text{ cm}} - \frac{1}{10.0\text{ cm}} = -\frac{2}{10.0\text{ cm}}$ $d_i = -5.00$ cm
Again, use $M = -d_i/d_o$ (Eq. 38.6) to find the magnification.	$M = -\frac{d_i}{d_o} = -\frac{-5.00\text{ cm}}{10.0\text{ cm}}$ $M = +0.500$
C Find the image distance when $d_o = 5.00$ cm.	$\frac{1}{d_i} = -\frac{1}{10.0\text{ cm}} - \frac{1}{5.00\text{ cm}} = -\frac{3}{10.0\text{ cm}}$ $d_i = -3.33$ cm
Find the magnification as before.	$M = -\frac{d_i}{d_o} = -\frac{-3.33\text{ cm}}{5.00\text{ cm}} = +0.666$

CHECK and THINK

As expected, the image distance is negative in all three cases, so the image is in front of the lens and virtual. Also, the magnification is positive and less than 1, so the image is upright and smaller than the object in all three cases. When the object distance equals the focal length (part B), the image is half the size of the object. When the object is farther away than the focal length (part A), the image is less than half the size of the object. When the object is closer than the focal length (part C), the image is more than half the size of the object.

38-7 Images Formed by Converging Lenses

In this section, we explore converging lenses using ray diagrams. The image produced by a converging lens may be real or virtual, upright or inverted, depending on the placement of the object. As in the case of diverging lenses, our goal here is to find the position, orientation, and magnification of a converging lens when an object is at some finite distance from it. There are three regions in which to place the object: (1) farther away from the lens than the focal point, (2) closer to the lens than the focal point, and (3) at the focal point. For each of these regions, we'll draw a ray diagram, including the primary rays, and then we'll connect our ray diagram to the lens equation.

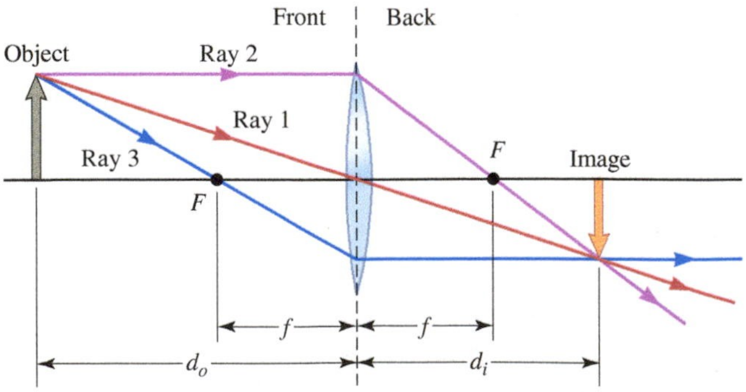

FIGURE 38.32 An object is farther from the converging lens than the focal point. The image is inverted, real, and behind the lens. In this figure, the image is smaller than the object.

Converging Lens: Object Farther than Focal Point

Consider an object that is farther away from the front of a converging lens than the focal point (Fig. 38.32). The three primary rays refract and cross behind the lens, forming a real image. If you put a screen where the rays cross, you would see an inverted image. So, the image is inverted, behind the lens, and real.

Let's connect our ray diagram (Fig. 38.32) to the thin-lens equation (Eq. 38.5). The focal length is positive because the lens is converging, and the object distance is positive because the object is real and in front of the lens. So the thin-lens equation becomes

$$\frac{1}{d_i} = \frac{1}{+|f|} - \frac{1}{+|d_o|}$$

Because the object is farther away from the lens than the focal point, $d_o > f$ and the first term on the right must be greater than the second term:

$$\frac{1}{|f|} > \frac{1}{|d_o|}$$

which means the image distance must be positive:

$$\frac{1}{d_i} > 0$$

A positive image distance (Table 38.3) means the image is real and behind the lens, in agreement with Figure 38.32.

The image may be smaller or larger than the object, depending on the object's position. To see why the object's position matters, imagine switching the situation shown in Figure 38.32 so that the arrow labeled "Image" is an object in front of the lens. Now the front of the lens is on the right, and the back of the lens is on the left. When you reverse the arrowheads on the primary rays, the rays come together at the arrow labeled "Object," which we now think of as the image of the arrow on the right. In this case, the image is larger than the object.

We can also do this switching mathematically. For a converging lens with an object farther from the lens than the focal point, the object distance, image distance, and focal length are all positive. So, if the object and image distances are switched in the thin-lens equation,

$$\frac{1}{d_o} + \frac{1}{d_i} = \frac{1}{d_i} + \frac{1}{d_o} = \frac{1}{f}$$

the equation still holds for the same focal length. We say the thin-lens equation is *symmetrical* with respect to such a switch of variables. However, switching these distances changes the magnification because $M = -d_i/d_o$ (Eq. 38.6) is not symmetrical. If $d_o > d_i$, as in Figure 38.32, $|M| < 1$ and the image is smaller than the object. If $d_o < d_i$, as if the arrowheads in Figure 38.32 were reversed, $|M| > 1$ and the image is larger than the object.

Converging Lens: Object Closer than Focal Point

Figure 38.33 shows an object that is closer to a converging lens than the focal point. Drawing the primary rays is a little trickier because to draw ray 3 you must start by extending the ray from the focal point on the front side. Of course, no ray comes from that focal point, so the portion of the ray from the focal point to the tip of the object is shown as a dashed line.

The three refracted rays do not cross behind the lens, so no real image forms. If you trace the three refracted rays backward, they intersect at a point in front of the lens. A virtual image forms at this point. The image is upright and larger than the object.

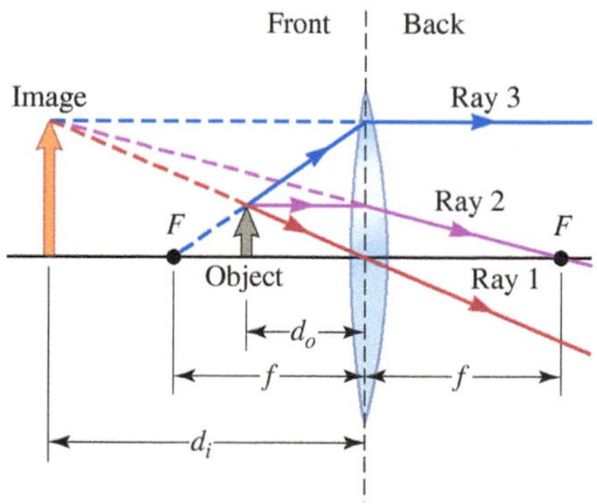

FIGURE 38.33 An object is closer to the converging lens than the focal point. The image is upright, virtual, and in front of the lens.

Now let's connect our ray diagram (Fig. 38.33) to the thin-lens equation (Eq. 38.5). The object distance is positive because the object is real and in front of the lens. The focal length is positive because the lens is converging. The object distance is less than the focal length: $d_o < f$. So,

$$\frac{1}{|f|} < \frac{1}{|d_o|}$$

The thin-lens equation becomes

$$\frac{1}{d_i} = \frac{1}{f} - \frac{1}{d_o} < 0$$

so the image distance is negative. According to the sign conventions in Table 38.3, the image is virtual and in front of the lens, in agreement with the ray diagram.

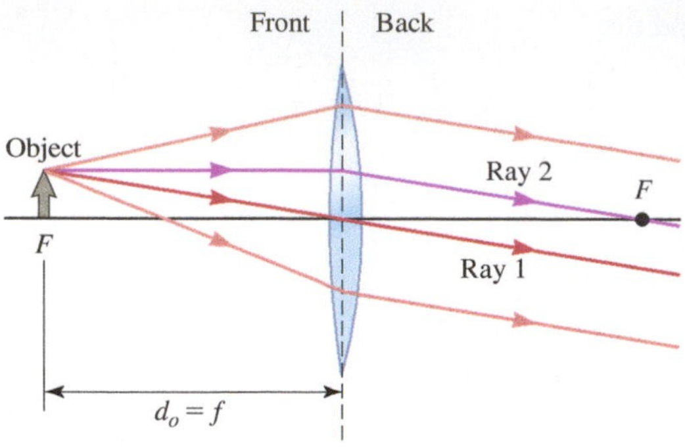

FIGURE 38.34 An object at the focal point does not form an image; the rays that emerge from the lens are parallel. Put another way, the image forms at infinity.

Converging Lens: Object at Focal Point

Finally, consider an object at the focal point of a converging lens, along with two primary rays (Fig. 38.34). We cannot include ray 3 because we cannot extend a ray through the focal point and the tip of the arrow. (The light would have to pass through the object itself.) However, we have included two other rays that also emerge from the tip of the object. All the refracted rays are parallel. Because they don't intersect, no real or virtual image forms.

Now let's apply the thin-lens equation (Eq. 38.5) to this situation by setting the object distance equal to the focal length ($d_o = f$):

$$\frac{1}{d_o} + \frac{1}{d_i} = \frac{1}{f} + \frac{1}{d_i} = \frac{1}{f}$$

The focal length cancels out:

$$\frac{1}{d_i} = 0$$

So the image distance must be infinite: $d_i \to \infty$. Our interpretation of this result is that the parallel refracted rays cross at infinity, which means the image forms at infinity. Compare this result to the situation when an object is placed at the focal point of a concave mirror (Section 37-6).

CONCEPT EXERCISE 38.5

Review the discussions of convex and concave mirrors in Sections 37-5 and 37-6. Is a converging lens more like a convex mirror or a concave mirror? Explain.

EXAMPLE 38.9 All Possible Regions for a Converging Lens

A converging lens has focal length $|f| = 10.0$ cm. Find the image distance and magnification if the object is

A 15.0 cm,
B 10.0 cm, and
C 5.00 cm in front of the lens.

INTERPRET and ANTICIPATE

For a converging lens, whether the image is upright or inverted, in front of or behind the lens, and larger or smaller than the object depends on where the real object is placed. If the object is farther away than the focal point, we expect an inverted real image to form behind the lens. That image may be larger or smaller than the object. If the object is at the focal length, we expect an image to form at infinity (no image). Finally, if the object is closer than the focal point, we expect an upright virtual image to form in front of the lens, larger than the object. The distances here are the same as in Example 38.8 for a diverging lens. To think about our results, we'll compare our answers in this example to what we found there.

SOLVE

Isolate the image distance on one side of the thin-lens equation (Eq. 38.5). Substitute for the focal length, which is positive because this is a converging lens.	$\frac{1}{d_i} = \frac{1}{f} - \frac{1}{d_o} = \frac{1}{10.0\,\text{cm}} - \frac{1}{d_o}$
A Find the image distance when $d_o = 15.0$ cm. This situation corresponds to Figure 38.32.	$\frac{1}{d_i} = \frac{1}{10.0\,\text{cm}} - \frac{1}{15.0\text{ cm}} = \frac{1}{30.0\,\text{cm}}$ $d_i = 30.0$ cm

Use $M = -d_i/d_o$ (Eq. 38.6) to find the magnification.	$M = -\frac{d_i}{d_o} = \frac{-30.0\text{ cm}}{15.0\text{ cm}} = -2.00$

CHECK and THINK

The positive image distance means the image is behind the lens and real, as in Figure 38.32. The negative magnification means the image is inverted. Because $|M| > 1$, the image is larger than the object.

B Find the image distance when $d_o = 10.0$ cm.	$\frac{1}{d_i} = \frac{1}{10.0\text{ cm}} - \frac{1}{10.0\text{ cm}} = 0$ $d_i \to \infty$

CHECK and THINK

When the object is at the focal point, the image forms at infinity. In this case, it does not make sense to calculate a magnification because no actual image forms.

C Find the image distance when $d_o = 5.00$ cm.	$\frac{1}{d_i} = \frac{1}{10.0\text{ cm}} - \frac{1}{5.00\text{ cm}} = -\frac{1}{10.0\text{ cm}}$ $d_i = -10.0$ cm
Again, use Equation 38.6 to find the magnification.	$M = -\frac{d_i}{d_o} = -\frac{-10.0\text{ cm}}{5.00\text{ cm}} = +2.00$

CHECK and THINK

In part C, we found that the image distance is negative, so the image is virtual and in front of the lens. The magnification is positive and greater than 1, so the image is upright and larger than the object, as in Figure 38.33.

Now let's compare our results for this converging lens to what we found for the diverging lens in Example 38.8. The diverging lens always produces an upright virtual image in front of the lens, and the image is always smaller than the object. When the object is at the focal point, there is nothing special about the image other than that it is exactly half the size of the object. By contrast, a converging lens does not produce an image if the object is at the focal point. If the object is farther than the focal point, a converging lens produces a real inverted image that may be larger or smaller than the object. Finally, if the object is closer than the focal point, the converging lens produces an upright virtual image that is always larger than the object. So, a converging lens can enlarge an image, whereas a diverging lens cannot. These facts will be important as we consider the applications of lenses in the case study.

38-8 The Human Eye

This chapter's case study is about devices with lenses that you use to see. Your vision begins with your eyes, so first we look at the how the human eye works. We then consider devices that use one lens, and finally devices that use two lenses.

Anatomy of the Human Eye

Figure 38.35 shows the anatomy of the human eye. The **sclera** is the white part of the eye; the **iris** is the colored part. The **pupil** is the dark spot in the center of the iris. It is actually a hole or aperture that allows light to pass into the eye's interior. The pupil's diameter changes in response to the amount of light. In bright light, the pupil's diameter may be as small as 2 mm, whereas in the dark, its diameter may

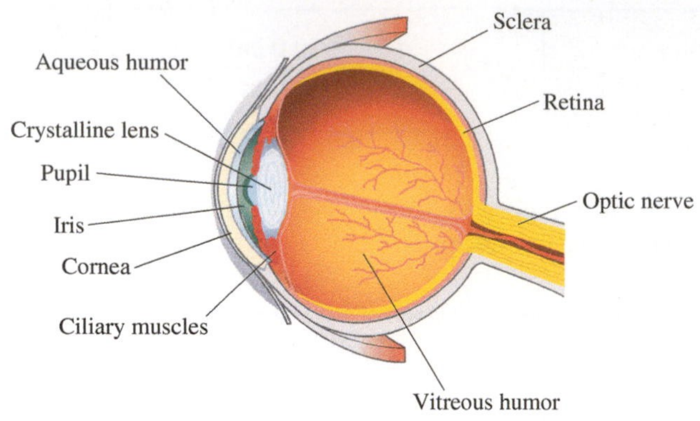

FIGURE 38.35 Anatomy of the human eye.

be 8 mm. The **cornea** is a clear protective covering over the iris and pupil that allows light to pass through. Behind the cornea is a compartment full of transparent fluid similar to water, called the **aqueous humor**. Behind this compartment is the **crystalline lens**, or simply the **lens**. Behind the lens is another compartment full of watery fluid called the **vitreous humor**. At the back of the eye is the light-sensitive **retina**. In response to light, the retina produces a signal that travels along the **optic nerve** to the brain.

Modeling the Eye

Now we develop a model for the optical system of the eye. Light incident in air first encounters the cornea. The index of refraction of the cornea is about 1.376. Next, the light enters the aqueous humor, a fluid with an index of refraction of 1.336. Then the light passes through the lens, which has an index of refraction of 1.410. And, finally, the light travels through the vitreous humor to the retina. The vitreous humor is very similar to the aqueous humor, so its index of refraction is also 1.336.

The light thus passes through four boundaries—(1) from air to cornea, (2) from cornea to aqueous humor, (3) from aqueous humor to lens, and (4) from lens to vitreous humor—on its way to the retina. The biggest difference in the indices of refraction is between the air ($n_{\text{air}} = 1.00026$) and the cornea ($n_{\text{cornea}} = 1.376$), so the greatest refraction occurs at the air–cornea boundary. (This explains why it is difficult for you to see underwater. The index of refraction of the water is close to the index of refraction of the cornea, so light is not refracted as much when you look through water as when you look through air.) Although the light passes through four media before reaching the retina, we can still model the eye as a single lens. (A more detailed model would include two lenses—the cornea and the crystalline lens—and would take into account the watery fluid surrounding the crystalline lens.)

Does the eye produce a real or virtual image? Refracted rays of light must make contact with the retina in order to stimulate its photo sensors and create an electrical signal that passes on to the brain. So a real image must be projected onto the retina. Because a diverging lens does not form a real image, we should model the eye as a converging lens and the retina as a screen where a real image forms. Now, is the image inverted or upright? From Figure 38.32, the real image formed by a converging lens is inverted. Objects do not appear upside down to you because the electrical signal that transmits the image is processed by your brain, which has learned to invert that image. As an analogy, consider a digital camera. The image that forms on the light sensor in the camera may be inverted, but the software in the camera displays an upright image to you.

How the Human Eye Focuses

The image distance in your eye is fixed because the distance between your lens and retina does not change. In a normal human eye, the image distance is about an inch (2.2 cm to 2.6 cm). Of course, people like to look at objects at various distances d_o. Because d_i is fixed, the focal length of the lens must be adjusted to form a real image on the retina.

How can the focal length of the lens be adjusted? The answer comes from the lens maker's equation:

$$\frac{1}{f} = (n - 1)\left[\frac{1}{r_1} - \frac{1}{r_2}\right] \tag{38.7}$$

We cannot change the index of refraction of our lens. However, the crystalline lens is flexible and is held in place by ligaments attached to the muscles of the **ciliary**, a ring-like structure surrounding it (Fig. 38.35). The contraction (or relaxation) of the ciliary muscles changes the shape (r_1 and r_2) and therefore the focal length of the lens.

When its ciliary muscles are relaxed, a normal eye is able to focus on objects at infinity. In practical terms, these are objects that are about 5 m or more from the eye.

The rays from such distant objects are parallel. So, when the eye focuses on a distant object, its (converging) lens forms an image on the focal plane and $d_i = f$. In a normally working eye, the focal plane is the retina, typically located 2.5 cm behind the lens (Fig. 38.36A). The *far-point distance* d_{far} is the maximum distance at which an eye can form an image of an object on its retina. The far point of a normal eye is at infinity, and its lens has focal length $f = 2.5$ cm when it focuses on distant objects.

There are two common disorders that alter the far-point distance. Figure 38.36B shows a myopic or nearsighted eye. In a myopic eye, the distance between the lens and the retina is farther than in a normal eye. Parallel rays from a distant object converge at a point in front of the retina. As a result, the person's far point is closer than infinity. In a hyperopic or farsighted eye (Fig. 38.36C), by contrast, the distance between the lens and the retina is shorter than in a normal eye. As a result, the image of a distant object forms behind the retina.

To see a close object, on the other hand, the eye's ciliary muscles must contract (Fig. 38.37). For example, a comfortable reading distance is $d_o = 25$ cm. The image distance is still fixed at $d_i = 2.5$ cm as before. By the thin-lens equation, the focal length must now be

$$\frac{1}{25\text{ cm}} + \frac{1}{2.5\text{ cm}} = \frac{1}{f}$$

$$f = 2.3\text{ cm}$$

which is shorter than the relaxed focal length ($f = 2.5$ cm). Contracting the ciliary muscles in order to look at close objects for a long time, such as when you read, can cause eyestrain. Some doctors recommend that you briefly focus on distant objects every time you reach the end of a page.

The closest distance at which an eye can focus is called the *near-point distance* d_{near}. As a person grows older, the eye sheds cells that build up on the crystalline lens, causing the lens to grow thicker and become harder to flex. Older people thus find it more difficult to focus on nearby objects. The near point moves farther away with increasing age, an effect called *presbyopia*. Table 38.5 lists typical near-point distances as a function of age. When you are in your twenties, your near-point distance is about 10 cm, so you can hold a menu 10 cm from your face and read it comfortably. By the time you are in your sixties, you must hold the menu 1 m to 2 m from your face in order to read it. Because your arms are simply not long enough, you must wear corrective lenses.

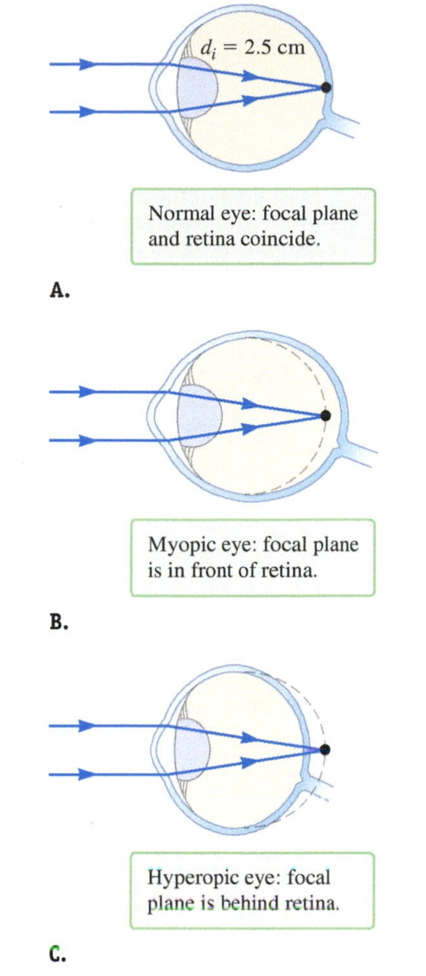
$d_i = 2.5$ cm
Normal eye: focal plane and retina coincide.
A.
Myopic eye: focal plane is in front of retina.
B.
Hyperopic eye: focal plane is behind retina.
C.

FIGURE 38.36 When an eye focuses on a distant object, the muscles are relaxed and the focal length of the lens is at its maximum. **A.** For a normal eye, the image is formed on the retina. **B.** For a myopic eye, the image is formed in front of the retina. **C.** For a hyperopic eye, the image is formed behind the retina.

TABLE 38.5 Typical near point as a function of age.

Age in years	Near point d_{near} in cm
10	8
20	10
30	13
40	20
50	42
60	100

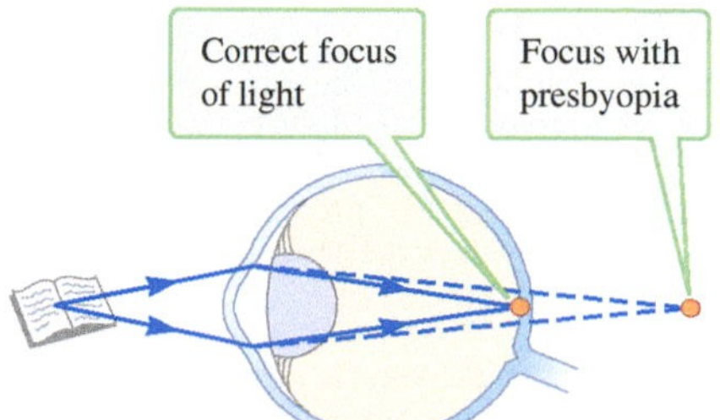
Correct focus of light
Focus with presbyopia

FIGURE 38.37 Reading text at a close distance requires the ciliary muscles to contract so that the focal length of the lens is shortened. As people age, it becomes more difficult to shorten that focal length. When an older person tries to read text up close, the image forms at a point behind her retina.

Corrective Lenses

We have discussed three problems with the human eye. Myopia occurs when the far-point distance d_{far} is too small, and the image of a distant object forms in front of the retina. Hyperopia and presbyopia occur when the near-point distance d_{near} is too great, and the image of a near object forms behind the retina (Fig. 38.36C)—because of either the shape of the eye (hyperopia) or the inflexibility of its lens

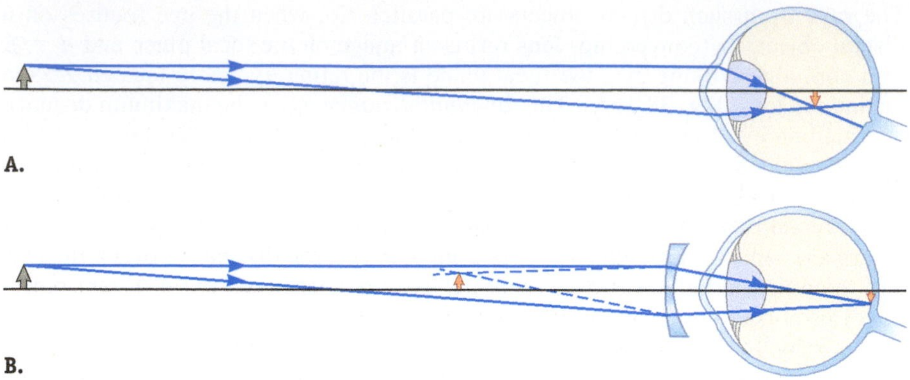

FIGURE 38.38 **A.** In a myopic eye, the image forms in front of the retina. The eye's lens over-refracts the rays. The result is that the far point is too close to the person's eye. **B.** A diverging corrective lens compensates for this over-refraction by spreading out the rays and producing an image closer than the object. The person looks at this closer image.

(presbyopia). All three problems can be fixed with corrective lenses such as eyeglasses or contact lenses.

In Figure 38.38A, an uncorrected myopic eye forms an image of a very distant object in front of the retina. In effect, the lens has over-refracted the rays. If the object were closer, the lens would refract the rays by the right amount and the image would form on the retina. A corrective lens is needed that will create an image of a very distant object closer to the eye, at the person's natural far point. Because the person will look at this image through the corrective lens, the image must be upright and virtual (Fig. 38.38B). A diverging lens always produces an upright virtual image that is closer to the lens than the object is (Fig. 38.31) and thus corrects for myopia. The diverging lens spreads out the rays to correct for the over-refracting converging lens of the myopic eye.

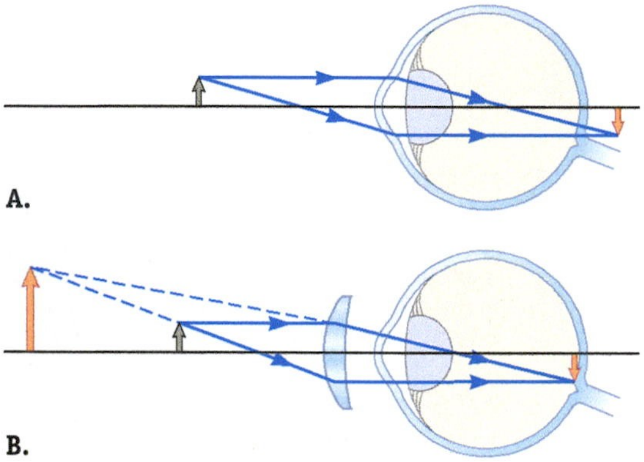

FIGURE 38.39 **A.** In a hyperopic or presbyopic eye, the image forms behind the retina because the eye's lens cannot refract the rays sufficiently. As a result, the near point is too far from the person's eye. **B.** A converging corrective lens compensates for this under-refraction by bringing the rays closer together, producing an image farther away than the object. The person looks at this far image.

In physics, *power* can also mean energy per unit time. Don't confuse the two usages.

In both a hyperopic and a presbyopic eye, the near-point distance is too great and the image forms behind the retina (Fig. 38.39A). In effect, the eye's lens does not refract the rays sufficiently. If the object were farther, the lens would refract the rays by the right amount and the image would form on the retina. A corrective lens must create an image that is farther from the eye. Because the person looks at this image through the lens, this image must be upright and virtual (Fig. 38.39B). A converging lens produces an upright virtual image that is farther than the object as long as the object is closer than the focal point (Fig. 38.33). The normal reading distance is 25 cm, so as long as the focal length of the corrective converging lens is greater than 25 cm, the lens produces an upright virtual image. By bending the rays together, the converging lens corrects the hyperopic or presbyopic eye's under-refraction of the rays, allowing an image to form on the retina.

Instead of specifying the focal length of a corrective lens, it is common to specify its *power*. The **power** D of a lens is the reciprocal of the focal length:

$$D = \frac{1}{f} \tag{38.8}$$

The SI unit of focal length is the meter, so the SI unit of power is m^{-1}, but the power of a lens is usually expressed in diopters, where

$$1 \text{ diopter} = 1 \text{ m}^{-1} \tag{38.9}$$

For example, a prescription for a corrective lens may be $D = +5$ diopters, corresponding to a focal length of $f = 0.2$ m. Because the power is positive, the focal length is positive, and this prescription is for a converging lens. If the power is negative, the lens is diverging.

EXAMPLE 38.10 CASE STUDY How to Read Contact Lens Prescriptions

Sophia and Ezra both wear contact lenses. Sophia's corrective lenses have a power of −4.50 diopters; Ezra's lenses have a power of +2.75 diopters.

A Determine whether each person has myopia or hyperopia. Then find the focal length of each person's corrective lenses.

B Assuming a nearsighted person can comfortably see infinitely distant objects when wearing corrective lenses, what is the nearsighted person's uncorrected far-point distance d_{far}? Give your answer in centimeters.

C Assume the farsighted person's corrective lenses allow him or her to just barely read at 25 cm. What is that person's near-point distance d_{near} without corrective lenses?

INTERPRET and ANTICIPATE

In the case of myopia, the person's far-point distance is too short; a nearsighted person can see things that are up close but not far away. So the corrective lenses must produce an image that is closer than the object. In the case of hyperopia, the person's near-point distance is too great; a farsighted person can see things that are far away but not up close. So the corrective lenses must produce an image that is farther than the object. With these ideas in mind, we can reason out whether the corrective lenses must be converging or diverging in each case and, from that, the sign of the focal length and the power. We expect the nearsighted person has a far point closer than 5 m and the farsighted person has a near point farther than 25 cm.

SOLVE

A A person with myopia needs diverging lenses to create an image that is closer than the object. A diverging lens has a negative focal length and power.	Sophia's prescription is for lenses that have a negative power; Sophia is myopic.
A person with hyperopia needs converging lenses to create an image that is farther away than the object. A converging lens has a positive focal length and power.	Ezra's prescription is for lenses that have a positive power; Ezra is hyperopic.
Now let's find the focal length of Sophia's contacts. The focal length is the reciprocal of the power (Eq. 38.8).	$D = \frac{1}{f}$ (38.8) $f = \frac{1}{D} = \frac{1}{-4.50\,\text{diopters}} = -0.222\,\text{m}$ $f = -22.2$ cm for Sophia
Find the focal length of Ezra's contacts in the same way.	$D = \frac{1}{f}$ $f = \frac{1}{D} = \frac{1}{2.75\,\text{diopters}} = 0.364\,\text{m}$ $f = 36.4$ cm for Ezra
B Next, we find Sophia's uncorrected far-point distance. Sophia's contacts produce an image at her far point. Assume the object is very far away, $d_o \to \infty$. In this case, the image distance is negative because the image is virtual and in front of the lens (Fig. 38.38B). Sophia's uncorrected far point is the absolute value of this image distance.	$\frac{1}{\infty} + \frac{1}{d_i} = \frac{1}{f}$ $d_i = f = -22.2$ cm $d_{\text{far}} = 22.2$ cm for Sophia

Example continues on page 1248 ▶

C Ezra's contact lenses allow him to read words that are 25 cm from his eyes, so this is the object distance d_o. These lenses produce an image located at his natural near point. The image distance is negative because the image is in front of the contact lens (Fig. 38.39B). This image is the object of his eye's lens. Ezra's uncorrected near-point distance is the absolute value of the image distance.

$$\frac{1}{d_o} + \frac{1}{d_i} = \frac{1}{f}$$

$$\frac{1}{d_i} = \frac{1}{f} - \frac{1}{d_o} = \frac{1}{36.4 \text{ cm}} - \frac{1}{25 \text{ cm}}$$

$$d_i = -79.8 \text{ cm}$$

$$d_{\text{near}} = 80 \text{ cm for Ezra}$$

CHECK and THINK

As expected, the nearsighted Sophia has a far-point distance less than 5 m. Her uncorrected eyes cannot focus on objects that are farther than about 22 cm. Also, as expected, the farsighted Ezra has a near-point distance greater than 25 cm. His uncorrected eyes cannot focus on objects that are closer than about 80 cm. If Ezra suffers from presbyopia rather than hyperopia, he is probably about 55 years old (Table 38.5).

38-9 One-Lens Systems

In this section, we look at two practical devices that require only a single lens. The first device is a basic camera. Cameras were originally used by artists, but other individuals quickly saw their value. Today, the camera plays an important role in astronomy, biology, police work, medicine, and communications. The second device is a **simple magnifier**, or just **magnifier**. In ordinary language, it is more commonly called a magnifying glass. You have probably used a magnifier to look at coins, stamps, fingerprints, or jewelry.

akg-images/Newscom

FIGURE 38.40 Daguerre's camera.

The Camera

The digital camera you carry in your pocket can trace its evolution back to a modified pinhole camera (Section 37-1). You can make a pinhole camera (a small camera obscura) by poking a small hole in a box. Light from outside the camera forms an inverted real image on a wall or screen inside the device. Artists, such as the scene painter Louis Daguerre (1787–1851), used the camera obscura to trace the image of a subject on paper or on another medium. Daguerre was also a physicist who wanted to find another way to preserve nature in an image. His primary contributions to photography were a photosensitive medium (iodized silver on a glass plate), a development process (exposure to mercury vapors), and—most important—a process for fixing the image on the medium in order to make it permanent (putting the plate in a salt solution). Today's film cameras contain film coated in a photosensitive medium, and part of the development process fixes the image so that further exposure to light does not cause the image to disappear.

Charles D. Winters/Cengage Learning

FIGURE 38.41 A contemporary camera has an adjustable aperture, much like the pupil of your eye.

Daguerre used a modified pinhole camera (Fig. 38.40). Its chamber had a somewhat large hole or aperture at one end, with a lens behind the hole to keep the image sharp. The image was projected onto a glass plate coated with iodized silver. The aperture in a camera is adjustable, so it can be widened or narrowed depending on the lighting conditions (Fig. 38.41). The photosensitive iodized silver is analogous to the photosensitive cells of the retina. A real image must form on the glass plate (or film or digital detector) in a camera, as on the retina. So a camera's lens must be converging, and the image formed on the plate or film is inverted. After the image is printed, we simply invert the photograph again in order to view it.

In the human eye, the image distance is fixed because the distance between the lens and retina does not change. You are able to focus on objects at various distances because your crystalline lens is flexible, able to change shape and therefore focal length. By contrast, the camera's lens is not flexible; its focal length is fixed by its shape (and the material it is made of). In order to form an image on the plate, then, the image distance must be flexible. As shown in Figure 38.40, Daguerre's camera consisted of two open-ended boxes, one holding the lens and the other the plate. Daguerre could slide the two boxes relative to each other to change the image distance, creating a sharp image. In a contemporary camera, the lens sits in a cylinder that moves in and out in order to adjust the image distance.

A contemporary camera usually has a lens made of several optical elements that collectively act like a single converging lens. These optical elements are designed to work together to reduce aberrations (defects that result in a poor image; Section 37-7). Aberrations are not a result of poor construction but are a natural consequence of how light is refracted through lenses. Like spherical mirrors, spherical lenses suffer from spherical aberration when the rays that reach them are not paraxial.

Lenses also suffer from **chromatic aberration**, which occurs because the index of refraction of light depends on its color. Violet light is refracted more than red light (Fig. 38.9), so when white light is refracted by a lens, the different colors that make up white light focus at different positions (Fig. 38.42A). When colors do not focus at exactly the same position, the resulting photograph shows a halo of colors. Figure 38.42B shows how chromatic aberration may be partially compensated for with a second lens. In this case, red and yellow light form an image at a single point, but blue light still does not focus at this point.

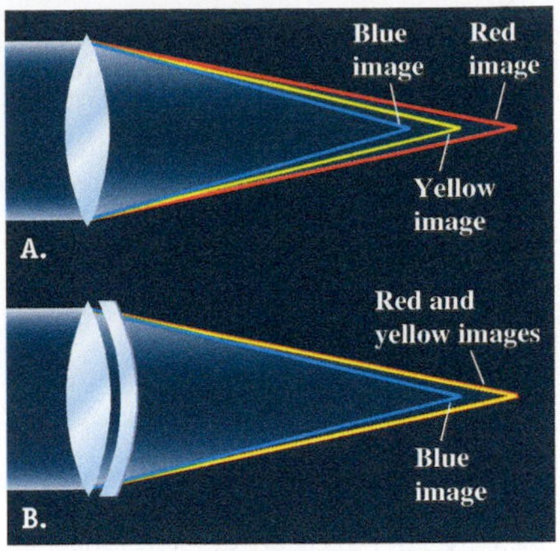

FIGURE 38.42 **A.** A single converging lens suffers from chromatic aberration. Blue light forms an image in front of yellow light, which forms an image in front of red light. **B.** A second lens is added, and the chromatic aberration is improved. Now the red and yellow images are formed at the same position.

EXAMPLE 38.11 Making Your Own Daguerre Camera

You can build a simple camera out of a pair of open boxes provided one box can easily slide inside the other. Wrap the seam between the boxes in dark fabric to keep the light out. Place light-sensitive paper at the back wall of one box, and poke a small hole in the front wall of the other box. Attach a lens in front of the hole.

Suppose you use a converging lens with a 30.0-cm focal length. Find the closest and farthest useful distances between the front and back walls of the two boxes. Be practical in your design. Assuming you will typically take pictures of objects that are 5 m from the camera, what is the typical useful distance between the front and back walls?

INTERPRET and ANTICIPATE

The thin-lens equation applies in this case. The lens is converging and the object is in front of it, so both the focal length and the object distance are positive. The image distance, which is also positive, is the distance between the front and back walls. So, for the first question, think about finding the full useful range of image distances. For the second question, solve for the image distance when the object distance is 5 m.

SOLVE

The image distance is shortest when the object is at infinity. In that case, the image distance equals the focal length. So, the closest useful distance between the front and back walls equals the focal length.

$$\frac{1}{d_o} + \frac{1}{d_i} = \frac{1}{f} \quad (38.5)$$

$$\frac{1}{\infty} + \frac{1}{d_i} = \frac{1}{f}$$

$$d_i = f = 30.0 \text{ cm}$$

(closest useful distance between walls)

Example continues on page 1250 ▶

A converging lens does not produce a real image if the object is at the focal point or closer (Figs. 38.33 and 38.34). Assume that, for practical purposes, you don't expect to photograph objects that are closer than 60.0 cm, or twice the focal length of the lens. (60.0 cm is about 2 ft—shorter than your arm. You probably won't take a picture of a friend from less than an arm's length away.)

$$\frac{1}{d_i} = \frac{1}{f} - \frac{1}{d_o}$$

$$\frac{1}{d_i} = \frac{1}{30.0\text{ cm}} - \frac{1}{60.0\text{ cm}} = \frac{1}{60.0\text{ cm}}$$

$$d_i = 60.0\text{ cm}$$

(farthest useful distance between walls)

But it seems reasonable to take pictures of objects that are about 5 m from the camera. We find that this image distance is very close to the focal length of the lens.

$$\frac{1}{d_i} = \frac{1}{f} - \frac{1}{d_o} = \frac{1}{30.0\text{ cm}} - \frac{1}{500\text{ cm}}$$

$$d_i = 32\text{ cm}$$

(typical useful distance)

CHECK and THINK

Now we see why early cameras were so large. For them to focus on close objects, the distance between the lens and the glass plate had to be large. Notice that the image distance nearly equals the focal length when the object is at 5 m. This gives us a practical measure for what we mean when we say an object is infinitely far. For a lens with a 30.0-cm focal length, an object at 5 m is, practically speaking, infinitely far away.

CASE STUDY The Magnifying Glass

Christian Musat/Shutterstock.com

A.

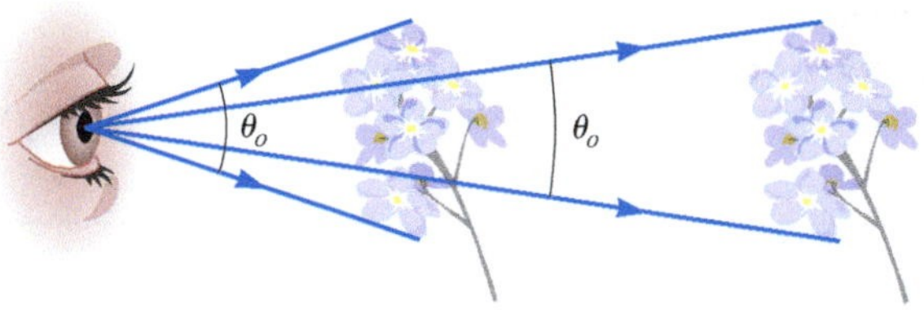

B.

FIGURE 38.43 A. A magnifier produces an upright virtual image. **B.** The angular size θ_o is the angle subtended by the object at your eye. When the flower is close to your eye, it subtends a larger angle, so it appears larger than it does when it is far from your eye.

Now let's turn our attention to the second example of a single-lens device—a magnifier (Fig. 38.43A). A magnifying lens differs from a camera lens in that when you look through a magnifying lens, you always see an enlarged image. This image must be upright and on the same side of the lens as the object. This image is virtual, but it cannot be made by a diverging lens because such a lens always produces an image that is smaller than the object. A converging lens produces a virtual upright image that is larger than the object when the object is closer to the lens than the focal point (Fig. 38.33).

An object looks bigger to you if its image fills up more space on your retina. Mathematically, we describe the apparent size of an object in terms of its angular size θ_o as defined in Figure 38.43B. An object that is close to your eye has a large angular size and appears large. However, you cannot focus on objects closer to your eye than your near point d_{near}. So there is a limit to the angular size of a particular object (Fig. 38.44A). If the object's height is h_o, the largest angle the object can subtend is

$$(\theta_o)_{\max} = \frac{h_o}{d_{\text{near}}} \qquad (38.10)$$

where $(\theta_o)_{\max}$ is small so that $\tan(\theta_o)_{\max} \approx \sin(\theta_o)_{\max} \approx (\theta_o)_{\max}$.

If you wish to see a larger image of the object, you can use a magnifier to make an enlarged image that is far from your eye. Ideally, the image forms at infinity so that your ciliary muscles can relax. To achieve this, the object must be placed at one focal point of the lens with your eye at the other focal point. In Figure 38.44B, we have drawn two rays that emerge from an object—one from the top and the other from the bottom. Both rays are parallel to the optical axis and bend toward the focal point of the lens. This figure is not a ray diagram; in a ray diagram, the rays emerge from a single point of an object. Compare Figure 38.44B to the ray diagram for an object at the focal point of a converging lens (Fig. 38.34).

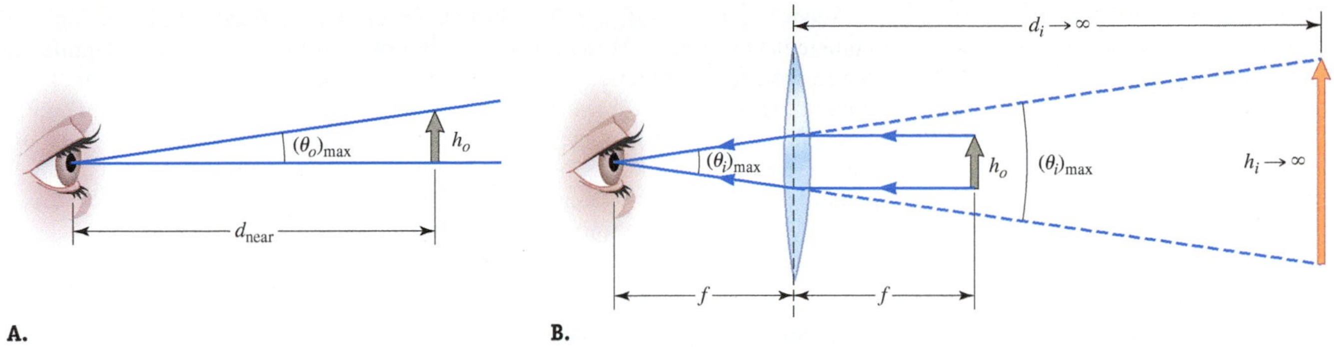

FIGURE 38.44 **A.** For an unaided eye, the maximum angular size of an object is seen when the object is at the person's near point. **B.** When you use a magnifier, the maximum angular size of the image is seen when the object is at one focal point of the converging lens and your eye is at the other focal point. Then the image is at infinity, so its height is infinite. For practical purposes, an image with infinite height at infinity is indistinguishable by a normal eye from a large image at a distance greater than 5 m whose rays are paraxial when they reach the eye.

As shown in the ray diagram, rays from a single point on the object, such as the tip, do not come together at the focus. Instead, the rays emerge from the lens parallel to one another; the image forms at infinity, and its height is infinite. Such an image cannot be drawn, so in Figure 38.44B we have drawn a large image at a great distance.

Because the rays in Figure 38.44B are from the top and bottom of the object, the angle between the refracted rays is the angular size of the image θ_i. In this ideal geometry, where the image is understood to have an infinite height and to be located at infinity, the angular size of the image must be at its maximum $(\theta_i)_{max}$. Assuming $(\theta_i)_{max}$ is small enough so that $\tan(\theta_i)_{max} \approx \sin(\theta_i)_{max} \approx (\theta_i)_{max}$, we can express the image's angular size as

$$(\theta_i)_{max} = \frac{h_o}{f} \tag{38.11}$$

The **angular magnification** m of a magnifier is the ratio of the image's angular size with the magnifier to the object's angular size without it:

ANGULAR MAGNIFICATION

Major Concept

$$m \equiv \frac{\theta_i}{\theta_o} \tag{38.12}$$

Using the maximum values (Eqs. 38.11 and 38.12) for these angles, we find

$$m = \frac{(\theta_i)_{max}}{(\theta_o)_{max}} = \left(\frac{h_o}{f}\right)\left(\frac{d_{near}}{h_o}\right)$$

$$m = \frac{d_{near}}{f} \tag{38.13}$$

Equation 38.13 gives the angular magnification for a simple magnifier. Because it is common to assume the near point of a person's eye is 25 cm, the angular magnification of a magnifier is often expressed as

$$m = \frac{25\text{ cm}}{f} \tag{38.14}$$

Throughout this textbook, we do not drop the term *angular*, but we usually drop the term *linear*.

It is important to distinguish between the angular magnification m and the (linear) magnification M, the ratio of image height to object height (Eq. 37.1). In the case of a magnifier, the image height is infinite, so the linear magnification is infinite. However, the image does not look infinitely large. Often the terms *angular* and *linear* are dropped when magnification is specified. When you buy a magnifier, the manufacturer usually reports the angular magnification as a number followed by ×, which stands for *times*. For example, binoculars that have a magnification of 50× produce an image whose angular size is 50 times larger than the angular size of the object.

According to $m = d_{near}/f$ (Eq. 38.13), the shorter the focal length, the higher the angular magnification. However, there is a practical limit to angular magnification due to spherical aberration. With no correction for spherical aberration, the highest angular magnification for a single-lens magnifier is about 4×. Such magnification is practical for many purposes, but not for all. Coin collectors and jewelers often require a higher magnification. Correcting for spherical aberration can produce a magnifier with about 20× angular magnification. If you need an even higher magnification, you must use a microscope (Section 38-10).

CONCEPT EXERCISE 38.6

CASE STUDY An Emergency Magnifier

You are in the woods taking photos with your camera when you come across an interesting and very rare bird's feather. Your camera has a removable lens. Can you use that lens as a magnifier to take a closer look at the feather? If so, explain what you need to do. If not, explain why not.

CONCEPT EXERCISE 38.7

CASE STUDY Focal Length and Angular Magnification

A magnifier has an angular magnification of 3×. What is the focal length of the lens?

38-10 Multiple-Lens Systems

In this section, we study two devices that require two lenses. The first is a **compound microscope**, or simply a **microscope**. The other is a **refracting telescope**. The purpose of a microscope is to produce a greatly enlarged image. Many people think a telescope serves the same purpose, but that is not the case. The purpose of a telescope is to gather a great deal of light from very distant objects. However, both devices are made of two converging lenses.

CASE STUDY The Compound Microscope

Suppose you are using a magnifier to see some fine detail on a small object but the image is simply not large enough. Fortunately, you have a second magnifier, so you can use both of them to see the detail. This is the basic idea behind a microscope (Fig. 38.45A). The object is placed on a slide and illuminated from below by a light source. Just above the object is the **objective lens**, or simply the **objective**. (As a memory aid, the *objective* lens is close to the *object*.) At the top of the microscope is a second lens called the **eyepiece**. You place your eye near the eyepiece.

Figure 38.45B shows a ray diagram for the two lenses that make up the microscope. The focal length of the objective lens is f_o, and the object distance is greater than the focal length: $d_o > f_o$. Because the objective is a converging lens and the object is farther away than its focal point, the lens forms a real inverted image at the image distance d_i (as in Fig. 38.32). To make sure this real image is larger than the object, the object must be close to the focal point of the lens. Then this enlarged real image becomes the object for the second lens—the eyepiece.

The job of the eyepiece is simply to act as a magnifier of the real image produced by the objective. Like any simple magnifier, the eyepiece should produce a virtual enlarged image. From Section 38-9, we know that a converging lens works as a

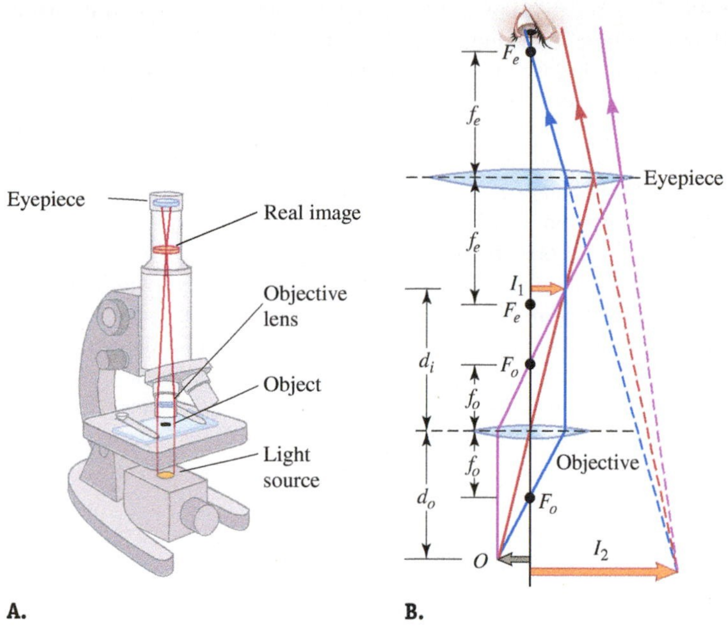

FIGURE 38.45 **A.** Parts of a compound microscope. **B.** A ray diagram for two lenses that make up the microscope. I_1 is the image produced by the objective; it is the object of the eyepiece. I_2 is the image that you observe through the eyepiece.

magnifier when the object is just inside its focal point (Fig. 38.44). So the real image produced by the objective lens must form just inside the focal point of the eyepiece.

The total magnification of a microscope depends on the linear magnification of the objective lens and the angular magnification of the eyepiece. The linear magnification of the objective is given by

$$M = \frac{h_i}{h_o} = -\frac{d_i}{d_o} \tag{38.6}$$

Because the object and image distances are both positive for the objective, the magnification is negative and the real image I_1 produced by the objective is inverted (Fig. 38.45B). The magnification is highest when the object distance is short and the image distance is long. Because we need the objective to produce a real image, however, the object cannot be closer to the lens than the focal point. So, in a well-designed microscope, the object distance is only slightly longer than the focal length of the objective. Let's make the approximation that the object distance equals the focal length of the objective lens. Then $M = -d_i/d_o$ (Eq. 38.6) becomes

$$M_o = -\frac{d_i}{f_o} \tag{38.15}$$

By contrast, the angular magnification of the eyepiece is given by $m_e = d_{\text{near}}/f_e$ (Eq. 38.13). The total angular magnification of the microscope is found by multiplying the linear magnification of the objective by the angular magnification of the eyepiece:

$$m_{\text{micro}} = M_o m_e = -\frac{d_i}{f_o}\frac{d_{\text{near}}}{f_e} \tag{38.16}$$

The negative sign means that the image you see through a microscope is inverted. But, if you imagine using a microscope to look at blood cells, for example, an inverted image is not a problem.

Often microscope manufacturers drop the negative sign in Equations 38.15 and 38.16. Also, they usually assume the near point is at 25 cm, so the angular magnification of a microscope is reported as

$$m_{\text{micro}} = (25\text{ cm})\frac{d_i}{f_o f_e} \tag{38.17}$$

Equation 38.17 shows that the angular magnification is increased by using an eyepiece and an objective with short focal lengths.

When you use a microscope, you must be able to adjust its angular magnification. Suppose you want to look at a particular blood cell. You start by studying an image with low angular magnification, so that you can see all the cells in the sample. Then you make sure the cell you are interested in is in the center of the image, so that when you look with a higher angular magnification, that cell remains in the image.

According to Equation 38.17, the magnification of a microscope can be controlled by using eyepieces and objectives of varying focal lengths. If the manufacturer supplies only one eyepiece, the magnification is controlled by switching the objective lenses. As shown in Figure 38.45A, several objective lenses are often arranged on a wheel. You control the magnification by choosing a particular objective; the objective with the shortest focal length produces the highest magnification.

We modeled the microscope as a system of two thin converging lenses. However, like modern camera lenses, each lens of a modern microscope usually consists of several optical elements.

CASE STUDY The Refracting Telescope

A microscope provides a greatly enlarged image of a small close object. Contrast this to an astronomical telescope, which provides an image of a large distant object. The main job of the telescope is to gather more light than you could with your unaided eyes. Both microscopes and refracting telescopes are made of two converging lenses, but the arrangement of these lenses determines which instrument results.

Figure 38.46A shows the major components of a refracting telescope. As in a microscope, the lens closer to the object is the objective and the lens closer to your eye is the eyepiece. An object examined through a telescope is far away, so rays from the object are parallel and come together at the focal point of the objective lens (Fig. 38.46B). The image formed at the focal point is very small. The second lens—the eyepiece—magnifies that image. The image produced by the objective becomes the object for the eyepiece, which produces an enlarged virtual image.

Ideally, the image you see through the eyepiece is at infinity so that your ciliary muscles can relax, which means the image produced by the objective should be at the focal point of the eyepiece. So the distance between the objective lens and the eyepiece should be the sum of their focal lengths: $f_o + f_e$ in Figure 38.46B.

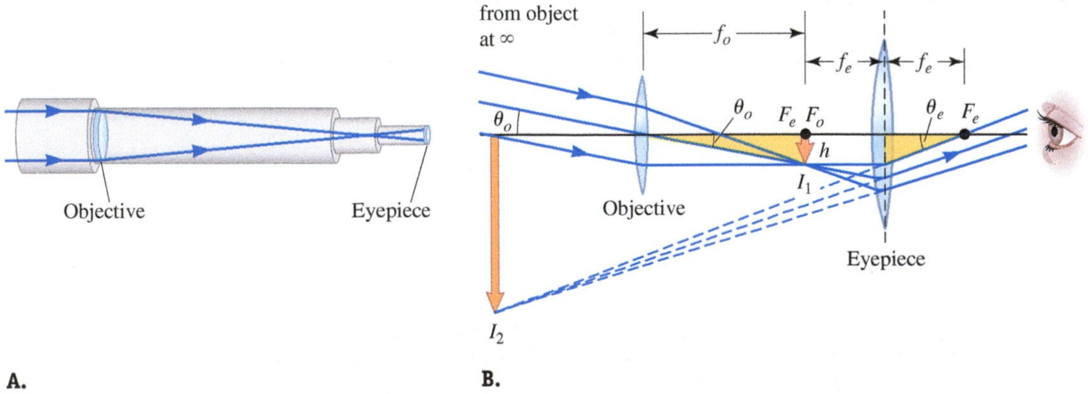

FIGURE 38.46 A. A refracting telescope has two lenses. **B.** A ray diagram for an astronomical telescope. I_1 is the image produced by the objective; it is the object of the eyepiece. I_2 is the image that you observe through the eyepiece.

Although the primary purpose of a telescope is to gather light and not to magnify, it does produce an enlarged image. This enlargement is useful when you want to see something far away, but not *astronomically* far. The total angular magnification of a telescope depends on the angular magnifications of the objective and the eyepiece. The angular size seen with the unaided eye is the same as the angle θ_o subtended by the object at the objective lens, and the angular size seen through the telescope is about equal to the angle θ_e subtended at the eyepiece (Fig. 38.46B). The angular magnification of the telescope is the angular size of the image seen with the telescope θ_e divided by the angular size of the object seen with the unaided eye θ_o:

$$m_{\text{tele}} = \frac{\theta_e}{\theta_o} \tag{38.18}$$

Using the two right triangles highlighted in Fig. 38.46B and some trigonometry, we find

$$\tan\theta_o = \frac{h}{f_o} \text{ and } \tan\theta_e = -\frac{h}{f_e}$$

The minus sign is introduced because the image seen through the eyepiece is inverted. (The image seen with the unaided eye is not inverted.) The angles are small, so we use the small-angle approximation for the tangent:

$$\theta_o \approx \frac{h}{f_o} \tag{38.19}$$

and

$$\theta_e \approx -\frac{h}{f_e} \tag{38.20}$$

Substitute Equations 38.19 and 38.20 into Equation 38.18 to find that h cancels out:

$$m_{\text{tele}} = -\frac{f_o}{f_e} \tag{38.21}$$

Now compare the angular magnification of a microscope, $m_{\text{micro}} = (25\text{ cm})(d_i/f_o f_e)$ (Eq. 38.17), to the angular magnification of a telescope (Eq. 38.21). For both instruments, the angular magnification is inversely proportional to the focal length of the eyepiece. However, for the telescope the angular magnification is directly proportional to the focal length of the objective, whereas for the microscope it is inversely proportional. Often a microscope has only one eyepiece, and the magnification is controlled by changing the objective. The opposite is true for a telescope. Usually the telescope manufacturer provides only one objective, and that lens is not easily removed. However, the eyepieces are relatively inexpensive and easy to change. An astronomer generally starts by using a low-magnification (long focal length) eyepiece to see a large field of view. When she knows that the object she wishes to see is in the center of the image, she may switch to a higher-magnification (short focal length) eyepiece to see details. The larger image is fainter, however, because the same amount of light is spread out over a larger area of her retina. As in the case of microscopes, the negative sign is often dropped when a telescope's angular magnification is reported.

The primary purpose of an astronomical telescope is to collect light from objects too distant to see with unaided eyes. We describe this property of telescopes as their *light-gathering power*, or LGP. The LGP of a telescope depends on its diameter. As an analogy, suppose that instead of collecting photons from a distant object, you wish to collect rain. You will collect more rain in the same amount of time if you use a bowl with a large diameter rather than a tall glass with a small diameter. Similarly, the light-gathering power of a telescope is proportional to the area A of its objective lens:

$$\text{LGP} \propto A$$

Because the cross section of the lens is a circle, the LGP is proportional to the square of the radius or the diameter of the lens:

$$\text{LGP} \propto \pi R^2 \propto \pi\left(\frac{d}{2}\right)^2$$

$$\text{LGP} \propto d^2 \tag{38.22}$$

The LGP of a telescope is so important that telescopes are usually described in terms of their objective lens diameter. So, an astronomer may say he is going to use the "8-inch," which means he will use a telescope whose objective lens is 8 inches in diameter.

Refracting telescopes are great fun and were very important until the early 1900s. Refracting telescopes are not generally used by today's professional astronomers, however, because very large telescopes are needed to see very faint objects. Large lenses are heavy and also suffer from chromatic aberration. It is expensive to build very large refracting telescopes that are corrected for chromatic aberration. Instead, astronomers use large reflecting telescopes.

CONCEPT EXERCISE 38.8

CASE STUDY How Good Is Your Telescope?

When you talk to an astronomer about his telescope, why doesn't it make sense to ask about its magnification? Why does it make more sense to ask about the size of the objective lens?

EXAMPLE 38.12 CASE STUDY Your Eye Versus Your Telescope

For your birthday, your very nice aunt gives you a 4-inch refracting telescope. The objective lens has a focal length of 880 mm. The telescope has two eyepieces with focal lengths of 10 mm and 50 mm.

A How does the light-gathering power of your telescope compare to the light-gathering power of your fully open pupil?

INTERPRET and ANTICIPATE

The LGP depends on the diameter. For a telescope, the diameter of the objective is all that matters. The phrase "4-inch refracting telescope" means the objective has a 4-inch diameter. By contrast, the diameter of a fully open pupil is about 8 mm (Section 38-8). Because the diameter of the telescope is so much larger than the diameter of your pupil, we expect the LGP of the telescope to be much greater than that of your eye. It is helpful to work in millimeters: 4 in. = 102 mm.

SOLVE

The ratio of the LGP of the telescope to the LGP of your eye is proportional to the ratio of their diameters squared (Eq. 38.22).

$$\frac{(\text{LGP})_{\text{tele}}}{(\text{LGP})_{\text{eye}}} = \left(\frac{d_{\text{tele}}}{d_{\text{eye}}}\right)^2 = \left(\frac{102\text{ mm}}{8\text{ mm}}\right)^2$$

$$\frac{(\text{LGP})_{\text{tele}}}{(\text{LGP})_{\text{eye}}} = 13$$

CHECK and THINK

As expected, the LGP of the telescope is much higher (13 times higher) than the LGP of your eye. So, when you use a telescope, you can see much fainter objects than you can without it.

B What is the telescope's highest angular magnification?

INTERPRET and ANTICIPATE

The magnification depends on the focal length of the objective and the focal length of the eyepiece you choose. There is only one objective, but you can use either the 10-mm or the 50-mm eyepiece. Notice that we refer to an eyepiece by its focal length, not its diameter. The eyepiece with the shorter focal length produces the higher angular magnification. So, to find the telescope's highest angular magnification, you would choose the 10-mm eyepiece.

SOLVE

Use Equation 38.21 to find the angular magnification. We have dropped the minus sign, as is often done.

$$m_{\text{tele}} = \frac{f_o}{f_e} \quad (38.21)$$

$$m_{\text{tele}} = \frac{880\text{ mm}}{10\text{ mm}} = 88\times$$

CHECK and THINK

The highest angular magnification you can achieve with this instrument is 88×. You might be able to get a higher magnification by using an eyepiece with an even shorter focal length, but that would make the image fainter. A telescope manufacturer often specifies the highest practical angular magnification. If the magnification is too high, the image is too faint to be useful.

SUMMARY

Underlying Principles

No new principles are introduced. This chapter uses geometric optics.

Major Concepts

1. The first part of the **law of refraction** states that the refracted ray, the incident ray, and the normal all lie in a single plane—the plane of incidence. The second part is **Snell's law**:

$$n_t \sin\theta_t = n_i \sin\theta_i \quad (38.1)$$

where θ_t and θ_i are the *t*ransmitted and *i*ncident angles (Fig. 38.1), and n_t and n_i are the indices of refraction for the transmitted and incident media, respectively.

2. **Total internal reflection** is possible if light is incident in a medium that has a higher index of refraction than that of the medium beyond the boundary. Then, no light is transmitted if the angle of incidence is larger than the critical angle as given by

$$\sin\theta_c = \frac{n_t}{n_i} \quad (38.2)$$

3. The spreading out of light by color due to differences in the index of refraction is called **dispersion**.

4. The **thin-lens equation** is identical to the mirror equation:

$$\frac{1}{d_o} + \frac{1}{d_i} = \frac{1}{f} \quad (38.5)$$

5. The **magnification** by a thin lens is the same as for a spherical mirror:

$$M = \frac{h_i}{h_o} = -\frac{d_i}{d_o} \quad (38.6)$$

6. The **lens maker's equation** is

$$\frac{1}{f} = (n-1)\left[\frac{1}{r_1} - \frac{1}{r_2}\right] \quad (38.7)$$

Equations 38.5, 38.6, and 38.7 use the sign conventions given in Table 38.2.

7. The **angular magnification** is given by

$$m \equiv \frac{\theta_i}{\theta_o} \quad (38.12)$$

Special Cases

1. When a real object is in front of a **diverging lens**, the image formed is in front of the lens and virtual. The image is upright and smaller than the object.
2. When a real object is in front of a **converging lens**, the image formed depends on the position of the object relative to the focal point.

a. A real inverted image is produced if $d_o > f$. If $d_o > 2f$, the image is smaller than the object. If $f < d_o < 2f$, the image is larger than the object.
b. An upright virtual image is produced if $d_o < f$. The image is always larger than the object.

Tools

The elements of a **ray diagram** (Fig. 38.31) are:

1. The **lens**, including a vertical midline
2. The **optical axis**
3. The **focal points** F drawn on both sides of the lens
4. An arrow representing the **object**
5. Two or three **primary rays** (described in Table 38.4) emerging from one point of the object, usually the tip of the arrow

PROBLEMS AND QUESTIONS

A = algebraic C = conceptual E = estimation G = graphical N = numerical

38-1 Law of Refraction

1. **N** The Sun appears at an angle of 53.0° above the horizontal as viewed by a dolphin swimming underwater. What angle does the sunlight striking the water actually make with the horizon?
2. **C** In this chapter, we studied the refraction of visible light. Have we ignored other parts of the electromagnetic spectrum because radiation at other frequencies does not refract? If so, explain why radiation of other wavelengths cannot refract. If not, give an example of radiation that refracts at other wavelengths.

Problems 3 and 4 are paired.

3. **N** A light ray is incident on an interface between water ($n = 1.333$) and air ($n = 1.0002926$) from within the water. If the angle of incidence in the water is 30.0°, what is the angle of the refracted ray in the air?
4. **N** A light ray is incident on an interface between water ($n = 1.333$) and air ($n = 1.0002926$) from within the air. If the angle of incidence in the air is 30.0°, what is the angle of the refracted ray in the water?

Problems 5 and 6 are paired.

5. **A** A mirror is above a swimming pool and perpendicular to the surface of the water. A narrow beam is incident on the mirror (Fig. P38.5). Find an expression for the angle of refraction in the water.
6. **N** A mirror is above a swimming pool and perpendicular to the surface of the water. A narrow beam is incident on the mirror (Fig. P38.5). If $\theta_i = 22.3°$, find the angle of refraction in the water.

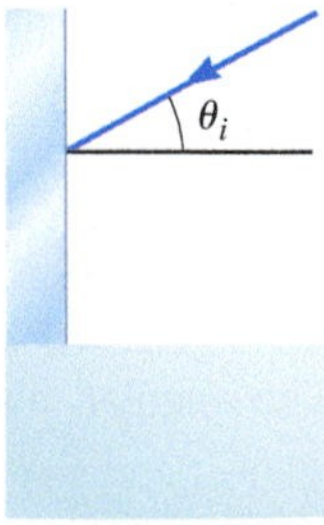

FIGURE P38.5 Problems 5 and 6.

7. **N** Figure P38.7 shows a monochromatic beam of light striking the right-hand face of a right-angle prism at normal incidence. If $\theta_t = 20.0°$, what is the index of refraction of the prism?

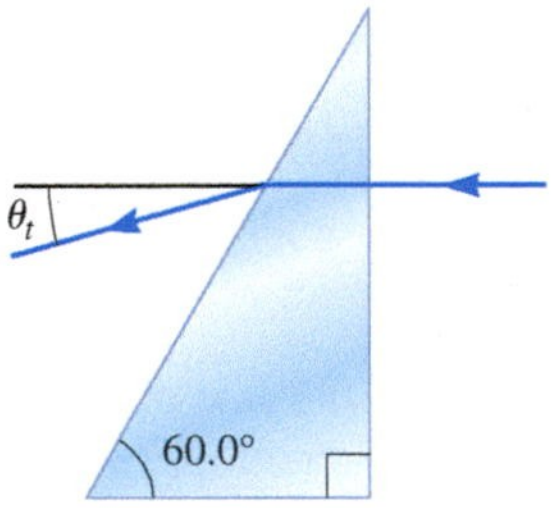

FIGURE P38.7

8. **N** A ray of light enters a liquid from air. If the angle between the incident and refracted rays is 150° and the angle between the reflected and refracted rays is 60°, find the refractive index of the liquid. Assume the refractive index of air is 1.00.
9. **N** A mirror is above a swimming pool and tilted toward the surface of the water. A narrow beam is parallel to the surface of the water and incident on the mirror (Fig. P38.9). The angle of incidence on the mirror's surface is 22.3°. Find the angle of refraction in the water.

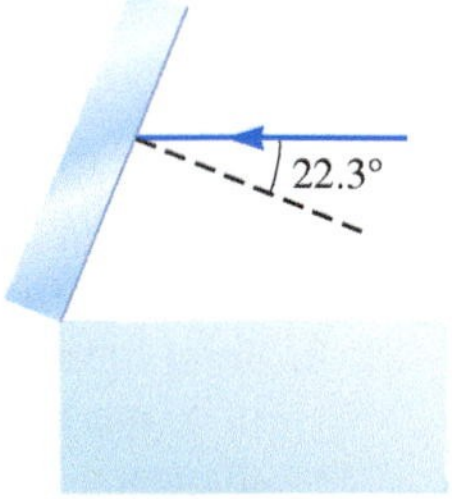

FIGURE P38.9

10. **N** Figure P38.10 on the next page shows a monochromatic beam of light of wavelength 575 nm incident on a slab of crown glass surrounded by air. Use a protractor to measure the angles of incidence and refraction. **a.** What is the speed of the beam of

light within the glass slab? **b.** What is the frequency of the beam of light within the glass slab? **c.** What is the wavelength of the beam of light within the glass slab?

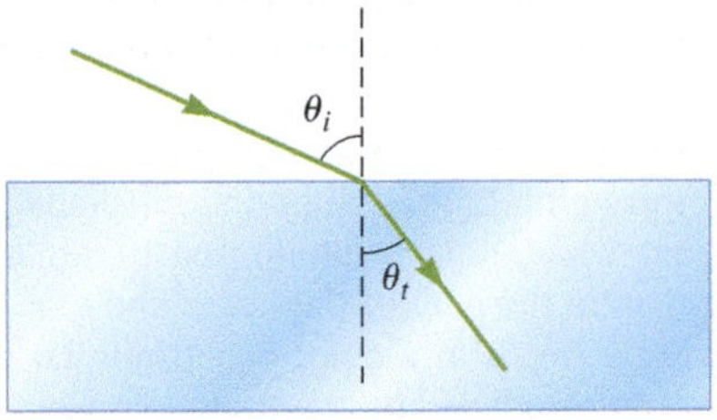

FIGURE P38.10

11. **N** Cone receptors in the human eye contain visual pigments that are sensitive to the color of light. In the 1800s, two physiologists, Thomas Young and Herman von Helmholtz, proposed that for each of the three fundamental color sensations in human vision (red, green, and blue), there is a different kind of color receptor. The excitation of a particular receptor leads to the sensation of the corresponding color. This theory, known as the Young–Helmholtz or trichromatic theory, is accepted today. The pigments are found in the aqueous humor inside the eyeball. The receptor responsible for red pigment shows peak sensitivity to light at a wavelength of about 580.0 nm in the aqueous humor. Using red light in air with a wavelength of 700.0 nm, determine the index of refraction of the aqueous humor and the speed and frequency of the light in this substance.

12. **C** You are camping in the woods and find yourself running low on food. You notice several fish in a small pool of clear water. After making a spear from a tree limb, you try to spear a meal by thrusting the spear into the water where you see a fish. Why is this a poor strategy for using the spear? Where should you target a particular fish if you hope to spear it?

Problems 13 and 107 are paired.

13. **N** A block is constructed from layers of cubic zirconia ($n = 2.14$), flint glass ($n = 1.80$), and quartz ($n = 1.54$) as shown in Figure P38.13. The block is surrounded by air. A ray of monochromatic light is incident on the cubic zirconia–flint glass interface with $\theta_i = 23.0°$. What is the refraction angle θ_t when the ray exits the slab at the bottom?

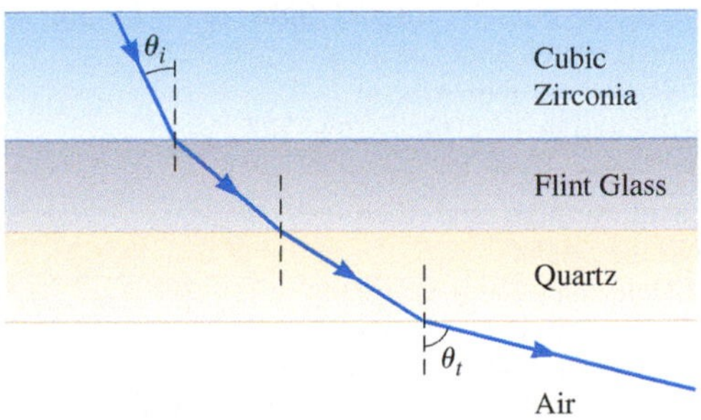

FIGURE P38.13 Problems 13 and 107.

38-2 Total Internal Reflection

14. **C** If you are underwater, can you look straight up and use total internal reflection to see your own face? Explain.

15. **N** What is the minimum value of the index of refraction of a right-angled, isosceles prism for which a light ray entering normally on one face will be totally internally reflected as shown in Figure P38.15? Assume the index of refraction of air is 1.00029.

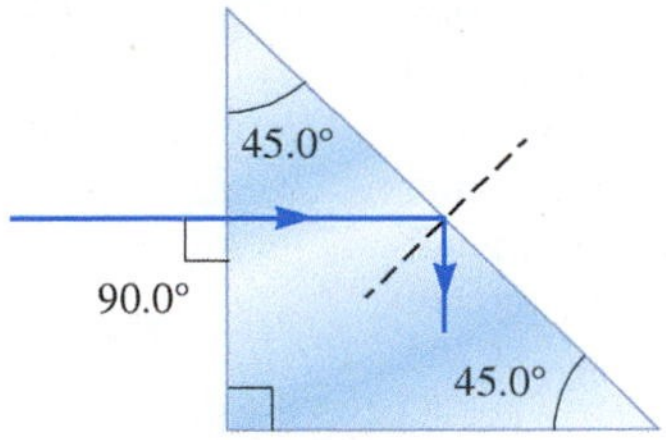

FIGURE P38.15

Problems 16 and 17 are paired.

16. **C** A fish is 3.25 m below the surface of still water (Fig. P38.16). You do not want the fish to see your fishing boat. Is it possible to place your boat so that total internal reflection keeps it hidden from the fish? If so, explain how this is done. If not, explain why not.

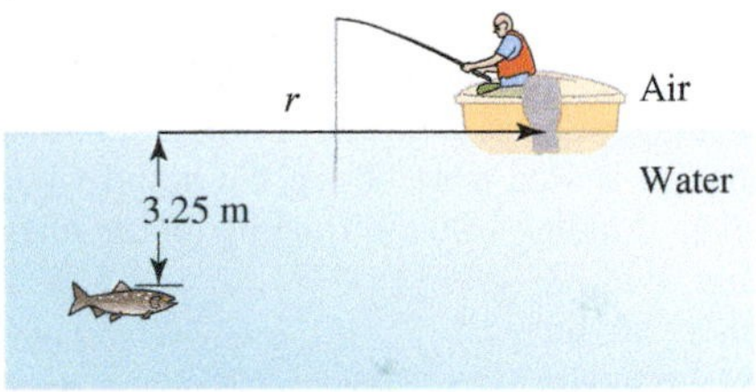

FIGURE P38.16 Problems 16 and 17.

17. **N** A fish is 3.25 m below the surface of still water. Because of total internal reflection, it is hidden from the view of a fisher in a boat on the water as long as the boat is outside a circle of radius r. The center of the circle is directly above the fish (Fig. P38.16). Find the minimum value of r.

18. **N** A beam of monochromatic light within a fiber optic cable is incident on one of the sides of the cable ($n = 1.485$) at an angle of incidence θ_i. Assume the fiber is surrounded by air ($n = 1.00029$). **a.** What is the critical angle for total internal reflection so that the beam stays within the fiber? **b.** What would the critical angle be if the fiber were completely immersed in water ($n = 1.33$)?

19. **N** A beam of light travels through a fiber with an index of refraction of 1.55. The fiber is surrounded by a medium with an index of refraction of 1.36. At what minimum angle of incidence is all the light reflected back into the fiber?

Problems 20 and 21 are paired.

20. Consider a light ray that enters a pane of glass with air on either side as shown in Figure P38.20. The light ray experiences refraction at the first interface when it enters the glass and again at the second interface when it exits the glass. Assume the index of refraction of the glass is 1.54.

a. **N** If $\theta_1 = 53.0°$, find θ_2, θ_3, and θ_4.

b. **C** Explain why it is impossible for a light ray striking the glass pane to experience total internal reflection at either interface.

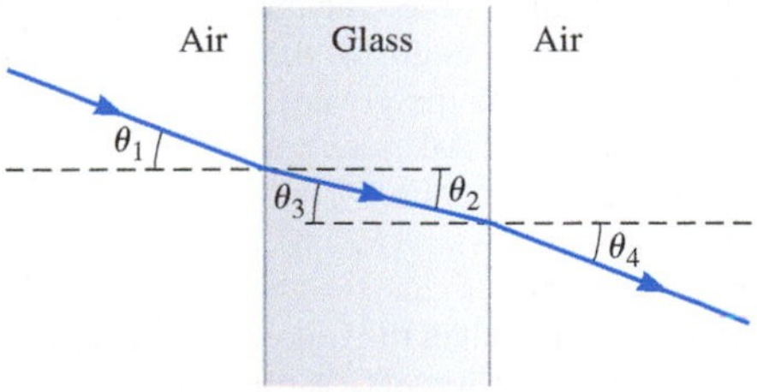

FIGURE P38.20

21. **N** Consider a light ray that enters a pane of glass with air on one side and water on the other side as shown in Figure P38.21. The light ray experiences refraction at the first interface when it enters the glass from the water and again at the second interface when it exits the glass into the air. Assume the index of refraction of the glass is 1.54. For a ray of light, find the angle of incidence θ_1 in the water such that the ray experiences total internal reflection when it strikes the glass–air interface on the other side.

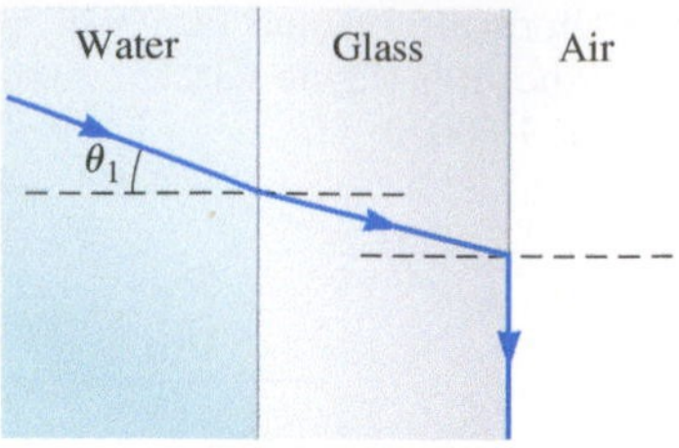

FIGURE P38.21

22. **N** Consider a beam of 486-nm blue light within a slab of material. The beam is incident on the slab's interface with air. What is the critical angle for total internal reflection if the slab is made of **a.** ice (n = 1.309), **b.** fluorite (n = 1.434), and **c.** diamond (n = 2.417)?

23. **N** A source of light S is placed at the bottom of a container holding a liquid that has an index of refraction of 1.67. A person is viewing the source from above the liquid while there is an opaque disk of radius 1.00 cm floating on the surface as shown in Figure P38.23. The center of the disk lies vertically above the source S. The liquid from the container is slowly drained out through a tap, while the center of the disk remains directly above the light source. What is the height of the liquid when the source can suddenly no longer be seen from above? Assume the index of refraction of air is 1.00029.

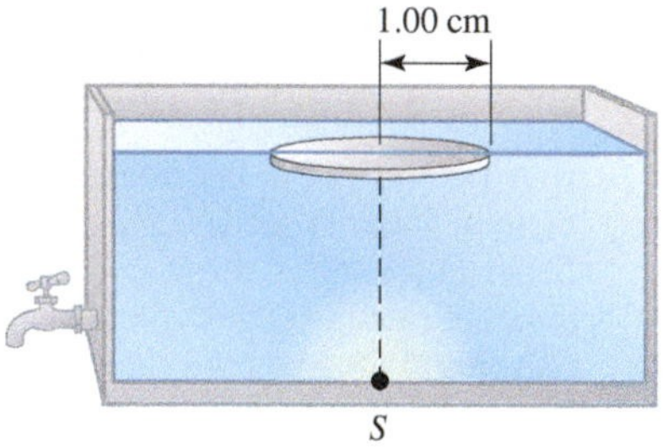

FIGURE P38.23

24. A light ray is traveling inside a block of glass. Assume the index of refraction of the glass is 1.54.

a. **N** If the medium outside the block of glass is air, find the critical angle for the glass–air interface.

b. **C** Any light ray that strikes the glass–air interface at an incident angle greater than the critical angle will experience total internal reflection. Will any ray that strikes the glass–air interface at an incident angle less than the critical angle necessarily experience transmission out of the block of glass on the other side? Explain.

25. **N** Figure P38.25 shows a beam of 656-nm red light striking the top surface of a slab of quartz (n = 1.54) surrounded by air at an angle of incidence θ_i. After refraction at the air–quartz interface, the beam strikes the right-hand edge of the slab at point P. What is the greatest angle θ_i for which total internal reflection will occur at point P?

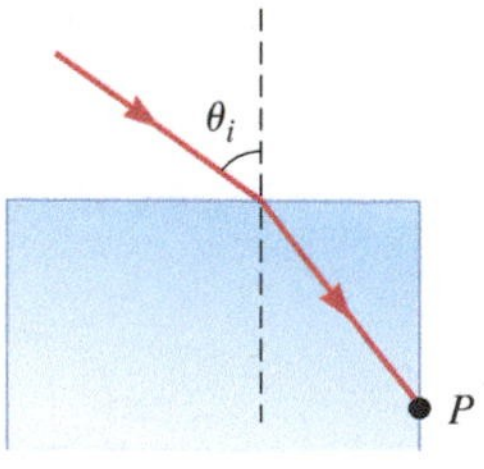

FIGURE P38.25

38-3 Dispersion

26. **N** Dispersion occurs when the index of refraction of a medium depends on the wavelength of incident light, resulting in different angles of refraction for each wavelength. A beam of white light is incident on a block of zinc crown glass with θ_i = 34.0°. The index of refraction of this glass for 589-nm yellow light is 1.528, while the index of refraction for 486-nm blue light is 1.517. **a.** What is the refraction angle θ_t for yellow light in the glass? **b.** What is the refraction angle θ_t for blue light in the glass?

27. **C** Isaac Newton, working in a dark room, let a beam of sunlight pass through a hole in the window shade. His goal was to show that white light is composed of all the colors. He placed a prism in the path of the beam and saw that the white light spread out to form a color spectrum. To prove that the colors were not produced by the prism, he used a screen with a small slit to select one color of light. He then placed a second prism in the path of this monochromatic (one-color) beam. **a.** What colors emerged from the second prism? **b.** Next, Newton removed the filter (the slit) and used the second prism to form a beam of white light (Fig. P38.27). Why was this step necessary?

FIGURE P38.27

28. **C** When UV and IR radiation pass from a vacuum into some medium, which one is refracted more by the change in media? Explain.

29. The wavelength of light changes when it passes from one medium into another. Suppose green light at 550.0 nm passes from air into glass (n = 1.5).

a. **N** What is its wavelength in glass?

b. **N** What is its frequency?

c. **C** Do you see the same color in both media?

30. **N** A new material is being considered for encapsulating and protecting a sensitive electronic device. If light of wavelength 500.0 nm in air (n = 1.00029) has a wavelength of 240.0 nm when it enters the new material, what is the index of refraction of the material?

31. Light is incident on a prism as shown in Figure P38.31. The prism, an equilateral triangle, is made of plastic with an index of refraction of 1.46 for red light and 1.49 for blue light. Assume the apex angle of the prism is 60.00°.

a. **C** Sketch the approximate paths of the rays for red and blue light as they travel through and then exit the prism.

b. **N** Determine the measure of dispersion, the angle between the red and blue rays that exit the prism.

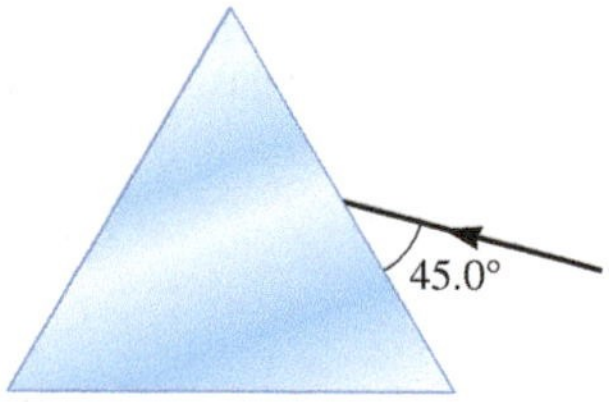

FIGURE P38.31

32. **C** Zak and Sallie are playing outside in a water sprinkler. Sallie can see a rainbow when she looks at the water, but Zak cannot. Where are the two children standing? Explain.

33. **A** Use geometry and Figure 38.13 (page 1225) to show these relationships: **a.** $\theta_2 + \theta_3 = 60°$, **b.** $\alpha = \theta_4 - 30°$, and **c.** $\theta_1 = 30°$.

34. **C** Light travels through air and enters a prism. How, if at all, do the speed, wavelength, frequency, and energy of the light change as it enters the prism?

38-4 Refraction at Spherical Surfaces

35. N A paperweight is made of a transparent material with an index of refraction of 1.75. The paperweight is a hemisphere of radius R. At its base is a flattened ladybug. Where is the image of the ladybug, and what is its magnification? Assume the index of refraction for air is 1.00.

36. A Use geometry and Figure 38.17 (page 1227) to show these relationships: **a.** $\theta_i = \alpha + \beta$ and **b.** $\theta_t = \beta - \gamma$.

37. N In a still pond, you see a fish whose distance from the surface of the water is 3.25 m. How far below the surface does the fish appear to you? If you were trying to catch the fish in a net, would you need to take the difference between the image distance and the object distance into account? For this problem, take the index of refraction for air to be 1.00 and for water to be 1.33.

38. N A Lucite slab ($n = 1.485$) 5.00 cm in thickness forms the bottom of an ornamental fish pond that is 40.0 cm deep. If the pond is completely filled with water, what is the apparent thickness of the Lucite plate when viewed from directly above the pond?

39. N Figure 38.20 (page 1231) showed an image produced by a vase of water projected onto a screen. Suppose the object is 1.0 m from surface 1. How far is the screen from surface 2, and what is the magnification of the final image?

40. N In Example 38.6, we found that a spherical vase of radius 6.0 cm full of water acts like a magnifying lens for an object 3.0 cm in front of it. Imagine taking the empty vase, capping the top to contain the air, and submerging it upside down in a pool to try to form a spherical "air lens" in a water environment. Determine the magnification of an object 3.0 cm in front of the spherical "air lens" surrounded by water to see whether the vase in these circumstances also acts like a magnifying lens.

41. N The end of a solid glass rod of refractive index 1.50 is polished to have the shape of a hemispherical surface of radius 1.0 cm. A small object is placed in air (refractive index 1.00) on the axis 5.0 cm to the left of the vertex. Determine the position of the image.

42. A Figure P38.42 shows a hemispherical material with radius of curvature R and refractive index 1.5. The flat portion is silvered, so it is a reflective surface. Find the necessary distance x of the silvered plane surface from the point P so as to form an image of a very distant object at that point.

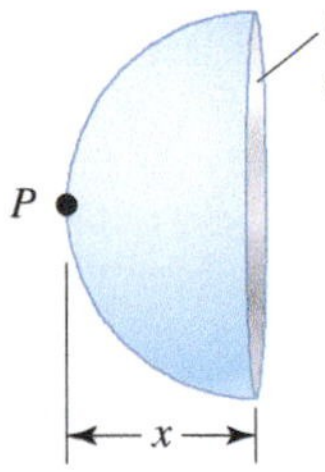

FIGURE P38.42

38-5 Thin Lenses

43. N Figure P38.43 shows a concave meniscus lens. If $|r_1| = 8.50$ cm and $|r_2| = 6.50$ cm, find the focal length and determine whether the lens is converging or diverging. The lens is made of glass with index of refraction $n = 1.55$. **CHECK and THINK:** How do your answers change if the object is placed on the right side of the lens?

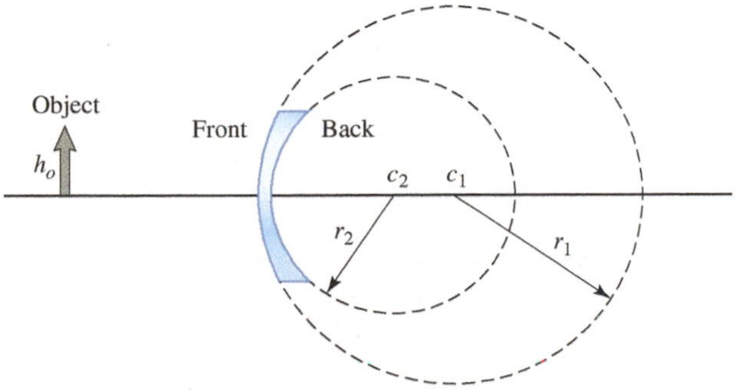

FIGURE P38.43

44. A Show that the magnification of a thin lens is given by $M = -d_i/d_o$ (Eq. 38.6). *Hint*: Follow the derivation of the lens maker's equation (page 1233) and start with a thick lens.

45. N A converging lens has a focal length of 30.0 cm and is made of glass ($n = 1.50$). If $|r_1| = |r_2| \equiv |r|$, what is $|r|$?

46. C Compare a converging lens to a diverging lens by filling in each blank with either **converging** or **diverging**: The focal length of a _______ lens is positive, and the focal point is behind the lens. Parallel rays form a real image at the focal point of a _______ lens. The focal length of a _______ lens is negative, and the focal point is in front of the lens. Parallel rays form a virtual image at the focal point of a _______ lens.

47. N A thin lens made of acrylic ($n = 1.49$) has $|r_1| = 17.5$ cm and $|r_2| = 7.5$ cm, where both surfaces of the lens have their radii of curvature on the same side of the lens. **a.** What is the focal length of this lens? **b.** Is the lens diverging or converging?

Problems 48 and 49 are paired.

48. N The radius of curvature of the left-hand face of a flint glass biconvex lens ($n = 1.60$) has a magnitude of 8.00 cm, and the radius of curvature of the right-hand face has a magnitude of 11.0 cm. The incident surface of a biconvex lens is convex regardless of which side is the incident side. What is the focal length of the lens if light is incident on the lens from the left?

49. N The radius of curvature of the left-hand face of a flint glass biconvex lens ($n = 1.60$) has a magnitude of 8.00 cm, and the radius of curvature of the right-hand face has a magnitude of 11.0 cm. The incident surface of a biconvex lens is convex regardless of which side is the incident side. What is the focal length of the lens if light is now incident on the lens from the right?

38-6 Images Formed by Diverging Lenses

50. C Devise an experiment that uses a small object (such as a coin) and a ruler to find the focal length of a diverging lens.

51. N A diverging lens with focal length $|f| = 15.5$ cm produces an image with a magnification of $+0.750$. What are the object and image distances?

52. N An object with a height of -0.050 m points below the principal axis (it is inverted) and is 0.150 m in front of a diverging lens. The focal length of the lens is -0.30 m. **a.** What is the image distance? **b.** What is the magnification? **c.** What is the image height?

53. N An object is placed 10.0 cm in front of a diverging lens of focal length -6.00 cm. Determine the image magnification and the image distance.

54. An object is placed 14.0 cm in front of a diverging lens with a focal length of -40.0 cm.

a. N What are the location and magnification of the image?

b. G Draw a ray diagram showing the locations of the object and the image by tracing at least two rays.

55. N An object is in front of a diverging lens with a focal length of -15.0 cm. The image seen has a magnification of 0.400. **a.** How far is the object from the lens? **b.** If the object has a height of -10.0 cm because it points below the principal axis (it is inverted), what is the image height h_i?

56. A student with a diverging lens is struggling to form an image on a screen. A small lightbulb is placed 25.0 cm in front of the lens, which has a focal length of 10.0 cm.

a. G Draw a ray-tracing diagram to find the position of the image formed by the lens, and determine why the student is unable to project an image onto the screen.

b. N Calculate the location and magnification of the image to confirm the accuracy of your drawing.

38-7 Images Formed by Converging Lenses

57. **N** An object is placed a distance of $4.00f$ from a converging lens, where f is the lens's focal length. **a.** What is the location of the image formed by the lens? **b.** Is the image real or virtual? **c.** What is the magnification of the image? **d.** Is the image upright or inverted?

58. A magnifying glass is a converging lens, which could be used to make a real image of an object. A small lightbulb is placed 25.0 cm in front of a convex lens with a focal length of 10.0 cm.
a. **G** Draw a ray-tracing diagram to find the position of the image formed by the lens.
b. **N** Calculate the location and magnification of the image to confirm the accuracy of your drawing.

59. **N** An object has a height of 0.050 m and is held 0.250 m in front of a converging lens with a focal length of 0.150 m. **a.** What is the magnification? **b.** What is the image height?

60. **E** Estimate the magnification and focal length of the pitcher of water shown in Figure P38.60.

Elizabeth Owens/American Association of Physics Teachers

FIGURE P38.60

61. **N** An object is in front of a converging lens with a focal length of 15.0 cm. The image seen has a magnification of -2.50. **a.** How far is the object from the lens? **b.** If the object has a height of -10.0 cm because it points below the principal axis (it is inverted), what is the image height h_i?

62. **C** Devise an experiment that uses a ruler to find the focal length of a converging lens.

63. **C** Explain the relative positions of the black and white backgrounds seen through the champagne glass shown in Figure P38.63.

Phuong Ha/American Association of Physics Teachers

FIGURE P38.63

38-8 The Human Eye

64. **E** Use a ruler and a small object (such as a coin) to find the near point of each of your eyes. Explain your procedure. How does your near point compare to the typical values in Table 38.5? Does each eye have the same near point? Estimate the focal length of each lens when you focus on an object at your near point. You may instead measure the near point of another person's eyes.

65. **C** Explain how wearing goggles while swimming enables you to see clearly given that your underwater vision is much less clear without goggles.

66. **C** CASE STUDY Two people are in the woods, and they want to start a fire by using their eyeglasses to focus sunlight onto some dry brush. One person is myopic, and the other is hyperopic. Whose glasses should they use, and why?

67. **C** CASE STUDY Take a look at the eyes of the cartoon character in Figure P38.67. Is she farsighted or nearsighted? Explain.

FIGURE P38.67

68. **C** CASE STUDY Ben Franklin invented bifocals (Fig. P38.68). The lower lenses are for reading, while the upper lenses are for looking at distant objects. What kind of lenses are in the bottom? What kind are in the top? Explain your answers.

Jeff Black

FIGURE P38.68

69. **N** CASE STUDY Susan wears corrective lenses. The prescription for her right eye is -7.75 diopters, and the prescription for her left eye is -8.00 diopters. Is she nearsighted or farsighted? If she is nearsighted, what is the far point for each of her eyes? If she is farsighted, what is the near point for each of her eyes? **CHECK and THINK:** In which eye does she have better vision?

70. **A** Fill in the missing entries in Table P38.70.

TABLE P38.70

	Convex mirror or diverging lens, $f < 0$	**Concave mirror or converging lens, $f < 0$**		
Object distance	$d_o > 0$	$d_o > 2f > 0$	$2f > d_o > f > 0$	$d_o < f > 0$
Image distance	$0 > d_i > f$		$d_i > 2f > 0$	
Magnification				$\|M\| > 1$
Real or virtual image	Virtual		Real	
Upright or inverted image		Inverted		

38-9 One-Lens Systems

71. N A converging lens is made such that both sides have a radius of curvature of 7.50 cm. The glass used has an index of refraction of 1.510 for red light and 1.530 for blue light, leading to chromatic aberration. **a.** What is the focal length of the lens for both red and blue light? **b.** If this lens is used to form an image of an object that is 10.0 cm in front of the lens, how far apart are the images formed for red and blue light?

72. C Is it possible to look through a magnifier and see a real image?

73. N In Example 38.11 (page 1249), we imagined making a camera from a single converging lens with a 30.0-cm focal length. If you wanted to use this lens as a magnifier, what would be the achieved magnification if you use the lens to create a virtual image by placing it 25.0 cm from an object, where you might have placed your eye to see the object as clearly as possible?

74. C Suppose you want to use a single lens to project a slide (a transparent film) onto a screen. Do you need a converging or diverging lens? Explain.

38-10 Multiple-Lens Systems

75. N An object 2.50 cm tall is 15.0 cm in front of a thin lens with a focal length of 5.00 cm. A thin lens with a focal length of −12.0 cm is placed 2.50 cm beyond this converging lens as shown in Figure P38.75. Determine the final image height and the position of the final image relative to the second lens.

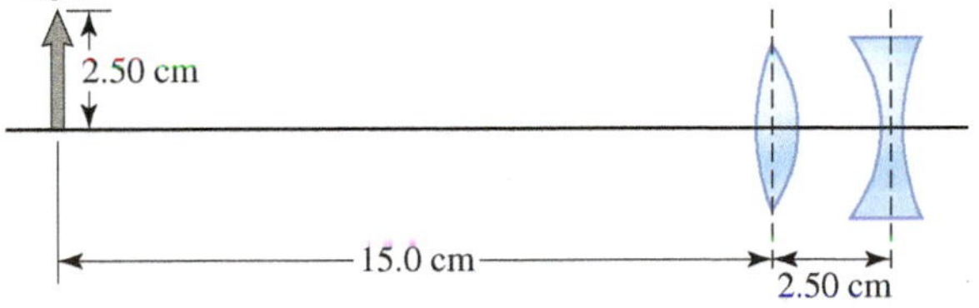

FIGURE P38.75

76. N Figure P38.76 shows an object placed a distance d_{o1} from one of two converging lenses separated by s = 1.00 m. The first lens has focal length f_1 = 22.0 cm, and the second lens has focal length f_2 = 45.0 cm. An image is formed by light passing through both lenses at a distance d_{i2} = 15.0 cm to the left of the second lens. **a.** What is the value of d_{o1} that will result in this image position? **b.** Is the final image formed by the two lenses real or virtual? **c.** What is the magnification of the final image? **d.** Is the final image upright or inverted?

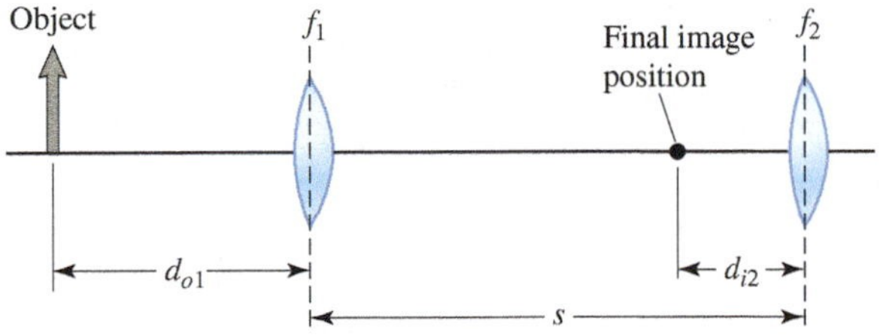

FIGURE P38.76

Problems 77 and 78 are paired.

77. N A system of lenses has light passing through a converging lens and then through a diverging lens. An object is 2.00 cm to the left of the converging lens, the lenses are 4.00 cm apart, and the focal length of the converging lens is 1.00 cm. **a.** Find the image distance for the image formed by the first lens. **b.** Find the focal length of the diverging lens if the final image position (after the light has gone through both lenses) is 0.50 cm to the left of the diverging lens.

78. G A system of lenses has light passing through a converging lens and then through a diverging lens. An object is 2.00 cm to the left of the converging lens, the lenses are 4.00 cm apart, and the focal length of the converging lens is 1.00 cm. Draw the ray diagram for the image formed by this system of lenses.

79. N An object is placed 22.0 cm to the left of a pair of lenses, with a converging lens of focal length +15.0 cm on the left, and a diverging lens of focal length −12.0 cm a distance of 40.0 cm to its right. **a.** What are the location and magnification of the final image formed by the two lenses? **b.** Is the final image formed by the two lenses upright or inverted?

Problems 80 and 81 are paired.

80. CASE STUDY A group of students is given two converging lenses. Lens A has a focal length of 12.5 cm, and lens B has a focal length of 50.0 cm. The diameter of each lens is 6.50 cm. The students are asked to construct a telescope from these lenses if possible, and they have this discussion:

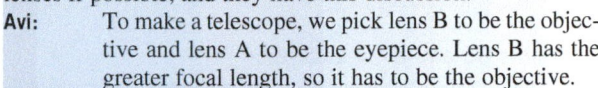

Avi: To make a telescope, we pick lens B to be the objective and lens A to be the eyepiece. Lens B has the greater focal length, so it has to be the objective.

Cameron: Both lenses have the same diameter—6.50 cm. It doesn't matter which is the objective.

Shannon: It does matter because the magnification depends on their relative focal lengths. We still want to get the best magnification.

a. C What do you think?

b. N If a telescope can be constructed from these two lenses, describe its design. What are its LGP and angular magnification? Compare the LGP to the value for your fully open pupil.

81. CASE STUDY A group of students is given two converging lenses. Lens A has a focal length of 12.5 cm, and lens B has a focal length of 50.0 cm. The diameter of each lens is 6.50 cm. The students are asked to construct a microscope from these lenses that has the same magnification as the telescope in Problem 80 if possible, and they have this discussion:

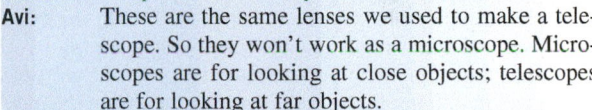

Avi: These are the same lenses we used to make a telescope. So they won't work as a microscope. Microscopes are for looking at close objects; telescopes are for looking at far objects.

Cameron: All you need for a microscope are two converging lenses. I think the difference from a telescope is just that the order of the lenses is switched. A microscope is just a backward telescope.

Shannon: I think the order of the lenses doesn't matter because the magnification is inversely proportional to both focal lengths. I think we have to adjust the distance between the lenses.

a. C What do you think?

b. N If a microscope can be constructed with these two lenses, describe its design. What is the minimum separation of the lenses? Where must you place the object?

82. N Microscope objectives and eyepieces are often labeled simply with a magnification. A 4× eyepiece and a 20× objective result in a total magnification of 80× for the system. Assuming these are simple converging lenses and the objective is brought to a distance of 8.5 mm from the specimen, what focal lengths are needed for the eyepiece and the objective to give the 80× system magnification?

83. N Two lenses are placed along the x axis, with a diverging lens of focal length −8.00 cm on the left and a converging lens of focal length 16.0 cm on the right. When an object is placed 10.0 cm to the left of the diverging lens, what should the separation s of the two lenses be if the final image is to be focused at $x = \infty$?

84. **C** CASE STUDY What is the purpose of the tube used in the construction of a microscope or a telescope?

General Problems

85. **N** A monochromatic beam of light is incident on a slab of unknown material at angle $\theta_i = 51.0°$ (Fig. P38.85). If the angle at which the beam exits the slab is $\theta_t = 69.0°$, what is the index of refraction of the slab?

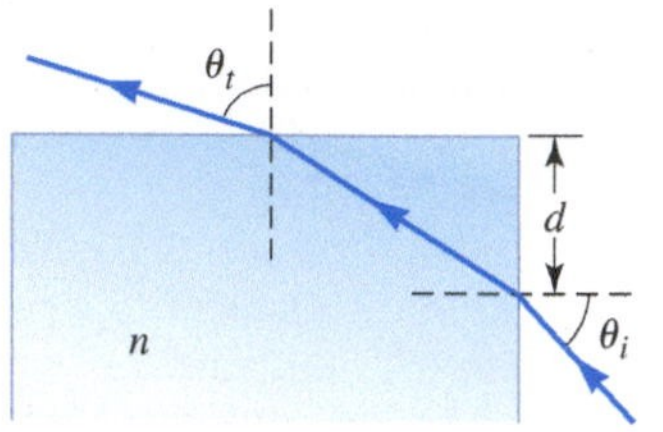

FIGURE P38.85

Problems 86 and 87 are paired.

86. **N** At a still mountain lake where the air temperature is 0.0°C, a plane sound wave with wavelength 602 mm is incident on the lake's surface with angle of incidence $\theta_i = 9.00°$. The lake's water is at 15.0°C. *Note*: Refer to Table 17.1 (page 498) for the speed of sound in air and in water at the given temperatures. **a.** What is the angle of refraction θ_t of the sound wave in the water? **b.** What is the wavelength of the sound wave in the water?

87. Consider the mountain lake in Problem 86. The angle of incidence of a beam of 656-nm red light on the lake's surface is $\theta_i = 9.00°$.

a. **N** What is the angle of refraction θ_t of the ray of light in the water?

b. **N** What is the wavelength of the ray of light in the water?

c. **C** Compare the answers to parts (a) and (b) to the corresponding parts of Problem 86. How do the behaviors of sound waves and light waves differ during refraction?

88. **N** A thick glass container ($n = 1.523$) with inner and outer vertical walls has its interior filled with water ($n = 1.333$) and has a quarter lying on the bottom. A ray of light with wavelength 486 nm in the water has reflected from the quarter and strikes the interface between the water and the inner wall of the glass at an angle of incidence of 15.0°. **a.** What is the frequency of the light while in the water? **b.** What is the light ray's transmitted angle when it finally leaves the glass, passing into the air?

89. **N** A Pyrex ($n = 1.47$) pan is filled with water, and a monochromatic beam of light is directed at the water–Pyrex interface. The beam is observed to reflect off the bottom of the Pyrex such that it hits the water–Pyrex interface at an angle of incidence of 24.3°. What is the angle of refraction of the beam of light as it emerges into the water?

Problems 90 and 91 are paired.

90. **A** Prove that when a wave moves from medium 1 into medium 2, $n_1\lambda_1 = n_2\lambda_2$.

91. **N** When a light wave moves from one medium into another, the frequency of the wave stays the same. If a light wave with a wavelength of 489 nm moving in air ($n = 1.0002926$) enters water ($n = 1.333$), find **a.** the frequency of the light in the air, **b.** the wavelength of the light in the water, and **c.** the speed of the light in the water.

92. **N** A beam of monochromatic light is incident at $\theta_i = 33.0°$ on a block of Lucite ($n = 1.495$) surrounded by air. **a.** What is the angle of refraction θ_t? **b.** What would the angle of refraction be if the block were made of flint glass with a 71% lead content ($n = 1.805$)?

Problems 93, 94, and 95 are grouped.

93. **N** Light in air is incident on diamond, with $\theta_i = 34.5°$. **a.** What is the angle of reflection? **b.** What is the angle of refraction in the diamond?

94. **N** Light in water is incident on diamond, with $\theta_i = 34.5°$. **a.** What is the angle of reflection? **b.** What is the angle of refraction in the diamond?

95. **N** Light in air is incident on cubic zirconia, with $\theta_i = 34.5°$. **a.** What is the angle of reflection? **b.** What is the angle of refraction in the cubic zirconia?

96. **N** A monochromatic beam of light is incident on a large rectangular sapphire crystal ($n = 1.760$) with $\theta_i = 27.0°$. **a.** What is the angle of refraction θ_t in the sapphire crystal? **b.** At what angle is the beam incident on the back surface of the sapphire crystal? **c.** At what angle is the beam refracted at the crystal–air interface?

97. **N** A beam of monochromatic light strikes the interface between air and safflower oil ($n = 1.466$) at angle θ_i and is refracted (Fig. P38.97). The beam next strikes the safflower oil–water interface at angle $\gamma = 17.0°$ and is refracted at angle θ_t. **a.** What is angle θ_i? **b.** What is angle θ_t?

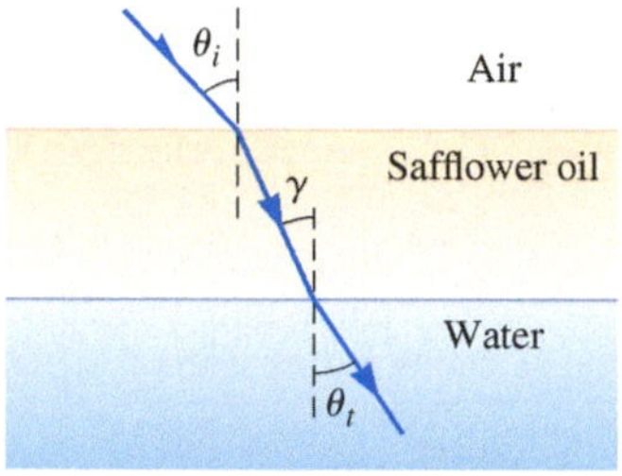

FIGURE P38.97

98. **A** **Fermat's principle of least time for refraction.** A ray of light traveling in a medium with speed v_1 leaves point A and strikes the boundary between the incident and transmitted media a horizontal distance x from point A as shown in Figure P38.98. The refracted ray travels with speed v_2 in the second medium, eventually reaching point B. The horizontal distance between points A and B is L. **a.** Calculate the time t required for the light to travel from A to B in terms of the parameters labeled in the figure. **b.** Now take the derivative of t with respect to x. What is the condition for which the ray of light will take the shortest time to travel from A to B?

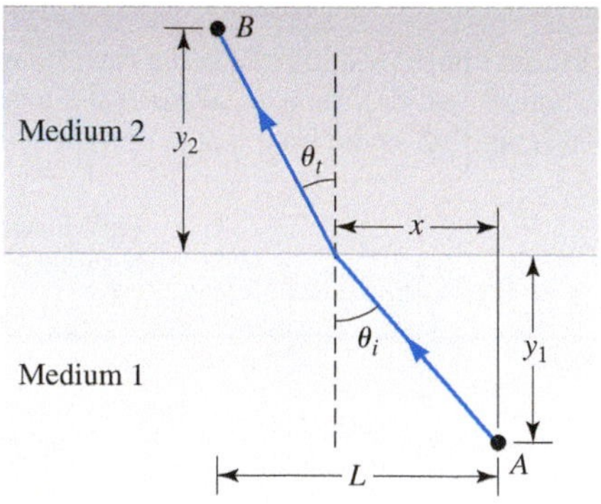

FIGURE P38.98

99. **N** A monochromatic beam of light is incident on a slab of Lucite ($n = 1.495$) surrounded by air. What is the incident angle θ_i that results in refraction angle $\theta_t = \theta_i/2$? *Hint*: Use the trigonometric identity $\sin 2\theta = 2 \sin\theta \cos\theta$.

100. **N** When the index of refraction of a medium depends on the wavelength of the incident light, each wavelength refracts at a different angle, causing dispersion that can be quantified as the difference in the refraction angles of two wavelengths (called the measure of dispersion). A beam of light containing red and yellow wavelengths strikes a block of heavy flint glass at $\theta_i = 66.0°$. If the index of refraction of the glass for 656-nm red light is 1.879 and the index of refraction for 589-nm yellow light is 1.919, what is the dispersion angle of the block of flint glass for these wavelengths?

101. **N** An outdoor hot tub is cylindrical with depth d and radius 1.25 m and is completely filled with water. In the morning, the bottom of the hot tub begins to receive sunlight when the Sun is 24.0° above the horizon. What is the depth d of the hot tub?

Problems 102 and 103 are paired.

102. **N** A quarter lies on the bottom of a glass ($n = 1.54$) that contains water ($n = 1.33$). Light rays bounce off the quarter in all directions. **a.** What is the critical angle for light traveling from the glass into the water? **b.** What is the critical angle for light traveling from the glass into the air ($n = 1.00029$)?

103. **N** A quarter lies on the bottom of a glass ($n = 1.54$) that contains water ($n = 1.33$). Light rays bounce off the quarter in all directions. What is the minimum angle of incidence for a ray of light from the quarter in the water that will result in the ray being at the critical angle when it arrives at the glass–air interface?

104. **N** Figure P38.104 shows a monochromatic beam of light incident on one end of a cylinder with index of refraction $n = 1.38$ at an angle of incidence θ_i. What is the maximum angle for which the beam of light will be totally internally reflected by the cylindrical wall and exit at the left-hand end?

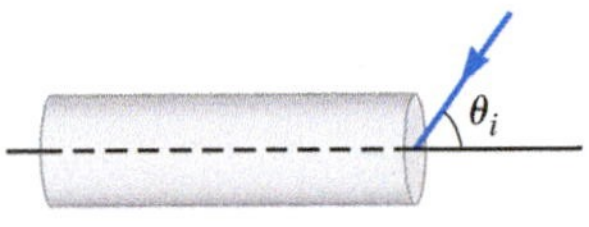

FIGURE P38.104

Problems 105 and 106 are paired.

105. **N** Curved glass–air interfaces like those observed in an empty shot glass make it possible for total internal reflection to occur at the shot glass's internal surface. Consider a glass cylinder ($n = 1.54$) with an outer radius of 2.50 cm and an inner radius of 2.00 cm as shown in Figure P38.105. Find the minimum angle θ_i such that there is total internal reflection at the inner surface of the shot glass.

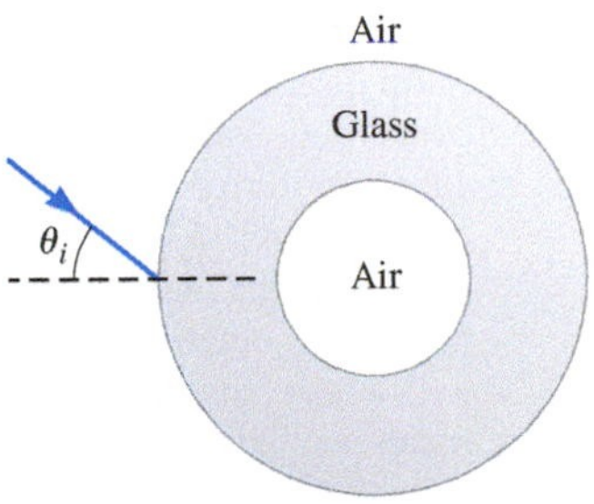

FIGURE P38.105 Problems 105 and 106.

106. **N** In Problem 105, we discovered how a light ray that enters a cylindrical glass container can experience total internal reflection at the internal glass–air interface when the outer angle θ_i is larger than a particular value. This phenomenon is very difficult to observe for thin containers because the outer angle is very close to 90°, so not much light experiences total internal reflection. However, we can observe this interesting example of total internal reflection in a thin-walled glass cylinder ($n = 1.54$) if we submerge it in water. Consider a cylinder with an outer radius of 2.50 cm and inner radius of 2.00 cm. Find the minimum angle θ_i at which total internal reflection is observed. The interior will still be air but the glass will be surrounded by water on the outside.

107. **N** A block is constructed from layers of cubic zirconia ($n = 2.14$), flint glass ($n = 1.80$), and quartz ($n = 1.54$) as shown in Figure P38.13. The block is surrounded by air. For what values of the incident angle θ_i does total internal reflection occur at the quartz–air interface?

108. **N** A hiker stands at the edge of a clear alpine lake that is 5.00 m deep. **a.** What is the apparent depth of the lake? **b.** Returning in the summer, the hiker finds the lake surface 2.00 m lower than before. What is the apparent depth of the lake now?

Problems 109 and 110 are paired.

109. **N** A light source forms parallel beams that strike the flat face of a transparent hemisphere of flint glass ($n = 1.65$) at normal incidence (Fig. P38.109). At what distance behind the hemisphere do the paraxial rays of the light source focus if the magnitude of the hemisphere's radius of curvature is 8.50 cm?

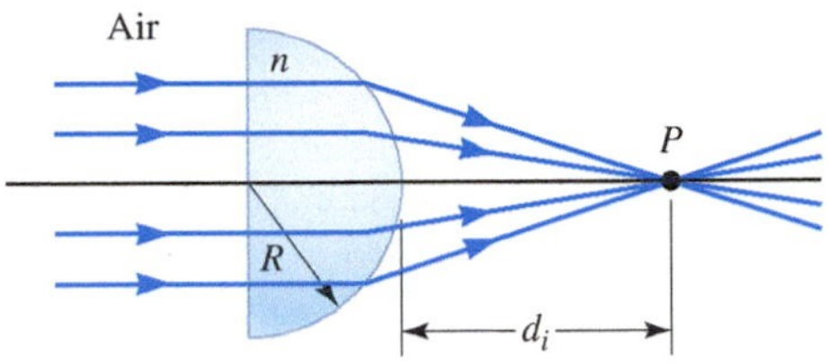

FIGURE P38.109

110. **N** A point source of light is placed at point P a distance d_o in front of the curved face of a transparent hemisphere of flint glass ($n = 1.65$) as in Figure P38.110. What should the object distance d_o be if the rays emerging from the flat surface of the hemisphere are to form an image at infinity? The radius of curvature is 8.50 cm.

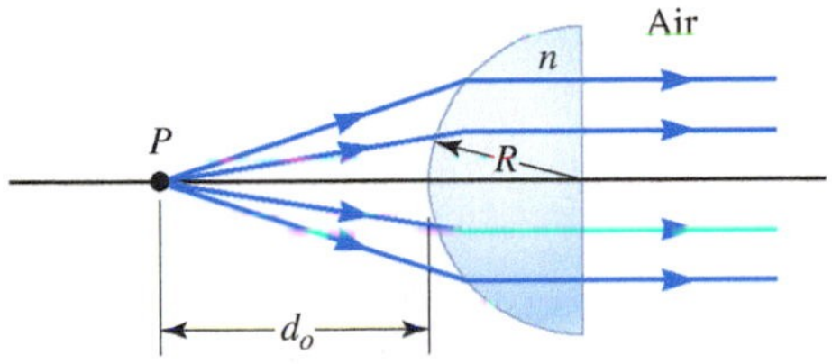

FIGURE P38.110

111. **N** A group of children watches a playful dolphin swim toward the underwater viewing window at an aquarium. If the dolphin's speed in the water is 1.00 m/s, what is its apparent speed as seen by the children?

112. **C** Return to the derivation of the lens maker's equation (page 1234), and explain why $d_{o2} = w - d_{i1}$ (Eq. 2) holds in all three cases shown in Figures 38.24, 38.25, and 38.26.

Problems 113 and 114 are paired.

113. An object is 20.0 cm to the left of a converging lens with a focal length of 30.0 cm. A second lens is placed 15.0 cm to the right of the converging lens. The second lens is a diverging lens with focal length −20.0 cm.

a. **N** Find the final image position.

b. **C** Describe where the final image is located.

114. **N** An object is 20.0 cm to the left of a converging lens with a focal length of 30.0 cm. A second lens is placed 15.0 cm to the right of the converging lens. The second lens is a diverging lens with a focal length of −20.0 cm. **a.** If the object has a height of 17.0 cm and is upright, find the magnification due to each lens and the magnification of the final image. **b.** What is the final image height?

115. **N** The magnification of an upright image that is 34.0 cm away from its object is 0.400. What type of lens is used to form the image, and what is its focal length?

116. **C** What is the physical difference between a real and a virtual image? Describe a method for using a converging lens to demonstrate the difference to a friend.

117. **N** The plano-convex lens made of glass ($n = 1.52$) shown in Figure P38.117 has a focal length of 20.0 cm. Calculate the radius of curvature of the curved surface of the lens.

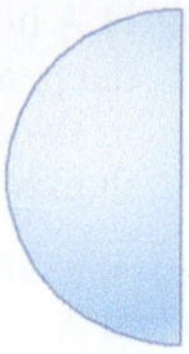

FIGURE P38.117

118. **C** You are handed two lenses, one of which is a diverging lens and the other a converging lens. Explain how you would tell the difference between these lenses if you were able to **a.** feel them with your hands, **b.** try to project an image of a lightbulb onto a piece of paper, or **c.** look through each of them at an object on your desk.

Problems 119 and 120 are paired.

119. **N** Consider a converging lens with a 16.0-cm focal length. How far is the object from the lens if a real image is formed **a.** 22.0 cm behind the lens and **b.** 42.0 cm behind the lens?

120. **N** Consider a converging lens with a 16.0-cm focal length. How far is the object from the lens if a virtual image is formed **a.** 22.0 cm in front of the lens and **b.** 42.0 cm in front of the lens?

121. **N** Consider a diverging lens with focal length $|f| = 14.0$ cm. **a.** What is the location of the image if $d_o = 28.0$ cm? What is the magnification of the image? Is the image real or virtual? Is it upright or inverted? **b.** What is the location of the image if $d_o = 6.00$ cm? What is the magnification of the image? Is the image real or virtual? Is it upright or inverted?

122. **A** In Example 38.7, we found that when a convex meniscus lens is flipped front to back, it remains a converging lens with the same focal length. Is this true of all lenses, or is it a particular feature of the convex meniscus lens? Starting with the lens maker's equation, determine which variables change when a lens is flipped and whether the results of Example 38.7 are true for all lenses.

123. **N** An object is placed 32.0 m in front of a converging lens with focal length $f = +18.0$ cm, forming a real image. The object is moved away from the lens with an initial speed of 3.00 m/s. **a.** With what initial speed does the image formed by the lens move? **b.** What is the direction of motion of the image—toward the lens or away from the lens?

Problems 124, 125, 126, and 127 are grouped.

124. Consider a prism with apex angle β. A light ray is incident at angle θ_1 as shown in Figure P38.124. The refractive index of the prism relative to air is n, and assume the index of refraction of the surrounding air is 1.

a. **A** Show that the total deviation α of a ray from its original path is given by

$$\alpha = \theta_1 - \beta + \sin^{-1}\left[\sin\beta\sqrt{n^2 - \sin^2\theta_1} - \cos\beta\sin\theta_1\right]$$

b. **G** Plot the angle of deviation α versus the angle of incidence θ_1 for $n = 1.5$ and the apex angle $\beta = 60.0°$.

c. **N** For what angle θ_1 is α minimized? What is the significance of this minimum?

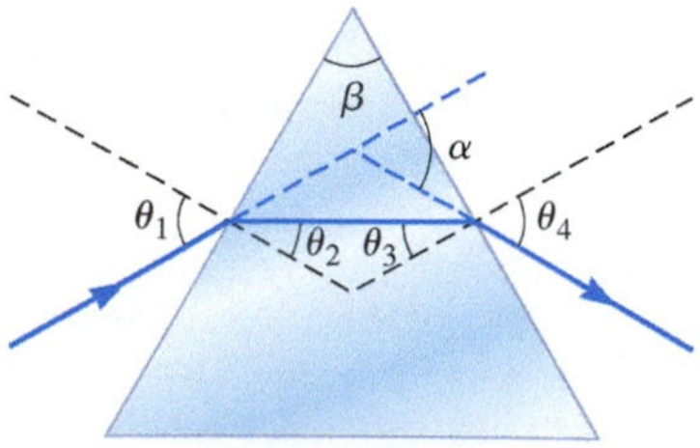

FIGURE P38.124

125. **A** In Problem 124, we found that the deviation α is minimized when the ray of light passes through the prism symmetrically.
a. Show that the minimum deviation α_{min} satisfies the equation

$$n = \sin\left(\frac{\beta + \alpha_{min}}{2}\right)\Big/\sin\left(\frac{\beta}{2}\right)$$

b. Assume the apex angle of the prism is small, so that $\sin\beta \approx \beta$. Obtain the minimum deviation angle.

126. **A** A combination of two prisms in which the deviation produced by the first prism is equal and opposite to that produced by the second prism is called a **direct vision prism**. This combination produces dispersion without deviation. If n_1 and n_2 are the refractive indices of the two prisms and β_1 is the apex angle of the first prism, determine the apex angle β_2 of the second prism so that the net deviation is zero. Assume the angles of the prisms are small, so that $\sin\beta_1 \approx \beta_1$ and $\sin\beta_2 \approx \beta_2$. *Hint*: You may use the result from part (b) in Problem 125.

127. **N** Light enters a prism of crown glass and refracts at an angle of 5.00° with respect to the normal at the interface. The crown glass has a mean index of refraction of 1.51. It is combined with one flint glass prism ($n = 1.65$) to produce no net deviation. **a.** Find the apex angle of the flint glass. **b.** Assume the index of refraction for violet light ($\lambda_v = 430$ nm) is $n_v = 1.528$ and the index of refraction for red light ($\lambda_r = 768$ nm) is $n_r = 1.511$ for crown glass. For flint glass using the same wavelengths, $n_v = 1.665$ and $n_r = 1.645$. Find the net dispersion.

128. **A** A converging lens with a focal length of 18.0 cm is placed at $y = 40.0$ cm on the y axis. An object originally at the origin is moved slowly along the y axis toward the lens until it is within 10.0 cm of the lens. What is the position of the image formed by the lens as a function of the position y of the object?

129. **N** An object is placed a distance of 10.0 cm to the left of a thin converging lens of focal length $f = 8.00$ cm, and a concave spherical mirror with radius of curvature +18.0 cm is placed a distance of 45.0 cm to the right of the lens (Fig. P38.129). **a.** What is the location of the final image formed by the lens–mirror combination as seen by an observer positioned to the left of the object? **b.** What is the magnification of the final image as seen by an observer positioned to the left of the object? **c.** Is the final image formed by the lens–mirror combination upright or inverted?

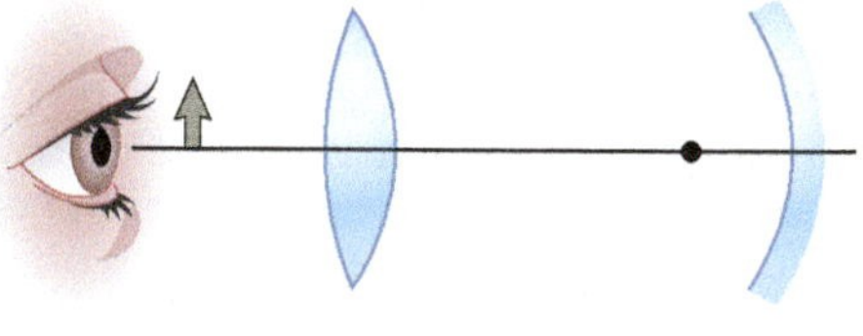

FIGURE P38.129

PART SIX
20th Century Physics

39 Relativity

Underlying Principles

1. Einstein's postulates of special relativity
2. Einstein's postulates of general relativity

Major Concepts

1. Lorentz transformations
2. Length contraction
3. Time dilation
4. Relativistic Doppler effect
5. Perpendicular velocity transformation
6. Parallel velocity transformation
7. Mass transformation
8. Relativistic momentum
9. Mass–energy equivalence principle
10. Curvature of space
11. Gravitational Doppler shift and gravitational time dilation

Special Cases

Galilean relativity

Key Questions

How is Einstein's theory of special relativity different from Galilean relativity? What are the strange consequences of special relativity?

How is general relativity different from special relativity?

We expect that some measurements should not depend on the observer. For example, your physics class is 50 minutes long, you are 5 feet 8 inches tall, and the speed of light is 3.00×10^8 m/s. The truth, however, is that observers need to agree about only one of these three observations—the speed of light. The duration of your physics class and your height depend on the velocity of the observer. It may seem crazy, but an observer moving at a very high speed will say that your physics class lasts longer than 50 minutes and you are shorter than 5 feet 8 inches!

The idea that you and a moving observer won't agree on something as fundamental as your height or the duration of your physics class may seem unbelievable. But, early in the 20th century, Albert Einstein (1879–1955) found that measurements of space and time depend on the motion of the observer. This led to some amazing discoveries, such as (1) the observed mass of an object depends on its speed, (2) the ultimate speed limit is the speed of light, and (3) mass and energy are equivalent. In this chapter, we explore Einstein's fundamental ideas and their consequences.

39-1 It's in the Eye of the Observer

In this textbook, we have studied physics in nearly chronological order—from Galileo's work on kinematics in the late 1500s to Maxwell's study of light in the middle 1800s. In this last part, we turn our attention to discoveries made in the early 1900s—a departure from classical physics referred to as *modern physics* or *20th-century physics*. Much of Einstein's work in developing his theories of relativity is actually based on a careful study of kinematics, so we must return to the beginning and study motion from the perspectives of different observers.

Experimenting in a Noninertial Reference Frame

Recall that we defined an inertial reference frame as one in which Newton's first law holds (Section 5-5). An inertial reference frame may be at rest or may move at constant velocity relative to the observer, but it cannot accelerate. Imagine for a moment that all scientists, including Newton, lived their entire lives on cruise ships that sometimes accelerated dramatically (Fig. 5.6, page 125). It would be difficult for these scientists to discover the law of inertia because sometimes they would observe the dramatic acceleration of objects that had no net force exerted on them.

The idea of living your entire life on a cruise ship may seem ridiculous, but it is actually closer to your experience than you might think. All human beings live in a *noninertial* reference frame. The Earth spins on its axis and orbits the Sun. The entire solar system orbits the center of the Milky Way galaxy, and the entire galaxy is falling toward the Andromeda galaxy. All this motion means that all of our laboratories are always accelerating. For most of this book, we could ignore the Earth's acceleration, but in sensitive experiments we cannot. One of Newton's great insights was to imagine a truly inertial reference frame. He had to infer the existence of such a frame and of the law of inertia by extrapolating from the results of experiments made on the Earth. His thought experiment led scientists to wonder whether such an inertial frame exists physically.

Looking for the Inertial Reference Frame

In the 19th century, scientists believed that light propagated in a medium known as the ether, which was also supposed to be at rest. So, perhaps the ether was the object that defined the one true inertial reference frame. If we could measure the speed of objects moving with respect to the ether, we would find that these objects obey Newton's laws. However, the Michelson–Morley experiment (case study in Chapter 36) failed to detect the motion of the Earth with respect to the ether, showing that either the ether is attached to the Earth or it does not exist.

The Aberration of Starlight

The possibility that the ether is attached to the Earth was ruled out by data collected even before Michelson and Morley conducted their experiment. An important piece of evidence to disprove that possibility comes from the observation of a star. The British astronomer James Bradley (1693–1762) observed the star Gamma Draconis (γ Dra) near the zenith for more than a year. He found that γ Dra appeared to move with respect to the zenith and that its greatest angular displacement (20.5 arcsec) occurred in September and March. This apparent motion of a star is known as **stellar aberration** and is a direct consequence of the Earth's orbital motion.

Let's start with the assumptions that the ether exists and that it is attached to the Earth. In this case, the velocity of the Earth with respect to the ether is zero (consistent with Michelson and Morley's results). If you want to observe a star that is directly overhead, you point your telescope straight up at the zenith (Fig. 39.1A). If the Earth, the ether, and the star are all at rest in the same reference frame, light from the star travels straight down the tube of your telescope to your detector at the bottom. You would say the star is at the zenith because that is the direction in which you pointed your telescope to collect the starlight.

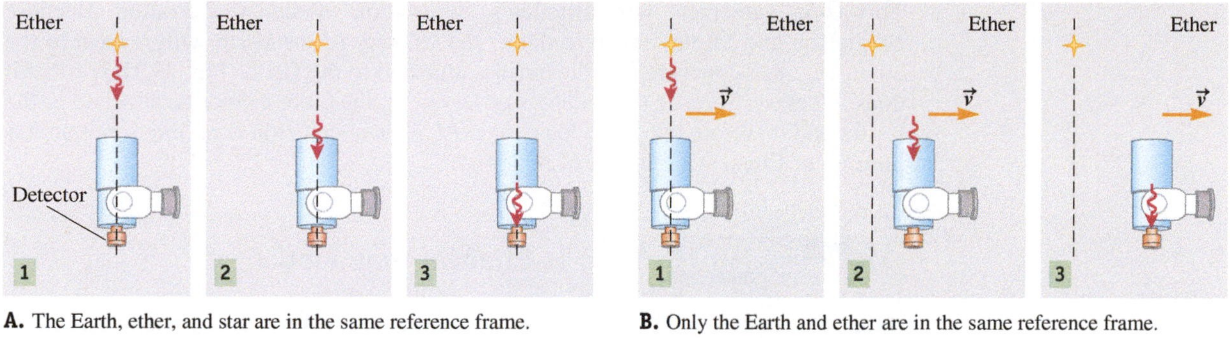

FIGURE 39.1 **A.** The Earth, the ether, and the star are all at rest in the same frame. **B.** The ether is attached to the Earth, and both move to the right relative to the star. In both cases, 1 light (shown as a photon) is emitted by the star, 2 the photon enters the telescope, and 3 the photon is detected. In both cases, the telescope is aimed straight up and the star is detected at the zenith. *Neither of these scenarios is true!*

Now suppose the ether and the Earth are at rest with respect to each other in the same reference frame, which is moving to the right with respect to the star (Fig. 39.1B). You point your telescope straight up at the star. Light from the star enters the aperture of your telescope. As the photon drifts down toward your detector, the telescope (attached to the Earth) and the ether move to the right, so the light must also move to the right as it drifts down. The result is that you detect the light. Again, you would claim the star is at the zenith because that is the direction in which you must point your telescope. But, if this were the case, if the ether were attached to the Earth, Bradley would not have observed the aberration of γ Dra. No matter what the time of year, he could point his telescope toward the zenith to see γ Dra. Instead, Bradley had to tilt his telescope by as much as 20.5 arcsec to see γ Dra at certain times of the year.

To see why, imagine the ether exists but is not attached to the Earth. Instead, the Earth moves through the ether at roughly the same speed ($v \approx 30$ km/s) at which it orbits the Sun. Suppose you wish to observe γ Dra in March, when the Earth's velocity is to the right with respect to the ether. The star is still at the zenith, but you must point your telescope at the angle θ in order to see it (Fig. 39.2A), so you believe this is the angular position of the star. In time t, the photon travels a vertical distance ct, while the telescope travels a horizontal distance vt (Fig. 39.2B). So, the (small) angle θ is approximately

$$\theta \approx \frac{vt}{ct} = \frac{v}{c} = \frac{30 \times 10^3\,\text{m/s}}{3.00 \times 10^8\,\text{m/s}} \approx 20.6\ \text{arcsec}$$

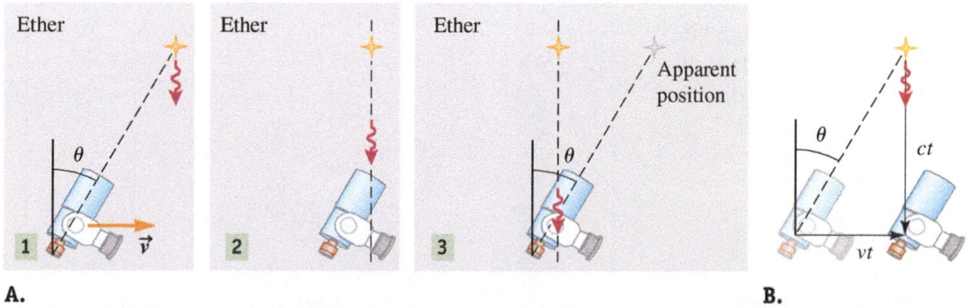

FIGURE 39.2 Suppose the Earth moves through the ether to the right at speed v. **A.** 1 A star emits a photon, 2 the photon enters the telescope, and 3 the photon is detected. If the Earth moves with respect to the ether, a star at the zenith appears to be at an angle θ with respect to the zenith. **B.** In time t, the photon travels distance ct and the detector travels distance vt. This scenario explains Bradley's observation, but the Michelson–Morley experiment requires that the ether does not exist.

This is in agreement with Bradley's observation of stellar aberration. Because Michelson and Morley failed to detect the velocity of the Earth with respect to the ether, we must conclude that the ether is attached to the Earth (Fig. 39.1B) or that it does not exist. Because of Bradley's observation, the ether cannot be attached to the Earth, and therefore the ether does not exist. This conclusion is an important underpinning of Einstein's theories of relativity.

CASE STUDY Truth Is Stranger than Fiction

Educated people acquire knowledge from many different sources. Sometimes, scientific discoveries inspire works of fiction, and we can learn about scientific breakthroughs through such works. Fiction is a great avenue for a first encounter with something new. It inspires our imagination, which makes the learning exciting. But there is a drawback to learning about science through fiction: Fiction writers are not obligated to be truthful. They can exaggerate or ignore a scientific truth in order to create their art. So, it can be hard for us to sort out what is true and what is made up.

In this chapter's case study, we use fiction to explore the strange consequences of Einstein's theory of relativity. If you first encountered these strange ideas in a work of fiction, you would probably think they were made up. But each one is a prediction that results from relativity and has been confirmed by observation. Einstein's theory of relativity demonstrates the old saying: *Truth is stranger than fiction.*

CONCEPT EXERCISE 39.1

Which of the following are (approximately) inertial reference frames? Explain your answer.

a. Your classroom
b. A railroad car moving at constant speed along a straight track
c. A car moving at constant speed around a turn
d. An airplane during takeoff

EXAMPLE 39.1 A Rotating Room

Aaron lives in a room that rotates with constant speed. The room has a coordinate system painted on its glass ceiling, and it is located entirely inside a stationary room in which Hannah lives (Fig. 39.3). Through the glass ceiling of his rotating room, Aaron can see a ball glued (off axis) to the ceiling of the stationary room. Draw the path of the ball as seen by Aaron through his rotating glass ceiling. What does he conclude about the law of inertia? A desk lamp in Aaron's room projects his coordinate system onto Hannah's ceiling.

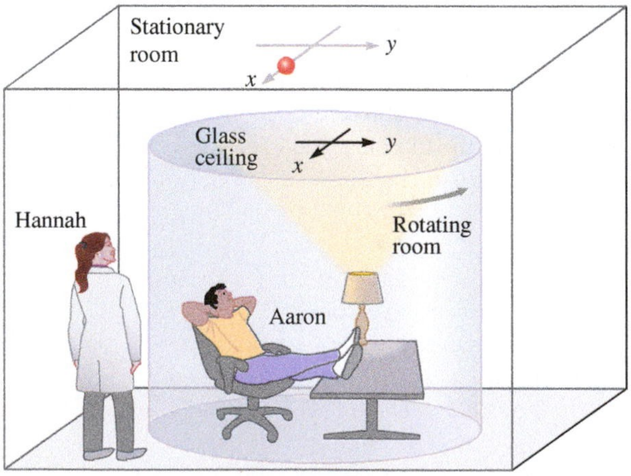

FIGURE 39.3 What path does the ball take in Aaron's frame?

INTERPRET and ANTICIPATE
Because it is accelerating, a rotating room is not an inertial reference frame (Section 5-5). So we expect to find that the ball, as observed from Aaron's frame, violates the law of inertia.

SOLVE

It is difficult to imagine immediately what Aaron sees from the rotating room. It is somewhat easier to imagine first what Hannah sees if Aaron's coordinate system is projected onto her stationary ceiling. To Hannah, the ball is at rest and the coordinate system rotates counterclockwise as shown at three time instants in Figure 39.4.

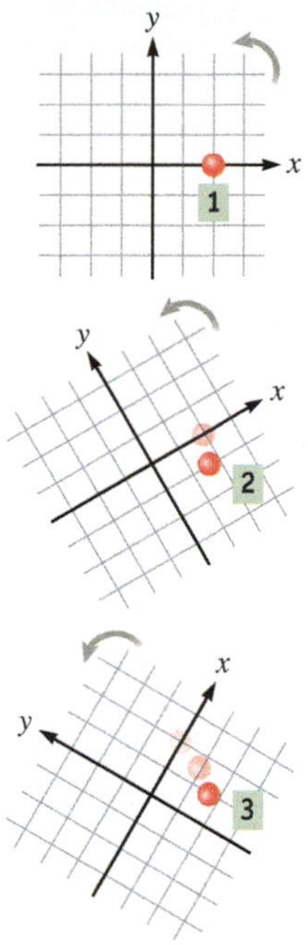

FIGURE 39.4 Hannah sees the ball remain fixed (bright red) and Aaron's coordinates moving (shown with images of the ball's previous position in Aaron's coordinates).

In the rotating room, Aaron sees a fixed coordinate system on his ceiling. He sees the ball moving clockwise in a circle around his coordinate system (Fig. 39.5). He cannot identify a source of centripetal force, however, or any net force on the ball. So, he concludes that the ball's motion violates the law of inertia.

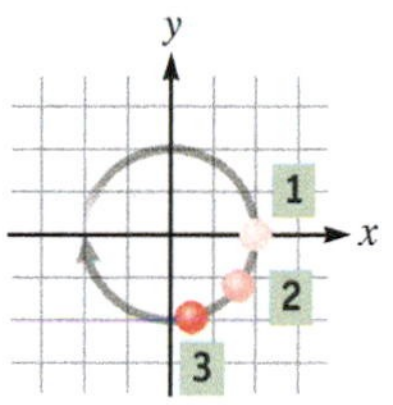

FIGURE 39.5 Aaron sees the ball moving in a circle.

CHECK and THINK

This is exactly what we would expect from an observer in an accelerating reference frame. Aaron sees an apparent violation of the law of inertia because, although the ball is stationary in an inertial frame, he is in a noninertial frame. If the ball had been glued to Aaron's glass ceiling, he would have seen the ball at rest and he would not have observed any violation of the law of inertia.

39-2 Special Case: Galilean Relativity

GALILEAN RELATIVITY

Special Case

Einstein developed much of his theory of relativity by thinking about the kinematics observed from different reference frames. His ideas about space and time are a departure from the classical ideas of space and time used until now in this textbook. To help distinguish the classical model of relative motion from Einstein's model, we'll refer to the classical model as **Galilean relativity.**

We'll start with a review of Galilean relativity as in Section 4-7, using mostly scalar components instead of vector ones and with less cumbersome notation. Consider two reference frames, one of which is at rest and is referred to as the *laboratory* frame. Unless otherwise specified, the other frame is moving at constant velocity $\vec{v}_{rel}$ relative to the laboratory frame. To distinguish between the two coordinate systems, a prime ($'$) is used on all the symbols for the moving frame, and no prime is used on the symbols for the laboratory frame (Fig. 39.6).

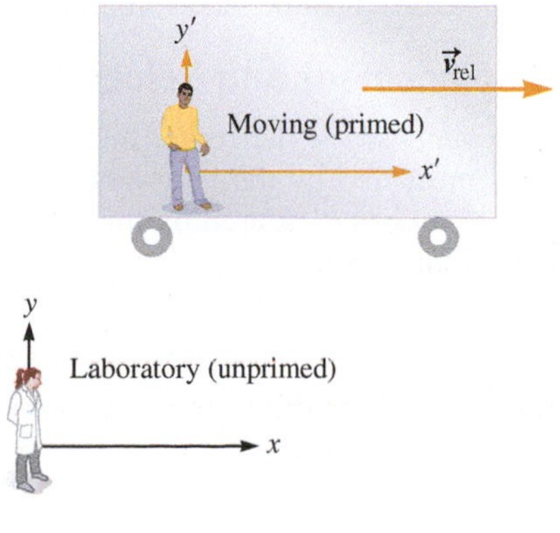

FIGURE 39.6 Two inertial frames with the laboratory frame at rest while the moving frame has velocity $\vec{v}_{rel}$ as measured by the observer in the laboratory frame. Each observer in each frame chooses a coordinate system such that he or she is at the origin. The two coordinate systems are parallel. We show them offset so that they don't overlap, but the primed frame is not actually located above the unprimed frame. In later figures, we just draw the coordinate systems; however, we will refer to the *laboratory observer* and the *primed* (or moving) *observer.*

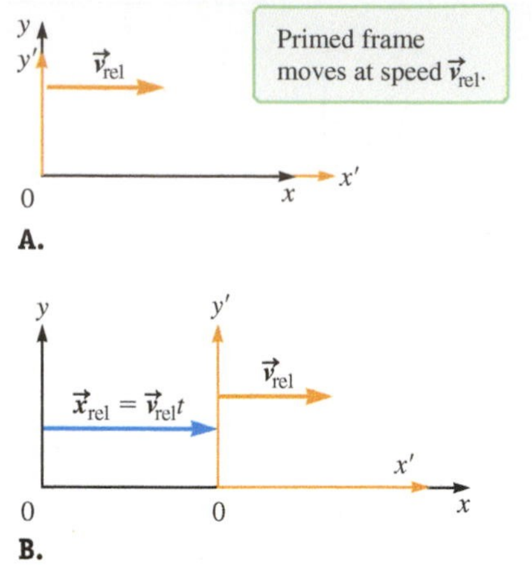

FIGURE 39.7 **A.** At $t = 0$, the two origins coincide. **B.** At a later time t, the observer in the laboratory frame finds that the moving frame's origin is at x_{rel}.

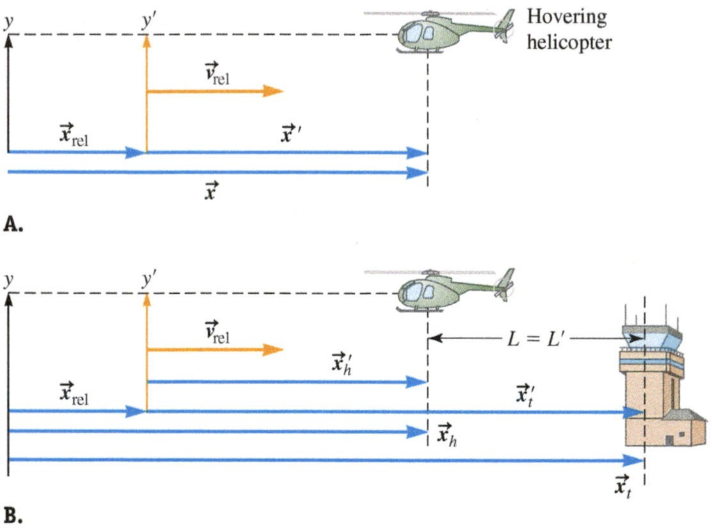

FIGURE 39.8 **A.** At t, both observers measure the position of a helicopter hovering relative to the laboratory. **B.** At t, both observers measure the distance $L = L'$ between a helicopter and a tower.

Position and Distance

Initially (at $t = 0$), the origins of two inertial frames overlap (Fig. 39.7A). The primed frame moves to the right at speed v_{rel}, and by time t, the observer in the laboratory frame finds that the primed frame's origin is at x_{rel} as shown in Figure 39.7B:

$$x_{rel} = v_{rel}t \tag{39.1}$$

Now suppose that at time t, both observers measure the position of a helicopter hovering in place relative to the laboratory (Fig 39.8A). They measure the same vertical position:

$$y = y' \tag{39.2}$$

However, they don't get the same horizontal position: $x = x' + x_{rel}$ (Eq. 4.45) and, using Equation 39.1, we find

$$x = x' + v_{rel}t \tag{39.3}$$

Equations 39.2 and 39.3 are called **transformation equations** because they transform the measurements made in one frame into the corresponding measurements made in the other frame. When a quantity is always the same in both frames, such as the vertical position of the helicopter, we say that quantity is *invariant*. Usually, we consider the simple situation in which one frame moves in the x direction relative to the other frame so that the perpendicular positions (y and z) are invariant.

Our next step is to find a transformation equation for a horizontal separation, such as between the helicopter and a tower (Fig. 39.8B). According to the laboratory observer, the separation is

$$L = x_t - x_h \tag{39.4}$$

where the subscript t stands for *t*ower and h for *h*elicopter. To the primed observer, the separation is $L' = x'_t - x'_h$. The transformation equation for the separation distance is found by substituting Equation 39.3 into Equation 39.4 twice: once for the tower's position and once for the helicopter's position:

$$L = (x'_t + v_{rel}t) - (x'_h + v_{rel}t) = (x'_t - x'_h) + (v_{rel}t - v_{rel}t)$$

The relative velocity is constant, and the measurements of the helicopter and tower's positions were made simultaneously (same t), so the second term is zero. The first term is the distance measured in the primed frame:

$$L = L' \tag{39.5}$$

Equation 39.5 shows that the distance or length is invariant under Galilean relativity. However, measurements of the endpoint positions (those of the helicopter and tower in this case) must be made simultaneously.

Displacement and Velocity

Now suppose the helicopter flies to the right from position x_i at time t_i to x_f at time t_f, as measured by the laboratory observer (Fig. 39.9). We can find the initial and final positions measured by the primed observer by using Equation 39.3 twice. For the initial position,

$$x_i = x'_i + v_{rel}t_i \tag{39.6}$$

and for the final position,

$$x_f = x'_f + v_{rel}t_f \tag{39.7}$$

The displacement measured by the primed observer is

$$\Delta x' = x'_f - x'_i \tag{39.8}$$

The displacement of the helicopter measured by the laboratory observer is

$$\Delta x = x_f - x_i \tag{39.9}$$

We don't expect the displacement to be invariant, however, because in the time between the two measurements, the primed frame has moved. We expect the displacement $\Delta x'$ measured by the primed observer to depend on the elapsed time. Relate the two displacements by substituting Equations 39.6 and 39.7 into Equation 39.9:

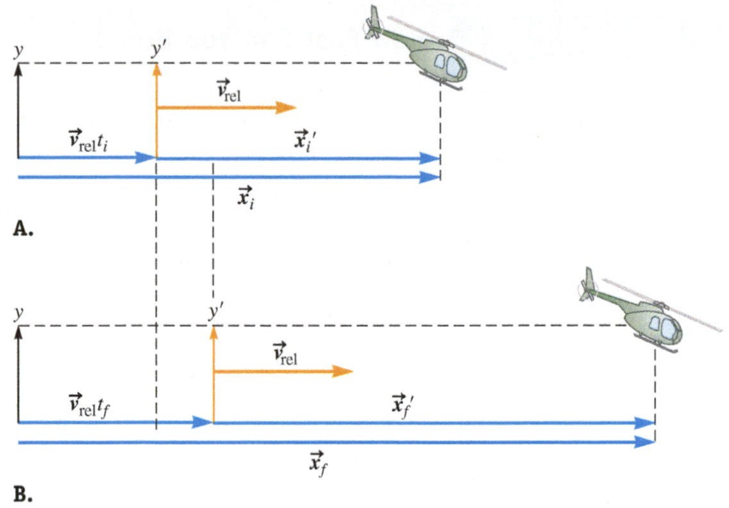

FIGURE 39.9 Both observers measure the position of a helicopter **A.** at t_i and **B.** at t_f.

$$\Delta x = (x'_f + v_{rel}t_f) - (x'_i + v_{rel}t_i) = (x'_f - x'_i) + v_{rel}(t_f - t_i)$$

The first term is the displacement as measured by the primed observer, so

$$\Delta x = \Delta x' + v_{rel}\Delta t \tag{39.10}$$

where $\Delta t = t_f - t_i$ is the elapsed time. Unlike a length, the displacement of the moving object is not invariant because it depends on the elapsed time as expected.

To find the velocity transformation equation, take the time derivative of Equation 39.3:

$$\frac{dx}{dt} = \frac{d}{dt}(x' + v_{rel}t) = \frac{dx'}{dt} + \frac{d(v_{rel}t)}{dt}$$

The relative velocity is a constant, so

$$v_x = v'_x + v_{rel} \tag{39.11}$$

where $v_x = dx/dt$ is the helicopter's velocity along the x direction measured in the laboratory frame and $v'_x = dx'/dt$ is its velocity along the same direction measured in the primed frame. Equation 39.11 shows that horizontal velocity is not invariant under Galilean relativity. However, the velocity in the perpendicular directions (y and z) is invariant: $v_y = v'_y$ and $v_z = v'_z$ (Problem 5).

Acceleration

Now suppose the helicopter is accelerating in some arbitrary direction. To find the transformation of acceleration in the x direction, take the time derivative of Equation 39.11:

$$\frac{dv_x}{dt} = \frac{dv'_x}{dt} + \frac{dv_{rel}}{dt}$$

The primed frame is an inertial frame, which means it is not accelerating. So v_{rel} is constant, $dv_{rel}/dt = 0$, and we find $dv_x/dt = dv'_x/dt$. Thus, acceleration in the parallel direction is invariant: $a_x = a'_x$. In Problem 5, you are asked to show that acceleration in the perpendicular directions is also invariant.

CONCEPT EXERCISE 39.2

Suppose the primed and laboratory observers want to measure the length of a rod that rests on the ground horizontally in the space between the helicopter and the tower (Fig. 39.8B). To derive the length transformation $L = L'$ (Eq. 39.5), we had to assume that the positions of the two ends were determined simultaneously. What happens to the length transformation equation if both observers measure the end below the helicopter at one time t_1 and the other end at a later time t_2?

EXAMPLE 39.2 How Fast Can You Bowl?

Aaron, a good bowler, is in the primed frame. He rolls a ball at 8.50 m/s (about 19 mph) in the x' direction. The bowling lane in the primed frame is 18.3 m (60 ft) long. The primed frame's speed with respect to Hannah in the laboratory frame is 25.0 m/s in the x direction. Assume the ball's speed is constant as it rolls down the lane.

A According to Hannah, what is the displacement of the ball?

INTERPRET and ANTICIPATE

During the time the ball rolls along the lane to the right (Fig. 39.10), Hannah sees Aaron's frame move to the right. So we expect her to see a greater displacement of the ball than he does.

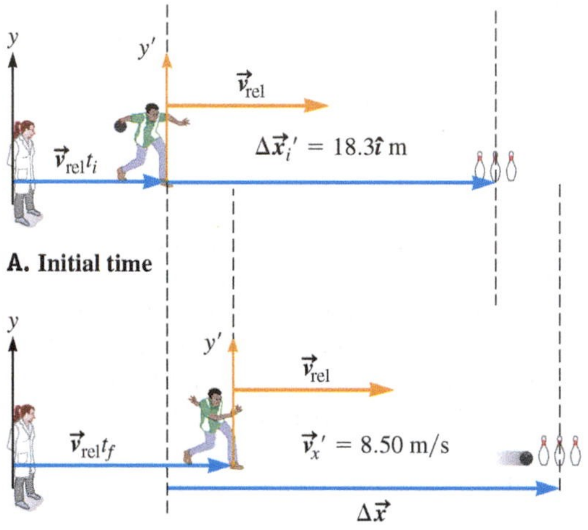

FIGURE 39.10

SOLVE

First, find the time the ball takes to travel down the lane. Because the ball's velocity is constant, we find the time interval by dividing the displacement by the ball's speed as observed by Aaron.

$$\Delta t = \frac{\Delta x'}{v_x'} = \frac{18.3\text{ m}}{8.50\text{ m/s}} = 2.15\text{ s}$$

Use Equation 39.10 to transform Aaron's observation of displacement into Hannah's observation of displacement.

$$\Delta x = \Delta x' + v_{rel}\Delta t \qquad (39.10)$$

$$\Delta x = 18.3\text{ m} + (25.0\text{ m/s})(2.15\text{ s})$$

$$\Delta x = 72.1\text{ m}$$

CHECK and THINK

As expected, the displacement measured by Hannah is greater than the displacement measured by Aaron.

B According to Hannah, what is the speed of the ball?

INTERPRET and ANTICIPATE

There are two ways to find the ball's constant speed in the laboratory frame. We can use the velocity transformation equation, or we can divide the ball's displacement by the time interval. We'll use the transformation equation and check our results by dividing the displacement by the time interval. Because Aaron's frame and the ball are both moving in the same direction, we expect that Hannah observes a higher speed than Aaron does.

SOLVE

The velocity transformation is given by Equation 39.11.

$$v_x = v_x' + v_{rel} \qquad (39.11)$$

$$v_x = 8.50\text{ m/s} + 25.0\text{ m/s} = 33.5\text{ m/s}$$

CHECK and THINK

To check the answer, divide the displacement measured by Hannah by the time interval of the ball's travel. Both methods give the same speed for the ball in the laboratory frame, and, as expected, Hannah finds that the ball is moving faster than the speed Aaron observes in his moving frame.

$$v_x = \frac{\Delta x}{\Delta t} = \frac{72.1\,\text{m}}{2.15\,\text{s}}$$

$$v_x = 33.5\ \text{m/s}\ \checkmark$$

The second method assumes the time interval is the same in both the primed and laboratory frames ($\Delta t = \Delta t'$). Although this assumption holds in Galilean relativity, it does not hold in Einstein's theory of relativity.

39-3 Postulates of Special Relativity

A **postulate** is a presupposition or condition that underlies a line of reasoning. The postulate of Galilean relativity is that the laws of mechanics (Newton's laws) hold in all inertial reference frames. You might expect that the laws of electricity and magnetism (Maxwell's equations) also hold in all inertial reference frames. However, near the end of the 19th century, it seemed that Maxwell's equations did not hold in all inertial frames; these laws seemed to be true in only one reference frame—the ether's frame. But, as described in Section 39-1, the observation of stellar aberration along with Michelson and Morley's experiment revealed that the ether does not exist. In the early 1900s (with no knowledge of Michelson and Morley's experiment), Einstein, a theoretical physicist, postulated that all the laws of physics (mechanics, electricity, and magnetism) are true in all inertial reference frames. With this presupposition in mind, Einstein published a paper in 1905 describing a new theory of relativity. This theory is known as *special relativity* because it is a restricted (or special) case of relativity, in which the reference frames are inertial (nonaccelerating). In the next decade, Einstein developed *general relativity*—an unrestricted theory that allows for the acceleration of reference frames.

In 1905, Einstein was working in the Swiss Patent Office in Bern.

Einstein's theory of special relativity has two postulates:

EINSTEIN'S POSTULATES OF SPECIAL RELATIVITY **Underlying Principle**

1. All the laws of physics are true in all inertial reference frames. This means there is no special frame, such as the ether's frame.
2. The speed of light in a vacuum has the same value c as measured by all observers, regardless of the observer's velocity. Put another way, the speed of light is invariant.

Although the second postulate may seem to come out of nowhere, it is the fundamental condition that ensures that the laws of electricity and magnetism hold in all inertial frames. To see why, recall that Maxwell derived the wave equation for light by starting with Faraday's law and Ampère–Maxwell's law. Faraday's law says that a changing magnetic field is the source of an electric field, and Ampère–Maxwell's law says that a changing electric field is the source of a magnetic field. Because the wave equation for light results from the combination of these two laws, a light wave needs both a changing magnetic field to create an electric field and a changing electric field to create a magnetic field.

Suppose an observer in the laboratory frame sees a light wave traveling at c. Einstein imagined observing the same light wave from a primed frame moving at a relative speed $v_{rel} = c$ in the same direction as the wave. According to Galilean relativity (Eq. 39.11), the light wave would have no speed in the primed frame: $v'_x = v_x - v_{rel} = c - c$. Such a primed (moving) observer would claim that the light wave does not exist and would not see a changing magnetic or electric field. Of course, this observation by the moving observer would violate Einstein's first postulate because, in the laboratory frame, a light wave is observed as predicted by Maxwell's equations, and so the wave must exist in all frames. In order for Maxwell's equations to hold in both frames, both observers must see a light wave moving at the speed of light.

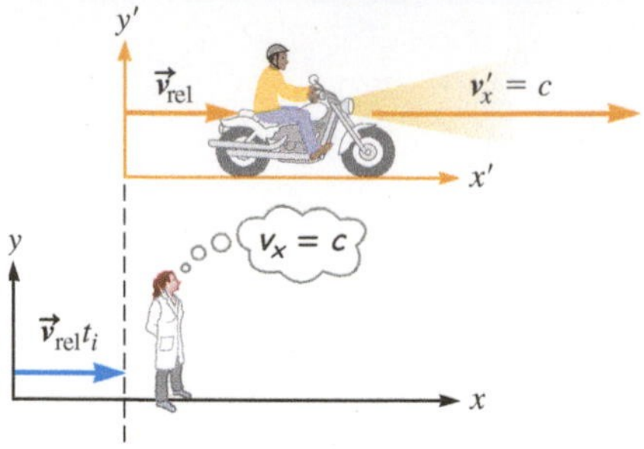

FIGURE 39.11 Aaron in the primed frame shines a light to the right. Observers in both the primed frame and the laboratory frame see the light moving at c.

Einstein's second postulate is consistent with Michelson and Morley's results (Section 36-6). Recall that in their experiment, light traveled along two perpendicular paths of equal length. Imagine two reference frames: each one moving with the light along each of the two paths. Their entire apparatus was on the moving Earth. But when they changed the orientation of this apparatus, they effectively changed the relative speed of the two frames. No matter what the orientation, no time delay was found between the two paths. Michelson and Morley found that the speed of light (in a vacuum) is constant, independent of the observer's speed.

Because our intuition is based on observing the motion of objects that have speeds much less than the speed of light, the second postulate leads to some counterintuitive results. For example, suppose Aaron in the primed frame turns on his headlight (Fig. 39.11). If Aaron could actually watch photons as they emerge from the headlight, he would see the light moving at speed $v'_x = c$. You might expect that Hannah in the laboratory frame would see the photons moving at $v_x = v_{rel} + c$. According to Einstein's second postulate, though, Hannah must still observe the light moving at $v_x = c$. The relative speed of the two frames does not matter. (Alternatively, you might think Hannah observes $v_x \approx c$ because the relative speed is low compared to c, so that when we add a low speed to the speed of light, we get approximately the speed of light. However, Einstein's second postulate is *not an approximation*; it says the speed of light is exactly c for all observers.)

39-4 Lorentz Transformations

One of the most fundamental ideas in Galilean relativity is that a time interval is invariant for all observers. So, one of the most counterintuitive implications of Einstein's second postulate is that time intervals are *not* invariant.

To see this, let's compare the motion of an object such as a helicopter (Fig. 39.9) to the motion of a photon. The displacement of the moving helicopter in the primed frame is $\Delta x'$ (Eq. 39.8), and its displacement as measured in the laboratory frame is greater: $\Delta x = \Delta x' + v_{rel}\Delta t$ (Eq. 39.10). Divide Equation 39.10 by Δt:

$$\frac{\Delta x}{\Delta t} = \frac{\Delta x'}{\Delta t} + v_{rel} \tag{39.12}$$

Now imagine replacing the helicopter moving to the right in Figure 39.9 with a photon. As in the case of the helicopter, the displacement of the photon in the primed frame is $\Delta x'$, and the displacement as measured in the laboratory frame is greater: $\Delta x > \Delta x'$. However, according to Einstein's second postulate, the speed of the photon as measured by either observer must be equal to c:

$$\frac{\Delta x}{\Delta t} = \frac{\Delta x'}{\Delta t'} = c$$

It is impossible for the time interval in the primed frame $\Delta t'$ to equal the time interval in the laboratory frame Δt because the displacements are not equal: $\Delta x' \neq \Delta x$. Furthermore, because the displacement is greater in the laboratory's frame, the time interval measured in that frame must be longer: $\Delta t > \Delta t'$. Compare this moving-photon analysis to the situation of the moving helicopter. According to Galilean relativity, the time interval of the helicopter's trip is the same in both frames, so Equation 39.12 shows that $\Delta x/\Delta t > \Delta x'/\Delta t'$; the helicopter's speed in the laboratory frame is higher than it is in the primed frame. By contrast, the speed of light is the same in both frames, and as a result the time interval for the photon's trip is longer in the laboratory's frame.

LORENTZ TRANSFORMATIONS
Major Concept

Because a time interval is *not* invariant in special relativity, we must come up with a transformation equation not only for position but also for time. These equations for transforming position and time are called the **Lorentz transformations**, named for the Dutch theoretical physicist Hendrik Lorentz. Lorentz did not start with Einstein's postulates; instead, he formulated these equations in an effort to understand Michelson and Morley's results. However, our arguments and derivations are based on Einstein's postulate that the speed of light is the same for all observers.

We can use the Lorentz transformations to derive other important equations. For convenience, these transformation equations along with the Galilean transformations are listed in Table 39.1.

TABLE 39.1 Galilean and Lorentz transformation equations.

Galilean (valid for low relative velocity)				Lorentz			
From primed to lab		**From lab to primed**	**From primed to lab**		**From lab to primed**	
$x = x' + v_{rel}t$	(39.3)	$x' = x - v_{rel}t$	$x = \gamma(x' + v_{rel}t')$	(39.15)	$x' = \gamma(x - v_{rel}t)$	(39.16)
$y = y'$	(39.2)	$y' = y$	$y = y'$	(39.13)	$y' = y$	(39.13)
$z = z'$		$z' = z$	$z = z'$	(39.14)	$z' = z$	(39.14)
$t = t'$		$t' = t$	$t = \gamma\left(t' + \frac{v_{rel}x'}{c^2}\right)$	(39.19)	$t' = \gamma\left(t - \frac{v_{rel}x}{c^2}\right)$	(39.20)
			where $\gamma = \frac{1}{\sqrt{1 - (v_{rel}/c)^2}} = \frac{1}{\sqrt{1 - \beta^2}}$			(39.17)

The Correspondence Principle

The **correspondence principle** is based on the observation that Galilean relativity seems to hold in circumstances when the relative speed between frames is low compared to the speed to light. According to the correspondence principle, the Lorentz transformations must be identical to the Galilean transformations when the relative speed is low. Sometimes a high relative speed is referred to as a **relativistic speed**, shorthand for saying the relative speed is high enough that we must take Einstein's theory of special relativity into account because Galilean relativity is not a good approximation. We will use the correspondence principle to check the Lorentz transformations.

Lorentz Transformations for *x*, *y*, and *z*

Let's take a look at the Lorentz transformations for position (Table 39.1). Figure 39.7 establishes our coordinate systems. Because the primed frame is moving at constant velocity in one dimension (along x), there is no difference between the Galilean and Lorentz transformations for the perpendicular components of positions:

$$y = y' \tag{39.13}$$

and

$$z = z' \tag{39.14}$$

These perpendicular components are invariant, but the x component of position is not. Instead, the Lorentz transformation for x is given by

$$x = \gamma(x' + v_{rel}t') \tag{39.15}$$

and the inverse transformation (from the laboratory frame to the primed one) is given by

$$x' = \gamma(x - v_{rel}t) \tag{39.16}$$

This dimensionless constant γ is sometimes called the **gamma factor**, or the **Lorentz factor**, and is given by

$$\gamma = \frac{1}{\sqrt{1 - (v_{rel}/c)^2}} = \frac{1}{\sqrt{1 - \beta^2}} \tag{39.17}$$

where

$$\beta \equiv \frac{v_{rel}}{c} \tag{39.18}$$

There are three important features of the Lorentz factor γ. First, it depends only on the relative speed of the primed observer. It is constant in time (no t dependence) and uniform in space (no x, y, or z dependence). Second, the Lorentz factor is greater than or equal to 1: $\gamma \geq 1$. Third, if the relative velocity is low compared to the speed of light, $\lim_{\beta \to 0} \gamma \to 1$.

Because $\gamma \to 1$ when the relative velocity is low, the Galilean transformation for x is a very good approximation of the Lorentz transformation at low relative speed, as expected by the correspondence principle. (In Problem 72, you will use Einstein's second postulate to confirm Equation 39.17 for the gamma factor.)

Lorentz Transformation for *t*

Though it is counterintuitive to think so, time also depends on the motion of the observer. The Lorentz transformation for time is given by

$$t = \gamma\left(t' + \frac{v_{\text{rel}}x'}{c^2}\right) \tag{39.19}$$

The inverse transformation from the laboratory frame to the primed frame is given by

$$t' = \gamma\left(t - \frac{v_{\text{rel}}x}{c^2}\right) \tag{39.20}$$

The Lorentz time transformation must be the same as the Galilean transformation when the relative velocity is low. When $v_{\text{rel}} << c$, the gamma factor is approximately 1. Also, the second term in Equations 39.19 and 39.20 must be approximately zero because $\lim_{v_{\text{rel}} \to 0} v_{\text{rel}}/c^2 \to 0$. So, for low relative speed, the Lorentz transformation for time is approximately the same as in Galilean relativity. (In Problem 14, you are asked to confirm Equations 39.19 and 39.20.)

Simultaneity Is Relative

A related corollary to Einstein's second postulate is that simultaneity is not invariant: Two events that occur simultaneously in one reference frame are not necessarily simultaneous in another frame moving with respect to the first frame. So when we say two events occur simultaneously, we must state in which frame these events are observed. The idea that simultaneity is not invariant seems counterintuitive, because in our everyday experience, in which relative speeds are low, simultaneous events occur at roughly the same time in both frames.

EXAMPLE 39.3 Displacement in the Lab Frame

The helicopter in Figure 39.9 is at x_i' at t_i' and it moves to x_f' at t_f', as seen by the primed observer. Find an expression for the displacement observed in the laboratory frame in terms of the displacement $\Delta x'$ and the time interval $\Delta t'$ measured in the primed frame. Use the Lorentz transformation.

INTERPRET and ANTICIPATE

The procedure for finding the displacement here is similar to the procedure in Example 39.2A, when we found a displacement under Galilean relativity. We expect the two results to be consistent when the relative velocity is low.

SOLVE

The displacement in the laboratory frame is the difference between the final position x_f and the initial position x_i.	$\Delta x = x_f - x_i$

To find the transformation between the primed frame and the laboratory frame, substitute the transformation for each position using $x = \gamma(x' + v_{\text{rel}}t')$ (Eq. 39.15).	$\Delta x = \gamma(x_f' + v_{\text{rel}}t_f') - \gamma(x_i' + v_{\text{rel}}t_i')$
Regroup the terms.	$\Delta x = \gamma[(x_f' - x_i') + v_{\text{rel}}(t_f' - t_i')]$
Write the result in terms of the displacement and the time interval in the primed frame.	$\Delta x = \gamma(\Delta x' + v_{\text{rel}}\Delta t')$ (39.21)

CHECK and THINK

Equation 39.21 is the transformation of displacement to the laboratory frame, and the result depends on the displacement and the time interval in the primed frame. Compare that to the displacement transformation under Galilean relativity, $\Delta x = \Delta x' + v_{\text{rel}}\Delta t$ (Eq. 39.10), which also depends on the displacement in the primed frame. In Galilean relativity, however, the time interval is the same in both frames, $\Delta t' = \Delta t$, so we are free to write the Galilean transformation in terms of Δt.

Check to see that the correspondence principle holds. At a low relative speed, the Lorentz factor is approximately 1 and time is invariant, so the time interval is the same in both frames.	$\lim_{v_{\text{rel}}\to 0} \Delta x = \lim_{v_{\text{rel}}\to 0} \gamma(\Delta x' + v_{\text{rel}}\Delta t')$ $\Delta x \to \Delta x' + v_{\text{rel}}\Delta t$ ✓

39-5 Length Contraction

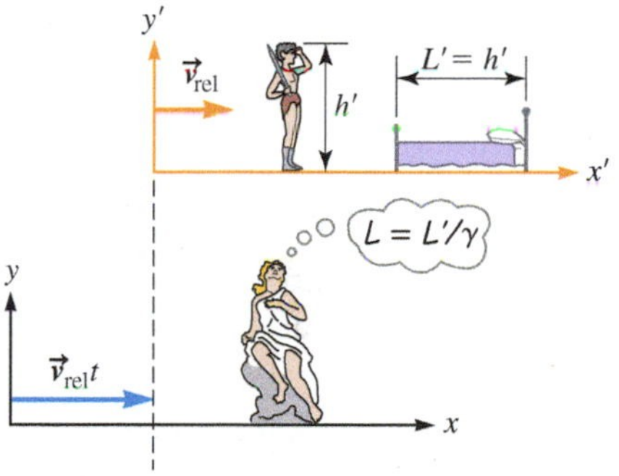

FIGURE 39.12 In the primed frame, the length of the bed equals Theseus's height. According to Ariadne in the laboratory frame, the bed is shorter than Theseus's height.

As discussed in the case study, the consequences of Einstein's theory of relativity may sound like science fiction or fantasy. For example, a consequence of the Lorentz transformation for position is that the length of an object depends on its motion relative to the observer. For a fictional take on this idea, think of the ancient Greek myth about a villain named Procrustes, who forced his victims to lie on a bed. If the victim was longer than the bed, Procrustes cut his victim down to make him fit. If the victim was shorter than the bed, Procrustes stretched his victim until he fit snugly in the bed.

Let's imagine a slight twist on this ancient myth. Suppose Procrustes's bed and our hero Theseus are in the primed frame moving at relative velocity v_{rel} with respect to the laboratory frame (Fig. 39.12). Theseus measures his height h' and the length of the bed L' in the primed frame, and he finds to his relief that they are equal: $L' = h'$. So Theseus believes he will fit in the bed. But Theseus's friend Ariadne is in the laboratory frame. Our challenge (in Example 39.4) is to see whether Ariadne agrees that Theseus will fit in the bed.

When you find the length of an object, you must measure the positions of its endpoints simultaneously. So, Theseus measures the positions of the bed's foot x_f' and head x_h' simultaneously. Then he subtracts to find the bed's length:

$$L' = x_h' - x_f' \tag{39.22}$$

Theseus makes his measurement in the primed frame, where the bed is at rest. A measurement of length made in the same frame in which the object is at rest is called the **proper length**. So L' is the bed's proper length.

Ariadne also measures the positions of the bed's foot x_f and head x_h simultaneously and then subtracts:

$$L = x_h - x_f \tag{39.23}$$

To see how her measurement compares to the proper length, we substitute Equation 39.16, $x' = \gamma(x - v_{\text{rel}}t)$, for the positions in Equation 39.22:

$$L' = \gamma(x_h - v_{\text{rel}}t) - \gamma(x_f - v_{\text{rel}}t)$$

Her measurements are made simultaneously, so the terms involving $v_{rel}t$ cancel:

$$L' = \gamma(x_h - x_f)$$

Substitute Equation 39.23:

LENGTH CONTRACTION
Major Concept

$$L' = \gamma L \tag{39.24}$$

or

$$L = \frac{L'}{\gamma} \tag{39.25}$$

So Ariadne's measurement made in the laboratory frame is the proper length divided by the Lorentz factor. Because $\gamma \geq 1$, Ariadne's measurement is, alarmingly, shorter than the proper length! It is always true that the proper length is the longest measurement of an object's length. Because an observer moving relative to an object always finds it has a shorter length than does an observer in the object's frame, we say that moving relative to an object causes an observed **length contraction**.

Equations 39.24 and 39.25 indicate that, in Einstein's theory of special relativity, length depends on the relative motion of the observer. However, length is invariant in Galilean relativity (Eq. 39.5), and according to the correspondence principle, this is what we expect to find at low relative speeds. When $v_{rel} \ll c$, the Lorentz factor approaches 1 ($\gamma \to 1$) and

$$L' = \gamma L \to L$$

as expected.

EXAMPLE 39.4 CASE STUDY Does Theseus Fit in the Bed?

Suppose Theseus's frame is moving at $0.6c$ relative to Ariadne's frame in Figure 39.12. Theseus measures his own height $h' = 2$ m (exactly), and he measures the length of the bed $L' = 2$ m. He is happy to find that his height equals the length of the bed. Theseus believes he is safe: Provided Procrustes is in his frame, Procrustes will not have to "adjust" Theseus's height.

A What is Theseus's height as measured by Ariadne?

INTERPRET and ANTICIPATE

Measuring Theseus's height means measuring the positions of his endpoints (feet and head) simultaneously. Because both are y coordinates and the y axes are perpendicular to the primed frame's velocity, we expect his height to be invariant.

SOLVE

Theseus measures the positions of his feet y'_f and head y'_h simultaneously and then subtracts to get his height h'.	$h' = y'_h - y'_f = 2$ m
Likewise, Ariadne measures the positions of Theseus's feet y_f and head y_h simultaneously and subtracts to get his height h.	$h = y_h - y_f$
Use the Lorentz transformation of y (Eq. 39.13) to transform Ariadne's measurement of Theseus's height.	$y = y'$ (39.13) $h = y'_h - y'_f$ $h = h' = 2$ m

CHECK and THINK

As expected, Ariadne and Theseus measure his height to be the same. In general, length measurements that are perpendicular to the frame's motion are invariant.

B What is the length of the bed as measured by Ariadne?

INTERPRET and ANTICIPATE

Because the bed is horizontal (parallel to the primed frame's motion), we expect Ariadne's measurement to differ from Theseus's measurement. Theseus is in the same frame as the bed, so he measures the bed's proper length. We expect Ariadne's measurement to be shorter than 2 m because she is not in the bed's frame.

SOLVE For problems involving special relativity, it often helps to find a value for the Lorentz factor first (Eq. 39.17).	$\gamma = \dfrac{1}{\sqrt{1-(v_{rel}/c)^2}} = \dfrac{1}{\sqrt{1-\beta^2}}$ (39.17) $\gamma = \dfrac{1}{\sqrt{1-0.6^2}} = 1.25$
Find Ariadne's measurement from Equation 39.25. Because we already found the Lorentz factor, simply divide Theseus's measurement by that value.	$L = \dfrac{L'}{\gamma}$ (39.25) $L = \dfrac{2\text{ m}}{1.25} = 1.6\text{ m}$

CHECK and THINK

As expected, Ariadne claims the bed is shorter than 2 m. To her alarm, it seems that Theseus won't fit in the bed (so Procrustes will cut him down to size). But Theseus thinks he will fit exactly. Who is right? Both are right. As long as Theseus remains vertical, they will both agree that his height is 2 m. However, Theseus finds the bed is 2 m long, while Ariadne finds it is only 1.6 m long. Still, Theseus believes he can safely lie in the bed—and he is right. As soon as he lies down, he will be horizontal, and Ariadne will find that his height is now 1.6 m—the same as the length of the bed. So neither observer will see the need for Procrustes to make any "adjustments" to Theseus. This is reassuring because it would violate the laws of physics to have Theseus cut down to size in one frame and remain intact in another frame.

39-6 Time Dilation

The Forever War by Joe Haldeman is a science fiction novel in which people must fight a war in a distant part of the galaxy. They must travel at nearly the speed of light to get to their battles. To these soldiers, each expedition lasts only a few months. However, when the solders return to the Earth, they find that many centuries have passed, and it is difficult for them to adjust to society in the future. This novel makes use of a key result of special relativity—namely, that the length of a time interval is relative, depending on the motion of the observer. This phenomenon is known as **time dilation**. In this section, we derive an expression for time dilation, and in Example 39.5, we apply it to Haldeman's characters.

DERIVATION The Time Dilation Equation

We will show that a time interval measured in the laboratory frame is given by

TIME DILATION

Major Concept

$$\Delta t = \gamma \Delta t' \qquad (39.26)$$

where $\Delta t'$ is known as the **proper time**. The proper time is defined as the time interval between two events that occur at the same position. Here, the proper time is measured in the primed frame.

Derivation continues on page 1282 ▶

Suppose in the primed frame Aaron fires a flashbulb at t_i'. A mirror is located directly above the bulb at h' (Fig. 39.13).

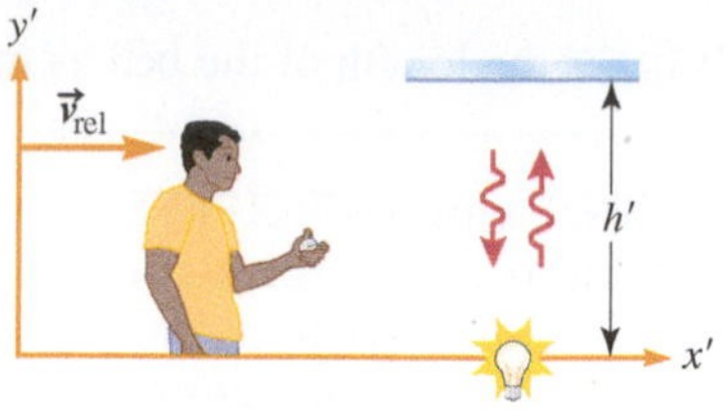

FIGURE 39.13 The observer in the primed frame sees a photon travel straight up and down.

Imagine Aaron can watch as a single photon travels up to the mirror, reflects from the mirror, and returns to the flashbulb at t_f'. The two events happen at the instant the photon leaves the bulb and the instant it returns to the bulb. Aaron measures the proper time between the two events because to him they both occur at the same position—the position of the bulb.

Proper time: $\Delta t' = t_f' - t_i'$

According to Aaron, the total path length is $2h'$. The proper time is the path length divided by the photon's speed c.

$$\Delta t' = \frac{2h'}{c}$$

$$h' = \frac{c\Delta t'}{2} \qquad (1)$$

However, in the laboratory frame, Hannah does not measure the proper time because the bulb is not in the same position when the photon returns to it as when the photon leaves it. Instead, she sees the photon move along two legs of a triangle, with the displacement Δx of the bulb forming the triangle's base (Fig. 39.14).

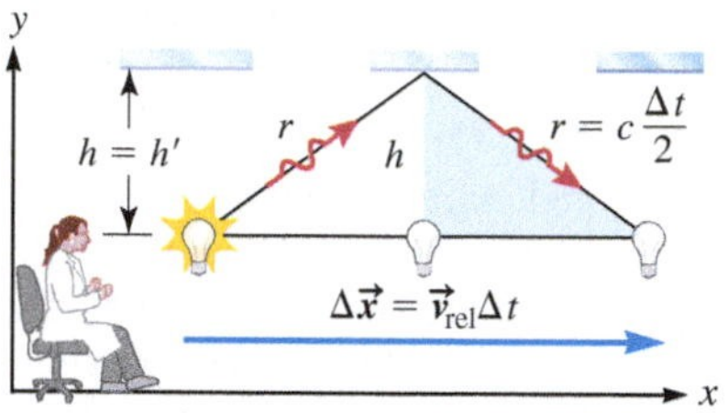

FIGURE 39.14 The observer in the laboratory frame sees the photon trace out two legs of a triangle.

The displacement of the bulb depends on the relative speed of the primed frame. To Hannah, the bulb moves in the positive x direction at speed v_{rel}. In the time Δt it takes the photon to move from the bulb, reflect from the mirror, and return to the bulb, the bulb has been displaced by Δx.

$$\Delta x = v_{\text{rel}}\Delta t$$

The photon travels along the legs of the triangle of length r at speed c. The time it takes the photon to go from the bulb to the mirror is equal to the time it takes to go from the mirror to the bulb. Each is half the total travel time.

$$r = c\frac{\Delta t}{2}$$

The height of the triangle is h. Apply the Pythagorean theorem to the right triangle shaded blue in Figure 39.14. The length of the horizontal leg of this right triangle is half the displacement of the bulb.

$$r^2 = h^2 + \left(\frac{\Delta x}{2}\right)^2$$

$$\left(c\frac{\Delta t}{2}\right)^2 = h^2 + \left(\frac{v_{\text{rel}}\Delta t}{2}\right)^2$$

$$h^2 = \left(c\frac{\Delta t}{2}\right)^2 - \left(\frac{v_{\text{rel}}\Delta t}{2}\right)^2 \qquad (2)$$

The height of the triangle equals the vertical distance between the mirror and the bulb. There is no length contraction in the perpendicular direction.

$$h = h' \qquad (3)$$

Square Equation (3) and substitute Equations (1) and (2).

$$h^2 = h'^2$$

$$\left(c\frac{\Delta t}{2}\right)^2 - \left(\frac{v_{\text{rel}}\Delta t}{2}\right)^2 = \left(\frac{c\Delta t'}{2}\right)^2$$

Solve for Δt^2.

$$\Delta t^2(c^2 - v_{\text{rel}}^2) = c^2\Delta t'^2$$

$$\Delta t^2 = \frac{c^2}{(c^2 - v_{\text{rel}}^2)}\Delta t'^2$$

Write the factor $c^2/(c^2 - v_{rel}^2)$ in terms of the gamma factor (Eq. 39.17).	$\frac{c^2}{(c^2 - v_{rel}^2)} = \frac{1}{1 - (v_{rel}/c)^2} = \gamma^2$
Solve for Δt.	$\Delta t^2 = \gamma^2 \Delta t'^2$ $\Delta t = \gamma \Delta t'$ ✓ (39.26)

COMMENTS

Because $\gamma \geq 1$, the time interval Δt measured in the laboratory frame is longer than the proper time $\Delta t'$. To help remember this, we sometimes use the phrase *moving clocks run slowly*. (The proper clock is in the primed frame, and the laboratory frame moves relative to that frame.) Now use the correspondence principle to check Equation 39.26. There is no time dilation in Galilean relativity. If the relative speed is low, then $\gamma \to 1$ and $\Delta t \to \Delta t'$ as expected.

EXAMPLE 39.5 CASE STUDY Back to the Future

In *The Forever War*, the soldiers are away from the Earth for about 2 months. When they return, they find that centuries have passed. Let's consider a slightly simpler problem based on those soldiers' experience. Suppose an observer in the primed frame claims that the time between two events at the same location is exactly 60 days (5.184×10^6 s). To keep this problem simple, you can imagine that the two events are the emission and return of the photon in Figure 39.13. An observer in the laboratory frame measures the time interval of the same two events as exactly 100 years (3.15576×10^9 s). What is the constant relative speed of the primed observer? Give your answer to eight significant figures in terms of the speed of light.

INTERPRET and ANTICIPATE

There is no time dilation if the relative speed is low. The time dilation in this problem is enormous, so we expect the relative speed to be very high.

SOLVE Solve Equation 39.26 for the Lorentz factor.	$\Delta t = \gamma \Delta t'$ (39.26) $\gamma = \frac{\Delta t}{\Delta t'}$
Substitute values. The proper time is the 60 days measured in the primed frame.	$\gamma = \frac{3.15576 \times 10^9 \text{ s}}{5.184 \times 10^6 \text{ s}} = 608.75$
Now solve $\gamma = \frac{1}{\sqrt{1 - \beta^2}}$ (Eq. 39.17) for β.	$1 - \beta^2 = \frac{1}{\gamma^2}$ $\beta = \sqrt{1 - \frac{1}{\gamma^2}}$
Substitute for γ.	$\beta = \sqrt{1 - \frac{1}{608.75^2}} = 0.99999865 = v_{rel}/c$ $v_{rel} = 0.99999865c$

Example continues on page 1284 ▶

CHECK and THINK

As expected, the primed frame's relative speed must be very relativistic—more than 99% of the speed of light. This is a real consequence of Einstein's theory of relativity. The only thing that makes this result fictional is that we have yet to come up with a spacecraft that can move at nearly the speed of light.

EXAMPLE 39.6 Muon Decay

Time dilation isn't found only in fiction; it has been confirmed by experiments involving subatomic particles known as muons. Muons can be created by high-energy particle accelerators, and they are unstable. When muons are at rest with respect to the laboratory, their half-life is roughly 2 μs. Muons are also made in the Earth's atmosphere by cosmic rays and subsequently move downward at roughly $0.98c$.

A How far can these muons move in the Earth's atmosphere during the time interval of their half-life?

INTERPRET and ANTICIPATE

We can consider the Earth and its atmosphere to be in the laboratory frame and the muons to be in the primed frame. The question is really asking us to find the distance the muons travel in the laboratory frame. We'll use a downward-pointing y axis because that is the direction in which the primed frame moves relative to the laboratory frame.

SOLVE

This is a straightforward application of kinematics. In the laboratory frame, the muons move downward at $0.98c$. We must find how far they travel in 2 μs.

$$\Delta y = v_{\text{rel}} \Delta t$$

$$\Delta y = (0.98)(3.00 \times 10^8\,\text{m/s})(2 \times 10^{-6}\,\text{s})$$

$$\Delta y = 588\,\text{m} \approx 6 \times 10^2\,\text{m}$$

CHECK and THINK

If we put muon detectors in the laboratory frame separated by $\Delta y = 600$ m, we would expect to find that only half the muons would survive to reach the lower detector. However, if you really did this, you would find that *more* than half the muons would survive to reach the second detector due to time dilation. The muons' half-life in their own frame is 2 μs. The detectors are in the laboratory frame, though, and we should *not* expect the time interval to be the same in both frames.

B What is the half-life of these muons observed in the laboratory frame?

INTERPRET and ANTICIPATE

The half-life measured when the muons are *at rest* in a laboratory is the proper time for their half-life. No matter how fast the muons are moving, in their own frame their half-life is 2 μs. But, because of the relative speed of the laboratory frame, the half-life must be longer in the laboratory frame.

SOLVE

Find the value of the Lorentz factor (Eq. 39.17).

$$\gamma = \frac{1}{\sqrt{1 - \beta^2}} \tag{39.17}$$

$$\gamma = \frac{1}{\sqrt{1 - 0.98^2}} = 5.0$$

Use the time dilation equation (Eq. 39.26) to find the half-life of a muon as observed in the laboratory frame.	$\Delta t = \gamma \Delta t'$ (39.26) $\Delta t = (5.0)(2\,\mu\text{s}) \approx 10\,\mu\text{s}$

CHECK and THINK

More than half the muons would survive to reach the lower detector because when muons move at $0.98c$ relative to detectors in the laboratory frame, their half-life is 10 μs, not 2 μs. So, if only half the muons are to reach the second detector, the detectors must be farther than 600 m apart to allow more time to pass.

C How far apart should the detectors be placed for half the muons produced at the higher detector to reach the lower detector?

INTERPRET and ANTICIPATE

This is similar to part A, except we must use the half-life we expect to observe in the laboratory frame.

SOLVE In the laboratory frame, the muons move downward at $0.98c$. Find how far they travel in 10 μs.	$\Delta y = v_{\text{rel}} \Delta t$ $\Delta y = (0.98)(3.00 \times 10^8\,\text{m/s})(1 \times 10^{-5}\,\text{s})$ $\Delta y = 2940\,\text{m} \approx 3000\,\text{m}$

CHECK and THINK

When the detectors are 3000 m apart, the lower detector finds half as many muons as the upper detector. You may be wondering what would happen if an observer moved with the muons. Would he also see half as many muons arrive at the second detector? The answer is yes. But, to this observer in the primed frame, the distance between the two detectors is contracted to 600 m, and the half-life of the muons is 2 μs. So the time interval between when the first detector encounters the muons and when the second detector encounters the muons is equal to their proper half-life in the primed frame.

39-7 The Relativistic Doppler Effect

You have probably noticed that as a police car using its siren passes you, you hear a change in the siren's frequency. This change in the frequency of sound waves is known as the **Doppler shift** (Section 17-9). The frequency and wavelength of light also depend on the relative motion of the source and the observer, so the color of an object (a light source) depends on the relative motion between the source and an observer. When a light source moves away from an observer, the wavelength is longer and the light looks redder. When a light source moves toward the observer, the wavelength is shorter and the light looks bluer.

DERIVATION The Relativistic Doppler Formula

We will show that the relativistic Doppler formula for change in wavelength is

$$\lambda = \sqrt{\frac{1 \mp \beta}{1 \pm \beta}}\,\lambda' \text{ (radial motion)} \quad (39.27)$$

RELATIVISTIC DOPPLER EFFECT

Major Concept

Choose the top signs if the source is moving toward the observer; choose the bottom signs if the source is moving away from the observer.

where λ' is the wavelength measured in the primed frame and λ is the wavelength measured in the laboratory frame. An observer in the primed frame sees that the frequency of light is f', the period (time interval between arriving wave fronts) is $T' = 1/f'$, and the wavelength is $\lambda' = c/f' = cT'$.

Derivation continues on page 1286

A monochromatic light source is stationary in the primed frame. In the laboratory frame, Hannah is standing in front of the light source and sees the light source approaching her (Fig. 39.15). We use the phrase *radial motion* to describe motion that is directly toward or away from the observer; we have chosen to consider motion toward the observer, and we'll generalize our results at the end of the derivation.

FIGURE 39.15 An observer in the laboratory frame sees a light source approaching her.

In the laboratory frame, the time between the emissions of each wave crest is T. But this is not the same as the period Hannah observes; because the source is moving toward her, the crests will reach her with a shorter period.

Figure 39.16 shows the motion of the source and the emission of its wave crests in the laboratory frame, similar to Figure 17.28 (page 510). At time 1, a wave crest is emitted. The source moves to the right at v_{rel}. By the time the second crest is emitted at time 2, the source has moved to $v_{rel}T$. By time 3, when a third crest is emitted, the second crest's radius has grown to cT and the first crest's radius has grown to $2cT$. The distance between the centers of the first and second crests is $v_{rel}T$. Along the direction of motion, the distance between the first and second crests (wave fronts) is the wavelength λ observed in the laboratory.

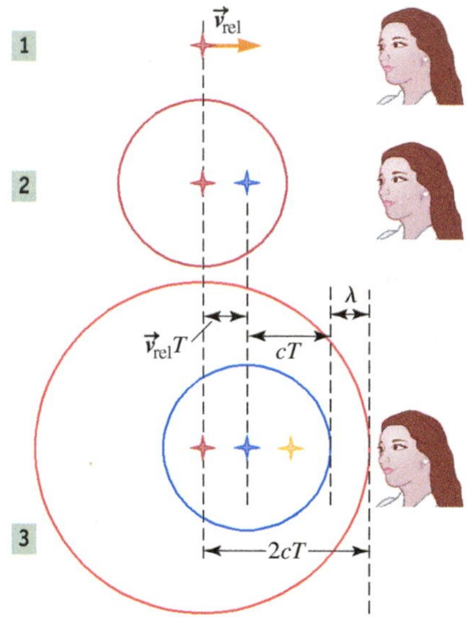

FIGURE 39.16 The colors of the source and wave crests have been chosen to help you connect the wave crest with the location of the source when the wave was emitted. For example, the large red wave crest shown at time 3 was emitted when the source was at the far left position at time 1.

Use Figure 39.16 to find an expression for the wavelength λ.

$$2cT = v_{rel}T + cT + \lambda$$

$$\lambda = cT - v_{rel}T = (c - v_{rel})T \quad (1)$$

The period T' measured in the primed frame is the proper time between the emissions of the crests. To find T, take time dilation (Eq. 39.26) into account.

$$T = \gamma T' \quad (39.26)$$

$$T' = \frac{T}{\gamma} \quad (2)$$

Write the period in the primed frame in terms of the wavelength observed in the primed frame.

$$\lambda' = \frac{c}{f'} = cT' \quad (3)$$

Substitute Equation (2) into Equation (3), and solve for T.

$$T = \frac{\gamma}{c}\lambda' \quad (4)$$

Substitute Equation (4) into Equation (1), and simplify by using $\gamma = \frac{1}{\sqrt{1-\beta^2}}$ and $\beta = \frac{v_{rel}}{c}$ (Eqs. 39.17 and 39.18).

$$\lambda = (c - v_{rel})\left(\frac{\gamma}{c}\lambda'\right) = \gamma(1 - v_{rel}/c)\lambda'$$

$$\lambda = \frac{(1-\beta)}{\sqrt{1-\beta^2}}\lambda' = \frac{(1-\beta)}{\sqrt{(1-\beta)(1+\beta)}}\lambda'$$

$$\lambda = \sqrt{\frac{1-\beta}{1+\beta}}\lambda'$$

We assumed Hannah is standing in front of the source, so that the source is moving toward her. If the source were moving away, the signs in front of β would be reversed. We write our final relativistic Doppler formula by writing these signs explicitly.

$$\lambda = \sqrt{\frac{1 \mp \beta}{1 \pm \beta}}\lambda' \ ✓ \quad (39.27)$$

The relativistic Doppler formula can also be written in terms of frequency (Problem 36).

$$f = \sqrt{\frac{1 \pm \beta}{1 \mp \beta}} f' \quad (39.28)$$

Choose the top signs if the source is moving toward the observer; choose the bottom signs if the source is moving away from the observer.

COMMENTS

Equations 39.27 and 39.28 are used when the source is moving radially toward or away from the observer. There is also a transverse Doppler shift associated with relative motion that is perpendicular to the observer's line of sight. This transverse Doppler shift is due entirely to time dilation.

CONCEPT EXERCISE 39.3

Show that Equation 39.27 predicts that when a light source is moving toward an observer, the source appears bluer, and when the source is moving away, it appears redder.

EXAMPLE 39.7 Quasars

In 1960, Thomas Matthews and Allan Sandage discovered a starlike object whose spectrum could not be identified with any known substance. Sandage said, "The thing is exceedingly weird." When other similar objects were discovered, they were called quasi-stellar radio sources, or quasars, because they are sources of radio emission that appeared like stars in our own galaxy. Today, we know that quasars are much more powerful than stars, and they are among the most distant objects we can observe in the Universe. One particular quasar emits light with a wavelength of 121.6 nm in its own frame. On the Earth, this light is observed to have a wavelength of 885.2 nm. The quasar is moving either toward or away from the Earth. Decide which, and then find the quasar's speed relative to the Earth.

INTERPRET and ANTICIPATE

First, decide what is in the primed frame and what is in the laboratory frame. We need to find the quasar's speed relative to the Earth. This suggests the quasar is in the primed frame and the Earth is in the laboratory frame.

SOLVE

The observer in the laboratory sees a longer wavelength than the one emitted in the primed frame. So the quasar must be moving away from the Earth, and we choose the bottom signs in Equation 39.27.

$$\lambda = \sqrt{\frac{1 + \beta}{1 - \beta}} \lambda' \quad (39.27)$$

Solving for β requires a few steps of algebra.

$$\left(\frac{\lambda}{\lambda'}\right)^2 = \frac{1 + \beta}{1 - \beta}$$

$$-\beta[(\lambda/\lambda')^2 + 1] = 1 - (\lambda/\lambda')^2$$

$$\beta = \frac{(\lambda/\lambda')^2 - 1}{(\lambda/\lambda')^2 + 1}$$

Example continues on page 1288 ▶

Substitute values. Use $\beta = v_{rel}/c$ (Eq. 39.18) to find the relative speed.	$\beta = \dfrac{\left(\dfrac{885.2\text{ nm}}{121.6\text{ nm}}\right)^2 - 1}{\left(\dfrac{885.2\text{ nm}}{121.6\text{ nm}}\right)^2 + 1} = 0.9630$ $v_{rel} = 0.9630c$

CHECK and THINK

When an object's speed is very relativistic, it is common to report that speed in terms of the speed of light, rather than in conventional units such as m/s. This quasar has a high recessional speed from the Earth. This fits with other observations made in astronomy: In general, the farther an object is from us, the faster it moves away. Quasars are very distant objects, so their relative speed is very high.

39-8 Velocity Transformation

Imagine that the primed observer and the laboratory observer measure the velocity of a moving object, while (as always) the primed observer is moving at $\vec{v}_{rel}$ along the x axis with respect to the laboratory. According to Galilean relativity, the two observers will agree on the component of the velocity perpendicular to $\vec{v}_{rel}$, while the parallel component's transformation is given by $v_x = v'_x + v_{rel}$ (Eq. 39.11).

The transformation of velocity is more complicated in special relativity, however. Because the observed velocity of an object depends on both its displacement and the time interval of measurement, a velocity transformation equation involves both length contraction and time dilation. There is no length contraction in the perpendicular directions, but there is time dilation, so the two observers will not measure the same perpendicular velocity. And, because there is length contraction in the parallel direction, the parallel velocity transformation involves both time dilation and length contraction. Our goal is to derive expressions for parallel and perpendicular velocity transformations. We can check our work in two ways. First, when the relative speed is low, our results should match the results of Galilean transformation. Second, if the moving object is a photon, we should find that both observers agree that its speed is c.

Problem 40 asks you to show that the transformation of a velocity component perpendicular to $\vec{v}_{rel}$ is given by

$$v_y = \frac{v'_y}{\gamma\left(1 + \frac{v_{rel}}{c^2}v'_x\right)} \tag{39.29}$$

PERPENDICULAR VELOCITY TRANSFORMATION ✪ Major Concept

and the inverse transformation is given by

$$v'_y = \frac{v_y}{\gamma\left(1 - \frac{v_{rel}}{c^2}v_x\right)} \tag{39.30}$$

We can write similar equations for the z component of velocity. The perpendicular velocity transformations depend on the parallel velocity component v_x.

DERIVATION Parallel Velocity Transformation

Here we show that the transformation of a velocity component parallel to $\vec{v}_{rel}$ is given by

$$v_x = \frac{(v'_x + v_{rel})}{\left(1 + \frac{v_{rel}}{c^2}v'_x\right)} \tag{39.31}$$

PARALLEL VELOCITY TRANSFORMATION ✪ Major Concept

Consider an object that moves parallel to the primed frame's motion, such as the helicopter in Figure 39.9. The x component of its velocity in both frames is the time derivative of its parallel position. Length contraction and time dilation mean that both the numerator and the denominator of this derivative differ in the two frames.

$$v_x = \frac{dx}{dt} \tag{1}$$

$$v_x' = \frac{dx'}{dt'} \tag{2}$$

First, we derive the transformation from the primed frame to the laboratory frame. Start by taking the differentials of Equations 39.15 and 39.19.

$$x = \gamma(x' + v_{\text{rel}}t') \tag{39.15}$$

$$dx = \gamma(dx' + v_{\text{rel}}dt') \tag{3}$$

$$t = \gamma\left(t' + \frac{v_{\text{rel}}x'}{c^2}\right) \tag{39.19}$$

$$dt = \gamma\left(dt' + \frac{v_{\text{rel}}dx'}{c^2}\right) \tag{39.32}$$

Substitute Equations (3) and 39.32 into Equation (1).

$$v_x = \frac{\gamma(dx' + v_{\text{rel}}dt')}{\gamma\left(dt' + \frac{v_{\text{rel}}dx'}{c^2}\right)}$$

The Lorentz factor cancels. Divide the numerator and denominator by dt'.

$$v_x = \frac{\left(\frac{dx'}{dt'} + v_{\text{rel}}\right)}{\left(1 + \frac{v_{\text{rel}}}{c^2}\frac{dx'}{dt'}\right)}$$

Use Equation (2) to write the expression in terms of the velocity measured in the primed frame.

$$v_x = \frac{(v_x' + v_{\text{rel}})}{\left(1 + \frac{v_{\text{rel}}}{c^2}v_x'\right)} \checkmark \tag{39.31}$$

Equation 39.31 is the parallel velocity transformation from the primed frame to the laboratory frame. Problem 38 asks you to show that Equation 39.33 is the corresponding transformation from the laboratory frame to the primed frame.

$$v_x' = \frac{(v_x - v_{\text{rel}})}{\left(1 - \frac{v_{\text{rel}}}{c^2}v_x\right)} \tag{39.33}$$

COMMENTS

Check to see whether Equation 39.31 matches the Galilean transformation in the case of low relative speed: $v_{\text{rel}}/c^2 \to 0$. Our result approaches Equation 39.11 (the Galilean transformation).

$$\lim_{v_{\text{rel}}\to 0} v_x = \frac{(v_x' + v_{\text{rel}})}{\left(1 + \frac{v_{\text{rel}}}{c^2}v_x'\right)} \to v_x' + v_{\text{rel}} \checkmark$$

Next, imagine the helicopter is replaced by a photon moving to the right at $v_x' = c$. As expected, both observers see the photon moving at c.

$$v_x = \frac{(c + v_{\text{rel}})}{\left(1 + \frac{v_{\text{rel}}}{c^2}c\right)}$$

$$v_x = \frac{(c + v_{\text{rel}})}{\left(1 + \frac{v_{\text{rel}}}{c}\right)} = \frac{c\left(1 + \frac{v_{\text{rel}}}{c}\right)}{\left(1 + \frac{v_{\text{rel}}}{c}\right)}$$

$$v_x = c \checkmark$$

EXAMPLE 39.8 A Photon Flies Straight Up

Suppose a photon flies straight up (along the y direction) at speed c in the primed frame, so $v_x' = 0$ and $v_y' = c$. As usual, the primed frame moves in the positive x direction at speed v_{rel} as observed from the laboratory frame. What is the speed of the photon observed in the laboratory frame?

INTERPRET and ANTICIPATE

Because of Einstein's second postulate, we expect the laboratory observer will observe the photon moving at c. This problem is tricky because the perpendicular velocity transformation depends on the horizontal motion of the photon. The situation is somewhat similar to the one in Figure 39.13, but there is no mirror, so the photon does not reverse direction. As in that case, because of the relative horizontal (parallel) motion of the primed frame, the laboratory observer sees the photon moving in both the y and x directions (as in Fig. 39.14). So we must find both v_x and v_y for the photon.

SOLVE

First, find the horizontal speed v_x in the laboratory frame (Eq. 39.31). The photon moves straight up in the primed frame, so $v_x' = 0$. The photon's horizontal speed in the laboratory frame equals the relative speed.

$$v_x = \frac{(v_x' + v_{rel})}{\left(1 + \frac{v_{rel}}{c^2} v_x'\right)} = \frac{(0 + v_{rel})}{\left[1 + \frac{v_{rel}}{c^2}(0)\right]}$$

$$v_x = v_{rel} \quad (1)$$

Next, find the photon's vertical speed in the laboratory frame (Eq. 39.29). Because the photon moves straight up at c in the primed frame, set $v_x' = 0$ and $v_y' = c$.

$$v_y = \frac{v_y'}{\gamma\left(1 + \frac{v_{rel}}{c^2} v_x'\right)} = \frac{c}{\gamma\left[1 + \frac{v_{rel}}{c^2}(0)\right]}$$

$$v_y = \frac{c}{\gamma}$$

Substitute for the gamma factor.

$$v_y = \frac{c}{\gamma} = c\sqrt{1 - \left(\frac{v_{rel}}{c}\right)^2} \quad (2)$$

The photon speed in the laboratory frame is found in the usual way from its x and y velocity components. Substitute Equations (1) and (2); then simplify.

$$v^2 = v_x^2 + v_y^2$$

$$v^2 = v_{rel}^2 + c^2\left[1 - \left(\frac{v_{rel}}{c}\right)^2\right]$$

$$v^2 = v_{rel}^2 + c^2 - v_{rel}^2 = c^2$$

$$v = c$$

CHECK and THINK

As expected, the speed of the photon as observed in the laboratory frame is c. We have confirmed that the perpendicular velocity transformation (Eq. 39.29) satisfies Einstein's second postulate.

39-9 Mass and Momentum Transformation

One strange consequence of special relativity is that the mass of an object also depends on the relative motion of the object and the observer. To see why this is true, we consider a pair of elastic collisions. Recall that in an elastic collision, both momentum and kinetic energy are conserved.

Figure 39.17 shows a collision between two perfectly elastic hockey pucks that are viewed in the same reference frame: a red puck of mass m_{rest} and a blue puck of mass m. Each puck is given the same initial speed v_{puck}. They collide elastically and each reverses direction, traveling afterward at the same speed v_{puck}. The change in momentum of the red puck is

$$\Delta\vec{p}_{\text{red}} = \vec{p}_f - \vec{p}_i = m_{\text{rest}}v_{\text{puck}}\hat{k} - \left(-m_{\text{rest}}v_{\text{puck}}\hat{k}\right) = 2m_{\text{rest}}v_{\text{puck}}\hat{k} \qquad (39.34)$$

The change in the blue puck's momentum is

$$\Delta\vec{p}_{\text{blue}} = -mv_{\text{puck}}\hat{k} - mv_{\text{puck}}\hat{k} = -2mv_{\text{puck}}\hat{k} \qquad (39.35)$$

Momentum is conserved, so we apply $\Delta\vec{p}_{\text{red}} = -\Delta\vec{p}_{\text{blue}}$ (Eq. 11.9) to Equations 39.34 and 39.35: $2m_{\text{rest}}v_{\text{puck}}\hat{k} = 2mv_{\text{puck}}\hat{k}$. We find that the pucks have the same mass:

$$m = m_{\text{rest}} \qquad (39.36)$$

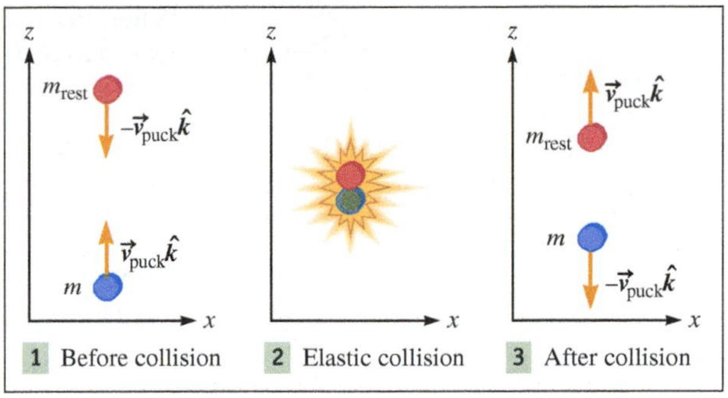

FIGURE 39.17 As observed in the laboratory frame, two pucks have the same initial speed. They collide, and afterward their speeds are unchanged.

This is not a surprising result; it is what we found for collisions in Section 11-5.

Now imagine that the blue puck and Paul are at rest in the primed frame, which moves to the right with speed v_{rel} relative to the laboratory frame. The red puck and Lil are in the laboratory frame. Paul and Lil agree to launch their pucks with the same speed v_{puck} parallel to the z axis, so that the two pucks collide. Of course, Paul must launch his puck *before* he is lined up with Lil's puck. As in the case of Figure 39.14, according to Lil in the laboratory frame, Paul's the blue puck traces out two legs of a triangle (Fig. 39.18). However, she sees her own red puck travel back and forth along the z axis at speed v_{puck}. She finds that the red puck's change in momentum is exactly what it was when both pucks were in the same frame:

$$\Delta\vec{p}_{\text{red}} = 2m_{\text{rest}}v_{\text{puck}}\hat{k} \qquad (39.34)$$

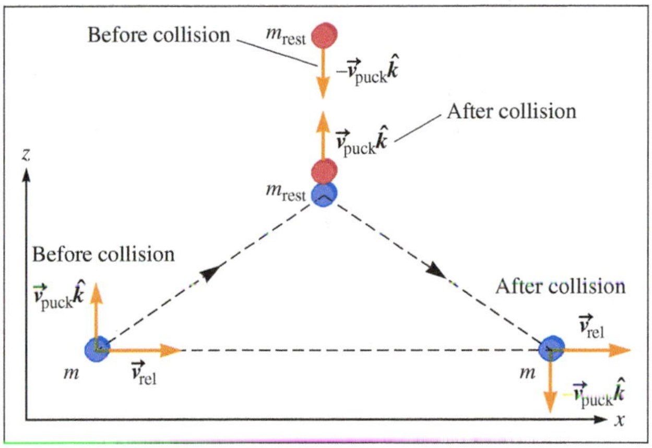

FIGURE 39.18 As seen from the laboratory frame, two pucks collide. In this frame, the red puck's velocity has no x component. In the primed frame (not shown), the blue puck's velocity has no x component.

In the primed frame, Paul launched the blue puck at v_{puck} in the z' direction. So, in his frame, the puck's initial velocity components are $v'_x = 0$ and $v'_z = v_{\text{puck}}$, while the final velocity components are $v'_x = 0$ and $v'_z = -v_{\text{puck}}$. With $v'_x = 0$ in Equation 39.31, $v_x = (v'_x + v_{\text{rel}})/(1 + v_{\text{rel}}v'_x/c^2)$ (parallel velocity transformation), Lil finds that the x component of the blue puck's velocity is $v_{\text{rel}}\hat{\imath}$, which is unchanged by the collision. So, to Lil, the x component of the blue puck's momentum is unchanged: $(\Delta p_{\text{blue}})_x = mv_{\text{rel}} - mv_{\text{rel}} = 0$. However, she sees the blue puck reverse its motion in the z direction, so she finds a nonzero change in its momentum along z. Before we can calculate this change in momentum, we need to know the speed Lil observes for the blue puck. Use Equation 39.29, $v_z = v'_z/[\gamma(1 + v_{\text{rel}}v'_x/c^2)]$ (perpendicular velocity transformation), with $v'_x = 0$ and $v'_z = v_{\text{puck}}$, to find the initial velocity along z seen in Lil's laboratory frame:

$$v_z = \frac{v_{\text{puck}}}{\gamma} \qquad (39.37)$$

Both observers agree that after the collision, the blue puck reverses its motion in the z direction with no change in speed. So Lil finds that the change in the blue puck's momentum is given by

$$\Delta\vec{p}_{\text{blue}} = \left(-m\frac{v_{\text{puck}}}{\gamma} - m\frac{v_{\text{puck}}}{\gamma}\right)\hat{k} = -2m\frac{v_{\text{puck}}}{\gamma}\hat{k} \qquad (39.38)$$

According to the conservation of momentum, $\Delta\vec{p}_{\text{red}} = -\Delta\vec{p}_{\text{blue}}$ (Eq. 11.9), and using Equations 39.34 and 39.38, we find

$$2m_{\text{rest}}v_{\text{puck}}\hat{k} = -\left(-2m\frac{v_{\text{puck}}}{\gamma}\hat{k}\right)$$

which reduces to

MASS TRANSFORMATION
Major Concept

$$m = \gamma m_{\text{rest}} \qquad (39.39)$$

Some physicists do not like to define *rest* mass; instead, they use the term *mass* (an invariant quantity). Then only other quantities such as momentum and energy are affected by relative motion.

When the two pucks were in the same reference frame, we found that their masses were equal (Eq. 39.36). However, Equation 39.39 says that the mass of an object depends on its relative motion. The **rest mass** m_{rest} of any object must be measured in the object's frame. Because $\gamma \geq 1$, if the mass is measured in a frame in which the object is moving, its mass will be greater than its rest mass.

An implication of the experiment in Figure 39.18 is that we must modify how we express the momentum of an object. Suppose that in the laboratory frame, we observe a particle moving at velocity $\vec{v}$. If we think of the particle as being at rest in the primed frame, its momentum as observed in the laboratory frame is

RELATIVISTIC MOMENTUM

Major Concept

$$\vec{p} = m\vec{v} = \gamma m_{\text{rest}}\vec{v} \tag{39.40}$$

The Lorentz factor in this case depends on the speed v of the particle:

$$\gamma = \frac{1}{\sqrt{1-(v/c)^2}} = \frac{1}{\sqrt{1-\beta^2}} \tag{39.41}$$

The conservation of momentum holds in all inertial reference frames as long as we use Equation 39.40 for momentum.

The Ultimate Speed Limit

Often in works of science fiction, there is some mention of traveling at or faster than the speed of light. According to the special theory of relativity, though, the speed of light is the ultimate speed limit. Nothing can travel faster than light. Specifically, any object that has mass must travel at less than the speed of light. To see why, substitute Equation 39.41 into Equation 39.39:

$$m = \frac{m_{\text{rest}}}{\sqrt{1-(v/c)^2}} = \frac{m_{\text{rest}}}{\sqrt{1-\beta^2}} \tag{39.42}$$

As $\beta \to 1$ (or, equivalently, as $v \to c$), the object's mass approaches infinity: $m \to \infty$. Because it makes no sense to have an object with infinite mass (as measured in any reference frame), we conclude that any object that has mass cannot exceed the speed of light. Put another way, the object would have infinite momentum, and it would take an infinite amount of energy to accelerate the object to the speed of light. So, the speed of light is the ultimate speed limit and is reached only by massless particles such as photons.

EXAMPLE 39.9 CASE STUDY Try Working Out

In Example 39.5, we estimated that the characters in *The Forever War* move at a very high speed ($\beta = 0.99999865$). If a typical soldier's mass is 85.0 kg in his own frame, what is his mass according to an observer in the laboratory frame?

INTERPRET and ANTICIPATE

The smallest possible mass of the soldier is his rest mass. We expect his mass observed in the laboratory frame to be greater than his rest mass.

SOLVE

From Example 39.5, we know the Lorentz factor.	$\gamma = 608.75$
Use Equation 39.39 to find the mass observed in the laboratory frame.	$m = \gamma m_{\text{rest}}$ (39.39) $m = (608.75)(85.0\,\text{kg})$ $m = 5.17 \times 10^4\,\text{kg}$

CHECK and THINK

The soldier's weight as observed in the laboratory frame is nearly 60 tons! Of course, in his own frame, his mass is still 85.0 kg.

39-10 Newton's Second Law and Energy

You have rearranged Newton's second law countless times to find an object's acceleration: $a = F/m$. However, according to special relativity, the mass of an object depends on its motion relative to an observer. Consider the seemingly simple scenario of observing an object that is initially at rest. You and the object are in the laboratory frame, and you find the object's mass equals its rest mass m_{rest}. Then a constant force is applied to the object, so the object accelerates while you remain in the laboratory frame. As the speed of the object increases, you see its mass increase: $m > m_{\text{rest}}$. By Newton's second law, then, its acceleration must decrease. However, a constant force should result in a constant acceleration, and there is a simple way around this contradictory observation. We just need to express Newton's second law in terms of momentum, $\vec{F}_{\text{tot}} = d\vec{p}/dt$ (Eq. 10.2), rather than in terms of mass and acceleration. This expression of Newton's second law holds even when objects are moving at relativistic speeds. The familiar expression $\vec{F}_{\text{tot}} = m\vec{a}$ is an approximation that holds when the object's speed is low enough that its mass equals its rest mass.

Just as the familiar form of Newton's second law is an approximation, the familiar form of kinetic energy, $K = \frac{1}{2}mv^2$, is also an approximation. In this section, we derive a more general equation for kinetic energy that applies in the case of relativistic motion. We also examine Einstein's most famous equation, $E = mc^2$.

DERIVATION Relativistic Kinetic Energy

We will show that the relativistic kinetic energy is given by

$$K = mc^2 - m_{\text{rest}}c^2 \tag{39.43}$$

for a particle of rest mass m_{rest} whose mass is m when it is observed in some frame other than its own.

The classical kinetic energy of a particle depends on both its speed and its mass. According to special relativity, the mass of a particle depends on its speed. So, to find the change in a particle's kinetic energy, we must take into account both its change in speed and its change in mass. Throughout this derivation, we observe a particle from the laboratory frame.

Suppose a constant force in the x direction does work on a particle, with no potential energy or thermal energy involved, so that the work done by the force goes only into changing the particle's kinetic energy (Fig. 39.19).

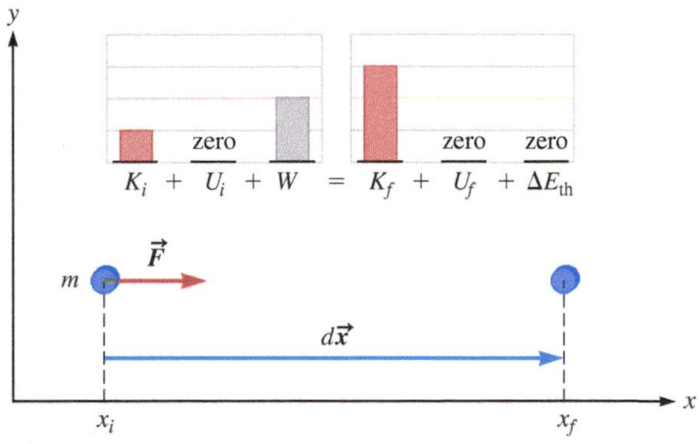

FIGURE 39.19 A constant force does work on a particle. The particle moves from x_i to x_f, and its kinetic energy increases.

When the particle moves a short distance dx, the work done is the force times this distance (Eq. 9.1). According to the work–kinetic energy theorem (Eq. 9.5), this work equals the increase in its kinetic energy dK.

$$W = \vec{F} \cdot d\vec{x} = dK \tag{1}$$

The force is parallel to the displacement, so the dot product becomes $F\,dx$. This is the only force exerted on the particle, so substitute Newton's second law, $F = dp/dt$ (Eq. 10.2), into Equation (1). Simplify using $v = dx/dt$.

$$dK = \frac{dp}{dt}dx = dp\frac{dx}{dt}$$

$$dK = v\,dp$$

Derivation continues on page 1294 ▶

In the laboratory frame, we can express the momentum as $p = mv$ as long as we keep in mind that $m = \gamma m_{\text{rest}}$ (Eq. 39.39).	$dK = v\,d(mv)$
When we take the differential of mv, we must use the product rule because both m and v change in this case.	$dK = v(m\,dv + v\,dm)$ $dK = mv\,dv + v^2\,dm$ (2)
To take into account the change in the particle's mass, start with Equation 39.42 and isolate m^2c^2.	$m = \dfrac{m_{\text{rest}}}{\sqrt{1-(v/c)^2}}$ (39.42) $m^2 = \dfrac{m_{\text{rest}}^2}{1-(v/c)^2} = \dfrac{c^2}{c^2-v^2}m_{\text{rest}}^2$ $m^2c^2 = m^2v^2 + m_{\text{rest}}^2c^2$
Differentiate using the product rule. The last term is zero because both c and m_{rest} are constants.	$d(m^2c^2) = d(m^2v^2) + d(m_{\text{rest}}^2c^2)$ $2mc^2dm = 2m^2v\,dv + 2mv^2\,dm + 0$
Cancel $2m$ from each term.	$c^2dm = mv\,dv + v^2\,dm$ (3)
The right sides of Equations (2) and (3) are identical, so we can set these equations equal to each other.	$dK = c^2dm$ (4)
Equation (4) shows that a change in mass dm results in a change in kinetic energy dK. We can integrate to find an expression for the kinetic energy. The lower limit on the integrals comes from the observation that the kinetic energy is zero when the particle is at rest in the laboratory frame, in which case the particle's mass equals its rest mass.	$\int_0^K dK = \int_{m_{\text{rest}}}^{m} c^2dm = c^2m\Big\vert_{m_{\text{rest}}}^{m}$ $K = mc^2 - m_{\text{rest}}c^2$ ✓ (39.43)

COMMENTS

Equation 39.43 shows that the relativistic kinetic energy depends on the change in mass, $m - m_{\text{rest}}$. In Problem 52, you will show that this reduces to the familiar $K = \frac{1}{2}m_{\text{rest}}v^2$ in the case of a slow-moving particle.

The Famous Equation $E = mc^2$

Equation 39.43, $K = mc^2 - m_{\text{rest}}c^2$, has important implications for the natures of mass and energy. Because we derived this equation for the special case of a particle, we didn't consider potential energy or any internal energy. However, if the system had been more complicated, we would have found that Equation 39.43 holds for a change in energy of any form. Conceptually, a change in energy of any form is proportional to a change in mass. If a system's mass increases, its energy increases. Just as James Joule concluded that heat is a form of energy because it can increase a system's thermal energy (Section 21-1), we can conclude that mass is a form of energy.

This is a major change in our understanding of mass. So far, we have thought of mass as inertia (Section 5-4). The more mass an object has, the more it resists a change in velocity. Now we can think of mass as a form of energy. This idea is known as the **mass–energy equivalence principle**. Mathematically, we write that the total energy E of a system is

MASS–ENERGY EQUIVALENCE PRINCIPLE ✪ Major Concept

$$E = mc^2 \tag{39.44}$$

where m is the observed mass of the system, $m = \gamma m_{\text{rest}}$ (Eq. 39.39). This principle allows us to write Equation 39.43 in a more general form by substituting Equation 39.44 and rearranging to find

$$E = mc^2 = m_{\text{rest}}c^2 + K \tag{39.45}$$

Equation 39.45 says that the total energy of a system is equal to its **rest mass energy** ($m_{rest}c^2$) plus its kinetic energy. Even if the system is at rest ($K = 0$), it still has energy associated with its rest mass.

There is another major consequence of Equation 39.44. Our previous most general statement of the conservation of energy principle was that the total energy of the universe must be conserved. Now, the mass–energy equivalence principle leads us to broaden the principle of energy conservation to include mass. In its broadest terms, the conservation of mass and energy says that **the total mass plus energy of the universe is constant.** In practical terms, the total mass and energy of an isolated system must be conserved. So, if an isolated system loses mass, it must gain energy in some other form, such as thermal energy.

CONCEPT EXERCISE 39.4

You have found a cold bowling ball at rest with effectively no thermal energy. If the system consists of only the ball, estimate its total energy

a. according to classical mechanics and
b. according to special relativity.

EXAMPLE 39.10 Fusion in the Sun

Sunlight is the ultimate source of energy on the Earth, and the Sun generates the energy for this light through nuclear fusion reactions in its core. For now, we need to know that the net (overall) fusion reaction in the Sun is

$$4\text{H} \rightarrow \text{He}$$

In this nuclear reaction, four hydrogen nuclei are fused together to make a single helium nucleus. A hydrogen nucleus is only one proton, and a helium nucleus is made up of two protons and two neutrons.

A Find the mass of the four hydrogen nuclei. Assuming the mass of the helium nucleus equals the mass of its four particles, find its mass. Report your answers to five significant figures.

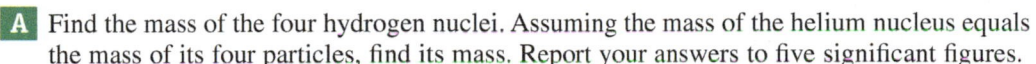

INTERPRET and ANTICIPATE

We must calculate the total mass of four protons and of a helium nucleus (two protons plus two neutrons). Because the neutron is slightly more massive than the proton, we expect the helium nucleus to be slightly more massive than the four protons.

SOLVE

Look up the masses of the particles. Retain six significant figures to avoid rounding errors.	$m_p = 1.67262 \times 10^{-27}\,\text{kg}$ $m_n = 1.67493 \times 10^{-27}\,\text{kg}$
Multiply the proton's mass by 4 to find the mass of four protons.	$4m_p = 4(1.67262 \times 10^{-27}\,\text{kg})$ $4m_p = 6.6905 \times 10^{-27}\,\text{kg}$
Add the mass of two protons to the mass of two neutrons to find the mass of a helium nucleus.	$m_{\text{He}} = 2m_p + 2m_n$ $m_{\text{He}} = 2(1.67262 \times 10^{-27}\,\text{kg}) + 2(1.67493 \times 10^{-27}\,\text{kg})$ $m_{\text{He}} = 6.6951 \times 10^{-27}\,\text{kg}$

Example continues on page 1296 ▶

CHECK and THINK

Our answer for the mass of the helium nucleus is greater than our answer for the mass of four protons. However, experiments show that the mass of a helium nucleus is *less* than the mass of four protons. Helium has less mass than the sum of its parts because some mass is converted to energy in binding the four particles together.

B A helium nucleus actually has a smaller mass than you found in part A; its mass is 6.643×10^{-27} kg. When four hydrogen nuclei are fused into a single helium nucleus, mass is lost. This loss is known as a **mass deficit**. According to the principle of conservation of mass and energy, this mass deficit requires an increase of another form of energy. Find the energy associated with this mass deficit in joules.

INTERPRET and ANTICIPATE

The mass deficit is the difference between the mass of the helium nucleus and the mass of the four protons. In order for mass and energy to be conserved, this mass must be converted to another form of energy. We can find the amount of energy using the famous mass–energy equivalence principle.

SOLVE

To find the mass deficit, subtract the (actual) mass of the helium nucleus from the mass of the four protons.

$$\Delta m = 4m_p - m_{\text{He}}$$

$$\Delta m = (6.6905 \times 10^{-27}\,\text{kg}) - (6.643 \times 10^{-27}\,\text{kg})$$

$$\Delta m = 4.75 \times 10^{-29}\,\text{kg}$$

To find the amount of energy equivalent to this mass deficit, multiply by c^2 (Eq. 39.44).

$$\Delta E = \Delta mc^2 \qquad (39.44)$$

$$\Delta E = (4.75 \times 10^{-29}\,\text{kg})(3.00 \times 10^{8}\,\text{m/s})^2$$

$$\Delta E = 4.28 \times 10^{-12}\,\text{J}$$

CHECK and THINK

The mass deficit due to the fusion of four protons into a helium nucleus is equivalent to a small amount of energy released by the Sun in the form of light.

C The Sun emits $P = 3.84 \times 10^{26}$ W. Assume all the energy associated with the mass deficit in part B goes into the Sun's power. How many fusion reactions take place per second in the Sun's core?

INTERPRET and ANTICIPATE

Because each fusion reaction releases very little energy and because the Sun's power is so great, we expect to find a great number of reactions per second.

SOLVE

Divide the Sun's power by the energy generated in each reaction to get the number n of reactions per second.

$$n = \frac{P}{\Delta E} = \frac{3.84 \times 10^{26}\,\text{J/s}}{4.28 \times 10^{-12}\,\text{J/reaction}}$$

$$n = 8.97 \times 10^{37}\,\text{reactions per second}$$

CHECK and THINK

As expected, the fusion reaction rate in the Sun is very high. The tiny mass deficit in each of these reactions supplies energy to generate sunlight, which is then transferred throughout the solar system. This sunlight can be seen even in distant parts of the galaxy, and it contributes to the glow of the Milky Way galaxy seen in distant parts of the Universe.

39-11 General Relativity

Einstein was not satisfied with restricting his theory of relativity to the special case of reference frames moving at constant velocity. In the decade that followed his groundbreaking work on special relativity, he developed a general theory of relativity that incorporated accelerating reference frames. In these last two sections, we take a brief look at general relativity, which includes a new theory of gravity.

Like special relativity, Einstein's general relativity is based on two postulates. The **first postulate of general relativity** is similar to the first postulate of special relativity: The laws of nature are the same in all reference frames, including accelerating ones. Before stating the second postulate, we must review a few things about mass, inertia, and gravity.

EINSTEIN'S FIRST POSTULATE OF GENERAL RELATIVITY

Underlying Principle

The Principle of Equivalence

For most of this book, we have taken for granted that the inertial mass of an object is equivalent to its gravitational mass. Although this may seem obvious, it was a problem for Newton and other scientists, including Einstein.

The mass of either of the two objects in the law of universal gravity, $F_G = GM_{\text{grav}}m_{\text{grav}}/R^2$ (Eq. 7.4), is referred to as the object's **gravitational mass**. This is the property of the objects that creates a gravitational force between them. From Section 5-4, the mass in Newton's second law ($\vec{F}_{\text{tot}} = m_{\text{inert}}\vec{a}$) is the **inertial mass** of an object. Experimental evidence supports the idea that the gravitational mass of any object equals its inertial mass (Section 7-3). But, until Einstein, no one knew why they should be equal, so no one could explain why the gravitational force between two particles should depend on their inertial masses rather than on some other fundamental property.

Einstein came up with an explanation for why the inertial and the gravitational masses are equivalent. He said that experimenting in a gravitational field (such as on the surface of the Earth) is equivalent to experimenting in an accelerating elevator that is located out in space, far from any gravitational field.

Figure 39.20 shows one such pair of experiments. Aaron and Hannah perform the same experiment in separate small rooms, neither of which has a view of the outside universe. Aaron is on the Earth, and Hannah is in an elevator that is far from any gravitational field but is accelerating upward at g. Aaron drops a ball, and, as you have probably done in your own laboratory, he measures the downward acceleration of the ball and finds it is g.

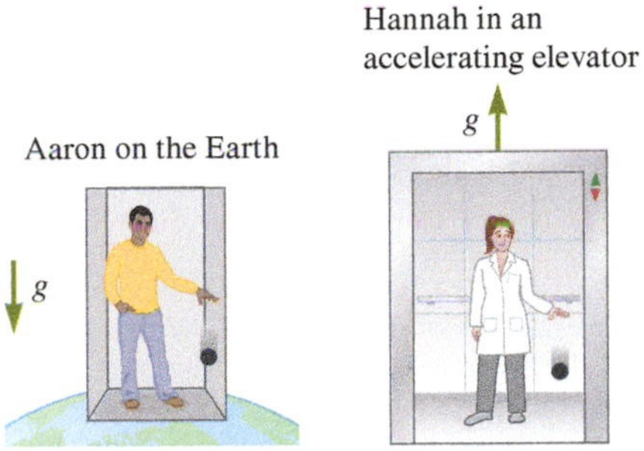

FIGURE 39.20 A. Aaron—an experimenter on the Earth—releases a ball that falls downward with an acceleration g. **B.** Hannah—an experimenter in an elevator accelerating upward at g, far from any gravitational field—releases a ball. She observes the ball moving downward with an acceleration g. Einstein concluded that gravity is equivalent to an accelerating reference frame.

Hannah in the elevator cannot tell that she is out in space. Her feet feel the normal force of the elevator, and it feels exactly like the normal force of the floor when she is standing on the Earth. Like Aaron, she drops a ball. She sees the ball fall to the floor of the elevator with an acceleration g. Because we can view this situation from an inertial frame outside the elevator, we see that in this frame, the ball hovers in place. There is no net force on the ball, so according to Newton's first law, the ball remains at rest. And, in our inertial frame, the floor of the elevator accelerates upward toward the ball. While we can see all of this, Hannah in the elevator cannot, so she cannot tell whether she is on the Earth or in an accelerating elevator. Likewise, Aaron cannot tell whether he is on the Earth or whether his small room is really an accelerating elevator.

In fact, Einstein said no experiment can be done to tell the difference between being in an accelerating frame and being in an equally strong gravitational field. This leads to the **second postulate of general relativity**: A gravitational field is equivalent to an accelerating reference frame without gravity. This second postulate is called the **principle of equivalence**, and it leads to some mind-blowing consequences.

PRINCIPLE OF EQUIVALENCE (EINSTEIN'S SECOND POSTULATE OF GENERAL RELATIVITY)

Underlying Principle

Gravity Is an Illusion

The first consequence has to do with gravity. According to Einstein, gravity is an illusion. An object with mass does not create a gravitational field; rather, it distorts or *curves* space. Humans see space as having three dimensions: (1) left and right, (2) forward and back, and (3) up and down. When Einstein says that an object with mass curves space, he means that the three-dimensional space we are used to thinking about is bent in a fourth spatial dimension.

CASE STUDY Flatlanders

No one—not even Einstein—can visualize four spatial dimensions. But, because visualizing a situation is an important part of understanding it, we'd like to have some picture of curved space in our heads. To create such a picture, we use an analogy with fewer dimensions. Instead of imagining the world through the eyes of three-dimensional beings, we imagine what two-dimensional beings would see. These two-dimensional beings are sometimes referred to as Flatlanders because of a novel by Edwin Abbott called *Flatland*. Flatlanders live on a two-dimensional surface. They describe the kinematics of everything in their universe in terms of just two dimensions: (1) left and right and (2) forward and back. Flatlanders may be able to imagine a third dimension (up and down) and work out the mathematical description of three-dimensional space, but they cannot picture it any more than we can picture four-dimensional space.

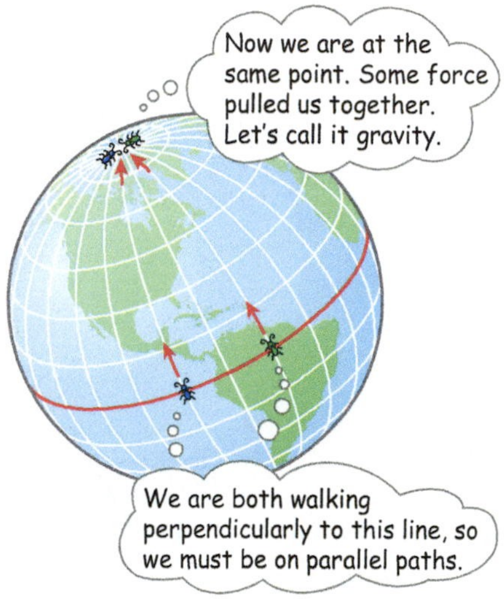
Now we are at the same point. Some force pulled us together. Let's call it gravity.
We are both walking perpendicularly to this line, so we must be on parallel paths.

FIGURE 39.21 Flatlanders believe their universe is two-dimensional. When they walk on parallel paths, they find that they are drawn together. They conclude that a force must have pulled them together. We can see that their paths simply cross in three-dimensional space.

Suppose Flatlanders live on the two-dimensional surface of a smooth sphere. You can easily picture them living on such a sphere, but they would have no way to picture the curvature of their two-dimensional universe in three-dimensional space.

Now imagine two Flatlanders start at the equator and walk north on two different lines of longitude (Fig. 39.21). They see the equator as a straight line, and they see their two paths as perpendicular to this line, so they say their paths are parallel to each other. They know parallel lines never cross, so they expect to remain the same distance apart. However, when they reach the North Pole, they find they have arrived at the same position. They conclude that they were drawn away from their parallel straight-line paths by some attractive force that they call *gravity*. As three-dimensional beings, we could explain to the Flatlanders that (1) no force was involved, (2) their paths crossed because they were really walking along great circles (lines of longitude) on the surface of a three-dimensional sphere, and (3) the force they experienced and called "gravity" is an illusion because of their inability to picture three-dimensional space.

Einstein argued that we are like the Flatlanders. Gravity is an illusion; we invented it because, as three-dimensional beings, we cannot picture four spatial dimensions. Gravity is just our explanation for the curved motion of objects, such as the orbital motion of the Moon around the Earth. Although we cannot picture a fourth spatial dimension, it exists, and our three-dimensional universe is curved in this four-dimensional space due to the presence of massive objects such as the Earth. According to general relativity, the Earth does *not* exert a gravitational force on the Moon; instead, the Earth curves space and the shape of the Moon's path is a natural consequence of that curvature.

The idea that gravity is an illusion also explains why the gravitational mass is the same as the inertial mass. Or, said more precisely, if gravity is an illusion, the question is moot. There is no such thing as two different kinds of mass. Objects have only one kind of mass, and that mass curves space.

CURVATURE OF SPACE
Major Concept

You might feel uncomfortable with the idea that gravity is an illusion. However, we can still use Newton's law of universal gravity in most practical situations, such as when calculating the orbits of satellites. Gravity may be an illusion, but Newton's description of this illusion works very well. Also, even when we talk about the theory of general relativity, we do not abandon the terms *gravity* and *gravitational*. They are shorthand for the illusion or effect we perceive due to the presence of an object with mass.

Time Dilation and the Gravitational Doppler Shift

Consider an elevator in free space (Fig. 39.22) accelerating upward at g. On the floor of the elevator is a monochromatic light source emitting light at frequency f_{emt}. A detector on the ceiling of the elevator measures the frequency of the light that enters it. At some instant, the source emits a photon. In the time it takes the photon to cross from the floor to the ceiling, the elevator has moved upward. As we have seen before in both Galilean and special relativity, when the source and detector move toward or away from each other, there is a Doppler shift (Sections 17-9 and 39-7). Further, we saw that relative motion also produces time dilation (Section 39-6). According to the principle of equivalence, the acceleration due to gravity near a massive object is equivalent to an accelerating reference frame in empty space. So, the source on an elevator can be replaced by a source in the gravitational field of a massive object, and then the detector, located far from the object, must still measure a *gravitational* Doppler shift and *gravitational* time dilation. The **gravitational Doppler shift** and **gravitational time dilation** are given by

GRAVITATIONAL DOPPLER SHIFT AND GRAVITATIONAL TIME DILATION ✪ Major Concepts

$$\frac{f_{obs}}{f_{emt}} = \frac{\Delta t_{emt}}{\Delta t_{obs}} = \left(1 - \frac{2GM}{c^2 r}\right)^{1/2} \tag{39.46}$$

where the subscript "obs" indicates the *observed* measurement made by the detector and the subscript "emt" indicates the frequency emitted (or the time interval measured) in the source's frame. The mass of the object is M, and the source's distance from the object is r. The detector is assumed to be infinitely far from the object.

The negative sign in Equation 39.46 indicates that the observed frequency is lower than the emitted frequency. Expressed in terms of wavelength, the observed wavelength is longer than the emitted wavelength. So the observed light is redder than the emitted light. We say the light has been *gravitationally red-shifted*. Further, Equation 39.46 indicates that time passes more slowly near the massive object.

It is often convenient to approximate Equation 39.46. If $2GM/c^2 r << 1$, then

$$\frac{f_{obs}}{f_{emt}} = \frac{\Delta t_{emt}}{\Delta t_{obs}} \approx 1 - \frac{GM}{c^2 r} \tag{39.47}$$

Further, if the source and the detector are relatively close together and the gravitational field g is nearly uniform, it is convenient to write the Doppler shift as

$$\frac{f_{obs} - f_{emt}}{f_{emt}} = \frac{\Delta f}{f_{emt}} \approx -\frac{gh}{c^2} \tag{39.48}$$

where h is the distance between the source and the detector. (You can derive this equation in Problem 82.)

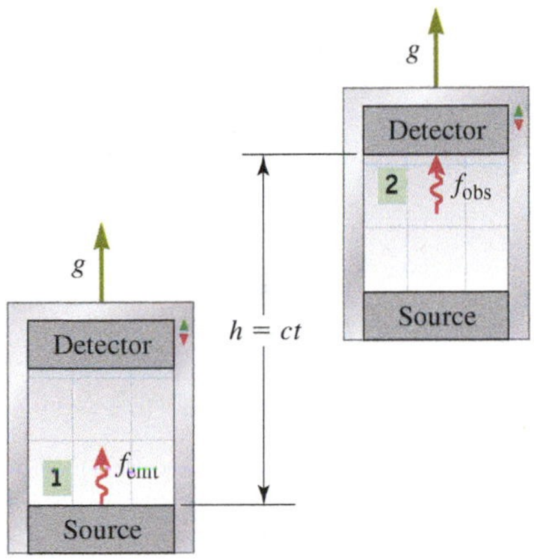

FIGURE 39.22 An elevator in free space accelerates upward at g. **1** A photon is released at $t = 0$. **2** The photon is detected at a later time t.

EXAMPLE 39.11 Harvard's Test of the Gravitational Doppler Shift

In 1960, a test of the gravitational Doppler shift was conducted at Harvard University. A γ–ray photon was emitted from the bottom of a tower of height 22.6 m. The photon was then detected at the top of the tower. The experimental result was $\Delta f/f_{emt} = -(2.57 \pm 0.26) \times 10^{-15}$. Does this agree with the prediction of general relativity?

INTERPRET and ANTICIPATE

We must calculate $\Delta f/f_{emt}$ based on the tower's height and the gravitational field of the Earth. Because the tower is on the Earth's surface, g is nearly uniform over the height of the tower: $g = 9.81 \text{ m/s}^2$.

SOLVE

Substitute into Equation 39.48.

$$\frac{\Delta f}{f_{emt}} \approx -\frac{gh}{c^2} \tag{39.48}$$

$$\left(\frac{\Delta f}{f_{emt}}\right)_{pred} \approx -\frac{(9.81 \text{ m/s}^2)(22.6 \text{ m})}{(3.00 \times 10^8 \text{ m/s}^2)^2}$$

$$\left(\frac{\Delta f}{f_{emt}}\right)_{pred} \approx -2.46 \times 10^{-15}$$

Example continues on page 1300 ▶

To make the comparison between the predicted value and the measured value, find the minimum and maximum of the measurement using the given experimental error.

$$-2.83 \times 10^{-15} < \left(\frac{\Delta f}{f_{\text{emt}}}\right)_{\text{meas}} < -2.31 \times 10^{-15}$$

The predicted value falls within the experimental range.

CHECK and THINK

The prediction and the measurement are consistent. This result is one confirmation of the gravitational Doppler shift. Such confirmations are comforting because there are so many counterintuitive consequences of general relativity. In the next section, we look at another test.

39-12 Gravitational Lenses and Black Holes

One of the most mind-blowing consequences of the principle of equivalence is that light is bent or refracted by a massive object, much as a lens refracts light. To see why this happens, consider Aaron in free space in an elevator accelerating upward at g. The elevator has large windows on opposite walls. A light source and a target are stationary outside the elevator, on opposite sides (Fig. 39.23A). The source emits light that hits the target. Hannah is in the frame of the source and target; to her, the light's path is horizontal. To Aaron, however, the light must move downward to hit the target. When the light was emitted, the source was lined up with his shoulders. In the time the light takes to cross the elevator and arrive at the target, the elevator moves upward, and the target ends up aligned with Aaron's feet. So Aaron must see the light fall from the height of his shoulders to the height of his feet as it moves across the elevator. The light's motion is analogous to that of a tennis ball launched horizontally at shoulder height on the Earth: The ball falls to the ground in a parabolic arc (Section 4-5). So the path of the light seen by Aaron is also a parabola (Fig. 39.23B).

According to the principle of equivalence, the accelerating elevator is the same as a gravitational field produced by an object with mass. So the curved path the light takes near a massive object is a natural consequence of the shape of space. No force is exerted on the light.

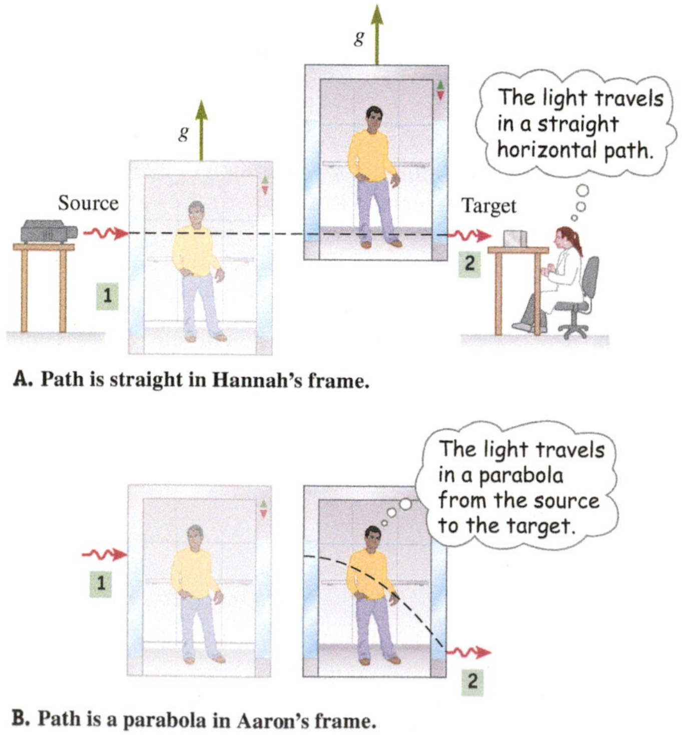

FIGURE 39.23 An elevator in free space accelerates upward at g. The elevator has windows on both sides. **A.** Light travels horizontally from the source to the target, passing through the elevator as seen in the laboratory frame. **B.** But in the moving frame, the light's path is a parabola.

Using an Eclipse to Test General Relativity

The bending of light near a massive object is not predicted by Newton's law of universal gravity. According to Newton's law, only objects that have mass can be affected by gravity. Because light has no mass, it should move in a straight line through a gravitational field. An observation that light is refracted by a massive object would be a confirmation of Einstein's theory of general relativity and a refutation of Newton's theory of gravity.

In 1919, a total solar eclipse was observed in Brazil and Principe in the Gulf of Guinea. During the eclipse, astronomers were able to take pictures of stars whose light passed near the Sun. They found that the positions of the stars appeared shifted by 1.64 arcsec compared to their normal positions. This shift was due to the refraction of starlight by the Sun's gravitational field, a confirmation of Einstein's theory of general relativity.

Gravitational Lenses

Since 1919, refraction of light by a massive object has been observed in many circumstances. The massive object that causes the light to refract is called a **gravitational lens** and is always located between the source and the observer. Rays from the source are refracted by the lens and come together at the observer's position. By tracing the rays back along straight paths, we find the image formed by the lens. If the source is directly behind the lens and the lens is symmetrical, the image is a ring (Fig. 39.24). Einstein described such a ring in 1936. The angular radius θ_r (in radians) of the ring depends on the mass M of the gravitational lens:

$$\theta_r = \sqrt{\frac{4GM}{c^2}\left(\frac{d_{\text{src}} - d_{\text{lens}}}{d_{\text{src}}d_{\text{lens}}}\right)} \tag{39.49}$$

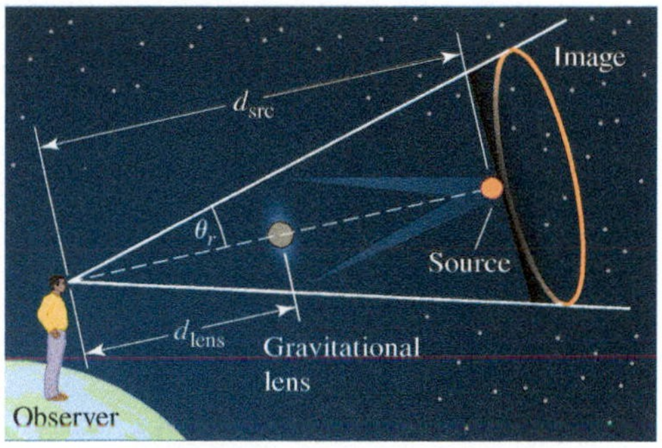

FIGURE 39.24 Any massive object between the source and the observer may refract the source's light and act as a gravitational lens. If the source is directly behind the lens, it is possible for the image to be a ring.

where d_{src} and d_{lens} are the distances to the source and to the lens, respectively. By observing the radius of such a ring, astronomers are able to determine the mass of the gravitational lens. Often the gravitational lens is a galaxy, so this allows astronomers to measure the masses of entire galaxies. Such observations indicate the presence of dark matter (case study in Chapter 7).

Black Holes

In a star like the Sun, the inward pull of its own gravity is balanced by the thermal pressure that results from nuclear fusion taking place in its core. But all stars eventually run out of the reactants needed for fusion. When that happens, gravity causes the star to collapse. The Sun will form a white dwarf—an object about the size of the Earth. The inward pull of gravity will eventually be balanced by the pressure exerted by a sea of electrons in the white dwarf. For a star that is several times as massive as the Sun, the electron pressure is not enough to keep it from collapsing under its own weight. A very massive star is eventually crunched into a single point known as a *singularity*. Anything that gets too close to a singularity cannot escape. By "too close" we mean inside a spherical region called the **event horizon** (Fig. 39.25). The singularity plus the event horizon are known as a **black hole**.

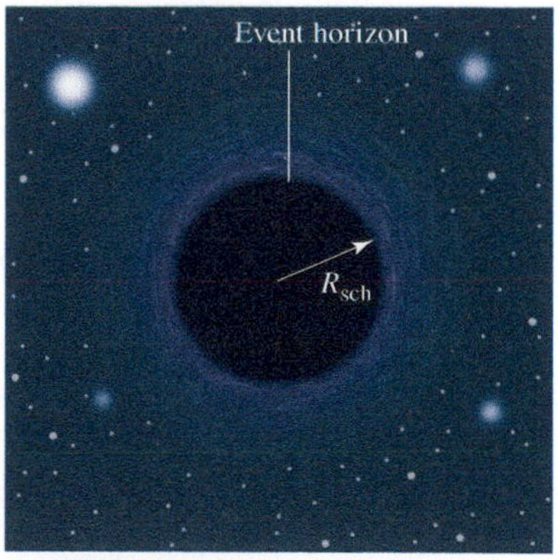

FIGURE 39.25 Light that passes inside the event horizon never escapes. The radius of the event horizon is called the Schwarzschild radius.

These dense objects are called *black holes* because anything, even light, that enters the event horizon cannot escape. The radius of the event horizon is called the **Schwarzschild radius** R_{sch} (Fig. 39.25). The Schwarzschild radius depends on the mass of the black hole. The correct way to find an expression for the Schwarzschild radius is to use general relativity to take into account the curvature of space and time dilation at the event horizon.

However, we can also find the correct expression by using classical mechanics. The escape speed from an object of mass M is $v_{\text{esc}} = \sqrt{2GM/R}$ (Eq. 8.17). At the event horizon, the escape speed equals the speed of light. We can find the Schwarzschild radius by setting v_{esc} equal to c:

$$c = \sqrt{\frac{2GM}{R_{\text{sch}}}}$$

So the Schwarzschild radius is given by

$$R_{sch} = \frac{2GM}{c^2} \quad (39.50)$$

Often, black holes are several times the mass of the Sun ($M_\odot$) and have Schwarzschild radii of just a few kilometers, so it is helpful to rewrite Equation 39.50 in the form

$$R_{sch} \approx 3\frac{M}{M_\odot}\text{km} \quad (39.51)$$

where M is measured in solar mass units and the Schwarzschild radius is in kilometers.

CONCEPT EXERCISE 39.5

The Sun will not form a black hole. But if a black hole were discovered to have as great a mass as the Sun, what would its Schwarzschild radius be in kilometers?

SUMMARY

Underlying Principles

1. **Einstein's postulates of special relativity** All the laws of physics are true in all inertial reference frames. The speed of light c in a vacuum is invariant.
2. **Einstein's postulates of general relativity** The laws of nature are the same in all reference frames, including accelerating ones. A gravitational field is equivalent to an accelerating reference frame without gravity. This second postulate is called the **principle of equivalence**.

Major Concepts

1. **Lorentz transformations** for position and time measured in two different reference frames hold in special relativity (Table 39.1, page 1277). These transformations are written in terms of the Lorentz (gamma) factor:

$$\gamma = \frac{1}{\sqrt{1 - (v_{rel}/c)^2}} = \frac{1}{\sqrt{1 - \beta^2}} \quad (39.17)$$

where

$$\beta \equiv \frac{v_{rel}}{c} \quad (39.18)$$

2. **Length contraction** is described by

$$L = \frac{L'}{\gamma} \quad (39.25)$$

The object's proper length is measured in its own frame and is its longest observable length.

3. **Time dilation** is given by

$$\Delta t = \gamma \Delta t' \quad (39.26)$$

Moving clocks run slowly. The proper time is the time interval between two events that occur at the same position.

4. The **relativistic Doppler effect** is given by

$$\lambda = \sqrt{\frac{1 \mp \beta}{1 \pm \beta}}\lambda' \quad (39.27)$$

and

$$f = \sqrt{\frac{1 \pm \beta}{1 \mp \beta}}f' \quad (39.28)$$

Choose the top signs if the source is moving toward the observer; choose the bottom signs if the source is moving away from the observer.

5. The **perpendicular velocity transformation** is given by

$$v_y = \frac{v_y'}{\gamma\left(1 + \frac{v_{rel}}{c^2}v_x'\right)} \quad (39.29)$$

with a similar equation for z; this transforms a component of an object's velocity that is perpendicular to the relative motion between reference frames.

6. The **parallel velocity transformation** is given by

$$v_x = \frac{(v_x' + v_{rel})}{\left(1 + \frac{v_{rel}}{c^2}v_x'\right)} \quad (39.31)$$

This relationship holds for the component of an object's velocity that is parallel to the relative motion between reference frames.

7. **Mass transformation** is given by

$$m = \gamma m_{rest} \quad (39.39)$$

where the rest mass m_{rest} must be measured in the object's frame.

8. **Relativistic momentum** for a particle observed with velocity $\vec{v}$ in the laboratory frame is given by

$$\vec{p} = m\vec{v} = \gamma m_{rest}\vec{v} \quad (39.40)$$

where

$$\gamma = \frac{1}{\sqrt{1 - (v/c)^2}} \quad (39.41)$$

9. According to the **mass–energy equivalence principle**, mass is a form of energy given by

$$E = mc^2 \quad (39.44)$$

10. Gravity is an illusion. An object that has mass does not create a gravitational field. Instead, the object **curves space**.

11. The **gravitational Doppler shift** and **gravitational time dilation** are given by

$$\frac{\Delta t_{emt}}{\Delta t_{obs}} = \frac{f_{obs}}{f_{emt}} = \left(1 - \frac{2GM}{c^2 r}\right)^{1/2} \quad (39.46)$$

Special Cases

Galilean relativity: The Galilean transformations hold when the relative speed is low (Table 39.1, page 1277). The relativistic transformations must be identical to the Galilean transformations in this special case.

PROBLEMS AND QUESTIONS

A = algebraic **C** = conceptual **E** = estimation **G** = graphical **N** = numerical

39-1 It's in the Eye of the Observer

1. **C** A simple pendulum surrounded by pegs is known as a Foucault pendulum (Fig. P39.1). Imagine watching such a pendulum placed at the Earth's North Pole. As the pendulum swings back and forth, it gradually hits each of the pegs. Explain why it hits all the pegs in 24 h. *Hint*: What would happen if the pendulum were in an inertial reference frame?

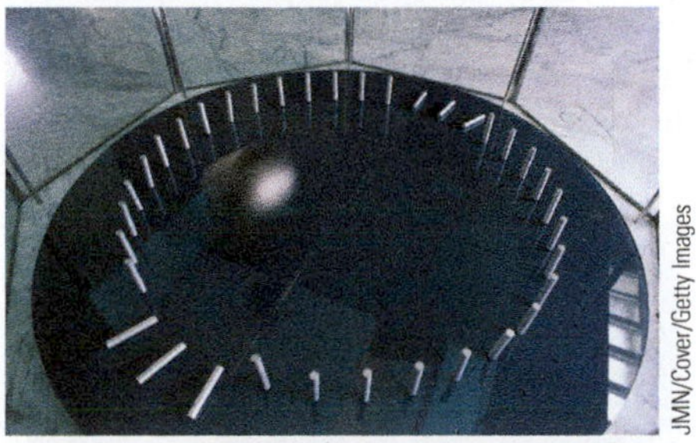

FIGURE P39.1

39-2 Special Case: Galilean Relativity

2. **N** Desmond and Lilani are traveling on different high-speed trains on tracks that run parallel to each other. Lilani's train is traveling west at 305 km/h, and Desmond's train is traveling east at 275 km/h. **a.** How far does Desmond travel in 30.0 s according to a stationary farmer near the tracks? **b.** How far does Lilani travel in 30.0 s according to a stationary farmer near the tracks? **c.** How far does Desmond travel in Lilani's frame of reference in 30.0 s? **d.** How far does Lilani travel in Desmond's frame of reference in 30.0 s?

Problems 3 and 4 are paired.

3. **N** Carrying a suitcase in an airport, you walk at about 2.0 mph. A fast-moving sidewalk (conveyor belt) moves at about 9.00 km/h and is 0.50 km long. **a.** If you stand still on the moving sidewalk, how long will it take you to cover the 0.50 km? **b.** If you walk next to the conveyer belt, how long will it take you to walk 0.50 km? **c.** If, instead, you walk on the sidewalk at your usual pace, how long will it take you to cover this distance? (Report your answers to the nearest second.)

4. **N** In an airport terminal, there are two fast-moving sidewalks (9.0 km/h); one carries its passengers south, and the other carries its passengers north. Each sidewalk is 0.50 km long. At the instant a woman steps onto the north end of the southbound sidewalk, a man steps onto the south end of the northbound sidewalk. He stands still with respect to the sidewalk, while she walks south at 5.0 km/h. **a.** How long after stepping onto the

sidewalks do they pass each other? (Report your answer to the nearest second.) **b.** How far does each person travel in that time? (Report your answer in kilometers.)

5. **A** A primed frame is moving in the positive x direction relative to the laboratory frame. Use $y = y'$ (Eq. 39.2) to show that both v_y and a_y are invariant under Galilean relativity.

Problems 6 and 7 are paired.

6. **N** Jason is driving north on a highway at 60.0 mph when he sees a car ahead. Kevin is driving that car north at 53.0 mph. **a.** What are the magnitude and direction of Kevin's velocity according to Jason? **b.** How far does Kevin travel in Jason's frame of reference in the time it takes Jason to travel 3.0 mi?
7. **N** Jason is driving north on a highway at 60.0 mph when he sees a car ahead. Kevin is driving that car south at 53.0 mph. **a.** What are the magnitude and direction of Kevin's velocity according to Jason? **b.** How far does Kevin travel in Jason's frame of reference in the time it takes Jason to travel 3.0 mi?

39-3 Postulates of Special Relativity

8. **C** Suppose there are two observers. Observer A is traveling toward a light source at $0.5c$, and observer B is traveling away from the same light source at $0.5c$. The source emits a brief pulse. **a.** What is the speed of the light pulse measured by each observer? **b.** At the instant the pulse is emitted, the two observers are the same distance from the source. Which observer will see the pulse first, as determined by a person in the frame of the source? Explain.
9. **C** One reference frame is moving relative to another, and it is said that both reference frames are inertial. The first postulate of special relativity is that the laws of physics hold in all inertial reference frames. What does it mean to say that a particular reference frame is "inertial"? In other words, what is required for a reference frame to be called "inertial"?

39-4 Lorentz Transformations

10. **N** A particle is at a point $x' = 12.0$ m, $y' = 4.00$ m, $z' = 6.00$ m at time $t' = 4.00 \times 10^{-4}$ s as measured in the primed frame. The particle is at rest in the primed frame. Assume the origins of the primed and unprimed frame were coincident at time $t = 0$. What are the coordinates (x, y, z, t) as measured in the laboratory frame if the particle is moving along the x and x' axes with speed **a.** $v_{rel} = 400.0$ m/s, **b.** $v_{rel} = -400.0$ m/s, and **c.** $v_{rel} = 2.0 \times 10^8$ m/s?
11. **N** In a stationary reference frame S, a green laser located at $x_1 = 1.00$ m is switched on at $t_1 = 5.00$ ns, and 11.0 ns later, a blue laser is switched on at $x_2 = 4.00$ m. A second reference frame S′ is moving in the positive x direction so that the origins of the two reference frames coincide at time $t = 0$ as measured in both reference frames. To an observer in S′, both lasers are switched on at the same location. **a.** What is the relative speed between the S and S′ reference frames? **b.** What is the x coordinate of the two lasers in the S′ reference frame when they are switched on? **c.** What is the time interval between the two lasers being switched on in the S′ reference frame?

Problems 12 and 13 are paired.

12. **C** The primed frame has speed $0.75c$ in the x direction according to an observer in the laboratory frame. An observer in the primed frame sees a balloon hovering at $(x', y') = (6.0$ m, 12.0 m). What can you say about the position of the balloon as seen in the laboratory frame?
13. **N** The primed frame has speed $0.75c$ in the x direction according to an observer in the laboratory frame. At $t = t' = 0$, the origins of the two frames coincide. At $t' = 16.0$ s, an observer in the primed frame sees a balloon hovering at $(x', y') = (6.0$ m, 12.0 m). **a.** What can you say about the position of the balloon as seen in the laboratory frame? **b.** How much time has elapsed in the laboratory frame when the observer in the primed frame makes this observation?
14. **A** Starting with the Lorentz transformations relating x and x', confirm **a.** Equation 39.19, $t = \gamma(t' + v_{rel}x'/c^2)$, and **b.** Equation 39.20, $t' = \gamma(t - v_{rel}x/c^2)$.

Problems 15 and 16 are paired.

15. An electron is moving with a speed of 6.75×10^7 m/s as measured by an observer in a laboratory.
    a. **N** What is the Lorentz factor for the electron moving in the laboratory frame of reference?
    b. **C** Which frame of reference (the electron's or the laboratory's) would we label the x' frame and which the x frame?
16. **N** An electron is moving with a speed of 6.75×10^7 m/s as measured by an observer in a laboratory. The electron is at rest in the primed frame of reference and at its origin. If we assume the origins of the reference frames of the electron and the laboratory are coincident at $t = t' = 0$ and the electron is moving along the $+x$ axis, what is the position of the electron in the primed frame of reference when the position in the unprimed frame is **a.** 1.0 mm, **b.** 1.0 cm, **c.** 1.0 m, and **d.** 1.0 km?

39-5 Length Contraction

17. **N** A spaceship is moving at a relativistic speed. The length of the spaceship is measured to be exactly half its proper length. Determine the speed of the spaceship (in units of the speed of light, c) relative to the laboratory frame.
18. **C** In a factory, a rod lies on a conveyor belt so that its long axis is parallel to the direction of motion. A factory worker is watching the rod on the belt when the belt stops. Does the worker see the rod's length grow or shrink? Assume the worker can observe the change in length due to the velocity of the rod.
19. **N** CASE STUDY A space traveler aboard a spacecraft measures its length and finds it is 15.0 m long. An observer in another inertial reference frame sees that the spacecraft is 7.50 m long. What is the relative speed of the spacecraft in this frame?
20. **N** What is the speed with which a ruler 12.0 in. (30.48 cm) long, as measured in its own reference frame, must be moving if its apparent length is 20.0 cm?
21. **N** A meterstick in the primed frame makes a 45.0° angle with the x' axis. The meterstick is at rest in the primed frame. As seen by an observer in the laboratory frame, the primed frame's relative speed in the x direction is $0.975c$. What is the length of the meterstick as observed in the laboratory frame?
22. **N** An astronaut on the Moon observes a UFO flying by with a velocity of $0.75c$ as shown in Figure P39.22. **a.** If the long side of the UFO, when at rest, has a length of 265 m, what is the length of this side as perceived by the astronaut if it flies by as shown in case 1? **b.** What is the perceived length of the long side if it flies by the astronaut as shown in case 2? Explain.

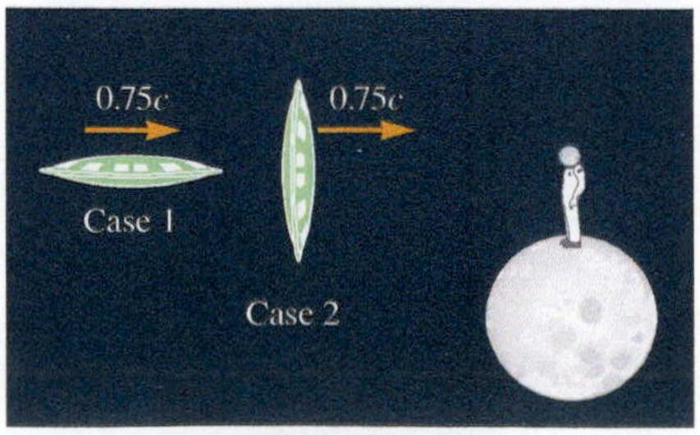

FIGURE P39.22

23. **N** As Brent's spacecraft overtakes Jasmin's identical spacecraft at high speed, Brent notifies Jasmin that his spacecraft is 35.0 m long; Jasmin's spacecraft appears to be 31.0 m long according to his observations. **a.** What is the length of Jasmin's spacecraft in her frame of reference? **b.** What is the length of Brent's spacecraft as observed by Jasmin? **c.** What is the relative speed between their spacecrafts?

Problems 24 and 25 are paired.

24. **N** A starship is 1025 ly from the Earth when measured in the rest frame of the Earth. The ship travels at a speed of $0.80c$ on its way back to the Earth. What is the distance traveled as measured by the crew of the starship?

39-6 Time Dilation

25. **N** A starship is 1025 ly from the Earth when measured in the rest frame of the Earth. The ship travels at a speed of $0.80c$ on its way back to the Earth. **a.** How much time will it take for the ship to return to the Earth as measured by the Earth observer? **b.** How much time will it take for the ship to return to the Earth as measured by the crew of the ship?

26. **N** Two atomic clocks are carefully synchronized on the Earth (considered an inertial frame). One remains in a laboratory at the airport. The other is carried on an airplane traveling at twice the speed of sound. When the plane returns, according to the laboratory clock, 3.50 h have passed. According to the clock that traveled on the plane, how much time has elapsed? What is the difference between the two clocks? Ignore the effects of acceleration on the clocks. *Hint*: Work to 10 significant figures throughout solving the problem, until expressing the time difference, and use the binominal theorem (Appendix A) to find the Lorentz factor.

27. **N** What is the speed with which a clock must be moving with respect to an observer at rest if its second hand sweeps through 45.0 s during each 60.0 s measured by the observer's clock?

28. **N** A neutron outside of a nucleus is an unstable particle. It decays in roughly 10.0 min. How fast would a neutron need to travel in order to be stable for 11.0 min as seen by an observer in the laboratory frame? How far would the neutron travel in that time according to the laboratory observer?

29. **A** The primed observer measures a time interval of $\Delta t'$, and the laboratory observer measures $\Delta t = 2\Delta t'$. What is the primed frame's relative speed?

30. **N** An astronaut takes her grandfather clock with her on her spaceship. Later, her ship is traveling by the Earth at a speed of $0.65c$. According to the astronaut, the pendulum on the clock has a frequency of 1.00 Hz. What is the frequency of the pendulum according to an observer on the Earth?

31. **N** Atomic clocks operate by measuring the resonances of certain atoms, most commonly cesium-133 (which is used to define the second). An atomic clock is held stationary on the ground, while a second identical atomic clock moves with a constant speed of 350.0 m/s for 2.00 h as measured by the stationary clock. What is the time difference between the two clocks after this 2.00-h interval?

32. **N** It takes 1.0 h to bake a casserole in a frame moving at a speed of 2.0×10^8 m/s. How long is the baking time as measured by a cook in the laboratory frame?

33. **N** Scientists are conducting timed experiments on two spacecraft and on the Earth. The experimenters on the Earth observe spacecraft A moving with speed $0.400c$ away from the Earth and toward spacecraft B, which is moving with speed $0.710c$ toward the Earth. The experiment requires a 2.00-h incubation period as measured by the crew on spacecraft A. **a.** How much time has elapsed on spacecraft B when the spacecraft A clock shows an elapsed time of 2.00 h? Assume that the relative speed between the two spacecraft is $0.864c$ (you will learn why in section 39-8). **b.** How much time has elapsed for the experimenters on the Earth when the spacecraft A clock shows an elapsed time of 2.00 h?

39-7 The Relativistic Doppler Effect

34. **C** A spiral galaxy is seen edge-on. From this perspective, half the disk moves toward the observer and half moves away. Describe the relative color of the two sides as seen by the observer.

35. **N** A supernova explosion sends out a spherical shell of hydrogen gas. The shell expands at the rate of 1000.00 km/s. The gas is transparent, so we can see the front of the shell moving toward us and the back moving away. In the frame of the gas, the hydrogen gives off radio waves with a frequency of 1420.04 MHz. What are the lowest and highest frequencies observed on the Earth?

36. **A** Beginning with Equation 39.27, derive Equation 39.28:

$$f = \sqrt{\frac{1 \pm \beta}{1 \mp \beta}}f' = \sqrt{\frac{c \pm v_{\text{rel}}}{c \mp v_{\text{rel}}}}f'$$

37. **N** A source of hydrogen gas gives off Hα (Fig. 36.19A, page 1166). The line is emitted with a wavelength of 656 nm and observed with a wavelength of 652 nm. What is the constant speed of the source relative to the observer? Is the source moving toward or away from the observer?

39-8 Velocity Transformation

38. **A** Show that Equation 39.33,

$$v_x' = \frac{(v_x - v_{\text{rel}})}{\left(1 - \dfrac{v_{\text{rel}}}{c^2}v_x\right)}$$

is the x velocity transformation from the laboratory frame to the primed frame.

39. **N** As measured in a laboratory reference frame, a linear accelerator ejects a proton with a speed of $0.780c$. Moments later, a muon is ejected at a speed of $0.920c$ as measured in the laboratory reference frame. What is the speed of the proton in a reference frame where the velocity of the muon is zero?

40. **A** Show that Equation 39.29,

$$v_y = \frac{v_y'}{\gamma\left(1 + \dfrac{v_{\text{rel}}}{c^2}v_x'\right)}$$

is the transformation of a velocity component perpendicular to $\vec{v}_{\text{rel}}$ and Equation 39.30,

$$v_y' = \frac{v_y}{\gamma\left(1 - \dfrac{v_{\text{rel}}}{c^2}v_x\right)}$$

is the inverse transformation from the laboratory frame to the primed frame.

41. **N** The components of the velocity of an electron measured in the primed frame are $v_x' = 5.00 \times 10^7$ m/s, $v_y' = 4.00 \times 10^7$ m/s, and $v_z' = 3.00 \times 10^7$ m/s. Determine the magnitude of the velocity in the laboratory frame. Assume the relative velocity between the primed and the laboratory frames is $2.00 \times 10^7\hat{\imath}$ m/s.

Problems 42, 43, and 44 are grouped.

42. **N** In a laboratory, two particles are fired sequentially. Particle 1 has speed $0.80c$ and is traveling to the east. A moment later, particle 2 is launched with speed $0.70c$ and is also traveling to the east. What is the speed of particle 2 as measured in the reference frame of particle 1?

43. **N** In a laboratory, two particles are fired sequentially. Particle 1 has speed $0.80c$ and is traveling to the east. A moment later, particle 2 is launched with speed $0.70c$ but is traveling to the west. What is the speed of particle 2 as measured in the reference frame of particle 1?

44. **N** In a laboratory, two particles are fired sequentially. Particle 1 has speed $0.80c$ and is traveling to the east. A moment later, particle 2 is launched with speed $0.70c$. **a.** What is the speed of particle 1 as measured in the reference frame of particle 2, if particle 2 is also traveling to the east? **b.** What is the speed of particle 1 as measured in the reference frame of particle 2, if particle 2 is traveling to the west?

45. **N** CASE STUDY In a fictional spaceport, a fast-moving sidewalk moves at about $0.90c$. The sidewalk is 0.50 ly long. A man gets on the sidewalk, and a woman standing near the sidewalk watches him. In each case, find his speed in her frame in terms of c. **a.** He is at rest with respect to the sidewalk. **b.** He walks at 2.2 m/s with respect to the sidewalk. **c.** He uses fictional roller blades so that his speed relative to the sidewalk, and in the same direction, is $0.50c$.

39-9 Mass and Momentum Transformation

46. **N** What is the momentum of a proton ($m_p = 1.67 \times 10^{-27}$ kg) that has a speed of **a.** $0.0500c$, **b.** $0.550c$, and **c.** $0.950c$?

Problems 47 and 48 are paired.

47. **N** The rest mass of a baseball is 0.145 kg. Find the mass observed if the ball is moving at **a.** 90.0 mph (40.2 m/s)—a fast pitch, **b.** 250.0 km/s—the speed of the Sun around the center of the Milky Way galaxy, and **c.** $0.850c$ with respect to the laboratory.

48. **N** The rest mass of a baseball is 0.145 kg. Find the momentum observed if the ball is moving at **a.** 90.0 mph (40.2 m/s)—a fast pitch, **b.** 250.0 km/s—the speed of the Sun around the center of the Milky Way galaxy, and **c.** $0.850c$ with respect to the laboratory.

49. **N** Consider a proton moving with speed $0.00100c$. How fast does an electron have to move to have the same momentum as this proton?

Problems 50 and 51 are paired.

50. **N** In transmission electron microscopy (TEM), electrons are accelerated to have kinetic energies of hundreds of thousands of electron volts (1 eV $= 1.602 \times 10^{-19}$ J). Suppose an electron has a kinetic energy of 300.0 keV. **a.** What is the mass of the electron according to an observer in the lab? **b.** What is the momentum of the electron according to an observer in the lab? Answer using three significant figures.

39-10 Newton's Second Law and Energy

51. **N** In transmission electron microscopy (TEM), electrons are accelerated to have kinetic energies of hundreds of thousands of electron volts (1 eV $= 1.602 \times 10^{-19}$ J). Suppose an electron has a kinetic energy of 300.0 keV. **a.** What is the rest mass energy of the electron? **b.** What is the total energy of the electron? **c.** What is the speed of the electron? Answer using three significant figures.

52. **A** Show that $K = mc^2 - m_{\text{rest}}c^2$ (Eq. 39.43) reduces to the familiar $K = \frac{1}{2}m_{\text{rest}}v^2$ in the case of a slow-moving particle.

53. **N** The *top quark*, an elementary particle and a fundamental constituent of matter, was discovered in 1995 at Fermilab in Chicago. It has a mass of 172.9 GeV/c^2. What is the mass of the top quark in kilograms?

54. **N** Consider a proton moving with speed $0.2000c$. How fast would an electron have to move to have the same kinetic energy as this proton?

55. **N** A 15.00-kg satellite orbits the Earth at 8314 m/s. Calculate the satellite's kinetic energy using special relativity. How does your answer compare to what you find if you use classical physics?

56. **E** Assume the Sun's luminosity (the same as its power) is constant over its entire 10-billion-year life. Estimate the amount of mass it must consume in nuclear fusion to put out the required amount of energy. Give your answer in kilograms and in solar masses.

57. **N** Consider an electron moving with speed $0.980c$. **a.** What is the rest mass energy of this electron? **b.** What is the total energy of this electron? **c.** What is the kinetic energy of this electron?

58. **N** VY Canis Majoris, the largest known star in the Milky Way galaxy, is 4.30×10^5 times more luminous than the Sun. If this star has an average power output of 1.65×10^{32} W, how many kilograms of its mass are converted to energy per second?

39-11 General Relativity

Problems 59 and 60 are paired.

59. **N** Hα ($f_{\text{emt}} = 4.57 \times 10^{14}$ Hz) is emitted from the surface of the Sun. What is the frequency of the light when it is observed far from the Sun?

60. **N** Hα ($f_{\text{emt}} = 4.57 \times 10^{14}$ Hz) is emitted from the surface of a neutron star. Its radius is 6.96×10^4 m and its mass is 2 solar masses. What is the frequency of the light observed far from the neutron star?

61. **N** According to an observer on the Earth, a solar flare lasts for 3.25 min on the surface of the Sun. What is the lifetime of the flare as observed from the Sun?

62. **N** Imagine a star with a mass that is 1000 times that of the Sun and a radius that is 1/100 the radius of the Sun. **a.** If the star emits red light with a frequency of 3.2×10^{14} Hz, what is the frequency observed far away from the star? **b.** A dark spot forms on the surface of the star and lasts for 13.5 days as observed from the star. How long does the spot last according to observers far from the star?

39-12 Gravitational Lenses and Black Holes

63. **N** The Milky Way galaxy has a supermassive black hole at its center. Its mass is 3.6 million solar masses. What is its Schwarzschild radius? Give your answer in kilometers and in Earth radii.

64. **N** When a 10-solar-mass star runs out of reactants in its core, it sheds its outer layers in a supernova explosion, and the core collapses to form a black hole of about 4.0 solar masses. What is its Schwarzschild radius?

65. **N** A particular Einstein ring has an angular diameter of 2.1 arcsec. The distance to the source (a quasar) is 2.3×10^{26} m, and the distance to the gravitational lens (a galaxy) is 3.3×10^{25} m. What is the mass of the lens? Give your answer in kilograms and in solar masses. Compare your answer to the Andromeda galaxy, whose mass is about 710 billion solar masses.

General Problems

66. **N** At a distance of 6.00×10^2 ly, Kepler-22b is the nearest habitable planet found orbiting another star as of 2012. What would the speed of a spacecraft traveling to this planet have to

be so that the distance to the star is 1.50×10^2 ly in the reference frame of the people on the spacecraft?

67. **N** An event has a duration of 1.00 ms in the laboratory frame. How much time passes during this event for an observer moving past the lab at a speed of $0.733c$?

68. **N** The energy released in the first peacetime nuclear weapon test at Bikini Atoll in 1946 was approximately 6.3×10^{16} J. What mass of plutonium-239 would have the same rest mass energy as the amount of energy released in this explosion?

69. **N** While her spacecraft is traveling from the Earth to the Alpha Centauri star system at a speed of $0.910c$, Lea exercises by doing midair cartwheels, each taking 4.20 s to complete. According to an observer on the Earth, how long does each of Lea's cartwheels take to complete?

70. **C** Consider Example 39.1 (page 1270). Imagine one more scenario in which the ball hangs from the glass ceiling by a string. Aaron would see the string make an angle with respect to the vertical. What would Aaron conclude about Newton's first law?

71. **N** Joe and Moe are twins. In the laboratory frame at location S_1 (2.00 km, 0.200 km, 0.150 km), Joe shoots a picture for a duration of $t = 12.0\ \mu$s. For the same duration as measured in the laboratory frame, at location S_2 (1.00 km, −0.200 km, 0.300 km), Moe also shoots a picture. Both Joe and Moe begin taking their pictures at $t = 0$ in the laboratory frame. Determine the duration of each event as measured by an observer in a frame moving at a speed of 2.00×10^8 m/s along the x axis in the positive x direction. Assume that at $t = t' = 0$, the origins of the two frames coincide.

72. **A** Start with the Lorentz transformation for x and x', and show that the Lorentz factor is given by Equation 39.17:

$$\gamma = \frac{1}{\sqrt{1 - (v_{rel}/c)^2}} = \frac{1}{\sqrt{1 - \beta^2}}$$

Hint: Imagine that at $t = t' = 0$, the origins of both coordinate systems overlap (Fig. 39.7A) and at that instant the observer in the primed frame fires a flashbulb, sending a photon to the right parallel to the x and x' axes.

73. **N** On its way back to the Earth, an interplanetary spaceship with a proper length of 180.0 m is observed by flight controllers to pass a stationary beacon on the Moon in 0.550 μs. What is the speed of the spacecraft in the Earth's reference frame?

74. **N** What is the speed of a proton that has a momentum 4.00 times greater than its classical momentum?

75. **N** A light in a spaceship moving at a relativistic speed appears red (wavelength 675 nm) to the passengers in the ship. But it appears yellow (wavelength 575 nm) to an observer on the Earth. What is the speed (in terms of the speed of light c) of the spaceship? Is it coming toward the Earth or moving away from the Earth?

76. **N** The Sun radiates energy with a power of about 3.8×10^{26} W. **a.** What is the rest mass energy of the Sun? **b.** How much mass is "lost" per second by the Sun if the power radiated is due to the conversion of mass to energy? **c.** Given your answer to part (b), what percentage of the Sun's mass is radiated each second?

77. **N** Far into the future, trains on the Earth may travel at speeds approaching the speed of light. Imagine such a train traveling at $0.875c$ and having a proper length of 145 m. The first of several tunnels along the train's path is 80.0 m long. Is the train ever completely in the tunnel according to an observer standing on a trackside platform near the tunnel? If so, by how much? If not, how much longer than the tunnel is the train?

78. **N** In December 2012, researchers announced the discovery of ultramassive black holes, with masses up to 40 billion times the mass of the Sun (seen as the bright spot at the center of the galaxy near the center of Fig. P39.78). **a.** What is the Schwarzschild radius of a black hole that has a mass 40 billion times that of the Sun? **b.** Suppose this black hole is 1.3 billion ly from the Earth. What is the angular radius of a galaxy that is 1.7 billion ly behind it, as viewed from the Earth?

FIGURE P39.78

Problems 79 and 80 are paired.

79. **A** A rectangular tank filled with a liquid of density ρ' is in a moving frame with its edges parallel to the coordinate axes. What is the density of the liquid as measured in the laboratory frame? Assume the relative speed along the x axis between the two frames is v_{rel}.

80. **A** A rectangular tank filled with a liquid of density ρ' is in a moving frame with its edges parallel to the coordinate axes. If the liquid in the tank is evaporating, show that in both the laboratory and the primed frames, the rate at which mass is being lost from the tank by evaporation is the same.

81. **N** How much work is required to increase the speed of a helium nucleus ($m_{He} = 6.64 \times 10^{-27}$ kg) from $0.750c$ to $0.990c$?

82. **A** Show that the gravitational Doppler shift formula is approximately given by Equation 39.48:

$$\frac{f_{obs} - f_{emt}}{f_{emt}} = \frac{\Delta f}{f_{emt}} \approx -\frac{gh}{c^2}$$

Hint: Start by considering the scenario illustrated in Figure 39.22.

83. **N** A heavy nucleus breaks up into two pieces while in motion. The first piece, of rest mass $m_1 = 1.45\ \text{MeV}/c^2$, is observed to have momentum $\vec{p}_1 = -10.0\ \hat{\imath}\ \text{MeV}/c$, and the second piece, of rest mass $m_2 = 0.850\ \text{MeV}/c^2$, is observed to have momentum $\vec{p}_2 = 5.75\ \hat{\jmath}\ \text{MeV}/c$ by an observer in the laboratory frame. **a.** What is the speed of the first piece, as viewed by an observer in the lab? **b.** What is the speed of the second piece, as viewed by an observer in the lab?

84. **N** In radioactive decay, a parent nucleus decays into one or more daughter nuclei. During one such reaction, the parent nucleus with rest mass 8.99×10^{-27} kg decays into two daughter nuclei with rest masses m_1 and m_2 moving with velocities $v_1 = -0.915c\hat{\imath}$ and $v_2 = 0.785c\hat{\imath}$. What are the rest masses of the two daughter nuclei?

85. **N** A train moving with constant velocity $0.775c$ approaches a station where a large mirror is hung next to the tracks. A playful boy on the train aims a laser pointer at the mirror when the train is 10.0 km from the station, as measured by observers at the station, and the beam is reflected by the mirror back toward the train. **a.** According to the boy, how much time elapses from when he switches on the laser pointer until the beam gets back to him? **b.** According to observers at the station, how much time elapses from when the boy switches on the laser pointer until the beam gets back to him?

40 The Origin of Quantum Physics

Key Questions

How is quantum physics different from classical physics?

What is the wave-particle duality?

Underlying Principles

1. Planck's **quantum theory** and model of black-body radiation
2. **Wave-particle duality** of light and matter

Major Concepts

1. Black body
2. Photon
3. Compton shift
4. De Broglie wavelength

By the late 1800s, many physicists thought that all the major questions of nature had been answered by Newton's laws of mechanics and Maxwell's equations for electricity and magnetism. Even controversial topics, such as the nature of light, had been worked out. They were satisfied that light is a wave, as predicted by Maxwell's equations and shown experimentally by Young and Arago. They believed that only a few minor questions remained to be answered and that experimentalists would have nothing else to do but make more accurate measurements, such as measuring the speed of light to six significant figures.

But as we saw in Chapter 39, they couldn't have been more wrong. Michelson and Morley didn't believe that their experiment would lead to groundbreaking results. They thought that they were just measuring the speed of the Earth with respect to the ether. They certainly didn't expect that their failure to find the ether would be explained by Einstein's theory that observations of space and time depend on the motion of the observer.

Those who thought that the major questions of nature had been answered were in for another big surprise in the early 1900s. In trying to understand something as seemingly mundane as the glow given off by a warm object, such as a wood-burning stove, another radical departure from classical physics was discovered. Today, we call this new departure the theory of *quantum physics* (or *quantum mechanics*).

40-1 Another Modern Idea

The basic idea behind quantum physics is that energy is not necessarily continuous; instead, energy comes in small packets called *quanta*. We have already seen a similar idea in our study of electromagnetism. Electric charge is quantized in that the elementary charge e is carried by either a proton ($+e$) or an electron ($-e$). An object that is positively charged carries an excess number n of protons, and its charge is $q = ne$, where n is an integer. Because it is impossible to find an object with a fractional number of excess electrons or protons, an object cannot have a charge of $\pm 1.75e$, for example. Since the elementary charge is so small ($e = 1.6 \times 10^{-19}$ C), the charge of a macroscopic object (at even a few nanocoulombs) appears continuous: The addition or subtraction of a single elementary charge has very little effect compared to the total charge.

It has been discovered that protons are not fundamental particles; instead they are made up of quarks. Quarks have a fractional charge (Chapter 43).

The same is true for quantized energy. When we consider the energy of a baseball in flight, for example, to the batter the energy seems not quantized but continuous. Quantum physics is most applicable on the submicroscopic scale of atoms. So perhaps it is not surprising that quantum physics plays an important role in many of today's electronic devices, two of which are discussed in this chapter's case study.

CASE STUDY Digital Cameras and Electron Microscopes

In this chapter, we'll return to the controversy over whether light should be modeled as a wave or a particle, and we'll find that it should be modeled as both. In addition, matter (such as electrons, protons, and even people) can be modeled both as waves and as particles. It might seem that how we choose to model a physical phenomenon is only important to theorists, who need words to describe the phenomenon. However, models lead to practical devices as well. For example, your digital camera is based on modeling light as a particle. By contrast, the high-powered electron microscope used in fields such as medicine, biology, and crystallography is based on modeling the electron as a wave.

CONCEPT EXERCISE 40.1

a. An atom is missing one electron. What is its net charge?
b. A pith ball's charge is 9 nC. How many electrons is it missing?
c. Use these two examples to explain why charge seems continuous on the macroscopic level of a pith ball but quantized on the level of a single atom.

40-2 Black-Body Radiation and the Ultraviolet Catastrophe

BLACK BODY ✪ Major Concept

The first steps leading to quantum mechanics were taken by the German physicist Max Karl Ernst Ludwig Planck (1858–1947). Like scientists before him, Planck was interested in the thermal radiation given off by an idealized object known as a **black body**. A black body is both a perfect emitter and a perfect absorber of radiation. So, for example, when a black body is cooler than its surroundings, by comparison it appears dark or black because it absorbs radiation of every color (or frequency).

Although no real object is a black body, many objects are well modeled as such. For example, consider a small hole in a cavity of an opaque vessel, such as an old-fashioned keyhole in a closet door. (The keyhole, not the closet, is the black body.) Any light that passes into the hole is reflected off the interior surfaces many times, but very little, if any, light may escape the hole (Fig. 40.1). So the hole has effectively absorbed all the radiation impinging on it and is well modeled as a black body.

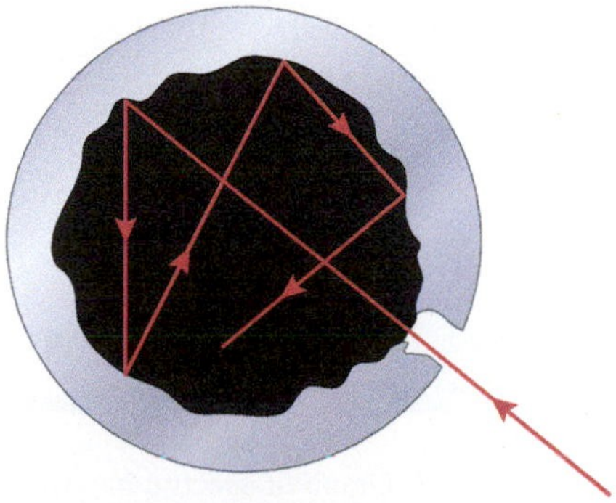

FIGURE 40.1 The hole in a cavity is well modeled as a black body.

A black body is also a perfect emitter, giving off electromagnetic waves at all frequencies. The power P emitted as thermal radiation (another name for black-body

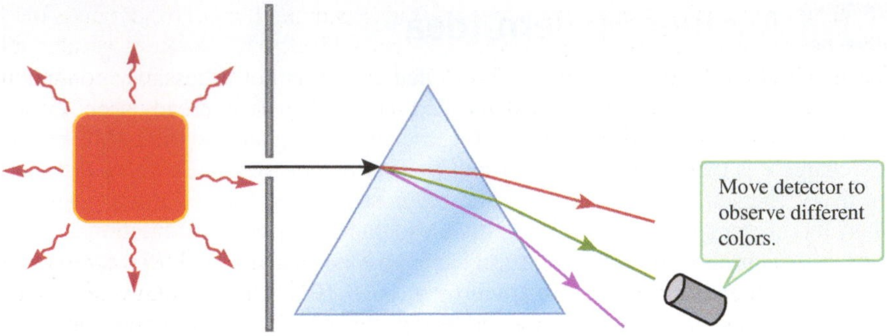

FIGURE 40.2 A prism is used to disperse the light from a hot lump of coal. By moving the detector, the intensity at each color may be observed.

Recall σ is Stefan–Boltzmann's constant and is found on the inside back cover.

radiation) depends on the temperature T and surface area A of the object: $P = \sigma\varepsilon AT^4$ (the Stefan–Boltzmann equation with $P = Q/\Delta t$, Eq. 21.34). The emissivity ε in this equation is a measure of a substance's ability to emit or absorb radiation. So a black body is an ideal object whose emissivity is one: $\varepsilon = 1$. A real object's emissivity is always between 0 and 1, but some objects and materials, such as coal ($\varepsilon = 0.95$), come very close to the ideal. For example, charcoal is well modeled as a black body, and Figure 21.30 (page 642) shows that when charcoal is cooler than its surroundings, it appears black. When it is hotter than its surroundings, however, it appears to glow brightly. For the rest of this section, we'll model such real objects as black bodies. (Often, real objects that are nearly black bodies are referred to as "gray bodies." However, in this book we'll use the term *black body* with the understanding that no real object is an ideal black body.) When an object is in thermal equilibrium with its surroundings, the thermal power it absorbs must equal the thermal power it emits.

When a black body gives off radiation, the emitted spectrum contains a complete rainbow of colors. (The white light source in Fig. 36.16 is well modeled as a black body, but the compact fluorescent bulb in Fig. 36.17 is not.) The intensity (power per unit area) emitted at each color depends on the black body's temperature. Suppose you wished to measure the intensity of radiation emitted by a black body, such as a lump of hot coal, as a function of wavelength. The hot coal emits visible light. So you might place the lump of coal behind a small slit, so that you get a nice beam of light. Then you could use a device such as a prism or a diffraction grating to disperse the light into its color components (Fig. 40.2). A detector could measure the intensity of the radiation it receives through the prism. By moving the detector to different positions, you could measure the intensity of light as a function of color or wavelength. To be more precise, the detector is placed at one position at a time and has a finite width, so it measures the intensity of light at λ (due to the position of the detector) with range of wavelengths $d\lambda$ (due to the detector's width). So the detector measures the intensity per unit wavelength band. There are several names for intensity per wavelength; in this book we use the term *spectral intensity* and the symbol I_λ

A **black-body curve** is a graph of emitted spectral intensity I_λ versus wavelength. Figure 40.3 shows several black-body curves for temperatures ranging from 6000 K to 12,000 K. Notice that the warmest black body (which gives off the most power) has the shortest peak wavelength. The peak wavelength λ_{max} of the black-body curve is given by Wien's law:

$$\lambda_{max} T = 2.898 \times 10^{-3}\,\text{m}\cdot\text{K} = 2.898 \times 10^{6}\,\text{nm}\cdot\text{K} \qquad (40.1)$$

where T is the black body's temperature in kelvins. Figure 40.3 also shows a warm black-body curve with peak intensity in the green part of the visible spectrum. Because the curve is fairly flat, this black body looks nearly white—something like the Sun. Compare this curve for a warm black body to that shown for a hot black body, for which the most intense color is indigo. If you were to look at such a black body, it would look bluish, like the hot star Rigel in the constellation Orion (Fig. 40.4).

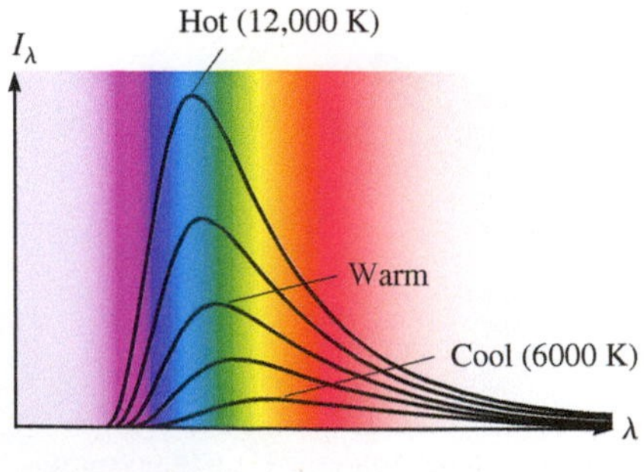

FIGURE 40.3 Graph of spectral intensity versus wavelength for five black bodies of different temperatures. All the curves have the same shape, differing only in scale.

The total intensity over all wavelengths (the area under the curve) depends on the temperature of the black body, so that a warmer black body emits a greater total intensity. (Often "total" is dropped from the term *total intensity* when the meaning can be understood from the context.) The total intensity is found by integrating the spectral intensity over all wavelengths:

$$I = \int_0^\infty I_\lambda d\lambda \qquad (40.2)$$

Notice that the symbol for total intensity I has no subscript. The total intensity depends on the absolute temperature of the object. We find the total intensity of a thermal radiator by dividing the power it emits (Eq. 21.34) by its surface area:

$$I(T) = \varepsilon\sigma T^4 \qquad (40.3)$$

As usual, for a black body the emissivity $\varepsilon = 1$.

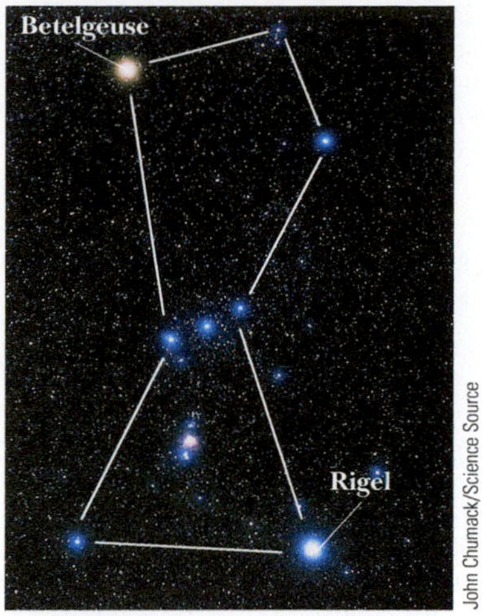

FIGURE 40.4 The constellation Orion has two very bright stars—Rigel and Betelgeuse. Rigel's peak wavelength is actually in the ultraviolet; it looks blue to us because our eyes are not sensitive to UV.

Modeling Black-Body Curves

Before Planck, scientists noticed that all black-body curves (Fig. 40.3) had the same shape (with different scalings). But no one came up with a model for matter and radiation that would reproduce the observed curves, although two important attempts were made. One attempt was made by the German physicist Wilhelm Wien (1864–1928), who noticed that black-body curves look much like the Maxwell–Boltzmann distribution for the speeds of particles in gas (Fig. 20.8, page 589). Wien reasoned that the temperature of a black body is related to the motion of the particles (such as the atoms or molecules) inside the black body. Think of it this way: An object that is in thermal equilibrium with its surroundings absorbs as much thermal energy as it emits. The thermal energy the object absorbs causes an increase in its particles' energy (atoms and molecules). Of course, the particles in a solid object oscillate around their equilibrium positions, and when their energy increases, their oscillations become more vigorous. (We measure this as an increase in the object's temperature.) The particles are made up of electrons and protons. These charged particles (typically electrons) are accelerated by the oscillations, and accelerating charged particles give off radiation. This emitted radiation may keep the black body in thermal equilibrium if the power emitted equals the power absorbed. The important point here is that the motion of the particles in a black body is responsible for the observed black-body radiation, so it made some sense for Wien to think that the Maxwell–Boltzmann distribution is related to the black-body curves. Using the Maxwell–Boltzmann distribution

$$f(v) = 4\pi\left(\frac{m}{2\pi k_B T}\right)^{3/2} v^2 e^{-(mv^2/2k_B T)} \qquad (20.19)$$

as a guide, Wien derived a black-body intensity formula that was a good fit for the data at short wavelengths (Fig. 40.5). (At this point, you may wish to review Section 20-4 on the Maxwell–Boltzmann distribution function.)

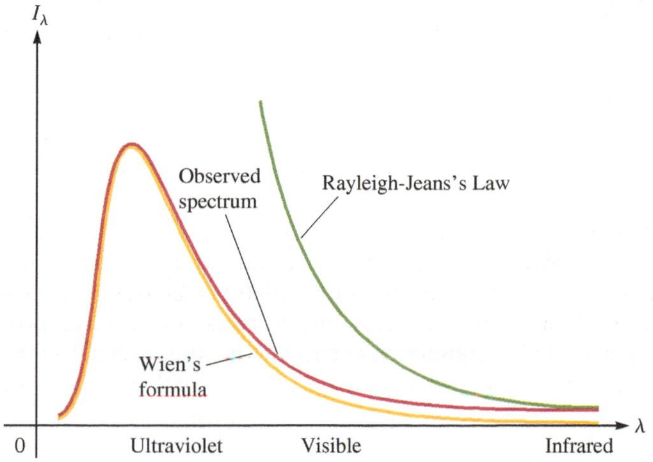

FIGURE 40.5 Neither Wien's formula nor Rayleigh–Jeans's law fits the black-body curve observed for an object at a given temperature. Wien's formula is best at short wavelengths, and Rayleigh–Jeans's law at long wavelengths.

Black-body curves may be observed by heating up an oven with a very small hole in one of its sides. The hole is similar to the one shown in Figure 40.1, but now the cavity (an oven) is hotter than its surroundings. The oven may be heated to different temperatures, and light emitted from the hole is detected using a prism (Fig. 40.2) to construct a black-body curve. Perhaps thinking about such an oven inspired the British scientists Lord Rayleigh (1842–1919) and Sir James Jeans (1877–1946) to come up with some ideas for black-body radiation. Instead of thinking about the atoms that make up the black body, as Wien did, Rayleigh focused on the electromagnetic waves in the cavity. He reasoned that the radiation inside the oven reflects from the walls and sets up standing waves. Like a wave on a string, each standing wave has a fundamental mode and an infinite number of higher harmonic modes. The spectral intensity is measured per waveband $d\lambda$. So Rayleigh found an expression for the number of modes per volume with wavelengths between λ and $\lambda + d\lambda$. According to classical mechanics, each mode contributes an average energy of $k_B T$. From the number density of modes per waveband and the average energy per mode, the energy density per waveband in the cavity can be found. And from this average energy density, Rayleigh–Jeans's law for the spectral intensity as a function of wavelength

$$I_\lambda = \frac{2\pi c}{\lambda^4} k_B T \tag{40.4}$$

is found. Rayleigh–Jeans's law is shown in Figure 40.5.

Take a close look at Figure 40.5, and you will see that Rayleigh–Jeans's formula misses one of the most important features of the actual black-body curves—the turnover. A black-body curve has a peak, but Rayleigh–Jeans's curve continues to climb at short wavelengths. To see what this means, imagine that you want to cook over a campfire. A campfire's radiation peaks somewhere in the infrared or visible red part of the spectrum. According to Rayleigh–Jeans's formula, though, even the most modest campfire would give off high-intensity radiation in the ultraviolet, X-ray, and gamma-ray range. You would suffer much more than a bad sunburn from such radiation. Because everyone knew that black-body radiation did not follow this intensity distribution, Rayleigh–Jeans's formula was said to result in an **ultraviolet catastrophe**.

Planck's Solution

In 1900, the German physicist Max Planck solved the ultraviolet catastrophe empirically by adjusting Wien's formula to fit the black-body curves at both long and short wavelengths. **Planck's formula** for the spectral intensity I_λ of a black body is:

$$I_\lambda(\lambda,T) = \frac{2\pi hc^2}{\lambda^5(e^{hc/\lambda k_B T} - 1)} \tag{40.5}$$

Planck's formula depends on the temperature T of the black body. For any single curve in Figure 40.3, the temperature is constant. Each curve in Figure 40.3 is a graph of I_λ versus wavelength λ and is described by Equation 40.5 for a particular T. There are three physical constants in Planck's formula: the speed of light c, Boltzmann's constant k_B, and Planck's constant h with the value

$$h = 6.626 \times 10^{-34}\,\text{J}\cdot\text{s} = 4.136 \times 10^{-15}\,\text{eV}\cdot\text{s}$$

At first, Planck's formula was merely an empirical fit to the data. He knew that the formula described the observed black-body curves, but he didn't have a physical model or theory to explain why the formula worked. He wanted to start from some fundamental principle and derive Equation 40.5.

So Planck—like Wien—started with the kinetic theory of gases. He found that he could derive his formula if he (1) modeled the particles in a black body as oscillators and (2) postulated that the energy of each oscillator is **quantized**, existing in discrete bundles instead of being continuous. In this postulate, the smallest possible amount of energy E_{min} is called a **quantum of energy** and is proportional to the frequency of the oscillator:

$$E_{\text{min}} = hf \tag{40.6}$$

If a molecule in the black body oscillates at frequency f, then its energy is an integer multiple of E_{min}:

$$E = nE_{\text{min}} = nhf \qquad n = 1, 2, 3 \ldots \tag{40.7}$$

PLANCK'S QUANTUM THEORY

Underlying Principle

The idea that this energy is quantized is called **Planck's quantum theory,** and it marks a radical departure from the classical idea of energy. According to the classical model, an oscillator can have any energy. According to the quantum model, by contrast, an oscillator's energy comes in discrete bundles. However, no one realized that Planck had made such a radical statement until Einstein used the idea to explain the photoelectric effect (see Section 40-3).

To illustrate Planck's quantum theory, consider an analogy with two types of radios. The old radio in Figure 40.6A is tuned by turning a knob. In theory, the knob can be set to any position, and so the radio can be tuned to any frequency in its range. This is analogous to the classical model of an oscillator. According to classical mechanics, an oscillator can have any energy, just as the old radio can be tuned to any frequency. The radio in Figure 40.6B is a more contemporary digital radio. It is tuned by pressing a button. Each time the button is pressed the frequency is adjusted by discrete amount. So the frequency of a digital radio can only be set to certain discrete frequencies. This is analogous to Planck's quantum theory for an oscillator. According to this theory, an oscillator's energy can only take on certain discrete values.

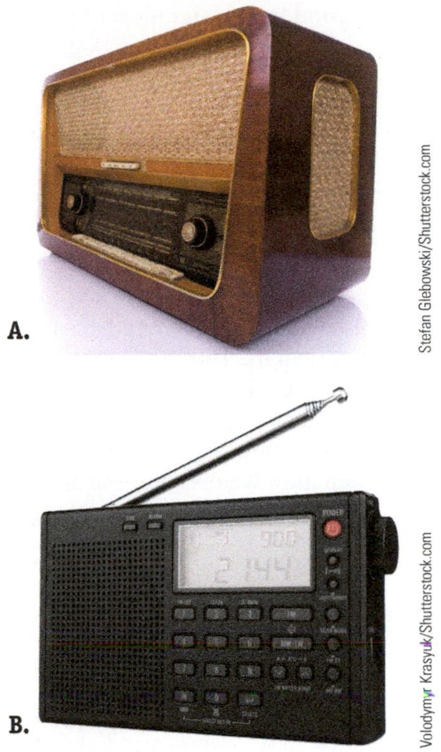

FIGURE 40.6 **A.** An analog radio can be tuned to any frequency. **B.** A digital radio can only be tuned to discrete frequencies.

CONCEPT EXERCISE 40.2

Planck's idea that energy is quantized is a radical departure from classical physics. You may find the radio analogy (Fig. 40.6) useful, but like all analogies this one is not perfect. Describe the ways in which this analogy fails and the ways in which it succeeds.

EXAMPLE 40.1 How hot is that flame?

We normally associate the color blue with cold objects and the color red with hot ones, but this association probably arises from our thinking of water as blue and cool and fire as red and hot. In reality, a blue black body is hotter than a red one. Take a close look at the flame in Figure 40.7, and you will see the flame is blue near the source and red on the far end. Model the flame as a black body, and estimate the temperature at each end. In the check and think step, consider that a candle's flame has a temperature of about 1000°C.

FIGURE 40.7

INTERPRET and ANTICIPATE

We expect to get a higher temperature for the blue part of the flame.

SOLVE

From the photos and Figure 34.11 (page 1101), we estimate the wavelengths based on the color of each flame. (It is okay to keep an extra significant figure for this step.)

Blue: $\lambda_{\text{blue}} = 475\,\text{nm}$

Red: $\lambda_{\text{red}} = 675\,\text{nm}$

Assume that the observed color is near the peak of the black-body spectrum for each flame; then the temperature can be estimated from Wien's law. Our estimates are only good to two significant figures.

$$\lambda_{max} T = 2.898 \times 10^6\,\text{nm}\cdot\text{K} \tag{40.1}$$

$$T = \frac{2.898 \times 10^6\,\text{nm}\cdot\text{K}}{\lambda_{max}}$$

$$\text{Blue: } T_{blue} = \frac{2.898 \times 10^6\,\text{nm}\cdot\text{K}}{475\,\text{nm}} \approx 6.1 \times 10^3\,\text{K}$$

$$\text{Red: } T_{red} = \frac{2.898 \times 10^6\,\text{nm}\cdot\text{K}}{675\,\text{nm}} \approx 4.3 \times 10^3\,\text{K}$$

CHECK and THINK

We found that the blue part is hotter than the red part. This result is satisfying because the blue part is closer to the source, and we expect the flame to be hotter near the source. However, there must be something wrong because the temperatures we found (about 5000°C) are considerably greater than the 1000°C we were expecting for a candle's flame. The problem is that we assumed the flame could be modeled as a black body, but a candle's flame is more complicated than a simple black body. The color (and temperature) variation depends on the chemical reactions that occur in various parts of the flame. In Section 42-3, we see how the structure of atoms affects the radiation we observe; not all radiation is well-modeled as black body radiation.

EXAMPLE 40.2 Cosmic Background Radiation

As we learned in Section 34-5, radiation was generated early in the history of the Universe. This radiation may be observed today. Our observations show that this radiation is well modeled as a black body with a temperature of 2.725 K. Sketch a graph of spectral intensity as a function of wavelength for this cosmic background radiation.

INTERPRET and ANTICIPATE

This problem gives us a chance to use Planck's formula $I_\lambda(\lambda,T) = \dfrac{2\pi hc^2}{\lambda^5(e^{hc/\lambda k_B T} - 1)}$ (Eq. 40.5).

It is always helpful to plot complicated functions, as we are doing here.

SOLVE

Substitute all the constants into Planck's formula, as well as the given temperature. A spreadsheet program is helpful. A few values have been provided so that you can check your own results.

λ (m)	I_λ (W/m³)
1.00×10^{-4}	3.47×10^{-19}
2.00×10^{-4}	3.57×10^{-9}
3.00×10^{-4}	3.24×10^{-6}
4.00×10^{-4}	6.38×10^{-5}
5.00×10^{-4}	2.97×10^{-4}

Plotting spectral intensity on the vertical axis in milliWatts per meter cubed and wavelength on the horizontal axis in millimeters (Fig. 40.8) produces a curve that looks very much like the black-body curves in Figure 40.3.

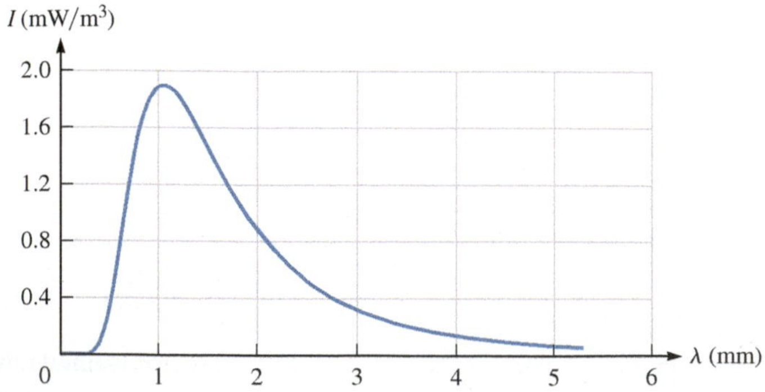

FIGURE 40.8

CHECK and THINK

To check our work, let's use Wien's law (Eq. 40.1) to find the peak wavelength for a black body with a temperature of 2.725 K.

$$\lambda_{max} T = 2.898 \times 10^{-3} \text{ m} \cdot \text{K} \quad (40.1)$$

$$\lambda_{max} = \frac{2.898 \times 10^{-3} \text{ m} \cdot \text{K}}{T} = \frac{2.898 \times 10^{-3} \text{ m} \cdot \text{K}}{2.725 \text{ K}}$$

$$\lambda_{max} = 1.06 \times 10^{-3} \text{ m} = 1.06 \text{ mm}$$

The peak wavelength we just found is consistent with our graph, which you can see peaks at just a little more than 1 mm. Further, we learned in Section 34-5 that the cosmic background radiation is observed in the microwave part of the electromagnetic spectrum. Compare Figure 40.8 to Table 34.2, and you will see that our entire graph falls in the microwave part of the spectrum. Finally, we note that cosmologists measure the background radiation, and then fit the spectrum to Planck's function to arrive at the corresponding black-body temperature of 2.725 K. (This is the reverse of what we did in this example.) Often, this temperature is referred as *the temperature of the Universe*, which may be a little misleading because not everything in the Universe is at the same temperature.

EXAMPLE 40.3 The Ultraviolet Catastrophe

Find the total intensity of a black body as predicted by Rayleigh–Jeans's law (Eq. 40.4).

INTERPRET and ANTICIPATE

We know that Rayleigh–Jeans's law predicts at short wavelengths a higher spectral intensity than is observed (Fig. 40.5). So we expect that the total intensity will be greater than what is observed.

SOLVE

The total intensity comes from integrating the spectral intensity over all wavelengths (Eq. 40.2).

$$I = \int_0^\infty I_\lambda d\lambda \quad (40.2)$$

Substitute Rayleigh–Jeans's formula (Eq. 40.4).

$$I = \int_0^\infty \left(\frac{2\pi c}{\lambda^4} k_B T \right) d\lambda$$

Pull out the constants, integrate, and substitute limits.

$$I = 2\pi c k_B T \int_0^\infty \frac{d\lambda}{\lambda^4} = -\frac{2\pi c k_B T}{3} \frac{1}{\lambda^3} \bigg|_0^\infty$$

$$I = \frac{2\pi c k_B T}{3} \left(\frac{-1}{\infty} + \frac{1}{0} \right) \rightarrow \infty$$

CHECK and THINK

We just found that the total intensity predicted by Rayleigh–Jeans's formula is infinite, which, as expected, is greater than the total intensity observed in a black body. Had we integrated Planck's law instead, we would have found that the total intensity is given by $I(T) = \sigma T^4$ where $\sigma = \frac{2\pi^5 k_B^4}{15 h^3 c^2}$, a finite value consistent with Equation 40.3, when $\varepsilon = 1$.

EXAMPLE 40.4 A Star's Radius

Have you ever wondered how astronomers measure the size of distant stars? They are too far away to resolve into a disk, so their radius is inferred from their luminosity and temperature. The star's temperature is easy to estimate from its spectrum, which looks like a black-body spectrum, but with some particular dark lines due to photon absorption in the star's atmosphere. These lines are used to estimate the temperature. The star's luminosity is another name for its total emitted power. Luminosity is harder to estimate than temperature because it comes from measuring the star's brightness and estimating its distance. Consider a star that has the same temperature as the Sun but is 1.0×10^4 times brighter than the Sun. Model the star and the Sun as black bodies, and find the star's radius in terms of the Sun's radius.

INTERPRET and ANTICIPATE

The luminosity or power emitted by a star is easily found from $P = \sigma \varepsilon A T^4$ (Eq. 21.34). By comparing the luminosity of the star to that of the Sun, we can find the ratio of their radii.

SOLVE

We are modeling the star and the Sun as black bodies, so we set the emissivity ε to 1.	$P = \sigma A T^4$
Both the star and the Sun are spheres. Substitute the surface area of a sphere into the luminosity expression.	$P = 4\pi R^2 \sigma T^4$
Write an expression for the ratio of the star's power to that of the Sun. We have used the usual symbol ⊙ to stand for the Sun, and the symbol * to stand for the star.	$\frac{P_*}{P_\odot} = \frac{4\pi R_*^2 \sigma T_*^4}{4\pi R_\odot^2 \sigma T_\odot^4}$ $\frac{P_*}{P_\odot} = \left(\frac{R_*}{R_\odot}\right)^2 \left(\frac{T_*}{T_\odot}\right)^4$
Use the fact that the star has the same temperature as the Sun to write an expression for the ratio of their radii.	$\frac{P_*}{P_\odot} = \left(\frac{R_*}{R_\odot}\right)^2$ $\frac{R_*}{R_\odot} = \sqrt{\frac{P_*}{P_\odot}}$
Substitute values.	$\frac{R_*}{R_\odot} = \sqrt{1.0 \times 10^4} = 1.0 \times 10^2$ $R_* = 1.0 \times 10^2 R_\odot$

CHECK AND THINK

This very bright star is 100 times larger than the Sun. Such a large star is called a supergiant. Supergiants are "dying stars," which means they have run out of the fuel in their core and are quickly fusing other material before their final demise.

40-3 The Photoelectric Effect

Einstein's other 1905 paper was about Brownian motion.

For Einstein, 1905 was an *annus mirabilis*—a wonderful year—because in that year he published four major papers. Two of these were on special relativity (Sections 39-3 through 39-10), and another was on the quantization of light energy. This paper revived the age-old debate over whether light is best modeled as a wave or as a particle. In the early 1900s, it was generally agreed that the debate had been settled. Maxwell's equations predicted that light is a wave. Interference experiments (Young's double slit) and diffraction experiments (Arago's bright spot in an object's shadow) confirmed the wave model. However, Einstein found that light can also be modeled as a particle known as a *photon*. In this section, we'll explore how Einstein reasoned that light can also be modeled as a particle, and we'll describe the experimental confirmation of that model.

In his paper *Über einen die Erzeugung und Verwandlung des Lichtes betre fenden heuristischen Gesichtspunkt* [*On a Heuristic Viewpoint Concerning the Production and Transformation of Light*], Einstein reasoned that

Similar experiments were conducted before Einstein's theoretical work.

1. according to Planck's quantum theory, a vibrating molecule's energy is quantized;
2. when such a molecule gives off light, it loses energy equal to the light's energy;
3. the energy the molecule loses must be quantized: $\Delta E = nhf$; so
4. the light's energy must be quantized.

As an example, suppose that the vibrating molecule initially had 10 quanta of energy $E_i = 10hf$. Then the molecule gives off light and loses energy so that it has seven quanta of energy $E_f = 7hf$. The light's energy must be equal to the energy lost by the molecule, so the light has three quanta of energy $3hf$.

This argument has broader implications for light. Because all light comes from the motion of charged particles, light's energy must depend on the energy lost by those particles. If the particle's energy is quantized, then the light's energy is quantized (discrete), and it is convenient to think of light as little bundles of energy called **photons**. A photon's energy depends on the frequency of light:

PHOTON Major Concept

$$E = hf \tag{40.8}$$

When a photon is absorbed by an atom (or a molecule), the atom's energy increases by $E = hf$ and the photon ceases to exist.

A test to see if light can be modeled as photons is made possible by a process called the **photoelectric effect**. When light shines on a metallic surface, electrons are sometimes released. You can think of it this way: An orbiting electron is bound to the nucleus of an atom like objects on the Earth are bound to its surface by gravity. If the light delivers enough energy to the electron, the electron may escape the pull of the nucleus. By conducting an experiment that involves the photoelectric effect, we can determine whether the light delivers energy in continuous amounts, the way a wave would, or whether it delivers energy in discrete amounts, as photons.

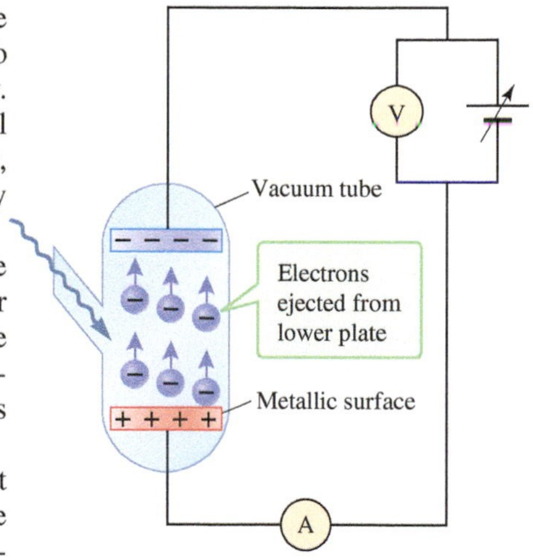

FIGURE 40.9 An experimental setup for testing the photoelectric effect.

Figure 40.9 shows the basic photoelectric effect experiment. The circuit is made up of a capacitor, a variable DC power supply, and a couple of meters. The capacitor is charged so that the upper plate is negative and the lower plate is positive. The capacitor is fully charged when the potential across its plates equals the power supply's terminal voltage, and there is no current in the circuit. The capacitor's plates are enclosed in a vacuum tube with opaque walls.

A small transparent hole in the tube allows light from a monochromatic light source to strike only the lower (positive) plate, freeing some of its electrons. The free electrons are repelled by the upper (negative) plate and attracted to the lower (positive) plate, so most of the electrons quickly fall back to the lower plate. However, a few free electrons with high kinetic energy K make it to the negative plate. When an electron crosses from the lower plate to the upper plate, it loses kinetic energy, and the system (electric field plus electron) gains potential energy. The kinetic energy lost by the electron and gained as the system's potential energy is $|\Delta K| = |\Delta U| = |q\Delta V|$ (Eq. 26.7).

Electrons that make it across the gap in the capacitor will then travel through the circuit's wires and through the ammeter on their way to the lower plate. Their passage through the ammeter means that a current will be detected. The number of electrons that can make it through the gap to the negative plate depends on the potential difference across the capacitor. If that potential difference is low, then many electrons can get through, but if the potential difference is high, then very few can make it. Because the potential difference across the capacitor equals the power supply's voltage, the current through the ammeter will depend on that voltage.

The key to testing the photon model is to adjust the power supply's voltage so that the current just barely stops. Starting with a low voltage, we slowly turn the voltage knob up until the ammeter reads zero current. This voltage is called the *stopping potential* V_0. When the capacitor's potential difference equals the stopping potential, the free electrons with the most kinetic energy K_{max} stop just short of crossing from the lower plate to the upper plate. These electrons are ejected from the lower plate with a kinetic energy equal to the potential energy gained by the system:

$$K_{max} = eV_0 \tag{40.9}$$

Now let's think about a single electron bound to the nucleus of an atom on the surface of the lower plate. Such a bound electron must gain a certain minimum amount of energy called the **work function** W_0 in order to be freed. If the light delivers an amount of energy exactly equal to the work function, then the free electron will have no kinetic energy and will quickly fall back to the lower plate. If the light delivers an amount of energy E_{light} equal to the work function W_0 plus the kinetic energy K_{max} required to get from the lower plate almost to the upper plate, then this electron is one of the few that almost makes the trip across the gap:

$$E_{light} = K_{max} + W_0 \tag{40.10}$$

The work function W_0 is the minimum energy required to free an electron. But most electrons are more tightly bound and require more energy. After they are freed, these tightly bound electrons don't have enough kinetic energy to make it across the gap. So Equation 40.10 holds only for the mostly loosely bound electrons in the lower plate's surface.

Now we come to the measurement that tests Einstein's photon theory against the wave theory. The kinetic energy of the most energetic electrons is calculated from the stopping potential using $K_{max} = eV_0$ (Eq. 40.9), which is related to the energy delivered by the monochromatic light as:

$$eV_0 = K_{max} = E_{light} - W_0 \tag{40.11}$$

The light source can be tuned to adjust the frequency and the intensity independently. Although adjusting f or I may change the delivered energy E_{light}, the work function W_0 depends only on the type of substance that the plate is made of. The wave model and the photon model of light differ in predicting what happens as the light intensity and frequency are varied. In the wave model, the electrons crossing the gap (as measured by the current) depend on the intensity, but not on the frequency of the light. The photon model makes the opposite prediction: The electrons crossing the gap depend on the frequency, not the intensity, of the light. Table 40.1 compares the predictions made by the two models.

The American physicist Robert Andrews Millikan (1868–1953) was convinced that Einstein's photon theory was wrong. He believed that the particle-or-wave

TABLE 40.1 **Results of photoelectric effect experiments as predicted by the wave and photon models of light.**

Wave Model	Photon Model
K_{max} is proportional to I but not affected by f. The energy delivered by an electromagnetic wave is proportional to the wave's intensity, but **not** its frequency (Section 34-6). As the light intensity I is increased, E_{light} and therefore K_{max} will increase, but increasing f will have no effect.	**K_{max} is proportional to f but not affected by I.** Light's energy depends on its frequency, **not** on its intensity: $E = hf$ (Eq. 40.8). Assuming the frequency is above the threshold ($f > W_0/h$), increasing the light's frequency increases the kinetic energy of the electrons. However, increasing the light's intensity increases the number of electrons released but not the kinetic energy of each electron.
At low light intensity, energy is delivered slowly, so it takes some time before a single electron gains sufficient energy to cross the gap. Hence, there will be a time delay before electrons arrive at the upper plate.	**Light intensity does not affect the time for electrons to arrive at the upper plate.** There will in effect be no time delay. As long as a single photon has enough energy to liberate a single electron with enough kinetic energy to cross the capacitor gap, then it will do so, as it does when the intensity is high.
At very low frequency, current will be observed through the ammeter in Figure 40.9. The frequency of the light does not affect the ejection of electrons in the wave model.	**At very low frequency, even if there is no potential difference across the capacitor, there will be no current through the ammeter.** If the light energy is lower than the work function $E_{light} < W_0$, then no electrons will be ejected. This condition corresponds to $f < W_0/h$.

debate had been settled and that light clearly should be modeled as a wave. He spent nearly a decade building and conducting the photoelectric experiment. His results (Fig. 40.10) were consistent with the predictions made in the column on the right in Table 40.1, supporting Einstein's photon model. Despite these results, initially Millikan wasn't convinced. As late as 1927, he put out a textbook that only mentioned Einstein's theories as part of his biographical notes (and that included the ether). We might feel some sympathy for Millikan because it is difficult to accept a new theory. So it may help us to know that other experiments also confirmed the photon model. The next section describes another such historic experiment.

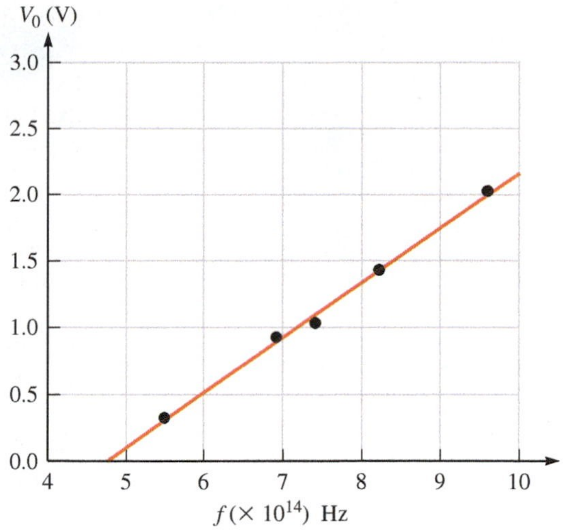

FIGURE 40.10 Stopping potential V_0 (see Fig. 41.5) as a function of frequency based on Millikan's 1916 results (*A Direct Photoelectric Determination of Planck's "h," Physical Review* 7, pp. 355–388). The vertical axis is proportional to K_{max} ($K_{max} = eV_0$). So this linear fit shows that K_{max} is proportional to f as predicted by the photon model (Table 40.1).

CONCEPT EXERCISE 40.3

Millikan used the data in Figure 40.10 to measure Planck's constant. Use the data to estimate h to two significant figures and compare your answer to the accepted value $h = 6.6 \times 10^{-34}\,\text{J}\cdot\text{s}$.

EXAMPLE 40.5 The Energy Range of Visible Photons

We are able to see light in the wavelength range from about 400 nm to about 700 nm. What is the corresponding range in the energy of photons that we can detect with our eyes? Report your answer to one significant figure in joules and in electron-volts.

INTERPRET and ANTICIPATE

An electromagnetic wave's frequency can be found from its wavelength. The photon's energy is directly proportional to the wave's frequency.

SOLVE

Find the frequencies using Equation 34.20. (It is okay to keep an extra significant figure in this step.)

$$\lambda f = c \tag{34.20}$$

$$f = c/\lambda$$

$$\text{Red: } f_{red} = \frac{3.00 \times 10^8\,\text{m/s}}{700 \times 10^{-9}\,\text{m}} = 4.3 \times 10^{14}\,\text{Hz}$$

$$\text{Violet: } f_{violet} = \frac{3.00 \times 10^8\,\text{m/s}}{400 \times 10^{-9}\,\text{m}} = 7.5 \times 10^{14}\,\text{Hz}$$

Find the corresponding energies from the frequencies.

$$E = hf \tag{40.8}$$

$$\text{Red: } E_{red} = (6.626 \times 10^{-34}\,\text{J}\cdot\text{s})(4.3 \times 10^{14}\,\text{Hz}) = 3 \times 10^{-19}\,\text{J}$$

$$E_{red} \approx 2\,\text{eV}$$

$$\text{Violet: } E_{violet} = (6.626 \times 10^{-34}\,\text{J}\cdot\text{s})(7.5 \times 10^{14}\,\text{Hz}) = 5 \times 10^{-19}\,\text{J}$$

$$E_{violet} \approx 3\,\text{eV}$$

CHECK and THINK

So we are able to detect photons with energies between approximately 2 and 3 eV.

FIGURE 40.11 A charge-coupled device (CCD) based on modeling light as photons. A CCD creates a digital image by collecting electrons that are excited into a higher energy state by the photons that strike the detector.

CASE STUDY Digital Cameras

As difficult as it may be to accept that light is modeled as particles, the design of today's digital cameras is based on the photon model of light. Some digital cameras use a semiconductor detector known as a charge-coupled device (CCD), such as the one shown in Figure 40.11. A CCD is made of a two-dimensional array of picture elements known as *pixels*. (Your camera's CCD may have around a million pixels.) When a photon strikes a pixel, an electron is excited into the semiconductor's conduction band. (Electrons in the conduction band are "free" to move throughout the conductor, much as conduction electrons are free to move throughout a conductor. See Chapter 28.) These electrons are collected and processed to form an image. The number of electrons collected from each pixel is proportional to the number of photons striking it.

Digital cameras are convenient and a lot of fun because you can see the image just moments after taking a picture. These cameras also play a very important role in contemporary astronomy. Early astronomers had to sketch what they saw, so the human eye was the original astronomical detector. But the eye can only detect about 1 out of 100 photons; we say that the human eye has a quantum efficiency of 1%. In the mid-1800s, astronomers began to take photographs using photographic plates, which are about equally as good at detecting photons as the human eye. But since the late 1900s, CCD cameras have been commonly used in astronomy. Not only do CCDs have almost 100% quantum efficiency, they also offer other advantages:

1. They are sensitive over a wide range of photon energies, from the infrared to X-rays.
2. They have a wide dynamic range, meaning that they can differentiate between very faint and very bright objects.
3. They have a linear response, meaning that if the number of photons doubles, the signal doubles.

CCDs are so important in astronomy today that it is probably fair to say that all contemporary astronomical images you have seen were taken with a CCD. So not only was the photon model of light revolutionary, but CCDs designed on the basis of that model have revolutionized our view of the universe.

40-4 The Compton Effect

One of the most important experiments confirming Einstein's photon model was carried out in 1923 by the American physicist Arthur Holly Compton (1892–1962). Four years later, Compton shared the Nobel Prize with the British physicist Charles Wilson (1869–1959) for this experiment, which discovered what is now known as the **Compton effect**. Compton was investigating the scattering of X-rays from elements such as carbon. He found two surprising results: (1) The scattered X-rays have a longer wavelength (lower frequency) than the original X-rays, and (2) the wavelength (or frequency) of the scattered X-rays depends on their scattering angle. These results are surprising because according to classical physics, scattering does not change the frequency of electromagnetic radiation.

Compton's work confirmed the photon model because he could explain his two experimental results if he treated the scattering event as an elastic collision between an incoming X-ray photon carrying a discrete amount of momentum and energy, and a free electron initially at rest. The electrons may be scattered at high speeds, so Compton needed the results from special relativity to analyze their motion after the collision. Provided the X-rays were treated as particles (photons), Compton could apply conservation of energy and momentum, as for any other elastic collision, to derive an expression for the change in the X-ray's wavelength.

DERIVATION The Relationship between Relativistic Energy and Momentum

Compton needed a useful relationship between total relativistic energy and momentum:

$$E^2 = p^2c^2 + m_{\text{rest}}^2c^4 \tag{40.12}$$

We will show that this relationship is true.

We can express any particle's momentum in terms of its energy. Start by writing the famous Equation 39.44 in terms of the rest mass and the Lorentz factor γ. (Use Equation 39.39, $m = \gamma m_{\text{rest}}$.)	$E = mc^2$ (39.44) $E = \gamma m_{\text{rest}}c^2$
We need to use two algebraic tricks. First, square both sides.	$E^2 = \gamma^2 m_{\text{rest}}^2 c^4$
Second, add $\gamma^2 m_{\text{rest}}^2 c^2 v^2$ and subtract $\gamma^2 m_{\text{rest}}^2 c^2 v^2$ on the right side.	$E^2 = \gamma^2 m_{\text{rest}}^2 c^4 + \gamma^2 m_{\text{rest}}^2 c^2 v^2 - \gamma^2 m_{\text{rest}}^2 c^2 v^2$ $E^2 = \gamma^2 m_{\text{rest}}^2 c^2(c^2 + v^2 - v^2)$ $E^2 = \gamma^2 m_{\text{rest}}^2 c^2 v^2 + \gamma^2 m_{\text{rest}}^2 c^2(c^2 - v^2)$
Use $\gamma = \frac{1}{\sqrt{1 - (v/c)^2}}$ (Eq. 39.41) to rewrite the second term on the right so that the Lorentz factor cancels out.	$E^2 = \gamma^2 m_{\text{rest}}^2 c^2 v^2 + \gamma^2 m_{\text{rest}}^2 c^4[1 - (v/c)^2]$ $E^2 = \gamma^2 m_{\text{rest}}^2 c^2 v^2 + \gamma^2 m_{\text{rest}}^2 c^4\left(\frac{1}{\gamma^2}\right)$ $E^2 = \gamma^2 m_{\text{rest}}^2 v^2 c^2 + m_{\text{rest}}^2 c^4$ (1)
Use the magnitude of relativistic momentum, $p = \gamma m_{\text{rest}} v$ (Eq. 39.40) to rewrite the first term on the right in Equation (1).	$\gamma^2 m_{\text{rest}}^2 v^2 c^2 = (\gamma m_{\text{rest}} v)^2 c^2 = p^2c^2$ $E^2 = p^2c^2 + m_{\text{rest}}^2 c^4$ ✓ (40.12)

COMMENTS

Equation 40.12 is another expression for momentum and energy that holds under special relativity for any particle. Compton used it to derive an expression for the momentum of a photon.

Starting with Equation 40.12, we can find a simple expression for the momentum of a photon by noting that its rest mass is zero.

$$E_{\text{photon}}^2 = p_{\text{photon}}^2 c^2 + 0 = p^2c^2$$

$$E_{\text{photon}} = p_{\text{photon}}c \tag{40.13}$$

Solving for momentum and substituting $E_{\text{photon}} = hf$ (Eq. 40.8) for the energy of a photon results in

$$p_{\text{photon}} = \frac{E_{\text{photon}}}{c} = \frac{hf}{c}$$

Since $f = c/\lambda$ (Eq. 34.20),

$$p_{\text{photon}} = \frac{h}{\lambda} \tag{40.14}$$

According to Equation 40.14, a photon's momentum depends only on its wavelength (or frequency). With this equation in hand, we are ready to examine the Compton effect.

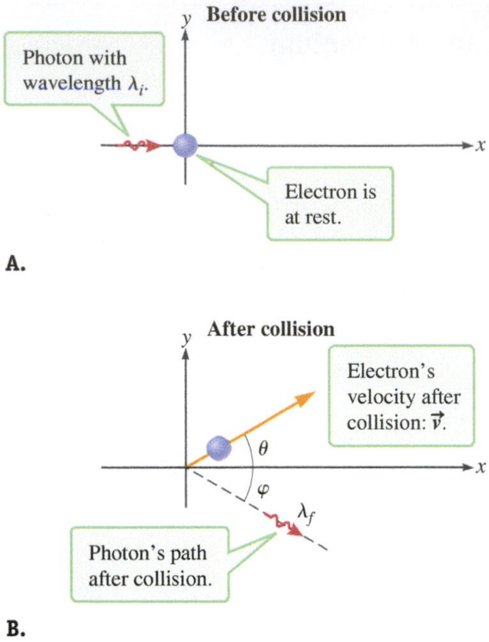

Figure 40.12 illustrates the event that produces the Compton effect. A photon of wavelength λ_i collides with an electron that is initially at rest (Fig. 40.12A). After the collision, the electron moves off at angle θ with respect to the x axis, while the scattered photon moves off at angle φ with respect to the x axis, with wavelength $\lambda_f \geq \lambda_i$ (Fig. 40.12B).

We seek to understand both of Compton's results. (1) Why does the photon's wavelength increase (and frequency decrease)? (2) Why does the scattered photon's wavelength (and frequency) depend on the scattering angle? As stated earlier, we find the answers by modeling light as a photon (carrying a discrete amount of momentum and energy) and then applying the principles of conservation of energy and momentum. The result will be an expression for the photon's change in wavelength $\Delta\lambda = \lambda_f - \lambda_i$, known as the *Compton shift.*

Our procedure resembles the analysis of two-dimensional elastic collisions in Section 11-6, but using special relativity to calculate the electron's energy and momentum. Because the photon has no mass, we must use $p = h/\lambda$ (Eq. 40.14) for its momentum.

FIGURE 40.12 The Compton effect. **A.** Before the collision, the electron is at rest and the photon approaches with wavelength λ_i along the x axis. **B.** After the collision, both the photon and the electron are in motion and the photon's wavelength has changed to λ_f.

DERIVATION The Compton Shift

COMPTON SHIFT

Major Concept

We will show that a photon scattering from an electron (Fig. 40.12) has a change in wavelength known as the ***Compton shift***, given by:

$$\Delta\lambda = (\lambda_f - \lambda_i) = \lambda_C(1 - \cos\varphi) \qquad (40.15)$$

where λ_C is the **Compton wavelength** of a free electron:

$$\lambda_C \equiv \frac{h}{m_e c} = 2.426 \times 10^{-12}\,\text{m} = 2.426\,\text{pm} \qquad (40.16)$$

and m_e is the rest mass of an electron.

CONSERVATION of ENERGY

Begin with a bar chart (Fig. 40.13) for the two particles (photon and electron). The photon has no rest mass energy, so we need only one bar to represent its energy. The two bars for the electron represent its rest mass energy and kinetic energy.

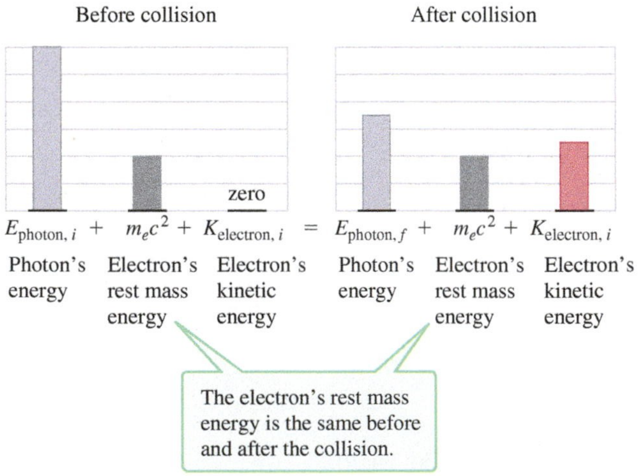

FIGURE 40.13

The electron's rest mass energy is the same before and after the collision, so the conservation of energy condition involves the photon's energy E_{photon} and the electron's kinetic energy K_{electron}. We assume the electron is initially at rest ($K_{\text{electron},\,i} = 0$).

$$E_{\text{photon},\,i} + K_{\text{electron},\,i} = E_{\text{photon},\,f} + K_{\text{electron},\,f}$$

$$E_{\text{photon},\,i} = E_{\text{photon},\,f} + K_{\text{electron},\,f} \qquad (1)$$

Use $E = hf$ (Eq. 40.8) for the photon's total energy, then substitute $f = c/\lambda$ (Eq. 34.20).	$E_{\text{photon}} = hf = \dfrac{hc}{\lambda}$ (40.17)
Substitute Equation 40.17 for the photon's energy before and after the collision, and $K_{\text{electron}} = mc^2 - m_e c^2$ (Eq. 39.43) for the electron's kinetic energy after the collision into Equation (1). Then, substitute $m = \gamma m_{\text{rest}}$ (Eq. 39.39) for the scattered electron's observed mass.	$\dfrac{hc}{\lambda_i} = \dfrac{hc}{\lambda_f} + (mc^2 - m_e c^2) = \dfrac{hc}{\lambda_f} + (\gamma m_e c^2 - m_e c^2)$ $\dfrac{hc}{\lambda_i} = \dfrac{hc}{\lambda_f} + (\gamma - 1)m_e c^2$ (2)

CONSERVATION of MOMENTUM

As in Section 11-6, apply conservation of momentum separately in the x and y directions for the two-dimensional collision in Figure 40.12.

Before the collision, the photon's momentum is in the positive x direction. After the collision, both the photon and the electron have momentum in the positive x direction.	Along x: $p_{\text{photon},i} = p_{\text{photon},f}\cos\varphi + p_{\text{electron},f}\cos\theta$ (3)
Before the collision, neither particle has momentum in the y direction. After the collision, the photon has momentum in the *negative* y direction, whereas the electron has momentum in the *positive* y direction.	Along y: $0 = -p_{\text{photon},f}\sin\varphi + p_{\text{electron},f}\sin\theta$ (4)
Substitute $p_{\text{photon}} = h/\lambda$ (Eq. 40.14) and $p_{\text{electron}} = \gamma m_e v$ (Eq. 39.40) into Equations (3) and (4).	$\dfrac{h}{\lambda_i} = \dfrac{h}{\lambda_f}\cos\varphi + \gamma m_e v\cos\theta$ (5) $0 = -\dfrac{h}{\lambda_f}\sin\varphi + \gamma m_e v\sin\theta$ (6)

The next steps involve many lines of algebra to eliminate v, γ, and θ from Equations (2)–(6). We'll do some of the work here and leave most of the steps for homework. The goal is to arrive at a single expression for $\Delta\lambda = \lambda_f - \lambda_i$ as a function of the scattering angle φ.

In Problem 83, you will eliminate θ from Equations (5) and (6) and arrive at Equation (7).	$\gamma^2 m_e^2 v^2 = \dfrac{h^2}{\lambda_i^2} - \dfrac{2h^2}{\lambda_i\lambda_f}\cos\varphi + \dfrac{h^2}{\lambda_f^2}$ (7)
In Problem 28 you will show that $\gamma^2 v^2 = c^2(\gamma^2 - 1)$. Use this equation to eliminate v from Equation (7).	$c^2(\gamma^2 - 1)m_e^2 = \dfrac{h^2}{\lambda_i^2} - \dfrac{2h^2}{\lambda_i\lambda_f}\cos\varphi + \dfrac{h^2}{\lambda_f^2}$ (8)
Now we must eliminate γ, which can be done by starting with Equation (2). The result is an expression for $\gamma^2 - 1$ (Problem 32).	$(\gamma^2 - 1) = \dfrac{h^2}{m_e^2 c^2}\left(\dfrac{1}{\lambda_i^2} - \dfrac{2}{\lambda_i\lambda_f} + \dfrac{1}{\lambda_f^2}\right) + \dfrac{2h}{m_e c}\left(\dfrac{1}{\lambda_i} - \dfrac{1}{\lambda_f}\right)$ (9)
In Problem 84 you will substitute Equation (9) into Equation (8) and arrive near our goal of finding an expression for the Compton shift.	$(\lambda_f - \lambda_i) = \dfrac{h}{m_e c}(1 - \cos\varphi)$
The constant $h/m_e c$ is the **Compton wavelength** λ_C of a free electron.	$\Delta\lambda = (\lambda_f - \lambda_i) = \lambda_C(1 - \cos\varphi)$ ✓ (40.15)

COMMENTS

The wave model for light predicts that when light scatters, $\Delta\lambda = 0$, so that the light does not change in wavelength (or frequency). But by using the photon model, we found a very different prediction for $\Delta\lambda$. The Compton shift (Eq. 40.15) predicts that scattered light should have a longer wavelength (that is, be redder) than incoming light. Additionally, the shift depends on the scattering angle φ.

The equation $\Delta\lambda = \lambda_C(1 - \cos\varphi)$ mathematically describes what Compton observed. Because $-1 \le \cos\varphi \le 1$, the Compton shift is always positive or zero: $\Delta\lambda \ge 0$. The wavelength of the scattered photon is the same or longer than that of the incoming photon. Put in terms of frequency and energy, the scattered photon has a frequency that is the same or lower, and an energy that is the same or lower, than the incoming photon.

To help visualize how the shift in wavelength depends on φ, Figure 40.14 diagrams three different scattering angles. The incoming photon and the electron are not shown. The smallest scattering angle (Fig. 40.14A) corresponds to a photon with the smallest Compton shift, the shortest wavelength, the highest frequency, and the most energy after the event. The greatest scattering angle (Fig. 40.14C) corresponds to the greatest Compton shift, the longest possible wavelength, the lowest frequency, and the least amount of energy after scattering.

Remember that the wave model predicts that all scattered radiation will have the same frequency (and wavelength) as the incoming radiation. By contrast, the Compton shift equation $\Delta\lambda = \lambda_C(1 - \cos\varphi)$ (Eq. 40.15) comes from modeling electromagnetic radiation as a photon with energy and momentum. Because Compton's experiment is well described by Equation 40.15, which was derived using the photon model, Compton's experimental results support the photon model for light.

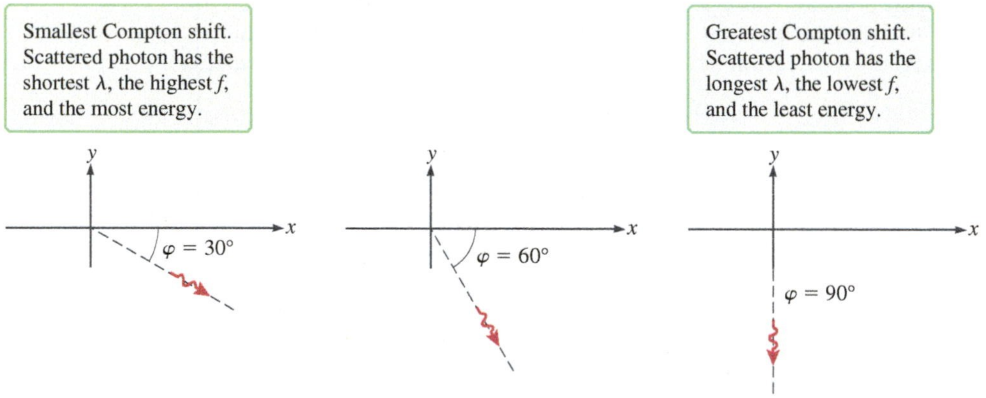

A. $\Delta\lambda = \left(1 - \frac{\sqrt{3}}{2}\right)\lambda_C = 0.13\lambda_C$ **B.** $\Delta\lambda = \left(1 - \frac{1}{2}\right)\lambda_C = 0.50\lambda_C$ **C.** $\Delta\lambda = (1 - 0)\lambda_C = \lambda_C$

FIGURE 40.14 Three different Compton scattering angles φ are shown. Compare the wavelengths, frequencies, and energies of the scattered photons.

CONCEPT EXERCISE 40.4

In Figure 40.14, the three electrons are not shown. However, you can infer their change in energy from the photon's Compton shift. Which of the three electrons gains the most energy?

EXAMPLE 40.6 Plotting the Compton Shift

Graphs usually clarify complicated mathematical expressions. Make a graph of the relative Compton shift $\Delta\lambda/\lambda_C$ versus the scattering angle φ for $0 \le \varphi \le 180°$. In the **CHECK and THINK** step, comment on what is happening physically when $\varphi > 90°$.

INTERPRET and ANTICIPATE

From Figure 40.14, we expect the Compton shift to be smallest for small scattering angles.

SOLVE

Start with Equation 40.15 and isolate $\Delta\lambda/\lambda_C$ on one side.

$$\frac{\Delta\lambda}{\lambda_C} = (1 - \cos\varphi)$$

Substitute values from 0 to 180° for the scattering angle and plot the results (Fig. 40.15). A spreadsheet program may be helpful.

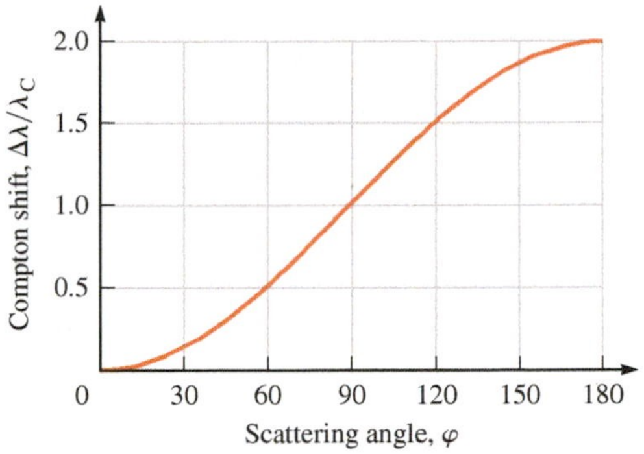

FIGURE 40.15

CHECK and THINK

As expected, the Compton shift increases as the scattering angle increases. When the scattering angle is greater than 90°, the scattered photon's motion begins to have a backward component. Consider the case when $\varphi = 180°$, so that the photon travels on a path back to its source. These photons have the greatest Compton shift, so they have the longest possible wavelength, lowest possible frequency, and least possible amount of energy after the scattering.

EXAMPLE 40.7 What Happens to the Electron?

The Compton shift equation tells us about the photon that is scattered by the electron, but what about the electron? Suppose an X-ray photon with a wavelength of 0.2400 nm is scattered off an electron that is initially at rest, as in Figure 40.12. The scattered X-ray beam is observed at an angle of 60.0°: down and to the right, as shown. Find the speed and momentum of the electron after the collision.

INTERPRET and ANTICIPATE

The X-ray photon loses energy and momentum as a result of its collision with the electron. The electron gains this energy and momentum. We need to find the electron's speed and momentum. Because speed is a scalar, we don't need to find the direction. So the easiest way to find the electron's speed is from its energy. Getting the electron's momentum is a little trickier because momentum is a vector quantity.

SOLVE

Start with the Compton shift to find the wavelength of the scattered photon.

$$\frac{\Delta\lambda}{\lambda_C} = \frac{\lambda_f - \lambda_i}{\lambda_C} = (1 - \cos\varphi) \qquad (40.15)$$

$$\lambda_f = \lambda_i + \lambda_C(1 - \cos\varphi)$$

$$\lambda_f = 0.2400 \text{ nm} + 0.002426 \text{ nm}(1 - \cos 60.0°)$$

$$\lambda_f = 0.2412 \text{ nm}$$

Next use $E = hc/\lambda$ (Eq. 40.17) to find the change in the photon's energy.

$$\Delta E = E_f - E_i = h(f_f - f_i)$$

$$\Delta E = hc\left(\frac{1}{\lambda_f} - \frac{1}{\lambda_i}\right)$$

Example continues on page 1326 ▶

The energy lost by the photon goes into the kinetic energy of the electron. (Remember, the electron is initially at rest.)	$K_e = -\Delta E = hc\left(\frac{1}{\lambda_i} - \frac{1}{\lambda_f}\right)$

Let's start by assuming the speed is nonrelativistic. If we find a speed that is close to the speed of light, we will start over, taking special relativity into account.

$$\frac{1}{2}m_e v_e^2 = hc\left(\frac{1}{\lambda_i} - \frac{1}{\lambda_f}\right)$$

$$v_e = \sqrt{\frac{2hc}{m_e}\left(\frac{1}{\lambda_i} - \frac{1}{\lambda_f}\right)}$$

$$v_e = \sqrt{\frac{2(6.626 \times 10^{-34}\,\text{J}\cdot\text{s})(2.998 \times 10^8\,\text{m/s})}{9.101 \times 10^{-31}\,\text{kg}}\left(\frac{1}{0.2400\,\text{nm}} - \frac{1}{0.2412\,\text{nm}}\right) \times 10^9\,\text{nm/m}}$$

$$v_e = 3.008 \times 10^6\,\text{m/s}$$

CHECK and THINK

The speed we found is two orders of magnitude lower than the speed of light. We don't have to take special relativity into account.

SOLVE

The photon's momentum is initially in the x direction, and its magnitude is given by $p_{\text{photon}} = \frac{h}{\lambda}$ (Eq. 40.14).

$$\vec{p}_i = \frac{h}{\lambda_i}\hat{\imath}$$

We find the photon's change in momentum, remembering that momentum is a vector. Find the components separately. As shown in Figure 40.12, the photon's y momentum change is negative.

$$\Delta p_x = p_f \cos\varphi - p_i = h\left(\frac{1}{\lambda_f}\cos\varphi - \frac{1}{\lambda_i}\right)$$

$$\Delta p_x = (6.626 \times 10^{-34}\,\text{J}\cdot\text{s})\left(\frac{1}{0.2412}\cos 60.0^\circ - \frac{1}{0.2400}\right) \times 10^9\,\text{nm/m}$$

$$\Delta p_x = -1.3875 \times 10^{-24}\,\text{kg}\cdot\text{m/s}$$

$$\Delta p_y = p_f \sin\varphi = -\frac{h}{\lambda_f}\sin\varphi$$

$$\Delta p_y = -\frac{6.626 \times 10^{-34}\,\text{J}\cdot\text{s}}{0.2412 \times 10^{-9}\,\text{m}}\sin 60.0^\circ$$

$$\Delta p_y = -2.3789 \times 10^{-24}\,\text{kg}\cdot\text{m/s}$$

Because momentum is conserved, the loss in the photon's momentum must equal the gain in the electron's momentum. The electron is at rest before the collision, so the gain in momentum is the electron's momentum after the collision.

$$p_{\text{electron}} = (1.3875\hat{\imath} + 2.379\hat{\jmath}) \times 10^{-24}\,\text{kg}\cdot\text{m/s}$$

CHECK and THINK

According to our expression for momentum, the electron moves up and to the right, consistent with Figure 40.12. Other than how we calculated the photon's energy and momentum, this problem is much like the two-dimensional collision problems we solved in Chapter 11 for particles. Put in simple terms, the electron cannot tell if it collided with another particle or with light. This example further shows why the Compton effect supports the photon model. We normally think that particles can collide but waves cannot. (Waves interfere; particles collide.)

40-5 Wave-Particle Duality

Let's review the debate over whether light should be modeled as a particle or a wave. In the 1600s, Newton was the strongest advocate of the particle model, whereas Huygens was the strongest advocate of the wave model. Because Newton's other theories (his three laws of dynamics and his law of universal gravity) were so successful, it was difficult for scientists to think that Newton could be wrong about light. Not until the late 1800s, after Young's interference experiment and Arago's diffraction experiment, was the wave model accepted.

Since Newton's and Huygens's time, scientists had believed that light must be modeled either as a wave or as a particle, but not as both. So in the 1800s, when the wave model was widely accepted, the particle model was widely rejected. Scientists believed that Young's and Arago's experiments disproved the particle model. However, the photoelectric effect and the Compton effect, observed in the early 1900s, could not be explained by the wave model.

Today we know of many experiments showing that light behaves as a wave and many other experiments showing that light behaves as a particle. So to have a full understanding of light, we must accept both models. This two-part description of light is known as **wave-particle duality**. Sometimes we must model light as a wave and at other times as a stream of photons.

WAVE-PARTICLE DUALITY OF LIGHT

Underlying Principle

How do we know when to model light as a wave and when to model it as a photon? To explain the interference pattern produced by Young's double-slit experiment (Fig. 35.10, page 1128), we considered the path taken by two light waves that emerged from the two slits (Figs. 35.13 and 35.14, page 1129). Then we used the principle of superposition to find places of constructive and destructive interference. In general, we model light as a wave in any experiment where the *propagation* of light plays the key role. In these interference experiments, we consider the paths taken by the light waves and use the principle of superposition.

By contrast, in the Compton effect, light is modeled as a particle (a photon) that collides with another particle (an electron). The photon carries momentum and energy, which are conserved in its collision with the electron. In general, we model light as photons in any experiment where its *interaction* with matter plays a key role. These scattering experiments require that light be modeled as a particle having energy given by $E = hf = hc/\lambda$ (Eq. 40.17) and momentum given by $p = hf/c = h/\lambda$ (Eq. 40.14). In fact, these equations link the wave model of light (its frequency and wavelength) to the particle model (its energy and momentum).

We return to wave-particle duality in Chapter 41.

On the macroscopic level, we see some objects, such as basketballs and airplanes, that are best modeled as particles. We also experience phenomena, such as sound, that are best modeled as waves. The particle model and the wave model seem completely distinct. So when scientists were arguing about how to model light, it seemed that they were really arguing about its fundamental nature. Is light made up of a bunch of objects such as tiny balls, or is it a disturbance like a sound wave? When we conclude that our experiments show that we must model light *both* as a stream of particles and as a wave, how can we know what light is? The duality of light seems unsatisfactory.

Light's existence as both a wave and a stream of particles is dissatisfying because it doesn't fit our everyday experience. We can hold the basketball that we model as a particle in our hands, and we can watch it translate from place to place. By contrast, modeling sound as a wave is a little harder because we ordinarily can't see sound waves. However, we can make an analogy with water waves. By watching the water surface moving up and down in a water wave, we can imagine air molecules moving back and forth in a sound wave in a similar fashion. But it is even harder to digest wave-particle duality because we have no single macroscopic analogy. Instead, we must sometimes visualize light as a stream of tiny particles, and at other times as a waving disturbance. This is our best current description of light's nature.

Wave-particle duality is also counterintuitive because waves and particles are so different. For example, picture a water wave in a large pool; the wave spreads to take up the whole pool. If a second wave is introduced, the result is the superposition of both waves, which overlap and occupy the entire pool. By contrast, a basketball sitting on the floor of the court occupies a small place. If a second ball is placed on the floor of the court, you can be certain that its position is different from the position of the first ball. Two particles cannot occupy the same position; they would collide. So when we say that light is both a wave and a particle, we mean that sometimes we'll use the principle of superposition when two light waves overlap (as in Young's experiment), and at other times we'll treat light as composed of particles that can collide with other particles (as in Compton's experiment).

CONCEPT EXERCISE 40.5

What do we mean when we say that $E = hf = hc/\lambda$ (Eq. 40.17) "links" the wave model of light to the particle model? (How does the energy carried by a photon differ from the energy carried by a wave in the classical model?)

40-6 The Wave Properties of Matter

Our story of quantum physics continues with a French prince, Louis Victor Pierre Raymond de Broglie (1892–1987), who was originally a historian. He became interested in science during World War I. After the war, he earned his doctorate in physics in 1924. His doctoral thesis was revolutionary—in it, de Broglie proposed that the wave-particle duality extended beyond light. In fact, his thesis was so revolutionary that his professor wasn't sure what to make of it. So he sent a copy to Einstein, who said that de Broglie might be on to something.

WAVE-PARTICLE DUALITY FOR MATTER

Underlying Principle

As we have seen, exploiting the symmetry of nature sometimes leads to the discovery of new laws. De Broglie thought that if light behaves both as a wave and as a particle, then other things that we traditionally model as particles, such as electrons, protons, and even basketballs, might also be modeled as waves.

De Broglie said that the wavelength associated with such a particle comes from solving $p = h/\lambda$ (Eq. 40.14) for λ:

$$\lambda = \frac{h}{p} \tag{40.18}$$

DE BROGLIE WAVELENGTH

Major Concept

The wavelength of a particle is called the **de Broglie wavelength**. The numerator is Planck's constant, which is a very small number. The denominator is large for macroscopic objects such as basketballs, so the wavelength of macroscopic objects is too small to be detected. Therefore, we don't notice the wave properties of such objects, and it is best just to model them as particles—as we have been doing.

If the particle is small, however, such as an electron, then it doesn't have a lot of mass, and its momentum is low too. In the case of a microscopic particle, the de Broglie wavelength is not negligible, and the particle's wave nature cannot be ignored.

Einstein pointed out that if de Broglie is correct, then electrons will show interference and diffraction patterns, similar to the interference and diffraction patterns observed by Young and Arago for light. The de Broglie wavelength of an electron is on the order of 10^{-10} m, or about 5000 times shorter than the wavelength of visible light. In order to observe an electron diffraction pattern like the one produced when light passes through a grating, you need a grating with rulings on the order of 10^{-10} m apart. Fortunately, the array of atoms in a solid can have spacings on this order of magnitude.

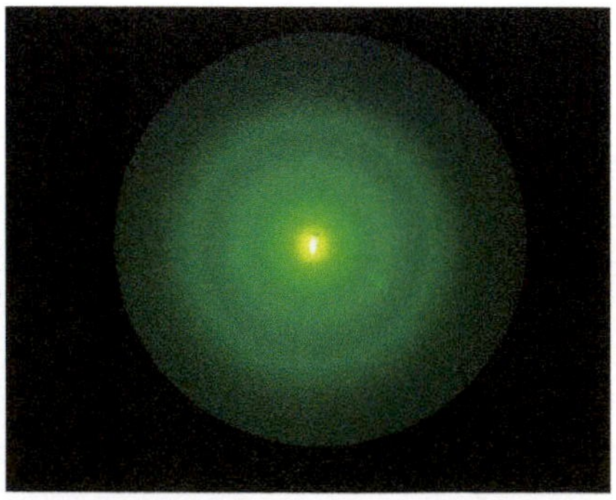

Andrew Lambert Photography/Science Photo Library

FIGURE 40.16 An electron diffraction pattern obtained from a thin film of polycrystalline graphite consists of concentric rings. In general the diffraction pattern observed depends on the structure of the sample; many observed patterns are more complicated that the simple concentric rings shown here.

In 1927, Clinton Davisson (1881–1958) and Lester Germer (1896–1971) were working in their New York City laboratory. They had been scattering electrons from the surface of a nickel crystal, when by accident they discovered that the scattered electrons form a pattern of peaks and valleys. Figure 40.16 shows the electron diffraction pattern obtained in a contemporary experiment involving graphite instead of nickel. The underlying concept of this experiment is the same as that of Davisson and Germer's: The scattered electrons form a pattern like the diffraction pattern seen when light passes through a grating or a circular aperture (Fig. 36.2, page 1156). If you model the electrons as waves and the nickel crystal as a diffraction grating, then the pattern you see can be interpreted as a diffraction pattern. The spacing of fringes in the pattern can be used to find the electron's wavelength. When Davisson and Germer made this calculation, they found that the electron's wavelength is in agreement with the de Broglie wavelength.

Since this experiment in the 1920s, many others have been carried out showing that protons, neutrons, and other subatomic particles are well modeled as waves with wavelengths given by the de Broglie relation (Eq. 40.18). One of these experiments was carried out by Sir George Paget Thomson (1892–1975)—the son of J. J. Thomson. Since J. J. Thomson is famous for showing that the electron is a particle, it is ironic that his son successfully showed that the electron is a wave.

Wave-particle duality applies to objects that we traditionally model as particles, as well as to light. On the macroscopic scale, we continue to model basketballs as particles, but on the microscopic scale, we often need to model objects such as electrons, protons, and neutrons as waves instead of as particles. Sometimes we use the term **matter wave** to describe the wave associated with objects traditionally modeled as particles.

EXAMPLE 40.8 Your de Broglie Wavelength

A jogger with mass 65.0 kg is jogging at 4.00 m/s. What is the jogger's de Broglie wavelength?

INTERPRET and ANTICIPATE

We model the jogger as a matter wave whose wavelength depends on his momentum. Since a person's jogging speed is much less than the speed of light, we do not need to use special relativity.

SOLVE

Find the magnitude of the momentum using Equation 10.1.

$$p = mv \qquad (10.1)$$

$$p = (65.0\,\text{kg})(4.00\,\text{m/s})$$

$$p = 260\,\text{kg}\cdot\text{m/s}$$

Substitute this momentum into Equation 40.18 to find the de Broglie wavelength of the jogger.

$$\lambda = \frac{h}{p} \qquad (40.18)$$

$$\lambda = \frac{6.626 \times 10^{-34}\,\text{J}\cdot\text{s}}{260\,\text{kg}\cdot\text{m/s}}$$

$$\lambda = 2.55 \times 10^{-36}\,\text{m}$$

CHECK and THINK

The wavelength of a jogger is about 26 orders of magnitude smaller than the diameter of a single atom ($\sim 10^{-10}$ m). This is much too small to be detected, so the wave properties of a jogger can be safely ignored. This is why we are able to model objects as particles on the macroscopic scale.

EXAMPLE 40.9 An Electron's de Broglie Wavelength

In a laboratory experiment, an electron has a speed of 3.00×10^6 m/s. What is its de Broglie wavelength?

INTERPRET and ANTICIPATE

As in the previous example, find the electron's de Broglie wavelength from its momentum. The electron's speed is $0.01c$—much less than the speed of light. Once again, we don't need special relativity.

SOLVE

Find the magnitude of the momentum using Equation 10.1.

$$p = mv \qquad (10.1)$$

$$p = (9.11 \times 10^{-31}\,\text{kg})(3.00 \times 10^6\,\text{m/s})$$

$$p = 2.73 \times 10^{-24}\,\text{kg}\cdot\text{m/s}$$

Example continues on page 1330 ▶

Substitute this momentum into Equation 40.18 to find the electron's de Broglie wavelength.	$\lambda = \frac{h}{p}$ (40.18) $\lambda = \frac{6.626 \times 10^{-34}\,\text{J}\cdot\text{s}}{2.73 \times 10^{-24}\,\text{kg}\cdot\text{m/s}}$ $\lambda = 2.42 \times 10^{-10}\,\text{m} = 0.242\,\text{nm}$

CHECK and THINK

The electron's de Broglie wavelength is about the size of an atom or of the space between atoms in a crystal. So when a beam of electrons is aimed at a crystal, the electrons form a diffraction pattern such as the one shown in Figure 40.16. In this case, the electrons behave like a wave, and we must use the wave model. In order to understand the structure of atoms (Chapter 42), we must model their electrons as matter waves.

CASE STUDY The Electron Microscope

De Broglie's idea that electrons may be modeled as waves may seem theoretical, with no practical applications. However, less than a decade after he put forth this idea, the first primitive electron microscope was built. An electron microscope uses electron waves in much the same way that an optical microscope uses light waves (Section 36-9). However, the optical lenses are replaced by magnetic (or electrostatic) fields that exert forces on the electrons, bringing them into focus.

There are two basic designs for an electron microscope (Fig. 40.17). In both designs, a beam of electrons is generated by an electron gun composed of a hot, V-shaped negative plate and a disk-shaped positive plate. Electrons are accelerated by the potential difference between the two plates. In a transmission electron microscope, the electron beam passes through the sample, and its image is projected onto a screen at the other end of the device (Fig. 40.17A). In a scanning electron microscope, the electron beam liberates electrons from the sample, and these electrons are then collected and used to form a three-dimensional image of the sample (Fig. 40.17B).

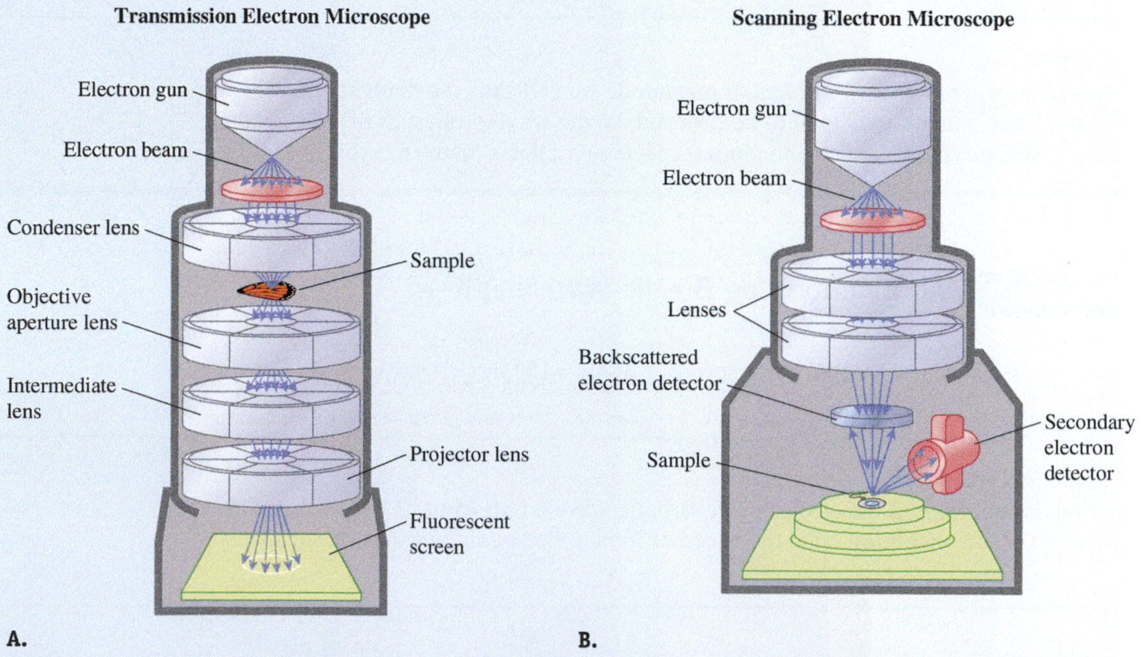

FIGURE 40.17 Two electron microscope designs. In both designs, an electron gun generates a beam. **A.** In a transmission electron microscope, a series of lenses cause the electron beam to pass through the sample and form an image on a screen. **B.** In a scanning electron microscope, the electron beam liberates electrons from the sample. Some are scattered backward toward the electron gun, and others are scattered (roughly) perpendicular to the original beam. The two scattered beams of electrons are used to generate a three-dimensional image.

You have probably seen a photograph that has been overmagnified; the image looks like many individual squares, and it is hard to discern its subject. The useful magnification of an image is limited by how many individual pixels or resolution elements went into creating the image. For a microscope, the useful magnification depends on its resolution. Recall that the diffraction-limited resolution of an instrument such as a microscope is proportional to the wavelength (Eq. 34.4). To construct a high-resolution, greatly magnified image, you should illuminate the sample with short wavelengths. As illustrated in the next example, it is more practical to do this with electron waves than with light.

EXAMPLE 40.10 CASE STUDY Comparing a Conventional Microscope to an Electron Microscope

In this example, we compare a conventional microscope that illuminates a sample with an electromagnetic wave to an electron microscope (of either design) that uses an electron wave. Suppose you wish to image a strand of DNA, which has a width of roughly 2 nm. You'd like a high-resolution, greatly magnified image, and you estimate that you require a wavelength of about 0.02 nm, so that there will be roughly 100 pixels or resolution elements across the DNA's width.

A Find the energy (in eV) of the electromagnetic wave used to illuminate the DNA in a conventional microscope. In the **CHECK and THINK** step, discuss the practical limitations of such a device.

INTERPRET and ANTICIPATE

Use the estimated wavelength to find the energy of the photons in the electromagnetic wave your microscope would employ.

SOLVE

The energy of the photons depends on their frequency, according to Equation 40.8. Their frequency is related to their wavelength by $\lambda f = c$ (Eq. 34.20). Convert the energy to eV.

$$E = hf \tag{40.8}$$

$$E = h\frac{c}{\lambda} = (6.63 \times 10^{-34}\ \text{J}\cdot\text{s})\left(\frac{3.00 \times 10^{8}\ \text{m/s}}{2 \times 10^{-11}\ \text{m}}\right)$$

$$E = 9.9 \times 10^{-15}\ \text{J}$$

$$1\ \text{eV} = 1.60 \times 10^{-19}\ \text{J}$$

$$E = 6 \times 10^{4}\ \text{eV}$$

CHECK and THINK

As shown in Table 34.2 (page 1101), electromagnetic radiation with a wavelength of 2×10^{-11} m is in the X-ray or gamma-ray band. Such high-energy radiation would pose a threat to the researchers, their microscope, and their sample. Using X-rays is not practical in a high-resolution microscope.

B Find the kinetic energy (in eV) of the electron wave used in an electron microscope. Assume that the electrons are not relativistic. In **CHECK and THINK**, compare your results to those in part A.

INTERPRET and ANTICIPATE

Find the momentum of the electrons from their wavelength, and find their kinetic energy from their momentum.

SOLVE

Use Equation 40.18 to find the electron momentum.

$$\lambda = \frac{h}{p} \tag{40.18}$$

$$p = \frac{h}{\lambda}$$

Example continues on page 1332 ▶

Write the expression for kinetic energy in terms of the momentum. Substitute values, then convert the result to eV.

$$K = \frac{1}{2}m_e v^2 = \frac{(m_e v)^2}{2m_e}$$

$$K = \frac{p^2}{2m_e} = \frac{1}{2m_e}\left(\frac{h}{\lambda}\right)^2$$

$$K = \frac{1}{2(9.11 \times 10^{-31}\ \text{kg})}\left(\frac{6.63 \times 10^{-34}\ \text{J}\cdot\text{s}}{2 \times 10^{-11}\ \text{m}}\right)^2$$

$$K = 6.0 \times 10^{-16}\ \text{J} = 4 \times 10^3\ \text{eV}$$

CHECK and THINK

The energy required by the electron microscope is about 15 times lower than that required by the conventional microscope. It will be easier to power an electron microscope and to control the high-energy electrons with magnetic (or electrostatic) fields.

Although the resolution of an electron microscope is not limited significantly by diffraction, like a conventional microscope, it is subject to spherical aberration, distortion, chromatic aberration, and other imaging defects. So even an electron microscope cannot achieve the resolution of 0.02 nm that we would like here. The best resolution achieved by an electron microscope is typically between 0.1 and 0.5 nm, and its highest useful magnification is typically between 10^4 and 10^5. Figure 40.18 is an image of a bundle of strands DNA taken with a transmission electron microscope. The bar in the bottom left corner is the scale; it represents 20 nm. This amazing image shows six DNA stands wrapped around a seventh strand. The inset shows a close-up of the helical structure, with red arrows pointing to the individual turns of the helix. The width of the DNA is 2 nm, so assuming that the resolution is 0.2 nm, we find that there are about 10 resolution elements across the width of a single strand of DNA. Without such detailed resolution, the helical structure would not be seen, the image would look like a smooth line without any structure.

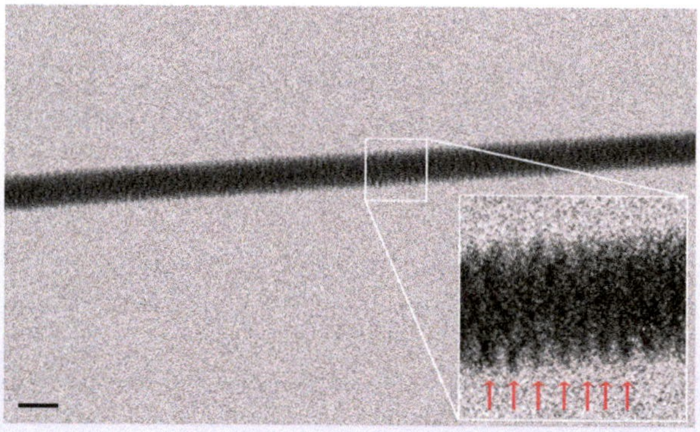

FIGURE 40.18

Professor Enzo Di Fabrizio, IIT/Science Source

SUMMARY

Underlying Principles:

1. According to **Planck's quantum theory**, the smallest possible amount of energy E_{min} (a **quantum of energy**) for an oscillator of frequency f is:

$$E_{min} = hf \quad (40.6)$$

Planck modeled the particles in a black body as oscillators and concluded that the energy of each oscillator is **quantized** and given by:

$$E = nE_{min} = nhf \quad n = 1, 2, 3 \ldots \quad (40.7)$$

Planck's formula for the intensity I of a black body's **radiation** is:

$$I(\lambda, T) = \frac{2\pi hc^2}{\lambda^5\left(e^{hc/\lambda k_B T} - 1\right)} \quad (40.5)$$

where **Planck's constant** is $h = 6.626 \times 10^{-34}\ \text{J}\cdot\text{s}$.

2. **Wave-particle duality**: A wave model and a particle model must be applied to both light and matter.

Major Concepts

1. In the photon model, light is modeled as bundles of energy or particles known as **photons**. A photon's energy depends on the frequency of the light:

$$E_{\text{photon}} = hf = \frac{hc}{\lambda} \quad (40.17)$$

A photon's momentum is given by:

$$p = \frac{hf}{c} = \frac{h}{\lambda} \quad (40.14)$$

2. When a photon is scattered (Fig. 40.12), it has a change in wavelength known as the **Compton shift**, given by:

$$\Delta\lambda = (\lambda_f - \lambda_i) = \lambda_C(1 - \cos\varphi) \quad (40.15)$$

where λ_C is the **Compton wavelength** of a free electron,

$$\lambda_C \equiv \frac{h}{m_e c} = 2.426 \times 10^{-12}\,\text{m} = 2.426\,\text{pm} \quad (40.16)$$

3. The **de Broglie wavelength** associated with a particle is

$$\lambda = \frac{h}{p} \quad (40.18)$$

PROBLEMS AND QUESTIONS

A = algebraic C = conceptual E = estimation G = graphical N = numerical

40-1 Another Modern Idea

1. **C** Many of the images that we see today are composed of many small dots of color, or pixels. Normally, we don't notice that the images are made out of discrete pixels. Why do the images seem continuous? What would happen if the size of the pixels increased?

40-2 Black-Body Radiation and the Ultraviolet Catastrophe

2. **N** The surface temperature of the Sun is about 5.8×10^3 K. Model the Sun as a black body; find the peak wavelength of its radiation.
3. **N** A stove top burner operates at a temperature of 350.0°F (Fig. P40.3). Modeling the burner as a black body, what is the peak wavelength emitted by the burner?

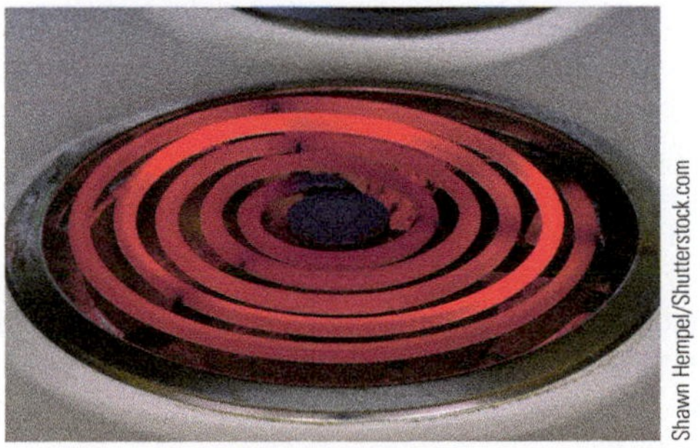

FIGURE P40.3

Problems 4 and 5 are paired.

4. **C** Samantha tells George that stars that appear yellow are really hot and that the color is indicative of temperature. George says he's heard that red stars are often very big and should thus be hotter than yellow ones, which are often smaller. Samantha tells George that he is wrong about red stars being hotter and observes that some stars can even be blue. She says these would be even hotter than the yellow ones. George insists that she has it backwards because blue things should be cold. Who is correct and why?
5. **N** Calculate the peak wavelength and identify the color of the following stars based on their different temperatures. Treat them as black bodies: **a.** 4.00×10^3 K, **b.** 5.30×10^3 K, **c.** 6.40×10^3 K.
6. **E** Model yourself as a black body and estimate your peak wavelength.
7. **N** What is the temperature of a black body with a wavelength distribution peaking at 628 nm?
8. **A** Show that Planck's constant h has the same dimensions as angular momentum.
9. **N** An incandescent lightbulb filament reaches a temperature of around 4.14×10^3 K. Assuming that it radiates as a black body, at what wavelength is the peak intensity of the black-body radiation? In what region of the electromagnetic spectrum is this radiation?
10. **N** What is the frequency of the light associated with the peak wavelength emitted by a black body with a temperature of 4.83×10^3 K?
11. **N, C** A molecule vibrates with a minimum frequency of 9.50×10^{13} Hz. What is the minimum quantum of energy that this molecule can have? Can this molecule vibrate at 12.5×10^{13} Hz? If not, why not? If so, what is its energy when it vibrates at this frequency?
12. A black body has a temperature of 3575 K.
    - **a.** **N** What is the peak wavelength of this black body?
    - **b.** **N** What is the intensity of the light from the black body at this wavelength?
    - **c.** **C** How would the answer to part (b) compare to the intensity of other wavelengths from the black body?
13. **N** The Sun, which can be modeled as a black body, radiates 3.85×10^{26} J of energy from its surface each second. The Sun's radius is 6.9551×10^5 km.
    - **a.** What is the black-body temperature of the Sun's surface?
    - **b.** What is the peak wavelength of the radiation emitted by the Sun?

14. **C** When you look at a flame or a fire, you may notice that different regions appear to glow with different colors (Fig. P40.14). In terms of black-body radiation, explain why different regions within the flame or fire appear to be different colors.

FIGURE P40.14

40-3 The Photoelectric Effect

15. **N** The work function of silicon is $W_0 = 4.85$ eV.
 a. What is the maximum wavelength of incident light for which photoelectrons will be released from silicon?
 b. What is the minimum frequency of incident light, called the cutoff frequency, for which photoelectrons will be released from silicon?
 c. What is the maximum kinetic energy of photoelectrons emitted by silicon if 7.00-eV photons strike its surface?

16. A material has a work function of 2.3 eV.
 a. **N** What frequency of light is necessary to eject electrons from its surface?
 b. **C** Assuming light of the frequency you found in part (a), does the number of ejected electrons increase if the intensity of the light is increased? How might you explain this in terms of the particle representation of light?
 c. **N** What is the stopping potential for this material if it is illuminated by light with a frequency of 6.50×10^{14} Hz?
 d. **N, C** What is the maximum kinetic energy of ejected electrons if light with a frequency of 3.50×10^{14} Hz is incident on this material? Explain.

Problems 17 and 18 are paired.

17. **N** The work function of gold is 5.1 eV. What is the minimum frequency of light that will produce the photoelectric effect in gold?

18. **G** The work function of gold is 5.1 eV. Plot the maximum speed of the ejected electrons from gold versus the frequency of incident photons for photon energies between 0 and 10 eV.

19. **N** Far ultraviolet light with a wavelength of 175 nm is incident on an unknown solid surface, which releases photoelectrons with a maximum speed of 7.70×10^5 m/s.
 a. What is the work function of the unknown solid?
 b. What is the minimum frequency of incident light for which photoelectrons will be released from this surface?

20. **N** For each entry in Table P40.20, find the photon's energy in eV.

TABLE P40.20

Color	Wavelength λ (nm)	Frequency f ($\times 10^{12}$ Hz)	Energy E (eV)
Red	620–780	380–480	
Orange	600–620	480–500	
Yellow	575–600	500–522	
Green	500–575	522–600	
Blue	450–500	600-670	
Indigo	420–450	670–710	
Violet	400–420	710–750	

Problems 21 and 22 are paired.

21. **N** A strange metallic rock is found and is being tested. Suppose that light with a frequency of 7.50×10^{14} Hz is incident upon the rock and a stopping potential of 1.00 V is needed to reduce the electron current to zero in a photoelectric experiment.
 a. What is the maximum kinetic energy of an electron ejected by this light from this material?
 b. What is the work function of this material?

22. **N** For the situation in Problem 21, what is the minimum frequency of light for which electrons are still ejected from the surface of this material?

23. **N** Iron has a work function of 4.50 eV. What is the minimum frequency of light needed to eject electrons from iron?

24. **C** What has more energy, an X-ray photon or a radio photon? Explain.

25. **N** CASE STUDY Larger diameter telescopes are more desirable, in part, because they are better at gathering photons. The light-gathering power of a telescope is proportional to the area of its primary mirror. Large telescopes are expensive, so they are generally owned and operated by large institutions. However, the quantum efficiency of CCDs means that small institutions with small telescopes are also equipped to detect faint astronomical objects. Consider a small telescope equipped with a CCD. To make a concrete comparison, consider the unlikely scenario of a large 5.00-m diameter telescope equipped with a photographic plate. Suppose that in the same time exposure, the two telescopes detect equally faint objects. If the photographic plate's quantum efficiency is 1.5% of the CCD's, what is the diameter of the small telescope?

26. **C** The work functions for sodium and platinum are 2.28 eV and 6.35 eV respectively. The same light source with a sufficient frequency to free electrons from each metal is shined on each surface.
 a. How does the number of electrons freed from each metal differ?
 b. How does the maximum kinetic energy of the electrons freed from each metal differ?

40-4 The Compton Effect

27. **N** Suppose that the ratio of the wavelengths of two photons is $\lambda_1/\lambda_2 = 4.00$. What is the ratio of their momenta, p_1/p_2?

28. **A** The derivation of the Compton shift in Section 40-4 takes many lines of algebra. You are asked to provide many of the missing steps. Show that $\gamma^2 v^2 = c^2(\gamma^2 - 1)$.

29. **N** X-rays of wavelength 1.250 Angstroms (10^{-10} m) incident on an aluminum target undergo Compton scattering.
   a. For a photon scattered at 90° relative to the incident beam, what is the wavelength of the scattered photon?
   b. What are the magnitude of the momentum and the energy of this scattered photon at this angle?

Problems 30 and 31 are paired.

30. **A** A photon undergoes a shift in wavelength due to a collision with an electron so that its wavelength changes by 10.0%. Find an expression for the scattering angle in terms of the initial wavelength of the photon.
31. **N** For the photon in Problem 30, what is the scattering angle if the initial wavelength was 15.0 pm?
32. **A** The derivation of the Compton shift in Section 40-4 takes many lines of algebra. You are asked to provide many of the missing steps. Show that

$$(\gamma^2 - 1) = \frac{h^2}{m_e^2c^2}\left(\frac{1}{\lambda_i^2} - \frac{2}{\lambda_i\lambda_f} + \frac{1}{\lambda_f^2}\right) + \frac{2h}{m_ec}\left(\frac{1}{\lambda_i} - \frac{1}{\lambda_f}\right).$$

33. **N** What is the shift in wavelength of X-rays that are scattered from a target at an angle of 62.0° to the incident beam?
34. **A** Show that $\lambda_C \equiv h/(m_ec)$ has the dimensions of length.

Problems 35, 36, and 37 are grouped.

35. **N** Suppose that you perform the Compton experiment with X-rays of wavelength 1.984×10^{-11} m. What is the wavelength (in picometers) of the scattered X-rays observed at **a.** 0°, **b.** 45° and, **c.** 180°?
36. **N** Suppose that you perform the Compton experiment with X-rays of wavelength 1.984×10^{-11} m. What is the frequency (in exaHertz) of the scattered X-rays observed at **a.** 0°, **b.** 45° and, **c.** 180°?
37. **N** Suppose that you perform the Compton experiment with X-rays of wavelength 1.984×10^{-11} m. What is the energy (in keV) of the scattered X-rays observed at **a.** 0°, **b.** 45.00° and, **c.** 180.0°?
38. **N** An X-ray with a frequency of 8.7700×10^{16} Hz undergoes a Compton scattering process and is detected at an angle of 45.000° relative to its original velocity.
   a. What is the wavelength of the photon after it has scattered?
   b. What is the kinetic energy of the electron after its collision with the X-ray?
39. **N** A beam of 75.0-keV X-rays is Compton scattered by a target in such a way that the scattered X-rays are at an angle of 44.0° to incident beam.
   a. What is the wavelength for the scattered X-rays?
   b. What is the energy of the scattered X-rays?
   c. What is the energy of the electrons scattered by the incident X-rays?
40. **C** According to the theory behind Compton scattering, is there a difference expected for scattering from targets made from two different metals? Why or why not?
41. **N** An X-ray with a frequency of 4.500×10^{17} Hz undergoes a Compton scattering process. What is the wavelength of the X-ray that is scattered at **a.** 30.00°, **b.** 60.00°, and **c.** 90.00°?

40-5 Wave-Particle Duality

42. **C** Wave-particle duality is hard to accept. Let's work through an analogy that might make the duality more comprehensible. You know you cannot be both a human being and a chimpanzee. However, you know that you *can* be both a physics student and a server in a restaurant. Explain why you cannot be both a human being and a chimpanzee, but you can be both a student and a server. Use this as an analogy to explain why light can be both a particle and a wave.
43. **C** For each of the following experiments, or cases, identify whether the particle or wave nature of light would be observed: **a.** light is incident on a metal giving rise to free electrons, **b.** light is incident on a single slit giving rise to an interference pattern on a screen, **c.** light is incident on a pair of closely spaced slits giving rise to an interference pattern on a screen, and **d.** x-rays are scattered elastically by free electrons.
44. **C** We know that the intensity of light from a point source decreases as a function of distance r with a dependence of $1/r^2$. Explain that inverse square dependence using the photon model.

40-6 The Wave Properties of Matter

45. **N** Consider a photon and a proton carrying equal momenta of 2.13×10^{-25} kg · m/s.
   a. What is the wavelength of a photon with this momentum?
   b. What is the speed of a proton with this momentum?

Problems 46 and 47 are paired.

46. **E** Determine an estimate for the wavelength of a fastball thrown by a major league baseball pitcher.
47. **C** The space between slats on a fence is about 1.5 cm. Would it be possible to observe the diffraction of a baseball thrown at the fence? Explain why or why not.
48. **C** A number of times throughout this textbook, we have said that by exploiting this symmetry nature new physical laws are proposed. For examples see Chapter 34 (page 1085) and Section 40-6. What do we mean when we say that nature is symmetric? How does the term compare and contrast to our common usage?
49. **N** If an electron has a kinetic energy 3.40 eV, what is the de Broglie wavelength of the electron?
50. **C** Compare and contrast an electron with a photon.
51. **C** In a transmission electron microscope, it is possible to control the energy of the electrons when imaging. High-resolution imaging usually requires 300 keV electrons. A researcher might describe 100 keV electrons as being "too fat" for high-resolution imaging. What does the researcher mean when she says this?
52. **E** Similar to X-ray and electron diffraction, neutron diffraction can occur if the de Broglie wavelength of the neutrons is comparable to the interatomic spacing of the target. Approximately what speed would be required to produce neutron diffraction effects?
53. **N** A proton is observed with a speed of 7.98×10^5 m/s. What is the de Broglie wavelength of this proton?
54. **N** An alpha particle (a helium nucleus) moves with a speed of 2.46×10^5 m/s. What is the wavelength of this particle?
55. **N** A stream of neutrons moves with a speed of 1.23×10^3 m/s and is incident on two rows of atoms within a sample material with a spacing of 0.35 nm between the rows. The neutrons undergo diffraction as they pass between the rows of atoms. What is the width of the central bright region on a screen that is 2.4 cm away from the sample material?
56. **C** In Example 40.8 (page 1329), we found that a jogger's wavelength is around 10^{-36} m, and in Example 40.9 we found the wavelength of an electron is about 0.2 nm. Why isn't it possible to detect the wave properties of a jogger, whereas it is possible to detect the wave properties of an electron?

57. **N** Suppose that an electron and a proton are traveling at the same speed. What is the ratio λ_p/λ_e of their wavelengths?

58. **N** Find the wavelength and kinetic energy of each of the following objects, assuming they move with a speed of 1.23×10^6 m/s: **a.** an electron, **b.** a dust particle ($m = 1.54 \times 10^{-10}$ kg), and **c.** a baseball ($m = 0.145$ kg).

Problems 59 and 60 are paired.

59. **N** Find **a.** the energy of a photon with a wavelength of 3.56 nm, and **b.** the kinetic energy of an electron with a de Broglie wavelength of 3.56 nm. **c.** Compare your results by finding the ratio of your answers (part (a)/part(b)).

60. **N** Find **a.** the energy of a photon with a wavelength of 3.56 fm, and **b.** the kinetic energy of an electron with a de Broglie wavelength of 3.56 fm. **c.** Compare your results by finding the ratio of your answers (part (a)/part(b)).

General Problems

61. **N** What is the stopping potential for a samarium (Sm) surface with work function $W_0 = 2.70$ eV illuminated with 405-nm blue light?

62. **N** An electron moves with a speed of $0.60c$.
- **a.** Find the wavelength of the electron semiclassically by using $p = mv$ for the momentum of the electron.
- **b.** Find the wavelength of the electron using special relativity to express the momentum of the electron.
- **c.** Calculate the percent difference between your answers for parts (a) and (b).

Problems 63 and 64 are paired.

63. **N** A researcher fires a stream of X-rays with a frequency of 2.50×10^{18} Hz at a target where they undergo Compton scattering. A circular detector is set up to detect X-rays at all possible scattering angles. What are **a.** the maximum and **b.** the minimum wavelengths that are detected?

64. **N** Consider the stream of X-rays and the circular detector in Problem 63.
- **a.** An X-ray scatters, resulting in the maximum possible wavelength being detected. What is the magnitude of the momentum of the electron scattered during this process?
- **b.** An X-ray scatters, resulting in the minimum possible wavelength being detected. What is the magnitude of the momentum of the electron scattered during this process?

65. **N** How many photons are emitted in 3.0 ms by a HeNe laser ($\lambda = 633$ nm) if the laser has a power of 15 mW?

66. **C** CASE STUDY You are helping a friend buy a new digital camera. The salesperson insists that the more expensive camera is better because it has 2 megapixels. Your friend has limited funds and only plans to print relatively small (8×10 in) images. How do you advise your friend? Explain.

67. **N** A proton has a de Broglie wavelength of 74.50 fm. What is its speed?

68. **N** A cup containing 5.67 kg of water is placed in a microwave oven. The oven creates microwaves with a frequency of 2.45 GHz. The initial temperature of the water is 20.0°C, and the specific heat capacity of water is 4186 J/kg · K.
- **a.** How much energy is necessary to raise the water to its boiling point?
- **b.** How many microwave photons are necessary to bring the water to its boiling point, assuming all of the energy from each photon is absorbed by the water?

69. **N** An electric tea kettle is plugged in without water in the kettle, and the heating element of the kettle reaches a temperature of 175.0°C. What is the peak wavelength of the radiation from the heating element?

70. **N** What are the wavelength and the energy in electron volts of a photon with a frequency of **a.** 892 MHz, **b.** 14.8 GHz, and **c.** 7.99 THz (terahertz = 10^{12} Hz)?

71. **N** CASE STUDY In Example 40.10 (page 1331), we found that the kinetic energy of the electrons in an electron microscope is 4 keV. What potential difference must be provided by the electron gun (Fig. 40.17)?

Problems 72 and 73 are paired.

72. **N** Light of ever-increasing frequency is incident on a metal until electrons are observed to be freed from the surface. If free electrons are first observed when the light has a frequency of 5.45×10^{14} Hz, what is the work function of the metal?

73. **N** For the light incident on metal in Problem 72, what stopping potential is necessary to completely stop free electrons created by the light when the light's frequency is **a.** 7.45×10^{14} Hz and **b.** 9.45×10^{14} Hz?

74. **N** Electron microscopes have a much greater resolving power than optical microscopes because electrons have wavelengths that are much shorter than visible light wavelengths. What energy must electrons have to image atoms with a transmission electron microscope operating at a wavelength of 0.0300 nm?

Problems 75, 76, and 77 are grouped.

75. **N** If 5% of the power radiated by a 100-W bulb is emitted as visible light at 550 nm, determine the number of photons that are emitted per second at this wavelength. Answer with two significant figures.

76. **E** Estimate the number of photons that are emitted by the Sun per second.

77. **N** CASE STUDY Let's imagine replacing the electrons in an electron microscope (Fig. 40.17) with protons. If you desired the same resolution as in Example 40.10, what would be the necessary kinetic energy of the protons?

78. **N** A free electron initially at rest is scattered by a photon with an energy of 1.24 MeV in such a way that the scattering angle of the photon and the electron are equal (Fig. P40.78).
- **a.** What is the scattering angle φ?
- **b.** What are the energy and momentum of the photon after it is scattered?
- **c.** What are the kinetic energy and momentum of the electron after it is scattered?

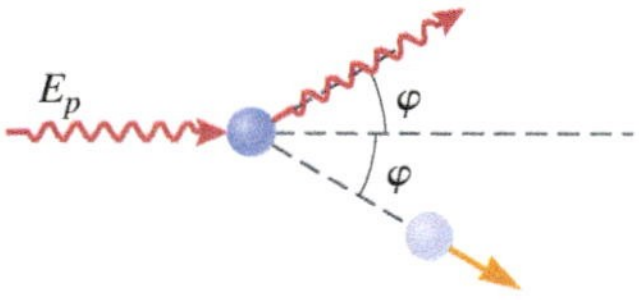

FIGURE P40.78

79. **C** Why can't we use $p = \gamma m_{\text{rest}}c^2$ to find the magnitude of a photon's momentum? *Hint*: What does this expression predict for photon's momentum?

Problems 80 and 81 are paired.

80. **C** A solar sail is a proposed propulsion device for space travel. The sail intercepts photons from the Sun similar to how a sail on a ship intercepts the wind on the ocean (Fig. P40.80). Should the side of the sail that faces the Sun be reflective or absorptive, assuming the goal is to get the most momentum possible from each photon collision with the sail? Explain.

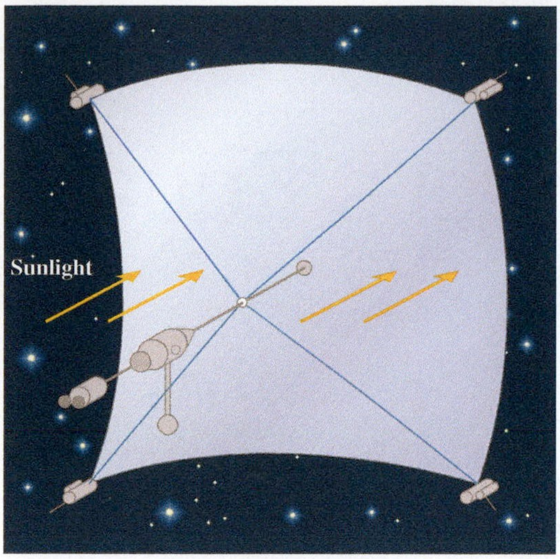

FIGURE P40.80

81. **N** Suppose you have a solar sail with an area of 1.00×10^4 m. In space near the Earth, the intensity of the sunlight is approximately 1.40 kW/m^2. Suppose the light consists entirely of green photons ($\lambda = 535$ nm).

a. If the sail absorbs the photons as they strike the sail, what is the change in momentum of the solar sail each second?

b. If the sail reflects the photons as they strike the sail, what is the change in momentum of the solar sail each second?

82. **N** A free electron initially at rest recoils with a speed of 1.750×10^6 m/s during the Compton scattering of an X-ray photon. The scattering angle of the photon is 21.50°.

a. What is the wavelength of the incident X-ray photon?

b. What is the scattering angle of the electron?

83. **A** The derivation of the Compton shift in Section 40-4 takes many lines of algebra. Start with Equations (5) and (6) from the derivation (page 1322), and eliminate θ from these equations to derive:

$$\gamma^2 m_e^2 v^2 = \frac{h^2}{\lambda_i^2} - \frac{2h^2}{\lambda_i \lambda_f}\cos\varphi + \frac{h^2}{\lambda_f^2}$$

This is Equation (7) in the derivation.

84. **A** The derivation of the Compton shift in Section 40-4 takes many lines of algebra. Substitute Equation (9) into Equation (8) from the derivation (page 1322), and simplify your expression until you derive the Compton shift expressed as:

$$\lambda_i \lambda_f \left(\frac{1}{\lambda_i} - \frac{1}{\lambda_f}\right) = (\lambda_f - \lambda_i) = \frac{h}{m_e c}(1 - \cos\varphi)$$

41 Schrödinger's Equation

Key Questions

If a particle is a wave, then what is doing the "waving"?

In what fundamental way are the predictions made by quantum mechanics different from those made by classical mechanics?

What does quantum mechanics predict for the energy and position of a trapped particle?

Underlying Principles

1. Schrödinger's equation
2. Bohr's correspondence principle
3. Heisenberg's uncertainty principle

Major Concepts

1. Probability waves and probability density
2. Normalization condition
3. Boundary conditions
4. Barrier tunneling

Special Cases:

1. Particle in an infinite square well
2. Particle in a finite square well
3. Quantum simple harmonic oscillator

Tools

Energy-level diagram

You and your friend are up late studying for physics finals. Suddenly, he jumps up and says that he can walk through walls! You think he must be really worried about his exams tomorrow; but then he explains himself. He says that atoms are mostly empty space. The nucleus is a tightly concentrated ball in the center of the atom, and the electrons hover around it in a kind of cloud. So both he and the wall are mostly empty space. He should be able to pass his atoms through the wall's atoms like a comb passes through hair.

You laugh and say that he is wrong. When you press yourself against the wall, it exerts a normal force on you, owing to the way the atoms in your body interact with those in the wall. But he replies that your view is a *classical* view of the world, where everything is knowable and predictable. In modern physics, he says, nothing is knowable. You can only say that things are probable or improbable. So there really is a small probability that he can walk through walls.

Believe it or not, your friend is basically right. According to quantum physics, everything is subject to probability. However, probability matters a lot more on the microscopic scale. It plays a very important role for a single small particle, such as an electron; but on the macroscopic scale, probabilities calculated by quantum physics are essentially indistinguishable from the values determined by classical physics. So can your friend walk through walls? The probability of doing such a thing is so very low that you can safely say *no*—and go back to studying.

41-1 The New Quantum Theory

At the end of the previous chapter, we learned that de Broglie came up with a wave model for matter, such as an electron. This model marks the end of the first phase of the development of the *old* quantum physics theory. The old quantum theory was a great departure from classical physics, but it left many questions unanswered. Why should energy be quantized instead of continuous? And if an electron is a wave, what is doing the "waving"? (We could ask this about any particle.)

The *new* quantum theory was first developed by three independent theorists—Werner Heisenberg (1901–1976), Paul Dirac (1902–1984), and Erwin Schrödinger (1887–1961). Whole textbooks and courses could be devoted to the work of any one of these individuals; in this book, we primarily focus on Schrödinger's contribution.

The Austrian Erwin Schrödinger took a copy of de Broglie's thesis on a vacation during the winter of 1925. Schrödinger was bothered by the ad hoc nature of Planck's quantum theory (Section 40-2). He liked de Broglie's *matter wave* model, but he believed that de Broglie's theory was too vague. While he was on his vacation, he came up with a wave equation for these matter waves. This equation is known as **Schrödinger's (nonrelativistic) time-dependent wave equation**; in one dimension, Schrödinger's equation is:

SCHRÖDINGER'S TIME-DEPENDENT EQUATION **Underlying Principle**

$$-\frac{h^2}{8\pi^2 m}\frac{\partial^2\Psi(x,t)}{\partial x^2} + U(x,t)\Psi(x,t) = \frac{ih}{2\pi}\frac{\partial\Psi(x,t)}{\partial t} \tag{41.1}$$

where U is potential energy. Schrödinger's equation is a wave equation for $\Psi(x, t)$, called the **wave function**. (The symbol Ψ is the uppercase Greek letter psi.) Compare Schrödinger's equation to the classical wave equation:

In Equation 41.1, h is Planck's constant, $i \equiv \sqrt{-1}$, and $\Psi(x, t)$ may be a complex quantity.

$$\frac{\partial^2 y(x,t)}{\partial x^2} = \frac{1}{v_x^2}\frac{\partial^2 y(x,t)}{\partial t^2} \tag{17.33}$$

The two equations are similar in that both involve partial derivatives with respect to x and t. Schrödinger's equation is more complicated than the classical wave equation because a classical wave only carries energy and momentum. Schrödinger's equation describes the propagation of a matter wave, which includes all the things we normally associate with a particle: energy, momentum, mass, and sometimes charge.

Schrödinger's equation is an underlying principle of quantum physics, much as Newton's laws are principles of classical physics. This equation will be our focus in this chapter. Although we cannot derive it, we'll show why it is plausible, and then we'll apply it to a number of situations. Schrödinger's equation makes predictions that are counterintuitive, and even seem impossible—such as the one in this chapter's case study.

CASE STUDY The Sun

The Sun generates energy through the nuclear fusion of hydrogen into helium (Example 39.10) in a three-part chain reaction. The first reaction is the most difficult because it requires fusing two hydrogen nuclei—that is, two protons—together. By Coulomb's law (Section 23-5), two protons repel one another, and the closer they get together, the greater their mutual repulsion. Nothing we have studied so far can explain how two protons can be smashed together. In fact, according to classical physics the reaction shouldn't take place—but we know it does. If it didn't, there would be no sunlight. We'll see how what seems to be impossible becomes possible in quantum physics.

41-2 A Trapped Particle

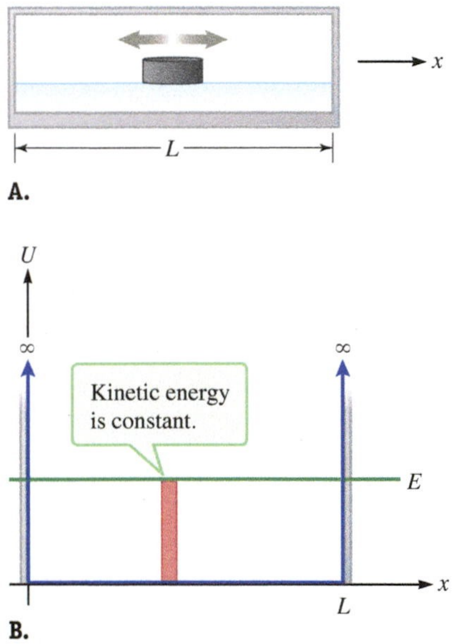

FIGURE 41.1 **A.** A hockey puck on ice is trapped inside a long narrow box. The puck can only move in one dimension, along the x axis. Only the short walls perpendicular to the x axis can limit the puck's one-dimensional motion. **B.** The potential energy curve is zero in the box and infinite at the walls.

Throughout our discussion of quantum mechanics, we'll use a puck as an example of a particle. This is simply a visual aid; you should imagine that the puck is so small that quantum-mechanical effects are important.

Let's start by studying a simple situation using a *quasi-quantum* approach. Consider a hockey puck trapped in a long narrow box, so that it can only move back and forth along a single line (Fig. 41.1A). The surface underneath the puck is ice. We model this as an ideal system consisting of the puck and the walls of the box, in which there are no dissipative forces. No work is done and no energy is lost by heat—so mechanical energy is conserved. Our goal is to use a quasi-quantum approach to find the system's energy, allowing us to see how problems in quantum mechanics are characterized. For this simple example, we'll get the same results we'll find later using Schrödinger's equation (Section 41-5). We call this situation "a particle in a one-dimensional box," because the puck can only move along one dimension. In this chapter we only consider the one-dimensional Schrödinger equation, so this problem is as general as we need. For this one-dimensional problem, we can ignore the two long walls. Because we only need to concern ourselves with the short walls at each end, we'll refer to these simply as "the walls." (The puck is never in contact with the ceiling, and there is no net vertical force exerted on the puck.)

When the puck is moving between the walls of the box, its kinetic energy is constant. Also, no net force is exerted on the system, so the system's potential energy is set to zero. In contrast, when the puck encounters a wall, it experiences a normal force that causes its momentum to be reversed instantaneously, without any change in magnitude. So at the walls, the system's potential energy is infinite. In quantum mechanics we work with potential energy instead of forces. Figure 41.1B shows an energy graph for this system. Between the walls, the system's total energy is finite—that is, too low for the puck to be found outside the box—so the walls are at turning points. If this system were being modeled according to classical mechanics, the puck could be at rest or moving at any speed within the box, and the system's energy E could have any value. But our quasi-quantum approach will show that the system's energy is restricted to certain values.

De Broglie proposed that a particle should be modeled as a wave with a wavelength given by $\lambda = h/p$ (Eq. 40.18). Because the puck is trapped in a box of length L, let's model the puck as a standing wave on a string of length L fixed at both ends, like a guitar string (Section 18-5, page 534). Recall that a standing wave on a string has a wave function of the form given by Equation 18.6, $y(x, t) = [2y_{\text{max}} \sin(kx)] \cos(\omega t)$. We can separate this wave function's time part (the cosine term) from its space part (the sine term). We are only interested in the space part, which we write as

$$y(x) = A \sin kx \tag{41.2}$$

where A is the amplitude of the space part. Because the ends of the guitar string are fixed, they cannot oscillate, so the standing wave that forms on the string must have a node on each end. Therefore, only standing waves with wavelengths given by

$$\lambda_n = 2\frac{L}{n} \qquad (n = 1, 2, 3 \ldots) \tag{18.9}$$

will fit on the string. This restriction on the wavelength can also be expressed as a restriction on the angular wave number $k = 2\pi/\lambda$ (Eq. 17.5) as

$$k_n = \frac{n\pi}{L} \qquad (n = 1, 2, 3 \ldots) \tag{41.3}$$

We say that the wavelength and angular wave number are *quantized,* meaning that only certain separate values are allowed. So the family of possible standing waves is given by

$$y_n(x) = A_n \sin k_n x = A_n \sin \frac{n\pi x}{L} \tag{41.4}$$

where n is referred to as the "harmonic number" and each possible standing wave y_n is called the "nth harmonic." The amplitude of each harmonic is A_n.

Unless otherwise specified, n is a counting number: $n = 1, 2, 3 \ldots$

Now let's go back to modeling the trapped hockey puck. In Equation 41.4, y is the vertical position of the string at the horizontal position x. When we say that the puck is modeled as a wave on a string, we don't mean that its vertical position is waving. We really don't know what is waving (although we'll get a better idea of that in the next section); we just mathematically represent whatever is waving as ψ, and write (by analogy with Eq. 41.4):

$$\psi_n(x) = A_n \sin k_n x = A_n \sin \frac{n\pi x}{L} \tag{41.5}$$

Conceptually, Equation 41.5 says that the trapped puck is modeled as a family of standing waves. Instead of referring to n as the harmonic number, it is called the **quantum number**, and instead of referring to each possible standing wave ψ_n as the "nth harmonic," we refer to each ψ_n as a **quantum stationary state**, or simply a **state**. The lowest state is called the **ground state**, corresponding to $n = 1$.

The angular wave number k_n and the wavelength λ_n of the waves that represent the puck are quantized, just as they are for a standing wave on a string. Further, because de Broglie proposed that the momentum of a matter wave depends on its wavelength, the puck's momentum is quantized:

$$p_n = \frac{h}{\lambda_n} = \frac{nh}{2L} \tag{41.6}$$

Because the system's potential energy is zero inside the box (the only place the puck can be found), its mechanical energy is equal to the puck's kinetic energy. If the puck's mass is m, then the system's energy is $E = p^2/2m$. Since the puck's momentum is quantized, the system's energy is quantized:

$$E_n = \frac{(nh/2L)^2}{2m} = n^2\left(\frac{h^2}{8mL^2}\right) \tag{41.7}$$

Contrast this quasi-quantum analysis of a puck in a box with a classical analysis. According to classical mechanics, the puck can have any momentum and the system can have any energy. However, when we model the puck as a standing wave, we find it can only have certain momenta (Eq. 41.6), and the system can only have certain energies (Eq. 41.7). This sounds a lot like Planck's quantum hypothesis (Section 40-2): Planck proposed that the particles inside a black body have quantized energy levels. He didn't have an explanation for why this is true, but he found that it fit the data. De Broglie's model is a little better because it is based on the idea that nature is symmetric; because light is modeled as both a wave and a particle, matter should be modeled as both as well. The quantization of energy levels is a natural consequence of modeling a particle as a wave. However, de Broglie's model is not completely satisfying because we don't have a reason for modeling the trapped puck as a string fixed on two ends. Part of our goal in this chapter is to

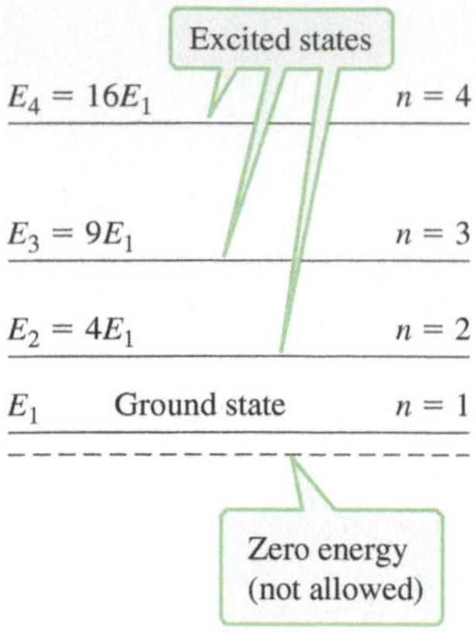

FIGURE 41.2 An energy-level diagram shows the relative energies possible for a system. In this case, the system is a particle in a box with infinitely high walls.

ENERGY-LEVEL DIAGRAM Tool

achieve a more complete understanding of quantum mechanics—then we will return to this trapped particle.

Energy Levels

There are a few more things we can learn from our quasi-quantum approach. First, because the lowest allowed value for the quantum number n is 1, the puck's lowest allowed momentum is $p_1 = h/2L$, which means the puck cannot be at rest. Second, the minimum value of the system's energy is $E_1 = h^2/8mL^2$. Both of these facts are true for all quantum-mechanical systems in which a particle is trapped, and the system's minimum energy E_1 is called its **zero-point energy** or **ground-state energy**. Third, if such a system is to gain or lose energy, it must do so in whole quantized steps, so that the system is in one of the states with an energy given by Equation 41.7.

The energy of a quantum system is often best represented visually by an **energy-level diagram**. The energy of each state is represented by a horizontal line labeled with its quantum number n, and the relative spacing of the lines represents the energy between each state. Figure 41.2 shows the energy levels for a particle trapped in a box (Eq. 41.7). For this system, the energy levels get farther apart for higher quantum numbers, which we can best see by writing the energy levels in terms of the zero-point energy: $E_n = n^2E_1$.

EXAMPLE 41.1 A Hockey Puck in a Box

Suppose that a hockey puck's mass is 0.160 kg and it is in a box of length 6.35 m (Fig. 41.1). Find the system's zero-point energy and the energies of the first three excited states. In **CHECK and THINK**, comment on the difference between the third excited state and the ground state.

INTERPRET and ANTICIPATE

A hockey puck in a box is a classical (macroscopic) system. So we expect to find results that seem familiar.

SOLVE

Find the zero-point energy by setting $n = 1$ in Equation 41.7.

$$E_n = n^2\left(\frac{h^2}{8mL^2}\right) \qquad (41.7)$$

$$E_1 = \frac{h^2}{8mL^2} = \frac{(6.63 \times 10^{-34}\ \text{J}\cdot\text{s})^2}{8(0.160\ \text{kg})(6.35\ \text{m})^2}$$

$$E_1 = 8.52 \times 10^{-69}\ \text{J}$$

The first three excited states have quantum numbers 2, 3, and 4. Write the energy of the excited states in terms of E_1.

$$E_n = n^2E_1$$

$$E_2 = 2^2E_1 = 3.41 \times 10^{-68}\ \text{J}$$

$$E_3 = 3^2E_1 = 7.66 \times 10^{-68}\ \text{J}$$

$$E_4 = 4^2E_1 = 1.36 \times 10^{-67}\ \text{J}$$

CHECK and THINK

The lowest energy level is not zero, but it is so small that we would not distinguish it from zero. Also, keep in mind that we derived the energy levels for the ideal situation in which there are no dissipative forces, and that any real hockey puck would experience dissipative forces. Furthermore, the energy difference between the third excited state and the ground state is about 10^{-67} J. Again, this is too small to be distinguished from zero.

41-3 The Double-Slit Experiment Revisited: Probability Waves

Schrödinger was able to apply his equation to a number of situations, but he did not come up with a physical interpretation of the wave function Ψ. In other words, he was unable to answer the question "*What is doing the waving?*" To answer that question, we must think more about wave-particle duality, and return to the double-slit experiment.

When coherent light passes through two narrow slits, an interference pattern of bright and dark fringes is seen on a screen (Fig. 35.10, page 1128). The pattern is well described by the wave model for light (Section 35-1). We've also seen that objects that we have classically modeled as particles, such as electrons, can produce diffraction patterns (Fig. 40.16). The fringe pattern in part E of Figure 41.3 looks much like the interference pattern created by light passing through two slits, but in this case it was formed by electrons. What is really remarkable about Figure 41.3 is that the fringe pattern is built up one electron at a time. (Photons will also create an interference pattern that is built up one photon at a time.) How can an *interference* pattern be created when each particle (a photon or electron) encounters the slit by itself, with no other particles to interfere with it?

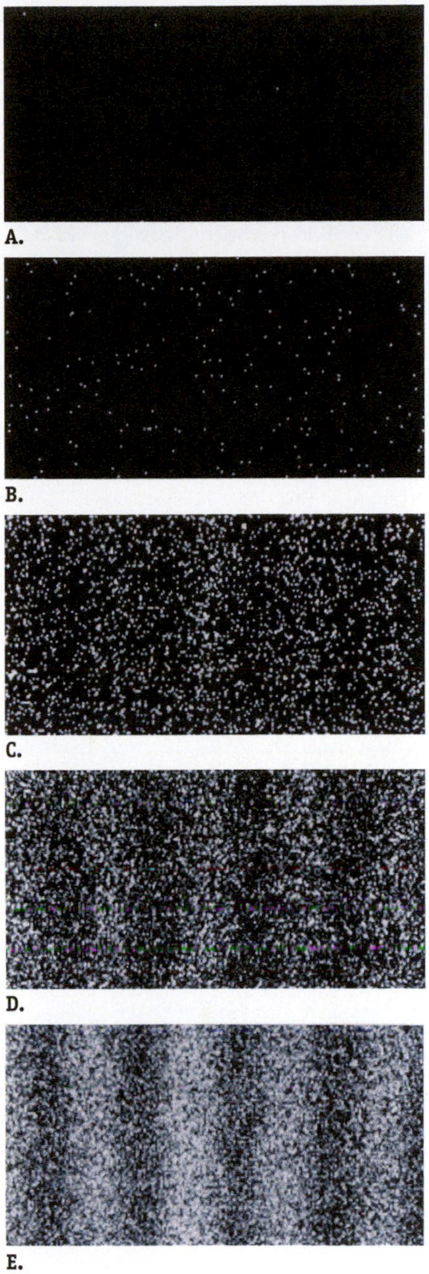

FIGURE 41.3 When individual electrons are successively incident on two slits, an interference pattern of bright and dark fringes builds up. Individual electrons alone do not reveal the pattern; see progression from part **A.** to part **E.**

This question was answered by the German physicist Max Born (1882–1970) who, in doing so, came up with an interpretation for Ψ. He said that for matter waves there is no *physical* interpretation of the wave function Ψ itself, but we can think of Ψ as a **probability wave**. (It makes sense that there is no physical interpretation of Ψ because Ψ involves imaginary numbers.) Furthermore, he said that we don't actually detect Ψ; instead we detect $|\Psi|^2$, which is the **probability density** of finding the particle at a particular location and time. As we already stated, we will mostly consider one-dimensional situations, and so $|\Psi|^2$ is probability per unit length.

PROBABILITY WAVE AND PROBABILITY DENSITY ✪ Major Concept

Some authors refer to a probability wave as a wave function.

As described in Section 41-4, $|\Psi|^2 = \Psi \cdot \Psi^*$.

To understand the idea of a probability wave, let's return to the familiar interference pattern created by photons that are incident on a double slit. We know that one way to model light is as an electromagnetic wave, and that the bright and dark fringes are *not* a direct measure of the light's electric (or magnetic) field. Instead, the brightness of the fringes is a measure of the light's intensity, and intensity is proportional to the amplitude of the electric (or magnetic) field: $I \propto E^2 \propto B^2$ (Eq. 34.30). So the fringes are measuring E^2 (or B^2), but not E (or B) directly. When we see a bright fringe, we infer that amplitude of the electric field (and magnetic field) is great.

Now let's apply the photon model to the double-slit experiment. Imagine that we use narrow detectors, labeled A and B, to count the arrival of photons at two locations on the screen (Fig. 41.4). Let's also imagine that each detector makes a unique sound when it detects a particle: A chirps and B clicks. Because the individual photons arrive at random times, we hear chirps and clicks from the detectors at random time intervals. We cannot predict *when* a particle will arrive at either detector. However, detector A is located on a bright fringe, whereas detector B is located somewhere between a bright fringe and a dark fringe. The bright fringes are bright because they receive many more photons per area than do the other regions of the screen. That is, the probability of a particle arriving at detector A is much greater than the probability of it arriving at B—which means that in the same time interval we expect to hear more chirps from A than clicks from B. We have just reasoned that it is more likely that a particle will be detected at a bright place than at a dark place. Mathematically, we can say that the probability density (here, probability per unit area) of detecting a photon at a particular location on the screen is proportional to intensity, I. Because $I \propto E^2 \propto B^2$, we can argue that the probability density is proportional to E^2 (or B^2). Put simply in informal terms, the "things" that do the waving in an electromagnetic wave are E and B, and the probability density is proportional to their *squares*.

In Born's interpretation, however, Ψ is the "thing" that waves. Ψ is analogous to the electric (or magnetic field) of a light wave, but it has no physical interpretation.

$|\Psi|^2$ is analogous to E^2 (or B^2).

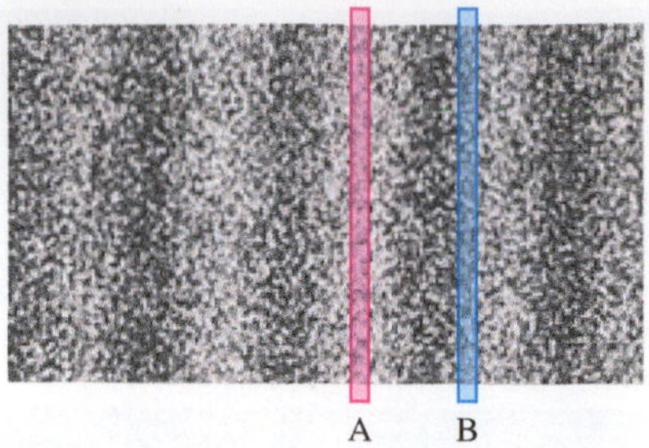

FIGURE 41.4 Detector A is located on a bright fringe; detector B is located between a bright and a dark fringe. (Demonstration of single-electron buildup of an interference pattern Tonomura, A.; Endo, J.; Matsuda, T.; Kawasaki, T.; Ezawa, H. American Journal of Physics, Volume 57, Issue 2, pp. 117–120 (1989). Reprinted courtesy of the Central Research Laboratory, Hitachi, Ltd., Japan.)

So by analogy with the intensity, the probability density is proportional to $|\Psi|^2$. Using this interpretation, we can understand how the interference pattern can be built up one electron (or one photon) at a time. A source generates a particle (an electron or a photon). Associated with *each* particle is a probability wave function Ψ, which we cannot measure directly. This wave travels from the source to the detector (the screen). When the wave arrives at the detector, it is detected as a particle (an electron or a photon). So the pattern on the screen is a measure of the probability density $|\Psi|^2$ of the particle arriving at a particular location. Of course, each particle arrives at some particular location and does not create the entire pattern by itself. The entire pattern is created only when a large number of particles have arrived (Fig. 41.3).

The idea that we can only measure the probability density $|\Psi|^2$ is a major departure from classical physics. According to classical physics, if we know the energy and momentum of a particle at some particular time, we can determine whether or not it will arrive at some detector. (This is just like calculating whether a basketball will drop through the basket.) According to classical mechanics, your answer is completely determined: Its probability is 100%.

Probability only enters into classical physics when we try to study a system with too many particles to count practically. If we are interested in whether a gas molecule will pass through a hole in the wall of its container, we answer with a probability only because there are too many particles to calculate the trajectory of each particle based on each particle's energy and momentum.

The probability that enters into quantum physics is different in nature. Quantum mechanics says that all of our calculations result in probabilities and that none of our answers are 100% determined. So instead of asking, "Will the basketball drop through the basket?" or "Will the gas molecule pass through the hole?" we should ask, "What is the probability of the basketball dropping through the basket or the molecule through the hole?" Quantum mechanics says that this is not a matter of the basketball player's skill or of having too many particles to keep track of. Probability is a fundamental principle of nature.

The idea that nature works in probabilities was very controversial when Born first proposed it. It took about 10 years to become generally accepted, and some scientists—such as Schrödinger, Bohr, and Einstein—never really accepted this interpretation of nature. Although the idea may be hard to accept, the probabilities calculated by using Schrödinger's equation work well to describe many experiments and physical phenomena that cannot be explained by the deterministic interpretation of classical mechanics.

CONCEPT EXERCISE 41.1

Three students were asked why the temperature of a gas is proportional to the *average* kinetic energy of the gas particles as opposed to the *exact* kinetic energy of the particles. (You may wish to review Chapter 20.) Read the discussion and decide who you think is correct. Explain your answer.

Avi: According to Born's interpretation, nature works in probabilities. We can only calculate averages and not exact values.

Cameron: We cannot observe the exact microscopic kinetic energy of every particle in a gas because there are too many particles. We are forced to work with averages when we have so many particles.

Shannon: We use an average because the gas molecules are very tiny, and quantum mechanics applies to tiny particles. That's why we are forced to use Born's interpretation.

EXAMPLE 41.2 Throwing Dice

A die is a cube whose six sides are marked by dots; one side has one dot, another two dots and so forth. If you roll a single die, the probability of rolling a 4 is 1/6 because there are six sides and only one has exactly four dots. If you roll two dice simultaneously, what is the probably of rolling a 4?

INTERPRET and ANTICIPATE

This is a classical mechanical problem, and if we knew exactly the initial conditions of the dice, we could use dynamics and kinematics to determine how the dice land. The only reason we can apply probability is that we don't know exactly how the two dice were tossed. This problem gives us a chance to review probability. The probability is found by determining how many possible ways the two dice may land such that the total is 4 out of all the possible ways the dice may land.

SOLVE

Make a sketch (Fig. 41.5), to find all ways the two dice may land. Imagine one die is blue and the other is red. The sketch shows that if blue die is a 1, the red die may be any value 1 through 6 for six possible combinations. The same reasoning can be used if the blue die is 2 for another six combinations, and so forth.

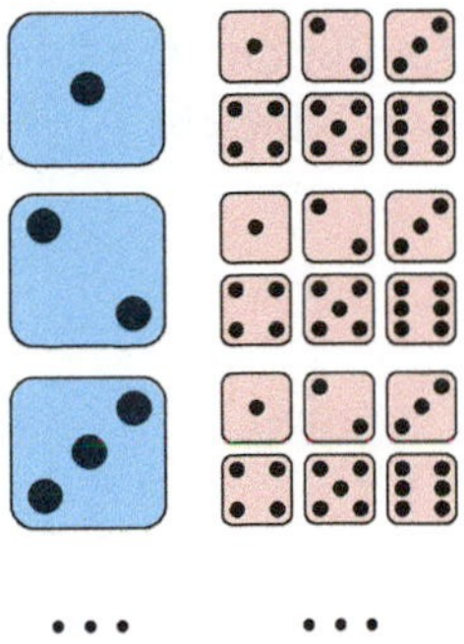

FIGURE 41.5

This sketch (Fig. 41.5) only shows the first three possible values for blue die, but there are six. The sketch shows 18 possible combinations, but that is only half of the total possible combinations.

$$N_{tot} = 36$$

Next make another sketch (Fig. 41.6) to consider all the ways the two dice may be tossed so they total 4.

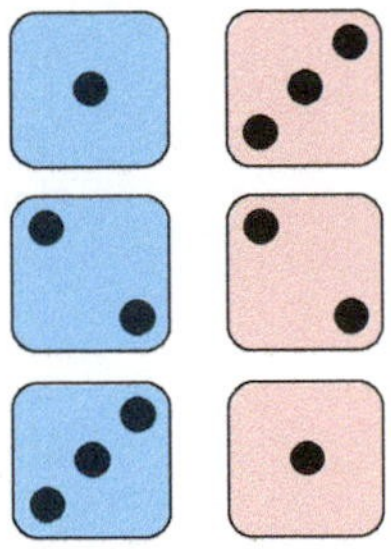

FIGURE 41.6

From Figure 41.6, we see there are three ways to roll a 4.

$$N = 3$$

The probability of rolling a 4 is the number of ways to roll a 4 divided by the number of possible ways the dice may land.

$$P = \frac{N}{N_{tot}} = \frac{3}{36} = \frac{1}{12}$$

$$P \approx 8\%$$

CHECK and THINK

So the probability of rolling a 4 when you roll just one die is twice as likely as when you roll two dice.

41-4 Schrödinger's Equation

With Born's probability interpretation in hand, we now turn our attention back to Schrödinger's equation. When we apply Schrödinger's equation to a situation, our goal is to come up with an expression for Ψ, which we can then use to find the probability density $|\Psi|^2$—a quantity that we can measure.

We must begin with a simplification. In Section 41-1, we presented the time-dependent version of Schrödinger's equation:

$$-\frac{h^2}{8\pi^2 m}\frac{\partial^2\Psi(x,t)}{\partial x^2} + U(x,t)\Psi(x,t) = \frac{ih}{2\pi}\frac{\partial\Psi(x,t)}{\partial t} \tag{41.1}$$

In our study of classical waves, we learned that standing-wave solutions are somewhat easier to study because we can separate the space and time dependencies into two factors, such as $y(x, t) = [2y_{\max} \sin(kx)] \cos(\omega t)$ in Equation 18.6. The same is true for the solutions to Schrödinger's equation; standing-wave solutions can be expressed as $\Psi(x, t) = \psi(x)e^{-i\omega t}$, where the lowercase Greek letter psi ψ represents the space part of the solution and $\omega = 2\pi E/h$. Finding general solutions for the time-dependent partial differential equation adds a layer of complication we do not explore in this book. Here we only consider standing-wave solutions for the time-independent, one-dimensional Schrödinger equation:

TIME-INDEPENDENT SCHRÖDINGER'S EQUATION ❶ **Underlying Principle**

$$-\frac{h^2}{8\pi^2 m}\frac{d^2\psi(x)}{dx^2} + U(x)\psi(x) = E\psi(x) \tag{41.8}$$

where E is the mechanical energy of the system.

We don't need partial derivatives in the time-independent equation.

Equation 41.8 holds for any system in which the potential energy $U(x)$ only depends on space (and not time) and the mechanical energy E is a constant. You will derive this time-independent equation in Problem 41.11. Conceptually, you can think of Schrödinger's equation as describing a particle modeled as a matter wave; the particle is part of a system whose mechanical energy is E and whose potential energy $U(x)$ does not change with time, but varies from place to place. From now on, when we refer to the Schrödinger equation, you can assume we mean the time-independent version (Eq. 41.8) unless otherwise specified.

A Plausible Equation

Schrödinger's equation is a departure from classical physics, so it cannot be derived from the principles of classical physics. However, we can show that it seems plausible. The argument that follows is a test that is just a little more thorough than checking the equation's dimensions. Recall that Schrödinger was looking for a theory to support de Broglie's matter wave model. Assume that $\psi(x)$ is a simple sine wave with a wavelength given by the de Broglie wavelength $\lambda = h/p$ (Eq. 40.18):

$$\psi(x) = \psi_{\max} \sin kx = \psi_{\max} \sin\frac{2\pi x}{\lambda}$$

Take the second derivative with respect to x:

$$\frac{d^2\psi}{dx^2} = -\psi_{\max}\frac{4\pi^2}{\lambda^2}\sin\frac{2\pi x}{\lambda} = -\frac{4\pi^2}{\lambda^2}\psi \tag{41.9}$$

Use $K = p^2/2m$ to write the de Broglie wavelength in terms of the particle's (nonrelativistic) kinetic energy:

$$\lambda = \frac{h}{p} = \frac{h}{\sqrt{2mK}}$$

Substitute into Equation 41.9:

$$\frac{d^2\psi}{dx^2} = -4\pi^2\frac{2mK}{h^2}\psi = -\frac{8\pi^2 mK}{h^2}\psi \tag{41.10}$$

Finally, substitute $K = E - U$ and rearrange:

$$\frac{d^2\psi}{dx^2} = -\frac{8\pi^2 m(E - U)}{h^2}\psi$$

$$-\frac{h^2}{8\pi^2 m}\frac{d^2\psi(x)}{dx^2} + U(x)\psi(x) = E\psi(x) \checkmark$$

We have *not derived* Schrödinger's equation; instead we have simply shown that it is plausible that the equation describes a particle modeled as a matter wave with a de Broglie wavelength in a system with potential energy U.

Finding Solutions

Now we're ready to explore solutions to Schrödinger's equation. Each solution ψ depends on a particular form of the potential energy $U(x)$. Once we arrive at a solution, $|\psi|^2$ is the probability density of finding the particle at some position x. Because we are working with one-dimensional systems, the probability density is the probability per unit length, and the probability of finding the particle in a region dx is $|\psi|^2 dx$. Of course, the particle must be found somewhere along the x axis; mathematically, we express this commonsense idea as

NORMALIZATION CONDITION

Major Concept

$$\int_{-\infty}^{\infty} |\psi(x)|^2 dx = 1 \quad 41.11$$

Equation 41.11 is known as the **normalization condition**, and it is required in order to interpret $|\psi|^2$ as a probability density. In practice, we must measure the positions of a large number of identical systems with the same wave function ψ in order to come up with an average value for the position x. The average position of x is called the **expectation value** of x, and is given by:

The symbol $\langle q \rangle$ means the expectation value of the quantity q. Don't be alarmed by this terminology: it is the same as the familiar phrase *average* value of q.

$$\langle x \rangle \equiv \int_{-\infty}^{\infty} x|\psi(x)|^2 dx \quad (41.12)$$

The process of solving Schrödinger's equation is mathematical. From our experience with physics, we know that often there are solutions that work mathematically but have no physical meaning. In order for a solution ψ to be physically meaningful so that we can interpret $|\psi|^2$ as a probability density, we require $\psi(x)$ to meet the following conditions.

1. $\psi(x)$ and $d\psi(x)/dx$ must be a continuous, single-valued function for all points along x. (Otherwise we couldn't take the second derivative $d^2\psi(x)/dx^2$ in Schrödinger's equation.)
2. $\psi(x)$ must equal zero if x is in a region where the particle cannot be found.
3. $\psi(x)$ must meet the normalization condition (Eq. 41.11). This is same as requiring that $\psi(x) \rightarrow 0$ as $x \rightarrow \pm\infty$ in order for the integral to converge.

In general, the solution $\psi(x)$ to Schrödinger's equation may be a complex number. Then the probability density is $|\psi|^2 = \psi \cdot \psi^*$, where ψ^* is called the **complex conjugate**. The complex conjugate is found from ψ by replacing i with $-i$. For example, if $\psi = \psi_{max} e^{ikx}$, then its complex conjugate is $\psi^* = \psi_{max} e^{-ikx}$, and $|\psi|^2 = \psi^2_{max} e^{ikx} e^{-ikx} = \psi^2_{max}$. Much of the rest of this chapter is devoted to $\psi(x)$ and then to finding the probability density ψ^2 for a number of special cases. Our major goal is to find the energy levels of each system and present these on an energy-level diagram.

One way that we will check our results is with **Bohr's correspondence principle**. According to this principle, in the limit of very large quantum numbers, the classical calculation and the quantum calculation must yield the same results. This principle is much like what we do in special relativity, when we check to see that at low relative speed our results are consistent with Galilean relativity.

BOHR'S CORRESPONDENCE PRINCIPLE

Underlying Principle

CONCEPT EXERCISE 41.2

Come up with at least one example from the previous 40 chapters where a mathematical solution does not make sense physically. *Hint*: You are likely to find an example from as early as Chapter 2.

EXAMPLE 41.3 Average Position for a Particle in an Infinite One-Dimensional Well, Classical Approach

Consider the hockey puck trapped between the walls of the one-dimensional box in Figure 41.1. As described in Section 41-2, when the puck encounters a wall, it reverses direction instantaneously and there are no dissipative forces. So the puck's kinetic energy is constant and nonzero. Use classical mechanics to find the (average position) expectation value of x for a puck in an infinite well of width L. (In Example 41.4, we will take a quantum-mechanical approach to answer the same question.)

INTERPRET and ANTICIPATE

The expectation value is another way of saying the average value. Because the puck slides back and forth without losing speed, you might expect that its average position is in the center of the box.

SOLVE

The puck doesn't change speed, and so it is equally likely to be anywhere inside the box. So the probability density $|\psi|^2$ is a constant between the walls of the box and zero outside the box as shown in Figure 41.7.

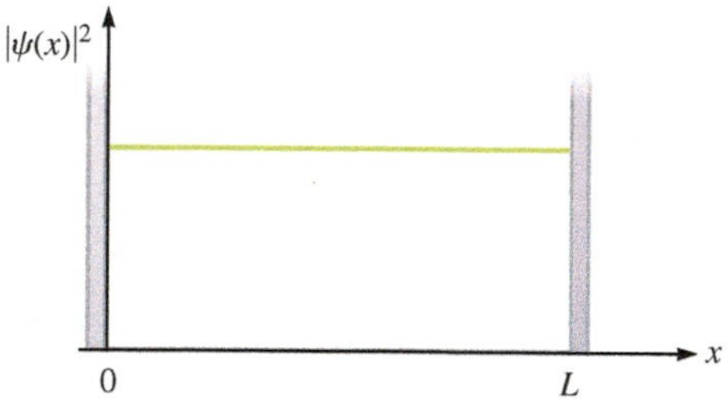

FIGURE 41.7

Of course, the puck must be located somewhere along the x axis inside the box. To express this commonsense idea, we apply the normalization condition (Eq. 41.11). In this case, we know $|\psi|^2$ is a constant between $x = 0$ and L and zero elsewhere.

$$\int_{-\infty}^{\infty} |\psi(x)|^2 dx = 1 \tag{41.11}$$

$$\int_{-\infty}^{0} |\psi(x)|^2 dx + \int_{0}^{L} |\psi(x)|^2 dx + \int_{L}^{\infty} |\psi(x)|^2 dx = 1$$

$$0 + |\psi(x)|^2 \int_{0}^{L} dx + 0 = |\psi(x)|^2 L = 1$$

$$|\psi(x)|^2 = \frac{1}{L} \text{ (for } 0 < x < L)$$

Now we are ready to find the expectation value of x using Equation 41.12. (Keep in mind that the integrals from $x = -\infty$ to 0 and from $x = L$ to ∞ are zero because $|\psi|^2$ is zero in these regions. These two integrals are not shown here.)

$$\langle x \rangle = \int_{-\infty}^{\infty} x|\psi(x)|^2 dx \tag{41.12}$$

$$\langle x \rangle = \int_{0}^{L} x|\psi(x)|^2 dx = \int_{0}^{L} x\frac{1}{L} dx = \frac{1}{L}\left(\frac{x^2}{2}\right)\Bigg|_{0}^{L}$$

$$\langle x \rangle = \frac{L}{2}$$

CHECK and THINK

We found that the average position is in the center of the box as expected. In Example 41.4, we'll find the same thing using quantum mechanics.

41-5 Special Case: A Particle in an Infinite Square Well

Now we are ready to return to a particle trapped in a one-dimensional box, like the puck in Figure 41.1. This time, however, we'll solve Schrödinger's equation to find the system's energy levels. We'll gain a deeper insight into quantum mechanics because we'll also find expressions for where the particle is most likely to be found. Because we know that the quantum nature of a hockey puck in a box is difficult to observe, it is better to imagine a much smaller particle, such as a proton or an electron, that is "trapped in a box" owing to some electric potential energy well.

We use the term *potential energy well* to describe the general shape of a potential energy curve that rises on two sides, known as its **boundaries.** Figure 41.1B is a good illustration for a particle in an infinite *square* well. The system consists of the (unknown) source of potential energy and a particle that is trapped between $x = 0$ and $x = L$. The potential energy is $U = 0$ for $0 < x < L$ and $U = \infty$ for $x \leq 0$ and for $x \geq L$.

BOUNDARY CONDITION

Major Concept

The particle cannot be outside the box, so $\psi(x) = 0$ for $x \leq 0$ and for $x \geq L$. The value that $\psi(x)$ must have at a boundary is known as the **boundary condition.** We'll use boundary conditions to help make a general solution to Schrödinger's equation fit a particular situation. The strict boundary conditions in this special case correspond to our physical intuitions about an infinite square well. But when we solve Schrödinger's equation in the more general case of a *finite* square well using more relaxed boundary conditions, we'll find that our intuition that the particle cannot be outside the infinite square well is correct.

Inside the well $0 < x < L$, $U(x) = 0$, and Schrödinger's equation is

$$-\frac{h^2}{8\pi^2 m}\frac{d^2\psi(x)}{dx^2} = E\psi(x)$$

It is convenient to combine the constants as $k^2 \equiv 8\pi^2 mE/h^2$; then we write Schrödinger's equation as:

$$\frac{d^2\psi(x)}{dx^2} + k^2\psi(x) = 0 \tag{41.13}$$

The general solution to Equation 41.13 may be written as:

$$\psi(x) = A\sin kx + B\cos kx \tag{41.14}$$

We apply the boundary condition $\psi(0) = 0$:

$$\psi(0) = A\sin 0 + B\cos 0 = 0$$

and we find that $B = 0$. So now our solution is a little more specific; there is no cosine term. We apply the other boundary condition $\psi(L) = 0$:

$$\psi(L) = A\sin kL = 0$$

and we find that this boundary condition holds if k is quantized:

$$k_n = \frac{n\pi}{L} \quad (n = 1, 2, 3 \ldots) \tag{41.15}$$

where we start with $n = 1$. Although $n = 0$ also fits the boundary condition, it cannot be normalized, Put another way, allowing $n = 0$ is the same as saying that the wave does not exist. So we reject this physically unacceptable solution.

Without finding our complete solution to Schrödinger's equation, we have already found that its energy levels are quantized. To see this, combine $k^2 \equiv 8\pi^2 mE/h^2$ with Equation 41.15:

PARTICLE IN AN INFINITE SQUARE WELL

Special Case

$$E_n = \left(\frac{h^2}{8\pi^2 m}\right)\left(\frac{n^2\pi^2}{L^2}\right) = n^2\left(\frac{h^2}{8mL^2}\right) \tag{41.16}$$

This is exactly what we found using our quasi-quantum approach (Eq. 41.7). Now, however, we have a better understanding of how we got to these quantized levels. Schrödinger's equation describes the mechanical energy of the system. The potential energy is zero in the box, and the particle cannot be found outside the box. When we express the boundary conditions mathematically, we find that the only solution to Schrödinger's equation requires the energy levels to be quantized. This is a general truth about trapped particles: The energy of of a trapped particle comes in discrete values.

Normalizing the Wave Function

Let's complete our solution by normalizing $\psi(x)$. So far, $\psi(x) = A \sin k_n x$ inside the well, where k_n is given by Equation 41.15. When we normalize the wave function, we'll find A. Conceptually, the normalization condition says that the particle must exist somewhere, and mathematically we have $\int_{-\infty}^{\infty} |\psi(x)|^2 dx = 1$ (Eq. 41.11). Substitute $\psi(x)$ into the normalization condition:

$$\int_{-\infty}^{\infty} |A \sin k_n x|^2 dx = 1$$

We break this integral into three parts, corresponding to potential energy values to the left of the well, inside the well, and to the right of the well:

$$\int_{-\infty}^{0} |A \sin k_n x|^2 dx + \int_{0}^{L} |A \sin k_n x|^2 dx + \int_{L}^{\infty} |A \sin k_n x|^2 dx = 1$$

We have already established that the particle cannot be outside the box, so the first and last integrals are zero:

$$\int_{0}^{L} |A \sin k_n x|^2 dx = 1$$

Using

$$\int \sin^2 ax\, dx = \frac{x}{2} - \frac{\sin 2ax}{4a} + C$$

from Appendix A to solve this integral, we find that $A^2(L/2) = 1$ (Problem 41.79). So $A = \sqrt{2/L}$ and

$$\psi(x) = \sqrt{\frac{2}{L}} \sin k_n x \tag{41.17}$$

Interpreting the Wave Function

Equation 41.17 is the solution to Schrödinger's equation for a particle in an infinite square well. (Of course, outside the well $\psi = 0$.) Although we have no physical interpretation for ψ, $|\psi|^2$ is the probability density, given by:

$$|\psi|^2 = \frac{2}{L} \sin^2 k_n x \tag{41.18}$$

Figure 41.8 shows the probability density function plotted for four different states.

For a moment, let's think classically about the hockey puck oscillating back and forth between the walls of its box (Fig. 41.1). You might ask, "At any instant, where is the puck most likely to be in the box?" The answer has to do with the puck's speed as it moves back and forth. If the puck maintains its speed, then at any instant it is equally likely to be anywhere inside the box (Example 41.3 and Problem 41.14). Of course, this is true for the puck in the box whether the system's energy is great or small.

By analogy, Equation 41.18 and Figure 41.8 are attempts to answer the question, "At any instant, where is the particle most likely to be in the well?" Perhaps the results for the first quantum states are the most surprising because these show that the particle is most likely to be in one, two, or three particular regions in the box. Only for a high quantum number state are the maxima so closely spaced that it becomes equally likely to find the particle anywhere in the box.

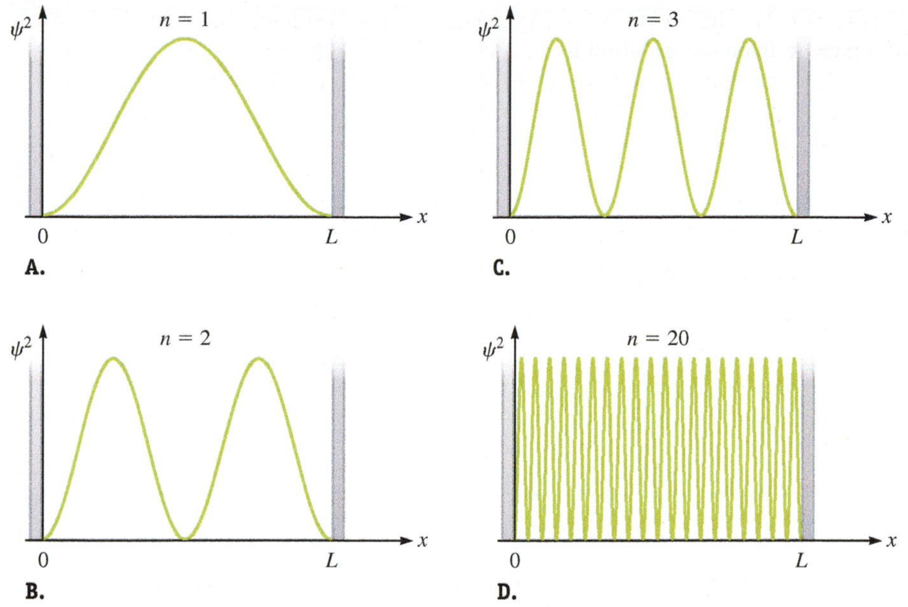

FIGURE 41.8 The probability densities for a particle in a box for quantum states **A.** $n = 1$, **B.** $n = 2$, **C.** $n = 3$, and **D.** $n = 20$.

However, we need to be careful with our analogy. Our model of a puck sliding back and forth between the walls of a box is insufficient for a particle in quantum mechanics. Instead, we model a particle as a wave whose probability density is shown in Figure 41.8. Instead of visualizing a solid particle in motion, it is better to think of (the probability distribution of) the particle as a cloud. In its ground state, the cloud is most concentrated in the center of the box and thinly spread over the rest. In its $n = 2$ state, the cloud is concentrated in two regions and so on. For a very high quantum number, the cloud is well distributed throughout the box.

CONCEPT EXERCISE 41.3

Does Figure 41.8 show the motion of a trapped particle? If so, describe this motion in words. If not, describe what the figure is really showing about the particle.

EXAMPLE 41.4 Average Position for a Particle in an Infinite One-Dimensional Well, Quantum Approach

Find the expectation value of x for a particle in an infinite well of width L. This time apply the solution we found to Schrödinger's equation for this situation.

INTERPRET and ANTICIPATE

This problem is the same as Example 41.3, but here we take a quantum-mechanical approach. In the end we'll compare our results.

SOLVE

Start with Equation 41.12 for the expectation value of x. Change the limits of integration, since the particle cannot be found outside the box.

$$\langle x \rangle = \int_{-\infty}^{\infty} x|\psi(x)|^2\,dx \qquad (41.12)$$

$$\langle x \rangle = \int_{0}^{L} x|\psi(x)|^2\,dx$$

Substitute $\lvert\psi\rvert^2 = \frac{2}{L}\sin^2 k_n x$ (Eq. 41.18), which came from our solution to Schrödinger's equation.	$\langle x\rangle = \int_0^L x\left(\frac{2}{L}\sin^2 k_n x\right)dx$
This integral can be solved by parts.	$\langle x\rangle = \frac{2}{L}\left(\frac{x^2}{4} - \frac{x\sin 2k_n x}{4k_n} - \frac{\cos 2k_n x}{8k_n^2}\right)\Big\rvert_0^L$
Substitute the limits and use $k_n = n\pi/L$ (Eq. 41.15). Notice that $\sin 2n\pi = 0$ and $\cos 2n\pi = 1$. (In quantum mechanics the energy is quantized.)	$\langle x\rangle = \frac{2}{L}\left(\frac{L^2}{4} - \frac{\cos 2k_n L}{8k_n^2} + \frac{1}{8k_n^2}\right)\Big\rvert_0^L$ $\langle x\rangle = \frac{2}{L}\left(\frac{L^2}{4} - \frac{\cos 2n\pi}{8k_n^2} + \frac{1}{8k_n^2}\right)$ $\langle x\rangle = \frac{L}{2}$

CHECK and THINK

Many times quantum mechanics seems counterintuitive, but not this time. Here the average position is in the center of the box, just as you'd expect (and found in Example 41.3).

EXAMPLE 41.5 A Ground-State Energy Free Particle

Take the limit of $E_1 = h^2/8mL^2$ (Eq. 41.16 with $n = 1$) to find the zero-point energy of a free particle.

INTERPRET and ANTICIPATE

The key to this problem is knowing what limit we must take in order to go from the energy expression for a trapped particle to that for a free particle. There are only two parameters we can change: m and L. Changing the mass does not set a particle free, but increasing the width of the well does. An infinitely wide well means that the particle can go anywhere.

SOLVE Take the limit as $L \to \infty$.	$\lim_{L\to\infty} E_1 = \lim_{L\to\infty} \frac{h^2}{8mL^2} \to 0$

CHECK and THINK

When a particle is free (not part of any system), it can have any energy value. So its minimum energy is zero and it may be at rest. Only a confined particle is required to have nonzero energy and to be in motion.

41-6 Special Case: A Particle in a Finite Square Well

In the previous section, we applied quantum mechanics to a particle trapped in a square well with infinite sides. Now we consider a particle in a finite square well. Astonishingly, we'll find that particle may be found slightly outside of the box! This is something like watching a hockey puck ooze through the walls of its box. This bizarre behavior is predicted by the solution to Schrödinger's equation.

The finite square well looks much like the infinite square well; as in that case, the potential energy is $U = 0$ for $0 < x < L$ inside the well (Fig. 41.9). The difference is that the potential energy is finite and uniform outside the well: $U = U_0$ for $x \leq 0$ and for $x \geq L$. We are interested in a trapped particle, so we will only consider a system for which $E < U_0$.

Schrödinger's equation for the region inside the well is the same as it was in the case of an infinite well:

$$\frac{d^2\psi(x)}{dx^2} + k^2\psi(x) = 0 \tag{41.13}$$

where $k^2 \equiv 8\pi^2 mE/h^2$. The general solution must be the same as in the infinite well:

$$\psi(x) = A\sin kx + B\cos kx \tag{41.14}$$

But now the boundary conditions are different. We no longer require that $\psi(x) = 0$ for $x \leq 0$ and for $x \geq L$. Instead, we only need $\psi(x)$ and $d\psi/dx$ to be continuous functions at the boundaries. (Later in this chapter we will show that these boundary conditions are sufficient to produce the same results as in the case of the infinite square well when $U_0 \to \infty$.) With these more relaxed boundary conditions, $B \neq 0$. Both terms in Equation 41.14 hold inside the well, and together they describe an oscillating function.

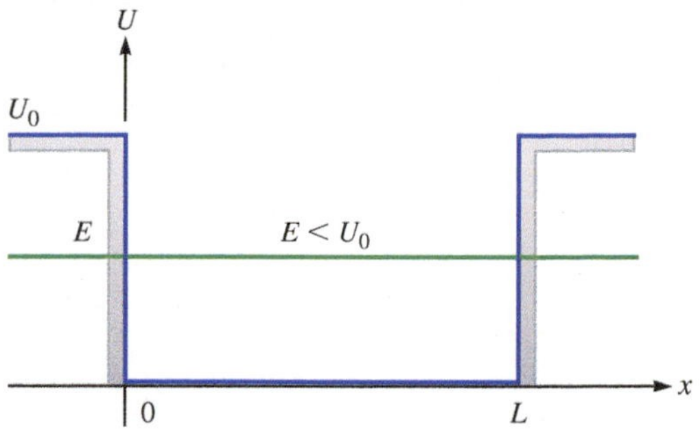

FIGURE 41.9 The potential energy is zero inside the well and finite outside.

Outside the well, Schrödinger's equation is:

$$-\frac{h^2}{8\pi^2 m}\frac{d^2\psi(x)}{dx^2} + U_0\psi(x) = E\psi(x)$$

We group terms to write:

$$\frac{d^2\psi(x)}{dx^2} - \frac{8\pi^2 m}{h^2}(U_0 - E)\psi(x) = 0$$

and define $\kappa^2 \equiv (8\pi^2 m/h^2)(U_0 - E)$, so that we have:

$$\frac{d^2\psi(x)}{dx^2} - \kappa^2\psi(x) = 0 \tag{41.19}$$

for Schrödinger's equation outside the well. But we are only considering the case of a particle trapped in the well, such that $E < U_0$ or $\kappa^2 > 0$. The general solution to Equation 41.19 in this case is:

$$\psi(x) = \alpha e^{-\kappa x} + \beta e^{\kappa x} \tag{41.20}$$

In the region to the left of the well, where $x < 0$, we require $\psi(x) \to 0$ as $x \to -\infty$, and so we must set $\alpha = 0$ in this region and the solution reduces to $\psi_{\text{left}}(x) = \beta e^{\kappa x}$. In the region to the right of the well, where $x > L$, we require $\psi(x) \to 0$ as $x \to \infty$, and so we must set $\beta = 0$ in this region and the solution reduces to $\psi_{\text{right}}(x) = \alpha e^{-\kappa x}$.

In principle, we solve for A, B, α, and β by ensuring that $\psi(x)$ and $d\psi/dx$ are continuous at both boundaries. In practice, this is a tedious process, and unnecessary for the points we wish to make here. Instead we'll present the solution graphically for the first three quantum states in Figure 41.10A. Like the infinite square well (Fig. 41.10B), the solution inside the finite square well is sinusoidal. The major difference is that outside the well, the solution is a decaying exponential function. However, you can see a more subtle difference if you carefully compare the solutions in parts A and B of the figure: The wavelengths inside the finite square well are slightly

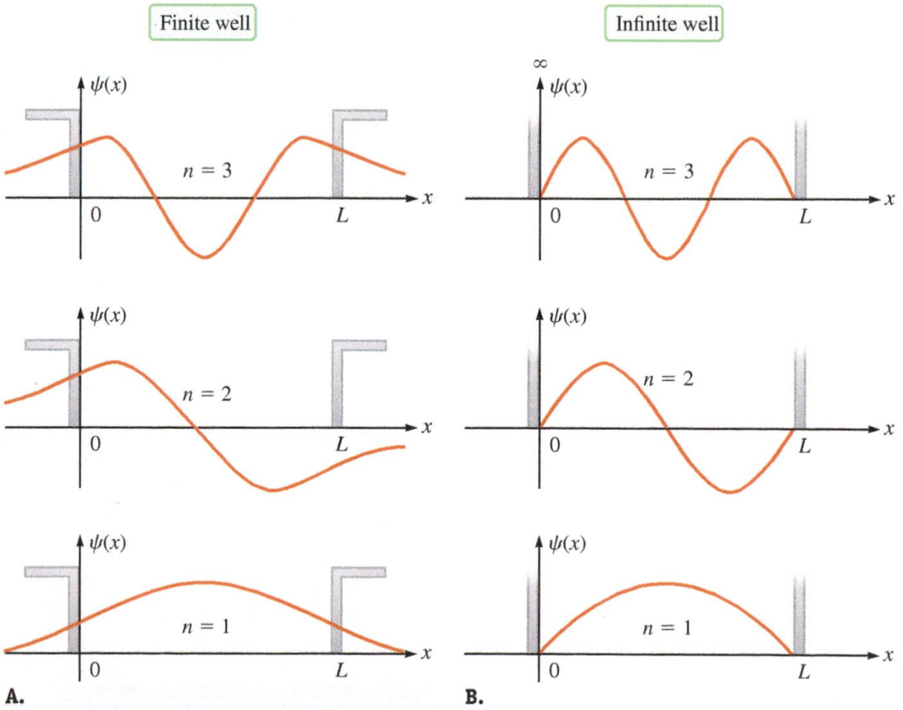

FIGURE 41.10 **A.** The wave functions for a finite square well extend beyond the boundaries of the well. **B.** The wave functions for an infinite square well are zero beyond the boundaries.

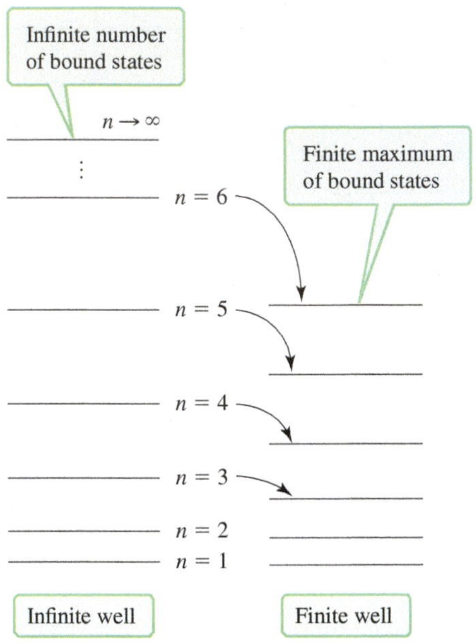

FIGURE 41.11 Compare the energy levels for two square wells of the same width. The well with a finite height has lower energy levels.

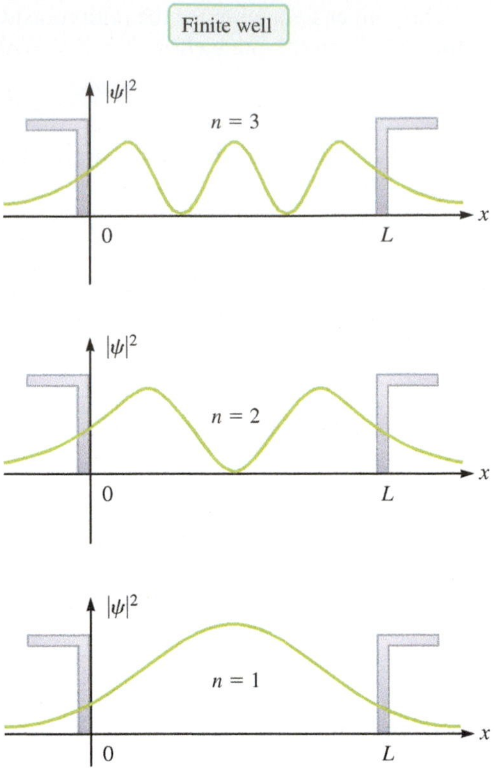

FIGURE 41.12 The probability densities for the first three quantum states of a particle in a finite square well.

longer than the wavelengths inside the infinite square well. Because $E_n = p_n^2/2m$ and $p_n = h/\lambda_n$ (Eq. 41.6), the longer wavelengths for the finite square well mean that the energy levels there are a little lower (Fig. 41.11). Also, if the system's energy is greater than U_0, the particle is free. So there are a finite number of bound energy levels in the case of a finite square well.

PARTICLE IN A FINITE SQUARE WELL
Special Case

Of course, the most amazing difference between the infinite and the finite square wells is that in the case of the finite well, the particle has a nonzero probability of being found outside the well (Fig. 41.12). According to classical physics, it would be impossible for the particle to be found outside the well. Because the system's energy E is less than the potential energy U_0, a particle outside the well would have negative kinetic energy. So the region beyond the boundaries of the well is called the *classically forbidden region.* The probability of finding the particle beyond the boundaries is small and drops off as the distance beyond the boundaries increases. The **penetration distance** Λ depends on κ:

$$\Lambda_n = \frac{1}{\kappa_n} = \frac{h}{2\pi\sqrt{2m(U_0 - E_n)}} \tag{41.21}$$

The penetration distance is a measure of how far the particle may be reasonably found in the classically forbidden region. (When $U_0 \to \infty$, that is, when the finite square well approaches an *infinite* square well, the penetration distance $\Lambda_n \to 0$ in agreement with our intuition that the particle cannot be found outside an infinite well.) This quantum-mechanical possibility, that a particle can be outside the boundaries of an finite square well, plays an important role in our case study.

CONCEPT EXERCISE 41.4

Use cloud imagery to describe Figure 41.12.

EXAMPLE 41.6 The Penetration Distance of Particle in a Finite Square Well

When a particle in a finite square well is in its ground state, its penetration distance is Λ_1. The well's height is $U_0 = 5.55$ eV. When the same system is in an excited state with an energy of 5.35 eV, its penetration distance is $4\Lambda_1$. What is the system's ground-state energy (in eV)?

INTERPRET and ANTICIPATE
The penetration distance depends on the system's energy (Eq. 41.21). We expect the ground-state energy to be lower than the excited state's energy.

SOLVE
We are given that when the system is in the excited state, the penetration distance is four times longer than when it is in the ground state.

$$\Lambda_x = 4\Lambda_1 \qquad (1)$$

Substitute Equation 41.21 into Equation (1) and simplify.

$$\frac{h}{2\pi\sqrt{2m(U_0 - E_x)}} = \frac{4h}{2\pi\sqrt{2m(U_0 - E_1)}}$$

$$\frac{1}{\sqrt{(U_0 - E_x)}} = \frac{4}{\sqrt{(U_0 - E_1)}}$$

Solve for E_1.

$$E_1 = 16E_x - 15U_0$$

$$E_1 = 16(5.35 \text{ eV}) - 15(5.55 \text{ eV})$$

$$E_1 = 2.35 \text{ eV}$$

CHECK and THINK
As expected, the ground-state energy is less than the excited state's energy.

41-7 Barrier Tunneling

We have just found that a trapped particle has a nonzero probability of being found outside its confines (Fig. 41.12). This amazing result contradicts classical mechanics, which is based on our everyday experience. However, our lives depend daily on reactions in the Sun that require particles to "slip out of their containers." As an analogy, let's consider a more ordinary experience from a classical perspective.

Isaac is playing an unusual game of catch with his mother. Isaac is at the bottom of a slide and his mother is behind the slide; Isaac rolls a ball up the slope (Fig. 41.13). Ideally, the ball makes it over the top of the slide and falls into his mother's hands. This is a problem that we can analyze with classical physics. Consider a system that consists of the ball of mass m and the Earth, and use conservation of energy. We'll ignore the rolling kinetic energy of the ball and any energy lost due to dissipative forces. If we assign the zero point of gravitational potential energy to the point at the bottom of the slide where Isaac releases the ball, then initially the only energy in the system is the ball's kinetic energy K_i. The potential energy at the top of the slide is mgy, where y is the height of the slide. So if the ball's kinetic energy at the bottom is greater than the potential energy at the top ($K_i > mgy$), then the ball will make it over the top and to Isaac's mother. If the ball's initial kinetic energy is less, so that the potential energy at its peak $K_i < mgy$, then the ball will get partway up the slide before rolling back down to Isaac. (Of course, if $K_i = mgy$, the ball will reach the top of the slide and stop there.) Classical physics is completely deterministic. If the ball has enough energy at the bottom

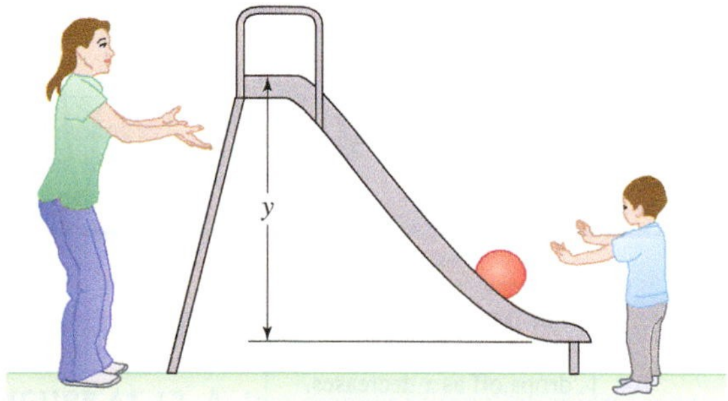

FIGURE 41.13 Isaac's ball will make it to his mother if the ball's kinetic energy at the bottom is greater than the system's potential energy when the ball is at the top of the slide.

CASE STUDY Hydrogen Fusion in the Sun

Barrier tunneling happens all the time inside the Sun. The Sun generates the energy for its light (the ultimate source of all energy on the Earth) through nuclear fusion reactions in its core. From Example 39.10 (page 1295), the net fusion reaction in the Sun is

$$4\text{H} \rightarrow \text{He}$$

This is a nuclear reaction in which four hydrogen nuclei are fused together to make a single helium nucleus. A hydrogen nucleus is just a proton, whereas a helium nucleus is made up of two protons and two neutrons. Fusion in the Sun is a three-step chain reaction. In the first step, two protons must come together.

In classical physics, there are two problems with this first step. The first of them applies to any nucleus. According to Coulomb's law, the protons are repelled by one another with a force that grows stronger the closer together they are. So how can protons stay together in a nucleus? Why are many nuclei stable? The answer is that another force is involved called the *strong nuclear force* (or simply the *strong force*), a short-range attractive force that is stronger than Coulomb's repulsive force inside the nucleus. When the protons are very close together in the nucleus, we can ignore the Coulomb force.

The strong force explains why nuclei are stable. But there is a second problem: Outside the nucleus, Coulomb's repulsive force dominates. How can two protons ever get close enough together to form a nucleus? The answer has to do with barrier tunneling.

To keep things simple, imagine that one proton is at rest at the origin, $x = 0$. Another proton approaches it from the positive x direction. Figure 41.18 shows the potential energy of the two-proton system. When the protons are farther apart than x_0, the Coulomb force dominates and the electric potential energy falls off as shown in the right part of the graph. When the protons are closer together than x_0, the strong force dominates and the potential energy is a flat-bottomed well. In order for the protons to stick together, then, they must get closer together than x_0.

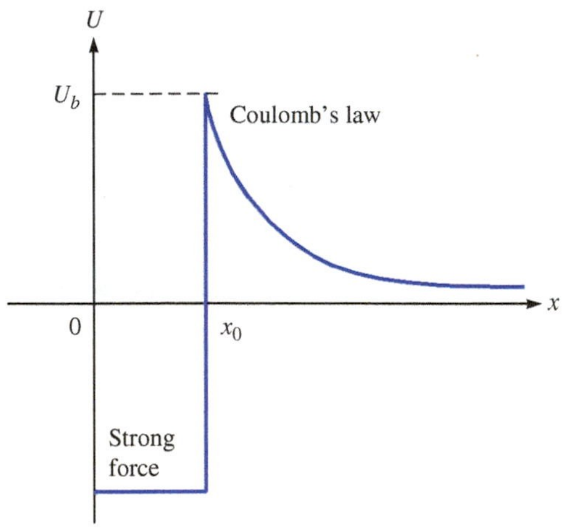

FIGURE 41.18 According to quantum mechanics, it is possible for two protons to come together even if the kinetic energy is less than peak of the potential energy curve $K_i < U_b$.

The electric potential curve forms a barrier shaped something like a playground slide. According to classical physics, the protons will get together if the moving proton's initial kinetic energy (when it is far away) is greater than the peak of the barrier's potential energy, $K_i > U_b$. In this case the barrier's potential energy is $U_b = 700$ keV. We can find the proton's initial kinetic energy from the temperature in the core of the Sun: about 1.5×10^7 K, so that the average kinetic energy of the protons is about 1 keV (Problem 41.1). Since K_i is about 700 times smaller than U_b, according to classical physics there is no chance the protons will get together.

You might argue that we have used the *average* kinetic energy for the protons. Of course, some protons have more kinetic energy, and perhaps some of them have more than 700 keV. The Maxwell–Boltzmann speed distribution equation (Eq. 20.19, page 589) predicts that the probability of a proton having enough kinetic energy to cross over the top of the barrier is less than 10^{-200}.

EXAMPLE 41.7 CASE STUDY The Probability of Proton Tunneling in the Sun

Equation 41.22 is for a rectangular barrier. Although the barrier in Figure 41.18 is not rectangular, use Equation 41.22 to make an order-of-magnitude estimate of the probability of tunneling required for two protons to fuse together in the Sun. (Don't be surprised if your estimate is off.) Note that in Figure 41.18, $x_0 \sim 10^{-15}$ m (approximate diameter of a proton).

INTERPRET and ANTICIPATE

We expect to find a small but non-zero result.

SOLVE

We need the barrier potential energy and the initial kinetic energy in SI units.

$$U_b = 700\,\text{keV} \sim 10^{-13}\,\text{J}$$
$$K_i = 1\,\text{keV} \sim 10^{-16}\,\text{J}$$

It is also helpful to find their difference.

$$U_b - K_i \sim (10^{-13}\,\text{J} - 10^{-16}\,\text{J})$$
$$U_b - K_i \sim 10^{-13}\,\text{J}$$

The strong force is important at x_0, so take this as the left edge of the barrier.

$$x_0 \sim 10^{-15}\,\text{m}$$

At the right edge of the barrier, the proton's initial kinetic energy equals the system's potential energy given by Coulomb's law.

$$\frac{ke^2}{x} = K_i$$
$$x = \frac{ke^2}{K_i} = \frac{(9 \times 10^9\,\text{N}\cdot\text{m}^2/\text{C}^2)(1.6 \times 10^9\,\text{C})^2}{10^{-16}\,\text{N}\cdot\text{m}}$$
$$x \sim 10^{-12}\text{m}$$

If we simply subtract the left edge x_0 from the right edge x, we get an overestimate for the width of the barrier because the barrier is not a rectangle.

$$x - x_0 = (10^{-12}\,\text{m} - 10^{-15}\,\text{m}) = 10^{-12}\,\text{m}$$

A slightly better estimate for the average width of the barrier is about an order of magnitude smaller.

$$w \sim 10^{-13}\,\text{m}$$

Now we can use Equation 41.23 to estimate γ. (The subscript p on the m stands for proton.)

$$\gamma^2 \equiv \frac{32\pi^2 m_p w^2 (U_b - K_i)}{h^2}$$
$$\gamma^2 \sim \frac{(300)(10^{-27}\,\text{kg})(10^{-13}\,\text{m})^2(10^{-13}\,\text{J})}{(7 \times 10^{-34}\,\text{J}\cdot\text{s})^2} \sim 600$$
$$\gamma \sim 25$$

The probability of tunneling comes from Equation 41.22

$$P_{\text{tunnel}} \approx e^{-\gamma}$$
$$P_{\text{tunnel}} \sim e^{-25} \sim 10^{-11}$$

CHECK and THINK

We modeled the barrier as rectangular. A better model finds a higher probability of about $P_{\text{tunnel}} \sim 10^{-10}$. It means that if 10^{10} pairs of protons attempt to tunnel, only one pair is successful. This might sound like a very low probability. However, it is a high enough probability to account for the 10^{38} fusion reactions (see Example 39.10, page 1295) that take place every second in the core of the Sun.

41-8 Special Case: Quantum Simple Harmonic Oscillator

QUANTUM SIMPLE HARMONIC OSCILLATOR ▶ Special Case

In classical mechanics, we found the simple harmonic oscillator to be a good model for a number of complex systems. Its role is equally important in quantum mechanics, and there is a particularly important reason for studying it. Planck's quantum hypothesis was that particles in a black body can be modeled as oscillators with the energy of each oscillator given by:

$$E = nE_{\text{min}} = nhf \qquad (n = 1, 2, 3\ldots) \qquad (40.7)$$

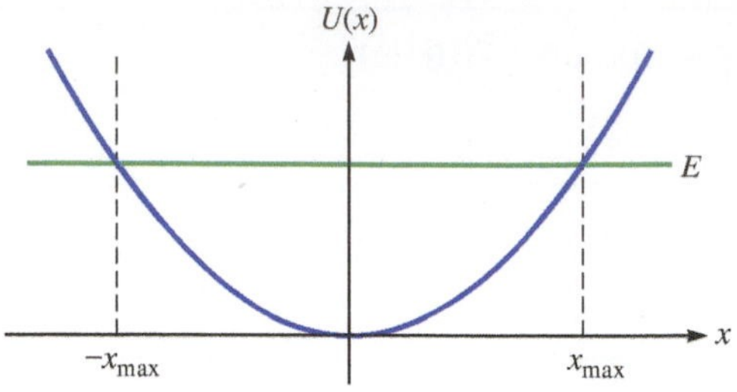

FIGURE 41.19 A simple harmonic oscillator's energy diagram.

where f is the frequency and $E_{min} = hf$. In Planck's model, the energy levels are evenly spaced. By contrast, when a particle is confined to either an infinite or a finite well, we have seen that the system's energy is quantized as in Planck's hypothesis, but the energy levels are not evenly spaced. Because the energy levels in a simple harmonic oscillator are evenly spaced, it makes a much better model for the molecules in a black body than for the particles in a well.

Our system consists of a particle connected to a spring that is fixed on its other end, so that the particle oscillates back and forth along an x axis. The spring obeys Hooke's law, so the magnitude of force on the particle is $F_H = kx$. The system's potential energy is $U(x) = \frac{1}{2}kx^2$. The natural frequency of the system is $\omega = \sqrt{k/m}$, and we use this equation to write the potential energy as $U(x) = \frac{1}{2}m\omega^2x^2$. The system's energy is E, and classically the particle oscillates between $\pm x_{max}$ (Fig. 41.19).

Schrödinger's equation for the simple harmonic oscillator is

Here k is the spring constant in Hooke's law.

$$-\frac{h^2}{8\pi^2 m}\frac{d^2\psi(x)}{dx^2} + \frac{1}{2}m\omega^2x^2\psi(x) = E\psi(x) \tag{41.24}$$

The x^2 in the second term makes this a somewhat difficult equation to solve, so we will just give the solution for the ground state:

$$\psi_1(x) = A_1 e^{-(x/x_0)^2} \tag{41.25}$$

where $x_0^2 = h/\pi m\omega$ and A_1 is a constant that is found by normalization. In Problem 41.36, you will confirm that Equation 41.25 is a solution to Schrödinger's equation, and you will find that the ground-state energy is $E_1 = h\omega/4\pi$.

We can write the ground-state energy in terms of the frequency f as $E_1 = \frac{1}{2}hf$. The solutions for the excited states show that the energy is quantized and given by:

$$E_n = \left(n - \frac{1}{2}\right)hf \tag{41.26}$$

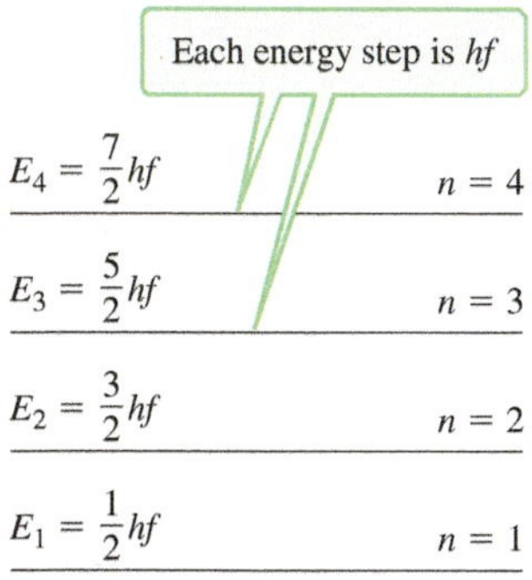

FIGURE 41.20 An energy-level diagram for a simple harmonic oscillator. The energy levels are evenly spaced and much like those in Planck's model for a black body.

According to Equation 41.26, the energy levels of a simple harmonic oscillator are evenly spaced (Fig. 41.20). Compare these energy levels to those proposed by Planck (Eq. 40.6); the difference between adjacent levels is $\Delta E = hf$ in both cases. The only difference is that Planck's minimum energy $E_{min} = hf$ is twice the ground state energy $E_1 = \frac{1}{2}hf$ of the simple harmonic oscillator. Like any trapped particle, according to quantum mechanics our particle attached to a spring cannot be at rest; the system's minimum energy cannot be zero.

EXAMPLE 41.8 In Which State Is the Oscillator?

A simple harmonic oscillator in its ground state has a frequency of 3.450 MHz. Now its energy is found to be 2.400×10^{-26} J. What state is it in?

INTERPRET and ANTICIPATE

The frequency of the ground state is the frequency in the excited state. So we can use the given frequency and the energy of the excited state to find which state the oscillator is in.

SOLVE

Solve Equation 41.26 for n.

$$E_n = \left(n - \frac{1}{2}\right)hf \tag{41.26}$$

$$n = \frac{E_n}{hf} + \frac{1}{2}$$

Substitute values. The answer must be an integer, so in this case we round up to the nearest whole number.

$$n = \frac{(2.400 \times 10^{-26}\ \text{J})}{(6.626 \times 10^{-34}\ \text{J}\cdot\text{s})(3.450 \times 10^6\ \text{Hz})} + \frac{1}{2}$$

$$n = 10.9988 = 11$$

CHECK and THINK

To check, let's find the difference between adjacent levels: $\Delta E = hf = (6.626 \times 10^{-34}\ \text{J} \cdot \text{s})(3.450 \times 10^{6}\ \text{Hz}) = 2.286 \times 10^{-27}$ J. So the system's energy is about $2.400 \times 10^{-26}\ \text{J}/2.286 \times 10^{-27}\ \text{J} = 10.5$ times greater than the difference between adjacent energy levels. The ground-state energy ($n = 1$) is $0.5\ hf$, and the system is 10 levels above the ground state, so $n = 1 + 10 = 11$, as we found.

EXAMPLE 41.9 Modeling Iodine

A number of times throughout this book, we have modeled molecular bonds as tiny springs connecting atoms. (For examples, see Figures 5.15, 14.21, and 21.25.) In particular, a vibrating diatomic molecule may be modeled as a simple harmonic oscillator (Fig. 41.21). In this problem, we connect the quantum-mechanical simple harmonic model to the classical model we have used for a molecular bond. An iodine molecule consisting of two iodine atoms is observed to have equally spaced energy levels of 4.24×10^{-21} J. Model the molecule as a particle-spring system, with one end of the spring fixed and the other attached to the oscillating particle. The particle's mass is 1.053×10^{-25} kg. (This the *reduced* or effective mass of the iodine molecule.) Find the spring constant.

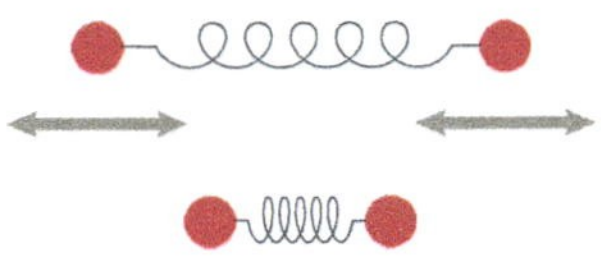

Vibrate in and out.

FIGURE 41.21

INTERPRET and ANTICIPATE

Modeling the iodine molecule as a quantum simple harmonic oscillator is reasonable because the observed energy spacing is uniform as predicted by Equation 41.26. The frequency of a classical object-spring oscillator (Section 16-5) depends on the mass and spring constant. So, making the connection between the classical and quantum models means finding the frequency of the quantum oscillator and from that frequency finding the spring constant.

SOLVE

The difference in adjacent energy levels depends on frequency.

$$\Delta E = hf \quad (1)$$

The frequency of an object-spring oscillator is given by Equation 16.26 (modified by using $\omega = 2\pi f$.)

$$f = \frac{1}{2\pi}\sqrt{\frac{k}{m}} \quad 16.26$$

Combine Equations (1) and 16.26 to arrive at the spring constant.

$$\frac{\Delta E}{h} = \frac{1}{2\pi}\sqrt{\frac{k}{m}}$$

$$k = 4\pi^2 m\left(\frac{\Delta E}{h}\right)^2$$

$$k = 4\pi^2(1.053 \times 10^{-25}\ \text{kg})\left(\frac{4.24 \times 10^{-21}\ \text{J}}{6.626 \times 10^{-34}\ \text{J} \cdot \text{s}}\right)^2$$

$$k = 1.70 \times 10^{2}\ \text{N/m}$$

CHECK and THINK

In this problem we have come full circle on our models of molecular bonds. We see why it was reasonable to model such bonds by tiny springs throughout this book. However, we must remember there is a major difference between a classical and quantum simple harmonic oscillator: In a classical oscillator the mechanical energy can have any value, but in a quantum oscillator the energy can only have certain discrete, evenly spaced values.

41-9 Heisenberg's Uncertainty Principle

Quantum mechanics is a great departure from classical mechanics largely because classical mechanics is deterministic and quantum mechanics is probabilistic. According to classical mechanics, both the location and momentum of a particle may be precisely determined. By contrast, quantum mechanics tells us that no matter how carefully we conduct an experiment, we cannot precisely determine a particle's position and momentum simultaneously. To see why quantum mechanics gives us such a different picture of nature, let's consider a free particle.

A Free Particle

So far we have only considered particles that have been trapped in potential wells. Now we'll consider a free particle traveling in the positive x direction that is not part of any system, so that $U(x) = 0$. Our plan is to solve for both the time-independent and the time-dependent wave functions. We then find the probability density. For a free particle, Schrödinger's equation is given by

$$-\frac{h^2}{8\pi^2 m}\frac{d^2\psi(x)}{dx^2} = E\psi(x) \tag{41.27}$$

The total energy of a free particle is its kinetic energy, which we express in terms of its momentum $E = p^2/2m$ and substitute into Equation 41.27:

$$-\frac{h^2}{8\pi^2 m}\frac{d^2\psi(x)}{dx^2} = \frac{p^2}{2m}\psi(x)$$

Rearrange this equation to isolate the second derivative on the left side:

$$\frac{d^2\psi(x)}{dx^2} = -\frac{4\pi^2 p^2}{h^2}\psi(x) \tag{41.28}$$

Use the de Broglie wavelength $\lambda = h/p$, and $\lambda = 2\pi/k$ to write the angular wave number in terms of momentum:

$$k = \frac{2\pi p}{h} \tag{41.29}$$

Substitute Equation 41.29 into Equation 41.28:

$$\frac{d^2\psi(x)}{dx^2} = -k^2\psi(x) \tag{41.30}$$

Mathematically, Equation 41.30 is identical to the differential equation for the position y of a simple harmonic oscillator, Equation 16.47: $d^2y(t)/dt^2 = -(k/m)y(t)$. Unfortunately, the solution given by Equation 16.3, $y(t) = y_{max}\cos(\omega t + \varphi)$, is not general enough in this case. As before, we won't solve this differential equation; instead, we assert that the most general solution is

$$\psi(x) = \psi_0 e^{ikx} + \psi_1 e^{-ikx} \tag{41.31}$$

where ψ_0 and ψ_1 are arbitrary constants (Problem 41.40).

Although we are not normally interested in the time-dependent Schrödinger equation in this book, in this one case, the solution is interesting. We can easily find the time-dependent solution from $\Psi(x, t) = \psi(x)e^{-i\omega t}$; we just multiply the solution for $\psi(x)$ in Equation 41.31 by $e^{-i\omega t}$:

Remember $\omega = \frac{2\pi E}{h}$

$$\Psi(x, t) = \psi_0 e^{i(kx-\omega t)} + \psi_1 e^{-i(kx+\omega t)} = \Psi_0(x, t) + \Psi_1(x, t) \tag{41.32}$$

In Chapter 17, we wrote the wave function of a wave traveling in the positive x direction as Equation 17.4, $y(x, t) = y_{max}\sin(kx - \omega t)$, and in the negative x direction as Equation 17.9, $y(x, t) = y_{max}\sin(kx + \omega t)$. But using the sine function is an arbitrary choice in representing a wave. The key is that a function of the form $F(kx \pm \omega t)$ represents a wave traveling in either the positive or the negative x direction. $\Psi_0(x, t) = \psi_0 e^{i(kx-\omega t)}$ represents a wave traveling in the positive x direction, and $\Psi_1(x, t) = \psi_1 e^{-i(kx+\omega t)}$ represents a wave traveling in the negative x direction. We are interested in a free particle traveling in the positive x direction, so we focus our attention on $\Psi_0(x, t) = \psi_0 e^{i(kx-\omega t)}$.

Let's find this particle's probability density:

$$|\Psi_0|^2 = \left(\psi_0 e^{i(kx-\omega t)}\right)\left(\psi_0 e^{-i(kx-\omega t)}\right) = \psi_0^2 e^{[i(kx-\omega t)-i(kx-\omega t)]} = \psi_0^2 e^0$$

$$|\Psi_0|^2 = \psi_0^2$$

We just found that the probability density is a constant that is independent of x. This means that the free particle is equally likely to be found anywhere along the x axis. The uncertainty Δx in the particle's position is infinite; the particle's position is *completely* uncertain. However, the particle's momentum is given by $p = hk/2\pi$ (Eq. 41.29), and because the angular wave number k is precisely known, the free particle's momentum is precisely known.

Uncertainty Is Inherent

Our inability precisely to know both the position and the momentum of a free particle is an example of Heisenberg's uncertainty principle. This uncertainty has nothing to do with the human ability to design an experiment; it is fundamental to nature. Our intuition that we can simultaneously measure position and momentum to an unlimited precision is based on our classical model of a particle, but the quantum model of a "particle" is of a probability *wave*. No forces are exerted on a free particle; its momentum is a constant that we can theoretically measure to any level of precision so that there is no inherent uncertainty in momentum $\Delta p = 0$. But because it is possible to measure its momentum precisely, when we do so, the idea of the particle having a position loses its meaning; the particle can be anywhere.

A free particle is an extreme example of precisely knowing a particle's momentum but having no way of knowing its position. What about a particle that is trapped? In such a case, we may know something about both its momentum and its position. But the more precisely we can know its position, the less precisely we can know its momentum.

To come up with a mathematical relationship for these uncertainties, let's think about how we could make such measurement. Suppose you wish to measure the position of a puck: You would probably reflect light off the puck to see it. Because of diffraction, the measurement of the puck's position is limited by the wavelength of the light used. So the uncertainty in the position measured is $\Delta x \approx \lambda$. You could, of course, improve the measurement by using very short wavelength radiation (or even an electron beam); but no matter how you design your experiment, there will be an inherent uncertainty in your position measurement. Next, you would like to measure the puck's momentum. You do this by measuring its position at two different times, calculating its velocity, and then multiplying by its mass. If the photons you use to illuminate the puck at these times have wavelength λ, then their momentum is h/λ. So the photons' interaction with the puck introduces an uncertainty in the puck's momentum: $\Delta p \approx h/\lambda$. If we multiply the uncertainty in the position by the uncertainty in the momentum, we find:

$$\Delta x \Delta p \approx (\lambda)\left(\frac{h}{\lambda}\right) \approx h \tag{41.33}$$

The German physicist Werner Heisenberg (1901–1976) worked this relationship out more carefully. According to **Heisenberg's uncertainty principle**, the standard deviation in position Δx and the standard deviation in momentum are given by:

HEISENBERG'S UNCERTAINTY PRINCIPLE **Underlying Principle**

$$\Delta x \Delta p \geq \frac{h}{4\pi} \tag{41.34}$$

Equivalently, Heisenberg's uncertainty principle may be expressed in terms of the standard deviation in energy and the standard deviation in time:

$$\Delta E \Delta t \geq \frac{h}{4\pi} \tag{41.35}$$

The equalities in Equations 41.34 and 41.35 are rarely realized. To make estimates, we usually use Equation 41.33 for position and momentum, and

$$\Delta E \Delta t \approx h \tag{41.36}$$

for energy and time.

Conceptually, $\Delta E \Delta t \approx h$ tells us that as the time taken to make an energy measurement increases, the uncertainty inherent in the result decreases. This equation has an even bolder implication. It says that the conservation of energy principle can be violated as long as the time period of this violation is short enough to be consistent with Equation 41.35.

Heisenberg's uncertainty principle gives us another way to understand the requirement that trapped particles can never be at rest; that is, their zero-point energy is not zero. The position of a particle trapped in a potential well has a relatively small uncertainty: within a box of width L, so we can say that $\Delta x \approx L$ (Fig. 41.1). Such a particle cannot be at rest because then its momentum would be precisely known, $\Delta p = 0$. Then $\Delta x \Delta p \approx (L)(0) = 0$, which violates Heisenberg's uncertainty principle.

Heisenberg's uncertainty principle also sheds light on the reactions in the core of the Sun (CASE STUDY). Fusion in the Sun requires protons to bind together, and for this to happen, quantum tunneling is required. Here is another way to look at tunneling. A proton may be characterized by its wavelength λ, and the uncertainty in its position must be at least on the order of its wavelength. As long as the barrier is no wider than a few times this wavelength, then tunneling is possible because the proton's position uncertainty includes both sides of the barrier. In other words, the proton is already on the other side of the barrier. It is like finding Isaac's ball is already on both sides of the slide (Fig. 41.13).

EXAMPLE 41.10 Science Fiction

As we saw in Chapter 39's CASE STUDY (page 1270), science fiction writers often use real physics in their stories. Let's test an idea a writer had. Feeding people on deep space missions is challenging. In a sense, the mission would need to somehow pack all the nutrition the crew needs for the entire voyage. The writer's solution to this problem is based on Heisenberg's uncertainty principle. According to the principle, the conservation of energy can be violated for a brief period of time given by $\Delta E \Delta t \approx h$. So a meal with energy ΔE could be created "out of nothing" and last a short time Δt. The real trick is that the person would need to eat the meal in that short time. Estimate Δt.

INTERPRET and ANTICIPATE

The key to this problem is estimating the energy created out of nothing. This estimate comes from Einstein's famous equation $E = mc^2$ (Eq. 39.44). We find the mass of a typical meal, then the energy that must be created.

SOLVE

Let's say the meal is a quarter-pound hamburger. The mass including the bun, French fries, and drink may be about 0.2 kg. We find the energy that must be created.

$$\Delta E = mc^2 \quad (39.44)$$

$$\Delta E = (0.2\text{ kg})(3 \times 10^8\text{ m/s})^2$$

$$\Delta E = 1.8 \times 10^{16}\text{ J}$$

According to Heisenberg's uncertainty principle, the mass can be created out of nothing as long as it disappears again in a short time given by Equation 41.36.

$$\Delta E \Delta t \approx h \quad (41.36)$$

$$\Delta t \approx \frac{h}{\Delta E} = \frac{6.63 \times 10^{-34}\text{J} \cdot \text{s}}{1.8 \times 10^{16}\text{ J}} = 3.68 \times 10^{-50}\text{ s}$$

$$\Delta t \approx 4 \times 10^{-50}\text{ s}$$

CHECK and THINK

This is a very short time. As a comparison, it takes about 300 milliseconds to blink your eye. So in a blink of an eye about 10^{45} such meals could come into existence and vanish. The writer should try to think of another solution for feeding the fictional crew. In case you are worried that we haven't done a good job of making our estimate because the person really only needs the chemical energy stored in the food (the calories), and not the mass, try Problem 41.80.

EXAMPLE 41.11 Hawking Radiation Is NOT Science Fiction

In the previous example, we imagined a whole meal being created from nothing. The result was that such a meal could only exist for a negligible amount of time.

But don't think that such events are impossible. The British physicist Stephen Hawking has a theory that predicts radiation from black holes. Recall from Section 39-12 that nothing can escape from the event horizon of a black hole. However, Hawking theorizes that black holes can radiate; we call this radiation *Hawking radiation.* His idea is that particle and antiparticle pair are created out of nothing near the event horizon of a black hole. Normally, when a particle and antiparticle are created, they soon come together and annihilate one another. But if in their short lifetime one of the particles falls inside the event horizon, the other particle will escape.

In principle we could see this escaped particle as Hawking radiation from the black hole. This radiation carries energy away from the black hole. Thus the black hole effectively loses energy and its mass is reduced, and we say the black hole *evaporates*. This evaporation process is very slow for solar mass black holes; such a black hole would take over 10^{67} years to evaporate. The universe is only 13.7 billion years old, and so no solar mass black holes have evaporated. However, much lower mass black holes known as *primordial black holes* may have had enough time to evaporate. Such small black holes would produce a final burst of high-energy Hawking radiation. Although no such radiation has been observed yet, it is theoretically possible.

Consider the creation of an electron-position pair near a black hole. Similar to the writer's idea about the meal in Example 41.10, one of the particles must be consumed by the black hole in their brief lifetime. Estimate the lifetime of such an electron-positron pair.

INTERPRET and ANTICIPATE

Like the previous example, we use Einstein's famous equation to estimate the energy that must be created.

SOLVE

The mass of electron is the same as positron's mass.

$$\Delta E = mc^2 \quad (39.44)$$

$$\Delta E = 2m_e c^2 = 2(9.11 \times 10^{-31}\ \text{kg})(3.00 \times 10^8\ \text{m/s})^2$$

$$\Delta E = 1.64 \times 10^{-13}\ \text{J}$$

According to Heisenberg's uncertainty principle, the mass can be created out of nothing as long as it disappears again in a short time given by Equation 41.36.

$$\Delta E \Delta t \approx h \quad (41.36)$$

$$\Delta t \approx \frac{h}{\Delta E} = \frac{6.63 \times 10^{-34}\text{J}\cdot\text{s}}{1.64 \times 10^{16}\ \text{J}}$$

$$\Delta t \approx 4.04 \times 10^{-21}\ \text{s}$$

CHECK and THINK

This is still a very short time. However, it is 10^{29} longer than the length of time the meal would survive. According to quantum mechanics, such particle–antiparticle pairs are constantly created in empty space. And according to Hawking, there is a nonzero probability that in some cases one of the particles falls into the black hole and the other is set free.

SUMMARY

Underlying Principles

1. **Schrödinger's equation** describes the propagation of a matter wave, and a matter wave carries all the things we normally associate with a particle—energy, momentum, mass, and sometimes charge.

Schrödinger's (nonrelativistic) time-dependent wave equation is

$$-\frac{h^2}{8\pi^2 m}\frac{\partial^2\Psi(x,t)}{\partial x^2} + U(x,t)\Psi(x,t) = \frac{ih}{2\pi}\frac{\partial\Psi(x,t)}{\partial t} \quad (41.1)$$

Underlying Principles cont'd

The time-independent, one-dimensional Schrödinger equation is

$$-\frac{h^2}{8\pi^2 m}\frac{d^2\psi(x)}{dx^2} + U(x)\psi(x) = E\psi(x) \quad (41.8)$$

2. According to **Bohr's correspondence principle** in the limit of very large quantum numbers, the classical calculation and the quantum calculation must yield the same results.
3. According to **Heisenberg's uncertainty principle**, the standard deviation in position Δx and the standard deviation in momentum Δp are given by:

$$\Delta x \Delta p \geq \frac{h}{4\pi} \quad (41.34)$$

and the standard deviation in energy ΔE and the standard deviation in time Δt by:

$$\Delta E \Delta t \geq \frac{h}{4\pi} \quad (41.35)$$

Major Concepts

1. According to Born, there is no *physical* interpretation of the wave function Ψ itself, but we can think of Ψ as a **probability wave**. We don't actually detect Ψ. Instead we detect $|\Psi|^2$, which is the **probability density** of finding the particle at a particular location and time.
2. The **normalization condition** mathematically expresses this commonsense idea that the particle must be located somewhere along the x axis as

$$\int_{-\infty}^{\infty} |\psi(x)|^2 dx = 1 \quad (41.11)$$

the **normalization condition**, and it is required in order to interpret $|\psi|^2$ as a probability density.
3. The value that $\psi(x)$ must have at a boundary is known as the **boundary condition**. The boundary conditions are used to make a general solution to Schrödinger's equation fit a particular situation.
4. **Barrier tunneling** describes the prediction made by quantum mechanics that a particle may be found on the other side of a potential energy barrier that would be insurmountable according to classical mechanics. The probability that a particle with energy K_i will tunnel through a rectangular barrier of height U_b is given by:

$$P_{\text{tunnel}} \approx e^{-\gamma} \quad (41.22)$$

where

$$\gamma^2 \equiv \frac{32\pi^2 m w^2 (U_b - K_i)}{h^2} \quad (41.23)$$

and w is the width of the barrier.

Special Cases

1. Schrödinger's equation for a **particle in an infinite square well** is

$$\frac{d^2\psi(x)}{dx^2} + k^2\psi(x) = 0 \quad (41.13)$$

and the specific solution is $\psi(x) = \sqrt{\frac{2}{L}}\sin k_n x$ (Eq. 41.17). The system's energy levels are

$$E_n = n^2\left(\frac{h^2}{8mL^2}\right) = n^2 E_1 \quad (41.16)$$

2. Schrödinger's equation for a **particle in a finite square well** is the same as that for an infinite well (Eq. 41.13). However, outside the well, Schrödinger's equation is

$$\frac{d^2\psi(x)}{dx^2} - \kappa^2\psi(x) = 0 \quad (41.19)$$

where $\kappa^2 = (8\pi^2 m/h^2)(U_0 - E)$. The solutions for the first three quantum states are presented graphically in Figure 41.10A.

The energy levels are a little lower in the finite well than in the infinite well (Fig. 41.11), and if the system's energy is greater than U_0, the particle is free.

A particle in a finite well may be found outside the well. The **penetration distance** Λ depends on κ:

$$\Lambda_n = \frac{1}{\kappa_n} = \frac{h}{2\pi\sqrt{2m(U_0 - E_n)}} \quad (41.21)$$

3. Schrödinger's equation for the **quantum simple harmonic oscillator** is:

$$-\frac{h^2}{8\pi^2 m}\frac{d^2\psi(x)}{dx^2} + \frac{1}{2}m\omega^2 x^2\psi(x) = E\psi(x) \quad (41.24)$$

where the ground-state solution is:

$$\psi_1(x) = A_1 e^{-(x/x_0)^2} \quad (41.25)$$

The system's ground-state energy is $E_1 = h\omega/4\pi$ or $E_1 = \frac{1}{2}hf$, and the energies of the excited states are:

$$E_n = \left(n - \frac{1}{2}\right)hf \quad (41.26)$$

Tools

In an **energy-level diagram**, the energy of each state is represented by a horizontal line labeled with its quantum number n, and the relative spacing of the lines represents the change in energy between each state (Figs. 41.2, 41.11, and 41.20).

PROBLEMS AND QUESTIONS

A = algebraic C = conceptual E = estimation G = graphical N = numerical

41-1 The New Quantum Theory

1. **E** CASE STUDY The Sun's core temperature is about 1.5×10^7 K. Show that this means the average kinetic energy of the protons in the core is about 1 keV (order of magnitude only).
2. **C** In what way is de Broglie's matter wave theory ad hoc (make-shift)? In what way is Schrödinger's equation ad hoc, and in what way is it not ad hoc?

41-2 A Trapped Particle

Problems 3 and 4 are paired.

3. **N** Suppose that a hockey puck's mass is 0.160 kg, and it is in a box of length 6.35 m (Example 41.1). What excited state would give an energy of 8.52 J?
4. **C** Why do the energy levels of a hockey puck trapped in a box seem to be continuous?
5. **N** A particle of mass m is trapped in a one-dimensional box of length L. What combination of mL^2 is required for the zero-point energy to be 25.0 eV?
6. **N** Suppose we wanted to try to observe the quantization of energy for a 1.00-g peanut trapped in a box. If our detector has a resolution such that it is able to measure the energy difference between the ground state and first excited state, as long as the difference is greater than 0.00100 eV, what is the maximum length L of the box? Note that we, of course, cannot have a box this small. Thus, we do not observe the quantization of peanuts.
7. **N** A proton is trapped in a one-dimensional box of length L. If the zero-point energy is to be 25.0 eV, what is L?

41-3 The Double-Slit Experiment Revisited: Probability Waves

8. **E** Suppose that 6.5×10^6 particles created the interference pattern in Figure 43.4. There is a 13% probability of a particle being detected by device A and a 2.5% probability of a particle being detected by B. **a.** Estimate the number of particles detected by each detector. **b.** How many more particles were detected by A than by B?

Problems 9 and 10 are paired

9. Consider again the throwing of the dice in Example 41.2. In quantum mechanics, we will often speak of the current state of the system given a set of possible states. The sum of the faces of the two dice that are face-up after being thrown (or rolled) can be thought of as representing the state of the dice. This means the possible total, or state of the dice, could be 2, 3, 4, and so on, with 12 being the maximum value.
   a. **N** How many possible ways are there for the state of the dice to equal 10?
   b. **N** How many possible ways are there for the state of the dice to equal 6?
   c. **C** Which of these two states is more likely to occur? Explain your answer.
10. Consider the throwing of the dice in the previous problem. The number of possible ways a state can be achieved is called the *multiplicity* of the state and can be used to calculate the probability of the state occurring.
    a. **N** What is the multiplicity of the state where the total of the dice is equal to 7?
    b. **N** What is the total number of possible states of the dice when they are thrown? Be sure to consider that each possible state may have a multiplicity greater than one.
    c. **N** What is the probability of finding the dice in the state where their total is 7 after being thrown?
    d. **C** Is 7 the most likely result? Explain your answer.

41-4 Schrödinger's Equation

11. **A** Substitute $\Psi(x, t) = \psi(x)e^{-i\omega t}$ (where $\omega = 2(\pi)E/h$) into Equation 41.1, the time-dependent Schrödinger equation:

$$-\frac{h^2}{8\pi^2 m}\frac{\partial^2\Psi(x, t)}{\partial x^2} + U(x, t)\Psi(x, t) = \frac{ih}{2\pi}\frac{\partial\Psi(x, t)}{\partial t}$$

and derive the time-independent version, Equation 41.8:

$$-\frac{h^2}{8\pi^2 m}\frac{\partial^2\psi(x)}{\partial x^2} + U(x)\psi(x) = E\psi(x)$$

12. **C** A chapter of the Society of Physics Students at the United States Naval Academy designed a T-shirt with a pun based on quantum mechanics (Fig. P41.12). Explain the meaning of the T-shirt.

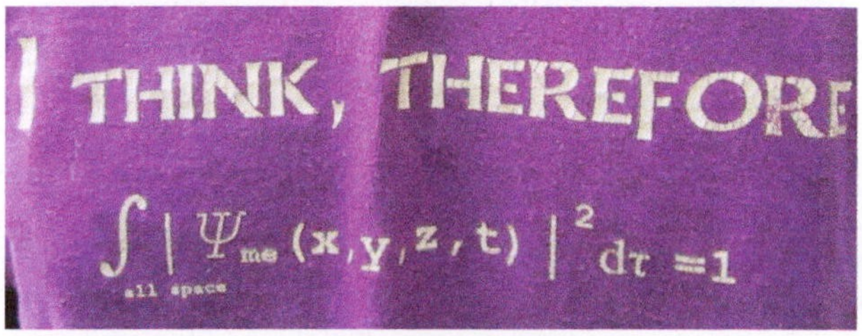

Debora Katz

FIGURE P41.12

41-5 Special Case: A Particle in an Infinite Square Well

13. **N** A neutron is bound in an infinite square well with a width of 3.25 nm. Find the energies of the ground state and the first three excited states.

14. **C** We imagine that a puck is moving in a box with very stiff walls, so that the puck reverses direction without any time delay. Of course, in a real box the puck would take a moment to reverse direction. Where would you most likely find the puck in such a real situation? How does your answer change depending on the length of the box?

15. **N** A proton is bound in an infinite square well. A photon with a frequency of 2.34×10^{15} Hz is emitted when the proton makes a transition from the $n = 6$ excited state to the ground state. What is the width of the square well?

16. **G** The wave function at a particular moment in time for an electron in an infinite one-dimensional box of length L, as in Section 41-5, is

$$\psi(x) = \sqrt{\frac{2}{L}}\sin\left(\frac{2\pi x}{L}\right) \text{ for } 0 < x < L$$

Sketch the probability density for the electron in the box.

Problems 17 and 18 are paired

17. **N** An infinite square well of width L is used to confine a particle. The well extends from $x = 0$ to $x = L$. Assume the particle is in the ground state.

a. What is the probability of detecting the particle between $x = 0$ to $x = L$?

b. What is the probability of detecting the particle between $x = 0$ to $x = 0.50L$?

18. **N** An infinite square well of width L is used to confine a particle. The well extends from $x = 0$ to $x = L$. Assume the particle is in the first excited state ($n = 2$).

a. What is the probability of detecting the particle between $x = 0$ to $x = L$?

b. What is the probability of detecting the particle between $x = 0$ to $x = 0.50L$?

Problems 19, 20, and 21 are grouped.

19. **N** An electron is trapped in an infinite well of width 2.50 nm. What is its zero-point energy? Give your answer in joules and electron-volts.

20. **C** An electron is trapped in an infinite well of width 2.50 nm. Initially, the system is in the $n = 3$ quantum state; it then jumps to the $n = 5$ quantum state. Did the system lose or gain energy? Explain.

21. **N** An electron is trapped in an infinite well of width 2.50 nm. Initially, the system is in the $n = 3$ quantum state; it then jumps to the $n = 5$ quantum state. Determine the amount of energy lost or gained by the system. Give your answer in electron-volts.

22. **A** Write the wave functions for the stationary ground state and each of the first three stationary excited states for a particle in an infinite square well of width L_0.

23. **N** The ground state ($n = 1$) of a proton confined to an infinite square well of width L is 3.75 keV.

a. What is the width L of the infinite square well?

b. What is the energy difference between the $n = 3$ excited state and the ground state for this proton?

Problems 24 and 25 are paired.

24. **E** A proton in the nucleus of an atom can be modeled crudely as a proton trapped in an infinite well with a width on the order of 10^{-14} m. What is the order of magnitude of its zero-point energy? Give your answer in joules and electron-volts.

25. **E** An electron bound in an atom can be modeled crudely as an electron trapped in an infinite well with a width on the order of 10^{-10} m. What is the order of magnitude of its zero-point energy? Give your answer in joules and electron-volts. If you worked Problem 24, compare your results.

Problems 26 and 27 are paired.

26. **N** An electron is in an infinite potential well with a width of 1.50 nm.

a. What is the wavelength of the photon emitted when the electron transitions from the fourth excited state, $n = 5$, to the ground state, $n = 1$?

b. What is the maximum wavelength of a photon capable of transitioning the electron from the ground state to the third excited state, $n = 4$?

27. **N** A proton is in an infinite potential well with a width of 1.50 nm.

a. What is the wavelength of the photon emitted when the proton transitions from the fourth excited state, $n = 5$, to the ground state, $n = 1$?

b. What is the maximum wavelength of a photon capable of transitioning the proton from the ground state to the third excited state, $n = 4$?

41-6 Special Case: A Particle in a Finite Square Well

28. **C** A particle is in the ground state of a finite square well with energy $E_1 = 5.42$ eV. The walls of the well have a height 10.62 eV. If the particle absorbs a photon with an energy of 6.00 eV, what can we say about the state of the particle? How is this different than if the particle were in an infinite square well with the same ground-state energy and absorbed the same photon?

29. **N** An electron is bound in the ground state of a finite square well with $U_0 = 75$ eV.

a. How much energy is required to free the electron from the well if the ground-state energy is 3.0 eV?

b. If this transition is accomplished through the absorption of one photon of light, what is the maximum wavelength of that photon?

Problems 30 and 31 are paired.

30. **C** A particle is bound in the ground state of a finite square well. An external potential is applied such that the walls of the well increase in height, or the well becomes deeper. Describe what happens to the ground state energy of the particle as the wall height increases. What happens to the penetration distance?

31. **N** A proton is bound in a finite square well such that its energy E is 10.0% of the height of the well, U_0. Find the penetration distance of the particle when the height of the well is **a.** 5.0 eV, **b.** 50.0 eV, and **c.** 500.0 eV.

32. **A** Find the dimensions of the quantity in Equation 41.21,

$$\Lambda_n = \frac{h}{2\pi\sqrt{2m(U_0 - E_n)}}$$

Problems 33 and 34 are paired.

33. **N** A proton is in a finite square well with $U_0 = 35.5$ MeV. It is found that $\Lambda_2 = 2.35 \times 10^{-15}$ m. What is E_2? Give your answer in joules and in electron-volts.

34. **N** For the proton in Problem 33, given your answer, place a limit on the system's zero-point energy.

35. **N** The width of a finite square well can be varied such that the ground-state energy remains unchanged as we vary the depth of the square well. Suppose this is done, where the ground state energy of a proton is 3.50 eV and the various depths of the well are **a.** 10.0 eV, **b.** 100.0 eV, and **c.** 1000.0 eV. Find the penetration distance for the proton in the ground state for each well depth.

41-7 Barrier Tunneling

36. **E** CASE STUDY The probability of two protons tunneling in the Sun's core is $P_{\text{tunnel}} \sim 10^{-10}$. This means that out of each 10^{10} pairs of protons, one pair tunnels successfully. Make an estimate showing that this probability can account for the 10^{38} fusion reactions that take place each second in the Sun's core. *Hint:* Assume that the Sun's core contains about 40% of its mass.

37. **N** CASE STUDY Use the Maxwell–Boltzmann speed distribution equation (Eq. 20.19, page 589) to find the probability that a proton has enough kinetic energy to cross over the top of the barrier in Figure 41.18. *Hint*: The Sun's core temperature is about 1.5×10^7 K, so the average kinetic energy of the protons in the core is about 1 keV (order of magnitude only). In this case the barrier's potential energy is $U_b = 700$ keV.

Problems 38, 39, and 40 are grouped.

38. **N** An electron with a kinetic energy of 45.34 eV is incident on a square barrier with $U_b = 54.43$ eV and $w = 2.400$ nm. What is the probability that the electron tunnels through the barrier?

39. **N** An electron with a kinetic energy of 45.34 eV is incident on a square barrier with $U_b = 54.43$ eV and $w = 2.400$ pm. What is the probability that the electron tunnels through the barrier?

40. **N** A proton with a kinetic energy of 45.34 eV is incident on a square barrier with $U_b = 54.43$ eV and $w = 2.400$ pm. What is the probability that the proton tunnels through the barrier?

41. **N** A proton is approaching a potential barrier with a height of 10.24 eV and a width of 4.560 pm. If there is a 25.00% chance of the proton tunneling through the barrier, what is the de Broglie wavelength of the proton?

42. **A** Derive an expression that relates the rate of change of the probability of tunneling for a particle incident on a potential barrier with height U_b and width w, to the rate of change of the kinetic energy of the particle.

43. **N** In a quantum wire, an electric current can be controlled by varying the height of a potential barrier. Suppose that the electrons each have a kinetic energy of 8.24 eV and that 1.33×10^4 electrons pass through the wire every second when there is no potential barrier. If the width of the barrier is 50.00 pm, what barrier energy height would cause the current to become 1.00% of its initial value?

41-8 Special Case: Quantum Simple Harmonic Oscillator

44. **A** Confirm that Equation 41.25, $\psi_1(x) = A_1 e^{-(x/x_0)^2}$, where $x_0^2 = h/(\pi m\omega)$ and A_1 is a constant that is found by normalization, is a solution to Schrodinger's equation (Eq. 41.24):

$$-\frac{h^2}{8\pi^2 m}\frac{d^2\psi(x)}{dx^2} + \frac{1}{2}m\omega^2 x^2\psi(x) = E\psi(x)$$

In doing so, also show that the ground state energy is $E_1 = h\omega/4\pi$.

45. **N** The energy difference between two adjacent energy levels in a quantum simple harmonic oscillator is 3.58 eV. What is the energy of the ground state?

46. **A** Assume an electron is in the ground state of a quantum simple harmonic oscillator potential well with a ground-state energy of 4.65 eV. Write the wave equation for the stationary ground state. You do not need to determine the normalization constant, A_1.

47. **N** A particle is in the $n = 5$ excited state of a quantum simple harmonic oscillator well. A photon with a frequency of 1.25×10^{15} Hz is emitted as the particle moves to the $n = 3$ excited state. What is the minimum photon frequency required for this particle to make a quantum jump from the ground state of this well to the $n = 4$ excited state?

48. **N** A quantum simple harmonic oscillator has an angular frequency $\omega = 6.43 \times 10^{15}$ rad/s. Find the energy of the ground state and the first three excited states of the oscillator.

41-9 Heisenberg's Uncertainty Principle

49. **A** Substitute Equation 41.31, $\psi(x) = \psi_0 e^{ikx} + \psi_1 e^{-ikx}$, into Equation 41.30, $d^2\psi(x)/dx^2 = -k^2\psi(x)$, to show that it is a solution.

50. **A** Show that a free particle that is traveling in the negative x direction is equally likely to be found anywhere along the x axis.

51. **N** A hard-luck baseball player who can't seem to hit the fastball of an all-star pitcher claims that his difficulties rest in the physics of the ball being pitched. "Given the speed of the ball being 103 mph, the minimum uncertainty in the position of the ball is just too large to be able to know where and when to swing," he says. Assuming the mass of the baseball is 0.145 kg, find the minimum uncertainty in the position of the baseball. Is the player's claim legitimate?

52. **E** CASE STUDY In Chapter 42 we study atoms, and part of our case study is about hydrogen. Let's use Heisenberg's uncertainty principle to estimate the ground-state energy of hydrogen. In our model, the electron is confined in a one-dimensional well with a length about the size of hydrogen, so that $\Delta x = 0.0529$ nm. Estimate Δp, and then assume that the ground-state energy is roughly $\Delta p^2/2m_e$. Give your answer in joules and electron-volts.

Problems 53 and 54 are paired.

53. **N** An alpha particle is a helium nucleus consisting of two protons and two neutrons. It is moving with a speed of 1.10×10^3 m/s.
a. What is the momentum of the alpha particle?
b. If there is a 10.0% uncertainty in the momentum of this alpha particle, what is the minimum uncertainty in the position of the alpha particle?

54. **N** **Review** An alpha particle is a helium nucleus consisting of two protons and two neutrons. It is moving with a speed of 1.10×10^8 m/s.
a. What is the momentum of the alpha particle?
b. If there is a 10.0% uncertainty in the momentum of this alpha particle, what is the minimum uncertainty in the position of the alpha particle?

Problems 55 and 56 are paired.

55. **N** **Review** The lifetime of a muon in its own frame of reference is measured to be 3.10×10^{-6} s. What is the minimum uncertainty in the energy of the muon?

56. **Review** The lifetime of a muon moving relative to an observer is measured to be 15.6×10^{-6} s.
a. **N** What is the minimum uncertainty in the energy of the muon according to the observer?
b. **C** Consider the muon from Problem 55 and the uncertainty in its energy. Explain how the energy can be more certain for this moving muon than for a muon that is at rest for the same observer.

General Problems

57. **A** An infinite square well of width L confines an electron that transitions from the $n = 1$ ground state to the $n = 4$ state after absorbing a photon with wavelength λ_A. Assume the photon has just enough energy to accomplish this transition.
   a. What is the width L of the well confining the electron in terms of the mass of the electron, m_e, h, λ_A, and c?
   b. The electron next transitions from the $n = 4$ to the $n = 2$ state. What is the wavelength λ_B of the photon emitted during this transition in terms of λ_A?
58. **C** Although there is no physical interpretation for the wave function $\psi(x)$, it does have units. Consider the form of wave function for the case of the infinite square well, $\psi(x) = (\sqrt{2/L})\sin(k_n x)$. What are the SI units of the wave function?
59. **N** An infinite square well of width $L = 155$ pm is used to confine an electron. The well extends from $x = 0$ to $x = L$. Assume the electron is in the ground state.
   a. What is the probability of detecting the electron between $x = 0$ to $x = 0.50L$?
   b. What is the probability of detecting the electron between $x = 0$ to $x = 0.25L$?
60. **N** An infinite square well of width $L = 155$ pm is used to confine an electron. The well extends from $x = 0$ to $x = L$. Assume the electron is in the first excited state ($n = 2$).
   a. What is the probability of detecting the electron between $x = 0$ to $x = 0.50L$?
   b. What is the probability of detecting the electron between $x = 0$ to $x = 0.25L$?

Problems 61, 62, and 63 are grouped.

61. **N** A proton is trapped in an infinite well of length 0.350 nm. Find the system's ground-state energy. Give your answer in joules and electron-volts.
62. **N** A proton is trapped in an infinite well of length 0.350 nm. Find the energy of the first four excited states ($n = 2$ to $n = 5$). Give your answers in joules and electron-volts.
63. **N** A proton is trapped in an infinite well of length 0.350 nm. Initially the system is in its ground state. It then absorbs a photon with an energy of 1.34×10^{-2} eV. What state is the system in as a result?
64. **C** A free particle with kinetic energy 185 eV travels in a region of space where $U = 0$. It passes over a square potential well of depth 152 eV.
   a. What happens to the total energy of the particle as it passes over the well?
   b. What happens to the kinetic energy of the particle as it passes over the well?
65. **C** Consider the quantum simple harmonic oscillator potential well. Would you expect there to be a penetration distance into the walls of the harmonic potential well? Explain your answer.

Problems 66 and 67 are paired

66. **N** A hockey puck ($m = 0.160$ kg) is placed in a two-dimensional box where the distance between one set of walls is 0.500 m and the distance between the other set of opposing walls is 0.250 m. Assume the walls in each dimension represent an infinite square well in each case. In this two-dimensional case, the state of the puck would be represented by the combination of standing waves in each of the two dimensions.
   a. What is the zero-point energy of the puck in each dimension?
   b. The actual zero-point energy is the sum of the zero point energy in each dimension. Find this zero-point energy for the puck in the two-dimensional box.
67. **N** A hockey puck ($m = 0.160$ kg) is placed in a three-dimensional box where the distance between one set of walls is 0.500 m, the distance between another set of opposing walls is 0.250 m, and the distance between the remaining set of walls is 0.350 m. Assume the walls in each dimension represent an infinite square well in each case. In this three-dimensional case, the state of the puck would be represented by the combination of standing waves in each of the three dimensions.
   a. What is the zero-point energy of the puck in each dimension?
   b. The actual zero-point energy is the sum of the zero-point energy in each dimension. Find this zero-point energy for the puck in the three-dimensional box.

Problems 68 and 69 are paired.

68. **C** A particle is in an infinite square well that has its width halved. Is the percentage change in the third excited state energy (E_4) the same as the percentage change in the ground-state energy E_1? Explain your answer.
69. **C** Consider two adjacent energy levels for a particle in an infinite square well, n and m, where $n > m$. When the width of the well is halved, does the energy difference $E_n - E_m$ stay the same, increase, or decrease? Explain your answer.
70. **A, C** The width of a potential barrier decreases at a constant rate (dw/dt) until the width becomes zero. Derive an approximate expression for the rate of change of the probability (dP_{tunnel}/dt) that a particle with mass m and kinetic energy K_i tunnels through the barrier with height U_i. Interpret the meaning of your answer.
71. **N** When transitioning from the $n = 4$ excited state to the ground state ($n = 1$), an isolated atom emits a 567-nm photon. When transitioning from the $n = 3$ excited state to the ground state, the same atom emits a 626-nm photon. What is the wavelength of the photon emitted by this atom when it transitions from the $n = 4$ excited state to the $n = 3$ excited state?
72. **A** In this chapter, you have seen the solution for the time-independent, one-dimensional Schrödinger equation for an infinite square well with a width from $x = 0$ to $x = L$. Though much more difficult to derive, Schrödinger's equation can be solved for an infinite square well that is symmetric about $x = 0$, running from $x = -L/2$ to $x = L/2$. In this case, the normalized results for the wave function are $\psi(x) = (\sqrt{2/L})\sin(k_n x)$ when n is an odd number, and $\psi(x) = (\sqrt{2/L})\cos(k_n x)$ when n is an even number. Show that $\psi(x) = (\sqrt{2/L})\cos(k_n x)$ is a solution for the time-independent, one-dimensional Schrödinger equation.
73. **N** A particle incident on a potential barrier has a probability $P_{\text{tunnel}} = 50\%$ of tunneling through the barrier.
   a. If the width of the barrier is halved, what is the new probability of the particle tunneling through the barrier?
   b. If, instead, the width of the barrier is doubled, what is the new probability of the particle tunneling through the barrier?
74. **C** You may often encounter derivations, models, or solutions that claim to use a "semiclassical" approach. Describe what is meant by "semiclassical" in these instances.
75. **C** Shannon and Cameron have been studying all morning for tomorrow's physics final. Avi has slept in.

Cameron: Avi, so good of you to join us. Do you already know everything in all 43 chapters?

Avi: No, but I don't have to. All I need to know is what Heisenberg said, *nothing is certain, and anything is possible.* I can ace the exam without knowing anything else.

Shannon: Let me tell you just one thing about Heisenberg's uncertainty principle that might even help you tomorrow...

Complete Shannon's thought: What does Heisenberg's uncertainty principle really say?

Problems 76, 77 and 78 are grouped

76. **A** Although each stationary state represents a single-state solution to Schrödinger's time-independent equation for a particle in an infinite square well, a more general solution might include a mix of states, where there might be a different probability of finding the particle in each state. Suppose a linear combination of the ground state and the first excited state describes the current state of the particle, $\psi(x) = A_1 \sin(k_1 x) + A_2 \sin(k_2 x)$. The coefficients of each stationary state, A_1 and A_2, are related to the probability of each state occurring ($|A_1|^2$ and $|A_2|^2$ respectively). When a particle is measured, this kind of mixed state will "collapse" into one of the possible states – in this case, either the ground state or the first excited state. Substitute each individual state into the left side of the time-independent Schrödinger's equation (Eq. 41.8), with $U(x) = 0$ and show that in each case, $k_n = \sqrt{8\pi^2 m E_n / h^2}$.

77. **N** Consider the mix of stationary states described in Problem 76. If there is a 1 in 4, or 25%, chance of finding the particle in the first excited state ($n = 2$), what is the probability of finding the particle in the ground state?

78. **N** Consider the mix of stationary states described in Problem 76. Suppose there is a 1 in 4, or 25%, chance of finding the particle in the first excited state ($n = 2$).

a. What is the value of A_2?

b. What is the value of A_1?

79. **A** Start with the normalization condition $\int_0^L |A \sin k_n x|^2 dx = 1$ and show $A = \sqrt{2/L}$. Therefore, the wave function for a particle trapped in an infinite square well is given by $\psi(x) = \sqrt{\frac{2}{L}} \sin k_n x$ (Eq. 41.17).

80. **E** Suppose the science fiction writer in Example 41.10 only wishes to produce 2000 calories of pure energy (with no mass) for a crew member to consume. (We won't worry how the body can absorb the energy without the usual metabolic process.) Use Heisenberg's uncertainty principle to estimate how long such energy could be created from nothing before vanishing.

42 Atoms

Key Questions

How do we best model an atom?

What observations support our model?

Underlying Principles

Pauli's exclusion principle

Major Concepts

1. Thomson's plum pudding model
2. Rutherford's solar system model
3. Bohr's atomic model
4. De Broglie's model applied to an atom
5. Schrödinger's equation applied to an atom
6. Periodic table of the elements
7. Zeeman effect

In Chapter 41, we considered particles under ideal conditions, such as when they are trapped in an infinite well. Exploring those ideal conditions enabled us to tackle the mathematics more readily. So you might think that quantum mechanics is a theory having little to do with anything practical. In this chapter, we take the opposite approach and explore something very practical—atoms. Almost everything is made of atoms. (Some exceptions include plasmas, white dwarfs, and neutron stars.) In this chapter we'll use quantum mechanics to explain the structure of atoms. Quantum mechanics explains the organization of the periodic table of the elements and shows why some types of atoms are highly reactive and others are not reactive at all. Although the mathematics involved in modeling atoms with quantum mechanics is beyond the scope of this book, we'll imitate our practice in the previous chapter by deducing the steps that are required without actually performing all the calculations. However, you can expect to understand and use elsewhere the results derived here from quantum mechanics.

42-1 Early Atomic Models

You have probably encountered a solar system model of the atom from your early school days (Fig. 42.1). At the center of the atom, there is a nucleus consisting of protons and neutrons. Because it consists only of positive and neutral particles, this nucleus is positively charged. Negatively charged electrons orbit the nucleus, much as the planets orbit the Sun. The atom as a whole is neutral because the negative charge of all the electrons has the same magnitude as the positive charge of the nucleus. But because this solar system model does not take into account the wave properties of subatomic particles, it is an incorrect description of the atom. In this chapter, we take a closer look at the structure of the atom, one that goes beyond the solar system model. We begin with a brief history of atomic models.

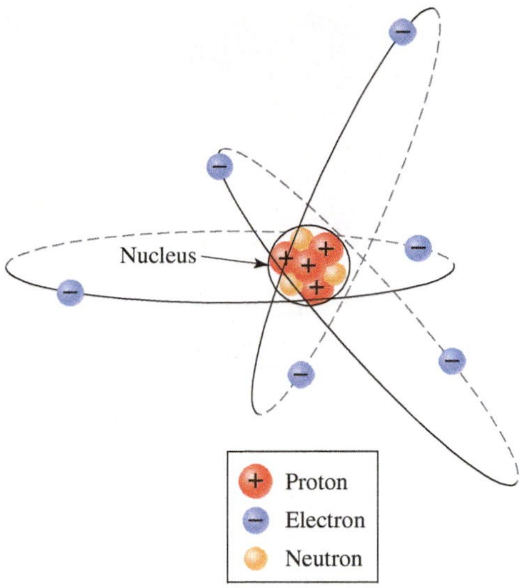

FIGURE 42.1 Solar-system model for an atom.

The word *atom* comes from the Greek word *atomos*, which means *undivided*; so, an atom was thought to be an *indivisible* particle. The nature of matter was debated by ancient (fifth century BCE) Greek philosophers. The dominant idea at the time was that matter is continuous, but a few philosophers believed that matter was made up of small, discrete, indivisible particles called atoms. These atoms were thought to come in different sizes and shapes, which would determine the properties of matter on the macroscopic scale. Nevertheless, these atoms were also thought to be fundamental particles with no internal structure. This ancient atomic view of matter was largely ignored or rejected until the 1600s, when the modern scientific method was developed.

Evidence in favor of the modern atomic model was based on the study of gases (the kinetic theory of gases; Chapter 20) and chemical reactions. By the latter half of the 1800s, over 63 elements were identified and arranged by mass to form the beginning of the **periodic table** (Appendix B). By then scientists also knew that the chemical properties of an element correlate with its place in the periodic table. So the molecules that could form out of various elements were determined by the place of these elements in the periodic table. At first scientists didn't know what forces bind the atoms together to form molecules. Later, scientists such as Sir Humphry Davy (1778–1829) and his assistant Michael Faraday (1791–1867)—whose law is one of Maxwell's equations (Chapter 34 and page 1093)—discovered, through experimentation, that molecules are held together by electrical forces.

The idea that atoms have no internal structure persisted until the late 1800s, when the work of the British physicist Sir Joseph John Thomson (1856–1940) first showed that atoms are made up of smaller particles. Physicists at that time were trying to understand an experiment involving a capacitor in a vacuum tube (Fig. 42.2). The capacitor is charged, and then the tube is evacuated. As the air is removed from the tube, a blue glow appears between the capacitor's plates. As more air is removed, the glow turns pink. Physicists at the time called this glow (actually glowing particles) *cathode rays*. The nature of cathode rays was controversial. It was known that cathode rays could be deflected by a magnetic field, but in early experiments they were not deflected by an electric field. If cathode rays were made up of charged particles, then an electric field should deflect them.

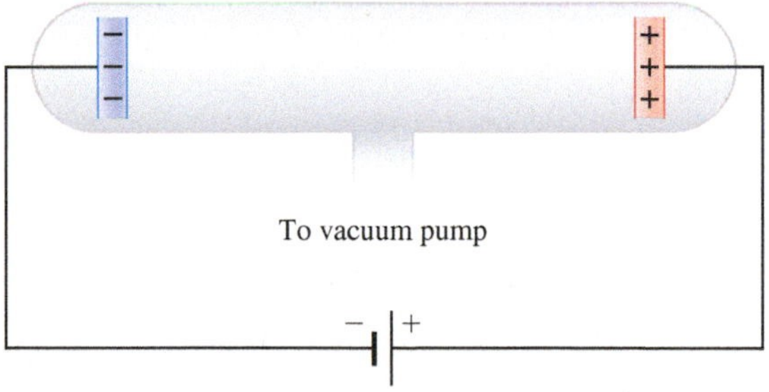

FIGURE 42.2 A capacitor in a tube can be used to observe several phenomena, such as cathode rays or the emissions from a gas such as hydrogen.

At first, Thomson believed that cathode rays were charged particles and were not deflected by an electric field because the tube still contained too much air. By 1897, Thomson developed a much better vacuum system, and showed that cathode rays are indeed deflected by an electric field, as would be expected for a stream of negatively charged particles. Thomson concluded that the negative particles—electrons—emerge from the atoms that make up the metal of the capacitor plates. He then went on to change the type of metal used for the plates, with similar results. So he concluded that electrons are present in all atoms.

THOMSON'S PLUM PUDDING MODEL
Major Concept

Based on his experiments, Thomson developed a model for the atom. Thomson thought that the atom consists of two parts: a positive homogeneous sphere and electrons embedded in this positive sphere. This was known as the plum pudding

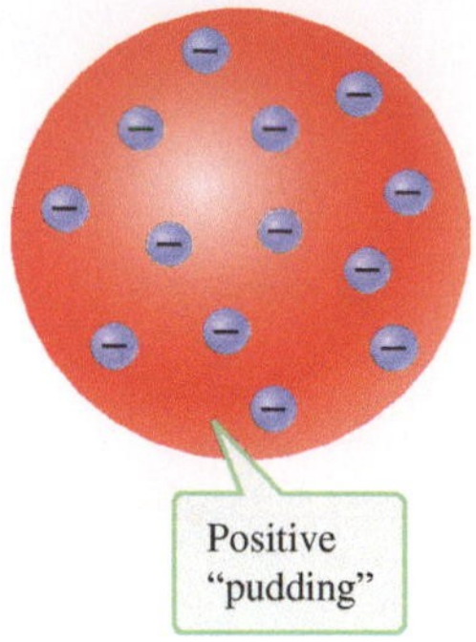

FIGURE 42.3 In the plum pudding model of an atom, negatively charged electrons are embedded in a positive "pudding." There is no nucleus, and the electrons do not orbit.

model (Fig. 42.3). Since you have probably never seen a plum pudding, it might be better to think of it as the chocolate chip ice cream model. The scoop of vanilla ice cream represents the positive sphere, and the chocolate chips represent the embedded electrons.

Although Thomson's model is consistent with his experiment, it cannot explain the spectral lines we observe from atoms (Section 36-4). Any acceptable atomic model must be able to explain the spectral lines. One part of this chapter's case study focuses on the spectrum produced by the simplest atom—hydrogen.

CONCEPT EXERCISE 42.1

What is misleading about the term *atom*? Come up with a term to replace *atom*. Explain your choice.

CASE STUDY Part 1: Lasers

As discussed in the opening paragraph, our focus in this chapter is atoms. Using quantum mechanics to understand atoms deepens our understanding of the universe and leads to practical devices. So the case study in this chapter is divided into two parts. One part is about a practical device—the laser. The other part is about hydrogen—the simplest and most abundant element in the universe. Lasers have become ubiquitous. You or a friend may have a laser on your key ring; your professor is likely to use a laser pointer during a lecture. You probably used a laser in your laboratory to produce interference or a diffraction pattern such as those shown in Figures 35.10, 35.16, 36.2, and 36.5. Lasers are used in medicine to correct vision, to remove tattoos, and to treat some cancers. In the future, lasers may be used to detect gravity waves from distant astronomical events such as the merging of black holes (Fig. 42.4). By the end of this chapter, you will understand the basic operation of a laser.

FIGURE 42.4 General relativity predicts that a gravity wave is triggered by the merging of two massive objects such as black holes or neutron stars. These gravity waves, like waves on the surface of water, travel out into space and cause distant objects to oscillate. Detectors on Earth use lasers to search for the slight oscillations, which would confirm the existence of gravity waves. As of press time, no gravity waves have been detected, but you may wish to check with such organizations as LIGO (Laser Interferometer Gravitational-Wave Observatory).

CASE STUDY Part 2: The Importance of the Hydrogen Spectrum

Because hydrogen is the most abundant element in the universe, astronomers spend a lot of time observing hydrogen, especially the hydrogen spectrum (Fig. 42.5). This spectrum has much to teach us: for example, it is used to find the temperature and size of stars, the mass of the Milky Way galaxy, and the velocity of distant quasars. Each kind of atom has a unique internal structure that determines its unique spectrum. In this case study, we'll focus on the structure of hydrogen atoms to see how that structure is responsible for hydrogen's spectrum. We'll end this part of the case study by seeing if observations of the Sun's spectrum can be used to measure the Sun's magnetic field.

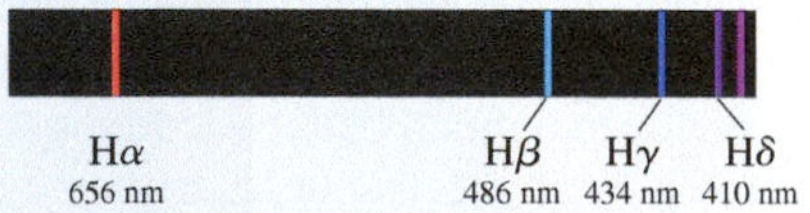

FIGURE 42.5 Hydrogen produces several colored lines in the visible part of the electromagnetic spectrum. The wavelength (in nanometers) of four dominant lines is provided on the figure. Hydrogen is the most abundant element in the Universe and so plays a major role in astronomy.

EXAMPLE 42.1 **CASE STUDY** **The Hydrogen Spectrum**

As shown in Figure 42.5, hydrogen produces four lines in the visible part of the spectrum: Hα (656 nm), Hβ (486 nm), Hγ (434 nm) and Hδ (410 nm). Find the energy associated with the photons that produce each of these lines.

INTERPRET and ANTICIPATE

A photon's energy is proportional to its frequency. The red line (Hα) has a lower frequency than the violet line (Hδ), so we expect the red line to be produced by photons that are less energetic than the ones that produce the violet line.

SOLVE

Use $f = c/\lambda$ (Eq. 34.20) to find the frequency of each line.

Line	f ($\times 10^{14}$ Hz)
Hα	4.57
Hβ	6.17
Hγ	6.91
Hδ	7.32

Use $E = hf$ (Eq. 40.8) to find the energy of each photon.

Line	E ($\times 10^{-19}$ J)
Hα	3.02
Hβ	4.09
Hγ	4.58
Hδ	4.85

The energies are very small. When the energy is so low, it is customary to work in electron volts (eV) instead of in SI units. The conversion factor is $1\ \text{eV} = 1.6 \times 10^{-19}$ J.

Line	E (eV)
Hα	1.89
Hβ	2.56
Hγ	2.86
Hδ	3.03

CHECK and THINK

As expected, the red line (Hα) is produced by photons having less energy than those producing the violet line (Hδ). In this chapter's case study, we'll show why these four unique lines are produced by hydrogen.

42-2 Rutherford's Model of the Atom

Thomson's plum pudding model only survived until about 1911, when Ernest Rutherford (1871–1937) designed an experiment that refuted it. Rutherford was born and began his education in New Zealand. In 1895 he won a scholarship, and decided to continue his education at the Cavendish laboratory in Cambridge, where Thomson was working. Thomson was known as a great teacher and mentor, and Rutherford flourished in Cambridge.

By studying the natural radioactive decay of uranium, Rutherford discovered that it emits two kinds of radiation. One kind is better than the other at penetrating a substance such as aluminum foil. Rutherford named the less-penetrating radiation **alpha rays**, and the more-penetrating radiation **beta rays**. Today we know that alpha rays are helium nuclei. Since helium nuclei consist of two protons and two neutrons, they have a charge of $+2e$ and are about 7000 times more massive than an electron.

Today we use the term *ray* to mean a stream of particles. For example, an alpha ray is a stream of alpha particles (helium nuclei).

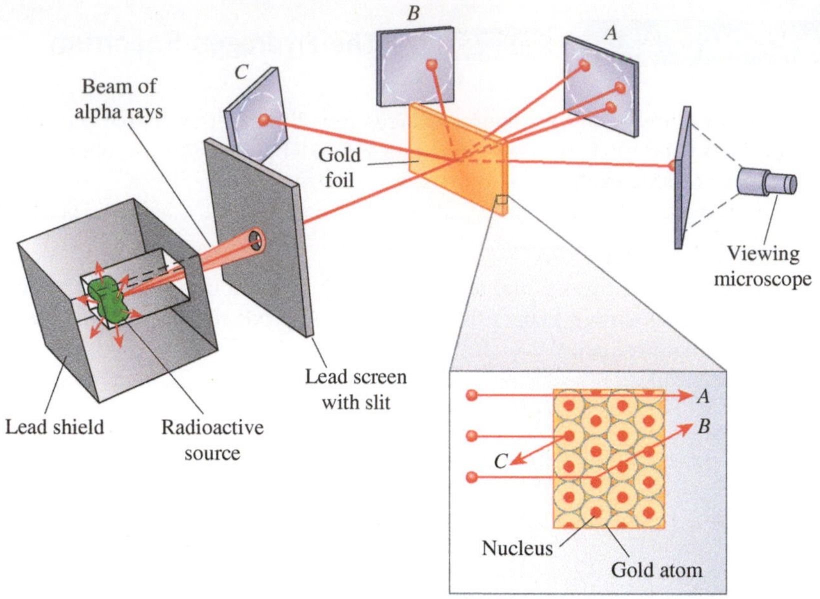

FIGURE 42.6 Rutherford's experiment aimed a beam of alpha rays at a sheet of gold foil. If the beam hits the screen at A, an alpha particle experiences little or no deflection because it passes through the gold foil far from any atomic nucleus. If it hits the screen at B, the alpha particle is somewhat deflected by passing close to an atomic nucleus. If it hits the screen at C, the alpha particle must hit the atomic nucleus nearly head-on. Rutherford was surprised to detect the scattered beam at all angles around the foil. Rutherford's experiment was carried out by Hans Geiger and by an undergraduate named Ernest Marsden.

Rutherford aimed alpha rays at a piece of gold foil (Fig. 42.6). A screen coated with zinc sulfide was used to detect the scattered alpha rays. When the zinc sulfide is struck by an alpha particle, it emits light, which the experimenter observes with a small microscope. Rutherford moved his screen to a number of positions around the foil in order to see if the foil would deflect the alpha rays. If the plum pudding model was correct, then the alpha rays would not be deflected very much by the foil. Electrons in the foil (consisting of plum pudding gold atoms) could not cause much of a deflection because they are 7000 times less massive than alpha rays. These electrons would be accelerated by the alpha rays owing to the electromagnetic force, but the alpha rays themselves would only experience a negligible acceleration. By contrast, the positive "pudding" in the gold atoms would be too spread out to cause a significant net force on the alpha rays. The slight deflection caused by the many gold atoms in the foil would nearly cancel out, so that most of the alpha rays would be detected near point A in Figure 42.6.

Rutherford did find that many of the alpha particles were not deflected by the foil and arrived near point A. However, a significant number were greatly deflected and were detected at positions B and C. A few even backscattered off the foil and were detected near their source! Rutherford was surprised by his results; he said

> It was quite the most incredible event that has ever happened to me in my life. It was almost as incredible as if you fired a 15-inch shell at a piece of tissue paper and it came back and hit you.

RUTHERFORD'S SOLAR SYSTEM MODEL
✪ Major Concept

Because the alpha particles were deflected at all angles, including 180°, Rutherford concluded that the positive part of the atom is concentrated at its center. His model of the atom consisted of a massive positively charged nucleus at the center surrounded by orbiting electrons, so that most of the atom is empty space. Rutherford's model is basically the solar system model that we discussed earlier (Fig. 42.1). This model accounts for the backscattering results as follows. When an alpha particle passes through atoms without coming close to a nucleus, it experiences almost no deflection and is detected at position A. But if the alpha particle comes close to a positive nucleus, then according to Coulomb's law the alpha particle will experience a repulsive force. The closer the alpha particle comes to a nucleus, the greater the force. This repulsive force causes the alpha particle to be deflected and in some cases even to go backward.

In Rutherford's model, the electrons must orbit the nucleus. If they did not orbit, then the attractive force between the nucleus and the electron would cause the electron to fall into the nucleus. If that were the case, then matter would be unstable. (As we'll see in the next sections, quantum mechanics does not require electrons to orbit in order for atoms to be stable.)

EXAMPLE 42.2 Orbiting Electrons in Hydrogen

Let's examine Rutherford's model more closely. Consider hydrogen, the simplest atom, with a nucleus consisting of a single proton that is orbited by a single electron. Model the nucleus as a sphere with a radius of roughly 10^{-15} m, and assume the electron orbits the center of the nucleus in a circle with a radius of roughly 10^{-10} m. Find the magnitude of the force exerted on the electron by the nucleus and the electron's speed.

INTERPRET and ANTICIPATE

We could have done this problem in Chapter 23. Coulomb's law gives the force exerted on the electron, and because no other forces are present, this is the centripetal force. We can find the electron's speed from its centripetal acceleration.

SOLVE

Use Coulomb's law (Eq. 23.3) to find the force on the electron. As usual, quantity r is the distance from the center of the sphere to the particle; in this case r is the radius of the electron's orbit. For hydrogen, the nucleus and the electron have the same magnitude of charge, e.

$$F_E = k\frac{|q_1 q_2|}{r^2} \tag{23.3}$$

$$F_E = k\frac{e^2}{r^2}$$

$$F_E = (8.99 \times 10^9\ \mathrm{N \cdot m^2/C^2})\frac{(1.60 \times 10^{-19}\ \mathrm{C})^2}{(10^{-10}\ \mathrm{m})^2}$$

$$F_E = 2.3 \times 10^{-8}\ \mathrm{N}$$

This is the centripetal force, so the electron's speed is found using $F_c = mv^2/r$ (Eq. 6.7).

$$F_c = F_E = m_e\frac{v^2}{r}$$

$$v = \sqrt{\frac{rF_E}{m_e}} = \sqrt{\frac{(10^{-10}\ \mathrm{m})(2.3 \times 10^{-8}\ \mathrm{N})}{9.11 \times 10^{-31}\ \mathrm{kg}}}$$

$$v = 1.6 \times 10^6\ \mathrm{m/s}$$

CHECK and THINK

The electron's orbital speed is about 0.5% the speed of light, so there is no need to use special relativity. Our estimate of this speed seems reasonable (not too fast, not too slow), which partly explains why the solar system model of the atom along with classical physics explains so many phenomena. However, Rutherford's solar system model cannot account for the hydrogen spectrum (Fig. 42.5), the subject of part 2 of our CASE STUDY.

42-3 Bohr's Model and Atomic Spectra

Rutherford's solar system model encountered problems immediately. First, since the electrons orbit around the nucleus, they are accelerating. Accelerating charged particles give off radiation, so the electrons should be giving off radiation continually. Second, since radiation carries energy away from its source, the electrons should lose energy. You would expect the electrons eventually to spiral into the nucleus, and matter would be unstable. However, no such radiation is observed, and matter is stable. Finally, each element is known to give off a unique set of discrete spectral lines. How can Rutherford's model of the atom explain these observations? A complete model of the atom must be able to account for the observed spectral lines.

The Hydrogen Spectrum

Let's begin by considering in some detail what researchers in the late 19th century were able to observe. One way to view the spectrum emitted by a gas is to use a tube like the one in Figure 42.2. Instead of maintaining a vacuum in the tube, replace the air with the rarefied gas that you wish to study, such as hydrogen. The capacitor is then connected to a high-voltage power supply. This excites the gas, the gas radiates, and the spectrum can be seen by using a diffraction grating (Section 36-4). From Figure 42.5, we see that the resulting visible spectrum of hydrogen consists of four lines. The wavelengths of these lines (656 nm, 486 nm, 434 nm, and 410 nm) may seem arbitrary or random. However, in 1885 a Swiss high school mathematics teacher named Johann Jakob Balmer (1825–1898) discovered their pattern. He realized that the wavelengths of the four lines could be found from the formula

$$\frac{1}{\lambda} = R\left(\frac{1}{2^2} - \frac{1}{n^2}\right) \quad (n = 3, 4, 5, 6) \tag{42.1}$$

where R is called the Rydberg constant and has the value:

$$R = (1.0973731568539 \pm 0.0000000000055) \times 10^7\,\text{m}^{-1} \tag{42.2}$$

Today we refer to the series of hydrogen lines that fits Equation 42.1 as the **Balmer series.**

Four other hydrogen line series were discovered and named for the researchers who discovered them. The wavelengths of the lines that make up these other series fit an equation (known as the **Rydberg formula**) similar to Equation 42.1:

$$\frac{1}{\lambda} = R\left(\frac{1}{m^2} - \frac{1}{n^2}\right) \tag{42.3}$$

(In Eq. 42.3, $m < n$; you can remember that the variables are in alphabetical order, from low values to high.) The value of m is a constant (integer) for each series, and the lowest value of n is $(m + 1)$ for that series. For example, in the Balmer series $m = 2$, and the lowest value of n is 3. The other series and their values of m are listed in Table 42.1. Equation 42.3 is an empirical fit; it was first stated without a physical model to explain it. That physical model was later developed by Bohr.

TABLE 42.1 The hydrogen spectral line series.

Name of series	m
Lyman	1
Balmer	2
Pashen	3
Brackett	4
Pfund	5

CONCEPT EXERCISE 42.2

CASE STUDY **The Balmer Series**

Show that the Balmer formula correctly gives the four observed wavelengths Hα (656 nm), Hβ (486 nm), Hγ (434 nm) and Hδ (410 nm) in Figure 42.5.

Bohr's Atomic Model

BOHR'S ATOMIC MODEL
✪ Major Concept

Niels Hendrik David Bohr (1885–1962) was part of a very scientific family. Not only did he win the Nobel Prize in physics in 1922, his son Niels Aage Bohr also won the Nobel Prize in physics in 1975. Bohr worked with Rutherford for four years. For a moment, try to imagine being in Bohr's position in 1911. Rutherford's experiment had showed that the atom's positive charge is concentrated in a nucleus and that most of the atom is empty space. If you wanted to explore this model, you would probably focus your attention on the simplest atom—hydrogen. Bohr knew that the visible hydrogen spectrum showed a small number of lines whose wavelengths were given by Balmer's formula (Eq. 42.1). Bohr also knew that Planck had solved the "ultraviolet catastrophe" by hypothesizing that the energy of each oscillator is *quantized.* In fact, Bohr and Planck were in similar positions. Planck had discovered an empirical fit to the black-body curve, and he wanted to develop a physical model for black bodies (Section 40-2). Bohr knew that the hydrogen spectrum was described by Balmer's empirical fit, and he wanted a physical model for hydrogen that would allow him to derive that empirical fit from basic physical principles.

Bohr recognized the problems with Rutherford's atomic model, but he set out to fix these problems without completely overturning it. In both Rutherford's model and Bohr's model, the positive nucleus is at the center of the atom, and the electrons orbit this nucleus like the planets orbit the Sun.

However, Bohr's model differs from Rutherford's in two major ways, known as **Bohr's postulates**. His **first postulate** is that *the orbit of the electron is restricted to one of a number of allowed circular paths.* In making this postulate, Bohr was guided by Planck's success in modeling black bodies, which led Bohr to hypothesize that the angular momentum of the electron orbit is quantized. Specifically, Bohr proposed that the only allowable orbits have an angular momentum L given by:

$$L = n\left(\frac{h}{2\pi}\right) = n\hbar \qquad (n = 1, 2, 3 \ldots) \tag{42.4}$$

where $\hbar = h/2\pi = 6.58211814 \times 10^{-16}\ \text{eV}\cdot\text{s}$. The integer n is called the **principal quantum number**.

Contrast this to the orbits of planets and other objects in the solar system. Any object orbiting the Sun can have any orbital path with any angular momentum. Comet and asteroid orbits have a wide range of semi-major axes, and thus a wide range of angular momenta. Bohr's first postulate is equivalent to saying that most of these paths are impossible and that all objects orbiting the Sun must follow certain allowed paths.

Bohr's second postulate is that *electrons in orbit around the nucleus do not radiate.* An electron only radiates when it moves from one allowed orbit to another allowed orbit that has a lower energy. Because the electron orbits are quantized, an electron cannot be found moving in between allowed orbits. We must imagine that the electron *jumps* discontinuously from one allowed orbit to another.

Although Bohr's first postulate is based on Planck's work on black body radiation, his second postulate involves a unique departure from classical physics. According to classical physics, a charged particle moving in a circular orbit must radiate. Bohr could not use classical physics to explain why the electrons orbiting the nucleus don't radiate, but this must be true if Rutherford's solar system model is basically correct.

Bohr's two postulates "fixed" the problems with Rutherford's model. First, an electron in a particular orbit does not radiate continually, which explains why we don't observe this sort of radiation coming from matter. Second, because an orbiting electron is not continually losing energy through radiation, it does not spiral into the nucleus. This fixes the problem with the stability of matter. Third, the atom only emits radiation when the electron jumps from one orbit to a lower energy orbit. When that happens, the atom loses a specific amount of energy in the form of a photon that has a specific wavelength. So we only observe photons of certain wavelengths (which in the case of hydrogen are given by Balmer's fit); as a result, the observed spectrum is made up of a few specific lines.

These two postulates were the only two departures Bohr made from classical physics. He was able to derive Balmer's empirical fit using classical physics and these postulates. We'll work through Bohr's derivation here in order to answer part of the question posed by our CASE STUDY—why does hydrogen produce just a handful of lines in the visible spectrum (Fig. 42.5)?

DERIVATION Spectral Lines of Hydrogen

We will use Bohr's model to derive the Rydberg formula,

$$\frac{1}{\lambda} = R\left(\frac{1}{m^2} - \frac{1}{n^2}\right) \tag{42.3}$$

which was originally found empirically for the hydrogen spectrum.

In Bohr's model, the nucleus is a proton with charge $+e$. The electron has charge $-e$, and its orbit is a circle of radius r. The electron's mass is given by its rest mass m_e.

As in Example 42.2, find the speed squared of the orbiting electron by equating the electric force (Eq. 23.3) exerted by the nucleus on the electron with the centripetal force (Eq. 6.7).	$F_E = F_c$ $k\dfrac{e^2}{r^2} = m_e\dfrac{v^2}{r}$ $v^2 = \dfrac{ke^2}{m_e r}$	(1)
Next we apply Bohr's first postulate: The angular momentum is quantized and given by Equation 42.4, where n is an integer.	$L = n\left(\dfrac{h}{2\pi}\right)$ where $n = 1, 2, 3\ldots$	(42.4)
The magnitude of the angular momentum is given by $L = rp \sin\varphi$ (Eq. 13.25) and $p = mv$ (Eq. 10.1). Since the electron's velocity is perpendicular to its orbital radius, $\sin\varphi = 1$.	$L = rm_e v \sin\varphi$ $L = rm_e v$	(2)
Set Equation 42.4 equal to Equation (2) and solve for v. Remember n is an integer, so we find that the electron's speed is also quantized, with allowed speeds given by Equation (3).	$L = n\left(\dfrac{h}{2\pi}\right) = rm_e v$ $v = \dfrac{nh}{2\pi m_e r}$	(42.5) (3)
Substitute Equation (3) into Equation (1).	$\dfrac{n^2h^2}{4\pi^2 m_e^2 r^2} = \dfrac{ke^2}{m_e r}$	
Solve for r. We find that r is also quantized.	$r = \dfrac{n^2h^2}{4\pi^2 k m_e e^2}$	(4)
The smallest possible orbital radius is known as the **Bohr radius** r_B, and we find it by substituting $n = 1$ as well as all of those constants into Equation (4) (Problem 42.9). The allowed radii are given by Equation 42.7.	$r_B = 5.29 \times 10^{-11}\,\text{m}$ $r = n^2 r_B$	(42.6) (42.7)
Now consider the energy of the atom, a system consisting of the nucleus and an electron. The nucleus is at rest, so the system's kinetic energy is due only to the electron's orbital speed. From Example 42.2, speed is not relativistic. Substitute Equation (1) for v^2.	$K = \dfrac{1}{2}m_e v^2$ $K = \dfrac{1}{2}m_e\left(\dfrac{ke^2}{m_e r}\right)$ $K = \dfrac{1}{2}\dfrac{ke^2}{r}$	(5)
The electric potential energy of the nucleus–electron system is given by $U_E = kQq/r$ (Eq. 26.3). By the usual convention, the potential energy is zero when the electron is infinitely far from the nucleus. For all other positions, the potential energy is negative.	$U_E = -\dfrac{ke^2}{r}$	(42.8)

Add Equations (5) and 42.8 to find the atom's total mechanical energy.	$E = K + U_E = \frac{1}{2}\frac{ke^2}{r} - \frac{ke^2}{r}$ $E = -\frac{1}{2}\frac{ke^2}{r}$
Substitute Equation (4) for r. The resulting Equation 42.9 says that the atom's energy is quantized, depending on n. Because of the negative sign, the higher the value of n, the greater (less negative) the atom's energy.	$E = -\frac{1}{2}ke^2\left(\frac{4\pi^2 k m_e e^2}{n^2 h^2}\right)$ $E = -\frac{2\pi^2 k^2 m_e e^4}{h^2}\frac{1}{n^2}$ (42.9)
Now we apply Bohr's second postulate: The atom only emits radiation when the electron jumps from its orbit to a lower energy orbit. The higher energy orbit has a greater quantum number n, and the lower energy orbit has a lower quantum number m.	$\Delta E = E_n - E_m$ $\Delta E = -\frac{2\pi^2 k^2 m_e e^4}{h^2}\left(\frac{1}{n^2} - \frac{1}{m^2}\right)$ $\Delta E = \frac{2\pi^2 k^2 m_e e^4}{h^2}\left(\frac{1}{m^2} - \frac{1}{n^2}\right)$ (6)
When the atom loses energy ΔE, it emits radiation in the form of a photon whose energy, according to Einstein, is $E = hc/\lambda$ (Eq. 40.17). Set hc/λ equal to Equation (6) and find the wavelength of the photon emitted by solving for $1/\lambda$.	$hf = \Delta E$ $\frac{hc}{\lambda} = \frac{2\pi^2 k^2 m_e e^4}{h^2}\left(\frac{1}{m^2} - \frac{1}{n^2}\right)$ $\frac{1}{\lambda} = \left(\frac{2\pi^2 k^2 m_e e^4}{h^3 c}\right)\left(\frac{1}{m^2} - \frac{1}{n^2}\right)$
In Problem 42.8 you will show that the constant $2\pi^2 k^2 m_e e^4/h^3 c$ is equal to the Rydberg constant R (Eq. 42.2).	$\frac{1}{\lambda} = R\left(\frac{1}{m^2} - \frac{1}{n^2}\right)$ ✓ (42.3)

COMMENTS

We have shown that Equation 42.3 follows from Bohr's two postulates and classical physics. We can explain the visible hydrogen spectrum in terms of the Bohr atomic model, in which the hydrogen atom consists of a proton, which is the nucleus and an electron in orbit around the nucleus. The electron orbit is quantized: Only certain orbital radii (Eq. 42.7) or energies (Eq. 42.9) are allowed. The atom only radiates when the electron jumps from one orbit to another orbit with lower energy. The energy of the photon given off by the atom equals the energy difference between these two orbits. Because only certain orbital energies are allowed, only photons of certain specific energies can be emitted. The energy of the emitted photons corresponds to their wavelength or the color we see; thus only lines of those colors are found in the hydrogen spectrum.

Energy Levels in Hydrogen

In following Bohr's derivation, we found that the hydrogen atom's energy is quantized, as given by Equation 42.9. The absolute value of this energy is small, so it is more convenient to work in electron volts (eV) instead of in SI units (joules). Substituting values of the constants in Equation 42.9, we find that the energy levels of the hydrogen atom are given by:

$$E_n = -\frac{13.6\,\text{eV}}{n^2} \qquad (n = 1, 2, 3 \ldots) \qquad (42.10)$$

It is also convenient to express Planck's constant in eV instead of in joules: $h = 4.1356 \times 10^{-15}$ eV·s

All atoms and molecules have quantized energy levels, although Equation 42.10 only holds for the hydrogen atom. The energy levels of any atom or molecule are best visualized with an energy-level diagram (Section 41-2). Figure 42.7 is an energy-level diagram for hydrogen. The quantum number of the lowest energy level is $n = 1$; this level is known as the **ground state**. The atom cannot have less energy

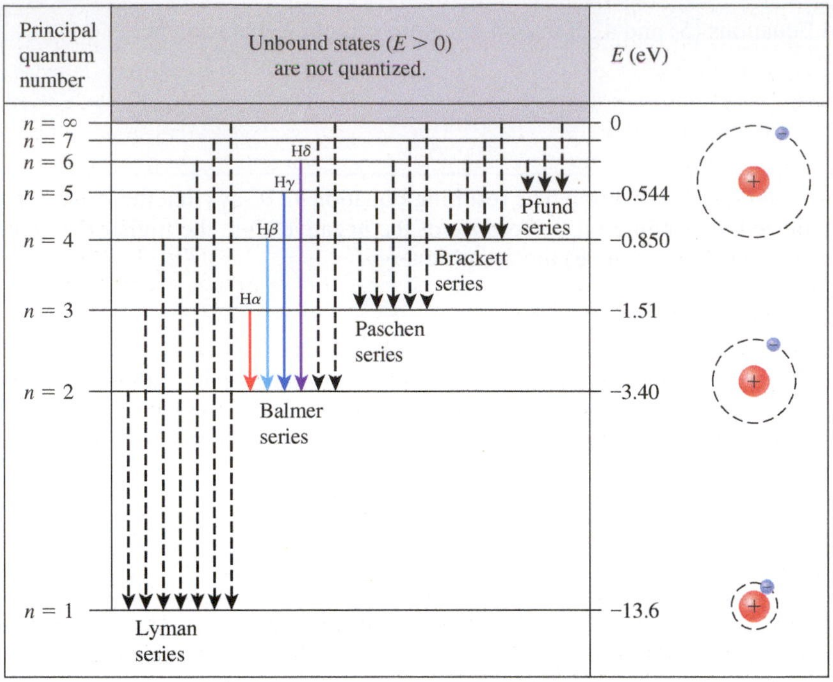

FIGURE 42.7 Energy-level diagram for hydrogen (not to scale). According to Bohr's model, the energy of a hydrogen atom increases as the size of the electronic orbit increases. The lines are labeled by Greek letters in Greek alphabetical order, starting with the lowest energy photon. Bohr's model is based on Rutherford's solar system model. The size of the electron's orbit corresponds to the atomic energy, as shown here for three levels.

than its ground-state energy. For hydrogen, the ground-state energy is $E_1 = -13.6$ eV. Equation 42.10 can be written in terms of hydrogen's ground state energy:

$$E_n = \frac{E_1}{n^2} \tag{42.11}$$

Because the other energy levels are found by dividing by n^2, these energy levels are closer together for high quantum numbers than they are for low quantum numbers. The energy is negative, as is our usual convention for any *bound* system, such as the electrons bound to a nucleus to form an atom.

Energy-level diagrams are particularly useful for trying to visualize the processes that change an atom's energy state. For example, when a hydrogen atom loses energy, its electron moves into a lower orbit—one that is smaller and has less energy. Such a transition is represented by a vertical, downward-pointing arrow on an energy-level diagram. The arrow must begin on one energy level and end on a lower energy level. When an atom loses energy, it emits a photon whose energy equals the difference between the two energy levels. For example, the vertical arrow labeled Hα in Figure 42.7 begins on the $n = 3$ level and ends on the $n = 2$ level, representing the electron's transition from the larger $n = 3$ orbit to the smaller $n = 2$ orbit. The amount of energy lost by the hydrogen atom is the difference between the two levels:

$$\Delta E = E_2 - E_3 = -3.40\text{ eV} - (-1.51\text{ eV})$$

$$\Delta E = -1.89\text{ eV}$$

The energy lost by the atom takes the form of a photon of energy 1.89 eV (Example 42.1), and wavelength:

$$\lambda = \frac{hc}{E} = \frac{(4.1356 \times 10^{-15}\text{ eV}\cdot\text{s})(2.9979 \times 10^8\text{ m/s})}{1.89\text{ eV}}$$

$$\lambda = 6.56 \times 10^{-7}\text{m} = 656\text{nm}$$

This is the wavelength observed for Hα. Figure 42.7 shows the transitions corresponding to all the hydrogen series listed in Table 42.1. The Balmer series (the visible light series shown in Fig. 42.5) corresponds to transitions *to* the $n = 2$ level.

Excited Atoms and Ions

You can imagine that when a hydrogen atom is in its ground state, its electron is in the closest orbit around the nucleus. The atom is stable and no photons are emitted. In order for an atom to emit a photon, it must first be in some energy level above the ground state. When an atom has more than its ground-state energy, we say that it is **excited**. Excitation requires that energy be transferred from the environment to the atom, usually by one of two mechanisms: (1) a collision with another object, such as another atom, a molecule, or even a free electron; or (2) absorption of a photon.

Because the atom's energy is always quantized, energy transferred to the atom must also be quantized (unless the atom becomes ionized). If a hydrogen atom is in its ground state, the environment could transfer 13.6 eV − 3.40 eV = 10.2 eV to the atom, placing the atom in its $n = 2$ state. Ground-state hydrogen can absorb other amounts of energy as long as that energy is the difference between some energy level and the ground state. But it cannot absorb 1 eV, for example, because there is no available energy level at −12.6 eV.

However, the atom can absorb more than 13.6 eV. In that case, the electron is freed from the atom, and the atom becomes an ion. So the energies listed in Figure 42.7 can also be seen as the minimum amounts of energy the atom needs in each state in order to become an ion. They are called the **binding energy** or **ionization energy** for that state. If a hydrogen atom is in its $n = 2$ state, for example, it needs 3.40 eV in order to become an ion. If an atom gains an amount of energy that is greater than its binding energy, then the electron carries off the excess as kinetic energy. For example, if a ground-state hydrogen atom gains 15.6 eV, it becomes an ion, and the freed electron's kinetic energy is 15.6 eV − 13.6 eV = 2.0 eV.

Bohr's model works very well for the hydrogen atom and is even a good model for ions that resemble hydrogen (that is, those that have only one remaining electron). Bohr's model is not good for other atoms, but his two postulates were important steps toward discovering better models for atoms and molecules.

CASE STUDY Part 2: The Sun's Hydrogen Spectrum

You have probably seen the spectrum produced by sunlight as it passes through a prism; it appears to be a complete rainbow of colors. However, only an ideal black-body spectrum shows all the colors (Fig. 42.8A). A black-body spectrum is known as a ***continuum spectrum***.

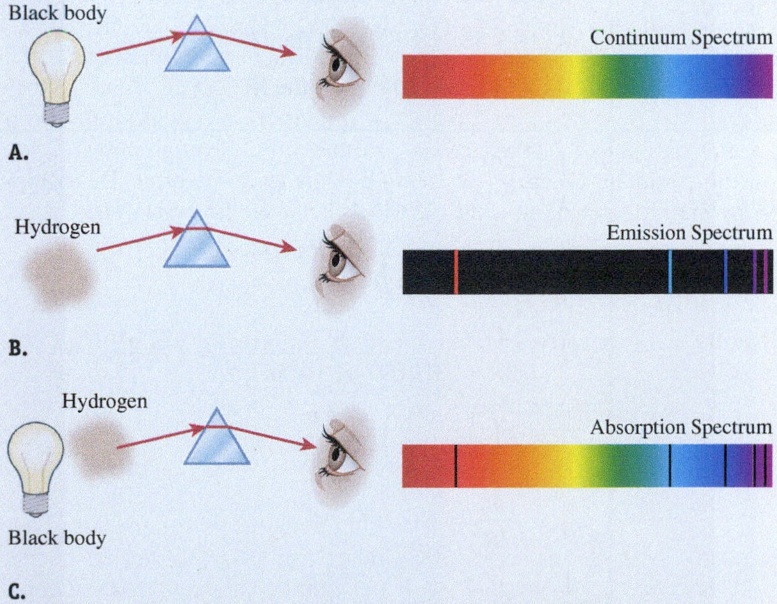

FIGURE 42.8 **A.** A black body is represented by the incandescent bulb. Its spectrum contains all the colors. **B.** A hydrogen emission spectrum is observed. **C.** A hydrogen absorption spectrum is observed when the light from a black body passes through hydrogen gas.

The Sun is not an ideal black body, so what colors does its spectrum consist of? The Sun is mostly made of hydrogen, so you might expect its spectrum to show the four bright Balmer lines in Figure 42.8B. Such a spectrum is known as an **emission spectrum** because these photons are emitted by atoms. However, what is actually observed looks like a continuum spectrum minus the Balmer lines (Fig. 42.8C). Such a spectrum is called an **absorption spectrum**.

To see why we observe an absorption spectrum from the Sun (and most other stars), imagine that some hydrogen gas (in a low energy state) is illuminated by white light from a black body (Fig. 42.8C); the hydrogen absorbs the photons whose energy corresponds to the difference between two hydrogen energy levels. The rest of the photons pass through the gas. So if you look at the spectrum from the perspective shown in Figure 42.8C, you will see the black-body spectrum, minus the photons that have been absorbed by hydrogen.

When we observe the Sun, we are looking at its outermost layer. The layers below radiate like a black body, and the outermost layer of hydrogen absorbs the photons that have energies corresponding to energy-level differences in the hydrogen atom. Normally, when we pass sunlight through a prism, we don't notice the few missing absorption lines. However, these absorption lines are present and with more sensitive equipment, astronomers regularly observe these lines to determine physical properties (such as temperature and magnetic field strength) of the Sun and other stars.

EXAMPLE 42.3 CASE STUDY Other Hydrogen Lines

Use the information in Figure 42.7 to find the wavelength of the first (lowest energy) line of the Lyman, Paschen, and Brackett series. In what part of the electromagnetic spectrum are these lines found?

INTERPRET and ANTICIPATE

Find the wavelength of the emitted photons from the energy differences shown in Figure 42.7. The greater the energy difference, the greater the energy of the photon. High-energy photons have short wavelengths. Of the three series, the Lyman transition has the greatest energy difference, so we expect its photon will have the shortest wavelength. Check the results by using the empirical fit (Eq. 42.3).

SOLVE

Find the energy difference for each of the lines. The subscripts L, P, and B refer to the first line of the *L*yman, *P*aschen and *B*rackett series, respectively.

$$\Delta E_{\text{L}} = E_2 - E_1 = -3.40 \text{ eV} - (-13.6 \text{ eV}) = 10.2 \text{ eV}$$

$$\Delta E_{\text{P}} = E_4 - E_3 = -0.850 \text{ eV} - (-1.51 \text{ eV}) = 0.66 \text{ eV}$$

$$\Delta E_{\text{B}} = E_5 - E_4 = -0.544 \text{ eV} - (-0.850 \text{ eV}) = 0.306 \text{ eV}$$

In each case, the energy lost by the atom equals the photon energy. Use $E = hc/\lambda$ (Eq. 40.17) to find the photon's wavelength.

$$\lambda = \frac{hc}{E} = \frac{(4.1356 \times 10^{-15} \text{ eV}\cdot\text{s})(2.998 \times 10^{8} \text{ m/s})}{E} = \frac{1.240 \times 10^{-6} \text{ eV}\cdot\text{m}}{E}$$

$$\lambda_{\text{L}} = \frac{1.240 \times 10^{3} \text{ eV}\cdot\text{nm}}{10.2 \text{ eV}} = 122 \text{ nm}$$

$$\lambda_{\text{P}} = \frac{1.241 \text{ eV}\cdot\mu\text{m}}{0.66 \text{ eV}} = 1.88\,\mu\text{m}$$

$$\lambda_{\text{B}} = \frac{1.241 \text{ eV}\cdot\mu\text{m}}{0.306 \text{ eV}} = 4.06\,\mu\text{m}$$

Use Figure 34.11 and Table 34.2 (page 1101) to identify the part of the electromagnetic spectrum in which each line is found.

$$\lambda_{\text{L}} = 1.22 \times 10^{-7} \text{ m in the UV}$$

$$\lambda_{\text{P}} = 1.88 \times 10^{-6} \text{ m in the IR}$$

$$\lambda_{\text{B}} = 4.05 \times 10^{-6} \text{ m in the IR}$$

CHECK and THINK

As expected, the Lyman line has the shortest wavelength. We can check our results using Equation 42.3.

$$\frac{1}{\lambda} = R\left(\frac{1}{m^2} - \frac{1}{n^2}\right) \tag{42.3}$$

$$\frac{1}{\lambda_{\text{L}}} = R\left(\frac{1}{1^2} - \frac{1}{2^2}\right) = \frac{3R}{4}$$

$$\lambda_{\text{L}} = \frac{4}{3R} = \frac{4}{3(1.097 \times 10^{7} \text{ m}^{-1})} = 122 \text{ nm} \checkmark$$

$$\frac{1}{\lambda_{\text{P}}} = R\left(\frac{1}{3^2} - \frac{1}{4^2}\right) = \frac{7R}{144}$$

$$\lambda_{\text{P}} = \frac{144}{7R} = 1880 \text{ nm} \checkmark$$

$$\frac{1}{\lambda_{\text{B}}} = R\left(\frac{1}{4^2} - \frac{1}{5^2}\right) = \frac{9R}{400}$$

$$\lambda_{\text{B}} = \frac{400}{9R} = 4050 \text{ nm} \checkmark$$

EXAMPLE 42.4 Tungsten

Although Bohr's model doesn't work well for complicated atoms, it does fit ions that have been stripped of all but one electron. Consider such an ion of tungsten. If the ion goes from its first excited state ($n = 2$) down to the ground state, what is the energy of the photon emitted?

INTERPRET and ANTICIPATE

We start by modifying the expression (Eq. 42.9) for hydrogen's energy levels to find an expression for ionized tungsten's energy levels. Then we can find the energy between the first excited state and the ground state.

SOLVE

Revisit the derivation of the hydrogen spectral lines (Eq. 42.3). We started by applying Coulomb's law to the single proton and electron in a hydrogen atom. When we apply Coulomb's law to ionized tungsten, we must take into account that it has Z protons, where Z is its atomic number. So the speed of the electron (Eq. 1 in the derivation) increases by a factor of Z.

For hydrogen

$$F_E = k\frac{q_1 q_2}{r^2} = k\frac{e^2}{r^2}$$

For other ions (with one electron) and Z protons

$$F_E = k\frac{q_1 q_2}{r^2} = k\frac{Ze^2}{r^2}$$

$$v^2 = \frac{kZe^2}{m_e r} \tag{1}$$

This higher speed means that the kinetic energy is higher by a factor of Z.

For hydrogen

$$K = \frac{1}{2}m_e v^2 = \frac{1}{2}\frac{ke^2}{r}$$

For other ions (with one electron) and Z protons

$$K = \frac{1}{2}\frac{kZe^2}{r} \tag{2}$$

Continue to revisit the derivation, and you find that the radius of the electron's orbit (Eq. 4 in the derivation) is reduced by this factor of Z.

For hydrogen

$$r = \frac{n^2h^2}{4\pi^2 k m_e e^2}$$

For other ions (with one electron) and Z protons

$$r = \frac{n^2h^2}{4\pi^2 k m_e Z e^2} \tag{3}$$

There is another modification we must make to the derivation. The potential energy of the nucleus-electron system is modified by a factor of Z.

For hydrogen

$$U_E = k\frac{Qq}{r} = -\frac{ke^2}{r}$$

For other ions (with one electron) and Z protons

$$U_E = -\frac{kZe^2}{r} \tag{4}$$

Combine equations (2) through (4) to find an expression for the energy levels in an ion with only one electron remaining.

$$E = K + U = -\frac{1}{2}\frac{kZe^2}{r}$$

$$E = -\frac{1}{2}kZe^2\left(\frac{4\pi^2 k m_e Z e^2}{n^2h^2}\right) = -\frac{2\pi^2k^2m_eZ^2e^4}{h^2}\frac{1}{n^2}$$

Compare the expression here to Equation 42.9 for hydrogen's energy levels, and you see that they differ by a factor of Z^2.

For hydrogen

$$E = -\frac{2\pi^2k^2m_e e^4}{h^2}\frac{1}{n^2} \tag{42.9}$$

We find a convenient expression for the energy levels in the ion by multiplying Equation 42.10 by Z^2.	$$E_n = -\frac{13.6\text{ eV}}{n^2} \quad (42.10)$$ For other ions (with one electron) and Z protons $$E_{\text{ion}} = Z^2 E_{\text{Hydrogen}} = -\frac{Z^2\,(13.6\text{ eV})}{n^2} \quad (5)$$
From the periodic table (Appendix B), tungsten's atomic number is 74, which we substitute in Equation (5).	$$E_{\text{Tungsten}} = -\frac{74^2\,(13.6\text{ eV})}{n^2} = -\frac{74.5\text{ keV}}{n^2} \quad (6)$$
We use Equation (6) to write an expression for the difference in the tungsten ion's energy levels, which corresponds to Equation 4 in the derivation for the hydrogen spectral lines. Then we substitute for the energy levels. The photon carries away energy equal to the difference in these levels.	$$\Delta E_{\text{Tungsten}} = 74.5\text{ keV}\left(\frac{1}{m^2} - \frac{1}{n^2}\right)$$ $$\Delta E_{\text{Tungsten}} = 74.5\text{ keV}\left(\frac{1}{1^2} - \frac{1}{2^2}\right) = 55.9\text{ keV}$$ $$E_{\text{photon}} = 55.9\text{ keV}$$

CHECK and THINK

When tungsten is stripped of all but one electron, it has 74 protons and 1 electron. We expect that when the ion is in its ground state, it should be much more tightly bound than when a hydrogen atom is in its ground state. This is exactly what we find when we compare Equation (6) in this example for the tungsten ion to Equation 42.10 for hydrogen. The numerator in these equations indicates the amount of energy it would take to free the electron from its ground state. In the case of hydrogen, this is 13.6 eV. For tungsten the energy requirement is roughly 74500 eV—a factor of $74^2 = 5476$ times greater.

42-4 De Broglie's Theory and Atoms

DE BROGLIE'S MODEL APPLIED TO AN ATOM Major Concept

Although Bohr's postulates work well at creating a model for hydrogen, Bohr did not provide a theoretical basis for these postulates that would explain *why* the energy levels in an atom are quantized. This question is at least partially answered by de Broglie's idea that particles, such as electrons, are also waves (Section 40-6). Bohr did not have the benefit of de Broglie's thinking because de Broglie wrote his thesis about a decade after Bohr came up with his model of the atom.

An electron confined inside a hydrogen atom by the electric potential of its nucleus is like a particle confined in a potential well (Chapter 41). In both situations, the system's energy is quantized and the lowest energy level is nonzero. We modeled a particle confined in an infinite square well as a standing wave on a string with both ends fixed. We can use a similar model for an electron confined in a hydrogen atom.

First, write the de Broglie wavelength (Eq. 40.18) explicitly in terms of the speed v: $\lambda = h/p = h/(m_e v)$. Solve for $m_e v$:

$$m_e v = \frac{h}{\lambda} \quad (42.12)$$

Keep in mind that L is used to represent both angular momentum and length. You may use dimensional analysis if you get confused.

Then substitute Equation 42.12 into Equation 42.5 for the angular momentum, $L = n(h/2\pi) = rm_e v$:

$$L = n\left(\frac{h}{2\pi}\right) = r\left(\frac{h}{\lambda_n}\right)$$

Solve for the wavelength:

$$\lambda_n = \frac{2\pi r}{n} \quad (42.13)$$

Equation 42.13 shows that the wavelengths are quantized, and it also gives us a new model for the atom. Compare Equation 42.13 to Equation 18.9, $\lambda_n = 2L/n$, and you'll see that an electron confined in an atom behaves like a standing wave on a string of length $L = \pi r$ (or like a particle confined in a box of this length). Bohr's model predicts the hydrogen spectral lines, so the Rydberg formula derivation still holds—but with a new physical model behind it. Instead of interpreting r as the radius of the electron's orbit, we relate r to the length of the "box" in which the electron is confined. So we can replace the solar system model of the atom with the model of a standing wave on a string.

Because the numerator in Equation 42.13 is $2\pi r$—the circumference of a circle—we can think that the electron is confined in a circular "box" of radius r. However, a circular "box" is not the best visual analogy. Instead we consider the analogy of a standing wave on a string, and imagine a thin circular wire of radius r (Fig. 42.9). Bohr's model correctly gives the energy levels of the hydrogen atom (Fig. 42.7), but these levels are no longer said to correspond to the size of the electron's orbit. Instead, in de Broglie's model, the energy levels correspond to (circular) standing waves, with wavelengths given by Equation 42.13.

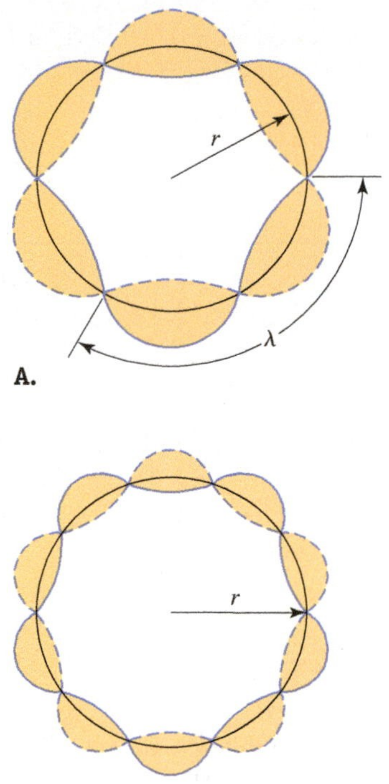

FIGURE 42.9 An electron confined in an atom is like a standing wave on a circular wire. Here we show two wavelengths: **A.** $\lambda_3 = 2\pi r/3$ and **B.** $\lambda_5 = 2\pi r/5$.

CONCEPT EXERCISE 42.3

In what fundamental way is de Broglie's model of hydrogen different from Bohr's model? In what fundamental way is de Broglie's model of hydrogen the same as Bohr's model?

EXAMPLE 42.5 De Broglie's Model of Hydrogen

If a hydrogen atom's energy is -1.51 eV, what is the de Broglie wavelength of the atomic electron?

INTERPRET and ANTICIPATE

Use the energy-level diagram to find the principal quantum number. The quantum number determines the radius of the circular wire. (In Bohr's model, this is the radius of the electron's orbit.) The wavelength is determined by the quantum number and the radius of the circular wire.

SOLVE

The energy-level diagram (Fig. 42.7) shows which quantum number corresponds to an energy of -1.51 eV.	$n = 3$
Find the orbital radius from Equation 42.7, where r_B is the Bohr radius and $n = 3$.	$r = n^2 r_B$ (42.7) $r = (3^2)(5.29 \times 10^{-11}\text{ m}) = 4.76 \times 10^{-10}\text{ m}$
The de Broglie wavelength is given by Equation 42.13, $\lambda_n = 2\pi r/n$.	$\lambda_3 = \dfrac{2\pi(4.76 \times 10^{-10}\text{ m})}{3}$ $\lambda_3 = 9.97 \times 10^{-10}\text{ m}$ $\lambda_3 = 0.997\text{ nm}$

CHECK and THINK

This problem required a reinterpretation of Bohr's model. Instead of thinking of r as the radius of the electron's orbit, we think of it as the radius of the wire to which the "waving" electron is confined.

42-5 Schrödinger's Equation Applied to Hydrogen

Neither de Broglie's nor Bohr's model provides a theoretical basis for why the energy levels of hydrogen are quantized. However, such a basis comes from Schrödinger's equation. In Chapter 41, we learned that the procedure for applying Schrödinger's equation is to (1) substitute the potential energy function, (2) solve for a general expression for ψ, (3) apply the boundary conditions, and (4) interpret $|\psi|^2$ as the probability density. When we applied Schrödinger's equation to a trapped particle, we found that the energy levels are quantized. Because the electron is trapped in the potential well created by the proton, we expect that solving Schrödinger's equation for the hydrogen will yield quantized energy levels. And in particular, when we apply the boundary conditions, we will find an expression for its energy levels.

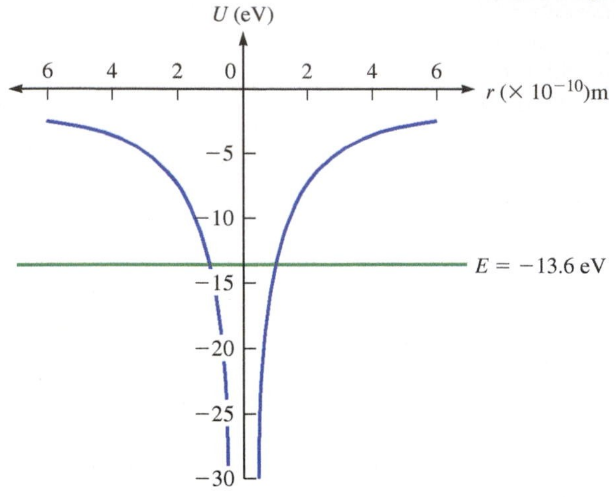

FIGURE 42.10 The energy graph for a hydrogen atom in its ground state. To show the spherical symmetry of the function, we have plotted (positive) r on both sides of the vertical (energy) axis.

The potential energy for a hydrogen atom is given by $U_E = -ke^2/r$ (Eq. 42.8) and is plotted in Figure 42.10. The atom and its potential energy function are three dimensional, requiring a three-dimensional version of Schrödinger's equation, which is beyond the scope of this book. However, finding the solution to a three-dimensional version of Schrödinger's equation follows the same steps as solving the one-dimensional version, and we can understand the solution and use it without doing the complicated derivation.

Hydrogen's Quantum Numbers

SCHRÖDINGER'S EQUATION APPLIED TO AN ATOM Major Concept

When we solved the one-dimensional Schrödinger's equation, we applied one set of boundary conditions and found that one quantity—energy—is quantized and described by one quantum number. When Schrödinger's equation is solved and boundary conditions are applied in three dimensions, three quantities are quantized and are described by three quantum numbers.

The first quantum number is the **principal quantum number** n found in Bohr's model. This quantum number describes the quantization of energy; for hydrogen we have $E_n = E_1/n^2$ (Eq. 42.11), where hydrogen's ground-state energy is $E_1 = -13.6$ eV. The principal quantum number can be any positive integer: $n = 1, 2, 3 \ldots$

The second quantum number is the **orbital quantum number** ℓ. It describes quantization of the magnitude of the electron's orbital angular momentum L. The orbital quantum number is an integer whose range depends on the principal quantum number such that $\ell = 0, 1, 2, \ldots (n - 1)$. The magnitude of the electron's orbital angular momentum is:

$$L = \sqrt{\ell(\ell + 1)}\hbar \qquad (42.14)$$

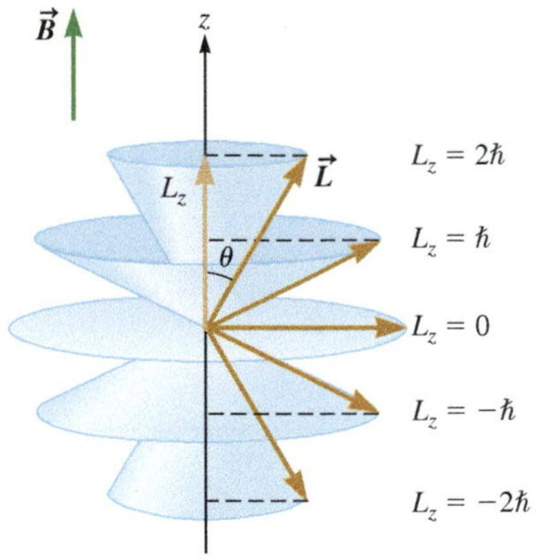

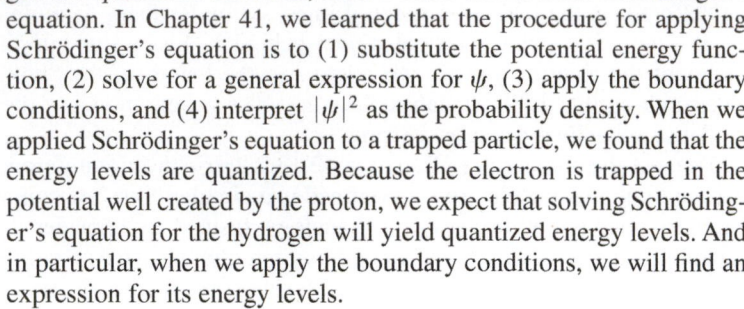

FIGURE 42.11 The magnetic field is in the z direction. When $\ell = 2$, $m = -2$, -1, 0, 1 and 2, corresponding to the five L_z listed here. The components of the angular momentum that are perpendicular to z are not known, so the angular momentum vector for any one value of m lies on the surface of a cone (or disk in the case of $m = 0$.)

The third quantum number is the **orbital magnetic quantum number**, or simply the **magnetic quantum number** m. It describes quantization of the direction of the orbital angular momentum. The magnetic quantum number is an integer whose range depends on the orbital quantum number such that $m = -\ell, -(\ell - 1), \ldots, 0, \ldots +(\ell - 1), +\ell$. According to the uncertainty principle, the direction of $\vec{L}$ is only partially measurable; we can measure only one component, and then the others become uncertain. Traditionally, we imagine that the atom is in a magnetic field pointing in the positive z direction; then L_z is the component of $\vec{L}$ that we measure (Fig. 42.11). This z component is quantized and given by

$$L_z = m\hbar \qquad (42.15)$$

Don't confuse m for magnetic quantum number with m for mass. It helps to keep in mind that the quantum numbers are unitless.

The vector $\vec{L}$ can be anywhere on the surface of a cone that makes an angle θ with the z axis. This angle is quantized, and from Figure 42.11, we find

$$\cos\theta = \frac{L_z}{L} \qquad (42.16)$$

Traditionally, the states with the same principal quantum number n are considered part of the same **shell**. You may have run across this terminology in chemistry. These shells are designated by the uppercase letters K, L, M, and so forth alphabetically, corresponding to $n = 1, 2, 3 \ldots$ Further, these shells are divided into **subshells** based on orbital quantum number ℓ. These subshells are labeled with lowercase letters corresponding to particular values of ℓ (Table 42.2). This practice allows for a compact notation. For example, if hydrogen has $n = 3$ and $\ell = 2$, then we write its state as $3d^1$, where the superscript 1 tells us that one electron occupies this subshell.

TABLE 42.2 Subshell labels

ℓ	Lowercase letter label
0	*s*
1	*p*
2	*d*
3	*f*
4	*g*
5	*h*

After *f*, the labels proceed alphabetically.

CONCEPT EXERCISE 42.4

True or false? For an electron in an atom such as hydrogen, the angular momentum vector $\vec{L}$ can never be in the $\pm z$ direction if the magnitude of z component is given by $L_z = m\hbar$ (Eq. 42.15). Explain your answer.

Hydrogen's ground state

If hydrogen is in its ground state, then $n = 1$, and the maximum and only value of its orbital quantum number is $\ell = 1 - 1 = 0$ in this case. Substituting $\ell = 0$ in Equation 42.14, the electron's orbital angular momentum in the ground state is $L = \sqrt{0(0+1)}\hbar = 0$. Contrast this with Bohr's model, in which the ground-state electron has an angular momentum $L = \hbar$ (Eq. 42.4 with $n = 1$). To make this contrast more vivid, consider the classical "orbit" of any particle, such as a planet, that has no angular momentum. Such a particle travels on a straight line toward (or away) from the central object, eventually colliding with this object (or escaping from it). The classical system just described is unstable. However, hydrogen does exist in its ground state with no orbital angular momentum, and according to quantum mechanics such an "orbit" is stable.

You may notice something strange in this description. We went to great lengths in the previous section to discredit the solar system model and replace it with a standing-wave model. However, we now refer to the electron's angular momentum, implying that we are treating the electron as a particle in orbit around the nucleus. This is a common practice in the description of atoms. The solar system model is so familiar that we don't entirely abandon it; instead, we treat it as an analogy. Our terminology including the word "orbit" is based on that analogy, but remember that it is only an analogy. We find a more appropriate visual image for our description of the electron, after we consider the solutions to Schrödinger's equation.

We don't need to solve the three-dimensional version of Schrödinger's (time-independent) equation ourselves to appreciate and work with the solutions. In general, these three-dimensional solutions depend on three Cartesian coordinates x, y, and z. However, because of the spherical symmetry, the solutions are easier to find and understand using the spherical coordinates r, θ, and φ. We are only interested in the part of the solution that depends on r, the radial distance from the origin. This is called the *radial solution*, but from brevity we'll just refer to it as the *solution*.

The ground state ($n = 1$) solution for hydrogen is

$$\psi_{1s}(r) = \frac{1}{\sqrt{\pi r_B^3}} e^{-r/r_B} \tag{42.17}$$

where r_B is the Bohr radius.

As with any wave function, there is no physical interpretation for ψ, but we can interpret $|\psi|^2$ as the probability density. (Since an atom is three dimensional, this is the probability per unit volume.)

Figure 42.12 illustrates the probability density for an electron in its ground state in hydrogen. The cloud gives us a way to visualize the probability density. The probability density is greatest where the cloud is thickest. This figure stands in sharp contrast to the solar system model. The electron is not a particle in orbit around the nucleus. Instead, the electron is represented by a cloud that is thick near the nucleus and more diffuse at greater distances. Remember that in the ground state, the electron has no angular momentum. In the solar system model an electron with no angular momentum, trapped in the potential well of the nucleus,

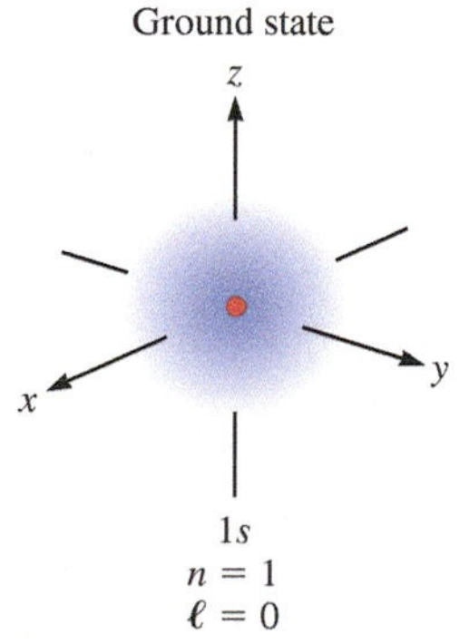

FIGURE 42.12 The probability density for a hydrogen atom in the ground state is spherically symmetric. The electron cloud is densest in the center, meaning that the electron is most likely to be found in the center.

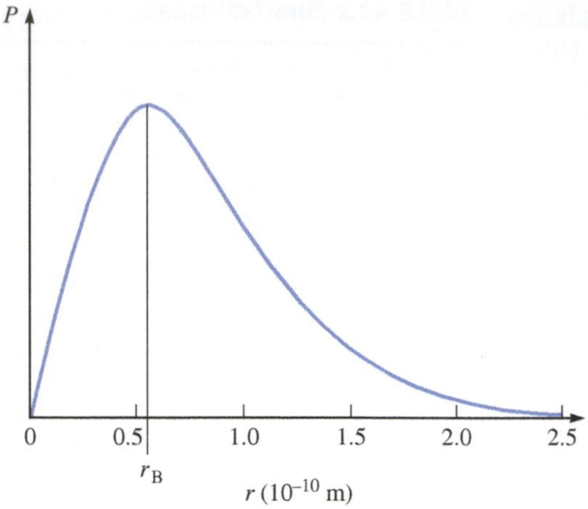

FIGURE 42.13 The radial probability per unit length for the electron in the ground state of the hydrogen atom. The peak occurs at the Bohr radius.

would end up in the nucleus. Contrast this to the quantum-mechanical model. In the ground state, the electron has no angular momentum, but the electron's position is uncertain. So we cannot say that the electron is in the nucleus. Instead we say that the probability density is greatest in the center of the atom.

To find the most likely location for the electron, we need the *radial* probability density $P(r)$, not the *volume* probability density $|\psi|^2$. The two probability densities are related by

$$P(r)dr = |\psi|^2 dV \tag{42.18}$$

The volume element dV is the volume of a spherical shell of thickness dr, so $dV = 4\pi r^2\, dr$. So the radial probability density is

$$P(r) = 4\pi r^2 |\psi|^2 \tag{42.19}$$

The radial probability density allows us to determine the probability $P(r)$ of finding the electron as function distance r from the nucleus. Problem 42.64 asks you to show that

$$P(r) = \frac{4}{r_B^3} r^2 e^{-2r/r_B} \tag{42.20}$$

which is plotted in Figure 42.13. The peak of the graph is located at the Bohr radius. So when hydrogen is in its ground state, the electron is most likely to be found at the Bohr radius—and we can see why Bohr's model predicts the behavior of hydrogen so well. However, quantum mechanics gives a probabilistic interpretation of Bohr's model for hydrogen; allows us to model more complicated atoms; explains the arrangement of atoms in the periodic table; and explains observed lines in hydrogen's (and other atom's) spectrum that cannot be accounted for by the Bohr model.

Hydrogen in an Excited State

So far we have considered hydrogen in its ground state. The wave functions and possibilities are more complicated for the excited states. For example, when hydrogen is in an excited state with $n = 2$, there are two possible values for ℓ and three possible values for m; these are listed in Table 42.3. So there are four possible states in which $n = 2$ (not taking the next quantum number m_s into account). These four states have the same principal quantum number n and make up a single shell with the same energy given by $E_2 = -13.6/(2^2)$ eV (Eq. 42.10 with $n = 2$). The energy of other atoms beyond hydrogen depends on both n and ℓ.

In principle, these four $n = 2$ states are described by four different wave functions and four separate probability densities. However, the probability densities are identical for two of these states: $n = 2$, $\ell = 1$ and $m = +1$ or -1. For hydrogen in the 2s state ($n = 2$, $\ell = 0$) the solution is

$$\psi_{2s}(r) = \frac{1}{\sqrt{2\pi r_B^3}}\left(\frac{1}{2} - \frac{r}{4r_B}\right)e^{-r/2r_B} \tag{42.21}$$

The other two wave functions depend on the azimuthal angle measured from the z axis in the xy plane and the polar angle measured from the z axis. So, to keep the mathematics of our discussion manageable, we omit the other two wave functions.

TABLE 42.3 Quantum numbers for hydrogen in the $n = 2$ state.

n	$\ell = 0, 1, 2 \ldots (n-1)$	$m = 0, \pm1, \pm2 \ldots \pm\ell$
2	0	0
2	1	+1
2	1	0
2	1	−1

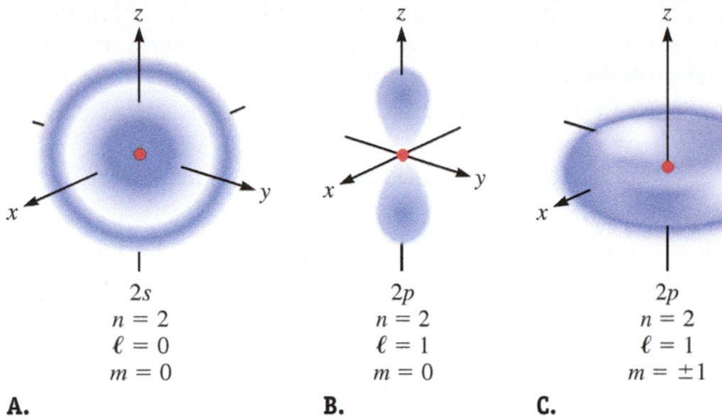

FIGURE 42.14 **A.** The probability density for a hydrogen atom in the $n = 2$, $\ell = 0$ state is spherically symmetric. Notice the white spherical shell region where $|\psi|^2 = 0$. **B.** The probability density in the $n = 2$, $\ell = 1$, $m = 0$-state is not spherically symmetric. **C.** The probability density in the $n = 2$, $\ell = 1$, $m = \pm 1$ state.

However, Figure 42.14 shows $|\psi|^2$ for all four states. The probability density for $n = 2$, $\ell = 0$ is made up of two components (Fig. 42.14A). One component is a sphere that is densest in the center. The second component is a spherical shell, and there is a spherical gap between the two components. Overall, the probability density for this state is spherically symmetric, much like the probability density for $n = 1$, $\ell = 0$ (Fig. 42.12). This is a general property of wave functions: All quantum states with $\ell = 0$ have spherically symmetric wave functions.

For more on symmetry see p. 756.

By contrast, the probability density for $n = 2$, $\ell = 1$, $m = 0$ (Fig. 42.14B) has two separate lobes and is symmetric, but not spherically symmetric. The probability density for $n = 2$, $\ell = 1$, $m = \pm 1$ (Fig. 42.14C) looks like a donut, again symmetric but not spherically symmetry. The donut display of this state is often shown in physics textbooks. In chemistry books, this same state is separated into two components, as shown in Figure 42.15. The display traditionally used in chemistry textbooks is helpful to chemists who are interested in studying how atoms bond, and the stretched out displays help us to understand in which directions these bonds are likely to form.

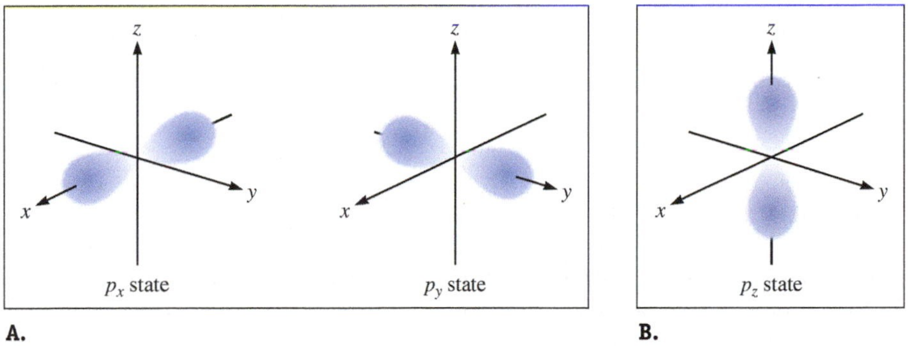

FIGURE 42.15 **A.** In chemistry books the probability density for the $n = 2$, $\ell = 1$, $m = \pm 1$ state is divided into p_x and a p_y component. **B.** Then the probability density of the $n = 2$, $\ell = 1$, $m = 0$ is referred to as the p_z. This is identical to Figure 42.14B.

EXAMPLE 42.6 Radial Probability density for Hydrogen in 2*p* states

Figure 42.16 shows the radial probability density for hydrogen in both the $n = 2$, $\ell = 0$ and $n = 2$, $\ell = 1$ subshells. The radial probability density for the $\ell = 1$ subshell peaks at $r_{max} = 4r_B$. The radial probability density for the $\ell = 0$ peaks at a greater distance; find this r_{max} in terms of the Bohr radius r_B.

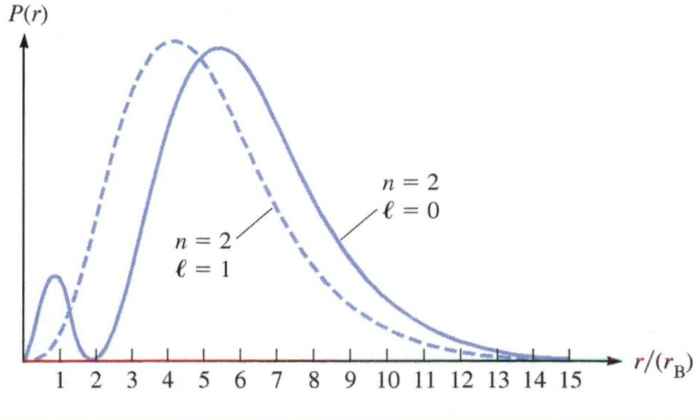

FIGURE 42.16

Example continues on page 1392 ▶

INTERPRET and ANTICIPATE

We know the wave function for this state. To find the radial probability density $P(r)$, we'll need to square this wave function and multiply the result by $4\pi r^2$. To find the maximum, we use the fact that the derivative of a function is zero at its extrema.

SOLVE

We substitute the wave function (Eq. 42.21) into Equation 42.19 for the radial probability density.

$$P(r) = 4\pi r^2 |\psi|^2 \tag{42.19}$$

$$P(r) = 4\pi r^2 \left[\frac{1}{\sqrt{2\pi r_B^3}}\left(\frac{1}{2} - \frac{r}{4r_B}\right)e^{-r/2r_B}\right]^2$$

We need a few steps of algebra. Because we will be taking the derivative and setting it to zero, we can make the algebra a little easier if we replace the equals with a proportionality and drop the leading constants.

$$P(r) \propto \left[\left(\frac{1}{2} - \frac{r}{4r_B}\right)r\right]^2 e^{-r/r_B}$$

$$P(r) \propto \left[\frac{1}{4}\left(2 - \frac{r}{r_B}\right)r\right]^2 e^{-r/r_B} \propto \left[\left(2 - \frac{r}{r_B}\right)r\right]^2 e^{-r/r_B}$$

$$P(r) \propto \left(2r - \frac{r^2}{r_B}\right)^2 e^{-r/r_B}$$

Now take the derivative with respect to r and set the derivative to zero.

$$\frac{dP}{dr} \propto 2\left(2r - \frac{r^2}{r_B}\right)\left(2 - \frac{2r}{r_B}\right)e^{-r/r_B} - \frac{1}{r_B}\left(2r - \frac{r^2}{r_B}\right)^2 e^{-r/r_B} = 0$$

$$2\left(2 - \frac{2r}{r_B}\right) - \frac{1}{r_B}\left(2r - \frac{r^2}{r_B}\right) = 0$$

$$4 - \frac{4r}{r_B} - \frac{2r}{r_B} + \frac{r^2}{r_B^2} = 0$$

$$4 - \frac{6r}{r_B} + \frac{r^2}{r_B^2} = 0$$

Finally, use the quadratic formula to solve for r in terms of the Bohr radius r_B. We choose the + sign. The – sign gives the smaller peak shown in Figure 42.16.

$$r = \frac{6r_B \pm \sqrt{36r_B^2 - 4(4r_B^2)}}{2} = 5.24r_B$$

CHECK and THINK

Compare the answer here to the ground-state level of hydrogen (Fig. 42.13). It makes sense that when hydrogen is in a higher energy state, the electron is most likely to be found farther from the nucleus.

42-6 Magnetic Dipole Moments and Spin

You may be wondering why m is called the *magnetic* quantum number. The reason has to do with how a material becomes magnetized. So we'll digress from our focus on hydrogen and turn our attention to atoms in general. From Section 30-6, a material becomes magnetized partly from the orbital motion of the electrons in its atoms. The magnetic moment of a current loop is given by $\vec{\mu} = (I\pi R^2)\hat{\jmath}$ (Eq. 30.14), and in Problem 42.40 you will show that the magnetic moment of a charged particle of mass m in a circular orbit is given by

$$\vec{\mu}_{\text{orbit}} = \frac{q\vec{L}}{2m} \tag{42.22}$$

Because the magnitude of $\vec{L}$ is quantized, the magnitude of the magnetic moment is quantized. Furthermore, because the direction of $\vec{L}$ is quantized, the direction of the magnetic moment is quantized. For an atom in an external magnetic field pointing in the z direction, this quantization is best expressed as

$$(\mu_{\text{orbit}})_z = -m\mu_{\text{Bohr}} \tag{42.23}$$

where $\mu_{\text{Bohr}} = e\hbar/2m_e = 9.274 \times 10^{-24}$ J/T is the **Bohr magneton**, m is the magnetic quantum number, and m_e is the mass of an electron. The negative sign in Equation 42.23 indicates that the electron carries a negative charge and its magnetic moment points in the opposite direction as its orbital angular momentum. The magnetic quantum number m in Equation 42.23 determines the component of the magnetic moment that is antiparallel to the external magnetic field, and so justifies calling m the magnetic quantum number.

Recall that the magnetization of a material also depends on the spin of the electrons in its atoms (Section 30-6). Again, the term *spin* gives us a mental picture of an object, such as a planet, that spins as it orbits. Although this is a useful image, it is not a good description for the actual motion of the electrons. First, classical mechanics cannot even account for the spin of an electron because particles have no spatial extent and cannot rotate. Second, quantum mechanics abandons the particle model in favor of a wave model for the electron, and it is hard even to come up with a way to visualize a "spinning" probability wave. Instead, spin is in an intrinsic property of a particle, much like charge is an intrinsic property.

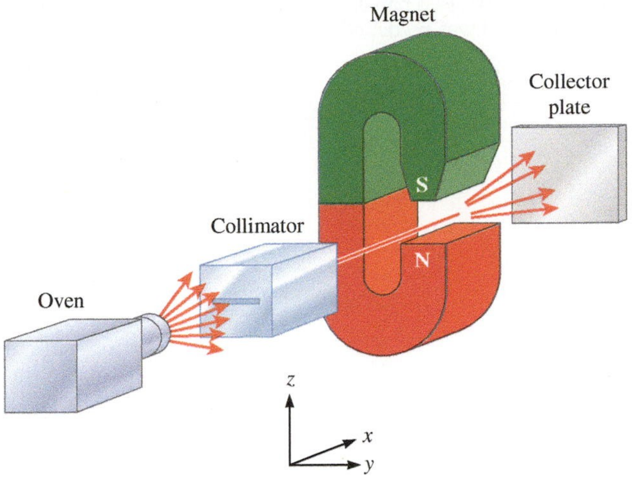

FIGURE 42.17 In the Stern–Gerlach experiment, atoms are released by an oven, and a collimator creates a narrow beam that passes through a nonuniform magnetic field. The magnetic field deflects the beam before encountering the collector plate.

Stern–Gerlach Experiment

How do we know that particles such as electrons have this intrinsic property? The answer comes from experimental evidence. In 1920, the German physicists Otto Stern (1888–1969) and Walther Gerlach (1889–1979) performed an experiment that provided that evidence. Figure 42.17 shows the basics of their experiment. An oven is used to vaporize silver. (A similar experiment was conducted by other researchers about seven years later using hydrogen.) A collimator creates a beam, which then passes through a nonuniform magnetic field. Let's model atoms in the beam as small dipole magnets with magnetic moment $\vec{\mu}$. We studied what happens when a magnetic dipole is in a *uniform* magnetic field; we found there is a net torque but no net force (Section 30-13). Now we must consider what happens when a magnetic dipole is in a *nonuniform* magnetic field. The result is a net force (and a net torque). The magnitude of the force depends both on the magnetic dipole moment and on the gradient of the magnetic field. The Stern–Gerlach experiment is set up so there is only a gradient in the z direction (Fig. 42.17). As a result, the atoms may be deflected in the positive or negative z direction based on their magnetic moment. Figure 42.18 shows that if the magnetic moment is along the x axis the atom is not deflected, but some atoms are deflected upward and others downward.

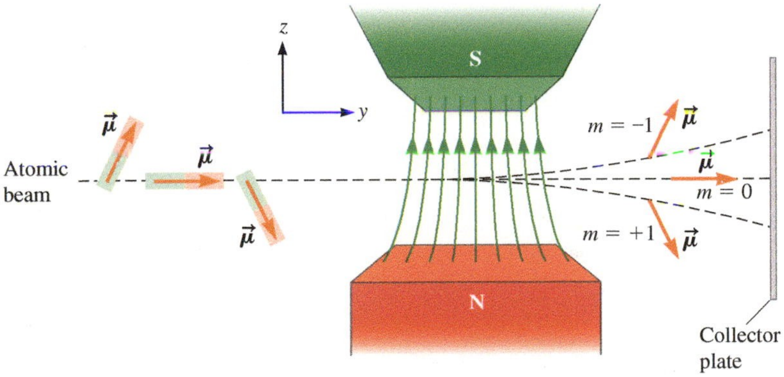

FIGURE 42.18 Magnetic dipoles are deflected by a nonuniform magnetic field. Remember you can think of a magnetic dipole as a small bar magnet as shown on the left side of the figure with the magnetic dipole vectors. For clarity only the magnetic dipole vectors are shown on the right side of the figure.

According to classical mechanics, atoms can have any magnetic moment, oriented in any direction. So when the beam passes through a magnetic field, atoms should be deflected up and down, landing on the collector plate at a wide range of positions. The data on the collector plate should completely fill the region over which the beam is deflected (Fig. 42.19A).

Compare this to the quantum-mechanical prediction. If the atom has $\ell = 1$, then there are three possible values for m: $m = 0$ and $m = \pm1$. This means there are only three discrete orientations for the magnetic moment. So there are only three possible ways each atom may be deflected, and the data on the collector plate should show three different bands (Fig. 42.19B). Because the number of values m may take

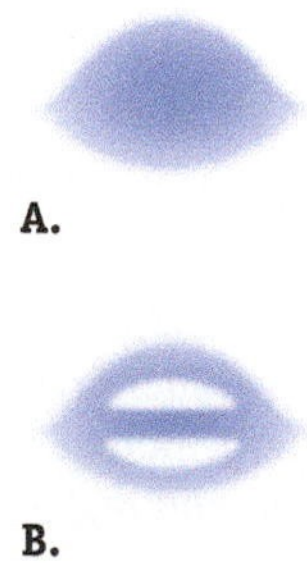

FIGURE 42.19 **A.** Classical prediction. **B.** Quantum prediction.

FIGURE 42.20 Two bands observed when only one band was predicted.

$(2\ell + 1)$ is an odd number, quantum mechanics predicts that an odd number of bands should be observed on the collector plate. In the Stern–Gerlach experiment using silver and in the subsequent experiment using hydrogen, the atoms where in the ground state with $\ell = 0$. Quantum mechanics predicts that one band should be observed. However, two bands were observed (Fig. 42.20). This data suggested that the atom must have a nonzero magnetic moment. So there must be another source (other than orbital angular momentum of the electron) for that magnetic moment. This other source was attributed to the intrinsic spin of the electron. Further, it was affirmed that spin must be quantized in order to form separate bands.

Spin

The spin angular momentum $\vec{S}$ of any electron (trapped or free) has a magnitude

$$S = \sqrt{s(s+1)}\hbar \tag{42.24}$$

where s is called the **spin quantum number**. For an electron, $s = 1/2$, and so $S = \sqrt{3}\hbar/2$. The magnetic moment due to an electron's spin is $\vec{\mu}_{\text{spin}} = q\vec{S}/m$. (Notice the similarity between this and Equation 42.22, $\vec{\mu}_{\text{orbit}} = q\vec{L}/2m$.) Because the magnitude of the spin angular momentum is single-valued, the magnitude of the spin magnetic moment is also single-valued, and for an electron is given by $\mu_{\text{spin}} = \sqrt{3}\hbar e/2m_e$.

We cannot measure $\vec{S}$ or $\vec{\mu}_{\text{spin}}$, but we can measure one of their components. As in the case of $\vec{L}$ and $\vec{\mu}_{\text{orbit}}$, we choose to measure the z component. The z component of $\vec{S}$ is quantized:

$$S_z = m_s\hbar \tag{42.25}$$

where m_s is the **spin magnetic quantum number**, having only one of two values: either $\pm 1/2$. If $m_s = +1/2$, we say the particle is *spin up*, and if $m_s = -1/2$, we say the particle is *spin down*. Table 42.4 summarizes the five quantum numbers. Because S_z is quantized, the z component of the spin magnetic moment is quantized:

$$(\mu_{\text{spin}})_z = -2m_s\mu_{\text{Bohr}} \tag{42.26}$$

The spin magnetic moment in Equation 42.26 and the orbital magnetic moment in Equation 42.23, $(\mu_{\text{orbit}})_z = -m\mu_{\text{Bohr}}$, obey very similar relations.

TABLE 42.4 Allowed values for quantum numbers.

Name	Symbol	Allowed Values
Principal	n	1, 2, 3, …
Orbital	ℓ	0, 1, 2, … $(n - 1)$
Orbital magnetic	m	0, ±1, ±2, … $\pm\,\ell$
Spin	s	½
Spin magnetic	m_s	±½

EXAMPLE 42.7 Excited States of Hydrogen

Suppose a hydrogen atom's energy is -1.51 eV. How many separate states are available to it? Take into account the possible values of ℓ, m, and m_s.

INTERPRET and ANTICIPATE

This example is much like Example 42.3, where we used the energy-level diagram in Figure 42.7 to find that the principal quantum number in this case is $n = 3$. Now we must find the possible values for ℓ, m, and m_s.

SOLVE

From Table 42.4, use $\ell = 0, 1, 2, \ldots (n-1)$ to find the values for the orbital quantum number when $n = 3$.	$\ell = 0, 1, 2$

Consulting Table 42.4 again, use $m = 0, \pm 1, \pm 2, \ldots \pm \ell$ to find the values for the orbital magnetic quantum number for each possible value of ℓ. Count the number of states for each unique value of ℓ and m.

ℓ	m	Number of states
0	0	1
1	$-1, 0, 1$	3
2	$-2, -1, 0, 1, 2$	5

For each state, the electron can be either spin up ($m_s = 1/2$) or spin down ($m_s = -1/2$). Add up the number of unique states specified by ℓ and m, and then multiply by 2.

Number of states $= 2(1 + 3 + 5) = 18$

CHECK and THINK

In all 18 of these states, hydrogen has the same energy E (-1.51 eV). So the 18 states form a single shell. This shell is broken into three subshells for the three values of l. Each subshell has the same energy and the same orbital angular momentum L, but within the subshells for $\ell = 1$ and $\ell = 2$ there is variation in the orbital magnetic moment $(\mu_{\text{orbit}})_z$. In addition, in each state the spin magnetic moment $(\mu_{\text{spin}})_z$ is either up or down.

42-7 Other Atoms

Hydrogen is the simplest atom; its atomic number $Z = 1$ means that it has one proton in the nucleus, and because it is neutral, it must also have one electron. All other atoms have multiple electron clouds surrounding a nucleus (consisting of both neutrons and protons). So in other atoms the potential well that confines any one electron is more complicated, and solving Schrödinger's equation is more difficult. Typically, it is solved numerically with a computer. Fortunately, we don't need to consider such solutions to understand the structure of atoms. We can use our understanding of hydrogen instead.

Because hydrogen has only one electron, we need only one set of four quantum numbers (n, ℓ, m, and m_s) to specify its state (Table 42.4). (We don't bother to list s because it is single-valued.) For other atoms, we need to specify these four quantum numbers for each of their electrons. For hydrogen, the energy is determined by n alone; so the energy is the same for each shell (Fig. 42.21). For other atoms, the energy depends on n and ℓ; so the energy is the same for each *sub*shell (Fig. 42.22).

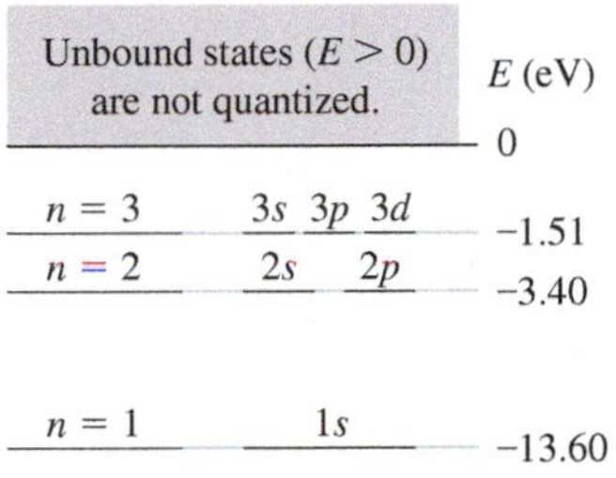

FIGURE 42.21 Compare the energy-level diagram labeled simply in terms of n to the one labeled in terms of n and ℓ for the first three energy levels in hydrogen.

The Pauli Exclusion Principle

The state with the lowest energy is the ground state. You might expect that in the ground state of a multielectron atom, all the electrons are in the 1*s* subshell. For example, carbon has an atomic number of 6, so it has six protons and six electrons. If all six electrons were in the 1*s* subshell, then on average they would be very close to the nucleus. Because the electrons are attracted to the protons, we would expect the electrons to be drawn close to the nucleus, making carbon a small atom. You might even argue that a carbon atom should be smaller than a hydrogen atom with its electron in the 1*s* subshell because there is a greater attractive force between six protons and six electrons than there is between one proton and one electron. However, we observe the opposite: Hydrogen atoms are smaller than carbon atoms. The answer to this puzzle is that not all the electrons in carbon can occupy the lowest subshell.

The idea that not all the electrons can be in the lowest subshell is part of a broader principle formulated by the theoretical physicist Wolfgang Pauli (1900–1958). According to the **Pauli's exclusion principle,** *no two electrons confined in the same well can have the same set of values for their quantum numbers.* We have expressed Pauli's exclusion principle in terms of electrons, but it applies equally well to any particles (such as protons and neutrons) which have $s = 1/2$. In practice, Pauli's exclusion principle means that in an atom, no two electrons can have exactly the same set of the four quantum numbers n, ℓ, m and m_s.

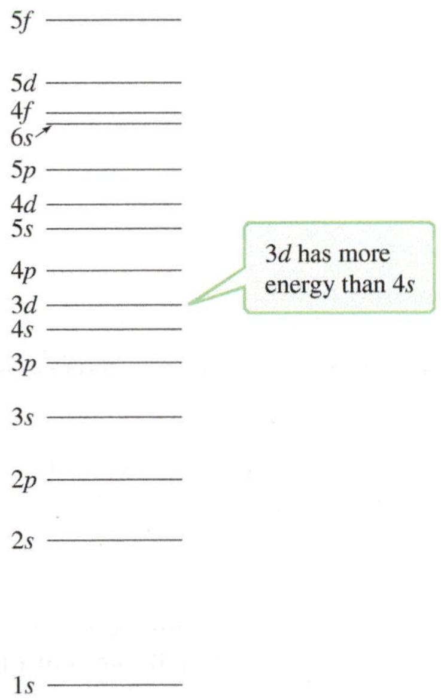

FIGURE 42.22 The energy levels in multielectron atoms depend on both n and ℓ. Notice that in some cases a state with a higher quantum number n has less energy than a state with lower n.

PAULI'S EXCLUSION PRINCIPLE

Underlying Principle

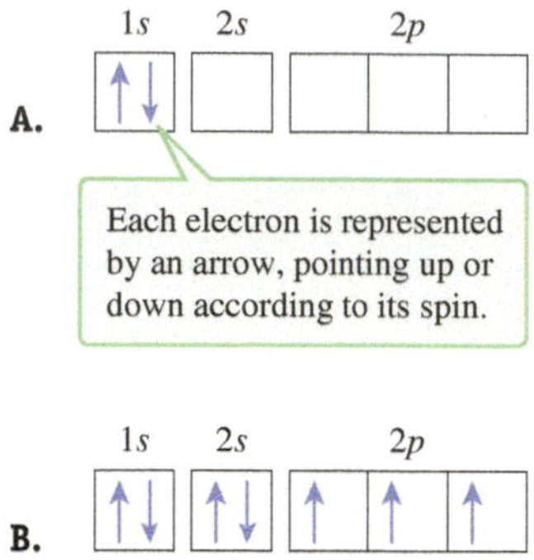
1s
2s
2p
A.
Each electron is represented by an arrow, pointing up or down according to its spin.
1s
2s
2p
B.

FIGURE 42.23 Each box represents one orbital; each orbital can hold up to two electrons. **A.** The two electrons in He are both in the 1*s* subshell. One is spin up and the other is spin down. This state is written in the compact form as $1s^2$. **B.** The seven electrons in N fill 1*s* and 2*s*, but only half of 2*p*. Following Hund's rule, all the electrons in 2*p* are spin up. This state is written as $1s^22s^22p^3$.

Pauli's exclusion principle limits the number of electrons that can be in any one subshell. For example, any electron in the 1*s* subshell has three quantum numbers in common $(n, \ell, m) = (1, 0, 0)$. So two electrons can be in this subshell because one can be spin up with $m_s = 1/2$, and one can be spin down with $m_s = -1/2$. Put another way, each subshell is divided into *orbitals,* and each orbital may have up to two electrons of opposite spin.

When an atom is in its ground state, the electrons fill the lowest-energy subshells first while obeying Pauli's exclusion principle. Let's consider two examples. Helium has two electrons, and in the ground state both are in the 1*s* subshell with $(n, \ell, m) = (1, 0, 0)$: One is spin up, so the other must be spin down (Fig. 42.23A). Nitrogen has seven electrons. Two of these electrons are in the 1*s* state, just as they are in helium. Then two electrons are in the 2*s* subshell, with quantum numbers $(n, \ell, m) = (2, 0, 0)$: One is spin up and the other spin down. Finally, the last three electrons are in the 2*p* subshell with $(n, \ell) = (2, 1)$. Here there are three possible values for m $(-1, 0, 1)$ and two values for $m_s\,(-1/2, 1/2)$; so there are six different combinations of quantum numbers, but only three electrons. How do we find their quantum numbers? The answer is that we follow **Hund's rule**, which says that *every orbital in a subshell is singly occupied with one electron before any one orbital is doubly occupied, and all electrons in singly occupied orbital have the same spin.* This means that the maximum number of electrons have unpaired spins. Hund's rule is based on experimental evidence, which shows that the maximum number of unpaired spins corresponds to the most stable state with the lowest energy level. So, remaining electrons in nitrogen have the same m_s, say spin up, but three different values for m (Fig. 42.23B). A listing of the subshells occupied by the electrons in an atom is called the **electron configuration**. The electron configuration for helium is $1s^2$ and for nitrogen it is $1s^22s^22p^3$.

EXAMPLE 42.8 The Maximum Number of Electrons in Each Subshell

Argue that Pauli's exclusion principle means that the maximum number of electrons allowed in each subshell is

$$N = 2(2\ell + 1) \qquad (42.27)$$

Check this for the 2*p* subshell.

INTERPRET and ANTICIPATE

A subshell is characterized by a particular value of n and ℓ. Table 42.4 will help us to determine the possible quantum numbers for a subshell, and Pauli's exclusion principle says there can only be one electron with each unique set of quantum numbers.

SOLVE

Because $m = 0, \pm 1, \pm 2, \ldots \pm \ell$, for each value of ℓ, there are $2\ell + 1$ possible values of m.	The range in m allows for $2\ell + 1$ electrons in a subshell.
For each value of m, the electron can be either spin up ($m_s = 1/2$) or spin down ($m_s = -1/2$). The spin magnetic quantum number doubles the number of electrons (found above) that can possibly be in the subshell.	$N = 2(2\ell + 1)$ ✓ (42.27)

CHECK and THINK

The 2*p* subshell has $\ell = 1$, so we expect $N = 2(2 \cdot 1 + 1) = 6$ electrons. Check this by listing all the unique combinations of quantum numbers. (Since $n = 2$ and $\ell = 1$ for the 2*p* subshell, we don't bother to list them separately here.)	$(m, m_s) = (-1, -1/2)$ $(m, m_s) = (-1, +1/2)$ $(m, m_s) = (0, -1/2)$ $(m, m_s) = (0, +1/2)$ $(m, m_s) = (1, -1/2)$ $(m, m_s) = (1, +1/2)$ 6 combinations

42-8 Organizing Atoms

By the mid-1800s, over 60 elements had been discovered and studied. Chemists knew their relative masses, their chemical activity, and some of their other physical properties. However, scientists didn't know why there are different types of elements, or why they have different properties. One way to answer such a question is to look for patterns or correlations among these properties. In 1869, a chemist from Siberia, Dmitri Mendeleev (1834–1907), found that if he arranged the elements by atomic mass in a grid, certain properties recurred periodically. Mendeleev's table is the basis of our periodic table; he successfully predicted the existence of undiscovered elements that filled in "blank" places in the grid. However, Mendeleev's table was not completely useful because atomic mass is not always well correlated with an element's chemical properties. The British physicist Henry Moseley (1887–1915) later showed that atomic number is better correlated with chemical properties, so it is better to arrange the elements by their atomic number, as we do in today's periodic table.

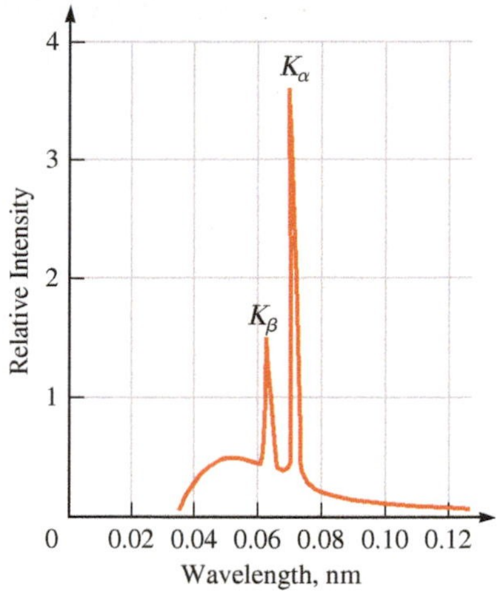

FIGURE 42.24 X-ray spectrum of molybdenum. The spectrum looks like a blackbody spectrum with two tall spikes—called lines. The K_α line comes from the transition from $n = 2$ to $n = 1$. The K_β line comes from the transition from $n = 3$ to $n = 1$. Moseley measured the wavelength of the K_α line for about 40 elements to arrive at his plot similar to the one shown in Figure 42.25.

Moseley's Experiment

Moseley devised an experiment that revealed new discoveries about the internal structure of 38 elements. He placed a sample in an evacuated tube and fired a beam of electrons at it. When an energetic electron collided with an atom, one of its low-energy electrons was freed so that a "hole" was opened in a shell near the atom's nucleus. An electron from a higher shell then dropped into this hole, releasing energy in the form of an X-ray. Moseley studied the X-ray generated by an electron that drops from the $n = 2$ shell to the $n = 1$ shell, which he labeled as K_α (Fig. 42.24). He measured the frequency and therefore the energy of this emission line for approximately 40 elements and found that the X-ray energy was correlated with each element's atomic number Z, not with its atomic mass (Fig. 42.25). Moseley's law is an empirical fit given by:

$$f(Z) = C(Z - 1)^2$$

where $C \approx 2.47 \times 10^{15}$Hz. Today we know that arranging the periodic table by atomic number leads to a better prediction of the behavior and properties of atoms.

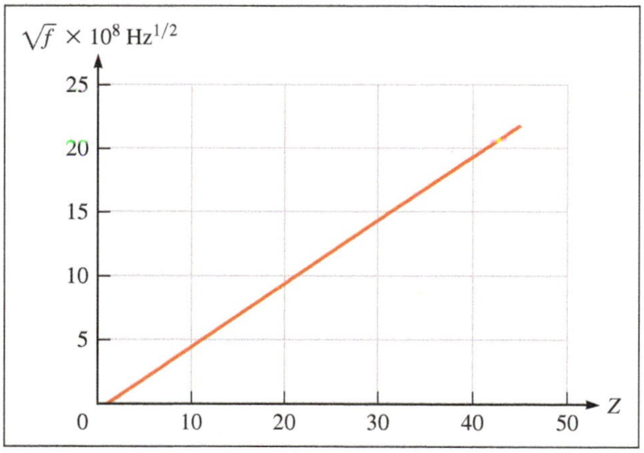

FIGURE 42.25 When trying to fit data, it is often best to choose axes that produce a straight line rather than a curve as Moseley did when he plotted his data. Similarly, this graph has been constructed to display Moseley's law as a straight line. To do this, the vertical axis is $\sqrt{f}$ on the graph here.

Shielding

In a moment we'll consider other evidence that supports the ordering of the elements in the periodic table. But now we'll digress to consider why an X-ray emission from an atom yields more information about the atom's nuclear charge Z than optical emission does. X-ray emission lines are produced by atomic transitions between shells near the nucleus, such from the $n = 2$ shell to the $n = 1$ shell. By contrast, optical lines come from transitions that involve higher-numbered shells farther from the nucleus, and the electrons that occupy such distant shells are *shielded* from the nucleus by other, closer electrons. Shielding means that the effective charge that attracts these outer electrons to the nucleus is smaller than for "core" electrons, and the Coulomb force is weaker. So when an outer electron makes a transition to another (lower) outer shell, the photon released has relatively low energy. Such low-energy photons may be observed in the optical, but not in the X-ray part of the electromagnetic spectrum. This idea of shielding also helps explain the chemical behavior of various types of elements in the periodic table.

Atomic Size and Ionization Energy

The outermost electrons known as the *valence electrons* in an atom are primarily responsible for determining how atoms make contact or bond together. The radius of an atom is measured by how close together atoms must come to one another to form bonds or to make nonbonded contact. The **first ionization energy** is the energy required to remove the first electron from a neutral atom in its ground state. Figure 42.26A shows atomic radius as a function of atomic number, whereas Figure 42.26B shows the first ionization energy as a function of atomic number. Both graphs display a *general* trend and a *periodic* trend.

First, the radius graph (Fig. 42.26A) shows that overall a greater atomic number means a larger atom. We explain this *general* trend in radius by noting that electrons in an atom fill their energy levels, from the lowest to the highest, while obeying Pauli's exclusion principle. (Not all the electrons can be in the ground state, for example.) Atoms with large Z have a great number of electrons, some of which must occupy outer shells. Such atoms have a large radius.

Second, the *general* trend in the ionization energy graph (Fig. 42.26B) shows that overall a greater atomic number means a lower ionization energy. We can explain this trend in terms of shielding. If an atom has a great number of electrons, then a few outer electrons are greatly shielded by many core electrons. These outermost electrons are weakly attracted to the nucleus, which means the atom is large and one of the outermost electrons is relatively easily removed. Thus, the ionization energy is low.

A regular *periodic* pattern within the overall trend also occurs in both graphs. The radius graph shows that the atomic radius decreases periodically with increasing atomic number within a certain range of Z values. The ionization graph shows that ionization energy increases periodically with increasing Z within each similar range. We can also explain these periodic trends in terms of shielding. Core electrons are

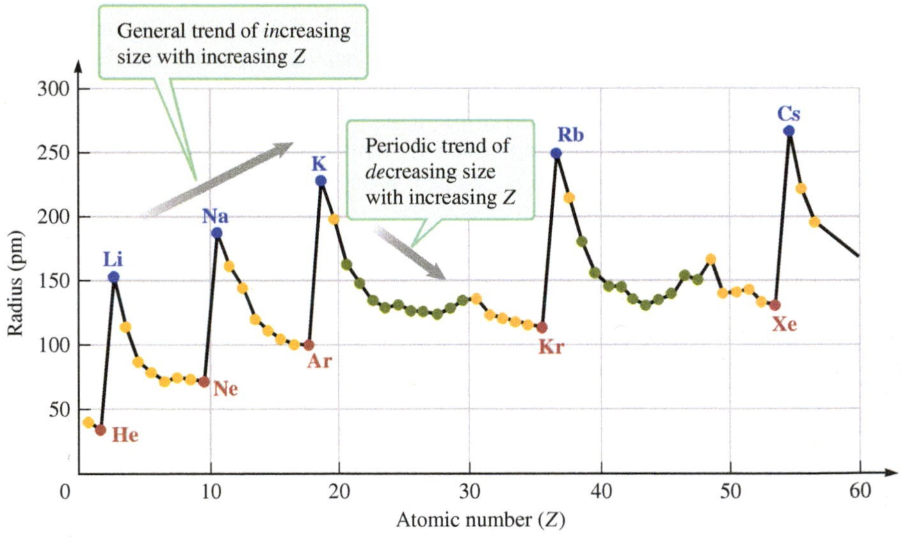

A. Atomic Radii

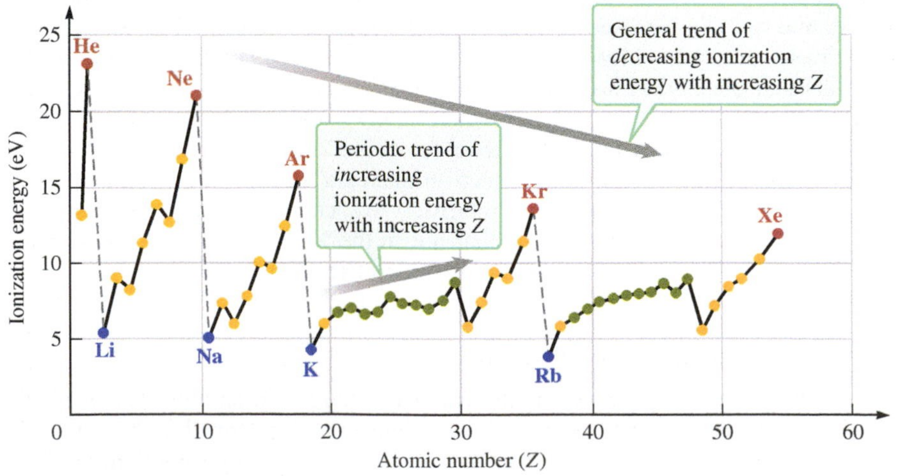

B. First Ionization Energies

FIGURE 42.26 **A.** Plotting the radii of atoms as a function of atomic number shows a regular periodic pattern. **B.** Plotting the first ionization energy as a function of atomic number shows a periodic repetition. In both graphs alkali metals are labeled in blue and noble gases in red.

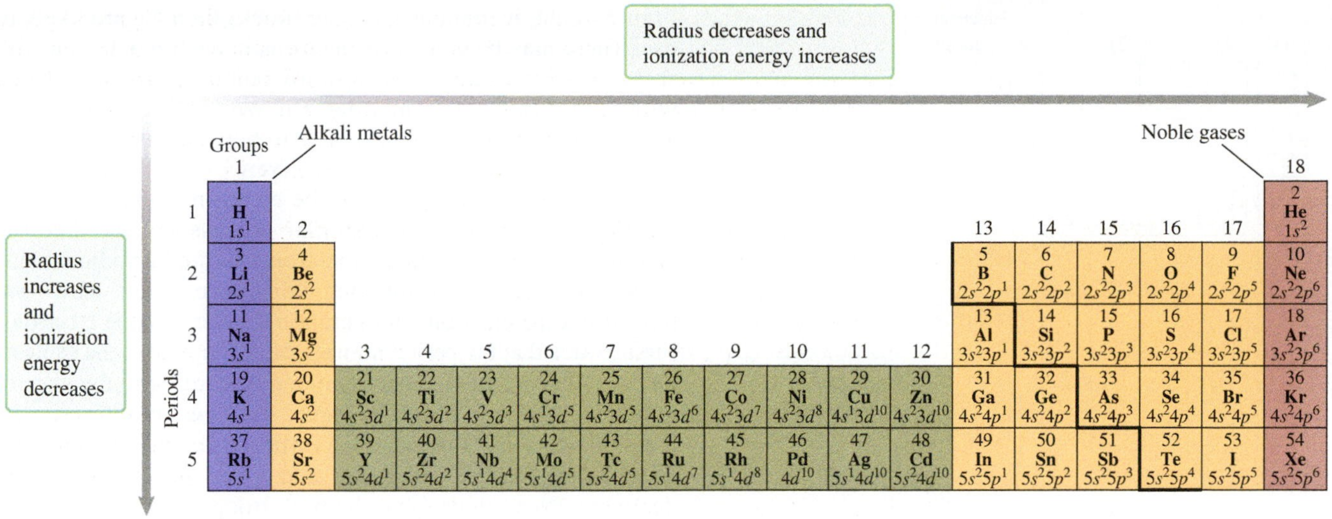

FIGURE 42.27 The first 54 elements in the periodic Table help us to see how the table is constructed. The complete periodic table is in Appendix B. Here we show 5 periods and 18 groups.

very effective at shielding outer electrons, but the valence electrons in the same subshell are not very effective at shielding one another. So, if the valence electrons fill a subshell, then there is only a little shielding due to the core electrons, and these outer electrons experience a great Coulomb attraction to the nucleus, and the atom is small. It is also difficult to remove an electron from such an atom, so the ionization energy is great. For example, neon (Ne) has 10 electrons, with electron configuration $1s^22s^22p^6$ indicating that all of its subshells are filled. Neon is relatively small (Fig. 42.26A) and it has a relatively large ionization energy (Fig. 42.26B). By contrast, if an atom has just one more electron than is required to fill a subshell, this outer electron is greatly shielded and weakly attracted to the nucleus. Such an atom is relatively large and has a relatively low ionization energy. For example, helium is the smallest atom because its $1s$ subshell is filled and it only has two electrons. Lithium (Li) has just one more electron, so its electron configuration is $1s^22s^1$. The core electrons in the $1s$ subshell shield the valence electron in the $2s$, so this outer electron is weakly bound. As a result, lithium has a lower ionization energy and is larger than helium.

PERIODIC TABLE OF THE ELEMENTS

Major Concept

The Periodic Table

Moseley's experiment, along with the graphs of radius and ionization energy as functions of atomic number, gives us good reasons to expect that if the elements are ordered by atomic number, we'll find trends that explain their chemical behavior. The elements in the modern periodic table are ordered by their atomic number into rows known as *periods* and columns known as *groups* (Fig. 42.27). An element's chemical behavior—how reactive it is—results from the configuration of its valence electrons. The valence electrons are important in chemical bonding because they are the most weakly attracted to the atom, and so they are the easiest to lose or share in a chemical reaction. These tend to be the outermost electrons (in s or p subshells), but for some elements the outermost d subshell electrons also act as valence electrons. The electron configuration of an element is related to its placement on the periodic table, and its placement on the table is related to its chemical and physical properties.

The structure of an element is given by its electron configuration in the ground state. To find an element's electron configuration, first find its atomic number on the periodic table. Then fill in its shells and subshells starting with the lowest energy, while following Pauli's exclusion principle and Hund's rule. The energy levels are shown in Figure 42.22. It is helpful to represent each energy level with a series of empty boxes in which arrows representing electrons are drawn. Figure 42.28 is such a blank diagram for all the energy levels in

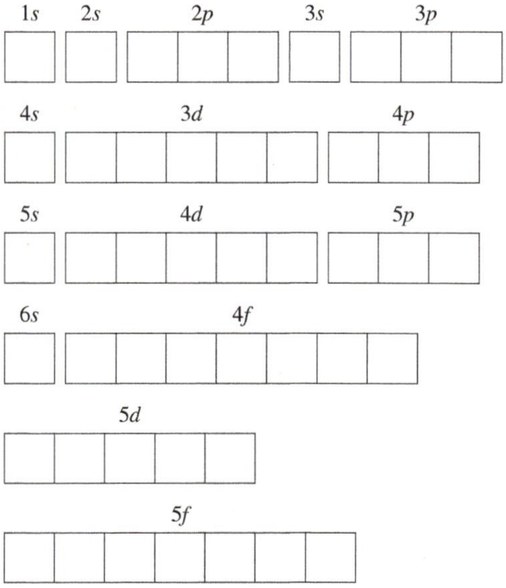

FIGURE 42.28 A blank diagram such as this can help you to find the electron configuration of an atom. Start by filling in electrons from the lowest energy on the left. Follow Pauli's exclusion principle, which says that any one box can have up to two electrons in it: one with spin up, and the other with spin down. Also, follow Hund's rule, which says that you must fill a subshell with single electrons before adding a second electron to a box.

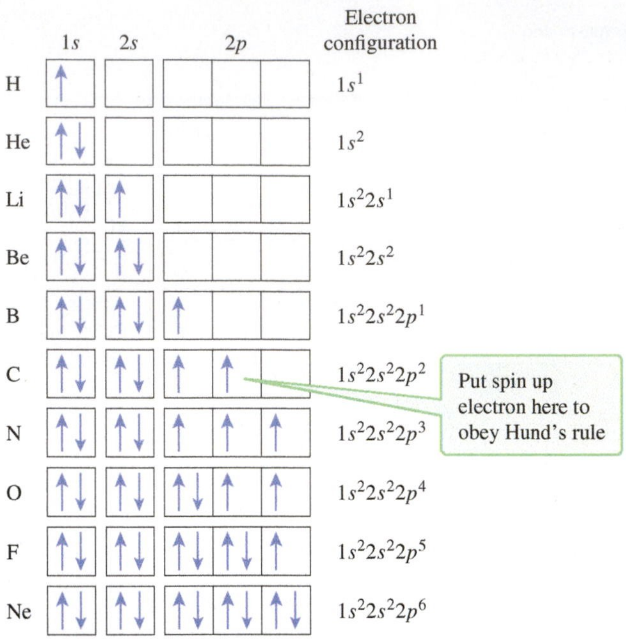

FIGURE 42.29 Use a diagram like that in Figure 42.28 to find the electron configuration of an atom. Here is the electron configuration for the first 10 elements in the periodic table.

Figure 42.22; it contains far more blocks then we are likely to use. There may be up to two electrons in each block: one with spin up represented by an upward-pointing arrow, and one with spin down represented by a downward-pointing arrow. The number of blocks corresponds to half the number of electrons allowed in each subshell, given by $(2\ell + 1)$. Figure 42.29 shows how to determine the electron configuration for the first 10 elements. By a similar process, we can find the electron configuration of all the atoms in the periodic table.

Beneath the symbol for each element in Figure 42.27, we have listed the electron configuration for the valence electrons. First, notice that the period number of an element is the same as its highest principal quantum number. According to the general trends in the graphs shown in Figure 42.26, as we consider elements down a particular group (vertical column), we find that their atomic radius increases and their ionization energy decreases. Next, notice that the first group is the alkali metals and the last group is the noble gases. So according to the periodic trends in the graphs, as we consider elements across a period from left to right, we find that their atomic radius decreases and their ionization energy increases.

We can account for the chemical behavior of the elements by means of their placement on the periodic table. For example, the alkali metals are in the first group, and their valence electrons all have configurations of the form ns^1. These metals have one valence electron that is well shielded by the core electrons in the lower subshells. This electron is weakly attracted to the nucleus, so in chemical reactions, alkali metals tend to lose their valence electron. For another example, consider the noble gases in the last group. Their valence electron configurations are all of the form ns^2np^6. These eight valence electrons fill the atom's outer subshell, making them very stable, so noble gases are generally nonreactive. In general, elements in a group tend to show similar chemical behavior.

EXAMPLE 42.9 Silicon's Electron Configuration

Find the electron configuration for silicon (Si).

INTERPRET and ANTICIPATE

According to the periodic table, Si has 14 electrons. We expect its configuration to be longer than the configuration of neon, which has only 10 electrons.

SOLVE

Fill in a diagram (Fig. 42.30) following these three rules: (1) start with the lowest energy subshell, (2) obey Pauli's exclusion principle, and (3) obey Hund's rule.

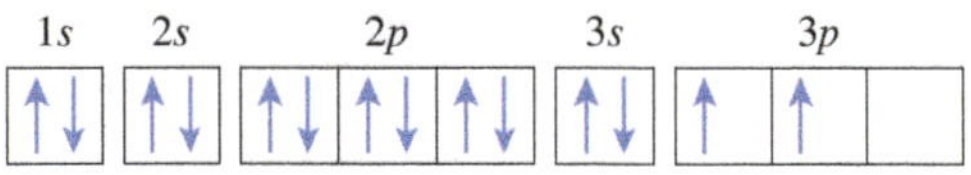

FIGURE 42.30

Write the electron configuration based on this figure.

$1s^2 2s^2 2p^6 3s^2 3p^2$

CHECK and THINK

As expected, the electron configuration for silicon is longer than that for neon ($1s^2 2s^2 2p^6$).

42-9 The Zeeman Effect

ZEEMAN EFFECT Major Concept

In 1902, the Dutch physicists Pieter Zeeman (1865–1943) and Hendrik A. Lorentz (1853–1928) shared the Nobel Prize in physics for what is known as the **normal Zeeman effect**. (Lorentz is the theorist for whom the transformation equations in Chapter 39 are named.) Zeeman was an experimentalist who discovered in 1896 that the spectral lines of certain atoms split into three lines when a sample is placed in a magnetic field. The Zeeman effect confirms that angular momentum is quantized.

To see how, we'll consider the hydrogen atom, though our arguments can be applied to more complicated atoms. The energy levels E_n we have discussed so far have been for hydrogen without the presence of an external field; our system consisted of the atom's electron and nucleus. Because of the electron's orbital angular momentum $\vec{L}$, the electron has a magnetic moment given by $\vec{\mu}_{\text{orbit}} = q\vec{L}/2m$ (Eq. 42.22). Now let's consider a hydrogen atom in a uniform external magnetic field pointing in the z direction. In this case, our system consists of the electron, the nucleus, and the external field. The external magnetic field adds a potential energy term U to the energy E_n, given by $U = -\vec{\mu}_{\text{orbit}} \cdot \vec{B}$ (Eq. 30.48). Because the angular momentum is quantized in both magnitude and direction, so is the magnetic moment. And because the magnetic moment is quantized, so is this additional potential energy.

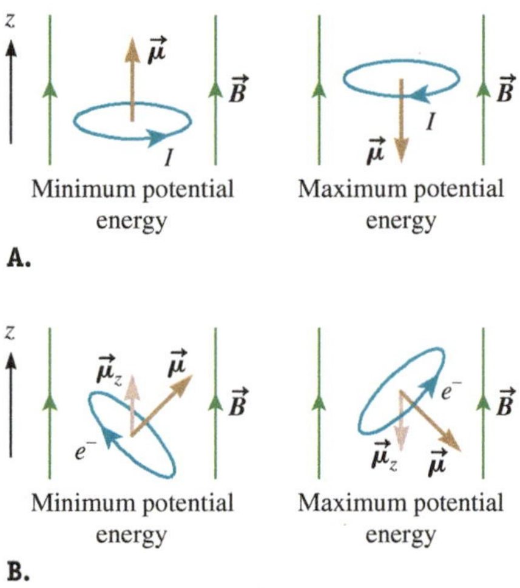

FIGURE 42.31 **A.** The extrema of the potential energy for a current loop-magnetic field system. **B.** The same for an atom in a magnetic field.

Contrast these ideas with a classical interpretation. If a current loop is placed in an external magnetic field, the potential energy of the loop-field system depends on the direction of the magnetic moment with respect to the magnetic field (Fig. 42.31A). The minimum potential energy occurs when the magnetic moment is aligned with the magnetic field: then $U_{\min} = -\mu B$. The maximum potential occurs when the magnetic moment is antiparallel to the field: then $U_{\max} = +\mu B$. As a result, the system's potential energy may be anywhere in the range from $U_{\min}$ to $U_{\max}$.

For the system consisting of an atom in an external magnetic field, not only is its potential energy quantized, but $U_{\min}$ and $U_{\max}$ are restricted by quantization of the direction of angular momentum and magnetic moment. According to Concept Exercise 42.4, for a magnetic field in the z direction, the angular momentum $\vec{L}$ can never be in the $\pm z$ direction. The electron's magnetic moment is antiparallel to its orbital angular momentum. Thus, the magnetic moment cannot be parallel (or antiparallel) to the external magnetic field. Instead, the z component of magnetic moment is given by $(\mu_{\text{orbit}})_z = -m\mu_{\text{Bohr}}$ (Eq. 42.23) where m is the magnetic quantum number. The maximum and minimum potential energy are determined by this z component, rather than by the magnitude of the magnetic moment (Fig. 42.31B).

When there is no external magnetic field, the energy of a subshell depends on the quantum numbers n and ℓ, so we label it $E_{n\ell}$. All the orbitals of that subshell have the same energy, independent of the orbital quantum number m. The energy of the subshell is said to be *degenerate.* But when an atom is in an external magnetic field pointing in the z direction, the additional potential energy is found by substituting Equation 42.23 into Equation 30.48:

$$U = -\vec{\mu}_{\text{orbit}} \cdot \vec{B} = -(\mu_{\text{orbit}})_z B_z = -(-m\mu_{\text{Bohr}})B_z = m\mu_{\text{Bohr}}B$$

where $B_z = B$. The energy of an orbital $E_{n\ell m}$ in a particular subshell depends not only on n and ℓ, but also on m:

$$E_{n\ell m} = E_{n\ell} + m\mu_{\text{Bohr}}B \tag{42.28}$$

We say that the external magnetic field removes the degeneracy, so that each orbital has its own distinct energy.

The energy of the subshells of hydrogen only depends on the principal quantum number n. In the case of hydrogen, Equation 42.28 becomes:

$$E_{nm} = E_n + m\mu_{\text{Bohr}}B$$

$$E_{nm} = -\frac{13.6\text{ eV}}{n^2} + m\mu_{\text{Bohr}}B \tag{42.29}$$

We can use Equation 42.29 to construct an energy-level diagram for hydrogen in an external magnetic field pointing in the z direction. For the ground state, $n = 1$ and $m = 0$, so $E_{10} = -13.6$ eV. Because there is only one orbital, the ground state in this case is exactly the same as the ground state when there is no external magnetic field. However, the $2p$ state has three orbitals, with $m = -1$, 0 and $+1$, and three corresponding energies:

$$E_{2(-1)} = -\frac{13.6\text{ eV}}{2^2} + (-1)\mu_{\text{Bohr}}B = -3.4\text{ eV} - \mu_{\text{Bohr}}B$$

$$E_{20} = -3.4\text{ eV} + (0)\mu_{\text{Bohr}}B = -3.4\text{ eV}$$

$$E_{21} = -3.4\text{ eV} + (1)\mu_{\text{Bohr}}B = -3.4\text{ eV} + \mu_{\text{Bohr}}B$$

The $m = 0$ orbital has the same energy as if there were no external magnetic field, whereas the $m = -1$ orbital has a lower energy and the $m = 1$ orbital has a higher energy. For easy comparison, Figure 42.32 shows the energy levels computed here for hydrogen both with and without an external magnetic field. (In Problems 42.44 and 45, you will be asked to calculate the energy of other orbitals.) Figure 42.32A shows that if a hydrogen atom undergoes a transition from either the $2s$ or $2p$ subshell to the ground state, a Lyman α photon (of energy 10.2 eV) is released if there is no external magnetic field. However, if there is an external magnetic field, the $2p$ subshell has three distinct energies, one for each orbital. So there are three distinct possible transitions from the $2p$ subshell to the ground state, and the Lyman α line is split into three closely spaced lines (Fig. 42.32B). The energies of the corresponding photons are given by:

$$E_{2(-1)} - E_1 = (-3.4\text{ eV} - \mu_{\text{Bohr}}B) - (-13.6\text{ eV}) = 10.2\text{ eV} - \mu_{\text{Bohr}}B$$

$$E_{20} - E_1 = -3.4\text{ eV} - (-13.6\text{ eV}) = 10.2\text{ eV}$$

$$E_{21} = (-3.4\text{ eV} + \mu_{\text{Bohr}}B) - (-13.6\text{ eV}) = 10.2\text{ eV} + \mu_{\text{Bohr}}B$$

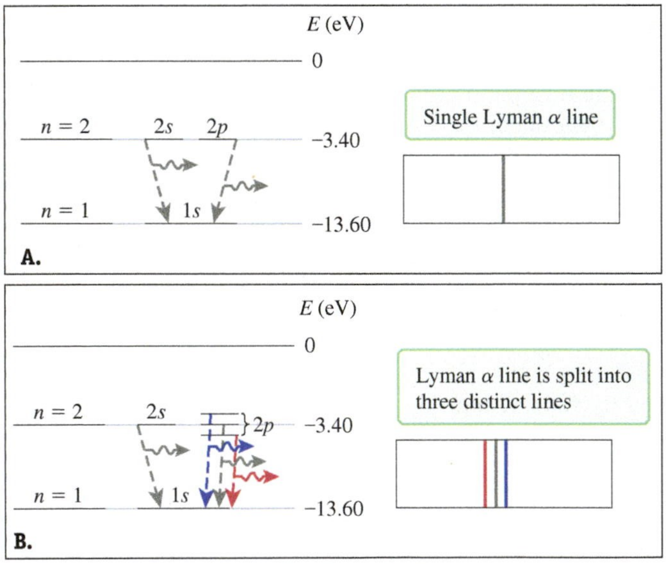

FIGURE 42.32 **A.** Energy-level diagram for the first two shells of hydrogen when no external magnetic field is present. A transition from $n = 2$ to $n = 1$ results in a single Lyman α line. **B.** The same applies when an external magnetic field points in the z direction. Notice that the $2p$ orbital energies are divided into three levels. Now the Lyman α line is split into three lines corresponding to $m = -1$, 0, and 1. (Energy levels are not drawn to scale, and line colors are illustrative only.) The wavy arrows represent the photons that are emitted.

So in general, the energy of the emitted photon is given by

$$E = E_0 - \Delta m \mu_{\text{Bohr}}B \tag{42.30}$$

where E_0 is the energy of the corresponding photon in the absence of an external magnetic field, and $\Delta m = m_f - m_i$ is the difference in the orbital magnetic quantum number between the two states. Not all transitions are equally probable, however. Although we won't show the details, applying Schrodinger's equation reveals that the most probable transitions have $\Delta\ell = \pm 1$ and $\Delta m = 0, \pm 1$; these are known as **allowed transitions**. So in the normal Zeeman effect, a single spectral line may be split into three lines, one for each allowed value of Δm. Transitions with very low probability, by contrast, are known as **forbidden transitions**. Although forbidden transitions are not important under normal conditions on the Earth, forbidden transitions are observed from astronomical sources. Observations of forbidden lines (produced by forbidden transitions) in astronomical sources tell us a great deal about the conditions in these sources such as their temperature and density.

Zeeman's observation that spectral lines divide into several closely spaced lines in the presence of an external magnetic field confirms the theory that the angular momentum vector $\vec{L}$ is quantized. Let's review the argument. Quantization of $\vec{L}$ means that the magnetic moment $\vec{\mu}_{\text{orbit}}$ is quantized, so the angle between $\vec{\mu}_{\text{orbit}}$ and the external field $\vec{B}$ is also quantized. The additional potential energy of the atom-external field system $U = -\vec{\mu}_{\text{orbit}} \cdot \vec{B}$ depends on this relative angle, so it too must be quantized. When an atom is in an external magnetic field, the energy of a particular orbital is divided into discrete energy levels. Zeeman detected these discrete energy levels when he saw one spectral line split into distinct spectral lines if an external magnetic field was applied.

EXAMPLE 42.10 CASE STUDY The Zeeman Effect in the Sun's Hydrogen Spectrum

In 1908, George Hale was the first person to exploit the Zeeman effect to measure the magnetic field in sunspots. Although Hale observed heavy elements such as iron, we'll consider how we might observe the Zeeman effect in the Sun's hydrogen spectrum.

A Assume that the magnetic field strength in a typical sunspot is 0.1 T, and that only allowed transitions occur under the conditions in the Sun. Estimate the magnitude of the difference in wavelength $\Delta\lambda$ between either of the two Hγ lines that result from the normal Zeeman effect and the original ($B = 0$) central line. What is $\Delta\lambda/\lambda_0$, where $\lambda_0 = 434.047$ nm is the wavelength of original line?

INTERPRET and ANTICIPATE

The difference in energy between the central line and either of the other two lines comes from Equation 42.30. This energy difference corresponds to a wavelength difference, which we can derive using calculus.

SOLVE

The allowed transitions have $\Delta m = 0, \pm 1$. Start with Equation 42.30 to write an expression for the energy difference between the central line and the higher energy line generated by $\Delta m = 1$.

$$E = E_0 + \Delta m \mu_{\text{Bohr}} B \quad (42.30)$$

$$dE = E_0 - E = (1)\mu_{\text{Bohr}} B \quad (1)$$

Write the energy of a photon in terms of its wavelength by using $f = c/\lambda$ (Eq. 34.20) and $E = hf$ (Eq. 40.8).

$$E = hf = \frac{hc}{\lambda} \quad (2)$$

Differentiate Equation (2) with respect to wavelength λ.

$$\frac{dE}{d\lambda} = -\frac{hc}{\lambda^2} \quad (3)$$

Eliminate dE from Equations (1) and (3) to arrive at an expression for $d\lambda$.

$$d\lambda = -\frac{\mu_{\text{Bohr}} B}{hc}\lambda^2$$

Use this expression to find the magnitude of the finite difference $\Delta\lambda$ and $\Delta\lambda/\lambda_0$.

$$\Delta\lambda = \frac{\mu_{\text{Bohr}} B}{hc}\lambda^2$$

$$\Delta\lambda = \frac{(9.274 \times 10^{-24} \text{J/T})(0.1 \text{ T})}{(6.626 \times 10^{-34} \text{ J}\cdot\text{s})(3.00 \times 10^8 \text{ m/s})}(434 \times 10^{-9} \text{ m})^2$$

$$\Delta\lambda = 8.8 \times 10^{-13} \text{ m} \approx 9 \times 10^{-4} \text{ nm}$$

$$\frac{\Delta\lambda}{\lambda_0} = \frac{8.8 \times 10^{-4} \text{ nm}}{434.047 \text{ nm}} \approx 2 \times 10^{-6}$$

CHECK and THINK

The Zeeman effect is very subtle. Next we'll consider the quality of the diffraction grating required.

B Hale reported that to see the Zeeman effect with his diffraction grating, he had to use the third-order part of the spectrum. If we wish to see the splitting of Hγ, with a resolving power that is three times the minimum required in the third order, how many rulings will our grating need?

INTERPRET and ANTICIPATE

This part of the problem requires a review of Section 36-5. The minimum resolving power needed is given by $R = \lambda_{\text{av}}/\Delta\lambda$ (Eq. 36.20), where λ_{av} is the average wavelength of two lines that are just barely separated, and $\Delta\lambda$ is the difference in their wavelength. For these lines, $\lambda_{\text{av}} \approx \lambda_0$, so $R = \lambda_0/\Delta\lambda$.

SOLVE The minimum resolving power needed is the inverse of the expression we found in part A.	$R_{\text{min}} = \dfrac{\lambda_0}{\Delta\lambda} = \dfrac{1}{2 \times 10^{-6}}$
The resolving power is related to the number of rulings according to Equation 36.21. (In this equation, m refers to the order number, not the quantum number.)	$R = Nm$ (36.21)
We'd like R to be three times the minimum required resolving power, and like Hale we plan to use the third-order part of the spectrum.	$R = Nm = 3R_{\text{min}}$ $N = \dfrac{3R_{\text{min}}}{m} = \dfrac{3}{3}\left(\dfrac{1}{2 \times 10^{-6}}\right)$ $N = 5 \times 10^5$ rulings

C Hale also reported that his grating had 567 rulings per millimeter. Assuming the same for our grating, what is the angular separation between the two lines?

INTERPRET and ANTICIPATE

This part of the problem also requires information from Section 36-5. The dispersion is given by Equation 36.19,

$$D = \frac{\Delta\theta}{\Delta\lambda} = \frac{m}{d\cos\theta}$$

which we can solve for $\Delta\theta$. First, find the angular position of the third-order spectrum, using $d\sin\theta = m\lambda$(Eq. 36.15).

SOLVE Find the distance between rulings, using the given 567 rulings per millimeter.	$d = \dfrac{1\text{ mm}}{567\text{ rulings}} = 1.76 \times 10^{-3}\text{ mm}$ $d = 1.76 \times 10^3\text{ nm}$
Both lines have very nearly the same wavelength, so their angular position is roughly the same.	$d\sin\theta = m\lambda$ $\theta = \sin^{-1}\left(\dfrac{m\lambda}{d}\right) = \sin^{-1}\left(\dfrac{3(434.047\text{ nm})}{1.76 \times 10^3\text{ nm}}\right)$ $\theta = 47.7°$
Now use Equation 36.19 for the dispersion. Substitute $m = 3$ (third-order part of spectrum) and information found previously.	$D = \dfrac{\Delta\theta}{\Delta\lambda} = \dfrac{m}{d\cos\theta}$ $\Delta\theta = \dfrac{m\Delta\lambda}{d\cos\theta} = \dfrac{3(8.8 \times 10^{-4}\text{ nm})}{(1.76 \times 10^3\text{ nm})(\cos 47.7°)}$ $\Delta\theta = 2.2 \times 10^{-6}$ rad
CHECK and THINK To see if this angular separation is sufficient to resolve the two lines, compare it to either line's half-width. The angular separation between the two lines is about three times greater than their half-width. Lines are said to be barely resolved if the angular separation equals the half-width. In this case, the lines are better than barely resolved.	$\Delta\theta_{hw} = \dfrac{\lambda}{Nd\cos\theta}$ (36.16) $\Delta\theta_{hw} = \dfrac{434.047\text{ nm}}{(5 \times 10^5)(1.76 \times 10^3\text{ nm})(\cos 47.7°)}$ $\Delta\theta_{hw} \approx 7 \times 10^{-7}$ rad

Although hydrogen is the most abundant element in the Universe and is very often observed in astronomy, there are reasons to observe other elements. Hale observed the Zeeman effect in the spectra of heavier elements in order to estimate the magnetic field strength at various depths within sunspots. He found that the magnetic fields in sunspots are about three to four orders of magnitude greater than the Earth's magnetic field. The more we learn about atoms in the laboratory, the more we can learn about the Universe.

42-10 Practical Devices

In this chapter so far, we've applied the theory of quantum mechanics to understand atoms better. In Example 42.10, we saw that this deeper understanding of atoms leads us to a deeper understanding of the Universe. In this final section we will see how a deeper understanding of atoms also leads to the development of practical devices.

The Cesium Clock

Now that we are nearly at the end of this textbook. We are in a position to explain statements we made near the very beginning of the book. In Section 1-5, we learned that the time standard is determined by a Cesium clock (Fig. 1.5, page 7). Now we can describe the basic operation of such a clock. A sample of Cesium gas is heated, so that its atoms are in an excited state. The atoms then lose energy and return to their lower energy state. Of course, in losing energy the atoms must emit photons. Each photon's energy is determined by the energy difference in the two states ΔE. Their frequency is given by $f = \Delta E/h$ (Eq. 40.8). For the two energy levels exploited in a Cesium clock, this frequency is very high: 9,192,631,770 Hz. The frequency of the emitted light is the basis for the definition of a second. You can think of the frequency as 9,192,631,770 "ticks" per second. Then the second is accuracy defined as duration of 9,192,631,770 ticks. We could not have built such a precise timepiece without a quantum-mechanical model of the atom because this level of precision requires knowing that atomic energies are quantized and that an atom emits a photon with energy equal to the difference in two energy levels whenever it transitions to a lower energy level.

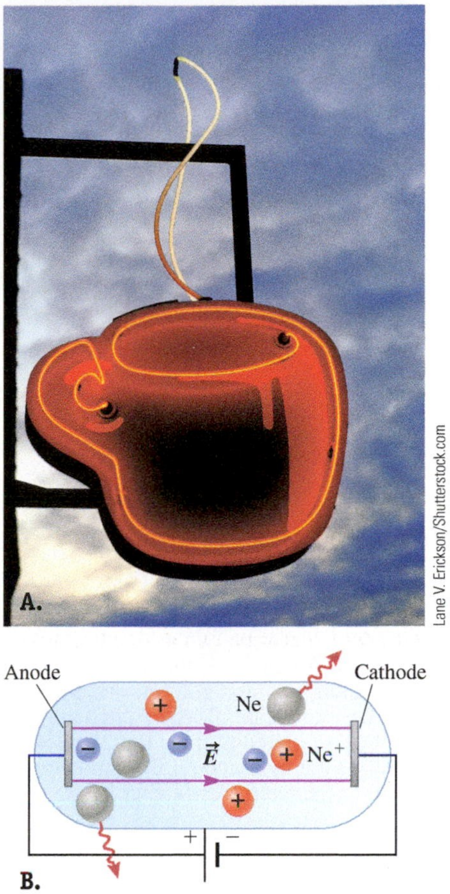

FIGURE 42.33 **A.** A neon sign glows red. **B.** Gas in a neon bulb is excited by an electric current. Photons of a few particular frequencies (colors) are released with the atoms' transition to a lower energy state.

Fluorescent Bulbs

We can also use our deeper understanding of atoms to see why the spectrum produced by a fluorescent bulb is made up of a discrete number of emission lines (Fig. 36.17, page 1166). First, let's look at a neon bulb, such as those used in a red neon sign (Fig. 42.33A). The neon gas is confined in a tube with metal plates near its ends (Fig. 42.33B). These plates are connected to a power supply, so that an electrical current can be passed through the gas. The current ionizes the gas, producing some free electrons and ions. So now there are free electrons, ions, and neutral atoms in the tube. These numerous particles undergo many collisions, which cause many of the neutral atoms to be in an excited state. But then these excited atoms quickly transition back down to a lower energy state and release a photon. Each photon's energy is determined by the energy difference between the two states ΔE, and its frequency is given by $f = \Delta E/h$ (Eq. 40.8).

Of course, not all the neon atoms will be in the same excited state or will transition to the same lower energy state. But because only a few possible transitions are available to the neon in the tube, the emitted photons can only have a few possible frequencies. As a result, the light produced by such a neon tube is limited to just a few spectral lines. In the case of neon, this limitation makes the light reddish. A red light may be desirable in many situations, but often we would prefer a white light. In such cases, fluorescent bulbs use mercury, which produces lines in the ultraviolet. The tube of a mercury bulb is coated with a material that absorbs the ultraviolet light. The atoms in the coating are in an excited state. Then they transition to a lower energy state, emitting photons at various frequencies in the visible spectrum. These photons blend together to form nearly white light.

CASE STUDY Lasers

As we mentioned when we first introduced the case study for this chapter, lasers have many practical everyday uses. They have also played a major role in our understanding of physics. You have probably used a laser in your laboratory to re-create experiments such as Young's double-slit experiment (Fig. 35.10, page 1128). Here we consider the basic operation of lasers.

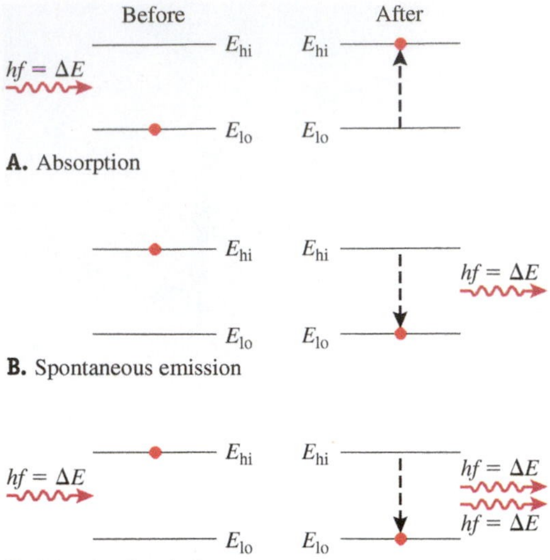

FIGURE 42.34 **A.** An atom can absorb a photon if the photon's energy equals the difference between two available energy states. **B.** An excited atom can spontaneously emit a photon and transition to a lower energy state. **C.** An excited atom can be stimulated to transition to a lower energy state by a photon with an energy equal to the energy difference of the two states.

To understand lasers, we need to think about ways that photons and atoms can interact. As we have been discussing, when a photon encounters an atom, it can be absorbed if the energy of the photon equals the energy difference between two of the atom's energy states (Fig. 42.34A). We have also discussed *spontaneous emission*, which occurs when an atom in a high-energy state transitions to a lower one spontaneously and emits a photon whose energy equals the energy difference between the two atomic states (Fig. 42.34B).

There is another way for photons and atoms to interact. If an atom is in a high state with energy E_{hi} when a photon encounters it, it is possible for the photon to *stimulate* the atom to transition to a lower state with energy E_{lo} (Fig. 42.34C). This **stimulated emission** occurs when the photon's energy equals the difference between the two energies: $E_{photon} = E_{hi} - E_{lo}$. In such a case, the atom transitions to a lower state and *releases a photon that is identical to the original stimulating photon.* So now there are two photons where before there had been one. Lasers are based on stimulated emission.

Under normal conditions, stimulated emission is rare because it requires an atom to be in an excited state when a photon of just the right energy encounters it. To justify this claim, we consider **Boltzmann's distribution law**, which relates the number density of atoms in the high-energy n_{hi} to the number density of atoms in the lower energy state n_{lo}:

$$n_{hi} = n_{lo}e^{-\Delta E/k_B T} \tag{42.31}$$

where $\Delta E = E_{hi} - E_{lo}$ and T is the temperature in kelvin. (We won't derive Boltzmann's distribution, but in Problem 42.80 you will show that Boltzmann's distribution law can be used to derive the Maxwell–Boltzmann distribution we studied in Chapter 20.) The energy difference ΔE is positive, so all the terms in the exponential function are positive. The negative sign in the exponential function means that $n_{hi} < n_{lo}$. In fact, under normal conditions there are very few atoms in the higher energy state. If such an atom were to spontaneously transition to the lower energy state and release a photon, this photon would be very unlikely to encounter another atom in the higher energy state. This photon is much more likely to encounter an atom in the lower energy state and get absorbed.

Laser is an acronym for *l*ight *a*mplification by *s*timulated *e*mission *r*adiation. Lasers depend on stimulated emission. We have just argued that under normal conditions stimulated emission is rare because there are more atoms in the lower energy state than in the higher energy state $n_{hi} < n_{lo}$. So the first step in designing a laser is to create a **population inversion**; that is, we need to have more atoms in the higher energy state than in the lower one $n_{hi} > n_{lo}$.

But a population inversion can only be created if the atoms have an available **metastable state** above their ground state. (We touched on such a state in Section 42-8 when we discussed forbidden transitions.) An atom can exist in a metastable state for a relatively long time because the transition to a lower state has a very low probability. An atom in a normal energy state may exist in that state for roughly 10^{-8} s before spontaneous emission occurs, but an atom in a metastable state can remain in that state for orders of magnitude longer making it available for stimulated emission to cause it to transition to a lower energy state.

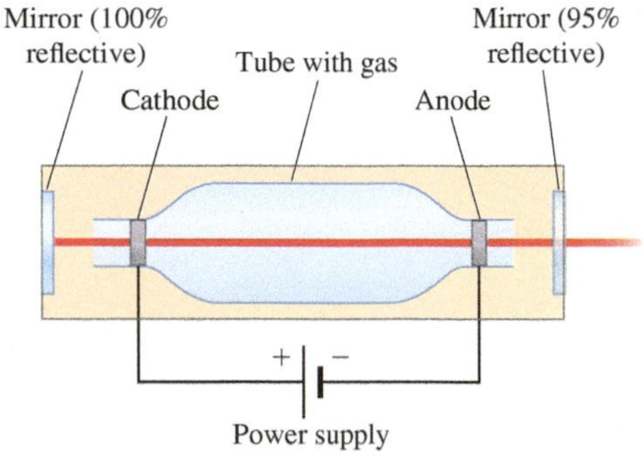

FIGURE 42.35 A schematic of a basic laser design.

So the laser must **pump** the atoms up to a metastable state. We'll start by describing how this is done in a pulsed laser. In a laser that uses a gas, the pumping process is similar to the one found in a neon bulb. An electric current is passed through the gas, creating some ions and free electrons (Fig. 42.35). Collisions between the atoms, ions, and electrons put the atoms into a higher energy state. So far, this is exactly like the gas in the neon bulb. The difference is that the atoms in the laser have a metastable state available to them. (In the neon gas, the atoms spontaneously emit photons.)

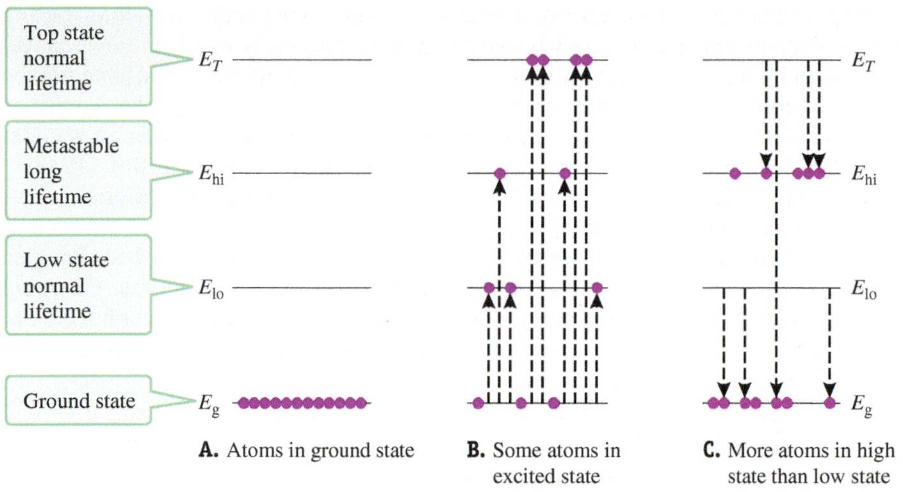

FIGURE 42.36 **A.** Before pumping, the atoms are in the ground state. **B.** Soon after pumping atoms occupy all four states. **C.** After a short time, spontaneous emission causes the atoms' transition to lower states. Atoms in the metastable state do not spontaneously transition in this short time. So there is a population inversion between the high and low states.

Figure 42.36 shows the energy levels of atoms in a laser. The dots represent atoms in their various energy states. These atoms have a metastable state with energy E_{hi}. Below the metastable state are two other states: One with energy E_{lo}, which we'll call the low state and another which is the ground state, with energy E_g. Above the metastable state is another energy state, which we'll call the top state, with energy E_T. Before the current is turned on, the atoms are in their ground state (Fig. 42.36A). But once pumping begins, atoms are in all four possible states (Fig. 42.36B). After a short time (about 10^{-8} s), the atoms in the top state and those in the middle state spontaneously transition to a lower energy state. However, those atoms in the metastable state do not transition. In addition, some atoms that were in the top state will transition to the metastable state. There will be very few atoms, if any, in the low state because these atoms would have transitioned to the ground state. The result is a population inversion between atoms in the higher metastable state and those in the low state (Fig. 42.36C).

In about a millisecond, a few atoms in the metastable state will spontaneously transition to the low state, releasing a photon of energy $E_{photon} = E_{hi} - E_{lo}$. Because of the population inversion, a spontaneously released photon is now *likely* to encounter an atom in the metastable state, causing it to release an identical photon. To ensure that these photons stimulate other atoms, mirrors are used to send them back and forth through the gas many times (Fig. 42.35). One of these mirrors is partially transparent, so that some of the resulting light can escape in a beam. In a *pulsed* laser, there is a burst of laser light, and then the pumping must begin again.

Because the photons produced by stimulated emission are identical, they have the same frequency, are in phase, are similarly polarized, and travel in the same direction. The result is laser light that is coherent (in phase) and monochromatic. The reflections back and forth across the tube ensure that the laser light diverges little. (Photons traveling along other paths do not stimulate many atoms.) These properties make laser light particularly useful in our laboratory experiments.

In your laboratory, you have probably used a helium–neon laser. Such a laser is a *continuous* laser rather than a pulsed laser of the kind we just described. The workings of such a laser are similar to the pulsed laser. An electric current is applied to the gas, which is a mixture of 10% He and 90% Ne. The helium is excited by this current to a 2*s* state. The excited helium then collides with neon atoms in the ground state. Because the energy of helium's 2*s* state is nearly the same as neon's 3*s* state, these collisions efficiently cause the neon to transition to its 3*s* state; this is a metastable state for neon. The neon atoms will stay in these metastable states until they undergo stimulated emission. These lasers are continuous rather than pulsed because the collision process between the helium and neon can maintain the population inversion.

Experiments like Young's done nearly two hundred years ago—long before lasers were invented in the 1960s—used sunlight (which is much more inconvenient than lasers). But lasers are **not** just more convenient to use. They have played a critical role in aiding scientific progress. For example, in 1997 the American physicist Steven Chu (1948–), who later became the Secretary of Energy under President Obama, shared the Nobel Prize with William Daniel Phillips (1948–) and Claude Cohen-Tannoudji (1933–) for their work involving lasers. As we know, atoms are always in motion owing to their thermal energy. But a physicist would like to slow atoms down to study them better. Chu used six lasers aimed toward the center of sodium gas to slow down and trap sodium atoms. He could cool the atoms down to a few hundred microkelvin for about half a second. So our knowledge of atoms allows us to build lasers, which allows us to trap atoms to study them better!

EXAMPLE 42.11 CASE STUDY Got to Pump

In order for a helium–neon laser to operate, the neon atoms must be in an excited state. The wavelength of light from a helium–neon laser is 633 nm. At normal room temperature, what is the ratio of the number density of neon atoms in the high-energy state to that of the atoms in the low-energy state?

INTERPRET and ANTICIPATE

We can use the wavelength of the emitted light to find the energy difference between the two energy states that emitted that light. Then we can use Boltzmann's distribution law to find the ratio.

SOLVE

The energy of a photon can be found by combining $f = c/\lambda$ (Eq. 34.20) and $E = hf$ (Eq. 40.8).	$E_{\text{photon}} = hf = \dfrac{hc}{\lambda}$
The energy of the photon must equal the difference in the energy of the two states.	$E_{\text{photon}} = E_{\text{hi}} - E_{\text{lo}} = \Delta E$ $\Delta E = \dfrac{hc}{\lambda} = \dfrac{(6.63 \times 10^{-34}\,\text{J}\cdot\text{s})(3.00 \times 10^{8}\,\text{m/s})}{633 \times 10^{-9}\,\text{m}}$ $\Delta E = 3.14 \times 10^{-19}\,\text{J}$
To use Boltzmann's law (Eq. 42.31), we must also estimate room temperature. Typically, room temperature is about 20° C = 293 K. Rearrange to find the ratio and substitute values.	$n_{\text{hi}} = n_{\text{lo}} e^{-\Delta E/k_B T}$ (42.31) $\dfrac{n_{\text{hi}}}{n_{\text{lo}}} = e^{-\Delta E/k_B T}$ $\dfrac{n_{\text{hi}}}{n_{\text{lo}}} = \exp\left(\dfrac{-3.14 \times 10^{-19}\,\text{J}}{(1.38 \times 10^{-23}\,\text{J}\cdot\text{K})(293\,\text{K})}\right)$ $\dfrac{n_{\text{hi}}}{n_{\text{lo}}} = 1.78 \times 10^{-34}$

CHECK and THINK

We just found a very small ratio. To make this more understandable, if there were 1 mole of neon gas, and so there were 6.0×10^{23} atoms, then no atom would likely be in the high-energy state. Now we see why pumping and the existence of a metastable state are so important. Stimulated emission requires atoms in the higher energy state. Atoms must be pumped to the higher state and remain there until they can be simulated to emit photons.

SUMMARY

Underlying Principles

According to **Pauli's exclusion principle,** *no two particles with $s = 1/2$ confined in the same well can have the same set of values for their quantum numbers.* In practice, this means that in an atom, no two electrons can have exactly the same set of the four quantum numbers n, ℓ, m, and m_s.

Major Concepts

1. **Thomson's plum pudding model** for an atom consists of two parts: a positive homogeneous sphere and electrons embedded in this positive sphere (Fig. 42.3). Although Thomson's model is consistent with Thomson's experiments on cathode rays, it cannot explain the spectral lines that we observe from atoms (Section 36-4).
2. In **Rutherford's solar system model**, the positive part of the atom is concentrated at the center and surrounded by orbiting electrons. Most of the atom is empty space (Fig. 42.1). This model accounts for Rutherford's experimental results. However, in this model the electrons must orbit the nucleus in order for the atom to be stable, a requirement contradicted by quantum mechanics.
3. As in Rutherford's model, in **Bohr's atomic model** the positive nucleus is at the center of the atom and the electrons orbit that nucleus much as the planets orbit the Sun. The two major differences with Rutherford's model are known as **Bohr's postulates.**
   a. **Bohr's first postulate** is that only certain electron orbits are allowed, with angular momentum L given by:
   $$L = n\left(\frac{h}{2\pi}\right) = n\hbar \qquad (n = 1, 2, 3 \ldots) \quad (42.4)$$
   where $\hbar = h/2\pi = 6.58211814 \times 10^{-16}\ \text{eV}\cdot\text{s}$. The integer n is called the **principal quantum number.** The energy levels of hydrogen are then:
   $$E_n = -\frac{13.6\,\text{eV}}{n^2} \qquad (n = 1, 2, 3 \ldots) \quad (42.10)$$
   b. **Bohr's second postulate** is that electrons in orbit around the nucleus do not radiate. An electron only radiates when it moves from one possible orbit to another possible orbit that has lower energy. For hydrogen, the wavelength of the radiated photon is given by Rydberg's formula
   $$\frac{1}{\lambda} = R\left(\frac{1}{m^2} - \frac{1}{n^2}\right) \quad (42.3)$$
4. In applying **de Broglie's model** to the atom, we model the electrons as circular standing waves with quantized wavelengths given by:
   $$\lambda_n = \frac{2\pi r}{n} \quad (42.13)$$
   Like Bohr's model, de Broglie's model successfully predicts the hydrogen spectrum.
5. When **Schrödinger's equation** is applied to an atom and the boundary conditions are applied in three dimensions, the result is three quantities (energy and the magnitude and direction of the orbital angular momentum) that are quantized and described by three quantum numbers. In addition, **spin**—an intrinsic property of a particle—is also quantized in magnitude and direction, as described by two more quantum numbers. Table 42.4 summarizes all five quantum numbers.
6. The **periodic table** (Appendix B) is the ordering of the elements by atomic number into periods and groups. The chemical and physical properties of the elements are correlated with their place in the table.
7. In the normal **Zeeman effect**, the spectral lines of certain atoms split into three lines when the sample is in an external magnetic field B. The energy of the emitted photon is given by
   $$E = E_0 - \Delta m \mu_{\text{Bohr}} B \quad (42.30)$$
   where E_0 is the energy of the corresponding photon in the absence of an external field and the difference in the orbital magnetic quantum number is $\Delta m = 0, \pm 1$ for allowed transitions.

PROBLEMS AND QUESTIONS

A = algebraic C = conceptual E = estimation G = graphical N = numerical

42-1 Early Atomic Models

Problems 1 and 4 are paired.

1. **C** One thing we know about atoms is that they are stable; that is, they don't collapse. What accounts for an atom's stability in the plum pudding model?
2. **E Review** Estimate the number of atoms that exist in a cubic centimeter of a solid material. This process will be aided by choosing an element and examining its density in the solid phase.
3. **C** Suppose you cut a wooden log or an iron bar in half repeatedly. Based on the natural models of the ancient Greeks, how would these two experiments end?

42-2 Rutherford's Model of the Atom

4. **C** One thing we know about atoms is that they are stable; that is, they don't collapse. What accounts for an atom's stability in the solar system model?
5. **C** Why was Rutherford so surprised by his experimental results?
6. **C** Explain this sentence: "You are mostly empty space."
7. **E** In the development of atomic models, it was realized that the atom is mostly empty. Consider a model for the hydrogen atom where its nucleus is a sphere with a radius of roughly 10^{-15} m, and assume the electron orbits in a circle with a radius of roughly 10^{-10} m. In order to get a better sense for the emptiness of the atom, choose an object and estimate its width. This object will be your "nucleus". How far away would the "electron" be located away from your "nucleus?"

42-3 Bohr's Model and Atomic Spectra

8. **N** Show that the constant $(2\pi^2k^2m_ee^4)/(h^3c)$ is equal to the Rydberg constant $R = 1.10 \times 10^7\ \text{m}^{-1}$ (Eq. 42.2). To keep this task manageable, work with four significant figures, and report your answer to three significant figures.
9. **N** In the Bohr model of hydrogen, an electron makes a transition from the $n = 5$ excited state to the $n = 3$ excited state. What is the change in the orbital radius of the electron?
10. **N** Show that the constant $(2\pi^2k^2m_ee^4)/h^2$ is equal to -13.6 eV (Eq. 42.9). To keep this task manageable, work with four significant figures, and report your answer to three significant figures.
11. **N** According to Bohr's model, what is the diameter of hydrogen in its ground state?
12. **C** Suppose that an electron in the ground state of a hydrogen atom absorbs a photon with 15.0 eV of energy. Is the electron still bound to the nucleus? What is the energy of the electron after it absorbs this photon? What kind of energy does the electron then possess? Explain your answer.
13. **N** Consider photons incident on a hydrogen atom.
    a. A transition from the $n = 3$ to the $n = 7$ excited state requires the absorption of a photon of what minimum energy?
    b. A transition from the $n = 1$ ground state to the $n = 5$ excited state requires the absorption of a photon of what minimum energy?

Problems 14, 15, and 16 are grouped.

14. **A** Use a solar system model to come up with an expression for the speed of an electron orbiting in the ground state of hydrogen. Express your answer in terms of the Bohr radius, the mass of the electron, and the elementary electric charge.
15. **A** Use Bohr's model to come up with an expression for the speed of an electron orbiting in the ground state of hydrogen. Express your answer in terms of the fundamental constants. Your final expression should not involve the mass of the electron.
16. **C** Using Bohr's model for hydrogen, you can come up with a speed for the orbiting electron that does not involve the electron's mass. Explain this result, present an example of a similar situation, and explain what it means for the atom's energy.
17. **N** An atom in an excited state can, on average, exist in that state for 10^{-8} s. If a hydrogen atom is in the $n = 2$ state, about how many orbits will it undergo before the atom returns to the ground state? Assume Bohr's model for hydrogen.
18. **N** A hydrogen atom transitions from the $n = 6$ excited state to the $n = 2$ excited state, emitting a photon. **a.** What is the energy, in electron volts, of the photon emitted by the hydrogen atom? **b.** What is the wavelength of the photon emitted by the hydrogen atom? **c.** What is the frequency of the photon emitted by the hydrogen atom?
19. **N** What is the maximum photon wavelength that would free an electron in a hydrogen atom when it is in the $n = 6$ excited state?

Problems 20, 21, and 22 are grouped.

20. **N** The Balmer series consists of the spectral lines from hydrogen for an electron making a transition from an excited state to the $m = 2$ state. The Lyman series consists of the spectral lines from hydrogen for an electron making a transition from an excited state to the $m = 1$ state. Determine the wavelengths of the first four spectral lines of the Lyman series ($n = 2, 3, 4$, and 5).
21. **N** The Balmer series consists of the spectral lines from hydrogen for an electron making a transition from an excited state to the $m = 2$ state. The Paschen series (or Bohr series) consists of the spectral lines from hydrogen for an electron making a transition from an excited state to the $m = 3$ state. Determine the wavelengths of the first four spectral lines of the Paschen series ($n = 4, 5, 6$, and 7).
22. **N** The Balmer series consists of the spectral lines from hydrogen for an electron making a transition from an excited state to the $m = 2$ state. The Brackett series consists of the spectral lines from hydrogen for an electron making a transition from an excited state to the $m = 4$ state. Determine the wavelengths of the first four spectral lines of the Brackett series ($n = 5, 6, 7$, and 8).

Problems 23, 24, and 25 are grouped.

23. **C** The Bohr model for the hydrogen atom can be extended to cover other atoms when they are stripped free of all but one electron. When this occurs, the energy levels for the single electron in an atom with atomic number, Z, are given by $E_n = (-13.6\ \text{eV})Z^2/n^2$ (see Example 42.4). As we examine atoms with greater atomic number, the ground-state energy becomes more negative. Explain why this would be expected and what it means regarding the requirements for freeing the electron in the ground state.
24. **N** The Bohr model for the hydrogen atom can be extended to cover other atoms when they are stripped free of all but one electron. When this occurs, the energy levels for the single electron in an atom with atomic number, Z, are given by

$E_n = (-13.6 \text{ eV})Z^2/n^2$ (see Example 42.4). Calculate the electron energy for the first five energy levels ($n = 1$ to $n = 5$) of ionized lithium (Li^{++}).

25. **N** The Bohr model for the hydrogen atom can be extended to cover other atoms when they are stripped free of all but one electron. When this occurs, the energy levels for the single electron in an atom with atomic number, Z, are given by $E_n = (-13.6 \text{ eV})Z^2/n^2$ (see Example 42.4). What wavelength of photon would be emitted if an electron transitions from the $n = 4$ excited state to the ground state in ionized lithium (Li^{++}).

42-4 De Broglie's Theory and Atoms

26. **N, C** Compare the circular standing waves of an electron in a hydrogen atom (Fig. 42.9, page 1387) to the standing waves on a guitar string (Fig. 18.25, page 535). How many nodes and antinodes exist in the third harmonic for the guitar string? How many nodes and antinodes exist for the third quantum state of an electron in hydrogen? Comment on your findings.

27. **N** The angular momentum of an electron in a hydrogen atom is $2h/\pi$. **a.** What is the radius of the electron's orbit? **b.** What is the de Broglie wavelength of the electron?

28. **C** An electron in a hydrogen atom undergoes a transition such that its de Broglie's wavelength doubles. What is the relationship between the initial and final energy levels occupied by the electron? Explain your answer.

29. **N** A hydrogen atom is in its $n = 6$ state. Find the de Broglie wavelength of its electron.

42.5 Schrödinger's Equation Applied to Hydrogen

30. **N** An electron in a hydrogen atom is in a state with a principal quantum number of 4. How many possible subshells could the electron occupy?

Problems 31 and 32 are paired.

31. **N** If the state of hydrogen is given by $n = 3$ and $\ell = 2$, find the possible values of m, L and L_z.

32. **G** If the state of hydrogen is given by $n = 3$ and $\ell = 2$, find the possible values of m, L, and L_z. Assume that the vector $\vec{L}$ lies in the x–z plane and sketch its possible values (magnitude and direction). Indicate angles and magnitudes.

Problems 33 and 35 are paired.

33. **N** If hydrogen is in its $n = 3$ state, what are the possible magnitudes of the electron's orbital angular momentum L?

34. **N** An electron in a hydrogen atom is in a state with $n = 4$. Find the orbital angular momentum and z component of the orbital angular momentum for each possible value of ℓ and m for the electron.

42.6 Magnetic Dipole Moments and Spin

35. **N** If hydrogen is in its $n = 3$ state, what are the possible values of $(\mu_{\text{orbit}})_z$ in terms of the Bohr magneton?

Problems 36 and 37 are paired.

36. **N** An electron in a hydrogen atom is in a state with a principal quantum number of 3 and an orbital quantum number of 2. What is the number of possible states for the electron? Take into account the possible values of m, and m_s.

37. **N** An electron is bound to a hydrogen atom. What is the number of possible states for the electron in each of the following cases? Take into account the possible values of m, and m_s. **a.** $n = 4$ **b.** $n = 5$ **c.** $n = 6$.

38. **A** Show that an electron's spin magnetic moment is a constant, $\mu_{\text{spin}} = \sqrt{3}\hbar e/2m_e$, regardless if it is bound or free.

39. **N** An electron in a hydrogen atom is in a state with $n = 3$. Find the orbital angular momentum and the z component of the orbital magnetic moment for each possible value of ℓ and m for the electron.

40. **A** The magnetic moment of a current loop is given by $\vec{\mu} = (I\pi R^2)\hat{\jmath}$ (Eq. 30.14). Show that the magnetic moment of a charged particle in a circular orbit of radius R is given by $\vec{\mu}_{\text{orbit}} = (q\vec{L})/(2m)$.

42.7 Other Atoms

Problems 41 and 42 are paired.

41. **C** Is it possible for the z component of the spin magnetic moment of an electron to equal its z component of the orbital magnetic moment? If so, what value(s) of m make(s) this possible? If not, explain why not.

42. **N, C** Is it possible for the spin magnetic moment of an electron to cancel its orbital magnetic moment? If so, what value(s) of m make this possible? If not, explain why not.

43. **N** Find the maximum number of electrons in each of the following subshells: **a.** s **b.** p **c.** d **d.** f **e.** g

44. **C** Your friend tells you a story about her two cousins, Paul and Paula. She jokes about how just when one leaves the other seems to always show up, and that it doesn't seem that they can ever be in the same place twice. In light of the physics class you share, she jokes that they must be obeying Pauli's exclusion principle. Is this an accurate comparison? In what way is she using the analogy correctly? In what way is the analogy not correct? Give an example.

45. **C** When a shell is full, does it always have an even number of electrons? Explain.

42.8 Organizing Atoms

46. **C** How do you account for the fact that lithium (Li) and sodium (Na) exhibit similar chemical behavior?

47. **N** What is the ground-state electron configuration of iodine (I)?

48. **C** Do you expect carbon (C) and calcium (Ca) to exhibit similar chemical behavior? Explain.

Problems 49 and 50 are paired.

49. **N** Write the ground-state configurations for the first three noble gases on the periodic table: helium (He), neon (Ne), and argon (Ar).

50. **C** Consider the electron configurations of the noble gases in the rightmost column on the periodic table. What aspect of their electron configuration makes them so nonreactive?

51. **N** What atom's ground-state electron configuration is given by $1s^2 2s^2 2p^6 3s^2 3p^6 4s^2 3d^8$?

42.9 The Zeeman Effect

52. **G** Make an energy-level diagram for the first three shells of hydrogen when it is in an external magnetic field B pointing in the z direction. Include those same shells for the case of no external field for comparison. Draw a distinct diagram for each subshell, labeling n, ℓ, m.

53. **N** A hydrogen atom is exposed to a magnetic field of 25.00 T. An electron is in the state $n = 5$, $\ell = 1$. Find each of the possible wavelengths of the photon emitted when this electron transitions directly from its current state to the ground state of the atom.

54. **A** Calculate the energy levels for hydrogen's 3d subshell when hydrogen is in an external magnetic field B pointing in the z direction. Express your answer in terms of the Bohr magneton.

Problems 55 and 56 are paired.

55. **N** A hydrogen atom with an electron in an excited state with $n = 3$, $\ell = 2$ undergoes a single transition to the state $n = 2$, $\ell = 1$, $m = 0$ while in an external magnetic field with magnitude 4.000 T. How many possible wavelengths could be observed for the emitted photon?

56. **N** A hydrogen atom with an electron in an excited state with $n = 3$, $\ell = 2$ undergoes a single transition to the state $n = 2$, $\ell = 1$, $m = 0$ while in an external magnetic field with magnitude 4.000 T. What is the maximum possible wavelength of an observed photon from this transition?

57. **N** An electron in a hydrogen atom is in the $n = 5$ shell and $\ell = 4$subshell. What is the difference between the highest and lowest possible energies of the electron when the atom is exposed to a 4.000 T magnetic field?

58. **C** For states of the electron in the hydrogen atom with $n > 1$, is it safe to say that there is at least one subshell that would show three distinct energies when the atom is exposed to an external magnetic field? Explain.

General Problems

59. **N** An isolated hydrogen atom is found to be in the $n = 5$ excited state. **a.** What is the radius of the Bohr orbit of the electron in this atom? **b.** What is the de Broglie wavelength for this electron?

60. **N** The Balmer series consists of the spectral lines from hydrogen for an electron making a transition from an excited state to the $m = 2$ state. The Pfund series consists of the spectral lines from hydrogen for an electron making a transition from an excited state to the $m = 5$ state. Determine the wavelengths of the first four spectral lines of the Pfund series (n = 6, 7, 8, and 9).

Problems 61 and 62 are paired.

61. **N** The magnitude of the orbital angular momentum is $L = \sqrt{30}\hbar$. What is ℓ?

62. **A** The magnitude of the orbital angular momentum is $L = \sqrt{30}\hbar$. What are the possible values of m?

63. The Bohr model for the hydrogen atom can be extended to cover other atoms when they are stripped free of all but one electron, as in Example 42.4, $E_n = (-13.6\text{ eV})Z^2/n^2$. An electron in the $n = 3$ state of a single-electron Lithium atom (Li^{++}) makes a transition to the ground state by emitting a photon.
   a. **N** What is the wavelength of the emitted photon?
   b. **C** If this emitted photon were absorbed by an electron in the ground state of a hydrogen atom, would it be enough to free the electron? Explain your answer.

64. **A** To find the most likely location for the electron, we need the *radial* probability density P, not the *volume* probability density $|\psi|^2$. The two probability densities are related by $P(r)dr = |\psi|^2 dV$. The volume element dV is the volume of a spherical shell of thickness dr, so $dV = 4\pi r^2 dr$. Show that $P(r) = (4/r_B^3)r^2 e^{-2r/r_B}$.

65. **N** An electron in a hydrogen atom makes a transition from the ground state to the $n = 4$ state after absorbing a photon **a.** What is the minimum energy of the photon? **b.** What is the frequency of the photon with this minimum energy?

66. **C** Covalent bonds form between atoms or molecules that share electrons and often allow each atom to fill its outer shell via the sharing process. Consider hydrochloric acid HCl. Which shell is filled for each atom when they share electrons?

Problems 67 and 68 are paired.

67. **C** An electron in a hydrogen atom is in the $n = 5$ shell. Is there a subshell in which the electron could have five distinct energies when the atom is exposed to an external magnetic field? If so, which subshell? Explain. If not, why not?

68. **N** An electron in a hydrogen atom is in the $n = 5$ shell and $\ell = 1$subshell. Find the three orbital energies of the electron, when an external magnetic field of 2.00 T is applied.

Problems 69, 70, and 71 are grouped.

69. **N** An imaginary atom has only one electron, and its energy levels are given by $E_n = (-19.5\text{ eV})/n^2$ where $n = 1, 2, 3 \ldots$. Find the first five energy levels for this atom.

70. **N** An imaginary atom has only one electron, and its energy levels are given by $E_n = (-19.5\text{ eV})/n^2$ where $n = 1, 2, 3 \ldots$. If the atom goes from the $n = 3$ state to the $n = 1$ state, what is the frequency of the emitted photon?

71. **N** An imaginary atom has only one electron, and its energy levels are given by $E_n = (-19.5\text{ eV})/n^2$ where $n = 1, 2, 3 \ldots$. Initially, the atom is in its ground state. If the atom absorbs a photon with a frequency of 4.42×10^{15} Hz, in what state is the atom now?

72. **N** An electron is bound in a hydrogen atom such that it is in the $n = 6$ shell. What is the de Broglie wavelength of this electron?

73. Consider an electron in a hydrogen atom in the $n = 2$, $\ell = 1$, $m = 1$ excited state.
   a. **N** What magnitude of an applied magnetic field would cause the energy of this electron to be equal to zero?
   b. **C** Given that the energy of this electron would be zero under these conditions (if the electron does not undergo a transition to another state), what would be true about the electron?

74. **C** A common saying is that "you can be two places at the same time." How might this colloquialism be reworded to describe Pauli's exclusion principle?

Problems 75, 76, and 77 are grouped.

75. **C** Suppose an electron in a hydrogen atom transitions from its current state, n, to the state, $n/2$. What, if any, limitations can you place on the value of n? Are there any values not allowed?

76. **A** Suppose an electron in a hydrogen atom transitions from its current state, n, to the state, $n/2$. Use Eq. 42.2 to express the wavelength of the resulting emitted photon in terms of R, and n.

77. **N** Consider the result for Problem 76. Find the wavelength of the emitted photons for the cases where **a.** $n = 2$, **b.** $n = 6$, and **c.** $n = 12$.

Problems 78 and 79 are paired.

78. **C** An electron in a hydrogen atom is to make a transition from the $n = 4$ excited state to the ground state. There are three ways the electron could make this transition from the $n = 4$ state to the ground state by emitting photons. Describe each of the ways the electron could accomplish this feat.

79. **N** An electron in a hydrogen atom is to make a transition from the $n = 4$ excited state to the ground state. There are three ways the electron could make this transition from the $n = 4$ state to the ground state by emitting photons. Determine the energy of the photons that would be emitted in each way the electron can make the transition.

80. **A** Start with Boltzmann's distribution $n_{hi} = n_{lo}e^{-\Delta E/k_BT}$ (Eq. 42.31) and derive Maxwell–Boltzmann's equation $f(v) = 4\pi\left(\frac{m}{2\pi k_BT}\right)^{3/2} v^2 e^{-(mv^2/2k_BT)}$ (Eq. 20.19).

Nuclear and Particle Physics

43

Underlying Principles

1. Models of nuclei
2. Strong nuclear force
3. Weak nuclear force
4. Electroweak theory
5. QED and QCD
6. The standard model

Major Concepts

1. Nuclear radius
2. Nuclear decay
3. Mass deficit and binding energy
4. Fusion and fission
5. Absorbed, equivalent, and effective dose
6. Bosons and fermions
7. Leptons and hadrons

Special Cases

1. Gamma rays
2. Alpha particles (and rays)
3. Beta particles (and rays)
4. Electron capture

Key Questions

How do we model the nucleus and subatomic particles?

What are nuclear reactions?

What is radioactivity, and how does it affect living organisms?

Here's something odd. According to physics, there are four fundamental forces; but aside from a few brief mentions, for the past 42 chapters we discussed only two of these four forces. How did we avoid the last two forces for so long, and why do we need to discuss them now? The answer is that both of these remaining forces are *nuclear* forces. So far we have treated the nucleus as a positively charged particle, but in this chapter we explore nuclear structure and behavior. Along the way, we'll learn where we all came from, how to measure the age of fossils, and the basic principles of nuclear reactors. We'll see how all of this is organized in a well-confirmed theory known as the *standard model*.

43-1 Describing the Nucleus

A nuclear species is referred to as a *nuclide*; however, we will typically use the terms *nucleus* (singular) and *nuclei* (plural).

You already have a good idea of what a nucleus is: an object at the center of an atom, made up of protons and neutrons. You also know that the nucleus was discovered in Rutherford's scattering experiment (Section 42-2). Rutherford concluded that the nucleus is about 10^4 times smaller than the atom.

Most of an atom's mass is concentrated in its nucleus, which is composed of particles called **nucleons** (either protons or neutrons). Although neutrons are slightly more massive than protons, each nucleon's mass is about 1800 times the mass of an electron. Both kinds of nucleons have spin ½, so both obey Pauli's exclusion principle (Section 42-7). The number of neutrons in a nucleus is known as the **neutron number** N. The number of protons in the nucleus is called the **atomic number** Z. (For a neutral atom, Z is also the number of electrons.) The total number of nucleons is called the **mass number** A, and $A = N + Z$. The simplest nucleus is that of hydrogen, consisting of a single proton. All other nuclei have both protons and neutrons. The most massive naturally occurring nucleus is that of uranium, with 92 protons and 146 neutrons. Heavier nuclei have been created artificially.

Nuclei are represented by their chemical symbol X (taken from the periodic table, Appendix B) and by a preceding superscript and subscript, as in ${}^{A}_{Z}X$. The superscript A is the mass number, and the subscript Z is the atomic number. For example, the most massive naturally occurring uranium *isotope* (more on this term below) may be written as ${}^{238}_{92}U$. Often the subscript is dropped because the chemical symbol indicates the atomic mass number, making the subscript redundant. (Uranium is the 92nd element in the periodic table.) There is no need to indicate the neutron number because it can always be found from A and Z: $N = A - Z$. (Uranium has $238 - 92 = 146$ neutrons.) We often speak of an isotope using its name and mass number: for example, we call ${}^{238}_{92}U$ "uranium-238."

An element is characterized by its atomic number Z; that is, its chemical properties are determined by its number of electrons. An **isotope** is a species of an element (with the same Z) that differs in its neutron number, and therefore its mass number. Because all isotopes of an element have the same number of protons (and of electrons), they occupy the same place on the periodic table, have the same electron configuration, and therefore the same chemical behavior. For example, deuterium is an isotope of hydrogen with one proton and one neutron in its nucleus. The atom still has one electron, though, and like hydrogen, deuterium bonds with oxygen to form (heavy) water. As another example, uranium-235 and uranium-238 have the same chemical properties and so cannot be separated by a chemical reaction. However, because their masses are different, they can be separated mechanically with, for example, a centrifuge.

Although the chemical behavior of isotopes of a particular atom is the same, the stability of its nucleus depends on the isotope. A periodic table is not useful for showing the differences in isotopes, so we need another sort of chart. Typically, we plot Z as a function of N, creating an **isotope chart** (Fig. 43.1). There are over 3000 known isotopes, but only about 266 are stable and occur naturally. The stable isotopes are shown in blue in Figure 43.1. The other isotopes are radioactive, meaning that they **decay** or break apart by emitting a particle or radiation. Most isotopes are radioactive, and will decay (sometimes repeatedly) until the end result is a stable isotope. The stable isotopes tend to form a thin line in the middle of the chart. The *small* stable isotopes (with N less than about 20) tend to fall on the $N = Z$ line, so small stable isotopes have roughly equal numbers of protons and neutrons. However, the larger stable isotopes fall below the $N = Z$ line, meaning that they tend to have fewer protons than neutrons. Stable as well as unstable isotopes occur naturally.

The **natural abundance** of an isotope is the number of a naturally occurring isotope relative to the total number of naturally occurring isotopes of that element. For example, chlorine comes in two naturally occurring isotopes, and the natural abundance of ${}^{35}Cl$ is about 76% and that of ${}^{37}Cl$ is about 24%. The isotopes of chlorine have different numbers of nucleons, so they must have different masses. The atomic mass reported for an element, for example, on the periodic table, is the weighted average of the atomic masses of the naturally occurring isotopes.

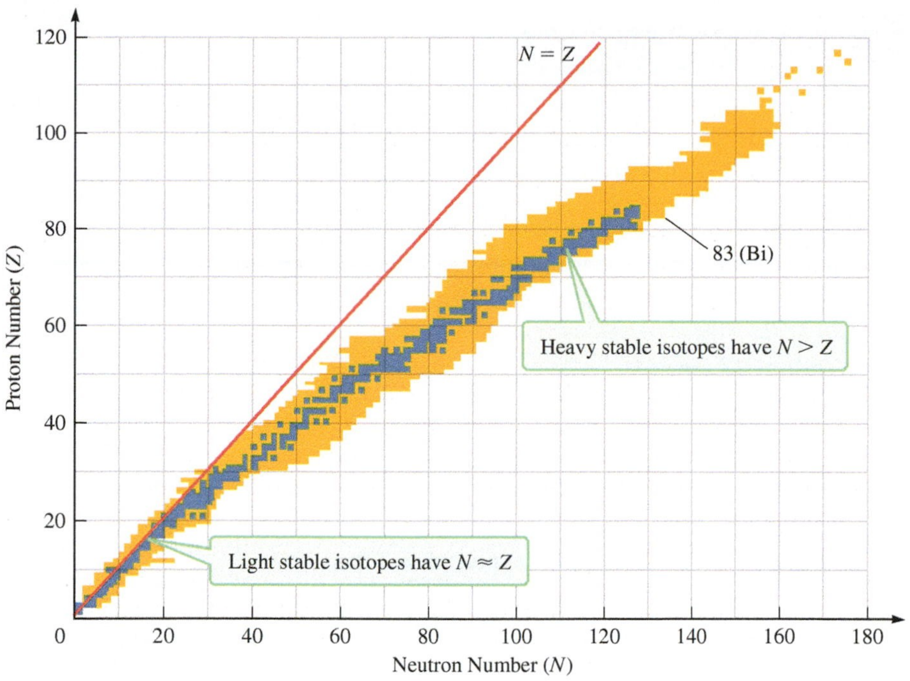

FIGURE 43.1 Isotopes are shown on a graph of Z as a function of N. The stable isotopes are in blue. The red line indicates isotopes for which the atomic number equals the neutron number. This line is well above the high-N isotopes but passes through the low-N ones. There are no stable nuclei for $Z > 83$ (bismuth). This type of graph is sometimes called a *Segrè chart*. Notice that we have plotted Z on the vertical axis and N on the horizontal, but you may encounter the graph with these axes reversed.

Atomic masses are often reported in atomic mass units (u); the mass of ^{12}C is defined as exactly 12 u. So the mass number A gives an approximate mass in atomic mass units. For example, the atomic mass of chorine is (roughly) given by: $m_{Cl} \approx 0.76(35\text{ u}) + 0.24(37\text{ u}) \approx 35.5\text{ u}$, which is close to the value of 35.453 u found in the periodic table. In Appendix B, conversion factors are generally given to five significant figures, but you may need more significant figures to convert from atomic mass units to kilograms:

$$1\text{ u} = 1.66053879 \times 10^{-27}\text{ kg} \tag{43.1}$$

Often we will need to know the mass to five or so significant figures, so estimating mass from A is not sufficient.

Nuclear Radius

The radius of a nucleus is best measured in femtometers (1 fm = 10^{-15} m). A nucleus is not necessarily spherical. Nevertheless, experiments have shown that we can model some nuclei as spheres with an effective radius given by:

$$R = r_0 A^{1/3} \tag{43.2}$$

where $r_0 = 1.2$ fm and A is the mass number. The volume of a sphere is $V = 4\pi R^3/3$, so the volume of a nucleus is proportional to A. From this we can conclude that nucleons are tightly packed into the nucleus and are incompressible. Also, because the volume is proportional to the number of nucleons, the density of nuclear matter is nearly independent of the type of nucleus (Example 43.1).

A femtometer is also known as a *fermi* (both abbreviated "fm"), named for the Italian physicist Enrico Fermi (1901–1954).

Some nuclei produced in laboratories are not well modeled by Equation 43.2; these nuclei are larger than what the model predicts.

NUCLEAR RADIUS ✪ **Major Concept**

CASE STUDY **Part 1: Where did you come from?**

Your body is made of many different elements. Much of it is water, made of hydrogen and oxygen. Your blood contains iron and your thyroid contains iodine. In this case study, we explore the origins of these four elements. We'll find that some elements in your body are as old as the Universe itself (13.7 billion years), whereas other elements were made in a supernova explosion over 4.5 billion years ago. In learning where the elements in your body came from, we'll learn also about the history of the Universe.

CONCEPT EXERCISE 43.1

We concluded that because the volume of a nucleus is proportional to the number of nucleons, the nucleons must be tightly packed and incompressible. Use the analogy of packing marbles tightly into a flexible bag to explain why nucleons must be incompressible. What would happen if the marbles were replaced by highly compressible foam balls (like small Nerf® balls)?

EXAMPLE 43.1 The Density of Nuclear Material

Estimate the density of nuclear material.

INTERPRET and ANTICIPATE

Because both the volume of a nucleus and its mass are proportional to A, we expect to find that its density is a constant, independent of A.

SOLVE

Find the density from the mass and the volume.

$$\rho = \frac{m}{V} \quad (1.1)$$

The mass is approximately given by A in atomic mass units. Substitute $R = r_0 A^{1/3}$ (Eq. 43.2) for r in the volume expression.

$$\rho \approx \left(\frac{A}{\frac{4}{3}\pi R^3}\right) = \left(\frac{A}{\frac{4}{3}\pi r_0^3 A}\right)$$

$$\rho \approx \frac{3(1\text{ u})}{4\pi r_0^3} \approx \frac{3(1.66 \times 10^{-27}\text{ kg})}{4\pi(1.2 \times 10^{-15}\text{ m})^3}$$

$$\rho \approx 2 \times 10^{17}\text{ kg/m}^3$$

CHECK and THINK

As expected, our answer does not depend on A, so it is constant for all nuclei. Our result is much denser than familiar objects such as the Earth ($\rho_\oplus \approx 5500\text{ kg/m}^3$) or water ($\rho_{\text{water}} \approx 1000\text{ kg/m}^3$), but this isn't surprising because we know that atoms and therefore objects are mostly empty space.

43-2 The Strong Force

STRONG NUCLEAR FORCE
Underlying Principle

You already know that protons in a nucleus experience a repulsive force due to Coulomb's law, and from the case study in Chapter 41 (page 1358), another force binds the protons together in the nucleus. This force, called the **strong force** or the **strong nuclear force**, is one of the four fundamental forces of physics. The strong force has a short range, acting over distances of roughly 1 fm or less. Within this short range, the strong force is about 100 times stronger than the Coulomb force. In this section, we take a closer look at the strong force, which is essential in modeling the energy levels of a nucleus.

So far, we have treated protons and neutrons as elementary particles. However, unlike electrons, protons and neutrons are actually made up of other particles known as **quarks**. Quarks come in six types, with the whimsical names *up, down, strange, charm, bottom* and *top.* A proton, for example, is made of two up quarks and one down quark; a neutron is made of two down quarks and one up quark. The strong force binds the quarks together to form these nucleons. Because protons are made of quarks, and because the strong force is an attractive force between quarks, protons are bound together in the nucleus by the strong force. The same can be said for neutrons: They are also bound together by the strong force. Electrons, which are elementary particles that are *not* made up of quarks, do not experience the strong force.

Protons experience a Coulomb repulsive force, but neutrons do not. This difference explains why light stable isotopes have nearly equal numbers of protons and neutrons, whereas heavy stable isotopes have a greater number of neutrons than protons (Fig. 43.1). Although the nucleons are always tightly packed in the nucleus, protons on the opposite side of a large nucleus may be several fermis apart. At this relatively large distance, the strong force is negligible and the net force on such protons is repulsive due to the dominance of the Coulomb force. But the neutrons in a large nucleus experience no such repulsion. They are attracted to and attract neighboring nucleons, so they keep Coulomb repulsion from splitting the nucleus. In a small nucleus, one neutron for each proton is sufficient for stability. However, larger nuclei need more neutrons than protons to ensure that the strong force dominates over the Coulomb repulsion.

CASE STUDY **Part 2: Making Protons and Neutrons**

According to the **Big Bang model**, the Universe began in a very hot and very dense state about 13.7 billion years ago. Since that time, the Universe has been expanding, cooling down, and becoming less dense. Cosmologists can model the state of the early Universe and make predictions about what can be observed today. There have been some minor and major adjustments to the model, but observational evidence supports the basic idea that the Universe was once very hot and very dense. Conditions in the (relatively) early Universe have been reproduced in high-energy accelerators, with results that support the Big Bang model.

So what was the very early Universe like? In those extreme conditions, the particles, atoms, and molecules that we are familiar with could not exist. Instead the early Universe was a sea of elementary particles, including quarks. As the Universe cooled, the strong force bound quarks together to form protons and neutrons. Because a proton is the nucleus of hydrogen, hydrogen was likely the first element to form in the Universe. This kind of hydrogen is called *primordial* hydrogen.

CONCEPT EXERCISE 43.2

When we consider the attraction between the Earth and Moon, we only need to take gravity into account, and we can ignore the Coulomb force they exert on one another because they are electrically neutral. We can also ignore the strong force. Why?

43-3 Models of Nuclei

MODELS OF NUCLEI
Underlying Principle

We have been using the language of particles to describe the nucleus. However, these "particles" must be modeled as probability waves governed by Schrödinger's equation. We won't solve this equation for a nucleus; instead we'll use our work on trapped particles and atoms to model the nucleus.

The Shell Model

The first step in applying Schrödinger's equation is to find the potential energy of the system. Nuclear potential energy is complicated because the system involves many nucleons exerting both attractive and repulsive forces on one another. The American physicist Maria Goeppert-Mayer (1906–1972) and the German physicist J. Hans D. Jensen (1907–1973) came up with a clever way to think about the potential energy of nucleons: the **shell nuclear model**, which is based on Bohr's model of hydrogen. According to this model, each nucleon can be treated as a particle in a system whose remainder is the rest of the nucleus. The potential energy of the system then results

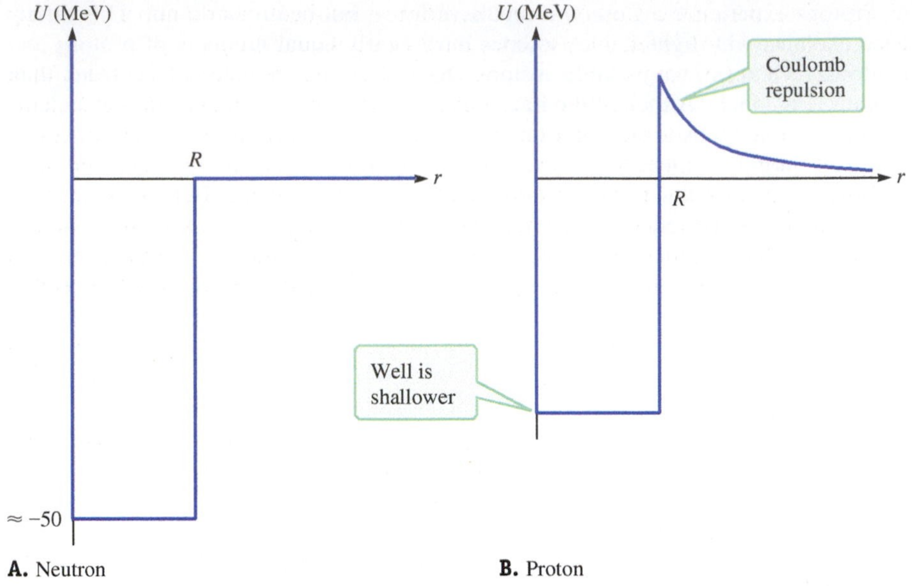

FIGURE 43.2 Schematic potential energy curves for nucleons. **A.** A neutron only experiences an attractive force due to the other nucleons in the nucleus, so the potential energy curve is a simple well. **B.** A proton experiences an attractive force due to all the other nucleons, but also a repulsive force due to the other protons; so the potential energy curve contains both a well and a barrier.

For brevity, we'll drop the full name of each system, and just refer to the proton system and the neutron system, or even just to the proton and the neutron.

from the average force experienced by this nucleon due to the rest of the nucleus. For a neutron, that average force is simply attractive, but for a proton, the force is attractive for short distances and repulsive for longer distances.

Figure 43.2 shows rough potential energy curves as a function of r, where r is the distance between the nucleon and the center of the nucleus, and R is the radius of the nucleus given by $R = r_0 A^{1/3}$ (Eq. 43.2). The potential energy curve for a system consisting of a neutron and the remainder of the nucleus is a simple well (Fig. 43.2A) with a depth of roughly 50 MeV for all nuclei. By contrast, the potential energy curve for a system consisting of a proton and the remainder of the nucleus is a somewhat shallower well with a positive barrier (Fig. 43.2B). Figure 43.2B is a crude diagram; the depth and shape of the proton potential energy curve depend on the mass of the nucleus because protons in a more massive nucleus experience a greater Coulomb repulsion than those in a low-mass nucleus.

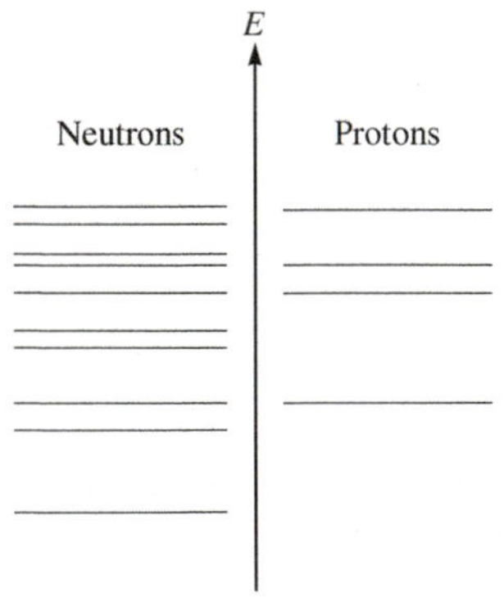

FIGURE 43.3 Energy levels for the nucleus are quantized, with the difference in energy levels on the order of several MeV. The neutron system energy levels are more closely spaced than those of the proton system. The lowest energy for protons is greater than the lowest energy for neutrons.

When we apply Schrödinger's equation to a trapped particle, such as a nucleon bound to a nucleus, the system's energy is quantized (Section 41-5). The spacing of energy levels depends on the depth and shape of the potential energy well. So we expect the energy levels in the proton system to be different from those in the neutron system. For high-mass nuclei, the proton energy levels are farther apart than are the neutron energy levels (Fig. 43.3). However, for low-mass nuclei ($Z \approx 8$ or less), Coulomb repulsion of the protons is weak compared to the strong force; therefore, in such nuclei the proton system and neutron system have similar potential wells and thus their energy levels are more similarly spaced.

Describing the structure and behavior of a nucleus is much like describing the structure and behavior of an atom: In both cases we do so by finding which energy levels are occupied. Because nucleons have spin ½, they obey Pauli's exclusion principle (Section 42-7). Just as when finding the electron configuration of an atom, we find the configuration of a nucleus by filling up energy levels starting with the lowest level while obeying Pauli's exclusion principle. In Chapter 42, we filled boxes in a diagram with upward- and downward-pointing arrows representing electrons with a particular set of quantum numbers (Fig. 42.23, page 1396). We can use a similar diagram here (Fig. 43.4). Each box on the left can be filled by up to two arrows (one up and the other down), representing two neutrons, while each box on the right can be filled by up to two arrows, representing protons. The vertical spacing on both sides of Figure 43.4 is equal for our convenience; remember that proton energy levels are actually farther apart than those of neutrons. Examine Figure 43.3 to see why $N > Z$ for stable heavy nuclei. Stable nuclei tend to have their energy levels filled, from the lowest level up to some highest level, for both protons and neutrons. If the highest proton energy level is roughly equal to the highest neutron energy level, then there must be more neutrons to fill the greater number of energy levels below this top level, so $N > Z$ for a heavy stable nucleus.

As in the case of atoms, the nuclear energy levels are referred to as **shells**. Filled shells produce a stable, tightly bound system. For example, noble gases are atoms with a particular number of electrons (2, 10, 18, 54 ...) that allows the atomic electron shells to be completely filled (Section 42-8). Recall, for example, that helium is a noble gas and its 1*s* subshell is completely filled by its two electrons. Also recall that lithium has just one more electron, so its electron configuration is $1s^2 2s^1$; lithium's valence electron is outside of a completely filled shell and is weakly bound. As a result, lithium has a lower ionization energy than helium. Helium is more stable than lithium.

Similarly, it is easier to remove a nucleon that is outside a filled nuclear shell than it is to remove one from a completely filled shell. In nuclear physics, the particular number of protons or neutrons that completely fills a nuclear shells is known as a **magic number**. The magic numbers are 2, 8, 20, 28, 50, 82, 126.... so a nucleus that has either an atomic number Z or a neutron number N equal to a magic number will have a filled shell and be tightly bound and stable. For example, ^{18}O ($Z = 8$) and ^{92}Mo ($N = 50$) are very stable. For some nuclei, both Z and N are magic numbers; these are referred to as **doubly magic** nuclei. For example, ^{4}He ($Z = N = 2$), ^{40}Ca ($Z = N = 20$) and ^{208}Pb ($Z = 82$, $N = 126$) are doubly magic. In fact, the helium nucleus particle is so tightly bound that we often model it as a single particle called an *alpha particle*. Further, it is impossible to form a stable nucleus by adding another neutron or proton to an alpha particle.

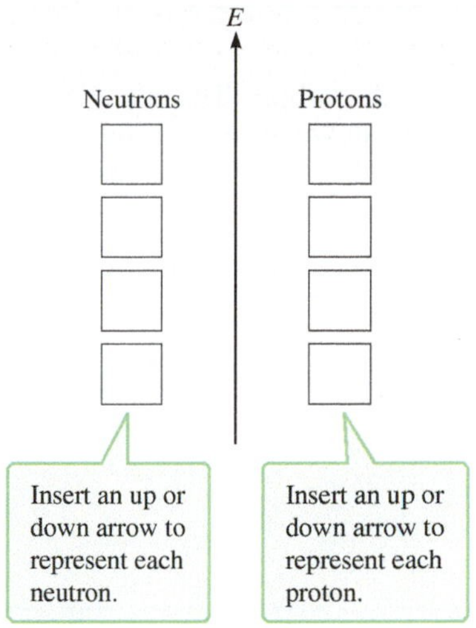
E
Neutrons
Protons
Insert an up or down arrow to represent each neutron.
Insert an up or down arrow to represent each proton.

FIGURE 43.4 Use this diagram to find a nucleus's occupied energy levels. Up to two nucleons can be in each box. A proton and a neutron in the same nucleus can have the same set of quantum numbers because they are in different potential energy wells.

Other Models

Before the shell model, Niels Bohr and the American physicist John Wheeler (1911–2008) came up with the **liquid drop model** for the nucleus. According to this model, nucleons move around the nucleus much like molecules moving in a drop of liquid, and there are frequent collisions between the nucleons. The liquid drop model does a good job of accounting for **fission**, the splitting of a nucleus into smaller nuclei. Fission can be spontaneous, but it can also be triggered by the absorption of a particle such as a proton or neutron into the original nucleus (Fig. 43.5). According to the liquid drop model, when such a particle is absorbed, it quickly shares its energy with the nucleons that were already present in the nucleus. The nuclear shape is altered from nearly spherical to a two-lobed form. The nucleus then divides at the waist, forming two daughter nuclei and releasing energy (and a few nucleons).

Today's nuclear models are a combination of the shell model and the liquid drop model. In such a combined model, the *core* consists of filled shells that contain a magic number of nucleons. The core is modeled as a liquid drop. The outer nucleons fit the shell model with its quantized states. They also interact with the core nucleons, causing the core to vibrate, rotate, and distort.

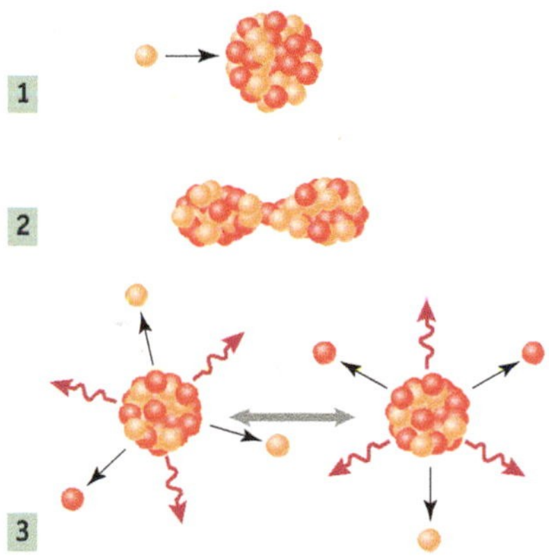
1
2
3

FIGURE 43.5 **1.** A particle, such as a neutron, is absorbed by a nucleus. **2.** The neutron's energy is shared by the other nucleons; the nucleus's shape is distorted as a result. **3.** The nucleus breaks into pieces; some nucleons are emitted and energy is released.

EXAMPLE 43.2 Carbon and Nitrogen

The most abundant isotope of carbon is Carbon-12, which has six protons and six neutrons, and the most abundant isotope of nitrogen is Nitrogen-14. However, for this problem consider Nitrogen-12, which has seven protons and five neutrons. Fill in energy-level diagrams (Fig. 43.4) for both nuclei. In the **CHECK and THINK** step, answer the question: Which nucleus has more energy?

INTERPRET and ANTICIPATE

Start at the lowest energy level and fill in protons and neutrons, while following Pauli's exclusion principle.

Example continues on page 1420 ▶

SOLVE

Start with carbon. The six neutrons fill three boxes (spin up and spin down in each box), and likewise for the six protons (Fig. 43.6).

Follow the same procedure for nitrogen (Fig. 43.7).

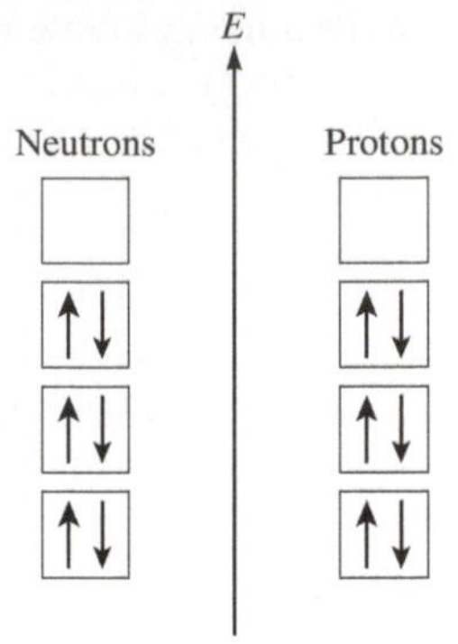

FIGURE 43.6 Nuclear energy levels in Carbon-12.

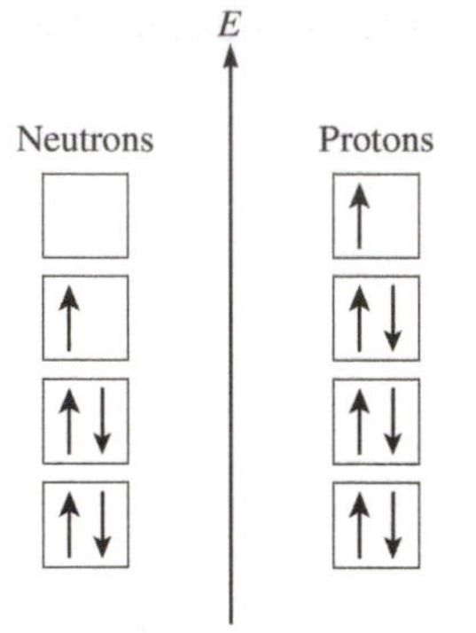

FIGURE 43.7 Nuclear energy levels in Nitrogen-12.

CHECK and THINK

Because the last proton in nitrogen is in the fourth box, nitrogen has more energy than carbon, whose highest energy nucleons are only in the third box. However, this isotope of nitrogen is unstable, so one of its protons will transform into a neutron through a process known as *inverse beta decay* (Section 43-5).

43-4 Radioactive Decay

In 1903, Antoine-Henri Becquerel (1852–1908) shared the Nobel Prize in physics with Pierre (1859–1906) and Marie Curie (1867–1934). Becquerel discovered spontaneous radioactivity, and the Curies investigated the resulting radiation. Becquerel had learned of Wilhelm Röntgen's discovery of X-rays and was investigating whether there is a connection between visible light and invisible radiation when he serendipitously discovered that uranium gives off penetrating radiation. Marie Curie named this phenomenon **radioactivity**. Ernest Rutherford then showed that radioactive substances emit more than one kind of radiation. Some radiation more readily penetrates a substance such as aluminum foil than other radiation. Rutherford named the less penetrating radiation **alpha rays** and the more penetrating radiation **beta rays**. In 1900, the French chemist Paul Villard (1860–1934) discovered an even more penetrating form of radiation known as **gamma rays**. When these three types of rays pass through a magnetic field, only the alpha and beta rays are deflected. From the direction of the deflection, we know that alpha rays are positively charged and beta rays are negatively charged. Gamma rays are not deflected, so they must be neutral.

ALPHA AND BETA RAYS
Special Case

The term *ray* is somewhat misleading. Today we know that X-rays and gamma rays are both photons (electromagnetic radiation), and that alpha rays and beta rays are also particles. **Alpha particles** are helium nuclei consisting of two protons and two neutrons. **Beta particles** are electrons. We continue to use the somewhat archaic term *ray*, but remember we mean a beam of particles. In this section we explain the origins of gamma rays and alpha particles. In Section 43.5, we'll take up beta particles.

Gamma Rays

GAMMA RAYS Special Case

The production of gamma rays is much like that of visible light or of X-rays emitted by an atom. An excited atom emits a photon when it transitions to a lower energy level. The energy of the photon equals the difference between the two energy levels. For example, the Balmer series of lines are due to photons in the visible part of the spectrum (Section 42-3). (And, as we saw in Section 42-8, atoms may generate X-rays.)

The same is true for the origin of nuclear gamma rays, except that a nucleus, instead of an atom, transitions to a lower energy level. The difference in nuclear energy levels is on the order of hundreds of keV to a few MeV. This is about 10^5 times the energy levels in the atom, so the photons released when a nucleon drops to a lower energy level are about 10^5 times more energetic. As a result, gamma rays are *typically* emitted.

Alpha Particles

ALPHA PARTICLES Special Case

Nucleons within a nucleus occasionally form themselves into a collection of two protons and two neutrons. This is an **alpha particle** or, equivalently, a helium nucleus. We already know that a helium nucleus is very stable; now we'll consider it as a particle on its own. We can model a nuclear system as having two components: an alpha particle and the rest of the nucleus. The alpha particle is positively charged, so it experiences Coulomb repulsion due to the other protons and strong force attraction due to the other nucleons. So the alpha particle is in a potential well much like the one shown in Figure 43.2B. The system's energy level is well below the peak of the potential energy curve (Fig. 43.8), so the alpha particle can only leave the nucleus by tunneling out of it. This is the reverse of the process that brings two protons together in the Sun, as discussed in the case study in Chapter 41 (Section 41-7). The probability of the alpha particle tunneling out of the nucleus depends on the height and width of the barrier.

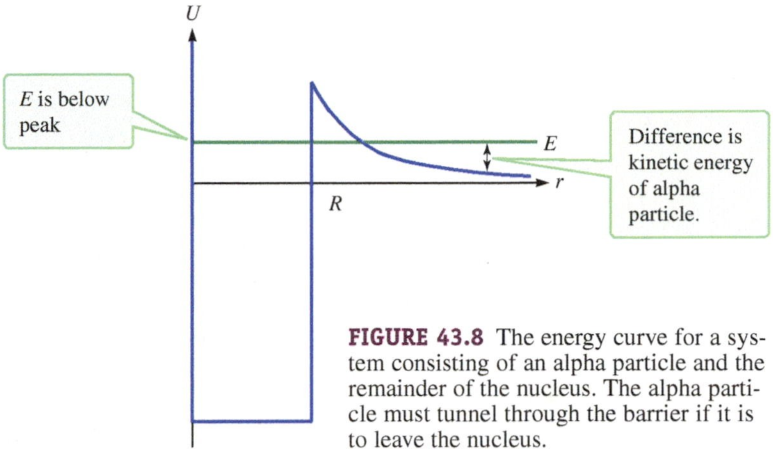

FIGURE 43.8 The energy curve for a system consisting of an alpha particle and the remainder of the nucleus. The alpha particle must tunnel through the barrier if it is to leave the nucleus.

The process of emitting an alpha particle is called **alpha decay**. The original nucleus is called the *p*arent nucleus and is symbolized by ${}^{A}_{Z}\text{P}$, where the superscript and subscript stand for the mass number A and the atomic number Z, respectively. When the parent nucleus undergoes alpha decay, it changes into a new nucleus called the *d*aughter nucleus, symbolized by D. We write alpha decay as:

$$ {}^{A}_{Z}\text{P} \rightarrow {}^{A-4}_{Z-2}\text{D} + {}^{4}_{2}\text{He} + \text{energy} \qquad (43.3)$$

P stands for *p*arent and D for *d*aughter.

The energy released in this reaction is (mostly) in the form of kinetic energy of the escaped alpha particle (Fig. 43.8).

Decay Rate

NUCLEAR DECAY Major Concept

When a parent nucleus decays by either giving off an alpha particle (this section) or beta particle (Section 43-5), it doesn't vanish; instead, it becomes a daughter nucleus. So when there is a radioactive substance made from a number of such parent nuclei, the substance itself doesn't vanish; instead, over time it turns into a sample of the daughter substance. According to quantum mechanics, the decay process is probabilistic. We cannot know precisely when a particular nucleus will decay. Instead, we can say that each nucleus has a probability $\eta\, dt$ of decaying in time dt, where the lowercase Greek letter eta (η) is the **decay constant** with dimensions $[\![\text{T}^{-1}]\!]$. In a sample of N nuclei, the average number decaying in time dt is $dN = N\eta\, dt$. Thus, the number of parent nuclei in the sample changes at a rate of

$$\frac{dN}{dt} = -\eta N \qquad (43.4)$$

The letter N stands both for the neutron number and for the number of parent nuclei. You must discern the meaning from the context.

where the negative sign indicates that the number of parent nuclei is decreasing. The absolute value of the decay rate is called the **activity** a, so $a = -dN/dt$. In Problem 26 you will integrate Equation 43.4 to show that the number of parent nuclei decreases exponentially:

$$N = N_0 e^{-\eta t} \qquad (43.5)$$

where N_0 is the number of parent nuclei at time $t = 0$.

The activity as a function of time comes from taking the derivative of Equation 43.5:

$$a = a_0 e^{-\eta t} \qquad (43.6)$$

where $a_0 = \eta N_0$ is the activity at $t = 0$ (Problem 27). The SI unit of activity is the **becquerel** (Bq); however, activity is often reported in **curies** (Ci), where

$$1\,\text{Ci} = 3.7 \times 10^{10}\,\text{Bq} = 3.7 \times 10^{10}\,\text{decays/s}$$

It is often convenient to replace the decay constant η with either the half-life $T_{1/2}$ or the time constant τ. The **half-life** is the time it takes the number of nuclei in the sample to decrease to half that number, so at $t = T_{1/2}$

$$\frac{N}{N_0} = \frac{1}{2} = e^{-\eta T_{1/2}}$$

In Problem 28, you will show that the half-life is given by

$$T_{1/2} = \frac{\ln 2}{\eta} \approx \frac{0.693}{\eta} \tag{43.7}$$

The half-life of a small sample of isotopes is given in Table 43.1. The **time constant** is defined as $\tau \equiv 1/\eta$, so at $t = \tau$ the number of parent nuclei in the sample is given by:

$$N = N_0 e^{-\eta/\eta} = \frac{N_0}{e} \approx 0.368 N_0$$

This means that when $t = \tau$, the number of nuclei remaining is $0.368N_0$. Each radioactive isotope is characterized by its own decay constant, or equivalently, by its half-life or by its time constant. The number of parent nuclei in the same isotope at time t may be written in terms of its half-life $T_{1/2}$ (Problem 29) or time constant τ as

$$N = N_0\left(\frac{1}{2}\right)^{t/T_{1/2}} = N_0 e^{-t/\tau} \tag{43.8}$$

TABLE 43.1 Half-life for selected nuclei

Isotope	Half-life	A	Z	Decay mode
^{12}B	0.0202 s	12	5	Beta decay
^{12}N	11.00 ms	12	7	Inverse beta decay and alpha decay
^{14}C	5730 yr	14	6	Beta decay
^{16}N	7.13 s	16	7	Beta decay
^{40}K	1.25×10^9 yr	40	19	Beta decay
^{57}Co	271 d	57	27	Electron capture
^{131}I	8.04 d	131	53	Beta decay
^{222}Rn	3.82 d	222	86	Alpha decay
^{226}Ra	1600 yr	226	88	Alpha decay
^{238}U	4.46×10^9 yr	238	92	Alpha decay

Radioactive Dating

When a sample contains parent nuclei that decay with some known half-life, it is possible to measure the relative amount of parent nuclei to find the sample's age. Different isotopes are used, depending on the approximate age of the sample. For example, potassium is found in rocks and soil, and the isotope ^{40}K has a half-life of 1.25×10^9 yr. It decays to a stable isotope of argon, ^{40}Ar. By measuring the ratio of ^{40}K/^{40}Ar, it is possible to estimate the ages of rock and soil samples that formed between roughly 10^5 and 4.5×10^9 (age of the Earth) years ago.

Carbon dating is another example of radioactive dating. A naturally occurring stable isotope of carbon is ^{12}C. Radioactive ^{14}C with a half-life of 5730 ± 40 yr is produced in the Earth's atmosphere when cosmic rays collide with nitrogen. A constant ratio of ^{14}C/^{12}C is maintained in the atmosphere due to the balance between radioactive decay and the production of new ^{14}C. Both isotopes of carbon have the same chemical behavior, so both form the carbon dioxide molecule CO_2. All living organisms absorb CO_2, so all living organisms maintain ^{14}C and ^{12}C in the constant ratio of $^{14}C/^{12}C = 1.3 \times 10^{-12}$, found in the atmosphere. When these organisms die, they stop absorbing CO_2, and any absorbed ^{14}C decays without being resupplied. The ratio $^{14}C/^{12}C$ decreases over time in the organism's remains, owing to the decay of ^{14}C. Carbon dating is used to measure ages between about 500 and 50,000 yr.

CONCEPT EXERCISE 43.3

A group of college students discusses a homework problem.

Shannon: It says that iodine-131 has a half-life of 8 days. We have to find what fraction of the sample is still iodine-131 after 16 days.

Cameron: That's one I can answer. None. In the first 8 days half the sample decays, and in the second 8 days the other half decays.

Shannon: No. The answer depends on how much iodine-131 was in the sample to begin with. You assumed the whole sample was iodine-131 at the beginning of the 16 days. I don't think they gave us enough information to do the problem numerically. I think it is a conceptual problem, and we are just meant to explain that not enough information is given.

Avi: I don't think this is a conceptual problem that we do in our heads. I think we can come up with a numerical answer. I agree that the answer isn't zero because every eight days the amount of iodine-131 is halved; but that doesn't mean it is all gone in 16 days.

Shannon: But that means I'm right about needing to know how much iodine-131 there was at the beginning of the 16 days.

With whom do you agree, and why?

EXAMPLE 43.3 An Old Rock

Another isotope used in radioactive dating is ^{238}U, which decays into ^{206}Pb and has a half-life of 4.46×10^9 yr. Suppose a rock contains a ratio of lead-206 to uranium-238 nuclei of $N_{Pb}/N_U = 0.65$. How old is the rock? (Assume the rock was originally all uranium-238.)

INTERPRET and ANTICIPATE

Because a substantial amount of the uranium has decayed into lead, we expect that the age of the rock will be close to the uranium half-life of several billion years.

SOLVE

At $t = 0$, the sample contained N_0 nuclei of uranium. Today's number N_U is given by Equation 43.8.

$$N_U = N_0\left(\frac{1}{2}\right)^{t/T_{1/2}} \quad (43.8)$$

The number of lead nuclei N_{Pb} in the rock today is the number of original uranium nuclei minus the number of uranium nuclei today.

$$N_{Pb} = N_0 - N_U \quad (1)$$

Substitute Equation 43.8 into Equation (1).

$$N_{Pb} = N_0 - N_0\left(\frac{1}{2}\right)^{t/T_{1/2}}$$

$$N_{Pb} = N_0\left[1 - \left(\frac{1}{2}\right)^{t/T_{1/2}}\right] \quad (2)$$

Example continues on page 1424 ▶

Use Equations 43.8 and (2) to write a ratio for N_{Pb}/N_U.	$\frac{N_{Pb}}{N_U} = \frac{N_0[1 - (\frac{1}{2})^{t/T_{1/2}}]}{N_0(\frac{1}{2})^{t/T_{1/2}}}$ $\frac{N_{Pb}}{N_U} = \left(\frac{1}{2}\right)^{-t/T_{1/2}} - 1$
Substitute values and solve for t.	$0.65 = \left(\frac{1}{2}\right)^{-t/T_{1/2}} - 1$ $\left(\frac{1}{2}\right)^{-t/T_{1/2}} = 1.65$ $t = \frac{T_{1/2} \ln 1.65}{\ln 2} = \frac{(4.46 \times 10^9 \text{ yr}) \ln 1.65}{\ln 2}$ $t = 3.2 \times 10^9 \text{ yr}$

CHECK and THINK

As expected, our answer is several billion years. When working such a problem, you shouldn't expect to find rocks that are older than the age of the Earth (about 4.5 billion years).

43-5 The Weak Force

BETA PARTICLE Special Case

Beta decay is another way for an unstable isotope to decay. In **beta decay**, a neutron becomes a proton, and in the process an electron is emitted. The electron is called a **beta particle** in this context. Consider ^{12}B, which has seven neutrons and five protons (Fig. 43.9). If ^{12}B undergoes beta decay, its neutron number is reduced by 1, its atomic number increases by 1, and its mass number remains unchanged. Because the result has a new atomic number, it is a different element: in this case ^{12}C. Carbon-12 is more tightly bound than boron-12, so it is stable. To explain beta decay, we must discuss an elementary particle known as a **neutrino**, and the fourth fundamental force: the **weak nuclear force**.

Neutrinos and the Weak Force

Rutherford named beta rays in the early 1900s, so in a sense beta decay was discovered then. However, later studies found problems with the theory of beta decay. It seemed in particular that energy was not conserved by beta decay; instead, it appeared that energy was lost. The supporting data was so convincing that Bohr proposed that perhaps energy was really not conserved in this situation, violating one of the most fundamental principles of physics.

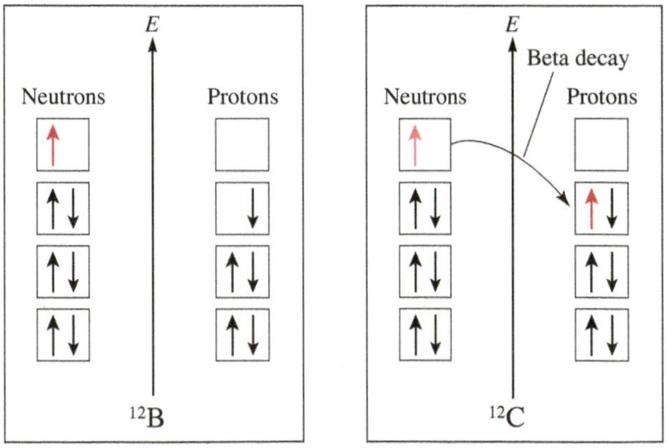

FIGURE 43.9 In the beta decay of ^{12}B, one of boron's neutrons becomes a proton. The result is carbon. Both nuclei have the same mass number A.

Pauli and Fermi responded with a solution to the problem, proposing the existence of another particle emitted during beta decay that carries the missing energy. Fermi called this particle a *neutrino*. He reasoned that the particle must be neutral, or the reaction would have a charge imbalance, and that the particle must have low mass, or it would have been easy to discover. So he gave it a name that means "little neutral one" in Italian. The neutrino also has an antiparticle, known as an **antineutrino**. The antineutrino was discovered in the mid-1950s by a team of American physicists led by Frederick Reines (1918–1998) and Clyde Cowan (1919–1974), both shown in Figure 43.10. (Reines shared the Nobel Prize with Martin Perl (1927–2014) for this work in 1995; the Nobel Prize is not awarded posthumously, so Cowan was not eligible. Perl was awarded the prize for his discovery of the tau lepton.) The symbol for the neutrino is the lowercase Greek letter nu (ν), whereas the antineutrino is designated by a bar over the symbol ($\bar{\nu}$). So we write beta decay as

$$n \rightarrow p + e^- + \bar{\nu} \qquad (43.9)$$

The number of particles and antiparticles is balanced. There is one particle on the left. On the right there are two particles and one antiparticle; one particle is balanced by one antiparticle. The net result is a matter particle on each side of the reaction.

WEAK NUCLEAR FORCE

Underlying Principle

The beta decay reaction (Eq. 43.9) says that a neutron is transmuted into a proton. This almost sounds like alchemy, a pseudoscience based on the medieval belief that cheap metal could be transmuted into gold. But beta decay has been observed, and alchemy has not. So we need an explanation for how neutrons can be transformed into protons. The answer involves another fundamental force of nature—the **weak force** (or the **weak nuclear force**), which is short-ranged (about 10^{-3} fm) and is exerted on quarks and neutrinos. It transforms a neutron into a proton at the level of quarks, so in particular it turns a down quark into an up quark. (It can also turn an up quark into a down quark, which transforms a proton into a neutron.) The weak force is so named because it is about 10^7 times weaker than the strong force. Any process that converts one type of quark into another involves the weak force.

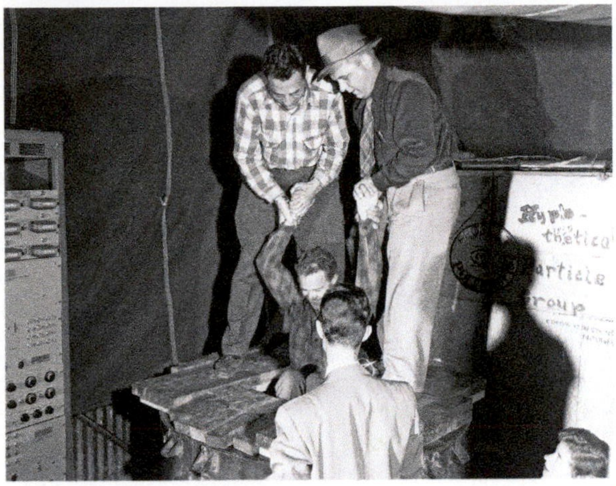

Historical/Corbis

FIGURE 43.10 Frederick Reines and Clyde Cowan worked on the Poltergeist project—an experiment to confirm the existence of the neutrino. Here Reines (left) and Cowan (in hat) demonstrate their methods in searching for the existence of the neutrino, using one of their colleagues.

The neutrino is a neutral particle, so the electromagnetic force does not act on it. In addition, because it is not made of quarks, the strong force does not act on neutrinos. Neutrinos also have a very low mass, so they are only weakly affected by gravity. The only force that really affects neutrinos is the weak force. For this reason, neutrinos are extremely difficult to detect: They hardly interact with matter. Their mean free path through water is thousands of light-years. So although they are the most numerous particles in the Universe (see the CASE STUDY at the end of this section), billions of them pass through your body every second without leaving a trace.

Beta Decay, Inverse Beta Decay, and Electron Capture

Now let's examine three reactions that involve the weak force and neutrinos. In beta decay a neutron is transmuted into a proton, so the beta decay reaction for a parent nucleus with N neutrons and Z protons is:

$$^{A}_{Z}\mathrm{P} \rightarrow {}^{A}_{Z+1}\mathrm{D} + e^- + \bar{\nu} \qquad (43.10)$$

The beta decay reaction is also referred to as β^- decay because a negative beta ray (electron) is released.

The daughter nucleus has one more proton and one fewer neutron than its parent, but both nuclei have the same mass number A. Beta decay from a given type of isotope always yields the same amount of energy, but that energy can be split in an infinite number of ways between the emitted electron and antineutrino.

The weak force can also transmute a proton into a neutron, as given by:

$$p \rightarrow n + e^+ + \nu \text{ (inside nucleus)} \qquad (43.11)$$

Such a reaction is called **inverse beta decay**. Free protons do not decay, but a proton inside a nucleus can decay. For a parent nucleus with N neutrons and Z protons, the inverse beta decay reaction is:

$$^{A}_{Z}\mathrm{P} \rightarrow {}^{A}_{Z-1}\mathrm{D} + e^+ + \nu \qquad (43.12)$$

The inverse beta decay reaction is also referred to as β^+ decay because a positive beta particle (positron) is released.

The daughter nucleus has one fewer protons, one additional neutron, and the same mass number as its parent. In the case of inverse beta decay, a **positron** e^+ and a neutrino ν are released. A positron is the **antiparticle** of an electron; it's like the electron except that it has a positive charge $+e$. In Equation 43.12 there is one particle on the left. On the right there are two particles and one antiparticle; one particle is balanced by one antiparticle. So the net result is that there is one matter particle on each side of the reaction.

Nuclei that are rich in neutrons are likely to undergo beta decay, and atoms that are rich in protons are likely to undergo inverse beta decay. So beta decay and inverse beta decay act to stabilize nuclei and to even out the numbers of protons and neutrons. The half-lives of beta decay and inverse beta decay for nuclei vary greatly, ranging from a few minutes to tens of millions of years. A free neutron (outside the nucleus) will undergo beta decay in a short time ($T_{1/2} \approx 10.2$ min). However, free protons do not spontaneously undergo inverse beta decay into neutrons.

ELECTRON CAPTURE **Special Case**

Finally, inner electrons have a nonzero probability of being in the nucleus (Section 42-5). So it is possible that one of these electrons can be *captured* by a proton in the process called **electron capture**. This process is more likely to happen in high Z atoms because their inner electron shells are more tightly bound to the nucleus. The reaction for electron capture is:

$$^{A}_{Z}\text{P} + \text{e}^{-} \rightarrow {}_{Z-1}^{\ \ A}\text{D} + \nu \quad (43.13)$$

Electron capture is also called *K* capture because the innermost electrons are in the *K* shell.

The daughter nucleus has one more neutron and one fewer proton than its parent. After the inner electron is captured, an outer electron will fall into the resulting gap, releasing an X-ray photon. In principle, we can observe electron capture by observing the released neutrino in Equation 43.13. However, because of the difficulty observing neutrinos, we infer that electron capture has taken place when we observe the released X-ray photon.

Uranium: A Naturally Occurring Radioactive Element

The most abundant naturally occurring radioactive element on the Earth is uranium, $^{238}_{92}\text{U}$. Uranium decays in a series of reactions into lead, $^{206}_{82}\text{Pb}$. Figure 43.11 illustrates this series of reactions on a portion of a **Segrè chart**. The horizontal axis shows the neutron number and the vertical axis the atomic number. Alpha decay decreases both the neutron number and the atomic number by two, so on the chart it is represented by a diagonal line: two boxes down and two boxes to the left. For example, $^{238}_{92}\text{U}$alpha decays into $^{234}_{90}\text{Th}$ according to

$$^{238}_{92}\text{U} \rightarrow {}^{234}_{90}\text{Th} + {}^{4}_{2}\text{He}$$

as represented by the blue arrow between $^{238}_{92}\text{U}$ and $^{234}_{90}\text{Th}$. Beta decay lowers the neutron number by one and increases the atomic number by one, so it is represented on the chart by a diagonal line: one box to the left and one box up. For example, $^{234}_{90}\text{Th}$ beta decays into $^{234}_{91}\text{Pa}$ according to:

$$^{234}_{90}\text{Th} \rightarrow {}^{234}_{91}\text{Pa}^{*} + \text{e}^{-} + \bar{\nu}$$

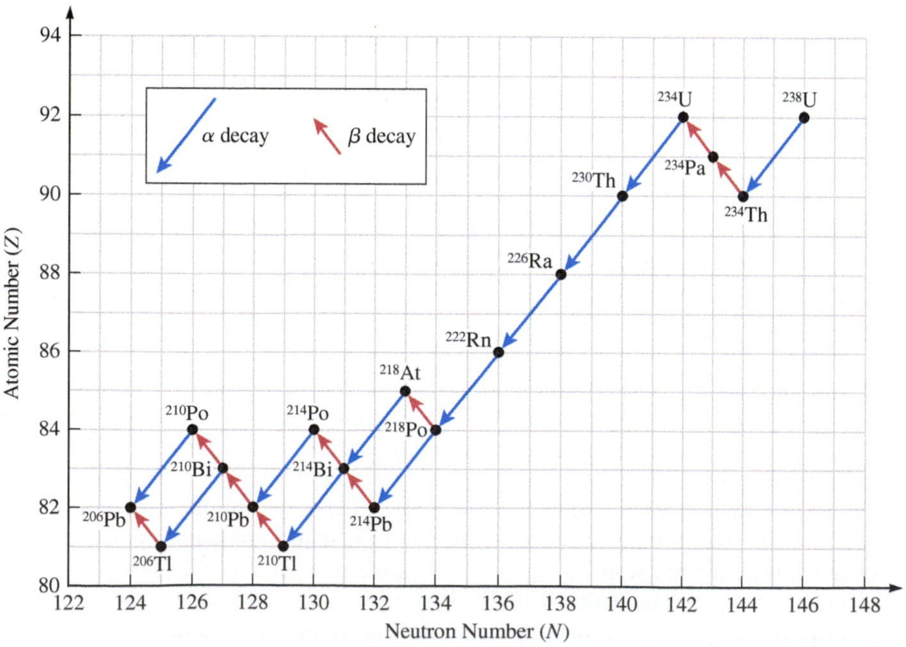

FIGURE 43.11 Uranium-238 is unstable and decays into lead-206, which is stable. Radon is part of this series of reactions. Radon gas is found in some homes, and is a hazard to human health.

as represented by the red arrow between $^{234}_{90}\text{Th}$ and $^{234}_{91}\text{Pa}$. After this beta decay, $^{234}_{91}\text{Pa}^*$ is an in excited state, indicated by the asterisk * next to its abbreviation. By emitting a gamma ray, $^{234}_{91}\text{Pa}^*$ drops down to its ground state:

$$^{234}_{91}\text{Pa}^* \rightarrow {}^{234}_{91}\text{Pa} + \gamma$$

The energy released due to the decay of uranium in the Earth heats up the Earth's interior. This process is partially responsible for plate tectonics, which in turn causes volcanic activity and earthquakes.

CONCEPT EXERCISE 43.4

Positron emission tomography (PET) is a technique used to see inside a living body. When a positron encounters an electron, the two annihilate and give off two gamma photons that are used to create the image. Fluorine-18 is used as the source of positrons. The fluorine decays to oxygen-18. Which type of decay process (beta, inverse-beta, or electron capture) produces the necessary positrons? Write down the equation for decay of fluorine-18 to oxygen-18.

CONCEPT EXERCISE 43.5

If ^{12}N (Fig. 43.7) undergoes inverse beta decay, what element is the daughter nucleus?

CASE STUDY Part 3: Making Hydrogen and Helium

Quarks in the early Universe were bound together to form protons and neutrons. Other high-energy particles were present too, such as electrons, positrons, neutrinos, and antineutrinos. These particles were kept in thermal equilibrium by the weak force through a few reactions:

$$\begin{gathered} \text{p} + \text{e}^- \rightleftarrows \text{n} + \nu \\ \text{p} + \bar{\nu} \rightleftarrows \text{n} + \text{e}^+ \end{gathered} \tag{43.14}$$

where the double arrow $\rightleftarrows$ means that the reactions happened in both directions. So protons were converted into neutrons, and neutrons into protons. When the Universe was just tens of seconds old, the positrons were annihilated by reactions with electrons. The number of electrons was greatly reduced, but not to zero. Likewise, the antineutrinos were annihilated by reactions with neutrinos; the number of neutrinos decreased, but not to zero. The remaining neutrinos still exist today and are believed to be the most numerous particles in the Universe. Without a great number of electrons, positrons, neutrinos, and antineutrinos, the reactions in Equation 43.14 could not maintain equilibrium. Instead, free neutrons began to undergo beta decay into protons (Eq. 43.9), with no compensating reaction to restore protons into neutrons.

The half-life of a free neutron via beta decay is about 10 minutes. However, neutrons in a nucleus have much longer half-lives. So many of the neutrons survived by combining with a proton, forming an isotope of hydrogen known as deuterium, $^{2}_{1}\text{H}$. In the first few minutes of the early Universe, much of the deuterium that formed was used up in making helium nuclei, $^{4}_{2}\text{He}$. A few other low-mass elements formed in trace amounts, but only hydrogen and helium nuclei were produced in great abundance as the Universe cooled. Once the temperatures were low enough, electrons began to bind to these nuclei to form atoms. Because no other nuclei were available, only hydrogen and helium atoms were present. So where did the iron in your blood and the iodine in your thyroid come from? We will explain that in Section 43-8.

EXAMPLE 43.4 CASE STUDY Evidence in Favor of the Big Bang Model

One convincing way to confirm the Big Bang model is by calculating the expected amount of primordial helium as a ratio $N_{He}/(N_H + N_{He})$. Helium began to form after sufficient time had passed for many neutrons to undergo beta decay. Roughly speaking, the theoretically calculated ratio of neutrons to protons was $N_n/N_p \approx 1/7$ when helium began to form. Assume that all the available neutrons went into helium and the remaining protons went into hydrogen, and estimate the ratio $N_{He}/(N_H + N_{He})$. Express your answer as a percentage.

INTERPRET and ANTICIPATE

A single-helium nucleus contains two protons and two neutrons. If we assume that all the available neutrons go into helium, then an equal number of protons also go into helium. The remainder becomes hydrogen. So if the number of neutrons equaled the number of protons, there would be no hydrogen at all.

SOLVE

Every helium nucleus contains 2 neutrons, so the number of helium nuclei that can form is half the number of neutrons. Since one proton must go into helium for every neutron that does so, the number of hydrogen nuclei is the number of protons minus the number of neutrons.

$$\frac{N_{He}}{N_{He} + N_H} = \frac{\frac{1}{2}N_n}{\frac{1}{2}N_n + (N_p - N_n)}$$

$$\frac{N_{He}}{N_{He} + N_H} = \frac{\frac{1}{2}N_n}{N_p - \frac{1}{2}N_n}$$

Divide the numerator and denominator by N_p.

$$\frac{N_{He}}{N_{He} + N_H} = \frac{\frac{1}{2}(N_n/N_p)}{1 - \frac{1}{2}(N_n/N_p)}$$

Substitute $\frac{N_n}{N_p} \approx \frac{1}{7}$.

$$\frac{N_{He}}{N_{He} + N_H} = \frac{\frac{1}{2}\left(\frac{1}{7}\right)}{1 - \frac{1}{2}\left(\frac{1}{7}\right)} = \frac{1}{13}$$

$$\frac{N_{He}}{N_{He} + N_H} \approx 8\%$$

CHECK and THINK

This result is very close to the observation that about 90% of the ordinary matter in the Universe is primordial hydrogen and about 10% is primordial helium. More careful calculations have been done that include other elements created in the early Universe, and careful observations have confirmed these calculations. Although there have been great changes to our understanding of the evolution of the Universe, these fundamental ideas about how elements formed, taken from the original Big Bang theory, have changed very little and are well supported by observations.

43-6 Binding Energy

FUSION AND FISSION

Major Concept

Nuclear reactors on the Earth supply us with some of the electrical power we need to run useful devices, such as our refrigerators, TVs, and computers. The energy produced by these reactors comes from nuclear *fission* reactions that break up large nuclei into smaller ones. The Sun is the ultimate source of energy in the solar system, and its core is essentially a nuclear reactor. However, the energy generated by the Sun's core is the result of *fusion* reactions that combine small nuclei into larger ones. In the following two sections, we take a closer look at fission and fusion reactions. In this section, we'll explain why both fission reactions and fusion reactions can release energy. First, we need to consider the energy stored in a nucleus.

MASS DEFICIT AND BINDING ENERGY

Major Concept

Mass Deficit and Binding Energy

Imagine assembling a bound system, such as a nucleus, from free particles such as nucleons. If you could do this, you would find that the mass of the bound system is *less* than the mass of the individual free particles. The resulting system has a **mass deficit** Δm, given by:

$$\Delta m = \sum m_i - M \qquad (43.15)$$

where $\sum m_i$ is the sum of the free particles' individual (rest) masses and M is the (rest) mass of the bound system. According to Einstein's theory of relativity, mass is a form of energy, as given by $E = mc^2$ (Eq. 39.44). The mass deficit that results when free particles are combined into a bound system is a form of energy known as the **binding energy** E_B. Mathematically, the binding energy is given by

$$E_B = \Delta mc^2 \qquad (43.16)$$

Some authors define binding energy with a negative sign: $E_B = -\Delta mc^2$, but in this book there is no negative sign in Equation 43.16.

Energy is required to break apart the bound system of the nucleus. The binding energy is the amount of energy, supplied by an outside source, required to pull the particles out of the bound system and separate them into free particles at rest. If more energy than the binding energy is supplied, the separated free particles would have some nonzero kinetic energy.

In some *fusion* reactions, the binding energy may be released. The released energy may go into some combination of particles, photons, and their kinetic energy. To understand this process, let's use an analogy. Suppose an asteroid is initially at rest and far (but not infinity far) from the Earth and then falls to the Earth. The asteroid–Earth system loses gravitational potential energy, and gains kinetic energy. Figure 43.12 shows the familiar energy bar chart for such a process. Let's think about this familiar process in a slightly different way. We can say that, initially, the system had energy stored as gravitational potential energy. Think of the asteroid's fall to Earth in terms of it "fusing" with the Earth to form a new complex body. (Keep in mind that this is an analogy. The asteroid was initially bound to the Earth.) Some of the stored energy is released in the process of "fusing" the asteroid and the Earth; this process is represented by the kinetic energy bar in Figure 43.12. Often we think of this released energy as going into increasing the asteroid's kinetic energy, but we can improve our analogy somewhat if think about other places where this released energy may go. For example, the released energy goes into increasing the atmosphere's thermal energy, light is emitted, and particles of debris fly into the air.

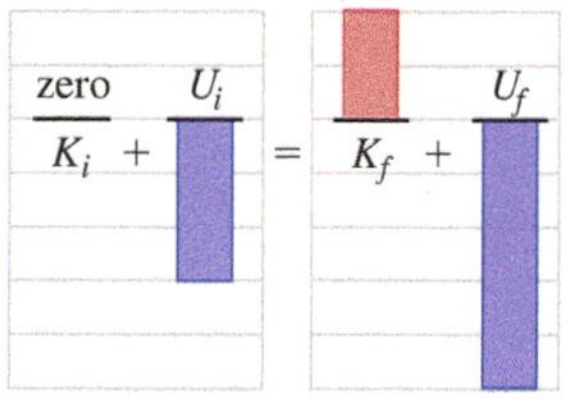

FIGURE 43.12 Energy bar chart for an asteroid-Earth system.

Our language here is somewhat loose. Remember: energy describes the motion, configuration and mass of a system. Energy is not a substance.

A similar process happens during the nuclear fusion of small nuclei. When such small reactants (free particles) are fused to form a larger product, the nucleons are bound together by the strong force, and an amount of energy equal to the binding energy must be added to separate the nucleus back into its original reactants. Such fusion reactions release energy. Consider two free particles having mass m_1 and m_2. Each particle stores energy in the form of mass, and according to Einstein's famous equation, these stored energies are m_1c^2 and m_2c^2. So the total energy stored in their mass is given by $(m_1 + m_2)c^2$. Figure 43.13A illustrates these ideas about the energy stored in their mass. Now suppose the two particles fuse together. In some fusion reactions, the reactants (the free particles) store more energy in their individual masses than the product stores in its mass M. This difference in stored mass energy is the binding energy (Fig. 43.13B), and the binding energy is released in the fusion reaction. Analogous to the energy released when an asteroid falls to Earth, the energy released by a fusion reaction can go into kinetic energy, but it can also go into new particles and photons.

But the analogy with gravity only goes so far. Gravity is an attractive force, and gravitational "fusion" always increases a system's binding energy. You would have to put energy into the bound asteroid-Earth system to separate them; separating the two is analogous to fission. So for gravity, "fusion" always releases stored energy and "fission" always requires a supply of energy. However, in the nucleus there are two competing forces—the attractive strong nuclear force and the repulsive Coulomb force. So some nuclear *fusion* reactions release energy and others require a supply of energy. Similarly, some nuclear *fission* reactions release energy and other require it.

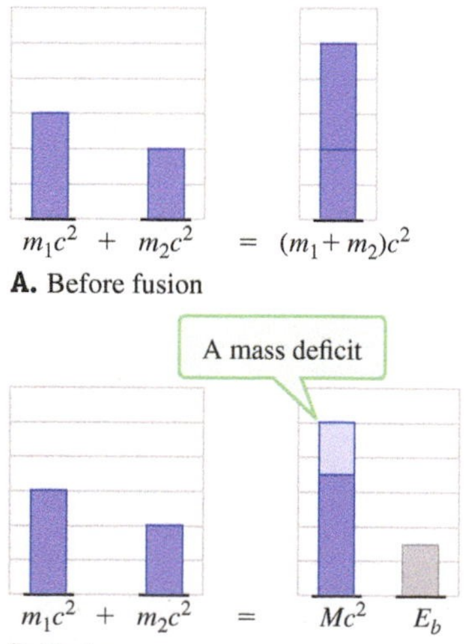

FIGURE 43.13 **A.** The total rest mass energy of two particles. **B.** In this fusion reaction, the sum of the rest mass energy of the reactants is greater than the rest mass energy of the product: $(m_1 + m_2)c^2 > Mc^2$.

The repulsive force tends to be more important in larger nuclei. If a large nucleus undergoes a *fission* reaction that breaks it into two smaller nuclei, the products may have a greater binding energy than the original nucleus. In this case, it would take energy from an outside source to fuse these smaller nuclei back together to form a larger nucleus.

Putting this together, in either a fusion or fission reaction, energy may be released or be required. If there is a positive mass deficit such that the product has less mass than the sum of the reactant's masses, then energy is released. In other words, energy is released when the reaction's products have a greater binding energy than its reactants. The following may help you to remember: *A reaction that increases binding energy releases energy.* Similarly, energy must be supplied by an external source when the reaction's products have less binding energy (more mass) than its reactants. *A reaction that decreases binding energy requires energy.*

Convenient Units of Mass and Binding Energy

We use the term *pseudo-unit* here because MeV/c^2 is a combination of unit for energy and the speed of light constant.

Because of the intimate connection between mass and energy, it is often convenient to express rest mass in the *pseudo-units* of MeV/c^2. It is also common to express the rest mass of a nucleon in atomic mass units. The translation between these two sets of units can be found as follows. The conversion factor between atomic mass units and kilograms comes from Equation 43.1 ($1\ \text{u} = 1.66053879 \times 10^{-27}$ kg), and the rest mass energy of 1 u is found by applying $E = mc^2$ (Eq. 39.44):

$$E = (1\ \text{u})c^2 = (1.66053879 \times 10^{-27}\,\text{kg})(2.99792458 \times 10^{8}\,\text{m/s})^2$$
$$= 1.492417837 \times 10^{-10}\,\text{J}$$

This is more conveniently expressed in MeV:

$$E_B = \frac{1.492417837 \times 10^{-10}\,\text{J}}{1.602176565 \times 10^{-13}\,\text{J/MeV}} = 931.494\,\text{MeV}$$

So 1 u of mass is equivalent to 931.494 MeV (rounded to six significant figures for convenience) of energy, and the translation can be written as:

It is common to drop the "/c^2" and report rest mass in units of MeV.

$$1\ \text{u} = 931.494\,\text{MeV}/c^2 \tag{43.17}$$

For convenience, the rest mass energies of the proton, the neutron, and a few nuclei are provided in Table 43.2.

The Binding Energy Curve

To compare the properties of various nuclei, it is helpful to calculate the binding energy per nucleon ε by dividing the binding energy of a nucleus by its mass

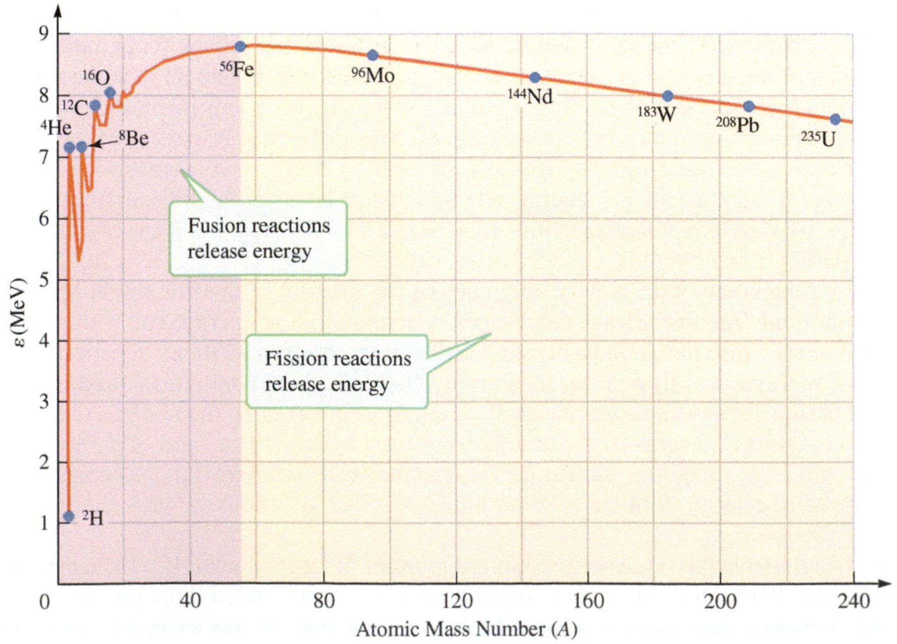

FIGURE 43.14 Binding energy per nucleon ε as a function of mass number A. The curve peaks at iron ^{56}Fe .

number: $\varepsilon \equiv E_B/A$. A graph of ε as a function of A for the known nuclei forms the **binding energy curve** (Fig. 43.14). This curve rises steeply from deuterium up to around oxygen. Then it peaks at $A \approx 56$, corresponding to iron, after which it becomes relatively flat for heavier nuclei. The flat shape of the binding energy curve is evidence that the strong force is short-ranged, only extending far enough that a nucleon is bound by the few nucleons that are relatively close to it and not by those farther away.

Because the curve is very flat, we should more precisely say that iron is *roughly* at the peak. We won't worry about such fine distinctions in our language here.

A nucleus with high binding energy per nucleon is tightly bound; it will take considerable energy per nucleon to separate it into free particles. Iron is the most tightly bound nucleus, but there are other local peaks in the binding energy curve, most notably ${}^4_2$He, ${}^8_4$Be, ${}^{12}_6$C, and ${}^{16}_8$O. These nuclei are more tightly bound than are their neighbors on the binding energy curve. We have already seen that such nuclei are particularly stable, having the same number of protons and neutrons. Each can be modeled as one or more alpha particles.

The binding energy curve explains why some fusion and fission reactions release energy and why others require energy. *When nuclear reactions result in a product that has a greater binding energy and so is more tightly bound than its reactants, energy is released.* (Recall our analogy: Energy is released when an object such as an asteroid falls to the Earth.) Otherwise, energy would be required to make the reaction possible.

Because iron is at the peak of the binding energy curve, both the fusion of nuclei that are lighter (to the left of the peak) and the fission of nuclei that are heavier (to the right of the peak) release energy. So the fission of nuclei lighter than iron into even lighter nuclei requires the input of energy. Likewise, the fusion of nuclei heavier than iron into even heaver nuclei also requires the input of energy.

TABLE 43.2 Rest mass of selected particles and nuclei

Nucleon or Nucleus	m (in MeV/c^2)
Proton	938.27
Neutron	939.57
Deuteron ${}^2_1$H	1875.612859
Helium ${}^4_2$He	3727.379
Carbon ${}^{12}_6$C	11177.9
Oxygen ${}^{16}_8$O	14899.2
Iron ${}^{56}_{26}$Fe	52103.06
Copper ${}^{63}_{29}$Cu	58603.84
Gold ${}^{197}_{79}$Au	183433.33
Lead ${}^{208}_{82}$Pb	193687.68
Uranium ${}^{238}_{92}$U	221696.64

Energy Released by Reactions

The energy released in a nuclear reaction, including the decay processes we have discussed, depends on the change in mass $\Delta M = M_f - M_i$, where M_i is the total mass of the reactants before the reaction (such as nuclei, electrons, and positrons), and M_f is the total mass of the products that result from the reaction. The energy released is then,

In this book we ignore the mass of neutrinos, but strictly speaking they should be taken into account.

$$E_{\text{released}} = -\Delta M c^2 \qquad (43.18)$$

This quantity is positive as long as the mass of the reactants before the reaction is greater than the mass of the products that result. If this isn't the case, then energy must be provided to make the reaction possible. So a spontaneous reaction will occur only if $\Delta M > 0$. For example, alpha decay is possible when the mass of the original parent nucleus is greater than the sum of the masses of the daughter nucleus and helium-4 (the alpha particle). When using Equation 43.18, you may need to know the mass of the electron or the positron in atomic mass units; each of them has mass $m_e = 5.48579909 \times 10^{-4}$ u.

CONCEPT EXERCISE 43.6

Why are the units MeV/c^2 referred to as *pseudo*-units? Why call Equation 43.17 a translation instead of a conversion factor? *Hint*: Why did we consider 1 kg = 2.2 lb to be a translation instead of a conversion factor (Section 5-7, page 131)?

CONCEPT EXERCISE 43.7

The ground-state energy of hydrogen is −13.6 eV. What is its binding energy? What is another name for binding energy in this case?

EXAMPLE 43.5 Binding Energy of a Deuteron

A deuteron is the nucleus of deuterium (an isotope of hydrogen, ${}^{2}_{1}H$), consisting of one proton and one neutron. Use the information in Table 43.2 to find the binding energy of a deuteron. Use no more than six significant figures in your work.

INTERPRET and ANTICIPATE

The mass of a deuteron is less than the mass of a proton plus a neutron. The mass deficit is proportional to the binding energy.

SOLVE

First, find the sum of the masses of the proton and neutron.

$$m_p + m_n = (938.27 + 939.57)\ \text{MeV}/c^2$$
$$m_p + m_n = 1877.84\ \text{MeV}/c^2$$

Next, find the mass deficit (Eq. 43.15).

$$\Delta m = \sum m_i - M \qquad (43.15)$$
$$\Delta m = (m_p + m_n) - m_{\text{deuteron}} = 1877.84\ \text{MeV}/c^2 - 1875.61\ \text{MeV}/c^2$$
$$\Delta m = 2.23\ \text{MeV}/c^2$$

Multiply by c^2 to find the binding energy. Now we see why these pseudo-units are so convenient—the c^2 cancels out.

$$E_B = \Delta mc^2 = (2.23\ \text{MeV}/c^2)c^2$$
$$E_B = 2.23\ \text{MeV}$$

CHECK and THINK

This may not seem like much energy at first, but compare it to the energy required to ionize hydrogen (13.6 eV). The binding energy of a nucleus is about 10^5 times greater than the binding energy of an atom.

EXAMPLE 43.6 What Happens to Cobalt-57?

The rest mass of ${}^{57}_{27}Co$ is 56.936296 u, and the rest mass of ${}^{57}_{26}Fe$ is 56.935399 u. Can ${}^{57}_{27}Co$ undergo spontaneous electron capture to become ${}^{57}_{26}Fe$? If so, how much energy is released? Give your answer in MeV to three significant figures. To keep this simple ignore the neutrino.

INTERPRET and ANTICIPATE

We need to know if the mass of the reactants is greater than or less than the mass of the products. It is helpful to write the electron capture reaction: ${}^{57}_{27}Co + e^- \rightarrow {}^{57}_{26}Fe + \nu$.

SOLVE

First, find the sum of the masses of the electron and ${}^{57}_{27}Co$. Use $m_e = 5.48579909 \times 10^{-4}$ u for the electron's mass.

$$m_e + m_{Co} = 5.48579909 \times 10^{-4}\ \text{u} + 56.936296\ \text{u}$$
$$m_e + m_{Co} = 56.936845\ \text{u}$$

Compare the mass of the reactants to that of the products.

$$m_e + m_{Co} > m_{Fe}$$

Electron capture is possible.

Find the difference in mass and multiply by c^2 to find the energy released.

$$\Delta M = 56.936845\ \text{u} - 56.935399\ \text{u} = 1.4460 \times 10^{-3}\ \text{u}$$
$$E_{\text{released}} = \Delta Mc^2 = 1.4460 \times 10^{-3}\ \text{u}\left(\frac{931.494\ \text{MeV}/c^2}{1\ \text{u}}\right)c^2$$
$$E_{\text{released}} = 1.35\ \text{MeV}$$

CHECK and THINK

This released energy goes into the total energy of the neutrino and the kinetic energy of the iron. The neutrino travels outward without interacting further, but the iron can interact with other particles, sharing this energy.

43-7 Fission Reactions

FISSION Major Concept

Perhaps you remember the Great Sendai Earthquake on March 11, 2011. It began off the northern coast of Japan's main island. Later it caused large tsunamis, triggering the second worst nuclear accident at a nuclear power plant. The power plant—Fukushima Daiichi—melted down and released radiation because the tsunami damaged the backup generators and the loss of power meant the cooling systems could not operate. You might think that such tragedies mean that nuclear power plants are too dangerous. In fact, some nations (such as Germany) shut down many of their power plants after the meltdown at Fukushima Daiichi. However, when you consider how carbon emission causes climate change (CASE STUDY in Chapter 20), you might consider nuclear power plants a "greener" source of energy than power plants that use fossil fuels. So other nations (such as France and the United States) are keeping their existing nuclear plants running, and some nations (such as China) are building many new ones. In this section, we consider the fission reactions and some of the practical concerns about generating energy with these reactions.

Germany's carbon emission per capita is actually pretty low.

In a fission reaction, a heavy nucleus splits into smaller daughter nuclei. From the previous section, we know that energy is released if the combined mass of the daughter nuclei is less than that of the parent nuclei. In other words, there must be a positive mass deficit (an increase in binding energy) in order to release energy in the fission reaction.

Although other parent nuclei are possible, we'll consider uranium-235 to make our discussion concrete. The fission reaction is triggered by the absorption of a neutron (Fig. 43.15). Because neutrons are neutral, they don't experience a Coulomb repulsion and can easily penetrate a nuclei. The addition of the neutron means that uranium-235 becomes uranium-236:

$$ {}^{1}_{0}\text{n} + {}^{235}_{92}\text{U} \rightarrow {}^{236}_{92}\text{U}^{*} $$

Although the half-life of uranium-235 is many millions of years, the half-life of uranium-236 is very short and it only lasts for about a picosecond before it splits into two smaller daughter nuclei and releases more neutrons. There are many possible daughter nuclei. For example,

$$ {}^{236}_{92}\text{U}^{*} \rightarrow {}^{141}_{56}\text{Ba} + {}^{92}_{36}\text{Kr} + 3\left({}^{1}_{0}\text{n}\right) $$

Or, as another example,

$$ {}^{236}_{92}\text{U}^{*} \rightarrow {}^{147}_{57}\text{La} + {}^{87}_{35}\text{Br} + 2\left({}^{1}_{0}\text{n}\right) $$

In these two reactions we see that 2 or 3 neutrons are released. In fact, on average 2.5 neutrons are released. It make sense that a net number of neutrons are released in a fission reaction because, in general, larger nuclei require a higher neutron-to-proton ratio for stability (Fig. 43.1), so the smaller daughter nuclei don't require as many neutrons (per proton) and these extra neutrons are released. These released neutrons can trigger more fission reactions, leading to a chain reaction (Fig. 43.15).

The daughter nuclei typically have atomic mass numbers between 70 and 160 (Fig. 43.16). So, they are about half the size of the uranium nucleus. The most probable values are 96 and 135. The binding

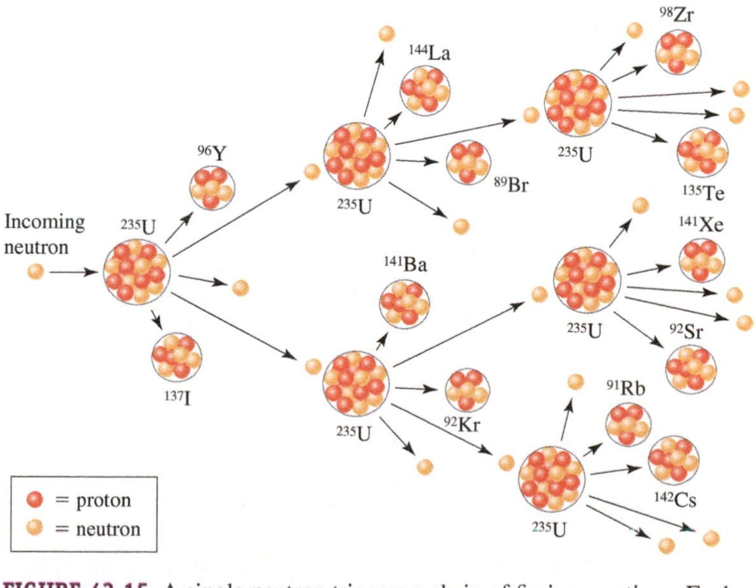

FIGURE 43.15 A single neutron triggers a chain of fission reactions. Each fission reaction releases more neutrons that trigger more fission reactions.

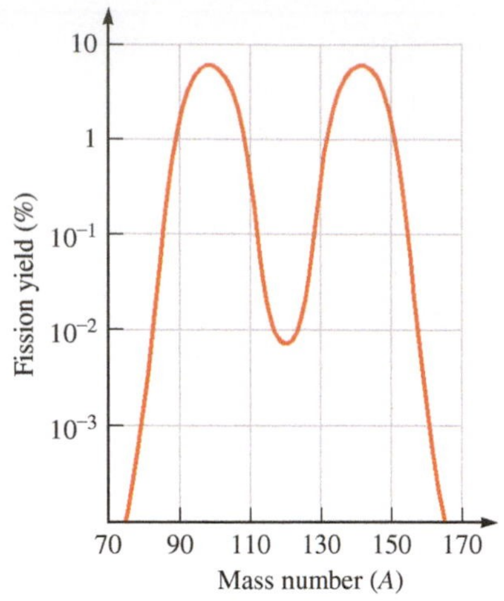

FIGURE 43.16 When uranium-235 undergoes neutron-induced fission, many possible daughter nuclei are produced. This figure roughly shows the distribution of these daughter nuclei as a function of mass number *A*.

energy per nucleon for uranium-235 is $\varepsilon_U \approx 7.6$ MeV. The daughter nuclei are more tightly bound, with a binding energy per nucleon that is nearly 1 MeV greater; so about 1 MeV per nucleon is released by each fission reaction. Uranium-235 has 235 nucleons, so each fission reaction releases nearly 235 MeV. A better estimate is about 200 MeV released per fission reaction. This released energy goes into producing gamma rays and into the kinetic energy of the reactants.

Furthermore, the daughter nuclei are generally not stable because they have a high neutron-to-proton ratio. So either they undergo beta decay (a neutron is transformed into a proton), or a neutron is emitted. This beta decay releases energy but not as much as the fission reaction itself. So, on the one hand, these daughter nuclei can contribute to energy production, but on the other hand they can slow the chain reaction by capturing neutrons, which would otherwise be contributing to it.

Nuclear Power Plants

Nuclear fission reactions have been put to practical use in generating energy that is used to power ships, submarines, spacecraft, cities and towns. Let's take a look at some of the practical aspects of a nuclear power plant used to produce electricity. The power plant must be able to sustain and control fission chain reactions such as those shown in Figure 43.15. The first challenge is finding the right fuel source. For example, the natural abundance of uranium-235 is 0.7%, and the rest (99.3%) is uranium-238. Uranium-238 tends to absorb neutrons without a subsequent fission reaction. So naturally occurring uranium is difficult to use as a fuel source in a power plant. One solution to this difficulty is to *enrich* the uranium fuel, such that uranium-235 achieves an artificial abundance of a few percent. Although many power plants use other fuel sources, we'll discuss enriched uranium to make our discussion specific. But the general ideas apply to other power plant designs as well.

Figure 43.17 shows the basic design of a nuclear power plant. Figure 43.17A shows the reactor core. The reactor contains a *moderator*—often liquid water. One of the moderator's jobs is to slow down the free neutrons. Slower neutrons are more likely to interact with nuclei and cause a fission reaction. *Fuel rods* (comprised of enriched uranium) are placed in the moderator, and these rods may be replaced when the fuel is exhausted. In addition to the fuel rods, there are *control rods* made of a material such as cadmium or boron, which is good at absorbing free neutrons. Both the control rod and the fuel rods may be inserted or removed to control the rate of fission reactions. For example, if the reaction rate is too high, inserting control rods to absorb neutrons can slow down the reaction rate.

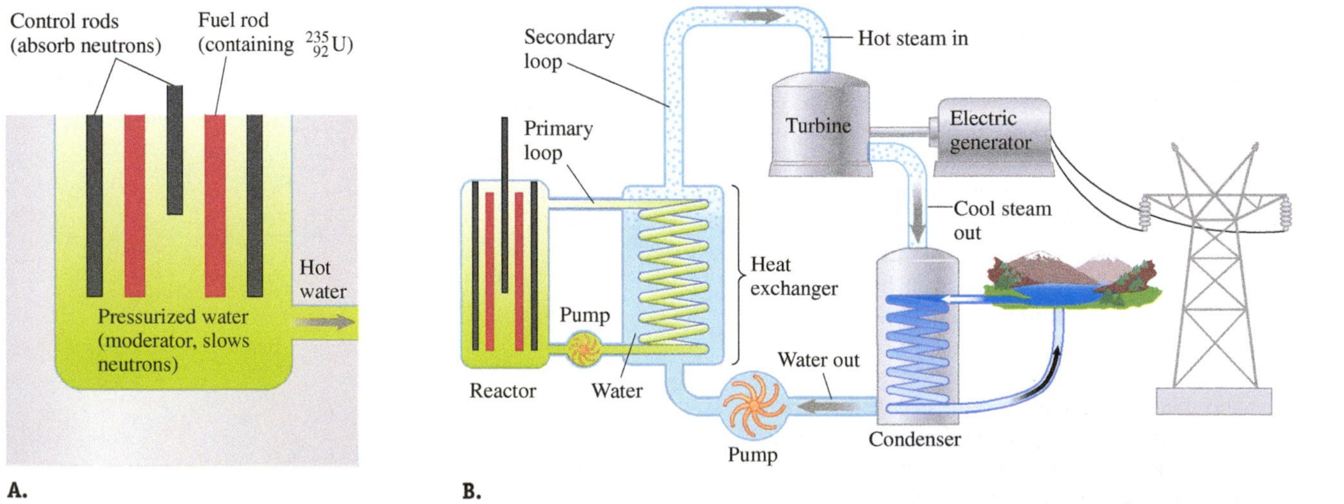

FIGURE 43.17 **A.** Inside the reactor core are fuel rods and control rods in a water bath. These rods are removable. **B.** The water is heated and pumped into a heat exchanger. The hot water heats water in a secondary loop. The water in the secondary loop becomes steam, which turns the turbines that generate electricity. The steam loses energy and cools. The cool steam is condensed by cold water.

Let's take a close look at the moderator's other job. For the purposes of our discussion, we'll assume the moderator is water. The fission reactions cause this water to heat up, and the circulation of this water results in the production of electricity. The water in the primary loop is kept at high pressure so that it doesn't become steam. This hot liquid water from the reactor circulates through its own closed primary loop so that it doesn't contaminate water in a secondary loop, which contains water as shown in Figure 43.17B. The hot water from the reactor heats the water in the secondary loop in the heat exchanger. The result of our discussion does not depend on the fact that the power plant has a nuclear reactor; it could also be a fossil fuel power plant. The water in the secondary loop boils into steam. The steam drives the blades of the turbine. In Section 32-6, we learned that the basis of an AC electrical generator involves turning coils of wire in a magnetic field (Fig. 32.21, page 1027). So the rotational motion of the turbine is used to turn wire coils and produce electricity. Next, cool water is pumped through a third loop in the condenser. This cool water is used to condense the steam in the secondary loop back into liquid water so that it can circulate back through the heat exchanger. As a result, the water in the third loop is heated. Typically, power plants are built near a natural body of water, and the water in the third loop comes from that natural source. The result is that power plants heat the natural body of water, which can have harmful consequences for organisms that live in that water. So, many newer power plants do not return hot water directly to the natural body of water.

Nuclear fission power plants have other environmental and public health risks. When the fuel rods are no longer effective, they are replaced. However, the "spent" rods are still radioactive and considered toxic for centuries. Safely storing these spent rods is a major challenge. In a disaster, the chain reaction in the reactor could run away leading to a meltdown, and even an improperly functioning reactor can leak radioactive material into the atmosphere or water. In addition, transporting both nuclear fuel and spent fuel rods has the potential for harming the environment if an accident occurs en route.

Nuclear Fission Bombs

In principle the physics of a fission bomb and of a fission power plant are the same. However, in a nuclear power plant, the goal is to achieve a sustainable and controlled fission chain reaction so that the energy is released slowly over a long period of time. In a fission bomb, the goal is to have a runaway chain reaction so that the energy is released very quickly. As before, we consider uranium-235 to be the fuel. In a power plant, uranium must be enriched so that the abundance of uranium-235 is about 3%, but for a bomb the abundance of uranium-235 must be about 90%.

In a power plant, the enriched uranium is shaped into pellets that are about the size and shape of an eraser at the end of your pencil (Figure 43.18). These pellets are stacked to form fuel rods that are several meters tall. The shape and size of the fuel matters because nuclei near the surface are likely to release neutrons that escape rather than trigger further fission reactions. So a small sample with a large surface area-to-volume ratio may not be able to sustain the chain reaction for very long. The shape with the smallest surface area-to-volume ratio is the sphere. A bomb requires a low surface area-to-volume ratio; this means a relatively large sample of enriched uranium shaped into a sphere. The minimum mass of nuclear fuel required to sustain a chain reaction is called the *critical mass*. For uranium, the critical mass is about 50 kg. In a bomb, the uranium is kept in two or more separate pieces, each below the critical mass. (This is known as a subcritical mass.) When the nuclear bomb is detonated, those pieces are brought together to form a single object whose mass is greater than the critical mass. (This is called a supercritical mass.) These pieces must be brought together quickly, and a conventional bomb or explosion is used. Figure 43.19 shows the schematic of the bomb (known as *Little Boy*) dropped over Hiroshima near the end of World War II. The uranium was shaped into two hemispheres and kept separate until detonation. A conventional explosion inside Little Boy forced the two hemispheres together to form a single sphere of roughly 64 kg (about 14 kg above the critical mass).

Pyatakov Sergey/RIA Novosti EyePress/Newscom

FIGURE 43.18 Nuclear fuel pellets.

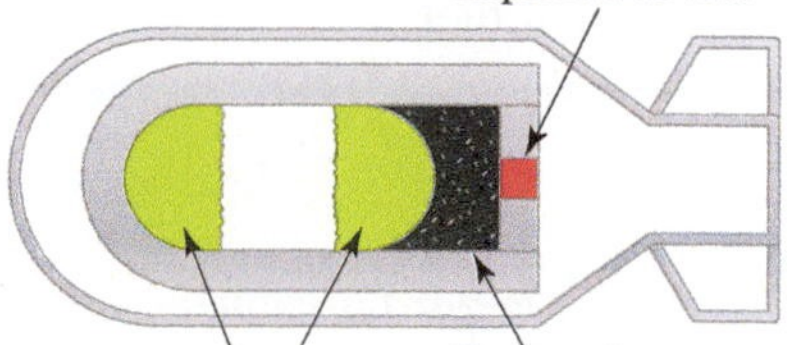

FIGURE 43.19 In Little Boy—the bomb dropped over Hiroshima—a conventional explosion brought the two subcritical hemispheres together to form one supercritical sphere of nuclear fissile material.

EXAMPLE 43.7 Uranium versus Fossil Fuel

Figure 43.18 shows pellets of enriched uranium used in a nuclear power plant. Estimate the total energy released by just one of the pellets. In the **CHECK and THINK** step, compare your answer to energy released by same mass of fossil fuel such as coal or gasoline. *Hint*: See Table 9.1 (page 265).

INTERPRET and ANTICIPATE

We can use the photo to estimate the mass of single pellet. Once we know the mass of a whole pellet, we'll estimate the mass of uranium-235. From this we can estimate the number of uranium-235 nuclei. We know that each fission reaction releases about 200 MeV per nuclei. We'll assume all the nuclei undergo fission to find the total energy released.

SOLVE

The pellets look fairly light. Perhaps each one is the mass of a few quarters. From Appendix B, we find that the mass of a single quarter is about 6 g. So let's assume a single pellet is about 20 g. We'll also assume that it has been enriched so that the abundance of uranium-235 is about 3%. We'll assume the pellet is comprised of uranium isotopes of roughly the same mass. (That is not really true. The uranium isotopes have slightly different masses and the pellets are not pure uranium, but this is an estimate and such details don't make a huge difference.)

$$m_{235} = 0.03m_{\text{pellet}} = 0.03(20\text{g}) = 0.6 \text{ g}$$

Next we need the number of uranium-235 nuclei. The molar mass of uranium-235 is 235 g, which means 1 mole of uranium-235 has a mass of 235 g. We calculate the number of nuclei using Avogadro's number N_A found on the back inside cover.

$$N = m_{235}\frac{N_A}{235 \text{ g}}$$

$$N = (0.6 \text{ g})\frac{6.02 \times 10^{23} \text{ nuclei}}{235 \text{ g}}$$

$$N = 1.5 \times 10^{21} \text{ nuclei}$$

Assume every uranium-235 nuclei undergoes a fission reaction and the each reaction releases 200 MeV of energy.

$$E_{\text{released}} = (200 \text{ MeV/nuclei})N$$

$$= (200 \text{ MeV/nuclei})(1.5 \times 10^{21})$$

$$E_{\text{released}} = 3 \times 10^{23} \text{ MeV} = 5 \times 10^{10} \text{ J}$$

CHECK and THINK

First, to check our answer we can use the fact that Table 9.1 says that 1 lb of uranium-235 contains 3.7×10^{13} J. 1 lb is the weight of 453 g. So, according to Table 9.1, our 0.6 g of uranium-235 contains $0.6 \text{ g}\left(\frac{3.7 \times 10^{13} \text{ J}}{453 \text{ g}}\right) = 5 \times 10^{10}$ J exactly as we estimated. Next, Table 9.1 says that 1 lb of coal contains 1.6×10^7 J. So 20 g of coal contains $20 \text{ g}\left(\frac{1.6 \times 10^7 \text{ J}}{453 \text{ g}}\right) = 7 \times 10^5$ J. Our results are amazing; a pellet of uranium releases tens of thousands of times more energy than an equal amount of coal. If you would like, you can repeat this comparison for gasoline, which you will also find in Table 9.1. The results are similar; fission releases millions of times more energy than burning fossil fuel.

FUSION ✪ Major Concept

43-8 Fusion Reactions

We've just looked at the practical application of *fission* reactions. As described in Section 43-7, the *fusion* of nuclei that are lighter than iron (to the left of the peak) releases energy. The practical application of fusion reactors on Earth could meet our energy needs without many of the complications of fission reactors. For example, fission reactors require the mining and enrichment of fuel. Fusion begins with light elements such as hydrogen or deuterium (an isotope of hydrogen with one extra neutron). Such light elements can easily be found in water. Fusion produces very few radioactive byproducts. So there are fewer environmental concerns.

© ITER Organization, 2015

FIGURE 43.20 ITER construction (May 2015).

However, many obstacles must be overcome before fusion reactors power our towns and cities. Several experimental fusion reactors have been built on Earth, which have overcome many of these obstacles. One in particular, known as ITER (International Thermonuclear Experimental Reactor) (pronounced "eater"), is currently being built in the south of France (Figure 43.20). The name, ITER, comes from the Latin word for "the way," and the goal of ITER is to generate 10 times more energy by fusion than is required to run the reactor. Perhaps, ITER will make its goal by 2019.

Fusion in the Sun's Core

Many great technological advancements often come from observing nature. So we turn our attention to the fusion reactor in the core of our Sun. The reactions in the Sun's core that fuse hydrogen into helium (CASE STUDY Chapter 41) are:

$$\begin{aligned} &{}^1_1\text{H} + {}^1_1\text{H} \rightarrow {}^2_1\text{H} + e^+ + \nu \\ &{}^2_1\text{H} + {}^1_1\text{H} \rightarrow {}^3_2\text{He} + \gamma \\ &{}^3_2\text{He} + {}^3_2\text{He} \rightarrow {}^4_2\text{He} + {}^1_1\text{H} + {}^1_1\text{H} \end{aligned} \tag{43.19}$$

According to these reactions, two protons (hydrogen nuclei) fuse together to form deuterium, and then deuterium and a proton fuse to make an isotope of helium. Because the reaction chain begins with the fusion of two protons, it is called the **proton-proton chain**. Finally, two isotopes of helium fuse to form helium-4. Helium-4 is at a local peak on the binding energy curve (Fig. 43.14); it has more binding energy per nucleon than the hydrogen and helium isotopes from which it is formed.

Astronomers often refer to the proton–proton chain as the *pp* chain, without even cracking a smile.

Notice that the third reaction requires two helium-3 nuclei, which means the first two reactions must occur twice as often as the third reaction:

$$\begin{aligned} &2({}^1_1\text{H} + {}^1_1\text{H} \rightarrow {}^2_1\text{H} + e^+ + \nu) \\ &2({}^2_1\text{H} + {}^1_1\text{H} \rightarrow {}^3_2\text{He} + \gamma) \\ &{}^3_2\text{He} + {}^3_2\text{He} \rightarrow {}^4_2\text{He} + {}^1_1\text{H} + {}^1_1\text{H} \end{aligned}$$

Also notice that the third reaction releases two protons that are free to undergo fusion. The net result is that four protons fuse to form helium-4, gamma rays, positrons, and neutrinos:

$$4({}^1_1\text{H}) \rightarrow {}^4_2\text{He} + 2e^+ + 2\nu + 2\gamma \tag{43.20}$$

Compare Equation 43.20 to the one given in Example 39.10 (page 1295) and you will see we've included more detail here, but the essence of that example still applies.

Each complete *pp* chain releases 4.28×10^{-12} J (26.7 MeV) and there are about 8.97×10^{37} reactions per second occurring in the Sun to maintain its power output.

In the case study of Chapter 41 (page 1358), we learned that the first reaction cannot be explained by classical physics. The two protons must fuse and be held together by the strong force, but according to classical physics, Coulomb repulsion would prevent them from coming close enough for the strong force to bind them. According to quantum mechanics, however, there is a nonzero probability that they will tunnel through the repulsive barrier and fuse (Example 41.7, page 1358). The probability of tunneling increases with temperature and density; the Sun's core is very hot (1.5×10^7 K) and very dense (1.5×10^5 kg/m^3). Such a high temperature and density requirements make the *pp* chain impractical on Earth. These conditions are met in the Sun because of the great gravitational force that crushes the core into a very dense and very hot sphere.

Fusion on Earth

Using fusion reactors to meet our energy needs is an important goal. Here we look at some of the practical concerns of achieving that goal with a focus on ITER because it is currently under construction.

First, we need to think about the possible fusion reactions. As described above, the *pp* chain is not a great option because of the difficulty of fusing two protons. It is much easier to fuse isotopes of hydrogen. The most efficient fusion reaction possible on Earth begins with deuterium and tritium:

$$^2_1\text{H} + {}^3_1\text{H} \rightarrow {}^4_2\text{He} + {}^1_0\text{n} \tag{43.21}$$

This reaction releases an energy of 17.59 MeV. Deuterium can be distilled from water and so is easily available. Tritium is radioactive with a short half-life (12.3 yr), and it only occurs in a trace amount in nature. So tritium must be manufactured; this is known as *breeding*. There are about 20 kg of tritium available worldwide, some of which ITER will use in its early stages. However, tritium is bred from lithium, which is plentiful in the Earth's crust. So in later stages, ITER will breed tritium using lithium. So the fuels for fusion are deuterium and lithium, both of which are abundant, easy to obtain, and safe to handle. (Compare this to the fuel required for fission—enriched uranium.)

Second, the deuterium and tritium are both positively charged. So like the first reaction in the *pp* chain, there is a repulsive barrier and the particles must tunnel through this barrier. Fortunately tunneling can be achieved on Earth without requiring the enormous gravitational force exerted on the Sun's core. All that is required is a temperature that is about 10 times higher than the Sun's core temperature, and this is achievable on Earth. However, achieving such a high temperature is not easy. ITER will use three techniques to achieve this high temperature.

Before discussing ITER's heating techniques, we must look at a third concern—confinement. As the fuel temperature is increased, it becomes a very hot plasma. Like a gas, a plasma fills its container, but there are no walls that can be built that would withstand the high temperature of the plasma. The plasma must not come in contact with the walls of its container. Instead, it is confined by a magnetic field. ITER will use a tokamak chamber (see Fig. P31.48, page 1007). It is a donut-shaped container with a strong magnetic field produced by superconductors.

Now we can continue our discussion of ITER's three techniques for heating the plasma. The first involves the tokamak's magnetic field. The tokamak uses a changing magnetic field, and from Faraday's law, we know that a changing magnetic field produces an electric field. This electric field creates a current in the plasma. As we've seen before, a current heats a conductor through collisions (Chapter 25). Similarly the plasma is heated, however, this will not achieve the required temperature. So ITER will use two external heating methods; these are the second and third techniques. The second technique involves injecting the plasma with high-energy neutral deuterium. Outside the tokamak, charged deuterium is accelerated and then neutralized before being injected into the plasma. The deuterium transfers energy to the plasma through collisions. This increases the temperature but still not high enough for fusion. The third technique is similar to the way a microwave oven operates.

Energy is transferred by high-frequency electromagnetic waves to the plasma by matching the resonance frequency of the ions and of the electrons. Once fusion begins, ITER's goal is to achieve a *burning plasma*. In such a burning plasma, the temperature is maintained by the fusion reaction itself, and then the external heating methods can be phased out.

If ITER meets its goals, then we may be on our way to achieving fusion-powered cities, towns, cars and trains. Such an energy source may be important to preventing further climate change while sustaining our way of life. Fusion emits no pollution or greenhouse gas. Its end product—helium—is an inert, nontoxic gas. There is no possibility of runaway reaction because the conditions for fusion are too difficult to maintain: if the conditions are altered, the plasma cools and fusion stops. So keep your fingers crossed, and check the ITER website for further developments.

CASE STUDY Part 4: Where Did Oxygen and Iron Come From?

We've seen that small nuclei—primarily those of hydrogen and helium—formed in the early universe. As the Universe continued to expand and cool, these nuclei bound with electrons to form atoms. This doesn't explain where the larger elements, such as oxygen and iron, came from. The water (H_2O) in your body and the iron in your blood cannot be primordial, so how were their elements made?

As the Universe continued to expand and cool, matter clumped together due to gravitational attraction, forming clouds of hydrogen (and helium) gas. (One of the contemporary problems physicists face is to explain how that clumping took place on timescales that match our observations.) The clouds of gas then collapsed due to gravity, forming galaxies full of stars. The core of every living star is a nuclear fusion reactor, and the primary fuel in such a reactor is hydrogen (Eq. 43.19). This hydrogen must be hot enough for quantum tunneling to occur (case study, Chapter 41, page 1358), so only the hydrogen in the core of a star undergoes fusion. Other elements are not able to undergo fusion at all because they require even higher temperatures in order to achieve quantum tunneling.

But eventually, there isn't enough hydrogen to fuel the reactions in a star's core. At this point, the star begins to collapse under its own weight. The temperature near the core increases, making the fusion of heavier elements possible. In stars several times more massive than the Sun, this process continues until the fusion reactions produce iron. Because iron is at the top of the binding energy curve (Fig. 43.14), more fusion reactions would require the addition of energy instead of releasing energy. Fission reactions could release energy, but they require elements heavier than iron—and none existed in the earliest stars. (Even later generations of stars do not have enough large nuclei to make fission possible.) So the early heavy stars ran out of fuel and died in explosions known as supernova events. However, before they died, they produced elements that fall between helium and iron on the binding energy curve—elements including oxygen. Before our Sun was born, one such heavy star exploded, and material from that star went to form everything in our solar system. This answers our question about oxygen, but we still need to find out where even heavier elements, such as the iodine in your thyroid, came from.

CASE STUDY Part 5: Where did Iodine Come From?

In the core of even the most massive stars, fusion can only produce elements up to iron. Where did the heavier elements, such as the iodine in your thyroid, come from? (Insufficient iodine can lead to goiters; see Fig. 43.21.) It may seem impossible that such heavy elements exist because they have less binding energy than iron, and so making them through fusion requires energy. Where does the energy come from, and what reactants fuse together?

The answer is that in a supernova event, much gravitational potential energy is released and a great number of neutrons are produced. These neutrons penetrate nearby nuclei, creating products that are heavier than iron. Let's consider this process in a little more detail.

First, we need a source of free neutrons. These are produced in nuclear reactions between relatively small nuclei, such as:

$$^{4}_{2}\text{He} + ^{13}_{6}\text{C} \rightarrow ^{16}_{8}\text{O} + \text{n}$$

$$^{16}_{8}\text{O} + ^{16}_{8}\text{O} \rightarrow ^{31}_{16}\text{S} + \text{n}$$

Next, the free neutrons made in such reactions penetrate nearby nuclei. Because the neutron is electrically neutral, there is no Coulomb barrier to overcome, and no quantum tunneling is necessary. The daughter nucleus that results has a higher mass number than the parent, but no change in atomic number:

$$^{A}_{Z}\text{P} + \text{n} \rightarrow ^{A+1}_{Z}\text{D} + \gamma$$

Thus, the daughter in these reactions is an isotope of the parent. Such neutron penetration can happen in the relatively normal conditions inside a star or during a supernova event.

FIGURE 43.21 An iodine deficiency in the thyroid gland can cause goiters. Iodine added to our salt helps to prevent this condition.

Under the normal conditions inside a star, neutrons are generated and absorbed by nuclei at a slow rate. When a nucleus captures a neutron under these slow conditions, it undergoes beta decay before another neutron can be captured. Of course, beta decay decreases the neutron number and increases the atomic number. So the nucleus is no longer an isotope of the original parent; instead it becomes a new element. This is called the ***s*-process**, where *s* stands for slow.

Under the extraordinary conditions during a supernova event, neutrons are produced and absorbed rapidly. Under these conditions, a nucleus will likely absorb many neutrons before undergoing beta decay. So the daughter nucleus may be a higher mass number isotope of the parent. After the period of rapid neutron penetration, these isotopes are generally unstable and undergo either alpha or beta decay. This is called the ***r*-process**, where *r* stands for rapid.

Today's isotopes beyond iron were generally formed in either the *s*-process or the *r*-process, though some isotopes can be formed in both. How was iodine formed? To answer such a question, astronomers measure the abundance of various elements in stars, while laboratory scientists measure the properties of various nuclei and theorists model the conditions in stars and in supernova events. These models must match the astronomers' observations and the experimentalists' data. Some elements, such as copper, silver, and gold, were most likely produced in the *s*-process. Other elements, such as radium, uranium, plutonium, and thorium, were almost certainly produced in the *r*-process. In general, elements as massive as bismuth ($^{209}_{83}\text{Bi}$) can be produced in the *s*-process, and heavier elements in the *r*-process. So in principle, iodine ($^{127}_{53}\text{I}$) can be produced in either process, but theoretical models and experimental data suggest that it was most likely produced in the *r*-process.

Perhaps one day a child will ask you, "Where did I come from?" You might answer, "You were made by the cosmos. You began as hot particles soon after time began. You grew in a star that exploded and gave birth to our Sun and to our planet, and to everyone you know, including you." One of the main purposes of these case studies is to connect physics to the human experience. With this final case study we see that physics can even explain how the human experience came to be.

EXAMPLE 43.8 Deuterium versus Fossil Fuel

In seawater there is about 1 deuterium atom for every 6400 hydrogen atoms. If the deuterium from 1 gal of such water were used in the fusion reaction given in Equation 43.21, estimate how much energy would be released? Don't worry about details such as obtaining the tritium. In the **CHECK and THINK** step compare your answer to energy released by the same volume of fossil fuel such as coal or gasoline. *Hints*: See Table 9.1 (page 265) and Table 15.1 (page 420); also 1 m^3 = 264 gal. Report your final answer to two significant figures.

INTERPRET and ANTICIPATE

We need to find the number of deuterium atoms in a gallon of seawater. We don't need to worry about the details such as ionizing and heating the deuterium. We'll assume that the number of deuterium nuclei available for the fusion reaction equals the number of deuterium atoms. Finally, we'll use the fact that each fusion reaction releases energy equal to 17.59 MeV.

SOLVE

First, we need the mass of 1 gallon of seawater. The density of seawater is 1025.18 kg/m³ (Table 15.1).	$m_{\text{water}} = 1\text{ gal}\left(\dfrac{1\text{ m}^3}{264\text{ gal}}\right)\left(\dfrac{1025\text{ kg}}{\text{m}^3}\right) = 3.88\text{ kg}$
Next we need the number of water molecules in the gallon. The molar mass of H_2O is 18 g, which means 1 mole of water has a mass of 18 g. (Of course, heavy water is slightly heavier, but we are making an estimate to two significant figures, and we don't need to be concerned about such details.)	$N_{\text{molecules}} = m_{\text{water}}\dfrac{N_A}{18\text{ g}}$ $N_{\text{molecules}} = m_{\text{water}}\dfrac{6.02 \times 10^{23}\text{ molecules}}{18\text{ g}}$ $N_{\text{molecules}} = (3.88 \times 10^3\text{ g})\dfrac{6.02 \times 10^{23}\text{ nuclei}}{18\text{ g}}$ $N_{\text{molecules}} = 1.3 \times 10^{26}\text{ molecules}$
There are two hydrogen atoms in each molecule of water. So we'll multiply the number of molecules by 2 to find the number of hydrogen atoms.	$N_H = 2N_{\text{molecules}} = 2.6 \times 10^{26}$
We need the number of deuterium atoms. Because there is 1 deuterium atom for every 6400 hydrogen atoms, we divide by 6400. As described, this is also the number of nuclei that undergo fusion.	$N_D = \dfrac{N_H}{6400} = \dfrac{2.6 \times 10^{26}}{6400} = 4.1 \times 10^{22}\text{ nuclei}$
Assume deuterium nuclei undergo a fusion reaction and that each reaction releases 17.59 MeV of energy.	$E_{\text{released}} = (17.59\text{ MeV/nuclei})N_D$ $E_{\text{released}} = (17.59\text{ MeV/nuclei})(4.1 \times 10^{22}\text{ nuclei})$ $E_{\text{released}} = 7.1 \times 10^{23}\text{ MeV} = 1.1 \times 10^{11}\text{ J}$

CHECK and THINK

According to Table 9.1, 1 gal of gasoline contains 1.3×10^8 J. Fusing the deuterium in 1 gal of seawater releases about a thousand times more energy than burning 1 gal of gasoline, and fusion doesn't produce greenhouse gases.

43-9 Human Exposure to Radiation

Throughout your life, you are exposed to both artificially-generated and naturally-occurring radiation. This radiation includes the X-rays and gamma rays that are part of the electromagnetic spectrum, as well as alpha rays, beta rays, and other emitted particles. Artificial sources of radiation include nuclear power plants, medical examinations and procedures, fallout from past nuclear tests of weapons, and accidents such as the 1986 disaster at the Chernobyl nuclear power plant in Ukraine. Natural sources of radiation include cosmic rays from space and radioactive elements in the Earth. Radiation may break apart molecules and atoms, which can then damage living cells. The good news is that life on Earth has evolved in an environment with a low level of radiation so that living tissue can repair itself. However, living organisms cannot repair all the damaged cells, and damaged cells can produce tumors, cause cancer, and destroy bone marrow—to name just a few health problems. In this section, we explore the effects of nuclear radiation on living organisms, with a focus on how this radiation affects humans. We'll also learn how the exposure to nuclear radiation is quantified.

DNA (deoxyribonucleic acid) contains the genetic information that is transmitted during cell reproduction.

DNA is arranged into chromosomes in the cell. During cell reproduction, each chromosome is disassembled.

First, how does radiation damage tissue? The radiation transfers energy to the atoms and molecules that make up the tissue. (To make our discussion simpler, we'll refer to the molecules, but the same description could be applied to the atoms.) This energy either excites or ionizes the molecules. If a molecule is particularly important for life such as DNA, the damage may be detrimental to the tissue. The ionization of other molecules may create *free radicals.* Free radicals are neutral atoms with an unpaired electron. Such free radicals are highly reactive, and they can diffuse into the organism, setting off chemical reactions in critical sites such as in organs. These chemical reactions may involve important biological systems such as chromosomes. In such cases, the reactions can kill cells or cause genetic mutations. So some of the damage done by radiation exposure may be apparent only a few hours after exposure as critical biological systems are destroyed. Other consequences such as cancer or transferring genetic defects may take years or generations to appear.

The extent of the damage that occurs depends on the type of radiation. It is helpful to consider four types: alpha rays, beta rays, neutrons, and photons (such as gamma rays). Alpha particles transfer energy directly to living tissues by colliding with electrons in the tissue. These collisions excite the molecules in the tissue and can ionize them. The alpha particles are easily stopped by the skin and don't usually penetrate past the dead layer of skin cells. So normally, alpha particles do little damage.

If, however, the source of alpha particles is ingested or inhaled, they can do damage. The series of reactions in Figure 43.11 includes radon $^{222}_{86}\text{Rn}$. Radon is a colorless, odorless, radioactive noble gas, with a half-life of 3.82 days. Radon gas is found in some homes. Although its half-life is less than four days, it is continually replenished by the decay of $^{226}_{88}\text{Ra}$(radium), which is found in the rock and soil on which some homes are built. Radon is a particularly dangerous radioactive element because, as a gas, it can be inhaled into the lungs of a home's inhabitants. Once inside the lungs, radon undergoes alpha decay. The reactions in Figure 43.11 may take place inside the lungs, releasing more alpha (and beta) particles into the body.

According to the U.S. Surgeon General's office, as many as 20,000 deaths due to lung cancer are caused by radon annually. Smokers are at a higher risk than nonsmokers because radon decay particles attach to tobacco leaves and lodge in smoke particles in the lungs and bronchi. According to the U.S. Environmental Protection Agency, 1 in 15 homes in the United States has such a high radon level that the family living there should take action. Recommended actions include ventilation to remove radon gas from the home, and sealing the floors and walls to prevent more radon from leaking in.

Next, let's consider how beta particles affect living tissue. Alpha particles are heavy, and though they collide, they travel in nearly straight lines through tissue. However, beta particles are electrons. Electrons are much lighter, and so when they collide, they are easily scattered and travel in a zigzag path through tissue. Beta particles don't lose much energy with each collision, so they penetrate farther than alpha particles. It would take a few millimeters of metal to stop beta particles with an energy of 1 MeV, whereas alpha particles of that energy are stopped by a few 10^{-2} millimeters of skin tissue. Beta particles also lose energy by radiating photons, and these photons may be absorbed elsewhere in the organism. So the energy from beta particles is deposited over a greater volume than the energy of the heavier alpha particles.

Now we consider how neutrons deposit energy in tissues. Of course, neutrons are neutral, and so they penetrate and interact with nuclei. The most common element in a living organism is hydrogen because living organisms contain a lot of water. So, a neutron is most likely to interact with a proton in hydrogen. When a high-energy neutron (more than 1 keV) collides with a proton (in hydrogen), the neutron loses a large fraction of its energy to the proton. The proton then travels through tissue, ionizing atoms and molecules as it loses energy.

Finally, we consider the effect of photons on tissue. Photons transfer energy to electrons through the Compton effect (Section 40-4). This transfer is most dramatic for photons with energies between 40 keV and a few tens of MeV. The energy may be deposited very deep in the tissue. Gamma rays with energies of a few MeV are very penetrating.

Dosimetry

Now that we've discussed how *too much* radiation can damage living tissue, we want to know: How much is *too much*? To answer this question, we first need to know how radiation exposure is measured.

Radiation dosimetry is the branch of medical physics that involves the calculation and measurement of ionizing radiation absorbed by tissue. In this subsection we learn about the many quantities that are defined and measured in this branch of physics. The first quantity is the **absorbed dose** D; it is the energy delivered to tissue per unit mass of the tissue. The SI unit of dose is called the **gray** and abbreviated Gy: 1 Gy = 1 J/kg. However, an earlier system of dosimetry units is still in popular use today. In that system, the absorbed dose D is measured in rad, which stands for *r*adiation *a*bsorbed *d*ose. The conversion factors between the two systems are:

ABSORBED DOSE Major Concept

$$1 \text{ rad} = 0.01 \text{ J/kg} = 0.01 \text{ Gy}$$

It is also helpful to use the mass **average absorbed dose** D_{av} for an organ or a particular tissue such as the lungs or bone marrow. The average absorbed dose is the total energy E transferred to the organ (or tissue) divided by its mass m:

$$D_{av} = \frac{E}{m} \tag{43.22}$$

Living tissue is equipped to repair itself. So if the radiation dose is delivered slowly, it is possible that the tissue will repair itself before permanent damage is done. This means that it is important to quantify not just the dose of radiation absorbed, but also the rate at which radiation is absorbed. This is particularly important when radiation is used as a therapy to fight disease such as cancer. The large dose of radiation required to kill the cancer cells is given in several smaller doses spread over time so that the healthy cells have time to repair themselves between treatments.

The degree of permanent injury also depends on the degree of ionization. Because the four types of radiation (alpha rays, beta rays, neutrons, and photons) differ in their ability to ionize molecules, the same dose of each type of radiation will injure a particular tissue differently. To quantify the impact different types of radiation have on living tissue, a dimensionless quantity called the **radiation weighting factor** Q is introduced. For convenience, the quality factor scale is set such that X-ray radiation has $Q = 1$. So, for example, radiation with $Q = 5$ is equivalent to X-ray radiation with a 5 times higher dose. The values of Q come from experimentation, and some values are provided in Table 43.3. The effect of ionizing radiation on living tissue depends on both the dose and the type of radiation. This is quantified by the **equivalent dose** H, which is given by

The radiation weighting factor was formerly called the quality factor; hence, we use Q in this textbook. Q is based on the relative biological effectiveness (RBE)—the empirically determined ratio of the biological effectiveness of one type of radiation to X-ray radiation. Q is then an average arrived at by the consensus of regulators, industries, and governments.

$$H = QD \tag{43.23}$$

EQUIVALENT DOSE Major Concept

The SI unit for H is the sievert (Sv), which is a joule per kilogram. Although Q is dimensionless, it is reported in sieverts per gray. When $Q = 1$, the equivalent dose in sieverts equals the absorbed dose in gray. There is an older system of units still in use today. The equivalent dose H in that system is in rem, which stands for *r*oentgen *e*quivalent for *m*an and 1 rem = 0.01 Sv. (The roentgen is named for Wilhelm Röntgen, 1845–1923, who discovered the X-ray.) One roentgen is 2.58×10^{-4} C/kg.

TABLE 43.3 Some radiation weighting factors

Type of radiation	Q (Sv/Gy)
α particles	20
Neutrons (in range 100 keV – 2 MeV)	20
Neutrons (in range 10-100 keV or 2 MeV – 20 MeV)	10
Neutrons (of less than 10 keV or greater than 20 MeV)	5.0
Photons (of 30 keV or more)	1.0
β particles (of 30 keV or more)	1.0
β particles (of less than 30 keV)	1.7

Another factor that must be taken into account is the part of the organism that is exposed. Some parts are more sensitive to radiation than others. For example, the lungs are 12 times more sensitive than the skin. The **tissue weighting factor** w is a dimensionless quantity that takes the sensitivity of tissues into account. The human body is divided into 15 parts. Each part has a tissue weighting factor. The sum of all 15 tissue weighting factors is 1.00 (Table 43.4). The **effective dose** D_{eff} takes the tissue weighting factor into account and is given by

EFFECTIVE DOSE Major Concept

$$D_{eff} = wH \tag{43.24}$$

Now that we know how exposure to radiation is quantified, we can consider our exposure to radiation. Table 43.5 gives the average annual exposure to both natural and artificial sources of radiation both in the United States and worldwide. The typical natural radiation dose equivalent in the United States is roughly 3 mSv annually, and the average annual dose equivalent is roughly another 3 mSv. So Americans are exposed to about 6 mSv annually, about double the average worldwide. Perhaps this is due to American's access to medical procedures. For example, a full mammogram exam with a total of six images has a dose equivalent of about 8 mSv, and the average annual dose due to medical procedures in the United States is 3 mSv. (This average includes other procedures including diagnostic and treatment procedures.) The American Medical Association recommends that women over 40 have an annual mammogram, but in the United Kingdom the National Health Service recommends that woman between the ages of 40 and 70 get a mammogram every three years. To help make sense of how all this exposure to radiation affects our health, we end with this statistic: On average 5% of the people exposed to an effective dose of 1000 mSv will contract a fatal cancer as a result.

TABLE 43.4 Tissue weighting factors

Organ or tissue	w
Bladder	0.04
Bone surface	0.01
Brain	0.01
Breast	0.12
Colon	0.12
Esophagus	0.04
Gonads	0.08
Liver	0.04
Lung	0.12
Red bone marrow	0.12
Salivary glands	0.01
Skin	0.01
Stomach	0.12
Thyroid	0.04
Remainder of the body	0.12
Total	1.00

The data is from ICRP Publication 103 37 (2007).

TABLE 43.5 Average annual human exposure to radiation in mSv

Source	U.S.	Worldwide
Inhalation of air (primarily radon)	2.28	1.26
Ingestion of food and water	0.28	0.29
Ground	0.21	0.48
Cosmic rays	0.33	0.39
Total natural sources	3.10	2.40
Medical	3.00	0.60
Other	0.14	0.012
Total artificial sources	3.14	0.612

EXAMPLE 43.9 Are bananas safe to eat?

Potassium K is an important mineral in the human body. Your muscles need potassium to operate, potassium keeps the sodium levels in your body under control, and all of your cells need potassium to function normally. The average adult needs 4700 mg of potassium per day. Many people try to meet this need by eating bananas. A single typical banana has a mass of 125 g, about 450 mg of which is potassium, nearly 10% of your daily requirement. However, potassium-40 is a radioactive isotope of potassium with a natural abundance of 0.012%. Potassium-40 has a half-life of 1.25×10^9 years, and it radiates gamma and beta rays. The average energy released with each decay is about 0.5 MeV. The body takes around 50 hours to expel digested food after it has been consumed. Assume the radiation from the potassium is mostly absorbed by the tissue and organs along the digestive tract: the esophagus, liver, salivary glands and stomach. Further assume the combined mass of all of these tissues is about 2 kg. Estimate the absorbed dose, the equivalent dose, and the effective dose delivered to these tissues. Report your estimates to one significant figure.

INTERPRET and ANTICIPATE

The absorbed dose D is the energy delivered to tissue per unit mass of the tissue. We know the mass of the tissue, so we just need to find the energy delivered. We also know the energy released by each decay. So we must find the number of decays in 50 hours to find the total energy delivered to the tissues. Once we know the absorbed dose, the effective dose D_{eff} is found from the equivalent dose and the tissue weighting factor.

SOLVE

To find the number of decays, we must first find the initial number N_0 of potassium-40 nuclei when the banana first entered the body. The banana contains 0.45 g of potassium. The molar mass of potassium is 39 g, and the natural abundance of potassium-40 is 0.012%.

$$N_0 = (\text{fraction of } {}^{40}\text{K}) \times \frac{(\text{mass of K ingested})}{(\text{molar mass of K})} \times N_A$$

$$N_0 = 0.00012 \frac{0.45 \text{ g}}{39 \text{ g/mol}} 6.02 \times 10^{23} \text{ nuclei/mol}$$

$$N_0 = 8.3 \times 10^{17} \text{ nuclei}$$

From the half-life, we find the number of decays that occur in the 50 hours in which the potassium is in the body. The number of parent nuclei as a function of time is given by Equation 43.8. The number of decays is the initial number of parent nuclei minus the number of parent nuclei remaining after 50 hours. The half-life must be converted from years to hours (1.09×10^{13} h); the units have been left off the exponent for clarity.

$$N = N_0\left(\frac{1}{2}\right)^{t/T_{1/2}} \quad (43.8)$$

$$N_{decays} = N_0 - N$$

$$N_{decays} = N_0\left[1 - \left(\frac{1}{2}\right)^{t/T_{1/2}}\right]$$

$$N_{decays} = 8.3 \times 10^{17}\left[1 - \left(\frac{1}{2}\right)^{50/1.09 \times 10^{13}}\right]$$

$$N_{decays} = 2.7 \times 10^{6}$$

Each decay releases 0.5 MeV of energy. So we find the total energy delivered by multiplying the number of decays by 0.5 MeV.

$$E = (0.5\,\text{MeV})N_{decays} = (0.5\,\text{MeV})(2.7 \times 10^6)$$

$$E = 1.3 \times 10^6 \text{ MeV} = 2 \times 10^{-7} \text{ J}$$

The absorbed dose D is found by dividing the delivered energy by the mass of the tissue.

$$D = \frac{E}{m} = \frac{2 \times 10^{-7}\text{J}}{2 \text{ kg}} = 1 \times 10^{-7}\text{Gy}$$

The radiation weighting factor Q is 1 Sv/Gy for both photons and beta particles with energy above 30 keV (Table 43.3). So the equivalent dose (Eq. 43.23) is numerically equal to the absorbed dose.

$$H = QD = (1 \text{ Sv/Gy})(1 \times 10^{-7}\text{Gy})$$

$$H = 1 \times 10^{-7}\text{Sv}$$

The effective dose takes into account the sensitivity of the particular organs and tissues that absorb the radiation. We add the tissue weighting factors for these organs and tissues (Table 43.4). Then find D_{eff} using Equation 43.24. (The subscripts stand for *esoph*agus, *liver*, *sal*ivary *gl*an*d*s and *st*o*m*a*ch*.)

$$w = w_{esoph} + w_{liver} + w_{sal\ gld} + w_{stmch}$$

$$w = 0.04 + 0.04 + 0.01 + 0.12 = 0.21$$

$$D_{eff} = wH \quad (43.24)$$

$$D_{eff} = (0.21)(1 \times 10^{-7}\text{Sv}) = 2 \times 10^{-8} \text{ Sv}$$

CHECK and THINK

If you search the Internet, you are likely to find the vague statement that the radiation dose of a banana is 0.1 μSv. Perhaps the statement is referring to the equivalent dose H, in which case the statement exactly matches what we found. Your search will also reveal an informal unit for measuring a radiation dose—the banana equivalent dose (BED). The idea of this informal unit is to help make radiation doses more understandable. For example, the maximum permitted radiation leakage for a nuclear power plant is 2500 BED. We all feel safe eating a banana, but how about 2500 bananas?

THE STANDARD MODEL
Underlying Principle

43-10 The Standard Model

We've reached the last section of this textbook. Here we introduce one final theory—known as the *standard model.* The **standard model** is a theory that lists the particles that make up that the Universe, explains their properties, and describes the forces by which they interact. So one of the wonderful things about saving the standard model for the end of this book is we get to review many forces and particles we have studied so far, and order them in a logical scheme.

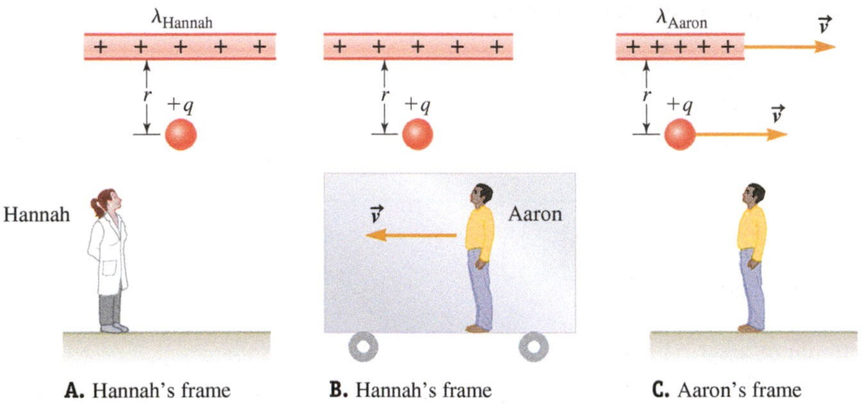

FIGURE 43.22 **A.** Hannah in the laboratory frame observes a particle with charge $+q$ near a charged rod with linear charge density λ_{Hannah}. **B.** *According to Hannah,* Aaron observes the same particle and rod from a frame moving at constant velocity to the left. **C.** To Aaron the rod and the particle seem to move to the right at constant velocity. To Aaron the rod is shorter, and the linear charge density is greater $\lambda_{\text{Aaron}} > \lambda_{\text{Hannah}}$.

Electricity, Magnetism, and Relativity

We begin by taking another look at the electric force, the magnetic force, and special relativity. Hannah and Aaron both observe the force between a positively charged particle and a nearby positively charged rod (Fig. 43.22). Hannah is in the laboratory frame, which is at rest with respect to the particle and the rod. Aaron is in a frame moving to the left at constant velocity. So he sees the rod and the particle moving to the right.

Hannah observes that force on the particle is repulsive and so points straight down (Fig. 43.23A). To Hannah, the magnitude of the electric field produced by the rod is

$$E_{\text{Hannah}} = \frac{1}{2\pi\varepsilon_0}\frac{\lambda_{\text{Hannah}}}{r} \tag{25.13}$$

and the electric force on the charged particle is $F_{\text{Hannah}} = qE_{\text{Hannah}}$.

Aaron sees the rod and the particle moving to the right. Because of this relative motion, there is a length contraction (Section 39-5), and to Aaron the rod appears shorter. Because for him the rod is shorter, Aaron sees the same amount of charge spread over a shorter rod, and therefore, a greater linear charge density. So Aaron claims that the electric field points downward but has a greater magnitude than that observed by Hannah.

However, Einstein postulated that all laws of physics are true in all inertial reference frames. Put simply, according to Einstein, Aaron and Hannah must agree on the force exerted on the charged particle; if they disagree, something else must be going on. But there is something else going on. Aaron sees the charged rod is moving; to him the charged rod is producing a current. According to the simple right-hand rule for currents (Fig. 30.11, page 939) the magnetic field is directed into the page at the location of charged particle. According to Equation 30.17 $\vec{F} = q(\vec{v} \times \vec{B})$, the magnetic force on the charged particle is upward toward the rod. The downward electric force and the upward magnetic force observed by Aaron give a net force on the particle that exactly matches the electric force observed by Hannah (Fig. 43.23B).

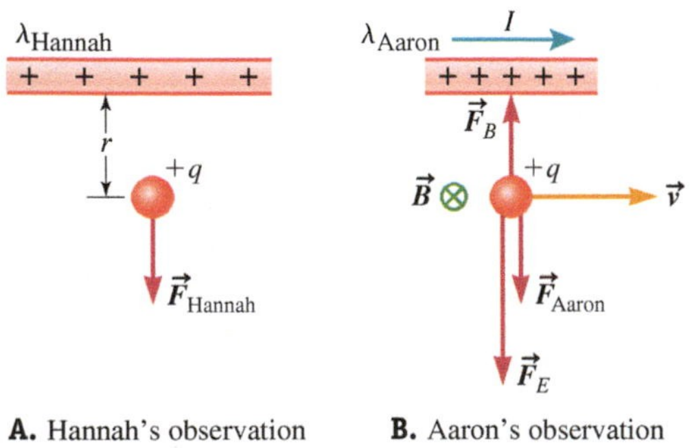

FIGURE 43.23 **A.** Hannah observes only a downward-pointing electric force. **B.** Aaron observes a downward-pointing electric force and an upward-pointing magnetic force. The net force he observes is exactly equal to the net force Hannah observes.

This thought experiment illustrates the idea behind unified forces. Two forces are unified when they are understood as two aspects of the same underlying force. Here, a net force on a charged particle can be understood in terms of the electric force alone, or in terms of a combination of the electric and magnetic forces, depending on the relative motion between the particle and the observer. So we say the two forces are aspects of a single underlying *electromagnetic force.*

Quantum Electrodynamics (QED)

QED **Underlying Principle**

Several times throughout this textbook, we have mentioned that Maxwell's equations give a unified theory of electricity and magnetism. Furthermore, this unified theory explains a phenomenon—light—that neither theory alone can explain. The American physicists Richard Feynman (1918–1988) and Julian Schwinger (1918–1994), and the Japanese physicist Sin-Itiro Tomonaga (1906–1979), shared the Nobel Prize in 1965 for developing a theory known as **quantum electrodynamics** (QED). This theory combines electromagnetism with quantum mechanics. According to QED, photons are more than just particles that make up the phenomenon of light. They play a role in all phenomena that involve the electromagnetic force.

To understand the role played by photons, we consider a crude analogy. Figure 43.24A shows Sallie and Pete playing a game of catch. Pete has just thrown the ball. He pushes on the ball, and by Newton's third law the ball pushes on him: he experiences a recoil. When Sallie catches the ball she exerts a force on the ball, and the ball exerts a force on her. We can think of the ball as a particle that carries a force between Pete and Sallie.

According to QED, when two charged particles exert forces on one another, a photon is exchanged (Fig. 43.24B). The photons carry the force from one particle to the other, much as the ball carries a force between Pete and Sallie. Both Pete and Sallie feel pushed away from each other, so the ball carries a repulsive force. This analogy is crude because it cannot account for an attractive force, but according to QED, both attractive and repulsive forces are attributed to the exchange of photons. We'll soon learn that other forces are accounted for by the exchange of other particles. The particles, such as photons, that are exchanged when forces are exerted are referred to as "messenger particles," "force-carrying particles," or "exchange particles." The standard model includes a number of other exchange particles, and we say that a particular force is "mediated" by its particular exchange particles. But next, we must digress to learn about the particles that make up the standard model.

A. **B.**

FIGURE 43.24 **A.** Pete throws a ball to Sallie. **B.** Two charged particles interact. In this analogy, the ball is like a photon carrying a force from Pete to Sallie.

A Zoo of Particles

BOSON AND FERMIONS **Major Concept**

The standard model classifies particles into several different categories. Some categories overlap, and you can find particles in more than one category. In this section we'll discuss these categories and some of the particles in the standard model. Many of the particles will be familiar to you, but there are a few new ones as well.

Like the electromagnetic force, each fundamental force is thought to be propagated by exchange particles. The exchange particles are **bosons**. Bosons do not obey Pauli's exclusion principle (Section 42-6), so they can have the same quantum numbers. Electrons are an example of another type of particle, called **fermions**. Fermions do obey the Pauli exclusion principle and cannot have the same quantum numbers. We saw that, in the case of electrons, Pauli's exclusion principle explains why atoms take up space (not all the electrons can be in the atom's ground state.) Put simply, solid objects, such as the chair you are sitting on, must be made of fermions because they obey Pauli's exclusion principle and take up space. Because bosons do not obey Pauli's exclusion principle, however, they can be packed together without taking up space. Bosons also differ from fermions in their spins. All known fermions, such as electrons, are spin 1/2 particles, whereas bosons have integer spin. For example, the photon has a spin of 1.

Just as the photon is a boson that mediates the electromagnetic force, there are eight **gluons** that are bosons that mediate the strong force. There are also three bosons that mediate the weak force. All of these bosons have been observed in a laboratory and are part of the standard model. In addition, there is a theory that gravity should also be mediated by a boson. This boson is called the graviton, but its existence hasn't been confirmed by experimentation. The standard model does not account for gravity, so the graviton isn't properly part of the standard model either. Table 43.6 summarizes all of these bosons.

TABLE 43.6 Bosons

Force	Name or symbol	Mass (GeV/c^2)	Charge	Spin
Electromagnetic	Photon	0	0	1
Strong nuclear	(8) Gluons	0	0	1
Weak nuclear	W^+	80.4	+1	1
	W^-	80.4	−1	1
	Z	91.2	0	1
Gravity	Graviton	0	0	2

LEPTONS AND HADRONS

Major Concepts

The standard model also divides particles into **leptons** and **hadrons**. Leptons are fundamental particles that cannot be subdivided. We are already familiar with two of them: the electron and the neutrino. In fact, there are six leptons and six antiparticles, called **antileptons**. The neutrino we are familiar with from inverse beta decay (Eq. 43.12) is actually called an **electron neutrino**. The right part of Table 43.7 summarizes the six leptons. Notice that they fall into three pairs, such that each particle is paired with a neutrino. Leptons are subject to the weak force but not the strong force. Charged leptons are further subject to the electromagnetic force. To help understand reactions involving leptons, it is convenient to assign each lepton a lepton number of +1 and each antilepton a lepton number of −1.

Hadrons, in contrast, are made of quarks. They are further subdivided into **baryons** and **mesons**. Baryons are made of three quarks. Familiar examples of baryons are protons and neutrons. Mesons are made of two quarks. Both quarks and leptons are fermions, so they obey Pauli's exclusion principle. Quarks are charged and have a charge of either $\pm e/3$ or $\pm 2e/3$. Table 43.7 summarizes the fermions. All hadrons are composed of quarks, so they are subject to the strong force. Also, because quarks are charged, they are subject to the electromagnetic force. Lastly, they are subject to the weak force. Table 43.8 provides examples of baryons and mesons. Each meson is made of a quark and an antiquark. A bar over the symbol for a quark denotes an antiquark; for example, $\bar{s}$ is a strange antiquark. A meson or baryon's charge is the sum of the charge of its quarks. For example, a neutron is made of two down quarks and one up quark, and its charge is $2(-e/3) + 2e/3 = 0$ as expected. Each quark has a baryon number of +1/3, and each antiquark has a baryon number of −1/3. So the baryon number for all the mesons is 0. However, the baryon number for a proton or a neutron is +1, and the baryon number for an antiproton or an antineutron is −1.

TABLE 43.7 Fermions

Quark	Symbol	Charge	Mass (GeV/c^2)
Up	u	$+2e/3$	0.002
Down	d	$-e/3$	0.005
Charm	c	$+2e/3$	1.5
Strange	s	$-e/3$	0.15
Top	t	$+2e/3$	172
Bottom	b	$-e/3$	4.7

Lepton	Symbol	Charge	Mass
Electron	e	$-e$	0.511 MeV/c^2
Electron neutrino	ν_e	0	Between 0.05 and 2 eV/c^2
Muon	μ	$-e$	106 MeV/c^2
Muon neutrino	ν_μ	0	Less than 0.19 MeV/c^2
Tau	τ	$-e$	1.78 GeV/c^2
Tau neutrino	$\nu\tau$	0	Less than 18 MeV/c^2

TABLE 43.8 Some baryons and mesons

Baryon	Symbol	Quarks
Proton	p	uud
Neutron	n	udd
Omega	Ω^-	sss
Lambda	Λ	sud

Meson	Symbol	Quarks
Pion	π^+	$u\bar{d}$
Phi	ϕ	$s\bar{s}$
Kaon (plus)	K^+	$u\bar{s}$
Kaon (minus)	K^-	$\bar{u}s$

Forces in the Standard Model

ELECTROWEAK THEORY AND QCD

Underlying Principles

Now let's see what the standard model says about how all these particles interact. First, the standard model is a combination of two theories. One theory is the **electroweak theory**—the unification of the electromagnetic force and the weak force (Fig. 23.2, page 684). Because we have studied the electromagnetic force and the weak force, we can surmise that the electroweak force acts on charged particles, leptons, and quarks. The other theory describes how the strong force is mediated by gluons and acts on quarks. This part of the standard model is called **quantum chromodynamics** (QCD). So the standard model accounts for the behaviors involving all the forces except gravity. Here we briefly look at some of these behaviors.

No one has found a free particle with a fractional charge. This means that no one has found a free quark. According to the standard model, quarks are confined in hadrons because the energy required to separate quarks *grows* as their separation increases. At a large enough separation, this energy is large enough to produce a new quark–antiquark pair, which binds to form a hadron.

Quarks are fermions, and they obey Pauli's exclusion principle. So you might be wondering how two up quarks can exist in a single proton. The answer is that quarks have another intrinsic quantum property, called *color*. Quarks come in three colors: red, green, and blue. Of course, these are not actual colors. You can think of color much like spin. Spin is an intrinsic quantum property, but quantum particles are not really spinning like a top. Just as two electrons can be in the same subshell because one can be spin up with $m_s = 1/2$, and one can be spin down with $m_s = -1/2$, two up quarks can be in the same proton, as long as they are different colors.

The standard model makes a prediction that helps us understand some of the reactions we have studied. According to QCD, baryon number is conserved during a reaction. The standard model also requires that the lepton number be conserved. As an illustration, consider this reaction (inside a nucleus): $p \rightarrow n + e^+ + \nu$ (Eq. 43.11). We start with a baryon number of $+1$ for the proton. After the reaction, the baryon number is $+1$ for the neutron. Before the reaction, the lepton number is 0. After the reaction, the lepton numbers are -1 for the positron and $+1$ for the neutrino, which sum to 0. Both the baryon number and the lepton number are conserved in this reaction. The standard model also requires that the total charge of quarks confined in a hadron be either $+e$, $-e$ or 0. Check Table 43.8, and you will find that those hadrons meet the charge requirement.

The Higgs Boson

As we have mentioned several times, the electroweak theory unifies the electromagnetic and the weak forces. The Nobel Prize was awarded in 1979 to the American physicists Sheldon Glashow (1932–) and Steven Weinberg (1933–), and the Pakistani physicist Abdus Salam (1926–1996), for developing the electroweak theory. Their electroweak theory explains phenomena that cannot be accounted for by either the electromagnetic theory or the weak theory alone. Perhaps more amazingly, the electroweak theory predicted the existence of three bosons: W^+, W^-, and Z (Table 43.6). The existence of these bosons was later confirmed by experimentation.

The standard model also predicts the existence of another boson, known as the **Higgs boson**. The Higgs boson is named for the British physicist Peter Higgs (1929–), one of the theorists who predicted its existence. Take a look at Table 43.6, and you might be struck by the fact that for 3 out of the 4 fundamental forces, the force-carrying bosons are massless, whereas the force-carrying bosons for the weak force have mass. In the 1960s, particle theorists were disturbed by this too, and they wanted the standard model to account for this discrepancy. According to the particle physics theories at that time, all force-carrying particles should be massless. So the question they asked was "what could cause these force-carrying bosons to acquire mass?" For that matter, they wanted to account for why other massive particles have mass. Their answer predicted the Higgs boson. Unlike the other bosons we've discussed, the Higgs boson has a spin of 0.

According to the standard model, the Higgs boson is a very massive (found to be 125 GeV/c^2) neutral particle that decays very rapidly. Because the Higgs is so massive, it requires a lot of energy to create it. Because the Higgs is neutral and decays rapidly, once created, it is hard to detect.

The Large Hadron Collider (LHC) is an experiment at CERN, the European Organization for Nuclear Research, located on the French-Swiss border. CERN operates the world's largest and most powerful particle physics laboratory. The LHC was designed to find the Higgs boson by smashing together very high-energy protons. After the protons collide, a plethora of particles emerge. By studying these particles, physicists look for evidence of the Higgs boson. On July, 4, 2012, two research groups who independently analyzed the data from the LHC announced that they had found evidence for the Higgs boson. Soon after the announcement, champagne corks popped as physicists celebrated the power of theorists to make predictions and the power of experimentalists to discover evidence testing those predictions.

CONCEPT EXERCISE 43.8

On an exam, you've been asked to write down the expression for beta decay. You remember the daughter nucleus has an extra proton and there are two products. You know one is either an electron or a positron and the other is either a neutrino or an antineutrino: ${}^{A}_{Z}\text{P} \rightarrow {}_{Z+1}^{\;\;\;A}\text{D} + \underline{\text{e}^- \text{ or e}^+} + \underline{\nu \text{ or } \overline{\nu}}$. Use conservation principles to figure out which product belongs on each line. Explain.

A Final Word

Congratulations! You have just finished reading this entire, enormous book. But studying physics does not end when you close this book, not even after you ace your final exam. Physics takes a long time to digest fully. It is the sort of subject you need to live with. So my advice is: Keep thinking about physics. Keep your book. Keep your notes. Keep your old homework and tests. You will find times when you will look at all of these. Next semester, you may be asked to solve a homework problem similar to something you saw in this class. Perhaps one day, you will need to decide whether to buy a diesel car. Before seeing the dealer, it will help to reread Chapter 22 on engines. Or you may read that there are more neural connections in your brain than atoms in the universe, and you may want to look up a few facts in Appendix B to see if that makes sense. Maybe you will just be sentimental one day, and take a look at your old college work. Rutherford (the inventor of the solar system model of the atom) once took a look at his old notes. When he did he said,

> I've just finished reading some of my early papers, and you know, when I'd finished I said to myself, "Rutherford, my boy, you used to be a damned clever fellow."

You are clever too. Don't throw your work away, or sell your used book for the price of a pizza. Keep connecting physics with your life. Thank you for reading this book.

Debora M. Katz

SUMMARY

Underlying Principles

1. **Models of nuclei**
   a. In the **shell nuclear model**, each nucleon can be treated as a particle in the system, while the other part of the system is the rest of the nucleus. In this model the system's energy is quantized. In general, proton energy levels are farther apart than are neutron energy levels.
   b. In the **liquid drop model**, nucleons move around the nucleus much like molecules moving in a drop of liquid, and collide frequently.

   Today's nuclear models are a combination of the shell model and the liquid drop model, in which the core is modeled as a liquid drop and the outer nucleons fit the shell model with its quantized states.
2. The **strong nuclear force** is one of the four fundamental forces of physics; it binds quarks together to form nucleons and holds the nucleus together. The strong force has a short range, acting over distances of roughly 1 fm or less. At this short range, the strong force is about 100 times stronger than the Coulomb force.
3. The **weak nuclear force** is another of the four fundamental forces of physics; at the level of quarks it transforms a neutron into a proton or a proton into a neutron. The weak force has a short range and is about 10^7 times weaker than the strong force.
4. **Electroweak theory** unifies the electromagnetic and the weak forces. The electroweak theory explains phenomena that cannot be accounted for by either the electromagnetic theory or the weak theory alone. Perhaps more amazingly, the electroweak theory predicted the existence of three bosons: W^+, W^- and Z. The existence of these bosons was later confirmed by experimentation.
5. **QED and QCD**
   a. **Quantum electrodynamics** (QED) is a confirmed theory combining electromagnetism with quantum mechanics. According to QED, photons are force carrying particles that play a role in all phenomena that involve the electromagnetic force.
   b. **Quantum chromodynamics** (QCD) is another confirmed theory that describes how the strong force is carried by gluons and acts on quarks.
6. **The standard model** is a combination of the electroweak theory and QCD. So the standard model accounts for the behaviors involving all the forces except gravity. Put simply the standard model lists the particles that make up that the universe, explains their properties and describes the forces by which they interact.

Major Concepts

1. The effective **nuclear radius** is based on a spherical model and is given by:

$$R = r_0 A^{1/3} \tag{43.2}$$

   where $r_0 = 1.2$ fm and A is the mass number.
2. **Nuclear decay:** The parent nucleus of a radioactive substance may **decay** into a daughter nucleus, so over time a sample of radioactive material is transformed into of the daughter nuclei. The number of parent nuclei decreases exponentially:

$$N = N_0 e^{-\eta t} \tag{43.5}$$

   where N_0 is the number of parent nuclei at time $t = 0$. The decay constant η is more conveniently expressed in terms of the half-life $T_{1/2}$:

$$T_{1/2} = \frac{\ln 2}{\eta} \tag{43.7}$$

   or in terms of the **time constant**, defined as $\tau \equiv 1/\eta$.
3. A bound system has a **mass deficit** Δm given by:

$$\Delta m = \sum m_i - M \tag{43.15}$$

   where $\sum m_i$ is the sum of the free particles' individual masses, and M is the mass of the bound system. The **binding energy** is the amount of energy supplied by an outside source required to pull the particles out of the bound system and separate them into free particles at rest. Mathematically, the binding energy is

$$E_B = \Delta mc^2 \tag{43.16}$$

   Putting this together, if there is a positive mass deficit such that the product has less mass than the sum of the reactant's masses, then binding energy *increases* and energy is released.
4. In a **fusion** reaction lighter nuclei combine to form a heavier product. In a **fission** reaction a heavy nucleus divides in to two (or more) fragments.
5. a. The **absorbed dose** D is the energy delivered to tissue per unit mass of the tissue. The SI unit of dose is called the gray and abbreviated Gy: 1 Gy = 1 J/kg.
   b. The **equivalent dose** is given by

$$H = QD \tag{43.23}$$

Major Concepts (Continued)

where Q is a dimensionless quantity called the **radiation weighting factor** Q, which quantifies the impact different types of radiation have on living tissue

c. The **effective dose** D_{eff} takes the tissue weighting factor into account and is given by

$$D_{\text{eff}} = wH \quad (43.24)$$

6. **Bosons** (Table 43.6) do not obey Pauli's exclusion principle, but **fermions** (Table 32.7) do. So bosons can have the same quantum numbers, but fermions cannot have the same quantum numbers. Further, all known fermions, such as electrons, are spin ½ particles, whereas bosons have integer spin.
7. **Leptons** are fundamental particles that cannot be subdivided. There are six leptons and six antileptons (Table 43.7). **Hadrons**, on the other hand, are made of quarks. Hadrons are further subdivided into **baryons** and **mesons.** Baryons are made of three quarks, and mesons are made of two quarks (Table 43.8). Both quarks and leptons are fermions.

Special Cases

1. **Gamma rays** are photons that are emitted when a nucleon drops from a high energy level to a lower energy level.
2. **Alpha rays** are also called **alpha particles**, and are equivalent to helium nuclei. The process of emitting an alpha particle is called **alpha decay**:

$$^{A}_{Z}\text{P} \rightarrow {}^{A-4}_{Z-2}\text{D} + {}^{4}_{2}\text{He} + \text{energy} \quad (43.3)$$

3. **Beta rays, or beta particles**, is another name for electrons. The process of emitting a beta particle is called **beta decay**, in which a neutron in the parent nucleus becomes a proton:

$$^{A}_{Z}\text{P} \rightarrow {}_{Z+1}^{A}\text{D} + e^- + \bar{\nu} \quad (43.10)$$

In **inverse beta decay, a proton in** the parent nucleus becomes neutron:

$$^{A}_{Z}\text{P} \rightarrow {}_{Z-1}^{A}\text{D} + e^+ + \nu \quad (43.12)$$

4. In **electron capture**, an inner atomic electron is captured by a nuclear proton. So the nucleus gains a neutron and loses a proton:

$$^{A}_{Z}\text{P} + e^- \rightarrow {}_{Z-1}^{A}\text{D} + \nu \quad (43.13)$$

PROBLEMS AND QUESTIONS

A = algebraic C = conceptual E = estimation G = graphical N = numerical

43-1 Describing the Nucleus

1. **C** A nucleus is symbolized by $^{202}_{87}\text{Fr}$.
a. What is the name of the element? **b.** How many protons are in the nucleus? **c.** How many neutrons are in the nucleus?
2. **E** One way to probe the nucleus is to bombard a sample with high-energy electrons. To learn about the nuclear structures in a sample, the de Broglie wavelengths of these electrons would need to be a little smaller than a nuclear radius. Estimate the energy of such electrons. Give your answer in electron-volts.
3. **C** An isotope of lead has 132 neutrons. Write the symbol for this isotope in the form $^{A}_{Z}\text{X}$.
4. **N** The element nitrogen (N) has two stable isotopes—one with an atomic mass number of 14 and the other with an atomic mass number of 15. **a.** How many neutrons are in each of these nuclei? **b.** If the natural abundance of ^{15}N is 0.37%, what is the natural abundance of ^{14}N?

Problems 5 and 6 are paired.

5. **N** The lightest naturally occurring element on Earth is hydrogen (H), and the heaviest is uranium (U). **a.** What is the nuclear radius of hydrogen? **b.** What is the nuclear radius of uranium?
6. **E, N** In order to get a sense of scale of the size difference between hydrogen and uranium nuclei, assume that a baseball represents the size of the hydrogen nucleus. In this scaled system, what is the diameter of the uranium nucleus?
7. **N** Find the nuclear radius of ^{40}Ca in fermis.

43-2 The Strong Force

Problems 8 and 9 are paired.

8. **N, C** Consider two protons that are separated by 1.5 fm. What is the magnitude of the Coulomb repulsive force between them? Assume that these protons are one another's nearest neighbors and that the strong force between them is about 2000 N. Will these protons hold together or fly apart?
9. **N, C** Consider two protons that are separated by 7.0 fm. What is the magnitude of the Coulomb repulsive force between them? Assume that these protons are on opposite sides of a nucleus and that the strong force on their nearest neighbors is about 2000 N, but nearly zero between nucleons on opposite sides of the nucleus. Comment on the stability of this nucleus.
10. **C** Could an isotope of helium with two protons and no neutrons exist? Why or why not?
11. **C** Protons in a nucleus experience four fundamental forces: Coulomb repulsion, the weak nuclear force, the strong force attraction, and gravitational attraction. Why did we ignore the gravitational force in our discussions?

43-3 Models of Nuclei

Problems 12, 13, and 52 are grouped.

12. **G, C** Fill in an energy-level diagram (Fig. P43.12) for $^{16}_{8}O$. Based on your diagram, do you expect $^{16}_{8}O$ to be stable? Explain.

13. **G, C** Fill in an energy-level diagram (Fig. P43.12) for $^{19}_{8}O$. Based on your diagram, do you expect $^{19}_{8}O$ to be stable? Explain.

14. **C** Express the symbol for each of the first five doubly magic isotopes in the form $^{A}_{Z}X$, beginning with helium.

15. **C** Consider the energy-level diagram for a particular nucleus shown in Figure P43.15. Express the symbol for this isotope in the form $^{A}_{Z}X$.

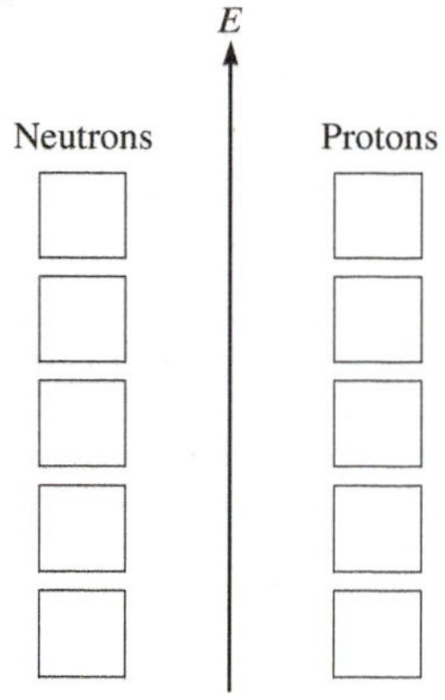

FIGURE P43.12 Problems 12 and 13.

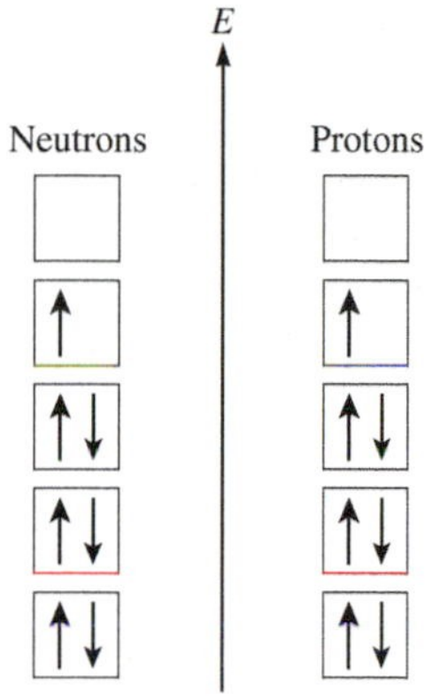

FIGURE P43.15

16. **C** There are only two stable nuclei ($^{1}_{1}H$ and $^{3}_{2}He$) that have more protons than neutrons. Why does $Z > N$ in general cause a nucleus to be unstable? Why do you suppose that these two nuclei are exceptions to the general rule?

17. **C** Consider the energy-level diagram for a particular nucleus shown in Figure P43.17. Is this nucleus stable? Why or why not?

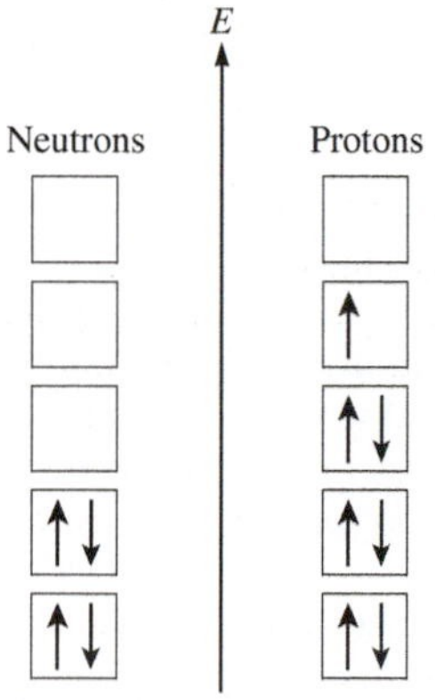

FIGURE P43.17

43-4 Radioactive Decay

18. **N** The half-life of ^{14}C is 5730 yr, and a constant ratio of $^{14}C/^{12}C = 1.3 \times 10^{-12}$ is maintained in all living tissues. A fossil is found to have $^{14}C/^{12}C = 2.65 \times 10^{-13}$. How old is the fossil?

19. **N** At $t = 0$, a radioactive sample has exactly 20,000 radioactive nuclei. How many radioactive nuclei does it have after two half-lives have passed?

Problems 20 and 21 are paired.

20. **N** The number of radioactive isotopes in a sample is found to drop to 22.50% of its original value in 924.5 s. What is the decay time constant of the isotope?

21. **N** The number of radioactive isotopes in a sample is found to drop to 22.50% of its original value in 924.5 s. What is the half-life of the isotope?

Problems 22 and 23 are paired.

22. **N** In the Fukushima Daiichi nuclear disaster (March 2011), one of the major radioactive contaminants released was iodine-131. The half-life of iodine-131 is 8.04 days. **a.** What is the decay constant in s^{-1} for iodine-131? **b.** What is the decay time constant in seconds for iodine-131?

23. **N** During the disaster at Fukushima, an estimate for the activity of iodine-131 released in the event was 355 PBq (Petabecquerels, 10^{15} Bq). The half-life of iodine-131 is 8.04 days. **a.** Assuming this was the initial activity, what will be the activity 2 days later? **b.** How many days must pass for the activity to drop to 1.00% of this initial value?

24. **N** Nitrogen-16 has a half-life of 7.13 s and decays to oxygen-16. If half of the ^{16}N nuclei in a sample have decayed, how much time has passed since the sample was created?

25. **C** Why don't we use carbon dating to measure the age of fossils that are roughly a hundred million years old?

26. **A** Integrate $dN/dt = -\eta N$ (Eq. 43.4) to show that the number of parent nuclei decreases exponentially according to $N = N_0 e^{-\eta t}$ (Eq. 43.5).

27. **A** Show that $a = a_0 e^{-\eta t}$(Eq. 43.6), where $a_0 = \eta N_0$ is the activity at $t = 0$. Start with $N = N_0 e^{-\eta t}$ (Eq. 43.5).

28. **A** Show that the half-life is given by Equation 43.7, $T_{1/2} = (\ln 2)/\eta$.

29. **A** Show that $N = N_0(1/2)^{t/T_{1/2}}$ (Eq. 43.8).

43-5 The Weak Force

30. **E** CASE STUDY Before free neutrons in the early Universe had much time to beta decay, the ratio of neutrons to protons was $N_n/N_p \approx 1/5$. At this point, what was the abundance of helium $N_{He}/(N_H + N_{He})$? Express your answer as a percentage.

31. **C** Carbon-14 beta decays into another isotope as follows: $^{14}_{6}C \rightarrow {}^{?}_{?}D + e^- + \bar{\nu}$. What is the daughter nucleus? Be sure to include the missing numbers.

32. **C** Use Equation 43.10 to express the beta decay process of each of the following parent isotopes: **a.** $^{60}_{26}Fe$, **b.** $^{10}_{4}Be$, **c.** $^{129}_{52}Te$.

33. **C** Use Equation 43.13 to express the electron capture process of each of the following parent isotopes: **a.** $^{188}_{77}Ir$, **b.** $^{127}_{54}Xe$ **c.** $^{57}_{27}Co$.

34. **C** Iodine undergoes a decay process as follows: $^{124}_{?}I \rightarrow {}^{?}_{?}Te + e^+ + \nu$. Rewrite this reaction, filling in the missing numbers. What sort of decay is this?

35. **C** Oxygen undergoes the following reaction: $^{15}_{?}O + e^- \rightarrow {}^{15}_{?}? + \nu$. Rewrite this reaction, filling in all the missing information. What sort of reaction is this?

36. **C** Use Equation 43.12 to express the inverse beta decay process of each of the following parent isotopes: **a.** $^{23}_{12}Mg$, **b.** $^{40}_{19}K$, **c.** $^{15}_{8}O$.

37. C Oxygen undergoes the following reaction: $^{19}_{?}O \rightarrow ^{19}_{?}F + ? + ?$. Rewrite this reaction, filling in all the missing information. What sort of reaction is this?

43-6 Binding Energy

38. N In one example of nuclear fusion, deuterium (^2H) fuses with tritium (^3H) to form an alpha particle and a neutron. The rest mass energies of the deuterium and the tritium are 1875.62 MeV and 2808.92 MeV, respectively, whereas the rest mass energies for the alpha particle and the neutron are 3727.38 MeV and 939.57 MeV, respectively. **a.** What is the energy released in this fusion reaction? **b.** What is the mass deficit in this reaction?

39. N According to Table 43.2, the rest mass of iron $^{56}_{26}Fe$ is 52103.06 MeV/c^2. Find its binding energy and its binding energy per nucleon. Check your answer with Figure 43.14.

40. The mass of a proton and that of a neutron are not exactly the same. The mass of a proton is 1.6726×10^{-27} kg, whereas the mass of a neutron is 1.6749×10^{-27} kg

a. N Calculate the rest mass energy of each nucleon.

b. N A neutron can decay into a proton. The decay also results in the creation of other particles. What must be the total rest mass energy of these particles?

c. C Compare the rest mass you found in part (b) to the rest mass energy of an electron (or positron).

41. N Uranium $^{238}_{92}U$ has a binding energy of 1802 MeV. What is its mass deficit in atomic mass units?

42. N According to Figure 43.11, uranium-238 alpha decays into thorium-234. Calculate the energy released in this alpha decay. The rest masses of $^{238}_{92}U$, $^{234}_{90}Th$, and $^{4}_{2}He$ are 238.05079 u, 234.04363 u, and 4.00260 u ,respectively.

43. N Use the binding energy curve (Fig. 43.14) to find the binding energy per nucleon for $^{4}_{2}He$. Then find the binding energy for the whole nucleus.

44. N Consider the masses of ^{12}C (12.0000 u), ^{13}C (13.0034 u), and ^{11}C (11.0114 u). What is the binding energy per nucleon for each of these isotopes?

45. N Consider the masses of ^{6}Li (6.015122 u) and ^{7}Li (7.016004 u). What is the binding energy per nucleon for each of these isotopes?

43-7 Fission Reactions

46. C In each of the following cases, state whether or not the fission depicted is possible and explain why or why not.

a. $^{236}_{92}U \rightarrow ^{141}_{56}Ba + ^{92}_{36}Kr + ^{1}_{0}n$

b. $^{239}_{94}Pu + ^{1}_{0}n \rightarrow ^{148}_{58}Ce + ^{89}_{36}Kr + 3(^{1}_{0}n)$

c. $^{235}_{92}U + ^{1}_{0}n \rightarrow ^{141}_{56}Ba + ^{92}_{36}Kr + 2(^{1}_{0}n)$

Problems 47, 48, and 49 are grouped

47. C Complete the following reaction by replacing $^{A}_{Z}X$ with the appropriate isotope: $^{239}_{94}Pu + ^{1}_{0}n \rightarrow ^{148}_{58}Ce + 3(^{1}_{0}n) + ^{A}_{Z}X$.

48. N Consider the fission reaction from the previous problem, where $^{239}_{94}Pu$ (239.05216 u) undergoes fission into $^{148}_{58}Ce$ (147.9242 u) and $^{A}_{Z}X$ (88.91764 u). Compute the binding energy of **a.** $^{239}_{94}Pu$, **b.** $^{148}_{58}Ce$, and **c.** $^{A}_{Z}X$.

49. N Consider the fission reaction in Problem 47 and the atomic masses given in Problem 48. **a.** Determine the energy released when the reaction occurs once. **b.** How many of these reactions must occur to equal the total output energy of the Sun each day, about 2.1×10^{44} MeV.

50. C Suppose that an operator at a nuclear power plant loses the ability to control the movement of both the fuel and control rods in a fission reactor. What are the possible, disastrous consequences, and explain how they would occur?

43-8 Fusion Reactions

51. C In each of the following cases, state whether or not the fusion depicted is possible and explain why or why not.

a. $^{1}_{1}H + ^{3}_{1}H \rightarrow ^{3}_{2}He + \gamma$

b. $^{4}_{2}He + ^{4}_{2}He + ^{4}_{2}He \rightarrow ^{12}_{6}C$

c. $^{2}_{1}H + ^{3}_{1}H \rightarrow ^{4}_{2}He + ^{1}_{0}n$

Problems 52 and 53 are paired

52. N The mass of tritium is 2808.9261 MeV/c^2. Use this and Table 43.2 to compute the mass deficit in the fusion reaction depicted in Equation 43.21, $^{2}_{1}H + ^{3}_{1}H \rightarrow ^{4}_{2}He + ^{1}_{0}n$.

53. N The mass of tritium is 2808.9261 MeV/c^2. Use this and Table 43.2 to verify that the energy released in the fusion reaction depicted in Equation 43.21, $^{2}_{1}H + ^{3}_{1}H \rightarrow ^{4}_{2}He + ^{1}_{0}n$ is 17.59 MeV.

Problems 54 and 55 are paired

54. N A fusion process called the *triple-alpha process* involves the fusing of three alpha particles, resulting in $^{12}_{6}C$. Although there is actually an intermediate step in the process, determine the mass deficit for the combination of three alpha particles resulting in $^{12}_{6}C$. Express your answer using the units MeV/c^2.

55. E, N A fusion process called the *triple-alpha process* involves the fusing of three alpha particles, resulting in $^{12}_{6}C$. Although there is actually an intermediate step in the process, determine an estimate for the energy released in this process.

43-9 Human Exposure to Radiation

56. C The quality factor for alpha radiation is much greater than that of X-rays, gamma radiation, or beta radiation. What is the meaning behind the difference in these values? In other words, why is the quality factor so much higher for alpha radiation?

57. N During a chest X-ray a patient is exposed to a dose equivalent of 0.3 mSv. The X-ray has an energy of 25 keV. If the portion of the chest that absorbed the radiation has a mass of 2.5 kg, how many X-ray photons were absorbed?

58. C How does the activity, a, affect the average absorbed dose, D_{av}? Suppose that you have a substance that emits alpha radiation of a particular energy. How will the activity affect the average absorbed dose? For example, will the effect be linear?

Problems 59 and 60 are paired.

59. N While walking outside of New Vegas, you pick up a piece of metal and are exposed to alpha radiation with a quality factor of 20. The dose absorbed by your body is 30.0 rad. **a.** What was the absorbed dose expressed in the units of grays (Gy)? **b.** What was the dose equivalent expressed in units of sieverts (Sv)? **c.** What was the dose equivalent expressed in units of rem?

60. C While walking outside of New Vegas, you pick up a piece of metal and are exposed to alpha radiation with a quality factor of 20. The dose absorbed by your body is 30.0 rad. Should you be worried about your health? Explain your answer.

61. N Beginning in the early 1920s, people sometimes exposed themselves to X-rays at the shoe store in search of a better-fitting shoe. The X-ray fluoroscope (Figure P43.61) was installed in shoe stores and offered customers the opportunity to see their feet in their shoes to observe the fit. Though primarily a gimmick, the machines remained in use in some stores until some time in the 1970s, at which point many states had banned their use. On average, during a single viewing, a user would be exposed to an equivalent dose of about 13 rem. The quality factor of X-rays is 1.0. What was the dose to which people were exposed when using the X-ray fluoroscope. Express you answer in both Gy and rad.

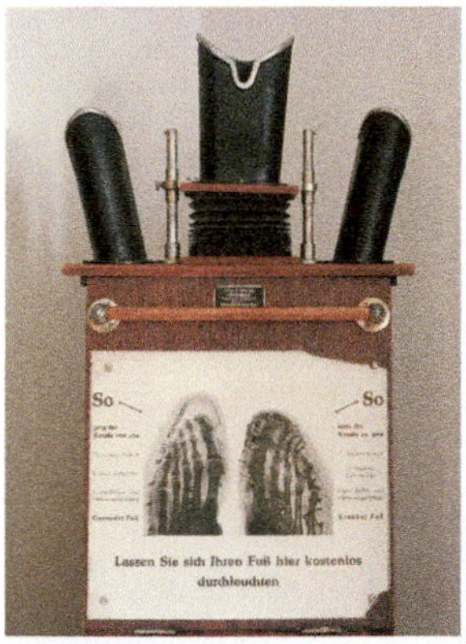

akg-images/Newscom

FIGURE P43.61

43-10 The Standard Model

62. **C** Given that the electron is a fermion, should the positron (e^+) be a fermion or a baryon? Justify your answer by considering a reaction that involves a positron.
63. **C** Is the following reaction possible: ${}^1_1p + {}^1_0n \rightarrow {}^1_1p + \mu^- + \mu^+$? Consider the baryon number before and after the reaction number. Explain your answer.
64. **C** **a.** Prior to the discovery of the electron antineutrino, its existence was predicted because of the varied amount of kinetic energy of the electron resulting from neutron decay. Given what you have learned about lepton number in this chapter, why is the following reaction dissatisfying: ${}^1_0n \rightarrow {}^1_1p + e$? **b.** Include an electron antineutrino in this reaction such that the issue you identified in part (a) is resolved. Write the resulting reaction.
65. **N** The neutron is composed of an up quark and two down quarks. How many color combinations are possible for the two down quarks?

General Problems

66. **N** A bone found in a crypt is being dated. If 25% of the original amount of carbon-14 has decayed, what is the age of the bone? The half-life of ${}^{14}C$ is 5730 yr.
67. **C** A Bose–Einstein condensate is a form of matter consisting of bosons, all occupying the same ground state. Could a form of matter consisting of fermions occupying the same ground state be created? Explain your answer.
68. **C** The isotope ${}^{19}_8O$ is unstable. What do you expect will happen to make a more stable nucleus?
69. **N** The carbon cycle is a series of reactions beginning with the fusion of ${}^{12}_6C$ and ${}^1_1p$, whereby the end result is ${}^{12}_6C$ and ${}^4_2He$. If the addition of a proton is involved at three more points during the cycle, how many of these protons must also decay into positron and neutrino pairs?
70. **N** The element neon (Ne) has three stable isotopes with atomic mass numbers of 20, 21, and 22. The ratios of the natural abundances are ${}^{20}Ne/{}^{21}Ne = 335.11$ and ${}^{20}Ne/{}^{22}Ne = 9.78$. Find the natural abundance of each isotope.
71. **N** What atomic nucleus would have a volume that is three times greater than the volume occupied by a carbon-12 nucleus? Assume that the nuclei are spheres. Make your selection by comparing to atomic masses as listed on the periodic table of elements.
72. **N** Consider the masses of ${}^{16}O$ (15.994915 u), ${}^{17}O$ (16.999132 u), and ${}^{18}O$ (17.999160 u). What is the binding energy per nucleon for each of these isotopes?

Problems 73 and 74 are paired.

73. **N** A sample material is observed to have an activity of 1.45×10^9 decays/s. Express the activity in **a.** Bq, and **b.** Ci. **c.** If the material has a half-life of 30.07 yr, how long will it take for the activity to decrease by 10.0%? **d.** How long will it take for the activity to decrease by 20.0%?
74. **N** Consider the sample material in Problem 73. **a.** What is the number of radioactive nuclei when the activity is 1.45×10^9 decays/s? **b.** How long will it take for 10% of the nuclei to decay?
75. **N** The mass of ${}^{107}Ag$ is 106.9051 u. What is the binding energy of this silver nucleus?
76. One of the few stable isotopes with more protons than neutrons is ${}^3He$ (3.016 u).
    - **a.** **C** Why is this such a rarity? Why are there not more stable isotopes of elements with more protons than neutrons?
    - **b.** **N** What is the binding energy per nucleon of ${}^3He$?

Problems 77 and 78 are paired

77. **N** The Chernobyl power station was capable of producing 4.00×10^3 MW of electrical power, prior to the disaster at one of its four reactors in 1986. Assuming that each fission reaction produces about 200 MeV and that the power plant is about 30.0% efficient, determine the number of fission events that must occur each second in the operation of the power plant. Retain three significant figures in your answer.
78. **N** The Chernobyl power station was capable of producing 4.00×10^3 MW of electrical power, prior to the disaster at one of its four reactors in 1986. Assuming that each fission reaction produces about 200 MeV and that the power plant is about 30.0% efficient, determine the mass of ${}^{235}_{92}U$ that must be used in operating the plant for 1.00 hours. Retain three significant figures in your answer.

Problems 79 and 80 are paired.

79. In nuclear fission, ${}^{235}U$ interacts with a neutron and could split into ${}^{90}Rb$, ${}^{143}Cs$, and three neutrons, releasing energy.
    - **a.** **C** Express this nuclear reaction as an equation with ${}^{235}U$ and the neutron it absorbs on the left side of the equation.
    - **b.** **N** What is the binding energy of ${}^{235}U$ (235.0439 u)?
    - **c.** **N** ${}^{90}Rb$ (89.9148 u)?
    - **d.** **N** ${}^{143}Cs$ (142.9278 u)?
80. **N** In nuclear fission, ${}^{235}U$ (235.0439 u) interacts with a neutron and could split into ${}^{90}Rb$ (89.9148 u), ${}^{143}Cs$ (142.9278 u), and three neutrons, releasing energy. What is the energy released in this reaction?

Mathematics

APPENDIX

A

This appendix is not meant to serve as a tutorial or review. It is a short list of useful formulas. A more comprehensive list may be found in various handbooks such as the *Handbook of Chemistry and Physics* (Boca Raton, FL: CRC Press, published annually). These handbooks are available in hardcover and as e-books, and they make great birthday presents for science and engineering students.

A-1 Algebra and Geometry

Quadratic formula: If $ax^2 + bx + c = 0$, then $x = \dfrac{-b \pm \sqrt{b^2 - 4ac}}{2a}$

Factorial notation: $n! = n(n-1)\cdots 2 \cdot 1$

Binomial theorem: $(1+x)^n = 1 + \dfrac{nx}{1!} + \dfrac{n(n-1)x^2}{2!} + \cdots \quad (x < 1)$

Commonly used approximation: $(1+x)^n \approx 1 + nx \ (x \ll 1)$

Exponential expansion: $e^x = 1 + x + \dfrac{x^2}{2!} + \dfrac{x^3}{3!} + \cdots$

Logarithms (any base):

$$\log(x)^n = n\log x$$
$$\log(AB) = \log A + \log B$$
$$\log(A/B) = \log A - \log B$$

Logarithms (base 10):

$$10^{\log x} = x$$
$$\log 10^x = x$$

Natural logarithms (base *e*):

$$e^{\ln x} = x$$
$$\ln e^x = x$$

Area of common shapes:

Rectangle
$A = \ell w$

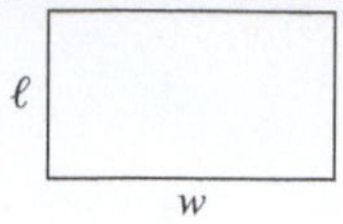

Circle
$A = \pi r^2$

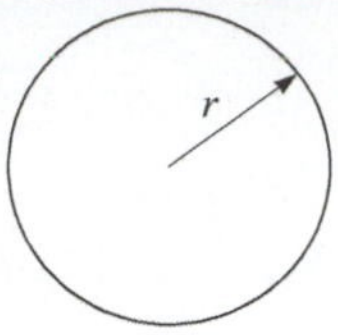

Parallelogram
$A = bh$

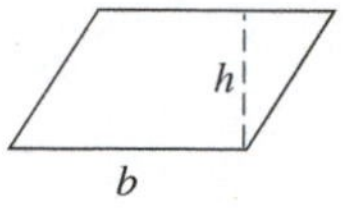

Ellipse
$A = \pi ab$

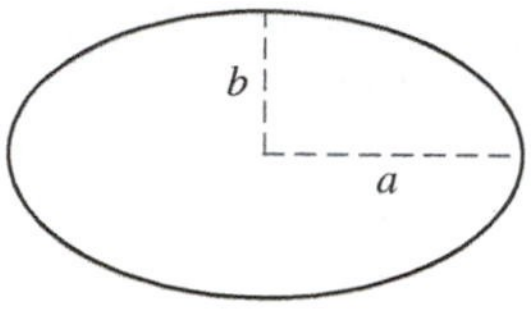

Triangle
$A = \frac{1}{2}bh$

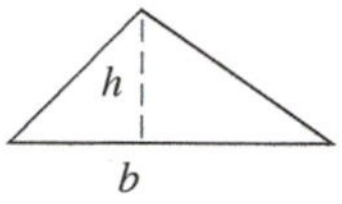

Equation of a straight line of slope *m* and *y* intercept *b*: $y = mx + b$

Equation of a parabola: $y = ax^2 + bx + c$

Equation of a circle: $x^2 + y^2 = r^2$

Equation of an ellipse: $\left(\frac{x}{a}\right)^2 + \left(\frac{y}{b}\right)^2 = 1$

Circumference of a circle: $c = 2\pi r$

Volume and surface area of common solids:

Rectangular box $V = \ell wh$ $A = 2(\ell w + \ell h + wh)$

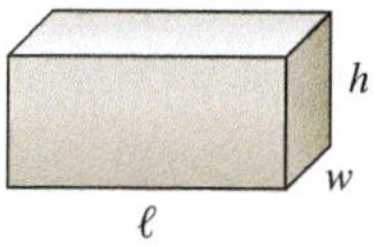

Right circular cylinder $V = \pi r^2 h$ $A = 2\pi r^2 + 2\pi rh$

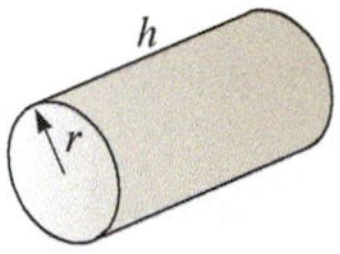

Sphere $V = \frac{4}{3}\pi r^3$ $A = 4\pi r^2$

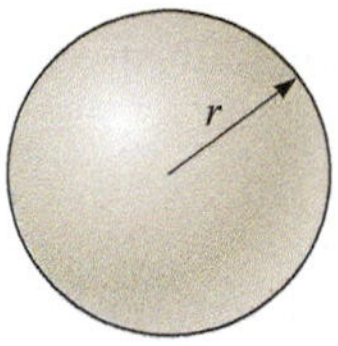

A-2 Trigonometry

Pythagorean theorem (applied to a right triangle): $x^2 + y^2 = r^2$

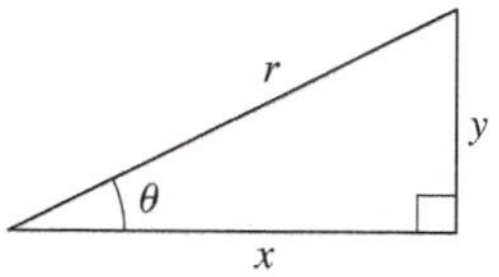

Trigonometric functions (applied to a right triangle):

$$\sin\theta = \frac{y}{r} \qquad \csc\theta = \frac{1}{\sin\theta} = \frac{r}{y}$$

$$\cos\theta = \frac{x}{r} \qquad \sec\theta = \frac{1}{\cos\theta} = \frac{r}{x}$$

$$\tan\theta = \frac{y}{x} \qquad \cot\theta = \frac{1}{\tan\theta} = \frac{x}{y}$$

Commonly used approximations:

$$\sin\theta \approx \tan\theta \approx \theta \text{ for small } \theta \text{ in radians}$$

$$\cos\theta \approx 1 \text{ for small } \theta$$

Trigonometric identities:

$\sin(-\theta) = -\sin\theta$ $\cos(-\theta) = \cos\theta$ $\tan(-\theta) = -\tan\theta$	$\sin 2\theta = 2\sin\theta\cos\theta$ $\cos 2\theta = \cos^2\theta - \sin^2\theta$
$\sin(90° - \theta) = \sin\left(\frac{\pi}{2} - \theta\right) = \cos\theta$	$\sin(\alpha \pm \beta) = \sin\alpha\cos\beta \pm \cos\alpha\sin\beta$ $\cos(\alpha \pm \beta) = \cos\alpha\cos\beta \mp \sin\alpha\sin\beta$
$\cos(90° - \theta) = \cos\left(\frac{\pi}{2} - \theta\right) = \sin\theta$	
$\sin(90° + \theta) = \sin\left(\frac{\pi}{2} + \theta\right) = \cos\theta$	$\cos\theta = \sin(\theta + 90°)$
$\cos(90° + \theta) = \cos\left(\frac{\pi}{2} + \theta\right) = -\sin\theta$	
$\tan\theta = \frac{\sin\theta}{\cos\theta}$ $\sin^2\theta + \cos^2\theta = 1$	$\sin\alpha \pm \sin\beta = 2\sin\frac{1}{2}(\alpha \pm \beta)\cos\frac{1}{2}(\alpha \mp \beta)$ $\cos\alpha + \cos\beta = 2\cos\frac{1}{2}(\alpha + \beta)\cos\frac{1}{2}(\alpha - \beta)$ $\cos\alpha - \cos\beta = -2\sin\frac{1}{2}(\alpha + \beta)\sin\frac{1}{2}(\alpha - \beta)$

A-3 Calculus

Derivatives

In this section, f, g, u, and v are functions of x; a, b, C, and n are constants.

Derivative of $f(x)$: $\dfrac{df}{dx} = \lim\limits_{\delta x \to 0} \dfrac{f(x + \delta x) + f(x)}{\delta x}$

Derivative of a constant: $\dfrac{dC}{dx} = 0$

The power rule: $\dfrac{dx^n}{dx} = nx^{n-1}$

The derivative of a sum: $\dfrac{d}{dx}[f(x) + g(x)] = \dfrac{df}{dx} + \dfrac{dg}{dx}$

The product rule: $\dfrac{d}{dx}[f(x)g(x)] = g\dfrac{df}{dx} + f\dfrac{dg}{dx}$

Special case of the product rule: $\dfrac{d}{dx}[Cf(x)] = C\dfrac{df}{dx}$

The chain rule: $\dfrac{df}{dx} = \dfrac{df}{du}\dfrac{du}{dx}$

Second derivative: $\dfrac{d^2f}{dx^2} = \dfrac{d}{dx}\left(\dfrac{df}{dx}\right)$

Derivatives of special functions:

$\frac{d}{dx}\ln x = \frac{1}{x}$	$\frac{d}{dx}\sin x = \cos x$
$\frac{d}{dx}e^x = e^x$	$\frac{d}{dx}\cos x = -\sin x$
$\frac{d}{dx}e^u = e^u\frac{du}{dx}$	$\frac{d}{dx}\tan x = \sec^2 x$

L'Hôpital's rule: If the limit of the numerator $f(x)$ and of the denominator $g(x)$ of a fraction both approach zero or both approach infinity, the limit of the fraction is indeterminate in the form of type $0/0$ or ∞/∞. In such cases, it may be possible to find the limit using L'Hôpital's rule:

$$\lim\frac{f(x)}{g(x)} = \lim\frac{df/dx}{dg/dx}$$

Essentially, L'Hôpital's rule replaces the limit of a fraction with the limit of a new fraction, where the numerator and denominator are the derivatives of their original counterparts.

Integrals

Indefinite integral or antiderivative: $\int df(x) = f(x) + C$

Indefinite integral of a constant: $\int a dx = ax + C$

The power rule: $\int x^n dx = \frac{x^{n+1}}{n+1} + C \; (n \neq -1)$

Indefinite integral of a sum: $\int [f(x) + g(x)]dx = \int f(x)\,dx + \int g(x)\,dx$

Indefinite integrals of particular functions:

$\int \frac{1}{x}dx = \ln\lvert x\rvert + C$	$\int(\sin ax)dx = -\frac{1}{a}\cos ax + C$	$\int\frac{dx}{\sqrt{x^2+a^2}} = \ln\lvert x + \sqrt{x^2+a^2}\rvert + C$
$\int e^x\,dx = e^x + C$	$\int(\cos ax)\,dx = \frac{1}{a}\sin ax + C$	$\int\frac{dx}{\sqrt{a^2-x^2}} = \sin^{-1}\frac{x}{\lvert a\rvert} + C$
$\int e^{-ax}\,dx = -\frac{1}{a}e^{-ax} + C$	$\int(\tan ax)\,dx = -\frac{1}{a}\ln(\cos ax) + C$	$\int\frac{dx}{x^2+a^2} = \frac{1}{a}\tan^{-1}\frac{x}{a} + C$
	$\int\sin^2 ax\,dx = \frac{x}{2} - \frac{\sin 2\,ax}{4a} + C$	$\int\frac{dx}{(x^2+a^2)^{3/2}} = \frac{x}{a^2\sqrt{x^2+a^2}} + C$
	$\int\cos^2 ax\,dx = \frac{x}{2} + \frac{\sin 2\,ax}{4a} + C$	$\int\frac{x\,dx}{(x^2+a^2)^{3/2}} = -\frac{1}{\sqrt{x^2+a^2}} + C$

***U* substitution integration method:** $\int f(x)\frac{df}{dx}dx = \int u\,du$, where $u \equiv f(x)$ and $du = \frac{df}{dx}dx$

Integration by parts: $\int udv = uv - \int vdu$

Definite integrals and the Fundamental Theorem of Calculus:
If $F(x)$ is continuous on the interval from $x = a$ to b, then

$$\int_a^b F(x)\,dx = f(x)\rvert_a^b = f(b) - f(a)$$

where $f(x)$ is the antiderivative of $F(x)$ and a and b are known as the ***limits***.

Average value of a function in the interval from $x = a$ to $x = b$: $f_{av} = \frac{1}{b-a}\int_a^b f(x)\,dx$

Definite integrals of particular functions:

$\int_0^\infty x^n e^{-ax}\,dx = \frac{n!}{a^{n+1}}$	$\int_0^\pi (\sin^2 ax)\,dx = \frac{\pi}{2}$
$\int_0^\infty e^{-ax^2}\,dx = \frac{1}{2}\sqrt{\frac{\pi}{a}}$	$\int_0^\pi (\cos^2 ax)\,dx = \frac{\pi}{2}$
$\int_0^\infty xe^{-ax^2}\,dx = \frac{2}{a}$	$\int_0^\infty e^{-ax}(\sin nx)\,dx = \frac{n}{a^2+n^2}\ (a > 0)$
$\int_0^\infty x^2 e^{-ax^2}\,dx = \frac{1}{4}\sqrt{\frac{\pi}{a^3}}$	$\int_0^\infty e^{-ax}(\cos nx)\,dx = \frac{a}{a^2+n^2}\ (a > 0)$

A-4 Propagation of Uncertainty

Physics is an experimental science that relies on measurements. All measurements have uncertainty or error. We often seek quantities that result from the combination of uncertain measurements, so the resulting quantity is also uncertain. One way to estimate the uncertainty δQ in a quantity q calculated from uncertain measurements is to find the best estimate Q and then the extreme possible values of q.

For example, if $q = 2(a + b)/c$, $a = 5.2 \pm 0.2$, $b = 7.5 \pm 0.3$, and $c = 57.6 \pm 0.5$, the best estimate of q is

$$Q = \frac{2(5.2 + 7.5)}{57.6} = 0.44$$

The maximum value of q is

$$Q_{max} = \frac{2(5.4 + 7.8)}{57.1} = 0.46$$

where we chose the maximum possible values in the numerator and the minimum possible value in the denominator. Similarly, we find that the minimum value of q is

$$Q_{min} = \frac{2(5.0 + 7.2)}{58.1} = 0.42$$

The quantity must fall between its minimum and maximum values: $0.42 \le q \le 0.46$. So,

$$q = Q \pm \delta Q = 0.44 \pm 0.02$$

where $\delta Q = Q_{max} - Q = Q - Q_{min}$.

We apply this technique to come up with three rules for propagating uncertainty.

Sums and Differences

If $q = a + b$, $a = A \pm \delta A$, and $b = B \pm \delta B$, then

$$q = Q \pm \delta Q = (A + B) \pm (\delta A + \delta B)$$

If $q = a - b$, $a = A \pm \delta A$, and $b = B \pm \delta B$, then

$$q = Q \pm \delta Q = (A - B) \pm (\delta A + \delta B)$$

When measured quantities are added or subtracted, their errors add.[1]

[1]If the original uncertainties δA and δB are random and independent, then $\delta A + \delta B$ is an overestimate of the propagated error.

Products, Quotients, and Powers

If $a = A \pm \delta A$, the *fractional uncertainty* in a is $\delta A/|A|$. If $q = ab$, $a = A \pm \delta A$, and $b = B \pm \delta B$, then $Q = AB$, and the fractional uncertainty in q is approximately

$$\frac{\delta Q}{|Q|} \approx \frac{\delta A}{|A|} + \frac{\delta B}{|B|} \tag{1}$$

and

$$q = Q \pm \delta Q \approx AB \pm |AB|\left(\frac{\delta A}{|A|} + \frac{\delta B}{|B|}\right)$$

If $q = a/b$, $a = A \pm \delta A$, and $b = B \pm \delta B$, then $Q = A/B$. The fractional uncertainty in q is given by Equation (1), and

$$q = Q \pm \delta Q \approx \frac{A}{B} \pm \left|\frac{A}{B}\right|\left(\frac{\delta A}{|A|} + \frac{\delta B}{|B|}\right)$$

If $q = a^n$ and $a = A \pm \delta A$, then $Q = A^n$, and the fractional uncertainty in q is

$$\frac{\delta Q}{|Q|} \approx n\frac{\delta A}{|A|}$$

and

$$q = Q \pm \delta Q \approx A^n \pm |A^n|\left(n\frac{\delta A}{|A|}\right)$$

When measured quantities are multiplied or divided, their fractional errors add.

Multiplication by an Exact Number

If $q = ab$, a is exact, and $b = B \pm \delta B$, then

$$q = Q \pm \delta Q = aB \pm |a|\delta B$$

Reference Tables

APPENDIX

B

B-1 Symbols and Units

Prefixes for powers of 10

Name	Abbreviation	Value
yocto	y	10^{-24}
zepto	z	10^{-21}
atto	a	10^{-18}
femto	f	10^{-15}
pico	p	10^{-12}
nano	n	10^{-9}
micro	μ (Greek letter "mu")	10^{-6}
milli	m	10^{-3}
centi	c	10^{-2}
deci	d	10^{-1}
deka	da	10^{1}
hecto	h	10^{2}
kilo	k	10^{3}
mega	M	10^{6}
giga	G	10^{9}
tera	T	10^{12}
peta	P	10^{15}
exa	E	10^{18}
zetta	Z	10^{21}
yotta	Y	10^{24}

Greek alphabet

Name	Uppercase	Lowercase	Name	Uppercase	Lowercase	Name	Uppercase	Lowercase
Alpha	A	α	Iota	I	ι	Rho	P	ρ
Beta	B	β	Kappa	K	κ	Sigma	Σ	σ
Gamma	Γ	γ	Lambda	Λ	λ	Tau	T	τ
Delta	Δ	δ	Mu	M	μ	Upsilon	Υ	υ
Epsilon	E	ε	Nu	N	ν	Phi	Φ	φ
Zeta	Z	ζ	Xi	Ξ	ξ	Chi	X	χ
Eta	H	η	Omicron	O	o	Psi	Ψ	ψ
Theta	Θ	θ	Pi	Π	π	Omega	Ω	ω

SI base units

Dimension	SI unit	Symbol	Definition
Time	second	s	1 second is the duration of 9,192,631,770 periods of the radiation (corresponding to the transition between hyperfine levels of the ground state) of the cesium-133 atom.
Length	meter	m	1 meter is the distance light travels through empty space in 1/299,729,458 second.
Mass	kilogram	kg	1 kilogram is the mass of a prototype (a particular platinum-iridium cylinder).
Thermodynamic temperature	kelvin	K	1 kelvin is the fraction 1/273.16 of the thermodynamic temperature of the triple point of water.
Amount of substance	mole	mol	1 mole is the amount of a substance of a system that contains as many elementary entities as there are atoms in 0.012 kg of carbon-12.
Electrical current	ampere	A	1 ampere is the constant current that, if maintained in two straight parallel conductors of infinite length, of negligible circular cross section, and placed 1 m apart in a vacuum, would produce between these conductors a force per unit length equal to 2×10^{-7} N/m.
Luminous intensity	candela	cd	1 candela is the luminous intensity, in a given direction, of a source that emits monochromatic radiation of frequency 540×10^{12} Hz and that has a radiant intensity in that direction of 1/683 watt per steradian.

Source: Adapted from "Definitions of the SI base units," National Institute of Standards and Technology. See http://physics.nist.gov/cuu/Units/current.html.

Symbols and abbreviations for units

Unit	Symbol	Unit	Symbol
ampere	A	light-year	ly
atmosphere	atm	liter	L
atomic mass unit	U	meter	m
British thermal unit	Btu	mile	mi
calorie	cal	miles per hour	mph
coulomb	C	millimeter of mercury (torricelli)	mm Hg (torr)
day	d	minute	min
degree Celsius	°C	mole	mol
degree Fahrenheit	°F	newton	N
electron volt	eV	ohm	Ω
farad	F	pascal	Pa
foot	ft	pound	lb
gallon	gal	pounds per square inch	psi
gauss	G	radian	rad
gram	g	revolution	rev
henry	H	revolutions per minute	rpm
hertz	Hz	second	s
horsepower	hp	tesla	T
inch	in.	volt	V
joule	J	watt	W
kelvin	K	weber	Wb
kilocalorie	Cal	yard	yd
kilogram	kg	year	yr
kilowatt-hour	kWh		

B-2 Conversion Factors

Length

	meter	cm	km	in.	ft	mi
1 meter	1	10^2	10^{-3}	39.37	3.281	6.214×10^{-4}
1 centimeter	10^{-2}	1	10^{-5}	0.3937	3.281×10^{-2}	6.214×10^{-6}
1 kilometer	10^3	10^5	1	3.937×10^4	3281	0.6214
1 inch	2.540×10^{-2}	2.540	2.540×10^{-5}	1	8.333×10^{-2}	1.578×10^{-5}
1 foot	0.3048	30.48	3.048×10^{-4}	12	1	1.894×10^{-4}
1 mile	1609	1.609×10^5	1.609	6.336×10^4	5280	1
1 angstrom	10^{-10}	10^{-8}	10^{-13}	3.937×10^{-9}	3.281×10^{-10}	6.214×10^{-14}
1 AU	1.496×10^{11}	1.496×10^{13}	1.496×10^8	5.890×10^{12}	4.908×10^{11}	9.296×10^7
1 nautical mile	1852	1.852×10^5	1.852	7.291×10^4	6076	1.151
1 light-year	9.461×10^{15}	9.461×10^{17}	9.461×10^{12}	3.725×10^{17}	3.104×10^{16}	5.878×10^{12}
1 parsec	3.086×10^{16}	3.086×10^{18}	3.086×10^{13}	1.215×10^{18}	1.012×10^{17}	1.917×10^{13}
1 yard	0.9144	91.44	9.144×10^{-4}	36	3	5.682×10^{-4}

Mass

	kilogram	g	slug	u
1 kilogram	1	10^3	6.852×10^{-2}	6.022×10^{26}
1 gram	10^{-3}	1	6.852×10^{-5}	6.022×10^{23}
1 slug	14.59	1.459×10^4	1	8.786×10^{27}
1 atomic mass unit	$1.6605402 \times 10^{-27}$	1.661×10^{-24}	1.138×10^{-28}	1

Force

	newton	dyne	lb	oz	ton
1 newton	1	10^5	0.2248	3.597	1.124×10^{-4}
1 dyne	10^{-5}	1	2.248×10^{-6}	3.597×10^{-5}	1.124×10^{-9}
1 pound	4.448	4.448×10^5	1	16	5×10^{-4}
1 ounce	0.2780	2.780×10^4	6.250×10^{-2}	1	3.125×10^{-5}
1 ton	8.896×10^3	8.896×10^8	2000	3.2×10^4	1

Pressure

	pascal	atm	Torr (mm Hg)	psi	dyne/cm^2
1 pascal	1	9.869×10^{-6}	7.501×10^{-3}	1.450×10^{-4}	10
1 atm	1.013×10^5	1	760	14.70	1.013×10^6
1 Torr	1333	1.316×10^{-2}	1	0.1934	1.333×10^4
1 psi	6.895×10^3	6.805×10^{-2}	51.71	1	6.895×10^4
1 dyne/cm^2	0.1	9.869×10^{-7}	7.501×10^{-4}	1.405×10^{-5}	1

Energy

	joule	erg	ft·lb	cal	eV
1 joule	1	10^7	0.7376	0.2389	6.242×10^{18}
1 erg	10^{-7}	1	7.376×10^{-8}	2.389×10^{-8}	6.242×10^{11}
1 ft·lb	1.356	1.356×10^7	1	0.3238	8.464×10^{18}
1 cal	4.184	4.184×10^7	3.088	1	2.612×10^{19}
1 eV	1.602×10^{-19}	1.602×10^{-19}	1.182×10^{-19}	3.827×10^{-20}	1

B-3 Some Astronomical Data

Object	Symbol	Rotation period (hh:mm:ss.s or days)	Mass ($\times 10^{24}$ kg)	Equatorial radius ($\times 10^6$ m)	Free-fall acceleration near surface (m/s^2)	Escape speed (km/s)	Blackbody temperature (K)
Sun	⊙	$\approx$ 25 to 36 days[1]	1.9891×10^6	695.51	274	618	5777
Mercury	☿	58.65 days	0.3302	2.4397	3.7	4.3	440.1
Venus	♀	243 days	4.87	6.052	8.9	10.36	184.2
Earth	⊕	23:56:4.1	5.9736	6.378136	9.81	11.186	254.3
Moon	☾	27.3 days	0.07	1.738	1.6	2.38	270.7
Mars	♂	24:37:22.6	0.64	3.397	3.7	5.03	210.1
Ceres	⚳	09:04:19	9.6×10^{-4}	0.48			239
Jupiter	♃	9:50:30	1900	71.493	24.8	59.5	110.0
Saturn	♄	10:14:00	569	60.268	10.4	35.5	81.1
Uranus	⛢	17:14:00	87	25.559	8.87	21.3	58.2
Neptune	♆	16:03:00	103	24.764	11.2	23.5	46.6
Pluto	♇	6.387 days	0.01	1.135	0.58	1.2	37.5
Eris			$\approx 10^{-2}$	1.2			30

[1]The Sun is gaseous and does not rotate as a solid body; its period near the equator is shorter than at the poles.

Orbital parameters for objects that orbit the Sun

Object	Orbital period (days or years)	Semimajor axis (AU)	Eccentricity
Mercury	87.969 days	0.387	0.2056
Venus	224.701 days	0.723	0.0067
Earth	365.26 days	1.000	0.0167
Mars	1.8808 years	1.524	0.0935
Ceres	4.603 years	2.767	0.097
Jupiter	11.8618 years	5.204	0.0489
Saturn	29.4567 years	9.5482	0.0565
Uranus	84.0107 years	19.201	0.0457
Neptune	164.79 years	30.047	0.0113
Pluto	247.68 years	39.482	0.2488
Eris	559 years	67.89	0.4378

Some natural satellites

Satellite	Planet	Orbital period (days)	Semimajor axis (10^6 m)	Mass (10^{22} kg)	Radius (10^6 m)
Moon	Earth	27.322	384.4	7.349	1.7371
Io	Jupiter	1.769	421.6	8.932	1.8216
Europa	Jupiter	3.551	670.9	4.800	1.5608
Ganymede	Jupiter	7.155	1070.4	14.819	2.6312
Callisto	Jupiter	16.689	1882.7	10.759	2.4103
Titan	Saturn	15.945	1221.8	13.455	2.575
Triton	Neptune	5.877	354.8	2.14	1.3534

B-4 Rough Magnitudes and Scales

Numbers

Quantity	Approximate value or order of magnitude
Number of atoms in the Earth	10^{50}
Number of atoms in a 70-kg person	7×10^{27}
Number of cells in a person	5×10^{13}
Number of mobile phones in U.S.	330 million
Number of mobile phones worldwide	5 billion
Number of dogs in U.S.	78 million
Number of dogs in Italy	8 million
Population of the Earth	7 billion
Population of students at University of CA	220,000
Population of U.S.	300 million
Population of China	1.3 billion
Population of New York City	8 million
Population of Annapolis, MD	38,000
Population of Chesterton, IN	13,000
Population of Morris, MN	5,000
Veterans in U.S.	23 million
Percentage of people in U.S. under age of 18	24%
Money spent in film investments in U.S.	$15 billion
Money spend in film investments in India	$200 million

Sizes: Lengths, diameters, areas, and volumes

Quantity	SI or metric units	U.S. customary units
Area of a $1 bill	100 cm^2	17 $in.^2$
Area of a typical college campus	15 km^2	6.5 mi^2
Area of a cell phone	30 cm^2	5 $in.^2$
Area of continents	1.5×10^{14} m^2	6×10^7 mi^2
Area of oceans	3.6×10^{14} m^2	1.4×10^8 mi^2
Area of palm	40 cm^2	6 $in.^2$
Area of U.S. land	9×10^{12} m^2	3.5×10^6 mi^2
Average human stride	1 m	1 yd
Diameter of a hydrogen atom	10^{-10} m	
Diameter of a pollen grain	10–100 μm	
Diameter of a proton	10^{-15} m	
Diameter of a U.S. nickel	2.121 cm	0.835 in.
Diameter of the Milky Way galaxy	10^{21} m	10^5 ly
Height of a typical adult human	2 m	5–6 ft
Height of a typical story	3 m	10 ft
Length of a house fly	0.5 cm	0.2 in.
Length of a human thumb	5 cm	2 in.
Length of a match stick	5 cm	2 in.
Size of a living cell	10 μm	
Size of the smallest visible dust particle	0.1 μm	
Thickness of a human hair	50 μm	
Thickness of a U.S. nickel	1.95 mm	
Width of human finger	1–3 cm	

Speeds

	SI or metric units	U.S. customary units
Top speed of a car	200 km/h	120 mph
Top speed of a *typical* bicycle	50 km/h	30 mph
Walking	1.3 m/s	3 mph
Running or jogging	15 km/h	6-min. mile (10 mph)
Commercial airplane cruising speed	250 m/s	550 mph
Speed of a snail	1 mm/s	2–3 inch/min
Speed of a cheetah	28 m/s	62 mph
Speed of a rifle bullet	700 m/s	1600 mph

Weights and masses

	SI or metric units	U.S. customary units
Mass of a U.S. nickel	5.000 g	3×10^{-4} slug (approx)
Weight of a car	10000–20000 N	1–2 tons
Mass of a car	1000–2000 kg	70–140 slug
Weight of a physics book	50 N	10 lb
Mass of a physics book	5 kg	0.4 slug
Weight of a U.S. quarter	6×10^{-2} N	0.2 oz
Mass of a U.S. quarter	6 g	4×10^{-4} slug
Mass of the Milky Way galaxy	10^{42} kg	10^{41} slug
Mass of an elephant	5×10^{3} kg	340 slug
Mass of a frog	100 g	7×10^{-3} slug
Mass of a house fly	8–20 mg	(5–14) $\times 10^{-4}$ slug
Weight of an adult human	500–1000 N	110–200 lbs
Mass of an adult human	50–100 kg	4–7 slug

Times, ages, periods, frequencies, and angular momentum

Quantity	Convenient units
Resting heart beat	60–80 per min
Age of the Universe	14 billion years
Age of human written history	10^4 years
Age of the Earth	4.5 billion years
Age of oldest fossil	2.7 billion years
Time for light to travel from the Sun to the Earth	10 min
Time for light to cross the diameter of a proton	3.3×10^{-24} s
Period of Halley's comet	2.4×10^9 s
Period of a typical x-ray	10^{-19} s
Time for light to travel from nearest star	4.3 years
Angular speed of record turntable	33 rpm
Angular momentum of record (33 rpm)	6 mJ · s
Angular momentum of electric fan	1 J · s
Angular momentum of Frisbee	0.1 J · s
Angular momentum of helicopter rotor (320 rpm)	5×10^4 J · s

Periodic Table of the Elements

Key:
Symbol — Ca
Atomic number — 20
Atomic mass† — 40.078
Electron configuration — $4s^2$

Group I	Group II	Transition elements						
H 1 1.007 9 $1s$								
Li 3 6.941 $2s^1$	Be 4 9.0122 $2s^2$							
Na 11 22.990 $3s^1$	Mg 12 24.305 $3s^2$							
K 19 39.098 $4s^1$	Ca 20 40.078 $4s^2$	Sc 21 44.956 $3d^14s^2$	Ti 22 47.867 $3d^24s^2$	V 23 50.942 $3d^34s^2$	Cr 24 51.996 $3d^54s^1$	Mn 25 54.938 $3d^54s^2$	Fe 26 55.845 $3d^64s^2$	Co 27 58.933 $3d^74s^2$
Rb 37 85.468 $5s^1$	Sr 38 87.62 $5s^2$	Y 39 88.906 $4d^15s^2$	Zr 40 91.224 $4d^25s^2$	Nb 41 92.906 $4d^45s^1$	Mo 42 95.94 $4d^55s^1$	Tc 43 (98) $4d^55s^2$	Ru 44 101.07 $4d^75s^1$	Rh 45 102.91 $4d^85s^1$
Cs 55 132.91 $6s^1$	Ba 56 137.33 $6s^2$	57–71*	Hf 72 178.49 $5d^26s^2$	Ta 73 180.95 $5d^36s^2$	W 74 183.84 $5d^46s^2$	Re 75 186.21 $5d^56s^2$	Os 76 190.23 $5d^66s^2$	Ir 77 192.2 $5d^76s^2$
Fr 87 (223) $7s^1$	Ra 88 (226) $7s^2$	89–103**	Rf 104 (261) $6d^27s^2$	Db 105 (262) $6d^37s^2$	Sg 106 (266)	Bh 107 (264)	Hs 108 (277)	Mt 109 (268)

*Lanthanide series	La 57 138.91 $5d^16s^2$	Ce 58 140.12 $5d^14f^16s^2$	Pr 59 140.91 $4f^36s^2$	Nd 60 144.24 $4f^46s^2$	Pm 61 (145) $4f^56s^2$	Sm 62 150.36 $4f^66s^2$
**Actinide series	Ac 89 (227) $6d^17s^2$	Th 90 232.04 $6d^27s^2$	Pa 91 231.04 $5f^26d^17s^2$	U 92 238.03 $5f^36d^17s^2$	Np 93 (237) $5f^46d^17s^2$	Pu 94 (244) $5f^67s^2$

Note: Atomic mass values given are averaged over isotopes in the percentages in which they exist in nature.
† For an unstable element, mass number of the most stable known isotope is given in parentheses.

			Group III	Group IV	Group V	Group VI	Group VII	Group 0
							H 1 1.007 9 $1s^1$	**He** 2 4.002 6 $1s^2$
			B 5 10.811 $2p^1$	**C** 6 12.011 $2p^2$	**N** 7 14.007 $2p^3$	**O** 8 15.999 $2p^4$	**F** 9 18.998 $2p^5$	**Ne** 10 20.180 $2p^6$
			Al 13 26.982 $3p^1$	**Si** 14 28.086 $3p^2$	**P** 15 30.974 $3p^3$	**S** 16 32.066 $3p^4$	**Cl** 17 35.453 $3p^5$	**Ar** 18 39.948 $3p^6$
Ni 28 58.693 $3d^84s^2$	**Cu** 29 63.546 $3d^{10}4s^1$	**Zn** 30 65.41 $3d^{10}4s^2$	**Ga** 31 69.723 $4p^1$	**Ge** 32 72.64 $4p^2$	**As** 33 74.922 $4p^3$	**Se** 34 78.96 $4p^4$	**Br** 35 79.904 $4p^5$	**Kr** 36 83.80 $4p^6$
Pd 46 106.42 $4d^{10}$	**Ag** 47 107.87 $4d^{10}5s^1$	**Cd** 48 112.41 $4d^{10}5s^2$	**In** 49 114.82 $5p^1$	**Sn** 50 118.71 $5p^2$	**Sb** 51 121.76 $5p^3$	**Te** 52 127.60 $5p^4$	**I** 53 126.90 $5p^5$	**Xe** 54 131.29 $5p^6$
Pt 78 195.08 $5d^96s^1$	**Au** 79 196.97 $5d^{10}6s^1$	**Hg** 80 200.59 $5d^{10}6s^2$	**Tl** 81 204.38 $6p^1$	**Pb** 82 207.2 $6p^2$	**Bi** 83 208.98 $6p^3$	**Po** 84 (209) $6p^4$	**At** 85 (210) $6p^5$	**Rn** 86 (222) $6p^6$
Ds 110 (271)	**Rg** 111 (272)	**Cn** 112 (285)	113†† (284)	**Fl** 114 (289)	115†† (288)	**Lv** 116 (293)	117†† (294)	118†† (294)

Eu 63 151.96 $4f^76s^2$	**Gd** 64 157.25 $4f^75d^16s^2$	**Tb** 65 158.93 $4f^85d^16s^2$	**Dy** 66 162.50 $4f^{10}6s^2$	**Ho** 67 164.93 $4f^{11}6s^2$	**Er** 68 167.26 $4f^{12}6s^2$	**Tm** 69 168.93 $4f^{13}6s^2$	**Yb** 70 173.04 $4f^{14}6s^2$	**Lu** 71 174.97 $4f^{14}5d^16s^2$
Am 95 (243) $5f^77s^2$	**Cm** 96 (247) $5f^76d^17s^2$	**Bk** 97 (247) $5f^86d^17s^2$	**Cf** 98 (251) $5f^{10}7s^2$	**Es** 99 (252) $5f^{11}7s^2$	**Fm** 100 (257) $5f^{12}7s^2$	**Md** 101 (258) $5f^{13}7s^2$	**No** 102 (259) $5f^{14}7s^2$	**Lr** 103 (262) $5f^{14}6d^17s^2$

††Elements 113, 115, 117, and 118 have not yet been officially named. Only small numbers of atoms of these elements have been observed.
Note: For a description of the atomic data, visit *physics.nist.gov/PhysRefData/Elements/per_text.html.*

Answers to Concept Exercises and Odd-Numbered Problems

CHAPTER 1: Concept Exercises

1.1 The puns are **a.** a megaphone **b.** a piccolo **c.** 2 kilomockingbirds **d.** a microfiche **e.** a terrapin

1.2 To find the mass of the raisins in kilograms,

$$42.5\text{ g}\left(\frac{1\text{ kg}}{1000\text{ g}}\right) = 4.25 \times 10^{-2}\text{ kg}$$

1.3 To find the dimensions of energy,

$$[\![E]\!] = [\![m]\!][\![c^2]\!] = \text{M}\left(\frac{\text{L}}{\text{T}}\right)^2$$

1.4 Only (**b**) may represent energy. Use dimensional analysis to find the dimensions of each quantity.

a. $[\![Q]\!] = \text{M}\frac{\text{L}}{\text{T}}$

b. $[\![Q]\!] = \text{M}\left(\frac{\text{L}}{\text{T}}\right)^2$, which may represent energy because it has the correct dimensions.

c. $[\![Q]\!] = \text{M}\frac{\text{L}}{\text{T}^2}$ **d.** $[\![Q]\!] = \text{M}\frac{\text{L}^2}{\text{T}}$

1.5 $[\![\rho]\!] = \frac{\text{M}}{\text{L}^3}$, and the SI units are kg/m^3.

1.6 a. Exact. Numbers that are not measured but are instead derived mathematically are exact **b.** 3 **c.** Ambiguous: 2 or 3. The final zero may be significant, or it may just be a placeholder. Scientific notation avoids such ambiguity. **d.** 3 **e.** 2 **f.** 2

1.7 a. 9.95×10^1 **b.** 3.6×10^4 **c.** -7.3 **d.** -3.90

CHAPTER 1: Problems and Questions

1. 2.6×10^9 s
3. 1.4×10^{17} s
5. a. 5.3×10^{-3} m **b.** 0.12892 kg **c.** $3.57 \times 10^{-5}\text{ m}^3$ **d.** $6.57 \times 10^4\text{ kg/m}^3$
7. $1.93 \times 10^4\text{ kg/m}^3$
9. 1.5×10^{11} m
11. $2 \times 10^2\text{ bloobits/bot}^3$
13. 2.8×10^{-3} m/s
15. a. 1.10×10^3 gal/min **b.** 69.4 L/s **c.** 576 s
17. $2.12 \times 10^4\text{ kg/m}^3$
19. a. $\frac{\text{M}\cdot\text{L}}{\text{T}^2}$ **b.** $\frac{\text{M}\cdot\text{L}}{\text{T}^2}$ **c.** $\frac{\text{M}\cdot\text{L}}{\text{T}}$ **d.** $\frac{\text{M}\cdot\text{L}^2}{\text{T}}$
21. Both quantities have dimensions $(\text{M}\cdot\text{L}^2)/\text{T}^2$
23. $f \propto \sqrt{k/m}$
25. a. $\text{M}\cdot(\text{L/T}^2)$ **b.** $\frac{\text{kg}\cdot\text{m}}{\text{s}^2}$ **c.** kg/s^2
27. 3
29. a. 4 **b.** 3 **c.** 2 **d.** 3
31. a. 8.65 **b.** 177 **c.** 25.891
33. −6.7
35. a. $1.0868 \times 10^{21}\text{ m}^3$ **b.** $5.50 \times 10^3\text{ kg/m}^3$
37. a. The result should be reported as 1.2 ± 0.1 g instead. **b.** The reported result makes sense. **c.** The reported result makes sense. **d.** The reported result contains only one digit to the right of the decimal, whereas the reported uncertainty would suggest knowledge of the value out to two digits to the right of the decimal. Either the uncertainty has been underestimated, or the measured value has been reported erroneously.
39. Answers may vary. Assume that a student studies 2 hr a week for each 1 hr of class time each week. If we assume that a student has a total of 6 hr of class time each week, then in a given 15-week semester, t = (2 study hr/class hr) × (6 class hr/week) × (15 week) = 180 study hr. We have kept 2 significant figures because we are estimating and have a leading "1" in the answer. If there are 50 students in your physics class, then an estimate for the total time spent studying by the entire class would be t_{tot} = (180 study hr/student) × (50 students) = 9000 study hr.
41. about 2 m
43. $5 \times 10^{-3}\text{ m}^3$ (lungs) and $2 \times 10^{-3}\text{ m}^3$ (stomach). Answers may vary by a factor of two or so.
45. 10^{14} cells
47. 15 days. Answers may vary by a factor of two or so.
49. 1.83×10^3
51. a. $3.58 \times 10^3\text{ m}^3$ **b.** 4.24×10^4 N
53. 9.13 in.
55. Both quantities have dimensions $\text{M}\,(\text{L/T}^2)$.
57. 1.50×10^3 ft
59. a. 205.9 g and 204.7 g **b.** (205.3 ± 0.6) g
61. 4.2 ly, 4.0×10^{16} m

CHAPTER 2: Concept Exercises

2.1 The spacing between the dots reveals how the particle is moving.

Case 1. Even spacing between dots: Particle is moving at constant speed.
Case 2. Dots get farther apart: Particle is speeding up.
Case 3. Dots get closer together: Particle is slowing down.
Case 4. Dots get farther apart: Particle is speeding up.
Case 5. All dots lie on top of one another: Particle is at rest.

2.2 All four vectors have the same magnitude, 2.4 m. To find the components, look for the unit vector. For example, if $\hat{\imath}$, the vector component is $\vec{x} = x\hat{\imath}$, where x is the scalar component. The vector components and scalar components are as follows.

a. $\vec{x} = -2.4\hat{\imath}$ m; $x = -2.4$ m
b. $\vec{z} = -2.4\hat{k}$ m; $z = -2.4$ m
c. In this case, the vector component $\vec{z}$ is given; $\vec{z} = -2.4\hat{k}$ m; $z = -2.4$ m
d. In this case, the scalar component x is given; $\vec{x} = -2.4\hat{\imath}$ m; $x = -2.4$ m

2.3 If the displacement of every point on the object is the same, the object is undergoing purely translational motion and may be modeled as a particle.

Case 1. A person on a Ferris wheel may be modeled as a particle.
Case 2. A person on a loop-the-loop roller coaster may not be modeled as a particle because the person flips upside down as he moves along the track.
Case 3. The tire on a bicycle may not be modeled as a particle.
Case 4. The person on a bicycle may be modeled as a particle.

2.4 For each course, divide the displacement by 2 hours to get the magnitude of the runner's average velocity. For the course in Figure 2.14A, we find 10 mph, and for that in Figure 2.14B, we find 6 mph. In both cases, the magnitude of the average velocity is less than the average speed of 13.1 mph because the magnitude of the displacement is shorter than the distance traveled by the runner. Average velocity is not a particularly useful quantity for describing the motion of a marathon runner.

2.5 It refers to magnitude of the displacement.

2.6 Because the tiger's displacement is zero, his average velocity is zero. The tiger's average speed is the total distance $(100 \times 20 \text{ m})$ traveled divided by 2 hours: $\frac{2000 \text{ m}}{2 \text{ h}} \times \frac{1 \text{ h}}{60 \text{ min}} \times \frac{1 \text{ min}}{60 \text{ s}} = 0.3 \text{ m/s}$. If the tiger remains in one place, both his average speed and average velocity are zero.

2.7 **a.** Plot acceleration on the vertical axis and time on the horizontal one. **b.** The slope of the tangent line for each point on the position-versus-time graph gives the velocity, and the slope on the velocity-versus-time graph gives acceleration. **c.** Position-, velocity- and acceleration-versus-time graphs are a concave-up parabola, an upward-sloping line and a horizontal line above the time axis, respectively.

CHAPTER 2: Problems and Questions

1. No; the path is nearly circular when viewed from the north celestial pole, indicating two-dimensional motion.

3. **a.** $x_A = -6$ m, $x_B = 0$, $x_C = 4$ m **b.** $z_A = 0$, $z_B = 6$ m, $z_C = 10$ m

5. **a.** $\vec{v} = 35.0\hat{\jmath}$ m/s, $v_y = 35.0$ m/s, $v = 35.0$ m/s **b.** $\vec{v} = 53.0\hat{\imath}$ m/s, $v_x = 53.0$ m/s, $v = 53.0$ m/s **c.** $\vec{v} = -3.50\hat{k}$ m/s, $v_z = -3.50$ m/s, $v = 3.50$ m/s **d.** $\vec{v}_x = -5.30\hat{\imath}$ m/s, $v_x = -5.30$ m/s, $v = 5.30$ m/s

7. **a.** Less than **b.** Speeding up **c.** Less than because the slope of the line drawn from $t = 0$ to 60 s for B is less than the slope of the similar line for A.

9. **a.**

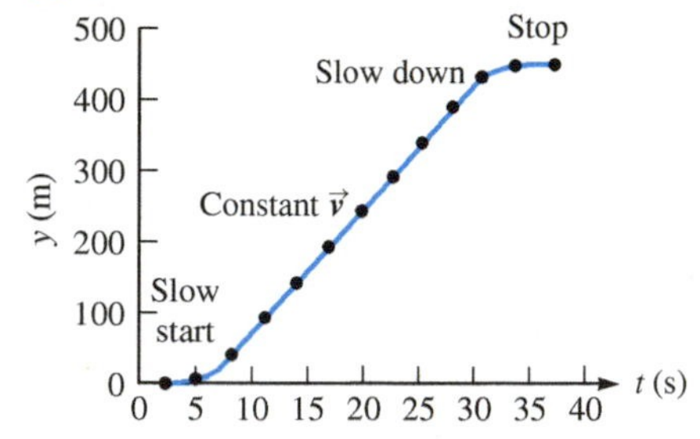

b. The elevator's speed gradually increases for the first 5 seconds, is nearly constant for about 30 seconds, and gradually decreases over the last 5 seconds. **c.** 17 m/s; from about 5 s to about 30 s **d.** Make sure that the acceleration is gradual at the beginning and end.

11. Never

13. 1.00×10^2

15. **a.** 6.44×10^5 m **b.** $2.99 \times 10^5\hat{\imath}$ m

17. **a.** $14.5\hat{\jmath}$ cm **b.** $14.5\hat{\jmath}$ cm

19. **a.** Zero; the initial and final positions are equal. **b.** 15.7 m; distance is the whole path length, but displacement only depends on the initial and final positions. **c.** There are many answers; here are two: the atomic bonds in a solid and the muscles in an animal are sometimes modeled by springs.

21. **a.** $9.0\hat{\jmath}$ m/s **b.** 0

23. **a.** The estimate depends on the distance between you and the lamp, but you should find an answer on the order of nanoseconds. **b.** 5.00×10^2 s **c.** The photon moves a very great distance in a very short amount of time. The jogger's speed is low enough to be measured with ordinary metersticks (or measuring tape) and stopwatches.

25. **a.** About 3×10^3 m **b.** About 9×10^{-6} s. Yes; the light travel time is much less than the sound travel time. **c.** The maximum difference in measurements would be approximately 200 m, or about one-tenth of a mile. Depending on the needed level of accuracy, it can probably be neglected.

27. 5.1×10^6 yr

29. **a.** $+L/t_1\hat{\jmath}$, **b.** $-L/t_2\hat{\jmath}$ **c.** 0 **d.** $2L/(t_1 + t_2)$

31. **a.** It is speeding up. **b.** 0, $-7.58\hat{\jmath}$ m/s, $-2.27 \times 10^2\hat{\jmath}$ m/s **c.** 0, 7.58 m/s, 2.27×10^2 m/s

33. **a.** $4.2\hat{\jmath}$ m/s **b.** $6.0\hat{\jmath}$ m/s **c.** 3.7 s

35. **a.** $-22.5\hat{k}$ m/s, $-52.5\hat{k}$ m/s **b.** $-37.5\hat{k}$ m/s

37. 1.33×10^4 m/s^2

39. **a.** No, the acceleration is not necessarily zero when the velocity is zero. **b.** A ball thrown vertically upward comes to a stop momentarily at the top of its path, but it moves back down again. So, its acceleration at the top cannot be zero.

41. It must be positive.

43. **a.** The derivative of v is negative, and so it is slowing down. **c.** No and no **d.** A motor must be used to produce a force that cancels the drag force from the water, thus producing zero acceleration and constant velocity.

45. **a.** $-(492.6 \text{ m} \cdot \text{s})\left(\frac{1}{t + 2.0 \text{ s}}\right)^3\hat{k}$ **b.** Slowing down. The velocity and the acceleration have opposite signs. **c.** No; it is always moving forward, although its speed becomes vanishingly small as time goes on.

47. 27.1 m/s^2

49. **a.** $31.2\hat{\imath}$ m/s **b.** $-5.55\hat{\imath}$ m/s^2

51. **a.** 30.4 m/s **b.** 2.56 m/s^2

53. **a.** $36.8\hat{\imath}$ m **b.** 36.8 m **c.** $36.8\hat{\imath}$ m **d.** 38.2 m

55. **a.** $307\hat{\imath}$ ft **b.** $116\hat{\imath}$ ft/s **c.** $24.0\hat{\imath}$ ft/s^2

57. 2.31 s

59. **a.** 44.1 m/s, upward **b.** 99.3 m

61. **a.** $-23\hat{\jmath}$ m/s **b.** $-9.2\hat{\jmath}$ m/s^2

63. 1.2 s and 3.7 s

65. **a.** 46.3 s **b.** 2.16 km **c.** 206 m/s, downward

67. About 4 m/s^2

69. 16 m

71. About 0.008 s (Answers within a factor of two are probably okay.)

73. **a.** $-3.04\hat{\jmath}$ m **b.** $5.00\hat{\jmath}$ m/s

75. 455 m east

77. **a.** 13.0 AU/yr **b.** 5.00 AU/yr and 21.0 AU/yr **c.** 0.375 yr

79. **a.** $34.0\hat{k}$ m/s **b.** $39.5\hat{k}$ m/s

81. There are many good choices; here are some tips for choosing good coordinate systems. (1) It is best to place an axis along the direction of motion. (2) It is sometimes helpful to choose an origin so that the position of the particle is always positive. (3) Choose positive as the direction in which the moving particle speeds up. (4) When the motion is caused by a spring, it is common to put the origin at the spring's relaxed position; because the motion switches direction, it doesn't matter which direction is chosen to be positive. Use these tips to check your coordinate system and see if you can make improvements.

83. $\vec{v}_y = \lim_{\Delta t \to 0} \frac{\Delta \vec{y}}{\Delta t} = \frac{d\vec{y}}{dt} = \frac{dy}{dt}\hat{\jmath}$,

$\vec{v}_z = \lim_{\Delta t \to 0} \frac{\Delta \vec{z}}{\Delta t} = \frac{d\vec{z}}{dt} = \frac{dz}{dt}\hat{k}$

85. $\vec{a}_y = \frac{dv_y}{dt}\hat{\jmath}$, $\vec{a}_z = \frac{dv_z}{dt}\hat{k}$

87. $\vec{a}_x = 0.0157\hat{\imath}$ m/s²; 92.6 s; $\vec{a}_x = -0.0157\hat{\imath}$ m/s². ThrustSSC's acceleration is nearly 400 times greater than Jeantaud's, and Jeantaud's timed mile took about 20 times longer than ThrustSSC's.

CHAPTER 3: Concept Exercises

3.1 Using any graphical method, we find
a. $\vec{A} - \vec{B} = 0$ **b.** $\vec{B} - \vec{A} = 0$ **c.** $\vec{A} - \vec{C} \neq 0$
d. $\vec{C} - \vec{A} \neq 0$ **e.** $\vec{A} + \vec{C} = 0$

3.2 **a.** Vector addition is commutative, so $A + C = C + A$ for any two vectors; in this case $\vec{A} + \vec{C} = \vec{C} + \vec{A} = 0$. **b.** The resultant of $\vec{A} - \vec{C}$ points to the right, and the resultant of $\vec{C} - \vec{A}$ points to the left, so they are not equal. **c.** Part (b) shows that vector subtraction is not commutative.

3.3 **a.** 9.81 cm **b.** (9.81 cm)/2 = 4.90 cm

3.4 Only case 2 shows a right-handed coordinate system.

3.5 **a.** $\alpha + \theta = 270°$
$\alpha = 270° - \theta = 270° - 215°$
$\alpha = 55°$

b. $\vec{B} = B\cos\theta\hat{\imath} + B\sin\theta\hat{\jmath}$
$\vec{B} = B\cos 215°\hat{\imath} + B\sin 215°\hat{\jmath}$
$\vec{B} = -0.82B\hat{\imath} - 0.57B\hat{\jmath}$

$\vec{B} = -B\sin\alpha\hat{\imath} - B\cos\alpha\hat{\jmath}$
$\vec{B} = -B\sin 55°\hat{\imath} - B\cos 55°\hat{\jmath}$
$\vec{B} = -0.82B\hat{\imath} - 0.57B\hat{\jmath}$ as before.

CHAPTER 3: Problems and Questions

1. **a.** 7.20 cm **b.** 1 cm → 200 m/s **c.** No; they both seem to fit comfortably on the page and are large enough.

3. $A\sqrt{2}$ or $B\sqrt{2}$

5. 24 units in the negative x direction

7. The magnitude is correct, but the direction is reversed.

9. The direction is correct, but the magnitude is half as long as it should be.

11. $\frac{1}{3}\vec{A} - \vec{B}$

13. B > C > A

15. **a.** 4.5×10^{-5} m/s (Answers may vary slightly due to measurement error.) **b.** The direction is the same as the displacement vector.

17. **a.** He is farthest from the starting point in the final drill. He is closest to his starting point in the second drill. **b.** He is farther than d in all three cases. **c.** 120°

19. The painting won't fit in the gallery on any wall.

21. 9.02 km

23. **a.** 4.24 blocks at 45.0° north of east **b.** 10.00 blocks

25. 3.47 mi south and 4.14 mi west

27. 36.9 at 63.4° clockwise from the x axis

29. $B_x = -11.79$, $B_y = 15.47$

31. 60°

33. **a.** $\vec{A} = -4.00\hat{\imath} - 2.00\hat{\jmath}$ **b.** 4.47 at 207° **c.** $\vec{B} = -2.00\hat{\imath} + 2.00\hat{\jmath}$

35. $(v_{1x}, v_{1y}) = (20.90, 12.07)$ m/s,
$(v_{2x}, v_{2y}) = (0, 8.96)$ m/s,
$(v_{3x}, v_{3y}) = (-13.25, 0)$ m/s,
$(v_{4x}, v_{4y}) = (-7.648, -21.01)$ m/s

37. $(-35.76\hat{\imath} + 61.94\hat{\jmath})$ m

39. **a.** $\vec{A} = (15.0\hat{\imath} - 6.00\hat{\jmath} - 3.00\hat{k})$ m **b.** $\vec{B} = (5.00\hat{\imath} - 2.00\hat{\jmath} - 1.00\hat{k})$ m **c.** $\vec{C} = (-45.0\hat{\imath} + 18.0\hat{\jmath} + 9.00\hat{k})$ m

41. **a.** 19° below the horizontal direction **b.** 14° below the horizontal direction; she is safe.

43. 65.6° north of east (or 24.4° east of north)

45. −0.740 m

47. $-18.5\hat{\imath} + 36.0\hat{\jmath} - 4.00\hat{k}$

49. $\sqrt{d_1^2 + d_2^2 + 2d_1d_2\cos\theta}$

51. $C_x = 6.10$ m, $C_y = -10.1$ m

53. 147°

55. 334 m

57. The spider's displacement is between d and $2d$. Add the horizontal displacements to find $1.7d$.

59. $(3.4\hat{\imath} + 1.3\hat{\jmath})$ m

61. $R_x = 6.58$ units, $R_y = -21.5$ units

63. 55.8 at 234°

65. Greater than 5, and $90° > \theta > 0$. The x component of each vector must be equal to 5.

67. The angle between $\vec{A}$ and $\vec{B}$ must be 60°.

69. **a.** The vectors must point in opposite directions. **b.** $s < 0$

71. **a.** 1.7 **b.** Each of the components must be proportional with the same proportionality constant m (given the function in the problem), but that is not the case here.

73. 72°

75. The direction of the instantaneous velocity at both times is up and to the right (clockwise). The direction of the average velocity is down and to the left (counterclockwise).

CHAPTER 4: Concept Exercises

4.1 **a.** Two dimensions **b.** Two dimensions
c. Two dimensions **d.** One dimension

4.2 **a.** Two dimensions
b. The particle slows down from A to G and then speeds up to L.

4.3 Because the initial and final positions are the same for the ball as they are for the cart, the displacement of the cart is equal to the displacement of the ball, $\Delta\vec{r} = 0.75\hat{\imath}$ m.

4.4 To answer these questions, we need to know the difference between range and displacement.

a. Because the balloon returns to the ground, its vertical displacement is zero, which means that its displacement equals the range. In this case, $\Delta r = R = 300$ yd.

b. The vertical displacement is not zero in this case. From the figure, we see that the horizontal displacement in this case is greater than the range.

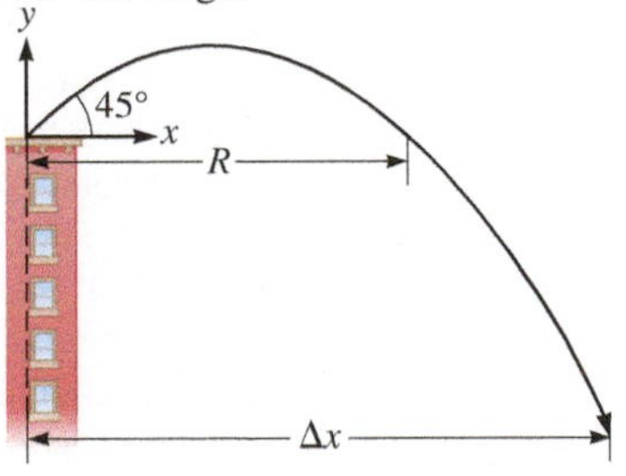

4.5 From $R = (v_0{}^2/g)\sin 2\theta$ (Eq. 4.28), we see that the range of a device depends on g. The free-fall acceleration is lower on the Moon than it is on the Earth. So, the range of the device on the Moon is longer that it is on the Earth.

4.6 **a.** This solution is a straightforward application of $v = 2\pi r/T$ (Eq. 4.30):

$$v = \frac{2\pi(0.5 \text{ m})}{12 \text{ s}} = 0.26 \text{ m/s}$$

b. This solution is a straightforward application of $\omega = 2\pi/T$ (Eq. 4.33):

$$\omega = \frac{2\pi \text{ rad}}{12 \text{ s}} = 0.52 \text{ rad/s}$$

c. In 4 s, the particle has completed only $\frac{4}{12} = \frac{1}{3}$ of its circular path. That means that it has swept out only one third of 2πrad: $\theta = \frac{1}{3}2\pi = \frac{2}{3}\pi$.

Because we are looking at a time that is one third of a revolution, the particle has traveled one third of the circle's circumference: $s = \frac{1}{3}(2\pi r) = \frac{2}{3}\pi r = \frac{2}{3}\pi(0.5 \text{ m}) = 1.0 \text{ m}$ (to 2 significant figures). It is the same as simply using $s = \theta r$ with $\theta = \frac{2}{3}\pi$.

4.7 Because we know the distance between the two cities as measured in the frame of the ground, we need the groundspeed to find the flight time.

CHAPTER 4: Problems and Questions

1. **a.** One dimension **b.** One dimension **c.** One dimension because the train moves in a single straight line in all cases

3. Neither player has an advantage as long as the ship moves with a constant velocity.

5. **a.** Two dimensions **b.** Points near the bottom of each arc (DEF and JKL) **c.** Points near the top of each arc (AB, GHI, and MN)

7. **a.** 5.00 m

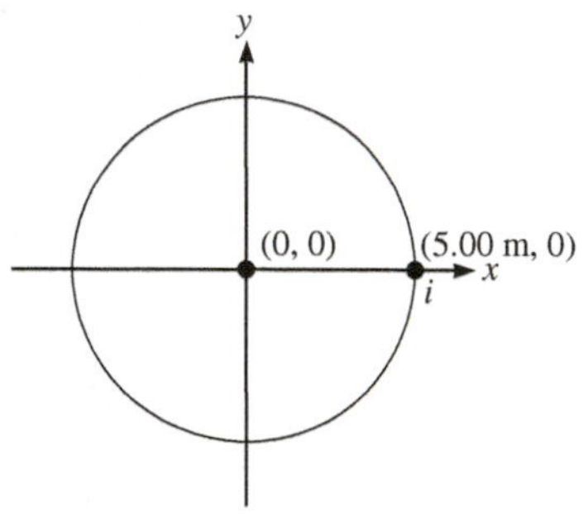

b. $x = (5.00 \text{ m}) \cos[(\pi \text{ rad/s})t]$

c.

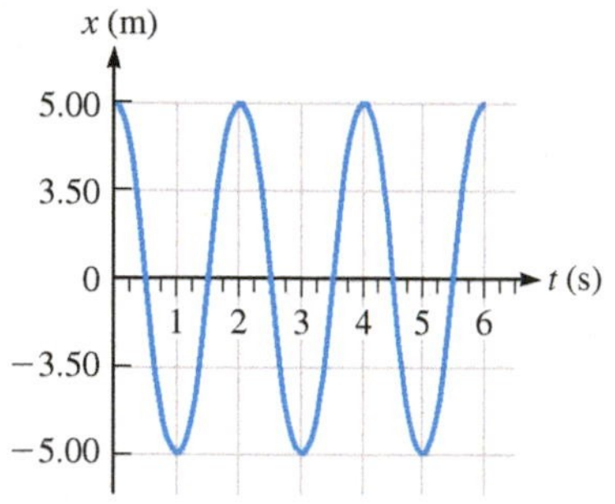

d. The graph also repeats such that x is between −5.00 m and +5.00 m.

9. **a.** 3.59 m **b.** No change

11. **a.** $(1.25t^2\hat{\imath} + 3.00t\hat{\jmath})$ m **b.** $(2.50t\hat{\imath} + 3.00\hat{\jmath})$ m/s **c.** $(11.3\hat{\imath} + 9.00\hat{\jmath})$ m **d.** 8.08 m/s

13. $90° < \theta \leq 180°$

15. **a.** $\vec{v}_i = (15.72\hat{\imath} - 7.88\hat{\jmath})$ m/s **b.** $\vec{a} = (-7.88\hat{\jmath} + 11\hat{k})$ m/s^2

17. $2b/3c$

19. $\vec{v} = A\omega\{[\cos(\omega t) - \omega t \sin(\omega t)]\hat{\imath} + [\sin(\omega t) + \omega t \cos(\omega t)]\hat{\jmath}\}$

21. **a.** 2.50 m/s **b.** −56.7°

23. 2.4×10^2 m

25. $x = 72.5$ m, $y = 2.95$ m

27. **a.** 22 m above the ground **b.** No

29. 15.1 m/s

31. **a.** 26.5 m **b.** 87.0 m **c.** 2.05 s

33. **a.** 7.8 m **b.** $(8.2\hat{\imath} - 5.6\hat{\jmath})$ m/s **c.** 1.58 m

35. 323.7 m/s^2

37. **a.** 7.40×10^2 m/s^2 **b.** Greater with the longer string **c.** 6.32×10^2 m/s^2

39. 2:1

41. 15.8 m/s^2

43. **a.** 2.6×10^{-6} rad/s **b.** 2.6×10^{-3} m/s^2

45. **a.** 27.0 km/h (west) **b.** 27.0 km/h (east) **c.** 43.3 s

47. **a.** While riding the train, pitch the ball in the same direction as the train's velocity. **b.** Reference frame fixed to the Earth; no

49. 5 min

51. 90°

53. 15.6° south of east

55. **a.** 1.34 h **b.** 1.15 h **c.** 1.24 h

57. **a.** 10.1 m/s^2 at 13.4° west from the vertical **b.** 9.81 m/s^2 vertically downward

59. 90°

61. $\sqrt{2gh}/\tan\theta$

63. 2.4 m from the end of the track

65. A parabola

67. $\vec{v} = A\hat{\imath} + (B - 2Ct)\hat{\jmath}$; $\vec{a} = -2C\hat{\jmath}$

69. 82.9°

71. 566 m/s

73. 3.34 m above the floor

75. Because the relative motion between the Earth and Mars is zero at those times

77.

a. $\left(2.00 + \frac{2\sqrt{3}}{3}t^{3/2}\right)\hat{\imath} + \left(7.00 - \frac{t^2}{2}\right)\hat{\jmath}$

b. $\left(2.00t + \frac{4\sqrt{3}}{15}t^{5/2}\right)\hat{\imath} + \left(7.00t - \frac{t^3}{6}\right)\hat{\jmath}$

79. **a.** 5.5 m **b.** $(3.2\hat{\imath} - 15\hat{\jmath})$ m/s

81. **a.** Circular **b.** $(\vec{v}_P)_M = -R\omega\sin(\omega t)\hat{\imath} + R\omega\cos(\omega t)\hat{\jmath}$ **c.** $(\vec{a}_P)_\oplus = (\vec{a}_P)_M = -R\omega^2\cos(\omega t)\hat{\imath} - R\omega^2\sin(\omega t)\hat{\jmath}$ **d.** The acceleration should be equal as found in part (c).

CHAPTER 5: Concept Exercises

5.1 **a.** Because friction cannot be completely eliminated, any sliding object has at least one force acting on it. So friction made it appear that rest was a natural state and that it takes a force to maintain motion. **b.** Newton's first law implies rest is **not** a natural state and replaces the idea of a "natural state of rest" with "constant velocity." **c.** No force is required to maintain constant velocity.

5.2 Avi's first statement is false. No force throws you through the windshield. Cameron's first statement is true. The passengers are already in motion, and (according to Newton's first law) it would take a force to stop them (when the vehicle stops suddenly). Avi's second statement is true. Cameron's second statement is true. Shannon's underlined statement is false. A seat can exert a force on you.

5.3 **Case 1.** Source: glove. Direction: up. Contact force.
Case 2. Source: freight train. Direction: right. Contact force.
Case 3. Source: Earth. Direction: toward Earth. Field force.

5.4 Identify which sources are outside the system in each case, and then decide whether these sources exert an external force on the system. **a.** The spring scale, and the Earth. The elevator car and the cable are external, but are not in contact with the system, so they cannot exert a force on the system. **b.** The elevator car, and the Earth.

The cable is not in contact with the system, so it cannot exert a force on the system. **c.** The Earth and the cable are outside the system, and they both exert forces on it.

5.5 **a.** A person with "a lot of inertia" can't seem to get moving on tasks, analogous to an object with a lot of mass that is difficult to accelerate from rest. **b.** It is just as difficult to stop an object with a lot of mass as it is to get it to start moving, but we don't say that a person that is working on task has a lot of inertia. **c.** In our everyday language, *massive* means heavy or large, such as "I have a massive amount of homework."

5.6 The only inertial frame is **a,** an airplane cruising in a straight path at constant speed. The others are accelerating frames.

5.7 **a.** Newton's first law states that acceleration requires at least one force; and is used to identify inertial frames. Newton's second law mathematically connects net force with inertia (mass) and acceleration. **b.** According to the second law, if the total force is zero, the object does not accelerate. This statement is consistent with the first law. It is likely that Newton stated the first law separately from the second law to address the notion of a *natural state*, which had been discussed for centuries before he published the *Principia*.

5.8 The apparent weight of the bunch of bananas does not change when it is hung from scale 2, so $\Delta y_2 = \frac{1}{3}\Delta y_1$.

5.9 Because the magnitude of the tension is the same all along the rope, in all three cases the magnitude of the tension force from each rope on Rochelle is 15 N and the rope pulls Rochelle. **a.** Its direction is toward Buddy. **b.** Its direction is toward the pole. **c.** Rochelle's left hand is pulled toward Joe, and her right hand is pulled toward Buddy.

5.10

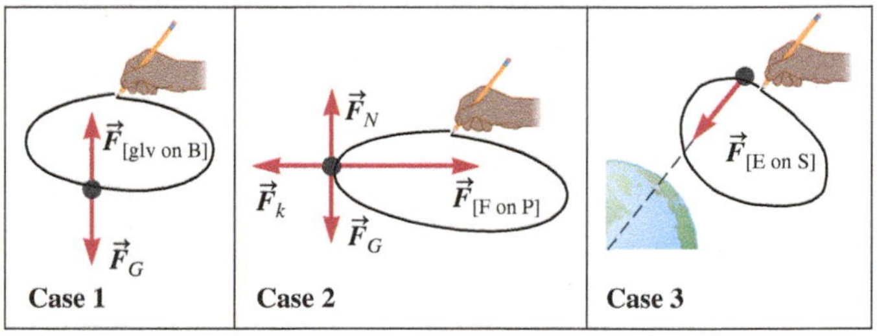

5.11 The magnitude of the force exerted on the child is the same as the magnitude of the force exerted on the Earth:

$$m_E a_E = m_c a_c$$

$$\frac{a_E}{a_c} = \frac{m_c}{m_E}$$

The Earth accelerates, but because its mass is so very much larger than the mass of the child, the Earth's acceleration is negligible.

5.12 Shannon is correct. According to Newton's third law, each train exerts a force on the other, with both forces of the same magnitude. (According to Newton's second law, the passenger train was accelerated more because it is lighter than the freight train.)

CHAPTER 5: Problems and Questions

1. It is easier to lift the beach ball because it is lighter. Other properties such as its size and shape aren't important.

3. The light block has less mass and so is easier to accelerate than the heavy block. The same push will make the light block go faster initially and take longer to stop.

5. Choice (b) is correct. The hero's inertia keeps him moving at a constant velocity with little force affecting him as he travels.

7. **a.** No force is necessary. **b.** A force must cause the deceleration. **c.** A force is needed to cause the change in direction.

9. 17.98 N

11. $(-2.32\hat{\imath} + 23.57\hat{\jmath})$ N

13. The fish exerts a contact force. It pushes on the water, and the water pushes back. The fish would also exert a gravitational force on the surrounding water, but the magnitude would be much less than the contact force.

15. The ball has the same mass in each instance and therefore the same inertia whether it is at rest or rolling.

17. The heavy man is more massive and thus has more inertia.

19. The wind causes the deflection. The cart is moving at a constant velocity and is thus an inertial frame.

21. 3.13 kg

23. 769 N

25. **a.** $\vec{F}_{\text{tot}} = (-1.63\times10^6\hat{\imath} + 7.59\times10^6\hat{\jmath})$ N **b.** $\vec{a} = (-641\hat{\imath} + 2.98\times10^3\hat{\jmath})\text{m/s}^2$

27. **a.** 0.294 **b.** 0.966 m/s²

29. **a.** $\vec{a} = (6.089\hat{\imath} - 6.595\hat{\jmath})$ m/s² **b.** 8.976 m/s² at $-47.28°$

31. **a.** $F_x = 0.0800$ N, $F_y = 0.480$ N **b.** 0.487 N

33. At $t = b/3c$

35. 1.5×10^{-2} N

37. 1.7 lb/in. and 3.0×10^2 N/m

39. $1.20 \times 10^3\hat{\jmath}$ N, upward

41. 124 N and 2.00 m/s²

43. $\dfrac{\mu_k mg}{\mu_k \sin\theta + \cos\theta}$

45. 0.340

47. 38.6 N, 53.9 N, 3.07 m/s²

49. 6.89 m/s²

51. **a.**

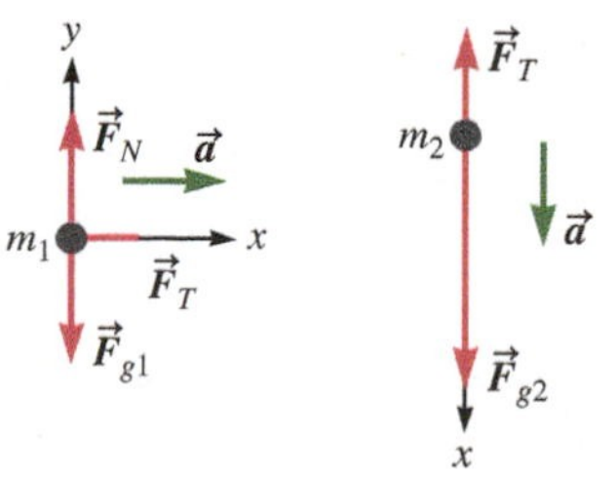

b. 21.8 N **c.** 7.25 m/s²

53. By Newton's third law, you push on the ground, and the ground pushes back with a force equal in magnitude.

55. **a.** and **b.**

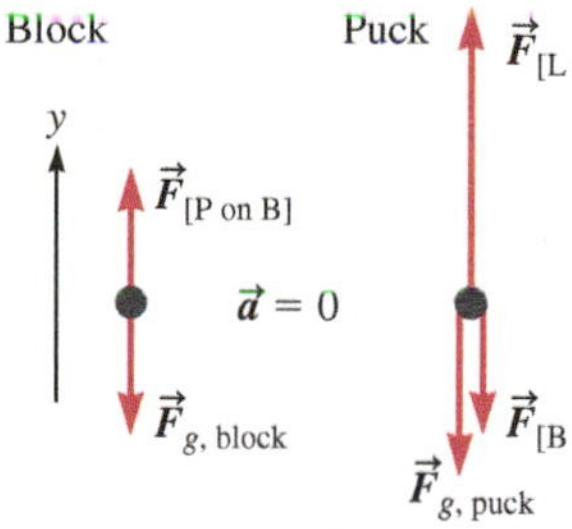

c. (1) The normal force of the puck on block and the force of the block on the puck; (2) the gravitational force of the Earth on the block and the gravitational force of the block on the Earth, (3) the gravitational force of the Earth on the puck and the gravitational force of the puck on the Earth, and (4) the normal force of the frozen lake on the puck and the force of the puck on the frozen lake

57. **a.** When she pushes the backpack, the backpack exerts an equal and opposite force on her according to Newton's third law. So, when she forces the backpack away from the craft, the backpack forces her toward the craft. **b.** 7.04 m/s² **c.** 2.78 m/s²

59. The force is directed toward you, or pointing upward from the ground toward you. The Earth is so massive that the effect of the gravitational force you exert on the Earth is not observed.

61. By Newton's third law, each train exerts a force of equal magnitude on the other.
63. 3.63×10^3 N to the right
65. 47.9 N
67. **a.** 3.13×10^{-15} N **b.** The force is 1.70×10^{12} greater than the muon's weight.
69. North
71. **a.** $F_{pull} = 95.0$ N, $F_g = 147$ N, $F_N = 147$ N **b.** 9.75 m/s
73. $\tan^{-1}\left[\dfrac{mg - F_N}{\mu_k F_N}\right]$
75. **a.**

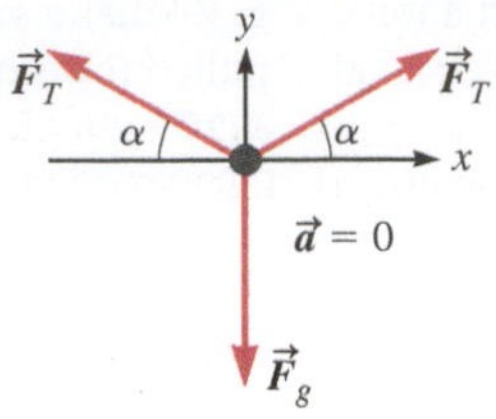

b. 552 N
77. $F_{T1} = 549$ N, $F_{T2} = 1.00 \times 10^3$ N, $F_{T3} = 1.23 \times 10^3$ N
79. **a.**

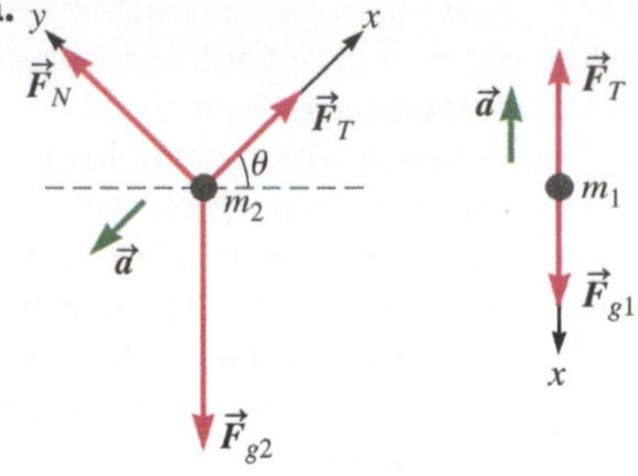

b. 0.701 m/s² **c.** 42.0 N **d.** 2.10 m/s
81. **a.** 46.7° **b.** 1.01 N
83. **a.** 61.3 N **b.** 1.50×10^2 N

CHAPTER 6: Concept Exercises

6.1 **a.** Avi is correct. Air does exert a force (the drag force) on the skydivers. It may be helpful to think of an extreme example of two skydivers, one in free fall and the other with an open parachute. The skydiver with the open parachute falls much more slowly than the skydiver in free fall. If the first skydiver to leave the plane is the one with the open parachute, the skydiver in free fall would be able to catch up. The same is true of all the other skydivers in the photograph. **b.** Shannon suggests that the plane does not fly along a level path but instead points downward so that each diver leaves the plane at a lower altitude than the previous diver. The problem here is that the skydivers are (nearly) in free fall. If the plane were to keep near them, it would also be in free fall. People on a free-falling plane would feel weightless. **c.** Cameron is correct; we often study physics under ideal conditions. Doing so helps make the problems simpler, but physical laws and principles govern more complicated problems as well. It is true that in some situations, people cannot be modeled as particles, but that is not an issue in this Case Study. The skydivers' motion is nearly translational. The reason this case is more complicated is due to the air. In many situations, we can ignore the medium in which a particle moves, but not in the case of skydivers.

6.2 Panel 1. Aluminum on steel has a higher coefficient of static friction than Teflon on steel (Table 6.1), so it requires a greater tension force to move the box with the aluminum side down.

Panel 2. The area of the object does not change the maximum static friction. Therefore, this comparison is the same as the one made in Panel 1.

Panel 3. The coefficient of static friction is the same in both cases, but a greater normal force is exerted on the box with extra weight so a greater tension force is required for motion to occur.

6.3 **a.** Because the particle is at rest and no force is applied horizontally, $F_s = 0$. **b.** The tension is less than the maximum static friction force. Because the object is at rest, $F_s = 5$ N. **c.** The tension equals the maximum static friction force $F_s = 15$ N. **d.** The object moves when the tension exceeds $F_{s,\,max} = 15$ N.

6.4 When we push a heavy sofa, we increase the force we apply until the sofa begins to move. The force we apply must be at least equal to $F_{s,\,max}$. Once the sofa begins to slide, kinetic friction takes over. Kinetic friction is weaker than $F_{s,\,max}$ and therefore weaker than the force we are applying at the moment the sofa begins to move. Therefore the sofa accelerates. We usually reduce our applied force in order to match kinetic friction. The sofa then moves at a constant speed.

6.5 We lubricate engines, bicycle chains, and pocket knives to reduce friction because kinetic friction causes abrasions and wears out the equipment.

6.6 a. Parked: b. Rolling: c. Sliding:

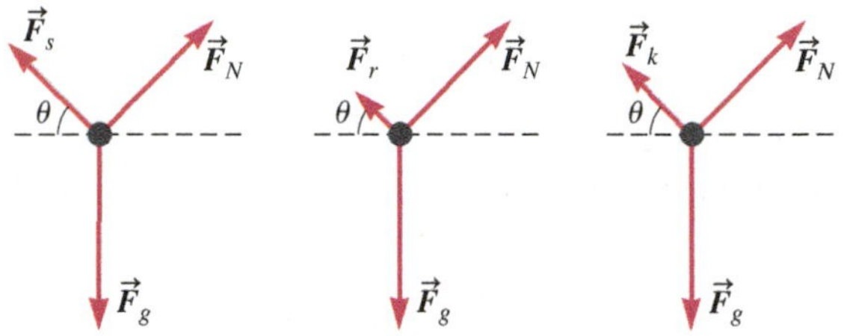

6.7 Three forces act on you as shown in the free-body diagram: gravity, the normal force, and the force of static friction. This friction force, which is exerted on you by the floor, is the force that moves you. Newton's third law says that the harder you press on the floor, the harder the floor presses back on you. To walk, you must apply a horizontal component of force to the floor. If the floor is slippery, static friction is smaller. If the horizontal force you apply to the floor exceeds the maximum value of the static friction force, your foot slips. On icy days, you might notice that people's steps are more vertical.

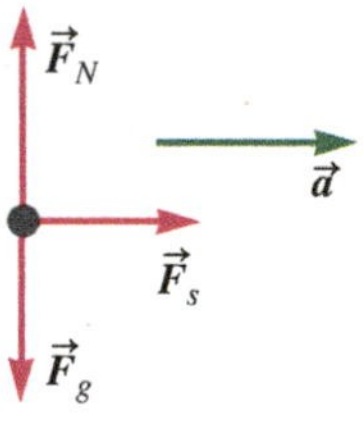

6.8 The cross-sectional areas are
Panel 1: $A = wd$, Panel 2: $A = \pi R^2$, Panel 3: $A = 2Rh$, and Panel 4: $A = \pi R^2$

6.9 Not all objects reach their terminal speed before they land. For example, if you drop a ball from a sufficiently low height, it will not reach terminal speed before it hits the ground.

6.10 **a.** An object falling through a vacuum is in free fall. Its acceleration is constant, which makes Graph 3 the best representation. Its speed continues to increase as best represented by Graph 4. **b.** An object falling through air starts in free fall, and then drag reduces its acceleration. Graph 2 best represents its acceleration. Initially, the object is at rest and its speed increases until it reaches its terminal speed, a condition best represented by Graph 6.

6.11 The force responsible for the centripetal force is circled in each free-body diagram.

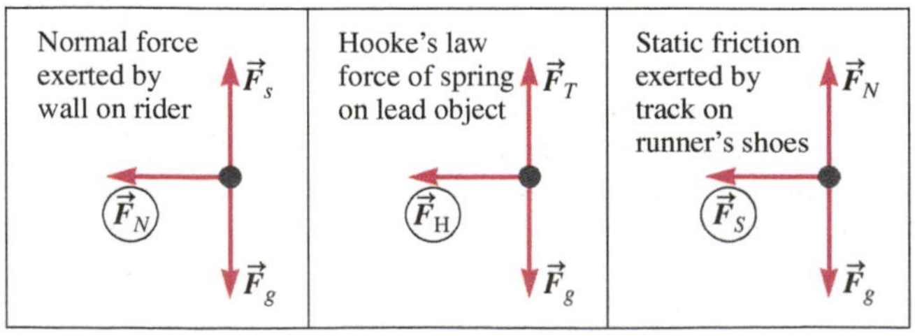

CHAPTER 6: Problems and Questions

1. Consider one example. In Chapter 5, Problem 3 we are asked to consider two blocks on ice. Of course, there is friction between the ice and the blocks, but it is fairly weak. So the problem is asking us to assume that there is no friction.
3. No; air still exerts a force on the skydivers.
5. a. They are equal but in opposite directions. **b.** The normal force will also be reduced. **c.** The static friction force will also be reduced because the normal force is reduced.
7. Static friction between the rod and each wall is strong enough to counteract the gravitational force on the rod. No tension is involved because the rod presses on the wall and the wall presses back. Perhaps the rod should be called a compression rod instead.
9. a. 6.52 N **b.** 8.4 N (The box won't slip because its weight is less than the maximum value of static friction.)
11. a. 249 N **b.** 374 N
13. a. 385 m **b.** 48.9 m
15. a. $\mu_S g$ **b.** It will not change.
17. 1.7×10^3 N
19. $\tan^{-1} \mu_S$, independent of the mass of the rider
21. No; the maximum static friction force is not enough to keep them from sliding given the angle of the hill because the downhill component of the gravitational force (186 N) is greater than the maximum value of static friction (118 N).
23. The block will remain motionless because $F_{s,\,\max} = 10.2$ N is greater than $mg \sin \theta = 9.81$ N.
25. a. 0.387 s **b.** 56.4°
27. a. $a_x = -\mu_k g \hat{\imath}$ **b.** $\dfrac{v_i^2}{2\mu_k g}$
29. 1.47 m/s^2. The result will be the same if the rider's mass is increased because acceleration doesn't depend on mass.
31. 32°
33. -2.7 m/s^2
35. a. 118 m/s **b.** 52.7 m
37. 24 m/s
39. a. $v_{A,\,\min} = 32.2$ m/s, $v_{A,\,\max} = 103$ m/s, $v_{B,\,\min} = 30.1$ m/s, $v_{B,\,\max} = 78.2$ m/s, $v_{C,\,\min} = 29.0$ m/s, $v_{C,\,\max} = 70.0$ m/s **b.** Skydiver C leaves first at minimum speed, then skydiver B jumps headfirst until catching up with skydiver C, and skydiver A jumps headfirst last. Both skydivers B and A should flatten out as they each catch up with skydiver C. **c.** 84 s. If we take the acceleration time into account, the wait is longer because terminal speed is not reached immediately.
41. 0.460 m/s
43. a. 0° **b.** 8.18°
45. The sphere, like an aerodynamic car, does not have sharp edges and would have a lower drag coefficient.
47. 11.6 N upward
49. 4.90 m/s^2
51. a. 9.05 m/s **b.** 0 and 136 m/s^2 **c.** 3.02 m/s^2
53. $v_{\text{ucm}} = \sqrt{F_T R/m}$
55. 340 N toward the center of the circular motion
57. 4.8×10^{30} N
59. a. Between 2.26 m/s and 37.6 m/s **b.** 0.404
61. 0.148
63. a. 0.296 **b.** 0.217
65. $\tan^{-1}(1/\mu_s)$
67. $\dfrac{2\pi d \cos\theta}{\sqrt{gd \sin\theta}}$
69. a. 1.89 N **b.** 4.48 m/s
71. $\dfrac{2\pi d \cos\theta \sqrt{\cos\theta + \mu_s \sin\theta}}{\sqrt{gd\cos\theta(\sin\theta - \mu_s\cos\theta)}}$
73. a. 38 m/s **b.** 4.4×10^3 N
75. a. 1.75 m/s **b.** $F_1 = 3.70 \times 10^2$ N and $F_2 = 455$ N
77. $\dfrac{v_i}{1 + v_i k t}$
79. a. 9.06×10^{22} m/s^2 inward **b.** 8.26×10^{-8} N inward **c.** 5.16×10^{-9} N inward
81. a. 3.70×10^3 m/s **b.** 1.75 h

CHAPTER 7: Concept Exercises

7.1 Tycho worked to eliminate procedural errors from his observations. We should evaluate our laboratory procedures. Often, redesigning an experiment leads to better results.
7.2 The orbits A and C are possible. Orbit A is circular with the Sun at the center. A circle is a special ellipse in which the two foci are the same point. Orbit C is an ellipse with the Sun at one focus. Orbit B is not possible because the Sun is not at a focus.
7.3 a. According to Kepler's third law, the periods are equal because the semimajor axis of the orbits are equal. (For a circle, the radius becomes the semimajor axis.) **b.** The speed of planet A is constant because its orbit is circular. The speed of planet B varies, being fastest near perihelion. Planet B's speed is higher than planet A's speed whenever planet B is closer to the Sun than planet A.
7.4 a. The semimajor axis is found from Kepler's third law $T^2 = a^3$:

$$a = T^{\frac{2}{3}} = 500^{\frac{2}{3}} = 63.0\,\text{AU}$$

b. The aphelion distance comes from $a = \dfrac{r_P + r_A}{2}$ (Eq. 7.1):

$$r_A = 2a - r_P = 2(63.0\,\text{AU}) - 0.5\,\text{AU} = 125.5\,\text{AU}$$

c. The comet moves very slowly when it is near aphelion. It is faster near perihelion. Therefore, the comet spends most of its time in the outer solar system.
7.5 Kepler's contemporaries maintained that the planets moved in uniform circular motion. Kepler hypothesized that planetary motion was nonuniform and elliptical.

CHAPTER 7: Problems and Questions

1. These terms give the impression that the Sun moves around the Earth. However, we know that heliocentric model has been verified.
3. 2.80×10^2 yr
5. Anywhere between 0 and 10 cm. The length of the semi-major axis does not depend on the distance between the foci.
7. a. 3.84×10^8 m **b.** 2.1×10^7 m **c.** A small version of the sketch should look similar to this drawing. A larger scale shows the Earth's offset better.

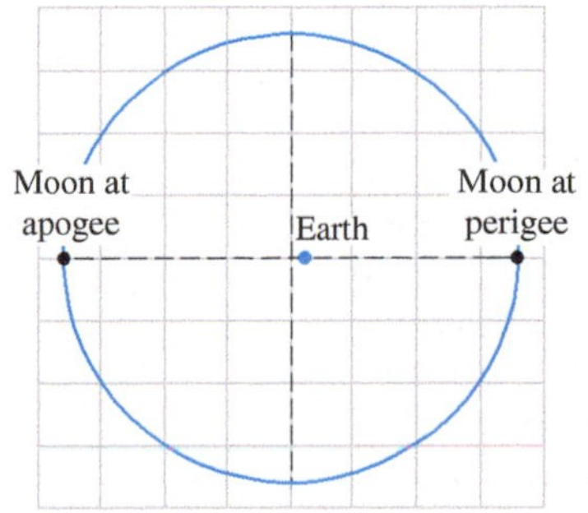

9. a. 1.5 **b.** 1.2 AU and 1.8 AU **c.** 0.3 AU
11. 4.97 days
13. a. 7.5×10^{36} kg **b.** 1.5×10^6 m/s. The Earth's speed is about 2% of this star's speed. This star's speed is about 0.5% of the speed of light.
15. a. 480 AU **b.** 884 AU
17. a. 1.99×10^{20} N **b.** 40 N
19. 1.96×10^{-8} N
21. a. 4.27×10^{-8} N toward the 750-kg sphere **b.** 2.97 m from the 750-kg sphere

23. **a.** 2.83×10^3 N toward the Moon **b.** 2.83×10^3 N toward the LRO
25. 4.66×10^4 m/s
27. **a.** $T_{outer}/T_{inner} = 1.8$ **b.** 7.9 h **c.** 4.74×10^8 m
29. 3.62×10^{22} N at perihelion, 3.44×10^{22} N at aphelion
31. 1.18 m/s² toward the Earth
33. $\frac{Gm}{L^2}\left(\sqrt{2} + \frac{1}{2}\right)$ at an angle of 45° to the horizontal
35. **a.** No, unless you and your dog have the same mass (which is not likely) **b.** The field is still the same. The source of the field (the Earth) has not changed, nor has the location under consideration.
37. **a.** 5.90×10^{-3} m/s²
b.

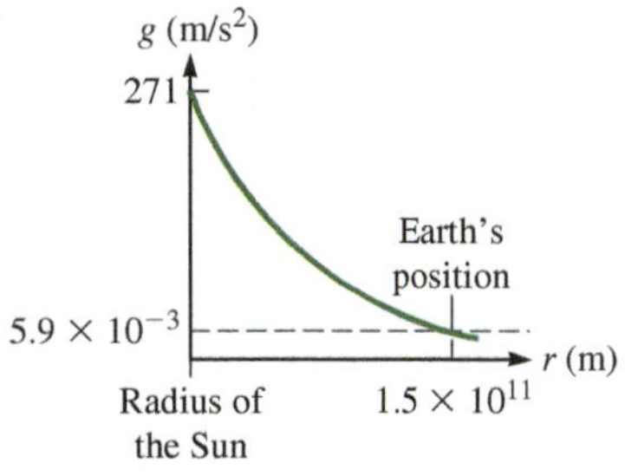

c. 3.53×10^{22} N
39. The mass of the Sun is much larger than the mass of the Earth. When calculating the gravitational force on each object, you are comparing a weaker field strength acting on a large mass and a stronger field acting on a smaller mass.
41. $(-4 \times 10^{-12}\,\hat{\imath} + 1 \times 10^{-11}\,\hat{\jmath})$ m/s²
43. No. To remain above the same point, the satellite must orbit in the same direction as the Earth, which is only possible at the equator.
45. **a.** $-9.24\hat{r}$ m/s², which is about 94% of $\vec{g}$ at the surface of the Earth. **b.** In essence, astronauts in orbit are in a continued state of free fall.
47. **a.**

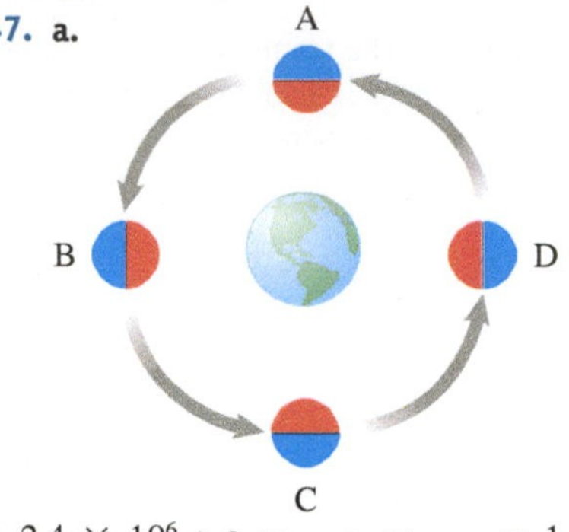

b. 2.4×10^6 s **c.** $w_{pole} - w_{equator} = 1.2 \times 10^{-5}$ N **d.** 3.4×10^{-2} N is the difference on the Earth, so the difference on the Moon is about 2800 times less than that on the Earth.
49. **a.** 360° **b.** 0°
51. 1.52 h
53. 1.00×10^{30} kg and 3.00×10^{30} kg
55. **a.** 3.44×10^{-5} m/s² **b.** 3.22×10^{-5} m/s² **c.** 3.32×10^{-5} m/s²
d.

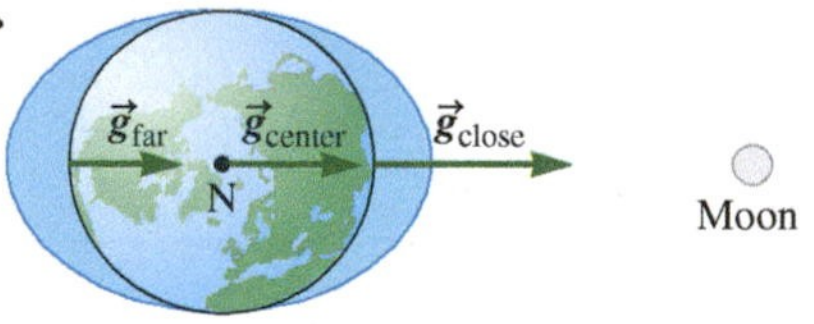

e. The two bulges in the figure show that there are two high tides a day. These bulges form as a result of the gradient in Moon's gravitational field across the Earth. The side closest to the Moon feels the greatest force and is pulled the most in the general direction of the Moon. The side farthest from the Moon feels the weakest force and is not pulled nearly as much as the center of the Earth or the close side.
57. **a.** $-1.07 \times 10^{-2}\,\hat{r}$ N/kg **b.** 7.15×10^{13} N
59. 1.9 rpm, 0.20 rad/s
61. No; the satellite must be in orbit around the center of the Earth.
63. **a.** 2.3 **b.** 1.2 **c.** With a 20% difference in speed, it would not be proper to approximate the orbit as circular. Also, the acceleration differs by about 120%.
65. **a.** 3.748×10^8 s (approximate) and 3.746×10^8 s (proper). So, the difference is 2×10^5 s, or about 0.05%. **b.** Somewhat valid. Equation 7.6 is most valid when the masses are very different in magnitude. The mass of the Moon is about 1.2% that of the Earth. We might expect a 1% difference between the two calculations. **c.** Not valid. Assuming the masses of the asteroids are quite similar, the difference between these two calculations might be on the order of 30%.
69. **b.** $-4.0 \times 10^2\,\hat{r}$ m/s² **c.** This difference is 40 times greater than the gravitational acceleration on the Earth. You are being "spaghettified."

CHAPTER 8: Concept Exercises

8.1 **a.** Use Kepler's third law to find Comet Halley's semimajor axis from its period given in the case study:

$$T^2_{[yr]} = a^3_{[AU]} \quad (7.2)$$

$$a = t^{2/3} = (76\,\text{yr})^{2/3} = 17.9 \approx 18\,\text{AU}$$

b. Comet Halley's aphelion distance can be found using

$$a = \frac{r_P + r_A}{2} \quad (7.1)$$

$$r_A = 2a - r_p = 35.2\,\text{AU}$$

which is quite close to the measured aphelion distance.

8.2 At the top of its one-dimensional flight, the ball momentarily stops. Therefore, its kinetic energy is momentarily zero.
8.3 Scalars can be negative; an example is a negative temperature. Kinetic energy cannot be negative because neither mass nor speed (squared) can be negative.
8.4 The change in gravitational potential energy only depends on the vertical displacement. The ball has the same vertical displacement in both cases 1 and 2, so $\Delta U_1 = \Delta U_2$, and because that vertical displacement is negative, $\Delta U_1 = \Delta U_2 < 0$. In case 3, the ball's vertical displacement is zero, so $\Delta U_3 = 0$, which means that the system's potential energy did not change.
8.5

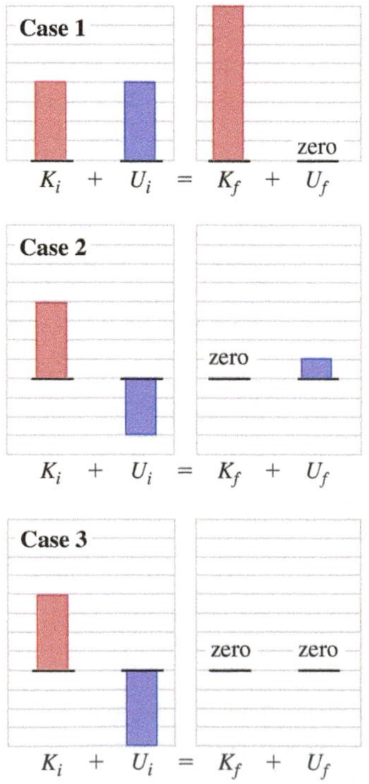

CHAPTER 8: Problems and Questions

1. The comet is slowing down just before it reaches aphelion because the angle between the gravitational force and the velocity will be greater than 90°. It is speeding up just after passing aphelion because the angle between the gravitational force and the velocity will be less than 90°.
3. **a.** 55 J **b.** 43 J
5. 167 J
7. **a.** 48.5 J **b.** 17.6 J
9. 2.3 m/s
11. 2.8×10^2 J
13. Case 1: $-mgh$; Case 2: $-\frac{1}{2}mgR$; Case 3: 0
15. −25.8 J
17. **a.** $U_T = 717$ J, $U_B = 0$, $\Delta U = -717$ J **b.** $U_T = 0$, $U_B = -717$ J, $\Delta U = -717$ J
19. 0.34 J
21. **a.** -5.3×10^{33} J **b.** -1.1×10^{27} J **c.** There is significantly more negative potential energy in the Earth–Sun system than in the Earth–Mars system. Thus, the force on the Earth due to the Sun is greater than that due to Mars. The Sun's mass is the dominant factor.
23. **a.** 0.0300 J **b.** 0.0675 J **c.** Elastic potential energy becomes kinetic energy when the spring relaxes. Because more energy is stored when the car is pulled back farther, the final speed of the car is greater in that case.
25. 0.29 J
27. $-mg\ell \sin\theta + \frac{1}{2}k\ell^2$
29. 30.8 m/s
31. 1.16×10^6 m
33. 4.15×10^4 m/s
35. Mechanical energy is conserved for the watermelon–Earth system. The initial and final total energies are the same regardless of which direction the watermelon is tossed.
37. **a.** 7.25 m **b.** 21.0 m/s
39. **a.** 20 J and 5 J, respectively **b.** 4 m/s and 2 m/s, respectively
41. 1.09×10^4 m/s
43. 4.5 m/s
45. 4.9 m/s
47. 15.5 m
49. 25.0 m
51.

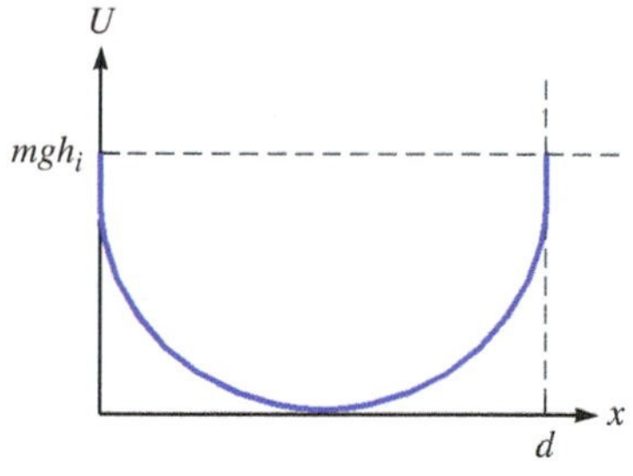

53.

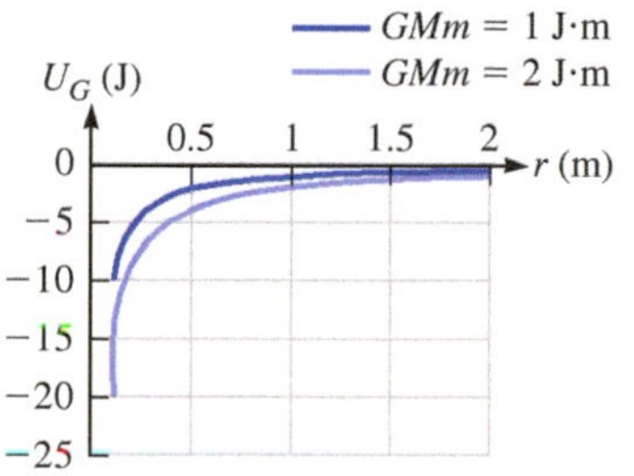

55. **a.**

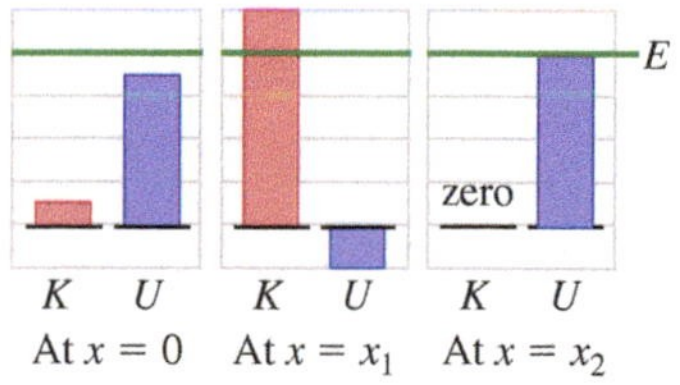

b. x_1 **c.** The particle cannot get there because the total energy is not sufficient.
57. **a.** 13.4 J **b.** 5.9 J **c.** The particles will not touch because the system does not have enough energy.
61. **a.** 8.4×10^4 m/s **b.** 3.0×10^8 m/s **c.** His speed is a factor of 3000 or so less than the speed of light in part (a) and approximately equal to the speed of light in part (b). He would have to go faster than light if he were to orbit any closer (which is theoretically impossible).
63. **a.** 0.989 **b.** 2540 yr **c.** -3.11×10^{22} J
65. **a.** $2\sqrt{gR}$ **b.** 1.47 N downward
67. -1.41×10^{32} J and 1.41×10^{32} J, respectively
69. $\sqrt{\dfrac{2gH}{1 + m_1/m_2}}$
71. **a.** 9.13 m/s **b.** 3.9 m
73. **a.** 238 km **b.** −500 MJ **c.** 1000 MJ
75. **a.** 110 m **b.** 29.4 m/s²
77. 1.95 m/s
79. 1.54 kg
81. We are looking for locations where the derivative of U as a function of position is equal to 0. This would indicate locations where the net force on the object could be 0, and could thus be in equilibrium. The object could be in equilibrium where the potential energy curve is at its peak. A small displacement in either direction would destroy the equilibrium, however, because there will be a force on the object that increases in magnitude as the object moves. The force drives the object away from the equilibrium point in this case. This makes the equilibrium an unstable one because small displacements from the equilibrium point do not drive the object back toward the location where there is no net force.
83. **a.** -3.07×10^{34} J **b.** -2.71×10^{34} J

CHAPTER 9: Concept Exercises

9.1 Work (area under curve) is positive in the first two cases and negative in the last case. Work is positive when the force is parallel to the displacement. So, in the first two cases, the force is parallel to the displacement, and in the last case, the force is antiparallel to the displacement.
9.2 **a.** The normal force on her feet accelerates her upward.
b. Because the point of application is not displaced (when the normal force is being applied), the force does zero work on her.
9.3 **a.** The normal force on his feet accelerates him.
b. This force does positive work on the man because the point of application is displaced in the same direction as the force is applied.
9.4 Gravity is a conservative force, so the change in the gravitational potential energy depends only on the initial and final configurations. Because the Earth–book system is restored to its initial configuration when the book is returned to Avi, there is no change in gravitational potential energy.
9.5 Thermal energy depends on the total path length. In other words, each time the book slides across the table, the thermal energy of the book–tabletop system increases. This answer is in contrast to Concept Exercise 9.4, which concerns gravitational potential energy that is unchanged when the system is restored to its original configuration.
9.6 There are 720 hours in 30 days, and 100 W is 0.1 kW. So, in 30 days, her lightbulb requires a total of $(720\ \text{hr})(0.1\ \text{kW}) = 72$ kWh, and her cost is $(72\ \text{kWh})(\$0.0472/\text{kWh}) = \$3.40$.

CHAPTER 9: Problems and Questions

1. **a.** Earth and satellite **b.** Earth, atmosphere, and plane **c.** Earth, truck, road, and air **d.** Earth, person, and floor
3. No. The force is always perpendicular to the direction of a rider's motion.
5. −12 J
7. 1.60×10^4 J
9. Both objects require forces that point to the west. Assuming the objects are restricted

to move in the east–west direction, the first object must have work performed on it to bring it to a stop. This amount of work will be negative. Once that occurs, both objects require the same magnitude of work to get them moving to the west at 25 m/s. Because it first needs to be stopped, the first object requires a net amount of work equal to zero, whereas the second object experiences a positive amount of net work.

11. 7.19

15. **a.** 0° **b.**

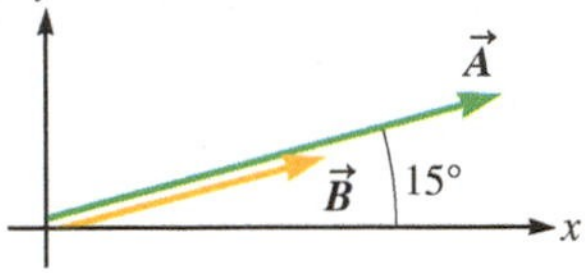

17. **a.** 7.9 J **b.** 7.9 J **c.** 7.9 J

21. −23

23. 1.3 J

25. 2.7×10^3 J

27. **a.** 75.0 J **b.** −75.0 J **c.** 0

29. 1.166×10^7 J

31. $-(\frac{1}{2})(k_1 + k_2)x_f^2 + (k_1 + k_2)[\ell\sqrt{x_f^2 + \ell^2} - \ell^2]$

35. The sled is speeding up (accelerating). $\vec{F}_P$ in the figure represents the force applied by Paul, and $\vec{F}_S$ represents the static friction force.

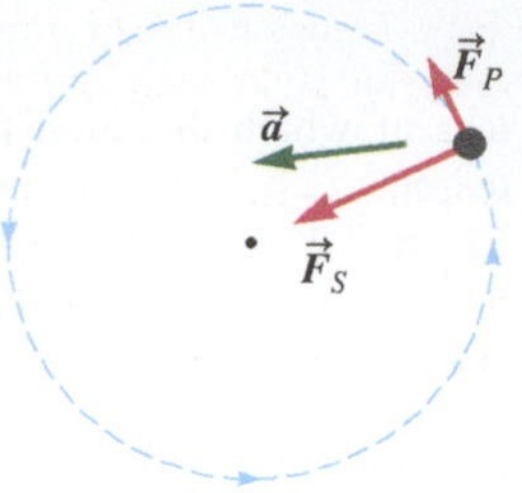

37. **a.**

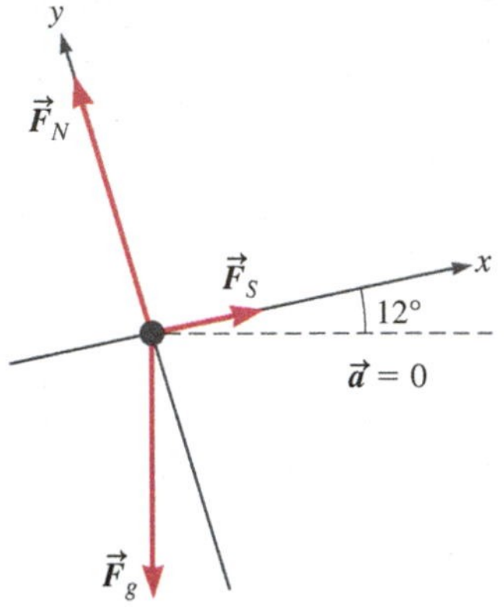

b. 14 J

39. **a.** 5.00 cm **b.** 2.40 J

41. **a.** 11.3° **b.** Both surfaces are warmed as the thermal energy increases due to kinetic friction, so it is best to include both the block and the incline in the system. **c.** 20.3 J

43. 1.9×10^6 J

45. $\sqrt{\dfrac{m_b v_b^2 - 2\Delta E_{th}}{m_b + m_{wb}}}$

47. **a.** 39.8 m/s **b.** 5.94×10^5 J **c.** 0.350

51. **a.** To account for the increase in thermal energy and the changes in gravitational potential energy, the system of choice is the Earth-box-track-spring surface. **b.** 8.6 m **c.** The block slides a shorter distance on this surface because friction acts throughout the entire motion.

53. 0.623 m

55. 1.9×10^2 N

57. 3.75 m

59. 7.3 N

61. 8.3×10^4 W

63. 6.28×10^4 W

65. 978 W

67. **a.** 90° **b.** 120° **c.** 86°

69. **a.** 1.60×10^3 J **b.** 491 J **c.** 1.11×10^3 J **d.** 9.42 m/s

71. 1.50×10^8 J

73. **a.** 37.9 J **b.** 44.6 N

75. **a.** 100 J **b.** −100 J **c.** 102 J

77. **a.** 3.3 m/s **b.** 1.8 N

79. **a.** 1.21×10^5 J **b.** 24.1 m/s

81. 99.4 s

83. 7.84 m

85. **a.** $K = 36t^2 + 72t^3 + 36t^4$ **b.** $a = 6.00 + 12.0t$ and $F = 12.0 + 24.0t$ **c.** $P = 72.0t + 216t^2 + 144t^3$ **d.** 5040 J

CHAPTER 10: Concept Exercises

10.1 Thrusters are not just used to launch spacecraft; they are used to maneuver spacecraft in space. (Check out NASA's website.)

10.2 Set the origin of the coordinate system to the center of the Earth. Then Equation 10.5 becomes

$$\vec{r}_{CM} = \frac{1}{M}\sum_{j=1}^{n} m_j\vec{r}_j = \frac{1}{M_\oplus + M_{Moon}}(M_{Moon} r_{Moon})\hat{r} = 4.67 \times 10^6\,\hat{r}\text{ m}$$

The center of mass of the Earth–Moon system is inside the Earth, $r_{CM} < R_\oplus = 6.37 \times 10^6$ m. For most problems, assuming the center of mass of the Earth–Moon system is at the center of the Earth is a good approximation.

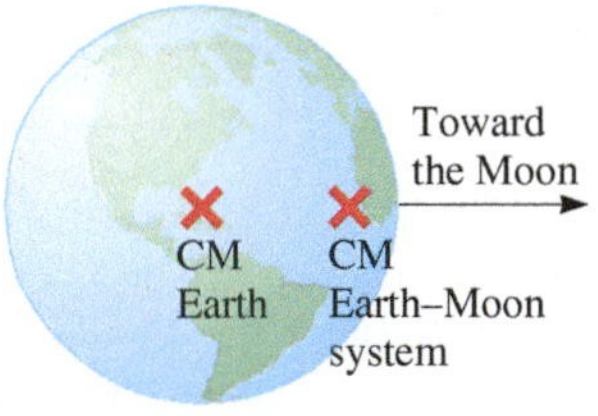

10.3 The center of mass moves closer to the trunk.

10.4 The center of mass is in the center of the can, so $(x_{CM}, y_{CM}, z_{CM}) = (0, 7.5, 0)$ cm.

10.5 When a cannonball is fired, the cannon must move backward for the momentum of the ball–cannon system to be conserved. The backward motion of the cannon is called *recoil*. The ropes are used to restrain recoil and therefore prevent damage to the ship or crew.

CHAPTER 10: Problems and Questions

1. Neither (b) nor (c) can be answered using the conservation of energy principle because both ask about direction. Energy is a scalar and therefore does not contain directional information.

3. **a.** $78\hat{\imath}$ kg · m/s **b.** $-1.7\hat{\imath}$ kg · m/s

5. $(4.64\hat{\imath} - 0.640\hat{\jmath})$ kg · m/s

9. 1.78×10^{29} kg · m/s

11. **a.** 2.35 kg **b.** 17.0 m/s

13. 13.7 N

15. (0.67 m, 0.29 m)

17. **a.** $3.76\hat{\imath}$ m **b.** $4.02\hat{\imath}$ m. The two answers do not agree because there is more mass to the right of the average position of the students.

19. **a.** 0.25 m from the center in the direction of the sister. **b.** She moved 0.43 m closer to the center of the see-saw.

21. $(0, 6.51 \times 10^{-12}$ m)

23. 4.56×10^5 m from the center of the Sun

25. 2.0×10^4 m from the center of Jupiter

27. $-60\hat{\imath}$ kg · m/s

29. 1.0 m/s²

31. Equal. The net change in momentum of the system is zero.

33. $v/2$

35. $x = 1.00$ m

37. 3.73×10^{-23} m/s

39. $(5.30\hat{\imath} - 2.52\hat{\jmath})$ m/s

41. **a.** There is no net external force on the system. **b.** $-1.01\hat{\imath}$ m/s

43. **a.** Yes; 2.56×10^{-2} m/s toward the shore. **b.** The forces acting on the child are a gravitational force between the child and the

Earth, a normal force on the child from the icy surface, and the force of the ball pushing on the child as he throws it. From Newton's third law, as the child exerts a force and throws the ball, the ball exerts an equal and opposite force on the child.
45. $(4.04\hat{\imath} - 2.67\hat{\jmath})$ m/s
47. 3.2 m/s
49. 13.4 m/s
51. **a.** 20.2% **b.** 3.73×10^5 kg
53. When Shannon throws the ball, the ground exerts static friction on Shannon. Friction creates a net external force, so momentum is not conserved. When analyzing the raft rocket in Section 10-6, we ignored the drag of the water on the boat. If we take that into account, then momentum is not conserved. Because the drag on a boat is not very large, the boat will still accelerate in response to the water being launched, but its acceleration will be less than in a frictionless case.
55. 4.60 m/s
57. 362 kg/s
59. 20.4 N perpendicular to the wall
61. $(1.2 \times 10^{-2}\hat{\imath} + 1.2 \times 10^{-2}\hat{\jmath})$ kg · m/s
63. 2.7 m/s
65. $(1.26 \times 10^3\hat{\imath} - 4.30 \times 10^2\hat{\jmath})$ N
67. **a.** $(6.00\hat{\imath} - 11.6\hat{\jmath})$ m **b.** $(5.58\hat{\imath} - 8.15\hat{\jmath})$ m/s **c.** $(19.5\hat{\imath} - 28.5\hat{\jmath})$ kg · m/s
69. 5.55×10^{-3} m/s
71. $(2/3)L$
73. $(4/7)L$
75. **a.** $v_{sled} = 3.36$ m/s **b.** $v_{pack} = 1.64$ m/s in the opposite direction to the sled's velocity relative to the ground
77. Gravity pulls straight down at the center of mass. Your body shifts so that the normal force exerted on the bottom of your foot is directly below your center of mass.
79. 0.107 m/s and 0.214 m/s in opposite directions

CHAPTER 11: Concept Exercises

11.1 The car leaves skid marks when the brakes have locked and kinetic friction decelerates the car. If the car comes to a stop, we know its final speed (zero). If it collides with something before coming to a stop, we may be able to use information about the collision to find the speed of the car at the end of the skid marks (that is, the speed just before impact). If we know the car's speed at impact, we can use the length of the skid marks and kinematics (Chapter 2) to find the speed of the car just as the brakes were locked.
11.2 By bending her legs as she lands on the ground, the gymnast increases the time Δt over which her momentum is reduced to zero. Increasing Δt reduces the average force she experiences from the ground.
11.3 Newton's third law
11.4 The two trains are stuck together after the collision; therefore, the collision is completely inelastic. Equations 11.5 through 11.9 apply (conservation of momentum and Newton's third law).
11.5 From Example 11.6, we know that $\Delta v_S = 2v_P$. We find Venus's speed from its orbital information, assuming its orbit is nearly circular:

$$v_P = \frac{2\pi r}{T} = \frac{2\pi(1.08 \times 10^{11}\text{ m})}{1.94 \times 10^7\text{ s}} = 3.50 \times 10^4\text{ m/s}$$

$$\Delta v_S = 2v_P = 7.00 \times 10^4\text{ m/s} = 70\text{ km/s}$$

11.6 **a.** The eight-ball will sink if

$$\theta_2 = \tan^{-1}\left(\frac{0.50\text{ m}}{0.75\text{ m}}\right) = 34°$$

which is the angle shown. So, the eight-ball is sunk.
b. The cue ball will sink if

$$\theta_1 = \tan^{-1}\left(\frac{0.50\text{ m}}{0.75\text{ m}}\right) = 34°$$

For an elastic collision between pool balls, we have (from Eq. 11.30):

$$\theta_1 = 90° - \theta_2 = 90° - 34° = 56°$$

so the player does not lose the game.

CHAPTER 11: Problems and Questions

1. When the force between two interacting objects is a field force, the objects do not need to touch to collide. The force exerted between a spacecraft and a planet is gravity (a field force), but the truck and the car interact through the normal force and friction.
3. 3.9×10^2 N. Adding padding does not change the impulse, but it does increase the time needed for the man's head to stop. A longer stopping time means a weaker average force.
5. **a.** 8.11 kg · m/s upward **b.** 54.1 N upward
7. **a.** 11 kg · m/s to the left **b.** 0.23 s
9. **a.** $-748\hat{\imath}$ kg · m/s **b.** $-4.98 \times 10^3\hat{\imath}$ N
11. 17.2 N
13. Equal. Both objects experience forces of the same magnitude for the same amount of time.
15. **a.** 0 **b.** 0 **c.** 0.900 m
17. **a.** 5.55×10^{-3} m/s **b.** 5.55×10^{-3} m/s
19. $\dfrac{M^2v^2}{2m^2\mu_k g}$
21. 3.3 m/s
23. **a.** 2.50 m/s **b.** 0.319 m
25. 1.3×10^3 N
27. **a.** 2.50 m/s to the left **b.** 2.96×10^5 J
29. $\dfrac{(m + M)}{m}\sqrt{2\mu_k gD}$
31. 207 m/s
33. **b.** 0
35. $v_1 = 6.80$ m/s and $v_2 = 2.80$ m/s, both to the right
37. Particle 1 must be much more massive than particle 2. Particle 2's velocity is in the same direction as particle 1's initial velocity, and its speed is about twice that of particle 1.
39. $v_1 = 11.7$ m/s to the left and $v_2 = 6.26$ m/s to the right
41. $\vec{v}_{1f} = -6.2\hat{\imath}$ m/s and $\vec{v}_{2f} = -8.7 \times 10^{-2}\hat{\imath}$ m/s
43. **a.** $\dfrac{4m_1^2}{(m_1 + m_2)^2}h_1$ **b.** 1.78 m
45. $\vec{v}_{1i} = 0.647\hat{\imath}$ m/s and $\vec{v}_{2i} = 0$; $\vec{v}_{1f} = 1.22\hat{\imath}$ m/s and $\vec{v}_{2f} = 0.793\hat{\imath}$ m/s
47. 25.6 m
49. $h_1 = 2.94$ m and $h_2 = 5.22$ m
51. $6.93\hat{\jmath}$ m/s
53. $(1.5\hat{\imath} + 2.0\hat{\jmath})$ m/s
55. **a.** 4.44 m/s at 2.38° west of south **b.** 2.86×10^4 J **c.** More energy is lost in this collision than in Problem 54. In a completely inelastic collision, the energy lost is maximized.
57. 50.0 m/s at an angle of 143° relative to the direction of A
59. 0
61. $(2.10\hat{\imath} + 1.35\hat{\jmath})$ m/s
63. 141 m/s
65. **a.** 2.72 m/s at 34.1° north of east (b) 677 J
67. **a.** $(v_{1i}/5)\hat{\imath} + (4v_{1i}/15)\hat{\jmath}$ **b.** $\frac{5}{13}$
69. 1.3 kg
71. 1.1×10^2 N
73. **a.** 1.37 m/s to the left **b.** No. As long as there is no net external force on the system, the total momentum is conserved regardless of the order of the collision events.
75. 5.8×10^5 m/s
77. 9.8 m
79. 0.059 m/s in the direction the ball was traveling initially
81. **a.** $v_m = \sqrt{2}v_i$ and $v_{3m} = \left(\sqrt{\tfrac{2}{3}}\right)v_i$ **b.** 215°

CHAPTER 12: Concept Exercises

12.1 The rotation axis for each object is shown in the figure.

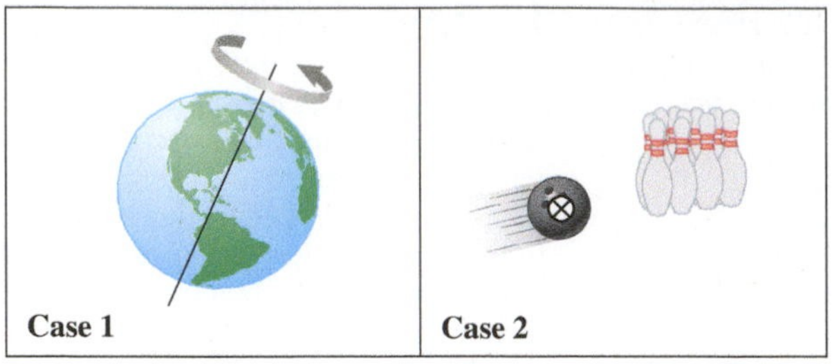

Case 1. The rotation axis of the Earth is fixed in the Earth's center-of-mass frame.

Case 2. The bowling ball's rotation axis is fixed in a frame that translates to the right at the same speed as the ball's center of mass.

12.2 Solve Equation 12.1 for s:

$$s = r\theta_{[\text{rad}]}$$

In Equation 12.1, θ is the ratio of two lengths, so θ has no units. (The "radian" is dimensionless.) We are given θ in degrees, however. To convert from an angle measured in degrees to one measured in radians, recall that there are 360° in a circle and that the ratio of the circumference c of a circle to its radius r is 2π:

$$\frac{c}{r} = 2\pi \text{ rad} = 360°$$

So, if θ is measured in degrees instead of radians, we can find s from

$$s = r\theta_{[\text{deg}]}\left(\frac{2\pi \text{ rad}}{360°}\right)$$

$$s = (0.25\text{ m})(15.5°)\left(\frac{2\pi \text{ rad}}{360°}\right) = 6.8 \times 10^{-2}\text{m}$$

If θ is measured in radians, Equation 12.1 takes its simplest form without the factor $(2\pi \text{ rad}/360°)$. We say that θ is most naturally measured in radians.

12.3 Use the values of θ from Table 12.1 to mark off the angular position of the reference line for each given time.

a.

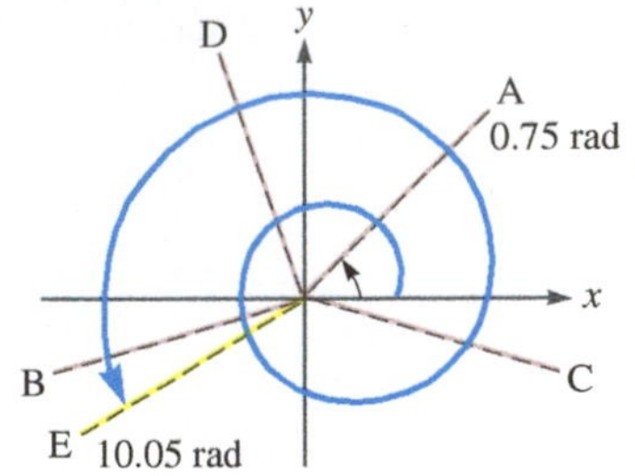

b. The angular displacement from A to E is given by Equation 12.2:

$$\Delta\theta = \theta_E - \theta_A = 10.05 \text{ rad} - 0.75 \text{ rad}$$

$$\Delta\theta = 9.30 \text{ rad}$$

The positive angular displacement indicates that the bottle's rotation is counterclockwise.

12.4 Use Equation 12.3 and the angular displacement found in Concept Exercise 12.3 to find the average angular speed of the bottle:

$$\omega_{av} = \frac{\Delta\theta}{\Delta t} = \frac{9.3 \text{ rad}}{8\text{ s}} = 1.2 \text{ rad/s}$$

Notice that we kept an extra significant figure because the first digit is 1.

12.5 The rotation is counterclockwise when viewed from $+z$, so $\vec{\omega}$ points in the positive z direction. We know that the bottle is slowing down because the reference lines get closer together for later time intervals. Because the bottle is slowing down, the angular acceleration must point in the opposite direction of the angular velocity. In this case, $\vec{\alpha}$ points in the negative z direction.

12.6 Imagine the force exerted on each dumbbell. The moment arm is the component of $\vec{r}$ that is perpendicular to the force $\vec{F}$.

Case 1. The moment arm is the forearm from the elbow to the hand.

Case 2. The moment arm is the whole arm from the shoulder to the hand.

Case 3. There is (almost) no moment arm. The force on the dumbbell is nearly vertical along the person's arm. So, there is no perpendicular component of $\vec{r}$.

CHAPTER 12: Problems and Questions

1.

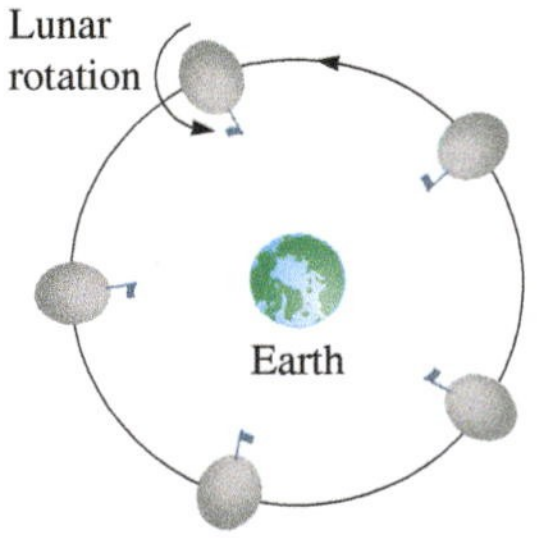

The same side of the Moon always faces the Earth; so, as the Moon orbits the Earth, it rotates. Our sketch shows the Moon as an exaggerated oval to emphasize the rotation. A rotating object cannot be modeled as a simple particle in translational motion.

3. The ice skater is rotating. The axis of rotation is vertical, passing through his foot. Each off-axis point on the skater's body moves in a circle around this axis.

5. a. 13 rad/s **b.** 25 rad **c.** 4.0 rev

7. a. 3.13 rad/s **b.** 3.08 rad/s^2

9. a. 1.76×10^{-4} rad/s **b.** Because of its angular speed, Jupiter bulges at the equator.

11. a. $\Delta\vec{\theta} = -\pi/2\hat{k}$ rad **b.** $\vec{\omega} = -1.7 \times 10^{-3}\hat{k}$ rad/s **c.** 0

13. ~10^7 rev/yr

15. a. 189 rad/s **b.** 2.40×10^{-9} rad/s^2 **c.** 1.9×10^3 s

17. a. $\theta(0) = 2.00$ rad, $\theta(1.50) = 5.45$ rad **b.** $\omega(0) = 0.800$ rad/s, $\omega(1.50) = 3.80$ rad/s **c.** $\alpha = 2.00$ rad/s^2. It is the same at both times and is constant.

19. $\vec{\alpha} = 1.62 \times 10^{-8}\hat{k}$ rad/s^2, $\vec{\omega} = -1.69 \times 10^{-3}\hat{k}$ rad/s

21. a. 17.0 rad/s^2 **b.** 1.63 rev

23. a. $\vec{\alpha} = -2.22\hat{k}$ rad/s^2 **b.** 10.9 s

25. a. No. The derivative of the angular velocity will be a function of time. **b.** $\alpha(t) = (9.34 \text{ rad/s}^2) + (4.48 \text{ rad/s}^3)t$

27. $v_{large}/v_{small} = 2.40$

29. a. 0.453 rad/s **b.** 10.9 m/s^2 toward the center of the circular track

31. 9.2 m/s^2

33. a. All points on the 40-m blade meet this condition. Both blades have the same angular speed, so any point on the 40-m blade has the same angular speed as the point at the end of the 20-m blade. **b.** The midpoint at 20.0 m would have the same translational speed as the point at the end of the 20-m blade. The translational speed at a point depends on both the angular speed and the distance from the center of rotation.

35. **a.** 12.0 rad/s **b.** 2.40 m/s **c.** 28.8 m/s^2 at 1.19° with respect to the radial direction **d.** 24.6 rad

37. 52.3 m/s^2

39. The axis of rotation is vertical and passes through the hinges. By putting the doorknob on the opposite edge, the torque we exert has the longest moment arm. Assuming we exert a perpendicular force, the force we exert is the minimum needed to rotate the door open or closed.

41. $\tau_1 = 4.53 \times 10^2$ N · m, $\tau_2 = 2.60 \times 10^2$ N · m, $\tau_3 = 0$

43. 0.891 N · m

45. $\tau_5 = \tau_3 < \tau_2 < \tau_1 < \tau_4$

47. **a.** 0 **b.** $225\hat{k}$ **c.** $-225\hat{k}$

49. 30.0°

51. $(5\hat{\imath} - 10\hat{\jmath} + 5\hat{k})$N · m

53. **a.** $-53\hat{k}$ N · m **b.** $-2.2\hat{k}$ rad/s^2

55. **a.** 4.5×10^2 N · m **b.** 4.2×10^2 N · m **c.** $\alpha_a = 2.7$ rad/s^2, $\alpha_b = 2.6$ rad/s^2 **d.** Case 1 is more effective because the resulting angular acceleration is greater and will cause the disk's angular speed to increase more rapidly.

57. **a.** $\omega_0 = 4.00$ rad/s, $b = 0.277$ s^{-1} **b.** 0.277 rad/s^2 **c.** 1.72 rev

59. **a.** 2.54 s **b.** 11.3 rad

61. **a.** 26.2 rad **b.** increase by a factor of nine

63. **a.** They are the same. **b.** Assuming there is nonzero angular acceleration, the tangential acceleration at H will be half that at R because $a_T = r\alpha$. **c.** Assuming H is not the same distance as R from the center, the answer to (a) will remain unchanged, but we can only say that the tangential acceleration is less at H than at R for (b). **d.** In this case, both the angular and tangential accelerations would be the same and are equal to zero.

65. **a.** 61.4 rev **b.** 24.6 rev/s

67. $a_{ES} = 3.4 \times 10^{-15}$ m/s^2, $a_{NO} = 3.0 \times 10^{-15}$ m/s^2

69. 248 rad

71. $\tau_1 = F_1R$, $\tau_2 = (1/2)F_1R$, $\tau_3 = 2F_1R$

73. $-(5/2)F_1R\hat{k}$

75. 5.25 N

77. 0.522

79. 39.1 N · m

81. 44.6 N

83. **a.** 0.625 rad/s **b.** 0.125 m/s **c.** 0.833 s **d.** 0.144 rev

CHAPTER 13: Concept Exercises

13.1 Mechanical energy is conserved under two conditions: (1) No energy is lost or gained by the system, and (2) there are no changes to the system's internal energy. If the system is isolated, its energy (mechanical and internal) is conserved. When Chris defined his system to include the ball, the track, and the Earth, he ensured that the system is isolated. There is nothing left in the environment that could do work on the system. Chris has every reason to believe that energy is conserved. In fact, it is conserved.

13.2 Use the right-hand rule to find the direction of $\vec{L}$.

Case 1. The angular momentum is into the page, in the negative z direction: $\vec{L} = -(rp \sin\varphi)\hat{k}$.

Case 2. The angular momentum is zero because $\vec{r}$ is parallel to $\vec{p}$.

Case 3. The angular momentum is out of the page, in the positive z direction: $\vec{L} = (rp \sin\varphi)\hat{k}$.

13.3 Set $\dfrac{d\vec{L}_{tot}}{dt} = \sum \vec{\tau}_{ext}$ (Eq. 13.27) and $\vec{\tau}_{tot} = I\vec{\alpha}$ (Eq. 12.24) equal to each other:

$$\frac{d\vec{L}_{tot}}{dt} = \sum \vec{\tau}_{ext} = I\vec{\alpha}$$

So, a change in angular momentum $d\vec{L}_{tot}$ is proportional $\vec{\alpha}$, and the two vectors must point in the same direction.

13.4 When the helicopter is on the ground with its rotors off, its total angular momentum is zero. The rotors rotate in opposite directions so that the total angular momentum of the system remains constant at zero. If one of the rotors stopped operating, the body of the helicopter would have to rotate (in the opposite sense as the functioning rotor) for the total angular momentum of the system to remain constant.

13.5 If no net external torque acts on the disk, its angular momentum (a vector) is conserved, always pointing in the same direction. Imagine arranging the gyroscope so that the angular momentum vector points at Polaris, the North Star. The computers on a space telescope can then use the direction of Polaris as a reference direction to find other astronomical objects. To precisely aim the telescope at particular objects, space telescopes use several gyroscopes with angular momentum vectors pointing in different directions.

CHAPTER 13: Problems and Questions

1. **a.** The kinetic energy of the sled increases. Because the system is not completely frictionless, the thermal energy of the ice and sled increases. So, it is possible to pull the sled at constant speed. **b.** Same as part (a), but the energy also goes into increasing the gravitational potential energy of the sled–Earth system. In fact, it may be possible to pull the sled at constant speed even if there were no friction between the sled and the snow. **c.** The rotational kinetic energy of the pulley increases. If there is friction or another dissipative force, the work you do may also increase the thermal energy of the system, and it is possible to rotate the pulley at constant angular speed. **d.** The translational kinetic energy of the center of mass and the rotational kinetic energy of the wheels both increase. If there are dissipative forces, it is possible to move the cart with constant angular speed and constant center-of-mass speed.

3. **a.** The observer sees the center of mass translate and the rotation of the Frisbee. So, the system has both translational and rotational kinetic energy. **b.** The dog runs under the Frisbee and so does not observe its translational motion. The dog would only observe the system's rotational kinetic energy. **c.** The ant is at rest on the Frisbee and so does not observe its motion. According to the ant, the system has no kinetic energy.

5. 1.36 kg · m^2

7. 51.6 rev

9. $\frac{7}{5}MR^2$

11. $(3\pi/2)R^6$

13. **a.** $0.267\hat{k}$ rad/s^2 **b.** $16.0\hat{k}$ rad/s **c.** 32.0 m/s

15. **a.** 6.08 s **b.** 40.5 rev

17. 303 J

19. 6.40×10^2 J
21. 23.7 J
23. 6.89×10^{-5} J
25. -3.39×10^{-2}J
27. a. 0.41 kg · m² **b.** 38 J
29. a. The disk reaches the bottom first because the ratio of its rotational inertia to its mass is smaller than for the hoop; this result is independent of the radius. **b.** $v_{disk} = \sqrt{4gh/3}$, $v_{hoop} = \sqrt{gh}$
31. 1.9×10^{-4} J
33. a. 0.66 m/s **b.** 0.78 m/s
35. In the case of a rolling ball, the entire system rotates. No part of it is in pure translational motion. In the case of a cart on a track, however, most of the system's motion is translational; only the wheels, which have very low rotational inertia, are rotating.
37. 28 J
39. a. 1.98×10^4 J **b.** 4.40×10^3 W
41. 20 N · m
43. 627 kg · m²/s
45. 31.2 kg · m²/s
47. 3.26 kg · m²/s
49. 2.88 rad/s
51. Your friend is incorrect. Your rotational inertia will increase when you extend your arms, and your angular speed will slow due to conservation of angular momentum. If the experiment were repeated but with the motions reversed (arms begin extended and are later drawn in), you would observe an increase in angular speed after you change your rotational inertia. The effect would be magnified by increasing the mass of the objects held in your hands.
53. a. 1.1×10^3 kg · m²/s **b.** 1.7×10^3 J **c.** 4.9 rad/s
55. 3.60×10^{34} kg · m²/s
57. $(-13.8t^2)\hat{\imath} + (-2.38t)\hat{\jmath} + (6.26)\hat{k}$
59. 1.31 m
61. 3.65×10^{28} J
63. a. 25.0 J **b.** 12.5 J **c.** 37.5 J
65. $0.6H$
67. 1×10^{-3} W
69. $-79.0\hat{k}$ kg · m²/s
71. a. 1.80 kg · m²/s **b.** 3.15 kg · m²/s
73. a. 1.98 kg · m²/s **b.** 1.46 s
75. a. $mv\ell$ **b.** $M/(M + m)$
77. (1.0 m, 1.7 m)
79. a. $10F/7M$ **b.** $3F/7$ to the right **c.** $\sqrt{20Fd/7M}$
81. $MRgt \sin\theta$

CHAPTER 14: Concept Exercises

14.1 If you displace the duck up or down slightly, it will return to its original position. So, it is in stable equilibrium in the vertical direction. If you move the duck from side to side, however, it will not return to its original position, illustrating neutral equilibrium in the horizontal direction.
14.2 Like the duck, if you move the boat from side to side, you will see that it is in neutral equilibrium horizontally. If you press the boat downward far enough that water gets in, however, the boat will sink; so, the boat is unstable in the vertical direction.
14.3 a. You just attach one cable to a point directly above the center of mass in the middle of the plane.
b. The center of mass will be directly below the cable. By picking the point of attachment to be off the midline, you can cause the plane to be tilted.
14.4 a. The pivot should be in the middle of the seesaw.
b. The pivot should be closer to the heavier person.
14.5 Because Young's modulus is much greater for steel, if you can just barely eyeball the stretch in the rubber rod, your eyes will not be able to detect the stretch in the steel cable.

CHAPTER 14: Problems and Questions

1. Yes, the ball is in a stable, static equilibrium because if it were displaced by a small amount, it would return or pass through the equilibrium position.
3. Yes, the system is in a neutral, static equilibrium because if you displaced the system from equilibrium, it would stay in the new position.
5. A and G would be locations of neutral equilibrium; C and E would be locations of stable equilibrium; and B, D, and F would be locations of unstable equilibrium.
7. $(-3.3 \times 10^2\hat{\jmath} + 2.8 \times 10^3\hat{k})$ N · m
9. $F_{Ay} = F_{By} = \frac{1}{2}mg$ and $F_{Ax} = -F_{Bx}$. Each neighboring block supports half the weight of the keystone.
11. You lean toward the side of your supporting foot to keep from falling. This leaning is necessary to ensure your center of mass is directly above your foot so that the net torque on your body is zero.
13. $\tan^{-1}(W/H)$
15. $\vec{R} = (-131.8\hat{\imath} + 538.0\hat{\jmath} + 582.9\hat{k})$ and $R = 804.1$
17. a. $-559\hat{k}$ N · m **b.** The torque is unaffected. **c.** The torque will increase if the y component of the force is increased.
19. $\vec{r} = 0.868\hat{k}$ m
21. 72.0 N · m
23. $F_B = 119$ N and $F_J = 106$ N
25. a. 0.198 **b.** The coefficient of static friction between rubber and concrete is approximately 1.0, so rubber ladder tips on dry concrete would be safe.
27. $F_1 = 234$ N, $F_2 = 209$ N, and $F_3 = 286$ N
29. 2.0×10^2 N
31. $F_{upper} = F_{lower} = 65$ N, but in opposite directions
33. 242 N
35. $F_A = 9.95 \times 10^3$ N and $F_B = 8.90 \times 10^3$ N
37. 1.09×10^2 N
39. 2.0×10^{-4}
41. 8.53×10^{-4} m
43. 1.07×10^{-5} m
45. a. 8.62 m **b.** 1.82×10^3 N
47. 8.06×10^{-3} m
49. 0.039 m
51. 6.30×10^{-6}
53. 5.75×10^{-6}
55. 1.88×10^{-5} m
57. 1.3 m
59. (0.467 m, 1.48 m)
61. $(-MR^2\alpha/2)\hat{k}$
63. $|\vec{\tau}_{F_g}| = |\vec{F}_g||\Delta\vec{r}| \sin\theta = 0$, so $\sin\theta = 0$, and $q = 0$
65. 2.85×10^{-6}
67. 11.7°
69. $F_P = Wmg/2h$
71. 667 Pa
73. a. 7.78×10^6 N **b.** 7.78×10^6 N
75. 12.8°, $F_A = 666$ N, $F_T = 927$ N
77. 0.29 m from the left side of the cabinet
79. 17.7 N
81. 1.6×10^{-3} m
83. $F_{vertical} = 2.55 \times 10^3$ N and $F_{horizontal} = 8.29 \times 10^2$ N
85. $(F_{T1}L_2 - F_{T2}L_1)/(F_{T1} - F_{T2})$

CHAPTER 15: Concept Exercises

15.1 The net force on the airplane must be zero because the airplane is not accelerating. Air drag must be balanced by the thrust of the plane's engines, and gravity must be balanced by an upward force exerted by the air. The free-body diagram for an airplane flying at constant velocity is shown here.

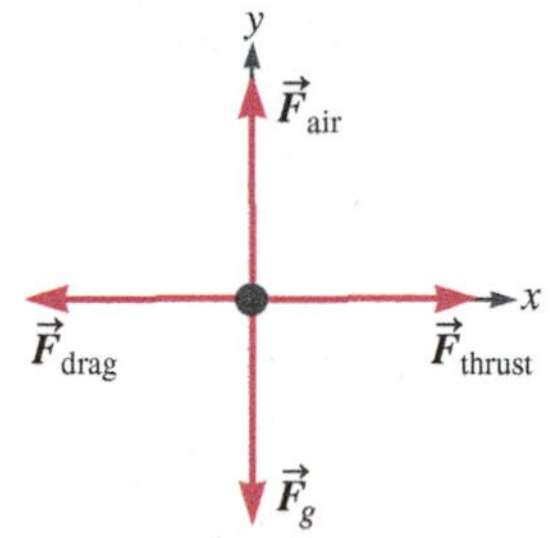

15.2 Consistent with the physics use of the word *pressure*: *His mom and dad are pushing him, and he is feeling pressure from both parents.* Pressure is a scalar, so pressure applied does not cancel out. Instead, it adds! So, the son feels squeezed like a balloon under water.

Inconsistent with physics use of the word *pressure*: *Peer pressure is just one form of negative pressure on him.* The word *negative* here seems to imply that peer pressure pushes in particular direction (the wrong way?). In physics, pressure is a scalar. A better analogy may be with the word *force*: *His peers force him down* is analogous to *gravity exerts a downward force on the apple*. In both *peer force* and *gravity*, that force is always in the same direction.

15.3 Estimate the depth of the pool to be about 2 m. The pressure at the bottom of the pool is given by Equation 15.6:

$$P = \rho g y + P_0 = (1000 \text{ kg/m}^3)(9.81 \text{ m/s}^2)(2 \text{ m}) + 1.01 \times 10^5 \text{ Pa}$$
$$P = 1.21 \times 10^5 \text{ Pa}$$

The difference in pressure is

$$P - P_0 = \rho g y = (1000 \text{ kg/m}^3)(9.81 \text{ m/s}^2)(2 \text{ m})$$
$$P - P_0 = 1.96 \times 10^4 \text{ Pa}$$

15.4 If the same pool only contains air, the pressure at the bottom is

$$P = \rho g y + P_0 = (1.29 \text{ kg/m}^3)(9.81 \text{ m/s}^2)(2 \text{ m}) + 1.01 \times 10^5 \text{ Pa}$$
$$P = 25.3 \text{ Pa} + 1.01 \times 10^5 \text{ Pa} = 1.01 \times 10^5 \text{ Pa}$$

There is only a very slight difference in air pressure between the bottom and top of the "empty" pool, so it is safe to ignore this pressure difference and assume the pressure is constant.

15.5 a. We simply substitute the densities into Equation 15.10:

$$\frac{V_{below}}{V_{obj}} = \frac{1000 \text{ kg/m}^3}{1250 \text{ kg/m}^3} = 0.8$$

which means that about 80% of the sunbather's body is below the surface of the water and about 20% of her body is above the surface.

b. We see from Figure 15.12B that the sunbather's head, shoulders, forearms, hands, knees, and toes are above the water. The question is: Does that amount to 20%, or one-fifth, of the sunbather's volume? It is helpful to sketch a woman and break her volume into roughly five pieces. As shown in the sketch, it seems about right that one-fifth of the sunbather's body is above water.

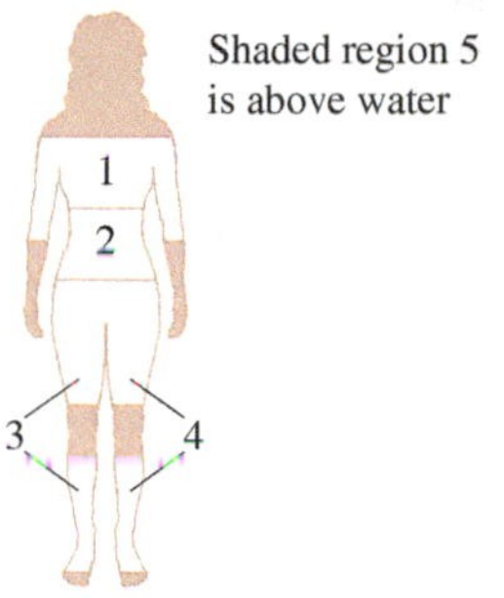

15.6 Many people use the words *light* and *heavy* to describe density. So, the phrase *It is lighter than air* should be translated as *It is less dense than air*. It isn't clear if Cameron's statement is correct because we don't know if the volumes being compared are the same.

CHAPTER 15: Problems and Questions

1. a. ~4×10^{17} kg/m^3 **b.** The density of an atomic nucleus is about 10^{13} times greater than the density of osmium, which shows that an atom is mostly empty space. Solids and liquids, as well as gases, are mostly empty space.
3. 1.25 m^3
5. a. 6.51×10^6 Pa **b.** 6.51×10^4 N
7. Because there are many nails in a bed of nails, the total area of their contact is fairly large. If half the person's cross-sectional area is in contact with the nails, the pressure exerted by the nails isn't very great. In fact, when the person stands on one foot, the floor probably exerts a greater pressure because the area of one foot is smaller than the area covered by the nails.
9. a. 1.10×10^8 Pa **b.** 0.050 **c.** Water is compressible, but the strain near atmospheric pressures can be ignored. At about 1000 times atmospheric pressure, the volume change of water is only about 5%.
11. a. 2.0×10^{-5} **b.** 4.8×10^{-4}
13. 1.03×10^4 Pa
15. a. 196 N **b.** 3.23×10^4 Pa
17. a. 4.53×10^7 N **b.** 1.13×10^7 N **c.** 4.24×10^6 N
19. 750 kg/m^3
21. 1875.0 kg/m^3
23. 0.106
25. a. 0.0610 m **b.** 2.37×10^{-3} m
27. $(F_B)_{air} = 5.7$ N and $(F_B)_{water} = 4.4 \times 10^3$ N
29. a. The object will sink ($F_g > F_B$). **b.** 1.1 m/s^2 downward
31. 54.1 kg/m^3
33. a. 1.35×10^{-3} m^3 **b.** 3.15 kg
35. 1.52 m
37. 3.4×10^5 Pa. The absolute pressure in the tire is only about three times more than atmospheric pressure, so gauge pressure is very convenient when you want to know the pressure inside your tire.
39. 1.01×10^5 Pa
41. 9.13×10^4 Pa
43. a. 60.0 N **b.** 0.0500 m
45. a. The fluid slows down as it moves in the positive x direction.

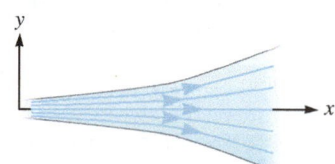

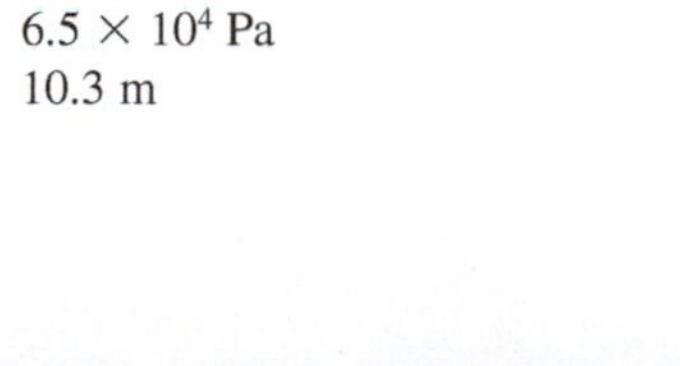

b. $v_0/[1 + (C/B)x^2]^2$
47. a. 8.22 m/s **b.** 58.8 m/s
49. 18 m/s
51. 6.5×10^4 Pa
53. 10.3 m

55. **a.** 1.14 m/s **b.** 5.97 m/s
57. 4.30×10^5 Pa
59. 46 s
61. 1.80×10^8 Pa
63. The surrounding air exerts pressure on all sides of the balloon, but the pressure on the bottom of the balloon is slightly greater than at the top of the balloon. So, the air exerts a net upward (buoyant) force. Because the balloon is full of low-density hot air, the downward gravitational force on it is relatively small. Thus, the net force (buoyant force minus gravity) on the balloon is upward.
65. $\vec{F}_B = 4.2 \times 10^2 \hat{\jmath}$ N
67. **a.** 2.67×10^{-4} m^3/s **b.** 4.22×10^{-4} m^3/s
69. 0.25 m/s^2 upward
71. 363 kg
73. 2.14×10^6 kg
75. **a.** The pressure should be highest at the bottom of the siphon. **b.** 9.91×10^5 Pa or 9.78 atm
77. **a.** 3.66×10^5 N **b.** 3.05×10^5 N upward **c.** 3.11×10^4 kg
79. **a.** -0.027 L **b.** 1.06×10^3 kg/m^3 **c.** Even at these great depths where the pressure is very high compared to atmospheric pressure, the volume of the water is only changed by about 2.7%. It is acceptable in most circumstances to consider water to be incompressible.
81. 25:21

CHAPTER 16: Concept Exercises

16.1 Read the times for each given position off the position-versus-time graph (Fig. 16.5A). Then read the corresponding velocity and acceleration off parts B and C of Figure 16.5 for those times.

Position y	Times (s)	Velocity v_y (m/s)	Acceleration a_y (m/s^2)
0	0.5, 1.5, and 2.5	1.3, −1.3, and 1.3	0 for all three times
0.4 m	1.0, 3.0	0 for both times	−3.8 m/s^2 for both times
−0.4 m	0, 2.0	0 for both times	+3.8 m/s^2 for both times

16.2

a. Crall has set the initial time to the moment when the disk is at $y_i = 0$. The initial phase comes from Equation 16.4:

$$\varphi = \cos^{-1} 0 = \pm\frac{\pi}{2}$$

Because the disk's position is increasing at point 2, we choose the negative sign. The disk's position, velocity, and acceleration according to Crall are

$$y(t) = (0.4 \text{ m}) \cos\left(3.1t - \frac{\pi}{2}\right)$$

$$v_y(t) = -(1.3 \text{ m/s}) \sin\left(3.1t - \frac{\pi}{2}\right)$$

$$a_y(t) = -(3.8 \text{ m/s}^2) \cos\left(3.1t - \frac{\pi}{2}\right)$$

Whipple has set the initial time to the moment when the disk is at $y_i = y_{max}$. The initial phase comes from Equation 16.4:

$$\varphi = \cos^{-1}\left(\frac{y_{max}}{y_{max}}\right) = \cos^{-1} 1 = 0$$

According to Whipple, the disk's position, velocity, and acceleration are given by

$$y(t) = (0.4 \text{ m}) \cos(3.1t)$$
$$v_y(t) = -(1.3 \text{ m/s}) \sin(3.1t)$$
$$a_y(t) = -(3.8 \text{ m/s}^2) \cos(3.1t)$$

b. The only difference is the initial phase φ, which reflects the observers' different choices of "when to start the clock," giving a different position of the disk at the chosen initial time.

16.3

	Angular frequency ω (rad/s)	Period T (s)	Initial phase φ	Amplitude y_{max} (m)
a.	14.5	0.433	0	0.75
b.	14.5	0.433	$\pi/2$	5.17×10^{-2}
c.	0.75	8.4	$\pi/2$	26

16.4 Using Equation 16.26,

$$\omega_s = \sqrt{\frac{k}{m}} = \sqrt{\frac{0.987 \text{ N/m}}{0.100 \text{ kg}}} = 3.1 \text{ rad/s}$$

$$T = \frac{2\pi}{\omega_s} = \frac{2\pi \text{ rad}}{3.1 \text{ rad/s}} = 2.0 \text{ s}$$

These results fit Crall and Whipple's data.

16.5

θ (degrees)	θ (rad)	$\left\lvert\frac{\sin\theta - \theta}{\sin\theta}\right\rvert$
0	0	0
±5°	$\pm 8.7 \times 10^{-2}$	1.3×10^{-3}
±10°	±0.17	4.8×10^{-3}
±15°	±0.26	1.1×10^{-2}

When the angle is between −15° and +15°, the difference between sine of the angle and the angle itself is about 1% or less.

16.6

a. $\alpha(t) = -\alpha_{max} \cos(\omega t + \varphi)$
b. $\omega_{max} = \theta_{max}\omega$
c. $\alpha_{max} = \theta_{max}\omega^2$
d. In all three equations for parts (a) through (c), ω without a subscript is angular frequency. In part (b), ω_{max} is the maximum angular speed.

CHAPTER 16: Problems and Questions

1. **a.** At 0.5 s and 2.5 s, $y = 0$ and $a = 0$ **b.** At 1.5 s, $y = 0$ and $a = 0$ **c.** At 0, 1.0, 2.0 and 3.0 s, $y = -0.4, 0.4, -0.4$, and 0.4 m, respectively, and $a = 3.8, -3.8, 3.8$ and -3.8 m/s^2, respectively.
3. $3.5\hat{\jmath}$ m/s^2
5. **a.** −0.0383 m, 0.751 m/s, 4.14 m/s^2 **b.** −0.0817 m, 0 m/s, 8.84 m/s^2 **c.** 0.0813 m, 0.0916 m/s, −8.79 m/s^2

7. **a.** 0.10 m **b.** 0.20 s **c.** 37°
9. $T/6$
11. **a.** 0.188 m/s **b.** 0.880 m/s^2
c. $x = (4.00 \text{ cm})\cos(4.69t)$,
$v = -(18.8 \text{ cm/s})\sin(4.69t)$,
$a = -(88.0 \text{ cm/s}^2)\cos(4.69t)$
13. The shadow of the ball will oscillate back and forth horizontally in simple harmonic motion with the angular frequency given in the problem statement. In general, the position of the object as a function of time would be given by $x(t) = (L \sin\theta)\cos(\omega t + \varphi)$.
15. **a.** 0.654 s **b.** $\vec{v} = 2.88\hat{\jmath}$ m/s
c. $\vec{a} = -27.7\hat{\imath}$ m/s^2
17. $v_y(t) = y_{max}\omega \cos(\omega t)$,
$a_y(t) = -y_{max}\omega^2 \sin(\omega t)$
19. Yes; $\pi L/\sqrt{3gR}$
21. **a.** 2.61 Hz **b.** 0.140 m on the opposite side of the equilibrium position **c.** 0.876 m/s away from the point where the block was released
23. 73.8 kg
25. **a.** $\frac{1}{2}\left(M + \frac{m}{3}\right)v^2$ **b.** $\frac{1}{2\pi}\sqrt{\frac{k}{M + m/3}}$
27. **a.** When the elevator is accelerating upward, the apparent weight of the pendulum bob is increased to $m(g + a)$, which means that there is a stronger restoring force and a shorter period. **b.** The period of a mass hanging by a spring only depends on m and k, so the acceleration of the elevator has no effect on the period.
29. **a.** 0.777 s **b.** 0.106 m/s **c.** 0.856 m/s^2
31. **a.** 1.09 m/s **b.** 5.25 rad/s^2 **c.** It is 1.08 m/s using the conservation of energy approach, so the answers match very well.
33. **a.** The center of mass moves towards the axis of rotation, so the period decreases. **b.** The center of mass moves toward the axis of rotation, so the period decreases. **c.** The center of mass moves away from the axis of rotation, so the period increases.
35. $2\pi\sqrt{\frac{3a^2 + b^2}{2ga}}$
37. 1.45 s
39. **a.** 4.7 m and 1.1×10^3 kg · m^2 **b.** 4.8 s **c.** 60 s
41. **a.** 1.3×10^{-3} N · m **b.** 4.9 s
43. **a.** 3.70×10^3 N/m **b.** 5.9 m/s **c.** 5.8 m/s
45. **a.** 0.332 m **b.** 6.17 m/s
47. $K = 1.7$ J and $U = 1.7$ J
49. **a.** 4.09×10^4 kg/s **b.** 0.701 s
51. The oscillator is no longer critically damped. If the mass is increased the term k/m decreases. The term $(b/2m)^2$ decreases even more. So $\omega_D = \sqrt{\frac{k}{m} - \frac{b^2}{4m^2}}$ is real and the oscillator is underdamped.
55. If the ball were filled with BBs, increasing its mass but not changing the drag force (which depends on the ball's shape and cross-sectional area), the time constant would increase. Yes, the ball would swing longer and presumably hold the cat's attention.
57. 0.307 m
59. The block starts at the right and oscillates back and forth. Numbers below the line give the block's position in meters.

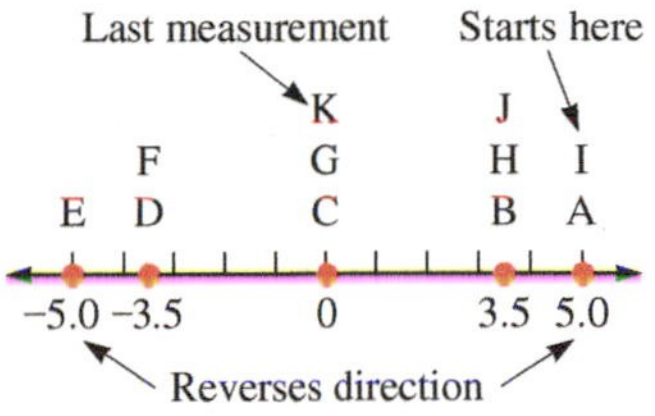

61. The block is moving in the *negative x* direction, so the velocity is negative. For the first two times the acceleration is in the negative x direction and for the last two times the acceleration is the positive x direction. There is no acceleration at the 0.50 s.
63. **a.** 2.47 N/m **b.** 12.3 N
c. $v_{max} = 22.2$ m/s and $a_{max} = 98.7$ m/s^2
65. **a.** 30.8 J **b.** $K = 30.8\sin^2(3.14t)$ and $U = 30.8\cos^2(3.14t)$
c.

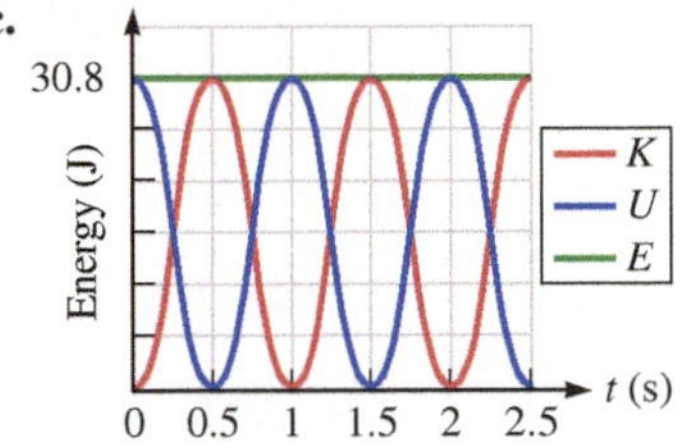

67. **a.** $z(t) = 1.40\cos(6.40\pi t - \pi/2)$ **b.** 28.1 m/s **c.** 566 m/s^2 **d.** 44.8 m
69. **a.** 2.24 s **b.** 13.1 J **c.** 55.2°
71. The period of a simple pendulum is proportional to the square root of its effective length (from the top to the center of mass of the sphere). Initially, the center of mass of the water-filled sphere is at its center. As the water flows out through the hole at the bottom of the sphere, the center of mass of the sphere moves downward. Thus, the effective length of the pendulum increases, so its period increases, reaching a maximum value. The period goes back to the original period when the sphere becomes empty.
73. $2\pi\sqrt{L/(g\cos\alpha)}$
75. $2\pi\sqrt{\frac{2R^2 + 5L^2}{5gL}}$
77. **a.** 0.822 m/s **b.** The speed found in part (a) is that of block A and block B at the equilibrium point. Beyond this point, block B moves with the constant speed of 0.822 m/s, whereas block A starts to slow down due to the restoring force of the spring. When block A returns to the equilibrium point, its speed will be equal to that of block B.
79. **a.** 5.03 Hz **b.** 8.55%

CHAPTER 17: Concept Exercises

17.1 When you check your pulse, you are measuring the period of your heart's vibrational motion. In physics, a pulse is a discrete disturbance in a medium.
17.2 The horizontal axis is x on the graph of the profile and t on the position-versus-time graph.
17.3 **a.** $S_{max} = 0.75$m, $k = 0.30$ rad/m, $\omega = 655$ rad/s

b. $v_x = \frac{\omega}{k} = \frac{655 \text{ rad/s}}{0.30 \text{ rad/m}} = 2.2 \times 10^3$ m/s

17.4 No. Avi's model is not a good model for sound. Sand is shot out of a sandblaster. Particles move outward. No air molecules are shot out of a speaker, however. In fact, the air molecules oscillate back and forth, staying relatively close to their original positions. Only the wave pattern moves outward, not the air molecules.
17.5

$$v_s = \sqrt{\frac{B}{\rho}} = \sqrt{\frac{2.2 \times 10^9 \text{ Pa}}{10^3 \text{kg/m}^3}} = 1.5 \times 10^3 \text{ m/s}$$

which is consistent with the speed of sound (1496.7m/s) in water given in Table 17.1.
17.6 Like any wave, sound transmits energy. So, the speaker's energy is transmitted to the hair cells in your ear through sound, and that energy is converted into motion of those hairs. The hairs oscillate back and forth. Imagine wiggling a toothpick back and forth. You can do it for a while, but if you do it for too long or if you wiggle too much, the toothpick breaks.

CHAPTER 17: Problems and Questions

1. No. The dog is modeled as a particle moving from one end to the other end of the pool.
3. We need only to multiply the original wave function by 2, so the new wave function is $y(x, t) = 2/[(x - v_x t)^2 + 1]$.
5. 4 m/s along the negative x axis
7. a. $\vec{v} = -3.5\hat{\imath}$ m/s
b. $y(x, t) = 7.5/[(x - 3.5t)^2 + 0.5]$
9. 0.125 m
11. The frequency and angular frequency of the wave will increase. Since the speed of the wave is constant, the wave number must increase, while the wavelength decreases.
13. 2.14×10^{-4} m
15. a. 0.667 m **b.** 8.00 s **c.** 0.0833 m/s **d.** $-0.0350\hat{\jmath}$ m/s **e.** $-0.0140\hat{\jmath}$ m/s^2
17. a. 10 cm **b.** 3.5 Hz **c.** $y(x, t) = (10 \text{ cm}) \sin[(0.39 \text{ cm}^{-1})x + (22 \text{ rad/s})t]$
19. a. 1.21 s and 0.604 m **b.** $S(0) = 0$, $S(T/4) = 0.850$ m, $S(T/2) = 0$, $S(3T/4) = -0.850$ m, $S(T) = 0$ **c.** $x(0) = 0.302$ m, $x(T/4) = 1.15$ m, $x(T/2) = 0.302$ m, $x(3T/4) = -0.548$ m, $x(T) = 0.302$ m
21. 189 m/s
23. 55.1 N
25. a. $y(x, t) = (1.00 \times 10^{-2} \text{ m}) \sin[37.1x - (2.04 \times 10^3)t]$ **b.** 0.0645 kg/m
27. 275 N
29. 9.9×10^{-2} Pa; about 3000 times greater than the pressure amplitude in air
31. a. 0.0137 m **b.** 4.60×10^2 rad/m **c.** $S(x, t) = (1.57 \times 10^{-6}) \sin[(4.60 \times 10^2)x - (1.58 \times 10^5)t]$
33. 2.0×10^{11} Pa
37. 0.0439 m
39. 0.769 J
41. 1.1 W/m^2
43. a. 6.2×10^2 W/m^2 **b.** 14 W/m^2
45. 10 dB
47. The flute is twice as intense as the cello.
49. 38°
51. 26°
53. The speed of sound in the thinner air above is different from the speed of sound in denser air below. As the sound wave travels, it refracts (changes direction). To an observer, the sound will seem to have come from a different direction.
55. 8.4×10^2 Hz
57. a. 754 Hz **b.** 657 Hz **c.** 825 Hz **d.** 601 Hz
59. 1.3×10^3 Hz
61. a. The frequency increases and the wavelength decreases as the source approaches. **b.** The frequency decreases and the wavelength increases as the source moves away.
63. 1.4
65. 2.55 km
67. a. 2.00 m **b.** 3.75 Hz **c.** 0.600 m **d.** 7.50 m/s
69. 0.043 dB
71. a. 0.0450 kg/m **b.** 23.3 m/s
73. a. 2.93 W **b.** 1.17 J
75. a. 0.0578 s **b.** 0.361 m
77. a. 1.38×10^3 Hz **b.** 1.51×10^3 Hz
79. 22.0 m
81. $\sqrt{F_T/[\rho\pi(2.50x - 0.200)^2]}$

CHAPTER 18: Concept Exercises

18.1

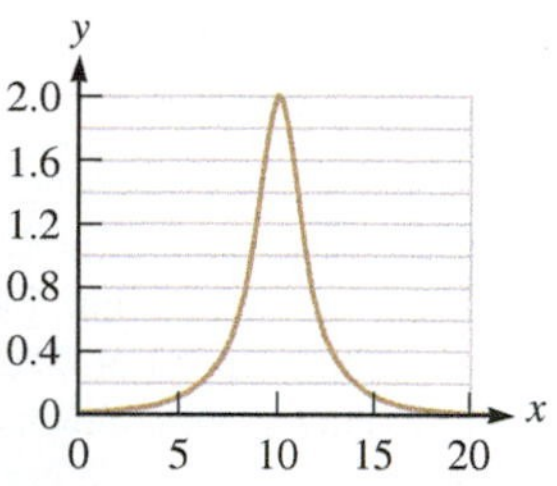

18.2

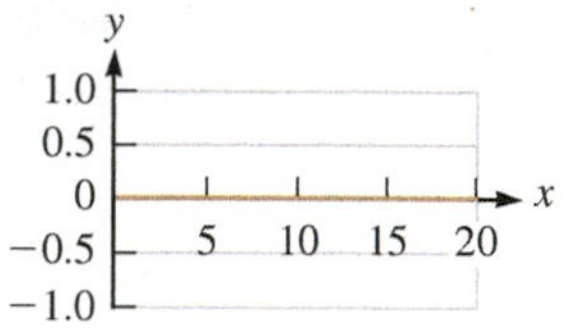

18.3

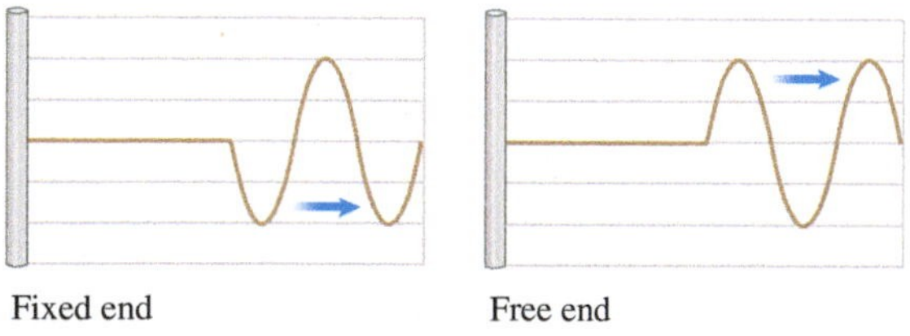

18.4

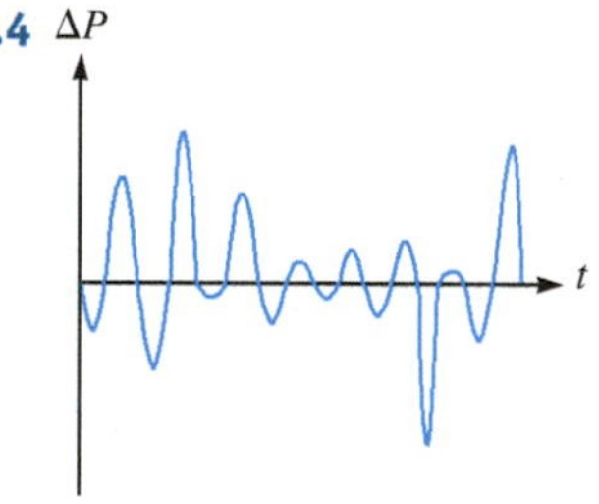

18.5 Because she hears a beat frequency of 2 Hz, we know from $f_{beat} = |f_2 - f_1|$ (Eq. 18.20) that her guitar either sounded at 440 Hz + 2 Hz = 442 Hz or at 440 Hz – 2 Hz = 438 Hz. Tightening the string made the beats disappear, and, according $v = \sqrt{F_T/\mu}$ (Eq. 17.11), an increase in the tension increases the speed of the wave. Finally, because $f_n = n(v/2L)$ (Eq. 18.10), the frequency increases as the speed increases. So, the guitar must have been at too low a frequency before it was tuned. The guitar was initially sounding an A at 438 Hz, and tightening the string increased the note's frequency to 440 Hz.

CHAPTER 18: Problems and Questions

1. a. Wave 1 propagates in the positive x direction, whereas wave 2 propagates in the negative x direction.
b. $y(x, t) = 4 \cos 3t \cos 10x$ **c.** $4 \cos 10x$
3. a. Wave 1 has an angular frequency of $\omega_1 = 3.2$ rad/s, whereas wave 2's angular frequency is $\omega_2 = 3.4$ rad/s. Additionally, the angular wave number of wave 1 is 8 rad/m and wave 2's is 6 rad/m.
b. $y(x, t) = 5.0 \sin(7x - 2.8t) \cos(x - 0.4t)$
c. $5.0 \cos(x - 0.4t)$
5. a. 0.251 m **b.** -0.0979 m **c.** -0.374 m
7. a. $y(x, t) = \dfrac{0.43}{(x + 13.6t)^2 + 1}$
b. $y(x, t) = -\dfrac{0.43}{(x + 13.6t)^2 + 1}$
9. a. 6 **b.** $11\ell/v_s$ **c.** 30 m
11. An animal could emit a sound and listen for the reflected sound, or echo. After learning to interpret the time between emission and echo as a relative distance in its environment, the animal can determine the location of nearby objects.
13. b. $A(0) = 2y_{max}$, $A(\pi/2) = \sqrt{2}\,y_{max}$, $A(\pi) = 0$

15. The sound will be louder where constructive interference occurs. If we assume the sound from the drum radiates equally in all directions, waves spread outward from the drum with two-dimensional cylindrical symmetry, traveling at the speed of sound in air v_s. The waves reflect off the wall and encounter the next outbound sound wave. If the time between drum strikes is equal to the time it takes the first wave to reach the outer edge of the monument, the reflected wave and the next generated sound wave will meet at a point on a circle with radius equal to half the monument radius, which will be a location of constructive interference. Other rates of striking could also be considered, resulting in different locations where constructive interference would occur.
17. **a.** 0.898 m **b.** 414 m/s
19. **a.** 3.5 cm **b.** Starting from the left end ($x = 0$) of the string, the nodes are at $x = 0$, 15 cm, 30 cm, 45 cm, and 60 cm. **c.** 0
21. **a.** 2.92×10^{-2} s **b.** 1.51 m **c.** 34.0 m
23. $y(x, t) = [6.90 \sin(1.01x)]\cos(0.184t)$
25. 22 Hz
27. 0.690 m
29. **a.** 1.20 m **b.** 6.24×10^2 m/s to 6.60×10^2 m/s
31. **a.** 1 **b.** $1/\sqrt{2}$
33. There is no change; the string is not moving at a node.
35. **a.** 47.6 N **b.** 392 Hz
37. **a.** The wavelength in air of the sound produced by the string is longer because the wave speed is greater. **b.** 0.351
39. $\Delta f_{\text{short}} = 19$ Hz and $\Delta f_{\text{long}} = 1.2$ Hz
41. $n(4.53 \times 10^{-3})$ Hz, where $n = 1, 2, 3, \ldots$
43. 1.27 m, 1.59 m, and 1.91 m
45.

Free at both ends	Fixed at one end (from Example 18.6)
$f_1 = 2140$ Hz	$f_1 = 1070$ Hz
$f_2 = 4280$ Hz	$f_3 = 3210$ Hz
$f_3 = 6420$ Hz	$f_5 = 5350$ Hz
$f_4 = 8560$ Hz	$f_7 = 7490$ Hz
$f_5 = 10{,}700$ Hz	$f_9 = 9630$ Hz
$f_6 = 12{,}840$ Hz	$f_{11} = 11{,}770$ Hz
$f_7 = 14{,}980$ Hz	$f_{13} = 13{,}910$ Hz
$f_8 = 17{,}120$ Hz	$f_{15} = 16{,}050$ Hz
$f_9 = 19{,}260$ Hz	$f_{17} = 18{,}190$ Hz
f_{10} is inaudible	$f_{19} = 20{,}330$ Hz

The rod that is free at both ends has a slightly smaller range, but the difference between adjacent resonant frequencies in both cases is 2140 Hz.
47. 3/2
49. **a.** The harmonic wavelengths will remain unchanged. **b.** The frequencies will increase when the speed of sound in the medium increases.
51. 210 Hz, which is an audible low tone
53. 8.65 Hz
55. 445 Hz
57. 3 Hz
59. 4.6°C
61. 9.19 cm
63. **a.** Wave 1 travels in the negative x direction, and wave 2 travels in the positive x direction. **b.** 0.75 m **c.** 0.300 s
65. **a.** 4.24 m **b.** 7.50 Hz **c.** 1.00 m
67. 0.32 m and 0.95 m
69. **a.** The pipe is closed at one end. **b.** 76 Hz
71. 24.8 Hz
73. **a.** 0.249 m **b.** 0.497 m
75. 0.60 m
77. **a.** 1.18 m **b.** 4.14 m
79. **a.** 192 Hz or 200 Hz **b.** 198 Hz **c.** The tension must be reduced by 2.01%.

CHAPTER 19: Concept Exercises

19.1 Use Equation 19.2 to convert from Fahrenheit to Celsius degrees and then Equation 19.1 to convert from Celsius degrees to kelvins. You should become comfortable thinking about temperature using all three scales.

a. The temperature range in Gdansk is –3.9°C to 26°C or 269 K to 299 K.

b. A very cold day (0°F) is –18°C or 255.15 K, and a very hot day (100°F) is 38°C or 311.15 K.

c. A high fever (105°F) is 41°C.

19.2 A thermometer reading indicates the temperature of the thermometer. If the thermometer is in thermal equilibrium with some object, the object and the thermometer are at the same temperature, and the reading on the thermometer reports the temperature of the object.

19.3 If the potential energy curve is symmetric (as in Fig. 8.36C), the molecules move just as far outward as they move inward when the temperature increases, and the average distance between molecules remains unchanged. Therefore, the material would not expand at higher temperatures.

19.4 Find the changes in length and radius from Equation 19.4, using the coefficient of linear expansion for steel in Table 19.1. The length of the steel rod increases by

$$\Delta L = L_0 \alpha \, \Delta T = (1 \text{ m})(13.0 \times 10^{-6} \text{ K}^{-1})(1 \text{ K})$$
$$\Delta L = 13.0 \times 10^{-6} \text{ m} = 13.0 \,\mu\text{m}$$

The factor $\alpha \Delta T$ is the same for the radius of the rod, but because the radius is initially 1 cm, it changes by only 0.130 μm:

$$\Delta r = r_0 \alpha \, \Delta T = (1 \text{ cm})(13.0 \times 10^{-6} \text{ K}^{-1})(1 \text{ K})$$
$$\Delta r = 13.0 \times 10^{-6} \text{ cm} = 0.130 \,\mu\text{m}$$

Because the change in radius is 100 times smaller than the change in length, the change in the cross-sectional area is 10,000 times smaller than the change in length. It is reasonable to model the expansion of the rod as a one-dimensional expansion.

19.5 You should cool the peg, perhaps with liquid nitrogen. The peg will shrink and therefore fit into the hole. When the peg warms up again, there will be a very tight fit.

19.6

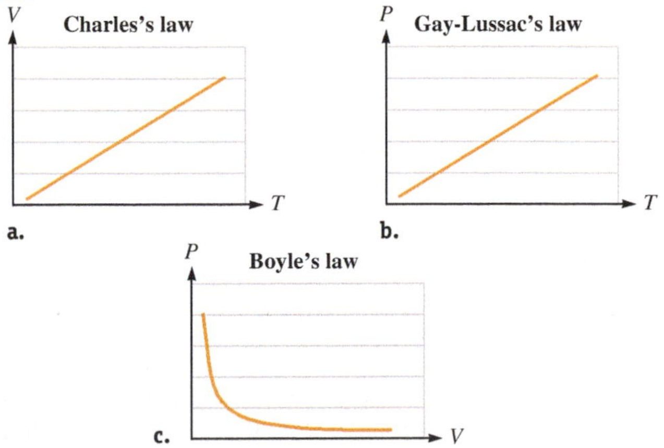

19.7 Label the mass of one molecule of O_2 as m_{O_2} and that of one He atom as m_{He}. The entire mass of the oxygen in the balloon is the number of oxygen molecules multiplied by the mass of each molecule:

$$M_{O_2} = N_{O_2} m_{O_2}$$

and the entire mass of the helium in the balloon is the number of helium atoms multiplied by the mass of each atom:

$$M_{He} = N_{He} m_{He}$$

According to Avogadro's law, because the gases have the same volume and are at the same temperature and pressure, they must contain the same number of particles:

$$N_{O_2} = N_{He}$$

So, the ratio of the balloon's masses is the same as the ratio of the particle's masses:

$$\frac{M_{O_2}}{M_{He}} = \frac{m_{O_2}}{m_{He}} = \frac{5.3 \times 10^{-26}}{6.6 \times 10^{-27}} = 8.0$$

Because the balloons have the equal volumes, the ratio of their densities is also just the ratio of the particle masses:

$$\frac{\rho_{O_2}}{\rho_{He}} = \frac{M_{O_2}/V}{M_{He}/V} = \frac{m_{O_2}}{m_{He}} = 8.0$$

CHAPTER 19: Problems and Questions

1. **a.** 24°C, 297 K **b.** 82°C to 88°C, 355 K to 361 K **c.** −89°C, 184 K
3. 2.6×10^3 °F to 9.5×10^4 °F
5. $T_1 = -40$°F $= -40$°C $= 233$ K
$T_2 = 0$°F $= -18$°C $= 255$ K
$T_3 = 32$°F $= 0$°C $= 273.15$ K
7. −460°F
9. 40°C
11. **a.** No. Your hand is at 98°F, and the ice is at 32°F. **b.** Yes. They are likely in thermal equilibrium with the freezer and thus with each other by the zeroth law of thermodynamics. **c.** No. Your bathing suit is in contact with your body, which is not in thermal equilibrium with the ocean. **d.** Yes. Your bathing suit and the water have had a sufficient amount of time to reach the same temperature.
13. **a.** The metal expands more than the glass because its thermal coefficient of expansion is greater. **b.** If they are made of the same metal, they will expand by the same amount, and the lid will not loosen.
15. 6.19×10^{-3} m
17. **a.** 472°C **b.** 0.035 cm
19. 8.3×10^{-4} °C^{-1}
21. 30.53 m
23. **a.** 11.896 m **b.** 1.01×10^8 Pa
25. $\sigma_{com} \approx 4\sigma$. The compressive strength of cement is greater than the thermal stress by a factor of four. This result suggests that thermal stress is not the cause of the cracking.
27. 1.14×10^5 Pa
29. 1.62×10^7 Pa
31. 3.39 L
33. 1.18×10^6 Pa
35. The pressure and number of particles remains constant, whereas the temperature and the volume must be decreasing.
37. 1.09 atm
39. 0.405
41. **a.** 9.92×10^{14} **b.** 1.65×10^{-9} mol
43. 1×10^{-13} Pa to 4×10^{-10} Pa, 1×10^{-18} atm to 4×10^{-15} atm. Pressure in this cloud is about the same a good laboratory vacuum.
45. 1.23×10^{17}
47. Yes. As in any lab, the bulb should be placed in contact with a triple-point cell and adjusted until the temperature reads 273.15 K.
49. $\dfrac{L_1 L_2}{L_2 - (L_2 - L_1)\cos\theta}$
51. The temperature measured is too low because the height of the mercury is too low. To use a constant-volume thermometer, the volume of the sample must remain constant. By failing to lift the right tube, the researcher has allowed the sample to expand.
53. 284 K
55. **a.** 111 K **b.** −260°F
57. 1.26×10^{-4} s
59. 5.95×10^{-4} m^3
61. **a.** 6.19 mol **b.** 3.73×10^{24} molecules
63. $V_0 \Delta T[\beta_\ell - 3\alpha_g]$
65. 12.0 MPa
67. 8.82 atm
69. **a.** 2.2 mol **b.** 0.071 kg
71. Your idea is a theoretical model because you have a *theory* or *explanation* for the colony's growth; that is, each cell divides into two daughter cells and so on.
73. **a.** $-0.845V_i$ **b.** 7.39×10^5 Pa
75. 315.8 K
77. 70°C
79. **a.** 3.95×10^{-4} kg/m **b.** 26.3 N **c.** 23.5 N **d.** 185 Hz
81. 1.2×10^5 Pa

CHAPTER 20: Concept Exercises

20.1 If all the numbers are equal, the rms value and the average are equal. For example, we can find the average and rms values of the set of six numbers {3, 3, 3, 3, 3, 3}:

$$\text{average} = \frac{3 + 3 + 3 + 3 + 3 + 3}{6} = 3$$

$$\text{RMS} = \sqrt{\frac{3^2 + 3^2 + 3^2 + 3^2 + 3^2 + 3^2}{6}} = \sqrt{\frac{54}{6}} = 3$$

If the numbers are not all equal, the rms weights the larger numbers; in that case, the rms is greater than the average value.

20.2 The average kinetic energy is proportional to temperature; so, if the temperature doubles, that kinetic energy also doubles. The rms speed is proportional to the square root of the temperature; so, if the temperature doubles, the rms speed goes up by $\sqrt{2}$.

20.3 Because the temperatures of the gases are equal, the average kinetic energies of their particles are equal: $K_{av,1} = K_{av,2}$. The rms speed depends on the mass of the particles, and we find

$$\frac{v_{rms,2}}{v_{rms,1}} = \frac{\sqrt{3k_BT/m_2}}{\sqrt{3k_BT/m_1}} = \sqrt{\frac{m_1}{m_2}} = \sqrt{\frac{4m_2}{m_2}} = 2$$

So, the rms speed of particles in gas *B* is twice that of particles in gas *A*.

20.4 Yes, fanning helps. The smoke or other noxious particles take a long time to mix into the air by diffusion. Usually, convection currents speed up that process. Fanning the air also helps by creating a breeze. Once the smoke is mixed into the air, it is less noticeable than before.

20.5 **a.** Above the critical pressure, water must be a blended state in which there is no distinction between liquid and gas. **b.** Between 1 atm and the critical pressure, water at 100°C is a liquid. **c.** Below 1 atm, water at 100°C is a vapor.

20.6 Because the Earth's temperature increases due to global warming, we would expect the evaporation rate to increase.

CHAPTER 20: Problems and Questions

1. The gas particles are in motion in random directions, and they do not interact with one another. Therefore, their motion is only altered by the walls of the container and the gas must fill the container.
3. The Earth's atmosphere is not contained by walls, but rather by the Earth's gravity. The pressure is greater at sea level than on the tops of mountains because gravity causes a higher density of particles near the surface, and therefore a higher pressure.
5. **a.** 5.80 m/s **b.** 6.57 m/s
7. The magnitude of the average velocity is less than the average speed, which is less than the rms speed: $v_0/2 < 3v_0/2 < 1.58v_0$.
9. **a.** 889 m/s **b.** 513 m/s
11. $T_{avg} = -0.18°C$, $T_{rms} = 1.3°C$
13. 724 m/s
15. 1.35×10^7 Pa
17. **a.** 7.52×10^{-21} J **b.** $v_{rms, Ne} = 670$ m/s and $v_{rms, Kr} = 329$ m/s
19. 3.32×10^{-21} J
21. 3.74×10^5 N
23. C is the most probable.

Letter	Number of boxes	Letter	Number of boxes
A	2	E	4
B	3	F	2
C	7	G	1
D	6	H	1

25. 2.03×10^{-25} kg, 122 u. This is heavier than any single atom on the periodic table, so only molecules of at least this mass would be retained.
27. **a.** $v_{mp} = 390$ m/s, $v_{av} = 440$ m/s, and $v_{rms} = 480$ m/s **b.** We expect N_2 molecules to be faster because their mass is lower: $v_{mp} = 420$ m/s, $v_{av} = 470$ m/s, and $v_{rms} = 510$ m/s.
29. $v_{rms} = 511$ m/s and $f(v_{rms}) = 0.00181$, $v_{mp} = 417$ m/s and $f(v_{mp}) = 0.00199$. The most probable speed is indeed at the peak of the distribution function. Because the function is not symmetric, the rms velocity is somewhat higher than the most probable speed.
31. **a.** $\sqrt{k_B T/m}$ **b.** $\sqrt{(\pi k_B T)/(2m)}$
33. 1.02×10^{-8} m
35. 1.3×10^{-7} m
37. **a.** 2.47×10^8 m^{-3} **b.** 5.69×10^9 m in the vacuum chamber, about 10^{17} times longer than in air.
39. **a.** Monica's hydrogen will win. Assuming the gases are at the same temperature, the molecules have the same average kinetic energy, but hydrogen molecules have less mass. Thus, the hydrogen will move faster, resulting in a higher diffusion rate. **b.** The diffusion rates differ by a factor of 4.
41. 2.33×10^{-10} m
43. 2.26×10^7 Pa
45. **a.** Imagine drawing a horizontal line at a pressure of 1 atm; this line does not cross the liquid region at any temperature. As the temperature of solid carbon dioxide (dry ice) is increased, the line crosses into the gas region without ever becoming liquid. The temperature at which this transition occurs appears to be around −75°C. **b.** Imagine drawing a vertical line at a temperature of 20°C to intersect the gas/liquid line. Tracing from this point to the pressure axis shows that the transition occurs at a pressure of around 50 atm.
47. **a.** The water would go from ice to liquid water and then to vapor. **b.** There would be no change.
49. 8.6×10^4 Pa
51. 2.1×10^5 Pa
53. 65%
55. **a.** Increase by a factor of $\sqrt{3}$ **b.** Decrease by a factor of $1/\sqrt{2}$
57. **a.** $(96mv^2)/V$ **b.** $8mv^2$
59. 1.12 mol
61. 5.0×10^5 Pa
63. **a.** 831 m/s **b.** 1.97×10^4 N
65. 8×10^{11} s
67. 1300 m/s, higher than the rms speed of a nitrogen molecule in the atmosphere
69. **a.** $\dfrac{dT}{dt} = \left(\dfrac{v_{av}\pi m}{4k_B}\right)\left(\dfrac{dv_{av}}{dt}\right)$ **b.** 0.735 m/s^2
71. **a.** 2.49×10^{-7} m **b.** 7.50×10^{-10} s
73. **a.** 2.37×10^4 K **b.** 4.72×10^3 K **c.** Although the sea-level temperature of the atmosphere is low, solar energy heats the upper atmosphere such that the thermosphere, at an altitude of 180 km, has a temperature exceeding 2200 K. Thus, a small fraction of the helium gas in the upper atmosphere always has a velocity that exceeds Earth's escape speed and is lost to space.

CHAPTER 21: Concept Exercises

21.1 **a.** Incorrect. Heat is not contained in Texas. The high temperature in Texas means that the air has a lot of thermal energy. **b.** Incorrect. Kinetic energy is converted into thermal energy due to friction between the table and the book. **c.** Incorrect. Heat and work are both methods of transferring energy. One does not become the other.
21.2 **a.** *Heat* is transferred from the air in the room to the water. **b.** If the Earth is not in the system, it does *work* on the ball. When the ball hits the floor, the normal force distorts the ball, which can be modeled as a spring. Kinetic energy is converted into elastic potential energy and thermal energy. **c.** As in part (b), the Earth does *work* on the system. Friction between the atmosphere and the meteor converts mechanical energy into thermal energy.
21.3 Many possible terms would be more helpful than the traditional terms *heat capacity*, *specific heat*, and *molar specific heat*. For example, the constant of proportionality in the equation for static friction (Eq. 6.1) is called the *coefficient of static friction*. How about replacing *heat capacity* with *heat coefficient*? Then, specific heat could be called the *mass heat coefficient*, and molar specific heat could be called the *molar heat coefficient*. Unfortunately, the traditional terms are in standard use.
21.4 Average kinetic energy only depends on the gas temperature, so the average kinetic energy does not change. Thermal energy depends on temperature and the amount of gas. Doubling the number of particles doubles the thermal energy.
21.5

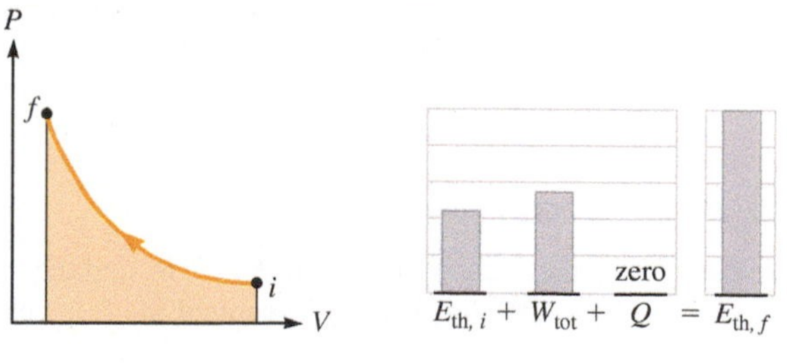

21.6

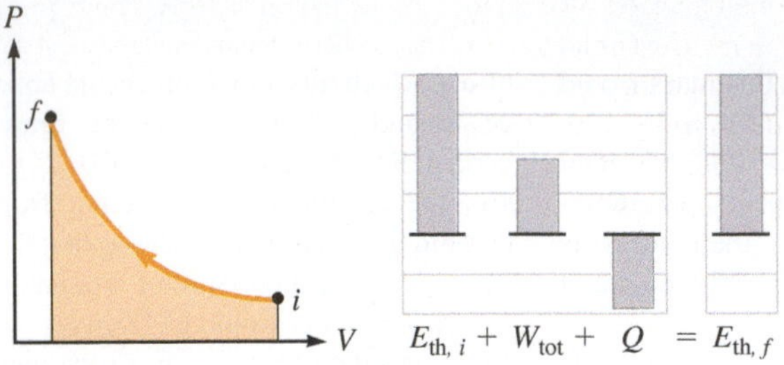

21.7 The volume decreases, so the work done by the gas is negative in both adiabatic compression and isothermal compression. The environment (the piston) does positive work on the gas. Because the area under the adiabatic path in the figure here is greater than the area under the isothermal path, the work done by the piston is greater in the adiabatic process.

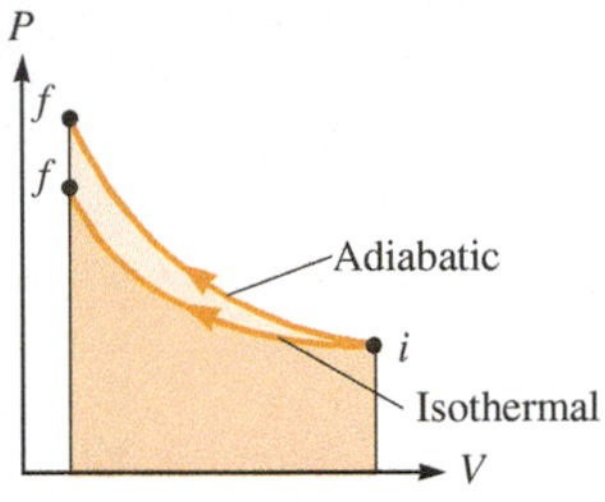

21.8 **a.** According to the *PV* diagram, the area under the expansion path is less than the area under the contraction path. Because the gas does positive work during expansion and negative work during contraction, the net work done by the gas is *negative* for this cycle. **b.** Because the gas does negative work during this cycle, we can conclude that the piston does *positive* work on the gas. Energy enters the system through work. **c.** Because the thermal energy is constant in a cyclic process, the amount of heat that flows into the system must equal the amount of energy the system loses through work. So, the net flow of heat is from the system to the environment.

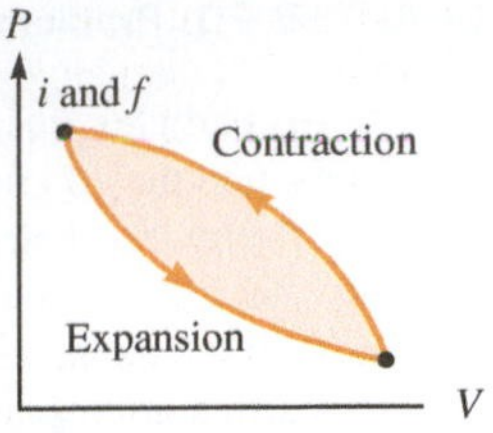

21.9 Changes in state variables only depend on the endpoints in a *PV* diagram. The method of energy transfer depends on the thermodynamic process and therefore depends on the path. So, (e) through (g) depend on the path, and (a) through (d) depend on the endpoints.

CHAPTER 21: Problems and Questions

1. Heat is the energy transferred from a system to the environment or vice versa due to a temperature difference between the two. Perhaps *thermotransportation energy* is a good term because it describes the temperature difference between the system and environment and the concept of energy transfer.
3. Energy is transferred by heat in both cases. The energy is transferred from the water or oven to the steaks because the steaks are at the lower temperature in each case.
5. The word *isolated* suggests that Bobby is cut off, remote, or secluded. The word *insulated* suggests that he is protected or shielded. *Isolated* means that the system cannot exchange energy through work, whereas *insulated* means that the system cannot exchange energy in the form of heat.
7. **a.** 8.380×10^5 J **b.** 1.90×10^3 steps
9. Energy is transferred by heat from the engine to the hood when the engine is hotter than the hood. During the day, sunlight will also transfer energy by heat from the Sun to the hood. The air around the hood will transfer energy to it by heat if the air is at a higher temperature than the hood. So if you are traveling on a cold winter night, you would conclude that the energy was only transferred from the engine to the hood.
11. -15.0 J. The energy enters by work and leaves by heat.
13. 0 J. The energy enters by work, but no energy is exchanged by heat.
15. 81.5°C
17. Yes. Thermal energy depends on n and T. So, if n is larger in one case (as it is for the bathtub), then T can be lower and the system can still have more thermal energy.
19. **a.** 82°C **b.** 88°C It would be better to use a steel cup than a glass cup.
21. 22.9°C
23. -7.77×10^3 J. The energy transferred by heat is leaving the system.
25. 5.1°C; cold liquid
27. 1.38×10^5 J
29. **a.** 2.83×10^4 J **b.** 1.88×10^4 J **c.** 9.47×10^3 J. Results are approximate because the value for c is for *solid* copper, and we used it for *liquid* copper in part (a). It takes more energy to melt the copper than it does to melt the silver.
31. **a.** 6.66×10^5 J **b.** 4.51×10^6 J
33. -6.50×10^2 kJ
35. $7.5P_0V_0$
37. **a.** -105.0 J **b.** 105.0 J **c.** Energy enters the system because the work done by the gas is negative.
39. 3.2×10^2 J
41. **a.** 0.00172 m^3 **b.** 391 K
43. **a.** 7.8×10^2 K **b.** 2.2×10^4 J **c.** -1.5×10^4 J
45. **a.** -1.05×10^3 J **b.** 1.05×10^3 J
47. **a.** $T_B = 1.20 \times 10^2$ K and $T_C = 1.64 \times 10^2$ K **b.** 1.00×10^3 J **c.** 5.02×10^3 J
49. 2.9×10^2 J
51. 5.2×10^4 J
53. **a.** 3.2 **b.** 2.6. Because some energy goes into internal degrees of freedom for nonmonatomic gases, the final pressure of the diatomic gas is lower.
55. **a.** The gas contracted. **b.** 0.486
57. 6.1×10^{-5} m^2
59. 6.67×10^{-4} m/s
61. 363 K
63. **a.** Approximately 9 h **b.** The water will not get much hotter than the air temperature due to convection and thus will never boil.
65. 6.07×10^2 J/(kg · K)
67. **a.** 9.53×10^8 J **b.** -8.46×10^8 J
69. **a.** 0°C **b.** 0.065 kg
71. 1.06 kg
73. 297 K
75. $2k_BT$
77. **a.** -0.279 J **b.** 47.4 kJ **c.** 47.4 kJ
79. 11.3 times faster
81. 0.157 kg
83. 0.386°C
85. **a.** 2.25×10^4 W **b.** 4.86×10^8 J

CHAPTER 22: Concept Exercises

22.1 The back of the truck is at a height $h \approx 1$ m, and the mass of an elephant is about 6000 kg. The work required to lift the elephant into the back of the truck is approximately

$$W_{\text{required}} = mgh = (6000\text{ kg})(9.81\text{ m/s}^2)(1\text{ m})$$
$$W_{\text{required}} \approx 6 \times 10^4\text{ J}$$

A perfect engine would require the same amount of heat, $Q_h = 60{,}000$ J. A real engine would require more heat because it would lose energy (for example, due to friction between the piston and the cylinder).

22.2 **a.** It is possible that the water has undergone a nearly reversible cycle. The temperature could be kept near the boiling point. No work is involved, so no friction dissipates energy. Finally, the process could be done very slowly. **b.** The wood has not undergone a reversible cycle. The chemical reactions that burn the wood cannot be reversed.

22.3 No. Opening the refrigerator door allows the hot and cold reservoirs to come into thermal contact. As they approach the same temperature $T_c \to T_h$, the coefficient of performance approaches infinity, $\kappa \to \infty$. The second law of thermodynamics does not allow the coefficient of performance to reach infinity.

Here's another way to think about it. Usually, a refrigerator removes heat from the cold interior and deposits heat outside. If the door is open, there is no distinction between the inside and the outside. So, where does the extracted heat go?

22.4 **a.** You know that a smashed car does not spontaneously roll back from a brick wall and return to perfect condition. Thus, the entropy of the Universe must increase when a car crashes into a wall. **b.** Ice cubes don't spontaneously form in cups of water, so the entropy of the Universe must increase when ice melts in a cup of water. **c** and **d.** You can imagine watching a puddle of water freeze or melt in reverse. It is possible that the entropy of the Universe does not change during these processes, but any change in the water's entropy must be balanced by a change in the heat reservoir's entropy. When the ice melts, its entropy increases. If the heat reservoir—the Earth and surrounding air—has an equal decrease in entropy, it is possible that the entropy of the Universe is unchanged.

22.5 **a.** The probability is found by dividing the number of microscopic states for the 50 heads/50 tails macroscopic state by the total number of microscopic states:

$$\frac{1.0 \times 10^{29}}{1.27 \times 10^{30}} = 0.079$$

So, there is a 7.9% chance of finding 50 heads and 50 tails.

b. Now we need to find the total number of microscopic states that are possible for all the macroscopic states that produce between 45 and 55 heads:

$$[2(6.14) + 2(7.35) + 2(8.44) + 2(9.32) + 2(9.89) + 10.1] \times 10^{28} = 9.24 \times 10^{29}$$

As in part (a), the probability is found by dividing this result by the total number of possible microscopic states:

$$\frac{9.24 \times 10^{29}}{1.27 \times 10^{30}} = 0.728$$

So, there is about a 73% chance of finding between 45 and 55 heads.

CHAPTER 22: Problems and Questions

1. The potato would get hotter, and the lake would cool. The temperature of each would get farther and farther apart, and equilibrium would not be reached if that were true.

3. The engine would do more work in the adiabatic process, B to C in Figure 22.2, if the process occurred at a higher pressure. Likewise, if the work done by the environment in the process from D to A occurred at a lower pressure, the net amount of work done by the engine would also increase. The area enclosed by the cycle on the PV diagram is the net work performed by the engine. These changes would increase that area.

5. **a.** 0.300 **b.** 70.0%

7. 1.9×10^5 W

9. **a.** 5.861×10^4 J **b.** 0.7835

11. 0.635

13. **a.** 0.58 **b.** 0.76

15. 0.5000

17. **a.** 707.3 K **b.** 7.556

19. **a.** 7.71×10^7 J **b.** 3.75×10^7 J

21. 4.3×10^4 J

23. **b.** The difference in efficiencies is $(T_D - T_A)/T_C$. This makes sense because the temperature at D should be greater than that at A, which means the difference is positive. It makes sense that the Carnot efficiency is greater than the Otto efficiency.

25. **a.** 999 K **b.** Yes. The intake temperature must be at least 143 K.

27. 1.33

29. 16.2

31. **a.** 52.5 J **b.** 262.5 J

35. 101 J/K

37. **a.** 0 **b.** -0.23 J/K

39. -152 J/K

41. **a.** 4.40×10^4 J **b.** -1.10×10^5 J **c.** -271 J/K

43. **a.** 25.7°C **b.** $\Delta S_{cold} = 8.98$ J/K, $\Delta S_{hot} = -8.40$ J/K **c.** 0.58 J/K

45. 14.1 J/K

47. **a.** -0.183 J/K **b.** Yes. You can watch this process in reverse without it appearing odd or improper. **c.** 0.183 J/K

49. The neatly arranged drawer is like a cool system because entropy is low. The messy basket is like a hot system because entropy is high. When you add one more piece of laundry to the basket, the disorder does not increase much compared to the amount of disorder that was already present. When you add a messy item to the drawer, the relative disorder increases significantly. Likewise, when you add a little heat to a cool system, the relative disorder increases significantly as opposed to when you add the same amount of heat to an already hot system.

51. She can decrease the entropy of the clothes as long as there is a corresponding increase in entropy of greater magnitude in the surrounding environment. When she uses her muscles to fold the laundry, she will produce excess heat that increases entropy.

53. A local decrease in entropy is possible as long as there is a corresponding increase in entropy somewhere else that is equal or greater in magnitude. Evolution does not violate the second law because the Earth is not a closed system, nor is any localized system on Earth a closed system. The Earth radiates energy back into space. This is one example of a process where there can be a nearby increase in entropy.

55. 1/36

57. 2.16×10^{-21} J/K

59. It is very likely that the number of heads will be near 50 because the curve has a broad hump near 50 but falls off quickly for values greater than 60 and less than 40.

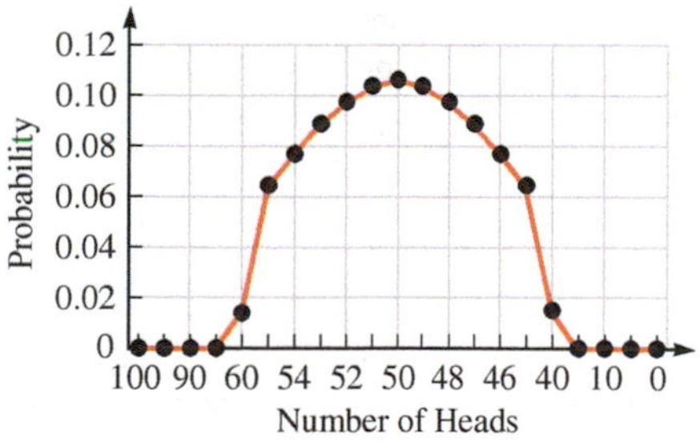

61. 289 K

63. Boiling would lead to a larger change in entropy. Each process occurs at a constant temperature such that $\Delta S = Q/T$. The boiling temperature is higher than the melting temperature (373 K versus 273 K). The heat added depends on the latent heat of fusion for melting and the latent heat of vaporization for boiling, however, and the latter is much higher (2260 J/g versus 333 J/g). The latent heat is almost seven times larger for boiling, whereas the boiling temperature is not even twice as large as the melting temperature; therefore, heat exchange is the dominant factor and the change of entropy must be higher for boiling, even though boiling occurs at a higher temperature.

65. **a.** $-1.75 \times 10^{-3}\ K^{-1}$ **b.** $8.98 \times 10^{-4}\ K^{-1}$
67. **a.** 934 J/K **b.** 123 J/K **c.** 9.37×10^3 J/K
69. **a.** 968 J **b.** 582 J
71. -15.9 J/K
73. **a.** -411 J **b.** 2.78×10^3 J **c.** 3.19×10^3 J **d.** 0.129 **e.** 0.667
75. **a.** -10.9 J/K **b.** -15.3 J/K
77. **a.** 48.3 J/K **b.** 1.43×10^4 J
79. 28.8 J/K
81. **a.** 2×10^{20} **b.** $S_{A_1} = 3.18 \times 10^{-22}$ J/K, $S_{A_2} = 3.27 \times 10^{-22}$ J/K, $S_A = 6.45 \times 10^{-22}$ J/K

CHAPTER 23: Concept Exercises

23.1 Because electrons are transferred from the glass to the silk, the silk has the same number of excess electrons as the number of excess protons in the glass. So the charge of the silk is negative as given by Equation 23.2:

$$q = -Ne = -(3.33 \times 10^{11})(1.60 \times 10^{-19}\ C)$$
$$q = 5.33 \times 10^{-8}\ C = 53.3\ nC$$

23.2 **a.** The red rod is rubbed with the red cloth, so they have charges of equal magnitude and opposite sign. We represent this fact by drawing the same number of signs on the red rod and the red cloth, three plus signs on the rod and three minus signs on the cloth. **b.** The green rod has the greatest positive charge because it is shown with the greatest number of plus signs—five in this figure.

23.3 To determine whether two charged objects are attracted to or repelled by each other, remember that *opposites attract*. The two rods are attracted to each other because the glass rod is positively charged and the amber rod is negatively charged. The two cloths are attracted to each other because the silk is negatively charged and the wool is positively charged. The silk is repelled by the amber rod because both are negatively charged. The wool and the glass rod are mutually repelled because both are positively charged.

23.4 **a.** Yes. Because A is repelled by B, we know that A and B must have charges of the same sign. Suppose both are positive. Because A is attracted to C, we know that A and C have charges of the opposite sign; if A is positive, then C is negative. So B must also be attracted to C.
b. No. If A is attracted to B, then A and B must have charges of opposite sign; if A is positive, then B is negative. In addition, if A is attracted to C, then A and C must also have charges of opposite sign; if A is positive, then C is negative. That means B is repelled by C because B and C are both the same sign—negative.

If you ever discover three charged objects such that A is attracted to B and both are attracted to C, you will be dealing with charge that doesn't come in only two types—positive and negative. It must come in three types.

23.5 **a.** No. The metal ball may be either positive or neutral. If the metal ball is positive, it is attracted to the negative rod because *opposites attract*. If the ball is neutral, it is attracted to the charged rod because the rod causes the ball to be polarized, and it is attracted to the rod just like the can in Figure 23.16B (page 693).
b. Yes. The metal ball must have a negative charge. If the ball were positive or neutral, it would be attracted to the rod as explained in part (a). The only way for the ball to be repelled is if the ball and the rod have charges of the same sign—in this case, negative.

23.6 **a.** Silk is an insulator. The silk rope is used so that the boy is not grounded. If he were grounded, he would discharge and the fun would be over. A metal chain would ground the boy by providing a pathway for electrons to go from the boy through the chain, through the building, and into the Earth.
b. The boy is charged, so he can induce dipoles in the paper. Like the comb in Figure 23.5C (page 686), the boy is able to attract paper and turn the pages of a book without touching it.
c. The woman is about to provide a pathway from the boy to the Earth; in other words, she is about to ground him. If the boy has built up a large charge, when the woman gets very close to him, the air will act as a conductor, momentarily completing the pathway to ground even before she makes contact. When that happens, there will be a spark as in the case of a lightning strike (but smaller!).

CHAPTER 23: Problems and Questions

1. A contact force is exerted when the source and subject touch; a field force does not require contact. The only field force presented in the first 22 chapters is gravity. All other forces (normal, friction, drag, tension, spring, buoyant) are contact forces and are manifestations of the electromagnetic field force. So, in a sense, all these "contact" forces are field forces.
3. 2.2×10^{11}
5. 6.25×10^{18}
7. **a.** The rod has lost mass. Electrons have been stripped from the rod, leaving behind a net positive charge. **b.** 2.60×10^{-16} kg
9. **a.** 2.63×10^{13} **b.** 1.81×10^{-12} **c.** 2.39×10^{-17} kg
11. Yes. The glass and the plastic are attracted to each other because they have opposite charges. The silk and the wool must also have opposite charges and be attracted because each was electrically neutral, like the glass and the plastic, but became charged when they exchanged electrons with those objects.
13. The technician wants to be protected and prevent electric charge from transferring from the lines to his body. Because rubber is an insulator, charges that are transferred to the rubber will stay at or near the point where they contact the rubber and not move through the technician and to the ground.
15. 9.54×10^{12} electrons/m^2
17. The charged insulator's net charge remains unchanged, $+30.0\ \mu C$, while the other maintains a net charge of 0. It does matter how the contact is made. If either of the insulators is moved while they are in contact such that a friction force acts on each insulator over a short distance, charges can be transferred from one to the other.
19. The charge is free to move once it is on the conductor. The charge spreads out uniformly across the surface of the conductor.
21. **a.** 4.92×10^{-8} N **b.** Attractive
23. 2.4×10^{-5} C
25. $r_i/2$
27. **a.** $6.66 \times 10^{-6}\hat{\imath}$ N **b.** $-6.66 \times 10^{-6}\hat{\imath}$ N
29. **a.** $d\sqrt{2}$ **b.** $d/\sqrt{2}$
31. **a.** 9.21×10^{-8} N **b.** 4.16×10^{42}
33. $\pm 4.3 \times 10^{-13}$ C
35. 2.28×10^{-9}
37. $-9.66 \times 10^{-4}\hat{\imath}$ N
39. $(-758\hat{\imath} - 2.35 \times 10^3\hat{\jmath})$ N
41. **a.** $-1.02 \times 10^{-7}\hat{\jmath}$ N **b.** $1.05 \times 10^{-7}\hat{\jmath}$ N **c.** $-2.74 \times 10^{-9}\hat{\jmath}$ N
43. 1.45 N, 46.7° below the $-x$ axis
45. $(\sqrt{2} - 1)d$
47. 4.55×10^{-7} C
49. -1.5×10^{-9} C
51. $-q((1 + 2\sqrt{2})/4)$
53. 1.33 m from the 8.00-nC sphere

55. $\sqrt{(4L^2mg\tan\theta\sin^2\theta)/5k}$
57. 2.80×10^{23}
59. 1.90×10^{-7} C
61. $(1.38\hat{\imath} - 1.93\hat{\jmath})$ N
63. a. $8.79\hat{\imath}$ N **b.** $21.6\hat{\imath}$ N
65. $kq^2/(4L^2g\tan\theta\sin^2\theta)$
67. 1.12×10^{-7} C
69. 2.28×10^{-5} C
71. 0.21 m
73. a. $(2kQqh)/(h^2 + a^2)^{3/2}\hat{\jmath}$
75. $(-0.273\hat{\imath} - 0.273\hat{\jmath} - 0.273\hat{k})$ N
77. $kq^2/(mgr^2)$

CHAPTER 24: Concept Exercises

24.1 According to $\vec{E}(r) = (kQ_S/r^2)\hat{r}$ (Eq. 24.3), the electric field vectors point radially. If the source is positive, the electric field vectors point outward, and if the source is negative, the electric field vectors point inward. Also according to Equation 24.3, the electric field is stronger near the source; the field drops off as $1/r^2$ as r increases. These statements are consistent with Figure 24.4C, which shows outward-pointing electric field vectors that are longer near the source.
24.2 To find the magnitude of the electric field at the position of a charged particle, take the limit $\lim_{r\to 0} E(r) = \lim_{r\to 0}(kQ_S/r^2) \to \infty$. The electric field due to a charged particle at its exact location approaches infinity.
24.3 The lines in cases 2 and 4 cannot represent electric field lines because they do not originate or terminate on charged particles.
24.4 a. The particle on the left is positively charged because the electric field lines originate from it. The particle on the right is negatively charged because the electric field lines terminate on it. **b.** The electric field is weakest in region C because the electric field lines in that region have the lowest density. **c.** The electric field is strongest in region A because the electric field lines in that region have the highest density.
24.5 The positive charge is due to 10 protons and the negative charge is due to 10 electrons, so $Q = 10e$. The dipole moment is $p = Qd = 10ed = 6.2 \times 10^{-30}$ C · m.
24.6 a. The surface charge density is

$$\sigma = \frac{Q}{A} = \frac{Q}{2\pi RL}$$

The small amount of charge dq in a length dx of the rod is

$$dq = \sigma dA = \sigma(2\pi R)dx$$

$$dq = \frac{Q}{2\pi RL}(2\pi R)dx = \frac{Q}{L}dx$$

b. The linear charge density is

$$\lambda = \frac{Q}{L}$$

The small amount of charge dq in a length dx of the rod is

$$dq = \lambda dx = \frac{Q}{L}dx$$

This is exactly what we found in part (a). We conclude that we can ignore the radius of a real rod and model the rod as an infinitely thin line.

CHAPTER 24: Problems and Questions

1. The symbol for electrostatic force is F_e and it has units of N. The symbol for electrostatic field is E and it has units of N/C. The force requires a source and a subject, whereas the field requires only a source.
3. 1.32×10^5 N/C
5. a. $-1.03 \times 10^5\hat{\jmath}$ N/C
b. $-5.54 \times 10^{-4}\hat{\jmath}$ N
7. 5.92×10^5 C
9.

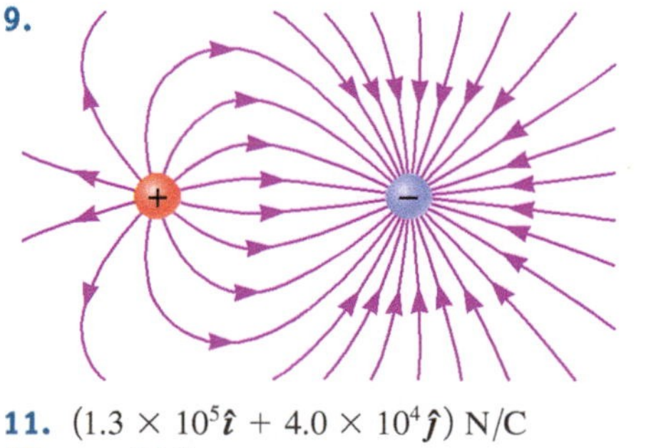

11. $(1.3 \times 10^5\hat{\imath} + 4.0 \times 10^4\hat{\jmath})$ N/C
13. $x = 12.8$ m
15. $x = -0.0986$ m
17. $-6.14 \times 10^{13}\hat{\imath}$ N/C
19. $-1.03 \times 10^5\hat{\imath}$ N/C
21. $n = 1$: $E_P \approx 2.81 \times 10^3$ N/C,
$n = 2$: $E_P \approx 3.40 \times 10^3$ N/C,
$n = 4$: $E_P \approx 3.64 \times 10^3$ N/C,
$n = 8$: $E_P \approx 3.72 \times 10^3$ N/C
23. $(k\lambda L)/(x^2 - xL)$
27. $y = R/\sqrt{2}$ and $\vec{E} = (2kQ)/(R^2 3\sqrt{3})\hat{\jmath}$
29. $E_{\text{disk}} = E_{\text{particle}} = 6.0 \times 10^3$ N/C
31. $3.95 \times 10^2\hat{\imath}$ N/C
33. 428 N/C
35. $5.31 \times 10^8\hat{\jmath}$ N/C
37. $-4.73 \times 10^5\hat{\jmath}$ N/C
39. 3×10^{-7} C. This is about 10 times less than the charge in the case of Wilson's rod.
41. 0.77 m
43. 3.27×10^3 N/C, upward
45. 1.81×10^{-18} C/m^2
47. 2.1×10^5 m/s
49. 4.21×10^{-7} C/m
51. a. $(-347\hat{\imath} - 635\hat{\jmath})$ N/C
b. $(-6.94 \times 10^{-7}\hat{\imath} - 1.27 \times 10^{-6}\hat{\jmath})$ N
53. a. -1.10×10^{-8} C **b.** 7.46×10^{-3} N
55. The water molecules reorient themselves as they fall due to the electric field created by the excess charge on the comb. The part of the water molecule that is attracted toward the comb rotates to be closer, while the other side is rotated away from the comb. The attracted portion of the molecule is closer to the comb, so the magnitude of the electrostatic force on that portion is greater than that on the other side of the molecule that is repelled. The water then is attracted to the comb due to the net force on the molecules.
57. a. 4.4×10^{-29} J **b.** 5.0×10^{-28} J **c.** -4.4×10^{-29} J
59. $x = 0.60$ m
61. a. 0 **b.** $kQ/(2\sqrt{2}R^2)$
63. $-2kQ/\pi R^2\hat{\jmath}$
65. $-4kq/5\hat{\imath}$
67. a. 5.86×10^6 N/C
b. 1.10×10^7 N/C
c. 5.56×10^6 N/C **d.** 1.21×10^5 N/C
69. 9.78×10^4 N/C
71. $(3.71\hat{\imath} + 111\hat{\jmath})$ N/C
73. 4.20×10^6 N/C, to the left
75. 0.743 N
77. a. $\sqrt{2}kQ/(4\ell^2)\hat{\jmath}$
b. $kQ/(\sqrt{2}\ell^2)\hat{\jmath}$ **c.** The ratio is 1/2.
79. a. $0.628kQ/\ell^2\hat{\jmath}$ **b.** $kQ/(\sqrt{2}\ell^2)\hat{\jmath}$
c. 0.888

CHAPTER 25: Concept Exercises

25.1 a. Other uppercase letters that have the same symmetry as **A** are **M**, **T**, **U**, **V**, **W**, and **Y**. **b.** Other uppercase letters that have the same symmetry as **Z** are **N** and **S**. **c.** No other uppercase letters have the same symmetry as **O**. **B**, **C**, **D**, **E**, **K**, and **R** are all symmetrical with respect to a 180° rotation around the x axis. **H**, **I**, and **X** are symmetrical with respect to 180° rotations around all three axes. **F**, **G**, **J**, **L**, **P**, and **Q** are not symmetrical.

25.2 *Electric force* is the field force exerted between charged objects. It may be repulsive or attractive. The SI units are newtons. *Electric field* is the electric force per unit charge. It can be found by imagining the electric force on a positive test charge. The SI units are newtons per coulomb. *Electric flux* is analogous to the flow of a fluid through a loop. The electric flux is represented by the number of electric field lines that penetrate an area. The SI units are newtons · meter squared per coulomb.

25.3 Take the dot product to find the electric flux $\Phi_E = \vec{E}\cdot\vec{A}$ (Eq. 25.3):

$$\Phi_E = (15\hat{\imath} + 25\hat{\jmath})\cdot(0.65\hat{\imath} + 0.35\hat{\jmath})\,\text{N}\cdot\text{m}^2/\text{C} = (15)(0.65) + (25)(0.35)\,\text{N}\cdot\text{m}^2/\text{C} = 19\,\text{N}\cdot\text{m}^2/\text{C}$$

25.4 Choices b and c are correct ways to express Gauss's law. Choice a is not correct because it is only an expression for flux.

25.5 In cases 1 and 2, the lumpy surface and the sphere both have +10 mC inside, so according to $\Phi_E = q_{\text{in}}/\varepsilon_0$ (Eq. 25.7), the net electric flux is

$$\Phi_E = \frac{10\times10^{-3}\,\text{C}}{8.85\times10^{-12}\,\text{C}^2/\text{N}\cdot\text{m}^2} = 1.1\times10^9\,\text{N}\cdot\text{m}^2/\text{C}$$

In case 3, there is no charge in the box, so according to Gauss's law, the electric flux through the box is zero: $\Phi_E = q_{\text{in}}/\varepsilon_0 = 0$.

25.6 To show that Equation 25.13 is consistent with Equation 24.15, take the limit of Equation 24.15 in the case that the line becomes very long, $\ell \gg y$:

$$\lim_{\ell\gg y}\vec{E} = \lim_{\ell\gg y}\frac{kQ}{y}\frac{1}{\sqrt{\ell^2+y^2}}\hat{\jmath} = \frac{kQ}{y}\frac{1}{\ell}\hat{\jmath} = \frac{k2\lambda\ell}{y}\frac{1}{\ell}\hat{\jmath}$$

where $\lambda = Q/2\ell$ for a rod of length 2ℓ.

$$\lim_{\ell\gg y}\vec{E} = \frac{k2\lambda}{y}\hat{\jmath} = \frac{1}{4\pi\varepsilon_0}\frac{2\lambda}{y}\hat{\jmath} = \frac{1}{2\pi\varepsilon_0}\frac{\lambda}{y}\hat{\jmath}$$

This is for a point on the y axis. In general, we can write $y \to r$ and $\hat{\jmath} \to \hat{r}$ for a point at some distance r from the rod. Then Equation 24.15 is the same as Equation 25.13 in the limit that the rod becomes infinitely long:

$$\lim_{\ell\to\infty}\vec{E} = \frac{1}{2\pi\varepsilon_0}\frac{\lambda}{r}\hat{r} \checkmark$$

25.7 No. The sphere in Figure 25.23 cannot be a conductor in electrostatic equilibrium. If it were a conductor, the excess charge would quickly move to the surface. The sphere must consist of either an insulator or many individual charges, such as the protons in a nucleus.

CHAPTER 25: Problems and Questions

1. We proceed letter by letter. **Z** has the same symmetry as **N** and **S**, but not **C** or **B**. Thus, only **NUT** and **SUE** are possible answers. **A** has the same symmetry as **U** in each of those words, but the symmetry of **K** matches only the symmetry of **E**, not **T**. Thus, the answer is (b), **SUE**.

3. The source inside must be negative because it attracts the positively charged ball.

5. The box contains both positively and negatively charged objects. The negatively charged object is near the attractive face and the positively charged object is near the repulsive face. The net charge in the box may be zero, positive, or negative depending on the magnitudes of the charged objects near each face.

7. a. The net electric flux is zero. The number of electric field lines that enter and leave the box is the same. **b.** The net electric flux is positive. More field lines leave the box than enter it.

9. 1.39×10^7 N/C

11. a. 0 **b.** $4.18\times10^4\ \text{N}\cdot\text{m}^2/\text{C}$

13. $2.62\times10^3\ \text{N}\cdot\text{m}^2/\text{C}$

15. $655\ \text{N}\cdot\text{m}^2/\text{C}$

17. $\Phi_E = \pi r^2 E\sin\omega t$

19. $\Phi_{E,1} = 0$, $\Phi_{E,2} = -3.05\times10^6\ \text{N}\cdot\text{m}^2/\text{C}$, $\Phi_{E,3} = -4.58\times10^6\ \text{N}\cdot\text{m}^2/\text{C}$, $\Phi_{E,4} = -3.05\times10^6\ \text{N}\cdot\text{m}^2/\text{C}$

21. $\Phi_{E,\text{A}} = 904\ \text{N}\cdot\text{m}^2/\text{C}$, $\Phi_{E,\text{B}} = 0$, $\Phi_{E,\text{C}} = 0$, $\Phi_{E,\text{D}} = 1.47\times10^3\ \text{N}\cdot\text{m}^2/\text{C}$

23. $q/8\varepsilon_0$

25. $\Phi_{E,\text{A}} = Q/\varepsilon_0$, $\Phi_{E,\text{B}} = -Q/\varepsilon_0$, $\Phi_{E,\text{C}} = 0$, $\Phi_{E,\text{D}} = Q/\varepsilon_0$

27. a. $8.81\times10^3\ \text{N}\cdot\text{m}^2/\text{C}$ **b.** $4.41\times10^3\ \text{N}\cdot\text{m}^2/\text{C}$

29. $\rho(r^2-a^2)/(2\varepsilon_0 r)\hat{r}$

31. a. 0 **b.** 4.50×10^4 N/C **c.** 1.80×10^3 N/C

33. $(\lambda r)/(2\pi R^2\varepsilon_0)\hat{r}$

35. $\vec{E}_A = -2.1\times10^5\hat{\imath}$ N/C, $\vec{E}_B = 4.6\times10^5\hat{\imath}$ N/C, $\vec{E}_C = -4.8\times10^5\hat{\imath}$ N/C

37. $-3.04\times10^{-4}\ \text{C/m}^2$

39. 0, 8.81×10^9 N/C, 1.76×10^{10} N/C, 7.83×10^9 N/C

41. 0

43. a. $cR^5/(5\varepsilon_0 r^2)$ **b.** $cr^3/(5\varepsilon_0)$

45. 5.54×10^3 N/C

47. a. $\vec{E}_A = -4.07\times10^6\hat{\imath}$ N/C, $\vec{E}_B = 1.36\times10^6\hat{\imath}$ N/C, $\vec{E}_C = 4.07\times10^6\hat{\imath}$ N/C **b.** $\vec{F}_A = 6.51\times10^{-13}\hat{\imath}$ N, $\vec{F}_B = -2.17\times10^{-13}\hat{\imath}$ N, $\vec{F}_C = -6.51\times10^{-13}\hat{\imath}$ N

49. a. $\sigma_{\text{upper}} = 3.98\times10^{-7}\ \text{C/m}^2$ and $\sigma_{\text{lower}} = -3.98\times10^{-7}\ \text{C/m}^2$ **b.** $Q_{\text{upper}} = 4.34\times10^{-8}$ C and $Q_{\text{lower}} = -4.34\times10^{-8}$ C

51. a. $\rho x/\varepsilon_0$ **b.** 6.78×10^3 N/C

53. 8.3×10^9

55. 1.12×10^4 N/C, 1.80×10^{-15} N

57. $Q_{\text{inner}} = -38.3\ \mu\text{C}$, $Q_{\text{outer}} = 38.3\ \mu\text{C}$

59. a. -24.6 mC **b.** 0 **c.** $-4.89\times10^{-2}\ \text{C/m}^2$ **d.** 0

61. $0.72\ \text{N}\cdot\text{m}^2/\text{C}$

63. a. 0 **b.** 3.62×10^6 N/C

65. $-0.120\ \text{N}\cdot\text{m}^2/\text{C}$

67. a. -1.46×10^{-7} C **b.** 1.68×10^{-7} C **c.** $Q_{\text{inner}} = 1.46\times10^{-7}$ C, $Q_{\text{outer}} = 2.2\times10^{-8}$ C

69. a. 0 **b.** 1.65×10^6 N/C **c.** 3.30×10^6 N/C **d.** 3.66×10^5 N/C

71. 0

73. 5.0×10^{-11} C

75. a. $2\pi AR^2$ **b.** $AR^2/(2\varepsilon_0 r^2)$ **c.** $A/(2\varepsilon_0)$

77. $E_{\text{outside}} = 5.6\times10^5$ N/C

79. $5.45\times10^4\ \text{N}\cdot\text{m}^2/\text{C}$

CHAPTER 26: Concept Exercises

26.1 a. Gravitational potential. Gravitational force depends on the mass of *both* the source and the subject, whereas gravitational field depends *only* on the mass of the source. Similarly, gravitational potential energy depends on the mass of *both* the source and the subject, whereas gravitational potential depends *only* on the mass of the source. **b.** Gravitational potential. Gravitational force is a *vector* that depends on the mass of both the source and the subject, whereas gravitational potential energy is a *scalar* that depends on the mass of both the source and the subject. Similarly, gravitational field is a *vector* that depends only on the mass of the source, whereas gravitational potential is a *scalar* that depends only on the mass of the source.

26.2 a. The sphere must be positively charged. Because the electric potential energy is positive, the source must have charge of the same sign as the subject (a proton).
b. $U_E = -15.0 \times 10^{-20}$ J

26.3 Start with $U_E = kQq/r$ (Eq. 26.3):

$$U_E = k\frac{(10e)(-10e)}{r} = -100k\frac{e^2}{r} = -100(8.99 \times 10^9 \text{ N}\cdot\text{m}^2/\text{C}^2) \times \left[\frac{(1.60 \times 10^{-19}\text{ C})^2}{3.9 \times 10^{-12}\text{ m}}\right]$$

$$U_E = -5.90 \times 10^{-15}\text{ J}$$

The negative sign means that energy is transferred from the system to the environment in bringing the particles together from infinity, and it would take positive work to separate the charged particles of a dipole.

26.4 Map A → Landscape 4, a simple symmetrical hill
Map B → Landscape 3, a hill that is steep on one side
Map C → Landscape 2, a sharp drop-off or cliff
Map D → Landscape 1, two hills

26.5 Only *voltage* is a synonym. It is a common term used for electric potential.

26.6 The landscape would look like two ramps meeting at a thin, long peak, something like the peaked roof line shown here.

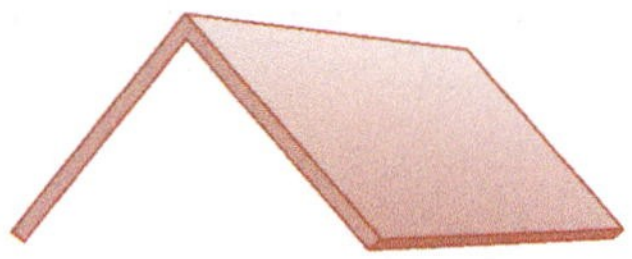

26.7 No. You cannot use an EKG to plot E versus position because an EKG is a graph of V versus time. You need a graph of V versus *position* because E comes from the spatial derivative of V. Remember that E is the slope only on a V-versus-*position* graph.

CHAPTER 26: Problems and Questions

1. When a vector, like force, is negative, its direction is opposite the direction that is defined to be positive. When the potential energy is negative, it means the potential energy is less than that of some reference configuration where the potential energy is zero.

3. While all of the quantities change as the planet moves, the gravitational potential and the field at the location of the planet do not depend on the mass of the planet. They are properties of the space affected by the star. The other quantities depend on the mass of the planet, its distance from the star, and its speed at any moment in time.

5.

Term	Mathematical symbol	Vector or scalar?	Source and subject or source only?	SI units
Electrostatic force	$\vec{F}_E$	vector	both	N
Electric field	$\vec{E}$	vector	source only	N/C or V/m
Electric potential energy	U_E	scalar	both	J
Electric potential	V	scalar	source only	V

7. -5.6×10^{-6} J
9. a. $U_E = kq^2/r$ **b.** $U_E = -kq^2/r$ **c.** $U_E = kq^2/r$
11. 8.81×10^{-8} J
13. 3.8×10^{-11} m
15. a. 58.2 J **b.** The particles will fly apart and their kinetic energies will increase. The kinetic energy of the system will eventually equal 58.2 J.
17. 2.05×10^{-5} J
19. a. 84.4 V **b.** The potential is higher at $y = 2.00$ cm.
21. a. 2.91 m **b.** 8.27×10^{-7} C
23. -1.5 eV, -2.4×10^{-19} J
25. 13.8 V
27. a. -8.21×10^5 V **b.** -6.65×10^5 V
29. 90.0 V
31. a. 2.2×10^4 V **b.** 4.3×10^4 V **c.** 4.3×10^4 V
33. 1.68×10^3 V
35. 2.22 J
37. -1.07×10^{-6} C
39. -1.07×10^7 V
41. a. $k\lambda \ln\left[(a + \sqrt{a^2 + y^2})/(-a + \sqrt{a^2 + y^2})\right]$
b. $kq\lambda \ln\left[(a + \sqrt{a^2 + y^2})/(-a + \sqrt{a^2 + y^2})\right]$
45. 111 V
47. $(Aa^2/2) - (Bb^3/3)$
49. a. $V = RE$ **b.** 4.8×10^5 V
51. $(V_0 e^{-ar}/r)(a + 1/r)$
53. a. -5.50 V, 12.5 V, 30.5 V **b.** 4.50 V/m in the $-y$ direction
55. a. $\vec{E} = -(8xz + 2y^2)\hat{\imath} - (4xy - 8z^2)\hat{\jmath} - (4x^2 - 16yz)\hat{k}$
b. 58.8 V/m
57. a.

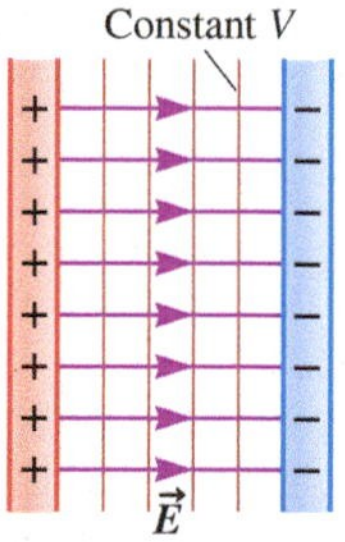

b. 0
59. a. 0, 3.45×10^4 V **b.** 1.57×10^5 V/m, 3.45×10^4 V **c.** 3.04×10^4 V/m, 1.52×10^4 V
61. a. -0.766 J **b.** -1.29×10^5 V
63. 23.4 m/s and 6.63 m/s
65. a. -125 V **b.** -125 V **c.** No. The electric force is a conservative force; thus, only the initial and final positions of a particle's movement determine the change in electric potential.
67. Yes. The sign of the charged particle does matter. In either case, the change in electric potential is the same (positive), but the change in electric potential energy is not. If the particle has a net negative charge, the

change in electric potential energy is negative according to $\Delta U_E = q\Delta V$. Likewise, if the particle has a net positive charge, the change in electric potential energy is positive.

69. 46.9 J

71. -2.69×10^4 V

73. $2\pi k\sigma[1 - x/(\sqrt{R^2 + x^2})]\hat{\imath}$

75. **a.** $V = Er\ln(R/r)$ **b.** -6.3×10^3 V

77. $(2\pi ka/3)\{\sqrt{R^2 + x^2}[R^2 - 2x^2] + 2x^3\}$

79. $2kq/3$

CHAPTER 27: Concept Exercises

27.1 Consider the spring, the object, and the Earth to be the system. Assume the object undergoes purely translational motion, and ignore air resistance. Then the potential energy stored by the spring at maximum compression equals the gravitational potential energy when the object is at its maximum height:

$$\frac{1}{2}ky^2 = mgy_{\max}$$

$$y = \sqrt{\frac{2mgy_{\max}}{k}} = \sqrt{\frac{2(0.09\,\text{kg})(9.81\,\text{m/s}^2)(120\,\text{m})}{1000\,\text{N/m}}}$$

$$y = 0.46\,\text{m} \approx 0.5\,\text{m}$$

This means that even a stiff spring has to be longer than 0.5 m (or about the length of your forearm) in order to compress it far enough.

27.2 With $Q = C\Delta V$ (Eq. 27.1), the slope of each graph (Fig. 27.8, page 832) is $1/C$. Find the slope for capacitor A:

$$\frac{1}{C_A} = \frac{30\,\text{V}}{3\,\text{nC}} = \frac{30\,\text{V}}{3 \times 10^{-9}\,\text{C}} = 1 \times 10^{10}\,\text{V/C}$$

Take the reciprocal of the slope to find the capacitance:

$$C_A = \frac{1}{1 \times 10^{10}\,\text{V/C}} = 1 \times 10^{-10}\,\text{F} = 100\,\text{pF}$$

Repeat this process for capacitor B:

$$\frac{1}{C_B} = \frac{30\,\text{V}}{1.5\,\text{nC}} = \frac{30\,\text{V}}{1.5 \times 10^{-9}\,\text{C}} = 2 \times 10^{10}\,\text{V/C}$$

$$C_B = 5 \times 10^{-11}\,\text{F} = 50\,\text{pF}$$

27.3 **a.** From the graph for capacitor B, when $Q = 0.5$ nC, the potential difference between the plates is $\Delta V = 10$ V. Combine $Q = C\Delta V$ with $U_E = Q^2/2C$ to find $U_E = \frac{1}{2}Q\Delta V$. Substitute values:

$$U_E = \tfrac{1}{2}(0.5\,\text{nC})(10\,\text{V}) = 2.5 \times 10^{-9}\,\text{J} = 2.5\,\text{nJ}$$

b. From Concept Exercise 27.2, $C_A = 100$ pF. Find the charge from $U_E = Q_A^2/2C_A$:

$$Q_A = \sqrt{2C_A U_E} = \sqrt{2(100\,\text{pF})(2.5\,\text{nJ})}$$

$$Q_A = 7.1 \times 10^{-10}\,\text{C} = 0.7\,\text{nC}$$

Find the potential difference from $U_E = \frac{1}{2}C_A\Delta V_A^2$:

$$\Delta V_A = \sqrt{\frac{2U_E}{C_A}} = \sqrt{\frac{2(2.5\,\text{nJ})}{(100\,\text{pF})}} = 7\,\text{V}$$

Use the graph (Fig. 27.8) to confirm this.

27.4 First consider connecting the terminals of a battery with a simple wire (Fig. 27.12B, page 835). Because one terminal is positive and the other is negative, there is an electric field between the terminals, and electrons are forced to move from the negative terminal toward the positive terminal. Now consider charging a capacitor with a battery (Fig. 27.14). An electric field forces electrons to move from the negative terminal toward the positive terminal. Because of the gap in the capacitor, electrons build up on one plate and an excess of protons (lack of electrons) builds up on the other plate. When the potential difference across the capacitor equals the terminal potential, $V_C = \mathcal{E}$, the potential of the negative capacitor plate equals the potential of the battery's negative terminal. Once the negative capacitor plate and the negative battery terminal are at the same potential, there is no electric field between those locations and electrons are no longer forced to move through the wire.

27.5 The **I**-shaped conductor remains neutral. Because electrons are free to flow inside the conductor, they are repelled by the negatively charged bottom plate of the capacitor and attracted to the positively charged top plate. The net result is that the bottom plate of the **I**-shaped conductor has excess positive charge $q_1 = +Q$, and its top plate has excess negative charge $q_2 = -Q$. The capacitor plates are not changed by the insertion of the **I**-shaped conductor, so the potential difference between the top and bottom plates is still the terminal potential of the battery: $V_C = \mathcal{E}$.

27.6 The six capacitors are in parallel, so we find the capacitance of each individual capacitor from $C_{eq} = \Sigma_{i=1}^{N} C_i$ (Eq. 27.8):

$$C_{eq} = C + C + C + C + C + C = 6C$$

$$C = \frac{C_{eq}}{6} = \frac{94\,\mu\text{F}}{6} = 16\,\mu\text{F}$$

27.7 Find the energy density from $u_E = \frac{1}{2}\kappa\varepsilon_0 E^2$ (Eq. 27.22) with $\kappa = 1$ (because the tube is evacuated):

$$u_E = \tfrac{1}{2}(8.85 \times 10^{-12}\,\text{C}^2/\text{N}\cdot\text{m}^2)(5 \times 10^6\,\text{N/C})^2 = 1.1 \times 10^2\,\text{J/m}^3$$

CHAPTER 27: Problems and Questions

1. Air resistance dissipates the mechanical energy, so more potential energy would be needed initially to get the ring to reach the same final height.

3. *Leyden jar* has the advantage of sounding like a storage device, but the disadvantage of making charge seem like a liquid contained in the jar. *Capacitor* and *condenser* do not sound like storage devices, although *capacitor* might remind you of the word *capacity*, perhaps suggesting that it has the ability to store charge. Similarly, *condenser* might remind you that charges are "condensed," or brought together on the plates.

5. 2.9×10^{-4} C

7. **a.** $Q_1 = 8.00 \times 10^2\,\mu$C, $Q_2 = 2.00 \times 10^2\,\mu$C **b.** 20%

9. **a.** 1.82×10^{-4} J **b.** 1.30×10^{-3} J

11. **a.** 4.69×10^{-3} C **b.** 2.93×10^{-4} F

13. 0.150 J

15. 9.002 V

17. **a.** 10.1 μF **b.** 2.34 μF

19. 15 V

21. $C_1 = 0.364\,\mu$F, $C_2 = 0.366\,\mu$F

23. $186C/241$

25. $1.00 \times 10^2\,\mu$F and 9.84 μF

27. 8.22 μF

29. 22.00 pF, 44.00 pF, and 66.00 pF

31. **a.** 1.69×10^{-12} F **b.** 1.52×10^{-11} C **c.** 3.91×10^3 V/m

33. $4\pi\varepsilon_0 r_{in}$

35. 3.56×10^{-5} m

37. **a.** Charge is halved. **b.** Energy is halved.

39. 1.88×10^3 V

41. 5.6×10^6 m^2

43. **a.** 2.71×10^5 V **b.** $r_{in} = 1.53 \times 10^{-2}$ m and $r_{out} = 2.18 \times 10^{-2}$ m
45. 2.39×10^{-9} C
47. **a.** 2.28×10^{-10} C **b.** 1.02×10^{-11} F **c.** 22.3 V **d.** 1.40×10^{-8} J
49. 35.0
51. **a.** 0.333 J **b.** 851 V
53. $C = 3C_0$, $Q = 3Q_0$, $V = V_0$, $U = 3U_0$
55. **a.** 1.59×10^{-8} J **b.** 4.00×10^3 V/m **c.** 946 V/m **d.** 3.05×10^3 V/m **e.** 3.77×10^{-9} J
57. 1.25×10^4 s
59. **a.** 1.27 J/m³ **b.** 1.08×10^9 J/m³
61. $Q/(\kappa\varepsilon_0\pi R^2)$
63. 2.0
65. 2.2×10^{-11} C/m
67. **a.** 1.20×10^{-8} J **b.** 4.00 V
69. **a.** 2.4 μF **b.** 9.72×10^{-5} J **c.** 2.16×10^{-5} C **d.** $U_A = 1.94 \times 10^{-5}$ J, $U_B = 7.78 \times 10^{-5}$ J
71. 2.12×10^{-8} C
73. 16.5 mJ. To achieve the highest terminal potential, connect the batteries in series. To achieve the greatest capacitance, connect the capacitors in parallel.

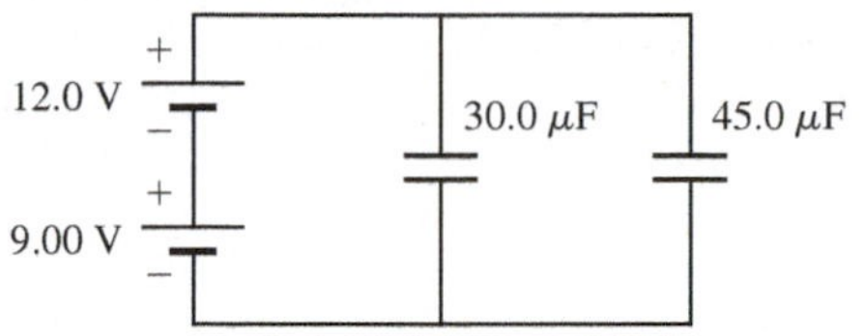

75. **a.** 2.91×10^{-6} F **b.** $Q_A = 3.50 \times 10^{-5}$ C, $Q_B = 2.61 \times 10^{-5}$ C, $Q_C = 8.93 \times 10^{-6}$ C, $Q_D = 8.93 \times 10^{-6}$ C
77. **a.** 3.51×10^{-12} F **b.** 4.21×10^{-11} C **c.** 8.28×10^3 V/m **d.** 7.32×10^{-8} C/m²
79. 8.30 μF and 4.70 μF
81. 34.6 μF
83. **a.** $U_1C_1C_2/(C_1^2 + 2C_1C_2 + C_2^2)$
85. **a.** $V_1 = V_4 = 6.67$ V, $V_2 = V_3 = 3.33$ V, $Q_1 = Q_2 = 40.0\,\mu$C, and $Q_3 = Q_4 = 26.7\,\mu$C **b.** $V_1 = V_3 = 5.33$ V, $V_2 = V_4 = 4.67$ V, $Q_1 = 32.0\,\mu$C, $Q_2 = 56.0\,\mu$C, $Q_3 = 42.7\,\mu$C, and $Q_4 = 18.7\,\mu$C
87. **a.** $0.5C$ **b.** $0.6C$ **c.** $0.62C$

CHAPTER 28: Concept Exercises

28.1 Imagine a single conduction electron in a swarm of other electrons, with each exerting a repulsive force on that single electron in different directions and with different magnitudes. If you sum all those random force vectors, the net force is zero (or nearly so).
28.2 If the current is constant, Equation 28.2 becomes $q = I\int_0^t dt = It$. The time in each case is 1 s, so the charge that passes through the cross section in each case is shown in the table.

Device	Charge in 1 s
a. Flashlight	$q = (0.5\text{ A})(1\text{ s}) = 0.5$ C
b. Starter motor	$q = (200\text{ A})(1\text{ s}) = 200$ C
c. Computer circuit	$q = (1 \times 10^{-12}\text{ A})(1\text{ s}) = 1 \times 10^{-12}\text{ C} = 1$ pC

28.3 Yes. Considering the actual conduction electrons moving in the negative x direction ($\vec{v}_{drift} = -v_{drift}\hat{\imath}$) and carrying charge $-e$, we find

$$\vec{J} = n(-e)(-v_{drift})\hat{\imath} = nev_{drift}\hat{\imath}$$

This result still shows the current density pointing in the same direction as the electric field—in this case, the positive x direction. The magnitude of the current density is also the same because the drift speed of electrons is equal to the drift speed of the imaginary positive particles: $v_{drift} = v_d$.
28.4 When the filament breaks, the circuit is open. Because air is normally an insulator, there can be no current in the network. The resistance measured between the ends of the filament is infinite (or at least very high).
28.5 The potential difference across the broken filament is $\mathcal{E}$, the same as the potential across the battery because no current is in the wire or the (broken) filament.
28.6 For a broken filament, $I = 0$, $R \to \infty$, and $\Delta V = \mathcal{E}$. Any of the equations for power gives $P = 0$; for example, $P = I\Delta V = 0\mathcal{E} = 0$.

CHAPTER 28: Problems and Questions

1. Metal 1 is the better conductor because there are fewer collisions as the electrons drift through the metal.
3. The cooler penny is the better conductor because electrons undergo fewer collisions with vibrating ions as they drift through the metal.
5. Answers will vary but should be on the order of 10^5 m/s.
7. 0.48 A, to the left
9. $I = -q_0\omega \sin \omega t$
11. **a.** The current oscillates back and forth, changing direction. Its maximum value in either direction is $q_0\omega$. **b.** Yes. This occurs when $\sin \omega t = 0$, when $\omega t = n\pi$ ($n = 0, 1, 2, 3, \ldots$), or when $t = n\pi/\omega$ ($n = 0, 1, 2, 3, \ldots$).
13. 1.7×10^{-3} m/s
15. 2.2×10^2 C
17. **a.** 56 A **b.** 1.8×10^5 A/m²
19. 1.04×10^5 s
21. 6.25 mA, 4.00×10^3 A/m²
23. 4.70×10^{-5} m/s
25. $2.94 \times 10^7\ \Omega^{-1} \cdot \text{m}^{-1}$
27. 1.15×10^{-2} V/m
29. **a.** decreases **b.** increases **c.** no change **d.** no change **e.** decreases. An increase in temperature means the atoms vibrate more and so have a larger effective cross-sectional area. An increased area decreases the time between collisions, increases the rate of collision, and so decreases the conductivity.
31. 6.8×10^8 A
33. 7.6×10^7 A
35. 0.564 A
37. 15.0 m
39. $1.50 \times 10^4\ \Omega$
41. **a.** Conductivity is a microscopic or materials property. **b.** Resistivity, the inverse of conductivity, is also microscopic. **c.** Conductance depends on both microscopic properties (through the conductivity) and macroscopic properties (the length and cross-sectional area of the conductor). **d.** Resistance is the inverse of conductance and, therefore, also depends on microscopic and macroscopic properties.
43. 30.0%
45. 1.52 m
47. **a.** 4.27 V/m from left to right **b.** 0.168 Ω **c.** 19.0 A from left to right **d.** 2.69×10^8 A/m²
49. No. The ratio of the potential difference to the current is not constant.
51. The word *rule* may be better because Ohm's law, like Hooke's law, is an empirical law that does not hold for all materials. It works well for modeling the behavior of many materials, but it is not universally true.
53. **a.** 0.4545 A **b.** 484.0 Ω
55. **a.** 7.083 A **b.** 16.94 Ω
57. 1.82 Ω
59. **a.** $2I_0$ **b.** $4V_0I_0$ **c.** $4V_0I_0$

61. 335 s
63. 2.50 W
65. 0.31 C
67. **a.** 24.0 Ω **b.** 2.40×10^3 Ω
69. $v_d = E/(\rho n e)$
71. **a.** 14.4 W and 10.3 W **b.** the 250.0-Ω resistor
73. 0.708
75. 4.16 Ω
77. $I/[ne\pi(r_b^2 - r_a^2)]$
79. 149°C
81. 1.19×10^9 s

CHAPTER 29: Concept Exercises

29.1 Volts

29.2 The voltmeter in experiment 2 will read +1.45 V, as it does in experiment 1. You might have thought the voltage would be negative because of the negative sign in the resistor rule (Eq. 29.1), but that sign holds only when the current is from the black lead to the red lead. In this circuit, the current is clockwise, so the current is from point *b* (red) to point *a* (black). Thus, the potential difference measured ($V_{\text{red}} - V_{\text{black}} = V_b + V_a$) is positive.

29.3 **a.** Start by solving $\mathcal{E}_1 - IR_1 - \mathcal{E}_2 - IR_2 = 0$ for current:

$$I = \frac{\mathcal{E}_1 - \mathcal{E}_2}{R_1 + R_2}$$

When battery 2's emf is nearly 0 ($\mathcal{E}_2 \approx 0$), the current is very high and the bulb is bright. As battery 2's emf increases, the current decreases and the bulb becomes fainter. In fact, if $\mathcal{E}_2 = \mathcal{E}_1$, the current is 0 and the bulb is not lit. So, when the indicator bulb goes out, the battery is recharged. **b.** The term *recharge* is misleading because it seems to imply that the function of an emf device is to supply charged particles. Of course, the emf device supplies energy, not charge. A better term may be *re-energize*.

29.4 In the pegboard analogy, a resistor is like a pegboard and the current is like rubber balls falling through and hitting the pegs. If you imagine a ball falling through two small pegboards in series, as in the accompanying figure, the resistance to the ball's motion is equal to the resistance the ball would encounter if it fell through a larger pegboard that is the sum of the two small pegboards. So, it makes sense to add resistances in series.

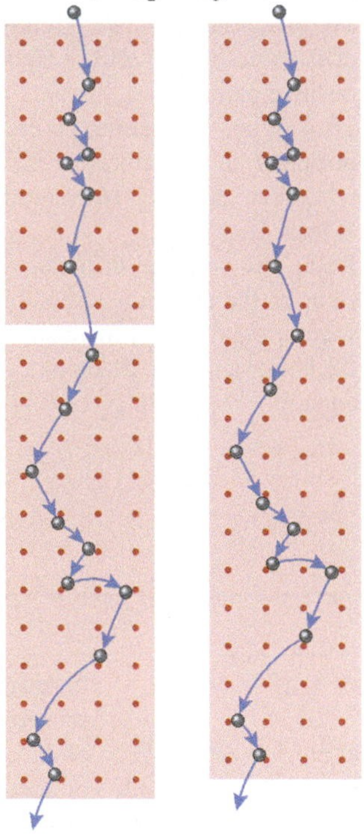

29.5 **a.** According to the emf rule, the potential difference is positive: $\Delta V = \mathcal{E}$. **b.** Whether we measure the potential difference across resistor 1 or resistor 2, the current is from the red lead to the black lead. According to the resistor rule, the potential difference is positive. So, we have $\Delta V = \mathcal{E} = I_1R_1 = I_2R_2$. It makes sense that the potential difference across all three circuit elements is the same because the voltmeter can give only one reading. When the voltmeter is connected to points *a* and *b*, it measures the voltage across all three circuit elements at once.

29.6 Applying the junction rule to either the junction at *a* or the junction at *b* gives $I_0 = I_1 + I_2$. From Concept Exercise 29.5, we have $I_1R_1 = I_2R_2$.

If the resistances are equal, then $R_1 = R_2$ and the current in branch 1 must equal the current in branch 2: $I_1 = I_2$. So, $I_1 = I_2 = \frac{1}{2}I_0 = \frac{1}{2}(3.0\text{ A}) = 1.5$ A.

29.7 Start with

$$\frac{1}{R_{\text{eq}}} = \frac{1}{R_1} + \frac{1}{R_2}$$

(Eq. 29.7 for two resistors in parallel). Then

$$\frac{1}{R_{\text{eq}}} = \frac{R_2}{R_1R_2} + \frac{R_1}{R_1R_2} = \frac{R_1 + R_2}{R_1R_2}$$

and

$$R_{\text{eq}} = \left(\frac{R_1 + R_2}{R_1R_2}\right)^{-1} = \frac{R_1R_2}{R_1 + R_2}$$

CHAPTER 29: Problems and Questions

1.

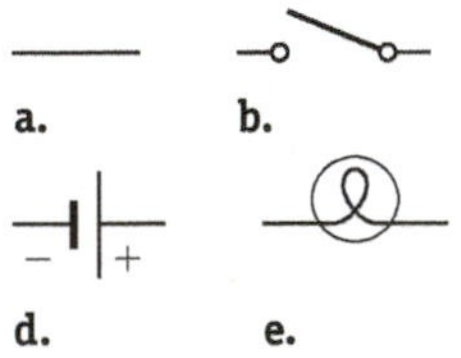

3. True. The magnitude may change, but the sign of each terminal is fixed for a DC emf.
5. **a.** 15 V **b.** 0 **c.** −15 V **d.** 0
7. **a.** decreases **b.** decreases **c.** decreases
9. **a.** 1.67 V **b.** 3.34 V **c.** 46.5 mA **d.** 93.0 mA
11. **a.** 12.0 Ω **b.** 2.32 Ω
13. 0
15. **a.** 0.339 A **b.** $\Delta V_1 = 4.54$ V, $\Delta V_2 = 6.94$ V, $\Delta V_3 = 3.32$ V **c.** $P_1 = 1.54$ W, $P_2 = 2.35$ W, $P_3 = 1.12$ W, $P_{\mathcal{E}} = 5.01$ W **d.** $I = 0.277$ A, $\Delta V_1 = 3.71$ V, $\Delta V_2 = 5.67$ V, $\Delta V_3 = 5.42$ V, $P_1 = 1.03$ W, $P_2 = 1.57$ W, $P_3 = 1.50$ W, $P_{\mathcal{E}} = 4.09$ W
17. **a.** The water is moving uphill, probably due to a pump of some kind. **b.** It is gradually flowing downhill, with a slight decrease in height. **c.** The water undergoes a large decrease in height as it flows downhill. **d.** The water is more or less back at the level where it started, undergoing a slight decrease in level as it finally returns to where it started.
19. 436 Ω
21. **a.** 9.6 Ω **b.** 11 Ω **c.** No. The same current would flow through each if they were in series, and there would be no means to select the desired resistance.
23. $67R/4$
25. **a.** 9.00 mA **b.** 18.0 mA
27. $I_2 = 7.0$ A, $I_5 = I_6 = 10.0$ A, $I_7 = 18.0$ A

29. **a.** $I_1 = (6/11)I_0$, $I_2 = (3/11)I_0$, $I_3 = (2/11)I_0$ **b.** The same as part (a) because they are the same circuit.
31. $21R/164$
33. The current is greater in B because the equivalent resistance of the monks is higher in A. (The monks are in series in A.) The current is also greater through each monk in B because each is in parallel with the Volta pile.
35. $11R/6$
37. 9.600 A
39. **a.** $I_0 = 0.917$ A, $I_1 = 0.500$ A, $I_2 = 0.250$ A, $I_3 = 0.167$ A **b.** same as in part (a)
41. **a.** $I_1 = 1.51$ A, $I_2 = 0.386$ A, $I_3 = 1.89$ A **b.** $P_1 = 31.8$ W, $P_2 = 1.19$ W, $P_3 = 35.8$ W
43. **a.** $I_1 = I_2 = I_3 = 0.133$ A, $I_4 = 0$ **b.** $I_1 = 0.179$ A, $I_2 = I_3 = 0.124$ A, $I_4 = 0.0552$ A
45. **a.** 14.7 Ω **b.** $I_{20.0} = 0.326$ A, $I_{5.00} = 1.30$ A, $I_{9.00} = 1.63$ A, $I_{2.00} = 1.40$ A, $I_{12.0} = 0.233$ A
47. 12.38 V
49. **a.** 338 Ω **b.** 1.50×10^2 W **c.** $\Delta V_1 = \Delta V_2 = \Delta V_3 = 113$ V
51. **a.** 15.0 V **b.** 0.0003 A. This current is 0.06% of the current through the 30.0-Ω resistor.
53. 6.2×10^{-5} s. This is nearly the same as the 5.0×10^{-5} s firing time.
55. **a.** 0.949 A **b.** −0.500 A **c.** 3.90 μC
57. **a.** 1.26×10^{-3} s **b.** 72.0 μC **c.** 68.4 μC
59. **a.** 5.25×10^{-4} s **b.** 3.64×10^{-4} s **c.** 0.429 A
61. 666 Ω and 154 Ω
63. 9.00 h
65. **a.** 7.00 A **b.** 4.00 A **c.** 13.0 V
67. 0.0738 A
69. **a.** 2.00 A **b.** 8.0 V
71. **a.** $I_{50} = 7.85$ A, $I_{90} = 4.63$ A, $I_{120} = 1.31$ A, $I_{175} = 1.91$ A **b.** 334 V
73. 33.3 Ω
75. **a.** 3.75×10^{-4} s **b.** 2.60×10^{-4} s **c.** 0.267 A
77. **a.** 2.37 A **b.** 1.26 V
79. 1.57×10^6 Ω
81. 0
85. $2\mathcal{E}/R$
87. $3\mathcal{E}/2R$
89. $\mathcal{E}/R$
91. $\mathcal{E}/R$
95. The minus sign indicates that the capacitor is losing charge as it discharges. That is, the current removes charge from the capacitor.

CHAPTER 30: Concept Exercises

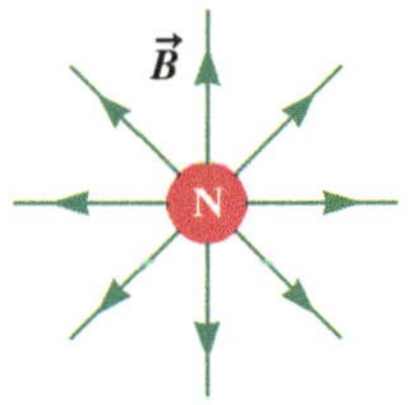

30.1 An isolated north pole is a monopole, and so far there have been no confirmed monopole discoveries. However, as shown here, the magnetic field of an isolated north magnetic pole would look like the electric field due to a positive point charge.
30.2 Consider the upward current in Figure 30.11 (page 939), the result of electrons moving downward. To find the magnetic field direction due to electron motion, use a simple left-hand rule: Point your left thumb in the direction of electron motion, and then your fingers wrap counterclockwise around the wire.
30.3 The constant $\mu_0/4\pi$, the current I, and the length $d\ell$ are the same for both segments. Because $\varphi_{\text{red}} + \varphi_{\text{blue}} = 180°$, we have $\sin\varphi_{\text{red}} = \sin\varphi_{\text{blue}}$. With all else being equal, because $r_{\text{red}} > r_{\text{blue}}$, the magnetic field dB due to the blue segment is stronger than that due to the red segment at point P.
30.4 Avi is correct. The magnetic force is perpendicular to both the magnetic field $\vec{B}$ and the subject's velocity $\vec{v}$. Because these vectors are both in the plane of the page, the magnetic force $\vec{F}_B$ is either into or out of the page. Use the right-hand rule to find the direction of $\vec{v} \times \vec{B}$; your thumb points into the page. Because the particle is positively charged, this is the direction of the magnetic force.
30.5 Because the radius of the spiral is larger near the equator than it is near the poles, we know the magnetic field is weaker in the region near the equator than it is near the poles.
30.6 From $\vec{F}_B = q(\vec{v} \times \vec{B})$ and the right-hand rule for cross products, the magnetic force is in the positive x direction.
30.7 Again use $\vec{F}_B = q(\vec{v} \times \vec{B})$ and the right-hand rule. Due to the negative charge, flip over your thumb. The magnetic force is to the right in the positive x direction.
30.8 Currents in the same direction attract; opposite currents repel.

CHAPTER 30: Problems and Questions

1. Magnets either repel or attract each other depending on the orientation of their poles. In the metaphor, the hands are pulling on each other in opposite directions, but if the magnets were actually pulling each other, they would be attracting each other. The instructor might describe two magnets that are repelling each other.
3. **a.** Gravity. The neutron has no net electric charge but has mass. **b.** Gravity and the electric force. The proton has a net positive charge and mass. **c.** All three: gravity, the electric force, and the magnetic force. A moving charge, or current, creates a magnetic field.
5. **a.** Yes. **b.** Yes. The line through the axis has no point of origination or termination and must form a closed loop.
7. By convention, the magnetic field lines point out of the north end of a bar magnet and into the south end, which is not consistent with the drawing. The lines on the right should not cross. The lines on the left should loop back through the bar magnet, and there should be no termination points for lines, as seen on the coin.
9. **a.** $\hat{\jmath}$ **b.** $-\hat{\jmath}$ **c.** $-\hat{k}$ **d.** $-\hat{k}$
11. **a.** The current should flow downward along the wire from the top toward the bottom. **b.** The current should flow along the wire from the upper left to the lower right. **c.** The current should flow along the wire into the page.
13. $(\mu_0 ev)/(4\pi r^2)$
15. 5.12×10^{-21} T into the page
17. 2.00×10^{-6} T
19. $\vec{B}_A = 0$, $\vec{B}_B = [(2\mu_0 I)/(3\pi r)]\hat{\jmath}$
21. 5.28×10^{-5} T
23. 3.00 A out of the page or 9.00 A into the page
25. 4.80×10^{-4} T
27. $B_1 : B_2 = 4:1$
29. **a.** $1.02 \times 10^{-5}\hat{k}$ (out of the page) **b.** 0.225 m
31. 1.59×10^{-2} A·m²
33. 8.07×10^{-10} T
35. Both poles should be attracted. Each pole will cause the magnetic moments of nearby atoms in the refrigerator to align against the field of the bar magnet. Thus, each pole will be attracted to the anti-aligned magnetic domains that form in the refrigerator near each pole.
37. **a.** $\vec{F}_1 = -1.90\hat{k}$ N and $\vec{F}_2 = (1.90\hat{\imath} - 3.04\hat{\jmath})$ N **b.** $\vec{F}_1 = 1.90\hat{k}$ N and $\vec{F}_2 = (-1.90\hat{\imath} + 3.04\hat{\jmath})$ N

39. $(-0.109\hat{\imath} + 0.326\hat{\jmath} - 0.716\hat{k})$ N
41. **a.** negative **b.** 3.19×10^{-7} C
43. **a.** 4.61×10^{-13} N **b.** 5.06×10^{17} m/s^2
45. **a.** 90° **b.** 2.17×10^{-4} T
47. **a.** 0.104 m **b.** 4.57×10^{6} Hz
49. $r_{He} = 2r_p$
51. 2.1×10^{5} m/s
53. 2.41×10^{-6} V
55. **a.** 0.535 A **b.** 6.21×10^{28} m^{-3}. This is very close to the value for tungsten.
57. $-61.6\hat{\imath}$ N
59. $(-8.86\hat{\imath} - 5.85\hat{\jmath} - 2.34\hat{k})$ N
61. **a.** 7.94 N **b.** 7.94 N **c.** 11.2 N
63. **a.**

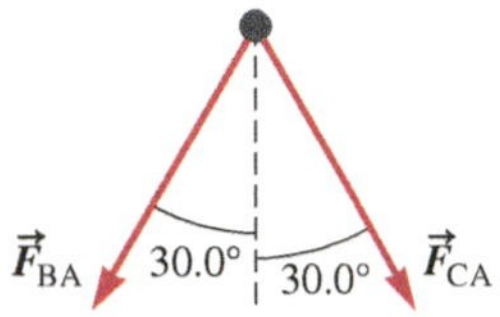

b. $-3.60 \times 10^{-6}\hat{\jmath}$ N/m
65. **a.** $(1.04 \times 10^{-6}\hat{\imath} - 5.39 \times 10^{-6}\hat{\jmath})$ N/m
b. $(5.19 \times 10^{-6}\hat{\imath} + 1.80 \times 10^{-6}\hat{\jmath})$ N/m
67. $[(\mu_0 I_1 I_2)/(2\pi r_1) + (\mu_0 I_3 I_2)/(2\pi r_2)]\hat{\imath}$
69. 2.16 A · m^2
71. If the magnetic field were uniform, the magnitude of the magnetic torque on the loop would change as it rotated. With the radial field, the magnetic torque on the loop is constant, and the loop will rotate until the torque from the spring is equal and opposite to the magnetic torque. If the loop were in a uniform field, it would attempt to rotate until the magnetic torque was 0.
73. **a.** 33.4 N · m **b.** The loop will rotate so as to align the magnetic moment with the B field, counterclockwise as seen looking down from a position on the positive z axis.
75. The two curves appear to approach the same curve above $y = 0.08$ m and diverge fairly significantly below $y = 0.05$ m, which is equal to the diameter of the loop. Therefore, at points more than one or two loop-diameters from the current loop, we are far enough away that the approximation is quite good.

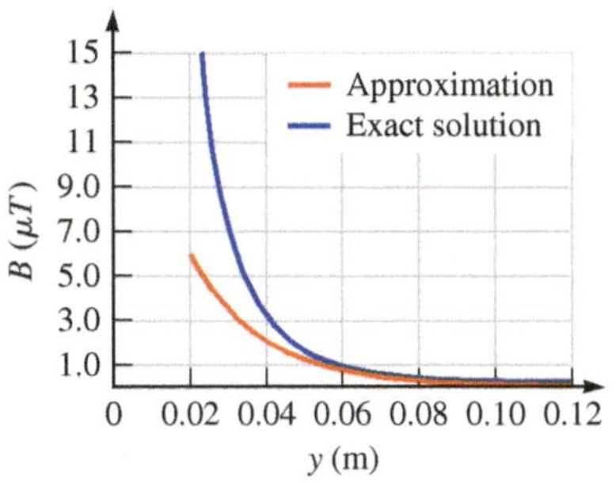

77. **a.** $3.13 \times 10^{-6}\hat{k}$ T
b. $(2.80 \times 10^{-6}\hat{\imath} - 0.800 \times 10^{-6}\hat{\jmath})$ T
79. $\left(\dfrac{N\mu_0 I}{2\pi R}\right) \sin(\pi/N)\,\hat{k}$
81. $2.31 \times 10^{-5}\hat{k}$ T
83. **a.**

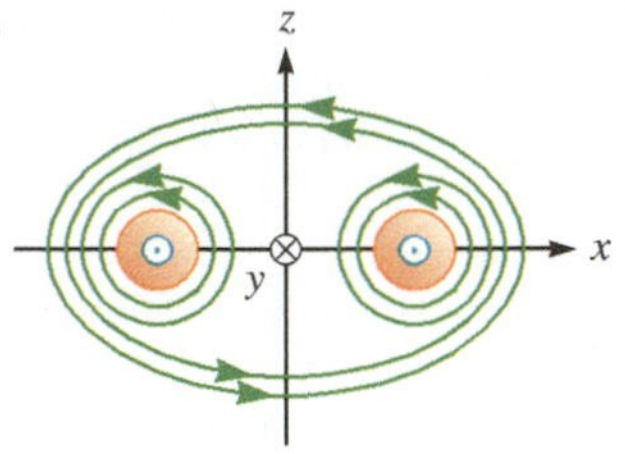

b. 0
c. $\dfrac{-(2.00 \times 10^{-6})z}{(1.21 \times 10^{-2}) + z^2}\,\hat{\imath}$ T, where z is in meters
85. **a.** out of the page **b.** toward the right **c.** toward the bottom of the page
87. $q(v_x B_y - v_y B_x)\hat{k}$
89. **a.** 0 **b.** 7.91×10^{-14} N
91. $f \approx -\left(\dfrac{\mu_0 I I_0}{\pi d^2}\right)x$ and $\omega = \sqrt{\dfrac{\mu_0 I I_0}{\pi \lambda d^2}}$
93. **a.** $I = \left(\dfrac{mg}{B\ell}\right)\tan\theta$ from Q to P **b.** No, $2g \sin\theta$
95. 37 m

CHAPTER 31: Concept Exercises

31.1 Here is one possible experimental procedure:

1. With the bar magnet far away, find the direction of the Earth's magnetic south pole with the compass. This is the reference position.
2. Set the bar magnet perpendicular to the original position of the compass needle. If you want the compass needle to deflect toward the east, place the bar magnet to the west of the compass and point the bar magnet's north end eastward as in the accompanying figure.
3. Measure the distance between the bar magnet's pole and the compass needle.
4. Measure the deflection of the needle, and use $B = B_{\oplus} \tan\theta$ (Eq. 31.1) to find B.
5. Move the compass eastward or westward, and then repeat steps 3 and 4 until you have covered the region of space you are interested in.

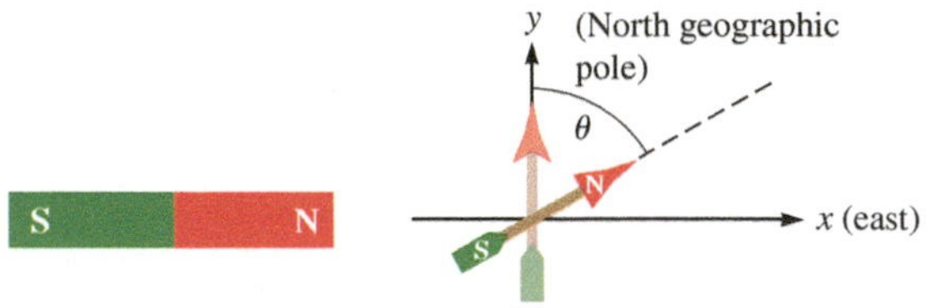

31.2 No. The integral in $\Phi_B = \int \vec{B} \cdot d\vec{A}$ (Eq. 31.2) is *not* taken over a closed (Gaussian) surface. Gauss's law for magnetism applies only to the flux through a Gaussian surface. The magnetic flux through an open surface may be nonzero.

31.3 The two Ampèrian loops are identical. Wrap the fingers of your right hand in the direction of the path; then your thumb points in the positive y direction.

Case 1: This wire penetrates the film twice: once carrying current downward in the negative y direction, and once upward in the positive y direction. The total current through the loop is $I_{thru} = I - I = 0$.

Case 2: This wire penetrates the film three times: twice carrying current downward in the negative y direction, and once upward in the positive y direction. The total current through the loop is $I_{thru} = I - 2I = -I$.

31.4 We need to find the graph of B versus r that represents Equation 30.9, which says that as $r \to 0$, $B \to \infty$, and as $r \to \infty$, $B \to 0$. Only graph 3 is consistent with the two extremes.

31.5 The deflection outside the solenoid is 0, independent of the current. Inside the solenoid, the current is found by combining $B = B_{\oplus} \tan\theta$ (Eq. 31.1) with $B = \mu_0 n I$ (Eq. 31.6), which yields $I = (B_{\oplus}/\mu_0 n) \tan\theta$. So, if we wish to measure $10° < \theta < 80°$, then 1.4×10^{-3} A $< I < 4.5 \times 10^{-2}$ A.

CHAPTER 31: Problems and Questions

1. 0.38%
3. The ball is attracted to the rod and swings toward it. The electric field is stronger near the rod, so the angle the thread makes with the vertical is larger for locations nearer the rod.
5. **a.** 0.17 A **b.** 1.4×10^{2} A
7. **a.** $B_x A_x$ **b.** $B_y A_y$ **c.** 0
9. 62.6°

11. 3×10^{-5} Wb, -3×10^{-5} Wb, 9×10^{-5} Wb, -9×10^{-5} Wb, 0, and 0. The sum of the magnetic flux through all the surfaces of the cube is 0, and thus Gauss's law for magnetism holds.

13. a. Yes. Any Ampèrian loop (closed path) drawn will satisfy the condition. **b.** No. There is no way to find a loop through which a net current will flow.

15. a. Yes. Any loop in the plane of the magnetic field will work because the area vector will be perpendicular to the magnetic field. **b.** Yes. Any loop whose area vector is not perpendicular to the magnetic field will work. For example, a loop that is perpendicular to the magnetic field has an area vector that is parallel to the magnetic field and will have a nonzero magnetic flux through it.

17. The contributions for sides 1, 2, 3, and 4 are 0.0313 T·m, −0.0438 T·m, −0.0313 T·m, and 0.0438 T·m, respectively. The net current must be zero.

19. 0.2500 A

21. $(1/2)\mu_0 J$

23. $B_{\text{in}} = \dfrac{\mu_0 I r^3}{2\pi a^4}$ and $B_{\text{out}} = \dfrac{\mu_0 I}{2\pi r}$

25. 2.00 m

27. a. For $r < r_1$, $B = \dfrac{\mu_0 I}{2\pi r_1^2} r$; for $r_1 < r < r_2$, $B = \dfrac{\mu_0 I}{2\pi r}$; for $r_2 < r < r_3$, $B = \dfrac{\mu_0 I}{2\pi r}\left[1 - \dfrac{r^2 - r_2^2}{r_3^2 - r_2^2}\right]$; and for $r_3 < r < r_4$ (really in the entire region $r_3 < r$), $B = 0$. **b.** For the long, straight wire, $B = (\mu_0 I)/(2\pi r)$ away from the wire, but for the coaxial cable, $B = 0$ outside the wire. So there is no need to worry about B interfering with outside devices in the case of the coaxial cable.

29. a. At $r = r_1$ **b.** 1.30×10^{-4} T

31. $\mu_0 c\left[\dfrac{r_A^2}{3} - \dfrac{R_1 r_A}{2} + \dfrac{R_1^3}{6 r_A}\right]$

33. 8.0×10^5

35. 0.612 A

37. a. 9.1×10^{-3} T **b.** There is no change.

39. $N = 9.78 \times 10^3$ and $\ell = 7.28 \times 10^2$ m

41. 9.47 cm

43. $n\mu_0 I_s I_w$

45. 2.0×10^1 A

47. a. 0.290 T **b.** 0.194 T

49. 0.230 m

51. 2.88 cm

53. 2.88×10^{12} N/(C·s)

55. The current is decreasing, so the magnitude of the magnetic field is also decreasing.

57. 1.76×10^{-7} T

59. a. -2.51×10^4 Wb **b.** 0

61. 3.98×10^5

63. 3.46×10^{-3} T

65. a. $1.60 \times 10^{-5}\hat{k}$ T **b.** $-1.20 \times 10^{-5}\hat{k}$ T

67. For loop a, $-4\pi \times 10^{-7}$ T·m; for loop b, $20\pi \times 10^{-7}$ T·m; for loop c, $16\pi \times 10^{-7}$ T·m; for loop d, 0; for loop e, $4\pi \times 10^{-7}$ T·m

69. a. 6.44×10^{-6} Wb **b.** 3.22×10^{-6} Wb

71. $\mu_0 n I_1 I_2 s^2$

73. $\vec{B}_{\text{inside}} = -\mu_0 J z \hat{\jmath}\ (-L < z < L)$, $\vec{B}_{\text{below}} = \mu_0 J L \hat{\jmath}\ (z > L)$, $\vec{B}_{\text{above}} = -\mu_0 J L \hat{\jmath}\ (z < -L)$

75. $\mu_0 \varepsilon_0 \omega^2 I R^2/4$

CHAPTER 32: Concept Exercises

32.1 Actually, we did. We need to find the magnetic flux through the rectangular loop by breaking the integral in Equation 31.2 into two pieces. One piece is integrated over the cross-sectional area of the solenoid A_{sol}; the other piece covers the area inside the loop minus the area of the solenoid $(A_{\text{loop}} - A_{\text{sol}})$:

$$\Phi_B = \int \vec{B} \cdot d\vec{A} = \int_{\text{sol}} \vec{B} \cdot d\vec{A} + \int_{(\text{loop}-\text{sol})} \vec{B} \cdot d\vec{A}$$

The second integral over $(A_{\text{loop}} - A_{\text{sol}})$ is 0 because the magnetic field is 0. The magnetic field is uniform over the cross-sectional area of the solenoid and $\vec{B}$ is parallel to $d\vec{A}$, so the first integral is

$$\Phi_B = \int_{\text{sol}} B\,dA = B\int_{\text{sol}} dA = BA_{\text{sol}}$$

So $A = A_{\text{sol}}$ in Equation 32.1.

32.2 Case 1: The angle φ changes, so an emf is induced in the loop.

Case 2: The magnetic field is strongest near the bar magnet. As the magnet swings back and forth, the magnetic flux through the loop alternately decreases and increases. An emf is induced in the loop.

Case 3: The magnet does not oscillate. The magnetic field, the loop's area, and the angle between the magnetic field and the area vector are all constant. So there is no change in the magnetic flux, and no emf is induced in the loop.

32.3 Case 1: The magnetic field $\vec{B}_{\text{sol}}$ is decreasing, so the loop's magnetic field $\vec{B}_{\text{loop}}$ must point in the same direction as $\vec{B}_{\text{sol}}$.

Case 2: The loop is tilted, so $\vec{B}_{\text{loop}}$ points either up and to the left or down and to the right. The magnetic field $\vec{B}_{\text{sol}}$ is decreasing, so the loop's magnetic field $\vec{B}_{\text{loop}}$ must have a component in same direction as $\vec{B}_{\text{sol}}$. Therefore, $\vec{B}_{\text{loop}}$ points up and to the left.

Case 3: There is no magnetic flux through the loop. When $\vec{B}_{\text{sol}}$ changes, there is no changing flux and no current induced in the loop, so there is no magnetic field $\vec{B}_{\text{loop}}$.

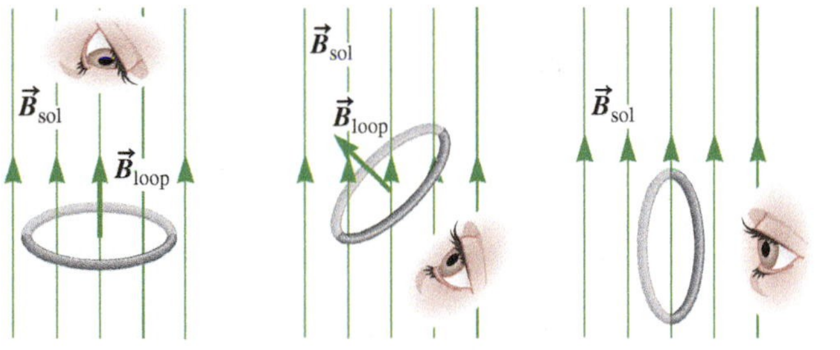

Case 1 Case 2 Case 3

32.4 The figure shows the loop's magnetic field in each case. To find the direction of the induced current, point your right thumb in the direction of $\vec{B}_{\text{loop}}$.

Case 1: The magnetic field $\vec{B}_{\text{loop}}$ points upward. Your right fingers wrap counterclockwise from the observer's perspective.

Case 2: The loop is tilted, and $\vec{B}_{\text{loop}}$ points up and to the left. Your right fingers wrap clockwise from the observer's perspective.

Case 3: There is no induced current.

32.5 The sliding magnet speeds up. Consider the sliding magnet to be the system; then both you and the fixed magnet are outside the system. Both you and the fixed magnet do positive work on the system. No friction or other forces act, so the total work done by external forces must equal the change in kinetic energy:

$$W_{\text{tot}} = \Delta K$$

Because $W_{\text{tot}} > 0$, $\Delta K > 0$, and the sliding magnet speeds up.

32.6 This device is the opposite of a magnetic brake. The relative motion of the magnet induces eddy currents in the disk. The direction of the eddy currents is given by Lenz's law; the eddy currents create their own magnetic fields that oppose the change. In this case, a region of the disk is magnetized such that it is attracted to the magnet. It might help to imagine sliding the magnet in Figure 32.18 (page 1025); how would the two loops respond in that case?

32.7 No. The current in the bulb varies sinusoidally according to $I = I_{max} \sin \omega t$ (Eq. 32.14). So, when the current is zero, the bulb is momentarily off. The fluorescent lights in your room actually flash on and off at 60 Hz, or $\omega = 2\pi(60 \text{ Hz}) = 377$ rad/s, because the coil in the power company's AC generator rotates at angular speed $\omega = 377$ rad/s. You don't see the light going on and off because your eyes cannot detect such a rapid change. AC generators have a similar effect on incandescent bulbs; they flicker slightly but don't completely flash on and off.

32.8 The peak voltage is found by multiplying the rms voltage by $\sqrt{2}$:

$$\mathcal{E}_{max} = \sqrt{2}\mathcal{E}_{rms} = \sqrt{2}(240 \text{ V})$$
$$\mathcal{E}_{max} = 340 \text{ V}$$

32.9 **a.** If a battery is used, the magnetic flux does not change, so there is no emf induced in the secondary coil of a transformer and no output current. **b.** If a DC generator is used, the magnetic flux increases and decreases. The result is that AC current is produced by the secondary coil; this would need to be converted back to DC for Edison's DC system.

CHAPTER 32: Problems and Questions

1. **a.** 5.29×10^{-4} Wb **b.** No. There is no induced current because the magnetic flux is not changing in the problem, as far as we can tell.
3. BA_{loop}
5. 8.45×10^{-2} V
7. **a.** 0 **b.** $-2BLW\omega/\pi$
9. 2.01 V
11. **a.** Increasing **b.** 2.26×10^{-4} V
13. **a.** 8.87×10^{-7} Wb **b.** 6.72×10^{-7} V
15. 0
17. As the magnet approaches the loop, the flux through the loop increases. The loop's magnetic field must point in the direction opposite the bar magnet's field, or to the left. Point your right thumb toward the left; then your fingers and the induced current wrap clockwise from the perspective of the observer. Now, as the magnet moves away from the loop, the magnetic flux decreases, so the loop's magnetic field must point in the same direction as the bar magnet's field, or to the right. Now point your right thumb toward the right in the direction of the loop's magnetic field: your fingers and the induced current wrap counterclockwise from the perspective of the observer.
19. 3.7×10^{-4} A, clockwise
21. **a.** 7.61 T/s **b.** an increase
23. B_0Lv_y/R, clockwise
25. 0.135 V
27. $-0.228 \sin(67.0\pi t)$
29. No. There will be no change in the magnetic flux through conductor A.
31. **a.** Loop B creates a magnetic field that is in the direction opposite that of loop A, so it is repelled away from loop A. **b.** Loop B creates a magnetic field in the same direction as that of loop A, so it is attracted toward loop A.
33. An electric current was created in the pipe as the magnet fell because of the changing magnetic flux as the magnet moves. Thus, there is electrical energy that accounts for the "missing energy."
35. **a.** 22.5 m/s **b.** No. The direction of the induced current depends on the direction of the bar's movement because the magnetic flux will either be increasing or decreasing.
37. 143 A
39. **a.** 0.176 V **b.** The bottom end of the bar, or the end with the lowest y value
41. 1.71×10^3 rev/min
43. 8.39×10^{-3} m^2, 9.16×10^{-2} m. The coil is about 9 cm $\times$ 9 cm, similar to the size of a cell phone.
45. **a.** $-114 \cos(93.2t)$ **b.** 114 V
47. **a.** 0.505 V **b.** 0°
49. The current in a generator switches direction and thus is an AC current. The output of the generator can be turned into a DC current through the use of a commutator, but in reality, the current in a generator is AC initially.
51. Yes. You would need a frequency of only about 0.8 Hz.
53. **a.** 43.8 V **b.** 0.250
55. 0.5. The average power is not 0.
57. **a.** 16.8 V **b.** 2.8 W
59. **a.** 2.18 A **b.** 28.4 W
61. **a.** 1.5 V **b.** −1.5 V
63. 11.1 m
65. 0.17 V
67. 21.6 A
69. 11 V
71. **a.** 4.73 N **b.** 5.91 W
73. 20.4 C
75. **a.** 2.19×10^{-6} Wb **b.** 1.09×10^{-5} V **c.** counterclockwise
77. 4×, or quadrupled
79. 1.34 A

CHAPTER 33: Concept Exercises

33.1 Yes. When the switch is moved back to position a, the magnetic flux through the circuit increases from zero. An emf is induced in the circuit (Faraday's law), and the induced emf opposes the flux change (Lenz's law). So, the induced emf is in the direction opposite the battery's emf—counterclockwise, producing a counterclockwise induced current.

33.2 Assume the initial inductance is L_i and the initial number of turns is N_i. The number of turns increases to $N_f = 1.1N_i$. The inductance increases according to Equation 33.5, $L = \mu_0 N^2 A/\ell$:

$$L_f = \mu_0 \frac{A}{\ell} N_f^2 = \mu_0 \frac{A}{\ell}(1.1N_i)^2 = 1.2\left(\mu_0 \frac{A}{\ell} N_i^2\right)$$
$$L_f = 1.2\,L_i$$

So the inductance increases by 20%.

33.3 **a.** The current is a constant (not necessarily 0). **b.** The current may be decreasing as it goes from the black lead to the red lead. It is also possible that the current is increasing and going from the red lead to the black lead.

33.4 Mathematically, the differential equations for RL and RC circuits are identical. Compare Figures 29.39 and 33.9; the circuits look the same except the capacitor has been replaced by an inductor. Compare the charge on the capacitor in the RC circuit to the current in the RL circuit. In an RC circuit, when the switch is at A, the charge on the capacitor builds up just as the current in an RL circuit builds up. When the switch is at B in the RC circuit, the charge on the capacitor decays, similar to the decay of the current in the RL circuit.

33.5 Equation 27.22 ($u_E = \frac{1}{2}\kappa\varepsilon_0 E^2$) is analogous to Equation 33.17. In both cases, the stored energy density is proportional to the square of the field.

33.6 Mathematically, y is analogous to Q and $v = dy/dt$ is analogous to $I = dQ/dt$. On the graph (Fig. 16.5, page 453), we see that the relative phase shift between y and v is 90° just like the relative phase shift between Q and I (Fig. 33.13, page 1060). Also, the amplitude of the velocity is given by $v_{max} = \omega y_{max}$ (Eq. 16.7) and the amplitude of the current is given by $I_{max} = \omega Q_{max}$ (Eq. 33.23).

33.7 Shannon is correct. The current and voltage are out of phase, so you can't just divide the voltage at any instant by X_C to find the current at that instant. In Figure 33.20 (page 1065), initially the voltage is zero and the current is at its maximum. You cannot divide zero voltage by X_C to find the current. However, the maximum voltage divided by X_C is the maximum current.

CHAPTER 33: Problems and Questions

1. a. The short, wide inductor has the greater inductance per unit length because inductance per unit length is proportional to A, not length ℓ. **b.** They could have the same inductance if the product $A\ell$ is the same in each case.
3. Option (a) would increase the inductance. The inductance of a solenoid is given by $L = \mu_0 n^2 \ell A$, where n is the number of turns per unit length. If we compressed the inductor to a smaller length, keeping the same number of turns, $n^2\ell = N^2/\ell$ would increase and therefore the inductance would increase.
5. a. 0.750 V **b.** -3.33×10^2 A/s **c.** In both cases, the bulb would become dimmer and stop glowing as the current gets smaller. In (a), the current slowly gets smaller, so the bulb's brightness decreases slowly compared to case (b).
7. 2.53×10^{-3} H
9. a. 0.324 s **b.** 2.40 s
11. a. $4.61L/R$ **b.** $L < 0.016$ H
13. a. 632 Ω **b.** 3.16×10^{-3} s
15. a. $\mathcal{E}\tau/eR$ **b.** $\left(\frac{\mathcal{E}\tau}{R}\right)\left(\frac{e-1}{e}\right)$ **c.** More charge passes through the resistor (in one time constant) when the current (and the energy stored in the inductor) is decreasing than when the current (and the energy stored in the inductor) is increasing. This is confirmed by comparing the areas under the curves for I versus t in each case.
17. a. $\frac{\mathcal{E}^2\tau}{R}\left[\frac{2}{e} - \frac{1}{2e^2} - \frac{1}{2}\right]$
b. $\frac{\mathcal{E}^2\tau}{2R}\left(1 - \frac{1}{e^2}\right)$ **c.** We find that the resistor dissipates more energy when the current is decreasing. The current is higher on average when the switch is moved to b than when the switch is at a, and the power dissipated by the resistor depends on the current squared.
19. 3.2×10^{-2} J
21. 0.975 H
23. a. 3.70×10^{-2} C **b.** 1.88 A **c.** $3.70 \times 10^{-2} \cos(50.8t)$ **d.** $-1.88 \sin(50.8t)$
25. 0.73 A
31. 8.8×10^{-8} F
33. a. 87.0 V **b.** 123 V **c.** 261 W **d.** 4.24 A
35. a. 1.4 A **b.** 1.4×10^2 V **c.** 2.0×10^2 V **d.** $(2.0 \times 10^2\text{ V}) \sin[(120\pi\text{ rad/s})t]$
37. a. 1.16×10^3 rad/s **b.** 6.23×10^{-4} s
39. a. 0.398 Ω **b.** 171 A **c.** 242 A
41. a. 424 Ω **b.** 1.49×10^3 V **c.** 4.95 A and 2.10×10^3 V **d.** $(2.10 \times 10^3\text{ V}) \sin[(377\text{ rad/s})t]$
43. The current through the capacitor leads the potential difference across the capacitor in an AC circuit by $\pi/2$ rad, so the current has a value of zero. Given that the circuit has only a source emf and a capacitor, both the potential difference across the capacitor and the source will be identical.
45. a. We need to use the RC filter as a low-pass filter, so we want to connect the detector (voltmeter) across the capacitor. **b.** 1.27×10^{-5}. Our answer is encouraging, but really we need to compare this to the natural signal received. Because that source is much farther away, its maximum emf (at the telescope) will be much smaller. So our result doesn't tell us whether there will be interference.
47. a. 47.0 Ω **b.** 2.55 A **c.** 3.61 A
49. a. 2.4×10^2 Ω **b.** 0.063 A **c.** 0.089 A and 21.2 V **d.** $(21.2\text{ V}) \sin[(4.40 \times 10^2\text{ rad/s})t]$
51. The current through the inductor lags the potential difference across the inductor in an AC circuit by $\pi/2$ rad, so the potential difference across the inductor is zero. Given that the circuit has only a source emf and a capacitor, both the potential difference across the capacitor and the source will be identical.
53. a. −6.91 A **b.** −4.83 A **c.** 6.91 A
55. 0.082 Hz
57. a. 0.20 mH and 31 nF **b.** 1.3×10^5 Hz
59. a. 8.41 Ω **b.** −3.66 Ω
61. a. 2.4×10^2 Ω **b.** $V_R = 2.00 \times 10^2$ V, $V_L = 4.1 \times 10^2$ V, $V_C = 849$ V **c.** −66° or −1.1 rad **d.** $\mathcal{E} = (4.8 \times 10^2\text{ V}) \sin[(120\pi\text{ rad/s})t - 1.1\text{ rad}]$ and $I = (2.00\text{ A}) \sin[(120\pi\text{ rad/s})t]$
63. a. 0.374 A and 33.6° **b.** 92.5 V and 90.0° **c.** 99.9 V and 0° **d.** 26.1 V and −90.0°
65. a. 3.28×10^2 Ω **b.** 5.72×10^{-2} A **c.** 8.28°
67. 5.74×10^{-5} J
69. a. 9.5 V **b.** 3.8 A in the upward direction in Figure P33.69
71. a. 26.2 V **b.** 1.70×10^2 V **c.** 116 V
73. a. 0.148 V **b.** 0.592 V **c.** 1.00 s
75. a. 123 Ω **b.** 354 Ω **c.** 4.60×10^2 Ω **d.** 398 Ω **e.** −30.2°
77. a. 6.08 A/s **b.** 0.181 A/s
79. 373 mA
81. $I_{max,R} = 8.49$ A, $I_{max,C} = 12.8$ A, $I_{max,L} = 2.25$ A
83. 3.47×10^{-5} F and 1.47×10^{-6} F

CHAPTER 34: Concept Exercises

34.1 Avi makes a better argument. Little g is the acceleration due to gravity near the surface of the Earth. An alien society would find a value for g near the surface of their own planet. Unless their planet happened to be very similar to the Earth, they would not find the same value for g. Big G is the constant of proportionality in Newton's law of universal gravity. It is true that its value depends on the system of units we choose, and it even depends on how we define mass and distance. So the aliens' exact value of G would not match our value. However, an alien society would have to discover some version of G that would work in their system of units and measurements. We could find a way to convert between the two systems, and then we would discover that G is fundamentally the same on both planets.

34.2 In each case, determine the sign of the magnetic flux. Then remember Lenz's law when determining the sign of the right side of Faraday's law, Equation 34.6:

$$\oint \vec{E} \cdot d\vec{\ell} = -\frac{d}{dt}\int \vec{B} \cdot d\vec{A}$$

If the right side is positive, the electric field must be in the same direction as the path. (Negative means opposite direction.) **a.** Same direction. **b.** Same direction. **c.** Opposite direction.
34.3 a. The amplitude of the electric field is the number in front of the sine function: $E_{max} = 0.75$ V/m. The amplitude of the magnetic field is given by Equation 34.23: $B_{max} = E_{max}/c = 2.5 \times 10^{-9}$ T.

b. The angular wave number is the value in front of x: $k = 0.30$ rad/m. Find the wavelength from Equation 17.5: $\lambda = 2\pi/k = 21$ m. **c.** The propagation speed is the speed of light c. Because there is a negative sign in the middle of the sine function's argument, the wave is moving in the positive x direction: $\vec{v} = 3.00 \times 10^8\,\hat{\imath}$ m/s. **d.** Find the angular frequency from $\omega/k = c$ (Eq. 34.20): $\omega = ck = 9.0 \times 10^7$ rad/s. Then find the frequency, $f = \omega/2\pi = 1.4 \times 10^7$ Hz, and the period, $T = 1/f = 7.0 \times 10^{-8}$ s.

34.4 Infrared radiation has a lower frequency than red light, so you might guess that *infrared* means "lower than red." (The prefix *infra* comes from the Latin word for "beneath.") Ultraviolet radiation has a higher frequency than that of violet light, so you might say that *ultraviolet* means "higher than violet." (The prefix *ultra* comes from the Latin word for "beyond.") Originally, "infrared" was called "ultrared."

34.5 The frequency can be found from Equation 34.20: $f = c/\lambda = 1.4 \times 10^9$ Hz $= 1.4$ GHz. From Table 34.2, this is in the radio or microwave part of the spectrum. Most astronomers do not distinguish between radio and microwave; they use the term *radio* for both.

34.6 Apply the right-hand rule to the cross product in $\vec{S} \equiv (1/\mu_0)(\vec{E} \times \vec{B})$ (Eq. 34.27). Point the fingers of your right hand in the direction of the electric field, and close your hand so that you push $\vec{E}$ into $\vec{B}$. Your thumb then points in the positive x direction, the direction of the Poynting vector.

34.7 *Unpolarized* light, shown in Figure 34.22, has electric fields oscillating in all possible planes (perpendicular to the direction of propagation). *All polarized* may be a better term because it might help us think of unpolarized light as *not polarized in any one particular plane.*

CHAPTER 34: Problems and Questions

1. Light is sometimes modeled as a particle, known as a photon. We think of each photon as a ball that possesses both energy and momentum. Light is modeled as a wave when it displays wave properties, such as interference.
3. $E(0.001\text{ s}) = 0.056$ V/m, $E(0.01\text{ s}) = 0.18$ V/m
5. **a.** $B = \mu_0 nCt$, $E = (1/2)\mu_0 nCa$ **b.** 4.7×10^{-6} N/C
7. $E = [(-B_0R)/(2\tau)]e^{-t/\tau}$. Yes.
9. **a.** 2.99×10^{10} V·m/s **b.** 0.265 A
11. 4.59×10^{-18} T to the right
15. $(2.87 \times 10^8\hat{\jmath} - 15.3 \times 10^8\hat{k})$ m/s^2
17. **a.** 1.50×10^{-3} m **b.** 3.75×10^{-9} m
19. 3.90×10^{-7} m
21. $f_{UVA} = 7.5 \times 10^{14}\text{ Hz} \rightarrow 9.4 \times 10^{14}$ Hz, $f_{UVB} = 9.4 \times 10^{14}\text{ Hz} \rightarrow 11 \times 10^{14}$ Hz
23. 7.41×10^{14} Hz
25. 1.4 μT
27. **a.** 2.03×10^2 m **b.** 9.30×10^6 rad/s **c.** 3.10×10^{-2} m^{-1} **d.** 9.0×10^{-3} V/m
29. **a.** 20.4 V/m **b.** 4.14×10^5 Hz **c.** 725 m
31. 1.4×10^{-7} T. This magnetic field is 0.0027 times the magnetic field of the Earth.
35. No. In fact, radio waves pass through you undetected all the time. You can hear a sound wave because your ears are able to detect pressure waves in the surrounding medium.
37. Both waves travel at the speed of light in a vacuum. Only their frequency and wavelength are different.
39. A wavelength measurement of a wave that is blue-shifted appears shorter than the actual wavelength. Because the speed of the light is not affected by the measurement, the measured frequency thus appears higher than the actual frequency of the wave.
41. **a.** 0.0083 W/m^2 **b.** 1.2×10^{-6} W/m^2
43. It must be about 4×10^7 times stronger.
45. 4.38×10^{-4} W/m^2
47. **a.** 1.67×10^{-11} T **b.** 1.11×10^{-16} J/m^3
49. **a.** $\hat{\imath}$ **b.** $\hat{\jmath}$ **c.** $\hat{k}$ **d.** $-\hat{\jmath}$
51. $VI/(2\pi RL)$
53. **a.** 3.33×10^{-6} Pa **b.** 1.33×10^{-7} m/s^2
55. 5.66×10^{-4} Pa
57. **a.** 4.53×10^{-6} Pa **b.** 9.07×10^{-6} Pa
59. **a.** $1.72 \times 10^{-9}\hat{k}$ kg·m/s **b.** $1.72 \times 10^{-9}\hat{k}$ N
61. 496 W/m^2
63. **a.** 352 W/m^2 **b.** 0
65. 889 W/m^2
67. **a.** $-\hat{\imath}$ **b.** 1.4×10^4 m^{-1}, 4.5×10^{-4} m, 6.7×10^{11} Hz **c.** $E(x,t) = (12\text{ V/m})\sin[(1.4 \times 10^4\text{ rad/m})x + (4.2 \times 10^{12}\text{ rad/s})t]$
69. 1.53×10^{-9} T
71. 224 m^2
73. **a.** $\hat{k}$ **b.** 6.61×10^{14} Hz, 4.15×10^{15} rad/s, 1.38×10^7 m^{-1} **c.** the x direction **d.** $B = (1.67 \times 10^{-11}\text{ T})\sin[(1.38 \times 10^7\text{ m}^{-1})z - (4.15 \times 10^{15}\text{ rad/s})t]$
75. 5×10^{-4} m/s^2
77. $E = (1.44 \times 10^4\text{ V/m})\sin[(1.01 \times 10^7\text{ m}^{-1})z - (3.02 \times 10^{15}\text{ rad/s})t]$, $B = (4.80 \times 10^{-5}\text{ T})\sin[(1.01 \times 10^7\text{ m}^{-1})z - (3.02 \times 10^{15}\text{ rad/s})t]$
79. **a.** 0.00349 N **b.** 1.11×10^{-5} m/s^2
81. 0
83. **a.** 2.61×10^3 V/m **b.** 1.08×10^3 m^{-1} **c.** 3.25×10^{11} rad/s **d.** the xy plane **e.** $9.03 \times 10^3\hat{\imath}$ W/m^2 **f.** 6.02×10^{-5} Pa **g.** 6.98×10^{-2} m/s^2

CHAPTER 35: Concept Exercises

35.1 If Newton had clearly observed a diffraction pattern, he might have abandoned his particle theory because the particle model cannot explain a diffraction pattern. Diffraction patterns are often very faint and may easily go unnoticed. We can see diffraction patterns under everyday circumstances if we deliberately look for them. Look at a lightbulb through your fingers when they are nearly touching, and you may be able to see one or two dark fringes.

35.2 As long as you are on one side of the sandbag and the source of the paint is on the other side, you do not get hit. Because the paint is made up of particles, it is absorbed by the sandbag. If you try to hide behind a rock from the water waves, you oscillate up and down because the waves are diffracted around the rock.

35.3 Moving the speaker would cause a reversal. The places where there was destructive interference would now have constructive interference, and the places where there was constructive interference would now have destructive interference. So there would be destructive interference at point C and constructive interference at D (Fig. 35.8).

35.4 The location of the bright fringes depends on the wavelength of light. In a beam of light made up of all colors, there are only certain screen positions where a particular color interferes constructively and forms a bright fringe. Because these places are different for each wavelength, the fringes are separated by color. The central maximum is white, however, because all colors experience constructive interference in the middle of the screen.

35.5 According to Figure 35.21, if the slit's width is very narrow (close to the wavelength of the light), the first dark fringes are far from the central maximum. Often we are not interested in light that falls far from the central maximum. For example, in Figure 35.9, light passes through a single slit before encountering the double slits. Only the light from the central maximum illuminates the double slits, so we are not concerned about the positions of the first dark fringes due to diffraction at the single slit.

35.6 The graph in Figure 35.25C corresponds to the light with the shortest wavelength ($\lambda = w/6$). The light with the longest wavelength produces the pattern in Figure 35.25A; its wavelength is $\lambda = w$. If light of all three wavelengths illuminated the same slit, the light with the longest wavelength (Fig. 35.25A) would produce the broadest central maximum.

CHAPTER 35: Problems and Questions

1. Model the paint as particles because the artwork has crisp, well-defined lines. There is no diffraction or bending of the paint.
3. $\lambda/4$
5. 2.08×10^{-6} m
7. 1.35×10^{-2} m
9. 31
11. 0.0222 m
13. **a.** 3.80 cm **b.** 39.3 cm **c.** The fringes are not evenly spaced. Fringes that occur at small angles appear equally spaced, but the angular separation between adjacent maxima increases as n increases.
15. 1.3×10^{-4} m
17. 8.86×10^{-5} m
19. 8.16×10^{-7} m
21. 5.33×10^{-7} m
23. 454.6 nm
25. 1.41×10^{-4} m
27. **a.** 3.58 cm **b.** 2.98 cm
29. 7.09×10^{-3} m
31. 0.5
33. Either 1.080×10^{-6} m or 2.700×10^{-7} m; no, the light is not visible in either case.
35. 5.81×10^{-7} m
37. 34
39. 1.05 rad, or 60.0°
41. **a.** 6.47 rad **b.** $I = 0.991 I_{max}$
43. $1.305 d_{green}$
45. 0.257%
47. 1.98×10^{-6} m
49. B, A, then C
51. 4.50×10^{-2}, 1.62×10^{-2}, and 8.27×10^{-3}
53. 0.0456
55. **a.** 4.50% **b.** 1.62%
57. 6.27×10^{-2}
59. three
61. The light from the two bulbs is not coherent. Even if the light were monochromatic, the phase difference would not be constant.
63. 4.73×10^{-4} m
65. ± 0.32 m
67. All orders are missing except the central bright fringe ($n = 0$).
69. 5.91 m
71. 4.15×10^{-7} m
73. 5.0×10^{-4} m
75. 3.03×10^{-3} m
77. 156° or 2.72 rad
79. **a.** 29.2° **b.** 5.43×10^{-2} m
81. 2.52×10^{-6} W/m^2
83. 2.80×10^{-4} m
85. 3.75 m
87. **a.** 6.03×10^{-3} m **b.** 1.21×10^{-2} m
89. $(I_0/9)\{1 + 8\cos^2[(\pi/\lambda)d \sin\theta]\}$

CHAPTER 36: Concept Exercises

36.1 According to Rayleigh's criterion, $\sin\theta_{min} = 1.22\lambda/d$ (Eq. 36.3), a telescope with a larger diameter d has better resolution. So the minimum angular separation between two different sources that are barely resolved is smaller for the larger telescope.
36.2 The conditions for constructive and destructive interference depend on the wavelength or color of light. If the condition for a bright fringe at some location on the film is met for one particular wavelength—say, red light—then it is not met for other wavelengths. The rainbow effect is due to the condition for constructive interference being met by only one color over some small band on the film.
36.3 The pattern produced by a diffraction grating is much like the interference pattern produced by two slits. Because so many rulings (slits) are involved, it might be better to call it an *interference* grating. Of course, each slit also produces a diffraction pattern. So you might argue that the best name would be a *diffraction and interference* grating. Such a long name would probably be shortened, however.
36.4 To find the dispersion $D \equiv \Delta\theta/\Delta\lambda$ (Eq. 36.18), we need $\Delta\theta$ and $\Delta\lambda$ (Table 36.2):

$$D = \frac{(0.334 - 0.246)\text{ rad}}{(6.56 - 4.86) \times 10^{-7}\text{m}} = 5.18 \times 10^5 \text{ rad/m}$$

Because these numbers apply to both gratings, both have the same dispersion.

36.5 For Young's experiment, there are two slits, so $N = 2$. For the first-order lines, $R = (2)(1) = 2$. Consider a red line of wavelength 650 nm and a violet line of wavelength 400 nm. Their average wavelength is 525 nm, and $\Delta\lambda = (650 - 400)\text{nm} = 250$ nm. By Equation 36.20, the resolving power required to just barely separate them is

$$\frac{\lambda_{av}}{\Delta\lambda} = \frac{525\,\text{nm}}{250\,\text{nm}} = 2.1$$

This is about equal to the resolution of Young's double-slit experiment. So the lines are just barely resolved at the opposite ends of the spectrum. The other colors are blended together, forming nearly white light in between.

CHAPTER 36: Problems and Questions

1. As the size of the aperture decreases, the diameter of the central maximum (the Airy disk) gets larger.
3. 295 m
5. 13.8 arcmin
7. **a.** 8.13×10^{-5} rad **b.** We would choose the shortest wavelength of visible light possible—so, violet with a wavelength of about 400 nm. **c.** 6.51×10^{-5} rad **d.** 4.89×10^{-5} rad
9. 1.2×10^2 m

11. 1.5×10^{-4} m
13. 1.16×10^{-7} m
15. **a.** $\lambda/4$ **b.** 9.96×10^{-8} m
17. $w_{green} = 9.35 \times 10^{-8}$ m, $w_{red} = 1.19 \times 10^{-7}$ m
19. 1.42×10^{-7} m
21. If we slowly decrease the thickness of the oil, green no longer exhibits constructive interference and is not seen. The observed color changes gradually to colors that have a shorter and shorter wavelength (from green to blue to violet).
23. **a.** 6.50×10^{-7} m **b.** 4.88×10^{-7} m
25. 1.59×10^5 rulings/m
27. 1.29×10^{-6} m
29. **a.** 4.36×10^3 slits/cm **b.** 9
31. 2.78×10^5 rulings/m
33. 9.76°
35. **a.** 4 **b.** 9
37. $D_1 = 6.85 \times 10^5$ rad/m, $D_2 = 1.82 \times 10^6$ rad/m
39. **a.** 58 **b.** 29 **c.** 20
41. 3.56×10^5 rad/m
43. 6.00×10^3
45. 3162 fringes. Each fringe viewed indicates the mirror has moved 316.3 nm.
47. **a.** 2.5×10^{-9} m **b.** 8.3×10^{-18} s
49. The component in part C is more flat. Part B includes a series of curved lines, which represent regions that are not linearly increasing in gap thickness as one moves from left to right across the flat. Thus, the component in part B is more curved.
51. 2.94×10^{-7} m
53. 1.21×10^{-5} m
55. 1.29
57. 4.62 m
59. 3.57×10^{-7} m
61. 1.04×10^3 m
63. **a.** 1.85×10^{-10} m **b.** 4
67. 3.00×10^{-3} m
69. 6.19×10^{-7} m

CHAPTER 37: Concept Exercises

37.1

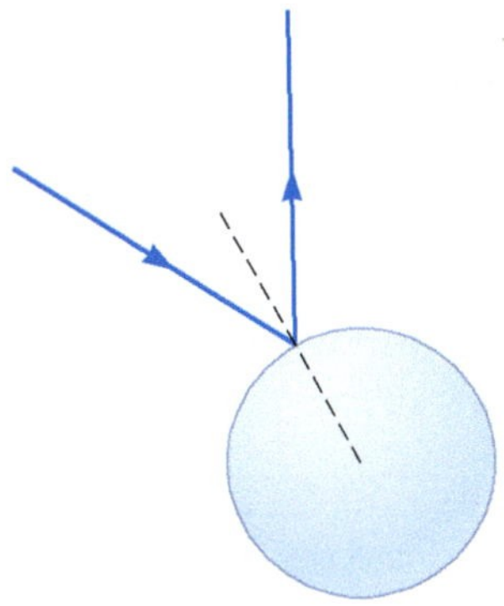

37.2 Answers will vary. Two possible sets of alternative terms for *virtual image* and *real image* are (1) *nonprojectable image* and *projectable image* and (2) *diverging image* and *converging image*. The first set of terms helps us to remember that a virtual image cannot appear on a screen, but a real one can. The second set helps us to remember that a virtual image is formed by rays that appear to diverge from each point that makes up the image, whereas each point that makes up a real image is the result of rays that converge at that point.

37.3 The spherical mirror is held in the artist's hand. It looks like the diameter of the sphere is about the same as the span of the artist's fingers, or about 9 in. or 23 cm. Then the radius of curvature is about 12 cm. The focal length is half the radius of curvature. This is a convex mirror, so according to the third sign convention, the focal length and radius of curvature are negative:

$$f = \frac{r}{2} = \frac{-12\,\text{cm}}{2} = -6 \text{ cm}$$

37.4 When traced back behind the mirror, all four rays intersect, so they produce the same results.

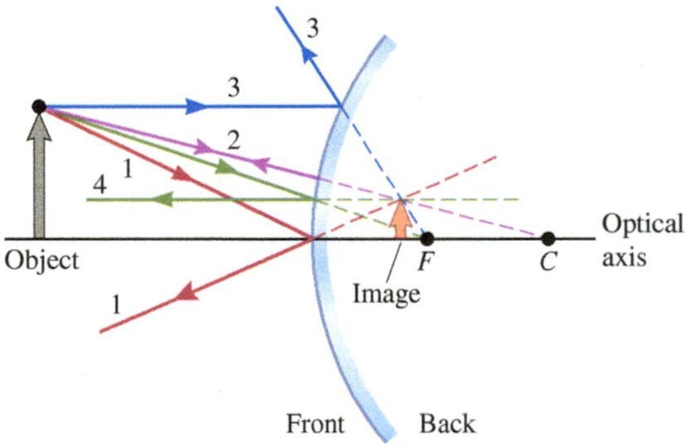

37.5

Plane mirror	$\|d_i\| = \|d_o\|$	$M > 0$ (upright image)	$\|M\| = 1$
Convex mirror with $d_o < \|f\|$	$\|d_i\| < \|d_o\|$	$M > 0$ (upright image)	$\|M\| < 1$ In fact, $0.5 < \|M\| < 1$
Convex mirror with $d_o > \|f\|$	$\|d_i\| < \|d_o\|$	$M > 0$ (upright image)	$\|M\| < 1$ In fact, $\|M\| < 0.5$
Concave mirror with $d_o < \|f\|$	$\|d_i\| > \|d_o\|$	$M > 0$ (upright image)	$\|M\| > 1$
Concave mirror with $\|r\| > d_o > \|f\|$	$\|d_i\| > \|d_o\|$	$M < 0$ (inverted image)	$\|M\| > 1$
Concave mirror with $d_o > \|r\|$	$\|d_i\| < \|d_o\|$	$M < 0$ (inverted image)	$\|M\| < 1$

37.6 Rays from distant objects are parallel and focused at the focal point. If the secondary mirror were missing, the image of distant stars would form 57.6 m in front of the primary mirror (more than four times the length of the HST's tube). One of the roles of the secondary mirror is to "fold" the optics. Light is not focused at the secondary mirror; instead, that mirror helps to focus the light through a hole and onto a plane behind the primary mirror.

37.7 The image of the four people is upright. According to the ray diagrams in Figures 37.33 and 37.34 (page 1200–1201), if the image is upright, it must also be virtual. The people must be standing at a position closer than the focal point F, so $d_o < f$ (or $d_o < 57.6$ m for the HST).

CHAPTER 37: Problems and Questions

1. 1.5 m
3. **a.** -0.20 **b.** $d_i = (0.20)d_o$
5. 4.4 m
7. No. The rest of the electromagnetic spectrum reflects from surfaces, too. For example, radio waves reflect from a parabolic dish in a radio telescope.
9. 67.7°
11. $\theta_r = 45.0° - \theta_i$
13. 4
15. 67.5°
17. 45.0°
19. 8.5 ft
21. **a.** 4.0 ft **b.** 10.0 ft
23. 9.62 m
25. 82.5 cm
29. -1.11 cm
31. 1.20 m
33. -75.0 cm
35. -17.5 cm
37. 2.66 cm
39. **a.** $d_i = 23.7$ cm, -0.395, real, inverted **b.** $d_i = 34.0$ cm, -1.00, real, inverted
41. A convex mirror is not able to form a real image. The image must be projected onto the CCD screen. A concave mirror is capable of reflecting the light so that it passes through the location of the image and onto the CCD.
43. 1.43 cm
45. **a.** 2.10 m in front of the mirror **b.** 0.977 cm
47. 26.3 cm
49. 20.0 cm and 40.0 cm
51. **a.** 5.24 cm **b.** 2.10
53. 4.3 m
55. **a.** -3.27 cm **b.** 0.182 **c.** upright **d.** 0.636 cm
57. **XOS ᗡN∀ XOℲ**
59. **XOƧ ᗡИA XOꟻ**
61. $2f$ or $4f$
63. **a.** 23° **b.** 5.7° **c.** 1.1° **d.** $(1.1 \times 10^{-5})°$
65. 10.0 cm
67. -6.0 cm
69. -8.96 cm
71. $(2n)D$
73. **a.** $\dfrac{\sqrt{x^2 + y_1^2} + \sqrt{(L - x)^2 + y_2^2}}{v}$
b. You should end up having derived the law of reflection, $\theta_i = \theta_r$. This is the condition for travel in the shortest time.

CHAPTER 38: Concept Exercises

38.1 The correct figure is sketch 1 because the angle of reflection equals the angle of incidence. Also, because the light travels from a medium with a low index of refraction to a medium with a high index of refraction, the beam is bent *toward* the normal, so the refracted angle must be smaller than the incident angle.

38.2 No. Because air has a lower index of refraction than water, when light goes from air into water, it cannot be totally reflected. Some of the light from your face is transmitted into the water.

38.3 The angle of refraction is zero in this case, which means the light does not bend in the glass and is a single white beam. The red light is still faster than the violet light, so in principle the leading edge of the beam should be red. But in practice this effect is too small to notice.

38.4 A diverging lens is like a convex mirror. Both produce virtual upright images, no matter where the real object is placed.

38.5 A converging lens is like a concave mirror. For both, whether the image produced is virtual or real, upright or inverted, depends on the placement of the real object.

38.6 Yes. Usually, the focal length of a camera lens is fairly short. In order to use the lens as a magnifier, you need to hold the feather closer to the lens than its focal point. So you should hold the lens very close to the feather, probably just a centimeter or so away.

38.7 Because we assume the near point is at 25 cm, the angular magnification is $m = 25 \text{ cm}/f$ (Eq. 38.14). Therefore, $f = 25 \text{ cm}/3 = 8.3$ cm.

38.8 To an astronomer, a telescope's LGP is more important than its magnification. If you know that a telescope has a large diameter, it can gather a lot of light and detect very faint objects.

CHAPTER 38: Problems and Questions

1. 36.7°
3. 41.8°
5. $\theta_t = \sin^{-1}\left(\dfrac{n_{\text{air}}}{n_{\text{water}}}\sin(90° - \theta_i)\right)$
7. 1.53
9. 32.3°
11. 1.207, 2.49×10^8 m/s, 4.29×10^{14} Hz
13. 56.7°
15. 1.41
17. 3.69 m
19. 61.3°
21. 48.6°
23. 1.34×10^{-2} m
25. All angles of incidence will cause the effect, so as long as $\theta_i < 90°$.
27. **a.** The second prism bent the monochromatic light's path, but no additional colors were seen. (See the figure.) So, Newton could conclude that prisms do not create color; prisms just spread the colors out. **b.** Using a second prism demonstrates that when all the colors come together, you get white light. Without this demonstration, you can say that white light may be separated into colors, but you cannot say that the color spectrum may be combined to make white light.

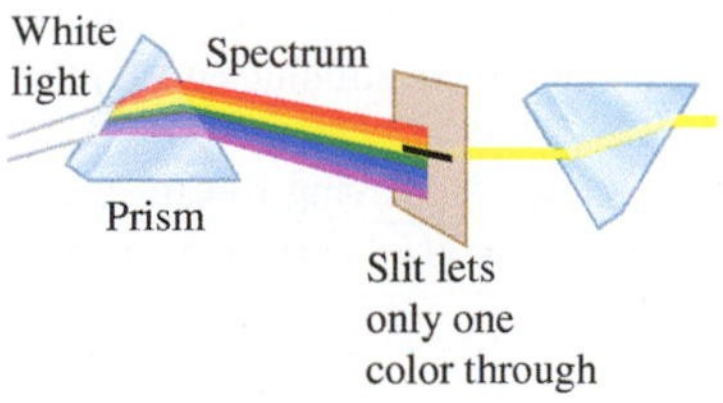

29. a. 3.7×10^{-7} m **b.** 5.5×10^{14} Hz **c.** Yes. The frequency is the same in both media.

31. a.

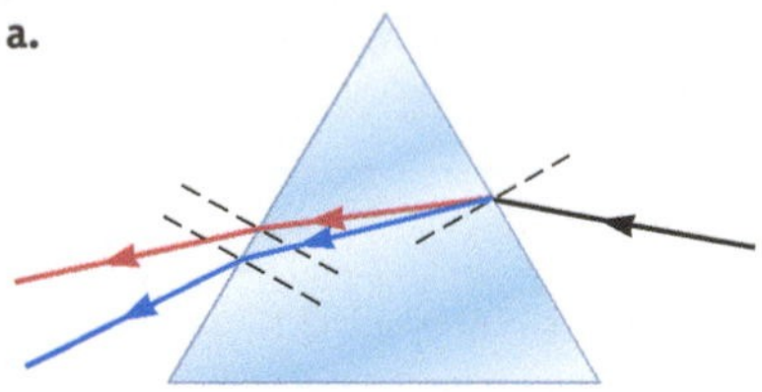

b. 2.65°

35. $d_i = -R$, $M = 1.75$

37. 2.44 m. This may be an issue when trying to catch the fish with a net if the net is not very big.

39. 7.5 cm, −0.13

41. 5.0 cm

43. −50.2 cm, diverging. The answers will not change if the object is placed on the other side of the lens.

45. 30.0 cm

47. a. −26.8 cm **b.** diverging

49. 7.72 cm

51. $d_o = 5.17$ cm, $d_i = -3.88$ cm

53. $M = 0.375$, $d_i = -3.75$ cm

55. a. 22.5 cm **b.** −4.00 cm

57. a. $1.33f$ **b.** real **c.** −0.333 **d.** inverted

59. a. −1.50 **b.** −0.075 m

61. a. 21.0 cm **b.** 25.0 cm

63. The glass and its contents act as a converging lens. The image we see is real and inverted.

65. Goggles enable you to maintain the air–eye interface so that your eye focuses as it normally does.

67. Her eyes are larger when viewed through her glasses, so their lenses act as magnifying glasses for us. She must be wearing converging lenses, with a focal length greater than the distance between her eyes and her glasses. She must be farsighted.

69. Susan's better vision is in her right eye, and she is nearsighted; $d_{\text{far, right}} = 12.9$ cm, $d_{\text{far, left}} = 12.5$ cm

71. a. $f_{\text{red}} = 7.35$ cm, $f_{\text{blue}} = 7.08$ cm **b.** 3.58 cm

73. 6.00

75. $h_i = -2.14$ cm, 8.57 cm behind the second lens

77. a. 2.00 cm **b.** −0.667 cm

79. a. 17.6 cm to the right of the diverging lens, $M = -5.29$ **b.** inverted

81. a. All three students have made statements that are at least partially correct. You can *in principle* construct a microscope with these lenses, but their separation will not be the sum of their focal lengths. The resulting size may be impractical. **b.** The distance between the two lenses must be at least 112.5 cm. That is more than a meter. If we now found the position of the object, we would see that the microscope would have to be more than 2 m long. So these lenses cannot be used to construct a practical microscope given these size constraints.

83. 11.6 cm

85. 1.22

87. a. 6.74° **b.** 4.92×10^{-7} m **c.** The light wave slows down as it moves from air into water, but the sound wave speeds up by a large factor. The light wave bends toward the normal and its wavelength shortens, but the sound wave bends away from the normal and its wavelength increases.

89. 27.0°

91. a. 6.13×10^{14} Hz **b.** 3.67×10^{-7} m **c.** 2.25×10^8 m/s

93. a. 34.5° **b.** 13.6°

95. a. 34.5° **b.** 15.4°

97. a. 25.4° **b.** 18.8°

99. 83.25°

101. 2.66 m

103. 48.8°

105. 53.1°

107. $\theta_i \geq 27.9°$

109. 13.1 cm

111. 0.750 m/s

113. a. 15.8 cm to the left of the diverging lens **b.** The final image is to the left of the lenses and is virtual.

115. diverging, −37.8 cm

117. −10.4 cm

119. a. 58.7 cm **b.** 25.8 cm

121. a. $d_i = -9.33$ cm, $M = 0.333$, virtual, upright **b.** $d_i = -4.20$ cm, $M = 0.700$, virtual, upright

123. a. 9.60×10^{-5} m/s **b.** toward the lens

125. b. $\alpha_{\min} = \beta(n - 1)$

127. a. 7.85° **b.** 0.007°

129. a. 11.3 cm to the right of the mirror **b.** −9.00 **c.** inverted

CHAPTER 39: Concept Exercises

39.1 Your classroom and the railroad car are approximately inertial frames because neither is accelerating (except for the acceleration of the Earth itself). The car making a turn has centripetal acceleration, and the airplane taking off must increase its speed, so neither of these is an inertial frame.

39.2 The primed observer measures a shorter length because in the time it takes to make the measurement, the primed frame moves toward the second end of the rod. If we use the tower and helicopter notation from Equation 39.4, the transformation equation is

$$L = x_t - x_h = (x'_t - x'_h) + v_{\text{rel}}(t_2 - t_1)$$
$$L = L' + v_{\text{rel}}(t_2 - t_1)$$

where $(t_2 - t_1)$ is the time between the two measurements. If the measurements are made simultaneously, then $t_2 = t_1$ and $L = L'$, as in Equation 39.5.

39.3 If the source is moving toward the observer, we choose the top signs in $\lambda = \lambda'\sqrt{(1 \mp \beta)/(1 \pm \beta)}$ (Eq. 39.27), and we find $\sqrt{(1 - \beta)/(1 + \beta)} < 1$. So $\lambda < \lambda'$, and the observer in the laboratory frame sees a bluer color than was emitted. If the source is moving away from the observer, we choose the bottom signs, and we find $\sqrt{(1 + \beta)/(1 - \beta)} > 1$. So $\lambda > \lambda'$, and the observer sees a redder color.

39.4 a. The ball has no energy. It isn't moving. It is cold and has almost no thermal energy. It has no potential energy because there is only one particle in the system.
b. The ball has no potential energy and essentially no thermal energy, but it has energy associated with its mass. A bowling ball's rest mass is about 6 kg; then its total energy is approximately $E = m_{\text{rest}}c^2 \approx 5 \times 10^{17}$ J (its rest mass energy). There is a considerable amount of energy stored in any massive object.

39.5 According to $R_{\text{sch}} \approx 3M/M_{\odot}$ km (Eq. 39.51), with $M/M_{\odot} = 1$, the radius would be 3 km—the distance of an easy jog.

CHAPTER 39: Problems and Questions

1. The Earth's motion causes the table of pegs to rotate under the pendulum. If the pendulum were in an inertial frame, only two pegs would be knocked over.
3. **a.** 2.0×10^2 s **b.** 5.6×10^2 s **c.** 1.5×10^2 s
7. **a.** 113.0 mph to the south **b.** 5.7 mi
9. The frame must not be accelerating.
11. **a.** 2.73×10^8 m/s **b.** -0.873 m **c.** 4.58×10^{-9} s
13. **a.** $x = 5.4 \times 10^9$ m and $y = 12.0$ m **b.** 24 s
15. **a.** 1.03 **b.** Although both frames could be labeled the x' frame, the convention used in this textbook is to identify the laboratory frame as the unprimed frame and the frame moving with the object as the primed frame. Realize, however, that from the electron's point of view, the laboratory is moving relative to it, and it would be fair to reverse the answers!
17. $(\sqrt{3}/2)c$
19. $(\sqrt{3}/2)c$
21. 0.724 m
23. **a.** 35.0 m **b.** 31.0 m **c.** $0.464c$
25. **a.** 1.3×10^3 yr **b.** 7.7×10^2 yr
27. $0.661c$
29. $(\sqrt{3}/2)c$
31. 4.90×10^{-9} s
33. **a.** 3.97 h **b.** 2.18 h
35. 1415.31 MHz and 1424.78 MHz
37. 1.83×10^6 m/s toward the observer
39. $0.496c$
41. 8.50×10^7 m/s
43. $0.96c$
45. **a.** $0.90c$ **b.** $0.90c$ **c.** $0.97c$
47. **a.** 0.145 kg **b.** 0.145 kg **c.** 0.275 kg
49. $0.877c$
51. **a.** 8.20×10^{-14} J or 512 keV **b.** 1.30×10^{-13} J or 812 keV **c.** $0.7762c$
53. 3.078×10^{-25} kg
55. 5.184×10^8 J whether using special relativity or classical physics
57. **a.** 8.20×10^{-14} J **b.** 4.12×10^{-13} J **c.** 3.30×10^{-13} J
59. 4.57×10^{14} Hz. We do not observe any gravitational shift, to three significant figures.
61. 3.25 min
63. 1.1×10^7 km = 1700 $R_\oplus$
65. 3.4×10^{41} kg = $1.7 \times 10^{11} M_\odot$. This is slightly smaller than the mass of the Andromeda galaxy.
67. 1.47×10^{-3} s
69. 10.1 s
71. 16.1 μs for Joe's camera and 16.1 μs for Moe's camera
73. $0.737c$
75. $0.159c$ toward the Earth
77. Yes. The trackside observer measures the length of the train to be 70.2 m, so the train is measured to fit inside the tunnel with 9.8 m to spare.
79. $\dfrac{p'}{1 - (v_{rel}/c)^2}$
81. 3.33×10^{-9} J
83. **a.** $0.990c$ **b.** $0.989c$
85. **a.** 23.7 μs **b.** 37.6 μs

CHAPTER 40: Concept Exercises

40.1 **a.** The atom is missing one electron, so it is positively charged, with $q = +e$.
b. The pith ball's charge is positive, so the ball is missing

$$n = \frac{9 \times 10^{-9}\text{C}}{1.6 \times 10^{-19}\text{C/electron}} = 5.6 \times 10^{10} \text{ electrons}$$

In the case of the pith ball, one electron fewer would make no difference to its total charge (to two significant figures). If you wanted the pith ball to have 1% more charge (9.09×10^{-9} C instead of 9.00×10^{-9} C), you would simply need to subtract more electrons (until 5.7×10^{10} were missing). So it seems that the pith ball can have any charge. However, in the case of a singly ionized atom, if there were one electron fewer, then its net charge would double. So you cannot increase the atom's charge by just 1%.

40.2 According to Planck's quantum theory, a system's energy comes in discrete bundles. So the idea that a digital radio can only be set to certain discrete frequencies helps us to imagine a quantity whose values are not continuous. For example, you may be able to set your digital radio to 88.1 KHz and to 88.3 KHz, but not to 88.2 KHz. Perhaps a quantum oscillator can have an energy of 88.1 eV or 88.3 eV, but not 88.2 eV. The analogy fails in that a radio is designed by a human engineer. There is no law that prevents the engineer from designing the radio so that you can tune it to 88.2 KHz. However, quantum mechanics tells us that nature sets the discrete energy levels in an oscillator.

40.3 Combine $K_{max} = eV_0$ and $K_{max} = E_{light} - W_0$ (Eqs. 40.9 and 40.11) with $E_{light} = hf$. Then the linear fit shown on the graph is $V_0 = \frac{h}{e}f - \frac{W_0}{e}$ and its slope is h/e. We estimate the slope $h/e = 4.1 \times 10^{-15}$ V/Hz. Solve for h to find $h = (4.1 \times 10^{-15}\text{ V/Hz})e = 6.6 \times 10^{-34}$ J·s. Our estimate is in agreement with the accepted value. However, estimating the slope may be difficult; don't worry if your estimate is within about 3% of the accepted value.

40.4 The electron that corresponds to Figure 40.14C gains the most energy because the scattered photon has the greatest Compton shift and loses the most energy.

40.5 In classical physics, frequency, wavelength, and amplitude are properties of waves, not particles. The equation $E = hf = hc/\lambda$ links the wave properties of a photon to the amount of energy it carries as a particle, and that energy does *not* depend on amplitude. However, in the classical model the energy carried by a wave *does* depend on the wave's amplitude (Section 17-6).

CHAPTER 40: Problems and Questions

1. The pixels are very small, so when viewed on our size scale they seem continuous. Adding or subtracting a few pixels out of millions would not make much of a difference. If pixels were larger and more comparable to our size scale, we would see adjacent squares of various colors, and the detail of the image would be lost.
3. 6.443×10^{-6} m
5. **a.** 7.25×10^{-7} m, red **b.** 5.47×10^{-7} m, green **c.** 4.53×10^{-7} m, blue
7. 4.61×10^3 K
9. 7.00×10^{-7} m, red visible light
11. 6.29×10^{-20} J. No, it cannot. Any other frequency must be a whole-number multiple of the minimum frequency.
13. **a.** 5.78×10^3 K **b.** 5.01×10^{-7} m
15. **a.** 2.56×10^{-7} m **b.** 1.17×10^{15} Hz **c.** 2.15 eV
17. 1.2×10^{15} Hz
19. **a.** 5.40 eV **b.** 1.31×10^{15} Hz
21. **a.** 1.00 eV **b.** 2.10 eV
23. 1.09×10^{15} Hz
25. 0.61 m

27. 0.250
29. a. 1.274×10^{-10} m **b.** 1.560×10^{-15} J and 5.200×10^{-24} kg · m/s
31. 67.6°
33. 1.29×10^{-12} m
35. a. 19.84 pm **b.** 20.55 pm **c.** 24.69 pm
37. a. 62.54 keV **b.** 60.38 keV **c.** 50.25 keV
39. a. 1.72×10^{-11} m **b.** 72.0 keV **c.** 3.0 keV
41. a. 6.670×10^{-10} m **b.** 6.679×10^{-10} m **c.** 6.691×10^{-10} m
43. a. particle **b.** wave **c.** wave **d.** particle
45. a. 3.11×10^{-9} m **b.** 127 m/s
47. Recall that the wavelength of the wave must be comparable to the width of the opening in order to observe diffraction. A typical wavelength of a thrown baseball would be about 1×10^{-34} m. This is far less than the spacing between the slits, and so we do not expect to observe diffraction.
49. 6.66×10^{-10} m
51. The lower energy electrons will have less momentum than the higher energy electrons, which means that their wavelength is greater than that of the higher energy electrons. Because the wavelength of the lower energy electrons is greater, one might describe them as being "fatter" than the higher energy electrons.
53. 4.96×10^{-13} m
55. 0.11 m
57. 5.446×10^{-4}
59. a. 5.58×10^{-17} J **b.** 1.90×10^{-20} J **c.** 2.94×10^{3}
61. 0.364 V
63. a. 1.25×10^{-10} m **b.** 1.20×10^{-10} m
65. 1.4×10^{14}
67. 5.317×10^{6} m/s
69. 6.467×10^{-6} m
71. 4×10^{3} V
73. a. 0.827 V **b.** 1.65 V
75. 1.4×10^{19} s^{-1}
77. 2 eV
79. The photon's momentum would be infinite and the rest mass would be zero.
81. a. 4.67×10^{-2} kg · m/s **b.** 9.33×10^{-2} kg · m/s

CHAPTER 41: Concept Exercises

41.1 Cameron is correct. We relate temperature to average kinetic energy because of the great number of particles.
41.2 Answers will vary. One common occurrence in Chapter 2 is solving a quadratic equation for time and arriving at two solutions. One solution often corresponds to a time *before* the problem began, so this solution is not physically reasonable.
41.3 Figure 41.8 does not show the motion of the particle. Each panel is fixed in time, showing us the probability distribution of the trapped particle for a particular quantum state.
41.4 Imagine the particle as a cloud. The cloud is completely contained by the infinite well, but leaks out of the finite well. In both cases, the cloud's concentration in the well depends on the quantum state. For example, in the ground state the cloud is most concentrated in the center and we are most likely to find the particle there. However in the $n = 2$ state, the cloud is concentrated in two regions between the center and each boundary.

CHAPTER 41: Problems and Questions

1. $K_{av} = 1.9 \times 10^{3}$ eV $\approx$ 1 keV
3. 3.16×10^{34}
5. 1.37×10^{-50} kg · m^2
7. 2.86×10^{-12} m
9. a. 3 **b.** 5 **c.** It is more likely that a total of 6 would turn up on the dice because there are more possible ways to make a total of 6 with the dice.
13. $E_1 = 1.94 \times 10^{-5}$ eV, $E_2 = 7.76 \times 10^{-5}$ eV, $E_3 = 1.74 \times 10^{-4}$ eV, $E_4 = 3.10 \times 10^{-4}$ eV
15. 2.72×10^{-11} m
17. a. 1 **b.** 0.50
19. 9.64×10^{-21} J or 0.0602 eV
21. 0.964 eV
23. a. 2.34×10^{-13} m **b.** 3.00×10^{4} eV
25. 10^{-18} J or 10 eV
27. a. 5.68×10^{-4} m **b.** 9.09×10^{-4} m
29. a. 72 eV **b.** 1.7×10^{-8} m
31. a. 2.1×10^{-12} m **b.** 6.80×10^{-13} m **c.** 2.15×10^{-13} m
33. 5.08×10^{-12} J or 3.17×10^{7} eV
35. a. 1.79×10^{-12} m **b.** 4.64×10^{-13} m **c.** 1.44×10^{-13} m
37. 10^{-235}
39. 0.9591
41. 9.692×10^{-12} m
43. 89.2 eV
45. 1.79 eV
47. 1.88×10^{15} Hz
51. 7.91×10^{-36} m, so the player's claim is not legitimate.
53. a. 7.36×10^{-24} kg · m/s **b.** 7.16×10^{-11} m
55. 1.06×10^{-10} eV
57. a. $\sqrt{(15\lambda_A h)/(8m_e c)}$ **b.** $5\lambda_A/4$
59. a. 0.50 **b.** $(\pi - 2)/(4\pi)$
61. 2.68×10^{-22} J or 1.67×10^{-3} eV
63. $n = 3$
65. Yes. The potential barrier does not have an infinite height except as x goes to infinity, so there must be some penetration distance into the walls of the well.
67. a. 1.37×10^{-66} J, 5.49×10^{-66} J, 2.80×10^{-66} J **b.** 9.66×10^{-66} J
69. The energy difference will increase. As the width is decreased, the energy levels themselves increase in value, but the higher-energy levels will increase more than the lower levels, when we look at Eq. 41.16. Both are quadrupled when the width is halved, but the n level was greater than the m level, so the spread is greater.
71. 5.91×10^{-6} m
73. a. 71% **b.** 25%
75. Heisenberg's uncertainty principle is much more limited than Avi's summary suggests. According to Heisenberg's uncertainty principle, we cannot simultaneously measure a particle's position and momentum to an unlimited precision. Avi's misconception is common. Many people think that physicists believe that the universe is completely uncertain. Understanding that quantum mechanics is probabilistic is not the same as saying that we cannot make predictions. If Avi doesn't review for the final exam, you know with great certainty what is going to happen.
77. 75%

CHAPTER 42: Concept Exercises

42.1 The word "atom" comes from the Greek word for *undivided.* So the term gives the impression that an atom's structure is fixed; but we know that an atom's electrons can move around in the atom and that an atom can even gain or lose electrons. Of course, other terms will vary, but here is a line of thought: Perhaps a better term would be based on the idea that atoms are the neutral elements that make up all matter.

42.2 Use Equation 42.1 for each value of n:

Line	n	$\frac{1}{\lambda}=R\left(\frac{1}{2^2}-\frac{1}{n^2}\right)$ (42.1)	$\frac{1}{\lambda}$ (m^{-1})	λ (nm)
Hα	3	$\frac{1}{\lambda}=R\left(\frac{1}{2^2}-\frac{1}{3^2}\right)$	1.52×10^6	656
Hβ	4	$\frac{1}{\lambda}=R\left(\frac{1}{2^2}-\frac{1}{4^2}\right)$	2.06×10^6	486
Hγ	5	$\frac{1}{\lambda}=R\left(\frac{1}{2^2}-\frac{1}{5^2}\right)$	2.30×10^6	434
Hδ	6	$\frac{1}{\lambda}=R\left(\frac{1}{2^2}-\frac{1}{6^2}\right)$	2.44×10^6	410

42.3 De Broglie's model offers some explanation for why hydrogen's energy is quantized: Quantization is a natural consequence of modeling the electron as a standing wave. By contrast, Bohr can only say that he hypothesizes that the energy is quantized because that worked well for Planck's model of a black body. Both models are ad hoc in nature. De Broglie cannot say why the electron should be modeled as a standing wave, and Bohr cannot say why his postulates are true. For a deeper understanding we must turn to Schrödinger's equation.

42.4 True. If $\vec{L} = \pm m\hbar\hat{k}$, its magnitude would be $L = \sqrt{\ell(\ell+1)}\,\hbar = m\hbar$, where m cannot be zero (or the angular momentum would be zero). This would require $m = \sqrt{\ell(\ell+1)}$, but this is not possible because both m and ℓ are nonzero integers.

CHAPTER 42: Problems and Questions

1. The plum pudding model is static. The electrons are embedded in a plum pudding of positive charge. In an odd sense, the electrons are fixed in place by something akin to a normal force and static friction, much like the chocolate chips are held in place in a cookie.

3. Based on the model of the ancient Greeks, the final result is the same in each case. Both objects would eventually be atoms, as you make the final cut between two atoms. However, the ancient Greeks also had the notion that these atoms would somehow be different from one another, as the iron and wood are different from each other.

5. The expectation was that the α particles should mostly pass through the gold foil, so the large deflections that were observed were surprising. If an α ray passed near the outside of the plum pudding sphere, you would expect a slight deflection due to the Coulomb force. However, if the α ray were to pass through the plum pudding, you would expect almost no deflection because only the charge that is in the spherical region inside the α ray's path could deflect it. But Rutherford observed large deflections, indicating that the positive charge in the pudding was concentrated and had a mass comparable to or higher than the mass of the alpha particles.

7. My dog is about 1 m long, so if he is my nucleus, the radius is 0.5 m. Then, the corresponding electron would be located at 5×10^4 m, or about 30 mi away.

9. -8.46×10^{-10} m

11. 1.06×10^{-10} m

13. **a.** 1.23 eV **b.** 13.1 eV

15. $2\pi ke^2/h$

17. 8×10^6

19. 3.28×10^{-6} m

21. 1.88×10^{-6} m, 1.28×10^{-6} m, 1.09×10^{-6} m, and 1.01×10^{-6} m

23. This is expected because there are more protons in the nucleus of larger atoms. This means the electron will be more attracted to the nucleus because the amount of the potential energy between an electron and the nucleus will increase. So, it will take more energy to free the electron from the atom.

25. 1.09×10^{-8} m

27. **a.** 8.46×10^{-10} m **b.** 1.33×10^{-9} m

29. 1.99×10^{-9} m

31. $m = 0, \pm 1, \pm 2$, $L = \sqrt{6}\hbar$, $L_z = 0, \pm\hbar, \pm 2\hbar$

33. $0, \sqrt{2}\hbar, \sqrt{6}\hbar$

35. $0, \pm\mu_{Bohr}, \pm 2\mu_{Bohr}$

37. **a.** 32 **b.** 50 **c.** 72

39. For $\ell = 0$, $L = 0$ and $(\mu_{orbit})_z = 0$; For $\ell = 1$, $L = \sqrt{2}\hbar$ and $(\mu_{orbit})_z = 0, \pm\mu_{Bohr}$; For $\ell = 2$, $L = \sqrt{6}\hbar$ and $(\mu_{orbit})_z = 0, \pm\mu_{Bohr}, \pm 2\mu_{Bohr}$

41. It is possible. Because the spin quantum number must be $\pm 1/2$, the orbital magnetic moment could equal the spin magnetic moment is if $m = \pm 1$. This can be seen by examining Eq. 42.23 and Eq. 42.26.

43. **a.** 2 **b.** 6 **c.** 10 **d.** 14 **e.** 18

45. Yes. This is because for every possible magnetic quantum number, there are two possible spin quantum numbers. Thus, when a shell is full, the total possible states, as dictated by the orbital quantum numbers and the magnetic quantum numbers, will be multiplied by 2 to account for the two possible spin states in each case.

47. $1s^2 2s^2 2p^6 3s^2 3p^6 4s^2 3d^{10} 4p^6 5s^2 4d^{10} 5p^5$

49. He: $1s^2$, Ne: $1s^2 2s^2 2p^6$, Ar: $1s^2 2s^2 2p^6 3s^2 3p^6$

51. Ni

53. 9.503×10^{-8} m, 9.504×10^{-8} m, 9.505×10^{-8} m

55. 5

57. 1.852×10^{-3} eV

59. **a.** 1.32×10^{-9} m **b.** 1.66×10^{-9} m

61. 5

63. **a.** 1.14×10^{-8} m **b.** Yes. The energy of this photon is 109 eV, which is much greater than the magnitude of the ground-state energy in hydrogen, 13.6 eV.

65. **a.** 12.8 eV **b.** 3.08×10^{15} Hz

67. Yes, the $\ell = 2$ subshell. The magnetic field will remove the degeneracy of each subshell. The $\ell = 2$ subshell has five degenerate states with $m = 0, \pm 1, \pm 2$.

69. -19.5 eV, -4.88 eV, -2.17 eV, -1.22 eV, -0.780 eV

71. $n = 4$

73. **a.** 5.87×10^4 T **b.** This would mean the electron is no longer bound to the nucleus, and would be free, if in that state.

75. Because the states must be whole numbers, n must be an even-numbered state. Thus, in this scenario, n cannot be odd, but $n/2$ could be odd.

77. **a.** 1.22×10^{-7} m **b.** 1.09×10^{-6} m **c.** 4.38×10^{-6} m

79. $n = 4 \rightarrow n = 1$: 12.8 eV, $n = 4 \rightarrow n = 3 \rightarrow n = 1$: 0.661 eV and 12.1 eV, $n = 4 \rightarrow n = 2 \rightarrow n = 1$: 2.55 eV and 10.2 eV, $n = 4 \rightarrow n = 3 \rightarrow n = 2 \rightarrow n = 1$: 0.661 eV, 1.89 eV, and 10.2 eV

CHAPTER 43: Concept Exercises

43.1 If you put several marbles in a flexible bag and tie it off, leaving no excess room, the bag will have a nearly spherical shape. Each marble takes up a certain fixed (constant) volume, so if you increased the number of marbles in the bag, the bag's volume would increase proportionally to the number of marbles added. If each marble were replaced by a foam ball, then the volume of these balls would not be constant. You could add more foam balls without increasing the volume of the bag, as long as you squeezed the foam balls further, reducing their individual volumes.

43.2 The strong force acts only over distances of roughly 1 fm or so. The Earth and Moon are much too far apart for the strong force to play any role.

43.3 Avi is correct. The students can use Equation 43.8, $N = N_0(\frac{1}{2})^{t/T_{1/2}}$. The problem asks for the fraction remaining, that is N/N_0, so they don't need to know N_0. The answer is not zero because every eight days *half* of the parent nuclei still remain. The number of parent nuclei goes to zero as time goes to infinity.

43.4 Inverse beta decay because it produces positrons: $^{18}_{9}\text{Fl} \rightarrow {}^{18}_{8}\text{O} + e^+ + \nu$.

43.5 The daughter nucleus becomes carbon. Compare Figures 43.6 and 43.7. If one of nitrogen's protons becomes a neutron through inverse beta decay, it will have six protons and six neutrons, exactly as carbon-12 does.

43.6 We use the term *pseudo*-units because MeV/c^2 are not true units, since they involve the constant c (the speed of light). We refer to 1 u = 931.4947 MeV/c^2 (Eq. 43.17) as a translation instead of a conversion factor because we are translating between two quantities, mass and energy, much as we did when translating between mass and weight in the expression 1 kg = 2.2 lb.

43.7 E_B = 13.6 eV because that is the amount of energy an external source must add to break up the hydrogen atom into free proton and a free both at rest. This is also called the ionization energy.

43.8 The correct reaction is $^{A}_{Z}\text{P} \rightarrow {}_{Z+1}^{A}\text{D} + e^- + \bar{\nu}$. You reason that charge must be conserved. Because the daughter nucleus has an extra proton, one of the products must be an electron, so there is no net gain in positive charge. You further reason that both baryon number and lepton number must be conserved. There are A baryons in the parent nucleus and A in the daughter, and the baryon number is conserved. The electron is a lepton. Because there are no leptons on the left of the reaction, there must be no net leptons on the right. The missing particle must be an antilepton to cancel out the electron's lepton number, in this case an antineutrino.

CHAPTER 43: Problems and Questions

1. **a.** Francium **b.** 87 **c.** 115

3. $^{214}_{82}\text{Pb}$

5. **a.** 1.2×10^{-15} m **b.** 7.4×10^{-15} m

7. 4.1 fm

9. 4.7 N. The nucleus should be stable because the attraction inward due to nearest neighbors is far greater than a Coulomb repulsion outward. Although there will be additional Coulomb repulsion due to nearer protons, the distance between a proton on the edge and a nearer neighbor would need to be orders of magnitude less than the distance between the two protons on the edge, which is not possible given their own dimensions.

11. We ignored it because the protons have low mass and because G, the universal gravitation constant, is small. The gravitational attraction between the protons is much weaker (about 10^{38} times weaker) than the Coulomb repulsion. It contributes very little to the stability of the nucleus. Instead, the strong nuclear force keeps the protons bound together.

13. We do not expect $^{19}_{8}\text{O}$ to be stable because there is considerably more energy in its neutron system than in its proton system.

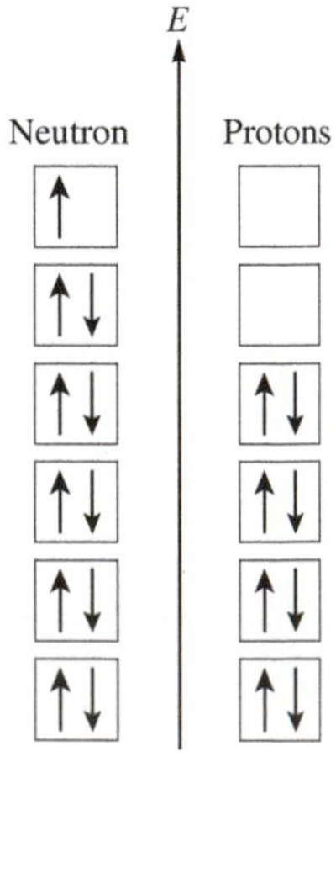

15. $^{14}_{7}\text{N}$

17. No. There are more protons than neutrons, and this is not one of the two exceptions to that general rule, H and He.

19. 5000

21. 429.6 s

23. **a.** 2.99×10^{17} Bq **b.** 53.4 days

25. The half-life of carbon-14 is roughly 6000 years. If a sample is a hundred million years old, it has lasted about 17,000 times carbon-14's half-life. So, very little carbon-14, if any, would remain in the sample. It would be very difficult to measure such a trace amount.

31. $^{14}_{7}\text{N}$

33. **a.** $^{188}_{77}\text{Ir} + e^- \rightarrow {}^{188}_{76}\text{Os} + \nu$
b. $^{127}_{54}\text{Xe} + e^- \rightarrow {}^{127}_{53}\text{I} + \nu$
c. $^{57}_{27}\text{Co} + e^- \rightarrow {}^{57}_{26}\text{Fe} + \nu$

35. $^{15}_{8}\text{O} + e^- \rightarrow {}^{15}_{7}\text{N} + \nu$, electron capture.

37. $^{19}_{8}\text{O} \rightarrow {}^{19}_{9}\text{F} + e^- + \bar{\nu}$, beta decay.

39. 479.06 MeV and 8.5546 MeV per nucleon

41. 1.935 u

43. ~7 MeV per nucleon and ~28 MeV

45. **a.** 5.0783 MeV per nucleon **b.** 5.3892 MeV per nucleon

47. $^{239}_{94}\text{Pu} + {}^{1}_{0}\text{n} \rightarrow {}^{148}_{58}\text{Ce} + 3({}^{1}_{0}\text{n}) + {}^{89}_{36}\text{Kr}$

49. **a.** 179.65 MeV **b.** 1.2×10^{42}

51. **a.** No, the total number of nucleons is not conserved or accounted for. **b.** Yes, the total number of nucleons is conserved or accounted for. **c.** Yes, the total number of nucleons is conserved or accounted for.

55. 4.237 MeV

57. 1.9×10^{11} photons

59. **a.** 0.300 Gy **b.** 6.00 Sv **c.** 6.00×10^2 rem

61. 0.13 Gy or 13 rad

63. No, it is not. The baryon number before the reaction is 2, and the baryon number after the reaction is 1. The baryon number should be conserved.

65. 6

67. No. Fermions cannot occupy the same state.

69. Two protons must decay. A total of four protons are added during the cycle, but the final product consists of only two extra protons. Thus, two of them must decay during the cycle.

71. Chlorine-36, Cl.

73. **a.** 1.45×10^9 Bq **b.** 0.0392 Ci **c.** 4.57 yr **d.** 9.68 yr

75. 891.43 MeV

77. 4.17×10^{20}

79. **a.** $^{235}_{92}\text{U} + {}^{1}_{0}\text{n} \rightarrow {}^{143}_{55}\text{Cs} + 3({}^{1}_{0}\text{n}) + {}^{90}_{37}\text{Rb}$ **b.** 1737.37 MeV **c.** 758.10 MeV **d.** 1150.62 MeV

Index

A

B

C

D

J

M

S

T

U

X

Y

Z

Some Astronomical Data

Quantity	Convenient units
Earth	
Mass of the Earth	$M_\oplus = 5.9736 \times 10^{24}$ kg
Radius of the Earth	$R_\oplus = 6.378 \times 10^6$ m
Orbital speed of the Earth	$v_\oplus = 30$ km/s
Year (measured with respect to fixed stars)	$T_\oplus = 365.25$ days
Sun	
Mass of the Sun	$M_\odot = 1.9891 \times 10^{30}$ kg
Radius of the Sun	$R_\odot = 6.9551 \times 10^8$ m
Effective temperature of the Sun's surface	$T_\odot = 5777$ K
Luminosity of (power emitted by) the Sun	$P_\odot = 3.839 \times 10^{26}$ W
Distance of the Sun from the Milky Way galaxy's center	$r_{\odot\,\mathrm{MW}} = 8.5$ kpc
Speed of the Sun around the Milky Way galaxy's center	$v_\odot = 220$ km/s
Moon	
Mass of the Moon	$M_{\mathrm{Moon}} = 7.35 \times 10^{22}$ kg
Radius of the Moon	$R_{\mathrm{Moon}} = 1.738 \times 10^6$ m
Distance between the Moon and the Earth	$r_{\oplus\mathrm{Moon}} = 3.84 \times 10^8$ m
Month (measured with respect to fixed stars)	$T_{\mathrm{Moon}} = 27.3$ days

Some Physical Data

Quantity	Convenient units
Air (dry, at 1 atm and 20°C except where noted)	
Density	1.21 kg/m^3
Specific heat (at −5°C)	2.108×10^3 J/(kg · K)
Molar specific heat (at −5°C)	37.6 J/(mol · K)
Ratio of specific heats (constant pressure/constant volume)	1.40
Speed of sound	343 m/s
Electrical breakdown strength	3×10^6 V/m
Effective molar mass	2.89×10^{-2} kg/mol
Water (at 1 atm)	
Density of water (at 4°C)	1000 kg/m^3
Density of ice (at 0°C)	9.167×10^2 kg/m^3
Melting temperature	273.15 K
Boiling temperature	373.15 K
Speed of sound in water (at 20°C)	1481 m/s
Specific heat (at room temperature)	4.187×10^3 J/(kg · K)
Molar specific heat (at room temperature)	75.4 J/(mol · K)
Heat of fusion	3.33×10^5 J/kg
Heat of vaporization	2.256×10^6 J/kg
Thermal conductivity	0.56 W/(m · K)
Emissivity	0.67
Index of refraction ($\lambda = 589$ nm, green)	1.33
Molar mass	1.80×10^{-2} kg/mol
Subatomic masses	
Electron mass	$9.10938291 \pm 0.00000040 \times 10^{-31}$ kg
Proton mass	$1.672621777 \pm 0.000000074 \times 10^{-27}$ kg
Neutron mass	$1.674927351 \pm 0.000000074 \times 10^{-27}$ kg
Hydrogen mass	$1.673532499 \pm 0.00000013 \times 10^{-27}$ kg

Source: Physical data and constants can be found at http://physics.nist.gov/cgi-bin/cuu/Category?view=html&All+values.x=115&All+values.y=7.

Some Astronomical Data

Object	Symbol	Rotation period (hh:mm:ss.s or days)	Mass ($\times 10^{24}$ kg)	Equatorial radius ($\times 10^6$ m)	Free-fall acceleration near surface (m/s^2)	Escape speed (km/s)	Blackbody temperature (K)
Sun	☉	$\approx$ 25 to 36 days[1]	1.9891×10^6	695.51	274	618	5777
Mercury	☿	58.65 days	0.3302	2.4397	3.7	4.3	440.1
Venus	♀	243 days	4.87	6.052	8.9	10.36	184.2
Earth	⊕	23:56:4.1	5.9736	6.378136	9.81	11.186	254.3
Moon	☾	27.3 days	0.07	1.738	1.6	2.38	270.7
Mars	♂	24:37:22.6	0.64	3.397	3.7	5.03	210.1
Ceres	⚳	09:04:19	9.6×10^{-4}	0.48			239
Jupiter	♃	9:50:30	1900	71.493	24.8	59.5	110.0
Saturn	♄	10:14:00	569	60.268	10.4	35.5	81.1
Uranus	⛢	17:14:00	87	25.559	8.87	21.3	58.2
Neptune	♆	16:03:00	103	24.764	11.2	23.5	46.6
Pluto	♇	6.387 days	0.01	1.135	0.58	1.2	37.5
Eris			$\approx 10^{-2}$	1.2			30

[1]The Sun is gaseous and does not rotate as a solid body; its period near the equator is shorter than at the poles.

Orbital parameters for objects that orbit the Sun

Object	Orbital period (days or years)	Semimajor axis (AU)	Eccentricity
Mercury	87.969 days	0.387	0.2056
Venus	224.701 days	0.723	0.0067
Earth	365.26 days	1.000	0.0167
Mars	1.8808 years	1.524	0.0935
Ceres	4.603 years	2.767	0.097
Jupiter	11.8618 years	5.204	0.0489
Saturn	29.4567 years	9.5482	0.0565
Uranus	84.0107 years	19.201	0.0457
Neptune	164.79 years	30.047	0.0113
Pluto	247.68 years	39.482	0.2488
Eris	559 years	67.89	0.4378

PEDAGOGICAL COLOR CHART

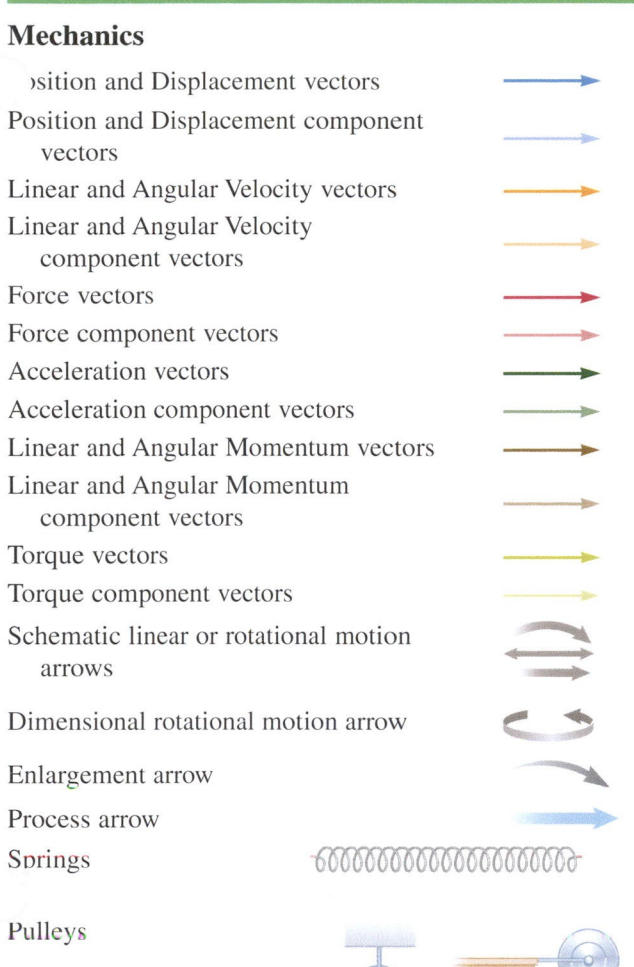

Mechanics

- Position and Displacement vectors
- Position and Displacement component vectors
- Linear and Angular Velocity vectors
- Linear and Angular Velocity component vectors
- Force vectors
- Force component vectors
- Acceleration vectors
- Acceleration component vectors
- Linear and Angular Momentum vectors
- Linear and Angular Momentum component vectors
- Torque vectors
- Torque component vectors
- Schematic linear or rotational motion arrows
- Dimensional rotational motion arrow
- Enlargement arrow
- Process arrow
- Springs
- Pulleys

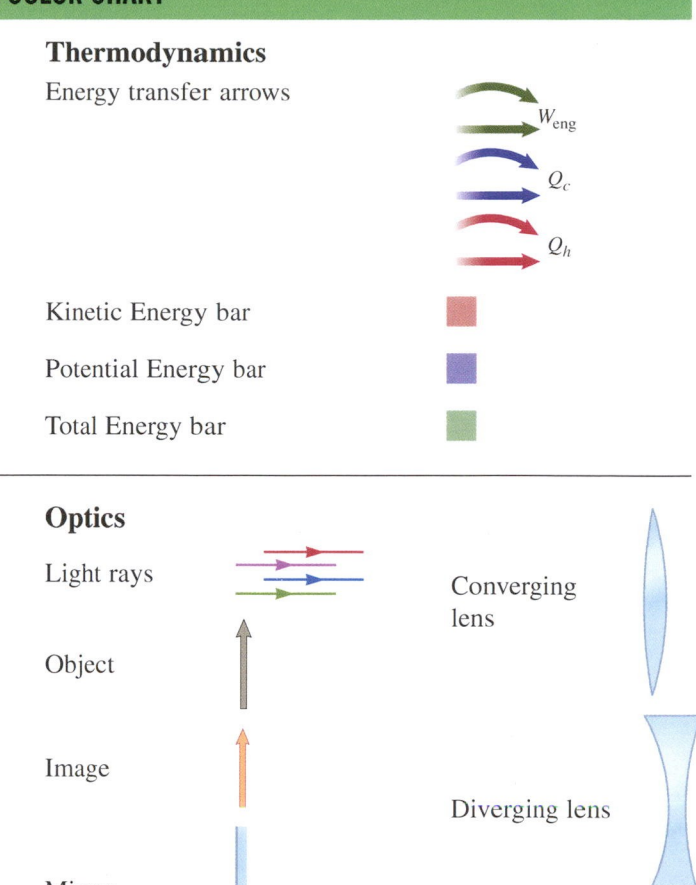

Thermodynamics

- Energy transfer arrows (W_{eng}, Q_c, Q_h)
- Kinetic Energy bar
- Potential Energy bar
- Total Energy bar

Optics

- Light rays
- Object
- Image
- Mirror
- Curved mirror
- Converging lens
- Diverging lens

Electricity and Magnetism

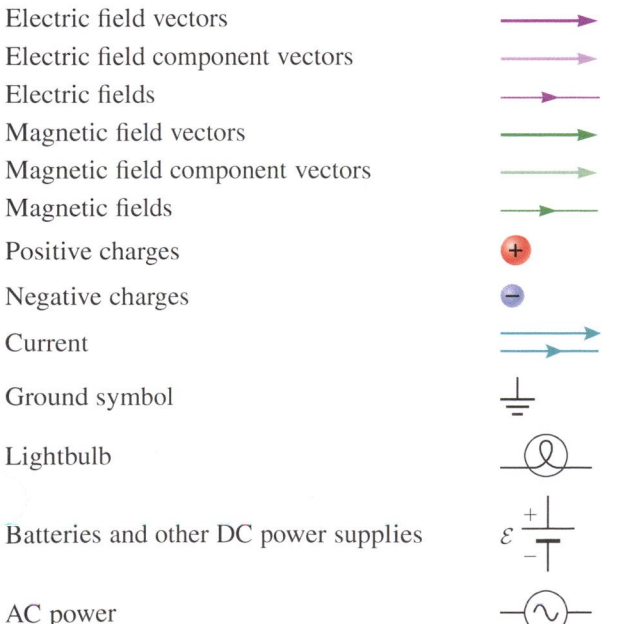

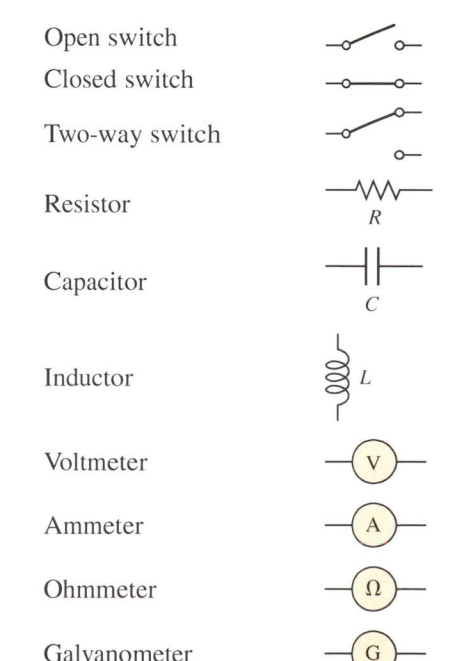

- Electric field vectors
- Electric field component vectors
- Electric fields
- Magnetic field vectors
- Magnetic field component vectors
- Magnetic fields
- Positive charges (+)
- Negative charges (−)
- Current
- Ground symbol
- Lightbulb
- Batteries and other DC power supplies (ℰ, +, −)
- AC power
- Open switch
- Closed switch
- Two-way switch
- Resistor (R)
- Capacitor (C)
- Inductor (L)
- Voltmeter (V)
- Ammeter (A)
- Ohmmeter (Ω)
- Galvanometer (G)

Symbols and abbreviations for units

Unit	Symbol	Unit	Symbol
ampere	A	light-year	ly
atmosphere	atm	liter	L
atomic mass unit	U	meter	m
British thermal unit	Btu	mile	mi
calorie	cal	miles per hour	mph
coulomb	C	millimeter of mercury (torricelli)	mm Hg (torr)
day	d	minute	min
degree Celsius	°C	mole	mol
degree Fahrenheit	°F	newton	N
electron volt	eV	ohm	Ω
farad	F	pascal	Pa
foot	ft	pound	lb
gallon	gal	pounds per square inch	psi
gauss	G	radian	rad
gram	g	revolution	rev
henry	H	revolutions per minute	rpm
hertz	Hz	second	s
horsepower	hp	tesla	T
inch	in.	volt	V
joule	J	watt	W
kelvin	K	weber	Wb
kilocalorie	Cal	yard	yd
kilogram	kg	year	yr
kilowatt-hour	kWh		

Greek alphabet

Name	Uppercase	Lowercase	Name	Uppercase	Lowercase	Name	Uppercase	Lowercase
Alpha	A	α	Iota	I	ι	Rho	P	ρ
Beta	B	β	Kappa	K	κ	Sigma	Σ	σ
Gamma	Γ	γ	Lambda	Λ	λ	Tau	T	τ
Delta	Δ	δ	Mu	M	μ	Upsilon	Υ	υ
Epsilon	E	ε	Nu	N	ν	Phi	Φ	φ
Zeta	Z	ζ	Xi	Ξ	ξ	Chi	X	χ
Eta	H	η	Omicron	O	o	Psi	Ψ	ψ
Theta	Θ	θ	Pi	Π	π	Omega	Ω	ω

Conversion Factors

Length

	meter	cm	km	in.	ft	mi
1 meter	1	10^{2}	10^{-3}	39.37	3.281	6.214×10^{-4}
1 centimeter	10^{-2}	1	10^{-5}	0.3937	3.281×10^{-2}	6.214×10^{-6}
1 kilometer	10^{3}	10^{5}	1	3.937×10^{4}	3281	0.6214
1 inch	2.540×10^{-2}	2.540	2.540×10^{-5}	1	8.333×10^{-2}	1.578×10^{-5}
1 foot	0.3048	30.48	3.048×10^{-4}	12	1	1.894×10^{-4}
1 mile	1609	1.609×10^{5}	1.609	6.336×10^{4}	5280	1
1 angstrom	10^{-10}	10^{-8}	10^{-13}	3.937×10^{-9}	3.281×10^{-10}	6.214×10^{-14}
1 AU	1.496×10^{11}	1.496×10^{13}	1.496×10^{8}	5.890×10^{12}	4.908×10^{11}	9.296×10^{7}
1 nautical mile	1852	1.852×10^{5}	1.852	7.291×10^{4}	6076	1.151
1 light-year	9.461×10^{15}	9.461×10^{17}	9.461×10^{12}	3.725×10^{17}	3.104×10^{16}	5.878×10^{12}
1 parsec	3.086×10^{16}	3.086×10^{18}	3.086×10^{13}	1.215×10^{18}	1.012×10^{17}	1.917×10^{13}
1 yard	0.9144	91.44	9.144×10^{-4}	36	3	5.682×10^{-4}

Mass

	kilogram	g	slug	u
1 kilogram	1	10^{3}	6.852×10^{-2}	6.022×10^{26}
1 gram	10^{-3}	1	6.852×10^{-5}	6.022×10^{23}
1 slug	14.59	1.459×10^{4}	1	8.786×10^{27}
1 atomic mass unit	$1.6605402 \times 10^{-27}$	1.661×10^{-24}	1.138×10^{-28}	1

Force

	newton	dyne	lb	oz	ton
1 newton	1	10^{5}	0.2248	3.597	1.124×10^{-4}
1 dyne	10^{-5}	1	2.248×10^{-6}	3.597×10^{-5}	1.124×10^{-9}
1 pound	4.448	4.448×10^{5}	1	16	5×10^{-4}
1 ounce	0.2780	2.780×10^{4}	6.250×10^{-2}	1	3.125×10^{-5}
1 ton	8.896×10^{3}	8.896×10^{8}	2000	3.2×10^{4}	1

Pressure

	pascal	atm	Torr (mm Hg)	psi	dyne/cm^2
1 pascal	1	9.869×10^{-6}	7.501×10^{-3}	1.450×10^{-4}	10
1 atm	1.013×10^{5}	1	760	14.70	1.013×10^{6}
1 Torr	1333	1.316×10^{-2}	1	0.1934	1.333×10^{4}
1 psi	6.895×10^{3}	6.805×10^{-2}	51.71	1	6.895×10^{4}
1 dyne/cm^2	0.1	9.869×10^{-7}	7.501×10^{-4}	1.405×10^{-5}	1

Energy

	joule	erg	ft·lb	cal	eV
1 joule	1	10^{7}	0.7376	0.2389	6.242×10^{18}
1 erg	10^{-7}	1	7.376×10^{-8}	2.389×10^{-8}	6.242×10^{11}
1 ft·lb	1.356	1.356×10^{7}	1	0.3238	8.464×10^{18}
1 cal	4.184	4.184×10^{7}	3.088	1	2.612×10^{19}
1 eV	1.602×10^{-19}	1.602×10^{-19}	1.182×10^{-19}	3.827×10^{-20}	1

Physical Constants

Quantity	Symbol	Convenient units
Avogadro's number	N_A	$6.02214129 \pm 0.00000027 \times 10^{23}$
Bohr magneton	$\mu_{Bohr} = e\hbar/2m_e$	$9.27400968 \times 10^{-24}$ +/− 0.00000020 J/T
Bohr radius	r_B	0.52917721067 +/− $0.00000000012 \times 10^{-10}$ m
Boltzmann's constant	k_B	$1.3806488 \pm 0.0000013 \times 10^{-23}$ J/K
Compton wavelength of an electron	λ_C	$2.4263102389 \pm 0.0000000016 \times 10^{-12}$ m
Coulomb constant	$k = \dfrac{1}{4\pi\varepsilon_0}$	$8.99 \times 10^9\ N \cdot m^2/C^2$
Elementary (electric) charge	e	$1.602176565 \pm 0.000000035 \times 10^{-19}$ C
Permeability of free space	μ_0	$4\pi \times 10^{-7}\ T \cdot m/A$
Permittivity of free space	ε_0	$1/\mu_0 c^2 = 8.854187817... \times 10^{-12}\ C^2/N \cdot m^2$
Planck's constant	h	$6.62606957 \pm 0.00000029 \times 10^{-34}\ J \cdot s$
Rydberg's constant	R	$10973731.568539 \pm 0.0000000000055\ m^{-1}$
Speed of light (in a vacuum)	c	2.99792458×10^8 m/s (exact)
Stefan-Boltzmann's constant	σ	$(2\pi^5 k_B^4)/(15c^2h^3) = 5.670373 \pm 0.000021 \times 10^{-8}\ W/(m^2 \cdot K^4)$
Universal gas constant	R	$8.314472 \pm 0.000015\ J/(mol \cdot K)$
Universal gravitational constant	G	$6.67384 \pm 0.00080 \times 10^{-11}\ N \cdot m^2/kg^2$

Source: Physical data and constants can be found at http://physics.nist.gov/cgi-bin/cuu/Category?view=html&All+values.x=115&All+values.y=7.

Prefixes for Powers of 10

Name	Abbreviation	Value
yocto	y	10^{-24}
zepto	z	10^{-21}
atto	a	10^{-18}
femto	f	10^{-15}
pico	p	10^{-12}
nano	n	10^{-9}
micro	μ (Greek letter "mu")	10^{-6}
milli	m	10^{-3}
centi	c	10^{-2}
deci	d	10^{-1}
deka	da	10^{1}
hecto	h	10^{2}
kilo	k	10^{3}
mega	M	10^{6}
giga	G	10^{9}
tera	T	10^{12}
peta	P	10^{15}
exa	E	10^{18}
zetta	Z	10^{21}
yotta	Y	10^{24}